DICTIONARY OF SCIENTIFIC BIOGRAPHY

PUBLISHED UNDER THE AUSPICES OF
THE AMERICAN COUNCIL OF LEARNED SOCIETIES

The American Council of Learned Societies, organized in 1919 for the purpose of advancing the study of the humanities and of the humanistic aspects of the social sciences, is a nonprofit federation comprising forty-five national scholarly groups. The Council represents the humanities in the United States in the International Union of Academies, provides fellowships and grants-in-aid, supports research-and-planning conferences and symposia, and sponsors special projects and scholarly publications.

MEMBER ORGANIZATIONS

AMERICAN PHILOSOPHICAL SOCIETY, 1743
AMERICAN ACADEMY OF ARTS AND SCIENCES, 1780
AMERICAN ANTIQUARIAN SOCIETY, 1812
AMERICAN ORIENTAL SOCIETY, 1842
AMERICAN NUMISMATIC SOCIETY, 1858
AMERICAN PHILOLOGICAL ASSOCIATION, 1869
ARCHAEOLOGICAL INSTITUTE OF AMERICA, 1879
SOCIETY OF BIBLICAL LITERATURE, 1880
MODERN LANGUAGE ASSOCIATION OF AMERICA, 1883
AMERICAN HISTORICAL ASSOCIATION, 1884
AMERICAN ECONOMIC ASSOCIATION, 1885
AMERICAN FOLKLORE SOCIETY, 1888
AMERICAN DIALECT SOCIETY, 1889
AMERICAN PSYCHOLOGICAL ASSOCIATION, 1892
ASSOCIATION OF AMERICAN LAW SCHOOLS, 1900
AMERICAN PHILOSOPHICAL ASSOCIATION, 1901
AMERICAN ANTHROPOLOGICAL ASSOCIATION, 1902
AMERICAN POLITICAL SCIENCE ASSOCIATION, 1903
BIBLIOGRAPHICAL SOCIETY OF AMERICA, 1904
ASSOCIATION OF AMERICAN GEOGRAPHERS, 1904
HISPANIC SOCIETY OF AMERICA, 1904
AMERICAN SOCIOLOGICAL ASSOCIATION, 1905
AMERICAN SOCIETY OF INTERNATIONAL LAW, 1906
ORGANIZATION OF AMERICAN HISTORIANS, 1907
AMERICAN ACADEMY OF RELIGION, 1909
COLLEGE ART ASSOCIATION OF AMERICA, 1912
HISTORY OF SCIENCE SOCIETY, 1924
LINGUISTIC SOCIETY OF AMERICA, 1924
MEDIAEVAL ACADEMY OF AMERICA, 1925
AMERICAN MUSICOLOGICAL SOCIETY, 1934
SOCIETY OF ARCHITECTURAL HISTORIANS, 1940
ECONOMIC HISTORY ASSOCIATION, 1940
ASSOCIATION FOR ASIAN STUDIES, 1941
AMERICAN SOCIETY FOR AESTHETICS, 1942
AMERICAN ASSOCIATION FOR THE ADVANCEMENT OF SLAVIC STUDIES, 1948
METAPHYSICAL SOCIETY OF AMERICA, 1950
AMERICAN STUDIES ASSOCIATION, 1950
RENAISSANCE SOCIETY OF AMERICA, 1954
SOCIETY FOR ETHNOMUSICOLOGY, 1955
AMERICAN SOCIETY FOR LEGAL HISTORY, 1956
AMERICAN SOCIETY FOR THEATRE RESEARCH, 1956
SOCIETY FOR THE HISTORY OF TECHNOLOGY, 1958
AMERICAN COMPARATIVE LITERATURE ASSOCIATION, 1960
AMERICAN SOCIETY FOR EIGHTEENTH-CENTURY STUDIES, 1969
ASSOCIATION FOR JEWISH STUDIES, 1969

DICTIONARY OF SCIENTIFIC BIOGRAPHY

CHARLES COULSTON GILLISPIE
Princeton University

EDITOR IN CHIEF

Volume 15

Supplement I

ROGER ADAMS – LUDWIK ZEJSZNER

TOPICAL ESSAYS

CHARLES SCRIBNER'S SONS · NEW YORK

Copyright © 1970, 1971, 1972, 1973, 1974, 1975, 1976, 1978, 1980
American Council of Learned Societies.
First publication in an eight-volume edition 1981.

Library of Congress Cataloging in Publication Data

Main entry under title:

Dictionary of scientific biography.

"Published under the auspices of the American Council
of Learned Societies."
Includes bibliographies and index.
1. Scientists—Biography.   I. Gillispie, Charles
Coulston.   II. American Council of Learned Societies
Devoted to Humanistic Studies.
Q141.D5   1981   509'.2'2 [B]   80-27830
ISBN 0-684-16962-2 (set)

ISBN 0-684-16963-0 Vols. 1 & 2   ISBN 0-684-16967-3 Vols. 9 & 10
ISBN 0-684-16964-9 Vols. 3 & 4   ISBN 0-684-16968-1 Vols. 11 & 12
ISBN 0-684-16965-7 Vols. 5 & 6   ISBN 0-684-16969-X Vols. 13 & 14
ISBN 0-684-16966-5 Vols. 7 & 8   ISBN 0-684-16970-3 Vols. 15 & 16

Published simultaneously in Canada
by Collier Macmillan Canada, Inc.
Copyright under the Berne Convention.

7 9 11 13 15 17 19   B/C   20 18 16 14 12 10 8 6

Printed in the United States of America

Editorial Board

Editorial Staff

Panel of Consultants

Georges Canguilhem
University of Paris

Pierre Costabel
École Pratique des Hautes Études

Alistair C. Crombie
University of Oxford

Maurice Daumas
Conservatoire National des Arts et Métiers

Allen G. Debus
University of Chicago

Marcel Florkin
University of Liège

John C. Greene
University of Connecticut

Mirko D. Grmek
Archives Internationales d'Histoire des Sciences

A. Rupert Hall
Imperial College of Science and Technology

Mary B. Hesse
University of Cambridge

Brooke Hindle
Massachusetts Institute of Technology

Joseph E. Hofmann
University of Tübingen

Reijer Hooykaas
State University of Utrecht

Michael A. Hoskin
University of Cambridge

E. S. Kennedy
Brown University

Sten H. Lindroth
University of Uppsala

Robert K. Merton
Columbia University

John Murdoch
Harvard University

Shigeru Nakayama
University of Tokyo

Charles D. O'Malley
University of California, Los Angeles

Derek de Solla Price
Yale University

J. R. Ravetz
University of Leeds

Maria Luisa Righini-Bonelli
Istituto e Museo di Storia della Scienza, Florence

Duane H. D. Roller
University of Oklahoma

Karl Eduard Rothschuh
University of Münster/Westphalia

S. Sambursky
Hebrew University, Jerusalem

Giorgio de Santillana
Massachusetts Institute of Technology

Aydin Sayili
University of Ankara

Robert E. Schofield
Case Western Reserve University

Christoph J. Scriba
Technical University, Berlin

Nathan Sivin
Massachusetts Institute of Technology

Bogdan Suchodolski
Polish Academy of Sciences

René Taton
École Pratique des Hautes Études

J. B. Thornton
University of New South Wales

Richard S. Westfall
Indiana University

W. P. D. Wightman
King's College, Aberdeen

L. Pearce Williams
Cornell University

A. P. Youschkevitch
Academy of Sciences of the U.S.S.R.

Contributors to Supplement I

The following are the contributors to Supplement I. Each author's name is followed by the institutional affiliation at the time of publication and the names of the articles written for this volume. The symbol † means that the author is deceased.

MARK B. ADAMS
University of Pennsylvania
VAVILOV

MICHELE L. ALDRICH
Aaron Burr Papers
PEALE

G. C. ANAWATI
Institut Dominicain d'Études Orientales, Cairo
ḤUNAYN IBN ISḤĀQ; IBN SĪNĀ

RICHARD P. AULIE
Loyola University of Chicago
EDWARD JOHN RUSSELL

A. CLIFFORD BARGER
Harvard Medical School
CANNON

JOHN J. BEER
University of Delaware
BESSEMER

SAUL BENISON
University of Cincinnati
CANNON

WILLIAM A. BLANPIED
American Association for the Advancement of Science
BHABHA; JAGADISCHANDRA BOSE; SATYENDRANATH BOSE

FRANCK BOURDIER
École Pratique des Hautes Études
DUMÉRIL

IVOR BULMER-THOMAS
MENELAUS OF ALEXANDRIA; THEODORUS OF CYRENE

WERNER BURAU
University of Hamburg
GRASSMANN

WILLIAM F. BYNUM
University College London
DALE

ROBERT S. COHEN
Boston University
ENGELS; MARX

WILLIAM COLEMAN
Johns Hopkins University
GEGENBAUR

ALBERT B. COSTA
Duquesne University
ARFVEDSON; BEHREND

PIERRE COSTABEL
École Pratique des Hautes Études
POISSON

ERIKA CREMER
BODENSTEIN

HOWARD A. CRUM
University of Michigan Herbarium
EVANS

STANISŁAW CZARNIECKI
Polish Academy of Sciences
STASZIC; ZEJSZNER

PER F. DAHL
Brookhaven National Laboratory
COLDING

JERZY DOBRZYCKI
Institute for the History of Science, Polish Academy of Sciences, Warsaw
HOËNÉ-WROŃSKI

ARTHUR DONOVAN
West Virginia University
CRAWFORD

J. G. DORFMAN†
ARTSIMOVICH

THE EDITORS
COLUMBUS

FRANK N. EGERTON III
University of Wisconsin-Parkside
FOREL; FUCHS

VĚRA EISNEROVÁ
VEJDOVSKÝ

FRANÇOIS ELLENBERGER
University of Paris-South
BOURGUET

JAMES ELLINGTON
University of Connecticut
PAUL CARUS

JOSEPH EWAN
Tulane University
ENGELMANN; HITCHCOCK; MERRILL

JOAN M. EYLES
FEATHERSTONHAUGH

VICTOR A. EYLES
FEATHERSTONHAUGH

VERA N. FEDCHINA
Academy of Sciences of the U.S.S.R.
MIKLUKHO-MAKLAY

ROBERT FOX
University of Lancaster
LAPLACE

F. FRAUNBERGER
RAMSAUER

JOSEPH S. FRUTON
Yale University
BERGMANN; HARDY

A. E. GAISSINOVITCH
Academy of Sciences of the U.S.S.R.
WOLFF

GERALD L. GEISON
Princeton University
CUSHNY; MELLANBY

KARL-ERNST GILLERT
Robert Koch-Institut des Bundesgesundheitsamtes
WASSERMANN

R. J. GILLINGS
The University of New South Wales
THE MATHEMATICS OF ANCIENT EGYPT

CHARLES C. GILLISPIE
Princeton University
LAPLACE

LOREN GRAHAM
Columbia University
BOGDANOV

IVOR GRATTAN-GUINNESS
LAPLACE

REESE E. GRIFFIN, JR.
HOLBROOK

A. T. GRIGORIAN
Academy of Sciences of the U.S.S.R.
ANDRONOV; BUNYAKOVSKY; GALERKIN; IOFFE

M. D. GRMEK
Académie Internationale d'Histoire des Sciences
NICOLLE

JEAN-CLAUDE GUÉDON
University of Montreal
BÉCHAMP

C. DORIS HELLMAN†
PEURBACH

ARMIN HERMANN
University of Stuttgart
BORN

C. C. HEYDE
Scientific and Industrial Research Organization
BIENAYMÉ

ERWIN N. HIEBERT
Harvard University
NERNST; OSTWALD

CONTRIBUTORS TO SUPPLEMENT I

DONALD R. HILL
AL-JAZARĪ

FREDERIC L. HOLMES
University of Western Ontario
DALTON

SALLY SMITH HUGHES
BEIJERINCK

TATYANA D. ILYINA
Soviet Academy of Sciences
MIKLUKHO-MAKLAY

ALBERT Z. ISKANDAR
The Wellcome Institute for the History of Medicine
ḤUNAYN IBN ISḤĀQ; IBN SĪNĀ

JEAN JOLIVET
École Pratique des Hautes Études
AL-KINDĪ

PHILLIP S. JONES
University of Michigan
KARPINSKI

R. V. JONES
University of Aberdeen
BOYS

DAVID JORAVSKY
Northwestern University
MICHURIN

GÜNTHER KERSTEIN
CRAMER; GEHLEN; HERMBSTAEDT

DAVID A. KING
Smithsonian Institution
Project in Medieval Islamic Astronomy, American Research Center in Egypt
AL-KHALĪLĪ

ROBERT E. KOHLER
University of Pennsylvania
ADAMS; KENDALL

HANS-GÜNTHER KÖRBER
Zentralbibliothek des Meteorologischen Dienstes der DDR, Potsdam
OSTWALD

YVES LAISSUS
Muséum National d'Histoire Naturelle
LA CONDAMINE

DELFINO LAURI
PAGANO

JOHN LAW
University of Keele
BRAGG

HENRY M. LEICESTER
University of the Pacific
FISCHER; KARRER

RICHARD LEMAY
The Graduate School and University Center of the City University of New York
GERARD OF CREMONA

CAMILLE LIMOGES
University of Montreal
DAUBENTON

FLOYD G. LOUNSBURY
Yale University
MAYA NUMERATION, COMPUTATION, AND CALENDRICAL ASTRONOMY

VICTOR A. McKUSICK
Johns Hopkins Hospital
BRÖDEL

MICHAEL McVAUGH
University of North Carolina
RUSCELLI

KARL MÄGDEFRAU
University of Munich
CAMERARIUS; CESALPINO

JOSEPH M. MALINE
COPE

YVES MARQUET
University of Dakar
IKHWĀN AL-ṢAFĀʾ

S. R. MIKULINSKY
Institute of the History of Science, Moscow
DYADKOVSKY

LETTIE S. MULTHAUF
NANSEN

SHIGERU NAKAYAMA
University of Tokyo
JAPANESE SCIENTIFIC THOUGHT

J. D. NORTH
University of Oxford
HENRY CHAMBERLAINE RUSSELL

A. LEO OPPENHEIM†
MAN AND NATURE IN MESOPOTAMIAN CIVILIZATION

JANE OPPENHEIMER
Bryn Mawr College
DOHRN

RICHARD A. PARKER
Brown University
EGYPTIAN ASTRONOMY, ASTROLOGY, AND CALENDRICAL RECKONING

H. W. PAUL
University of Florida
MEYERSON

MOGENS PIHL
University of Copenhagen
CHRISTIANSEN

DAVID PINGREE
Brown University
DOROTHEUS OF SIDON; HISTORY OF MATHEMATICAL ASTRONOMY IN INDIA

EMMANUEL POULLE
École Nationale des Chartes
FINE; FUSORIS

DEREK DE SOLLA PRICE
Yale University
VITRUVIUS POLLIO

ROSHDI RASHED
Centre National de la Recherche Scientifique
AL-KINDĪ

GUENTER B. RISSE
University of Wisconsin-Madison
HOUSSAY

GLORIA ROBINSON
Yale University
JULIUS VICTOR CARUS

ALAN J. ROCKE
University of Wisconsin-Madison
BARGER; JACOBS

JACQUES ROGER
University of Paris
BOUCHER DE PERTHES

SYDNEY ROSS
Rensselaer Polytechnic Institute
FREUNDLICH

DOROTHEA RUDNICK
CHAMISSO

GERHARD RUDOLPH
University of Kiel
JULIUS BERNSTEIN

A. S. SAIDAN
Jordanian University
AL-BAGHDĀDĪ

JULIO SAMSÓ
Universidad Autónoma de Barcelona
AL-BIṬRŪJĪ

CHARLES B. SCHMITT
Warburg Institute
BORRO; CAMPANELLA

CHRISTOPH J. SCRIBA
Technical University, Berlin
GRASSMANN

E. SENETA
Australian National University, Canberra
BIENAYMÉ

JACQUES SESIANO
DIOPHANTUS OF ALEXANDRIA

C. P. SNOW
BERNAL

FRANS A. STAFLEU
University of Utrecht
ENGLER; KUNTH; KUNTZE

ARTHUR R. STEELE
University of Toledo
PAVÓN Y JIMÉNEZ

NANCY STEPAN
Yale University
CRUZ

NOEL M. SWERDLOW
University of Chicago
PEURBACH

FERENC SZABADVARY
Technical University, Budapest
CLASSEN

CONTRIBUTORS TO SUPPLEMENT I

KENNETH L. TAYLOR
University of Oklahoma
DELAFOSSE

G. J. TOOMER
Brown University
CAMPANUS OF NOVARA; DIOCLES; HERACLIDES PONTICUS; HIPPARCHUS; VITRUVIUS POLLIO

HENRY S. TROPP
Humboldt State University
ECKERT

T. VON UEXKÜLL
University of Ulm
UEXKÜLL

PETER W. VAN DER PAS
BERNARD ALBINUS; BERNARD SIEGFRIED ALBINUS; CHRISTIAAN BERNARD ALBINUS; FREDERIK BERNARD ALBINUS; BIDLOO

B. L. VAN DER WAERDEN
University of Zurich
MATHEMATICS AND ASTRONOMY IN MESOPOTAMIA

J. VERNET
University of Barcelona
MUTIS Y BOSSIO

ALEKSANDR VOLODARSKY
BUGAEV

P. J. WALLIS
University of Newcastle Upon Tyne
GREEN

JOHN WARD-PERKINS
VITRUVIUS POLLIO

J. H. WOLF
MURALT

A. P. YOUSCHKEVITCH
Academy of Sciences of the U.S.S.R.
SERGEY NATANOVICH BERNSTEIN; DA CUNHA

BRUNO ZANOBIO
University of Pavia
PAGANO; PENSA; PIANESE

DICTIONARY
OF
SCIENTIFIC BIOGRAPHY

DICTIONARY OF SCIENTIFIC BIOGRAPHY

ADAMS – ZEJSZNER

ADAMS, ROGER (*b.* Boston, Massachusetts, 2 January 1889; *d.* Urbana, Illinois, 6 July 1971), *organic chemistry*.

Adams has a special place in the history of American chemistry as a scientific organizer. As head of the department of chemistry at the University of Illinois, and through his close contacts with the chemical industry and the many committees on which he served, he fostered the rapid growth of chemistry in the United States and its integration with industry and government.

Adams was educated at Harvard University, receiving the B.A. in 1909 and the Ph.D. in 1912, under C. L. Jackson. He then spent a year in Germany, studying with Otto Diels and Richard Willstätter – the leading exponents, respectively, of synthetic organic chemistry and the chemistry of natural products. Adams' own interests continued in these traditional lines of structure, stereochemistry, and synthetic methods, little influenced by the increasingly important applications of physical chemistry to organic reaction mechanisms in the 1920's and 1930's.

In 1913 Adams returned to Harvard as a postdoctoral fellow with Jackson (then emeritus), apparently intending to pursue a career in research – which at the time meant an industrial career. An unexpected vacancy in 1914 led to his becoming an instructor in organic chemistry at Harvard. In 1916 he accepted an offer from William A. Noyes of an assistant professorship at the University of Illinois. He was promoted to professor in 1919, and in 1926 succeeded Noyes as director of the department of chemistry and chemical engineering, which he headed for nearly thirty years. Noyes had made Illinois the leading school of organic chemistry in the United States, where physical chemistry generally took first place. Under Adams, Illinois expanded and became the most prolific source of chemists for the chemical industry.

The phenomenal growth of American chemistry during Adams' lifetime was largely a result of the demand for research chemists in industry, following World War I. With this new social relation came a marked change in the style of chemists. Noyes, for example, was an exponent of "pure" chemistry. He represented the generation of chemists, imbued with the ideal of German university science, that won chemistry a place as an academic discipline in American universities, in the face of strong traditions of "applied" science. Noyes's call in 1907 to Illinois, a land grant college with a very heavy emphasis on applied, technological science, was a measure of the spreading influence of the pure science ideal. Applied chemistry and chemical engineering continued to be important at Illinois under the leadership of Samuel W. Parr. But Parr and Noyes represented two distinctly separate traditions.

In the 1920's and 1930's this sharp distinction disappeared. University chemistry reestablished strong links to industrial institutions. Industries demanded more chemists; and, more important, demanded Ph.D.'s with academic style and training. Industrial work became attractive for chemists with academic tastes, as researchers or consultants. Adams did not initiate this trend but he did facilitate it, hastening and smoothing the meshing of the academic and industrial worlds.

During World War I, Adams turned a summer project for producing chemicals for classroom use into pilot-plant production of organic chemicals for war and industrial use, to replace lost German sources. This operation was continued after the war, to introduce students to industrial operation. Bulletins on synthetic methods developed at Illinois were issued, and in 1921 they became the academic monograph series Chemical Syntheses, which Adams edited for nineteen years.

Adams was impressed by the promise of organized science during the war. In the 1920's the demand for industrial chemists led to a rapid expansion of the department and to changes in the chemical curriculum, especially after Adams assumed chairmanship of the department. The many specialized courses typical of Noyes's period were eliminated from both undergraduate and graduate curricula. Emphasis was put on fundamentals, on the assumption that specific problem areas were best learned in industrial laboratories. Curricula were upgraded, and a Ph.D. program in chemical engineering was initiated. While the emphasis was heavily on training for industrial research, the immediately vocational aspect was dropped. The same training was given to prospective industrial researchers and university teachers. Under Adams' direction, industrial training became more scientific, and academic training more attuned to industrial needs. The conflicting traditions of Noyes's generation merged. In 1954 Adams epitomized this new mode of chemistry:

> The duty and one of the primary responsibilities [of universities] is to train chemists for industry. If the trained chemists from the universities should not be able to find jobs, it would not be long before the number of students in chemistry would dwindle to a trickle and scientific research, which in this country stems in large measure from the efforts of graduate students, would likewise diminish rapidly in volume. Industry, on the other hand, has the university as its only source of scientific personnel. The universities and industry are mutually interdependent.

This symbiotic relationship became the dominant pattern in American chemistry. As Adams' students entered positions of influence, demand for Illinois chemists increased. The department became an assembly line for the production of industrial chemists.

Adams served as a director of the American Chemical Society in 1930–1935 and 1944–1950, and as president in 1935. He was a member of the Executive Committee of the American Association for the Advancement of Science from 1941 to 1945 and from 1947 to 1951, and was president in 1950. He was also a director of the Council for Agricultural and Chemurgic Research, an organization devoted to the application of chemistry to agricultural industry.

Adams had a role in virtually every major development in the national organization of science in America from World War I into the era of government patronage. His involvement with government dates from 1918, when he served in the Chemical Warfare Service. Elected to the National Academy of Sciences in 1929, he served on the Council (1931–1937), as chairman of the Chemical Section (1938–1941), and on the National Research Council Fellowship Board (1928–1941). Adams was a member of the short-lived Science Advisory Board in 1934–1935, an early attempt at advising the president on science. From 1941 to 1946 he served on the National Defense Research Committee, which was responsible for organizing war research in chemistry and chemical engineering. From 1954 to 1960 he was a member of the board of directors of the National Science Foundation. Adams actively represented the interests of chemistry in the expanding national science system.

Adams was also influential in relationships with private patrons. In 1950 he oversaw the transfer of the income of Universal Oil Products to the American Chemical Society, as the Petroleum Research Fund. Three years later he chaired an advisory committee of the Sloan Foundation that initiated a program to aid basic research in the physical sciences.

Adams' research likewise reveals his organizational style. His main interest was in methods of organic synthesis, and he contributed many standard recipes to *Organic Syntheses* and to *Organic Reactions*, which he also helped to found. His synthetic work focused on aromatic compounds, important in the dye industry. The "Adams catalyst," a colloidal platinum oxide, became standard for hydrogenations. Probably his most influential academic work was his systematic investigation, in the late 1920's and 1930's, of the stereochemistry of substituted biphenyl and biaryl compounds, which, owing to sterically hindered rotation, can be resolved into optical isomers. This work raised basic theoretical questions concerning the relationship between steric and electronic effects, which were then attracting keen interest among physical organic chemists. Adams' approach owed less to these theoretical concerns than to more classical problems of organic synthesis and stereochemistry.

Adams' best-known work on natural products is that on the marijuana alkaloids, which he took up in the late 1930's and 1940's at the request of the Narcotics Bureau. He isolated and synthesized tetrahydrocannabinol and several of its analogues. He also elucidated and synthesized chaulmoogric and hydnocarpic acids, the active principles of chaulmoogra oil, a folk remedy for leprosy; gossypol, the toxic agent in cottonseed oil; and the *Senecio* and *Crotalaria* alkaloids, which poison grazing cattle. Adams synthesized a number of local anesthetics, such as Butyn. A characteristic feature of all this work is its relevance to the practical concerns of the drug industry.

ALBINUS

Adams received ten honorary degrees, twenty-four medals and awards from American and foreign scientific societies, and honorary membership in nine chemical societies and the National Academy of Sciences. A truer measure of his historical importance, however, is the number of his students in high positions in the academic world and in industrial chemistry, and the degree to which the style and institutional structure of American chemistry approaches his vision.

BIBLIOGRAPHY

I. Original Works. Adams' writings include "Bacteriological Action of Certain Organic Acids Toward *Mycobacterium leperae* and Other Acid-Fast Bacteria," in *Journal of Pharmacology and Experimental Therapy*, **45** (1932), 121–162; "The Stereochemistry of Diphenyls and Analogous Compounds," in *Chemical Reviews*, **12** (1933), 161–385; "Marihuana," in *Harvey Lectures*, **37** (1941–1942), 168–197; and "Universities and Industry in Science," in *Industrial and Engineering Chemistry*, **46** (1954), 506–510, the Perkin Medal address.

Several boxes of Adams' personal papers have been deposited at the University of Illinois archives.

II. Secondary Literature. See Nelson J. Leonard, "Roger Adams," in *Journal of the American Chemical Society*, **91** (1969), *a–d*; and C. S. Marvel, "Roger Adams," in *Yearbook. American Philosophical Society* for 1974, 111–114. *University of Illinois, Special Circular of the Department of Chemistry 1916–1927* (Urbana, 1927); and *University of Illinois, Department of Chemistry, Developments During the Period 1927–1941* (Urbana, 1941), contain useful historical material, policy statements, and complete bibliographies of department publications to 1940.

Robert E. Kohler

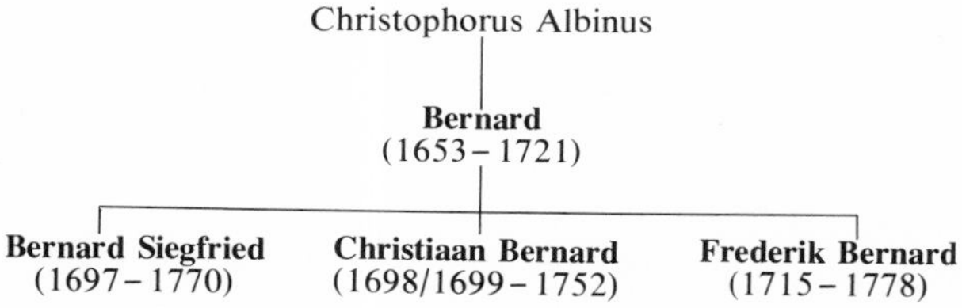

The Albinus family. Names in boldface are discussed in articles.

ALBINUS, BERNARD (*b.* Dessau, Anhalt, Germany, 7 January 1653; *d.* Leiden, Netherlands, 7 September 1721), *medicine.*

The family name originally was Weiss; the poet-historian Petrus Weiss (1534–1598), a granduncle of Bernard's, latinized the name to Albinus. Bernard's father, Christophorus Albinus, was mayor of Dessau. Bernard was first educated by a private tutor and later attended the public school, where Heinrich Alers was his principal teacher. When Alers was appointed teacher at the Bremen Athenaeum, Bernard followed him there. On 26 April 1675 he matriculated as a student of medicine at the University of Leiden, where he received the M.D. degree on 12 May 1676; the subject of his dissertation was catalepsy. Thereafter he traveled for four years in the Netherlands and France to learn the latest developments in anatomy, surgery, and medicine. Soon after his return to Dessau, he was appointed professor of medicine at the University of Frankfurt-an-der-Oder, where he started teaching on 13 January 1681. He was awarded a Ph.D. degree on 16 April 1681.

Bernard's career as a professor was interrupted twice, first when he was appointed court physician to the elector of Brandenburg, Friedrich Wilhelm, a post he held until the elector's death on 29 April 1688, and for the second time when he was appointed court physician to Elector Friedrich (later Frederick I of Prussia) in 1697. In 1702 he was invited to teach theoretical and practical medicine at the University of Leiden, where he began work on 19 October 1702. The teaching of Bernard Albinus and of Hermann Boerhaave, who had been appointed one year earlier, laid the foundation for the fame of Leiden as a world center for the study of medicine. In 1696 Bernard married the daughter of a Frankfurt professor, Susanna Catherina Ring. They had seven daughters and four sons, three of whom became known as anatomists.

BIBLIOGRAPHY

I. Original Works. Bernard Albinus' writings include *Dissertatio epistolica de elephantia Javae nova quam disputationis publicae loco, resp. Joh. Chr. Mentzel* (Frankfurt-an-der-Oder, 1683)—H. Sallander, in *Bibliotheca Walleriana* (Stockholm, 1955), nos. 323–335, lists 12 more Frankfurt dissertations, dated between 1684 and 1696; *Oratio de ortu et progressu medicinae; cum in Academia Lugduno-Batava medicinae theoretico-practicae professionem, die xix Octobris MDCCII, auspicaretur* (Leiden, 1702); *Oratio de incrementis et statu artis medicae seculi decimi septimi, dicta at diem vidus Februar. MDCCXI, cum magistratu academico se abdicaret* (Leiden, 1711); and *Oratio in obitum Johannis Jacobi Rau* (Leiden, 1719).

II. Secondary Literature. For a discussion of the Albinus family, see A. Hirsch, "Bernhard Albinus," *Allgemeine deutsche Biographie*, I (1875), 221–222; and J. E. Kroon, "Enkele aanteekeningen uit de geschiedenis van een beroemd anatomen geslacht," in *Bijdragen tot de geschiedenis der geneeskunde*, **3** (1923), 328–330.

Works on Bernard Albinus are the following, listed chronologically: H. Boerhaave, *Oratio academica de vita et obitu viri clarissimi Bernhardi Albini, ex decreto Magnifici Rectoris et Senatus Academici habita xxii Sept. anni MDCCXXI* (Leiden, 1721); F. Chaussier and N. P. Adelon, "Bernard Albinus," in *Biographie universelle ancienne et moderne*, I (1843), 346; A. J. van der Aa, "Bernardus Albinus," in *Biografisch woordenboek der Nederlanden*, I (n.d.), 156–157; and E. D. Baumann, "Bernard Albinus," in *Nieuw Nederlandsch biografisch woordenboek*, IV (1918), 21–22.

PETER W. VAN DER PAS

ALBINUS, BERNARD SIEGFRIED (*b.* Frankfurt-an-der-Oder, Germany, 24 February 1697; *d.* Leiden, Netherlands, 9 September 1770), *anatomy, medicine*.

Bernard Siegfried, the eldest son of Bernard, matriculated at Leiden on 16 September 1709, at the age of twelve. He studied under his father, H. Boerhaave, G. Bidloo, and J. J. Rau. In 1718 he began a study trip but was soon called back to become lecturer in comparative anatomy, relieving Rau, who was ill and died soon afterward. He began teaching on 2 October 1719, before he had been awarded the M.D. degree—which he received *honoris causa* shortly thereafter. After his father's death, Bernard Siegfried was appointed to succeed him, mainly on the recommendation of Boerhaave. He started his duties on 19 November 1721, teaching anatomy and surgery.

Although Bernard Siegfried became a professor at the uncommonly early age of twenty-four, the university never had cause to regret his appointment. He soon became the leading anatomist of his time; with Boerhaave he made the University of Leiden the world center of medical education. In order to supply his pupils with study material, he reissued (with Boerhaave) the anatomical atlas of Vesalius and, later, the anatomical works of Fabrici (1737) and of Eustachi (1744). He also edited the complete works of William Harvey (1736). Not satisfied with these, for their time excellent, anatomical works, he started publishing his own plates, on the human bones (1726), on the human muscles (1734), and on the development of the human skeleton (1737). These works were, however, only a forerunner of greater achievements.

In 1745 Bernard Siegfried's duties were partially lightened by the appointment of his youngest brother as lecturer in anatomy and surgery; starting 30 August 1745 he taught only general medicine and physiology. This easier schedule enabled him to embark on the great project at which he had been working for many years, a definitive anatomical atlas. He had devised a method for accurately rendering the proportions of a human skeleton or a muscle man—which method was, however, not original; Albrecht Dürer had already used it. (See the figure "Der Zeichner des liegenden Weibes," in Dürer's *Underweysung der Messung mit dem Zirkel und Richtscheyt, in Linien, Ebnen und gantzen Corporen* [Nürnberg, 1625].) Bernard Siegfried engaged Jan Wandelaar, one of the best draftsmen and engravers available, and spent some 30,000 guilders of his own money in the preparation of the work. In 1747 his work on the human skeleton and muscles was published in thirty-five sheets; in 1748, his illustration of the gravid uterus in nine sheets; and in 1753, an atlas of the human bones, drawn separately, in thirty-four sheets. These plates, supreme examples of anatomical illustration, have never been equaled in excellence. Between 1754 and 1768, his studies on human physiology were published in eight volumes.

After his death, Bernard Siegfried's anatomical preparations were sold at auction. They were bought by the University of Leiden after his widow, Clara Magdalena du Peyrou, had declined an offer of Empress Catherine II of Russia.

Bernard Siegfried was, without doubt, the most important member of the Albinus family. Besides his anatomical work, he made several medical discoveries, among them that the Haversian canals serve to carry blood vessels. He determined the existence of the hymen, which had been a subject of controversy for centuries, in an original way (E. D. Baumann, *Medisch historische Studiën* [privately printed, 1935], 16).

BIBLIOGRAPHY

I. ORIGINAL WORKS. Writings by Bernard Siegfried are *Oratio inauguralis de anatome comparata* (Leiden, 1719); *Oratio qua in veram viam, quae ad fabricae humani corporis cognitionem inquiritur ducat* (Leiden, 1721); *Andreae Vesali invictissimi Caroli V Imperatoris Medici, Opera omnia anatomica et chirurgica* (Leiden, 1725), with H. Boerhaave, partial English trans. by B. Farrington in *Transactions of the Royal Society of South Africa*, **19** (1930), 49–78; *Index supellectilis anatomicae, quam academiae batavae, qua Leidae est, legavit vir clarissimus Jo. Jac. Rau* (Leiden, 1725); *De ossibus corporis humani libellus* (Leiden, 1726; Vienna, 1746, 1757, 1759); *Historia musculorum corporis hominis* (Leiden, 1734; 2nd ed., Frankfurt, 1784); *Jac. Douglas, bibliographiae anatomicae specimen, sive ca-*

talogus omnium pene auctorum, qui ab Hippocrate ad Harvaeum rem anatomicam ex professo, vel obiter scriptis illustrarunt (Leiden, 1734), the 2nd ed. of Douglas' book, edited by Albinus; and *Gul. Harvaeus, Exercitatio anatomica de motu cordis et sanguinis in animalibus. Cui acced. ejusdem auctoris exercitationes duae anatomicae de circulatione sanguinis ad Johannem Riolanum, filium* and *Gul. Harvaeus, Exercitationes de generatione animalium* (Leiden, 1736), both vols. reissued under the title *Opera, sive exercitatio anatomica de motu cordis et sanguinis in animalibus. Atque exercitationes duae anatomicae de circulatione sanguinis ad Johannem Riolanum, filium. Tumque exercitationes de generatione animalium* (Leiden, 1737; 3rd ed., 1753).

Also see *Dissertatio de arteriis et venis intestinorum hominis* (Leiden, 1736); *Dissertatio secunda, de sede et caussa coloris Aethiopum et caeterorum hominum* (Leiden, 1737); *Fabricius ab Aquapendente, Opera omnia anatomica et physiologica* (Leiden, 1737); *Icones ossium foetum humani, accedit osteogoniae brevis historia* (Leiden, 1737); *Explicatio tabularum anatomicarum Barth. Eustachii* (Leiden, 1744; 2nd ed., 1761); *Tabulae sceleti et musculorum corporis humani* (Leiden, 1747), translated as *Tables of the Skeleton and Muscles of the Human Body* (London, 1749; 2nd ed., 1777); *Tabulae VII uteri mulieris gravidae, cum jam parturiret mortuae* (Leiden, 1748), with Appendix (Leiden, 1751); *Tabulae ossium humanorum* (Leiden, 1753); *Academicarum annotationum libri VIII, cont. anatomica, physiologica, zoographica, phytographica, pathologica etc.* (Leiden, 1754–1768); *Tabula vasis chyliferi, cum vena azyga, arteriis intercostalibus, aliisque vicinis partibus* (Leiden, 1757); and *De sceleto humano liber* (Leiden, 1762).

II. SECONDARY LITERATURE. See the following, listed chronologically: P. Camper, *Epistola ad anatomicorum principem magnum Albinum* (Groningen, 1767); (E. Sandifort), ["Obituary B. S. Albinus"], *Natuur en geneeskundige bibliotheek*, **7** (1770), 671; "Levensbeschrijving van den Hoog Edelen Hooggeleerden Heer Bernhard Siegfried Albinus," *ibid.*, 673–682; (J. N. S. Allamand), "Éloge historique de Mr. Albinus," in *Bibliothèque des sciences et des beaux-arts*, **36** (1771), 416–465; E. Sandifort, in *Natuur- en geneeskundige bibliotheek*, **8** (1771), 707; F. Chaussier and N. P. Adelon, "Bernard Sifroy Albinus," in *Biographie universelle ancienne et moderne*, I (1843), 346–347; A. J. van der Aa, "Bernhard Siegfried Albinus," in *Biografisch woordenboek der Nederlanden*, I (n.d., Haarlem), 157–160; and H. Halbertsma, *Oratio de Albini anatomiae tractandae methodo comparata cum ea, quam nostra tempora sibi desposcunt* (Leiden–Amsterdam, 1848; 2nd ed., Leipzig, 1931).

Also see G. C. B. Suringar, "De school van Bernard Siegfried Albinus," in *Nederlandsch tijdschrift voor geneeskunde*, **2** (1867), 1–21; D. Lubach, "Bernhard Siegfried Albinus," in *Album der natuur*, **36** (1887), 1–16; and "Nog iets betreffende B. S. Albinus," *ibid.*, 190; E. D. Baumann, "Bernard Siegfried Albinus," in *Nieuw Nederlandsch biografisch woordenboek*, IV (1918), 22–24; M. Villaret and F. Montlier, "Un essai d'anatomie 'dynamique' au XVIIIe siècle; les *Tabulae sceleti et musculorum* d'Albinus," in *Bulletin de la Société française d'histoire de la médecine*, **15** (1921), 205–212; J. C. van der Klaauw, "A Letter of B. S. Albinus From Leiden to R. Nesbitt in London," in *Janus*, **35** (1931), 217–220; W. R. le Fanu, "More Letters From B. S. Albinus to Robert Nesbitt," *ibid.*, **36** (1932), 1–26; J. A. J. Barge, "Het geneeskundig onderwijs aan de Leidsche Universiteit in de 18de eeuw," in *Bijdragen tot de geschiedenis der geneeskunde*, **14** (1934), 1–12; and E. D. Baumann, *Drie eeuwen Nederlandsche Geneeskunde* (Amsterdam, 1950), 250.

PETER W. VAN DER PAS

ALBINUS, CHRISTIAAN BERNARD (*b.* Berlin, Germany, 20 March 1698/19 March 1699; *d.* Utrecht, Netherlands, 5 April 1752), *anatomy*.

Christiaan Bernard, Bernard Albinus' second son, matriculated at the University of Leiden at the age of twelve. He received doctorates in philosophy and medicine on the same day, 31 July 1722. In 1733 he was appointed extraordinary professor of anatomy and surgery at the University of Utrecht, assuming his duties on 20 September of that year. He became ordinary professor on 17 January 1724.

Christiaan is mainly known as the editor of the second English edition of William Cowper's anatomical atlas. When he was elected magistrate of Utrecht in 1747, he resigned his professorate.

BIBLIOGRAPHY

I. ORIGINAL WORKS. Christiaan Bernard's works are *Specimen inaugurale [anatomicum] exhibens novam tenuium hominis intestinorum descriptionem* (Leiden, 1722; repr. 1724); *Oratio inauguralis de anatome prodente errores in medici* (Utrecht, 1723), also published as *De anatome errores detegente in medicina* (Utrecht, 1723); and W. Cowper, *The Anatomy of Humane Bodies, With Figures Drawn After the Life by Some of the Best Masters in Europe etc. Revised and Published by C. B. Albinus* (Leiden, 1737).

II. SECONDARY LITERATURE. See A. J. van der Aa, "Christiaan Bernard Albinus," in *Biografisch woordenboek der Nederlanden*, I (n.d.), 160; and E. D. Baumann, "Christiaan Bernard Albinus," in *Nieuw Nederlandsch biografisch woordenboek*, IV (1918), 23.

PETER W. VAN DER PAS

ALBINUS, FREDERIK BERNARD (*b.* Leiden, Netherlands, 20 June 1715; *d.* Leiden, 23 May 1778), *anatomy*, *medicine*.

Frederik Bernard, Bernard Albinus' fourth and youngest son, was born six years before his father's death. Nothing appears to be known about his early years; he matriculated as a student of literature at Leiden on 15 June 1731. He studied mathematics; philosophy (physics) under W. J. 'sGravesande; botany under A. van Royen; and medicine under Boerhaave, H. D. Gaubius, and his older brother. He was promoted to doctor of philosophy and of medicine on the same day, 22 December 1740. Thereafter he settled in Amsterdam as a physician.

On 9 August 1745 Frederik was appointed lecturer in anatomy and surgery at the University of Leiden, thereby lightening the burden of his older brother Bernard Siegfried. On 6 December 1747 he became ordinary professor. After his brother's death he turned over the teaching of anatomy and surgery to E. Sandifort, and taught human physiology himself. The oration with which he opened the physiology course (30 April 1771), *De ambulatione, eaque utili, et necessaria, et jucunda*, was ridiculed in an anonymous Latin poem and was defended in a pamphlet by P. van Schelle. Frederik's best-known work is *De natura hominis libellus* (1775), a book on human physiology based on the posthumous papers of his brother and on his own research.

BIBLIOGRAPHY

I. ORIGINAL WORKS. Frederik's writings are *De meteoris ignitis* (Leiden, 1740); *Dissertatio medica inauguralis de deglutitione* (Leiden, 1740); *De amoenitatibus anatomicis* (Leiden, 1745); *De causis dissensionum inter anatomices* (Leiden, 1748); *De praestantia chirurgiae* (Leiden, 1755); *De amictus noxis* (Leiden, 1767); *De ambulatione, eaque utili, et necessaria, et jucunda* (Leiden, 1771); *Supellex anatomica Bern. Siegfr. Albini* (Leiden, 1775); and *De natura hominis libellus* (Leiden, 1775), translated into Dutch as *Beschrijving van 's menschen natuur* (Middelburg, 1780).

II. SECONDARY LITERATURE. See the following, listed chronologically: P. van Schelle, *Tegen den maker van het laffe en eerlooze schimpdicht op F. B. Albinus* (Leiden, 1771); Coquebert de Thaizy, "Frédéric Bernard Albinus," in *Nouvelle biographie universelle ancienne et moderne*, I (1843), 347; A. J. van der Aa, "Frederik Bernard Albinus," in *Biografisch woordenboek der Nederlanden*, I (n.d.), 160–162; and E. D. Baumann, "Frederik Bernard Albinus," in *Nieuw Nederlandsch biografisch woordenboek*, IV (1918), 24–25.

PETER W. VAN DER PAS

ANDRONOV, ALEKSANDR ALEKSANDROVICH (*b.* Moscow, Russia, 11 April 1901; *d.* Gorky, U.S.S.R., 31 October 1952), *theoretical physics.*

The son of an office worker, Andronov graduated from secondary school in 1918 and immediately went to work in a factory. He subsequently joined a military supply detachment, with which he traveled to the Urals. In 1920 he enrolled at the electrotechnical faculty of the Moscow Higher Technical School, transferring to Moscow University three years later. He graduated from the physical-mathematical faculty in 1925; his specialty was theoretical physics. As a student he had shown great interest in theoretical mechanics and had begun to teach mechanics and theoretical physics at the Second Moscow University (now the V. I. Lenin Pedagogical Institute) in 1924. From 1926 to 1929 he was a postgraduate student at Moscow State University.

L. I. Mandelshtam played a major role in Andronov's development as a physicist. Under his guidance Andronov did his first scientific research, which was related to the theory of the scattering of light by a fluctuating liquid surface, and undertook a series of studies on perturbation theory. Even after Andronov had achieved recognition and a school had evolved around him, he continued to work closely with Mandelshtam.

Although Andronov was trained as a theoretical physicist, his major field of activity was rather removed from what theoretical physicists usually studied. Attracted by atomic physics, he began to study statistical quantum physics in the late 1920's. By 1930 his theoretical work was concentrated on the generation of oscillations, a subject to which great importance had been given by the development of the electron tube. Almost all of his subsequent research was a development of the ideas presented in his dissertation, which dealt with extremal Poincaré cycles and oscillatory theory.

In 1931 Andronov moved to Gorky and worked at the university there for the rest of his life. He was also involved in the activities of the Institute of Automation and Telemechanics of the Academy of Sciences of the U.S.S.R. and became an active member of the Academy in 1946.

Andronov's basic works concerned oscillation theory and automatic control theory. They laid the foundations for the subsequent mathematical treatment of oscillatory processes in nonlinear automatic systems that were based on the work of Poincaré and on Lyapunov's fundamental research on the theory of stability of motion. Especially fruitful were the ideas that Andronov developed in 1928 on the ways of examining undamped oscillations in nonlinear automatic systems. Andronov called such oscillatory processes autooscillations, a term that has become generally accepted, and his works led to the development of the basic methods of the theory of nonlinear oscillations. After Andronov related nonlinear oscillations to the qualitative theory of differential equations, there was an intensive development both of the mathematical methods and of the physical applications of the theory of nonlinear oscillations. Andronov's book on the theory of oscillations (1937), written with S. E. Khaykin, has become the basic work in the Soviet Union on nonlinear perturbation theory and has been used extensively in training specialists in this branch of radiophysics.

Andronov's school contributed substantially to the development of the theory of automatic control. His works in this area are the direct continuation and completion of Vyshnegradsky's *O regulyatorakh pryamogo deystvia* ("On Regulators of Direct Action," 1876). He also wrote on the works of Maxwell, Vyshnegradsky, and A. Stodola in the regulation of machines.

Many of Andronov's students continued his research in the theory of oscillations, the dynamics of machines, and the qualitative theory of differential equations, and developed his scientific ideas, rendering them accessible to a wide circle of physicists and mathematicians.

BIBLIOGRAPHY

I. ORIGINAL WORKS. Andronov's writings include *Teoria kolebany* ("Theory of Oscillations"; Moscow–Leningrad, 1937), written with S. E. Khaykin; and *Sobranie trudov* ("Collected Works"; Moscow, 1956).

II. SECONDARY LITERATURE. See the obituary in *Vestnik Akademii nauk SSSR*, **22**, no. 12 (1952), 81; and *Pamyati A. A. Andronova* ("Recollections of A. A. Andronov"; Moscow, 1955), a collection that includes an annotated list of Andronov's works.

A. T. GRIGORIAN

ARFVEDSON, JOHANN AUGUST (*b.* Skagerholms-Bruk, Skaraborgs-Län, Sweden, 12 January 1792; *d.* Hedensö, Sweden, 28 October 1841), *chemistry.*

Arfvedson was educated at home until 1806, when he enrolled in the mining course at Uppsala. He served as a secretary for the Royal Bureau of Mines at Stockholm; and early in 1817 he entered Berzelius' laboratory, where he began studies on the chemical analysis of minerals. Before the end of that year Arfvedson had completed research on the oxides of manganese, determining the composition of manganous oxide (MnO) and manganosic oxide (Mn_3O_4). Berzelius recognized the outstanding abilities of his student and allowed him to work on the analysis of petalite, a lithium aluminum silicate discovered at Utö, Sweden, in 1800 by José Bonifácio de Andrada e Silva, the Brazilian scientist and statesman.

Arfvedson isolated a new alkali from petalite, one of lower equivalent weight than any of the alkalies then known. He and Berzelius named it "lithia" (lithium oxide), since it had been discovered in the mineral kingdom. Arfvedson prepared many of the compounds of the alkali but never succeeded in decomposing it. In 1818 Humphry Davy obtained a minute amount of lithium by the electrolysis of lithium carbonate. In 1822 W. T. Brande obtained larger amounts and named the new metal "lithium."

In 1819 Arfvedson purchased an estate at Hedensö and pursued his investigations in mineral analysis there. In an 1822 study on the composition of chrysoberyl he isolated another new base (beryllium hydroxide) but mistook it for silica. He attempted to obtain uranium metal by the reduction of uranosic oxide. He did not succeed but did isolate a new oxide, uranous oxide (UO_2).

Arfvedson's business interests in several industrial plants left him little time for chemical research after 1822. In 1821 he became a member of the Royal Swedish Academy of Sciences, which in the last year of his life awarded him its gold medal for the discovery of lithium.

BIBLIOGRAPHY

I. ORIGINAL WORKS. Arfvedson's report of the discovery of lithia appears in "Undersökning af några vid Utö—jemmalms-brott träffade fossilier, och ett eget deri funnet eldfast alkali," in *Afhandligar i fysik, kemi, och mineralogi*, **6** (1818), 145–172; and "Analyses de quelques minéraux de la mine d'Utò en Suède, dans lesquels on a trouvé un nouvel alcali fixe," in *Annales de chimie*, 2nd ser., **10** (1819), 82–107.

An account of the discovery of lithium, with a bibliography of Arfvedson's published works, is in Mary Elvira Weeks, *Discovery of the Elements*, 6th ed. (Easton, Pa.,

1956), 484–488, 495–503. Bibliographies are in Poggendorff, I, 59; and Royal Society *Catalogue of Scientific Papers*, I, p. 89.

II. SECONDARY LITERATURE. There is an obituary notice in *Kungliga Svenska vetenskapsakademiens handlingar* (1841), 249–255, and a biographical sketch by H. G. Söderbaum in *Svenskt biografiskt lexikon*, II (Stockholm, 1920), 165–166. Also see J. R. Partington, *A History of Chemistry*, IV (London, 1964), 152.

ALBERT B. COSTA

ARTSIMOVICH, LEV ANDREEVICH (*b.* Moscow, Russia, 25 February 1909; *d.* Moscow, U.S.S.R., 1 March 1973), *physics*.

The son of a professor of statistics, Artsimovich graduated in 1928 from the Belorussian State University at Minsk and became laboratory assistant, then scientific associate, at the Leningrad Physical-Technical Institute, directed by A. F. Joffe. In his first scientific research, which he did with A. I. Alikhanov, he investigated the complete internal reflection of X rays from thin layers (1933). In 1934–1935 he studied the properties of the recently discovered neutron, work carried out by a group of physicists headed by I. V. Kurchatov. The research that Artsimovich did with Kurchatov, G. D. Latyshev, and V. A. Khramov (1935) was the first to show clearly that the cross section of the capture of slow neutrons by protons is comparatively very large. In his subsequent work (1936) with Alikhanov and A. I. Alikhanyan, R. S. Shankland's conclusions regarding the possibility of the violation of the laws of conservation of energy and momentum in the Compton effect were experimentally tested. The work, which was done in record time, showed the strict correctness of the laws of conservation during the electron-positron annihilation. On the strength of this evidence, Shankland's conclusions were refuted.

Artsimovich's further experimental research was devoted to the processes of interaction of fast electrons with matter, a problem that in the mid-1930's had been very little studied. The experimental data on the bremsstrahlung and angular distribution of electrons were divided into two categories by means of data derived from theoretical calculation. Through Artsimovich's research, extensive factual material was obtained regarding the dependence of the intensity of the bremsstrahlung and the complete loss of energy on the energy of the primary electrons. Having carefully analyzed this material, Artsimovich showed (in works published with Khramov [1938] and in an article published with I. I. Perrimond [1946]) that the data of contemporary quantum mechanical theory on the passage of fast electrons through a substance agree, within the limits of accuracy of the experiments, with the results of experimental research.

During World War II, Artsimovich studied electronic optics, particularly the theory of chromatic aberration of electron-optical systems (1944); and his experimental and theoretical research on electron-optical transformers had important practical results. In 1945 Artsimovich and I. Y. Pomeranchuk made a theoretical study of radiation loss in the betatron. This work made it possible to compute the maximum energy that could be imparted to electrons by acceleration in a given betatron (1946).

After the war Artsimovich solved several important problems of scientific technology. Under his direction a way was found to construct a safe plant for the electromagnetic separation of isotopes. While working in this area Artsimovich produced a detailed analysis of the conditions of nonaberrational focusing of wide-angled ion beams (for currents of the order of amperes) in axial-symmetric magnetic fields. His proposed construction of an ion optics source (1957) has been applied in contemporary systems, thus ensuring the possibility of obtaining pure, stable isotopes.

From the 1950's Artsimovich was increasingly attracted to the difficult problem of creating a controlled thermonuclear reactor, a project that required the organization of multifaceted experimental and theoretical research. Under Artsimovich's direction, a group of physicists began with the study of powerful pulsed discharges in rarefied deuterium. They succeeded in obtaining, although only for very short intervals, highly ionized plasma heated to several million degrees. In 1952 this research led to the discovery that a powerful pulsed discharge in deuterium at low pressure is the source of neutrons and "hard" X rays. They also discovered paramagnetic properties of gas-discharge plasma under pressure from the longitudinal magnetic field. These investigations showed that neutrons discovered in these experiments are released not by thermonuclear processes but by a specific acceleration process (1958). At the Second International Conference on the Peaceful Uses of Atomic Energy at Geneva (1958), Artsimovich gave a detailed report on Soviet studies of methods of obtaining controlled thermonuclear reactions.

Under Artsimovich's directorship of the scientific section of the I. V. Kurchatov Institute of Atomic Energy, large and complex plants for continuation of this research were built. At plants of the "Tokamak" (an acronym of "toroid camera with magnetic field") type, Artsimovich and his students achieved important results; in particular the lifetime of high-temperature plasma was sharply increased. This research culminated in a thermonuclear reaction, which strengthened the Soviet position in this field of physics.

In 1930 Artsimovich began teaching at the Leningrad Polytechnical Institute, and later he lectured at Leningrad University. After the war, he taught atomic and nuclear physics at the Moscow Physical-Engineering Institute and at Moscow State University. In 1946 Artsimovich became corresponding member, and in 1953 active member, of the Academy of Sciences of the U.S.S.R.; from 1957 he was academician-secretary of the section of general physics and astronomy. He was also a member of its Presidium and chairman of the National Committee of Soviet Physicists. Artsimovich was awarded the State Prize in 1953 and the Lenin Prize in 1958; was awarded the title Hero of Socialist Labor; and received six orders. He was honorary member of many foreign academies, and as a member of the Standing International Committee of the Pugwash Movement he made a great contribution to the fight of scientists for peace and disarmament.

BIBLIOGRAPHY

I. ORIGINAL WORKS. Artsimovich's writings include *Upravlyaemye termoyadernye reaktsii* ("Controlled Thermonuclear Reactions"; Moscow, 1961); *Shturm termoyadernogo sinteza* ("Assault on the Thermonuclear Synthesis"; Moscow, 1962); *Eksperimentalnye issledovania na ustanovkakh Tokamak* ("Experimental Research at 'Tokamak' Plants"; Moscow, 1968), written with G. A. Bobrovsky, E. P. Gorbunov, *et al.*; *Elementarnaya fizika plazmy* ("Elementary Plasma Physics"), 3rd ed. (Moscow, 1969), also in English trans. (New York, 1965); *Zamknutye plazmennye konfiguratsii* ("Closed Plasma Configurations"; Moscow, 1969); and *Dvizhenie zaryazhennykh chastits v elektricheskom i magnitom polyakh* ("Motion of Charged Particles in Electric and Magnetic Fields"; Moscow, 1972), written with S. Y. Lukyanov.

II. SECONDARY LITERATURE. There is a biography in *Great Soviet Encyclopedia*, II (1976), 384. See also A. I. Alikhanov, "Lev Andreevich Artsimovich (k 50-letiyu so dnya rozhdenia)" (" . . . Artsimovich [for His Fiftieth Birthday]"), in *Uspekhi fizicheskikh nauk*, **67**, no. 2 (1959), 367–369, with portrait and selected bibliography.

J. G. DORFMAN

AL-BAGHDĀDĪ, ABŪ MANṢŪR ᶜABD AL-QĀHIR IBN ṬĀHIR IBN MUḤAMMAD IBN ᶜABDALLAH, AL-TAMĪMĪ, AL-SHAFIᶜĪ (*b.* Baghdad; *d.* 1037), *arithmetic.*

The last two names indicate the tribe from which Abū Manṣūr was descended and the religious school to which he belonged. Born and raised in Baghdad, he left with his father for Nīshāpūr (or Nīsābūr), taking with him great wealth that he spent on scholars and scholarship. Riots broke out in Nīshāpūr, and he moved to the quieter town of Asfirāyīn. His departure was considered a great loss to Nīshāpūr. In his new home, he continued to pursue learning and to propagate it. He is reported to have lectured for years in the mosque, on several subjects, never accepting payment. Although he was one of the great theologians of his age, and many works are attributed to him, none has been studied scientifically. We are concerned here with two works on arithmetic.

The first is a small book on mensuration, *Kitāb fiʾl-misāḥa*, which gives the units of length, area, and volume and ordinary mensurational rules.

The second, *al-Takmila fiʾl-ḥisāb*, is longer and far more important. In the introduction Abū Mansūr notes that earlier works are either too brief to be of great use or are concerned with only one chapter (system) of arithmetic. In his work he therefore seeks to explain all the "kinds" of arithmetic in use.

The Islamic world knew three arithmetical systems: finger reckoning, the sexagesimal scale, and Indian arithmetic. Not long after the last was introduced, Greek mathematical writings became accessible and the works of Euclid, Nicomachus, and others were made known. All these elements underwent a slow unification. Abū Mansūr presents them at an intermediary stage in which each system still had its characteristics preserved but was already enriched by concepts or schemes from other systems.

Abū Manṣūr conceived of seven systems. The first two were the Indian arithmetic of integers and that of fractions. The third was the sexagesimal scale, expressed in Hindu numerals and treated in the Indian way.

The fourth was finger reckoning. Two works on Arabic finger reckoning before the time of Abū Mansūr are extant: the arithmetic of Abuʾl-Wafāʾ and that of al-Karajī (known also as al-Karkhī). Both works devote the most space to explaining a cumbersome and complicated fractional system that lacks the idea of the unrestricted common fraction. This system does not appear in the work of Abū Manṣūr, who seems to prefer the Indian system. His finger reckoning is confined to concepts lacking in Indian arithmetic, such as shortcuts, and to topics taken from Greek mathematics, such as the summation of finite series. He provides rules for the summation of the general arithmetic, and some special geometric progressions, as well as the sequences r^2, r^3, r^4 $(2r)^2$, $(2r-1)^2$, and polygonal numbers. These rules are expressed in words and assume that the number of terms in each case is ten, a Babylonian practice presented in the works of Diophantus.

Abū Manṣūr's next two systems are the arithmetic of irrational numbers and the properties of numbers. In the first of these, Euclid's rules of the irrationals in Book X of the *Elements* are given on a numerical basis. In the second the Pythagorean theory of numbers is presented with an improvement upon Nicomachus: To determine whether n is prime, test it for divisibility by primes $\leq \sqrt{n}$. Perfect numbers, such as 6, 28, 496, and 8,128, end in 6 or 8; but there is no perfect number between 10^5 and 10^6. The first odd abundant number is 945.

This part of Abū Manṣūr's work is ten chapters long, but some folios of the manuscript are missing; only the first three chapters and a few lines of the last are extant. The latter contain an attempt to divide a cube into several cubes by using the relation $3^3 + 4^3 + 5^3 = 6^3$.

The last of Abū Manṣūr's seven systems, business arithmetic, begins with business problems and ends with two chapters on curiosities that would find a place in any modern book on recreational problems or the modulo principle. One example is given here because it is found in Greek, Indian, and Chinese sources: Your partner thinks of a number not greater than 105. He casts out fives and is left with a; he casts out sevens and is left with b; he casts out threes and is left with c. Calculate $21a + 15b + 70c$; cast out 105's, and the residue is the number. The explanation shows that the author was quite familiar with the modulo concept.

Abū Mansūr's work also seems to solve a problem encountered by historians of medieval mathematics. Latin arithmeticians of the early Renaissance were divided into abacists and algorists. The exact significance of each name was unknown. It has recently been learned that Hindu-Arabic arithmetic required the use of the abacus, thus abacists were those who used the Hindu-Arabic system, and algorists must have adhered to the older system. This agrees with the fact that a work by Prosdocimo de Beldamandi containing an outspoken denunciation of the abacus is called *Algorithmus*. "Algorist" and *Algorithmus* come from the name of al-Khwārizmī, the first Muslim to write on Indian arithmetic. His work in Arabic is lost, but we have the *Algoritmi de numero indorum*, believed to be a translation of it.

But why should those who did not use the abacus be called algorists? This question can be answered as follows: Arabic biographers attribute to al-Khwārizmī a book called *Kitāb al-jamʿ waʾl-tafrīq*, now lost. It has been commonly accepted that this was the Arabic title of al-Khwārizmī's work on Indian arithmetic. Abū Manṣūr, however, refers to this book in his *al-Takmila*, and once he quotes methods from it. These methods follow typical finger-reckoning schemes, which indicates that this book of al-Khwārizmī's was of the finger-reckoning type. It seems that those who followed this book of al-Khwārizmī's were called algorists and those who followed his work on Indian arithmetic were the abacists.

A. S. SAIDAN

BARGER, GEORGE (*b.* Manchester, England, 4 April 1878; *d.* Aeschi, Switzerland, 6 January 1939), *organic chemistry.*

Barger was the eldest son of Gerrit Barger, a Dutch engineer, and Eleanor Higginbotham, an Englishwoman. After receiving his initial schooling at Utrecht, he studied at University College, London, and King's College, Cambridge, from which he graduated in 1901 with first-class honors in both chemistry and botany. In 1903 he accepted a position as chemist at the Wellcome Physiological Research Laboratories. Barger taught at Goldsmith's College (1909–1913) and at the Royal Holloway College (1913–1914). He spent the war years on the staff of the Medical Research Committee, and in 1919 he was appointed the first professor of chemistry in relation to medicine at Edinburgh. He held this post until 1937, when he accepted the regius professorship of chemistry at Glasgow.

Barger's scientific work centered on the isolation, structure determination, synthesis, and pharmacological importance of alkaloids and of naturally occurring amino acid degradation products. His first major research project was an analysis of the physiologically active components of ergot. In 1906 he and F. H. Carr announced the first isolation of an active ergot alkaloid, which they named "ergotoxine" (later shown to be a mixture of three alkaloids). Subsequently, Barger and his associates isolated and identified tyramine, histamine, and other amino acid derivatives from ergot extracts.

In 1910 he and the physiologist Henry Dale published a detailed investigation into the pharmacological activity of these and other amines, in a partially successful attempt to correlate activity with molecular structure. They termed these substances "sympathomimetic" amines, since their action resembled that of the sympathetic nervous system. Barger also played an essential role, principally in an advisory capacity, in C. R. Harington's synthesis of the thyroid hormone thyroxine (1927). Harington persuaded his former teacher to agree to joint publication of the final stage of the work.

Although his work often had important biochemical and pharmacological significance, Barger always considered himself to be fundamentally an organic chemist. A skillful experimenter, cautious toward hypotheses, he maintained a strictly mechanistic philosophy.

BIBLIOGRAPHY

I. Original Works. Barger wrote four books: *The Simpler Natural Bases* (London, 1914); *Some Applications of Organic Chemistry to Biology and Medicine: The George Fisher Baker Non-Resident Lectureship in Chemistry at Cornell University* [1927–1928] (New York, 1930); *Ergot and Ergotism: Based on the Dohme Lectures Delivered in Johns Hopkins University, Baltimore* [1928] (London, 1931); and *Organic Chemistry for Medical Students* (London, 1932; 2nd ed., 1936). In addition, Barger was author or coauthor of well over 100 scientific publications. The majority of these appeared in *Journal of the Chemical Society*; papers dealing with pharmacological aspects frequently appeared in *Biochemical Journal* or *Journal of Physiology*. A complete list of Barger's publications is in Dale's obituary notice (see below).

II. Secondary Literature. The best sources for information concerning Barger's life are biographical memoirs written by two of his collaborators: H. H. Dale, in *Obituary Notices of Fellows of the Royal Society of London*, **3** (1940), 63–85; and C. R. Harington, in *Journal of the Chemical Society* (1939), 715–721, repr., with minor changes, in A. Findlay and W. H. Mills, eds., *British Chemists* (London, 1947), 419–431. Poggendorff (VIIb, 241) lists several obituaries. The best summaries of Barger's scientific work are his own monographic treatments, which deal with his fields of specialization in a thorough and impartial manner. The two biographical memoirs also contain discussions of Barger's scientific career from the point of view of his associates. For a historical discussion of the significance of Barger and Dale's 1910 paper on the sympathomimetic amines, see B. Holmstedt and G. Liljestrand, *Readings in Pharmacology* (New York, 1963), 169–201.

Alan J. Rocke

BÉCHAMP, PIERRE JACQUES ANTOINE (*b.* Bassing, Moselle, France, 16 October 1816; *d.* Paris, France, 15 April 1908), *chemistry, biochemistry.*

Béchamp, the son of a miller, left France at an early age to live in Bucharest, where he studied pharmacy, probably at St. Sava College. In his late teens he moved to Strasbourg, where he was an apprentice in a pharmacy and obtained the title of pharmacist in 1843. He quickly abandoned pharmacy, however, to resume his education. Early in 1851 he was named *professeur agrégé* by a jury including Louis Pasteur. Béchamp taught at the Faculty of Science of Strasbourg (1853–1854), where he succeeded Pasteur, and then at the Strasbourg School of Pharmacy (1854–1856). In 1853 he earned his doctorate in the physical sciences and in 1856 he received the doctorate in medicine with an important thesis on albuminoid substances.

From 1856 until 1876 Béchamp taught medical chemistry at the Faculty of Medicine of Montpellier. He resigned his post there to become dean of the Free (Catholic) Faculty of Medicine in Lille. He retired from the latter post in 1886 amid bitter controversy. With his son, Béchamp resumed the pharmaceutical trade in Le Havre. The accidental death of his son, no mean chemist himself, led Béchamp to move to Paris, where the generous hospitality of Charles Friedel provided him with a small laboratory. There he carried out experiments until 1899.

Béchamp made discoveries in numerous fields. His doctoral thesis of 1856 led to a voluminous treatise (1884). Through a skillful, systematic use of the optical activity of albuminoid substances, Béchamp was able to distinguish a large number of complex compounds that his predecessors, relying on more standard analytical methods, had failed to discover. He also developed a cheap industrial process to produce aniline (1852) and thereby greatly contributed to the emergence of the synthetic dye industry. For this particular work the Société Industrielle de Mulhouse awarded Béchamp the Daniel Dollfus Prize (1864) jointly with W. H. Perkin, A. W. von Hofmann, and E. Verguin. He also identified the parasitic nature of two silkworm diseases and, in this context, anticipated Pasteur's results. The two became bitter rivals over that matter.

Béchamp's theory of life constituted the main thrust of his activity, however, and also brought about a never-ending string of disputes, with Pasteur in particular. He accepted neither spontaneous generation nor the parasitic theory of disease and of fermentation. He claimed instead that all life was derived from small, subcellular "molecular granulations," otherwise known as *microzymas*. Béchamp, mistakenly, was fond of comparing his project with that of Lavoisier and, indeed, the *microzymas* did play a role analogous to chemical elements. They differed from the elements, however, in that they were considered as the ultimate foundations of living structures, while for Lavoisier the elements existed only as experimental constructs.

When living organisms die, Béchamp argued, they revert to inert chemical substances and *microzymas*, the latter being essentially eternal. The *microzymas* also act as the organizing principle of living things. The way in which organization proceeds, however, depends as much on the chemical substances actually present and on the circumstances as on the nature of the *microzymas*. In other words, *microzymas* have no specificity of action, as do microbes in Pasteur's theory; on the contrary, they are quite polyvalent.

The strength of Béchamp's theory rested on its ability to use a single principle to explain a wide range of phenomena; with it he could account for such contradictory results as those adduced by Pasteur and Pouchet in the course of their famous controversy on spontaneous generation. But this was also the weakness of Béchamp's theory: it explained too much without lending itself to experimental testing. Its empirical foundation rested on two general statements: the microscope reveals molecular granulations; and whenever (and apparently only when) molecular granulations appear, life processes occur.

Of course the claim to have discovered the basic, material site of life, couched in scientific terms and bolstered by the scientific method, had direct import for extrascientific debates. It is no surprise, therefore, that Béchamp's theory was upheld or attacked by Catholics, evolutionists, and materialists. But being at the center of too many controversial issues is not always the best way to pursue a successful career, especially if one is confined to provincial teaching positions; and Béchamp ended his life in nearly complete isolation. He died ignored by most and eulogized as a scientific martyr by a few.

BIBLIOGRAPHY

I. ORIGINAL WORKS. There are 245 papers by Béchamp listed in the Royal Society *Catalogue of Scientific Papers*, I, 226–228; VII, 113–116; IX, 156–157; XII, 60; and XIII, 378–380. A bibliography of his works is appended to the unsigned obituary notice in *Moniteur scientifique* and that by Delassus (see below).

II. SECONDARY LITERATURE. The literature on Béchamp is not scarce, but little is dispassionate. Most articles or texts concerning him are extravagant in their praises or in their criticisms; as such they must be read with caution. They include the unsigned "Antoine Béchamp," in *Revue des questions scientifiques* (Brussels), 3rd ser., **13** (1908), 345–347; "Antoine Béchamp," in *Moniteur scientifique* (Quesneville), 4th ser., **22** (Dec. 1908), 790–800, with comprehensive bibliography of his works; *Inauguration du monument du Professeur Béchamp à Bassing (Moselle) le 18 septembre 1927* (Metz, 1927); and "La théorie de Béchamp s'oppose-t-elle à celle de Pasteur?," in *Est-Matin* (Mulhouse, 8 Nov. 1947); P. Bachoffner, "Considérations sur la vie et l'oeuvre du Professeur Béchamp de l'École de pharmacie de Strasbourg," in *Bulletin de la Société de pharmacie de Strasbourg*, **3**, no. 1 (1960), 25–30; J. Bucquoy, "Décès de M. Béchamp, correspondant national," in *Bulletin de l'Académie de médecine*, 3rd ser., **59** (1908), 520–521; Philippe Decourt, "Béchamp et Pasteur—Introduction," in *Archives internationales Claude Bernard*, no. 1 (Oct. 1971), 39–42; "Sur une histoire peu connue: La découverte des maladies microbiennes. Béchamp et Pasteur," *ibid.*, no. 2 (1972), 23–131; and "La justice et la vérité," *ibid.*, no. 3 (1973), 27–31; A. Delassus, "M. le Professeur Antoine Béchamp," in *Journal des sciences médicales de Lille*, **31**, no. 1 (1908), 444–448; Hector Grasset, *L'oeuvre de Béchamp* (Paris, 1911; 2nd ed., 1913); E. Douglas Hume, *Béchamp or Pasteur? A Lost Chapter in the History of Biology* (Chicago, 1923); M. Javillier, "Pierre Jacques Antoine Béchamp," in *Bulletin des sciences pharmacologiques*, **15** (1908), 281–284; L. Ménard, "Un émule de Pasteur: Béchamp," in *Cosmos*, n.s. **58**, no. 1215 (9 May 1908), 509–511; P. Pagès, "Antoine Béchamp. Sa vie, son oeuvre," in *Monspeliensis Hippocrates*, **2**, no. 3 (Mar. 1959), 13–29; Robert Pearson, *Pasteur Plagiarist, Impostor! The Germ Theory Exploded* (Denver, Colo., n.d. [1942]); Aurore Valérie (pseudonym of Madeleine Renault), *Béchamp et l'évolution européenne* (Paris, 1958); and *De Béchamp à Lazzaro Spallanzani. Essai historique des phénomènes d'oxydation* (Paris, 1963); and Martial Villemin, "Un savant lorrain méconnu: Antoine Béchamp (1816–1908) adversaire de Pasteur," in *Académie et société lorraines des sciences. Bulletin trimestriel*, **11** (1972), 276–284.

JEAN-CLAUDE GUÉDON

BEHREND, ANTON FRIEDRICH ROBERT (*b.* Harburg [now part of Hamburg], Germany, 17 December 1856; *d.* Hannover, Germany, 15 September 1926), *chemistry.*

Behrend was a law student at Freiburg im Breisgau (1876–1877) before enrolling at Leipzig (1877–1881) as a chemistry student. After receiving the Ph.D. he served as assistant in the physical chemistry institute at Leipzig until 1887. He became *Privatdozent* in chemistry in 1885 and professor of chemistry four years later. In 1895 Behrend moved to the Technische Hochschule in Hannover, where he was professor of organic chemistry (1897–1925) and of physical chemistry (1897–1911). He received gold medals at the expositions in Chicago (1893) and St. Louis (1904) and an honorary degree from the Technische Hochschule of Danzig (1924). He died during the 1926 typhus epidemic in Hannover.

Behrend was best known for his research in synthetic organic chemistry, especially as it related to biologically important materials. He provided the first synthesis of uric acid (1888) to reveal that it is a purine derivative. He condensed urea with ethyl acetoacetate to form β-uramidocrotonic ester. By means of a series of steps he converted this compound into isodialuric acid, which he condensed with urea to form uric acid. In 1882 Emil Fischer had ventured structural formulas for uric acid and other purines. Behrend's synthesis helped to confirm Fischer's formulas for the purine group. Behrend also synthesized isobarbituric acid and developed a general method for preparing 5-aminouracils.

Behrend made several important contributions to carbohydrate chemistry. In 1904 he established the existence of two forms of d-glucose (α- and β-glucose). He developed a method for preparing the β-form (1910) and investigated the isomeric phenylhydrazones of glucose (1910). These researches provided proof of a ring structure in d-glucose.

BIBLIOGRAPHY

I. ORIGINAL WORKS. Important papers by Behrend include "Über synthetische Versuche in der Harnsäurereihe," in *Berichte der Deutschen chemischen Gesellschaft*, **21** (1888), 999–1001, written with Oscar Roosen; "Synthese der Harnsäure," in *Justus Liebigs Annalen der Chemie*, **251** (1889), 235–236, written with O. Roosen; "Über die Birotation der Glykose," *ibid.*, **331** (1904), 359–382, written with Paul Roth; and "Über Diphenylhydrazone der Glykose," *ibid.*, **377** (1910), 189–220, written with Willy Reinsberg.

II. SECONDARY LITERATURE. There is a biographical sketch of Behrend and a bibliography of published papers by A. Skita in *Berichte der Deutschen chemischen Gesellschaft*, **59A** (1926), 158–164. His works are listed in Poggendorff, IV, 90–91; V, 84–85; and VI, 162. There is a biographical article by Georg Lockemann in *Neue deutsche Biographie*, II (Berlin, 1955), 11.

ALBERT B. COSTA

BEIJERINCK, MARTINUS WILLEM (*b.* Amsterdam, Netherlands, 16 March 1851; *d.* Gorssel, Netherlands, 1 January 1931), *microbiology, botany*.

Beijerinck was the youngest of four children of Derk Beijerinck, a tobacco dealer, and Jeannette Henriëtte van Slogteren. After going bankrupt when Martinus was two, the father became a railway clerk, which necessitated moving his family several times. When Beijerinck graduated from secondary school at Haarlem in 1869, he had decided upon botany as a career but could not afford a university education. With financial assistance from an uncle, he settled upon a three-year course at the Polytechnical School in Delft. He received a diploma in chemical engineering in July 1872, and in October enrolled in the doctoral program in botany at the University of Leiden. Beijerinck supported himself by teaching at the Agricultural School in Warffum and later at a secondary school in Utrecht. In 1876 he was appointed to the Agricultural School in Wageningen, where he taught and did botanical research. A year later he received the doctorate for his dissertation on the morphology of plant galls. Beijerinck was elected to the Royal Netherlands Academy of Sciences in 1884. In the same year he accepted a position as bacteriologist at the Dutch Yeast and Spirit Factory in Delft, although he had no experience in that field. In 1895 he became professor of microbiology at the Polytechnical School in Delft, where he founded the Laboratory for Microbiology. Beijerinck, who never married, remained at Delft for the next twenty-six years, leading a solitary life devoted to scientific pursuits.

Beijerinck received many honors, including the Order of the Dutch Lion (1903) and the Leeuwenhoek Medal of the Royal Netherlands Academy of Sciences (1905), as well as several awards from abroad, and was a member of several foreign scientific and medical societies. On his seventieth birthday in 1921, Beijerinck was presented with five volumes of his scientific papers, a publication financed through contributions from friends, former students, and the Dutch government. He reluctantly retired a few months later to the town of Gorssel, in eastern Holland, where he lived with his two sisters. He died of cancer at the age of seventy-nine.

Beijerinck wrote over 140 papers on botany, microbiology, chemistry, and genetics. His earliest research was on gall formation. A paper of 1882, in which he described and illustrated the biology of gall wasps and the development and structure of galls, still serves as a basic reference in cecidiology. In a paper of 1896 he compared gall formation with normal ontogeny and made the precocious suggestion that both processes are controlled by enzymes. Thus led to the subject of ontogeny, he concluded in 1917 that the development of higher plants and animals is determined by a series of "growth enzymes" that become active in fixed succession. Some of the many other botanical problems that he investigated were phyllotaxy, regeneration, and the genesis of adventitious organs.

Beijerinck's first major achievement in microbiology was the cultivation and isolation in 1888 of *Rhizobium leguminosarum*, a bacillus that fixes free nitrogen and causes the formation of nodules on the roots of Leguminosae. In 1889 he published a paper on auxanography, a method for culturing microorganisms that is based on the diffusion of nutrients through a solid gel. He also investigated luminous bacteria; obtained the first pure cultures of green algae, Zoochlorellae, and the gonidia of lichens; and made thorough studies of several yeasts and of the butyric acid and butyl alcohol bacteria.

One of Beijerinck's greatest accomplishments was the development, simultaneously with Winogradsky, of the technique of enrichment culture. Beijerinck had observed that most microorganisms occur in most natural materials, but in numbers too small to be studied. By transferring these materials to an artificial medium adapted to the specific nutritional requirements of the microorganism under study, he could accumulate the microorganism to the extent that it could be isolated in pure culture. Using enrichment cultures, Beijerinck was able to isolate numerous highly specialized microorganisms, many for the first time: sulfate-reducing bacteria, urea bacteria, oligonitrophilous microorganisms, denitrifying bacteria, lactic and acetic acid bacteria. Of note is his characterization of a new group of nitrogen-fixing bacteria, *Azotobacter*, which Winogradsky had previously isolated but had failed to recognize as nitrogen-fixing. In addition Beijerinck named a new genus, *Aerobacter*, of which he distinguished four different species, and also wrote several papers on microbial variation.

Intermittently from 1885 to 1900 Beijerinck did research on tobacco mosaic disease. This disease, which today is known to be caused by a virus, at that period was assumed, like all infectious diseases, to be microbial in origin. By 1898, however, Beijerinck had been led by several observations to conclude that he was dealing with a unique type of pathogen. First, he found that sap from plants infected with tobacco mosaic disease did not lose infectivity after passage through a filter impervious to microorganisms. Furthermore, the agent could be precipitated by alcohol, a property not normally associated with living organisms. Second, he observed that a filtered extract from infected plants could diffuse in a solid agar medium. To Beijerinck this meant that the agent had to be "fluid" or nonparticulate, since the capacity to diffuse was then believed to be a means of distinguishing molecular substances from larger, supposedly nondiffusing particles and cells. Third, he noted that the pathogen was unable to reproduce outside the host and seemed to multiply only in parts of the plant undergoing rapid cell division. Reluctant to accept the idea of an actively self-reproducing molecule, he suggested that replication might occur passively by incorporation of the pathogen into the reproductive machinery of the host cell. On the basis of these observations, Beijerinck concluded that tobacco mosaic disease was caused by a *contagium vivum fluidum*, a term coined to convey his concept of a living infectious agent in fluid (noncellular) form. This was a revolutionary idea at a time when life, as evidenced by reproduction, and cellularity were thought to be inextricably interconnected.

Both Ivanovsky, a Russian botanist, and Beijerinck have been credited with the "discovery" of the virus. There is no doubt that in 1892 Ivanovsky was first to report the filterability of an infectious agent now recognized to be a virus, yet he continued to believe that it had to be a type of bacterium. Beijerinck, in contrast, recognized the causal agent of tobacco mosaic disease as a completely new type of infectious agent, and suggested that other diseases might be caused by similar entities. His concept of the *contagium vivum fluidum* was virtually forgotten until 1935, when Stanley's crystallization of the tobacco mosaic virus revived interest in Beijerinck's self-reproducing, noncellular, infectious fluid.

Taciturn and somewhat acerbic, Beijerinck was nonetheless respected for the breadth of his scientific knowledge, his exceptional observational powers, and the originality of his research. His insatiable scientific curiosity often induced him to abandon a research problem once he had answered the major questions, leaving the exploration of finer details to others. Perhaps Beijerinck's most lasting achievement was to establish, with Winogradsky, the importance of microorganisms in the cycles of matter, and to demonstrate how each is specialized to perform a particular kind of chemical transformation.

BIBLIOGRAPHY

I. Original Works. Beijerinck's scientific papers are reproduced in *Verzamelde geschriften van M. W. Beijerinck*, G. van Iterson, Jr., L. E. den Dooren de Jong, and A. J. Kluyver, eds., 5 vols. (The Hague, 1921). The papers are arranged chronologically by date of original publication and appear in the language in which they were first published: Dutch, German, English, or French. Vol. VI (The Hague, 1940) contains papers by Beijerinck that appeared after 1921, indexes to all 6 vols., and biographical material.

II. Secondary Literature. *Verzamelde geschriften*, VI, contains three long biographical articles in English: L. E. den Dooren de Jong, "Beijerinck, the Man"; G. van Iterson, Jr., "Beijerinck, the Botanist"; and A. J. Kluyver, "Beijerinck, the Microbiologist," as well as 13 pls. The appendixes include the *stellingen* or propositions accompanying Beijerinck's doctoral dissertation (in Dutch), the names of Beijerinck's assistants while he was teaching, a list of publications by Beijerinck's collaborators at the Laboratory for Microbiology in Delft (1895–1921), a list of doctoral dissertations prepared under his direction, five papers and addresses given by Beijerinck's associates on important occasions in his life (in Dutch), an abstract from the lecture given by Beijerinck on his retirement in 1921 from the Delft Polytechnical School (in Dutch), an interview with Beijerinck in 1928 (in Dutch), and a list of obituary articles.

Sally Smith Hughes

BERGMANN, MAX (*b.* Fürth, Germany, 12 February 1886; *d.* New York, N.Y., 7 November 1944), *biochemistry, organic chemistry.*

The son of a prosperous coal merchant, Bergmann came of a Jewish family that had lived in Fürth for many generations. After completing his secondary schooling there, he took his first degree in 1907 at the University of Munich and then enrolled in the chemical department of the University of Berlin, headed by Emil Fischer. Working under Ignaz Bloch, Bergmann received the Ph.D. in 1911 with a dissertation on acyl polysulfides and became an assistant to Fischer.

During World War I, Bergmann was exempted from military service because of his position with Fischer, and was closely associated with his chief's research. In 1920 he was appointed *Privatdozent* at the University of Berlin and head of the department of chemistry of the Kaiser Wilhelm Institute for Textile Research. A year later he became director of the newly established Kaiser Wilhelm Institute for Leather Research in Dresden. Upon Hitler's rise to power in 1933, Bergmann moved to the United States, where he continued his work at the Rockefeller Institute until his death.

Bergmann's scientific research shows considerable continuity. During his association with Fischer, he made several basic contributions to carbohydrate, lipide, tannin, and amino acid chemistry, including the elucidation of the structure of glucal and the development of new methods for the preparation of α-monoglycerides. While at Dresden he created one of the leading laboratories for protein chemistry and attracted numerous young chemists from other countries. Among them was Leonidas Zervas, who was associated with Bergmann during the years 1926–1933 and joined him in the United States for two years (1935–1937) before returning to his native Greece. Bergmann, Zervas, and their associates made numerous contributions to the chemistry of amino acids and proteins. The most important of these came in 1932, when they devised the "carbobenzoxy" method for the synthesis of peptides. This marked a new era in protein chemistry, since it opened an easy route to the preparation of peptides that had hitherto been difficult or impossible to synthesize.

At the Rockefeller Institute, Bergmann directed the work of his laboratory along two lines. One was the use of the carbobenzoxy method for the synthesis of peptides to be tested as possible substrates for proteolytic enzymes, such as pepsin. This work, largely pursued by Joseph S. Fruton, led in 1936–1939 to the discovery of the first synthetic peptide substrates for these enzymes, thus opening the way for the study of their specificity. The second line of work was directed to the development of new methods for the quantitative analysis of the amino acid composition of proteins. With Carl Niemann, Bergmann proposed in 1938 that the arrangement of amino acids in a protein chain is periodic; although this hypothesis was later shown to be an oversimplification, it stimulated great experimental activity with proteins. In Bergmann's laboratory Stanford Moore and William H. Stein began work that led them, after World War II, to the accurate quantitative determination of the amino acid composition of proteins.

BIBLIOGRAPHY

I. Original Works. Bergmann published about 330 scientific and technical articles, and held 29 patents. His most important articles were "Über ein allgemeines Verfahren der Peptid-synthese," in *Berichte der Deutschen chemischen Gesellschaft*, **65** (1932), 1192–1201, written with L. Zervas; "Newer Biological Aspects of Protein Chemistry," in *Science*, **86** (1937), 187–190, written with C. Niemann; and "The Specificity of Proteinases," in *Advances in Enzymology*, **1** (1941), 63–98, written with J. S. Fruton. A list of Bergmann's publications is given in the biographical article by B. Helferich in *Chemische Berichte*, **102** (1969), i–xxvi.

II. Secondary Literature. In addition to the article by Helferich (see above), important obituary notices are by H. T. Clarke in *Science*, **102** (1945), 168–170; by C. R. Harington in *Journal of the Chemical Society* (1945), 716–718; and by A. Neuberger in *Nature*, **155** (1945), 419–420.

JOSEPH S. FRUTON

BERNAL, JOHN DESMOND (*b.* Nenagh, County Tipperary, Ireland, 10 May 1901; *d.* London, England, 15 September 1971), *crystallography, molecular biology*.

Desmond Bernal's first Christian name was John but was never used by his intimates.

Like almost everything else about him, his family origins were unusual. His father was what used to be called a squireen, somewhere between a farmer and a Catholic Irish squire. His mother was an American, educated at Stanford, who wrote some interesting journalism and had considerable resemblances to a Henry James expatriate heroine.

There were, as happened throughout Bernal's life, legends about this heredity, for he was a mythopoeic character about whom stories, and inaccurate statements of fact, massively accumulated. For private circulation there once appeared a loving document about him entitled *The Irish Jew*. In fact, Bernal is sometimes a Jewish name in Spain. It is possible that a Jewish forebear, conceivably a Marrano, had once settled in Ireland. If there were any Jewish genes at all in the family, however, they must have arrived many generations back. To anyone interested in social niceties, Bernal was by birth an Irish Catholic gentleman who had an upper-class English education and spoke all his life with the accent of a privileged Englishman.

Bernal was the oldest of three children. He was much attached to his mother but, so it seems, on bad terms in his boyhood with his father. There was a certain amount of money in the family. How much is hard to guess; but later he possessed some capital and with characteristic abandon and generosity was ready to give it away. He was sent away to boarding schools in England, which would have been beyond the means of an ordinary Irish farmer.

His first school, where he went at the excessively early age of ten, was Stonyhurst, the Jesuit establishment. He stayed only two years and, as usual, legends have collected. One is that, aged eleven, he was dissatisfied with the scientific education, decided that he could design a better one for himself, and accordingly departed. Another is more reliable, since it came directly from Bernal. At this stage, and until his first year at Cambridge, he was a passionate Irish nationalist and an even more passionate and fervent Catholic. Since the adoration of God was the first duty of a devout Catholic, he originated in his dormitory a Society of Perpetual Adoration. He was no doubt as eloquent and persuasive as he later showed himself, and thus wretched small boys were organized into watches. In relays there were two or three on their knees throughout each night. This must have presented a nice problem in situation ethics to sensible Jesuit priests.

After an interval in Ireland, Bernal was dispatched again to England to another public (public in the English sense, meaning private) school, Bedford. Bedford was a middle-of-the-road, middle-class, Protestant school; but it was competent and apparently met young Bernal's requirements as to scientific tuition. He must have had to cope with some aggression from other boys, for English public schools at the time of the First World War weren't gentle places and he already looked very odd—and, by the standards of his colleagues, actually was very odd. Still, Bernal was quite fearless and much stronger physically than anyone expected. So he survived; and his teachers probably recognized his ability from the start, which would have been hard not to do. Whatever the demerits of English education, talent normally is spotted very early. Bernal's was evident from his teens on, and even his enemies and detractors (he acquired both) couldn't doubt it. At eighteen he won a major open scholarship in mathematics to Emmanuel College, Cambridge.

Presumably Bernal opted for Emmanuel on advice from his school. It wasn't a lucky choice. In the seventeenth century Emmanuel had been one of the two great Puritan colleges, and had provided a high proportion of the first New England divines. By Bernal's time it had become more commonplace, and wasn't well adapted to deal with a remarkably nonconforming intellect that had some of the marks of genius. He entered the college as a believing Catholic but also as a believing Marxist. Since he couldn't reconcile the two faiths, the Catholicism had to go, although some friends thought they detected the effects long afterward. Bernal was always a man of faith; and all he did—in science, in politics, in his campaigns for peace and the perfectibility of humankind (in Marxist terms)—was suffused by something indistinguishable in spirit from religious emotion. Some of the more intellectual colleges would have realized what they had been presented with, and would have clung on to him whatever the cost. Emmanuel didn't, although forty years later it made the most handsome amend in its power by electing him to an honorary fellowship.

As it happened, Bernal's undergraduate record gave the college some excuse. His academic career was ludicrously checkered for someone who, only a few years afterwards, was to be described as the last man alive who knew and understood the whole of science. Part I of the mathematical tripos is an easy examination, and Bernal was a good natural mathematician, and well-trained. He ingeniously got a second class, which has never been explained. He then took part I of the natural sciences tripos in chemistry, mineralogy, geology. He was already committed to crystallography, and that was an appropriate combination. There he got a first class. He proceeded to part II, physics, and got another second class. Here, though, there is a creditable explanation. Bernal had become obsessively absorbed with crystallography (as he was to become obsessively absorbed in so many other projects later), and spent three-quarters of his time deriving the 230 space groups by means of Hamiltonian quaternions. This was an astonishing piece of work for an undergraduate—or, as far as that goes, for anyone else. It has never been published, on the grounds of the vast expense. It would have won him a fellowship at any of the great colleges—Trinity, St. John's, King's—which would have laughed off the consequent second class. It did win him a research post in W. H. Bragg's laboratory at the Royal Institution, and the immediate and undeviating support of both the Braggs.

At the Royal Institution, Bernal began his career as a professional scientist. Within a very short time he was known to be, with the Braggs, the most profound crystallographer in England. Among his less obvious gifts he was a first-rate experimenter, and his analysis of the structure of graphite was a classical piece of distinctly laborious work. As a more characteristic exercise of pure intellect, he created a diagram for interpreting X-ray photographs, now known as the Bernal chart.

In 1927, when he was twenty-six, Bernal was recalled to Cambridge. Arthur Hutchinson, the professor of mineralogy, had persuaded the university to establish a department of crystallography and to advertise for an assistant director of research (which meant, in the odd Cambridge terminology, the man in charge). Bernal applied. Hutchinson used to tell the story of the interview. During the formal questions Bernal either answered in monosyllables or sat mute, looking sullen (he could be curiously inept on such occasions). Then Hutchinson, as chairman, asked him what he would do if he was put in charge of the department. Bernal threw back his splendid head, the marvelous eyes sparkled, and he talked for half an hour—talked like the Pied Piper of Hamelin, whom he sometimes reminded one of, bewitching a good many men, even more women, and this hard-boiled academic interview board. After that, no more questions. Bernal was asked to withdraw, was called back in three minutes, and was given the job. From then on, Hutchinson said, being the most modest of men, that his only duty as Bernal's nominal superior was to act as a combination of nurse and housemaid, cleaning up after him.

Cambridge from 1927 to 1937 was the most creative period of Bernal's scientific life, in the specific sense of what he personally achieved. It was also the period when the personality of Bernal stamped itself on the English intellectual world, and soon on a world wider than that. At the Royal Institution in London he had already become something of a legend, on the fringe of what he called "low Bloomsbury." In many of his tastes he was, and remained, a child of the twenties. His writers were Donne and Joyce, he proclaimed, and though he went on reading widely, they continued to mean most to him. In inks of various colors he drew labyrinthine charts, one of his delights, to define the erotic combinations of the whole of Bloomsbury, and took a not undistinguished part in them himself. He married only once, and the marriage lasted till he died; but his amorous career bore a marked resemblance to that of H. G. Wells—who, incidentally, was very fond of him. As with Wells, nearly all of Bernal's attachments were with women of character and high intelligence. What isn't always true in such a life, none of those who loved him has in retrospect been known to say an unloving word.

In Cambridge, all that was pretty open. Bernal's

active Communism was quite open. That did him a little practical harm. Despite his reputation as one of the great scientists of the future, no college made him a fellow, although there were attempts to persuade them. That didn't matter very much. The scientific community gave him justice all through his career. He was elected at thirty-six—unusually young—to the Royal Society; and conservative elder statesmen of science, thoroughly disapproving of his extramural activities, came to Christ's College to celebrate the occasion, just to recognize that he was one of the new glories of English science. An attractive oddity, someone said fairly audibly.

It was at this time that he did his major pioneer work, which effectively started molecular biology. He began, with his pupils Dorothy Crowfoot Hodgkin, Isidore Fankuchen, and others, to take X-ray photographs of biologically important molecules, sterols, amino acids, proteins, nucleoproteins (e.g. the tobacco mosaic virus). He was in no sort of doubt that in the geometry and physical structure of such molecules must lie some of the explanations of the origin of life, and the way the living process works—as became magnificently clear twenty years later, with the Watson-Crick revelation of the structure of DNA. Molecular biology, as developed by Crick and his generation, is one of the greatest of scientific triumphs. Bernal was the founding father. In the thirties, he already foresaw a great deal of the future.

Two questions require some sort of answer. One isn't important; but even Alan Mackay, a colleague of Bernal's in later years and the scientist who has written most penetratingly about him, has probably given too much weight to an oversimplification. This is that, right from the beginning of his scientific life, Bernal was dominated by one purpose and one theme, namely the search for the origin of life itself—"biopoesis," to use his own term. Bernal certainly came to believe this. He was a man of cosmos-embracing mind, with great love for systems, who somehow never created a system himself. In the most brilliant part of his scientific career, and probably right up to the end, the origin of life was his major preoccupation. But it wasn't always so, and one has to be cautious about a rationalization after the event. Some people must remember his conversation in his first years at the Cambridge department of crystallography—conversation, scientific and otherwise, of which no one who heard it ever expects to hear the like again. But the conversation between, say, 1928 and 1931 was not about the genesis of molecular biology. Bernal could turn his mind to almost any scientific topic, especially if there was a practical outcome, however remote. With quantum physics he was an interested spectator, but it didn't grip him, and he played no part. On the other hand, he was passionately involved in the structure of metals and the prospect of making entirely new materials.

The second question is perhaps more significant. If he had been so utterly dedicated to molecular biology, it is hard to believe that Bernal wouldn't, after the war, have polished off the structure of the nucleic acids. That couldn't have been done in the thirties, for the techniques weren't developed. In the late forties they were, and no man alive was so well equipped to solve the problem single-handed. But he hadn't the single-mindedness. Other concerns, to him more imperative, had supervened. Further, all through his life he had a curious lack of the artistic impulse to perfect a piece of work and sign his own name underneath. He started so many things, and stayed to finish only a few. Others could do the final work. He was part of a collective enterprise. In some depths of his temperament he was self-centered, but also he was the most unselfish of men. It was that combination, as rare as the somewhat similar one of Einstein's, which made him different from most of the human species.

After the rise of Nazism, Bernal's most imperative concern became politics, and remained so until his death. He believed, with his religious intensity, that human intelligence and the goodwill of the people could, and would, produce a desirable life for human beings everywhere. In that respect he was a totally committed Marxist, and would cheerfully have died for his belief—and in fact, though not in battle, did so.

Bernal wasn't innocent. He knew a lot about human frailty; but he couldn't entertain the thought that men in power, even if they had arrived at power through impeccable Marxist channels, didn't necessarily behave with the sweetness that he would have shown himself. It was on that matter that he and the present writer couldn't remain in complete communication, though the friendship stayed intact. He couldn't imagine that the Stalin purges weren't precisely what they were officially said to be.

It is a mistake to think that he was absolutely uncritical. He was deeply torn about the Hungarian rising in 1956 and about Czechoslovakia in 1968. But with all the resources of one of the most subtle and labyrinthine minds of the twentieth century, he tried to elaborate at least a partial justification for Lysenko.

The war found Bernal at his happiest and best. True, he was in a position of some ethical delicacy during the months of the Nazi-Soviet pact. That was managed with characteristic ingenuity, however. He devoted himself to protecting the civilian population from bombing, which no one could object to. He was one of the most valuable of all wartime scientists. Bernal loved being set finite problems. This was science with a purpose—such as he had called for in his great testament and polemic *The Social Function of Science* (1939). This was how science ought to be used in peace. More than many scientists, he liked being instructed about what the job was and ordered to get it done.

He was also abnormally brave. About his entire wartime activities, a new wave of mythopoeia set in. Some were literally true. He really did dicker with an unexploded bomb at Liverpool Street Station—an unjustifiable risk that would have been stopped if there had been a responsible person around—and then ceremoniously declare: "Mr. Stationmaster, now you can announce that your station is open."

It is worth remembering that no one in the English official world ever worried that he was an overt Communist. As he said later, for wartime purposes he was regarded as respectable. After the Germans invaded Russia, he threw all his energies into aggressive war. He became an adviser to Lord Louis Mountbatten on special operations, went to the Quebec Conference, helped plan the invasion of Europe.

About that last, some of the mythopoeia went wild. He applied both science and scholarship to working out the condition of the Norman beaches. No one else could have done it so well. But the story that he had gone across the Channel at night, weeks before the invasion, to inspect the beaches for himself, is nonsense. He was brave enough, perhaps foolhardy enough. But it would never have been permitted. No one could have risked his being captured. He knew far too much.

War over, he returned to the professorship of physics at Birkbeck College, London University, to which he had been appointed in 1937, succeeding his great contemporary P. M. S. Blackett. There he did fine work himself, notably on the structure of liquids, even when crippled by illness; and he inspired other fine work in a small but jovially worshipping department. Both at Cambridge and Birkbeck his relations with colleagues and pupils were singularly happy. He was utterly devoid of envy, and good work done by anyone gave him joy. His method of presiding over research, although it would have made a tidy-minded administrator blink (it wasn't very close to his own theoretical model of planning), somehow worked. Nevertheless, he wasn't often to be found in the laboratory. He was wearing himself out as a world figure.

The honors were flowing in, not as many as lesser men receive, but enough. The Royal Society in 1946 gave him one of their highest rewards, the Royal Medal. He was still involved in British governmental committees on building, a throwback to his early war work. That he could do in his spare time, but he had very little. He was absorbed, body and soul, in trying to prevent another war.

By this time (1946 onwards) he had become—what was already partially true in the late thirties—the most respected and loved of Western intellectual Communists. This was partly due to his scientific eminence, which was unassailable, and at least as much to his character. People accepted that he was not only a great man but a good one. It helped that he had no solemnity at all, was remarkably amusing, and could charm birds off any kind of perch. So he was the first person in demand for the great Communist causes. Communists believed in the danger of another war. So did he. He threw himself into the campaign without a particle of reserve. He became chairman of the Presidential Committee of the World Council of Peace (Communist-organized), and that was only one of his duties.

This meant that he was in something like continuous motion, scurrying from Birkbeck, the most recent scientific paper being corrected in the car, to London Airport, off to one of the socialist capitals, Moscow, Warsaw, Bucharest, or Peking. This might have been endured by a man capable of more relaxation, but he had no idea of looking after himself. He was astonishingly careless of his own life. Bernal's friends looked on with horror. Gentle and unassuming, he was also rock-obstinate to any advice. His friends guessed that, though his muscles were strong, he probably wasn't organically tough. There were ominous clinical signs years before his first stroke.

He enjoyed the fuss and flurry of action. He had reveled in it during the war, and did so again. He positively enjoyed multinational committees. He was certain that it was all worthwhile; anything that made war even slightly less likely must be worthwhile. Friends such as Blackett, as experienced in military-political thinking as he was but much more skeptical, couldn't accept that it was. For objective reasons, Blackett argued, major war was most improbable. Fringe campaigns couldn't affect the chances by .001 percent. Meanwhile, Bernal was throwing away years of his life.

He had his first stroke in 1963, in an aircraft returning from one of his missions. That didn't stop him. He continued with his travels until he was finally immobilized by other strokes and a rare and terrible combination of pathologies. He fought against the spread of paralysis with a flawless stoical courage that was agonizing to see.

For some years before his death, Bernal had lost almost all muscular movement. He had been, more than most, an active, often restless man. He could scarcely speak audibly, even with amplifiers, except to those of his nearest connections who could catch his tone of voice. He had been the most brilliant talker of his time. His intellect was untouched almost to the end. He continued to think and work. It was the last thing left to him. He died, at the age of seventy, on 15 September 1971.

BIBLIOGRAPHY

I. Original Works. Bernal's writings include *The World, the Flesh and the Devil* (1929; Bloomington, Ind., 1969); *The Social Function of Science* (1939; Cambridge, Mass., 1967); *The Freedom of Necessity* (1949); *The Physical Basis of Life* (1950); *Science and Industry in the Nineteenth Century* (1953; Bloomington, Ind., 1969); *Science in History* (1954; 2nd ed., 1957; 3rd ed., 1965; ill. ed., 1969); *World Without War* (1958; 2nd ed., 1961), also in abridged paperback form as *Prospect of Peace* (1960) that was translated into French as *La paix, pourquoi faire?* (1965); *Origin of Life* (1967); and *The Extension of Man—Physics Before the Quantum* (1972).

Bernal's writings on science and the social relations of science, together with notes and any other relevant material, were bequeathed to Birkbeck College and are in its main library. His correspondence, notes on committee work, and biographical material are at Cambridge University library.

II. Secondary Literature. Obituaries include unsigned ones in *Nature*, **235** (28 Jan. 1972); *Scientific World*, **16**, no. 1 (1972); and the *Times* of London (16 Sept. 1971); Pierre Biquard, in *La vie culturelle* (17 Sept. 1971), 10; C. H. Carlisle, "J. D. Bernal—an Appreciation," in *Bulletin. Institute of Physics* (London), **22** (1971), 685; J. G. Crowther, "John Desmond Bernal—an Appreciation," in *New Scientist* (23 Sept. 1971), 666; Jack Legge, in *Australian Left Review* (May 1971), 31–33; Joseph Needham, "Desmond Bernal—a Personal Recollection," in *Cambridge Review* (19 Nov. 1971), 33–35; Linus Pauling, "Bernal's Contribution to Structural Chemistry," in *Scientific World*, **16**, no. 2 (1972); and Alden Whitman, "A Natural Philosopher," in *New York Times* (16 Sept. 1971), 46.

Also see Sir Lawrence Bragg, in *Chemistry in Britain* (4 Apr. 1970), 149; Maurice Goldsmith and Alan Mackay, eds., *Science of Science* (London, 1964), esp. C. P. Snow, "J. D. Bernal, a Personal Portrait," 19–29; *McGraw-Hill Encyclopaedia of Science and Technology* (New York, 1972); "Profile of J. D. Bernal," in *New Scientist*, **4** (1958), 164–165; and W. L. and C. K. Schultz, "John Desmond Bernal—Evangelist of Science," in *Journal of the American Society for Information Science*, **21**, no. 2 (1970), 142–144.

C. P. Snow

BERNSTEIN, JULIUS (*b.* Berlin, Germany, 8 December 1839; *d.* Halle, Germany, 6 February 1917), *physiology.*

Bernstein was the son of Aron Bernstein, a Jewish theologian, author, and politician from Danzig. The father himself established a reputation as a popularizer of science, and it was under his influence that Bernstein early became interested in the natural sciences. While still a schoolboy he sought to solve physical and technological problems. In 1858 he began to study medicine at the University of Breslau, where his admiration for Rudolf Heidenhain turned his interest to physiology. Through the intervention of his friend Ludimar Hermann, Bernstein entered the laboratory of Emil du Bois-Reymond at Berlin, and he earned his medical degree there in 1862 with a dissertation on invertebrate muscle physiology. In 1864 Bernstein went to Heidelberg as assistant to Hermann von Helmholtz, whose sober outlook, dominated by the ideals of mathematical physics, corresponded well with Bernstein's own temperament as a researcher. When Helmholtz was called to the chair of physics at Berlin in 1871, Bernstein became acting head of the Heidelberg Physiological Institute, but in the same year returned to Berlin. In 1872 Bernstein succeeded Friedrich Goltz as professor of physiology at the University of Halle, where he remained until his death forty-five years later.

Despite considerable difficulties, Bernstein succeeded in establishing a new physiology institute at Halle, but not a modern institution with separate departments for research and teaching. He made effective use of the very modest means at his disposal. Although it was not his ambition to form a school, he attracted the attention of many physiologists through the originality of his writings. His scientific interests and skills transcended physiology to include mathematics and astronomy.

Bernstein married the daughter of H. Levy, a Russian military physician. She took an interest in his work and shared his love of music. Their son, the mathematician Felix Bernstein, became well known in medicine for his computations of the inheritance of blood groups. Bernstein died of pneumonia at the age of seventy-seven.

During his long career Bernstein carried out important research. As creator of the "membrane theory of excitation," he has a claim to a major place in the development of modern physiology, a fact that seems not to have been sufficiently appreciated during his lifetime. His works treated such broad areas of physiology as bioelectricity, structural problems of contractile substances, cardiac and circulatory physiology, reproduction and growth, secretion and resorption, respiration, sensory physiology, and toxicology. He was also concerned with the teaching of natural sciences. Evidence of his outstanding gifts in this area is his *Lehrbuch der Physiologie des tierischen Organismus* (1894; 3rd ed., 1908). Used in the training of physicians, it also influenced many young researchers in their choice of specialization.

Of particular interest are Bernstein's contributions to the physiology of irritable structures—that is, general nerve and muscle physiology and membrane physiology. His research in this field—to which he devoted the majority of his 135 publications—may be divided into two phases. In the earlier one, that of classical electrophysiology, he was concerned above all with perfecting techniques of stimulation and measurement. For example, an experiment demonstrating the simultaneity of the propagation of the excitation (wave) and the alteration of the electrical potential in nerves (1867) belongs to this period. So does his determination (the first) that synapse time for the neuromuscular junction is 0.3 milliseconds (1882). Bernstein also very ingeniously measured the velocity, form, and course of the excitation wave with the help of his differential rheostat. With this apparatus he was able—by employing periodic stimulation and by recording point by point small portions of the current curve—to reconstruct the entire course of the curve with the aid of his slowly reacting galvanometer. This painstaking technique permitted him to describe the real time course of the excitation process. In 1912 Bernstein recommended the use of the inertialess cathode ray for the record of bioelectric activity, a suggestion that was taken up in physiological research, about a decade later, by Erlanger and Gasser.

Recognizing that no further explanation of the nature of the excitation process could be expected from existing views and conventional methods, Bernstein turned to ideas and techniques of modern electrochemistry, molecular physics, and thermodynamics. The second phase of his research began with his application of these experimental tools to unresolved problems. Bernstein's essential conceptual breakthrough was the assumption that even before stimulation, the cell membrane includes an electrically charged double layer of ions. According to this view, those ions that do not pass through are decisive in determining the differences in electrical potential between the two sides of the membrane. This idea was developed in the notion proposed by R. Höber (1873–1953) of selective permeability of the membrane and in his remarks on the role of salts in affecting the porosity of the membranes.

According to Bernstein, upon stimulation, the membrane's permeability increases as a result of a chemical change, so that the excited portion becomes negatively charged with respect to the unexcited portion. That is, the membrane's potential decreases upon stimulation. This alteration of the living substance propagates in the fiber in a wavelike manner and has a certain duration at every point along its course. It is the origin of the electric excitation wave (*Aktionspotential*, in Hermann's terminology). "A consequence of this theory would be that the negative deflection (or oscillation) must attain a maximum, which would be given by the strength of the membrane's potential, so that the latter could not be reversed during stimulation" (Bernstein, *Elektrobiologie*, 105). According to H. Grundfest, Bernstein thus sacrificed on the altar of his theoretical views the "overshoot" of the axon spike that he had observed as early as 1871.

It is worth noting, however—as E. Abderhalden stressed—that from the perspective of the history of science, Bernstein completed the conceptual and methodological restructuring of his research at an age when most scientists are either reluctant to seek new paths or incapable of following them. His strict adherence to the methods of exact research, whether in classical electrophysiology or in the new thermodynamic treatment of bioelectrical phenomena, demonstrated the futility of speculative nineteenth-century *Naturphilosophie* and argued strongly for what he himself called the "mechanistic theory of life" (1890).

BIBLIOGRAPHY

I. ORIGINAL WORKS. A chronologically ordered list of Bernstein's publications (135 titles, plus 72 titles of works by his students) is in the article by A. von Tschermak (see below), 79–89.

Bernstein published the following books: *Untersuchungen über den Erregungsvorgang im Nerven- und Muskelsystem* (Heidelberg, 1871); *Die fünf Sinne des Menschen* (Leipzig, 1876; 2nd ed., 1889); trans. as *The Five Senses of Man* (New York, 1876); *Die mechanistische Theorie des Lebens, ihre Grundlagen und Erfolge* (Brunswick, 1890); *Lehrbuch der Physiologie des tierischen Organismus* (Stuttgart, 1894; 3rd ed., 1908); and *Elektrobiologie. Die Lehre von den elektrischen Vorgängen im Organismus auf moderner Grundlage dargestellt* (Brunswick, 1912).

His journal articles include "Über den zeitlichen Verlauf der negativen Schwankung," in *Pflügers Archiv für die gesamte Physiologie des Menschen und der Tiere*, **1** (1868), 173–207; "Über den zeitlichen Verlauf des Polarisationsstromes," in *Annalen der Physik und Chemie*, 6th ser., **155** (1875), 177–211; "Über Ermüdung und Erholung des Nerven," in *Pflüger's Archiv*, **15** (1877), 289–327; "Die Erregungszeit der Nervenendorgane in den Muskeln," in *Archiv für Anatomie, Physiologie und wissenschaftliche Medizin* (1882), 329–346; "Neue Theorie der Erregungsvorgänge und elektrische Erscheinungen an der Nerven- und Muskelfaser," in *Untersuchungen aus dem physiologischen Institut der Universität Halle*, **1** (1888), 27–104; "Über das Verhalten der Kathodenstrahlen zueinander," in *Annalen der Physik*, n.s. **62** (1897), 415–424; "Chemotropische Bewegung eines Quecksilbertropfens. Zur Theorie der amöboiden Bewegung," in *Pflüger's Archiv*, **80** (1900), 628–637; "Untersuchungen zur Thermodynamik der bioelektrischen Ströme," *ibid.*, **92** (1902), 521–562, repr. in W. Blasius, I. Boylan, and K. Kramer, eds., *Founders of Experimental Physiology* (Munich, 1971); and "Kontraktilität und Doppelbrechung des Muskels," in *Pflüger's Archiv*, **163** (1916), 594–600.

Bernstein wrote brief evaluations of the scientific work of or, in some cases, obituaries of Paul du Bois-Reymond (1889), Helmholtz (1895, 1904), Carl Ludwig (1895), Emil du Bois-Reymond (1897), Frithiof Holmgren (1897), and Heidenhain (1897). He also became involved in polemics with Setchenov, Preyer, Engelmann, and Hermann. His "Errinerungen an das elterliche Haus" was printed in MS form in 1913.

II. SECONDARY LITERATURE. See E. Abderhalden, "Dem Andenken von Julius Bernstein gewidmet," in *Medizinische Klinik*, **13**, no. 9 (1917), 260–261; H. Grundfest, "Julius Bernstein, Ludimar Hermann and the Discovery of the Overshoot of the Axon Spike," in *Archives italiennes de Biologie*, **103** (1965), 483–490; K. E. Rothschuh, *Geschichte der Physiologie* (Berlin–Göttingen–Heidelberg, 1953), trans. as *History of Physiology*, G. B. Risse, ed. (Huntington, N.Y., 1973); G. Rudolph, "Julius Bernstein (1839–1917)," in J. W. Boylan, ed., *Founders of Experimental Physiology. Biographies and Translations* (Munich, 1971), 249–257; A. von Tschermak, "Julius Bernsteins Lebensarbeit–zugleich ein Beitrag zur Geschichte der neueren Biophysik," in *Pflüger's Archiv*, **174** (1919), 1–89; and F. Verzár, "Erinnerungen an Julius Bernstein 1910/1911" (unpublished; for access to this paper write to G. Rudolph, Institut für Geschichte der Medizin und Pharmazie der Universität Kiel).

GERHARD RUDOLPH

BERNSTEIN, SERGEY NATANOVICH (*b.* Odessa, Russia, 5 March 1880; *d.* Moscow, U.S.S.R., 26 October 1968), *mathematics.*

Bernstein was the son of Natan Osipovich Bernstein, lecturer in anatomy and physiology at the Novorossysky University in Odessa. After graduating from high school in 1898, he studied at the Sorbonne and the École d'Électrotechnique Supérieure in Paris; in 1902–1903 he also studied at Göttingen. He defended his doctoral dissertation at the Sorbonne in 1904 and, after returning to Russia in 1905, defended his master's thesis (1908) and his doctoral dissertation at Kharkov (1913) because scientific degrees awarded abroad did not entitle one to a university post in Russia. From 1907 to 1933 Bernstein taught at Kharkov University, first as a lecturer and then as a professor after 1917, laying the foundations of a mathematical school that included N. I. Akhiezer and V. L. Goncharov. During this period Bernstein frequently gave series of lectures and presented reports abroad; in 1915 he participated in the Second All-Russian Congress of High School Teachers; and in 1930 he organized the First All-Union Mathematical Congress, held in Kharkov. He was director of the Mathematical Research Institute in 1928–1931 and was one of the leaders of the Kharkov Mathematical Society from 1911. In 1925 he was elected member of the Academy of Sciences of the Ukrainian S.S.R.

In 1933 Bernstein began lecturing at the University of Leningrad and the Polytechnical Institute, while he worked also in the Mathematical Institute of the U.S.S.R. Academy of Sciences, of which he had been elected corresponding member in 1924 and member in 1929. He moved to Moscow in 1943, continuing his work at the Mathematical Institute. He edited Chebyshev's *Polnoe sobranie sochineny* ("Complete Works"; 1944–1951) and in his later years prepared a four-volume edition of his own writings.

Bernstein was a member of the Paris Académie

des Sciences (elected corresponding member in 1928 and foreign member in 1955), doctor *honoris causa* of the universities of Algiers (1944) and Paris (1945), and honorary member of the Moscow (1940) and Leningrad (1966) mathematical societies. In 1942 his scientific achievement was recognized with the State Award, first class; earlier he had received awards from the Belgian Academy of Sciences and from the Paris Academy of Sciences for the book based on his lectures at the Sorbonne in 1923.

In his work Bernstein united traditions of Chebyshev's St. Petersburg mathematical school with western European mathematical thought, particularly that of France (Picard, Vallée-Poussin) and Germany (Weierstrass, Hilbert). Three fields dominated his work: partial differential equations, theory of best approximation of functions, and probability theory.

Bernstein's doctoral dissertation at the Sorbonne held the solution of Hilbert's nineteenth problem (1900), particular cases of which had been treated shortly before by Picard and others. Bernstein's result read: If a solution $Z\ (x,y)$ of an analytical differential equation $F(x,y,z,p,q,r,s,t) = 0$ of elliptical type $4\frac{\delta F}{\delta r}\cdot\frac{\delta F}{\delta t} - \left(\frac{\delta F}{\delta s}\right)^2 > 0$, where $p = \frac{\delta z}{\delta x}$, $q = \frac{\delta z}{\delta y}$, $r = \frac{\delta^2 z}{\delta x^2}$, $s = \frac{\delta^2 z}{\delta x \delta y}$, $t = \frac{\delta^2 z}{\delta y^2}$ possesses continuous derivatives up to the third order (inclusive), the solution is analytical. In his master's thesis Bernstein solved Hilbert's twentieth problem, demonstrating the possibility of an analytical solution of Dirichlet's problem for a wide class of nonlinear elliptical equations. A number of theorems on the differential geometry of surfaces followed, particularly on the theory of minimal surfaces. The above-mentioned studies were further advanced by Bernstein and many others, and are still being developed.

Another area of Bernstein's work concerned the theory of the best approximation of functions, to which he contributed new ideas and that he called the constructive theory of functions (1938). In this theory, calculation and investigation of functions are carried out with the notion introduced by Chebyshev in 1854 of the best approximation of a given function $f(x)$ by means of a polynomial $g_n(x)$ of a given degree n or by means of some other relatively simple function $g(x)$ depending on the finite number of parameters. If parameters are selected so that on a segment (a,b) the value of max $|f(x) - g_n(x)| = E_n[f(x)]$ is minimal, the function $g_n(x)$ is called the best approximation of a function $f(x)$. Chebyshev and his immediate successors were more interested in finding the polynomial of best approximation $g_n(x)$ when n is given, than in exploring the general functional properties of the quantity $E_n[f(x)]$. In 1885 Weierstrass demonstrated that any function $f(x)$ continuous on a segment (a,b) can be expanded into a uniformly convergent series of polynomials so that $\lim_{n\to\infty} E_n[f(x)] = 0$, whatever the continuous function $f(x)$ is.

The point of departure for Bernstein was the problem of estimating the order of the quantity E_n for $f(x) \equiv |x|$ on a segment $(-1,+1)$, posed by Vallée-Poussin. In 1911 Bernstein demonstrated that the best approximation $E_{2n}(|x|)$ by means of a polynomial of degree $2n$ lay between $\frac{\sqrt{2}-1}{4(2n-1)}$ and $\frac{2}{\pi(2n+1)}$, and that there existed $\lim_{n\to\infty} 2n\,E_{2n}(|x|) = \lambda$, with $0.278 < \lambda < 0.286$, so that for sufficiently great values of n the inequality $\frac{0.278}{2n} < E_{2n} < \frac{0.286}{2n}$ holds. Bernstein also studied asymptotic values of the best approximation for various classes of functions and established the closest relations between the law of decrease of the quantity $E_n[f(x)]$ and the analytical and differential properties of the function $f(x)$. These relations were the subject of his report at the Fifth International Congress of Mathematicians, held at Cambridge in 1912. Bernstein later continued his studies in that area, solving problems relevant to the theory of interpolation, methods of mechanical quadratures, and the best approximation on an infinite axis. In 1914 he introduced an important new class of quasi-analytical functions.

Almost simultaneous with the theory of best approximation Bernstein approached probability theory. In 1917 he suggested the first system of axioms for probability theory; he later conducted a number of fundamental studies relevant to the generalization of the law of large numbers, the central limit theorem, Markov chains, the theory of stochastic processes, and applications of probability theory to genetics. All these works by Bernstein greatly influenced the advance of contemporary probability theory.

BIBLIOGRAPHY

I. Original Works. Bernstein's writings were collected as *Sobranie sochineny akademika S. N. Bernshteina* ("Collected Works . . ."), 4 vols. (Moscow,

1952–1964). They include "Sur la nature analytique des solutions des équations différentielles aux dérivées partielles du second ordre," in *Mathematische Annalen*, **59** (1904), 20–76, his doctoral diss. at the Sorbonne; "Issledovanie i integrirovanie differentsialnykh uravneny s chastnymi proizvodnymi vtorogo poryadka ellipticheskogo tipa" ("Research on the Elliptical Partial Differential Second-Order Equations and Their Integration"), in *Soobshcheniya i protokoly Kharkovskago matematicheskago obshchestva*, **11** (1910), 1–164, his master's thesis, also in *Sobranie sochineny*, III; *Teoria veroyatnostey* ("Probability Theory"; Kharkov, 1911; 4th ed., Moscow–Leningrad, 1946); *O nailuchshem priblizhenii nepreryvnykh funktsy posredstvom mnogochlenov dannoy stepeni* ("On the Best Approximation of Continuous Functions by Means of Polynomials of Given Degree"; Kharkov, 1912), his Russian doctoral diss., also in *Sobranie sochineny*, I, and in French as "Sur l'ordre de la meilleure approximation des fonctions continues par des polynomes de degré donné," in *Mémoires de l'Académie royale de Belgique. Classe des sciences. Collection*, 2nd ser., **4** (1912), 1–103; and *Leçons sur les propriétés extrémales et la meilleure approximation des fonctions analytiques d'une variable réelle* (Paris, 1926), based on lectures delivered at the Sorbonne in 1923.

II. SECONDARY LITERATURE. See N. I. Akhiezer, *Akademik S. N. Bernshtein i ego raboty po konstruktivnoy teorii funktsy* ("Academician S. N. Bernstein and His Research on the Constructive Theory of Functions"; Kharkov, 1955); A. N. Kolmogorov, Y. V. Linnik, Y. V. Prokhorov, and O. V. Sarmanov, "Sergey Natanovich Bernshtein," in *Teoria veroyatnostei i ee primenenie*, **14**, no. 1 (1969), 113–121; A. N. Kolmogorov and O. V. Sarmanov, "O rabotakh S. N. Bernshteina po teorii veroiatnostey" ("S. N. Bernstein's Research on Probability Theory"), *ibid.*, **5**, no. 2 (1960), 215–221; *Matematika v SSSR za sorok let* ("Forty Years of Mathematics in the U.S.S.R."), 2 vols. (Moscow, 1959), see index; *Matematika v SSSR za tridtsat let* ("Thirty Years of Mathematics in the U.S.S.R."; Moscow–Leningrad, 1948), see index; I. Z. Shtokalo, ed., *Istoria otechestvennoy matematiki* ("History of Native Mathematics"), II and III (Kiev, 1967–1968), see index; and A. P. Youschkevitch, *Istoria matematiki v Rossii do 1917 goda* ("History of Mathematics in Russia Before 1917"; Moscow, 1968), see index.

A. P. YOUSCHKEVITCH

BESSEMER, HENRY (*b.* Charlton, Hertfordshire, England, 19 January 1813; *d.* Denmark Hill, London, England, 15 March 1898), *technology.*

Of nineteenth-century inventors, few were more celebrated than Henry Bessemer, whose famed converter helped transform the "iron age" of the early industrial revolution into the present "age of steel." By training he was not an iron man; but since childhood he had been familiar with metal processing and fabrication in his father's type foundry at Charlton, where he attended school until he was seventeen. His father, Anthony Bessemer, had learned engineering by apprenticeship and had become a successful inventor and entrepreneur. The son charted a similar course. In London he acquired training in machine design and a fair command of chemistry.

Bessemer's first invention was a way of casting metal replicas of such delicate objects as flowers. To improve their attractiveness, he bronzed their surface by electroplating (a very early application of that process). In his next invention he collaborated with his fiancée, Anne Allen, in devising a method for canceling stamps by perforation. Other inventions followed (1833–1838) that yielded only modest returns but demonstrated Bessemer's versatility: a better way to make graphite pencil lead; a way of casting type under pressure; a process for making imitation Utrecht velvet; machinery for embossing cardboard.

About 1840 Bessemer began searching for a mechanical way to manufacture gold paint pigment (called bronze powder in the trade). He devised a sequence of six machines that made tiny particles of brass foil virtually automatically. The process, kept completely secret for forty years, supplied the world with gold paint at a fraction of its former cost, yet still so profitably that Bessemer was freed from financial worry. By 1848 he could attempt more complex innovations. An improved process for extracting sugar from cane won him a gold medal. In the early 1850's he took out many patents on railway wheels, axles, brakes, and other equipment. He sold a machine for rolling plate glass for £6,000.

The Crimean War stimulated Bessemer to experiment with elongated artillery shot having a grooved surface to induce rotation when fired from a smooth-bore cannon. Existing cast-iron guns were, however, too weak for his shot. Hence he sought a mix of iron and steel that was stronger than cast iron. While melting down a mixture of steel and pig iron in a furnace equipped with air blast to enhance combustion, he noticed that air alone can remove carbon from molten pig iron, thus yielding malleable iron without use of the then-standard, and very costly, puddling process. On 17 October 1855 Bessemer patented this process of blowing air through molten pig iron and hastened to design crucibles with tuyeres that soon evolved into tiltable, bottom-blown "Bessemer converters," patented on 5 February 1856. In

August he made public his new process in a paper read before the British Association for the Advancement of Science.

Bessemer at once received requests for licenses from numerous ironmakers. The licensees, however, soon discovered that malleable iron and steel made by Bessemer's process were brittle when cold and unworkable when hot. Stunned and embarrassed, Bessemer spent the next two years locating the problems and saving his invention. The main problem was contaminating phosphorus in the pig iron that "Bessemerization" failed to remove. Most British iron ores contain phosphorus, but by luck Bessemer had used pig made from phosphorus-free Blaenavon ore in his initial experiments.

Vain attempts to dephosphorize iron between 1856 and 1858 forced Bessemer to restrict his process to refining only phosphorus-free pig from Sweden. In 1859 he began operating his own steel plant at Sheffield, where he further improved his process by learning how to add spiegeleisen (a mixture of iron, carbon, and manganese) to the refined iron in the converter after the "blow." The carbon in spiegeleisen recarburizes the molten iron to the desired grade of steel, and its manganese removes oxygen bubbles left from the air blast. These bubbles make steel red-short—brittle when worked hot. Robert Mushet, an English metallurgist, called Bessemer's attention to the fact that manganese could remove trapped oxygen in the form of manganic oxide slag. Mushet was unable to patent his valuable idea, however, and later had to be content with a small pension that Bessemer voluntarily paid him.

Although Bessemer could make high-grade steel, his process was less suited for that purpose than was the competing open hearth being perfected in the 1860's. Bessemer steel proved ideal for rails, structural steel, and plate. As he grew rich, Bessemer left the perfection of his process to others, such as Sidney Thomas and Percy Gilchrist, who in 1878 solved the problem of removing phosphorus from pig iron.

The last thirty years of Bessemer's life were spent in busy leisure that included some inventing. A costly attempt to build a seagoing ship with a pivoted saloon to keep the passengers level and free from seasickness came to naught. He received many honors before his death at age eighty-five.

BIBLIOGRAPHY

Bessemer's *Autobiography* (London, 1905; new ed., 1924) is almost the only source used by biographers. Unfortunately it does not contain a bibliography or list of Bessemer's 114 patents.

Robert F. Mushet, *The Bessemer-Mushet Process* (London, 1883), tries to sustain Mushet's claim that his suggestion of adding spiegeleisen saved the Bessemer process from oblivion. *Journal of the Iron and Steel Institute*, **183** (1956), 179–195, contains five useful articles evaluating Bessemer's invention and commemorating the centenary of his process. No full-length biography exists. Of the shorter life sketches, that in the *Dictionary of National Biography*, supp., 185–191, is best in describing the inventor's personal life. The controversies over priority with the American inventor William Kelly and the claims of Robert Mushet are objectively evaluated in Philip W. Bishop, "The Beginnings of Cheap Steel," in *Bulletin. United States National Museum*, no. 218 (1959), 27–47. Several books on the history of iron and steel tell Bessemer's story. A good one is J. C. Carr and W. Taplin, *History of the British Steel Industry* (Oxford, 1962), 19–30.

JOHN J. BEER

BHABHA, HOMI JEHANGIR (*b.* Bombay, India, 30 October 1909; *d.* Mont Blanc, France, 24 January 1966), *physics*.

Bhabha was the son of Jehangir Bhabha, a barrister, and the former Meherbai Framji Panday, both members of the small but prosperous and influential Parsi community of Bombay. He was also connected to the Tata family through the marriage of his paternal aunt, Meherbai, to Sir Dorab Tata. He attended Cathedral High School in Bombay from 1916 to 1925, then began preparations for the Senior Cambridge Examination at Elphinstone College and the Royal Institute of Science in that city. In 1927 Bhabha entered Gonville and Caius College, Cambridge, where he studied mechanical engineering with the expectation of an industrial career in the Tata family's business empire. Bhabha preferred mathematics and theoretical physics, however; and after receiving first-class marks in 1930, he was accepted as a research student at the Cavendish Laboratory, still directed by Ernest Rutherford. His activities were centered there until 1939. He also traveled extensively on the Continent, however, particularly between 1934 and 1936, spending time with Enrico Fermi's group in Rome, with Wolfgang Pauli's group in Zurich, and at Niels Bohr's institute in Copenhagen. He was awarded the Ph.D. from Cambridge in 1935.

Bhabha was in India on a holiday at the outbreak of World War II in September 1939. Since he was unable to return to England, he accepted a readership created for him at the Indian Institute of Science at Bangalore, which owed its existence in part to the munificence of the Tata family. C. V. Raman was still a member of the physics department, and is said to have had a significant influence on Bhabha's work there. Bhabha was promoted to professor of cosmic ray physics in 1941, the same year in which he was elected fellow of the Royal Society. It is M. G. K. Menon's opinion that Bhabha found his mission in life during his years at Bangalore, since it was there that he became aware of the role he could play in the development of modern science and technology in India. In March 1944 he proposed to the chairman of the Sir Dorab Tata Trust the establishment of an institution that would be devoted to advanced research and teaching in physics, particularly cosmic ray and nuclear physics, and mathematics. The institute, named the Tata Institute for Fundamental Research, was founded in June 1945 at the Indian Institute of Science in Bangalore, with Bhabha as its director. In December 1945 it was relocated in Bombay.

From its inception the Tata Institute was conceived by Bhabha to be not only a first-rate center for basic research, but also an incubator where viable new types of industrial enterprises could be nurtured. His vigorous advocacy of a nuclear-powered electrical system for India led to the creation of the Indian Atomic Energy Commission in 1948 and to Bhabha's appointment as its chairman. Most of the commission's early research and development activities were carried out at the Tata Institute. When the commission was reorganized in 1954 as the Department of Atomic Energy, Bhabha was named its secretary with direct responsibility to the prime minister. During that same year he was also appointed director of the Atomic Energy Research Center, which was being constructed at Trombay, a northern suburb of Bombay.

Bhabha was a staunch advocate of international cooperation in science, and entered into agreements with Canada, Great Britain, France, and the United States for assistance in developing the Indian atomic energy program. He was the unanimous choice of the scientific advisory committee to the secretary-general of the United Nations for the post of president of the First International Conference on Peaceful Uses of Atomic Energy, held at Geneva in 1955. He became a governor of the International Atomic Energy Agency at Vienna, established as a result of that conference, as well as a member of the scientific advisory committee to its secretary-general. On 24 January 1966 an airplane carrying him to a meeting of that committee crashed on Mont Blanc. A year after his death the Trombay Center was rededicated as the Bhabha Atomic Research Center.

In 1966 Werner Heitler stated that the twenty papers Bhabha had published prior to his return to India in 1939 were sufficient in themselves to have earned him a lasting reputation as a theoretical physicist. Bhabha continued to publish prolifically until he became secretary of the Department of Atomic Energy. Of the sixty-six papers he wrote alone or with others between 1933 and 1954, sixty-two are either original contributions to theoretical physics or review articles on the current state of cosmic ray or nuclear physics. He published no papers in pure physics after 1954, although he invariably participated in the Wednesday afternoon theoretical physics seminar at the Tata Institute whenever he was in Bombay.

Bhabha's Cavendish period coincided with the early years of activity in a field that later emerged as high-energy physics. In 1932 John Cockcroft and E. T. S. Walton achieved the first disintegrations of nuclei with electrostatically accelerated particles. In the same year the positive electron, or positron, was discovered by Carl Anderson at the California Institute of Technology, thus providing spectacular confirmation of the relativistic electron theory of Paul Dirac, Bhabha's tutor in mathematics at Caius College. Shortly afterward Patrick M. S. Blackett and G. P. S. Occhialini, working at the Cavendish, used cloud chamber techniques to demonstrate that electron-positron pairs are produced by the interaction of high-energy gamma rays with matter. They also showed that primary gamma rays of sufficiently high energy can produce secondary electrons and positrons that interact with matter to produce additional gamma rays. The latter can then interact with matter to yield additional electron-positron pairs. Thus a primary gamma ray can dissipate its energy in the form of a shower.

The interpretation of shower phenomena required the development of methods for the detailed analysis of the successive interactions of gamma rays, electrons, and positrons with the atoms constituting bulk matter. Bhabha's first paper, published in 1933, was concerned with the absorption of high-energy gamma rays by matter. In 1935 he derived a correct expression for the cross section (probability) of the scattering of positrons by electrons, a process subsequently known as Bhabha scattering.

During the 1930's cosmic rays provided the only

source of high-energy particles. Thus Bhabha's research inevitably turned to the phenomenological interpretation of cosmic ray interactions. In 1937 he wrote a classic paper with Heitler on the theory of electron- and gamma-ray-induced cosmic ray showers. This paper also demonstrated that the very penetrating component of cosmic ray showers observed at ground level and underground could not be composed of electrons. The particles were later (1946) identified as muons (mu-mesons).

James Chadwick's discovery of the neutron in 1932 led inevitably to speculations about the character of the force that binds protons and neutrons into stable nuclei. In 1935 Hideki Yukawa proposed a model that vested the nuclear force in the exchange of an unstable particle between the nuclear constituents. These hypothetical particles, with about one-sixth the proton mass, became known as mesatrons, and later as mesons. Yukawa's model assumed scalar mesons; that is, mesons with zero intrinsic angular momentum, or spin. Bhabha, among others, developed an exchange model for vector mesons with one unit of spin, in which the unit is Planck's constant, h, divided by 2π. Yukawa's meson, now called the pi-meson, was first positively identified by Occhialini and Cecil F. Powell in a series of cosmic ray experiments in 1947. It soon became evident, however, that this particular meson could not account for all details of the nuclear interaction. With the discovery of several types of heavier mesons of both the scalar and vector types in the early 1960's, Bhabha's conviction that the meson exchange force need not be as simple as Yukawa had first hypothesized was in part vindicated.

Yukawa's meson theory made it clear that cosmic ray phenomena could not be described exclusively in terms of electromagnetic interactions, as had been tacitly assumed prior to 1935. During the late 1930's a good deal of Bhabha's theoretical work dealt with nuclear forces and cosmic rays, and reflected the complications that had been introduced by meson theory. Bhabha was also the first to point out, in a 1938 letter to *Nature*, that the lifetimes of fast, unstable cosmic ray particles would be increased because of the time-dilation effect that follows as a consequence of Einstein's special theory of relativity. The verification of this effect by means of cosmic ray experiments is often cited as one of the most straightforward pieces of experimental evidence supporting special relativity.

Bhabha's published contributions to phenomenological cosmic ray theory continued through 1954. According to Menon, however, Bhabha derived his greatest sense of intellectual achievement from a series of highly abstract papers published during his years at Bangalore and his first years at the Tata Institute. These dealt with the classical theory of point particles moving in a general field, and with relativistic wave equations for particles having half-odd integral intrinsic spins greater than one-half. To Menon, these abstract papers exhibited Bhabha's fascination with the aesthetic beauty of exact mathematical solutions.

Bhabha had a deep involvement with both music and art. As a boy he had learned to appreciate classical Western music by listening to the extensive record collections of his grandfather and aunt, and extended that appreciation during his years in Europe. According to Menon, he began to immerse himself in classical Indian music only during his years in Bangalore. Bhabha also became a serious artist while a student at Cambridge. Over the years the subject matter of his paintings changed from figures to landscapes and then to abstracts. Later, when he had less time for painting, he turned to figure drawing in charcoal and pencil. Bhabha became a major patron of contemporary art in India, purchasing paintings and sculpture for himself, for the Tata Institute, and for the Trombay Atomic Research Center.

The similarities between Bhabha and Jawaharlal Nehru have frequently been noted. Both were born into upper-class, Westernized families; both spent long periods in England during their formative years and traveled extensively in Europe; both discovered and came to appreciate Indian culture after they had learned to feel comfortable with high European culture; both became convinced at an early stage of the importance of science and technology for the development of independent India. The fact that Bhabha was able to convince the prime minister within a year of independence that the government should embark on an atomic energy program and that he, Bhabha, should have undisputed authority over its development was no doubt due in large part to the similarities in their backgrounds and aspirations. Nehru came to rely heavily on Bhabha's scientific advice. Indeed, by 1952 they met weekly whenever both were in India. Bhabha's success in combining in his own person both the scientific and the political direction of the Indian atomic energy program was so persuasive that his successor as secretary of the department of atomic energy, Vikram Sarabhai, was also a physicist who had spent his early professional years in pure research, and had subsequently become a junior colleague of Bhabha.

Bhabha never married. He was president of the Indian National Science Congress in 1951, and in 1954 was awarded the Padma Bhushan by the government of India. He was elected a foreign associate of the U.S. National Academy of Sciences in 1963.

BIBLIOGRAPHY

I. ORIGINAL WORKS. Bhabha's more significant scientific writings include "Zur Absorption der Hohenstrahlung," in *Zeitschrift für Physik*, **86** (1933), 120; "Passage of Very Fast Protons Through Matter," in *Nature*, **134** (1934), 934; "Electron-Positron Scattering," in *Proceedings of the Royal Society*, **A154** (1935), 195; "Passage of Fast Electrons Through Matter," in *Nature*, **138** (1936), 401, written with W. Heitler; "Passage of Fast Electrons and the Theory of Cosmic Showers," in *Proceedings of the Royal Society*, **A159** (1937), 432, written with W. Heitler; "Penetrating Component of Cosmic Radiation," *ibid.*, **A164** (1938), 257; "Theory of Heavy Electrons and Nuclear Forces," *ibid.*, **A166** (1938), 501; "Classical Theory of Mesons," *ibid.*, **A172** (1939), 384; "Elementary Heavy Particles With Any Integral Charge," in *Proceedings of the Indian Academy of Sciences*, **A11** (1940), 347, 468; "General Classical Theory of Spinning Particles in a Maxwell Field," in *Proceedings of the Royal Society*, **A178** (1941), 273, written with H. C. Corben; "General Classical Theory of Spinning Particles in a Meson Field," *ibid.*, 314; "On the Theory of Point Particles," *ibid.*, **A183** (1944), 134, written with Harish-Chandra; "Relativistic Equations for Particles of Arbitrary Spin," in *Current Science* (Bangalore), **14** (1945), 89; and "On the Fields and Equations of Motion of Point-Particles," in *Proceedings of the Royal Society*, **A185** (1946), 250, written with Harish-Chandra.

II. SECONDARY LITERATURE. See Sir John Cockcroft, "Homi Jehangir Bhabha," in *Proceedings of the Royal Institution of Great Britain*, pt. I, **188** (1966), 411–422; Werner Heitler, "Bhabha's Work on High Energy Physics," in *Science Reporter* (Oct. 1966), 449; M. G. K. Menon, "Homi Jehangir Bhabha," in *Proceedings of the Royal Institution of Great Britain*, pt. I, **188** (1966), 423–438; and Lord Penney, "Homi Jehangir Bhabha," in *Biographical Memoirs of Fellows of the Royal Society*, **13** (1967), 35–52.

WILLIAM A. BLANPIED

BIDLOO, GOVARD (*b.* Amsterdam, Netherlands, 21 March 1649; *d.* Leiden, Netherlands, 30 April 1713), *anatomy, biology.*

Bidloo was the son of Govert Bidloo and Maria Feliers, who were Dutch Baptists. Little is known about his education; but he must have received the traditional classical instruction, for at the age of twenty-three he translated a Latin anatomical treatise by Ruysch into Dutch. In 1670 he was apprenticed to a surgeon in Amsterdam and was obliged to attend Ruysch's anatomy lessons and Gerard Blasius' botany lessons at the Hortus Medicus. It is very probable that Bidloo also attended the lectures on medicine presented by Blasius at the Athenaeum Illustre in Amsterdam, for on 2 May 1682 he matriculated at the University of Franeker, where he received the M.D. three days later, after defending his dissertation, *De variis anatomico-medicis positionibus*. In the *Album promotorum* of Franeker, it is noted following his name: "According to the decision of the Senate, Mr. Goverd [*sic*] Bidloo was allowed to give a solemn promise instead of the customary (Hippocratic) oath, on account of his religion." Bidloo was appointed professor of anatomy at The Hague on 24 January 1688. Soon afterward he was invited to give weekly anatomy lessons in Rotterdam as well.

Bidloo's abilities were recognized by *stadhouder* William III,[1] who, in 1690, appointed him "superintendent-general of all physicians, apothecaries and surgeons of the military hospitals of the Netherlands." This was the time of the War of the Grand Alliance (1689–1699), which was fought largely in the Spanish Netherlands. Bidloo visited the various battlefields and made great improvements, especially in the provision of pharmaceutics. In May 1692 he was given the additional duty of supervising the British hospitals.

On 1 February 1694, Bidloo succeeded the celebrated anatomist Anton Nuck as professor of medicine and surgery at the University of Leiden. This appointment was probably due to the influence of William III, for it was customary to appoint only members of the Dutch Reformed Church to teaching positions at the universities. His teaching was frequently interrupted by calls to England from William III (1696, 1699). On 22 October 1701, Bidloo was appointed physician in ordinary to William, and he was among the doctors who attended him during the illness of which he died in 1702. Bidloo wrote an account of William's sickness and death. During his last stay in England, Bidloo was elected a fellow of the Royal Society (21 January 1701).

Because of his service to William III, Bidloo had badly neglected his teaching duties at Leiden, for which dereliction he was officially reprimanded in 1697. After William's death, however, Bidloo became a very good teacher and developed a large practice as well. He allowed some of his students to watch his more difficult operations, thus anticipating the technique of bedside teaching for which Boerhaave later made Leiden famous.

Bidloo is remembered not only as an anatomist and a surgeon but also as a poet and a playwright. He took a lively interest in the politics of his time and often used his poetical talents for political satire.[2]

Bidloo's chief work was his anatomical atlas, which he started in 1676 and must have completed before May 1682: for in that month Anton Nuck wrote a letter to the Royal Society "that Mons. Bidloo, a skilful chirurgeon of Amsterdam, had newly shewed him above 100 anatomical figures of the parts of a man as big as the life, ingraven on copper, with a description of the parts, but not of their use."[3] The 105 plates had been drawn by Gerard de Lairesse, one of the finest artists of his time, and had been etched in copper by Pieter and Philip van Gunst. The book—the first large-scale anatomical atlas since Vesalius' *De humani corporis fabrica* (1543)—was published with a Latin text in 1685 and with a Dutch text in 1690. Although some of Bidloo's contemporaries criticized the work on minor points, it was, and still is, generally admired. The book is of additional interest for showing many microscopical drawings of human organs. One of Bidloo's critics, his former teacher Ruysch, incorporated his criticisms in a series of ten letters addressed to various anatomists. Bidloo answered Ruysch's objections in a sarcastic pamphlet, and Ruysch replied in the same vein. The latter wrote six more letters but did not further attack Bidloo. The text of Bidloo's atlas was its weakest point, for he intended to present only a brief description of the plates and refrained from writing an anatomical treatise.

In 1698, *The Anatomy of Humane Bodies,* signed by William Cowper, was published at Oxford.[4] This publication is perhaps the most flagrant case of plagiarism in the history of medicine. All of Bidloo's 105 plates were used; nine more, of doubtful value, were added. Bidloo's text was not translated but was replaced by a new, more elaborate one. The cartouche in the engraved frontispiece was skillfully altered by covering the title of Bidloo's book with that of Cowper's. The only plate in Bidloo's atlas not used in this production was Bidloo's portrait, which was replaced by a large portrait of Cowper. Bidloo's name did not appear on the title page, and in the text it was accompanied only by criticism of his work.

Although aware that Cowper was working on this atlas, Bidloo had been led to believe that it was a translation of his work. He offered Cowper alterations of his text in several letters, which were never answered. When Cowper's book finally reached Bidloo, he wrote a pamphlet, addressed to the Royal Society, in which he exposed the fraud. This booklet ended as follows: "Take action and destroy, deprive this man of the title of honor of your Society. . . . Take away, noble judges, the foreign feathers which this impudent fellow shows off. . . . the ambitious miserly scoundrel and thief of literary knowledge. . . ." Cowper's defense was somewhat lame. He denied having received any letters from Bidloo and disputed Bidloo's right to the plates, claiming that they had been designed not by Bidloo but by Jan Swammerdam—and even if Bidloo had had a hand in them, they had been drawn by Lairesse. The statement that the plates were Swammerdam's was, of course, preposterous; in addition, it was as bad to plagiarize Swammerdam as it was to plagiarize Bidloo.

This was a painful matter for the Royal Society, especially since the book had been published by Smith and Walford, its official printers. Hans Sloane, the second secretary of the Society, was directed by the Council to write to Bidloo "that the Society are not erected for determining controversies, but promoting naturall and experimentall knowledge, which they will do in him or anybody else."[5]

Of Bidloo's other anatomical work, only his work on the nerves will be mentioned. In his *Opera omnia* he proved that the nerves are not hollow tubes, as had been believed since the time of Galen, but are taut, transparent fibrous threads. Consequently, the animal spirits that the nerves were supposed to conduct did not exist.

Bidloo is remembered as a biologist for his admirable work on the liver fluke (*Fasciola hepatica*). He described his work on this parasite in a letter to Leeuwenhoek, who had it published. It is sometimes found bound in a collection of Leeuwenhoek's letters.

NOTES

1. William III of Orange (1650–1702) was *stadhouder* of Holland from 1672. He and his wife Mary were crowned king and queen of England in 1689.
2. See *Nederduitsche en latijnsche keurdichten, bijeen verzamelt door de liefhebberen der oude Hollandsche vrijheit* (Rotterdam, 1710).
3. See T. Birch, *The History of the Royal Society of London*, IV (London, 1757), 151.
4. William Cowper was admitted as a barber-surgeon in London on 9 March 1691. His *Myotomia reformata* was published in 1694. In 1696 he was admitted as a fellow of the Royal Society. In addition to the book cited above and his

edition of Bidloo's atlas, he published a number of papers in the *Transactions of the Royal Society*.

5. See C. R Weld, *A History of the Royal Society*, I (London, 1848), 352.

BIBLIOGRAPHY

I. ORIGINAL WORKS. Bidloo's writings include *Ontdecking der klapvliezen, in de water en melkvaten, nevens eenige seldsame anatomische opservatien. In 't latijn beschreven door F. Ruysch, vertaelt door G. Bidloo. Alwaer bygevoegt zijn vijf briefsgewijse aenmerckingen van Wigger de Vogel* (Amsterdam, 1672); *Anatomia humani corporis, centum et quinque tab. per artificissime G. de Lairesse delin., illustrata* (Amsterdam, 1685), translated into Dutch as *Ontleding des menschelijken lichaams, uitgebeeld naar het leven in 105 afteekeningen door G. de Lairesse* (Amsterdam, 1690; reiss. with new title page, Utrecht, 1728; 1734); *Dissertatio [Oratio] de antiquitate anatomes* (Leiden, 1694); *Oratio in funere Pauli Hermanni* (Leiden, 1695); *Disputatio medica inauguralis de vera medicinae cognitione, resp. Andreas Lundelius* (Leiden, 1696); *Vindiciae quarundam delineationum anatomicarum contra ineptas animadversiones Fred. Ruyschii* (Leiden, 1697); *Observatio medico-anatomica de animalculis in ovino aliorumque animantium hepate, detegendis. Resp. Henricus Snellen* (Leiden, 1698), translated into Dutch as *Brief van G. Bidloo aan Antony van Leeuwenhoek; wegens de dieren welke men zomtijds in de lever der schapen en andere beesten vind* (Delft, 1698), English trans. in S. Hoole, *The Select Works of Antony van Leeuwenhoek* (London, 1800), 144–145; *Gulielmus Cowper, criminis literarii citatus, coram tribunali nobiliss: ampliss: Societatis britanno-regiae* (Leiden, 1700); *Verhaal der laatste ziekte en het overlijden van Willem de III^de, Koning van Groot Brittanje enz.* (Leiden, 1702); *Exercitationum anatomico-chirurgicarum, Decas I* (Leiden, 1704); *Exercitationum anatomico-chirurgicarum, Decas II* (Leiden, 1715); *De oculis et visu variorum animalium observationes physico-anatomicae* (Leiden, 1715); and *Opera omnia anatomico-chirurgica, edita et inedita* (Leiden, 1715).

II. SECONDARY LITERATURE. The letters in which Ruysch criticized Bidloo were published in a series of pamphlets: J. Gaubius, *Epistolae problematicae ad Fr. Ruyschium, et hujus responsiones* (Amsterdam, 1696–1704), repr. in F. Ruysch, *Opera omnia anatomico-medico-chirurgica, huc usque edita* (Amsterdam, 1737), translated into Dutch as *Alle de ontleed- genees- en heelkundige werken van Frederik Ruysch* (Amsterdam, 1744). See also F. Ruysch, *Responsio ad Godefridi Bidloi libellum, cui nomen vindicarum inscripsit* (Amsterdam, 1697; 2nd ed., 1738), Dutch trans. in F. Ruysch, *Ontleed-genees- en heelkundige werken,* I, 439–483. W. Cowper's works are *The Anatomy of Humane Bodies, With Figures Drawn After the Life by Some of the Best Masters in Europe, and Curiously Engraven in 114 Copperplates. Illustrated With Large Explications Containing Many New Anatomical Discoveries and Chirurgical Observations. To Which Is Added an Introduction Explaining the Animal Economy* (Oxford, 1698; 2nd ed., C. B. Albinus, ed., Leiden, 1737); and Εὐχαριστια *in qua dotes plurimae et singularis Godefridi Bidloo M.D. et in illustrissima Leydarum Academia anatomiae professoris celeberrimi, perita anatomica, probitas, ingenium, elegantiae latinatis, lepores, candor, humanitas, ingenuitas, solertia, verecundia, humilitas, urbanitas, etc., celebrantur et ejusdem citationi humilime respondetur* (London, 1701; repr. London, 1702).

See also the following, listed chronologically: J. P. Elias, *Overzicht van de geschiedenis der geneeskunde in Rotterdam* (Rotterdam, 1912), 33–35; J. van der Hoeven, "Een nieuw liedeken van de klagende lijken op 't Cartuysers kerkhof," in *Bijdragen tot de geschiedenis der geneeskunde*, **3** (1923), 335–336; E. D. Baumann, "Govard Bidloo," in *Nieuw nederlandsch biografisch woordenboek*, VIII (1930), 104–108; J. A. J. Barge, "Het geneeskundig onderwijs aan de Leidsche Universiteit in de 18de eeuw," in *Bijdragen tot de geschiedenis der geneeskunde,* **14** (1934), 1–22; B. A. G. Veraart, "G. Bidloo's verhaal van de laatste ziekte en het overlijden van Willem III, Koning van Engeland," *ibid.*, 203–210; F. Beekman, "Bidloo and Cowper, Anatomists," in *Annals of Medical History*, n.s. **7** (1935), 113–129; J. van Ditmar, "Dood van Stadhouder-koning Willem III," in *Bijdragen tot de geschiedenis der geneeskunde*, **17** (1937), 165–166; A. J. P. van den Broek, "Bidloo en Cowper," *ibid.*, **22** (1942), 72–77; W. Vasbinder, *Govard Bidloo en William Cowper* (Utrecht, 1948); E. D. Baumann, *Drie eeuwen nederlandsche geneeskunde* (Amsterdam, 1951), 246–249; R. Herrlinger, "Bidloo's 'Anatomia'; Prototyp barocker Illustration?" in *Gesnerus*, **23** (1966), 40–47; and N. Moore, "William Cowper," in *Dictionary of National Biography*, IV, 1313–1314.

PETER W. VAN DER PAS

BIENAYMÉ, IRÉNÉE-JULES (*b.* Paris, France, 28 August 1796; *d.* Paris, 19 October 1878), *probability, mathematical statistics, demography, social statistics.*

Bienaymé's secondary education began at the *lycée* in Bruges and concluded at the Lycée Louis-le-Grand, Paris. He took part in the defense of Paris in 1814 and enrolled at the École Polytechnique the following year; the institution was closed briefly in his first year, however. Bienaymé became lecturer in mathematics at the military academy at St.-Cyr in 1818, leaving in 1820 to enter the Administration of Finances. He soon became inspector and, in 1834, inspector general. At about this time he became active in the affairs of the Société

Philomatique de Paris. The revolution of 1848 led Bienaymé to retire from the civil service, and he received a temporary appointment as professor of the calculus of probabilities at the Sorbonne. Despite his retirement, he had considerable influence as a statistical expert in the government of Napoleon III; he was attached to the Ministry of Commerce for two years and was praised in a report to the Senate in 1864 for his actuarial work in connection with the creation of a retirement fund. At the Paris Academy, to which he was elected in 1852, he served for twenty-three years as a referee for the statistics prize of the Montyon Foundation, the highest French award in that field.

Bienaymé was a founding member of the Société Mathématique de France (president in 1875, life member thereafter), corresponding member of the St. Petersburg Academy of Sciences and of the Belgian Central Council of Statistics, and honorary member of the Association of Chemical Conferences of Naples. He became an officer of the Legion of Honor in 1844. Bienaymé had a considerable knowledge of languages; he translated a work by Chebyshev from Russian and at his death was preparing an annotated translation of Aristotle from classical Greek.

Laplace's *Théorie analytique des probabilités* (1812) was Bienaymé's guiding light; and some of his best work, particularly on least squares, elaborates, generalizes, or defends Laplacian positions. Bienaymé corresponded with Quetelet, was a friend of Cournot's, and was on cordial terms with Lamé and Chebyshev. His papers were descriptive, his mathematics laconic; and he had a penchant for controversies. No sooner was he elected to the Academy than he locked horns with Cauchy over linear least squares. In 1842 he had attempted to criticize Poisson's law of large numbers (pertaining to inhomogeneous trials); this invalid criticism was not published until 1855. He also sent a group of communications to the Academy criticizing the Metz Mutual Security Society. Bertrand was to take blistering exception to his style.

Bienaymé did not publish until he was in his thirties, and only twenty-two articles appeared. Almost half are in a now obscure source, the scientific newspaper-journal *L'Institut, Paris*, and were reprinted at the end of the year of their appearance in the compendium *Extraits des Procès-Verbaux de la Société philomatique de Paris*.

The early writings lean to demography; "De la durée de la vie en France" (1837) discusses the life tables of Antoine Deparcieux and E. E. Duvillard de Durand, both of which were widely used in France. Its major object is to present overwhelming evidence against the continued use of the Duvillard table, employment of which by insurance companies had been to their undoubted financial advantage, since the mortality rates that it predicted were much more rapid than was appropriate for France at that time. Apart from writings on the stability of insurance companies, Bienaymé's other direct contribution to the social sciences concerned the size of juries and the majority required for conviction of the accused. The jury system in France had been in a state of flux; its revision was based largely on interpretations of results obtained by Laplace, whose conclusions were later rejected by Poisson in *Recherches sur la probabilité des jugements . . .* (1837); Bienaymé naturally sided with Laplace. Some other papers by Bienaymé have demographic or sociological motivation but involve major methodological contributions to probability or mathematical statistics.

The stability and dispersion theory of statistical trials is concerned with independent binomial trials, the probability of success p_i in the ith trial in general depending on i. The subject forms the early essence of the "Continental direction" of statistics, which is typified by Lexis, Bortkiewicz, A. A. Chuprov, and Oskar Anderson. One of its major aims is to test for, and typify, any heterogeneity in the p_i's (the homogeneous, or Bernoulli, case is that of all p_i's being equal). Bienaymé introduced a physical principle of *durée des causes*, with which he showed that the proportion of successes exhibits more variability than in the homogeneous case (this fact might therefore explain such manifested variability). His reasoning here was not understood until much later. In the context of a set of Bernoulli trials in which there is a specified number of successes, divided into two successive blocks of trials, Bienaymé in 1840 showed an understanding of the important statistical concept of sufficiency, now attributed to Fisher (1920). These contributions established him, after Poisson, as a founder of the "Continental direction."

In linear least squares one is concerned with the estimation of (in modern notation) an $r \times 1$ vector, β, of unknowns from a number N of observations Y, related linearly to β but subject to error ϵ: $Y = A\beta + \epsilon$. Bienaymé extended Laplace's asymptotic treatment of the system as $N \to \infty$; but the only essential originality is in a largely unsuccessful attempt to find a simultaneous confidence region for all the coefficients β_i, $i = 1, \cdots, r$. Nevertheless, "Mémoire sur la probabilité des erreurs" contains a deduction of an almost final form of the continuous chi-squared density, with n degrees of freedom, for the sum of squares of n independent and identically distributed $N(0,1)$ random variables. The impassioned defense of least squares against Cauchy ("Considérations à l'appui de la découverte de Laplace," 1853) contains three important incidental results, the most startling of which is the Bienaymé-Chebyshev inequality,

$$P(|\bar{X}-EX| \geq \epsilon) \leq \text{var } X/(\epsilon^2 n),$$

proved by the simple argument still used. Chebyshev obtained it by a more difficult means in 1867 and published his results simultaneously in Russian and French. This was juxtaposed with a reprint of Bienaymé's paper and in 1874 Chebyshev himself credited Bienaymé with having arrived at the inequality via the "method of moments," the discovery of which he ascribed to Bienaymé.

Perhaps the most startling of Bienaymé's contributions to probability is a completely correct statement of the criticality theorem for simple branching processes. His "De la loi de la multiplication et de la durée des familles" (1845) anticipated the partly correct statement of Galton and Watson by some thirty years, and predated the completely correct one of Haldane (until recently thought to be the first) by over eighty years. This work may have been stimulated by L. F. Benoiston de Châteauneuf. Of only slightly less significance is a remarkably simple combinatorial test for randomness of observations on a continuously varying quantity. This method involves counting the number of local maxima and minima in the series; Bienaymé stated in 1874 that the number of intervals, complete and incomplete, between extrema in a sequence of N observations is (under assumption of randomness) approximately normally distributed about a mean of $(2N - 1)/3$ with variance $(16N - 29)/90$. This result, which is describable in modern terms as both a nonparametric test and a limit theorem, is technically complex to prove even by modern methods.

A sophisticated limit theorem proved (but not rigorously) by Bienaymé is the following: If the random variables Θ_i, $i = 1, \cdots, m$ satisfy $0 \leq \Theta_i \leq 1$, $\Sigma\Theta_i = 1$, and the joint probability density function of the first $(m - 1)$ is const. $\Theta_1^{x_1} \Theta_2^{x_2} \cdots \Theta_m^{x_m}$ where $x_i \geq 0$ is an integer, then if $V = \Sigma\gamma_i\Theta_i$, as $n \to \infty$

$$\sqrt{n}\,(V-\bar{\gamma})/\sqrt{\Sigma(\gamma_i-\bar{\gamma})^2 x_i/n} \to N(0,1),$$

where $\bar{\gamma} = \Sigma\gamma_i x_i/n$, $n = \Sigma x_i$, $r_i = x_i/n = \text{const.}_i > 0$. With a more general distribution for the Θ_i, this result was reobtained in 1919 by Von Mises, who regarded it as a *Fundamentalsatz.*

Finally, there is the algebraic result announced by Bienaymé in 1840: let $\{a_i\}$, $\{c_i\}$ be sets of positive numbers. Then

$$\left(\sum_{i=1}^{n} c_i a_i^m \Big/ \sum_{i=1}^{n} c_i\right)^{1/m}$$

is nondecreasing in $m \geq 0$. In a probabilistic setting this result is credited to Lyapounov, the mathematical attribution being to O. Schlömilch (1858). The result contains the Cauchy inequality and the earlier inequality between the arithmetic and geometric means, which it also complements by another consequence:

$$n^{-1}(a_1 + \cdots + a_n) < (a_1^{a_1} \cdots a_n^{a_n})^{(a_1 + \cdots + a_n)^{-1}}.$$

Bienaymé was far ahead of his time in the depth of his statistical ideas. Because of this and the other characteristics of his work, and his being overshadowed by the greatest figures of his time, his name and contributions are little known today.

BIBLIOGRAPHY

I. ORIGINAL WORKS. Bienaymé's writings include "De la durée de la vie en France depuis le commencement du XIXe siècle," in *Annales d'hygiène publique et de médecine légale*, **18** (1837), 177–218; "De la loi de multiplication et de la durée des familles," in *Société philomatique de Paris. Extraits des procès-verbaux des séances*, 5th ser., **10** (1845), 37–39, also in *L'Institut, Paris*, no. 589 (1845), 131–132, and repr. by Kendall in 1975 (see below); *De la mise à l'alignement des maisons* (Paris, 1851), a satirical dialogue that demonstrates his abhorrence of the legal system; "Mémoire sur la probabilité des erreurs d'après la méthode des moindres carrés," in *Journal des mathématiques pures et appliquées*, **17** (1852), 33–78, repr. in *Mémoires presentés par savants étrangers à l'Académie des sciences*, **15** (1868), 615–663; *Notice sur les travaux scientifiques de M. I.-J. Bienaymé* (Paris, 1852), with a partial bibliography; and "Considérations à l'appui de le découverte de Laplace sur la loi de probabilité dans la méthode des moindres carrés," in *Comptes rendus . . . de l'Académie des sciences*, **37** (1853), 309–324, repr. in *Journal des mathématiques pures et appliquées*, 2nd ser., **12** (1867), 158–176.

II. SECONDARY LITERATURE. See A. Gatine, "Bienaymé," in *École polytechnique. Livre du centenaire 1794–1894*, III (Paris, 1897), 314–316; J. de la Gournerie, "Lecture de la note suivant, sur les travaux de M. Bienaymé," in *Comptes rendus . . . de l'Académie des sciences*, **87** (1878), 617–619; C. C. Heyde and E. Seneta, "The Simple Branching Process, a Turning Point Test and a Fundamental Inequality: A Historical

Note on I.-J. Bienaymé," in *Biometrika*, **59**, no. 3 (1972), 680–683; and "Bienaymé," in *Proceedings of the 40th Session of the International Statistical Institute* (Warsaw, 1975); D. G. Kendall, "Branching Processes Since 1873," in *Journal of the London Mathematical Society*, **41** (1966), 385–406; and "The Genealogy of Genealogy: Branching Processes Before (and After) 1873," in *Bulletin of the London Mathematical Society*, **7** (1975), 225–253; M. G. Kendall and A. Doig, *Bibliography of Statistical Literature*, III (Edinburgh, 1968), the most extensive bibliography; H. O. Lancaster, "Forerunners of the Pearson χ^2," in *Australian Journal of Statistics*, **8** (1966), 117–126; R. von Mises, *Mathematical Theory of Probability and Statistics*, edited and complemented by Hilda Geiringer (New York, 1964), 352–357; and L. Sagnet, "Bienaymé (Irénée-Jules)," in *Grande encyclopédie, inventaire raisonné des sciences, des lettres et des arts*, VI (Paris, n.d. [1888]), 752.

C. C. HEYDE
E. SENETA

AL-BIṬRŪJĪ AL-ISHBĪLĪ, ABŪ ISḤĀQ, also known as **Alpetragius** (his surname probably derives from Pedroche, Spain, near Cordoba; *fl.* Seville, *ca.* 1190), *astronomy, natural philosophy.*

The outstanding astronomer among the Spanish Aristotelians, al-Biṭrūjī may have been from Los Pedroches (Bitrawsh), Córdoba province. In his only known work, *Kitāb fī'l-hay'a* ("Book of Astronomy"), he says that he is a pupil of Ibn Ṭufayl (*d.* 1185), who was already dead by the time the book was finished (I, 61; II, 49).[1] Since Michael Scot completed the Latin translation of al-Biṭrūjī's work as *De motibus celorum circularibus* in 1217, the *Kitāb* must be dated between those two years. According to Yahūda ibn Solomon Kohen of Toledo, al-Biṭrūjī died in 1217; the date is doubtful, however, because of the coincidence with the translation by Michael Scot. The *Kitāb fī'l-hay'a* was translated into Hebrew by Moses ibn Tibbon in 1259; in 1247 Yahūda ibn Solomon Kohen had produced an abridged version. The Hebrew text was translated into Latin by Qalonymos ben David.

According to al-Biṭrūjī, Ibn Ṭufayl expounded an astronomical system that differed from Ptolemy's and did not use eccentrics or epicycles. Although he promised to develop the system in a book, Ibn Ṭufayl seems not to have done so. In view of the agreements between al-Biṭrūjī's astronomical system and the much less elaborate ideas of Ibn Rushd, F. J. Carmody suggests that both were derived from the work of Ibn Ṭufayl.

Al-Biṭrūjī considered Ptolemy's system to be mathematical, not physical, with a recognizable exactitude and precision valuable to the astronomical computer (I, 59–60; II, 37–41). Therefore all the parameters in the *Kitāb fī'l-hay'a* are derived from the *Almagest.* Al-Biṭrūjī was familiar with Jābir ibn Aflaḥ's criticisms of certain defects in the Ptolemaic system (I, 113; II, 269; I, 122; II, 309) and of the problem of the order of the spheres of the inferior planets (I, 53; II, 5; I, 124, 125; II, 315, 321). Jābir's *Iṣlāḥ al-Majisṭī* is also one of the ways through which the sine theorem was introduced into Spain (I, 98; II, 207).

Nevertheless, the main "defect" that al-Biṭrūjī found in the Ptolemaic system was the incompatibility of its basic principles with Aristotle's physical concepts. If the source of all motion in the universe is the Prime Mover, situated in the ninth sphere, it would be absurd to suppose that the Prime Mover transmits motions in opposite directions to the different spheres—the diurnal motion from east to west, the movements of longitude from west to east (see I, 53–57; II, 5–29). One must accept that the motion of the ninth sphere—the fastest, strongest, and simplest—is transmitted to the lower spheres, which are slower in proportion to their distance from the Prime Mover. Precession of the sphere of the fixed stars and the movements of longitude of the planetary spheres constitute a sort of slowing down (*taqṣīr, incurtatio*) that affects the diurnal motion, the transmission of which from the ninth sphere is explained by using the theory of impetus (I, 78; II, 137). Thus Saturn would be the fastest-moving planet and the moon the slowest (see I, 63–68; II, 57–91).

These ideas were not original with al-Biṭrūjī. Lucretius (*De rerum natura*, vv. 621 ff.) attributes them to Democritus, and Alexander of Aphrodisias credits them to the Pythagoreans. Theon of Alexandria took them up again in his commentary on the *Almagest* (I, 7), as did Ibn Rushd. The motion transmitted from the ninth sphere rebounds in the sublunar world and is transmitted to the element fire, producing the shooting stars(?) (*shihāb* or *ashbāh al-kawākib*); it is then transmitted to the elements air and water, in the latter case producing the tides and waves (I, 64; II, 63–69). Another anti-Ptolemaic argument of Aristotelian origin used by both Ibn Rushd and al-Biṭrūjī is the *horror vacui* necessarily arising from the movements of the eccentric spheres (I, 61; II, 47–49). Al-Biṭrūjī also held antiempirical views: he did not trust human senses, given the distance between the observer and the spheres, and put his faith in human reason (I, 66; II, 79–81).

On these bases al-Biṭrūjī elaborated a system that sought only to give qualitative explanations, a limitation of which he was quite aware (I, 76; II, 127–129; I, 154; II, 427–429). He postulated the existence of a ninth sphere, the Prime Mover, the discovery of which he attributed to the "modern" astronomers. This sphere moves around the poles of the equator, from east to west, completing one revolution in twenty-four hours (I, 66; II, 77; I, 76; II, 129). Next is the eighth sphere, that of the fixed stars, which is moved by the motion of the ninth; its poles (those of the ecliptic) describe two small circles around the poles of the universe, since they participate in the diurnal movement of the ninth sphere.

With this scheme al-Biṭrūjī constructed a model of variable precession. B. R. Goldstein has calculated that in it, if k is the mean value of precession, the maximum value is 1.1 k and the minimum is 0.9 k. This indicates that al-Biṭrūjī did not accept the theory of trepidation, although he spoke of "accession" (*iqbāl*) and "recession" (*idbār*); but here the sense is that "accession" refers to precession at an amount greater than the mean value, and "recession" refers to precession at an amount less than the mean value.

Like al-Zarqālī,[2] al-Biṭrūjī, when presenting a brief history of the theories of precession and trepidation, attributed to Theon of Alexandria the belief in a combination of the trepidation of the equinoxes along an arc of eight degrees with the Ptolemaic precession of one degree in one hundred years. This may have been a forerunner of the models of variable precession that were used in Spain by the Alfonsine astronomers and in the Arab world by Naṣīr al-Dīn al-Ṭūsī and Quṭb al-Dīn al-Shīrāzī, and whose echo reaches Copernicus.

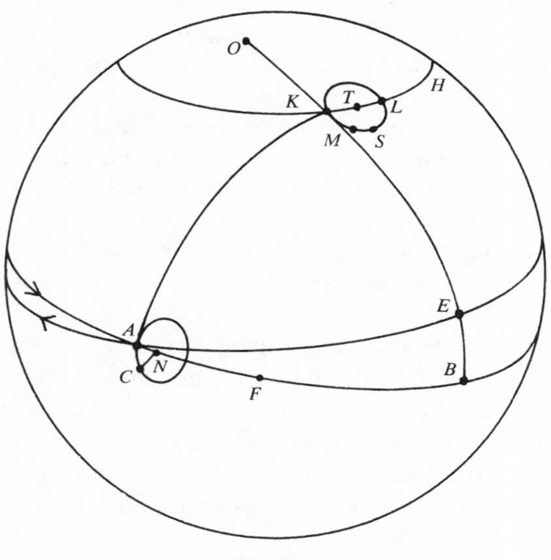

FIGURE 1

On the other hand, the fixed stars were subject to diurnal motion and to precession that moved them, so that they did not describe a perfect circle but a curve called *lawlab ḥalazūnī* by al-Biṭrūjī and, traditionally, interpreted as a spiral. Plato (*Timaeus* 39A) had spoken of the spiral motion of the heavenly bodies, but the source used by al-Biṭrūjī probably was Aristotle. The first mentions in *Kitāb fī'l-hay'a* (I, 61; II, 49; I, 62; II, 51, 53) of the *lawlabī* motion are citations of, or references to, *De caelo* II, 8, in which Aristotle states that the heavenly bodies have two types of motion—δίνησις and κύλισις—the first of which suggests the concept of a vortex and could be interpreted by the commentators as a spiral. Theon of Alexandria (commentary on the *Almagest*, I, 2) also spoke of spiral motion of the heavenly bodies, as did Ibn Rushd and Albertus Magnus, both of whom attributed the concept to Aristotle.

The *lawlabī* motion also affected the planets (I, 101; II, 219–221). The planetary models were based on the following schema (see Figure 1): Planet A, according to al-Biṭrūjī, always was ninety degrees from its pole K, moving with the motion of K. The movement in longitude occurred in a circle parallel to the equator, the polar deferent HT, which is the circle traced by the poles of the ecliptic around the poles of the universe. The motion in anomaly takes place on the polar epicycle KSL, which has a radius equal to the maximum latitude of the planet and a center T that moves on circle HT. Al-Biṭrūjī stated that the motions of T on HT and of the planetary pole on KSL are in the same direction (I, 110; II, 257) and, speaking of Jupiter (I, 117; II, 289), added that T moves in the direction of signs, thus abandoning the basic principle of only one movement, the diurnal, from east to west.

With this model al-Biṭrūjī could justify the variations in the velocity of the planet, which—in the case of Saturn, for example—would be average when its pole was at K and L; maximum when the pole was approximately at S; decreasing when the pole was between S and L; and at its stationary point, beginning afterwards to retrograde, when the pole was between L and K. The lunar model was very similar to the planetary one, although al-Biṭrūjī added some modifications in order to account for its variable motion and lack of stationary points and retrogradation.

An important aspect of al-Biṭrūjī's planetary theory is his discussion of the order of the inferior planets. After presenting the history of the question, he gave the order as moon, Mercury, sun, Venus, Mars, and so on: Mercury was slower than the sun, which was slower than Venus. He rejected the objections made to the traditional order (moon, Mercury, Venus, sun) based on the fact that the transits of Mercury and Venus across the sun are not visible. He pretended that Mercury and Venus have their own light and do not receive it from the sun, as the moon does (I, 123–125; II, 313–323). Therefore their transits cannot be perceptible.

As for the solar model, al-Biṭrūjī began by considering an epicyclic model similar to the planetary ones; but he rejected it because the radius of the epicycle would have to be extraordinarily small in order not to produce a perceptible latitude. He solved the problem by moving Ptolemy's eccentric to the north pole of the equator. Then the pole of the solar sphere would describe a circle around the pole of the universe, and the sun would always remain (in theory) ninety degrees from its pole, which would be moving twice as fast as the sun.

E. S. Kennedy states: "Al-Biṭrūjī's system is a clever adaptation, with Ptolemaic parameters, of a device invented by Eudoxus (*fl. ca.* 360 B.C.) and incorporated into Aristotelian cosmology."[3] He thus demonstrates that the Eudoxian model for Saturn is equivalent to al-Biṭrūjī's. B. R. Goldstein, on the other hand, rejects any Eudoxian influence and favors that of al-Zarqālī: al-Biṭrūjī cites a work by the latter, *Fī ḥarakat al-iqbāl wa'l-idbār* ("On Accession and Recession"), that can be identified with the *Treatise on the Movement of the Fixed Stars*. In the *Treatise* al-Zarqālī justifies the positions of the equinoxes with a polar deferent and an epicycle. Goldstein considers this the inspiration for al-Biṭrūjī, who substituted a planet for the equinox.

Al-Biṭrūjī's astronomical system spread through much of Europe in the thirteenth century. William the Englishman cited it; and Grosseteste referred to it in several works, even plagiarizing from it in his refutation of the Ptolemaic system. In the second half of the century there were disputes between supporters of Ptolemy and Aristotelian defenders of al-Biṭrūjī; Albertus Magnus spread al-Biṭrūjī's ideas in simplified form, although he ultimately preferred the Ptolemaic system. His *De caelo*, or a similar work, may have been the source in which Dante found the ideas of al-Biṭrūjī; the same may be true of Vincent of Beauvais in his *Speculum naturale*. Richard of Middleton also chose the Ptolemaic system, rejecting the views of Ibn Rushd and al-Biṭrūjī. Roger Bacon, in his *Communia naturalium*, expounded al-Biṭrūjī's system in detail and compared it with Ptolemy's: parts of this work have been brought together in *Liber tertius Alpetragii in quo tractat de perspectiva*. In his *Opus maius* Bacon discussed al-Biṭrūjī's theory of the tides. Others who considered the problem and opted for the Ptolemaic system were Bernard of Verdun, Giles of Rome, Pietro d'Abano, and John of Jandun.

The fourteenth century saw the definite return of the Ptolemaic theory, although Henry Bate of Malines toward the end of the previous century and Henry of Hesse sought to construct astronomical systems without eccentrics and epicycles.[4] The system of Henry of Hesse, however, seems to be related not to al-Biṭrūjī's but to that of Abū Ja'far al-Khāzin (died between 961 and 971), although the connection is not clear.

Among Hebrew authors, besides the abridgment of the *Kitāb* by Yahūda ibn Solomon Kohen (1247), Isaac Israeli of Toledo (*fl.* 1310) seems to refer to al-Biṭrūjī in his *Liber Jesod Olam* when speaking of *ha-īsh ha-mar'īsh* ("the man [whose theory] shook [the world]"), who lived until 1140 and opposed the Ptolemaic system. Isaac Israeli seems quite skeptical, however, about the reality of the new system, since it cannot be verified. Levi ben Gerson, in his *Sefer Tekunah*, discusses al-Biṭrūjī's model for Saturn, introducing corrections that lead Goldstein to believe that the discussion is based on memory and that Levi did not have al-Biṭrūjī's text before him. In general, Levi ben Gerson rejects the system for not agreeing with observed reality and indicates that various philosophical arguments of al-Biṭrūjī's are refuted in other chapters of his *Milḥamot Adonai*.

The diffusion of al-Biṭrūjī's ideas continued in the fifteenth and sixteenth centuries. At the end of the fifteenth, the astrologer Simon de Phares cited him (but had not read his work) and attributed fantastic ideas to him. Regiomontanus wrote a brief work on al-Biṭrūjī's errors, in which he used both astronomical arguments—parallax, the impossibility of explaining total and annular solar eclipses if the moon is always at the same distance from the earth—and astrological ones. In some instances his criticism reflects misunderstanding of al-Biṭrūjī's system; for instance, he states that the latter placed Mercury and Venus above the sun. In the sixteenth century Copernicus (*De revolutionibus*, I, 10) cited his system in connection with theories of the order of the inferior planets.

NOTES

1. The references in parentheses are to Goldstein's English trans. (I) with Arabic text (II).
2. J. M. Millás Vallicrosa, *Estudios sobre Azarquiel* (Madrid–Granada, 1943–1950), 275–276.
3. *Speculum*, **29** (1954), 248.
4. Claudia Kren, "Homocentric Astronomy in the Latin West: The *De reprobatione eccentricorum et epiciclorum* of Henry of Hesse," in *Isis*, **59** (1968), 269–281.

BIBLIOGRAPHY

I. ORIGINAL WORKS. Bernard R. Goldstein has edited, with English trans., the Arabic text of *Kitāb fī'l-hay'a* (facs. of MS Escorial 963 with variants from MS Istanbul Seray 3302) and the Hebrew trans. by Moses ibn Tibbon: *Al-Biṭrūjī: On the Principles of Astronomy* . . ., 2 vols. (New Haven–London, 1971). The Latin trans. by Michael Scot has been edited by Francis J. Carmody as *Al-Biṭrūjī De motibus celorum* . . . (Berkeley–Los Angeles, 1952). The Latin version by Qalonymos ben David was published with Sacrobosco's *Sphere* as *Alpetragii arabi planetarum theorica physicis rationibus probata nuperrime latinis litteris mandata a Calo Calonymos hebreo napolitano* (Venice, 1531).

II. SECONDARY LITERATURE. See F. J. Carmody, "Regiomontanus' Notes on al-Biṭrûjî's Astronomy," in *Isis*, **42** (1951), 121–130; and "The Planetary Theory of Ibn Rushd," in *Osiris*, **10** (1952), 556–586; J. L. E. Dreyer, *A History of Astronomy from Thales to Kepler* (repr. New York, 1953), 264–267, 278; P. Duhem, *Le système du monde*, II (Paris, 1914), 146–156; Léon Gauthier, "Une réforme du système astronomique de Ptolémée tentée par les philosophes arabes du XIIe siècle," in *Journal asiatique*, **14** (1909), 483–510, reproduced, with simplifications and additions, in Gauthier's *Ibn Rochd (Averroès)* (Paris, 1948), 113–127; Bernard R. Goldstein, "On the Theory of Trepidation According to Thābit b. Qurra and al-Zarqāllu and Its Implications for Homocentric Planetary Theory," in *Centaurus*, **10** (1964), 232–247; and "Preliminary Remarks on Levi ben Gerson's Contributions to Astronomy," in *Proceedings of the Israel Academy of Sciences and Humanities*, **3** (1969), 239–254; E. S. Kennedy, review of Carmody's *De motibus*, in *Speculum*, **29** (1954), 246–251; Bruno Nardi, "Dante e Alpetragio," in *Giornale Dantesco*, **29** (1926), 41–53; E. Rosen, "Copernicus and al-Biṭrūjī," in *Centaurus*, **7** (1961), 152–156; and George Sarton, *Introduction to the History of Science*, II (Baltimore, 1931), 399–400, also 18, 620, 749–750, 757, 925, 930, 937, 952, 956, 964, 990, 995; and III (Baltimore, 1947–1948), 440, 539, 544.

JULIO SAMSÓ

BODENSTEIN, MAX (*b.* Magdeburg, Germany, 15 July 1871; *d.* Berlin, Germany, 3 September 1942), *chemistry.*

Bodenstein was the son of Franz Julius Bodenstein, a brewery owner, and Elise Meissner, the daughter of August Christian Meissner, a district court judge. His father's business awakened an interest in chemistry, which he soon began to study in a sort of alchemical workshop erected in the cellar of the private house attached to the brewery. Second, the roof of the brewery tempted him to practice mountain-climbing techniques, which he later put to use on the glaciers forming the roof of the Swiss Alps. In short, he became a chemist and an Alpinist. At age seventeen Bodenstein entered the University of Heidelberg. After six semesters he began his doctoral work under Victor Meyer. At first the subject, the thermal decomposition of hydrogen iodide, did not appeal to him; he would have preferred preparing organic compounds and colorful substances. He ably conducted the experiments involved, which required great technical skill (including glassblowing) and interpreted the results in accord with the very "new" kinetic theory of gases. Everything seemed to agree perfectly. Even though the interpretation has subsequently been altered, his work stands as a milestone in the study of reaction kinetics.

Bodenstein was awarded the doctorate *summa cum laude* in 1893. For the next two years he continued his training under Liebermann at Berlin-Charlottenburg and under Nernst at Göttingen. He also spent one year in the army, serving in a cavalry regiment. In 1896 he returned to Heidelberg and married Marie Nebel. Bodenstein's chief scientific interest was still reaction kinetics, and in 1899 he published his *Habilitationsschrift*, "Über die Gasreaktionen in der chemischen Kinetik," a work that brought him to the attention of the scientific world.

From 1900 to 1906 Bodenstein worked with Ostwald's group of researchers at Leipzig, where he investigated the kinetics of catalytic processes. He was the first to suggest that in heterogeneous catalysis it is not the concentration of the substance in the reaction vessel that is decisive, but rather the concentration at the surface of the catalyst. And as early as 1906 he applied the method later named for Langmuir and Hinshelwood to the heterogeneous dissociation of antimony hydride. In the same year Nernst summoned Bodenstein to Berlin as extraordinary professor and department head at the institute of which he became director many years later at the end of his career. While at Berlin he studied catalysis in flowing systems, as well as the kinetics of gas reactions (hydrogen bromide dissociation). A notable result from this period was the designation of diffusion and reaction as possible steps determining the velocity. The dimensionless quotient obtained by multiplying the velocity of flow with the pipe length and dividing by the diffusion constant is today known as the Bodenstein number.

In 1908 Bodenstein accepted a professorship at the Technische Hochschule in Hannover. There, during a balloon flight that took place on an exceptionally stormy day and ended with a dramatic landing, he met Walter Dux; on the way back to Hannover they worked out a plan for the experiments on the photochemical chlorine hydrogen reaction that later became famous. The dissociation of hydrogen bromide had been shown to be far more complicated than the simple proportionality relationships that held for hydrogen iodide. The study of the photochemical chlorine hydrogen reaction resulted in a further surprise: the velocity was found to be proportional to the square of the chlorine concentration and inversely proportional with the oxygen concentration. Through the concept of a chain reaction Bodenstein explained this law and, simultaneously, the fact that the photochemical yield exceeded the Einstein law of equivalents by a factor of 10^4. This accomplishment was the start of a long struggle to determine the nature of the chain propagators involved, their formation and deformation. Finally, more than a decade after Bodenstein's work with Dux, the question was settled by postulating the existence of the "atomic chain" originally proposed by Nernst.

By that time Bodenstein had returned to the University of Berlin, where he had become Nernst's successor as head of the Institute of Physical Chemistry in 1923. Although reluctant to leave Hannover, he could not refuse a call to a university with a physics faculty that included Planck, Einstein, von Laue, and Nernst. In the succeeding years Bodenstein was awarded many honors: membership in the Prussian Academy of Sciences, an honorary doctorate from the Technische Hochschule in Hannover, an honorary doctorate of science from Princeton University, and honorary membership in the Chemical Society (London) and in other academies and chemical and physical-chemical societies, in Germany and abroad.

Until the end of his life Bodenstein was occupied with "Abschlussarbeiten am Chlorknallgas." He originally believed he had been particularly fortunate to begin his research on kinetic reactions with the obviously "simple" case of hydrogen iodide. But the more scientists studied chain reactions, the more they found that even very simple-seeming gross reaction equations can result from complicated chain mechanisms, and they were no longer sure whether hydrogen iodide formation is really a reaction between hydrogen and iodine molecules. A remark that Bodenstein made late in his life may be interpreted as indicating that he had come to believe it necessary to revise his original views: "I would like to investigate hydrogen iodide once again with the aid of modern techniques." He did not live to do so. It was only in 1967 that this reaction, too, was shown by John H. Sullivan to take place through the intermediary of halogen atoms.

Bodenstein always regarded his work on "Chlorknallgas" as his most important scientific achievement. We must agree with him when we consider everything that has emerged from the research he and his school devoted to that reaction: the concept of the chain reaction, the explanation of why the law of equivalents is violated in photochemical reactions, the computation of rate equations from a reaction mechanism scheme, the theoretical calculation of the chain length, the idea of explosions produced by chain branching, the explanation of the sensitivity of a chain reaction to impurities and wall influences and the quantitative control of these disturbances, the importance of the threefold collision for kinetics, the determination of the absolute values of the velocity constants of all individual reactions in the formation of hydrogen chloride and of phosgene, and the explanation of the inhibition of reactions during strong drying as resulting from the introduction of impurities that break the chains. The knowledge gained in studying the chlorine hydrogen reaction has been applied to many other chain reactions.

Bodenstein died following a brief illness. He was engaged in research until the end of his life, not only guiding the work of students but also conducting experiments himself.

BIBLIOGRAPHY

The obituary notice by M. von Laue, in *Jahrbuch der Preussischen Akademie der Wissenschaften* for 1946–1949, 127–139, includes a bibliography of about 200 papers by Bodenstein. Among them are "Gasreaktionen in der chemischen Kinetik," in *Zeitschrift für physikalische Chemie*, **29** (1899), pt. 1, 167–158, pts. 2–3, 295–333; "Photochemische Kinetik des Chlorknallgases," *ibid.*, **85** (1913), 297–397; "Abschlussarbeiten am Chlorknallgas," pt. 1 in *Sitzungsberichte der Preussischen Akademie der Wissenschaften zu Berlin*,

Phys.-math. Abt., 1936, 2–18; and pts. 2–3 in *Zeitschrift für physikalische Chemie*, **B48** (1941), 239–288, and "100 Jahre Photochemie des Chlorknallgases," in *Berichte der Deutschen chemischen Gesellschaft*, **75A** (1942), 119–125.

ERIKA CREMER

BOGDANOV (MALINOVSKY), ALEKSANDR ALEKSANDROVICH (*b.* Tula, Russia, 22 August 1873; *d.* Moscow, U.S.S.R., 7 April 1928), *philosophy, medicine.*

Bogdanov was the second of six children of Aleksandr Malinovsky, a schoolteacher and later school inspector. An outstanding student at the Tula Gymnasium, where he received state scholarships, Bogdanov early developed an interest in the natural sciences. In 1891 he entered Moscow University, where at first his scientific studies were quite general, including mathematics, physics, chemistry, and biology. Bogdanov soon became involved in student political activities, initially of the non-Marxist populist variety; in 1894 he was arrested and exiled to Tula. There he became active in factory workers' discussion groups, began to study Marxism, and lectured workers on popular science and radical politics. A belief in the validity of a union of science and revolutionary activity remained a characteristic of his entire career.

In 1895 Bogdanov resumed his university studies, this time at the Medical Faculty of Kharkov University, where he spent four academically successful and politically active years. Soon after graduation he was again arrested, held for a time in Moscow, and then exiled for three years to northern Russia, where he worked as a psychiatrist. In 1903 he joined the Bolshevik wing of the Russian Social Democratic Labor Party; by 1905–1907 he was second only to Lenin in the party. His intellectual unorthodoxy, particularly his criticism of philosophical materialism, soon led to a split with Lenin. In 1911 Bogdanov left organized politics. During World War I he served as a military physician. Following the Revolution he was the theoretician of the "Proletarian Culture" movement, a group calling for the creation of a distinct revolutionary culture. After 1918 he was a member of the Socialist (later Communist) Academy and was a major promoter of universities for students from the lower classes and the *Great Soviet Encyclopedia.*

Bogdanov's last years were spent organizing scientific research, particularly in blood transfusion, a practice that had become standard during World War I. He set up the first large-scale research center for the study of blood transfusion in Soviet Russia and tirelessly propagandized the value of cooperative blood transfusion for "rejuvenation." He often carried out experiments on himself, and his death followed an exchange of blood with a young student who had had both malaria and tuberculosis. He made observations of his condition until just before his death.

Bogdanov's voluminous writings can be divided into five categories: political economy, historical materialism, philosophy, science, and proletarian culture. Only philosophy and science are discussed here, although his influence in the other areas was probably greater. His most important scientific work was not that on blood transfusion but his attempt to create a "universal organizational science."

In his two major theoretical works, *Empiriomonizm* (1904–1907) and *Vseobshchaya organizatsionnaya nauka* (Tektologia) ("Universal Organizational Science [Tectology]"; 1913–1929), Bogdanov presented a major effort to harmonize Marxism and modern science. In the first he maintained that the key to an understanding of knowledge lay in the principles of its organization, not in a search for "reality" or "essence." Neither materialism nor idealism, therefore, was an appropriate or useful epistemological position. Bogdanov followed Ernst Mach and Richard Avenarius in denying the dualism of sense perceptions and physical objects, but he believed that they had not gone far enough in explaining the existence of the subjective and objective realms of experience. Bogdanov attempted to unite these realms in a new philosophical system, empiriomonism, by deriving the physical world from "socially organized experience" and the mental world from "individually organized experience." The two worlds revealed two different "biological-organizational tendencies." Why, asked Bogdanov, do men differ so radically about the second realm, the sphere of individually organized values? The answer, he thought, was that men are torn apart by class, race, linguistic and national conflicts, by specialization arising from technical knowledge, and by relations of dominance and subordination of all kinds. If these conflicts could be overcome, he continued, a new consciousness would emerge as a result of which men would be in much greater agreement about values than ever before.

In developing his concept of "tectology," Bogdanov sought to find, through structural analogies and models, the organizational principles that would unite "the most disparate phenomena" in the organic and inorganic worlds under one conceptual scheme. All objects that exist, he wrote, can be distinguished in terms of the degree of their organization. Entities on higher levels of organization possess properties that are greater than the sum of their parts. Living beings and automatic machines are dynamically structured complexes in which "bi-regulators" provide for the maintenance of order. Recent commentators on Bogdanov have cited this apparent prefiguring of the concept of cybernetic feedback. Bogdanov called for the application of "bi-regulation" and degree of organization in his "universal organizational science," which would embrace the biological and social worlds in the way that mathematics had described classical mechanics.

After the 1920's Bogdanov's work was largely ignored in the Soviet Union, no doubt a result both of Lenin's earlier criticism of his epistemology and of a turn away from theoretical concerns to the practical tasks of industrialization. In the 1960's a group of Soviet philosophers attempted to revive interest in Bogdanov by relating his work to the later ideas of Norbert Wiener, W. Ross Ashby, and Ludwig von Bertalanffy. Thus far, however, Bogdanov has remained little-known both in the Soviet Union and abroad, and no critical evaluation of his scientific work exists.

BIBLIOGRAPHY

I. ORIGINAL WORKS. Bogdanov's works number in the hundreds, but no complete bibliography has been published. His major writings in science and philosophy are *Empiriomonizm*, 3 vols. (Moscow – St. Petersburg, 1904 – 1907); and *Vseobshchaya organizatsionnaya nauka (Tektologia)* ("Universal Organizational Science [Tectology]"), 3 vols. (Moscow – St. Petersburg, 1913 – 1929). His best-known work in political economy is *Kratky kurs ekonomicheskoy nauki* (Moscow, 1897). On historical materialism, see *Nauka ob obshchestvennom soznany* (Moscow, 1914). Proletarian culture is treated in *O proletarskoy kulture* (Leningrad – Moscow, 1924), a collection of his articles from 1904 – 1924, and in his two utopian novels, *Krasnaya zvezda* (St. Petersburg, 1908) and *Inzhener Menni* (Moscow, 1913). A short autobiography was published in *Entsiklopedichesky slovar*, XLI (Moscow, 1926), 29 – 34.

II. SECONDARY LITERATURE. The major biography is Dietrich Grille, *Lenins Rivale: Bogdanov und seine Philosophie* (Cologne, 1966), which contains the most complete bibliography of Bogdanov's works yet published, pp. 253 – 259. For a discussion of his work on logic, see V. F. Asmus, "Logicheskaya reforma A. Bogdanova," in *Vestnik kommunisticheskoy akademii*, no. 22 (1927), 122 – 144. A discussion of his work in hematology is A. A. Belova, "Odna iz pamyatnykh stranits proshlogo," in *Problemy gematologii i perelivania krovi* (1967), no. 6, 54 – 57. An article calling for recognition of Bogdanov's contribution to systems theory is M. I. Setrov, "Ob obshchikh elementakh tektologii A. Bogdanova, kibernetiki i teorii sistem," in *Uchenye zapiski kafedr obshchestvennykh nauk vuzov goroda Leningrada* (1967), no. 8, 49 – 60. Lenin's criticism of Bogdanov is in his *Materializm i empiriokrititsizm* (Moscow, 1909), esp. 265 – 273. English-language discussions devoted primarily to Bogdanov's political concerns include Peter Scheibert, "Lenin, Bogdanov and the Concept of Proletarian Culture," in Bernard W. Eissenstat, ed., *Lenin and Leninism* (Lexington, Mass., 1971), 43 – 57; and S. V. Utechin, "Philosophy and Society: Alexander Bogdanov," in Leopold Labedz, ed., *Revisionism: Essays on the History of Marxist Ideas* (New York, 1962), 117 – 125; and "Bogdanovism," in his *Russian Political Thought: A Concise History* (New York – London, 1963), 207 – 213.

LOREN GRAHAM

BORN, MAX (*b*. Breslau, Germany [now Wrocław, Poland], 11 December 1882; *d*. Göttingen, Germany, 5 January 1970), *theoretical physics*.

Born was the son of Gustav Born, a professor of anatomy, and Margarethe Kauffmann, who came from a family of Silesian industrialists. Born attended the König-Wilhelms-Gymnasium in Breslau and began his university studies in that city in 1901. At first he attended lectures on a wide variety of subjects, but he soon discovered that his interest lay in mathematics and the exact sciences.

After three semesters at Breslau and two summer semesters at Heidelberg and Zurich, in 1904 Born entered the University of Göttingen, where he immediately established a close relationship with Hilbert. Born was given the honor of preparing the lecture transcripts for the mathematics reading room, and in this capacity he became Hilbert's private assistant in 1905. Born never let these manuscripts, which are written in an exceptionally clear hand, out of his possession; they later accompanied him when he emigrated and when he returned to Germany. The extensive knowledge of mathematics that Born acquired in preparing Hilbert's lectures became one of his greatest assets.

Even before the publication of Einstein's papers, Born had been introduced to the problems of relativity in Hermann Minkowski's seminar on the electron theory. There he became familiar with the works of Fitzgerald, Lorentz, Larmor, Poincaré, and others, as well as with Minkowski's own ideas (first published in 1907).

At the same time, in the winter semester of 1904–1905, Born attended the seminar on selected topics in the theory of elasticity given by Felix Klein in collaboration with Ludwig Prandtl, Carl Runge, and Woldemar Voigt. His report on the stability of elastic wires and tapes impressed Klein, who persuaded the philosophical faculty to set this problem as the subject of a prize competition. Born's decision not to submit an entry but to turn his attention, instead, to relativity angered Klein, who was virtually all-powerful at Göttingen. Born found himself obliged, against his will, to write a paper for the contest. This circumstance also determined the subject of his dissertation, for papers accepted in the prize competition automatically qualified as dissertations.

After receiving his doctorate in 1907, Born returned to Breslau. His plans to learn scientific experimentation from Lummer and Ernst Pringsheim were not realized; and he decided to pursue the questions raised by Minkowski. In 1908 he sent the manuscript of a paper to Minkowski, who immediately invited him to come to Göttingen as an associate. What should have been a fruitful collaboration ended with Minkowski's sudden death at the beginning of 1909. It fell to Born to put Minkowski's scientific papers in order, and from them he was able to produce a paper "Eine Ableitung der Grundgleichungen für die elektromagnetischen Vorgänge in bewegten Körpern vom Standpunkte der Elektronentheorie (Aus dem Nachlass von Hermann Minkowski bearbeitet von Max Born)," in *Mathematische Annalen*, **68** (1910), 526–551.

When Born presented his own, independently developed ideas on the self-energy of the relativistic electron to the Mathematische Gesellschaft in Göttingen, Klein interrupted him so often that he became completely confused. But Hilbert and Runge took his part and arranged for him to speak again at a later session. This time it went well, and Born was invited by the theoretical physicist Woldemar Voigt to use this work for his habilitation.

In 1909 Born went to Cambridge, to J. J. Thomson and Joseph Larmor, "in order to learn something about the electron at the source." After several months he returned to Breslau, where Stanislaus Loria acquainted him with the works of Einstein. Born wrote, "Although I was thoroughly familiar with the notions of relativity and the Lorentz transformations, Einstein's train of thought came as a revelation to me."

In 1912 Born accepted an invitation from Michelson to lecture on relativity theory at the University of Chicago. The following year he married Hedwig Ehrenberg, the daughter of Viktor Ehrenberg, a professor of law in Leipzig. They had two daughters and one son, Gustav, who, like his paternal grandfather, became a biologist.

While Born's initial scientific works concerned relativity theory, the second major group of his publications was inspired by Einstein's work of 1907 on specific heat. In 1912 Born and Theodor von Kármán published "Über Schwingungen in Raumgittern," which contains a theoretical foundation for the deviations from Einstein's quantum formula that had been experimentally ascertained by Nernst and his collaborators. Einstein considered only the case of crystals with a single frequency. Born and Kármán, however, showed that it is not the atoms of a crystal that should be considered as independent resonators but, rather, the principal oscillations, which lie in a determined frequency range. Their paper was recorded as "received on 20 March 1912" by the *Physikalische Zeitschrift*. A few days earlier, Peter Debye had reported similar results (reached, however, by an entirely different approach) to the March session of the Swiss Physical Society. Born and Kármán did not learn of these findings until after the publication of their own paper. Scientists generally preferred Debye's theory, which was easier to work with and clearer from a physical point of view. This theory fails at very low temperatures, however, whereas that of Born and Kármán gives the correct observed values in this region as well.

Born undertook to erect a unified crystal physics on atomic foundations. This work soon exceeded the normal compass of a journal article and appeared in 1915 as *Dynamik der Kristallgitter*. The physical nature of the forces between the atoms in the lattice was unknown at the time (except in the case of the ionic crystals); it was only decades later that the question was elucidated with the help of the quantum theory. Nevertheless, by assuming very small deviations of the "crystal particle" from the position of rest—an assumption that, independently of the universally valid force law, always leads to a linear relationship between displacement and force—it was possible to derive the essential elastic, thermal, electric, and optical properties of the crystals. Born provided a comprehensive exposition of lattice dynamics, in brief, through *Dynamik der Kristallgitter* and his article in *Encyclopädie der mathematischen Wissenschaften*, which appeared in 1923 as a book entitled *Atomtheorie des festen Zustandes*. Simultaneously, he laid one of the cornerstones of solid-state physics.

Born also devoted a considerable part of his later work to these problems. In *Handbuch der Physik* he and Maria Göppert-Mayer reported on developments up to 1933. After World War II he collaborated with Kun Huang, professor at the University of Peking, on a new version of this work, based completely on quantum theoretical foundations, *Dynamical Theory of Crystal Lattices* (1954).

In 1915 Born was called to Berlin as extraordinary professor, to relieve Max Planck of his lecture obligations. "It is really an uncommon stroke of good luck for me," he wrote to Johann Jakob Laub in Buenos Aires, "to be appointed in the middle of the war. . . . I have been assigned to the army service corps (telegraph, etc.) . . . Germany's power is great and her cause is good; we are happy to be her sons." Born was able to spend most of his time in the army in Berlin, in the office of the ordnance testing commission. Despite the difficulties and privations of the war and of the postwar period, his years in Berlin (1915–1919) were, in his own judgment, among the best of his life. He was close to Planck and Einstein, and with the latter he began a deep friendship to which both have left ample testimony.

In 1919 Born and Max von Laue exchanged their teaching positions. Born assumed Laue's post at Frankfurt, where he had a small laboratory at his disposal. Here his assistant, Otto Stern, working with Walther Gerlach, conducted experiments that later became famous on the directional quantization of atoms in the magnetic field. Stern began in 1919 with a direct demonstration of Maxwell's law of velocity distribution. "This gave me the idea," Born stated, "of making other quantities appearing in the gas theory, which up to now have been established only indirectly, accessible to measurement by means of molecular rays." This intention was the origin of a series of experimental works, above all a direct measurement of the free path length of gas molecules, which he did with his assistant Elisabeth Bormann.

Two years later Born was offered a post at the University of Göttingen, where more than a decade earlier he had made a strong personal impression by earning the doctorate with a prize work and by his habilitation. In negotiations with the farseeing Prussian minister of education, Carl Becker, Born succeeded in having James Franck appointed at Göttingen at the same time as himself. Since it now included Hilbert, Born, Franck, and Robert Pohl—all of whom were deeply immersed in current scientific problems—the Göttingen faculty covered the whole range from mathematics, through theoretical physics, to experimental physics. The preconditions were thereby created for productive collaborative efforts.

At first Born continued his investigation of the dynamics of the crystal lattice, but later regretted having devoted so much time to it. (He encouraged his students to choose dissertation topics in this field until 1926.) After the "Bohr-Festspiele"—a major lecture series that Bohr gave at Göttingen in June 1922—questions concerning the quantum theory became paramount. In the seminar on the structure of matter, which was attended by mathematicians as well as physicists, the structure of the atom henceforth held the center of attention.

Through his "correspondence principle," Bohr had explained that any future "quantum theory" must differ fundamentally from classical physics but that a relationship nevertheless exists between them and that, accordingly, the task is to find the "transition." Thus Born, inspired by Bohr's remarks, sought to establish a "quantum mechanics." The new term first appeared in "Über Quantenmechanik," in *Zeitschrift für Physik* (1924). Born believed that in substance this work was only a precursor employing a method already known (by Kramers, Heisenberg, and others) in which certain differential operators of classical mechanics were replaced by difference operators. The resulting second-order perturbation formula for the energy was in full agreement with the one later developed in quantum mechanics. The paper "Zur Quantenmechanik aperiodischer Vorgänge," in *Zeitschrift für Physik*, **33** (1925), 479–505, written with Jordan, utilized this same method to calculate the absorption and emission of a resonator in the radiation field, with the goal of overcoming the contradiction between classical field theory and quantum hypothesis in the derivation of Planck's radiation formula. In the course of this investigation it was found that the quantum physics pertained not to individual states but, rather, to pairs of states, to which a "transition amplitude" must be assigned.

Building on this foundation, the twenty-four-year-old Heisenberg, who was Born's assistant, succeeded in cutting the Gordian knot in a manuscript he gave to Born in July 1925 with a view toward publication. Entitled "Über quantentheoretische Umdeutung kinematischer und mechanischer Beziehungen," this paper contains the conceptual foundations of matrix mechanics.

Born immediately recognized the work's far-reaching significance. Heisenberg's multiplication rule left him no peace, and after eight days of intensive thinking and examining, he suddenly remembered an algebraic theory he had learned at Breslau from his teacher Rosanes. Such quadratic schemata, well-known to mathematicians, are called "matrices" when they are in conjunction with a definite multiplication rule. He applied this rule to Heisenberg's quantum condition and found that the latter agreed with the quantities appearing in the diagonal. It was easy to guess what the remaining quantities must be—null—and he was immediately confronted with the formula $pq - qp = h/2\pi i$. This meant that the coordinates q and the impulses p could not be represented by numerical values but, rather, through symbols the product of which is dependent on ordering—they do not commute. This result convinced Born that they were on the right path. Yet a large portion of it had only been guessed at: the disappearance of the nondiagonal members in the above expression. In this connection Born enlisted the aid of his student Pascual Jordan, and in a few days they succeeded in showing that Born had guessed correctly.

Born and Jordan published "Zur Quantenmechanik" (1925) as a sequel to Heisenberg's work. After this joint paper had been sent to *Zeitschrift für Physik*, Born traveled with his family to the Engadine. After returning in October, he continued his collaborative study with Jordan. At first they maintained a lively correspondence with Heisenberg, who was with Bohr in Copenhagen; Heisenberg then returned to Göttingen before the end of the month and participated in writing a definitive account of the new ideas. They had to work quickly, since Born had to leave Göttingen at the beginning of November to give a series of lectures at the Massachusetts Institute of Technology. This was the origin of the *Dreimännerarbeit* ("Zur Quantenmechanik II"), written shortly before Born's departure for the United States.

The new matrix mechanics was at first applied only to the very simplest cases, such as the harmonic and the anharmonic oscillator; but by October 1925 Wolfgang Pauli had calculated the complete hydrogen spectrum. Physicists finally had at their disposal the long-sought new method for computing the atom's stationary states, which had so long resisted an exact mathematical treatment. As Pauli had shown in the first nontrivial example, the results were correct. Still, in its original formulation matrix mechanics was suited to the description only of periodic processes.

In the summer of 1925 Norbert Wiener gave a well-received lecture at Göttingen; and as a result he was invited by Born, Hilbert, and Richard Courant to work at the university during the summer session of 1926. Before that, however, in the winter of 1925–1926, Born had been a guest lecturer at the Massachusetts Institute of Technology, where he began a highly productive collaboration with Wiener. At the latter's suggestion, "matrix" was replaced by the general concept of an operator. Wiener has said that Born had some doubts; Hilbert was consulted and gave his approval. The idea was elaborated in "Eine neue Formulierung der Quantengesetze für periodische und nicht periodische Vorgänge" (1926).

Very soon after publication of Erwin Schrödinger's works on wave mechanics, Born recognized—despite Heisenberg's and Pauli's objections to its basic conceptions—that the new theory was acceptable from a mathematical point of view; and he used Schrödinger's method of treating atomic scattering processes. Applied to a standard scattering problem with known interaction—the scattering of a particle in an external field—the quantum theory permitted an exact calculation only in principle; except in special cases the basic differential equations could not be solved. With "Quantenmechanik der Stossvorgänge" (1926) Born elaborated the basis of the "Born approximation method" for carrying out the actual computations; the method has since grown steadily in importance.

Rutherford's formula for the scattering of α particles by atomic nuclei, which he derived in 1911 from nonrelativistic mechanics and confirmed experimentally, was derived as the first Born approximation by Born's student Gregor Wentzel. It is characteristic of the entire method that it yields the statistics of the scattering process directly without having to introduce physically meaningless—because not measurable—quantities. If the question is treated by means of classical mechanics, it is necessary to work with the concept of the collision parameter and then to ascertain its mean value. From the formalism Born was also able to derive the existence of energy thresholds, a typically quantum mechanical effect in inelastic scattering processes. He thereby established a connection between the theory and the experiments of Franck and Hertz.

With Born's method it became possible to provide a systematic treatment of a large portion of the atomic scattering processes. This point can be illustrated, for example, by the great extent to which N. F. Mott and H. S. W. Massey draw on Born's work in their *Theory of Atomic Collisions* (1933). The fact that the Born approximation converges to a steadily increasing degree with the growth of energy of the colliding particles makes the method very useful in modern high-energy physics. Further, quantum electrodynamics, which is so far the only area of elementary particle theory that has received a satisfactorily coherent elucidation, is based on a version of Born's approximation method. Richard Feynman accomplished this in a clear and elegant form with his "Feynman diagrams."

Born's "Quantenmechanik der Stossvorgänge" furnished the basis not only of the approximation method named for him but also of the statistical interpretation of the quantum theory, for which he was awarded the Nobel Prize in 1954. Born considered the square of the absolute value of the amplitude of the scattered wave $|f(k,\Theta)|^2 d\Omega$ as the probability that the scattered particle is deflected through the angle in the solid angle $d\Omega$. The next step was to conceive of the square of the absolute value of the wave amplitude $|\Psi(x,y,z)|^2 dV$ as a quantity indicating the probability that the particle will be encountered in the region of space dV about the point (x,y,z). The current density to be assigned according to the continuity equation, to the volumetric density $|\Psi|^2$ gives the probability per unit area that after the collision the scattered particle will pass through a surface perpendicular to the direction of impact.

In order to obtain additional verification for this conception, Born prepared two studies (the second with Vladimir Fok) in which he considered the influence of external forces on the state of a physical system. Thus, in the summer of 1926 Born complemented the computational schema of the quantum theory with precise physical representations. The interpretation of the new calculus, further elaborated by Heisenberg and Bohr, culminated in the "Copenhagen interpretation."

Born's works found worldwide recognition, and gifted young researchers flocked to work under him. The "Born school" at Göttingen was as important to the flowering of theoretical physics as the schools of Bohr at Copenhagen and of Arnold Sommerfeld at Munich. Born's students and associates included Max Delbrück, Walter Elsasser, Enrico Fermi, Vladimir Fok, Yakov Frenkel, Maria Göppert-Mayer, Werner Heisenberg, Walter Heitler, Friedrich Hund, Pascual Jordan, Theodor von Kármán, John von Neumann, Lothar Nordheim, J. Robert Oppenheimer, Wolfgang Pauli, Léon Rosenfeld, Edward Teller, Victor F. Weisskopf, Norbert Wiener, and Eugene P. Wigner.

From his lectures Born produced the lengthy textbook *Optik* (1933), which was photostatically copied and widely used in the United States. He often returned to optics later in his career and in 1959 published *Principles of Optics*, written with Emil Wolf and limited to optics in the strict sense.

Through his contributions to crystal physics, matrix mechanics, the statistical interpretation of the quantum theory, and the creation of his approximation method in quantum mechanics, Born showed himself to be one of the most capable physicists of his day. All this counted for nothing in 1933. On 25 April he was placed on "leave of absence," in conformity with stipulations of the "Gesetz zur Wiederherstellung des Berufsbeamtentums" (law for the restoration of the professional civil service), which was directed against Jewish civil servants. He left Germany on 15 May and spent several months in Wolkenstein in the South Tyrol. On 25 June, at a conference in Zurich, he met Patrick M. Blackett, who invited him to Cambridge. He was appointed to the Stokes lectureship in applied mechanics, a minor post compared with his position at Göttingen. The Rockefeller Foundation gave him an additional stipend for research.

During his stay in Wolkenstein, Born had the use neither of his own books nor of a public library. Under this constraint he turned to nonlinear electrodynamics, on which very little literature existed. He discovered a deviation from Maxwell's theory, in which the energy of a point charge is finite. He worked out this idea at Cambridge with Leopold Infeld, and their account of the phenomenon is known as the Born-Infeld theory.

Finding it necessary to earn additional money, Born published *Atomic Physics* (1935), based on his 1932 lectures at the Technische Hochschule in Berlin-Charlottenburg. The German edition, *Moderne Physik* (1933), had been suppressed by the Nazis.

In October 1936 Born was named Tait professor of natural philosophy at the University of Edinburgh, succeeding Sir Charles G. Darwin. Gradually he gathered a circle of students, including a number of Chinese (Peng, Cheng, L. M. Yang, and, above all, Kun Huang). Born now had heavy teaching responsibilities, and he and his associates published many works based on the subjects he taught. Thus Born again became the leader of a large group of researchers. Nevertheless, as he himself stated, Edinburgh, where he was obliged to teach many basic courses, was not what Göttingen had been in its glorious period.

Besides Klaus Fuchs (who later became famous as an "atom spy"), Born's most important student at Edinburgh was Herbert S. Green. After World War II he and Born jointly published many works on the statistical mechanics of condensed systems. These publications, along with the virtually contemporaneous ones of J. E. Mayer, J. G. Kirkwood, J. Yvon, and G. E. Uhlenbeck, established the rational theory of liquids. Born and Green provided an overall picture of the subject in *A General Kinetic Theory of Liquids* (1949).

In 1953 Born became professor emeritus at the University of Edinburgh. He returned to Germany the following year and retired with his wife to Bad Pyrmont, where he spent the rest of his life. When his name became known to the public through the awarding of the Nobel Prize, Born discovered a new mission: to draw attention to the dangers confronting man in the atomic age. On 15 July 1955, at a meeting of Nobel laureates on Mainau Island in Lake Constance, he signed, along with fifteen of his colleagues, a statement condemning the development of atomic weapons. Eventually fifty-one Nobel laureates signed the statement. In the debate over equipping the German army with atomic weapons (April 1957), he joined other leading German physicists in drawing up the "Appeal of the Göttingen 18," which strongly influenced public opinion in the Federal Republic of Germany.

Born's later essays on the history of science are, in part, extracts from the posthumously published autobiography (1975), written in two parts in 1944–1948 and 1961–1963. He wrote the latter in English so that it could be read by his grandchildren, who had become British citizens. The essays and autobiography reflect the profound change in Born's views on science and politics. From a loyal German citizen he had become a cosmopolite with British citizenship who viewed Germany's postwar problems (such as the partition of the country) with the cool detachment of a foreigner.

To ease his disposition after the depressing political analyses, he translated into English the humorist Wilhelm Busch. Finally, after some hesitation, he published his correspondence with Einstein, accompanied by an extensive commentary. Altogether Born published more than twenty books; and his bibliography lists more than 300 articles in physics journals, including those written with his students and friends.

Born was awarded the Grosses Bundesverdienstenkreuz and nine honorary doctorates. A member of numerous scientific societies, he also received the Stokes Medal (Cambridge, 1936), the Macdougall-Brisbane Prize (Edinburgh, 1945), the Max Planck Medal of the Deutsche Physikalische Gesellschaft, and the Hughes Medal of the Royal Society.

BIBLIOGRAPHY

I. Original Works. The most important of Born's more than 300 articles are reprinted in his *Ausgewöhlte Abhandlungen*, 2 vols. (Göttingen, 1963), with a bibliography in II, 695–706. The papers on quantum mechanics are reprinted in *Zur statistischen Deutung der Quantentheorie*, 2 vols. (Stuttgart, 1962); and in *Zur Begründung der Matrizenmechanik* (Stuttgart, 1962), both of which are in the series Dokumente der Naturwissenschaft.

His most important books are *Dynamik der Kristallgitter* (Leipzig, 1915); *Der Aufbau der Materie* (Berlin, 1922), also in English (London, 1923); *Die Relativitätstheorie Einsteins und ihre physikalischen Grundlagen* (Berlin, 1923), also in English (London, 1924); *Vorlesungen über Atommechanik* (Berlin, 1925); *Probleme der Atomdynamik*, 2 vols. (Berlin, 1926), also in English (Cambridge, Mass., 1926); *Optik. Ein Lehrbuch der elektromagnetischen Lichttheorie* (Berlin, 1933); *Atomic Physics* (Glasgow, 1935); *The Restless Universe* (Glasgow, 1935); *Experiment and Theory in Physics* (Cambridge, 1943); *A General Kinetic Theory of Liquids* (Cambridge, 1949); *Natural Philosophy of Cause and Chance* (Oxford, 1949); *Dynamical Theory of Crystal Lattices* (Oxford, 1954); *Principles of Optics* (London, 1959), written with Emil Wolf; *Einstein's Theory of Relativity* (New York, 1962); *Von der Verantwortung des Naturwissenschaftlers* (Munich, 1965); and *Der Luxus des Gewissens. Erlebnisse und Einsichten im Atomzeitalter* (Munich, 1969); and *The Born-Einstein Letters*, Irene Born, trans. (London–New York, 1971), covering the period 1916–1955.

Autobiographical writings are *Physics in My Generation* (London, 1956), a collection of autobiographical papers; "Recollections," in *Bulletin of the Atomic Scientists*, 3 pts., **21**, nos. 7–9 (Sept.–Nov. 1965); and *Mein Leben. Die Erinnerungen des Nobelpreisträgers* (Munich, 1975), a German trans. of his autobiography.

II. Secondary Literature. See Viktor Frenkel, "Max Born," in *Ideen des exakten Wissens* (1972), 289–298; Armin Hermann, "Max Born. Eine Biographie," in Born's *Zur statistischen Deutung der Quantentheorie*, with bibliography 120–130; and Friedrich Herneck, "Max Born," in *Bahnbrecher des Atomzeitalters. Grosse Naturforscher von Maxwell bis Heisenberg* (Berlin, 1965), 336–355.

Armin Hermann

BORRO, GIROLAMO, also known as **Borri** and **Hieronymus Borrius** (*b.* Arezzo, Italy, 1512; *d.* Perugia, Italy, 26 August 1592), *natural philosophy, methodology of science.*

Little is known of Borro's early life, except that he was born at Arezzo, the son of Mariano Borro. He studied theology, philosophy, and medicine, perhaps at Padua, although there is no record of his having received a degree there. About 1537 Borro entered the service of Cardinal Giovanni Salviati, whom he served as a theologian for sixteen years. After teaching for a short period at the University of Perugia (about 1538), Borro lived in Venice, Padua, and Rome. In 1548 he was in Paris, where he engaged in public disputations in the presence of King Henry II and perhaps also taught at the university, although published records do not indicate this. In 1550, still in the service of Salviati, he returned to Rome to participate in the conclave that resulted in the election of Pope Julius III. After Salviati's death in 1553, Borro began lecturing on philosophy at the University of Pisa, where he remained for six years.

Borro left Pisa in 1559 and we know relatively little about his life for the next sixteen years. In 1561 his first published work appeared, *Del flusso e reflusso del mare*, a treatise which attempts to explain the motion of the tides by appealing to Aristotelian principles. In 1567 he was implicated in the third heresy trial of Pietro Carnesecchi, which took place in Rome and resulted in Carnesecchi's execution. Borro returned to Pisa in 1575, teaching natural philosophy once more until 1586. During that period he published *De motu gravium et levium* (1575) and *De peripatetica docendi atque addiscendi methodo* (1584), an exposition of scientific method according to Aristotelian principles. Borro was absent from Pisa in 1582–1583, when he again had difficulties with the Roman Inquisition, but was freed through the intercession of Pope Gregory XIII.

Borro's years at Pisa were marred by polemics and personal quarrels with colleagues, including Francesco de'Vieri, Francesco Buonamici, and Andrea Camuzio, who apparently reacted violently to his abrasive personality. His opponents finally persuaded the university to dismiss him in 1586. Borro is reported to have returned to the University of Perugia (although his presence there is as yet undocumented), where he continued teaching philosophy until his death at the age of eighty.

Borro was lecturing on natural philosophy at the University of Pisa when Galileo was a student there. It may well be that Galileo heard lectures containing material similar to what had been published in Borro's *De motu gravium et levium*. In any case, Borro's work was in Galileo's library and is referred to specifically in his *De motu* (Edizione Nazionale, I, 333), written at Pisa about 1590. Galileo also knew Borro's work on tides, which he owned in the 1577 edition; and Borro's teaching on the subject is cited in the *Dialogo sopra i due massimi sistemi*.

Borro represents the conservative type of Peripatetic philosophy, against which Galileo began rebelling early in his life. On the whole he was a rather backward-looking Aristotelian: there is only scanty evidence that he was acquainted with the Greek text of Aristotle, he made little use of newly discovered ancient commentaries on Aristotle (such as those of Philoponus), and he was not receptive to the new approaches to the study of philosophy brought by the Renaissance. Moreover, he strongly rejected the application of mathematical methods to the study of natural philosophy, as well as all varieties of Platonism. In short, his basic approach to the study of the natural world was deeply rooted in the Western medieval Peripatetic tradition influenced by the Latin translations of Ibn Rushd's commentaries.

Emphasizing an experiential approach to the study of natural philosophy (as opposed to a mathematical one), Borro was responsible for several interesting and perhaps significant suggestions. In *De motu gravium et levium* (pp. 214–217) he described a protoexperiment performed at his home with the aid of students. In order to resolve a dispute over whether a body of heavy material will fall faster than one of light material, he "took refuge in experience [*experientia*], the teacher of all things." He described how lead and wood balls were dropped from a "high window" to resolve this question: "The lead descended more slowly, namely [it descended] above the wood, which had fallen first to the ground; however many times we were all there waiting for the result of this occurrence, we saw the latter [the wood] fall downward [before the lead]. Not only once but many times we tried it with the same results." The inconclusiveness and somewhat puzzling nature of his results parallel those described by Galileo in *De motu* (I, 333–337). Thus, Galileo's putative Leaning Tower experiment was anticipated some years earlier by one of the most conservative of his own Peripatetic teachers of natural philosophy.

BIBLIOGRAPHY

I. Original Works. Published works, all relatively rare, are *Del flusso e reflusso del mare* (Lucca, 1561; rev. and enl. ed., Florence, 1577; 2nd rev. ed., Florence, 1583), which is accompanied by other writings; *De motu gravium et levium* (Florence, 1575; repr. 1576); and *De peripatetica docendi atque addiscendi methodo* (Florence, 1584). His printed correspondence with Pietro Aretino is listed by Stabile (see below). There is also a letter addressed to him in *Delle lettere del sig. Bonifatio Vannozzi*, I (Venice, 1606), 227–228. In addition to the MSS listed by Stabile, the following should be noted: Florence, Biblioteca Nazionale, II.V.168, fols. 70–74, an oration on the death of Pietro Calefati; and Magl. IX.25, epigrams dedicated to Borro. For an edition and discussion of an unpublished text, see C. B. Schmitt, "Girolamo Borro's *Multae sunt nostrarum ignorationum causae* (MS. Vat. Ross. 1009)," in *Philosophy and Humanism: Renaissance Essays in Honor of Paul Oskar Kristeller* (New York, 1976), 448–462.

II. SECONDARY LITERATURE. The most important biographical study is the article by G. Stabile in *Dizionario biografico degli Italiani*, XIII (1971), 13–17, which has an extensive bibliography. Earlier works that are still useful include A. Fabroni, *Historia Academiae Pisanae*, II (Pisa, 1792; repr. Bologna, 1971), 281–282, 341–346, 469; and U. Viviani, *Medici . . . della provincia aretina* (Arezzo, 1923), 103–109.

For Borro's intellectual and scientific contributions, see the following, listed chronologically: P. Duhem, *Études sur Léonard de Vinci*, III (Paris, 1913), 205–207; L. Olschki, *Geschichte der neusprachlichen wissenschaftlichen Literatur*, II (Leipzig–Florence–Rome–Geneva, 1922; repr. Vaduz, Liechtenstein, 1965), 253–256; *Opere di Galileo Galilei*, A. Favaro, ed., XX (Florence, 1939), 97, 398; E. A. Moody, *Studies in Medieval Philosophy, Science, and Logic* (Berkeley–Los Angeles–London, 1975), 203–286; N. W. Gilbert, *Renaissance Concepts of Method* (New York, 1960), 186–192; E. Garin, *Scienza e vita civile nel Rinascimento italiano* (Bari, 1965), 123–126, 141–142; C. Vasoli, *Studi sulla cultura del Rinascimento* (Manduria, 1968), 341–342; W. A. Wallace, *Causality and Scientific Explanation*, I (Ann Arbor, Mich., 1972), 149–150, 178; and C. B. Schmitt, "The Faculty of Arts at Pisa at the Time of Galileo," in *Physis*, **14** (1972), 243–272.

CHARLES B. SCHMITT

BOSE, JAGADISCHANDRA (*b.* Mymensingh, India [now Nasirabad, Bangladesh], 30 November 1858; *d.* Giridih, India, 23 November 1937), *physics, plant physiology*.

Bose was the son of Bhagawanchandra Bose, a functionary in the Indian civil service, who was a deputy magistrate at Mymensingh, in the Dacca district of Bengal, at the time of his son's birth. His mother was Bamasundari Bose. The family were members of the Kayastha caste, though Jagadischandra later joined the Brahmo Samaj. Bose's early childhood was spent in Faridpur, where he attended a vernacular school established by his father. In 1869 he was enrolled in St. Xavier's School in Calcutta, and in 1875 he passed the entrance examinations for St. Xavier's College, where he received the B.A. in 1879. At St. Xavier he was deeply influenced by the professor of physics, the Belgian Jesuit Eugène LaFont, who had become renowned in Calcutta for his public lecture demonstrations.

Bose traveled to England in the late spring of 1880 to begin medical studies in London. Recurrent attacks of a fever (probably kala azar) contracted during his late teens, however, made those studies too taxing and he decided to pursue pure science instead. He entered Christ's College, Cambridge, in January 1881, taking the natural science tripos and receiving the B.A. in 1884. In the same year he also was awarded the B.Sc. by London University. His teachers at Cambridge included Sydney Vines in botany, Francis Darwin in plant physiology, and Lord Rayleigh in physics. Following his return to India in 1884, Bose was appointed officiating professor of physics in the Imperial Education Service at Presidency College, Calcutta. Three years later he received a lifetime appointment as professor of physics at the college, a post he held until his retirement from the service in 1915. On 30 November 1917 he presided over the dedication of the Bose Research Institute in Calcutta. Modeled on the Royal Institution, London, the establishment was endowed from the proceeds of Bose's own investments, from private benefactions, and from public donations. He retained the directorship of the institute until his death, although from 1931 he lived in virtual retirement.

Bose has generally been regarded as the first modern Indian scientist to establish an international reputation. The first phase of his research was concerned with the generation, reception, and properties of radio waves with wavelengths of about one centimeter. He was awarded the D.Sc. from London University in 1896 for a dissertation based on this research. His first paper, "On the Polarisation of Electric Rays by Double Refracting Crystals," was published in the *Journal of the Asiatic Society of Bengal* in May 1895. A second paper, "On the Determination of the Index of Refraction of Sulphur for the Electric Ray," was submitted to the Royal Society by Lord Rayleigh and was published in its *Proceedings* in October 1895. During that year, Bose demonstrated the capabilities of centimeter waves for wireless telegraphy, transmitting them over a distance of seventy-five feet. In 1896–1897 the government of Bengal sent him on a nine-month lecture tour of Europe. Most of his time was spent in England, where he repeated his wireless telegraphy demonstration; he also visited France and Germany.

Bose's attempts to improve his transmitting and receiving apparatus resulted in the observation that after being exposed for a time to centimeter radiation, most materials experience a decreasing ability to respond to further irradiation but ultimately recover their former sensitivity if left in a quiescent state. In a paper read at the 1900 International Congress of Physics in Paris, Bose likened the electrically induced fatigue in inorganic matter to muscle fatigue in living tissue. His further research on the responses of a wide variety of plants to electrical and mechanical stress led him to posit, in 1901, that vegetable matter might be regarded as a connecting link between animal and inorganic matter. Although the ingenuity of his physiological experiments was generally appreciated, his conclusions and several of his observations were disputed by a group of plant physiologists, including J. S. B. Sanderson, and several of his papers were rejected by the Royal Society.

From about 1902 until his retirement from active scientific work some thirty years later, Bose's research was devoted primarily to studies of plant responses to a broad range of stimuli. These investigations were carried out with a number of ingenious and highly sensitive instruments of his own design. Although the precise measurements that Bose made with these instruments were generally regarded as valid, the body of his research was not taken seriously by the majority of contemporary plant physiologists. Bose's stated ambition during the latter part of his career was to comprehend the lives of plants in their entirety. Ultimately he expressed the conviction that animal and vegetable life exhibited a fundamental unity, and also stated that science had provided him with the means to comprehend the truths inherent in the classical philosophy of India.

Bose maintained a close and lasting friendship with Rabindranath Tagore, whom he met after the poet had dedicated a poem in Bengali to him on the occasion of his return from Europe in 1897. He became a leading exponent of Tagore's Bengali renaissance movement serving from 1911 to 1913 as president of the Bangiya Sahitya Sammilan (Bengali Literary Society).

Bose was made Companion of the Indian Empire in 1903 and Companion of the Star of India in 1912, and was knighted in 1916. In 1920 he was elected fellow of the Royal Society, in 1926 was appointed to the Intellectual Cooperative Committee of the League of Nations, and in 1927 was elected president of the Indian Science Congress Association.

BIBLIOGRAPHY

I. Original Works. Bose's major books are *Response in the Living and Non-Living* (London, 1902); *Plant Response: As a Means of Physiological Investigation* (London, 1906); *Comparative Electrophysiology. A Physico-Physiological Study* (London, 1907); *Researches on Irritability of Plants* (London, 1913); *Life Movements in Plants*, 4 vols. (Calcutta, 1918–1921); *The Ascent of Sap* (London, 1922); *Physiology of Photosynthesis* (London, 1924); *Nervous Mechanisms of Plants* (London, 1926); and *Plant Autographs and Their Revelations* (London, 1927).

His early papers are in *Collected Physical Papers* (London, 1927), with a foreword by J. J. Thomson.

II. Secondary Literature. See Patrick Geddes, *The Life and Work of Sir Jagadis C. Bose* (London, 1920); Monoranjon Gupta, *Jagadischandra Bose: A Biography* (Bombay, 1964); and Meghnad Saha, "Sir Jagadis Chunder Bose," in *Obituary Notices of Fellows of the Royal Society of London*, **3** (1939–1941), 3–12, with portrait.

William A. Blanpied

BOSE, SATYENDRANATH (*b.* Calcutta, India, 1 January 1894; *d.* Calcutta, 4 February 1974), *physics.*

Bose was the son of Surendranath Bose, an accountant in the executive engineering department of the East India Railways, who later was a founder of the Indian Chemical and Pharmaceutical Works. His mother was Amodini Raichaudhuri. The family were members of the Kayastha caste. Bose began his primary education in the local English-language schools. The upsurge of Bengali nationalism that followed Lord Curzon's decision to divide the province of Bengal into two administrative units, however, convinced his father to send him to a Bengali-language secondary school in 1907. Two years later Bose enrolled as an undergraduate in Presidency College, Calcutta, where his teachers included Jagadischandra Bose in physics and mathematics, and Prafullachandra Ray in chemistry. Meghnad Saha, Jranchandra Ghosh, and Jnanendranath Mukherjee were among his classmates. He received the M.Sc. in mathematics in 1915, ranking first in his class.

Sir Asutosh Mookerjee, vice-chancellor of Calcutta University, had inaugurated the University College of Science in 1914. This institution, funded largely through the endowments of Sir Taraknath Palit and Dr. Rashbehari Ghose, was the first college in India to offer advanced studies in science. In 1915 Bose and Saha, among others, suggested that Mookerjee build upon the existing postgraduate chemistry curriculum at the college by instituting courses in mathematics and physics. Mookerjee agreed to their request and also obtained for them stipendiary scholarships and funds for procuring scientific journals and laboratory apparatus. When the physics department was organized in 1917, they were appointed lecturers. A year later Chandrasekhara V. Raman, then a civil servant in the Indian Finance Department, joined the department as Palit Professor of Physics.

Bose left Calcutta in 1921 to become Reader in Physics at the newly established University of Dacca in East Bengal. In July 1924 he sent a short manuscript entitled "Plancks Gesetz und Lichtquantenhypothese" to Albert Einstein for criticism and possible publication. Einstein himself translated the paper into German and had it published in the *Zeitschrift für Physik* later that year. He added a note that stated: "In my opinion Bose's derivation of the Planck formula signifies an important advance. The method used also yields the quantum theory of the ideal gas as I will work out in detail elsewhere."

Einstein's enthusiastic endorsement of his work enabled Bose to obtain a two-year paid study leave from Dacca University, which he spent in France and Germany. During his year in France he was guided in his studies by Paul Langevin and was in close contact with Maurice and Louis de Broglie. Late in 1925 he had a brief but reportedly cordial meeting in Berlin with Einstein, and in the early summer of 1926 he heard Max Born's lectures at Göttingen on the new matrix mechanics of Werner Heisenberg. Later that summer Bose returned to Dacca as professor and head of the physics department. He held these posts until 1945, when he returned to Calcutta University as Khaira professor of physics and, from 1952 to 1956, as dean of the Faculty of Sciences. Following his retirement from Calcutta, he served for three years as vice-chancellor of Visva-Bharati University, an institution in West Bengal that had been established by Rabindranath Tagore. He relinquished that position in 1959 upon his appointment as a national professor by the government of India. Bose was president of the National Institute of Sciences of India in 1949–1950, and from 1952 to 1958 served in the upper house of the Indian parliament. He was awarded the Padma Vibhushan by the government of India in 1954, and was elected fellow of the Royal Society in 1958. Bose married Ushabala Ghosh in 1914, and was the father of two sons and five daughters.

Bose's twenty-six original scientific papers, published between 1918 and 1956, include contributions to statistical mechanics, the electromagnetic properties of the ionosphere, the theories of X-ray crystallography and thermoluminescence, and unified field theory. Two of his first four papers were investigations of the equation of state for gases, written with Saha. In 1919 the Calcutta University Press published a two-volume edition of Einstein's collected papers on the special and on the general theories of relativity, translated into English by Saha and Bose, respectively. Bose's first paper on quantum theory (1920) demonstrated that the empirical formulas for the line spectra of the alkali atoms are derivable from the Bohr-Sommerfeld quantization rules, and included the assumption that the effective potential in which the valence electrons of these atoms move can be expressed as the superposition of potentials due to a point charge and an electric dipole.

Bose is known outside India primarily for his first paper in the *Zeitschrift für Physik* (1924) in which he succeeded in deriving the Planck blackbody radiation law without reference to classical electrodynamics. Einstein's generalization of Bose's method led to the first of two systems of quantum statistical mechanics, known as the Bose-Einstein statistics. Paul Dirac later coined the term "boson" for particles that obey these statistics.

Planck's radiation law, derived in 1900, relates the electromagnetic energy density in equilibrium with a blackbody, or ideal radiator, at an absolute temperature, T, to the radiation frequency, ν. Using modern notation:

$$\frac{E(\nu)d\nu}{V}=\frac{8h\nu^3 d\nu}{c^2}\,\frac{1}{\exp\left(\frac{h\nu}{kT}\right)-1},$$

where V is the volume of the radiator, c the speed of light, k the Boltzmann constant, and h Planck's constant. Planck based his derivation on a model in which the radiation emitted and absorbed by the blackbody is in equilibrium with a set of charged oscillators. He took the laws of classical electrodynamics as valid but assumed in addition that each oscillator could emit and absorb only in quanta proportional to its frequency of oscillation. That is, $E = nh\nu$, where n is any integer.

Planck regarded his quantum hypothesis as an ad hoc assumption to be grafted onto the inviolable body of classical electrodynamics. In contrast, Einstein, in his "photoelectric effect" paper of 1905, used general thermodynamic arguments to show that electromagnetic radiation could be regarded as having an atomic or quantum structure. Thus, he argued in effect that electromagnetic radiation in equilibrium with matter could be regarded as a gas similar in some respects to an ordinary gas the quanta of which are atoms or molecules. The zero-rest-mass quanta of the electromagnetic field are now called photons, a term introduced in 1926.

Bose's 1924 paper showed that the Planck law was completely consistent with Einstein's quantum gas model. His derivation followed a general procedure introduced by Boltzmann for determining the equilibrium energy distribution of the microscopic entities that constitute a macrosystem. The procedure begins by enumerating all the possible, distinguishable microstates of the entities, where each such state is defined by a set of coordinates and momenta. That is, each possible state of a single entity is specified by a point in six-dimensional phase space the axes of which correspond to the three spatial coordinates and the three components of momentum. Each possible state of the system is specified by a distribution of such phase points. Bose's innovation was to assume that two or more such distributions that differ only in the permutation of phase points within a subregion of phase space of volume h^3 (where h is Planck's constant) are to be regarded as identical. Thus, in effect he asserted that two truly identical photons cannot be distinguished even in principle. This method of counting has the effect of enhancing the populations of lower-energy photon states at the expense of those of higher energy, and leads to the correct Planck distribution law.

The assumption that the region h^3 sets a limit on the distinguishability of two photons appears to have been completely ad hoc and was arrived at, according to Bose's later recollection, in the course of preparing a lecture on the Planck law for a postgraduate physics class at Dacca. It can easily be shown to be consistent with the uncertainty principle of Werner Heisenberg, announced in 1927, and, more fundamentally, with either the matrix mechanics of Heisenberg (1925) or the wave mechanics of Schrödinger. In July 1924 Einstein had already generalized Bose's results to particles of nonzero rest mass the total number of which is conserved; and in January 1925 he showed that such a gas would, under conditions of extreme temperature and pressure, exhibit marked deviations in behavior from that of a classical ideal Maxwell-Boltzmann gas. The latter paper also showed that Bose's assumption is consistent with the relationship between the wavelength and momentum of a particle, $\lambda = h/p$, which Louis de Broglie had hypothesized in 1923. Thus, by generalizing Bose's theory, Einstein completed the formulation of the first of two types of quantum statistics. In 1926 Enrico Fermi derived a second system of quantum statistics, now called the Fermi-Dirac statistics, in which it is assumed that each subvolume h^3 in phase space can be occupied by no more than one point, consistent with the exclusion principle enunciated by Wolfgang Pauli in 1925.

Bose's first paper in *Zeitschrift für Physik* was followed by another that was also translated by Einstein and published during 1924. In it Bose provided a general statistical treatment of emission and absorption processes for electromagnetic radiation in equilibrium with matter. This paper was accompanied by a note by Einstein expressing serious doubts about the method. In January 1925 Bose wrote to Einstein from Paris that he was working on a paper he felt would remove these doubts. But it seems never to have been completed.

Bose's next published scientific contribution consisted of two papers on mathematical statistics (1936). Two works on the electromagnetic properties of the ionosphere were published in 1937 and 1938, respectively. A paper on the mathematical properties of the Lorentz group appeared in 1939. Two more mathematical works, one on the inhomogeneous Klein-Gordon equation and one on an integral equation for the hydrogen atom, were published in 1941. Most of his published theoretical work between 1943 and 1950 was on X-ray crystallography and thermoluminescence, both areas in which experimental groups were active at Dacca. Bose's last six scientific papers, published between 1953 and 1955, were on unified field theory, a topic on which he and Einstein exchanged at least one letter in 1953.

Bose is reputed to have been a devoted and inspiring teacher. His ability to deliver polished lectures without notes was legendary and considered phenomenal even in India, where professors take considerable pride in that accomplishment. This skill was aided by a remarkable memory which he developed as a schoolboy, partly because of his exceedingly weak vision.

Born and educated in an era when Rabindranath Tagore was presiding over the Bengali cultural renaissance, Bose remained devoted to that movement throughout his life. In 1948 he founded the

Bangiya Bijnam Parishad, or Science Association of Bengali, as a means of popularizing science in his native language. Like Tagore, he loved poetry, which he read and quoted not only in Bengali and Sanskrit, but also in English and French, both of which he spoke fluently.

BIBLIOGRAPHY

I. ORIGINAL WORKS. Bose's papers include "On the Influence of the Finite Volume of Molecules on the Equation of State," in *Philosophical Magazine*, 6th ser., **36** (1918), 199–202, written with M. N. Saha; "On the Equation of State," *ibid.*, **39** (1920), 456, written with M. N. Saha; "On the Deduction of Rydberg's Law From the Quantum Theory of Spectral Emission," *ibid.*, **40** (1920), 619–627; "Plancks Gesetz und Lichtquantenhypothese," in *Zeitschrift für Physik*, **26** (1924), 178–181; "Wärmegleichgewicht im Strahlungsfeld bei Anwesenheit von Materie," *ibid.*, **27** (1924), 384–390; "Anomalous Dielectric Constant of Artificial Ionosphere," in *Science and Culture*, **3** (1937), 335–351, written with S. R. Khastgir; "On the Total Reflection of Electromagnetic Waves in the Ionosphere," in *Indian Journal of Physics*, **12** (1938), 121–144; "On an Integral Equation Associated With the Equation for Hydrogen Atom," in *Bulletin of the Calcutta Mathematical Society*, **37** (1945), 51–61; and "Solution d'une équation tensorielle intervenant dans la théorie du champ unitaire," in *Bulletin de la Société mathématique de France*, **83** (1955), 81–88.

Five letters from Bose to Einstein (1924–1926) and a copy of a 1953 letter from Einstein to Bose are among the Einstein papers at the Institute for Advanced Study, Princeton, New Jersey.

II. SECONDARY LITERATURE. See William A. Blanpied, "Satyendranath Bose: Co-Founder of Quantum Statistics," in *American Journal of Physics*, **40** (1972), 1212–1220; Nirendranath Roy, "Professor Satyendranath Bose," in A. K. Datta and Asima Chatterjee, eds., *Satyendranath Bose 70th Birthday Commemoration Volume* (Calcutta, 1964), 6–12; and Jagadish Sharma, "Satyendra Nath Bose," in *Physics Today*, **27**, no. 4 (Apr. 1974), 129–131.

WILLIAM A. BLANPIED

BOUCHER DE CRÈVECOEUR DE PERTHES, JACQUES, (*b.* Rethel, Ardennes, France, 10 September 1788; *d.* Abbeville, France, 5 August 1868), *prehistory, human paleontology.*

Boucher de Crèvecoeur was descended on his father's side from an old family of the Champagne district and through his mother, Stéphanie de Perthes, from an uncle of Joan of Arc. The only male heir of his mother's family, he took its name in 1818 and thus became Boucher de Perthes. His father, Jules Armand Guillaume Boucher de Crèvecoeur, had been a financier before the Revolution; under the Empire he became director of customs at Abbeville. A good naturalist and a corresponding member of the Institute for the subject of botany, he knew Lamarck, Cuvier, and Alexandre Brongniart.

Boucher, who was only a mediocre student, joined the Customs Administration, in which he spent his entire career. In 1825 he was appointed director of customs at Abbeville, where he spent the rest of his life.

At a very young age Boucher de Perthes began to write on literary, philosophical, economic, and social questions. In *Opinion de M. Cristophe* (4 vols., 1825–1834), which is written in a humorous style, he defended political and commercial liberty, the dignity of work and industry, and the abolition of the death penalty. Other works dealt with the condition of women and popular education. He was one of the first to suggest the holding of universal expositions. It seems that he was close to the Saint-Simonians. He defined his most important philosophical work, *De la Création* (5 vols., 1838–1841) as an "essai sur l'origine et la progression des êtres." Inspired, it appears, by Buffon, Charles Bonnet, and Lamarck, but also by German *Naturphilosophie*, the work presents a theory of the universal evolution of nature—a rather confused theory composed of diverse elements, but typical of French romanticism.

After moving to Abbeville, Boucher de Perthes became president of its local learned society, the Société d'Émulation (founded in 1798). The exceptional richness of the nearby beds in the lower valley and estuary of the Somme led its members to take a particular interest in archaeology and paleontology. Numerous mammalian fossils had been exhumed and sent to Cuvier; and in 1835 a young physician, Casimir Picard, began to discover and identify axes made of polished stone.

Following Picard's death in 1841, Boucher de Perthes continued his investigations, on which he had been collaborating since 1837. The results of this research, presented in 1840 to the Société d'Émulation, were published in 1846 as *De l'industrie primitive, ou des arts à leur origine.* Boucher de Perthes carefully distinguished two categories of objects, according to the site at which they were discovered. The first group, consisting of axes of hewn or polished stone and tools made of stone or horn, came from the beds near the surface of the earth, peat bogs, and the bed of the

Somme. They were attributed to a "Celtic" people, supposedly the immediate predecessors of the Gauls. The others, hewn stone axes and tools, came from much older "diluvial" terrain at Menchecourt, l'Hôpital, and Moulin-Quignon. These beds, the "diluvium" of Cuvier and Brongniart, were generally considered to be the remnants of a geological catastrophe, a gigantic "deluge" dating from the end of the Tertiary. They were soon called "terrains de transport" and were later interpreted as resulting from the glaciations of the Pleistocene. In these beds the stone axes and tools were mixed with the mammalian fossil bones that Cuvier had identified as being of extinct species: *Elephas primigenius*, rhinoceros, and so on. From them Boucher de Perthes deduced the existence of "homme antédiluvien," contemporary with the extinct species and therefore much older than any known human population.

In order to arrive at these conclusions, it was necessary to define man by work and by tools, and not only by reason and language; to identify hewn flints as objects made by man; and to employ a rigorous stratigraphic method. All this was new at the time, and the conclusions seemed shocking: Boucher de Perthes convinced almost nobody. Reissued in 1847 under the new title *Antiquités celtiques et antédiluviennes*, his book had no success. Superficially criticized in the press, it was ignored by official science. The Académie des Sciences named a commission to verify the facts of the case; but the commission took no action, and nothing more was said about the matter. It should be noted, however, that Boucher de Perthes made the mistake of including in his collection of axes and tools a much more questionable group of small stones on which he saw "objects of art," "symbols," and antediluvian "hieroglyphics." Moreover, his philosophical-historical interpretation of antediluvian man was quite bold. But the main cause of the resistance among scientists was their loyalty to Cuvier's teaching. All the previous discoveries in this field—those of John Frere (1799), Ami Boué and Crahay (1823), Paul Tournal and Julien de Christol (1826 and 1829), and especially those of Schmerling (published at Liège in 1833 and 1834)—had encountered the same resistance, as had those of J. MacEnery and Robert Godwin-Austen in England, where Buckland played the role that Cuvier did in France. None of this work was adequately appreciated until Boucher de Perthes had achieved his success.

Meanwhile, Boucher de Perthes continued his research, encouraged by the discoveries of F.-A. Spring in Belgium (1853) and the finds made at St.-Acheul, near Amiens (1856). The second volume of *Antiquités celtiques et antédiluviennes* (1857) enjoyed no more success than the first; but it did inspire Joseph John Prestwich and Hugh Falconer, who were pursuing analogous investigations in England, to come to Abbeville in 1858 and 1859. Soon convinced of the value of Boucher de Perthes's work, they encouraged other English scientists, including Charles Lyell, to join them. All of them corroborated Boucher de Perthes's assertions. The Académie des Sciences was finally obliged to break its silence. It sent Albert Gaudry to Abbeville to examine the question and, upon his favorable report, converted to Boucher de Perthes's views. In *De l'homme antédiluvien et de ses oeuvres*, a little book written with great restraint, Boucher de Perthes recounted his work, his trials, and his ultimate triumph (1860). He regretted only that skepticism was still voiced regarding the art objects, symbols, and hieroglyphics.

Although it appeared that actual fossil remains of this hypothetical ancient man had yet to be found, a great many of them had already been discovered; but these finds had all been challenged. Those of Spring had been made in a grotto, as had those of Lartet at Aurignac (1860). On 23 March 1863, in the quarry at Moulin-Quignon, Boucher de Perthes found a human tooth and some axes, and on 28 March, three teeth, a jawbone, and an axe. The discovery caused much excitement; but it also unleashed a violent campaign in the English press, where Boucher de Perthes was accused of being the author, or else the very naive victim, of a crude fabrication. The French scientists, now supporters of Boucher de Perthes, and the English scientists, who had become skeptical, gathered in Paris and then in Abbeville from 9 to 18 May 1863. They concluded that the discovery was authentic. Even so, public opinion, especially in England, was not satisfied; the work of Falconer, Lartet, Vibraye, and others multiplied the discoveries, however, and very quickly established the antiquity of man beyond what even Boucher de Perthes had been able to conceive.

Boucher de Perthes related all these events in the third volume of *Antiquités* (1864). He pursued his excavations, accumulating more or less authentic human fossils, and continued to write (notably *Des outils de pierre* [Paris, 1865]). He was elected honorary president of the Société d'Émulation in 1866, only a few years before a disease brought an end to his life.

Boucher de Perthes was not the first to discover

hewn flints, or even human fossils; and he owed part of his ultimate success to the unusual richness of the beds he excavated. Still, he far surpassed his predecessors in the rigorous use of the stratigraphic method and the skillful application of the techniques of excavation. His other strength was never to let himself be discouraged by the resistance to the idea of the great antiquity of the human species. His influence on anthropology was considerable. Although an amateur whose work was sometimes of uneven quality, he deserves recognition as the man who won acceptance for the idea of a human paleontology.

BIBLIOGRAPHY

Boucher's most important works are cited in the text. A complete bibliography would include more than thirty titles.

Studies devoted to Boucher include Leon Aufrère, *Essai sur les premières découvertes de Boucher de Perthes et les origines de l'archéologie primitive (1838–1844)* (Paris, 1936); and *Figures de préhistoriens* (Paris, 1939); and Victor-Amédée Meunier, *Les ancêtres d'Adam. Histoire de l'homme fossile* (Paris, 1875).

JACQUES ROGER

BOURGUET, LOUIS (*b.* Nîmes, France, 23 April 1678; *d.* Neuchâtel, Switzerland, 31 December 1742), *archaeology, philology, philosophy, biology, geology, crystallography, physics.*

Bourguet was the eldest son of Jean Bourguet, a rich wholesale merchant in Nîmes, and of Catherine Rey, both Huguenots. In 1685, the year of the revocation of the Edict of Nantes, Bourguet's family fled to Switzerland, seeking refuge first in Lausanne and then in Zurich. Bourguet enthusiastically began his studies at the College of Zurich in 1688 but was compelled the following year to accompany his father to Casteseigna, in the Grisons, to establish a stocking and muslin factory. Upon his return in 1690, Bourguet was obliged to enter the family firm. Thus began a conflict between his scholarly vocation and business that did not end until 1716. He read widely in archaeology, numismatics, and philology and began assembling a large collection of medals, antiquities, and rare books, notably in Oriental and Slavic languages.

Between 1697 and 1715 Bourguet was often in Italy on business with his father or other members of his family. Until 1705 he stayed principally in Verona, Venice, and Bolzano, where he studied Hebrew and the Mishnah with a rabbi. Between these trips he lived in Zurich until the French refugees were expelled and his family was separated; he then lived in Bern and finally in Neuchâtel, where he became a citizen in 1704. During this period Bourguet began to correspond with J.-J. Scheuchzer and a number of other scholars. In 1707 he went to Rome where he copied ancient inscriptions (including the Iguvine Tables), studied the ancient monuments, and established contact with the leading scholars and Jesuit missionaries.

Toward the end of 1708, upon his return to Neuchâtel, Bourguet took up the study of geology. In 1709 he explored the Jura Mountains near Neuchâtel, collected fossils, and examined mineral springs, as well as cave stalactites and grottoes. This new interest undoubtedly reflected the influence of Scheuchzer, whom Bourguet greatly admired. During a trip to Italy in 1710, Bourguet befriended Vallisnieri at Padua and Zannichelli at Venice. He accompanied them on a geological expedition in the area around Verona and Vicenza, where they observed the fossiliferous strata of the Alpine foothills, especially those containing *pierres lenticulaires* (nummulites and related forms, the nature of which was then a subject of debate). In 1711, a few months after returning to Neuchâtel, Bourguet and his wife moved to Venice, where they remained until 1715. During these intellectually decisive years, guided by Jakob Hermann (then professor at Padua) and probably by Bernardino Zendrini, Bourguet began to study Leibniz' infinitesimal calculus and astronomy. He further expanded his already extensive correspondence, particularly with Leibniz, who esteemed him highly from the time they began corresponding about the binary calculus of the Chinese, an exchange of letters that lasted from 1707 until 1716.

Bourguet discussed geology and biology with Vallisnieri, the famous student of Malpighi. He studied the viviparous plant lice and through microscopic study verified the existence of Leeuwenhoek's *vers séminaux* (spermatozoa), but he questioned their role in reproduction. In addition, he assembled a collection of fossils with the aid of Vallisnieri, Gian Girolamo Zannichelli (to whom he communicated his enthusiasm), and Giuseppe Monti of Bologna. At Bologna oceanographic studies were much in favor, and Bourguet, with Monti and Jean Daniel Geissel, explored the nearby Apennines. Without adopting the very limited diluvial views of his Italian friends, Bourguet shared their conviction that the fossils they found were of organic and marine origin; he felt an urgent need to combat the outdated theses put forward by Karl Lang (1708), against which he directed "Dissertation sur les pierres figurées" (1711), which was not published.

For unknown reasons, Bourguet's life underwent a major change at the end of 1715. Leaving Venice and Italy for good, he moved to Neuchâtel and did not leave Switzerland again except for a trip to Holland in 1725 to resolve some printing problems. He abandoned commerce and set about finding employment. His health deteriorated in 1718, and in 1721 his financial problems became acute following bankruptcies at Basel and Geneva. In order to obtain money to live he gradually sold his collections, retaining only his Bibles in nearly fifty languages.

In 1717 Bourguet's friends urged him to seek nomination to the vacant chair of law at Lausanne, but he did not receive the post. It was not until October 1731 that, at the age of fifty-three, he was finally named to a professorship of philosophy and mathematics created for him at Neuchâtel. Bourguet's teaching duties, which he continued to fulfill until his death, consisted of private courses in logic, philosophy, natural law, and natural sciences, and public lectures on philosophy, history, alchemy, minerals, meteors, true and false miracles, and the formation of the earth.

Bourguet spent liberally to establish and maintain two periodicals. With several friends, he founded a scholarly review, *Bibliothèque italique* (1728), designed to present the results of Italian science to a French general audience. In 1732 Bourguet created another review with a group of Swiss collaborators. Devoted to literary and historical as well as scientific subjects, it first appeared as the *Mercure suisse* and later as the *Journal helvétique*. Bourguet expended great energy on this monthly publication, which survived its founder and played a role in the cultural awakening of French-speaking Switzerland. While satisfying the reader's taste for lighter material, it also published, for example, a series of articles debating the merits of Leibniz' system and the notion of preestablished harmony, of which Bourguet, like Christian Wolff, was a fervent and well-informed advocate.

At Neuchâtel, Bourguet expanded his circle of correspondents to more than seventy persons. Through his exchange of letters he set forth the major controversies of the time and played an important role in the diffusion of ideas, as in the debates in geology over the historicity of the Flood and in biology over the preformation of germ cells. He was unaware that in according priority to strictly interpreted biblical theses (especially those in Genesis 1:1–11:9 and II Peter 3:5–7), he was involving himself in serious errors.

Encyclopedic in his erudition, Bourguet, whom his friends called the Pliny of Neuchâtel, became a member of the Berlin Academy of Sciences and the Etruscan Academy of Cortona. He read or spoke a great many languages, knowledge of which he acquired from the daily reading of a translation of the Bible, assisted, where possible, by a grammar. In 1733, for example, he learned Basque and, about 1735, the Celtic languages. Throughout these studies he was motivated by the hope of establishing the filiation and harmony of the world's languages.

Works. Bourguet's surviving works include two books: *Lettres philosophiques*, which has as an appendix the *Mémoire sur la théorie de la terre*, written in 1723 and published in 1729; and *Traité des pétrifications* (1742), several sections of which are by Pierre Cartier or are collaborative efforts. There are also numerous articles and news columns, signed either with Bourguet's name, with such pseudonyms as Philalèthe or Philanthrope, or with initials, or published anonymously, and a large number of manuscripts.

Lettres philosophiques. This book, which Bourguet had published in 1729 at Amsterdam, is a unified work (except for the pages on the theory of the earth) presented in the form of four letters to Scheuchzer. Taking as his point of departure the demonstration of the animal origin of belemnites and *pierres lenticulaires*, Bourguet compares the elementary processes of the mineral world (crystallization) and of the living world (generation, assimilation, growth).

Crystal growth is a product of the mineral mechanism, whereas the growth of plants and animals is an effect of the organic mechanism. Bourguet makes a highly instructive comparison of the fine structures of a stalactite and a belemnite. The internal complexity of the latter is sufficient proof that it is a product of the living world and not of the mineral. That it was once in a marine environment is attested by various details, such as incrusting organisms. Similarly, the organic origin of the *pierres lenticulaires* (nummulites and similar forms) can be clearly seen in their structure. Bourguet thought they should be grouped with the *couvercles* or opercula of the Ammonoidea, an order that then customarily included the ammonites, nautiluses, spirulas, and even the small spiral Foraminifera discovered shortly before by Beccari. He also showed why the entrochites must be placed among the *étoiles de mer arbreuses* (crinoids) and not among the zoophytes.

Bourguet studied stalactites and described their internal radial structure, as well as their final cleavage into small crystalline rhomboids; their genesis obeys the same laws of crystallization as soluble salts and natural crystals (rock crystals). Their deviations revealed to him that they are accidental productions of nature in contrast with the unvarying symmetry and organization of living creatures, which are products of a final purpose. Bourguet's attempt to explain crystallization was innovative. First, he asserted that a crystal is created not from water but in water that carries a multitude of extremely small particles, the fusion of which creates a crystal or concretion. To each crystalline substance there correspond particles of specific, determinate shape where particular laws of growth govern the external form. Each corpuscle, endowed with a specific dynamism or vital activity, contributes to a solid body as it coalesces with other particles. The corpuscles group together according to divine plan but may form nothing more advanced than a crystal. This notion derives from the concept of the chain of immaterial beings that Bourguet adapted from Cudworth and Grew and modified to fit a Leibnizian context. Even if the actual microscopic order escapes our senses, everything is organized in matter, which we perceive only as a whole, as if it were a crowd or a distant maneuvering army. All organization in the universe was originally and simultaneously created by God.

Since Bourguet had no notion of evolution, he did not consider the question of the increasing organization and complexity of the biosphere. He merely related the problem to the origin of the universe, thereby rendering it inaccessible to human reason. Thus, he adhered firmly to the theory of the *emboîtement des germes*, or "encasement" theory of germ cells (particularly as formulated by Malebranche) and consequently denied any role in reproduction to the *vers séminaux* (spermatozoa) and to pollen. Bourguet believed that embryonic processes did not signify a genuine formation but rather the appearance of previously invisible, preformed organs. This conception fitted a mathematical image of convergent and infinite series but, as Firmin Abauzit pointed out, it was not physically feasible—even for the short biblical chronology—as soon as it was assumed that matter is composed of finite corpuscles.

Bourguet seems to have established direct contact with Buffon through Jean Bouhier, and according to Jacques Roger they began to discuss the problem of generation in 1733. However, the hypothesis of organized organic molecules later adopted by Buffon was a regression from Bourguet's early notions on the *mécanisme organique*. Bourguet held that there are neither special organic molecules nor occult vital forces. The elementary corpuscles circulate ceaselessly and by their combinations constitute the infinitely complex tissues of organisms. He compared the growth of a crystal with that of a shell and with more advanced structures (an ear of grain, a feather, and so on). Both shell and crystal are merely concretions assembled in layers, differing only in that the productive mechanism of the former is regulated by final causes. The mechanism is called organic "because it acts by means of an organized body, without which it would not exist." In more advanced structures, growth occurs by the addition of molecules selected from the general stream of corpuscles according to the local needs of each site of the organism. Only that which is already organized can organize, for it is the visible expression of an immaterial being, endowed with a perfect, specific power of organization. Bourguet made this postulation in the spirit of Leibniz and in an attempt to reconcile various contemporary (nonmaterialistic) doctrines. Thus the Organic Mechanism was simply the combination of the motion of an infinity of differing moving particles, in conformity with particular systems, which are predetermined by God. Each system is bound to a single dominant monad, to which the other participant monads are subordinated. That which we name Matter is nothing but the totality of countless actions and reactions that bound every creature to others.

Traité des pétrifications. This book (1742), a collaborative work done mainly by Bourguet with Pierre Cartier (a minister) and other friends, consists of two parts—a collection of rather polemical letters or articles dealing with subjects encountered in the *Lettres philosophiques* and an atlas illustrating the bulk of the fossil animals discovered by the authors in Switzerland or drawn from the works of Lang and Scheuchzer. The atlas, with its succinct explanatory text, its extensive bibliography of European paleontology (enlarged and modernized by Guettard for the second edition of 1778), and its list of the world's recorded fossiliferous localities, was the first of its kind to be published in French. Bourguet sought to arrange all mineral and organic species in a single chain, a series characterized by imperceptible gradations leading from the simplest (mineral) products of nature to the most complex: a concrete representation of the Great Chain of Being.

In his classification of the mineral kingdom (partly based on John Woodward and partly original—see *Bibliothèque italique*, **2** [1728], 99–131), Bourguet distinguished the clays (including soils, chalks, etc.), the stones, and the metals (including the minerals). In 1728 he divided the stones according to their origin: 1) sand, gravel, and pebbles; 2) bitumen, flint, marble, etc., formed through a kind of fusion; and 3) salts, crystals, granite, etc., formed by aqueous crystallization. In 1742 he rejected any fusion in favor of a strict neptunism that attributed even granite and metamorphic rock to the great *changement* wrought by the Flood. Still, he allowed that some rocks dated from the first formation of the earth and that some were still being formed. Despite his erroneous conceptions, Bourguet was more closely allied to Hutton than to Werner in that he posited that the same kinds of rock can date from different cycles and that certain rocks form only as a result of a *grand changement* (in fact, orogeny) or of processes different from the sedimentation currently occurring in the depths of the sea (for example, flint, fossil pyrites).

The *Traité* presents the fossil animals in four classes: 1) the *champignons* (sponges and certain corals), madrepores, and coralloids; 2) our bivalves and brachiopods (poorly distinguished); 3) our gasteropods and cephalopods, plus the spiral worms (Serpulidae) and the *pierres lenticulaires*; 4) our echinoderms and crinoids, the dentalia, our crustaceans, the actual or alleged fish teeth (glossopetrae), belemnites, and the race of men that, according to Scheuchzer, witnessed the Flood. Despite its rudimentary character, the *Traité* was an indispensable manual of paleontology that diffused widely the competing theories.

Mémoire sur la théorie de la terre consists of a methodological and historical introduction; an objective, descriptive section; a theoretical portion comprising twenty-six propositions that seek to present a hypothesis for the development of the earth; and a conclusion containing an outline of a future work. Bourguet believed that the solid mass of the earth, continuous from the mountains to the sea floors, is formed of heterogeneous materials arranged in beds and strata of varying thickness. Quarries, mines, and natural sections through the mountains reveal the existence of strata extending for great distances in every direction. The hardest rocks (sandstone, conglomerates, *pierres grises*, slate, limestone, marble, granite, and porphyry) compose the mountains, the beds of which may be horizontal, vertical, inclined, or curved in concave or convex arcs. Bourguet found that each bed preserves a constant thickness despite its various inflections but is often broken by vertical or inclined joints and cracked in all directions. Some beds have remained horizontal over great distances but are not always ordered according to density, as postulated by Woodward. Fossil shells and other animal and plant remains exist in an infinity of strata of earth and rock from the summits of mountains to the depths of quarries. Bourguet noted that shells are invariably filled with the same material surrounding them, that their origin is organic, and that they had had a period of marine existence.

Bourguet emphasized the order evident in the disposition and structures of the mountains. It was known that order existed on a global scale, since mountains were found to be grouped into chains generally displaying a preferred orientation, some parallel to the meridian, others to the equator. Further, inside a given chain—and here Bourguet was proud of having made what he considered an important discovery—advanced thrusts lie perpendicular to each side of the secondary formation, whether the latter was north-south or east-west. When two secondary ridges are parallel, the advanced thrusts of one correspond to the reentrant angles of the one facing it, in the fashion of fortifications. This geometric disposition can occasionally be seen at lower levels, in traces of riverbeds, on the plain, and along rocky marine shorelines. For Bourguet, it was the *mot de l'énigme* and the key to the theory of the earth. Concerning the destruction of the earth, Bourguet presented the evidence of erosion and fluvial transport, including coastal alluvial deposits, but assigned little significance to it. Nor did he (again like Buffon and De Maillet) attach much importance to the uplift and subsidence of islands and mountains through the action of volcanoes and earthquakes.

Theory of the Earth. In speaking of a "theory of the earth" Bourguet intended a rational reconstruction of its history and dynamic causality, which he claimed to deduce solely from his observations. He declared that all parts of the earth as it is presently constituted are more or less coeval, dating from the time of the Flood. This deluge was not a supernatural miracle but the amplification of normal processes and the exceptional convergence, ordained by God, of ordinary causes. In rough outline, Bourguet's scheme is at first close to that proposed by Woodward. But, influenced by Scheuchzer, Bourguet elaborated a genuine theory of orogeny (however warped to fit orthodox diluvianism), which he linked to the dynamics of terrestrial rotation. His theory encompassed the phenomena of sedimentation fed by attacks on old outcrops, the disturbance and raising of originally horizontal strata by a coordinated motion, and the circulation of mineralizing solutions.

In Bourguet's view, mountains were formed only near the end of the *grand changement* (or renovation of the earth by the Flood), when the flat-lying, concentric new strata, rebuilt from the dissolution of the ancient world, began to harden. The acceleration of the earth's rotation then induced an upward thrust, or dynamic *direction d'élévation.* He explained the differing local arrangement of the strata by the variable degree of consolidation at the start. Furthermore, he attributed their "more or less regular conformity" to the earth's motion and to the general and particular orientation of the chain being formed. Such considerations (unfortunately overcondensed) point directly to the ideas of H. B. de Saussure. For Bourguet, as for Buffon after him, mountains and all their topographic features were carved by the retreating sea (hence their zigzag pattern and ripplelike marks); but Buffon denied or ignored Bourguet's assertions both of the diluvial catastrophe and of the linked orogeny.

Bourguet rejected both the mineral generation *in situ* of fossils and any notion that the seas had washed over the face of the earth. He also examined the theory proposed by Henri Gautier in *Nouvelles conjectures sur le globe de la terre* (1721). According to this theory (which is very close to Hutton's), rain and rivers had leveled the mountains over long periods of time, and the material that was thus removed and deposited in the seas later formed the strata of new mountains erected by lateral upheaval. Bourguet discarded this theory because of the immense periods of time that it implied. Thus, it appears that Bourguet unwittingly eased the way for Hutton, Playfair, and Lyell by rescuing from oblivion the main ideas of this forgotten predecessor and demonstrating that one must either accept the biblical chronology or a very long uniformitarian one.

Humanities. In the eyes of his contemporaries Bourguet's greatest achievement was his philological investigation of the Etruscan alphabet. He was guided in this work by his conviction in the common origin of all peoples, of their languages, and of their alphabet, which he believed derived from ancient Hebrew. He worked in this perspective from 1704 to 1709 on his "Histoire critique de l'origine des lettres," of which only the outline was published. The material assembled on this matter is preserved in the manuscript division of the Bibliothèque Nationale in Paris. Bourguet also sought to discover affinities between the Amerindian and Asiatic languages on the basis of his notions of the late dispersion and migration of the peoples in the relevant areas. He also wrote on the traditions and antiquities of China, challenging their presumed great age and isolation.

BIBLIOGRAPHY

I. Original Works. Bourguet published two books: *Lettres philosophiques sur la formation des sels et des crystaux, et sur la génération et le mechanisme organique des plantes et des animaux, à l'occasion de la pierre belemnite et de la pierre lenticulaire, avec un mémoire sur la théorie de la terre* (Amsterdam, 1729; 2nd ed., 1762); and, anonymously, *Traité des pétrifications* (Paris, 1742; 2nd ed., enl. and rev., with author's name, 1778); another ed. of the *Traité* is cited under the title *Mémoires pour servir à l'histoire naturelle des pétrifications dans les quatre parties du monde; avec figures, et divers indices aussi méthodiques que nécessaires* (The Hague, 1742).

Among the most important of Bourguet's many articles are "Plan abrégé de l'histoire critique de l'origine des lettres," in *Histoire critique de la république des lettres*, **1** (1712), 300–303; "Lettre . . . à Monsieur Antoine Vallisnieri . . . sur la gradation et l'échelle des fossiles," in *Bibliothèque italique ou histoire littéraire de l'Italie*, **2** (1728), 99–131; "Lettre sur deux prétendues inscriptions étrusques," *ibid.*, **3** (1728), 174–204; "Éloge historique de Mr. Antoine Vallisnieri," *ibid.*, **5** (1729), 46–73; "Éloge historique de Mr. Jean-Jérôme Zannichelli," *ibid.*, **6** (1729), 152–169; "Litanies pélasges, des anciens habitants del'Italie," *ibid.*, **14** (1732), 1–52; "De Bononiensi scientarum et artium instituto atque academia commentarii – Mémoires de l'Institut et de l'Académie des sciences et des arts de Bologne," *ibid.*, 78–125; "Suite de l'extrait des *Mémoires de l'Académie des sciences de l'Institut de Bologne*," *ibid.*, **15** (1732), 126–161; "Lettre sur l'alphabet étrusque," *ibid.*, **18** (1734), 1–62, translated into Italian – along with "Lettre sur deux prétendues inscriptions étrusques" – by A. degli Abati-Olivieri as *Spiegazione di alcuni monumenti degli antiche Pelasgi. Trasportata del francese, con alcune osservazioni sovra i medesimi* (Pesaro, 1735); "Analyse du discours inaugural de M. Bourguet sur l'histoire abrégée de la philosophie et des mathématiques," in *Mercure suisse* (*Journal helvétique*) (Dec. 1732), 65–67; "Lettre sur les noyés par un philanthrope . . . adressée à l'Académie des sciences de Berlin," *ibid.* (Nov. 1733), 70–84; "Mémoire sur les progrès du Christianisme dans les Indes orientales," *ibid.* (July 1734), 82–88; and "Conclusion de la dispute sur les noyés," *ibid.*, 88–93.

See also "Discours sur les phénomènes, que les an-

BOURGUET

ciens regardaient comme miraculeux," in *Mercure suisse* (*Journal helvétique*) (Jan. 1735), 100–113; "Lettre . . . sur la jonction de l'Amérique à l'Asie," *ibid.* (July 1735), 67–97; "Particularités intéressantes sur la colonie protestante de Herrenhut, et sur les missions du Groenland, et de la Côte de Coromandel," *ibid.* (Sept. 1735), 49–66; "Lettre à Mr. Engel . . . sur la jonction de l'Amérique avec l'Asie," *ibid.* (Feb. 1736), 53–62; "Lettre sur l'office de bibliothécaire . . . à Monsieur Engel, bibliothécaire à Berne," *ibid.* (July 1736), 64–83, extracts of which are reproduced in Godet (see below); "Lettre à Monsieur C. . . . servant de réponse à l'épître adressée à Mr. le professeur B. . . . dans le *Mercure* de septembre 1735, p. 96," *ibid.* (Jan. 1737), 90–106, on the errors of the philosophers and a eulogy of Wolff; "À Mr. de la Faie . . ., à l'occasion d'un extrait des *Lettres philosophiques* de M. B. inséré dans les *Mémoires de Trévoux*," *ibid.* (Dec. 1740), 531–549, in which Bourguet records his change of opinion on the belemnites and corrects Castel's errors of interpretation; and "Lettre à Monsieur Seigneux de Correvon . . . sur deux ouvrages envoyés de Rome à l'auteur, qui traitent de quelques monuments antiques," *ibid.* (Nov. 1737), 33–59, a review of *Spiegazione di alcuni monumenti* and a critique of Olivieri's commentary.

Further works are "Lettre à Mr. Meuron . . . sur la philosophie de Mr. le baron de Leibnitz," in *Mercure suisse* (*Journal helvétique*) (May 1738), 333–419, notably on binary arithmetic and the *Théodicée*; "Seconde lettre à Mr. Meuron . . . sur la philosophie de Mr. le baron de Leibnitz," *ibid.* (July 1738), 15–36, a refutation of the criticisms of Crousaz; "Lettre à Monsieur Meuron . . . sur les hypothèses concernant l'union de l'âme avec le corps," *ibid.* (Dec. 1738), 521–556, on preestablished harmony and against Locke; "Lettre à Mr. Roques . . . servant de réponse aux quatre lettres qui ont paru de lui dans le *Journal helvétique*, contre le système de Mr. de Leibnitz," *ibid.* (Aug. 1739), 49–84; "Lettre à Mr. Roques . . . sur les idées innées et leur développement," *ibid.* (Mar. 1740), 204–223, a comparison of the development of innate ideas and of organic bodies; and "Lettre à Monsieur Ostervald . . . sur la conversion des églises du Comté de Northampton, dans la Nouvelle Angleterre," *ibid.* (Nov. 1740), 427–435.

Additional writings are "De crystallorum generatione," in *Acta Academiae naturae curiosorum*, **4** (1730), app., 7–46; "Dissertatio de fatis philosophiae inde ejus natalibus ad nostra usque tempora," in *Tempe helvetica*, **1**, pt. 2 (1735), 129–160; "Dissertatio de vero atque genuino juris naturalis studii usu," *ibid.*, **3**, pt. 1 (1738), 9–41; "Epistolae Ludovici Bourgueti . . . ad primorum Evangelii apud Malabares praeconum ex ipsis autographis descriptae," in J. G. Schellhorn, *Amoenitates historiae ecclesiasticae et litterariae* . . . (Frankfurt–Leipzig, 1737–1738), 710–754; and "Dissertazione di Lodovico Bourguet, professore in Neufchastel sopra l'alfabeto etrusco," in *Saggi di dissertazioni accademiche pubblicamente lette nella nobile Accademia etrusca dell'antichissima città di Cortona*, **1** (1742), 1–23. It should be noted that a number of Bourguet's articles appeared either anonymously or pseudonymously.

Bourguet left numerous MSS. Among the collections containing principally reading notes and bibliographic information are "Livre d'extraits et de remarques" (Zurich, 1695–1696), now at the municipal library of Neuchâtel MS 1241, a compilation of reading notes; "Miscellanea ad theoriam telluris pertinentia," municipal library of Neuchâtel MS 1242, which contains Bourguet's bibliographic sources in geology, long extracts from works by various writers, a fragment of a narrative of his trip with Monti in the region of Bologna, and a memoir entitled "Quelques observations sur les coquillages qu'on trouve dans la terre et sur le Déluge"; "Livre d'extraits . . . relatifs à la théologie, aux langues, à l'épigraphie, à la numismatique sémitiques; aux langues de l'extrême-orient, chinois, siamois, etc., à l'histoire de l'écriture et de l'imprimerie; catalogue des livres qui peuvent servir pour l'histoire critique de l'origine des lettres; plan pour la même histoire," Bibliothèque Nationale, Paris, MS f.fr. 12809; and "Recueil des alphabets et des caractères de toutes les langues," Bibliothèque Nationale, Paris, MS n.a.f. 891, an incomplete dummy of the planned book.

Other MSS are of lectures or articles. These include "Dissertation sur les monuments de l'ancienne Égypte et les labyrinthes," municipal library of Neuchâtel MS 1250; "Discours sur les phénomènes prétendus miraculeux, 31 déc. 1734," Neuchâtel MS 1248, text of a public lecture summarized in *Mercure suisse* (Jan. 1735), 100–113; "Cours de philosophie, du 19 déc. 1731 au 12 mai 1734," Neuchâtel MS 1244, the texts of eleven lectures, those of particular interest being the ones of 7 Apr. and 12 May 1734, on the formation of the earth; "Cours de philosophie, Neuchâtel, du 19 mai 1734 au 15 avril 1735," Neuchâtel MS 1245, beginning with the text of the "Discours sur l'origine des pierres" (dated 15 Apr. 1735), which bears interesting variants and corrections, followed by a portion of a long letter from John Toland, without indication of the recipient, containing critical marginal notations in Bourguet's hand; an incomplete MS on the *secours aux noyés*, 28 Aug. 1734, Neuchâtel MS 1251; "De fatis philosophiae. Leçon du 19 déc. 1731," Neuchâtel MS 1247; and "Cours de philosophie en latin (Praelectiones XV); Cours sur la providence, 2 juillet 1733," Neuchâtel MS 1243.

Finally, Bourguet's MSS include numerous letters and copies of letters. His principal correspondents have been named by L. Favre (1866), who gives some representative extracts, as well as by L. Isely (1903–1904, 1904), and P. Bovet (1904). Only a few of the letters can be cited here. On the apologetic aim of the *Mémoire sur la théorie de la terre*, with several clarifications and an elucidation of the meaning of the word *miracle*, see "Lettres à Monsieur Polier" of 30 Mar. and 24 Apr. 1726, 26 Sept. 1727, 7 and 21 Feb., 4 Mar., 7 Apr., and 30 May 1728, Neuchâtel MS 1261; on the objections of Vallisnieri and Firmin Abauzit to Bourguet's account of the Flood, and on Rizzetti's theory of the change experienced by the polar axis, see "Lettres à Monsieur Abauzit" of 28 Feb. and 29 Apr. 1722; on the weight engendered by the rotation (*turbination*) of the heavenly bodies, see the letter to Abauzit of 15 Mar. 1724, Neuchâtel MS 1259; on the reconciliation of the churches, the history of the missions, the study of languages, and the edition of the letters of Leibniz, see the letters of Bourguet included in "Correspondance du président Jean Bouhier," MSS Bibliothèque Nationale, Paris, f. 24409 and f. 24464. See also the letters to Caze (MSS Neuchâtel and cantonal library of Lausanne), to du Lignon (MS Neuchâtel), to Castel (MS Neuchâtel), to Antoine de Jussieu (MS Neuchâtel), and to G. de Seigneux (MS Neuchâtel). Other MSS are at Basel, Lausanne, and Nîmes. A complete catalog of Bourguet's letters has not yet been compiled.

II. SECONDARY LITERATURE. The principal biographical source is the anonymous obituary "Abrégé de la vie de M. Bourguet, professeur en philosophie et en mathématiques à Neuchâtel, décédé le 31 décembre 1742," in *Mercure suisse* (*Journal helvétique*) (Feb. 1743), 184-195; (Mar. 1743), 295-306; and (Apr. 1743), 368-376. Nothing of substance is added by the biographical accounts of E. and É. Haag, in *La France protestante*, II (Paris, 1847; facs. repr., Geneva, 1966), 484-486; of M. Nicolas, *Histoire littéraire de Nîmes*, II (Nîmes, 1854), 76-83; or of the *Biographie universelle ancienne et moderne*, V (Paris, 1812), 384-385. Two remarkable studies of Bourguet's life and work are F. A. Jeanneret and J. H. Bonhôte, "Louis Bourguet," in *Biographie neuchâteloise*, I (Le Locle, 1863), 59-80, with bibliography; and L. Favre, "Inauguration de l'Académie de Neuchâtel, discours de M. Louis Favre," in *Musée neuchâtelois*, **3** (1866), 288-310. These works furnish much important new information but provide little analysis of Bourguet's printed works.

More recently, P. Bovet, in "Le premier enseignement de la philosophie à Neuchâtel, 1731," in *Musée neuchâtelois*, **41** (Sept.-Oct. 1904), 195-210, gives details on Bourguet's professorial activity; and in "Louis Bourguet, son projet d'édition des oeuvres de Leibniz," in *IIe Congrès international de philosophie* (Geneva, 1904), 252-263, recounts Bourguet's efforts to edit Leibniz's correspondence for publication. L. Isely published three articles on the Leibniz-Bourguet correspondence: "Leibniz et Bourguet, Correspondance scientifique et philosophique, 1707-1716," in *Bulletin de la Société neuchâteloise des sciences naturelles*, **32** (1903-1904), 173-214; *Cinq lettres inédites de Bourguet* (La Chaux-de-Fonds, 1904); and *Les origines de la théorie des fractions continues. Leibniz et Bourguet. Correspondance scientifique et philosophique* (Geneva, 1904).

Recent works include H. Perrochon, "Un homme du XVIIIe siècle: Louis Bourguet," in *Vie, art, cité*, **1** (1951), 34-38, esp. on the role played by the *Mercure suisse*; and M. Godet, "L'office de bibliothécaire, comment on le concevait à Neuchâtel il y a deux siècles," in *Nouvelles de l'Association des bibliothécaires suisses*, 3rd ser., **18**, no. 1 (1942), 1-12; and "Au temps de la Respublica litterarum: Jacob Christophe Iselin et Louis Bourguet," in *Festschrift Karl Schwarber* (Basel, 1949), 117-127.

See also N. Broc, *Les montagnes vues par les géographes et les naturalistes de langue française au XVIIIe siècle* (Paris, 1969), 49-50, 126, 147-148, which underestimates Bourguet's pioneering role as a challenger of Gautier's views; J. Roger, *Les sciences de la vie dans la pensée française du XVIIIe siècle* (Paris, 1963), 376-378, which, in a brief aside, places Bourguet in the context of contemporary thought; B. Sticker, "Leibniz et Bourguet. Quelques lettres inconnues sur la théorie de la terre," in *Actes du XIIe Congrès international de l'histoire des sciences* (Paris, 1971), 143-147; and F. B. C. Ullrich, "Scipione Maffei e la sua corrispondenza inedita con Louis Bourguet," which is *Memorie del R. Istituto veneto di scienze, lettere ed arti*, Classe di scienze morale, lettere ed arti, **34**, no. 4 (1969); and "Johann III Bernoulli ed il carteggio Bourguet," in *Rivista storica svizzera*, **19** (1969), 356-370.

Bourguet's crucial role in the history of crystallography is well treated by H. Metzger, *La genèse de la science des cristaux* (Paris, 1918), 44-52, 151-155; K. B. Bork, "The Geological Insights of Louis Bourguet (1678-1742)," in *Journal of the Scientific Laboratories of Dennison University*, **55** (1974), 49-77, is the first study to treat Bourguet's work in geology and paleontology in any real detail—it also offers much new information. See also F. Ellenberger, "De Bourguet à Hutton: Une source possible des thèses huttoniennes; originalité irréductible de leur mise en oeuvre," in *Comptes rendus . . . de l'Académie des sciences*, **275** (1972), ser. D, 93-96; "Un précurseur méconnu de James Hutton: L'ingénieur Henri Gautier (1660-1737)," *ibid.*, **280** (1975), vie acad., 165-169; and "À l'aube de la géologie moderne: Henri Gautier (1660-1737). Première partie," in *Histoire et nature*, **7** (1975), 3-58; deuxième partie, *ibid.*, **9** (in press, 1976).

During his lifetime and shortly after his death, Bourguet was cited in the *Journal des savants* (1742), 643-652, and especially in the *Journal de Trévoux, ou Mémoires pour servir à l'histoire des sciences et des beaux-arts* (facs. repr., Geneva, 1968), which published two detailed critical reviews by L. B. Castel of the *Lettres philosophiques*: **30** (1730), 1739-1750 (repr. ed. 442-444); and **40** (1740), 1636-1665 (repr. ed. 420-428).

Some years after Bourguet's death, Buffon referred to him in his *Théorie de la terre* (1749), arts. V, VIII, and IX, but his citations are very sparse considering the extent to which he drew on Bourguet's work. Desmarest devoted an article to Bourguet in the *Encyclopédie méthodique*, I (1795), 28-34, reproducing, although in somewhat altered form, a portion of the articles of the *Mémoire sur la théorie de la terre* dealing with geological phenomena. Bourguet's ideas on *pierres lenticulaires* are carefully discussed and criticized by Guettard, "Sur les pierres lenticulaires ou numismales . . . ," in *Mémoires sur différentes parties de la physique, de l'histoire naturelle, des sciences et des arts*, II (Paris, 1774), 197–207.

FRANÇOIS ELLENBERGER

BOYS, CHARLES VERNON (*b.* Wing, Rutlandshire, England, 15 March 1855; *d.* St. Mary Bourne, Andover, Hampshire, England, 30 March 1944), *physics*.

Boys was the son of Charles Boys, rector of Wing, Rutlandshire, and Caroline Goodrich Dobbie. In 1869 he enrolled at Marlborough College, where he was attracted to the teaching of G. F. Rodwell, the first teacher of science to be appointed at the school and a fellow of the Royal Society. Boys later dedicated his book *Soap-Bubbles and the Forces Which Mould Them* (1890) to Rodwell in appreciation of "the interest and enthusiasm which his advent and his lectures awakened in the author, upon whom the light of science then shone for the first time." In 1873 he entered the Royal School of Mines, London, and in 1881 he became a demonstrator in physics under Frederick Guthrie, who was professor at the Normal School of Science (renamed the Royal College of Science in 1890 and later incorporated in Imperial College). For a short period after Guthrie's death in 1886, Boys took charge of the physics department and was promoted to assistant professor in 1889. One of those to whom he lectured was H. G. Wells, who commented in *An Experiment in Autobiography* that while he had found Guthrie uninspiring, Boys was too fast: "Boys shot across my mind and vanished from my ken with a disconcerting suggestion that there was a whole dazzling universe of ideas for which I did not possess the key." Wells described Boys as "one of the worst teachers who had ever turned his back upon a restive audience, messed about with the blackboard, galloped through an hour of talk, and bolted back to the apparatus in his private room." Despite these defects Boys later became an able expositor, both as a lecturer at the Royal Institution (where soap bubbles were the subject of his Christmas lectures in 1889) and as an expert witness in the law courts.

In 1887 Boys published a preliminary note on his "radiomicrometer," which was designed to detect infrared radiation and in which he incorporated a blackened thermojunction into the suspension of a moving-coil galvanometer. This obviated the need for an external circuit and for conducting leads from fixed terminals to the moving-coil suspension. This instrument was probably more sensitive than any preceding infrared detector. Boys calculated that in combination with a condensing lens of one-inch diameter it could detect the heat from a candle at a distance of 1,530 feet, and with it he detected radiant heat from the surface of the moon. Despite its ingenuity and sensitivity, the radiomicrometer did not come into general use; it required both a steady horizontal platform and the proximity of the observer, and so the conventional thermopile or bolometer in combination with a remote galvanometer proved more convenient.

In the torsion head of the radiomicrometer there was no need for a conducting suspension, and Boys therefore experimented with fibers of various insulating materials. After trying glass, he proceeded to fused quartz, which he found to have a low expansion coefficient and a remarkably low hysteresis. One of his methods of drawing very thin fibers (one micron or less in diameter) was to fuse a portion of a quartz rod of which one end was fixed and the other was attached to the bolt of a crossbow, which was then discharged, drawing the rod into a fiber before it had time to cool. Fused quartz has since become one of the basic materials of experimental physics and has been used in such experiments as those of Beth (1936) and Holbourn (1936) for measuring the angular momentum of a photon. (Holbourn was the first winner of the C. V. Boys Prize, which the Physical Society founded in Boys's memory.)

Noting that the combination of mechanical strength with low hysteresis in the suspension would facilitate a much improved version of Cavendish's experiment (1798) to measure the constant of gravitation, Boys carried out the measurements between 1890 and 1895. Conditions at South Kensington were not good enough, because of vibration due to traffic, and he finally worked in the cellars of the Clarendon Laboratory at Oxford. The experiment was one of the classics of experimental physics and was particularly notable for its discussion to determine the optimum size of the apparatus, especially as regards the effects of temperature gradients and convection currents in the air surrounding the suspension. He gave dimensional arguments to show that the effects of convection currents were likely to depend on a high power, perhaps the fifth or the seventh, of the linear dimensions, and that the apparatus should therefore be as small as possible. Boys gave his final result for the constant of gravitation as 6.6576 $\times 10^{-8}$ in cgs units ($6.6576 \times 10^{-11} Nm^2kg^{-2}$) estimating that the fourth figure should not be more than 2 in error. The currently accepted best estimate for *G* lies a little (perhaps 0.5 percent) above Boys's upper limit, but the effective precision of his measurement has not yet been substantially improved.

The measurement of the constant of gravitation was the peak of Boys's achievement in pure science. In 1897, just as the experiment was ending, he became a metropolitan gas referee, resigning his post as assistant professor and setting up an office and laboratory in London; thenceforth he was primarily involved in applied physics. He had already begun (in 1893) to develop a lucrative practice as expert witness in patent cases; he himself applied for eighty-seven patents between 1881 and 1939 and took fifty-three of them to the stage of complete specification. These patents covered a wide range of devices in connection with mechanics, measurement, electricity, and gas metering; in the last area his work on instruments for measuring the calorific value of coal gas was especially important.

Boys prided himself on his invention of the rotating lens camera for photographing lightning flashes. He built the camera in 1900 and carried it about with him for twenty-eight years, until he at last had an opportunity to use it while visiting A. L. Loomis at Tuxedo Park, New York. He concluded that the flash started at the ground, and his work led to that of Schonland in South Africa. He also developed a system for the spark photography of flying bullets, obtaining widely reproduced pictures of shock waves; but, as he was careful to emphasize, he had been anticipated in the initiation of this development by Ernst Mach.

In 1892 Boys married Marion Amelia Pollock; they had two children. In private life he was somewhat unconventional, for example drinking his tea out of a saucer if it was too hot, and covering a dirty shirt when he was about to appear as an expert witness in a lawsuit by inserting a sheet of plain foolscap under his tie and waistcoat. Like some other physicists, most notably R. W. Wood and George Gamow, Boys delighted in practical jokes, in which the victim might equally be a cat invading his garden or a solemn academic colleague. The consequent strain on his domestic life led to a divorce in 1910.

At the same time, he was a much loved figure in scientific circles in London. In 1935 he was knighted, and the Royal Society Club celebrated his eightieth birthday with a dinner. With the deterioration of his eyesight, he retired to St. Mary Bourne, Andover; a year before his death he published a paper on an instrument for drawing ellipses. One of his earliest papers had been on the garden spider (*Nature*, **23** [1880], 149–150), and he retained his biological interests throughout his life. In his garden he became more interested in weeds than cultivated plants, summarizing his experiences in a small book, *Weeds, Weeds, Weeds* (London, 1937).

Among his honors were a Royal Medal (1896) and the Rumford Medal (1924) of the Royal Society (to which he had been elected in 1888), the Duddell Medal of the Physical Society (1925), and the Elliot Cresson Medal of the Franklin Institute, Philadelphia (1939). He was president of the Physical Society of London (1917–1918); his presidential address contained some characteristic advice to experimenters, quoting a dictum of Fresnel: "If you cannot saw with a file or file with a saw, you will be no good as an experimentalist." But for all his pragmatism, Boys had a strong sense of elegance, and spoke, for example, with humorous scorn of "any instrument designer who would lower himself to make use of a *cam*." His work on gravitation in particular showed his exemplary sense of the many factors in instrument design, his persistence in tracing experimental difficulties to their basic causes, and his patience in advancing the experimental techniques of his time to the limit of the possible.

C. T. R. Wilson (of the cloud chamber), whose investigation of the electrical conduction in the air surrounding a gold-leaf electroscope led to the discovery of cosmic rays, paid this tribute to Boys's experiments on the insulating power of quartz: "These experiments gave very strong indications that some at least of the leakage of electricity from a charged body suspended in a closed vessel is not through the insulating support but by conduction through the air." A great experimenter himself, Wilson also said of Boys: "His delight was in designing, constructing and manipulating apparatus for physical measurements of the highest accuracy, and in overcoming experimental difficulties which to most would have seemed insuperable. He was a really great experimenter, and his methods of working were original and often unconventional."

BIBLIOGRAPHY

Boys's memoirs include "The Radiomicrometer," in *Philosophical Transactions of the Royal Society*, **180A** (1890), 159–186; "Quartz Fibres," in *Nature*, **42** (1890), 604–608; and "On the Newtonian Constant of Gravitation," in *Philosophical Transactions of the Royal Society*, **186A** (1896), 1–72.

On Boys and his work, see R. A. S. Paget, "Sir Charles Boys, F.R.S.," in *Proceedings of the Physical Society*, **56** (1944), 397–403; Lord Rayleigh, "Charles Vernon Boys," in *Obituary Notices of Fellows of the Royal Society of London*, **4** (1944), 771–788, with portrait; and C. T. R. Wilson, "Sir Charles Vernon Boys, F.R.S.," in *Nature*, **155** (1945), 40–41.

R. V. JONES

BRAGG, SIR WILLIAM LAWRENCE (*b.* Adelaide, Australia, 31 March 1890; *d.* Ipswich, England, 1 July 1971), *X-ray crystallography, physics.*

Bragg was the eldest son of the physicist Sir William Henry Bragg and Lady Gwendoline Bragg, granddaughter of Sir Charles Todd, the postmaster general and government astronomer of South Australia. Bragg was educated at St. Peter's College, Adelaide, and at the University of Adelaide, where he studied mathematics. He continued his studies at Trinity College, Cambridge, where he was admitted in 1909; but after one year, at the suggestion of his father, he transferred to physics. After obtaining first class honors in the natural science tripos in 1912, he started research under J. J. Thomson. This work, however, was interrupted by the outbreak of World War I, when Bragg volunteered for service. After some time in a horse artillery battery, he was transferred to act as technical adviser on sound ranging at the Map Section, G.H.Q., where his knowledge of physics was partly responsible for the considerable success of the sound-ranging techniques. In 1915 he was jointly awarded the Nobel Prize for physics with his father for work on the X-ray determination of crystal structures.

After the war Bragg returned briefly to Trinity College as a lecturer, but in 1919 he was appointed Langworthy professor of physics at the University of Manchester, in succession to Rutherford. During this happy and successful period he was made a fellow of the Royal Society, and in 1921 married Alice Hopkinson, the daughter of a doctor, by whom he had four children. In 1937 Bragg moved for a year to the National Physical Laboratory as director, but after Rutherford's premature death, he was invited to Cambridge as Cavendish professor of experimental physics. He stayed at Cambridge (with some interruptions for wartime liaison work with Canada and the United States) until 1954, when he moved to the Royal Institution, London, as Fullerian professor of chemistry and as director of the Davy-Faraday Research Laboratory—a post at one time held by his father. At the Royal Institution he sponsored and conducted crystallographic research and also devoted himself with energy and talent to another of his lifelong interests—the popularization and teaching of science, including the history of science. He retired in 1966 but maintained an active interest in both crystallography and scientific popularization until shortly before his death.

Bragg started his first important work as a result of the claim—made in 1912 by Friedrich, Knipping, and Laue—that they had observed the diffraction of X rays by a crystal. William Henry Bragg, who advocated a corpuscular theory of X rays, was greatly interested in Laue's work, despite the fact that the observed effect was explained in terms of, and strongly supported Barkla's alternative wave theory of X radiation. Lawrence and his father discussed Laue's findings in 1912, and William Henry developed the theory that the diffraction effect might be explicable as the shooting of corpuscles down avenues between lines of atoms in crystals.[1] Bragg seems to have found this suggestion unconvincing, although he was careful not to contradict his father in public, and after further study of Laue's paper, came to the conclusion that this was indeed a diffraction effect, but that in its application to ZnS Laue's explanation was incorrect and unnecessarily complex. Part of the problem concerned Laue's suggestion that ZnS was a simple cubic system. As a result he had found a number of unexpected gaps in the diffraction pattern. He had explained that these gaps derived from the absence of particular wavelengths in the incident beam, or perhaps from fluorescence. Bragg suggested[2] that ZnS should be seen as face-centered cubic, rather than as simple cubic. He was then able to show that as a result the diffraction pattern was entirely explicable as having arisen from the diffraction of white X radiation through a three-dimensional grating. No special assumptions about the characteristics of the incident beam were required.

In his paper Laue had calculated the conditions for diffracted intensity maxima for the simple cubic system where the incident beam was parallel to one side of the cell. The path differences of diffracted rays in the three dimensions were thus represented by the three expressions $a \cos \Theta_1$, $a \cos \Theta_2$, and $a(1 - \cos \Theta_3)$, where a equals the length of the side of the unit cell; and intensity maxima were achieved when the following conditions were satisfied: $\alpha = h_1 \frac{\lambda}{a}$, $\beta = h_2 \frac{\lambda}{a}$, $1 - \gamma = h_3 \frac{\lambda}{a}$, where $\alpha = \cos \Theta_1$; $\beta = \cos \Theta_2$; $\gamma = \cos \Theta_3$; and

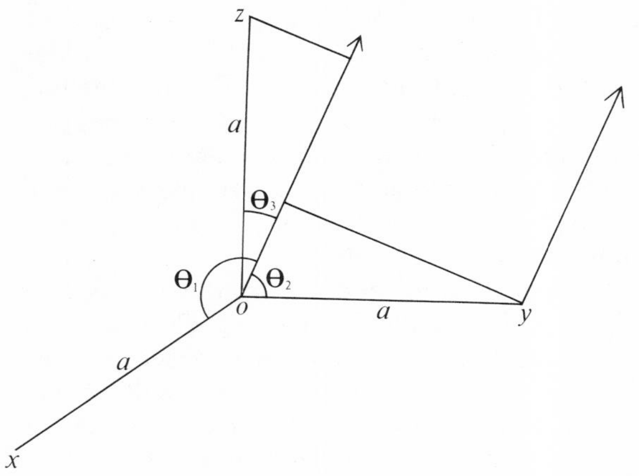

FIGURE 1. Laue's analysis of crystal diffraction.

h_1, h_2, and h_3 represented an integer equivalent to the order of the diffracted beam in each dimension. (See Figure 1.)

While Bragg concurred with Laue's identification and treatment of the problem, he nonetheless, in an impressive display of physical insight, reconceptualized the effect as that of the reflection of X rays off crystal planes, and formulated this in the expression $\lambda = 2d \cos \Theta$, which showed the relationship between angle of incidence Θ, wavelength λ, and distance between parallel atomic planes d. (This expression in its more usual form $n\lambda = 2d \sin \Theta$, where n is an integer corresponding to the order of refraction and Θ corresponds to the glancing angle—[the "Bragg angle"]—became universally known in the community of crystallographers as "Bragg's law." See Figure 2.)

This reworking was of far-reaching significance, for when compared with Laue's expression, Bragg's law (and the notion of reflection) rendered the process of diffraction easier to visualize and simplified calculation—advantages that were particularly important in the early development of X-ray crystallography, although Laue's expression

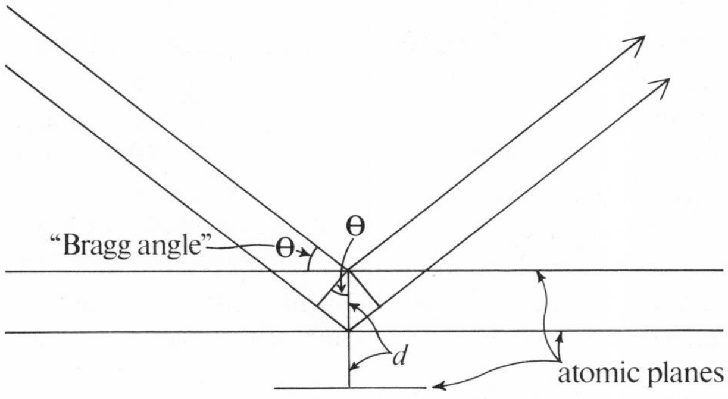

FIGURE 2. Bragg's analysis of X-ray reflection off atomic planes.

proved more appropriate for later quantitative work.

It was thus immediately clear that the crystal "manufactured" its own monochromatic X rays. The notion of reflection also explained why Laue had found that diffracted spots were circular when the photographic plate was close to the crystal, but became elliptical when the plate was more distant. Moving in a cone from the source, the X rays, once reflected, tended to converge in one plane.

With this work, Bragg's father abandoned the corpuscular theory, and father and son embarked on a brilliant period of intense collaboration, interrupted only by the outbreak of World War I. Although much of their work was published jointly, they brought rather different interests and skills to the collaboration. William Henry Bragg brought a deep concern with physical issues and great skill as an experimenter—most particularly in the form of the X-ray diffractometer. Lawrence Bragg brought a rapidly developing knowledge of and sensitivity for the crystalline state, and the powerful notion of reflection from atomic planes.

In a joint paper read in April 1913[3] the Braggs described the ionization spectrometer and the observed relative intensities of the different "orders" of diffracted X rays, when these were reflected off "normal" crystal planes. They calculated the length of side of the elementary cube of NaCl but were unable to derive a final value for λ, because it was not clear whether the distance between reflecting planes represented that between two identical or successive planes.

This problem was solved in a preliminary but spectacular fashion in a paper read by Bragg in June 1913.[4] He devised a simple method for projecting and indexing reflections, which he used to show that there were systematic differences between such simple cubic structures as KCl, such face-centered cubic structures as KBr, and NaCl which appeared to be intermediate between the other two structures. He explained this by suggesting that the scattering power of atoms varied in proportion to atomic weight. Thus, in the case of KCl, the atoms were of approximately equal scattering power, and this was reflected in the simple cubic lattice to which both, as it were, contributed. This was not the case for KBr where the lattice was defined by the heavier Br atom. NaCl was an intermediate case, reflecting the greater but not predominant scattering power of the Cl atom.

Lawrence Bragg then calculated a figure proportional to the number of molecules in each scattering center, and found the value of KCl to be half that of the other molecules considered. This sug-

BRAGG

gested strongly that in each of these other cases a single molecule was associated with each scattering center, while in the case of KCl an atom alone was so associated. With this knowledge the calculation of a tentative wavelength for X rays became possible.

In July 1913 the Braggs published the structure of diamond, showing the carbon atoms to be tetrahedral,[5] although this was largely the work of William Henry, and in November, Lawrence Bragg outlined further crystallographic work.[6] The importance of accurate intensity measurements induced him to abandon photographic methods and concentrate on the diffractometer. The importance of comparing the relative intensities of the different orders of characteristic X rays in determining the space group of the crystal had been made clear in the paper described above. Now Lawrence Bragg showed that the further quantitative study of intensities could be used to determine the position of those atoms the positions of which were not fixed with reference to symmetry considerations. In this paper he resolved the structures of FeS and $CaCO_3$—both with a single such parameter, and the first such structures to be solved. It may be suggested fairly that this set of papers constituted a "charter" for the later development of X-ray crystallography.

After World War I Lawrence Bragg continued this work. His general strategy was to develop the power of X-ray methods for crystal structure determination, whether by direct or indirect means. His first contribution was to publish a list of atomic radii,[7] which, however, were calculated from an incorrect baseline, and required later correction. The aim of this work was to set limits to possible atomic packing arrangements, and hence reduce the number of potential solutions of unknown structures with several parameters.[8] His second venture, with James and Bosanquet,[9] was more successful and involved work on absolute intensity measurements. They derived empirical *f*-curves—the ratio of the amplitude scattered by an atom at different angles to that scattered by a single classical electron—which were then used by Hartree in comparisons with theoretical *f*-curves derived from the Bohr model of the atom, and Schrödinger's later wave mechanics. In a crystallographic context, the work made it possible to check C. G. Darwin's prewar theory of the strength of reflections from perfect and mosaic crystals.

Bragg and his collaborators next turned to the determination of more difficult structures—the silicates—despite the many doubts about the possibility of solving structures with more than one or two parameters. In a few years, however, this work proved spectacularly successful, and it was shown that silicate structures depended on the ratio of silicon to oxygen atoms.[10] To a considerable extent this work depended on Bragg's almost intuitive sense of what constituted a plausible structure—a "good engineering job." At the same time he also encouraged work on metals, although his personal contribution was limited to the theoretical analysis of order-disorder phenomena.[11]

In his later years Bragg became more involved with administration and less concerned with day-to-day scientific work, although his scientific powers were in no way diminished. His continuing commitment to the development of X-ray crystallography was clearest in his support for and participation in the work of that small group of crystallographers (notably M. F. Perutz) who started work on the structure of globular proteins in the late 1930's and pushed on, despite great difficulties, to achieve spectacular success in the late 1950's.[12]

Bragg is therefore associated with the entire history and development of X-ray crystallography, and his major contributions can be seen as spanning a number of fields. The work before 1914 was seminal for X-ray crystallography, but also constituted a major contribution to both physics (X-ray wavelengths) and inorganic chemistry (continuous networks of ions rather than matched pairs in NaCl). The later work on absolute intensities laid a sound methodological basis for the development of quantitative X-ray crystallography, and was important for debates on atomic structure. The successful study of silicates not only revolutionized the basis of mineralogy, but was used by Pauling in his work on structural chemistry. The work on metals required fundamental rethinking in metal chemistry, while X-ray crystallography, and hence Bragg's work, contributed a vital element to the growth of molecular biology both directly and indirectly.

Bragg's importance for science may thus be judged in several ways. First, he was able to bring together elements from diverse fields. (It cannot, perhaps, be emphasized too strongly that at the roots of X-ray crystallography lies a synthesis of both great power and imagination.) Secondly, he possessed physical insight to an unusually high degree, combined with the happy knack of being able to present his ideas precisely but simply to both professional and lay audiences. Thirdly, his unusual degree of commitment and energy led him to press forward with the development of X-ray crystallography, where a less dedicated man might reasonably have considered the difficulties insuperable. Fourthly, Bragg was a great scientific organizer. He communicated his enthusiasm to those around him, whether in science or industry, and much work that does not bear his name can, as a result, be seen as having arisen from his inspiration.

NOTES

1. W. H. Bragg, "X-rays and Crystals," in *Nature*, **90** (1912), 219.
2. W. L. Bragg, "The Diffraction of Short Electromagnetic Waves by a Crystal," in *Proceedings of the Cambridge Philosophical Society. Mathematical and Physical Sciences*, **17** (1912), 43–57.
3. W. H. Bragg and W. L. Bragg, "The Reflection of X-rays by Crystals," in *Proceedings of the Royal Society*, **88A** (1913), 428–438.
4. W. L. Bragg, "The Structure of Some Crystals as Indicated by Their Diffractions of X-rays," *ibid.*, **89A** (1913), 248–277; this calculation is also used in a paper submitted at the same time by W. H. Bragg: "The Reflection of X-rays by Crystals (II)," *ibid.*, 246–248.
5. W. H. Bragg and W. L. Bragg, "The Structure of the Diamond," *ibid.*, 277.
6. W. L. Bragg, "The Analysis of Crystals by the X-ray Spectrometer," *ibid.*, 468–489.
7. W. L. Bragg, "The Arrangement of Atoms in Crystals," in *Philosophical Magazine*, 6th ser., **40** (1920), 169–189.
8. The problem was that it was not possible to move from diffraction data to a structure, but only backwards, deducing diffraction patterns from a supposed structure. Trial-and-error methods were therefore used, and it was important to limit the range of possible trials, which otherwise became very large.
9. W. L. Bragg, R. W. James, and C. H. Bosanquet, "Über die Streuung der Röntgenstrahlen durch die Atome eines Kristalles," in *Zeitschrift für Physik*, **8** (1921), 77–84.
10. See, for instance, W. L. Bragg and B. E. Warren, "The Structure of Diopside $CaMg(SiO_3)_2$," in *Zeitschrift für Kristallographie* . . . , **69** (1928), 168–193; and W. L. Bragg, "Atomic Arrangement in the Silicates," in *Transactions of the Faraday Society*, **25** (1929), 291–314.
11. W. L. Bragg and E. J. Williams, "The Effect of Thermal Agitation on Atomic Arrangements in Alloys," in *Proceedings of the Royal Society*, **145 A** (1934), 699–730; and W. L. Bragg and E. J. Williams, "The Effect of Thermal Agitation on Atomic Arrangements in Alloys (II)," *ibid.*, **151A** (1935), 540–566.
12. The first such structure was determined in 1957 by J. C. Kendrew, another of Bragg's collaborators; J. C. Kendrew *et al.*, "A Three-Dimensional Model of the Myoglobin Molecule," in *Nature*, **181** (1958), 662–666.

BIBLIOGRAPHY

I. Original Works. There is no complete bibliography of Bragg's published work, although when the biographical memoir of the Royal Society is published this should include the customary bibliography.

Bragg's works include *X-rays and Crystal Structure* (London, 1915), written with W. H. Bragg; *The Crystalline State*, I, *General Survey* (London, 1933); *Old Trades and New Knowledge* (London, 1933); and *Electricity* (London, 1936). He was editor of *The Crystalline State*, II–IV (London, 1948, 1953, 1965). Later works were "The Development of X-ray Analysis," in *Proceedings of the Royal Society*, **262A** (1961), 145–158; "Personal Reminiscences," in P. P. Ewald, ed., *Fifty Years of X-ray Diffraction* (Utrecht, 1962), 531–539; *The Crystalline State, vol. IV: Crystal Structures of Minerals* (London, 1965), written with C. F. Claringbull; "Manchester Days," in *Acta crystallographica*, **26A** (1970), 173–177; and *The Development of X-ray Analysis*, D. C. Phillips and H. Lipson, eds. (London, 1975).

II. Secondary Literature. On Bragg and his work, see P. P. Ewald, ed., *Fifty Years of X-ray Diffraction* (Utrecht, 1962), which contains a wealth of historical and biographical detail by and about crystallographers, including Bragg; and *Acta crystallographica*, **26A** (1960), 171–196, with assessments and appreciations by H. Lipson, I. Naray-Szabo, M. F. Perutz, D. C. Phillips, and J. Thewlis.

By permission of the family, extensive archival material has been deposited at the Royal Institution. A catalogue is in preparation.

JOHN LAW

BRÖDEL, MAX (*b.* Leipzig, Germany, 8 June 1870; *d.* Baltimore, Maryland, 26 October 1941), *medical illustration, anatomy.*

Brödel was the son of Louis Brödel and Henrietta Frenzel. He attended public schools in Leipzig and studied from 1885 to 1890 at the Leipzig Academy of Fine Arts. During vacations he worked at the Leipzig Anatomical Institute for His, C. W. Braune, and Werner Spalteholz and later at the Leipzig Physiological Institute for Carl Ludwig and his pupils. There he met Franklin P. Mall and William H. Welch, professors at Johns Hopkins. After fulfilling his military service obligation (1890–1892), Brödel worked in Leipzig as a freelance anatomical and physiological illustrator.

At Johns Hopkins he started work in 1894 with Howard A. Kelly, professor of gynecology, at the university's hospital. The association with Johns Hopkins continued for the rest of his career. In 1910, at the instigation of Brödel's friend and collaborator, Thomas S. Cullen, and with the benefaction of Henry Walters (the founder of the Walters Art Gallery in Baltimore), the department of art as applied to medicine was established for Brödel at Johns Hopkins.

Brödel revolutionized medical illustration by his contributions and also taught a group of leading artists who have continued his work. Brödel was an accomplished pianist and, with Henry L. Mencken, a member of the famous Saturday Night Club.

In 1902, he married Ruth Huntingdon, of Sandusky, Ohio, a biomedical artist who graduated from Smith College. They had four children.

Close attention to anatomic detail was the essence of his method. He wrote: "The artist must first fully comprehend the subject matter from every standpoint; anatomical, topographical, histological, pathological, medical and surgical. . . . A clear and vivid picture always must precede the actual picture on paper. The planning of the picture, therefore, is the all important thing, not the execution" ("Remarks . . . ," 71).

Brödel became an expert anatomist. Dissections of the kidney and vascular injections led Brödel to point out a relatively avascular plane and to suggest that the kidney be opened along this line when explored for stone. He also developed a suture for attaching a prolapsed kidney. He is eponymically memorialized in Brödel's line and Brödel's suture.

Brödel's best-known works included the illustrations for Howard A. Kelly's *Operative Gynecology* (1898), Cullen's *Cancer of the Uterus* (1900) and some for *Atlas of Human Anatomy* (1935). Among his nonmedical illustrations are two famous cartoons: "The Welch Rabbits," showing William H. ("Popsy") Welch standing with cigar in one hand and reins leading to his students, pictured as rabbits, in the other; and "The St. John's Hopkins Hospital," in which William Osler, adorned with halo and wings, is above the Johns Hopkins Hospital dome.

BIBLIOGRAPHY

Articles by Brödel are "Medical Illustration," in *Journal of the American Medical Association*, **117** (1941), 668–672; and "Remarks at Testimonial Dinner to Howard Atwood Kelly on his Seventy-fifth Birthday," in *Bulletin of the Johns Hopkins Hospital*, **53** (1933), 71.

Secondary literature includes T. S. Cullen, "Max Brödel," in *Bulletin of the Medical Library Association*, **33** (1945), 5–29; V. A. McKusick, "Brödel's Ulnar Palsy, With Unpublished Brödel Sketches," in *Bulletin of the History of Medicine*, **23** (1949), 469–479; and Judith Robinson, *Tom Cullen of Baltimore* (London–New York, 1949).

VICTOR A. MCKUSICK

BUGAEV, NICOLAY VASILIEVICH (*b.* Dusheti, near Tiflis [now Tbilisi], Russia, 14 September 1837; *d.* Moscow, Russia, 11 June 1903), *mathematics.*

The son of a military physician, Bugaev was educated in Moscow, where he was sent at the age of ten and where, while a student, he had to work as a tutor. In 1855 he entered the Physical-Mathematical Faculty of Moscow University. After graduating in 1859, he studied at the Engineering Academy in St. Petersburg but returned to Moscow in 1861 and from then on devoted himself to mathematics. In 1863 Bugaev defended his master's thesis, on constructing a general theory of the convergence of infinite series, then for two and a half years continued his education abroad, studying at Berlin under Kummer and Weierstrass and at Paris under Liouville.

In 1866 Bugaev defended his doctoral dissertation, which dealt with numerical identities related to the properties of the symbol *e*. The following year he became professor at Moscow University, serving as dean of the Physical-Mathematical Faculty from 1886 until 1903.

Many of Bugaev's pupils, including N. Y. Sonin and D. F. Egorov, became prominent scholars. In the last quarter of the nineteenth century he exerted considerable influence upon the work of his faculty and the Moscow Mathematical Society, of which he was cofounder, vice-president (1886), and president (1891). Bugaev was one of the most regular contributors to *Matematicheskii sbornik*, the society's journal, founded in 1866. Bugaev's proposal, that contributions to this journal by Russian authors always be published in Russian, led to the development of Russian mathematical terminology. As a proponent of the dissemination of mathematical knowledge and of the application of mathematics to technology, Bugaev was founder of the Society for Dissemination of Technological Knowledge in the late 1860's and headed its educational branch. He was elected a corresponding member of the St. Petersburg Academy of Sciences in 1898, and was an honorary member of numerous Russian and West European societies.

Bugaev's research was concerned mainly with mathematical analysis and number theory. In number theory Bugaev deduced many identities important in various problems of applied mathematics. Using elliptic functions, he proved formulas of number theory that had been given without proof by Liouville. He also published articles on algebra, the theory of algebraic functions, and the theory of ordinary differential equations.

BIBLIOGRAPHY

I. ORIGINAL WORKS. Bugaev's 76 books, articles, and reviews are listed in *Matematicheskii sbornik*, **25**, no. 2 (1905), 370–373. Works published before 1890 are also listed in F. A. Brockhaus and I. A. Efron, eds., *Entsiklopedichesky slovar*, IVa (St. Petersburg, 1891), 827–829. A synopsis of Bugaev's writings published before 1892 is in F. Y. Shevelev, ed., "Kratkoe obozrenie uchenykh trudov professora N. V. Bugaeva (sostavlennoe im samim)" ("A Short Review of Scientific Works by Professor N. V. Bugaev [Compiled by Bugaev Himself]"), in *Istoriko-matematicheskie issledovaniya* (1959), no. 12, 525–558.

II. SECONDARY LITERATURE. Various aspects of Bugaev's work are treated in a group of articles in *Matematicheskii sbornik*, **25**, nos. 1–2 (1904–1905). Of especial interest are L. K. Lakhtin, "Nicolay Vasilievich Bugaev (biografichesky ocherk)" (". . . Bugaev [a Biographical Essay]"), in no. 2, 251–269; and "Trudy N. V. Bugaeva v oblasti analiza" ("Bugaev's Works in the Area of Analysis"), *ibid.*, 322–330; and A. P. Minin, "O trudakh N. V. Bugaeva po teorii chisel" ("On Bugaev's Works in Number Theory"), *ibid.*, 293–321.

Bugaev's work is also described in general histories of mathematics: L. E. Dickson, *History of the Theory of Numbers*, 2nd ed., 3 vols. (Washington, D.C., 1934), see index; *Istoria otechestvennoy matematiki* ("National History of Mathematics"), II (Kiev, 1967), 297–299; A. A. Kiselev and E. P. Ozhigova, "K istorii elementarnogo metoda v teorii chisel" ("On the History of the Elementary Method in the Theory of Numbers"), in *Actes du XIe Congrès international d'histoire des sciences*, III (Wrocław–Warsaw–Cracow, 1968), 244–249; and A. P. Youschkevitch, *Istoria matematiki v Rossii do 1917 goda* ("History of Mathematics in Russia Before 1917"; Moscow, 1968), 483–489.

ALEKSANDR VOLODARSKY

BUNYAKOVSKY, VIKTOR YAKOVLEVICH (*b.* Bar, Podolskaya gubernia [now Vinnitsa oblast], Russia, 16 December 1804; *d.* St. Petersburg, Russia, 12 December 1889), *mathematics, mechanics.*

Bunyakovsky was the son of Colonel Yakov Vasilievich Bunyakovsky. After a basic education at home, he completed his studies abroad, receiving the doctorate in mathematical sciences at Paris in 1825. He returned the following year to St. Petersburg, where he subsequently began his scientific research and teaching. For many years Bunyakovsky lectured on mathematics and mechanics at the First Cadet Corps (later the Naval Academy) and at the Communications Institute. From 1846 to 1880 he was a professor at St. Petersburg University.

Bunyakovsky's scientific work was done at the St. Petersburg Academy of Sciences, of which he was named adjunct in mathematics (1828), extraordinary academician (1830), and ordinary academician (1841). He was elected vice-president in 1864 and retained the post for twenty-five years.

Of Bunyakovsky's approximately 150 published works in mathematics and mechanics, a monograph on inequalities relating to integrals in finite intervals (1859) is particularly well known. In this work he first stated the important integral inequality named for him:

$$\left[\int_a^b f(x)\,\phi(x)\,dx\right]^2 \leqslant \int_a^b f^2(x)\,dx \int_a^b \phi^2(x)\,dx.$$

Rediscovered and published by Hermann Schwarz in 1884, it is now often known as the Schwarz inequality. Bunyakovsky produced many works on number theory and in particular solved a series of specific equations and gave a new proof for the law of quadratic reciprocity.

Some of Bunyakovsky's results were included in P. Bachmann's *Niedere Zahlentheorie*, and about forty references to his original results appear in L. E. Dickson's *History of the Theory of Numbers*. His contributions to number theory include a work (1846) in which he gave an original exposition of this science and of its application to insurance and demography.

Bunyakovsky's works also deal with geometry. In 1853 he critically examined previous attempts to prove Euclid's fifth postulate concerning parallel lines and attempted a proof himself—unaware of the significance of Lobachevsky's non-Euclidean geometry. Active in disseminating mathematical knowledge in Russia, he also contributed substantially to the enrichment of Russian mathematical terminology.

Bunyakovsky's works on applied mechanics and hydrostatics are also of interest. To commemorate fifty years of his research and teaching, the St. Petersburg Academy in 1875 issued a medal and established a prize bearing his name for outstanding work in mathematics.

BIBLIOGRAPHY

I. ORIGINAL WORKS. Bunyakovsky's major writings are "Du mouvement dans la machine d'Atwood, en ayant égard à l'élasticité du fil," in *Mémoires de l'Académie impériale des sciences de St.-Pétersbourg*, 6th ser., **2** (1833), 179–186; *Leksikon chistoy i prikladnoy matematiki* ("Lexicon of Pure and Applied Mathematics"; St. Petersburg, 1837); *Osnovania matematicheskoy teorii veroyatnostey* ("Foundations of the Mathematical Theory of Probability"; St. Petersburg, 1846); "Note sur le maximum du nombre des positions d'équilibre d'un prisme triangulaire homogène plongé dans un fluide," in *Bulletin de la classe physico-mathématique de l'Académie impériale des sciences de St-Pétersbourg*, **10**, no. 4 (1852), 49–58; *Parallelnye linii* ("Parallel Lines"; St. Petersburg, 1853); *O nekotorykh neravenstvakh, otnosyashchikhsya k opredelennym integralam ili integralam v konechnykh raznostyakh* ("On Certain Inequalities Relating to Definite Integrals or Integrals in Finite Intervals"; St. Petersburg, 1859); "Sur les planimètres libres," in *Bulletin de l'Académie des sciences de St.-Pétersbourg*, 3rd ser., **11** (1860), 567–573; and *O samoschetakh i o novom ikh primenenii* ("On Computing Machines and New Uses for Them"; St. Petersburg, 1876).

II. Secondary Literature. See K. A. Andreev, *V. Y. Bunyakovsky* (Kharkov, 1890); L. E. Dickson, *History of the Theory of Numbers*, 3 vols. (Washington, D.C., 1919–1923), see index; I. G. Melnikov, "O rabotakh V. Y. Bunyakovskogo po teorii chisel" ("Bunyakovsky's Works on Number Theory"), in *Trudy Instituta istorii estestvoznaniya i tekhniki. Akademiya nauk SSSR*, **17** (1957), 270–286; *Opisanie prazdnovania doktorskogo yubileya vitse-prezidenta Akademii nauk akademika V. Y. Bunyakovskogo 19 maya 1875 g.* ("Description of the Celebration of the Doctoral Jubilee of the Vice-President of the Academy of Sciences, Academician Bunyakovsky, 19 May 1875"; St. Petersburg, 1876); V. E. Prudnikov, *V. Y. Bunyakovsky, ucheny i pedagog* (". . . Scientist and Teacher"; Moscow, 1954); and A. P. Youschkevitch, *Istoria matematiki v Rossii do 1917 goda* ("History of Mathematics in Russia Before 1917"; Moscow, 1968), esp. 296–302.

A. T. Grigorian

CAMERARIUS (CAMERER), RUDOLPH JAKOB (*b.* Tübingen, Germany, 12 February 1665; *d.* Tübingen, 11 September 1721), *medicine, botany.*

Camerarius was the older son of Elias Rudolph Camerarius (1641–1695), professor of medicine, and grandson of Johann Rudolph Camerarius (1618–1675), a physician. After earning his bachelor's (1679) and master's (1682) degrees, he traveled through Germany, Holland, England, France, and Italy from 1685 to 1687. Following his return to Tübingen, he received the doctorate at the university in 1687 and in 1688 was appointed extraordinary professor of medicine there and director of the botanical gardens. Upon the death of his father in 1695, Camerarius succeeded him as full professor. He died in 1721 of tuberculosis.

Camerarius' most important scientific achievement was the experimental demonstration of the sexuality of plants. This fundamental biological question had been debated by Theophrastus and, later, by Cesalpino (1583), Malpighi (1675), and Grew (1682), among others; but no clear conclusion had been reached. Camerarius was the first to prove by numerous experiments that sexuality must be attributed to plants. He reported his findings on the subject in the customary form of a long letter, *Epistola . . . de sexu plantarum*, which he sent in 1694 to Michael Bernhard Valentini, professor of medicine at Giessen, and which was subsequently printed several times. In it Camerarius first described, with great clarity, the structure of the flower in its various forms. Next he carefully compared plant and animal reproduction, drawing upon his own observations and experiments, some of the results of which he had already published. He had observed, for example, that a female mulberry tree (*Morus*) bears fruit even though no male tree is near it. The berries, however, contain only empty seeds, which he compared to the unfertilized "wind eggs" of the chicken.

After making this discovery, Camerarius conducted his first experiment with another dioecious plant, *Mercurialis annua* (1691), setting two female plants in flowerpots and isolating them from other plants. They produced fruit, but it soon dried up and contained no seeds. He repeated the same experiment with two other dioecious plants: *Spinacia oleracea* and *Cannabis sativa.* Turning his attention to the monoecious plants *Ricinus communis* and *Zea mays*, Camerarius cut off the male inflorescences below the opening of the anthers; in these cases, too, only empty fruit was produced. From these experiments Camerarius concluded that the anthers represent the male organs of the plant and the ovary, with style and stigma, the female organs. He considered those plants in which male and female organs coexist in the same flower to be hermaphroditic, and compared them with the snails, the hermaphroditism of which had been discovered shortly before by Swammerdam. Camerarius' assumption that the hermaphrodite plants, in contrast with the snails, fertilized themselves, later proved to be incorrect; Joseph Gottlieb Kölreuter (1761) and Christian Konrad Sprengel (1793) showed that allogamy predominates among hermaphrodites.

Camerarius further displayed his critical powers as a researcher by raising several objections to his view. He pointed out, for example, that *Lycopodium* and *Equisetum* form only flowers with many anthers, but that "they lack a female sex." The reproduction of these plants was not elucidated until a century and a half later, by Wilhelm Hofmeister. Camerarius also saw a serious challenge to his views in the fact that during his experiments, an ear of corn on a castrated plant had produced several fertile kernels, even though the intact plants were far away. In addition, he had encountered a similar phenomenon in experiments on *Cannabis*. Valentini wrote to Camerarius, in his response to the *Epistola*, that these last cases can be attributed to pollen having been carried by the wind over long distances.

Camerarius' experiments were confirmed at the Berlin Botanical Gardens in 1749. There, Johann Gottlieb Gleditsch induced a female palm tree (*Chamaerops humilis*) to produce seeds by dusting it with pollen obtained from a male specimen in the Leipzig Botanical Garden. Gleditsch repeated this experiment in 1750 and 1751 with equal success.

Camerarius had shown that ripe seeds form only if the stigma has been covered with pollen. The fact that the pollen also plays a role in the structure of the subsequent generation was demonstrated, however, by Kölreuter (1761) through the creation of numerous hybrids, among which could be detected various combinations of characteristics derived from the male or from the female parent. Nevertheless, many botanists continued to deny the sexuality of plants until the beginning of the nineteenth century. Finally, the works of Karl Friedrich von Gaertner (1844, 1849), who carried out over nine thousand hybridization experiments, overcame the last doubts on the matter.

In addition to the *Epistola . . . de sexu plantarum*, Camerarius published about thirty other, mostly briefer, works on botany. In the most notable of these, *De convenientia plantarum in fructificatione et viribus* (1699), he showed that plants displaying similar flower structures possess similar healing properties (*vires*) and, therefore, are constituted of similar substances. This notion, which Linnaeus, referring to Camerarius, later discussed in his *Philosophia botanica* (1751), has been thoroughly confirmed by modern chemotaxonomical methods.

BIBLIOGRAPHY

I. ORIGINAL WORKS. *Epistola ad D. Mich. Bern. Valentini de sexu plantarum* (Tübingen, 1694) was also published in Johann Georg Gmelin, *Sermo academicus de novorum vegetabilium . . . exortu* (Tübingen, 1749), 83–148; and in Johann Christian Mikan, *Opuscula botanici argumenti* (Prague, 1797), 43–117. It was translated into German, with extensive commentary and complete bibliography, by Martin Möbius as no. 105 in Ostwald's Klassiker der Exakten Wissenschaften (Leipzig, 1899). Also see *De convenientia plantarum in fructificatione et viribus* (Tübingen, 1699).

II. SECONDARY LITERATURE. See Alexander Camerarius, "Memoria Camerariana comprehendens programma funebre B. Rud. Jac. Camerarii," in *Acta Physico-Medica Exhibentia Ephemerides,* **1** (1727), app., 165–183; L. W. O. Camerer and J. F. W. Camerer, *Geschichte der Tübinger Familie Camerer 1503–1903* (Tübingen, 1903); A. Hirsch, ed., *Biographisches Lexikon der hervorragendsten Ärzte aller Völker und Zeiten,* I (1929), 808–809, and supp. (1935), 155; K. Mägdefrau, *Geschichte der Botanik* (Stuttgart, 1973), 108–111; J. Mayerhöfer, *Lexikon der Geschichte der Naturwissenschaften,* I (Vienna, 1958), 595; J. Sachs, *Geschichte der Botanik* (Munich, 1875), 416–421; and *World Who's Who in Science* (Chicago, 1968), 291.

KARL MÄGDEFRAU

CAMPANELLA, TOMMASO (*b.* Stilo, Calabria, Italy, 5 September 1568; *d.* Paris, France, 21 May 1639), *natural philosophy.*

Campanella, the son of Geronimo and Catarinella Martello Campanella, showed remarkable intellectual ability at a very early age. In 1582, inspired by the lives of Thomas Aquinas and Albertus Magnus, he entered the Dominican order, where his name was changed from Giovan Domenico to Tommaso, by which he has been known since. After his early philosophical studies, based on Aristotle, Campanella moved to the Dominican house at Cosenza in 1588 to study theology. There he read Telesio's *De rerum natura* in the 1570 edition, a work that remained a dominant influence for the remainder of his life. Telesio's work led him to publish *Philosophia sensibus demonstrata* (written 1589, published 1591) in response to G. A. Marta's *Pugnaculum Aristotelis adversus principia B. Telesii* (Rome, 1587). In 1589 Campanella went to Naples, where he met Giambattista della Porta and a Jewish astrologer named Abraham. He became active in della Porta's group, participating in its protoexperimentalism; through Abraham he became acquainted with the astrological and pseudoscientific traditions, which were important factors in his later thought. During this time he put forward his Telesian views more and more openly.

In May 1592 Campanella was denounced to the Inquisition for heresy and was confined to the Convent of San Domenico. Thus began a long series of imprisonments, trials, tortures, and other punishments that ended only with his release in 1629. His internment, in Naples (1599–1626) and in Rome (1626–1629), was sometimes under the most brutal conditions, while at other times he was allowed a certain degree of freedom for writing, reading, having visitors, and teaching. During these years Campanella continued his studies as far as was possible, and a number of manuscripts were smuggled out of prison to be published in Germany. They could not be issued in Italy, owing to the regulations imposed by the Inquisition.

After Campanella had had several years of partial freedom in Rome, the Inquisition discovered a plot in Naples (1634) by one of Campanella's followers and Campanella was implicated. He fled Rome before he could be arrested, reaching Aix-en-Provence in November 1634; there he was enthusiastically received by Peiresc and Gassendi, who had known of him and his work for some years. A few months later Campanella arrived in Paris. There Cardinal Richelieu came to his aid and helped him in various ways for the rest of his life, and he was also received at the court of Louis XIII. Finally accorded the honor he had been denied in Italy for so long, Campanella spent his last years writing and preparing for the press numerous works, which could be published in France with official approval. He died in the Dominican convent on Rue St.-Honoré and was buried in the adjoining Church of the Annunciation, which was destroyed during the Revolution (1795).

Campanella's writings cover a wide range of subjects. Although perhaps best known for work on political philosophy (*Civitas solis*) and metaphysics, he also wrote on most of the other branches of the comprehensive seventeenth-century philosophy *cursus*. The early attraction to Telesian science was never lost, and was reinforced by an attachment to Renaissance Neoplatonism and contact with della Porta's brand of experimentalism and magic. Although trained as a Dominican, Campanella very early recognized the limitations of Aristotelian philosophy and science, and a strong anti-Aristotelian tendency can be seen throughout his works.

Telesio's *De rerum natura* must be seen as the foundation stone of Campanella's thought (although not the only one). From it he derived a number of key principles, such as the reduction of active forces operating in the universe to those of heat and cold, and the conception of a void space in which natural events take place. Although Telesio's empirical view was still emphasized and reinforced by his contacts with the Neapolitan protoexperimentalists, Campanella also grafted strong theological and metaphysical elements onto it. Like Ficino and others before him, he viewed man as composed of the triad body-*spiritus*-soul. This position gave him scope to place man in a macrocosm-microcosm framework in which the use of magic and astrology could play a central interpretive role. Although scholars disagree somewhat on the precise extent to which magic influenced Campanella's thought, there is no doubt that it had a significant place. Like Ficino, but unlike Telesio, his world view allowed a key position to astrology and pseudoscience, as evidenced by works such as *De sensu rerum et magia* (1620) and *Astrologicorum libri VII* (1629). The latter work shows that Campanella acted as a consultant for Pope Urban VIII in astrological matters.

In addition to his interests in the astrological side of stellar phenomena, Campanella was a supporter of the Copernican system as defended by Galileo. In several works, principally the *Apologia pro Galilaeo* (1622), he sided with the embattled Florentine, not only in supporting the Copernican theory but also on broader issues, including those involving the competing claims of religion and science. Like Galileo, Campanella held that natural truth was not revealed in Scripture, but in the physical world. Thus the study of natural phenomena was seen as an important step toward theological understanding. Science and theology were to be clearly distinguished, however; both led to an understanding of God, who had revealed Himself in two books (*codices*), Nature and Scripture. While Galileo was essentially satisfied with an understanding of natural, physical reality, Campanella endeavored to go beyond this and to find the ultimate metaphysical truth of things. Again like Galileo, Campanella stressed—unfortunately for him, counter to the dominant opinion of his fellow Italians—the necessity of a *libertas philosophandi*. He was perhaps the first to use this particular formulation, later popularized by Spinoza and others.

Campanella's writings encompass a very broad range of scientific topics, and he must be viewed as one of the important systematizers of the seventeenth century. Iconoclastic and antitraditional, he attempted to provide a new system of natural knowledge based on the empirical, Neoplatonic, and astrological traditions at his disposal. While scientific knowledge played a more important role in his thought than most interpreters have admitted, it was basically subjugated to his very strong metaphysical and theological orientation, which aimed for ultimate causal explanations of physical phenomena. Thus Campanella's importance in the history of science was not so much through his own scientific discoveries as through his animistic, yet empirical, interpretation of the world, which influenced a number of contemporaries and successors.

BIBLIOGRAPHY

I. ORIGINAL WORKS. Campanella's output was voluminous. Owing to the circumstances of his life, many of his writings were destroyed; others were rewritten several times in various forms; some were circulated or even published without his knowledge; and others remained hidden until recent times. All of this makes for a very confusing state of affairs that is impossible to cover briefly yet in detail. For a thorough treatment see L. Firpo, *Bibliografia degli scritti di Tommaso Campanella* (Turin, 1940), and his biographical article cited below.

Among the works most important for the aspects of Campanella's work discussed in this article are *Philosophia sensibus demonstrata* (Naples, 1591); *Prodromus philosophiae instaurandae* (Frankfurt, 1617); *De sensu rerum et magia* (Frankfurt, 1620; Paris, 1637), also in Italian as *Del senso delle cose e della magia*, A. Bruers, ed. (Bari, 1925); *Apologia pro Galilaeo* (Frankfurt, 1622), repr. with intro. and Italian trans. by S. Femiano (Milan, 1971); *Astrologicorum libri VII* (Lyons, 1629; Frankfurt, 1630); *Medicinalium juxta propria principia libri VII* (Lyons, 1635); *Philosophia realis* (Paris, 1637); *Lettere*, V. Spampanato, ed. (Bari, 1927); *Mathematica*, R. Amerio, ed., in *Archivium Fratrum Praedicatorum*, V (1935), 194–240; and *Epilogo magno*, C. Ottaviano, ed. (Rome, 1939).

II. SECONDARY LITERATURE. The literature on Campanella is vast, and the reader is referred to the following bibliographies by L. Firpo: "Campanella nel settecento," in *Rinascimento*, **4** (1953), 105–154; "Cinquant'anni di studi sul Campanella (1901–1950)," *ibid.*, **6** (1955), 209–348; *Campanella nel secolo XIX* (Naples, 1956), repr. from *Calabria nobilissima*, **6–10** (1952–1956); and "Un decennio di studi sul Campanella (1951–1960)," in *Studi secenteschi*, **3** (1960), 125–164. The best and most recent brief account of his life and activities, with an excellent bibliographical survey, is Firpo's article in *Dizionario biografico degli Italiani*, XVII (1974), 372–401. Still fundamental are L. Amabile, *Fra Tommaso Campanella, la sua congiuria, i suoi processi e la sua pazzia*, 3 vols. (Naples, 1882); and *Fra Tommaso Campanella ne'castelli di Napoli, in Roma e in Parigi*, 2 vols. (Naples, 1887). Also see L. Firpo, *Ricerche campanelliane* (Florence, 1947). Of the vast interpretive literature, the more important studies include the following, listed chronologically: L. Blanchet, *Campanella* (Paris, 1920); G. di Napoli, *Tommaso Campanella, filosofo della restaurazione cattolica* (Padua, 1947); A. Corsano, *Tommaso Campanella* (Bari, 1961); N. Badaloni, *Tommaso Campanella* (Milan, 1965); and G. Bock, *Thomas Campanella: politisches Interesse und philosophische Spekulation* (Tübingen, 1974). A number of publications appeared in connection with the 400th anniversary of his birth, including *Tommaso Campanella (1569–1638)* (Naples, 1969).

CHARLES B. SCHMITT

CAMPANUS OF NOVARA

The documentation on the life of Campanus assembled by Francis Benjamin in *Campanus of Novara and Medieval Planetary Theory, Theorica planetarum*, Francis S. Benjamin and G. J. Toomer, eds. (Madison, Wis., 1971), 3–11, has been augmented by some important new documents published or noted by Agostino Paravicini Bagliani, "Un matematico nella corte papale del secolo XIII: Campano da Novara (†1296)," in *Rivista di storia della chiesa in Italia*, **27** (1973), 98–129, summarized by the same author in "Campano da Novara," in *Dizionario biografico degli Italiani*, **17** (1974), 420–424. These supplement, but do not essentially change, the picture of Campanus' life outlined in *DSB*, III, 23–24. They reveal that Campanus was a member of the well-known Panibada family of Novara (in his will is a bequest of fifty pounds to "his relative Francinus Panibada"), thus confirming the conjecture of L. A. Cotta, *Museo Novarese* (Milan, 1701), 87–89. They add the names of Nicholas III and Martin IV to the list of popes whom Campanus is known to have served. It now seems probable that he was chaplain to all twelve popes from Urban IV to Boniface VIII. To the long list of benefices already known to have been bestowed on Campanus in France (Paris, Rheims, and Saugeans [for which "Savines" in Benjamin, 5–7, and *DSB*, III, 24, is an error]), Italy (Domodossola), England (Felmersham), and Spain (Toledo) must be added an archdeaconry at Guadalajara.

The most important new document is Campanus' will (printed by Paravicini Bagliani, "Un matematico," 124–127). The legacies in this fully confirm the impression we have from other sources that Campanus was a wealthy man at his death. He instructed that he be buried in the Church of the Holy Trinity at Viterbo. Since the will, obviously made on his deathbed, was drawn up on 9 September 1296, the date of his death must fall between that day and 17 September, the date of a letter written by Boniface VIII announcing his death.

G. J. TOOMER

CANNON, WALTER BRADFORD (*b.* Prairie du Chien, Wisconsin, 19 October 1871; *d.* Franklin, New Hampshire, 1 October 1945), *physiology.*

Cannon was the only son of Colbert Hanchett Cannon and Sarah Wilma Denio. His father, a railroad worker, and his mother, a schoolteacher, were descendants of early American settlers. The Denio line began with the marriage of Abigail Stebbins to a French-Canadian *coureur de bois*, Jacques de Noyen, in Deerfield, Massachusetts, in 1704; the first of the Cannons (Samuel Carnahan) sailed to Boston from Ulster Province, Ireland, in 1718. Cannon's birthplace was the site of Fort Crawford, where, in the 1820's, William Beaumont made his classic observations on gastric function in his patient with a permanent gastric fistula, Alexis St. Martin. Cannon attended primary and grammar schools in Milwaukee, Wisconsin, and St. Paul, Minnesota. When Cannon was fourteen, his father, disappointed by his lack of diligence, put him to work in a railroad office, where he remained for two dull years. When he returned to high school, he completed the four-year course in three, was valedictorian and editor of the school paper, and then took a postgraduate year to prepare for college examinations. An extraordinary teacher of English literature at St. Paul High School, Miss M. J. Newson, persuaded Cannon to apply to Harvard College, and helped secure financial aid for him. On his departure for Cambridge in 1892 Cannon was given $100 by his father; this, plus a later gift of $80, was the sole help his father was able to provide during the rest of his educational career. His expense accounts were meticulously kept to the penny, as were the accounts relating to income from monitoring, tutoring, and, later, student assistantships.

Cannon's undergraduate days at Harvard were full and exciting. He was particularly attracted to the biological sciences and did extremely well in his course work. He took no mathematics while in college, however, and only two courses in chemistry. These decisions he later regretted. He graduated *summa cum laude* in 1896 with twenty-two courses to his credit. Two were research courses; and because of extra work done and the advanced character of some of it, Cannon was granted the M.A. degree in 1897. Among his most influential teachers were the philosopher and psychologist William James, the psychologist Hugo Münsterberg, the zoologist George H. Parker, and Charles B. Davenport, with whom he wrote his first research paper, "On the Determination of the Direction and Rate of Movement of Organisms by Light." While a student at Harvard College he became interested in neurology and psychology, interests that were continued in medical school.

Cannon may have preferred to study at the Johns Hopkins Medical School; but when his letter to Dean William H. Welch concerning financial assistance went unanswered, he enrolled in the Harvard Medical School. His studies with Davenport had provided his first insight into the nature of research, and led him into the physiology laboratory when he was a first-year student of medicine in the autumn of 1896. Henry P. Bowditch, then professor of physiology, suggested that Albert Moser, a second-year student who had likewise expressed a wish to undertake investigation, and Cannon should use the recently discovered X rays to examine the mechanism of swallowing. Thus, on 9 December 1896, less than a year after Röntgen made his announcement at Würzburg, Cannon first used X rays for the study of gastrointestinal motility. Cannon described the first demonstration, which was made on apparatus supplied by Ernest Amory Codman and set up in the small prosector's room in the anatomy department of the Medical School at the corner of Boylston and Exeter Streets.

> It was thought best to try first a small dog as a subject, and I was commissioned to get a card of globular pearl buttons for the dog to swallow. Dr. [Thomas] Dwight, Professor of Anatomy, Dr. Bowditch, Dr. Codman and I were the only witnesses. We placed a fluorescent screen over the dog's esophagus, and with the greenish light of the tube shining below we watched the glow of the fluorescent surface. Everyone was keyed up with tense excitement. It was my function to place the pearl button as far back as possible in the dog's throat so that he would swallow it. Nothing was seen! As intensity of our interest increased someone exploded: "Button, button, who's got the button?" We all broke out in a sort of hysterical laughter.[1]

Several days later, recalling Röntgen's observations that salts of heavy metals obstructed the passage of X rays, Cannon and Moser watched the passage of gelatin capsules filled with bismuth subnitrate in the frog and the goose. At the meeting of the American Physiological Society in Boston, 29 December 1896, the phenomena of deglutition, as exhibited by the goose when swallowing, were informally demonstrated to the members by means of the Röntgen rays. This was the first public demonstration of movements of the alimentary tract by the use of the new method.

As a student Cannon was an innovator not only in research but also in medical education. In the late 1890's he and his fellow students were subjected to four hours of continuous lectures, from two to six o'clock, five days a week. Little wonder that he questioned whether the didactic lecture was "the most satisfactory and effectual method of drilling the mind to careful thought in diagnosis and prognosis," or whether lectures were the most economical use of the students' time in "these days of the crowded curriculum." Either the student sat as a passive listener, he noted, or took notes so assiduously that he had no time to think.

Cannon had envied the eagerness and zest with which his roommate, a Harvard law student, and his classmates discussed cases and their implications, a method of instruction introduced by Christopher C. Langdell. He wrote:

> Undoubtedly the most brilliant example thus far of the use of cases in education is seen in the study of law. The change from text-book to the case system wrought out in the Harvard Law School has been called America's greatest contribution to educational reform. The newer method has roused an ardor and keenness of interest among students as was never known before. They learn their law not by dreary grubbing at text-books or lecture notes, but by vigorously "threshing out a case" with one another. And for its method and for its results the Harvard Law School is regarded by competent observers as perhaps the foremost centre of legal education in the English-speaking world.[2]

In 1900 Cannon adapted the case system for medicine, with printed data from actual case histories gathered from the various hospitals. These data, he suggested, were to be studied and analyzed by the students, and then discussed in detail at a conference with the instructor. He added, "To these cases the students can bring all their knowledge of anatomy, physiology, pathology and therapeutics, these subjects, which are now more like separate packets in the mind than related parts of a single system"—an early plea for integrated medical education. Much to the credit of the faculty, the case method was quickly and enthusiastically adopted by the various departments of the Medical School.

During his last year of medical school, Cannon was an instructor in zoology at Harvard College and taught comparative vertebrate anatomy, a course in which he had been an undergraduate assistant. When Cannon graduated, William Townsend Porter, who had assumed more and more responsibility for the teaching of physiology as Bowditch neared retirement, urged President Charles Eliot to make Cannon an instructor of physiology. Cannon accepted the position, and was promoted to assistant professor in 1902 with Porter's strong backing. The students were most appreciative of Cannon's teaching ability and, at the same time, became increasingly critical of Porter's. In 1904 a group protested the large number of failures in physiology, which they attributed to Porter. Thus, when Cannon received a call from Cornell Medical School in 1906, President Eliot decided to have him succeed Bowditch as George Higginson professor and chairman of the department; Porter became professor of comparative physiology. Cannon's appointment caused a rift with Porter and they remained estranged until 1937, when Cannon, as a member of the Council of the American Physiological Society, helped arrange for Porter to be named honorary president of the Society on the occasion of its fiftieth anniversary. Cannon remained chairman of the physiology department for thirty-six years and retired in 1942.

When Cannon joined the department of physiology, he continued his studies on gastrointestinal motility, using the fluoroscope as his principal tool. This pioneering research, done with a series of collaborators, examined the nature of swallowing, gastric peristalsis, the time of passage for different foodstuffs out of the stomach into the duodenum, control of the pylorus, and peristalsis of the small intestine. In 1911 Cannon summarized this experimental work in *The Mechanical Factors of Digestion.*

Cannon's early physiological research not only laid the groundwork for the development of gastrointestinal radiology but had other consequences as well. In 1908, following a savage antivivisectionist attack on the Rockefeller Institute, the American Medical Association organized a special Defense Committee in Support of Medical Research and, because of his growing eminence as an experimental physiologist, appointed Cannon as chairman. In so doing it put Cannon in the forefront of the struggle against antivivisection—a struggle in which he directed the strategy for the next eighteen years. The tenets that guided Cannon in that struggle were made explicit by him in an article in the *Journal of the American Medical Association* soon after he took office. They were not solely a defense. They were, in fact, a ringing affirmation of the right of medical research and experimentation.

> In the first place the investigators object to any step tending to check the use of animals for medical research. They maintain that such interference is not justified by the present treatment of the experimental animal. They declare that the imagined horrors of medical research do not exist. . . . Only the moral degenerate is capable of inflicting the torment that the anti-vivisectionist imagines. No one who is acquainted with the leaders in medical research who are responsible for the work done in the laboratories can believe for a moment that they are moral degenerates. The medical investigators further maintain that judgement should be based on knowledge not ignorance. They rightly insist that their critics are ignorant—ignorant of the conditions of medical research and ignorant of the complex relations of the medical sciences to medical and surgical practice and they contend that these critics in their ignorance are endeavoring to stop that experimental study of physiology and pathology.[3]

While Cannon's research did not develop in a linear fashion, it did develop logically, often on the basis of chance observations that other investigators might have ignored. During his investigations on the mechanical factors of digestion, Cannon noticed that the movement of the stomach and intestines of his experimental animals ceased when they were emotionally excited. These observations stimulated Cannon to study the effect of strong emotions on bodily functions and disease states—studies that led to an examination of the sympathetic nervous system. Between 1911 and 1915 Cannon developed the concept of the emergency function of the sympathetic nervous system, which he later synthesized in his *Bodily Changes in Pain, Hunger, Fear and Rage*. Despite the general acceptance of Cannon's research by physiologists, some of his concepts stirred controversy. George N. Stewart and Julius M. Rogoff, for example, sharply attacked Cannon's view of the emergency functioning of the adrenal medulla, maintaining that the organ secreted its hormone continuously. Despite the vigor of the attack, Cannon was ultimately able to sustain his view that the secretion of the adrenal medulla was much increased under emergency situations, such as those created by pain, cold, emotional stress, asphyxia, or injury.

In the fall of 1916, before the United States entered World War I, the National Research Council named Cannon a member of a committee on traumatic shock. Later he joined the Harvard University Hospital Unit. On his way to France in May 1917, he stopped in London and arranged with Fletcher, first secretary of the Medical Research Committee, to join the group of physicians and surgeons of the British Expeditionary Forces who were dealing with shock cases at the Casualty Clearing Station at Béthune. The following summer the Medical Research Committee formed a Shock Committee and E. H. Starling, H. H. Dale, W. M. Bayliss, T. R. Elliott, F. A. Bainbridge, John Fraser, and W. B. Cannon were named members. Cannon worked with Bayliss in his London laboratory in the winter of 1917–1918, and later was director of a surgical research laboratory attached to the Central Medical Department of the American Expeditionary Forces at Dijon. Initially Cannon and his associates in the field concentrated their therapeutic efforts on treating the acidosis that accompanies shock. Later they recognized that the acidosis was merely a secondary phenomenon, the result of the inadequacy of tissue perfusion. In 1923 Cannon summarized his wartime experience in *Traumatic Shock*.

After the war Cannon returned to investigations on the sympathetic nervous system and developed the denervated heart as an indicator of sympathetic activity. In 1921, at the same time that Otto Loewi was demonstrating the transmission of peripheral effects of the vagus nerve by a chemical mediator, Cannon and Joseph Uridil reported the acceleration of the denervated heart when the hepatic nerves were stimulated, even after the adrenal glands were extirpated. As Henry Dale noted, "Similar effects at a distance, transmitted by the circulation, were later recognized by Cannon and his colleagues as the result of stimulating other sympathetic nerves; and it seems clear that he had not been far from the discovery which later gained Loewi the Nobel Prize."[4]

Cannon's extraordinary manual dexterity and surgical skill enabled him to remove the entire sympathetic nervous system, a procedure that gradually led to the development of his ideas concerning the maintenance of steady states in the internal environment or fluid matrix of the body. In a series of studies continuing for almost a decade, Cannon showed that the function of the autonomic nervous system is the maintenance of a uniform condition in the body fluids, an elaboration of Claude Bernard's concept of the constancy of the *milieu intérieur*. Cannon later employed the specific designation "homeostasis" (from the Greek *homeos*, "like or similar" and *stasis* "condition") for these states. It is not clear when the concept of self-regulation of physiological processes first occurred to Cannon. In the preface to his *Wisdom of the Body* (1932), which summarized his investigations of homeostasis, Cannon noted:

> That relation was only slowly disclosed. Indeed, not a few researches on the service of the autonomic in providing for stability of the organism had been completed and published before the connection of that system with regulatory arrangements was clearly understood. We found that we had long been working on the role of the autonomic system in maintaining steady states without realizing that we were doing so! New facts already discovered took on new significance.

The studies that Cannon began with M. A. McIver and S. W. Bliss on the sympathetic and adrenal mechanism for mobilizing sugar in hypoglycemia induced by insulin (1923) may well have been seminal for the development of the concept of homeostasis. In an extension of this work published in *American Journal of Physiology* (1924), Cannon and his associates ended their paper with a theme that was later repeated for other physiological functions: "The mechanism here described is another remarkable example of autonomic adjustments within the organism when there is a disturbance endangering its equilibrium." At a meeting of the Congress of American Physicians in 1925, Cannon further elaborated this theme by presenting six propositions regarding physiological factors maintaining steady states in the body. It was not until the following year, however, that Cannon gave a name to his concept, in a modest paper entitled "Physiological Regulation of Normal States: Some Tentative Postulates Concerning Biological Homeostatics."[5] In 1929 Cannon published an exhaustive analysis of the problem of homeostasis in *Physiological Reviews*. This paper not only examined the homeostatic regulation of water and sodium chloride balance, glucose, protein, fat, and calcium but also reviewed the role of the autonomic nervous system in homeostasis and the homeostatic functions of hunger, thirst, maintenance of body neutrality, and uniform temperature. Cannon's closing statement, "that regulation in the organism is the central problem of physiology," was to have a profound influence on biological research.

Ironically, save for a popularization of the concept of homeostasis in *Wisdom of the Body* (1932) and the chapter "Ageing of Homeostatic Mechanisms" for E. V. Cowdry's *Problems of Ageing* (1939), Cannon did little more research on homeostasis. Instead, beginning in the 1930's, he and his collaborators, who included Z. M. Bacq and Arturo Rosenblueth, devoted themselves to the study of chemical transmission of nerve impulses and made pioneering advances in the field. The early synthesis of this research by Cannon and Rosenblueth, *Autonomic Neuro-Effector Systems* (1937), with its theoretical hypothesis of the existence of two sympathins, one excitatory and the other inhibitory, was critically received. Today, on the basis of Ulf von Euler's research, we know that they were in fact dealing with epinephrine and norepinephrine.

Despite his deep involvement in science, Cannon believed strongly that the scientist was also a citizen and, as a citizen, had an obligation to defend freedom. In his view freedom was an essential element for productive scholarship. His belief in the brotherhood of man and the universality of science took Cannon to the Peking Union Medical College in 1935, and subsequently involved him in the formation of the Medical Bureau to Aid Spanish Democracy, as well as in the organization of the American-Soviet Medical Society, the American Bureau for Medical Aid to China, and the United China Relief. Since these activities engendered much political controversy, they deserve fuller examination.

Cannon's work on digestion was responsible for his long and continuing interest in Russia, especially in the school of physiology sired by Pavlov. Initially Cannon and Pavlov came to know each other because of the similarity of the research in which they engaged. At the beginning of their friendship, they exchanged letters and papers but did not meet. In 1923, before an International Physiological Congress in Scotland, Pavlov, in response to Cannon's written entreaties, visited the United States. In 1929 Pavlov returned to the United States to attend the International Physiological Congress, held in Boston. Although there were many extraordinary events at this Congress, including Cannon's display of animals that had survived complete sympathectomy, the hero of the Congress was certainly Pavlov. In 1935 the two physiologists cemented their friendship further when Cannon visited Moscow to attend the fifteenth International Physiological Congress. On this occasion Cannon dominated the Congress, not because of the paper he presented but because of his prefatory statement on the relation of freedom to scientific research:

> During the last few years, how profoundly and unexpectedly the world has changed. Nationalism has become violently intensified until it is tainted with bitter feeling. Governments whose strength seemed deeply rooted in fixed traditions have vanished like phantoms only to be replaced by strange new forms and agents. The world wide economic depression has

CANNON

greatly reduced the material support for scholarly efforts. In consequence lameness is already at hand and paralysis is threatening. Creative investigators of high international repute have been degraded and subjected to privations. Some universities have been closed. Others have been deprived of their ideal social function of providing a sanctuary for scholars where the search for truth is free and untrammeled, and where novel ideas are welcome and evaluated. As scientific investigators commonly associate with universities, these conditions have serious meaning for all of us. That raises questions which are insistent and searching. What is the social value of the physiologist or the biochemist? What conditions promote and what hinder his best uses to society? Do these conditions now prevail? If not what care should be taken to protect them lest they be lost or forgotten?

The prescience and wisdom of Cannon's remarks have been forgotten, and he is often portrayed as naive in political matters. Cannon well understood the Soviet Union as a social experiment; and while he applauded Soviet governmental support of science, he did not overlook the harsher aspects of Soviet politics. When someone questioned some of the commendatory public statements he made about the Soviet Union in 1935, Cannon replied.

> During my stay in the Soviet Union I found myself oscillating between admiration for the ideas and achievements of the group in control of the vast social experiment which is being tried there and a sickening horror and repulsion because of tales which I heard from American and British sources of gross cruelties and injustices perpetrated on persons who were not given any reason for the treatment they received. There is no question whatever regarding the verity of these reports. . . . It is inconceivable to me that the unjust treatment of human beings should go on in such a harsh and cruel form as it takes. The extraordinary fact is that the present rulers are using the same sort of treatment against which their predecessors protested with utter disregard of their safety and indeed of their lives, in order to overthrow it.[6]

It was the comment of a critical friend who not only continued amicable relations with Soviet physiologists but also sought to improve relations between the Soviet Union and the United States through wider exchange of information. In 1937 he was offered the presidency of the American Russian Institute, which he refused. While Cannon's refusal was in large measure based on the limited time at his disposal, it is also clear that he wished to avoid daily political polemics. Nevertheless, he continued to express himself vigorously and critically on the Soviet Union in conferences, articles in popular magazines, and book reviews. With the outbreak of World War II, Cannon joined Hugh Cabot in organizing the Massachusetts chapter of the Committee for Russian War Relief—an act that brought harsh criticism from conservative members of the Boston medical community, who whispered darkly of Communist sympathies of the two independent members. In 1943, at the request of Henry Sigerist, Cannon became president of the American-Soviet Medical Society, the purpose of which was to organize the exchange and dissemination of medical information between the two countries.

In many respects Cannon's interest in Spain was motivated by the same concerns that shaped his interest in Russia: an abiding concern with the development of science and a humanitarian's ideal of aiding the weak and oppressed. Cannon recognized that one of the major forces of oppression and direct threat to scientific development was the rising tide of Nazism.

Cannon had strong ties with the medical and scientific community in Spain. Some colleagues, like R. Carrasco-Formiguera and Jaime Pi Suñer, had trained with him in physiology at Harvard. Others, like Georges Marañon and Juan Negrín, he met, and established friendships with, during his visit to Spain in 1930. That visit marked the beginning of Cannon's direct interest in Spain. When Alfonso XIII was overthrown in 1931, Cannon saw that political victory as a triumph of freedom and republican ideals, and wrote to Negrín that he hoped this new turn in Spanish politics augured well for Negrín's concept of a new national university. When Franco began a counterrevolution against the Spanish Republic in 1936, Cannon joined a group of distinguished American physicians in organizing the Medical Bureau to Aid Spanish Democracy. The purpose of this new organization, as Cannon saw it, was perhaps best expressed in a letter he wrote to A. S. Begg of the Boston University School of Medicine in November 1937:

> As you probably know there is an organization known as the Medical Bureau to Aid Spanish Democracy. Since near the beginning of the year I have been National Chairman of the organization. Its sponsors include [Anton] Carlson of Chicago, Haven Emerson of Columbia, Adolph Meyer and [Henry] Sigerist of Johns Hopkins, [Frederick A.] Coller and [Louis Harry] Newburgh of the University of Michigan, William H. Park of New York, [Joseph] Erlanger, Evarts Graham and Leo Loeb of Washington University, C. E. A. Winslow of Yale, Florence Sabin of the Rockefeller Institute and others. My inter-

CANNON

est in the organization was really based on an experience which I had in 1930 when I met the university group in Barcelona and Madrid and heard open talk of a republic in Spain. I was entertained by Juan Negrín then Professor of Physiology at the Medical School in Madrid. Ever since then I have kept in touch with him and now, as you know, he is Premier of Spain. The Medical Bureau has no political aims, it is not a propagandist organization, its purposes are strictly humanitarian. You know as well as I do that war is worse than pestilence and all sorts of privations and miseries which involve not only combatants but innocent non combatants—old men, women and children. . . . Of course in addition to these distresses there are the dreadful wounds of warfare with the long attention which they often demand of the victims of war. It is in an effort to lessen these miseries of pains I have attempted to serve the Medical Bureau.

Cannon's activities on behalf of the Medical Bureau embroiled him in some of his most bitter political controversies. The Spanish Civil War divided Americans, and there can be little doubt that many in the United States were deeply affected by the common charge that the Loyalist government of Negrín was Communistic. Some newspapers and journals in New England declared that the Medical Bureau "was a communist organization with headquarters in Russia" and that Cannon himself was not only a Communist but godless as well. Despite the obloquy heaped on him, Cannon continued to work tirelessly on behalf of medical aid to Loyalist Spain, organizing the medical community, speaking at innumerable rallies, and using the prestige of his name to collect medical supplies and, more especially, to persuade drug companies to send insulin and nicotinic acid to Spain to aid victims of diabetes and pellagra. When in 1937 the State Department blocked nurses and physicians from offering their services to the Loyalist government by requiring them to swear, before receiving their passports, that they would not enter Spain, Cannon joined in a protest of intellectuals that led to a change in that policy.

With Franco's victory, Cannon directed his efforts to rescuing scientists and physicians who fought for the Republic, and to finding posts for them in American and South American universities. As the 1930's drew to a close, these activities broadened to include physicians and scientists ousted or imprisoned by the Nazis in Germany and Austria, and Chinese scientists uprooted by the war with Japan. It was not only compassion that moved Cannon; he saw the rescue of the scientific victims of fascism as necessary for the further development of science and the preservation of its humanistic ideals.

With the advent of World War II, Cannon once more devoted himself to the problems of shock, this time not as an investigator but as chairman of the Committee on Shock and Transfusion of the National Research Council. In this capacity he saw Edwin J. Cohn and his associates develop methods of blood fractionation and blood preservation that later proved to be extraordinarily important in the prevention and treatment of shock.

In 1942 Cannon retired from Harvard Medical School. Although he subsequently was visiting professor at the New York University Medical School (1944) and did research with Arturo Rosenblueth in Mexico City (1945), he was in failing health. His illness was directly related to his early research. In 1931 it was discovered that Cannon was suffering from mycosis fungoides, a neoplastic disease that was the result of overexposure to the X rays he had used in research on digestion. He bore his affliction with great stoicism. On 1 October 1945, only eighteen days short of his seventy-fourth birthday, he succumbed to the disease.

When Cannon died he had received many honors. Although he never was awarded a Nobel Prize, his work was recognized by election to a number of learned societies, including the American Philosophical Society, the National Academy of Sciences, and the Royal Society. He was a valued member and officer of more than a score of professional associations and received numerous medals, honorary degrees, and lectureships. He was treasurer of the American Physiological Society from 1905 to 1912, and president from 1914 to 1916. Honors and recognition aside, one of the best evaluations of Cannon's career was made by Chandler McC. Brooks, Kuyomi Koizumi, and James O. Pinkston on the centenary celebration of his birth, held at the Downstate Medical Center of the State University of New York in 1971–1972:

Cannon was one of the first to use X-rays, but no one considered him to be a radiologist. He made major discoveries in the field of gastroenterology, but he was not considered a gastroenterologist. He contributed as much to our knowledge of the autonomic nervous system as any man, but he was not thought of as a neurophysiologist. In much of his work on transmitters he employed drugs to facilitate and drugs to block chemical actions, yet he was not a pharmacologist. Certainly he did much research relative to the function of the endocrine glands, but he is not considered to have been an endocrinologist. His work on emotional expression has been of great interest to psychologists and behaviorists, yet he was not classified as either. His studies of homeostasis, reactions to stress, the means used by the body to maintain required balances as well as his studies of traumatic shock qualify him to be a physician of major attainment. He was always considered to be a physiologist, no other title would suffice.[7]

NOTES

1. Letter from Walter B. Cannon to John F. Fulton, 16 Apr. 1942.
2. W. B. Cannon, "The Case Method of Teaching Systematic Medicine," in *Boston Medical and Surgical Journal*, **142** (1900), 31–36, 563–564.
3. W. B. Cannon, "The Opposition to Medical Research," in *Journal of the American Medical Association*, **51** (1908), 635–646.
4. H. H. Dale, "Walter Bradford Cannon," in *Obituary Notices of Fellows of the Royal Society of London*, **5** (1947), 407–423.
5. W. B. Cannon, "Physiological Regulation of Normal States: Some Tentative Postulates Concerning Biological Homeostatics," in *Jubilee Volume for Charles Richet* (Paris, 1926), 91–93.
6. Letter from Walter B. Cannon to Harry Freeman, 20 Sept. 1935.
7. C. McC. Brooks, K. Koizumi, and J. O. Pinkston, eds., *The Life and Contributions of Walter Bradford Cannon* (New York, 1975), xx.

BIBLIOGRAPHY

I. ORIGINAL WORKS. A complete bibliography of Cannon's work is in *The Life and Contributions of Walter Bradford Cannon . . .* (New York, 1975), 73–94. Cannon and his collaborators were prolific writers. Fortunately, at frequent intervals Cannon summarized the work he and his associates had been doing, and tried to assess the significance of the studies in relation to integrative physiology and medicine. Among his books were *The Mechanical Factors of Digestion* (London, 1911); *Bodily Changes in Pain, Hunger, Fear and Rage* (New York–London, 1915); *Traumatic Shock* (New York–London, 1923); *The Wisdom of the Body* (New York, 1932); *Digestion and Health* (New York, 1936); *Autonomic Neuro-Effector Systems* (New York, 1937), written with A. Rosenblueth; *The Body as a Guide to Politics* (London, 1942); *The Way of an Investigator* (New York, 1945); and *The Supersensitivity of Denervated Structures* (New York, 1949), written with A. Rosenblueth.

The Walter B. Cannon Archive in the Countway Library of the Harvard Medical School is the major repository of the letters, diaries, notebooks, MSS, and memorabilia. The collection, which contains 164 boxes of correspondence, is a valuable source of information on the development of physiology in the first half of the twentieth century. It also reflects Cannon's deep involvement in numerous scientific organizations and his role as a concerned and courageous citizen.

II. SECONDARY LITERATURE. No full-length biography of Cannon has been published. *The Life and Contributions of Walter Bradford Cannon (1871–1945)* (see above), which contains papers delivered at a centennial symposium on Cannon's influence on the development of physiology in the twentieth century, is an extensive review of the significance of his scientific contributions by former colleagues and others in his field of interest, and includes personal reminiscences by his son, Bradford Cannon. The esteem in which he was held by his contemporaries at Harvard and by others is evident in the commemorative volume *Walter Bradford Cannon: Exercises Celebrating Twenty-five Years as George Higginson Professor of Physiology, Oct. 15, 1931* (Cambridge, Mass., 1932). Biographical sketches of Cannon, listed chronologically, include W. J. Meek, "An Appreciation of Walter B. Cannon," in *Texas Reports on Biology and Medicine*, **11** (1953), 24–45; J. Mayer, "Walter Bradford Cannon. A Biographical Sketch," in *Journal of Nutrition*, **87** (1965), 1–8; and J. Garland, "Walter Bradford Cannon: George Higginson Professor of Physiology," in *Harvard Medical Alumni Bulletin* (Sept.–Oct. 1971), 4–8. Of the many obituaries published, Henry H. Dale, "Walter Bradford Cannon," in *Obituary Notices of Fellows of the Royal Society of London*, **5** (1947), 407–423, presents the most detailed critique of Cannon's physiological contributions, while R. M. Yerkes, "Walter Bradford Cannon: 1871–1945," in *Psychological Review*, **53** (1946), 137–146, assesses the impact of Cannon's views on psychology, particularly on the topic of emotions.

SAUL BENISON
A. CLIFFORD BARGER

CARUS, JULIUS VICTOR (*b.* Leipzig, Germany, 23 August 1823; *d.* Leipzig, 10 March 1903), *zoology.*

A comparative anatomist, Carus contributed to zoology as a historian and bibliographer; and his translation and promulgation of the works of Charles Darwin were influential upon scientific thought in Germany.

Carus came of a family that had included several professors at the University of Leipzig; his grandfather was Friedrich August Carus, professor of philosophy, and his father, Ernst August Carus, a doctor of medicine and surgeon, was *professor extraordinarius* there. His mother, the former Charlotte Agnes Eleonore Küster, was the daughter of a gynecologist.

After graduating from the Nicolai Gymnasium, Carus matriculated in 1841 at the University of Leipzig, where he studied medicine and natural history. When his father was appointed as professor and as director of a surgical clinic at Dorpat in 1844, Carus accompanied the family, planning to continue his clinical studies there. At the university his enthusiasm was kindled by the lectures he attended in comparative anatomy, physiology, and embryology; and he was especially influenced by Karl Bogislaus Reichert.

Carus returned to Leipzig in 1846 and was an assistant physician at the St. George Hospital while he completed his medical training. He received his doctorate in medicine and surgery in 1849 and decided to follow his inclination to study comparative anatomy. Using a travel stipend given by the Medical Faculty at Leipzig, he spent some time at Würzburg, where the histologist Albert von Kölliker taught, and with Karl Theodor von Siebold at Freiburg im Breisgau. Also in 1849 he published a paper on the alternation of generations and became conservator of the Museum of Comparative Anatomy at Oxford, where he remained until 1851. While at Oxford he concentrated on zoology and was in touch with Richard Owen, among others. He visited Sicily in the summer of 1850; in later years he returned to the Mediterranean several times, to study marine fauna at Messina and Naples, and published *Prodromus faunae Mediterraneae* (1885–1893).

After leaving Oxford, Carus returned to Leipzig, where he lectured on comparative anatomy. He was appointed *professor extraordinarius* of comparative anatomy in 1853 and was director of the zootomic museum. He was married in that year to Sophie Catherine Hasse; three daughters and a son were born to them. Carus' wife died in 1884, and in 1886 he married Alexandra Petroff of St. Petersburg, who bore him one son.

Over the years Carus lectured on comparative anatomy, embryology, and histology. During the summers of 1873 and 1874 he lectured at Edinburgh in place of C. Wyville Thomson, who was then on the *Challenger* expedition.

Within a few years of the publication of Darwin's *On the Origin of Species*, discussion of the new theory had become a part of Carus' lectures; but it was through his translations that he furthered evolutionary theory most actively. Carus was careful and painstaking; in his correspondence with Darwin he brought to the latter's attention minor inaccuracies. He improved on the earlier translation of the *Origin* by H. G. Bronn and, although he was not favorably impressed by Darwin's theory of pangenesis, translated into German *The Variation of Animals and Plants Under Domestication*, in which the theory is presented. It is not surprising that Carus became interested in the propagation of Darwin's works, for in his own *System der thierischen Morphologie* (1853) he had stated that the aim of zoology is the explanation of animal forms and, in a passage that has been cited frequently since then, had maintained the direct connection of presently living organisms with the earliest created forms. But this was not a concept that Carus expressed consistently or developed at that time, even though morphological relationships were important to him in tying together a wealth of scattered facts and details. Nevertheless, Darwin acknowledged these "views on the genealogical connection of past and present forms" in a letter to Carus in 1866, and remarked that he wished he had known of them when he wrote his historical sketch.

Carus also made available to German readers works by Thomas Huxley, Lionel Beale, and George Henry Lewes; thus he was important in linking zoological thought in Germany with contemporary English thought on the subject.

His predilection for scholarship that connected facts related to zoology, rather than for original research, led Carus to write a history of zoology (1872) to which other writers have often referred. His fluency in English was enhanced by his stay at Oxford and gave him the necessary expertise for his translations. Also, his work in a special division of the library at the University of Leipzig (1852–1859) augmented his income while providing experience that aided him in his extensive endeavors in zoological bibliography. With Wilhelm Engelmann he edited *Bibliotheca zoologica*, which proved to be an invaluable reference work; and his interest was reflected in the bibliographical section of the *Zoologischer Anzeiger*, the journal he founded in 1878 and edited until just before his death in 1903. Morphological systematization remained Carus' enduring interest. In the highly detailed *Handbuch der Zoologie*, which he wrote with A. Gerstaecker, he strove for a complete ordering and related present to fossil forms. It was necessary, he found, to introduce new nomenclature in various groups. Similarly, in his work on the Mediterranean fauna he organized and made available for other scientists a wide range of material.

Carus' scholarly contributions were recognized with honorary degrees conferred by Oxford, Edinburgh, and Jena; he belonged to learned societies in Germany and England and received several decorations. Although as a zoologist he was not long interested in research, his history of the science and his systematic and bibliographic compilations enlarged and enriched the literature.

BIBLIOGRAPHY

I. ORIGINAL WORKS. A comprehensive list of Carus' many writings, including his translations, appears in the concluding section of Taschenberg's article in *Leopoldina* (see below). They include *Zur nähern Kenntniss des Generationswechsels* (Leipzig, 1849); *System der thierischen Morphologie* (Leipzig, 1853); *Bibliotheca zoologica*, 2 vols., edited with Wilhelm Engelmann and published as a supp. to the latter's *Bibliotheca historico-naturalis* (Leipzig, 1861); *Handbuch der Zoologie*, 2 vols. (Leipzig, 1863–1875), written with A. Gerstaecker; *Geschichte der Zoologie bis auf Joh. Müller und Charl. Darwin* (Munich, 1872; repr. New York–London, 1965); and *Prodromus faunae Mediterraneae*, 2 vols. (Stuttgart, 1885–1893).

II. SECONDARY LITERATURE. See Karl von Bardeleben, "Julius Victor Carus," in *Anatomischer Anzeiger*, **23** (1903), 111; Max Beier, "Carus," in *Neue deutsche Biographie*, III (Berlin, 1957), 161; C. Chun, "Julius Victor Carus," in *Berichte über die Verhandlungen der Königlich Sächsischen Gesellschaft der Wissenschaften zu Leipzig*, Math.-phys. Kl., **55** (1903), 423–428; "Julius Victor Carus," in *Proceedings of the Linnean Society of London*, sess. 115 (1903), 28; Otto Taschenberg, "Julius Victor Carus," in *Leopoldina*, **39** (1903), 50–64, 66–73; and "Zur Erinnerung an Julius Victor Carus," in *Zoologischer Anzeiger*, **26** (1903), 473–483; and J. A. T. [J. Arthur Thomson], "Julius Victor Carus (1832–1903)," in *Nature*, **67** (1903), 613–614.

GLORIA ROBINSON

CARUS, PAUL (*b.* Ilsenburg, Germany, 18 July 1852; *d.* La Salle, Illinois, 11 February 1919), *philosophy of science.*

Carus received his Ph.D. from the University of Tübingen in 1876 after studying philosophy, classical philology, and natural science. He next was an instructor at the military academy in Dresden, but he was forced to leave because of religious views he had expressed in pamphlets. He went to England for a while and eventually moved to New York, where he published *Monism and Meliorism* (1885). This book aroused the interest of Edward Carl Hegeler, a German chemist in La Salle, Illinois. He had founded a periodical called *The Open Court* and soon invited Carus to become its editor. In 1888 another and more technical journal, to be devoted to the philosophy of science, was founded under the name of *Monist*; Carus was the editor of this as well. He also published philosophical and scientific classics (including Ernst Mach's *Science of Mechanics*) that are still widely used. In 1888 he married Hegeler's daughter Mary, who assisted him in editing the two journals and, after his death, continued to edit them until her death in 1936. Their daughter has continued to operate the Open Court Publishing Company, which publishes the Carus lectures, given by distinguished philosophers at meetings of the American Philosophical Association.

Carus wrote hundreds of articles and scores of books, which are largely collections of these articles. His mission in all his writing was to conciliate science and religion. Between 1890 and 1900 he elaborated a philosophy that he sometimes called "monism" and at other times "religion of science." He was convinced that science provided the key to all human problems. Religion was in a state of change, in which it would overcome its superstitions and find a solid foundation in the scientific *Weltanschauung*.

The key tool for constructing this "religion of science" was provided by the German mathematician Hermann Grassmann, who developed vector analysis. He had created a general science of pure forms that, unlike the empirical sciences, was not limited by the three-dimensional spatial world. His vector geometry could be expanded to unlimited dimensions. Carus' account of the nature of the world and our knowledge of it is a generalization of this method of geometry. There are forms in the world that belong to everything. Form is the quality not only of mind but of all reality. There is a one-to-one correspondence between knowledge and the external world because all elements of objective reality are inseparably united with the corresponding elements of subjective reality. According to his monism, all things are one by virtue of certain eternal laws that exist in things and are discovered by the inquirer. Such laws of nature are dependent on a single law—God.

Carus rejected Kant's position that the laws of nature depend on the mind of the knower; knowledge is the grasping of forms that are in the world. Hence Carus may be called a realist. He repudiated not only idealism but also materialism, claiming that every part of the world is both material (and acts according to the laws of matter) and spiritual (and acts according to the laws of mind). Thus he avoided the bifurcation of nature into minds and bodies in a way that is suggestive of Whitehead, who developed his process philosophy some years later. Human minds and souls differ from the rest of the world only in degree, not in kind. Carus' monism regarded the world as a living actuality that in its evolution from lower to higher forms naturally evolved ever-higher souls and, accordingly, raised the subjectivity of atomic life to the intellectuality of a human being.

Carus was somewhat disturbed by Einstein's relativity, not because it destroyed classical notions of absolute space, time, and motion but because the language of relativity sounded too much like the language of paradox and even of mysticism. This new physics, he thought, manifested a subjectivist tendency. It even implied the relativity of truth itself, as had the philosophies and theories of Bergson, Freud, Nietzsche, and the pragmatists.

BIBLIOGRAPHY

I. ORIGINAL WORKS. Carus wrote hundreds of articles that appeared in the journals *Open Court* and *Monist.* He also wrote many books, most of which are collections of his articles. They include *Monism and Meliorism* (New York, 1885); *Fundamental Problems* (Chicago, 1889); *The Soul of Man* (Chicago, 1891); *Religion of Science* (Chicago, 1893); *Primer of Philosophy* (Chicago, 1896); *Surd of Metaphysics* (Chicago, 1903); *Whence and Whither* (Chicago, 1903); *God* (Chicago, 1908); *Truth on Trial* (Chicago, 1911); and *Principle of Relativity* (Chicago, 1913).

II. SECONDARY LITERATURE. For biographical information, see *Dictionary of American Biography*, III, 548–549; and William Ellery Leonard, "Paul Carus," in *Dial*, **66** (1919), 452–455. For studies of his works, see William H. Hay, "Paul Carus: A Case-Study of Philosophy on the Frontier," in *Journal of the History of Ideas*, **17** (1956), 498–510; Donald Harvey Meyer, "Paul Carus and the Religion of Science," in *American Quarterly*, **14** (1962), 597–607; and James Francis Sheridan, "Paul Carus: A Study of the Thought and Work of the Editor of the Open Court Publishing Company" (Ph.D. diss., Univ. of Illinois, Urbana), also on microfilm (Ann Arbor, Mich., 1957).

JAMES ELLINGTON

CESALPINO, ANDREA (or **Andreas Caesalpinus**) (*b.* Arezzo, Italy, 6 June 1519; *d.* Rome, Italy, 23 February 1603), *medicine, botany, philosophy.*

Cesalpino studied philosophy and medicine at Pisa, where he received the doctorate in 1551. Four years later he succeeded his teacher Luca Ghini as professor of medicine and director of the botanical garden at Pisa. In 1592 he was called to Rome as physician to Pope Clement VIII and, simultaneously, professor at the Sapienza, where he taught until his death.

In his philosophical views, which he set out in *Quaestionum peripateticarum . . .* (1571) and which formed the framework of his medical and botanical works, Cesalpino was a follower of Aristotle, although he partially reformulated the latter's theories. He retained the favor of the church and of the pope and was acquitted of the accusation of heresy made against him.

Cesalpino's most important medical studies concern the anatomy and physiology of the movement of the blood. He gave a good description of the cardiac valves and of the pulmonary vessels connected to the heart, as well as of the minor blood circulation; he also recognized that the heart is the center of the circulation of the blood and accepted the existence of the traditional synanastomoses of the arteries with the veins. He did not, however, discover the major blood circulation (first demonstrated in 1628 by William Harvey). Cesalpino also published several works on practical medicine, which contain his observations on diseases of the heart and chest, syphilis, and other ailments.

Cesalpino's principal contribution to science lies in botany. Whereas such contemporary botanists as Brunfels, Bock, Leonhart Fuchs, Mattioli, and Tabernaemontanus merely described and illustrated a great number of plants in their *Kräuterbücher*, Cesalpino wrote the first true textbook of botany, *De plantis libri XVI* (1583). The first book of this text is of outstanding historical importance. Here, in thirty pages of admirably clear Latin, Cesalpino presented the principles of botany, grouping a wealth of careful observations under broad categories, on the model of Aristotle and Theophrastus.

Cesalpino considered the portion of the plant between the roots and the shoots—which he called the "heart" (*cor*)—to be the seat of its "soul" (*anima*), although he added that the soul is present throughout the plant. The task of the roots is to draw nourishment from the ground, and that of the shoots is to bear seeds. The leaves protect the shoots and the fruit from sunlight; they fall off in autumn, when the fruit is ripe and the shoots are developed. Cesalpino's description of the tendrils on the shoots and leaves, the climbing petioles of the Clematis, the anchoring roots of the *Hedera*, the secretion of the nectar from the blossoms, and many other phenomena testify to extraordinary skill in observation.

The parts of the plant, Cesalpino asserted, exist

either "for a purpose" (*alicuius gratia*) or "out of [inner] necessity" (*ex necessitate*); with this distinction he anticipated the concepts of adaptive characteristics and organizational characteristics. Cesalpino considered the fruit to be the most important part of the plant and, accordingly, made it the basis of his system of the plant kingdom. In this system the perianth and the stamens serve only to protect the young fruit; for in his opinion plants do not possess sexuality. He called the outer covering of the fruit the *pericarpium*. Among fruits he distinguished "racema" (*Vitis*), "juba" (*Milium*), "panicula" (*Panicum*), and "umbella" (*Ferula*).

Like Aristotle, Cesalpino divided plants into four "genera": *Arbores* (trees), *Frutices* (shrubs), *Suffrutices* (shrubby herbs), and *Herbae* (herbs). Trees possess a single stem, whereas shrubs have many thin stems. Shrubby herbs live for many years and often bear fruit, but herbs die after formation of the seeds. The distinction among species should be made, he held, only according to similarity and dissimilarity of forms; "unessential features" (*accidentia*), such as medicinal use, practical application, and habitat, must not be considered.

The remaining fifteen books are devoted to the classification and description of plants. The trees and shrubs are divided according to whether their fruits are single, bipartite, tripartite, quadripartite, or multipartite; herbs and shrubby herbs are classified in the same manner. Books 2–15 deal with the flowering plants, and book 16 treats those plants that form no seeds: ferns, duckweed, mosses, lichens, fungi, and algae. The latter groups arise, Cesalpino believed, from "putrefaction" (*ex putretudine*). Among the ferns, however, the leaves form on their underside a "down" (*lanugo*) out of which the new plants emerge and which thus corresponds to the seeds.

Cesalpino was the first to elaborate a system of the plants based on a unified and coherent group of notions. Not content to confine himself to describing plants, he also set forth the basic elements of general botany. By paying scant attention to the medicinal uses of plants—which were of crucial importance to his contemporaries—he raised botany to the level of an independent science.

Cesalpino exerted little influence on contemporary botanists; but later botanists, especially Joachim Jungius and Linnaeus, valued his work highly. The latter called Cesalpino the "first true systematizer" (*primus verus systematicus*) and named a plant genus after him (*Caesalpinia*).

BIBLIOGRAPHY

I. ORIGINAL WORKS. Cesalpino's writings include *Quaestionum peripateticarum libri V* (Venice, 1571); *Daemonum investigatio peripatetica, in qua explicatur locus Hippocratis si quid divinum in morbis habetur* (Florence, 1580); *De plantis libri XVI* (Florence, 1583); *Quaestionum medicorum libri II* (Venice, 1593; 1604); *De metallicis libri III* (Rome, 1596); *Katoptron, sive speculum artis medicae Hippocraticum* (Rome, 1601); *Appendix ad libros de plantis* (Rome, 1603); and *Praxis universae artis medicae* (Tarvisio, 1606).

II. SECONDARY LITERATURE. U. Viviani, *Vita ed opere di Andrea Cesalpino* (Arezzo, 1923), contains a list of works on Cesalpino up to 1916 and eight portraits of him. On Cesalpino as a physician, see G. P. Arcier, *The Circulation of the Blood and Andrea Cesalpino* (New York, 1945); and A. Hirsch, ed., *Biographisches Lexikon der hervorragendsten Ärzte aller Zeiten und Völker*, I (1929), 866–868, and supp. (1935), 166. Cesalpino as a botanist is discussed in K. Mägdefrau, *Geschichte der Botanik* (Stuttgart, 1973), 37–38, 41–43; L. C. Miall, *The Early Naturalists* (London, 1912), 36–39; E. Rádl, *Geschichte der biologischen Theorien*, 2nd ed., I (Leipzig, 1913), 122–126; J. Sachs, *Geschichte der Botanik* (Munich, 1875), 45–62, 487–490; and W. Zimmermann, *Evolution* (Freiburg-Munich, 1953), 125–130.

KARL MÄGDEFRAU

CHAMISSO, ADELBERT VON (*b.* Ante parish, Marne, France, *ca.* 27 January 1781; *d.* Berlin, Germany, 21 August 1838), *natural history, botany.*

Adelbert von Chamisso is known—if at all—by most readers as the creator of Peter Schlemihl, the man who bartered away his shadow. In Germany he is still a beloved poet whose verses were set to music by composers such as Schumann and Grieg. For scientists he is the naturalist who explored Pacific shores many years before Darwin, bringing back rich botanical collections, incidentally naming the California poppy and leaving his own name to an Alaskan island.

Chamisso was a younger son in an aristocratic provincial French family of the Champagne, just west of the Argonne forest. His father, Count Louis-Marie de Chamisso de Boncourt, was descended from established minor nobility with traditions of loyal military service to the crown; his mother, born Marie Anne Gargam, from wealthy bourgeois. The boy was christened Louis-Charles Adélaïde on 31 January 1781; in later life he always signed himself Adelbert. His early years were spent at the Château de Boncourt.

In 1792 the family abandoned Boncourt and dispersed northward. Not until 1796 were they enabled, by courtesy of the Prussian crown, to reunite and settle in Berlin. Adelbert, who had been painting china to help support the family, first became a court page and then, in 1798, entered military school. He was commissioned lieutenant in the Prussian army in 1801. His regiment remained in Berlin until the campaign of 1805–1806, when it occupied Hameln and, after the defeat at Jena, was surrendered to the Dutch. Chamisso, tortured by the possibility of fighting his own countrymen, was relieved to be released to France.

Between 1806 and 1812, in the uneasy peace and subsequent rearming of Germany, Chamisso, more displaced than ever, unsuccessfully sought employment in France (1806–1807 and 1810–1811), before returning to Berlin. Originally his ambitions were all literary; in 1811, however, at Coppet on Lake Geneva as guest of Madame de Staël, he began intensive botanizing. In the fall of 1812 he returned to Berlin to enter the new university and for the next three years followed medical and scientific courses.

In 1815 Chamisso obtained a berth as naturalist on the brig *Rurik*, commanded by Otto von Kotzebue and sent under Russian flag to explore in the Pacific. Setting off via Plymouth, they crossed the Atlantic and rounded Cape Horn. Despite very difficult conditions, Chamisso was able to collect and observe, notably in the Marshall and Hawaiian islands, Kamchatka, the Aleutians, and California. The expedition returned via the Philippines, the Cape of Good Hope, and London, arriving safely at Saint Petersburg in the fall of 1818.

Chamisso returned to Berlin with botanical collections and notebooks, received an honorary doctorate, and was appointed adjunct curator of the Royal Botanical Gardens in 1819. The rest of his life was spent as a scientific official, succeeding to the curatorship of garden and herbarium in 1833 upon the departure of his chief and good friend, D. F. L. von Schlechtendal. He was elected to the Berlin Society of Friends of Natural History, the Leopoldina, the Imperial Society of Naturalists of Moscow, and in 1835 to the Berlin Academy of Sciences. During this period he had great success as a poet and popular writer. In 1832, with Gustav Schwab, he undertook editing the *Deutscher Musenalmanach*, essentially as an elder statesman of German poetry.

Chamisso married Antonie Piaste in 1820, settling into a cheerful domestic life and raising seven children. In 1833 he suffered a severe pulmonary illness from which he never fully recovered. Working intensively between bouts of fever, he died in 1838 at the age of fifty-seven.

The foregoing outline requires commentary. In childhood his first interests had been those of the naturalist. In Berlin, exposed for the first time to cosmopolitan scholarship, he soon became a child of the age, an egalitarian disciple of Rousseau, with literary ambitions in the German Romantic vein. He made military service tolerable by using his leisure for study, reading, and learning to compose in German. He was welcomed in literary circles, where he made warm friendships that attached him to Germany. In his twenties he wrote much, essayed many genres, but finished little and published almost nothing that he later cared to acknowledge.

Chamisso's early impulse toward natural science was evidently catalyzed into action by various circumstances. Most notable were a visit in 1810 with Alexander von Humboldt to Paris and the hospitality and stimulating conversation of Madame de Staël at Chaumont-sur-Loire and later in her exile at Coppet. He once attributed his firm direction toward botany at Coppet to a chance remark in a friend's letter. Surely however the companionship of Auguste de Staël, elder son of his hostess and an accomplished botanist, as well as the richly diversified terrain, served to fix his attention not only on plants but on their distribution.

That his decision, once taken, resolved his inner uncertainties was soon clear. In the summer of 1813, perhaps bored with unrelieved botany, he composed his masterpiece of fantasy and self-reconciliation, *Peter Schlemihl's wundersame Geschichte*. First published in 1814, the tale was an immediate success, and was widely translated. On returning to Berlin from the Pacific, no longer an enemy alien but an author and scientist, Chamisso continued to intertwine poetic fantasy with meticulous science, without ever confusing the two.

In his scientific role, also, Chamisso presents contrasting aspects. The bulk of his technical publications were in botanical taxonomy, written jointly with Schlechtendal and to be found in *Linnaea*, the journal launched by the latter in 1826. A few geological articles appeared in appropriate journals. Of his equally few zoological publications, one contained a real discovery, that of alternation of generations in salps. *De Salpa* (Berlin, 1819) includes an accurate identification of the aforesaid phenomenon, which was only later to be recognized in other animal and in most plant groups. The publication was a special one, illustrated with beautiful lithographs of Chamisso's accurate and finely executed drawings. In the preface, tribute is paid to the collaboration of the zoologist J. F. Eschscholtz, ship's doctor on the *Rurik*. Chamisso always felt himself something of an amateur, seeking criticism and assurance from others of more academic experience.

By contrast to his dry, formal, and diffident approach to technical science, Chamisso's narrative accounts of the *Rurik* voyage show him at his best, an active, keen, and humorous observer in fields as various as biogeography, ecology, meteorology, oceanography, linguistics, and ethnology—not to mention botany, zoology, and geology. He was the first to propose floating fruits as agents for populating islands and to ascribe the coloration of seawater to pigmented microorganisms. He was also a born ethnologist whose geniality transcended communication barriers. He not only joined natives in their ceremonies and games but conscientiously collected vocabularies. His last project, unfinished at his death, was a dictionary of the Hawaiian language.

BIBLIOGRAPHY

I. ORIGINAL WORKS. See Günther Schmid, *Chamisso als Naturforscher. Eine Bibliographie* (Leipzig, 1942), for the numerous technical and scientific works. The literary works, poetry and prose, were collected as *Adelbert von Chamissos Werke*, Julius Hitzig, ed., 6 vols. (Leipzig, 1836–1839), begun under Chamisso's supervision—the 5th ed., Friedrich Palm, ed. (Berlin, 1864), contains some additional material; *Werke in zwei Bänden*, Ulrike Wehres and Wolfgang Deninger, eds. (Zurich, 1971), is less complete. *Peter Schlemihl's wundersame Geschichte* (Nuremburg, 1814), has been reprinted numerous times, recently with concluding essay by Thomas Mann, Insel Taschenbuch no. 27 (Frankfurt, 1973).

II. SECONDARY LITERATURE. See Louis Choris, *Voyage pittoresque autour du monde* (Paris, 1820); E. du Bois-Reymond, *Adelbert von Chamisso als Naturforscher* (Leipzig, 1889); Werner Feudel, *Adelbert von Chamisso. Leben und Werk* (Leipzig, 1971); Otto von Kotzebue, *A Voyage of Discovery Into the South Sea and Beering's Straits*, H. E. Lloyd trans., 3 vols. (London, 1821); August Carl Mahr, *The Visit of the "Rurik" to San Francisco in 1816* (Stanford, Calif., 1932); M. Möbius, "Chamisso als Botaniker," in *Botanisches Zentralblatt, Beihefte*, **36**, Abt. 2 (1918), 270–306; René Riegel, *Adalbert de Chamisso. Sa vie et son oeuvre* (Paris, 1934); and D. F. L. von Schlechtendal, "Den Andenken an Adelbert von Chamisso als Botaniker," in *Linnaea*, **13** (1839), 93–112.

DOROTHEA RUDNICK

CHRISTIANSEN, CHRISTIAN (*b.* Lónborg Jutland, Denmark, 9 October 1843; *d.* Copenhagen, Denmark, 28 November 1917), *physics.*

Christiansen was the son of Mads Peter Christiansen, a landowner and member of parliament, and Ane Marie Mortensdatter. He obtained the M.S. at the University of Copenhagen in 1866 and then worked for several years as scientific assistant, and subsequently as instructor, at the Polytechnical Institute and the Agricultural College of Copenhagen. In 1886 he was appointed professor of physics at the University of Copenhagen and the Polytechnical Institute.

In 1870 Christiansen discovered the anomalous dispersion of light in solutions of the dyestuff fuchsine. In 1872 he constructed the water-jet pump to produce low pressures. His other experimental work included research on the optical properties of crystals (with Haldor Topsøe), the properties of heat conduction, heat radiation, the movement of air currents through small openings, and friction electricity.

An outstanding teacher, Christiansen exerted great influence on his students, who included Martin Knudsen and Niels Bohr. He himself was strongly influenced by Ludvig Lorenz and played a major role in establishing Lorenz' position in the Danish scientific community. Under Lorenz' influence Christiansen laid the foundation for the study of theoretical physics at the University of Copenhagen; and his textbook in this field was translated into German, English, and Russian.

BIBLIOGRAPHY

Christiansen's *Indledning til den matematiske Fysik*, 2 vols. (Copenhagen, 1887–1890) was translated into German by J. Müller as *Elemente der theoretischen Physik* (Leipzig, 1894; 4th ed., 1921), and appeared in English trans. by W. F. Magie as *Elements of Theoretical Physics* (London, 1897). His other works are listed in Poggendorff, III, 269–270; IV, 247; and V, 221–222.

On his life and work, see H. M. Hansen, in *Dansk biografisk leksikon*, V (Copenhagen, 1934), 195–198; the obituary by Martin Knudsen, in *Festskrift udgivet af Københavns universitet* (1918), 119–122; and K. Prytz, in *Kongelige Danske Videnskabernes Selskabs Oversig-*

ter (1917–1918), 31–54, which includes a complete bibliography of his works.

MOGENS PIHL

CLASSEN, ALEXANDER (*b.* Aachen, Germany, 13 April 1843; *d.* Aachen, 28 January 1934), *analytical chemistry.*

Classen, who came from a wealthy family, studied at Giessen and Berlin before building a private laboratory in Aachen. After working independently for three years as an analytical chemist, he obtained an appointment to teach analytical chemistry at the Technische Hochschule in Aachen. He was named professor of inorganic chemistry there in 1883, and in 1894 he became director of the Electrochemical Institute, where he remained until his retirement in 1914.

Classen devoted the greater part of his work to electrogravimetric analysis. The chemical effect of electric current had been observed in 1800, and in that year William C. Cruikshank had reported that metal is deposited at the negative pole. Yet, remarkably, the electrical deposit of metal was first employed for analytical purposes—by Oliver Wolcott Gibbs—only much later, in 1864. Although several works had already been published on the subject, Classen was really the first to make a thorough study of it. The method that he developed and refined over several decades became standard practice in analytical chemistry.

Before Classen's work, the only studies that had been made concerned the nature of the electrolytic solution. Classen was the first to examine the influence of the current and applied voltage, and he also introduced measuring devices into the circuit. Further, he substituted storage cells for galvanic cells. The first to discover the advantages of using warm solutions, he was able, by combining this technique with efficient mixing of the solution, to develop rapid methods of electrolysis. Classen wrote the first book on electrogravimetry, *Quantitative Analyse auf elektrolytischem Weg*, a work of fundamental importance in the development of analytical chemistry. It went through many editions and did much to spread the use of the gravimetric method. The Electrochemical Institute became a center for the study of this field, and technical personnel came from throughout the world to attend special classes on electrogravimetry.

BIBLIOGRAPHY

Classen's books include *Quantitative Analyse auf elektrolytischem Weg* (Brunswick, 1881; 6th ed., 1920); *Ausführliches Lehrbuch der Chemie,* 3rd ed. (Brunswick, 1895), written with Henry Roscoe and Carl Schorlemmer; *Ausgewählte Methoden der analytischen Chemie* (Brunswick, 1901); and *Theorie und Praxis der Massanalyse* (Leipzig, 1912). He also revised the 7th ed. of F. Mohr, *Lehrbuch der chemisch-analytischen Titriermethode* (Brunswick, 1896).

A secondary source is F. Szabadváry, *History of Analytical Chemistry* (Oxford, 1966), 315.

FERENC SZABADVÁRY

COLDING, LUDVIG AUGUST (*b.* Holbaek, Denmark, 13 July 1815; *d.* Copenhagen, Denmark, 21 March 1888), *engineering, physics.*

Colding was the second son of Andreas Christian and Anna Sophie Colding. His father, a former sea captain and officer in the privateering service of the Danish navy, unwisely embarked on a new career as farmer and manager of an estate near Copenhagen at about the time of Ludvig's birth. There the boy was raised under somewhat strained economic circumstances, as a result of difficult times following Denmark's involvement in the Napoleonic wars and particularly through the elder Colding's evident lack of talent for agriculture. The deeply religious household reflected his mother, daughter of a prominent clergyman, and left a permanent stamp on the boy. His elementary schooling was rather irregular, although eventually he served an apprenticeship under a well-known craftsman in Copenhagen, on the advice of the physicist Hans Christian Oersted, an old family acquaintance. In 1836 he became journeyman, and the following year enrolled at the Polytechnic Institute in Copenhagen with an engineering career in mind, again following Oersted's suggestion. After his graduation in 1841, Colding held several modest teaching and tutorial positions before being offered his first important appointment in 1845, as inspector of roads and bridges in Copenhagen. The post enabled him to marry Henriette Louise Lange, a cousin.

Oersted's influence on his education and, indeed, on his subsequent career and philosophy of life began when Colding was a student at the Copenhagen Polytechnic. There he had easy access to Oersted, who, as director of the institution and professor of physics, took considerable interest in him and aided him. The strength of the curriculum in those years lay in basic and theoretical subjects, although practical application of fundamental knowledge was not neglected, in line with Oersted's own philosophy. Oersted had himself been instrumental in founding the Institute in 1829 on the pattern of the École Polytechnique.

Commencing with his appointment as inspector in 1845, Colding held a succession of technical and administrative positions with the municipal government of Copenhagen, which culminated in his being named to the specially created post of state engineer for Copenhagen in 1857. During these years he oversaw the planning of public housing and parks, tramways and railways, roads and bridges, and harbor and canal projects. He also supervised major improvements in city services, including the introduction of gas illumination and the overhauling of desperately inadequate water and sanitation systems. In addition, he was consultant on many projects in other cities, and acquired a high reputation both within and outside Denmark.

Despite the burden of his official responsibilities, Colding found time to pursue an impressive range of scientific subjects. Some of this research was of a basic nature, while other portions concerned applied aspects, many of them evolving naturally from engineering problems encountered during his regular activities. His work covered fluid dynamics, hydrology, oceanography, meteorology, electromagnetism, and especially thermodynamics. From 1869 he was professor at the Polytechnic in addition to his official duties. He was a member of numerous technical commissions and scientific societies, and received many honors at home and abroad, including the Cross of Honor of the Men of Dannebrog, awarded on his retirement in 1886.

A concept of special and abiding importance to Colding was the notion of a "principle of the imperishability of the forces of nature," and it is this phase of his scientific and philosophical work that is likely to be of principal interest to scholars outside Denmark. Aside from the general intellectual milieu fostered at the Polytechnic, several specific factors are discernible in the background of this idea. By Colding's own account the earliest hint of the idea occurred to him in 1839, while still a student, when he was led to contemplate "d'Alembert's principle concerning lost forces." The starting point for these speculations was, however, as much of a religious and metaphysical nature:

> My first thought concerning the imperishability of the forces of nature I have, as I have mentioned earlier, borrowed from the view that the forces of nature must be related to the spiritual in nature, to the eternal reason as well as to the human soul. Thus it was the religious philosophy of life that led me to the concept of the imperishability of forces. By this line of reasoning I became convinced that just as it is true that the human soul is immortal, so it must also surely be a general law of nature that the forces of nature are imperishable (L. A. Colding, *Naturvidenskabelige Betragtninger . . . , 155).*

Oersted's personal influence on Colding's religiophilosophical development is clear and was repeatedly acknowledged by Colding himself. Moreover, his preoccupation with a fundamental underlying harmony or unity in the material world was a recurrent and dominant contemporary philosophical view, voiced especially in *Naturphilosophie*, of which Oersted was a firm adherent.

In 1839 Colding finally had the opportunity to assist Oersted in a series of fundamental experiments on the compression of water, a subject that had occupied Oersted off and on for many years. The results were puzzling at the time, and only in the following decade was the underlying effect—the increase in temperature of a liquid when it is adiabatically compressed—recognized as central to the concept of energy and its conservation.

The energy concept was vaguely recognized by 1840, having essentially been anticipated but not published or expounded publicly by Sadi Carnot. The first explicit suggestion of an invariable relationship between heat and mechanical work was put forth by J. R. Mayer in Germany in 1842 and a year later, independently, by J. P. Joule in England and by Colding. The circumstances surrounding this nearly simultaneous enunciation of the principle of the conservation of the "forces" of nature, and its manifold sources, have received considerable attention from historians of nineteenth-century science. Mayer, a physician, was first guided to the principle through metabolic observations on the relationship between heat produced by the human organism and the work expended in producing it. Joule, in contrast, was a physical scientist and brilliant experimenter. Originally concerned with improving the efficiency of electric motors, he employed his gift for precision measurements to yield a long series of determinations of the "mechanical equivalent" of heat with steadily increasing accuracy. Colding appears to have compounded the energy principle from a complex merger of metaphysical speculation and experiment. In 1847 appeared the classic treatise by Helmholtz, which treated the interconversion of forms of energy from a very general point of view. It had considerable influence on the mature formulation of the energy concept, although the systematic development of thermodynamics came in the second half of the century, mainly by R. Clausius, W. J. Rankine, and William Thomson.

Colding's first paper on the subject, "Nogle Saetninger om Kraefterne" ("Theses Concerning Force"), was read before the Danish Society of Sciences in 1843, although its publication was delayed until 1856. In it he first summarized available data on friction and compressibilities of various materials indicating that the heat evolved is dependent on the compression or friction. Next, he reported on an experiment of his own in which the mechanical work required to move a weighted sled over a track of metal rails was correlated with the linear expansion of rails and sled runners, as a result of the frictional heating. He concluded that ". . . the quantities of heat evolved are, in every case, proportional to the lost moving forces," although he neglected to condense these preliminary results in a numerical coefficient of proportionality. Consequently, "When a force seems to disappear it merely undergoes a transformation, whereupon it becomes effective in other forms."

On Oersted's recommendation, a more elaborate version of this experiment, supported by the Danish Society of Sciences, was subsequently carried out by Colding. The new results were reported at the 1847 meeting of Scandinavian scientists in Copenhagen, and published in the first of two companion papers in 1850, "Undersøgelse om de almindelige Naturkraefter og deres gjensidige Afhaengighed og isaerdeleshed om den ved visse faste Legemers Gnidning udviklede Varme" ("Investigation Concerning the Universal Forces of Nature and Their Mutual Dependence and Especially Concerning the Heat Evolved From the Friction of Certain Solid Bodies"). From these measurements he adopted a value for the mechanical equivalent of heat approximately 14 percent lower than the modern accepted value (by this time Joule was within 1 percent of the accepted value). An alternative derivation of this constant yielding (fortuitously, as it turned out) essentially the same value, based on Oersted's experiments on the compression of water and on the then available (albeit not very accurate) value for the specific heat of air, appeared in his second paper of 1850, "Om de almindelige Naturkraefter og deres gjensidige Afhaengighed" ("On the Universal Forces of Nature and Their Mutual Dependence"), which is largely theoretical and devoted to a mathematical treatment of thermodynamic relations. A considerably more accurate value (only 3 percent in error) was presented in a fourth paper on the subject (1852), dealing with practical problems of steam engineering: "Undersøgelse over Vanddampene og deres bevaegende Kraft i Dampmaskinen" ("Investigation Concerning Steam and Its Motive Power in Steam Engines").

The philosophical and religious side of Colding's thesis was first elaborated in his cumbersome treatise of 1856, "Naturvidenskabelige Betragtninger over Slaegtskabet mellem det aandelige Livs Virksomheder og de almindelige Naturkraefter" ("Scientific Reflections on the Relationship Between the Intellectual Life's Activity and the General Forces of Nature"), his last paper on this topic to the Danish Society of Sciences; it was published on the occasion of his being elected member of the Society. It stressed his own philosophical convictions and dwelt at length on the relationship between the material and the spiritual in nature, clearly echoing the intellectual and aesthetic influence of Oersted and in a tone reminiscent of Kant and Schelling. This paper is also of interest for Colding's discussion of Joule's experiments, with which he was acquainted by then, and for his passionate priority dispute with Mayer.

Colding returned to thermodynamics and the mechanical theory of heat in an 1863 manuscript, submitted in response to a prize competition on the theme announced that year by the French Academy, "Samlet Fremstilling af Naturkraefternes gjensidige Afhaengighed med Anvendelse paa den mekaniske Varmetheori" ("Unified Presentation of the Mutual Dependence of the Forces of Nature With Application to the Mechanical Theory of Heat"). It failed to win the prize, however, and was not published. His last and only paper on energy published abroad was a memoir in *Philosophical Magazine* (1864); there was also a translation of his second paper of 1850 in *Philosophical Magazine* (1871). The former work, "On the History of the Principle of the Conservation of Energy," is of value chiefly as a summary of his various contributions to the subject.

Among his prolific research the energy concept appears to have most occupied Colding's mind. Unfortunately, however, this aspect of his scientific productivity went rather unnoticed in his time, even in Denmark; and today Colding remains a somewhat neglected pioneer in the research through which the first law of thermodynamics came to be recognized. His lasting scientific contributions are found elsewhere, particularly in his numerous hydrodynamic and meteorological works. He was, in fact, largely responsible for the establishment of a meteorological institute in Denmark. His major achievement, however, was undoubtedly the extraordinary number of municipal projects started on his initiative and completed in full or in part during his forty years of service to the Danish community.

BIBLIOGRAPHY

I. ORIGINAL WORKS. Colding's writings include "Undersøgelse om de almindelige Naturkraefter og deres gjensidige Afhaengighed og isaerdeleshed om den ved visse faste Legemers Gnidning udviklede Varme," in *Det Kongelige Danske Videnskabernes Selskabs Skrifter*, 5th ser., **2** (1850), 122–146; "Om de almindelige Naturkraefter og deres gjensidige Afhaengighed," *ibid.*, 167–188; "Undersøgelse over Vanddampene og deres bevaegende Kraft i Dampmaskinen," *ibid.*, **3** (1852), 1–35; "Naturvidenskabelige Betragtninger over Slaegtskabet mellem det aandelige Livs Virksomheder og de almindelige Naturkraefter," in *Oversigt over det Kongelige Danske Videnskabernes Selskabs Forhandlinger*, no. 4–6 (1856), 136–168; "Nogle Saetninger om Kraefterne," supp. to *ibid.*, no. 8 (1856), 1–20; and "On the History of the Principle of the Conservation of Energy," in *London, Edinburgh, and Dublin Philosophical Magazine and Journal of Science*, **27** (1864), 56–64.

A copy of the MS "Samlet Fremstilling af Naturkraefternes gjensidige Afhaengighed med Anvendelse paa den mekaniske Varmetheori" is in the possession of Torben Holck Colding, in Copenhagen.

II. SECONDARY LITERATURE. See Theodore M. Brown, "Resource Letter EEC-1 on the Evolution of Energy Concepts from Galileo to Helmholtz," in *American Journal of Physics*, **33** (1965), 759–765; Torben Andreas Colding, *Professor L. A. Coldings Naturvidenskabelige Betragtninger* (Copenhagen, 1924); Per F. Dahl, "Ludvig A. Colding and the Conservation of Energy," in *Centaurus*, **8** (1963), 174–188; and *Ludvig Colding and the Conservation of Energy Principle; Experimental and Philosophical Contributions* (New York–London, 1972), which includes English translations of Colding's papers relevant to energy conservation; Erwin N. Hiebert, *Historical Roots of the Principle of Conservation of Energy* (Madison, Wis., 1962); Thomas S. Kuhn, "Energy Conservation as an Example of Simultaneous Discovery," in Marshall Clagett, ed., *Critical Problems in the History of Science* (Madison, Wis., 1959), 321–356; and Asger Lomholt, *Fortegnelse over Det Kongelige Danske Videnskabernes Selskabs Publicationer 1742–1930; Saertryk af Oversigt over Det Kgl. Danske Videnskabernes Selskabs Forhandlinger 1929–1930* (Copenhagen, 1930), with an index of Colding's scientific papers published in the journals of the Royal Danish Society of Sciences on 221–222 and a list of Oersted's papers on 361–364.

Also see Vilhelm V. Marstrand, "Ingeniøren og Fysikeren Ludvig August Colding," in *Ingeniørvidenskabelige Skrifter*, no. 20 (1929); Hans Christian Oersted, *H. C. Oersted, Scientific Papers*, Kirstine Bjerrum Meyer, ed., 3 vols. (Copenhagen, 1920), with two essays on Oersted's work by the editor, I, xiii–clxvi; III, xi–clxvi; George Sarton, "The Discovery of the Law of Conservation of Energy," in *Isis*, **13** (1929), 18–49; and Povl Vinding, "Colding, Ludvig August," in *Dansk Biografisk Leksikon*, V (Copenhagen, 1934), 377–383.

PER F. DAHL

COLUMBUS, CHRISTOPHER (*b.* Genoa, Italy, 26 August/31 October 1451; *d.* Valladolid, Spain, 20 May 1506), *exploration.*

Columbus was the eldest son of Domenico Colombo and Susanna Fontanarossa. Although there has been some debate about the site of his birth, several documents in the State Archives at Genoa confirm that city as his place of origin. Moreover, Columbus' will of 22 February 1498 exhorts his eldest son to "make every effort . . . for the good, honor and increase of the city of Genoa, where . . . I was born." Christopher was the eldest of five children in this family of rather humble economic status. The other children were Giovanni Pellegrino (who died young), Bartolomeo, Jacopo (known later, in Spain, as Diego), and Bianchinetta. Bartolomeo and Diego accompanied Columbus on his voyages, the former displaying a forceful and energetic character that contrasted with Christopher's indecision and his often excessive submissiveness under harsh circumstances. His two brothers proved very valuable to Columbus: in Haiti, Bartolomeo quelled a native rebellion; and in 1509 Diego replaced Nicolás de Ovando as governor of Santo Domingo.

Nothing certain is known about Columbus' early years. According to a passage in the log of the first ocean crossing, he first went to sea at the age of eighteen; and in 1472 he referred to himself as a "Genoese wool draper." Shortly afterward, in 1473, Columbus and his father moved to Savona, from which port Columbus made voyages on behalf of Genoese firms.

One of the many problems in modern Columbian literature is the alleged dependence of Columbus' voyage plan on similar views held by the Florentine Paolo dal Pozzo Toscanelli (1397–1482). Although a letter from Toscanelli to Ferdinando Martini has survived in a copy made by Columbus, there is no evidence of any direct correspondence between Columbus and Toscanelli. It is unreasonable to suppose that Columbus would have said that a letter was addressed to him when it was clearly headed "Ferdinando Martini canonico ulixiponensi Paulus physicus [dixit]." Moreover, Toscanelli favored a route on the Lisbon parallel, whereas Columbus held to that of the Canaries, fourteen degrees farther south; and when he reached Hispaniola he was convinced, after having traveled sixty degrees west, that he had arrived at Cipango (Japan). According to Toscanelli's map, he would have been forty degrees away.

The principal theoretical assumptions drawn from classical sources (Aristotle, Strabo, Pliny the Elder, Seneca, Marinus of Tyre, Ptolemy), Hebrew (Esdras), Arab (al Maʾmun), and European sources (from Marco Polo to d'Ailly and Pius II) that made possible the discovery of the New World were two major but fortunate errors: an exaggerated extension of the inhabited landmass eastward and a considerable reduction in the terrestrial meridian, which was estimated to be about one-fourth less than it actually is. Columbus correlated Toscanelli's data with ancient and medieval sources and arrived at a colossal miscalculation. As Samuel Eliot Morison (*The European Discovery of America*, p. 30) has indicated, the distance from the Canaries to Japan via Antilia, which Toscanelli estimated at 3,000 nautical miles (and Columbus whittled down to 2,400), is actually about 10,000 miles between their respective meridians, measured on latitude 28° north.

Toscanelli's Canaries-to-Quinsay route of 5,000 miles (reduced by Columbus to 3,550) is actually about 11,766 nautical miles by air. Columbus seems to have reduced the length of a degree of longitude by one-quarter, stretching Ptolemy's estimate of the length of the Eurasian continent (Cape St. Vincent to eastern Asia) from 180 degrees to 225 degrees, adding 28 degrees for the discoveries of Marco Polo and 30 degrees for his estimated distance from the east coast of China to the east coast of Japan. He also saved another 9 degrees of westing by starting his ocean crossing from the outermost of the Canary Islands. This left only 68 degrees of ocean to cross before reaching Japan, yet Columbus reduced this figure as well. Arguing that the medieval calculators used too long a degree of longitude, he proposed to cross on latitude 28° north, where he thought the degree measured only forty nautical miles; thus he estimated that he had only 2,400 miles of water to traverse. In other words, his figures placed Japan in relation to Spain about where the West Indies actually are.

Columbus' miscalculations should not be construed as a lack of the necessary nautical and cosmographical training. His observations on magnetic declination, its variation, and the daily movement of the lodestar around the pole reveal that he was a very competent navigator.

In order to place the evolution of Columbus' plan within its proper time frame, one would have to know details of his life and work in Portugal; yet even the date of his arrival there is unknown. We know only that in July 1479 he was in Genoa, about to depart for Lisbon. Columbus apparently went to Madeira and then to Porto Santo, where he married Felipa Perestrello e Moniz. Once established in Madeira, he sailed with the Portuguese as far as Mina (Elmina, on the Gulf of Guinea), thus obtaining valuable maritime experience.

There appears to be little basis for the traditional notion that Columbus submitted his plan for a voyage of discovery to the Portuguese king, John II (1481–1495), who rejected it. In 1485 or 1486 Columbus moved to Spain, but little is known of his activities there. He returned to Portugal in the fall of 1488 and was present, on 2 December, at Bartolomeu Dias' return from his southern exploration of 1487–1488. Although Dias had reached the southern tip of Africa and opened a new sea route to Asia, his voyage of 6,300 miles still left him far short of China. Doubtless this circumstance encouraged Columbus to place even greater reliance on his own views.

Nevertheless, Columbus was compelled to wait for favorable political and economic conditions in Spain, to which he had returned by 1492; on 17 April of that year he received the title of *almirante mayor del mar oceano* and was granted the viceroyalty and governorship of any lands he might discover. Two brothers, Martín Alonso and Vicente Yáñez Pinzón, wealthy and expert ship outfitters, organized the expedition and prepared the flagship, the *Santa Maria*, at their own expense. Columbus' first transatlantic voyage set sail on 3 August 1492 from Palos with the *Santa Maria*, the *Pinta*, the *Niña* (totaling 450 tons) and with a letter from the Spanish sovereigns addressed to the grand khan of China. He touched land on 12 October on a little island in the Bahamas that was called Guanahaní by the natives. Christened San Salvador by Columbus, it was later renamed Watling Island by the British. He then sailed to the northern coast of Cuba, which he mistakenly took for Zipango (Cipango), still convinced that he would soon reach Marco Polo's Quinsay (Hangchow). Sailing along the coast of Cuba, he came to believe that he had reached Cathay and dispatched an embassy to deliver Ferdinand and Isabella's letter to the grand khan. The mission soon aborted, and he turned his attention to the large island of Babeque (Great Inagua Island), where the natives had assured him gold was to be found.

Moving west, Columbus touched the northwestern tip of Haiti, established a settlement on its north shore, and traded with the natives, who, he was sure, would lead him to gold. Unfortunately, the *Santa Maria* was lost through carelessness on Christmas night, and Columbus was obliged to postpone his departure for Spain. Leaving forty-eight of his companions at a fortress he had established—Villa de la Navidad, on Hispaniola—and charging them to study the island's inhabitants and produce, Columbus left for Spain on 3 January 1493, convinced that he had reached Asia. Severe storms nearly ended the return voyage, but he reached Palos on 13 March, after stopping in Lisbon to confer with the Portuguese king. Although all of Spain welcomed him, the voyage had made the international situation increasingly precarious, for Columbus' route had taken him through Portuguese waters, thereby violating the Treaty of Toledo (1480). Through remarkable diplomatic skill Columbus managed to overcome the difficulty.

Six months later a new expedition was outfitted with fourteen caravels and about 1,400 men. The voyage began on 25 September 1493 and again proceeded toward the islands on the southern edge of the Caribbean Sea. Columbus discovered the Leeward Islands and Puerto Rico on his way to Haiti. At Hispaniola he found that all forty-eight men he had left at Villa de la Navidad were dead—their greed had moved the once friendly natives to murder. Columbus set sail again, discovering Jamaica (May 1494) and skirting the southern coast of Cuba. Returning to Haiti shortly afterward, he found the colony in confusion. Word of the colonists' discontent had reached Spain, and Juan de Aguado had been dispatched in June 1495 to learn the reasons for the situation and to take the necessary measures. Before Aguado reached Haiti, however, Columbus returned to Spain (11 June 1495), leaving his brother Bartolomeo in charge. Columbus succeeded in winning the court's confidence and had his privileges confirmed; in addition, he obtained the right to transmit titles and rights to his descendants.

Despite this vote of confidence, Columbus' prestige had diminished. Disappointment was only too evident in Spain, where great hopes had been frustrated by the low level of profit that the distant posts had yielded. In order to outfit a third voyage, Columbus had to sign on convicts, and the fleet was reduced to eight caravels (fitted out by Amerigo Vespucci on behalf of the house of Berardi). Two vessels left in January 1498; but the other six, with Columbus, did not sail from Sanlúcar until 30 May. Taking three caravels, Columbus followed a more southerly route than those previously adopted and reached Trinidad on 31 July. Subsequently he sailed across the mouth of the Orinoco, thus fully meriting recognition—often mistakenly denied him—as the discoverer of the American continent.

Columbus headed north to Haiti and landed on the south coast of Santo Domingo, to which Bartolomeo had transferred the island's seat of government. Rebellion and intrigue had left the colony in such wretched condition that Columbus felt unable to settle matters without harsh disciplinary action and the interference of the Spanish government. The Spanish court immediately sent Francisco de Bobadilla to act as royal commissioner. He reached Santo Domingo on 23 August 1500 and was shocked to find that Columbus was making frequent use of the gallows. He put Columbus in chains and sent him back to Spain (November 1500) with his brothers Bartolomeo and Diego. Despite his disgrace, the Spanish sovereigns received Columbus and granted him permission for a fourth voyage, although they stripped him of his governorship of Hispaniola.

Spain's precarious political and economic situation quelled much of the enthusiasm for Columbus' explorations, especially in view of the successful expedition (1497) of Vasco da Gama, who returned to Lisbon in 1499 after having reached the southern tip of India. Nevertheless, Columbus, taking his brother Bartolomeo and his thirteen-year-old son, Fernando, sailed from Seville with a fleet of four caravels on 3 April 1502, still in search of a passage to the Indian Ocean. He stopped briefly at Santo Domingo to replace a damaged caravel; but Nicolás de Ovando, his successor as governor, refused his request for aid and denied him permission to land. Setting off again, Columbus sailed south of Jamaica and reached the Gulf of Darien. His discovery there of a Mayan canoe persuaded him that he was on the brink of finding a civilization more advanced than that of the natives previously encountered; and he sailed further south, convinced that he would soon reach the long-sought passage to India. He discovered Honduras, Nicaragua, Costa Rica, Panama, and Colombia, from which hostile natives and malaria forced him to retreat. Columbus took refuge in Jamaica, his vessels unseaworthy and his crew on the verge of mutiny. Two of his officers, Diego Méndez and Bartolomeo Fieschi, outfitted a canoe and courageously paddled the 108 miles to Santo Domingo. It was nearly a year before they were granted permission by Ovando to outfit a ship, which rescued Columbus and his men on 28 June 1504. Returning to Spain broken and ill, Columbus died ignorant of the extent of his discoveries.

BIBLIOGRAPHY

Columbus is credited with a number of writings, which are listed in J. H. Vignaud, *Histoire critique de la grande entreprise de Christophe Colomb*, 2 vols. (Paris, 1911), I, 18, 21, 352–353, 547–548, 602, 679; and II, 6, 208, 242. There is still no adequate critical ed. to remove apocryphal material.

Much of our information on Columbus derives from two questionable sources: *Historie della vita et dei fatti dell'ammiraglio don Cristoforo Colombo*—first published (Venice, 1571) under the name of Fernando, Columbus' natural son, by Alfonso Ulloa (thirty-two years after the presumed, and still missing, original); and *Historia de las Indias*, by Bartolomeo de Las Casas (1474–1566), who began the work in 1527 and completed the final draft in 1553; it was finally published at Madrid in 1875–1876. The authorship and history of these two sources has long been debated. Henri Harrisse (1872) revealed many inconsistencies and contradictions in the *Historie* and concluded that it could not have been written by Fernando Columbus. His argument has been largely discredited, and Alberto Magnaghi has shown in the compilation of the *Historie* the responsibility of an anonymous author, probably Luis Colón, a descendant of Columbus who was exiled to Oran by Charles V. Various legendary details were inserted into the Columbian tradition: Columbus' having graduated from the University of Pavia, the mention of other admirals in his family, and an account of a voyage to Tunis that portrays Columbus as a nautical buffoon and a shameless inventor of fairy tales. He also was reputed to have made an equally fanciful voyage to Iceland and beyond in 1477. One must be wary in using these sources, for dates and personages often are confused.

The Columbian bibliography is constantly growing, keeping pace with the critical study of the many complex questions that arise from tradition and from a facile appeal to innovative views that are not always adequate to the complexity and seriousness of the problems treated.

The following, cited in chronological order, offer useful guidance and background material for further research: J. B. Muñoz, *Historia del nuevo mundo* (Madrid, 1793); G. B. Spotorno, *Codice diplomatico Colombo-americano* (Genoa, 1823); M. Fernández de Navarrete, *Colección de viajes y descubrimientos que hicieron por mar los españoles desde fines del siglo XV*, 5 vols. (Madrid, 1825–1828); Washington Irving, *History of the Life and Voyages of Columbus* (London, 1828); A. von Humboldt, *Examen critique de l'histoire de la géographie du Nouveau Continent*, 5 vols. (Paris, 1836–1839); H. Harrisse, *Fernand Colomb, sa vie, ses oeuvres. Essai critique* (Paris, 1872); *Christophe Colomb, son origine, sa vie et ses voyages* (Paris, 1884); Justin Winsor, *Christopher Columbus and How He Received and Imparted the Spirit of Discovery* (Boston, 1891); H. Harrisse, *Christophe Colomb devant l'histoire* (Paris, 1892); C. de Lollis, *Cristoforo Colombo nella leggenda e nella storia* (Milan, 1892); Italian Ministry of Education, *Raccolta di documenti e studi pubblicati dalla R. Commissione colombiana*, 15 vols. (Rome, 1892–1894); H. Harrisse, *The Discovery of North America* (Paris, 1897); J. Boyd Teacher, *Christopher Columbus, His Life, His Work, His Remains, as Revealed by Original Printed and Manuscript Records* (New York, 1903); G. Nunn, *The Geographical Conceptions of Columbus* (New York, 1924); G. Pessagno, "Questioni colombiane," in *Atti della Società ligure di storia patria*, **53** (1926), 539–691; L. Olschki, "Herman Pérez de Oliva's 'Yistoria de Colón,'" in *Hispanic American Historical Review*, **23** (1943), 165–196; Martin Torodash, "Columbus Historiography Since 1939," *ibid.*, **46** (1966), 409–428; and Antonio Rumeu de Armas, *Hernando Colón, historiador del descubrimiento de América* (Madrid, 1973).

Noted for novelty of research and for the use of innovational and thorough criticism are the writings of Alberto Magnaghi, which led to a more accurate presentation of Columbus as man, sailor, and discoverer. See, for example, his "I presunti errori che vengono attribuiti a Colombo nella determinazione della latitudine," in *Bollettino della Società geografica italiana*, **64** (1928), 459–494, 553–582; "Ancora dei pretesi errori che vengono attribuiti a Colombo nella determinazione delle latitudini," *ibid.*, **67** (1930), 457–515; "Questioni colombiane," in *Annali della istruzione media*, **6** (1930), 691–515; "Incertezze e contrasti delle fonti tradizionali sulle osservazioni attribuite a C. Colombo intorno ai fenomeni della declinazione magnetica," in *Bollettino della Società geografica italiana*, 6th ser., **10** (1933), 595–641; "Di una recente pubblicazione italiana su Cristoforo Columbo," in *Atti dell'Accademia delle scienze* (Turin), Classe di scienze morali, storiche e filologiche, **74** (1938–1939), 69–141; "La nuova storia della scoperta dell'America," in *Miscellanea della Facoltà di lettere e filosofia dell'Università di Torino* (1933), 1–111.

The most useful works in English are those of Samuel Eliot Morison: *The Second Voyage of Christopher Columbus From Cádiz to Hispaniola and the Discovery of the Lesser Antilles* (Oxford, 1939); *Admiral of the Ocean Sea: A Life of Christopher Columbus*, 2 vols. (Boston, 1942), in 2 eds.—in 2 vols. with copious footnotes and other scholarly apparatus and in 1 vol. for the general public; *Christopher Columbus, Mariner* (Boston, 1955; New York, 1956); *Journals and Other Documents on the Life and Voyages of Christopher Columbus*, S. E. Morison, ed. and trans. (New York, 1963);

and *The European Discovery of America: The Southern Voyages 1492–1616* (New York, 1974).

THE EDITORS

COPE, EDWARD DRINKER (*b.* Philadelphia, Pennsylvania, 28 July 1840; *d.* Philadelphia, 12 April 1897), *vertebrate paleontology, zoology.*

A pioneer in the development of American vertebrate paleontology, Cope gained notoriety for his disputes with Othniel C. Marsh and fame as the leading theorist of the neo-Lamarckian movement in American biology. Cope's father, Alfred Cope, was descended from a prominent, wealthy, and well-established Quaker merchant family; his mother, Hannah Edge Cope, died when he was three years of age.

Cope's formal education began in 1849, when he was enrolled in the Friends Select School in Philadelphia. Four years later he entered the West-town Boarding School, another Quaker institution, just outside of Philadelphia, where he studied intermittently until 1859. Cope's schooling, which emphasized the classics, did not kindle his interest in natural history. From 1854 to 1860, however, his father tried to persuade him to follow a career in agriculture by sending him to the farms of relatives during the summer months; it was there that Cope's interest in natural history developed. In 1859 he began his lifelong, all-consuming endeavor—the study of natural history—by recatologing without remuneration the herpetological collection at the Academy of Natural Sciences of Philadelphia.

During the early 1860's, Cope's interests in natural history took him to the Smithsonian Institution, the Museum of Comparative Zoology, and the prominent museums of natural history in Europe. Except for a one-year appointment at Haverford College (1866), Cope spent all of his time doing research or preparing his work for publication. His inherited wealth enabled him to support his wife and daughter while pursuing his interests in natural history, almost at will, until 1880. He was originally interested in the reptiles, amphibians, and fish of North America, but by 1866 he had begun to conduct research on the fossil remains of these organisms' ancestors. By 1870 Cope was studying and publishing works on the mammals of North America, both contemporary and extinct. His first theoretical work on evolution appeared in 1868; three years later he began a long career in the popular presentation of evolutionary ideas and the facts of natural history.

Cope's work, particularly his studies of the cold-blooded vertebrates, quickly established his reputation. Despite his aggressive and at times abrasive personality, he was admitted to the leading American scientific societies: the Academy of Natural Sciences of Philadelphia (1861), the American Philosophical Society (1866), and the National Academy of Sciences (1872). Later he continued to receive recognition for his work in natural history: the Bixby Medal of the Geological Society of London (1879), the Hayden Medal of the Academy of Natural Sciences (1891), and the presidency of the American Association for the Advancement of Science (1896).

Cope made his most lasting contributions in the field of vertebrate paleontology. When he began his work, the parameters of vertebrate paleontology had been broadened by the rapid western expansion of the United States. Prior to this time, most of vertebrate fossil materials from western America consisted of bone fragments, hastily collected by explorers. After the Civil War the pressures of westward expansion led to the development of an extensive transcontinental railroad system, which provided less costly transportation for tons of newly discovered fossils intact within their matrix. In order to facilitate the development of western lands, the government authorized and funded surveys to investigate and make geographical and geological maps of the wilderness; there were four such expeditions between 1866 and 1879. These expeditions provided not only access to the rich and previously unknown fossil fields but also money for the expensive monographs and plates in which the fossils were described and illustrated.

From 1871 to 1879 Cope spent eight months of each year with one of the United States geological surveys. During this time he visited or discovered fossil fields in Colorado, Kansas, Montana, New Mexico, South Dakota, Texas, and Wyoming. On these travels Cope obtained the materials and knowledge of natural history that served him for the rest of his life. His prodigious discoveries led him to work quickly—at times too rapidly—a trait that led to inaccuracies in his work, as even his friends admitted. The geological surveys published Cope's two most significant works: "The Vertebrata of the Cretaceous Formations of the West" (1875) and "The Vertebrata of the Tertiary Formations of the West, Book I" (1884), also known as "Cope's Bible." These writings represented the first comprehensive descriptions of vertebrates from the early Eocene formation. Cope thus reoriented naturalists' understanding of animal life by pushing the origin of the "age of the mammals" further back in time.

Ironically, it was a dispute with Marsh over material collected during one of the geological surveys that resulted in Cope's losing the privilege of access to government facilities. The dispute began in 1872 when Marsh, also a vertebrate paleontologist, challenged the validity of Cope's fossil discoveries around Fort Bridger, in southwestern Wyoming. Cope and Marsh clashed repeatedly over the next twenty years, trading accusations of unethical conduct, attempting to undermine each other's scientific achievements, and trying to limit each other's access to government resources. This controversy frequently has been described as the result of the collision of two unusually volatile and ambitious personalities in the same field at the same time. (Commentators have felt that only this coincidence could explain the twenty years of bitter rivalry and conflict.) Although the characterization is accurate, it has tended to obscure the nature of the dispute: involved were priority in scientific discoveries and access to publication facilities. The controversy drastically altered Cope's life after 1879, when Marsh became chief vertebrate paleontologist of the newly consolidated United States Geological Survey. From then on, all of Cope's affiliations and privileges associated with the geological survey ceased.

This appointment coincided with another misfortune for Cope: he lost his entire inheritance of $250,000 in one of the numerous mining hoaxes of the Gilded Age. These misfortunes forced him to reduce his research on western fossil fields. Even though he still received valuable material from professional fossil hunters, most of his time was spent either in trying to find a publisher for the discoveries he had made during the 1870's or in seeking a way to support his family. His deteriorating financial situation compelled him to undertake popular scientific lecture tours during the 1880's and to sell his invaluable fossil collections to the American Museum of Natural History in 1894. Fortunately, the University of Pennsylvania appointed him to a teaching post in geology in 1889. In 1895 the university promoted him to the chair of zoology and comparative anatomy, a post he retained until his death.

Despite the unsavoriness of the Cope-Marsh controversy and the coincidence of unhappy events in Cope's private life, his reputation in vertebrate paleontology remains secure. Even his methods for preserving fragile fossil bones during transportation from the field to the museum are still in use. In addition, he and Marsh share the distinction of having discovered the first complete remains of the giant dinosaurs, the reptiles that roamed the earth during the Cretaceous period. Of equal importance were Cope's contributions to the discussions on the question of evolution. In the last quarter of the nineteenth century, Cope was the leading theorist of an uniquely American neo-Lamarckian system of thought.

Cope's theory of evolution was articulated in 1866, when Alpheus Hyatt, another American biologist, published his thoughts on evolution. Both scientists relied upon the concepts of acceleration and retardation and the phenomenon of embryological recapitulation. Cope, however, played a larger role in the neo-Lamarckian movement because of the greater number and wider range of his publications on evolution. He argued that if organisms of different genera went through nearly identical stages of early development, they were closely related and descended from the same racial stock. In this relation, which he called parallelism, an evolved series of genera emerged as structural changes took place, through time, by means of acceleration and retardation. Acceleration referred to the appearance of a character (or set of characters) earlier in an organism's development than its appearance in the organism's ancestors. Cope argued that this stage left time in the organism's development for the emergence of newly adaptive characters, which, if they appeared before reproduction, could be transmitted to the organism's offspring. Retardation was the reverse process, the slowing down of an organism's development. In this manner, a less highly developed character (or set of characters) was passed on to the offspring. By closely examining the embryological development of an organism, Cope believed he could trace its life history and phylogeny.

Cope developed his theory of evolution most fully in "On the Origins of Genera" (1868) and "The Method of Creation of Organic Forms" (1871). In the former work he revealed his adherence to typological classification when he attacked Darwin's theory of natural selection. He argued that the theory of natural selection might explain the evolution of the characteristics that define a species, but it could not explain the appearance of well-adapted, apparently nonfortuitous variations of the higher characteristics of classification (such as characters of genera, family, or order). Cope merely proposed that these variations could be understood through his concepts of acceleration and retardation, but he did not explain how these processes took place. In "The Method of Creation of Organic Forms" he admittedly relied upon Herbert Spencer's *Principles of Biology* (1864–1867) to explain the phenomena of variations. This explanation entailed the application of neo-Lamarckian concepts of use and disuse and of the inheritance of acquired characteristics.

A careful examination of Cope's evolutionary thought would reveal a certain ambiguity in his explanations of inheritance. Although by 1871 Cope defended the inheritance of acquired characteristics, at some points he argued that there were laws of matter (in his earlier work said to be divinely inspired) that foreordain the direction of evolution; at other points he maintained that some organisms possess an ability to direct their evolution. Cope resolved this paradox by asserting that the more highly developed organisms, those with intelligence, can control their environment, and, hence, their evolutionary development. Man, the most highly developed organism, possessed the power to control his destiny. Although his final work on evolution, *The Primary Factors of Organic Evolution* (1896), was a defense of acquired characteristics, throughout his work there is a strain of orthogenesis, which perhaps reflects the legacy of his religious background.

Although Cope published after 1872 more than fifty articles and books on evolution, his thought did not change significantly after that year. In response to the repeated attacks of the neo-Darwinians, especially after August F. L. Weismann's *Germ-Plasm* (English translation, London, 1893), however, Cope attempted to explain with greater clarity the effects of use and disuse upon an evolving organism. He stressed the importance of the environment in evolution and adhered, with greater determination, to the belief in the inheritance of acquired characteristics.

During the 1870's and 1880's, the Cope-Hyatt school of evolutionary thought was the most active and representative of the American biological community, Asa Gray and March notwithstanding. For example, Cope had the support of, and later owned, the first scientific journal devoted exclusively to biology, *American Naturalist*. By the 1890's, however, he and his supporters in the United States were on the defensive, because of the arguments advanced against the inheritance of acquired characteristics by the neo-Darwinians and those who supported a mutation theory of evolution. Yet many naturalists, as opposed to experimental biologists and biometricians, continued to hold evolutionary concepts similar to Cope's until the 1930's and the advent of the modern synthesis in evolution theory. If nothing else, it can be said that the strength and endurance of Cope's evolutionary thought highlights the weaknesses associated with the theory of natural selection before the advances of population biology and genetics.

BIBLIOGRAPHY

I. ORIGINAL WORKS. The most important MS collection of Cope material, including correspondence and diaries, is in the Osborn Library at the American Museum of Natural History, New York. The Quaker Collection at Haverford College, Haverford, Pa., contains pertinent material from Cope's childhood and his early years as a scientist. The American Philosophical Society's Cope collection includes his field diaries and material concerning his work at the society. The Academy of Natural Sciences also possesses material pertaining to his life and work in Philadelphia.

A chronological bibliography of Cope's more than 1,500 publications follows Henry Fairfield Osborn, "Edward Drinker Cope," in *Biographical Memoirs. National Academy of Sciences*, **13** (1930), 129–317. Cope's most important works include "On the Origins of Genera," in *Proceedings of the Academy of Natural Sciences of Philadelphia*, **20** (1868), 242–300; "The Method of Creation of Organic Forms," in *Proceedings of the American Philosophical Society*, **12** (1871), 229–263; "The Vertebrata of the Cretaceous Formations of the West," in *Report of the U.S. Geological Survey of the Territories*, II (Hayden, Ariz., 1875); "Consciousness in Evolution," in *Penn Monthly*, **6** (1875), 560–575; "The Relation of Animal Motion to Animal Evolution," in *American Naturalist*, **12** (1878), 40–48; "The Vertebrata of the Tertiary Formations of the West, Book I," in *Report of the U.S. Geological Survey of the Territories*, III (Hayden, Ariz., 1884); *The Origin of the Fittest* (New York, 1886): "The Batrachia of North America," in *Bulletin. United States National Museum*, no. 34 (1889); *The Primary Factors of Organic Evolution* (Chicago, 1896); and "The Crocodilians, Lizards and Snakes of North America," in *Report of the United States National Museum* (1898), 153–1270.

II. SECONDARY LITERATURE. Henry Fairfield Osborn, *Cope: Master Naturalist* (Princeton, N.J., 1931), is the unsurpassed guide to Cope's life, and contains a large portion of his correspondence. See also Edward Pfeifer, "The Genesis of American Neo-Lamarckism," in *Isis*, **56** (1965), 156–167; Elizabeth Noble Shor, *The Fossil Feud Between E. D. Cope and O. C. Marsh* (Hicksville, N.Y., 1974); and George Stocking, "Lamarckianism in American Social Science: 1890–1915," in *Journal of the History of Ideas*, **23** (1962), 239–256.

JOSEPH M. MALINE

CRAMER, JOHANN ANDREAS (*b.* Quedlinburg, Germany, 14 December 1710; *d.* Berggiesshübel, near Dresden, Germany, 6 December 1777), *chemistry*.

Cramer's father, who was leaseholder of the state ironworks in Quedlinburg, introduced him to metallurgy, in which he took an immediate interest.

When Cramer was fourteen, his father died; he was then sent to the Johanneum in Hamburg, where he received an excellent grounding in natural science.

Upon completing secondary school, Cramer began to study medicine but became dissatisfied with the subject. He took up legal studies, although he also attended lectures on chemistry and metallurgy. He had little professional success as a lawyer, and his frequently brusque manner intimidated many clients. As a result he had much free time, which he used to visit mines and foundries in the Harz region and to broaden his knowledge of analytical chemistry. This hobby so captivated Cramer that, at the age of twenty-four, he gave up his legal practice and entered the University of Helmstedt with the intention of becoming a chemist and metallurgist. After concluding his studies, he went to Leiden to lecture on analytic chemistry.

At the same time Cramer worked on a textbook, the first of its kind, which was published in 1737. Entitled *Elementa artis docimasticae*, the profusely illustrated work encompassed the entire art of assaying in two parts, one theoretical and one practical. In the preface he referred to the works of Agricola, Lazarus Ercker, and Stahl. All the instruments and apparatus of contemporary analytical chemistry were depicted and described exactly.

In the *Elementa*, Cramer first described the use of the blowpipe in smelting small amounts of substances and in analyzing them. The sample was heated to glowing over charcoal, and in many cases borax beads were also utilized. The blowpipe, which was made of copper, included a bulge to collect the saliva secreted during the heating.

In 1738 and 1739 Cramer made a long trip through England to learn more about the subject, and he gave lectures in London. After further travels he returned home and in 1743 was appointed director of the Brunswick Mining and Metallurgy Administration in the Harz mountains. In addition to his official duties, he taught chemical analysis, eventually gathering a school around himself. His impetuous nature brought him many enemies, however, and as a result he lost his post for a time. Nevertheless, Cramer declined an offer from Russia and worked for only a short time in Prussia. He was asked to return to his previous post but became disgusted with it because of the intrigues in which it involved him. In 1774 he accepted a new position in Saxony, and from there he went for two years to Hungary. He returned, ill, to Germany in 1777 and died of dropsy at the end of that year.

BIBLIOGRAPHY

Cramer's main work is *Elementa artis docimasticae* (Leiden, 1737; 4th ed., 1744), translated into German by Christlieb Ehregott Gellert as *Anfangsgründe der metallurgischen Chemie*, 2 vols. (Leipzig, 1751–1755).

There is a biography in J. R. Partington, *A History of Chemistry*, II (London, 1961), 710–711; and III (London, 1962), 36.

GÜNTHER KERSTEIN

CRAWFORD, ADAIR (*b.* Ireland, 1748; *d.* Lymington, England, 29 July 1795), *physiology, physics, chemistry.*

Little is known about Crawford's family background or personal life. His father was a nonsubscribing Presbyterian minister in Crumlin, County Antrim. After obtaining the M.D. degree, he is said to have had a very successful practice in London. His career evidently flourished both as a result of his own achievements and through the intervention of his many influential friends. He was appointed a physician at St. Thomas's Hospital, London; was elected to the Royal Society of London and that of Edinburgh, the Royal Irish Academy of Dublin, and the American Philosophical Society; and became professor of chemistry at the Royal Military Academy, Woolwich. Perhaps his emphasis upon experimentation and his use of phlogistic arguments owed something to the influence of his friends Joseph Priestley and Richard Kirwan. He died shortly after poor health had forced him into retirement.

From 1764 to 1776 Crawford studied arts, theology, and medicine at the University of Glasgow. In the course of his medical studies he heard William Irvine lecture on his new theory of heat, and Crawford later joined the chemical society that Irvine organized in Glasgow. Irvine had spent several years as Joseph Black's pupil and assistant during the period in which Black was developing his theory of latent heat. Irvine's particular task had been to measure the specific heats of various substances. Unlike Black, whose theory proceeded from the assumption that the absorption of heat is a chemical process, Irvine viewed heat as being physically contained in substances having varying capacities for it. He further argued that his theory of capacities enabled him to determine the absolute zero point of temperature and the absolute quantity of heat in a body. Irvine was appointed lecturer in chemistry at the University of Glasgow in 1770, and by 1775 he was presenting a fully developed account of his new theory in his lectures. It was this theory that captured Crawford's attention in 1776, and it was as a champion of Irvine's theory that he made his mark as a scientist.

In the summer of 1777 Crawford performed the first experiments designed to determine the specific heats of gases. He believed that respiration, which causes a chemical change in air, also changes the air's capacity for heat. According to this application of Irvine's theory, the air that supports respiration has a greater capacity for heat—that is, a higher specific heat—than the air given off by respiration. Thus, in breathing, a certain quantity of "absolute heat" is transferred to the body and, Crawford concluded, is the source of the body's heat. In 1777 Crawford enrolled in the University of Edinburgh medical school, where he read an account of his experiments to the student medical society; and the following year he described his work to Thomas Reid and several other professors in Glasgow. He then published it as *Experiments and Observations on Animal Heat, and the Inflammation of Combustible Bodies; Being an Attempt to Resolve These Phaenomena Into a General Law of Nature* (London, 1779).

To establish the truth of his theory, Crawford put forward three fundamental propositions. The first states that the atmospheric air inhaled into the lungs contains more absolute heat than the air exhaled from the lungs. His experiments on the specific heats of gases were designed to verify this proposition. The second asserts that the arterial blood leaving the lungs contains more absolute heat than the venous blood pumped to the lungs. The truth of this proposition is proved, he argued, by the calorimetric experiments on blood described in his book. The third proposition states that a body's capacity for heat is reduced by the chemical fixation of phlogiston and is increased by the separation of phlogiston.

On the basis of his third proposition Crawford constructed a general theory of combustion. He began his argument by demonstrating that metallic calxes, which are products of combustion and are, according to the phlogiston theory, dephlogisticated, have higher heat capacities than do the metals from which they are formed. Then, to explain how respiration causes a release of "absolute heat," he appealed to the analogy between combustion and respiration and suggested that during respiration a double decomposition takes place in the lungs, atmospheric air and venous blood yielding phlogisticated air and arterial blood [(air + heat) + (blood + phlogiston) → (air + phlogiston) + (blood + heat)]. Apparently Crawford was not aware that this view of respiration obscures the distinction between Black's chemical theory of heat and Irvine's physical theory of capacities.

Crawford's book attracted considerable attention. Its attempt to explain animal heat in physicochemical terms was not especially novel or shocking; but in providing the first published account of Irvine's theory of capacities, it seriously challenged the chemical theories of heat advanced by Black and Lavoisier. The scientific "intelligencer" J. H. de Magellan provided a lengthy description of Crawford's experiments in his *Essai sur la nouvelle théorie du feu élémentaire, et de la chaleur des corps* (London, 1780) and claimed: "We owe the birth of this branch of natural philosophy to the publication of this excellent work by Dr. Adair Crawford." After visiting Paris in 1781, J. A. Deluc reported: "Dr. Crawford's theory on the phenomena of heat . . . was then in great agitation." The challenge posed by Crawford's theory was soon taken up by Lavoisier and Laplace. In 1782 they began a series of experiments on the specific heats of gases and heats of combustion that led them to conclude, in their "Mémoire sur la chaleur," that no reliable figure for absolute zero could be established and that Irvine's theory therefore remained a highly dubious hypothesis.

A greatly enlarged but not fundamentally altered second edition of Crawford's book was published in 1788. By then his critics had come to realize that the available experimental data on the specific heats of gases were too imprecise to permit a resolution of the central theoretical issue. Irvine's theory remained influential in Great Britain, however, largely through the advocacy of John Dalton, who admitted that he was "overawed by the authority of Crawford." But in 1812 the prize-winning work of François Delaroche and J. E. Bérard provided experimental values for the specific heats of gases and finally rendered the theory of capacities untenable.

BIBLIOGRAPHY

I. ORIGINAL WORKS. In addition to the two editions of his book mentioned in the text, Crawford published two papers in *Philosophical Transactions of the Royal Society*: "Experiments on the Powers That Animals, When Placed in Certain Circumstances, Possess of Producing Cold," **71** (1781), 479–491; and "Experiments and Observations on the Matter of Cancer, and on the Aerial Fluids Extricated From Animal Substances by Distillation and Putrefaction; Together With Some

Remarks on Sulphureous Hepatic Air," **80** (1790), 391–426. The archives of the Royal Society contain an MS that Crawford read on 6 Dec. 1787, "Experiments and Observations on the Stability of Heat in Animals." "On the Medicinal Properties of Muriated Barytes," which contains the first description of strontium, was published in *Medical Communications* (of the Society for Promoting Medical Knowledge), **2** (1790), 301–359; and *An Experimental Enquiry Into the Effects of Tonics, and Other Medicinal Substances, on the Cohesion of the Animal Fibre* was edited by his younger brother, Alexander Crawford (London, 1816). An MS note and letter by Crawford are in G. W. Corner and William Goodwin, "Benjamin Franklin's Bladder Stone," in *Journal of the History of Medicine*, **8** (1953), 359–377.

II. SECONDARY LITERATURE. Everett Mendelsohn, *Heat and Life, the Development of the Theory of Animal Heat* (Cambridge, Mass., 1964), discusses Crawford's place in the history of physiology and gives references to translations of his book and the writings of his critics. Crawford's theory of combustion is analyzed in J. R. Partington and Douglas McKie, "Historical Studies on the Phlogiston Theory. III. Light and Heat in Combustion," in *Annals of Science*, **3** (1938), 337–371; and in J. R. Partington, *A History of Chemistry*, III (London, 1962), 156–157. Crawford's contributions to calorimetry are examined in Douglas McKie and Niels H. de V. Heathcote, *The Discovery of Specific and Latent Heats* (London, 1935), 126–129, 136. For an account of the heat theories of Black and Irvine, see A. L. Donovan, *Philosophical Chemistry in the Scottish Enlightenment: The Doctrines and Discoveries of William Cullen and Joseph Black* (Edinburgh, 1975), 222–249, 265–276. On Crawford's influence on Lavoisier, see Robert J. Morris, "Lavoisier and the Caloric Theory," in *British Journal for the History of Science*, **6** (1972), 1–38. On the study of the specific heats of gases in France, see Robert Fox, *The Caloric Theory of Gases From Lavoisier to Regnault* (Oxford, 1971).

ARTHUR DONOVAN

CRUZ, OSWALDO GONÇALVES (*b*. São Luís de Paraitinga, São Paulo, Brazil, 5 August 1872; *d*. Petrópolis, Rio de Janeiro, Brazil, 11 February 1917), *public health, medicine*.

Cruz's contributions to public health and medicine were closely intertwined. As the "sanitizer" of Rio de Janeiro, he rid that city of yellow fever and bubonic plague; and as director of what became known as the Oswaldo Cruz Institute, he created the first important center for medical research in Brazil. His career is thus of considerable interest to scholars concerned with the growth of science in developing countries.

Cruz was the son of Bento Gonçalves Cruz, a physician who was active in public health work for the imperial government. In 1877 the family moved to Rio de Janeiro, where Cruz attended medical school, completing the M.D. requirements in 1892 with a thesis on waterborne bacteria. In 1896 Cruz was able to continue his studies in experimental medicine. He went to Paris, where he worked at several institutions, the most important being the Pasteur Institute. Cruz also specialized in the clinical field of urology; but, as he wrote to a friend, he detested clinical medicine and planned to use his training in microbiology, pathology, histology, and chemistry to set up a laboratory in Brazil where he would perform medical diagnoses. Cruz returned to Rio de Janeiro in the fall of 1899. A highly trained medical scientist and nationalist, he believed firmly that science could play an important role in his native country.

Shortly after his return he was appointed to the staff of a small laboratory on the outskirts of the city. The laboratory had not been founded as a center of research. The municipal authorities in Rio had opened the laboratory to produce plague vaccine and serum to fight an epidemic that had spread from Santos. A little later the federal government took control and renamed it the Serum Therapy Institute of Manguinhos. By 1902 Cruz had become director.

Until 1903 the institute was a classic crisis laboratory, that is, one designed to respond to a specific challenge. Such laboratories often have overly restricted functions, budget, and staff, and cannot expand their activities once the immediate medical crisis has passed. The Serum Therapy Institute escaped this fate when Cruz took charge of Brazil's first large-scale and systematic sanitation campaign in Rio de Janeiro. From 1903 to 1909 the institute was an integral part of the campaign.

Francisco de Paula Rodrigues Alves was elected president of Brazil in 1902. He had campaigned on the need to reform Brazilian culture and to eliminate the epidemic diseases that had given Rio de Janeiro the reputation of being one of the least healthy cities in South America. In 1903 he entrusted the supervision of sanitation to Cruz, who proposed an ambitious program against bubonic plague, smallpox, and yellow fever, basing the extinction of yellow fever on the recent work of the Reed Commission in Havana. Cruz organized the yellow fever prophylaxis service, the purpose of which was to police zones of infection, eradicate the *Aedes aegypti* mosquito, and identify, register, and isolate yellow fever victims.

Although Cruz enjoyed the political support of Alves, many factors contributed to making the campaign controversial. Opposition to compulsory vaccination was strong and although it became mandatory by law in 1904, many people were not vaccinated and in 1908 the city experienced one of its worst outbreaks of smallpox. The campaigns against yellow fever and the plague, however, were successful. As early as 1906 Cruz reported that yellow fever no longer existed in epidemic form in Rio.

In 1903 Cruz had asked the Brazilian Congress to make his institute a "Pasteur Institute," where preparation of vaccines and serums, teaching, and research would be combined. His proposal was rejected as costly and unnecessary. As director of the nation's public health program, however, Cruz was able to steer money, materials, and people to Manguinhos, and in 1907 with the sanitation campaign already a success, the Congress granted his request. The Serum Therapy Institute became the Institute of Experimental Pathology, and its budget was tripled. That year Cruz was honored with the gold medal at the International Congress on Hygiene and Demography at Berlin, and in 1908 the institute was renamed the Cruz Institute by presidential decree.

Although it is clear that the sanitation campaign had a decisive impact on the institute, its survival after Cruz resigned from the directorship of public health in 1909, and its growth into a productive research institution, must be viewed as the result of his shrewd institution-building.

Several factors explain his success. As a nationalist trained at a first-rate European institution of medical science, he enjoyed great scientific prestige. As the director of one of the most influential government agencies of the day, Cruz was able to attract technicians and students to the sanitary sciences. His decision to use the Serum Therapy Institute as a strategic arm of the public health campaign meant that technicians came to the institute to prepare vaccines and serums, physicians came to carry out research, and medical students came to prepare their theses.

Cruz published thirty-six scientific papers during his lifetime—some describing new species of mosquitoes or diseases caused by protozoa, others concerning measures of prophylaxis against epidemic disease—but scientific writing slowly took second place to organization and teaching. With his young staff he was stern but supportive, stressing the value of both basic and applied research as well as practical hygiene. He was excellent in directing the staff toward fruitful problems for research. To encourage familiarity with foreign work in the field, Cruz instituted regular seminars during which scientific articles in foreign journals were discussed. He created the first library in Latin America that specialized in microbiology, and he supervised the production of glass equipment for the institute.

From among the doctors and students recruited and trained from 1903, Cruz chose the first official staff in 1907. All of the eight original staff members were Brazilians, and five—Carlos Chagas, Henrique da Rocha Lima, Artur Neiva, Henrique de Beaurepaire Aragão, and Antônio Cardoso Fontes—achieved fame in medical research. In 1908 Rocha Lima left Brazil to continue his career in Germany. His place in pathology was taken by Gaspar Vianna, who studied yellow fever and leishmaniasis until his accidental death in 1914. The distinguished protozoologist and entomologist Adolfo Lutz joined Cruz in 1908. In 1909 the informal methods of training were replaced by a formal course in microbiology, based on the *cours de microbie technique* taught by Émile Roux at the Pasteur Institute. That year the Cruz Institute also began publication of *Memorias do Instituto Oswaldo Cruz*, long regarded by medical scientists as one of the few significant Latin American medical journals.

Although Cruz is not remembered for original contributions to medicine, he established the first solid institutional base for research in Brazil and assembled a team of outstanding medical researchers. In 1908 the staff began the move to the new and imposing laboratories of the Oswaldo Cruz Institute, where research on yellow fever, malaria, leishmaniasis, and other tropical diseases flourished.

After 1908 Cruz was increasingly incapacitated by renal disease and early in 1916 he retired to Petrópolis, outside Rio de Janeiro. He was appointed mayor, but continued illness soon led him to relinquish his official duties. He died the next year.

BIBLIOGRAPHY

I. ORIGINAL WORKS. Cruz's papers have been republished as *Opera omnia* (Rio de Janeiro, 1972); this vol. includes his two surveys of the Amazon, *Considerações gerais sôbre as condições sanitárias do Rio Madeira* (Rio de Janeiro, 1910); and *Relatório sôbre as condições médico-sanitárias do Valle do Amazonas* (Rio de Janeiro, 1913). A collection of documents concerning the sanitary campaign, and letters from Cruz to his colleague Henrique da Rocha Lima (1901–1915), have been published in connection with the centennial celebrations of Cruz's birth; see Edgard Cerqueira Falcão, *Oswaldo Cruz: Monumenta histórica*, 3 vols. (São Paulo, 1971–1973). Twenty-five files containing documents, many of them the official notes exchanged between the institute and the government (and in Cruz's hand), are at the Museum of the Oswaldo Cruz Institute.

II. SECONDARY LITERATURE. The standard biography, E. Sales Guerra, *Osvaldo Cruz* (Rio de Janeiro, 1940), and the biography by Clementino Fraga, *Vida e obra de Osvaldo Cruz* (Rio de Janeiro, 1972), are fairly useful but have no references. More important are the accounts of Cruz's work by colleagues at the institute. See especially Henrique de Beaurepaire Aragão, *Oswaldo Cruz e a escola de Manguinhos* (Rio de Janeiro, 1945) and *Notícia histórica sôbre a fundação do Instituto Oswaldo Cruz* (Rio de Janeiro, 1950); Ezequiel Caetano Dias, *O Instituto Oswaldo Cruz: Resumo histórico 1899–1918* (Rio de Janeiro, 1918) and *Traços biográficos de Oswaldo Cruz* (Rio de Janeiro, 1945); and Henrique da Rocha Lima, "Com Oswaldo Cruz em Manguinhos," in *Ciência e cultura*, **4**, no. 1–2 (1952), 15–38.

A good analysis of the institute's scientific work is Olympio da Fonseca, "A escola de Manguinhos: Contribuição para o estudo do desenvolvimento da medicina experimental no Brasil," in Falcão, *op. cit.*, pp. 24–128. On the public health campaign, see Octavio G. de Oliveira, *Oswaldo Cruz e suas atividades na direção da saúde pública brasileira* (Rio de Janeiro, 1955); and Donald B. Cooper, "Oswaldo Cruz and the Impact of Yellow Fever in Brazilian History," in *Bulletin of the Tulane Medical Faculty*, **26** (1967), 49–52. Much of the original sanitation legislation organized by Cruz is repr. in the work prepared during his directorship of public health: Placido Barbosa and Cassio Barbosa de Rezende, *Os serviços de saúde pública no Brasil, especialmente na cidade do Rio de Janeiro de 1808 à 1907 (Esboço histórico e legislação)*, 2 vols. (Rio de Janeiro, 1909).

Two sources in English are Nancy Stepan, "Initiation and Survival of Biomedical Research in a Developing Country: The Oswaldo Cruz Institute of Brazil, 1900–1920," in *Journal of the History of Medicine and Allied Sciences*, **30** (1975), 303–325; and *Beginnings of Brazilian Science: Oswaldo Cruz Medical Research and Policy, 1890–1920* (New York, 1976).

NANCY STEPAN

CUNHA, JOSÉ ANASTÁCIO DA (*b.* Lisbon, Portugal, 1744; *d.* Lisbon, 1 January 1787), *mathematics.*

The son of Lorenzo da Cunha, a painter, and his wife, Jacinta Ignes, da Cunha learned grammar, rhetoric, and logic at the Lisbon school of the Congregation of the Oratory; he also studied mathematics and physics on his own. At the age of nineteen he volunteered as a lieutenant in the artillery and spent nearly ten years at Valença do Minho. At this time Portugal was experiencing antifeudal and anticlerical reforms, which were carried out by the marquis of Pombal, minister of King Joseph I. Da Cunha belonged to a group of freethinking intellectuals who supported Pombal and disseminated ideas of the Enlightenment; and he became known as a progressive thinker, talented poet, and author of an original memoir on ballistics. In 1773 Pombal appointed da Cunha as a geometry professor at the Faculty of Mathematics of Coimbra University. Da Cunha's university career was short. In 1777, after the death of Joseph I, reactionaries returned to power and Pombal was dismissed, then exiled. In the same year da Cunha was arrested and imprisoned by the Inquisition. Charged with supporting heretical doctrines, in October 1778 he was sentenced by the General Council of the Inquisition in Lisbon to three years in prison. Freed at the beginning of 1781, da Cunha, under the protection of a high official, obtained a mathematical professorship at the College of São Lucas and resumed his scientific research. His health had been weakened in jail, however, and he died before his forty-fourth birthday.

Da Cunha's main scientific work was *Principios mathemáticos*, published serially beginning in 1782 and as a complete book in 1790. Intended to be a textbook, this work is a concise encyclopedia of mathematics in twenty-one parts that embrace all basic branches of the science, from geometry and arithmetic to the solution of differential equations and problems in the calculus of variations.

Excessive conciseness was the pedagogical deficiency of this exposition, which contained many fresh and fruitful ideas. The most striking feature was da Cunha's tendency to rigorous exposition of mathematics in general and of the calculus in particular. Needless to say, not all of his attempts in this direction were successful.

In book IX of *Principios* da Cunha presented a new theory of the exponential function that anticipated the methods of the modern theory of analytic functions and that was based on the use of solely convergent series (a very uncommon restriction at that time). The convergence of series in question was tested by comparing the given series with a convergent geometric series with each term greater than the series. Let a be a (positive) number $a = 1 + c + \frac{c^2}{2!} + \frac{c^3}{3!} + \cdots$, which series is seen to converge for all values of c. Then the exponential function a^x to the base a is defined for all values of x as the sum of the series $1 + cx + \frac{(cx)^2}{2!} + \frac{(cx)^3}{3!} + \cdots$; to that end da Cunha demonstrated that every positive number a may be represented by the series

$1 + c + \frac{c^2}{2!} + \frac{c^3}{3!} + \cdots$. From this definition the laws of exponents were derived. The binomial theorem—the power-series expansion of the function $(1 + x)^n$—was obtained in a very ingenious way: da Cunha represented both $1 + x$ and $(1 + x)^n$ in the form of the exponential expansions, a device also used in the modern theory of complex functions.

In book XV, devoted to the elements of the calculus, the fundamental concepts were those of infinitely great and infinitely small variables; the concept of limit was not explicitly used. Following Leibniz' notations, da Cunha profited to some extent from Newton's terminology; for instance, for the differential he adopted the symbol d but called it, as did some other mathematicians, "fluxion," a word that Newton used to designate the velocity of change of the variable or fluent. The definition of "differential" given by da Cunha was remarkable. During the eighteenth century the differential of a function, $y = f(x)$, was generally understood to be, and was defined as, its infinitely small increment, $dy = \Delta y$; but in practice dy was calculated as a part of the increment linear with respect to Δx, a distinction that was one of the sources of paradoxes and endless discussions. The definition proposed by da Cunha legitimated the procedures of differential calculus and was equivalent to one introduced in the nineteenth century following the works of Cauchy: If the increment $\Delta y = f(x + \Delta x) - f(x)$ can be represented as $\Delta y = A\Delta x + \epsilon\Delta x$, where A does not depend on Δx and ϵ approaches zero together with Δx, then $A\Delta x$ is called the differential of function $y, dy = A\Delta x$. In this way da Cunha deduced some formulas of the differential calculus.

Da Cunha was one of the main precursors of the reform of the foundations of infinitesimal calculus, initiated in the first decades of the nineteenth century. Neither the Portuguese nor the French edition of his *Principios* had a wide circulation, however, and they did not greatly influence the development of mathematics. Da Cunha's manuscripts on the problems of mathematics and its foundations are briefly mentioned in the preface to the French edition of *Principios*, but their subsequent fate is unknown.

BIBLIOGRAPHY

I. Original Works. Da Cunha's main writings are *Principios mathemáticos para instrucçao dos alumnos do Collegio de São Lucas, da Real Casa Pio do Castello de São Jorge* (Lisbon, 1790), French trans. by J. M. d'Abreu as *Principes mathématiques de feu Joseph-Anastase da Cunha* (Bordeaux, 1811; repr. Paris, 1816); and *Ensayo sobre os principios da mechanica obra posthuma de J. A. da Cunha, dada à luz por D. D. A. de S. C. possuidor do MS autographo* (London, 1807; Amsterdam, 1808).

II. Secondary Literature. See the following, listed chronologically: T. Braga, *Historia da Universidade de Coimbra*, III (Lisbon, 1898), 500–619; F. G. Teixeira, *História das matemáticas em Portugal* (Lisbon, 1934), 255–256; J. V. Gonçalves, "Análise do livro VIII dos *Principios mathemáticos* de José-Anastácio da Cunha," in *Congresso do mundo português*, I (n.p., 1940), 123–140; and A. P. Youschkevitch, "J. A. da Cunha et les fondements de l'analyse infinitésimale," in *Revue d'histoire des sciences et de leurs applications*, XXVI (1973), 3–22.

A. P. Youschkevitch

CUSHNY, ARTHUR ROBERTSON (*b.* Fochabers, Morayshire, Scotland, 6 March 1866; *d.* Edinburgh, Scotland, 25 February 1926), *pharmacology, physiology.*

Cushny was the fourth son of Rev. John Cushny, of the manse of Speymouth, and his wife, Catherine Ogilvie Brown. After attending Fochabers Academy, he went to the University of Aberdeen, where he graduated M.A. in 1886. The year before he had entered Marischal College, the medical school of Aberdeen, from which he graduated Bachelor of Medicine and Master of Surgery in 1889. His interest in pharmacology and physiology was aroused by John Theodore Cash, who taught physiology at Aberdeen. Upon his medical graduation, Cushny won the university's George Thompson fellowship, which he used to study on the Continent. He went first to Bern, where he worked in the physiological laboratory of Hugo Kronecker. After about a year, during which he seems also to have studied at Würzburg, Cushny went to Strasbourg to work with pharmacologist Oswald Schmiedeberg. He stayed there for nearly three years, during the last two of which he held an assistantship in Schmiedeberg's laboratory.

In the spring of 1893 J. J. Abel, himself a student of Schmiedeberg's, came to Strasbourg and persuaded Cushny to accept the chair of pharmacology at the University of Michigan, which Abel was giving up to go to Johns Hopkins. Cushny taught at Michigan from 1893 to 1905. In the latter year he went to University College, London, as the first occupant of a new chair in pharmacology and materia medica. Hitherto the latter subject had been taught at University College by one of the physicians attached to the affiliated hospital. The regulations for the new chair, the first of its kind in England, were designed to ensure an emphasis on experimental pharmacology and to release its occupant from clinical duties. Cushny was initially housed in one small, poorly lit room; but a gift from Andrew Carnegie permitted the construction of a new and well-equipped institute of pharmacology, which was opened in 1912. He remained at University College until 1918, when he succeeded Sir Thomas Fraser in the chair of materia medica at the University of Edinburgh. Simultaneously a rearrangement was made in the duties attached to the chair. As at University College, Cushny was relieved of immediate clinical duties, these falling instead to a professor in clinical therapeutics. It was arranged, however, that the occupants of the experimental and of the clinical chairs should work in close association.

Cushny was elected to fellowship of the Royal Society of London in 1906 and of the Royal Society of Edinburgh in 1919. His death in 1926 was attributed to cerebral hemorrhage. He was survived by a daughter and by his wife, Sarah Firbank, an Englishwoman whom he met in Strasbourg and married in the United States in 1896.

Cushny made important contributions to three separate problems: the action and therapeutic application of digitalis, the secretion of urine by the kidney, and the action and general biological relationships of optical isomers. The study of digitalis came first, both chronologically and in terms of practical importance. Cushny began to study the effects of drugs on the heart while working under Schmiedeberg, who had attracted considerable attention for his own work on the effects of muscarine and other poisons on the frog's heart. One of Cushny's first major papers, published while he was at Strasbourg, concerned the same problem.[1] Although he had also begun to study digitalis at Strasbourg, his first publications in this area did not appear until after he had gone to the University of Michigan.

In 1897 Cushny published a major paper on the effects of the digitalis series on the circulation in mammals. He emphasized that the usual method of studying this problem rested on estimates of the blood pressure in the systemic arteries. Rejecting this method as "entirely unsuited" to the question at hand, he used instead the myocardiograph and cardiometer recently introduced to physiology by Charles Smart Roy and J. G. Adami. The measurements permitted by these instruments gave Cushny a new insight into the effects of digitalis on the mammalian heart. He found that the effects could be divided into two main stages. In the first stage, the heartbeat slowed, the ventricular contraction became more complete, and the systolic pressure increased. The second stage was marked by an accelerated rhythm due to increased irritability in the heart muscle, and by periodic changes in the strength of the ventricular contractions.

Cushny suggested that these effects resulted from a dynamic interplay between two antagonistic powers possessed by the digitalis series. On the one hand, digitalis exerted a marked inhibitory action on the heart, probably by directly stimulating the vagus nerve as well as by stimulating the central inhibitory apparatus in the medulla oblongata. On the other hand, digitalis also exerted a direct tonic effect on the cardiac musculature. These two actions were difficult to separate, and neither seemed to be confined entirely to either of the two stages. Since, however, a slower ventricular rhythm was the first dramatic indication of digitalis action, Cushny concluded that the nervous, inhibitory action predominated at first. He believed that the tonic, muscular action developed more gradually and persisted longer, so that it predominated by the time the accelerated rhythm of the second stage appeared.

Cushny then sought to relate these experimental results to the therapeutics of digitalis. He emphasized that the beneficial effects of digitalis, both as a diuretic and in heart disease, depended mainly on its direct action on cardiac muscle, and particularly on its capacity to promote the general circulation by increasing the strength and completeness of the ventricular contractions. Thus, he argued, digitalis increases the flow of urine not by any direct effect on the kidney but, rather, by increasing the blood flow in the renal arteries and thus allowing the kidney to filter a larger quantity of urinary fluid from the blood plasma. By the same token, it was the capacity of digitalis to increase the strength of ventricular contractions that made it "practically specific" in patients with incomplete systole of the ventricle.

Cushny also emphasized how cautiously the results of animal experiments should be extended to human patients. In particular, his experimental animals had received their digitalis by direct intravenous injection, whereas patients took the drug orally. By the time it was absorbed from the stomach, Cushny suggested, digitalis probably affected human hearts chiefly through its tonic muscular action rather than its nervous inhibitory action. He conceded, however, that this conclusion required further demonstration.

In this paper of 1897, Cushny defined the central issues that would guide all of his later work on digitalis. He revealed throughout an unusual concern with the clinical and therapeutic aspects of heart disease and digitalis. In 1899, in an important paper on the interpretation of pulse tracings, Cushny complained that clinicians had failed to keep pace with the tremendous advance in knowledge about the physiology of the mammalian heart. In this paper he sought "to fill in one hiatus existing between clinical observation and physiological experiment." He had already noted in his paper of 1897 that lethal doses of digitalis tended to produce irregularities in the dog's heart during the second stage of the drug's action. In the meantime he had conducted purely physiological experiments in an attempt to elucidate these irregularities.[2] Very often, he had observed, his digitalized dogs displayed an auricle in which the fibers beat rapidly but not as a coordinated whole (auricular fibrillation). These conditions—auricular fibrillation and general cardiac irregularity—could be produced by rapid electrical stimulation of the auricle as well as by lethal doses of digitalis.

Cushny now began to suspect that the auricle had been inadequately appreciated as a possible factor in the cardiac irregularities observed clinically. He was particularly struck by the apparent similarity between auricular fibrillation in dogs and an extreme irregularity observed frequently in the diseased human heart. This condition, known clinically as delirium cordis, was attributed either to paralysis of the auricle or, more commonly, to ventricular abnormalities; but Cushny thought it possible that auricular fibrillation was responsible instead. He found that the clinical sphygmogram in cases of delirium cordis paralleled exactly that obtained from dogs with auricular fibrillation; and he pointed out that this fibrillation, like delirium cordis, could persist for a long time without proving fatal. While cautiously resisting the conclusion that "the clinical delirium cordis is identical with the physiological delirium auriculae," he did insist that "the resemblance is certainly striking."

In 1906, independently of H. E. Hering (1903), Cushny and Charles W. Edmunds published a diagnosis of auricular fibrillation in the case of a diseased human heart they had been observing since late 1901.[3] Their diagnosis was at first doubted by clinicians, who found it hard to believe that a patient could survive so long with a fibrillating auricle. By 1910, however, Thomas Lewis and others had applied the electrocardiogram to a large number of diseased human hearts and had found that auricular fibrillation was frequently involved. Almost paradoxically, moderate doses of digitalis proved to be of special benefit in cases of auricular fibrillation; and in much of his later work on digitalis, Cushny sought to explain why this should be so. From the beginning he seemed to suspect that the muscular action of digitalis was the main reason for its therapeutic value in auricular fibrillation, despite suggestive evidence that its inhibitory action must be responsible.

Clearly the most beneficial effect of digitalis in such cases was to slow the ventricular beats so that the chamber could recuperate and generate a stronger beat each time. This reduction in ventricular rate was normally ascribed to inhibitory vagus action; but Cushny eventually concluded that it ought, rather, to be ascribed to the capacity of digitalis directly to impair the muscular fibers of the atrioventricular bundle. He emphasized this view in his monograph *The Action and Uses in Medicine of Digitalis and Its Allies* (London, 1925). This important work also bore witness to Cushny's long experience in the administration and assay of the digitalis series and in the use of digitalis in cases of heart disease other than auricular fibrillation. In many of these other cases, Cushny emphasized, the drug's inhibitory action probably played an important role.

From the outset of his work on digitalis, Cushny must have encountered and considered the problem of urinary secretion. Digitalis had originally been introduced into the materia medica not as a remedy for heart disease but chiefly because of its marked diuretic properties. In his first major paper on urinary secretion (1902), however, Cushny traced his serious study of the question to an isolated observation that did not seem to fit the prevailing theory of kidney action. According to this theory, due largely to Rudolf Heidenhain, urine was formed in two stages: water and inorganic salts, taken from the blood plasma, were secreted by the cells of the glomerular capsule into the urinary tubules; then urea and uric acid were secreted into this water by the cells of the epithelium lining the tubules. In its general features this theory represented a return to the view William Bowman had set forth in 1840, when he established the connection between the glomerular capsule and the urinary tubule. Heidenhain had, however, added impressive experimental evidence to Bowman's histological work; and in 1883 he had gathered all the evidence in a large and cogent treatise.[4]

The observation that first aroused Cushny's suspicion followed the injection of phlorhizin into rabbits. This procedure produced glycosuria, as expected, but Cushny also noticed a concomitant increase in the concentration of chloride ions in the urine. This chloride increase was hard to explain on the basis of Heidenhain's theory, which assumed that the concentration of inorganic ions in urine must depend essentially on their concentration in the circulating blood. Cushny decided to investigate more closely the fate of inorganic salts in the body. He injected isoanionic amounts of sodium chloride and sodium sulfate into the bloodstream and found that the two salts eventually appeared in the urine in concentrations that varied independently of their concentration in the blood. He argued that these results could best be explained by a modified version of Carl Ludwig's filtration-reabsorption theory. According to that theory, which had been eclipsed by Heidenhain's, urine was formed in the following way: First, all of the constituents of urine, including urea and uric acid, were filtered out of the blood plasma in the glomerular capsule; second, the dilute fluid thus produced was concentrated into urine by a simple process of diffusion—more specifically, by the diffusion of most of the water in the filtrate through the tubular epithelium. Thus Ludwig had emphasized the physical forces involved in urinary secretion, while Heidenhain emphasized the unknown "vital" forces possessed by secretory cells.

To explain his results with sodium chloride and sodium sulfate, Cushny adopted Ludwig's theory with one fundamental modification concerning the events taking place in the urinary tubules after filtration. Simple diffusion could not explain the differential concentration of inorganic salts in the urine. Cushny therefore proposed that the epithelial cells of the tubule possessed a capacity for differential reabsorption—or, to put it another way, that the chloride and sulfate ions possessed differential powers of permeating those cells.

Later in the same year (1902) Cushny published a paper confirming this suggestion. When the outflow from one ureter was obstructed, he found, absorption in the tubules was differential, with the water and chlorides returning to the blood much more readily than did the sulfates, phosphates, urea, and pigment. In 1904 he argued that the formation of an acid urine from an alkaline plasma, a phenomenon typical during saline diuresis, could also be explained by this modified filtration-reabsorption theory. Not until 1917 did Cushny publish another major paper on urinary secretion. This time he directly criticized one of Heidenhain's major experimental arguments for his theory. In 1874 Heidenhain had injected indigo carmine into rabbits after experimentally abolishing urinary secretion. He found that the dye subsequently accumulated in the cells of the tubular epithelium. From this he concluded that under normal circumstances, the dye would have been secreted into the urine by the epithelial cells. He argued by analogy that urea, which, like carmine, is a nitrogenous substance, must also be secreted by the epithelial cells. Without disproving Heidenhain's conclusions about indigo carmine, Cushny destroyed the analogy. He abolished urinary filtration in a rabbit, using Heidenhain's method. Two hours later he removed the kidney, macerated it, and subjected it to careful chemical analysis. No urea could be detected, and therefore could not have accumulated in the cells of the tubular epithelium. Its total absence indicated that it had no access to the kidney once its normal passage through the glomerular filter was abolished.

Also in 1917 Cushny published his monograph *The Secretion of Urine*, in which he advocated the filtration-reabsorption theory. He was now able to undermine the second major argument for Heidenhain's theory—that the tubular epithelium was incapable of absorbing the large quantity of water that would be required by the filtration-reabsorption theory. By quantitative determinations of the rate of absorption in the urinary tubule of a cat, Cushny showed that it was in fact adequate to meet Heidenhain's objection. He modestly called the filtration-reabsorption theory the "modern theory," although he had, since 1902, probably done more than anyone else to establish it. He believed that this theory, unlike Heidenhain's, would pave the way for new research; and he hoped that his book might serve "as an advance post from which others may issue against the remaining ramparts of vitalism." He insisted that differential absorption, while it too involved unknown and therefore "vital" forces, was an essentially automatic action. The epithelial cells of the urinary tubule did not really possess powers of discrimination. Each substance in the filtrate possessed its own constant "threshold" of permeability (urea and uric acid were "nonthreshold" substances and therefore were never absorbed), and "blind" physicochemical forces determined the composition of the exudate returned to the circulating blood.

In 1921 Cushny and a co-worker admitted that the modern theory probably emphasized too exclusively the role of the blood supply in diuresis. Finding that the action of caffeine and certain other diuretics was independent of changes in the

CUSHNY

blood supply, they concluded that these substances might reduce the resistance to filtration by a specific action on the cells of the glomerular capsule. Cushny incorporated this and other changes into the second edition of his monograph (1926). Subsequent research has confirmed Cushny's theory in its essentials, although many details remain in dispute.[5]

The pharmacological action of optical isomers was the third major area in which Cushny made fundamental contributions. The problem had attracted great interest ever since Louis Pasteur, in the 1840's, had discovered optical isomers in the tartrates. Cushny felt, however, that much of the literature on the subject was distinguished more by zeal than by sound judgment. Beginning in 1904, he published seven major papers in which he carefully and quantitatively investigated the pharmacological action of several pairs of optical isomers and their optically inactive racemic forms. His first paper can be taken as illustrative. He compared the effects of inactive atropine with those of its isomer, levorotatory hyoscyamine. In frogs, he found, the two isomers were equally poisonous to many organs, such as the heart and muscle, but atropine was a more powerful stimulant for the central nervous system. In mammals, on the other hand, the two isomers acted equally on the central nervous system, but the levorotatory form acted almost exactly twice as powerfully on the terminations of the autonomic nervous system. By this work, Cushny provided the first decisive evidence that a pair of optical isomers could differ in their action on the cells of higher organisms.

In his later work Cushny demonstrated more fully just how complex the pharmacological action of optical isomers could be. In many cases neither of a pair of isomers exerted any effect; while in many other cases, either of the isomers would yield the same effect. Any one isomer might produce entirely different effects on different tissues in the same organism, or on the same tissues in different organisms. Just before his death Cushny collected his own work and that of others into a critical monograph, issued posthumously under the title *Biological Relations of Optically Isomeric Substances* (1926). In keeping with his own experience, broad generalizations are rare. He even hesitated to endorse the general belief that the isomers occurring naturally are more active than their synthetic counterparts. He applied equal caution to the fundamental question of the relation between chemical structure, physical properties, and pharmacological action. On balance, Cushny supposed that such physical properties as solubility, volatility, and diffusibility played a more important pharmacological role than did chemical structure in general or alleged specific "receptor substances" in particular.[6] But on this issue, too, he displayed considerable theoretical reserve. "It may be surmised," he wrote, "that in pharmacological activity the relation between the living cell and the drug is seldom purely chemical or purely physical and that the proportion that these forces bear to each other varies from instance to instance." He did, however, agree with Pasteur that there was a close connection between optical activity and life. Earlier, in 1919, he had described optical activity as "the most persistent evidence" and "the most definite physical characteristic" of life.[7]

Cushny published papers on various topics other than digitalis, urinary secretion, and optical isomerism, including two on the pharmacology of the respiratory center and two on the hepatic cirrhosis produced in cattle by the *Senecio* alkaloids. Besides the monographs on each of his three major interests, he also wrote *A Textbook of Pharmacology and Therapeutics* (1899), which reached its thirteenth edition in 1947. Here, as elsewhere, Cushny revealed his concern for the practical applications of his own work and of pharmacology in general. Because of this concern he often worked in close cooperation with clinicians, and he deplored the general lack of communication between experimental pharmacologists and clinicians.[8] But neither did he minimize the importance of animal experimentation or basic research. Indeed, convinced that therapeutics had become fundamentally and irreversibly dependent on animal experiments, he produced a pamphlet in support of vivisection. And in his book on optical isomers, after giving several examples of the practical benefits of research pursued out of mere "idle curiosity," he wrote that "Saul was not the last who, going forth to seek his father's asses, found a kingdom."

NOTES

1. "Ueber die Wirkung des Muscarins auf das Froschherz," in *Archiv für experimentelle Pathologie und Pharmacologie*, **31** (1893), 432–453.
2. "On the Effects of Electrical Stimulation on the Mammalian Heart," in *Journal of Physiology*, **21** (1897), 213–230, written with S. A. Matthews.
3. "Paroxysmal Irregularity of the Heart and Auricular Fibrillation," in *Studies in Pathology Written . . . to Celebrate the Quatercentenary of Aberdeen University*, William Bulloch, ed. (Aberdeen, 1906), 95–110; also in *American Journal of the Medical Sciences*, n.s. **133** (1907), 66–77.
4. Rudolf Heidenhain, "Physiologie der Absonderungs-

organge," in Ludimar Hermann, ed., *Handbuch der Physiologie*, V (1883), 279–373.

5. See F. R. Winton, ed., *Modern Views on the Secretion of Urine* (London, 1956), the Cushny memorial lectures, esp. 246. For another attempt to place Cushny's work on urinary secretion in historical perspective, see John F. Fulton and Leonard G. Wilson, eds., *Selected Readings in the History of Physiology*, 2nd ed. (Springfield, Ill., 1966), 347–378.
6. See John Parascandola, "Arthur Cushny, Optical Isomerism, and the Mechanism of Drug Action," in *Journal of the History of Biology*, **8** (1975), 145–165; and "The Controversy Over Structure-Activity Relationships in the Early Twentieth Century," in *Pharmacy in History*, **16** (1974), 54–63.
7. See Abel, "Arthur Robertson Cushny and Pharmacology," 280.
8. See esp. "A Plea for the Study of Therapeutics," in *Proceedings of the Royal Society of Medicine*, **4** (1910–1911), Therapeutical and Pharmacological Section, 1–12.

BIBLIOGRAPHY

I. Original Works. For the titles and dates of publication of Cushny's three monographs and his textbook, see the text above. His most important papers on digitalis and heart action are "On the Action of Substances of the Digitalis Series on the Circulation in Mammals," in *Journal of Experimental Medicine*, **2** (1897), 233–299; "On the Interpretation of Pulse-Tracings," *ibid.*, **4** (1899), 327–347; "Paroxysmal Irregularity of the Heart and Auricular Fibrillation," in *American Journal of the Medical Sciences*, n.s. **133** (1907), 66–77; "The Therapeutics of Digitalis and Its Allies," *ibid.*, **141** (1911), 469–485; and "Irregularity of the Heart and Auricular Fibrillation," *ibid.*, 826–837.

His major papers on urinary secretion are "On Diuresis and the Permeability of the Renal Cells," in *Journal of Physiology*, **27** (1902), 429–450; "On Saline Diuresis," *ibid.*, **28** (1902), 431–457; "On the Secretion of Acid by the Kidney," *ibid.*, **31** (1904), 188–203; "The Excretion of Urea and Sugar by the Kidney," *ibid.*, **51** (1917), 36–44; and "The Action of Diuretics," *ibid.*, **55** (1921), 276–286, written with C. G. Lambie.

His major papers on optical isomers are "Atropine and Hyoscyamines; a Study of the Action of Optical Isomers," in *Journal of Physiology*, **30** (1904), 176–194; "The Action of Optical Isomers. II. Hyoscines," *ibid.*, **32** (1905), 501–510, written with A. R. Peebles; "The Action of Optical Isomers. III. Adrenalin," *ibid.*, **37** (1908), 130–138; "Further Note on Adrenalin Isomers," *ibid.*, **38** (1909), 259–262; "On Optical Isomers. V. The Tropeines," in *Journal of Pharmacology and Experimental Therapeutics*, **13** (1919), 71–93; "On Optical Isomers. VI. The Tropeines," *ibid.*, **15** (1920), 105–127; and "On Optical Isomers. VII. Hyoscines and Hyoscyamines," *ibid.*, **17** (1921), 41–61.

J. J. Abel (see below) gives a bibliography of eighty-five works by Cushny, with incomplete pagination and minor ambiguities in dating. Although nearly complete, it does omit "Paroxysmal Irregularity of the Heart and Auricular Fibrillation" (1907).

Several bound volumes of Cushny's research notebooks and lecture notes are in the MSS Division, Edinburgh University Library.

II. Secondary Literature. The best and most complete account of Cushny's life and work is J. J. Abel, "Arthur Robertson Cushny and Pharmacology," in *Journal of Pharmacology and Experimental Therapeutics*, **27** (1926), 265–286, repr., without bibliography, in *Science*, **63** (1926), 507–515. Also of special interest is Helen MacGillivray, "A Personal Biography of Arthur Robertson Cushny, 1866–1926," in *Annual Review of Pharmacology*, **8** (1968), 1–24. For other obituary notices, see G. B., in *Proceedings of the Royal Society of Edinburgh*, n.s. **46** (1926), 354–356, repr. in *Edinburgh Medical Journal*, n.s. **33** (1926), 247–249; Henry H. Dale, in *Archives internationales de pharmacodynamie et de thérapie*, **32** (1926), 3–8, with bibliography identical to that given by Abel; H. H. D[ale], in *Proceedings of the Royal Society*, **B100** (1926), xix–xxvii; John D. Comrie, in *Medical Life*, **33** (1926), 245–248; J. F. F[ulton], in *Dictionary of American Biography*, 2nd ed., III, 6–7; Hans H. Meyer, in *Naunyn-Schmiedebergs Archiv für experimentelle Pathologie und Pharmakologie*, **113** (1926), i–iv (after 128); and J. A. Gunn, in *Dictionary of National Biography, 1922–1930*, 234–235. Also see B. Holmstedt and G. Liljestrand, *Readings in Pharmacology* (New York, 1963), 261–268, which includes a considerable selection from Cushny's book on optical isomers; and Parascandola (see n. 6), who provides the best available analysis of Cushny's work on the pharmacological action of optical isomers.

Gerald L. Geison

DALE, HENRY HALLETT (*b.* London, England, 9 June 1875; *d.* Cambridge, England, 23 July 1968), *physiology, pharmacology.*

Son of a London businessman, Dale received his education at Tollington Park College, London; the Leys School, Cambridge; and Trinity College, Cambridge. He earned first-class honors in the natural sciences tripos in 1898, then succeeded Ernest Rutherford in the Coutts-Trotter studentship at Trinity. At Cambridge he came under the influence of the great physiologists of the "Cambridge School," including Michael Foster, W. H. Gaskell, J. N. Langley, and H. K. Anderson. Dale began clinical training at St. Bartholomew's Hospital in 1900, receiving his B. Chir. in 1903 and proceeding to the M.D. in 1909. He continued his physiological studies at London under Starling and Bayliss from 1902 to 1904, first as George Henry Lewes Student, then as Sharpey Student with the department of physiology of University College. He also studied for several months with Paul Ehrlich at Frankfurt.

DALE

Against the advice of his colleagues, Dale in 1904 accepted a position at the Wellcome Physiological Research Laboratories, a post that paid a salary adequate to allow him to marry Ellen Harriet Hallett; they had a son and two daughters. At the Wellcome he began, at Sir Henry Wellcome's suggestion, his investigations into the physiological actions of ergot. This work eventually led him, through a series of fortuitous discoveries, to his two major research interests:

> These two lines of enquiry have led, on the one hand, by way of studies which involved the specific actions of adrenaline and of acetylcholine, to a widening application of the conception of a chemical phase in the transmission of excitation from nerve-fibre endings to responsive cells; and, on the other hand, by way of studies of the actions of histamine and of its distribution in the animal body, to evidence for its contribution to local and general reactions, by which the organism as a whole and its separate tissues respond to various chemical, immunological, or physical assaults upon the integrity of their living cells (*Adventures in Physiology*, x).

Dale left the Wellcome Laboratories in 1914 to become a member of the scientific staff of the Medical Research Committee (or, as it was called after 1920, the Medical Research Council); from 1928 until 1942 he served as the first director of the organization into which the Medical Research Council evolved, the National Institute for Medical Research. From 1942 to 1946 Dale was resident director of the Royal Institution of Great Britain, acting as chairman of the Scientific Advisory Committee to the War Cabinet in 1942–1947. He received virtually every honor England could bestow on a man of science. Elected a fellow of the Royal Society in 1914, he served as secretary from 1925 to 1935, received its Copley Medal in 1937, and was its president from 1940 to 1945. The Royal College of Physicians made him a fellow in 1922. He served as president of the British Association in 1947, of the Royal Society of Medicine 1948–1950, and of the British Council 1950–1955. Dale was knighted with the Grand Cross Order of the British Empire in 1943, and in the next year the Order of Merit was bestowed on him. He shared the Nobel Prize for medicine or physiology in 1936 with his friend Otto Loewi. The Dale Medal was struck by the Society for Endocrinology in 1959, to be awarded annually; and the Royal Society established the Henry Dale professorship in 1961. The latter was endowed by the Wellcome Trust, which Dale had served as chairman from 1938 to 1960.

Dale never wrote a book; in his later life, however, two collections of his papers were published: *An Autumn Gleaning* (1954), a selection of his addresses and occasional papers; and *Adventures in Physiology* (1953), a collection of his scientific papers with a valuable retrospective comment by Dale. In the latter volume Dale frequently draws attention to the key role of serendipity in his work. His early researches into the physiological actions of ergot failed in their explicit goal of clarifying the pharmacologic and therapeutic properties of the drug. A set of fortunate circumstances, however, caused him to use a spinal animal, which had just been given several doses of a preparation of ergot, to measure the strength of a newly extracted batch of adrenaline. (A spinal animal is one in which the brain has been separated from the spinal cord.) He rejected the first set of data, which was exactly opposite to the expected results; but when the same sequence of events recurred the next week, Dale pursued the anomalous data, thereby discovering the important phenomenon of adrenaline reversal. The present-day use of phentolamine in the diagnosis of pheochromocytoma is based on Dale's investigations. Two years later, in 1909, while using posterior pituitary extracts as a control for his ergot investigations, he first identified the oxytocic effects of the extracts when he incidentally noted their tonic actions on the pregnant cat uterus.

Dale first witnessed the use of the horn of the cat uterus as an isolated experimental organ in 1907, when Ferdinand Kehrer employed it at a physiological congress in Heidelberg to demonstrate the immediate and quite powerful tonic effects on the uterus of an ergot extract, results that suggested to Dale the presence of an unidentified contaminant. In collaboration with George Barger, a colleague at the Wellcome Laboratories whose name is frequently associated with Dale's early work, Dale in 1910 successfully identified the contaminant as histamine. In the same year Dankwart Ackermann at Würzburg produced histamine by decarboxylation of histidine. Over the next few years Dale, working first with Barger and then with P. P. Laidlaw, A. N. Richards, J. H. Burn, and others, subjected histamine to an intensive pharmacologic investigation. As early as 1911 Dale and Laidlaw pointed out the similarities between the effects of histamine poisoning and anaphylactic shock. Furthermore, in 1911 Dale and Barger showed that histamine is a naturally occurring substance by identifying it in the intestinal mucosa of the ox. They minimized the importance of their findings, however, for they could not exclude the presence of the histamine as secondary to bacterial decarboxylation of histidine. Consequently, the nexus between histamine and anaphylaxis was not established for another fifteen years.

In the meantime, in a series of classic experiments Dale attacked the problem on two fronts: his work with isolated guinea pig uteri (also begun through a chance observation) established that the anaphylactoid antibodies are attached to cells rather than circulating; and his research on the pharmacology of histamine yielded precise data on the mechanism of "histamine shock," as he called it. Through the simple expedient of carefully measuring the volume of a limb of an experimental animal into which histamine was being injected, correlated with serial hematocrit determinations, Dale showed that histamine causes the loss of plasma fluid into the tissues, to produce edema, along with vasodilation, resulting in marked hemoconcentration, stagnation of the blood in the dilated vessels, decreased blood return to the heart, decreased cardiac output, and shock. His work also demonstrated the importance of chemical and humoral factors in the control of the circulation, especially at the micro level. Two Croonian lectures delivered by Dale summarize his work during this period: "The Biological Significance of Anaphylaxis," before the Royal Society in 1919, and "Some Chemical Factors in the Control of the Circulation," before the Royal College of Physicians in 1929; both are reprinted in *Adventures in Physiology*.

In 1927 Dale, collaborating with C. H. Best, H. W. Dudley, and W. V. Thorpe, conclusively proved that histamine is normally present in animal tissues, being particularly rich in the lung. In his 1929 Croonian lectures he summarized the overwhelming evidence in favor of histamine release as the primary cause of the physiological features of anaphylaxis. He also identified the "H-substance" of Sir Thomas Lewis—the chemical that produces the characteristic erythema, wheal, and flare of the Lewis "triple-response" when cells of the skin are injured—as probably histamine itself. Dale's lectures stimulated an intense interest in the physiological properties of histamine, an interest unabated today.

Dale's work with histamine not only clarified several poorly understood phenomena; it also served as a pivotal point for his other primary research interests: histamine, in common with adrenaline, vasopressin, and acetylcholine, is a vasoactive substance. It has already been mentioned that early in his career Dale had noted that some preparations of ergotoxin reverse the action of adrenaline, and had separated the oxytocic from the vasopressive qualities of pituitary extract. (The vasopressive property was discovered by Oliver and Schäfer in 1895.) Acetylcholine is also vasoactive and, like histamine, first came to Dale's attention as an active contaminant of ergot extracts. The intense depressor (vasodilatory) effect of acetylcholine was first reported by Reid Hunt in 1906. Dale published a comprehensive review of the physiological properties of the substance in 1914, noting its muscarinic action, paralyzed by atropine, and its nicotinic action, paralyzed by excess of nicotine. He called attention to the striking similarities between the actions of acetylcholine and those produced by stimulation of the parasympathetic nervous system.

Dale's lifelong friend T. R. Elliott had shown in 1904 that many of the actions of the sympathetic branch of the autonomic nervous system could be duplicated by the injection of adrenaline; he had even suggested that the effects of the sympathetic system might be mediated by the release of adrenaline at the nerve endings. Elliott did not follow up his notion, although in 1907 W. E. Dixon extended Elliott's concept by postulating that a muscarine-like substance is liberated when the vagus is stimulated. But the parallels between nerve stimulation and muscarine or adrenaline injection, although striking, were not exact; and there the matter rested until Dale reopened the door with his 1914 paper on acetylcholine. The main difficulty in accepting a physiological role for acetylcholine was that its natural occurrence in animal tissue had not been proved.

The beautifully simple experiments of Otto Loewi, working with vagal and sympathetic stimulation of the isolated frog heart in 1921 and the succeeding years, made chemical mediation of nerve impulses seem highly probable. In 1929 Dale and H. W. Dudley—while assaying for histamine—unexpectedly found high concentrations of acetylcholine in the spleen of the ox and horse, thus proving its natural occurrence. Over the next seven years Dale and his colleagues at the National Institute for Medical Research, including W. Feldberg, J. H. Gaddum, M. Vogt, and G. L. Brown, greatly extended the concept of chemical transmission. They showed that acetylcholine is liberated not only at parasympathetic postganglionic endings but also at some sympathetic postganglionic and all preganglionic endings. They also experimentally proved that acetylcholine is released at nerve endings in voluntary muscles. In 1933 Dale introduced the useful terms "cholinergic" and "adrenergic" to refer to fibers releasing acetylcholine and an adrenaline-like compound (later shown to be noradrenaline) at their endings.

Dale and Otto Loewi shared the 1936 Nobel Prize for medicine or physiology for "their discoveries relating to the chemical transmission of nerve impulses." It served as a fitting climax to Dale's outstanding career as an investigator; after 1936 other obligations commanded a larger portion of his time. He continued as director of the National Institute for Medical Research until 1942, devoting much energy to the planning of the new facilities at Mill Hill, a project largely interrupted by the war and consequently erected under the direction of Dale's successor, Sir Charles Harington.

In his later years Dale became a spokesman for British men of science, a role for which he was qualified not only by the force of his character but also by virtue of the active concern he had shown for the social consequences of science and medicine. Dale and Thorvald Madsen of Copenhagen were largely responsible for the adoption of an international scheme of standardizing drugs and antitoxins. At the 1925 conference of the Health Organization of the League of Nations, meeting in Geneva, Dale prevailed upon his colleagues to accept international standards for insulin and pituitary extract; and other drugs soon were agreed upon. He was also instrumental in the passage of the Therapeutic Substances Act in England. His concern for preserving the apolitical nature of science and for the peaceful use of nuclear energy was noteworthy.

BIBLIOGRAPHY

I. ORIGINAL WORKS. *Adventures in Physiology* (London, 1953; repr. 1965) contains thirty of Dale's major scientific papers, with comment on each paper by Dale. *An Autumn Gleaning* (London, 1954) reprints an additional fourteen of his occasional lectures and addresses. The two volumes together give an extraordinary picture of Dale, both as a man and as a scientist. *Adventures in Physiology* contains a bibliography of Dale's published writings through 1953. Of his subsequent papers the most important are "Autobiographical Sketch," in *Perspectives in Biology and Medicine*, **1** (1957–1958), 125–137; "Sir Michael Foster K.C.B., F.R.S., a Secretary of the Royal Society," in *Notes and Records. Royal Society of London*, **19** (1964), 10–32; and the intros. to *Collected Papers of Paul Ehrlich*, edited by F. Himmelweit, with the assistance of Martha Marquardt, 3 vols. (London–New York, 1956–1960). Dale wrote the obituary notices in *Biographical Memoirs of Fellows of the Royal Society* of T. R. Elliott (1961) and Otto Loewi (1962), which are also of interest.

II. SECONDARY LITERATURE. Several short assessments of Dale's work appeared during his lifetime. See *British Medical Journal* (1955), **1**, 1355–1361, 1378–1379, for articles by several authors. Also see James F. Riley, "Histamine and Sir Henry Dale," *ibid.* (1965), **1**, 1488–1490. The most perceptive obituaries are those of W. Feldberg, in *British Journal of Pharmacology and Chemotherapy*, **35** (1969), 1–9; and in *Biographical Memoirs of Fellows of the Royal Society*, **16** (1970), 77–173. Dale's work is put into perspective in Charles Singer and E. Ashworth Underwood, *A Short History of Medicine*, 2nd ed. (Oxford, 1962), 555–577. There is also sound historical treatment in L. S. Goodman and A. Gilman, eds., *The Pharmacological Basis of Therapeutics*, 3rd ed. (New York, 1965), esp. chs. 21, 23, and 29.

WILLIAM F. BYNUM

DALTON, JOHN CALL (*b.* Chelmsford, Massachusetts, 2 February 1825; *d.* New York, New York, 12 February 1889), *medicine, physiology.*

Dalton was the eldest of four sons of John Call Dalton, a prominent physician, and of Julia Ann Spalding. He graduated from Harvard College in 1844 and from the Harvard Medical School in 1847. During the next three years he probably practiced medicine in Boston. In 1848 he reported, to the Boston Society for Medical Observation, the case of a child with lead poisoning who subsequently recovered; and in 1849 he published a description of a human fetus with a malformed cranium. By means of postmortem examinations of human ovaries, Dalton established that the corpus luteum of pregnancy is distinguishable from, and persists longer than, the corpus luteum produced during the normal menstrual cycle. His essay on this subject won a prize from the American Medical Association in 1851.

By then Dalton had decided to study experimental physiology. In 1850 he visited Paris, where he attended the lectures of Claude Bernard, just as Bernard was emerging as the foremost physiologist in France. Dalton was deeply impressed by Bernard, whom he regarded afterward as his mentor. After his return to America, Dalton gave up medical practice to devote all of his time to physiology. No American had done so before, and his colleague Silas Weir Mitchell therefore later called him "our first professional physiologist." He was professor of physiology at the University of Buffalo Medical School from 1851 to 1854, then held a similar position at the University of Vermont until 1859, when he accepted the chair of physiology at Long Island College Hospital. In 1854 he lectured on physiology in a course entitled "Physiology and Pathology" that was given at the College of Physicians and Surgeons of New York. In 1855 the College made physiology independent of pathology, through the creation of a new chair of "physiology and microscopic anatomy." Dalton was the first person appointed to this chair.

Undoubtedly inspired by Bernard's brilliant vivisection demonstrations, Dalton introduced into American teaching the practice of illustrating physiology lectures with experiments on live animals. During the first five years of his teaching career he repeated many of the experimental investigations recently made in Europe, confirming all of the major results through which Bernard had discovered the action of the pancreatic juice on fats, the production of sugar by the liver, and the effects on body temperature of sectioning the sympathetic nerves. While examining the larynxes of etherized dogs during the winter of 1853–1854, Dalton observed that the glottis is opened and closed synchronously with the respiratory movements, through the action of the posterior cricoarytenoid muscles on the vocal cords. Since this observation was contrary to statements in François Longet's recent textbook of physiology, Dalton published his finding.

Over the next four years he focused his research on digestion, taking up problems that were then at the center of attention of French and German physiologists. Using the gastric fistula method devised by René Blondlot and perfected by Bernard, Dalton investigated the action of gastric juice in 1854. Among the conclusions he reached was that "true" gastric juice cannot be obtained by irritating the stomach walls with indigestible substances, a method that some investigators had recommended to obtain it in a pure state; the secretion of the juice is excited only by the kinds of food it normally digests. He also found, in support of Bernard's view, that starch is not converted to sugar in the stomach and that the action of saliva on starch does not continue in the presence of gastric juice. During this investigation, however, Dalton noticed that the presence of glucose could not always be detected, because gastric juice interfered with the standard Trommer's test for that sugar. Later he discovered that organic substances in such fluids as gastric juice, saliva, pancreatic juice, bile, and blood serum interfered with the test for starch, blocking its color reaction with iodine. The difficulty could be overcome, he reported in 1856, by adding a little nitric acid prior to adding the iodine.

Between 1855 and 1857 Dalton carried out sixty-seven experiments on the chemical and physiological properties of bile. He confirmed Adolph Strecker's identification of glycocholate and taurocholate of soda as the two characteristic ingredients of ox bile. In opposition to Strecker's opinion, however, Dalton found in the bile of other animals substances that, although similar, were not identical with these compounds. In order to determine at what time, after feeding, the bile is secreted most abundantly, he established permanent duodenal fistulas in dogs. Collecting the intestinal fluids at various intervals after a given meal, he estimated the quantity of bile poured into the intestine, as well as its proportion to the total quantity of the intestinal fluids, by means of Pettenkofer's test for the two characteristic bile compounds. Dalton concluded that the largest quantity of bile is secreted during the first hour, a lower but uniform quantity over the next eighteen hours, and that some bile is present at all times. Taking up the long-debated question of whether bile is a simple excretion or plays a role in digestion, he showed that the characteristic biliary compounds disappear during their passage through the intestine; they are, therefore, probably reabsorbed into the blood. He was unable, however, to detect them in the blood of the portal vein, so he had to assume that the compounds underwent some chemical change before being absorbed.

In 1859 Dalton published *A Treatise on Human Physiology*, designed for "students and practitioners of medicine." Judged by his stated objective—to communicate, "in a condensed form," the recent progress of physiology—his text was uneven. Although quite current for those areas in which he had been especially interested, such as digestion and nutrition, his discussion of other topics was less so. His chapter on the circulation made no mention of the important quantitative methods of the Ludwig school; and that on the electrical phenomena of nerve conduction relied on the work of Longet and Matteucci rather than the later, more rigorous investigations of du Bois-Reymond. Dalton dwelt at disproportionate length on those topics about which he had previously published his own investigations, while treating other equally important subjects rather cursorily. His inclusion of the stages of embryological development as an integral part of physiology had become somewhat old-fashioned by 1859. Nevertheless, Dalton's gift for lucid, vivid description and his ability to present complex issues simply and clearly made his book eminently readable. Perhaps reflecting the dual nature of his professional position, he grounded his discussions of functions in detailed descriptions of the microscopic anatomy of the structures involved. His extensive reliance on experiments that he had repeated himself, his own chemical analyses, and illustrations drawn from his own anatomical specimens or histological preparations gave his discussions an authoritative tone. In America his textbook, which, according to Mitchell, was "without a rival," went through seven editions by 1882.

Dalton's *Treatise* reveals that, despite his admiration for Bernard, he did not always follow the latter's views. Although he adopted Bernard's interpretation of the functions of the pancreas and liver, he also utilized extensively investigations such as those of Bidder and Carl Schmidt, which Bernard considered nearly useless. For his account of the nervous system Dalton relied heavily on Flourens and Longet, who had been at odds with Bernard and his teacher Magendie. (Dalton made no mention of Magendie, even in discussions of discoveries usually associated with him.)

In his *Treatise* Dalton reported that he had frequently repeated Flourens's and Longet's experiments on the removal of the cerebral hemispheres of birds. Like them, he had found that such birds could survive for long periods, that they were capable of movements and of receiving sensations, but that they had lost memory and "the power of forming mental associations." Dalton also repeated many times the celebrated experiments in which Flourens had removed portions of the cerebellum of pigeons. Like Flourens, he found that the more extensive the portion of the cerebellum he excised, the more irregular and uncoordinated the posture, gait, and movements of the head, neck, and wings became. In January 1859, after demonstrating these phenomena to a medical class, he noticed an effect that had escaped Flourens. The pigeon, which he afterward found to have lost two-thirds of its cerebellum, survived for sixteen days, during which time it gradually recovered its muscular coordination. After attaining similar results in further experiments, Dalton concluded in 1861 that one must either suppose that the irregularity of motions results from the sudden injury to the cerebellum as a whole rather than from the loss of a part of its substance, or else that the remaining portions of the cerebellum gradually become able to take the place of the lost parts.

When the Civil War broke out, Dalton volunteered his services and was appointed surgeon to the Seventh Massachusetts Regiment. He accompanied it to Washington in 1861 and was then commissioned surgeon of a brigade of volunteers serving under Brigadier General Egbert Viele. Dalton was with this brigade during the expedition to Georgia on which it captured Port Royal Island and Fort Pulaski. He became medical director of the brigade in 1862 and thereafter supervised the treatment of sick and wounded soldiers. By 1863 he was back in New York often enough to carry on his regular physiology lectures, but he still undertook several missions to transport rebel prisoners. Chronic health problems, aggravated by malaria, induced him in 1864 to resign his commission. In the same year he was named professor of physiology at the College of Physicians and Surgeons of New York and was elected to the National Academy of Sciences.

In June 1871 Dalton read to the New York Academy of Medicine a paper entitled "Sugar Formation in the Liver," in which he defended Bernard's belief that the liver normally produces sugar. In 1854 another of Bernard's former students, Frederick Pavy, had found no sugar in blood removed through a catheter from the right heart of an unanesthetized animal, whereas blood taken from the right heart after the animal was killed did contain sugar. From this and related evidence he concluded that the formation of glucose from glycogen, which Bernard had demonstrated to take place in isolated livers, was only a postmortem phenomenon. During the 1860's a number of people attempted to settle the question of whether the liver produces sugar in life, or only after death, by analyzing pieces of liver for sugar as quickly as possible after removing the tissue from a living animal. Some of them found small quantities, others found none, so that the issue remained in dispute in 1869, when Dalton took it up. He undertook a thorough review of the question, beginning with an assessment of the relative sensitivity of the different tests for glucose. He found that Fehling's solution was capable of detecting the smallest quantities. Then, devising a special grinder that enabled him to reduce a portion of a liver to a pulp and immerse it in alcohol within ten seconds after removing it from an animal, he was able to establish that in all cases there was an unmistakable glucose reaction. He concluded that sugar is a normal ingredient of hepatic tissue. In his *Leçons sur le diabète* (1877) Bernard used this "extremely careful work" of Dalton as a central element in his own refutation of Pavy's criticisms.

In 1874 Dalton returned to his earlier interest in the constitution of bile, using the methods of spectroscopic analysis, which were then being applied widely to animal substances. He discovered a hitherto unnoticed absorption band in the red, characteristic of all bile having a greenish tint.

After these two papers Dalton did little original experimental work. Ill health, and his appointment as vice-president of the New York Academy of Medicine (1874–1877) and as president of the College of Physicians and Surgeons (1884), left him little time for research. In his last years, however, he did write a monumental textbook on the topography of the brain, for which he prepared all the specimens and drawings himself and closely supervised the production of photographs. He wrote essays on the experimental method in physiology as a defense against the antivivisectionist movement. Dalton also produced books on the history of the discovery of the circulation and on the history of the College of Physicians and Surgeons, as well as historical lectures on spontaneous generation, "Galvani and Galvanism in the Study of the Nervous System," "Buffon and Bonnet in the Eighteenth Century," and "Nervous Degenerations and the Theory of Sir Charles Bell" (the last three in *The Experimental Method in Medical Science*). Dalton's use of primary sources, his perceptive insights, and his clear definition of issues make his historical writings still valuable.

The many physiological experiments that Dalton performed during his years of active investigations produced no major original contributions. Most of his results merely confirmed those achieved previously by European scientists; a few provided useful emendations or extensions of earlier discoveries. According to S. Weir Mitchell, Dalton's reputation in America exceeded what "his discoveries justified," because his singleness of purpose, his recognition that physiology must henceforth be taught through experimental demonstrations rather than from texts, and his skill in teaching and writing enabled him to open a new era in American physiology. It was as a teacher, therefore, that Dalton played his most effective role; but it was because he devoted himself persistently to his experiments as well that he could convey to his students a realistic understanding of the nature of a rapidly developing science. Through these means he did as much as any man to prepare for the spread of physiological research from the European centers where it had originated to the American centers where it was already beginning to thrive by the time he died.

BIBLIOGRAPHY

I. Original Works. Dalton's books include *On the Corpus Luteum of Menstruation and Pregnancy* (Philadelphia, 1851); *A Treatise on Human Physiology* (Philadelphia, 1859; 7th ed., 1882); *A Treatise on Physiology and Hygiene; for Schools, Families, and Colleges* (New York, 1868; repr. 14 times between 1869 and 1890); *Experimentation on Animals, as a Means of Knowledge in Physiology, Pathology, and Practical Medicine* (New York, 1875); *The Experimental Method in Medical Science* (New York, 1882); *Doctrines of the Circulation; a History of Physiological Opinion and Discovery, in Regard to the Circulation of the Blood* (Philadelphia, 1884); *Topographical Anatomy of the Brain* (Philadelphia, 1885); *History of the College of Physicians and Surgeons in the City of New York* (New York, 1888); and *John Call Dalton M.D., U.S.V.* (Cambridge, Mass., 1892), a narrative of Dalton's military experience.

Dalton's principal scientific articles are "Some Account of the Proteus anguinus," in *Edinburgh New Philosophical Journal,* **60** (1853), 332–340; "On the Movements of the Glottis in Respiration," in *American Journal of the Medical Sciences*, **28** (1854), 75–79; "On the Gastric Juice and Its Office in Digestion," *ibid.*, 313–320; "On the Decomposition of Iodide of Starch by the Animal Fluids," *ibid.*, **31** (1856), 326–330; "On the Constitution and Physiology of the Bile," *ibid.*, **34** (1857), 305–323; "On the Cerebellum, as the Centre of Co-ordination of the Voluntary Movements," *ibid.*, **41** (1861), 83–88; "Anatomy of the Placenta," in *Transactions of the New York Academy of Medicine*, **2** (1863), 33–50; "On the Rapidity and Extent of the Physical and Chemical Changes in the Interior of the Body," *ibid.*, 51–76; "Sugar Formation in the Liver," in *New York Medical Journal*, **14** (1871), 15–33; "On the Spectrum of Bile," *ibid.*, **19** (1874), 579–598; and "On the Form and Topographical Relations of the Corpus Striatum," in *Brain*, **3** (1881), 145–159. A collection of letters written by Dalton and his brothers during the Civil War is in *Proceedings of the Massachusetts Historical Society*, **56** (1923), 354–495.

II. Secondary Literature. A biographical notice and bibliography appeared in *Proceedings of the American Academy of Arts and Sciences*, **24** (1889), 445–447. S. Weir Mitchell's memoir, in *Biographical Memoirs. National Academy of Sciences*, **3** (1895), 179–185, is a good general description of Dalton's life and personality. Articles on Dalton in *Dictionary of American Biography*, V (New York, 1930), 40; and in *National Cyclopaedia of American Biography*, X (New York, 1909), 500, rely heavily on Mitchell's memoir.

Frederic L. Holmes

DAUBENTON, LOUIS-JEAN-MARIE (*b.* Montbard, France, 29 May 1716; *d.* Paris, France, 1 January 1800), *medicine, anatomy, mineralogy, zootechny.*

Daubenton was the son of Jean Daubenton, a notary, and of Marie Pichenot. Intended for the priesthood by his family, he became a novice at the age of twelve and studied at the Jesuit *collège* in Dijon, then with the Dominicans in that city. Sent to Paris to study theology at the Sorbonne, he became interested in medicine—apparently without his father's knowledge—and attended the anatomy lectures at the Jardin du Roi, as well as the botany lectures. After his father's death in 1736, Daubenton completed his medical training and took his M.D. at Rheims in 1741.

Daubenton then returned to Montbard to practice medicine. But Buffon, also a native of Montbard and *intendant* of the Jardin du Roi since 1739, summoned him to Paris at the end of 1742 and had him named *garde et démonstrateur* of the natural history collection at the Jardin du Roi on 12 June 1745. Daubenton lodged there and drew a salary of 500 livres. (His salary reached 4,200 livres by 1790.) Buffon, who was treasurer of the Académie Royale des Sciences, procured several additional sums for Daubenton after having arranged his appointment as *adjoint botaniste* on 19 March 1744. Daubenton had become known to the Academy the previous year with a memoir on a method of classifying the shellfish. He was promoted to *associé botaniste* on 13 August 1758, *associé anatomiste* on 29 May 1759, and to *pensionnaire anatomiste* on 16 May 1760. He retained the last post when the Academy was reorganized in 1785; and on 20 November 1795 he was named *membre résident* of the anatomy and zoology section of the First Class of the Institut National. During his career he also became a member of all the major foreign academies and of the Société de Médecine and the Société d'Agriculture.

Besides teaching at the Jardin du Roi, Daubenton held the chair of natural history at the Collège Royal (1778) and of rural economy at the veterinary school at Alfort (1783). (The latter post brought him a salary of 6,000 livres.) Daubenton wrote several articles for the *Encyclopédie* of Diderot and d'Alembert. He also supervised part of the *Encyclopédie méthodique* devoted to the animal kingdom. In *an* III he lectured on natural history at the newly founded École Normale Supérieure.

In 1790, at the request of the Assemblée Nationale, the officers of the Jardin du Roi presented a project for the reorganization of their institution; his colleagues chose Daubenton, rather than La Billardière, Buffon's successor as *intendant*, to preside over the meeting. After La Billardière's resignation at the end of 1791, and until the nomination of Henri Bernardin de Saint Pierre on 1 July 1792, Daubenton served as *de facto intendant*. He played a major role in the reform project and, after the creation of the Muséum National d'Histoire Naturelle by the decree of 10 June 1793, he was elected director by his colleagues (9 July 1793). Daubenton chose not to present himself at the end of his term in office; he wished to avoid a precedent for a return to the authoritarianism instituted by Buffon.

Daubenton married Marguerite Daubenton, a first cousin, on 21 October 1754; they had no children. His contemporaries described him as circumspect and meticulous, living a well-regulated life and keeping his distance from the political dissensions of the age. The time he did not devote to science was spent reading literature with his wife, who enjoyed a certain success as a novelist (*Zélie dans le désert*, 2 vols. [1786–1787], which went through twenty-one reprints by 1861). Daubenton referred to this pastime as "putting his mind on a diet."[1] Although he gave the impression of being physically frail, he enjoyed good health throughout his life, except for attacks of gout in his old age.

Named to the Sénat Conservateur in 1799, Daubenton suffered an attack of apoplexy at the first session and died on 1 January 1800. His colleagues at the museum organized an impressive funeral for him, and he was buried at the Jardin des Plantes.

Daubenton's work in anatomy consisted mainly in his share in the preparation of the great *Histoire naturelle*, undertaken by Buffon at the suggestion of the minister of state, Jean Maurepas. When he came to the Jardin du Roi in 1745, Daubenton's first task was to organize the natural history collection, which was simply a set of herbal and pharmaceutical samples, to which minerals, rocks, shells, and Tournefort's botanical collection had been added. Daubenton had to refine methods of preserving the specimens, to prepare and classify them, and to examine the many items sent by correspondents of the Jardin du Roi and by travelers. In addition, as curator he was obliged to be available as a guide twice a week and to provide impromptu lectures for visitors.

At the start, to judge by the prospectus[2] and the address to the king that open its first volume (1749), Buffon and Daubenton were officially on equal footing with regard to authorship of the *Histoire naturelle*. It is evident from reading Buffon's text, however, that he considered his collaborator's descriptions to be a subordinate undertaking, designed to furnish material for his own expositions. Nevertheless, Daubenton considered his work valuable in its own right, as "the sole means that leads to positive knowledge of the animal economy."[3]

The limits of the *Histoire* imposed certain constraints on Daubenton. Unable to provide exhaustive descriptions, he concentrated on the skeleton, the principal internal organs, the brain, the sexual organs, and the embryo and its enveloping membranes. Each description concluded with measurements of various parts of the organs. In order to facilitate the making of comparisons, the descriptions were set out in a uniform manner—a new procedure at the time. Nevertheless, Daubenton made no attempt to present a general synthesis of his some 200 detailed descriptions, which were published in volumes III through XV (1749–1767) of the work. This reticence prompted Peter Camper to comment, "Daubenton's modesty did not permit him to know all the discoveries that he made." Nevertheless, Daubenton's descriptions constitute an important innovation. Cuvier wrote of their author (upon whose work he drew abundantly in preparing his lectures on comparative anatomy) that he was "the first who united anatomy and natural history in a continuous manner."[4]

In dealing with certain parts of the body, particularly the skeleton (thereby laying the foundation for the work of his student Vicq d'Azyr), Daubenton was careful always to use the same terms to designate corresponding parts of the various quadrupeds.[5] In this way he made it easier to distinguish the structural similarities and differences among the groups that make up the quadrupeds. Thus, he wrote:

> . . . in describing the horse, we have, so to speak, in large part described the ass; all that is left is to state the resemblances and to furnish evidence of the differences that we have observed between these two animals. But to the same degree that the description of the ass is related to that of the horse, that of the bull is independent of it [,] for the bull resembles the horse only in being a quadruped.[6]

Daubenton seemed to adhere to the taxonomic definitions established by Buffon but tended to lack the latter's boldness in making generalizations. While his caution was sometimes justified, some authors have maintained that it verged on a total lack of concern with theoretical questions. For example, in comparative anatomy he did not accord special importance to repeated similarities, being satisfied merely to mention them without drawing the possible taxonomic implications. As a result, he never grasped the principle of the subordination of characters.[7]

Daubenton did, however, display a degree of intellectual independence from Buffon, whom he sometimes corrected (notably concerning the existence of mammae in the horse). He eventually called Buffon to task for his position on this question, accusing him of having criticized Linnaeus' method without really having understood it.[8]

Daubenton's skill in anatomy was quickly recognized and continued to be admired in the new ways he found to demonstrate his virtuosity. For example, he appears to have been the first to apply the methods of comparative anatomy to fossil forms (1762). Also, with the publication of *Mémoire sur les différences de la situation du grand trou occipital dans l'homme et dans les animaux* (1764), he brilliantly displayed his mastery of the comparative method. He showed not only that the position of the occipital foramen varies among the quadrupeds, apes, and man but also that the angle of inclination of this orifice with respect to a line passing from the root of the nose to the posterior side of the occipital foramen becomes increasingly smaller as man is approached in the series. This, he thought, explained the uniqueness of man's erect posture. Examining transitional types in his "Introduction à l'histoire naturelle des animaux" for the *Encyclopédie méthodique*, Daubenton cast doubt on certain generally accepted ideas concerning the series of living animals. Specifically, he attacked the notion of the zoophyte and rejected certain transitional forms that previously had been accepted as occurring between various groups of animals.

Daubenton ceased to work in close collaboration with Buffon around 1766, when the latter decided to eliminate Daubenton's descriptions from subsequent editions of the *Histoire naturelle*. This decision was perhaps motivated by financial considerations. In 1764 Buffon had bought back the publication rights to the *Histoire*; and Daubenton's contributions, which were very technical and rendered the volumes quite large, had little sales appeal to the general public. The reason, however, may simply have been jealousy on Buffon's part, since Daubenton had received much praise from his fellow scientists. In any case, the volumes on the birds and mineralogy suffered from the absence of Daubenton's participation.

At almost the same time, as he later reported in *Instruction pour les bergers* (1782), Daubenton was commissioned to investigate "by a series of well-conceived and carefully executed experiments the most favorable natural arrangement for improving wool. M. M. Trudaine informed me of this project in 1766 and asked me to conduct all the experiments I believed necessary to find a good way of improving wool: I felt prepared to undertake this task . . .; I was encouraged in this matter by the observations I had been making for twenty years on the structures of animals." In all his research Daubenton was guided by the hope of using scientific knowledge of nature to benefit mankind. Accordingly, until the Revolution he pursued his research on the sheep with great energy. He traveled in Spain to study the breeding of merino sheep and in France to learn current breeding practices. Beginning in 1767, he presented many memoirs on his research to the Académie des Sciences and the Société Royale de Médecine.

Daubenton erected a sheepfold near Montbard in 1767, and he had a second one built at the Alfort veterinary school when he began teaching there in 1783. He also pursued his wool experiments at the Jardin du Roi. The study of the mechanism of rumination led him to formulate more precisely his ideas on how the flocks should be fed. He studied the resistance of sheep to drops in temperature and insisted on the necessity of continual folding of wool-bearing animals and on the disadvantages of keeping them in sheds. Following the English example, he undertook crossbreeding experiments and recommended the use of rams of higher race. He used the microscope to examine the fineness of the wools obtained and in 1784 published "Sur le premier drap de laine superfine du crû de France."

Two years earlier, after fifteen years of research, Daubenton had published *Instruction pour les bergers et les propriétaires de troupeaux*, which dealt with the physiology, therapeutics, and surgery of wool-bearing animals, as well as with pastures, races, and breeding methods. The book, which made him famous, was soon translated into several foreign languages; and in *an* III the Convention ordered it to be reprinted at state expense. Excerpts were often published, and it was as a "shepherd" that Daubenton obtained his certificate of good citizenship in 1793. Encouraged by his results, Daubenton decided about 1790 to conduct experiments with a view toward acclimatizing in France a number of exotic animals, such as the tapir, the peccary, and the zebra; but nothing came of this project.

Despite his extensive work on breeding, Daubenton continued his research on mineralogy, a subject that he had been teaching at the Jardin du Roi since 1745. He presented papers on mineralogical questions to the Académie des Sciences and the Société de Médecine, and wrote many articles on mineralogical topics for the *Encyclopédie*. The first edition of his *Tableau méthodique des minéraux* appeared in 1784. While they display no great originality, Daubenton's teaching and writings on mineralogy contributed significantly to the spreading of current knowledge of the subject. It was, moreover, under Daubenton's patronage that Haüy began his career. Upon the creation of the Muséum d'Histoire Naturelle, Daubenton was given the chair of mineralogy, which he held until his death.

Daubenton also taught botany. He published a number of works on this subject and made several discoveries in plant physiology (the way in which the trunks of palm trees grow and the existence of tracheae in the bark of trees). In accord with Buffon's instructions, he supervised the writing of the "Discours préliminaire" for Lamarck's *Flore française* (3 vols., 1777–1778), which was written by Lamarck himself and was put into final form by Haüy. Daubenton investigated several medical questions and published *Mémoire sur les indigestions* (1785). He planned to work with Malesherbes on an annotated edition of the works of Pliny, but the project was never carried out.

Daubenton's official collaborators at the Jardin du Roi included his cousin Edmé Daubenton, who occupied the specially created post of *sous-démonstrateur* (from 1767); Lacépède, who replaced Edmé in 1784, and, from 1787, Faujas de Saint-Fond, who was in charge of correspondence.

NOTES

1. P. A. Cap, *Le Muséum d'histoire naturelle* (Paris, 1854), 75.
2. See *Journal des sçavans* (Oct. 1748), 639.
3. See Daudin, *De Linné à Jussieu . . .*, 154.
4. *Leçons d'anatomie comparée*, III (1805), xix.
5. "Description du cheval," in *Histoire naturelle*, IV, 338.
6. "Description du taureau," *ibid.*, 474.
7. Daudin, *op. cit.*, 134–135.
8. See *Séances des écoles normales recueillies par des sténographes*, I (Paris, 1802), 295.

BIBLIOGRAPHY

I. Original Works. Daubenton's most important publication is the *Histoire naturelle, générale et particulière, avec description du Cabinet du Roy*, 15 vols. (Paris, 1749–1767), on which he collaborated with Buffon for the part on quadrupeds.

Daubenton's memoirs published by the Académie Royale des Sciences include "Mémoire sur des os et des dents remarquables par leur grandeur," in *Histoire de l'Académie royale des sciences . . .* (1762), 206–229; "Mémoire sur le méchanisme de la rumination et sur le tempérament des bêtes à laine," *ibid.* (1768), 389–398; "Sur le premier drap de laine superfine du crû de France," *ibid.* (1784), 76–84; "Observations sur l'organisation et l'accroissement du bois," *ibid.* (1790), 665–675; "Observations sur les caractères génériques en histoire naturelle," in *Mémoires de l'Institut national des sciences et arts. Sciences mathématiques et physiques,* **1** (1796), 387–396; "Plan des expériences qui se font au Jardin des Plantes sur les moutons et d'autres animaux domestiques," *ibid.*, 377–386; and "Moyen d'augmenter la production du bled sur le sol de la République française par le parcage des moutons et la suppression des jachères," *ibid.*, 397–404.

Among the memoirs he presented to the Société Royale de Médecine is "Mémoire sur le régime le plus nécessaire aux troupeaux, dans lequel l'auteur détermine par des expériences ce qui est relatif à leur aliments et à leur boisson," in *Histoire de la Société royale de médecine* (1777–1778), 570–578. To the *Encyclopédie* of Diderot and d'Alembert he contributed the articles "Animalcule," "Botanique," and "Cornaline." Daubenton also wrote the articles "Quadrupèdes" and "Règne animal," as well as the "Introduction à l'histoire naturelle" in the part of vol. I of *Encyclopédie méthodique* entitled "Histoire naturelle des animaux." See also "Observations sur la division générale et méthodique des productions de la nature," in *Magasin encyclopédique*, **3** (1796), 5–10.

Daubenton's books include *Instruction pour les bergers et les propriétaires de troupeaux* (Paris, 1782)—on the various eds. of this work, see J. B. Huzard, *Notice historique et bibliographique sur les éditions et les traductions de l'Instruction pour les bergers* (*an* X and subsequent eds.); *Tableau méthodique des minéraux, suivant leurs différentes natures et avec les caractères distinctifs, apparents ou faciles à reconnaître* (Paris, 1784 and many subsequent eds.); and *Mémoires sur les indigestions, qui commencent à être plus fréquentes pour la plupart des hommes, à l'âge de quarante à quarante cinq ans* (Paris, 1785; 2nd ed., 1798).

MSS by Daubenton or concerning him are mainly in Paris, at the Bibliothèque Centrale du Muséum National d'Histoire Naturelle, the Archives Nationales, and the archives of the Académie des Sciences.

II. Secondary Literature. See A. Albrier, "La famille Daubenton, notice historique et généalogique," in *Revue historique nobiliaire et biographique*, **11** (1874), 152–181; G. Cuvier, "Éloge historique de Daubenton," in *Recueil des éloges historiques lus dans les séances publiques de l'Institut royal de France*, I (Strasbourg–Paris, 1819), 37–80; P.-M.-J. Flourens, "Notice sur Daubenton," in *Recueil des éloges historiques lus dans les séances publiques de l'Académie des sciences*, III (Paris, 1862), 313–328; E. T. Hamy, "La mort et les funérailles de L. J. M. Daubenton (1800)," in *Les débuts de Lamarck* (Paris, 1908), 236–254; B. Lacépède, "Discours sur la vie et les ouvrages de Daubenton," in Daubenton's *Instruction pour les bergers*, 3rd ed. (Paris, *an* X), 1–38; L. Roule, *Daubenton et l'exploitation de la nature* (Paris, 1925); and P. Vangeon, "La vie laborieuse du docteur Daubenton," in *Bulletin de la Société archéologique et biographique du canton de Montbard*, no. 31 (1934), 1–14, and no. 32 (1935), 1–9. The best account of Daubenton's scientific work is H. Daudin, *De Linnée à Jussieu. Méthodes de la classification et idée de série en botanique et en zoologie (1740–1790)* (Paris, 1926), esp. ch. 3.

Camille Limoges

DELAFOSSE, GABRIEL (*b.* Saint-Quentin, France, 24 April 1796; *d.* Paris, France, 13 October 1878), *mineralogy, crystallography.*

The son of a provincial magistrate, Delafosse graduated from schools in Saint-Quentin and Rheims and became a scholarship student at the École Normale Supérieure. Upon completion of his studies there in 1816, he went to work for Haüy, whose mineralogy courses he had attended at the Muséum d'Histoire Naturelle. He assisted Haüy in editing and publishing *Traité de cristallographie* (1822) and the posthumous last three volumes of the second edition of the *Traité de minéralogie* (1822–1823). Delafosse's appointment as assistant naturalist in 1817 began his lifelong association with the Muséum, where he became a professor in 1857, replacing Dufrénoy. He was also associated with the Paris Faculté des Sciences from 1822, obtaining a professorship of mineralogy in 1841, as successor to Beudant. For many years he taught at the École Normale Supérieure (1826–1857). His election to the section of mineralogy in the Académie des Sciences occurred in 1857. He was a founding member of the Société Géologique de France.

As Haüy's pupil, Delafosse was a member of the second generation of the creators of modern crystallography. Like his teacher he was guided by a deep commitment to symmetry as the basic principle for the investigation of crystals. Unlike Haüy, however, Delafosse distinguished clearly between the integrant or subtractive molecule, a geometrical invention, and the chemical molecule, without discarding the belief that a crystal's exterior symmetry must be governed by the internal symmetry of the molecular arrangement. Haüy's subtractive molecule might then correspond to groups of several chemical molecules, and cleavage would proceed by cutting through subtractive molecules (instead of through the interstices between them) and along networks of chemical molecules forming a surface. By this means Delafosse was able to bring hemihedral properties of crystals into accord with crystallographic theory. In modern terminology the hemihedralism was ascribed to a geometric polarity of the chemical molecule.

Seeking to relate the structure and chemical composition of crystals, Delafosse investigated plesiomorphism, or similarity in crystal form unaccompanied by chemical identity. His method of mineral classification was based on both chemical composition and crystal form. He also studied relations between the morphology of crystals and their physical properties, such as electricity, light, and heat. Among his students who extended such studies fruitfully was Louis Pasteur.

BIBLIOGRAPHY

I. ORIGINAL WORKS. Delafosse's most comprehensive work was *Nouveau cours de minéralogie comprenant la description de toutes les espèces minérales avec leurs applications directes aux arts*, 3 vols. and atlas (Paris, 1858–1862), drawn from his courses at the École Normale, the Muséum, and the Sorbonne, and published after he had attained recognition as a major figure in mineralogy. His doctoral dissertation at the Paris Faculté des Sciences, *De la structure des cristaux considérée comme base de la distinction et de la classification des systèmes cristallins* (Paris, 1840), set forth his theory of crystal structure. He linked the form and physical properties of crystals in "Recherches sur la cristallisation, considérée sous les rapports physiques et mathématiques," in *Mémoires présentées par divers savants à l'Académie royale des sciences de l'Institut de France*, **8** (1843), 641–690; chemical composition was related to crystal morphology in "Mémoire sur une relation importante qui se manifeste, en certains cas, entre la composition atomique et la forme cristalline, et sur une nouvelle application du rôle que joue la silice dans les combinaisons minérales," *ibid.*, **13** (1852), 542–579.

His *Mémoire sur le plésiomorphisme des espèces minérales* (Paris, 1851) was extracted in *Comptes rendus . . . de l'Académie des sciences*, **32** (1851), 535–539, as were his "Mémoire sur la structure des cristaux et ses rapports avec les propriétés physiques et chimiques," **43** (1856), 958–962, and "Sur la véritable nature de l'hémiédrie et sur ses rapports avec les propriétés physiques des cristaux," **44** (1857), 229–233. Earlier articles include "Observations sur la méthode générale du Rév. W. Whewell pour calculer les angles des cristaux," in *Annales des sciences naturelles*, **6** (1825), 121–126; and "Mémoire sur l'électricité des minéraux," in *Annales des mines*, **3** (1818), 209–226.

Delafosse contributed articles to Charles d'Orbigny's *Dictionnaire universel d'histoire naturelle*, 13 vols. and atlas in 3 vols. (Paris, 1841–1849), of which he was an editor. He was one of the principal editors of Baron de Férussac's *Bulletin des sciences naturelles et de géologie* (1824–1831), a section of *Bulletin universel des sciences et de l'industrie*. He prepared a review of the development of mineralogy in France during this time: *Rapport sur les progrès de la minéralogie* (Paris, 1867).

In accord with his deep interest in teaching, Delafosse published a number of didactic works in science, including *Précis élémentaire d'histoire naturelle, à l'usage des collèges et des maisons d'éducation*, 4 vols. (Paris, 1830), republished in many subsequent eds.; *Notions élémentaires d'histoire naturelle*, 3 vols. (Paris, 1835–1836), also republished many times; and *Leçons d'histoire naturelle* (Paris, 1838).

Delafosse also prepared a *Notice sur les travaux scientifiques de M. Delafosse* (Paris, 1851).

II. SECONDARY LITERATURE. There is no biography of Delafosse. Biographical information comes mainly from such contemporary eulogies as Ch[arles] Friedel, "La vie et les travaux de Delafosse," in *Revue scientifique*, no. 21 (1878), 481–484; Ed[mond] Hébert, *Notice sur Gabriel Delafosse* (Versailles, 1879), extracted from *Compte rendu annuel de l'Association des anciens élèves de l'École normale* and based largely on Friedel's sketch; Ed[ouard] Jannettaz, "Notice nécrologique sur M. G. Delafosse," in *Bulletin de la Société géologique de France*, ser. 3, **7** (1878–1879), 524–530; A. L. O. Legrand Des Cloizeaux's note on Delafosse in *Comptes rendus . . . de l'Académie des sciences*, **87** (1878), 569–570; and an unsigned biographical sketch in *l'Année scientifique et industrielle*, **22** (1878), 494–495.

KENNETH L. TAYLOR

DIOCLES (*fl. ca.* 190 B.C.), *mathematics*.

The few facts known about Diocles' life are derived entirely from his one surviving work, *On Burning Mirrors* (Περὶ πυρίων). His date can be determined approximately from his acquaintance with the mathematician Zenodorus, who is known to have lived in the early second century B.C. This date accords well with the terminology and treatment of conic sections in Diocles' work, which shows little or no trace of influence by Apollonius' *Conics*, since it makes him an exact contemporary of Apollonius. Diocles mentions only mathematicians contemporary with or earlier than Archimedes (except for Zenodorus). He was living in Arcadia when Zenodorus visited him.

Until the recent discovery of the Arabic translation of *On Burning Mirrors*, it was lost except for three excerpts in the commentary by Eutocius on Archimedes' *Sphere and Cylinder*. Since Eutocius did not quote Diocles verbatim, but reformulated his proofs (for instance, introducing references to Apollonius' *Conics*), modern inferences about Diocles' date and place in the history of the theory of conics are misleading and usually wrong. The following account is based on the Arabic text.

The work consists of an introduction and sixteen

propositions, of which numbers 6, 9, and 14 are spurious (probably interpolated in the Arabic transmission). The title *On Burning Mirrors* is somewhat misleading, as it applies only to the first five propositions. Numbers 7 and 8 deal with a problem in Archimedes' *Sphere and Cylinder*, and propositions 10–16 with the problem of "doubling the cube." The book as a whole has no unity except that it deals with "higher geometry"; as is natural for a Hellenistic Greek work, much of it is concerned with conic sections.

Diocles starts from two problems, the first posed by one Pythion (otherwise unknown) to Conon of Samos: What mirror surface will reflect the sun's rays to the circumference of a circle? The second was posed by Zenodorus to Diocles: What mirror surface will reflect the sun's rays to a point? Diocles says that the second problem was solved by Dositheus (well-known as a correspondent of Archimedes). This implies that the focal property of the parabola was recognized by about the middle of the third century B.C. Diocles indicates, however, that he himself is the first to give a formal proof of the property. After an obscure but historically interesting discussion of the application of burning mirrors to sundial construction, he proves (prop. 1) the focal property of the parabola, and shows how Pythion's and Zenodorus' problems can be solved by suitable rotation of the parabola.

In propositions 2 and 3 Diocles shows that it is useless to construct a spherical burning mirror from an arc greater than 60°, and proves that all rays reflected from such an arc will pass through a section less than 1/24 the diameter of the mirror.

Propositions 4 and 5 are of great historical interest. They address the problem of constructing a parabolic mirror of given focal length. Diocles' solution is as follows (see Figure 1). If the given focal length is AB, complete the square $ABEF$, extend AF to K so that $FK = AF$, join KE and produce it to meet AB produced in R. Take arbitrary points D,G on AB, draw DH,GZ parallel to AF, and produce them to meet KE in L,M. Then, with center A and radius DL, draw a circle to cut DL in N and (on the other side in) Φ. Similarly, with center A and radius GM, draw a circle to cut GM in Θ and Ψ. Make $AX = AK$. Then points $K, N, \Theta, B, \Psi, \Phi, X$ lie on a parabola with A as focus. This construction is equivalent to the construction of the parabola from focus and directrix—as is obvious if, like Diocles, we complete the square $ARSK$ and drop onto SR the perpendiculars LQ, NO, MC, ΘP. For $NA = LD$, by construction, and $NO = LQ = LD$, so $NA = NO$, or the distance of the point N from A (the focus) is equal to its vertical distance from the line SR (the directrix). Similarly for the other points K, Θ, and so on. It is noteworthy that Diocles proves not only this, but also that a curve so generated is indeed a parabola (according to the classical Greek definition, by the equivalent of the relationship $y^2 = px$). The obvious inference is that Diocles himself discovered the focus-directrix property of the parabola.[1]

In propositions 7 and 8 Diocles discusses a problem arising out of Archimedes' *Sphere and Cylinder* II, 4: to divide a sphere in a given ratio. The problem involves, in modern terms, a cubic equation, which Diocles solves by the intersection of a hyperbola and an ellipse. His solution was already known from Eutocius, who also gives solutions by Dionysodorus and (possibly) Archimedes that likewise employ the intersection of two conics.

The rest of the book is devoted to the problem of doubling the cube, to which much attention was paid by Greek mathematicians from the fifth century B.C. on. Like everyone else in antiquity, Diocles in fact solves the equivalent problem of finding two mean proportionals between two given magnitudes.[2] His first solution, employing the intersection of two parabolas, was already known (in an altered form) from Eutocius; but since Eutocius did not mention the author, in modern times it has been almost universally misattributed to Menaechmus.

The second solution is both more interesting and more influential (see Figure 2). In a circle, with diameters AB, GD intersecting at right angles, there are marked off from D equal small arcs DZ, ZH, $H\Theta$ . . . and DN, NS, SO . . . on the other side of D. Drop onto AB the perpendiculars ZK, HL, ΘM . . . and join BN, BS, BO. . . . Mark the points P, Q, R . . . where BN cuts ZK. . . . Then it can be shown that

$$\frac{AK}{KZ}=\frac{KZ}{KB}=\frac{KB}{KP}.$$

That is, KZ and KB are two mean proportionals between AK and KP. Similarly for point Q, LH and LB are two mean proportionals between AL and LQ, and so on. The points P, Q, R . . . are joined in a smooth curve $DPQRB$, which can be used to find two mean proportionals between any two magnitudes, and to solve related problems, as Diocles demonstrates at length in propositions 11–16.

Despite his contributions to the theory of conics,

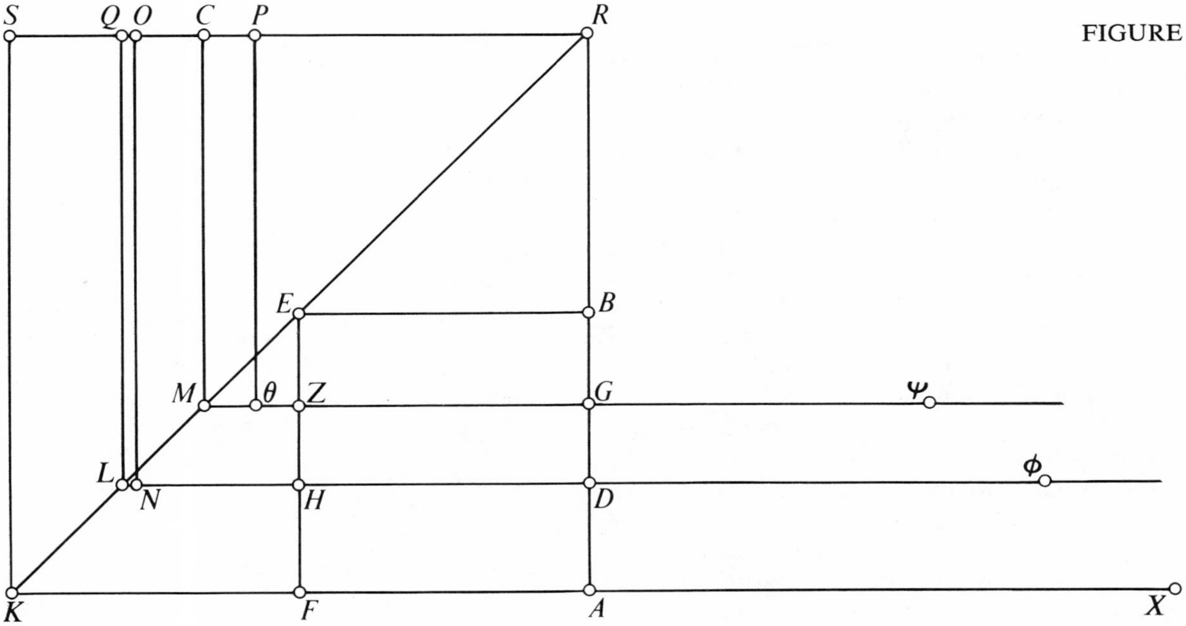

FIGURE 1

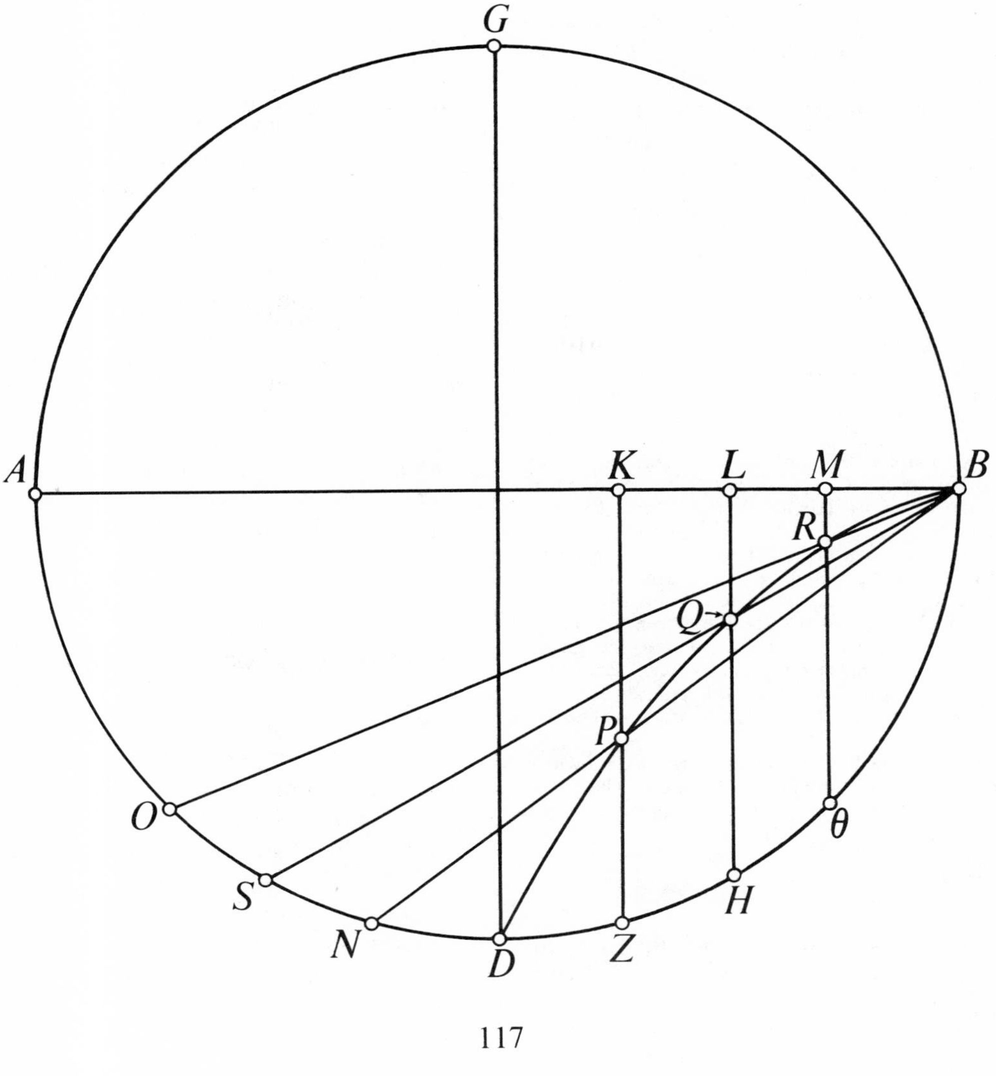

FIGURE 2

there is no mention of Diocles in surviving Greek mathematical works until late antiquity. In the sixth century his work was used by Eutocius and, about the same time, by the unknown author of the "Bobbio Mathematical Fragment."[3] It seems likely that Diocles had a considerable indirect influence on medieval discussions of the parabolic burning mirror. There is only one known explicit reference to his work in Islamic literature.[4] It is, nevertheless, very probable that it is one of the sources of Ibn al-Haytham's *On Paraboloidal Burning Mirrors*[5], which was well-known not only in the Islamic world but also in the West after its translation into Latin. Diocles was known by name in the West, however, only through the extracts in Eutocius, whose commentary on Archimedes attracted the attention of mathematicians of the late sixteenth and seventeenth centuries particularly for its discussion of curves used by the ancient geometers to solve the problem of doubling the cube. Among these was Diocles' curve (see Figure 2 above), part of the discussion on which had been excerpted by Eutocius. This curve was dubbed "cissoid" in the seventeenth century.[6] It was discussed by some of the most notable mathematicians of that time, including Fermat, Descartes, Roberval, Huygens, and Newton. To them we owe the generalization of the curve, the discovery of its infinite branch, and the revelation of many of its beautiful properties.

NOTES

1. The extension to all three conic sections is found in Pappus, VII, 312–318, Hultsch ed., II, 1004–1014. It was probably made in the later Hellenistic period. The argument that it was known as early as Euclid cannot be sustained: see Toomer ed. of Diocles, 17.
2. The problem of doubling the cube had been reduced to finding two mean proportionals between two lines, one of which was double the other, by Hippocrates of Chios (late fifth century B.C.)
3. The author mentions a work "On the Burning Mirror," which he attributes to Apollonius. For arguments that this is in fact Diocles' work, see Toomer ed. of Diocles, 20–21.
4. In the encyclopedic work by al-Akfānī (fourteenth century), Sprenger ed., 82. All other references known to me are derived from Eutocius, whose commentary on Archimedes' *Sphere and Cylinder* was also translated into Arabic.
5. For arguments in favor of this see Toomer ed. of Diocles, 22.
6. Because it was identified with a class of curves known as κισσοειδής ("ivy-shaped") from ancient sources. The identification is almost certainly wrong, as I have argued in my ed. of Diocles, 24.

BIBLIOGRAPHY

The Arabic trans. of Diocles' *On Burning Mirrors* is edited by G. J. Toomer, with English trans. and commentary (New York, 1976), as Sources in the History of Mathematics and the Physical Sciences, no. 1. This includes the extracts by Eutocius, which are found in his commentary on Archimedes' *Sphere and Cylinder*, bk. II: *Archimedis Opera Omnia*, J. L. Heiberg, ed., 2nd ed., III (Leipzig, 1915), 66–70, 82–84, 160–174. My intro. to the ed. (1–17) gives a full discussion of the evidence for the date of Diocles and his place in the history of the theory of conics. On the date of Zenodorus, see G. J. Toomer, "The Mathematician Zenodorus," in *Greek, Roman and Byzantine Studies*, **13** (1972), 177–192. Pappus' discussion of the focus-directrix properties of the three conic sections is in *Pappi Alexandrini Collectionis quae supersunt*, F. Hultsch, ed., 3 vols. (Berlin 1875–1878), II, 1004–1014. On the solutions to Archimedes' problem, see T. L. Heath, *A History of Greek Mathematics*, II (Oxford, 1921), 45–49. On the history of the problem of doubling the cube, see *ibid.*, I, 244–270. The "Bobbio Mathematical Fragment" is edited by J. L. Heiberg in his *Mathematici graeci minores*, which is Kongelige Danske Videnskabernes Selskab, Historisk-filologiske Meddelelser, **13**, no. 3 (Copenhagen, 1927), 87–92.

The work of al-Akfānī is edited by A. Sprenger, *Two Works on Arabic Bibliography*, which is Bibliotheca Indica, VI (Calcutta, 1849), 14–99. The passage referring to Diocles is translated by Eilhard Wiedemann in *Aufsätze zur Arabischen Wissenschaftsgeschichte*, **I** (Hildesheim, 1970), 119–120. Ibn al-Haytham's work on the parabolic burning mirror is printed as the third treatise in his *Majmūʿ al-Rasāʾil* (Hyderabad, 1938). A German trans. from the Arabic, together with the medieval Latin trans., was published by J. L. Heiberg and E. Wiedemann, "Ibn al-Haiṯams Schrift über parabolische Hohlspiegel," in *Bibliotheca mathematica*, 3rd ser., **10** (1910), 201–237. For a modern mathematical treatment of the cissoid and an account of the discoveries of its properties by seventeenth-century mathematicians, see F. Gomes Teixeira, *Traité des courbes spéciales remarquables planes et gauches*, **I**, which is his *Obras sobre mathematica*, IV (Coimbra, 1908), 1–26; and Gino Loria, *Spezielle algebraische und transzendente ebene Kurven*, 2nd ed., I (Leipzig–Berlin, 1910), 36–51.

G. J. TOOMER

DIOPHANTUS OF ALEXANDRIA (*fl.* A.D. 250), *mathematics.*

Since the publication of Kurt Vogel's article "Diophantus of Alexandria" in volume IV of the *Dictionary*, our knowledge of the *Arithmetica* has been increased significantly by the discovery of four new books in Arabic translation.

The manuscript, which seems to be a unicum, is Codex 295 of the Shrine Library in Meshed. It contains eighty leaves, numbered in recent times as pages; and each page contains twenty lines of

DIOPHANTUS OF ALEXANDRIA

text. The manuscript can be dated, for it is indicated at the end that it was completed on Friday 3 Safar 595 A.H. (4 December 1198). The translation itself is attributed to Qusṭā ibn Lūqā and is no doubt the one alluded to by Ibn al-Nadīm in his *Fihrist*.[1] Unlike the Greek text that we know, the Arabic version is entirely rhetorical and extremely prolix. Thus it is difficult to infer the form of the Greek text lying behind the translation; it appears that the translator was responsible for the general prolixity. This verbosity is exemplified by extraneous computations designed to determine whether the initial conditions of the problems are met, and by detailed analyses of the problems. These additions, completely absent from the Greek text that we know, justify the title "commentary" (*tafsīr*) given to Qusṭā's translation by Ibn al-Nadīm.

The Arabic translation contains four books (*maqālāt* = βιβλία), labeled IV to VII. There is indeed evidence suggesting that the Arabic books were intended to follow the Greek books I–III; in the Arabic books all the methods or results needed, whether explicitly cited or not, are found in the first three (Greek) books, while methods used in the last three Greek books (among them the use of the second-degree equation) are totally absent from the Arabic books. In early Islamic time, the order of the books is confirmed by the order of the problems considered in al-Karajī's *Fakhrī*: problems borrowed from books I–III are immediately followed by problems from (Arabic) book IV. Indeed, a marginal gloss in the Paris manuscript of the *Fakhrī* even states indirectly that the order is (Greek) book III–(Arabic) book IV.[2] This shows that (Greek) books IV–VI, although their content must be drawn from the original work of Diophantus, are the result of a later recension that changed the numbering, and perhaps the order of contents, of Diophantus' work.

Book IV. Book IV begins with an introduction stating that, subsequent to the consideration of problems involving linear numbers (x, y) and plane numbers (x^2, y^2, xy) as found in books I–III, the problems to come will involve solid numbers and their associations with numbers of the two preceding powers. The problems, graduated in difficulty, aim at "experience and practice"; this suggests that one objective of this part of the *Arithmetica* was to extend and consolidate previously acquired knowledge. Next the author defines the names of the third and higher powers (up to the sixth); incidentally, this indicates that the similar part of the Greek introduction is a later interpolation. These definitions are extended in problem IV.29 with the introduction of the eighth and ninth powers. Finally, Diophantus, as in the Greek introduction,[3] reminds the reader that a problem must be brought to an equality between a power of the unknown (*naū* = εἶδος) and a number (hence, no equation of the second degree will occur in the Arabic books).

The forty-four problems of book IV cover seventy-three pages of the manuscript; most of them have been appropriated by al-Karajī in the last section of the *Fakhrī*. They are presented in order of increasing difficulty, but one can easily recognize well-ordered and interdependent groups of problems (in contrast with the extant Greek text), of which typical representatives are the following.

(IV.1) $b^3 + a^3 = \square$.

Taking $a = x$, $b = mx$, we have $(m^3+1)x^3 = \square$.

$$\text{With } \square = (nx)^2\text{: } x = \frac{n^2}{m^3+1}.$$

Diophantus chooses $m = 2$, $n = 6$, and obtains $x = 4$.

(IV.6) $b^3 \cdot a^2 = \square$.

If $a = x$, $b = mx$, then $m^3x^5 = \square$.

If we assume $\square = (nx)^2$, m and n will have to underlie a condition for the rationality of the solution. Thus we take $\square = (nx^2)^2$, and $x = n^2/m^3$. For $m = 2$, $n = 4$: $x = 2$.

(IV.20) $\begin{cases} k \cdot a^3 = \square \\ l \cdot a^3 = \sqrt{\square}. \end{cases}$

Here $k = 200$, $l = 5$ are given numbers that must obey the condition $k/l^2 =$ a cube.

If $a = x$, we have $kx^3 = (lx^3)^2 = l^2x^6$. Hence

$$x = \sqrt[3]{\frac{k}{l^2}} = 2.$$

(Actually, this problem had been solved in I.26.)

(IV.31) $(b^2)^2 - (a^3)^3 = \square$.

If $a = x$, $b = mx^2$, then $m^4x^8 - x^9 = \square$.

Assuming $\square = (nx^4)^2$, one obtains $m^4x^8 - x^9 = n^2x^8$, or $x = m^4 - n^2$.

For $m = 2$, $n = 2$ $(m^2 > n)$: $x = 12$.

Here, as in many problems of the Arabic books, the high powers lead to large numbers $(b^2)^2$, $(a^3)^3$ as results. This is one of the salient characteristics of the Arabic books.

(IV.34) IV.34 to IV.44 are systems of two equations.

$$\begin{cases} a^3 + b^2 = \square \\ a^3 - b^2 = \square'. \end{cases}$$

If $a = x$, $b = 2x$, the system becomes

$$\begin{cases} x^3 + 4x^2 = \square \\ x^3 - 4x^2 = \square'. \end{cases}$$

DIOPHANTUS OF ALEXANDRIA

Diophantus gives two methods of solution. One is the method of the double equation (II.11), based on the identity

$$\left\{\frac{1}{2}\left(\frac{u}{v}+v\right)\right\}^2-\left\{\frac{1}{2}\left(\frac{u}{v}-v\right)\right\}^2\equiv u.$$

If $\square-\square'=8x^2$ replaces u, and $2x$ replaces v, then

$$\square=\left\{\frac{1}{2}\left(\frac{8x^2}{2x}+2x\right)\right\}^2 \text{ and } \square'=\left\{\frac{1}{2}\left(\frac{8x^2}{2x}-2x\right)\right\}^2$$

will both lead to $x^3=5x^2$, or $x=5$.

We can also avoid the method of the double equation:

With $\square=(mx)^2$, the first equation will yield $x^3+4x^2=m^2x^2$, or $x=m^2-4$.

With $\square'=(nx)^2$ $(n<m)$, the second equation will yield $x^3-4x^2=n^2x^2$, or $x=n^2+4$.

Hence $m^2-n^2=8$. How to find two square numbers having a given difference has been shown in II.10. Thus we set $m=n+h$, and choose a suitable value of the parameter h. $h=1$ leads to $x=16\frac{1}{4}$ (while $h=2$ would give the solution found by the method of the double equation).

(IV.43) $\begin{cases}(a^3)^3+k\cdot(b^2)^2=\square\\(a^3)^3-l\cdot(b^2)^2=\square'.\end{cases}$

$k=1\frac{1}{4}$ and $l=\frac{3}{4}$ are given numbers.

If $a=x$, $b=2x^2$, the system becomes

$$\begin{cases}x^9+20x^8=\square\\x^9-12x^8=\square'.\end{cases}$$

Setting $\square=(mx^4)^2$, $\square'=(nx^4)^2$ $(m>n)$, we obtain $x=m^2-20=n^2+12$, or $m^2-n^2=32$, of which a solution, found by II.10, is $m^2=36$, $n^2=4$. Hence $x=16$.

Book V. Book V contains sixteen problems, covering pages 73–97 of the manuscript. The existence of three groups of problems (V.1–6, V.7–12, V. 13–16) is easily recognized.

(V.1) $\begin{cases}(b^2)^2+k\cdot a^3=\square & k=4\\(b^2)^2-l\cdot a^3=\square' & l=3.\end{cases}$

If $b=x$, the system becomes

$$\begin{cases}x^4+4a^3=\square\\x^4-3a^3=\square'.\end{cases}$$

Taking $a^3=r\cdot x^4$ and $\square=m^2x^4$, $\square'=n^2x^4$ $(m>n)$, we obtain

$$\begin{cases}x^4(1+4r)=m^2x^4\\x^4(1-3r)=n^2x^4.\end{cases}$$

Then $r=\dfrac{m^2-1}{4}=\dfrac{1-n^2}{3}$ or $\dfrac{m^2-1}{1-n^2}=\dfrac{4}{3}$.

How to solve the general problem

$$\frac{m_1^2-p_1^2}{p_1^2-n_1^2}=\frac{k}{l}$$

is known from II.19. A solution for $k=4$, $l=3$ is $m_1^2=81$, $p_1^2=49$, $n_1^2=25$. The norm $p^2=1$ gives the particular solution of our problem:

$$m^2=\frac{81}{49},\ n^2=\frac{25}{49}.$$

Hence $r=\dfrac{8}{49}$ and $a^3=\dfrac{8}{49}x^4$.

We may now set $a=mx$—for instance, $a=2x$; then $x=49$.

(V.8) $\begin{aligned}a-b&=k\\a^3-b^3&=l.\end{aligned}$

k, l are given numbers, fulfilling the condition

$$\frac{4l-k^3}{3k}=\text{a square.}$$

Such a pair is $k=10$, $l=2{,}170$.

Write $a+b=2x$, so that $a=x+\dfrac{k}{2}$, $b=x-\dfrac{k}{2}$ (see I.27 ff.).

Hence $a^3-b^3=l=\dfrac{k^3}{4}+3kx^2$, and $x=\dfrac{1}{2}\sqrt{\dfrac{4l-k^3}{3k}}$.

With $k=10$, $l=2{,}170$: $x=8$.

V.8 and V.7 are found in the extant Greek text as the first two problems of "book IV," without the given condition and with different parameters. These certainly found their way from some commentary into the Greek text; for such additions, when introduced, tend to be located at the beginning or at the end of books (see II.1–5, derived from I.31–34; or III.20–21, derived from II.15–14 respectively). One should further note that in the group of problems V.7–12, Diophantus does not state explicitly the condition of positivity for b, which in the present case would amount to $l>k^3$.

(V.15) $\begin{cases}k\cdot a^2-l=m+n & k=9,\ l=18,\text{ given numbers.}\\a^3+m=⧉\\a^3-n=⧉'\end{cases}$

Write $a=x$ and take $⧉=(x+p)^3$, $⧉'=(x-q)^3$, $(q>p$; see below), so that $m=(x+p)^3-x^3$ and $n=x^3-(x-q)^3$.

Then $kx^2-l=m+n=3x^2(p+q)-3x(q^2-p^2)+(p^3+q^3)$.

One way of obtaining an equality between two terms is to eliminate the multiples of x^2 by setting the condition $k/3=p+q$.

Then $x=\dfrac{l+p^3+q^3}{3(q^2-p^2)}$.

DIOPHANTUS OF ALEXANDRIA

With $k=9$, $l=18$, and choosing $p=1$, $q=2$, we obtain $x=3$.

Book VI. Book VI runs from page 97 to page 130 of the manuscript and contains twenty-three problems of various kinds—sometimes complementary to those of the preceding Arabic books, sometimes of a new type. The last (from VI.17 on) are solved by ad hoc methods, without aiming at generality of the solution.

(VI.1) $(a^3)^2 + (b^2)^2 = \square$.

This problem has already been solved in IV.25 by putting $a = x$ and $b = mx$; but Diophantus here imposes the condition $a = mb$ (m integral). Taking $m = 2$ and $b = x$, we have $64x^6 + x^4 = \square\ (= [nx^3]^2)$.

Hence $x^4 = (n^2 - 64)x^6$, or $n^2 - 64 =$ a square $= p^2$, say, so that $n^2 - p^2 = 64$, the solution of which is found by means of II.10. Then

$$p^2 = 36 \text{ and } x = \frac{1}{p} = \frac{1}{6}.$$

(VI.12) $$\begin{cases} a^2 + \dfrac{a^2}{b^2} = \square \\ b^2 + \dfrac{a^2}{b_2} = \square' \end{cases} \quad \text{with } a > b.$$

Write $b = x$. Assuming $\dfrac{a^2}{b^2} = \dfrac{9}{16}x^2$, the second equation will yield $x^2 + \dfrac{9}{16}x^2 = \dfrac{25}{16}x^2$ and will be satisfied. Since $a^2 = \dfrac{9}{16}x^4$, the first equation yields

$$\frac{9}{16}x^4 + \frac{9}{16}x^2 = \square.$$

Taking $\square = m^2x^4 : m^2 - \dfrac{9}{16} = \dfrac{9}{16}\dfrac{1}{x^2} \equiv p^2$, or $m^2 - p^2 = \dfrac{9}{16}$, to be solved by II.10.

Since $a^2 > x^2$, $\dfrac{9}{16}x^2 > 1$, or $\dfrac{16}{9} \cdot \dfrac{1}{x^2} < 1$, and therefore $\dfrac{9}{16}\dfrac{1}{x^2} < \dfrac{9}{16}\dfrac{9}{16}$, we have the condition $p^2 < \dfrac{81}{256}$. (On account of an incidence of a homoeoteleuton, the condition in the text has been inappropriately reduced to $p^2 < 1$.)

Taking $m = p + h$ and choosing the acceptable value $h = \dfrac{1}{2}$, we obtain $p^2 = \dfrac{25}{256}$ and $x = \dfrac{12}{5}$.

(VI.19) $a^2 \cdot b^2 \cdot c^2 - (a^2 + b^2 + c^2) = \square$

We choose $a^2 = 1$; then $b^2c^2 - 1 - b^2 - c^2 = \square$, or $c^2(b^2 - 1) - (b^2 + 1) = \square$. In order to equate this to a square, we can make $b^2 - 1 =$ a square, taking for instance, $b^2 = \dfrac{25}{16}$. With $c^2 = x^2$ as unknown: $\dfrac{9}{16}x^2 - \dfrac{41}{16} = \square = \left(\dfrac{3}{4}x - \dfrac{1}{4}\right)^2$, say.

Then $\dfrac{6}{16}x = \dfrac{42}{16}$ and $x = 7$.

(VI.23) $$\begin{cases} \dfrac{k^2}{a^2} + \dfrac{k^2}{b^2} = \square \\ a^2 + b^2 + k^2 = \square' \end{cases} \qquad k^2 = 9, \text{ given number}$$

Lemma: If $u^2 = u_1^2 + u_2^2$, then $\dfrac{c^2}{u_1^2} + \dfrac{c^2}{u_2^2} =$ a square for any c^2.

(For we have $\dfrac{c^2}{u_1^2} + \dfrac{c^2}{u_2^2} = \dfrac{c^2(u_1^2 + u_2^2)}{u_1^2u_2^2} = \dfrac{c^2u^2}{u_1^2u_2^2} =$ a square.)

Thus the first equation will be identically satisfied if we write $a^2 = \dfrac{16}{25}x^2$, $b^2 = \dfrac{9}{25}x^2$.

The second equation yields $x^2 + 9 = \square' = (x + 1)^2$, say. Hence $x = 4$.

Book VII. Book VII (pp. 130–159) begins with a small introduction which states that the (eighteen) problems in it will not depart from the type seen in books IV and V, although being of a different kind; indeed, one solves them mostly with ad hoc methods, and the aim of book VII is to provide deeper insight and practice in the technique of mathematical devices.

(VII.2) This problem illustrates best the very particular way of getting a solution.

$$(a^2)^3 \cdot (b^2)^3 \cdot (c^2)^3 = \square^2$$

Since $64 = (2^2)^3$, we put $(a^2)^3 = \dfrac{1}{64}$, $(b^2)^3 = 64$, so that there will remain, with $c = x$:

$$x^6 = \square^2, \text{ or } x^3 = \square.$$

Taking $\square = (2x)^2$, we obtain $x = 4$.

(VII.7) $$\begin{cases} (a^3)^2 = a_1 + a_2 + a_3 \\ a_1 + a_2 = \square \\ a_1 + a_3 = \square' \\ a_2 + a_3 = \square''. \end{cases}$$

Diophantus gives two methods of solution. In the first we take $a = x$ as unknown, so $x^6 = a_1 + a_2 + a_3$. If $a_i = a_i' \cdot x^4$, the problem will be reduced to that of finding three numbers a_1', a_2', a_3' such that their sum is a square (namely, x^2) and such that the sum of any two is a square.

But, as stated in the text, this has already been solved in III.6, with the solution

$$a_1' = 80, a_2' = 320, a_3' = 41.$$

Hence $x^2 = 441$ and $x = 21$.

Second, we write $(a^3)^2 = 64$. Again using the result of III.6, we multiply each part by $\frac{64}{441}$, which will give a solution to our problem.

This second method, imposing a value a priori, often permits the avoidance of large numbers in the solutions. Diophantus uses it in the problems immediately following VII.7.

$$\text{(VII.13)}\begin{cases} a^2 = a_1 + a_2 + a_3 & a^2 = 25, \\ a^2 + a_1 = \square & \text{given number} \\ a^2 + a_2 = \square' \\ a^2 + a_3 = \square''. \end{cases}$$

Adding the last three equations, we obtain

$$3a^2 + a_1 + a_2 + a_3 = 100 = \square + \square' + \square''.$$

Thus we are led to the division of 100 into three squares, each of which is larger than 25. We choose $\square = 36$ and apply II.8 to the remainder 64, a procedure that yields

$$\square' = 33\frac{471}{841}, \square'' = 30\frac{370}{841}.$$

These three numbers, when decreased by 25, give the magnitudes for which we are looking.

$$\text{(VII.15)}\begin{cases} a^2 = a_1 + a_2 + a_3 + a_4 & a^2 = 25, \\ a^2 + a_1 = \square & \text{given number} \\ a^2 + a_2 = \square' \\ a^2 - a_3 = \square'' \\ a_2 - a_4 = \square'''. \end{cases}$$

We search for some square $u^2 = x^2$, $u^2 = u_1 + u_2 + u_3 + u_4$, fulfilling conditions similar to those required for a^2, but with $u^2 \neq 25$.

The four latter conditions ($u^2 + u_1 = \square_1$, and so on) will be satisfied identically if we take

$u_1 = 2x + 1$, $u_2 = 4x + 4$, $u_3 = 2x - 1$, $u_4 = 4x - 4$.

Then the first equation will give

$$u^2 = x^2 = 12x; \text{ hence } x = 12.$$

Now, since $a^2 : u^2 = 25 : 144$, the parts of a^2 will be $a_i = \frac{25}{144} u_i$.

Then the text remarks that the same method can be used to solve the system

$$\begin{cases} a^2 = \sum_{k=1}^{8} a_k \\ a^2 + a_i = \square_i & i = 1, \cdots, 4 \\ a^2 - a_j = \square_j & j = 5, \cdots, 8. \end{cases}$$

Indeed, the method is generally valid for an even number $2n$ of parts, of which n are additive and n subtractive.

NOTES

1. See K. Vogel, "Diophantus," *DSB*, IV, 117; Ibn al-Nadīm, however, speaks of the translation of three and a half books.
2. See F. Woepcke, *Extrait du Fakhrî*. The gloss in question is in MS Paris BN Arabe 2959, on the margin of fol. 98r, on which begins the fifth section of the *Fakhrî*; it states that part of the problems of the fourth section and the whole of the fifth section "are taken from the books of Diophantus, in the order." (Woepcke took the gloss to refer to the third and fourth sections; see *Extrait du Fakhrî*, 22.)
3. See Tannery, *Diophanti Opera*, I, 14 (def. XI).

BIBLIOGRAPHY

See J. Sesiano, "The Arabic Text of Books IV to VII of Diophantus' Ἀριθμητικά in the Translation of Qusṭā ibn Lūqā," ed., with trans. and commentary (Ph.D. diss., Brown University, 1975); or Springer-Verlag edition in the Sources in the History of Mathematics and Physical Sciences series (Berlin, 1977). A very summary outline of the problems is R. Rashed, "Les travaux perdus de Diophante," in *Revue d'histoire des sciences*, **27** (1974), 97–122 (bk. IV), and **28** (1975), 3–30 (bks. V–VII). An Arabic version of this summary, together with the Arabic text, has been published by R. Rashed (Cairo, 1977). See my review of it in *Isis*, **68**, no. 244 (1977). See also F. Woepcke, *Extrait du Fakhrî, traité d'algèbre* . . . (Paris, 1953). The bibliography given by K. Vogel must be completed by a Russian trans. with commentary on the Greek books: Diofant Aleksandrysky, *Arifmetika i kniga o mnogougolnykh chislakh*, translated by I. N. Veselovsky, commentary by I. G. Bashmakova (Moscow, 1974); also translated by L. Boll, as *Diophant und diophantische Gleichungen* (Stuttgart, 1975). Forty-four problems of book IV are discussed by E. S. Stamatis, in *Platon*, **28** (1976), 121–133.

JACQUES SESIANO

DOHRN, FELIX ANTON (*b.* Stettin, Germany [now Szczecin, Poland], 29 December 1840; *d.* Munich, Germany, 26 September 1909), *zoology*.

Dohrn, named Felix Anton for his godfather Felix Mendelssohn-Bartholdy but always called Anton, was the third of three sons born to Carl August and Adelheid Dietrich Dohrn; an older sister, Anna, was born before the parents were married. The Dohrn family was prosperous financially, talented musically, and socially and intellectually prominent in Stettin and in Germany more generally. Yet its members were highly individualistic rather than establishmentarian, and for a number of generations sons rebelled against their fathers. Anton's paternal grandfather, Heinrich Dohrn, trained as a surgeon, established a sugar refinery that proved durable and lucrative. Anton's father, a natural son later legitimized, first reacted against family traditions but later became interested in natural history through Alexander von Humboldt, a family friend. He became an entomologist of distinction and editor of the influential *Stettiner entomologische Zeitschrift*. Anton was not the only one of his sons to become a naturalist; the oldest son, Heinrich, for some time followed his father's career of entomology but eventually became a member of the Reichstag; the other son, Wilhelm, became a landowner.

Dohrn became attracted to entomology through his father and published his first article at the age of sixteen. He attended, as was then customary in Germany, a number of universities, including Königsberg, Bonn, Berlin, and Jena, where he studied with Gegenbaur and Haeckel. His studies were, however, intermittent, interrupted by military service and by extensive travels, and also as a result of his own self-doubts. The zoology that he learned in the 1860's failed to excite him, and at one time he decided to become a bookdealer. But Darwin's *On the Origin of Species* rekindled his interest, and he was habilitated in Jena in 1868.

Dohrn's dissertation dealt with the anatomy of hemipterans, but after its completion he became more interested in crustaceans than in insects. In his zoological studies he carried out combined morphological and embryological investigations in an attempt to elucidate the phylogeny of arthropods. He then began to study homologies between arthropods and vertebrates, especially with regard to head and appendages; and this led him to speculation on the origin of vertebrates. He ultimately attempted to show that they had been derived from annelids.

But it was not through his scientific investigations that Dohrn left his greatest mark. His primary contribution was the establishment of the Zoological Station in Naples; it was not only the first laboratory set up specifically for marine studies but also the first institute formally organized for the sole pursuit of research and the prototype of those that followed. Once Dohrn was habilitated, the thought of narrow academic life did not appeal to him; he also did not wish to remain in Jena near Gegenbaur and Haeckel, the fixity of whose ideas now was repellent to him. In the autumn of 1868 he visited Messina, Sicily, where he took some portable aquariums; by using them he was able to follow continuous embryological and larval transformations during crustacean development that had never before been observed. This was a decisive factor in the development of his idea of establishing a research laboratory for the study of marine organisms. In 1865 Dohrn had studied marine animals in Helgoland with Haeckel and others; and from that time on, he was aware of the importance of their investigation for the new comparative physiology and embryology, and of the necessity for organized laboratories in which to study them. Public exhibits in aquariums were beginning to be set up in several European cities, including London, Hamburg, and Berlin; and techniques were being developed for keeping aquatic organisms alive in such facilities.

Dohrn began to formulate the plans for his new venture during the winter of 1870. In the autumn of 1871 he settled permanently in Naples. He maintained his permanent residence there until he died. In February 1874 the Zoological Station was opened. The years between 1870 and 1874 were highly turbulent for Dohrn. Obtaining the land on which the station was to be built from the municipality of Naples, in the Villa Reale, the most beautiful park in the city, in order to establish what then seemed to be a German institution, presented only one of many problems. Success depended in part on the beauty of the sketches, made by Dohrn himself, for the proposed building. Securing the permission to start construction, then to continue it after the height of the building was found to have been miscalculated because of an architect's surveying error, caused more serious worries. Subject to periodic depressions, Dohrn also had to combat ill health. The most compelling concern was the source of funds to erect, and then to maintain and administer, the station.

Dohrn's father was at first totally unsympathetic to the enterprise, viewing the plan as utopian and hopelessly impractical, and considering investment in it as the wildest of gambles. He ultimately became convinced of its worth, however, after having received letters in support of it from Charles Darwin and Karl Ernst von Baer. No one could write more feelingly than Baer; when he had gone to Trieste in 1846 to attempt to return to the embryology he had abandoned twelve years before, he lost his experimental material when it was thrown away by the maid who cleaned his hotel room, which was the only place he had to keep it. Carl August Dohrn's financial support was the first given to the laboratory. After long, complicated, and difficult negotiations, the next funds received were granted by the Prussian Academy of Sciences.

Earlier in his thinking, Dohrn had the original and imaginative idea that entrance fees paid to see the exhibits in the public aquarium on the ground floor of the building might support the research in the laboratories on the floors above. When it became evident that this was infeasible, the plan for incorporating a public aquarium in the building was nonetheless maintained; effective methods of circulating well-aerated seawater in the tanks were developed, and the aquarium was for many decades one of the finest in the world. In his next plan for research support, Dohrn, mindful that corporations and wealthy private donors guaranteed annual fees for the support of beds in hospitals, for the use of individuals designated by them, conceived the idea that governments or institutions might support work tables in the Zoological Station, to be used by investigators of their choosing. Translated into reality, this brilliant notion not only ensured support for the station but also made the laboratory a truly international venture. While early support was governmental or through academies of science, eventually universities or such agencies as national research councils supported tables. The American Association of University Women for many years supported a table for American women to use.

The tables in the station were not bare; when Theodor Boveri spoke, in 1910, about their establishment, he compared them with the "Table, set yourself" of the fairy tale. Each had a freshwater and a seawater aquarium; well-stocked communal supply rooms for chemicals, glassware, and apparatus were established; and keen-eyed collectors and willing laboratory assistants provided for all needs of the investigators. When the station opened, there were ten tables; by 1910 the number had risen to eighty used at a single time. A library was set up that became one of the richest biological collections in the world. The station published both its own periodical and an extensive series of monographs describing the exceptionally rich flora and fauna of the Gulf of Naples; by 1972 thirty-nine monographs had appeared. Dohrn had originally thought that he might set up several marine stations—a plan he did not carry through—but Naples served as a model for other marine laboratories soon established elsewhere. It was also a model for the many research institutions that became so important for the progress of science in the century to follow. Thus the laboratory became, in the words of E. B. Wilson, "a potent force in the progress of biological science throughout the world."

Progress in the study of morphology, taxonomy, life history, comparative embryology, and many other aspects of marine organisms was given enormous impetus by the availability of invertebrates and vertebrates captured alive, and of facilities to examine and work with them under optimum laboratory conditions. Investigations on the distribution of marine forms that now would be classified under the rubric of ecology were abundant, and important pioneering experimental investigations were carried out at the station. Especially significant were investigations in comparative physiology and experimental embryology, which became the foundations of whole new sciences.

On 3 June 1874, shortly after the station was opened, Dohrn married Marie von Baranowska, whom he had first met in Messina in 1868. They had five children: a daughter, who died in infancy, and four sons: Boguslav, Wolf, Reinhard, and Harald. Boguslav studied physiology and medicine, but ultimately became a property owner. Wolf had literary interests, and both he and Harald had political inclinations; Harald was arrested by Nazi storm troopers and was shot without a trial. Reinhard, like his father, became a zoologist; and after his father's death, which terminated his directorship of the station, Reinhard carried on the administrative responsibilities of the station with energy and imagination.

Music, as well as science and public affairs, was an integral part of the life of the Dohrns. The home of Heinrich Dohrn, Anton's paternal grandfather, was a musical center in Stettin. Anton's parents were both musical as were his brothers. Wilhelm Dohrn's son (Anton's nephew) became music director in Breslau. Anton's sister Anna was the grandmother of the conductor Wilhelm Fürtwangler. Besides Mendelssohn, many other great musicians frequented the Dohrn household: Joseph Joachim, Johann Nepomuk Hummel, and Jenny Lind among them. The Dohrns also numbered others of note among their circle: Eleonora Duse, Fridtjof Nansen, Rudolf Binding, Werner Siemens, and Ernst Abbe. Dohrns had talent for friendship as well as for science and art. Anton's friendship with the sculptor Adolf Hildebrand and the painter Hans von Marées had great consequence for the station: Hildebrand designed the fine façade, and Marées painted in one of the public rooms a group of frescoes that soon became as famous among artists as did the laboratories of the station among zoologists. Music, art, and literature were not peripheral to science in the life of these truly cosmopolitan spirits, but an integral part of it. The success of the station is a monument to international cooperation on the highest planes of combined intellectual and cultural endeavor.

BIBLIOGRAPHY

A complete bibliography of Dohrn's published works (80 items) is in A. Kühn, "Anton Dohrn und die Zoologie seiner Zeit," in *Pubblicazione della Stazione zoologica di Napoli*, supp. (1950). This is an extensive and intensive analysis of Dohrn's scientific work. The definitive biography of Dohrn is Theodor Heuss, *Anton Dohrn* (Berlin–Zurich, 1940; 2nd ed., enl., Stuttgart–Tübingen, 1948). The 2nd ed. reprints a touching essay by Margret Boveri, "Das Haus am Rione Amedeo," describing life in the Dohrn household; it was written after the house had been destroyed by Allied bombs. Its destruction marked the end of an era in European science and culture.

Naturwissenschaften, **28**, no. 51 (1940), commemorated the hundredth anniversary of Dohrn's birth; it includes, among a number of other articles, a reprint of the perceptive memorial address on Dohrn delivered by Theodor Boveri at the International Zoological Congress in Graz, previously published separately by Boveri as *Anton Dohrn. Gedächtnisrede, gehalten auf dem Internationalen Zoologen-Kongress in Graz am 18. August 1910* (Leipzig, 1910). An excellent brief evaluation of Dohrn is E. B. Wilson, "The Memorial to Anton Dohrn," in *Science*, **34** (1911), 632–633.

JANE OPPENHEIMER

DOROTHEUS OF SIDON (*fl.* Egypt[?], first century), *astrology.*

Dorotheus was one of the most influential astrologers of antiquity. His theories form the basis of the astrological treatises of Firmicus Maternus, Hephaestio of Thebes, and Rhetorius of Egypt; and he became one of the chief authorities for Arabic astrologers beginning with Māshā'allāh, ʿUmar ibn al-Farrukhān al-Ṭabarī, and Abū Maʿshar. Through Firmicus and the translations of Arabic texts Dorotheus' doctrines also profoundly influenced the astrology of the later Middle Ages and Renaissance in the Latin West. Unfortunately, we know virtually nothing of his person, although Symon de Phares[1] in the late fifteenth century identified him with the Dorotheus, son of Nathanael, whom Josephus (*Ant. Iud.* 20, 14) mentions as a member of an embassy in A.D. 51; there is nothing to substantiate this legend.

The work of Dorotheus was originally published as five books in hexameters; the first two on the basic reading of a horoscope, the third and fourth on methods of continuous astrology (prorogation and the revolution of the years of nativities), and the fifth on catarchic (concerning initiatives) and interrogational astrology. In these last three books Dorotheus was particularly innovative.

But these Greek hexameters do not survive save in the citations of Byzantine astrologers, in particular Hephaestio of Thebes. We must rather study Dorotheus in an Arabic version made in about 800 by ʿUmar ibn al-Farrukhān al-Ṭabarī on the basis of a lost Pahlavi translation, which Ibn al-Nadīm, following Ibn Nawbakht,[2] dates in the time of Ardashīr I (226–240) and Shāpūr I (240–271). ʿUmar's Pahlavi original, however, represents a redaction of about A.D. 400, when various insertions into the Dorothean texts were made. These insertions include passages from the Pahlavi translation of Vettius Valens' *Anthologies*, definitions of the Indian *navāṃśas* or ecliptic arcs of 1/9 of a sign, and two horoscopic examples, one of which can be dated 20 October 281, the other 26 February 381. The latter is of some significance as it proves that in Sassanian Iran at this time an astronomer would give a date in the era of Diocletian and would employ Ptolemy's table of oblique ascensions. In the first book Dorotheus gives a series of horoscopic examples, which can be dated between 7 B.C. and A.D. 43. These allow us to fix his date and to gain some added insight into the methods of computing planetary longitudes employed in the fifty years after the beginning of our era.

NOTES

1. E. Wickersheimer, ed., *Recueil des plus célèbres astrologues et quelques hommes doctes* (Paris, 1929), pp. 44, 134.
2. D. Pingree, *The Thousands of Abū Maʿshar* (London, 1968), p. 10.

BIBLIOGRAPHY

Dorotheus' work on astrology in ʿUmar's translation survives in two manuscripts: Yeni Cami 784 and Berlin or. oct. 2603; an edition with translation and Greek and Latin fragments by D. Pingree was published at Leipzig in 1976. Aside from this, there exist a large number of texts and opinions falsely attributed to Dorotheus; unfortunately, this was not fully realized by V. Stegemann in his incomplete *Die Fragmente des Dorotheos von Sidon*, 2 pts. (Heidelberg, 1939–1943); Stegemann also did not know of ʿUmar's translation.

DAVID PINGREE

DUMÉRIL, ANDRÉ-MARIE-CONSTANT (*b.* Amiens, France, 1 January 1774; *d.* Paris, France, 14 August 1860), *zoology, herpetology, entomology.*

As a child Duméril showed a lively interest in zoology, collecting insects and improvising lectures for his schoolmates. He wished to become a physician, but as the eighth child of a poor minor official he could not immediately satisfy this desire. Finally, in 1795, with state support, he began his medical studies in Paris, where he became friendly with the enthusiastic young men who were soon to fill the newly reorganized Académie des Sciences: M.-F.-X. Bichat, G. Dupuytren, A. Richerand, É. Geoffroy Saint-Hilaire, Alexandre Brongniart, A.-P. de Candolle, G. Cuvier, and J.-B. Biot.

Duméril's skill in preparing anatomical demonstrations, the clarity of his exposition, and his enthusiasm brought him rapid success. Named *prosecteur* in 1795 and then chief assistant in practical anatomy in 1799, he became professor of medicine in 1801 and retained this post until 1857. From 1812 to 1852 he was *médecin des hôpitaux*, and also in 1812 he inherited the extensive practice of his father-in-law, who was a well-known physician. Methodical and tireless, Duméril combined the careers of physician and naturalist; his rise in the latter area, however, was somewhat slower, but was supported by Cuvier. The two had a powerful common bond—Protestantism. Cuvier, who already had many responsibilities and lacked expert knowledge of human anatomy, arranged for Duméril to become his substitute in 1800. Duméril then became editor of the first two volumes of Cuvier's *Leçons d'anatomie comparée* (1800). In 1803, probably at Cuvier's suggestion, Lacépède, overburdened with official duties, chose Duméril as his substitute in the chair of zoology specializing in reptiles and fishes at the Muséum d'Histoire Naturelle. To fulfill his responsibilities as a professor, Duméril was obliged to compensate quickly for his scanty knowledge of these subjects. Again with the help of Cuvier, Duméril obtained a contract to write *Traité élémentaire d'histoire naturelle* (1804, 1807), the official manual for the *lycées,* to which he added, for university students, *Zoologie analytique . . .* (1806). In both works his methodical approach harmonized well with his gifts as a popularizer.

Cuvier, whose opinions had become very orthodox—following Napoleon's restoration of the official recognition of religion to his own advantage (1804)—was certainly less than enthusiastic when Duméril advocated the vertebral theory of the cranium around 1803 and again in 1805 and indicated in 1812 that he viewed the cyclostomes (for example, the lamprey) as intermediary between the invertebrates and the vertebrates. In a letter to Agassiz, Duméril asserted that despite appearances, he stood opposed to Cuvier in this matter.

Duméril did not become titular professor until 1825, following the death of Lacépède. In 1832 Duméril engaged as *aide-naturaliste* Gabriel Bibron, who shared his ardent interest in natural science. They undertook the publication of the nine-volume *Erpétologie générale,* which had reached its sixth volume at the time of Bibron's premature death in 1848. Duméril then called upon his son, Auguste, to assist him in completing the work.

After studying medicine, Auguste Duméril (1812–1870) entered P. Flourens's laboratory of comparative physiology at the Muséum around 1840. More original and less inclined to classification than his father, Auguste was especially interested in the development of the human fetus, in odors, glandular secretions, and in the chemical substances that alter the internal temperature of higher vertebrates. He conducted an experiment in which he revived a batrachian that he had previously frozen. At the Muséum, Constant Duméril had created the first *ménagerie* of reptiles (*vivarium*). There, in 1865, Auguste observed the transformation of an axolotl (larval salamander) into an ambystoma (a terrestrial salamander); this transformation proved that under certain circumstances a living creature can breed in its larval stage. Duméril's observation gave rise to the notion of neoteny, which is of great importance in general biology.

Still, after 1848, Auguste Duméril's essential scientific activity consisted in helping his father, whom he succeeded as professor at the Muséum in 1857. Following his father's death in 1860, he continued the latter's work, as well as that of Cuvier on the fishes (for example, in 1870 he gave a description of the dipneusts). Volumes VII and IX of the *Erpétologie générale* are largely his work. Altogether, the nine volumes describe 120 species of chelonians, 445 of saurians, 528 of ophidians, and 218 of batrachians, or a total of 1,311 species—more than twice the number contained in Blasius Merrem's classic *Versuch eines Systems der Amphibien* (1820). The necessary sequel to the major works of Lamarck and Latreille on the invertebrates and of Cuvier on the fishes, the *Erpétologie générale* constitutes a crucial stage in the development of descriptive and systematic zoology.

Duméril produced an enormous body of work in

entomology, ranging from hundreds of articles for the *Dictionnaire des sciences naturelles* (1816–1830, 1840), published by Levrault under Cuvier's direction, to the 1,334 pages of his own *Entomologie analytique* (1860). An assessment of this somewhat neglected and essentially classificatory part of his literary production would require much further study.

BIBLIOGRAPHY

I. ORIGINAL WORKS. The bibliography of Duméril's writings (193 works) in the *éloge* in A. Moquin-Tandon, *Discours prononcé par . . . Moquin-Tandon* (Paris, 1861), should be supplemented by the article in *Annales de la Société entomologique de France*, 3rd ser., **8** (1860), 647–662, which lists 82 items, principally on entomology; and by the Royal Society *Catalogue of Scientific Papers*, II, 392–395, with 75 references.

Duméril's works include *Traité élémentaire d'histoire naturelle* (Paris, 1804)—a special printing was presented in the form of questions and answers, and a 2nd ed. (1807) contained illustrations; *Zoologie analytique . . . à l'aide de tableaux synoptiques* (Paris, 1806), German trans. by L. F. Froriep (Weimar, 1806); *Dissertation sur la famille des poissons cyclostomes* (Paris, 1812), a résumé of one of Duméril's *thèses de sciences; Erpétologie générale*, 9 vols. and atlas (Paris, 1835–1854), written with G. Bibron and Auguste Duméril; "Ichtyologie analytique . . . à l'aide de tableaux synoptiques," which is *Mémoires de l'Académie des sciences de l'Institut impérial de France*, **27**, pt. 1 (1856); and "Entomologie analytique," 2 pts., *ibid.*, **31** (1860).

II. SECONDARY LITERATURE. A. Moquin-Tandon's *éloge* (see above), which draws on Duméril's extensive personal papers; and P. Flourens, *Éloge historique de A. M. C. Duméril* (Paris, 1863), which contains severe but carefully disguised criticism; were summarized by J. Guibé, "A.M.-C. Duméril le père de l'erpétologie. . . ," in *Bulletin du Muséum national d'histoire naturelle*, **30**, no. 4 (1958), 329–341. See also De Caraven, *A. M. C. Duméril médecin et son temps* (n.p., n.d.), with genealogical tables, which is well documented concerning Duméril's medical career.

On Duméril's collaborators, see C. Duméril, "Allocution sur la tombe de G. Bibron," in *Revue et magasin de zoologie*, 2nd ser., **1** (1849), 589–592; and P. Gervais, "Discours . . . sur la tombe de . . . A. Duméril," in *Bulletin de Muséum national d'histoire naturelle*, **7** (1871), 15–24, with bibliography of 50 works by A. Duméril.

FRANCK BOURDIER

DYADKOVSKY, IUSTIN EVDOKIMOVICH (*b.* Dyadkovo, Ryazan gubernia, Russia, 12 June 1784; *d.* Pyatigorsk, Russia, 2 August 1841), *medicine, philosophy.*

The son of a sexton, Dyadkovsky received his primary education at the Ryazan religious seminary. In 1809 he entered the Moscow Medical-Surgical Academy, and after graduating in 1812 he was retained to prepare for a scientific career. He served for two years in the Moscow militia and returned in 1814 to the academy, where he was named assistant to the professor of botany and pharmacology. Two years later he was awarded the M.D. degree for a dissertation on the mode of action of medicine in the human body. He was professor at the academy from 1824 and, from 1831, simultaneously professor at Moscow University, at various times teaching botany, pharmacology, general pathology, and therapy. His dismissal from the university in 1835 for atheism terminated his teaching and scientific career.

An outspoken opponent of *Naturphilosophie* and vitalism, Dyadkovsky believed that the movement of matter underlies all natural phenomena; matter is the source of development of everything that exists and alone explains all observed phenomena. At the same time he defended the thesis of the unity of matter and motion, striving to free the theory of matter both from mechanistic concepts and from the speculations of the dynamists, particularly the followers of Schelling. In his opposition to trivial empiricism and rationalism, he held that knowledge must come from an organic combination of experimental research and theoretical generalizations made in accord with materialistic ideas about nature. The organic world, according to Dyadkovsky, evolved from the inorganic through natural transformations under specific conditions. Thus, in 1816—long before scientific recognition of the theory of evolution—he supported the idea of the transformation of species under the influence of food, climate, and way of life.

Following Diderot, Dyadkovsky held that the basis of perception lies in the property of matter to correspond, in one form or another, to influences from the external world; and he considered this property to be the essential quality of matter. Gradually, by means of the increasing complexity of organic matter, there arose the capacities of perception and then of thought, for which he considered the brain to be the sole organ.

According to Dyadkovsky, life is a continuous physicochemical process influenced by the interaction of the organism and its environment. Thus, he challenged the vitalistic denial of the applicability of the laws of physics and chemistry to physiological phenomena, believing that physicochemical study of living bodies and their processes forms the very basis of physiology. In opposition to the idealists, he stated that "experience is the sole source of our knowledge" and argued for the transformation of physiology into an experimental science based on precise methods of research. Transcending simplified mechanistic concepts of the organism, he refuted the absolute distinction between living and nonliving, emphasizing the qualitative difference between living and inorganic bodies.

The essence of Dyadkovsky's work was his attempt to discover the general theoretical bases of medicine, without which, he believed, it would never rise to the level of a science. The most important area of his investigation was his theory of the nervous system and its role in the life of an organism, and he believed that the nervous system and the brain govern every vital activity of man. Especially concerned with mental illness, he considered the necessity of substituting physiological research for abstract speculation about it and saw its cause exclusively in the disturbance of the normal functioning of the nervous system and brain. The origin of this disturbance was, in turn, to be found in external living conditions. The nervous system itself formed the basis of his classification of illnesses (1833). A merely mechanical summary of symptoms, according to Dyadkovsky, does not give any indication of the nature of the illness, which can be understood only by examining the organism as a whole in relation to living conditions and the environment.

Dyadkovsky was popular both as a practicing physician and as a theoretician. His friends and patients included the great Russian writer N. V. Gogol, the philosopher P. Y. Chaadaev, and the historian T. N. Granovsky.

BIBLIOGRAPHY

I. Original Works. Only eight of Dyadkovsky's scientific writings were published during his lifetime. They include *De modo quo agunt medicamenta in corpus humanum* (Moscow, 1816), also in Russian (Moscow, 1845), his doctoral diss.; and *Systema morborum* (Moscow, 1833). His lectures on private therapy were published posthumously by his student K. V. Lebedev, as *Prakticheskaya meditsina* ("Practical Medicine"), 2 vols. (Moscow, 1845–1847). Dyadkovsky's major writings were collected in his *Sochinenia* ("Works"; Moscow, 1954). Many of his MSS are at the Lenin Library, Moscow.

II. Secondary Literature. See A. G. Lushnikov, *I. E. Dyadkovsky i klinika vnutrennikh bolezney pervoy poloviny XIX v.* ("Dyadkovsky and the Clinic of Internal Disease in the First Half of the Nineteenth Century"; Moscow, 1953); S. R. Mikulinsky, *I. E. Dyadkovsky. Mirovozzrenie i obshchebiologicheskie vzglyady* (". . . World View and General Biological Views"; Moscow, 1951), with detailed bibliography; and S. L. Sobol, "I. E. Dyadkovsky—russky materialist-biolog nachala XIX v." (". . . Russian Materialist-Biologist of the Early Nineteenth Century"), in *Trudy Instituta istorii estestvoznaniya. Akademiya nauk SSSR*, **5** (1953), 145–190.

S. R. Mikulinsky

ECKERT, WALLACE JOHN (*b.* Pittsburgh, Pennsylvania, 19 June 1902; *d.* Englewood, New Jersey, 24 August 1971), *celestial mechanics, computation.*

Raised on a dairy farm in Albion, Pennsylvania, Eckert was the second of four sons born to John Eckert and Anna Heil. He received his A.B. from Oberlin College in 1925, his M.A. from Amherst in 1926, and his Ph.D. in astronomy from Yale in 1931. He joined the department of astronomy at Columbia University as assistant instructor in 1926 and rose through the ranks to become professor of celestial mechanics.

Eckert was familiar with Comrie's adaptation of commercial machines to perform scientific computation, and at Columbia he applied this technique to astronomical calculations. In 1928, using punched-card equipment donated by Thomas J. Watson, Sr., of IBM, Ben Wood founded the Columbia University Statistical Bureau. From 1929 to 1933 Eckert used Wood's facility for the interpolation and reduction of astronomical data, and the numerical solution of planetary equations. In 1933, with the encouragement of Wood and the philanthropy of Watson, Eckert established the T. J. Watson Astronomical Computing Bureau, which was operated as a joint effort of Columbia University, the American Astronomical Society, and IBM. His equipment consisted of an IBM 601 multiplying punch, a credit-balancing accounting machine, and a summary punch controlled by a pluggable relay box taken from Wood's statistical tabulator. This gave him the capability of mechanical reading, writing, and arithmetic for scientific computation.

From 1940 to 1945 Eckert was the director of the U.S. Nautical Almanac Office at the Naval Observatory in Washington, D.C. It was there that he introduced machine methods in the production of the observatory's *American Ephemeris and Nautical Almanac*. In 1940, as director of the U.S. Nautical Almanac Office, he designed and developed the *American Air Almanac*, which began continuous publication in 1941 and proved to be a vital navigational aid during World War II.

Watson invited Eckert to join IBM in 1944 as director of a newly created department of pure science. Eckert proposed that the goals of the department could best be accomplished by the establishment of a research center at Columbia University. Watson agreed, and in March 1945 Eckert returned to Columbia as director of the Watson Scientific Computing Laboratory. In the same year Eckert began working on the logical design of a large-scale general-purpose computer. Under the engineering supervision of Frank Hamilton, IBM's Selective Sequence Electronic Computer (SSEC) was completed and dedicated in January 1948. Using the SSEC, Eckert, Dirk Brouwer (Yale), and G. M. Clemence (U.S. Naval Observatory) computed precise positions of Jupiter, Saturn, Uranus, and Pluto at forty-day intervals over a span of four centuries (1653–2060). In the early 1950's Eckert supervised the design and construction by IBM of the Naval Ordnance Research Calculator (NORC), which was the world's most powerful electronic computer when completed in 1954.

Eckert made some of his most significant contributions in predictive data and theory related to the orbital motion of the moon. In the first two decades of this century E. W. Brown of Yale had laboriously computed the predicted coordinates of the position of the moon. He subsequently published a lunar ephemeris and developed a basic theory and procedure for predictive computation. At the Columbia Computing Laboratory, Eckert used his machine capability to check Brown's tables and found them to be error-free. Later, first on the SSEC and then on the NORC, Eckert returned to Brown's theory to improve the predictive accuracy of data concerning the motion of the moon. This work, which revived Brown's theory and resulted in significantly improved lunar data, provided the basis for the orbital calculations of various NASA moon programs.

Eckert retired from IBM in 1967 and from Columbia University in 1970. He was awarded an honorary doctorate of science by Oberlin College (1968), and in 1968–1969 he was a visiting professor of astronomy at Yale University. Other honors included the James Craig Watson Medal of the National Academy of Sciences (1966) and appointment as an IBM fellow (1967).

BIBLIOGRAPHY

I. Original Works. Eckert's publications include "The Computation of Special Perturbations by the Punched Card Method," in *Astronomical Journal*, **44** (1935), 177–184; "The Astronomical Hollerith-Computing Bureau," in *Publications of the Astronomical Society of the Pacific*, **49** (1937), 249–253; *Punched Card Methods in Scientific Computation* (New York, 1940); "Air Almanacs," in *Sky and Telescope*, **4** (1944), 5–8; "The Printing of Mathematical Tables," in *Mathematical Tables and Other Aids to Computation*, **2** (1947), 197–202, written with R. F. Haupt; "Punched Card Techniques and Their Application to Scientific Problems," in *Journal of Chemical Education*, **24** (1947), 54–58; and "The IBM Selective Sequence Electronic Calculator," *IBM Publications No. 52-3927-2* (1948).

Other works include *Coordinates of the Five Outer Planets, 1653–2060*, vol. XII of *Astronomical Papers Prepared for the Use of the American Ephemeris and Nautical Almanac* (Washington, D.C., 1951), written with D. Brouwer and G. M. Clemence; *Faster, Faster* (New York, 1955), written with Rebecca Jones; "Computing in Astronomy," in Preston C. Hammer, ed., *The Computing Laboratory in the University* (Madison, Wis., 1957), 43–50; "Planetary Motions and the Electronic Calculator," in Harlow Shapley, ed., *Source Book in Astronomy 1900–1950* (Cambridge, Mass., 1960), 93–102, written with G. M. Clemence and D. Brouwer; "On the Motions of the Perigee and Node and the Distribution of Mass in the Moon," in *Astronomical Journal*, **70** (1965), 787–792; "Transformations of the Lunar Coordinates and Orbital Parameters," *ibid.*, **71** (1966), 314–332, written with M. J. Walker and Dorothy Eckert; "The Literal Solution of the Main Problem of the Lunar Theory," *ibid.*, **72** (1967), 1299–1308, written with D. Eckert; "The Motion of the Moon" (IBM Watson Laboratory, Columbia University, N.Y., 1967), a fourteen-page MS; and *The Solution of the Main Problem of the Lunar Theory by the Method of Airy* (Washington, D.C., 1970), written with H. F. Smith, Jr.

II. Secondary Literature. Jean Ford Brennan, *The IBM Watson Laboratory of Columbia University: A History* (Armonk, N.Y., 1971), contains considerable information on Eckert's accomplishments from 1928 to 1970. See also Eleanor Krawitz, "The Watson Scientific Computing Laboratory," in *Columbia Engineering Quarterly* (Nov. 1949); Paul Herget, "The Minor Planet Center at the Cincinnati Observatory," in *Bulletin of the Cincinnati Historical Society*, **24** (1966), 175–187, which discusses Eckert's contributions; "Wallace Eckert," a six-page unpublished document (signed by Martin C. Gutzwiller, Watson Laboratory, 18 Sept. 1969), which discusses Eckert's pioneering role in scientific calculations; "Most Precise Computation of Moon's Orbit Completed," a fourteen-page IBM press release dated 14 Apr. 1965; Thomas Watson, Sr.'s speech at the NORC dedication (2 Dec. 1954), 7–11, which contains a discussion of the early vision and accomplishment of Eckert and Ben Wood; and "Giant Calculator Joins the Navy," in *Business Machines* (23 Dec. 1954), 4–9, a description of NORC, with photographs of the computer and dignitaries, including John von Neumann, who also gave an address on that occasion (see A. H. Taub, ed., *John von Neumann: Collected Works*, V [New York, 1963], 238–247).

See also J. Ashbrook, "A Great American Astronomer," in *Sky and Telescope*, **42** (1971), 207; "W. J. Eckert: In Memoriam," in *Celestial Mechanics*, **6** (1972), 2–3; William M. Freeman, "Dr. Wallace Eckert Dies at 69; Tracked Moon with Computer," in *New York Times* (25 Aug. 1971), 41; and *IBM News*, **8**, no. 18 (Sept. 1971), which contains an unsigned obituary notice. Each of these obituaries includes a photograph.

Other sources include Eckert's widow, Mrs. Dorothy Applegate Eckert, who granted a personal interview and also helped in my perusal of her husband's papers, on many of which she had collaborated; and the IBM Corporation, which made available copies of corporate publications, a thirty-five-item bibliography, two taped interviews with Eckert conducted by Larry Saphire on 11 July 1967 and 20 July 1967, and access to the exhibit on Eckert prepared by the office of Charles Eames.

HENRY S. TROPP

ENGELMANN, GEORGE (*b.* Frankfurt am Main, Germany, 2 February 1809; *d.* St. Louis, Missouri, 4 February 1884), *botany*.

Engelmann was the ablest and best-known scientist in the Mississippi Valley for the fifty years during which important expeditions explored and surveyed the West. More than anyone else he was responsible for persuading Henry Shaw to establish a botanic garden in St. Louis, not merely for flower beds but as a scientific center. Since its founding in 1859 the Missouri Botanical Garden, its proper name, has become an international center for research and teaching in systematic botany. With exceptional library and herbarium resources, the *Annals* of the garden and its graduates occupy positions on all continents. Besides practicing medicine Engelmann studied many taxonomically difficult plant groups and published fundamental revisions.

The eldest child of George and Julia May Engelmann, who conducted a school for girls in Frankfurt am Main, Engelmann wrote that he "became greatly interested in plants" at the age of fifteen. He entered the University of Heidelberg as a scholarship student and was influenced there by the work of Alexander Braun and Karl Schimper. Because of his liberal political sympathies he moved to the University of Berlin in 1829 and then to the University of Würzburg, where he received the M.D. degree in 1831 with a thesis on plant monstrosities, *De antholysi prodromus*, published in 1832. Later that year, on a trip to Paris, Engelmann incidentally met Louis Agassiz and other scientists; and in September he set out for the United States. He landed at Baltimore and reached St. Louis the following February. For two years Engelmann lived on the farm of a relative twenty miles from St. Louis and studied the local flora.

In December 1835 he began medical practice in St. Louis and was immediately successful. He did not, however, neglect his scientific interests: in 1836 he started meteorological observations that he continued until his death, launched a German-language paper, *Das Westland*, and began studying plant groups overlooked by other botanists. After meeting Asa Gray in 1840, Engelmann operated a clearinghouse in St. Louis for botanists who, with their own funds or as members of government expeditions, were collecting plants in the West. Engelmann's service resulted in a flow of new information to study centers in the East. He encouraged, among others, Ferdinand Lindheimer in Texas, August Fendler in New Mexico and Central America, and J. C. Frémont in the West.

Engelmann's first botanical recension (1842), which treated the genus *Cuscuta*, attracted international commendation. His *Cactaceae of the Boundary* (1859), illustrated by Paul Roetter of St. Louis, was based on his "Synopsis of the Cactaceae of the United States" (1856) and established fundamental criteria for the classification of the cacti. His monographs on conifers, grapes, dodders, mistletoes, yuccas, rushes, and quillworts remain essential references.

Oligotropic pollination studies began with Engelmann's findings on the requisite role of the moth *Tegeticula (Pronuba) yuccasella* in effecting seed set in *Yucca*; this work was later elaborated by C. V. Riley (1892). Engelmann published "Diseases of Grapes" (1873), which John A. Stevenson characterized as the "only [paper] of scientific significance in the plant disease field prior to 1870." The first monograph of a genus of Pteridophyta published in America was Engelmann's "Genus *Isoëtes* in North America" (1882).

On 11 June 1840, Engelmann married his cousin Dorothea Horstmann during his first visit to Germany. Their only child, George Julius Engelmann, became a distinguished physician. After his wife's death Engelmann accompanied Charles Sprague Sargent on a four-month field survey, during which, according to Charles White, Engelmann's "pluck, good nature, good spirits, and good fellowship" struck Sargent "most forcibly." Asa Gray said in his obituary of Engelmann: "Not very many of those who could devote their whole time to botany have accomplished as much."

BIBLIOGRAPHY

I. ORIGINAL WORKS. C. S. Sargent lists about 100 titles in "Botanical Papers of George Engelmann," in *Botanical Gazette*, **9** (1884), 69–74; William G. Bek adds about a dozen titles in *Missouri Historical Review*, **23** (1929), 189–195. Articles written with Asa Gray are indexed in *American Journal of Science*, 3rd ser., **36** (1888), app., 1–42. Engelmann's scattered publications were published as *Botanical Works of the Late George Engelmann Collected for Henry Shaw*, William Trelease and Asa Gray, eds. (Cambridge, Mass., 1887). Thousands of letters from 509 correspondents; 60 volumes of notes and drawings; his books, often with marginalia; his extensive herbarium; and theses of European colleagues, often privately printed and rare in the United States, are preserved at the Missouri Botanical Garden, St. Louis. A few of Engelmann's books are in the Hitchcock-Chase Library at the Smithsonian Institution. There are 532 Engelmann letters (1840–1884) at the Gray Herbarium, Harvard University; a selection, in abstract, was published by Jane Loring Gray, *Letters of Asa Gray* (Boston–New York, 1893), *passim*.

II. SECONDARY LITERATURE. No biography of Engelmann has been published. Biographical sketches, largely repetitive, are based on the obituary by Asa Gray, in *Proceedings of the American Academy of Arts and Sciences*, **19** (1884), 516–522; repr. verbatim in *American Journal of Science*, 3rd ser., **28** (1884), 61–67; in Engelmann's *Botanical Works*, iii–vi; and in *Scientific Papers of Asa Gray*, C. S. Sargent, ed., II (Boston–New York, 1889), 439–446. Sargent adds a personal estimate in *Science*, **3** (1884), 405–408; and in his *Silva of North America*, VIII (Cambridge, Mass., 1895), 84. Ignatz Urban, in *Berichte der Deutschen botanischen Gesellschaft*, **2** (1884), xii–xv; and Charles A. White, in *Biographical Memoirs. National Academy of Sciences*, **4** (1902), 1–21, provide incidental facts.

For Engelmann's role in the introduction of American grape stocks to combat the *Phylloxera* epidemic in Europe, see John A. Stevenson, "Beginnings of Plant Pathology in America," in C. S. Holton *et al.*, eds., *Plant Pathology, Problems and Progress* (Madison, Wis., 1959), 14–23. Edgar Anderson relates Engelmann's role in the development of the Missouri Botanical Garden, in *Washington University Magazine*, **39**, no. 3 (1969), 38–43. The omission of the intended map of distribution of cacti in *Cactaceae of the Boundary* is discussed by L. E. Newton, in *National Cactus and Succulent Journal*, **17**, no. 3 (1962), 43–45; and Engelmann's pioneer work on cacti is analyzed by Larry W. Mitich, in *Excelsa*, no. 4 (1974), 31–39. For bibliographic notes on his *Cuscuta* papers, see F. A. Stafleu, *Taxonomic Literature* (Utrecht, 1967), 132. C. V. Riley, "The Yucca Moth and Yucca Pollination," in *Report. Missouri Botanical Garden*, **3** (1892), 99–158, is placed in modern perspective by Michael Proctor and Peter Yeo, *Pollination of Flowers* (London, 1973), 316–318.

JOSEPH EWAN

ENGELS, FRIEDRICH (*b*. Barmen [now part of Wuppertal], Prussian Rhineland, 28 November 1820; *d*. London, England, 5 August 1895), *history, sociology, political science, economics, philosophy, history and philosophy of science and technology, military science*.

Friedrich Engels was the eldest of eight children of Friedrich Engels, senior, a third-generation conservative cotton mill owner, and his wife, Elizabeth van Haar, daughter of a gymnasium principal of Dutch origin. The father held to strictly orthodox German Pietism, side-by-side with enjoyment of theater, fine arts, and music (he played the bassoon and the cello)—his orthodoxy had been tempered toward an unprejudiced and critical outlook by his many trips to England and elsewhere in Europe. The mother was lively, loving, imaginative, well-read, and known for her sense of humor, laughing "until the tears ran down her cheeks" even in old age. Engels was devoted to her; typically, in 1869, when he was financially able to retire from his eighteen-year business career in the Manchester textile mills, he wrote: "Dear Mother, Today is my first day of freedom and I can put it to no better use than by writing to you first thing." Without accepting his politics, Engels' mother always believed in his integrity.

To his father, Engels was more distant, critical, cooler; indeed, the elder Engels was soon disturbed by the strong tendency of his adolescent son toward conflict with the established order, with convention, and with religion. In later years father and son overcame that pattern of paternal disappointment and filial resentment; their business relations became cordially collaborative and reflected correct mutual respect. Friedrich Engels, senior, died in 1860; Elizabeth Engels, in 1873. Typically, Engels resolved a dispute with his brothers over their father's estate by renouncing his full claim as eldest son under law as well as by family tradition, in order to avoid embittering his mother. He wrote to her: "I can get a hundred other businesses, but never again a mother."

In 1834, Engels was sent to a better gymnasium at nearby Elberfeld, where he studied Latin, Greek, French, mathematics, natural sciences, geography, and philosophy. He was especially stimulated by one teacher, a Dr. Clausen, and immersed himself in German literature and history. Engels admired Clausen as the only one who knew "how to awaken the feeling for poetry in the students . . . the feeling that otherwise would have had to grow stunted under the Philistines of the Wupper Valley." Engels' heroes then were Siegfried, William Tell, and Faust.

But other institutions educated him too. During these school years, Engels, unlike his fellow students, was increasingly affected by the miserable conditions of life among workers in the Barmen factories and artisans in their home shops; by smoky, airless workrooms; child labor from the age of six; jobless and homeless laborers, often ruined by cheap gin. In short, he was overwhelmed by the condition of the poverty-stricken working class in the prosperous, industrialized Wupper Valley. Engels came to feel a deep conflict with his politically and religiously conservative home, and even more with the inhumane hypocrisy of the social-economic system all about him.

Strongly encouraged by his father to enter the business world rather than attend a university to study law, as he had wished for a time, Engels left the gymnasium a year before graduating. He worked for a year at his father's firm, then for about two years as a clerk for a wholesale merchant at Bremen. Next he volunteered his year of military service in an artillery guards' regiment at Berlin, then went to England in November 1842 to work in his father's textile mill at Manchester.

At school and in Bremen, Engels had tried his hand at literary and critical essays, poetry, and translations; and he enthusiastically espoused the rationalist, anti-authoritarian, and humanist views of the "Young Germany" literary movement. He published several dozen writings under pseudonyms; the first under his name was a romantic translation of a Spanish poem in honor of Gutenberg, in which the poet Manuel José Quintana praised the inventor for bringing mankind reason and truth—and thereby peace and freedom—through the printing press. Among his unsigned pieces were two vividly critical "Letters from Wuppertal," written at the age of nineteen, which were published in a Hamburg paper; they anticipated the analytic and factual style of his lifework by their exposé of the linked Pietist egoism and mass distress in his home town, and they settled his account with that religion, which he saw as bigoted, obscurantist, and the denial of human creativity and decency. While still a clerk, Engels was deeply affected by David Friedrich Strauss's just-published *Life of Jesus*, and soon wrote to a close school friend: "What science rejects—the development of which now includes the whole of church history—should no longer exist in life either" (*Collected Works,* II, letter to F. Graeber, 12–27 July 1839, p. 457). Strauss led him to Hegel: first the upward spiral of the *Philosophy of History*, and then the deeper waters of the logical treatises and the *Phenomenology of Mind*, and soon back into the writings of Kant and other predecessors.

During his military year (1841), Engels lived privately and plunged into life among university students and an informal club of young Left Hegelians called "The Free"; he attended Schelling's inaugural lecture, was outraged by Schelling's rejection of reason, science, and progress, and proceeded to publish three anonymous pamphlets in defense of progressive Hegelianism and against Schelling. That year he was profoundly influenced by Ludwig Feuerbach's works, which confirmed both his total breakaway from his Christian upbringing, and his abiding belief in the humanist basis and purpose of any reasonable ethical principles. And then came a decisive turn: he was impressed by Moses Hess, who brought the French socialism of Saint-Simon into the Left Hegelians' discussions; it was Hess whom Engels called the first communist among them, the first to show that human liberation required communism. Hess also pointed away from Germany and toward England, which also meant to Engels a turn away from philosophy to practice, from theory to action.

Engels' activities accelerated. In 1842 he wrote for Marx's newspaper, the *Rheinische Zeitung*, and the two met briefly at Cologne. While in England for the next two years, Engels continued his commercial training at the family firm and investigated English social and economic conditions. He also developed his association with the German émigré League of the Just, began his long friendship with the Chartist leader G. J. Harney, and wrote for English Chartist and Utopian socialist papers. During this period Engels published his first economic work—the extraordinary anticipation of Marx's later critique of economic categories, *Outlines of a Critique of Political Economy*. He also fell in love with a young Irish working-class woman, Mary Burns, with whom he lived until her death in 1863.

During his return journey to Barmen in August 1844, Engels stopped for ten days in Paris, to see Marx. The visit was decisive for both; their intimate friendship and collaboration began then. Engels returned to Barmen to visit his family, to plan his

ENGELS

business career, to undertake political activity with Hess and other socialists in the Rhineland, and to write *The Condition of the Working Class in England.*

This pioneering book in social science, a work of first-hand investigation combined with thorough analysis of reports and secondary sources, both historical and empirical, was extraordinarily moving both in detail and in its overall portrait of the new urban industrial society of Manchester. Engels set forth the rapid pace of change in England during the previous sixty years that had brought new ways of production, new ways of living, new classes of people; most important, he described the extraordinary upheaval in human relations within the new industrial working class, the proletariat, which "was called into existence by the introduction of machinery" (*Collected Works*, IV, 321). Machinery, the division of labor, and water power (especially in the form of steam) were "the three great levers . . . busily putting the world out of joint." They also were the causal technology that served the main force of development, which he saw to be capitalist concentration and centralization of people and economic power.

Thereafter Engels continued his work on concrete social matters, both contemporary and historical: *The Peasant War in Germany* (1850); twenty articles on the 1848 revolution in Germany for the *New York Tribune* (1851–1852)—signed by Marx; and six articles on housing (1872). He also produced philological studies on the history of German dialects; several dozen significant articles and pamphlets on tactical and political-military subjects; and essays and notes on the history of science and technology (which were used in his longer theoretical works and in *Dialectics of Nature*).

In 1844–1845, Engels had quickly joined Marx in a number of projects, the first being their sarcastic critique of "speculative" Left Hegelians, *The Holy Family* (1845). It was followed by *The German Ideology* (unpublished until 1932), which set forth an early major statement of their theory of historical materialism; initiated their conception of sociological analysis; extended their separately developed concepts of economic laws, categories, and socioeconomic formations; and provided various suggestive *aperçus* and theoretical starting points for understanding language, the evolution of materialism, the social foundations of the psychology of thinking and of emotions, and the contributions of technology to social development. Engels later said that he mainly listened and questioned while Marx wrote; and Marx himself commented, "We abandoned the manuscript to the gnawing criticism of the mice, all the more willingly as we had achieved our main purpose—self-clarification."

In 1848 came their famous *Manifesto of the Communist Party*, written by Marx but influenced by Engels, and anticipated in many aspects by Engels' *Principles of Communism* (1847). In sharp and spare language the *Manifesto* combined empirical sociological investigation with passionate exposition of their view of the social crisis and its resolution: "In place of the old bourgeois society with its classes and class antagonisms, we shall have an association in which the free development of each individual is the condition for the free development of all." Engels aided Marx with more than 170 articles as European correspondent for the *New York Tribune* and nine, chiefly on military subjects, for the *New American Cyclopedia* (1857–1862). (After Marx's death in 1883, Engels put the incomplete manuscripts in order and prepared the second and third volumes of Marx's *Capital* for publication [1885, 1894], as well as some eighteen new editions of Marx's other works. He had collaborated for decades in the research and thinking that had gone into Marx's research for *Capital.*)

Engels returned to Manchester in 1850, after six years of activity among revolutionary and working-class organizations, during which he served in the unsuccessful German uprisings of 1849. He resumed his managerial post at the family cotton mill, becoming a partner in 1864, and also continued his studies, his writing, an immense correspondence with Marx and others, and his political associations, notably the International Working Men's Association. After his retirement in 1869, Engels settled in London to continue scientific work, writing, political activity, and work with Marx. He wrote *Anti-Dühring* (1876–1878), the most popularly effective of all his and Marx's expositions of their world view; *The Origin of the Family, Private Property and the State* (1884), based in part on Marx's anthropological notes; and *Ludwig Feuerbach and the Outcome of Classical German Philosophy* (1886); he also drafted parts of his proposed work on the philosophy of science, a "dialectic of nature."

Engels had lived happily with Lizzie, the sister of Mary Burns, after Mary's death; they were formally married the day before Lizzie died in 1878. In his last years his housekeeper was Louise Kautsky, who had been married to Karl Kautsky (a leader of the German socialist movement). When she married Ludwig Freyberger, a medical doctor who attended Engels, he shared a new residence with them.

ENGELS

As social scientist and political figure, Engels was primarily Marx's collaborator, beloved friend, support and disciple, research assistant and critic. Perhaps he also was Marx's window to the actualities of Victorian England. Good-willed and good-humored, healthy and vigorous, he was a confident man of great physical and mental energy, quick at languages, at reading, at personal relationships, and at the immense tasks of self-imposed studies. He saw his own achievement accurately, for he knew how very much of his work in politics and in science (and not least in his hypotheses and projects for research into the natural sciences and their history) remained programmatic. He was, however, immensely pleased by what he had accomplished. He knew Marx's quality better than anyone, of course: "Marx stood higher, saw further and took a wider and quicker view than all the rest of us. Marx was a genius; the rest of us were talented at best" ("Ludwig Feuerbach and the Outcome of Classical German Philosophy," in *Marx-Engels Selected Works*, II [Moscow, 1950], 349).

Engels and Science. A remarkable autodidact in the full breadth of the natural sciences as much as in philosophy, history, and the social sciences, Engels wrote on science in several distinct modes: (1) popular exposition of natural science in support of his and Marx's general outlook upon society; (2) investigation of the history of the sciences, especially their social history, as supportive research for the Marxist theory of historical materialism; (3) reports and commentary on scientific developments for himself and for Marx in their congenial division of intellectual labors; (4) general criticism of the uses of science in practical and in ideological developments; (5) philosophical and methodological interpretations of the sciences.

Engels' appreciation of science began early, as is shown in this passage from his first economic work, written in 1843, when he was twenty-three:

> . . . a third element which, admittedly, never means anything to the economist—science—whose progress is as unlimited and at least as rapid as that of population. What progress does the agriculture of this century owe to chemistry alone—indeed to two men alone, Sir Humphry Davy and Justus Liebig! . . . And what is impossible to science? (*Outlines of a Critique of Political Economy; Collected Works*, III, 440).

This appreciation continued; in 1894, at the end of his life, Engels wrote to a correspondent his now well-known comment on the place of science in modern history, and on the correct understanding of the history of science:

> What we understand by the economic relations, which we regard as the determining basis of the history of society, is the manner and method by which men in a given society produce their means of subsistence and exchange the products among themselves (in so far as division of labor exists). Thus the *entire technique* of production and transport is here included. . . . Further included in economic relations [is] the *geographical basis* on which they operate. . . .
>
> If, as you say, technique largely depends on the state of science, science depends far more still on the *state* and the *requirements* of technique. If society has a technical need, that helps science forward more than ten universities. The whole of hydrostatics (Torricelli, etc.) was called forth by the necessity for regulating the mountain streams of Italy in the 16th and 17th centuries. We have known anything reasonable about electricity only since its technical applicability was discovered. But unfortunately it has become the custom in Germany to write the history of the sciences as if they had fallen from the skies (letter to H. Starkenburg, 25 January 1894; *Selected Correspondence* [Moscow–London, 1956], no. 234, pp. 548–549).

Engels and Marx had a common outlook upon science as a historical phenomenon, so that any account of Marx's thought on science is largely an account of Engels' thought as well. First, we summarize their joint view of the history of science, and then Engels' work on the history of nature as well as his understanding of the philosophy of science. Then we set forth these matters in some detail.

Engels' and Marx's principal contribution to understanding natural science was an analysis of its changing, that is, historical nature. They went further, to conclude that science not only is in process, as part of the larger historical social process, but also has a fundamentally social character and impact, and must be understood as such. By the mid-nineteenth century, the role and impact of science upon social life, both actual and prospective, by direct means and through technology, may have needed no argument; nevertheless, Marx and Engels saw insufficient recognition of that impact in the scholarly works of contemporary historians and political economists. This failure occurred despite the explicit hopes of scientists and philosophers from the time of Galileo and Bacon to Diderot, Kant, and Goethe for a benevolent impact of science upon human life.

Moreover, study of the converse effect, the impact of the social order upon science, was even more neglected; and to Engels such study promised to provide a more profound understanding of science. Here Engels included several distinct aspects of science: the sociology and political economy of scientists; provision and selection of problems for scientific investigation; material tools and techniques for investigating by observation and experiment; ideas, metaphors, heuristics, and mathematical and abstract (and perhaps other) models for use in concepts, theories, and general hypotheses; methodology and classification of the sciences, including social impact upon the development of epistemology and upon the criteria of scientific explanation. These external factors were to be taken along with the internal factors of thought and discovery in order fully to comprehend the cognitive, as well as the instrumental, nature of science. Engels disclaimed any simple and distinct separation of the internal and the external factors.

For Engels, science evidently was complex. An essential part of the productive basis of society, as both contributor and recipient, science was also seen by Engels as part of the cultural and conceptual superstructure of society. At times it was allied with religion, philosophy, fine art, or other cultural tendencies; at other times, opposed. It was never stable, never completed. Science was in part self-developing through rational argument, discursive critique, and either deliberate or chance enlargement of its empirical foundations; but—of great importance—in substantial part science developed in response to command, purchase, opportunity, or social values. Science was therefore surely both ideological and objective. Science was ideological in that it reflected, utilized, projected, and carried out social purposes; but just as surely science was objective in its cognitive achievements. It attained truths about the physical, chemical, astronomical, and biological processes of nature, and about the political economy of societies in their historical development. In doing so, it served the most advanced stage of the human project to master nature.

Both in attaining truths and in serving to dominate nature, science was, for Marx and Engels, a source of individual fulfillment: in satisfying the human capacity to take pleasure in analytic understanding and in the combination of individual achievement with social cooperation, a particularly satisfying form of the cooperative division of labor that typifies all social and scientific advance. Essential for the liberation of mankind, but limited by its entrenchment within the ironic progress and conflicts of the bourgeois epoch, science was, when they wrote, the achievement and servant of capitalism.

Nature and Science. The historical quality of science had another significance for Engels. Not only must science be understood through its history and its social context; nature too required such understanding. Deeply influenced by his grasp of Hegel's logic, Engels saw the essential criterion for understanding nature in a cosmic dialectic of development, from a stellar and galactic universe of physical and chemical evolutionary processes, through the biochemical processes that are the genesis of living matter, to the biological evolution of all life forms, and to the specific historical evolution of the human species, with its particular bodily and mental changes and its social development due to its primary character in labor. The meaning and distinctive significance of "dialectic" for Engels is still disputed, but he stated the minimal and necessary property in 1859: "What distinguished Hegel's way of thinking from that of all other philosophers was the enormous historical sense upon which it was based" (*On Karl Marx's "A Contribution to the Critique of Political Economy," Das Volk*, 20 August 1859; trans. in *Selected Works*, I [1935], 367, originally published separately at London in 1895).

For Engels, the idea of a full continuity of inanimate, animate, and human nature (individual and social) had already been well established by the sciences of his day. Human nature was embedded within the larger natural history; and the larger process of nature likewise would have a genuine dialectical development, its levels of beings coming into existence where they did not exist before, with a striking ability to generate novel phenomena that had their own characteristic properties and activity. Engels saw three great instances of genuine natural history: our sun with its planets had a beginning (the Kant-Laplace hypothesis), and no doubt all other astronomical and galactic structures had too; life had a chemical genesis (Wöhler and subsequent biochemistry); our species had its biological origin, as did all the others (Darwin). So, to the history of ideas of nature, Engels added the idea of the history of nature. He believed that the study of the history of nature would best be undertaken by "Hegel's way of thinking," applied within the sciences of nature. To write such an interpretive synthesis of the sciences in order to demonstrate such a dialectic of nature remained a cherished goal for Engels, to which he turned initially in 1873, again in 1875–1876, and later in 1882; but aside from a popular presentation in his *Anti-Dühring*, the project remained drafts, notes, fragments, and a plan. His notes were published in 1927 and are now known as his *Dialectics of Nature*.

The History of Science. At the age of twenty-three, Engels wrote: ". . . a single achievement of science like James Watt's steam-engine has brought in more for the world in the first fifty years of its existence than the world has spent on the promotion of science since the beginning of time" (*Outlines of a Critique of Political Economy; Collected Works*, III, 428). He soon realized that to understand nineteenth-century capitalist society, he must understand not only Watt and his steam engine, and others he cited—Berthollet, Davy, Liebig, Edmund Cartwright—but the general history of science and technology as well.

Thus, for Marx and Engels, science was one among many social activities, arising from society and developing as other social activities did—in part autonomously, in part closely coupled to a component of the overall structure. Science was, in their view, characteristically a fusion of several distinct activities: (1) observation and its systematic development through exploration, instrumentation, measurement, experimentation, mathematical analysis, and the accumulated learning of craft and technology; (2) explanatory thinking through increasingly abstract principles, using correlations, hypotheses, analogies, metaphors, pictorial and abstract models, and logical demonstrations; (3) presentation of an increasingly concrete conceptual rendering of the objects of investigation, together with critical testing in social practice and in subsequent theoretical practice as well.

As thoughtful human activity, science was a peculiar form of consciousness, since scientific consciousness was not only cognitive but also ideological in being responsive to social interests and powers; likewise, science was not only instrumentally practical but also a meditative and intrinsically a joyful activity, an essential contribution toward a creative outlook upon the world. Scientific consciousness, Engels said, changes with changing times, just as religious, political, aesthetic, and other forms of mental life do. Like culture in general, science itself can be investigated scientifically; and as part of human history, science should be explained by Marx's theory of historical materialism. Marx's compact summary applied, then, to science as a component of intellectual life (1859):

> The mode of production of material life conditions the general process of social, political, and intellectual life. It is not the consciousness of men that determines their existence, but their social existence that determines their consciousness (*Preface* to *A Contribution to the Critique of Political Economy* [New York, 1970], 20–21).
>
> *The successive development* of the separate branches of natural science should be studied.—First of all, *astronomy*, which, if only on account of the seasons, was absolutely indispensable for pastoral and agricultural peoples. Astronomy can only develop with the aid of *mathematics*. Hence this also had to be tackled.—Further, at a certain stage of agriculture and in certain regions (raising of water for irrigation in Egypt), and especially with the origin of towns, big building structures and the development of handicrafts, *mechanics* also arose. This was soon needed also for *navigation* and *war*.—Moreover, it requires the aid of mathematics and so promotes the latter's development. Thus, from the very beginning the origin and development of the sciences has been determined by production (*Dialectics of Nature* [Moscow, 1954], 247).

Social existence is fundamental to science, then, at least at the beginning. For Marx and Engels, the main content of their materialist understanding of society had been tentatively settled early in their work, and with increased conviction as they undertook further studies. There were three central and related components: (1) Societies are historically evolved systems, the structures of which depend essentially on the social relations among men and women; these relations primarily serve the material functions of production and reproduction. The human relations are within, and between, social classes; social systems are hierarchical structures with a basis in the production and reproduction needed to satisfy material and related needs, and with a political-cultural superstructure. (2) While the social systems are self-preserving with respect to their external natural and human environment and to their internal activities carried on by individuals and classes, this stabilizing function is countered by internal conflicts and tensions, the Marxian "contradictions," which tend to undermine and propel the systems. (3) The historical quality pervades all social phenomena, and must characterize any adequate explanation of either the material forces and the social relations of production in the base, or of the political and cultural realities of the superstructure.

Marx and Engels observed that science functions in the economic basis, but is situated as a cognitive and partly autonomous activity within the cultural superstructure. As a component of the base, scientists have their special tasks in the historically developed division of labor. Under nineteenth-century capitalism, the overall scientific enterprise increasingly and massively served as one of the productive forces of society. As a component of the superstructure, changing science reflected and, in a complex way, mediated, and contributed to, the structure, values, and human relations of changing society. It did so in the course of the investigation and interpretation of the forces of nature.

To Marx and Engels, a theory of economic determinism, and a fortiori any blunt technological determinism, was a misunderstanding of their views. Direct social causation of scientific activities and ideas by economic or technological factors is a crude reductionism, a primitive "mechanical" determinism. What they intended was clear and univocal to them, although disputed by contemporaries and subsequent Marxist thinkers: to wit, an association of scientific achievements with the economic structure and technical processes of their time, but an association of which the feedback modes of reciprocal conditioning and influencing of science and society set subtle research problems before the historians of science and of society. Engels wrote about the danger of simplification, of a "vulgar" Marxism, explicitly, some years after Marx died:

> . . . According to the materialist conception of history, the ultimately determining element in history is the production and reproduction of real life. More than this, neither Marx nor I have ever asserted. . . . The economic situation is the basis, but the various elements of the superstructure—political forms of the class struggle and its results . . . and even the reflexes of all these actual struggles in the brains of the participants, political, juristic, philosophical theories, religious views and their further development into systems of dogmas—also exercise their influence upon the course of the historical struggles and in many cases preponderate in determining their *form*. There is an interaction of all these elements in which, amid all the endless host of accidents (that is, of things and events whose inner interconnection is so remote, or so impossible of proof that we can regard it as non-existent, as negligible) the economic movement finally asserts itself as necessary. Otherwise the application of the theory to any period of history would be easier than the solution of a simple equation of the first degree. . . .
>
> Marx and I are ourselves partly to blame for the fact that the younger people sometimes lay more stress on the economic side than is due to it. We had to emphasize the main principle *vis-à-vis* our adversaries, who denied it, and we had not always the time, the place or the opportunity to give their due to the other elements involved in the interaction. But when it came to presenting a section of history, that is, to making a practical application, it was a different matter and there no error was permissible . . . (letter of 21–22 September 1890, to J. Bloch; *Selected Correspondence*, no. 214, pp. 498–499).

The Marxist heuristic for historians of science, then, included a spectrum of social, economic, and cultural influences upon science. (See, however, L. Colletti, *From Rousseau to Lenin*, 63–72, for the argument that this is not Marx's own view—and, indeed, that in this respect Engels differs.) The blunt and promising power of the Marxist approach was the complex externalism, seeking explanation of scientific activities in the multiple social contexts of those activities. Whatever subtlety would be found was due to the complex nature of those social contexts. Furthermore, their view of the social determination of ideas, including scientific ideas, took Marx and Engels beyond any externalist account of science that would be limited to a commonsense list—problem choice, resource allocation, professional codes, social roles, public relations, personnel recruitment and training, industrial and military exploitation, religious and aesthetic influence, and the like—however illuminating such an account would be. Marx and Engels also looked for social influences upon the content of science, upon the factual findings, concepts and theories, explanatory criteria, and epistemology.

Applying historical materialism to a variety of cultural phenomena, including religion, philosophy, and the natural and the social sciences, Marx and Engels began to formulate the research problem of locating and understanding the social determinants of truth seeking and truth attainment. Their work was to be a major source of the sociology of knowledge (for Karl Mannheim, Max Scheler, and others). As a rather general method, they looked, in the case of science, for differing sciences—which means differing accounts of nature—and different practical activities and interactions with natural materials in differing historical epochs. Nature as understood and science as developed in an aristocratic and pastoral society contrast with nature as understood and science as practiced in an industrial, urban, and competitively individualist society. Engels was amused by projections of social norms upon nature in the content of science, as in the nineteenth-century English adaptive transition from a religion of cosmic harmony to a theology of progress through struggle and the "survival of the fittest."

But such a general assessment of social context provides no specific account of any particular scientific development. What it can do is to suggest the problem of the historical periodization of science, and the related problem of the systematic classification of the sciences—but again without specificity. In treating these two problems, Engels analyzed the sciences first with respect to their historical stages, and separately with respect to their differing, specialized subject matter (which he took to be identified by particular forms of motion—or, more properly, of energy). Periodization of science largely followed Marx's account of the stages of historical formations, as in his striking remarks in *The Poverty of Philosophy* about the societies of feudal lords and industrial capitalists as "given" by the hand mill and the steam mill; but neither Marx nor Engels wrote at length on it. There are extended fragments on classification in Engels' notes. (See Kedrov's full exposition.)

It is likely that Engels understood the strength, subtlety, complexities—even the difficulties—of a Marxist historiography of science. In his view, all intellectual and institutional formations in the social superstructure had their own "internal" histories once they had been initiated, their own "forces" influencing the development of the base and the rest of the superstructure, their relative autonomy. Just how far that independence might go could only be determined concretely, case by case, by the scientist and the historian of science after him; just how personally idiosyncratic and autonomous an individual experimenter or theorist might be, going against the dominant patterns of scientific activity (whether they in turn were set by internal or external factors); just where and how the requirements set by the productive base of society finally prevailed or how the established norms of the cultural superstructure fixed limits upon scientific thinking; just how, and how much, the received and accustomed patterns of thought and explanation might be stretched—these, and a host of similar questions, remain for the historian.

Given the prevailing histories of ideas at his time, Engels surmised that the first Marxist analyses of the history of science would have to be blunt, only a first approximation, stressing economic and technical necessities. One of the first pioneering efforts, B. Hessen's "The Social and Economic Roots of Newton's *Principia*" (1931), deliberately emphasized "the complete coincidence of the physical thematics of the period, which arose out of the needs of economics and technique, with the main contents of the *Principia*, which in the full sense of the word is a survey and systematic resolution of all the main group of physical problems" (Hessen, p. 176). But Engels and Marx, as Hessen also recognized, would in addition have looked for internal factors at work upon Newton and upon his scientific context: it is the balance of all factors that only the historian's genius can establish. How far autonomous superstructure may stray from the constraining influences of the base is too abstract a question to be put in general terms; but Engels, at any rate, would not agree with those historians for whom even the historical genesis, as well as the major properties, of modern science are fully independent, "the fruit of intellectual mutation" (Hall, "Merton Revisited" [1963]).

Engels did write of reciprocal relations between science and philosophy, as well as between science and (independent) technology, so that the actual sources of ideas must be a matter of inquiry. Moreover, for Engels the social context of science included ideas, such as the metaphysical or other presuppositions about nature, where metaphysics in turn was the object of investigation with respect to its social role and ideological interests. Likewise, the history of the continuum of interactions between science and technology, near to common ground in Engels' time, as he saw it, was construed within the same theory of social contexts.

If Engels was to avoid simple reduction to either economic need or technological implication, what was to remain of the distinctively Marxist explanation of science? A reply can be drawn from *The German Ideology*. The economic basis of society comprises productive forces and the social relations of those who take part in the working of those forces. Out of these social relations of production comes the consciousness that characterizes the given times; and the general prevailing state of consciousness provides the framework for ideas and other cultural activities. But all these relations are reciprocal, asymmetric, shifting; moreover, the study of culture in society should be the study of an entire social system, not of pairs of separate and independent entities that interact externally. Even so fundamental an analytic distinction as that of basis and superstructure would be open to qualification, as may be seen in the discussion of "ideology" and in the consideration of the formal process whereby ideas and other ideological factors, such as values, arise. In 1893, Engels wrote to the literary critic and historian Mehring:

> . . . one more point is lacking, which, however, Marx and I always failed to stress enough in our writings and in regard to which we are all equally guilty. That is to say, we all laid, and *were bound* to lay, the main emphasis, in the first place, on the *derivation* of political, juridical, and other ideological notions, and of actions arising through the medium of these notions, from basic economic facts. But in so doing we neglected the formal side—the ways and means by which these notions, etc., come about—for the sake of the content. This has given our adversaries a welcome

ENGELS

opportunity for misunderstandings and distortions. . . .

Ideology is a process accomplished by the so-called thinker consciously, it is true, but with a false consciousness. The real motive forces impelling him remain unknown to him; otherwise it simply would not be an ideological process. . . . The historical ideologist (historical is here simply meant to comprise the political, juridical, philosophical, theological—in short, all the spheres belonging to *society* and not only to nature) thus possesses in every sphere of science material which has formed itself independently out of the thought of previous generations and has gone through its own independent course of development in the brains of these successive generations. True, external facts belonging to one or another sphere may have to exercise a codetermining influence. . . .

. . . It is the old story: form is always neglected at first for content. . . .

Hanging together with this is the fatuous notion of the ideologists that because we deny an independent historical development to the various ideological spheres which play a part in history we also deny them any *effect upon history*. The basis of this is the common undialectical conception of cause and effect as rigidly opposite poles, the total disregarding of interaction . . . once an historic element has been brought into the world by other, ultimately economic causes, it reacts, can react on its environment and even on the causes that have given rise to it (letter of 14 July 1893; *Selected Correspondence*, no. 232, pp. 540–542).

The "ways and means" by which ideas come about will be similar to the ways in which they function: they mediate among the factors of human life, among needs, values, techniques, social contexts, and nature. Marx and Engels saw that scientific ideas, and the differentiated forms of scientific rationality, support developing capitalist society at each level, sustaining both the accelerating technologies and the appropriate ideology for the capitalist division of labor. In *Capital* and other writings, Marx had examined these aspects of the functions of science in bourgeois society, notably in his analysis of machinery and through his elucidating conception of the effects of fetishism of commodities and reification of human relations: things were seen to be human, while relations among human beings were reified—that is, seen and treated as relations among inanimate things, which is to say treated as "objective," outside human control. To Marx and Engels, this delusive and indeed false objectivity carried the ideological burden of a spurious science, spurious because it brought a dangerous analogy of impersonal and objective laws into the ways people relate to one another.

Darwin's work signaled the ambiguities of science to Marx and Engels. They welcomed Darwin for his insight into the nature of life, for taking the principal scientific step toward understanding nature as historical and toward the defeat of fixity in scientific ideas and static metaphysical conceptions. Marx wrote to Lasalle: "Darwin's book is very important and serves me as a natural-scientific basis for the class struggle in history . . . despite all deficiencies [in Darwin's argument], not only is the death-blow dealt here for the first time to 'teleology' in the natural sciences, but its rational meaning is empirically explained . . ." (16 January 1861). Engels observed: "The Darwinian theory [is] to be demonstrated as the practical proof of Hegel's account of the inner connection between necessity and chance" (*Dialectics of Nature*, p. 402). But he also noted the social content of Darwin's theory:

> . . . Darwin did not know what a bitter satire he wrote on mankind, and especially on his countrymen, when he showed that free competition, the struggle for existence, which the economists celebrate as the highest historical achievement, is the normal state of the *animal kingdom*. Only conscious organization of social production . . . can lift mankind above the rest of the animal world as regards the social aspect, in the same way that production in general had done this for mankind in the specifically biological aspect . . . (*Dialectics of Nature*, p. 49).

Indeed, Engels went on to criticize the adoption of Darwinism by the "bourgeois Darwinists":

> The whole Darwinist teaching of the struggle for existence is simply a transference from society to living nature of Hobbes's doctrine of *bellum omnium contra omnes* and of the bourgeois-economic doctrine of competition together with Malthus's theory of population. When this conjurer's trick has been performed . . . the same theories are transferred back again from organic nature into history and it is now claimed that their validity as eternal laws of human society has been proved . . . (letter of 12 November 1875 to P. Lavrov; *Selected Correspondence*, no. 153, pp. 366–370).

Marx and Engels had argued against Malthus for decades, notably in Engels' *The Condition of the Working Class in England* (1845), in which he bitterly pointed out that "poverty, misery, distress and immorality" prevail as a result of economic and political forces, not of any alleged natural necessity of a biological law of overpopulation.

Could it be that Darwin's work was free of ideology, and that social Darwinism was quite simply spurious science? Could the Marxist historian distinguish sharply between the social conditioning of Darwin's own work and the social forces that affected the various receptions of his theories? Engels did not, in the end, clarify the issue. He did, however, clearly maintain that one must hold fast to the Darwinian achievement, whatever its social and personal circumstances. He paid the highest compliment to Marx's lifework in his funeral oration for his comrade in seeing Marx's work on social nature as the true parallel to Darwin's work on the origin and development of biological species. Moreover, evolutionary biology was, for Engels, the decisive connection between nature as a whole and mankind. Indeed, the species-specific properties of man—to labor, and thus through productive activity to dominate and transform nature, and thereby also to make and transform himself—were necessarily to be understood also as evolved natural properties, a materialist account of the idealist notion of nature transforming itself through the mediation of our novel species.

To Engels, it was inevitable that scientific truths would have ideological impacts and exploitation. Science existed in society, subject to social uses and understandings. When would it be opposed to the ruling class? Engels, throughout his writings, argued that the working-class interest lay in realistic judgment and illusion-free knowledge, ranging from economics to medicine, from history to chemistry—and, hence, that science was already an implicit ally, in method if not entirely in choice of investigation. Going further, the political interest of the modern industrial working class was to bring about a new society that had no ruling class, neither class distinctions nor exploitation. The implication for science was that precisely such working-class interest required an end to delusions about nature and about social nature, an end to metaphysical abstractions and to the ideological fantasies that flourish in the consciousness of class societies. Science could be taken over into the working class, he thought, and it would be science at its best. A later Marxist (Neurath) remarked hopefully that the proletariat would be the bearer of science without metaphysics. But Engels saw that science, like all social activities, would have its struggles—those about its findings, those about its direction, and those within its thought.

In the debates over Darwinism, Engels saw science being used to support a false consciousness through religious, political, philosophical, and even juridical interpretation and adaptation of scientific results; not only reconciliation between scientific novelty and existing social relations, but also support of the ideology of the ruling class. Of course, for Marx and Engels the struggle took place within social theory and social science as much as within the ideological marketplace. They never rejected the need for a genuine scientific analysis of human society, for knowledge of objective laws of social development—only the covert distortion of that analysis to fit the special interests of those who benefited from it. But such a genuine science of human society, in Engels' analysis, must treat the essential novelty of the human species. Here the role of consciousness was central—the subjective feelings, intentions, thoughts, emotions, the mental life that emerged with mankind in the long course of evolution from objective chemical and biological nature.

To explain the strikingly novel phenomenon of the human mind in scientific terms without reductionist mechanisms was of greatest importance to Marx and Engels. Marx saw the causal link between nature and this new natural entity, man, to be labor. Labor was the chief explanatory category for understanding the genesis and history of man. Engels set forth his tentative account of this process in an essay entitled "The Part Played by Labor in the Transition From Ape to Man" (1876), with a sketch of the early inventions and uses of tools in the formation of new human physiological features and new social relations:

> . . . the hand is not only the organ of labor, *it is also the product of labor* . . . [furthermore, the transformed larynx and the mouth developed when] men in the making arrived at the point where *they had something to say* to each other . . . first labor, after it and then with it, articulate speech—these were the two most essential stimuli under the influence of which the brain of the ape gradually changed into that of man . . . the reaction on labor and speech of the development of the brain and its attendant senses, of the increased clarity of consciousness, power of abstraction and of judgment, gave an ever-renewed impulse to the further development of both labor and speech . . . the animal merely *uses* external nature, and brings about changes in it simply by his presence [while] man by his changes makes it serve his ends, *masters* it. . . .

Engels concluded with a warning:

> . . . at every step we are reminded that we by no means rule over nature like a conqueror over a foreign people, like someone standing outside nature—but that we, with flesh, blood, and brain, belong to nature, and exist in its midst, and that all our mastery of it consists in the fact that we have the advantage over all

ENGELS

> other creatures of being able to know and correctly apply its laws. . . . Let us not, however, flatter ourselves overmuch on account of our human conquests over nature. For each such conquest nature takes its revenge on us (as printed in *Dialectics of Nature,* 230 ff.).

Scientific ideas were both classless and class-situated for Engels. With Marx, he had written decisively about the role of ruling-class interests in the thought of an epoch:

> The ideas of the ruling class are in every epoch the ruling ideas; i.e. the class which is the ruling *material* force of society, is at the same time its ruling *intellectual* force. The class which has the material means of production at its disposal, has control at the same time over the means of mental production, so that thereby, generally speaking, the ideas of those who lack the means of mental production are subject to it. The ruling ideas are nothing more than the ideal expression of the dominant material relationships . . . grasped as ideas; hence of the relationships which make the one class the ruling one, therefore, the ideas of its dominance. . . . If now . . . we detach the ideas of the ruling class from the ruling class itself and attribute to them an independent existence . . . if we thus ignore the individuals and world conditions which are the source of the ideas, we can say, for instance, that during the time that the aristocracy was dominant, the concepts honor, loyalty, etc., were dominant, during the dominance of the bourgeoisie the concepts of freedom, equality, etc. . . . increasingly abstract ideas hold sway, ideas which increasingly take on the form of universality. For each new class which puts itself in the place of one ruling before it, is compelled . . . to represent its interest as the common interest of all the members of society, that is, expressed in ideal form: it has to give its ideas the form of universality, and represent them as the only rational, universally valid ones (*German Ideology*, 61 – 62).

If this analysis bears upon scientific thought too, could Engels take a uniformly optimistic view of science as socially progressive? At times he equivocated: science and its interpretation were not automatically progressive, and science should not itself be a fetish; but there was, in Engels' view, a particular optimistic world outlook based upon his critical interpretation of the ensemble of specialized sciences, of their methods of thought, and of their findings about the world. His dialectics of nature, and his project for placing radical social analysis and revolutionary practice within such a cosmology of emergent evolution, have been subject to vigorous debate on several counts (philosophical, scientific, political, ideological).

The question has also been pursued of whether Marx differed from Engels with respect to the place and role of science in human affairs, and in his conception of nature. For both, science as a social ideology in bourgeois society tended both to conceal social realities and to provide methods and findings to reveal those realities; Marx's chief work was deliberately a critique of the previous works on the science of political economy, in part for their ideological, nonscientific character. Critique of what claimed to be social science was different from critique of natural science, for the critique of the "science" of political economy required a critique of the whole system of social reality. Furthermore, successful science in their time was limited by its analytic, specialized approach, which needed to be supplemented by a critical awareness of the whole that had been subjected to analytic abstraction, by what Marx intended to be a unity in concrete exposition.

Science, and all scientific theories of nature, were irrevocably social, and must be examined as such. Engels and Marx well understood Protagoras: man is the measure of all things. Their gloss was on "measure," which was through human labor and was a transforming of "all things." In that materialist sense, labor was constitutive of all things. For Marx, every practical or cognitive encounter with nature—to understand, to change and transform, even to obliterate—was mediated by human consciousness in its historically determined state; and that state of consciousness had arisen along with the social metabolism that was the dialectic of the human species with nature through socially organized labor. Such mediation through consciousness, Engels' "ways and means," seemed to Marx inescapably to be through the several, and often conflicting, ideological presuppositions of class societies. Engels never denied this. Neither Marx nor Engels wrote about the nature of science in a future classless society, in a society with a minimum of necessary labor (but see *Capital*, III) and without the exquisite specialized division of labor that thus far had characterized industry and science alike; nor did they describe the classless society in any other respect.

Marx's dialectic within social nature arose fully from within his critical examination of the social science of his time: economics, political theory, and related social thought; by contrast, while Engels' dialectic accepted that, his writings carried it further, to interpret the theories and empirical findings of the natural sciences. This interpretive process was from without, external to the scientific activities. Whether Engels' dialectic was external to the object of such scientific investigations—to nature—is an open question for philosophers and scientists; but it was not posed, nor was it implicitly answered, by Marx. At any rate, nature is dialectical by bringing men into existence, since for Marx human beings were at once forces of nature and social subjects. For Marx too the dialectic of labor was a natural process. But Marx set his limit in 1845 in the second of his *Theses on Feuerbach:* "The dispute over the reality or non-reality of thinking which is isolated from practice is a purely *scholastic* question."

Practical social mediation applies to all knowledge; this is the beginning of Marxist epistemology. The social context is the same for ideas of nature and for ideas of society. Analysis of the social context of ideas, by the Marxist history of science, is joined with analysis of content, because social context, being mediated by practice and by cultural forms, is in any case constitutive of ideas and of their objects. "One can look at history from two sides and divide it into the history of nature and the history of men. The two sides are, however, inseparable; the history of nature and the history of men are dependent on each other so long as men exist" (*German Ideology, Collected Works*, V, 28).

The historian of science is laboring, too, working in a determinate time and place, bringing understanding to the works of scientists and to their lives in history. The historian is also embedded in the values, metaphors, criteria, conflicts, and false consciousness of his time. He must, in Engels' mode, both praise and query science. He should puzzle out the dialectical interaction of anthropomorphic projection onto nature and objective knowledge of nature—meditate, for example, upon Darwin's great metaphor (Young, "Darwin's Metaphor") and ask: Does Nature select?

History of Nature and Philosophy of Science. Engels wrote of his dialectical conception in sweeping but somewhat heuristic terms: ". . . the dialectic is nothing more than the science of the general laws of motion and development of nature, human society, and thought." Throughout his notes, and in his *Anti-Dühring* and other publications, he gave illustrations from the natural sciences, the social sciences, philology, philosophy, and even mathematics. Engels took Hegel's logic as the most suggestive result of classical philosophical investigations, in contrast with the inadequacies, as he saw them, of empiricism and its purely inductive generalizations. But Engels himself believed that scientific investigators abstract dialectical laws (for the greater part unknowingly) from "the history of nature and human society." He agreed with his understanding of Hegel, that there are primarily three such laws: transformation of quantity into quality, and vice versa; interpenetration of opposites; and negation of the negation.

Engels was writing in broad strokes, in an exuberant and serious critical style. Is it popular science or philosophical critique? He plunged into his examples, as in his criticism of Haeckel's claims on behalf of induction:

> The concepts with which induction operates: species, genus, class have been rendered fluid by the theory of evolution and so have become *relative*: but one cannot use relative concepts for induction.
>
> . . . Light corpuscles and caloric were the results of induction. Where are they now? Induction taught us that all vertebrates have a central nervous system differentiated into brain and spinal cord, and that the spinal cord is enclosed in cartilaginous or bony vertebrae—whence indeed the name is derived. Then *Amphioxus* was revealed as a vertebrate with an undifferentiated central nervous strand and *without* vertebrae . . . (*Dialectics of Nature*, 303).

Engels' work is a lively exploitation of nineteenth-century science. He praises the method of abstraction, so far as it goes, but not its product; and he adopts Hegel's remark "We can eat cherries and plums, but not *fruit*, because no one has so far eaten fruit as such." Engels also adds that the eternal abstract laws of nature become historical and, hence, concrete: "That water is fluid from 0°–100°C is an eternal law of nature, but for it to be valid, there must be (1) water, (2) the given temperature, (3) normal pressure. On the moon there is no water, in the sun only its elements, and the law does not exist for these heavenly bodies" (*Dialectics of Nature*, 316). Moreover, whole bodies of science are exclusively geocentric: "The moon has no atmosphere, the sun one of glowing metallic vapors; the former has no meteorology, that of the latter is quite different from ours." And "The geocentric standpoint in astronomy is prejudiced and has rightly been abolished. But as we go deeper in our investigations, it comes more and more into its own. The sun, etc., *serve* the earth (Hegel). Anything other than geocentric physics, chemistry, biology, meteorology, etc. is impossible for us, and these sciences lose nothing by saying that they hold good for the earth and are therefore only relative. If one takes that seriously and demands a centerless science, one puts a stop to *all* science" (*Dialectics of Nature*, 317).

Engels' range was great, for he was sensitive to puzzles, polarities, and contrasts:

ENGELS

> On simple and compound: "An animal is neither simple nor compound."
>
> On linguistic polarization: ". . . Frankish is a dialect that is *both* High German *and* Low German."
>
> On chance and necessity: "As long as we are not able to show on what the number of peas in the pod depends, it remains just a matter of chance, and to say that the case was foreseen already in the primordial constitution of the solar system does not get us a step further. . . . The *one* [individual] pea-pod would provide more causal connections for following up than all the botanists in the world could solve. . . . Hence chance is not explained here by necessity, but rather necessity is degraded to the production of what is merely accidental."
>
> On nature at large: "The whole of nature accessible to us forms a system, an interconnected totality of bodies, and by bodies we understand here all material existences extending from stars to atoms, indeed right to ether particles, in so far as one grants the existence of the last named." . . . "Among the Greeks—just because they were not yet advanced enough to dissect, analyze nature—nature is still viewed as a whole . . . the universal connection of natural phenomena."

As his examples and briefly stated observations accumulate, the chief characteristics of Engels' approach come into focus.

First, while scientific methods are said to be implicitly dialectical and may be understood as such, Engels' objective materialism goes considerably further, beyond method to the hypothesis of a universal dialectical content in all that exists. For Engels this old ontological theme, held by thinkers from Heraclitus to Hegel, requires materialist meaning and empirical demonstration; but his own articulation is substantially and admiringly—often literally—that of Hegel: "There is no motion without matter, so also there is no matter without motion" (Hegel, *Philosophy of Nature*, sec. 261) becomes "Matter without motion is just as inconceivable as motion without matter" (*Dialectics of Nature*, 86).

Second, dialectical thought, or the dialectical method, applies to the development of concepts. It is carried out by the thinker, who is, however, merely bringing conceptual development into awareness. It also applies to the historical evolution of cultural forms, including scientific ideas; but in either realm, the dialectical process presupposes the working out, the "unfolding," of tensions, contradictions, and oppositions within the meaningful content of the concepts and the cultural forms. Evidently only man, among the thinking animals, can think dialectically; this is so, as Engels sees it, because only man's brain, developed in size and function, makes it possible for him to abstract, to universalize, to engage in self-critical analysis of the concepts that have been formed by that ability to generalize, and to perform what Marx called "universal labor."

Third, dialectics of a subject matter clearly describes and requires indeterminateness, vagueness, transitional states and critical points, the simultaneous and reconciled presence of polar opposites. On the other hand, the everyday life of crafts and technology, and the specialized sciences for the most part also—Engels calls it "for daily use, for scientific retail"—can get along with the approximations of fixed, at times naively metaphysical, notions and categories that have only relative validity and artificial, even arbitrary and conventional, standing.

Fourth, whatever the subject matter, a dialectical theory requires interaction forces, an idea supported by Engels' optimistic reading of contemporary science, which he saw as engaged in seeking and understanding "universal reciprocal interactions."

Fifth, Engels argues for the power of hypothetical thinking, which he sees too little respected by empiricist philosophers and historians of science:

> The form of development of natural science, in so far as it thinks, is the *hypothesis*. A new fact is observed which makes impossible the previous method of explaining the facts belonging to the same group. From this moment onwards new methods of explanation are required—at first based on only a limited number of facts and observations. Further observational material weeds out these hypotheses, doing away with some and correcting others, until finally the law is established in pure form . . . (*Dialectics of Nature*, 318–319).

It is easy to see Engels' desire for a sensible fusion of the strikingly imaginative achievements of modern scientific thought (hypotheses), based on empirical findings, with the flexibility, dynamics, and self-critical analytic power of dialectical logic. But he had other goals to pursue in his reflections on science and nature. In view of one penetrating theme of historical materialism—that all knowledge, without exception, is a function of human labor, socially situated—how was Engels to understand the continuing component of materialism? Human labor is both biological and sensuous activity and socially mediated throughout. Engels sought a materialist explanation of the objective source of the active, sensuous, laboring qualities; and he pressed for causal (and no doubt dialectical) hypotheses to account for *praxis* and morality.

This revival of some goals of French materialism

ENGELS

was deliberate, but not reductionist. Materialist explanation is not biologism; but, similarly, a socially mediated biology should not eliminate biologically given natural limits—that is, it must not be Utopian and idealist. For Engels this problem of material causation of human actions bore upon human choice and moral questions, since human decisions are the end results of multiple causes—the vector sum of social, biological, and other forces. At one point he complained, "*Ideal* driving forces are recognized but . . . the investigation [is] not carried further back behind these into their motive causes." However inadequately developed materialist psychology must have seemed to him at that time, Engels believed in the determinism of a materialist account of subjective events—acts of will, cognitive processes, emotional dynamics, evaluating and choosing values, evaluating and choosing the means. Human subjectivity is educable, and thereby responsible—often unpredictable in the individual case, not because there are no determining causes but because there are too many! Marx's delight in Darwin, after all, was in Darwin's causal account of *telos*.

The draft of *Dialectics of Nature*, together with *Anti-Dühring* and the book on Feuerbach, sketch out Engels' concerns with science, which are deep and subtle despite his powerfully clear and simple style:

(1) Nonhuman nature is not fixed in its states of existence, but historical; not merely moving, but genuinely changing, since novel entities and properties come into existence.

(2) Even novelty is relative (to its causal origins).

(3) Scientific thinking about nature plausibly introduces useful but ultimately superficial abstractions, essences, fixed species, unchanging atoms, static instincts; and these must be criticized as only relatively valid.

(4) The knowledge of various levels of nature cannot, in their separate qualities, show them to be accountable for each other; neither mechanism, nor vitalism, nor a sociocentric account of all human knowledge will suffice.

(5) Philosophical presuppositions and interpretations of science are unavoidable and only foolishly denied: philosophical clues, heuristics, and anticipations have occurred throughout European history.

(6) The natural environment sets ultimate boundaries to mankind, limits to social and individual growth, pleasure, and age.

(7) Natural and human history alike bring dreadfully painful as well as progressive qualities to evolutionary advances—as portrayed in Engels' vivid account of the original initiation of class society out of so-called gentile society, in *The Origin of the Family* (indebted to the research of L. H. Morgan). Engels went so far as to say, "Each advance in organic evolution is at the same time a regression, fixing *one-sided* evolution and excluding evolution along many other directions."

(8) The relation of man to nature is active, through labor, but also passive; the causal influences of nature upon man stimulate action, no doubt, but set his abilities and inabilities. Here, for example, Engels thought of heredity and of thermodynamics, but his point was general.

(9) The achievement of science must be winnowed into what is objective and what is not. Part of the superstructure yet part of the base, transcending historical limits of class societies, science gives cognitive and practical mastery that cannot be wholly understood as ideological; it must also be understood that the scientific achievements of slave, feudal, and capitalist societies are not slave, feudal, or capitalist truths.

(10) Hence the history of science is not simply a sector of the history of culture; science is qualitatively different from religion or jurisprudence, for example.

In his final contemplation of man within nature, Engels' view might seem to be unresolved. The natural limitations of disease, accident, and span of life are eternally there, to be treated and coped with. No doubt a socialist society will no longer let the improvements due to science cause alienation and exploitation. But scientific coping with natural limits causes no revolutionary transformation in the laws of nature, and apparently cannot. This is no justification for being anti-science, in Engels' view. Revolutionary action is wholly a matter of human history. In the end it seems clear and not troublesome to him that communism cannot defeat the inexorable qualities of the natural order.

Engels writes tentatively of the origin of our "island universe," of the solar systems of 20 million stars and their gradual, though certain, extinction. He does not know whether the ashes of our planetary system will be the raw material for a new one; but he remarks that either there is a divine creator or there is the possibility of a new planetary system arising in accordance with the "*nature inherent* in moving matter, and the conditions for which, therefore, must also be reproduced by matter, even if only after millions and millions of years and more or less by chance, but with the necessity that is also inherent in chance" (*Dialectics of Nature*, 52). Speculating upon the mutual collapse of the heavenly bodies and the resultant vast temperature increase, Engels sees as yet unknown forces responsible for energy storage, for recombination of novel forms of motion, and for the "reconversion of extinct suns into incandescent vapor," with new galaxies, suns, and planets once more evolving.

Like his contemporary T. H. Huxley, and Arthur Eddington, fifty years later, Engels wrote a scientific epilogue to Mephistopheles:

> . . ."all that comes into being deserves to perish." Millions of years may elapse, hundreds of thousands of generations be born and die, but inexorably the time will come when the declining warmth of the sun will no longer suffice to melt the ice thrusting itself forward from the poles; when the human race, crowding more and more about the equator, will finally no longer find even there enough heat for life; when gradually even the last trace of organic life will vanish; and the earth, an extinct frozen globe like the moon, will circle in deepest darkness . . . instead of the bright, warm solar system with its harmonious arrangement of members, only a cold, dead sphere will still pursue its lonely path through universal space. And what will happen to our solar system will happen sooner or later to all the other systems of our island universe; it will happen to all the other innumerable island universes, even to those the light of which will never reach the earth while there is a living human eye to receive it.
>
> And when such a solar system has completed its life history and succumbs to the fate of all that is finite, death, what then? Will the sun's corpse roll on for all eternity through infinite space, and all the once infinitely diversely differentiated natural forces pass for ever into one single form of motion, attraction? (*Dialectics of Nature*, 49–50).

But no, Engels sees a scientific truth in ancient wisdom. The mechanism of chance and the emergence of structures is the materialist source of a dialectic in nature, the eternal source of the renewal of life. Engels concludes the sole completed section of his *Dialectics of Nature* with serene cosmic optimism:

> It is an eternal cycle in which matter moves, a cycle that certainly only completes its orbit in periods of time for which our terrestial year is no adequate measure . . . a cycle in which every finite mode of existence of matter, whether it be sun or nebular vapour, single animal or genus of animals, chemical combination or dissociation, is equally transient, and wherein nothing is eternal but eternally changing, eternally moving matter and the laws according to which it moves and changes. But however often, and however relentlessly, this cycle is completed in time and space; however many millions of suns and earths may arise and pass away, however long it may last before, in one solar system and only on *one* planet, the conditions for organic life develop; however innumerable the organic beings, too, that have to arise and to pass away before animals with a brain capable of thought are developed from their midst, and for a short span of time find conditions suitable for life, only to be exterminated later without mercy—we have the certainty that matter remains eternally the same in all its transformations, that none of its attributes can ever be lost, and therefore, also, that with the same iron necessity that it will exterminate on the earth its highest creation, the thinking mind, it must somewhere else and at another time again produce it (*Dialectics of Nature*, 54).

BIBLIOGRAPHY

I. Original Works. For information on the published and unpublished writings of Marx and Engels see the article on Karl Marx elsewhere in this volume. In addition see the extensive bibliography by L. Stohr, *Friedrich Engels*, which is Bibliographische Kalendarblatter, 30. Sonderblatt (Berlin [Stadtbibliothek], 1970), 1–72.

A chronological list of the principal works by Engels, including those written with Marx, includes the following. The date of composition is indicated in parentheses. *Letters From Wuppertal* (1839); *Outlines* [Umrisse] *of a Critique of Political Economy* (1843); *The Holy Family* (1844), written with Marx; *The Condition of the Working Class in England* (1844–1845); *The German Ideology* (1845–1846), written with Marx; *Principles of Communism* (1847); *Manifesto of the Communist Party* (1847–1848), written with Marx; *The Peasant War in Germany* (1850); *Dialectics of Nature* (1873–1883; incomplete); *Herr Eugen Dühring's Revolution in Science*, known as *Anti-Dühring* (1876–1878); *The Origin of the Family, Private Property and the State* (1884); *Ludwig Feuerbach and the Outcome of Classical German Philosophy* (1886); population studies, especially on Malthus, collected in *Marx and Engels on Malthus*, translated and edited by D. I. and R. L. Meek (London, 1953); and military studies, collected in *Engels as a Military Critic*, W. O. Henderson and W. H. Chaloner, eds. (London, 1959).

II. Secondary Literature. The vast literature on Karl Marx usually extends to important aspects of the life and work of Friedrich Engels. Biographical studies of Engels include the standard and pioneering work of Gustav Mayer, *Friedrich Engels, eine Biographie*, 2 vols., 2nd ed. (The Hague, 1934), which has a drastically edited English trans. in 1 vol. by G. and H. Highet and R. H. S. Crossman, *Friedrich Engels, a Biography* (London–New York, 1936); Auguste Cornu, *Karl Marx und Friedrich Engels, Leben und Werk*, 3 vols. (Berlin, 1954–1962), a detailed work on the early Marx and Engels; H. Gemkow *et al., Friedrich Engels. Eine Biographie* (Berlin, 1970; English trans., Berlin, 1972); H. Hirsch, in *Selbstzeugnissen und Bilddokumenten*

(Reinbek bei Hamburg, 1968); and a long "Profil" (Wuppertal, 1970); L. F. Ilyichov *et al., Frederik Engels. Biografia* (Moscow, 1970; English trans., Moscow, 1974); and W. O. Henderson, *The Life of Friedrich Engels*, 2 vols. (London, 1976). Also see Steven Marcus, *Engels, Manchester and the Working Class* (New York, 1974), a study of the young Engels; and M. Klein *et al.*, eds., *Geschichte der Marxistischen-Leninistischen Philosophie in Deutschland* (Berlin, 1969). A major and invaluable portion of Yvonne Kapp, *Eleanor Marx*, II, *The Crowded Years 1884–1898* (London, 1976), is devoted to Engels.

Among the many commentaries, expositions, and debates concerning Engels' work, the following (as well as those listed in the Marx article) bear particularly in whole, or at least in significant parts, on his analysis of nature, science, methodology, and the history of science: F. Adler, "Friedrich Engels und die Naturwissenschaft," in O. Jenssen, ed., *Marxismus und Naturwissenschaft* (Berlin, 1925), 146–177; M. Adler, *Engels der Denker* (Berlin, 1920); F. Baptiste, *Studien zu Engels' 'Dialektik der Natur'* (Bonn, 1971); J. D. Bernal, *The Social Relations of Science* (London, 1939); and *Science and Industry in the Nineteenth Century* (London, 1953); E. Bloch, *Das Materialismusproblem, Gesamtausgabe,* VII (Frankfurt, 1972); F. Borkenau, *Der Übergang vom feudalen zum bürgerlichen Weltbild* (Paris, 1934; repr. Berlin, n.d. [*ca.* 1970])—see critique by H. Grossmann, "Die gesellschaftlichen Grundlagen der mechanistischen Philosophie und die Manufaktur," in *Zeitschrift für Sozialforschung*, **4** (1935), 161–231 (also bound in the Berlin reprint); R. S. Cohen, "Dialectical Materialism and Carnap's Logical Empiricism," in P. A. Schilpp, ed., *The Philosophy of Rudolf Carnap* (La Salle, Ill., 1963), 99–158; Lucio Colletti, *From Rousseau to Lenin: Studies in Ideology and Society* (London, 1972), trans. of *Ideologia e società* (Rome, 1969); *Marxism and Hegel* (London, 1973), trans. of pt. II of *Il Marxismo e Hegel* (Bari, 1969); and "Marxism and the Dialectic," in *New Left Review,* no. 93 (1975), 3–29; S. Coontz, *Population Theories and the Economic Interpretation* (London, 1957); A. Deborin, "Materialistische Dialektik and Naturwissenschaft," in *Unter dem Banner des Marxismus*, I (Berlin–Frankfurt, 1925–1926), 429 ff.; also in Russian, "Materialisticheskaya dialektika i estestvoznanie," in *Voinstvooyushchy materialist*, no. 5 (1925); and in A. Deborin, *Dialektika i estestvoznanie* (Moscow–Leningrad, 1929–1930), 22–25; repr. from 1925 German trans. in A. Deborin, N. Bukharin, *et al., Kontroversen über dialektischen und mechanistischen Materialismus,* O. Negt, ed. (Frankfurt, 1969), 93–134; G. Della Volpe, *Logica come scienza storica* (Rome, 1969); and E. Fiorani, *Friedrich Engels e il materialismo dialettico* (Milan, 1971).

Also see J. Habermas, "Technology and Science as 'Ideology,'" in *Toward a Rational Society* (London–Boston, 1971), 81–122; A. R. Hall, "Merton Revisited, or Science and Society in the Seventeenth Century," in *History of Science*, **2** (1963), 1–16; B. Hessen, "Social and Economic Roots of Newton's 'Principia,'" in *Science at the Crossroads* (London, 1931), 149–212 (repr. 1971), also repr. separately, R. S. Cohen, ed. (New York, 1971); D. C. Hodges, "Engels' Contribution to Marxism," in R. Miliband and J. Saville, eds., *Socialist Register 1965* (London, 1965), 297–310; S. Hook, "Dialectic and Nature," in his *Reason, Social Myths and Democracy* (New York, 1940), 183–226; J. H. Horn, *Wiederspiegelung und Begriff* (Berlin, 1958); Z. A. Jordan, *The Evolution of Dialectical Materialism* (London–New York, 1967); B. M. Kedrov, *Über Engels Werk 'Dialektik der Natur'* (Berlin, 1954), trans. from the Russian (Moscow, 1950); and *Klassifizierung der Wissenschaften*, I (Cologne–Moscow, 1975), Russian ed., including vol. II (Moscow, 1961); M. Klein and H. Ley, eds., *Friedrich Engels und modern Probleme der Philosophie des Marxismus* (Berlin, 1971), Russian ed., M. T. Jowtschuk, ed. (Moscow, 1971); K. Korsch, *Marxism and Philosophy* (New York–London, 1970; 1st German ed., Berlin, 1923); W. Leiss, *The Domination of Nature* (New York, 1972); H. Ley, *Friedrich Engels' philosophische Leistung und ihre Bedeutung für die Auseinandersetzung mit der bürgerlichen Naturphilosophie* (Berlin, 1957); G. Lichtheim, "Engels" and "Dialectical Materialism," in *Marxism: An Historical and Critical Study* (London, 1961), 234–258; S. Lilley, "Social Aspects of the History of Science," in *Archives internationales d'histoire des sciences*, **28** (1949), 376–443; E. Lucas, "Marx' und Engels' Auseinandersetzung mit Darwin," *International Review of Social History*, **9** (1964), 433–469; and G. Lukacs, *Geschichte und Klassenbewusstsein. Studien über marxistische Dialektik* (Berlin, 1923), translated by Rodney Livingstone as *History and Class Consciousness* (London–Cambridge, Mass., 1971).

Additional works are S. F. Mason, *A History of the Sciences* (New York, 1962); H. Mehringer and G. Mergner, eds., *Debatte um Engels*, 2 vols. (Reinbek bei Hamburg, 1973); M. Merleau-Ponty, "Marxisme et philosophie," in his *Sens et non-sens* (Paris, 1948), 253–277; J. Needham, *Time the Refreshing River* (London, 1948); relevant chapters newly edited as *Moulds of Understanding*, Gary Werskey, ed. (London, 1976); *The Grand Titration* (London–Toronto, 1969); H. Pelger, ed., *Friedrich Engels 1820–1970*, edited proceedings and documents from the International Conference in Wuppertal, 25–29 May 1970 (Hannover, 1971); G. Prestipino, *Natura e società: Per una nuova lettura di Engels* (Rome, 1973); K. Reiprich, *Die philosophisch-naturwissenschaftlichen Arbeiten von Karl Marx und Friedrich Engels* (Berlin, 1969); J.-P. Sartre, "Materialisme et revolution," in his *Situations*, I (Paris, 1947), 135–225, trans. as "Materialism and Revolution," in J.-P. Sartre, *Literary and Philosophical Essays* (London, 1955), 185–239; Alfred Schmidt, "Toward a Critique of Engels's Dialectics of Nature," in his *The Concept of Nature in Marx* (London, 1971), 51–62, trans. of the rev. German ed. (Frankfurt, 1971), 45–58;

S. Timpanaro, *On Materialism* (London–Atlantic Highlands, N. J., 1975), trans. of *Sul materialismo* (Pisa, 1970); R. M. Young, "The Historiographic and Ideological Contexts of the 19th Century Debate on Man's Place in Nature," in M. Teich and R. M. Young, eds., *Changing Perspectives in the History of Science* (London–Dordrecht–Boston, 1973), 344–438; and "Darwin's Metaphor: Does Nature Select?" in *Monist*, **55** (1971), 442–503; and E. Zilsel, "The Sociological Roots of Science," in *American Journal of Sociology*, **47** (1942), 544; and "The Genesis of the Concept of Physical Law," in *Philosophical Review*, **51** (1942), 245–279.

See the coordinate article on Karl Marx in this volume of the *Dictionary*.

ROBERT S. COHEN

ENGLER, HEINRICH GUSTAV ADOLF (*b*. Sagan, Silesia, Germany [now Zagań, Poland], 25 March 1844; *d*. Berlin-Dahlem, Germany, 10 October 1930), *botany*.

Engler was the son of August Engler, a merchant, and Pauline Scholtz. At an early age his mother took him to Breslau, Germany (now Wrocław, Poland), where he completed a classical education at the Magdalenen Gymnasium. During his years at the Gymnasium, he accompanied Rudolf von Uechtritz on botanical field trips. Engler later worked on the difficult and variable genus *Saxifraga* for his dissertation, for which he obtained a doctorate under Goeppert at the University of Breslau. After some years of teaching at Breslau, Engler became curator of the botanical collections at the Botanische Staatsanstalt in Munich, then under the supervision of Naegeli. Under the latter's critical guidance, Engler reached maturity in his systematic work. After completing his monograph on *Saxifraga*, he began work on the tropical families Olacaceae, Icacinaceae, and the especially difficult Araceae for Martius and Eichler's international team effort, the *Flora Brasiliensis* (1840–1906). With these wider studies Engler developed the systematic theory and methodology with which he would influence and dominate almost a half-century of plant taxonomy: a basically comparative morphological method was supplemented with elements from phytogeography, anatomy, embryology, and even—although more modestly—from phytochemistry. While conducting these studies Engler came to accept the Darwinian ideas on evolution as a background explanation of organic diversity. In 1878 he was offered the chair of systematic botany at the small (250 students) University of Kiel. Engler enthusiastically accepted this chance to pursue a university career, although he was not particularly gifted either as a teacher or an orator. His very light teaching duties left him ample time to develop his original ideas as well as his organizational skill. In 1880 he founded the *Botanische Jahrbücher für Systematik, Pflanzengeschichte und Pflanzengeographie*, of which he edited and published sixty-two volumes in the course of nearly fifty years. For most of that period *Engler's Jahrbücher*, as it was usually called, was the world's foremost journal of plant systematics and phytogeography.

While at Kiel, Engler wrote *Versuch einer Entwicklungsgeschichte der Pflanzenwelt . . .* (1879–1882), which brought him widespread fame. Essentially, the book is the first attempt at a genetic and historical theory of the origin of the floristic diversity of the Northern Hemisphere. Engler proved that during the Tertiary period there existed in this hemisphere a somewhat homogeneous flora, which, modified by later glaciation, resulted in the present diversity of the floras of Europe, Eastern Asia, and Pacific and Atlantic North America.

In 1884 Engler was recalled to the University of Breslau to succeed Goeppert as professor of systematic botany and director of the botanical garden. His colleague from earlier Breslau days, Ferdinand Julius Cohn, was in charge of physiological botany; and it was Cohn who was mainly responsible for Engler's remarkably broad interest in botany, comprising the cryptogams and also the higher plants. The Breslau period was significant because of Engler's idea of producing an encyclopedia of the descriptive plant sciences. With Karl A. E. Prantl from Aschaffenburg, who was primarily in charge of the cryptogams and who died in 1893 in the early stages of the work, Engler began *Die natürlichen Pflanzenfamilien*, which was to become the last complete systematic and monographic survey of the plant kingdom from the generic level up to the present day. Between March 1887 and September 1915, 248 installments appeared, written by numerous collaborators. The basis of all treatments was clearly provided by Engler, who published several versions of his "Principien der systematischen Anordnung" in various volumes of the *Pflanzenfamilien* and also in the nine editions of his highly successful one-volume abridgment, *Syllabus der Pflanzenfamilien*.

Realizing that the theory of evolution provided the background for an understanding of organic diversity, Engler stressed that in actual practice there were insufficient historical and experimental data for a fully justified phylogenetic arrangement.

Believing that it would be wrong to assume *per se* a monophyletic origin of plant life, he pointed out that parallel developments must have taken place at different times and places. Therefore, the actual basis for the system elaborated in the *Pflanzenfamilien* remained that outlined during his years at Munich: morphology, phytogeography, and related subdisciplines had to provide the basic data for what would remain a subjective assessment of formal or genetic relationships. Premature evolutionary speculations had to be avoided. First, priority had to be given to the assembling of basic data. The Darwinian and Mendelian ideas on the origin of diversity, however, should never be lost sight of when trying to explain diversity. The Engler system of plant classification came to its full development in the *Pflanzenfamilien* and the *Syllabus*. Essentially the system recognized, contrary to earlier systems, the great diversity of what had so far been called "cryptogams," and attributed high and equal rank to the various groups of algae, fungi, mosses, and ferns. The phanerogams, called Embryophyta siphonogama, were on an equal footing with twelve divisions of former cryptogams. The arrangement of the flowering plants still put the more highly specialized monocotyledons before the dicotyledons, which were later shown to contain all ancestral forms. Nevertheless, Engler's system remained for many years the most widely accepted classification of all plants.

Through his many publications and clearly stated theoretical assumptions Engler had already won worldwide fame by 1889, when he was chosen to succeed Eichler as professor of botany and director of the botanical garden at Berlin. It was there that his great organizational gifts found their scope in the Second Reich environment of economic and colonial as well as cultural and scientific expansion. In the large new botanical garden, which was set up and developed between 1895 and 1910 in nearby Dahlem, Engler was able to illustrate his ideas on phytogeography. To continue the expansion of his work and ideas, a botanical museum and a staff of collaborators were at his disposal. The cryptogamic volumes of the *Pflanzenfamilien* began to appear, and material from Africa directed research on phanerogams increasingly toward the tropical regions of this continent.

During his Berlin period Engler produced his most important publication as sole author, the various volumes of *Die Pflanzenwelt Afrikas*. This work was part of a series of volumes on the main phytogeographic regions of the world: *Die Vegetation der Erde*. Engler also started a monographic survey of the plant world down to the species level, *Das Pflanzenreich*. After his retirement in 1921, he started a second edition of his *Natürlichen Pflanzenfamilien*.

As a scientist and organizer Engler dominated the *Engler-Zeit*, an entire era in systematic botany, in which the descriptive activities in the discipline reached an all-time high.

BIBLIOGRAPHY

A complete list of Engler's publications is in Ludwig Diels, "Zum Gedächtnis von Adolf Engler," in *Botanische Jahrbücher für Systematik, Pflanzengeschichte und Pflanzengeographie*, **64** (1931), i–lvi.

Engler's works include *De genere Saxifraga* (Halle, 1866); *Monographie der Gattung Saxifraga* (Breslau, 1872); *Versuch einer Entwicklungsgeschichte der extratropischen Florengebiete der nördlichen Hemisphäre* (Leipzig, 1879–1882); *Versuch einer Entwicklungsgeschichte der Pflanzenwelt insbesondere der Florengebiete, seit der Tertiärperiode*, 2 vols. (Leipzig, 1879–1882); *Die natürlichen Pflanzenfamilien* (Leipzig, 1887–1915; 2nd ed., Leipzig, 1924–1960), written with Karl A. E. Prantl, as editor and contributor; *Syllabus der Vorlesungen über specielle und medicinisch-pharmaceutische Botanik* (Berlin 1892; 9th ed., 1924); *Die Pflanzenwelt Ost-Afrikas* (Berlin, 1895); *Das Pflanzenreich* (Berlin, 1900–1953), as editor and contributor; and *Die Pflanzenwelt Afrikas* (Leipzig, 1910–1915), which is pt. 9 of his *Die Vegetation der Erde*, 15 vols. (Leipzig, 1896–1923), written with O. Drude.

No formal biography of Engler exists. Diels's obituary, cited above, is the most complete and independent evaluation of Engler as a scientist; the many other published obituaries add little to the basic picture. For a review of secondary literature on Engler's major publications, see F. A. Stafleu, *Taxonomic Literature* (Utrecht, 1967), 133–149.

FRANS A. STAFLEU

EVANS, ALEXANDER WILLIAM (*b.* Buffalo, New York, 17 May 1868; *d.* New Haven, Connecticut, 6 December 1959), *botany.*

Alexander Evans died, at the age of ninety-one, the undisputed leader in two unrelated areas of botanical research. After achieving eminence as a hepaticologist, he turned to lichenology and made contributions of equal importance to that field. His mysterious withdrawal from hepaticology, at an advanced age, and his youthful approach to a new interest have continued to charm and intrigue practitioners of both specialties.

Evans was the youngest of seven children of

EVANS

William A. and Maria Ives Beers Evans. His father, a planing mill operator, died in 1880, whereupon the family moved to New Haven. Evans attended public high school there and continued his education at Yale University, where he received the Ph.B. (1890), M.D. (1892), Ph.D. (1899), and D.Sc. degrees (1947).

After medical school Evans served a two-year internship at the New Haven City Hospital, but he never practiced medicine. As early as 1888, his first year in college, he had been influenced by Daniel Cady Eaton to take up the study of bryophytes. During undergraduate and medical training he added more than twenty liverworts and a good number of *Sphagna* to the known flora of Connecticut and also found, in New Hampshire, a curious moss that Eaton described as new. (This moss, *Bruchia longicollis*, has not been collected since then.) From 1891 to 1893 Evans published four papers on liverworts (of Hawaii, Connecticut, Virginia, West Virginia, and Patagonia), including excellent descriptions and illustrations of seven new species. He demonstrated such competence that it is not surprising that he decided to give up medicine altogether. In the fashion of the time, he went to Germany in the fall of 1894 to study botany at the universities of Berlin and Munich.

In the following spring Evans was summoned by cablegram to return to Yale to fill the vacancy left by the death of Eaton. He served there for the rest of his life, as instructor (1895–1901), assistant professor (1901–1906), Daniel Cady Eaton professor (1906–1936), and professor emeritus (1936–1959). He was chairman of the department of botany for many years.

On his return Evans prepared himself to meet the requirements of a Ph.D. degree. Tradition has it that as sole preceptor of botany, he rejected his first dissertation (a revision of North American species of *Frullania*), whereupon he successfully defended a second, entitled "The Hawaiian Hepaticae of the Tribe Jubuloideae."

For years Evans spent part of each summer in Europe, visiting herbaria and searching for literature. He attended several international botanical congresses at Paris (1900), Vienna (1905), and Brussels (1910).

Evans was not skilled as an administrator, and as a teacher he was not able to handle large groups of uninterested students. He worked well, however, with individual students who showed motivation; and two of his graduate students, Margaret Fulford and Hempstead Castle, became hepaticologists of note.

Evans' taxonomic acumen was based on a complete familiarity with plants. Extensive field study gave authority to his monumental publications on *Cladonia*, to many papers on liverworts, and to a book, *The Bryophytes of Connecticut* (1908), written with George Nichols. The last, modest in size and format (and selling for thirty cents), appeared at a time when there were no convenient means of identifying bryophytes in any part of North America. It is still useful because of well-constructed keys to genera and species representative of most of eastern North America. Evans also collected extensively elsewhere in New England and along the eastern seaboard, and made short visits to Cuba, Jamaica, Puerto Rico, Panama, Venezuela, and Colombia.

Evans' work on hepaticology, by no means outmoded now, is admirable because of his effective use of morphological detail in interpreting relationships. As a teacher he had offered instruction, at one time or another, in virtually every discipline of plant science; but it was to morphology that he was most devoted. A product of the classical school of morphology, he contributed much in the way of structural and developmental detail, and gave phylogenetic meaning to a mass of information already in the literature. His account of reduction as a prevailing trend in liverwort evolution and his scheme of classification based on that trend reflect a thoroughly modern point of view.

In 148 articles on liverworts, Evans described seven genera and more than 135 species new to science. His papers were thorough and even encyclopedic in documentation, yet concise and readable. (After Evans' retirement, M. L. Fernald, editor of *Rhodora* and master craftsman at writing, published a lament that few botanists could write as well as the "late" Alexander Evans—who was active in publication long after Fernald's demise.) Evans' taxonomic contributions were enhanced by unusually graceful and accurate drawings.

Most of Evans' papers were floristic accounts—more than half on North America, a quarter on Latin America, and a goodly number on the exotic floras of Hawaii, Sumatra, Japan, and elsewhere. His "Notes on New England Hepaticae" (1902–1923) and "Notes on North American Hepaticae" 1910–1923) are among the most informative reference sources available to American workers. His "Hepaticae of Puerto Rico" (1902–1911) is essentially a monograph of the Lejeuneaceae. Evans contributed monographs on numerous families of thalloid liverworts to the *North American Flora* and published many generic revisions—of

Marchantia, Dumortiera, Riccardia, Metzgeria, Symphyogyna, Hymenophytum, Acromastigum, Herberta, Frullania, and *Lejeunea.*

Evans intended to go a further step and prepare a comprehensive liverwort flora of North America. That project was never completed because of the rash and precipitate appearance of T. C. Frye and Lois Clark's *Hepaticae of North America*, a warts-and-all synthesis from the literature, and even more because of Evans' late-born interest in *Cladonia.* He began to collect lichens in Connecticut in 1924, methodically filling in county and township records. His specimens, mostly named by G. K. Merrill, were recorded in *Catalogue of Connecticut Lichens* in 1926, with additions in 1927. Evans produced a *Cladonia* flora of the state in 1930; and for the next thirty years he specialized in that large and difficult genus and only rarely bothered with other lichens or hepatics.

Evans was not interested in popularizing *Cladonia*, attractive as the genus is, and his papers can be appreciated only by a specialist. Although he described fewer than ten new species, he made *Cladonia* the world's best-known lichen genus. His major achievement was the successful use of chemistry to define species, employing microchemical methods that Yasuhiko Asahina had only recently applied to *Cladonia* in Japan. Heinrich Sandstede, in a supplement to E. A. Vainio's monograph on *Cladonia*, had given notice of these methods in 1938; but Evans, in 1943, explained and exemplified them in a particularly effective way. These microchemical methods provided the means to extract substances from mere bits of thalli and to identify them microscopically by their crystalline structure, thus enabling a taxonomist to demonstrate distinctive substances quickly and easily, without needing the knowledge or methods of the organic chemist. Evans had a narrow concept of species, based on familiarity with plants gained from unhurried study. His knowledge of *Cladonia* was so intimate that he could recognize many taxa that less critical workers might have overlooked, and in some instances he was able to correlate morphological subtlety with chemical difference. His conclusions regarding chemical taxa were so logical and so convincing that chemotaxonomy has become essential in the study of *Cladonia* and has been applied with outstanding success to numerous other genera of lichens.

Evans published twenty-five papers on lichens, including important accounts of the *Cladoniae* of Connecticut, New Jersey, Vermont, Florida, and the Carolinas. With P. R. Burkholder and others he wrote two papers reporting antibiotic properties of lichens—another area of pioneering activity.

Evans' *Cladonia* herbarium of nearly 40,000 specimens is housed at the Smithsonian Institution, and his hepatic herbarium of more than 30,000 specimens is kept at Yale.

Evans was on the editorial board of *Bulletin of the Torrey Botanical Club* from 1907 to 1934 and served as editor from 1914 to 1924. He was associate editor of *Bryologist* for many years, secretary of the Connecticut Academy of Arts and Sciences (1897–1903), vice-president of the Botanical Society of Ameria (1911), a fellow of the American Association for the Advancement of Science, and honorary member of the Sullivant Moss Society. In 1947 he received an honorary D.Sc. degree from Yale University. Evans was invited to serve as honorary president of the section on lichenology at the Eighth International Botanical Congress meeting at Paris in 1954, but was unable to attend. On the occasion of the golden jubilee of the Botanical Society of America, in 1956, he was given a certificate of merit as one of fifty who had made outstanding contributions to botanical science.

Evans married Phoebe Whiting of New Haven in 1914. They had three daughters: Margaret, Janet, and Allison. (His daughter Margaret was killed, reportedly by an estranged husband, in the Evans' home in the summer of 1954. Difficult as this was, and at his age, Evans found strength to bear his grief and continue his work with quiet dignity.)

In October 1959, at the age of ninety-one, he fell and broke his hip. He submitted to an operation and was making good progress toward recovery when he contracted pneumonia and died. He was survived by his wife, two daughters, and seven grandchildren. A private service was held in his home and a memorial service at St. John's Episcopal Church.

Lewis Anderson wrote in reminiscence:

> His quiet and genuine modesty, his sense of humility, his continual and almost overpowering regard for others, his brilliant but often subtle sense of humor, and his tremendous personal integrity formed the hard core of his lovable personality. These traits seemed almost to be etched in his face in later years. Somewhat heavy of jowl and for many years afflicted with a slight but very noticeable palsy he had a way of looking at one so that his face would literally light up, and his eyes would twinkle and somehow communicate a mischievous quality that one cannot forget [*Bryologist*, **63** (1960), 87–88].

Evans was neither dynamic nor forceful. He had a mild voice, a gentle manner, and a warm personality, as well as strong cultural interests and a complete indifference to financial concerns. His dignity sometimes expressed itself in incongruity. He rode a bicycle but wore a sober hat in keeping with his age and position. In the field he was careful to wear a jacket and tie in combination with disreputable pants and shoes. Evans was "half decent" in the laboratory too. His articles were models of precision and his herbaria carefully tended, but his office was a shambles. As an old man he had astonishing endurance. In the field, even though he was decidedly paunchy and forbidden by his doctor to collect in hilly country, he could make his way surely over any other terrain, unmindful of fatigue or difficulties underfoot. His palsy made it difficult and tiring to use a microscope, but he worked—productively—six or seven hours nearly every day until the end of his life.

BIBLIOGRAPHY

Bibliographies listing Evans' work in bryology have been provided by Schuster and in lichenology by Hale (see below).

For secondary literature, see L. E. Anderson, "Personal Reflections on Alexander W. Evans," in *Bryologist*, **63** (1960), 84–88, with photograph; M. E. Hale, "Alexander W. Evans and Lichenology," *ibid.*, 81–83, with bibliography; and "Alexander William Evans," in *Bulletin of the Torrey Botanical Club*, **87** (1960), 354–356, with photograph; H. A. Miller, reminiscences, in *Revue bryologique et lichénologique*, n.s. **29**, nos. 1–2 (1960), 140; G. E. Nichols, "Alexander William Evans, Hepaticologist," in *Annales bryologici*, **11** (1938), 1–5; J. R. Reeder, "Alexander William Evans (1868–1959)," in *Taxon*, **9** (1960), 168–169, with photograph; and R. M. Schuster, "Alexander W. Evans—an Appreciation," in *Bryologist*, **63** (1960), 73–81, with photographs and bibliography; and "Alexander W. Evans (1868–1959)," in *Revue bryologique et lichénologique*, n.s. **29**, nos. 1–2 (1960), 132–139, with photograph and bibliography.

HOWARD CRUM

FEATHERSTONHAUGH, GEORGE WILLIAM (*b.* London, England, 9 April 1780; *d.* Le Havre, France, 26 September 1866), *geology*.

Featherstonhaugh was born shortly after his father had died suddenly at the age of twenty-three. He is said to have been related to Sir Henry Featherstonhaugh, second baronet, of Uppark, Sussex, the only son of Sir Matthew Fetherstonhaugh of Northumberland. Featherstonhaugh's mother, Dorothy, alarmed by the Gordon riots, took her infant son to Scarborough, Yorkshire, where her parents, George and Ann Simpson, resided. The boy was brought up there and received an excellent classical education at Stepney Hall, a private academy near Scarborough. He did not enter a university, but at the age of twenty-one began to travel in Europe, visiting France, Italy, and other countries. In 1806 he went to the United States, where he made many friends. He was about to return to England when he met, and subsequently married, Sarah, daughter of the late James Duane, a wealthy landowner who had been mayor of New York City and a federal judge.

Featherstonhaugh then began to farm his wife's property at Duanesburg, near Schenectady, and developed a great interest in agriculture. In 1809 he was elected to the American Philosophical Society; but there is little on record about the next fifteen years of his life, except that in 1819 he became corresponding secretary of the newly established New York Board of Agriculture.

In 1825, with his friend Stephen van Rensselaer (patron of Amos Eaton and a principal figure in the development of geology in the United States), Featherstonhaugh promoted a railway (opened in 1831) from Albany to Schenectady. To improve his knowledge of railway construction and to consult with the railway engineer George Stephenson, in September 1826 he returned to England, accompanied by his wife. There he became acquainted with the leading geologists and displayed an active interest in geology. At Scarborough, where his mother still resided, Featherstonhaugh met the geologist William Smith and took lessons from him in the recognition of different strata and the fossils found in them. He visited Scotland and met Jameson but lost all confidence in him after discovering his outdated Wernerian views. He also met Murchison and, with him, attended some of Buckland's lectures at Oxford. With Buckland he visited Mantell's museum at Lewes and later studied the fossils of the Greensand (Lower Cretaceous). On 17 June 1827 he accompanied Sedgwick, Murchison, and others on a visit to the tunnel under the Thames then being constructed by Marc Brunel. He and Sedgwick were the first to descend in a diving bell to examine the river bottom, returning to the surface, according to Murchison, with "red faces and staring eyes."

In August 1827 Featherstonhaugh went to Paris and spent several weeks with the leading French geologists, particularly Cuvier and the Brongniarts; at the same time he studied the Tertiary beds of the Paris basin. On his return to England, he went to Dorset and Devon with Buckland, and later saw Cornish geology under the guidance of Sir John St. Aubyn, a keen mineralogist.

Shortly after his departure from England at the end of November 1827, he was elected a fellow of the Geological Society of London.

A few months after his return to the United States, his wife died. Featherstonhaugh then abandoned agriculture and left Duanesburg to live in Philadelphia. Only a year later his house in Duanesburg, Featherston Park, was destroyed by fire. In 1828 he sent Murchison a letter "On the Series of Rocks in the United States," comparing the succession with that of England; the letter was read to the Geological Society of London on 2 January 1829 and was published in the *Proceedings*. In it Featherstonhaugh stated his opinion that the American coal measures were analogous to the English ones and were not post-Liassic, as Amos Eaton maintained.

During 1829 Featherstonhaugh gave a series of public lectures on geology at the New York Lyceum of Natural History; these were followed by a similar course in Philadelphia. He had returned to America with a very large collection of minerals, fossils, and recent shells (in a letter to Mantell in July 1827 he stated that a large number had been packed, but about 4,000 still awaited cataloging and packing). Most of these were distributed to museums and societies, and he was soon begging his friends in England to send him more.

Featherstonhaugh felt the lack of an American journal devoted to geology, and in July 1831 began to issue *Monthly American Journal of Geology and Natural Science*. He both edited and contributed extensively to it. Although the journal was well supported, with President Jackson and many military men heading the list of subscribers, the bankruptcy of his publisher after only eight issues had appeared meant that Featherstonhaugh had to finance the rest of the volume himself, and a second one was not commenced.

Between 1832 and 1838 Featherstonhaugh carried out a number of field investigations, some at his own expense, others sponsored officially. He spent the summer and fall of 1833 examining the base of the Cretaceous beds along the "fall line," as far as Virginia. In 1834 he received government instructions to examine the minerals and geology of the Ozark Mountains in northern Arkansas and southern Missouri. With this appointment Featherstonhaugh became the first United States government geologist, receiving six dollars a day and traveling expenses. In his report (1835) he claimed to have traveled 4,600 miles in the time allowed—less than six months. In July 1835 he was instructed to go to Green Bay, Wisconsin, on Lake Michigan, and examine the country to the west as far as the headwaters of the Minnesota River. (The assistant geologist on this trip was W. W. Mather, who, however, was not mentioned in the final report.) In his report Featherstonhaugh also described the rocks he had observed on his way from Washington to Green Bay. In the summer of 1837 he made another lengthy journey to examine various mineral deposits, including the lead mining area near Galena, Illinois (in company with the geologist Richard Cowling Taylor), the Missouri iron deposits, and the gold diggings in Georgia and the western Carolinas. This journey is described in the second part of his book *A Canoe Voyage up the Minnay Sotor* (1847).

In 1835 Featherstonhaugh was elected a fellow of the Royal Society; later that year he showed he had not forgotten the town where he grew up, sending a donation of £25 to the Scarborough Philosophical Society. A few years later he sent a collection of American minerals and fossils, as well as freshwater shells from American rivers, for the society's museum.

In September 1838 Featherstonhaugh went to Canada, where he became involved in diplomatic affairs and had discussions with the governor-general, Lord Durham. The dispute over the boundary between Maine and New Brunswick had brought Britain and the United States near war, and it was not to be settled for some years. In February 1839 Featherstonhaugh arrived in England and was immediately involved in discussions with Lord Palmerston, the foreign secretary. He accepted an invitation to serve as commissioner in the boundary dispute and in July returned to the United States with R. Z. Mudge as associate commissioner and a small party of surveyors. He spent three months, often under extremely arduous conditions, examining the area involved. At the end of December he returned to England to make his report. He was accompanied by his wife (he had remarried in 1831) and a daughter. He never returned to the United States; but his eldest son, James, helped to complete some unfinished work on the New Brunswick boundary.

Featherstonhaugh continued to work at the Foreign Office in London; and he attended several meetings of the British Association, reading a paper at the York meeting in 1844. In 1843 he was asked to return to the United States as boundary commissioner but, being over sixty and not in good health, he refused. In 1844 Featherstonhaugh was appointed British consul at Le Havre, a post he held until his death. As consul he was responsible, in 1848, for smuggling out of France the fleeing king, Louis Philippe, and his queen, passing them off as his uncle and aunt, "Mr. and Mrs. Smith."

A detailed assessment of Featherstonhaugh's contributions to American geology has yet to be made, but certainly the fund of information and large collections with which he returned from England in 1828 must be taken into consideration. It was a critical time, when Amos Eaton was attempting to correlate the American and British strata, but with little knowledge of the characteristic fossils of either. Samuel Morton and Vanuxem were on the right track, and the firsthand account of British strata and fossils brought by Featherstonhaugh to Philadelphia would have been very valuable. But Featherstonhaugh's scathing criticism of Eaton's terminology so infuriated the latter that he widely expressed his contempt for the Anglo-American geologist, and any cooperation between them became impossible. By maintaining a regular correspondence with English geologists, particularly Murchison, Featherstonhaugh kept in touch with new discoveries; and his official reports contain evidence of his wide reading.

(Although some members of the Featherstonhaugh family in England pronounce the name "Fanshaw," others [including his American descendants] pronounce it in full. It should be noted that Murchison's wife always wrote to him as "Dear Mr. Featherstone," and his house at Duanesburg was called "Featherston Park.")

BIBLIOGRAPHY

I. ORIGINAL WORKS. Featherstonhaugh's published geological work is not large. His paper "On the Series of Rocks in the United States" is in *Proceedings of the Geological Society of London*, **1** (1834), 91–93. Max Meisel, *A Bibliography of American Natural History*, II (New York, 1926), 259–260, 526–530, 542, 563, 569, lists papers and reports published by Featherstonhaugh, including those in *Monthly American Journal of Geology and Natural Science* (July 1831–June 1832; facs. repr., New York, 1969). His two official reports are *Geological Report of an Examination Made in 1834 of the Elevated Country Between the Missouri and Red Rivers* (Washington, D.C., 1835) and *Report of a Geological Reconnaissance Made in 1835, From the Seat of Government, by Way of Green Bay and the Wisconsin Territory, to the Côteau de Prairie* (Washington, D.C., 1836). There is much geology in his two popular works, *Excursion Through the Slave States From Washington on the Potomac to the Frontier of Mexico; With Sketches of Popular Manners and Geological Notices*, 2 vols. (London–New York, 1844), and *A Canoe Voyage up the Minnay Sotor; With an Account of the Lead and Copper Deposits in Wisconsin*, 2 vols. (London, 1847; facs. repr., St. Paul, Minn., 1970). His last geological paper seems to have been "On the Excavation of the Rocky Channels of Rivers by the Recession of Their Cataracts," in *Report of the British Association for the Advancement of Science*, **14** (1845), sec. 2, 45–46.

Unpublished MS sources are numerous and have been drawn on for the above account. There is a large collection of letters to Featherstonhaugh from British geologists, and some from him, in the University of Cambridge Library. These include over eighty from Murchison and his wife, dated 1827–1861. Others to Murchison are in the library of the Geological Society of London, and three to Mantell are in the Alexander Turnbull Library, Wellington, New Zealand. Over 100 letters to H. S. Fox, British envoy at Washington, written 1838–1843, are in the Bodleian Library, Oxford, and throw much light on Featherstonhaugh's work as a boundary commissioner. Some of his MSS are in the collections of the American Philosophical Society, and others remain with his descendants (James Featherstonhaugh, of Duanesburg, N.Y., has correspondence and day journals).

II. SECONDARY LITERATURE. There is no full-length biography of Featherstonhaugh. W. H. G. Armytage, "G. W. Featherstonhaugh, F.R.S., 1780–1866, Anglo-American Scientist," in *Notes and Records. Royal Society of London*, **11**, no. 2 (1955), 228–235, provides a useful account, valuable for the number of references in American literature to Featherstonhaugh and his activities. Other biographical accounts are G. W. White, in the facs. repr. of *Monthly American Journal of Geology* (New York, 1969), xi–xix, with a bibliography; and the introduction by W. E. Lass in the facs. repr. of *A Canoe Voyage . . .* (St. Paul, Minn., 1970).

JOAN M. EYLES
VICTOR A. EYLES

FINE, ORONCE (*b.* Briançon, France, 20 December 1494; *d.* Paris, France, 6 October 1555), *astronomy, mathematics, cosmography.*

Fine (Orontius Finaeus Delphinatus) was born in the Dauphiné but spent his scientific career at Paris.[1] His father, François Fine,[2] had attended the University of Paris[3] and practiced medicine in Briançon. Upon his father's death Fine was sent to

Paris and was confided to the care of Antoine Silvestre, regent of the Collège de Montaigu and later of the Collège de Navarre. Although he earned his Bachelor of Medicine degree in 1522,[4] his career developed outside the university; in 1531 he was appointed to the chair of mathematics at the recently founded Collège Royal, where he taught until his death.

From 1515 Fine edited astronomical and mathematical writings for printers in Paris and abroad. Among them were Peuerbach's *Theoricae planetarum*, Sacrobosco's *De sphaera* (1516), and Gregor Reisch's *Margarita philosophica* (1535), as well as a tract by his grandfather, Michel Fine, on the plague (1522). He also was responsible for an edition of Euclid's *Elements*, of which he published only the first six books (the manuscript of the seventh book, prepared for the printer, is extant).

Fine's first book (1526) was a treatise on the equatorium, an instrument designed to determine the true positions of the planets. In this work Fine exploited the possibilities of curves traced by points (the diagrams of the equations of center), used to facilitate the placement, with respect to the equant, of the mean apsidal line (*auge*) on the epicycle. These curves, drawn on the basis of lists of the equations of center and of the proportional minutes furnished by the Alphonsine Tables, were a very ingenious innovation. At the same time Francisco Sarzosa composed a treatise on the equatorium with the same innovations.[5] It is difficult to believe that their research was independent, but it is now impossible to establish the proper priority.

Fine wrote four other treatises on the equatorium that are extant in manuscript at the library of the University of Paris (Univ. 149); three treatises are little more than outlines, and the fourth treatise describes an instrument similar to Apian's *Astronomicum Caesareum*, with the planetary instruments bound as a book (each of them simply reproduces the geometric decomposition of the Ptolemaic theory of epicycles).

Fine's further works on astronomical instruments include treatises on the new quadrant (1527, 1534) and on the astrolabe (incomplete manuscript layouts are in Paris lat. 7415 and Univ. 149). These are not innovative and offer only the standard university account. Fine also inserted a treatise on the new quadrant at the end of his work on gnomonics, *De solaribus horologiis et quadrantibus*, which first appeared in 1532 as the concluding section of his *Protomathesis*. The latter consists of four parts, each with its own title page and each separately reprinted: *De solaribus horologiis* has a separate title page dated 1531, but it is not known to exist separately, and it is unlikely that it was distributed by itself. Among the many types of sundials described in this book are a multiple dial and a *navicula*.[6] A very rare ivory *navicula* signed "Opus Orontii F. 1524"—the only scientific instrument certainly attributable to Fine, and perhaps the only one he ever constructed—is in the private Portaluppi collection at Milan.

Besides treatises on instruments, Fine's astronomical work included theoretical writings of a popular nature. These were presented at the two levels of traditional instruction: the elementary one represented by Sacrobosco's *De sphaera*, and the higher one of epicyclic astronomy. The *Cosmographia*, an elementary manual, was first published in Latin as the third part of the *Protomathesis* (1532), with a separate title page dated 1530, and was reprinted several times, both in Latin and French. It includes the description of the fixed celestial sphere used for reference, the essential ideas concerning the astronomy of the *primum mobile* (right and oblique ascensions and the duration of diurnal arcs), and a few brief notions of astronomical geography (climates and terrestrial longitudes and latitudes); but it contains no information on the motions of the planets. The latter were discussed in the *Théorique des cieux* (published anonymously in 1528), which gives a detailed exposition of the Alphonsine epicyclic theory, the first one in French. The brief *Canons . . . touchant l'usage . . . des communs almanachs* (1543) is a succinct explanation of an almanac computed for the meridian of Tübingen (undoubtedly by Johann Stöffler, which exists in editions dated 1531 and 1533).

Although Fine's interest in astronomy extended to astrology, he wrote only minor works on that subject. The *Almanach* of 1529, actually a calendar giving the dates and hours of the new moons in the nineteen-year cycle and the duration of the diurnal arcs, included a short commentary on medical astrology. *De XII coeli domiciliis* (1553) was a complete theory of the celestial houses, important for casting horoscopes. In this work Fine adopted the definition of the houses advocated by Campanus, for whom the divisions of the celestial vault, for a given horizon, are constructed on the equal divisions of its first azimuth and converge to the south and to the north of the horizon; following this definition, the one usually employed on astrolabes, the lines of the celestial houses were projected onto the astrolabe in accordance with circles passing through the points of intersection of the horizon and of the meridian. Fine also gave an original definition

FINE

of the unequal hours, however, which were no longer the equal divisions of the diurnal arc and of the nocturnal arc, but the equal divisions of the ecliptic computed, at each moment, from its intersection with the horizon.

The result, for the construction of the tympana of the astrolabes, was a highly original plotting of the hourly lines—a geometric locus of these equal divisions when the ecliptic turns about the axis of the earth according to the daily movement. Although no astrolabe is known to have been constructed on the basis of this definition, it did find an application in the astrolabic dial of the planetary clock at the Bibliothèque Ste. Geneviève. The dial expressly refers to Fine, and it has therefore been assumed since the seventeenth century that the clock itself was his work. This is highly improbable.[7] It is virtually certain that the clock dates from the fifteenth century and that about 1553 one of its panels, containing the dial of the hours and the astrolabic dial, was replaced by Fine with a panel designed to illustrate his new conception of the unequal hours. The level of technical competence displayed by the mechanism of this dial is very low, for the *araignée* (the stereographic projection of the celestial vault) completes its revolution in a mean solar day and not in a sidereal day.

While the third and fourth parts of the *Protomathesis* dealt with astronomy, the first two treated arithmetic and geometry. *De arithmetica practica*, the first part, is Fine's only work on arithmetic. In accordance with the traditional schema of medieval arithmetic, the various operations carried out on the numbers were enumerated and described following a plan that distinguished whole numbers, common fractions (*fractiones vulgares*), and natural or sexagesimal fractions. The latter were of particular interest to practitioners of Alphonsine astronomy, since they were the basis of their preferred mathematical tool. Fine made it easier to work with these fractions by providing a *tabula proportionis* (so called because of its aid in computing proportional parts of the equation of the argument), or multiplication table in sexagesimal numeration, similar to the same table by John of Murs or Bianchini. The last book of the *De arithmetica*, on ratios and proportions, developed theorems established by Euclid and Ptolemy.

The two books on geometry (dated 1530 in the *Protomathesis*) treated the subject at a more elementary level. After stating the definitions of plane and solid figures, borrowed from the *Elements*, as well as the Euclidean postulates, Fine discussed the measurement of length, height, and width in the tradition of the treatises on practical geometry, of which one of the most popular aspects was geometrical canons for the use of the astrolabe. To this end he treated the geometric square, the quadrant, the cross-staff (Jacob's staff), and the mirror. The calculation of surfaces and volumes, which was the complement of the measurement of lengths, included that of circular surfaces and volumes. For the latter, Fine computed the ratio of the circumference of the circle to its diameter as 22/7.

Returning to the ratio of circumference to diameter in *De quadratura circuli* (1544), Fine offered what he believed to be a more precise value of π: $\frac{22+2/9}{7}$. In *De circuli mensura*, which follows *De quadratura*, he reduced that ratio to 47/15. Finally, in the posthumous *De rebus mathematicis* (1556), he increased the value slightly to one between the two preceding ones: $3+\frac{11}{78}$. These attempts to determine the true figure were but one aspect of his efforts to solve the quadrature of the circle, for which he examined several solutions. None of them was satisfactory; and Fine was vehemently attacked by some of his contemporaries, notably Pedro Nuñez Salaciense, in *De erratis Orontii Finaei* (1546), and Johannes Buteo (Jean Borrel), in *De quadratura circuli* (1559). It must be acknowledged that Fine's arrogance about his own accomplishments undoubtedly made his errors of logic all the more intolerable to his opponents.

Fine's work in trigonometry scarcely went beyond what was necessary to establish a table of sines: three chapters of book II of *De geometria*, included in the *Protomathesis*, and *De rectis in circuli quadrante subtensis* and *De universali quadrante*, both published in 1542 as appendixes to the first reprinting of *De mundi sphaera* (which had been included in the *Protomathesis*). Although the works of Regiomontanus and Copernicus on this subject were printed during Fine's lifetime, his writings fell entirely within the Ptolemaic tradition. For example, he limited himself to demonstrating the properties that allow successive evaluations of the half chords of arcs starting from the half chords of some other noteworthy arcs. Also, the table of sines that he constructed for intervals of fifteen minutes and a radius of sixty units is very similar to that (for example) of Fusoris.[8] Nevertheless, Fine indicated how his sines, expressed in sexagesimal notation, may be transformed into those given by Regiomontanus, which were calculated with a radius of 60,000 units.

The universal quadrant described in 1542 was the

trigonometric quadrant deriving from the eleventh-century *quadrans vetustissimus*. This earlier instrument had been described and commented upon by Apian in his *Instrumentum . . . primi mobilis* (1534). Fine dealt only with the strictly trigonometric uses of the quadrant, determination of the right and versed sines of a given arc—or vice versa—and the products or ratios of two sines. Virtually ignoring the application of its properties to astronomical calculations—a task carried out by J. Bonie and by B. de Solliolis in works that Fine owned[9]—Fine did no more than enumerate these possibilities.

In his Latin thesis of 1890, L. Gallois dealt only with Fine's cartography: a large map of France on four sheets and two cordiform world maps, one of the eastern hemisphere and the other, doubly cordiform, of the northern and southern hemispheres. Gallois held that the world maps were original creations and provided the source of the similar maps executed by Schöner and Apian.[10] This hypothesis is unlikely; and in the absence of an established chronology of these maps, it may be supposed that the relations of dependence were in fact the reverse, for Fine's usual procedure was to elaborate his astronomical works on the basis of the writings of others. This was undoubtedly the case with his map of France, but the scarcity of the surviving documents does not allow its genesis to be reconstructed. Fine's map of France does not truly comprise the grid of the parallels and meridians but, rather, transfers the schema to the margins; the longitudes are computed there from *l'extremité occidentale du monde*, as in the Alphonsine Tables.

Fine's scientific work may be briefly characterized as encyclopedic, elementary, and unoriginal. It appears that the goal of his publications, which ranged in subject from astronomy to instrumental music, was to popularize the university science that he himself had been taught. In this perspective, it is perhaps his works in French (such as *Théorique des cieux*) or the French translations of his works first published in Latin (for instance, *Canons et documents très amples touchant l'usage des communs almanachs* and the *Sphère du monde*) that best illustrate his scientific career.

NOTES

1. There is disagreement as to whether the last letter of his name should be accented. Citing the Latin form Finaeus, bookkeeping records in which the name is spelled Finee, and the bad rhyme of Finé with Dauphiné and *affiné* made by André Thevet, L. Gallois (*De Orontio Finaeo*, 2) opted for the pronunciation Finé. This is the one that has generally been accepted, despite the objections of Dauphinois scholars, who, citing local usage—which ought to decide the question—prefer Fine. The form Finée probably resulted from rendering the Latin form into French. As for Thevet's rhymes, which are very late (1584), their significance is diminished by the fact that a contemporary and close friend of Fine's, Antoine Mizauld, rhymed Fine with *doctrine* in his verses. (See MS Paris fr. 1334, fol. 17.) The date of his birth is specified in an autograph note in MS Paris lat. 7147, fol. ii.
2. See E. Wickersheimer, *Dictionnaire biographique des médecins en France au moyen âge* (Paris, 1936), 553 and 154.
3. There are two MSS of a course on Aristotle given by Jean Hannon and copied by François Fine in 1472–1473 (Paris lat. 6436, 6529); see *Catalogue des manuscrits en écriture latine portant des indications de date, de lieu ou de copiste*, II (Paris, 1962), 341, 353; pl. cli.
4. *Commentaires de la Faculté de médecine de l'Université de Paris*, II, 1516–1560, M.-L. Concasty, ed. (Paris, 1964), 50b, 54a. This is in the series Collection de Documents Inédits sur l'Histoire de France.
5. E. Poulle and Fr. Maddison, "Un équatoire de Franciscus Sarzosius," in *Physis*, **5** (1963), 43–64.
6. D. J. de Solla Price, "The Little Ship of Venice, a Middle English Instrument Tract," in *Journal of the History of Medicine and Allied Sciences*, **15** (1960), 399–407; and Fr. Maddison, *Medieval Scientific Instruments and the Development of Navigational Instruments in the XVth and XVIth Centuries* (Coimbra, 1969), 14. This book is Agrupamento de Estudos de Cartografia Antiga, XXX.
7. D. Hillard and E. Poulle, "Oronce Fine et l'horloge planétaire de la Bibliothèque Sainte-Geneviève"; and E. Poulle, "Les mécanisations de l'astronomie des épicycles, l'horloge d'Oronce Fine," in *Comptes rendus des séances de l'Académie des inscriptions et belles-lettres* (1974), 59–79.
8. E. Poulle, *Un constructeur d'instruments astronomiques au XVe siècle, Jean Fusoris* (Paris, 1963), 75–80. This work is Bibliothèque de l'École Pratique des Hautes Études, IVe sect., fasc. 318.
9. E. Poulle, "Théorie des planètes et trigonométrie au XVe siècle d'après un équatoire inédit, le sexagenarium," in *Journal des savants* (1966), 129–161, esp. 131–132.
10. L. Gallois, *Les géographes allemands de la renaissance* (Paris, 1890), 92–97. This work is Bibliothèque de la Faculté des Lettres de Lyon, XIII.

BIBLIOGRAPHY

I. ORIGINAL WORKS. The list of books published by Fine is difficult to establish, for it involves sorting out many reprintings, some of them only partial, and a number of translations. There are four contemporary lists, three of them inserted in his eds. of Euclid's *Elements* of 1536, 1544, and 1551; the fourth was included by Antoine Mizauld in his ed. of Fine's *De rebus mathematicis* in 1556. All of these lists, however, pose problems. That drawn up by L. Gallois in *De Orontio Finaeo gallico geographo* (Paris, 1890), 71–79, is incomplete and has been superseded by those of R. P. Ross, in his unpublished doctoral dissertation, "Studies on Oronce Fine (1494–1555)" (Columbia University, 1971); and of D. Hillard and E. Poulle, in "Oronce Fine et l'horloge planétaire de la Bibliothèque Sainte-Geneviève," in *Bibliothèque d'humanisme et renaissance*, **33** (1971), 311–351, see 335–351. The latter list is numbered and indexed, and includes MSS. One should consult the latest findings concerning the bibliography in R. P. Ross, "Oronce Fine's Printed Works: Additions to Hillard and Poulle's Bibliography," in *Bibliothèque d'humanisme et renaissance*, **36** (1974), 83–85.

II. SECONDARY LITERATURE. Gallois's *De Orontio Finaeo* has become quite dated and presents an extremely limited picture of Fine's work. Ross's "Studies on Oronce Fine (1494–1555)" deals only with the mathematical works, among which those on astronomy are not included. It does, however, contain a recent bibliography. An overall account of Fine's work is in the exposition catalog of the Bibliothèque Ste.-Geneviève, *Science et astrologie au XVIe siècle, Oronce Fine et son horloge planétaire* (Paris, 1971). See also Richard P. Ross, "Oronce Fine's 'De sinibus libri II': The First Printed Trigonometric Treatise of the French Renaissance," in *Isis*, **66** (1975), 378–386.

EMMANUEL POULLE

FISCHER, HANS (*b.* Höchst am Main, Germany, 27 July 1881; *d.* Munich, Germany, 31 March 1945), *chemistry*.

Fischer's father, Eugen, was a dye chemist in the large dye works of Meister, Lucius, and Brüning in Höchst; and while the boy was still young, the father became director of the laboratory of the Kalle Dye Works in Biebrich, near Wiesbaden. Young Fischer thus was early acquainted with the chemistry of pigments. He began the study of medicine and chemistry at Marburg, where, under the influence of Theodor Zincke, he decided to specialize in chemistry. After graduating in 1904 he continued his medical studies at Munich and spent a year in Berlin as assistant to Emil Fischer (no relation), working on peptides and sugars.

In 1910 Fischer went to a medical clinic in Munich, where he began to study the constitution of the bile pigment bilirubin; and his *Habilitationsschrift* (1912) was based on these studies. He accepted the chair of medical chemistry at the University of Innsbruck, as successor to Adolf Windaus, in 1916; and two years later he assumed a similar position in Vienna. During World War I and the reconstruction period that followed, he was unable to carry on any continuous research. In 1921, however, Fischer became head of the Institute of Organic Chemistry at the Technische Hochschule of Munich, succeeding Heinrich Wieland. He remained there for the rest of his life, and it was there that he conducted his most important studies.

In 1935 Fischer married Wiltrud Haufe, who was thirty years younger than he. It was a happy marriage. Although he suffered from tuberculosis at the age of twenty and had had a kidney removed in 1917, he was active in mountaineering and skiing and enjoyed long automobile trips. During World War II, Fischer became severely depressed, especially when his institute was almost totally destroyed by bombing. Convinced that his lifework had been shattered, he committed suicide in March 1945.

When Fischer began his research in Munich, he continued his earlier work on bile pigments and related substances. The general outlines of porphyrin chemistry were known, and in 1912 W. Küster had proposed a formula for the porphyrin ring that was essentially correct. It was Fischer's task to establish the accuracy of the earlier work and to carry it to its logical conclusion. He recognized that the significant feature of bilirubin was its content of four pyrrole rings and that the pigment itself was a degradation product of hemin, the active portion of the hemoglobin molecule. In addition, he soon learned that in the pathological condition known as porphyria, several other porphyrin ring compounds were excreted, including coproporphyrin and uroporphyrin. Other porphyrins were found in a wide variety of natural sources. Chlorophyll, the green pigment of leaves, was also a porphyrin, differing from hemin in its content of magnesium instead of iron. Fischer found that the various porphyrins were distinguished from one another chiefly by the presence of differing substituents on the pyrrole rings.

He therefore began a systematic study of a number of synthetic pyrrole derivatives and became an authority on pyrrole chemistry, writing the definitive monograph on this subject (1934–1940). While working on the syntheses of various porphyrins, Fischer developed microanalytical techniques so successfully that more than 60,000 microanalyses were carried out in his laboratory. He found reactions by which two pyrrole rings could be combined into dipyrrylmethenes, which in turn could be combined to form porphyrin rings. In 1926 he carried out the first synthesis of a porphyrin, and in 1929 he was able to prepare hemin. For the latter achievement he received the 1930 Nobel Prize in chemistry. It was only later that Fischer finally determined the structure of bilirubin, the compound with which he had begun his research. Since hemin was closely related to chlorophyll, he next studied the plant pigment. He identified the pyrrole rings of chlorophyll but died before completing its synthesis, which was accomplished in 1960 at Munich and, independently, at Harvard.

The author of over 300 papers, Fischer also received the Davy Medal of the Royal Society and an honorary doctorate from Harvard.

BIBLIOGRAPHY

I. ORIGINAL WORKS. Fischer's chief book is *Die Chemie des Pyrrols*, 3 vols. (Leipzig, 1934–1940), written with Hans Orth. An extensive review of his own work is "Hemin und Porphyrine," in *Verhandlungen der Deutschen Gesellschaft für innere Medizin*, **45** (1933), 7–27.

II. SECONDARY LITERATURE. Fischer's scientific work is reviewed by F. Baumgärtel, "Zur Erforschung der Blut- und Gallenfarbstoffe. Hans Fischer zum Gedächtnis," in *Medizinische Klinik*, **42** (1947), 31–34. Biographical details are given by Karl Keile, "Das Lebenswerk Hans Fischers," in *Naturwissenschaften*, **33** (1946), 289–291; H. Wieland, "Hans Fischer und Otto Hönigschmid zum Gedächtnis," in *Angewandte Chemie*, **62** (1950), 1–4—the portion relating to Fischer has been translated by Ralph Oesper in E. Farber, ed., *Great Chemists* (New York, 1961), 1527–1533; and Alfred Treibs, "Hans Fischer, 1881–1945," in *Chemie in unserer Zeit*, **1** (1967), 58–61.

HENRY M. LEICESTER

FOREL, FRANÇOIS ALPHONSE (*b*. Morges, Switzerland, 2 February 1841; *d*. Morges, 8 August 1912), *limnology, earth sciences.*

Forel was the eldest of three children born to François Marie Étienne Forel and Adélie Morin Forel, and their only child to reach adulthood. His father, a distinguished jurist and historian of Switzerland, encouraged his son's scientific interests. Forel began his studies in Morges and continued them at the Académie de Genève, from which he graduated in science. For two years he studied medicine at Montpellier and then completed his medical degree at the University of Würzburg. After teaching there briefly, in 1870 he joined the faculty of the Académie de Lausanne, where he taught general anatomy and physiology for twenty-five years. In 1872 he married Fanny Elizabeth Mathilde Monneron; they had three daughters and a son.[1] Darwinism undermined Forel's Christian faith, but he viewed this as a private matter not to be imposed upon his devout family.

Forel was trained primarily in zoology, but his strong interest in Lake Geneva led him to broaden his research to include physical as well as biological problems. His physical research was facilitated by a close association with his teacher (and later colleague) Charles Dufour, whose own research was in physical geology and meteorology. The breadth of Forel's investigations may have prevented him from making a major contribution in any one area, but it prepared him for making his unique contribution—the founding of limnology.

Forel's earliest important research was on the profundal fauna of Lake Geneva, which he accidentally discovered while attempting to learn whether the lake bottom had ripple marks indicating bottom waves. Sars, Lindström, W. B. Carpenter, T. H. Huxley, and Pourtalès had already studied profundal faunas of the oceans; but Forel was the first to discover their existence in deep lakes. He studied the profundal species in Lake Geneva and then made comparative studies in lakes Constance, Neuchâtel, and Zurich. He also investigated the light, currents, temperature, organic and inorganic matter in the profundal waters, and the soil characteristics of the bottoms. This series of studies culminated in *La faune profonde des lacs suisses* (1884).

Forel's other zoological investigations included studies of both the zooplankton and the littoral faunas of Swiss lakes. He was particularly interested in the causes of the daily vertical migration of Crustacea. His hypothesis that they are controlled by currents turned out to be less significant than August Weismann's that light is the major controlling factor. Forel was also interested in discovering the probable evolutionary history of the littoral, pelagic, and profundal faunas.

In 1869 Forel began studying the seiches of Lake Geneva. This type of wave had first been detected on Lake Constance in 1549 and had been studied at Lake Geneva by several investigators. Forel measured the magnitude and duration of seiches at various points on the lake and then compared his data with other data derived from experimental models. In 1828 J. R. Merian had developed the mathematics for describing waves in rectangular tanks, and Karl von der Mühl and Sir William Thomson helped Forel to apply Merian's formula to lake seiches. Forel then turned to a study of causes and concluded in 1878 that the main ones are wind, rain, and variations in atmospheric pressure.

Forel's other investigations in physical science usually had some connection with lake phenomena. His studies of earthquake measurement were related to seiche measurement, for earthquakes cause seiches. His studies in meteorology were connected with his interest in variations of the level of Lake Geneva, as was his interest in glaciology. With Dufour he investigated the relationships between glaciers and atmospheric moisture, those between glaciers and loss of water to the lake (1871).

After devoting most of his life to the study of Lake Geneva, Forel decided to write a detailed monograph of its characteristics; his three-volume *Le Léman* (1892–1894) is probably the most detailed study ever written about a single lake. He devoted the first two volumes to its physical features—geography, hydrography, geology, climate, hydrology, hydraulics, temperature, optics, acoustics, and chemistry. The third volume was concerned mainly with biology, although it also included history, navigation, and fishing. It is a model limnological reference work, but Forel realized it could not serve as an introductory text. He therefore wrote the first limnology text, *Handbuch der Seenkunde: Allgemeine Limnologie* (1901). In most respects it follows the organization of *Le Léman*, but treats the subjects more briefly and more generally. Forel thus fulfilled the requirements that seem necessary for the founding of a new science: a reference monograph based upon a series of reliable studies and a text that served to introduce students to the problems of the science.

NOTES

1. I am indebted to Forel's grandson, the Reverend François Forel, for genealogical information.

BIBLIOGRAPHY

I. Original Works. Forel compiled his own bibliography (288 titles), which was published with his obituary by Henri Blanc, "Le professeur Dr. François Alphonse Forel, 1841–1912," in *Actes de la Société helvétique des sciences naturelles*, **95** (1912), 110–148.

II. Secondary Literature. There are surveys of Forel's work by Blanc (see above) and by Frank N. Egerton III, "The Scientific Contributions of François Alphonse Forel, the Founder of Limnology," in *Schweizerische Zeitschrift für Hydrologie*, **24** (1962), 181–199. Kaj Berg has discussed Forel's concept of limnology and his influence in "The Content of Limnology Demonstrated by F.-A. Forel and August Thienemann on the Shore of Lake Geneva," in *Proceedings of the International Association of Theoretical and Applied Limnology*, **11** (1951), 41–57. There is a useful collection of commemorative articles on the various aspects of Forel's work in *Bulletin de la Société vaudoise des sciences naturelles*, **49** (1914), 291–341. This society also published a memorial issue on the fiftieth anniversary of Forel's death, with articles illustrating the continuing influence of his work on modern Swiss investigations in limnology and glaciology: *ibid.*, **68** (1963), 189–229. B. H. Dussart has provided a comprehensive survey and bibliography of the limnology of large Swiss lakes in "Les grands lacs d'Europe occidentale," in *Année biologique*, 4th ser., **2** (1963), 499–572.

C. H. Mortimer has reviewed the history of lake hydrodynamics from Forel's time to the present in "Lake Hydrodynamics," in *International Association of Theoretical and Applied Limnology, Mitteilungen*, **20** (1974), 124–197. For placing Forel's geological work in context, there is a useful collection of brief articles edited by Johann-Christian Thams, *The Development of Geodesy and Geophysics in Switzerland* (Zurich, 1967).

Forel's pleasant personality and his religious outlook have been described by his cousin, August Forel, in *Out of My Life and Work*, Bernard Miall, trans. (New York, 1937), 83.

Frank N. Egerton III

FREUNDLICH, HERBERT MAX FINLAY (*b.* Berlin-Charlottenburg, Germany, 28 January 1880; *d.* Minneapolis, Minnesota, 30 March 1941), *colloid and interface science.*

Freundlich was the older brother of Erwin Freundlich. At the age of nineteen he transferred from the University of Munich to that of Leipzig, in order to study chemistry under Wilhelm Ostwald. After graduating in 1903, he remained as Ostwald's chief assistant, later private assistant, until 1911, when he became associate professor of physical chemistry at the Technische Hochschule in Brunswick. During World War I, Freundlich's knowledge of adsorption enabled him to find agents for gas masks. He was noticed by Fritz Haber, wartime chief of the Chemical Warfare Service, who found a post for him after the war at the Kaiser Wilhelm Institute for Physical Chemistry and Electrochemistry in Berlin-Dahlem. From 1919 until 1933 Freundlich's laboratory was one of the world's chief centers of research in colloid and interface science. He followed Haber's example of linking his fundamental research to industrial processes (such as brewing and ore flotation) and to technically important systems (such as rubber, paint, oil, and detergents). In 1933 Freundlich left Germany to escape Nazi harassment. A five-year research appointment at University College, London, was created for him by Imperial Chemical Industries. He was elected an honorary fellow of the Chemical Society (1938) and a foreign member of the Royal Society (1939). At the end of his tenure in London he accepted a special research professorship at the University of Minnesota but served there for only two years before he died of a coronary thrombosis.

Freundlich's most important contributions are on the flocculation of colloidal dispersions by electrolytes, particularly on the significance of the zeta potential and his demonstration that it is not the same as the ordinary electrode (Nernst) potential; on the viscosity and elasticity of colloidal dispersions and the phenomenon of thixotropy (a term he coined); on the effects of ultrasonic vibrations on colloidal stability; and on the biological applications of the principles of colloid science, many of which he himself had developed or amplified.

Freundlich was highly cultivated and an accomplished musician; his temperament combined the artist and the scientist. A lively and engaging conversationalist, he wrote books and papers that are models of organization and forceful exposition.

BIBLIOGRAPHY

I. ORIGINAL WORKS. Freundlich's major book is *Kapillarchemie* (Leipzig, 1909; 4th ed., 2 vols., 1930–1932). The English trans. of the 3rd ed. is *Colloid and Capillary Chemistry* (London, 1926); the 4th German ed., twice as long as the 3rd ed., has not yet been translated. A complete list of Freundlich's 8 books and 243 papers is in the obituary notice by Donnan (see below).

II. SECONDARY LITERATURE. Obituaries are F. G. Donnan, in *Obituary Notices of Fellows of the Royal Society of London*, **4** (1942), 27–50; R. A. Gortner and K. Sollner, in *Science*, **93** (1941), 414–416; H. S. Hatfield, in *Nature*, **147** (1941), 568–569; E. K. Rideal, *ibid.*, 568; and J. Traube, *ibid.*, **148** (1941), 18.

SYDNEY ROSS

FUCHS, LEONHART (*b.* Wemding, Germany, 17 January 1501; *d.* Tübingen, Germany, 10 May 1566), *medicine, botany.*

Fuchs's father and paternal grandfather were *Burgermeister* of Wemding. His father, Hans Fuchs, married Anna Denteni in 1490; they had two sons and a daughter, Leonhart being the youngest child. When Leonhart was four years old his father died; and his grandfather, Johann Fuchs, assumed responsibility for the family. On walks through the countryside Johann taught his grandson the names of flowers and evidently imparted a lasting interest in them.

At the age of ten Fuchs was sent to Heilbronn to prepare for a university education. Progressing well, he transferred in the following year to the Marienschule in Erfurt; and in the fall of 1515 he entered the University of Erfurt. After only three semesters he took the baccalaureate examination and received the degree in 1517. Returning to Wemding, Fuchs opened a school that prospered; but in 1519 he enrolled at the University of Ingolstadt and studied for a master's degree under the controversial linguist and humanist Johann Reuchlin; he also came under the sway of Luther's writings. Fuchs later applied their reformist outlook to his own medical and scientific contributions. After receiving the master's degree in 1521, he studied medicine at Ingolstadt and received the doctorate in 1524.

For two years Fuchs practiced medicine at Munich. In that city he met and married Anna Friedberger; they had four sons and six daughters. He returned to Ingolstadt in 1526 as a professor of medicine but resigned two years later to become court physician to a fellow Lutheran, George von Brandenburg, margrave of Ansbach. There Fuchs gained wide respect in 1529 for his treatment of an epidemic that has been identified with the English sweating sickness. He received another appointment as professor of medicine at Ingolstadt in 1533, but opposition to his Lutheran religion prevented his assuming the position.

Instead, Fuchs became professor of medicine (1535) at Tübingen, where he remained for the rest of his life. He exerted a strong influence at the university and was elected rector in 1536 and in 1540. He also wrote the statutes of the medical faculty, issued in 1539. While on the faculty he continued to practice medicine.

In attempting to reform medicine, Fuchs emphasized the importance of relying upon the ancient Greek authorities rather than upon later authors, just as Protestant leaders emphasized the importance of the Bible, rather than later authors and traditions, as the source of Christianity. He was active in the movement to publish new and more accurate editions of the Greek texts and the Latin translations based upon them. One of the editors of a Greek edition of Galen's works (Basel, 1538), he translated both Hippocratic and Galenic medical texts and also the pharmaceutical work of Nicolaus Myrepus Alexandrinus.

Fuchs's reforming zeal led him into many controversies. Sometimes, as in his castigations of the hack writer Walther Hermann Ryff, his barbs were clearly deserved; but sometimes he seemed to side with Greek authors merely because they were ancient and their critics were less venerable. His bias nevertheless suited the polemical atmosphere of his times, and he became a very successful author—due to his organizing ability rather than to originality. Some contemporary complaints against him of plagiarism have been sustained, and other instances have more recently come to light. His great ability to organize knowledge is illustrated in his medical textbooks, beginning with the *Compendiaria* (1531). He frequently revised and enlarged them, and his *Institutionum medicinae* was reprinted as late as 1618.

Fuchs is best known for his pharmaceutical herbal, *De historia stirpium* (1542); the text included his own observations but was mostly a compilation of the work of other authors. The impressive illustrations came mainly from original drawings and woodcuts made under his supervision. They were often more spectacular than those that Otto Brunfels had published in his *Herbarum vivae eicones* (1530–1536) and were subsequently borrowed to illustrate many other herbals.

Fuchs was not so blinded by reverence for the Greeks, however, that he lacked interest in new knowledge. Of the 487 species and varieties included in *De historia stirpium*, over 100 were recorded for the first time in Germany. Among the species that had never before been illustrated were foxglove (*Digitalis purpurea*) and Indian corn (*Zea mays*). In 1543 Fuchs issued a German edition of the *Historià*, with some revisions in the text and six additional illustrations. Later he aspired to publish one or two additional herbal volumes but was unable to obtain the necessary funds.

BIBLIOGRAPHY

I. Original Works. There is a very good bibliography of Fuchs's works in Eberhard Stübler, *Leonhart Fuchs, Leben und Werk* (Munich, 1928), 115–133. Another good list is Richard J. Durling, *A Catalogue of Sixteenth Century Printed Books in the National Library of Medicine* (Bethesda, Md., 1967), 199–204, including eds. omitted by Stübler. There are two modern facs. eds. of the *New Kreüterbüch* (Leipzig, 1938; Munich, 1964), the German trans. of *De historia stirpium*. The Latin MS is extant and has been described by Kurt Ganzinger, "Ein Kräuterbuchmanuskript des Leonhart Fuchs in der Wiener Nationalbibliothek," in *Sudhoffs Archiv für Geschichte der Medizin*, **43** (1959), 213–224. Also extant are 25 wood engravings meant to illustrate a second volume of the *New Kreüterbüch*, preserved at the Botanical Institute of the University of Tübingen, and 173 printed illustrations, preserved in the herbarium of Félix Platter at the University of Bern. Some of the plants that Fuchs's artists used for their drawings have been preserved in the herbarium of Leonhard Rauwolf at the Rijksherbarium, Leiden. These specimens have been discussed by Kurt Ganzinger, "Rauwolf und Fuchs, Ein Beitrag zur Geschichte der Botanik im 16. Jahrhundert," in *Veröffentlichungen der Internationalen Gesellschaft für Geschichte der Pharmazie*, **22** (1963), 23–33.

Fuchs's outline of eye diseases has been republished in modern times in German, in Latin, and in English translation. See Edward Pergens, "Leonhard Fuchs' alle Kranckheydt der Augen (1539)," in *Zentralblatt für praktische Augenheilkunde*, **23** (1899), 199–203, 231–238; and Karl Sudhoff, "Des Leonhart Fuchs, Professors in Tübingen, 'Tabelle der Augenkrankheiten,' im lateinischen Originalwortlaut von 1538 bekanntgegeben," in *Archiv für Augenheilkunde*, **97** (1926), 493–501. In 1936 the British Optical Association published as separate charts a facs. of the original Latin outline and also an English trans.

Most of Fuchs's writings except for *Historia stirpium* were published in his *Operum . . . omnia*, 3 vols. (Frankfurt am Main, 1566–1567; 2nd ed. in 1 vol., 1604).

II. Secondary Literature. The standard contemporary source is the obituary by Fuchs's colleague, Georg Hizler: *Oratio de vita et morte clarissimi viri, medici et philosophi praestantissimi, D. Leonharti Fuchsii, artis medendi in Academia Tubingensi professoris doctissimi* (Tübingen, 1566), repr. in Fuchs's *Opera* (1566, 1604). Stübler (see above) is the best and most detailed modern survey of Fuchs's entire career and works. Horst Rudolf Abe has provided additional details about Fuchs's early studies at Erfurt in "Zur Datierung des Erfurter Universitätsaufenthaltes von Leonhart Fuchs," in *NTM/Schriftenreihe für Geschichte der Naturwissenschaften, Technik und Medizin*, **9** (1972), 56–61. Further information concerning Fuchs's life and work from 1541 to 1565 is available from his letters to the Camerariuses, which Gerhard Fichtner has discussed in "Neues zu Leben und Werk von Leonhart Fuchs aus seinen Briefen an Joachim Camerarius I. und II. in der Trew-Sammlung," in *Gesnerus*, **25** (1968), 65–82.

Luigi Samoggia has documented Fuchs's poorly acknowledged indebtedness to Leoniceno's treatise on the errors of Pliny (1492), as revealed in Fuchs's *Errata recentiorum medicorum* (1530; rev. and retitled *Paradoxorum medicinae*, 1535), in "Le ripercussioni in Germania dell' indirizzo filologico-medico Leoniceniano della scuola ferrarese per opera di Leonardo Fuchs," in *Quaderni di storia della scienza e della medicina*, **4** (1964), 3–41. Samoggia has also discussed Fuchs's borrowings from Giovanni Manardo's *Epistolarum medicinalium* when writing his own *Compendiaria* (1531), *Paradoxorum medicinae* (1535), and *Tabula oculorum morbos comprehendens* (1538), in "Manardo e la scuola umanistica filologica tedesca con particolare riguardo a Leonard Fuchs," in *Atti del Convegno internazionale per la celebrazione del V centenario della nascità di*

Giovanni Manardo, 1462–1536 (Ferrara, 1963), 241–251.

W. P. D. Wightman has discussed the understanding of scientific method in Fuchs's *Compendiaria* (1531; rev. and retitled *Methodus*, 1541) in *Science and the Renaissance: I, An Introduction to the Study of the Emergence of the Sciences in the Sixteenth Century* (Edinburgh–New York, 1962), 213–217. Wightman has also discussed Fuchs's methodological predecessors, Giovanni Battista da Monte and Manardo, in "*Quid sit Methodus*? 'Method' in Sixteenth Century Medical Teaching and 'Discovery,'" in *Journal of the History of Medicine and Allied Sciences*, **19** (1964), 360–376.

Georg Harig has discussed Fuchs's pharmacological theory, pointing out its ultimate origin in the writings of Galen, and possibly those of Oribasius and several Arabic authors, in "Leonhart Fuchs und die theoretische Pharmakologie der Antike," in *NTM/Schriftenreihe für Geschichte der Naturwissenschaften, Technik und Medizin*, **3** (1966), 74–104. The likelihood of Fuchs's indebtedness to the pharmacological writings of late medieval European physicians, such as Arnald of Villanova and Bernard of Gordon, has not yet been studied, however.

Van Schevensteen has discussed the relationship between the writings of Fuchs on eye diseases and the work of an anonymous contemporary in "À propos de traités d'ophtalmologie parus à Strasbourg au début du XVIe siècle," in *Janus*, **28** (1924), 1–20. J. H. Sutcliffe has clarified Fuchs's understanding of eye diseases and has explained the policy of the English translation in "The *Tabula oculorum morbos comprehendens* of Leonhart Fuchs," in *Dioptric Review*, **38** (1936), 347–353.

Although Fuchs acknowledged that he used Vesalius' *Fabrica* and *Epitome* when writing his own *De humani corporis fabrica ex Galeni & Andrea Vesalii libris concinnatae epitome*, 2 vols. (Tübingen, 1551), he nevertheless plagiarized Vesalius in it. See C. D. O'Malley, *Andreas Vesalius of Brussels, 1514–1564* (Berkeley, Calif., 1965), 245–248.

Thomas A. Sprague and E. Nelmes have published a very useful analysis of Fuchs's contributions to botany, "The Herbal of Leonhart Fuchs," in *Botanical Journal of the Linnean Society of London*, **48** (1931), 545–642. Heinrich Marzell published a useful intro. to the *New Kreüterbüch*, which appeared with the facs. ed. of 1938, and also separately as *Leonhart Fuchs und sein New Kreüterbüch (1543)* (Leipzig, 1938). Agnes Arber has placed Fuchs's herbal in the perspective of contemporary botany in *Herbals, Their Origin and Evolution: A Chapter in the History of Botany, 1470–1670*, 2nd ed. (Cambridge, 1938). Edward Lee Greene has discussed Fuchs's use of botanical terminology and nomenclature in *Landmarks of Botanical History*, 2nd ed., Frank N. Egerton III, ed., I (in press), ch. 6. Helen A. Choate has translated Fuchs's glossary in "The Earliest Glossary of Botanical Terms: Fuchs 1542," in *Torreya*, **17** (1917), 186–201. The claim for Fuchs's originality cannot be sustained, however, because he drew heavily upon Jean Ruelle's terminology in *De natura stirpium* (1536), as Greene discovered—see *Landmarks* of *Botanical History*, II, ch. 16.

A. H. Church has discussed the relative merits of the illustrations in the herbals of Brunfels and of Fuchs, and has indicated the illustrations used by Fuchs that were plagiarized from Brunfels: "Brunfels and Fuchs," in *Journal of Botany*, **57** (1919), 233–244.

Other useful contributions to an understanding of Fuchs's activities as a botanist are Stübler, *op. cit.*, ch. 5; Ganzinger, "Ein Kräuterbuchmanuskript . . ." (1959) and "Rauwolf und Fuchs . . ." (1963); and F. W. E. Roth, "Leonhard Fuchs, ein deutscher Botaniker, 1501–1566," in *Beihefte zum Botanischen Zentralblatt*, **8** (1898), 161–191.

FRANK N. EGERTON III

FUSORIS, JEAN (*b.* Giraumont, Ardennes, France, *ca.* 1365; *d.* 1436), *astronomy.*

Fusoris is the only medieval maker of astronomical instruments about whom there is considerable information on both his life and his work. The son of a pewterer, he left his native region to study at the University of Paris, where he earned master's degrees in arts and medicine and a bachelor's degree in theology. Named successively canon of Rheims (1404), of Paris (1411), and of Nancy, and curate of Jouarre-en-Brie, Fusoris resided in Paris until 1416. He directed a large workshop for the manufacture of astronomical instruments, and his clients included the king of Aragon, the duke of Orléans, the antipope John XXIII, the king of Navarre, and the bishop of Norwich. His relations with the last, however, proved to be compromising when the Hundred Years' War with England resumed in 1415. Fusoris was charged with high treason but, in the absence of conclusive evidence against him, he was only sentenced to exile in Mézières-sur-Meuse, where he had spent his youth. His name reappears in documents in 1423, at which time he had a commission from the chapter of Bourges to build an astronomical clock. He also established trigonometric tables for King Charles VII and, in 1432, wrote a treatise on cosmography for the chapter of Metz.

Much of Fusoris's work has survived, and comprises both brass instruments and books. His workshop must have produced a great many astrolabes, for at least eighteen examples still exist—a truly unique case for western medieval astrolabes. None of them is signed, but all display a characteristic error: the presence of the star Cornu Arietis (β Arietis) in the southern hemisphere. Fourteen of these instruments have already been listed;[1] others are at the Collegium Maius in Cracow, the Bayerisches Nationalmuseum in Munich (this instrument is from the Bassermann-Jordan collection),[2] the Astronomisch-Physikalisches Kabinett in Kassel, and the Nordiska Museet in Stockholm. This list is probably not definitive and other astrolabes may be added to it.

Fusoris built at least one equatorium, which he sold for 400 écus to the bishop of Norwich. It consisted of several "instruments"—one for each planet—each of which bears two brass disks: the "mother," which is analogous to that of an astrolabe but without an armil, and the equant disk. The equant of the instrument for Mars has been found; it was reused as a clock face in the seventeenth century.[3]

Other instruments made in Fusoris's workshop include dials, armillary spheres, and clocks; they seem not to have survived. The astronomical clock made for the chapter of Bourges has been preserved, however, although in poor condition. Its representation of the motions of the sun and the moon in front of an astrolabic dial is excellent, for the relationships of the teeth of the gear train yield an astonishingly exact periodicity for the solar and lunar motions.

Fusoris's written work includes texts in both Latin and French. Among the latter are a treatise on the uses of the astrolabe, dedicated to Pierre of Navarre, comte de Mortain, and a treatise on its construction, as well as one on cosmography written for the canons of Metz. Fusoris was perhaps also the author of an elementary treatise on arithmetic and geometry, preserved in two manuscripts. All these texts are pedagogical and without great originality, but they attest the author's concern with presenting difficult material to the layman in a clear manner.

The Latin texts are much more technical. The treatise on the equatorium describes the construction and use of the *instruments des sept planètes* sold to the bishop of Norwich. The device is a geometric equatorium, similar in its operating principle to that of Campanus of Novara: each "instrument" reproduces very truly the geometric decomposition of the motion of one planet. In Fusoris's device, however, the graduations of the equant and epicycle in degrees are replaced by a movable chronological graduation.

Toward the end of his life, Fusoris undertook to reconstruct the astronomical tables, but he was able to execute only the trigonometric part of his project, which he dedicated to King Charles VII. The establishment of his trigonometric table, which contains only sines and chords, was based on the theorem that every chord subtended by a median arc of two arcs of which the chords are known is equal to the product of half the sum of these chords and a coefficient that depends only on the difference of the two arcs and not on the arcs themselves. The chords and, as a result, the sines were therefore calculated gradually, starting from the chords of certain noticeable arcs. The smallest arc to which this procedure could be applied in order to obtain its trigonometric lines is one of 45′; the lines of the arcs of 15′ and 30′ were simply taken to be equal, respectively, to one-third and two-thirds of those of 45′. In accordance with the Ptolemaic tradition, the quantities were expressed as sexagesimal fractions of the radius, with a precision, depending on the angles, of the third, the fourth, or the sixth sexagesimal division of the sixtieth of the radius.

Still surviving are Fusoris's notes on gnomonics, concerning the construction of the traveler's clock and the tracing of horizontal sundials for the latitude of Paris. Also preserved are Fusoris's annotated drawings of an astronomical clock that, according to his pupil Henri Arnaut of Zwolle, was built for the duke of Burgundy.

Fusoris appears to have been a highly skilled practitioner of the astronomy of his time, and he provides excellent testimony to the solid training in science given by the University of Paris at the end of the fourteenth century.

NOTES

1. E. Poulle, *Un constructeur d'instruments astronomiques*, 20–21. The astrolabe in the Whitney Warren collection (Gunther's no. 193B) is no longer at the Smithsonian Institution; the astrolabe of the Conservatoire des Arts et Métiers, Paris, is not from the collections of the École d'Horlogerie, but is a gift from Gaston Brière, who obtained it from Vallet de Viriville.
2. This instrument bears a date, 1447, which seems to have been added subsequently.
3. A. Simoni, "Un fortunato rinvenimento," in *La clessidra*, **27**, no. 9 (1971), 25–27.

BIBLIOGRAPHY

The documents relating to Fusoris's trial were published, with a long intro., by L. Mirot as "Le procès de maître Jean Fusoris, chanoine de Notre-Dame de Paris, 1415–1416; épisode des négociations franco-anglaises durant la guerre de Cent ans," in *Mémoires de la Société de l'histoire de Paris et de l'Ile-de-France*, **27** (1900), 137–287. Fusoris's scientific work is set forth and commented on in E. Poulle, *Un constructeur* *d'instruments astronomiques au XVe siècle, Jean Fusoris* (Paris, 1963; = Bibliothèque de l'École Pratique des Hautes Études, IVe section, fasc. 318); the book has editions of the treatises on the uses of the astrolabe and on its construction, as well as of the treatise on the equatorium, of the notes on gnomonics and of the intro. to the astronomical tables.

The following references supplement those given by Poulle. There is an Italian trans. of the treatise on the use of the astrolabe: Vatican Reg. 1732, fols. 44–59v. Fusoris's star table—which is part of the treatise on the construction of the astrolabe but can be found separately in Salamanca 2621 (see Poulle, *op. cit.*, 17–18)—is also in Wolfenbüttel, 2816, fol. 139, where there is the same reference to Fusoris and to the year 1428. Besides the two Paris MSS (Poulle, *op. cit.*, 42, 125), the treatise on the equatorium exists in Florence Magl. XX. 53, fols. 1v–34 (the beginning of the first chapter is missing).

An ed. of Fusoris's treatise on cosmography (preserved in Paris MS fr. 9558, fols. 7–20) has been prepared by L.-O. Grundt as a doctoral dissertation for the University of Bergen, but it has not been published.

EMMANUEL POULLE

GALERKIN, BORIS GRIGORIEVICH (*b.* Polotsk, Russia, 4 March 1871; *d.* Moscow, U.S.S.R., 12 June 1945), *mechanics, mathematics.*

Galerkin was born into a poor family. He received his secondary education at Minsk, and in 1893 he entered the Petersburg Technological Institute. During his studies there Galerkin had to support himself first by giving private lessons and then, from the year 1896, by working as a designer.

In 1899, after graduating from the Technological Institute, Galerkin entered the Kharkov Locomotive Building Mechanical Plant. In 1903 he moved to St. Petersburg and started work in the Northern Mechanical and Boiler Plant as manager of the technical section. He quickly became known in engineering circles.

From 1909 to 1914 Galerkin studied Russian and foreign factories and engineering installations. He visited Germany, Sweden, Switzerland, Belgium, and Austria, becoming acquainted with the outstanding examples of foreign technology. In 1909 Galerkin was invited to teach at the Petersburg Polytechnical Institute. He was chosen head of the department of structural mechanics there in 1920. From 1923 to 1929 he was dean of the structural engineering department. During this period Galerkin was professor of the theory of elasticity at the Leningrad Institute of Communications Engineers and professor of structural mechanics at Leningrad University. From 1940 to 1945 he headed the Institute of Mechanics of the Soviet Academy of Sciences.

In 1928 Galerkin was elected a corresponding member of the Soviet Academy of Sciences, and in 1935 an active member. In 1934 he was awarded the title of honored scientist and technologist. In 1942 he received the title of state prize laureate.

Galerkin's scientific work was devoted to difficult problems in the theory of elasticity and structural mechanics. His first scientific work, *Teoria prodolnogo izgiba . . .* ("Theory of Longitudinal Curvature . . ."), appeared in 1909. Galerkin extended the theory of longitudinal curvature, created by Leonhard Euler, to multistage uprights formed by the joining of a series of vertical and horizontal rods.

In his second work, *Izgib i szhatie* ("Curvature and Compression," 1910), Galerkin investigated the curvature of a rod strengthened at one end by the action of force applied, parallel to the axis, to the free end with eccentricity (or force applied at any angle to the axis). Galerkin's early works in the area of longitudinal curvature opened broad possibilities for the application of the theory of longitudinal curvature to the calculation of the stability of bridges, the frames of buildings, and similar systems.

From 1915 to 1917, in connection with the beginning of the use of beamless floors in industrial and civil construction, Galerkin made his first profound research in the theory of the curvature of thin plates. He devoted many years of his life to its development. Galerkin's many works in this field were generalized in the monograph *Uprugie tonkie plity* ("Elastic Thin Plates"), published in 1933. In 1915 Galerkin proposed a method for the approximate integration of differential equations that was widely used for the solution of problems in mathematical physics and technology. It became known as Galerkin's method.

A series of works by Galerkin on the theory of torsion and the curve of prismatic rods, published in 1919–1927, developed an interesting problem in the theory of elasticity. In 1930–1931 his fundamental works on the theory of elasticity, which contain a general solution of three-dimensional problems, appeared.

Galerkin's scientific research in the theory of casing (1934–1945) revealed its broad application in industrial construction. His works in the field constitute a new direction in this important area.

Galerkin was a consultant in the planning and building of many of the Soviet Union's largest hydrostations. In 1929, in connection with the building of the Dnepr dam and hydroelectric station, Galerkin investigated stress in dams and breast walls with trapezoidal profile. His results were used in planning the dam. For many years Galerkin was head of the All-Union Scientific Engineering-Technical Society of Builders.

The Soviet government established prizes in Galerkin's name for distinguished work in the theory of elasticity, structural mechanics, and the theory of plasticity, as well as stipends for graduate students.

BIBLIOGRAPHY

I. ORIGINAL WORKS. Galerkin's writings were brought together in *Sobranie sochineny* ("Collected Works"), 2 vols.: I, *Issledovania po stroitelnoy mekhanike teorii uprugosti, teorii obolochek* ("Research in Structural Mechanics, the Theory of Elasticity, and the Theory of Casing," Moscow, 1952); II, *Raboty po teorii uprugikh plit* ("Works in the Theory of Elastic Plates," Moscow, 1953).

II. SECONDARY LITERATURE. On Galerkin or his work see "Akademik Boris Grigorievich Galerkin. K 70-letiyu so dnya rozhdenia i 45-letiyu nauchnoy deyatelnosti" ("Academician Boris Grigorievich Galerkin. On the Seventieth Anniversary of His Birth and the Forty-Fifth Anniversary of the Beginning of His Scientific Career"), in *Izvestiya Akademii nauk SSSR, Otdelenie tekhnicheskikh nauk*, no. 4 (1941), pp. 115–120; *Boris Grigorievich Galerkin. K 70-letiyu so dnya rozhdenia i 45-letiyu inzhenernonauchnoy, pedagogicheskoy i obshchestvennoy deyatelnosti* ("Boris Grigorievich Galerkin. On the Seventieth Anniversary of His Birth and the Forty-Fifth Anniversary of the Beginning of His Engineering, Teaching and Public Career," Leningrad–Moscow, 1941); A. Joffe, A. Krylov, and P. Lazarev, "Zapiska ob uchenykh trudakh professora B. G. Galerkina" ("A Note on the Scientific Works of Professor B. G. Galerkin"), *ibid., Otdelenie fiziko-matematicheskikh nauk*, nos. 8–10 (1928), pp. 616–618; A. N. Krylov *et al.*, "Akademik B. G. Galerkin. (K 70-letiyu so dnya rozhdenia)" ("Academician B. G. Galerkin. [On the Seventieth Anniversary of His Birth]"), in *Vestnik Akademii nauk SSSR*, no. 4 (1941), pp. 91–94; V. V. Sokolovsky, "O zhizni i nauchnoy deyatelnosti akademika B. G. Galerkina" (" On the Life and Scientific Career of Academician B. G. Galerkin"), in *Izvestiya Akademii nauk SSSR, Otdelenie tekhnicheskikh nauk*, no. 8 (1951), pp. 1159–1164; and V. V. Sokolovsky and G. S. Shapiro, "Metody B. G. Galerkina v teorii uprugosti" ("B. G. Galerkin's Methods in the Theory of Elasticity"), in *Yubileyny sbornik, posvyashchenny tridtsatiletiyu velikoy oktyabrskoy sotsialisticheskoy revolyutsii* ("Jubilee Collection, Dedicated to the Thirtieth Anniversary of the Great October Socialist Revolution"), pt. 2 (Moscow–Leningrad, 1947); and "Boris Grigorievich Galerkin (1871–1945)," in *Materialy i konstruktsii v sovremennoy arkhitekture* ("Materials and Construction in Contemporary Architecture"), pt. 2 (Moscow, 1948).

A. T. GRIGORIAN

GEGENBAUR, CARL (*b.* Würzburg, Germany, 21 August 1826; *d.* Heidelberg, Germany, 14 June 1903), *comparative anatomy and morphology, zoology.*

Beginning his academic career as a medical student, Gegenbaur quickly became a leading invertebrate zoologist and, soon thereafter, emerged as the foremost vertebrate morphologist of the Darwinian era. Assiduity, thoroughness, and a keen insight into formal relations characterize his scientific endeavors. Through his efforts in the classroom and laboratory and, above all, his series of closely reasoned and richly documented monographs, Gegenbaur largely established the priority of comparative anatomy in the task of phylogenetic reconstruction and determined the centrality of this science within the biological curriculum, a position it has lost only with the rise since 1940 of molecular biology.

Gegenbaur's parents, Franz Joseph Gegenbaur and Elisabeth Karoline Roth, belonged to well-established Roman Catholic families of southern Germany. His father's responsibilities as an official in the Bavarian tax administration led to numerous moves and varied educational experiences during Carl's younger years. In 1838, however, rigorous training began, first at the Würzburg Lateinschule and then at that city's Gymnasium. The severity and narrowness of spirit of these schools decided Gegenbaur's lifelong aversion to ecclesiastical practices and personnel. Nonetheless, in these schools he received exemplary training in the classics and, no doubt, in mental discipline. His hours out of school were devoted to impassioned self-instruction in all aspects of natural history, based largely on walking tours of the Franconian countryside.

There followed a conventional sequence of university training and service. Gegenbaur entered the University of Würzburg in 1845, completed the preparatory *biennium philosophicum* in 1847, and received the medical diploma in 1851. He immediately began to travel, visiting Johannes Müller in Berlin and, upon his advice, staying briefly at Helgoland to study the North Sea fauna. "For the first time [I saw] the sea!"[1] he later exclaimed, and that experience was decisive. The following year

(1852), with Rudolf Albert von Koelliker and Heinrich Müller, he began an eighteen-month research period at Messina. Returning to Würzburg, he presented his *Habilitationsschrift* (1854) and became a *Privatdozent* in zoology. Gegenbaur was called to Jena in 1856 as extraordinary professor of zoology in the medical faculty. He remained there, occupying various chairs (ultimately that of anatomy) until 1873, when he accepted the chair of anatomy recently vacated by his father-in-law, Friedrich Arnold, at Heidelberg. There Gegenbaur found himself once more in the familiar landscape of south Germany and in regular communication with an intimate friend from his Jena days, the philosopher Kuno Fischer. The Heidelberg years were especially devoted to the teaching of human anatomy, direction and improvement of the university's important Anatomy Institute, and continued investigation of vertebrate morphology, a subject begun in earnest at Jena.

Gegenbaur presents a marvelous example of the better qualities of the higher professoriate during the golden years of the German universities. Absolute integrity, vast strength of mind and will, and a deep sense of responsibility characterized him. He attracted numerous outstanding students, including Oscar and Richard Hertwig, Max Fürbringer, T. W. Engelmann, Hans Gadow, and Giovanni Battista Grassi. These students were immediately exposed to the intensive research training that had marked Gegenbaur's own education at Würzburg. Intellectually, Gegenbaur sought breadth, but only insofar as he could personally master in depth the subject matter before him; it was partly for this reason that at Jena he sought—successfully—to remove instruction in physiology from the responsibilities of the chair of anatomy.

The emphasis within Gegenbaur's scientific studies shifted during his career. His anatomical investigations may be distributed into three periods. At Würzburg and during the early years at Jena his attention was devoted primarily to the life cycles and morphology of different stages of various marine animals. About 1860 he began the intensive comparative examination of vertebrate musculature, osteology, and neural structures upon which his great contemporary reputation so largely rested. This work continued during the third period, that at Heidelberg, and included much effort directed toward specifying in detail the anatomical bases for regarding man as an animal. During this period he also published several general and critical essays on the state of the anatomical art.

The materials and instruments of research were always an intrinsic component of the anatomist's concern. Gegenbaur was responsible for the care and augmentation of prepared and preserved specimens at both Jena and Heidelberg, a duty assumed with utmost seriousness. The research on invertebrates was pursued, as far as possible, with living specimens, thus necessitating study periods by the sea. At Würzburg, Gegenbaur had had the extraordinary good fortune to have received his scientific training from perhaps the finest group of research biologists then active in Germany: Rudolf Virchow, Franz Leydig, and Koelliker. He early learned microscopical techniques and the significance of the cell for organic structure and development. This was a critical lesson, for the formal and temporal priority of the cell provided the base point for interpretation in literally the whole of his scientific work.

Besides materials and instruments, Gegenbaur devoted much care to means of publication. His approximately 160 publications include both books and articles. Already an active participant in the meetings and publications of the Physikalisch-medizinische Gesellschaft in Würzburg, he assumed primary responsibility for issuing the *Jenaische Zeitschrift für Medizin und Naturwissenschaft* in 1864. This responsibility was relinquished when Gegenbaur went to Heidelberg; but soon thereafter (1875) he began his own journal, *Morphologisches Jahrbuch*, the outstanding vehicle for publication in comparative anatomy over several generations.

Fürbringer provides a detailed commentary on the scope and value of Gegenbaur's diverse contributions to zoology and morphology.[2] This work often was both descriptive and interpretive. Especially notable were the tracing of life cycles of free-swimming marine invertebrates, particularly of heteropods and pteropods (1855), and three lengthy monographs on the homologies of the vertebrate appendages and of the cranial bones (1864, 1865, 1872). Gegenbaur's interests and orientation are strikingly similar to those of his contemporary T. H. Huxley, who also turned from invertebrate to vertebrate morphology and utilized both in the evolutionary campaign. Better-known are Gegenbaur's interpretive manuals and student textbooks, in which first principles and superbly marshaled evidence are combined to exhibit the achievements and promise of comparative anatomy. Of these works the most significant is doubtless the second, much revised edition (1870) of *Grundzüge der vergleichenden Anatomie*, in which Gegenbaur explains how he, and fellow comparative anato-

GEGENBAUR

mists, may—and most certainly should—transfer their allegiance from the traditional, static type concept to the new evolutionary interpretation announced by Charles Darwin.

A Quest for Generality. Gegenbaur was well aware of the propensity of anatomists to accumulate facts and to designate the ensuing mass of incidental description a science. He condemned this practice. "It is evident . . . that mere description of isolated observed facts cannot be the highest goal of science."[3] Indeed, the major goal of comparative anatomy in nineteenth-century Germany was to escape the apparent mindlessness of traditional anatomical practice. Generalization was sought; and that generalization could only be the product of judgment, of a positive act of the mind as it reflects on organic similarities and dissimilarities. Gegenbaur and his contemporaries, most notably his colleagues at Jena, Matthias Schleiden and Ernst Haeckel, were thoroughly imbued with the epistemological demands of the post-Kantian generation. There had been diverse reactions to the speculative excesses of the *Naturphilosophen*, including not only banal empiricism and programmatic physicochemical reductionism but also an effort, based ultimately on Kant, to find a dependable manner of moving from seemingly bare fact or facts to increasingly broader generalization. The understanding itself would play a positive role in the creation of general propositions, but its lead was always subject to the rigorous control of empirical evidence.

The mind thus imposed its own order on nature, Gegenbaur held. But in doing so, neither willful imagination nor directionless intuition deserved serious consideration. At the heart of the matter, the guidance and control of the reflective mind, lay the generally accepted factual wealth of the science and, above all, comparison. The adjective "comparative" in the term "comparative anatomy" was most certainly a working word. Comparison, wrote J. V. Carus, is the "base point" of the science, not as mere procedure but "as the form of observation appropriate to morphological relationships."[4] Obviously these relationships themselves (as such, being the very essence of things, they could not, of course, be demonstrated) could well anticipate the comparative assessment by which they were revealed. Before 1859 those of organic form were expressed by the concept of animal types to which the perceived diversity of organisms could be reduced; after 1859 many morphologists, including Gegenbaur, explained all relationship by means of common descent.

But whatever interpretation was imposed, the objective of comparative anatomy as a science was the pursuit of such generalizations. "The combinatory power of the mind," wrote Gegenbaur, "enters at precisely that point where sensory perception reaches its limit."[5] Man's mind is truly an "ordering mind,"[6] and it generates those categories—species, genera, types—that simple sensory experience alone would fail to disclose. Gegenbaur's emphasis on generalization reached through comparison therefore served a double purpose: it exhibited structural similarity and dissimilarity among organisms, thereby revealing broader patterns of relationship, and it directed attention to the comparative procedure itself, the indispensable basis of a genuine science of form.

Any consideration of Gegenbaur's general outlook on the sciences requires at least brief notice of his passionate allegiance to the Old Catholic position. In the midst of the many conflicts dividing nineteenth-century Germany (Gegenbaur, incidentally, favored unification and spoke approvingly of Bismarck), that provoked by religious dispute within the Roman Catholic Church was prominent. Pressed by growing unbelief and increasing secularism, after mid-century the papacy and its defenders mounted an ever more determined effort to reassert universal authority over matters both temporal and spiritual. The movement culminated in the definition of papal infallibility at the Vatican Council (1869–1870). A furious reaction, led by Ignaz von Döllinger, immediately arose in Germany, creating a small but autonomous assemblage of German and Dutch Roman Catholics known thenceforth as "Old Catholics."

Virtually all commentators on Gegenbaur's career remark upon his keen sympathies for the traditional German forms of church organization and practice. None, however—and Gegenbaur provides no more in his autobiography—offers even minimal exposition of his views. It is clear from the autobiography, nonetheless, that Gegenbaur held the church militant in total contempt. Its spokesmen—the clergy and, above all, the Jesuits—he obviously loathed. Church control over the mind—he had reported its effects upon himself in the Jesuit schools of Würzburg—was surely even a greater affront to him than its loudly proclaimed pretensions to social control and political power.

There is no witness to Gegenbaur's adherence to or neglect of expected practices of worship (his second marriage was outside the church). Certainly, however, church doctrine—which, in contrast to that of various Protestant sects, was remarkably discrete—had absolutely no bearing upon Gegenbaur's acceptance and propagation of the descent theory. That acceptance occurred probably early in the 1860's; at least by the following decade he was insisting to his Heidelberg students that man was but another animal. Unlike his friend Haeckel, Gegenbaur did not shrilly preach his rejection of accepted Christian teaching regarding the uniqueness of man, but preach he did—and with powerful effect. The many editions of his *Lehrbuch der Anatomie des Menschen* conveyed above all the following single message: "The human organism is not isolated in nature, but is only one member of an endless series within which knowledge of the whole will illuminate that of the individual."[7] Gegenbaur's proud assertion of *Lehrfreiheit* tolerated no opposition from ultramontane or other sources.

The Central Generalization: Descent Theory. Prior to the 1860's Gegenbaur freely accepted the type concept as the basis for interpreting the facts of zoology. At Jena, however (presumably in 1861 or 1862), he discovered Darwin and the descent theory. Over the next several years the descent theory achieved full hegemony over Gegenbaur's zoological and anatomical outlook. In this conversion Gegenbaur had the constant and enthusiastic support of his intimate friend Haeckel. The latter, attracted to study at Jena by Gegenbaur, soon joined the faculty there and in 1862 launched the first of his notorious series of polemical lectures on Darwinism. Throughout the 1860's, while Haeckel was composing the text of the *Generelle Morphologie* (1866), his preeminent contribution to discussion of the general issues evoked by Darwin, Gegenbaur was preparing his own response to the new doctrine, first fully expressed in the second edition of *Grundzüge der vergleichenden Anatomie* (1870).

Despite the great differences in the emotional character and public style of the two men, they were joined by close personal and intellectual bonds. They shared a lively skepticism regarding the truth—or at least the wide application—of the doctrines of revealed religion, and harbored a splendid disdain for the clergy. They agreed that an explanation of the origin and interrelationship of organic forms—that is, of the structural patterns of animals and plants—taken both in particular and in general, was the foremost problem facing the zoologist or morphologist. By so defining the issue, they emphasized that morphology, the science of organic form, and not zoology, with its traditional emphasis upon description and classification, was of paramount concern to the post-Darwinian biologist. Morphology (itself built upon comparative anatomy and descriptive embryology) would not, obviously, altogether replace zoology; it was deemed, nonetheless, the more significant science because it promised to explain, at a seemingly more basic level, the otherwise incidental facts of zoology.

The type concept, the predominant morphological and zoological generalization of the first half of the nineteenth century, was the creation of a sizable group of Continental biologists, notably Goethe, Cuvier, and K. E. von Baer. According to these authors, the great diversity of form manifested in the animal world could be reduced to a small number of units or even a single unit, called types. Each type exhibited a unique set of characteristics. Its many members, however dissimilar they might at first glance appear to be, nonetheless presented common elements in the essential features of their organization; and each member, by sharing these same elements, stood absolutely apart from members of all other types.

The mature type concept of the 1840's and 1850's assured the morphologist and zoologist that the animal kingdom was neither an uncontrollable chaos of altogether separate (and therefore unclassifiable) species—the view entertained by the taxonomic nominalist—nor were animals to be gathered together as simple permutations on a single structural theme—the opinion voiced by the still lively school of the *Naturphilosophen* and against the claims of which the morphological typologists were most firmly decided. It was in this milieu that Gegenbaur received his intellectual formation. Until the 1860's he expressed full agreement with the premise of many typologists that the recognition and ever closer specification of the formal limits of the individual types was the principal objective of the science. Speculation regarding the putative origin of these types and their genealogical connections (the latter an inadmissible hypothesis) was excluded from consideration; morphology must attend to formal relationships only.

But the relationships between those types (Gegenbaur recognized seven: Protozoa, Coelenterata, Echinodermata, Vermes, Arthropoda, Mollusca, and Vertebrata) were indeed suggestive and lent themselves easily, when that option became clearly visible, to an alternative explanation. Gegenbaur's acceptance of Darwinian descent theory was very carefully circumscribed. The theory of natural selection and its associated themes, particularly the laws of variation and heredity and the fact of adaptation, lay beyond his concern. Analysis of the dynamics of the evolutionary process he altogether ignored. It was, rather, the evolutionary product, the great diversity of organisms produced in time, that commanded attention. Gegenbaur sought to explore the pertinence of the vast fund of morphological fact for the demonstration of the likelihood of the occurrence of evolutionary change and to expand that fund as far as possible in support of the new interpretation. His turn to descent theory was, as he recognized, an act of judgment, the deliberate adoption of a new interpretive stance to obtain a better grasp of the order of organic nature. Unlike Haeckel, Gegenbaur refused to assert that organic descent was a proven or even a demonstrable fact. It was, instead, a leading idea, a guide for the reason in its encounter with the myriad and apparently unrelated facts of natural history. As such, it mediated between pure reason and bare empiricism and, in so doing, enabled morphologist and zoologist better to apprehend the overall temporal and formal patterns of nature.

Concretely expressed, Gegenbaur's adherence to descent theory led in 1870 to an explicitly genealogical formulation of the relationship between the types. From the lowly Protozoa had arisen, in time, the entire array of living and extinct animal forms. Man stood at the apex of this ascent; other groups of animals (still designated "types") were at lower levels along the main evolutionary trunk or, more commonly, were branches from that trunk. "We recognize," Gegenbaur wrote, "that there exists a connection between the separate lines of descent. . . . [W]e can deal with the relations between types in a manner no different from that with which we treat of divisions within the types, that is, [treat them as expressions of] genealogical diversification."[8] The revised edition of the *Grundzüge*, as well as both editions (1874, 1878) of its simplified offspring, *Grundriss der vergleichenden Anatomie*, described the comparative morphology and presumed evolutionary connections of the invertebrate as well as vertebrate types. Gegenbaur's final synthetic work, *Vergleichende Anatomie der Wirbelthiere* (1898–1901), however, placed overwhelming emphasis upon the vertebrates. This work thus is a fair representation of the emphasis of the morphologist's later years. Through Gegenbaur's influence and that of contemporary morphologists of like mind, such as Robert Wiedersheim, comparative vertebrate anatomy began increasingly to represent the *summa* of post-Darwinian investigation of the major lines of evolutionary descent and to provide the indispensable foundation for the introduction of the serious student to biological inquiry, a movement well advanced by the 1890's.

An important reason for Gegenbaur's overriding concern with the priority of comparative anatomy in evolutionary reconstruction (phylogeny) was his doubts concerning the scope and accuracy of alternative approaches. Paleontology, he conceded, provided direct access to the history of life. Regrettably, however, the fossil record seemed altogether too fragmentary to allow the re-creation of what was, above all else, a quite literal continuity: the course and changes of living forms over vast periods of time. He appears to have expected more from what Haeckel had in 1866 dramatized as the biogenetic law ("Ontogeny is the summary and rapid recapitulation of Phylogeny") and that over the several preceding decades had received much sympathetic attention. If the developing embryo of a presumably higher form—a vertebrate, for example—did indeed display, in strictly comparable temporal and spatial order, the evolutionary history that lay behind its own existence, then an altogether extraordinary instrument of study was at hand. One need only carefully observe the emerging complexity of such an embryo in order to discover facts ("stages" of development) that would allow one to fill in those gaps inevitably present in the known fossil record.

Originally agreeable to the idea, by the 1880's Gegenbaur was vigorous in pointing out the pitfalls facing the unrestrained recapitulation theorist. Most important, he emphasized that caenogenesis (another of Haeckel's neologisms) was a common and perhaps ubiquitous phenomenon. Caenogenesis referred to new evolutionary acquisitions by the embryo itself and thus dealt with parts or processes that, when manifest in the embryo, could not be accepted as reliable indicators of the ancestral condition or conditions from which the parent stock of the embryo had arisen. Basically, Gegenbaur's criticism cast all in doubt, for how was one to decide whether a given part or process was caenogenetic or palingenetic (that is, truly indicative of ancestral conditions)? Here, once again, was a problem demanding the most exquisite exercise of judgment. Gegenbaur suggested that probably the best standard for venturing such judgments was provided by comparative anatomy, the science that could at least provide some reliable representation of the forms and interrelations of form toward which any known embryo was developing. The damage, however, had been done, for Gegenbaur was persuaded that ontogeny presents "no true likeness" of phylogeny.[9]

This negative assessment of the recapitulation theory led to a further conclusion. Just as paleontology had displayed shortcomings in its fitness for phylogenetic studies, so did the great expectations placed on embryology fail. From this fact Gegenbaur drew great comfort: surely comparative anatomy must assume leadership in the high function of re-creating the history of life. Comparative anatomy therefore commanded priority of place among all the special disciplines dealing with the descent theory. In 1858 the chair of zoology at Jena (which included anatomy among its responsibilities) was divided. Gegenbaur, who had demanded this action, retained the chair of zoology and comparative anatomy; a new chair devoted exclusively to physiology was occupied by Albert von Bezold. This event marked an important moment in the development of the natural sciences within the German university organization. Physiology, long tied to anatomy (as *anatomia animata*), was given independence and could develop according to its own needs and opportunities. Upon the death of Johannes Müller (1858), the most important anatomical chair in Germany, that at Berlin, was split following the Jena example. Anatomical responsibilities fell to Karl Bogislaus Reichert and physiological ones to Emil du Bois-Reymond.

But, to Gegenbaur, anatomy also had been liberated. On diverse grounds he had been seeking, and throughout his career continued to seek, a rationale and appropriate institutional foundations for the autonomy of anatomical investigation. Physiology, increasingly an experimental science, had now received its rights. Descent theory presented a host of novel possibilities for investigation; and anatomy, especially comparative anatomy, appeared to be uniquely suited to this task. Gegenbaur continued the abiding interest of German morphologists in form itself—that is, in the interrelations between the purely spatial disposition of the parts of organisms. These several motives lay behind Gegenbaur's unrelenting defense of the rights of anatomy and his claims for its central role in biological research and generalization.

NOTES

1. Gegenbaur, *Erlebtes und Erstrebtes*, p. 58.
2. M. Fürbringer, "Carl Gegenbaur," in *Heidelberger Professoren*, 426–450; also in *Gesammelte Abhandlungen*, I, iii–xxii.
3. Gegenbaur, "Condition and Significance of Morphology," 40.
4. J. V. Carus, *System der thierischen Morphologie* (Leipzig, 1853), 30.
5. Gegenbaur, *Grundzüge*, 2nd ed., 76.
6. *Ibid.*, 72.
7. Gegenbaur, *Lehrbuch*, 4th ed. (1890), 2.
8. Gegenbaur, *Grundzüge*, 2nd ed., 77.
9. Gegenbaur, "Cänogenese," 496.

BIBLIOGRAPHY

I. ORIGINAL WORKS. An exemplary bibliography (with subject divisions) of Gegenbaur's publications is M. Fürbringer, "Systematisches Verzeichnis der Veröffentlichungen von Carl Gegenbaur," in *Heidelberger Professoren aus dem 19. Jahrhundert*, II (Heidelberg, 1903), 455–466; it reappears in modified form in Gegenbaur's *Gesammelte Abhandlungen*, III, 575–597. The anatomist's principal monographs are *Untersuchungen über Pteropoden und Heteropoden. Ein Beitrag zur Anatomie und Entwicklungsgeschichte dieser Thiere* (Leipzig, 1855); *Grundzüge der vergleichenden Anatomie* (Leipzig, 1859; 2nd ed., rev., 1870); *Untersuchungen der vergleichenden Anatomie der Wirbelthiere: Erstes Heft. Carpus und Tarsus* (Leipzig, 1864), *Zweites Heft. Schultergürtel der Wirbelthiere und Brustflosse der Fische* (Leipzig, 1865), *Drittes Heft. Das Kopfskelet der Selachier, Ein Beitrag zur Erkenntniss der Genese des Kopfskeletes der Wirbelthieres* (Leipzig, 1872); *Grundriss der vergleichenden Anatomie* (Leipzig, 1874; 2nd ed., 1878); *Lehrbuch der Anatomie des Menschen* (Leipzig, 1883; 7th ed., 2 vols., 1898–1899); and *Vergleichende Anatomie der Wirbelthiere mit Berücksichtigung der Wirbellosen*, 2 vols. (Leipzig, 1898–1901).

Gegenbaur's articles have been collected and published as *Gesammelte Abhandlungen von Carl Gegenbaur*, M. Fürbringer and H. Bluntschli, eds., 3 vols. (Leipzig, 1912). Those dealing with matters of general biological interest include *De animalium plantarumque regni terminis et differentiis* (Leipzig, 1860), a rare 16-page oration also in *Gesammelte Abhandlungen*, II, 1–13; "Die Stellung und Bedeutung der Morphologie," in *Morphologisches Jahrbuch*, **1** (1875), 1–19, trans. as "The Condition and Significance of Morphology," in *The Interpretation of Animal Form*, W. Coleman, ed. and trans. (New York, 1967), 39–54; "Einige Bemerkungen zu Götte's 'Entwicklungsgeschichte der Unke als Grundlage einer vergleichenden Morphologie der Wirbelthiere,'" in *Morphologisches Jahrbuch*, **1** (1875), 299–345; "Cänogenese," in *Anatomischer Anzeiger*, **3** (1888), 493–499; and "Ontogenie und Anatomie in ihren Wechselbeziehungen betrachtet," in *Morphologisches Jahrbuch*, **15** (1889), 1–9.

II. SECONDARY LITERATURE. There is no major biography of Gegenbaur. His autobiography, *Erlebtes und Erstrebtes* (Leipzig, 1901), provides an invaluable account of the Würzburg and Jena years. Dependent upon the autobiography, but also presenting much additional information, are two accounts by M. Fürbringer: "Carl Gegenbaur," in *Heidelberger Professoren*, II, 389–466, and, in briefer form, "Carl Gegenbaur," in *Anatomischer Anzeiger*, **23** (1903), 589–608. Also useful is E. Göppert, "Gegenbaur, Karl," in *Biographisches Jahrbuch und deutscher Nekrolog*, **8** (1903), 324–339 (Gegenbaur did not adopt the spelling Karl).

On Gegenbaur as investigator and teacher, see Friedrich Maurer, "Carl Gegenbaur, Rede zum Gedächtnis seines 100. Geburtsjahres," in *Jenaische Zeitschrift für Medizin und Naturwissenschaft*, **55** (1926), 501–518; and E. Göppert, "Karl Gegenbaurs genetische Methode im anatomischen Unterricht," in *Anatomischer Anzeiger*, **63** (1927); and "Friedrich Maurer und der Kreis um Carl Gegenbaur," in *Anatomischer Anzeiger*, **85** (1938), 313–331. E. S. Russell offers extended discussion of nineteenth-century morphological interpretations in *Form and Function* (London, 1916); on Gegenbaur and Haeckel see esp. 246–267; see also the brief comment by William Coleman, *Interpretation of Animal Form* (New York, 1967), xi–xxx. For a description and assessment of why and how Gegenbaur converted to the descent theory in the 1860's, consult William Coleman, "Morphology Between Type Concept and Descent Theory," in *Journal of the History of Medicine*, **31** (1976), 149–175.

On comparative anatomy since Cuvier, the basic source remains Wilhelm Lübosch, "Geschichte der vergleichenden Anatomie," in Bolk *et al.*, eds., *Handbuch der Anatomie der Wirbelthiere*, I (Berlin, 1931), 3–76; Lübosch must be approached with care, however, for he presents an unannounced apology for idealistic or "pure" morphology as well as an ill-disguised nationalistic bias. A classic article on problems central to Gegenbaur's inquiry that also provides valuable historical insight is Hans Spemann, "Zur Geschichte und Kritik des Begriffs der Homologie," in C. Chun and W. Johannsen, eds., *Allgemeine Biologie*, in the series Kultur der Gegenwart (Leipzig, 1915), 63–86. Georg Uschmann, *Geschichte der Zoologie und der zoologischen Anstalten in Jena, 1779–1919* (Jena, 1959), 27–62, describes Gegenbaur's work at Jena in great detail and provides an insight into how and when Gegenbaur and Haeckel discovered Darwin. On descent theories in Germany before *The Origin of Species*, see the indispensable article by Owsei Temkin: "The Idea of Descent in Post-Romantic German Biology: 1848–1858," in Bentley Glass *et al., Forerunners of Darwin: 1745–1859* (Baltimore, 1959), 323–355.

WILLIAM COLEMAN

GEHLEN, ADOLF FERDINAND (*b.* Bütow, Pomerania [now Bytów, Poland], 15 September 1775; *d.* Munich, Germany, 15 July 1815), *pharmacy, chemistry.*

The last decades of the eighteenth century and the first of the nineteenth century constituted a high point in the development of pharmacy in Germany. The apothecaries themselves made a particularly vigorous contribution to the scientific advancement of their field, and many of them worked with great enthusiasm in areas far beyond the narrow confines of their profession. Gehlen. the son of an apothecary, grew up in this fruitful period. Like his younger brother, he was destined for a career in pharmacy. At school in his native city he acquired a very good knowledge of the classical languages, then received his practical training under the direction of Karl Gottfried Hagen, the court apothecary in Königsberg. After three years Gehlen entered the University of Königsberg, where he studied modern languages and subjects related to pharmacy, so that he was later able to correspond—in eight languages—with scientists throughout the world. He graduated with the M.D. degree.

As a child, Gehlen suffered from an ear ailment that left him hard of hearing. Since it would therefore have been difficult for him to wait on customers in an apothecary shop, he sought to establish contacts in Berlin, where he could occupy himself primarily with laboratory research and writing. After working briefly with Klaproth, who helped him improve his knowledge of chemistry, Gehlen was hired by Valentin Rose, another former student of Hagen's. Together they edited the first six volumes of the *Neues Berliner Jahrbuch für die Pharmazie* (1803–1806). At the same time Gehlen took over the editorship of the *Allgemeines Journal der Chemie* from Alexander Nikolaus Scherer, when the latter was appointed professor at the University of Dorpat. After publishing six volumes from 1803 to 1806 under the title *Neues allgemeines Journal der Chemie*, he decided to expand its scope and renamed it *Journal für Chemie, Physik und Mineralogie*. Despite the broadening of subject matter promised by the title, the nine volumes that appeared until 1810 remained devoted almost exclusively to chemical topics. The only change was the inclusion of articles on the physical aspects of chemistry, a field then growing steadily in importance.

Gehlen quickly revealed how well suited he was to being an editor. He was very strict in selecting articles, accepting only those that corresponded to the title of the journal and avoiding any overlap with other periodicals. Whereas most editors of the time, in an effort to offer as much material as possible, set scarcely any boundaries between their own areas and the other sciences, Gehlen published only material pertaining to chemistry. He was also, to a large extent, successful in convincing other editors to follow his policy, and in a few cases he even drew up contracts establishing the editorial limits between his journal and others. He concluded such an agreement with Johann Barthelomäus Trommsdorff, editor of the *Journal der Pharmacie*. L. W. Gilbert, too, indicated his readiness to limit as much as possible the inclusion of chemistry articles in his *Annalen der Physik*. The cooperation went so far that the editors sent each other material they had received but thought more suited to the other's journal.

Gehlen displayed considerable skill in enlisting a large circle of distinguished collaborators, including Hermbstädt, Klaproth, and Trommsdorff. Most of the articles he published were original, although he also included important foreign works that he himself had translated. Each volume of the journal contained an extensive bibliography, as well as a lively correspondence section that was much appreciated by its readers.

The importance of the articles that Gehlen wrote at this time did not lie in any wholly new discoveries or concepts. Rather, his strength was always in his ability to make critical evaluations and to point out ways in which known results could be further improved and submitted to careful testing. For example, he refuted the assertion made by Fourcroy and Vauquelin that formic acid is a mixture of acetic and malic acids, and he was also one of the first to recognize the toxicity of hydrocyanic acid.

In 1806 Gehlen went to Halle, where he devoted his time exclusively to academic concerns. He wrote a dissertation in the natural sciences and qualified as a *Privatdozent*. He was engaged by J. C. Reil to lecture on zoochemistry at the clinical institute.

The following year Gehlen accepted an appointment as *akademischer Chemiker* at the Bavarian Academy of Sciences, of which he automatically became a member. Following the accession to the throne of Maximilian I, a more liberal spirit reigned in Bavaria. It thus became possible to attract even foreign scientists to the capital, especially since the university science faculty was at that time the center of some very advanced experimental research. Gehlen's activities at the Academy absorbed all his energies. He was required to study the Bavarian mining and metallurgical industries, as well as its glass and porcelain manufacturing. Moreover, since the king took a strong personal interest in the encouragement of agriculture, Gehlen was often obliged to travel around the country gathering information, to analyze soil samples, and to write reports and scientific papers on the subject. In order to have the time for all these duties, he gave up his editorial activities in 1810 and was succeeded by Johann Christoph Schweigger.

In 1807 Gehlen was promised a laboratory, but construction was delayed for eight years by unfavorable economic conditions. As a result, he was obliged to erect a laboratory in his own house. Gehlen's favorite field of study was still pharmacy. With J. A. Buchner he founded the Pharmazeutischer Verein in Baiern, which, going beyond the usual professional concerns, set itself broad social aims as well. Gehlen published a plan proposing old-age insurance for apothecary's assistants, and, with the support of Bucholz and Trommsdorff, he called for the establishment of a fund for this purpose. It existed for years as the Bucholz-Gehlen-Trommsdorff Stiftung. Gehlen frequently helped capable young apothecaries to enroll in universities or assisted them in other ways. In 1815 he and Buchner began publication of *Reportorium für die Pharmacie*. Finally, in that same year, construction was started on the long-promised laboratory.

Unfortunately, Gehlen was not able to benefit from the new laboratory. Having undertaken a major work on the compounds of arsenic without being aware of the toxicity of arsenic hydride, he had inhaled a large quantity of that substance; he succumbed, after much suffering, to arsenic hydride poisoning. With his death chemistry lost one of the most famous scientists of the age. The apothecaries of Bavaria erected a monument to him, and his friend Georg F. C. Fuchs named a silicon compound gehlenite in his honor.

BIBLIOGRAPHY

I. ORIGINAL WORKS. Gehlen's writings include "Bemerkungen über die Ätherarten," in *Neues allgemeines Journal der Chemie*, **2** (1804), 206–227; "Über den bässeschen Salzäther und über das Verhältniss der Acidität der Essigsäure zu ihrem spezifischen Gewicht," *ibid.*, **5** (1805), 689–695; "Beiträge zur wissenschaftlichen Begründung der Glasmacherkunst," in *Denkschriften der Königlichen Bayerischen Akademie der Wissenschaften zu München* (1809–1810), 197–224; *Fassliche Anleitung zu der Erzeugung und Gewinnung des Salpeters zunächst für Landleute* (Nuremberg, 1812; 2nd ed., 1815); "Bemerkungen über die Eigenthümlichkeit der Ameisensäure," in *Journal für Chemie und Physik*, **4** (1812), 1–41; *Anleitung zum Bau der waldpflanze* (Munich, 1814); "Über das electrochemische System," *ibid.*, **12** (1814), 403–411; "Über das

Glasmachen ohne Pottasche vermittelst des Glaubersalzes," *ibid.*, **15** (1815), 89–107; and "Neue Bearbeitungsart des Arsenwasserstoffes," *ibid.*, 501–503. A list of Gehlen's writings is in *Denkmal auf dem Grabe des Adolf Ferdinand Gehlen* . . ., prepared by the Pharmaceutischer Verein in Baiern (Munich, 1820).

II. SECONDARY LITERATURE. See the obituaries by J. A. Buchner in *Reportorium für die Pharmacie*, **1** (1815), 435–446; and by F. von Schlichtegroll in *Denkschriften der Königlichen Bayerischen Akademie der Wissenschaften zu München* (1814–1815), xxix–xxxv.

GÜNTHER KERSTEIN

GERARD OF CREMONA (*b.* Cremona, Italy, *ca.* 1114; *d.* Toledo, Spain, 1187), *translation of scientific and philosophical works from Arabic into Latin.*

Gerard of Cremona was the most prolific translator of scientific and philosophical works from Arabic in the Middle Ages. He also has been credited with a few "original" works in the same fields, although these attributions are less certain. Through their abundance, subject matter, and quality, Gerard's translations made a decisive contribution to the growth of medieval Latin science. The impact of his work was felt well into the early modern period.

Life. Although Gerard's birth in Cremona is virtually certain, the site of his death continues to be disputed between Toledo, where he worked for the major part of his life, and Cremona, where C. A. Nallino, as recently as 1932[1] (followed by George Sarton), argued that he died. The dispute turns on two bits of evidence. The first, apparently contemporary, is a short eulogy inserted in manuscripts after the list of Gerard's translations compiled by his companions shortly after his death. In this eulogy, the last distich runs as follows:

> Hunc sine consimili genuisse Cremona superbit
> Toleti vixit, Toletum reddidit astris.[2]

Nallino argues that the final clause merely implies that Gerard, by his works, "extolled Toledo to the stars." This interpretation, however, forces inadmissible syntactical contortions upon the distich, which is of parallel construction. Toletum, like Cremona, is a nominative and the grammatical subject of the clause it introduces. Hence, the meaning of "Toletum reddidit astris" must be that the city of Toledo returned this illustrious man to the stars.

The second piece of evidence is a statement, dating from more than a century after Gerard's death, by the Italian chronicler Franciscus Pipinus,[3] a Dominican friar who wrote about 1300 and died in 1316. Pipinus asserted that Gerard was buried in the convent of Santa Lucia in Cremona, to which he had bequeathed his library. The sources invoked by the eighteenth-century scholar Francisco Arisi in his *Cremona literata* (1702)[4] testify that in his time there was absolutely no trace of Gerard's library in Cremona. Taking into consideration the eulogy and the fact that Pipinus never stated that Gerard actually died in Cremona, a plausible supposition would be that Gerard's body was returned to Cremona from Toledo to be reinterred, and that some kind of arrangement concerning his estate benefited a monastery in his native Cremona.

For further details concerning Gerard's life, as well as his activity as a translator in Toledo, the interpretation of manuscript evidence remains *grosso modo* where it was left by B. Boncompagni (1851) and F. Wüstenfeld (1877).[5] The principal evidence is a document in three parts, found inserted at the end of some copies of Gerard's translation of Galen's *Tegni* with the commentary of ʿAlī ibn Riḍwān. This document contains a short biography, a list of seventy-one works translated by Gerard, and the eulogy. An additional piece of evidence, important for the insight it provides into Gerard's intellectual interests and his way of life in Toledo, was added to the dossier by V. Rose in 1874.[6] In his *Philosophia*, the English scholar Daniel of Morley, who traveled to France and Spain in the late twelfth century, gives a detailed account of his lively encounter in Toledo with Gerard (whom, incidentally, he calls Gerard of Toledo) while attending Gerard's public lectures on Abū Maʿshar's *Great Introduction to the Science of Astrology*.[7]

Traditional interpretation has attributed all three parts of the biobibliography to Gerard's companions (*socii*) in Toledo. Although the bibliography may confidently be attributed to the *socii*, as the biography states,[8] there is evidence internal to the *vita* that intimates it was not drafted by the companions. Furthermore, the bibliography contains certain inferences in its manuscript transmission to suggest that at one time it existed independently. When it is accompanied by the biography, the link between the two is marked by a *vero* inserted in the title of the list ("Haec vero sunt nomina librorum quos transtulit"); this *vero* is absent from the title when the list appears independently, as in Oxford, MS All Souls 68, fol. 109,[9] and also in the labeling of the several partial lists, such as Ashmole 357, fol. 57v. But it is mostly the wording of the biography that raises doubts. It speaks of the authors of the list in the third person and in the past tense (*per socios ipsius diligentissime fuerunt connumerata*), and indicates the place in the corpus of Gerard's writings where this list has been appended: the end of Galen's *Tegni*, the last of his translations (*novissime ab eo translati*). In the original list without the biography, it should have been obvious where the list had been appended, without a need to specify it. The precision given by the biography sounds like the justification for having moved it from its original place, and the mention of Galen's *Tegni* reinforces this impression. Later, the *vita* recalls that it was the love of the *Almagest*, which he knew was not available in Latin, that led Gerard to Toledo (*Toletum perrexit*): "there [*ubi*], seeing the abundance of books in Arabic on every subject . . . he learned the Arabic language, in order to be able to translate. . . ." Also at the end of the *vita* and introducing the list of translations is the statement "These are the titles of the books translated by Master Gerard at Toledo."

These two citations indicate that the *vita* probably was not written in Toledo but more likely in Cremona, where Pipinus found it and transcribed portions of it in the late thirteenth century. This assessment clarifies the apparently contradictory passage at the beginning of the biography that praises the value of fame and simultaneously depicts Gerard as so humble that he refrained from signing his translations. Certainly this pompous exordium did not originate with the *socii*, who knew Gerard better. According to the discussion with Gerard reported by Daniel of Morley, Gerard was far from humble: "Ego qui loquor rex sum. . . . Cum vero ironice interrogarem ubi regnaret, respondit, in animo, quia nemini mortalium servirem. . . ."[10] In describing the translations, the *vita* characterizes the list as complete (*cuncta opera ab eo translata*) and summarizes it by following *ad litteram* the categories mentioned in the list at the opening of each group. This procedure suggests imitation and exaggeration. The eulogy is so closely patterned on the *vita* that the variants offered by the manuscripts of the eulogy can be selected from the *vita*. Of the two readings, *spirituali* and *spiritualis*, offered by the manuscripts in the third line of the eulogy Wüstenfeld chose the wrong one, *spiritualis*. The *vita* was stating clearly: *Carnis desideriis inimicando, solis spiritualibus adhaerebat*. In Wüstenfeld, the corresponding line in the eulogy reads *Voto carnali fuit hostis spiritualis* [*applaudens* . . .], where *spirituali applaudens* is obviously the correct reading.

Both textual and biographical clues reinforce the conviction that the *vita* could not have been written by Gerard's associates. This conclusion rests primarily on the absence of Gerard's signature to his translations and on the depiction in the *vita* of Gerard's activity in Toledo. In both cases, the *socii* appear to be denying themselves credit they might legitimately claim in the choice and work of the translations. A simple listing of the translations, probably included in the original document before the biographer's intervention, would not be subject to these difficulties of interpretation.

The date of Gerard's move from Italy to Toledo and the number of years he devoted to translations also are uncertain. The *vita* stresses that Gerard completed his education in the schools of the Latins before going to Toledo, but it also implies that he moved there very shortly afterward. Since there is no evidence that Gerard devoted to theology the many years necessary for a degree, one may estimate that his stay in the Western schools lasted until he was twenty-five or, at most, thirty years old. Thus, he must have reached Toledo by 1144 at the latest.[11] This would give him some forty-three years of activity in Toledo—a reasonable estimate that also renders his enormous output more credible.[12]

The only later date in his life that has hitherto been taken for granted is the year 1175, in which he was thought to have completed the translation of Ptolemy's *Almagest*.[13] But this date seems suspect: he would have taken more than thirty years to complete the translation of the work for which he had moved from his native country, despite the help he received with this translation from the Mozarab Galib (Galippus Mixtarabe). The date of 1175 is attested only for the transcription made at Toledo by Thaddeus of Hungary (one of Gerard's associates?) of the text of the *Almagest*; it does not follow that Gerard completed it then. On the contrary, as V. Rose has observed,[14] the use of native Spaniards as collaborators with foreign translators such as Gerard appears to have usually been limited in scope and duration. At any rate, Galib is never named as a collaborator on any other of Gerard's translations; and the assumption that translators usually worked in pairs[15] is an undue extrapolation from the very scanty occurrences. Rather than assume, as is easily done, that all translators worked in collaboration, the truth appears to be that the case was special and remarkable enough to be noted whenever it occurred. At any rate, in the case of Gerard, only Galib is mentioned as having collaborated on the *Almagest*; and the list drawn up by Gerard's associates makes no mention whatsoever of their direct collaboration in any translation.

What can these associates have been doing, if not helping in the translation? Several answers are possible without having to renounce the conviction that Gerard worked alone most of the time on all the translations ascribed to him (except the *Almagest*). The associates may have been engaged in tracing the many original copies of the Arabic texts to be used by Gerard. Ibn ʿAbdūn of Seville's early twelfth-century treatise of *ḥisba* (translated by E. Lévi-Provençal)[16] relates that a Muslim supervisor of the market in Andalusia was asked to forbid "the selling of Arabic books of science to the Christians" because the Christian translators allegedly attributed these works "to their bishops." This prohibition may not have been strictly enforced in Toledo; but surely, given the enormous number and scope of scientific works translated by Gerard, it is easy to imagine the tremendous task of assembling these works and their relevant sources. Gerard may have assigned this task to some of his associates; others may have been employed as editors, "proofreaders," or mere scribes, as Thaddeus of Hungary, mentioned above in relation to the *Almagest*, may have been.

Characteristics and Method of Gerard's Translations. Very few definitive characteristics have been isolated, despite the laudable attempts of a few scholars.[17] The difficulty lies in determining the proper method to be used in the search for identifying characteristics of style. We propose here a new method based on the close examination of one set of translations in Gerard's list and of some others that may be appended, despite their absence from the list. A provisional compilation of such new versions would include, without being exhaustive, the following translations (the number in square brackets is that on Wüstenfeld's list):

1 [1]. Aristotle, *Posterior Analytics* (James of Venice; "Io [annes]")

2 [4]. Euclid, *Elements* (Adelard of Bath; Hermann of Carinthia [?])

3 [5]. Theodosius of Bithynia, *Spherica* (Plato of Tivoli [?]; Robert of Chester [?])

4 [6]. Archimedes, *De mensura circuli* (Plato of Tivoli)

5 [13]. Al-Khwārizmī, *Algebra* (Robert of Chester)

6 [29]. Thābit ibn Qurra, *De motu accessionis* (John of Seville)

7 [21]. Al-Farghānī, *Liber continens capitula XXX* (John of Seville)

8 [33]. Pseudo-Aristotle, *Liber de causis* (Gundissalinus[?])

9 [34]. Aristotle, *Physics* (anonymous)

10 [42]. Al-Fārābī, *De scientiis* (Gundissalinus)

To these may be added some new versions not included in the list compiled by the *socii*:

11. Al-Khwārizmī, *Arithmetic* (John of Seville; Adelard of Bath [?])

12. Qusṭā ibn Lūqā, *De differentia spiritus et animae* (John of Seville)[18]

13. Al-Zarqālī, *Canones* (Robert of Chester)

14. Alcabitius (al-Qabīṣī) [?] (John of Seville)

15. Abū Maʿshar, *Liber maioris introductorii* (John of Seville)

The above list includes revisions of translations by earlier scholars and by others contemporary with Gerard in northern Spain. The manuscript tradition of Abū Maʿshar's still unpublished *Greater Introduction* reveals a revision—sometimes extensive, occasionally sporadic or nearly absent—that exhibits identifying characteristics observable in some of Gerard's other reworkings. On the other hand, Daniel of Morley provides testimony that Gerard delivered public lectures in Toledo on this very text. It had been translated in 1133 by John of Seville, whose approach to translation Gerard adopted.

A close examination of three of these reworkings through comparison with the Arabic original and with the earlier Latin translation sheds light on Gerard's strategy. He appears to have weighed the earlier Latin translation against the Arabic, retaining what seemed to him passable and simultaneously reducing the lexical usage to a standard one, substituting *generatio* for *effectus* or *constitutio* (Arabic *kawn*) and *corruptio* for *destructio* (Arabic *fasād*). Moreover, he restored all passages and constructions omitted in the earlier translation, often altering its vocabulary to make it better suited to Latin scientific usage. In this process of trial and practice, Gerard showed an increasing preference for the manner of John of Seville, from whom Gerard adopted many characteristic expressions. For example, in al-Farghānī's work, Gerard restored the chapter on comparative eras omitted by John, who had already treated this subject in another work. But in al-Fārābī's *De scientiis*, already translated by Gundissalinus, noting Gundissalinus' ruthless editing, Gerard completely restored the original and transformed Gundissalinus' Latin vocabulary. The form of Gerard's translations thus resembles that of John of Seville: closeness to the Arabic original, preservation as far as possible of the construction of the Arabic sentences, and scrupulous rendering of nearly every word contained in the Arabic. Undoubtedly, much of Gerard's knowledge of Arabic was acquired through working with the older translations, which directed him ultimately to the adoption of the word-for-word method of John of Seville.

Gerard's Translations and Works. Although the list drawn up by Gerard's *socii* has often been proved absolutely reliable with respect to the items it contains, it cannot be considered exhaustive. Modern scholarship concurs in this view, and a strong need has been felt to add to, rather than subtract from, the list.

Although Sarton's list in his *Introduction to the History of Science* (II, 339–344) may serve some purpose for its bibliographical information, it has the great inconvenience of rearranging the order of subject matter as well as the numbering from the *socii* list given by Wüstenfeld, which it seems preferable to preserve for clarity and historical accuracy.[19] Sarton's aim appears to have been to separate "classical" (Greek) authors and works in science and philosophy from Arabic commentaries on Greek and Arab authors in the same fields. The distinction is of questionable scholarly value, since the Arab authors depended heavily upon their Greek models, adding much commentary of their own. Moreover, Sarton thus provided an unwarranted basis for the mistaken historical notion (too easily adopted by users of the list, if not by Sarton himself) that the twelfth-century translations helped Western scholars to recover Greek science in its original form.

The interplay between Greek science and Arab scholarship is an important historical factor and should not be underemphasized. Gerard's transmission of this heritage to the West had an exclusively Arab garb[20] that directly reflected scholarship in the Arab world at the time of the Latins' contact. Furthermore, Gerard's influence perpetuated this interplay throughout most of the Middle Ages. Even the Greek interests revived by William of Moerbeke's translations in the late thirteenth century remained minor in comparison with the weight still carried by the Arab tradition (except in Moerbeke's translations of Aristotle, which were extensively used in the fourteenth century): the best proof is the influence on optics and on geometry exerted by Gerard's translations of Ibn al-Haytham (Alhazen) and of Euclid. Latin science and philosophy of the twelfth–fourteenth centuries was totally dependent on the Arabic tradition. This is in sharp contrast with the situation of scholarship in the Renaissance, which revived the original Greek texts. This crucial distinction should be borne in mind, for Sarton's rearrangement of the *socii* list mistakenly transfers to the twelfth century the humanists' fixation on Greek authors.

The *socii* list has been published and commented upon several times during the last century[21] and has recently become available in an English translation by Michael McVaugh with additional useful bibliographical annotations. We shall list here Gerard's works, both translations and original treatises, according to the order and categories of the *socii* list. Only such bibliographical information and criticism as can be added from recent scholarship are presented. For the Arabic background to the originals translated by Gerard, ample references are given to the recent works of two Orientalists: Fuat Sezgin, *Geschichte des arabischen Schrifttums*, and Manfred Ullmann, *Die Medizin im Islam* (1970) and *Die Natur- und Geheimwissenschaften im Islam* (1972).

Gerard's Translations. The first number is that of Wüstenfeld; the number from Sarton's list follows in square brackets.

A. Works on logic (3 works).

1 [1]. Aristotle, *Analytica posteriora*. There is a second, revised edition (1968) by L. Minio-Paluello and B. Dod in which Minio-Paluello shows that the original Arabic text used by Gerard was not the translation of Matta ibn Yūnus, but another, anonymous one. See also *Aristoteles latinus* (*AL*), I, 48, no. 13.

2 [2]. Themistius, *Commentarius super Posteriores analiticos*. See *AL*, I, 99, no. 97, and "Specimina," pp. 206–207. Edited by J. Reginald O'Donnell, in *Mediaeval Studies*, **20** (1958), 239–315.

3 [3]. Al-Fārābī, *De syllogismo*. Incipit *TK* 925: "Nostra in hoc [libro] intentio est famosas scientias. . . ." (*TK* is L. Thorndike and P. Kibre, *A Catalogue of Incipits of Mediaeval Scientific Writings in Latin*, 2nd ed. [Cambridge, Mass., 1963].) Probably this was some form of *Prior Analytics* commentary, according to Wüstenfeld, p. 59. It should be compared with al-Fārābī, *Short Commentary on Aristotle's Prior Analytics*, translated from the original Arabic with introduction and notes by Nicholas Rescher (Pittsburgh, 1963).

B. Works on geometry, mathematics, optics, weights, dynamics (17 works).

4 [24]. *XV Books of Euclid* (*Elements*), M. Clagett, "The Medieval Translations From the Arabic of the *Elements* of Euclid," in *Isis*, **44** (1953), 16–42; and *Archimedes in the Middle Ages*, I (Madison, Wis., 1964), p. 228, note, and index; A. A. Björnbo, "Gerhard von Cremonas Überset-

zung von Alkhwarizmis Algebra und von Euklids Elementen"; F. Sezgin, *Geschichte des arabischen Schrifttums* (*GAS*), V (Hajjaj's translation), 89–90, no. 4, 100–101, 103, 116, 117; H. L. L. Busard, "Über einige Euklid-Skolien . . .," in *Centaurus*, **18** (1974), 97–128. [Text as yet unpublished, but worked upon by Busard and J. Murdoch.] Books I and V have been edited in the dissertations of Sister Mary St. Martin Van Ryzin, O.S.F. ("The Arabic-Latin Traditions of Euclid's *Elements* in the Twelfth Century," University of Wisconsin, 1960, 321–399); and Thomas J. Cunningham ("Book V of Euclid's *Elements* in the Arabic-Latin Traditions," University of Wisconsin, 1972, 337–398), respectively.

5 [29]. Theodosius, *Spherica*. *GAS*, V, p. 155, 272; F. Carmody, *The Astronomical Works of Thabit ibn Qurra*, 2nd ed. (Berkeley, 1960), p. 22 (hereafter *Thabit* [1960]). Incipit *TK* 1523: "Sphera est figura corporea una quidem superficie. . . ." Also *ibid*.: "Sphera est figura solida una tantum superficie. . . ." Differences in wording do not necessarily imply a different version but, rather, probably scribal or reader's variants. F. Carmody, *Arabic Astronomical and Astrological Works in Latin Translation* (Berkeley, 1956), 22 (hereafter *AAAL*), ascribes the first incipit to Plato of Tivoli and the second to Gerard.

6[26]. Archimedes, *De mensura circuli*. M. Clagett, "Archimedes in the Middle Ages; the *De mensura circuli*," in *Osiris*, **10** (1952), 587–618; and *Archimedes*, I, esp. 30–58; *GAS*, V, 130–131, 289. [Clagett, 30, n. 3, erroneously gives no. 7 to this item in the list.]

7[37]. Aḥmad ibn Yūsuf, *De arcubus similibus*. F. Carmody, *AAAL*, no. 20, 2; and *Thabit* (1960), 231; *GAS*, V, 160; 288–290, 402. Edited by M. Curtze, in *Mitteilungen des Copernicus-Vereins für Wissenschaft und Kunst zu Thorn*, **6** (1887), 48–50; and more recently by H. L. L. Busard and P. S. van Koningsveld, in "Der *Liber de arcubus similibus* des Ahmed ibn Jusuf," in *Annals of Science*, **30** (1973), 381–406. Wüstenfeld places Aḥmad's death in 1002 (Busard [*loc. cit.*, p. 383] says 912), yet he was writing in 920 and probably died in 942.

8 [32]. Menelaus, *Sphaerica*. Carmody, *AAAL*, 22; and *Thabit* (1960), 221, spec. 15; *GAS*, V, 162, no. 3. Incipit *TK* 397: "Declarare volo qualiter faciam supra punctum datum. . . ." Francesco Maurolico's Latin text (Messina, 1558; Rome, 1587, with Clavius' *Astrolabium*) seems to contain another Latin version. At any rate, neither for this text, nor for Theodosius' *De habitationibus* (no. 26, below), which he also published with his own scholia, did Maurolico indicate the source of his Latin text.

9 [40]. Thābit ibn Qurra, *De figura alchata*. Carmody, *AAAL*, 121–122; and *Thabit* (1960), 159–164, three versions; the first one (*versio* F)—"Quod de figura nominata sectore . . ." (p. 159) corresponds closely to *TK* 1252. Also see *GAS*, V, 268.

10 [34]. Banū Mūsā, *Geometria*. Carmody, *AAAL*, 48–49; and *Thabit* (1960), 22; *GAS*, V, 246 ff. Also see M. Curtze, "Liber trium fratrum de geometria," in *Nova acta Academiae Caesarae Leopoldino Carolinae germanicae naturae curiosorum*, **49** (1885); and Clagett, *Archimedes*, I, 223–367.

11 [38]. Aḥmad ibn Yūsuf, *De proportione et proportionalitate*. Carmody, *AAAL*, 130–131; *GAS*, V, 289. Recent edition by D. Schrader (Ph.D. diss., University of Wisconsin, 1961).

12 [44]. *Liber iudei super X^m Euclidis*. *GAS*, V, 175, 287, 389. G. Junge, "Das Fragment der lateinishen Übersetzung des Pappus-Kommentars zum 10. Buche Euklids," in *Quellen und Studien zur Geschichte der Mathematik, Astronomie und Physik*, Abt. B, **3** (1934), 1–17, is a partial Latin version from Paris, MS BN lat. 7377A, which Junge thinks may be the translation by Gerard of Cremona.

13 [35]. Al-Khwārizmī, *De iebra et almucabala* (Algebra). G. Libri's edition (*Histoire des sciences mathématiques en Italie*, I [Paris, 1838], 253–297), is of a different text than the one published by B. Boncompagni in *Atti dell'Accademia pontificia dei Nuovi Lincei*, **4** (1851), 412–435. See A. A. Björnbo, "Gerhard von Cremonas Übersetzung von Alkwarizmis Algebra und von Euklids Elementen," in *Bibliotheca mathematica*, 3rd ser., **6** (1905), 239–248; L. Karpinski, *Robert of Chester's Latin Translation of the Algebra of al-Khowarizmi* (Ann Arbor, 1915), repr. as pt. I of Karpinski and J. G. Winter, *Contributions to the History of Science* (Ann Arbor, 1930); *GAS*, V, 239–240; Carmody, *AAAL*, 47–48.

14 [47]. *Liber de practica geometrie*. M. McVaugh suggests that this is identical with al-Karaji's *De mensuratione terrarum*, without specifying further (see no. [82]). The identification remains obscure and problematic; three works briefly described in *GAS*, V, 387–391, could qualify as the Arabic original: Abū ʿUthmān Saʿīd (387), Abbacus (388), and al-Karajī (389–391). We have

been unable to identify the incipit of Gerard's version included in the *socii* list.

15 [39]. Al-Nayrīzī, *Super Euclidem*. GAS, V, 283–284; Carmody, *Thabit* (1960), 22. Edited by M. Curtze in *Euclidis opera omnia. Supplementum* (Leipzig, 1899).

16 [25]. Euclid, *Data*. *GAS*, V, 116; Carmody, *Thabit* (1960), 22. See Shunturo Ito, "The Medieval Latin Translation of the Data of Euclid" (Ph.D. dissertation, University of Wisconsin, 1963), 19–20.

17 [53]. Tideus (Diocles), *De speculo*. Carmody, *AAAL*, 79; and *Thabit* (1960), 233 (spec.); *GAS*, V, 117. The original list has "Tideus, De speculis comburentibus," which represents a compression of two distinct works translated by Gerard: *Tideus de speculis*, edited in A. A. Björnbo–S. Vogl, "Alkindi, Tideus und Pseudo-Euklid"; and Ibn al-Haytham's *De speculis comburentibus*, edited by J. L. Heiberg and E. Wiedemann as "Ibn al-Haiṭams Schrift über parabolische Hohlspiegel," in *Bibliotheca mathematica*, 3rd ser., **10** (1909–1910), 201–237.

18 [54]. Al-Kindī, *De aspectibus* (optics). See Björnbo and Vogl, "Alkindi, Tideus und Pseudo-Euklid"; Carmody, *AAAL*, 79; and *Thabit* (1960), 231 (spec.); *GAS*, V, 117.

19 [not in Sarton]. *Liber divisionum*. Is this a work by Al-Baghdādī (Ibn Tahir)? See R. C. Archibald, *Euclid's Book on Divisions of Figures* (Cambridge, 1915); and *GAS*, V (1974), 387–388, 394–395.

20 [55]. Thābit ibn Qurra, *Liber Qarastonis*. Edited by M. Clagett and E. A. Moody, in *The Medieval Science of Weights* (Madison, Wis., 1952; repr. 1960). Previous information on this work by Thābit must now be checked against K. Jaouiche, "Le livre du Qarasṭun de Thabit ibn Qurra," in *Archive for History of Exact Sciences*, **13**, no. 4 (Nov. 1974), 325–347. Jaouiche's dissertation (Paris, 1972) on Thābit's *Qarastun* appears to be of importance (it includes one Arabic text of the *Qarastun*).

C. Works on astrology (astronomy) (12 works).

21 [36]. Al-Farghānī (Alfraganus). John of Seville's translation (edited by F. Carmody [Berkeley, 1943]) was entitled *Liber Alfragani in quibusdam collectis scientie astrorum et radicum motuum planetarum et est 30 differentiarum*. Gerard's version (edited by R. Campani [Florence, 1910]) was entitled *Liber de aggregationibus scientiae stellarum et de principiis coelestium motuum quem Ametus composuit filius Ameti qui dictus est Alfraganus 30 continens capitula*. The differences of vocabulary introduced by Gerard in his new versions appear strikingly in the titles. On al-Farghānī's work, see *GAS*, V, 259–260.

22 [33]. Ptolemy, *Almagest*. See P. Kunitzsch, *Der Almagest: Die Syntaxis mathematica des Claudius Ptolemäus in arabisch-lateinischen Überlieferung* (Wiesbaden, 1974).

23 [31]. Geminus of Rhodes [?], *Liber introductorius Ptolemei ad artem spericam*. The original text cannot have been written by Geminus, as M. McVaugh makes clear; however, K. Manitius (*Deutsche Literaturzeitung* [1899], col. 578) reminds his readers of the close parallels between this Latin version and Geminus' work, which he edited; see *GAS*, V, 157–158 (Aġāniyus). It also could be by Pappus of Alexandria (see GAS, V, 175, no. 2). The relation of Gerard's translation to the *Introduction à l'Almageste* by Eutocius presented by J. Mogenet (Brussels, 1956), who knew of a thirteenth-century(?) Latin translation (Mogenet, p. 38), is not yet clear. At any rate, Gerard's original must have been an Arabic text.

On the other hand, a Leningrad (Acad. Cod. AB-III; present shelf mark F.N.8) manuscript in its earlier (twelfth-century) portion contains, among works of Gerard of Cremona, an anonymous treatise beginning "Dividitur orbis signorum in 12 . . ." (fol. 25r) and containing the explicit (fol. 35va) ". . . incipit stella superbus oriri et accidit cum ea aqua. (R) Explicit quod abreviatum est de libro introductorii Ptholomei ad librum suum nominatum Almagesti." The incipit is strikingly similar to a work frequently ascribed to Ptolemy in medieval Latin manuscripts and beginning "Signorum alia sunt masculina . . .," easily confused with a work by Zahel (Sahl ibn Bishr) with a nearly identical incipit—and content, for that matter—translated by John of Seville. Paris BN codex lat. 16208 contains three different works with nearly the same incipit as above; one is ascribed to Zahel, another to Ptolemy, and the third has been tentatively assigned to Raymond of Marseilles (Mlle. M. T. d'Alverny, private communication). Perhaps item 23 in the *socii* list of Gerard's translations indicates this work; but we have our doubts.

24 [46]. Jābir ibn Aflah (Geber), *De astronomia libri IX* (Nuremberg, 1534). Carmody, *AAAL*, no. 35, 1 (p. 163); *GAS*, V, 53; R. P. Lorch, "Jābir ibn Aflaḥ and His Influence in the West" (Ph.D. dissertation, University of Manchester, 1970).

25 [80], Māshāʾallāh (Messahalla), *De elementis et orbibus celestibus*. Incipit TK 722: "Incipiam et dicam quod orbis est prescritus spericus. . . ." This corresponds to the incipit assigned by Albertus Magnus in his *Speculum astronomiae* to a

work by Messahalla, *De scientia motus orbis*, which Albert describes as a "late et compendiosius" treatment of the sphere according to the *Almagest*. In *Le système du monde* (II, 204–206) P. Duhem gives a summary analysis of this astronomical work by Māshāʾallāh in Gerard of Cremona's translation. Sarton, I, 531, calls it the most popular work by Māshāʾallāh in the Middle Ages, an evaluation that does not correspond to the vestiges of the manuscript tradition. There are literally hundreds of surviving manuscripts of the astrological works of Māshāʾallāh translated by John of Seville (see L. Thorndike, "The Latin Translations of Astrological Works by Messahalla," in *Osiris*, **12** [1956], 49–72), but only a few manuscripts of the *De elementis et orbibus* (see Carmody, *AAAL*, 32–33). Of Māshāʾallāh's astrological works translated by John of Seville, only his *Epistola de rebus eclipsium* has a slight astronomical background; this work, however, is in twelve chapters with the incipit "Quia Dominus altissimus fecit terram ad similitudinem spere . . ." (*TK* 1217), and thus would seem to correspond to article 2 in Ibn al-Nadīm's list in the *Fihrist*: "The Great Book of the 21 [12?] on Conjunctions, Religions and Sects." M. Ullmann, *Die Natur- und Geheimwissenschaften im Islam*, 304, supplies information on the works of Māshāʾallāh that is based on the research of L. Thorndike, E. S. Kennedy, and D. Pingree. In fact, Thorndike's studies on Māshāʾallāh's Latin manuscripts show the great complexity of identification by titles and manuscript ascriptions.

The translation of Māshāʾallāh's work by Gerard of Cremona is entitled *De elementis et orbibus celestibus* in the *socii* list and was so printed by Joachim Heller (Nuremberg, 1549): *De elementis et orbibus coelestibus liber antiquus ac eruditus Messahalae laudatissimi inter arabes astrologi*; this edition is not to be confused with the 1549 edition published at Nuremberg of other works of Māshāʾallāh translated by John of Seville under the title of *Messahallae libri tres*. Gerard's translation had already been published by J. Stabius in 1504, also at Nuremberg, under the title *De scientia motus orbis*. This was the title under which it was known to Albertus Magnus, who describes it in his *Speculum astronomiae* (on the *Speculum*, see the forthcoming edition by Paola Zambelli): "de eodem [material in the *Almagest*] agitur satis late et compendiosius in libro Messahalach *De scientia motus orbis* qui sic incipit: 'Incipiam et dicam quod orbis . . .'"—the incipit in our present work translated by Gerard of Cremona. The Arabic original seems to be the one listed as no. 8 in the *Fihrist* (see H. Suter, "Die Astronomen und Mathematiker," 5): "The book known as the 27"—it contains twenty-seven chapters. Ullmann gives no hint of the existence of this book by Māshāʾallāh either in Arabic or in Latin translation.

26 [30]. Theodosius, *De locis habitationibus*. Incipit *TK* 660 (and 684): "Illis ["In illis," 684] quorum habitationis loca sunt sub polo. . . ." This wording makes it slightly different from the text used by Maurolico, who added his own scholia and some ancient ones (Messina, 1558; Rome, 1587). The 1587 version is entitled *Autolyci De sphaera quae movetur* (see no. 30). Maurolico's text begins (prop. 1a): "Qui sub polo boreali habitant, iis quidem mundi hemisphaerium alterum idem semper conspicuum est. . . ." See Carmody, *Thabit* (1960), 219, spec. 7; *GAS*, V, 155–156.

27 [28]. Esculeus [Hypsicles], *De ascensionibus signorum*. Incipit *TK* 1449: "Si fuerint quotlibet quantitates quarum numeratio. . . ." Carmody, *Thabit* (1960), 22, and spec. p. 201; *GAS*, V, 143–145. See V. de Falco and M. Krause, eds. and trans., "Hypsikles, Die Anfangszeiten der Gestirne," *Abhandlungen der Akademie der Wissenschaften zu Göttingen*, 3rd ser., no. 62 (1966), with intro. by O. Neugebauer, which contains a Latin version by Gerard incomplete in the footnote section; reviewed by P. Kunitzsch in *Zeitschrift der Deutschen morgenländische Gesellschaft* (ZDMG), **118** (1968), 180–181.

28 [41]. Thābit ibn Qurra, *De expositione nominum Almagesti*. Edited by F. Carmody in *Thabit* (1960), 131–139.

29 [42]. Thābit ibn Qurra, *De motu accessionis et recessionis* (also known as *De motu octave spere*). Probably also translated by John of Seville; see J. Millás Vallicrosa, "Una obra desconocida," in *Osiris*, **1** (1936), 456–458; and "El liber de motu octave spere de Tabit ibn Qurra," in *al-Andalus*, **10** (1945), 89–108. Edited by F. Carmody in *Thabit* (1960), 102–113. See also Otto Neugebauer, "Thābit ben Qurra . . . 'On the Motion of the Eighth Sphere,'" in *Proceedings of the American Philosophical Society*, **106** (1962), 290–299. The incipit of this work of Thābit translated by Gerard is *TK* 661: "Imaginabor speram . . ." and is the same as in Carmody's edition. *TK* gives two other incipits for the same work: *TK* 106, "Annus itaque solaris vere loquendo," and *TK* 1703, "Vis motus et maxime corporum celestium . . .," the latter identified by *TK* as words of the prologue to a commentary on Thābit's work. The information given by Albertus Magnus in his *Speculum astronomiae* on the relation between the astronomical

GERARD OF CREMONA

works of Thābit, al-Zarqālī, al-Battānī, Jābir ibn Aflaḥ, John of Seville, al-Biṭrūjī, and Ptolemy's *Almagest* (and Hipparchus' precession) ought to bear on the solution of that question. We must, however, reserve for a separate work on the authenticity of the *Theorica planetarum* ascribed to Gerard of Cremona the consideration and interpretation of this evidence.

30 [23]. Autolycus, *De sphaera mota*. Incipit *TK* 1151: "Punctum equali motu dicitur moveri. . . ." The text published at Rome in 1587, *Autolyci De sphaera quae movetur liber*, together with *Theodosii Tripolitae De habitationibus liber* (bound with Christoph. Clavius, *Astrolabium* [Rome, 1593], in Columbia University Library copy) and given as from "Iosepho. Avria. Neapolitano Interprete," has a different incipit: "Hypotheses. I. Aequabiliter puncta ferri dicuntur. . . ." It probably is not the translation by Gerard, although the "interpreter" claims not to have translated it from the Arabic or Greek but to have extracted it from the Vatican Library. This printed text of the *De sphaera quae movetur* includes the scholia by Maurolico as does Theodosius' *De locis habitationibus* (see no. 26).

31 [50]. *Tabulae Jahen cum regulis suis*. See H. Hermelink, "Tabulae Jahen," in *Archive for History of Exact Sciences*, **2** (1964), 108–112. Hermelink establishes that this is the work entitled *Scriptum antiquum Saraceni cuiusdam, de diversarum gentium eris annis ac mensibus, et de reliquis astronomiae principiis*, published by J. Heller (Nuremberg, 1549). The author is Qadi Abū ʿAbdallāh ibn Muʿadh al-Jaihānī of Jaén (989–1079[?]), the second of the two Abū ʿAbdallāhs suggested by Sarton. A. I. Sabra ("The Authorship of the *Liber de crepusculis*, an Eleventh-Century Work on Atmospheric Refraction," in *Isis*, **58** [1967], 77–85) had already suggested this Qadi as author of the Jahen Tables as well as the author of no. 32 (below).

32 [56]. Abū ʿAbdallāh, Muhammad ibn Muʿadh *De crepusculis*, sometimes entitled *De ascensionibus nubium*. No longer to be ascribed to Ibn al-Haytham, as was demonstrated by Sabra in *Isis*; see *GAS*, V, 49, 364. Edited by Petri Nonii (Pedro Nuñez) Salaciensis as *De crepusculis liber. Item Alhacen Arabis vetustissimi de causis crepusculorum liber unus a Gerardo Cremonense jam olim latinitate donatus; nunc vero omnium primum in lucem editus* (Lisbon, 1541). Also published by F. Risner in *Opticae thesaurus Alhazeni* under Alhazen's name (Basel, 1572), 283–288. The 1541 edition includes an appendix by Gerard of Cremona in which he states that he has "omitted the terminal words of the Arabic original by which the author praises God in the manner of the Saracens" as being needless (see Wüstenfeld, p. 66). The incipit of the *De crepusculis* is *TK* 1021, "Ostendam quid sit crepusculum" and *TK* 1022, "Ostendere autem volo in hoc tractatu quid sit crepusculum."

D. Works on philosophy (11 works).

33 [9]. Pseudo-Aristotle, *De expositione bonitatis pure* (*Liber de causis*). *AL*, I, 94, no. IV; and spec. 196. See Bernard Carra de Vaux, "El 'liber de Causis' primis et secundis et de fluxu qui consequitur eas," in *al-Andalus*, **9** (1944), 419–440; M. Alonso Alonso, "El liber dẹ Causis," *ibid.*, 43–69, and **10** (1945), 345–382. Incipit *TK* 996: "Omnis causa primaria [primitiva] plus est influens. . . ."

34 [5]. Aristotle, *De naturali auditu* (*Physica*). See *AL*, I, 51, no. XV; and spec., 125–126 (this translation had little success, according to *AL*).

35 [4]. Aristotle, *Liber caeli et mundi*. See *AL*, I, 53, no. 18, and spec., 128–129 (the only popular translation of this text in the early thirteenth century, according to *AL*). See the study of its lexical particularities by I. Opelt, "Zur Übersetzungstechnik des Gerhard von Cremona," in *Glotta*, **38** (1959), 135–170.

36 [10]. Pseudo-Aristotle, *De causis proprietatum elementorum*. See *AL*, I, 91 (not to be confused with *Liber de causis*, no. 33 above).

37 [6]. Aristotle, *De generatione et corruptione*. Sybil D. Wingate, *The Medieval Latin Versions of the Aristotelian Scientific Corpus* . . . (London, 1931), 45–46; *AL*, I, 55, no. 21, and spec., 132–133 (little success in the Middle Ages, according to *AL*).

38 [7]. Aristotle, *Meteorologica*, books I–III. See Wingate, *Medieval Latin Versions*, 45–46; *AL*, I, 56, no. 23, and spec., 133; also II, 788; A. Pelzer, "Une source inconnue de Roger Bacon" (1919), reprinted and enlarged in a posthumous edition by A. Pattin and Émile van de Vyver, *Études d'histoire littéraire sur la scolastique médiévale*, Philosophes Médiévaux, no. 8 (Louvain–Paris, 1964), 241–271.

39 [11–14]. Alexander of Aphrodisias, *De tempore*, *De sensu*, and *De augmento*, to which Wüstenfeld adds *De intellectu*. Wüstenfeld states that Paris, MS BN lat. 6443 contains *De augmento*, *De tempore*, and *De intellectu* as items 19, 20, and 25 in the codex. The same portion of the manuscript, however, contains a *De intellectu* at fol. 195r (no. 21), immediately after the *De augmento* (no. 19) and the *De tempore* (no. 20); this *De*

GERARD OF CREMONA

intellectu is ascribed there to al-Kindī, while no. 18 of the codex (fol. 193r) is a *De unitate* ascribed to Alexander in the title (and in Sarton's list, no. 15) but to al-Kindī in the colophon, thus showing the complexity of identifying Gerard's translations through manuscript ascriptions. Similarly, a *De tempore* ascribed to al-Fārābī and said to be translated by Gerard in a London BM codex (Royal 12.C.XV, fol. 149) has been shown by A. Birkenmajer to be not by al-Fārābī but by Alexander of Aphrodisias.

On al-Kindī's *De intellectu*, see J. Jolivet, *L'intellect selon Kindi* (Leiden, 1971). *TK* lists three different incipits of a *De intellectu* or its equivalent ascribed to al-Kindī. *TK* 755 (and 756): "Intellexi [756, "Intelligo vel intellexi"] quod queris [scilicet] scribi tibi sermonem brevem" is identified as belonging to al-Kindī's *De intellectu*, but in a translation by John of Seville. *TK* 755—"Intellexi quod quesivisti de scribendo sermonem in ratione abbreviatum . . ."—is ascribed to al-Kindī with the title *Verbum de intentione antiquorum in ratione*; and Gerard is credited as translator. It was published in *Beiträge zur Geschichte der Philosophie des Mittelalters* (*Beiträge*), **2**, no. 5 (1897), 1–10. *TK* 1390 is "Scias quod videmus apud Aristotelem est tribus modis . . .," identified as a *De intellectu* and ascribed to Alexander (Alkindi) [*sic*]. The incipit of Alexander's *De sensu* is *TK* 1069; "Postquam consumavit Aristoteles in libro suo . . ."; it was published by Gabriel Théry in *Alexandre d'Aphrodise* (Kain [Belgium], 1926), 81–91 (see McVaugh). The incipit of Alexander's *De augmento* is *TK* 136: "Aristoteles dicit in libro de generatione et corruptione quod. . . ." Published by Théry, 99–100.

40 [19]. Al-Fārābī, *Commentary on the Physics of Aristotle*. The translation, long thought lost, was recovered by Alexander Birkenmajer, *Aus der Geisteswelt des Mittelalters. Studien und Texte Martin Grabmann zur Vollendung des 60. Lebensjahres . . . gewidmet* (Münster, 1935), text on 472–475.

41 [16]. Al-Kindī, *De quinque essentiis*. Edited by Albino Nagy, in *Beiträge*, **2**, no. 5 (1897), 28–40. The subject matter is nearly identical (the five essences are cause, matter, form, time, and space, paraphrased from Aristotle's *Metaphysics*) with Hermann of Carinthia's *De essentiis*, completed in 1143 and dedicated to Robert of Chester; publication of the latter was announced by Hermann directly to his "beloved" master, Thierry of Chartres. C. H. Haskins thought that the two works were totally different, failing to see that Hermann was merely trying to "Christianize" the Arab Peripateticist's new metaphysics of Neoplatonic bent.

The term *essentiae* had been thrown about among Latin scholars by Hermann, who sought to render the Arabic *wujūd* ("existent ones") as "the primary causes of all beings." This term, frequently used also by al-Fārābī in his *De scientiis* (translated by Gundissalinus and again by Gerard [see no. 42 below]), was treated quite differently in Latin by Gundissalinus and by Gerard. In chapter IV of the *De scientiis* (A. González Palencia, ed., *Al-Fārābī, Catálogo de las ciencias . . .*, 2nd ed. [Madrid, 1953], 106 ff.), for instance, we find two expressions in the Arabic rendered diversely by *esse* or *essentia* in the two translations; they are *qiwām* or *wujūd* (or *mawjūd, mawjūdāt*). *Qiwām*, meaning prop, support, basis, or sustenance, is employed six times in the chapter, always with reference to the mode of existence of an accident inherent in a subject. Although Gundissalinus always rendered it by *esse*,[22] Gerard always translated it as *essentia*. The root *wjd* occurs twenty-six times in al-Fārābī's *De scientiis*. In the twelve instances of the form *wujūd*, it is rendered as *esse* by both Gundissalinus and Gerard, except for one case when Gerard slips into *essentia*; Gerard's identification with Gundissalinus here is a good instance of what is involved in the "reworkings": the basic canvas of the older translation is altered only in specific circumstances of disagreement. Three times the root appears in the form *mawjūd*, also rendered as *esse* by both Gundissalinus and Gerard. When the root appears in the form *mawjūda* (four instances), to designate existent beings, it is rendered twice by *sint* in Gundissalinus (idiomatically closer to correct Latin usage), while Gerard uses *inventa* ("are found,"—that is, "found to exist"), precisely the term that was preferred by John of Seville in identical context. The two other instances of *mawjūda* are rendered by *essentia* and by *essentialis* in Gundissalinus, and each time as *existentia* by Gerard. In the final section of the chapter dealing with the subject matter of Aristotle's *Metaphysics*, for the first causes or principles of real being (the celestial bodies), the Arabic term most frequently used by al-Fārābī is *mawjūdāt*, the "existent" ones. The term occurs seven times in that section, always with the same connotation. Here, Gundissalinus uses *essentia* six times and Gerard has *existentia*; the seventh occurrence is rendered by *esse* in both Gundissalinus and Gerard.

These simple statistics tend to indicate that in

the early phase of translations of philosophical or astrological works (John of Seville is also heavily involved in the vagaries of this terminology) from the Arabic, the term *essentia* inherited from Boethius to designate the "essence" of a thing was imperceptibly transferred (by Gundissalinus and partly by Gerard) to designate "existent" beings, or the highest causes or principles of being in Peripatetic metaphysics. The direct role of the Arabic *falsafa* is understandable in al-Kindī's *De quinque essentiis* as well as in Hermann of Carinthia's *De essentiis* (published very inadequately by M. A. Alonso [Santander, 1946]). In *De quinque essentiis* al-Kindī reckoned the following five *essentie* to explain the entire chain of beings and causality therein: cause, matter, form, time, and space. The number of *essentie* corresponds to al-Fārābī's own number in chapter IV of his *De scientiis*, where he hesitates, however, between three and five, depending on the approach).

Incidentally, the idea of selecting a fixed number of *essentie* to explain the world of being seems to have originated with the Haranians or Sabaeans, whose philosophical elucubrations on that theme the *Fihrist* reports casually. Their approach is quite different from that of Aristotle in the *Metaphysics*, where the "unmoved movers," discovered through the astronomy of Eudoxus, may amount to anywhere between forty and fifty. Before the advent of the works of Aristotle in Arabic translations, the prism of Haranian and Hermetic speculation diffracted the problem-setting process among Arab philosophers. As the *failasūf al-ʿarāb* al-Kindī demonstrated in his *De quinque essentiis* how he reduced philosophical speculation to a strict Aristotelian framework, although the cast of a *De essentiis* left its mark on the format of Arabic philosophical speculation. So also did the cast work on Latin philosophical speculation in the twelfth century under the impact of the Arabs. Hermann of Carinthia counted six rather than five *essentie* by Christianizing the first one, *causa*, which he divided into two *essentie*: on the one hand, God as creator and First Cause, and, on the other hand, the celestial bodies as created but first active causes in the physical universe.

The most notable feature of Gerard of Cremona's reworking of Gundissalinus' translation of al-Fārābī's *De scientiis*, as explained above, was the distinct effort to depart from the recent practice of translating *mawjūdāt* by *essentie* and to stress, in the form of his *existentia*, the opposition between essence and existence, thus moving away from the earlier Boethian usage of *essentia*. It is, however, somewhat puzzling why, in translating al-Kindī's *De quinque essentiis*, Gerard retained the expression in the title. Either he translated this work very early in his career, possibly before his reworking of Gundissalinus' translation of al-Fārābī's *De scientiis*, or he may have sought to show the true origin of Hermann's *De essentiis*, a work already known in Spain and at Chartres at the time it was written (1143). It is not possible to state with any accuracy the further impact of Gerard's translation and of Hermann's original *De essentiis* upon the orientation of philosophical speculation among the Latins in the twelfth century, and particularly upon the series of treatises written by Latin scholars as *De principiis*, or *De sex principiis*.[23] The question, however, deserves greater attention in this expanded context.

42 [20]. Al-Fārābī, *De scientiis*, also translated by Dominicus Gundissalinus [Gundisalvi]. In supplying the bibliographical information concerning the edition of this translation by Gerard in A. González Palencia (Madrid, 1931; 2nd ed., 1953) —see no. 41—M. McVaugh omitted to state that in addition to the Arabic text and the Latin of Gerard, Palencia's edition also contains the text of Gundissalinus' translation from the edition of Guilelmus Camerarius (Paris, 1638) compared with that of L. Baur (*Beiträge*, **4**, nos. 2–3 [1903]) and a Castilian version by González Palencia. This surprising omission can create problems, since in a note by E. Grant to an extract from Gundissalinus' version that appears later in Grant's *Source Book of Medieval Science*, there is a reference to a Latin passage in Gerard that is actually to the portion of González Palencia's edition that contains Gundissalinus' translation. A later edition of Gundissalinus' *De scientiis* by Alonso[24] shows that this text is a compilation not only from al-Fārābī as a primary source but also from many other Latin sources. Gundissalinus was a recidivist in this kind of compilation that borders on plagiary.[25]

43 [17]. Al-Kindī, *De somno et visione*. Carmody, *AAAL*, 83; Albino Nagy, ed., *Beiträge*, **2**, no. 5 (1897), 12–27; *GAS*, III, 376.

E. Works on medicine (physica) (24 works).

44 [60]. Galen, *De elementis*. See *GAS*, III, 87; M. Ullmann, *Die Medizin im Islam*, 38, no. 4. Incipit: "Quoniam cum sit elementum minor pars." Richard J. Durling, "Corrigenda and Addenda to Diels' Galenica," 465; and H. Diels, "Die Handschriften der antiken Ärzte. Griechische Abt. I. Hippokrates und Galenos," 64.

45 [58]. Galen, *Expositiones super librum Ypocratis de regimine acutarum egritudinum*. *GAS*, III, 33, 118; Ullmann, 56, no. 61. See Diels,

102–103; Durling, 476. Printed in the *Articella* (Venice, 1513).

46 [61]. Pseudo-Galen, *De secretis*. *GAS*, III, 126, no. 91; Ullmann, 60, no. 103. See Wüstenfeld, 69, for manuscripts, incipits, and editions.

47 [62]. Galen, *De complexionibus*. Incipit: "Insignes antiqui medicorum et philosophorum." Durling, 472; Diels, 65.

48 [63]. Galen, *De malicia complexionis diverse*. *GAS*, III, 109; Ullmann, 39, no. 7. Incipit: "Malitia complexionis diverse quandoque." Durling, 466; Diels, 84.

49 [64]. Galen, *De simplici medicina I–V*. *GAS*, III, 109–110; Ullmann, 47, no. 49. Incipit: "Non mihi necesse est hic ostendere." Durling, 471; Diels, 97.

50 [65]. Galen, *De creticis diebus*. *GAS*, III, 96, no. 19; Ullmann, 43, no. 30. Incipit: "Ut egritudinum que paulatim non." Durling, 465; Diels, 91.

51 [66]. Galen, *De crisi*. *GAS*, III, 95, no. 18; Ullmann, 43, no. 29. Incipit: "Ego quidem non intendo." Durling, 464; Diels, 90.

52 [59]. Galen, *De expositione libri Ypocratis in pronosticatione*. *GAS*, III, 32, no. 3a, and 123, no. 74; Ullmann, 50, no. 59. See Diels, 107–108; Durling, 476. Printed in the *Articella* (Venice, 1513).

53 [57]. Pseudo-Hippocrates, *Liber veritatis Ypocratis*. Printed in 25 aphorisms in the ancient *Articella*. Modern edition by K. Sudhoff (1915). See Ullmann, 33–34.

54 [21]. Isḥāq al-Isrā'īlī, *De elementis*. English translation from the Arabic original by A. Altmann and S. M. Stern, *Isaac Israeli, a Neoplatonic Philosopher of the Early Tenth Century* (London, 1958). See Ullmann, 138.

55 [22]. Isḥāq al-Isrā'īlī, *Liber diffinitionum*. Edited by J. T. Muckle, in *Archives d'histoire doctrinale et littéraire du moyen âge (AHDLMA)*, **12–13** (1937–1938), 299–340, translated from the Arabic original in Altmann and Stern. See Ullmann, 138.

56 [71]. Al-Rāzī, *Liber Almansorius*, the shorter of his great medical compilations. It is dedicated in 903 to the ruler of Rayy, Abū Ṣāliḥ Manṣūr ibn Isḥāq ibn Aḥmad ibn Asad (see Bayard Dodge, ed. and trans., *The Fihrist of al-Nadīm*, II [New York, 1970], 704, n. 169), and not to the Abbāsid Caliph al-Manṣūr (754–775), as stated by M. McVaugh. See *GAS*, III, 275, 281–282; Ullmann, *Die Medizin . . .*, 132; and H. Schipperges, *Die Assimilation der arabischen Medizin durch das lateinische Mittelalter* (Wiesbaden, 1964), 92.

57 [72]. Al-Rāzī, *Liber divisionum continens CLIIIIor capitula cum quibusdam confectionibus ejusdem*. From Vatican MS lat. 2392 Boncompagni added *Almansoris* before *continens*, thus inviting double confusion with the *Liber Almansorius* (no. 56), translated by Gerard, and with the famous *Continens* (*al-Ḥāwī*) by al-Rāzī, translated in the thirteenth century by Faraj ben Salīm (Moses Farachi or Faragut) and printed many times (*GAS*, III, 280; Ullmann, 131). The Arabic original of Gerard's *Liber divisionum* seems to be the *Kitāb Taqsīm al-ʿilal* (or *al-tašjīr*) described in *GAS*, III, 284, no. 5; and Ullmann, 132. As indicated in the *socii* list, Gerard's translation contained 154 chapters, whereas both the Lyons, 1510, edition and the Basel, 1544, edition of al-Rāzī's works have a *Liber divisionum* in 159 chapters with the incipit "Ventilata fuit. . . ."

58 [73]. Al-Rāzī, *Liber introductorius in medicinam parvus*. This translation, probably of the *Kitāb al-mudḫal'ila ṭ-ṭibb*, is no. 6 in Sezgin's list (III, 284), and the eighth work entitled *Introductorium medicine* in the Venice, 1500, edition. The seventh work in the Lyons, 1510, edition is entitled (fol. 279v) *Liber introductorius parvus in medicinam Rasis*, with the incipit "Salvator excelsus et gloriosus. . . ."

The following translations (nos. 58a, 58b, 58c, 58d) were shown by L. Thorndike ("Latin Manuscripts of Works by Rasis at the Bibliothèque Nationale. Paris," in *Bulletin of the History of Medicine*, **32** [1958], 54–67) to be regularly present together in a good sampling of manuscript collections of works by al-Rāzī that were translated by Gerard of Cremona. The ascription to Gerard is not formally stated for each work, but each collection contains some general indication to that effect. The same collection of works likewise was frequently printed together (Milan, 1481; Venice, 1497; Venice, 1500; Lyons, 1510; Basel, 1544) and was ascribed to Gerard of Cremona. There thus exists a strong probability that they were translated as a group by Gerard, as Wüstenfeld (pp. 71–72) suggested in 1877, although the *socii* list did not include them by name. By omitting them from his list of Gerard's translations, Sarton indicates his tacit rejection of Wüstenfeld's hypothesis. Thorndike's "Latin Manuscripts . . .," however, seems to us to be nearly decisive in favor of Wüstenfeld's suggestion.

58a [74]. Al-Rāzī, *De iuncturarum egritudinibus* (or *doloribus*). The incipit Thorndike, "Latin Manuscripts . . .") is "Dixit Rasis. Volo in hoc capitulo dicere medicinas que necessarie sunt doloribus iuncturarum." In several of the Paris manuscripts studied by Thorndike, this translation is grouped with no. 58b under the joint title *Experimenta Rasis*. It is printed with the group in the five early editions listed in no. 58. In Sezgin's list (III, 288) this work is nos. 27 and 28: *Aujāʿ al-mafāṣil* and *Aujāʿ al-niqris*. Sezgin (III, 289) expresses doubt about the authenticity of an *Experimenta Rasis* (*Kitāb al-tajārib*; *GAS*, no. 34); although he knows Thorndike's "Latin Manuscripts," Sezgin does not bring it to bear on the authenticity of the *Experimenta Rasis*.

58b [not in Sarton]. Al-Rāzī, *De egritudinibus puerorum* (sometimes called *Practica puerorum* or *de cura* [*curis*] *puerorum*; see Paris, BN lat. 6893, fol. 283; lat. 7406, fol. 167; J. de Ketham, *Fasciculus medicine . . . Tractatus Rasis de egritudinibus puerorum et earum cura qui appellatur practica puerorum* [Venice, 1500]). The incipit begins "Sahafati. . . ." The work is often grouped in manuscripts with no. 58a under the title *Experimenta Rasis* and is published with it in the early editions of works by al-Rāzī (no. 58). Neither Ullmann, nor Sezgin's list compiled from earlier Arabic bibliographical lists of Ibn al-Nadīm, al-Bīrūnī, and Ibn Abī Uṣaybiʿa, contains a title that corresponds to this work translated by Gerard. It may, however, be an extract from a larger work; the matter requires further investigation.

58c [not in Sarton]. Al-Rāzī, *Antidotarium*. The incipit is "Dixi in hoc meo libro medicinas. . . ." The original Arabic of this translation may be the *Aqrābāḏīn al-kabīr*, as suggested by Sezgin (III, 283, no. 4) and Ullmann (*Die Medizin . . .*, 303). We must note, however, that Arabic bibliographical lists also mention a *Kitāb al-Qarābāḏīn aṣ-ṣaġīr* in four chapters by al-Rāzī (*GAS*, III, 292, l. 28; Ullmann, 103). Ullmann thought that this was the second *Antidotarium* printed in the Basel, 1544, edition (pp. 546–559) and also in the Venice, 1497, edition (pp. 95–98; "Cap. 1. De medicinarum"); while Sezgin observed that there was no manuscript trace of this work in Arabic. Perhaps one should consider the pseudo-Rāzī *Kitāb Nuzhat al-mulūk* (*GAS*, III, 291, no. 69; Ullmann, 135 and 332) as another *Antidotarium* by al-Rāzī, for Ullmann describes its contents as dealing with "the removal of all harmful effects of simple and composite poisons and drugs of animal, plant or mineral origin, or of poisoned clothes, drinks or foods, or of things handled by the hand such as snakes, scorpions and various animals." We have not ourselves established the necessary comparison of the two texts to decide the issue.

The Latin *Antidotarium Rasis* is present with nos. 56, 57, 58a, and 58b in seven of the eight Paris manuscripts examined by Thorndike; its incipit—"Dixi in hoc libro meo medicinas quarum necessitas est inseparabilis. . . . (R) Verba Abubecri de redactione librorum suorum (N) Iam pridem pervenimus ad expositionem relationis egritudinum que sunt a summitate capitis usque ad pedes . . ."—corresponds to the text on p. 452 of the Basel, 1544, edition, *De antidotis*. But in the Leiden, 1510, edition, fol. cclxviii verso, there is mention of an *Antidotarium Rasis* "in quo continentur compositiones plurium medicinarum ad diversas dispositiones . . . Cap. lum. De aptatione medicinarum ut sine horribilitate possint sumi. . . ." This last sentence shows that the extract with this incipit found in Paris, BN lat. 6893, by Thorndike, who was inclined to treat it as a separate work, probably is part of the short *Antidotarium*.

58d [not in Sarton]. Al-Rāzī, *De preservatione ab egritudine lapidis*. The work that in Paris, BN lat. 6893, follows the one beginning "De aptatione medicinarum . . ." (no. 58c) has the incipit "De lapide qui in renibus vel vesica formatur. . . ." The work printed with the incipit "Dixit Rasis. De lapide qui in renibus vel vesica formatur" in the Leiden, 1510, edition (fols. 278–279) under the title *Tractatus Rasis de preservatione ab egritudine lapidis*, may be a version by Gerard of Cremona of another well-known work by al-Rāzī. See *GAS*, III, 288, no. 26; Ullmann, 134.

A work in Latin translation entitled *Aphorismi Rasis*, in six chapters and containing well over 300 aphorisms, was printed among the works of al-Rāzī in the early editions. Wüstenfeld suggested Gerard as a translator of these, as well as of the other works by al-Rāzī in these editions, although they were not included in the *socii* list. Sezgin (III, 284, no. 7) identifies the Latin *Aphorismi Rasis* with an Arabic original *Kitāb al-Murshid* or *Kitāb al-Fuṣūl*, although he does not suggest Gerard as a translator. Ullmann (pp. 134–135) describes this *Kitāb al-Murshid* ("Der Führer") as a late work by al-Rāzī that consists of 377 aphorisms in 37 chapters, but he does not mention a Latin translation of it. The original Arabic was published by A. Zaki Iskandar in *Revue de l'Institut des manuscrits arabes* (Cairo), **7** (1961), 1–125. In Latin, the work was published sometimes with the other Rāzī translations by Gerard of Cremona, sometimes in a collection of *Aphorisms* headed by the *Aphorismi Raby . . . Moyses* (Maimonides). In all these editions it has approximately the same title, prologue, and incipit as the *De secretis medicine secundum*

GERARD OF CREMONA

Rasim studied by Thorndike (1958) in Paris, BN lat. 17847, where it begins: "Inquit Abubecri . . . Congregavi in divisionibus egritudinum et ostendi curas et causarum. . . ." This text, however, is not included among the tightly knit group of works by al-Rāzī translated by Gerard of Cremona. On the contrary, in Paris, BN lat. 17847, as observed by Thorndike, it is said to be translated by a "Magister Egidius apud sanctam habenam," whom Thorndike does not identify further. Thorndike's observation, based on manuscript tradition, does away with Wüstenfeld's hypothesis based on printed texts. It would seem, therefore, that there is positive proof that the *Aphorismi Rasis* was not translated by Gerard.

59 [77]. Ibn al-Wāfid (Abenguefit), *Liber medicinarum simplicium et ciborum*. Ibn al-Wāfid was a physician of Córdoba who died in 1075, according to Wüstenfeld (p. 72); a vizir at Toledo who died after 1068, according to Ullmann (*Die Medizin*, 210); and in 1068, according to Sezgin (III, 228). See *GAS*, III, 228–229; Ullmann, *Die Medizin*, 210 and 273; Max Meyerhoff, "Esquisse d'histoire de la pharmacologie et botanique chez les musulmans d'Espagne," in *al-Andalus*, **3** [1935], 13 ff. The *Liber medicinarum simplicium* was printed at Strasbourg in 1531 with Ibn Buṭlān's [Elluchasem Elimithar] *Tacuini sanitatis* (see Ullmann, *Die Medizin*, 157–158) and at Venice in 1558 with Mesuë.

60 [70]. Yaḥyā ibn Sarāfyūn (Filius Serapionis), *Breviarius Iohannis Serapionis tract. VII*, better known as *Practica Serapionis*. See *GAS*, III, 241, no. 2; Ullmann, 102–103. Printed at Venice in 1497 and reprinted several times. Ullmann gives a description of the contents from the Lyons, 1525, edition.

61 [75]. Abuʾl-Qāsim (Abulcasis) al-Zahrāwī, *De cirurgia, tres tractatus*. Of the thirty component parts of this enormous work, only the last part, dealing with surgery and its instruments in three books, was translated by Gerard. The original Arabic is profusely illustrated in manuscripts and was published with English translation by M. S. Pink and G. L. Lewis, *Albucasis on Surgery and Instruments. A Definitive Edition of the Arabic Text with English Translation and Commentary* (London, 1973). Also see *GAS*, III, 323–325, and V, 414 (*Nachträge*); and Ullmann, 149–151. The Latin was printed at Venice in 1487 with the *Chirurgia* of Guy de Chauliac.

62 [68]. Al-Kindī, *De gradibus*. A work on pharmacology. Latin text edited by Michael R. McVaugh, *Arnaldi de Villanova Opera medica omnia*, II, *Aphorismi de gradibus* (Granada–Barcelona, 1975), 263–295. *GAS*, III, 245; Ullmann, 302; Carmody, *AAAL*, 84–85.

63 [76]. Ibn Sīnā [Avicenna], *Canon*. Fifteen editions by 1500 and as many after that. First Arabic edition issued by the Medici Press (Rome, 1596). See Ullmann, 152–154. English extract in E. Grant, ed., *A Source Book in Medieval Science*.

64 [78]. Galen, *Tegni* [*Microtegni* or *Ars parva*], *cum expositione Ali ab Rodohan* (ʿAlī ibn Riḍwān). See *GAS*, III, 81; Ullmann, 45; Durling, 463; Diels, 61–63. It is at the end of this work that the manuscripts usually contain the *socii* list of Gerard's works that we are following here, after Wüstenfeld. Concluding the list of Gerard's medical translations and nearly at the end of the entire list, the *Tegni* with Ibn Riḍwān's commentary is named in the *vita* as among the last works translated by Gerard (*novissime ab eo translati*). This statement seems to imply a chronological order in the list, at least as far as the medical works are concerned. Durling, 463; Diels, 61–63.

F. Works on alchemy (3 works).

65 [84]. [Jābir ibn Ḥayyān], *Liber divinitatis de LXX*. Published by M. Berthelot in *Archéologie et histoire des sciences* (Paris, 1906). Renaldus Cremonensis (Wüstenfeld, 74) seems to be a misreading or a scribal misspelling of Gerardus Cremonensis. At least, the *socii* list is explicit about Gerard's translation of this work, and no other translator of it is known; a Renaldus Cremonensis as translator also is unknown.

66 [85]. Pseudo-Rāzī [?], *De aluminibus et salibus*. Incipit *TK* 677. Edited by R. Steele in *Isis*, **12** (1929), 14–42. See J. Ruska, *Das Buch der Alaune und Salze; ein Grundwerk der spätlateinischen Alchemie* (Berlin, 1935); and "Pseudepigraphe Rasis-Schriften," in *Osiris*, 7 (1939), 30–93. See Ullmann, *Die Natur- . . .*, 210–213, 228; *GAS*, IV, 282. Wüstenfeld, following Steinschneider, holds that because the words "apud nos in Yspania" appear in this text, the author cannot be the famous physician al-Rāzī, who never was in Spain. We must observe, however, from a limited practical experience with medieval alchemical manuscripts, that alchemists, who often were their own scribes, used these texts to inform each other of their accomplishments or of some other alchemist they knew or had heard about. These conditions may render invalid any criterion of textual authenticity based solely on such incidental remarks.

67 [86]. Pseudo-Rāzī, *Liber luminis luminum*. Incipit *TK* 290: "Cum de sublimi . . ." (*TK* says this translation is by Raymond of Marseilles). The original is ascribed to al-Rāzī in Gerard's translation. In the edition by B. Rhenanus, *Harmonia* (Wüstenfeld, p. 75), the author is named Rases Castrensis. This may very well be a misreading of a manuscript ascription to R[obert] Castrensis, or Robert of Chester (*ca.* 1140), who is known to have made alchemical translations from the Arabic. Citations of this kind in early printed editions cannot be decisive as to the authenticity of the manuscript tradition: only careful examination of the entire manuscript tradition could be conclusive in such cases.

G. Works on geomancy and divination.

68 [87]. *Liber geomantie de artibus divinantibus qui incipit: Estimaverunt Indi.* Also translated by Hugh of Santalla [?]. See Carmody, *AAAL*, 173. But is this a translation or an original work by Gerard? *TK* gives three different incipits of a geomancy ascribed to Gerard of Cremona, not specifying whether it is a translation or an original work, but with a question mark next to two of them.

TK 697: "In nomine illius qui maior est incipit geomancia . . ." Gerard of Cremona.

TK 1446: "Si de statu corporis questio proponatur utrum meliorari . . ." Gerard of Cremona (?), *Geomancia*.

TK 1461: "Si quis per artem geomanticam de preteritis, presentibus et futuris . . ." Gerard of Cremona (?), *Geomancy*.

On the other hand, the work beginning "Estimaverunt indi," stated in the *socii* list to be a translation by Gerard, is listed by both *TK* and Carmody, *AAAL*, 173, as translated by Hugh of Santalla. Hugh of Santalla is known exclusively as a translator from the Arabic; if he translated this geomancy, then the original must have existed in Arabic. On the other hand, if the *socii* list is to be believed—as surely it must, in view of the great reliability of the remainder of the list—then what happened to Gerard's translation, or original work? If the *Geomancy* currently ascribed to him is not his (printed at Paris 1661 in a French translation by "le sieur de Salerne"), there nevertheless are manuscript copies of a geomancy that clearly credit Gerard as author (Wüstenfeld, 75). This geomancy by Gerard often has been ascribed to the other Cremonensis, Gerard of Sabbioneta—but with no medieval authority, and on the same grounds as for the authorship of the *Theorica planetarum*.

69 [79]. *Liber Alfadhol i. est arab de bachi.* A book on lots and fates determined by questions and answers. The identity of the author indicated as "arab de bachi" (with variants ".i. tharab," or "de brachi," or "z d harab de bachi"; see Wüstenfeld, 75) should be read much more simply, it seems, as "an Arab from Balkh" (*arab de balchi*). Astrologers and occultists related to the city of Balkh were numerous in the classical period. On the other hand, some Latin manuscripts carry the additional identification of Alfodhol de Meregi, Alfodhol de Merengi, or Aralfodhol de Merengi. Wüstenfeld (p. 75) proposed to read it as a corruption of al-Nayrīzī (already corrupted into Tabrīzī Yazīdī in some important sources), a well-known commentator on the *Almagest* and on Euclid, because this author's name happened to be al-Faḍl; but the passage from al-Nayrīzī to de Merengi appears a bit farfetched. It seems much more natural to read *de niranji*, or *nayranji*, a frequently used term of Persian origin for books of magic, and magic is really the subject of this book. See M. Ullmann, *Die Natur- und Geheimwissenschaften im Islam*, 360, 362, 363, 366, 367, 375, 376, 393.

See P. Kunitzsch, "Zum 'Liber alfadhol,' eine Nachlese," in *ZDMG*, **118** (1968), 297–314, which is an appendix to a German edition and version by B. F. Lutz, *Das Buch 'Alfado,' Untersuchung und Ausgabe nach der Wiener Hds. 2804* (Heidelberg, 1967), with bibliography. Kunitzsch points out the existence of a two-fold Latin manuscript tradition, only one of which may be called a direct translation. He confesses to being unable to decide which of the two groups of text should be ascribed to Gerard of Cremona. The Arabic tradition also varies considerably, with fanciful attribution of authorship to al-Kindī or to Caliph Harun al-Rashid. No manuscript of the Arabic text known to Kunitzsch mentions al-Faḍl ibn Sahl al-Sarakhsī as the author, while the Latin text, presumably by Gerard, is specific on this name.

70 [83]. *Liber de accidentibus alfel.* A book of auguries and omens. See H. Suter, "Über einige noch nicht sicher gestellte Autorennamen in den Übersetzungen des Gerhard von Cremona," in *BM*, 3rd ser., **4** (1903), 25.

71 [45]. ʿArib (not Harib, as in McVaugh, which is a Jewish form) ibn Sa'd al-Kātib al-Qurtubī (secretary to Bishop Rabi ibn Zaid of Córdoba), *Liber anoe*. A work on calendar and agricultural usages, dedicated to al-Hakam II, caliph of Córdoba (961–976). Editions by G. Libri, *Histoire*

des sciences mathématiques en Italie, I (1838), 293–458 (correct McVaugh's 393); by R. P. A. Dozy, *Le calendrier de Cordoue de l'année 961* . . . (Leiden, 1873); and by Charles Pellat, *Le calendrier de Cordoue* (Leiden, 1961), with French translation.

H. Additional translations ascribed to Gerard and not on the *socii* list.

Both Boncompagni and Wüstenfeld agreed that medieval Latin tradition credited Gerard with more translations than are found in the *socii* list. Wüstenfeld's additions are included in Sarton's expanded list, with the exception of some works by Ibn Sīnā (Wüstenfeld, 78); likewise, most of Boncompagni's additions are included in Sarton's list, the only omissions being Ibn al-Haytham's *Perspectiva* (Boncompagni, 408–409) and the *Alchabitius* (Boncompagni, 443). The latter work had already been translated by John of Seville; and if Gerard really did translate it again, it must have been a reworking, as were so many other of his early efforts in translation.

The story of the medieval Latin *Alchabitius* is extremely involved, since this work came to be included in the curriculum of some teachers of astronomy-astrology-medicine;[26] hence its manuscript tradition is very complicated, and no conclusive statement on Gerard's possible contribution can yet be made. The omission of Ibn al-Haytham's *Perspectiva* from Sarton's list, however, is important; and it is difficult to understand Sarton's rationale in this case. Although it seems likely, it is by no means certain that Gerard was the translator of the *Perspectiva*. We shall give here in brief summary the additional items on Sarton's list, which includes nearly all ascriptions of Latin translations from the Arabic credited to Gerard of Cremona (Sarton's numbering is in square brackets).

[8]. *Liber lapidum*, an anonymous work quoted by Arnold of Saxony (ascribed to Aristotle).

[11, 13, 14]. These correspond to Wüstenfeld's nos. 39a, 39c, and 39d, respectively.

[15]. Alexander of Aphrodisias, *De unitate*. In Paris, BN lat. 6443, fol. 193r, it is ascribed to Alexander in the title but to al-Kindī in the colophon.

[18]. Al-Kindī, *De ratione*. See A Nagy, ed., *Beiträge*, **2**, no. 5 (1897), 2–11.

[27]. Apollonius, *De conicis*. Gerard unquestionably translated a fragment of book I of the *Conics*, which he used as introduction to his translation of Ibn al-Haytham's *De speculis comburentibus*. The fragment of the *Conics* translated by Gerard was published by Heiberg in his edition of the *Conics*, II (Leipzig, 1893), lxv–lxxx.

[43]. Abū Kāmil, *Liber de algebra et almucabala*. See L. Karpinski, *Robert of Chester's Latin Translation* . . ., 2nd ed. (Ann Arbor, 1930), 19–20; and M. Levey, *The Algebra of Abū Kāmil* (Madison, Wis., 1966), 9–10.

[48]. *Algorismus de integris*. This is perhaps a translation (probably a new version) of al-Khwārizmī's *Arithmetic* in one of the many Latin adaptations of this famous Arabic text (now lost). A. Allard of Tourpes, Belgium, is presently working on the problem of the medieval algorism; and his research may throw light on the various versions of Al-Khwārizmī's *Arithmetic* among Latins and Greeks.

[49]. *Liber co-aequationis planetarum*. It is unclear which of the several astronomical works ascribed to Gerard in manuscripts (see Wüstenfeld, 78–79) corresponds to this title given by Sarton, after Steinschneider ("Die europäischen Übersetzungen . . .").

[51]. Al-Zarqālī, *Canones*. Rules for the use of astronomical tables, compiled by the Toledan astronomer Zarqālī (Azarchel) about 1070. Zarqālī also drew up the Toledan Tables to accompany these *Canones*. The *Canones* of al-Zarqālī were very popular until they were superseded in the fourteenth century by the Alfonsine Tables.[27] Some manuscripts of the translation of the *Canones* named Gerard as the translator; it seems that he translated only the *Canones* and not the tables, although he may have composed tables of his own, either for the meridian of Toledo (Boncompagni, 445) or for the meridian of Cremona. See M. Reinaud, *Géographie d'Aboulféda*, I (Paris, 1848), ccxlvi–ccxlviii, according to whom MS Paris, BN lat. 7421, contains the *Theorica planetarum* ascribed to Gerard of Cremona (fol. 131) immediately after the tables of al-Zarqālī "translated by Gerard of Cremona" (fol. 100). On the Toledan Tables in general, see G. J. Toomer, "A Survey of Toledan Tables," in *Osiris*, **15** (1968), 5–174.

[52]. *Liber omnium sperarum caeli et compositionis tabularum*. The same authenticity problems as for no. 51. Wüstenfeld (p. 78) lists it as a translation under the title *De compositione sphaerae*. The possibility that this could be an obscure designation for the *Theorica planetarum*, which is now generally removed from Gerard's list of original works, will be discussed when we deal with Gerard's original works.

[67]. Galen, *Tegni*. Sarton lists separately this

work by Galen that is included, with ʿAlī ibn Riḍwān's commentary, as no. 64 in the *socii* list. He later lists the commentary by ʿAlī as [78].

[69] Ibn Māsawayh (Mesuë), *Aphorisms*. See Ullmann, *Die Medizin* . . ., 113; *GAS*, III, 233 (ascribed to Johannes Damascenus by Constantine the African). Edited by P. Sbath (Cairo, 1934).

[81]. [Alchandrus]. *Arcandam de veritatibus et praedicationibus astrologicis*. It is very doubtful that Gerard translated this simplistic astrological work of which some manuscript copies of the tenth and eleventh centuries exist (Paris, BN lat. 17868 [10c] and London, BM Add. 17808 [11c]); see Lynn Thorndike, *A History of Magic* . . ., I (New York, 1923), 710 ff. On *Arcandam*, see A. van de Vyver, "Les plus anciennes traductions latines . . .," in *Osiris*, **1** (1936), 658–691. Sarton probably confused or misread Alchandreus in Boncompagni's suggestion to add *Alchabitius* to the list of Gerard's translations.

[82]. Abhabuchri [Heus], *Liber in quo terrarum corporumque* [*caelestium*(?)] *continentur mensurationes Abhabuchri*. Wüstenfeld, 79. See H. Suter, "Über einige noch nicht sicher gestellte Autorennamen . . .," 19–20; *GAS*, V, 389–390; edited by H. L. L. Busard, "L'algèbre au moyen âge: Le 'Liber mensurationum' d'Abu Bekr," in *Journal des savants* (Apr.-June 1968), 66–124.

A last important addition may be the *Perspectiva* (*De aspectibus*) of Ibn al-Haytham. See A. Jourdain, *Recherches critiques sur les anciennes traductions latines d'Aristote* (Paris, 1843; repr., New York, 1960). Carmody, *AAAL*, 139–140; D. Lindberg, "Alhazen's Theory of Vision and Its Reception in the West," in *Isis*, **58** (1967), 321–341.

As already shown, the *socii* list is not exhaustive. Certainly the *socii* strove for completeness, but Gerard's long career and his reluctance to sign his work help to explain the limitations of their list. The *socii*, who apparently belonged to a circle of collaborators late in Gerard's life, simply were ignorant of the details of his early years as a translator. Gerard most certainly was not the originator of the list—not even in imitation of Galen at the end of the *Tegni*: it was the *socii* who found in Galen's practice the justification or inspiration to draw up the list as best they could.

Arab scholars interested in sciences seem to have created or inherited collections of basic works in the various branches of the *quadrivium* that students in each field had to master. The fact is fairly clear in the case of mathematical and astronomical works, as shown in the studies of M. Steinschneider ("Die mittleren Bücher . . .") and F. Carmody (*Thabit* [1960], 22). Similar pedagogical collections of required works in medicine probably were inherited by the Arabs from the *Summaria Alexandrinorum* (M. Ullmann, *Die Medizin* . . ., 65–67; *GAS*, III, 140–150), to which active physicians and writers like Ḥunayn ibn Isḥāq and al-Rāzī seem to have added significantly. It would seem that Gerard of Cremona sought out such collections, particularly in mathematics, astronomy, and medicine, in order to translate them as a corpus in each branch of the *quadrivium*; for the large number of his translations in those fields frequently agrees with the order of those Arab scientific collections.

Among the Latins, perhaps even directly under the influence of Gerard of Cremona, similar collections were valued and new ones, especially astrological translations from the Arabic, were assembled. In 1902 A. A. Björnbo discovered, in the Paris manuscript BN lat. 9335, which contains some twelve translations by Gerard of Cremona, the direct statement by Ḥunayn ibn Isḥāq on the subject of those "mittleren Bücher" (*BM*, 3rd ser., **3** [1902], 68). The following year, Björnbo published (*ibid.*, **4** [1903], 288–290) a very interesting specialized program of study in mathematics-astronomy that he was certain corresponded to some university curriculum, although he could not positively identify the university. The parallelism between this curriculum and the Ḥunayn extract is obvious, in the inspiration if not in the detailed series of works. Although medicine was not part of the *quadrivium*, it was in this field that Gerard produced the greatest number—and his best—works. He translated at least twenty-one medical writings, among them Ibn Sīnā's *Canon* and al-Rāzī's *Almansorius*. Consequently, his translations had an immeasurable impact upon Latin medicine of the Middle Ages, which profited greatly from the advanced state of medicine in medieval Islam. Second in importance in number and quality were his translations in geometry, mathematics, and astronomy, totaling some thirty works. Here again, Gerard's translations influenced the strivings of Latin scholars toward a scientific approach to knowledge of nature that subordinated philosophical and theological inclinations. Six additional works on geomancy and alchemy also contributed largely to the scientific orientation of the medieval West. Gerard's eleven translations of works in philosophy and three on dialectics appear to have had a rather minimal influence; they seem to have been selected for their relevance to the epistemology of natural science and to a scientific interpretation of the cosmos.

During the thirteenth century, the *parens scientiarum* of the era, the University of Paris, after the prohibition of Aristotle's work and of Arabic learning in 1210–1215, turned toward philosophical and theological speculation, although the consequent Scholasticism always had room for physical and cosmological considerations. The evolution of the university curriculum during the thirteenth and fourteenth centuries reveals the slow but sure penetration of many of Gerard's translations, which nourished the awakened interest in natural science until the end of the Middle Ages. Although the Renaissance infatuation with Greek texts at the expense of their Arab counterparts perhaps exerted a delaying action that began at the end of the fourteenth century, one can still observe among the luminaries of the "new science" at Oxford, Paris, and Padua a reliance on some of the texts produced by Gerard. Regiomontanus' scathing indictment of the *Theorica planetarum* reveals that it was still used largely as an introductory book by students of astronomy. Gerard may have written this text, although its authorship has been widely disputed. Still, the first Latin *Almagest* to be printed (Venice, 1515) was Gerard's, of which it seems that Copernicus soon procured a copy.

Gerard of Cremona's Original Works. There is still much uncertainty about the number, value, and even the existence of Gerard's original works. Aware of the incompleteness of the *socii* list and impressed by the number of manuscript ascriptions of works to Gerard, Wüstenfeld (pp. 79–80) added several allegedly original works, including two medical glosses on works by Isḥāq al-Isrā᾿īlī: his *Viaticum* and his *Diaetae universales* (see *GAS*, III, 296–297), and a *Summa de modo medendi et ordine curandi*. Nevertheless, Wüstenfeld held that Gerard did not compose the celebrated *Theorica planetarum* and the *Geomantia astronomica*, attributing both to Gerard of Sabbioneta on the strength of G. Tiraboschi's and Boncompagni's argumentation, which is based on three equally weak and unsound premises: the absence of this work in the *socii* list; the "doctrinal meagerness and linguistically inappropriate" style of the *Theorica*; and the arbitrary selection of the insignificant astrologer Gerard of Sabbioneta (near Cremona) as the probable author, without any medieval testimony for this. But this theory is severely flawed, for the original *socii* list never made any claim to completeness. Moreover, the list refers exclusively to "translations"—the primary role of Gerard in the service of his "beloved" Christendom. Occasional minor works composed by Gerard, especially in fields where his reputation as a translator was so eminent, were certainly of secondary importance to the *socii*—if they were even aware of their existence.

Olaf Pedersen holds (see E. Grant, *A Source Book in Medieval Science*) that the *Theorica planetarum Gerardi* dates from the middle of the thirteenth century. Yet, one little-observed manuscript of Spanish origin, now preserved in Leningrad at the Library of the Academy of Sciences, Codex XX, Ab-III (present shelf mark F°.N.8), already described by Sangin (*CCAG*, **12,** 205–229), contains translations of astronomical and astrological works by Gerard of Cremona and by John of Seville in its older portion that apparently dates from the late twelfth century (Sangin said thirteenth-fourteenth century; but we saw the codex in August 1974 and the first portion is definitely of the late twelfth century: on folio 77 it has a world horoscope dated 13 March 1178). On folios 13r–18r it contains the *Theorica planetarum*, here formally ascribed to Gerard of Cremona in a collection of translations by him and by John of Seville. This copy may date from the lifetime of Gerard of Cremona and may have originated in his circle. In 1959 Thorndike ("John of Seville," in *Speculum*, **34** [1959], 31–32) noted two fifteenth-century manuscripts of the *Theorica* bearing the ascription to John of Seville. There is, in fact, a distinct possibility that, like so many other cases of close relation between John's and Gerard's translations, the *Theorica planetarum* may have originated with John of Seville, whose style it matches perfectly, and was reworked in some fashion by Gerard of Cremona.

There is little to be added concerning the quality of Gerard's translations. Beginning with the Renaissance and through the early centuries of printing, the criticisms of his rendering of his Arabic models were many and harsh. These criticisms, however, rarely took into account the long period of manuscript transmission prior to the invention of printing that rendered Gerard's works so susceptible to scribal errors. Moreover, none of the critical comments has accused Gerard of mistranslating from the Arabic, simply because no scholar has compared his translations with the Arabic.

This double task, advocated by Wüstenfeld in 1877 (p. 80), has not yet even been attempted.

NOTES

1. C. A. Nallino, "Il Gherardo Cremonese autore della *Theorica planetarum* deve ritenersi esse Gherardo Cremonese da Sabbioneta," in *Atti dell'Accademia dei Lincei Rendiconti*. Cl. di sci. mor., stor. e fil., 6th ser., **8** (1932), 383–404; repr. in Maria Nallino, ed., *Raccolta di scritti editi e inediti*, VI (Rome, 1948), 304–320—see 307, n. 2.
2. Text in F. Wüstenfeld, "Die Übersetzungen arabischer Werke in das lateinische seit dem XI. Jahrhundert," 77; also in B. Boncompagni, "Della vita e delle opere di Gherardo Cremonese, traduttore del secolo duodecimo, e di Gherardo da Sabbionetta astronomo del secolo decimoterzo" in *Atti dell'Accademia pontificia dei Nuovi Lincei*, **4** (1851), 387–493, also published separately (Rome, 1851), 3–109. In his recent English trans., published in E. Grant, ed., *A Source Book in Medieval Science* (Cambridge, Mass., 1974), 35–38, Michael McVaugh omitted this section.
3. L. Muratori, *Rerum italicarum scriptores*, IX, 600.
4. Francisco Arisi, *Cremona literata, seu in Cremonenses doctrinis et literariis dignitatibus eminentiores chronologicae adnotationes*, I, 269–273.
5. K. Sudhoff, "Die kurze 'Vita' und das Verzeichnis der Arbeiten Gerhards v. Cremona, von seinem Schülern und Studien genossen kurz nach dem Tode des Meisters (1187) zu Toledo verabfasst," refers to Boncompagni's and V. Rose's studies but ignores that of Wüstenfeld. Sudhoff claims to have searched MSS of the trans. of the *Tegni* for new copies of the *vita* and list; he says he has found three in addition to the Vatican MS used by Boncompagni. Two of the new MSS are those used (and presumably discovered) by Wüstenfeld thirty-eight years earlier (MSS Leipzig 1119 and 1148). In presenting his "new" text from these four MSS, Sudhoff committed all the scribal errors rejected by Wüstenfeld. The fourth MS was cited by V. Rose in 1874.
6. V. Rose, "Ptolemaeus und die Schule von Toledo," in *Hermes*, **8** (1874), 327–349, esp. 347–349. The Erfurt MS of Sudhoff is cited on p. 334, n. 2.
7. See R. Lemay, *Abu Ma'shar and Latin Aristotelianism in the Twelfth Century* (Beirut, 1962), 315.
8. Since M. McVaugh's English trans. (1974) is now easily available, we shall refer to it as a rule and mark only occasionally the need for closer interpretation by reference to the Latin text of Wüstenfeld.
9. See Boncompagni, *op. cit.*, 397; sep. printing, p. 12.
10. See Rose, *op cit.*, 349. Also K. Sudhoff's ed. in *Archiv für die Geschichte der Naturwissenschaften und der Technik*, **8** (1918), 1–40, with better variants offered by A. Birkenmajer, *ibid.*, **9** (1920), 46–51.
11. S. D. Wingate, *The Mediaeval Latin Versions of the Aristotelian Scientific Corpus* (London, 1931), 46, suggests a year as early as 1134, which would give Gerard fifty-three years of activity in Toledo, a still more plausible situation. This early date, if it could be more securely based, would carry tremendous importance for Gerard's training in Arabic. The year 1133 is the date of the completion of John of Seville's translation of Abū Maʿshar's *Liber maioris introductorii*. John is known to have been active in translating, both on his own and for Archbishop Raymond, for at least a decade after that. This raises the strong possibility of direct contact between Gerard and John of Seville, his model as a translator.
12. In doubting Gerard's authorship of any number of translations because of this incredibly large output, Sarton did not consider the time factor. There is nothing inherently impossible in Gerard's direct authorship of so many translations if he had spent more than forty or even fifty years in that work. As his biography states, he was a very industrious worker throughout his stay in Toledo.
13. See Wingate, *op. cit.*, 46 and references.
14. *Hermes*, **8** (1874), 335 and 336, n. 1. We doubt, however, that this collaboration with a native Spaniard was necessarily done "in mündlichem Dictate," as Rose states, taking his example from Rudolf of Bruges. This process clearly prevented the Latin collaborator from working directly with the Arabic text. Such a situation, if probable for Rudolf, who made very few translations, is quite unthinkable for Gerard of Cremona.
15. See the extreme claim made by M. T. d'Alverny, "Deux traductions latines du Coran au moyen âge," *AHDLMA*, **22–23** (1947–1948), 69–131, esp. 85, n. 3, and 114.
16. *Séville musulmane au début du XIIe siècle: Le traité d'Ibn ʿAbdūn*, translated and annotated by E. Levi-Provençal (Paris, 1947), 128. See R. Lemay, *Abū Maʿshar*, 15, n. 1.
17. I. Opelt, "Zur Übersetzungstechnik des Gerhard von Cremona," in *Glotta*, **38** (1959), 135–170, is a valuable effort concentrating on a single work. Yet the results, embodied in a series of suggested criteria (pp. 138–151) for determining Gerard's characteristic manner, can only be weakened by the lack of comparative approach with other possible translations. This may be the case with the *De caelo et mundo*, the object of Opelt's attention. The weakness of Opelt's approach shows even more in the final glossaries: her glossary of Arabic terms contains only thirty entries, whereas the glossary of Greek-Latin terms (surely of little direct concern to Gerard) contains 600 terms. Gerard patiently and earnestly worked with all of his Arabic texts, an effort that does not appear in Opelt's analysis, at least not clearly enough. L. Minio-Paluello made a more direct examination of Gerard's technique in relation to the Arabic in *AL*, 2nd ed., L. Minio-Paluello and B. G. Dod, eds., I (Bruges–Paris, 1968), 1–4, containing Aristotle's *Analytica posteriora* [no. 1 on Wüstenfeld's list]; see esp. pp. li–lxv, which contain results indicating the soundness of this approach.
18. *AL*, I, 16 and 197–198; II, index, under Qusṭā ibn Lūqā; and M. A. Alonso, "Traducciones del Arabe al latín por Juan Hispano (Ibn Dawūd)," in *al-Andalus*, **17** (1952), 134–139. Alonso's information should always be used with caution. These studies conclude that there were two translations of this text, one surely by John of Seville and the other anonymous. E. Bertola, however, after a close comparison with several MSS of this Latin text, believes that there was only one translation, with scribal variants: E. Bertola, "Le traduzioni delle opere filosofiche arabo-giudaiche nei secoli XII e XIII," 269. The criteria proposed above for distinguishing Gerard's in relation to earlier translations seem to be directly applicable here. The difference in the two incipits of Qusṭā ibn Lūqā's Latin translations shows that John of Seville's version omitted the typically Muslim clause of "honorificet te Deus," which the other version restored. Since this is exactly Gerard's manner in his new versions, it would seem that the "anonymous" version should definitely be ascribed to him.
19. M. McVaugh is to be commended for having returned to the *socii* list in his recent English trans. of the *vita* and bibliography in E. Grant, ed., *op. cit.*, 35–38.
20. Minio-Paluello's hesitant suggestion that Gerard might have known Greek and done translation from this language in southern Italy before going to Toledo—"Note sull'Aristotele latino medievale," in *Rivista di filosofia neoscolastica (RFNS)*, **42** (1950), 227–228—was rejected by A. Mansion, *AL*, VII, 2 (1957), vii–viii.
21. See Wüstenfeld, *op. cit.*; Boncompagni, *op. cit.*; Sudhoff, *op. cit.* Now in English in M. McVaugh (see n. 19).
22. See Manuel Alonso Alonso, "Al-qiwām y 'al-aniyya' en las

traducciones de Gundisalvo," in *al-Andalus*, **22** (1957), 377–405.

23. Such as the work sometimes attributed under this title to Gilbert de la Porrée. See also H. Silverstein, "Liber Hermetis Mercurii triplicis de VI rerum principiis," in *AHDLMA*, **30** (1955), 217–302, for a parallel twelfth-century source from translations of Hermetic works.
24. *Domingo Gundisalvo, De Scientiis, texto latino . . .*, Manuel Alonso Alonso, ed. (Madrid–Granada, 1954). See J. T. Monroe, *Islam and the Arabs in Spanish Scholarship (Sixteenth Century to the Present)* (Leiden, 1970), 229.
25. Plagiarism was hinted at by Alonso when he edited *Hermann de Carinthia, De Essentiis* (Santander, 1946), in which he showed that Gundissalinus' *De processione mundi* took entire paragraphs from Hermann's work.
26. By Robert le Normand at Paris in 1358. See Henri Denifle and Émile Châtelain, *Chartularium Universitatis parisiensis. Auctarium*, I (Paris, 1889), col. 225. Other allusions to teachers of astrology at Paris are in *Chartularium*, III (1894), 265, 449; they concern Johannes Durand, "scolaris in medicina in secundo anno et [legens] Parisius astrologiam ex precepto domini regis." In Bologna the statutes of 1405 for the Faculty of Arts and Medicine also contain mention of *Alchabitius* as a text for students of medicine. See Malagola 276, translated in L. Thorndike, *University Records and Life in the Middle Ages* (New York, 1944), 279–282. It may be suspected that in addition to the original *Alchabitius*, either in the trans. of John of Seville or that of Gerard of Cremona, the students of medicine also used the commentary (*Liber isagogicus*) completed at Paris in 1331 by John of Saxony. See Simon de Phares, *Recueil des plus célèbres astrologues . . .*, Ernest Wickersheimer, ed. (Paris, 1929), 256. Simon recalls that in his youth (*ca.* 1460) he went to Paris "en la rue du Feurre [seat of the Faculty of Arts] ou je aprins *De spera* et mes introductoires de l'Acabice."
27. Originally composed at the court of Alfonso el Sabio in Toledo, *ca.* 1255–1260, but introduced after important modifications at the University of Paris about 1335. E. Rosen is preparing a study on the fate of the Alfonsine Tables in the medieval universities.

BIBLIOGRAPHY

The following abbreviations are used:

AHDLMA. Archives d'histoire doctrinale et littéraire du moyen âge.

AL. Aristoteles latinus. I, *Codices descripsit G. Lacombe in societatem operis adsumptis A. Birkenmajer, M. Dulong, Aet. Franceschini. Pars prior* (Rome, 1939); II, *Codices . . . supplementis indicibusque instruxit L. Minio-Paluello. Pars posterior* (Cambridge, 1947 ff.); III, *Codices. Supplementa altera edidit L. Minio-Paluello* (Bruges–Paris, 1961); IV, 1–4. *Analytica posteriora. 2 et 3 editio altera. Translationes Iacobi, anonymi sive "Ioannis," Gerardi et recensio Guillelmi de Moerbeka*, L. Minio-Paluello and Bernardus G. Dod, eds. (Bruges–Paris, 1968); VII, 2. *Physica. Translatio Vaticana*, A. Mansion, ed. (Bruges–Paris, 1957).

Beiträge. Beiträge zur Geschichte der Philosophie des Mittelalters; BM. Bibliotheca mathematica; Carmody, *AAAL.* F. Carmody, *Arabic Astronomical and Astrological Sciences in Latin Translation* (Berkeley, 1956); Carmody, *Thabit* (1960). F. Carmody, *The Astronomical Works of Thabit ibn Qurra* (Berkeley, 1941; 2nd ed., 1960); *RFNS. Rivista di filosofia neo-scolastica; TK.* L. Thorndike and P. Kibre, *A Catalogue of Incipits of Mediaeval Scientific Writings in Latin*, 2nd ed. (Cambridge, Mass., 1963); *ZDMG. Zeitschrift der Deutschen morgenländischen Gesellschaft.*

Works to be consulted are Francisco Arisi, *Cremona literata, seu in Cremonenses doctrinis et literariis dignitatibus eminentiores chronologicae adnotationes*, I, *Priscorum temporum monumenta complectens usque ad annum millesimum quingentesimum primum . . .* (Parma, 1702); E. Bertola, "Le traduzioni delle opere filosofiche arabo-giudaiche nei secoli XII e XIII," in *Studi di filosofia e di storia della filosofia in onore di Francesco Olgiati* (Milan, 1962), 235–270; A. Birkenmajer, "Eine wiedergefundene Übersetzung Gerhards von Cremona," in A. Lang, J. Lechner, and M. Schmaus, eds., *Aus der Geisteswelt des Mittelalters. Studien und Texte Martin Grabmann . . . gewidmet* (Münster, 1935), 472–481; A. A. Björnbo, "Ueber zwei mathematische Handschriften aus dem vierzehnten Jahrhundert," in *BM*, 3rd ser., **3** (1902), 63–75; and "Gerhard von Cremonas Uebersetzung von Alkhwarizmi's Algebra und von Euklides Elementen," *ibid.*, **6** (1905), 239–248; A. A. Björnbo and S. Vogl, "Alkindi, Tideus und Pseudo-Euklid. Drei optische Werke," *Abhandlungen zur Geschichte der mathematischen Wissenschaften mit Einschluss ihrer Anwendungen*, **26**, no. 3 (1912); B. Boncompagni, "Della vita e delle opere di Gherardo Cremonese, traduttore del secolo duodecimo, e di Gherardo da Sabbionetta, astronomo del secolo decimoterzo," in *Atti dell'Accademia pontificia dei Nuovi Lincei*, **4** (1851), 387–493, also separately printed (Rome, 1851), 3–109; H. L. L. Busard and P. S. Van Koningsveld, "Der *Liber de arcubus similibus* des Ahmed ibn Jusuf," in *Annals of Science*, **30** (1973), 381–406; and "Über einige Euklid-Skolien, die den Elementen von Euklid, übersetzt von Gerard von Cremona, angehängt worden sind," in *Centaurus*, **18** (1974), 97–128; H. Diels, "Die Handschriften der antiken Ärzte. Griechische Abt. I, Hippokrates und Galenos," *Abhandlungen der K. Preussischen Akademie der Wissenschaften*, Phil.-hist. Kl. (1905), no. 3; R. J. Durling, "Corrigenda and Addenda to Diels' Galenica," in *Traditio*, **23** (1967), 461–476; and C. H. Haskins, *Studies in the History of Medieval Science*, 2nd ed. (Cambridge, Mass., 1927).

Also to be consulted are G. Sarton, *An Introduction to the History of Science*, II (Baltimore, 1931), 338–349; Fuat Sezgin, *Geschichte des arabischen Schrifttums* (*GAS*), III, *Medizin-Pharmazie-Zoologie-Tierheilkunde bis ca. 430 H.* (Leiden, 1970); IV, *Alchimie-Chemie-Botanik-Agrikultur bis ca. 430 H.* (Leiden, 1971); and V, *Mathematik bis ca. 430 H.* (Leiden, 1974); M. Steinschneider, "Die mittleren Bücher der Araber und ihre Bearbeiter," in *Zeitschrift für Mathematik und Physik*, **10** (1865), 456–498; "Vite di matematici arabi . . . B. Baldi," in *Bollettino . . . Boncompagni*, **5** (1872), 427–460; and "Die europäischen Übersetzungen aus dem arabischen bis Mitte des 17. Jahrhunderts," *Sitzungsberichte der K. Akademie der Wissenschaften in*

Wien, Phil.-hist. Kl., **149**, no. 4 (1904), and **151**, no. 1 (1906); K. Sudhoff, "Die kurze 'Vita' und das Verzeichnis der Arbeiten Gerhards v. Cremona, von seinem Schülern und Studien genossen kurz nach dem Tode des Meisters (1187) zu Toledo verabfasst," in *Archiv für Geschichte der Medizin*, **8** (1915), 73–92; H. Suter, "Die Mathematiker und Astronomen der Araber und ihre Werke," *Abhandlungen zur Geschichte der mathematischen Wissenschaften*, **10** (1900); "Nachträge und Berichtigungen . . . ," *ibid.*, **14** (1902), 155–185; H. Suter, "Über einige noch nicht sicher gestellte Autorennamen in den Übersetzungen des Gerhard von Cremona," in *BM*, 3rd ser., **4** (1903), 19–27; L. Thorndike, *A History of Magic and Experimental Science*, II (New York, 1923), see index; "The Latin Translations of Astrological Works by Messahalla," in *Osiris*, **12** (1956), 49–72; and "John of Seville," in *Speculum*, **34** (1959), 20–38; Manfred Ullmann, *Die Medizin im Islam* (Leiden–Cologne, 1970); and *Die Natur- und Geheimwissenschaften im Islam* (Leiden, 1972); and F. Wüstenfeld, "Die Übersetzungen arabischer Werke in das lateinische seit dem XI. Jahrhundert," in *Abhandlungen der Gesellschaft der Wissenschaften zu Göttingen*, **22** (1877), 55–81.

RICHARD LEMAY

GRASSMANN, HERMANN GÜNTHER (*b.* Stettin, Pomerania [now Szczecin, Poland], 15 April 1809; *d.* Stettin, 26 September 1877), *mathematics.*

Life and Works. Grassmann came from a family of scholars. His father, Justus Günther Grassmann, studied theology, mathematics, and physics. He was a minister for a short time, then became a teacher of mathematics and physics at the Gymnasium in Stettin, where he did much to raise the level of education. He also wrote elementary mathematics textbooks and did research on problems in physics and crystallography. Grassmann's mother was Johanne Medenwald, a minister's daughter, from Klein-Schönfeld.

The third of twelve children, Grassmann received his earliest instruction from his mother and at a private school before attending the Stettin Gymnasium. He also learned to play the piano. At the age of eighteen he passed the final secondary school examination, ranking second. With his eldest brother, Gustav, Hermann studied theology for six semesters at the University of Berlin, where his teachers included August Neander and Friedrich Schleiermacher. At the same time he studied classical languages and literature and attended the lectures of August Böckh.

Grassmann returned to Stettin in the fall of 1830 and began intensive independent study of mathematics and physics. In December 1831, at Berlin, he took the examination for teaching at the Gymnasium level; the examiners stated, however, that Grassmann had to display greater knowledge of his subjects before he could be considered qualified to teach the higher grades. At Easter 1832 he obtained a post as assistant teacher at the Stettin Gymnasium, and two years later he passed the first-level theology examination given by the Lutheran church council of Stettin.

Grassmann did not become a minister, however. In the autumn of 1834 he was hired as senior master at the Gewerbeschule in Berlin, succeeding Jakob Steiner, who had been called to the University of Berlin. A year later Grassmann was appointed to the faculty of the newly founded Otto Schule in Stettin, where he taught mathematics, physics, German, Latin, and religion. Meanwhile he pursued his studies in theology, mathematics, and natural science. In 1839 he passed the second-level theology examination in Stettin and, the following year at Berlin, an examination in mathematics, physics, chemistry, and mineralogy that fully qualified him to teach all grades of secondary school.

The latter examination included a written portion to be done at home, the subject of which was the theory of the tides. This assignment proved decisive for Grassmann's career. In 1832 he had begun to work on a new geometric calculus. With its aid he was now able to give a simplified exposition of the mathematical developments in Lagrange's *Mécanique analytique* and to derive, in an original manner different from that of Laplace, the portions of the latter's *Mécanique céleste* relevant to the theory of the tides. Although Grassmann used the new methods only to the extent necessary to solve the problem at hand, he was undoubtedly aware of the far-reaching significance of his creation; and by 1840 he had decided to concentrate entirely on mathematical research.

At first, however, Grassmann devoted considerable effort to teaching. He wrote several brief textbooks for use in secondary school, some of which were frequently reprinted. They included *Grundriss der deutschen Sprachlehre* (1842) and *Leitfaden für den ersten Unterricht in der lateinischen Sprache* (1842). In collaboration with W. Langbein he published *Deutsches Lesebuch für Schüler von acht bis zwölf Jahren* (1846). Reorganizations of the Stettin school system led to his being transferred on several occasions (the Otto Schule, the Gymnasium, the Friedrich Wilhelm Schule; at the last he received the title *Oberlehrer* in May 1847). In 1852 Grassmann succeeded his father (who had

GRASSMANN

died in March of that year) as fourth-ranking teacher at the Stettin Gymnasium, a post that brought with it the title of professor.

Meanwhile, in the fall of 1843, Grassmann had completed the manuscript of the first volume of his chief work, *Die lineale Ausdehnungslehre*, which appeared the following year as *Die Wissenschaft der extensiven Grösse oder die Ausdehnungslehre*. Its fundamental significance was not grasped by contemporaries; and even a mathematician of the caliber of A. F. Möbius—whose own geometric research was to some extent related to Grassmann's—did not fully understand the author's intentions. Disapproving of the many new concepts and certain philosophical formulations, he declined to write a review of the book, and thus it was totally disregarded by the experts.

As an application of the *Ausdehnungslehre*, which is based on the general concept of connectivity, Grassmann published *Neue Theorie der Elektrodynamik* (1845), in which he replaced Ampère's fundamental law for the reciprocal effect of two infinitely small current elements with a law requiring less arbitrary assumptions. Thirty years later Clausius independently rediscovered Grassmann's law but acknowledged his priority as soon as he was apprised of it. In a series of articles published between 1846 and 1856 Grassmann applied his theory to the generation of algebraic curves and surfaces, in the hope that these papers, which were much less abstract than his book, would inspire mathematicians to read the *Ausdehnungslehre*. His hopes were not fulfilled.

On the other hand, Grassmann received rapid public recognition for a work submitted in 1846, at Möbius' suggestion, to the Fürstlich Jablonowsky'sche Gesellschaft der Wissenschaften in Leipzig. In it he solved the problem posed by the society of establishing the geometric characteristic, first outlined by Leibniz, for designating topological relations, without recourse to metric properties. His entry, *Geometrische Analyse*, was awarded the full prize and was published by the society in 1847. Möbius, who was one of the judges, justly criticized both the abstract manner in which Grassmann introduced his new concepts and his neglect of intuitive aids, defects that made the last part of the text particularly difficult to read. Accordingly, Möbius incorporated into the book his own essay, "Die Grassmann'sche Lehre von den Punktgrössen und den davon abhängigen Grössenformen," in which he explained Grassmann's *Scheingrössen* as abbreviated expressions of intuitively interpretable quantities.

In May 1847 Grassmann wrote to the Prussian ministry of education, requesting that he be considered for a post as university professor, in the event that one became available. The ministry thereupon requested an opinion of Grassmann's prize essay from E. E. Kummer, a mathematician at Breslau. Kummer's severe judgment that the work contained "commendably good material expressed in a deficient form" led to the rejection of Grassmann's application.

A man of broad interests and a strong sense of political responsibility, Grassmann participated in the political events leading to the Revolution of 1848. With his brother Robert, his scientific collaborator for many years, he founded the *Deutsche Wochenschrift für Staat, Kirche und Volksleben*, which was soon replaced by the daily *Norddeutsche Zeitung*. Advocates of a Germany united under Prussian leadership, the brothers hoped for the establishment of a constitutional monarchy, ruled by the king in cooperation with the Reichstag. Revolution and civil war, they contended, were not the proper means of winning greater freedom. In articles published in 1848 and 1849 Grassmann considered chiefly problems of constitutional law; but with the restoration he became increasingly dissatisfied and withdrew from the paper.

On 12 April 1849 Grassmann married Marie Therese Knappe, the daughter of a Pomeranian landowner. Of their eleven children, two died in early childhood and two others somewhat later. His sons Justus and Max became mathematics teachers at the Stettin Gymnasium; Ludolf, a physician; Hermann, professor of mathematics at the University of Giessen; and Richard, professor of mechanical engineering at the Technische Hochschule in Karlsruhe.

As a student Grassmann had been admitted to the Freemason lodge in Stettin, and from 1856 he held the post of treasurer. In 1857 he became a member of the board of directors of the Pommersche Hauptverein für die Evangelisierung Chinas. Founded in 1850, this society published a journal and occasionally sent missionaries to China. Under Grassmann's chairmanship it was unified with the Rheinische Missions-Gesellschaft in Barmen in 1873.

Soon after the political unrest of 1848–1849 had subsided, Grassmann began to study Sanskrit and then Gothic, Lithuanian, Old Prussian, Old Persian, Russian, and Church Slavonic—investigations that laid the foundations for his studies in comparative linguistics. Profiting from his acute sense of hearing and his capacity for making careful observations, Grassmann developed a theory of the physical nature of speech sounds (1854). He had recognized that each vowel sound arises through definite overtones that are specifically characteristic of it and planned to substantiate his theory experimentally, using a tone generator instead of the human voice. The device he intended to use, however, did not furnish sufficiently pure vibrations—that is, it was not sufficiently free of overtones.

In "Zur Theorie der Farbenmischung" (1853) Grassmann opposed certain conclusions that Helmholtz had drawn from experiments on the mixing of colors. According to Grassmann's theory, each color can be represented by a weighted point plotted on a circular surface; the position of the point indicates its hue and saturation, and the weight, its intensity. If the colors to be mixed are thus represented, then the center of gravity, in which the entire mass is seen as being concentrated, gives the intensity of the visual impression of the mixture. Helmholtz later acknowledged the correctness of the center-of-gravity construction but retained reservations concerning the circular form of the overall field.

Around the middle of 1854 Grassmann resumed work on the *Ausdehnungslehre*. He may have been stimulated in this effort by a remark made by Möbius, who informed him of two papers in which equations with several unknowns were solved by means of an approach incorporating *clefs algébriques* that corresponded to the one that Grassmann had developed in sections 45, 46, and 93 of his book. Indeed, Möbius insisted that Grassmann assert his priority against the authors of these papers, A. B. de Saint-Venant (in *Comptes rendus . . . de l'Académie des sciences*, **21** [1845], 620–625) and Cauchy (*ibid.*, **36** [1853], 70–76, 129–136). Grassmann subsequently addressed himself to the Paris Academy as well as to both authors, without charging plagiarism. Whether Cauchy, in particular, knew of the *Ausdehnungslehre* when he wrote his paper cannot be established.

Rather than writing a second volume of the *Ausdehnungslehre*, as he had originally intended, Grassmann decided to rework the text; and *Die Ausdehnungslehre. Vollständig und in strenger Form bearbeitet* was published at Berlin in 1862. The new version, unfortunately, fared no better than the first; Grassmann, in presenting a systematic foundation of his theory, had failed to see that a different approach was necessary for mathematicians to acquaint themselves with the new concepts. Given the failure of the first edition, it is all the more astonishing that Grassmann did not attempt to emphasize the advantages of his ideas by demonstrating them through specific examples. Again, no serious attempts were made by mathematicians to work through the peculiar terminology of this strange theory, and the revised edition was long ignored. Although the Leopoldina elected Grassmann to membership in 1864, it was in recognition of his achievements in physics—not in mathematics.

Disappointed by his continued lack of success, Grassmann gradually turned away from mathematical research. Besides a few minor articles, in 1865 he published a supplement to his *Lehrbuch der Arithmetik* (1860) entitled *Lehrbuch der Trigonometrie für höhere Lehranstalten* (a planned third part on geometry never appeared) and "Grundriss der Mechanik," in the *Stettiner Gymnasialprogramm* of 1867.

Grassmann also concentrated increasingly on linguistic research. Works on phonetics that were based on the historical study of language (1860, 1862) were followed by the important "Über die Aspiranten und ihr gleichzeitiges Vorhandensein im An- und Auslaute der Wurzeln" (1863), in which he formulated the law of aspirates that is named for him. A crucial contribution to the study of the Germanic sound shift, this law has become part of the science of comparative linguistics.

Linguistic research of a different kind forms the basis of *Deutsche Pflanzennamen* (1870), which Grassmann wrote with his brother Robert and his brother-in-law Christian Hess. The goal of the work was to introduce German names for all plants grown in the German-language area, terms as precise as the Latin and Greek forms that would be derived etymologically from the various Germanic languages. Grassmann hoped that the effort of collecting and explicating involved in the project would prove useful to both botanists (especially biology teachers) and linguists.

An achievement of a much higher order is represented by Grassmann's work on the Sanskrit language. Realizing that research in comparative linguistics must take the oldest Indic languages as a starting point, Grassmann began an intensive study of the hymns of the *Rig-Veda* about 1860. Unlike his mathematical work, which was greatly ahead of its time, these studies benefited from opportune timing. Theodor Aufrecht's authoritative text of the *Rig-Veda* and the St. Petersburg Sanskrit dictionary compiled by Otto von Böthlingk and W. R.

von Roth were also published at this time. Grassmann's complete glossary of the *Rig-Veda* was modeled on the Biblical concordances, and the entry for each word indicates the grammatical form in which it appears. Although criticized on points of detail, the six-part *Wörterbuch zum Rigveda* (1873–1875) was generally praised by specialists and remained the standard work for many years.

Although Grassmann had originally planned to follow the glossary with a grammar, he came to consider it more important to publish a translation of the hymns, believing it essential to their interpretation; it appeared in two parts as *Rig-Veda. Übersetzt und mit kritischen Anmerkungen versehen* (1876–1877). Alfred Ludwig's simultaneously published complete translation of the texts (1876), although far superior philologically to Grassmann's work, was presented in a German that was difficult to understand. Grassmann, in contrast, seeking to reproduce not only the meanings of the words but also the overall feeling of the original, retained a metrical form—at the sacrifice of a faithful rendering in certain passages.

Unlike his mathematical works, Grassmann's linguistic research was immediately well received by scholars. A year before his death he became a member of the American Oriental Society, and the Faculty of Philosophy of the University of Tübingen awarded him an honorary doctorate.

In the meantime, a few mathematicians had become aware of Grassmann's work. His election on 2 December 1871 as a corresponding member of the Göttingen Academy of Sciences encouraged him to publish some short mathematical papers during the last years of his life. In 1877 he prepared another edition of the *Ausdehnungslehre* of 1844 for publication; it appeared posthumously in 1878. Despite steadily increasing infirmity, he continued to work until succumbing to cardiac insufficiency.

Influence. Toward the end of his life, and even more after his death, a growing recognition of Grassmann's accomplishments was observed among specialists. Alfred Clebsch, the leading German mathematician of the time, for example, made a special effort to call attention to Grassmann during a memorial address that he delivered in honor of Julius Plücker in 1871. H. Hankel was the first to discuss the *Ausdehnungslehre* in a book: *Theorie der komplexen Zahlensysteme* (1867). The obituary of Grassmann by Sturm, Schröder, and Sohncke displays considerable appreciation for his work. At about the same time there appeared a short biography of him by V. Schlegel, who had previously used Grassmann's ideas in his *System der Raumlehre* (1872–1875).

The geometry that Grassmann established on the basis of the *Ausdehnungslehre* was later elaborated as *Punktrechnung*, first in Peano's *Calcolo geometrico* (1888). Grassmann's most gifted son, Hermann, wrote three textbooks on analytic geometry based on the approach of the *Ausdehnungslehre*; and in 1909 he applied the latter to the theory of gyroscopic motion. The other most important German advocates of the book were R. Mehmke and his student A. Lotze. French mathematicians were introduced to Grassmann's concepts by F. Caspary, who used them in his research on the generation of algebraic curves. The spread of Grassmann's ideas was further aided by the publication of a collected edition of his works and of papers written on the centenary of his birth. An early advocate of Grassmann's *Ausdehnungslehre* was the American physicist J. W. Gibbs (1839–1903). Well-known, among many other achievements, for his popularization of a vector algebra as it may be derived from Grassmann's and W. R. Hamilton's ideas, he emphatically preferred—in contrast to Tait's rival claims for Hamilton's quaternions—Grassmann's less limited concepts. Gibbs created what he called "dyadics," an operational approach that found favor with theoretical physicists until it was replaced by modern vector and tensor analysis. In this dyadics, a matrix in its operational application is understood as a sum of more simple operators, the so-called simple dyads. As Gibbs himself explained, the germ of these ideas must be seen in certain indeterminate products (*Lückenausdrücke*), which Grassmann introduced in a note at the end of the first edition of his *Ausdehnungslehre* of 1844. Thus the birth of linear matrix algebra, often associated with the publication of Cayley's classic "Memoir on the Theory of Matrices" in 1858, may be said to have occurred already in 1844. A century later H. G. Forder's *Calculus of Extension* (1941) testified to the continuing appeal of the *Ausdehnungslehre* among mathematicians in the English-speaking world.

The Ausdehnungslehre. The *Ausdehnungslehre* concerns geometric analysis, a border region between analytic geometry, which uses only the algebra of coordinates and equations, and synthetic geometry, which dispenses with all algebraic aids. The first to conceive of this kind of geometric analysis was Leibniz. In a letter of 1679 to Huygens (see Grassmann, *Werke*, I, 417) he wrote that in order to develop such an analysis, one should work directly with the symbols of geometric concepts—

such as points, straight lines, and planes—that is, without the intermediary of coordinates; further, loci and other properties should always be expressed through algebraic expressions written in the symbols of these basic concepts. Leibniz did not fully elaborate his idea, and systems of geometric analysis did not appear until the nineteenth century, after analytic and synthetic geometry had undergone considerable development. On this question one may consult three articles by H. Rothe, A. Lotze, and C. Betsch in the *Encyklopädie der mathematischen Wissenschaften* (1916–1923); the article by Lotze deals especially with the *Ausdehnungslehre*.

All systems of geometric analysis share the characteristic that one of their fundamental elements is the geometric addition of directed line segments, an operation borrowed from mechanics. Accordingly, one may place among these systems the description of Euclidean plane geometry by means of complex numbers that was formulated around 1800. Although known as the theory of the Gaussian plane, it was also elaborated independently of Gauss by Wessel and Argand. A purely geometric computation of line segments in the plane, in which even the complex numbers do not appear explicitly, had to await the appearance of the method of equipollences elaborated by Bellavitis in the 1830's. Another example of geometric analysis is Möbius' barycentric calculus (1827). The barycentric coordinates of Euclidean space are special systems of projective coordinates obtained with the aid of concepts drawn from point mechanics. More important in the present context, however, is the fact that in 1844, in a work devoted to establishing the barycentric calculus on a firmer basis, Möbius conceived of the line segment as the difference between two points, a notion that played an important role in the calculus of points based on the *Ausdehnungslehre*.

The theory of quaternions developed by W. R. Hamilton between 1843 and 1853 originated in an attempt to generalize the complex numbers in a manner that would preserve, if possible, all the laws of arithmetic. This generalization could be effected, however, only by giving up the commutative law of multiplication. Doing so gave rise to the system of quaternions named for Hamilton. In current terminology, this system is a skew field of a fourth-rank algebra over the field R of the real numbers.

During this same period, Grassmann developed his *Ausdehnungslehre*. Its algebraic entities are the extensive quantities. These consist, in the first instance, of quantities of the first rank of the base domain S_n^1; this, in modern terms, is an n-dimensional vector space over R, which is expressed here for the first time in such generality. Its base vectors are $e_1, \cdots, e_n$. In his attempt to find a suitable multiplication of two quantities of S_n^1, Grassmann proceeded differently from Hamilton. He did not seek to make S_n^1 into a ring but, instead, added to S_n^1 a domain $S^2{}_{\binom{n}{2}}$ of quantities of the second rank—that is, a vector space of dimension $\binom{n}{2}$ with base quantities e_{ij} $(1 \leq i < j \leq n)$. The product of two quantities of S_n^1—which he called the outer product—is so constructed that it lies in $S^2{}_{\binom{n}{2}}$. This outer multiplication is to be distributive, so it need be defined only for the e_i. If the multiplication is designated by brackets, then

$$[e_ie_j] = -[e_je_i] = e_{ij} \qquad (1 \leq i < j \leq n)$$

and

$$[e_ie_i] = 0 \qquad (1 \leq i \leq n).$$

Then, for arbitrary r where $1 \leq r \leq n$, Grassmann established the domain $S^r{}_{\binom{n}{r}}$ of the quantities of the r-th rank with base $e_{i_1 \ldots i_r}$ $(1 \leq i_1 < \cdots < i_r \leq n)$. By using the formulation $[e_{j_1} \cdots e_{j_r}] = +e_{i_1 \ldots i_r}$, or $= -e_{i_1 \ldots i_r}$, or $= 0$—according as $(j_1 \cdots j_r)$ is an even permutation of $(i_1 \cdots i_r)$ or an odd permutation of it, or whether the j_ν are not all different—the outer product of r base quantities of S_n^1 can be expressed as a quantity of $S^r{}_{\binom{n}{r}}$. At this stage one can immediately calculate the outer product of r arbitrary quantities of S_n^1. Grassmann called "simple" the special quantities of $S^r{}_{\binom{n}{r}}$ that arise in this manner. Since he set $e_{1 \ldots n} = 1$, he was again able to conceive S_n^1 as the scalar domain R.

Through reduction to the basic unities and use of the associative law, it can be shown that for arbitrary unities of S^r and S^s $[e_{i_1 \ldots i_r} e_{j_1 \ldots j_s}] = 0$, if i and j are not all different; that this expression equals $+1$ or -1 in the cases, respectively, that $(i_1 \cdots i_r j_1 \cdots j_s)$ is an even or an odd permutation of a combination $(n_1 \cdots n_r \cdots n_{r+s})$, where $1 \leq n_1 < \cdots < n_{r+s} \leq n$. From this result one may easily obtain, through distributive multiplication, the progressive product of the quantities $A^r \in S^r$ and $B^s \in S^s$. If one takes $e_{1 \ldots n}$ as an independent unity and adds quantities of arbitrary rank, which Grassmann avoided doing, one obtains, in modern terms, the Grassmann algebra of rank 2^n over S_n^1. In this algebra $[A^r, B^s]$ vanishes for $r + s > n$. Grassmann therefore constructed a "regressive" product of A^r and B^s. To this end he associated to each unity $e_{i_1 \ldots i_r}$ its supplement $| e_{i_1 \ldots i_r} = \pm e_{i_{r+1} \ldots i_n}$.

Here $(i_1 \cdots i_n)$ is a permutation of $(1 \cdots n)$, and equals $+1$ or -1 according as the latter is even or odd. From this it can be shown that the extension $|A^r$ for every $A^r \in S^r$ is a quantity of S^{n-r}.

The inner product of two quantities A^r, B^r of S^r is given by $P = [A \cdot |B]$, which Grassmann considered a scalar, since it is a quantity in S^n. At this point he could explain the notation of orthogonality and absolute value in S^r. He interpreted the regressive product of the quantities $A^r \in S^r$ and $B^s \in S^s$ (where $r + s > n$), for the unities ϵ^r, η^s, as that unity of rank $2n - (r + s)$ the supplement of which is the progressive product of the supplement of ϵ^r and η^s. This makes it easy to define the regressive product of arbitrary $A^r \in S^r$ and $B^s \in S^s$, where $r + s > n$.

On the basis of the preceding steps, Grassmann could obtain the outer products of arbitrarily many extensive quantities. These are "pure" if the multiplications to be performed are either all progressive or all regressive, and "mixed" if this is not the case. In general, a mixed product is neither commutative nor associative, although it does fulfill certain computational rules that Grassmann established. In addition to the outer products, Grassmann developed a multiplication that he termed "algebraic," which obeys the law $e_i e_j = e_j e_i$ for $i = 1 \cdots n$ and leads to what is today known as the polynomial ring.

Grassmann derived the ideas of his *Ausdehnungslehre,* as well as his new way of forming products, essentially from geometry, particularly from the geometry of n dimensions, which was then still in its infancy. In his calculus of points, the points of R_n, which are provided with weights (that is, numbers different from zero), are conceived as the fundamental domain S^1_{n+1} of a totality of extensive quantities. The points that are assigned weight 1 are also called "simple." $\lambda a + \mu b$ then yields all points of the straight line A spanned by the simple points a, b, those points having the weight $\lambda + \mu$. This is identical to the scheme developed by Möbius, where (λ, μ) are the barycentric coordinates of the relevant point of A. The only difference is that no weighted point of A corresponds to a-b but, rather, to the vector leading from b to a. In Grassmann's calculus, therefore, the vector space V_n over R appears as a subset of the domain S^1_{n+1} of the point range. If $a_0 \cdots a_k$ and $b_0 \cdots b_k$ are each linearly independent simple points that span the same $R_k \subset R_n$, then the outer products $[a_0 \cdots a_k]$ and $[b_0 \cdots b_k]$ do not vanish; and they differ by, at most, a scalar factor $\lambda \neq 0$. They are therefore associated with this R_k. Their components, related to a base $e_{i_0 \cdots i_k}$, are now called the Grassmann coordinates of R_k. They fulfill a system of quadratic relations that Grassmann gave for $k = 1$; and if they are interpreted as points of a projective space, they describe the manifold $G_{n,k}$ that is named for Grassmann.

If the simple quantities $[a_0 \cdots a_r]$ and $[b_0 \cdots b_s]$ are associated with the spaces R^a_r and R^b_s of R_n, and if these have no finite or infinite point in common, then the progressive outer product $[a_0 \cdots a_r b_0 \cdots b_s]$ corresponds to the connection space R_{r+s+1} of R^a_r and R^b_s. If R^a_r and R^b_s span the entire R_n, then the regressive product of $[a_0 \cdots a_r]$ and $[b_0 \cdots b_s]$ is associated with the intersection space $R^a_r \cap R^b_s$. In works directly inspired by Grassmann the perceptual spaces R_2, R_3 are treated in detail with these methods, as are the projective spaces P_2, P_3 over R. For example, in them the line vector (also known as the rod), which is bound to the connecting line A of a,b, is described by means of the outer product $[a,b]$ of the points a,b. Then $[abc]$ defines an oriented surface element bound to the plane containing a,b,c; its content can be either positive or negative, or zero when a, b, and c are collinear. The outer product of two free vectors yields a bivector.

Apart from details, the *Ausdehnungslehre* appears in retrospect as a very general and comprehensive treatise, the implications of which reached far beyond the "state of the art" in 1844 or even in 1862. In geometry, mathematicians were still thinking in terms of a "real" three-dimensional space and saw no need to occupy themselves with such a theory of "extended magnitudes" in a fictional n-dimensional space. Hamilton's goal had been to find a consistent algebra of rotations and vectors in three-dimensional space; when he reached it in his system of quaternions he was forced to sacrifice the traditional principle of commutativity for multiplication. Grassmann, on the other hand, had not only immediately considered manifolds of an arbitrary number of dimensions, but also had introduced new, seemingly artificial kinds of multiplication for its various types of elements. Nobody in his day could foresee that, in its general algebraic aspects, the *Ausdehnungslehre* did much more than merely accomplish for any finite number of dimensions what Hamilton's quaternions were designed to do for Euclidean space of three dimensions. Beyond that it anticipated (or even included) wide areas of modern linear algebra and of matrix, vector, and tensor analysis. Thus, three lines of later development may be distinguished in connection with Grassmann's principal work: first, the generalization of the geometrical concept of space (also anticipated at much the same time by Cayley and by Riemann); second, the influence on Gibbs and thus on the creation of vector analysis; and third, the important anticipation of fundamental parts of modern algebra, though this was not immediately noticed by the mathematical public.

Other Mathematical Works. Despite the long neglect of his ideas, Grassmann was always convinced of their importance. In several works he attempted to show how the theory of quaternions and invariant theory (then called modern algebra) can be understood on the basis of the *Ausdehnungslehre*. Still more important, however, are his writings on the "lineal" generation of algebraic entities, in which he also draws on his theory. This group of publications deals, for example, with the theory of constructing points of algebraic curves and surfaces by simply drawing straight lines and planes through given points, as well as with the determination of intersection points of known straight lines.

As early as 1721 Maclaurin had demonstrated that given the three points a, b, c and the two straight lines A, B of general position in the plane, the locus of the third vertices of all triangles the first two vertices of which lie, respectively, on A and B and the sides of which pass, correspondingly, through a, b, c is a conic section. In terms of the calculus of points, this statement means that the mixed outer product $(a \times AbB \times c)$ vanishes. Grassmann made the important discovery that in this way every plane algebraic curve C can be generated lineally. As a result, if C is of order n, it can be described by setting equal to zero an outer product in which, in addition to symbols for certain fixed points and straight lines, the expression for the variable point x of C appears n times. A cubic can be expressed, accordingly, as $(xaA) \cdot (xbB) \cdot (xcC) = 0$. This signifies that the locus of the point x of the plane is a cubic, if the line connecting x with three fixed points a, b, c cuts the three fixed straight lines A, B, C in three collinear points. Moreover, one can obtain every plane cubic in this manner. Grassmann was thus able to refute Plücker's assertion that curves higher than the second degree could be conceived only in terms of coordinate geometry. In writings collected in volume II, part 1, of the *Werke*, Grassmann considered, in particular, the lineal generations of plane cubics and quartics, as well as of third-degree spatial surfaces. (One of these generations bears his name.) He demonstrated that by setting equal to zero the products he designated as planimetric or stereometric, all these generations could be obtained from the *Ausdehnungslehre*.

A large portion of the *Ausdehnungslehre* is devoted to analysis. Grassmann treats functions of n real variables as functions of extensive quantities of a base domain S_n^1. Since he introduced a metric into S_n^1 in the form of the inner product, he was able to derive Taylor expansions, remainder formulas, and other items. His most important studies in analysis concern Pfaff's problem—that is, the theory of the integration of a Pfaffian equation

$$\omega = A_1(x_1 \cdots x_n)\,dx_1 + \cdots + A_n(x_1 \cdots x_n)\,dx_n = 0.$$

This question had interested leading nineteenth-century mathematicians both before and after Grassmann, especially Pfaff and Jacobi. Grassmann contributed the following important theorem: If one calls k the class of ω—that is, the minimum number of variables into which ω can be transformed—then, when $k = 2h$, ω can be transformed into the normal form

$$z_{h+1}\,dz_1 + \cdots + z_{2h}\,dz_h^1$$

and, when $k = 2h - 1$, into

$$p \cdot (dz_h + z_{h+1}\,dz_1 + \cdots + z_{2h-1}\,dz_{h-1}),$$

where p is a function of $z_1 \cdots z_{2h-1}$. Even these results, however, which appeared in the 1862 edition of the *Ausdehnungslehre* and surpassed Jacobi's achievements, obviously did not attract much attention. Recognition had to await their translation into the more customary language of analysis by F. Engel in his commentary on Grassmann's works.

The calculus of differential forms, which is based on Grassmann's outer multiplication, occupies a firm position in modern analysis. This calculus has enabled mathematicians to develop differential geometry in an elegant manner, as is particularly evident in the work of E. Cartan.

BIBLIOGRAPHY

Grassmann's writings were collected as *Mathematische und physikalische Werke*, F. Engel, ed., 3 vols. in 6 pts. (Leipzig, 1894–1911).

Biographical and historical works are E. T. Bell, *The Development of Mathematics* (New York–London, 1945), 198–206; M. J. Crowe, *A History of Vector Analysis* (Notre Dame, Ind.–London, 1967), ch. 3; A. E. Heath, "Hermann Grassmann. The Neglect of

His Work. The Geometric Analysis and Its Connection with Leibniz' Characteristic," in *Monist*, **27** (1917), 1–56; F. Engel, "H. Grassmann," in *Jahresberichte der Deutschen Mathematikervereinigung*, **18** (1909), 344–356; and "Grassmanns Leben," which is Grassmann's *Werke*, III, pt. 2 (1911); G. Sarton, "Grassmann—1844," in *Isis*, **35** (1944), 326–330; V. Schlegel, *Hermann Grassmann, sein Leben und seine Werke* (Leipzig, 1878); and *Die Grassmannsche Ausdehnungslehre. Ein Beitrag zur Geschichte der Mathematik in den letzten 50 Jahren* (Leipzig, 1896); and R. Sturm, E. Schröder, and L. Sohncke, "H. Grassmann. Sein Leben und seine mathematisch-physikalischen Arbeiten," in *Mathematische Annalen*, **14** (1879), 1–45, with bibliography of his works.

Works on geometrical analysis in general include H. Hankel, *Theorie der komplexen Zahlensysteme* (Leipzig, 1867); A. F. Möbius, *Der baryzentrische Kalkül* (Leipzig, 1827), also in his *Gesammelte Werke*, I (1885), 1–388; and "Über die Zusammensetzung gerader Linien und eine daraus entspringende neue Begründung des baryzentrischen Calculs," in *Journal für die reine und angewandte Mathematik*, **28** (1844), 1–9; and three articles collectively entitled "Systeme geometrischer Analyse," in *Encyklopädie der mathematischen Wissenschaften*, III, pt. 1 (1916–1923): by H. Rothe, 1277–1423; A. Lotze, 1425–1550; and C. Betsch, 1550–1595.

Works based on Grassmann's *Ausdehnungslehre* include F. Caspary, "Über die Erzeugung algebraischer Raumkurven durch veränderliche Figuren," in *Journal für die reine und angewandte Mathematik*, **100** (1887), 405–412; and "Sur une méthode générale de la géométrie, qui forme le lieu entre la géométrie synthétique et la géométrie analytique," in *Bulletin des sciences mathématiques et astronomiques*, 2nd ser., **13** (1889), 202–240; H. G. Forder, *Calculus of Extension* (Cambridge, 1941); H. Grassmann, Jr., "Über die Verwertung der Streckenrechnung in der Kreiseltheorie," in *Sitzungsberichte der Berliner mathematischen Gesellschaft*, **8** (1909), 100–114; and *Projektive Geometrie der Ebene, unter Verwendung der Punktrechnung dargestellt*, 2 vols. in 3 pts. (Leipzig, 1909–1923); F. Kraft, *Abriss des geometrischen Calculs nach H. G. Grassmann* (Leipzig, 1893); A. Lotze, *Die Grundgleichungen der Mechanik, neu entwickelt mit Grassmanns Punktrechnung* (Leipzig, 1922); and *Punkt- und Vektorenrechnung* (Berlin–Leipzig, 1929); R. Mehmke, *Vorlesungen über Punkt- und Vektorenrechnung*, I (Leipzig–Berlin, 1913); G. Peano, *Calcolo geometrico* (Turin, 1888); and V. Schlegel, *System der Raumlehre nach den Prinzipien der Grassmann'schen Ausdehnungslehre*, 2 vols. (Leipzig, 1872–1875). A bibliography was compiled by G. Peano: "Elenco bibliografico sull' 'Ausdehnungslehre' di H. Grassmann," in *Rivista di matematica*, **5** (1895), 179–182.

W. BURAU
C. J. SCRIBA

GREEN, GEORGE (*b*. Nottingham, England, July 1793 [baptized 14 July]; *d*. Sneinton, near Nottingham, 31 May 1841), *mathematics, natural philosophy*.

Although Green left school at an early age to work in his father's bakery, he had probably already developed an interest in mathematics that was fostered by Robert Goodacre, the leading private schoolmaster of Nottingham and author of a popular arithmetic textbook. Virtually self-taught, Green acquired his knowledge of mathematics through extensive reading. Many of the works he studied were available in Nottingham at the Bromley House Subscription Library, which he joined in 1823. By that time the family had moved to Sneinton, a suburb, where his father had established a successful milling business; Green used the top story of the mill as a study.

Green's most important work, *An Essay on the Application of Mathematical Analysis to the Theories of Electricity and Magnetism*, was published by subscription in March 1828. Apparently, almost all of the fifty-two subscribers were patrons and friends of Green's; a local baronet, Edward ffrench Bromhead of Thurlby, assisted Green later but was not an early promoter. Until other evidence is available, one can only conjecture that Green's supporters included some of the leading members of the Bromley House Library; the list of subscribers suggests only limited circulation outside Nottingham.

In the preface Green indicated that his "limited sources of information" preventing his giving a proper historical sketch of the mathematical theory of electricity, and indeed, he cites few sources. Among them are Cavendish's single-fluid theoretical study of electricity of 1771, two memoirs by Poisson of 1812 on surface electricity and three on magnetism (1821–1823), and contributions by Arago, Laplace, Fourier, Cauchy, and T. Young. The preface concludes with a request that the work be read with indulgence, in view of the limitations of the author's education.

The *Essay* begins with introductory observations emphasizing the central role of the potential function. Green coined the term "potential" to denote the results obtained by adding the masses of all the particles of a system, each divided by its distance from a given point. The general properties of the potential function are subsequently developed and applied to electricity and magnetism. The formula connecting surface and volume integrals, now known as Green's theorem, was introduced in the work, as was "Green's function," the concept now extensively used in the solution of partial differential equations.

Bromhead correctly surmised that Green's "publication must be a complete failure and dead born," but he was unaware that its significance would be appreciated later. Bromhead persuaded Green to matriculate at Caius College, Cambridge—a decision undoubtedly influenced by the death of Green's father in January 1829, for the subsequent sale of the family business afforded him the necessary financial backing. Before he could be admitted, however, he had to close the gaps in his classical education; but it is not known whether this was done at Nottingham, Cambridge, or both.

Green's second work, "The Laws of the Equilibrium of Fluids Analogous to the Electric Fluid," was read at the Cambridge Philosophical Society on 12 November 1832, so it is possible that he had already moved there. The paper was reportedly communicated by Bromhead, as was Green's next memoir, "Exterior and Interior Attractions of Ellipsoids of Variable Densities," read to the society the following May. Although not as significant as the *Essay*, both papers contain generalizations of his methods to cover an inverse nth power law of force and s dimensions.

Admitted to Caius College in October 1833, Green became scholar on 25 March 1834, after having submitted, again through Bromhead, a paper to the Royal Society of Edinburgh. In "Researches on the Vibration of Pendulums in Fluid Media," read in December 1833, Green obtained formulas, valid for small oscillations, for the effective increase of the mass of the pendulum due to the density of the surrounding fluid. In January 1837 Green received the B.A. as fourth wrangler—"to the disappointment of his friends." His attention may perhaps have been distracted by the demands of his own mathematical research.

"On the Motion of Waves in a Variable Canal of Small Depth and Width," read to the Cambridge Philosophical Society in May 1837, included the formula for determining the height of a wave,

$$\zeta \propto \beta^{-1/2}\gamma^{-1/4},$$

where β and γ represent the variable breadth and depth of the canal. A note to the paper, read in February 1839, commented on J. S. Russell's report (1837) to the British Association for the Advancement of Science.

Two papers followed in December 1837: "On the Reflexion and Refraction of Sound" and "On the Reflexion and Refraction of Light at the Common Surface of Two Non-Crystallized Media"; a supplement to the latter work, in which both papers were related, followed in May 1839. The first memoir simplified—and in one respect corrected—Poisson's memoir (1831); the second followed the work of Cauchy and Airy. In May 1839 Green read his second most important paper, "On the Propagation of Light in Crystallized Media," in which he used the vis viva theorem (conservation of mechanical energy) to simplify Cauchy's treatment.

This succession of works secured Green's election in October 1839 as Perse Fellow of Caius College, although he apparently made no significant contribution to academic life. It is reported that he set the problem papers for two college examinations but never lectured. In May 1840 he was in Nottingham; and it is doubtful that Green ever returned to Cambridge. His will, dated 28 July 1840, was probably written at Nottingham and confirms that he was in poor health, but no details of his illness are given. A codicil, added four months before his death, is his last known action. A locally published obituary, referring to Bromhead's support, concluded that "had his life been prolonged, he might have stood eminently high as a mathematician."

Only a few weeks before Green's death, William Thomson had been admitted to St. Peter's College, Cambridge. In a paper by Robert Murphy published in the *Transactions of the Cambridge Philosophical Society*, Thomson noticed a reference to Green's *Essay*, although Murphy did not mention any of his other works published in that journal. Thomson was unable to find a copy of the *Essay* until, just after receiving his degree in January 1845, his coach, William Hopkins, gave him three copies. Sixty years later Thomson recalled his excitement and that of Liouville and Sturm, to whom he showed the work in Paris in the summer of 1845. After returning to Cambridge, Thomson was responsible for republishing the work, with an introduction (1850–1854). Through Thomson, Maxwell, and others, the general mathematical theory of potential developed by an obscure, self-taught miller's son would lead to the mathematical theories of electricity underlying twentieth-century industry.

BIBLIOGRAPHY

I. ORIGINAL WORKS. *An Essay on the Application of Mathematical Analysis to the Theories of Electricity and Magnetism* (Nottingham, 1828) is extremely rare;

the total number of copies is estimated to have been less than 100. There are two facsimile reprints (Berlin, 1889; Göteborg, 1958) and a German trans. in *Ostwalds Klassiker der exakten Wissenschaften*, no. 61 (Leipzig, 1895). Thomson republished the *Essay*, with a brief biography of Green, a list of his writings, and a bibliography of eight "independent investigations on the subject of Green's *Essay*," in *Journal für die reine und angewandte Mathematik*, **39** (1850), 73–89; **44** (1852), 356–374; and **47** (1854), 161–221.

All but one of Green's subsequent writings were published in *Transactions of the Cambridge Philosophical Society*, **5–7** (1835–1842). Together with the *Essay*, they were edited for Caius College by N. M. Ferrers (London–Cambridge, 1871; facs. repr., Paris, 1903). The preface includes a brief biography that is chiefly a sketch of the contents of the papers; a few notes on particular points in them are collected in an appendix.

II. SECONDARY LITERATURE. The above account is based largely on H. G. Green, "A Biography of George Green," in *Studies and Essays . . . Offered in Homage to George Sarton* (New York, 1946), 545–594, which includes a complete bibliography. Adam W. Thomas, *A History of Nottingham High School 1513–1953* (Nottingham, 1958), is useful for Green's educational background; and A. R. Hall, *The Cambridge Philosophical Society: A History, 1819–1969* (Cambridge, 1969), provides details of the Society in whose *Transactions* most of Green's work was published. H. G. Green (see above) reveals the importance of the Bromley House Library in providing facilities for the young mathematician; and John Russell, *A History of the Nottingham Subscription Library* (Nottingham, 1916), ch. 5, indicates its keen interest in science in the 1830's. See also J. E. G. Farina, "The Work and Significance of George Green, the Miller-Mathematician, 1793–1841," in *Bulletin of the Institute of Mathematics and Applications*, **12**, no. 4 (1976), 98–105.

J. S. Russell's report was published as "Report of the Committee on Waves," in *Report of the British Association for the Advancement of Science*, **7** (1837), 417–496—see esp. 425 and 494; Poisson's memoir was "Sur le mouvement de deux fluides élastiques superposés," in *Mémoires de l'Académie des sciences*, 2nd ser., **10** (1831), 317–404.

P. J. WALLIS

HARDY, WILLIAM BATE (*b.* Erdington, England, 6 April 1864; *d.* Cambridge, England, 23 January 1934), *biology, colloid chemistry.*

After being educated at Framlingham College, Hardy entered Gonville and Caius College, Cambridge, in 1884, and was awarded a first class in the natural sciences tripos (zoology) in 1888. His association with his Cambridge college was a close one throughout his life; he was elected a fellow in 1892 and was a tutor from 1900 to 1918. In 1913 Hardy became university lecturer in physiology. He was elected to the Royal Society in 1902 and received its Royal Medal in 1926.

Hardy's exceptional ability as a scientific adviser to the government became evident after 1915 when, as biological secretary of the Royal Society, he organized a food committee that dealt with nutritional problems during World War I. He then became chairman of the Food Investigation Board (1917–1928) and director of food investigation (1917–1934) in the Department of Scientific and Industrial Research. Hardy also was chairman of the Advisory Committee on Fisheries Development (1919–1931), and was superintendent of the Low Temperature Research Station at Cambridge from 1922 to 1934. He was knighted in 1925, and at the time of his death was president of the British Association for the Advancement of Science.

Hardy began his scientific career as a histologist, and during the 1890's he performed a miscellaneous series of investigations in this field. By 1899, however, he had come to question the validity of the fixing and staining techniques used to reveal the details of cell structure; and he criticized these techniques in an important paper that marks his transition from a practicing biologist to a colloid chemist. During the succeeding twelve years he studied the colloidal properties of proteins, beginning with his discovery (in 1899) of the amphoteric behavior of albumin particles in an electric field. Although he initially doubted that proteins reacted stoichiometrically with acids and bases, by 1910 the weight of the evidence adduced by others led him to accept this view. In 1912 Hardy turned his attention to the properties of molecular films at air-liquid interfaces, and then to the study of the friction between surfaces and the nature of lubrication. As was the case with several of his contemporaries, notably Jacques Loeb, Hardy's scientific development exemplifies a transition from purely biological work to physicochemical investigations, in the search for a "physical basis of life." Until the end of his life, he continued to emphasize his biological orientation, especially in *To Remind* (1934).

Hardy's admirable personal qualities won him many friends and made his influence on British science a significant one.

BIBLIOGRAPHY

I. ORIGINAL WORKS. Hardy published about 60 scientific articles and lectures during the period 1891–1934.

The most important were "On the Structure of Cell Protoplasm," in *Journal of Physiology*, **24** (1899), 158–210; "On the Coagulation of Proteid by Electricity," *ibid.*, 288–304; "On Globulins (Croonian Lecture)," in *Proceedings of the Royal Society*, **79** (1907), 413–426; "The General Theory of Colloidal Solutions," *ibid.*, **A86** (1912), 601–610; and *To Remind: A Biological Essay* (Baltimore–London, 1934). *Collected Scientific Papers of Sir William Bate Hardy*, E. K. Rideal, ed. (Cambridge, 1936), was published under the auspices of the Colloid Committee of the Faraday Society.

II. SECONDARY LITERATURE. Obituary articles include I. Bircumshaw, in *Proceedings of the Physical Society*, **46** (1934), 902; F. G. Hopkins and F. E. Smith, in *Obituary Notices of Fellows of the Royal Society of London*, **1** (1932–1935), 327–333; F. G. Hopkins, in *Nature*, **133** (1934), 281–283; and T. M., in *Chemistry and Industry*, **12** (1934), 133–134. A biographical notice was written by A. V. Hill for *Dictionary of National Biography (1931–1940)*, 397–398. On the occasion of the centenary of Hardy's birth, a collection of tributes was edited by E. C. Bate-Smith, *Sir William Bate Hardy: Biologist, Physicist, and Food Scientist* (Cambridge, 1964). It contains an article by E. K. Rideal that also appeared in *Transactions of the Faraday Society*, **60** (1964), 1681–1687, and in *Proceedings of the Royal Institution of Great Britain*, **40** (1964), 178–185.

JOSEPH S. FRUTON

HERACLIDES PONTICUS (*b.* Heraclea Pontica [now Ereğli, Turkey], *ca.* 390 B.C.; *d.* Heraclea Pontica, after 339 B.C.), *astronomy, philosophy.*

Heraclides, son of Euthyphron, came from a noble and wealthy family of Heraclea Pontica, a Greek city on the south coast of the Black Sea. He traced his descent from one of the original founders of Heraclea. His birthdate can be inferred approximately from his relationship to various members of the Academy and from his statement that the destruction of the city of Helice in Achaea by an earthquake (373 B.C.) took place in his lifetime.[1] He came to Plato's Academy in Athens, some time before 360 B.C., if we can believe the story that Plato left the Academy in the charge of Heraclides when he went to Sicily.[2] Although counted as one of Plato's pupils (Heraclides himself said that Plato sent him to Colophon to collect the poems of Antimachus), he was apparently more closely associated with Speusippus, Plato's successor as head of the Academy. He also attended Aristotle's lectures.

Upon the death of Speusippus (339 B.C.), Heraclides was one of the candidates to succeed him, but Xenocrates won by a few votes, whereupon Heraclides returned to his native city, where he died some time later. The attempts to establish a *terminus post quem* for his death from his alleged mention of the cult of Sarapis[3] or of his pupil Dionysius[4] are unconvincing. Two different stories are connected with his death. According to one account, Heraclea was afflicted by a famine and sent envoys to the Delphic oracle to ask what to do. Heraclides bribed the ambassadors and the Pythia to pretend that the god had replied that the city would be relieved if Heraclides were honored with a gold crown while alive and a hero's cult after death. During the ceremony of bestowing the crown in the theater, Heraclides died of a stroke (or fell and hit his head on a step according to another version). The other account is even more implausible: Heraclides raised a tame snake and persuaded a friend to substitute the snake for his body when he died, so that people would think that he had become a god. Both stories may have been invented to match Heraclides' well-attested penchant for tall tales and his pretensions. We are told that he dressed richly, was very fat and stately, and was nicknamed Pompikos by the Athenians ("stately, magnificent") instead of Pontikos.

Heraclides' many books were greatly admired in antiquity both for style and content. Not a single work has survived, and of most we know only the title. Many of them were in the form of dialogues, as was common practice in the Academy. The subjects, which were typical of a fourth-century "philosopher," included ethics, literature, rhetoric, history, politics, and music. Although some of the works were very influential (for example, Heraclides' contribution to the legend of Pythagoras), they do not concern us here. A number of his works belong to a group called φυσικά, which is best translated "on the nature of things." These works too cannot be considered scientific but belong to the kind of prescientific speculation that characterized most early Greek philosophy. The following are some examples: each of the stars is a world of its own; the moon is earth surrounded by mist; and a comet is a high cloud reflecting light.[5] Heraclides' work "On Diseases" was more concerned with thaumaturgy (for example, a woman who lay apparently dead for seven days and was restored to life) than with medicine, to judge from the surviving fragments.[6]

In modern times Heraclides is famous chiefly for an astronomical theory that has been attributed to him, namely that the orbits of Venus and Mercury

HERACLIDES PONTICUS

have the sun as their center, while the sun in turn moves around the earth. Although there is no good reason to believe that Heraclides proposed such a theory, the attribution has become so much the received opinion that the theory commonly goes under the name of "the system of Heraclides Ponticus," and Heraclides is variously considered a precursor of Tycho Brahe, Aristarchus, or Copernicus. It is therefore appropriate to give some account, not only of the ancient evidence on the subject, but also of the numerous modern misunderstandings of that evidence.

The theory was indeed held in antiquity, but the contexts in which it occurs show that it arose at a much later stage of Greek astronomy, for reasons which were not operative at the time of Heraclides. We must start from Ptolemy's discussion of the order of the planets in *Almagest* IX, 1. He says there[7] that while all agree that all the planets lie between the sun and the fixed stars and that Mars, Jupiter, and Saturn lie (in ascending order) beyond the sun, there is disagreement about the position of Venus and Mercury. The "older astronomers" placed them below the sphere of the sun, while some of the later astronomers put them above the sphere of the sun. Ptolemy's account is fully confirmed by our fragmentary sources for pre-Ptolemaic astronomy.[8] It seems likely that the hypothesis that the orbits of Venus and Mercury encircle the sun (and that thus the two planets are sometimes above and sometimes below the sun) was introduced as a third choice. Moreover it seems highly probable that it was introduced after the development of the epicycle theory, according to which the mean motions of the sun, Venus, and Mercury are identical, that is, the centers of their epicycles lie on the same straight line. This is the form in which it is found in our most explicit source, Theon of Smyrna,[9] who says that according to this theory the epicycles of the sun, Venus, and Mercury have a common center. Thus the theory can hardly predate 200 B.C. Of the three sources who mention the theory only one, Macrobius, attributes it to a specific authority, namely "the Egyptians."

The only astronomical doctrine of Heraclides for which there is solid evidence is the rotation of the earth on its axis. This is attested by a number of sources,[10] from which it is abundantly clear that Heraclides proposed that the earth lies in the center of the universe and turns on its axis once a day. This is a simple variation of the common belief, canonized in Ptolemaic astronomy, that the earth is central and stationary, while the whole heavens revolve once a day. In *Almagest* I, 7, Ptolemy argues against the rotation of the earth (on purely physical grounds).[11] Although he mentions no names, it appears that the doctrine was fairly common. Heraclides is the earliest philosopher who is known beyond question to have held this opinion.[12] Unfortunately certain ambiguous expressions in the ancient descriptions of Heraclides' doctrine have misled some modern scholars into thinking that he held that the earth moves in a circular orbit (see below on Schiaparelli and van der Waerden). Examination of all the evidence shows that this is wrong.[13]

The only evidence concerning Heraclides' opinion on Venus is a passage of the commentary on Plato's *Timaeus* by Calcidius (fifth century A.D.), which I translate as follows:

> Finally Heraclides Ponticus, when he drew the circle of Venus, and also [the circle] of the sun, and assigned a single center to both circles, showed that Venus is sometimes above, sometimes below the sun. For he says that the sun and moon and Venus and all the planets, wherever each of them is, are [each] indicated by a single line drawn from the center of the earth through the center of the heavenly body. So there will be one line drawn from the center of the earth indicating the sun, and two other lines drawn to left and right of it, fifty degrees from the sun and a hundred degrees from each other. The eastern line indicates Venus when it is at greatest distance from the sun toward the east, and therefore has the name evening star ("Hesperus") because it appears in the east [sic] in the evening after sunset. The western line [indicates Venus] when it is at greatest distance from the sun toward the west and therefore is called the morning star ("Lucifer"). For it is obvious that it is called evening star when it is seen in the east [sic] following sunset, and morning star when it sets before the sun and rises again before the sun when the night is almost over.[14]

Seizing on the remark that Venus is "sometimes above, sometimes below the sun," modern scholars have concluded that Heraclides believed that the orbit of Venus (and, by analogy, that of Mercury) encircles the sun. The first to draw this conclusion was T. H. Martin in 1849.[15] Schiaparelli not only accepted Martin's conclusion[16] but even conjectured that Heraclides proposed the Tychonic theory, in which the orbits of all the planets encircle the sun, which in turn revolves about the central earth.[17] Since there is not a scrap of evidence that anyone in antiquity proposed the Tychonic theory, discussion of the point is idle. Schiaparelli further suggested that Heraclides anticipated Aristarchus in proposing a heliocentric system as at least a theoretical possibility.[18] The basis for this is the following passage in Simplicius, quoting Geminus (first century A.D.): "So someone comes forward and says [Heraclides Ponticus] that if the earth moves in a certain way and the sun stands still, the apparent anomaly of the sun can be represented."[19] The words "Heraclides Ponticus" are an intrusion into the syntax and sense of the sentence, and are obviously interpolated by a reader who wanted to explain the "someone," as was remarked by Tannery.[20] We can be sure that the interpolator, no doubt misled by the doxographical tradition that Heraclides assumed the axial rotation of the earth, was in error.

Yet another theory was attributed to Heraclides by van der Waerden,[21] according to which the sun, Venus, and the earth (in ascending order) all revolve around a common center. This is based largely on misinterpretation of diagrams in the manuscripts of Calcidius, explaining the maximum elongations of Venus from the sun according to the epicycle theory. In any case, it is flatly contradicted by the unanimous testimony that Heraclides put the earth in the center of the universe.

Although Calcidius was a bungler and certainly did not read Heraclides' work,[22] what he says in the passage translated above makes reasonable sense. It is an explanation of the fact that Venus appears as both a morning and evening star:[23] the sun and Venus both move on circles with the earth as "a single center." If one draws lines from that center to the sun and Venus, one finds that the line earth – Venus is sometimes to the left (to the east) of the line earth – Sun, and sometimes to the right (to the west). The only phrase inconsistent with this is the statement that Venus is "sometimes above, sometimes below the sun." If we interpret "above" and "below" to mean, not "farther from and nearer to the earth," but "to the west" and "to the east" of the sun, the inconsistency is removed. This was suggested by G. Evans[24] and was confirmed by O. Neugebauer, who pointed out that Calcidius' "superior/inferior" is simply a translation of the Greek *ἀνώτερον/κατώτερον*, which are found in works on "spherics" with exactly the meaning required here.[25] Thus the whole basis for attributing to Heraclides the theory that Venus revolves around the sun vanishes, and so does the influence on the development of ancient astronomy, which has often been attributed to him in modern times. Heraclides' only claim to a place in the history of astronomy is his assertion of the axial rotation of the earth, which won him mention as one of the ancient authorities for this in Copernicus' *De revolutionibus*.[26]

NOTES

1. Strabo, *Geography*, VIII, 384 (Wehrli, *Herakleides Pontikos*, fr. 46a).
2. Suidas, s.v. (Wehrli, fr. 2). If true, this must refer to Plato's third Sicilian journey (probably in 360 B.C.). The whole of "Suidas'" account, however, inspires little trust.
3. Plutarch, *Isis and Osiris*, 361e (Wehrli, fr. 139). Even if one emends the MS reading "Heraclitus" to "Heraclides," Plutarch may be referring to another "Heraclides Ponticus," a grammarian of a later period. Furthermore, the date of the foundation of the cult of Sarapis is greatly disputed.
4. Wehrli, fr. 12, with commentary on 62.
5. *Ibid.*, frs. 113a – b, 114a – c, 116.
6. *Ibid.*, frs. 76 – 89.
7. Heiberg, ed., *Syntaxis mathematica*, II, 206 – 207.
8. For details, see O. Neugebauer, *A History of Ancient Mathematical Astronomy*, II, 647 – 650, 690 – 693.
9. Theon of Smyrna, *Expositio rerum mathematicarum . . .*, Hiller, ed., 186 – 187. Of the other two sources, Macrobius, *Commentarii in Somnium Scipionis*, I, 19, 5 – 6, gives the same version in cruder language; while Martianus Capella, VIII, 857, simply says that Venus and Mercury have "the sun" as the center of their circles.
10. Wehrli, frs. 104 – 108.
11. Heiberg, ed., I, 24 – 26.
12. It is also ascribed to the obscure figure of "the Pythagorean Ecphantus" (Wehrli, fr. 104). In the 5th century B.C. Philolaus of Crotona had constructed a theory in which the earth not only rotates but moves about "the central fire"; this seems, however, to have been inspired more by mystical speculation than by astronomical considerations (see Kurt von Fritze, *Dictionary of Scientific Biography*, X, 589 – 591).
13. Best demonstrated by A. Pannekoek, "The Astronomical System of Herakleides," 375 – 379.
14. Calcidius, *Timaeus a Calcidio . . .*, CX, Waszink, ed., 157. The crucial first sentence is "Denique Heraclides Ponticus, cum circulum Luciferi describeret, item solis, et unum punctum atque unam medietatem duobus daret circulis, demonstrauit ut interdum Lucifer superior, interdum inferior sole fiat."
15. In his edition of Theon of Smyrna, 120, 426 – 428. Elaborated in *Mémoires de l'Académie des inscriptions et belles-lettres*, **30**, pt. 2 (1883), 21 – 43.
16. Schiaparelli, "I precursori di Copernico," 401 – 408.
17. Schiaparelli, "Origine del sistema eliocentrico," esp. 165 – 166.
18. *Ibid.*, 163 – 164.
19. Wehrli, fr. 110. On the date of Geminus, which is often wrongly stated to be the first century B.C., see Neugebauer, *History of Ancient Mathematical Astronomy*, II, 579 – 581.
20. *Mémoires scientifiques*, IX, 255 – 258. Elaborated by Heath, *Aristarchus*, 275 – 283.
21. "Die Astronomie des Herakleides von Pontos"; repeated in "Die Astronomie der Pythagoreer," 62 – 73.
22. Since much of what Calcidius says about astronomy is almost identical to passages in Theon of Smyrna, who is avowedly drawing on a certain Adrastus, it is likely that Adrastus is Calcidius' source here as elsewhere.
23. The discovery that the morning star and evening star are the same body was attributed to Pythagoras, but Heraclides may still have needed to explain it in the 4th century B.C.

24. Evans, "The Astronomy of Heracleides Ponticus," esp. 110–111.
25. Neugebauer, "On the Allegedly Heliocentric Theory of Venus," referring to Theodosius (1st century B.C.).
26. Zeller, ed., IV, 14.

BIBLIOGRAPHY

The chief source for Heraclides' life and works is his biography in Diogenes Laërtius, *Lives of the Philosophers*, V, 86–91 (Leipzig, 1884), 246–248. Like all of Diogenes' biographies, this is a mixture of puerile anecdotes and sound information derived from excellent authorities; it includes a list of the titles of Heraclides' works. The fragments relating to the life and works have been collected in Fritz Wehrli, *Herakleides Pontikos*, 2nd ed. (Basel–Stuttgart, 1969), which is *Die Schule des Aristoteles*, vol. VII. Wehrli also provides a useful commentary (somewhat muddled on astronomical matters), and should be consulted for further bibliography (esp. 57).

The astronomical theory, wrongly attributed to Heraclides, is in Theon of Smyrna, *Expositio rerum mathematicarum ad legendum Platonem utilium*, E. Hiller, ed. (Leipzig, 1878), 186–187; in Macrobius, *Commentarii in Somnium Scipionis*, I, 19, 5–6, J. Willis, ed. (Leipzig, 1973), 74; and in Martianus Capella, *De nuptiis philologiae et Mercurii*, VIII, 857, A. Dick, ed. (Leipzig, 1925), 450–451. Ptolemy's discussion of the order of the planets is in *Syntaxis mathematica*, J. L. Heiberg, ed., 2 vols. (Leipzig, 1898–1903), which is *Claudii Ptolemaei opera quae exstant omnia*, I, IX, 1, Vol. 2, 24–26. An extensive discussion of the ancient evidence on this topic is in O. Neugebauer, *A History of Ancient Mathematical Astronomy*, Studies in the History of Mathematics and the Physical Sciences, no. 1, 3 vols. (New York, 1975), II, 694–696.

The passage of Calcidius referring to Heraclides is *Timaeus a Calcidio translatus commentarioque instructus*, J. H. Waszink, ed., which is *Plato Latinus*, IV (London–Leiden, 1962), Commentarius CX, p. 157. It was first interpreted as making Venus' orbit heliocentric by T. H. Martin, in his edition of part of Theon of Smyrna, *Theonis Smyrnaei liber de astronomia* (Paris, 1849; repr. Groningen, 1971), 120–121, 426–428. See also T. H. Martin, "Mémoire sur l'histoire des hypothèses astronomiques chez les Grecs et les Romains," in *Mémoires de l'Académie des inscriptions et belles-lettres*, **30**, pt. 2 (1883), 1–43; and Giovanni Schiaparelli, "I precursori di Copernico nell'antichità," in his *Scritti sulla storia della astronomia antica*, I (Bologna, 1925), 363–458, originally published in *Memorie dell'Istituto lombardo di scienze e lettere*, **12** (1873). Schiaparelli developed his further hypotheses in "Origine del sistema planetario eliocentrico presso i Greci," in *Scritti . . .*, II (Bologna, 1926), 115–177, originally published in *Memorie dell'Istituto lombardo di scienze e lettere*, Classe di scienze matematiche e naturali, 3rd ser., **18** (1896–1900), 61–100. This was refuted by Paul Tannery, "Sur Héraclide du Pont," in *Revue des études grecques*, **12** (1899), 305–311, repr. in his *Mémoires scientifiques*, IX, J. L. Heiberg, ed. (Toulouse–Paris, 1929), 253–259.

A good account of the ancient evidence and modern discussions is T. L. Heath, *Aristarchus of Samos* (Oxford, 1913; repr., Oxford, 1959), 249–283. B. L. van der Waerden proposed his interpretation in "Die Astronomie des Heraklides von Pontos," in *Sitzungsberichte der Sächsischen Akademie der Wissenschaften zu Leipzig*, Math.-naturwiss. Kl., **96** (1944), 47–56, and repeated it (within a greatly expanded reconstruction of Pythagorean astronomy) in "Die Astronomie der Pythagoreer," *K. Verhandelingen der Nederlandse akademie van wetenschappen*, Afd. Natuurkunde, ser. A, **20**, no. 1 (1951). He was refuted by A. Pannekoek, "The Astronomical System of Herakleides," *ibid.*, ser. B, **55**, no. 4 (1952), 373–381. Godfrey Evans, "The Astronomy of Heracleides Ponticus," in *Classical Quarterly*, n.s. **20** (1970), 102–111, although occasionally confused and ignoring much of the modern literature, was the first to give a correct explanation of the Calcidius passage. A similar explanation, with the crucial terminological evidence, was given independently by O. Neugebauer, "On the Allegedly Heliocentric Theory of Venus by Heraclides Ponticus," in *American Journal of Philology*, **93** (1972), 600–601. See also his *A History of Ancient Mathematical Astronomy*, II (New York, 1975), 694–696.

Copernicus refers to Heraclides in *De revolutionibus orbium caelestium*, F. and C. Zeller, eds. (Munich, 1949), which is *Nikolaus Kopernikus Gesamtausgabe*, vol. II, in the Dedicatory Epistle, p. 5, and IV, p. 14.

G. J. TOOMER

HERMBSTAEDT, SIGISMUND FRIEDRICH (*b.* Erfurt, Germany, 14 April 1760; *d.* Berlin, Germany, 22 October 1833), *chemistry, technology.*

After attending St. Michaelis School and the Erfurt Gymnasium, Hermbstaedt studied medicine. Through contact with Wilhelm Bernhard Trommsdorff, however, he became so enthusiastic about chemistry that he decided to specialize in that field instead. He completed his training in chemistry under Trommsdorff's son Johann Bartholomäus and earned a doctorate in chemistry. To increase his knowledge of the subject, Hermbstaedt attended J. C. Wiegleb's school in Langensalza and also worked in Wiegleb's apothecary shop. He then became an assistant in the Ratsapotheke at Hamburg, owned by Johann Albert Reimarus; and in 1784–1785 he managed the Schwan-Apotheke in Berlin, which belonged to Valentin Rose the elder. During this period Hermbstaedt became friendly with the chemist Klaproth, whose lectures he attended at the Medical-Surgical College. At the same time, he was assisted in his career by Christian Gottlieb Selle, personal physician to the king of Prussia, and from privy councillor Christian Andreas Cothenius, the head of the Prussian medical administration.

When Valentin Rose the younger took over his father's apothecary in 1785, Hermbstaedt made an extended trip for professional purposes; it took him to the Harz Mountains, Thuringia, and the Saxon Erzgebirge. Especially important was his visit to Beckmann in Göttingen, who "awakened [his] taste for . . . technology and public finance and administration."

After returning to Berlin, Hermbstaedt founded *Magazin für die Technologie* (1788). Two years earlier he had completed a volume of *Physikalisch-chemische Versuche und Beobachtungen*, which was followed by a second volume in 1789. He also collaborated on the *Bibliothek der heuesten physikalischen, chemischen, metallurgischen . . . und pharmaceutischen Literatur*, a bibliographical work of which he compiled four volumes. In addition he reviewed books and translated foreign works.

In 1787 Hermbstaedt began to lecture privately, and he became an adviser to the Wegley chemical factories in Berlin. The following year he married Magdalene Rose, the daughter of his former employer, and founded a chemical boarding school for young men. Unfortunately, the firm of Wegley went out of business, and Hermbstaedt had to seek work elsewhere. He submitted an application to the king for the post of *Dozent*. To obtain it, however, he was required to take an examination and was "found not qualified."

Shortly thereafter it was discovered that the court apothecary had been embezzling funds. He was dismissed, and the king ordered that a successor be found. His advisers recalled Hermbstaedt, who was invited to take an examination again. This time he was "found very capable" and was immediately appointed court apothecary (1790). In addition, he was given the title of professor (1791). Hermbstaedt took up his duties with great zeal. To facilitate his lectures, he wrote the three-volume textbook *Systematischer Grundriss der allgemeinen Experimentalchemie* (1791), which was widely read. In this work he rejected Stahl's phlogiston theory; indeed, he was the first German chemist to do so. The following year he published an annotated German translation of Lavoisier's *Traité élémentaire de chimie*.

Hermbstaedt soon acquired an extraordinary reputation and was included among Germany's leading scholars. The number of his publications quickly reached one hundred. He developed an unusual ability to solve organizational problems and, accordingly, was named to the Obercollegium Sanitatis (1794) with the rank of *Obersanitätsrat*. In the following year he obtained a professorship in Berlin at the Pépinière, a school for army physicians. He was further honored by being named apothecary to the General Staff and charged with organizing the Prussian army's entire apothecary service. His skill in accomplishing this task can be reconstructed from the papers in the state archives in Merseburg, which include detailed questionnaires on every apothecary active at the time in Prussia.

In 1797 Hermbstaedt became assessor in the salt administration and, in 1808, full member of the Prussian Academy of Sciences, as well as member of the Technical Industrial and Trade Committee. Eager to encourage higher professional standards among apothecaries, he collaborated on the first Prussian pharmacopoeia (1799) and on the *Revidierte Apothekerordnung* (1801). As the burden of his manifold activities became too great, Hermbstaedt retired as court apothecary and then gave up his boarding school. The French conquest of Prussia terminated his duties as chief army apothecary.

Hermbstaedt's greatest achievement lay in the field of technology. Upon the creation of the University of Berlin in 1810, the founding committee voted unanimously to appoint him professor of technological chemistry. At his own request he lectured for as many as twenty-two to twenty-four hours a week. In his first semester he gave a general course entitled "Summary of Technology" and courses on "Technische, ökonomische und medizinische Warenkunde" (six hours) and on "Von den ökonomischen Gewerben." In the summer semester of 1811 he lectured on general technology according to Beckmann and on chemistry as related to dyeing.

As a member the Technical Industrial and Trade Commission, Hermbstaedt frequently met industrialists, including many from outside Prussia. Through the publication of his many textbooks, he performed a valuable service for Prussian industry. His *Grundriss der Technologie* (1814), which ran to almost 800 pages, was widely consulted by merchants, factory owners, and officials, who, confronted by the industrialization of Prussia, found themselves obliged to procure precise technical information. The court established in 1815 in Berlin to adjudicate disputes among factory owners and the government frequently drew upon Hermbstaedt's writings in reaching its decisions.

Hermbstaedt's activities were considerably reduced in 1813, when many of his students volunteered in the war of liberation against Napoleon. After the war the Prussian educational system was considerably expanded. Hermbstaedt was particularly eager to broaden the outlook of his students by taking them to visit factories. His attempt to institute practical sessions for students failed for lack of sufficient funds.

Hermbstaedt was able, however, to help in the realization of one important project in this area. In 1821 a group of officials on the state board of trade set up a Society for the Encouragement of Industry in Prussia; it succeeded in lifting the economy from the nadir it had reached around 1815. Hermbstaedt headed the organization's physics and chemistry group until his death.

BIBLIOGRAPHY

I. ORIGINAL WORKS. A fairly comprehensive list of Hermbstaedt's writings may be compiled from the Royal Society *Catalogue of Scientific Papers*, III, 314–317; and VII, 961, which lists over 90 memoirs; Poggendorff, I, 1082–1083; and Harnack (see below).

Hermbstaedt's early monographs include *Untersuchung des Milchzuckers* (Berlin, 1782); *Bernsteinsäure* (Berlin, 1784); *Handbuch der pharmazeutischen Praxis* ([Berlin], 1784); *Physikalisch-chemische Versuche und Beobachtungen*, 2 vols. (Berlin, 1786–1789); *Bibliothek der neuesten physikalisch-chemischen-metallurgischen-technologischen- und pharmaceutischen Literatur*, 4 vols. (Berlin, 1787–1795); *Systematischer Grundriss der allgemeinen Experimentalchemie zum Gebrauch seiner Vorlesungen*, 3 vols. (Berlin, 1791; 3rd ed., 5 vols., 1812–1826); *Grundriss der Experimentalpharmacie*, 2 vols. (Berlin, 1792–1793), 2nd ed., *Grundriss der theoretischen und experimentellen Pharmacie*, 2 vols. (Leipzig, 1806); *Katechismus der Apothekerkunst* (Berlin, 1792); *Rede über den Zweck der Chemie* (Berlin, 1792); and *Kurze Anleitung zur chemischen Zergliederung der Vegetabilien nach physikalisch-chemischen Grundsätzen* (1795–1799; 2nd ed., 1807), also in French trans. by Desertine.

Monographs published after 1800 include *Grundriss der Färbekunst* (Berlin, 1802); *Allgemeine Grundsätze der Bleichkunst* (Berlin, 1804); *Chemisch-technologische Grundsätze der Lohgerberei*, 2 vols. (Berlin, 1805–1807); *Grundriss der experimentellen Kameralchemie* (Berlin, 1808); *Die Wissenschaft des Seifesiedens* (Berlin, 1808); *Anleitung zur praktisch-ökonomischen Fabrikation des Zuckers aus den Runkelrüben* (Berlin, 1811); *Anleitung zur Fabrikation des Syrups und des Zuckers aus Stärke* (Berlin, 1814); *Chemische Grundsätze der Kunst Bier zu brauen* (Berlin, 1814); *Grundlinien der theoretischen und experimentellen Chemie* (Berlin, 1814); *Grundriss der Technologie* (Berlin, 1814; 2nd ed., 1829); *Anleitung zur Kunst wollene, seidene, baumwollene und leinerne Zeuge zu bleichen* (Berlin, 1815); *Gemeinnütziger Rathgeber für den Bürger und Landmann*, 3 vols. (Berlin, 1816–1819); *Chemische Grundsätze der Kunst Branntwein zu brennen* (Berlin, 1817); and *Chemische Grundsätze der Destillierkunst und Liqueurfabrikation* (Berlin, 1819).

His translations and editions include *Lavoisiers System der antiphlogistischen Chemie*, 2 vols. (Berlin, 1792); *Scheeles Sämtliche Werke*, 2 vols. (Berlin, 1793); and *Chaptals Chemie*, 2 vols. (Berlin, 1808).

Among the many periodicals to which Hermbstaedt contributed are *Journal für Lederfabrikanten und Gerber* (1802–1803); *Magazin für Färber, Zeugdrucker und Bleicher* (1802–1819); *Archiv für Agrikulturchemie* (1803–1817); *Bulletin des Neuesten und Merkwürdigsten aus der Naturwissenschaften* (1809–1813); and *Museum des Neuesten und Wissenswürdigsten aus dem Gebiet der Naturwissenschaften* (1814–1818).

His scientific contributions also appeared in *Berlinisches Jahrbuch der Pharmacie; Chemische Annalen; Journal der Pharmacie für Aerzte und Apotheker; Journal der Physik, Neueste Entdeckungen in der Chemie*; and *Physikalisch-chemische Versuche und Beobachtungen*. See esp. "Über die Erzeugung der Essigsäure," in *Abhandlungen der Preussischen Akademie der Wissenschaften* (1804–1811), 11–20; and "Über diabetischen Urin," *ibid.* (1814–1815), 53–62.

II. SECONDARY LITERATURE. See G. E. Dann, "Klaproths Wandlung zum Antiphlogistiker," in *Wissenschaftliche Zeitschrift der Karl-Marx-Universität, Leipzig*, **5** (1955–1956), 49–53; and *Martin Heinrich Klaproth* (Berlin, 1958), *passim*; A. Harnack, *Geschichte der Königlich Preussischen Akademie der Wissenschaften zu Berlin*, III (Berlin, 1900), 126–127, with bibliography of twenty-five of his papers published by the Prussian Academy of Sciences; *Haude- und Spenersche Zeitung* (24 Oct. 1833); *Intelligenzblätter der Allgemeinen Literatur-zeitung*, no. 98 (Dec. 1833), 794–795; I. Mieck, in *Technik-geschichte*, **32** (1965), 325–382, with bibliography and portrait; and G. E. Dann, in *Neue deutsche Biographie*, VIII (Berlin, 1969), 666–667.

GÜNTHER KERSTEIN

HIPPARCHUS (*b.* Nicaea, Bithynia [now Iznik, Turkey], first quarter of second century B.C.; *d.* Rhodes [?], after 127 B.C.), *astronomy, mathematics, geography.*

The only certain biographical datum concerning Hipparchus is his birthplace, Nicaea, in northwestern Asia Minor. This is attested by several ancient sources[1] and by Nicaean coins from the second and third centuries of the Christian era that depict a seated man contemplating a globe, with the legend ΙΠΠΑΡΧΟΣ. His scientific activity is dated by a number of his astronomical observations quoted in Ptolemy's *Almagest*. The earliest observation indubitably made by Hipparchus himself is of the autumnal equinox of 26/27 September 147 B.C.[2] The latest is of a lunar position on 7 July 127 B.C.[3] Ptolemy reports a series of observations of autumnal and vernal equinoxes taken from Hipparchus, ranging from 162 to 128 B.C., but it is not clear whether the earliest in the series had been made by Hipparchus himself or were taken from others; Ptolemy says only that "they seemed to Hipparchus to have been accurately observed." We can say, then, that Hipparchus' activity extended over the third quarter of the second century B.C., and may have begun somewhat earlier. This accords well with the calculations of H. C. F. C. Schjellerup and H. Vogt concerning the epoch of the stellar positions in Hipparchus' commentary on Aratus.[4]

It is probable that Hipparchus spent the whole of his later career at Rhodes: observations by him ranging from 141 to 127 B.C. are specifically attributed to Rhodes by Ptolemy. In Ptolemy's *Phases of the Fixed Stars*, however, it is stated that the observations taken from Hipparchus in that book were made in Bithynia (presumably Nicaea). We may infer that Hipparchus began his scientific career in Bithynia and moved to Rhodes some time before 141 B.C. The statement found in some modern accounts that he also worked in Alexandria is based on a misunderstanding of passages in the *Almagest* referring to observations made at Alexandria and used by or communicated to Hipparchus.[5] The only other biographical information we have is an utterly untrustworthy anecdote that Hipparchus caused amazement by sitting in the theater wearing a cloak, because he had predicted a storm.[6]

Hipparchus is a unique figure in the history of astronomy in that, while there is general agreement that his work was of profound importance, we are singularly ill-informed about it. Of his numerous works only one (the commentary on Aratus) survives, and that a comparatively slight one (although valuable in the absence of the others). We derive most of our knowledge of Hipparchus' achievements in astronomy from the *Almagest*; and although Ptolemy obviously had studied Hipparchus' writings thoroughly and had a deep respect for his work, his main concern was not to transmit it to posterity but to use it and, where possible, improve upon it in constructing his own astronomical system. Most of his references to Hipparchus are quite incidental; and some of them are obscure to us, since we cannot consult the originals. Some supplementary information can be gathered from remarks by other ancient writers; but for the most part they were not professional astronomers, and they frequently misunderstood or misrepresented what Hipparchus said (a typical example is the elder Pliny). Since the evidence is so scanty, the following account necessarily contains much that is uncertain or conjectural. Painstaking analysis of that evidence, however, which has begun only in recent years,[7] has revealed, and will reveal, a surprising amount. It also has demonstrated the groundlessness of the assumption, stated or tacit, of most modern accounts of Hipparchus: that everything in the *Almagest* that Ptolemy does not expressly claim as his own work is derived from Hipparchus. The truth is more complicated and more interesting. A further difficulty is that although we know the titles of a good many of Hipparchus' works, much of our information on his opinions and achievements cannot be assigned with certainty to any known title. Moreover, while we can make some inferences about the chronology of his work (it is certain, for instance, that his discovery of precession belongs to the end of his career), in general the dates and order of composition of his works are unknown. This discussion is therefore arranged by topics rather than by titles (the latter are mentioned under the topics that they are known or conjectured to have treated).

Mathematical Methods. In Greek astronomy the positions of the heavenly bodies were computed from geometrical models to which numerical parameters had been assigned. An essential element of the computation was the solution of plane triangles; Greek trigonometry was based on a table of chords. We are informed that Hipparchus wrote a work on chords,[8] and we can reconstruct his chord table. It was based on a circle in which the circumference was divided, in the normal (Babylonian) manner, into 360 degrees of 60 minutes, and the radius was measured in the same units; thus R, the radius, expressed in minutes, is

$$R = \frac{360 \cdot 60'}{2\pi} \approx 3438'.$$

This function is related to the modern sine function (for α in degrees) by

$$\frac{\text{Crd } 2\alpha}{2} = 3438 \sin \alpha.$$

Hipparchus computed the function only at inter-

HIPPARCHUS

vals of 1/48 of a circle (7-1/2°), using linear interpolation between the computed points for other values. Thus he was able to construct the whole table on a very simple geometrical basis: it can be computed from the values of Crd 60° (= R), Crd 90° (= $\sqrt{2}R$), and the following two formulas (in which d is the diameter of the base circle and s is the chord of the angle α):

(1) $$\text{Crd}\,(180°-\alpha)=\sqrt{d^2-s^2}.$$

(2) $$\text{Crd}\,\frac{1}{2}\alpha=\sqrt{\frac{1}{2}\,(d^2-d\sqrt{d^2-s^2})}\,.$$

The first is a trivial application of Pythagoras' theorem, and the second was already known to Archimedes.

This chord table survives only in the sine table commonly found in Indian astronomical works, with $R = 3438'$ and values computed at intervals of 3-3/4°, which is derived from it. But its use by Hipparchus can be demonstrated from calculations of his preserved in *Almagest* IV, 11. Otherwise, apart from a couple of stray occurrences of its use,[9] it vanishes from Greek astronomy, being superseded by Ptolemy's improved chord table based on the unit circle ($R = 60 = 1, 0$ in Ptolemy's sexagesimal system) and calculated to three sexagesimal places at intervals at ½°. The results of trigonometrical calculations based on Hipparchus' chord table, although less accurate than those based on Ptolemy's, are adequate in the context of ancient astronomy. The main disadvantage of its use, in contrast with Ptolemy's, is the constant intrusion of the factor 3438 in the calculations. It has the compensating advantage, however, that for small angles (up to 7-1/2°), the chord can be replaced by the angle expressed in minutes (in this respect it is analogous to modern radian measure), which greatly simplifies computations (for an example, see below on the distances of the sun and moon).

Given the chord function, Hipparchus could solve any plane triangle by using the equivalent of the modern sine formula:

$$\frac{\text{Crd}\ 2\alpha}{a}=\frac{\text{Crd}\ 2\beta}{b}\,.$$

No doubt, like Ptolemy, he usually computed with right triangles, breaking down other triangles into two right triangles. In the absence of a tangent function, he had to use the chord function combined with Pythagoras' theorem; but his methods were as effective as, if more cumbersome than, those of modern trigonometry. Particular trigonometrical problems had been solved before Hipparchus by Aristarchus of Samos (early third century B.C.) and by Archimedes, using approximation methods; but it seems highly probable that Hipparchus was the first to construct a table of chords and thus provide a general solution for trigonometrical problems.[10] A corollary of this is that, before Hipparchus, astronomical tables based on Greek geometrical methods did not exist. If this is so, Hipparchus was not only the founder of trigonometry but also the man who transformed Greek astronomy from a purely theoretical into a practical, predictive science.

In Greek astronomy most problems arising from computations of the positions of the heavenly bodies were either problems in plane trigonometry or could be reduced to such by replacing the small spherical triangles involved by plane triangles. The principal exception was those problems in which the earth is no longer treated as a point—that is, those in which the position of the observer on the earth must be taken into account, notably those concerned with parallax and rising times. Exact mathematical treatment of these requires spherical trigonometry. Since Hipparchus did treat these subjects, the question arises whether he used spherical trigonometry. In the *Almagest* spherical trigonometry is based on a theorem of Menelaus (late first century of the Christian era),[11] and there is no evidence for the existence of the trigonometry of the surface of the sphere before Menelaus. It was possible for Hipparchus to solve the problems he encountered in other ways, however, and there is considerable evidence that he did so.

The problem of the rising times of arcs of the ecliptic at a given latitude was usually connected in ancient astronomy with the length of daylight. The usual method of reckoning time in antiquity was to divide both the daylight and the nighttime into twelve hours. These "seasonal hours" (ὧραι καιρικαί) varied in length throughout the year, depending on the season and the latitude of the place. In order to convert them into hours of equal length ("equinoctial hours," ὧραι ἰσημεριναί), one needs to know the length of daylight on the date and at the place in question. This is given by the time it takes the 180° of the ecliptic following the longitude of the sun on that date to cross the horizon at that latitude (or, in spherical terms, the arc of the equator that rises with those 180°). For most purposes it is sufficient to know the rising times of the individual signs of the zodiac. This problem was solved in Babylonian astronomy by a simple arithmetical scheme that, although only approximately correct, produced remarkably good results.

If we number the rising times of the signs of the ecliptic, beginning with Aries, $\alpha_1, \alpha_2, \ldots, \alpha_{12}$, then, for the first six signs, it is assumed that the increment d between the rising time of a sign and the preceding sign is constant: $\alpha_2 = \alpha_1 + d$, $\alpha_3 = \alpha_2 + d = \alpha_1 + 2d$, and so on. The rising times of the signs of the second half of the ecliptic are equal to those of the corresponding signs of the first half according to the symmetry relations $\alpha_7 = \alpha_6$, $\alpha_8 = \alpha_5$, and so on. The length of the longest daylight, M, is the sum of the six arcs α_4 to α_9 inclusive; that of the shortest day, m, is the sum of the remaining six arcs. Thus, if the ratio M/m is given, the value of α_1 and d—and hence of all the rising times—can be computed arithmetically. For, in degrees of the equator, $M + m = 360°$, $M = 6\alpha_1 + 24d$, $m = 6\alpha_1 + 6d$. In Babylonian astronomy the ratio M/m was always taken as 3/2, which is approximately correct for Babylon. A contemporary of Hipparchus, Hypsicles of Alexandria, in his extant work Ἀναφορικός ("On Rising-times"), expounds exactly the same method, but for Alexandria, taking M/m as 7/5. It seems that Hipparchus extended the scheme to a number of different latitudes (probably the seven standard "climata" characterized by longest daylights extending from thirteen hours to sixteen hours at half-hour intervals, which Hipparchus used in his geographical treatise), for Pappus mentions a work by Hipparchus entitled "On the Rising of the Twelve Signs of the Zodiac" (Ἐν τῷ περὶ τῆς τῶν ιβʹ ζῳδίων ἀναφορᾶς) in which he proved a certain proposition "arithmetically" (δι' ἀριθμῶν).[12]

Other examples of the employment of arithmetical schemata for problems that would require spherical trigonometry if solved strictly can be inferred from Strabo's quotations from Hipparchus' geographical treatise. For instance, Hipparchus gave the following information: for the region where the longest daylight is sixteen hours, the maximum altitude of the sun above the horizon at winter solstice is nine cubits; for M = seventeen hours, it is six cubits, for M = eighteen hours, four cubits, and for M = nineteen hours, three cubits.[13] The altitudes form a series with constant second-order difference. Strabo also excerpted from Hipparchus' treatise the distance in stades between parallels with a given longest daylight. Here too the relationship is based on a constant difference, of third order.[14] Thus problems involving the relationship between geographical latitude and the length of daylight, which Ptolemy solved by spherical trigonometry at *Almagest* II, 2–5, were solved arithmetically by Hipparchus in every case for which there is evidence.

It does not seem possible, however, that Hipparchus solved arithmetically every problem that would normally require spherical trigonometry. Moreover, he himself said that he had written a work enabling one to determine "for almost every part of the inhabited world" which fixed stars rise and set simultaneously.[15] Theoretically one could solve such problems approximately by suitable manipulation of a celestial globe (see below on instruments). Hipparchus, however, stated that he had solved a particular problem of this type "in the general treatises we have composed on this subject [presumably the same as the above] geometrically" (διὰ τῶν γραμμῶν).[16] It is a plausible conjecture that at least one of the methods Hipparchus used for the solution of such problems was that known in antiquity as "analemma" and in modern times as descriptive geometry. This is best explained by an actual example, in which a numerical result given by Hipparchus is recomputed.[17] In the commentary on Aratus he said that a certain star is 27-1/3° north of the equator, and therefore (15/24 − 1/20 · 1/24) of the parallel circle through that star is above the horizon.[18] The horizon in question is Rhodes, with latitude $\phi = 36°$; the star's declina-

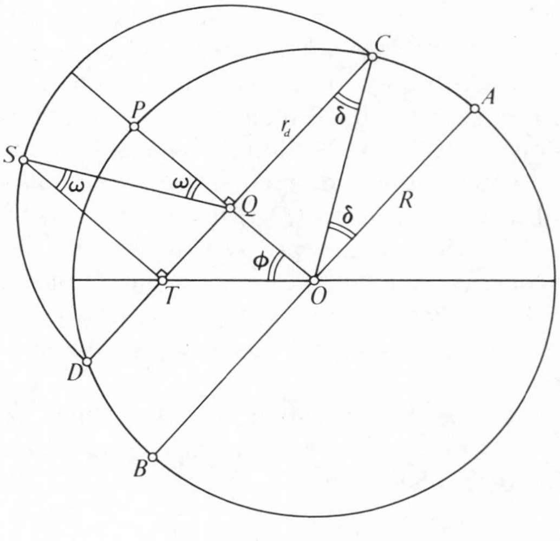

FIGURE 1

tion, δ, is 27-1/3°. In Figure 1 the circle *ACDB* represents the meridian at latitude 36°, *OT* the trace of the local horizon, *AB* the trace of the equator, and *CD* the trace of the parallel circle on which the star lies. Half of the parallel circle *CSD* is then drawn (rotated through 90° to lie in the plane of the meridian). It is required to find arc *SC*

HIPPARCHUS

(= $90° + \omega$), which is the portion of the parallel circle lying between meridian and horizon. The radius of the parallel circle, r_d, is computed by

$$r_d = CQ = \frac{1}{2}\,\text{Crd}\,(180° - 2\delta),$$

and

$$OQ = \text{Crd}\,2\delta.$$

Then

$$QT = \frac{\text{Crd}\,2\phi \cdot OQ}{\text{Crd}\,(180° - 2\phi)} = \frac{\text{Crd}\,2\phi\ \text{Crd}\,2\delta}{\text{Crd}\,(180° - 2\phi)},$$

and, in triangle STQ,

$$\text{Crd}\,2\omega = \frac{QT \cdot R}{SQ} = \frac{QT \cdot R}{r_d}$$

$$= \frac{\text{Crd}\,2\phi\ \text{Crd}\,2\delta \cdot 2R}{\text{Crd}\,(180° - 2\phi)\ \text{Crd}\,(180° - 2\delta)}$$

$$= \frac{4039 \cdot 3156 \cdot 2 \cdot 3438}{5559 \cdot 6105} \approx 2583,$$

giving $2\omega = 44\text{-}1/8°$ and the part of the circle above the horizon as 224-1/8° (Hipparchus' result was equivalent to 224-1/4°).

The method we assume here is not attested for Hipparchus, but his use of small circles (such as the parallel circle here) is alien to the spherical astronomy of the *Almagest*, which uses only great circles, while it is exactly parallel to the methodology of Indian astronomical texts in which the analemma is implicit.[19] On the other hand, for computation of simultaneous risings and settings of fixed stars, stereographic projection would be more convenient; and we shall see, in connection with the astrolabe, that there is evidence that Hipparchus used stereographic projection too.

Theory of the Sun and Moon. Geometrical models that would in principle explain the anomalistic motion of the heavenly bodies had been developed before Hipparchus. Ptolemy attests the use of both epicycle and eccentric by Apollonius (*ca.* 200 B.C.),[20] and it is clear from his discussion that Apollonius was fully aware of the equivalence between the two models. Hipparchus used both models, and is known to have discussed their equivalence. What was new was (in all probability) his attempt to determine numerical parameters for the models on the basis of observations. We are comparatively well-informed about his solar theory, since Ptolemy adopted it virtually unchanged. Having established the lengths of the four seasons, beginning with the spring equinox, as 94-1/2, 92-1/2, 88-1/8, and 90-1/8 days, respectively, and assuming a single anomaly (in Greek terms, an eccentric with fixed apogee or the equivalent epicycle model), he was able to determine the eccentricity (1/24) and the position of the apogee (Gemini 5-1/2°) from the first two season lengths combined with a value for the length of the year.[21] We do not know what value Hipparchus adopted for the latter (his treatise "On the Length of the Year," in which he arrived at the value 365-1/4 – 1/300 days, belongs to the end of his career, after his discovery of precession, whereas he must have established a solar theory much earlier), but for this purpose the approximate value of 365-1/4 days (established by Callippus in the fourth century B.C.) is adequate. Hipparchus presumably then constructed a solar table similar to (but not identical with) that at *Almagest* III, 6, giving the equation as a function of the anomaly; the astrological writer Vettius Valens claimed to have used Hipparchus' table for the sun.[22]

In attempting to construct a lunar theory Hipparchus was faced with much greater difficulties. Whereas (in ancient theory) the sun has a single anomaly, the period of which is the same as that of its return in longitude (one year), for the moon one must distinguish three separate periods: the period of its return in longitude, the period of its return to the same velocity ("anomalistic month"), and the period of its return to the same latitude ("dracontic month"). Related to the return in longitude and the solar motion is the "synodic month" (the time between successive conjunctions or oppositions of the sun and moon). Hipparchus enunciated the following relationships: (1) In 126,007 days, 1 hour, there occur 4,267 synodic months, 4,573 returns in anomaly, and 4,612 sidereal revolutions, less 7-1/2° (hence the length of the mean synodic month is 29; 31, 50, 8, 20 days); (2) In 5,458 synodic months there occur 5,923 returns in latitude.

According to Ptolemy, *Almagest* IV, 2, Hipparchus established these relationships "from Babylonian and his own observations." Had he in fact arrived at these remarkably accurate periods merely by comparison of eclipse data, he would indeed be worthy of our marvel. In fact, as F. X. Kugler showed,[23] all the underlying parameters can be found in Babylonian astronomical texts. The second relationship can be derived directly from those texts, while the first is a result of purely arithmetical manipulation of the following Babylonian parameters: (a) 251 synodic months = 269 anomalistic months; (b) 1 mean synodic month = 29; 31, 50, 8, 20 days; (c) 1 year = 12; 22, 8 synodic months. One finds the first relationship by

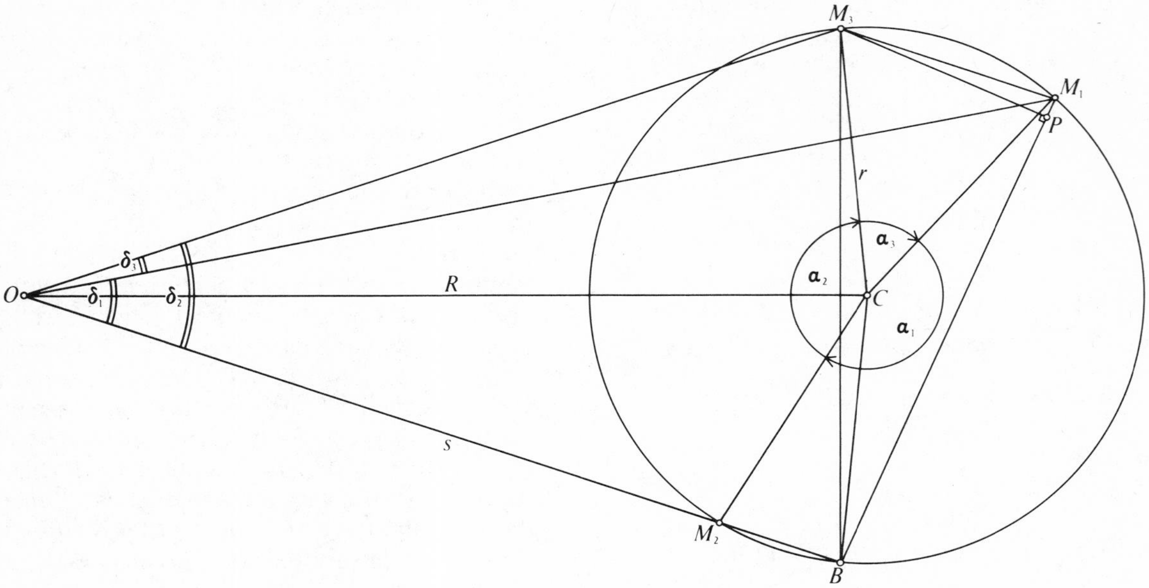

FIGURE 2

multiplying relationship (a) by 17. Hipparchus did this because he wanted to produce an eclipse period; and 17 is the smallest multiplier of (a) that will generate a period in which the moon can be near a node and the sun is in approximately the same position, at both beginning and end.[24] He wanted an eclipse period because he wished to confirm the Babylonian parameters by comparison of Babylonian and his own eclipse data; we can identify some of the eclipses that he used for this observational confirmation. Since one is the eclipse of 27 January 141 B.C., we have a *terminus post quem* for the establishment of the lunar theory.

The revelation of Hipparchus' dependence on Babylonian sources raises the question of what material was available to him, and in what form. We can infer that he possessed a complete or nearly complete list of lunar eclipses observed at Babylon since the reign of Nabonassar (beginning 747 B.C.). This list was available to Ptolemy. Hipparchus also had a certain number of Babylonian observations of planets (see below). Furthermore, he had access to some texts (apparently unknown to Ptolemy) that gave the fundamental parameters listed above. It seems highly improbable that he derived them from the type of texts in which they have come down to us, the highly technical lunar ephemerides. Rather, they must have been excerpted and translated by someone in Mesopotamia who was well acquainted with Babylonian astronomical methods. But when and how the transmission occurred is unknown. Hipparchus is the first Greek known to have used this material. Without it his lunar theory, and hence his eclipse theory, would not have been possible.

To represent the lunar anomaly Hipparchus used a simple epicycle model in which the center of the epicycle C moves at constant distance about the earth O with the mean motion in longitude, while the moon M moves about the center of the epicycle with the mean motion in anomaly. To determine the parameters of this model (that is, the size of the radius of the epicycle r relative to the radius of the deferent R), he devised an ingenious method involving only the observation of the times of three lunar eclipses. By calculating the position of the sun at the three eclipses (presumably from his solar table), he found the true longitude of the moon at the middle of each eclipse (180° away from the true sun). From the time intervals between the three eclipses he found the travel in mean longitude and mean anomaly between the three points. In Figure 2 points M_1, M_2, M_3 represent the positions of the moon on the epicycle at the three eclipses. The angles α at the center of the epicycle are given by the travel in mean anomaly (modulo 360°); the angles δ at the earth are the equational differences, found by comparing the intervals in mean longitude with the intervals in true longitude. Then, by solving a series of triangles, he found, first r in terms of $OB = s$, then M_2B in terms of s, and finally r in terms of R from

$$R^2 - r^2 = (R + r)(R - r) = s \cdot M_2O = s(s - M_2B).$$

Hipparchus performed this calculation twice, with two different sets of eclipses, using the epicycle model for one and the equivalent eccentric model for the other. Unfortunately the elegance of his mathematical approach was not matched by the accuracy of his calculations, so that carelessness in computing the intervals in time and longitude produced widely differing results from the two calculations, as Ptolemy demonstrates at *Almagest* IV, 11. The parameters that he found were $R:r$ = 3122-1/2:247-1/2 for the epicycle model, and $R:e$ = 3144:327-2/3 for the eccentric model.[25] We do not know how Hipparchus dealt with the discrepancy (he may have considered the possibility of a variation in the size of the epicycle), but we do know that he adopted the value 3122-1/2:247-1/2 (which is distinctly too small for the epicycle) in his work "On Sizes and Distances."

Ptolemy shows (*Almagest* V, 1–5) that a lunar model of the type developed by Hipparchus, with a single anomaly, works well enough at syzygies (oppositions and conjunctions) but that for other elongations of the moon from the sun, one must assume a second anomaly that reaches its maximum near quadrature (elongation of 90° or 270°). We can infer from material used here by Ptolemy that, toward the end of his career, Hipparchus had at least an inkling that his lunar theory was not accurate outside syzygies, and that he was systematically making observations of the moon at various elongations. Three of his latest observations that Ptolemy quotes (5 August 128 B.C., 2 May 127 B.C., and 7 July 127 B.C.) are of elongations of about 90°, 315°, and 45° respectively.[26] Hipparchus, however, seems never to have reached any firm conclusions as to the nature of the discrepancies with theory that he must have found; and it was left to Ptolemy to devise a model that would account for them mathematically.

In order to construct a theory of eclipses (one of the ultimate goals of his theory of the moon), Hipparchus had to take account of its motion in latitude. We have seen that he accepted a Babylonian parameter for the mean motion in latitude. He confirmed this by comparing two eclipses at as great an interval as possible.[27] He established the maximum latitude of the moon as 5° (*Almagest* V, 7) and devised a method of finding its epoch in latitude from an observation of the magnitude of an eclipse, combined with the data (obtained by measurement with the diopter; see below) that the apparent diameter of the moon at mean distance is 1/650 of its circle (that is, 360°/650) and 2/5 of the shadow.

Sizes, Distances, and Parallax of the Sun and Moon. These last data are connected with another topic on which Hipparchus wrote. In order to predict the circumstances of a solar eclipse, one must know the relative sizes and distances of the bodies concerned: sun, moon, and earth (for lunar eclipses it suffices to know the apparent sizes; but in solar eclipses parallax, which depends on the distances of the moon and sun, is very important). Hipparchus devoted a treatise "On Sizes and Distances" (Περὶ μεγέθων καὶ ἀποστημάτων),[28] in two books, to the topic. By combining the remarks of Ptolemy and Pappus,[29] we can infer that Hipparchus proceeded as follows. By measurement with the diopter he had established the following data:

(1) The moon at mean distance measures its own circle 650 times.

(2) The moon at mean distance measures the earth's shadow (at the moon) 2-1/2 times.

(3) The moon at mean distance is the same apparent size as the sun.

He also had established by observation that the sun has no perceptible parallax. But from this he could deduce only that the sun's parallax was less than a certain amount, which he set at seven minutes of arc.

In book I, Hipparchus assumed that the solar parallax was the least possible—that is, zero. He then derived the lunar distance from two observations of a solar eclipse (which can be identified as the total eclipse of 14 March 190 B.C.),[30] in which the sun's disk was totally obscured near the Hellespont and four-fifths obscured at Alexandria. The assumption that the sun has zero parallax means that we can take the whole shift in the obscured amount of the sun's disk (a fifth of its diameter) as due to lunar parallax. Assuming that the eclipse was in the meridian at both Alexandria and the Hellespont, we have the situation of Figure 3, where M represents the center of the moon, O the earth's center, H the observer at the Hellespont, A the observer at Alexandria, Z the direction of the zenith at the Hellespont, and POQ the equator of the earth, the radius of which is $r_{\oplus}$. Since Hipparchus knew ϕ_H and ϕ_A, the latitudes of the Hellespont and Alexandria (about 41° and 31° respectively), and could find the moon's declination δ at the time of the eclipse (about −3°) from his tables, he

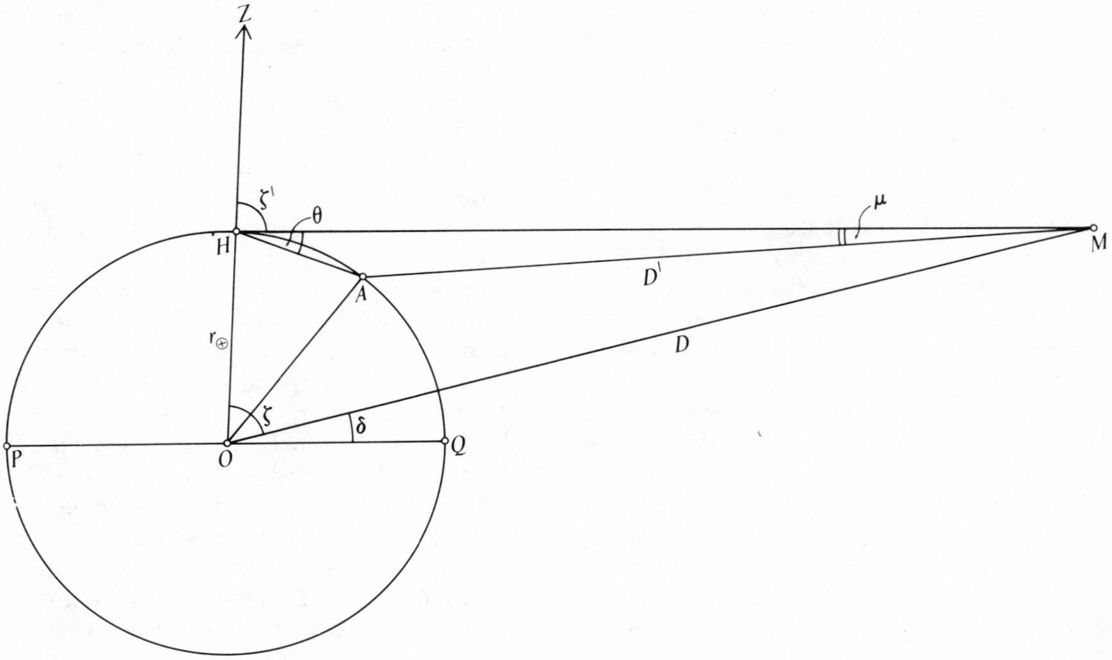

FIGURE 3

could calculate the distance of the moon (OM in Figure 3):

$$AH = r_{\oplus} \cdot \text{Crd}\,\widehat{AOH} = r_{\oplus} \cdot \text{Crd}(\phi_H - \phi_A) \approx r_{\oplus} \cdot \text{Crd}\,10°.$$

The zenith distance of the moon at H, ζ', is approximately equal to ζ; and $\zeta = \phi_H - \delta \approx 44°$.

$$\theta = 180° - \zeta' - \widehat{OHA} \approx 51°.$$

μ is one-fifth of the apparent diameter of the sun, or $21600'/(5 \cdot 650)$ (from [1] and [3] above). From AH, ϑ, and μ the triangle AHM is determined (in terms of $r_{\oplus}$); and we find

$$AM = D' \approx 70r_{\oplus}, \text{ and } D \approx D' + r_{\oplus} \approx 71r_{\oplus}.$$

This is the distance of the moon at the time of the eclipse. To find the least distance of the moon, we have to reduce it by one or two earth radii. Hipparchus found $71r_{\oplus}$ as the least distance. The small discrepancy is no doubt due to the approximations (in ϕ_H, ϕ_A, and ζ') made above. By applying the ratio $R{:}e$ = 3122-1/2:247-1/2, derived from his lunar model, Hipparchus found the greatest distance as $83r_{\oplus}$. The assumption that the eclipse took place in the meridian (which Hipparchus knew to be false) implies, however, that the distances must be greater than those computed (for as the moon moves away from the meridian, the angle ϑ, and hence D', increases), so that $71r_{\oplus}$ represents the minimum possible distance of the moon.

Whereas in book I, Hipparchus had assumed that the solar parallax was the least possible, in book II he assumed that it was the greatest possible (consistent with the fact that it was not great enough to be observed)—that is, 7′. This immediately gives the solar distance, for since the angle is small, we can substitute the angle for the chord and say that the sun's distance is $3438/7 \approx 490r_{\oplus}$. In Figure 4, S, M, O, U are the centers of the sun, moon, earth, and shadow, respectively, and $OU = OM$. From the similar triangles with bases UA, OB, and MD, it follows that

$$MD = 2OB - UA.$$

From (2) $$UA = 2\tfrac{1}{2}MC.$$

Therefore $$CD = MD - MC = 2OB - 3\tfrac{1}{2}MC$$

and $$OB - CD = 3\tfrac{1}{2}MC - OB.$$

From (1), and substituting angles for chords,

$$MC = \frac{1}{2} \cdot \frac{21600}{650} \cdot \frac{OM}{3438}.$$

And from the similar triangles OMC, OSE and OBE, CDE

$$\frac{OM}{OS} = \frac{OC}{OE} = \frac{OB - CD}{OB} = \frac{\dfrac{3\frac{1}{2} \cdot 21600\,OM}{2 \cdot 650 \cdot 3438} - OB}{OB}.$$

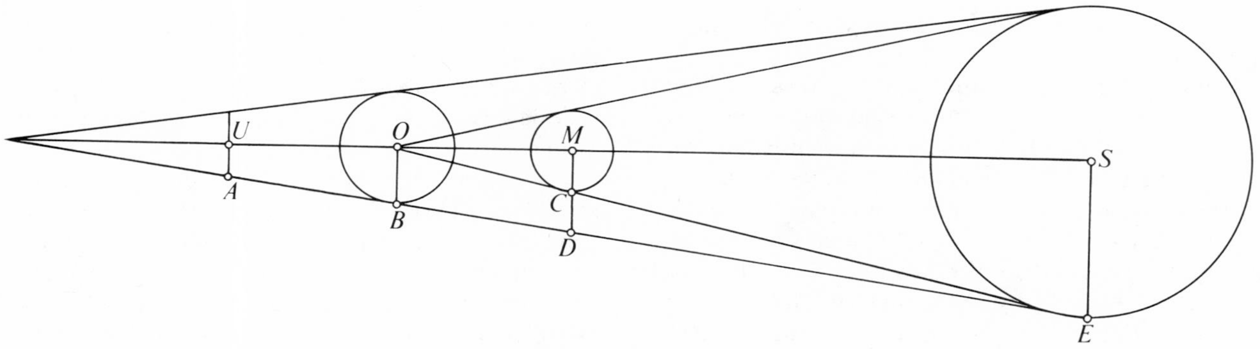

FIGURE 4.

Since $OS = 490\ OB$,

$$(4)\quad OM = \frac{490}{\dfrac{490 \cdot 3\frac{1}{2} \cdot 21600}{2 \cdot 650 \cdot 3438} - 1}\ OB \approx 67\tfrac{1}{5}\ OB.$$

Hipparchus found the mean distance of the moon as $67\text{-}1/3 r_{\oplus}$, and the distance of the sun as $490 r_{\oplus}$. These too were computed under an extremal assumption however: that the solar parallax was the maximum possible (the solar distance was the minimum possible). This in turn implies that the resultant lunar distance is a maximum, for it is easily seen from equation (4) that as the sun's distance (490 in the expression) increases, the moon's distance decreases. Moreover, it does not decrease indefinitely, for as the sun's distance tends to infinity, the expression in (4) tends to the limit $(2 \cdot 650 \cdot 3438)/(3\text{-}1/2 \cdot 21600) \approx 59 r_{\oplus}$. Although it is not explicitly attested, there is little doubt that Hipparchus realized this, and stated that the mean distance of the moon lay between the bounds 67-1/3 and 59 earth radii.[31] He thus arrived at a value for the lunar mean distance that was not only greatly superior to earlier estimates but was also stated in terms of limits that include the true value (about 60 earth radii). His methods must be regarded as a tour de force in the use of crude and scanty observational data to achieve a result of the right order of magnitude.

Ptolemy criticized Hipparchus' procedure (unjustifiably, if the above reconstruction is correct), on the ground that it starts from the solar parallax, which is too small to be observed.[32] He himself used Hipparchus' second method (with different data) but started from the lunar distance (derived from an observation of the moon's parallax), whence he found the solar distance. This modified procedure became standard in later astronomical works, and was used by Copernicus.

Hipparchus also computed the true sizes of sun and moon (relative to the earth), which are easily found from the distances and apparent diameters. Theon of Smyrna stated that Hipparchus said that the sun was about 1,880 times the size of the earth, and the earth 27 times the size of the moon.[33] Since a lunar distance of about 60 earth radii implies that the earth's diameter is about 3 times the size of the moon's, Hipparchus must have been referring to comparative volumes. Hence the sun's diameter was, according to him, about 12-1/3 times the earth's; and therefore (by the method of book II) the sun's distance was about 2,500 earth radii and the moon's about 60-1/2 earth radii. These may have been the figures he finally decided on for the purposes of parallax computation.

In any case, his investigation of the distances furnished Hipparchus with a horizontal parallax that, for the moon, was approximately correct. For computation of the circumstances of solar eclipses, and to correct observations of the moon with respect to fixed stars, he had to find the lunar parallax for a given lunar longitude and terrestrial latitude and time. This is a very unpleasant problem in spherical astronomy. We know that Hipparchus solved it; but we know very little about his solution, although it is certain that it was not carried out with full mathematical rigor. We can infer from a criticism made by Ptolemy[34] that Hipparchus wrote a work on parallax in at least two books. The details of Ptolemy's criticism are quite obscure to me; and the only safe inference is that in converting the total parallax into its longitudinal and latitudinal components, Hipparchus treated spherical triangles as plane triangles (some of these triangles are too large for the procedure to be justifiable, but one can apply the same criticism to Ptolemy).[35] The actual corrections for parallax that Hipparchus is known to have applied to particular lunar observations are, however, reasonably accurate.[36] One might guess that the methods of computing parallax found in Indian astronomical texts (which are quite different from Ptolemy's) are related to

Hipparchus' procedures, but at present this is mere speculation.

Eclipses. Mastery of the above topics enabled Hipparchus to predict eclipses of both sun and moon for a given place; and we may presume, although there is no specific evidence, that he compiled eclipse tables for this purpose. The elder Pliny tells us that he predicted eclipses of the sun and moon for a period of 600 years[37]; and this has been taken seriously by some modern scholars, who envisage Hipparchus producing a primitive "Oppolzer." Although utterly incredible as it stands, the story must have some basis; and O. Neugebauer makes the plausible suggestion that what Hipparchus did was to arrange the eclipse records available to him from Babylonian and other sources (which, as we have seen, covered a period of 600 years from Nabonassar to his own time) in a form convenient for astronomical use. If this was so, we can explain more easily how Ptolemy was able to select eclipses with very special circumstances to suit a particular demonstration[38]: he was using material already digested by Hipparchus.

Pliny provides another valuable piece of information about Hipparchus' work on eclipses. He says that Hipparchus discovered that lunar eclipses can occur at five-month intervals, solar eclipses at seven-month intervals, and solar eclipses at one-month intervals—but in the last case only if one eclipse is seen in the northern hemisphere and the other in the southern.[39] Hipparchus, then, had discussed eclipse intervals; and we see from Ptolemy's discussion of the topic at *Almagest* VI, 5–6, how Hipparchus must have approached the problem. Babylonian astronomers were aware that lunar eclipses could occur at intervals of five synodic months as well as the usual six. Ptolemy showed that five-month intervals are possible for lunar eclipses but seven-month intervals are not, that both five- and seven-month intervals are possible for solar eclipses, and that one-month intervals are not possible for solar eclipses "in our part of the world." Hipparchus must have proved all this, and also that solar eclipses at one-month intervals are possible for different hemispheres. The essential elements in the proof (besides the apparent sizes of sun, moon, and shadow) are parallax (for solar eclipses) and the varying motion of the moon in latitude in a true synodic month. It is to this topic that we must refer the work "On the Monthly Motion of the Moon in Latitude" (Περὶ τῆς κατὰ πλάτος μηνιαίας τῆς σελήνης κινήσεως), the title of which is preserved in the *Suda*.[40] Presumably Achilles Tatius is referring to Hipparchus' work on eclipse intervals when he names him as one of those who have written treatises on solar eclipses in the seven "climata,"[41] for the geographical latitude is one of the most important elements in the discussion.

Fixed Stars. Hipparchus devoted much time and several works to topics connected with the fixed stars. His observations of them must have extended over many years. On these points there is general agreement among modern scholars. On all other points confusion reigns, although Hipparchus' sole surviving work, "Commentary on the Phaenomena of Aratus and Eudoxus" (Τῶν Ἀράτου καὶ Εὐδόξου Φαινομένων ἐξήγησις), in three books, is concerned with the fixed stars. Nevertheless, it is possible, if we look at the evidence without formulating prior hypotheses, to come to some conclusions.

In the mid-fourth century B.C., Eudoxus wrote a pioneering work naming and describing the constellations. This is now lost (Hipparchus is the main source of our knowledge of it). In the early third century B.C., Aratus wrote a poem based on Eudoxus, called "Phaenomena," which became immensely popular and is still extant. Not long before Hipparchus a mathematician, Attalus of Rhodes, wrote a commentary on Aratus (now lost). Hipparchus' treatise is a critique of all three works. None of the three contained any mathematical astronomy, only descriptions of the relative positions of stars, simultaneous risings and settings, and the like; and much of Hipparchus' criticism in books 1 and 2 is of the same qualitative kind. But even from this, one can see that he had fixed the positions of a number of stars according to a mathematical system (he incidentally notes some polar distances, declinations, and right ascensions). The last part of book 2 and the whole of book 3 are devoted to his own account of the risings and settings of the principal constellations for a latitude where the longest day is 14-1/2 hours. Whereas his predecessors had merely reported the stars or constellations that rise and set together with a given constellation, Hipparchus gave the corresponding degrees of the ecliptic (for risings, settings, and culminations). At the end of book 3 is a list of bright stars that lie on or near twenty-four-hour circles, beginning with the hour circle through the summer solstice. Hipparchus says that the purpose of this is to enable one to tell the time at night when making astronomical observations.

In this treatise Hipparchus indicates the position of a star in various ways. We have already mentioned declination (which he calls "distance from the equator along the circle through the pole") and polar distance (the complement of declination). He frequently uses an odd form of right ascension. Thus he says that a star "occupies three degrees of Leo along its parallel circle." This means that each small circle parallel to the equator is divided into twelve "signs" of 30°, and thus the right ascension of the star is 123°. The assignment of the stars to the hour circles at the end of book 3 is also a form of right ascension. Besides these equatorial coordinates, we find in the section on simultaneous risings and settings a mixture of equatorial and ecliptic coordinates: Hipparchus names the point of the ecliptic that crosses the meridian together with a given star. In other words, he gives the point at which the declination circle through the star cuts the ecliptic. It is significant that this, the "polar longitude," is one of the standard coordinates for fixed stars in Indian astronomical texts. There are no purely ecliptic coordinates (latitude and longitude) in Hipparchus' treatise.

Far from being a "work of his youth," as it is frequently described, the commentary on Aratus reveals Hipparchus as one who has already compiled a large number of observations, invented methods for solving problems in spherical astronomy, and developed the highly significant idea of mathematically fixing the positions of the stars (Aristyllus and Timocharis recorded a few declinations in the early third century, but we know of nothing else before Hipparchus). There is no hint of a knowledge of precession in the work; but, as we shall see, that discovery falls at the end of his career. The treatise is probably subsequent to Hipparchus' catalog of fixed stars.

The nature of this catalog (presumably "On the Compilation of the Fixed Stars and of the [?] Catasterisms") to which the *Suda* refers[42] is puzzling. The worst excess of the modern notion of the strict dependence of the *Almagest* on Hipparchus is the belief that we can obtain Hipparchus' catalog simply by taking Ptolemy's catalog in *Almagest* VII and VIII and lopping 2-2/3° (to account for precession) off the longitudes. This was conclusively refuted by Vogt, who showed by a careful analysis of the coordinates of 122 stars derived from the commentary on Aratus that in almost every case there was a significant difference between Hipparchus' and Ptolemy's data. The most explicit statement about Hipparchus' star catalog is found in Pliny, who says that Hipparchus noticed a new star and, because it moved, began to wonder whether other fixed stars move; he therefore decided to number and name the fixed stars for posterity, inventing instruments to mark the positions and sizes of each.[43] This confused notice could be a description of a star catalog like Ptolemy's, but it could equally well refer to an account of the number and relative positions of the stars in each constellation (without coordinates). In fact Ptolemy, wishing to show that the positions of the fixed stars relative to each other had not changed since the time of Hipparchus, reported a number of star alignments that Hipparchus seemed to have recorded for just the purpose alleged by Pliny; to allow posterity to determine whether the fixed stars have a proper motion.[44] Naturally no coordinates were given.

Apparent excerpts from Hipparchus' star catalog that are found in late Greek and Latin sources merely give the total number of stars in each constellation. These suggest that Hipparchus counted a large number of stars,[45] but they do not prove that he assigned coordinates to all—or, indeed, to any. Ptolemy quoted the declinations of a few bright stars mentioned by Hipparchus; but in every case he compared the declination with that recorded by Aristyllus and Timocharis, in order to determine the precession. These declinations may well have been taken not from Hipparchus' catalog but from one of his works on precession. There is one text, however, that does appear to preserve coordinates from Hipparchus' star catalog. In late Latin scholia on Aratus[46] there are found, for the circumpolar constellations, polar distances and what must be interpreted as polar longitudes, which are approximately correct for the time of Hipparchus. Whether Hipparchus gave such coordinates for every star in the catalog or only for some selected stars remains uncertain. The only thing we know for certain about the catalog is that the coordinates employed were not latitude and longitude: Ptolemy (*Almagest* VII, 3) would not have chosen such a roundabout way (conversion of declinations) to prove that latitudes of fixed stars remain constant, had he been able to cite the latitudes from Hipparchus' catalog. It is a plausible conjecture that the star coordinates given in the commentary on Aratus are taken directly from the catalog; at the very least, they are a good indication of the ways in which the position of a star was described in that work.

Precession and the Length of the Year. Hipparchus is most famous for his discovery of the "precession of the equinoxes," the slow motion of the solstitial and equinoctial points from east to west through the fixed stars. This topic is intimately connected with the length of the year, for precession implies both that the coordinates of the fixed stars (such as right ascension and declination) change over a period of time, and also that the tropical year (return of the sun to the same equinox or solstice) is shorter than the sidereal year (return of the sun to the same star). We do not know what phenomenon first led Hipparchus to suspect precession, but we do know that he confirmed his suspicion by both the above approaches. According to Ptolemy, the first hypothesis that he suggested was that only the stars in the zodiac move.[47] Later, in "On the Change in Position of the Solstitial and Equinoctial Points" (*Περὶ τῆς μεταβάσεως τῶν τροπικῶν καὶ ἰσημερινῶν σημείων*), he formulated the hypothesis that all the fixed stars move with respect to the equinoxes (or, rather, vice versa). He supported this hypothesis in two ways. First he compared the distance of the star Spica (α Virginis) from the autumnal equinox in the time of Timocharis (observations of 294 and 283 B.C.) and his own time. Unfortunately this involved considerable uncertainty, since Timocharis' observations were simply occultations of Spica by the moon, which had to be reduced to longitudes by means of the lunar theory; and Hipparchus' own observations of elongations of Spica from the moon at two lunar eclipses led to results differing by 3/4°.[48] He concluded that the longitudes of Spica had increased by about 2° in the 160-odd years since Timocharis, but he was well aware that his data were too shaky to allow any confidence in the precise amount.

Hipparchus' other approach was to try to find the length of the tropical year. For this purpose he set out a series of equinox observations (listed by Ptolemy at *Almagest* III, 1) ranging from 162 to 128 B.C. (the latter date is probably close to that of the composition of the book, as is confirmed by Hipparchus' observation of the longitude of Regulus in the same year [*Almagest* VII, 2]). Unfortunately, these seemed to indicate a variation in the length of the year, which Hipparchus evidently was prepared to consider at this point. We do not know what conclusion, if any, he reached in this work; but he must already have assumed some value for the length of the sidereal year (see below), for otherwise there would have been no point in investigating the length of the tropical year in order to determine the amount of precession. Hipparchus reverted to the topic in a later work, "On the Length of the Year" (*Περὶ ἐνιαυσίου μεγέθους*). In this he came to more definite conclusions. He determined that the tropical and equinoctial points move at least 1/100° a year backward through the signs of the ecliptic, and that the length of the tropical year is 365-1/4 days, less at least 1/300 of a day.[49] These two formulations are of course related. The basis of his proof was comparison of the solstice observed by himself in 135 B.C. with those observed by Aristarchus in 280 B.C. and Meton in 432 B.C. Having thus found a (maximum) value for the tropical year, Hipparchus subtracted it from his value for the sidereal year, thus producing a (minimum) value for precession.[50] From his figures for the tropical year and precession we can deduce the value he assigned to the sidereal year: 365-1/4 + 1/144 days, a very accurate estimate. There is independent evidence that he used this value, and that it is of Babylonian origin. In "On the Length of the Year" Hipparchus also assumed (correctly) that precession takes place about the poles of the ecliptic, and not of the equator, but still expressed uncertainty on that matter. He also realized that one must define the solar year as the tropical year.[51]

Having established the length of the tropical year, Hipparchus wrote "On Intercalary Months and Days" (*Περὶ ἐμβολίμων μηνῶν τε καὶ ἡμερῶν*), in which he proposed a lunisolar intercalation cycle that was a modification of the Callippic cycle:[52] it contained 304 tropical years, 112 with 13 synodic months and 292 with 12, and a total of 111,035 days. This produces very good approximations to Hipparchus' values for both the tropical year and the mean synodic month. There is no evidence that it was ever used, even by astronomers.

Planetary Theory. Ptolemy tells us (*Almagest* IX, 2) that Hipparchus renounced any attempt to devise a theory to explain the motions of the five planets, but contented himself with showing that the hypotheses of the astronomers of his time could not adequately represent the phenomena. We may infer from Ptolemy's discussion that Hipparchus showed that the simple epicycle theory propounded by Apollonius produced constant arcs of retrogradation, whereas observation showed that they vary in length. Ptolemy further informs us that Hipparchus assembled observations of the planets "in a more convenient form." He mentions Hipparchus specifically in connection with an observation of Mercury of 262 B.C. according to the strange "Dionysian era,"[53] and it is probable that all observations in that era preserved in the *Almagest* are derived from his collection; the same may be true of Babylonian observations of Mercury and Saturn according to the Seleucid era. Ptolemy mentions no planetary observations made by Hipparchus himself, and it is probable that he made few. Ptolemy does, however, ascribe to him the following planetary period relations (*Almagest* IX, 3):

> Saturn in 59 years makes 57 revolutions in anomaly and 2 in the ecliptic.
> Jupiter in 71 years makes 65 revolutions in anomaly and 6 in the ecliptic.
> Mars in 79 years makes 37 revolutions in anomaly and 42 in the ecliptic.
> Venus in 8 years makes 5 revolutions in anomaly and 8 in the ecliptic.
> Mercury in 46 years makes 145 revolutions in anomaly and 46 in the ecliptic.

These are well-known relations in Babylonian astronomy, occurring in the so-called "goal-year" texts.

The only other indication of Hipparchus' interest in the planets is a passage in Ptolemy's *Planetary Hypotheses*[54] reporting an attempt to determine minimum values for the apparent diameters of the heavenly bodies. Hipparchus stated that the apparent diameter of the sun is thirty times that of the smallest star and ten times that of Venus.

Instruments. The only observational instrument that Ptolemy specifically ascribes to Hipparchus is the "four-cubit diopter" with which he observed the apparent diameters of the sun and moon and arrived at the estimates given above (*Almagest* V, 14). The instrument is described by Pappus and Proclus.[55] It consisted of a wooden rod of rectangular cross section some six feet in length. The observer looked through a hole in a block at one end of the rod, and moved a prism that slid in a groove along the top of the rod until the prism exactly covered the object sighted. The ratio of the breadth of the prism to its distance from the sighting hole gave the chord of the apparent diameter of the object.

As J. B. J. Delambre remarked,[56] certain observations of Hipparchus' seem to imply use of the "armillary astrolabe" described by Ptolemy at *Almagest* V, 1, and used by him to observe the ecliptic distance between heavenly bodies, particularly between sun and moon. These are observations of the elongation of the moon from the sun reported in *Almagest* V, 5. In particular, the observation of 7 July 127 B.C.[57] seems to indicate that Hipparchus observed not merely the elongation but also, and at the same time, the ecliptic position of the sun (which is given by the shadow on the armillary astrolabe). It seems odd, however, that Ptolemy should describe the instrument so carefully (and claim it as his own) if in fact it were old and well known. Furthermore, for one of the observations Ptolemy says Hipparchus used instruments (plural).[58] It is possible that Hipparchus used one instrument to determine the sun's position and another to measure the elongation, but use of the astrolabe cannot be excluded.

Hipparchus determined the times of equinox and solstice, and perhaps measured the inclination of the ecliptic, confirming Eratosthenes' estimation that the distance between the solstices was 11/83 of the circle.[59] For these observations he must, like Ptolemy, have measured the altitude or zenith distance of the sun at meridian transit. Much of his data on fixed stars also seem to imply observation of the meridian altitude. We can only speculate about the instrument(s) he used. One might conjecture that one was a primitive form of the diopter described by Hero, which could be used for both meridian and elongation observations.

An incidental remark of Ptolemy's implies that Hipparchus (like Ptolemy at *Almagest* VIII, 3) gave instructions for constructing a celestial globe and marking the constellations on it.[60] Such globes existed before Hipparchus, probably as early as the fourth century B.C., but more as artistic than scientific objects. Presumably Hipparchus mentioned the globe in connection with his star catalog. If the stars were marked on it according to a coordinate system, and if, like Ptolemy's, it were furnished with a ring indicating the local horizon to which its axis of rotation was inclined at the correct angle, it could be used not merely for demonstration purposes but also to read off simultaneous risings and settings directly.

There is evidence that Hipparchus invented the plane astrolabe. Our authority for this, Synesius, is late (A.D. 400) and his account confused. But the whole necessary theory of stereographic projection is set out in Ptolemy's *Planisphaerium*; and use of such projection would explain, among other things, how Hipparchus computed simultaneous risings and settings. The balance of probability is that Hipparchus used (and perhaps invented) stereographic projection. If that is so, there is no reason to deny his invention of the plane astrolabe, which provides an easy solution to the problems of the rising times of arcs of the ecliptic and simultaneous risings of stars at a given latitude.

Geography. Hipparchus wrote a work in (at least) three books entitled "Against the Geography of Eratosthenes" (Πρὸς τὴν Ἐρατοσθένους γεωγραφίαν). In the mid-third century Eratosthenes had given a description of the known world, in which he attempted to delineate crudely the main outlines of a world map by means of imaginary lines (including, but not restricted to, meridians and parallels of latitude) drawn through fixed points, and to establish distances along such lines in absolute terms (stades). This was related to astronomical geography in the sense that Eratosthenes had made an estimate of the circumference of the earth (252,000 stades), and had taken some account of the ratios of the shadow to the gnomon at various latitudes. Hipparchus' work was a detailed criticism of Eratosthenes' data, showing that the distances and relationships given were inconsistent with each other and with other geographical data. He evidently supplied numerous corrections and supplements to Eratosthenes; but, as we have seen, his astronomical geography is based on simple arithmetical schemes, and although he enunciated the principle that longitudes should be determined by means of eclipses,[61] this was merely an expression of an ideal. The only evidence that Hipparchus used astronomical observations to improve geographical studies is the statement of Ptolemy (*Geography* I, 4) that he gave the latitudes of "a few cities." Nor does he seem to have contributed anything to mathematical geography, which emerges from infancy only with the work of Marinus of Tyre and Ptolemy.

Other Works. Extant astrological works refer to Hipparchus as an authority on astrology; but we know almost nothing about the content of his writings on this subject, except that the topics attested are all well-worn astrological themes, such as "astrological geography," in which the various regions of the world are assigned to the influence of a zodiacal sign or part of a sign.[62] His work on weather prognostication from the risings and settings of fixed stars, known from numerous citations in Ptolemy's "Phases of the Fixed Stars," also belonged to a traditional genre of Greek "scientific" literature. Chance quotations mention a work on optics,[63] in which he endorsed the theory of vision that "visual rays" emanate from the eye; a book "On Objects Carried Down by Their Weight" (Περὶ τῶν διὰ βαρύτητα κάτω φερομένων);[64] and a work on combinatorial arithmetic. The last is almost the only reference we have to the topic in antiquity, but this merely illustrates how little we know of Greek mathematics outside the "classical" domain of geometry.

Both in antiquity and in modern times Hipparchus has been highly praised and misunderstood. Since Delambre modern scholarship has tended to treat Hipparchus as if he had written a primitive *Almagest*, and to extract his "doctrine" by discarding from the extant *Almagest* what are thought to be Ptolemy's additions. This unhistorical method has obscured Hipparchus' real achievement. Greek astronomy before him had conceived the idea of explaining the motions of the heavenly bodies by geometrical models, and had developed models that represented the motions well qualitatively. What Hipparchus did was to transform astronomy into a quantitative science. His main contributions were to develop mathematical methods enabling one to use the geometrical models for practical prediction, and to assign numerical parameters to the models. For the second, his use of observations and of constants derived from Babylonian astronomy was crucial (without them his lunar theory would not have been possible). His own observations were also important (he and Ptolemy are the only astronomers in antiquity known to have observed systematically). Particularly remarkable is his open-mindedness, his willingness to abandon traditional views, to examine critically, and to test by observation his own theories as well as those of others (Ptolemy's most frequent epithet for him is φιλαληθής, "lover of truth"). Remarkable too, in antiquity, was his attitude toward astronomy as an evolving science that would require observations over a much longer period before it could be securely established. Related to this were his attempts to assemble observational material for the use of posterity.

Hipparchus acquired a high reputation in antiquity (he was extravagantly praised by the elder Pliny, and later was depicted on the coins of his native city), but it seems likely that his works were little read. From Ptolemy's citations, one can see why. They were in the form of numerous writings discussing different, often highly specialized, topics. It seems that Hipparchus never gave an account of astronomical theory, or even part of it, starting from first principles (Ptolemy often inferred Hipparchus' opinion on a topic from incidental remarks). Hipparchus did not, then, construct an astronomical system: he only made such a system possible and worked out some parts of it. When Ptolemy, using Hipparchus' work as one of his essential foundations, did construct such a system, which became generally accepted, interest in Hipparchus' works declined further. It is not surprising that all are lost except the commentary on Aratus (which survived because of the popularity of Aratus' poem). Pappus, in the fourth century, still knew some works by Hipparchus, but there is no certain instance of firsthand knowledge of the lost works after him (the astrological treatise by Hipparchus referred to in Arabic sources was certainly pseudonymous). In Western tradition, then, the influence of Hipparchus was channeled solely through the *Almagest*. But there is much evidence that Indian astronomy of the *Siddhāntas* preserves elements of Hipparchus' theories and methods. These presumably come from some pre-Ptolemaic Greek astronomical work(s) based in part on Hipparchus (there are elements in Indian astronomy, notably in planetary theory, that are also of Greek origin but cannot have been derived from Hipparchus). The task of analyzing the Indian material from this point of view has hardly begun.

NOTES

1. For instance, *Suidae Lexicon*, Ada Adler, ed., II (Leipzig, 1931), 657, no. 521; Aelian, *De natura animalium*, R. Hercher, ed. (Leipzig, 1864), 175.
2. *Almagest* III, 1, Manitius ed., I, 142.
3. *Ibid.*, V, 5, Manitius ed., I, 274.
4. H. C. F. C. Schjellerup, "Recherches sur l'astronomie. . .," 38–39 (−140); H. Vogt, "Versuch einer Wiederherstellung . . .," cols. 25, 31–32 (different dates from different data: −138, −150, −130, −156).
5. The three eclipses of 201–200 B.C. (*Almagest* IV, 11, Manitius ed., I, 251–252) are obviously too early to have been seen by Hipparchus. For the equinox of 24 March 146 B.C. (*Almagest* III, 1, Manitius ed., I, 135), the observation made on the ring in Alexandria is stated to have differed by five hours from that made by Hipparchus himself.
6. See note 1. The text as printed by Hercher says that it was "Hieron the tyrant" who was amazed; presumably this is meant to be the tyrant of Syracuse in the early fifth century. This is, however, Valesius' emendation of "Nero the tyrant," which is equally impossible chronologically.
7. See the contributions of A. Aaboe, O. Neugebauer, N. Swerdlow, and G. J. Toomer in the bibliography. A notable exception from an earlier period is the article by Vogt in the bibliography.
8. A. Rome, *Commentaires . . . sur l'Almageste*, I (Rome, 1931), 451. The expression *ἡ πραγματεία τῶν ἐν κυκλῷ εὐθειῶν* (treatment of straight lines in a circle"; that is, chords) is not the title of the work, as is commonly supposed, but Theon's description of its contents, repeating verbatim the words Ptolemy applied to his own treatment (*Almagest* I, 9, Heiberg ed., I, 31, ll. 4–5). For a conjecture about the origin of Theon's incredible statement that Hipparchus' treatment was in twelve books see G. J. Toomer, "The Chord Table of Hipparchus . . .," 19–20.
9. Ptolemy, *Geography*, I, 20, C. F. A. Nobbe, ed., I (Leipzig, 1843), 42, ll. 26–30; Pappus, *Collectio* VI, F. Hultsch, ed., II (Berlin, 1877), 546, ll. 24–27. On this see G. J. Toomer, in *Zentralblatt für Mathematik*, **274** (1974), 7 (author's summary of article, "The Chord Table of Hipparchus . . .").
10. For a refutation of the argument of J. Tropfke that Archimedes had already constructed a chord table, see Toomer, "The Chord Table of Hipparchus . . .," 20–23.
11. See G. J. Toomer, "Ptolemy," *DSB*, XI, 189.
12. Pappus, *Collectio* VI, Hultsch ed., 600, ll. 10–11. The phrase "arithmetically" is implicitly opposed to "geometrically" (*διὰ τῶν γραμμῶν*), and in this context implies that Hipparchus was using not Euclidean geometry but Babylonian methods.
13. Strabo, *Geography* 75 and 135, H. L. Jones, ed., I, 282, 514–516. A "cubit" in the astronomical sense is 2°; it is known from cuneiform texts, and its use here and elsewhere by Hipparchus is another indication of his heavy debt to Babylonian astronomy.
14. See O. Neugebauer, *History of Ancient Mathematical Astronomy*, I, 304–306.
15. *Commentary on Aratus* II, 4, Manitius ed., 184. Ptolemy explains how to calculate simultaneous risings and settings trigonometrically at *Almagest* VIII, 5.
16. *Ibid.*, 2, Manitius ed., 150, ll. 14–17.
17. See Neugebauer, *op. cit.*, 301–302.
18. *Commentary on Aratus* II, 2, Manitius ed., 148–150.
19. See, for instance, O. Neugebauer and D. Pingree, *The Pañcasiddhāntikā of Varāhamihira (Kongelige Danske Videnskabernes Selskab. Historisk-Filosofiske Skrifter*, **6**, no. 1), II (Copenhagen, 1970), 41–42, 85. Use of the analemma is not attested before the first century B.C. (Diodorus of Alexandria) and is mainly, but not exclusively, associated with sundial theory in extant Greek texts (notably Ptolemy's *Analemma*). I believe that it was invented before Hipparchus for purely graphical purposes in sundial theory, and adapted by him to numerical solutions in spherical astronomy (whence it is found in Indian astronomy). But it must be admitted that this is all conjectural.
20. *Almagest* XII, 1. For a detailed discussion of the theories and their equivalence see Toomer, "Ptolemy," 190.
21. *Ibid.*, 190–191. It is likely that Hipparchus found only the first two season lengths by observation; the last two were then derived from the eccentric model after he had calculated the eccentricity.
22. Vettius Valens, *Anthologiae*, W. Kroll, ed. (Berlin, 1908), 354.
23. Kugler, *Die Babylonische Mondrechnung*, 21, 24, 40, 46, 108.
24. On all this see A. Aaboe, "On the Babylonian Origin of Some Hipparchian Parameters."
25. For details see Toomer, "The Chord Table of Hipparchus . . .," 7–16.
26. *Almagest* V, 3, Manitius ed., I, 266; V, 5, Manitius ed., I, 271, 274.
27. *Ibid.*, VI, 9, Manitius ed., I, 394–395.
28. On the exact title of the work see G. J. Toomer, "Hipparchus on the Distances of the Sun and Moon," 126, n. l.
29. *Almagest* V, 11, Manitius ed., I, 294–295; Rome, *op. cit.*, 67–68.
30. For the identification see Toomer, "Hipparchus on the Distances of the Sun and Moon," 131–138.
31. If so, it can hardly be a coincidence that Ptolemy establishes the mean distance of the moon as exactly $59r_{\oplus}$. See *ibid.*, 131, n. 25.
32. *Almagest* V, 11, Manitius ed., I, 294–295.
33. Theon of Smyrna, E. Hiller, ed. (Leipzig, 1878), 197.
34. *Almagest* V, 19, Manitius ed., I, 329–330.
35. Pappus' "explanation" of this passage, Rome, *op cit.*, 150–155, appears to be a completely fictitious reconstruction. I therefore ignore it here and in the account of Hipparchus' trigonometry (it is not credible that Hipparchus took the sides of large spherical triangles proportionate to the opposite angles).
36. Notably that for his observation of 2 May 127 B.C. (*Almagest* V, 5, Manitius ed., I, 271).
37. Pliny, *Naturalis historia* II, 54, Beaujeu ed., 24.
38. For instance, *Almagest* IV, 9; VI, 5.
39. Pliny, *op. cit.*, II, 57, Beaujeu ed., 25.
40. See note 1.
41. Maass, *Commentariorum in Aratum reliquiae*, 47.
42. See note 1. The Greek title is *Περὶ τῆς τῶν ἀπλανῶν ἀστέρων συντάξεως καὶ τοῦ †καταστηριγμοῦ†* (the last word should perhaps be emended to *καταστερισμου*).
43. Pliny, *op. cit.*, II, 95, Beaujeu ed., 41–42. The "new star" is commonly identified with a nova Scorpii of 134 B.C., alleged from Chinese records. The identification is of the utmost uncertainty (Pliny seems rather to refer to a comet).

HIPPARCHUS

44. *Almagest* VII, 1, Manitius ed., II, 5–8.
45. About 850, according to F. Boll, "Die Sternkataloge des Hipparch . . .," 194; but his calculations rest on no secure basis.
46. Maass, *op. cit.*, 183–189.
47. *Almagest* VII, 1, Manitius ed., II, 4. We do not know to what work of Hipparchus he refers.
48. *Ibid.* III, 1, Manitius ed., I, 137. Compare *ibid.*, VII, 2, Manitius ed., II, 12, where Hipparchus seems to have settled on a mean value between his two discrepant results.
49. *Ibid.*, III, 1, Manitius ed., I, 145 (this quotation is in fact from the later work "On Intercalary Months and Days"), and VII, 2, Manitius ed., II, 15.
50. That he did so is certain from his calculation of the precession over 300 years (that is, between his time and Meton's), which Ptolemy reports from this work at *Almagest* VII, 2, Manitius ed., II, 15.
51. Very interesting verbatim quotation of Hipparchus at *Almagest* III, 1, Manitius ed., I, 145–146.
52. See G. J. Toomer, "Meton," *DSB*, IX, 337–338. The details of the cycle are known from Censorinus, *De die natali* 18, O. Jahn, ed. (Berlin, 1845), 54.
53. *Almagest* IX, 7, Manitius ed., II, 134
54. Goldstein ed., 8. It is not clear whether the values for the apparent diameters of the other planets that Ptolemy adopts are also taken from Hipparchus.
55. Proclus, *Hypotyposis* . . ., Karl Manitius, ed. (Leipzig, 1909), 126–130; Rome, *op. cit.*, 90–92. The latter appears more reliable.
56. *Histoire de l'astronomie ancienne*, I, 117, 185; II, 185.
57. Manitius ed., I, 274–275.
58. That of 2 May 127 B.C., Manitius, ed., I, 271.
59. *Almagest* I, 12, Manitius ed., I, 44.
60. *Ibid.*, VII, 1, Heiberg ed., II, 11, ll. 23–24: *ταῖς κατὰ τὸν τοῦ Ἱππάρχου τῆς στερεᾶς σφαίρας ἀστερισμὸν διατυπώσεσιν.*
61. Strabo, *Geography* 7, H. L. Jones, ed., I, 22–24.
62. Hephaestion, *Apotelesmatica*, D. Pingree, ed., I (Leipzig, 1973), 4, 22.
63. *Doxographi Graeci*, H. Diels, ed. (Berlin, 1879), 404.
64. Simplicius. *Commentary on Aristotle's De caelo*, J. L. Heiberg, ed. (Berlin, 1894), which is *Commentaria in Aristotelem Graeca*, VII, 264–265.

BIBLIOGRAPHY

The only general account of Hipparchus that is at all adequate is O. Neugebauer, *A History of Ancient Mathematical Astronomy*, 3 vols. (New York, 1975) (which is Studies in the History of Mathematics and the Physical Sciences, no. 1), I, 274–343, on which I have drawn heavily. Still worth reading, despite their assumption of Ptolemy's servile dependence on Hipparchus, are J. B. J. Delambre, *Histoire de l'astronomie ancienne*, I (Paris, 1817), 106–189, and II (Paris, 1819) (the section on the *Almagest*); and Paul Tannery, *Recherches sur l'histoire de l'astronomie ancienne* (Paris, 1893). A. Rehm, "Hipparchos 18," in Pauly-Wissowa, *Real-Encyclopädie der classischen Altertumswissenschaft*, VIII, 2 (Stuttgart, 1913), cols. 1666–1681, is useful only for the references to Hipparchus in ancient sources that it assembles. No attempt can be made here to mention all the scattered references to Hipparchus in ancient sources (the more important are cited in text and notes). Our chief source, Ptolemy's *Almagest*, was edited by J. L. Heiberg, *Claudii Ptolemaei Opera quae extant omnia*, I, *Syntaxis mathematica*, 2 vols. (Leipzig, 1898–1903). Most of my references are to the German trans. by K. Manitius: *Ptolemäus, Handbuch der Astronomie*, 2nd ed., 2 vols., with corrections by O. Neugebauer (Leipzig, 1963). The references to Hipparchus by name can be found from the index in Manitius or that in J. L. Heiberg, *Ptolemaei Opera . . . omnia*, II, *Opera astronomica minora* (Leipzig, 1907).

Pliny the elder gives some important information in bk. II of his *Naturalis historia*, J. Beaujeu, ed. (Paris, 1950). For bibliography of the coins of Nicaea representing Hipparchus, see W. Ruge, "Nikaia 7," in Pauly-Wissowa, *Real-Encyclopädie*, XVII, 1 (Stuttgart, 1936), col. 237. Also see *British Museum Catalogue of Greek Coins, Pontus*, W. Wroth, ed. (London, 1889), 167 and pl. xxxiii.9; and *Sylloge nummorum graecorum, Deutschland, Sammlung v. Aulock, Pontus* (Berlin, 1957), nos. 548 (pl. 549), 717. There is a good photograph in Karl Schefold, *Die Bildnisse der antiken Dichter Redner und Denker* (Basel, 1943), 173, no. 30. Ptolemy's "Phases of the Fixed Stars" is printed in Heiberg's ed. of Ptolemy's *Opera astronomica minora*, 3–67. On Hipparchus' trigonometry, see G. J. Toomer, "The Chord Table of Hipparchus and the Early History of Greek Trigonometry," in *Centaurus*, **18** (1973), 6–28. On arithmetical methods for rising times, see O. Neugebauer in V. de Falco and M. Krause, *Hypsikles, Die Aufgangszeiten der Gestirne,* which is *Abhandlungen der Akademie der Wissenschaften zu Göttingen*, Phil.-hist. Kl., 3rd ser., no. 62 (1966), 5–17.

On arithmetical methods in Hipparchus' geography, see O. Neugebauer, *History of Ancient Mathematical Astronomy*, I, 304–306. On the "seven climata" see E. Honigmann, *Die sieben Klimata und die* ΠΟΛΕΙΣ ΕΠΙΣΗΜΟΙ (Heidelberg, 1929). On analemma methods, see Neugebauer, *History of Ancient Mathematical Astronomy*, II, 839–856. On Ptolemy's *Analemma* in particular, see P. Luckey, "Das Analemma von Ptolemäus," in *Astronomische Nachrichten*, **230** (1927), 17–46. On Hipparchus' equinox observations and solar theory see A. Rome, "Les observations d'équinoxes et de solstices dans le chapitre 1 du livre 3 du Commentaire sur l'*Almageste* de Théon d'Alexandrie," I and II, in *Annales de la Société scientifique de Bruxelles*, **57**, no. 1 (1937), 219–236, and **58**, no. 1 (1938), 6–26; and "Les observations d'équinoxes de Ptolémée," in *Ciel et terre*. **59** (1943), 1–15; and Viggo Petersen and Olaf Schmidt, "The Determination of the Longitude of the Apogee of the Orbit of the Sun According to Hipparchus and Ptolemy," in *Centaurus*, **12** (1968), 73–95.

The derivation of Hipparchus' lunar parameters from Babylonian astronomy was discovered by F. X. Kugler, in *Die Babylonische Mondrechnung* (Freiburg im Breisgau, 1900), and fully explained by A. Aaboe, "On the Babylonian Origin of Some Hipparchian Parameters," in *Centaurus*, **4** (1955), 122–125. On Hipparchus' derivation of the radius of the moon's epicycle, see G. J. Toomer, "The Chord Table of Hipparchus . . .," 8–16;

and "The Size of the Lunar Epicycle According to Hipparchus," *ibid.*, **12** (1967), 145–150. For an explanation of his method of finding the moon's epoch in latitude, see Olaf Schmidt, "Bestemmelsen af Epoken for Maanens Middelbevaegelse i Bredde hos Hipparch og Ptolemaeus," in *Matematisk Tidsskrift*, ser. B (1937), 27–32 (whence Neugebauer, *History of Ancient Mathematical Astronomy*, I, 313). Our most detailed information about "On Sizes and Distances" comes from Pappus' commentary on *Almagest* V, in A. Rome, ed., *Commentaires de Pappus et de Théon d'Alexandrie sur l'Almageste*, I (which is Studi e Testi, no. 54) (Rome, 1931), 67–68. This passage was edited and discussed by F. Hultsch, "Hipparchos über die Grösse und Entfernung der Sonne," in *Berichte der Königlich sächsischen Gesellschaft der Wissenschaft zu Leipzig*, Phil.-hist Kl., **52** (1900), 169–200; but Hultsch failed to understand the text and, misled by the size of the sun given by Theon of Smyrna, "emended" the distance 490 earth radii to 2,490. The correct explanation of Hipparchus' procedure in bk. II was given by N. Swerdlow, "Hipparchus on the Distance of the Sun," in *Centaurus*, **14** (1969), 287–305; and a reconstruction of the whole of Hipparchus' procedure by G. J. Toomer, "Hipparchus on the Distances of the Sun and Moon," in *Archive for History of Exact Sciences*, **14** (1975), 126–142. The commentary on Aratus was edited by Karl Manitius, *Hipparchi in Arati et Eudoxi Phaenomena commentariorum libri tres* (Leipzig, 1894), with German trans. and often misleading notes. A brilliant analysis of the fixed-star data contained therein was made by H. Vogt, "Versuch einer Wiederherstellung von Hipparchs Fixsternverzeichnis," in *Astronomische Nachrichten*, **224** (1925), 17–54. A comparison of the stars on the hour circles with modern data was made by H. C. F. C. Schjellerup, "Recherches sur l'astronomie des anciens I. Sur le chronomètre céleste d'Hipparque," in *Urania*, **1** (1881), 25–39.

On the "new star" allegedly observed by Hipparchus, see J. K. Fotheringham, "The New Star of Hipparchus and the Dates of Birth and Accession of Mithridates," in *Monthly Notices of the Royal Astronomical Society*, **79** (1918–1919), 162–167. The excerpt from Hipparchus' star catalog is published and discussed by F. Boll, "Die Sternkataloge des Hipparch und des Ptolemaios," in *Bibliotheca mathematica*, 3rd ser., **2** (1901), 185–195. Related texts are printed in E. Maass, *Commentariorum in Aratum reliquiae* (Berlin, 1898), 128, 137–139, 177–275. There may be some connection between Hipparchus' catalog and a list of stars in late Latin MSS published and discussed by W. Gundel, *Neue astrologische Texte des Hermes Trismegistos* (Munich, 1936), which is *Abhandlungen der Bayerischen Akademie der Wissenschaften*, Phil.-hist. Abt., n.s. **12**, but most of Gundel's conclusions are to be rejected. The evidence for the value adopted by Hipparchus for the length of the sidereal year is discussed by O. Neugebauer, "Astronomical Fragments in Galen's Treatise on Seven-Month Children," in *Rivista degli studi orientali*, **24** (1949), 92–94. On his intercalation cycle see F. K. Ginzel, *Handbuch der mathematischen und technischen Chronologie*, II (Leipzig, 1911), 390–391.

On Babylonian "goal-year" texts see A. Sachs, "A Classification of the Babylonian Astronomical Tablets of the Seleucid Period," in *Journal of Cuneiform Studies*, **2** (1948), 282–285. The section of Ptolemy's "Planetary Hypotheses" mentioning Hipparchus is found only in the ed. of the Arabic text by B. R. Goldstein, "The Arabic Version of Ptolemy's Planetary Hypotheses," in *Transactions of the American Philosophical Society*, n.s. **57**, pt. 4 (1967). The instruments mentioned are well described and illustrated by D. J. Price and A. G. Drachmann in *A History of Technology*, Charles Singer *et al.*, eds., III (Oxford, 1957), 586–614. On the four-cubit diopter also see F. Hultsch, "Winkelmessungen durch die Hipparchische Dioptra," in *Abhandlungen zur Geschichte der Mathematik*, **9** (1899), 193–209; on the armillary astrolabe see A. Rome, "L'astrolabe et le météoroscope d'après le commentaire de Pappus sur le 5e livre de l'*Almageste*," in *Annales de la Société scientifique de Bruxelles*, ser. A, **47**, no. 2 (1927), "Mémoires," 77–102, with good illustration on 78. On the history of the celestial globe in antiquity, see A. Schlachter, *Der Globus* (Leipzig–Berlin, 1927), which is ΣΤΟΙΧΕΙΑ, no. 8. There are good illustrations of the most famous surviving example, the globe borne by the Farnese Atlas, in Georg Thiele, *Antike Himmelsbilder* (Berlin, 1898), pls. II–VI. The (defective) text of Hero's work on the diopter is in *Heronis Alexandrini Opera quae supersunt omnia*, H. Schöne, ed., III (Leipzig, 1903), 188–315. On the evidence for the invention of stereographic projection and the astrolabe by Hipparchus, see O. Neugebauer, "The Early History of the Astrolabe," in *Isis*, **40** (1949), 240–256; and *A History of Ancient Mathematical Astronomy*, II, 868–869. The fragments of Hipparchus' geographical treatise are available in two eds. (both inadequate): H. Berger, *Die geographischen Fragmente des Hipparch* (Leipzig, 1869); and D. R. Dicks, *The Geographical Fragments of Hipparchus* (London, 1960). Our knowledge of the work is derived almost entirely from Strabo, cited here in the ed. by H. L. Jones, 8 vols., Loeb Classical Library (London–New York, 1917–1932).

On Hipparchus as an astrologer, see W. and H. G. Gundel, *Astrologumena* (Wiesbaden, 1966), which is *Sudhoffs Archiv . . .*, supp. 6, 109–110, with references to most relevant texts. The astrological work falsely ascribed to him in Arabic bibliographical works was called "The Secrets of the Stars" (*Kitāb asrār al-nujūm*): see Ibn al-Qifṭī, *Taʾrīkh al-ḥukamāʾ*, J. Lippert, ed. (Leipzig, 1903), 69; and Abuʾl Faraj (Bar Hebraeus), *Chronography*, E. A. Wallis Budge, trans., I (London, 1932), 29. The description of the contents (*ibid.*) proves that this work originated in Islam. On Hipparchus' arithmetical work, see (for what little is known) T. L. Heath, *A History of Greek Mathematics*, II (Oxford, 1921), 256. An inconclusive attempt to reconstruct his procedure was made by Kurt R. Biermann and Jurgen Mau,

"Überprüfung einer frühen Anwendung der Kombinatorik in der Logik," in *Journal of Symbolic Logic*, **23** (1958), 129–132. On combinatorial arithmetic in antiquity, see A. Rome, "Procédés anciens de calcul des combinaisons," in *Annales de la Société scientifique de Bruxelles*, ser. A., **50** (1930), "Mémoires," 97–104.

G. J. TOOMER

HITCHCOCK, ALBERT SPEAR (*b.* Owosso, Michigan, 4 September 1865; *d.* at sea, crossing North Atlantic, 16 December 1935), *botany*.

Hitchcock became the leading agrostologist in the United States and was acknowledged around the world. The son of Albert Hitchcock and Alice Martin Jennings, he was adopted by the J. Seabury Hitchcocks of St. Joseph, Missouri. At Iowa State Agricultural College, where he was profoundly influenced by C. E. Bessey, Hitchcock received his B.S.A. (1884) and M.S. (1886), and subsequently was instructor for three years. From 1889 to 1891 he assisted Trelease at the Missouri Botanical Garden in St. Louis, popularly known as Shaw's Gardens. As apprentice in the herbarium he gained experience in acting as a curator of historic collections, gathered information on the location of type specimens in European herbaria, and codified the concept of a "type specimen" as the base of a species. Its adoption was finally incorporated in the International Rules of Botanical Nomenclature at the Fifth International Botanical Congress in 1930, this leading to its universal acceptance.

Elected president of the newly founded Botanical Society of America in 1914, Hitchcock was named one of four editors for its journal. His writing was marked by clarity and precision, and his *Methods of Descriptive Systematic Botany* (1925) was widely followed. To encourage beginners he privately published *Field Work for the Local Botanist* (Washington, D.C., 1931). He botanized from Alaska to Argentina and from the Hawaiian Islands to the West Indies, in the Philippines, the Far East, Indochina, South Africa, and Europe, assembling at the National Herbarium, Washington, D.C., one of the world's largest and most comprehensive grass collections. These specimens served as documentation for monographs and grass floras of Central America, British Guiana, Ecuador, Peru, Bolivia, and the West Indies, and, most importantly, for his *Manual of the Grasses of the United States* (1935). This manual, prepared with the assistance of Agnes Chase, who joined his staff in 1907 and who prepared the second edition (1951), rested on the identity of valid species but also recorded all known synonyms. Although Hitchcock does not so state, his manual follows Bessey's phylogeny, which depends on the relative complexity of flower structure. Hitchcock genially contributed to the orderly handling of grass nomenclature for agronomists, foresters, and the general public.

As chairman of the executive committee of the Institute for Research in Tropical America (1920–1926) Hitchcock encouraged the establishment of Barro Colorado Island, Panama Canal Zone, as a field station.

In 1890 Hitchcock married Rania Belle Dailey of Ames, Iowa. The youngest of their five children, Albert Edwin Hitchcock, became a plant physiologist at Boyce Thompson Institute for Plant Research, Inc., Yonkers, New York.

Aimée Camus named *Hitchcockella*, a monotypic grass genus of Madagascar, in his honor. A. B. Rendle of the British Museum summed up: "Hitchcock was a grass enthusiast, and the results of his 34 years' work on the family are an invaluable legacy. A kindly and cheery soul."

BIBLIOGRAPHY

I. ORIGINAL WORKS. Hitchcock's most important work is *Manual of the Grasses of the United States*, United States Department of Agriculture Miscellaneous Publication 200 (Washington, D.C., 1935; 2nd ed., revised by Agnes Chase, 1951). *Methods of Descriptive Systematic Botany* (New York, 1925) is still useful. He published about 85 experiment-station bulletins and popular articles during his years at Kansas State University. A few, such as "Camping in Florida," in *Industrialist* (Nov. 1897), have historical value but have not been noticed in biographical accounts. Hitchcock's field records and memorabilia are in the botany division of the United States National Museum. Some correspondence is preserved in the special collections of the Kansas State University Library, Manhattan, Kans. An incomplete bibliography of his writings is in L. R. Parodi's sketch, in *Revista argentina de agronomia*, **3** (1936), 116–119, with portrait.

II. SECONDARY LITERATURE. No wholly satisfactory account of Hitchcock has been published. Frans Verdoorn's summary distributed as a broadside (Leiden, 1937), prepared for his proposed "Index botanicorum," is the fullest account. Also useful are the obituary notices by Agnes Chase, in *Science*, **83** (1936), 222–224, although, contrary to her statement, Hitchcock did not initiate the Barro Colorado Island plan—cf. F. M. Chapman, *My Tropical Air Castle* (New York, 1929); and by A. B. Rendle, in *Journal of Botany, British and Foreign*, **74** (1936), 54. See also M. L. Fernald, "Hitchcock's

Manual of the Grasses," in *Rhodora*, **37** (1935), 369–372; and H. B. Humphrey, *Makers of North American Botany* (New York, 1961), 109–111, to be read with caution–cf. J. Ewan's review of this book in *Rhodora*, **64** (1962), 186–190.

JOSEPH EWAN

HOËNÉ-WROŃSKI (or **HOEHNE**), **JÓZEF MARIA** (*b.* Wolsztyn, Poland, 23 August 1776; *d.* Neuilly, near Paris, France, 8 August 1853), *philosophy, mathematics.*

Hoehne was the son of Antoni Hoehne, the municipal architect of Poznań, and Elżbieta Pernicka. Educated in Poznań and Warsaw, he took part as a young artillery officer in the national uprising of 1794, commanding a battery during the siege of Warsaw by the Prussian army. In the same year he was taken prisoner by the Russian army, which he joined for a short period. He was released about 1797 and, living on money left to him by his father, spent the next few years studying philosophy at several German universities. In 1800 Hoehne settled in Marseilles, where he became a French citizen and addressed himself to scientific research. At first he took occasional jobs in scientific institutions, working at the Marseilles astronomical observatory and as secretary of the local medical association. In later years he earned his living by giving private lessons in science and philosophy. At various periods he was supported by patrons who had been converted to his philosophical doctrine and thus obtained the funds for his prolific publishing activity.

About 1810 Hoehne married Victoire Henriette Sarrazin de Montferrier, sister of the mathematician Alexandre Montferrier. At approximately the same time he adopted the surname Wroński, which he used alternatively with Hoehne; but most of his writings are signed Hoëné-Wroński, without a first name.

In 1810 Hoëné-Wroński moved to Paris and submitted to the Institut his first memoir on the foundations of mathematics, "Premier principe des méthodes analytiques." The paper received a rather sketchy review by Lacroix and Lagrange, and the ensuing polemic initiated by Hoëné-Wroński quickly led to a break in relations with the Institut. During his first years in Paris, he conducted intensive research in mathematical analysis, subsidized by the financier Pierre Arson, who was at first a devoted disciple. Their relations dissolved in a violent quarrel over financial arrangements that resulted in a trial, famous at the time, in 1819.

From 1820 to 1823 Hoëné-Wroński tried unsuccessfully to obtain the award of the British Board of Longitude for research on the determination of longitude at sea. He also failed in attempts to interest the Royal Society in his writings on hydrodynamics. In both cases Hoëné-Wroński became embroiled in polemics that quickly extended to extrascientific matters. He continued his mathematical research after returning to Paris, although his main interest had turned to the explication of his Messianic philosophy. In the 1830's Hoëné-Wroński investigated locomotion and sought to build vehicles that could compete both technically and economically with the newly developing railways, but the caterpillar vehicles that he designed did not progress beyond the model stage. His last years were spent in poverty.

Hoëné-Wroński's extant manuscripts and published writings cover a wide range of knowledge. His philosophy, which is central, forms the basis for reforming various branches of the exact and social sciences. Hoëné-Wroński's philosophical notions were formed under Kant's influence; and his first published work, *Philosophie critique découverte par Kant* (Marseilles, 1803), was the first exhaustive presentation of Kant's teachings in French. Hoëné-Wroński's philosophical system was based on the sudden revelation of the "Absolute," a concept never made precise, from which all aspects of existence evolve. This universal and rationalistic "absolute philosophy" could, according to its author, solve all theoretical and practical problems. Three main concepts constitute its framework: the "highest law," the foundation of reality independent of human influence; the "universal problem," man's supplementing of the Creation by introducing new realities; and the "final concordance," that harmony among various aspects of reality which is humanity's ultimate aim.

Hoëné-Wroński applied his philosophy to mathematics in a series of works that began with *Introduction à la philosophie des mathématiques.* In these writings rigorous mathematical proof retreated before arguments of the absolute philosophy–which, with the specific nomenclature introduced, made the reception and evaluation of his works difficult. Hoëné-Wroński criticized the standpoint taken by Lagrange in his *Théorie des fonctions analytiques*, disagreeing with both Lagrange's insufficient grounds for the use of the series development and his opposition to the introduction of infinite quantities in analysis. According to Hoëné-Wroński, the "highest law" in mathematics consisted in the development of any function in the series

$$F(x) = A_0\Omega_0(x) + A_1\Omega_1(x) + A_2\Omega_2(x) + \cdots,$$

where Ω_i denotes any function of the variable x. The "highest law" was to constitute the basis of the entire theory of differential equations. The lack of proof and imprecise range of applicability rendered its evaluation difficult; it is functional analysis that can determine the scope of Hoëné-Wroński's theorem. The determinants used to compute the coefficients A_i are known as Wronskians, a term introduced by Thomas Muir in 1882.

In 1812 Hoëné-Wroński published his universal solution of algebraic equations, *Résolution générale des équations de tous les degrés.* Althrough Ruffini's research had already demonstrated that this solution cannot be correct, it is applicable in particular cases. Several errors were found in Hoëné-Wroński's papers in other branches of sciences, for instance, in his treatment of the laws of hydrodynamics. His celestial mechanics—although based on a law allegedly more general than Newton's—was in fact equivalent to it. On the other hand, his method of resolving perturbative functions contained new ideas and was later found to be feasible.

A recurrent pattern in Hoëné-Wroński's relations with various institutions, both academic and social, indicates a marked psychopathic tendency: grandiose exaggeration of the importance of his own research, violent reaction to the slightest criticism, and repeated recourse to nonscientific media as allies against a supposed conspiracy. His aberrant personality, as well as the thesis of his esoteric philosophy (based on a revelation received on 15 August 1803 or, according to his other writings, 1804), tempt one to dismiss his work as the product of a gigantic fallacy engendered by a troubled and deceived mind. Later investigation of his writings, however, leads to a different conclusion. Hidden among the multitude of irrelevancies are important concepts that show him to have been a highly gifted mathematician whose contribution, unfortunately, was overshadowed by the imperative of his all-embracing absolute philosophy.

BIBLIOGRAPHY

I. Original Works. A bibliography of Hoëné-Wroński's writings and a catalog of MSS preserved at the library of the Polish Academy of Sciences at Kórnik is in S. Dickstein, *Hoene Wroński. Jego życie i prace* (". . . His Life and Works"; Cracow, 1896). Subsequent literature is in B. J. Gawecki, *Wroński i o Wrońskim. Katalog prac filozoficznych Hoene Wrońskiego oraz literatury dotyczącej jego osoby i życia* ("Wroński and About Wroński. Catalog of Philosophical Works by Hoëné-Wroński and of Works on the Man and His Philosophy"; Warsaw, 1958). An incomplete ed. of Hoëné-Wroński's philosophical works is F. Warrain, ed., *L'oeuvre philosophique de Hoene Wroński*, 2 vols. (Paris, 1933–1936). A selection of philosophical works published in Italian is *Collezione italiana degli scritti filosofici di Hoene Wronski*, 4 pts. (Vicenza, 1870–1878). The mathematical works appeared as J. M. Wroński, *Oeuvres mathématiques*, 4 vols. (Paris, 1925).

II. Secondary Literature. An introduction to Hoëné-Wroński's philosophy is P. d'Arcy, *Hoëné-Wroński, une philosophie de la création* (Paris, 1970). On the mathematical "supreme law" see S. Banach, "Über das 'loi suprême' von J. Hoene Wroński," in *Bulletin international de l'Académie polonaise des sciences et des lettres*, ser. A (1939), 1–10; and C. Lagrange, "Démonstration élémentaire de la loi suprême de Wroński," in *Mémoires couronnés . . . publiés par l'Académie royale des sciences . . . de Belgique*, **47**, no. 2 (1886). His astronomy is discussed in F. Koebcke, "Über Hoëné-Wroński's Überlegungen zur Himmelsmechanik," in *Acta astronomica*, ser. C, **3** (Feb. 1938), 73–81.

Jerzy Dobrzycki

HOLBROOK, JOHN EDWARDS (*b.* Beaufort, South Carolina, 30 December 1794; *d.* Norfolk, Massachusetts, 8 September 1871), *herpetology, ichthyology.*

Holbrook, son of Silas Holbrook, a schoolmaster, and Mary Edwards Holbrook, spent his childhood and received his early education at Wrentham, Massachusetts, his father's birthplace. He received his B.A. in 1815 from Brown University and his M.D. from the University of Pennsylvania in 1818. He then continued his medical studies in Boston, London, and Edinburgh. In 1820 Holbrook traveled to Paris, where he studied natural history with Cuvier and his associates at the Muséum d'Histoire Naturelle. He returned to the United States in 1822 and established a medical practice in Charleston, South Carolina. Holbrook joined the group of local physicians who in 1824 founded the Medical College of South Carolina, where he served as professor of anatomy until about 1854. In 1827 he married Harriott Rutledge, the daughter of a prominent South Carolina plantation owner. During the Civil War, Holbrook served as a medical officer in the Confederate Army and as chief medical examiner for the state of South Carolina. He was a member of the American Philosophical Society, the American Association for the Advancement of Science, the National Academy of Sciences, the Academy of Natural Sciences of Philadelphia, and several other scientific and professional organizations.

Although Holbrook was among the most respected medical men of South Carolina, it was in herpetology that he enjoyed an international reputation, secured by the publication of *North American Herpetology*, at the time the most accurate and comprehensive descriptive work on the American reptiles. He began work on *North American Herpetology* in 1826, but the first edition was not completed until 1840. During these fourteen years he collected and described the species he intended to cover. Much time was spent in preparing the hand-colored lithograph prints, by J. Sera, illustrating each of the 111 species. Holbrook, a perfectionist, insisted that descriptions and illustrations be taken from observation of live animals rather than deformed or discolored preserved specimens. A reorganized second edition, including descriptions and illustrations of thirty-six additional species, was published in 1842.

Following the completion of *North American Herpetology*, Holbrook visited Europe, where his work was well received. His colleagues at the Muséum d'Histoire Naturelle were especially impressed and asked him to reclassify the American reptiles in their collections, using the scientific nomenclature that appeared in *North American Herpetology*.

Satisfied that he had brought some order to herpetology in North America, Holbrook turned to ichthyology, hoping to accomplish the same for that field. He realized that the ichthyology of North America was too broad a topic and soon discovered that even the fish of the southern states were too numerous to describe by using his techniques of live observation. Finally he limited his studies to the fish of South Carolina, publishing *Ichthyology of South Carolina* in 1855. Again, detailed descriptions were accompanied by colored illustrations, with the addition of anatomical data. A second edition was published in 1860, after a fire destroyed the lithograph plates prepared for the first edition. Holbrook's ichthyological works were not as well received as *North American Herpetology*.

Holbrook is remembered not as a theoretical innovator but as a collector and describer of animals. On questions of taxonomy he generally adhered to the system prescribed by Cuvier and his followers, both preserving its good points and perpetuating its inaccuracies. He resisted many of the innovations in taxonomy introduced by his contemporaries, including those of his colleague and close friend Louis Agassiz. In *North American Herpetology*, for example, Holbrook minimized the importance of embryological data in developing his taxonomical scheme. For this reason he did not consider the amphibians a separate class of animals but placed them in an order, Batrachia, of the class Reptilia.

It was his great energy as a collector, his talent for meticulous observation and description, his fascination with his subjects, and his thorough scholarship that made Holbrook's works contributions of lasting value. Present-day naturalists recognize Holbrook as an important figure in the history of herpetology. Of the genera and species of reptiles, amphibians, and fish of which the discovery or earliest description is credited to Holbrook, many still carry the genus name *Holbrookia* or the species name *holbrookii*.

BIBLIOGRAPHY

I. Original Works. See *North American Herpetology; or, a Description of the Reptiles Inhabiting the United States*, 4 vols. (Philadelphia, 1836–1840; 2nd ed., 5 vols., Philadelphia, 1842)—the 1st ed. was recalled and is thus very rare; *Southern Ichthyology; or, a Description of the Fishes Inhabiting the Waters of South Carolina, Georgia and Florida* (New York–London, 1847), a work Holbrook planned to publish in installments but abandoned after the first installment; and *Ichthyology of South Carolina* (Charleston, S.C., 1855; 2nd ed., 1860), the ichthyological tract Holbrook managed to complete. He published only one journal article, "An Account of Several Species of Fishes Observed in Florida, Georgia, &c.," in *Journal of the Academy of Natural Sciences of Philadelphia*, **3** (1855), 47–58. All of the above works included lithograph plates prepared by Lehman and Duval Co. of Philadelphia. Unpublished Holbrook letters, which document his activities as a collector, are found at the Academy of Natural Sciences of Philadelphia, the American Philosophical Society, and the South Caroliniana Library of the University of South Carolina.

II. Secondary Literature. There are short biographical articles on Holbrook in *Appleton's Cyclopedia of American Biography; Dictionary of American Biography*, IX, 129–130; and *Dictionary of American Medical Biography*. Lengthier biographies include Louis Agassiz, "Eulogy on John E. Holbrook," in *Proceedings of the Boston Society of Natural History*, **14** (1872), 347–351; Theodore Gill, "Biographical Memoir of John Edwards Holbrook," in *Biographical Memoirs. National Academy of Sciences*, **5** (1903), 49–77; and Thomas Ogier, *A Memoir of Dr. John Edwards Holbrook . . .*

(Charleston, S.C., 1871). Theodore Gill also wrote two reviews of Holbrook's works: "Holbrook's *Ichthyology of South Carolina* Noticed," in *American Journal of Science and Art*, 2nd ser., **3** (1864), 89–93; and "First Edition of Holbrook's *North American Herpetology*," in *Science*, **17** (1903), 910–912. See also Joseph I. Waring, *A History of Medicine in South Carolina, 1670–1825* (Columbia, S.C., 1964); and Reese E. Griffin, Jr., "The Social Structure of Science in Nineteenth-century South Carolina" (M.A. thesis, Univ. of Pennsylvania, 1973).

REESE E. GRIFFIN, JR.

HOUSSAY, BERNARDO ALBERTO (*b.* Buenos Aires, Argentina, 10 April 1887; *d.* Buenos Aires, 21 September 1971), *physiology, pharmacology, medicine.*

Houssay was one of the most prominent and influential Latin American scientists of the twentieth century. His total dedication to the pursuit of knowledge and his untiring efforts to foster scientific and technical training among his compatriots received worldwide recognition. For more than twenty-five years his Institute of Physiology at the University of Buenos Aires was the scientific beacon for all of Latin America, and from its laboratories emerged disciples who now occupy prominent positions in scientific research and training throughout the continent.

The son of a French lawyer, Houssay was a precocious child who by the age of thirteen had already received his baccalaureate degree with honors from the Colegio Nacional de Buenos Aires. Thus in 1901 he was able to enter the School of Pharmacy of the University of Buenos Aires, from which he graduated first in his class at age seventeen. Houssay subsequently studied medicine at the University of Buenos Aires between 1904 and 1910; and his doctoral dissertation concerning the physiological activities of pituitary extracts, which won the school's highest award, was published in 1911.

While such academic achievements were being completed, Houssay was working as a hospital pharmacist to pay for his education and personal expenses. In 1908 he was named an assistant in the department of physiology of the Medical School, and was appointed to the chair of physiology at the School of Veterinary Science of the university the following year.

A measure of Houssay's versatility and capacity for work can be seen in his activities after graduation from medical school. He established a private practice and became chief of a municipal hospital service while continuing as full professor in the School of Veterinary Science and part-time substitute professor in physiology at the Medical School. Beginning in 1915, Houssay took on the additional duties of chief of the section of experimental pathology at the National Public Health Laboratories in Buenos Aires. In the latter capacity he studied the action of snake and insect bites on coagulation, and developed a protective serum against certain spider toxins.

In 1919 Houssay was appointed to the chair of physiology at the University of Buenos Aires School of Medicine, and promptly converted the department into a full-fledged Institute of Physiology capable of engaging in experimental investigations. For the next twenty-five years he developed the Institute into one of the most prestigious world centers of physiological research.

Houssay's devotion to academic and political freedom collided with the military dictatorship ruling Argentina after the 1943 revolution. Consequently he was stripped of his university posts and forced to continue his research in a private laboratory especially organized for him and his collaborators by the Sauberán Foundation.

A short-lived restoration of Houssay's academic position after the general amnesty of 1945 was followed by a second dismissal, ordered by the new government of Juan Perón. Despite numerous offers from other countries, Houssay remained in Argentina and was officially reinstated as director of the Institute of Physiology in 1955. He spent his last years directing the Argentine National Council for Scientific and Technical Research, which he had conceived and founded in 1957. This governmental organization sought to create new scientific careers, support research institutes, and stem the emigration of technical personnel.

Houssay was an outstanding, largely self-taught scientist who was influenced at the beginning of his career by Claude Bernard's applications of the scientific method to medical problems. His early interests in the physiology of the pituitary gland and systematic studies regarding the action of insulin eventually led to a recognition of the role played by the anterior lobe of the hypophysis in carbohydrate metabolism. For this work Houssay shared the Nobel Prize in physiology or medicine with G. T. and C. F. Cori in 1947. Before Houssay's research, it was commonly accepted that the posterior lobe of the hypophysis played a role in carbohydrate metabolism. After the discovery of insulin Houssay systematically studied the influence of endocrine glands on its activity. He soon discovered that hypophysectomized dogs were very sensitive to the hypoglycemic action of insulin.

By 1930, Houssay had proved the diabetogenic effect of extracts from the anterior lobe of the pituitary gland. Conversely, there was a remarkable decrease in the symptomatic severity of pancreatic diabetes after removal of the anterior pituitary lobe. The new vistas in endocrinological research opened by Houssay's attention to the anterior portion of the pituitary gland were momentous, leading to the discovery of a number of hormonal feedback mechanisms involving the thyroid, adrenals, and gonads.

With his disciples Houssay also studied the pancreatic secretion of insulin, the hormonal control of fat metabolism, and the factors regulating arterial blood pressure. Over 600 scientific papers and several books attest to the breadth as well as the depth of his research.

Houssay's activities were widely admired and recognized. A long-time member of the Argentine Academy of Medicine and founder of the Argentine Association for the Advancement of Science and the Argentine Biological Society, Houssay received many honors, including degrees from Paris, Oxford, Cambridge, and Harvard. In addition he was an associate foreign member of many scientific societies in the United States, Britain, Germany, France, Italy, and Spain.

BIBLIOGRAPHY

I. Original Works. Houssay's doctoral dissertation appeared as a book: *Estudios sobre la acción de los extractos hipofisarios. Ensayo sobre la fisiología del lóbulo posterior de la hipófisis* (Buenos Aires, 1911). Among his best-known books are *La acción fisiológica de los extractos hipofisarios* (Buenos Aires, 1918); *Tiroides e inmunidad, estudio crítico y experimental* (Buenos Aires, 1924), written with A. Sordelli; and *Acción de las tiroides sobre el metabolismo de los hidratos de carbono y en la diabetes* (Buenos Aires, 1945).

Most of Houssay's research papers were published for more than five decades in leading Argentine journals, such as *Revista de la Sociedad argentina de biologia, Revista de la Asociación médica argentina, Boletín de la Academia nacional de medicina de Buenos Aires*, and *Prensa médica argentina*. Others appeared in foreign publications, such as *Comptes rendus des séances de la Société de biologie* and *American Journal of Physiology*.

In 1935 Houssay delivered three Dunham lectures at Harvard, which were published: "What We Have Learned From the Toad Concerning Hypophyseal Functions," in *New England Journal of Medicine*, **214** (1936), 913–926; "The Hypophysis and Metabolism," *ibid.*, 961–971; and "Carbohydrate Metabolism," *ibid.*, 971–986. These papers were reprinted with four additional talks as a book, *Functions of the Pituitary Gland* (Boston, 1936).

A summary of Houssay's most important studies, also written in English, are "Advancement of Knowledge of the Role of the Hypophysis in Carbohydrate Metabolism During the Last Twenty-Five Years," in *Endocrinology*, **30** (1942), 884–897; and "The Hypophysis and Secretion of Insulin," in *Journal of Experimental Medicine*, **75** (1942), 547–566.

Houssay's Nobel Prize lecture, "The Role of the Hypophysis in Carbohydrate Metabolism and in Diabetes," was delivered on 12 December 1947 (in English) and reprinted in Nobel Foundation, *Nobel Lectures in Physiology or Medicine* (New York, 1964), 210–217. With a few disciples, Houssay wrote a compendium of physiology, *Fisiología humana* (Buenos Aires, 1950), which was translated into various languages and went through several eds. The English one was *Human Physiology*, translated by Juan and Olive T. Lewis (New York, 1951). A collection of his earlier writings and speeches on various subjects is *Escritos y discursos* (Buenos Aires, 1942).

II. Secondary Literature. While there is as yet no complete biographical study available, some information can be obtained from the obituaries written by disciples in a number of journals. The most comprehensive one is V. Foglia, "Bernardo Alberto Houssay (1887–1971)," in *Acta physiológica latino americana*, **21** (1971), 267–285. Shorter essays by the same author appeared in *Diabetes*, **22** (1973), 212–214; and *Acta diabetológica latina*, **8** (1971), 1209–1216.

A short biographic sketch of Houssay is appended to his Nobel Prize lecture, in Nobel Foundation, *Nobel Lectures in Physiology or Medicine* (New York, 1964), 218–219; and another, written by Juan T. Lewis, appeared in *Perspectives in Biology*, a collection of papers dedicated to Houssay on the occasion of his 75th birthday, C. F. Cori, V. G. Foglia, L. F. Leloir, and S. Ochoa, eds. (Amsterdam, 1963), vii–xiv.

A list of Houssay's writings until 1961, with data concerning his various activities, positions, and awards, is in Abel Sánchez Díaz, *Bernardo A. Houssay* (Buenos Aires, 1961).

Among the various public acknowledgments of which there are published proceedings are the commemoration of Houssay's 25th anniversary as an academician, *Libro jubilar del Profesor Dr. Bernardo A. Houssay 1910–1934* (Buenos Aires, 1935); and the celebrations of the Argentine Academy of Medicine on the occasion of Houssay's 80th birthday, *Homenajes al Dr. Bernardo A. Houssay* (Buenos Aires, 1967). A complete bibliography of his writings has been prepared and awaits publication in Argentina.

Guenter B. Risse

ḤUNAYN IBN ISḤĀQ AL-ʿIBĀDĪ, ABŪ ZAYD, known in the Latin West as **Johannitius** (*b.* near Ḥīra, Iraq, 808; *d.* Baghdad, Iraq, 873), *medicine, philosophy, theology, translation of Greek scientific works.*

Ḥunayn, a physician, philosopher, and theologian, was the most famous ninth-century translator of works from Greek antiquity into Arabic and Syriac. His *nisba*, al-ʿIbādī, is from an Arab tribe, al-ʿIbād, the members of which became Christian long before the rise of Islam and continued to belong to the Syrian Nestorian church. His father, Isḥāq, was a pharmacist (*ṣaydalānī*) at Ḥīra. While still young, Ḥunayn learned Arabic and Syriac, perfecting his knowledge of the former at Baṣra. Ibn Juljul incorrectly reports that in Baṣra, Ḥunayn met Khalīl ibn Aḥmad, founder of Arabic grammar; M. Plessner has shown, on chronological grounds, that this would have been impossible.[1]

Ḥunayn went to Baghdad to study medicine—a difficult undertaking, according to the picturesque and lively account given by Ibn Abī Uṣaybiʿa in his *ʿUyūn al-anbāʾ*. The teaching of medicine was then dominated by Yūḥannā ibn Māsawayh, originally from Jundīshāpūr. The physicians of that city, highly cultivated men who had long devoted themselves to medicine, felt contempt for the people of Ḥīra, who were concerned primarily with commerce and banking. Further, they were not happy to see these merchants' sons becoming interested in medicine.

Nevertheless, Ibn Māsawayh agreed to supervise Ḥunayn's studies and gave him a book on the various medical schools. Avid for knowledge, Ḥunayn never tired of raising questions (*ṣāḥib su'āl*); one day, when they became particularly urgent and difficult to answer, Ibn Māsawayh flew into a rage and brusquely rebuked his young student:

> What makes the people of Ḥīra want to study medicine? Go away and find one of your friends; he will lend you fifty *dirhems*. Buy some little baskets for a *dirhem*, some arsenic for three *dirhems*, and with the rest buy coins of Kūfa and of Qādisiyya. Coat the money of Qādisiyya with arsenic and put it in the baskets and stand by the side of the road crying: "Here is true money, good for giving alms and for spending. . . ." Sell the coins; that will earn you much more than the science of medicine.[2]

He then ordered Ḥunayn to leave his house.

Refusing to accept defeat, Ḥunayn resolved to pursue his vocation as a physician. He disappeared from Baghdad for several years, during which time he made a profound study of Greek, either in Alexandria or in the "*bilād al-Rūm*," probably in Byzantium. He was remarkably successful in learning the language; Ibn Abī Uṣaybiʿa recounts in detail that one evening, while visiting friends, Ḥunayn recited verses from Homer. He also translated a work by Galen for Jibrāʾīl ibn Bukhtīshūʿ. Ibn Māsawayh, who had sent him away, was forced to recognize his abilities. He was reconciled with Ḥunayn and accepted him as a disciple; the two became close collaborators.

Following the decision of al-Maʾmūn (*d.* 833) to have translations made of the works of Aristotle and of other classical authors, a cultural mission was sent to Byzantium to obtain manuscripts. Ḥunayn, who, according to Ibn Abī Uṣaybiʿa, possessed the best knowledge of Greek of anyone of his time, probably was a member of this mission. The rich lords followed the caliph's example and soon were competing to acquire manuscripts and have them translated.

Ḥunayn, his son Isḥāq, his nephew Ḥubaysh ibn al-Ḥasan al-Aʿsam, and another disciple, ʿĪsā ibn Yaḥyā, earned particular distinction as translators. Alone or in collaboration, Ḥunayn translated works of Plato and Aristotle, and of their commentators. Even more important were his translations of the major portion of the works of the three founders of Greek medicine, whose ideas were also central to the development of Arab medicine: Hippocrates, Galen, and Dioscorides.

Ḥunayn's translation methods were excellent and generally correspond to the standards of modern philology. Severe in his judgment of poor translations made by other writers, he had even more exacting standards for his own work. Referring to his translation of *De sectis*, he wrote:

> I translated it when I was a young man . . . from a very defective Greek manuscript. Later, when I was forty-six years old, my pupil Ḥubaish asked me to correct it after having collected a certain number of Greek manuscripts. Thereupon I collated these so as to produce one correct manuscript and I compared this manuscript with the Syriac text and corrected it. I am in the habit of proceeding thus in all my translation work.[3]

Ḥunayn made long journeys in order to find manuscripts, such as that of Galen's *De demonstratione*: "I sought for it earnestly and travelled in search of it in the lands of Mesopotamia, Syria, Palestine and Egypt, until I reached Alexandria, but I was not able to find anything, except about half of it at Damascus."[4]

ḤUNAYN IBN ISḤĀQ

Ḥunayn and his disciples strove to render the Greek text as clearly as possible. In fact, Ḥunayn may be considered one of the creators of the philosophical and scientific idiom of classical Arabic. In his study of Ḥunayn and his school, G. Bergsträsser emphasizes the superiority of Ḥunayn's versions: "The correctness is greater; nevertheless one is left with the impression that this is not the result of anxious effort, but of a free and sure mastery of the language. This is seen in the easier adaptation to the Greek original and the striking exactness of expression obtained without verbosity. It is all this that constitutes the famous *faṣāḥa* (eloquence) of Ḥunain."[5]

Ḥunayn soon became famous and participated in the scholarly meetings at which physicians and philosophers discussed difficult problems in the presence of Caliph al-Wāthiq (*d.* 847). Al-Mutawakkil (*d.* 861) named him head physician—thereby dismissing the Bukhtīshūʿ family from this post—after having assured himself of Ḥunayn's absolute loyalty. He asked Ḥunayn to prepare a poison to be used in eliminating a supposed enemy, encouraging him with both promises of rewards and with threats; Ḥunayn refused. His intense scientific activity and his favor with the caliph gave rise to jealousy among his colleagues and even among his friends and students. Bukhtīshūʿ ibn Jibrāʾīl, in particular, sought to turn al-Mutawakkil against him, exploiting Ḥunayn's iconoclastic views to this end. He induced Ḥunayn to spit on an icon in the presence of the caliph, who thereupon dismissed Ḥunayn from his post, confiscated his library, and had him imprisoned. The catholicos Theodosius treated Ḥunayn with equal severity, excommunicating him and dismissing him from his functions as deacon. But six months later the caliph fell ill and had to recall Ḥunayn from prison. He again granted him favors, and Bukhtīshūʿ was exiled.[6]

Ḥunayn retained his post until his death. Although his two sons, Dāʾūd and Isḥāq, both became physicians, only Isḥāq followed in his father's footsteps, devoting his efforts primarily to the translation of Greek philosophical works.

Ḥunayn's immense scientific activity consisted mainly of producing translations or revisions of earlier translations (into Arabic or Syriac) but also included a number of original works. Of the several ancient lists of these writings, the most extensive is that in Ibn Abī Uṣaybiʿa, *ʿUyūn al-anbāʾ*.[7] G. Furlani analyzed this list and attempted to draw up a classification scheme.[8] The list in the *Fihrist* is shorter.[9] An even shorter one is that of Ibn al-Qifṭī.[10] A complete but uncritical list of all the works was compiled by Lutfi M. Saʿdi.[11]

For convenience, Ḥunayn's works can be divided into those concerning medicine and those dealing with other subjects. Among the medical works are translations of ancient texts, summaries and paraphrases of these texts, and original treatises. For this group there is a *risāla* sent by Ḥunayn to Abuʾl Ḥasan ʿAlī ibn Yaḥya al-Munajjim, in which he indicates which works of Galen he has translated.[12] (For this area of his activity, consult the section of this article by A. Z. Iskandar.)

The nonmedical works are varied. The nature of some has been definitely established, while others are difficult to classify with certainty as a translation, a paraphrase, or an original work, since all we have is the title. Moreover, many of these works are known only from the lists of Arabic sources.[13] Here we shall present only a selection of Ḥunayn's works.

On his personal life there is *Kitāb ilā ʿAlī ibn Yaḥya*, written in response to the latter's suggestion that Ḥunayn become a Muslim. ʿAlī ibn Yaḥyā was secretary and friend of Caliph al-Mutawakkil. Ḥunayn also addressed his *risāla* on the translated works of Galen to him. *Risāla fīmā aṣābahu min al-miḥan wal-shadāʾid* ("Letter concerning Afflictions and Calamities") is partially quoted in Ibn Abī Uṣaybiʿa.[14] F. Rosenthal, however, contests its authenticity.

Ḥunayn's writings in philosophy comprise translations (into Arabic or Syriac) and original works: translations of Plato (the *Politics*, the *Laws*, and the *Timaeus*); translations of Aristotle (*Categories*, *De interpretatione*, *Analytica priora* and *Analytica posteriora*, *De anima*; fragments of the *Metaphysics*, and the *Ethics*, with Porphyry's commentary); *Kitāb fīmā yuqraʾu qabl kutub Aflāṭūn* ("What to Read Before the Books of Plato"); *Jawāmiʿ kitāb al-samāʾ wal-ʿālam* ("Compendium of the *De caelo et mundo*"); *Masāʾil muqaddama li-kitāb Furfūryūs al-maʿrūf bil-Madkhal* ("Introductory Questions to the Book of Porphyry Known as the Isagoge"), *Jawāmiʿ tafsīr al-qudamāʾ al-yūnāniyyīn li-kitāb Arisṭūṭālīs fiʾl-samāʾ wal-ʿālam* ("Compendium of the Commentary of the Ancient Greeks on Aristotle's Book *De caelo et mundo*"); *Sharḥ kitāb al-firāsa li-Arisṭū* ("Commentary on Aristotle's Book on Physiognomy");[15] *Masāʾil istakhrajahā min kutub al-manṭiq al-arbaʿa* ("Questions Extracted From the Four Books of Logic"); and a translation of the *Book of Dreams of Artemidorus* of Ephesus.[16]

To this list should be added some ten philosophical texts by Galen that Ḥunayn translated

ḤUNAYN IBN ISḤĀQ

into Syriac or Arabic, particularly *That Good People May Benefit From Their Enemies, The Prime Mover Is Immobile*, and *What Plato Says in the Timaeus.*[17] In addition, Ḥunayn supposedly translated from the original Greek the allegorical novel upon which Ibn Sīnā was to base his *Salāmān wa-Absāl.*[18]

Works on Arabic grammar and lexicography include *Kitāb fī aḥkām al-iʿrāb ʿala madhab al-yūnāniyyīn* ("The Rules of Desinential Syntax According to the Greeks"); *Kitāb fiʾl-naḥw* ("On Grammar"); and *Kitāb al-nuqaṭ*. Ḥunayn's grammar was mentioned by Elias of Nisibis in his *Majālis.* In the sixth *majlis* Elias speaks of the superiority of the Syriac language to Arabic.[19] The same chapter contains a passage in which Elias mentions a book by Ḥunayn that is not otherwise known, entitled *Kitāb al-nuqaṭ aʿnī nuqaṭ al-kitāb* ("Book of the Points, I Mean, Points of the Book"). As proof of the superiority of Syriac, Ḥunayn states that the Syrians, Greeks, and Iranians had many names for drugs (*al-ʿaqāqīr wal-adwiya*) and instruments (*ālāt*) that the Arabs did not have.[20]

Kitāb fī masāʾilihi al-ʿarabiyya ("Book on His Arabic Questions") is mentioned immediately after a book entitled *Kitāb fī asmāʾ al-adwiya al-mufrada ʿalā ḥurūf al-muʿjam* ("Names of Medicines Listed Alphabetically"). Ḥunayn may have wished to study, in a second book, the problems posed by the translations of terms for certain medicines.

Scientific subjects other than medicine were treated in the following works: *Kitāb khawāṣṣ al-aḥjār* ("On the Properties of Stones");[21] *Kitāb al-filāḥa* ("On Agriculture"); *Maqāla fīʾl-alwān* ("On Colors"); *Fīʾl-ḍawʾ wa ḥaqīqatih* ("On Light and on Its Nature");[22] *Maqāla fī tawallud al-nār bayn al-ḥajarayn* ("On the Generation of Fire Between Two Stones"); *Maqāla fiʾl-sabab al-ladhī min ajlihi ṣārat miyāh al-baḥr māliḥa* ("The Reason That the Waters of the Sea Are Salty"); *Jawāmiʿ li-Kitāb Arisṭū fiʾl āthār al-ʿulwiyya* ("Compendium of Meteorology");[23] *Maqāla fiʾl-madd wal-jazr* ("On Ebb and Flow"); *Maqāla afʿāl al-shams wal-qamar* ("On the Effects of the Sun and the Moon"); and *Maqāla fī qaws quzaḥ* ("On the Rainbow").

Religion is the subject of *Maqāla fī khalq al-insān wa annahu min maṣlaḥatihi wal-tafaḍḍul ʿalayhi an juʿila muḥtājan* ("On the Creation of Man and That It Is in His Interest and That It Is a Grace for Him to Have Been Created Needy") and *Kitāb fī idrāk ḥaqīqat al-adyān* ("How One Grasps the Truth of Religions"). Ibn Abī Uṣaybiʿa also cites *Kitāb kayfiyyat idrāk al-diyāna* ("How to Grasp Religion"),[24] which may be the same book. The manuscript tradition contains the title *Fī kayfiyyat idrāk ḥaqīqat al-diyāna* ("How to Understand the Truth of Religion"). Ḥunayn provides the criteria that make it possible to distinguish error from truth in religious matters. The true is distinguished from the false by the reasons that one adopts at the beginning. Six reasons lead people to accept falsehood: violence, misery and affliction from which one hopes to escape, the desire for glory and honors, the insidious words of a wily man, the listener's ignorance, and kinship between the preacher and his audience.

Four reasons lead a man to embrace the truth: miracles that surpass human power, discovering proofs of the truth of hidden things in the external and perceptible signs of religion, rational and irrefutable demonstration of the truth that one embraces, and the recognition of the authenticity of the origin of a religion by the successive phases of its development.

This treatise was preserved by Abuʾl Faraj Hibatullāh, known as Ibn al-ʿAssāl. L. Cheikho, who mentions Ḥunayn's book, notes that the Bibliothèque Orientale of the University of St. Joseph in Beirut possesses two manuscripts of the book by Ibn al-ʿAssāl: A, dating from the fifteenth century, and B, which is from the nineteenth century and is written in Karshūnī. The treatise by Ḥunayn that Ibn al-ʿAssāl reproduces is at the beginning of chapter 12 of the first part (A, pp. 233–238; B, pp. 93–96).[25]

Other religious works are *Maqāla fī dalālat al-qadar ʿala l-tawḥīd* ("How Divine Predetermination Is a Proof of the Unicity of God"); *Maqāla fiʾl-ājāl* ("On the Hour of Death")—the theologian Abū Isḥāq al-Muʾtaman ibn al-ʿAssāl mentions a *Kitāb al-ājāl*; and, according to a tradition reported by Masʿūdī,[26] an Arabic translation of the entire Old Testament based on the version in the Septuagint, which was considered one of the best in existence.

A miscellaneous work is *Nawādir al-falāsifa*, a collection of stories, letters, and apothegms of ancient Greek philosophers to which Ḥunayn added his own remarks. At a later period other books were compiled from extracts taken from this collection. There are two manuscripts of the Arabic text, still unpublished, and a medieval Hebrew translation published by A. Loewenthal.[27] In 1921 Karl Merkle made a thorough study of the work, in which he discussed the two manuscripts (Escorial and Munich), the title, and the authenticity of its attribution to Ḥunayn, which dates from the Middle Ages.[28] He also examined the translations: the Hebrew (done in 1200 by the Spanish poet Jehuda ben Salomo al-Kharīzī, the translator of Ḥarīrī), the Spanish (*Libro de los buenos proverbios*), and the German version prepared by Loewenthal. In addition, Merkle established the relationships between the *Ādāb al-falāsifa* and the Arabic sources, unpublished as well as published. Finally, he provided a translation, with commentary, of certain passages. The book is based on a similar Byzantine anthology and contains ancient elements.[29] The third part deals with the death of Alexander.[30] Most of Ḥunayn's translations into Syriac have been lost.[31]

NOTES

1. See M. Plessner, "Der Astronom und Historiker ibn Ṣā'id al-Andalusī und seine Geschichte der Wissenschaften," in *Rivista degli studi orientali*, **31** (1956), 235–257, esp. 244 ff.
2. A. Müller, ed., I (Cairo, 1882), 185 (hereinafter IAU).
3. Cited by M. Meyerhof in *The Book of the Ten Treatises*, xxiv.
4. *Ibid.*
5. *Ibid.*, xxv.
6. On the historical authenticity of this event and the reconciliation of facts reported in a contradictory fashion, see Yūsuf Ḥabbī, *Ḥunayn ibn Isḥāq* (Baghdad, 1974), 36–38 (in Arabic). See also F. Rosenthal, "Die arabische Autobiographie," in *Studia arabica*, **1** (1937), 15–19, which contests the authenticity of the account; B. Hemmerdinger, "Ḥunain ibn Isḥāq et l'iconoclasme byzantin," in *Actes du XIIe Congrès international des études byzantines*, II (Belgrade, 1964), 467–469, in which the author accepts Leclerc's view that Ḥunayn was influenced by the iconoclastic movement during his presumed stay in Byzantium; and G. Strohmaier, "Ḥunain ibn Ishāq und die Bilder," in *Klio*, **43–45** (1965), 525–533.
7. See IAU, I, 184–200.
8. G. Furlani, "Ḥunayn ibn Isḥāq," in *Isis*, **6** (1924), 287–292.
9. *Fihrist*, G. Flügel, ed., I (Leipzig, 1871), 294 ff.
10. *Ta'rīkh al-ḥukamā'*, J. Lippert, ed., I (Leipzig, 1903), 171 ff.
11. Lufti M. Sa'di, "A Biobibliographical Study of Ḥunayn ibn Isḥāq al-Ibadi . . .," in *Bulletin of the Institute of the History of Medicine*, **2** (1934), 409–446, which draws on the Arab sources and on the list made by Lucien Leclerc, *Histoire de la médecine arabe*, 145–152.
12. Compare the ed. and German trans. of this *risāla* by G. Bergsträsser, "Ḥunain ibn Isḥāq über die syrischen und arabischen Galenübersetzungen," which is *Abhandlungen für die Kunde des Morgenlandes*, **17**, no. 2 (1925); his "Neue Materialen zu Ḥunain ibn Isḥāq Galen-bibliographie," *ibid.*, 19, no. 2 (1932), also contains biographical details and information on his working methods.
13. For details concerning these works, see M. Steinschneider, "Die arabischen Übersetzungen aus dem griechischen," in *Zeitschrift der Deutschen morgenländischen Gesellschaft*, **50** (1896), also reprinted (Graz, 1960), 390 (index); G. Bergsträsser, *Ḥunain ibn Isḥāk und seine Schule* (Leiden, 1913); F. Peters, *Aristoteles Arabus* (Leiden, 1968); and 'Abdurraḥmān Badawi, *La transmission de la philosophie grecque au monde arabe*, Études de Philosophie Médiévale, no. 56 (Paris, 1968).
14. See IAU, I, 191–197.
15. On this book, see M. Grignaschi, "La 'physiognomonie' traduite par Ḥunayn ibn Isḥāq," in *Arabica*, **21**, fasc. 3 (1974), 287–291.
16. T. Fahd, ed. (Damascus, 1964).
17. See R. Walzer, "Djālīnūs," in *Encyclopedia of Islam*, new ed., II (Leiden, 1965), 402–403.
18. See A. F. Mehren, in *Muséon*, **4** (1885), 38 ff.; and C. A. Nallino, "Filosofia 'orientale' od 'illuminativa' d'Avicenna?" in *Rivista degli studi orientali*, **10** (1923–1925), 465.
19. See L. Cheikho, *Trois traités anciens* (Beirut, 1923), n. 59.
20. See L. Cheikho, *al-Mashriq*, II (Beirut, 1899), 373.
21. See J. Ruska, *Untersuchungen über das Steinbuch des Aristoteles* (Heidelberg, 1911).
22. This letter was edited by P. Cheikho, in *al-Mashriq*, II, 1105–1115.
23. This work has been edited with a trans. and notes by Hans Daiber, as *Ein Kompendium der Aristotelischen Meteorologie in der Fassung des Ḥunain ibn Isḥāq* (Oxford, 1975).
24. See IAU, 199.
25. See L. Cheikho, "Un traité inédit de Ḥonein," in *Orientalische Studien Theodor Noldeke zum siebzigstein Geburtstag (2. März 1906) gewidmet*, Carl Bezold, ed. (Giessen, 1906), with Arabic text and French trans. The text was reproduced in Cheikho, *Seize traités*, 2nd ed. (Beirut, 1911), 121–123; and *Vingt traités théologiques*, 2nd ed. (Beirut, 1920), 143–146. For other MSS of the text, see G. Graf, *Geschichte der christlich-arabischen Literatur*, II, 123–124.

    On Ḥunayn's religious attitude, see two recent studies: R. Haddād, "Ḥunayn b. Isḥāq, apologiste chrétien," in *Arabica*, **21** (1974), 292–302; and P. Nwyia, "Actualité du concept de religion chez Ḥunayn Ibn Isḥāq," *ibid.*, 313–317.
26. On this point, see *Kitāb al-tanbīh*, Michael Jan de Goeje, ed. (Leiden, 1894), 112.
27. Abraham Loewenthal, *Sefer mūsrē ha pīlōsōfīm* (Frankfurt, 1896), also in German trans. (Berlin, 1896).
28. K. Merkle, *Die Sittensprüche der Philosophen "Kitāb ādāb al-falāsifa" von Ḥonein ibn Isḥāq in der Überarbeitung des Muḥammed ibn 'Ali al-Anṣārī* (Leipzig, 1921).
29. See G. Strohmaier, "Zu einem weiberfeindlichen Diogenesspruch aus Herculaneum," in *Hermes*, **95** (1967), 253–255.
30. On this book, see Hartwig Derenbourg, "Les traducteurs arabes d'auteurs grecs et l'auteur musulman des Aphorismes des philosophes," in *Mélanges Henri Weil* (Paris, 1898), 117–124; M. Plessner, "Analecta to Ḥunain ibn Isḥāq 'Apopthegms of the Philosophers' and Its Hebrew Translation," in *Tarbīz*, **24** (1954–1955), 60–72, with summary in English, vi ff.; J. Kraemer, "Arabische Homerverse," in *Zeitschrift der Deutschen morgenländischen Gesellschaft*, **106** (1956), 292–302; and A. Spitaler, "Die arabische Fassung des Trost-briefs Alexanders an seine Mutter," in *Studi orientalistici in onore di Giorgio Levi della Vida*, II (Rome, 1956), 497 ff.
31. See A. Baumstark, *Geschichte der syrischen Literatur* (Bonn, 1922), 227–230. On the possibility of attributing certain of the fragments to Ḥunayn, see G. Furlani, "Bruchstücke einer syrischen Paraphrase der 'Elemente' des Eukleides," in *Zeitschrift für Semistik*, **3** (1924), 28; and J. Schleifer, "Zum syrischen Medizinbuch. II. Der therapeutische Teil," in *Rivista degli studi orientali*, **18** (1940), 341–372. Two more recent studies are A. Vööbus, "Discovery of New Syriac Manuscripts on Hunain," in *Ephrem-Hunayn Festival* (Baghdad, 1974), 525–528; and W. F. Macomber, "The Literary Activity of Ḥunayn b. Isḥāq in Syriac," *ibid.*, 554–570.

BIBLIOGRAPHY

In addition to the works cited in the notes, other Arab sources are Ibn Juljul, *Ṭabaqāt al-aṭibbā'*, F. Sayyid, ed. (Cairo, 1955), 68–72; Ibn Ṣā'id al-Andalusi, *Kitāb Ṭabaqāt al-Umam*, L. Cheikho, ed. (Beirut, 1912), 36 ff., French trans. by Régis Blachère as *Livres des catégories des nations* (Paris, 1935), 80 ff.; 'Alī ibn Zayd al-Bayhaqī, *Tatimmat ṣiwān al-ḥikma*, Mohammad Shāfi'ī, ed., I (Lahore, 1935), 3 ff.; Ibn Khallikān, *Vitae illustrium vivorum*, Ferdinand Wüstenfeld, ed., II (Göttingen, 1843–1871), 159 ff. (no. 208); Bar Hebraeus, *Chronicon ecclesiasticum*, Jean Baptiste Abbeloos and Thomas Joseph Lamy, eds., III (Louvain, 1875–1877), 197–200; *Chronicon syriacum*, Paul Bedjan, ed. (Paris, 1890), 162 ff.; and *Ta'rīkh Mukhtaṣar al-duwal*, Anṭūn Ṣalḥānī, ed. (Beirut, 1890), 250–253.

Modern works are L. Leclerc, *Histoire de la médecine arabe* (Paris, 1876; repr. New York, n.d.), 139–152, which is not a critical study; M. Steinschneider, *Die hebraischen Übersetzungen des Mittelalters* (Berlin, 1893; repr. Graz, 1956), 1055 (index); and *Die arabischen Übersetzungen aus dem griechischen* (Leipzig, 1896; repr. Graz, 1956), 390 (index); H. Suter, *Die Mathematiker und Astronomer der Araber* (Leipzig, 1900), 20–23; M. Steinschneider, "Die europäischen Übersetzungen aus dem arabischen," *Sitzungsberichte der Akademie der Wissenschaften in Wien*, Phil.-hist. Kl., **149**, no. 4; and **151**, no. 1 (1905), photo repr. (Graz, 1956), 98 (index); G. Bergsträsser, *Ḥunain ibn Isḥāk und seine Schule* (Leiden, 1913); G. Gabrieli, "Ḥunayn ibn Isḥāq," in *Isis*, **6** (1924), 282–292; M. Meyerhof, "New Light on Ḥunain ibn Isḥāq and His Period," *ibid.*, **8** (1926), 685–724; and "Les versions syriaques et arabes des écrits galéniques," in *Byzantion*, **3** (1926), 35–51; and G. Sarton, *Introduction to the History of Science*, I (Baltimore, 1927; repr. 1950), 611–613. A long biography of Ḥunayn and an analysis of his working methods are in the intro. to M. Meyerhof, *The Book of the Ten Treatises on the Eye Ascribed to Ḥunain ibn Is-ḥāq . . .* (Cairo, 1928).

See also C. Brockelmann, *Geschichte der arabischen Literatur*, I (Leiden, 1943), 224–227, and supp. I (Leiden, 1937), 366–369; F. Rosenthal's review of R. Walzer, ed., *Galen on Medical Experience*, in *Isis*, **36** (1945–1946), 253 ff.; and *The Technique and Approach of Muslim Scholarship* (Rome, 1947), *passim*; and G. Graf, *Geschichte der christlich-arabischen Literatur*, II (Vatican City, 1947), 122–129, which analyzes particularly the philosophical and theological works.

The most recent works are G. Strohmaier, "Ḥunain b. Ishak al-'Ibādī," in *Encyclopedia of Islam*, new ed., III, 578–581; *Arabica*, **21**, fasc. 3 (1974), 229–330, a special issue that contains ten articles prepared for the colloquium on Ḥunayn held at Paris during the 29th International Congress of Orientalists in July 1973; Yūsuf Ḥabbī, *Ḥunayn b. Isḥāq* (Baghdad, 1974), in Arabic; and *Ephrem-Hunayn Festival* (Baghdad, 1974); *Actes* of the Congress, which includes sixteen articles on Ḥunayn: four are in English (those by A. Vööbus, G. Strohmaier, W. F. Macomber, and D. M. Dunlop), and the rest in Arabic.

G. C. ANAWATI

ḤUNAYN THE TRANSLATOR

Ḥunayn's contributions to the art of physic consist of accurate translations, by him and the members of his school, of Greek medical and philosophical texts into Syriac and Arabic, and his own popular and concise books on medicine.[1] Here we shall survey his activities as a translator of Galen's works and attempt a critical study of his most popular book, *Questions on Medicine*.

In their role as mediator between Galen, on the one hand, and Syriac- and Arabic-reading physicians, on the other, the contributions of Ḥunayn and each member of his school of translation are detailed in *Ḥunayn ibn Isḥāq's Missive to 'Alī ibn Yaḥyā on Galen's Books Which, so Far as He [Ḥunayn] Knows, Have Been Translated and Some of Those Books Which Have Not Been Translated*.[2] An invaluable commentary on this work was written by M. Meyerhof.[3]

In his youth and mature years, Ḥunayn wrote his translations in his own hand; later his copyists al-Aḥwal and al-Azraq transcribed for him.[4] The latter wrote in large Kufic script, the lines far apart on thick paper of great size; hence the survival of many of these manuscripts until the time of Ibn Abī Uṣaybi'a (*d.* 1270),[5] who reports that Ḥunayn was paid in gold for his translations, weight for weight.[6]

Ḥunayn was forty-eight when he completed the first draft of his *Missive*, and about eight years later he made certain additions to its text.[7] Further posthumous textual interpolations suggest that it was brought up to date, possibly by Abu' l-Ḥasan 'Alī ibn Yaḥyā al-Munajjim (*d* 888/889), to whom the *Missive* was addressed.[8] The fact that Ḥunayn's *Missive*, which presents impressive information on his period and on some of his predecessors, was compiled—he says—after the loss of his own library, indicates dedication and a thorough knowledge of his work. He could easily remember such details as the contents and purpose of each of Galen's books, the names of persons to whom certain works were addressed, the number of treatises in each work, his age and experiences when translating certain books (including those he either retranslated or merely revised and corrected), previous translators (with fair evaluation of their achievements), the names of customers who requested Syriac and Arabic versions, the role of each member of his school of translation, and accounts of Greek manuscripts of Galen's works that survived in full or in part, as well as those lost in his time.[9]

The late Alexandrian school of medicine (sixth to seventh century) followed a certain order of reading Galen's books that was quite unlike Galen's own order, which is outlined in *On the Order of his books*.[10] This deviation from Galen's order is discussed in Ḥunayn's *Missive*, which also contains critical didactic remarks on textual alterations that were introduced by teachers of medicine in the late school of Alexandria. Not surprisingly, therefore, a few slips of the mind are encountered in the *Missive*. One is Ḥunayn's failure to mention—in the correct order—Galen's *On the Method of the Preservation of Health*.[11] According to Arabic manuscripts and bibliographical sources, it was the last book in the late Alexandrian corpus entitled *Summaries and Commentaries of Galen's Sixteen Books Which Were Read in Alexandria* (better known as the *Summaria Alexandrinorum*), in which sixteen of Galen's books appear in a fixed order of succession.[12] Furthermore, Ḥunayn concludes his statement about *On the Method of Healing* (which he wrongly gives as the last item of the *Summaria*) by writing: "Greek copies of this book are few because it was not among the books which were read in the school of Alexandria."[13] A few lines later he contradicts himself, saying: "These are the books to which reading was confined at the place of the teaching of medicine in Alexandria, and were read in the order I have cited them. [Students] gathered every day to read and understand a principal book, in the same way as our Christian companions assemble at present at the places of teaching known as 'Schools.'"[14] It seems that Ḥunayn's list of Galen's books was incomplete. Al-Rāzī (Rhazes, *d.* 925 or 935) wrote a book entitled *Kitāb fī istidrāk mā baqiya min kutub Jālīnūs mimmā lam yadhkurh Ḥunayn wa-lā Jālīnūs fī fihristih* ("An Appendix Containing Those Books of Galen Which Are Mentioned Neither by Galen in His Index, nor by Ḥunayn").[15]

Galen's books that appear in Ḥunayn's *Missive* number 129; of these, he translated about ninety from Greek into Syriac and about forty into Arabic. Some books were translated more than once from Greek into Syriac for Christian physicians, and many were further rendered into Arabic for Muslim doctors, book collectors, and patrons.[16] Ḥunayn's research technique and method of translation are exemplary, even by present standards. His two earliest translations were made from Greek into Syriac for Jibrāʾīl ibn Bukhtīshūʿ (*d.* 828/829).[17] The first, *On the Types of Fevers*,[18] completed when Ḥunayn was not yet seventeen, later displeased him—not because of its quality, but because he discovered gaps in the Greek text. He eventually filled these gaps through reading better manuscripts in order to prepare a reliable copy that he could transcribe for one of his sons.[19] *On the Natural Faculties*,[20] the second in Ḥunayn's long list of translated works, was made when he was seventeen. Again, lacunae in the text of Greek manuscripts used in an earlier Syriac translation led him to correct the same book twice, before rendering it into Arabic; hence a warning to prospective readers of *On the Natural Faculties* that three Syriac versions, all by Ḥunayn, exist.[21] At one stage, therefore, in the development of his technique, and probably that of his school of translations, Ḥunayn first established a reliable Greek text and then embarked on translating it.[22] This scholarly procedure does not seem to appear in any other known works written during—or before—Ḥunayn's time.

Those who undertook the "revision and correction" of texts were excellent translators who had mastered Greek, Syriac, and Arabic.[23] The method of revision is described in Ḥunayn's own words: at times, it sufficed to insert corrections in order to amend a previous translation; at others, it proved easier to make a fresh translation from the Greek. He gives interesting information about these alternative methods in his critical account of Sarjiyūs al-Raʾs ʿAynī's[24] (Sergius of Resaina, *d.* 536) translation of *On the Method of Healing*, of which the first six treatises (part one) were of a poorer quality than the remaining eight treatises (part two):

> Sergius translated this book into Syriac. His translation of the first six treatises was made when he was still weak and inefficient in translation. Further, he translated the remaining eight treatises after he had had [some] training and thus produced a better translation than that of the first [six] treatises. Salmawayh[25] compelled me to amend the second part of this [book]; he had hoped that this [method] would be easier and better than making another translation. He and I collated a part of treatise seven: he held the Syriac [copy]; I held the Greek; he read the Syriac to me, and wherever I encountered differences from the Greek, I told him, and he inserted the amendments. He found this [method] difficult, thus realizing that a new translation would have been quicker, better

ḤUNAYN IBN ISḤĀQ

and more consistent, and asked me to translate these treatises, which I completed when we were at Raqqa, during the conquests of al-Maʾmūn. He [Salmawayh] passed his copy on to Zakariyyā ibn ʿAbd Allāh known by the name al-Ṭayfūrī,[26] on his coming to Baghdad, to have it there transcribed. Fire broke out in the ship in which Zakariyyā traveled; the book was burned and the copy lost. A few years later, I translated the [same] book from the beginning for Bukhtīshūʿ ibn Jibrāʾīl.[27] I had several Greek copies of the last eight treatises. I collated them and authenticated one copy from which—to the best of my ability—I produced a well-investigated and eloquent translation. As to the first six treatises, I had only one manuscript; this was very defective, and accordingly I could not authenticate the text as I should have done. Later, I found a copy which I collated against the other and corrected whatever mistakes I could rectify, and would collate [the text] against a third copy, should I come across one. Greek copies of this book are few because it was not among the books which were read in the school of Alexandria. Ḥubaysh ibn al-Ḥasan translated this book from my Syriac versions [into Arabic] for Muḥammad ibn Mūsā. After he had completed his translation, he asked me to read through its last eight treatises and to correct any mistakes. I acquiesced to his [request] and did very well.[28]

A few examples will suffice to show the extent of Ḥunayn's involvement in translating Galen: the *Index* was translated into Syriac for Dāʾūd, a practitioner, and into Arabic for Muḥammad ibn Mūsā;[29] *On the Order of His Books* into Arabic for Aḥmad ibn Mūsā, but not into Syriac;[30] *On Sects* into Syriac when Ḥunayn was twenty years old, again (at Ḥubaysh's request) when he was forty, and, a few years later, into Arabic for Muḥammad ibn Mūsā;[31] *On the Art of Physic* into Syriac at the age of thirty for Dāʾūd, and later into Arabic for Muḥammad ibn Mūsā;[32] *On the Pulse, to Teuthras* and *To Glaucon, on Therapy* into Syriac for Salmawayh ibn Bunān, and later into Arabic for Muḥammad ibn Mūsā;[33] *On Anatomy of the Bones* into Syriac for Yūḥannā ibn Māsawayh, and into Arabic for Muḥammad ibn Mūsā;[34] *On Anatomy of the Muscles* and *On Anatomy of the Nerves* into Syriac for Yūḥannā ibn Māsawayh, but not into Arabic;[35] *On Anatomy of the Veins and Arteries* into Syriac for Yūḥannā ibn Māsawayh, and into Arabic for Muḥammad ibn Mūsā;[36] and *On the Elements According to Hippocrates* into Syriac for Bukhtīshūʿ ibn Jibrāʾīl and into Arabic for Abuʾ l-Ḥasan ʿAlī ibn Yaḥyā al-Munajjim.[37]

Ḥunayn described Syriac translations from Greek by Yūsuf al-Khūrī and Thiyūfīl al-Ruhāwī as poor and unreliable.[38] Al-Khūrī translated the first five treatises of *On Materia Medica* and Thiūfīl translated *On the Method of the Preservation of Health*.[39] Ḥunayn mentions an unacceptable Syriac translation of *On Ethics* by a Sabaean, Manṣūr ibn Athānās.[40] A translator from Greek and Syriac into Arabic, Ibrāhīm ibn al-Ṣalt,[41] is credited with three achievements: *On Tumors* (or *On Swellings*), into Arabic for Aḥmad ibn Mūsā;[42] *Galen's Abstract of His Book "On the Method of Healing,"* into Syriac;[43] and *On Receipts for an Epileptic Youth*, into Syriac and Arabic.[44] Thābit ibn Qurra al-Ḥarrānī (836–901),[45] a younger contemporary of Ḥunayn, translated *On the Good and the Bad [Kinds of] Humor* and *On His Own Opinion* into Arabic.[46] Ḥunayn believed that his contemporary Yaḥyā ibn al-Biṭrīq[47] might have translated *To Pison, on Theriac* into Arabic.[48] Ayyūb al-Ruhāwī[49] (Job of Edessa), also known by the name al-Abrash ("Spotted"), translated some thirty works from Greek into Syriac. In view of their poor quality, his translations of Galen's anatomical works either were carefully corrected or were altogether retranslated from Greek into Syriac by Ḥunayn; for example: *On Anatomical Procedures*,[50] *On All Dissension in Dissection*,[51] *On Anatomy of the Dead Animal*,[52] *On Anatomy of the Living Animal*,[53] *On Hippocrates' Knowledge of Anatomy*,[54] *On Anatomy of the Uterus*,[55] and *On Anatomy of the Eye*.[56] The names of two hitherto unknown translators appear in the *Missive*: Shamlī,[57] who translated *On the Good and the Bad [Kinds of] Humor*; and Ibn Shuhdī, from al-Karkh,[58] a suburb of Baghdad, who produced poor Syriac versions of *On Sects, On the Art of Physic*, and *On the Pulse, to Teuthras*.

Isḥāq ibn Ḥunayn. A prominent member of Ḥunayn's school was his son Isḥāq ibn Ḥunayn (*d.* 910),[59] who translated from Greek into Syriac and Arabic and also revised, against the original Greek, translations by his colleagues, particularly of those works that had been rendered from Syriac into Arabic. For example, he revised the last section of ʿĪsā ibn Yaḥyā's translation (from Syriac into Arabic) of *On Anecdotes of Prognosis*[60] against the Greek text.[61] Likewise, he checked Ḥubaysh's Arabic translation (from Ḥunayn's Syriac version) of *On Physical Exercise With the Small Ball*.[62] It should be noted, however, that Isḥāq's translations did not escape his father's criticism. Ḥunayn revised and corrected Isḥāq's Arabic translation (for ʿAlī ibn Yaḥyā) of *On the Number of Syllogisms*,[63] a book that Ḥunayn had earlier translated from Greek into Syriac.[64]

That Isḥāq was more interested in philosophy

ḤUNAYN IBN ISḤĀQ

than in medicine is substantiated by the great number of philosophical works that he translated into Arabic or Syriac.[65] They include *That the Prime Mover is Immobile*[66] and treatises 12 to 15 inclusive of Galen's *On Demonstration*,[67] of which the original Greek text was partly lost during Ḥunayn's lifetime.[68] Galen's books translated from Greek into Syriac by Isḥāq for Bukhtīshūʿ ibn Jibrāʾīl are *On the Order of His Books* and *On the Opinions of Erasistratus on Therapy*.[69] Isḥāq translated *On the Method of the Preservation of Health* into Arabic for ʿAlī ibn Yaḥyā, although it had previously been translated into Arabic by Ḥubaysh for Muḥammad ibn Mūsā.[70]

Isḥāq continued to work after his father's death. Toward the end of his life, Ḥunayn was engaged in an Arabic translation of *On the Parts of Medicine*,[71] of which he had earlier produced a Syriac version. He died after having completed more than half of the Arabic translation. After Ḥunayn's death, Isḥāq completed this unfinished work[72] and also translated *On the Organ of Smell* (into Arabic) and *On the Use of the Pulse*.[73] In addition to his fame as a translator-reviser, Isḥāq is credited with a long medicophilosophical bibliography and an interesting book entitled *History of Physicians*.[74]

Ḥubaysh ibn al-Ḥasan al-Aʿsam. Next to Ḥunayn, the most prolific translator of medical texts from Syriac into Arabic was his nephew Ḥubaysh ibn al-Ḥasan al-Aʿsam.[75] Nowhere in the *Missive* is his name mentioned in connection with Greek translations. That Ḥubaysh's translations were not made from the Greek may be construed from Ḥunayn's account of *That the Faculties of the Soul Follow the Constitution of the Body*,[76] which he translated from Greek into Syriac for Salmawayh. Ḥunayn's Syriac version was rendered into Arabic by Ḥubaysh for Muḥammad ibn Mūsā, who, for the purpose of checking, had read Ḥubaysh's Arabic translation while listening to Iṣṭafān ibn Bāsīl's Arabic rendering made directly from a Greek manuscript held in Iṣṭafān's hands; Iṣṭafān suggested amendments that were inserted in Ḥubaysh's translation.[77] Ḥubaysh translated some thirty-five medical works from Syriac into Arabic. Further, he rendered three Arabic versions of Galen's works into Syriac for Yūḥannā ibn Māsawayh: *On the Movement of the Chest and Lungs*[78] (from Iṣṭafān ibn Bāsīl's Arabic), *On the Voice*,[79] and *On Ethics*[80] (both from Ḥunayn's Arabic). Ḥunayn writes that he did not translate any of these three books into Syriac.[81]

The question has arisen as to who translated Galen's *On Anatomical Procedures* into Arabic. This work, in fifteen books, was translated into Syriac by Ayyūb al-Ruhāwī for Jibrāʾīl ibn Bukhtīshūʿ. A revised Syriac translation was made by Ḥunayn for Yūḥannā ibn Māsawayh. In the *Missive*, Ḥunayn does not mention any Arabic translation by him or by any of his assistants.[82] Ibn al-Nadīm writes: "I examined an Arabic translation by Ḥubaysh,"[83] and remarks that "to Ḥunayn's good fortune, Arabic translations by Ḥubaysh ibn al-Ḥasan al-Aʿsam, by ʿĪsā ibn Yaḥyā, and others are attributed to Ḥunayn. Should we consult Ḥunayn's *Missive* to ʿAlī ibn Yaḥyā on Galen's books, we would learn that Ḥunayn mostly translated into Syriac. He probably corrected, or read through, Arabic translations made by others."[84] M. Meyerhof gives an account of G. Bergsträsser's conclusion that the style and grammar of the Arabic version of *On Anatomical Procedures* indicate that Ḥubaysh was its translator into Arabic.[85] From numerous corrections in the text (all by Ḥunayn), Meyerhof believes that Ḥubaysh produced an Arabic version around the end of Ḥunayn's life and with his active collaboration.[86] The appearance of Ḥunayn's name as translator in three Arabic manuscripts of this book may be due to scribal error.[87] The fame of Ḥunayn probably would have overshadowed that of Ḥubaysh: their names in Arabic could be easily misread one for the other, especially if diacritical points are missing.[88]

Interpolations in Galen's text, preceded by the words "Ḥunayn said" (*qāl Ḥunayn*), appear occasionally in Arabic manuscripts of *On Anatomical Procedures*,[89] from which it may safely be concluded that Ḥunayn at least read the Arabic text. He probably wrote "marginal commentaries" (*ḥawāshī*) that were later incorporated in the text,[90] preserved by copyists as "marginal commentaries,"[91] or, in a few instances, written in the text and then deleted.[92]

The following passage from Galen's account of the eye (book X) deals with inaccurate terminology in the Greek manuscripts of *On Anatomical Procedures*. In his comment Ḥunayn includes a cross-reference to an ambiguous term that he had marked with a special sign in two places on the Arabic version:

> [Galen's text]: As to my statement about the "surface" (*saṭḥ*) [that is, the tissue] drawn around the posterior aspect of the crystalline humor (*al-ruṭūba al-shabīha biʾl-jalīd min khalfihā*), the matter here is as I have said [before]. This "surface" does not enclose it completely and it does not possess any smoothness which as we have indicated, is found in the arachnoidal tunic (*al-ṭabaqa al-shabīha bi-nasīj*

al-ʿankabūt) [anterior] to the pupil. The "surface" [tissue] of the vitreous humor (*al-ruṭūba al-shabīha biʾl-zujāj*) is not soft on either of its sides: neither on the side which faces the crystalline humor, nor on that which encounters the vitreous humor.[93]

Ḥunayn said: You should understand Galen's statement here about the [word] "surface," not according to the vocabulary of the geometricians (*al-muhandisūn*)—[implying] a length and breadth, yet without a depth—but according to the laity who usually call the visible part of a thing its "surface." This is because a "surface," according to the vocabulary of the geometricians, cannot be imagined as possessing two [lateral] sides; whereas, according to the laity, it is imagined with two sides, both of which being parts of the body. Accordingly, you find that when Galen used slightly more precise vocabulary, what he called here a "surface," he had called a "body," at the place [in the manuscript] where I have inserted the same sign which I have placed here.[94]

ʿĪsā ibn Yaḥyā. Another of Ḥunayn's assistants was his pupil ʿĪsā ibn Yaḥyā,[95] who specialized in rendering Syriac texts into Arabic. That he did not know Greek may be surmised from Ḥunayn's statement about *On Bloodletting*[96]—"I translated its second treatise [from Greek] into Syriac for ʿĪsā, and ʿĪsā rendered it into Arabic"[97]—and from his account of *On Antidotes*[98]—"This book, of which a Greek copy existed among my books, had not been previously translated. Later, with my help, Yūḥannā ibn Bukhtīshūʿ translated it into Syriac, and ʿĪsā ibn Yaḥyā rendered it from his [Bukhtīshūʿ's] translation into Arabic for Aḥmad ibn Mūsā."[99] Medical and philosophical works translated by ʿĪsā include *On the Differences Between Homogeneous Organs*,[100] *On the Arteries: Does Blood Run Naturally Within Them or Does It Not?*,[101] *On the Strength of Cathartic Drugs*,[102] *On Anecdotes of Prognosis*, *On Bloodletting* (in part), *On Marasmus*,[103] *On Antidotes*, *On Theriac to Pamphilianus*,[104] *Commentary on the Hippocratic Oath*,[105] *Commentary on the Book of "Humors,"*[106] *That the Excellent Physician Is a Philosopher*,[107] *On His Own Opinion*, *On Demonstration* (in part), *That Good People May Make Use of Their Enemies*,[108] and *That the Prime Mover Is Immobile*.

ʿĪsā translated into Arabic two of Ḥunayn's Syriac abstracts of Galen's books, *On the Black Bile*[109] and *On the Attenuating Regimen*,[110] and collaborated with Ḥunayn on an Arabic translation of Galen's *Commentary on Prognosis*[111] in which the text of Hippocrates was Ḥunayn's responsibility, and Galen's commentary was entrusted to ʿĪsā.[112] *On His Own Opinion*, Galen's last written book, according to al-Rāzī,[113] appeared in two Syriac and two Arabic translations. One Arabic version by ʿĪsā was twice revised, first by Isḥāq and then by Ḥunayn himself. "This book," Ḥunayn writes, "was translated by Job [of Edessa] into Syriac, and I [also] translated it into Syriac for my son Isḥāq. Thābit ibn Qurra translated it into Arabic for Muḥammad ibn Mūsā; ʿĪsā ibn Yaḥyā translated it into Arabic, Isḥāq collated this [last] translation against the original [Greek text], and I corrected it for ʿAbd Allāh ibn Isḥāq."[114]

Iṣṭafān ibn Bāsīl. Perhaps this study of Ḥunayn and his school of translation should be concluded with a mention of Iṣṭafān ibn Bāsīl,[115] the translator of Dioscorides' *Materia medica* from Greek into Arabic, a book that was revised by Ḥunayn.[116] Iṣṭafān, a translator and reviser, mastered Greek, Syriac, and Arabic. Among his translations from Greek into Arabic are *On the Movement of the Chest and Lungs*,[117] *On the Causes of Respiration*,[118] and *On the Use of Respiration*.[119] Iṣṭafān translated the third treatise of *On Bloodletting* into Arabic from a Syriac version originally made from the Greek by Sergius.[120] *On the Black Bile*, which had been translated into Syriac by Job of Edessa for Bukhtīshūʿ ibn Jibrāʾīl, was translated by Iṣṭafān into Arabic for Muḥammad ibn Mūsā; and *On Marasmus* was also translated by Iṣṭafān into Arabic and was partly corrected by Ḥunayn.[121]

ḤUNAYN THE PHYSICIAN

A case of professional jealousy is reported in which al-Ṭayfūrī,[122] or more likely Bukhtīshūʿ ibn Jibrāʾīl,[123] conspired against Ḥunayn. Ibn Abī Uṣaybiʿa asserts that Ḥunayn wrote a treatise entitled *On Misfortune and Hardships Which Befell Him at the Hands of His Adversaries, Those Renowned but Wicked Physicians of His Time*,[124] in which he complained of his relatives and most of the Christian physicians for whom he had made translations, and taught the art of physic. At times, when seriously ill, they consulted Ḥunayn for diagnosis, therapeutic recipes, and regimens. The "wicked physicians" were fifty-six selfish doctors who, at one time or another, were in the service of caliphs and exploited authority to rally support from the laity against Ḥunayn. Whenever Ḥunayn visited a patient, he was scoffed at and ridiculed by members of the medical profession, to such an extent that he had contemplated committing suicide. It was true, his adversaries argued, that he had translated medical books and was

ḤUNAYN IBN ISḤĀQ

rewarded generously, yet he was not a physician—just as a blacksmith could forge a beautiful sword without himself being a swordsman.[125]

Of Ḥunayn's medical books, *al-Masāʾil fiʾl-ṭibb* ("Questions on Medicine")[126] has been chosen for use in an assessment of his status as a physician and author, and to evaluate his impact on medical education.

Al-Masāʾil fiʾl-ṭibb is also known by the title *al-Madkhal fiʾl-ṭibb* ("Introduction to Medicine").[127] Some of its manuscripts have a more detailed title: *Masāʾil Ḥunayn ibn Isḥāq fiʾl-ṭibb liʾl-mutaʿallimīn maʿ ziyādāt Ḥubaysh tilmīdhih* ("Ḥunayn ibn Isḥāq's 'Questions on Medicine for Students,' With Additions by His Pupil Ḥubaysh").[128] Halfway through the text in a Bodleian Library manuscript of this book (Marsh 403), a marginal note (*ḥāshiya*) by a copyist who transcribed the entire work reads: "From here until the end of the book are additions by Ḥubaysh al-Aʿsam, a pupil of Ḥunayn ibn Isḥāq."[129] Ḥubaysh's contribution begins with a section on the periods of disease (*awqāt al-amrāḍ*), which include the onset (*al-ibtidāʾ*), increase (*al-tazayyud*), culmination (*al-muntahā*), and decline (*al-inḥiṭāṭ*).

The coauthorship of Ḥunayn and Ḥubaysh was known to Ibn Abī Ṣādiq al-Nīsābūrī (*d.* after 1068),[130] author of *Sharḥ masāʾil Ḥunayn ibn Isḥāq* ("Commentary on Ḥunayn ibn Isḥāq's 'Questions' "),[131] which was used in Syria as a medical textbook until the end of the thirteenth century. Manuscript Marsh 98 (Bodleian Library) has preserved a license (*ijāza*) written and signed by a famous physician and educator, Muwaffaq al-Dīn Yaʿqūb al-Sāmirī (*d.* 1282),[132] which runs as follows: "This part [book I] of the large '*Commentary on Ḥunayn's Questions*,' by the learned and eminent physician philosopher Ibn Abī Ṣādiq, was read before me by a student, Amīn al-Dawla Tādrus, son of shaykh Naṣr ibn Malīḥ. He read it in order to investigate its questions, and to understand and ascertain [its contents]. Written by Yaʿqūb al-Sāmirī, *al-mutaṭabbib*."[133]

Al-Nīsābūrī's introduction confirms the popularity of Ḥunayn's *Questions on Medicine* and elucidates its purpose:

> Said Abuʾl-Qāsim ʿAbd al-Raḥmān ibn ʿAlī ibn Abī Ṣādiq al-Nīsābūrī . . . experts on the art of physic have agreed that students of this science should begin by learning Ḥunayn ibn Isḥāq's book, the "*Questions*," which he wrote as an introduction to medicine for students. Accordingly, he excluded all that might be considered obscure, and wrote it in the form of questions and answers, so that every question might draw the attention of students to the purpose of each quest, which would be settled by providing an answer . . . .[134] I say: In this book, Ḥunayn's purpose was to compile summaries and commentaries of medicine, thus providing a course in its general rules and principles. It serves as a very useful introduction for beginners, who should become familiar with these [rules and principles] and should find it easy to understand more difficult and obscure matters which will arise later. Ḥunayn compiled material for this book on leaves in rough drafts; and during his lifetime he managed to re-write some final drafts. Later, Ḥubaysh ibn al-Ḥasan al-Aʿsam, his pupil and nephew, arranged the remaining leaves, and annexed his own additions to Ḥunayn's notes written for this book. Hence, the book is found entitled "*Ḥunayn's Questions, with Ḥubaysh al-Aʿsam's Additions*." Ḥunayn gave it the title "*Questions on Medicine*," because it is a record of medical questions.[135]

The text of al-Nīsābūrī's commentary is *mamzūj* (mixed), in that it provides passages from the *Questions on Medicine* preceded (in WMS. Or. 2) by the statement *qāl Ḥunayn* (Ḥunayn said), and continues with al-Nīsābūrī's commentary identified with the words *qāl al-mufassir* (the commentator said). This *mamzūj* pattern continues throughout the text, the cue *qāl Ḥunayn* being replaced by *qāl Ḥubaysh* from the beginning of Ḥubaysh's additions to the end of the book.[136] Except in the author's introductory note, MS 98 does not provide such a marked distinction between the contributions of Ḥunayn and Ḥubaysh; it merely gives *al-fiṣṣ* (extracts from the text), followed by *al-tafsīr* (commentary).

In his *Kitāb al-nāfiʿ fī kayfiyyat taʿlīm ṣināʿat al-ṭibb* ("Useful Book on the Method of Medical Education"), Abuʾl-Ḥasan ʿAlī ibn Riḍwān al-Miṣrī (*d.* 1061),[137] an Egyptian contemporary of al-Nīsābūrī, disapproves of the *Questions on Medicine*. Nevertheless, he writes a similar opinion of its purpose in his time,[138] says that Ḥunayn calls it *al-Madkhal ilā ṣināʿat al-ṭibb* ("Introduction to the Art of Physic"), and quotes from it.[139]

The *Questions on Medicine* is in the form of short questions, arising mostly from studies in Galen's *Art of Physic*, and straightforward answers that draw heavily on the *Summaria Alexandrinorum*. It was written as an introduction to the art of medicine—that is, the Galenic system of medicine including later commentaries and not merely, as is often claimed, an "Introduction to Galen's *Art of Physic*."[140] The opening passage runs as follows:

ḤUNAYN IBN ISḤĀQ

Into how many parts is medicine divided? Into two parts. What are these two [parts]? Theory and practice. And into how many parts is theory divided? Into three parts. And what are these? Investigation of the naturals, whereby pathological knowledge due to their alteration may be discovered; investigation of causes [of disease]; and investigation of symptoms. How many are the naturals? Seven. And what are these? The elements, the temperaments, the humors, the organs, the faculties, the actions, and the spirits. How many are the elements? Four. What are these? Fire, air, water, and earth. What is the strength of fire? Hot and dry. What is the strength of air? Hot and moist. What is the strength of water? Cold and moist. What is the strength of earth? Cold and dry. How many types are the temperaments? Nine. What are these? Eight are immoderate, and one is moderate. And of the eight which are immoderate, four are simple: hot, cold, moist, and dry; and four are combined: hot-dry, hot-moist, cold-dry, and cold-moist. How many are the humors? Four. What are these? Blood, phlegm, bile, and black bile. What is the strength of blood? Hot and moist. What is the strength of bile? Hot and dry. What is the strength of black bile? Cold and dry. What is the strength of phlegm? Cold and moist.[141]

Some manuscripts of the *Questions on Medicine* have excluded the "questions" altogether and merely provide subject matter in the form of subdivided presentation (*tashjīr*), as indicated by the title *Masāʾil Ḥunayn ibn Isḥāq ʿalā ṭarīq al-taqsīm waʾl-tashjīr* ("Ḥunayn ibn Isḥāq's 'Questions,' by Way of Subdivided Classification").[142]

Probably through their fame as translators of Galen's books, Ḥunayn and Ḥubaysh almost lost the *Questions on Medicine* to Galen. One manuscript (The British Library, London, Arundel Or. 10)[143] has a misleading title just before its opening passage: *Kitāb Īsāghūjī li-Jālīnūs tarjamat Ḥunayn ibn Isḥāq, yumtaḥanu bihi mutaʿallimū al-ṭibb* ("Galen's Book 'Isagoge,' Translated by Ḥunayn ibn Isḥāq, for Examining Students of Medicine"). This particular manuscript begins with the words *qāl Jālīnūs* (Galen said); and the questions and answers on the first few lines are prefixed by the word *qāl*, which also appears occasionally before some answers throughout the book.[144] Its text, however, represents the *Questions on Medicine*.

In the *Questions on Medicine*, Ḥunayn divides medicine into theory and practice,[145] and provides the classical definition of the three types of bodily condition: health, disease, and neutrality. Health is a condition of which the actions are natural; disease is a condition of which the actions are contranatural (against nature); and neutrality is a condition of nonhealth and nondisease, or a condition that is neither absolutely healthful nor absolutely unhealthful. Three examples of neutrality are cited: disability, such as blindness or lameness; marginal health of elderly people and convalescents, when the body is not entirely free from disease; and immoderate temperaments, when health and disease would be subject to seasonal changes and age. Each of the three conditions is related to the body, causes, and symptoms. Two genera of causes, natural and contranatural, are discussed in relation to health and disease; natural causes actively restore health in diseased bodies and preserve the health of people, whereas contranatural causes are either pathological, in that they bring about disease or prolong it, or are responsible for bringing about or promoting a neutral condition of nonhealth and nondisease.[146]

Six types of general causes, common to health and disease, are defined as the necessary causes (*al-asbāb al-ḍarūriyya*): ambient air (*al-hawāʾ al-muḥīṭ biʾl-abdān*), food and drink (*mā yuʾkal wa yushrab*), sleep and wakefulness (*al-nawm waʾl-yaqaẓa*), evacuation and congestion (*al-istifrāgh waʾl-iḥtiqān*),[147] movement and rest (*al-ḥaraka waʾl-sukūn*), and perturbation of the mind (*al-aḥdāth al-nafsāniyya*).[148] These six causes are called necessary because, of necessity, they will affect health in one of two ways: when used in proper quantity and quality, time and order, they will restore health and preserve it; if abused in quantity or quality, time or order, they will induce disease and prolong it.[149]

Origin of the Six Necessary Causes. The six necessary causes exist in three Arabic books of the *Summaria Alexandrinorum*,[150] in the British Library (London) MS Add. 23407:[151] *On Sects, On the Art of Physic*; and *To Glaucon, on Therapy*.[152] Ḥunayn translated *To Glaucon* into Arabic.[153] These three books of the *Summaria* could have been among Ḥunayn's sources of the six necessary causes, so clearly stated in Ḥunayn's section of the *Questions on Medicine*. The following are translations of extracts from MS Add. 23407:

The *Summaria Alexandrinorum*: Galen's book *On Sects, Arāsīs*.[154]
Commentary on Chapter Three; the causes which alter the body:

Some must, of necessity, alter the body, and these are six: first, air surrounding it; second, movement and rest; third, food and drink; fourth, sleep and wakefulness; fifth, evacuation and congestion;[155] and sixth, perturbation of the mind, such as distress, anxiety, fright, joy, and anger.

ḤUNAYN IBN ISḤĀQ

Other [causes] may alter the body, but not of necessity, as for example a sword, an arrow, a stone, and fire.[156]

The *Summaria Alexandrinorum*: Galen's small book *On the Art of Physic*, a summary and commentary.[157]
The causes which alter the body: some things alter the body of necessity, these are six genera; and some do not alter the body of necessity, such as harmful animals, stones, swords, and the like.

As to the six necessary causes, they are: air surrounding the body; a genus [including] things which are eaten and drunk; a genus, sleep and wakefulness; a genus, movement and rest of the whole of the body, or of some organs only; a genus, evacuation and constipation (*iḥtibās*); and a genus, perturbation of the mind, which includes joy, sadness, distress, jealousy, and fright. These six genera promote health, providing that they are properly preserved in quantity and quality in moderation; and cause disease if they become immoderate in quantity and quality, deviating to one of the extremes. In order to preserve the health of a person whose constitution and structure are good, these six genera must be kept moderate, I mean: moderation of air surrounding the body, moderation of food and drink, moderation of perturbation of the mind, moderation of sleep and wakefulness, moderation of movement and rest, and moderation of evacuation of superfluities.[158]

The *Summaria Alexandrinorum*: the first treatise of Galen's book *To Glaucon*.[159]
The necessary causes are: air which surrounds the body, food and drink which are taken in, conditions of sleep and wakefulness, movement and rest, whether the body is evacuated or constipated, and perturbation of the mind.[160]

Origin of the Term Nonnatural. It is surprising that although the six necessary causes constitute normal—or natural—everyday activities, they became known as the six nonnaturals, an expression that is mentioned in medical literature from the Renaissance until the early nineteenth century.[161] An attempt will be made here to show how the unusual expression "nonnaturals" superseded the meaningful term "necessary causes."

The term "nonnatural" has been traced to the Greek text and Latin translation of Galen's *On the Pulse, to Teuthras*, in a section on the "causes which alter the pulse";[162] and in Arabic the equivalent term, *laysa bi-ṭabīʿiyy*, appears in the same book, according to Ḥunayn's short account of the same section on *al-asbāb allatī tughayyir al-nabḍ*.[163] Further, Galen's six items that will of necessity alter the body have been located in his *Art of Physic*;[164] these probably were developed by the Alexandrian teachers of the sixth to seventh century or by some of their hitherto unknown predecessors, into the six necessary causes, which were borrowed by Ḥunayn in the *Questions on Medicine*.

The term "nonnatural" is mentioned in Ḥubaysh's section of the *Questions on Medicine*, where he outlines the scope of theory. Ḥunayn's six necessary causes differ from Ḥubaysh's nonnaturals in that, in Ḥubaysh's account, bathing replaces evacuation and marriage (or sexual intercourse) supersedes perturbation of the mind—the latter is mentioned merely as a possible additional item:

> Some doctors divide theory of medicine as follows: they say that it is divided into knowledge of the naturals (*al-umūr al-ṭabīʿiyya*), the nonnaturals (*al-umūr allatī laysat bi-ṭabīʿiyya*), and the contranaturals (*al-umūr al-khārija ʿanʾl-amr al-ṭabīʿiyy*). To the seven naturals which we mentioned earlier, they have added another four consequential and continuous matters, which are: age, color, complexion, and differences between male and female sexes. How many are the nonnaturals? Six. What are these? First, ambient air; second, movement and rest; third, bathing (*al-istiḥmām*); fourth, food and drink; fifth, sleep and wakefulness; and sixth, marriage (*al-nikāḥ*), and to these six, some have added another matter, perturbation of the mind.[165]

In some manuscripts of the *Questions on Medicine*, a marginal note considers "bathing and sexual intercourse" as means of "evacuation,"[166] thus reducing the nonnaturals mentioned by Ḥubaysh to the traditional necessary causes. Many Arabic works of Ḥunayn's successors devote much space to the six necessary causes,[167] on which monographs also have been written.[168]

Of particular interest is Ibn Hubal al-Baghdādī's (*d.* 1213) definition of the six necessary causes, in which he classifies them as "neither natural nor contranatural": "waʾl-nabdaʾ awwalā biʾl-asbāb al-ʿāmma allatī laysat bi-ṭabīʿiyya wa-lā khārija ʿanʾl-ṭabʿ, bal mushtaraka liʾl-ṣiḥḥa waʾl-maraḍ waʾl-ḥāla al-mutawassiṭa, wa hiya al-sitta al-ḍarūriyya"[169] ("First, we begin with the general causes that are neither natural nor contranatural; yet, they are common to health, disease, and the intermediate condition between health and disease—these are the six necessary [causes]").

On its own, the expression "laysat bi-tabīʿiyya" could be translated as "nonnatural"; but when it is

ḤUNAYN IBN ISḤĀQ

used in a clause of negative conjunction, "laysa bi-ṭabīʿiyya wa-lā khārija ʿanʾl-ṭabʿ," the translation should be "neither natural nor contranatural." The expression "nonnatural," therefore, is an abbreviation of "neither natural nor contranatural"; and the terms "natural," "contranatural," and "nonnatural" are derived from the classical definitions of health: "a condition of the body the actions of which are 'natural' "; of disease: "a condition of the body the actions of which are 'contranatural' "; and of neutrality: "a condition of 'nonhealth and nondisease.' "

Unlike the Arabic equivalent of "necessary causes," that of the "nonnaturals" is neither completely meaningful nor linguistically eloquent and consequently was not generally used in medieval medical terminology. Further research should be pursued in Greek and Latin sources, with a view to checking the validity of this assumption on the origin of the expression "nonnaturals" in Arabic literature.

Ḥunayn could not have produced accurate translations of Galen without careful studies of each text, thus learning much of the corpus, as is reflected in his own books. That he was consciously aware of minor details may be illustrated by his account of the seventeen books of Galen's *On the Compounding of Drugs*,[170] which Ḥunayn translated from Greek into Syriac for Yūḥannā ibn Māsawayh during the caliphate of al-Mutawakkil; Ḥubaysh later rendered it into Arabic from Ḥunayn's Syriac version for Muḥammad ibn Mūsā.[171] Its first seven treatises are entitled *On the Compounding of Drugs According to Genera*, known by the shorter transliterated title *Qāṭājānis*;[172] the remaining ten treatises constitute *On the Compounding of Drugs According to Affected Places*, also entitled *al-Mayāmir*.[173] Ḥunayn's statement that the seven treatises *Qāṭājānis* precede the ten treatises *al-Mayāmir* in *On the Compounding of Drugs* is derived directly from Galen.[174] This order is reversed in C. G. Kühn's edition: the *De compositione medicamentorum secundum locos*[175] (*al-Mayāmir*) comes first and the *De compositione medicamentorum per genera*[176] (*Qāṭājānis*) is second, an order that should be rectified in any forthcoming edition.

Like many authors, Ḥunayn did not make an original contribution to medicine. Nevertheless, he succeeded in presenting simplified abstracts, mainly for students, from the books of his predecessors. His translations rendered Greek medicine accessible to Arabic-speaking physicians and paved the way for medical education. They also preserved works that otherwise would have been totally lost.[177]

NOTES

1. Ibn al-Nadīm, *Kitāb al-Fihrist*, G. Flügel, ed., I (Leipzig, 1871), 294–295 (hereinafter *Fihrist*); Ibn Abī Uṣaybiʿa, *Kitāb ʿUyūn al-anbāʾ fī ṭabaqāt al-aṭibbāʾ* . . ., A. Müller, ed., I (Cairo, 1882), 184–200 (hereinafter IAU); al-Qifṭī, *Taʾrīkh al-ḥukamāʾ* . . ., J. Lippert, ed. (Leipzig, 1903), 171–177 (hereinafter Qifṭī); C. Brockelmann, *Geschichte der arabischen Litteratur*, I (Leiden, 1943), 224 (hereinafter *GAL*), and supp., I (Leiden, 1937), 366 (hereinafter *S*); Ibn Juljul, *Ṭabaqāt al-aṭibbāʾ waʾl-ḥukamāʾ* . . ., F. Sayyid, ed. (Cairo, 1955), 68–72; F. Sezgin, *Geschichte des arabischen Schrifttums, Medizin-Pharmazie-Zoologie-Tierheilkunde bis ca. 430 H.*, III (Leiden, 1970), 247–256 (hereinafter *GAS*); M. Ullmann, *Die Medizin im Islam*, Handbuch der Orientalistik, Abt. 1, supp. 6 (Leiden–Cologne, 1970), 115–119 (hereinafter Ullmann).
2. *Risālat Ḥunayn ibn Isḥāq ilā ʿAlī ibn Yaḥyā fī dhikr mā turjima min kutub Jālīnūs bi-ʿilmihi wa baʿḍ mā lam yutarjam*, ed. with a trans. by G. Bergsträsser as "Ḥunain ibn Isḥāq über die syrischen und arabischen Galen-übersetzungen," in *Abhandlungen für die Kunde des Morgenlandes*, **17**, no. 2 (1925). All references are to the Arabic text in this work (hereinafter Ḥunayn).
3. M. Meyerhof, "New Light on Ḥunain ibn Isḥāq and His Period," in *Isis*, **8**, no. 4 (1926), 685–724; an informative paper is by Lutfi M. Saʿdi, "A Biobibliographical Study of Hunayn ibn Is-haq al-Ibadi (Johannitius, 809–877 A.D.)," in *Bulletin of the Institute of the History of Medicine*, **2**, no. 7 (1934), 409–446.
4. Ḥunayn (15, ll. 17–20) wrote that he transcribed his revised Syriac translation of *On the Types of Fevers* for his son but did not mention the son's name. Yāqūt al-Ḥamawī (*d.* 1229) mentions Abuʾl-ʿAbbās Muḥammad ibn al-Ḥasan ibn Dīnār al-Aḥwal ("Cockeyed"), a great philologist, poet, author, and "copyist of Ḥunayn ibn Isḥāq the practitioner's translations of the sciences of the Ancients" (*kān warrāqā yuwarriq li-Ḥunayn ibn Isḥāq al-mutaṭabbib fī manqūlātihi li-ʿulūm al-awāʾil*). See Yāqūt al-Ḥamawī, *The Irshād al-arīb ilā maʿrifat al-adīb* . . ., D. S. Margoliouth ed., 2nd ed., VI (London, 1931), 482, ll. 16–17; 483, l. 7; *Fihrist*, I, 79.
5. IAU, I, 197, ll. 12–18; also I, 187, ll. 25–26.
6. *Ibid.*, I, 187, ll. 8–9.
7. See A. Z. Iskandar, "An Attempted Reconstruction of the Late Alexandrian Medical Curriculum," in *Medical History*, **20**, no. 3 (1976), 237, n. 11 (hereinafter "Alexandrian Medical Curriculum"). The Arabic titles of Galen's books that appear in this paper are given in transliteration, together with Latin translations and bibliographical references. Only those works that do not appear in the "Alexandrian Medical Curriculum" will be dealt with hereafter.
8. *Fihrist*, I, 143; Qifṭī, 117, ll. 19–20; 129, l. 1; 132, l. 2; IAU, I, 205–206.
9. Ḥunayn, 1–2.
10. "Alexandrian Medical Curriculum," 237, n. 13.
11. *Ibid.*, 239, n. 44.
12. On the *Summaries*, see *ibid.*, 237, nn. 15, 16; on the order of succession, *ibid.*, 237–239.
13. *On the Method*, see *ibid.*, 239, n. 43; the quotation is from Ḥunayn, 18, ll. 14–15.
14. The Arabic transliteration *askūl* is from the Syriac *skōlē*, derived from the Greek σχολή. See ʿAbdurraḥmān Badawi, *La transmission de la philosophie grecque au monde arabe*, Études de Philosophie Médiévale, no. 56 (Paris, 1968), 15. The quotation is from Ḥunayn, 18, ll. 19–22.

15. On al-Rāzī, see *GAL*, I, 267; *S*, I, 417; *GAS*, III, 274–294; Ullmann, 128–136. On his *Kitāb*, see *Fihrist*, 300, ll. 12–13; P. Kraus, *Épître de Bērūnī contenant le répertoire des ouvrages de Muḥammad b. Zakariyyā ar-Rāzī* (Paris, 1936), 21, no. 175; J. Ruska, "Al-Bīrūnī als Quelle für das Leben und die Schriften al-Rāzī's," in *Isis*, **5** (1923), 48, no. 175.
16. Ḥunayn, 4, no. 3; 6, no. 5; 6, no. 6; 7, no. 7; 9, no. 11.
17. Physician to the caliphs Hārūn al-Rashīd (786–809), al-Amīn (809–813), and al-Maʾmūn (813–833). Son of Bukhtīshūʿ ibn Jūrjīs (*d.* 801). See Qifṭī, 132; IAU, I, 127; *GAS*, III, 226; Ullmann, 109.
18. "Alexandrian Medical Curriculum," 239, n. 40.
19. Ḥunayn, 15, ll. 16–21.
20. "Alexandrian Medical Curriculum," 238, n. 23.
21. Ḥunayn, 11, ll. 2–10.
22. For example, see Ḥunayn's account of *On Sects* (Ḥunayn, 4–5, no. 3), and the section of this article by G. C. Anawati.
23. A study in the technique of translation and revision was published by ʿA. Badawi in *La transmission de la philosophie grecque . . .*," 15–34.
24. Qifṭī, 175, ll. 4–5; IAU, I, 109, ll. 25–26; G. Sarton, *Introduction to the History of Science*, I (Baltimore, 1927), 423 (hereinafter Sarton); *GAS*, III, 177; Ullmann, 100.
25. Salmawayh ibn Bunān, physician to Caliph al-Muʿtaṣim (833–842). See *Fihrist*, 296; Qifṭī, 207–208; IAU, I, 164–170; *GAS*, III, 227; Ullmann, 112.
26. *Fihrist*, I, 298; Qifṭī, 187–189; IAU, I, 157. See note 122.
27. Died in 870; physician to Caliph al-Mutawakkil (847–861). See Qifṭī, 102; IAU, I, 138; *GAS*, III, 243; Ullmann, 109.
28. Ḥunayn, 17, l. 15–18, l. 18.
29. *Fīnaks* = *Bīnaks* = *Fihrist kutubih* (Ḥunayn, 3, no. 1; *GAS*, III, 78, no. 1; Ullmann, 35, no. 1) = *De libris propriis*; Ḥunayn, 3, ll. 23–24. See C. G. Kühn, *Claudii Galeni opera omnia*, 20 vols. in 22 (Leipzig, 1821–1833) (hereinafter Kühn), XIX, 8–48; H. Diels, *Die Handschriften der antiken Ärzte . . . im Auftrage der Akademischen Kommission. I. Teil: Hippokrates und Galenos* (Berlin, 1905), 109 (hereinafter Diels); English trans. of extracts in A. J. Brock, *Greek Medicine, Being Extracts Illustrative of Medical Writers From Hippocrates to Galen* (London–Toronto, 1929), 174–181.
30. Ḥunayn, 4, ll. 7–9.
31. "Alexandrian Medical Curriculum," 238, n. 17; Ḥunayn, 5, ll. 1–9.
32. "Alexandrian Medical Curriculum," 238, n. 18; Ḥunayn, 6, ll. 2–6.
33. "Alexandrian Medical Curriculum," 238, nn. 19, 20; Ḥunayn, 6, ll. 14–19, and 7, ll. 14–16.
34. "Alexandrian Medical Curriculum," 238, n. 25; Ḥunayn, 8, ll. 9–12. On Yūḥannā ibn Māsawayh, also known as Mesuë (*d.* 857), physician to the caliphs from al-Maʾmūn to al-Mutawakkil, see *Fihrist*, I, 295–296; Qifṭī, 380–391; IAU, I, 175–183; Sarton, I, 574; *GAL*, I, 266; *GAS*, III, 231–236; Ullmann, 112–115.
35. "Alexandrian Medical Curriculum," 238, nn. 26, 27; Ḥunayn, 8, l. 22–9, l. 1, and 9, l. 6.
36. "Alexandrian Medical Curriculum," 238, n. 28; Ḥunayn, 9, ll. 13–15.
37. "Alexandrian Medical Curriculum," 238, n. 21; Ḥunayn, 10, ll. 4–7.
38. Ḥunayn, 30, ll. 1–2, and 39, ll. 7–8.
39. "Alexandrian Medical Curriculum," 251, n. 117; Ḥunayn, 30, ll. 1–2, and 39, ll. 7–8.
40. The work is entitled *Fiʾl-akhlāq* (Ḥunayn, 49, no. 119; Ullmann, 63, no. 113; an "Abridgement of Galen's Book 'On Ethics'" (*Mukhtaṣar min kitāb al-akhlāq li-Jālīnūs*), possibly by Abū ʿUthmān Saʿīd ibn Yaʿqūb al-Dimashqī, physician-translator and a contemporary of ʿAlī ibn ʿĪsā, was published with an intro. by P. Kraus, "Kitāb al-akhlāq li-Jālīnūs," in *Bulletin of the Faculty of Arts of the University of Egypt*, **5**, no. 1 [1937], 1–51 [Arabic]) = *De moribus*, lost in Greek. The name Manṣūr ibn Athānās is found in Ḥunayn, 49, ll. 6–9; IAU, I, 205, l. 17, gives the name "Manṣūr ibn Bānās," who knew Syriac better than Arabic.
41. Qifṭī, 39, l. 11; 98, l. 8; 130, l. 17; 131, l. 3; IAU, I, 205, l. 3.
42. *Fiʾl-awrām* = *Aṣnāf al-ghilaẓ al-khārij ʿan al-ṭabīʿa* (Ḥunayn, 31, no. 57; *GAS*, III, 111, no. 47; Ullmann, 43, no. 28) = *De tumoribus praeter naturam* (Kühn, VII, 705–732; Diels, I, 83); Ḥunayn, 31, l. 8.
43. *Kitābuhu alladhī ikhtaṣar fīhi kitābahu fī ḥīlat al-burʾ* = *Ikhtiṣār ḥīlat al-burʾ* = *Mukhtaṣar ḥīlat al-burʾ* (Ḥunayn, 34, no. 70; *GAS*, III, 115, no. 56; Ullmann, 59, no. 96); Ḥunayn, 34, l. 18.
44. *Fī ṣifāt li-ṣabiyy yaṣraʿ* (Ḥunayn, 35, no. 73; *GAS*, III, 116, no. 59; Ullmann, 46, no. 43; F. Heller, "Ueber Pathologie und Therapie der Epilepsie im Altertum," in *Janus*, **16** [1911], 589–605; Augusto Botto-Micca, "Il 'De puero epileptico' di Galeno," in *Rivista di storia critica delle scienze mediche e naturali*, **21**, no. 12 [1930], 149–169; O. Temkin, "Galen's 'Advice for an Epileptic Boy,'" in *Bulletin of the Institute of the History of Medicine*, **2**, no. 3 [1934], 179–189) = *Puero epileptico consilium* (Kühn, XI, 357–378; Diels, I, 96); Ḥunayn, 35 ll. 13–14.
45. *Fihrist*, I, 272; Qifṭī, 115–122; IAU, I, 215–220; Sarton I, 599–600; *GAS*, III, 260–263; Ullmann, 123.
46. "Alexandrian Medical Curriculum," 252, n. 123; *Fīmā yaʿtaqiduhu raʾyā* (Ḥunayn, 46, no. 113; Ullmann, 51, no. 64) = *De propriis placitis* (see note 113); Ḥunayn, 36, l. 8, and 47, ll. 1–2.
47. *Fihrist*, I, 290, l. 28–291, l. 1; Qifṭī, 131, ll. 9–10; IAU, I, 205; Sarton, I, 556–557; *GAS*, III, 225; Ullmann, 326.
48. *Fiʾl-tiryāq ilā Fīsun* = . . . *ilā Bīsun* = . . . *ilā Qayṣar* (Ḥunayn, 38, no. 83; *GAS*, III, 121, no. 68; Ullmann, 49, no. 51; E. Coturri, *De theriaca ad Pisonem. Testo latino, traduzione italiana ed introduzione*, Biblioteca della "Rivista di storia delle Scienze Mediche e Naturali, no. 8 (Florence, 1959); Arabic ed. with a German trans. by Lutz Richter-Bernburg, *Eine arabische Version der pseudogalenischen Schrift "De theriaca ad Pisonem"* (Göttingen, 1969), a Ph.D. diss. = *De theriaca ad Pisonem* (Kühn, XIV, 210–294; Diels, I, 99); Ḥunayn, 39, ll. 1–2.
49. *Fihrist*, I, 244, l. 12; IAU, I, 170; Sarton, I, 574; *GAS*, III, 230–231; Ullmann, 101.
50. "Alexandrian Medical Curriculum," 246, n. 87.
51. *al-Ikhtilāf alladhī kān bayn aṣḥāb al-tashrīḥ* = *Mā kān minʾl-ikhtilāf bayn aṣḥāb al-tashrīḥ* = *Fīmā waqaʿa minʾl-ikhtilāf fiʾl-tashrīḥ* (Ḥunayn, 21, no. 24; *GAS*, III, 133, no. 132). This book, lost in Greek, is cited by Galen in *On Anatomical Procedures*. For example, see Arabic text of University of California, Los Angeles, MS Ar. 90, pp. 11, ll. 13–14, and 300, l. 8. In his trans. from Greek, Charles Singer, in *Galen: On Anatomical Procedures . . .* (London, 1956), 9, and n. 34, writes that *De dissentione anatomica* cannot be identified. Margaret T. May gives references to *De dissentione anatomica* in *Galen: On the Usefulness of the Parts of the Body "De usu partium" . . .*, I (Ithaca, N.Y., 1968), 307, and also gives Galen's own references to *De dissentione anatomica* in the Greek text of *De anatomicis administrationibus*, I, 4, and VII, 11 (Kühn, II, 236, 625), and in *De ordine librorum suorum* (Kühn, XIX, 55).
52. *Fī tashrīḥ al-ḥayawān al-mayyit* = *Fī tashrīḥ al-mawtā* (Ḥunayn, 21, no. 25; *GAS*, III, 100, no. 22; Ullmann, 53, no. 74) = *De mortuorum dissectione* (not extant).
53. *Fī tashrīḥ al-ḥayawān al-ḥayy* = *Fī tashrīḥ al-aḥyāʾ* (Ḥunayn, 21, no. 26; *GAS*, III, 100, no. 23; Ullmann, 54, no. 75) = *De vivorum dissectione* (not extant).
54. *Fī ʿilm Buqrāṭ biʾl-tashrīḥ* = *Fiʾl-tashrīḥ ʿalā raʾyi Buqrāṭ* (Ḥunayn, 21, no. 27; *GAS*, III, 133, no. 133; Ullmann, 54, no. 76) = *De Hippocratis anatomice* (not extant).

ḤUNAYN IBN ISḤĀQ

Singer, in *Galen: On Anatomical Procedures . . .*, 1, 238 (n. 5) mentions *De Hippocratis et Erasistrati anatomice*, in three books. According to Ḥunayn, *Fī ʿilm Buqrāṭ biʾl-tashrīḥ*, addressed to Boëthus, is in five books; and *Fī ʿilm Arsisṭrāṭus fiʾl-tashrīḥ* ("On Erasistratus' Knowledge of Anatomy"), also addressed to Boëthus, is in three books. For references to *Fī ʿilm Arsisṭrāṭus fiʾl-tashrīḥ*, see Ḥunayn, 22, no. 28; *GAS*, III, 101, no. 24; Ullmann, 54, no. 77.

55. *Fī tashrīḥ al-raḥim* (Ḥunayn, 22, no. 31; *GAS*, III, 101, no. 26; Charles M. Goss, "On the Anatomy of the Uterus," in *Anatomical Record*, **144**, no. 2 [1962], 77–84; D. Nickel, *Galen über die Anatomie der Gebärmutter herausgegeben, übersetzt und erläutert*, Corpus Medicorum Graecorum, **5**, no. 2, pt. 1 [Berlin, 1971]) = *De uteri dissectione* (Kühn, II, 887–908; Diels, I, 68).
56. *Fī tashrīḥ al-ʿayn* (Ḥunayn, 23, no. 35; *GAS*, III, 101, no. 27) is a pseudo-Galenic work. Ḥunayn thinks that it probably was written by Rufus of Ephesus or by another. less competent author.
57. Ḥunayn, 36, l. 9; *Fihrist*, I, 244, l. 10.
58. G. Bergsträsser (Ḥunayn, 4, l. 20; 6, ll. 2, 14) gives the form Ibn Sahdā; Ibn al-Nadīm (*Fihrist*, I, 244, l. 14) gives Ibn Shuhdī al-Karkhī. See also Sarton, I, 573.
59. Isḥāq ibn Ḥunayn, a contemporary of al-Rāzī, is frequently quoted in *al-Ḥāwī* (see Ullmann's citations, 119, n. 8). *Fihrist*, I, 285 and 298; Qifṭī, 80; IAU, I, 200–201, and 188, ll. 23–28; *GAS*, III, 267–268; Ullmann, 119.
60. *Fī nawādir taqdimat al-maʿrifa* (Ḥunayn, 34, no. 69; *GAS*, III, 114, no. 55; Ullmann, 44, no. 34) = *De praenotione ad Posthumum* = *De praegnotione ad Epigenum* (Kühn, XIV, 599–673; Diels, I, 100). A critical ed. of the Greek text, with an English trans. by Vivian Nutton, is scheduled for Corpus Medicorum Graecorum.
61. Ḥunayn, 34, ll. 12–14.
62. *Ibid.*, 39, ll. 21–23; "Alexandrian Medical Curriculum," 252, n. 125.
63. *Fī ʿadad al-maqāyīs* = *Fī iḥṣāʾ al-qiyāsāt* (Ḥunayn, 51, no. 127). A MS of this book, now lost, was seen by Abuʾl-Futūḥ Aḥmad ibn Muḥammad ibn al-Sarī ibn al-Ṣalāḥ (*ca.* 1090–1153). See N. Rescher, *Galen and the Syllogism* (Pittsburgh, 1966), 76, ll. 20–21 (Arabic text).
64. Ḥunayn, 51, ll. 15–16.
65. For example, see *Fihrist*, I, 249, ll. 1, 7, 12, 15; 250, l. 1; 251, ll. 3, 11–12, 26; 252, l. 2; and M. C. Lyons, ed., *An Arabic Translation of Themistius' Commentary on Aristoteles "De anima,"* which is Oriental Studies II (London, 1973). Notes in the MS used in Lyons' ed. purport that the text is a second trans. by Isḥāq ibn Ḥunayn (*tarjamat Isḥāq ibn Ḥunayn al-thāniya*). See pp. vii, 42, 88, 169, 214.
66. *Fī annaʾl-muḥarrik al-awwal lā yataḥarrak* (Ḥunayn, 51, no. 125). For some of Isḥāq's activities as translator of philosophical works, see Badawi, *La transmission de la philosophie grecque . . .*, 17, 79, 80, 85.
67. *Fiʾl-burhān* (Ḥunayn, 47, no. 115; *GAS*, III, 70, 72; Ullmann, 62, no. 112) = *De demonstratione* (lost in Greek; fragments are extant in al-Rāzī's *Kitāb al-Shukūk ʿalā Jālīnūs* ["Dubitationes in Galenum"], in Baghdatli Vehbi, Istanbul, MS 1488, XXVI, fols. 231b–248b; Majlis Shūrā-yi Millī, Teheran, MS 3821, fols. 150b–185b; and Millī Malik, Teheran, MS 4554, XXIII; see *GAS*, III, 77; Ullmann, 67).
68. Ḥunayn, 47, l. 12–48, l. 8.
69. *Ibid.*, 4, ll. 7–9; *Fī afkār Arsisṭrāṭus fī mudāwāt al-amrāḍ* (Ḥunayn, 36, no. 77; *GAS*, III, 136, no. 141).
70. Ḥunayn, 39, ll. 10–11.
71. *Fī ajzāʾ al-ṭibb* (Ḥunayn, 31, no. 61; *GAS*, III, 112, no. 49; M. Lyons, *Galen "On the Parts of Medicine," "On Cohesive Causes," "On Regimen in Acute Diseases in Accordance With the Theories of Hippocrates," First Ed. of the Arabic Versions With English Translation; the Latin Versions of "On the Parts of Medicine" Ed. by H. Schoene, and "On Cohesive Causes" Ed. by K. Kalbfleisch, Reedited by J. Kollesch, D. Nickel, and G. Strohmaier*, Corpus Medicorum Graecorum, supp. Orientale, II [Berlin, 1969], 24–49, 113–129) = *De partibus artis medicativae* (Diels, I, 137).
72. Ḥunayn, 31, l. 20–32, l. 3.
73. "Alexandrian Medical Curriculum," 252, n. 126; Ḥunayn, 27, ll. 11–12; "Alexandrian Medical Curriculum," 252, n. 129; Ḥunayn, 25, ll. 13–15.
74. IAU, I, 201, ll. 26–32. The Arabic text of *History*, with an English trans., was published by F. Rosenthal, "Isḥāq b. Ḥunayn's Taʾrīkh al-aṭibbāʾ," in *Oriens*, **7** (1954), 55–80.
75. *Fihrist*, I, 297; Qifṭī, 177; IAU, I, 202; *GAS*, III, 265–266; Ullmann, 119.
76. *Fī anna quwā al-nafs tābiʿa li-mizāj al-badan* (Ḥunayn, 50, no. 123; *GAS*, III, 72, ll. 24–25; Ullmann, 39, no. 6; English trans. of extracts in Brock, *Greek Medicine . . .*, 231–244; H. H. Biesterfeldt, "Galens Traktat 'Dass die Kräfte der Seele den Mischungen des Körpers folgen' in arabischer Übersetzung," in *Abhandlungen für die Kunde des Morgenlandes*, **40**, no. 4 [1973]) = *Quod animi mores corporis temperamenta sequantur* (Kühn, IV, 767–822; Diels, I, 72).
77. Ḥunayn, 50, ll. 9–12.
78. "Alexandrian Medical Curriculum," 252, n. 130.
79. *Fiʾl-ṣawt* (Ḥunayn, 24, no. 38; *GAS*, III, 103, no. 30; Ullmann, 54, no. 79; A. Barduagni, "Galeno ascriptus liber 'De voce et anhelitu,'" in *Pagine di storia della medicina*, **9**, no. 6 [1965], 39–51) = *De voce et anhelitu* (Diels, I, 147), pseudo-Galenic; the original work is lost. Ḥunayn (24, ll. 16–24) translated *On the Voice* into Arabic for Muḥammad ibn ʿAbd al-Malik al-Wazīr, who made some textual alterations. Muḥammad ibn Mūsā read the copies of Ḥunayn and al-Wazīr and chose to transcribe Ḥunayn's text.
80. Ḥunayn, 49, ll. 13–14.
81. *Ibid.*, 23, l. 22–24, l. 4; 24, ll. 16–24; 49, ll. 10–15.
82. *Ibid.*, 20, ll. 6–8.
83. *Fihrist*, I, 290, ll. 4–5.
84. *Ibid.*, I, 289, ll. 15–18.
85. G. Bergsträsser, *Ḥunain ibn Isḥāḳ und sein Schule* (Leiden, 1913), 15–24, 28, 48.
86. Meyerhof, "New Light on Ḥunain ibn Isḥāq . . .," 693, no. 21.
87. Bodleian Library, MS Marsh 158, II, fols. 406b, l. 3; 423b, l. 17; 476a, l. 23–476b, l. 1; 540b, ll. 19–20; The British Library, MS Add. 23406, fols. 2b, l. 2; 75b, ll. 6–7; 94a, ll. 9–10; University of California, Los Angeles, MS Ar. 90, p. 161, l. 20.
88. G. Bergsträsser, "Ḥunain ibn Isḥāq über die syrischen und arabischen . . .," p. v; and A. Z. Iskandar, "Bibliographical Studies in Medical and Scientific Arabic Works: Galen's *Fī ʿAmal al-tashrīḥ* ("On Anatomical Procedures"), the Alexandrian Book Entitled *Fi ʾl-tashrīḥ ila ʾl-mutaʿallimīn* ("On Anatomy for Students"), and Rhazes' *al-Kāfī fiʾl-ṭibb* ("The Sufficient Book on Medicine")," in *Oriens*, **25–26** (1976), p. 134, nn. 5–8.
89. For example, see M. Simon, *Sieben Bücher Anatomie des Galen*, I (Leipzig, 1906), 50, l. 7; 58, l. 6; 63, l. 1; 74, l. 3; 109, l. 12; 117, l. 15; 118, l. 12; 122, l. 12; 131, l. 10; 147, ll. 4, 11; 156, l. 11; 158, l. 3; 187, ll. 9, 16; 203, l. 8; 206, l. 15; 304, l. 4; 305, l. 1; 306, l. 5.
90. For example, see MS Marsh 158, II, fols. 502a, ll. 19–21; 564b, ll. 20–22; 566a, l. 9; MS Add. 23406, fols. 103b, l. 20–104a, l. 2; 107a, ll. 9–14; MS Ar. 90, pp. 177, ll. 15–22; 187, l. 22–188, l. 3.
91. MS Ar. 90, pp. 398, margin; 400, margin.
92. MS Marsh 158, II, fols. 607b, ll. 8–10; 608b, ll. 17–18.

93. MS Marsh 158, II, fol. 562b, ll. 15–19; MS Add. 23406, fol. 167a, ll. 10–13; MS Ar. 90, pp. 298, l. 24–299, l. 3; Simon, *Sieben Bücher* . . ., I, 50, ll. 1–6; also see W. L. H. Duckworth, *Galen: "On Anatomical Procedures," the Later Books,* M. C. Lyons and B. Towers, eds. (Cambridge, 1962), 41.
94. MS Add. 23406, fols. 167a, l. 13–167b, l. 2 (Arabic text is partly corrupt); MS Ar. 90, p. 299, ll. 3–15; Simon, *Sieben Bücher* . . ., I, 50, l. 7–51, l. 4. According to MS Ar. 90, the last few lines of this passage read: "wa dhālik fiʾl-mawḍiʿ alladhī qad waḍaʿ tu ʿalayhi hadhihiʾl-ʿalāmati allatī waḍaʿtuhā hāhunā bi-ʿaynihā." Ḥunayn's comment on Galen's text does not appear in MS Marsh 158, II, or in Duckworth's trans.
95. ʿĪsā ibn Yaḥyā ibn Ibrāhīm. *Fihrist*, I, 297; IAU, I, 203, l. 8; 204, l. 1; Sarton, I, 613.
96. *Fiʾl-faṣd* (Ḥunayn, 34, no. 71; *GAS*, 115, no. 57; Ullmann, 59, no. 97) = *De venae sectione* (Kühn, XIX, 519–528; Diels, I, 112).
97. Ḥunayn, 35, ll. 3–4.
98. "Alexandrian Medical Curriculum," 252, n. 121.
99. Ḥunayn, 38, ll. 10–13.
100. *Fiʾikhtilāf al-aʿḍāʾ al-mutashābihat al-ajzāʾ* (Ḥunayn, 23, no. 33; *GAS*, III, 101, no. 25; Ullmann, 55, no. 80; Arabic text edited with a German trans. and indexes by G. Strohmaier, *Galen Über die Verschiedenheit der homoiomeren Körperteile, in arabischer Übersetzung zum erstenmal herausgegeben, übersetzt und erläutert,* Corpus Medicorum Graecorum, Supp. Orientale, **III** [Berlin, 1970]) = *De partium homoeomerium differentia.*
101. *Fiʾl-ʿurūq al-ḍawārib hal yajrī fīhāʾl-damm biʾl-ṭabʿi am lā* (Ḥunayn, 25, no. 43; *GAS*, III, 104, no. 34; Ullmann, 41, no. 19) = *An in arteriis natura sanguis contineatur* (Kühn, IV, 703–736; Diels, I, 71).
102. *Fī quwā al-adwiya al-mushila* (Ḥunayn, 26, no. 44; *GAS*, III, 105, no. 35) = *De purgantium medicamentorum facultate* (Kühn, XI, 323–342; Diels, I, 95).
103. *Fiʾl-dhubūl* (Ḥunayn, 35, no. 72; *GAS*, III, 116, no. 58; Ullmann, 43, no. 27; T. C. Theoharides, "Galen: 'On Marasmus,'" in *Journal of the History of Medicine and Allied Sciences*, **26**, no. 4 [1971], 369–390) = *De marasmo* = *De marcore* (Kühn, VII, 666–704; Diels, I, 83).
104. *Fiʾl-tiryāq ilā Bamfūliyānus* (Ḥunayn, 38, no. 82; *GAS*, III, 121, no. 67; Ullmann, 49, no. 52) = *De theriaca ad Pamphilianum* (Kühn, XIV, 295–310; Diels, I, 99).
105. *Tafsīr kitāb ʿahd Buqrāṭ* (Ḥunayn, 40, no. 87; *GAS*, III, 123, no. 70; Ullmann, 62, no. 111). This book was translated into Syriac by Ḥunayn, who wrote a commentary on its difficult passages, and also was translated into Arabic by Ḥubaysh (Ḥunayn, 40, 11. 2–4).
106. *Tafsīr kitāb al-akhlāṭ* [*li-Buqrāṭ*] (Ḥunayn, 42, no. 96; *GAS*, III, 123, no. 79; Ullmann, 62, no. 110) = *In Hippocratis de humoribus librum commentarii* (Kühn, XVI, 1–488; Diels, I, 103).
107. *Fī annaʾl-ṭabīb al-fāḍil faylasūf* (Ḥunayn, 44, no. 103; Ullmann, 38, no. 2; P. Bachmann, *Galens Abhandlung darüber, dass der vorzügliche Arzt Philosoph sein muss. Arabisch und deutsch herausgegeben* [Göttingen, 1965]) = *Quod optimus medicus sit quoque philosophus* (Kühn, I, 53–63; Diels, I, 59).
108. *Fī annaʾl-akhyār minʾl-nās qad yantafiʿūn bi-aʿdāʾihim* (Ḥunayn, 49, no. 121).
109. *Fiʾl-mirra al-sawdāʾ* (Ḥunayn, 32, no. 64; *GAS*, III, 113, no. 52; Ullmann, 40, no. 10) = *De atra bile* (Kühn, V, 104–148; Diels, I, 73).
110. *Fiʾl-tadbīr al-mulaṭṭif* (Ḥunayn, 36, no. 75; *GAS*, III, 117, no. 61; Ullmann, 47, no. 47; W. Frieboes and F. W. Kobert, "Galens Schrift 'Ueber die säfteverdünnende Djät.' Uebersetzt und mit Einleitung und Sachregister versehen," in *Abhandlungen zur Geschichte der Medicin,* no. 5[1903]) = *De victu attenuante* (Diels, I, 125).

ḤUNAYN IBN ISḤĀQ

111. *Tafsīr kitāb taqdimat al-maʿrifa* (Ḥunayn, 40, no. 91; *GAS*, III, 123, no. 74; Ullmann, 50, no. 59) = *In Hippocratis prognosticum commentarii* (Kühn, XVIIIB, 1–317; Diels, I, 107).
112. Ḥunayn, 40, l. 23–41, l. 2.
113. Millī Malik, Teheran, MS 4554, XXIII, p. [2], l. 22; Baghdatli Vehbi, Istanbul, MS 1488, XXVI, fol. [232a], ll. 22–23. See notes 46 and 67.
114. Ḥunayn, 46, l. 23–47, l. 3.
115. *Fihrist*, I, 244, l. 10; IAU, I, 204; Sarton, I, 613.
116. C. E. Dubler and E. Terés, *La "Materia médica" de Dioscórides, transmisión medieval y renacentista,* II (Tetuán–Barcelona, 1952–1957), Arabic text; on the revisions by Ḥunayn, see pp. 125, 236, 442.
117. Ḥunayn, 24, ll. 1–2.
118. "Alexandrian Medical Curriculum," 252, no. 127. See Ḥunayn, 24, ll. 8–9.
119. *Fiʾl-ḥāja ilaʾl-tanaffus* = *Fī manfaʿat al-tanaffus* = *Fī manfaʿat al-nafas* (Ḥunayn, 25, no. 42; *GAS*, III, 104, no. 33; Ullmann, 41, no. 17) = *De utilitate respirationis* = *De usu respirationis* (Kühn, IV, 470–511; Diels, I, 70).
120. Ḥunayn, 35, ll. 1–3.
121. *Ibid.*, 32, ll. 15–16; 35, ll. 8–10.
122. Qifṭī, 172, ll. 5–20; Ibn Juljul, *Ṭabaqāt* . . ., 69–70; IAU, I, 190, ll. 7–23. See note 26.
123. IAU, I, 190, ll. 25–30.
124. *Fīmā aṣābahu minʾ l-miḥan waʾl-shadāʾid min alladhīn nāṣabūh al-ʿadāwata min ashrāri aṭibbāʾi zamānih al-mash-hūrīn.* IAU, I, 190, l. 31–197, l. 12, presents long extracts from this lost work.
125. *Ibid.*, I, 190–192.
126. *GAL*, I, 224, nos. 1, 2; *S*, 1, 367, nos. 1, 2; *GAS*, III, 249, no. 1; Ullmann, 118, no. 5. For MSS see bibliography to this article.
127. IAU, I, 197, l. 24; and 202, ll. 5–6; H. P. J. Renaud, *Les manuscrits arabes de l'Escurial, décrits d'après les notes de H. Derenbourg*, II, no. 2 (Paris, 1941), 60, MS 853, I, fols. 1b–64a. See also notes 138, 139. Brockelmann (*GAL*, I, 224, nos. 1, 2) considers *al-Madkhal fiʾl-ṭibb* (Escorial MS 848, I [*sic*]) as a book different from *al-Masāʾil fiʾl-ṭibb*. MS collation, however, has shown that these are variant titles of the same book. Escorial MS 853, I, is defective. Its copyist missed a passage of about three lines (*isqāṭ*) that deals with the division of "theory of medicine," as well as short passages in other places.
128. J. Uri, *Bibliothecae Bodleianae codicum manuscriptorum orientalium*, I (Oxford, 1787), 140, no. 595 (MS Marsh 403, fols. 1a–62a; title appears in the closing passage of fol. 62a); A. Nicoll and E. B. Pusey, *Bibliothecae Bodleianae* . . ., II, 1 (Oxford, 1821–1835), 170–171, no. 195 (MS Marsh 16, II, fols. 56b–174b; title appears in the opening passage of fol. 56b).
129. MS Marsh 403, fol. 39a.
130. *GAL*, I, 638; *S*, I, 886.
131. Two MSS have been consulted: Bodleian Library, MS Marsh 98; and Wellcome Institute for the History of Medicine, WMS. Or. 2. See Uri, *Bibliothecae Bodleianae* . . ., I, 141, no. 600; and A. Z. Iskandar, *A Catalogue of Arabic Manuscripts on Medicine and Science in the Wellcome Historical Medical Library* (London, 1967), 179 (the first ten leaves of WMS. Or. 2 are badly damaged).
132. *GAL*, I, 648; *S*, I, 899; IAU, II, 272–273.
133. This license appears at the end of book I of al-Nīsābūrī's commentary, in MS Marsh 98, fol. 208a.
134. MS Marsh 98, fol. 1a, ll. 3–10; WMS. Or. 2, fol. 1b, ll. 2–8.
135. MS Marsh 98, fols. 1b, l. 8–2a, l. 7.
136. WMS. Or. 2, fols. 148a–264a.
137. *GAL*, I, 637; *S*, I, 886; Qifṭī, 443–444; IAU, II, 99–105; Ullmann, 158.
138. "Alexandrian Medical Curriculum," 239–240. In Ibn

Riḍwān's *Useful Book* . . ., the "second chapter [of treatise II] is on errors and mistakes made by Ḥunayn in his works; these errors are injurious to the art of medicine." This chapter consists of Ibn Riḍwān's belligerent criticism of Ḥunayn's *Questions on Medicine*. See Chester Beatty Library, Dublin, MS 4026, fols. 12b, l. 11–16b, l. 12 (A. J. Arberry, *The Chester Beatty Library. A Handlist of the Arabic Manuscripts*, V [Dublin, 1962], 9).

139. Chester Beatty Library, MS 4026, fols. 17a, ll. 1–2; 21a, l. 13; 22a, ll. 8, 12–13.
140. Renaud, *Les manuscrits arabes de l'Escurial* . . ., 60.
141. Escorial Library, MS 853, fol. 1b, ll. 2–13; Bodleian Library, MS Marsh 403, fol. 1a, ll. 3–18; and MS Marsh 16, II, fols. 56b, l. 4–57a, l. 12.
142. W. Ahlwardt, *Die Handschriften-Verzeichnisse der königlichen Bibliothek zu Berlin. Verzeichniss der arabischen Handschriften*, V (Berlin, 1893), 515 (MS 6258 = Spr. 1885, fols. 1–18); A. G. Ellis and E. Edwards, *A Descriptive List of the Arabic Manuscripts Acquired by the Trustees of the British Museum Since 1894* (London, 1912), 45 (MS Or. 5862, III, fols. 67b–88a; the title of this book appears on fol. 67b as *Masāʾil Ḥunayn ibn Isḥāq ʿalā ṭarīq al-taqsīm waʾl-tashjīr*); H. Ritter and R. Walzer, "Arabische Übersetzungen griechischer Ärzte in Stambuler Bibliotheken," in *Sitzungsberichte der Preussischen Akademie der Wissenschaften*, phil.-hist. Kl., **26** (1934), 827 (Aya Sofya, MS 3324). E. T. Withington, in *Medical History From the Earliest Times* . . . (London, 1894; repr. London, 1964), 386–396, gives an English trans. of Ḥunayn's *Introduction to Medicine* (*Isagoge*), under a general title—"The 'Galenic' System of Medicine"—adding that it contains "an excellent account of the Galenic and mediaeval medical theories." Withington does not specify the Latin source of his English trans. The Latin version used by Withington cannot have been made from the Arabic text of *al-Masāʾil fiʾl-ṭibb*; nevertheless, there are similarities and differences between Withington's English trans. and the Arabic text of *Masāʾil Ḥunayn ibn Isḥāq ʿalā ṭarīq al-taqsīm waʾl-tashjīr* (that is, *al-Masāʾil fiʾl-ṭibb*, from which the questions have been eliminated). It would be interesting to make a comparative study of the Arabic text of this book and the Latin text of the *Isagoge*. See also O. Temkin, *Galenism: Rise and Decline of a Medical Philosophy* (Ithaca, N.Y.–London, 1973), 104–107.
143. *Catalogus codicum manuscriptorum orientalium qui in Museo Britannico asservantur, pars secunda, codices arabicos amplectens*, II (London, 1846–1871), 456–457 (no. 984; MS Arundel Or. 10, III, fols. 28a–55a). A photostatic copy of this MS is at Dār al-Kutub al-Miṣriyya (Cairo), MS Ṭibb 1103.
144. MS Arundel Or. 10, fols. 28b, ll. 3–9; 29a, ll. 1–4; 33b, ll. 5–6.
145. Escorial Library, MS 853, fol. 1b, ll. 2–3; Bodleian Library, MS Marsh 403, fol. 1a, ll. 3–7; and MS Marsh 16, II, fol. 56b, ll. 4–9. The division of medicine into theory and practice appears in the *Summaria Alexandrinorum*, in a summary and commentary of Galen's book *On Sects*. See British Library, MS 23407, fol. 2b, ll. 2–15.
146. Escorial Library, MS 853, fols. 5b, l. 7–6b, l. 14; Bodleian Library, MS Marsh 403, fols. 5a, l. 9–6b, l. 3; MS Marsh 16, II, fols. 64a, l. 6–66a, l. 14.
147. In Arabic MSS the word *iḥtiqān* (congestion), not *imtilāʾ* (repletion), appears in the text—for example, in Escorial Library, MS 853, fol. 6a, l. 16; Bodleian Library, MS Marsh 403, fol. 6a, l. 6; and MS Marsh 16, II, fol. 65b, l. 9. The word *iḥtiqān* also appears in books by later writers who quote the six necessary causes (see note 167).
148. Escorial Library, MS 853, fol. 6a, ll. 14–16; Bodleian Library, MS 403, fol. 6a, ll. 3–6; and MS Marsh 16, II, fol. 65b, ll. 5–9.
149. Escorial Library, MS 853, fols. 6a, l. 16–6b, l. 3; Bodleian Library, MS 403, fol. 6a, ll. 7–10; and MS Marsh 16, II, fol. 65b, ll. 9–14.
150. "Alexandrian Medical Curriculum," 237–239, and nn. 15, 16.
151. *Catalogus codicum manuscriptorum orientalium qui in Museo Britannico* . . ., II, 629–630 (no. 1356).
152. MS Add. 23407, fols. 2b, l. 2–20b, l. 3; fols. 20b, l. 5–48a, l. 17; fols. 72b, l. 7–157a, l. 5.
153. *Ibid.*, fol. 129b, ll. 6–8 (at the beginning of treatise II).
154. *Ibid.*, fol. 2b, ll. 2–3.
155. See note 147.
156. MS Add. 23407, fol. 6a, l. 17–6b, l. 4.
157. *Ibid.*, fol. 20b, ll. 5–6.
158. *Ibid.*, fol. 42a, l. 16–42b, l. 12.
159. *Ibid.*, fol. 72b, ll. 7–8.
160. *Ibid.*, fol. 76b, ll. 8–11.
161. L. J. Rather, "The 'Six Things Non-Natural': A Note on the Origin and Fate of a Doctrine and a Phrase," in *Clio medica*, **3** (1968), 337–347; S. Jarcho, "Galen's Six Non-Naturals: A Bibliographic Note and Translation," in *Bulletin of the History of Medicine*, **44** (1970), 372–377; J. J. Bylebyl, "Galen on the Non-Natural Causes of Variation in the Pulse," *ibid.*, **45** (1971), 482–485; P. H. Niebyl, "The Non-Naturals," *ibid.*, 486–492; and C. R. Burns, "The Nonnaturals: A Paradox in the Western Concept of Health," in *Journal of Medicine and Philosophy*, **1**, no. 3 (1976), 202–211.
162. Bylebyl, "Galen on the Non-Natural Causes of Variation in the Pulse," 483; Niebyl, "The Non-Naturals," 486.
163. Ḥunayn, p. 6, ll. 11–12.
164. Rather, "The 'Six Things Non-Natural' . . .," 339 (quoting Lemnius' reference to *Ars medica*); and Jarcho, "Galen's Six Non-Naturals . . .," 376 (a trans. from the Latin text of *Ars medica*, in Kühn's ed.).
165. Escorial Library, MS 853, fols. 41b, l. 14–42a, l. 6; Bodleian Library, MS Marsh 403, fol. 42a, ll. 5–13; and MS Marsh 16, II, fol. 131a, l. 12–131b, l. 11; British Library, MS Or. 5725, fol. 69a, l. 10–69b, l. 9. The same account appears, in a tabulated form, in British Library, MS Or. 5862, fol. 80b (*Masāʾil Ḥunayn ibn Isḥāq ʿalā ṭarīq al-taqsīm waʾl-tashjīr*; see note 142).
166. For example, see Bodleian Library, MS Marsh 403, fol. 42a, ll. 13–14; and MS Marsh 16, II, fol. 131b, l. 9 (this note is incorporated in the text of the two MSS, with a marginal word, *ḥāshiya*, appearing in MS Marsh 403). Al-Majūsī (*d.* late tenth century; *GAL*, I, 273, and *S*, I, 423) also considers bathing, sexual intercourse, and micturition as means of evacuation. See *Kāmil al-ṣināʿa al-ṭibbiyya*, Būlāq, ed., I (Cairo, 1877), 153, ll. 3–4.
167. Al-Rāzī, in one MS of *al-Ḥāwī*, writes "qad intahaynā ilā ākhir al-kalām fiʾl-asbāb al-sitta al-ḍarūriyya al-mushtaraka liʾl-ṣiḥḥa waʾl-maraḍ" ("We have come to the end of the account of the six necessary causes that are common to health and disease"). See Wellcome Institute for the History of Medicine, WMS. Or. 123, fol. 18b, ll. 8–9. Some of al-Rāzī's accounts of the six necessary causes are preserved in this defective MS (Iskandar, *A Catalogue of Arabic Manuscripts on Medicine and Science* . . ., 3–26, 104–105). The six necessary causes are also mentioned by al-Majūsī in *Kāmil al-ṣināʿa* . . ., I, 152, l. 30–153, l. 3; by Ibn Sīnā (*d.* 1037; *GAL*, I, 589; and *S*, I, 812) in *al-Qānūn fiʾl-ṭibb*, Būlāq, ed., I (Cairo, 1877), 80, ll. 23–25; by Ibn Hubal al-Baghdādī (*d.* 1210; *GAL*, I, 646, and *S*, I, 895) in *al-Mukhtārāt fiʾl-ṭibb*, **I** (Hyderabad–Deccan, 1943), 105, ll. 7–10; and by Ibn al-Nafīs (*d.* 1288; *GAL*, I, 649, and *S*, I, 899) in *Mūjiz al-qānūn* (Lucknow, 1906), 9–12.
168. For example, Ibn Buṭlān of Baghdad (*d. ca.* 1066) wrote *Taqwīm al-ṣiḥḥa biʾl-asbāb al-sitta allatī lābudda li-kulli insān yuʾthir dawām ṣiḥḥatihi min taʿdīliha* ("Tables for the Preservation of Health, by Means of the Six Causes Which Must Inevitably Be Altered if a Person Wishes to

Remain Healthy"), *GAL*, I, 636, no. 1; *S*, I, 885, no. 1; Ullmann, 157. See also M. Azeez Pasha, "Yusrul ilaj (a Persian Medical Manuscript Compiled in India by Hekeem Hida-etullah, in 1731 A.D.)," in *Bulletin of the Institute of History of Medicine*, **3**, no. 3 (1973), 126–131, 201–206; and *Risāla mushtamila ʿalā mukhtaṣarāt al-maṭālib al-ṭibbiyya minʾ l-sitta al-ḍarūriyya* ("A Missive Comprising a Résumé of Medical Requirements [Fulfilled] Through the Six Necessary [Causes]"), by Muḥammad Muḥsin [?], in Wellcome Institute for the History of Medicine, WMS. Or. 114 (Iskandar, *A Catalogue of Arabic Manuscripts on Medicine and Science* . . ., 172).

169. *al-Mukhtārāt fiʾl-ṭibb* . . ., I, 105, ll. 8–9.

170. *Fī tarkīb al-adwiya* (Ḥunayn, 36, no. 79; IAU, I, 98, ll. 14–19; *GAS*, III, 118, no. 64; Ullmann, 48, no. 50).

171. Ḥunayn, 37, ll. 12–14.

172. *Fī tarkīb al-adwiya ʿalaʾl-jumali waʾl-ajnās* (Ḥunayn, 36, l. 19–37, l. 6; IAU, I, 98, ll. 16–17).

173. *Fī tarkīb al-adwiya bi-ḥasab al-mawāḍiʿ al-ālima* (Ḥunayn, 37, ll. 6–11; IAU, I, 98, ll. 17–19).

174. "Alexandrian Medical Curriculum," 251, n. 118.

175. Kühn, XII, 378–1007; XIII, 1–361.

176. *Ibid.*, XIII, 362–1058. This reversed order is preserved by Diels (I, 97–98), Sezgin (*GAS*, III, 118, no. 64), and Ullmann (48, no. 50).

177. For example, see notes 50 and 67; and "Alexandrian Medical Curriculum," 243, n. 76.

BIBLIOGRAPHY

I. ORIGINAL WORKS. Ḥunayn's works are listed below in alphabetical order of book titles.

1. *Fiʾl-aghdhiya* ("On Foods"), in Khudā Bakhsh, Patna Bankipore, MS 2, I, fols. 1–109 (Maulavī ʿAẓīmuʾd-Dīn Aḥmad and E. Denison Ross, eds., *Catalogue of the Arabic and Persian Manuscripts in the Oriental Public Library at Bankipore*, IV (Calcutta, 1910), 5–7; an Arabic ed., with a German trans., is in preparation by R. Degen.

2. *Kitāb al-ʿashar maqālāt fiʾl-ʿayn al-mansūb li-Ḥunayn ibn Isḥāq* . . . (M. Meyerhof, *The Book of the Ten Treatises on the Eye Ascribed to Hunain ibn Is-Hāq (809–877 A.D.)* . . . [Cairo, 1928]).

3. *Fī awjāʿ al-maʿida* ("On Disorders of the Stomach") = *Fī maʿrifat awjāʿ al-maʿida wa-ʿilājihā* ("On the Diagnosis and Therapy of Disorders of the Stomach"), in Aya Sofya, Istanbul, MS 3555, fols. 149b–156a (*GAL*, I, 225, no. 3); Escorial Library, MS 852, III, fols. 41a–68a (H. P. J. Renaud, *Les manuscrits arabes de l'Escurial* . . ., II, 2 [Paris, 1941], 59–60); University of California, Los Angeles, MS Ar. 98, I, pp. 1–47.

4. *Fiʾl-daghdagha* ("On Titillation"), in Aya Sofya, Istanbul, MS 3725, fols. 68a–72b (H. Ritter and R. Walzer, "Arabische Übersetzungen griechischer Ärzte in Stambuler Bibliotheken," in *Sitzungsberichte der Preussischen Akademie der Wissenschaften*, phil.-hist. Kl., **26** [1934], 827); Majlis-i Shūrā-yi Millī, Teheran, MS 1551 (Abdol Hossein Haeri, *Fihrist-i Kitābkhāna-i Majlis-i Shūrā-yi Millī*, IV (Teheran, 1956), 252–254).

5. *Fi ḥifẓ al-asnān wa-istiṣlāḥihā* ("On Hygiene of the Teeth, and Dental Repair") = *Qawl fī ḥifẓ al-asnān waʾl-litha wa-istiṣlāḥihā* ("Statements on Hygiene of the Teeth and Gums, and Dental Repair"), in Ẓāhiriyya Library, Damascus, MS 4516 (Ṣ. al-Munājjed, "Maṣādir jadīda ʿan tārīkh al-ṭibb ʿind al-ʿarab," in *Majallat Maʿhad al-Makhṭūṭāt al-ʿArabiyya*, **5**, no. 2 [1959], 294, no. 300); Bodleian Library, Oxford, MS Hunt 461, fols. 52b–63a (R. Degen, "Eine weitere Handschrift von Ḥunain ibn Isḥāqs Schrift über die Zahnheilkunde," in *Annali dell'Istituto orientale di Napoli*, n.s. **26** [1976], 236–243; J. Uri, *Bibliothecae Bodleianae codicum manuscriptorum orientalium*, I [Oxford, 1787], 140, no. 598).

6. *Jawāmiʿ arbaʿat ʿashar maqāla min kitāb Jālīnūs fī ḥīlat al-burʾ* ("Summary and Commentary of Fourteen Treatises of Galen's Book 'On the Method of Healing'"), in Chester Beatty Library, Dublin, MS 4001, II, fols. 9b–14 (A. J. Arberry, *The Chester Beatty Library. A Handlist of the Arabic Manuscripts*, V [Dublin, 1962], 1).

7. *Jawāmiʿ maʿānī al-khams al-maqālāt al-ūlā min kitāb Jālīnūs fī quwā al-adwiya al-mufrada mansūqa ʿalā ṭarīq al-masʾala waʾl-jawāb* ("Summary and Commentary of the First Five Treatises of Galen's Book on 'Materia Medica,' Presented in the Form of Questions and Answers"), in Nūr ʿUthmāniyya Library, Istanbul, MS 3555, fols. 1–217a (Ritter and Walzer, "Arabische Übersetzungen griechischer Ärzte in Stambuler Bibliotheken," 828).

8. *Kitāb al-karma* ("A Book on the Grapevine") = *Qawl Ḥunayn ibn Isḥāq fīmā dhakarahu Jālīnūs fiʾl-juzʾ al-rābiʾ min al-maqāla al-thāniya min kitābihi fī quwā al-adwiya waʾl-aghdhiya ʿalā ṭarīq al-masʾala waʾl-jawāb* ("Ḥunayn ibn Isḥāq's Statement on What Galen Had Mentioned in the Fourth Part of Treatise Two of His Book 'On the Virtues of Drugs and Foods,' in the Form of Questions and Answers"), in Aya Sofya, MS 3703, fols. 155a–202a (Ritter and Walzer, "Arabische Übersetzungen griechischer Ärzte in Stambuler Bibliotheken," 828).

9. *Fiʾl-laban* ("On Milk") = *Fī maʿrifat quwwat al-laban* ("On Estimating the Virtues of Milk"), in Āṣafiyya Library, Hyderabad, MS 360 (ʿAbbās . . . al-Kantūrī, *Fihrist-i kutub-i ʿarabī wa fārsī wa urdū makhzūna Kutubkhāna-i Āṣafiyya-i Sarkār-i ʿAlī*, II (Hyderabad, 1914–1915), *ṭibb yūnānī*, 936).

10. *Risālat Ḥunayn ibn Isḥāq ilā ʿAlī ibn Yaḥyā* . . . (see note 2).

11. *Kitāb al-Masāʾil fiʾl-ʿAyn* (P. Sbath, "Le livre des questions sur l'oeil de Ḥunaïn ibn Isḥāq," in *Mémoires présentés à l'Institut d'Egypte*, **36** (1938).
Bulletin de l'Institut d'Égypte, **17** (1934–1935), 129–138; P. Sbath and M. Meyerhof, "Le livre des questions sur l'oeil de Ḥunaïn ibn Isḥāq," in *Mémoires présentés à l'Institut d'Egypte*, **36** (1938).

12. *al-Masāʾil fiʾl-ṭibb* ("Questions on Medicine") = *Masāʾil Ḥunayn ibn Isḥāq fiʾl-ṭibb liʾl-mutaʿallimīn maʿ ziyādāt Ḥubaysh tilmīdhih* ("Ḥunayn ibn Isḥāq's 'Questions on Medicine for Students,' With Additions by His Pupil Ḥubaysh") = *al-Madkhal fiʾl-ṭibb* ("Introduction to Medicine") = *al-Madkhal ilā ṣināʿat al-ṭibb* ("Introduction to the Art of Physic"), in Aḥmadiyya

Library, Tunis, MS 5437 (Ṣ. al-Munajjed "Maṣādir jadīda . . .," 294, no. 303); Bodleian Library, MS Marsh 403, fols. 1a–62a (J. Uri, *Bibliothecae Bodleianae codicum manuscriptorum orientalium*, I, 140, no. 595); MS Marsh 16, II, fols. 56b–174b (A. Nicoll and E. B. Pusey, *Bibliothecae Bodleianae . . .*, II, 1, 1821–1835, 170–171, no. 195); British Library, MS Or. 5725, fols. 2a–100b, incomplete at both beginning and end (A. G. Ellis and E. Edwards, *A Descriptive List of the Arabic Manuscripts Acquired by the Trustees of the British Museum . . .*, [London, 1912], 45); Dār al-Kutub al-Miṣriyya, MS Ṭalʿat 511 (Munajjed, "Maṣādir jadīda . . .," 294, no. 303); Escorial Library, MS 853, I, fols. 1b–64a (Renaud, *Les manuscrits arabes de l'Escurial . . .*, 60–61); Forschungsbibliothek, Gotha, MS 2023, I, fols. 1–55; MS 2028, fols. 1–79; MS 2036, III, fols. 186a ff., incomplete (W. Pertsch, *Die arabischen Handschriften der herzoglichen Bibliothek zu Gotha*, IV [Gotha, 1883], 54–55, 73); al-Gharawiyya Library, Nejef, Iraq, MS 19 (Ḥ. ʿA. Maḥfūẓ, "Fihris al-Khizāna al-Gharawiyya biʾl-Najaf fī Mash-hadi Amīr al-Muʾminīn al-Imām ʿAlī ibn Abī Ṭālib ʿalayhi al-salām," in *Majallat Maʿhad al-Makhṭūṭāt al-ʿArabiyya*, **5**, no. 1 [1959], 25); Manisa Kitapsaray, Istanbul, MS 1779, I, fols. 1b–54a (A. Dietrich, "Medicinalia Arabica. Studien über arabische medizinische Handschriften in türkischen und syrischen Bibliotheken," in *Abhandlungen der Akademie der Wissenschaften zu Göttingen*, phil.-hist. Kl., 3rd ser., no. 66 [1966], 39); Saray Aḥmad III, Istanbul, MS 2131, I, fols. 1b–178b (*ibid.*, 41–42); al-Mārūniyya Library, Aleppo, MS 568 (S. Kataye, *Les manuscrits médicaux et pharmaceutiques dans les bibliothèques publiques d'Alep* [Aleppo, 1976], 355–358); University of Leeds, MS Arab. 265 (J. Macdonald, "Catalogue of Oriental Manuscripts, the University of Leeds, Department of Semitic Languages and Literatures," VI [1962, typescript], 16).

13. *Masāʾil Ḥunayn ʿalā ṭarīq al-taqsīm waʾl-tashjīr* ("Ḥunayn's 'Questions, by Way of Subdivided Classification' "). The MSS in which this title appears represent the *Questions on Medicine*, from which the questions have been omitted; the answers appear in tabulated forms in Aya Sofya, MS 3324 (Ritter and Walzer, "Arabische Übersetzungen griechischer Ärzte in Stambuler Bibliotheken," 827), see note 142; British Library, MS Or. 5862, III, fols. 67b–88a (Ellis and Edwards, *A Descriptive List of the Arabic Manuscripts Acquired by the Trustees of the British Museum . . .*, 45); Deutsche Staatsbibliothek, Berlin, MS 6258 = Spr. 1885, fols. 1–18 (W. Ahlwardt, *Die Handschriften-Verzeichnisse der königlichen Bibliothek zu Berlin . . .*, V [Berlin, 1893], 515).

14. *Fī tadbīr al-nāqih* ("On Treatment of Convalescents") = *Maqālat Ḥunayn ibn Isḥāq li-Abī Jaʿfar Muḥammad ibn Mūsā jamaʿa fīhā mā qālahu Jālīnūs fī tadbīr al-nāqih fī jamīʿi kutubihi allatī dhakar fīhā hādha al-bāb* ("Ḥunayn ibn Isḥāq's Treatise Addressed to Abū Jaʿfar Muḥammad ibn Mūsā, in Which Ḥunayn Collected From All of Galen's Books the Opinions of Galen on Treatment of Convalescents") = *Majmūʿ min aqwāl Jālīnūs wa-Abuqrāṭ fī tadbīr al-nāqih wa-man shābahah minʾl-mahzūlīn* ("A Collection of Statements by Galen and Hippocrates on the Treatment of Convalescents and Similar Emaciated Patients"), in Aya Sofya, MS 3590, fols, 137b–163b (Ritter and Walzer, "Arabische Übersetzungen griechischer Ärzte in Stambuler Bibliotheken," 828); University Library, Cambridge, MS Or. 1022 (A. J. Arberry, *A Second Supplementary Hand-List of the Muḥammadan Manuscripts in the University and Colleges of Cambridge* (Cambridge, 1952), 6, no. 30).

II. SECONDARY LITERATURE. The following works are listed chronologically: J. Freind, *The History of Physick . . .*, II (London, 1726), 17–19; L. Leclerc, *Histoire de la médecine arabe . . .*, I (Paris, 1876; repr. New York, 1971), 139–152; C. Prüfer and M. Meyerhof, "Die aristotelische Lehre vom Licht bei Ḥunain b. Isḥāq," in *Der Islam*, **2** (1911), 117–128; "Die Augenanatomie des Ḥunain b. Isḥāq," in *Archiv für Geschichte der Medizin*, **4** (1911), 163–190; and "Die Lehre vom Sehen bei Ḥunain b. Isḥāq," *ibid.*, **6** (1913), 21–33; al-Yāfiʿī, *Mirʾāt al-janān . . .*, II (Hyderabad, 1918–1919), 172; E. G. Browne, *Arabian Medicine* (Cambridge, 1921; repr. Cambridge, 1962), 24–26; G. Gabrieli, "Hunāyn ibn Ishāq," in *Isis*, **6** (1924), 282–292; D. Campbell, *Arabian Medicine and Its Influence on the Middle Ages*, I (London, 1926), 61–63; *Encyclopaedia of Islam*, II (Leiden–London, 1927), 336, also new ed., III (Leiden–London, 1971), 578–581; Y. A. Sarkis, *Muʿjam al-maṭbūʿāt al-ʿarabiyya . . .* (Cairo, 1928), 801–802; M. Meyerhof, "Science and Medicine," in Thomas Arnold and Alfred Guillaume, eds., *The Legacy of Islam* (Oxford, 1931), 316–320; Ḥājjī Khalīfa, *Kashf al-ẓunūn . . .*, II (Istanbul, 1943; repr. Teheran, 1967), 1668; Ibn Khallikān, *Wafayāt al-aʿyān . . .*, I (Cairo, 1948), 528–529; A. A. Khairallah, *Outline of Arabic Contributions to Medicine and the Allied Sciences* (Beirut, 1946), 44–48; I. al-Baghdādī, *Hadiyyat al-ʿārifīn . . .*, I (Istanbul, 1951), 339–340; and C. Elgood, *A Medical History of Persia and the Eastern Caliphate . . .*, (Cambridge, 1951), 104–113.

See also Kh. al-Ziriklī, *al-Aʿlām . . .*, 2nd ed., II (Cairo, 1954), 325; ʿU. R. Kaḥḥāla, *Muʿjam al-muʾallifīn . . .*, IV (Damascus, 1957), 87–88; M. Tayyab, "Al-ashr-ul-maqalat fīl ayn. The First Compilation on Ophthalmology in the World," in *Hamdard Medical Digest*, **2**, no. 8 (1958), 1–9; S. K. Hamarneh, "Bibliography on Medicine and Pharmacy in Medieval Islam. Mit einer Einführung Arabismus in der Geschichte der Pharmazie von Rudolf Schmitz," in *Veröffentlichungen der Internationalen Gesellschaft für Geschichte der Pharmazie*, n.s. **25** (1964), 60–61; *Index of Manuscripts on Medicine, Pharmacy, and Allied Sciences in the Ẓāhiriyah Library* (Damascus, 1969), 63–76, 227; and *Catalogue of Arabic Manuscripts on Medicine and Pharmacy at the British Library* (Cairo, 1975), 35–40; F. Rosenthal, *Das Fortleben der Antike im Islam* (Zurich, 1965), 20–23, 36–38; and *The Classical Heritage in Islam*, translated from the German by E. and

J. Marmorstein (Berkeley–Los Angeles, 1975), 7–8, 19–20; G. C. Anawati, "Science," in *The Cambridge History of Islam*, II (Cambridge, 1970), 768–769; R. Y. Ebied, *Bibliography of Mediaeval Arabic and Jewish Medicine and Allied Sciences* (London, 1971), 81; S. H. Nasr, *An Annotated Bibliography of Islamic Science . . .*, I (Teheran, 1975), 227–229; and "Ḥunayn ibn Isḥāq. Collection d'articles publiée à l'occasion du onzième centenaire de sa mort," in *Arabica*, **21**, no. 3 (1975), 229–330.

ALBERT Z. ISKANDAR

IKHWĀN AL-ṢAFĀʾ. The group of epistles entitled *Ikhwān al-Ṣafāʾ*, often called *Encyclopedia of the Brothers of Purity*, is a compendium of all the sciences known in the tenth century and the first complete exposition of the Ismaili philosophical system. The Shiites sought to prove that the imams—divinely appointed successors of Muḥammad in the line of ʿAlī—are alone entitled to rule the Muslim community, and that their authority should be seen in the context of a cosmic mission. Toward this end they adopted the Neoplatonic doctrine of emanation. The Ismaili further systematized and integrated with the Hermetic and astrology of the Ḥarrānians.

According to the Ismaili schema, God created the universe through a series of emanations originating with Himself, each giving rise directly to its successor. The series consists of the Intellect and its archetypes, the Soul, and prime matter. The Intellect educated the Soul through the aid of its archetypes; and the Soul, in turn, fragmented into innumerable faculties or souls, which gave prime matter its forms. First came the nine celestial spheres, by means of which the lower world is governed, and then the four elements, the lowest point reached by the souls in their descent. The souls then began a reascent, a sort of test in which they must progressively pass through all the stages through which they had descended. This process gave rise to the minerals, then the plants, then the animals, and finally man, who stands at the border of the two worlds. But the only souls that can cross this threshold into the celestial spheres are those that, purified by morality and knowledge, have been liberated from the prison of matter.

Just as there is a hierarchy of all beings, a hierarchy is established among human souls; the ignorant who are of good will remain at the same level, while the wicked fall to the rank of animals or even lower. This is why the very long life of the universe is divided into cycles of 7,000 years, which are determined by astrological cycles. Toward the end of a cycle, the souls that have not yet been chosen are judged. Those found unworthy of escaping from a cycle are reincarnated in the next and will again have to try their luck. In the long run, the most wicked will become demons tortured by carnal appetites.

But God, the theory promises, sustains men through the intermediary of the imams and, especially, by the prophets He sends. The first is Adam, and the sixth is Muḥammad, who will also be the *qāʾim* of the Resurrection in the seventh millennium. Within each millennium there are successive series of seven imams, who, in accordance with the astrological cycles, pass alternately from concealment to public view. The ascent of the souls takes place in the context of the Spiritual City (inspired by Plato's *Republic*—as seen in the light of Neoplatonism—like the Virtuous City of al-Fārābī). In absolute terms the Spiritual City is the archetypal soul of humanity (celestial Adam) and is therefore constituted of all the chosen souls of the past and of the future—notably the souls of the Messengers, who bring a new revealed Law, and of the imams.

On earth the City is ruled by the prophet sent by God or by the imam of the moment. He is assisted by the initiated, a mystical hierarchy through which the influx descends from on high and which consists of four ranks, corresponding to the levels of reason (instinctive or practical reason; acquired reason, demonstration, and free will; philosophy and inspiration; revelation). Opposed to it is an inverted hierarchy—that of the fiends, headed by the anticaliph and oriented toward the center of the earth. The *Ikhwān* present at length a theory—likewise inspired by Plato—of the hierarchy of the sciences. (It is the sciences that prepare the souls for rising through the sequence of stages.) The *Ikhwān* also offer a detailed theory of proselytism and of propaganda. The doctrine is identical with that of the proto-Ismaili, and the Spiritual City reflects the initial organization of the Ismaili *daʿwa*. The epistles were, in fact, one of the principal—if not the principal—means of spreading propaganda for the *daʿwa*.

The doctrine described here represents a new syncretism. Its chief component is an earlier Hellenistic syncretism in which the views of Aristotle, Euclid, Ptolemy, and others are subordinated to a mixture of Platonism and Neoplatonism recast in the form of a Pythagorean Hermeticism. To this amalgam are joined Hindu, Persian, and Christian elements. Finally, the whole is integrated with the doctrines of Islam.

IKHWĀN AL-ṢAFĀ᾿

The Ismaili attributed the epistles to the sixth imam, Jaʿfar al-Ṣādiq, or to three of his successors, the secret imams (765–909), as well as to four propagandists, including ʿAbd Allah ibn Maymūn al-Qaddāḥ. On the other hand, al-Tawḥīdī (*d.* 1023[?]; cited by F. H. Dieterici, *Die Philosophie der Araber*, 142) and the great Muʿtazilite judge of Rayy, ʿAbd al-Jabbār al-Hamadāni (936–1025; cited by S. M. Stern, "New Information About the Authors of the 'Epistles of the Sincere Brethren' "), give the names of several inhabitants of Baṣra who, they say, are the authors. Three of these names are mentioned by both Tawḥīdī and ʿAbd al-Jabbār; the judge Abū᾿l-Ḥasan ʿAlī ibn Hārūn al-Zanjānī (whom Tawḥīdī considers "the author of the doctrine" and a dangerous Ismaili), Abū Aḥmad al-Nahrajūrī, and al-ʿAwfī. Further, Tawḥīdī and ʿAbd al-Jabbār both cite Zayd ibn Rifāʿa. According to ʿAbd al-Jabbār, he is one of the authors; but Tawḥīdī sees him as only one of their friends. Only Tawḥīdī includes among the authors a certain Abū Sulaymān ibn Maʿshar al-Bustī al-Maqdisī (whom Tawḥīdī's teacher thought to be the principal author); and only ʿAbd al-Jabbār mentions a secretary and astrologer named Abū Muḥammad ibn Abī᾿l-Baghl.

In 1876 Dieterici noted several circumstances in support of his opinion that the epistles were written between 961 and 986. Specifically, he cited the presence of verses by Mutanabbī and Tawḥīdī's mention of the presumed authors in a conversation that he had with a vizier in 981. L. Massignon agreed with this dating, pointing out a mathematical fact and the presence of verses by Ibn al-Rūmī. One may also cite allusions to the polemic between Muʿtazilites and Ashʿarites, together with an explicit mention of these latter, as well as an allusion to the theory of modes proposed by Abū Hāshim al-Jubbā᾿ī. Nevertheless, several passages mentioning the conjunction of Saturn and Jupiter allude not to the date foreseen for the definitive triumph (about 1047) but, rather, to that which, 119 years earlier, inaugurates the period of ascension (about 928). And the success predicted is very probably the establishment of the Fāṭimid caliphate in Ifrīqiyya. These passages, therefore, should be dated earlier than 909. The writing may, however, have extended over a long period. Accordingly, there may well be some truth in the Ismaili tradition concerning the authorship of the epistles. Jaʿfar al-Ṣādiq may have been the originator of the doctrine, which was then developed by the secret imams. Finally, the great propagandists may have continued the writing under the auspices of the Fāṭimid caliphs of Ifrīqiyya. (One of the epistles purports to have been written by an imam, but otherwise the writers mention their common effort.)

The epistles probably did not receive their definitive form until after the conquest of Egypt (969), when the Ismaili were already active in Iraq in anticipation of the conjunction of 1047. There is evidence supporting the Mesopotamian origin of certain of the authors. The Basrians named by Tawḥīdī and ʿAbd al-Jabbār may therefore have been among the latter authors, although this is not certain. On the other hand, they were undoubtedly important propagandists who were at the height of their activity around 981 and who used the epistles, which were then completed, as an instrument of propaganda. The necessity of secret activity accounts for the inconsistency in the assertions of Tawḥīdī and of ʿAbd al-Jabbār (whom the propagandists had perhaps at first believed to be receptive to their ideas, the first as a philosopher and the second as a supporter of the idea of free will). A person working in secret must sometimes risk revealing himself while protecting his companions; it was natural that these people allowed some doubt to exist concerning the scope and the origin—whatever it may have been—of the epistles.

The tenth century has been called the century of Shiism. The justice of this assertion is attested by the activity of the Karmathians and of the Fāṭimids, the infiltration of the administration by the imamates, and the guardianship of the caliphate instituted by the Buwayhids. It was also the time in which al-Fārābī perhaps attempted (with greater success than Plato had with Dionysius the Younger of Syracuse) to convince the Hamdanid sovereign of Aleppo, Sayf al-Dawla, to make real his conception of the Virtuous City. The Ismaili Fāṭimids made it real in their organization, and the epistles undoubtedly played a not unimportant role in their historic successes.

BIBLIOGRAPHY

See F. H. Dieterici, ed., *Die Philosophie der Araber*, 16 vols. (Leipzig–Berlin, 1858–1891); and *Die Abhandlungen der Ichwān aṣ-ṣafā᾿ in Auswahl zum ersten Mal aus arabischen Handschriften herausgegeben* (Leipzig, 1886); Y. Marquet, *La philosophie des Ikhwān al-ṣafā᾿: De Dieu à l'homme* (Lille, 1973), a doctoral dissertation presented in 1971; and "La philosophie des Ikhwān al-ṣafā᾿: L'imām et la société," in *Travaux et documents du Département d'arabe, Faculté des lettres, Université de Dakar*, no. 1 (1973); S. M. Stern, "New Information About the Authors of the 'Epistles of the Sincere Brethren,' " in *Islamic Studies*, **3** (1964), 405–428; and

A. L. Ṭībāwī, "Jamāʿat Ikhwān aṣ-ṣafāʾ," in *Journal of the American University of Beirut* (1930–1931), 1–80; and "Ikhwān aṣ-ṣafāʾ and Their Rasāʾil," in *Islamic Quarterly*, **2** (1956), 28–46.

YVES MARQUET

IOFFE, ABRAM FEDOROVICH (*b.* Romna, Poltava gubernia, Russia, 29 October 1880; *d.* Leningrad, U.S.S.R., 14 October 1960), *physics, technology*.

The son of a merchant, Ioffe graduated from the St. Petersburg Technical Institute in 1902 with a degree in technical engineering. Strongly attracted to physics, however, he entered the University of Munich in 1902, studying at Röntgen's Physical Institute. He acted as Röntgen's assistant and received the Ph.D. *summa cum laude* in 1905. Declining Röntgen's suggestion that he continue to work at Munich, Ioffe returned to Russia the following year and assumed the post of senior laboratory assistant in the physics department of the St. Petersburg Polytechnical Institute. From 1906 to 1913 he lectured there in general physics, and from 1908 to 1914 he offered a course in thermodynamics at the Institute of Mines.

The desire to create a major Russian school of experimental physics guided Ioffe's activities in the physics section of the Russian Physics and Chemistry Society and in his own research. In 1913, after defending his master's thesis, he became professor at the Polytechnical Institute and, in 1914, assistant professor at St. Petersburg University as well. In 1915 he received a doctorate in physics for his detailed investigation of the elasticity and electrical characteristics of quartz. The physics seminar that he organized in 1916 at the Polytechnical Institute soon became the center of advanced physical thought in St. Petersburg. In 1918 Ioffe participated in the creation of the State Institute of Röentgenology and Radiology, heading the section of physics and technology. In 1919 he organized the Faculty of Physics and Mechanics at the Polytechnical Institute and served as chairman until 1948. In order to further the rapprochement of physics and industrial technology, Ioffe and his student N. N. Semenov founded an independent laboratory of physics and technology, which merged in 1929 with the State Institute of Physics and Technology to form the Physical and Technological X-ray Institute; Ioffe was director for twenty-five years.

From the outset work at the institute focused on two areas: the most advanced concepts of physics—atomic and nuclear physics and the physics of X rays—and elaborating scientific topics for application to industrial technology. Ioffe initiated or participated directly in studies of X rays, electronic phenomena, atomic nuclei, and the mechanical, electrical, and magnetic properties of solids, especially dielectric crystals.

At the State Institute of Physics and Technology, Ioffe continued the research on the conductivity of dielectric crystals that he had begun in Röntgen's laboratory; it was there that he had discovered the internal photoeffect in halite crystals that had been subject to X-radiation. This research led Ioffe to present his ideas concerning the role of ions in the interstices of a crystalline lattice. His physical representations formed the basis of the modern explanation of the electrical characteristics of dielectrics.

In 1930 Ioffe's main interest turned to semiconductors. He elaborated an extensive program for studying their conductivity, photoelectric properties, galvanomagnetic and thermoelectric phenomena, rectification effects, the nature of the barrier layer, and the influence of impurities. Two mechanisms of conductivity were subsequently revealed: by electrons and by holes; and methods were developed for manufacturing semiconductors having specified properties.

Ioffe's writings on the history of physics include a historical analysis of the development of ideas concerning the mechanical and electrical properties of solids (1938). Interspersed with his other comments on the work of former physicists are colorful reminiscences of Einstein, Planck, and Röntgen. The author of college textbooks, monographs on semiconductors and dielectrics, and works on the philosophical implications of contemporary physics, he also produced two major works on basic concepts in modern physics (1949) and on the physics of semiconductors (1957).

His students included three generations of Soviet physicists, among whom were P. L. Kapitsa, N. N. Semenov, I. V. Kurchatov, A. I. Alikhanov, I. K. Kikoin, V. N. Kondratiev, P. I. Lukirsky, I. V. Obreimov, and D. V. Skobeltsyn.

In 1918 Ioffe was elected an associate member and, in 1920, member of the U.S.S.R. Academy of Sciences. He was an honorary member of the American Academy of Arts and Sciences (1929), the National Academy of Sciences of India (1958), and the Accademia Nazionale dei Lincei (1959); and he received honorary doctorates from the University of California (1927), the University of

Paris (1946), and the University of Bucharest (1948).

BIBLIOGRAPHY

I. ORIGINAL WORKS. Ioffe's earlier writings include *Elastische Nachwirkung im kristallinischen Quarz* (Leipzig, 1906); *Elementarny fotoelektrichesky efekt. Magnitnoe pole katodnykh luchey* ("The Elementary Photoelectric Effect. The Magnetic Field of Cathode Rays"; St. Petersburg, 1913), his master's thesis; *Uprugie i elektricheskie svoystva kvartsa* ("The Elasticity and Electrical Characteristics of Quartz"; St. Petersburg, 1915), his doctoral diss.; *Lektsii po molekulyarnoy fizike* ("Lectures on Molecular Physics"; Petrograd, 1919); "Elektrizitätsdurchgang durch Kristalle," in *Annalen der Physik*, 4th ser., **72** (1923), 461–500, written with Röntgen; *Kurs fiziki* ("Course in Physics"; Moscow–Leningrad, 1927); *The Physics of Crystals*, L. B. Loeb, ed. (New York–London, 1928), also in Russian (Moscow–Leningrad, 1929); *Moya zhizn i rabota* ("My Life and Work"; Moscow–Leningrad, 1933), an autobiography; *Sur la distribution spectrale de l' effet photoélectrique dans l'oxyde cuivreux* (Paris, 1934); and *Semiconducteurs électriques* (Paris, 1935).

Later publications are *Osnovnye predstavlenia sovremennoy fiziki* ("Basic Concepts in Modern Physics"; Moscow–Leningrad, 1949); *Poluprovodniki v sovremennoy fizike* ("Semiconductors in Modern Physics"; Moscow–Leningrad, 1954); *Poluprovodniki* ("Semiconductors"; Moscow–Leningrad, 1955); *Poluprovodniki i ikh primenenie* ("Semiconductors and Their Application"; Moscow–Leningrad, 1956); *Poluprovodnikovye termoelementy* ("Semiconductor Thermoelements"; Moscow–Leningrad, 1956); *Semiconductors, Thermoelements and Thermoelectric Cooling* (London, 1957); *Halbleiter-Thermoelemente* (Berlin, 1957); *Fizika poluprovodnikov* ("The Physics of Semiconductors"; Moscow–Leningrad, 1954; 2nd ed., 1957), also in German (Berlin, 1958).

II. SECONDARY LITERATURE. On Ioffe and his work, see Y. I. Frenkel, "Akademik A. F. Ioffe," in *Vestnik Akademii nauk SSSR* (1940), no. 10, 72–77, published on his sixtieth birthday; *Sbornik, posvyashchenny 70-letiyu akademika A. F. Ioffe* ("Collection Dedicated to the Seventieth Birthday of Academician A. F. Ioffe"; Moscow, 1950), with contributions by various authors; I. K. Kikoin and M. S. Sominsky, "Abram Fedorovich Ioffe," in *Uspekhi fizicheskikh nauk*, **72**, no. 2 (1960), 307–321; and M. S. Sominsky, *A. F. Ioffe* (Moscow–Leningrad, 1964).

A. T. GRIGORIAN

JACOBS, WALTER ABRAHAM (*b.* Brooklyn, N.Y., 24 December 1883; *d.* Los Angeles, California, 12 July 1967), *organic chemistry.*

Jacobs was born and raised in Brooklyn, where his father, a tailor, encouraged him to study science. After receiving the bachelor's and master's degrees from Columbia University, he studied under Emil Fischer in Berlin, receiving the Ph.D. in 1907. He was then appointed fellow in chemistry at the Rockefeller Institute for Medical Research, where he remained until his retirement fifty years later.

Jacobs' first research, as assistant to the biochemist P. A. Levene, dealt with the chemistry of the nucleic acids. They identified the sugar units as D-ribose (1909), and determined several of the purine bases that occur in the RNA molecule. Levene and Jacobs designated the base-sugar unit a "nucleoside," corresponding with Levene's earlier designation of the base-sugar-phosphate unit as a "nucleotide."

The director of the institute, Simon Flexner, was greatly interested in Paul Ehrlich's discovery of the antisyphilitic Salvarsan, and in 1912 he assigned Jacobs to head a new division of chemotherapy. In 1919 Jacobs and his assistant Michael Heidelberger developed the trypanocidal Tryparsamide, which proved highly effective in the treatment of sleeping sickness, then endemic in Africa. The two chemists, with the biologists Wade H. Brown and Louise Pierce, were later honored by the Belgian government.

In 1922 Jacobs abandoned research on chemotherapeutic agents, and turned to structural investigations of pharmacologically significant natural products derived from plants. He and his co-workers established the general structural features of several of the cardiac glycosides derived from *Strophanthus*, squill, and digitalis; the nonsugar moieties (aglycones) were all shown to be steroids. The last twenty-five years of Jacobs' career were devoted to the study of alkaloids. He and Lyman Craig isolated lysergic acid from an ergot alkaloid in 1934; its structure was established in Jacobs' laboratory during the next eleven years, by degradation and synthetic studies. Jacobs' last researches concerned the elucidation of the general structures of several of the veratrine and aconite alkaloids.

BIBLIOGRAPHY

I. ORIGINAL WORKS. An extensive bibliography of over 200 of Jacobs' scientific publications is in Poggendorff, VI, 1210–1212, and VIIb, 2231–2233. Most of Jacobs' papers appeared in *Journal of Biological Chemistry* and *Journal of the American Chemical Society*. He was author or coauthor of several government reports, but no books.

II. SECONDARY LITERATURE. An obituary appeared in *New York Times* (14 July 1967), 31; and a biographical memoir by Lyman Craig was published in *Rockefeller University Review* (Nov.–Dec. 1967), 23–25. George W. Corner, *History of the Rockefeller Institute: 1901–1953* (New York, 1964), contains many references to Jacobs' scientific career. The nucleic acid research of Levene and Jacobs is discussed in Joseph S. Fruton, *Molecules and Life: Historical Essays on the Interplay of Chemistry and Biology* (New York, 1972), 204–207.

ALAN J. ROCKE

AL-JAZARĪ, BADĪᶜ AL-ZAMĀN ABŪʾL-ᶜIZZ ISMĀᶜĪL IBN AL-RAZZĀZ (*fl.* Diyār Bakr, 1206), *machinery, techniques of construction.*

All that we know of al-Jazarī's life is contained in the introduction to his work, *Kitāb fī Maᶜrifat al-ḥiyal al-handasiyya* ("Book of Knowledge of Mechanical Devices"). He tells us that at the time of writing his book he was in the service of Nāṣir al-Dīn, the Artuqid king of Diyār Bakr, and that he had spent twenty-five years with the family, having served the father and the brother of Nāṣir al-Dīn. The Artuqids were a Turcoman dynasty who maintained a precarious autonomy during the twelfth century in Mesopotamia. By 1181, however, they had become vassals of Saladin.[1]

The book, which al-Jazarī wrote at the command of Nāṣir al-Dīn, is divided into fifty chapters, grouped into six categories: I, water clocks and candle clocks (ten chapters); II, vessels and figures suitable for drinking sessions (ten chapters); III, pitchers and basins for phlebotomy and ritual washing (ten chapters); IV, fountains that change their shape and machines for the perpetual flute (ten chapters); V, machines for raising water (five chapters); and VI, miscellaneous (five chapters): a large ornamental door cast in brass and copper, a protractor, combination locks, a lock with bolts, and a small water clock.

The book is clearly written in straightforward Arabic; and the text is accompanied by 173 drawings, ranging from rudimentary sketches to full-page paintings. On these drawings the individual parts are in many cases marked with the letters of the Arabic alphabet, to which al-Jazarī refers in his descriptions. The drawings are usually in partial perspective; but despite considerable artistic merit, they seem rather crude to modern eyes. They are, however, effective aids to understanding the text.

It is apparent from statements made by al-Jazarī, in the introduction and elsewhere, that he took pride in belonging to an international fraternity of craftsmen and in continuing the work of his predecessors. Thus he acknowledges that his monumental water clock (chapter 1 of category I) is based upon the clock of pseudo-Archimedes.[2] Al-Jazarī also mentions the Banū Mūsā in connection with their work on fountains and cites the treatise of Apollonius of Byzantium on an automatic musical instrument.[3] Others, about whom we know little or nothing, are also named; yet elsewhere he refers to devices made by earlier workers whom he does not name. Clearly, then, many of al-Jazarī's machines were derived from earlier models, mainly Islamic, some of which were improved versions of devices described by such classical writers as Hero of Alexandria and Philo of Byzantium. Almost certainly there was also transmission from India and the Far East.[4] Al-Jazarī was therefore not primarily an inventor but an engineer who saw his task as the perfecting of earlier work. And, although many of his devices were designed for entertainment rather than for utility, he obviously took his work very seriously.

Of all al-Jazarī's complete machines, perhaps only the double-cylinder slot-rod pump driven by a paddle wheel (chapter 5 of category V) is a significant contribution to the history of machines. Water clocks, candle clocks, and trick vessels, the descriptions of which occupy about three-quarters of the book, have little importance in the subsequent development of mechanical technology. It is the individual components and the constructional techniques, described by al-Jazarī in such scrupulous detail, that are of far greater importance, since centuries later many of them entered the general vocabulary of European engineering. Among the most important of these components and techniques are conical valves, casting of brass and copper in closed mold boxes with greensand, static balancing of large pulley wheels, use of wooden templates, use of a paper model in design, calibration of orifices, lamination of timber to minimize warping, use of true suction pipes rather than drowned suction (in the pump mentioned above), tipping buckets that discharge their contents automatically after a set time, and segmental gears.

The driving mechanism for the monumental water clock (chapter 1 of category I) provides a typical example of al-Jazarī's methods and gives a good idea of the character of the book as a whole. The design is not original—al-Jazarī acknowledges its derivation from the water machinery in the "Ar-

AL-JAZARĪ

chimedes" clock, and a similar construction was used by Muḥammad al-Khurāsānī al-Saʿātī for the large water clock at the Jayrūn Gate in Damascus (described by his son Riḍwān).[5] Al-Jazarī, however, undoubtedly made improvements upon the other designs. For instance, the conical valve, as described by "Archimedes," is a rather crude device; and neither he nor al-Saʿātī made an accurate flow regulator.

The water in reservoir *A* (see Figure 1) sank at a constant rate, and the large float *B* operated most of the time-recording automata by the pull of the string attached to it. The static head was kept constant in the float chamber *D* by means of the conical valve plug on top of float *C*, which entered the valve seat *H* at the end of tap *F*. When water issued from orifice *G*, with tap *F* open, it flowed momentarily from the reservoir into the float chamber, whereupon valve *H* closed momentarily.

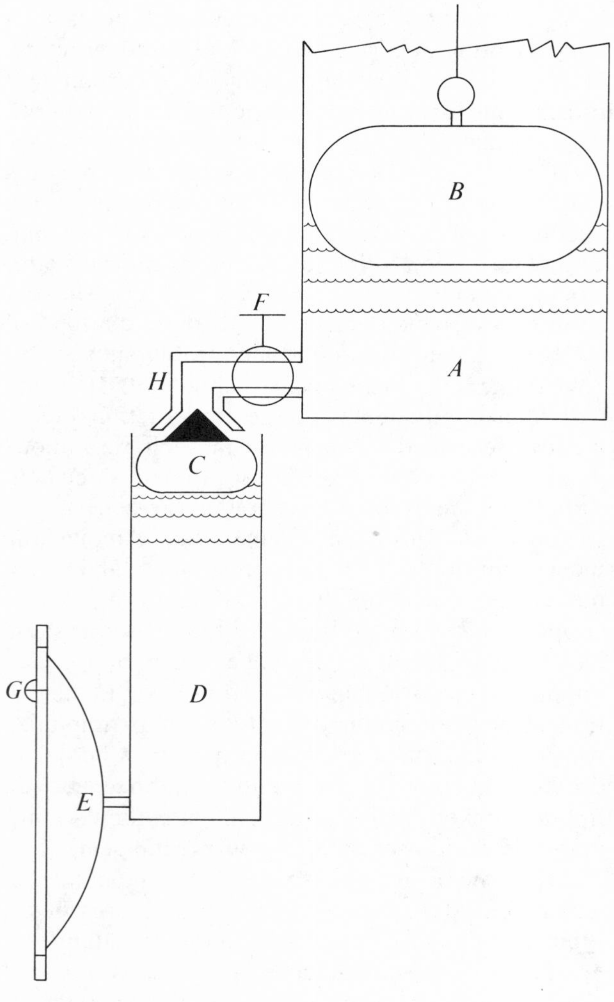

FIGURE 1 Al-Jazarī's water clock.

Thus there was only a very slight fluctuation in the static head in the float chamber. The clock recorded the passage of solar (or temporal) hours: the hours of daylight and the hours of darkness were divided by twelve, giving "hours" of varying length throughout the year, and from daytime to nighttime. The purpose of the flow regulator was to obtain these variations by altering the static head above the orifice. The system of pipes and channels inside the regulator was constructed in such a manner that the disk *E* carrying the orifice could be rotated through 360 degrees.

The reservoir (about 5.5 feet high by 1.0 foot in diameter), the float chamber (about 1.25 feet high by 3.5 inches in diameter), and the two floats were all made from sheet copper, soldered along the seams. Great care was taken to ensure that the vessels were of uniform cross section. The tap, valve seat, and valve plug were made from cast bronze. Plug and seat were ground together with emery on the lathe until the valve was watertight when closed, but the plug was free to slide easily out of the seat.

For the flow regulator al-Jazarī first describes how he tested various earlier designs—for example, an equally divided semicircle and an equally divided full circle[6]—but rejected them when he found that they were inaccurate. His own solution was to calibrate the orifice, a piece of drilled onyx, in the topmost position, for the required rate of flow for the day of the summer solstice. The other positions—zodiacal "houses," five-degree, and single-degree divisions—were determined by trial and error. These were engraved on an annulus surrounding the rotatable disk. A pointer attached to the disk (on a line passing through the orifice) extended over the annulus and was set to the appropriate degree on a given day, and to the diametrically opposed degree on that night. After manufacture, the pieces of equipment were carefully assembled on firm foundations and were brought to the vertical by using a plumb line.

The foregoing exemplifies al-Jazarī's main virtues: careful manufacture and assembly of components, and the ability to devise real improvements on the work of his predecessors. His main faults were a tendency to be inconsistent in his dimensions and some vagueness about the positioning of equipment. Nevertheless, taking drawings and text together, it can be said that he fulfilled his declared intention of describing the devices so that they could be reconstructed by a successor. Indeed, the clock described above was reconstructed in the Science Museum, London, for the 1976 World of

Islam Festival. It works perfectly, exactly in accordance with al-Jazarī's intentions.

Many of the components and techniques used by al-Jazarī and his Islamic congeners reappeared in Europe, apparently as reinventions, centuries later. Casting in closed mold boxes with greensand began in Europe about 1500.[7] The first mention of conical valves is by Leonardo da Vinci; and a float-controlled regulator for steam boilers, similar to the system described above for the water clock, was patented in England in 1784.[8] (It is important to distinguish between the use of automatically operating conical valves for feedback control and the simple lift-and-release "bath plug" type, which occurred as early as Hero of Alexandria[9] and was also used extensively by Islamic craftsmen, including al-Jazarī.)

It is hoped that further research will tell us whether European engineers had access to Islamic ideas; and if so, how they were transmitted. A book of the Banū Mūsā (*ca.* 850) was translated into Latin at the end of the twelfth century by Gerard of Cremona.[10] Versions of the book by al-Jazarī, and works by other Islamic engineers, may therefore also exist. It is possible, however, that by the thirteenth century, when the impetus of translation was slackening in Europe, ideas from Islam and other cultural areas were transmitted by travelers' reports and by personal contacts among craftsmen, rather than by written descriptions.

NOTES

1. See Claude Cahen, "Le Diyār Bakr au temps des premiers Urtuḳides," in *Journal asiatique*, **227** (1935), 219–276.
2. D. R. Hill, *On the Construction of Water-Clocks* (London, 1976), an annotated translation of Archimedes made from three Arabic MSS. E. Wiedemann and F. Hauser, "Uhr des Archimedes und zwei andere Vorrichtungen," in *Nova acta Academiae Caesarae Leopoldino-Carolinae germanicae naturae curiosorum*, **103**, no. 2 (1918), 164–202. This includes a German trans. made from two of the MSS.
3. F. Hauser, "Über das Kitāb al-Ḥiyal," in *Abhandlungen zur Geschichte der Naturwissenschaften und der Medizin* (Erlangen, 1922), nos. 89–94; E. Wiedemann, *Aufsätze zur arabischen Wissenschaftsgeschichte*, II (Hildesheim, 1970), 50–56, which includes a German trans. of the Apollonius MS.
4. In ch. 4 of category V, al-Jazarī refers to the chain of pots as a "Sindī wheel." Several times he mentions a metal that he calls *isfādruh*. According to al-Dimashqī (quoted by Wiedemann in *Aufsätze*, II, 120) this metal originated in China. At the present state of knowledge, other cultural borrowings cannot be identified with certainty.
5. E. Wiedemann and F. Hauser, "Über die Uhren in Bereich der islamischen Kultur," in *Nova acta Academiae Caesarae Leopoldino-Carolinae . . .*, **100** (1915), 167–272.
6. The pseudo-Archimedes used the semicircular regulator and mentions the full circle. Al-Saʿātī used the full circle. These designs incorporate two errors: that the variation of daylight throughout the year follows a sine curve, and that the rate of flow is proportional to the static head.
7. Cyril Stanley Smith, "The Early History of Casting, Moulds and the Science of Solidification," in W. W. Mullins and M. C. Shaw, eds., *Metal Transformations* (New York, 1968), 23. Al-Jazarī makes it clear that the technique was well-established in Mesopotamia during his lifetime.
8. F. M. Feldhaus, *Die Technik* (Wiesbaden, 1970), 499; Otto Mayr, "The Origins of Feedback Control," in *Scientific American*, **223** (Oct. 1970), 113.
9. *The Pneumatics of Heron of Alexandria*, facs. of 1851 Woodcroft ed., with intro. by Marie Boas Hall (London, 1971), 37.
10. The MS is Madrid, Biblioteca Nacional, 10010. This is a mathematical treatise, not the machine book mentioned in Note 3. An English version of the latter is being prepared by the present author.

BIBLIOGRAPHY

Al-Jazarī's book has not been published in Arabic, although a number of MSS are extant. There is, however, a full English ed.: Donald R. Hill, *The Book of Knowledge of Ingenious Mechanical Devices* (Dordrecht, Netherlands, 1974). This is a complete trans. of al-Jazarī's work, made mainly from MS Graves 27 at the Bodleian Library, Oxford. All the original illustrations are included as unaltered photographs with the text, and in addition there are thirty-two plates of the miniature paintings from two Istanbul MSS. Notes are provided for each chapter, in many cases with explanatory drawings; and the "General Notes" discuss al-Jazarī's machines, components, and techniques in a historical context. The MSS are described, and their locations and references given, on 3–6 (to which should be added Topkapi Seray, nos. A3350 and H414).

The only earlier work of any significance is a series of seven articles in German periodicals, written by E. Wiedemann and F. Hauser between 1908 and 1921. Partly translations and partly paraphrases, they deal with the complete book and are listed by Hill (see above) in the bibliography and by Coomaraswamy (see below) on 20–21.

Two of the five MSS in Istanbul were dated A.D. 1315 and A.D. 1354. Miniature paintings from both of these are in public and private collections, and some articles have been written about them from an artistic viewpoint. The most important of these are A. K. Coomaraswamy, *The Treatise of al-Jazarī on Automata* (Boston, 1924); and Freer Gallery (Washington, D.C.), folder sheets 30.71 and 32.19.

DONALD R. HILL

KARPINSKI, LOUIS CHARLES (*b.* Rochester, New York, 5 August 1878; *d.* Winter Haven, Florida, 25 January 1956), *history of mathematics, cartography*.

Karpinski was the son of Henry H. Karpinski,

who had emigrated from Warsaw, and of Mary Louise Engesser, from Gebweiler in Alsace. Karpinski completed high school in Oswego, New York, in 1894 and received a teacher's diploma from Oswego State Normal School in 1897. After teaching in the normal department of Berea College, he attended Cornell University, from which he received an A.B. in 1901.

From 1901 to 1903 Karpinski attended Kaiser Wilhelm American College of the University of Strasbourg), where after submitting the thesis "Über die Verteilungen der quadratischen Reste" to a committee composed of Heinrich Weber, Theodor Reye, and Ferdinand Braun, he received the doctorate of natural philosophy. From 1903 to 1904 Karpinski was in charge of physics and chemistry at Oswego Normal College and of arithmetic in its practice school. Besides teaching in summer schools at New York University in 1904 and at the Chautauqua Institution (1905–1908), Karpinski wrote several reviews and articles, and made speeches on pedagogic topics.

Karpinski became an instructor in mathematics at the University of Michigan in 1904. He was subsequently promoted to assistant professor (1910), associate professor (1914), and professor (1919), and became professor emeritus upon his retirement in 1948. In 1905 Karpinski married Grace Maude Woods of Lockport, New York; they had six children.

Karpinski's productive interest in the history of science dates from the academic year 1909–1910, which he spent as a Teachers College fellow and university extension lecturer at Columbia University. His collaboration with David Eugene Smith resulted in *The Hindu-Arabic Numerals* (1911), still the authoritative work on this topic. His early research was based on manuscript algorisms and algebras that he located in European libraries. He elucidated works by Jordanus de Nemore, Sacrobosco, Robert of Chester, al-Khwārizmī, and Abū Kāmil.

During a sabbatical year (1926–1927) Karpinski photographed manuscript maps pertaining to American history in archives in France, Spain, and Portugal. Copies of these manuscripts were sold to many American research libraries. This interest in maps led to other activities, one of which was the collecting of atlases and maps, which he sold to Yale University. This collection is now acknowledged as one of the bases of Yale's Thorne Collection of Cartography and Geography. He also prepared for the Michigan Historical Commission the *Bibliography of the Printed Maps of Michigan, 1804–1880* and the accompanying *Historical Atlas of the Great Lakes and Michigan*, both of which appeared in 1931. Evidence of his bibliographic interests appears in a number of journal articles and especially in his definitive *Bibliography of Mathematical Works Printed in America Through 1850*.

In 1928 Karpinski was elected a *membre effectif* of the Comité International d'Histoire des Sciences and in 1937 was the American representative to the Descartes Tercentenary held in Paris. He was a charter member of the Mathematical Association of America and of the History of Science Society, of which he was elected president in 1943. His interest in the history of mathematics and in photographs led to his preparation of the slides that were projected in the hall of science of the Century of Progress Exposition in Chicago in 1933. Karpinski was also an associate editor of *Scripta mathematica*, and under his direction seven students wrote theses in the history of mathematics.

BIBLIOGRAPHY

I. Original Works. Karpinski wrote nearly 200 publications; his original historical papers can largely be found in *American Mathematical Monthly, Archeion (Archivio di storia della scienza), Bibliotheca mathematica, Isis*, and *Scripta mathematica*. His books are *The Hindu-Arabic Numerals* (Boston, 1911), written with David Eugene Smith; *Robert of Chester's Latin Translation of the Algebra of Al-Khowarizmi, With an Introduction, Critical Notes, and an English Version*, University of Michigan Studies, Humanistic Series, XI, pt. 1 (New York, 1915); *Unified Mathematics* (Boston, 1918; rev., 1922), written with Harry Y. Benedict and John W. Calhoun; and *The History of Arithmetic* (Chicago, 1925; repr. New York, 1965).

Other works are *Nicomachus of Gerasa, Introduction to Arithmetic*, Martin Luther D'Ooge, trans., University of Michigan Studies, Humanistic Series, XVI (New York, 1926; Ann Arbor, 1936; 3rd ed., 1946), with studies of Greek arithmetic by Frank Egleston Robbins and L. C. Karpinski; *Bibliography of the Printed Maps of Michigan, 1804–1880* (Lansing, 1931), including discussions of Michigan maps and mapmakers by William Lee Jenks; *Historical Atlas of the Great Lakes and Michigan* (Lansing, 1931); *Bibliography of Mathematical Works Printed in America Through 1850* (Ann Arbor, 1940); and *Early Military Books in the University of Michigan Libraries* (Ann Arbor, 1941), written with Thomas M. Spaulding.

II. Secondary Literature. A picture of Karpinski as chairman of Section L and as vice-president of the

American Association for the Advancement of Science is in *Scientific Monthly,* **50** (Jan. 1940), 90. See also Phillip S. Jones, "Louis Charles Karpinski, Historian of Mathematics and Cartography," in *Historia mathematica,* **3** (1976), 185–202. Further data and memorabilia are in the Michigan Historical Collection, Bentley Historical Library, University of Michigan.

PHILLIP S. JONES

KARRER, PAUL (*b.* Moscow, Russia, 21 April 1889; *d.* Zurich, Switzerland, 18 June 1971), *chemistry.*

Karrer was the son of Paul Karrer, a Swiss dentist who practiced in Russia, and Julie Lerch Karrer. In 1892 the family returned to Switzerland, where Karrer was educated in the cantonal schools of Aargau. In 1908 he entered the University of Zurich and studied chemistry with Alfred Werner, whose lecture assistant he became. He began an independent study of organic arsenic compounds, and because of his interest in this field he went to Frankfurt in 1912 to work with Ehrlich. In 1914 Karrer married Helene Froelich, daughter of a director of a psychiatric clinic; they had three sons, one of whom died in childhood. The following year, after Ehrlich's death, Karrer became director of the chemical division of Georg Speyer Haus. In 1918 he accepted a call to the University of Zurich, where he succeeded Werner as professor of chemistry in 1919. In spite of many offers from other institutions, Karrer remained at Zurich for the rest of his life, serving as rector of the university from 1950 to 1952.

Karrer was a versatile organic chemist. The large number of students whom he attracted and the wide variety of problems that he attacked are attested by the more than 200 dissertations he directed, his more than 1,000 publications of all types, and his successful *Lehrbuch der organischen Chemie* (1928), which went through fourteen editions and was translated into seven languages.

When Karrer returned to Zurich, by undertaking study of the structures of amino acids, peptides, and proteins, he demonstrated that all these compounds had the same steric configuration. At the same time he studied a number of highly polymerized carbohydrates such as starch and cellulose. During the 1920's his attention was drawn to the pigments of plants, and he began the study of natural products, which occupied most of his career. After studying anthocyanin pigments, he turned his attention to the study of the carotenoids, a branch of organic chemistry with which his name is closely associated. By 1930 he had solved the long-puzzling problem of the structures of carotene and lycopene. During this work he did much to revive the almost forgotten contribution of Tsvet to chromatography; this not only made possible the isolation of a number of new carotenoids, but in its later developments became an important tool for many branches of chemistry.

The work on carotene led him to investigate the question of the nature of vitamin A, the configuration of which—closely resembling that of a part of the carotene molecule—he recognized in 1931. He was able to establish its structure before the pure substance had been isolated. He then studied other fat-soluble vitamins, which in part resembled vitamin A. In 1938 he established the formulas for α- and β-tocopherol (vitamin E), and in 1939 he isolated vitamin K. During the same decade he investigated the flavonoids, a class of yellow pigments. In 1935 he synthesized one of the most important of the flavins, riboflavin (vitamin B_2), and in addition investigated the chemistry of vitamin C and another B vitamin, biotin. In 1937 he received the Nobel Prize for chemistry for his "researches into the constitution of the carotenoids, flavonoids, and vitamins A and B." He shared the award with Walter N. Haworth, who had worked on the constitution of carbohydrates and vitamin C.

In 1942 Karrer contributed greatly to an understanding of the structure and function of nicotinamide-adenine dinucleotide (NAD), a coenzyme essential for the transfer of electrons in the energy system of the cell. In 1950 he accomplished the total synthesis of carotenoids. At the beginning of his career he had worked for a time on alkaloids, and after 1945 he resumed these studies, determining the structures of certain curare-like alkaloids. Karrer was honorary member of several scientific societies, from many of which he received medals. Reserved and retiring, Karrer led a quiet life in his home and garden on the Zurichberg. He refused to own an automobile, and upon his retirement in 1959 he burned most of his scientific correspondence.

BIBLIOGRAPHY

In addition to *Lehrbuch der organischen Chemie* (14th ed., Stuttgart, 1963), Karrer summarized his work on carotenoids in *Carotinoide* (Basel, 1948), written with E. Jucker. His biography, including references to his most important papers, is A. Wettstein, "Paul Karrer," in *Helvetica chimica acta*, **55** (1972), 317–328. An appreciative biography is C. H. Eugster, "Paul Karrer 1889–1971," in *Chemie in unserer Zeit*, **6** (1972), 146–153.

HENRY M. LEICESTER

KENDALL, EDWARD CALVIN (*b.* South Norwalk, Connecticut, 8 March 1886; *d.* Princeton, New Jersey, 4 May 1972), *endocrinology, biochemistry.*

The son of George Stanley Kendall and Eva Frances Abbott, Kendall obtained the B.S. (1908) and M.S. (1908) from Columbia University, where he received the Ph.D. in 1910 with a dissertation on the kinetics of pancreatic amylase. In 1910 he began work on thyroid extracts at Parke, Davis, and Company, as the sole Ph.D. chemist in the control laboratory, which was then little more than a testing department. From 1911 to 1913 he worked in the new chemical pathology laboratory at St. Luke's Hospital in New York City. Although encouraged in his attempts to isolate the active principle in thyroid extracts, Kendall came to feel increasingly marginal in the context of clinical medicine; and in 1914 he went to work in the biochemical section of the newly established research laboratories of the Mayo Clinic in Rochester, Minnesota, where he remained until his retirement in 1951. From 1921 to 1951 he was also professor of physiological chemistry at the Mayo Foundation and the University of Minnesota.

In the early twentieth century public concern with "the menace of the feebleminded" was focusing attention on disorders of the endocrine system. In 1914 the Mayos' faith in Kendall's work on the thyroid was rewarded with the first isolation of crystalline thyroxine. Kendall subsequently set out to elucidate its chemical structure, but his work ended in bitter disappointment in 1926, when C. R. Harington in England worked out the correct structure. Kendall's investigations of adrenal extracts were stimulated by Albert Szent-Györgyi, who came to the Mayo Clinic in 1929 to use Kendall's extensive equipment and rich source of adrenal glands (from Midwestern meat-packing companies) to extract vitamin C. In 1930 Kendall committed himself to the search for the hormones of the adrenal cortex, extracts of which alleviated Addison's disease.

As a growing number of diseases of growth and mental development had been traced to endocrinal disorders, the elucidation of the steroid structure and the identification of the sex hormones as steroids attracted leading organic chemists to the field in the early 1930's. Research on "cortin," the presumed adrenal "life maintenance hormone" in adrenal extracts, was pursued by many competing groups of endocrinologists and chemists, notably Tadeus Reichstein in Switzerland, as well as by drug manufacturers. In 1936 Kendall reported the identification of "a substance" with the physiological properties of adrenal extract. This substance, however, proved to be a mixture of different steroid hormones. In studying the complex chemistry of the corticosteroids, the organic chemists had the advantage; and in 1937 and 1943 Reichstein synthesized key members of this group. In his work on the physiological effects of the adrenal steroids, however, Kendall's clinical facilities gave him the edge: he could turn to experts in physiological and pathological research in well-equipped laboratories, and the interdisciplinary structure of the research institute was better adapted to physiological and clinical work than were chemical departments. In the long run the advantage was crucial. Initial results, however, were disappointing; by 1940 it had become clear that no single steroid hormone could duplicate all the physiological effects of crude adrenal extract in treating Addison's disease, and the promise of "cortin" faded.

In 1948 a search was begun for other clinical applications of corticosteroids. In 1949 Kendall and Philip S. Hench discovered that "compound E," later renamed "cortisone," alleviated the effects of rheumatoid arthritis. For this Kendall and Hench, with Tadeus Reichstein, were awarded the Nobel Prize in physiology or medicine in 1950.

Upon retirement in 1951 Kendall became research professor at Princeton University. There and in the laboratories of Merck and Company in Rahway, New Jersey, he continued work until his death on the nonsteroid components of adrenal extracts of possible therapeutic value.

BIBLIOGRAPHY

I. ORIGINAL WORKS. Kendall's papers include "The Isolation in Crystalline Form of the Compound Containing Iodin Which Occurs in the Thyroid: Its Chemical Nature and Physiological Activity," in *Transactions of the Association of American Physicians*, **30** (1914), 420–449; "The Identification of a Substance Which Possesses the Qualitative Action of Cortin," in *Journal of Biological Chemistry*, **116** (1936), 267–276; "The Effect of Hormone of the Adrenal Cortex (17-hydroxy-

11-dehydrocorticosterone: Compound E) and of Pituitary Adreno-corticotropin Hormone on Rheumatoid Arthritis: Preliminary Report," in *Proceedings of Staff Meetings of the Mayo Clinic*, **24** (1949), 181–197, written with H. L. Mason and C. S. Myers; and "The Development of Cortisone as a Therapeutic Agent," in *Nobel Lectures, Physiology or Medicine, 1942–1962* (Amsterdam, 1964), 270–288. Kendall's autobiography, *Cortisone* (New York, 1971), is a useful source. The Firestone Library, Princeton University, has a collection of personal papers. Columbia University Library has a taped interview in its series Nobel Laureates on Scientific Research.

II. SECONDARY LITERATURE. See Dwight Ingle, "Edward C. Kendall," in *Biographical Memoirs. National Academy of Sciences*, **47** (1975), 249–292, which contains a complete bibliography; and Hugh Taylor, "Edward Calvin Kendall, 1886–1972," in *Yearbook. American Philosophical Society* (1972), 216–220.

ROBERT E. KOHLER

AL-KHALĪLĪ, SHAMS AL-DĪN ABŪ ᶜABDALLĀH MUḤAMMAD IBN MUḤAMMAD (*fl.* Damascus, Syria, *ca.* 1365), *astronomy, mathematics.*

Al-Khalīlī (Suter, no. 418) was an astronomer associated with the Umayyad Mosque in the center of Damascus. A colleague of the astronomer Ibn al-Shāṭir, he was also a *muwaqqit*—that is, an astronomer concerned with *ᶜilm al-mīqāt*, the science of timekeeping by the sun and stars and regulating the astronomically defined times of Muslim prayer. Al-Khalīlī's major work, which represents the culmination of the medieval Islamic achievement in the mathematical solution of the problems of spherical astronomy, was a set of tables for astronomical timekeeping. Some of these tables were used in Damascus until the nineteenth century, and they were also used in Cairo and Istanbul for several centuries. The main sets of tables survive in numerous manuscripts, but they were not investigated in modern times until the 1970's.

Al-Khalīlī's tables can be categorized as follows: tables for reckoning time by the sun, for the latitude of Damascus; tables for regulating the times of Muslim prayer, for the latitude of Damascus; tables of auxiliary mathematical functions for timekeeping by the sun for all latitudes; tables of auxiliary mathematical functions for solving the problems of spherical astronomy for all latitudes; a table displaying the *qibla*, that is, the direction of Mecca, as a function of terrestrial latitude and longitude; and tables for converting lunar ecliptic coordinates to equatorial coordinates.

The first two sets of tables correspond to those in the large corpus of spherical astronomical tables computed for Cairo that are generally attributed to the tenth-century Egyptian astronomer Ibn Yūnus. They are recomputed for Ibn al-Shāṭir's parameters: 33;30° for the latitude of Damascus and 23;31° for the obliquity of the ecliptic. Al-Khalīlī does not mention any of his Egyptian predecessors. We know, however, that an elder colleague of his, the instrument maker al-Mizzī (*d. ca.* 1350; Suter, no. 406), who spent the first part of his life in Egypt and then moved to Damascus, had already compiled a set of hour-angle tables and prayer tables similar to those used in Egypt and based on 33;27° for the latitude of Damascus and 23;33° for the obliquity, a pair of parameters used by earlier Syrian astronomers. Al-Khalīlī's first and second tables were thus intended to replace al-Mizzī's set. These tables were used in Damascus until the nineteenth century. The Damascus *muwaqqit* Muḥammad ibn Muṣṭafā al-Ṭanṭāwī, who died in 1889, was one of the last to use them; he also converted the entries from equatorial degrees and minutes to equinoctial hours and minutes.

The third set of tables compiled by al-Khalīlī consisted of auxiliary tables for timekeeping by the sun and a table of the solar azimuth as a function of the solar meridian altitude and instantaneous altitude. The auxiliary tables, which contain over 9,000 entries, are intended specifically for facilitating the computation of the hour angle for given solar altitude and solar longitude, and any terrestrial latitude. They were plagiarized by later astronomers in both Egypt and Tunis.

Al-Khalīlī's fourth set of tables was designed to solve all the standard problems of spherical astronomy, and they are particularly useful for those problems that, in modern terms, involve the use of the cosine rule for spherical triangles. Al-Khalīlī tabulated three functions and gave detailed instructions for their application. The functions are the following (the capital notation indicates that the medieval trigonometric functions are computed to base $R = 60$, thus $\text{Sin}\,\vartheta = R\,\text{Sin}\,\vartheta$, and so on):

$$f_\varphi(\vartheta) = \frac{R\ \text{Sin}\ \vartheta}{\text{Cos}\ \varphi},\ g_\varphi(\vartheta) = \frac{\text{Sin}\ \vartheta\ \text{Tan}\ \varphi}{R},$$

and

$$G(x,y) = \text{arc Cos}\left\{\frac{R\,x}{\text{Cos}\ y}\right\},$$

computed for the domains

$\vartheta = 1°, 2°, \cdots, 90°$

$\varphi = 1°, 2°, \cdots, 55°$, as well as 21;30° (Mecca) and 33;30° (Damascus)

$x = 1, 2, \cdots, 59$
$y = 0°, 1°, \cdots, n(x)$,
where $n(x)$ is the largest integer such that

$$R\, x \leqslant \text{Cos}\, n(x).$$

The entries in these tables, which number over 13,000, were computed to two sexagesimal digits and are invariably accurate. An example of the use of these functions is the rule outlined by al-Khalīlī for finding the hour angle t for given solar or stellar altitude h, declination δ, and terrestrial latitude ϕ. This may be represented as

$$t(\text{h},\delta,\varphi) = G\{[f_\varphi(h) - g_\varphi(\delta)], \delta\},$$

and it is not difficult to show the equivalence of al-Khalīlī's rule to the modern formula

$$t = \text{arc cos}\left\{\frac{\sin h - \sin \delta \sin \varphi}{\cos \delta \cos \varphi}\right\}.$$

These auxiliary tables were used for several centuries in Damascus, Cairo, and Istanbul, the three main centers of astronomical timekeeping in the Muslim world.

Al-Khalīlī's computational ability is best revealed by his *qibla* table. The determination of the *qibla* for a given locality is one of the most complicated problems of medieval Islamic trigonometry. If (L, φ) and (L_M, φ_M) represent the longitude and latitude of a given locality and of Mecca, respectively, and $\Delta L = |\, L - L_M \,|$, then the modern formula for $q(L, \varphi)$, the direction of Mecca for the locality, measured from the south, is

$$q = \text{arc cot}\left\{\frac{\sin \varphi \cos \Delta L - \cos \varphi \tan \varphi_M}{\sin \Delta L}\right\}.$$

Al-Khalīlī computed $q(\varphi, L)$ to two sexagesimal digits for the domains $\varphi = 10°, 11°, \cdots, 56°$ (also 33;30°) and $\Delta L = 1°, 2°, \cdots, 60°$; the vast majority of the 2,880 entries are either accurately computed or in error by ±0;1° or ±0;2°. He states that he used the method for finding the *qibla* expounded by the late thirteenth-century Cairo astronomer Abū ʿAlī al-Marrākushī (Suter, no. 363); and it seems that he used his universal auxiliary tables to compute the *qibla* values, although they are generally more accurate than can be derived from the auxiliary tables in their present form. Several other *qibla* tables based on approximate formulas are known from the medieval period. Al-Khalīlī's table does not appear to have been widely used by later Muslim astronomers.

The last set of tables known to have been compiled by al-Khalīlī is for converting lunar ecliptic coordinates to equatorial coordinates, in order to facilitate computations relating to the visibility of the lunar crescent.

Al-Khalīlī wrote at least one treatise on the use of the quadrant with a trigonometric grid (*al-rubʿ al-mujayyab*), but his writings on this instrument have not yet been studied.

BIBLIOGRAPHY

I. ORIGINAL WORKS. MS Paris B.N. ar. 2558, copied in 1408, contains all of the tables in al-Khalīlī's major set (nos. 1, 2, 4, and 5). MS Berlin Ahlwardt 5753–6 (Wetzstein 1138) contains all but the hour-angle tables. MS Oxford Bodleian Seld. sup. 100 contains the prayer tables and the hour-angle tables; MS Oxford Marsh 39 contains the hour-angle tables. MSS Oxford Marsh 95 and Escorial ar. 931 contain only the universal auxiliary tables; and MS Gotha Forschungsbibliothek A1406, only the prayer tables. MSS Damascus Ẓāhiriyya 3116 and 10378 also contain tables from the corpus. MS Cairo Ṭalaat *mīqāt* 228 contains al-Khalīlī's universal auxiliary tables and hour-angle tables for Damascus, as well as an anonymous set of prayer tables for the latitude of Tripoli (Lebanon). MS Cairo Dār al-Kutub *mīqāt* 71M contains al-Khalīlī's tables of the hour angle and time since sunrise, plus tables of the solar azimuth computed by al-Ḥalabī (*d.* 1455; Suter, no. 434). Egyptian copies of the auxiliary tables exist in MSS Princeton Yahuda 861, 2 and Cairo Dār al-Kutub *mīqāt* 43M. MS Istanbul Aya Sofya 2590 consists of a recension of the auxiliary tables by the Ottoman astronomer Muḥammad ibn Kātib Sinān (*ca.* 1500; Suter, no. 455). Al-Ṭanṭāwī's prayer tables and hour-angle tables are extant in MSS Damascus Ẓāhiriyya 9233, and Cairo Taymūr *riyāḍiyyāt* 129 and Dār al-Kutub *mīqāt* 1007.

MS Dublin Chester Beatty 4091 is an apparently unique copy of al-Khalīlī's auxiliary tables for timekeeping by the sun and azimuth tables. Later Egyptian and Tunisian copies of the auxiliary tables are MSS Cairo Dār al-Kutub *mīqāt* 644 and Istanbul S. Esad Ef. Madresesi 119,2 and MS Cairo Dār al-Kutub *mīqāt* 689, respectively.

Al-Khalīlī's tables for crescent visibility are in the treatise entitled *ʿIqd al-durar*, by the later Egyptian astronomer Ibn al-Majdī (Suter, no. 432), and his criteria for crescent visibility are outlined in the writings of his nephew Sharaf al-Dīn al-Khalīlī (Suter, no. 427).

Al-Khalīlī's treatise on the quadrant with trigonometric grid is in MSS Cairo Dār al-Kutub *mīqāt* 138,9 and 201M. Another treatise of considerable interest on the same subject, preserved in MS Cairo Dār al-Kutub *mīqāt* 167M,8, may be by al-Khalīlī. A treatise describing a horizontal sundial is attributed in MS Princeton Yahuda 373, fols. 131v–135r, to Abū ʿAbd Allāh al-Khalīlī, and in MS Manchester 361G to the later Egyptian astronomer ʿAbd al-ʿAzīz ibn Muḥammad al-Wafāʾī (Suter, no. 437).

II. SECONDARY LITERATURE. See D. A. King, "Al-Khalīlī's Auxiliary Tables for Solving Problems of Spherical Astronomy," in *Journal for the History of Astronomy*, **4** (1973), 99–110; and "Al-Khalīlī's *Qibla* Table," in *Journal of Near Eastern Studies*, **34** (1975), 81–122; and H. Suter, *Die Mathematiker und Astronomen der Araber und ihre Werke* (Leipzig, 1900), 169.

DAVID A. KING

AL-KINDĪ, ABŪ YŪSUF YAʿQŪB IBN ISḤĀQ AL-ṢABBĀḤ (*b. ca.* 801; *d.* Baghdad, *ca.* 866), *philosophy, science.*

Ancient biobibliographers, as well as writers such as al-Jāḥiẓ, report many legends concerning the life of al-Kindī; but little certain, or even fairly reliable, information has come down to us. Even the years of his birth and death are not definitely known: it was only by collating various data that Muṣṭafāʿ Abd al-Rāziq was able to determine the years given above.[1] It has been established, however, that al-Kindī was descended from a noble branch of the Kinda tribe of Yemen and that he began his education in Kūfa, Iraq, completing it in Baghdad—both centers of intellectual activity. It was in Baghdad that al-Kindī came to the attention of Caliph al-Maʾmūn, who took him into his court and named him to the "Academy" of Baghdad—Dār al-Ḥikma—with the task of improving the often defective translations made from the Greek. Al-Maʾmūn's successor, al-Muʿtaṣim, chose al-Kindī as tutor to his son Aḥmad, on whose behalf al-Kindī wrote several philosophical essays.

Following the death of al-Muʿtaṣim, al-Kindī's relations with the court became less close, and they remained that way throughout the caliphate of al-Wāthiq. They improved when the latter was succeeded by al-Mutawakkil. Yet al-Kindī soon fell into disgrace, the victim of such rivals as the mathematicians Banū Mūsā and the astrologer Abū Maʿshar, and of his possible sympathies for the Muʿtazilites, who were persecuted by al-Mutawakkil. During the last years of his life, he remained in relative isolation.

The "first Arab philosopher," as he was commonly called by the bibliographers, al-Kindī participated in the expansion and dissemination of what might be called the contemporary encyclopedia of knowledge. In addition, he played an important role in the elaboration and definitive formulation of Arabic philosophical and, in some cases, scientific terminology. A particular aspect of his intellectual biography is therefore worth investigating: did he know Greek? Ancient biographers and bibliographers, such as Ibn Abī Uṣaybiʿa and Ibn al-Qifṭī, note that al-Kindī took part in an intense campaign to translate Greek philosophical and scientific works. Nevertheless, an examination of the works translated with his collaboration reveals that his role was less than that of a translator. In the case of certain works of Aristotle translated by Ḥunayn ibn Isḥāq, Abū Bishr Mattā, Qusṭā ibn Lūqā, Yaḥyā ibn ʿAdī, and others, as well as certain writings by Euclid, Ptolemy, and Eutocius, al-Kindī either corrected the Arabic text of an already completed translation, commented upon it, or summarized it. Consequently, we are led to believe that he did not know Greek well enough to translate directly from that language but that he did know enough of the rudiments to correct Arabic translations and, in some degree, to establish the terminology, particularly with respect to philosophical discourse.

Some fifteen philosophical works by al-Kindī have been preserved. Although they are often complex, they can be classified according to their main subjects.

Only four chapters of the *Book of First Philosophy* have survived. It begins with a defense of philosophy (in particular, of the *falsafa* inspired by the Greeks) and then discusses the difference between the sensible and the intelligible, methods of obtaining knowledge, and questions concerning eternity and the body. The last two parts develop a complete dialectic of the one and the many that leads to the designation of the Unique True Being (*al-wāḥid al ḥaqq*), the Creator. With this text may be grouped the *Letter on the True, First, and Perfect Agent*, which deals with the Creation and the hierarchy of causes, and *De quinque essentiis*, of which we have only a Latin translation; the "five essences" are matter, form, place, motion, and time.

Al-Kindī devoted three letters to demonstrating that the world is finite, not only in space but also in time. (On this point he obviously differed from the Greek philosophers.) These are the letters *On the Demonstration of the Finitude of the Corpus of the Universe*, *On the Quiddity of That Which Cannot Be Infinite*, and *On the Unity of God and the Finitude of the Corpus of the Universe*.

Two of al-Kindī's texts describe the universe according to its structure and the principal types of causality that obtain in it: *Book in Which the Efficient Proximate Cause of Generation and Corrup-*

tion Is Explained and *Letter in Which the Submission of the Outermost Body and Its Obedience to God Is Explained.* The latter is a physical and cosmological commentary on a passage of the Koran (LV, 6).

Al-Kindī wrote five very different works dealing with the soul and the intellect. *Letter on the Existence of Incorporeal Substances* demonstrates the existence of souls. *Discourse on the Soul*, written explicitly under the inspiration of "Aristotle, Plato, and other philosophers," describes in mystical and parenetic fashion, the relations of the soul to the body, as well as its fate. The enigmatic and very brief *Statement Concerning the Soul* apparently deals with the cosmic soul. *Letter on the Quiddity of Sleep and Dreams* gives a psychology and physiology of these phenomena. Finally, *Letter on the Intellect* presents a Neoplatonic interpretation of Aristotle's noetics.

Letter on the Method of Banishing Sadness recommends that the sufferer apply himself to the only enduring object—the world of the intellect. In this work al-Kindī maintains that sadness can be eliminated through dialectic and through behavior that is marked by resignation, prudence, and the avoidance of situations that might cause sadness—advice that is firmly in the tradition of the moralists of late antiquity. In addition to this letter, approximately 100 opinions and sayings, primarily concerning ethics, are attributed to al-Kindī by the *Muntakhab ṣiwān al-ḥikma* of Abū Sulaymān al-Sijistānī.

Letter on the Number of the Books by Aristotle and on What Is Needed to Learn Philosophy consists essentially of a catalog of Aristotle's writings, a program of studies, and a philosophical commentary on a passage of the Koran (XXXVI, 78–82).

Letter on the Definitions and Descriptions of Things poses difficult problems, for among its approximately 100 occasionally enigmatic definitions are some that do not accord with the rest of al-Kindī's known works.

The writings listed above constitute only a portion of the philosophical opus mentioned by the biobibliographers. Consequently, one cannot hope to give a complete or even adequately balanced account of al-Kindī's thought. Furthermore, his works, which are never very long, are composed mainly of elaborately presented arguments employing numerous concepts and are thus virtually impossible to summarize faithfully. Nevertheless, some of the major themes of the known works may be indicated.

The first sections of *First Philosophy* and *Letter on the Books by Aristotle* define al-Kindī's plan. The former work, dedicated to Caliph Muʿtaṣim, states that knowledge is built over the centuries through the cumulative efforts of many scholars and asserts the right to the truth, regardless of its source. This portion, which is obviously inspired by Aristotle, sometimes reproduces passages verbatim from his *Metaphysics*. The *Letter* is even more explicitly Aristotelian but specifies that the mathematical sciences are to be studied as preparation for acquiring knowledge of all other subjects. Here we have two traits characteristic of al-Kindī's thought and work: as a mathematician he often constructs long, closely reasoned arguments of the type found in geometry; and as a *faylasūf* he draws abundantly on Greek sources.

Al-Kindī rarely cites Greek authors other than Plato and Aristotle. He seems, moreover, to be inspired not directly by the former but, rather, by the Platonic tradition. In any case, he owes considerably more to Aristotle, at least in regard to the basic notions of philosophy: the concepts of act and potential, of matter and form, of substance and accident, the four causes, the various kinds of motion, and the fundamental principles of noetics, as well as the basic outlines of Aristotle's cosmology. A careful reading of the details, however, reveals other influences: Porphyry, the Alexandrian school of the sixth century (as it is known from the works of John Philoponus and David of Alexandria), Proclus, the Stoics, and probably the *Corpus Hermeticum.* From these sources al-Kindī borrowed certain concepts and themes in noetics, metaphysics, and ethics, such as the relations among intellects, the distinction of the sensible world from the intelligible world, the relation of the many to the one, and the salvation of the soul.

Al-Kindī organized these various elements into an overall pattern of his own invention. Opposing Aristotle, he maintained that the world is not of infinite duration. His speculation on the many and the one led him to posit the One True Being who is the cause of the existence of everything and, as such, is the Creator. The One can in no way be conceived of in the manner of ordinary objects. Al-Kindī explicitly denies the possibility of applying, in this case, the concepts inherited from Greek philosophy—the predicables, the categories, the soul, the intellect—for this being is "above the attributes that the heretics ascribe to him" (*jalla wa-taʿālā ʿan ṣifāt al-mulḥidīn*). The great living creature that is the outermost body (the first sphere) obeys God. On the other hand, the teaching of the philosophers is the same as that of the prophets. The only difference is that the latter proclaim all at once, in concise terms, and by the intervention of the divine will that which the former discover and set forth only through great effort and in the form of long treatises.

From what has been said, it can be seen that al-Kindī established the conceptual framework that remained, on the whole, characteristic of the *falsafa*. It was formed by the union of Greek philosophy, especially in its Neoplatonic version, and Islam. (More precisely, he drew on ideas inspired by the Muʿtazilite *kalām*.) In this scheme prophets and philosophers both teach a doctrine of purification and salvation; by observing this doctrine, the soul, which comes "from the light of God," can triumph over desire and ascend through the spheres to the "world of the intellect" (*ʿālam al-ʿaql*). The philosophical manifesto in *First Philosophy* concludes with a prayer to God. A close examination of al-Kindī's choice of words shows that he favored terms used in both religion and philosophy.

Al-Kindī was contemporary with the first generation of translators of Greek works into Arabic, including Usṭāt, Ibn al-Biṭrīq, and ʿAbd al-Masīḥ ibn Nāʿima, who, respectively, translated Aristotle's *Metaphysics*, *De caelo*, and the so-called *Theology of Aristotle* (which al-Kindī did not attribute to him). His vocabulary comprises many technical terms that have remained in use, as well as some that were not retained. Thus, to designate matter, ὕλη, he used *hayūlā* and *mādda*, as well as *ṭīn* and *ʿunṣur*; and within several pages he employed, to translate the Greek φθίσις, the words *naqṣ* and *ḍamr*. Al-Kindī took pleasure in exploiting the possibilities offered by verbal derivation. For example, starting from the pronoun *huwa* ("he" or "him"), normally used to designate an existing entity in its most elementary form, he constructed a series of terms that permitted him to express various stages and elements of a creationist ontology: *huwiyya* ("existence"), *tahawwī* ("existentiation"), *mutahawwī* ("existentiated"), and *mutahawwiyyāt* ("existentiated entities").

In this area, as in others, al-Kindī was an innovator but one with an archaic streak. He was soon supplanted by other philosophers and was henceforth cited primarily as a scientist. Thus his name does not appear in the list of *falāsifa* given by Ibn Khaldūn in his *Muqaddima*, and the few passages in this book where he is mentioned concern scientific questions. Al-Kindī is, of course, occasionally cited as a philosopher by Arab authors—for example in a few places in the *Tahdīb al-akhlāq* of Miskawayh and the *Kitāb al-saʿāda wa-l-isʿād* of Abū'l-Ḥasan al-ʿĀmirī. In addition, the Christian Yaḥyā ibn ʿAdī wrote a refutation of al-Kindī, and al-Sijistānī cites him at some length in his *Ṣiwān al-ḥikma*. Many of the definitions attributed to al-Kindī are found in the *Muqābasāt* of al-Tawḥīdī; and there is a *Refutation of al-Kindī the Philosopher* attributed to Ibn Ḥazm. The biobibliographers—Ibn al-Nadīm, Ibn Juljul, and Ibn al-Qifṭī—praise his great knowledge of philosophy. But these scattered traces do not amount to the genuine survival that would have resulted from the establishment of a school; and it appears that al-Kindī never founded one, although he has a direct disciple in Aḥmad ibn al-Ṭayyib al-Sarakhsī.

All things considered, the same may be said of al-Kindī's influence in the Latin West. A few of his treatises were translated in the twelfth century, Albertus Magnus cited him, a few other authors alluded to him, and Giles of Rome devoted a chapter of his *De erroribus philosophorum* to him. (The majority of the "errors" for which Giles reproaches him concern astrology and, more generally, the system of the world.) But this cannot be compared with the massive presence in the Latin Middle Ages of Ibn Sīnā, al-Ghazālī, Ibn Rushd, or even of al-Fārābī and Ibn Bājja.

Given the present state of knowledge it is difficult, if not impossible, to offer anything approaching a complete, systematic exposition of al-Kindī's scientific work. Encyclopedic in scope, it comprises writings on arithmetic, geometry, astronomy, music, medicine, pharmacology, and other fields. Among the titles cited by ancient and modern biographers are some that have not yet been found; and others have not appeared in a critical edition.[2] Except for a few short treatises, al-Kindī's scientific writings have not fared as well as his more abundant philosophical works. Moreover, the difficulty of presenting his ideas applies to his scientific as well as to his philosophical concepts; for, to a much greater degree than such successors as al-Fārābī, Ibn Sīnā, and Ibn Rushd, he followed an ancient tradition of basing his philosophical reflection on scientific investigation.[3] Consequently, a truly thorough assessment of the scope and limits of his contribution must await considerable further research.

Nevertheless, two abiding concerns may be detected in the corpus of al-Kindī's scientific writings. The first is that of the commentator, the transmitter of Hellenic scientific works, whose goal is to prepare his readers for the study of philosophy. The other is the completion and, if possible, the augmentation of the body of inherited scientific knowledge. Although sometimes separate, these two preoccupations are frequently commingled. *Kitāb fī l-ṣināʿat al-ʿuẓmā*[4] stems from the former concern; while *De aspectibus*, although a manual of ancient optics—primarily that of Euclid—is conceived with a view to extending and perfecting this older knowledge. Accordingly, if this twofold mission is ignored and attention is restricted to works designed mainly for pedagogical purposes (an error committed by certain historians), then their author is unjustly considered to be no more than a commentator of the Greek texts that inspired them. Nothing better expresses al-Kindī's intentions in this regard than his own words: "It is good . . . that we endeavor in this book, as is our habit in all subjects, to recall that concerning which the Ancients have said everything in the past, that is the shortest and easiest to adopt for those who follow them, and to go further in those areas where they have not said everything, according to the use of the [Arabic] language and the customs of the time, to the degree that we are able."[5]

Such was the project that al-Kindī sought to realize in the majority of his surviving scientific works. While nothing can be affirmed about him as a mathematician, since the most important texts have not been found, it can be said that he pursued this goal in his works on optics, pharmacology, and music.[6]

This statement is confirmed by his two principal works on optics. While in the *De aspectibus* he reiterates the same idea, in the *Burning Mirrors* he begins with a critique of Anthemius of Tralles and sets out to finish what the latter had left undone.[7] Yet, although his ultimate aim occasionally leads him to adopt a critical attitude toward the ancient authorities, al-Kindī remains basically committed to the optical ideas commonly held before Ibn al-Haytham. Essentially a geometer rather than an experimenter, al-Kindī is a prisoner of the traditional approach to the subject, in which no distinction is made between a theory of light and a theory of vision. That is, it was assumed that to see was to illuminate. His criticisms, therefore, are not elements of a reform but amendments to the optics of the geometers—principally Euclid. Nonetheless, his optical writings were read and studied by the Arab physicists as well as by Roger Bacon, John Pecham, and probably Robert Grosseteste.

In *De aspectibus* al-Kindī seeks to demonstrate what Euclid had postulated: the notion of the rectilinear propagation of light and the theory of emission. Although al-Kindī adopted the emissionist point of view, he attempted to demonstrate the rectilinear propagation of light rays on the basis of geometrical considerations about the shadows of opaque bodies exposed to luminous sources and about light passing through slits.[8] Then, in order to defend the emission theory, he gives new arguments against the ancient theories of vision, notably those of the intromission of forms and of the combined emission-intromission of forms. This critique ultimately amounted to showing the impossibility of reconciling the theory of the intromission of forms, the intromission of totalities not analyzable into their simple elements, and the fact that the perception of an object is a function of its localization in ordinary space. Al-Kindī notes in this connection that if the theory of the intromission of forms were correct, then a circle in the same plane as the eye would be perceived in all its circularity, which is not the case.[9]

Going still further, al-Kindī ultimately rejects the Euclidean theory of emission, amending it in order to make it conform to observed data. For example, he asserts that a visual cone is not formed of discrete rays, as Euclid stated, but appears as a volume of continuous radiations. Much more important than this modification is the idea on which it is based: rays are not geometrical lines but, rather, impressions produced by three-dimensional bodies; consequently, according to al-Kindī, the ray cannot be considered a one-dimensional geometrical line.[10] Rays are therefore three-dimensional and form a continuous radiant cone. To some degree this critique prepared the way for Ibn al-Haytham's distinction between light rays and the straight lines along which they are propagated.

Al-Kindī still had to explain how perception varies according to the region of the cone considered.[11] His position on this question differed from those held by both Euclid and Ptolemy. He assumed that from every point of the eye there emanate radiations along every straight line that can be drawn from these points and from those of the visual field. Thus, a visual cone emerges from every point of the eye.[12] To apply this idea, al-Kindī proceeds by analogy with light emanating from a luminous source and states more clearly than his predecessors—although in a less elaborate fashion than Ibn al-Haytham—the principle of rectilinear propagation.[13]

In the second optical treatise, on burning mirrors or "rays," al-Kindī first recalls Anthemius' report on how ships were set aflame by burning mirrors during a naval battle:

> Anthemius should not have accepted information without proof. . . . He tells how to construct a mirror from which twenty-four rays are reflected on a single point, without showing how to establish the point where the rays unite at a given distance from the middle of the mirror's surface. We, on the other hand, have described this with as much evidence as our ability permits, furnishing what was missing, for he has not mentioned a definite distance.[14]

Al-Kindī's demonstration is based on implicit knowledge of the law of reflection, which stipulates that the angles of incidence and reflection are equal, as well as that the incident ray, the normal, and the reflected ray are all in the same plane.

The desire to extend and improve upon the knowledge of antiquity can be detected in another field that al-Kindī investigated, pharmacology.[15] Besides a compilation of medical preparations, *Aqrābādhīn*, he wrote *Risāla fī maʿrifa quwwat al-adwiyat al-murakkaba*, which was translated into Latin as *De medicinarum compositarum gradibus investigandis*; in it he treats the composition of medicines from a Galenic point of view.[16] In the course of his study he raises the problem of the quantification of qualities and formulates a law the adaptation of which in physics yielded the law named for Bradwardine.

Like the ancient authors, al-Kindī held that the four qualities employed in ancient medical theory (heat, cold, dryness, and humidity) could assume four degrees of intensity. Each degree can be recognized by the effects observable in the patient, and the degrees are ordered along a scale of fixed units of sensation running from the smallest perceptible difference to (in some cases) the destruction of the patient's body. After recalling the opinions of the ancients, al-Kindī turns to completing their task: "They have not attempted to do as much for the compound medicines: they have not said that a certain compound medicine exists at such and such a degree of heat, cold, dryness, [and] humidity. Now, such knowledge is even more important and more valuable in the case of a compound medicine than in that of a simple medicine."[17] He then sets out to elaborate a theory that will enable him to extend to compound medicines a precise calculus based on an examination of the medicine's composition and on its effect upon the patient. Al-Kindī's fundamental notion is that "the faculties of the compound medicine necessarily increase or diminish according to the variations in the faculties of its components, and that its faculties cannot be reduced to those of one of its components to the exclusion of the others."[18] This conception is justified by an atomistic doctrine in which al-Kindī seeks "to represent the smallest possible part of the temperate disposition, which is indivisible because of its smallness. There ought to be in it as much heat as cold, since the overall temperate substance is composed of these parts."[19]

In his effort to broaden the scope of the earlier studies, al-Kindī was led to pose the problem of the quantification of the qualities. This occurred when he attempted systematically to link the degrees of intensity of the compound medicine with the numerical changes in the qualitative forces that produce them. Since the qualities remain separate in the medicine, the different parts have separate effects; and since the degree of the medicine's intensity is determined by the proportion of the qualities, it was possible for al-Kindī to formulate mathematical relationships between the increments in the number of parts—such as heat and cold—and the increments in the effects experienced. He states that the proportion 2:1 of hot parts to cold produces a heat effect of the first degree; the proportion 4:1, a heat effect of the second degree; the proportion 8:1, a heat effect of the third degree; and the proportion 16:1, a heat effect of the fourth degree. In another terminology, the degree of intensity, I, of a medicine is proportional to the logarithm of base 2 of the proportion of one quality to the other: $I \simeq \log_2 (\text{heat}/\text{cold})$.

The influence of this system during the Middle Ages seems to have been much greater among "physicists" than among physicians. Physicians such as Abu l-Qāsim al-Zahrāwī (Abulcasis) used al-Kindī's ideas;[20] but his works were difficult for the nonmathematicians among the physicians to interpret. (This opinion was expressed by Roger Bacon.)[21] The situation seems to have been different, however, for scientists, such as Bradwardine.[22]

The intention to advance ancient science is also evident in al-Kindī's four known works on music. Although he adopted "a system of nomenclature for the notes and tetrachords in the scale similar to that used in the old Greek theory,"[23] he used the letters of the alphabet to designate the notes of the scale—a procedure employed in Europe a century later. Al-Kindī's musical treatises, which are among the first works on the theory of music written in Arabic, paved the way for such major works as that by al-Fārābī.

Al-Kindī did not neglect other areas of knowledge. He studied optics, using an approach combining both physical and philosophical notions (especially the theory of color), and he investigated topics in geology, meteorology, geography, climatology, geophysics, astronomy, and astrology—considering the last as a science. But he went still further in his research, undertaking studies with a technological aim as well: the making of clocks, astronomical instruments, and even of objects such as swords.

Throughout his scientific writings, with varying success according to the subject involved, al-Kindī utilized the same approach: to work through the legacy of ancient science and then to transcend it in furtherance of his twofold aim of advancing both scientific pedagogy and research. His method always combined an empirical strain with a mathematical tendency that led him to seek geometrical or numerical relationships between phenomena. This is perhaps why his influence proved greater among philosopher-scientists and scientists than among the great metaphysicians.

NOTES

1. Muṣṭafā ʿAbd al-Rāziq, *Faylasūf al-ʿarab* . . . (Cairo, 1945), 17–20.
2. See especially Ibn al-Nadīm, *Kitāb al-Fihrist*, G. Flügel, ed., I (Leipzig, 1871), 255–261.
3. "The earliest philosophers in Islam were, like the first Greek thinkers, nature philosophers." J. de Boer, "Zu Kindī und seiner Schule," in *Archiv für Geschichte der Philosophie*, **13** (1900), 159. I. Madkour voices a similar idea: that al-Kindī "belongs principally to the physical science tendency that dominated Islamic philosophical speculation in its early stages." *La place d'al-Fārābī dans l'école philosophique musulmane* (Paris, 1934), 5. See also A. Cortabarria Beitia, "La classification des sciences chez al-Kindī," in *Mélanges de l'Institut dominicain d'études orientales* (Cairo), **11** (1972), 50–76.

   It is known that al-Kindī wrote a short work, not yet found, entitled *Risāla fī annahu lā tunāl al-falsafa illā bi-ʿilm al-riyāḍiyyāt* ("Treatise in Which It Is Shown That Philosophy Can Be Attained Only Through the Science of Mathematics"). See R. J. McCarthy, *Al-taṣānīf al-mansūba ilā faylasūf alʿarab*, 9.
4. See F. Rosenthal, "Al-Kindī and Ptolemy," in *Studi orientalistici in onore di G. Levi della Vida*, II (Rome, 1956), 436–456.
5. See al-Kindī, *Fī l-falsafat al-ūlā*, edited by ʿAbd al-Hādī Abū Ridā in his collection *Rasāʾil al-Kindī l-falsafiyya*, I, 103.
6. In *Kitāb al-Fihrist*, Ibn al-Nadīm gives ten titles of works on arithmetic by al-Kindī and twenty-two titles on geometry.
7. *De aspectibus* has been edited by A. A. Björnbo. See A. A. Björnbo and S. Vogl, "Alkindi und Pseudo-Euklid: Drei optische Werke," in *Abhandlungen zur Geschichte der mathematischen Wissenschaften*, **26**, no. 3 (1912), 3–41. The goal is stated on p. 3 of *De aspectibus*: "Oportet, postquam optamus complere artes doctrinales, et exponere in eo, quod antiqui praemiserunt nobis in eis, et augere, quod inceperunt et in quibus fuerunt nobis occasiones adipiscendi universas bonitates animales, ut de diversitatibus aspectus secundum nostrae possibilitatis mensuram universaliter et demonstrative loquamur. . . ."

   Kitāb Yaʿqūb ibn Isḥāq al-Kindī fī al-shuʿāʿāt ("Treatise on the Burning Mirror"), MS 2048 in Patna, India, has been photographically reproduced by the Institute of Arabic Manuscripts in Cairo as no. 3121 in its series. M. Y. Hāshimī claimed that this MS had been found and then lost. The MS does indeed exist—and he reproduced a photocopy not of the MS itself but of a transcription, without clearly stating that this was the case. In addition, the photocopy has some pages reversed. See Hāshimī, *Propagation of Ray: The Oldest Arabic Manuscript About Optics—"Burning Mirror" From Ya'kub ibn Isḥaq al-Kindī* (Aleppo, 1967). Jean Jolivet and Roshdi Rashed are currently preparing a critical ed. of this text.
8. "Quod vero videmus ex rectitudine finium umbrarum corporum in latitudine et luminibus per fenestras ingredientibus necessario ducit nos ad hoc, ut transitus radiorum procedentium a corporibus luminosis fiat secundum rectitudinem rectarum linearum." *De aspectibus*, p. 4.
9. *Ibid.*, props. 7–10.
10. *Ibid.*, props. 12–13.
11. *Ibid.*, p. 23.
12. *Ibid.*, pp. 24–25.
13. "Iam ergo exemplificavimus, qualiter quaeque pars corporis luminosi illuminet, quod ei obviat, scilicet a quo est possibile, ut ad ipsum producatur linea." *Ibid.*, p. 23.
14. *Fī l-shuʿāʿāt*, fol. 3
15. See the ed. of the Arabic text, with French trans., by Léon Gauthier in *Antécédents gréco-arabes de la psychophysique* (Beirut, 1939).
16. The Latin trans. was printed at Strasbourg in 1531. Sarton suggests that it was done by Gerard of Cremona and was used by Arnold of Villanova (*d.* 1311). See *Introduction to the History of Science*, 3 vols. (Baltimore, 1927–1931), II, 342.
17. Gauthier, *op. cit.*, 44.
18. *Ibid.*, 45
19. *Ibid.*, 56–57.
20. See S. K. Harmaneh and G. Sonnedecker, *A Pharmaceutical View of Abulcasis al-Zahrāwī in Moorish Spain* (Leiden, 1963), 61; "Al-Zahrāwī does not go into al-Kindī's complicated calculations and geometric proportions of degrees of faculties, yet he seems to apply such a system in determining the degrees of faculties expected from mixed drugs or diets containing more than two ingredients, and each with its own degree of humoral faculty or action."
21. *De erroribus medicorum*, which is *Opera hactenus inedita Rogeri Baconi*, R. Steele, ed., fasc. 9 (Oxford, 1928), 166–167.
22. See M. Clagett, *The Science of Mechanics in the Middle Ages* (Madison, Wis., 1959), 349; and, especially, M. McVaugh, "Arnold of Villanova and Bradwardine's Law," in *Isis*, **58** (1967), 56–64.
23. Youssef Shawqī, *Al-Kindī's Essay on Composition* (Cairo, 1969), 9.

BIBLIOGRAPHY

I. ORIGINAL WORKS. No existing ed. includes all of al-Kindī's known works. A group of scholars sponsored by the Centre National de la Recherche Scientifique is currently preparing a complete ed. of the available texts, with French trans. and notes. Some unpublished MSS have already been collected for this project.

Among the existing eds. the most useful are ʿAbd al-Hādī Abū Rīda, ed., *Rasāʾil al-Kindī l-falsafiyya*, 2 vols. (Cairo, 1950–1953); and Z. Yūsuf, ed., *Muʾallafāt al-Kindī al-mūsiqīyya* (Baghdad, 1962).

For a listing of individual texts, translations in various languages at different periods, specialized bibliographies, biobibliographical data, and both general and specialized studies, see R. J. McCarthy, *Al-taṣānīf al-mansūba ilā faylasūf al-ʿarab* (Baghdad, 1963); and N. Rescher, *Al-Kindī. An Annotated Bibliography* (Pittsburgh, 1964).

II. SECONDARY LITERATURE. Recent studies include M. Lewey, *The Medical Formulary or Aqrābādhīn of al-Kindī*, translated with a study of its materia medica (Madison, Wis., 1966); G. N. Atiyeh, *Al-Kindī: The Philosopher of the Arabs* (Rawalpindi, Pakistan, 1966); A. Cortabarria Beitia, "A partir de quelles sources étudier al-Kindī?" in *Mélanges de l'Institut dominicain d'études orientales* (Cairo), **10** (1970), 83–108; J. Jolivet, *L'intellect selon Kindī* (Leiden, 1971); David C. Lindberg, "Alkindī's Critique of Euclid's Theory of Vision," in *Isis*, **62** (1971), 469–489; M. T. d'Alverny and F. Hurdy, "Al-Kindī, *De radiis*," in *Archives d'histoire doctrinale et littéraire du moyen-âge*, **41** (1974), 139–260, which contains a bibliography that, in part, supplements that of Rescher; Alfred L. Ivry, *Al-Kindī's Metaphysics* (Albany, 1974), a translation of al-Kindī's treatise "On First Philosophy," with introduction and commentary; and Michael R. McVaugh, *Arnaldi de Villanova Opera medica omnia*, II, *Aphorismi de Gradibus* (Granada–Barcelona, 1975), esp. chs. 3 and 6, and app. 1.

JEAN JOLIVET
ROSHDI RASHED

KUNTH, CARL SIGISMUND (*b.* Leipzig, Germany, 18 June 1788; *d.* Berlin, Germany, 22 March 1850), *botany*.

Kunth was the son of a lecturer in English at Leipzig University and was educated at the Leipziger Rathsschule. In 1806 he was sent to Berlin to stay with his uncle Gotlob Christian Kunth, who had been tutor of Alexander and Wilhelm von Humboldt. The administrative post that Kunth was given at the Seehandlungs-Institut left him sufficient time to study botany at the University of Berlin under Carl Ludwig von Willdenow. Through his uncle Kunth became acquainted with Alexander von Humboldt, who was living in Paris during publication of the account of his South and Central American expedition and who invited Kunth to stay with him and publish the botanical results. Just before leaving Berlin, Kunth published his first *Flora berolinensis* (1813), essentially a purely floristic inventory. While he was in Paris, however, his botanical interest changed from floristics to analytical systematics.

The Paris group of botanists, which included René Desfontaines, A. L. de Jussieu, and Louis Claude Marie Richard, was involved in developing a more broadly based systematics, working toward a natural system of classification that could more effectively deal with the influx of the many new forms from tropical regions than could the old Linnaean system. Rather than accept the strict Linnaean hierarchy of characteristics of a few characters thought to be "essential," they took as a base a broad spectrum of morphological characters derived from the entire plant rather than from just the gross morphology of the flower. Kunth, when dealing with the 3,600 new species brought home by Humboldt and Bonpland, contributed significantly to this development by giving special attention to a minute analysis of floral structures and to a better understanding of the significance of vegetative characteristics.

The sumptuous way in which Humboldt presented the results of his expeditions enabled Kunth to publish his comprehensive analytical studies in great detail. He wrote the text for, and prepared most analytical drawings of the plates of, the seven volumes of the *Nova genera et species plantarum* (1815–1823), one of the sections of Humboldt and Bonpland's *Voyage aux régions équinoxiales du Nouveau Continent, fait en 1799 . . . 1804*. These volumes were followed by *Mimoses et autres plantes légumineuses du Nouveau Continent* (1819[–1824]), *Révision des graminées* (1829[–1834]), and *Synopsis plantarum* (1822–1825). These major works, describing in great detail and illustrating with beautiful plates thousands of taxa of New World plants, were written during nearly seventeen years of research in Paris, and drew upon not only the collections made by Kunth's patrons but also the other public and private herbaria that made Paris the world's capital of systematic botany during this period.

In 1829 Kunth became professor at the University of Berlin and head of the Berlin botanical garden. During his later years he published textbooks and works of taxonomic synthesis, such as *Enumeratio plantarum* (1833–1850), thus adding significantly to his impressive output. During his last years Kunth's work was impaired by bad health.

BIBLIOGRAPHY

I. ORIGINAL WORKS. Kunth's contributions to Humboldt and Bonpland's *Voyage aux régions équinoxiales du Nouveau Continent, fait en 1799 . . . 1804*, pt. 6, *Botanique*, are sec. 2, *Monographie des mélastomacées* (Paris, 1816–1823)—part of *Melastomae*, I, and the major part of *Rhexiae*, II; sec. 3, *Nova genera et species plantarum*, 7 vols. (Paris, 1815–1823[–1825], facs. repr. Weinheim, 1963; Amsterdam, 1974)—the text and the analytical parts of the plates; sec. 4, *Mimoses et autres plantes légumineuses du Nouveau Continent* (Paris, 1819[–1824]); sec. 5, *Synopsis plantarum*, 4 vols. (Paris, 1822–1825 [1826]); sec. 6, *Révision des graminées*, text and atlas (Paris 1829[–1834]), reiss. as *Distribution méthodique de la famille des graminées* (Paris, 1835–1837). Also published in Paris were *Malvaceae, Büttneriaceae, Tiliaceae* (1822) and *Terebinthacearum genera* (1824).

Other books are *Flora berolinensis* (Berlin, 1813); *Handbuch der Botanik* (Berlin, 1831); *Enumeratio plantarum*, 5 vols. and supp. (Stuttgart–Tübingen,

1833–1850); and *Lehrbuch der Botanik* (Berlin, 1847). *Flora berolinensis*, 2 vols. (Berlin, 1832), is a different work from the book with the same title of 1813.

For Kunth's many articles in scientific journals and periodicals, see *Annales des sciences naturelles* (Botanique), 3rd ser., **14** (1850), 95–106; and Stearn (see below), 149–151.

II. SECONDARY LITERATURE. There are two major contemporary sources of information: A. de Jussieu, *Notice sur la vie et les ouvrages de Charles Sigismund Kunth* (Paris, 1850), also in *Annales des sciences naturelles (Botanique)*, 3rd ser., **14** (1850), 76–106; and A. von Humboldt, *Beilage zum preussischen Staats-Anzeiger*, no. 128 (611) (9 May 1851), repr. in *Botanische Zeitung*, **9** (1851), 427–432, and in W. T. Stearn, *Humboldt, Bonpland, Kunth and Tropical American Botany* (Lehre, 1968), 143–148. For bibliographical details of Kunth's works, see F. A. Stafleu, *Taxonomic Literature* (Utrecht, 1967), 247–250.

FRANS A. STAFLEU

KUNTZE, CARL ERNST OTTO (*b.* Leipzig, Germany, 23 June 1843; *d.* San Remo, Italy, 28 January 1907), *botany*.

Otto Kuntze, essentially a self-taught scientist, received a rather limited primary and secondary education at a *Realschule* and a commercial school in Leipzig. His interest in botany was stimulated by Carl Otto Bulnheim at Leipzig and, after he had accepted an administrative post in Berlin at the age of twenty, by Alexander Braun and Paul Ascherson. His first publications, both published in 1867, were a pocket flora of Leipzig and a revision of the German brambles (*Rubus*).

After he established a factory for the manufacture of volatile oils and essences at Leipzig in 1868, Kuntze's financial situation improved so spectacularly that he could retire from business in 1873 and devote himself to travel and botany without further need to earn a living. He first went on a journey around the world from February 1874 until February 1876, traveling to the West Indies, northern South America and Central America, then via the United States (from New York to San Francisco) to Japan, China, southeast Asia, and India back to Europe through Aden and Egypt. The first published result of this trip was his original and stimulating *Reise um die Erde* (1881), in which Kuntze shows himself to be a highly critical and nonconformist observer. He made large ethnological and botanical collections, the former for the Museum für Völkerkunde at Leipzig and the latter constituting the nucleus of his sizable private herbarium, on the basis of which he elaborated his nomenclatural reforms in later years.

After the journey Kuntze chose botany as the main subject for somewhat belated university training at Leipzig and Berlin (1876–1878). He received a doctorate from the University of Freiburg im Breisgau for a thesis on *Cinchona* in June 1878. From then on, his time was spent in botanical studies at Leipzig, Berlin, Leiden, and Kew, interspersed with trips to Asiatic Russia and the Canary Islands. After his dissertation Kuntze elaborated his more general ideas on systematic botany in a second monograph on *Rubus*, treating mainly non-German material, by presenting theoretical considerations on the methodology of the recognition and description of species. The travelogue of 1881 was followed by *Phytogeogenesis* (1884), which offered a view of the origin of plant life. The main task ahead of Kuntze, however, was the publication of the botanical results of his journey around the world, which he presented in the first two volumes of his *Revisio generum plantarum* (1891).

While Kuntze was working on the naming of his plants, his highly legalistic mind had been troubled by many obvious defects in the current rules of botanical nomenclature, which had been set up by Alphonse de Candolle and were approved and accepted by the International Botanical Congress held at Paris in 1867. Thus far the community of scientists had silently interpreted several of the unsettled points or ambiguities of the rules in accordance with informal mutual understandings. Kuntze challenged most of these unwritten rules by means of a literal but sometimes very one-sided interpretation of the existing regulations. For instance, with respect to priority they provided that botanical nomenclature started with Linnaeus' writings—without specifying the starting-point publication. Kuntze accepted as such a starting point the *Systema naturae* of 1735 and the *Genera plantarum* of 1737, written when Linnaeus' proposed reform could not yet have been accepted by his contemporaries. Most botanists, however, had ignored all publications, including Linnaeus' own, issued before the *Species plantarum* of 1753.

Kuntze's choice thus resulted in the reinstatement of many neglected non-Linnaean generic names. As a further result of his nonconformist, often (admittedly) perfectly logical interpretation of the Paris rules of 1867, Kuntze found it necessary to change the names of some 25,000–30,000 names of species, often of very well-known plants.

General acceptance of Kuntze's rules as set out in his *Revisio* thus became impossible. The reaction of the botanical establishment to Kuntze's reforms was sharp and often vehement. Kuntze went on another major trip, this time to South America, from November 1891 until January 1893 and made again extensive plant collections.

The reactions against his proposed reforms, together with the new collections, formed the basis for the third volume of the *Revisio*, which appeared in three parts (1893, 1898) and added to the number of name changes. A chaotic state of botanical nomenclature resulted, partly because botanists were not entirely unanimous in their refutation of Kuntze's rules. These rules contained several elements—such as a true and strict adherence to priority as soon as the starting point had been agreed on—which made sense and appealed, for instance, to a large group of American botanists. The main issues were resolved by the Second International Congress of Botany held at Vienna in 1905, against the spirited opposition of Kuntze. The extent of the disturbance is illustrated by the fact that the last issues created by Kuntze were settled only at the Fifth International Congress of Botany held at Cambridge in 1930. Kuntze published numerous pamphlets in which he did not always refrain from acrimony and personal attacks. Even so, although very few of his changes were adopted, he had forced botanists to put their house in order with respect to botanical nomenclature.

The controversy created by Kuntze cast a shadow over his later years, and the positive points in the contributions to botany by this colorful although slightly quixotic scientist tended to be overlooked by his contemporaries. Kuntze knew his plants very well and was a scholarly and many-sided historian of botany in addition to being a legalistic reformer of nomenclature. In 1903 he published *Lexicon generum phanerogamarum* with T. E. van Post. He continued to travel and amass a large herbarium, which, except for the European material, is now incorporated in that of the New York botanical garden; the European material is at the Charleston (South Carolina) Museum. Kuntze did not attend the Vienna Congress of 1905, at which the results of so many of his labors were rejected, except for a brief but spectacular and dignified appearance at one of the nomenclature sessions, at which he denounced the right of the Congress to legislate on these matters. Shortly thereafter his health deteriorated, and he died at his Italian home in San Remo.

BIBLIOGRAPHY

I. Original Works. No complete published list of Kuntze's many writings exists, but he himself provided lists up to 1898 in his *Revisio*: I, clvi, and II, pt. 2, 202. Besides his books he published many journal articles and pamphlets, mainly on nomenclature. His main works were *Reform deutscher Brombeeren* (Leipzig, 1867); *Taschenflora von Leipzig* (Leipzig–Heidelberg, 1867); *Cinchona. Arten, Hybriden und Cultur der Chininbäume* (Leipzig, 1878); *Methodik der Speciesbeschreibung und Rubus* (Leipzig, 1879); *Reise um die Erde* (Leipzig, 1881; 2nd ed., 1888); *Phytogeogenesis* (Leipzig, 1884); "Monographie der Gattung Clematis," in *Verhandlungen des Botanischen Vereins der Provinz Brandenburg*, **26** (1885), 83–202; *Revisio generum plantarum vascularium omnium atque cellularium multarum secundum leges nomenclaturae internationales*, 3 vols. (Leipzig, 1891–1898); *Geogenetische Beiträge* (Leipzig, 1895); *Nomenclaturae botanicae codex brevis maturis* (Stuttgart, 1903); and *Lexicon generum planerogamarum inde ab anno 1737 . . .* (Stuttgart 1903; reiss. 1904), written with T. E. van Post.

Most of the nomenclatural writings are contained or summarized in the *Revisio* and the *Lexicon*. Polemical pamphlets included *Exposé sur les Congrès pour la nomenclature botanique* (Geneva, 1900); *Protest gegen den vollmachtswidrig arrangierten und wegen die vielen unregelmässigkeiten inkompetenten Nomenclatur-Kongress* (San Remo, 1905); and *Motivierte Ablehnung der angeblich vom Wiener Kongress 1905 angenommenen inkompetenten und fehlerreichen botanischen Nomenclaturregeln* (San Remo, 1907). Kuntze also published on geological, ethnographical, statistical, and legal subjects: for instance, *Motivirter Entwurf eines deutschen Gesundheits-Baugesetzes* (Leipzig, 1882).

II. Secondary Literature. There is no formal biography of Kuntze. Most of the literature deals with his nomenclatural reforms rather than with the man behind the polemics. I. Urban, *Symbolae antillanae*, III (Berlin, 1902), 70–71, and *Flora brasiliensis*, I, pt. 1 (1906), 36–38, gives the most personal information based upon private communication; J. H. Barnhart, *Bulletin of the Charleston Museum*, **9**, no. 8 (1913), 65–68, also provides some original details. The few formal obituaries, such as that by W. B. Hemsley, in *Bulletin of Miscellaneous Information. Royal Botanic Gardens, Kew* (1907), 100–101, are uninformative.

Frans A. Stafleu

LA CONDAMINE, CHARLES-MARIE DE (*b.* Paris, France, 27 January 1701; *d.* Paris, 4 February 1774), *mathematics, natural history.*

La Condamine came from an established, wealthy, and well-connected noble family. His father, district tax collector of the Bourbonnais, married in 1700, at the age of sixty. His mother, the former Marguerite-Louise de Chourses, daughter of a president of the Cours des Comptes of Montpellier, was about half his age. They had two children: Charles-Marie, the elder, and a daughter.

La Condamine completed his studies, with no marked enthusiasm, under the direction of the Jesuit fathers of the Collège Louis-le-Grand in Paris. There he had several remarkable teachers: the famous Père Poree, for the humanities; Père Brisson, in philosophy; and Père Castel, in mathematics. Lacking a pronounced vocation, La Condamine took up a military career when he left the *collège*. War broke out against Spain, and he joined the army of Roussillon commanded by the Maréchal de Berwick. Present at the siege of Rosas (1719), he distinguished himself by his contempt for danger. He soon found life in the army unsuited to his taste, however. La Condamine thereupon established contact with scientific circles in Paris, which were better able to satisfy his unquenchable curiosity, sometimes carried to the point of recklessness.

Through his new relations, La Condamine entered the Académie Royale des Sciences, as *adjoint-chimiste*, on 12 December 1730. But the Academy was no more able than the army to hold his interest for long, and in May 1731 he sailed on a naval ship for the commercial parts of the Levant, under the command of Duguay-Trouin. He thus came to Algiers, Alexandria, the coast of Palestine, Cyprus, and Smyrna, disembarking at Constantinople (October 1731), where he spent five months. After about a year, La Condamine returned to Paris and presented to the Academy, on 12 November 1732, his "Observations mathématiques et physiques faites dans un voyage de Levant en 1731 et 1732." Although the memoir did not consist entirely of new results, it was sufficient to earn him the reputation of a competent mathematician, an observant traveler, and a good storyteller. It is thus not astonishing that, a few months later, the Academy chose him to participate in the mission known as the *Académiciens du Pérou*.

The expedition to Peru, encouraged by the minister Maurepas, had as its goal the verification of Newton's hypothesis on the flattening of the terrestrial globe in the polar regions and, thereby, the resolution of the controversy regarding the form of the earth that was then dividing French scientists. Maupertuis, Clairaut, and Le Monnier went to Lapland to measure several degrees of meridian at the arctic circle, while Godin, Bouguer, and La Condamine were sent to Peru, territory belonging to Philip V of Spain, in order to make the same measurement in the vicinity of the equator. La Condamine left Paris on 14 April 1735 for La Rochelle, where he embarked with his two companions and the naturalist Joseph de Jussieu. They set sail for America on 16 May and made stops at Martinique (22 June to 1 July), Santo Domingo (11 July to 31 October), and Carthagena (16 November to 24 November). On 29 November they dropped anchor at Portobelo, Panama, arriving at the city of Panama a month later (29 December), after having traversed the isthmus. They set sail again on 22 February 1736; and on 10 March the mission finally disembarked at Manta, the port of the province of Quito. In order to reach the city of Quito, Godin and Bouguer sailed to Guayaquil. La Condamine went overland and did not rejoin the others in Quito until 4 June.

The arc of meridian that had been selected passes through a high valley nearly perpendicular to the equator, extending from Quito in the north to Cuenca in the south. Work had scarcely begun when tension arose between Louis Godin, the head of the mission, and Pierre Bouguer. From 3 October to 3 November, however, the team conducted the measurements of the base for the triangulation operations in the Yaruqui plain. This task completed, the members returned to Quito at the beginning of December. In the meantime, the financial aid expected from Paris had not arrived, and money was beginning to run short. La Condamine, who upon leaving France had provided himself with letters of credit addressed to banks in Lima, offered his assistance. He left Quito on 19 January 1737, in the midst of the rainy season and traveled the long and difficult journey to Lima, which he reached on 28 February. He extended his journey in order to observe, near Loja, the cinchona tree, which was still not well-known to Europeans. Having concluded his business with some trouble, at the offices of the Lima bankers, he headed back to Quito, arriving on 20 June, just in time to observe the solstice there.

By this time, Godin was working alone and refused to communicate any of his results to his colleagues. Consequently, La Condamine began to collaborate with Bouguer; and after two years of work that was often interrupted, the geometric measurement of the arc of the meridian, undertaken in a mountainous and difficult country, was completed in August 1739. It remained to make the astronomical measurement of the same arc by determining exactly the latitude of its two extremities. In the meantime, the misunderstanding between the scientists steadily worsened. Godin broke definitively with his colleagues and continued to work by himself. In December 1741, while verifying observations made jointly with La Condamine, Bouguer discovered a small error, which he corrected but which gave rise to a long dispute. La Condamine wished to recheck the observations that they had made together, but Bouguer refused. Henceforth each man pursued his measurements and observations independently. The work was finally completed in 1743. Leaving behind them a commemorative plaque on the wall of the Jesuit church in Quito (preserved at the Quito observatory) and two pyramids at the extremities of the base in the Yaruqui plain (soon destroyed by the Spaniards but reconstructed in 1836), the three scientists left for home, traveling separately.

La Condamine chose the longest and most dangerous route, the Amazon. Heading south from Tarqui, near Cuenca, on 11 May 1743 he reached the village of Jaen, having passed through Loja, Valladolid, and Loyola. On 4 July he set off from Jaen in a canoe and, descending the Chuchungas River, reached its confluence with the Rio Marañón, which is a source of the Amazon. Traveling down the Amazon took him more than two months. Although concerned primarily with astronomical observations and topographical details, La Condamine also observed the use of rubber by several tribes. His remarks on this subject, however, are far less interesting than his observations concerning the cinchona. On 19 September 1743 he reached the Atlantic at Pará. He left that city on 29 December and sailed to Cayenne, French Guiana, where he landed on 25 February 1744.

Unable to find a ship leaving for France, La Condamine spent five months in Cayenne. There he repeated Richer's experiments on the variation of weight at different latitudes (see Jean Richer, "Observations astronomiques et physiques faites en l'isle de Caïenne" [1679]) and made many observations on physics, natural history, and ethnology. He met the physician and naturalist Jacques-François Artur and the royal engineer François Fresneau. Later he presented the results of Fresneau's research to the Académie des Sciences in his memoir "Sur une résine élastique . . . nouvellement découvert à Cayenne" (26 February 1751). La Condamine finally left Cayenne on 22 August 1744, going first to Paramaribo, capital of Surinam. He embarked from there on 3 September for Amsterdam, where he arrived on 30 November. On 23 February 1745 he was back in Paris, after an absence of ten years, bringing with him copious notes and a collection of more than two hundred natural history specimens and various works of art, which he soon gave to Buffon for the royal Cabinet d'Histoire Naturelle. La Condamine's health had not been weakened to any serious degree, but he did suffer from a deafness that later worsened and from a growing lack of sensitivity in the extremities, especially his feet, no doubt caused by the rigors of the Andean climate.

The scientific result of the expedition was clear: the earth is indeed a spheroid flattened at the poles, as Newton had maintained. Bouguer and La Condamine were unable, however, to agree on the joint publication of their works. Their long quarrel continued through a series of memoirs that were essentially mutual refutations of no scientific value; it ceased only with the death of Bouguer in 1758. (Godin died in 1760.) The last survivor of the expedition, La Condamine, who was a less gifted astronomer than Godin and a less reliable mathematician than Bouguer, often received the major part of the credit, probably because of his amiable nature and his talent as a writer.

La Condamine returned from Peru with a project for a universal measurement of length, the unit of which would be the length of a pendulum beating once a second at the equator. Although Huygens had already suggested the idea in his *Horologium oscillatorium* (1673), La Condamine explained it more clearly in a memoir presented to the Academy in November 1747, which was read at a public meeting the following April. His proposal was not acted upon, but the idea remained under consideration and was taken up again by Turgot and, before the Constituent Assembly in 1790, by Talleyrand.

In his youth La Condamine had contracted smallpox, which perhaps led him to take such a resolute stand in the debate over inoculation. His role in this matter was that of a popularizer, and he played it with considerable talent. The clarity and grace of his style served him well, as did his good nature. Even in his polemical writings, whether in prose or in verse (see, for example, his *Mémoire pour servir à l'histoire des révolutions du pain mollet* [1768]), his tone remained measured and courteous. His other works on inoculation include three memoirs read before the Academy (in 1754, 1758, and 1765), as well as his *Lettres . . . à M. le Dr Maty sur l'état présent de l'inoculation en France* (1764) and a two-volume *Histoire de l'inoculation de la petite vérole* (1773). By the end of his life, the "Don Quixote of inoculation," as Louis Petit de Bachaumont called him, had seen

LA CONDAMINE

the triumph of the ideas he had defended with such passion.

La Condamine's poems, although skillfully fashioned, do not merit special attention, nor does his *Lettre critique sur l'éducation,* published anonymously in 1751. The *Lettre* contains some valid ideas, such as the utility of modern foreign languages and of the exact sciences (in the front rank of which the author places geometry); but they are joined with reflections characteristic of the period and that prefigure Rousseau's *Émile* (1762). For example, he writes: "That the child becomes virtuous by becoming reasonable; that to the degree that his ideas develop, he learns that virtue is only the perfection of reason, while waiting to be shown that religion is the perfection of virtue."

A member of the Académie Royale des Sciences since 1730, as well as foreign member of the academies of London, Berlin, St. Petersburg, and Bologna, La Condamine was elected to the Académie Française on 29 November 1760. Piron greeted his election with a biting epigram:

La Condamine est aujourd'hui
reçu dans la troupe immortelle.
il est bien sourd, tant mieux pour lui:
mais non muet, tant pis pour elle.

The admission ceremony took place on 12 January 1761. The new member's speech was well regarded and the short reply by Buffon, who welcomed him in the name of the members, was magnificently eloquent.

To the end of his life La Condamine displayed the traits that had characterized him since his youth. He was inquisitive, restless, jealous of his reputation, gay, loyal, and at once malicious and credulous—in sum, very charming. A lively pastel portrait of him by Maurice Quentin de la Tour appeared in the Salon of 1753. During his trip to Italy, La Condamine obtained from Pope Benedict XIV a dispensation that allowed him to marry his niece, Charlotte Bouzier d'Estouilly, in August 1756. He was then fifty-five, and she twenty-five. Their marriage was a happy one. La Condamine had many friends, the closest of whom was certainly the impetuous and anxious Maupertuis, who bequeathed him all his papers. La Condamine died of the effects of a hazardous hernia operation, which, in a final bout of curiosity, he decided to undergo despite the risk involved.

BIBLIOGRAPHY

I. ORIGINAL WORKS. La Condamine's principal books are *Relation abrégée d'un voyage fait dans l'intérieur de l'Amérique méridionale* . . . (Paris, 1745)—the text differs from that printed in the *Mémoires de l'Académie des sciences* in having a preface and many variations in wording—also published as *Nouvelle édition augmentée de la Relation de l'émeute de Cuenca au Pérou et d'une lettre de M. Godin des Odonais contenant la relation du voyage de Madame Godin, son épouse* . . . (Maastricht, 1778); *Lettre à Mme *** sur l'émeute populaire excitée en la ville de Cuenca . . . contre les académiciens des sciences envoyés pour la mesure de la terre,* 2 pts. (Paris, 1745–1746), most of which is concerned with judicial proceedings following the death of the surgeon Seniergues; *Journal du voyage fait par ordre du roi à l'équateur, servant d'introduction historique à la mesure des trois premiers degrés du méridien* (Paris, 1751); *Lettre critique sur l'éducation* (Paris, 1751), published anonymously; and *Mesure des trois premiers degrés du méridien dans l'hémisphère austral* . . . (Paris, 1751).

Also see *Supplément au Journal historique ou Voyage à l'équateur et au livre de la mesure des trois premiers degrés du méridien, servant de réponse à quelques objections* (Paris, 1752), of which there was also a *Seconde partie* . . . (Paris, 1754); *Lettre de M.D.L.C.* [de La Condamine] *à M *** sur le sort des astronomes qui ont eu part aux dernières mesures de la terre, depuis 1735 . . . 20 octobre 1773* (n.p., n.d.); *Lettres de M. de La Condamine à M. le Dr Maty sur l'état présent de l'inoculation en France* . . . (Paris, 1764); and *Histoire de l'inoculation de la petite vérole* . . . (Amsterdam, 1773).

La Condamine published many articles in the "Mémoires" of the *Histoire de l'Académie royale des sciences.* Earlier ones include "Sur une nouvelle espèce de végétation métallique," 1731 (Paris, 1733), 466–482 and pls. 28–29; "Observations mathématiques et physiques faites dans un voyage de Levant . . . ," 1732 (Paris, 1735), 295–322 and pls. 16–18; "Description d'un instrument qui peut servir à déterminer, sur la surface de la terre, tous les points, d'un cercle parallèle à l'équateur," 1733 (Paris, 1735), 294–301 and pls. 22–23; "Nouvelle manière d'observer en mer la déclinaison de l'aiguille aimantée," *ibid.*, 446–456 and pl. 26; "Recherches sur le tour. Premier mémoire . . . . Description et usage d'une machine qui imite les mouvements du tour," 1734 (Paris, 1736), 216–258 and pls. 13–19; "Recherches sur le tour. Second mémoire. . . . Examen de la nature des courbes qui peuvent se tracer par les mouvements du tour," *ibid.*, 295–340 and pls. 20–25; and "Addition au memoire . . . 'Nouvelle manière d'observer en mer la déclinaison de l'aiguille aimantée' . . . ," *ibid.*, 597–599.

Further articles are "Manière de déterminer astronomiquement la différence en longitude de deux lieux eu éloignés l'un de l'autre," 1735 (Paris, 1738), 1–11; "De la mesure du pendule à Saint-Domingue," *ibid.*, 529–544 and pl. 17; "Observations des degrés de hauteur du thermomètre, faites en 1736 . . . ," 1736 (Paris, 1739), 500–502; "Sur l'arbre du quinquina," 1738 (Paris, 1740), 226–243 and pls. 5–6; "Relation abrégée d'un voyage fait dans l'intérieur de l'Amérique méridionale . . . ," 1745 (Paris, 1749), 391–492 and pl. 8; "Extrait des opérations trigonométriques, et des observations astronomiques, faites pour la mesure des degrés du méridien aux environs de l'équateur," 1746 (Paris, 1751), 618–688 and pls. 43–44; "Nouveau projet d'une mesure invariable propre à servir de mesure commune à toutes les nations," 1747 (Paris, 1752), 489–514; "Mémoire sur une résine élastique, nouvellement découverte à Cayenne par M. Fresneau; et sur l'usage de divers sucs laiteux d'arbres de la Guiane ou France équinoctiale," 1751 (Paris, 1755), 319–333 and pls. 18–20; "Mémoire sur l'inoculation de la petite vérole," 1754 (Paris, 1759), 615–670; "Extrait d'un journal de voyage en Italie," 1757 (Paris, 1762), 336–410; "Second mémoire sur l'inoculation de la petite vérole . . . ," 1758 (Paris, 1763), 439–482; and "Suite de l'histoire de l'inoculation de la petite vérole. . . . Troisième mémoire," 1765 (Paris, 1768), 505–532.

La Condamine's papers and correspondence, formerly preserved at the chateau of Estouilly, near Ham (Somme), have been scattered. His dossier at the archives of the Academy of Sciences contains letters, notes concerning the mission to Peru (particularly on his quarrels with Bouguer), and statements favoring inoculation against smallpox. Many of the items are addressed to Grandjean de Fouchy, the perpetual secretary of the Academy since 1743. Other documents are at the Bibliothèque Nationale, MS department: Fr 11333; Fr 12222; Fr 22133; Fr 22135; NA Fr 3543, fol. 231; NA Fr 6197, fols. 9, 14, 22; NA Fr 3531, fol. 174; NA Fr 21015.

II. SECONDARY LITERATURE. See "Bicentenaire de la découverte du caoutchouc par La Condamine, 1736–1936," *Revue générale du caoutchouc*, **13**, no. 125 (Oct. 1936); Pierre Bouguer, *La figure de la terre, déterminée par les observations des Messieurs Bouguer & de La Condamine* . . . (Paris, 1749); M. J. Condorcet, "Éloge de M. de La Condamine," in *Histoire de l'Académie royale des sciences*, 1774 (Paris, 1778), 85–121; *Discours prononcés dans l'Académie françoise, le lundi 11 juillet 1774 à la réception de M. l'abbé Delille* (Paris, 1774), a eulogy of La Condamine; Victor Wolfgang von Hagen, *South America Called Them* . . . (New York, 1945), 3–85, translated as *Le continent vert des naturalistes* (Paris, 1948); Abbé Achille Le Sueur, *La Condamine, d'après ses papiers inédits* . . . (Paris, 1911); and M. C. Wolf. "Recherches historiques sur les étalons de l'observatoire," in *Annales de chimie et de physique*, 5th ser., **25** (1882), 5–112.

YVES LAISSUS

LAPLACE, PIERRE-SIMON, MARQUIS DE (*b.* Beaumont-en-Auge, Normandy, France, 23 March 1749; *d.* Paris, France, 5 March 1827), *celestial mechanics, probability, applied mathematics, physics.*

Laplace was among the most influential scientists in all history. His career was important for his technical contributions to exact science, for the philosophical point of view he developed in the presentation of his work, and for the part he took in forming the modern scientific disciplines. The main institutions in which he participated were the Académie Royale des Sciences, until its suppression in the Revolution, and then its replacement, the scientific division of the Institut de France, together with two other Republican foundations, the École Polytechnique and the Bureau des Longitudes. It will be convenient to consider the scientific life that he led therein as having transpired in four stages, the first two in the context of the old regime and the latter two in that of the French Revolution, the Napoleonic regime, and the Restoration.

The boundaries must not be taken more categorically than biography allows, but in the first stage, 1768–1778, we may see Laplace rising on the horizon, composing memoirs on problems of the integral calculus, mathematical astronomy, cosmology, theory of games of chance, and causality, pretty much in that order. During this formative period, he established his style, reputation, philosophical position, certain mathematical techniques, and a program of research in two areas, probability and celestial mechanics, in which he worked mathematically for the rest of his life.

In the second stage, 1778–1789, he moved into the ascendant, reaching in both those areas many of the major results for which he is famous and which he later incorporated into the great treatises *Mécanique céleste* (1799–1825) and *Théorie analytique des probabilités* (1812). They were informed in large part by the mathematical techniques that he introduced and developed, then or earlier, most notably generating functions, the transform since called by his name, the expansion also named for him in the theory of determinants, the variation of constants to achieve approximate solutions in the integration of astronomical expressions, and the generalized gravitational function that, through the intermediary of Poisson, later became the potential function of nineteenth-century electricity and magnetism. It was also during this period that Laplace entered on the third area of his mature interests, physics, in his collaboration with Lavoisier on the theory of heat, and that he became, partly in consequence of this association, one of the inner circle of influential members of the scientific community. In the 1780's he began serving on commissions important to the government and affecting the lives of others.

In the third stage, 1789–1805, the Revolutionary period and especially that of the Directory brought him to his zenith. The early 1790's saw the completion of the great series of memoirs on planetary astronomy and involved him centrally in the preparation of the metric system. More important, in the decade from 1795 to 1805 his influence was paramount for the exact sciences in the newly founded Institut de France; and his was a powerful position in the counsels of the École Polytechnique, which was training the first generation of mathematical physicists. The educational mission attributed to all science in that period of intense civic consciousness changed the mode of scientific publication from academic memoir to general treatise. The first four volumes of *Mécanique céleste* (Laplace himself coined the term), generalizing the laws of mechanics for their application to the motions and figures of the heavenly bodies, appeared from 1799 through 1805. The last parts of the fourth volume and the fifth volume, really a separate work that appeared in installments from 1823 to 1825, contain important material (on physics) not already included in the sequence of Laplace's original memoirs published previously by the old Academy.

Laplace accompanied both *Mécanique céleste* and *Théorie analytique des probabilités* by verbal paraphrases addressed to the intelligent public in the French tradition of *haute vulgarisation.* The *Exposition du système du monde* preceded the *Mécanique céleste* and appeared in 1796. The *Essai philosophique sur les probabilités*, published in 1814 as an introduction to the second edition of *Théorie analytique* and printed separately earlier in the same year, originated in a course of lectures at the École Normale in 1795.

The work of the fourth stage, occupying the period from 1805 until 1827, exhibits elements of culmination and of decline. It was then that the mature—perhaps the aging—Laplace, in company with Berthollet, formed a school, surrounding himself with disciples in the informal Société d'Arcueil. But the science that he set out to shape was not astronomy. The center of their interest, following Volume IV of *Mécanique céleste*, was in physics—capillary action, the theory of heat, corpuscular optics, and the speed of sound. The Laplacian school of physics has had a bad scholarly press since its identity was established, excessively so perhaps. But whatever else may be said about it, there can be no doubt about the encouragement that it gave to the mathematization of the science.

Beginning in 1810, Laplace turned his attention to probability again, moving back by way of error theory into the subject as a whole. Mathematically speaking, the *Théorie analytique des probabilités* (1812) may be said to belong to the previous phase of drawing together and generalizing the researches on special topics of his younger years. There were important novelties in the application, however, notably in the treatment of least squares, in the extension of probability in later editions to analysis of the credibility of witnesses and the procedures of judicial panels and electoral bodies, and in the increasing sophistication of the statistical treatment of geodesic and meteorological data.

A further preliminary word is in order about the circumstances of this article, for it is the work of many hands. Two successive agreements with individual contributors broke down, failure to honor the first having been responsible for the absence of an article from Volume VIII and similar failure in the case of the second for much delay in publication of the present volume. In what was becoming an extremity, there seemed no alternative but for the editor to enlist the collaboration of colleagues qualified in the several branches of science and to seek to combine their contributions into a coherent whole. Accordingly, Dr. Brian G. Marsden undertook to give an account of the *Mécanique céleste*, Dr. C. A. Whitney of the *Exposition du système du monde*, Dr. Ivo Schneider of the *Théorie analytique des probabilités*, and Dr. Robert Fox of Laplacian physics. All four have kindly and generously done so.

When I began writing the background, however, the account assumed a scale different from what we had imagined. At that stage, I also benefited from the investigations of members of my seminar at Princeton University, which was devoted in 1976–1977 to the career of Laplace. Papers by Messrs. George Anastaplo, Robert Bernstein, Chikara Sazaki, and Sherwin Singer were especially illuminating. Further, I have had invaluable guidance and criticism from two other colleagues, Dr. Stephen M. Stigler on the earliest memoirs and on statistical aspects in general, and Dr. Ivor Grattan-Guinness on mathematics, bibliography, style, organization—indeed, on everything. In the end, I have written Parts I, II, and III, together with Sections 25, 26, and 28 in Part IV. In composing the accounts of the respective treatises I have profited from the essays of Drs. Marsden, Whitney, and Schneider. Responsibility for errors or other shortcomings is, of course, not to be laid at their door. I have also compiled the Bibliography, with the able assistance of Mr. Joel Honig, whose editorial collaboration has been invaluable throughout. Dr. Robert Fox has written Sections 22, 23, 24, and 27 in Part IV, dealing with the Laplacian school of physics. Those four parts correspond to the stages of Laplace's career outlined above. In addition, Dr. Grattan-Guinness has written Part V (Section 29), on the history of the Laplace transform.

Before proceeding further, the reader would do well to turn to the Bibliography and familiarize himself with its several categories. The citations throughout have reference to titles that will be found in appropriate sections of the Bibliography as indicated; and its central section [I] constitutes a dual chronology of Laplace's work, by order of composition and by order of publication. References to the former sequence (Section I, Part 1) are given in italic: (*23*). References to the latter sequence (Section I, Part 2) are given by bracketing the italicized date. The several memoirs published in any given year are distinguished by letters of the alphabet, thus: [*1777a*]. Inevitably, the organization of the Bibliography reflects the complexity of that work, but I venture to hope that it may be a first approximation to that which does not yet exist, a complete bibliography of Laplace.

CHARLES COULSTON GILLISPIE

PART I: EARLY CAREER, 1768–1778

1. Youth and Education
2. Election to the Academy
3. Finite Differences, Recurrent Series, and Theory of Chance
4. Probability of Events and of Their Causes
5. Universal Gravitation
6. Cometary Distribution
7. Partial Differential Equations, Determinants, and Variation of Constants
8. The Figure of the Earth and the Motion of the Seas

PART II: LAPLACE IN HIS PRIME, 1778–1789

9. Influence and Reputation
10. Variation of Constants; Differential Operators
11. Probability Matured
12. Generating Functions and Definite Integrals
13. Population
14. Determination of the Orbits of Comets
15. Lavoisier and Laplace: Chemical Physics of Heat
16. Attraction of Spheroids
17. Secular Inequalities: Jupiter and Saturn; the Moons of Jupiter; Lunar Theory

PART III: SYNTHESIS AND SCIENTIFIC STATESMANSHIP

18. The Revolution and the Metric System
19. Scientific Work in the Early Revolution
20. *Exposition du système du monde*
21. *Mécanique céleste*

PART IV: LAPLACIAN PHYSICS AND PROBABILITY

22. The Velocity of Sound
23. Short-range Forces (Robert Fox)
24. The Laplacian School
25. Theory of Error
26. Probability: *Théorie analytique* and *Essai philosophique*
27. Loss of Influence (Robert Fox)
28. The Last Analysis

PART V: THE LAPLACE TRANSFORM

29. Laplace's Integral Solutions to Partial Differential Equations (Ivor Grattan-Guinness)

BIBLIOGRAPHY

PART I: EARLY CAREER, 1768–1778

1. Youth and Education. Genealogical records of the Laplace family in lower Normandy go back to the middle of the seventeenth century (J: Boncompagni, Simon). Laplace's father, Pierre, was a syndic of the parish, probably in the cider business and certainly in comfortable circumstances. The family of his mother, Marie-Anne Sochon, were well-to-do farmers of Tourgéville. He had one elder sister, also called Marie-Anne, born in 1745. There is no record of intellectual distinction in the family beyond what was to be expected of the cultivated provincial bourgeoisie and the minor gentry. One paternal uncle, Louis, an abbé although not ordained, is said to have been a mathematician and was probably a teacher at the *collège* (secondary school) kept at Beaumont-en-Auge by the Benedictines. He died in 1759, when his nephew was ten. Laplace was enrolled there as a day student from the age of seven to sixteen. Pupils usually proceeded to the church or the army; Laplace's father intended him for an ecclesiastical vocation.

In 1766 he went up to the University of Caen and matriculated in the Faculty of Arts, still formally a cleric. During his two years there he must have discovered his mathematical gifts, for instead of continuing in the Faculty of Theology, he departed for Paris in 1768. Apparently, he never took his M.A., although he may briefly have been a tutor in the family of the marquis d'Héricy and may also have taught at his former *collège*. The members of the faculty at Caen who opened his eyes to mathematics and their own to his talent were Christophe Gadbled and Pierre Le Canu. All that we know about them is that they were points of light in the philosophic and scientific microcosm of Caen, professors with the sense to recognize and encourage a gifted pupil.

On Laplace's departure for Paris at the age of nineteen, Le Canu gave him a letter of recommendation to d'Alembert, who immediately set him a problem and told him to come back in a week. Tradition has it that Laplace solved it overnight. Thereupon d'Alembert proposed another, knottier puzzle—which Laplace resolved just as quickly (J: Bigourdan, 381). The story may be apocryphal, but there is no doubt that d'Alembert was somehow impressed and took Laplace up, as he had other young men in the evening of his own career, although none of comparable merit mathematically. The next question was a livelihood, and d'Alembert himself answered to that necessity, securing his new protégé the appointment of professor of mathematics at the École Militaire. Imparting geometry, trigonometry, elementary analysis, and statics to adolescent cadets of good family, average attainment, and no commitment to the subjects afforded little stimulus, but the post did permit Laplace to stay in Paris.

2. Election to the Academy. It was expected of Laplace that he should concentrate his energies on making a mathematical reputation, and he won election to the Academy of Sciences on 31 March 1773, after five years in Paris. Condorcet had become acting permanent secretary earlier that month. Never, he wrote in the preface to the volume in which the first memoirs that Laplace published in Paris were printed (*SE*, **6** [1774], "Histoire," p. 19), had the Academy received from so young a candidate in so short a time as many important papers on such varied and difficult topics. On two previous occasions his candidacy had been passed over: in 1771, in favor of Alexandre Vandermonde, fourteen years his senior, and the following year in favor of Jacques-Antoine-Joseph Cousin, ten years older and a professor at the Collège Royal de France. Evidently, Laplace felt slighted, despite his youth. On 1 January 1773 d'Alembert wrote to Lagrange asking whether there was a possibility of obtaining a place in the Prussian Academy and a post at Berlin, since the Paris Academy had just preferred a person of markedly inferior ability (*Oeuvres de Lagrange*, XIII [1882], 254–256). The approach lapsed three months later when Laplace was chosen an adjunct member in Paris. Bigourdan's statement that he was admitted directly to the second rank of *associé* is incorrect (J: 384).

If the records are complete, the papers to which Condorcet was alluding numbered thirteen (*1–13*), presented in just under three years, beginning on 28 March 1770. The topics were extreme-value problems; adaptation of the integral calculus to the solution of difference equations; expansion of difference equations in a single variable in recurrent series and in more than one variable in recurro-recurrent series; application of these techniques to theory of games of chance; singular solutions for differential equations; and problems of mathematical astronomy, notably the variation of the inclination of the ecliptic and of planetary orbits, the lunar orbit, perturbations produced in the motion of the planets by the action of their satellites, and "the Newtonian theory of the motion of the planets" *(9)*. Of these papers, four were published *(1, 5, 8, 13)*. Laplace translated the first two into Latin and placed them in the *Nova acta eruditorum*, where the second ([*1771a*]) was printed before the first ([*1774a*]).

Perhaps it will be appropriate to call this pair of memoirs Laplace's juvenilia. The earlier, the first paper that he read before the Academy, was presented on 28 March 1770, five days after his twenty-first birthday, and was entitled "Recherches sur les maxima et minima des lignes courbes" (*1*). After a review of extreme-value problems, he proposed several improvements in the development that Lagrange had given to Euler's *Methodus inveniendi lineas curvas maximi minimive proprietate gaudentes* . . . (1744). One modification concerned Lagrange's finding, in a paper published in the *Mélanges . . . de la Société royale de Turin* (**2** [1760–1761]; *Oeuvres de Lagrange*, **1**, 335–362), that there was no need to follow Euler in assuming a constant difference. If the assumption was justified, the number of equations might be reduced by at least one, and otherwise the problem was unsolvable. Laplace found the same result by a method that his commissioners called "less direct, less rigorous in appearance, but simpler and fairly elegant" (*1*). In cases where a difference is not constant, the difficulty was shown to arise from a faulty statement of the problem. A variable was concealed that should have appeared in the function, and when it was identified the equations became determinate. If the solutions yield maximum values, the equations involve double curvature. Laplace further gave a general analytic criterion for distinguishing a true maximum or minimum from instances in which two successive values happen to be equal, and he appeared to have regarded this as his chief contribution.

It was, however, with the other Leipzig paper, "Disquisitiones de calculo integrale" ([*1771a*]), that Laplace made his debut in print. The subject is a particular solution for one class of differential equations. The method that he developed subsequently led to enunciation of a theorem, the statement of which he annexed without proof to his first memoir on probability ([*1774c*]; *OC*, VIII, 62–63), although it has nothing to do with that subject (see Section 7, Equation 23 below). Reworking this material in later years ([*1777a*]), Laplace repudiated this earliest publication, or very nearly so, apologizing for grave faults that he blamed on the printer ([*1774c*]; *OC*, VIII, 83). That was the only reference that he ever made to either of these youthful ventures into Latin.

3. Finite Differences, Recurrent Series, and Theory of Chance. Among his early interests, it turned out to be a memoir on the solution of difference equations that marks the beginning of one of the main sequences of his lifework. We may reasonably surmise that the applicability of such equations to problems of games of chance, and not any *a priori* penchant for that subject matter, was mainly responsible for the appearance given by the published record that probability attracted him more strongly than did celestial mechanics in this opening phase of his career. The appearance is misleading, or at least ironic. For chance was never mentioned in the titles of any of his investigations until February 1772, when the applicability had made itself evident (*10*). In 1771, at the outset of "Recherches sur le calcul intégral aux différences infiniment petites, et aux différences finies," he observed that the equations that he was studying turn up more frequently than any other type in applications of the calculus to nature. A general method for integrating them would be correspondingly advantageous to mechanics, and especially to "physical astronomy" ([*1771b*], p. 273). That science looms largest in the unpublished record (*1–13*); and judging from a report in the archives of the Academy, a paper (*9*) of 27 November 1771 may even have been the germ of *Mécanique céleste*.

Earlier in the same year Laplace had published "Recherches sur le calcul intégral aux différences infiniment petites, et aux différences finies," and since he was still only knocking at the door of the Academy, he sent it to the Royal Society of Turin for its *Mélanges*. It was almost surely the expansion of a paper on difference equations alone that he had read on 18 July 1770 in his second appearance before the Academy (*2*). He now reserved that topic for the second half of the memoir, where he proposed to adapt to the solution—or as he said in the looser terminology of the time, the "integration"—of difference equations a method that he was developing for infinitesimal expressions in the earlier version. He began by confirming in his own manner a theorem that Lagrange had recently proved concerning integration of equations of the following form ([*1771b*], p. 173):

$$X = y + H\frac{dy}{dx} + H'\frac{d^2y}{dx^2} + H''\frac{d^3y}{dx^3} + \cdots + H^{n-1}\frac{d(d^ny)}{dx^n}, \quad (1)$$

where X, H, H', H'', $\cdots$, are any function of x. Lagrange had shown that such equations can always be integrated if integration is possible in the homogeneous case when $X = 0$. His proof was of a type classical in analysis. It involved introducing a new independent variable, z; multiplying both sides of the equation by $z\,dt$; supposing integration of the resulting adjoint equation accomplished; and examining the steps needed to reduce the order one degree at a time until a solvable form should be reached. The proof worked for solving differential equations; but the procedure presupposed the validity of infinitesimal methods in analysis, and the operations were not applicable to the solution of difference equations.

Laplace's approach appears to be more cumbersome and turns out to be more general. Instead of introducing a multiplying factor and supposing the subsequent integration accomplished, which step restricted the method to the infinitesimal calculus, Laplace employed integrating factors obtained by substituting

$$\omega\frac{dy}{dx} + y = T, \quad (2)$$

where T and ω are functions of x. He then differentiated that expression successively n times. Of the resulting equations, he multiplied the first by ω', the second by ω'', the third by ω''', and so on; added the products to (1); and grouped the terms of the enormous resulting expression by orders of y. Manipulation then allowed for the determination of ω', ω'', ω''', $\cdots$, in terms of ω and H', H'', H''', $\cdots$. Thus, he could write equations equivalent to (1) in either finite or infinitesimal differences and evaluate them generally. Since his purpose was to extend a method from infinitesimal to finite analysis, his operations all conform to the rules of algebra, except that at the point where they involved differentiation, he justified the analogous step for finite differences.

The episode is largely typical of the relation of Laplace's point of departure to the work of elders and near contemporaries. There is the not quite ritual obeisance to a principle or practice, in this case the formulation of problems in terms of differential equations in general, attributed to d'Alembert, the patron. There is the tactful nod to a result found quite differently by Condorcet, the well-placed official. There is the pioneer analytical breakthrough achieved by Euler, although in restricted form. There is the formal mathematical theorem stated by Lagrange, emphasizing analyticity. There is, finally, the adaptation imagined and executed by Laplace, his motivation being the widest applicability to problems in the real world.

Although the memoir had opened with mention of mechanics and astronomy, Laplace introduced its raison d'être, the solution of difference equations, with a reminder that their calculus was the foundation of the entire theory of series ([*1771b*], p. 299). Coherently enough, therefore, he continued the discussion with a determination of the general term of series of the important class (*op. cit.*, p. 330)

$$y^x = A\phi^x y^{x-1} + {}'A\phi^x\phi^{x-1}y^{x-2} + {}''A\phi^x\phi^{x-1}\phi^{x-2}y^{x-3} + \cdots \quad (3)$$

wherein ϕ is a function of x. (In his early notation, the superscripts are often, as here, indices, not exponents.) In the simplest case, in which $\phi = 1$, Equation (3) reduces to the recurrent form,

$$y^x = Ay^{x-1} + {}'Ay^{x-2} + {}''Ay^{x-3} + \cdots. \quad (4)$$

The memoir ends with an application of the calculus of finite differences to a solution of this equation, and Laplace gives a method for determining the constants.

This finding that equations of the form (3) are always integrable became the starting point for the next memoir, dealing with what Laplace called "recurro-recurrent" series and their application to the theory of chance. Still not a member of the Academy, he submitted it for their judgment on 5 February 1772 and placed it in the *Savants étrangers* series ([*1774b*]). Recurrent series of the familiar form (3) were restricted to a single variable index, the definition being that "every term is equal to any number of preceding terms, each multiplied by a function of x taken at will" ([*1774b*]; *OC*, VIII, 5). While investigating certain problems in the theory of chance, so Laplace said a little later ([*1776a*, 1°]; *OC*, VIII, 71), he came upon equations in finite differences of another, novel type. They were the analogues in finite analysis of partial differential equations and gave rise to a complex set of series, the general term of which has two or more variable indices. In such series as (4), if ϕ is a function of x and n rather than of x alone, and if the integers 1, 2, 3, $\cdots$, are substituted for x and n, then for each value of n, a series results in which ${}^ny^x$ designates the term corresponding to the number x and n. The definition of a recurro-recurrent series is that ${}^ny^x$ is equal to any number of preceding terms, taken in rank or in order, in any number of such series, each multiplied by a function of x. Here is the example that Laplace displayed:

$$\begin{aligned} &{}^1y^x + A'\,y^{x-1} + B'\,y^{x-2} + \cdots + N = 0 \\ &{}^2y^x + A''{}^2y^{x-1} + B''{}^2y^{x-2} + \cdots + N'' \\ &= H''\,{}^1y^x + M''\,{}^1y^{x-1} + P''\,{}^1y^{x-2} + \cdots \\ &\cdots\cdots\cdots\cdots\cdots\cdots\cdots\cdots\cdots\cdots \\ &{}^ny^x + A^n\,{}^ny^{x-1} + B^n\,{}^ny^{x-2} + \cdots + N^n \\ &= H^n\,{}^{n-1}y^x + M^n\,{}^{n-1}y^{x-1} + \cdots. \end{aligned} \quad (5)$$

The value of ${}^ny^x$ must be determined, when A^n, B^n, $\cdots$, N^n, H^n, $\cdots$ are any functions of n; when also A'', B'', $\cdots$ $'''A$, $'''B$, $\cdots$ are what those functions become on substituting 1, 2, 3, $\cdots$ for n; and when, finally, A, B, $\cdots$, N are any constants.

The solution consists in showing that Equation (5) can always be transformed as follows (*OC*, VIII, 7):

$${}^ny^x = a^n\,{}^ny^{x-1} + b^n\,{}^ny^{x-2} + c^n\,{}^ny^{x-3} + \cdots + u^n, \quad (6)$$

where a^n, b^n, $\cdots$, u^n are functions of n and of constants that can be determined. This equation (6), in turn, is an expression for a recurrent series precisely of the type (3) that he had shown to be integrable in [*1771b*], his Turin memoir. A further problem of a third-degree equation in finite differences is then shown to be reducible successively to the forms (5) and (6), and Laplace went on to propose a general procedure for reducing an equation of any degree r to a lower degree, the requirement being that by the assumption made for the value of n, the equation of degree r become one of degree $r - 1$.

A passage included many years later in the *Essai philosophique sur les probabilités* (*OC*, VII, xxvi–xxvii) makes clearer to the noninitiate how he was visualizing these series. A recurrent series is the expression of a difference equation with a single variable index. Its degree is the difference in rank between its two extreme terms. The terms may be determined by means of the equation, provided the number of known terms equals its degree. These terms are in effect the arbitrary constants of the expression for the general term or (which comes to the same thing) of the solution of the difference equation. The reader is next to imagine a second series of terms arranged horizontally above the terms of the first series, a third series above the second, and so on to infinity. It is supposed that there is a general equation between the terms that are consecutive both horizontally and vertically and the numbers that indicate their rank in both directions. This will be an equation in finite partial differences, or recurro-recurrent. The reader is finally to imagine that on top of the plane containing this pattern of series there is another containing a similar pattern, and so on to infinity, and that a general equation relates the terms that are consecutive in the three dimensions with the numbers indicating their rank. That would be an equation in finite partial differences with three indices. Generally, and independently of the spatial model, such equations may govern a system of magnitudes with any number of indices. Some eight years later, Laplace replaced the use of recurro-recurrent series with the more efficient tools of generating functions ([*1782a*], see Section 12) for solving problems in finite differences, which he encountered mainly in the calculus of probability.

It is important biographically to notice his passing remark that investigations in the theory of chance had led him to the formulation of recurro-recurrent series. The latter part of the memoir illustrates how they might be applied to the solution of several problems concerning games of chance. In the first such example, two contestants, A and B, play a game in which the loser at each turn forfeits a crown to his opponent. Their relative skills are as a to b. At the outset A has m crowns, and B has n. What is the probability that the game will not end with x or fewer turns? Laplace found it by substituting values given by the conditions of the problem in a series of equations of the form (5), first for the case in which $a = b$, $m = n$, and n is even, and then for all possible suppositions about the parameters. This problem, like many others that Laplace adduced, appears in De Moivre, who seems to have furnished his first reading in the subject, but whose solutions were less direct (*Doctrine of Chances*, 3rd ed. [London, 1756], Problem LVIII, p. 191). A further example was suggested to Laplace by a bet made on a lottery at the École Militaire. What is the probability that all the numbers 1, 2, 3, · · ·, n will be taken after x draws? That, too, he found by formulating the problem in a recurro-recurrent series in two variable indices, observing that the approach could clearly have wide applicability in the theory of chance, where the most difficult problems often concern the duration of events.

Laplace chose precisely this juncture for defining probability. The statement occurs immediately after his development of the method of recurro-recurrent series and just before its application to the foregoing examples ([*1774b*]; *OC*, VIII, 10–11):

> The probability of an event is equal to the sum of each favorable case multiplied by its probability, divided by the sum of the products of each possible case multiplied by its probability, and if each case is equally probable, the probability of the event is equal to the number of favorable cases divided by the number of all possible cases.

The definition is noteworthy not for its content, which was standard, but for its location in the development of his work and for its phrasing. The wording should serve to temper the criticism often made of Laplace, particularly in respect to inverse probability (which, to be sure, he had not yet started), that he gratuitously assumed equal *a priori* probabilities or possibilities. It is true that he often did—although not, as will appear, when he had anything to go on. It is also interesting that in the passage immediately preceding the definition, Laplace should have written "duration of events" rather than "duration of play." For this was the first remark that he ever printed about the subject as a whole.

4. Probability of Events and of Their Causes. Historically and biographically, the instinctive choice of words is often more indicative than the deliberate. There is much anachronism in the literature, and it may be well, therefore, to take this, the juncture at which Laplace was entering upon one of the central preoccupations of his life, as the occasion to venture an observation about the early history of probability. For more substance has sometimes been attributed to it than the actual content warrants. It is true that adepts of the subject were much given to celebrating its applicability, but prior to Laplace what they were praising was a prospect rather than actual accomplishment—except in the theory of games of chance. Even there, most of the experiments were thought experiments. Jakob Bernoulli's *Ars conjectandi* (1713) is justly famous mathematically, although it is seldom mentioned that Part IV, headed "Usum et applicationem praecedentis doctrinae in civilibus, moralibus, & oeconomicis," contained simply the law of large numbers and otherwise remained uncompleted. As for the Dutch and English insurance industry in the eighteenth century, and the sale of annuities practiced by governing agencies, the tables on which actuarial transactions depended were numerically insecure. Risks were estimated empirically rather than calculated analytically.

The statement that games of chance provided the principal subject matter in which theorems could be demonstrated and problems solved mathematically will be confirmed by close attention to contemporary usage, which refers to *théorie des hasards*, or "theory of chance." The word "probability" was not used to designate the subject. That word appears in two ways, one more restricted and the other vaguer than in the post-Laplacian science. In the mathematical "theory of chance," probability was a quantity, its basic quantity, that which Laplace defines above. *Calcul des probabilités* refers to calculation of its amount for certain outcomes in given situations. The phrase *théorie des probabilités* rarely, if ever, occurs. The word "probability" had its second and larger sense, one that would have befitted a theory if any had existed, rather in the philosophic tradition started by Pascal's wager on the existence of God (for which see Hacking [L: 1975]). The changes rung on that idea belong to theology, epistemology, and moral philosophy, some pertaining to what is now called decision-making and others to political economy. Such is the discourse in Diderot's article "Probabilité" in the *Encyclopédie*. It was largely skepticism about the mathematical prospects for that sort of thing that inspired d'Alembert's overly deprecated hostility to the subject, and optimism about it that inspired Condorcet's overly celebrated enthusiasm. Laplace himself was clear about the difference, although in other terms. Indeed, he insisted upon it in the distinction between mathematical and moral expectation in one of the pair of papers to be discussed next, which between them did begin to join mathematical theory of games with philosophic probability and scientific methodology.

The bibliographical circumstances need to be discussed before the significance of these two papers can be fully appreciated. Both were composed before Laplace became a member of the Academy and were printed in successive volumes of the *Savants étrangers* series. The more famous was entitled "Mémoire sur la probabilité des causes par les événements" ([*1774c*]). The second, and lengthier, was delayed for two years and was then combined with an astronomical memoir under the title, "Recherches, 1°, sur l'intégration des équations différentielles aux différences finies, et sur leur usage dans la théorie des hasards. 2°, sur le principe de la gravitation universelle, et sur les inégalités séculaires des planètes qui en dépendent" ([*1776a*], 1° and 2°). As we shall see, this early coupling of probability with astronomy was no mere marriage of convenience; Laplace spent his entire professional life faithful to the pattern that it started. Before discussing that, however, we shall need to consider the way in which the aleatory part of the dual memoir completed his earlier application of recurro-recurrent series to solving problems in the theory of chance and at the same time complemented his new departure into the determination of cause. He submitted this resumption of his work on difference equations to the Academy on 10 March 1773 (*13*)—a marginal note dating it 10 February is in error.

The preamble of the paper on cause declares, and cross-references in both essays confirm, that they were conceived as companion pieces on the subject that Laplace was now beginning to call probability, the one breaking new ground in what was later called its inverse aspect, the other extending and systematizing a direct approach. It was, indeed, in these two writings that probability began to be broadened from the mathematics of actual games and hypothetical urns into the basis for statistical inference, philosophic causality, estimation of scientific error, and quantification of the credibility of evidence, to use terms not then coined. In preferring the word "probability" to suggest the wider scope that he was giving the subject itself as a branch of mathematics, Laplace may well have been following the precept of Condorcet, newly the acting permanent secretary of the Academy, who, in prefatory remarks to the volume in which the "Causes" paper appeared, praised it for its approach to predicting the probability of future events. "It is obvious," he wrote, "that this question comprises all the applications that can be made of the doctrine of chance to the uses of ordinary life, and of that whole science; it is the only useful part, the only one worthy of the serious attention of philosophers" (*SE*, **6** [1774], "Histoire," p. 18).

The memoir on cause opens with a preamble most of which might more appropriately have belonged to the concurrent piece on the solution of difference equations. Laplace referred readers to the latter memoir after reviewing what De Moivre and Lagrange had contributed to these problems and the way in which they had then involved him in the theory of chance. The present memoir had a different object, the determination of the probability of cause, given knowledge of events. Uncertainty, we are told, concerns both events and their causes (notice that at the outset Laplace took probability to be an instrument for repairing defects in knowledge). When it is given that an urn contains a set number of black and white slips in some definite ratio, and the probability is required of drawing a white one, then we know the cause and are uncertain about the event. But if the ratio is not given, and after a white slip is drawn the probability is required that it be as p is to q, then we know the event and are uncertain about the cause. All problems of theory of chance could be reduced to one of these two classes, and Laplace here proceeded to investigate problems of the second type. He began on the basis of a theorem that, like the definition of probability in the previous memoir, he enunciated verbally:

> If an event can be produced by a number n of different causes, the probabilities of the existence of these causes, reckoned from the event, are to each other as the probabilities of the event, reckoned from the causes; and the probability of each cause is equal to the probability of the event, reckoned from that cause, divided by the sum of all the probabilities of the event, reckoned from each of the causes ([*1774c*]; *OC*, VIII, 29).

In substance, this theorem is the same as that published in 1763, eleven years previously, by Thomas Bayes. Not only is it now named "Bayes's theorem" or "Bayes's rule," but the entire approach to probability and statistics depending on it is generally called Bayesian, at least in the literature influenced by the British tradition in analysis and philosophy of science since the early nineteenth century. That usage derives from the vindication by Augustus De Morgan and Boole of their obscure countryman's priority. Laplace did not mention Bayes in this memoir. Later he did refer to him in one sentence in the *Essai philosophique sur les probabilités* (*OC*, VII, cxlviii). It seems likely that in 1774 he had not read Bayes's paper in the *Philosophical Transactions of the Royal Society of London*; at this period leading Continental mathematicians seldom read or referred to their British contemporaries, and the statement and context of the relation in Bayes's piece are very different. On the other hand, he may have heard of it. Richard Price was known on the Continent, and especially among political theorists, including Condorcet. Moreover, in introducing the analysis of cause Laplace did not claim that it was an altogether new subject. He said it was novel "in many respects" and chimed in with Condorcet's prefatory remarks to the effect that the approach "the more merits being developed in that it is mainly from that point of view that the science of chance can be useful in civil life." It must also be said (once equity is served) that in the sequel the analysis of inverse probability derived from Laplace's memoir and further work and that Bayes remains one of those pioneers remembered only after the subject they intrinsically might have started had long been flourishing, in spite of their having been little noticed and thanks to work of others that did have consequence.

However that may be, Laplace proceeded from the statement of the theorem to an example. From an urn containing an infinite number of white and black slips in unknown ratio, $p + q$ slips are drawn, of which p are white and q black. What is the probability that the next slip will be white? The above theorem gives the following formula for the probability that x is the true ratio (*OC*, VIII, 30):

$$\frac{x^p\,(1-x)^q\,dx}{\int x^p\,(1-x)^q\,dx}, \tag{7}$$

where the integral is taken from $x = 0$ to $x = 1$; and Laplace calculated that the required probability of drawing a white slip on the next try is

$$\frac{p+1}{p+q+2}. \tag{8}$$

A second example from the same urn leads to the application that was the point of this analysis. What is the probability after pulling p white and q black slips in $p + q$ draws of then taking m white and n black slips in the next $m + n$ draws? For that probability, Laplace obtained the expression

$$\frac{\int x^{p+m}(1-x)^{q+n}\,dx}{\int x^{p}(1-x)^{q}\,dx}, \tag{9}$$

where the limits are again 0 to 1, and went on to ask the more significant question: how large would $(p + q)$ need to be, and how small would $(m + n)$ need to remain, in order to calculate the probability of the ratio of m to n on the basis of p and q? Clearly, the solution of that problem would give a basis for calculating the probability of a future event from past experience or for statistical inference (although Laplace never used that phrase). He did not try to solve the problem in general, pleading the lengthiness of the calculation. Instead, he proceeded to the demonstration of a limit theorem, which is nothing other than Bernoulli's law of large numbers, although Laplace did not call it that. In effect, he showed, in what he called a "curious proof," that the numbers $p + q$ can be supposed so large as to bring as close as you please to certainty the probability that the ratio of white slips to the total lies between $\frac{p}{p+q} + \omega$ and $\frac{p}{p+q} - \omega$, ω being less than any given magnitude.

Laplace had two changes to ring on one of the classic problems, the division of stakes in an interrupted game. First, he referred the reader to the companion memoir for deduction by the method of recurro-recurrent series of the canonical solution for the standard case, in which the relative skills of two players are given; and he also promised a general solution in the case of three or any number of players. So far as he knew, that problem had never been solved. (Laplace was not reticent about claiming credit when he believed himself to have done something altogether new, although Todhunter points out (L: [1865], 468) that this time he was wrong, for De Moivre had solved the problem.) Second, the case in which the relative skill of the players is unknown pertained to the probability of future events, and Laplace solved it for a two-man game, without attempting the generality of three or more players in this inverse example.

Altogether more significant is the next topic, first for its subject, which was the determination of the mean value among a series of observations; second for the area from which Laplace took it, which was astronomy; and third for the application, which was to the minimization of observational and instrumental error. In view of the whole development of Laplace's later career, this may well appear to be a highly indicative article, not least in his manner of introducing it. Two years previously, he said, he had worked out a solution for taking the mean value among several observations of the same phenomenon. Thinking it would be of little use, however, he decided to delete it from the memoir on recurro-recurrent series ([*1774b*]), where he had originally intended it as a postscript. He had since learned from the *Journal astronomique* that Daniel Bernoulli and Lagrange had both investigated the same problem ([*1774c*]; *OC*, VIII, 41–42). Their memoirs remained unpublished and he had never seen them. This announcement, together with what he now calls the utility of the matter, had led him to set out his own ideas. It would appear that he must have given them some further development, for the approach turned on treating the true value as the unknown cause of three observed values, taken for effects. Afterward, the mean value giving the minimum probability of error would be determined. It is evident from the outset that the observations Laplace had in mind were astronomical, for the point to be fixed was the time at which the event occurred.

Laplace now constructed a graph and a probability curve. In Figure 1, the line AB represents time; the points a, b, and c, the three successive observations of the phenomenon; and the point V its true instant. In Figure 2, the probability that an observation differs from the truth by the amounts Vp and Vp' is represented by the ordinates of the curve RMM'. The probability decreases with time according to an unknown law, which is expressed in the equation for the curve, $y = \omega(x)$.

Now then, either of two things may be intended in speaking of the mean to be taken among several such observations. One is the instant at which it is equally probable that the true time of the phenomenon occurs before or after it. That may be

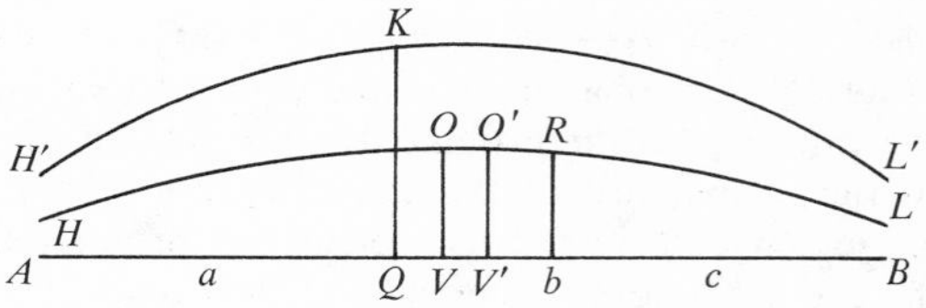

FIGURE 1

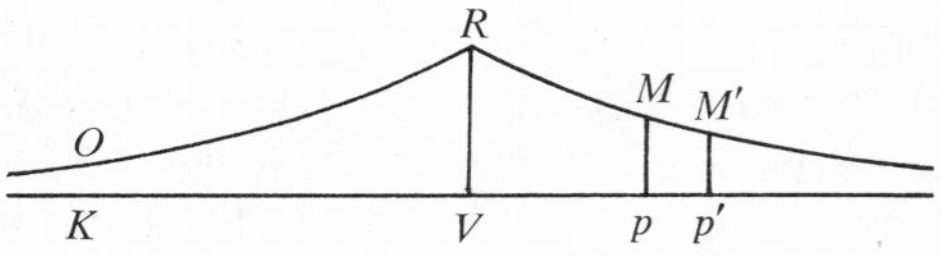

FIGURE 2

called the "probabilistic mean." The other is the instant at which the sum of errors to be incurred ("feared" in Laplace's terminology) multiplied by their probability is a minimum. That value may be called the "mean of error" or "astronomical mean," the latter because it is the term that astronomers ought to prefer. A lengthy calculation now shows that in fact the probabilistic and astronomical means come to the same value. Its determination in this memoir becomes too complicated to follow in detail. To find it, Laplace needed to identify the nature of the function $\phi(x)$. It is reasonable to suppose that the ratio of two consecutive infinitesimal differences is the same as that of the two corresponding ordinates, so that $d\phi/\phi$ is constant. Then the equation relating the ordinates to their infinitesimal differences is (*ibid.*; *OC*, VIII, 46):

$$\frac{d\phi(x+dx)}{d\phi(x)}=\frac{\phi(x+dx)}{\phi(x)}, \qquad (10)$$

whence

$$\frac{d\phi(x)}{dx}=-m\phi(x)\text{, and } \phi(x)=\beta e^{-mx}; \qquad (11)$$

m being constant, $\beta = m/2$ by symmetry, and

$$\phi(x)=\frac{m}{2}e^{-mx}. \qquad (12)$$

Thus Laplace did not here arrive at the famous least-square rule. His stated purpose was to convince astronomers that their normal practice of taking an arithmetical average was erroneous, although he had to acknowledge that his method was difficult to use. A further application of the principle of inverse probability calculates the probabilities that the different values of m are to each other as the probabilities that the values when obtained will be in the proportion of their respective distances. That calculation yielded a fifteen-degree equation. Resorting to approximation, therefore, Laplace gave a table for the correction that he advised applying to whichever of the two extreme observations was further from the middle one.

The final topic in the probability of cause may be thought equally significant in view of the importance that the finding assumed in its later application—and, indeed, in the philosophy of probability. It concerns the effect of inequalities in the prior probabilities that are unknown to the players (the examples come from theory of chance), or more largely to persons involved in the outcome of quantifiable events in the civic realm. In the normal theory, the assumption is that the various cases capable of producing an event are equally probable or, if not, that their relative probabilities are known. In physical fact, of course, everyone recognizes that there is no such thing as perfectly balanced coins or ideally symmetrical dice; but it was further assumed that the game is nevertheless fair, provided that both players are ignorant of the actual inequalities. On the contrary, Laplace found—and claimed to be the first to have noticed and demonstrated the fallacy ([*1774c*]; *OC*, VIII, 62)—that the latter assumption is valid only for situations involving simple probabilities. For example, if B should agree to give two crowns to A if A tosses heads on the first flip of a coin, then before the game begins, a fair division of the stakes would be 50–50 even if the coin is weighted, since neither knows which side is heavier. The assumption ceases to be valid, however, as soon as the game continues under the same rules, according to which B will then give A four crowns if a head turns up only at the second turn, six crowns at the third turn, and so on to x turns. Laplace pursued the matter in the case of a game of dice. A and B play, the rule being that if A throws a given side in n tosses of one dice, B pays him the sum a. What, then, should A forfeit to B if he fails? By the classical theory of chance, the amount equals the expectation of A, which is $a-(5^na/6^n)$. In fact, however, given any asymmetry at all, A's expectation will be less than that, for B has the advantage as the game continues. Determining the correct value gave Laplace a very lengthy calculation, issuing in a complicated function of the degree of asymmetry.

Laplace accompanied this discussion with a warning against excessively literal applications of mathematical theory to the actual world, physical or civic. In a somewhat unconvincing passage he claimed to have made experiments with English dice, the most regular he could obtain, only to find that they never fall true to theory in long runs. It is more interesting that this analysis contains the germs of two of the most characteristic techniques that Laplace employed in later work. Both pertain to inverse probability and the study of causes. First, the occurrence of seemingly aberrant patterns invites investigation of the source of departures from the results that equipossibility would entail. Second, the multiplication of observations provides the basis for calculating from experience the value of the prior probabilities when they are not equal. By the law of large numbers it can then be determined how large the sample must be in order to reduce the probability of error within prescribed limits.

There are textual grounds for thinking that the implications of asymmetries were an important factor in opening Laplace's eyes to the wider prospects for probability. He enlarged on them in a philosophical, or methodological, article ([*1776a*, 1°], no. XXV; *OC*, VIII, 144–153) strategically situated in the midst of the companion memoir on difference equations and theory of chance, to which we now turn, deferring for a moment the question of its combination with the first gravitational memoir. The *analyse des hasards* is there said to have objects of two kinds: first, the probability of happenings about which we are uncertain, whether in their occurrence or in their cause (it is worth noting that in the way he puts it here probability is still subsidiary to theory of chance); and second, the hopes that attach to their eventuality. Laplace acknowledged that questions could legitimately be raised about the very enterprise of applying mathematical analysis to situations of both sorts. In concerns of the former type, he attributed the difficulty to the mode of application, not to the definition. There is no ambiguity about the definition of probability itself, and no legitimate objections could be lodged against its calculus unless equal prior probabilities were to be assigned to cases that are not in fact equally probable. He had to admit that, unfortunately, all applications yet attempted to problems of civil life entailed precisely that fallacy. Laplace proceeded to illustrate it by a simplified repetition of the demonstration that the false supposition of symmetry in coins or dice is unfair to one of two players in any game involving compound probabilities.

The question of equipossibility was also involved in exposing the error committed by commentators who argued that a run of heads or tails is less likely than any alternation in a sequence of the same number of tosses of a coin. Proceeding from a mistaken notion of common sense, they supposed that each time a head turns up, the odds increase against another. In effect, they were saying that past events influence future ones. Laplace characterized that idea as "inadmissible" and proceeded to refute it in a general statement, his earliest, on the relation between regularity, chance, and causality. Why does so-called common sense give us to suppose that a sequence of twenty heads is not due to chance, whereas we would think nothing of any equally possible mixture of heads and tails in a total of twenty? The reason is that wherever we encounter symmetry, we intuitively take it for the effect of a cause acting in an orderly manner. If we were to come upon letters from a printer's font lying in the order I N F I N I T E S I M A L on some composing table, we should be disinclined to think the arrangement random, although if that were not a word in some known language, we should pay it no heed. Laplace even made a little calculation to show how our intuition that order bespeaks causality is itself conformable to probability. A symmetrical event has to be the result either of cause or of chance. Let $1/m$ be its probability in case it is due to chance, and $1/n$ its probability in case it is due to a regular cause. By his basic theorem (i.e., Bayes's rule) the probability of the cause will then be $\frac{1/n}{1/m + 1/n}$ or $\frac{1}{1 + n/m}$. The greater m in relation to n, the greater the probability that a symmetrical event is due to a regular cause.

In retrospect, it seems clear that an element of wanting it both ways was always lurking at the bottom of Laplace's outlook. On the one hand, the regularity of the universe as a whole bespeaks the rule of natural law. Order governs amid the infinity of its combinations. On the other hand, where there do appear to be disturbing factors, such as in the results of real games with imperfect dice or (to anticipate for a moment) in anomalies of planetary motion, indeterminacies in the shape of the earth, widely varying inclinations among the planes of cometary orbits, and inequality in the partition of births between boys and girls, such data call for identification of a particular cause. We do not seek out one that will be an exception to the larger realm of causality but, rather, one that, if properly calculated, will vindicate it. Thus, both apparent symmetry and exceptions to it bespeak causality, general or particular as the case may be, and the goal of analysis is to make the two cases one. Everything that a modern student wants to read into Laplace's outlook says that he should have attributed the larger regularity to randomness, and many things that he actually said point toward that anachronism. He did not, however. The notion of an order of chance would have been a contradiction in terms to Laplace ([*1776a*]; *OC*, VIII, 145):

> Before going further, it is important to pin down the sense of the words *chance* and *probability*. We look upon a thing as the effect of chance when we see nothing regular in it, nothing that manifests design, and when furthermore we are ignorant of the causes that brought it about. Thus, chance has no reality in itself. It is nothing but a term for expressing our ignorance of the way in which the various aspects of a phenomenon are interconnected and related to the rest of nature.

In regard to the second set of objections to

probability as a branch of analysis, namely that hopes, fears, and states of mind cannot be quantified, Laplace considered that reservations of this sort arose from a fundamental misunderstanding and not from a mere fallacy in the procedures. They could, therefore, be obviated by making clear distinctions in the definition of terms. Such had been the root of d'Alembert's resistance to the calculus of probability. In the future, practitioners of the science would feel obligated to him for having forced upon it clarification of its principles and recognition of its proper limitations. Laplace took the limits of probability to be identical with those of all the "physicomathematical sciences. In all our research, it is the physical cause of our sensations that is the object of analysis, and not the sensations themselves" (*ibid.*; *OC*, VIII, 147). It is obvious that out of estimates of the likelihood of future happenings come hopes and fears, and that the prospect for calculating probability was the reason that the science of chance had long been heralded for its potential utility in civil life. Taking advantage of that opportunity in the measure possible depended, however, upon seizing the distinction between *espérance morale* and *espérance mathématique*, between aspiration and expectation. (The French makes the contrast rhetorically more effective, but perhaps the English makes the difference more inescapable.) In the theory of chance, expectation is simply the product of the amount to be gained by the probability of winning it. It is a number, like probability itself—nothing more. No doubt something similar might be said about aspiration in the ordinary concerns of life, but only in a qualitative way; for that always depended on such indefinable factors that it was illusory ever to think of calculating it. Even Daniel Bernoulli's suggestion for measuring personal gain by the quotient of the value of the winnings (or other profit) divided by the total worth of the winner, although an ingenious idea, was one incapable of generalized mathematical application.

The article that makes this apology for probability in general opens with language famous from its reemployment almost verbatim nearly forty years later in the *Essai philosophique sur les probabilités* (*OC*, VII, vi). We give these thoughts of his youth as he set them down when he was twenty-six ([*1776a*]; *OC*, VIII, 144–145):

> The present state of the system of nature is evidently a consequence of what it was in the preceding moment, and if we conceive of an intelligence which at a given instant comprehends all the relations of the entities of this universe, it could state the respective position, motions, and general affects of all these entities at any time in the past or future.
>
> Physical astronomy, the branch of knowledge which does the greatest honor to the human mind, gives us an idea, albeit imperfect, of what such an intelligence would be. The simplicity of the law by which the celestial bodies move, and the relations of their masses and distances, permit analysis to follow their motions up to a certain point; and in order to determine the state of the system of these great bodies in past or future centuries, it suffices for the mathematician that their position and their velocity be given by observation for any moment in time. Man owes that advantage to the power of the instrument he employs, and to the small number of relations that it embraces in its calculations. But ignorance of the different causes involved in the production of events, as well as their complexity, taken together with the imperfection of analysis, prevents our reaching the same certainty about the vast majority of phenomena. Thus there are things that are uncertain for us, things more or less probable, and we seek to compensate for the impossibility of knowing them by determining their different degrees of likelihood. So it is that we owe to the weakness of the human mind one of the most delicate and ingenious of mathematical theories, the science of chance or probability.

Long afterward, Laplace observed, also in the *Essai philosophique* (*OC*, VII, 1xv), that his early investigations of probability were what had led him to the solution of problems of celestial mechanics in the first place. The historian is bound to temper respect for creative people with skepticism about their reminiscences of how they came to do their work. Nevertheless, it is at least interesting that Laplace's first general statement about nature and knowledge should have been consistent with the entire configuration of his *oeuvre*. That configuration had its origin in the relation of the two parts of the dual memoir under discussion. The structure of the first part is clear. It consists of thirty-five articles. Articles I–XXIV develop the integration of equations in finite differences by means of recurro-recurrent series. They constitute a comprehensive résumé of what Laplace and others had already done, incorporating many improvements. Article XXV then treats probability in general, opening with the sentences just quoted and concluding with a repetition of the exordium about the two classes of problems in probability from the companion "Mémoire sur la probabilité des causes par les événements" ([*1774c*]). The reader is directed there for inverse probability, while Laplace goes on in Articles XXVI–XXXV to apply the methods developed in the first twenty-four articles to the solution of problems of direct probability in the theory of games of chance.

Article XXXVI, although it continues the numbering, is subheaded "Sur le principe de la gravitation universelle, et sur les inégalités séculaires des planètes qui en dépendent." Matters of mathematical astronomy are treated throughout this second half of the dual memoir in a sequence of twenty-nine further articles, which we shall discuss in the next section. We do not know precisely when Laplace wrote them, but they must have been completed in the early months of 1774, for on 27 April he submitted another investigation of secular inequalities in planetary motion ([*1775a*]), which is described as a sequel (*15*), although it was printed earlier because of accidents in the academic publishing schedule. Almost certainly, therefore, he must have turned to Part 2° of the dual memoir immediately after completing the paper on the probability of causes ([*1774c*]) and the concurrent revision of Part 1°, which had occupied him between March and December 1773 (*13*). It also seems likely that the methodological Article XXV, with its philosophic propositions about the relation between probability and astronomy, would have been interpolated in the midst of the aleatory Part 1° during the time when Laplace was composing the astronomical Part 2°. As we shall see, passages introducing the latter are as prophetic, or rather programmatic, of his celestial mechanics as the paragraphs quoted above are of his epistemology. It would appear, therefore, that Laplace framed the main questions that his enormous lifework refined, extended, and largely answered, during a crucial period of about a year following his election to the Academy in March 1773.

5. Universal Gravitation. These questions were slightly but significantly different from what they are sometimes said to have been. A close reading of the astronomical part of the dual memoir, Laplace's first comprehensive piece on the mechanics of the solar system, serves to temper the conventional image of a vindicator of Newton's law of gravity against the evidence for decay of motion in the planets. Nothing is said about apparent anomalies gathering toward a cosmic catastrophe; on the contrary, the state of the universe is assumed to be steady. The problem is not whether the phenomena can be deduced from the law of universal gravity, but how to do it. Since that appeared to be impossible on a strict Newtonian construction of the evidence, Laplace proposed modifying the law of gravity slightly. He proceeded to try out the notion that gravity is a force propagated in time instead of instantaneously. Its quantity at a given point would then depend on the velocity of bodies as well as on their mass and distance. Even more interesting, the reasoning in this argument was not that of normal mathematical astronomy but was of the type that he brought to physics in other, much later writings. Lastly, in a problem that he did handle in the tradition of theoretical astronomy, namely the secular variations in the mean motions of Jupiter and Saturn, the conclusion is that the mutual attraction of the planets cannot account for them, contrary to what we expect from *Mécanique céleste*. Let us, therefore, examine these matters more fully.

As in other early papers, Laplace's point of departure was an analysis by Lagrange. In a memoir (*Oeuvres de Lagrange*, VI, 335–399) that had won the prize set by the Paris Academy for 1774, Lagrange had argued that it was impossible to derive from the theory of gravity an equation for the acceleration of the mean motion of the moon giving values large enough to agree with observation. Lagrange then wondered whether resistance of the ether might be slowing the rotation of the earth enough to resolve the apparent discrepancy. Laplace for his part took the question to be one involving the sufficiency of the law of gravity with implications for all of cosmology. (Even so he had expanded the significance of problems in the theory of chance and discussed probability in relation to all of knowledge.)

Announcing at the outset his intention of doing a series of memoirs on physical astronomy, he began by deriving general equations of motion for extended bodies referred to polar coordinates in a form especially adapted to analyzing problems of secular inequalities. Thereupon, he turned to the "principle of universal gravitation" in general, which he called the most incontestable truth in all of physical science. It rested, in his view, upon four distinct assumptions generally accepted among *géomètres*, or persons doing exact science. Given their importance in marking out the main lines along which Laplace developed his celestial mechanics, we shall state them in the form of a close paraphrase ([*1776a*, 2°]; *OC*, VIII, 212–213):

1. the force of attraction is directly proportional to mass and inversely proportional to the square of the distance;

2. the attractive force of a body is the resultant of the attraction of each of the parts that composes it;

3. the force of gravity is propagated instantaneously;

4. it acts in the same manner on bodies at rest and in motion.

The plan was to examine the respective physical consequences of these assumptions. Since Laplace continued that examination throughout his life, the tactics of this memoir in effect became the strategy of much of his subsequent research. We need to note, therefore, what he believed it was that followed physically from each of those four propositions.

The inverse-square law came first. Laplace asserted roundly that it was no longer permissible to doubt its applicability to the solar system. It is obvious to the modern reader that this law was the one ultimately served by his resolution of apparent irregularities in planetary motion. Only later, however, in a series of memoirs composed from 1785 through 1787, did he succeed in bringing off those investigations (see Section 17). Here at the outset, deductions from the inverse-square law engaged his interest in a different order of effects. In a speculative vein, he took issue with philosophers who doubted whether it holds for forces acting at very short ranges. Even though the radius of the earth was the smallest distance over which its validity had been confirmed by observation, analogy and canons of simplicity still gave reasonable grounds for supposing that the gravitational-force law obtains universally. If it be asked why gravity should diminish with the square of the distance rather than in some other ratio, Laplace—after duly objecting that such questions make mathematicians uncomfortable—would consent to say only that it is pleasing to think that the laws of nature are such that the system of the world would be the same whatever its size, provided the dimensions are increased or decreased proportionally.

It is perhaps somewhat surprising to find that in the first stage of Laplace's astronomical work, his analysis of the consequences following from the second of these assumptions, the principal of universal attraction, took precedence over the theory of planetary motion. Although a less famous topic, it may have been an even more fruitful preoccupation, if not necessarily for astronomy itself, then certainly for mathematical science in general; since the problem of the attraction of a spheroid was the source of the potential function (see Section 16). It may also be surprising to learn that in the later eighteenth century, skepticism persisted about whether the force of attraction really operates between all the particles of matter individually or only between centers of mass of macrocosmic bodies of which the shape and internal structure are governed by other, unknown laws. Again, Laplace saw no reason to suppose that there exists some least measure of distance below which analogy becomes an unconvincing mode of argument. For the present purpose, that did not matter, however, since this second assumption, unlike the inverse-square law of the intensity of short-range forces, could be subjected to analysis and the results submitted to the test of observation on the scale of terrestrial physics. D'Alembert had already derived the precession of the equinoxes and the nutation of the axis of the earth from the principle of universal attraction among all the particles of the globe, and he found the prediction confirmed by data. The tides were too complicated a phenomenon to tackle yet—although Laplace attacked them soon afterward in a major calculation (see Section 8).

There remained the shape of the earth. By Newtonian theory, it should be an ellipsoid with the polar and equatorial axes in a ratio of 229/230. Two sets of independently measured values existed. The first derived from geodesic surveys of the length of meridional arcs in Lapland by the Maupertuis expedition in 1735–1736, at the equator in Peru by the Bouguer-La Condamine expedition in 1736–1737, in France itself on the meridian of Paris by Lacaille in 1741–1742, and at the Cape of Good Hope again by Lacaille in 1751–1752. The other set of values derived from determinations of the length of the seconds pendulum at various latitudes during those expeditions and on other occasions. The geodesic data gave a flattening greater than an axial ratio of 229/230 would produce, and the pendulum yielded a smaller departure from the spherical than Newton required. Laplace considered it likely that simplifications introduced into the calculation were at fault rather than the theory itself. The most serious were the assumption of uniform density and the neglect of irregularities of the surface. However that might be, the failure of the earth's longitudinal profile to fit an elliptical curve would not disprove the mutual attraction of all its particles, unless it were shown that an ellipsoid was the only solid that could satisfy the equilibrium conditions for such a force, or else that every theoretically possible figure had been tried without satisfying the observations. Since neither of those propositions had ever been demonstrated, the principle of universal interparticle attraction held its ground. Thus, because of the relevance to that principle of the shape of celestial bodies considered as solids of revolution, Laplace entered upon this, another of his continuing preoccupations.

So much for the long-range importance of the first two gravitational assumptions in Laplace's work; the latter two, on the other hand, are interesting mainly for the discussion of them that he gave in the ensuing articles of the present memoir, and it was there that he least resembled the celestial dogmatist for which he has sometimes been taken. The third and fourth assumptions have a different standing from the inverse-square law and the principle of attraction. Anyone doing planetary astronomy would have thought it necessary to write down those first two principles in the axiomatic structure of a treatise. To most of Laplace's colleagues, on the other hand, the instantaneous propagation of gravity and its indifference to motion would have seemed prior, self-evident truths, like the rectilinear transmission of light or the three-dimensionality of space, something not usually needing statement. Not so the young Laplace, who stated these assumptions for the sake of taking issue with them. It is unreasonable, he observed immediately, to suppose that the power of attraction or any other force acting at a distance should be propagated instantaneously. Our sense is rather that it should correspond in its passage to all the intervening points of space successively. Even if communication should appear instantaneous, what really happens in nature may well be different, "for it is infinitely far from an unobservable time of propagation to one that is absolutely nil" (*ibid.*; *OC*, VIII, 220). (It later became a distinctive characteristic of Laplace's physics that the phenomena he analyzed should occur in the realm of the unobservable.) Thus, he would try what followed from the supposition that gravitation does take place in time. On the amount of time, however, he would disagree with Daniel Bernoulli, who in a piece on tidal motion had advanced the proposition that the action of the moon takes a day or two to reach the earth.

Broaching the matter in a more abstract manner, Laplace posited a corpuscle to be the bearer of gravitational force. In his analysis, the effect of weight in a particle of matter is produced by the impulse of such a gravitational corpuscle, infinitely smaller than the particle, moving toward the earth at some undetermined velocity. Given this model, the received hypothesis, according to which gravity has an identical effect whether bodies are at rest or in motion, is equivalent to supposing that velocity infinite. Laplace supposed it indefinite and determinable by observations, and his reasoning was similar to that employed to account for the aberration of light. Let us see how he set up the problem.

The calculation analyzes the motion of an infinitely small body p describing any orbit around S in the plane pSM. $Sp = r$; $\angle\ pSM = \phi$; and N is the corpuscle causing p to gravitate toward S.

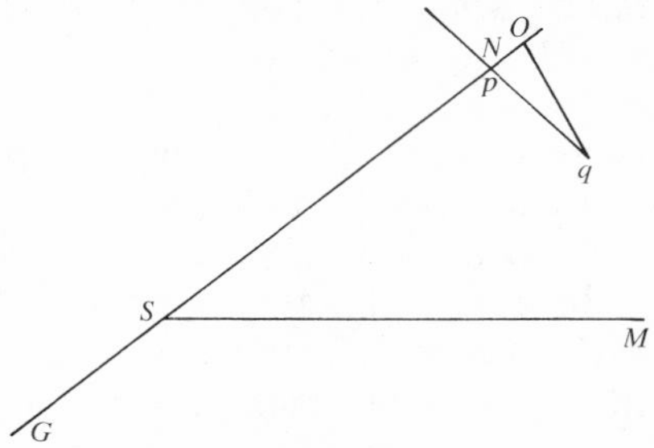

FIGURE 3

Resolving the forces and motions along the radius vector Sp and the perpendicular direction pq, Laplace introduced the expression θ/α for the measure of the distance described by the corpuscle N in time T. T and α are constants, α being an extremely small numerical coefficient and αe the ratio of the original eccentricity to the mean distance; t is time, and θ varies according to any function of the distance Sp. Calculation yields as equations of motion,

$$r = a\left(1 + \alpha e \cos nt - 2\frac{a\alpha}{\theta} Tn^2 t\right) \tag{13}$$

and

$$\phi = nt - 2\alpha e \sin nt + \frac{3}{2}\frac{a\alpha}{\theta} Tn^3 t^2, \tag{14}$$

where a and n are terms introduced in the change of variables

$$r = a(1 + \alpha y) \quad \text{and} \quad \phi = nt + \alpha x, \tag{15}$$

which refers the equations to rectangular coordinates for manipulation (*OC*, VIII, 224).

From the above equations, it appears that the mean motion of p is governed by a secular equation proportional to the square of the time. In the normal assumption of instantaneous gravitation, $a\alpha/\theta$ is infinitely small, and the secular equation vanishes. If that term were not zero, however, its effect would appear in the mean motion of planets and satellites. At this point, therefore, Laplace turned to the observations and particularly to the lunar tables, knowing that ancient and modern records of eclipses showed that the earth's satellite has been increasing its mean speed of revolution.

Laplace placed more confidence in Tobias Mayer than in any other practical astronomer. In his view, the acceleration of the mean lunar motion had amounted to one degree in 2,000 years. D'Alembert had shown that the acceleration could not be explained within the ordinary theory by any calculation involving the sun, earth, and moon alone. Lagrange had then shown that the nonsphericity of earth and moon, taken together with the influence of other planets, failed to account for the acceleration ("Sur l'équation séculaire de la lune" [1776], *Oeuvres de Lagrange*, VI, 335–399). It might be worthwhile, therefore, to try out the notion that gravitation takes time, which is to say that the term $a\alpha/\theta$ is not nil. To recall the conditions of the problem, θ/α is the space traversed by the gravitational corpuscle that is impelling the moon toward the earth during one revolution of the moon in its orbit. The velocity then is $\theta/\alpha T$. Substituting Mayer's values for the acceleration of the mean motion in these expressions, Laplace found that the velocity of the gravitational corpuscle is 7,680,000 times as great as the velocity of light. Since Mayer's values were thought to be off by 12′ in 2,000 years, the correct velocity was 6,400,000 times that of light.

Discussing this, which he admitted to be a conjecture, Laplace acknowledged that the Abbé Bossut's postulate of a very subtle fluid in space also permitted a calculation agreeing closely with the secular equation of the mean motion of the moon. What could that fluid be, however, unless it were light itself, emanating from the sun? If so, the consequences could be computed, for the orbit of the earth would have expanded and its mean motion would have been retarded. Unfortunately for Bossut's scheme, that has not happened. Its advocates might still say that sunlight has dilated the terrestrial atmosphere so as to produce the trade winds and has thus retarded the earth's rotation in another manner. Such a mechanism would indeed explain the apparent acceleration of the moon, but Laplace had also found, by a method that he promised to give elsewhere ([*1779b*]; see Section 8), that the earth's rotation cannot be detectably retarded by the friction of those winds. Thus, we are left with the force of gravity, "astonishing" in its activity but finite in velocity.

Following this analytic flight of fancy, Laplace "returned" to the normal assumptions about gravity and attacked traditional problems of mathematical astronomy concerning inequalities or variations in the elements of planetary motion. He thought of them in four main classes: positions of nodes and apsidal lines, eccentricities, inclinations of the orbital planes, and mean motions. The last were the most significant, and even they were less well-determined than Laplace could desire. On so signal a matter as the irregularities of the mean motions of Jupiter and Saturn, Euler and Lagrange disagreed, Euler having found them to be about equal and Lagrange very different (*Oeuvres de Lagrange*, I, 667). Laplace was worried about the applicability even of Lagrange's differential equations, for terms involving sines and cosines of very small angles with very small coefficients were dropped. Since the coefficients became large on integration, he feared lest the resulting formulas for determining the true movement of the planets would hold good for limited periods only. In his own calculation, Laplace took account of those terms and obtained formulas agreeing with Lagrange's in the values for apsides, eccentricities, and inclinations but differing drastically for the mean motions. Although his equations did contain terms proportional to the time and the square of the time, he would not claim that they succeeded in representing the true motions rigorously (*ibid.*; *OC*, VIII, 241). That question was interesting only mathematically and of no practical importance for astronomy over the historic span of recorded observations. More important were the results he obtained for Jupiter and Saturn.

Over the centuries most astronomers had come to consider that the observations show an acceleration for Jupiter and a much larger deceleration for Saturn. But when Laplace substituted values from Halley's tables in the expression that he had just derived for the secular equation of the mean motion of a planet, it reduced approximately to zero, leaving him to conclude that if an alteration existed in the mean motion of Saturn, it could not be caused by the influence of Jupiter (*ibid.*; *OC*, VIII, 252).

In Article LVIII (*OC*, VIII, 254–258), he applied d'Arcy's principle of areas to the same problem and obtained the same null results. It seemed unreasonable to suppose that so nearly complete a canceling out of positive and negative terms could be due to particular circumstances, and Laplace concluded that mutual gravitation between any two planets and the sun failed mathematically to account for any inequalities in their mean movements and that some other cause must be responsible for the observed anomalies. As candidate for the disturbing factor, he suggested—again by way of conjecture—the action of the comets.

He recognized that the conclusion was "contrary to what all mathematicians who have worked at the subject have hitherto supposed" (*ibid.*; *OC*,

VIII, 258). This statement will be equally surprising to students of the *Mécanique céleste*, still more so to those who know only the *Système du monde*, and most of all to those who are told in many textbooks that Laplace rescued the Newtonian planetary system from increasing instability by proving that precisely such mutual interactions do resolve the apparent inequalities that are actually periodic. That this finding, which he emphasized for its importance, should seem contrary to what he is most famous for, is explicable largely in consequence of the mathematics. He had not yet developed perturbation functions, and there was no provision in his expressions for long-term periodicity. Moreover, he nowhere worried about instability, unless that was what he had in mind in a passing remark that his conclusion would be less convincing if the secular inequalities increased proportionally to the square of the time, for that would mean that a continuously acting force was at work—and none other than universal gravitation was known.

The remaining articles of this, the first memoir on gravitation, will be somewhat anticlimactic to the modern reader, although there is no indication that Laplace thought them so. He considered that Euler in treating the secular inequalities in the motion of the earth had neglected significant terms and thus had given an incomplete analysis, particularly with respect to the eccentricity and apogee of the sun. The latter value, determined with precision from Halley's tables, could serve to establish the mass of Venus with sufficient accuracy for Laplace to calculate the share in the decrease of the obliquity of the ecliptic due to the action of the planets. In the brief Articles LXII and LXIII, he gave a method for determining that variation in general by means of analytic geometry (*ibid.*; *OC*, VIII, 268–272). They may possibly represent the memoir on that problem which Laplace had read before the Academy in November 1770, one of his earliest (*3*). In principle the method could have been applied to calculating the position of the equinoxes and hence the degree of that inclination at any time past or future, but the ancient observations were too imprecise to render the exercise worthwhile. Another conjecture about cometary influence speculates that it may be a factor offsetting the attraction of the planets in their influence on the decreasing inclination of the ecliptic. Finally, a postscript reviews the calculations of Lagrange that had left their author uncertain about the acceleration of the moon (*Oeuvres de Lagrange*, VI, 335–399). Laplace found, on the contrary, that they strengthen the case for it.

6. Cometary Distribution. In the same volume with the dual memoir on probability and gravitation discussed in the preceding two sections, Laplace published the tripartite "Mémoire sur l'inclinaison moyenne des orbites des comètes, sur la figure de la terre, et sur les fonctions" ([*1776b*]), the last work that he published in the *Savants étrangers* series. The three parts are entirely independent, and we shall discuss only the first here, deferring the others for consideration in connection with the further development that he gave to those respective topics (see Sections 8 and 10). Occupying over two-thirds of the entire memoir, it consists in an actual application of his work in probability to a study of the distribution of the orbital planes of comets in space.

Let us begin by recalling Laplace's remark about having been led into astronomy through probability (see Section 4). In construing what he meant, it will be helpful to distinguish between weak and strong interactions in the evolution of his interests. To the weak, or philosophical, sort belong the regulative remarks about order, causality, and knowledge; to the strong, or technical, belong the probabilistic analysis of the phenomena themselves and also of distributions and errors in the data, and more rarely the application of mathematical techniques conceived for one set of problems to another. The memoir on comets illustrates both aspects. Its motivation pertained to the weaker, philosophical or cosmological, aspect of probability and its execution, to the stronger.

Uncertainty still bedeviled the status of comets in the solar system. Did they fully belong to it or not? Like many others, Laplace was undecided. We have just seen, in Section 5, how at this early stage he invoked their action as a possible *deus ex machina* to explain the secular inequalities of the mean motions of Jupiter and Saturn, which appeared inexplicable on the basis of forces operating within the system. He opened the present memoir by pointing to the anomaly that their motions constitute if they do belong to the system of the world. All the planets and their satellites, some sixteen bodies in all, revolve in the same direction and almost in the same plane. The probability that a common cause lies behind the arrangement could be calculated and was approximately equal to certainty (although to the further question of what that cause might be, Laplace confessed that he had never found a convincing answer). The comets are quite another matter: they move in any direction in very eccentric orbits inclined at all angles to the plane of the ecliptic. Laplace's senior colleague, Dionis du Séjour, one of the few aristocratic amateurs who actually contributed to astronomy, had analyzed the problem of whether the cause of planetary motion—whatever it might be—had also produced the phenomena of comets. He calculated the mean inclination of the sixty-three known orbits and found it to be 46°16′. The departure from 45° was clearly insignificant. Moreover, the ratio of forward to retrograde motions was five to four, or nearly one to one. Dionis du Séjour had concluded, therefore, that the comets serve a principle of indifference characteristic of the universe at large, within which the causal system of the planets constitutes a distinctive ordering.

The conclusion was reasonable enough in Laplace's view, but a further calculation was needed to establish the degree of certainty that it held. Supposing the comets randomly projected into space, what are the probabilities that the mean inclination of their orbits, and also the ratio of clockwise to counterclockwise revolutions, will be contained within given limits? For if the value of the mean inclinations were $45° + \alpha$ and if there were very large odds (say a million to one), to bet that it should be less, it could plausibly be concluded that some particular cause does account for their moving in one plane rather than another. The same logic could be applied to the ratio of forward to retrograde motions; there the calculation is easy, involving only two possibilities. Not so the problem of the orbits. Perhaps the mention of a bet hints that he was thinking of it in terms that antedate the analysis of causes from events that he might have been expected to apply. At any rate, he went on to qualify the problem as one of the most complicated in the entire analysis of *hasards*, especially so since the goal was a general formula applicable to any number of comets.

The problem then posits an indefinite number of bodies randomly projected into space and revolving around the sun, and the probability is required that the mean inclination of the orbits with respect to a given plane lies between two limits. Laplace approached it by constructing probability curves for the simple cases, much as he had done in discussing the mean value of a series of astronomical observations in the memoir on probability of causes. Here, however, the mean is the arithmetical average, and the probability sought is direct, not inverse. The first case is that of two bodies only, and the probabilities are represented in Figure 4.

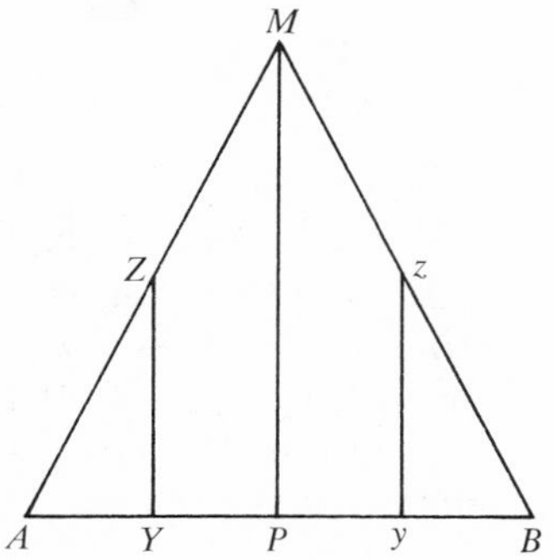

FIGURE 4

The line AB represents an inclination of 90°, and the ordinates determining the slope of the line $AZMzB$ are proportional to the probability that the corresponding abscissa measured along AB represents the mean inclination. Thus the probability that the mean inclination is AY, equal to x, is given by YZ; and Laplace set out the elementary reasoning to show that the required probability that the mean inclination lies between the limits Y and y will be equal to the area $YZMzy$ divided by the whole area AMB.

Laplace then turned to the case of three bodies (or comets) for which the curve is as shown in Figure 5:

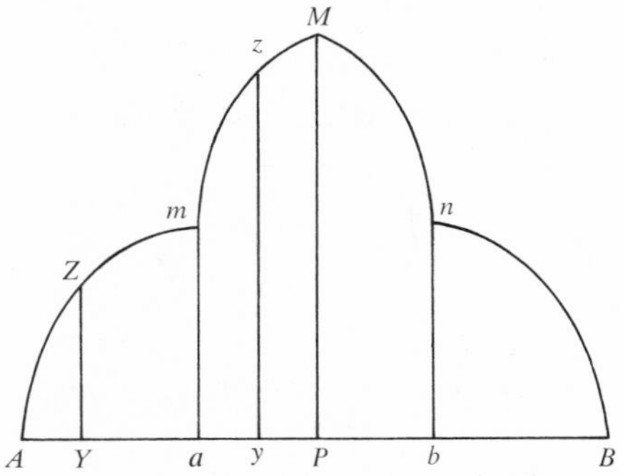

FIGURE 5

where $Aa = ab = bB$. Drawing on the above results for two bodies, he obtained the equations

$$ay = \frac{9}{2}x^2 \quad \text{and} \quad ay = \frac{1}{2}a^2 + 3az - 9z^2 \qquad (16)$$

for the curves AZm and mMn respectively. Again the required probability that the mean inclination falls between certain limits is given by the quotient of the area between those limits divided by the entire area under the curve AMB.

In a figure corresponding to a system of four comets, the equations for am and mM, obtained from the results for three comets, are respectively

$$a^2y=\frac{32}{3}x^3 \text{ and } a^2y=\frac{1}{6}a^3+2a^2z+8az^2-32z^3. \quad (17)$$

It would have been possible to continue inductively, dividing AB into five or more equal parts and obtaining the equations of the curves corresponding to n bodies from the results for $(n-1)$. To that end Laplace supposed the line AB of Figure 5 divided into n equal parts and sought an expression for the curve relative to the rth part, as shown in Figure 6:

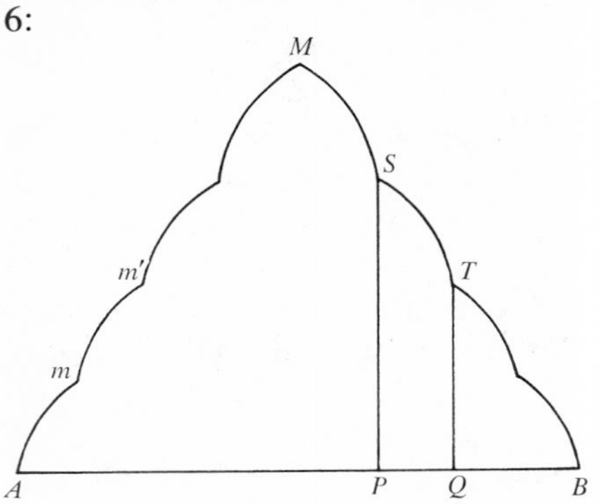

FIGURE 6

In finding it, he let $\frac{r-1}{n}a+z$ be the distance of one of its ordinates from A (z being less than a/n), and represented by ${}_ry_{n,z}$ the number of cases in which the mean inclination of n bodies is $\frac{r-1}{n}a+z$. That general expression could be given a value in the form

$$ {}_ry_{n,z}={}_rA_n z^{n-1}+{}_rB_n z^{n-2}+{}_rC_n z^{n-3} +\cdots+{}_rG_n z+{}_rH_n, \quad (18)$$

where ${}_rA_n$, ${}_rB_n$, $\cdots$, etc., are functions of r and n to be determined. Now it becomes evident what Laplace had been up to all the time. He could determine the value and solve the problem by a method that he had published in Article XII of the double memoir on difference equations and gravitation ([*1776a*, 1°]; *OC*, VIII, 97–102). Essentially, the problem was to find the general expression for quantities subject to a law governing their formation. In this instance, when $AB=a=90°$, and AB is divided into n equal parts, the method gives the equation for the curve corresponding to the rth part ([*1776b*]; *OC*, VIII, 299),

$$a^{n-2}y={}_ry_{n,z}. \quad (19)$$

The probability that the mean inclination of n orbits lies between any two points P and Q is then given by the quotient of the area AMB divided into the area $STQP$. In a complete application of the theory, the value for n would be 63, that being the number of comets for which astronomers had calculated the orbits. Leaving that computation to whoever wished to undertake it, Laplace calculated the probability that the mean inclination of twelve comets lies between 37-1/2° and 52-1/2° (for which case he divided the line AB into twelve equal parts of 7-1/2°). The resulting value is 0.339. Thus, the odds that the mean inclination is greater than 37-1/2° or less than 52-1/2° are 839 to 161, and the odds that it falls within those limits are 678 to 322. Referring to data, Laplace found that the mean inclination of the twelve most recently observed comets was 42°31′. There could be no reason to assign a cause tending to make these comets move in the plane of the ecliptic unless the contrary assumption of random projection resulted in enormous odds that the mean inclination should exceed 42°31′. In fact, the odds are less than six to one that it would exceed 37-1/2° and are very much smaller that it would exceed 42°31′. Hence, no such cause can be supposed to exist.

The important element here is not the support that this heaviest of probabilistic artillery brings to the position that Dionis du Séjour had already occupied in a comparatively straightforward and lightweight analysis. (Nor is it germane that no one had yet appreciated that the probability of an orbit is as the sine of the inclination, so that the mean should have been 60° rather than 45°.) What is interesting is Laplace's virtuosity in applying to the distribution of the comets a method for formulating in a general manner a problem encountered in the theory of chance.

7. Partial Differential Equations, Determinants, and Variation of Constants. The remaining mathematical investigations in Laplace's early period involved techniques of analysis developed in a group of three memoirs, two of them concerned also with astronomical problems. It will be convenient to begin with the piece that had no special astronomical relevance, "Recherches sur le calcul intégral aux différences partielles" ([*1777a*]), which, although published later than the other two, was started earlier. Laplace read a version in 1773 but did not submit it for publication until December 1776 (*19*). It consists of a reworking of his youthful *Acta eruditorum* paper ([*1771a*]), for which he had already apologized (see Section 2) in the paper on probability of causes ([*1774c*]). He had there appended a preliminary statement of the main theorem that he was now demonstrating. The long delay suggests that he may have had trouble with it. In any case, he had arrived at a general method (since called the cascade method) for solving linear partial differential equations by an approach that yielded the complete integral when it could be obtained and further served to identify equations that are insoluble. As originally printed, the memoir employs Euler's bracket notation, in which a general second-order linear partial differential equation is ([*1777a*]; *OC*, IX, 21)

$$0=\left(\frac{ddz}{dx^2}\right)+\alpha\left(\frac{ddz}{dxdy}\right)+\beta\left(\frac{ddz}{dy^2}\right)+\gamma\left(\frac{dz}{dx}\right) + \delta\left(\frac{dz}{dy}\right)+\lambda z+T, \quad (20)$$

α, β, γ, δ, λ and T being functions of x and y. Laplace then simplified the form by changing the variables x and y into others, ω and θ, which are functions of x and y, and by regarding z as a function of these new variables. He could thereby obtain the equation in a new form (*OC*, VIII, 22):

$$0=M\left(\frac{\partial^2 z}{\partial\omega\cdot\partial\theta}\right)+N\left(\frac{\partial z}{\partial\omega}\right)+L\left(\frac{\partial z}{\partial\theta}\right)+Rz+T'. \quad (21)$$

It followed that any second-order linear partial differential equation is reducible to the simple form

$$0=\left(\frac{\partial^2 z}{\partial\omega\cdot\partial\theta}\right)+m\frac{\partial z}{\partial\omega}+n\left(\frac{\partial z}{\partial\theta}\right)+lz+T. \quad (22)$$

Since m, n, and l are functions of ω and θ, ω and θ can be determined and substituted in (21), thus transforming it into (22), which in turn has all the generality of (20) but is simpler and easier to investigate. Laplace was now in a position to restate the theorem that is the heart of the matter. In cases where the complete integration of (22) is possible in finite terms (and those cases could always be identified), one of the two integral signs affecting the arbitrary functions $\phi(\omega)$ or $\psi(\theta)$ can be eliminated and the remaining integration performed. Moreover, the cases wherein such a solution is possible include almost all the problems encountered in mathematical physics. For that was the field in which Laplace expected the method to be applicable. Among the many uses of linear partial differential equations, he singled out for special mention the determination, whatever the state of a system, of infinitesimal oscillations among an infinite number of corpuscles interacting in any manner whatsoever ([*1777a*]; *OC*, IX, 7).

A paper started at about the same time, "Mémoire sur les solutions particulières des équations différentielles et sur les inégalités séculaires des planètes" ([*1775a*]), begins with the demonstration of a second theorem that Laplace had stated without proof in an appendix to the paper on probability of causes ([*1774c*]; *OC*, VIII, 64; cf. Section 2). It concerns Euler's discovery of singular solutions, which then appeared to be an anomaly in the theory of differential equations. Certain functions can satisfy a differential equation and still not be contained in its general integral, and the consequence is that integration of an expression is not necessarily tantamount to the complete solution of a problem. All the singular solutions also have to be determined, and the method that Euler gave was restricted to first-degree equations. Laplace went on to find further methods of determining special solutions. For example, given the differential equation $dy = p\ dx$, if $\mu = 0$ is a particular solution, then μ is a factor common to the expressions (*OC*, VIII, 339),

$$p+\frac{\dfrac{\partial^2 p}{\partial x\,\partial y}}{\dfrac{\partial^2 p}{\partial y^2}} \quad \text{and} \quad \frac{1}{\dfrac{\partial p}{\partial y}}, \quad (23)$$

and, reciprocally, any factor common to both expressions is a particular solution to the equation when it can be set equal to zero. He also obtained results for equations of higher degree.

In the case of this memoir, Condorcet observed (*HARS* [1772, part 1/1775], 70) that the two topics in the title are entirely distinct and have totally different objects. Formally, that is no doubt true; and indeed the mathematical part was submitted in July 1773 and the astronomical in December 1774 (*14* and *17*). More significant, however, is the evidence that throughout these years Laplace's mind was moving back and forth between problems of analysis and astronomy, and that the relation of emulation and competition with Lagrange embraced both domains. The astronomical section resumes his discussion (see Section 5) of the long-term invariance of the mean motions and hence the mean distances of the planets in respect to any gravitational interaction. (Although written earlier, this piece [*1776a*, 2°] was printed later.) Receipt of a manuscript memoir from Lagrange dealing with secular inequalities in the motions of the nodes and also in variations of orbital inclinations stimulated him to return to the subject after laying it aside in some frustration. By means of a transformation that Laplace greatly admired, Lagrange succeeded in reducing the problem to the integration of as many linear differential equations as there were unknowns and then to the determination of the constants, whatever the number of planets. Hastening to apply the transformation to his own formulas for secular inequalities of mean motion, Laplace succeeded in deriving the same equations. With considerable enthusiasm, he also tried whether he could determine the secular inequalities of the eccentricities and motions of aphelion by an analogous method—and happily he could. He apologized ([*1775a*]; *OC*, VIII, 355–356 n.) for rushing into print before Lagrange himself could publish. Between them, however, they had the makings of a complete and rigorous theory of all the secular inequalities, and Laplace for his part now proposed to draw together just such a general theory in a further study. Notice that this is the second intimation of the famous astronomical program eventually realized in the late 1780's (see Section 17) and first announced in the astronomical part of the dual probability-gravitation memoir ([*1776a*, 2°]) discussed above (see Section 5).

In the meantime, he had imagined a new method for approximating the solutions of differential equations in which problems of planetary motion were formulated. He gave a very summary idea of the technique in a hastily printed "Addition" ([*1775b*]; *OC*, VIII, 361–366) to the memoir ([*1775a*]) on singular solutions and secular inequalities, reserving its development in detail for the next volume of *Mémoires*. The essay "Recherches sur le calcul intégral et sur le système du monde" ([*1776c*]) constitutes the third item in this mainly mathematical series. Perhaps it was the most important, for the approach became a mainstay of Laplace's later work in planetary theory and thereby of positional astronomy in general. For readers unfamiliar with that calculus, Laplace explained how the complexity of planetary motions rules out all hope of achieving rigorous solutions to the problems they present. Called for instead were simple and convergent methods for integrating the differential equations by approximation. The most widely practiced technique, which Laplace attributed to d'Alembert, consisted in an adaptation of Newton's method for approximating the roots of polynomials. Since the values were known with only a small degree of uncertainty, accurate quantities could be substituted in their place and a very small indeterminate term added. Then the squares and higher powers of this new indeterminate might be neglected and the problem reduced to the integration of as many linear differential equations as there were variables.

All available methods for that, however, entailed one or both of two drawbacks. They were inapplicable to planets with more than one satellite and to the motion of two or more planets around the sun. Moreover, in some cases, notably that of the moon, the second approximation in the process of integration introduced secular terms with no physical meaning—then called "arcs of circles"—which spoiled the convergence. The method that he had just imagined avoided these inconveniences. It was specifically applicable to expressions in which the variables are functions of periodic quantities and also of other quantities that increase very slowly. That is precisely the situation in physical astronomy. The trick consisted in varying the arbitrary constants in the approximate integrals and then determining their values for a given time by integration. In the preface Condorcet despaired of conveying an idea of it verbally (*HARS* [1772, pt. 2/1776], 87–89). Agreeing that examples would convey a better notion than any generalities, Laplace produced a large number in a memoir of some 100 pages followed by several additions.

Andoyer does attempt a summary account, which may be of interest (J:[1922], pp. 54–55). Given a differential equation

$$\frac{d^2x}{dt^2}=P, \tag{24}$$

in which P, a function of x and of dx/dt, is periodic with respect to appropriate linear arguments in t, it is supposed that its general integral is obtained and that it takes the form

$$x=X+tY+t^2Z+\cdots. \tag{25}$$

The functions $X, Y, Z, \cdots$, depend on two arbitrary constants C_1 and C_2 and are periodic in t, as well as in any other arguments introduced by integration. Here Laplace restricted the conditions so that a function can be developed consistently with the above equation for x in only one way. The expression

$$x'=X'+(t-\theta)\,Y'+(t-\theta)^2\,Z'+\cdots \tag{26}$$

will then satisfy the equation, where θ is any arbitrary quantity and $X', Y', Z', \cdots$ are the forms taken by $X, Y, Z, \cdots$ when the constants C_1 and C_2 are replaced by similar variants, C'_1 and C'_2.

If C'_1 and C'_2 are now properly determined as functions of C_1, C_2, and θ, the expressions for x and x' become identical. Since x' will then be independent of θ, any value can be assigned to θ at will; and if θ is set equal to t, the expression for x reduces to

$$x = X', \tag{27}$$

when C'_1 and C'_2 are assigned corresponding values. If those are periodic, the manipulation will have eliminated the secular terms from the initial expression for x. In order to determine those values, it can be asserted that the value of x' becomes independent of θ, in such a way that

$$\frac{dx'}{dC'_1}\frac{dC'_1}{d\theta} + \frac{dx'}{dC'_2}\frac{dC'_2}{d\theta} + \frac{dx'}{d\theta} = 0. \tag{28}$$

That equation permits the determination of C'_1 and C'_2, given the further condition that for $\theta = 0$, these quantities reduce to C_1 and C_2.

Among his examples, Laplace applied the method to the determination of inequalities in eccentricities and the inclinations of the orbit; he derived anew his theorem from the probabilistic-astronomical memoir concerning the long-term invariance of the mean motions with respect to gravitational interactions ([*1776a*, 2°], Article LIX; *OC*, VIII, 258–263); and he tried out an analysis of a new sort of problem, the secular inequalities that a planet would exhibit if it moved in a resisting medium. His examples amount to the elements of just such a theory as he had promised in the previous memoir, and it was certainly the astronomical possibilities opened to him by the variation of constants that motivated the present paper. The reader who wishes an abbreviated recapitulation by Laplace himself may turn to Articles XVI–XIX of the great memoir on the "Théorie de Jupiter et de Saturne" ([*1788a*]; *OC*, XI, 131–140; see Section 17 below). Yet it is characteristic of Laplace that along the way he should have developed a method for solving problems capable of much more general applicability. Early in the course of the analysis ([*1776c*], Article IV; *OC*, VIII, 395–406), Laplace found himself confronting a set of linear equations from which n quantities had to be eliminated. The technique he there introduced involved an expansion since called by his name in the theory of determinants. Muir (N: [1906], I, 25–33) gives a summary of its importance in the development of that branch of mathematics.

8. The Figure of the Earth and the Motion of the Seas. Geophysical topics occupied Laplace in the final investigation that we wish to assign to his early period. The attribution is somewhat arbitrary, perhaps, since in *Mécanique céleste* (Volume II, Book IV) he did employ much the same method for the analysis of the tides. He there improved the calculations enough to make the difference a substantial one, however. At all events, in the program of the probability-gravitation memoir ([*1776a*, 2°], see Section 5) measurements of the figure of the earth constituted the one set of data by which the principle of universal interparticle attraction might be tested. Laplace first took up the matter in a single article of the tripartite memoir that opened with his analysis of the distribution of cometary orbits ([*1776b*], Article X; *OC*, VIII, 302–313). He conceived the problem to be one of determining the equilibrium conditions governing the form of planets considered as spheroids of revolution. Appropriately enough, the table of contents classifies the discussion under "Hydrostatics" rather than geodesy (*SE* [1773/1776], xv), for Laplace analyzed the same model that Newton had, a homogeneous fluid mass spinning on its axis with all the particles serving the inverse-square law of attraction among themselves. He considered Newton's assertion that such a body would be an ellipsoid to be very shaky, however. Although it was easy to show that an ellipsoid, or indeed some other solid of revolution, satisfied the conditions *a posteriori*, it was enormously difficult to determine *a priori* what those conditions must be.

Laplace undertook that task in a further brief analysis ([*1776d*]; *OC*, VIII, 481–501) tacked on to the memoir ([*1776c*]) on variation of constants in the integration of astronomical expressions. Although he failed to determine the equation for the curve described by a meridian, he did there discover the law of gravity in a generalized homogenous spheroid: on the surface of any such body in a state of internal gravitational equilibrium, the variation of gravity from the equator to the poles follows the same law as in an ellipsoid. Its force at any point on the surface is given by ([*1776d*]; *OC*, VIII, 493)

$$P = P'\ (1 + \frac{5}{4}\ \alpha m \cos^2 \theta)\,, \tag{29}$$

where P' is the value at the equator, αm is the ratio of centrifugal to gravitational force (α being infinitesimally small), and θ is the complement of the latitude.

Those two small pieces were forerunners of the "Recherches sur plusieurs points du système du monde" ([*1778a*], [*1779a*], [*1779b*]), which Laplace submitted to the Academy in three installments in 1777 and 1778 and which, taken together, comprise a treatise of over 200 pages in its *Mémoires*. The investigation had three objects: (1) the law of gravitational force on the surface of homogeneous spheroids in equilibrium; (2) the motion of the tides, together with the precession of the equinoxes and the nutation of the earth's axis; and (3) the oscillations of the atmosphere caused by the gravitational action of the sun and the moon. The first section is entirely formal and quite in the d'Alembert tradition, except that Laplace dispensed with the stipulation that the solid be generated by the revolution of a curve. He set up the problem by means of a diagram in which an element of surface fluid in the form of an infinitesimal parallelogram exerts a force upon the center of mass, and after eight pages of analysis he obtained the equation ([*1778a*]; *OC*, IX, 85)

$$P = P' + \frac{5}{4}\alpha f \cos^2 \theta - \frac{1}{2}\alpha V + \alpha R, \qquad (30)$$

where again θ is the complement of latitude, P' is a constant of integration, R is a centrally directed force at a point on the surface and hence a function of latitude and longitude, V is a term expressing the departure (assumed to be small) of the spheroid from the spherical, and αf is the centrifugal force at the equator. None of this would be very interesting, perhaps, except for what emerges as the motivation. Only at the end does it become evident why Laplace had reversed directions and determined the attraction of a surface element for the center. For he there adduçed a special case in which the forces are produced by the attraction of any number of bodies near to or far from the spheroid and subject to the conditions that they change its form negligibly and that they either participate in its rotation or are infinite in number and disposed in a uniform ring. Laplace frequently indulged in the practice of specifying some peculiarity of the world in highly abstract terms in order to make it appear to follow from a general analysis. This example is an instance, for the case is that of Saturn's rings. The conditions give a series expression for the last two terms in (30)

$$\sum \left(\frac{S}{2rs} - \frac{3}{2}Sb - \frac{3}{r^2 s}\frac{dr}{ds}\right), \qquad (31)$$

where S is the mass of a discrete body in an assembly constituting a ring, r its distance from the point M on the surface of the planet, and s the radius of the point M. If the shape and density of the ring, and its position relative to the axis of rotation of Saturn, were known, and also if the planet were supposed to be uniform in density, the law of gravity at the surface could be determined. For S would then be infinitesimal, and summing the series would become a simple exercise in integration. Laplace attempted no numerical solution, however, and the general significance of the analysis is its applicability to systems of particles external to the surface of an attracting spheroid.

It is the second topic in this series of memoirs that is the most interesting, as much for its subject matter as for the example it affords of Laplace's mode of thinking. The phenomenon analyzed is the tidal ebb and flow of the seas. In the program of the dual probability-gravitation memoir ([*1776a*, 1°], see Section 5) he had put that problem aside as too complicated for calculation. Now that he felt ready to take it on, the question (as always) was not whether the motion of the tides is subject to gravitational forces, which fact no one doubted, but how the law could be derived. He started with the assumption classically brought to the problem in the study it had received from Newton through Daniel Bernoulli and Maclaurin to d'Alembert, namely that the spheroid covered by a fluid should differ very little from a sphere; that the axis of rotation should be invariant; that the centrifugal force of the fluid particles and other forces, such as attraction of the stars, should be small compared to self-gravity; and that the sidereal motion should be slow compared to the speed of rotation of the sphere. To these *a priori* conditions, Laplace added two others that emerged in the course of his analysis and that are very interesting for the question they raise about what his sense actually was of the relation between the operations of analysis and physical facts.

The new conditions emerging from the analysis were that the seas be approximately uniform in depth and that the depth be about four leagues (twelve miles). If one were to read no further in Laplace than this lengthy memoir, one would be bound to suppose that he took these assertions to be statements about the way the world is made emerging from a mathematical analysis of tidal action under gravitational force. He acknowledged that if the case were different, his expressions would have been unmanageable. He repeatedly referred to it as "the case in nature," however, and nowhere stipulated that it was a simplification introduced in order to make the analysis possible at all. On the contrary, he took his predecessors to task for having simplified matters in a manner that was unfaithful to the facts. Beginning with Newton, the form they had investigated was that which would be assumed on a stationary planet by a thin covering of fluid in equilibrium under the gravitational pull of a stationary star. They then assumed that if the star were given a real or apparent motion around the planet, it would simply drag the bulge around without changing the form.

Laplace, for his part, would now give the problem a realistic analysis. The most obvious discrepancy with fact in the classical picture arose from the consequence it entailed that the two full tides occurring in a single day should differ markedly in height and most notably when the declination between the sun and moon was greatest. Actually, however, it was common knowledge at any seaside that consecutive high tides differ very little in level. A second difficulty was formal. The traditional analysis overlooked the variation of "the angular motion of rotation" ([*1778a*]; *OC*, IX, 90) of molecules of seawater with latitude, which would produce longitudinal displacements of the same order as those induced by the direct gravitational action of the sun and moon. From the frequency with which Laplace recurred to the former anomaly, it would appear to have been that which had fixed his attention on the problem at just this juncture. A letter of 25 February 1778 to Lagrange confirms that the lack of conformity between theory and observation over the inversion of the tides had struck him very forcibly (*Oeuvres de Lagrange*, XIV, 78–81). In his own analysis Laplace formed expressions containing terms for the actual momentum of fluid elements at all latitudes on a rotating globe. When he developed them, he was pleased to be able to argue that the theory yielded a convincing explanation for the near equality of consecutive high tides. Indeed, he said, it exhibited any number of reasons for it. What those expressions were like, we shall show in a moment.

Meanwhile, his analysis had led him to notice another deficiency in existing theory, this one affecting the precession of the equinoxes and the more recently discovered motion of nutation. Heretofore, calculations of these variations in the direction of the earth's axis had presupposed the action of the sun and moon on a solid globe. To be fully accurate, however, account would also need to be taken of the gravitational interaction of both bodies with the seas. If the sun and the moon were in the plane of the equator, the pull on the waters in the northern and southern hemispheres would be symmetrical and no problem would arise. In actuality, the inclination of the plane of the ecliptic is bound to result in uneven tidal distributions, and there can therefore be no warrant *a priori* for supposing that the unequal reaction of the waters in the two hemispheres will fail to produce disturbances in the direction of the polar axis. Quite the contrary.

As it happened, Laplace's calculations showed (he did not say to his surprise) that introducing these considerations produced no changes in the overall laws of precession and nutation. It turned out that only the ratio of the quantity of the nutation to that of precession is influenced by including terms for tidal asymmetry between the hemispheres. That composite effect is very small, moreover, when calculated on the basis of reasonable hypotheses concerning the depth and density of the seas. Nor was Laplace disappointed, for its amount came out to be proportional to the very small difference between consecutive high tides. Thus, the validity of the four-league uniform depth of the sea, which had been introduced to facilitate calculation of the tides, was confirmed by its applicability to the interrelation between the tides and the variations in direction of the earth's axis.

This piece of research makes a very nice example of the juncture at which Laplace came into the history of astronomical and geodesic problems, and equally of the approach that he brought to them. The first stage, the discovery of the laws themselves, had pertained to Newton for the most part; and the second, their analytic formulation, to the Bernoullis, Euler, Clairaut, and d'Alembert. Laplace's was a third stage, that of deriving interactions and interrelations and reducing or extending the scale between macrocosmic and microcosmic levels. Even so, the dynamical explanation of the tides belonged to Newton and its subsequent expression in terms of analytical mechanics to Laplace's predecessors. He cut into the theory by attacking the anomaly of the near equality of consecutive high tides and proceeded to derive from its resolution a relation to precession and nutation. When he corrected predecessors, it was seldom if ever over physical data: it was because their analyses did not account for the data. Thus, the existing theory of precession and nutation did not give the wrong result. But it ought to have done so, and discovering why it did not led Laplace to an unsuspected and higher-order relation between the two axial oscillations.

Here, as throughout, his own analyses bespeak an enormous fund of physical knowledge and acuteness of insight into the body of existing theory. He saw into the physical situation for consequences—for example, the asymmetrical gravitational pull on the seas in the two hemispheres—that others had either not noticed or had not been able to calculate. Then he would calculate them, in memoirs that run, like this one, into hundreds of pages of differentiation, integration, and approximation. Did he see how to make an analysis come out before deciding which phenomena to handle—

or the other way about? That question cannot be answered in the present state of scholarship, and probably the alternatives are two sides of the same coin. However that may have worked, one remark may be ventured with a certain confidence. His instinct for what was true and interesting in a set of physical relations was prompted by its accessibility to the kind of analysis in which he was a virtuoso. His knowledge of physics itself, however, was that of an omnivorous reader, an *érudit*, rather than an inquirer or discoverer.

A further feature of Laplace's mature technique first became fully apparent in these "Recherches," and that is the ingenuity with which he formulated expressions to contain terms representing the separate elements of a composite cycle of physical events. In this instance, the several cycles of oscillation—annual, diurnal, and semidiurnal—that when taken together produce the gross ebb and flow of the tides, are handled serially. Laplace himself did not use the term harmonic theory that has since been applied to the type of tidal analysis that he started here. The phrase is apt, however. In effect, he did treat the determination of the location of a molecule of seawater relative to its equilibrium position like that of the displacement of a point on a string vibrating so as to produce a beat of tones and overtones. The actual position at any moment is the linear resultant of the respective displacements suffered in consequence of oscillations of several frequencies. The technique appears to good advantage in the second installment of this series, submitted to the Academy on 7 October 1778, almost a year after the first. There Laplace went over the same material just summarized, cleaning up and ordering the analysis that he had spun out more or less as it came to him the first time around. Perhaps it will be useful to single out the salient features as they pertain to the elements of tidal oscillation. The problem is set up to consider the displacements of a fluid molecule, M, on the surface of the sea. The following are the parameters at the outset ([*1779a*]; *OC*, IX, 187–188):

θ — complement of the latitude; and
ω — longitude in relation to a meridian fixed in space

(after time t, θ becomes $\theta + \alpha u$; and ω becomes $\omega + nt + \alpha v$, where nt represents the rotational motion of the earth, and α is a very small coefficient);

αy — elevation of M above the surface in the equilibrium state it would have reached without the action of the sun and moon;
$\alpha\delta B$, $\alpha\delta C$ — components of the attraction for M of an aqueous spheroid of radius $(1+\alpha y)$, decomposed perpendicularly to the radius in the planes of the meridian and the parallel of latitude;
δ — density of seawater;
s — mass of the attracting star;
ν — complement of its declination;
ϕ — longitude reckoned on the equator from the fixed meridian;
h — distance from the center of the earth, the minor semiaxis of which is taken for unity, and we let $\frac{3s}{2h^3} = \alpha K$;
g — gravitational constant; and
$l\gamma$ — depth of the sea, l being very small and γ being any function of θ.

With these quantities, Laplace restated in terms of polar coordinates three equations that he had formulated in his initial analysis governing the motions of M, thus obtaining (*ibid.*; *OC*, IX, 188):

$$y = -\frac{l}{\sin\theta}\frac{\partial(\mu y \sin\theta)}{\partial\theta} - l\gamma\frac{\partial v}{\partial\omega} \qquad (32)$$

$$\frac{d^2u}{dt^2} - 2n\frac{dv}{dt}\sin\theta\cos\theta = -g\frac{\partial y}{\partial\theta} + \delta B + \frac{\partial R}{\partial\theta} \qquad (33)$$

$$\frac{d^2v}{dt^2}\sin^2\theta + 2n\frac{du}{dt}\sin\theta\cos\theta = -g\frac{\partial y}{\partial\omega} + \delta C\sin\theta + \frac{\partial R}{\partial\omega} \qquad (34)$$

where

$$R = K[\cos\theta\cos\nu + \sin\theta\sin\nu\cos(\phi - nt - \omega)]^2. \qquad (35)$$

(In these papers, R represents a force acting in a given direction to disturb the equilibrium of the particle M. Its derivatives are taken with respect to θ and ω; the terms K, ν, and ϕ are given by the law of motion of the attracting star as functions of time t [*ibid.*, nos. III, XXII, XXIV; *OC*, IX, 95, 187–189, 198–199; cf. *Mécanique céleste*, Book IV, nos. 1, 4; *OC*, II, 184, 195]). Equations 33, 34, and 35 were then combined and transformed by various manipulations into a single equation in rectangular coordinates, setting

$$\begin{aligned} y &= a\cos(it + s\omega + A) \\ y' &= a'\cos(it + s\omega + A) \\ u' &= b\cos(it + s\omega + A) \\ v &= c\sin(it + s\omega + A), \end{aligned} \qquad (36)$$

where i and s are any constant coefficients, and a, a', b, and c are functions of θ. Then, when $\sin\theta = x$, the following equation comprises the entire theory of the tides:

$$\begin{aligned} &ix^3 a\,(i^2-4n^2+4n^2x^2)^2 \\ &= -\,lgzix^2\,(1-x^2)\,(i^2-4n^2+4n^2x^2)\,\frac{\partial^2 a'}{\partial x^2} \\ &\quad + lgzix^3 \frac{\partial a'}{\partial x}\,(i^2+4n^2-4n^2x^2) \\ &\quad + lgsza'\,[(is+2n)\,(i^2-4n^2+4n^2x^2) \\ &\quad + 16n^3x^2\,(1-x^2)] \\ &\quad - lg\frac{\partial z}{\partial x}\,x\,(1-x^2)\,(i^2-4n^2+4n^2x^2) \\ &\quad \left(ix\frac{\partial a'}{\partial x}+2nsa'\right). \end{aligned} \qquad (37)$$

Fortunately, it did not have to be solved; it needed only to be satisfied, and that could be done piecemeal, introducing simplifications by dint of reasonable suppositions about the physical factors at work—as to equilibrium, fluid friction, resistance, and so on—and determining the coefficients i and s, and the function a' in terms of a.

To accomplish this task, Laplace grouped the different terms of the expression for R (35) according to the cycle that they represent in the total tidal flow. Thus:

$$\begin{aligned} R &= K[\cos\theta\cos\nu + \sin\theta\sin\nu\cos(\phi - nt - \omega)]^2 \\ &= K\cos^2\nu + \frac{1}{2}K\sin^2\theta\,(\sin^2\nu - 2\cos^2\nu) \\ &\quad + 2K\sin\theta\cos\theta\sin\nu\cos\nu\cos(nt+\omega-\phi) \\ &\quad + \frac{1}{2}K\sin^2\theta\sin^2\nu\cos(2nt+2\omega-2\phi). \end{aligned} \qquad (38)$$

Developing the values for R, Laplace obtained, in place of the three ranks of terms above, three series of the respective forms

$$\begin{aligned} &(1°)\ K' + K''\sin^2\theta\cos(it+A)\,, \\ &(2°)\ K'\sin\theta\cos\theta\cos(it+\omega+A)\,, \\ &(3°)\ K'\sin^2\theta\cos(it+2\omega+A)\,, \end{aligned} \qquad (39)$$

where K', K'', and A are any constant coefficients. These three classes of the expressions for R, and consequently for y, group terms in the following manner:

(1°) includes terms that are independent of ω. Their period is thus proportional to the time of revolution of the star in its orbit—annual in the case of the sun;

(2°) includes terms of which the period is approximately one day;

(3°) includes terms of which the period is half a day.

(By the conditions of the problem, i was small compared to n in terms of the first form; and if the orbit is assumed circular, $i = 0$ or $2m$, m being the mean motion; in terms of the second form, i differs very little from n, and this difference is 0 or $2m$ if the orbit is circular. In terms of the third form, i is very little different from $2n$, which difference is again 0 or $2m$ in the case of circular orbits.)

Examining now the terms of the first class (1° of Equation 39), which has the form

$$K' + K''\sin^2\theta\cos(it+A)\,, \qquad (40)$$

the condition that they govern the annual revolution gives values for the coefficients that need to be determined in order to satisfy (37), such that, for this part of the oscillation,

$$y = \frac{K\left(\cos^2\nu - \frac{1}{2}\sin^2\nu\right)}{6g\left(1-\frac{3\delta}{5\delta'}\right)}\,(1+3\cos 2\theta)\,, \qquad (41)$$

where δ' is the mean density of the earth, and $g = (4/3)\,\pi\delta'$. The expression is exact for the action of the sun, although less so for the moon, since the waters have less time to return to equilibrium in a month than in a year. But any errors are of little importance in the theory of the tides, since they affect only the absolute heights relative to the phases of the moon, and not the difference between high and low tide.

The depth of the sea figures in this analysis, its expression being $(1 + q^2 \sin\theta)$ in polar coordinates. Since $\gamma = 1 + (q/l)\sin^2\theta$, then $z = x + (q/l)x^3$ in rectangular coordinates, q being a very small constant coefficient of the order of l. This value is important for two reasons. Physically, the equilibrium considerations require that the mass of the seawater remain constant. Analytically—and this is more important, although it comes in rather unobtrusively—the determination of the values of a and a' to be substituted in (37) depended on assuming that q is approximately zero, and that the depth of the sea is constant. "This method," Laplace observed, "can thus serve to find the approximate values of a, given this hypothesis concerning the depth [of the sea], which we shall see in the following article is approximately that of nature" ([*1779a*]; *OC*, IX, 202).

It is the terms of the second class (2° of Equation 39) that are the most important for the inversion of the tides, their form being

$$K' \sin\theta \cos\theta \cos(it+\omega+A)\,, \tag{42}$$

and their period approximately a single day. Here the conditions are such that analysis yields the expression for the part of y corresponding to these terms

$$\frac{2q}{2qg\left(1-\frac{3\delta}{5\delta'}\right)-n^2} K' \sin\theta \cos\theta \cos(it+\omega+A)\,. \tag{43}$$

And now we begin to see the reason for all this. Laplace next sums all terms of this second form given by the development of R and designates by Y the corresponding expression for y:

$$Y=\frac{4Kq \sin\nu \cos\nu \sin\theta \cos\theta}{2qg\left(1-\frac{3\delta}{5\delta'}\right)-n^2}\cos(nt+\omega-\phi)\,. \tag{44}$$

The value of Y will be a maximum when $nt+\omega-\phi=0$ or π, which is to say, when the star that is the origin of the attracting force passes the meridian. The expression has a negative value when $nt+\omega-\phi=\pi$; and the difference between the maximum positive and negative values, which equals twice the former, will measure the difference between the two full tides of the same day. That difference will be proportional to the unwieldy coefficient of $\cos(nt+\omega-\phi)$ in (44). Since observation shows this difference to be extremely small, the value of q must be nil or almost nil. And finally, since the depth of the sea is $l+q\sin^2\theta$, it follows that "in order to satisfy the phenomena of ebb and flow of the tides, this depth must be approximately constant" (*ibid.*, *MARS* [1776/1779], 199–200). That is, it must be l, which result agrees with what he had already found. Thus, he could conclude—"without fear of observable error" (*ibid.*, 200)—that $Y=0$ and determine the values needed to satisfy (37).

Laplace went on to set out verbally what he had in mind as a physical picture. For this purpose, he designated by **u** and **U** the parts of u and v that correspond to the terms of the second class for the expression of R. It will be recalled that u and v are the variations respectively of θ, the complement of latitude, and of ω, the longitude. Thus defined,

$$\mathbf{u}=\frac{2K}{n^2}\sin\nu\cos\nu\cos(nt+\omega-\phi)\,, \tag{45}$$

and

$$\mathbf{U}=-\frac{2K}{n^2}\sin\nu\cos\nu\frac{\cos\theta}{\sin\theta}\sin(nt+\omega-\phi)\,. \tag{46}$$

Confining attention now to expressions of the form

$$2K\sin\nu\cos\nu\sin\theta\cos\theta\,(nt+\omega-\phi)\,, \tag{47}$$

and to the force they represent, it is (Laplace says) "easy to see" (*ibid.*, 200–202) that the molecules will slide over each other as if isolated, with no detectable loss from internal collisions in the system. We are to imagine a slice of this fluidity contained between two meridians and two parallels infinitely close together. Since the value of **u** is constant for all molecules located under the same meridian, the length of the slice remains constant in that direction. To the extent that the fluid flows toward the equator, however, the space enclosed between the two meridians increases. Since the slice thus widens along the parallels, the surface of the water would tend to descend if it were not for the latitudinal component of the velocity of the molecules, which tends to squeeze the two meridians together and correspondingly diminish the width. Again "it is easy," we are assured, to conclude that this diminution, resulting from the value for **U**, is compensated by the motion of the slice toward the equator. Therefore, the width actually remains constant along the parallels, and the height of the slice is not affected detectably by the motion of the fluid. That is the reason for which in a sea of uniform depth (such as would be constituted by the sum of all such slices taken side by side), the difference between consecutive high tides is almost undetectable. (It is essential to refer to the original printing for this passage, since *OC*, IX, 212, mixes up the notation.)

Now then, whatever may be thought of this argument, it can have referred only to the notion that Laplace had actually formed of physical reality. He was thus enlarging on the phenomenon, he said, because it was very important for the understanding of the tides and entirely contrary to received theory. He went on to illustrate the argument by data drawn from observations made at Brest and turned it against Daniel Bernoulli's idea that the earth rotates too rapidly for the tides to differ conformably with the results of theory. He himself had just shown that whatever the speed of rotation, consecutive high tides would be very unequal if the sea were not everywhere of approximately the same depth. "It seems to me," he concluded, "to result from these considerations that only an explanation founded on a rigorous calculation like this one . . . is capable of meeting all the objections that can be lodged against the principle of universal gravitation on this score" (*ibid.*; *OC*, IX, 214).

Terms of the form $K' \sin^2 \theta (\cos it + 2\omega + A)$ constitute the third class (3° of Equation 39) produced by the development of *R*. Their period is half a day, or that of a single tidal cycle. Examining these terms, Laplace substituted values for the observed tidal range, or difference between low and high tides, in various localities and determined that deriving the law of gravity from the data required assigning the figure of four leagues to the mean depth of the sea "on which only vague and uncertain conjectures have been formed until now" (*ibid.*; *OC*, IX, 216). We shall not excerpt that analysis, which is of the same type as the foregoing, nor summarize the calculation that exhibits the proportionality of the slight difference between daily high tides to the influence of the tides on the ratio of nutation to precession. We shall also pass over the third installment of this memoir, in which Laplace calculated much more succinctly that the gravitational effect of the sun and moon on the atmosphere is bound to cause oscillations comparable to tides but too slight to detect (see Section 28). More important, he redeemed the promise of an earlier memoir ([*1776a*, 2°]; see Section 5) and showed that these effects cannot be the source of the trade winds, recognizing that oversimplified assumptions, most notably that of constant temperature, reduced his results to a qualitative significance.

Instead, let us conclude this discussion of Laplace's youthful period with a further memoir, "Sur la précession des équinoxes" ([*1780c*]), which he read before the Academy on 18 August 1779. It opens with a simplified derivation of the results of his previous research on the relation of tidal action to precession and nutation, the consequence of which is now stated as a theorem (*ibid.*; *OC*, IX, 341–342):

> If the earth be supposed an ellipsoid of revolution covered by the sea, the fluidity of the water in no way interferes with the attraction of the sun and the moon on precession and nutation, so that this effect is just the same as if the sea formed a solid mass with the earth.

This theorem might well have been thought to hold true for any spheroid, observed Laplace, although he had been unable to demonstrate its generality by his previous analysis. The expressions that he formed there to represent the influence of the oscillations of the sea on precession could not be integrated generally. In the meanwhile, he had found a much simpler method, which formed the subject of the present memoir and which met this difficulty. Essentially it consisted of nothing other than the application to the problem of d'Arcy's principle of areas. Unlike the other basic principles, particularly vis viva conservation or least action, which are limited to gradual and continuous changes of motion, d'Arcy's principle holds good in cases of shock or turbulence, such as friction with the bed of the sea and resistance from the coasts in tidal movements. We shall not follow the analysis, except to note its crux, which was that the expression of the variation of the overall motion of the fluids covering a globe contains no terms dependent on the secular passage of time. Thus, both for mechanical and formal reasons, the new method might be extended to what Laplace now, not even a year later, called the case in nature, "in which the figure of the earth and the depth of the sea are very irregular, and the oscillations of the water are modified by a vast number of obstacles" (*ibid.*; *OC*, IX, 342).

Now that the seabed had suddenly—and one is tempted to say analytically—become very irregular instead of uniform, Laplace was led to speculate that the spinning earth may not be perfectly symmetrical. In that event, the resulting irregularities in angular momentum, added to the direct gravitational effects of the sun and the moon and the reaction of the seas to their tidal pull, might change the axis of rotation over very long periods. The poles would then migrate into other regions in the course of time—but this he would leave as a conjecture, worthy of the attention of mathematicians because of its difficulty and importance.

PART II: LAPLACE IN HIS PRIME, 1778–1789

9. Influence and Reputation. By the late 1770's Laplace had begun to win a reputation extending beyond the small circle of mathematicians who could understand his work, and by the late 1780's he was recognized as one of the leading figures of the Academy. On 15 May 1788 he was married to Marie-Charlotte de Courty de Romanges, of a Besançon family (J: Marmottan [1897], 7). She was twenty years younger than he, and they had two children. Laplace's son, Charles-Émile, born in 1789, followed a military career, became a general, and died without issue in 1874. The daughter,

Sophie-Suzanne, married the marquis de Portes and died in childbirth in 1813. Her child, a girl, survived and married the comte de Colbert-Chabannais. The living descendants derive from that marriage, having taken the name Colbert-Laplace.

We know practically nothing of Laplace's personal life in the years before his marriage, but there are a few indications of his effect on others. Not a single testimonial bespeaking congeniality survives. There are hints that the aged d'Alembert began to resent the regularity with which his recent protégé was relegating his own work to the history of rational mechanics. Overly elaborate tributes by Condorcet and Laplace himself bear the scent of mollification, notably in the prefaces to the "Recherches sur plusieurs points du système du monde," which we have just discussed. Anders Johan Lexell, a Swedish astronomer who spent part of the winter of 1780–1781 in Paris, wrote in a gossipy account of the Academy that Laplace let it be known that he considered himself the best mathematician in France. He also had extensive knowledge in other sciences, reported Lexell, but presumed too far upon it, "for in the Academy he wanted to pronounce on everything" (*Revue d'histoire des sciences*, **10** [1957], 148–166).

At about the same time, Laplace fell into a heated dispute with Jacques-Pierre Brissot, the future Girondist leader and a rising scribbler, over the optical experiments of Jean-Paul Marat, who was then seeking scientific recognition and entry to the Academy. In Brissot's dialogue on "academic prejudice" Laplace is the original of the Newtonian idolater, arrogant in his *fauteuil* and contemptuously spurning the aspiring "physicist" from the impregnable—and irrelevant—plane of mathematics (Brissot, *De la vérité* [Neuchâtel, 1782], 335).

In 1784 the government appointed Laplace to succeed Bezout as examiner of cadets for the Royal Artillery (J: Duveen and Hahn [1957]). The candidates had generally completed secondary school and a year or two of special preparatory school before going on to La Fère or one of the other artillery schools, or in some cases to Mézières for engineering. The responsibility was more serious than might at first glance be supposed. An individual report had to be written on each cadet, all of whom were of good family. The annual scrutiny brought Laplace, as it did Monge, the newly appointed examiner for naval cadets, into regular contact with ministers and high officers; and it introduced them to the practice of recruiting an elite by competitive examination that was later greatly expanded in scale and intensified in mathematical content by the procedures for selecting students to enter the École Polytechnique. (For Laplace's reports, see Bibliography, Section A). The government also named him to the most famous of the blue-ribbon commissions through which the Academy investigated and made recommendations on matters of civic concern in the last years of the old regime. Laplace was a member of the commission headed by Bailly to investigate the Hôtel Dieu, the major hospital in Paris, as well as hospital care in general. Calculations on the relative probabilities of emerging alive from its wards, and comparisons of the mortality there to other hospitals in France and abroad, must almost certainly have been his contribution to its report (*HARS* [1785/1788], 44–50).

Laplace was promoted to the senior rank of pensioner in the Academy in April 1785 in the vacancy created by the death of Le Roy. In this, technically the most proficient and productive period of his life, he pressed forward with all the topics that he had begun investigating in his youth, added physics to them, and achieved many of the results for which he is famous. From the published record, it appears that in the early 1780's his emphasis shifted from the problems of attraction, which were occupying him in 1777 and 1778 (*24, 26, 28*), to probability again, both in its calculus and now in its application to demography, and equally to the experimental and mathematical physics of heat. In the mid-1780's his interest again centered on attraction and the figure of the earth, and in the latter half of the decade it turned to planetary motions. It will be best to discuss these subjects in that sequence, beginning with a pair of relatively brief but important mathematical memoirs and continuing into probability. These dictates of convenience, however, must not obscure the evidence that, to an extraordinary degree, Laplace was able to hold and mature all these matters in his mind concurrently, readily turning from one to another. For even while occupied with probability about 1780, and with the Lavoisier collaboration simultaneously underway, he did the paper on precession already noticed and in 1781 drafted another, very important, memoir on the determination of cometary orbits, to be discussed in Section 14.

10. Variation of Constants; Differential Operators. The earlier of the two mathematical papers that preceded his resumption of probability pertained to astronomy and was entitled "Mémoire sur l'intégration des équations différentielles par approximation" ([*1780a*]). Laplace there simplified the technique of varying arbitrary constants in approximate solutions to differential equations of planetary motion in order to eliminate the troublesome secular terms that crept in and destroyed convergence. Having brought out the method by means of numerous examples in a very lengthy memoir ([1776c]; see Section 7), he now gave rules for applying it generally in problems of theoretical astronomy. Laplace's technical innovations were often thus introduced discursively and then shortened and pointed for restatement.

The latter paper, "Mémoire sur l'usage du calcul aux différences partielles dans la théorie des suites" ([*1780b*]), was the more important mathematically. It belongs to the nascent stages of the calculus of operations, as it was called in the nineteenth century, or, more recently, the calculus of differential operators. Laplace submitted it to the Academy on 16 June 1779 (*29*). Three years previously he had appended Articles XI and XII, "Sur les fonctions," to his memoir on the mean inclination of the comets ([*1776b*]) and there gave what he considered a general demonstration of a theorem that Lagrange had obtained by induction ("Sur une nouvelle espèce de calcul," *Oeuvres de Lagrange*, III, 441–476), Lagrange had developed the analogy between positive exponents and indices of differentiation, and reciprocally between negative exponents and indices of integration, and stated the following theorem. We give it in the notation of Laplace, who unlike Lagrange considered only a single variable:

$$\Delta^n u = \left(e^{\alpha \frac{du}{dx}} - 1\right)^n \quad \text{and} \quad \Sigma^n u = \frac{1}{\left(e^{\alpha \frac{du}{dx}} - 1\right)^n}. \qquad (48)$$

Laplace's 1776 proof depends on showing that in the expression for the developments

$$\alpha^n \frac{d^n u}{dx^n} = \Delta^n u + s\Delta^{n+1} u + s'\Delta^{n+2} u + \cdots \qquad (49)$$

and

$$\frac{1}{\alpha^n} \int^n u dx^n = \Sigma^n u + f\Sigma^{n-1} u + f'\Sigma^{n-2} u + \cdots \qquad (50)$$

the coefficients $s, s', \cdots, f, f', \cdots$ are constant and independent of α, thus depending only on n (*ibid.; OC*, VIII, 314–319). Lagrange's expressions are obtained if $u = e^x$, and the coefficients can thus be determined by the choice of the funtion u. Now, in the fullness of an entire memoir ([*1780b*]), Laplace dealt more generally with the possibility of moving back and forth between powers and indices of differentiation and integration, giving several alternative demonstrations of the Lagrange theorems and drawing out corollaries.

According to his own testimony, this research was one of the factors leading to the development of his theory of generating functions ([*1782a*]; *OC*, X, 2), which followed immediately after the memoir on probability ([*1781a*]) that he submitted to the Academy on 31 May 1780 (*33*). The draft was entitled "Mémoire sur le calcul aux suites appliqué aux probabilités," but as will appear in the next section, the treatment is far more comprehensive than that phrase would imply. For with principles or motifs, Laplace's pattern contrasts with that just noticed in the development that he often gave his technical innovations. Whereas the latter were abbreviated and focused after a lengthy initial statement, whole topics such as universal gravitation or inverse probability were first mentioned almost in passing and then, as the ideas continued germinating in his mind, were broadened to embrace entire domains of science and knowledge.

11. Probability Matured. The "Mémoire sur les probabilités" ([*1781a*])—the change of title epitomizes the process outlined above—gives the appearance of a finished piece of work. In a letter to Lagrange of 11 August 1780 he said that the principal object is the "method of going behind events to causes" (*Oeuvres de Lagrange*, XIV, 95). The preamble is more general. It reviews the entire status of inverse probability, pertaining as it does to a "very delicate metaphysics," the use of which is indispensable if the theory of probability is to be applied to life in society. The subject has two main aspects, closely related. In the first, the task is to calculate the probability of complex events compounded of elementary events of which the respective possibilities are unknown. In the second, the problem is to determine numerically the influence exerted by past events on the probability of future ones or, as we would say, to draw statistical inferences. In Laplace's view, the goal was to uncover the law that reveals the causes.

Before discussing problems of the former class, Laplace refined his epistemology slightly while distinguishing his procedure from previous practice in the theory of chance. In the traditional approach, the respective possibilities of simple events—that is, heads or tails in the tossing of a coin—had been determined in one of three ways: (1) *a priori*, by the assumption of equal possibility; (2) *a posteriori*, by repeated experiment; and (3) by whatever reasons we may have to judge the likely occurrence of the event. As to the last, if there is no reason to suppose player A more skillful than his opponent B, the probability assigned to his chance of winning is 1/2. The first approach gives the absolute possibility of events; the second—as he will show later—the approximate possibility; and the third, only their possibility relative to our information.

Since every event is actually determined by the general laws of the universe and is probable only relative to our knowledge, these distinctions may appear unreal. Still, they have their uses, and Laplace went on to consider a little more fully what they involve. Amid all the factors that produce some event, some are different every time, such as the precise movements of the hand in throwing dice. It is the overall effect of these factors that we call chance. Other factors are constant, such as the relative skill of the players or the weighting of the dice. Taken together, these are what constitute the "absolute possibility" (to which he intended to assimilate the approximate, although he did not say that at the outset). It is our greater or lesser knowledge of constant factors that constitutes their "relative possibility." These invariant factors do not suffice to produce the event. They must be joined to the operation of the chance or variable factors first mentioned and only increase the probability of events without determining their occurrence. Thus (if we have construed this passage correctly) Laplace considered at this juncture that the state of knowledge enters into the determination of probability at two levels: in what we know (relative possibility) of the absolute possibility (invariant factors) in events, and in our ignorance of the laws that will always appear to produce chance events. In this class of problems, then, the calculation of the probability of complex events, Laplace drew (as he had previously) upon models offered by games of chance and generalized the treatment that he had already given to estimating the influence of such unknown factors as slight asymmetries in coins or dice and differing habits or skills in players.

The development that he gave to the second topic, the estimation of causes from effects, broke new ground in the field of application. Indeed, it could be argued that social statistics as a mathematical subject had its beginning in this memoir. Laplace began the discussion by addressing himself to the first of two difficulties that impeded the application of Bayes's theorem to problems in the real world. (It is interesting that, although Laplace himself still ignored Bayes, whether in the French or English sense of the verb, Condorcet mentioned Bayes and Price in the summary of Laplace's memoir in his historical preface to the volume in which it appears [*HARS* (1778/1781), 43]. They are said to have stated the principle for determining causes from effects, but without any calculations.) The first impediment that Laplace identified was practical: experience was virtually never sufficiently extensive or controlled to yield reliable values for the *a priori* probabilities. The second impediment was analytical: in order to achieve numerical solutions, it was often necessary to integrate differential equations containing terms raised to very high powers. Laplace touched on that problem in the present memoir but reserved its full resolution for an even more technical companion paper that he appears to have been evolving concurrently.

To overcome the former, experiential difficulty, Laplace turned to the one subject on which statistically significant information had already been assembled—population. Population studies as a science owe much substance to the growing professionalism of eighteenth-century public administration. French parish registers in principle contained records of all births, marriages, and deaths. In 1771 the controller general of finance, the Abbé Terray, instructed all intendants in the provinces to have the figures for their generalities compiled annually and to report the results regularly to Paris in order that the government might have accurate information on the entire population. Turgot, a statesman close to the scientists in spirit and in program, was appointed controller general in August 1774. Through his influence, the Academy of Sciences became interested and published a summary of the figures for the city of Paris and the faubourgs covering the years 1709 through 1770. The record showed that in the last twenty-five years of that period, 251,527 boys and 241,945 girls had been born; and the ratio of approximately 105 to 101 remained virtually constant year by year. Figures also existed for London, and there too, more boys than girls had been born, although in the slightly greater ratio of 19 to 18.

Unlike imaginary black and white balls drawn from a hypothetical urn, the births of real children afforded a genuine numerical example, and Laplace seized on the opportunity to try out his technique for determining the limits of probability of future events based on past experience. Given p male and q female babies, the probability of the birth of a boy is $p/(p + q)$. Laplace next reiterated the theorem identical with Bernoulli's law of large numbers—again without mentioning Bernoulli (see Section 4)—to the effect that if P designates the probability that the chance of the birth of a boy is contained within the limits $p/(p + q) + \theta$ and $p/(p + q) - \theta$, where θ is a very small quantity, the difference between P and unity would vary inversely as the value of p and q. The latter quantity could be increased so that the difference between P and unity would be less than any given magnitude. To that end, he represented the definite integral for P by a highly convergent series, which on evaluation reduced to unity when p and q became infinite.

Readers may be interested in a précis of the calculation on which the above results depend ([*1781a*], no. XVIII; *OC*, IX, 422–429). What follows is a very close paraphrase in Laplace's notation. If x is the probability of the birth of a boy, and $(1 - x)$ that of a girl, then to determine the probability that x will fall within arbitrary limits became a problem of evaluating between those limits the definite integral

$$\int x^p (1-x)^q \, dx \qquad (51)$$

taken from $x = 0$ to $x = 1$, where p and q are very large numbers. Now, letting

$$y = x^p (1-x)^q, \qquad (52)$$

it followed that

$$y\,dx = \frac{x(1-x)}{p-(p+q)x}\,dy. \qquad (53)$$

If we set

$$p = 1/\alpha \quad \text{and} \quad q = \mu/\alpha, \qquad (54)$$

α being a very small fraction since p and q are very large, then (53) becomes

$$y\,dx = \alpha z\,dy, \qquad (55)$$

where

$$z = \frac{x(1-x)}{1-(1+\mu)x}. \qquad (56)$$

Thus, whatever the value of z,

$$\int y\,dx = C + \alpha y z \left\{ 1 - \alpha \frac{dz}{dx} + \alpha^2 \frac{d(z\,dz)}{dx^2} - \alpha^3 \frac{d[z\,d(z\,dz)]}{dx^3} + \cdots \right\}, \qquad (57)$$

where C is an arbitrary constant depending on the initial value of $\int y\,dx$.

The series (57) would no longer be convergent if the denominator of z in (56) were of the same order of magnitude as α, which would be the case when x only differed from $1/(1 + \mu)$ by a quantity of that order. The series was to be employed, therefore, only when that difference was very large with respect to α. But even that did not suffice. Since each differentiation increased the powers of the denominators of z and its derivatives by one, the term of which the coefficient was α^i had for denominator the term in z raised to the power z^{i-1}.

Thus, for the series to be convergent, α had to be not only much less than the denominator of z, but even much less than the square of the denominator. Under those conditions, the series (57) by very rapid approximations would give the value of the integral $\int y\,dx$ between the limits $x = 0$ and $x = 1/(1 + \mu) - \theta$ provided that α was much less than θ^2. If $x = 0$, then $y = 0$ and $z = 0$, and in that case (57) becomes

$$\int y\,dx = \frac{\alpha\mu^{q+1}[1-(1+\mu)\,\theta]^{p+1}\left(1+\frac{1+\mu}{\mu}\,\theta\right)^{q+1}}{\theta\,(1+\mu)^{p+q+3}} \times \left\{ 1 - \frac{\alpha[\mu + (1+\mu)^2\theta^2]}{\theta^2(1+\mu)^3} + \cdots \right\}. \qquad (58)$$

That series gave the limits between which the value of $\int y\,dx$ was contained, a value less than the first term in the curled brackets and greater than the sum of the first two terms (*OC*, IX, 425).

A similar demonstration shows that the series (57) also gave values of $\int y\,dx$ from $x = 1/(1 + \mu) + \theta$ to $x = 1$. In showing that such was the case, Laplace proved that the more p and q were increased, the more α was diminished, and that the difference between P and unity was proportional to α, so that by increasing p and q, and thus diminishing α, the difference could be reduced below any given magnitude.

It remained to draw upon the data. Calculating the probability that the possibility of a male birth in Paris was greater than 0.5, he found it to be less than unity by the fraction 1.1521×10^{-13}. He further calculated that the probability that in any year baby boys would fail to exceed baby girls in number was 1/259 in Paris, whereas in London it was 1/12,416. Although the ratio of male to female births was only slightly higher in London, the probability of male preponderance approached certainty at a drastically increasing rate as the proportion grew. It would have been reasonable to bet that boys would outnumber girls in any of the next 179 years in Paris and in any of the next 8,605 in London. Here, then, Laplace had worked out a method for finding numerical solutions to a type of problem the analytical solutions of which contained terms raised to such high powers that the expressions became impracticable when the numbers were substituted in the formulas. Moreover, his method was applicable to practical calculation of future events given the experience of the past. In short, not only did he have a method for statistical inference in hypothetical circumstances, but he actually drew inferences in the area of population.

12. Generating Functions and Definite Integrals. Of the next two memoirs to be discussed, the first, "Mémoire sur les suites" ([*1782a*]), was entirely mathematical. The second, "Mémoire sur les approximations des formules qui sont fonctions de très grands nombres" ([*1785a*]), was largely so. Both were motivated mainly, though not exclusively, by probabilistic concerns. The former, which introduced the theory of generating functions, was particularly important to Laplace. Years later, when composing the *Théorie analytique des probabilités*, he subordinated all of the analytical part to the theory of generating functions and represented the whole subject as their field of application; the first part of Book I is in the main a reprinting of this memoir, incorporating clarifications and simplifications in detail and several important additions. He defined generating functions in much the same words in the opening pages of the *Théorie analytique* and of the "Mémoire sur les suites" (*OC*, VII, 7; *OC*, X, 5):

> Let y_x be any function of x. If there be formed the infinite series, $y_0 + y_1t^1 + y_2t^2 + y_3t^3 + \cdots + y_xt^x + y_{x+1}t^{x+1} + \cdots + y_\infty t^\infty$, and u be designated the sum of that series, or (which comes to the same thing) the function of which the development forms the series, this function will be what I call the *generating function* of the variable y_x.

Thus, he explained, the generating function of a variable y_x is a function of t that, developed in powers of t, has that variable for the coefficient of t^x. Reciprocally, the corresponding variable of a generating function is the coefficient of t^x in the development of the function in powers of t. The exponent expressing the power of t then indicates the place that the variable y_x occupies in the series, which may also be extended indefinitely to the left according to negative powers of t.

Laplace regarded generating functions as something of a panacea for problems involving the development of functions in series and evaluation of the sums. Those procedures embraced nothing less than most of the possible applications of mathematics to nature. His memoir consists in showing how to apply the device to problems of interpolation both in convergent and recurrent series. As a corollary to the latter, he gave a method for solving the linear finite difference equations that express the relation between terms. From finite analysis he proceeded to the comparable infinitesimal expressions and from there to series involving two variables, recurro-recurrent as well as convergent in nature, observing by the way that he himself had initiated the theory of recurro-recurrent series. In the course of considering the solution of linear partial difference equations on which recurro-recurrent series depended, he had reduced the problem to one of infinitesimal differences by means of definite integrals involving a new variable. In effect, the technique results in finding discontinuous solutions to differential equations, the existence of which had emerged historically in the analysis of sound waves. Laplace reminded the reader that the integral of second-order partial differential equations contains two arbitrary functions in the form of differentials. Euler and Lagrange had been the ones to discover these mathematical objects in analyzing acoustical problems in which the movements of the air in transmitting sound were considered three-dimensionally. (This occasion was Laplace's first mention of the theory of sound.) It had been left to him, however, to find a general method for integrating second-order linear partial differential equations in the cases where that was possible and for identifying those cases in which it was not ([*1777a*]; see Section 7). Now he had found a method for integrating many such expressions by means of these definite integrals involving a new variable. The technique was further applicable to problems of vibrating cords, which in turn led to consideration of the employment of discontinuous functions in problems involving partial differences.

Thus, many of Laplace's existing analytical and probabilistic interests merged with certain of his future physical concerns in this memoir. What led him to the idea of generating functions may well have been further reflection on the calculus of differential operators (see Section 10). He mentioned at the outset how the relation between a generating function and its independent variable leads directly to the analogy between positive powers and derivatives and between negative powers and integrals. Lagrange had formulated that relation in a theorem that Laplace himself had demonstrated in a general manner. His calculus of generating functions is, in a sense, a calculus of exponents and characteristics (or operators) as well as of coefficients and variable quantities, and indeed he devoted Article X to deriving a theorem of Lagrange (*Oeuvres de Lagrange*, III, 450; cf. Section 10) in the form

$$'\Delta y_x = e^{\alpha \frac{dy_x}{dx}} - 1 \qquad (59)$$

by means of generating functions; the symbol $'\Delta y_x$ meaning that x varies by the quantity i.

An example will illustrate the method. Laplace analyzed the series formed by multiplying the terms of a convergent series by the terms of a geometric progression. The general term of the series produced would be $h^x y_x$ (y_x being the general term of the convergent series). Then u would be the sum of the infinite series

$$y_0 + y_1 ht + y_2 h^2 t^2 + y_3 h^3 t^3 + \cdots + y_\infty h^\infty t^\infty. \qquad (60)$$

It followed from the general analysis that he had already given of the relation between the coefficients of t and the corresponding powers of the variable that, in this example,

$$u\left(\frac{1}{t^i} - 1\right)^n = n\left[h^i\left(1 + \frac{1}{ht} - 1\right)^i - 1\right]^n. \qquad (61)$$

The coefficient of t^x on the left of that equation is the nth finite difference of $h^x y_x$, where x varies by the quantity i. Moreover, if the right-hand side of (61) is developed in powers of $\frac{1}{ht} - 1$, the coefficient of t^x in $u\left(\frac{1}{ht} - 1\right)^r$ will be $h^x \Delta^r y_x$, whatever the value of r. If the exponent n is negative instead of positive, the analysis will relate to integration instead of differentiation.

We do not know when Laplace resumed work upon the matter that had concerned him in a special case at the conclusion of the "Mémoire sur les probabilités" ([*1781a*]; see Section 11), namely numerical approximation of formulas that are functions of very large numbers; but four years had elapsed before he published his comprehensive treatment of the analytical aspects of the topic ([*1785a*]). It was chiefly, although not exclusively, in the theory of chance that analysis would often result in formulas impossible to use when large numbers were substituted in them. Only in two special cases, the product of the natural numbers 1, 2, 3, 4, ···, and determination of the middle term of a binomial raised to a high power, did techniques exist for readily achieving numerical solutions.

In the latter case, if the power were supposed even and equal to $2s$, that term would be, as everyone knew,

$$\frac{2s(2s-1)\,(2s-2)\,(2s-3)\cdots(s+1)}{1.2.3.4\cdots s}.$$

Even there, formulating the expression in numbers became difficult once s grew large. James Stirling, whose work Laplace consistently admired, had seen how to transform it into a series the sum of which was equivalent to $\sqrt{\pi}/2$ and which converged more rapidly as s grew larger. The transformation was remarkable in that it introduced a transcendent quantity into the investigation of purely algebraic quantities. It was still applicable only to special cases, however. Laplace himself had given a method in [*1781a*] for converting integrals of differential functions containing factors raised to very high powers into rapidly converging series. But he was busy with other matters, and only since then had further reflection shown him the way to extend the method generally to any functions involving very large numbers, thus reducing them to series that, even like the series of Stirling's theorem, become more convergent as the numbers increase.

Since the central difficulty lay in finding numerical solutions for complicated expressions containing many terms, the strategy had to be to transform such formulas into series that converged rapidly enough so that only the first few terms had to be considered. If each of them contained only a few factors, then it would not matter if they were raised to high powers, and resort to logarithms would yield solutions. On the face of it, there seemed no natural way to transform such complicated functions into convergent series. Reflecting, however, that differential expressions simple in form often yield just such functions on integration if the terms have large exponents, Laplace considered that complicated functions of any sort ought to be reducible to integrals of that type, which could then be transformed into convergent series. The problem had two aspects: on the one hand, to integrate by approximation differential equations involving functions that contain very large factors; and on the other, to convert the functions for which approximate values are required into integrals of this type.

Laplace expected that a solution to the former aspect would prove particularly valuable in estimating the probability of causes. The approximation involved series that complement each other: one type was to be employed for points far from the maximum value of the differential function, and the other was to be used for points close to it. The latter contained transcendent quantities, usually reducing to the form

$$\int e^{-t^2}\,dt. \qquad (62)$$

Since that integral, evaluated from $t = 0$ to $t = \infty$, equals $\sqrt{\pi}/2$, Stirling's theorem came out as a particular case of the general analysis.

As for transforming the approximate evaluation into the integration of differential expressions multiplied by factors raised to a high power, Laplace outlined a more indirect approach. The method consisted of representing complicated functions of s (where s is a large integer) by $y_s, y'_s, y''_s, \cdots$. These functions are to be supposed given by linear differential or difference equations of which the coefficients are rational functions of s. The equations are then manipulated so that

$$y_s = \int x^s \phi dx; y'_s = \int x^s \phi' dx; \cdots; \qquad (63)$$

and also so that they can be separated after integration by parts, one part coming under the integral sign and the other outside it. Equating the parts under the integral sign to zero then yields as many linear differential equations as there are variables $\phi, \phi', \phi'', \cdots$, which can thereby be determined as functions of x. Turning to the parts outside the integral sign, they also may be equated to zero. When the arbitrary constants of integration in the values for $\phi, \phi', \phi'', \cdots$, are eliminated, a definitive equation in x results. The roots of this equation then determine the limits between which the integrals $\int x^s \phi dx$, $\int x^s \phi' dx$, $\cdots$, are to be taken. It is very important to notice that the series obtained for $y_s, y'_s, \cdots$, hold good generally when the constants they contain change sign, because this circumstance greatly extends the applicability of the method, although as a result of such change of sign, the definitive equation in x, which gives the limits of the integrals, no longer has several real roots. The most serious difficulty to be overcome in applying this analysis arises from the nature of the differential equations in $\phi, \phi', \phi'', \cdots$, which often cannot be integrated. That obstacle might normally be surmounted through representing the functions $y_s, y'_s, \cdots$, in terms of multiple integrals such as $\int x^s x'^s \phi \, dx \, dx'$, $\int x^s x'^s \phi' \, dx \, dx'$, $\cdots$. The variables $\phi, \phi', \cdots$ might then be determined by equations of a lower order that could be integrated. All these possibilities being considered, Laplace claimed that the analysis might be employed generally for very complicated functions represented by ordinary or partial difference or differential equations—or for all the normal uses of analysis ([*1785a*]; *OC*, X, 213: cf. Section 29).

He reserved his own application for a sequel ([*1786b*]) to this memoir, which had grown very lengthy. Reading the continuation before the Académie des Sciences on 25 and 28 June 1785 (*49*), Laplace again reviewed the epistemological status of probability. It was there that he first employed phrasing of which the second sentence is famous from the *Essai philosophique* thirty years later (*OC*, X, 296; VII, viii):

> The word "chance" then expresses only our ignorance of the causes of the phenomena that we observe to occur and to succeed one another in no apparent order.
>
> Probability is relative in part to that ignorance and in part to our knowledge.

For the first sentence, however, Laplace substituted the following in the *Essai philosophique*: "The curve described by a simple molecule of air or of vapor is regulated in as certain a manner as the planetary orbits; there is no difference between them except that which our ignorance creates there" (cf. [*1810d*], p. 100).

We come now to the final technical factor that drew Laplace to the study of population statistics. In order to appreciate its role in his mathematical development, let us take stock of how far he had prevailed against difficulties still thwarting probabilistic analysis as he had outlined them at the beginning of the "Mémoire sur les probabilités" ([*1781a*]). He had there found in birth data a basis for determining the law according to which a succession of events gives access to knowledge of causes in a real case. In the intervening two memoirs he had developed generating functions and the technique for numerical evaluation of integrals of differential equations containing terms with very large exponents. It remained to put on a footing more realistic than the hypothetical case of lightly loaded dice the class of problems calling for the probability of complex events compounded from simple ones the respective possibilities of which are unknown quantities, and reciprocally for estimating the number of observations necessary in order that a predicted result might have a specified probability. It was in furnishing data for these calculations that the record of births, with its slight but known disproportion between male and female, proved to be such a valuable resource. Laplace devoted the remainder of the memoir to resolving from this point of view, and with the use of his improved techniques, the population problems that he had already handled ([*1781a*]; see Section 11), arriving in most instances at identical values.

13. Population. In 1786 Laplace finally addressed himself directly to demography, not merely as a convenient repository of problems and examples but as a subject in its own right. He then published a study of the vital statistics of Paris from 1771 to 1784, together with an estimate of the total population of France over a two-year period ([*1786c*]). As yet there was no actual census, which did not begin until 1801 under Napoleonic administration. Late eighteenth-century demography proposed to reach estimates of the population through determining the factor by which the average number of annual births was to be multiplied in order to approximate the total. Laplace's memoir consists of an application of the rule for predicting future events from the observation of those past. As always, direct numerical calculation was impractical. The problem was precisely the sort, however, that could be managed by the technique of predicting probable error and computing how far observations would need to be extended in order to reduce its range to specified limits. Samplings showed the number 26 to be the multiplier to be applied to the figure for annual births in order to give the population at any given time. Applying it to the average annual birth figure for the years 1781 and 1782 gave 25,299,417 for the population of France. In order to reduce to 1,000 to 1 the odds against making an error no greater than half a million in the estimate, the sampling that had established the factor of 26 would have needed to consist of 771,469 inhabitants. If the multiplier had been taken to be 26-1/2, then the figure for the population would have come to 25,785,944, and the sampling would have needed to be 817,219 in order to maintain the same odds against the same error. Faced with those results, Laplace recommended that the count be carried to 1,000,000 or even 1,200,000 in order to assure a degree of accuracy compatible with the importance of the information.

For the information was indeed important. Population, Laplace observed, is an index to national prosperity. A comparison of its variations in the light of antecedent events would serve as the most accurate measure of the effect of physical or moral agencies upon human welfare. Aware of the guidance that such information might provide to those responsible for public policy, the Academy had decided (Laplace noted) to insert each year in its published memoirs the summary of births, marriages, and deaths throughout the kingdom. In consequence of this decision, the Academy's last volumes of memoirs in the old regime contain annual installments of an "Essai pour connaître la population du royaume" (*MARS* [1783/1786], 703–718; [1784/1787], 577–592; [1785/1788], 661–689; [1786/1788], 703–717; [1787/1789], 601–610; [1788/1791], 755–767). Its purpose was to estimate the populations of municipalities and regions marked out on the Cassini map of France through multiplying the average number of annual births in each locality by a factor of 26. Citations to this compilation usually attribute it to the joint authorship of Condorcet, Laplace, and Dionis du Séjour, following therein the table of contents of the volumes themselves. That attribution is in error. Those three served merely as a commission to receive and communicate to the Academy the work of La Michodière, a magistrate who at the behest of the government was continuing research that he had undertaken at his own initiative some thirty years earlier in Auvergne and the Lyonnais, where he was then intendant.

There is no evidence that Laplace wrote further on probability prior to his series of lectures at the abortive École Normale in 1795 (see Section 20), one of which was the nucleus of the *Essai philosophique*, and (although there are probabilistic considerations in *Mécanique céleste*, notably in Book X) there is no evidence of his resuming sustained work until twenty-five years later, when he published a memoir ([*1810b*]) that derived the central-limit theorem for the reduction of error from his method for approximating the evaluation of formulas containing terms raised to very high powers (see Section 25). Let us turn back, therefore, to two sets of interests that occupied him concurrently with probability and with each other, cometary theory and his collaboration with Lavoisier in the chemical physics of heat.

14. Determination of the Orbits of Comets. Laplace's interest in the problem of the curves described by comets antedated the only publication he had yet devoted to these bodies, namely his probabilistic calculation that the cause behind the configuration of the planetary system fails to account for the distribution of the inclinations of the planes of cometary orbits ([*1776d*]; see Section 6). Earlier in 1776 he had involved himself in a controversy that became very acrimonious with Rudjer Bošcović, then resident in Paris. In 1771 Bošcović had presented to the Académie des Sciences a refinement of the method for determining cometary orbits that he had first advanced in 1746 and had advocated ever since ("De orbitis cometarum determinandis," *SE*, **6** [1774], 198–215).

In the case of comets, uncertainty about the nature of the conic sections they describe, combined with the irregularity of opportunities for observation, left a much wider gap between theoretical and practical astronomy than in the case of planets. In practice, the trajectories of the known comets had been mapped empirically, by fitting the curves to as many observations as could be recorded in the periods of visibility. Theoretically, on the supposition started by Newton (and widely adopted afterward) that the observable trajectory is a parabola approximately coinciding with a highly eccentric ellipse, three observations should suffice to determine the curve. The problem was how to arrive at it mathematically. At the very end of *De systemate mundi*, Newton gave an approximate solution that depended on considering a very short element of the trajectory as if it were a straight line that the comet traverses at constant velocity (*Principia mathematica*, Florian Cajori, ed., II [Berkeley, Calif., 1962], 619–626).

Bošcović's memoir consists in an analytical development of this, the standard approach, issuing in an elaborate sixth-degree equation of motion. According to Laplace's own account ([*1784b*]; *OC*, X, 93), it was on reading this memoir, published in 1774 in volume **6** of the *Savants étrangers* series—along with two of his own early papers ([*1774b*] and [*1774c*])—that he realized that the entire method involved a fatal fallacy. Treating the interval between the first and third observations as a first-order infinitesimal entailed neglecting second-order quantities that depend on the curvature of the orbit and on changes in velocity of the comet. At the same time, the position of the (supposedly) rectilinear fragment of trajectory that the comet is observed to describe has to be determined by second derivatives of its geocentric latitude and longitude. Thus, second-order quantities were both neglected and employed.

From the floor of the Academy, Laplace apparently read out a criticism of the Bošcović memoir, castigating all such procedures as "faulty, illusory, and erroneous." We do not know the precise date of his attack, which Bošcović took as a derogation of Newton and an insult to himself, demanding the appointment of a commission to adjudicate the dispute and the exchange with Laplace that ensued. The report of that commission, consisting of Vandermonde, d'Arcy, Bezout, Bossut, and Dionis du Séjour, is recorded in the minute book of the Academy (*PV* [5 June 1776], fols. 172–177). The members acknowledged that Laplace was right analytically, while deploring the abrasive manner in which he had couched his criticism and regretting that Bošcović had taken it personally. Both parties were counseled to bring their findings before the public rather than to quarrel in camera. Evidently, however, Laplace could not let the matter drop. On 19 June he read a further set of remarks (*ibid.*, fol. 191). We do not know what they contained, except that now he also had Lalande in his sights, and another commission was appointed to resolve the affair. Its report has not survived. Years later, in the opening paragraph of the "Mémoire sur la détermination des orbites des comètes" ([*1784b*]; *OC*, X, 93–94) Laplace recalled how he had proved that the method he was criticizing was so bad that it was capable, in an extreme instance, of reversing the apparent direction of a comet's motion from retrograde to direct.

Having refrained from pressing the investigation at that time, he now resumed it, stimulated by the recent work of Lagrange and Dionis du Séjour. As it was analytically impossible to operate with three widely separated observations, the standard method necessarily depended on three observations of positions that were fairly close together. Inevitably, therefore, small errors of observation would affect the results very considerably. Compensation of errors was then attempted, not by multiplying observations but by increasing the number of terms in the series that expressed the result, in order that it might approximate more closely to the truth. The technique was mathematically laborious and of little practical use. Seeking a simpler way to correct for observational error, Laplace saw that closely contiguous observations could serve that purpose if their number were increased. Standard methods of interpolation could then be applied to determine the observational data needed for a solution. The choice of parameters being arbitrary, he preferred to work with the geocentric longitude and latitude of the comet at a given moment and with the first and second derivatives of these quantities with respect to time. These data were the easiest to manipulate analytically, and he could obtain simple formulas that became more precise the larger the number and the greater the accuracy of the observations.

This approach to the determination of cometary orbits had the further advantage that observations separated by as much as 30° or 40° might be employed. In contrast to the established procedure, in which the analysis was an approximation and the observations had to be supposed perfectly exact, Laplace characterized his method as one in which the analysis is rigorous and the observations are acknowledged to be approximations. The second-order differential equations of motion of a comet around the sun at the focus of a conic section yielded directly a seventh-degree equation determining the distance of the comet from the earth (*ibid.; OC*, X, 110):

$$[\rho^2+2R\rho\cos\theta\cos(A-\alpha)+R^2]^3 \times(\mu R^2\rho+1)^2=R^6 \qquad (64)$$

where ρ is the comet-earth distance, α the geocentric longitude, θ the latitude, R the radius vector of the earth, r the radius vector of the comet, A the heliocentric longitude of the earth, and where $\mu\rho = d\rho/dt$. Since the theory holds for any conic section, the supposition that the orbit is a parabola and the major axis infinite yields a new sixth-degree equation for determining the distance of the comet from the earth (*ibid.; OC*, X, 121):

$$[\rho^2+2R\rho\cos\theta\cos(A-\alpha)+R^2] \times\left(m\rho^2+n\rho+\frac{1}{R^2}\right)^2-4=0 \qquad (65)$$

where, besides the above, m and n are abbreviations representing respectively

$$u^2+\left(\frac{d\theta}{dt}\right)^2+\left(\frac{d\alpha}{dt}\right)^2\cos^2\theta$$

and

$$\left(2u\cos\theta-2\frac{d\theta}{dt}\sin\theta\right) \times\left[(R'-1)\cos(A-\alpha)-\frac{\sin(A-\alpha)}{R}\right] +2\frac{d\alpha}{dt}\cos\theta\left[(R'-1)\sin(A-\alpha)+\frac{\cos(A-\alpha)}{R}\right],$$

and R' is the radius vector of the earth at longitude $90° + A$. It was possible to combine (64) and (65) and to obtain a linear equation for the distance and an equation of the conditions that the data must satisfy in the case of a parabolic orbit. But the calculation was difficult, and it was more direct to satisfy (64) or (65) by making trials with the data.

Since the problem of determining parabolic cometary orbits could be formulated in a system of equations exceeding the number of unknowns by one, there was a choice of ways to determine the distance of the comet from the earth. The important tactic was to select a method that would minimize the effect of observational error. The second derivatives of the geocentric latitude and longitude $(d^2\theta/dt^2)$ and $(d^2\alpha/dt^2)$ were the quantities most affected. Either but not both might be eliminated, and Laplace formulated two further sets of equations to be used alternatively, according to whether the second derivative of longitude or of latitude was the greater. The former ratio obtained in the case of comets of which the orbital plane was close to the ecliptic, and Laplace was pleased to discover that his first set of equations was nothing other than a translation into the language of analysis of what Newton had demonstrated synthetically in the *Principia mathematica*, Book III, Proposition XLI.

Although Laplace's method of determining cometary orbits was largely superseded by those of Olbers in the 1790's and of Gauss after 1801, its formulation marks an important stage in his career. It was the first piece of work that brought him into immediate contact with workaday observational astronomers. Following the analytical part of the memoir, the eighth article (*ibid.; OC*, X, 127–141) contains instructions for applying the method to comets themselves and even gives numerical examples. Prior to publication, Laplace sent a copy to the Abbé Pingré (*34*), who immediately applied it to computations in his monumental *Cométographie* (1783–1784). The memoir closes with a contribution by Méchain further illustrating the technique in the determination of a comet, the second that he discovered in the year 1781 (*ibid.; OC*, X, 141–146).

Laplace must have requested that computation from Méchain shortly after reading the draft of the analytical parts of the memoir before the Academy on 21 March 1781 (*34*). An entry in the *procès-verbaux* of 2 May (*35*) probably refers to that. Over three years elapsed before the memoir was published, after a delay rather longer than would be expected from the normal academic lag. The explanation may well be that Laplace became involved with William Herschel's discovery of an object that turned out to be the planet Uranus, initially taken for a comet. Herschel made the observation in Bath on 13 March 1781, and Laplace could scarcely have heard of it—and certainly cannot have been influenced in his treatment even if he had—when he delivered his paper eight days later. On 13 June, however, he read a note (*36*) to the Academy reporting that he had tried his general method on the new comet but found that neither it nor that of Lagrange—nor any other that he knew—was applicable to the present object, of which the apparent motion in latitude was almost indetectable relative to the motion in longitude (even so the notion of there being an unknown planet did not then occur to Laplace). He had been trying to find a new method by combining his equations in a different manner and believed that he had succeeded. He planned to read his investigations shortly but needed time to complete the calculations before the comet emerged from the sun, where it was presently hidden. On 28 July (*37*) he reported those calculations, still taking the object for a comet. Not until a year and a half later, on 22 January 1783 (*40*), did he make written reference to "Herschel's planet," the ephemerides of which he had been calculating in collaboration with Méchain. Their results showed it to be identical with the supposed star that Mayer had recorded in 1756 and that had mysteriously disappeared (*42*).

In the meantime, Laplace had become increasingly drawn into another aspect of empirical science, one involving hand as well as eye, and had entered into problems of physics in collaboration with Lavoisier.

15. Lavoisier and Laplace: Chemical Physics of Heat. Henry Guerlac has published a monograph (M:[1976]) on the investigations that Lavoisier carried out jointly with Laplace, beginning with phenomena of vaporization and evaporative cooling. The earliest record of Laplace's interest in the problem, or indeed in any physical experiment, is an entry (*22*) in the *procès-verbaux* of the Académie des Sciences. On 9 April 1777 at the Easter meeting, an occasion always open to the public, Laplace read a paper on "the nature of the fluid that remains in the receiver of the pneumatic machine." No trace of the text remains, but it is evident from the Lavoisier materials that the experiments pertained to determinations by the two colleagues of the effects of varying degrees of temperature and pressure on the vaporization of water, ether, and alcohol. Guerlac considers it virtually certain that Lavoisier instigated the trials.

Lavoisier and Laplace never published the memoir that they intended. They probably never wrote it, but their procedures and conclusion are evident from Lavoisier's other writings (M: Guerlac [1976], 198–199). The experiments gave Lavoisier the physical framework that he needed to come forward in November 1777 with a much rumored criticism of the phlogiston theory coupled with a statement of his own hypothesis on combustion and calcination. In a preface drafted in 1778 for a projected but phantom second volume of his *Opuscules* of 1774, Lavoisier wrote of his intention to practice so far as possible "la méthode des géomètres" (*ibid.*, 215–216).

As for Laplace, apart from the intrinsic interest of the work itself, collaboration with Lavoisier, almost six years his senior, offered him the chance to be more than a mathematician. It associated him with the one person who was clearly emerging as the scientific leader of the Academy in their generation, the newly appointed administrator of the Arsenal and reformer of the munitions industry, with influential connections in the worlds of government and finance.

That much transpired in 1777, and although Laplace and Lavoisier served together on occasional committees of the Academy and must have been in each other's company at its semiweekly meetings, they seem to have suspended active collaboration and to have resumed it only in 1781. During that summer they worked together to verify a design for fabricating a barometer with a flat meniscus imagined by Dom Casbois, a Benedictine in Metz (*ibid.*, 224). The occasion was Laplace's first recorded scrutiny of the problem of capillary action. Much more immediately, the barometric question led them during the winter of 1781–1782 to determinations of the thermal expansibility of glass as well as mercury and other metals (*38*). The motivation was both instrumental and theoretical. The experience stood them in good stead in the early 1790's, when both were serving on the commission charged with fabricating the standard weights and measures for the revolutionary metric system (see Section 18). Already, in their registers they were using decimal subdivisions of linear and gravimetric units. Along with these practical questions of measurement, the investigation set them both to thinking on heat capacities in general.

Their physical interests transcended heat. Electricity offered the companion example of a subtle fluid for which bodies have characteristic capacities. It was natural for Lavoisier and Laplace to explore the analogy between heat and electricity when Volta came to Paris early in 1782 for an extended visit. He brought with him an electroscope for detecting weak charges and was then harboring the theory that electrical charges in the atmosphere might be produced by vaporization. Lavoisier designed and had constructed a condenser with a marble plate to detect such charges, and Volta tried it in the company of Laplace. Conditions were bad and the experiment failed. Laplace and Lavoisier then tried it again on their own (*39*) and published a brief note on what they construed as positive effects ([*1784c*]).

Even then, they were planning the campaign of their experiments on heat, and Lavoisier must have commissioned construction of the famous ice calorimeters at much the same time. The *Mémoire sur la chaleur* first appeared in a separate printing ([*1783a*]). Laplace's influence was clearly paramount in its theoretical aspect, extending to the idea for measuring a quantity of heat by the amount of ice that it would melt and also to the choice of many experiments. The memoir consists of four articles. The first discusses the nature of heat and its quantification; the second, the determination of specific heats of selected substances and also certain heats of reaction; the third, theoretical consequences and a program for a chemical physics; and the fourth, the application of the techniques to the study of combustion and respiration.

Article I opens with a frequently paraphrased contrast between the fluid (soon called the "caloric") and the mechanical theories of heat, which hypotheses are said to be the only conceivable alternatives. It has generally been supposed that the contrast here was also between the opinions of the two authors, and there is no doubt that Lavoisier preferred the former. Moreover, Laplace would certainly have been the one to compose the passage elaborating the kinetic theory. According to that school, the quantity of heat in bodies is measured by the sum of the vis viva (mv^2) of the vibratory motions of their particles, and the conservation of heat in transfers is a form of the conservation of vis viva in gradual changes of motion. Robert Fox, however, has recently raised the question whether Laplace ever did adopt that position for himself (M: Fox [1971], 30). If that was his view at the time, he certainly changed it for the physics of his later years, when he consistently preferred the caloric model for analysis. In this connection, it is to be remarked that conservation of vis viva did not play an important part in his mechanics, any more than it had for Newton; and the body of the memoir consistently and naturally employs the vocabulary of the fluid theory. The contrast concludes with the authors' abstention from choosing between the alternatives. "Perhaps," it is said, "they both obtain at the same time" ([*1783a*]; *OC*, X, 153), an aside about which it may be worth remarking that in Laplace's final view caloric like other subtle fluids is itself particulate. In any case, the only admissible propositions are those that save the phenomena under both theories.

Foremost among those principles are conservation and reversibility in exchanges of heat within a system of bodies. Definitions are needed. *Chaleur libre*, or free heat, is that portion of the total heat contained in a body that may pass to another. Since different quantities of heat are required to raise the temperature of the same mass of different bodies equally, some unit must be designated with which to quantify these comparisons. The heat absorbed in raising one pound of water one degree makes a convenient amount, in terms of which the "specific heats" of other bodies may be expressed. Although probably not invariant with temperature, specific heats may be taken as nearly constant for the range between the freezing and boiling points of water (0° – 80° on the Réaumur scale).

It is to be noticed that the notion of specific heat is not quite the modern way of putting it, although it comes to the same thing when reduced to unit mass and referred to the amount for water as unit heat. A more serious reservation must be entered, however, lest this lead to anachronism. The free heat of a body is only that portion of its absolute quantity which may be exchanged with other bodies in consequence of differences of temperature, change of state, or chemical reaction. It is also the only manifestation of heat that is accessible to measurement, which, moreover, is in degrees of the thermometer. Here the reader must be especially careful to avoid thinking of the absolute quantity in terms of a hypothetical scale of absolute temperature, for there is no anticipation of Kelvin. The conception is that of the eighteenth-century theory of matter, which distinguishes between the electricity, heat, ether, or whatever, that is "fixed" in bodies, and the portion that may be disengaged in natural phenomena or in experiments.

These presuppositions in no way vitiated the design or execution of the experimental program. The problem of measuring specific heats in the simplest case of mixing two miscible substances was formulated algebraically, no doubt by Laplace. He let m and m' be the respective masses, a and a' the initial temperatures, and b the temperature resulting from mixture. Then the ratio of specific heats, q and q' is given by

$$\frac{q}{q'} = \frac{m'\,(b-a')}{m(a-b)}. \tag{66}$$

So straightforward an approach was inapplicable, however, to determinations of heat effects involving chemical combination, combustion and respiration, or change of state, the three most important and interesting processes. Faced with this limitation, the authors imagined a method of general applicability. The amount of ice melted in any process involving the evolution of heat could serve to measure the quantity. The notion is introduced by means of an image that bespeaks Laplace and mathematical modeling more obviously than does the elementary algebraic formulation for specific heats in mixture. We are to imagine a hollow sphere of ice with a shell thick enough to insulate the inner surface from the heat of the surroundings. Suppose a warm body were to be introduced in principle into the cavity. Its heat would melt away a portion of the inner surface until it had cooled to zero, and the weight of water would be proportional to the heat required to accomplish that effect.

Lavoisier commissioned the instruments and later named them calorimeters; two were made, each about three feet high. Air could be admitted into one for respiration experiments. A concentric nest of containers like an ice-cream freezer was packed with ice around a central receptacle. A basket could be suspended inside to hold the objects under study. Ice lining the inner shell served for the melting layer of the model. The water was run off through a petcock and weighed. The authors established first that the heat needed to melt a pound of ice will raise the temperature of a pound of water from 0° to 60° R. Here again, it must be emphasized that the concepts of intensity and quantity of heat were not yet differentiated. Ice, as they put it, "absorbs 60 degrees of heat in melting" (*ibid.; OC*, X, 167). Thereupon the remainder of their second, entirely experimental article reports the determination of certain specific heats, heats of reaction, and animal heats.

The third article is pure Laplace and exhibits both his capacity to analyze a set of physical phenomena in a highly abstract manner and also the limitations inherent in even the most sophisticated notion of a general theory of heat in the absence of any notion of thermodynamics. Not that Laplace fancied himself in a position to attain such a theory, although what he thought he needed was to know (1) whether specific heat increases with temperature at a uniform rate for all substances; (2) the absolute quantity of heat contained in bodies at given temperatures; and (3) the quantities of free heat given off or absorbed in chemical reactions. Lacking such data, which only a very elaborate program of investigation could work up, he would simply examine a few problems raised by experiments that the two authors had performed, beginning with the second topic just mentioned, the absolute or total quantity of heat in bodies.

Clearly, its amount is considerable, even at 0° on the thermometer. We wish to know that amount in degrees of the thermometer, but experiments such as those just tried on specific heat could form no basis for calculation, unless it was legitimate to suppose that the specific heats of bodies are proportional to their total heats. That would be a very risky relation to assume without examination. Moreover, there was unfortunately no way to get at it from the values determined by simple mixtures of substances at different temperatures; all that happens there is the exchange of heat. The case is comparable to local exchanges of motion, which tell the investigator nothing of the absolute motion of the earth through space.

There might be a deeper way, however, leading through chemistry. Since the heat of reaction is not the consequence of mere inequality of temperature, it might furnish a basis for relating change of temperature to absolute heat. As usual, Laplace's approach was analytic. He let x be the ratio of the absolute heat contained in water at 0° to the amount that can raise its temperature by 1° (note that he was not yet sufficiently prepared with his own definitions to call the latter "specific heat" and designate it unity). Then, the total heat contained in a pound of water at 0° would melt $x/60$ pounds of ice. Consider any two substances at 0°, m and n being their respective weights, and a and b the ratio of the heat contained in each to the heat contained in a pound of water. They react chemically, and the heat produced when the products are cooled down to 0° melts g pounds of ice. The heat of reaction alone is sufficient to melt y pounds (y being negative if the reaction is endothermic). Finally, c is the ratio of the heat contained in the mingled products of reaction to that in a pound of water. With these parameters, Laplace formulated and equated two expressions for the quantity of ice that would be melted by the residual free heat:

$$\frac{(ma+nb)x}{60}+y=\frac{(m+n)\ cx}{60}+g, \qquad (67)$$

whence

$$x=\frac{60(g-y)}{m(a-c)+n(b-c)}. \qquad (68)$$

Here x represents the number of degrees of heat contained by water at 0°. But how are numbers to be substituted for a, b, c, and y? Two hypotheses made evaluation possible. The first, conservation of free heat in chemical combinations, was generally admitted. The second, that the specific heats of substances are proportional to their absolute heat content, was precisely the risky assumption that could be tested from data tabulated in the preceding article. If it were correct, then $y=0$, and

$$x=\frac{60g}{m(a-c)+n(b-c)}. \qquad (69)$$

Those data contain values for the specific heats,

a, *b*, and *c*, in a few selected instances; and if the proportionality of specific heat to total heat was a justifiable hypothesis, all of the different cases ought to give the same value for *x*. Alas, they did not.

It is true—Laplace immediately went on (*ibid.; OC*, X, 178)—that a very small correction in each, no more than 1/40, would make them all satisfy the relation. Such a correction would be less than the margin of experimental error. Other considerations made such an exercise in curve-fitting unpromising, however. Endothermic reactions could not at all be accommodated, and neither could the phenomenon of the dilation of solids on heating. It was very probable that heat is fixed thereby, even as in change of state, although gradually and undetectably; and this reflection gave another reason to surmise that specific heats do increase with temperature but at a different rate for each substance.

Finally, Laplace turned to change of state itself, in order to consider what such episodes can reveal about equilibrium conditions in heat. Again, it was the analogy that is mechanical, rather than the theory of heat itself. Just as there are several positions of equilibrium for (say) a rectangular parallelepiped (resting on its side, balanced on one end, and so on), there may be several conditions around a change of state in which heat is in equilibrium, each involving different physical arrangements of the molecules and different distributions between the portions of the heat that are going into cooling and into freezing. As usual, Laplace formulated analytical expressions. He then adduced the example of super-cooling followed by a sudden crystallization creating a new equilibrium. We need not follow the algebra. The interesting feature is the glimpse that the discussion gives into Laplace's preoccupation with forces and structures at the molecular level. The mutual affinity of molecules of water draws them together on freezing and frees the heat that is keeping them apart. Thus, it seemed probable that their arrangement when frozen is that in which the force of affinity is at its most effective. Hence, it is natural that the surest means of inducing a supercooled sample to freeze is to introduce a bit of ice. The same holds true in all crystallizations.

More generally, and here Laplace laid down a program for research, study of the equilibrium between heat, which tends to separate molecules, and affinity, which draws them together, might well offer a method for comparing the intensities of these forces of affinity. For example, a certain mass of ice plunged into an acid would be melted to the point at which the acid was sufficiently weakened so that its attraction for the molecules of ice would be balanced by their mutual forces of adherence. That point would depend also on the temperature. The further below zero it fell, the higher would be the concentration at which the acid ceases to melt ice. It would thus be possible to construct a statical scale expressing the force of the affinity of the acid for water in terms of degrees of the thermometer. If this procedure were to be followed with solutions of every sort, the relative mutual affinities of all bodies could be stated numerically. But, our author breaks off, this is a major subject; another memoir would be devoted to it.

We shall not consider the final article, in which Lavoisier drew the threads together for the theory of combustion and respiration. Guerlac's monograph shows how it prepared the ground for the "Réflexions sur le phlogistique" (1786) and how, more generally, the further collaboration with Laplace styled the aspiration to make of chemistry in its revolution a mathematical science. Our concern is with Laplace. There are indications that just before undertaking this research, he had been somewhat on-again, off-again in his commitment. In a letter of 7 March 1782 he asked Lavoisier to release him from their agreement to work together (M: Guerlac, 240). On 21 August 1783 a covering letter to Lagrange (*Oeuvres de Lagrange*, XIV, 123–124), enclosing a finished copy of the memoir, was defensive about the time that he had invested in *physique*. Nevertheless, the passages on heat and affinity just discussed are evidence that the work had finally gripped his interest and carried him into the first stages of his physics of interparticulate forces. Only with Laplace, indeed, did chemical affinities begin to be seriously considered as physical forces of attraction. In the preface to the work to be discussed next, *Théorie du mouvement et de la figure elliptique des planètes* ([*1784a*]), Laplace referred to the experiments with Lavoisier and repeated the reflection that equilibrium between the attractive force of affinity and the repulsive force of heat might one day furnish analysis with the handle on chemistry that the discovery of gravitation had afforded to the mathematicians who had perfected astronomy since Newton. It also remained to determine the laws of force responsible for the physical effects of solidity in bodies, of crystallization, of the refraction and diffraction of light, and of capillary action.

Laplace continued working with Lavoisier into 1784, when the treatise just mentioned was published. They then began the program of research that he had imagined on affinity. They repeated, largely at his insistence, measurements of the heat generated by the combustion of charcoal, phosphorus, and other substances. The joint paper reporting those experiments was not written until 1793 and appeared after Lavoisier's execution ([*1793d*]). Laplace also worked with Lavoisier on the closely related combustion of hydrogen to produce water. He strengthened Lavoisier's hand in the battle against the phlogistonists by setting him straight on the source of hydrogen when acids act on metals. The gas comes from the acid and is not to be taken for phlogiston escaping from the metal (M: Guerlac, 265). Thereafter, however, his own affinities seem to have drawn him rather toward Berthollet in his chemical interests. But we must turn outward now, from particles to stars, from physics and chemistry to astronomy again.

16. Attraction of Spheroids. It is significant that the paragraph just discussed on physical and chemical forces should have figured in the opening of the *Théorie du mouvement et de la figure elliptique des planètes* ([*1784a*]), for that work is very revealing in other important ways of the continuity and persistence of Laplace's interests. It was his first separate publication, and the preface was his first writing suited to the comprehension of laymen. It consists of a summary view of the world in nontechnical language and might easily be taken for a prospectus of the *Exposition du système du monde*. He began by explaining how he came to write the book at the behest of an honorary member of the Academy, Jean-Baptiste-Gaspard Bochart de Saron, a magistrate of the Parlement of Paris and (although Laplace did not put it this way) one of several patrons of science who also contributed to its content. That Bochart de Saron should have commissioned the work and personally subsidized publication is the clearest evidence that Laplace's qualities had come to be recognized in high places. Long ago, he said, he had conceived the notion of drawing into a single work an exposition of the way in which planetary paths and figures follow mathematically from the law of gravity. It seems probable that he was referring here to one of the papers of his youth, "Une théorie générale du mouvement des planètes" (*9*), which he had submitted to the Academy in November 1771. But he would never have composed the treatise had Bochart de Saron not encouraged him on several occasions to show how the general properties of elliptical and parabolic motion may be derived from the second-order differential equations that determine the motion of celestial bodies.

That exposition occupies Part I of the treatise and, unlike the preface, is addressed to mathematically—and not merely verbally—literate readers. It has the quality of a textbook in the rational mechanics of the solar system limited to the principal motions of the celestial bodies and might appear to consist of a sketch for Books I and II of the *Mécanique céleste*. Besides that, Laplace explained certain of the techniques that he had imagined for achieving approximate solutions to equations that defied rigorous integration, and he also included a set of calculations to determine the orbit of Uranus. This topic bulks rather larger than its importance in the solar system would have warranted. It would appear that Laplace took the occasion to publish an analysis that he had submitted to the Academy on 22 January 1783 (*40*), in which he recognized that the object discovered by William Herschel in March 1781 was a planet and not a comet, as was first thought. Beyond that, he reprinted much of his memoir ([*1784b*]; see Section 14) on the determination of cometary orbits—or rather, preprinted it, since the memoir, although composed in 1781, came out some months later than the treatise. There is no other novelty in this first part, and Laplace simply hoped that the treatment would be pleasing to mathematicians and astronomers.

Part II, subtitled "De la figure des planètes" and addressed "uniquely to mathematicians" ([*1784a*], p. xxi), is quite another matter. Laplace there resumed investigating the laws of gravitational attraction of spheroids in a far more abstract manner than he had in his earlier pieces on the figure of the earth and on the oscillations of the tides ([*1778a*], [*1779a*], [*1779b*]); see Section 8). At the same time, the mathematical problems are of a more specific nature, and it is clear that he had matured his approach since the memoir on the precession of the equinoxes ([*1780c*]). In retrospect, the most interesting feature of the entire treatise is bound to be the emergence of the concept of what is now called potential, although that name was not given to the sum or integral of the action of the elements of an attracting body upon an external point until 1828, when George Green adopted Poisson's application of the expression for it to electrostatic and magnetic effects. There is no reason to suppose that here, at the outset, Laplace thought the notion more signal than other main aspects of his treatise.

In all these researches on attraction, Laplace played leapfrog with Legendre in a manner very like his relation of collaboration and competition with Lagrange over the integral calculus of planetary theory. In the present treatise, he referred ([*1784a*], pp. 96–97) to a formulation by Legendre that he had seen in the latter's "Attraction des sphéroïdes homogènes," which memoir was not published until 1785 (*SE*, **10** [1785], 411–434). Legendre for his part there attributed to a communication from Laplace in 1783 the notion that he developed, by way of generating functions, into the polynomials later named for Legendre; on this topic, see (N: Burckhardt [1908], pt. 5, 367–397). Legendre was not yet a member of the Academy, and the exchange must have occurred in connection with Laplace's preparation of a report on the memoir on behalf of Bezout, d'Alembert, and himself, who constituted the committee to which it was referred (*41*). Whatever these priorities may have been, Laplace in the treatise under discussion took as his point of departure the equation for a second-order surface where the origin of coordinates is the center of the spheroid,

$$x^2+my^2+nz^2=k^2, \tag{70}$$

m, n, and k being any constants. For the attraction that the enclosed solid exerts on an external point, he then formulated the integral (*ibid.*, 69)

$$V=\int\frac{dM}{(a-x')^2+(b-y')^2+(c-z')^2}, \tag{71}$$

where dM is the mass of a particle of the solid with coordinates x', y', and z', and a, b, and c are the coordinates of the external point. Then, when the attraction is decomposed along the three principal coordinates,

$$A=-\frac{\partial V}{\partial a};\quad B=-\frac{\partial V}{\partial b};\quad \text{and } C=-\frac{\partial V}{\partial c}. \tag{72}$$

Applying the analysis, Laplace began by looking backward to Newton's *Principia* rather than forward and by showing that the value of V is the same as if all the mass of the spheroid were concentrated at the center of gravity. In seeking to determine that value in a general manner capable of giving numerical solutions, Laplace turned to polar coordinates and expanded the transformed expression into an infinite series that could be evaluated by approximation. Wishing to achieve a rigorous solution, he in effect reverted to a strategy that he had employed in a fragmentary analysis of 1778, wherein he first investigated the attraction between a spheroid and an external point. At that time he had the rings of Saturn in mind and inverted the problem to consider the attraction of the point for the spheroid ([*1778a*]; *OC*, IX, 71–87; see Section 8). In the same manner, he now inverted the conditions and made the externally attracted point the origin of polar coordinates. Manipulating the resulting expressions yielded a theorem often called by his name: all ellipsoids with the same foci for their principal sections attract a given external point with a force proportional to their masses. The finding generalized a result already won by Maclaurin for the restricted case of particles located on the extension of the major axis. In modern terms, Laplace's theorem is said to assert that the potentials of confocal ellipsoids at a given point are proportional to their volumes. For the gravitational case, this would presuppose homogeneous density. His having conceived the theorem in terms of mass was not intended to obviate that restriction, however, since he did not then envisage any other application. It was simply consistent with his initial motivation, which was to give an up-to-date demonstration of the point-mass gravitational theorem. More significant, for the appreciation of Laplace himself, is to notice that his interest was first mathematical and second physical. His solution compared the attraction of the original ellipsoid to that of a new, confocal ellipsoid containing the attracted point in its surface. That is how he made the problem solvable, for rigorous integration was possible when a point lies on the surface of a figure. Physically, the procedure presented the further advantage of opening the possibility of distinguishing more directly between central attractions and perturbing forces in the motion of planets and their satellites.

Thereupon, Laplace turned to internal particles and found another theorem that surprised him more than anything so far: the attractive force exerted at any point within a homogeneous ellipsoidal shell is equal in all directions. For the component of such an attractive force parallel to the major axis, Laplace obtained the important definite integral (*ibid.*, 89),

$$A=\frac{2a\pi}{\sqrt{mn}}\int_0^1\frac{x^2\,dx}{\sqrt{\left(1+\frac{1-m}{m}x^2\right)+\left(1+\frac{1-n}{n}x^2\right)}}, \tag{73}$$

m and n (Equation 70) being positive for finite surfaces.

The rest of the treatise consists of similar mathematical investigations of particular problems concerning the equilibrium conditions and shape of rotating fluid masses. An important application to the moon shows that the difference in length between the earth-directed axis and the diameter of a spherical body of identical mass is four times the comparable elongation of the orthogonal axis in the orbital plane (*ibid.*, 116). The coincidence of the moon's periods of revolution and rotation is then deduced from that relation. The dependence of angular velocity on ellipticity forms another topic, and its application to the earth gives limits for the polar flattening. From his earlier essays, Laplace reiterated in simplified language the finding that, although it can be determined whether any given form for a solid (or, rather, a fluid) of revolution satisfies specified forces, it is impossible to determine in a general manner all the forms that do so. All these results appear in appropriate passages of Book III of *Mécanique céleste*, although he did not incorporate whole articles or sections from this treatise in his synthesis.

On 11 August 1784, less than a year after finishing the treatise just discussed (*47*), Laplace read the draft of a further memoir (*48*), most of which is reproduced with little change in *Mécanique céleste*, Book III, chs. 1–4. Entitled "Théorie des attractions des sphéroïdes et de la figure des planètes" ([*1785b*]), it contains the basic mathematical theory, which for Laplace meant formulation, of the subject. The entire memoir is an exercise in partial differential equations, manipulation of which enabled Laplace to solve problems by differentiation, series expansion, and analysis of coefficients when integration was impossible, as it normally was. The first of the memoir's five parts contains the simpler and more direct derivations that this method permitted for the main results of Part II of the immediately preceding *Théorie du mouvement et de la figure elliptique des planètes* ([*1784a*]). The rapid sequence fits the pattern in which Laplace often acted. Dissatisfied with his initial treatment of a subject as soon as he saw it in print, he would immediately set to work simplifying the analysis and generalizing the treatment.

That process marks a further stage in the development of what became the concept of potential. At the outset of his second section, Laplace gave it a formulation that provided him with a basic equation from which he could derive the whole theory of spheroidal attraction. The form will not be immediately recognizable to the modern reader, however, since the expression is in polar coordinates. He began by referring to an elementary observation of his earlier treatise, namely that if $V = \int dM/r$, differentiating V along a direction will give the attraction of a spheroid in that direction. In the equilibrium case, the attraction exerted on the particles of a planet takes that form, and Laplace proceeded to investigate V. With the origin of coordinates inside the spheroid,

a, b, c, are the coordinates of the point where the attraction is exerted;
x, y, z, the coordinates of a particle of the spheroid;
$r = \sqrt{a^2 + b^2 + c^2}$ the distance to the origin of the point attracted;
θ the angle that the radius r makes with the x-axis;
ω the angle formed by the intersection of the invariant plane in which the x-axis and y-axis lie with the plane that passes through the x-axis and the attracted point.

Designating by R the distance $\sqrt{x^2 + y^2 + z^2}$ from the origin to the attracting particle, and by θ' and ω' the values of θ and ω at the point occupied by that particle, Laplace obtained the following expression for V ([*1785b*]; *OC*, X, 362):

$$\int \frac{R^2\, dR\, d\omega'\, d\theta' \sin\theta'}{\sqrt{r^2 - 2rR[\cos\theta\cos\theta' + \sin\theta\sin\theta'\cos(\omega - \omega')] + R^2}} \tag{74}$$

in which the integral relative to R is taken from $R = 0$ to the value of R at the surface of the spheroid; that relative to ω' from 0 to 2π, and that relative to θ' from 0 to π.

Laplace's general mathematical virtuosity is nowhere more impressive than in this memoir, which also offers a particularly explicit example of the specific advantage that he could draw from his own mathematical innovations. At this point (and this may have been the breakthrough that made the approach possible at all), he applied an important finding of the memoir on generating functions ([*1782a*], Section XVIII; *OC*, *X*, 54–60, referring back to [*1777a*], Section V; *OC*, IX, 21–24). He had there shown that integrating second-order linear partial differential equations was often possible by means of—and only by means of—definite integrals of a form similar to the expression just given for V. In this instance, he found it easy to show that if $\cos\theta = \mu$, then differentiation produces the following partial differential equation (*ibid.; OC*, X, 362):

$$0 = \frac{\partial\left[1-\mu^2\right]\frac{\partial V}{\partial\mu}}{\partial\mu} + \frac{\frac{\partial^2 V}{\partial\omega^2}}{1-\mu^2} + r\frac{\partial^2(rV)}{\partial r^2}. \tag{75}$$

That equation is in fact equivalent to the modern expression for potential, $\Delta^2 V = 0$. Laplace never transformed it from spherical polar coordinates into Cartesian form in this memoir, however. Rather, he substituted for V in (74), so that he could write Legendre's equation (*ibid.; OC*, X, 375):

$$0=\left\{\frac{\partial\left[\left(1-\mu^2\right)\frac{\partial U^i}{\partial\mu}\right]}{\partial\mu}\right\}+\frac{\frac{\partial^2 U^i}{\partial\omega^2}}{1-\mu^2}+i(i+1)\,U^i, \quad (76)$$

where U^i is a polynomial function of μ, $\sqrt{1-\mu^2}$ sin ω, and $\sqrt{1-\mu^2}$ cos ω. In determining that function, and investigating the dependence of the variables on the angles ω and θ, sometimes called Laplace's angles, Laplace arrived at a formula for evaluating auxiliary factors ([*1785b*], 141)

$$\nu\lambda=2\frac{1\cdot3\cdot5\cdot\cdots\cdot(2i-1)}{2\cdot4\cdot6\cdot\cdots\cdot(i+n)\;2\cdot4\cdot6\cdot\cdots\cdot(i-n)}. \quad (77)$$

(An error arising from the assumption that $(i + n)$ is always even was corrected in *Mécanique céleste*, Book III, no. 15; *OC*, II, 42–43; cf. K: Todhunter, II, 57.) When the attracted point is internal, the expression for V has to be developed in a series of terms of ascending powers of r; and the analysis compounds the attraction of the sphere, on the surface of which the point lies, with that of the shell constituting the remainder of the spheroid.

Simplifying the problem to the case of almost spherical spheroids, Laplace in a third section achieved a general solution even on the supposition of heterogeneous density. Again he recurred to results he had won in an earlier piece ([*1776d*]), namely the equation for the attraction at the surface of a spheroid ([*1785b*]; *OC*, X, 372),

$$-a\frac{\partial V}{\partial r}=\frac{2}{3}\pi a^2+\frac{1}{2}V, \quad (78)$$

where a represents half the common diameter of the spheroid and the inscribed sphere, and r is the distance of the attracted point from the center of mass. Laplace reiterated that equation in *Mécanique céleste* (Book III, no. 10; *OC*, II, 30) and always accorded it great importance. The further analysis was now fairly simple. Again, he decomposed the attraction V into two components, the force exerted at the surface of a sphere and that exerted by a further shell of which the outer surface bounds the spheroid. According to the conditions of the problem,

$$r=a(1+\alpha y) \quad (79)$$

at the surface of the spheroid. Substitution in the basic expansion,

$$V=\frac{\mathrm{U}^0}{r}+\frac{U^1}{r^2}+\frac{U^2}{r^3}+\frac{U^3}{r^4}+\cdots, \quad (80)$$

and in (78) yielded the equation

$$4\alpha\pi a^2 y=\frac{\mathrm{U}'^{(0)}}{a}+\frac{3U^{(1)}}{a^2}+\frac{5U^{(2)}}{a^3}+\cdots. \quad (81)$$

Since y is also a polynomial function of μ, $\sqrt{1-\mu^2}$ cos ω, and $\sqrt{1-\mu^2}$ sin ω, then it could be thought of as a series of functions,

$$y=Y^0+Y^1+Y^2+Y^3+\cdots, \quad (82)$$

where Y^0, Y^1, Y^2, Y^3, $\cdots$ like U'^0, U^1, U^2, U^3, $\cdots$, serve the partial differential equation,

$$0=\frac{\partial\left[1-\mu^2\frac{\partial Y^i}{\partial\mu}\right]}{\partial\mu}+\frac{\frac{\partial^2 Y^i}{\partial\omega^2}}{1-\mu^2}+i(i+1)\,Y^i. \quad (83)$$

The functions Y and U are similar in form, and

$$U^i=\frac{4\alpha\pi}{2i+1}a^{(i+3)}Y^i. \quad (84)$$

Hence

$$V=\frac{4}{3}\pi\frac{a^3}{r}+4\alpha\pi\frac{a^3}{r}\left[Y^0+\frac{a}{3r}Y^1+\frac{a^2}{5r^2}Y^2+\frac{a^3}{7r^3}Y^3+\cdots\right]. \quad (85)$$

It thus proved possible to determine V by expanding y in a series of functions $Y^0 + Y^1 + Y^2 + Y^3 + \cdots$. Laplace then gave a method for evaluating V by analysis of the coefficients in the case of a spheroid the equation of whose surface, referred to Cartesian coordinates, is a polynomial function of the coordinates (*ibid.; OC*, X, 371–374).

Turning to the figure of the planets, Laplace treated the problem as a corollary of the foregoing analysis. In the simplified assumption that these bodies are homogeneous spheroids of revolution, he was finally able to demonstrate the long-sought theorem that, under the law of gravity, they can only be ellipsoids flattened at the poles. Proving that theorem enabled him to derive the law of attraction at the surface by an analysis *a priori*, which fully confirmed the validity of the above method of expansion of the function V into a series of what have been called Laplace's functions. The form in which he obtained it permitted comparison of the force of gravity, determined by means of the pendulum, with the value calculated for any point where the radius had been determined by geodesic measurements along the meridian. (It should be emphasized that this possibility helps to explain Laplace's strong preference in 1790 for basing the metric system on such a survey, instead of on a standard seconds pendulum [see Section 18]). Although existing data were less exact than he could have desired, they nevertheless permitted calculating that the figures assumed by the planets in the course of their rotations, even like their motions in orbit around the sun, confirm the principle of gravity with an overwhelming degree of probability. He had long since calculated ([*1779a*]; *OC*, IX, 269) that if the law of gravity were to satisfy the phenomena, the ellipticity of the earth had to be between the values 0.001730 and 0.005135. Since observation of the pendulum gives 0.0031171, it is well within the limits.

In the course of this analysis, Laplace demonstrated in passing a theorem, which assumed increasing importance in his later work, with respect to two functions of different orders of the type that he was here employing. If Y^i and $U^{i'}$ are polynomial functions of μ, $\sqrt{1-\mu^2}\sin\omega$, and $\sqrt{1-\mu^2}\cos\omega$, and if both satisfy partial differential equations of the form (83), then they have the property of orthogonality, namely that

$$\int_{-1}^{1}\int_{0}^{2\pi} Y^i\, U^{i'}\, d\mu\, d\omega = 0 \qquad (86)$$

([*1785b*]; *OC*, X, 389), if i and i' are different positive integers. Todhunter points out (K: [1873], II, no. 857, pp. 61–62) that although Laplace did not here consider the case in which i and i' are identical, there is an important equation in *Mécanique céleste* (Book III, no. 17; *OC*, II, 47) that does express all that this entails, namely

$$\iint Y'^{(i)}\, d\mu'\, d\omega'\, Q^i = \frac{4\pi\, Y^i}{2i+1}, \qquad (87)$$

where Q^i is a known function of μ and $\sqrt{1-\mu^2}\cos(\omega'-\omega)$, by means of which U^i may be calculated (*ibid.*, no. 9; *OC*, II, 26–27). Laplace did not write that equation down in his memoir; but he might well have done so, for it is implicit throughout the discussion of the figure of the planets.

Armed now with a general theory of the attraction of spheroids, which was precisely what he had lacked in his earlier investigation of tidal oscillations ([*1779a*]), Laplace returned to that subject and demonstrated conclusively that the equilibrium conditions entail periodicity and that the equilibrium will be stable only if the density of a layer of fluid is less than that of the spheroid it covers. There was nothing new in this, but it served to complete his theoretical investigation of the phenomena of spheroids of revolution serving the law of gravity.

Two short memoirs, on the shape of the earth ([*1786a*]) and the rings of Saturn ([*1789a*]), apply the theory of gravitational potential, or the attraction of a spheroid as Laplace always called it, to their respective topics. Like the problem of tidal oscillations, both represent earlier interests that Laplace had been unable to resolve in a conclusive manner for lack of an adequate theory. And although both pertain, again like the tides, to application rather than to innovation, both also contain—perhaps for that very reason—important clarifications and simplifications of the mathematical formalism. Thus, in the paper on the figure of the earth, he gave much greater prominence to (86), making it the basis of the analysis ([*1789a*]; *OC*, XI, 12).

That paper offers a particularly good example of Laplace's tendency to be preoccupied in the first instance with the analytical representation of the facts and only secondarily with the facts themselves, with how something could be the case, given what the case was. For he was not indifferent to the latter. In navigational and astronomical tables the apparent motions of the sun and moon are referred to the center of gravity of the earth; and it was incumbent, therefore, to know its precise location relative to the point of observation, particularly for the theory of the moon. If the flattening at the poles were 1/178, the lunar parallax would amount to 20″ at certain localities. Thus, it is again clear—to anticipate for a moment—that all these uncertainties about geodesic data were bound to reinforce the motivation to base the metric system on yet another survey of the meridian when the opportunity arose in the Revolution (see Section 18).

For the moment, however, the empirical state of the question was what it had been when Laplace in his initial programmatic dual memoir on probability and universal gravitation ([*1776a*]); see Section 5) first confronted the discrepancy between the shape of the earth calculated from measurements of the length of a seconds pendulum at different latitudes and that given by the classic surveys of the lengths of the arc in Lapland, Peru, the Cape of Good Hope, and France. From the former data, the ratio of minor to major axis came out to be 320:321; from the latter, 249:250. A brief calculation based on data from these surveys (apparently Laplace's first application of error theory to instrumental data) finds the probability negligible that the departure from the elliptical had resulted from observational error. The survey data also yielded a finite probability that the northern and southern hemispheres are not symmetrical, as well as a possibility that the earth might not be a solid of revolution at all.

As for the lengths of the radii determined by measurements of the seconds pendulum, they too failed to satisfy an ellipsoidal figure. That would have been troublesome to theory only if the earth were assumed to be of homogeneous density, for the differences did follow a regular pattern; and, most important, values for the force of gravity calculated from the length of the seconds pendulum approximated very closely to the law that its variation is proportional to the square of the sine of the latitude. This relation being of the utmost significance for the theory of the earth, Laplace proposed to combine it with the equilibrium conditions for the oscillations of the seas in order to draw from the combination the law of the variation of the radius of the earth.

He had already shown how the expression for the radius of any nearly spherical spheroid may take the form

$$1+\alpha(Y^0+Y^2+Y^3+Y^4\cdots). \qquad (88)$$

(In the case of the earth, the equilibrium conditions for the seas require that $Y^1=0$ and αY^0 be a constant (*ibid.; OC*, XI, 13). If the equilibrium is to be stable, it is a further condition that the axis of rotation be one of the principal axes of the earth. That requirement entailed the following form for Y^2:

$$H\left(\mu^2-\frac{1}{3}\right)+H^{IV}(1-\mu^2)\cos 2\omega, \qquad (89)$$

where the constants H and H^{IV}, depending on the physical constitution of the earth, are to be determined by observation (*ibid.; OC*, XI, 16).

These are equilibrium conditions that would hold for any celestial body covered by a fluid. In the case of the earth, measurements of the length of the seconds pendulum make it possible to calculate the values numerically. The constant H comes out approximately equal to 0.003111; H^{IV} is negligible relative to H; the quantity $Y^3+Y^4+\cdots$ is very small by comparison to Y^2, and so also is its first derivative by comparison to the first derivative of Y^2. Thus there would be no detectable error in calculating values for the radius of the earth, and also for its first derivative, from the formula

$$1-0.003111\left(\mu^2-\frac{1}{3}\right). \qquad (90)$$

What the surveys of the meridian show is that the same approximation is invalid for the second derivatives of the terrestrial radius. The reason for the discrepancy between the results of the two methods for calculating the length of the radius is that the function $Y^3+Y^4+\cdots$ becomes significant on a second differentiation. The expression for the radius of the earth has the form

$$1+\alpha H\left(\mu^2-\frac{1}{3}\right)+\alpha Y^i, \qquad (91)$$

where αY^i is the term representing the variation between the calculated value and the value that satisfies the law of the variation of gravity with the square of the sine of the latitude. Given this expression, the formula for the length of the seconds pendulum is (*ibid.; OC*, XI, 18):

$$l=L\left[1+\alpha\left(H+\frac{5}{2}\phi\right)\left(\mu^2-\frac{1}{3}\right)+(i-1)\,\alpha Y^i\right], \qquad (92)$$

where l and L are the lengths of the seconds pendulum corresponding to the force of gravity at two different latitudes. The formula for the degree of the meridian is

$$c+\frac{2}{3}\alpha cH-3\alpha cH\left(\mu^2-\frac{1}{3}\right)-i(i+1)\,\alpha cY^i$$
$$+\alpha c\frac{\partial(\mu Y^i)}{\partial\mu}-\alpha c\frac{\dfrac{\partial^2 Y^i}{\partial\omega^2}}{1-\mu^2}. \qquad (93)$$

We need not follow the calculation in detail in order to seize Laplace's explanation, which is that the term Y^i, representing the variation of the value for the radius calculated from observation from the value that satisfies the law of the square of the sine of the latitude, is differentiated once for the pendulum method and twice for the geodesic method. It becomes detectable only in the latter case. (On the significance of this paper in the history of error theory, see a remark in L: Plackett [1972], p. 239.)

The paper on the rings of Saturn was a much more provisional exercise. An "essay for a theory" Laplace called it, pending the development of telescopes powerful enough to reveal the correct number and dimensions of the rings. Like Galileo, Huygens, and many others, Laplace quite evidently found the phenomenon one of the most tantalizing in the whole field of astronomy. In his hands, of course, it was bound to take the form of a gravitational problem. He had first alluded to it in print in a special case discussed at the end of the opening section of his earliest major memoir on spheroid attraction ([*1778a*]; *OC*, IX, 86–87). There the rings are mentioned as the phenomenon instantiating the problem of the (reversed) attraction of an external point for a spheroid. The manuscript draft of that section, which he read on 22 January 1777 (*20*), closes with an undertaking to devote a future memoir to the problem of the figure of Saturn and the influence on it of the attraction of the rings. That was omitted in the printed text. Now that Laplace was finally redeeming the promise twelve years later, the problem was rather the figure of the rings themselves. There is still some appearance of hesitation. Laplace did not publish the memoir until 1789, five years after submitting the basic theory of spheroidal attraction ([*1785b*]) to the Academy (*48*).

Given the paucity of data, the discussion offers a particularly transparent example of recourse to a mathematical model. Instead of seeking the stability of the ring in any sort of mechanical connection between the particles, Laplace imagined its surface covered with an infinitely thin layer of a fluid in equilibrium under the influence of the forces at work. The shape of the ring would then be determined by the equilibrium conditions of the fluid. The most notable single feature of the memoir is the statement at the outset of the basic equation of spheroidal attraction theory, here referred to the rectangular coordinates that make it immediately recognizable for the first time as the potential function of later physics ([*1789a*]; *OC*, XI, 278):

$$0=\frac{\partial^2 V}{\partial x^2}+\frac{\partial^2 V}{\partial y^2}+\frac{\partial^2 V}{\partial z^2}. \qquad (94)$$

In the case of a spheroid of revolution, the equation becomes

$$0=\frac{1}{r}\frac{\partial V}{\partial r}+\frac{\partial^2 V}{\partial r^2}+\frac{\partial^2 V}{\partial z^2}, \qquad (95)$$

where $r^2 = x^2 + y^2$. If the body is a solid sphere, or a hollow sphere with a homogeneous shell, the equation becomes

$$0=\frac{2}{r'}\frac{\partial V}{\partial r'}+\frac{\partial^2 V}{\partial r'^2}, \qquad (96)$$

where $r' = \sqrt{(r^2 + z^2)}$.

Investigating the continuity of the ring, Laplace calculated that since the mass of Saturn must be much greater than that of the ring, and since a sphere exerts a much greater force on a particle at its surface than a flattened body of the same mass would do, the relation between the forces of attraction at a point on its inner circumference, its outer circumference, and the surface of Saturn are such that the ring must in fact be a series of concentric rings. He claimed to have been able to predict the discontinuity between the rings from the theory of gravity alone. He then showed how such a ring may be generated mathematically by a very flat ellipse of which the prolongation of the major axis passes through the center of Saturn and which revolves about that center in a plane perpendicular to its own. There is no point in following the formalism. Todhunter (L: [1865], II, 65–73) reproduced it more fully than was his wont, and Laplace himself omitted the derivations depending on the most questionable hypotheses from *Mécanique céleste*, where the topic occupies Book III, chapter 6 (*OC*, II, 166–177). It involves the analytic necessity of constructing another ellipse with an infinite major axis on the same equator as the generating ellipse. The two must exert an identical attraction at every point. The ring might then be invested with unequal dimensions in its different parts—and even with double curvature. It had to be so, for if it were circular and concentric with Saturn it would collapse into the planet. The most important findings are that the conditions for stability require that it revolve in the equatorial plane of Saturn, along with the four inner satellites, and that the center of gravity not coincide with the center of rotation. Abstract though the analysis was, Laplace pictured the rings physically as solid bodies, oscillating stably in asymmetrical rotation around the planet in such a way that their centers of gravity described elliptical orbits about the center of gravity of the planet even like normal satellites. With the delay in publication until 1789, the paper on Saturn's rings comes as an afterthought on the theory of attraction in the midst of the series of memoirs on planetary motion.

17. Secular Inequalities: Jupiter and Saturn; the Moons of Jupiter; Lunar Theory. Only in that series does vindication of the stability of the solar system, which became the central feature of the Laplace stereotype through its celebration in the nineteenth century, finally appear to be the central motivation. Laplace himself began insisting upon it in a sequence of five memoirs imparting his main discoveries in planetary theory. He composed them between November 1785 and April 1788 (*50, 52, 53, 55, 56*). Scientifically, however, they inaugurated an even lengthier preoccupation, for from then on his attention remained focused on problems of positional astronomy, practical as well as theoretical, for the next twenty years, through the completion of Volume IV of *Mécanique céleste* in 1805. A near coincidence marks the start of that orientation. He read the first of these pieces, "Un mémoire sur les inégalités séculaires des planètes" ([*1787a*]), on 23 November 1785 and one week later (*51*) read the last of his demographic papers ([*1786c*]; see Section 13).

The paper on the secular inequalities of the planets ([*1787a*]), a relatively brief piece by Laplace's standards, must be considered, together perhaps with "Théorie des attractions des sphéroïdes et de la figure des planètes" ([*1785b*]) and "Sur la probabilité des causes par les événements" ([*1774c*]), one of the really signal memoirs in his *oeuvre*. It opens with a serene and comprehensive survey of the history and state of the question of apparently cumulative discrepancies between the theoretical and observed positions of celestial bodies, and largely resolves them with a statement of two of his most famous determinations, the interdependence of the apparent acceleration of Jupiter's mean motion with the deceleration of Saturn's, and the rigorous necessity for the mathematical games played by the Jovian planets, the three inner satellites of Jupiter, in the figure dance of their revolutions around their parent. The second two memoirs in the sequence, actually a single enormous memoir with a sequel, on the theory of Jupiter and Saturn ([*1788a*] and [*1788b*]), contain calculations in detail. Only the further argument, that the apparent acceleration of the mean motion of our own moon over time depends upon the action of the sun compounded by variation of the eccentricity of the earth's orbit, would appear to have come to him after writing the covering paper on secular inequalities; for he announced it in the preamble to the Jupiter and Saturn memoir, which was presented on 6 May 1786 (*52*), reserving its development for a further memoir on lunar theory, the fourth of the series ([*1788c*]), submitted on 19 December 1787 (*55*).

Laplace began by explaining the distinction that astronomers habitually made between periodic and secular inequalities in the ellipticity of planetary orbits. The former depend on the positions of the planets in orbit, relative to each other and to their aphelions, and compensate for themselves in a few years time. They are to be considered as very small oscillations on either side of a point in motion on an ellipse described in consequence of the attractive force exerted by the sun alone. Secular inequalities are those that modify the elements of the orbits themselves—their inclinations, eccentricities, and longitudes—so slowly as to be undetectable in the course of a single revolution but that finally change the nature and position of the orbits (*ibid.; OC*, XI, 49). Their effects on the shape and position of the orbits become manifest only over centuries, hence the term "secular." The terminology may have become a touch inappropriate, since what Laplace was setting out to demonstrate precisely was that secular inequalities are themselves periodic, the periods occupying centuries. It is true that the term "secular" did not necessarily imply that very gradual inequalities need to be indefinitely cumulative, but such was certainly the apprehension. He himself spoke of them, perhaps inadvertently, as "accumulating ceaselessly" (*ibid.; OC*, XI, 49). Also, he was slightly less than candid—or at any rate clear—about the background of his own views on the matter. The most important secular inequalities would be those (if any such there were) that might change the mean motion of a planet and with it, by Kepler's third law, the mean distance from the sun; and practical astronomers were constrained to include a correction factor proportional to the square of the time in their expressions for the mean motions of Jupiter and Saturn. In writing now about his own early investigation of this problem ([*1776a*, 2°]; see Section 5), he claimed to have found that theory admitted no secular inequality in the mean motions or mean distances and to have concluded that such an inequality had been nonexistent or at least unobservable throughout the recorded history of astronomy. What he had actually said was not quite that. Rather, it was that theory proves the nonexistence of secular inequalities due to the interaction of any two mutually gravitating planets in historic time. He recalled that he had once thought to invoke the action of the comets to explain the undeniable speeding of Jupiter and slowing of Saturn. The analysis did not preclude that, although the physics did when he came to consider how slight the masses of the comets are. Moreover, whether the comets or some other agents were at work, Laplace had not at this early stage discussed or even mentioned the possibility of a long-term periodicity in these greater inequalities.

That opportunity to save the system was opened by the work of Lagrange. In a crucial memoir of 1783, "Sur les variations séculaires des mouvements moyens des planètes" (*Oeuvres de Lagrange,* V, 381–414), Lagrange investigated the question whether, even allowing unlimited time, the perturbing forces of other planets, themselves subject to variations in the elements of their orbits, could progressively alter the mean motions and mean distances of a planet and thus disturb its service to Kepler's third law. His analysis was more abstract than Laplace's had been, and he extended it to investigating the terms involving the squares of the eccentricities. There he did find a secular equation; but its maximum value was one-thousandth of a second, and its effect was negligible over all time, even in the case of the largest observed inequalities, those of Jupiter and Saturn. The question, therefore, was not whether those inequalities were to be explained by expressions involving lengthy periodicity, but how to do it. It was unthinkable that they should be random. This was ever the sort of problem at which Laplace excelled, and he set himself to examining the figures.

The first hope was that the mutual gravitation of the two planets might suffice to reveal periodicity. In planetary interaction over long periods, the sum of the masses of each planet divided by the major axis of the respective orbits is approximately constant. Thus, given Kepler's third law, if Saturn is slowed by Jupiter, Jupiter should be accelerated by Saturn. Their masses are respectively 1/1067.195 and 1/3358.40 that of the sun. It followed that the ratio of Jupiter's acceleration to Saturn's deceleration should have been approximately 7:3. Halley assumed 9°16′ for the deceleration of Saturn in two millennia. If that were correct, Jupiter's acceleration should have been 3°58′—which differed from the tables by only 9 arc minutes. Thus, near equality made it probable that their theory did indeed contain an inequality of very long term depending on their configuration and not on their positions individually. The problem was to find it.

That their mean motions were nearly commensurate was well known. Five times the mean motion of Saturn almost equals twice the mean motion of Jupiter. It was characteristic of Laplace that he saw how to exploit this fact. He suspected that the variations in inclination and eccentricity of the orbits were responsible for the changes in speed. Might not the differential equations of motion contain terms with an argument of $5n' - 2n$ (to use his designation for the two mean motions) that would become detectable on integration, even though they were involved with the cubes and third-order products of the eccentricities and inclinations? For the motions were not quite commensurate; actually $5n' - 2n$ came to about 1/74 of the mean motion of Jupiter. The differential equations for the longitude worked out so that on two integrations, the square of this quantity appeared in the divisors. Such terms were normally very minute, the order of magnitude being that also of cubes of eccentricities and inclination. In the case of Jupiter and Saturn, however, the smallness of the divisors rendered the terms detectable on the second integration. By a fortunate coincidence, the necessity to introduce the inequalities depending on the third and higher powers of eccentricities and inclinations also permitted neglecting the quantities that would have made their determination impossible. So theory could be compared to observation. The expressions for Saturn yielded a secular equation of approximately 46′50″ with a period of approximately 817 years; the theory of Jupiter contained an equation of 20′ with opposite sign and identical period. The ratio is about 3:7 ([*1787a*]; *OC*, XI, 49–56). Recalculating in the full Jupiter-Saturn memoir ([*1788a*]), Laplace found 48′44″ for Saturn and 20′49″ for Jupiter with a period of 929 years.

With these figures, and designating by nt and $n't$ the sidereal motions of Jupiter and Saturn respectively since the epoch of 1700, the longitudes will be given by the following formulas. For Jupiter:

$$nt + \epsilon + 20' \sin (5n't - 2nt + 49°8'40'') \; ; \quad (97)$$

and for Saturn,

$$n't + \epsilon' - 46'\, 50'' \sin (5n't - 2nt + 49°\, 8'\, 40''), \quad (98)$$

where ϵ and ϵ' are constants that depend on the longitudes of the two planets on 1 January 1700. In order to evaluate the acceleration and deceleration empirically, the mean motions determined over a very long period could be compared to the values obtained from recent data. The time elapsed between the opposition of Saturn recorded for 228 B.C. and that for 1714 should give the "true" mean motion.

If the deceleration given by the above theory was correct, the value computed for the mean motion between observations of 1595 and 1715 should be too small by 16.8″. In fact, the deceleration was 16″—and the fit between observation and theory was as good as the imprecision of sixteenth-century observation could allow. Agreement with Halley's tables was equally good for the acceleration of Jupiter, which comes out in the ratio of 7:3 to be approximately 7″. The changes of speed were at their maximum around A.D. 1580. Since then Saturn's deceleration and Jupiter's acceleration have been diminishing and the apparent mean motions coming closer to the "true" values taken over all time (*ibid.; OC*, X, 53–54). Laplace also showed his calculation to be conformable to a set of corrections to be applied over short periods that Lambert had published as "Résultat des recherches sur les irregularités du mouvement de Saturne et de Jupiter" in the *Nouveaux mémoires de l'Académie royale de Berlin* ([1773/1775], 216–221). His object had been an empirical determination of *la loi des erreurs* in Halley's tables, and it may be worth mentioning because it seems to be the first time that Laplace used the phrase "law of error."

Also contributing to determination of the inequalities of Jupiter and Saturn are terms due to the ellipticity of the orbits on the assumption that $5n' = 2n$ exactly, and others expressing perturbations due to variations in eccentricity of the orbits. Other mathematicians had calculated these effects, but with such discordant results that the whole procedure needed to be verified. That task Laplace deferred to a further memoir containing the calculations—here he gave only the results—in which all ancient and modern oppositions of both planets would be accommodated, and certain strange aberrations of Saturn explained. Centuries would have been required to accomplish that empirically: "Thus, on this point, the theory of gravity has moved ahead of observation" (*op. cit.; OC*, IX, 56).

The present memoir had two other objects, for which it does contain the analysis. They were equally integral to the system of the world. The first still concerned the uniformity of mean motion of celestial bodies, in the case of satellites rather than the primary planets, however, and specifically that of the moons of Jupiter. At issue was the invariance of their mean distances from a principal center of force. The second concerned the other variations of celestial orbits. The methods for determining the inclinations and eccentricities were very simple. But were these elements oscillating within certain, presumably narrow limits, or did permanent changes occur?

The motions of Jupiter's moons were a well-studied set of phenomena when Laplace took up the question of their stability over the span of astronomical time. They were most famous for the observations that had enabled Römer in 1676 to measure the velocity of light by recording the dependence of variations in the apparent time of Io's eclipses on the distance of Jupiter from the earth. More immediately, they were the main resort for navigators in determining longitude before the availability of reliable chronometers. The Swedish astronomer Pehr Wargentin had devoted much of his life to perfecting the tables, first published in 1746, which he revised constantly in correspondence with Lalande. In the theory of these bodies, the most important item by far was a memoir by Lagrange (*Oeuvres de Lagrange*, VI, 67–225), which had won the prize for 1766 set by the Académie des Sciences for a study of their inequalities. Lagrange calculated the effects on each of the four satellites of the oblateness of Jupiter, the action of the sun, and the action of the other three. He did not investigate the relation of the inequalities in longitude to the variations of eccentricity and inclination; and his results, however impressive analytically, were of no help in improving the tables. Moreover, he made a mistake. He assumed that the angle between the planes of Jupiter's equator and orbit was negligible.

Such, in summary, was the state of knowledge when Laplace took the problem in hand. The mean motion of the first satellite is about twice that of the second, which is about twice that of the third. Curiously enough, however, the difference between the mean motions of the first and second was incomparably closer to being precisely twice that between the mean motions of the second and third, than was the near doubling of mean motions in successive pairs. (The fourth was the odd moon out, its motions being incommensurable with the inner three; and Laplace neglected its influence here, as well as that of the sun.) If the tripartite relation were to prove really exact, so that

$$n + 2n'' = 3n', \qquad (99)$$

it would follow from this simple equation that the three moons could never be in simultaneous eclipse. In point of predictable fact, the data from Wargentin's tables show that such an eventuality could not occur in the next 1,317,900 years, and it would be entirely precluded by a modification in the annual motion of the second smaller than the margin of error in the tables. "Now," Laplace observed, "it may be laid down as a general rule that, if the result of a long series of precise observations approximates a simple relation so closely that the remaining difference is undetectable by observation and may be attributed to the errors to which they are liable, then this relation is probably that of nature" (*ibid.; OC*, XI, 57).

The remark offers an especially strategic example of the role of probability, causality, and error theory in his planning of an investigation. The equation (99) of the mean motions of Jupiter's moons could scarcely be thought the effect of chance; yet it was improbable that those bodies had originally been placed at just the distances from Jupiter that it requires. It was natural, therefore, to try out the hypothesis that their mutual attraction is the true cause. Thus, Laplace observed, in anticipation of a later investigation ([*1788c*]), the action of the earth on the moon had brought about the equality between its periods of revolution and rotation, which might have been very different at the beginning.

The object of the research, therefore, was to determine whether the mean motions of Jupiter's first three satellites became and remain stable by virtue of the law of universal gravitation. Laplace based the analysis on the motion of the second satellite. He set up general equations of motion of a system of mutually attracting bodies and applied them to the motions of all three satellites, showing how the principal inequalities of the second depend on the influence of the first and third. In his expressions,

t designates the time;

n, n' and n'', the mean motions of the first three satellites;

s, the quantity $n - 3n' + 2n''$;

$nt + \epsilon$, $n't + \epsilon'$, $n''t + \epsilon''$, the projections onto the same plane of the mean longitudes taken from the x-axis; and

V, the angle $(2n'' - 3n' + n)\,t + 2\epsilon'' - 3\epsilon' + \epsilon$, the mean longitudes being taken from a fixed point in Jupiter's orbit.

The tables call for s and V to be approximately constant in value, but Laplace could find no reflection of that fact in any terms that depended on the first powers of the perturbing masses. Accordingly, he proceeded to an examination of terms that depended on the squares and products of the masses. Multiplication of the expressions by the masses of the satellites, taken two at a time, did introduce into the values of s and V quantities proportional to time, giving for the variation of s,

$$\delta s = \alpha n^2 t \sin V, \tag{100}$$

where α stands for an immensely complicated function multiplied by $n^2 t \sin V$ in the equation ([*1787a*]; *OC*, XI, 77)

$$2\delta n'' - 3\delta n' + \delta n = \frac{3}{2} a n^2 t \sin V\alpha. \tag{101}$$

Laplace then eliminated from (100) the quantities that increase with time, using his method of varying the constants of integration (see Sections 7 and 10). He thus obtained two first-order differential equations among s, V, and t:

$$\frac{ds}{dt} = \alpha n^2 \sin V \tag{102}$$

and

$$dV = dt\,(2n'' - 3n' + n) = s\,dt, \tag{103}$$

whence

$$\frac{d^2 V}{dt^2} = \alpha n^2 \sin V. \tag{104}$$

Multiplying by dV and integrating gives

$$\frac{\pm dV}{\sqrt{\lambda - 2\alpha n^2 \cos V}} = dt, \tag{105}$$

where λ is an arbitrary constant. Of the three possibilities, first that λ is positive and $> \pm 2\alpha n^2$, second that α is positive and $\lambda < 2\alpha n^2$, and third that α is negative and $\lambda < -2\alpha n^2$, the second case entailed a periodic value for V oscillating around a mean of 180°; and the data showed this to be the case in nature, with the oscillations of very small amplitude.

Laplace drew four consequences from the analysis pushed thus far. The first was practical. Since both s and V are periodic, the relation (99), $n + 2n'' = 3n'$, is rigorously exact, and Wargentin's tables needed only slight corrections. The second was historical. It was not a necessary condition that the three satellites should originally have been placed at distances from Jupiter's center of mass that by Kepler's laws would result in mean motions satisfying (99). If they had merely been close to those distances (and Laplace calculated what the limits were), then their mutual attractions would bring them into that relation with each other. The third consequence concerned the stability of the system. There was no need to fear that the tables for the principal equation of the second satellite would be off, even after the lapse of many centuries. The fourth, and in Laplace's eyes the most important, raised problems for the future. Consideration of the angle V created a second condition that the tables had to satisfy, namely that

$$nt + 3n't - 2n''t = 180°, \tag{106}$$

where nt, $n't$, and $n''t$ are the respective mean longitudes.

The angle V is subject to a periodic inequality that Laplace likened to the oscillations of a pendulum, and that inequality affects the motions of the three satellites in varying degrees, according to the ratio of their respective masses and distances from the center of Jupiter. The mass of the second is adequately determined by the inequalities that it produces in the motions of the first—that is why Laplace could make its motion the basis of reference. The masses of the first and third were still unknown, however. All that could be said was that a relation exists that accounts for the inequalities of the second. That was how he was able to determine that the libration of V is contained within limits of 4-1/8 and 11-1/3 years. Both its zero point and amplitude are quantities to be determined by observation. To consider only the action of Jupiter's first three satellites on each other, their motion depends on no less than nine second-degree differential equations, the integrals of which contain eighteen arbitrary constants. The eccentricities, orbital inclinations, and positions of nodes and aphelia determine twelve of those constants. The mean motions and their epochs would give six others, without the two conditions to which these six arbitrary quantities are subject. That reduces them to four. To answer to that, the expression for V contains two arbitrary quantities. Since the tables do work well without regard to the periodic inequality of V, its amount must be small. But the uncertainties still prevailing over most of the elements of the theory of Jupiter's satellites made the determination very difficult, and Laplace left that question to the astronomers to resolve. He had shown them the two conditions that the tables must satisfy, namely (99) and (106). Concluding this discussion, he again recurred to the analogy with our own moon. Those conditions would still obtain if, like the moon, Jupiter's satellites showed an acceleration in the mean motions such that the mean distances progressively decreased. Even as the satellites drew in on Jupiter, the same relations would be preserved (*ibid.; OC*, XI, 60–61).

Thus, Laplace showed that the mean motions of Jupiter's moons are subject only to periodic inequalities, and he felt justified in drawing conclusions for the whole planetary system from this test case. Considering only the laws of universal gravity, the mean distances of celestial bodies from their principal centers of force are immutable. It by no means followed that the same is true of the other orbital elements, namely eccentricities, inclinations, and the positions of nodes and aphelia, all of which undergo continual change. Good methods for determining these quantities existed, subject to the hypothesis that the orbits differ little from the circular and that their planes are only slightly inclined to that of the ecliptic. Laplace himself had long since shown ([*1776c*]) that the eccentricities and inclinations of the orbits are bound to remain small under the operation of gravity, provided that only two planets are considered; and in 1782 Lagrange had further shown (*Oeuvres de Lagrange*, V, 211–345) that the same is true generally of the location of nodes and aphelia, given plausible hypotheses about the masses (over several of which some uncertainty did linger). But before the entire system of the world could be deduced from the law of gravity, it remained to demonstrate that the eccentricities and inclinations of any number of planets are contained within narrow limits. That was the final object of the present memoir. Analytically, what was required was to prove that the expressions for the secular inequalities of eccentricities and inclinations contain neither secular terms nor exponential terms. The analysis is uncharacteristically brief and occupies the final two chapters (*ibid.; OC*, XI, 88–92).

In those expressions,

$m, m', m'', \cdots$, designate the relative masses of the planets, referred to the sun as unity;
$a, a', a'', \cdots$, the semimajor axis;
$ea, e'a', e''a'', \cdots$, orbital eccentricity;
$V, V', V'', \cdots$, longitude of aphelion;
$\theta, \theta', \theta'', \cdots$, tangent of orbital inclination referred to a fixed plane; and
$I, I', I'', \cdots$, longitude of ascending node.

The quantities $e \sin V$, $e \cos V$, $e' \sin V'$, $e' \cos V'$, $\cdots$, $\theta \sin I$, $\theta' \sin I'$, $\theta \cos I$, $\theta' \cos I'$, $\cdots$, could then be given by linear differential equations with constant coefficients. Since the eccentricities and inclinations are very small, the system of equations for the eccentricities is independent of the system for the inclinations. The former system is the same as if the orbits were coplanar, and the latter is the same as if they were circular.

When the former system is integrated, each of the quantities $e \sin V$, $e \cos V$, $e' \sin V'$, $e' \cos V'$, $\cdots$ is given by the sum of a finite number of sines and cosines of angles proportional to time t. The numbers by which the times are to be multiplied to give the angles are the roots of an algebraic equation of a degree equal to the number of planets. Laplace calls that equation k. The same is true for the quantities of the second system, although the equation on which the angles depend is not the same. He calls that equation k', referring the reader to his discussion in [*1776c*] (*OC*, VIII, 406) and to Lagrange's "Théorie des variations séculaires des éléments des planètes" (1782) (*Oeuvres de Lagrange*, V, 211–344, esp. 249, 325).

If all the roots of k and k' are real and unequal, the values of the preceding quantities contain neither secular terms nor exponentials and thus remain within narrow limits. It is otherwise if some of the roots are imaginary, for then the sines and cosines change into secular terms or exponentials. But, in any event, the values of $e \sin V$, $e \cos V$, $e' \sin V'$, and $e' \cos V'$, $\cdots$ would always take the form shown in the four expressions

$$e \sin V = \alpha f^{it} + \beta f^{i't} + \cdots + \gamma t^r + \lambda t^{r-1} + \cdots + h$$
$$e \cos V = \mu f^{it} + \epsilon f^{i't} + \cdots + \phi t^r + \psi t^{r-1} + \cdots + l$$
$$e' \sin V' = \alpha' f^{it} + \beta' f^{i't} + \cdots + \gamma' t^r + \lambda' t^{r-1} + \cdots + h'$$
$$e' \cos V' = \mu' f^{it} + \epsilon' f^{i't} + \cdots + \phi' t^r + \psi' t^{r-1} + \cdots + l'. \quad (107)$$

(It should be explained that f stands for the e of later notation—that is the base of the natural system of logarithms—but i is not $\sqrt{-1}$.) In these expressions, the coefficients $\alpha, \beta, \mu, \epsilon, \cdots, \alpha', \beta', \mu', \epsilon'$, of the exponential terms are real quantities not subject to exponentials. They may, however, become functions of the arc t and of sines and cosines of angles proportional to that arc. As for the quantities $\gamma, \lambda, \phi, \psi, \cdots, h, l, \gamma', \lambda', \phi', \psi', \cdots, h', l', \cdots$, they are real, containing no exponentials or secular terms, and hence are either constant or periodic.

Laplace then supposes that, without regard to signs, $i > i'$, $i' > i''$, $\cdots$, when $e = (e \sin V)^2 + (e \cos V)^2$. It would then follow that

$$e^2 = (\alpha^2 + \mu^2)\, f^{2it} + \cdots + (\gamma^2 + \phi^2)\, t^{2r} + \cdots + h^2 + l^2. \quad (108)$$

Similarly,

$$e'^2 = (\alpha'^2 + \mu'^2)\, f^{2it} + \cdots + (\gamma'^2 + \phi'^2)\, t^{2r} + \cdots + h'^2 + l'^2, \quad (109)$$

and so on. These expressions give the values of the eccentricities, which hold good, however, only for a limited period of time. The precondition that they remain small, varying within narrow limits, is valid if—and only if—all the roots of the equation k are in fact real and unequal. It is very difficult to show this directly, and Laplace goes back to an expression for the eccentricities that he had obtained in discussing the general equations of motion of a system of mutually attracting bodies under the law of gravity (*ibid.; OC*, XI, 69):

$$c = m\sqrt{\frac{a\,(1-e^2)}{1+\theta^2}} + m'\sqrt{\frac{a'\,(1-e'^2)}{1+\theta'^2}} + m''\sqrt{\frac{a''\,(1-e''^2)}{1+\theta''^2}} + \cdots, \quad (110)$$

neglecting constant and periodic quantities of the degree m^2, where $a, a', a'', \cdots$ are the semimajor axes of the orbits of bodies m, m', m''; $ea, e'a', e''a'', \cdots$ are the eccentricities; and $\theta, \theta', \theta'', \cdots$ are the tangents of the inclinations; and c is an arbitrary constant.

If quantities of the degree e^4, $e^2\theta^2$, and θ^4 are ignored, then (110) becomes

$$c = m\sqrt{a} + m'\sqrt{a'} + \cdots - \frac{1}{2} m\,(e^2 + \theta^2)\sqrt{a} - \frac{1}{2} m'\,(e'^2 + \theta'^2)\sqrt{a} - \cdots. \quad (111)$$

Furthermore, since the mean distances of the planets from the sun are unaffected by their interactions,

$$m(e^2 + \theta^2)\sqrt{a} + m'\,(e'^2 + \theta'^2)\sqrt{a'} + \cdots = \text{constant}. \quad (112)$$

It had already been pointed out that the values of $e, e', e'', \cdots$ are independent of those for $\theta, \theta', \theta'', \cdots$, so that they will be the same as if the latter were null. Thus (112) reduces to

$$me^2\sqrt{a} + m'e'^2\sqrt{a'} + \cdots = \text{constant}, \quad (113)$$

which is the equation that must be satisfied by the values of $e, e', e'', \cdots$, after the lapse of any amount of time.

Substituting the general expression of those quantities as given in (107) into (113) gives

$$[m\sqrt{a}\,(\alpha^2 + \mu^2) + m'\sqrt{a'}\,(\alpha'^2 + \mu'^2) + \cdots] f^{2it} + \cdots$$
$$+ [m\sqrt{a}\,(\gamma^2 + \phi^2) + m'\sqrt{a'}\,(\gamma'^2 + \phi'^2) + \cdots]\, t^{2r} + \cdots$$
$$+ m\sqrt{a}\,(h^2 + l^2) + m'\sqrt{a'}\,(h'^2 + l'^2) + \cdots$$
$$= \text{constant} \quad (114)$$

This equation must hold whatever the value of t, and it is therefore essential to eliminate exponential powers of t and secular terms. To that end, the coefficient f^{2it} is equated to zero, so that

$$0 = m\sqrt{a}\,(\alpha^2 + \mu^2) + m'\sqrt{a'}\,(\alpha'^2 + \mu'^2) + \cdots. \quad (115)$$

Since $m\sqrt{a}, m'\sqrt{a'}, \cdots$ are positive quantities, and since $\alpha, \mu, \alpha', \mu', \cdots$ are real quantities, (115) will hold only on the supposition that $\alpha = 0$, $\mu = 0$, $\alpha' = 0$, $\mu' = 0$, $\cdots$. It follows that there are no exponential terms in the values of $e, e', e'', \cdots$.

Returning now to (114) and equating the coefficient t^{2r} to zero gives

$$0 = m\sqrt{a}\,(\gamma^2 + \phi^2) + m'\sqrt{a'}\,(\gamma'^2 + \phi'^2) + \cdots, \quad (116)$$

whence $\gamma = 0, \phi = 0, \gamma' = 0, \phi' = 0, \cdots$. Thus, neither do the values for $e, e', \cdots$, contain secular terms. They reduce, therefore, to periodic quantities of the form $\sqrt{h^2 + l^2}, \sqrt{h'^2 + l'^2}, \cdots$, and we know from (114) that these quantities serve the equation

$$\text{constant} = m\sqrt{a}\,(h^2 + l^2) + m'\sqrt{a'}\,(h'^2 + l'^2) + \cdots. \quad (117)$$

When the right-hand side of that equation is expanded in a series of sines and cosines, the coefficients of each sine and cosine vanish automatically.

As for the inclinations, the same reasoning may

be applied to the expressions for θ, θ', θ'', · · ·. Laplace thus satisfied himself analytically of the periodicity of variations in the orbital eccentricities and inclinations, thereby completing the mathematical demonstration that ". . . the system of the planets is contained within invariant limits, at least with respect to their action on each other" (*ibid.; OC*, XI, 92).

The calculations justifying the long-term interdependence of the inequalities of Jupiter and Saturn are enormously lengthier. Laplace deferred their publication to the "Théorie de Jupiter et de Saturne" ([*1788a*] and [*1788b*]), presented in two installments. Unlike the above, which was mainly a mathematical exercise, they involved him in detailed reference to the corpus of recorded astronomical observation and also in arriving at numerical solutions to his equations. He had tried, he said, to give his results a simple and convenient form. At the same time, he thought to improve on the existing practice of mathematicians, who limited themselves to the first powers of the eccentricities and inclinations in their calculations. He had come to recognize that the resulting approximations were insufficiently exact, and in working out the analytical theory of the mutual perturbations he took into account the inequalities that depend on the squares and higher powers of these variations, carrying the approximation up to the fourth power of the eccentricities.

The necessity to recur to the most ancient as well as the most recent data led Laplace into the way of certain historical reflections. It was a curious chance that the maximum acceleration of Jupiter and deceleration of Saturn, *ca.* 1560, should have largely coincided with the revival of astronomy in the generations of Copernicus and after. Since that time, the apparent values were approaching the true values of the mean motions. If astronomy had been reborn three centuries sooner, its practitioners would have observed the contrary phenomena. The motions that a culture attributes to the two planets might thus serve as an index of the period in which it was founded or at least sufficiently developed to support astronomical observation. Laplace would judge by data from Hindu astronomy that in India the mean motions had been determined at a time when Saturn appeared to be at its slowest and Jupiter at its fastest. Two of the principal Hindu astronomical eras appeared to fulfill these conditions, one centering around 3103 B.C. and the other around A.D. 1491 (*ibid.; OC*, XI, 178–179). For Laplace was as credulous in accepting historical fact as he was critical of astronomical information. He probably had the former from Bailly's *Traité de l'astronomie indienne et orientale* (1787).

Saturn being the more recalcitrant and less-studied planet, Laplace devoted the first installment of the memoir to calculating its theory in detail. The general strategy for finding its inequalities was to substitute its elements, together with those of Jupiter, into the formulas. Before that could be done, another inequality besides the long inequality deriving from the action of Jupiter had to be noticed. It was responsible for the slight discrepancy in what was only a near equality between twice the mean motion of Jupiter and five times that of Saturn, and its amount was about 10 arc minutes. If that equality had been exact, it would have coincided with the inequality due to the ellipticity of the orbit. A complete theory of Saturn had to provide for all these inequalities. Working out the mathematical basis in the first section of the memoir required varying the constants to eliminate terms introduced in successive integrations that increased with time and that would have precluded stability for the system if they referred to something real. Laplace included a more succinct account of the variation of constants than he had yet given (*ibid.; OC*, XI, 131–140).

Before Laplace could give his formulas numerical values, which task was carried out in the second section, he had to know what the arbitrary constants represented. In the expressions for the elliptical motion of the planets, these constants are the mean distance, inclination, the mean longitude at a given moment, the eccentricity, and the positions of aphelion and nodes. They come from the data of observation, which need to be corrected both for inaccuracies and for the effect of the perturbations to which the bodies are subject. The latter effects have to be known in advance from theory. For that purpose, the masses had to be known, taken relative to the sun as unity. In the cases of Jupiter and Saturn, the determination of their inequalities and of the elements of the orbits depend the one on the other in a reciprocal manner. The determination could therefore be accomplished by successive approximations. Halley's tables were the main source of data for Saturn. Dating back to 1719, they needed systematic correction, and Laplace had first of all to arrive at a formula, or "law of errors" (*ibid.; OC*, XI, 190), to be applied in using them. He then perfected the law by trying it on numerous oppositions of the planet. Jupiter presented less of a problem: Wargentin's tables were more up-to-date for that planet, and their author had allowed for the inequalities that were independent of eccentricity and also for those depending on the first power of that element. He had not known about the effect of the long inequality on the mean motion, eccentricity, and position of aphelion, and had bundled those effects into his terms for ellipticity, from which they could simply be subtracted in correcting the several elements. Laplace referred his numerical calculations to the beginning of the year 1750, mean time of Paris, choosing that date because it was in the midst of the period of good modern astronomical observations. Perhaps it will help the non-astronomer to visualize what he was about, if some numbers for Saturn are reproduced:

Mean longitude ϵ'	$7^s21°17'20''$
Mean longitude of aphelion ω'	$8^s28°7'24''$
Eccentricity e'	0.056263
Mean longitude of ascending node	$3^s21°31'17''$
Inclination of orbit to ecliptic (tan θ')	$2°30'20''$
Sidereal motion during 365-day common year, n	$12°12'46.5'' = 43966''.5$
Mean distance to sun (earth-sun = 1)	9.54007.

(118)

The mean sidereal motion of Jupiter (n') was $30°19'42''$, or $109,182''$. The problem was then to determine numerically the secular inequalities affecting the orbits of both planets, beginning with those dependent on the angle $5n't - 2nt + 5\epsilon' - 2\epsilon$. Calculation from the theory yielded the amount by which to correct the observed value for this class of inequality. Next came the periodic inequalities that are independent of eccentricities and inclinations, followed by those that derive from the eccentricity, and finally, those depending on the squares and higher powers of inequalities and inclinations. When the conversion from mean to true anomaly was included, Laplace could find the formula for determining the heliocentric longitude of Saturn in its orbit at any instant past or future.

These measures presupposed, however, that the above elements for Saturn in 1750 were exactly correct, together with the theory by which the inequalities were determined. No doubt the approximation was very close, but the values needed to be checked against the entire body of modern observations. To that end Laplace designated by $\delta\epsilon$, $\delta n'$, $\delta e'$, and $\delta\omega'$ the corrections that would need to be applied to the 1750 values for mean longitude, annual mean motion, eccentricity, and position of aphelion, respectively. To determine those quantities empirically, he drew on twenty-four oppositions of Saturn, choosing instances that were well distributed in time from 1591 to 1785 and that his prior investigation of the law of errors in Halley's tables had shown to be tolerably precise. Not that these could be supposed to represent the "true" longitudes—Halley's calculations on the basis of Flamsteed's observations had been executed in ignorance of the phenomena of aberration and nutation and of modern data on the location of certain reference stars. It would be well to recalculate them all. Pending that immense task, however, Laplace proposed to compensate for the lack of precision of the individual observations by increasing the number of them that he employed.

That procedure meant minimizing the error, and to that end Laplace wrote twenty-four linear equations of condition for the theory of the motion of Saturn. Here, for example, is the equation for the opposition observed on 25 December 1679 at 22h 39m Paris time, when the longitude was $3^s4°54'0''$:

$$0 = 3'9''.9 + \delta\epsilon' - 70.01\,\delta n' + 2\delta e'\,0.12591 - 2e'\,(\delta\omega' - \delta\epsilon')\,0.99204. \tag{119}$$

In an interesting article (L: [1975]) Stephen Stigler has recently characterized Laplace's solution of his system of twenty-four equations as an early example of a multiple linear regression. For Laplace combined them into a set of four equations by (1) summing all twenty-four; (2) subtracting the sum of the second twelve from the sum of the first twelve; (3) combining one set of twelve in the order: $-1+3+4-7+10+11-14+17+18-20+23+24$; and (4) combining the remaining twelve in the order: $+2-5-6+8+9-12-13+15+16-19+21+22$. He then solved these four equations to get

$$\begin{aligned} \delta\epsilon' &= 3'23''.544 \\ \delta\omega' &= 5'45'' \\ 2\delta e' &= 30''.094 \\ \delta n' &= 0''.11793. \end{aligned} \tag{120}$$

Applied to the elements for 1750, these corrections gave

$$\begin{aligned} \epsilon' &= 7^s21°20'44'' \\ \omega' &= 8^s28°13'19'' \\ e' &= 0.056336, \end{aligned} \tag{121}$$

and the value of $\delta n'$ showed that the mean sidereal motion of Saturn had to be increased by 1/9 second.

Armed with these corrections, Laplace gave simple formulas for finding the location of Saturn at any instant, having now brought theory and practice into conformity. Thus, the mean longitude

$n't + \epsilon'$ is given by taking the corrected mean longitude for 1 January 1750, $7^s 21°20'44''$ (121), and adding to it the angular distance traversed in the mean sidereal motion at a rate of 43966″.6 per year of 365 days for the number of days elapsed, instead of 43966″.5, as supposed in (118). A similar procedure was employed with the other elements of the orbit, and Laplace constructed a table not only of the twenty-four oppositions used in refining the data but of some forty-three extending from 1582 to 1786. In it he gave observed heliocentric longitudes together with the excess or residual differences between the observed value, the value that Halley had calculated in 1719, and the value that he had just calculated. Stigler points out that Laplace's solution—for the twenty-four equations that he linearized—had a residual sum of squares only 11 percent greater than a least-squares solution would have yielded and that the residual sum of squares from Halley's calculation was eighty times greater. The increase of Laplace's powers is no less striking in the contrast between this memoir and his early essays in error theory (see Section 4) than it is in astronomy itself. He concluded this part with an application of his formulas to the Ptolemaic data, the most ancient observation being that reported from the Babylonian legacy for 1 March 228 B.C.

The memoir had grown so lengthy that Laplace postponed Jupiter to the sequel. There he again turned the crank to accomplish the mutual reconciliation of the observations with the theory of the largest planet, in its interaction with Saturn as in its lesser inequalities ([*1788b*]; *OC*, XI, 226–239). More interesting for the history of his career is the evidence of his involvement in the activities of routine instrumental and positional astronomy. That had already begun, to be sure, in 1781 with the memoir on determination of cometary orbits and the discussion that followed Herschel's discovery of Uranus (see Section 14). With the Saturn memoir ([*1788a*]) the association became systematic. He read the first installment to the Academy on 10 May and the second on 15 July 1786 (*52, 53*). In the preamble to the second, he recognized that the theory of Saturn still contained three very small though detectable inequalities. Their sum came to less than one arc second, but they would need to be taken into account in the most exact and rigorous possible calculation. What was needed was that an astronomer familiar with such computations should rework all the oppositions of Jupiter and Saturn in the seventeenth and eighteenth centuries, calculating them directly from the perfected theory in each instance. Jean-Baptiste Delambre had come forward and volunteered to undertake that task—a *délicate et pénible* discussion, Laplace called it (*ibid.; OC*, XI, 211). The word drudgery might occur to others. For his part, he made the small corrections just mentioned in the theory, and Delambre proceeded to compare the formulas to a very large number of observations and to compile the results. His *Tables de Jupiter et de Saturne* were published in 1789, having been presented to the Academy on 26 April (*57*). Laplace called them the first tables ever to be based "on the law of gravity alone," since they depended on observation only for the data needed to determine the arbitrary constants introduced by the integration of the differential equations of planetary motion. One particular advantage of that austerity was the possibility that they afforded of deciding whether causes of any sort external to the solar system could disturb its motions. More mundanely, wrote Laplace of Delambre, "I acknowledge with pleasure that, if my research is useful to astronomers, the merit is owing principally to him" (*ibid.; OC*, XI, 211–212). From then on, Laplace in his astronomy was never without a calculator at his beck and call.

There remained the moon, the last member of the solar family whose apparent behavior failed to conform in all respects to the rule of universal gravity. Halley had discovered the acceleration of its mean motion, and since then astronomers had tried correcting for it by adding to the mean longitude a quantity proportional to the number of centuries elapsed since 1700. There was some disagreement among them on the exact rate of the increase in particular centuries, but none on its overall effect. Delambre had just confirmed that the secular motion was three or four minutes greater than in Babylonian times. As to the cause, the Academy had offered several prizes, but no one had been able to identify anything in the configuration or the motion of the earth or its satellite that would explain these variations in the lunar mean motion in a manner conformable to the law of gravity. Various ad hoc hypotheses had been invoked—resistance of the ether, the action of comets, the transmission of gravity with finite velocity. It seemed a pity to Laplace that the moon should still be a renegade, and he had decided to investigate the phenomenon, failing in his first few attempts and telling of his eventual success in "Sur l'équation séculaire de la lune" ([*1788c*]), the final substantive memoir in the seminal series on planetary motion. He had been led to his formula for the moon—so he recalled in later years—by the application that he had been able to make of his theory of Jupiter and Saturn to the problem of the moons of Jupiter ([*1799a*]; *OC*, XII, 193).

The cause, in a word, lies in the action of the sun combined with variations in the eccentricity of the earth's orbit owing to the action of the planets. The action of the sun—to take that first in a qualitative summary of the finding—tends to diminish the gravitational attraction of the earth for the moon and thus to expand its orbit. In itself that effect would slow the angular velocity and appear as a deceleration in the lunar mean motion. The solar action is strongest when the sun is in perigee, and the lunar orbit is then at its maximum, decreasing as the sun moves toward apogee. Thus, in the motion of the moon there is an annual equation that is identical in period with that of the apparent motion of the center of the sun, though opposite in sign. But the gravitational force exerted by the sun on the moon also varies with the changes in the terrestrial orbit that are caused by the resultant of the influence of the other planets. Over the long term, the major axis is fixed and the other elements—eccentricity, inclination, position of nodes and aphelion—change incessantly. The mean force exerted by the sun on the moon varies as the square of the eccentricity of the terrestrial orbit. This effect produces contrary variations in the motion of the moon, the periods of which are enormously longer, extending over centuries. Presently the eccentricity of the earth's orbit is decreasing, and the consequence is an acceleration of the moon. The motions of the lunar nodes and apogee are also subject to secular equations of a sign opposite to that of the equation of the mean motion: for the nodes the ratio to the equation of mean motion is 1:4 and for the apogee, 7:4.

It might also have been expected that variations in the position of the ecliptic would influence the action of the sun on the moon by changing the angle between the lunar orbit and the ecliptic. The sun keeps the angle between the lunar and terrestrial orbits constant, however, so that the declinations of sun and moon serve the same periodic law. Finally, Laplace calculated that neither the direct influence of planets on the moon, nor consequences arising from its figure, had any detectable effect on the mean motion. To Laplace, the most surprising finding was that the decrease in the eccentricity of the earth should have had an effect so much more evident in the motion of the moon than in the solar theory itself. The decrease in eccentricity of the earth's orbit came to less than 4 minutes since ancient times, while the mean motion of the moon had increased by more than 1-1/2°. The influences causing these variations clearly affect the lunar motion of revolution differently from the motion of rotation, and it might have been expected that they would disturb the identity of period between the two so that in time the moon would show the earth its other hemisphere. In fact, however—and this effect was the subject of Laplace's final calculation—the period of the moon's secular variation in mean motion of revolution was so long, and the earth's gravity by comparison so powerful a force, that the major axis of the lunar equator is always drawn toward the center of the earth, subject only to the slight libration that shows a tiny rim of the hidden hemisphere, now on one side and now on the other. The period of the moon's long inequality was the greatest that Laplace had yet studied, amounting to millions of years. Yet there could be no doubt of its existence, and he could assure his readers that the acceleration would be reversed one day, also by the force of gravity, and that the moon would retreat from the earth on its next beat. It would never come crashing down upon us, therefore, as it surely would do if ether resistance or a temporal transmission of gravity were the cause of its present approach. Acknowledging that his numerical results in these calculations were much less secure than in the theory of Jupiter and Saturn, he left it to posterity to perfect them. Posterity has obliged. In 1853 John Couch Adams estimated that barely half of the lunar acceleration can be explained by decrease in the earth's orbital eccentricity, and in 1865 Charles Delaunay attributed the remaining increase to slowing of the earth's rotation by tidal friction.

Laplace himself returned to the theory of the moon on several occasions prior to the publication of *Mécanique céleste* ([*1798c*], [*1799a*]; see Section 21). He there modified this analysis and its application more fully than he needed to do for his treatment of the other topics in what he had already developed into a body of celestial mechanics, dispersed amid these memoirs and lacking only the name. Since the last, and briefest, in the series, "Mémoire sur les variations séculaires des orbites des planètes" ([*1789b*]), merely gives a more general proof of the final item that he had demonstrated in the first ([*1787a*]), namely the periodicity of the variation of orbital eccentricities and inclinations, perhaps it will be fitting to conclude this account of the work that Laplace accomplished at the height of his powers with the peroration to the lunar memoir, presented in April 1788 (*56*), just one year prior to the onset of the French Revolution (*ibid.; OC*, XI, 248–249):

> Thus the system of the world only oscillates around a mean state from which it never departs except by a very small quantity. By virtue of its constitution and the law of gravity, it enjoys a stability that can be destroyed only by foreign causes, and we are certain that their action is undetectable from the time of the most ancient observations until our own day. This stability in the system of the world, which assures its duration, is one of the most notable among all phenomena, in that it exhibits in the heavens the same intention to maintain order in the universe that nature has so admirably observed on earth for the sake of preserving individuals and perpetuating species.

PART III: SYNTHESIS AND SCIENTIFIC STATESMANSHIP

18. The Revolution and the Metric System. There is no more instructive example of the effect that institutional developments may have, even in the most mathematical reaches of science, even upon the most insensitive of political temperaments, than the modification in the pattern of Laplace's career in consequence of the events following the French Revolution. The earliest recorded expression of his initiative in a reform was a proposal that he advanced—"anew," according to the minutes of the Academy—on 4 July 1789. It went dead against the current, not to say the torrent, of the times. Laplace wished the Academy to require an elementary knowledge of mathematics and physics of artisans qualified for a *brevet* in its licensed corps (*PV*, **108** [4 July 1789], fol. 184). A discussion was postponed to the next session, when his colleagues voted by a large majority to "require" nothing and resolved only that they would "prefer" to license those who knew these subjects (*ibid.* [8 July 1789], fols. 190–191). On 18 July, four days after the fall of the Bastille, Laplace read a paper on the inclination of the ecliptic (*58*). That autumn the Academy began considering a liberalization of its own structure and procedures to bring its regime into congruence with the emerging constitutional order. Laplace was appointed to a committee to review suggestions along these lines advanced by the duc de la Rochefoucauld (*ibid.* [23 Nov. 1789]), and he joined with Condorcet, Borda, Bossut, and Tillet in composing a memoir on the subject submitted on 1 March 1790 (*ibid.*, **109** [1790–1793], fol. 75). There is no evidence that he took further part in the reformation or political defense of the Academy, however, the leadership of which was left largely to Lavoisier, whose efforts proved increasingly forlorn as the extremists gained in power.

In the early years of the Revolution, the Academy was inundated with proposals for inventions and technical schemes of many sorts, and Laplace's committee work, like that of all his colleagues, became much more demanding. On 2 November 1791 he was elected to the panel of fifteen academicians who, together with a like number of representatives of inventors' societies, constituted a new Bureau de Consultation des Arts et Métiers intended to relieve the Academy itself of responsibility for advising government agencies on patents and technological policy in general. Prior to that, Laplace had taken one other initiative, together with Condorcet and Dionis du Séjour. Jointly, they framed a petition, subscribed to by others of their colleagues, and addressed it to the National Assembly, urging that body to instruct departmental authorities under the new system of local government to continue the population inquest begun by provincial intendants and published by the Academy (*ibid.*, **109** [11 Dec. 1790], fol. 257; see Section 13). The one revolutionary enterprise that consistently engaged Laplace's interest, however, was the preparation of the metric system, and there his part in the design was more decisive than has been appreciated.

Like other aspects of modern polity, the standardization of weights and measures in France had been urged in the programs of the Turgot ministry (1774–1776) and then frustrated until the Revolution made reform not only possible but imperative. It was also the sector in which science and civil life overlapped most extensively. We have seen the interest that Laplace was already taking in units during his collaboration with Lavoisier on the measurement of heat (see Section 15) and in geodesic surveying during his analysis of the figures of the planets (see Section 16). In the matter of the metric system, the scientists moved before the politicians did. On 27 June 1789 the Academy appointed a commission "for a piece of work on weights and measures"; the members were Lavoisier, Laplace, Brisson, Tillet, and Le Roy (*ibid.*, **108** [1789], fol. 170). No record remains of their deliberations or of a memoir of 14 April 1790 drawn up by Brisson on the relation of linear units to units of capacity and weight (*ibid.*, **109** [1790–1793], fol. 83). It is probable, however, that these early proposals revived a suggestion advanced by La Condamine on returning from the 1735–1736 expedition to Peru. His thought had been to take the length of the seconds pendulum at 45° latitude for the basic linear unit and to decimalize subdivisions. A pamphlet by one recalcitrant commissioner, Tillet, written together with a certain Louis-Paul Abeille on behalf of the Society of Agriculture, attacked precisely such a scheme (*Observations . . . sur l'uniformité des poids et mesures* [1790]). The authors urged limiting the reform to the standardization of existing units, pointing out that with a duodecimal basis the shopkeeper and engineer can manage quarters, thirds, and halves, and warning against sacrificing the daily convenience of farmer, merchant, and builder to the exigencies of a perfectionist science.

It may reasonably be surmised that the scientists, alarmed, felt the need for a spokesman skilled in the ways of the political world. The proposal that formed the basis of the initial metric law was adopted by the Constituent Assembly on 8 May 1790 (*Archives parlementaires*, 1st ser., **15** [1790], 438–443). Written by Talleyrand, it overwhelmed petty anxieties about habit with the grandeur of a universal reform drawing its units from nature (*Proposition . . . sur les poids et mesures* [1790]). The seconds pendulum was to be the linear basis, and gravimetric units were to be related to linear through the volume occupied by a unit weight of water at a given temperature. The last value had just been determined by Lavoisier, who almost surely had coached Talleyrand on these matters.

Thereupon, the Academy appointed a new commission on weights and measures consisting now of Laplace, Lagrange, Monge, Borda, and Condorcet. Less than a year later, on 25 March 1791, they made a different recommendation, which was enacted into law the very next day (*HARS* [1788/1791], 7–16). It provided for a new geodesic survey of the meridian of Paris, to be measured from Dunkerque to Barcelona, and for defining the still unnamed meter as the ten-millionth part of the quadrant. Its subdivisions and multiples would be decimal, and its length would be close to that of the seconds pendulum. The reason given for the change was that the length of the seconds pendulum depended upon a parameter in time, which was arbitrary, and upon the force of gravity at the surface of the earth, which was extraneous to the determination of a truly natural unit. In the words of Condorcet, who drafted the report:

> If it is possible to have a linear unit that depends on no other quantity, it would seem natural to prefer it. Moreover, a mensural unit taken from the earth itself offers another advantage, that of being perfectly analogous to all the real measurements that in ordinary usage are also made upon the earth, such as the distance between two places or the area of some tract, for example. It is far more natural in practice to refer geographical distances to a quadrant of a great circle than to the length of a pendulum . . . (*ibid.*, 9–10).

It has become conventional wisdom to dismiss this reasoning as a piece of grantsmanship (A. Favre, *Les origines du système métrique* [1931], 121–130). The argument is that eminent mathematicians and engineers must surely have understood that there is no such thing as a naturalistic metric and that any unit is based on a mere convention, the true meter being an agreed-upon stick. Their real motivation, it is said, was to mount another elaborate and costly scientific traverse at the expense of the state, with the immediate purpose of establishing the reputation of a new surveying instrument, an ingenious repeating circle, invented by one of their number, Borda. These dark thoughts were also noised at the time and were probably unfair, then as now. It seems likely that Laplace would have been instrumental in changing the commission's recommendation. He and Borda were the only members with any background in geodesy—Borda as a navigator and engineer, and Laplace as a theorist analyzing the figure of the earth. The law of 8 May 1790 had been passed in haste to forestall Tillet and the defenders of routine. The draft left it to the Academy to recommend an appropriate scale of subdivision. The importance of that was further emphasized in a companion decree passed the same day (*Archives parlementaires*, 1st ser., **15** [1790], 443), also calling for recommendations from the Academy for a new monetary system.

Evidently the vision of a universal decimal system, embracing not only ordinary weights and measures but also money, navigation, cartography, and land registry, unfolded before the commissioners as they explored the prospect in the summer and autumn of 1790. In such a system, it would be possible to move from the angular observations of astronomy to linear measurements of the earth's surface by a simple interchange of units involving no numerical conversions; from these linear units to units of area and capacity by squaring or cubing; from these to units of weight by taking advantage of the principle of specific gravity; and finally from weight to price by virtue of the value of gold and silver in alloys held invariant in composition through a rigorous fiscal policy. The seconds pendulum could never anchor that. The earth is round, and we fix our position by astronomical observation. The crux, therefore, was that linear units should be convertible to angular. At ordinary dimensions the curvature of the earth's surface would introduce no detectable error. In April 1795 Laplace put the point very simply in a lecture (see Section 20) before the École Normale:

> It is natural for man to relate the units of distance by which he travels to the dimensions of the globe that he inhabits. Thus, in moving about the earth, he may know by the simple denomination of distance its proportion to the whole circuit of the earth. This has the further advantage of making nautical and celestial measurements correspond. The navigator often needs to determine, one from the other, the distance he has traversed from the celestial arc lying between the zeniths at his point of departure and at his destination. It is important, therefore, that one of these magnitudes should be the expression of the other, with no difference except in the units. But to that end, the fundamental linear unit must be an aliquot part of the terrestrial meridian. . . . Thus, the choice of the meter was reduced to that of the unity of angles (*OC*, XIV, 141).

Laplace went on to justify taking decimal parts of the quadrant rather than of the whole circumference by reason of the role of the right angle in trigonometry. All the rhetoric about the universality of units taken from the earth itself thus turns out to have a perfectly sensible foundation. (The substance of the lecture was incorporated into the *Système du monde*, Book I, chapter 14; *OC*, VI, 64–86.) That only Laplace should have explained it clearly—then or now—gives further reason (in addition to the inherent probability, judging from the personalities involved) for thinking that he must have been the moving spirit in the design. Delambre confirmed the cogency of the reasoning in his authoritative account of the origins of the metric system (*Base du système métrique décimal*, 3 vols. [Paris, 1806–1810], III, 304). It is certainly indicative, moreover, that when the two surveying teams took the field in the summer of 1792, one for the southern and the other for the northern sector of the meridian, those chosen to head them should have been Laplace's first two calculators, Méchain, for cometary data (see Section 14), and Delambre, for the tables of Jupiter and Saturn (see Section 17).

A related project needs to be mentioned. The abolition of feudal landholding and the reorganization of local government made it essential to create a new land registry, or *cadastre*, for determinations of title and assessments of tax. The engineer in charge, Gaspard de Prony, was mandated by law to base the task upon the Cassini map of France, which was also the point of reference for the metric survey. He proposed to concert his efforts with those of the metric commission, construct his instruments on a decimal scale, and convert his units to the new basis as soon as it might be accurately known. His "Instruction" was submitted to the Academy and was largely approved in two reports, both drafted by Laplace. The first, which concerned methods, was also signed by Borda, Lagrange, and Delambre (*PV* [12 May 1792], fols. 147–151). In the second, Laplace took up units and proposed the names meter, decimeter, centimeter, and millimeter. The square on 100 meters was to have been an "are," divided into deciare, centiare, and milliare (*ibid.* [11 July 1792], fols. 205–207; the Bibliothèque Nationale has a *Recueil des documents . . . concernant le cadastre* [Lf^{158}**.236**]).

Laplace had an uncomfortable moment in connection with the design of the Revolutionary calendar, represented by its sponsors as an extension of the metric system to the realm of history and the arts (*Procès-verbaux du Comité d'instruction publique*, II [1894], 440–450, report of Romme, 20 Sept. 1793). Together with Lagrange and Lalande, he was consulted by the enthusiast who conceived it, Gilbert Romme. Lalande's notes of the encounter are in the archives of the Paris Observatory. According to him, Laplace refrained from bringing home to the zealots the incompatibility between their desire for a calendar that would embed the civil year in nature and the incommensurability between the day and the year entailing an unavoidable irregularity in the number of leap years per hundred years in centuries to come (Bibliothèque de l'Observatoire de Paris, B-5, 7). One of Laplace's political detractors (J: Merlieux) later called it apostasy in an astronomer that Laplace should have consented to draft the recommendation ([*P1805*]) on the strength of which Napoleon restored the Gregorian calendar effective 1 January 1806. In fact, astronomers require not a natural but an arbitrary and universal system of intercalation, and the Republican calendar conjured up instead the confusion of ancient Greek chronology, when particular cities intercalated a day or month named after some hero or victory. It may be true, however, that Laplace lacked the temerity to voice this sentiment in what was soon to be called the Year II.

As for the metric system itself, there is no doubt about his fidelity. Subsequent writings made a point of expressing angles in the form of decimal subdivisions of a right angle. After 1795 the metric system, the Observatory of Paris, and indeed all matters pertaining to navigation and official astronomy were placed under the administration of the new Bureau des Longitudes, organized under a law of 25 June 1795 (*Procès-verbaux du Comité d'instruction publique*, VI [1907], 321–327). Laplace served regularly as a member, often chose it as the forum to present appropriate papers, and published frequently in its journal, the *Connaissance des temps*. It is perhaps significant that his son in later years should have made himself something of a watchdog of the integrity of the metric system, as he did of the form of his father's work in general ("Notice sur le Général Marquis de Laplace," *OC*, I, v–viii).

Laplace took no other part in the affairs of the Revolution during the phase of radical republicanism. He had been among the scientists vilified by Marat in the diatribe *Les charlatans modernes* (1791), although not as virulently as Lavoisier. As power shifted to the left in the spring and summer of 1793, the Academy came under increasing pressure from the radicals, and it was suppressed by a decree of the National Convention on 8 August. Laplace appears to have withdrawn from participation as its political vulnerability increased. His attendance became increasingly infrequent in the latter part of 1792, and he was present for the last time on 21 December (*PV* [21 Dec. 1792], fol. 325). A provisional commission was left in charge of the metric system after the abolition of the Academy. On 23 December 1793 Laplace was purged from its membership, along with Lavoisier, Borda, Brisson, Coulomb, and Delambre, on the grounds that such responsibilities were to be entrusted only to those worthy of confidence "by their Republican virtues and hatred of kings" (*Procès-verbaux du Comité d'instruction publique*, III [1897], 239). By then the Terror was approaching its climax. Some time previously (we do not know precisely when) Laplace had decided to remove from Paris. He had then been married for five years, and his two children were infants. He and his wife took a lodging in Melun, thirty miles southeast of Paris, and remained there until he was recalled to Paris to participate in the reorganization of science that followed the fall of Robespierre and the Jacobin dictatorship in July 1794 (see Section 20).

19. Scientific Work in the Early Revolution. It is sometimes said that Laplace began writing the *Exposition du système du monde* and *Mécanique céleste* during this retreat at Melun. It may be so; there is no way of knowing. We do know that he had presented three further memoirs to the Academy before its situation deteriorated to the point that such works could scarcely have been received, even if they could have been composed. In July 1789 he communicated a memoir on the obliquity of the ecliptic (*58*), later combined with a miscellany of other topics ([*1793a*]); in April 1790 a further memoir on the satellites of Jupiter (*59*), to be followed by a sequel ([*1791a*] and [*1793b*]); and in December 1790 a study of tidal phenomena (*60*), largely in the port of Brest ([*1797a*]). The disparity of publication dates bespeaks the confusion of the circumstances.

All three memoirs exhibit the pattern of the investigations that Laplace put in hand during this second half of his career, alongside the magisterial treatises that remain its monument. On the whole the topics were not new to him, except for those in addition to heat that he took up in physics (see Sections 22–24, 27); neither, with the same exceptions, were the results. They contain no great surprises, nothing like the period of the long inequality of Jupiter and Saturn or the potential function. Instead, he returned to phenomena that he had already dealt with, usually with one or both of two purposes in mind: to give them a more detailed and general analysis and, where possible, to give the analysis numerical expression in actual instances. There was precedent, of course. The theory of Jupiter and Saturn had already issued in Delambre's tables (see Section 17), the cometary theory in the Abbé Pingré's *Cométographie* (see Section 14), and the probability of cause in population studies (see Section 13). It would therefore be difficult to say to what extent external pressure and opportunity favored the shift in emphasis that was occurring anyway in the natural evolution of his lifework. There can be no doubt about the pressures or the opportunities, however, and no reason to question their efficacy in this, as in any evolutionary process. Laplace was forty years old in 1789. Henceforth, his special investigations were conducted in regular and continuing interaction with practical astronomers, physicists, geodesists, meteorologists, and civil officials. The great treatises, on the other hand, owed much to the educational context, which is another, related story (see Section 20).

In the opening sentence of the voluminous "Théorie des satellites de Jupiter," Laplace took upon himself the challenge that he had posed to astronomers four years previously in concluding his discussion ([*1787a*]; see Section 17) of the pendulumlike libration of the three inner moons. He now intended to give "a complete theory of the perturbations that the satellites experience and to place before astronomers the resources that analysis can provide to perfect the tables of these stars" ([*1791a*]; *OC*, XI, 309). The earlier analysis had been cosmologically motivated. The moons of Jupiter were a test case for the stability of the planetary system, and the proof was limited to a relatively brief demonstration that, for the inner three, the two relations, $n - 3n' + 2n'' = 0$ and $nt - 3n't + 2n''t = 180°$, are rigorously exact. A more complete demonstration confirming those theorems occupies Articles XIII–XIV of the present memoir (*ibid.; OC*, XI, 369–387) concerning inequalities that depend on the squares and products of the perturbing forces. For the rest, he no longer restricted the problem to the three inner satellites, nearly coplanar and concentric with Jupiter and bound in their triune libration. He now included the outer moon, the orbit of which is more eccentric and a bit more inclined to the plane of Jupiter's equator. The calculations involve the interactions of all four together with the effects exerted by two other perturbing forces, those due to Jupiter's own oblateness and to the gravity of the sun, which is affected in its incidence by the angle between Jupiter's orbit and equator. The latter two factors made themselves felt mainly in the inequalities of motion of the fourth satellite.

Handling these complications one after another, Laplace drew successively on each of the main investigations that he had completed in planetary astronomy. The memoir amounts to a reprise of the entire subject put directly into practice within the compass of the Jovian system. In the overall strategy, he followed the model of the Jupiter-Saturn memoir ([*1788a*] and [*1788b*]). After first setting up general equations of motion for satellites, he then went through the calculation of the effect of each class of inequalities to which they are subject: those independent of eccentricity and inclination, those depending on eccentricity of orbit, those depending on the action of the sun, those appearing in the squares and products of expressions for the perturbing forces, and those depending on motion in latitude, where the angle of the planes of Jupiter's orbit and equator is significant. For the theory of the figure of Jupiter ([*1791a*]; *OC*, XI, 317) he drew on his work on the attraction of spheroids of revolution ([*1785b*]) and on its application to the figure of the earth ([*1786a*]). For the determination of inequalities independent of eccentricity and inclination ([*1791a*]; *OC*, XI, 329) he drew on the formula of the comparable section of the Jupiter-Saturn memoir ([*1788a*]). For the action of the sun on the motion of the satellites ([*1791a*]; *OC*, XI, 346) he drew on his discovery of the effect of variations in the eccentricity of the earth's orbit on the motions of our moon ([*1788c*]). In an interesting aside he explained how the data for Jupiter's moons were calculated from observation of the times and duration of eclipses ([*1791a*]); *OC*, XI, 361–369). The first three satellites disappear behind Jupiter on every revolution, and the fourth intermittently. All that the observer needs to report are the instants of disappearance and reappearance, and the information thus obtained is far more accurate than could be yielded by tracking the actual motions. The further possibility of multiplying observations indefinitely reduces the already small risk of instrumental error to zero.

Armed with this information from the Wargentin tables (see Section 17), Laplace could compute provisional numerical values for the mean motions and the mean distances from the center of Jupiter. For the distances, the most accurate method consisted in deriving the values for the three inner satellites from observed positions of the fourth, by means of Kepler's laws. With these quantities, he could calculate the flattening of Jupiter and the masses of the four satellites. Theoretically, determining those unknowns was the main object of the investigation. Practically, its purpose was to permit Delambre to construct tables for the satellites which, like his tables for Jupiter and Saturn, would be derived theoretically from the law of gravity. Observation would confirm the values and not be their source, as in the older tables. Laplace himself published a sample in the *Connaissance des temps* ([*1790a*]), the first of a series of contributions to practical astronomy in that journal. Delambre incorporated a more fully developed set, drawing on an enormous body of observations, in the third (1792) edition of Lalande's *Astronomie*.

Solving for the unknowns required manipulating a formidable array of equations. The three differential equations obtained at the outset govern the motion of each satellite; for four satellites there are twelve equations. Integration introduced twenty-four arbitrary constants, to be determined by deriving the elements of the orbits from observations of the eclipses. (Actually, the two relations of longitudes and mean motions among the inner three reduced the number of arbitrary constants to twenty-two, but the need to include terms for their libration added two others.) The indeterminate quantities—that is, the flattening of Jupiter and the masses of the four satellites—raised the number to twenty-nine. Five further items of observational data were required to make their determination possible: the principal inequality of the first satellite, the principal inequality of the second, the annual and sidereal motion of the apside of the fourth, the equation of the center of the third relative to the apside of the fourth, and finally the annual and sidereal motion of the orbital node of the second. It is a measure of Laplace's insight into the conditions of the problem that he could seize on these pieces of information as practically obtainable and analytically sufficient for a solution. Substituting those values in the analysis, he fixed the masses at 0.184113, 0.258325, 0.865185, and 0.5590808, each multiplied by 10^{-4}, the mass of Jupiter being taken for unity. As for the figure of Jupiter itself, the ratio of the minor to major axis came out to be 67:72 ([*1793b*]; *OC*, XI, 421). Formulas to compute the motion of each satellite in orbit then follow readily.

The first time that Laplace had run through these calculations, which appeared in a sequel ([*1793b*]) to the parent memoir, it had been with reference to a relatively small body of data, and he regarded his results as a first approximation to be corrected by a process of successive approximations that he set forth at the same time. All this was handed along to Delambre, and the original plan was that he would set to work preparing yet a further stage in the evolution of the tables toward perfection ([*1793a*]; *OC*, XI, 477–481). The preemption of Delambre's services by the survey of the meridian prevented that, and Laplace took back the material himself, substituting in the analysis the more exact and fuller data from the tables just mentioned in the Lalande compilation, which although still imperfect were far fuller than his own sketchy figures ([*1793b*]; *OC*, XI, 415–416). Delambre began work on the metric system in May 1792, and readying the calculations for publication must therefore have been occupying Laplace in the latter part of that year and perhaps into 1793.

The reflections with which he concluded the investigation are predictable in one respect. He held that the magnitude of the effect that the flattening of Jupiter has upon the inequalities of motion in its satellites proves that the attraction is compounded of the gravitational force exerted by every particle of the planet, that hypothesis having been assumed in the formulation of the equations of motion. A second rumination may be no more surprising in itself. It concerns the velocity of light, for which topic the moons of Jupiter had been instrumental long before Laplace's interest in them. He had Delambre calculate a value for aberration from figures for the eclipses of the first satellite. The results exactly confirmed Bradley's value drawn from the well-known method of direct observation of the fixed stars. That it should be so confirmed the uniformity of the velocity of light, at least within the dimensions of the diameter of the earth's orbit. Considering the configuration of his career, the conclusion that Laplace drew from this assertion is significant in a way different from the remark on gravity. Rather than celebrating a victory, it anticipates a battle that he would lose (*ibid.; OC*, XI, 473):

> That uniformity is a new reason for thinking that the light of the sun is an emanation from that body; for, if it were produced by the vibrations of an elastic fluid, there would be every reason to think that this fluid would be more elastic and denser on approaching closer to the sun, and that the velocity of its vibrations would thus not be uniform.

One further paper—or, more accurately, collection of short papers—appeared in the volume that the Academy managed to get printed in 1793 prior to its demise. "Sur quelques points du système du monde" ([*1793a*]) obviously consists of odds and ends from Laplace's worktable. The opening article, which concerns the investigation just discussed, was evidently written after completion of the analytical part ([*1791a*]) and before Laplace had decided to publish the numerical application himself ([*1793b*]). Thereafter, the two topics discussed most extensively are the variation in the obliquity of the ecliptic and geodesic data bearing on the figure of the earth.

Laplace had read a draft on the former topic in July 1789 (*58*). He reminds readers that the decrease in obliquity of the ecliptic was one of the best-studied celestial phenomena, that it occurs in consequence of the action of the other planets on the earth, and that in overall rate and period it is independent of the shape of the earth. Nevertheless, flattening at the poles and bulging at the equator do affect the action on the earth of the sun and moon, and the present analysis investigates the magnitude of these secondary effects in the variation of the plane of the ecliptic. They will appear in the rate of precession of the equinoxes and in the length of the year, and the chief purpose of the discussion was to correct the equations that astronomers employed for precession.

Two main points are to be noted. First of all, the results formed part of a work that he intended to publish on what he still called *astronomie physique* (*ibid.; OC*, XI, 483). This is the first indication since the general treatise on planetary motion ([*1784a*]; see Section 16) that *Mécanique céleste* was in gestation. Second, the mode of analysis recalls a technique that he had introduced in deriving general equations of motion of a system of mutually attracting bodies in the covering memoir on secular inequalities ([*1787a*], Article II; *OC*, XI, 69–70; see Section 17) and that he employed again in considering the motions of Jupiter's moons in latitude ([*1791a*], Article X; *OC*, XI, 347–361). He projected the orbits of planetary bodies onto a fixed plane, passing through the center of the sun, which served as the basis of a coordinate system for calculating angular momentum. Calculations of planetary motion in times past or future could always be reduced to such a plane invariant in space. It seems possible and even probable that this approach derives from his early enthusiasm for the application of d'Arcy's principle of areas to problems of planetary motion (see Sections 5 and 8). At any rate, Articles XXI–XXII ([*1793a*]; *OC*, XI, 547–553) develop it preparatory to a much more extensive use in *Mécanique céleste* (Book I, chapter 5; Book II, chapter 7; *OC*, I, 57–73, 309–345).

Laplace goes on to inquire whether the results of many meridional surveys, and also of determinations of the length of the seconds pendulum at various latitudes, could "without doing too much violence to the observations" ([*1793a*]; *OC*, XI, 493) be reconciled with the hypothesis that the earth is an ellipsoid. The method was simply to compare the data in the literature with the ellipsoidal requirement that the force of gravity at the surface vary with the square of the sine of the latitude. The answer in both cases was negative, although the data from measurements of the pendulum were about eight times closer to satisfying an ellipsoidal figure than were those from meridional surveying. It cannot be said that this finding carried Laplace much farther than his full-scale memoir on the figure of the earth ([*1786a*]; see Section 16), and perhaps the chief interest that it affords for the development of his work is the evidence that error theory was occupying a growing place in his thinking about physical problems. The form of the question was whether the discrepancies between the observed and calculated values exceeded what might be attributed to observational error. Calculating those limits represented another step in error theory itself ([*1793a*], Article XI; *OC*, XI, 506–509). It is worth mentioning that the degree measured by Mason and Dixon in Pennsylvania figures in these data.

For the rest, the memoir is a grab bag. One article points out that the earth cannot have taken form in the fluid state, since if it had, it could have assumed only an ellipsoidal figure under gravity. Another observes that the stability of the seas tolerates disturbances sufficiently great so that occasionally the highest mountains are submerged, a consideration that explains certain curious facts of natural history. There is new proof from the conservation of vis viva that the long-term equilibrium of the sea is stable, whatever its depth and whatever the law of rotation of the earth. Laplace gives yet another simplified and generalized demonstration of the variation of constants to eliminate secular terms introduced into the solution of integrals by the standard methods of approximation. Finally, there is a new and purely analytic formulation of the laws of motion of a system of any number of bodies attracting each other by any law whatever. Like the material on the variation of the ecliptic, much of this found its way into appropriate passages of *Mécanique céleste*.

Laplace returned to the problem of the tides in a memoir ([*1797a*]) first read in December 1790 (*60*). After remaining in manuscript for seven years, it was finally published together with other *Nachlässe* from the Academy in its posthumous volume. He may probably have touched it up during that long interval, for although none of the data refers to observations more recent than 1790, in recurring at the outset to the difficulty of the problem, he called it "the thorniest in all of celestial mechanics" (*ibid.; OC*, XII, 4). Thus, he casually launched that phrase in print two years before the appearance of the first two volumes that bear it for their title. The memoir has a very different quality from the enormous and intricate mathematical model that he had constructed some twenty years previously ([*1778a*], [*1779a*], [*1779b*]; see Section 8), which smells of the lamp and not of the brine. In the meantime, he had fixed his attention on the most considerable existing body of data, a corpus of tidal observations in the port of Brest dating from the early eighteenth century, which Jean-Dominique Cassini had found at the Paris Observatory among the papers of his grandfather, Jacques, intendants of that establishment each in his turn. In 1771 Lalande published this find in the second edition of his *Astronomie*. Jacques Cassini himself had drawn on the data for memoirs of no lasting value. Laplace took the occasion to remark how important it was in serious research on such topics to publish the original observations. The whole mass needed to be available before patterns could be discerned, trivial or accidental effects distinguished from fundamental rhythms, and causes assigned to the latter. When he examined the tables, he found one essential item of information missing; the collection contained no observations on the rate of the rise and fall of the tides at Brest. He therefore requested that detailed observations be made, evidently in the year 1790. He does not say who had carried out that commission; presumably, the naval authorities would have been responsible. It was a stroke of good fortune that Cassini had fixed on Brest; for Brittany juts into the sea, and the harbor itself had a long, narrow entrance to a large protected basin so that wave action and other irregular oscillations were damped. Few other locations could have been equally advantageous.

This memoir reads differently, somehow, from any of Laplace's previous works. It would abstract the discussion too far from his own career to say that he did not need to be the mathematician he was in order to compose it, but it can be said that the formulas were not beyond the grasp of anyone capable of doing astronomy or geodesy. It is tempting to infer that so direct a contact with the facts about something as tangible as the tides, all laden with seaweed washing in and out of a working harbor, had a chastening effect at least upon the writing. Whatever the truth may be, however, it is more enigmatic than that. For Laplace did mention, although in a somewhat subdued manner, that his earlier theory had predicted certain of the phenomena and that it had been deepened by others that he had thought irreconcilable on first learning of them ([*1797a*]; *OC*, XII, 21). The difficulty is that when the two memoirs are compared, the theory—unless Laplace meant the theory of gravity, a claim so broad as to be empty—appears to have been changed in certain respects that appear significant to the outside observer two centuries later.

It is natural simply to suppose that by 1790 he had learned to be clear about a physical picture that had been obscured in a thicket of calculation the first time around. Certainly his verbal account of the conclusions could hardly be clearer. Even now, it would be difficult to think of a better place to send a reader for a qualitative explanation of how the sun and moon contribute to the motions of the tides and of how the magnitude and incidence vary with the seasons and relative positions of those bodies and of the spinning earth. Why does the tide never fall to the lowest point called for in gravitational theory? Why does it always take a little less time to rise than to fall? Why is the magnitude of tidal effects greater the shallower and more extensive the oceanic area? How may local circumstances affect height and times? Why do tides as a rule run most swiftly in shallow bays and narrow passages? The reasons for these and many other effects are expounded with admirable lucidity.

Once again, as in the youthful analysis, there are three systems of tidal oscillation affecting the seas concurrently and superimposing their effects on the motions of individual particles of water. There, however, the identity of the two accounts ceases and merges into a resemblance wherein differences are at least as striking as similarities. In the earlier paper, the three cycles were accorded equal attention and were discussed in the order of the length of their annual, daily, and semidiurnal period. Now the order is reversed. The emphasis is also different, and—what is more surprising—so are important effects attributed to each system. In 1777, an account of the near equality of successive high tides (*23*), which the canonical Newtonian approach failed to give, was the starting point, if not quite the motivation, of the entire investigation. Moreover, Laplace then discussed it by means of a mathematical analysis of the terms in his expressions governing the middle set of oscillations, the period of which is one day. Physically, that analysis presupposed a uniform depth of the sea, and its success was said to confirm that hypothesis. Now, in 1790, nothing is said of the latter argument. Even more curious, the near equality of consecutive high tides is attributed to the physical conditions governing the first cycle discussed, the semidiurnal oscillation. What is attributed to the daily cycle is the small difference, rather than the near equality, between the two tides of the same day in times of syzygy.

Neither of these effects is now deduced in the first instance from the mathematics. The return of consecutive high tides to almost the same level is explained in terms of the equilibrium conditions affecting a single particle of seawater. Suppose the sun (or moon) is acting in the plane of the equator. The gravitational force that it exerts on a particle of water directly underneath will be slightly stronger than the force exerted on the center of the earth. Hence, the action of the sun will tend to separate the particle from the center of the earth. Twelve hours later, the particle is in opposition and the sun attracts the center of the earth more strongly than it does the particle. It will then tend to separate the center of the earth from the particle. Since the radius of the orbit is enormously greater than that of the earth, the two effects are virtually identical in magnitude. Thus it happens, generalizing over the whole ocean, that the seas return to the same state every twelve hours.

Oscillations of the second sort, with a period of one day, are also given a physical explanation. They arise because the attracting body does not normally act in the plane of the equator, and their amplitude is proportional to the product of the sine and cosine of the declination. This daily variation is now held responsible for the small difference in consecutive high tides. At Brest, the tide in the morning was about seven inches higher than in the evening during syzygies at the winter solstice. The magnitude was small in European latitudes. There might be places, however, where geographic conditions would be such that, in the semidiurnal cycle, a tidal crest coming from one direction would coincide with a trough coming from the other, so that the normal tides would annul each other. In such localities, the second system would produce the only tidal motion, and there would be one tide a day. Laplace understood the port of "Batsha" [Badong?]in Tonkin (Vietnam) to be such an instance (*ibid.; OC*, XII, 20).

Finally, Laplace discussed very briefly a system of oscillations like those that he had treated first, and at equal length with the others, in the early memoirs. They are no longer given the period of a year, however, but are simply said to be independent of the rotation of the earth and to result from the sharing of the seas in its other motions. Hence their period is very long though still finite, and their amplitude at Brest very small though still detectable. Centuries of observation would be needed to determine them precisely, after which time the values could be relied on to afford a valuable means of calculating the ratio of the mean density of the earth to that of the seas.

To what extent the theory that Laplace confirmed by the tidal data from Brest remained the theory that he had conceived in his youth is problematic, therefore. Fortunately the point is not one that needs to be settled. The burden of the new exercise would appear to have been descriptive rather than theoretical anyway. It culminates in a formula for finding the level of the tides at Brest at any instant by means of astronomical data (*ibid.; OC*, XII, 112), and the table with which it concludes was of a type that could have served the operations of any enlightened harbor master.

20. Exposition du système du monde. From the fall of Robespierre and the Jacobin regime on 27 July 1794 (9 thermidor an II) until 26 October 1795, France was governed by the Revolutionary Convention, purged of its radical elements and often said to have been reactionary. From then until the coup d'etat of 9 November 1799 (18 brumaire an VIII) that brought Napoleon Bonaparte to power, executive authority was vested in the collective hands of a Directory that never achieved stability, confronted by the survival of Jacobinism on the left and the revival of royalism on the right. Many judgments, mostly adverse, have been passed on the political tone of the period, but there is general agreement among historians about its importance in the institutionalization of modern French society. For science, the most signal instances are the first École Normale, which held classes for three months beginning on 21 January 1795; the École Polytechnique, which was given that name on 1 September 1795, having started classes on 21 December 1794 as the École Centrale des Travaux Publics; and the Institut de France, the scientific division of which began regular meetings on 27 December 1795. They were intended to be the apex of a system of primary and secondary education, trade schools, and medical schools in which science and systematic knowledge would largely replace the classics as the staple subject matter.

Conceptually, the dominant influence was the school of *idéologie*, certain of whose adepts had become administrators of science and culture in the government and who thought to implement a philosophy of science deriving from the Enlightenment. In their outlook the moral and civic function of science is to educate citizens in the order of nature and, by extension, of society. In practice, the Institute never became the quasi-ministry of education originally imagined. The planning for the École Normale was inadequate and it closed after three months, not to reopen until 1812. Napoleon dispersed the *idéologues*, and positivism predominated at the École Polytechnique after 1800. The practical effects were nonetheless decisive in the long run—and even in the short. The center of activity in science moved from academies, its home since the seventeenth century, to institutions of higher education, where it instilled the spirit of research. Scientists—Laplace among them—became educators and professors (cf. J: Lacroix [1828]).

There is no evidence that he was among the organizers or promoters of these enterprises but every indication that when called on to participate, he did so with alacrity and enthusiasm, naturally assuming a leading role. The first, or scientific, class of the Institute amounted to a reincarnation of the Academy decked out in national and republican garb, elitist rather than privileged, civic rather than royal. It so commanded attention in the world of learning that reference to the Institute brings to mind its doings rather than those of its fellow divisions concerned with social science and humanistic culture. At the organizational meeting on 27 December 1795, Laplace was elected vice-president and, on 26 April 1796, president. The office was mainly honorific, and more significant was his presence from the outset on a host of committees where policies were formulated and decisions preempted. To name only the most influential, he was a member of the committees that dealt with bylaws, with weights and measures, with finances, and with the specification of prize contests in physics as well as mathematics (*PVIF*, I [1910], 1, 30, 46, 410). At the end of the Institute's first year, Laplace was chosen to present before the joint meeting of the two legislative councils a formal address and résumé of the work accomplished ([*P1796*]). He took the occasion to exhort the legislators to support the implementation of the metric system.

Laplace's activities at the Institute were a case of new or increased prominence in a set of revised procedures, whereas his involvement with the École Polytechnique marked a new departure, for him and for scientific education at large. There, physics first came to be taught systematically as a mathematical subject to well-qualified and highly selected students. True, the graduates were intended to be engineers, but it cannot have been an accident that the most famous of them in the early nineteenth century were engaged in what is now called physics—Ampère, Sadi Carnot, Fresnel, Malus, and Poisson, to name only the best known. Laplace did not give a course before the Napoleonic period. The technique of teaching was more affected by Monge, the prime mover in the first foundation. Monge built upon the experience of the former Royal Engineering School at Mézières, where he had established his career. Laplace's post from 1795 to 1799 was examiner. It gave him power over content as well as standards. From 1797 until the end of 1799, Monge and Berthollet were largely absent with Napoleon, first in the Italian campaign and then in Egypt. In the interval Laplace became the predominant personality in the affairs of Polytechnique, and the experience that he gained there pertained to the professionalization of science.

The experience gained at the Ecole Normale, on the other hand, pertained rather to popularization. In addition to Laplace, the professors of exact science were Lagrange, Monge, Haüy, and Berthollet. Together with colleagues expounding natural history, geography, and political economy, they lectured in the auditorium of the Jardin des Plantes before audiences of more than 1,200 pupils. The students had been assembled in haste from all parts of the country with the notion that they would return to their own localities to impart what they had learned in a system of secondary schools through which science and learning would radiate. They ranged in age from extreme youth to near senility, and in ability from virtual illiteracy to the talent of the young Fourier, who moved to the staff of Polytechnique in 1795 as assistant lecturer. Laplace was named to a committee to select teachers for the Paris region (*Procès-verbaux du Comité d'instruction publique*, V [1904], 546), and he gave a course of ten lectures that was subsequently published ([*1800d*]). The first eight deal with elementary mathematics—arithmetic, algebra, plane geometry, trigonometry, and the simplest aspects of analytic geometry. The ninth describes the metric system. The tenth introduces probability, summarizing in nontechnical language the highlights of his earlier work in that field; he later enlarged and deepened it into the *Essai philosophique sur les probabilités*. In the opening sentences he explained that he was skipping to this because of its intrinsic interest and its relevance to many matters of great social utility. The program announced for his course had committed him to treat also of the differential and integral calculus, mechanics, and astronomy. Time did not permit, and he would simply refer his auditors to a book that he had in preparation, to be entitled *Description du système du monde*, in which he would give a nonmathematical account of all that had been discovered in these subjects (*ibid.; OC*, XIV, 146).

Whatever the comprehension of his lectures on the overcrowded benches of a noisy auditorium, the promised book proved to be one of the most successful popularizations of science ever composed. The impression it made is conveyed by an autograph on the flyleaf of a copy of the first edition presented by a graduate of the College of New Jersey to what is now the library of Princeton University:

> This treatise, considering its object and extent, unites (in a much higher degree than any other work on the same subject that we ever saw) clearness, order and accuracy. It is familiar without being vague; it is precise but not abstruse; its matter seems drawn from a vast stock deposited in the mind of the author; and this matter is impregnated with the true spirit of philosophy.

The *Système du monde* appeared in 1796; the two-volume work consists of five books. Book I begins with what any attentive observer may see if he will open his eyes to the spectacle of the heavens on a clear night with a view of the whole horizon. Book II, which is considerably shorter, sets out the "real" motions of planets, satellites, and comets and gives the dimensions of the solar system. Book III is a verbal précis of the laws of motion as understood in eighteenth-century rational mechanics, with special reference to astronomy and hydrostatics. In Book IV, Laplace in effect summarized his own work in gravitational mechanics. Much of it consists of simplification of the prefatory sections to the published memoirs. The topics are perturbations in planetary motions, the shape of the earth, the attraction of spheroids and the rings of Saturn, motions of the tides and atmosphere, the moons of Jupiter, precession, and lunar motions. Only Book V contains material that Laplace had not written up in technical form or presupposed. It gives an overview of the history of astronomy and concludes with the speculation since called the nebular hypothesis and another on the nature of the universe in outer space.

Laplace kept his book alive and abreast of his thinking and work throughout his life. A second edition was published in 1799 simultaneously with the first two volumes of *Mécanique céleste* (67), as a companion volume to that work and in identical format. The third edition appeared in 1808. For the commentary on the Republican calendar (Book I, chapter 3) it substitutes an explanation of why the Gregorian was more practical after all, despite its imperfections (*OC*, VI, 21–22). Laplace was by then deeply involved with the Arcueil group in problems of physics, most notably sound, capillary action, and refraction of light (see Sections 22 and 23). The fourth edition (1813) preceded by a few months the first printing (1814) of *Essai philosophique sur les probabilités* ([E]), which complemented it in giving intelligent laymen access to that subject. The fifth edition (1824) announced Laplace's intention to make molecular forces the basis of a theory of heat and gases (see Section 27), a project that he had apparently modified before preparing the sixth edition for a publication that was delayed until 1835, eight years after his death ([H]). We do not know what more he had in mind.

The work itself being a summary of the astronomical investigations that we have been discussing, it hardly seems practical or necessary to attempt a further epitome, the less so since it is readily available in one or another of the above editions and in translations in many libraries. It well repays perusal. Indeed, it may serve a purpose for modern students not unlike that which Laplace had in mind in writing for contemporaries, except that no one will need to feel edified and that by now it may be equally useful in a retrospective way to the scientifically initiated. It is a handbook of what was known of cosmology at the end of the eighteenth century. Perhaps it will be helpful to those who are not experts in the sciences concerned if the present author, who, although incapable of specializing in mechanics or its history, has had occasion to explore certain topics, singles out passages that he has found especially suggestive or illuminating. The discussion of the motion of a material point is a reminder that throughout the eighteenth century, force was taken to be proportional not to acceleration but to velocity, and that when Laplace and others spoke of the "force of a body" they meant the quantity of motion, mass times velocity, later called momentum (Book III, chapters 2, 3; *OC*, VI, 155–161, 173). For the equivalent of the quantity mass times acceleration, the more restricted term accelerative force was used. The importance to Laplace of d'Arcy's principle of areas and its equivalence to conservation of angular momentum emerges very explicitly in Book III, chapter 5 (*OC*, VI, 195–196). An especially felicitous analogy between the secondary and tertiary oscillations of a pendulum and the perturbations experienced by the planets makes it easier to see how he was envisioning and formulating problems of the latter sort (Book III, chapter 5; *OC*, VI, 190–193).

A chapter of "Reflections" on the law of gravity (Book IV, chapter 17) recapitulates the basic assumptions about the operation of this force that Laplace had stated as his point of departure in the dual probability-gravitation memoir ([*1776a*, 2°]; see Section 5). Now they are five instead of four, since the supposition that gravity is indifferent to the state of motion or rest of bodies, and that it acts instantaneously, has become two principles.

An interesting passage explains how his analysis of the secular inequalities of the moon had led him to change his mind on the latter point (*OC*, VI, 346). Finally, a passing observation is reminiscent of a saying of Einstein, to the effect that it is the laws of nature that are simple, not nature itself, which on the contrary is very complicated:

> The simplicity of nature is not to be measured by that of our conceptions. Infinitely varied in its effects, nature is simple only in its causes, and its economy consists in producing a great number of phenomena, often very complicated, by means of a small number of general laws (Book I, chapter 14; *OC*, VI, 65).

The remark is a reminder that Laplace, too, was a thinker about the world and not merely an indefatigable calculator or an overbearing dogmatist, although in the mix that made his personality those aspects may also have been combined.

Only in Book V did Laplace deal with matters on which he had not already published mathematical investigations. It consists of six chapters, the first five on the history of astronomy. He must have taken his history seriously, for he included it in all editions and published a revision separately ([*1821e*]). Laplace, however, was not the scholar that Delambre was, whose histories of ancient and modern astronomy continue to be valuable. Laplace's remarks on the great discoveries of the past are further evidence—if any is needed—that inventiveness in one discipline can accompany banality in another; and his treatment of the place of astronomy in the growth of knowledge is not so much warmed-over as it is cooled-down Condorcet, whose *Esquisse d'un tableau historique des progrès de l'esprit humain* had been posthumously published the previous year.

In the sixth, and last, chapter of Book V, Laplace introduced a speculation on the origin of the solar system and another on the nature of the universe beyond its confines. These concluding nineteen pages written for a popular audience have sustained a more continuing, although not a better-informed, commentary than all of Laplace's other pages put together. The former speculation, which has quite generally come to be misnamed the nebular hypothesis, was presented with the "misgivings" [*défiance*] that anything should arouse that is in no way the product of observation or calculation (B: [1796], II, 303). Perhaps it will not strain analogy (one of Laplace's favorite modes) too far to liken it to the Queries at the end of Newton's *Opticks*. It makes a curious commentary upon the history of science that the indulgence of exact minds in such flights of fancy should excite so much more interest—to be sure, it is a human interest—even among scientists in later times, than does the content, let alone the detail, of the work that gives them a claim on our attention in the first place. Laplace revised this concluding chapter for each edition of the *Système du monde*, as indeed he did other passages throughout the work, in the light of further reflection and of continuing astronomical discovery. S. L. Jaki has recently reviewed the successive modifications of his cosmogony (K: [1976]). We shall confine our attention to the first rendition and attempt to situate it in the context of Laplace's own thought.

In the century or more since the emergence of evolutionary modes of analysis and explanation, the Laplacian cosmogony along with the Kantian—it is very unlikely that Laplace knew of Kant in 1796—has conventionally been cited as an early instance, perhaps as marking the introduction, of a historical dimension into physical science. That attribution, indeed, has been its chief attraction. Unfortunately, however, it is also quite anachronistic. If the text itself is allowed to speak for Laplace, it will be altogether evident that evolutionary considerations in the nineteenth-century sense formed no part of his mentality. The conclusions that he had reached concerned stability; the evidence for that he had calculated, many and many a time. Although that was not the main burden of these passages, he again referred to it as a warranty for the care that nature has taken to ensure the duration of the physical universe, just as it has the conservation of organic species (for he did allude to them, in terms like Cuvier's). Clearly, it was not about the development of the solar system that he was thinking. It was about the birth.

If we were to find a phrase that would characterize what Laplace had in mind about that event, it would not be "nebular hypothesis." It would be "atmospheric hypothesis." And if, further, we were to identify the context in which he raised the question at all, it would not be the evolution or history of nature. It would be the probability of cause. The motifs are altogether familiar to the student of Laplace's own development. The reader is summoned to contemplate the whole disposition of the solar system. At the center spins the sun, turning on its axis every twenty-five and a half days. Its surface is covered by an "ocean" of luminous matter spotted with dark patches, some of which are the size of the earth. Above that zone is a vast atmosphere; how far it extends into space cannot be told. Beyond it turn the planets, seven of them, in almost circular orbits, with fourteen known satellites among them, all revolving almost in the same plane and in the same direction. Those whose rotation is observable—the sun, moon, five planets, and the rings and outer satellites of Saturn—also turn west to east on their respective axes. The question is, can such an arrangement be the effect of chance, or is the existence of a cause to be inferred? There are twenty-nine discrete movements in addition to the revolution of the earth around the sun. The earth's orbital plane serves as reference for determining whether the motion of other bodies is direct or retrograde. If any of the orbits fell outside a quadrant centered on the earth's orbital plane, the motions would appear retrograde. Now then, if the arrangement of the solar system were due to chance, the probability that at least one such inclination would exceed the quadrant is $1 - \frac{1}{2^{29}}$. Since that value amounts to virtual certainty, and since in fact no orbit does fall outside the quadrant, the arrangement cannot be the result of chance and must therefore bespeak a cause. Other appearances are no less remarkable, notably the very slight eccentricity of the orbits of all planets and satellites. Comets, on the other hand, are highly eccentric. They travel into regions still of the sun's dominion but far beyond the planetary sphere, in orbits inclined at all angles to the plane of the ecliptic.

What, then, can the cause be? It would need to explain five distinct sets of phenomena: (1) motion of the planets in the same direction; (2) motion of satellites in the same direction; (3) motion of rotation in the same direction; (4) small eccentricity of orbits for all the above; and (5) extreme eccentricity of cometary orbits.

The only modern writer Laplace had read who had tried to think seriously about the origin of the planets and satellites was Buffon. In his scheme, a comet had struck the sun and released incandescent matter that cooled and coalesced to become the planetary system. That hypothesis satisfied only the first among the above sets of phenomena. Laplace proposed to rise above that to the "true cause."

Whatever it was, it had to have included all the bodies. It had, therefore, to have been originally in the fluid state in order to have been expansible to the dimensions of the planetary system. It must, in a word, have surrounded the sun like an atmosphere. Might it not, indeed, have been the atmosphere of the sun, which in the course of contracting formed the planets by condensations in the plane of the solar equator at the successive limits of its extension? Similar processes centered on the planets could equally have produced the satellites. Such a mechanism would also account for the cometary appearances. The clue was in the absence of gradation between the near circularity of planetary and the extreme elongation of cometary orbits. The less eccentric comets, which would exhibit such a progression, had been drawn into the sun with the contracting atmosphere, leaving behind those describing the extreme trajectories. Hence, the appearance of chance in the distribution of their inclinations. Contraction of the solar atmosphere did not explain it—which was not to say that there was no other cause. It is often said that Laplace was mistaken in ruling the comets out of the solar system; only in respect to causality did he really do so, not physically.

So much for the origins of the solar system. It is true that Laplace mentioned nebulae in this chapter, but he did so in the course of the second speculation about the immensity of the universe beyond the solar system, and not in connection with contraction of the sun's atmosphere to form the latter. Large telescopes reveal great patches of undifferentiated light in the heavens. It is plausible to suppose that these *nébuleuses* without stars are really groups of very distant stars. This passage contains another conjecture that has recently been picked up in the light (or, perhaps, the dark) of black holes, rather than of evolution. The gravitational attraction of a star with a diameter 250 times that of the sun, if any such exists, would be so great that theoretically no light would escape from its surface, and it would be invisible by reason of its very magnitude ([*1799d*]).

The first edition of *Exposition du système du monde* (II, 312) closes with a panegyric of astronomy—and a political statement. The great merit of the science is that it

> . . . dissipates errors born of ignorance about our true relations with nature, errors the more damaging in that the social order should rest only on those relations. TRUTH! JUSTICE! Those are the immutable laws. Let us banish the dangerous maxim that it is sometimes useful to depart from them and to deceive or to enslave mankind to assure its happiness.

21. Mécanique céleste. Publication of the *Traité de mécanique céleste* was not only coincidentally but also circumstantially associated with the beginning of the Napoleonic regime. Laplace had first encountered the young Bonaparte at the École Militaire in Paris in September 1785 among the artillery cadets whom he examined that year in mathematics. On 25 December 1797 the Institute elected General Bonaparte, fresh from his victories in Italy, to the vacancy in the section of mechanics created by the exile of Lazare Carnot following the coup d'etat of fructidor (*PVIF* [21 brumaire an VI], I, 296). Laplace accompanied Berthollet in the ceremony of escorting the young general to take his seat. In October 1799, three weeks before the coup d'etat of 18 brumaire (9 November) that brought Napoleon to power as first consul, Laplace presented him with copies of the first two volumes of *Mécanique céleste*. The acknowledgement is famous. Bonaparte promised to read them "in the first six months I have free" and invited Laplace and his wife to dine the next day, "if you have nothing better to do" (*Correspondance de Napoléon Ier*, 27 vendémiaire an VIII [19 October 1799], no. 4384; VI [1861], 1). Laplace and Bonaparte were then serving on a commission together with Lacroix to report on an early mathematical memoir of Biot (*69*). Bonaparte never made the personal favorite of Laplace that he did of Monge and Berthollet, but in 1807 and 1808 his sister, Elisa, having been elevated to the rank of princess, took up Madame de Laplace and attached her as lady-in-waiting to her court in Lucca. Their correspondence offers a glimpse into the Napoleonic world of fashion (J: Marmottan [1897]).

On seizing power, Napoleon named Laplace minister of the interior. That ministry had responsibility for most aspects of domestic administration other than finance and police. Laplace lasted six weeks in the government, to be replaced by Napoleon's brother, Lucien. Napoleon's reminiscence at St. Helena is also famous. Laplace, he said, could never "get a grasp on any question in its true significance; he sought everywhere for subtleties, had only problematic ideas, and in short carried the spirit of the infinitesimal into administration" (*Correspondance de Napoleon Ier*, XXX [1870], 330). Thereupon, Napoleon saw value in Laplace as an ornament, though not as an instrument, of state. He appointed him to the senate and made him chancellor of that body in 1803, an office that Laplace enjoyed throughout the Consulate and Empire at an annual income of 72,000 francs. With other emoluments and honors, he "touched" (as the French has it) well over 100,000 francs a year and became a rich man. In 1805 Napoleon further named Laplace to the Legion of Honor, ennobled him the following year with the title of count of the empire, and in 1813 conferred on him the Order of La Réunion.

Laplace in return dedicated the third volume of *Mécanique céleste* (1802) and *Théorie analytique des probabilités* (1812) to Napoleon. The dedication in the latter is adulatory, even by sycophantic standards, and is not reproduced in the *Oeuvres complètes,* where the third edition occupies Volume VII. These apostrophes to power have incurred Laplace much odium since 1815 and have been taken by his detractors to epitomize a willingness to serve every set of masters in the state quite without regard to principle. It may have been so; his voice was rarely if ever raised in opposition to any action of any government in power. Fairness, however, requires the observation that his political conduct was no different from that of the scientific community as a whole. His eminence there exposed him to closer and more jealous scrutiny than has been directed at his colleagues, and his personality and influence may also have aroused greater hostility than was provoked by others. Fairness also requires recalling that the government of the restored monarchy showed no scruple in associating his reputation with its own, anticlimactic attempt at prestige.

In 1816 he was elected to the Académie Française, and in 1817 Louis XVIII elevated him in the peerage to the dignity of marquis. The reason for that was obvious, however, whereas the relation between Napoleon and the scientific community presents a problem that calls for further study and deeper insight. It was more than a straightforward matter of patronage, important though that was. Some special affinity was involved, comparable perhaps to the interdependence between artist and despot discerned by Jakob Burckhardt in *The Civilization of the Renaissance in Italy*. With all due allowances for differences in century and locus of talent, when this new cultural pact, which recruited scientists as courtiers, finds its analyst, he too may discover a clue to motivations in the illegitimacy of both parties with regard to traditional sources respectively of authority and of knowledge. The key to institutionalization, on the other hand, was the systematic need that authority, in the form of the modern state, and knowledge, in the form of science, were just then beginning to develop for each other in practical fact.

Reorganization of the École Polytechnique was the one important accomplishment that marked Laplace's tenure of the ministry of the interior. A law promulgated on 25 frimaire an VIII (16 December 1799) established a Conseil de Perfectionnement to oversee the curriculum and standards. The course was cut from three to two years and was made preparatory to the specialist schools, the École des Ponts et Chaussées, École des Mines, and École d'Artillerie, which became essentially professional schools at what would now be called the graduate level. Napoleon, well-disposed at first toward Polytechnique, was persuaded to provide adequate financing. Three members of the council were to be delegated from the Institute. Laplace, Berthollet, and Monge were chosen and reelected annually until Lagrange replaced Monge in October 1805 (*PVIF* [15 vendémiaire an XIV], III, 261). Laplace and Berthollet continued to serve throughout the Napoleonic regime, and, indeed, Laplace was commissioned by the government of the restored monarchy to oversee a further reorganization in 1816. His report at the end of the first year of the council's responsibility (24 December 1800) amounts to a catalog of courses and requirements and a prospectus of services to be expected of science by the state ([*P1800*]).

From this, the period of Laplace's greatest prominence, testimony remains of friendships no less than of enmities. They clustered around his work, naturally enough, and if the element of discipleship was predominant, that is not unusual in the lives of scientists. The most sympathetic personal recollection comes from Biot, who made it the subject of a reminiscence half a century later before the Académie Française. Biot had graduated from Polytechnique with the first class in 1797 and had received a post teaching mathematics at the École Centrale in Beauvais. It was common knowledge in scientific circles that Laplace was preparing *Mécanique céleste* for publication. Wishing to study the great work in advance, Biot offered to read proof. When he returned the sheets, he would often ask Laplace to explain some of the many steps that had been skipped over with the famous phrase, "It is easy to see." Sometimes, Biot said, Laplace himself would not remember how he had worked something out and would have difficulty reconstructing it. He was always patient in going back over these deductions and equally so with Biot's own early efforts. He encouraged him to present before the Institute the memoir (*69*) on the general method that Biot had conceived, in the isolation of Beauvais, for solving difference-differential equations. Only some time afterward did Laplace show Biot a paper he had put away in a drawer, a paper in which Laplace had himself arrived at much the same method years before.

Biot would often stay to lunch along with others of his age. After a morning of work, Laplace liked to relax in the company of students and young men at the beginning of their careers. In their mature years, they remained an entourage grouped around him like—Biot's phrase may be more revealing than he intended—"so many adopted children of his thought" (J: Biot [1850], 68). Madame de Laplace, still young and beautiful, treated them like a mother who could have been a sister. Lunch was frugality itself—milk, coffee, fruit. They would talk science for hours on end. Laplace would often ask them about their own studies and research, and tell them what he would like to see them undertake. He was equally concerned with practicalities of their future prospects and would point out opportunities. "He looked after us so actively," said Biot, "that we did not have to think of it ourselves" (*ibid.*, 69; cf. J: Lacroix [1828], *passim*). In 1800 Biot himself was appointed to the chair of mathematics at the Collège de France.

Among contemporaries, the friendship with Berthollet was the closest and most enduring of which record remains. They had begun to draw together in the mid-1780's, attracted to each other scientifically at least by their mutual interest in the physics of chemical forces. Both enjoyed greater prestige and influence at the Institute than either had achieved in the last years of the Academy, where Lavoisier had predominated. They were also close scientifically, at least after Berthollet's return from Egypt with Napoleon in 1799. Laplace contributed two notes on pressure-temperature relations in an enclosed gas to Berthollet's master treatise *Essai de statique chimique* ([*1803c*]). Berthollet then had a country house in the village of Arcueil, five miles south of Paris. He installed a chemical laboratory there and also a physical laboratory, and gathered around his work a younger group, the most notable among them being Gay-Lussac and Thenard. In 1806 Laplace bought the neighboring property. The transfer of his salon there, and their collaboration with Berthollet and his group, created a circle of mathematically and experimentally capable people under strong leadership who soon began informally calling themselves the Société d'Arcueil; its institutional history has been written by Maurice Crosland (J: [1967]). Laplace's part in the work of the group, which also included Bérard, Descotils, Biot, Arago, Malus, and Poisson on the side of physical sciences, occupies Sections 22–24 of the present article. One chronological fact is important to emphasize here. The activities of the Arcueil group clearly postdated Laplace's completion of his astronomical system with the publication of Volume IV of *Mécanique céleste* in 1805 (*83*). As will appear, he did do further astronomical work, but it was of an occasional nature, and Volume V comprises a series of addenda.

Before we proceed to a discussion of that treatise, two further memoirs on particular topics need to be noticed briefly. In January 1796 Laplace read before the Institute the draft of "Un mémoire sur les mouvements des corps célestes autour de leurs centres de gravité" ([*1798a*]) and had it revised and ready for publication less than a year later (*61*). His summary reflections on gravity in the *Système du monde* grouped the problems that the law presented into three categories—the motion of centers of gravity of celestial bodies about centers of force, the figures of the planets and the oscillations of the fluids that cover them, and the motion of bodies about their own centers of gravity (Book IV, chapter 17; *OC*, VI, 341). He did not say so, but his own work clearly had been addressed largely to problems of the first two types, and he had dealt with rotation only incidentally to analysis of precession, tidal motion, and the coincidence between the lunar periods of revolution and rotation. Now he proposed to give a complete analysis of motions of the last type.

In fact, the memoir is both more and less than that. The most important example, he says, is the earth. Precession of the equinoxes is produced by one of its motions, and we tell time by its rotation. First, however, he digressed to give an application that he had just developed of the generalized gravitational function (see Section 16, Equation 94) to the theory of perturbations in planetary motion. The approach came out of his studies of the moon. In analyzing its motion around the earth, he now treated its mass as infinitesimal and attributed to the earth a mass equal to the sum of the two masses. The new, and Laplace thought quite remarkable, equation of condition ([*1798a*]; *OC*, XII, 136) that permitted this gave a direct relation between parallactic inequality and inequalities of lunar motion in longitude and latitude. Moreover, it was easy to verify the theoretical values by observation, since the constants were given by the mean longitudes of the moon and of its perigee and nodes at a given time. More generally, the same equation was applicable to verifying the calculation of the perturbing influence exerted on one planet by another, whose own perturbation is ignored, which procedure was standard in astronomical practice. Laplace promised to develop this first-order theory further and kept his word in *Mécanique céleste* (Book II, nos. 14–15; *OC*, I, 163–170).

Coming back to the theory of rotation about centers of gravity, Laplace considered that the equations Euler had formulated in his *Mechanica* were the simplest and most convenient that he could use. In order to integrate them, terms needed to be expanded in series, and the whole art consisted in identifying those that on integration produced detectable quantities. The finding for the earth was that the only periodic variation in the position of the axis that needed to be considered was the so-called nutation, which depends on the longitude of the nodes of the lunar orbit. Two other axial wobbles, one much smaller and the other of much longer period, might be disregarded. Motions of the axis (and they are the main subject of this memoir) depend upon the shape of the earth, and the analysis led Laplace back to a review of his general memoir on attraction and the shape of spheroids ([*1785b*]; see Section 16).

He now found that the phenomena of nutation and also of precession confirmed the figure for the flattening, namely 1/320, given by the measurements of the length of the seconds pendulum at different latitudes. These results were in agreement and were much closer to satisfying an ellipsoidal figure than the curve constructed by the various surveys of arcs of the meridian. As was often the case when Laplace was changing his mind, he did not actually say that this was what he was doing. He said only that it had to be supposed that terms for the radius of the earth derived from geodesic surveys have less influence than those obtained from a seconds pendulum and that an ellipsoid is to be preferred for calculations of parallax, a figure flattened in the above degree and derived from measurements of the seconds pendulum and analysis of axial variation (*ibid.; OC*, XII, 131). He further gave a direct proof of the theorem that he had long since found indirectly ([*1780c*]; see Section 8), that precession and nutation have the same quantity as if the seas and the earth formed a solid mass. In the preliminary remarks, Laplace said that he was also extending the analysis to the variations in the direction of the lunar axis and in the inclination of the rings of Saturn. In fact, however, the memoir breaks off with an apology for its length and refers the reader to a further volume of the *Mémoires de l'Institut* for a continuation. That second installment never appeared, although what must probably have been the same thing was incorporated in appropriate passages of *Mécanique céleste*, Book V, chapters 2 and 3 (*OC*, II, 375–402). One other feature of this memoir is noteworthy. Laplace was now being assisted in his research by Alexis Bouvard, who had succeeded Delambre in the role of calculator and who performed all the work of calculation for *Mécanique céleste*.

Bouvard also made the calculations for a further investigation of lunar variations, the results of which Laplace read before the Institute on 20 April 1797 (*63*). The seminal series on planetary motions in the 1780's (see Section 17) had culminated in a paper on the secular inequalities of the moon ([*1788c*]), which Laplace had arrived at through the application to the moons of Jupiter of his approach to the theory of Jupiter and Saturn. He had there given a formula for determining the variations in mean motion, having found them to depend on variations of opposite sign in the eccentricity of the earth's orbit. The motions of nodes and apogee of the lunar orbit also exhibit secular inequalities. Laplace had then restricted his determination of those values to terms given by the first power of the perturbing force, although well aware that this was only half the story for the motion of the lunar apogee. The other half was expressed in terms dependent on the square of the perturbing force. Clairaut had discovered that this part was the resultant of the two large inequalities called variation and evection. The secular equation Laplace found for the motion of apogee, added to that for mean motion, gave a secular equation of the anomaly equal to 4.3 times that for mean motion. In like fashion, when he included terms depending on the square of the perturbing force in the secular equation of the nodes, he found its value to be 0.7 that of the mean motion, which amount was to be added to their mean longitudes. Thus, the motions of nodes and apogee decelerated when the mean motion of the moon accelerated, and the secular equations of the three effects were in the ratio 7:33:10 (*OC*, XII, 193–194).

These were large inequalities. One day they would produce changes in the secular motion of the moon equal to 1/40 of the circumference and up to 1/12 of the circumference in the secular motion of the apogee. Like the variations of the eccentricity of the earth's orbit, on which they depended, they were periodic; but the periods, which were enormously longer than any others that Laplace had yet identified, occupied millions of years. Slow though the changes were, they were sufficiently important to be incorporated in the tables and to appear in the comparison of ancient to modern observations. Laplace had Bouvard compare some twenty-seven eclipses recorded in antiquity, by Ptolemy and by the Arabs, to the figures in the tables, and he made no doubt of the importance of the acceleration of the motion of the lunar anomaly. He took the occasion to review the ancient corpus of lunar data as calculated by Ptolemy on the basis of the observations of Hipparchus and corrected by the further observations of al-Battānī.

Laplace's presentation of this research differed from his previous practice. He published the results in a brief paper in the *Connaissance des temps* ([*1798b*]), the almanac for practical astronomy and navigation, roundly recommending that astronomers increase the motion of the lunar anomaly by 8′30″ per century and apply a correction to its secular equation equal to 4.3 times the mean motion (*ibid.; OC*, XIII, 11). The details of the analysis he kept for publication in the *Mémoires* of the Institute ([*1799a*]). There he chided mathematicians for having been insufficiently scrupulous in examining which of the terms they might legitimately neglect in the successive integration of astronomical expressions (*ibid.; OC*, XII, 191–192). In this investigation, he found it best to follow d'Alembert's example and express the lunar coordinates in series of sines and cosines of angles depending on the true motion. In those expressions, he made the true longitude the independent variable, rather than the time, as he always did in his planetary theory. There would be an advantage, he thought, in constructing tables that would give time as a function of the true motion of the moon, since terrestrial longitudes were determined in practice on the basis of the time at which the moon was observed to be at some certain position in its motion in longitude.

The papers of this period are indicative of the pattern of Laplace's later work. Henceforth, he tended to divide his efforts between short communications giving the results or applications of his current investigations and the great treatises still to be compiled and issued. These brief reports appeared in the *Connaissance des temps* when they were astronomical; otherwise they were published in one of the other journals started in the 1790's concomitantly with the movement toward specialization in the sciences. Often, as will appear in the Bibliography, he published the same paper in several journals, sometimes with slight modifications. The day of the communication of scientific investigations through the medium of monographic research memoirs was, in any case, almost over. Those that Laplace had yet to publish show the tendency, already evident in the 1790's, to explore ever finer points of his earlier investigations. The first two volumes of *Mécanique céleste* were published in the same year, 1799, as the lunar analysis just discussed. In thus drawing together his science into the form of a treatise, Laplace like many of his colleagues was answering to another aspect of the evolution of science, the creation, actual or potential, by the new system of higher education of a truly scientific public within the larger audience that could be expected for his *Exposition du système du monde*.

Traité de mécanique céleste is a composite work. It has the aspects of a textbook, a collection of research papers, a reference book, and an almanac, and contains both theoretical and applied science. The first two volumes form a largely theoretical unit. Methodologically, their purpose is to reduce astronomy to a problem in mechanics, in which the elements of planetary motions become the arbitrary quantities. Phenomenologically, the purpose is to derive all the observed data from the law of gravity. The textbook character of the work is most apparent in Book I and, to a lesser degree, in Book II, which occupy the first volume.

Book I is a mathematical exposition of the laws of statics and dynamics in a development adapted to the formulation of astronomical problems. Laplace's normal practice in those investigations had been to open each memoir with a derivation of the laws of motion in a form suited to the particular set of problems. Here he arranged the same material systematically. The sequence was canonical: first the statics and dynamics of mass points, second of systems of bodies, and third of fluids; the point of view is d'Alembert's. Dynamical laws are derived from equilibrium conditions. Apart from the motivation, only two features appear to be distinctively Laplacian. In chapter 5, which is concerned with the general principles of mechanics, Laplace incorporated his concept of an invariant plane into the discussion of the principle of conservation of areas. His introduction of that idea ([*1793a*]; *OC*, XI, 547–553; see Section 19) had been the first published statement that a general work on physical astronomy was in preparation. He there emphasized the utility of such a plane in providing a frame of reference, fixed in space, to which calculations of planetary motion could be reduced in centuries to come. In the *Système du monde*, he had specified that a plane "that would always be parallel to itself" would pass through the center of the sun perpendicularly to the plane in which the sum of projected angular momentums of the bodies in the solar system is a maximum (Book III, chapter 5; Book IV, chapter 2; *OC*, VI, 198–199, 218–219). In later terminology, the reference plane is perpendicular to the total angular momentum vector of the system. Laplace had given the mathematical rule for finding it in the *Journal de l'École polytechnique* ([*1798c*]). Now, in *Mécanique céleste*, he moved the origin of coordinates from the sun to the center of the earth (Book I, no. 21; *OC*, I, 63–69), no doubt because in practice astronomers refer their observations of the motion of celestial bodies to the plane of the earth's orbit (cf. Book I, no. 60; *OC*, I, 337–338).

The second feature that one would not expect to find in a textbook of rational mechanics is the discussion in chapter 6 of the laws of motion of a system of bodies given any mathematically possible hypothesis concerning the relation of force to velocity (K: Vuillemin [1958]). In that apparent digression, Laplace may have been following Newton's example in certain propositions of Book II of the *Principia mathematica*. A completely abstract and general system of dynamics might be imagined in which the number of such relations involving no contradiction would be infinite. There are two laws of nature, however, that hold good as principles of dynamics only in the simplest case, that of force directly proportional to velocity. The first is the principle of rectilinear inertia; the second is the conservation of areas in angular motion. It is in this discussion that it becomes clearest how, in the astronomical application, Kepler's law of equal times in equal areas had become for Laplace a special case of the principle of conservation of areas, or of angular momentum. He was usually careful to point out, however, that an accurate determination of the masses of all the planets had yet to be achieved.

Taken in isolation, Book II might also appear to have been conceived as a manual in which the mathematically qualified student learns the analysis required for theoretical astronomy. In it the laws of motion are applied to deriving the law of gravity from phenomena and to calculating the displacements of celestial bodies. Here also Laplace is more concerned to impart techniques than results. As he moved beyond the differential equations of gravitational attraction (chapter 2) and of elliptical motion (chapter 3), however, the techniques became increasingly his own. Chapter 6, for example, on perturbation theory, generalizes the combination of perturbations in coordinates and in the orbital elements that he had evidently begun working out in the first gravitational memoir ([*1776a*, 2°], Article LXIII; *OC*, VIII, 241–246). Two bodies are assumed to move in coplanar, circular orbits with radii equal to the semimajor axis of the planetary orbits. A disturbing function is developed in sine and cosine series of the longitudinal difference between the bodies in orbit. The coefficients of the series are functions of the ratio of the semimajor axes, and Laplace established the analytical relationships among them and among their derivatives with respect to that ratio, finding expressions for which he could later give numerical values for all possible pairs of planets. The sequence of topics is also distinctively Laplacian; indeed, it is identical to that in the *Système du monde*. At the outset of Book II, Laplace says that he intends to give the mathematization of the phenomena that he had there described in detail. Even in this book, however, the treatment grows more specialized as he continues, and already he was incorporating blocs of material from earlier researches in the exposition. Passages from the memoir on cometary orbits ([*1784b*]), for example, reappear in chapter 4, on motion in very eccentric and parabolic orbits. Similarly, the reciprocity of the acceleration of Jupiter and deceleration of Saturn, and also the libration of the inner three satellites of Jupiter, are introduced to illustrate methodological points. The Jupiter-Saturn relation (chapter 8, no. 65) exemplifies the method of approximating periodic inequalities that appear in elliptical motion when it is legitimate to neglect terms involving squares or products of perturbing forces; the libration of the Jovian moons (chapter 8, no. 66) depends on inequalities that appear only in terms of the order of the squares of the perturbing masses.

Volume II continues and completes the mathematical analysis of the three main categories of phenomena outlined in the *Système du monde*. Having handled the motion of celestial bodies in translation in Book II, Laplace turned to the figure of the planets in Book III, to the motions of the seas and atmosphere in Book IV, and to rotational motion in Book V. Book III, nos. 8–15, on the attraction of spheroids, is a systematic reprinting, with some simplification of the mathematical development, of the material from his memoirs on the subject ([*1785b*], [*1786a*], [*1793a*]). He had promoted the statement of the most important equation, which gives the potential function (see Section 16, Equation 94), to the passage developing the basic differential equations governing the motions of mutually attracting bodies in Book II, no. 11 (*OC*, I, 153). For the rest, he repeated his discussion of the attraction exerted by spheroids of revolution on internal and external points, restated his theorem on the attractions of confocal ellipsoids, showed how to expand the expressions in series, and considered the cases of homogeneous and variable density (see Section 17). The third chapter brings in the demonstration that a liquid mass rotating under the influence of gravitational force will satisfy an ellipsoidal figure and that its axis of rotation will be in the direction that at the outset would have given it the maximum angular momentum (cf. [*1785b*]; [*1793a*], Article XV). The fourth chapter considers the spheroid covered with a layer of fluid and analyzes the equilibrium conditions (cf. [*1779a*], Article XXVIII; [*1785b*], Article XV); and the sixth and seventh discuss respectively the shape of the rings of Saturn (cf. [*1789a*]) and an equation governing the atmospheres of celestial bodies applied to the sun (cf. *Système du monde*, Book V, chapter 6; see Section 20).

The main novelty is the comparison (chapter 5) of spheroidal attraction theory with the results of geodesic surveys of meridional arcs. Laplace had introduced that topic in the miscellany published and largely lost to view in the waning days of the Academy ([*1793a*], no. 9; see Section 19). In *Mécanique céleste* he could draw on the data, not previously analyzed, from the Delambre-Méchain survey of the meridian from Dunkerque to Barcelona, on which the metric system was to have been based (see Section 18). He also, and perhaps more importantly, went further than he had previously done in applying error theory to the investigation of physical phenomena. An initial theoretical article (Book III, no. 38) develops the analytic geometry of geodesic lines and results in the following expressions applicable to the case of the earth. For the radius vector of an osculatory ellipsoid,

$$1-\alpha\sin^2\psi\,\{1+h\cos 2\,(\phi+\beta)\}; \qquad (122)$$

for the length of a meridional arc,

$$\epsilon-\frac{\alpha\epsilon}{2}\{1+h\cos 2\,(\phi+\beta)\}\{1+3\cos 2\psi-3\epsilon\sin 2\psi\}; \qquad (123)$$

and for the degree measured orthogonally to the meridian,

$$1^\circ+1^\circ\alpha\{1+h\cos 2\,(\phi+\beta)\}\sin^2\psi+4^\circ\alpha h\tan^2\psi\cos 2\,(\phi+\beta); \qquad (124)$$

where ψ is latitude, ϕ is the angle formed by intersection of the plane xz with the plane that includes the radius vector and the z axis, and β is a correction for the deviation of the true figure of the earth from an ellipsoid (*OC*, II, 133–134). Bowditch had to point out that Laplace erred in the calculation, and that his numerical application suffered from this as well as from several arithmetical mistakes (F: [1829–1839], II, 394, 412–416, 447, 459, 471).

Nevertheless, it was in the derivation of these expressions that spheroidal analysis was brought to bear on actual geodesic measurement. The science of geodesy was thereby moved a significant distance along the scale from the observational to the mathematical.

The method itself is more interesting than the results. It had two stages. The first (Book III, no. 39) involved estimates of observational error. The quantities $a^{(1)}, a^{(2)}, a^{(3)}, \cdots$, represent the lengths measured for a degree of the meridian in different latitudes; and $p^{(1)}, p^{(2)}, p^{(3)}, \cdots$, are the squares of the sines of the respective latitudes. On the supposition that the meridian describes an ellipse, the formula for a degree will be $z+py$. Designating the observational errors $x^{(1)}, x^{(2)}, x^{(3)}, \cdots$, Laplace wrote the following series of equations (*OC*, II, 135):

$$\begin{aligned} a^{(1)}-z-p^{(1)}y&=x^{(1)}, \\ a^{(2)}-z-p^{(2)}y&=x^{(2)}, \\ a^{(3)}-z-p^{(3)}y&=x^{(3)}, \\ \cdots\cdots&\cdots\cdots, \\ a^{(n)}-z-p^{(n)}y&=x^{(n)}, \end{aligned} \tag{125}$$

where n is the number of meridional degrees measured. The purpose is to determine y and z by the condition that the greatest of the quantities $x^{(1)}, x^{(2)}, x^{(3)}, \cdots, x^{(n)}$, shall have the least possible value. Laplace gave solutions for the cases of two, three, or any number of such equations of condition, pointing out that the method was applicable to any problem of the same type. He mentioned specifically the example of n observations of a comet, from which it would be required to determine (1) the parabolic orbit for which the largest error is smaller than in any parabola, and (2) whether the hypothesis of a parabolic trajectory can be reconciled with the observations in question. In the present, geodesic case the problem is to determine the ellipse for which the greatest deviation from measured values is a minimum.

The solution would reveal whether the hypothesis of an elliptical figure was contained within the limits of observational error. It would not, however, give the ellipse that the measured values themselves showed to be the most probable—the most probable ellipse, Laplace called it. Determining that figure was the object of the second stage of the analysis (*ibid.*, no. 40). Two conditions had to be satisfied, first that the sum of all the errors made in the surveys of entire arcs should be zero, and second that the sum of all the errors taken positively should be a minimum. In the preliminary version of this analysis ([*1793a*]) Laplace had attributed the idea for this approach to Bošković (K: Todhunter [1873], no. 962, II, 134), of whom he made no mention in *Mécanique céleste*. Having developed it, he proceeded to numerical calculations of the ellipticity of the earth, concluding from the data of the metric survey that it cannot be an ellipsoid and that the ratio of flattening of an osculatory ellipsoid is 1/250. The remainder of chapter 5 consists of calculations of the probable degree of error in the results of other surveys (Lapland, Peru, Cape of Good Hope, Pennsylvania), of the flattening of Jupiter, and of the length of the seconds pendulum at various latitudes.

A propos of the last topic, it is perhaps worth noting that as Laplace came to consider that measurements of the length of the seconds pendulum might be reconciled with an ellipsoidal figure ([*1798a*]; *OC*, XII, 131), his interest in them appears to have slackened. Instead, he increasingly turned his attention to the data from direct geodesic surveys, determining how much of the deviation of that figure was owing to observational error and how much of it to nature. For what the second stage in this investigation finds him estimating is how far nature itself departed from theoretically determined forms. In other words, Laplace was now applying error theory to an investigation of phenomena and not merely to the probability of cause. He had not yet arrived at the least-square rule, and Quetelet's notion that errors of observation and errors of nature may follow an identical distribution would have been foreign to Laplace (L: Gillispie [1963], 449). Both lay not far in the future along the same path, however.

Book IV, on the oscillations of the sea and the atmosphere (the last four of the 144 pages concern the atmosphere), contains less novelty in principle. Here, too, he first develops theory, which occupies three chapters, and compares it to the observations in a fourth. In his memoir on the tides at Brest ([*1797a*]) he had already revised the approach of his youthful investigations of the ebb and flow of the tides ([*1778a*], [*1779a*], [*1779b*]). The first chapter now gives the mathematical treatment which that revision had summarized. The second restates his two theorems, to the effect that the seas are in stable equilibrium if their density is less than the mean density of the earth, and conversely. The last chapters are largely a repetition of the Brest memoir on the influence of local conditions, illustrated by the same early eighteeenth-century data, the use of which is now tempered by incidental consideration of probable error in the observations. The second volume ends with Book V, on the rotation of celestial bodies. It is one of the shorter books of *Mécanique céleste*; and, like the memoir ([*1798a*]) hurried into print before publication of the work, has the appearance of an afterthought included for completeness.

Laplace had Volume III ready to present to the Institute in December 1802 (*80*). Three years of the Napoleonic consulate had elapsed since the publication of the first two volumes, which he had designated as Part I. The main purpose of Part II, he announced in the preface, was to improve the precision of astronomical tables. That motivation is consistent with the overall configuration of his career, at least in astronomy and probability. In both, the emphasis shifts to application, and only in physics did new theoretical problems engage his interest. The tendency was already evident in the internal sequence of particular investigations, and it would not force matters unduly to describe the first two volumes as representing largely the work of the early Laplace, and the second two that of the later Laplace. The transition occurred somewhere in the interval between 1790 and 1795, the years of the revolutionary liquidation of the old Academy and the quasi-technocratic reorganization of science at the Institute and related bodies. Nothing more than the kind of environmental conditioning that accompanies change of circumstance can be claimed for the coincidence, but influence is nonetheless important for being felt pervasively.

The third volume is entirely occupied by the theory of the planets in Book VI and of the moon in Book VII. In developing the general formulas and methods for planetary astronomy in Book II, Laplace had limited himself to expressions for inequalities in the motions that are independent of orbital eccentricities and inclinations or that depend only on the first power of those quantities. The precision was insufficient for accurate positional astronomy, however, and Book VI applied to all the planets the method employed for the theory of Saturn in the great Jupiter-Saturn memoir ([*1788a*]). Approximations were carried to the terms involving the squares and higher powers of these quantities and also to those depending on the squares of perturbing forces. Laplace then had Bouvard substitute the numerical values for each planet in these formulas, combined with the general formulas from Book II. The successive chapters then give numerical expressions for the radius vector and for its motion in longitude and latitude. It is in this book, and later in Book VIII (Volume IV) on the moons of Jupiter, that *Mécanique céleste* could serve the practical navigator and observer as the basis for an astronomical almanac. Bouvard was responsible for the enormous labor of numerical computation, for comparing the results with the findings of other astronomers, and for pinpointing the sources of disagreement. There might still be errors, Laplace acknowledged, but he was confident that they were too inconsiderable to vitiate the tables that might now be compiled.

Laplace himself had not previously worked on theories for Mercury, Venus, the earth, and Mars, for all of which the periodic inequalities are small and are now precisely given. Chapters 12 and 13, on Jupiter and Saturn, mainly repeat his classic work on their long periodic inequality. Although he had investigated the motion of Uranus as early as January 1783 (*40*), Chapter 14 is his first theoretical account of its motion. Chapter 15 formulates equations of condition for long-term periodic inequalities produced by the mutual perturbations of pairs of planets other than Jupiter and Saturn—earth-Venus, Mars-earth, Uranus-Saturn, Jupiter-earth—and shows how they corroborate the respective planetary theories. Finally, the masses of the planets and of the moon are calculated relative to the sun, the values for Saturn and Uranus still needing considerable refinement. In the preface, he mentioned the discovery of Ceres on the first day (Gregorian style) of the new century, followed by that of Pallas, but gave no detail.

Book VII is devoted to lunar theory. Its object is to exhibit in numerical detail the finding of the initial memoir on the moon ([*1788c*]) that all the inequalities of lunar motion, namely variation, evection, and the annual equation, result from the operation of universal gravity, and then to deduce from that law further explanations concerning finer points of the motion, and also of the parallax of the sun and moon and of the flattening of the earth. Laplace followed the practice of his second memoir ([*1799a*]) in taking true longitude rather than time for the independent variable in his differential equations of motion, which he had taken the precaution of adapting for the purpose in Book II (no. 15, Equation K). It will be recalled that he had published this paper only a few weeks before the first two volumes of *Mécanique céleste* itself, having worked out the analysis early in 1797 (*63*). In the meantime, the Austrian astronomer Johann Tobias Bürg had been investigating what appeared to be a periodic inequality in the motion of the lunar nodes with an interval of about seventeen years between the maximum positive values and about nineteen years between the maximum negative values. In 1800 the Institute awarded Bürg a prize for this research (*PVIF* [11 germinal an VIII], II, 129). He had already asked Laplace to investigate what the cause of these effects might be. Employing the appropriate equations from *Mécanique céleste* (Book II, no. 14; Book III, no. 35), Laplace analyzed the data in a memoir ([*1801b*]) read before the Institute in June 1800 (*71*). The episode illustrates that from the outset, *Mécanique céleste* was furnishing the apparatus for further research and calculation in both practical and theoretical astronomy. Laplace found that the effects result from a nutation in the lunar axis created by a variation in the inclination of the lunar orbit to the plane of the ecliptic. Its inclination is constant with respect to another plane passing through the equinoxes between the equator and the ecliptic. That angle would amount to 6.5″ on the assumption that the flattening of the earth is 1/334. Further comparison by Bouvard of Bürg's observations with those of Maskelyne indicated rather a figure of 1/314. In any case, the value was far from the fraction of 1/230 that spheroidal theory predicted for an earth of homogeneous density. Laplace was delighted that so minuscule an anomaly in the position of lunar nodes could thus confirm the direct measurements of geodesy on the shape of the earth and on conclusions to be drawn concerning its internal constitution. He incorporated the material in Book VII of *Mécanique céleste*, where it formed the major novelty, the bulk of the discussion being a recapitulation of his earlier research fortified by Bouvard's indefatigable calculation of numerical values for the formulas to serve in compiling precise tables.

For Volume IV, presented to the Institute in May 1805, over two years after Volume III, there remained the practical theory of the satellites of the outer planets, and also of the comets. Book VIII is almost entirely devoted to the moons of Jupiter. It consists of a revision of the calculations of the memoirs of the early 1790's ([*1791a*], [*1793b*]; see Section 19), which had issued in values for the masses of the four satellites relative to that of Jupiter and for the flattening of the latter. Laplace now gives greater numerical detail on the inequalities of the three inner satellites, concealed in the invariance of the libration that he had discovered in his first memoir on these bodies ([*1787a*]). He likens their lockstep to the observable libration of our own moon and compares other particularities of the motion of the Jovian satellites to the lunar evection, annual equation, and variation in latitude discussed in Book VII. (It is interesting that, having been led to his explanation of the apparent lunar acceleration by his first work on the moons of Jupiter, he was now illuminating finer points in their theory by analogy to the moon.) Chapter 7, giving numerical values for the various inequalities, is new. In deriving his formulas for the variations in radius vectors and longitudes, Laplace made several errors that were detected by Airy and corrected by Bowditch (F: Book VIII, no. 21, Vol. IV, 176–185). In all these formulas, Bouvard calculated the numerical values of the coefficients, although the tables to which Laplace refers navigators continue to be Delambre's (*OC*, IV, x–xi). The entire topic is presented as the confirmation by another method of the purely analytic demonstration in Book II, chapter 8 (*OC*, I, 346–395) of the theorems on the libration of the three inner moons. The difference in method consisted in the substitution of synodic for sidereal mean motions and longitudes, and of a moving for a stationary axis of rotation (*OC*, IV, viii). In general, Laplace made more than he had previously done of the spectacle offered by the Jovian satellites of a gravitational system in miniature, its elements oscillating about mean values at a higher rate than those of the whole slow-motion solar system.

The two brief chapters on the satellites of Saturn and Uranus that complete Book VIII are essentially a reprinting of a paper published in the *Mémoires de l'Institut* ([*1801a*]). The outermost satellite of Saturn and all six Uranian satellites that Herschel thought to have observed appeared to be inclined to the plane of the ecliptic at a much greater angle than other planetary bodies. Laplace's analysis demonstrates how that can follow from the weakening of the force of gravity given the distances and ratios of the masses. He acknowledged the data to be very uncertain, however, and indeed it has since been learned that the fifth and sixth moons of Uranus do not exist.

In Book IX, the shortest in the four main volumes, Laplace developed formulas for calculating cometary perturbations from the general equations of motion set forth in Book II, chapter 4 (*OC*, I, 210–254). The planetary formulas were inapplicable to orbits involving large eccentricities and inclinations. For comets, different formulas had to be applied to different parts of the same orbit. Laplace showed how to obtain numerical values for perturbations in orbital elements by means of generating functions. He would have liked to illustrate his techniques by calculating the elements of the curve to be described in the impending return of the comet last seen in 1759. Unfortunately, he was too busy and left the formulas to whoever wished to substitute the numerical data. He did complete two other examples. A chapter on the perturbation of comets that pass very close to a planet concludes that the gravity of Jupiter had drawn the perihelion of a previously invisible comet within range of sight in 1770 and then reversed the effect in 1779, after which year it never reappeared. A second calculation shows that the same comet produced no detectable change in the length of the sidereal year in 1770 despite its proximity to the earth. Laplace felt safe in concluding that the mass of comets is so small that they can have no influence on the stability of the solar system or on the reliability of astronomical tables.

Book X, subtitled "On Different Points Concerning the System of the World," contains largely new material and marks the shifting of Laplace's interest, as he was completing the fourth volume, to problems involving physics. Analysis of the effects of atmospheric refraction upon astronomical observation is the first topic. Laplace had already committed himself to a corpuscular emission theory of light in the remarks on aberration with which he concluded his numerical memoir on the moons of Jupiter ([*1793b*]; see Section 19). Now his derivation of the phenomena of atmospheric refraction presupposed that model, as did the investigations of 1808 in optics proper, discussed below in Section 24. The first problem in Book X is to find the law governing the dependence of refrangibility on the variation of atmospheric density with altitude and temperature. An elaborate analysis of the passage of light through a refracting medium yielded a formula (*OC*, IV, 269) that Laplace reckoned to be applicable when a star had risen to an elevation of 12° above the horizon. At angles greater than that, only the atmospheric pressure and temperature in the vicinity of the observer significantly affected the refraction, and these values could be read directly from the barometer and the thermometer.

In order to evaluate his expression numerically, Laplace needed to know the index of refraction of atmospheric air at a given temperature and pressure and the variation of its density respectively with pressure and with temperature. Delambre had determined the index of refraction for apparent elevations of 45° by observations of the least and greatest elevations of certain circumpolar stars at 0° with the barometer at 76 centimeters of mercury. As for the pressure-volume relations of atmospheric air, physicists were all agreed on the direct proportionality of density to pressure at constant temperature (Laplace did not call the law by the name of either Boyle or Mariotte). Despite many attempts at measurement, however, there was still no agreement about temperature-volume relations in gases, and Laplace engaged Gay-Lussac's assistance in examining the matter. For that purpose, Gay-Lussac calibrated a mercury thermometer against an air thermometer, took extraordinary precautions to dry the air and tubes composing the latter (for humidity was the main source of error), and found that at constant pressure of 76 centimeters of mercury, a unit volume of air at 0° expanded to a volume of 1.375 at 100°. Comparison of the two thermometers at intermediate temperatures argued for a linear expansion within that range. The final value represented the mean of twenty-five determinations, although Laplace did not say how the mean was calculated.

All this discussion, which somehow conveys a greater sense of enthusiasm than the preceding books of recapitulation and tabulation, bespeaks Laplace's growing interest in instrumentation, measurement, and the minimization of observational error. No one, he observed, had yet thought how to compensate for variations in humidity in measurements of atmospheric refraction. Small though he calculated the effect to be, he gave a correction table compiled on the reasonable hypothesis that the indices of refraction of air and water vapor are proportional to their densities. In a like, almost offhand manner he reported Gay-Lussac's ascension in a balloon to an altitude of over 6,500 meters, where the proportions of oxygen and nitrogen in the atmosphere turned out to be about the same as at the surface. As Laplace drew toward the intended conclusion of his treatise, the topics grew more recondite and more fanciful in the object—the effect of extreme atmospheric conditions on astronomical observations, the influence of differences in latitude on barometric measurements of altitude, the absorption of light by the atmospheres of earth and sun, and the influence of the earth's rotation on the trajectories of projectiles and on free fall from great heights (cf. [*1803b*]).

Before writing Book X, Laplace evidently intended to end it with calculations, which occupy chapter 7, contrasting consequences to be deduced from the wave theory and from the corpuscular theory of light. He there purported to show how the resistance of any ethereal medium supporting luminous oscillations would have entailed deceleration of planetary motion. The continuous impact of light corpuscles would, on the other hand, accelerate the planets, except that the effect is exactly compensated by the weakening of the sun's gravitational force through loss of mass. None of this disturbs stability, however. Since the mean motion of the earth shows no change over a 2,000-year span, Laplace calculated that the sun had not lost a two-millionth part of its substance in recorded history and that the effect of the impact of light particles on the secular equation of the moon is undetectable. In a way, it would have been fitting had this chapter been the last, for Laplace applied the calculation that he had just made to the gravitational force considered as the effect of a streaming of particles through space. Thus he would have emerged full circle from his celestial mechanics, coming out just where he went in with the first calculation of the youthful probability-gravitation memoir ([*1776a*, 2°]; see Section 5), except that now gravity is given a velocity of 1×10^9 times the speed of light, which is to say infinite.

That chance for symmetry (if such it may be called) disappeared with the publication of two further memoirs on the theory and tables of Jupiter and Saturn ([*1804a*] and [*1804b*]). Laplace immediately grafted them on to Book X, where they form the basis of chapters 8 and 9. The return to planetary astronomy was unconformable with the overall plan of the treatise, although it will already have been noticed that throughout its composition Laplace found ways to interpolate pieces of continuing research. He called chapter 8 "Supplément aux théories des planètes et des satellites," which may be bibliographically confusing since these chapters did appear in the first edition, unlike the true supplements to be mentioned in a moment. At any rate, in the interval since the publication of Book VI (Volume III, 1802), Bouvard had scrutinized all the oppositions of both planets observed at Greenwich and Paris since Bradley's time, and Laplace himself had reviewed the theory. The result was several new inequalities, and by taking them into account the agreement between his formulas and the observations was improved. The most signal advantage of the new data was that they permitted the first precise calculation of the mass of Saturn, hitherto known only roughly through the elongations of its satellites.

"Nothing more remains for me," wrote Laplace in the concluding sentence to the preface to Volume IV, "in order to fulfill the engagement that I undertook at the beginning of this work, but to give a historical notice of the works of mathematicians and astronomers on the system of the world: that will be the object of the eleventh and last book" (*OC*, IV, xxv). In fact, quite a lot remained, beginning with the studies of capillary action presented to the Institute on 28 April and 29 September 1806 (*86*, *87*) and separately printed as Supplements I and II to Book X. The years from 1806 through 1809 were evidently occupied with the further work in physics proper discussed in Sections 22–24 below, and those from 1810 through 1814 largely with probability (see Sections 25 and 26). Indeed, prior to 1819, by which time Laplace was seventy years old, he published only occasionally on problems of celestial mechanics, and these papers were on minor points. His only major addition in all that fifteen-year interval was a mathematical improvement in the method of calculating planetary perturbations, presented to the Bureau des Longitudes in August 1808 (*90*) as a supplement to Volume III of *Mécanique céleste*.

PART IV: LAPLACIAN PHYSICS AND PROBABILITY

22. The Velocity of Sound. Laplace apparently gained his first experience in physics in the experiments conducted jointly with Lavoisier (see Section 15). It will be recalled that in 1777 they investigated evaporation and vaporization, in 1781–1782 dilation of glass and metals when heated, and in 1782–1783 specific heats, heats of reaction, and animal heats. Apart from occasional collaboration with Lavoisier in the later 1780's, Laplace took little active interest in physics between 1784 and 1801. In the latter year he published a brief piece ([*1801d*]), which applies spheroidal attraction theory to analysis of the forces exerted by an infinitely thin layer of electrical fluid spread upon such a surface, and which may, therefore, have been a noteworthy link between gravitational theory and potential theory.

In 1802 Laplace made one of his enduring contributions to the science, in a paper that was written not by himself but by his young protégé Biot ("Sur la théorie du son," in *Journal de physique*, **55** [1802], 173–182). Using a knowledge of adiabatic phenomena, which had only recently become available in France (even though the heating and cooling associated with the rapid compression and expansion of a gas had been quite well known among British, Swiss, and German scientists since the 1770's), Laplace had suggested to Biot how the notorious discrepancy of nearly 10 percent between the experimental value for the velocity of sound in air and the calculated value using Newton's expression, $v = \sqrt{P/\rho}$, might be removed. According to Laplace, the discrepancy arose from Newton's neglect of the changes in temperature that occur in the regions of compression and rarefaction composing the sound wave. Hence Newton's assumption that $P \propto \rho$, which holds good only if isothermal conditions are maintained, was invalid.

Biot expressed the density of air at any point in a sound wave as $\rho' = \rho(1 + s)$, where ρ is the density of the undisturbed air and s the fractional change in density, taken as positive for compression. Where isothermal conditions were maintained, it followed simply that the pressure of the air could be similarly expressed as $P' = P(1 + s)$, where P is the pressure of the undisturbed air. However, if, as Biot supposed, heating and cooling occurred, respectively in the regions of compression and rarefaction, this equation could not hold. Making the reasonable but unproven assumption that the change in temperature was proportional to s, Biot arrived at the expression $P' = P(1 + s)(1 + ks)$, where k is a constant. Hence, assuming that s is small and neglecting the terms in s^2, Biot could show that the velocity of sound in air is

$$v = \sqrt{\frac{P}{\rho}(1+k)}\,.$$

The expression reduces to the more familiar

$$v = \sqrt{\gamma \frac{P}{\rho}}$$

if we substitute the modern term γ for Biot's constant $(1 + k)$.

Although Laplace's explanation of the discrepancy, as expounded by Biot, won immediate acceptance, replacing a variety of unsubstantiated proposals made during the eighteenth century by Lambert and Lagrange among others, the experimental evidence necessary for a rigorous proof became available only over the next twenty years. By 1807, however, Biot had made the important observation that sound waves could be transmitted through a saturated vapor ("Expériences sur la production du son dans les vapeurs," in *Mémoires de physique et de chimie de la Société d'Arcueil*, **2** [1809], 94–103). Thus he confirmed that some heating and cooling must occur, for if this were not so, condensation would take place in the regions of compression and the sound would not pass. Another major step forward came in 1816, when Laplace showed that Biot's constant $(1 + k)$ was equal to the ratio between the specific heat at constant pressure (c_p) and the specific heat at constant volume (c_v)—the ratio we now express as γ. Finally, in 1822, experiments by Gay-Lussac and J. J. Welter, for which Laplace was clearly the inspiration, yielded the first reliable, independently derived values for γ ("Sur la dilatation de l'air," in *Annales de chimie et de physique*, **19** [1822], 436–437). By observing the changes in temperature that occurred when air was suddenly allowed to enter a partially evacuated receiver (a method pioneered some years before by Clément and Desormes), Gay-Lussac and Welter arrived at a figure of 1.3748 for γ. This brought the theoretical value of v (337.14 meters per second) into good agreement with the currently accepted experimental figure.

How Laplace derived the correction factor of $\sqrt{\gamma}$ in Newton's expression for v was left unclear in the paper on the subject that he published as [*1816c*], but a reconstruction resting partly on [*1822a*] suggests that he began by showing that v must be equal to $\sqrt{dP/d\rho}$. In demonstrating that this quantity is in turn equal to $\sqrt{\gamma P/\rho}$, he assumed not only that the difference between c_p and c_v represents the heat required solely to bring about expansion in a gas expanding at constant pressure, but also that it is this same heat that causes heating when the gas is rapidly compressed. By this argument, a decrease in the volume of a unit mass of the gas from V_0 to $(V_0 - \Delta V)$ would release an amount of heat $(\Delta V/\alpha V_0)(c_p - c_v)$ which, in adiabatic conditions, would go to raise the temperature of the gas by $\Delta V/\alpha V_0\ \{(c_p - c_v)/c_v\}$, where α is the temperature coefficient of expansion. The effect of this rise in temperature would be to increase the pressure of the compressed gas by $P_0(\Delta V/V_0)\{(c_p - c_v)/c_v\}$, in addition to the increment in pressure that would be expected for an isothermal compression. Hence the total increase in pressure is given by

$$\left(\frac{\Delta P}{\Delta V}\right)_a = \frac{P_0}{V_0} + \frac{P_0}{V_0}\left\{\frac{c_p}{c_v} - 1\right\} = \frac{c_p}{c_v}\left(\frac{\Delta P}{\Delta V}\right)_i$$

where $\left(\frac{\Delta P}{\Delta V}\right)_i$ represents the pressure increment that would have been obtained under isothermal conditions. It followed simply that, when conditions were adiabatic, $\sqrt{dP/d\rho}$—Laplace's expression for the velocity of sound—was equal to

$$\sqrt{\frac{c_p}{c_v} \cdot \frac{P}{\rho}}\,.$$

It now remained to calculate a numerical value for c_p/c_v. At a time when no experimental data for c_v existed, this was no easy task, and in [*1816c*] Laplace was vague about the method that he had used. However, two attempts at a reconstruction (M: Finn [1964], 15; Fox [1971], 162–165), although differing on matters of detail, reveal quite clearly not only the unsatisfactory nature of the argument but also the importance for Laplace of the erroneous data concerning the specific heats of gases that had been published by Delaroche and Bérard in 1813. In particular, Laplace's argument rested squarely on their false observation that the specific heat of air decreases as it is compressed. Hence the similarity between Laplace's own figure for γ (1.5) and the figure of 1.43, required in order to secure exact agreement between prediction and observation, was quite fortuitous. Nevertheless, the plausibility of Laplace's treatment appears to have gone unquestioned, and the measurements of γ by Gay-Lussac and Welter in 1822 only confirmed a result that had already won general acceptance.

23. Short-range Forces. Nearly all of Laplace's work in physics from 1802 was characterized by an interest in what he saw as the outstanding problems of the Newtonian tradition; in this respect the attempt to correct Newton's expression for the velocity of sound was typical. From 1805, however, his interest in Newtonian problems assumed a more mathematical character. As we shall see, much of this later work was severely criticized. But, whatever its shortcomings—and it did little to enhance Laplace's reputation in his later life—it contained many results of enduring value, most notably perhaps in the theory of capillary action; and, even more important, it served to tighten the bond between mathematics and physics. This is not to imply that Laplace was in any sense the founder of mathematical physics—there were too many precursors in the eighteenth century for that claim to be sustained—but he did make a major contribution to the mathematization of a subject that had hitherto been predominantly experimental.

The increasingly mathematical thrust of Laplace's work in physics is very apparent in the studies of molecular physics that he pursued, and encouraged others to pursue, for the rest of his life. As early as 1783, when he composed *Théorie du mouvement et de la figure elliptique des planètes*, he elaborated a suggestion that he had advanced in the *Mémoire sur la chaleur* ([*1783a*]), written jointly with Lavoisier. It expressed his belief that optical refraction, capillary action, the cohesion of solids, their crystalline properties, and even chemical reactions were the results of an attractive force, gravitational in nature and even identical with gravity at bottom ([*1784a*], xii–xiii; see Section 16). Almost twenty years earlier, near the beginning of his career, he had remarked in the program of the dual probability-gravitation memoir ([*1776a*, 2°]; see Section 5) that analogy gives us every reason to suppose that gravity operates between all the particles of matter extending down to the shortest ranges. He repeated the detailed speculation in the first edition of *Exposition du système du monde* (B: [1796], II, 196–198) and looked forward to the day when the law governing the force would be understood and when "we shall be able to raise the physics of terrestrial bodies to the state of perfection to which celestial physics has been brought by the discovery of universal gravitation." In this comment, which is obviously reminiscent of the speculations on molecular forces in the Queries of Newton's *Opticks*, there lay the nucleus of a program that guided Laplace's own research in physics and that of several distinguished pupils until the 1820's.

Nevertheless, it was not until 1805 that Laplace began publishing on problems related to his program. By then, interest in molecular forces treated in the Newtonian manner had been greatly stimulated in France by Berthollet's work on chemical affinity, in particular by the *Essai de statique chimique* (1803), in which chemical reactions were explained in terms of short-range attractive forces, supposedly of a gravitational nature, of precisely the kind postulated by Laplace in his physics. It seems likely that Laplace was strongly influenced both by the *Essai* and by Berthollet himself, whose close friend he had been since the 1780's, and that the influence was mutual and reciprocal. In any event, Laplace's first work in molecular physics was published just two years after the publication of the *Essai*, in Book X of the fourth volume of the *Traité de mécanique céleste* (1805) (see esp. chapter 1, pp. 231–276; *OC*, IV, 233–277) and in two supplements to the book (*86, 87*) published in 1806 and 1807 (*OC*, IV, 349–417, 419–498).

It is a measure of Laplace's closeness to the Newtonian tradition that his first studies of molecular forces were concerned with optical refraction and capillary action. Both were manifestations of action at a distance on the molecular scale which had been of special interest to eighteenth-century Newtonians such as Clairaut and Buffon, as well as to Newton himself. (The work of Clairaut appears to have been especially important for Laplace. See, in particular, Clairaut's "Du système du monde dans les principes de la gravitation universelle," in *MARS* [1745/1749], 329–364. On page 338 of this article Clairaut ascribes "the roundness of drops of fluid, the elevation and depression of liquids in capillary tubes, the bending of rays of light, etc." to gravitational forces that become large at small [molecular] distances.) Both Clairaut and Buffon had raised major theoretical problems to which there was still no satisfactory answer by the end of the eighteenth century. In particular, although it was generally accepted that the force between the particles of ordinary matter (in the case of capillary action) and between the particles of ordinary matter and the particles of light (in the case of refraction) diminishes rapidly with distance, it had proved impossible to determine the law relating force and distance. Clairaut, for example, had tried to account for the intense short-range forces by suggesting that the law of gravitational force should contain a term inversely proportional to the fourth power of the distance, $1/r^4$ (*op. cit.*, 337–339). Buffon, by contrast, had upheld the $1/r^2$ law, although he had observed that such a law would be modified at short range by the shape of the particles of matter ("De la nature. Seconde vue" [1765], in his *Histoire naturelle, générale, et particulière*, 44 vols. [Paris, 1749–1804], XIII, xii–xv). Recognizing the intractability of the problem, Laplace proposed a much simpler solution that could be applied in all branches of molecular physics. In treatments that made good use of mathematical techniques developed in his earlier work on celestial mechanics, he showed that the precise form of the law was unimportant and that perfectly satisfactory theories could be given by simply making the traditional assumption that the molecular forces act only over insensible distances.

The treatment of refraction in Book X of the *Mécanique céleste* centered on the specific problem of atmospheric refraction, a matter of practical as well as theoretical concern to Laplace and his colleagues at the Bureau des Longitudes. The whole discussion was conducted in terms of the corpuscular theory of light, the truth of which was never questioned. According to Laplace, the path of a corpuscle of light passing through the successive layers of the earth's atmosphere is determined by the varying attractive forces exerted on it by the particles of air. The measure of these forces was what Laplace, following Newton, called the refracting force (*force réfringente*), a quantity equal to $(\mu^2 - 1)$, where μ is the refractive index of the air. In this analysis, $(\mu^2 - 1)$ is proportional to the increase in the square of the velocity of the incident corpuscles of light, and hence, in accordance with the normal laws of dynamics, it measures the force of attraction to which they are subject. In deriving his extremely complicated differential equation for the motion of light through the atmosphere, Laplace had to assume, not only the short-range character of the forces to which light corpuscles were subjected, but also that the refracting force was proportional to the density of the air, ρ. In order to integrate the equation, it was necessary to make further, speculative assumptions concerning the variation of ρ with altitude (the subject of a long and somewhat tentative section) and to allow for the effect on ρ of the air's humidity. The result was a method of calculating the magnitude of atmospheric refraction for which Laplace claimed complete reliability at any but small angles of elevation.

Despite subsequent refinements, much of Laplace's theory of capillary action, as expounded in the two supplements to Book X of the *Mécanique céleste*, survives in modern textbooks (a brief study of Laplace's theory appears in M: Bikerman [1975]). However, its roots lie as firmly in the vain quest for a comprehensive physics of short-range forces as do those of his work on refraction. As in the theory of refraction, it was crucial that the forces exerted by the particles of matter on one another could be ignored at any but insensible distances (although it was equally important that these distances be finite—an assumption that distinguished Laplace's theory from others in which adhesion was seen as the cause of capillary phenomena). Hence Laplace was glad to invoke Hauksbee's observation that the height to which a liquid rises in a capillary tube is independent of the thickness of the walls of the tube.

In each of the two supplements on capillary action Laplace presented a quite distinct version of his theory. In the first, he arrived at a general differential equation of the surface of a liquid in a capillary tube by considering the force acting on an infinitely narrow canal of the liquid parallel to the axis of the tube. In the second, he treated the equilibrium of the column of liquid in a capillary tube by considering the forces acting upon successive cylindrical laminae of the liquid parallel to the sides of the tube. The two versions were in no sense inconsistent, although in a number of applications the second version proved to be somewhat simpler and more fruitful.

Laplace was concerned, above all, to demonstrate the close agreement between his theory and experiment, and much of both supplements was devoted to this task. Among his most striking successes was a proof (obtained by solving the differential equation of the liquid surface in the first version of his theory) that the elevation of a liquid in a capillary tube is very nearly in inverse proportion to the tube's diameter. Using the same version of the theory, he also showed that the insertion of a tube of radius r_1 along the axis of a hollow tube of slightly larger radius r_2 causes the liquid between the tubes to rise to a height equal to that to which it would rise in a circular capillary tube of radius $(r_2 - r_1)$; thus he confirmed a well-known observation made but not explained by Newton. Among the other classic problems treated in the supplements were the behavior of a drop of liquid in capillary tubes of various shapes (including conical tubes), the rise of liquids between parallel or nearly parallel plates, the shape of a drop of mercury resting on a flat surface, and the force drawing together parallel plates separated by a thin film of liquid.

The importance, for Laplace's theory, of the short-range character of the molecular forces cannot be overstressed. Small terms involving the square of the distance were repeatedly ignored, and it is no coincidence that in his concluding remarks to the second supplement Laplace reiterated his belief in the identity of the forces at work in optical refraction, capillary action, and chemical reactions. In accordance with his belief that capillarity is a consequence of intermolecular action at a distance (albeit at a very small distance), he tried to determine the relative magnitude of the attractive force between the particles composing the liquid (F_1) and the force between the particles of the liquid and those of the tube (F_2). Neglecting variations in density near the surface of the liquid and the walls of the tube, he showed that if $F_2 > F_1/2$, the surface must be concave; otherwise it must be convex, being, in the limiting case of $F_2 = 0$, a convex hemisphere.

Even as the supplements on capillary action were being written, the comparison between theory and observation was being carried still further in experiments, performed at Laplace's request, by Gay-Lussac, Haüy, and Jean-Louis Trémery (the experiments are reported in [*1806a*]; see *OC*, IV, 403–405). These experiments gave the theory added plausibility, as Laplace himself was always ready to observe; and they certainly helped it to survive the criticism of Laplace's only contemporary rival in the treatment of capillarity, Thomas Young. By comparison, Young's theory ("An Essay on the Cohesion of Fluids," in *Philosophical Transactions of the Royal Society* [1805], 65–87), which was based on the concept of surface tension rather than intermolecular attraction, was obscure and unmathematical. Yet Laplace's theory was not without fault, and its author led the way in making modifications. In a paper that he read to the Academy of Sciences in September 1819, the theory was refined to take account of the effect of heat in reducing the attractive force between the particles of a liquid ([*1819h*]); the net attractive force was now taken as the difference between the innate attraction (the only force considered in the supplements to the *Mécanique céleste*) and a repulsive force that was supposed to be caused by the presence of heat. An even more important modification was made in 1831 in Poisson's *Nouvelle théorie de l'action capillaire*, in which Poisson remedied one of the most obvious weaknesses in Laplace's theory by taking account of the variations in density near the surfaces of the liquid and the material of the capillary tube.

24. The Laplacian School. From the time the studies of refraction and capillary action appeared until 1815, Laplace exerted a dominating influence on French physics. Its extent is equally apparent from the problems that younger men were encouraged to investigate (either directly or through prize competitions), from the nature of their answers (which with remarkable frequency served to endorse and extend the Laplacian program), and from educational syllabuses and textbooks, which seldom departed from Laplacian orthodoxy on such matters as the physical reality of the imponderable fluids of heat, light, electricity, and magnetism. Yet at no time was Laplace's control total. There were always those in France who worked outside the Laplacian tradition or even in opposition to it. For example, the paper on the distribution of heat in solid bodies that Fourier read to the First Class of the French Institute in 1807 shows no sign of Laplace's influence, either in its positivistically inclined physics or in its mathematical techniques. (See J: Grattan-Guinness with Ravetz [1972]. A revised version of the paper of 1807 was submitted in 1811 for the prize competition of the First Class. It won the competition but was not published until 1824–1826; see *MASIF*, **4** [1819–1820/1824], 185–555; and **5** [1821–1822/1826, 153–246.) Indeed, Fourier's treatment stands in marked contrast with Laplace's own discussion of the problem, incorporated as a "Note" to his paper on double refraction read to the Institute in January 1809 ([*1810a*], 326–342; *OC*, XII, 286–298). In this note Laplace set up a model for heat transfer by reference to the molecular radiation of caloric over insensible distances. Even Poisson's papers of 1811–1813 on electrostatics ("Sur la distribution de l'électricité à la surface des corps conducteurs," in *MI*, **12**, pt. 1 [1811/1812], 1–92; *ibid.*, pt. 2 [1811/1814], 163–274) owed more to Coulomb than to Laplace, although Poisson was close to Laplace at this time and Laplace would certainly have approved of his treatment, in particular his use of the two-fluid theory of electricity. Such instances of non-Laplacian physics leave no doubt that a fruitful union of the mathematical and experimental approaches to physics would have occurred in early-nineteenth-century France quite independently of Laplace. But the fact remains that Laplace did more than any of his contemporaries to foster that union and, at least in the short term, to determine the character of the work that emerged from it.

The years of Laplace's greatest influence in physics were also those in which his personal standing was at its height, outside the scientific community as well as within it; and he seized every opportunity of furthering his scientific interests. It was a simple matter for him to direct the attention of gifted young graduates of the École Polytechnique, such as Gay-Lussac, Biot, Poisson, and Malus, to problems of his own choice, often in return for help in advancing their careers in the teaching institutions of Paris or at the Bureau des Longitudes. And, once he had become Berthollet's next-door neighbor at Arcueil and the joint patron of the Société d'Arcueil in 1806, he could offer his protégés the additional attractions of access to Berthollet's private laboratory and an association with the elite of Parisian science in the weekend house parties that were a feature of Arcueil life until 1813 (J: Crosland [1967], *passim*). The work of Gay-Lussac, Haüy, and Trémery on capillary action (1806), of Biot on the transmission of sound in vapors (1807), and of Arago and Biot on the polarization of light (1811–1812) was very obviously a result of direct influence of this kind.

It was equally important for the course of French physics that Laplace wielded extraordinary power in the First Class of the French Institute. Here his ability to dictate problems and solutions was no less apparent than in the more intimate atmosphere of Arcueil. It was Laplace, for instance, who persuaded the First Class to engage Biot and Arago on the experimental investigation of refraction in gases, which they described to the class in March 1806 ("Mémoire sur les affinités des corps pour la lumière, et particulièrement sur les forces réfringentes des différens gaz," in *MI*, **7**, pt. 1 [July 1806], 301–387). It is a measure of his influence that their results, although obviously applicable to the practical problems of astronomical refraction as well, were presented in the context of a highly theoretical discussion of short-range molecular forces and of the affinities between the particles of the eight gases examined and the corpuscles of light. Biot and Arago, in fact, adopted Laplace's analysis of refraction without question. Although they measured refractive index (μ) in their experiments, they presented their results in terms of refractive power (Laplace's *pouvoir réfringent*), that is, the quantity $(\mu^2 - 1)/\rho$; and they provided the experimental evidence—conspicuously lacking in Book X of the *Mécanique céleste*—that, for any one gas, $\mu^2 - 1$ is proportional to ρ. It is also a mark of their allegiance to the prevailing orthodoxy of Arcueil—although in this case Berthollet was as much the inspiration as Laplace—that they speculated confidently on the analogy that they supposed to exist between chemical affinity and affinity for light, as measured by the refractive power. Such an analogy was consistent with Laplace's view that both types of affinity were gravitational in origin, so that it was highly satisfactory to be able to show that the order in which substances appeared in the two tables of affinity was very roughly similar.

Laplace also used his position at the Institute to good effect in the system of prize competitions. There is little doubt that he was chiefly responsible for the setting of the competition for a mathematical study of double refraction, which was announced in January 1808 and won by Malus, a recent recruit to the Arcueil circle, in January 1810. (For a detailed study of this competition, see M: Frankel [1974], 223–245.) The intention in setting this subject was clearly that Laplace's theoretical treatment of ordinary refraction, as given in the *Mécanique céleste* in 1805, should be extended to embrace double refraction as well; and to this extent the competition was a success.

Double refraction had never been satisfactorily explained either in the corpuscular theory or in Huygens' wave theory. Among the corpuscularians, Newton's brief analysis of the phenomenon in terms of the two "sides" of a ray of light (*Opticks*, 4th ed. [1730], Query 26) was still endorsed in textbooks but vaguely and without conviction; Haüy's *Traité élémentaire de physique* (2nd ed. [Paris, 1806], II, 334–355) is a good illustration of this. Huygens' explanation, as given in the *Traité de la lumière* (1690), not only had obvious weaknesses, particularly in its inability to explain the phenomena associated with crossed double-refracting crystals, but was also too closely allied to the wave theory to carry conviction at a time when—especially in France—the corpuscular theory was dominant.

It seems likely that the immediate stimulus for the competition on double refraction was the news of Wollaston's experimental confirmation of Huygens' construction for the ordinary and extraordinary rays. Huygens had used his wave theory and his notion of secondary wavelets to show that the wave surface of an extraordinary ray was an ellipsoid of revolution, while that of an ordinary ray was a sphere. He had then deduced the properties of the ellipsoid and had established laws governing the path of the extraordinary ray at different angles of incidence. Although Huygens confirmed these laws experimentally, his method, as described in the *Traité*, was obscure; hence the need for Wollaston's systematic confirmation ("On the Oblique Refraction of Iceland Crystal," in *Philosophical Transactions of the Royal Society* [1802], 381–386), described before the Royal Society in June 1802 but, because of the war, inaccessible in France until 1807.

Wollaston's paper appeared as an impressive endorsement of Huygens' construction and, by implication (for Wollaston did not endorse the wave theory), as a challenge to the corpuscularians. The challenge was one that Laplace could not resist, and his enthusiasm for a competition that was clearly intended to yield a corpuscularian counterpart to Huygens' wave theory of double refraction was only heightened by the availability of a candidate of impeccable credentials in Malus. The latter's analytical skills and commitment both to the corpuscular theory and to the doctrine of short-range forces were already apparent in the "Traité d'optique," which he read to the First Class of the Institute in April 1807 (published in *Mémoires présentés à l'Institut national . . . par divers savans . . . Sciences mathématiques et physiques*, **2** [1811], 214–302); hence the endorsement of these principles in his prize-winning paper was, we may assume, no more than the fulfillment of Laplace's expectations.

Laplace followed the course of the competition, and Malus's work in particular, very closely. A report on the paper of December 1808 in which Malus announced his discovery of polarization (*91*) reflects Laplace's admiration not only for the discovery but also for Malus's experimental confirmation of Huygens' law of extraordinary refraction (*OC*, XIV, 322). (When Laplace presented his report to the First Class of the Institute, on 19 December 1808, Malus's confirmation of Huygens' law was still unpublished. See M: Frankel [1974], 233.) At last it was established beyond doubt that any corpuscular theory of double refraction would have to be consistent with the law, and almost immediately Laplace showed how this might be achieved, in the "Mémoire sur le mouvement de la lumière dans les milieux diaphanes," which he read to the First Class of the Institute in January 1809 ([*1810a*]).

Laplace began by asserting, quite dogmatically, that Huygens' wave theory was inadequate for the explanation of double refraction and that the way ahead lay in devising a new explanation in terms of short-range molecular forces. His own, corpuscular theory rested on the principle of least action and on an arbitrary but plausible assumption concerning the relationship between the velocity of light inside a crystal (v), the velocity of light outside the crystal (c), and the angle (V) between the ray inside the crystal and the crystal's optic axis. Presumably by analogy with Snell's law for ordinary refraction, for which $v^2 = c^2 + a^2$, Laplace put, for the extraordinary ray, $v^2 = c^2 + a^2 \cos^2 V$. Assuming the truth of this equation, he derived expressions relating the direction of the extraordinary ray to the angle of incidence of the ray entering the crystal and the orientation of the optic axis.

Laplace's treatment was remarkably like Malus's. The expressions by which they both described the path of the extraordinary ray were similar in form and, by an adjustment of constants, could even be made identical. Moreover, in arriving at his expression, Malus, like Laplace, leaned heavily on the principle of least action, although he derived the dependence of v on the orientation of the extraordinary ray by assuming, as Laplace had not done, the truth of Huygens' law. The similarity between the two papers was such that the possibility of plagiarism cannot be ruled out. Malus felt that by the time Laplace wrote his paper of January 1809, he already knew the essentials of Malus's theory, which were almost certainly available by late 1808. And, whether or not it was intentional, Laplace's paper certainly had the effect of diminishing Malus's achievement in providing a corpuscular theory of double refraction.

Taken together, the papers of Laplace and Malus could be passed off as yet another triumph for corpuscular optics: now no one could doubt that Huygens' law was consistent with the doctrine of short-range forces and the materiality of light. Yet there were weaknesses. As Young observed, the ellipsoid of revolution, which represented a wave front in Huygens' theory, was reduced in Laplace's paper to a mathematical construct without physical significance (see Young's unsigned review of Laplace's memoir, in *Quarterly Review*, **2** [1809], 337–348, esp. 344); and it was by no means obvious that there really existed molecular forces with the special directional properties required in order to explain extraordinary refraction. But in the period of Laplacian domination of French physics, such objections were readily overlooked.

Laplace's involvement in the prize competition on the specific heats of gases, which was set in January 1811, was equally apparent though less direct. The winning entry, by Delaroche and Bérard ("Mémoire sur la détermination de la chaleur spécifique des différens gaz," in *Annales de chimie*, **85** [1813], 72–110, 113–182), made a decisive contribution to Laplacian physics, and we may be sure that Laplace's wishes were prominent both in the setting of the competition and in the adjudication. The first aim was the acquisition of reliable data in a notoriously uncertain branch of experimental physics. But a quite explicit subsidiary purpose was to decide whether it was possible for some caloric to exist in a body in a combined, or latent, state (that is, without being detected by a thermometer) or whether (as William Irvine, Adair Crawford, and John Dalton had supposed) all of the caloric was present in its "sensible," or free, state and therefore as a contribution to the body's temperature. In their *Mémoire sur la chaleur* ([*1783a*]) Lavoisier and Laplace had provided strong evidence against Irvine's theory, so that it was predictable enough that the winners, Delaroche and Bérard, should use their measurements of the specific heats of elementary and compound gases to endorse what was clearly the Laplacian view. As a result of experiments that they performed entirely at Arcueil, they firmly upheld the distinction between latent and sensible caloric.

So in the competitions on double refraction and the specific heats of gases Laplace was well served, but not all of his attempts to use the system of prize competitions were so successful. The competition on the distribution of heat in solids, which Fourier won in 1811, departed significantly from Laplace's approach to the problem (although, unlike Lagrange, he found much to admire in Fourier's paper); and the competition set in 1809 on the theory of elastic surfaces was for Laplace a failure. Before the closing date for this competition, Laplace added elastic surfaces to the list of phenomena that might be explained in terms of short-range molecular forces ([*1810a*], p. 329), but no entries of sufficient merit were received. The prize was eventually awarded in January 1816 to Sophie Germain, whose paper broke pointedly with the theory of elastic surfaces treated in the Laplacian manner, which Poisson had presented before the First Class of the Institute in 1814. (Germain's paper was published in an enlarged and modified form as *Recherches sur la théorie des surfaces élastiques* [Paris, 1821]. Poisson's paper appeared as "Mémoire sur les surfaces élastiques," in *MI*, **13**, pt. 2 [1812/1816], 167–225; see esp. 171–172 and 192–225.)

25. Theory of Error. The activity at Arcueil was at its height from 1805 through 1809, after which interval of preoccupation with problems of physics and younger physicists (see Sections 22, 23, 24), Laplace turned back to probability for the intensive effort that culminated in the production of the *Théorie analytique des probabilités* in 1812 and the companion *Essai philosophique sur les probabilités* in 1814. The prelude to these works consists in a pair of memoirs ([*1810b*] and [*1811a*]) and a supplement ([*1810c*]) to the earlier paper, together with a "Notice sur les probabilités," published anonymously in the *Annuaire publié par le Bureau des longitudes* ([*1810d*]) and since forgotten. Laplace presented the first of these papers before the Institute on 9 April 1810 (*95*). We have already seen how prior to that the analysis of probable error had assumed increasing importance for him in Book III of *Mécanique céleste*, particularly in the comparison of spheroidal attraction theory to geodesic data in chapter 5, and how the problem of correcting for instrumental error in physical observations had concerned him in Book X, mainly in relation to barometric and thermometric data (see Section 21).

The greatest novelty in the two analytical papers of 1810 and 1811 is the derivation from what is now called the central limit theorem of the least-square rule for determining the mean value in a series of observations. Legendre had published the rule in 1805 as a method for resolving inconsistencies between linear equations formed with astronomical data. His procedure was not a probabilistic one. In 1809 Gauss did derive the least-square law from an analysis of what is now called the normal distribution (L: Eisenhart [1964]; Plackett [1972]; Sheynin [1977]). It is sometimes said that this opportunity was what drew Laplace back to a preoccupation with the whole theory of probability in these, his advancing years, after the lapse of a quarter century since his initial immersion in its theory, definition, and application to population problems. That is uncertain, however, since it was only in the addendum ([*1810c*]; *OC*, XII, 353) to the earlier memoir that he first mentioned least squares. Moreover, he accompanied these memoirs with the popular "Notice sur les probabilités" (cf. Delambre, in *MI* [1811/1812],

"Histoire," i–ii) in which Laplace enlarged on the lecture he had given before the École Normale in 1795 ([*1800d*]). Most of the passages from this notice he then incorporated verbatim in the methodological and actuarial sections of the first edition of the *Essai philosophique* ([E]). In the *Annuaire* of the Bureau des Longitudes they serve to introduce a set of tables of mortality and feature a verbal statement of the central limit theorem (*op. cit.*, 110–111). Perhaps, therefore, it will be prudent to report his own account of the route that he had followed in finally arriving at that theorem.

The opening mathematical paper ([*1810b*]) in the pair under discussion has exactly the same title, "Mémoire sur les approximations des formules qui sont fonctions de très grands nombres," as did the important sequence twenty-five years before ([*1785a*] and [*1786b*]) that had moved probability from analysis of games of chance to population studies (see Section 12). In Laplace's own recollection, it was the lengthy repetition of events encountered in theory of probability that had initially brought home to him the inconvenience of evaluating formulas into which the numbers of these events had to be substituted in order to achieve a numerical solution. He had then attacked the difficulty by seeking a general method for accomplishing transformations of the type that Stirling had discovered for reducing the middle term of a binomial raised to a high power to a convergent series. The method he had given ([*1781a*] and [*1785a*]; see Section 12) transformed the integrals of linear differential or difference equations, whether partial or ordinary, into convergent series when large numbers were substituted in terms under the integral sign, the larger the numbers the more convergent the series. Among the formulas he could thus transform, the most notable was that for the finite difference of the power of a variable. In probability, the conditions of the problem often required restricting consideration to the positive values even though the variable decreases through zero into the negative range.

That was the case in the analysis of the probability that the mean inclination of any number of cometary orbits is contained within a given range. It now appears that Laplace had felt dissatisfied with everything he had so far tried on that problem. He had started it with his youthful analysis of the orbital inclinations of comets in order to determine whether the distribution bespeaks the same cause as the nearly coplanar arrangement of the planets ([*1776b*]; see Section 6). Apparently, he took up the question again soon afterward, for he now says that the problem can be resolved by a method that he had given in his first comprehensive memoir on probability ([*1781a*]; see Section 11), namely that the required probability could be expressed by the finite difference of the power of a uniformly decreasing variable in a formula where the exponent and the difference are the same as the number of orbits. Unfortunately, a numerical solution was unobtainable in practice, and it is reasonable to surmise that this was the reason that he failed to include the problem in the printed memoir ([*1781a*]). The obstacle stopped him for a long time, he acknowledges. Finally, and this would appear to be the background of the 1810 memoir, he had resolved the difficulty by approaching the problem from another point of view. Assuming in general that the prior probabilities (he now says *facilités*) of inclination serve any law whatever, he succeeded in expressing the required probability in a convergent series. For he had come to see the problem as identical with those in which the probability is required that the mean error in a large number of observations falls within certain limits, which question he had discussed at the end of "Mémoire sur les probabilités" ([*1781a*], Article XIII, 30–33]), although without numerical examples. (He does not say so here, but the point of view was also akin to that taken in his calculation that the error in estimating the population of France from a given sample would be within certain limits [*1786c*]; see Section 13.) He could then show that if the observations are repeated an indefinite number of times, their mean result converges on a limit such that, if an equal interval on either side be made as small as one pleases, the probability that the result will be contained therein can be brought so close to certainty that the difference is less than any assignable magnitude. If positive and negative errors are equipossible, this mean term is indistinguishable from the truth. Because the methods he had so far discussed were indirect, Laplace considered it preferable to find a direct approach to evaluating the finite differences of the higher powers of the variable; and he proceeded to apply to error theory a technique that he had published the previous year in *Journal de l'École polytechnique* ([*1809e*]; *OC*, XIV, 193). The memoir contains refinements of his earlier work on generating functions and on the use of definite integrals for solving certain classes of linear partial differential and difference equations that could not be integrated in finite terms ([*1782a*]). He applied to the present purpose an analysis involving the reciprocity of real and imaginary results, which he had introduced in the first memoir on approximate solutions to formulas containing very large numbers ([*1785a*]; cf. Section 29).

Thus, it was by way of analyzing the distribution of cometary orbits that Laplace came to the central limit theorem, having perfected methods for evaluating the mathematical expressions developed years before. The opening articles argue the old proposition that neither the mean inclination of the cometary orbits (by now ninety-seven were known, all of which he could now include in the computation), nor the proportion of direct to retrograde motions, can be supposed to result from the same cause as the arrangement of the planetary system. Article VI changes the problem of mean inclination into the problem that the mean error of a number n of observations will be contained within the limits $\frac{\pm rh}{\sqrt{n}}$, where r is the sum of the errors. At first, Laplace assumes the equipossibility of error in the interval h, but he goes on to the general case in which the error distribution follows any law, and obtains for the required probability the formula (*OC*, XII, 325):

$$\frac{2}{\sqrt{\pi}}\sqrt{\frac{k}{2k'}}\int e^{-\frac{k}{2k'}r^2}\,dr. \qquad (126)$$

When $\phi\left(\frac{x}{h}\right)$ is the probability of the error $\pm x$, k is

$$\int_{-h/2}^{h/2}\phi\frac{x}{h}\,dx,$$

and k' is

$$\int_{-h/2}^{h/2}\frac{x^2}{h^2}\phi\left(\frac{x}{h}\right)dx.$$

Later in 1809 (we do not know precisely when), Laplace composed a brief supplement ([*1810c*]) in which he returned to the choice of the mean in a series of observations, which task had first attracted him to error theory in the memoir on the probability of cause ([*1774c*], Article V; see Section 4). He invokes the procedure that he had imagined crudely there and more abstractly in the closing passages of the "Mémoire sur les probabilités" ([*1781a*], Articles XXX–XXXIII). A curve of probability might be constructed, for which the abscissa defines the "true" instant of the observation—presumably astronomical—and the ordinate is proportional to the probability that the value is correct. The problem is to find the point on the x-axis at which the departure *à craindre* from the truth is a minimum. Now then, just as in the theory of probability, the loss "to be feared" is multiplied by its probability and the product summed, so in error theory the amount of each error, regardless of sign, is to be multiplied by its probability and the product summed. Supposing that n observations of one sort, with equal possibility of error, result in a mean value of A; that n' of another sort, following a different law of error, result in a mean of $A + q$; that n'' of yet another sort and another distribution result in a mean of $A + q'$, and so on; and that $A + X$ is the mean to be preferred among all these results. From Equation 126 Laplace then derives the proposition that the required value of X will be that for which the function (*OC*, XII, 353)

$$(pX)^2+[p'(q-X)]^2+[p''(q-X)]^2+\cdots \qquad (127)$$

is a minimum, $p, p', p'', \cdots$, representing the greatest probabilities of the results given by the observations $n, n', n'', \cdots$. That expression gives the sum of the squares of each result multiplied respectively by the greatest ordinate in its curve of error. It is clear from Laplace's comment that the novelty was not in the least-square property itself. He considered its status merely hypothetical, however, when the mean depended on a few observations or on an average among a number of single observations. It became valid generally only when each of the results among which it gave the mean itself depended on a very large number of observations. Its basis had to be statistical (a word that he did not employ), and only then could it be derived from the theory of probability and employed whatever the distributions of error in instruments or observations.

Daniel Bernoulli, Euler, and Gauss are mentioned in this note, albeit rather vaguely, but not Legendre. Hard feelings about priorities in the matter of least squares had meanwhile arisen between Legendre and Gauss. Delambre, now a permanent secretary of the scientific division of the Institute, tried to make peace (*MI*, **12** [1811/1812], "Histoire," i–xiii). His contemporary account reaches the same conclusion that Laplace himself arrived at when reviewing the origin of least squares in *Théorie analytique des probabilités* (*OC*, VII, 353). Gauss had indeed had the idea first and made use of it in private calculation, but Legendre had come upon it independently and published it first. Delambre also reports Gauss's attribution of inspiration to a theorem that he had found in Laplace, namely that the value of the integral $\int e^{-t^2}\,dt$ taken from $-\infty$ to $+\infty$ is $\sqrt{\pi}$ (cf. Section 12, Equation 62). In fact, so Legendre informed Gauss in their exchange of reproach, the theorem belonged to Euler

(M: Plackett [1972], 250). Otherwise, Laplace escaped unscathed on the fringe of this dispute, never having claimed the least-square rule itself but only the generality that it could assume in virtue of his derivation of it from the probability of cause. His initial procedure would now be called Bayesian.

Laplace followed a different procedure in the second of these papers, "Mémoire sur les intégrales définies . . ." ([*1811a*]). He opened it also with a historical resume, recalling (what he had never claimed at the time) how the companion discoveries of generating functions ([*1782a*]) and approximations for formulas containing very large numbers ([*1785a*]; see Section 12) were really complementary aspects of a single calculus. He reminded his readers that the object of the former was the relation between some function of an indeterminate variable and the coefficients of its powers when the function is expanded in a series, and that generating functions had initially proved most valuable in solving difference equations in which problems of the theory of chance were formulated. The object of the latter, on the other hand, was to express variables that occur in difference equations in the form of definite integrals to be evaluated by rapidly convergent approximations. It turns out, however, that the quantity under the integral sign in such cases is nothing other than the generating function of the variable expressed by the definite integral. That is how the two theories merged in a single approach that he would henceforth call the calculus of generating functions, and that he also thought of as the exponential calculus of differential operators (*caractéristiques*) (*ibid.; OC*, XII, 360). It is valid both in infinitesimal and finite analysis. When a difference equation is expanded in powers of the difference taken as indeterminate but infinitesimal, and higher-order infinitesimals are held to be negligible relative to those of some lower order, the result is a differential equation. But the integral of that equation is also the integral of the difference equation wherein the infinitesimal quantities are similarly held to be negligible relative to the finite quantities. Justifying the neglect of such terms led Laplace into one of his few discussions of the foundations of the calculus and defenses of its rigor.

When first presenting the approximations for formulas containing large numbers ([*1785a*]), Laplace had shown how to evaluate several classes of the definite integrals that he employed in terms of transcendent quantities. The method depended on the passage from the real to the imaginary and resulted in series of sines and cosines. These were special cases, however, and only now was he in a position to give a direct and general method for evaluating any such expressions. That topic occupies the opening article, after which Laplace illustrated the method in three representative problems of the theory of probability. The first concerns a case of duration of play in theory of games of chance. The second is an urn problem. Two vessels contain n balls each. Of the total $2n$, half are black and half white. Each draw consists of taking a ball from each urn and placing it in the other. What is the probability that after r draws there will be x white balls in urn A? Mathematically, Laplace considered this the most interesting of his illustrations of the newly named calculus of generating functions, for his solution involved—so he claimed—the first application of partial differential equations to infinitesimal analysis in the theory of probability (*ibid.; OC*, XII, 361–362).

Nevertheless, he reserved the fullest treatment for the third problem and republished the articles containing its resolution in *Connaissance des temps* ([*1811b*]). He thus emphasized the method because of its widespread utility in the theory of chance, its object being the mean to choose among the results given by different sets of observations. For a modern mathematician's account of this second, strongly featured presentation of the method of least squares, see Sheynin's excellent article (L: [1977]), nos. 5.1–5.2). The following is a paraphrase of Laplace's own précis, which he intended for readers unversed in probability ([*1811a*]; *OC*, XII, 362). The idea was to employ the totality of a very large number of observations to correct several elements that are already approximately known. Each observation is a function of these elements, and their approximate value could be substituted in that function. Those values were to be modified by small corrections, which constituted the unknowns. The function was then expanded in a series of powers of those corrections. Squares, products, and higher powers are neglected, and the series is equated to the observed value. Those steps gave an initial equation of condition between the corrections to be applied to the elements. A second equation could be found from a second observation, a third from a third, and so on. If the observations had each been precise, only one equation apiece would have been needed. But since they were subject to error, the effect of which was to be minimized, a very great number of them had to be taken in order that the errors might compensate each other in the values deduced from the total number. There was the core of the problem. How were the equations to be combined?

Here was the point at which probability entered the procedure. Any mode of combination would consist in multiplying each equation by a particular factor and summing all the products. The respective systems of factors employed would yield systems of definitive equations between the corrections of the elements, and as many equations would be needed as there were elements to correct in solving them. It was obvious that the crux consisted in choosing factors such that the probability of error, positive or negative, should be a minimum for each element. Again defining mean error as the product of the amount of each error multiplied by its probability, Laplace then would show that in each equation of condition, when the respective coefficients are varied so that it can be set equal to zero, the sum of the squares is a minimum. As introduced by Legendre and Gauss, the method was limited (in Laplace's view) to finding the definitive equations needed for a solution. In his derivation, it also served to determine the corrections. Thus, once again, this time employing what is now called a linear regression rather than a Bayesian approach, Laplace claimed that a derivation of the method of least squares from theory of probability promoted it from the status of a rule of thumb to that of a mathematical law.

26. Probability: Théorie analytique and Essai philosophique. In the preamble to the "Mémoire sur les intégrales définies" just discussed, Laplace wrote: "The calculus of generating functions is the foundation of a theory that I propose to publish soon on probability" ([*1811a*]; *OC*, XII, 360). Good as his word, he presented the first part of *Théorie analytique des probabilités* to the Institute on 23 March 1812 and the second part on 29 June (*98, 99*). There is a minor bibliographical puzzle here. The complete treatise is a quarto volume of 464 pages in the first edition and is divided into two books, Book I consisting of a "Première partie" and a "Seconde partie." The significance of the partition will be clarified in a moment. What is unclear is whether Laplace had Book I, Part I printed first, and then Book I, Part II together with all of Book II; or whether, after the three-month interval, it was Book II that he saw through the press. The latter conjecture seems more logical, although the *procès-verbaux* of the Institute are confirmed by the "Avertissement" to the second edition, presented on 14 November 1814 (*104*). In either case, the general scheme is similar to that of *Mécanique céleste*. Book I is devoted to the mathematical methods. In Book II, occupying two-thirds of the volume, they are applied to the solution of problems in probability. There is some difference in the relation of the organization to the sequence of events in Laplace's career, however. In the field of probability, his resolution of a rather larger proportion of the problems than in celestial mechanics had preceded his development of the mathematical techniques incorporated in the finished treatise. That was notably the case in the areas of games of chance and probability of cause and, to a degree, in demography. The areas of application that he explored later tended to be in the realm of error theory, decision theory, judicial probability, and credibility of witnesses.

The first edition contains a brief introduction (pp. 1–3) that was eliminated in the second and third (1820) in favor of the *Essai philosophique*. It is worth notice, nevertheless, for the interest that Laplace claimed for the work in bringing it before the public:

> I am particularly concerned to determine the probability of causes and results, as exhibited in events that occur in large numbers, and to investigate the laws according to which that probability approaches a limit in proportion to the repetition of events. That investigation deserves the attention of mathematicians because of the analysis required. It is primarily there that the approximation of formulas that are functions of large numbers has its most important applications. The investigation will benefit observers in identifying the mean to be chosen among the results of their observations and the probability of the errors still to be apprehended. Lastly, the investigation is one that deserves the attention of philosophers in showing how in the final analysis there is a regularity underlying the very things that seem to us to pertain entirely to chance, and in unveiling the hidden but constant causes on which that regularity depends. It is on the regularity of the mean outcomes of events taken in large numbers that various institutions depend, such as annuities, tontines, and insurance policies. Questions about those subjects, as well as about inoculation with vaccine and decisions of electoral assemblies, present no further difficulty in the light of my theory. I limit myself here to resolving the most general of them, but the importance of these concerns in civil life, the moral considerations that complicate them, and the voluminous data that they presuppose require a separate work.

Laplace never wrote that separate work, although the thought of it may well have been what led him to expand his old lecture for the École Normale into the *Essai philosophique* two years later.

In conformity with the program announced the

preceding year ([*1811a*]), the general subtitle of Book I is "Calcul des fonctions génératrices." It consists almost entirely of a republication, with some revision, of the two cardinal mathematical investigations of the early 1780's. The "Mémoire sur les suites" ([*1782a*]), on generating functions themselves, has now become the basis of its first part, and that on the approximation by definite integrals of formulas containing very large numbers ([*1785a*] and [*1786b*]), the basis of its second part. (For a more detailed and mathematical summary of the latter than is given in our Section 12, see L: Todhunter [1865], nos. 956–968.) The introduction reiterates what Laplace had first stated in the memoir of the preceding year ([*1811a*]), that the two theories are branches of a single calculus, the one concerned with solving the difference equations in which problems of chance events are formulated, the other with evaluating the expressions that result when events are repeated many times.

Laplace says in the introduction that he is now presenting these theories in a more general manner than he had done thirty years before. The chief difference in principle is that the new calculus is held to have emerged along the main line of evolution of the analytical treatment of exponential quantities. In an opening historical chapter, Laplace traces the lineage back through the work of Lagrange, Leibniz, Newton, and Wallis to Descartes's invention of numerical indices for denoting the operations of squaring, cubing, and raising magnitudes to higher integral powers. The principal difference in practice between the two earlier memoirs and their revision in Book I is that Laplace omitted certain passages that had come to appear extraneous in the interval and gave greater prominence to others that now appeared strategic. The most important omission is three articles from the "Mémoire sur les suites" on the solution of second-order partial differential equations, which were important for problems of physics but not for theory of games of chance ([*1782a*], Articles XVIII–XX; *OC*, X, 54–70). On the other hand, Laplace gave greater emphasis than in the earlier memoirs to the passage from the finite to the infinitesimal and also from real to imaginary quantity. He now develops as an argument what he had merely asserted in the immediately preceding memoir on definite integrals ([*1811a*]), namely that rigor is not impaired by the necessity of neglecting in appropriate circumstances infinitesimal quantities relative to finite quantities and higher-order infinitesimals relative to those of lower order. It is in support of this proposition that he brings in the solution to the problem of vibrating strings from [*1782a*]. It affords a convincing example that discontinuous solutions of partial differential equations are possible under conditions that he specifies (*OC*, VII, 70–80). He attached even greater importance to the fertility that he increasingly found in the process of passing from real to imaginary quantity and discussed those procedures in the transitional section between the first and second parts. The limits of the definite integrals to be converted into convergent series are given by the roots of an equation such that when the signs of the coefficients are changed, the roots become imaginary. But it was precisely this property that led Laplace to the values of certain definite integrals that occur frequently in probability and that depend on the two transcendental quantities π and e. His early methods had been *ad hoc* and indirect, but since that time he had perfected direct methods for evaluating such integrals in a general manner, a procedure that (he acknowledged) Euler had arrived at independently (*OC*, VII, 88).

In Book II, subtitled "Théorie générale des probabilités," Laplace turns from the calculus to probability itself. Indeed, it is fair to say that he constituted the subject, drawing together the main types of problems from the theory of chance already treated by many mathematicians, including himself, in a somewhat haphazard manner, and rehandling them in tandem with problems from the new areas of application in philosophy of science, astronomy, geodesy, instrumentation, error, population, and the procedures of judicial panels and electoral bodies. Unlike the two parts of Book I and much of *Mécanique céleste*, Book II is more than a republication of earlier memoirs with minor and incidental revision. Material from earlier work is incorporated in it, to be sure, but it is revised mathematically and fortified with new material. What is carried over without significant change from the earliest memoirs ([*1774c*] and [*1776a*]) is the point of view from which the subject as a whole is treated and the spirit in which the various topics are approached. It has been said that *Théorie analytique des probabilités* is unsystematic, rather a collection of chapters that might as well be separate than a treatise in the usual sense. Perhaps so. Its organization certainly recapitulates the evolution of the subject matter rather than some logical system within it. It is also difficult to imagine its serving either as a textbook, as the first two volumes of *Mécanique céleste* could do, or as a work of reference, in the way that the third and fourth volumes really did do. Its relation to the subject was different. Rather than drawing together the lifework of a leading contributor to a vast and classical area of science, it was the first full-scale study completely devoted to a new specialty, building out from old and often hackneyed problems into areas where quantification had been nonexistent or chimerical. Later commentators have also sometimes castigated the obscurity and lack of rigor in many passages of the analysis. Once again, it may be so. It is constitutionally and temperamentally very difficult, however, for many mathematicians to enter sympathetically into what was once the forefront of research. Important parts of *Mécanique céleste* were also in the front lines, of course—but that was the location of *Théorie analytique des probabilités* as a whole. What no one has denied is that it was a seminal if not a fully systematic work.

The first chapter gives the general principles and opens with the famous characterization of probability as a branch of knowledge both required by the limitation of the human intelligence and serving to repair its deficiencies in part. The subject is relative, therefore, both to our knowledge and to our ignorance of the laws of a determined universe. After stating the definition of probability itself and the rule for multiplying the probabilities of independent events, Laplace includes as the third basic principle a verbal statement of his theorem on the probability of cause, still without mentioning Bayes. Thereupon, he takes the example of the unsuspected asymmetries of a coin to consider the effect of unequal prior probabilities mistakenly taken for equal. Finally, he distinguishes between mathematical and moral expectation. The basic content of these matters was drawn from the companion memoirs ([*1774c*] and [*1776a*, 1°]) composed thirty-nine years previously in 1773 (see Section 4).

The actual problems discussed in the early chapters also consist in part of examples reworked from these and the other early papers on theory of chance. (For a useful mathematical summary of many of them in modern terminology, see L: Sheynin [1976].) Laplace solved them by using generating functions and arranged them, not for their own sake, but to illustrate the typology of problems in probability at large, interspersing new subject matter where the methodology made it appropriate. Chapter 2, which is concerned with the probability of compound events composed of simple events of known probability, is much the most considerable, occupying about one-quarter of Book II. In a discussion of the old problem of determining the probability that all n numbers in a lottery will turn up at least once in i draws when r slips are chosen on each draw, he adduced the case of the French national lottery, composed of ninety numbers drawn five at a time. Laplace went on to other classic problems in direct probability: of odds and evens in extracting balls from an urn, of extracting given numbers of balls of a particular color from mixtures in several urns, of order and sequence in the retrieval of numbered balls, of the division of stakes and of the ruin or victory of one of a pair of gamblers in standard games.

Perhaps it will be useful to trace the sequence in one set of problems as an illustration of how Laplace made the connections between topics. He imagines (no. 13) an urn containing $n+1$ balls numbered 0, 1, 2, 3, $\cdots$ n. A ball is taken out and returned, and the number noted. What is the probability that after i draws, the sum of the numbers will be s? If t_1, t_2, t_3, $\cdots$ t_i are the numbers of balls taken in the first, second, third, etc. draws, then as long as t_2, t_3, $\cdots$ t_i do not vary,

$$t_1+t_2+t_3+\cdots+t_i=s. \qquad (128)$$

Only that one combination is possible. But if different numbers are taken, so that t_1 and t_2 are varied simultaneously and are capable of taking any value beginning at zero and continuing indefinitely, then (128) will be given by the following number of combinations:

$$s+1-t_3-t_4\cdots-t_i \qquad (129)$$

since the limits of t_1 are zero, which would give

$$t_2=s-t_3-t_4\cdots-t_i,$$

and $s-t_3-t_4\cdots-t_i$, which would give $t_2=0$. Negative values are excluded. By like reasoning, Laplace finds that the total number of combinations that can give Equation 128 on the supposition of the indefinite variability of t_1, t_2, t_3, $\cdots$ t_i, always greater than zero, is

$$\frac{(s+i-1)(s+i-2)(s+i-3)\cdots(s+1)}{1\cdot2\cdot3\cdots(i-1)}. \qquad (130)$$

By the conditions of the problem, however, these variables cannot exceed n, and the probability of any particular value of t_1 from zero to n is $\frac{1}{n+1}$. Since the probability of t_1 equal to or greater than $(n+1)$ is nil, it may be represented by the expression $\frac{1-l^{n+1}}{n+1}$ provided that l = unity. Now then, on condition that l be introduced only when t_i has reached the limit $n+1$, and that it be equal to unity at the end of the operation, any value of t_1 can be represented by $\frac{1-l^{n+1}}{n+1}$. The same is true for the other variables. Since the probability of (128) is the product of the probabilities of the values $t_1, t_2, t_3, \cdots$, its expression is $\left(\frac{1-l^{n+1}}{n+1}\right)^i$. The number of combinations given by that equation multiplied by their respective probabilities is then

$$\frac{(s+1)(s+2)\cdots(s+i-1)}{1\cdot 2\cdot 3\cdots(i-1)}\left(\frac{1-l^{n+1}}{n+1}\right)^i. \qquad (131)$$

In expanding that function, l^{n+1} is to be applied only to combinations in which one variable is beginning to exceed n; l^{n+2} only to combinations in which two of the variables begin to exceed n, and so on. Thus, if it is supposed that t_1 has grown larger than n, then by setting $t_1 = n+1+t'_1$, Equation 128 becomes

$$s-n-1=t'_1+t_2+t_3+\cdots, \qquad (132)$$

where t'_1 increases indefinitely. If two variables, t_1 and t_2, exceed n, then setting $t_1=n+1+t'_1$, and $t_2=n+1+t'_2$, Equation 128 becomes

$$s-2n-2=t'_1+t'_2+t_3+\cdots. \qquad (133)$$

The purpose of this manipulation is to decrease s in the function (130) by $n+1$, relative to the system of variables $t'_1, t_2, t_3 \cdots$, to decrease s by $2n+2$, relative to the system $t'_1, t'_2, t_3, \cdots$, and so on. In expanding (131) in powers of l, s is to be decreased by the exponent indicating the power of l, and when $l=1$, the function (131) becomes

$$\frac{(s+1)\ (s+2)\cdots(s+i-1)}{1\cdot 2\cdot 3\cdots(i-1)\ (n+1)^i}$$
$$-\frac{i(s-n)\ (s-n+1)\cdots(s+i-n-2)}{1\cdot 2\cdot 3\cdots(i-1)\ (n+i)^i}+\frac{i(i-1)}{1\cdot 2}$$
$$\times\frac{(s-2n-1)\ (s-2n)\cdots(s+i-2n-3)}{1\cdot 2\cdot 3\cdots(i-1)\ (n+1)^i}-\cdots, \qquad (134)$$

which series is continued until one factor, $(s-n)$, $(s-2n-1)$, $(s-3n-2), \cdots$, becomes zero or negative in value.

The formula (134)—to change the problem now—will give the probability of throwing any number s in tossing i dice each with $n+1$ sides, the smallest number on any side being 1. If s and n are infinite numbers, (134) becomes the following expression:

$$\frac{1}{1\cdot 2\cdot 3\cdots(i-1)\,n}\left\{\left(\frac{s}{n}\right)^{i-1}-i\left(\frac{s}{n}-1\right)^{i-1}+\frac{i\,(i-1)}{1\cdot 2}\left(\frac{s}{n}-2\right)^{i-1}\cdots\right\} \qquad (135)$$

This expression, proceeds Laplace—affording his reader not so much as a new paragraph to draw breath—can be employed to determine the probability that the sum of the inclinations of orbits to the ecliptic will be contained within given limits on the assumption of equipossibility of inclination between 0° and a right angle. If a right angle, $\frac{1}{2}\pi$, is divided into an infinite number n of equal parts, and s contains an infinite number of these parts, then if ϕ is the sum of the inclinations of the orbits,

$$\frac{s}{n}=\frac{\phi}{\frac{1}{2}\pi}. \qquad (136)$$

Multiplying (136) by ds, or $\frac{n d\phi}{\frac{1}{2}\pi}$, and integrating from $\phi-\epsilon$ to $\phi+\epsilon$, gives

$$\frac{1}{1\cdot 2\cdot 3\cdots i}\left\{\begin{aligned}&\left(\frac{\phi+\epsilon}{\frac{1}{2}\pi}\right)^i-i\left(\frac{\phi+\epsilon}{\frac{1}{2}\pi}-1\right)^i\\&\qquad+\frac{i(i-1)}{1\cdot 2}\left(\frac{\phi+\epsilon}{\frac{1}{2}\pi}-2\right)^i-\cdots\\&-\left(\frac{\phi-\epsilon}{\frac{1}{2}\pi}\right)^i+i\left(\frac{\phi-\epsilon}{\frac{1}{2}\pi}-1\right)^i\\&\qquad-\frac{i(i-1)}{1\cdot 2}\left(\frac{\phi-\epsilon}{\frac{1}{2}\pi}-2\right)^i+\cdots\end{aligned}\right\}. \qquad (137)$$

Formula 137 expresses the probability that the sum of the inclinations of the orbits is contained within the limits $\phi-\epsilon$ and $\phi+\epsilon$.

We shall not follow Laplace into yet another calculation (his last on this phenomenon) that the orbital arrangement of the planets results from a single cause and that the comets escape its compass, nor from that back to a variation on the original problem, in which any number of balls in the urn may be designated by the same integer, nor even into his derivation by the same method that the sum of the errors in a series of observations will be contained within given limits. Suffice it to indicate the sequence and the virtuosity it bespeaks. But we must notice his earliest venture, this late in life, into judicial probability. Imagine a number i of points along a straight line, at each of which an ordinate is erected. The first ordinate must be at least equal to the second, the second at least equal to the third, and so on; and the sum of these i ordinates is s. The problem is to determine, among all the values that each ordinate can assume, the mean value. For that quantity in the case of the rth ordinate, Laplace obtains the expression (*OC*, VII, 276)

$$\frac{s}{i}\left(\frac{1}{i}+\frac{1}{i+1}+\cdots+\frac{1}{r}\right). \tag{138}$$

Suppose now, however, that an event is produced by one of the i causes A, B, C, · · ·, and that a panel of judges is to reach a verdict on which of the causes was responsible. Each member of the panel might write on a ballot the various letters in the order that appeared most probable to him. Formula 138 will now give the mean value of the probability that he assigns to the rth cause (in this application s must amount to certainty and have the value of 1). If all members of the tribunal follow that procedure, and the values for each cause are summed, the largest sum will point to the most probable cause in the view of that panel.

Laplace hastened to add that since electors, unlike judges, are not constrained to decide for or against a candidate but impute to him all degrees of merit in making their choices, the above procedure may not be applied to elections. He proceeded to outline a probabilistic scheme for a preferential ballot that would produce the most mathematically exact expression of electoral will. Unfortunately, however, electors would not in fact make their choices on the basis of merit but would rank lowest the candidate who presented the greatest threat to their own man. In practice, therefore, preferential ballots favor mediocrity and had been abandoned wherever tried.

The third chapter deals with limits, in the sense in which the idea figures in the frequency definitions of the discipline of probability that have developed out of it (L: Molina [1930], 386). In his own terminology, his concern was with the laws of probability that result from the indefinite multiplication of events. No single passage is as clear and definite as his derivation of the central limit theorem in the memoir on approximating the values of formulas containing large numbers ([*1810b*]) that had brought him back to probability several years before, but the examples he adduces are much more various. He begins with a conventional binomial problem. The probabilities of two events a and b are respectively p and $1-p$. The probability that a will occur x times and b will occur x' times in $x+x'$ tries is given by the $(x'+1)$th term of the binomial $[p+(1-p)]^{x+x'}$. Laplace calculates the sum of two terms that are symmetrical on either side of the middle term of the expansion of the binomial. The formula is $\frac{2}{\sqrt{\pi}}\frac{\sqrt{n}}{\sqrt{2\pi xx'}}e^{\frac{-nl^2}{2xx'}}$, where $n=x+x'$.

When $t=\frac{l\sqrt{n}}{2xx'}$, the sum of all such pairs is (*OC*, VII, 283–284)

$$\frac{2}{\sqrt{\pi}}e^{-t^2}\,dt+\frac{\sqrt{n}}{\sqrt{2\pi xx'}}e^{-t^2}. \tag{139}$$

Discussing this formula and the reasoning, Laplace points out that two sorts of approximations are involved. The first is relative to the limits of the *a priori* probability (*facilité*) of the event a, and the second to the probability that the ratio of the occurrences of a to the total number of events will be contained within certain limits. As the events are repeated, the latter probability increases so long as the limits remain the same. On the other hand, so long as the probability remains the same, the limits grow closer together. When the number of events reaches infinity, the limits converge in a point and the probability becomes a certainty. Just as he had done in his earliest general memoir on probability ([*1781a*]), Laplace turned to birth records to illustrate how the ratio of boys to girls gives figures from experience for prior probabilities, or *facilités*.

The latter part of the discussion contains another of the many passages scattered throughout his writings that have led modern readers to feel that Laplace must somehow have had an inkling (or perhaps a repressed belief) that random processes occur in nature itself and not merely as a function of our ignorance. The mathematical occasion here is the use, started in ([*1811a*]), of partial differential equations in solving certain limit problems (see Section 25). The concluding example turns on a ring of urns, one containing only white and another only black balls, and the rest mixtures of very different proportions. Laplace proves that if a ball is drawn from any urn and placed in its neighbor, and if that urn is well shaken and a ball drawn from it and placed in the next further on, and so on an indefinite number of times around the circle, the ratio of white to black balls in each urn will eventually be the same as the ratio of white to black balls in all of them. But what Laplace really thought to show by such examples was the tendency of constant forces in nature to bring order into the most chaotic systems.

By comparison to the early probabilistic memoirs of the 1770's and 1780's, the fourth chapter on probability of error certainly represents the most significant development in the subject as a whole. It contains, of course, a derivation (no. 20) of the least-square law for taking the mean in a series of observations, which is given by essentially the same method as in [*1811a*], although in a more detailed and abstract form. There is much more to the discussion of error theory than that, however (see L: Sheynin [1977], no. 6, 25–34, who gives Laplace's formulations in modern notation). The chapter opens (no. 18) with determinations that the sum of errors of a large number of errors—equivalent to the distribution of sums of random variables—will be contained within given limits, on the assumption of a known and equipossible law of errors. It continues (no. 19) with the probability that the sum of the errors (again amounting to random variables), all considered as positive, and of their squares and cubes, will be contained within given limits. This is equivalent to considering the distribution of the sum of moduli. That leads to the problem of correcting values known approximately by the results of a great number of observations, which is to say by least squares, first in the case of a single element (no. 20) and then of two or more elements (no. 21). Mathematically, this involves a discussion of linearized equations with one unknown and with two unknowns, respectively. Laplace includes instructions on application of the analysis to the correction of astronomical data by comparison of the values given in a number of tables. He then considers the case in which the probability of positive and negative error is unequal, and derives the distribution that results (no. 22). The next-to-last section (no. 23) deals with the statistical prediction of error and methods of allowing for it on the basis of experience. At least, that seems to be a fair statement of what Laplace had in mind in speaking of "the mean result of observations large in number and not yet made," on the basis of the mean determined for past observations of which the respective departures from the mean are known (*OC*, VII, 338). The chapter closes with a historical sketch of the methods used by astronomers to minimize error up to the formulation of least squares, in which account Laplace renders Legendre and Gauss each his due (see Section 25).

In the fifth chapter, Laplace discussed the application of probability to the investigation of phenomena themselves and of their causes, wherein it might serve to establish the physical significance of data amid all the complexities of the world. The approach offers practical instances of his sense of the relativity of the subject to knowledge and to ignorance, to science and to nature. In the analysis of error, it is the phenomena that are considered certain, whereas here the existence and boundaries of the phenomena themselves are the object of the calculation. The main example is the daily variation of the barometer, which long and frequent observation shows to be normally at its highest at 9:00 A.M. and lowest at 4:00 P.M., after which it rises to a lower peak at 11:00 P.M. and sinks until 4:00 A.M. Laplace calculated the probability that this diurnal pattern is due to some regular cause, namely the action of the sun, and determined its mean extent. He then raised a further question which, for lack of data, he could not resolve mathematically here, but which is interesting since it came to occupy the very last calculation of his life (see Section 28). For in theory, atmospheric tides would constitute a second and independent cause contributing to the daily variations of barometric pressure. He referred his readers to his discussion of that hypothetical phenomenon, a corollary to the treatment of oceanic tides, in *Mécanique céleste*, Book IV (*OC*, II, 310–314). The detection of such a small effect was not yet possible, although Laplace expressed his confidence that observations would one day become sufficiently extensive and precise to permit its isolation.

In short, it was calculations of this sort that Laplace had in mind when he claimed, as he here remarked again, that on the cosmic scale probability had permitted him to identify the great inequalities of Jupiter and Saturn, even as it had enabled him to detect the minuscule deviation from the vertical of a body falling toward a rotating earth. He even had hopes for its calculus in physiological investigations, imagining that application to a large number of observations might suffice to determine whether electrical or magnetic charges have detectable effects upon the nervous system, and whether animal magnetism reflects reality or suggestibility. In general—he felt confident—the same analysis could in principle be applied to medical and economic questions, and even to problems of morality, for the operations of causes many times repeated are as regular in those domains as in physics. Laplace had no examples to propose, however, and closed the chapter with a mathematical problem extraneous in subject matter but not in methodology. The problem had been imagined by Buffon in order to show the applicability of geometry to probability (L: Sheynin [1976], 152). It consists of tossing a needle onto a grid of parallel lines, and then onto a grid ruled in rectangles, of which the optimal dimensions relative to the length of the needle constitute the problem. Laplace adapted it to a probabilistic, or in this instance a statistical, method for approximating to the value of π. It would be possible, he points out, although not mathematically inviting, to apply a similar approach to the rectification of curves and squaring of surfaces in general.

Chapter 6, "On the Probability of Causes and Future Events, Drawn From Observed Events," is in effect concerned with problems of statistical inference. In practice, the material represents a reworking of his early memoir on probability of cause ([*1774c*]) and of the application of inverse probability by means of approximations of definite integrals to calculations involving births of boys and girls and also to population problems at large ([*1781a*], [*1786b*], and [*1786c*]). He now had figures for Naples as well as for Paris and London. In calculating the probable error in estimates of the population of France based on the available samples, he made use of the partial census which, at his request, the government had instituted in 1801.

The next, very brief, chapter also starts with old material, to which he gave a new turn. He recurs to his own discovery ([*1774c*]) of the effect of inequalities in the prior probabilities that are mistakenly supposed to be equal (see Section 4). He always attached great importance to that finding, so much so that he alluded to it in the opening chapter of Book II, where definitions were laid down. The significance was that it brought out the care that needed to be taken when mathematical calculations of probability were applied to physical events. In effect, there are no perfect symmetries in the real world, and what Laplace was saying was that allowance has to be made for slight deviations of parameters from assumed values in making predictions. He discussed the problem in the same connection in which he had started it, in relation to the unfairness to one of two players of unsuspected asymmetries in a coin to be tossed. That could be mitigated, he now suggested, by submitting the chance of asymmetry itself to calculation. He let the probability of throwing heads or tails be $\frac{1 \pm \alpha}{2}$, where α represents the unknown difference between the prior probabilities of throwing one or the other. The probability of throwing heads n times in a row will then be (*OC*, VII, 410–411)

$$\frac{(1+\alpha)^n+(1-\alpha)^n}{2^{n+1}}; \qquad (140)$$

and a player who bets on heads or tails consecutively will have an advantage over one who bets on an alternation. Instead of that, it will be fairer to try tossing two similar coins simultaneously n times running. Then the true value of the probability that the two coins will fall the same way is

$$\frac{1}{2^{n+1}} \; [(1+\alpha\alpha')^n+(1-\alpha\alpha')^n], \qquad (141)$$

which is closer to the equipossible $\frac{1}{2^n}$ than is (140).

In chapters 8, 9, and 10, all of them quite brief, Laplace took up life expectancy, annuities, insurance, and moral expectation (or prudence). We do not know where he obtained his information, but it is reasonable to suppose that some of it must have been derived from occasional service on commissions appointed to review writings in this area and various actuarial schemes submitted to the government. The *procès-verbaux* of the Academy in its last years and of the Institute contain record of his having thus been called on from time to time. Moreover, the "Notice sur les probabilités" ([*1810d*]) was published as a rationale of the application that it was legitimate to make to tables of mortality. This piece, it will be recalled (see Section 25), was an expansion of his École Normale lecture of 1795 ([*1800d*]). It was then further expanded to become the first edition of the *Essai philosophique* and concludes with a summons to governments to license and regulate underwriters of insurance, annuities, and tontines, and to encourage investment in soundly managed associations. For an insurance industry and a literature did exist, although Laplace does not refer either to actual practice or to authorities. Comparison of his chapters with both would be required before a judgment could be made of what his contribution may have been.

Mathematically, his model for calculations of the "mean duration," both of life and marriage, is error theory. Given the tables of mortality covering a large population, a value for the mean length of life may be taken and the probability calculated that the mean life of a sample of stated size will fall within given limits. Calculation of life expectancy at any age follows directly. Laplace also gave a calculation for estimating the effect of smallpox on the death rate and of vaccination on life expectancy. The conclusion is that the eradication of smallpox would increase life expectancy by three years, if the growth in the population did not diminish the improvement by outrunning the food supply. Laplace did not give the data or provide numerical examples here, as he had done in his population studies. The succeeding chapter on annuities and tontines is equally abstract and gives expressions for the capital required to create annuities on one or several lives, for the investment needed to build an estate of given size, and for the advantages to be expected from participation in mutual benefit societies.

In the tenth chapter, with which Laplace concluded the first edition, he softened the asperity he had once expressed about Daniel Bernoulli's calculation of a value for moral expectation (see Section 4) in distinguishing that notion from mathematical expectation ([*1776a*, 1°], Article XXV). Bernoulli had proposed that prospective benefits, in practice financial ones, may be quantified as the quotient of their amount divided by the total worth of the beneficiary. Laplace now adopted that principle as a useful guide to conduct—without attributing it this time to Bernoulli. In infinitesimal terms, where x represents the fortune and dx the increment, its benefit will be $\frac{k\,dx}{x}$. If y represents the moral fortune corresponding to the physical value (*OC*, VII, 441),

$$y = k \log x + \log h, \tag{142}$$

where h is an arbitrary constant to be determined by the ratio of a value of y corresponding to a value of x. But perhaps it will not be necessary to follow the calculation in order to be convinced. Laplace concluded that in the most mathematically advantageous games of chance the odds are always unfavorable over time, and that diversification is a prudent practice in the investment of wealth. On matters of this sort, he wrote more persuasively in the ordinary language of the *Essai philosophique* than in the mathematics of the *Théorie analytique*.

Laplace must have continued straight on to expand the "Notice sur les probabilités" ([*1810d*]) into the *Essai philosophique*. He presented the first edition to the Institute in February 1814 (*102*), a year and a half after finishing the *Théorie analytique*. In August of that year he also read a memoir on the probability of testimony (*103*). Inclusion of these two pieces, the first as the introduction and the second as a new concluding chapter 11, together with three minor mathematical additions (*OC*, VII, 471–493), marks the difference between the first edition of *Théorie analytique des probabilités* and the second, completed by November 1814 (*104*). The two pieces have in common the tendency to move the subject further in the direction of civic relevance, the one in expounding it for laymen, the other in extending the application to concerns of life in society.

The *Essai philosophique sur les probabilités* has certainly had a longer life and almost certainly a larger number of readers than any of Laplace's other writings, including its counterpart in celestial mechanics, *Exposition du système du monde*. The reason for its continuing—indeed, its growing—success has clearly been the importance that probability, statistics, and stochastic analysis have increasingly assumed in science, social science, and philosophy of science. Inevitably, Laplace's technical writings have come to have the same sort of relation to the later development of the discipline of probability that, for example, Newton's *Principia mathematica* had to the later science of mechanics. Even if there were no other reason, that would suffice to explain why most readers who wish to repair to the fountainhead of what is often called the subjective interpretation of probability, in contrast to the frequency view, have recourse to the *Essai philosophique*. But there is a complementary reason, and that is the extreme difficulty of many parts of the *Théorie analytique*.

Given the accessibility of the *Essai philosophique* in many editions and languages, a summary scarcely seems necessary. The work itself is a summary. Instead, a reservation may be ventured, although somewhat hesitantly. If the two famous popularizations are compared in point of intrinsic merit, it is possible to consider the *Exposition du système du monde* the better book. At least, it conveys its subject with altogether greater clarity. For, once the reader is past the epistemological opening passages in the *Essai philosophique*, Laplace's paraphrase of the mathematics of *Théorie analytique* is not very easy to follow. No doubt the subject matter lent itself less well to verbal summary than did astronomy or geodesy. But it may also be worth noting that the order of composition was reversed. Laplace wrote the *Système du monde* as a book in its own right before he compiled the *Mécanique céleste*. The former was an outline or a prospectus for the latter. In the case of the *Essai philosophique*, he wrote all but the epistemological and actuarial sections after the *Théorie analytique*. In its first edition, it was a précis and bears the same relation to the treatise that the initial sections in many of his memoirs do to them, that of a preface written last. Moreover, there is nothing in the epistemology that he had not already said in principle in his lecture at the École Normale in 1795 ([*1800d*]; see Section 20) or indeed in the prolegomena of the youthful probability-gravitation memoir ([*1776a*]; see Section 4). It is true that in the later editions of the *Essai philosophique* he did enlarge upon topics likely to interest a wider public and that his observations on the credibility of witnesses and on the procedures of legislative assemblies and of judicial panels make more comprehensible the calculations of the supplementary material on the same topics added to the second and third editions of the *Théorie analytique*.

Let us consider briefly the approach of the eleventh chapter of *Théorie analytique* (*OC*, VII, 455–470), on the probability of testimony. The inevitable urn containing numbered slips is the model analyzed. A slip is drawn, and a witness reports the number to be n. Is he telling the truth? It was Laplace's idea to apply inverse probability to such problems, taking the statement for an event and estimating the probability that it was caused by the truthfulness of the witness. There are four possibilities: (1) he is neither lying nor making a mistake; (2) he is not lying and is mistaken; (3) he is lying and not mistaken; (4) he is both lying and mistaken. With these alternatives, Laplace employs a Bayesian analysis, first for this problem and then for a series of more complicated instances involving several occurrences witnessed by more than one observer. The corresponding discussion in the fourth edition (*117*) of the *Essai philosophique* (*OC*, VII, lxxix–xc) gives numerical examples and goes on to expose what he considered the fallacy in Pascal's wager on the existence of God. The probability of truthfulness in the witnesses who promise infinite felicity to believers is infinitely small. There is also a daunting estimate of the decay of reliability of historical information with the passage of time.

More interesting is the concluding article in the treatise. There the judgment rendered by a tribunal deciding between contradictory assertions is assimilated to the reports of several witnesses about the drawing of a numbered slip from an urn containing only two. The panel consists of r judges, and p is the probability that each will render a true judgment. The probability of the soundness of a unanimous verdict will then be $\frac{p^r}{p^r+(1-p)^r}$. The value of p is given by the proportion of unanimous verdicts, denoted by i, to the total number of cases n. Since $p^r + (1-p)^r = \frac{i}{n}$, or very nearly so, solving that equation will give the probability p of the veracity of each judge. Laplace then showed that if the tribunal consists of three magistrates (*OC*, VII, 470),

$$p=\frac{1}{2}\pm\sqrt{\frac{4i-n}{12n}}. \tag{143}$$

Choosing the positive root on the assumption that each judge has a greater propensity for truth than error, Laplace calculated the case of a court half of whose verdicts are unanimous. The probability of veracity in each judge will then be 0.789, and the probability that a verdict sustained on appeal is just will be 0.981 if the finding is unanimous, and 0.789 if the vote is divided. The greater the number of judges and the more enlightened they are, the better the chance that justice will be done—for to do Laplace himself justice, he acknowledged the artificiality of these calculations in introducing the topic in *Essai philosophique*, claiming only that they might provide guidance to common sense (*OC*, VII, lxxix).

Laplace extended the application of inverse probability to the analysis of criminal procedures in a supplement to the *Théorie analytique*, the first of four, composed in 1816 (*107*). For this purpose a condemnation is an event, and the probability is required that it was caused by the guilt of the accused. As always in the probability of cause, prior probabilities had to be known or assumed, and Laplace again supposed that the probability of a truthful juror or judge lies between 1/2 and 1. In a panel of eight members of whom five suffice to convict, the probability of error came to 65/256. He felt that the English jury system with its requirement for unanimity weighted the odds too heavily against the security of society, but that the French criminal code was unjust to the accused. By one provision, if a defendant was found guilty by a majority of 7 to 5 in a court of first instance, it required a vote of 4 to 1 to overturn the verdict in a court of appeal, since a majority of only 3 to 2 there still left a plurality against him in the two courts taken together. That rule was as offensive to common sense as to common humanity (*OC*, VII, 529). In view of the severe strictures that have been passed upon Laplace's political conduct, it should be noted that he took the trouble of publishing a pamphlet expounding on mathematical grounds the urgency of reforming these savage provisions ([*P1816*]); and in the definitive, fourth edition of *Essai philosophique* he gave his considered opinion that a majority of 9 out of 12 for conviction gave the most even balance between the interests of society in protection and in equity (*OC*, VII, xcxix).

Between 1817 and 1819, Laplace investigated the application of probability to sharpening the precision of geodesic data and gathered these studies for publication as the second and third supplements to the third edition (1820) of the *Théorie analytique* (*112, 118*). In recent years scholars concerned with the history of statistics have been especially interested in these two pieces, which Laplace himself saw in the context of his theory of error. His intention was to improve on the method of least squares in the minimization of instrumental and observational error. That had also been his motivation in the opening articles of the First Supplement, where before taking up judicial probabilities, he further developed what he called the "most advantageous" method of combining equations of condition formed from observations of a single element, like those exemplified in his initial justification of the least-squares method ([*1811a*]; see Section 22; cf. *Théorie analytique*, Book II, no. 21; *OC*, VII, 327–335). He then applied the method to estimating the probable error in Bouvard's recent, highly refined calculations of the masses of Uranus, Saturn, and Jupiter (*OC*, VII, 516–520).

In the Second Supplement, Laplace turned to geodesy and compared the results of his method with the so-called method of situation of Bošković (*OC*, VII, 531–580). Stigler considers that this discussion contains the earliest instance of a comparison of two well-elaborated methods of estimation for a general population, in which the conditions that make one of them preferable are specified. He is particularly enthusiastic about the growing statistical sophistication, as he sees it, of Laplace's later work in probability and argues that the analysis here is strikingly similar to that which led R. A. Fisher to the discovery of the concept of sufficiency in 1920 (L: [1973], 441–443; cf. Sheynin [1977], 41–44). In the Third Supplement, Laplace reports the result of applying his method to the extension of the revolutionary Delambre-Méchain survey of the meridian from a base in Perpignan to Formentera in the Balearic Islands. The data were the discrepancies between 180° and the sums of the angles measured for each triangle. There were only twenty-six triangles in the Perpignan-Formentera chain, however, and Laplace preferred to estimate the law of error on the basis of all 700 triangles in the original survey. He could then calculate the probabilities of error of various magnitudes in the length of the meridian by the formulas already developed for his modified, or most advantageous, method of least squares. In a further article on a general method for cases involving several sources of error, Laplace obtained a paradigm equation (129) for formulating equations of condition that relate values for observed elements to the error distributions involved. The equation served him in his later investigation of variations of the barometer as evidence for lunar atmospheric tides, the last he ever undertook (see Section 28).

In Laplace's own life and career, perhaps it is appropriate to see the curve breaking over in 1820 with the publication of the third, and definitive, edition of *Théorie analytique des probabilités*. In 1825 he did compose a fourth supplement, containing a minor modification to the theory of generating functions. By then he was showing signs of age. The minutes of the Institute record that the work was presented by Laplace and by his son.

27. Loss of Influence. It will be recalled (see Section 24) that in January 1816 Sophie Germain won the Institute's prize set in 1809 for the theory of elastic surfaces. Her paper broke sharply with the approach of the Laplacian school, and her successful challenge was one of the first signs that Laplace's power was beginning to wane. But it was by no means the only sign. The slackening of corporate research activity at Arcueil after 1812 and the cessation of regular meetings in 1813 did not augur well, and after 1815 Laplace became an increasingly isolated figure in the scientific community of the Restoration, particularly in the realm of physics. His personal reputation also suffered from the readiness of his accommodation to yet another change of political regime. In the Senate he voted for the overthrow of Napoleon in favor of a restored Bourbon monarchy in 1814. Conveniently absenting himself from Paris during Napoleon's temporary return to power in the Hundred Days—an episode that clearly embarrassed him—he remained loyal to the Bourbons until his death, becoming a bête noire of the liberals, most notably on his refusal in 1826 to sign a declaration of the Académie Française supporting the freedom of the press.

Perhaps the most serious assault on Laplace's physics came with the development of Fresnel's wave theory of light and its championing by Arago, a former member of the Arcueil circle, between 1815 and the early 1820's. In the face of this challenge, the Laplacian position was represented, rather typically, by Laplace's disciples. Thus when Fresnel won the competition of the Academy of Sciences for a study of the theory of diffraction in 1819 ("Mémoire sur la diffraction de la lumière," in *MASIF*, **5** [1821–1822/1826], 339–475), his theory was measured not against any of Laplace's writings but against the corpuscular theory of diffraction which Biot and Pouillet had devised, probably at Laplace's instigation, in 1816 (the theory is expounded in Biot's *Traité de physique expérimentale et mathématique*, 4 vols. [Paris, 1816], IV, 743–775); and it was Biot, not Laplace, who continued the open resistance to the wave theory into the 1820's. Similarly, Laplace did not respond publicly to the growing support for the chemical atomic theory in France after 1815, even though the theory was inconsistent with Berthollet's chemistry of affinities and the whole notion of short-range chemical forces. Nor did he react to the criticism of the caloric theory that was explicit in the work of Petit and Dulong on specific heats (1819) and clearly implied in Fourier's *Théorie analytique de la chaleur* (1822). (See A. T. Petit and P. L. Dulong, "Recherches sur quelques points importans de la théorie de la chaleur," in *Annales de chimie et de physique*, **10** [1819], 395–413, esp. 396–398 and 406–413. The criticisms of Petit and Dulong are discussed in Robert Fox, "The Background to the Discovery of Dulong and Petit's Law," in *British Journal for the History of Science*, **4** [1968–1969], 1–22, esp. 9–16.)

Despite the criticism to which his style of physics was subjected, Laplace never admitted defeat. By the early 1820's few could share his apparently unswerving belief in the physical reality of the imponderable fluids of heat and light, yet between 1821 and 1823 he developed the most elaborate version of his caloric theory of gases. He expounded the theory first in papers published chiefly in the *Connaissance des temps* and then, in a definitive and modified form, in Book XII of the fifth volume of the *Mécanique céleste* (*OC*, V, 97–160).

In all the versions of his theory, Laplace leaned heavily on the treatment of gravitation that he had published in the first volume of the *Mécanique céleste* in 1799. In particular, his expressions for the gravitational forces between spherical bodies proved to be readily applicable to the repulsion that, in accordance with the standard Newtonian model of gas structure, he supposed to exist between the particles of a gas; the modification that the force between the particles was not only repulsive but also inversely proportional to the distance between them was easily made.

In his first paper on caloric theory in the *Connaissance des temps* ([*1821d*]), which was the published version of one that he read to the Academy of Sciences in September 1821 (*123*), Laplace considered the equilibrium of a spherical shell taken at random in a gas. Invoking the condition that the force between the particles is effective only at short range, he showed that the pressure of the gas, P, is proportional to ρ^2c^2, where ρ is its density and c the quantity of heat contained in each of its particles. The argument had a highly speculative cast and rested on the unfounded assumption that the repulsive force between any two adjacent gas particles is proportional to c^2. Even more suspect was Laplace's model of dynamic equilibrium, in which, when the temperature is constant, the particles of a gas constantly radiate and absorb caloric at an equal rate. Postulating the simplest of mechanisms for the process, Laplace pictured the radiation from any particle as resulting from the mechanical detachment of the particle's own caloric by incident radiant caloric, the density of which, $\pi(u)$, was taken as a function of the temperature, u, alone. If the fraction of incident caloric absorbed was put equal to q (a constant depending solely on the nature of the gas) and if it was assumed (quite gratuitously) that the quantity of caloric detached was proportional both to c and to the total "density" of caloric in the gas, ρc, it followed that

$$\rho c^2 = q\pi(u). \qquad (144)$$

Since for Laplace $P \propto \rho^2c^2$, Boyle's law was an immediate consequence of this equation, as was Dalton's law of partial pressures. It also followed that since the function $\pi(u)$ was independent of the nature of the gas, all gases expand to the same extent for a given increment in temperature, as Dalton (1801) and Gay-Lussac (1802) had observed. The obvious next step, of assuming $\pi(u)$ to be proportional to the (absolute) temperature as measured on the air thermometer, was not taken in [*1821d*] but appeared very soon afterward in a paper published in the November issue of the *Annales de chimie et de physique* ([*1821a*]), and it was axiomatic in the definitive version presented in Book XII of the *Mécanique céleste* (*OC*, V, 125).

The fact that Laplace's theory was consistent with the main gas laws lent it obvious plausibility, but one prediction in particular raised difficulties. It followed from equation (144) that the isothermal compression of a gas to, for example, one-half of its original volume would cause c to decrease by a factor $\sqrt{2}$. Qualitatively, a reduction in the value of c was perfectly consistent with the phenomenon of adiabatic heating, but Laplace showed that a decrease by a factor $\sqrt{2}$ in isothermal compression was too great to account accurately for the error in Newton's expression for the velocity of sound; in fact, it led to the impossibly high figure of 2 for γ. Laplace's immediate solution was to suggest that the alternate compressions and rarefactions occurred slowly enough for there to be some heat exchange with the surroundings, but by December 1821 he had abandoned this hypothesis. In a paper submitted in that month to the Bureau des Longitudes and published the following year in the *Connaissance des temps* ([*1822a*]) he argued that heat exchange did not occur but that the heat "expelled" in excess of that required to reconcile the theoretical and experimental values for the velocity of sound merely becomes latent and so ceases to contribute to the interparticle force. Hence c now represents not the total quantity of heat in a particle of gas but only that part of it that is "sensible" or free; the "total heat" of a particle is Q or $(c + i)$, i being its latent or combined heat.

It was a simple matter for Laplace to restate his fundamental expression for the velocity of sound ($\sqrt{dP/d\rho}$) in a form that involved c, and he did this in [*1822a*]. Assuming $P \propto \rho^2 c^2$, it followed that the velocity of sound is equal to

$$\sqrt{\frac{2P}{\rho}\left(1+\frac{\rho}{c}\frac{dc}{d\rho}\right)};$$

and this in turn implied that

$$1+\frac{\rho}{c}\frac{dc}{d\rho}$$

is equal to $\gamma/2$. In a supplement to this paper, which dates almost certainly from early 1822 ([*1822b*]), Laplace also showed how he could express γ in an equation involving Q, ρ, and P. To do this, he assumed that Q was a function of any two of P (that is, $\rho^2 c^2$), ρ, and the absolute temperature, so that in adiabatic conditions ($\Delta Q = 0$)

$$dP\left(\frac{\partial Q}{\partial P}\right)_\rho + d\rho\left(\frac{\partial Q}{\partial \rho}\right)_P = 0, \qquad (145)$$

and

$$\gamma = -(\rho/P)\{(\partial Q/\partial \rho)_P/(\partial Q/\partial P)_\rho\}. \qquad (146)$$

In these ways Laplace showed how the investigation of c and Q, both quantities that were intimately related to his speculations on the state of caloric in bodies and not susceptible to a direct test, could proceed simply through the measurement of γ.

The experiments to determine γ that Gay-Lussac and Welter conducted in 1822 therefore assumed a heightened theoretical significance. Particularly fruitful for Laplace's purpose was the observation that γ remained very nearly constant over a wide range of temperature and pressure. When the condition γ = constant was introduced into (145), it followed, by integration, that Q must be a function of $P^{1/\gamma}/\rho$ or, inserting the absolute temperature u, of $uP^{\left(\frac{1}{\gamma}-1\right)}$. By postulating the simplest possible relationship between Q and $P^{\left(\frac{1}{\gamma}-1\right)}$—proportionality—Laplace could then show that

$$Q = KuP^{\left(\frac{1}{\gamma}-1\right)}, \qquad (147)$$

where K is an unknown constant, determined by the nature of the gas. In the *Mécanique céleste* (C: V, 128; *OC*, V, 143), Equation (147) appeared as

$$Q = F + KuP^{\left(\frac{1}{\gamma}-1\right)}, \qquad (148)$$

but for Laplace's purposes the two expressions were equivalent. One important result in particular followed with either (147) or (148). This was that the ratio between the volume specific heats of any gas at two different pressures P_0 and P_1, but at the same temperature, is equal to $(P_1/P_0)^{1/\gamma}$. In fact, the volume specific heats in such circumstances should be in the ratio P_0/P_1, but the close agreement with Delaroche and Bérard's erroneous results for the variations of specific heat with pressure only endorsed Laplace's conclusion and the assumptions from which it was derived.

To all appearances, Laplace had secured a major triumph in giving at least one branch of the caloric theory the quantitative character that it had always conspicuously lacked. Yet in important respects the triumph was illusory. There was no independent evidence to confirm his assertions concerning the state of caloric in gases; they were clearly determined by the requirement that the deductions made from them should agree with the gas laws—for discussion of this point, see (M: Fox [1971], 173–174). And logically, despite his elaborate expressions involving the hypothetical entities c and i, much of Laplace's argument rested on far simpler premises than he intimated. In 1823 this point was made implicitly but unmistakably by Laplace's most loyal pupil, Poisson ("Sur la vitesse du son," in *Connaissance des temps* [1826/1823], 257–277; and "Sur la chaleur des gaz et des vapeurs," in *Annales de chimie*, 2nd ser., **23** [1823], 337–352), who reviewed and extended several aspects of Laplace's theory of heat. Most of his results had already been obtained by Laplace, but new ground was also broken. In the first paper Poisson derived the now familiar expressions for adiabatic changes in volume, $TV^{\gamma-1}$ = constant, and PV^γ = constant; and in the second he arrived at the false conclusion that the principal specific heats of a gas, c_p and c_v, are equal to $BP^{\left(\frac{1}{\gamma}-1\right)}$ and $\frac{1}{\gamma}BP^{\left(\frac{1}{\gamma}-1\right)}$, where B is a constant. No reader could miss the point that all this was achieved without any mention of Laplace's mechanisms. Poisson merely assumed, as any supporter of the caloric theory would have done, that the heat content of a gas was a function of its pressure and density.

Possibly the most telling evidence of Laplace's diminished status in physics toward the end of his life is the almost total indifference with which his work on caloric was received. It aroused neither overt opposition nor support except, somewhat ambiguously, from Poisson, and it stimulated no further research. Yet in 1824 Laplace appeared as confident as ever that a physics based on imponderable fluids and short-range molecular forces could be achieved. In that year, in the "Avertissement" to the fifth edition of the *Système du monde*, he wrote that he intended to make molecular forces the subject of a special supplement. As it happened, the intention was never fulfilled (see Bibliography, Section H). The physics of the sixth edition of the *Système du monde* (which appeared in 1827, the year of Laplace's death, and again, as a quarto, in 1835) was virtually identical to that of the fourth edition (1813).

So in his last years there were few who endorsed Laplace's approach to physics. The great days of Arcueil were now a distant memory, and the once loyal disciples were no longer involved in the problems that their master had identified. In 1822 Biot virtually retired from the scientific community, following his conflict with Arago over the wave theory of light and his defeat by Fourier in the election for one of the two posts of permanent secretary of the Academy of Sciences. It was left for Poisson to carry on Laplacian physics into the late 1820's and 1830's. But even he was far from being an uncritical admirer. In his *Nouvelle théorie de l'action capillaire* (1831) he criticized and corrected a number of shortcomings in Laplace's theory of capillarity; and in the *Théorie mathématique de la chaleur* (1835), despite a prefatory discussion of the properties of caloric, he totally ignored—as he had in 1823—the model that Laplace had perfected in the early 1820's.

28. The Last Analysis. On the publication of Volume IV of *Mécanique céleste* in 1805, Laplace had undertaken to complete the original plan with an eleventh and final book giving an account of the work of predecessors and contemporaries in the science of astronomy. Instead, he had become immersed in physics and in probability, and by the time in the 1820's when he got down to the histories he had promised, his own work was already beginning to be history. Meanwhile, and notably in 1819 and the early 1820's, he had published other investigations on particular topics of celestial mechanics and decided to collect these pieces and append them to the historical summaries of the areas they concerned. Thus the intended Book XI became Books XI through XVI. Laplace had them printed as they were completed beginning in March 1823 (*125, 126, 129–131, 133*) and assembled them as the fifth volume at the end of 1825. The historical notices are more detailed than the concluding chapters of *Système du monde* and are written in a matter-of-fact rather than an inspirational vein. It would be interesting to know at what point in the great investigations of his own career he had studied the works of his predecessors, for it is clear that he knew them well. His histories are still worth consulting in one respect, for his own sense of what he himself had contributed to the several topics. It is doubtful that many of the new investigations assembled for Volume V made much difference to the further development of celestial mechanics; on the whole, their day was past. Book XI is said to be about the figure and rotation of the earth (cf. [*1818b*], [*1819a*], [*1819b*], [*1820e*], [*1820f*], [*1820g*]); Book XII, about the attraction and repulsion of spheres and the motion of elastic fluids, though it really contains the comprehensive development of Laplace's caloric theory of heat and gases discussed in Section 27 (cf. [*1820h*], [*1821d*], [*1822a*], [*1822b*], [*1822d*]); Book XIII about the oscillation of fluids surrounding the planets (cf. [*1815c*], [*1819d*], [*1819e*], [*1820d*], [*1821a*], [*1823a*]); Book XIV about the rotation of heavenly bodies about their centers of gravity (cf. [*1809d*], [*1824a*]); Book XV about the theory of planetary and cometary motions (cf. [*1813a*], [*1819c*], [*1821c*]); and Book XVI about satellite theory (cf. [*1809c*], [*1812a*], [*1820a*], [*1820b*], [*1820c*], [*1820i*], [*1821b*]). The comprehensive titles are a bit misleading. Most of Volume V consists of minor emendations to the data and improvements on fine points of the analysis of particular phenomena treated under those headings in the four main volumes.

Several novelties and peculiarities are worth signaling, however. Inevitably, Laplace had been in touch with the work of Cuvier, whose position and influence in the biological sciences during the Napoleonic period and afterward paralleled his own eminence in mathematical quarters. The first substantive chapter of Book XI consists of calculations purporting to show how the depth and configuration of the seas can be reconciled with the geological evidence for catastrophic inundations and extreme climatic changes in the history of the earth. Chapter 4 also concerns the theory of the earth, in respect to its cooling. As far back as 1809, Laplace had learned of Fourier's investigations of the diffusion of heat (J: Grattan-Guinness [1972], 444–452) and had referred in print to Fourier's pathbreaking but still unpublished paper of 1807 ([*1810a*]; *OC*, XII, 295). Despite the difference in their approaches (which he did not mention), he opened his chapter on the heat of the earth by expressing his pleasure that Fourier's two fundamental equations (*OC*, V, 82–83),

$$\frac{\partial^2 V}{\partial x^2}+\frac{\partial^2 V}{\partial y^2}+\frac{\partial^2 V}{\partial z^2}=k\frac{\partial V}{\partial t}; \qquad (149)$$

which expresses the diffusion of heat inside the earth, or any comparable body, where V is the heat of any point; and

$$-\frac{\partial V}{\partial r}=fV-fl, \qquad (150)$$

which expresses the transmission of heat through the surface, were simply modifications respectively of his general equation (94) on the attraction of spheroids (the potential function), and of his own Equation 2 in Book III, no. 10 of *Mécanique céleste* (*OC*, II, 30). Laplace did not allude to the ambivalence in the background of his further relations with Fourier and with his own disciple, Poisson, bested in the rivalry between the two (J: Grattan-Guinness [1972], 462–463). He went on to analyze the temperature gradient beneath the surface and the rate of cooling of the earth in expressions analogous to those that he had developed years before ([*1785b*]) for spheroidal attraction theory. His purpose was to estimate whether the shrinkage of the earth on cooling was sufficient to decrease the angular velocity and thus to alter the length of the day detectably. While acknowledging that his parameters were hypothetical, he felt safe in concluding that the effect, if it existed at all, did not amount to 0.01 seconds since the time of Hipparchus.

In Book XIII, Laplace thought to redeem a promise, or an assurance, other than historical. It will be recalled that the atmospheric tides had interested him since the earliest memoirs on the ebb and flow of the seas ([*1779b*]; see Section 8). He recurred to the subject briefly and inconclusively in *Mécanique céleste*, Book IV, and more confidently in Book II of *Théorie analytique des probabilités*. Discussing the significance of data in chapter 5, where the most important example came from barometric readings, Laplace there predicted that one day records would be sufficiently full and accurate to permit detecting the gravitational influence of sun and moon among the other, in effect much larger, causes that determine atmospheric pressure (see Section 23). Now, in 1823, taking his point of departure from his previous analysis of mean sea level, he set out to apply his own version of the least-squares analysis to the detection of significance in the variations of the barometer that could be correlated with the relative positions of earth, sun, and moon ([*1823a*]). The third supplement (*118*) to *Théorie analytique des probabilités*, published in 1820, again described his least-squares method—now somewhat modified—as the "most advantageous." As in the original distinction from Legendre's approach ([*1811a*]), Laplace meant by "most advantageous" that he combined the equations of condition for the unknowns so as to determine the most probable values (cf. Section 25). In an excellent recent discussion, Stigler (L: [1975]) describes the modification as weighted least squares. Laplace now did have access to a series of barometric measurements recorded at the Observatory in Paris—quite probably at his instigation—three times a day, at 9:00 A.M., noon, and 3:00 P.M. from 1 October 1815 through 1 October 1823. He published his findings in *Connaissance des temps* ([*1823a*]) and reprinted them with little change in Volume V (*OC*, V, 184–188, 262–268).

Laplace's idea was to determine whether the gravitational influence can be detected by comparing variations in barometric pressures over those eight years in the four days surrounding syzygies (when the sun, moon, and earth are in line) and quadratures (when they make a right angle). From the conditions, he formed the linear equations

$$x\cos(2iq)+y\sin(2iq)=E_i;$$

$$y\cos(2iq)-x\sin(2iq)=F_i, \qquad (151)$$

in which x and y are the unknowns to be estimated; i indicates the day of each set of data and has the value -1, 0, $+1$, or $+2$; q represents the synodic motion of the moon; and E_i and F_i are computed from the data by the following formulas:

$$E_i=A_i''-A_i+B_i-B_i'';$$

$$F_i=\{2A_i'-(Ai+A_i'')-2B_i'+(Bi+B_i'')\}\left(1+\sin\frac{q}{4}\right). \qquad (152)$$

In these expressions, A_i denotes the eight-year

mean of the 9:00 A.M. measurements for the ith day after syzygy; B_i for the ith day after quadrature; A_i' and B_i' the means for the noon values; and A_i'' and B_i'' for the 3:00 P.M. values. Thus, E_i is the mean barometric change between 9:00 A.M. and 3:00 P.M. for the ith day after syzygy, minus the same change for the ith day after quadrature; while F_i is proportional to the difference between the mean rates of change for those days. These expressions had to be combined by the modification that he had brought to his "most advantageous" method in the Third Supplement to the *Théorie analytique des probabilités* (*OC*, VII, 608–616). He multiplied each of the four equations for E by a factor of three and by the corresponding coefficient of x, and multiplied each of the equations for F by the corresponding coefficient of y, obtaining

$$x\,(8+\Sigma\cos 4iq)+y\Sigma(\sin 4iq) = 3\Sigma E_i\cos 2iq-\Sigma F_i\sin 2iq$$

$$y\,(8-\Sigma\cos 4iq)+x\Sigma(\sin 4iq) = 3\Sigma E_i\sin 2iq+\Sigma F_i\cos 2iq. \tag{153}$$

Substituting the data from the observations and calculating x and y (0.10743 and −0.017591 respectively), Laplace found the range of the lunar atmospheric tide to be 0.05443 millimeters of mercury and the time of the maximum tide in syzygy to be 18 minutes and 36 seconds after 3:00 P.M. There then remained the problem of determining the probability that these observations really did exhibit the existence of a lunar atmospheric tide. For it was not enough to compare the variations at syzygy and quadrature with those assumed to follow from irregular or random causes, as he had just done. Unless the probability of error in the conclusion is contained within very narrow limits, it might be that the data exhibit only the overall effects of irregular causes, a fallacy to which the science of meteorology was prone. In what amounted to the application of a central limit theorem (the phrase was never his), Laplace calculated the probability that chance alone would produce a variation in the means no greater than that indicated by the eight-year accumulation of observations. The value was 0.843, an unconvincing figure in the tidal quest. All that could be said was that there would be "some implausibility" (*OC*, V, 268) in attributing the variation to chance alone. To increase to near certainty the probability that tides are responsible, the very small effects detected would need to rest on something like nine times the 1,584 thrice-daily readings, or approximately 40,000 observations.

In the next four years Bouvard continued the program of recording barometric readings at the Paris Observatory, which data he compared with a comparable series assembled on similar principles by Ramond at Clermont-Ferrand, in the Puy-de-Dôme. Bouvard then recalculated the whole corpus, an enormous labor that led Laplace to take up his theory of lunar atmospheric tides yet again. In this, his last paper ([*1827b*]), composed in his seventy-eighth year, he modified once more his method of calculating probable error. Whether or not he had in the interval become aware of the importance of independence as between E_i and F_i in Equation 152—and Stigler thinks he may well have revised his approach for just that reason (L: [1975], 503)—his new calculation did combine the equations of condition in such a way that terms depending on the same measurements served to determine only one unknown. Moreover, in explaining it, he now put a distance between his approach and that of least squares, for which several mathematicians have given proofs that were "not at all satisfactory" (*OC*, V, 491). The art consists in choosing the factors by which the equations of condition formed from the data are to be multiplied in combining them into a system of final equations. Laplace now compares his procedure with that followed in planetary astronomy. In order to correct the elliptical elements, observed longitudes are equated to theoretical longitudes, each modified by the relevant correction. A large number of equations of condition are thus formed. Each is multiplied by the coefficient of the initial correction. Adding all of them gives the first final equation. The same procedure is followed for each successive correction, until there are as many final equations as there are corrections. But the longitude does not depend on a single observation. It is derived from two observations by different instruments, one giving right ascension and the other declination. The law of errors may not be the same for both or have the same effect on the longitude. It was in determining the most advantageous factors (given all these complications) that his method was superior to least squares, and he referred to the general expression that he had given for it in the Third Supplement to the *Théorie analytique des probabilités* (*OC*, VII, 612):

$$l^{(i)}x+p^{(i)}y+q^{(i)}z+\cdots = a^{(i)}+m^{(i)}\gamma^{(i)}+n^{(i)}\lambda^{(i)}+r^{(i)}\delta^{(i)}+\cdots, \tag{154}$$

where l, p, q, a, m, n, and r are coefficients given by the conditions; γ, λ, and δ are errors in the observations arising from different circumstances; and x, y, and z are unknowns to be estimated.

Obviously, his equations of condition (152) for the variations in the lunar tide had been formed in that mold, and now, four years later, he combined them in a manner conformable to the astronomical illustration just given of the "most advantageous" method. Having thus revised the method, Laplace let Bouvard perform the calculation, which yielded values of 0.031758 for x and 0.01534 for y, and a difference of 0.01763 millimeters of mercury for the range of the atmospheric tide. Again, the probability had to be calculated that the results show the existence of a regular cause, in this case the gravitational pull of the moon, rather than mere chance. If the value for x were the effect of chance alone, the probability that it would fall within the limits ± 0.031758 would be $\frac{1}{\sqrt{\pi}}\int g\, e^{-g^2l^2} \cdot dl$, where g is given by observation, and the integral is evaluated between those same limits. That works out to be only 0.3617, and it would again need to be very close to unity to be convincing evidence of the existence of a lunar tide. The value for y gave even a lower probability, and the detectability of a lunar tide in Paris had, therefore, still to be considered moot despite the additional data (*OC*, V, 500).

Corollary information proved more amenable to the search for causes. Ramond in his observations in Clermont-Ferrand had discovered—and Bouvard in Paris had confirmed—that the daily variation of the barometer between 9:00 A.M. and 3:00 P.M. varied with the seasons, the mean increase being 0.557 millimeters of mercury in the three months from November through January and 0.940 in the following quarter. In the remaining two quarters, the values were intermediate between those extremes. Did these differences result from cause or chance? Again, only probability could decide, and on making the calculation Laplace found that the two values just cited do argue the existence of a regular cause with a high degree of probability, but that the intermediate values and the annual mean of 0.762 can reasonably be attributed to the effects of chance. Bouvard had also noticed that the mean variation is positive in every month of the year, and in a corollary calculation Laplace found the pattern probabilistically predictable. Readers interested in comparing Laplace's capabilities with the resources of modern statistical science may wish to consult the evaluations of Laplace's handling of this problem by Stigler (L: [1975], 509–515) and by Sheynin (L: [1977], 56–58). The manuscript for this paper, published posthumously ([*1827b*]) with another on elliptical motion and the calculation of planetary distances, was found among Laplace's papers after his death and was incorporated in further printings of *Mécanique céleste*, Volume V, as a supplement.

Fourier in his éloge (K: [1835], xii) recalled that Laplace retained his extraordinary memory to a very advanced age. He always ate and drank very lightly and showed no sign of enfeeblement before his last two years. Magendie was his doctor and Bouvard was with him constantly at the end. There is a bust of him by Houdon, and Guérin did his official portrait as president of the senate in 1803. Most likenesses show a thin, pointed face with narrow lips and prominent nose. Later portraits suggest a slight tendency to dewlaps in his final years. Laplace was buried at Père Lachaise. The monument erected to him there was moved to Beaumont-en-Auge in 1878, when his remains were transferred to St. Julien de Mailloc, a small village in the canton of Orbec in Calvados.

PART V: THE LAPLACE TRANSFORM

29. Laplace's Integral Solutions to Partial Differential Equations. Laplace's name is most widely used today by mathematicians when referring to the "Laplace transform" method of solving differential, difference, and integral equations. Thus it is meet to outline here the contexts in which it arises in Laplace's writings and its later development into a systematized theory.

To begin at the end: the modern definition of the Laplace transform $\bar{f}$ of f is

$$\bar{f}(s) = \int_0^\infty e^{-su} f(u)\, du, \quad \mathrm{Re}\,(s) > 0. \qquad (155)$$

The essence of Laplace transform theory is to convert the given f-problem via (155) to a problem in $\bar{f}$, solve that, and then convert back to the f-solution. The theory itself includes addition, convolution, and shifting theorems, results on transforms of derivatives and integrals, and especially the inverse theorem by means of which we have a general rule to get back to f from $\bar{f}$:

$$f(x) = \frac{1}{2\pi\sqrt{-1}} \int_C \bar{f}(s)\, e^{sx}\, ds, \qquad (156)$$

where C is a certain kind of contour in the s-plane. Note that the problem context mentioned at the beginning, and the various theorems stated after (155), are needed to characterize the theory; integral forms similar to (155) and (156) are not sufficient on their own to earn the name of "Laplace transform."

Let us now trace some of this history. Historians of the calculus are accustomed to finding seeds of later ideas in Euler, and such is the case here. In a paper published in 1744 he examined

$$z=\int X(x)\,e^{ax}\,dx \tag{157}$$

as a possible form of solution of differential equations,[1] and a more extensive discussion was given in 1753.[2] However, his preference was for functional or, to a lesser extent, power-series solutions of differential equations (and sometimes combinations of both), although iterative indefinite integrals of a function would also sometimes be used.[3] Euler's ideas to date on solutions of differential equations were outlined in detail in *Institutiones calculi integralis* (1768–1770); integral solutions of the form of (157) were given some space, as was the similar form[4]

$$\int X(x)\,x^{\lambda}\,dx. \tag{158}$$

The influence of Euler on Lagrange was very profound; and of particular interest for the subsequent influence on Laplace is Lagrange's 1773 paper on finding the mean of a set of observations (of which Laplace was aware, as we saw in Section 4). Lagrange considered a few special discrete and continuous distributions, and evaluated the probability of errors falling within given limits. This involved him in "Laplace-transform"-looking integrals such as

$$\int X(x)\,e^{-\alpha x}\,a^{x}\,dx, \tag{159}$$

where X is a rational function, and their conversion into infinite series.[5] Although we can interpret the results in terms of modern Laplace transform theory, Lagrange, like Euler, did not see his results in quite that way.

The first significant signs of Laplace's interest in such expressions occur in his "Mémoire sur les suites" ([*1782a*]). There his solution method by successive approximations (see Section 12 above) involved formulas such as

$$u=\sum_{r=1}^{\infty} a_r(s,s_1)\int^{(r)}\phi(s)\,(ds)^r+\sum_{r=1}^{\infty} b_r(s,s_1)\int^{(r)}\psi(s_1)\,(ds_1)^r, \tag{160}$$

where ϕ and ψ are "arbitrary" (*OC*, X, 54). This kind of result was already given in his first presentation in [*1777a*] of the method (*OC*, IX, 24 ff.); but later in [*1782a*] he came close to a Laplace transform of the form (158) for integral values of μ, for he used iterative integration by parts to obtain

$$\int \phi(z)\,z^{\mu}\,dz= C+\sum_{r=0}^{\mu}(-1)^r\,\mu\,P\,r\,\phi_{r+1}(z)\,z^{\mu-r}, \tag{161}$$

where ϕ_r is the rth indefinite integral of ϕ (*OC*, X, 66). However, the purpose was not to explore the properties of (161) for themselves but to obtain this integral solution of a general linear second-order partial differential equation:

$$u=(2x)^{-m/2}\left\{\int^{x+at}\daleth\left(\frac{x+at-z}{2x}\right)\phi(z)\,dz+\int_0^{x-at}\daleth\left(\frac{x-at-z}{2x}\right)\psi(z)\,dz\right\} \tag{162}$$

(*ibid.*, 68), where ⅃ is the solution of a certain second-order ordinary differential equation.

More promising material for our purpose occurs in Laplace's paper [*1785a*] on approximating to functions of very large numbers. He took the linear difference equation

$$S(s)=\sum_r A_r(s)\,\Delta^r y(s), \tag{163}$$

(where Laplace's s is real) and applied two transforms to it:

$$y(s)=\int x^{s}\,\phi(x)\,dx, \tag{164}$$

akin to (161) and now sometimes called the "Mellin transform," which proved particularly useful when $S \not\equiv 0$; and the "almost-Laplace" transform

$$y(s)=\int e^{-sx}\,\phi(x)\,dx, \tag{165}$$

which is helpful when $S\equiv 0$ (*OC*, X, 212, 236–248). In the course of using these two transforms he derived a few basic properties; for example, from (165) he had a theorem on (forward) differences:

$$\Delta^r y(s)=\int e^{-sx}\,(e^{x}-1)^r\,\phi(x)\,dx \tag{166}$$

(*ibid.*, 236); while (164) provided a similar result for differences,

$$\Delta^r y(s)=\int x^{s}\,(x-1)^r\,\phi(x)\,dx, \tag{167}$$

and also one for derivatives,

$$\frac{d^k y(s)}{dx^k}=\int x^s (\log x)^k \phi(x)\, dx \qquad (168)$$

(*ibid.*, 242–247; see also 278–291). "In many circumstances," Laplace commented prophetically, "these forms [(165)] . . . are more useful than the preceding ones [(164)]" (*ibid.*, 248). It is worth noting also that earlier parts of this paper feature integrals the integrands of which involve e^{-t^2}, for the Laplace transform of the error function is of some importance in certain applications.[6] Later in the paper Laplace urged general solutions to differential equations using the form of (164) or (165) (*ibid.*, 253).

It is this work that is normally cited as the origin of the term "Laplace transform"; but we must also look at the effect on Laplace of Fourier's 1807 monograph on heat diffusion. Fourier had found the "diffusion equation," in forms such as

$$\frac{\partial^2 y}{\partial x^2}=\frac{\partial y}{\partial t}, \qquad (169)$$

to represent the physical phenomenon, and the "Fourier series" solution form

$$y=\sum_{r=0}^{\infty}(a_r \cos rx + b_r \sin rx)\, e^{-r^2 t} \qquad (170)$$

to solve it.

A strong controversy ensued at the Institut de France about this work, partly because solutions of the form of (170) had been considered and rejected in the eighteenth century.[7] A particularly relevant criticism to our current discussion is that the initial condition function f is not explicitly encased in the solution (170)—as it is in integral and functional solutions—but appears only in the integrals that define its coefficient; to the mathematical mind of the time, the explicit involvement of f helped to justify the generality of any solution. Lagrange, still alive, remained opposed to (170), as did Poisson and Biot; but Laplace accepted Fourier's results. In [*1810a*] he constructed a Newtonian intermolecular force model to obtain the heat transfer term in (169) (*OC*, XII, 293). More significantly, in [*1809e*] he published a miscellany paper on analytical methods that not only related to Fourier's work but also developed techniques presented in "Mémoire sur les suites" ([*1782a*]) and the paper ([*1785a*]) on very large numbers.

As a special case of advancing the results of [*1782a*], Laplace considered Fourier's diffusion equation (169) and solved a problem that is conspicuous by its absence from Fourier's 1807 monograph: to solve (169) for an *infinite* range of values of x, where the periodicity of the trigonometric functions rules out (170). Poisson, already aware in 1806 of the trend of Fourier's work, had offered this power-series solution[8] of (169):

$$y=\sum_{r=0}^{\infty}\frac{x^{2r}}{(2r)!}f^{(r)}(t)+\sum_{r=0}^{\infty}\frac{x^{2r+1}}{(2r+1)!}g^{(r)}(t)\,. \qquad (171)$$

In [*1809e*] Laplace used the same type of solution of (169). Applying the initial condition

$$y=\phi(x)\,,\ \text{when } t=0,\ -\infty\leq x\leq\infty, \qquad (172)$$

he obtained

$$y=\sum_{r=0}^{\infty}\frac{t^r}{r!}\phi^{(2r)}(x)\,, \qquad (173)$$

and then submitted it to an ingenious manipulation. The result

$$\int_{-\infty}^{\infty} z^{2r} e^{-z^2}\, dz=\frac{(2r)!}{4^r r!}\sqrt{\pi} \qquad (174)$$

—proved in [*1785a*] see *OC*, X, 269, where "t_{2r}" is misprinted from the original for "t^{2r}"; I have used z for t in (174), although Laplace did not explicitly recall his proof here—and the obvious

$$\int_{-\infty}^{\infty} z^{2r+1} e^{-z^2}\, dz=0 \qquad (175)$$

converted (173) to

$$y=\frac{1}{\sqrt{\pi}}\int_{-\infty}^{\infty}\sum_{r=0}^{\infty}\frac{(2z\sqrt{t})^r}{r!}\phi^{(r)}(x)\, e^{-z^2}\, dz. \qquad (176)$$

The integrand of (176) contains a Taylor expansion, so that we have finally the integral solution

$$y=\frac{1}{\sqrt{\pi}}\int_{-\infty}^{\infty}\phi(x+2z\sqrt{t})\, e^{-z^2}\, dz \qquad (177)$$

(*OC*, XIV, 184–193).

This solution form preserved the tradition of containing the initial condition function ϕ explicitly. Fourier himself now realized that an integral solution would work for an infinite range of values of x, and by ingenious if unrigorous manipulations of infinitesimals he very quickly found the "Fourier transform" and its inverse. They took forms such as

$$\underline{f}(q)=\sqrt{\frac{2}{\pi}}\int_0^{\infty} f(u)\cos qu\, du \qquad (178)$$

and

$$f(x)=\sqrt{\frac{2}{\pi}}\int_0^\infty \underline{f}(q)\cos qx\,dq, \qquad (179)$$

and led to double-integral solutions of the diffusion equation (169) in which, as in (177), the initial condition function f is encased:[9]

$$y=\frac{2}{\pi}\int_0^\infty \int_0^\infty f(u)\cos qu \cos qx\, e^{-q^2 t}\,du\,dq. \qquad (180)$$

Thus two new integral (and the Fourier series) solutions of linear partial differential equations were produced in a very short time. During the next decade an intense development of these methods occurred in a variety of physical contexts. Fourier integrals were the most popular, but Poisson used Laplace's form (177) whenever he could. Because Laplace himself was not prominently involved, we shall not pursue the details here;[10] but several of his results of this period are worthy of notice. He used Fourier integrals in [*1810b*], which continued the purpose of [*1785a*] on very large numbers (*OC*, XII, 334–344). In [*1811a*] he returned to another old interest, finding the mean of a set of observations, and derived more Fourier integrals from integrals such as

$$\int_0^\infty e^{-ax}x^{-\omega}e^{\sqrt{-1}\,rx}\,dx \qquad (181)$$

(*OC*, XII, 363; compare the results in [*1785a*] at *OC*, X, 264). Elsewhere in [*1811a*] he may possibly have revealed another influence from Fourier, for he showed that the set of functions

$$\left\{\frac{(-2)^i}{1\cdot 3\cdots(2i-1)\sqrt{\pi}}\int_{-\infty}^{\infty} e^{-s^2}(\mu+s\sqrt{-1})^{2i}\,ds,\quad i=1,2,\cdots\right\} \qquad (182)$$

was orthogonal over $(-\infty, \infty)$ with respect to the weighting function $e^{-\mu^2}$ (*ibid.*, 382)—an analysis that corresponds closely to Fourier's 1807 demonstration[11] of the orthogonality of the (misnamed) "Bessel functions" $\{J_0(a_i x)$; the $\{a_i\}$ are the roots of a certain transcendental equation$\}$ with respect to the weighting function x—and an orthogonality expansion similar to Fourier's (*ibid.*, 384). This work constitutes Laplace's anticipation of a form of the "Hermite polynomials" of half a century later (see L: Molina [1930]); the expansion is now misnamed "the Gram-Charlier expansion." Finally in this association of Laplace with Fourier, we recall from Section 28 Laplace's use of Fourier analysis in estimating the age of the earth.

Laplace's *Théorie analytique des probabilités*, which appeared in editions in 1812, 1814, and 1820, also deserves mention; for, as in his earlier work on mathematical probability (see Sections 4 and 11) he included, especially in Book I, part 1, treatments of the generating function in the form

$$\sum_r p_r t^r \qquad (183)$$

rather than as what we now call the moment generating function of the distribution function of the discrete random variable t:

$$\sum_t e^{\lambda t} f(t). \qquad (184)$$

He also used a continuous analogue of (183),

$$y(x)=\int t^{-x}\,T(t)\,dt, \qquad (185)$$

which harks back to his (164), and at one point adapted his earlier (168) to define the then rather novel fractional derivative of $y(x)$:

$$d^i y(x)/dx^i=\int t^{-x}(\log 1/t)^i\,T(t)\,dt, \qquad (186)$$

where i is *not* necessarily an integer (*OC*, VII, 86). He also used a continuous version of (184),

$$y(x)=\int e^{-xt}f(t)\,dt, \qquad (187)$$

where an inverse transform is attempted (*OC*, VII, 136), and also in treating again the "Hermite polynomials" (in Book II, chapter 3). Various other of his earlier results were given an airing: in Book I, part 2, some Fourier integrals were evaluated, and the use of (164) and (165) in solving difference equations and in developing asymptotic theory was again dealt with. Laplace also made some use of characteristic functions for discrete distributions (for example, in Book I, part 1 and Book II, chapter 4—and already in [*1810b*] at *OC*, XII, 309 for a uniform distribution).

However, the full relationship between mathematical probability and harmonic analysis does not seem to have been grasped at this time. For example, we now know that the distribution function of a sum of independent random variables is the convolution of the component distribution functions, and that the Laplace transform converts the convolution into the product of its transformed distributions; yet this was not shown, although Poisson was especially aware of the significance of convolutions in Fourier integral solutions of differential equations.[12] Again, the existence of the integral in (185) is better secured by the use of complex variables in defining the characteristic function of f:

$$X(\lambda)=\int_{-\infty}^{\infty} e^{\sqrt{-1}\,\lambda t} f(t)\,dt; \qquad (188)$$

but this move had to wait for a time, although Cauchy was very adept at handling complex variable forms of the Fourier integral (178).[13]

Thus we see a number of opportunities that later generations were to grasp. It is appropriate here only to outline the later development of the Laplace transform, which has been our principal theme. The first systematic study of its basic properties was carried out in an 1820's manuscript by Abel, first published in 1839:[14] he started from generating functions, and noted the multivariate transform

$$\phi(x, y, z, \cdots)=$$
$$\int e^{xu+yv+zp+\cdots} f(u, v, p, \cdots)\,du\,dv\,dp\cdots. \qquad (189)$$

Meanwhile, in 1833 Robert Murphy had taken the transform in the form

$$g(x)=\int t^{x-1}\phi(t)\,dt, \qquad (190)$$

where ϕ is a rational function, and made explicit this formula for the inverse:[15]

$$\phi(t)=\frac{1}{t}\left(\text{coefficient of}\,\frac{1}{x}\,\text{in}\,\frac{g(x)}{t^x}\right). \qquad (191)$$

It was the difficulty of finding a general formula for the inverse, as well as the difficulty of solving integral equations of any kind, that prevented the Laplace transform from revealing its power for so long;[16] Fourier's quick success in obtaining his inverse transform (179) was crucial to the much more rapid development of his methods. It should be pointed out that when the phrase "Laplace transformation" is found in nineteenth-century mathematical literature, the reference is normally either to Laplace's method of reducing partial differential equations or to his method of cascades (compare Sections 12 and 7 respectively). When progress did come, it was through the aid of Fourier analysis. For example, in 1859 Riemann converted the transform.

$$g(s)=\int_0^{\infty} h(x)\,x^{-s}\,d(\log x)\,,\ s=a+b\sqrt{-1}, \qquad (192)$$

to a sum of Fourier sine and cosine integrals, and he inverted both, by means of the appropriate versions of (179), to end up with[17]

$$h(y)=\frac{1}{2\pi\sqrt{-1}}\int_{a-\infty\sqrt{-1}}^{a+\infty\sqrt{-1}} g(s)\,y^s\,ds. \qquad (193)$$

Dini studied such inversions, in contour form, in 1880, and again Fourier analysis provided the means.[18] Laplace transform theory had to await such events as Poincaré's 1885 analysis of asymptotic solutions to differential equations,[19] and especially the development of Heaviside's operational calculus in the 1890's.[20] Modernly recognizable proofs of the inversion formula were produced early in the twentieth century,[21] and they helped substantially in the exegesis of Heaviside's ideas in the 1920's with the "operational calculus" and its applications. J. R. Carson's work was especially significant in systematizing the operational calculus, based on the Laplace transform, for its use in electrical circuit theory.[22] Thus textbooks on both theory and applications began to appear in the 1930's and early 1940's.[23] Then, with the theory well launched, some of its development was classified during World War II, for it was used in problems such as waveguide design for radar systems. A politically acute man like Laplace would have appreciated that.

NOTES TO SECTION 29

1. See esp. art. 6 of L. Euler, "De constructione aequationum" (1744), in *Opera omnia*, 1st ser., XXII, 150–161.
2. See arts. 6 ff. of L. Euler, "Methodus aequationes differentiales . . ." (1753), in *Opera omnia*, 1st ser., XXII, 181–213.
3. See, for example, esp. art. 28 of L. Euler, "Recherches sur l'intégration de l'équation . . ." (1766), in *Opera omnia*, 1st ser., XXIII, 42–73. On these and other references, see (N: Petrova [1975]).
4. *Institutiones calculi integralis*, II (1769), chs. 3–4 and 5 respectively; in *Opera omnia*, 1st ser., XII.
5. See esp. arts. 37–42 of J. L. Lagrange, "Mémoire sur l'utilité de la méthode. . ." (1773), in *Oeuvres*, II, 171–234.
6. This is well conveyed in, for example, Francis D. Murnaghan, *The Laplace Transformation* (Washington, D.C., 1962).
7. For further details on these matters, see (J: Grattan-Guinness with Ravetz [1972]).
8. S.-D. Poisson, "Mémoire sur les solutions. . .," in *Journal de l'École polytechnique*, cahier 13, **6** (1806), 60–116, esp. 109–111.
9. Fourier's most detailed treatment is in arts. 342–385 of his *Théorie analytique de la chaleur* (1822), in *Oeuvres*, I, 387–448.
10. Some hints are given in n. 7, chs. 21 and 22; and in chs. 6–10, *passim*, of (N: Burkhardt [1908]).
11. See (J: Grattan-Guinness with Ravetz [1972]), ch. 16, note 7.
12. See esp. S.-D. Poisson, "Mémoire sur la théorie des ondes," in *Mémoires de l'Académie royale des sciences de l'Institut de France*, **1** (1816), 71–186.
13. See esp. A.-L. Cauchy, "Sur les intégrales des équations linéaires. . ." (1823), in *Oeuvres*, 2nd ser., I, 275–357; and its continuation by operational means in "Sur l'analogie

des puissances et des différences" (1827), *ibid.*, 2nd ser., VII, 198–254. The form of (188) received some attention in the context of probability in S.-D. Poisson, *Recherches sur la probabilité des jugements. . .* (Paris, 1837), ch. 4.

14. N. H. Abel, "Sur les fonctions génératrices et leurs déterminantes," in *Oeuvres complètes*, B. Holmboe, ed. (1839), II, 77–88—he slightly misstates (189); *Oeuvres complètes*, L. Sylow and S. Lie, eds. (1881), II, 67–81.
15. R. Murphy, "On the Inverse Method of Definite Integrals . . .," in *Transactions of the Cambridge Philosophical Society*, **4** (1833), 353–408; see 362.
16. Literature on this history includes Harry Bateman, *Report on the History and Present State of the Theory of Integral Equations* (London, 1911); and Hans Hahn, "Bericht über die Theorie der linearen Integralgleichungen," in *Jahresbericht der Deutschen Mathematiker-Vereinigung*, **20** (1911), 69–117.
17. B. Riemann, "Ueber die Anzahl der Primzahlen unter einer gegebenen Grösse" (1859), in *Gesammelte mathematische Werke*, 2nd ed. (Leipzig, 1892), 144–155; see 149.
18. See esp. arts. 62–88 *passim*, of Ulisse Dini, *Serie di Fourier*. . . (Pisa, 1880), in his *Opere*, IV.
19. Henri Poincaré, "Sur les équations linéaires . . ." (1885), in *Oeuvres*, I, 226–289. Amusingly, Poincaré misnames it "Bessel transformation" throughout the paper and puts a corrective note at the end. The *Oeuvres* edition silently makes the correction, although the original page numbers are incorrectly given.

    The use of the term "Laplace transform" may stem from Boole's advocacy of the form $\int e^{ux} V(x)$ to solve differential equations in his *A Treatise on Differential Equations*, 2nd ed. (Cambridge, 1865), ch. 18, where its advocacy also in J. Petzval's *Integration der linearen Differentialgleichungen* . . ., 2 vols. (Vienna, 1853–1859), I, 38–119, 328–395 (see also II, 369–379) is mentioned. Notice, however, that in ch. 17 Boole uses the phrase "Laplace transformation" to refer to Laplace's method of cascades—and Petzval uses "Laplace's integral" to refer (I, 96) *only* to the equation $\int_0^\infty e^{-x^2}\,dx = \frac{1}{2}\sqrt{\pi}$, which is a special case of (174).
20. Oliver Heaviside, *Electromagnetic Theory*, 3 vols. (London, 1893–1912).
21. See esp. H. M. MacDonald, "Some Applications of Fourier's Theorem," in *Proceedings of the London Mathematical Society*, 1st ser., **35** (1902–1903), 428–443; H. Mellin, "Abriss einer einheitlichen Theorie. . .," in *Mathematische Annalen*, **68** (1910), 305–337; and T. J. I'A. Bromwich, "Normal Coordinates in Dynamical Systems," in *Proceedings of the London Mathematical Society*, 2nd ser., **15** (1916), 401–448.
22. See esp. J. R. Carson, *Electric Circuit Theory and the Operational Calculus* (New York, 1926; repr. New York, 1953). The book is essentially a reprint of his articles with the same title in *Bell System Technical Journal*, **4** (1925), 685–761; and **5** (1926), 50–95, 336–384; see also his "The Heaviside Operational Calculus," *ibid.*, **1** (1922), pt. 2, 43–55; and "Notes on the Heaviside Operational Calculus," *ibid.*, **9** (1930), 150–162.
23. In German, see esp. G. Doetsch, *Theorie und Anwendung der Laplacesche Transformation* (Berlin, 1937; repr. New York, 1943). In English, for theory see D. V. Widder, *The Laplace Transform* (Princeton, 1946); and for applications, H. S. Carslaw and J. C. Jaeger, *Operational Methods in Applied Mathematics* (Oxford, 1941). For bibliography, see M. F. Gardner and J. L. Barnes, *Transients in Linear Systems Studied by the Laplace Transformation* (New York, 1942), 359–382.

I am indebted to Jock MacKenzie and Stephen Stigler for advice on this article.

I. GRATTAN-GUINNESS

BIBLIOGRAPHY

This bibliography is divided into two major sections and is arranged under the following headings:

ORIGINAL WORKS
- A. Correspondence
- B. *Exposition du système du monde*
- C. *Traité de mécanique céleste*
- D. *Théorie analytique des probabilités*
- E. *Essai philosophique sur les probabilités*
- F. Translations of *Mécanique céleste*
- G. *Oeuvres de Laplace* (1843–1847)
- H. *Oeuvres complètes de Laplace* (1878–1912)
- I. Individual Memoirs
    1. Order of Composition
    2. Order of Publication

SECONDARY LITERATURE
- J. Biographical and General
- K. Celestial Mechanics
- L. Probability
- M. Mathematical Physics
- N. Mathematics

The key to abbreviations will be found in Section I.

ORIGINAL WORKS

A. Correspondence. Laplace conducted an extensive correspondence with leading scientists throughout his lifetime; and it must have been important to him as a record of his thoughts, for he kept copies of his own letters. Unfortunately, it was consumed, along with all the other papers in the possession of the family, by a fire that swept through the château of Mailloc in Normandy in 1925. The property was then owned by his great-great grandson, the comte de Colbert-Laplace. In a letter to Karl Pearson [J] he tells of his intention to publish the Laplace-Lagrange correspondence and of his grandmother's recollections of her grandfather. Also, F. N. David has stated that much personal and scientific material, including all correspondence with English scientists, was destroyed during the British bombardment of Caen in 1944, but no details are given and no authority is cited ("Some Notes on Laplace," in J. Neyman and L. M. LeCam, eds., *Bernoulli, Bayes and Laplace* [Berlin–Heidelberg–New York, 1965], 30–44). In any case, the losses are irreparable, for with one exception (*17a*) the editors of the *Oeuvres complètes*, 14 vols. (Paris, 1878–1912), failed to include anything not already in print.

A few of Laplace's other letters have been printed in editions devoted to his correspondents. There are fourteen letters to Lagrange and twelve from Lagrange in *Oeuvres de Lagrange*, XIV (Paris, 1892); and several others in *Oeuvres de Lavoisier, Correspondance*, René Fric, ed., 3 fascs. to date (Paris, 1955–1964). Occasional documents are to be found at the Bibliothèque de l'Institut de France, MS 2242, and next door at the Archives de l'Académie des Sciences. From the former repository R. Taton has published a few letters and other pieces illustrating Laplace's relations with Lacroix from 1789 until 1815, "Laplace et Sylvestre-François Lacroix," in *Revue d'histoire des sciences*, **6** (1953), 350–360; and from the latter R. Hahn has summarized a fragment on theology, "Laplace's Religious Views," in *Archives internationales d'histoire des sciences*, **8** (1955), 38–40. Yves Laissus has published letters to Alexis Bouvard (20 Feb. 1797) and J.-B. Delambre (29 Jan. 1798), in *Revue d'histoire des sciences*, **14** (1961), 285–296.

In 1886 Charles Henry published six letters from Laplace to Condorcet and d'Alembert, of which the originals are in the Condorcet papers at the Institut de France, and four together with a fragment on the orbits of comets from the papers of the Abbé A.-G. Pingré in the Bibliothèque Sainte-Geneviève, in *Bollettino di bibliografia e storia delle scienze matematiche e fisiche*, **19** (1886), 149–178. These documents are reprinted in *OC*, XIV, 340–371. See below, (*34*).

Two letters by Laplace as minister of the interior, dated 17 frimaire and 26 frimaire an VIII (8 Dec. and 17 Dec. 1799), the latter accompanied by a report, appear in the *Moniteur universel* (an VIII), no. 78 (18 frimaire), 307–308; and no. 87 (27 frimaire), 343–345.

There are scattered autographs of Laplace in other collections. For example, there are reports on his examination of artillery cadets in 1784, 1785, and 1786, and his recommendations for reform of the system in July and August 1789, at the Archives de la Guerre, X^{D} 249 (Vincennes). There is fragmentary correspondence with Fourier in the Fourier papers, Bibliothèque Nationale, fonds français, MSS 22501, 68–74; and 22529, 122–124. The Library of the American Philosophical Society, Philadelphia, has a few letters to him from Madame de Laplace and their son, and several from Laplace to Giovanni Fabroni in Florence. The Library of the Royal Society of London has a dozen letters to English colleagues, six to Charles Blagden, one to Thomas Young, and five to J. F. W. Herschel. But such glimpses are fleeting, and to a very large extent the history of Laplace's work must be reconstructed from the internal evidence of the published writings, controlled as to chronology by the records of the Académie des Sciences.

It will be convenient to give first the major treatises in the form in which they were published during Laplace's lifetime, followed by details of the translations, the two collected editions, and the individual memoirs. The place of publication is Paris, unless indicated otherwise.

B. Exposition du système du monde, 2 vols. (an IV [1796]); 2nd ed. (an VII [1799]); 3rd ed. (1808); 4th ed. (1813); 5th ed. (1824). An edition printed by de Vroom in Brussels in 1826 and 1827 appears to be a reprint of the 5th ed., even though the latter printing is called *sixième édition* on the title page. It seems probable that the true 6th ed., for which Laplace was reading proof at the end of his life, was delayed until 1835, when it appeared in Paris, prefaced by the *éloge* delivered by Fourier on 15 June 1829 before the Institut de France [J]. This edition occupies vol. VI of [G] and [H] below. For the differences between it and the 5th ed., see the discussion under [H].

C. Traité de mécanique céleste: vols. I and II (an VII [1799]); vol. III (an XI [1802]); vol. IV (an XIII [1805]); vol. V (1823–1825). A 2nd ed. was published in 4 vols. (1829–1839). This work occupies vols. I–V of [G] and [H] below.

D. Théorie analytique des probabilités (1812); 2nd ed. (1814); 3rd ed. (1820). With 3 supps. published in 1816, 1818, and 1820, respectively, this work, with a 4th supp. (1825), occupies vol. VII of [G] and [H] below. A facsimile of the 1st ed. was published by Éditions Culture et Civilisation (Brussels, 1967).

E. Essai philosophique sur les probabilités (1814), originally published as the "Introduction" to the 2nd ed. of [D] above; 2nd ed. (1814); 3rd ed. (1816); 4th ed. (1819); 5th ed. (1825). It is included with [D] above in vol. VII of both [G] and [H] below. A facsimile of the 1st ed. was published by Éditions Culture et Civilisation (Brussels, 1967).

F. Translations of Mécanique céleste: J. C. Burckhardt published a German trans. of vols. I and II as *Mechanik des Himmels*, 2 vols. (Berlin, 1800–1802). There are two early English translations of Book I and Books I and II respectively: by John Toplis (London–Nottingham, 1814) and by Henry Harte, 2 vols. (Dublin, 1822–1827). Both were entirely superseded by the splendid work of Nathaniel Bowditch, *Mécanique céleste by the Marquis de Laplace, Translated With a Commentary*, 4 vols. (Boston, 1829–1839). Bowditch's commentary in the footnotes is an indispensable vade mecum for the study of Laplace, explaining and filling out the demonstrations, and containing a great body of historical as well as mathematical and astronomical elucidation. Bowditch did not translate Volume V of *Mécanique céleste*.

G. Oeuvres de Laplace, 7 vols. (1843–1847), reprints [B], [C], and [D] above, with [E] included as the intro. to [D]. Its publication, initiated by Laplace's widow, was eventually subsidized by the state.

H. Oeuvres complètes de Laplace, 14 vols. (1878–1912), was financed by a bequest from Laplace's son, General the Marquis de Laplace, who died on 7 October 1874. His will entrusted the task to the Académie des Sciences. There is correspondence in the Laplace dossier in its archives concerning the arrangements between his niece, the comtesse de Colbert-Laplace, the permanent secretaries, and the publisher, Gauthier. General Laplace expressly directed that the edition was to contain neither commentary nor extraneous elements. He based this injunction on what he took to be the wishes of his father, who had often said in the last months of his life that no corrections should be made in the works of savants after their death; modifying their writings in any way could only distort the record of their initial thoughts and be prejudicial to the history of science.

In only one respect (apart from correction of typographical errors) was the new edition to depart from [G] in its printing of the major treatises in the first seven volumes. Before his death, Laplace had begun correcting the proofs for a 6th ed. of *Exposition du système du monde* [B]. His intention had been to return to the 4th ed. and restore its chapters 12 ("De la stabilité et de l'équilibre des mers"), 17 ("Réflexions sur la loi de la pesanteur universelle"), and 18 ("De l'attraction moléculaire"), which he had omitted from the 5th ed. In the foreword to the projected 6th ed. Laplace said that he intended a separate work bringing together "the principal results of the application of analysis to the phenomena due to molecular actions distinct from universal attraction, which had just been much extended." Since he did not have the time for that, his son considered it consistent with his principles to restore these chapters from the 4th ed.

For the rest, Volumes VIII–XII contain the individual memoirs published by the Académie des Sciences through 1793 and by the Institut de France after 1795. Volume XIII contains writings published in *Connaissance des temps* from 1798 until Laplace's death, and Volume XIV, memoirs reprinted from *Journal de l'École polytechnique* (most notably his course on mathematics given at the École Normale in 1795); articles from *Journal de physique, Annales de physique et chimie, Journal des mines*, and so on; scattered items of correspondence; and fragments concerning annuities, rents, and matters of public interest. There is a very inadequate "Table analytique."

Despite their instructions, the editors did modernize much of Laplace's notation, for which reason it is preferable to have recourse to the original printing whenever possible, particularly in the case of the earlier memoirs. Since that is possible only in large research libraries, it has seemed practical to give quotations from the original sources and to make page references to the *Oeuvres complètes* whenever the work in question is contained in them. Unfortunately, however, the edition is not exhaustive; the editors simply went through the journals mentioned above and reprinted what they found seriatim. Thus, they missed Laplace's earliest papers [*1771a, 1771b, 1774a*] published in Leipzig and Turin, as well as one major treatise, *Théorie du mouvement et de la figure elliptique des planètes* (1784), and other, lesser writings. It is quite possible that further publications that escaped their net may also have eluded us.

I. Individual Memoirs. For details of separate printings and also for later editions and translations of the *Système du monde* and *Essai philosophique sur les probabilités*, the reader will do well to have recourse to the Library of Congress, *National Union Catalog*, and the British Museum, *General Catalogue of Printed Books*, as well as to the Bibliothèque Nationale, *Catalogue général des livres imprimés*. The last named contains a phantom that should here be exorcised. It lists among the writings of Laplace *Essai sur la théorie des nombres. Second supplément* (1825). The volume corresponding to that call number (V. **7051**) has "Laplace" pencilled on the flyleaf but is in fact Legendre's 2nd supp. to his *Essai sur la théorie des nombres* (1808). Another copy is bound with the 2nd ed. of that work (1825).

What follows is (1) a chronology of Laplace's work in the order of the composition of particular writings, and (2) a bibliographical listing of his memoirs in the order of publication. As will become evident, the two sequences are not everywhere the same. Prolonged immersion in these confusing details has convinced us that any redundancy in this double listing will be more than compensated by the facility it creates for keeping the problem of tracing the development of Laplace's research distinct from the problem of tracing the history of the influence of his publications. The cross-references make it possible to relate the two sequences at any juncture.

The first list has been established from the records of the Académie des Sciences. The *procès-verbaux*, or minutes of the semiweekly meetings of that body, were transcribed into a register that is conserved in the Archives de l'Académie des Sciences in the Institut de France. Before Laplace was elected to membership, the papers that he submitted were referred to commissions for evaluation, according to the normal practice. The record of these reports is maintained in a separate register (which, however, is not complete); and many of the reports themselves remain in the archives, classified by date. The sequence of these early communications, (*1*)–(*13*) in our numbering, was investigated by Stephen M. Stigler for "Laplace's Early Work: Chronology and Citations," in *Isis,* **69,** no. 247 (June 1978). He has most generously communicated to us a preprint together with notes of his researches in the archives in the summer of 1976. The MSS of the papers themselves were normally returned to the contributor and have not been conserved in the archives. In rare instances, the text was transcribed in the *procès-verbaux*. After Laplace's election to the Academy, the only record we have consists in most instances of the original title of his communications and the date on which he read or simply submitted them. In certain instances he formally requested recognition of priority—*pour prendre date.* Unless otherwise indicated, the dates and titles below (through 1793) are from the *procès-verbaux*, which carry through to the suppression of the Academy on 8 August of that year by the Revolutionary Convention, and thereafter from the published *procès-verbaux* of the Institut de France.

Since the foundation of the Institut de France in 1795, the Académie des Sciences has formed one of its constituent bodies, as its Classe des Sciences Physiques et Mathématiques from 1795 to 1814 and under the name Académie des Sciences since the Restoration. For the period 1795–1835 its *procès-verbaux* have been published together with the reports of committees, which are even more valuable. In the Academy before 1793 and in the Institute after 1795, Laplace served on many commissions concerned with evaluating the work of others or with special projects. We have included in this listing only the reports for which he was primarily responsible as author or spokesman. His other involvements in the affairs of the Institute were manifold, particularly in relation to the prize programs. They may be followed by means of the indexes to each volume of the *procès-verbaux*.

As for the second listing, that of publications, a word is needed about the organization and dating of the Academy's memoirs. Under the old regime, an annual volume was published under the general title *Histoire et mémoires de l'Académie royale des sciences de Paris*. The two sections are separately paginated and for this reason are cited separately (see the table of abbreviations below). The *Histoire* consists of announcements—prizes, distinguished visitors, works received, and so on—together with an abstract of the memoirs prepared by the permanent secretary. The *Mémoires*, which constitute the bulk of each volume, are the scientific papers themselves published by the members. Confusion in dating easily arises, because publication was always two to four years in arrears. Thus the volume for the year 1780 appeared in 1784 and contained memoirs submitted in 1783—or any time between the nominal and publication dates. Our method is to indicate that volume as *MARS* (1780/1784), with the latter date that of publication, and to specify memoirs in the bibliography by the date of publication. The *Connaissance des temps* presents the reverse problem. The volume of this almanac "for 1818," for example, contained the ephemerides for that year but appeared in 1815. Thus, it is cited *CT* (1818/1815), with the latter date still that of publication.

Journals proliferated after 1795, and Laplace then fell into the practice of publishing short papers and abstracts of his long memoirs in the *Connaissance des temps* and other periodicals mentioned in the list of abbreviations. More often than not, he would publish the same piece, sometimes with minor modifications, in several journals. The Laplace entry in the Royal Society of London, *Catalogue of Scientific Papers, 1800–1863*, III (1869), cites all the journals in which each piece appeared, as well as the translations that appeared in Germany and in Britain. We have limited ourselves to the most readily accessible journal; and when more than one is indicated, we have cited the number of the memoir (indicated by *RSC*) as listed in the catalog. The editors took the ostensible date of the various journals at face value and thus cannot always be relied on for their dating.

In addition, Laplace published a few reports concerned with the administration of the Institute and École Polytechnique. Also, the opinions that he delivered in his political capacity as a senator in the Napoleonic regime and a member of the Chambre des Pairs after the Restoration were printed in the proceedings of these bodies. Although we are not noticing his rare interventions in debates, it has seemed practical to include these episodic political and administrative pieces in the chronology of publications, and to distinguish them from the scientific writings by a "*P*" placed before the date.

The following abbreviations are employed:

AC	*Annales de chimie*
AP	*Archives parlementaires*
BSPM	*Bulletin de la Société philomathique de Paris*
CT	*Connaissance des temps*
CX	*Correspondance de l'École polytechnique*
HARS	*Histoire de l'Académie royale des sciences de Paris* (the preface to the corresponding volume of *MARS* below)
JP	*Journal de physique*
JX	*Journal de l'École polytechnique*
MARS	*Mémoires de l'Académie royale des sciences de Paris*
MASIF	*Mémoires de l'Académie royale des sciences de l'Institut de France* (**1** [1816/1818]–. . .).
MI	*Mémoires de l'Institut national des sciences et arts; sciences mathématiques et physiques*, 14 vols. (thermidor an VI [1798]–1818), for the Directory and the Napoleonic period
OC	*Oeuvres complètes de Laplace* [H]
PV	"Registre des procès-verbaux des séances de l'Académie royale des sciences de Paris"
PVIF	Académie des sciences, *Procès-verbaux des séances . . . depuis la fondation de l'Institut jusqu'au mois d'août 1835*, 10 vols. (Hendaye, 1910–1922)
RSC	Royal Society of London, *Catalogue of Scientific Papers (1800–1863)*, III (London, 1869), 845–848
SE	*Mémoires de mathématique et de physique, présentés . . . par divers sçavans* (often cited *Savants étrangers*), 11 vols. (1750–1786)

I, 1. Individual Memoirs by Order of Composition

(1) 28 Mar. 1770. "Recherches sur les maxima et minima des lignes courbes." Referees: Borda and Condorcet. Report, 28 Apr. 1770, printed in Bigourdan (J: [1931]). Published as [*1774a*] below.

(2) 18 July 1770. "Sur quelques usages du calcul intégral appliqué aux différences finies." Referees: Borda and Bossut. Report, 1 Sept. 1770, is in the archives of the Académie. The last paragraph is printed in Bigourdan (J: [1931]). An early draft of [*1771b*].

(3) 28 Nov. 1770. "Sur une méthode pour déterminer la variation de l'écliptique du mouvement des noeuds et de l'inclinaison de l'orbite des planètes." Referees: Condorcet and Bossut. Report, 12 Dec. 1770, is in the archives of the Académie.

(4) Date unrecorded, but probably Dec. 1770. "Sur la détermination de la variation de l'inclinaison et les mouvements des noeuds de toutes les planètes et principalement la variation de l'obliquité de l'écliptique." Referees: Dionis du Séjour, Bezout, and Condorcet. Report, 9 Jan. 1771, is in the archives. An entry in *PV* for that same date refers to a "Suite" to a memoir with an almost identical title to *(4)* and names d'Alembert and Condorcet as referees. This is probably an error, since *(4)* was itself a sequel to *(3)*; and it is more likely that d'Alembert and Condorcet were referees for *(5)*.

(5) 19 Jan. 1771. "Sur le calcul intégral." No referees recorded in *PV*, and the archives contain no report; but

see *(4)* above. The memoir was translated into Latin and was published in *Nova acta eruditorum*, as per [*1771a*] below.

(6) 13 Feb. 1771. "Sur le calcul intégral, les suites récurrentes, et la détermination de l'orbite lunaire." Referees: Condorcet and Bossut. Report, 20 Mar. 1771, is in the archives. The register of reports further specifies the topics as "1° Sur l'intégration de l'équation linéaire d'un ordre quelconque. 2° Sur une généralisation de la méthode qu'il a déjà employé pour les séries récurrentes. 3° Sur une application des formules de son mémoire sur l'obliquité de l'écliptique à des équations de l'orbite lunaire."

(7) 4 May 1771. "Sur les perturbations du mouvement des planètes causées par l'action de leurs satellites." Referees: Bezout and Bossut. Report, 15 May 1771, is in the archives.

(8) 17 May 1771. "Sur le calcul intégral appliqué aux différences finies à plusieurs variables." Referees: Borda and Bossut. There is a report in the archives, but it is dated the same day as the memoir and is by Condorcet. It describes the memoir as an extension of a previous work, which must be *(2)*, on which Borda and Bossut had reported. Evidently a revised draft of [*1771b*].

(9) 27 Nov. 1771. "Une théorie générale du mouvement des planètes." Referees: d'Alembert, Bezout, and Bossut. The report, 29 Jan. 1772, is in the archives. It describes the paper as concerned with "la théorie Newtonienne des planètes" and is sufficiently full to permit the conclusion that Laplace expanded this memoir to constitute his first general treatise on celestial mechanics, [*1784a*] below.

(10) 5 Feb. 1772. "Sur les suites récurrentes appliquées à la théorie des hasards." Referees: Dionis du Séjour and Le Roi. Report, 26 Feb. 1772, is in the archives. Probably a draft of [*1774b*]. The clerk failed to record the term "recurro-recurrent," although the referees allude to it.

(11) 2 May 1772. "Recherches pour le calcul intégral." Referees: Le Roi and Condorcet. The report, 6 May 1772, is in the archives and identifies the topic as "sur les solutions des équations différentielles non comprises dans l'intégral général." The subject of singular solutions occupies the first part of [*1775a*]. This paper, *(11)*, was probably combined with *(12)* and was superseded by *(14)*.

(12) Date unrecorded in *PV*. "Nouvelles recherches sur les intégrales particulières." Referees: Le Roi and Condorcet. Report, 30 May 1772, is in the archives.

(13) 10 Mar. 1773. "Recherches sur l'intégration des équations différentielles aux différences finies et sur leur application à l'analyse des hasards." Reading continued 17 Mar. Referees: Le Roi, Borda, and Dionis du Séjour. Report, 31 Mar. 1773, is in the archives. Revised version presented "pour retenir date," 7 Dec. 1773. Published as [*1776a*, 1°] below. This was the last memoir that Laplace submitted prior to his election to the Académie on 31 Mar., the very date of the report. The referees were dazzled and wound up their account with the following judgment: "Tel est le mémoire. . . . Nous en avons dit beaucoup de choses avantageuses dans le courant de notre rapport, et nous sommes persuadés que le petit nombre de savants qui le liront en porteront le même jugement, et nous croions même qu-ils ajouteront à nos éloges. Enfin nous ne craignons pas d'avancer que cet ouvrage donne dès à présent à son auteur un rang très distingué parmi les géomètres."

(14) 14 July 1773. "Recherches sur les solutions particulières des équations différentielles." Reading continued 21 July. Published with *(15)* and *(17)* as [*1775a*].

(15) 27 Apr. 1774. "Une suite du mémoire sur les équations séculaires des planètes." The reference is probably to an early draft of *(17)* and hence of the second part of [*1775a*]. The memoir of which it is said to be the sequel was probably [*1776a*, 2°].

(16) 31 Aug. 1774. "Sur le calcul intégral." Probably the draft of the first part of [*1776c*].

(17) 17 Dec. 1774. "Sur les inégalités séculaires des planètes." Presented "pour retenir date." This memoir must have become the second part of [*1775a*] and may have been a revision of *(15)*.

(17a) 6 Sept. 1775. Report (with d'Alembert, Borda, Bezout, and Vandermonde) on Dionis du Séjour, *Essai sur les phénomènes relatifs aux disparitions périodiques de l'anneau de Saturne* (Paris, 1776); *OC*, XIV, 333–339. It is unclear why the editors of *OC* saw fit to include this one among Laplace's reports from the *registres* of the old Académie—and nothing else.

(18) 28 Feb. 1776. Began reading "Un mémoire sur les nombres." The only trace that exists of what was probably this piece is a remark in a letter from Lagrange thanking Laplace for sending him several memoirs and observing about one of them, "Votre démonstration du théorème de Fermat sur les nombres premiers de la forme 8n + 3 est ingénieuse . . ." (Lagrange to Laplace, 30 Dec. 1776, *Oeuvres de Lagrange*, XIV, 67). See also Laplace's reply, 3 Feb. 1778, acknowledging receipt of a further communication from Lagrange on Fermat's theorem (*ibid.*, 74). The only other recorded involvement of Laplace with number theory was his service as referee, together with Bezout, on two committees reviewing works by the Abbé Genty on prime numbers, 23 Aug. 1780 and 18 July 1781 (*PV*, **99**, fols. 219–220; and **100**, fol. 155); and on a third with Lagrange and Lacroix, 26 Mar. 1802 (*PVIF*, II, 485). See *(78)* below.

(19) 4 Dec. 1776. There is no record in the *PV*, but a marginal note in "Recherches sur le calcul intégral aux différences partielles" [*1777a*] states that the memoir was submitted on this date, having been read in 1773.

(20) 22 Jan. 1777. "Recherches sur la loi de la pesanteur à la surface d'un sphéroïde homogène en équilibre." Laplace deposited the text on 25 Jan. and observed at the outset that this paper was a continuation of what he had begun in [*1776d*]. It is transcribed in full in *PV*, **96**, fols. 17–25; and is virtually identical with the opening section of [*1778a*]; see *OC*, IX, 71–87. The printed version omits an undertaking that Laplace placed at the end of the manuscript memoir. He there proposed in a further memoir to investigate the figure of Saturn and

the law of gravity resulting from the action of its rings. That intention he fulfilled in [*1789a*].

(21) 8 Mar. 1777. "Recherches sur le milieu qu'il faut choisir entre les résultats de plusieurs observations." This memoir is transcribed in *PV*, **96**, fols. 122–142. It has never been published. Arrangements are in hand to print it in a forthcoming issue of *Revue d'histoire des sciences et de leurs applications*.

(22) 9 Apr. 1777. "Sur la nature du fluide qui reste dans le récipient de la machine pneumatique." No trace remains of this piece.

(23) 7 May 1777. "Un mémoire sur les oscillations des fluides qui recouvrent les planètes." On 31 May, Laplace read an addition to this memoir, which he is recorded as having withdrawn. After revision, it was combined with *(20)* in [*1778a*].

(24) 15 Nov. 1777. A marginal note in [*1778a*] records this date for submission of the memoir, combining *(20)* and *(23)*, the first installment of the investigation of tidal phenoma, "Recherches sur plusieurs points du système du monde."

(25) 18 Feb. 1778. "Un mémoire sur le calcul intégral." Probably a draft for all or part of [*1780a*].

(26) 13 May 1778. "Recherches sur les ondes, pour servir à son mémoire imprimé dans le volume de 1775." The reference is to [*1778a*], and this investigation of wave motion was printed as the final article (XXXVII) of the second sequel, [*1779b*]; see *OC*, IX, 301–310.

(27) 7 Oct. 1778. A marginal note in [*1779a*] records the submission of the memoir. No mention in *PV*. This was the first sequel on tides and on the motion of the earth.

(28) 25 Dec. 1778. A marginal note in [*1779b*] specifies the date of submission. See also *(26)*, which was appended. This was the second sequel on tides and on the motion of the earth.

(29) 16 June 1779. A marginal note gives the date for submission of "Mémoire sur l'usage du calcul aux différences partielles dans la théorie des suites" [*1780b*].

(30) 7 July 1779. "Une addition à son mémoire actuellement sous presse sur la précession des équinoxes." The reference is to [*1780c*]. Laplace apparently read the addition before reading the memoir. See *(31)*.

(31) 18 Aug. 1779. "Un mémoire sur la précession des équinoxes." The draft of [*1780c*].

(32) 1 Sept. 1779. "Un écrit où il répond à quelques objections faites contre son mémoire sur la précession des équinoxes, imprimé en 1776" [sic].

(33) 31 May 1780. "Mémoire sur le calcul aux suites appliqué aux probabilités." Draft of [*1781a*], submitted for publication on 31 July according to a marginal note.

(34) 21 Mar. 1781. Read "Un mémoire sur la détermination des orbites des comètes." A draft of the analytical part (Articles I–VII) of [*1784b*]. Laplace evidently revised the calculations *(36)* after learning of Herschel's "comet." It is also likely that the draft read to the Academy did not contain the instructions for application in Article VIII, an early version of which Laplace communicated to the Abbé Pingré no later than November 1782. It was published with fragments of correspondence by Henry (A: [1886]; *OC*, XIV, 355–368) and is virtually identical with Article VIII of the published memoir ([*1784b*]; *OC*, X, 127–141.

(35) 2 May 1781. "Une application de sa méthode à la comète qui paroît actuellement."

(36) 13 June 1781. "Un mémoire sur une méthode de calculer l'orbite des comètes." The dossier for this session includes a note containing the calculations in Laplace's hand.

(37) 28 July 1781. "Des eléments de la comète de M. Herschel, déterminés par un nouveau calcul." The numerical data are transcribed in *PV*, **100**, fols. 160–161.

(38) 21 Dec. 1781. An entry records on behalf of Lavoisier and Laplace the date on which they deposited the description of a new "pyromètre" by means of which the elongation of solid bodies under the influence of heat could be measured to an accuracy of "0.01 lignes," which is to say about 0.001 inches. Accompanying the account of the instrument, which had been constructed, was a series of experiments on the dilation of glass and metals. These must certainly have been among the earlier experiments carried out in the garden of the Arsenal in 1781 and 1782 that were published after Lavoisier's execution, "De l'action du calorique sur les corps solides, principalement sur le verre et sur les métaux, et du rallongement ou du raccourcissement dont ils sont susceptibles . . ." [*1793c*], in *Oeuvres de Lavoisier*. II (1862), 739–759. This instrument may well have been the one for which Biot reconstructed the design in his *Traité de physique* (1816), aided by the recollections of Laplace and Madame Lavoisier (*Oeuvres de Lavoisier*, II, 760).

(39) 2 Mar. 1782. An entry records on behalf of Lavoisier and Laplace a series of experiments already begun that show that substances passing from the liquid to the gaseous state and from the gaseous to the liquid state emit negative or positive electricity. These experiments were clearly the basis of [*1784c*].

(40) 22 Jan. 1783. "Un mémoire sur la planète d'Herschel." The gist of this memoir, together with the results of the calculations in *(42)*, was evidently incorporated in [*1784a*], pt. I, nos. 14–17, pp. 28–59; but it does not appear that Laplace ever published *(35)*, *(36)*, *(37)*, and *(42)* per se. According to an annotation on a handlist of Laplace's memoirs contained in the dossier at the Académie des Sciences concerning publication of *OC*, there was a memoir on "Eléments de la nouvelle planète Ouranus," in the *Mémoires de l'Académie impériale et royale des sciences et belles-lettres de Bruxelles* in 1788. In fact, there is nothing by Laplace in that collection, but **5** (1788), 22–48, does contain a memoir by F. von Zach (or "de Zach," as he is called there) entitled "Mémoire sur la nouvelle planète Ouranus," presented on 20 May 1785. Laplace and Zach were in frequent correspondence, and Laplace supplied him with the elements of the orbit, pp. 43–44.

(41) 15 Mar. 1783. Report (with Bezout and d'Alem-

bert) on Legendre's memoir on "Attraction des sphéroïdes homogènes." Transcribed in *PV*, **102**, fols. 85-87. A review of the state of the problem. The Legendre memoir was published in *SE*, **10** (1785), 411-434.

(42) 21 May 1783. "Une note d'où il résulte que d'après ses calculs, et ceux de M. Méchain, la planète Herschel est la même chose qu'une étoile observée par Mayer et qui ne se retrouve plus." Data given in *PV*, **102**, fols. 119-120.

(43) 24 May 1783. "Un mémoire sur l'attraction des sphéroïdes elliptiques." Presented in addition on 31 May. This piece almost certainly constituted part II of [*1784a*].

(44) 18 June 1783. "Un mémoire, fait conjointement avec M. Lavoisier, sur une nouvelle méthode de mesurer la chaleur." Laplace is recorded as having read the memoir, published as [*1783a*].

(45) 25 June 1783. Lavoisier and Laplace announced that they had repeated the combustion of combustible air (hydrogen) combined with dephlogisticated air (oxygen) in the presence of several observers, and obtained pure water. This demonstration, which occupied the public meeting of the Academy for St. Martin's Day, took place exactly one week after Laplace had read their joint *Mémoire sur la chaleur* [*1783a*] (*PV*, **102**, fol. 104 [18 June 1783], fol. 144 [25 June 1783]). Lavoisier then published this and other experiments, "Mémoire dans lequel on a pour objet de prouver que l'eau n'est point une substance simple . . . ," in *MARS* (1781/1784), 468-494; also *Oeuvres de Lavoisier*, II (1862), 334-359.

(46) 23 July 1783. Laplace read experiments done in England on the freezing of mercury (*PV*, **102**, fol. 159). It is not clear whose experiments these were, although Laplace's interest in them almost certainly pertained to the work on heat that he continued with Lavoisier through 1783 into 1784, which is reported in "Mémoire contenant les expériences faites sur la chaleur, pendant l'hiver de 1783 à 1784, par. P. S. de Laplace & A. L. Lavoisier." Published after Lavoisier's death—see *(38)*—and in *Oeuvres de Lavoisier*, II (1862), 724-738.

(47) 3 Dec. 1783. Laplace requested the appointment of a commission to review his treatise on the motion and figure of the planets [*1784a*]. The referees, Dionis du Séjour, Borda, and Cousin, reported on 31 Jan. 1784, recommending publication under the *privilège* of the Académie, as was necessary for a work issued independently of the *MARS*. The report is given in *PV*, **103**, fols. 72-76.

(48) 11 Aug. 1784. "Un mémoire sur l'équilibre des fluides sphéroïdes." The draft of [*1785b*].

(49) 25 June 1785. A memoir "Sur les probabilités," intended as sequel to his "mémoire de 1782." The reference is to [*1785a*], the memoir on approximate solutions to problems involving functions containing terms raised to very high powers, and the present memoir was the draft of its sequel [*1786b*]. Laplace continued the reading on 28 June.

(50) 19 Nov. 1785. "Un mémoire sur les inégalités séculaires des planètes." Condorcet recorded the memoir on this date, and Laplace read it on 23 Nov. This is the draft of [*1787a*].

(51) 30 Nov. 1785. "Un mémoire sur la population de la France." The draft of [*1786c*].

(52) 6 May 1786. Laplace presented a theorem on the motions of Jupiter and Saturn "pour retenir date." On 10 May he read a draft of the first memoir on the theory of Jupiter and Saturn [*1788a*].

(53) 15 July 1786. "Un 2e mémoire sur la théorie de Jupiter et de Saturne." The draft of [*1788b*].

(54) 21 July 1787. On behalf of a committee consisting also of Cousin and Legendre, Laplace read a *compte-rendu* of R.-J. Haüy, *Exposition raisonée de la théorie de l'électricité et du magnétisme, d'après les principes d'Aepinus* (Paris, 1787). The account is very appreciative, and the report is transcribed in *PV*, **106**, fols. 290-293.

(55) 19 Dec. 1787. "Un mémoire sur l'équation séculaire de la lune." A first draft of [*1788c*].

(56) 2 Apr. 1788. "Un mémoire sur l'équation séculaire de la lune." Probably a revision rather than an extension of *(55)*, for the memoir [*1788c*] is a brief one.

(57) 26 Apr. 1789. A commission composed of Lagrange, Lalande, and Méchain presented a *compte-rendu* of tables for Jupiter and Saturn prepared by Delambre, which compare the predictions from Laplace's theory of the two planets to the record of observations. The report is transcribed in *PV*, **108**, fols. 92-99.

(58) 18 July 1789. "Un mémoire sur l'inclinaison de l'écliptique." This is probably the draft of Articles II-VII of [*1793a*].

(59) 17 Apr. 1790. "Un mémoire sur la théorie des satellites de Jupiter." The draft of [*1791a*].

(60) 15 Dec. 1790. Laplace began "Un mémoire sur le flux et le reflux de la mer." Delayed in publication [*1797a*].

(61) 21 Jan. 1796 (1 pluviôse an IV). Laplace presented "Un mémoire sur les mouvements des corps célestes autour de leurs centres de gravité." *PVIF*, I, 6. Draft of [*1798a*]. Readied for publication on 5 Jan. 1797 (16 nivôse an V).

(62) 7 Oct. 1796. (16 vendémiaire an V). Report (with Lagrange) on two memoirs by Flaugergues, "De l'aberration de la lumière" and "Du phénomène de l'apparence de l'étoile sur le disque de la lune dans les occultations." *PVIF*, I, 114-115.

(63) 20 Apr. 1797 (1 floréal an V). Read a "Mémoire sur les équations séculaires du mouvement des noeuds et de l'apogée de l'orbite lunaire et sur l'aberration des étoiles." *PVIF*, I, 203. Draft of [*1798b*].

(64) 10 Jan. 1798 (21 nivôse an VI). Read the preamble of a "Mémoire sur les équations séculaires du mouvement de la lune, de son apogée et de ses noeuds." *PVIF*, I, 330. Text of [*1799a*].

(65) 21 Dec. 1798 (1 nivôse an VII). Laplace, Lacépède, and Fourcroy submitted a report on the ques-

tions to be presented to the Institut d'Égypte. The report was combined with those from the 2nd and 3rd classes and was printed in *Histoire de la classe des sciences morales et politiques de l'Institut de France*, **3** (prairial an IX [May–June 1801]), 5–19.

(66) 5 Apr. 1799 (16 germinal an VII). A report (with Lagrange) on several memoirs of Parceval on "le calcul intégral aux différences partielles." *PVIF*, I, 546–547.

(67) 7 Sept. 1799 (21 fructidor an VII). Presented copies of *Exposition du système du monde* and *Mécanique céleste*. *PVIF*, I, 619.

(68) 2 Nov. 1799 (11 brumaire an VIII). A report (with Coulomb and Lefèvre-Gineau—it is not clear that Laplace was the author) on a memoir by Libes on the role of caloric in elasticity. *PVIF*, II, 21–22.

(69) 12 Nov. 1799 (21 brumaire an VIII). A report (with Napoléon Bonaparte and Lacroix) on a memoir of Biot on "les équations aux différences mêlées." *PVIF*, II, 30–32. See also the joint report with Prony (read by the latter on 6 frimaire an VIII [27 Nov. 1799], *PVIF*, II, 45–48), on Biot's memoir "Considérations sur les intégrales des équations aux différences finies," *PVIF*, II, 45–48, published in full, *MI*, **3** (prairial an IX [May–June 1801]), "Histoire," 12–21.

(70) 2 Mar. 1800 (11 ventôse an VIII). Read a "Mémoire sur le mouvement des orbites des satellites de Saturne et d'Uranus." *PVIF*, II, 118. Abstracted [*1800b*]. Draft of [*1801a*].

(71) 15 June 1800 (26 prairial an VIII). Read a "Mémoire sur la théorie de la lune." *PVIF*, II, 177. Draft of [*1801b*].

(72) 20 June 1800 (1 messidor an VIII). Announced Bouvard's application of a new equation contained in the previous memoir *(71)* to observations by Maskelyne, yielding a flattening of 1/314. *PVIF*, II, 179.

(73) 2 Dec. 1800 (11 frimaire an IX). Proposed continuing in the *Mémoires de l'Institut* the notes on the French population contained in the final volumes of the Académie under the old regime. *PVIF*, II, 274.

(74) 10 June 1801 (21 prairial an IX). Read a "Mémoire sur la théorie de la lune." *PVIF*, II, 359. Not separately published. May have been incorporated in *Mécanique céleste*, Book VII (Volume III, 1802).

(75) 12 Nov. 1801 (21 brumaire an X). Report (with Delambre) on a memoir "Sur la théorie de Mars" by Lefrançois-Lalande. *PVIF*, II, 426–429.

(76) 26 Jan. 1802 (6 pluviôse an X). Read a "Mémoire sur une inégalité à longue période, qu'il vient de découvrir dans le mouvement de la lune. . . ." *PVIF*, II, 457. Not separately published at this time. May have been incorporated in *Mécanique céleste*, Book VII (Volume III, 1802). Laplace returned to this topic in [*1811c*] and [*1812a*]. See also *(94)*.

(77) 12 Mar. 1802 (21 ventôse an X). Read a "Mémoire sur la théorie lunaire." *PVIF*, II, 476. Not separately published. May have been incorporated in *Mécanique céleste*, Book VII (Volume III, 1802).

(78) 27 Mar. 1802 (6 germinal an X). Report (with Lagrange and Lacroix) on a memoir of Genty on number theory. *PVIF*, II, 485.

(79) 17 Nov. 1802 (26 brumaire an XI). Announces measures that the government has adopted to resume making exact estimates of the size of the population by taking samples in various regions and calculating the factor by which the annual number of births is to be multiplied. *PVIF*, II, 595.

(80) 29 Dec. 1802 (8 nivôse an XI). Presented Volume III of *Traité de mécanique céleste*. *PVIF*, II, 606.

(81) 2 May 1803 (12 floréal an XI). Read a set of observations on the tides, upon which a committee consisting of himself, Levêque, and Rochon was appointed. *PVIF*, II, 659. See [*1803a*].

(82) 12 Sept. 1803 (25 fructidor an XI). Read a "Mémoire sur les tables de Jupiter et la masse de Saturne." *PVIF*, II, 703. Draft of [*1804a*].

(83) 27 May 1805 (7 prairial an XIII). Presented Volume IV of *Mécanique céleste* (the minute mistakenly says Volume XIV). *PVIF*, III, 216.

(84) 14 Oct. 1805 (22 vendémiaire an XIV). Read a "Mémoire sur la diminution de l'obliquité de l'écliptique." *PVIF*, III, 262. There is no record of publication.

(85) 23 Dec. 1805 (2 nivôse an XIV). Read a "Mémoire sur les tubes capillaires." *PVIF*, III, 293. Published in part [*1806a*].

(86) 28 Apr. 1806. Presented "Théorie de l'action capillaire." *PVIF*, III, 344. Originally issued (1806) as a separate booklet under the above title, this piece was incorporated in later printings of *Mécanique céleste*, Volume IV, as the first Supplement to Book X. *OC*, IV, 349–417. Abstracted [*1806c*].

(87) 29 Sept. 1806. Read a "Suite à sa théorie de l'action capillaire." *PVIF*, III, 431. Abstracted [*1807a*]. A printed copy, separately issued although sometimes bound with *(86)*, was presented to the Institut on 6 July 1807. *PVIF*, III, 353. This piece was incorporated in later printings of *Mécanique céleste*, Volume IV, as the second Supplement to Book X. *OC*, IV, 419–498.

(88) 24 Nov. 1806. Read a memoir on "L'adhésion des corps à la surface des fluides." *PVIF*, III, 451. Printed in part as [*1806d*].

(89) 21 Mar. 1808. Presented a copy of the 3rd edition of *Exposition du système du monde*. *PVIF*, IV, 36.

(90) 17 Aug. 1808. Presented *Supplément au Traité de mécanique céleste* to the Bureau des Longitudes. This appendix concerns the theory of planetary perturbations developed in Books II and VI, and was incorporated in later printings of Volume III (*OC*, III, 325–350). Laplace also presented a copy to the Institut on 26 Sept. 1808. *PVIF*, IV, 106. The last section (5) corrects an error in the sign in the expressions for the fifth power of the eccentricities and inclinations of the orbits in the theory of the inequalities of Jupiter and Saturn, *OC*, III, 349–350.

(91) 19 Dec. 1808. Report (with Haüy, Chaptal, and Berthollet) on the memoir of Malus, "Sur divers phénomènes de la double réfraction de la lumière." La-

place's report is printed in *PVIF*, IV, 145–147; and in *OC*, XIV, 321–326. *RSC* 22.

(92) 30 Jan. 1809. Read a memoir on "La loi de la réfraction extraordinaire de la lumière dans les milieux transparents." *PVIF*, IV, 159. Abstracted in [*1809b*]. Published as [*1810a*].

(93) 10 Apr. 1809. Reported (with Lacroix) on a memoir of Poisson on "La rotation de la terre." *PVIF*, IV, 190–192.

(94) 18 Sept. 1809. Read a memoir on "La libration de la lune." *PVIF*, IV, 253. This memoir was probably combined with *(76)* in [*1811c*] and [*1812a*].

(95) 9 Apr. 1810. Read a report on "Les probabilités." *PVIF*, IV, 341. A draft of [*1810b*].

(96) 21 May 1810. Reported (with Biot and Arago) on a memoir by Daubuisson on "La mesure des hauteurs par le baromètre." *PVIF*, IV, 350–352.

(97) 29 Apr. 1811. Read a memoir on "Les intégrales définies." *PVIF*, IV, 475. A draft of [*1811a*].

(98) 23 Mar. 1812. Presented the "première partie" of *Théorie analytique des probabilités. PVIF*, V, 34.

(99) 29 June 1812. Presented the "seconde partie" of *Théorie analytique des probabilités. PVIF*, V, 69. The first edition was thus issued in at least two installments.

(100) 24 May 1813. Presented the 4th edition of *Exposition du système du monde. PVIF*, V, 214.

(101) 2 Aug. 1813. Read a memoir on "Les éléments des variations des orbites planétaires."*PVIF*, V, 235. There is no record of publication.

(102) 14 Feb. 1814. Presented *Essai philosophique sur les probabilités* (1814). *PVIF*, V, 316. Incorporated with *(104)* as its Introduction. The version printed in *OC*, VII, v–cliii, is *(117)*, the 4th ed. (1819).

(103) 8 Aug. 1814. Read a memoir on "La probabilité des témoignages." *PVIF*, V, 386. Incorporated in *(104)*, the 2nd ed. of *Théorie analytique des probabilités*, Book II, chapter 11; *OC*, VII, 455–485.

(104) 14 Nov. 1814. Presented the 2nd ed. of *Théorie analytique des probabilités*, incorporating *(102)* and *(103)*, together with three minor additions. *PVIF*, V, 422. *OC*, VII, 471–493.

(105) 10 July 1815. Read a memoir on "Les marées." *PVIF*, V, 527. A draft of [*1815c*].

(106) 18 Sept. 1815. Read a memoir on "Les probabilités, dans lequel il détermine la limite de l'erreur qui peut rester après qu'on a déterminé les valeurs les plus probables des inconnues." *PVIF*, V, 554. The draft of [*1815a*].

(107) 26 Aug. 1816. Presented a supplement to *Théorie analytique des probabilités. PVIF*, VI, 73. Incorporated in the 3rd ed. (1820) as Supplement I. *OC*, VII, 497–530. See *(118)*.

(108) 28 Oct. 1816. Read a "Note sur la pendule." *PVIF*, VI, 108. Probably the draft of [*1816a*].

(109) 25 Nov. 1816. Read "Note sur l'action réciproque des pendules et sur la vitesse du son dans les diverses substances." *PVIF*, VI, 113. A draft of [*1816b*].

(110) 23 Dec. 1816. Read a "Note sur la vitesse du son." *PVIF*, VI, 131. Related to [*1816c*] and [*1816d*]. See the marginal note in the latter.

(111) 4 Aug. 1817. Read a memoir on "L'application du calcul des probabilités aux opérations géodésiques." *PVIF*, VI, 208. A draft of [*1817a*].

(112) 2 Feb. 1818. Presented a second Supplement to *Théorie analytique des probabilités. PVIF*, VI, 263. The supplement consists of *(111)* together with additions and was incorporated in the 3rd ed., *Théorie analytique des probabilités* (1820). *OC*, VII, 531–580. See *(118)*.

(113) 18 May 1818. Read a "Mémoire sur la rotation de la terre." *PVIF*, VI, 316. The draft of [*1819a*].

(114) 3 Aug. 1818. Read a memoir "Sur la figure de la terre et la loi de la pesanteur à sa surface." *PVIF*, VI, 350. Abstracted in [*1818b*] and published in [*1819b*].

(115) 26 May 1819. Read a memoir "Sur la figure de la terre" before the Bureau des Longitudes; see the marginal note in *CT* (1822/1820), 284. The draft of [*1820e*].

(116) 13 Sept. 1819. Read a memoir "Considérations sur les phénomènes capillaires." *PVIF*, VI, 487. Presumably the draft of [*1819h*].

(117) 25 Oct. 1819. Presented the 4th ed. of *Essai philosophique sur les probabilités. PVIF*, VI, 504. Replaced the 1st ed. *(102)* as Introduction to *Théorie analytique des probabilités (104)* in the 3rd ed. (1820). *OC*, VII, v–cliii.

(118) 20 Dec. 1819. Read a "Mémoire sur l'application du calcul des probabilités aux opérations géodésiques," *PVIF*, VI, 515. The draft of what became the 3rd Supplement to the 3rd ed. of *Théorie analytique des probabilités* (1820). *OC*, VII, 581–616. Abstracted in [*1819f*] and printed as [*1819g*].

(119) 19 Jan. 1820. Read a memoir "Sur les inégalités lunaires dues à l'aplatissement de la terre" before the Bureau des Longitudes. See the marginal note in [*1820a*].

(120) 28 Feb. 1820. Presented the 3rd ed. of *Théorie analytique des probabilités*, containing the revised Introduction *(117)* and three supplements *(107)*, *(112)*, and *(118)*.

(121) 29 Mar. 1820. Read "Sur le perfectionnement de la théorie et des tables lunaires" before the Bureau des Longitudes. See the marginal note in [*1820b*].

(122) 12 Apr. 1820. Read "Sur l'inégalité lunaire à longue période, dépendante de la différence des deux hemisphères terrestres" before the Bureau des Longitudes. See the marginal note in [*1820c*].

(123) 10 Sept. 1821. Read a "Mémoire sur l'attraction des corps sphériques et sur la répulsion des fluides élastiques." *PVIF*, VII, 222.

(124) 12 Dec. 1821. Presented a memoir on elastic fluids and the speed of sound to the Bureau des Longitudes. See the marginal note in [*1822a*].

(125) 17 Mar. 1823. Presented Book XI of *Mécanique céleste. PVIF*, VII, 457. *OC*, V, 6–96.

(126) 21 Apr. 1823. Presented Book XII of *Mécanique céleste. PVIF*, VII, 480. *OC*, V, 99–160.

(127) 8 Sept. 1823. Presented a memoir "Sur le flux et le reflux de la mer." *PVIF,* VII, 538. Published as [*1823a*].

(128) 5 Jan. 1824. Presented the 5th ed. of *Système du monde. PVIF*, VIII, 3.

(129) 9 Feb. 1824. Presented Book XIII of *Mécanique céleste. PVIF*, VIII, 24. *OC*, V, 164–269.

(130) 26 July 1824. Presented Book XIV of *Mécanique céleste. PVIF*, VIII, 117. *OC*, V, 273–323.

(131) 13 Dec. 1824. Presented Book XV of *Mécanique céleste. PVIF*, VIII, 162. *OC*, V, 327–387.

(132) 7 Feb. 1825. Laplace, together with his son, presented a fourth supplement to *Théorie analytique des probabilités. PVIF*, VIII, 182. *OC*, VII, 617–645.

(*133*) 16 Aug. 1825. Presented Volume V of *Mécanique céleste*, Book XVI being the final book. *PVIF*, VIII, 261. *OC*, V, 389–465.

(134) 23 July 1827. The Academy received a Supplement to *Mécanique céleste*, Volume V; the MS had been found among Laplace's papers after his death. *PVIF*, VIII, 571. *OC*, V, 469–505.

I, 2. Individual Memoirs by Order of Publication

[*1771a*] "Disquisitiones de calculo integrale," in *Nova acta eruditorum, Anno 1771* (Leipzig, 1771), 539–559; not in *OC*. A draft was presented to the Academy on 19 Jan. 1771; see (*5*) above.

[*1771b*] "Recherches sur le calcul intégral aux différences infiniment petites, et aux différences finies," in *Mélanges de philosophie et de mathématiques de la Société royale de Turin, pour les années 1766–1769* (*Miscellanea Taurinensia*, IV), date of publication not given but probably 1771, pp. 273–345. A typographical error numbers pp. 273–288 as 173–188. Laplace read an early draft to the Academy on 18 July 1770 and a revised version on 17 May 1771 (*2*) and (*8*) above. Not in *OC*. I am indebted to Mr. George Anastaplo for a discussion of this memoir.

[*1774a*] "Disquisitiones de maximis et minimis, fluentium indefinitarum," in *Nova acta eruditorum, Anno 1772* (Leipzig, 1774), 193–213. Not in *OC*. First draft read to the Academy on 28 Mar. 1770 (*1*) above. I am indebted to Mr. Chikara Sazaki for a discussion of this memoir.

[*1774b*] "Mémoire sur les suites récurro-récurrentes et sur leurs usages dans la théorie des hasards," in SE, **6** (1774), 353–371; *OC*, VIII, 5–24. A draft was presented on 5 Feb. 1772. Referees: Dionis du Séjour and Le Roi, who reported on 26 Feb. 1772 (*10*) above.

[*1774c*] "Mémoire sur la probabilité des causes par les événements," in *SE*, **6** (1774), 621–656; *OC*, VIII, 27–65. *PV* contains no mention of this memoir. It seems probable, however, that it was composed between March and December 1773, concurrently with the revision of [*1776a*, 1°].

[*1775a*] "Mémoire sur les solutions particulières des équations différentielles et sur les inégalités séculaires des planètes," in *MARS* (1772, pt. 1/1775), 343–377; *OC*, VIII, 325–366. Laplace read the first part (*OC*, VIII, 325–354) on 14 and 21 July 1773 (*14*), having started the topic in a paper of 2 May 1772 (*11*). The second part (*ibid*., 354–366) was registered on 17 Dec. 1774 (*17*).

[*1775b*] "Addition au mémoire sur les solutions particulières . . .," in *MARS* (1772, pt. 1/1775), 651–656; *OC*, VIII, 361–366.

[*1776a*] "Recherches, 1°, sur l'intégration des équations différentielles aux différences finies, et sur leur usage dans la théorie des hasards. 2°, sur le principe de la gravitation universelle, et sur les inégalités séculaires des planètes qui en dépendent," in *SE* (1773/1776), 37–232; *OC*, printed as two memoirs, VIII, 69–197, 198–275. A note printed in the margin (*SE* [1773/1776], 37) gives 10 Feb. 1773 as the date on which 1° was read. The register of *PV* gives 10 Mar., with reading continued on 17 Mar. See (*13*) above. We do not know when Laplace readied 2° for publication, although it must have been before 27 Apr. 1774, when he presented a further memoir called a "Suite"; see (*15*) and (*17*). As printed, 2° probably contains elements from (*3*), (*4*), (*6*), (*7*), and (*9*) above, but most of it goes beyond anything suggested by these titles.

[*1776b*] "Mémoire sur l'inclinaison moyenne des orbites des comètes, sur la figure de la terre, et sur les fonctions," in *SE* (1773/1776), 503–540; *OC*, VIII, 279–321. There is no record of when Laplace presented these topics.

[*1776c*] "Recherches sur le calcul intégral et sur le système du monde," in *MARS* (1772, pt. 2/1776), 267–376; *OC*, VIII, 369–477. Laplace read the first part on 31 Aug. 1774 (*16*).

[*1776d*] "Additions aux recherches sur le calcul intégral et sur le système du monde," in *MARS* (1772, pt. 2/1776), 533–554; *OC*, VIII, 478–501.

[*1777a*] "Recherches sur le calcul intégral aux différences partielles," in *MARS* (1773/1777), 341–402; *OC*, IX, 5–68. A marginal note says that this memoir was read in 1773 and was submitted for publication on 4 Dec. 1776 (*19*).

[*1778a*] "Recherches sur plusieurs points du système du monde," in *MARS* (1775/1778), 75–182; *OC*, IX, 71–183. A combination of (*20*) and (*23*), the latter revised and submitted on 15 Nov. 1777 (*24*).

[*1779a*] "Recherches sur plusieurs points du système du monde" (Suite), in *MARS* (1776/1779), 177–267; *OC*, IX, 187–280. A marginal note specifies that this continuation of [*1778a*], with which the articles are numbered consecutively, was submitted on 7 Oct. 1778.

[*1779b*] "Recherches sur plusieurs points du système du monde" (Suite), in *MARS* (1776/1779), 525–552; *OC*, IX, 283–310. Submitted on 25 Dec. 1778. See *(26)* and (*28*).

[*1780a*] "Mémoire sur l'intégration des équations différentielles par approximation," in *MARS* (1777/1780), 373–397; *OC*, IX, 357–379. There is no record of anything to which this is more likely to have corresponded than (*25*), read on 18 Feb. 1778.

[*1780b*] "Mémoire sur l'usage du calcul aux differ-

ences partielles dans la théorie des suites," in *MARS* (1777/1780), 99–122; *OC*, IX, 313–335. A marginal note dates the submission 16 June 1779 (*29*).

[*1780c*] "Mémoire sur la précession des équinoxes," in *MARS* (1777/1780), 329–345; *OC*, IX, 339–354. A marginal note is confirmed by the entry in *PV* (*31*) that Laplace read this memoir formally on 18 Aug. 1779, even though he had already read an "addition" to it on 7 July (*30*). Apparently, it elicited some criticism and discussion (*32*).

[*1781a*] "Mémoire sur les probabilités," in *MARS* (1778/1781), 227–232; *OC*, IX, 383–485. On 31 May 1780 Laplace read a "Mémoire sur le calcul aux suites appliqué aux probabilités" (*33*) and submitted the memoir for publication on 19 July, according to a marginal note.

[*1782a*] "Mémoire sur les suites," in *MARS* (1779/1782), 207–309; *OC*, X, 1–89. Not mentioned in *PV*.

[*1783a*] *Mémoire sur la chaleur*, written with Lavoisier. This separate printing issued from the Imprimerie Royale. The memoir as published by the Académie a year later is in *MARS* (1780/1784), 355–408; *OC*, X, 149–200; also in *Oeuvres de Lavoisier*, J. B. Dumas, ed., II (Paris, 1862), 283–333. I am indebted to Mr. Robert Bernstein for discussion of problems arising from the physical chemistry of this memoir.

[*1784a*] *Théorie du mouvement et de la figure elliptique des planètes* (Paris, 1784). Not in *OC*. Part I of this treatise probably represents a revision and expansion of (*9*), together with material on comets from (*35*), (*36*), (*37*), (*40*), and (*42*), and other up-to-date matter. Part II almost surely consists of a memoir "Sur l'attraction des sphéroïdes elliptiques" (*43*) that Laplace read on 24 and 31 May 1783 and no doubt draws also on the Legendre memoir that he discussed in (*41*). Laplace requested a commission of review on 3 Dec. 1783 (*47*). A finished copy was presented to the Académie on 24 Feb. 1784 (*PV*, **103**, fol. 37).

[*1784b*] "Mémoire sur la détermination des orbites des comètes," in *MARS* (1780/1784), 13–72; *OC*, X, 93–146. The same title as (*34*), probably enlarged.

[*1784c*] "Mémoire sur l'électricité qu'absorbent les corps qui se réduisent en vapeurs," written with Lavoisier, in *MARS* (1781/1784), 292–294; *OC*, X, 203–205; also in *Oeuvres de Lavoisier*, II, 374–376.

[*1785a*] "Mémoire sur les approximations des formules qui sont fonctions de trés grands nombres," in *MARS* (1782/1785), 1–88; *OC*, X, 209–291. Not mentioned in *PV*.

[*1785b*] "Théorie des attractions des sphéroïdes et de la figure des planètes," in *MARS* (1782/1785), 113–196; *OC*, X, 341–419. Read on 11 Aug. 1784 (*48*).

[*1786a*] "Mémoire sur la figure de la terre," in *MARS* (1783/1786), 17–46; *OC*, XI, 3–32. Not mentioned in *PV*.

[*1786b*] "Mémoire sur les approximations des formules qui sont fonctions de très grands nombres" (Suite), in *MARS* (1783/1786), 423–467; *OC*, X, 295–338. Laplace read the draft on 25 and 28 June 1785 (*49*).

[*1786c*] "Sur les naissances, les mariages et les morts à Paris, depuis 1771 jusqu'en 1784, et dans toute l'étendue de la France, pendant les années 1781 et 1782," in *MARS* (1783/1786), 693–702; *OC*, XI, 35–46. Read on 30 Nov. 1785 (*51*).

[*1787a*] "Mémoire sur les inégalités séculaires des planètes et des satellites," in *MARS* (1784/1787), 1–50; *OC*, XI, 49–92. Recorded on 19 Nov. 1785 and read on 23 Nov. (*50*).

[*1788a*] "Théorie de Jupiter et de Saturne," in *MARS* (1785/1788), 33–160; *OC*, XI, 95–207. Laplace read the draft on 10 May 1786 (*52*).

[*1788b*] "Théorie de Jupiter et de Saturne" (Suite), in *MARS* (1786/1788), 201–234; *OC*, XI, 211–239. Read on 15 July 1786 (*53*). This memoir is part III of [*1788a*].

[*1788c*] "Sur l'équation séculaire de la lune," in *MARS* (1786/1788), 235–264; *OC*, XI, 243–271. The draft was read on 19 Dec. 1787 (*55*) and was revised on 2 Apr. 1788 (*56*).

[*1789a*] "Mémoire sur la théorie de l'anneau de Saturne," in *MARS* (1787/1789), 249–267; *OC*, XI, 275–292.

[*1789b*] "Mémoire sur les variations séculaires des orbites des planètes," in *MARS* (1787/1789), 267–279; *OC*, XI, 297–306.

[*1790a*] "Sur la théorie des satellites de Jupiter," in *CT* (1792/1790), 273–286. Not in *OC*.

[*1791a*] "Théorie des satellites de Jupiter," in *MARS* (1788/1791), 249–364; *OC*, XI, 309–411. Read on 17 Apr. 1790 (*59*). I am indebted to Mr. Sherwin Singer for a discussion of this memoir.

[*1793a*] "Sur quelques points du système du monde," in *MARS* (1789/1793), 1–87; *OC*, XI, 477–558. On 18 July 1789 Laplace read a paper on the inclination of the plane of the ecliptic (*58*), which may probably have been the draft of Articles II–VII; *OC*, XI, 481–493.

[*1793b*] "Théorie des satellites de Jupiter" (Suite), in *MARS* (1789/1793), 237–296; *OC*, XI, 415–473. Not mentioned in *PV*.

[*1793c*] "De l'action du calorique sur les corps solides, principalement sur le verre et sur les métaux. . .," in *Oeuvres de Lavoisier*, II (1862), 739–759, written by Laplace and Lavoisier. For the original printing, undated but *ca.* 1803, see *Dictionary of Scientific Biography*, VIII, 87. The year assigned here is that in which Lavoisier composed the report of his experiments with Laplace in the winter of 1781–1782 (*38*) and also those reported in [*1793d*]. Not in *OC*.

[*1793d*] "Mémoire contenant les expériences faites sur la chaleur pendant l'hiver de 1783 à 1784," in *Oeuvres de Lavoisier*, II (1862), 724–738, written by Laplace and Lavoisier. See [*1793c*]. Not in *OC*.

[*P1796*] *Discours prononcé aux deux conseils . . . au nom de l'Institut national des sciences et des arts*, 17 Sept. 1796 (1er jour complémentaire an IV). Report on the first year of the Institut. Not in *OC*.

[*1797a*] "Mémoire sur le flux et le reflux de la mer," in *MARS* (1790/1797), 45–181; *OC*, XII, 3–126. Laplace began the reading on 15 Dec. 1790 (*60*).

[1797b] "Sur le mouvement de l'apogée de la lune et sur celui de ses noeuds," in *BSPM*, **1** (1797), 22–23. Not in *OC*. *RSC* 1.

[1797c] "Sur les équations séculaires du mouvement de la lune," in *BSPM*, **1** (1797), 99–101. Not in *OC*. *RSC* 2.

[1798a] "Mémoire sur les mouvements des corps célestes autour de leurs centres de gravité," in *MI*, **1** (an IV [1795–1796]/thermidor an VI [July–Aug. 1798]), 301–376; *OC*, XII, 129–187. Presented on 21 Jan. 1796 (*61*).

[1798b] "Sur les équations séculaires des mouvements de l'apogée et des noeuds de l'orbite lunaire," in *CT* (an VIII [1799–1800]/pluviôse an VI [Jan.–Feb. 1798]), 362–370; *OC*, XIII, 3–14. Read to the Institut on 20 Apr. 1797 (*63*). The calculations were deferred to *[1799a]*.

[1798c] "Mémoire sur la détermination d'un plan qui reste toujours parallèle à lui-même, dans le mouvement d'un système de corps agissant d'une manière quelconque les uns sur les autres et libres de toute action étrangère," in *JX*, **2**, 5e cahier (prairial an VI [May–June 1798]), 155–159; *OC*, XIV, 3–7.

[1798d] "Sur les plus grandes marées de l'an IX," in *CT* (an IX [1800–1801]/fructidor an VI [Aug.–Sept. 1798), 213–218; *OC*, XIII, 15–19.

[1799a] "Mémoire sur les équations séculaires des mouvements de la lune, de son apogée et de ses noeuds," in *MI*, **2** (an VII [1798–1799]/fructidor an VII [Aug.–Sept. 1799]), 126–182; *OC*, XII, 191–234; *OC*, XII, 191–234. Read on 10 Jan. 1798 (*64*). The calculations for *[1798b]*.

[1799b] "Sur la mécanique," in *JX*, **2**, 6e cahier (thermidor an VII [July–Aug. 1799]), 343–344; *OC*, XIV, 8–9.

[1799c] "Sur quelques équations des tables lunaires," in *CT* (an X [1801–1802]/fructidor an VII [Aug.–Sept. 1799]), 361–365; *OC*, XIII, 20–24.

[1799d] *Allgemeine geographische Ephemeriden*, **4**, no. 1 (July 1799), 1–6. Laplace supplied the editor, F. X. von Zach, with a proof for the statement in *Exposition du système du monde* (B: [1796], II, 305) that a luminous body 250 times larger than the sun and of comparable density to the earth would exert an attractive power that would prevent the light rays from escaping from its surface and that the largest luminous bodies in the universe may thus be invisible. The proof is translated as Appendix A of S. W. Hawking and G. F. R. Ellis, *The Large-scale Structure of Space-time* (Cambridge, 1973), 365–368. I owe this reference to the kindness of John Stachel. Not in *OC*. Laplace eliminated this passage from later editions of *Système du monde*.

[1800a] "Sur l'orbite du dernier satellite de Saturne," in *BSPM*, **2** (1800), 109; not in *OC*.

[1800b] "Sur les mouvements des orbites des satellites de Saturne et d'Uranus," in *CT* (an XI [1802–1803]/messidor an VIII [June–July 1800]), 485–489; not in *OC*. Read on 1 Mar. 1800 (*70*); an abstract of *[1801a]*.

[1800c] "Sur la théorie de la lune," in *CT* (an XI [1802–1803]/1800), 504–506; not in *OC*. An abstract of *[1801b]* on a periodic inequality in the nutation of the lunar orbit.

[1800d] "Leçons de mathématiques professées à l'École normale en 1795," in *Séances de l'École normale* (an VIII [1799–1800]), I, 16–32, 268–280, 381–393; (pt. 2), 3–23, 130–134; II, 116–129, 302–318; III, 24–39; IV, 32–70, 223–263; V, 201–219; VI, 32–73; reprinted in *JX*, **2**, 7e and 8e cahiers (June 1812), 1–172; *OC*, XIV, 10–177, *RSC* 8.

[P1800] Rapport sur la situation de l'École polytechnique, 24 Dec. 1800 (3 nivôse an IX). Submitted to the minister of the interior on behalf of the Conseil de Perfectionnement. Not in *OC*.

[1801a] "Mémoire sur les mouvements des orbites des satellites de Saturne et d'Uranus," in *MI*, **3** (prairial an IX [May–June 1801]), 107–127; *OC*, XII, 237–253. Read on 1 Mar. 1800 (*70*).

[1801b] "Mémoire sur la théorie de la lune," in *MI*, **3** (prairial an IX [May–June 1801]), 198–206; *OC*, XII, 257–263. Read on 15 June 1800 (*71*). *RSC* 11.

[1801c] "Sur la théorie de la lune," in *CT* (an XII [1803–1804]/fructidor an IX [Aug.–Sept. 1801]), 493–501; not in *OC*. On an inequality in the lunar parallax; an abstract of *[1801b]*.

[1801d] "Sur un problème de physique, relatif à l'électricité," in *BSPM*, **3**, no. 51 (prairial an IX [May–June 1801]), 21–23. Not in *OC*. A concluding editor's note reads: "Nous devons au C. Laplace cette application à l'électricité, des formules relatives à la théorie de la figure de la terre."—I.B.

[1803a] "Sur les marées," in *BSPM*, **3**, no. 74 (floréal an XI [Apr.–May 1803]), 106; not in *OC*. *RSC* 13. See (*81*).

[1803b] "Mémoire sur le mouvement d'un corps qui tombe d'une grande hauteur," in *BSPM*, **3**, no. 75 (prairial an XI [May–June 1803]), 109–115; *OC*, XIV, 267–277.

[1803c] Two notes on the relation of pressure to temperature among the molecules of an enclosed gas, contributed to Claude-Louis Berthollet, *Essai de statique chimique*, 2 vols. (1803), I, 245–247, n. 5; I, 522–523, n. 18; *OC*, XIV, 329–332.

[1804a] "Sur les tables de Jupiter et sur la masse de Saturne," in *CT* (an XIV [1805–1806]/nivôse an XII [Dec. 1803–Jan. 1804], 435–440; *OC*, XIII, 25–29. Laplace read a draft at the Institut on 12 Sept. 1803 (*82*). *RSC* 16.

[1804b] "Sur la théorie de Jupiter et de Saturne," in *CT* (an XV [1806–1807]/frimaire an XIII [Nov.–Dec. 1804], 296–307; *OC*, XIII, 30–40. *RSC* 20.

[P1805] "Rapport sur le projet . . . portant rétablissement du calendrier grégorien." Sénat conservateur, 9 Sept. 1805 (22 fructidor an XIII), in *AP*, VIII, 722–723. Not in *OC*.

[1806a] "Sur la théorie des tubes capillaires," in *JP*, **62** (Jan. 1806), 120–128; *OC*, XIV, 217–227. An abstract of (*85*), read on 23 Dec. 1805. *RSC* 17.

[*1806b*] "Sur l'attraction et la répulsion apparente des petits corps qui nagent à la surface des fluides," in *JP*, **63** (Sept. 1806), 248–252; *OC*, XIV, 228–232. Read in part on 29 Sept. 1806 (*87*). *RSC* 18.

[*1806c*] "Sur l'action capillaire," in *JP*, **63** (Dec. 1806), 474–477; also *CX*, **1** (1804–1808), 246–256; *OC*, XIV, 233–246. An abstract of (*86*), read on 28 Apr. 1806. *RSC* 15.

[*1806d*] "De l'adhésion des corps à la surface des fluides," in *JP*, **63** (Nov. 1806), 413–418; *OC*, XIV, 247–253. An abstract of (*88*), read on 24 Nov. 1806. *RSC* 19.

[*1807a*] "Supplément à la théorie de l'action capillaire," in *JP*, **65** (July 1807), 88–95; not in *OC*. An abstract of (*87*), read on 29 Sept. 1806.

[*1809a*] "Mémoire sur la double réfraction de la lumière dans les cristaux diaphanes," in *BSPM*, **1**, no. 18 (Mar. 1809), 303–310; *OC*, XIV, 278–287. An abstract of (*92*), read on 30 Jan. 1809. *RSC* 23.

[*1809b*] "Sur la loi de la réfraction extraordinaire de la lumière dans les cristaux diaphanes," in *JP*, **68** (Jan. 1809), 107–111; *OC*, XIV, 254–258. An abstract of the passages of [*1810a*] (*92*) that deal with the application of the principle of least action to double refraction, in consequence of forces of attraction and repulsion acting at undetectable distances. *RSC* 23.

[*1809c*] "Sur l'anneau de Saturne," in *CT* (1811/July 1809), 450–453; *OC*, XIII, 41–43. *RSC* 24.

[*1809d*] "Mémoire sur la diminution de l'obliquité de l'écliptique qui résulte des observations anciennes," in *CT* (1811/July 1809), 429–450; *OC*, XIII, 44–70. *RSC* 30.

[*1809e*] "Mémoire sur divers points d'analyse," in *JX*, **8**, 15e cahier (Dec. 1809), 229–265; *OC*, XIV, 178–214.

[*1810a*] "Mémoire sur le mouvement de la lumière dans les milieux diaphanes," in *MI*, **10** (1809/1810), 300–342; *OC*, XII, 267–298. Read at the Institut on 30 Jan. 1809 (*92*). The date is mistakenly given as 1808 in *OC*, XII, 267n. *RSC* 27.

[*1810b*] "Mémoire sur les approximations des formules qui sont fonctions de très grands nombres, et sur leur application aux probabilités," in *MI*, **10** (1809/1810), 353–415; *OC*, XII, 301–353. Read on 9 Apr. 1810 (*95*). Delambre summarized this paper in *MI*, **11** (1810/1811), "Histoire," iii–v.

[*1810c*] "Supplément au mémoire sur les approximations des formules qui sont fonctions de très grands nombres," in *MI*, **10** (1809/1810), 559–565; *OC*, XII, 349–353.

[*1810d*] "Notice sur les probabilités," in *Annuaire publié par le Bureau des longitudes* (1811/1810), 98–125. Not in *OC*.

[*1810e*] "Sur la dépression du mercure dans une tube de baromètre, due à sa capillarité," in *CT* (1812/July 1810), 315–320; *OC*, XIII, 71–77. Bouvard published corrections to the table of data, which are printed in *OC*, XIII, 334–341. *RSC* 31.

[*1811a*] "Mémoire sur les intégrales définies, et leur application aux probabilités, et spécialement à la recherche du milieu qu'il faut choisir entre les résultats des observations," in *MI*, **11** (1810/1811), 279–347; *OC*, XII, 357–412. Read on 29 Apr. 1811 (*97*). There is a prefatory summary by Delambre, in *MI*, **12** (1811/1812), "Histoire," i–xiii, reviewing the state of the question of least squares.

[*1811b*] "Du milieu qu'il faut choisir entre les résultats d'un grand nombre d'observations," in *CT* (1813/July 1811), 213–223; *OC*, XIII, 78. *RSC* 32.

[*1811c*] "Sur l'inégalité à longue période du mouvement lunaire," in *CT* (1813/July 1811), 223–227; *OC*, XIII, 79–84. Continued in [*1812a*]. See (*76*) and (*94*).

[*1812a*] "Sur l'inégalité à longue période du mouvement lunaire," in *CT* (1815/Nov. 1812), 213–214; *OC*, XIII, 85–87. A continuation of [*1811c*]. See (*76*) and (*94*). *RSC* 33.

[*1813a*] "Sur les comètes," in *CT* (1816/Nov. 1813), 213–220; *OC*, XIII, 88–97.

[*P1814*] Debate on a proposal to authorize the exportation of grain, Chambre des Pairs, 8 Nov. 1814, in *AP*, XIII, 470–471. Laplace's opinion supported free trade. Not in *OC*.

[*1815a*] "Sur l'application du calcul des probabilités à la philosophie naturelle," in *CT* (1818/1815), 361–377; *OC*, XIII, 98–116. Read before the Institut on 18 Sept. 1815 (*106*). *RSC* 34.

[*1815b*] "Sur le calcul des probabilités appliqué à la philosophie naturelle," in *CT* (1818/1815), 378–381; *OC*, XIII, 117–120. Supplement to [*1815a*]. *RSC* 34.

[*1815c*] "Sur le flux et reflux de la mer," in *CT* (1818/1815), 354–361; not in *OC*. Read on 10 July 1815 (*105*). *RSC* 35 mistakenly identifies this with [*1820a*], of which it was a preliminary abstract.

[*1816a*] "Sur la longueur du pendule à secondes," in *AC*, **3** (1816), 92–94. An excerpt from [*1817a*]. Read on 28 Oct. 1816 (*108*).

[*1816b*] "Sur l'action réciproque des pendules et sur la vitesse du son dans les diverses substances," in *AC*, **3** (1816), 162–169; *OC*, XIV, 291–296. Read before the Académie on 25 Nov. 1816 (*109*). *RSC* 36.

[*1816c*] "Sur la transmission du son à travers les corps solides," in *BSPM* (1816), 190–192; *OC*, XIV, 288. An abstract of [*1816b*]. See (*110*). *RSC* 40.

[*1816d*] "Sur la vitesse du son dans l'air et dans l'eau," in *AC*, **3** (1816), 238–241; *OC, XIV*, 297–300. Read at the Académie on 23 Dec. 1816 (*110*). *RSC* 37.

[*P1816*] "Sur une disposition du code d'instruction criminelle." Pamphlet, 15 Nov. 1816. Bibliothèque Nationale Fp. **1187**. On the mathematical equity of majorities required to find an accused guilty, in court of first instance and on appeal. Proposed modification of Article 351 of Code d'Instruction Criminelle as unfair to the accused. Not in *OC*.

[*1817a*] "Sur la longueur du pendule à secondes," in *CT* (1820/1817), 265–280; *OC*, XIII, 121–139. A fuller text than [*1816a*], with which *RSC* 38 mistakenly identifies it. Read on 28 Oct. 1816 (*108*).

[*1817b*] "Addition au mémoire précédent sur la lon-

gueur du pendule à secondes," in *CT* (1820/1817), 441–442; *OC*, XIII, 140–142.

[*P1817*] Recommendation on the cadastre, Chambre des Pairs, debate on the budget of 1817, 21 Mar. 1817, in *AP*, 2nd ser., XIX, 506–507; *OC*, XIV, 372–374. Urged completion of cadastre tied to geodesic survey as the basis for a fair assessment of the land tax.

[*1818a*] "Application du calcul des probabilités aux opérations géodésiques," in *CT* (1820/1818), 422–440. Excerpted in *AC, BSPM*, and *JP*. *RSC* 41. Read before the Académie on 4 Aug. 1817 (*111*). Incorporated, with additions (*112*), in *Théorie analytique des probabilités*, 3rd ed. (1820), as Supplement 2; *OC*, VII, 531–580.

[*1818b*] "Sur la figure de la terre, et la loi de la pesanteur à sa surface," in *AC*, **8** (1818), 313–318; *CT* (1821/1819), 326–331. Abstracted from [*1819b*]. Read on 3 Aug. 1818 (*114*). *RSC* 44.

[*1819a*] "Sur la rotation de la terre," in CT (1821/1819), 242–259; *OC*, XIII, 144–164. Read before the Academy on 18 May 1818 (*113*). *RSC* 43.

[*1819b*] "Mémoire sur la figure de la terre," in *MASIF*, **2** (1817/1819), 137–184; *OC*, XII, 415–455. Read on 3 Aug. 1818 (*OC* mistakenly has 4 Aug.). See (*114*). *RSC* 42.

[*1819c*] "Sur l'influence de la grande inégalité de Jupiter et de Saturne, dans le mouvement des corps du système solaire," in *CT* (1821/1819), 266–271; *OC*, XIII, 175–180.

[*1819d*] "Sur la loi de la pesanteur, en supposant le sphéroïde terrestre homogène et de même densité que la mer," in *CT* (1821/1819), 284–290; *OC*, XIII, 165–172.

[*1819e*] "Addition au mémoire précédent," in *CT* (1821/1819), 353; *OC*, XIII, 173–174.

[*1819f*] "Mémoire sur l'application du calcul des probabilités aux observations et spécialement aux opérations du nivellement," in *AC*, **12** (1819), 337–341; *OC*, XIV, 301–304. An abstract of the 3rd Supplement of *Théorie analytique des probabilités*, 3rd ed. (1820); *OC*, VII, 581–616. Read on 20 Dec. 1819 (*118*).

[*1819g*] "Application du calcul des probabilités aux opérations géodésiques de la méridienne de France," in *BSPM* (1819), 137–139; reprinted in *CT* (1822/1820), 346–348. Incorporated in the 3rd Supplement of *Théorie analytique des probabilités*, 3rd ed. (1820); *OC*, VII, 581–585. Cf. *OC*, XIII, 188, for correction of an error of calculation. Read on 20 Dec. 1819 (*118*). *RSC* 47.

[*1819h*] "Considérations sur la théorie des phénomènes capillaires," in *JP*, **89** (Oct. 1819), 292–296; XIV, 259–264. Read on 13 Sept. 1819 (*116*). *RSC* 45.

[*P1819*] Recommendation on the lottery, Chambre des Pairs debate on the budget of 1819, 16 July 1819, in *AP*, 2nd ser., XXV, 683–684; *OC*, XIV, 375–378. Urged suppression of the lottery on the grounds that it was inappropriate to raise public funds by encouraging illusions among the citizens.

[*1820a*] "Sur les inégalités lunaires dues à l'aplatissement de la terre," in *CT* (1823/1820), 219–225; *OC*, XIII, 189–197. Read on 19 Jan. 1820 (119).

[*1820b*] "Sur le perfectionnement de la théorie et des tables lunaires," in *CT* (1823/1820), 226–231; *OC*, XIII, 198–204. Read on 29 Mar. 1820 (*121*). *RSC* 49.

[*1820c*] "Sur l'inégalité lunaire à longue période, dépendante de la différence des deux hémisphères terrestres," in *CT* (1823/1820), 232–239; *OC*, XIII, 205–212. Read on 12 Apr. 1820 (*122*).

[*1820d*] "Mémoire sur le flux et le reflux de la mer," in *MASIF*, **3** (1818/1820), 1–90; *OC*, XII, 473–546. An expansion of [*1815b*].

[*1820e*] "Addition au mémoire sur la figure de la terre, inséré dans le volume précédent," in *MASIF*, **3** (1818/1820), 489–502; *OC*, XII, 459–469. The preceding memoir was [*1819b*]. Also printed under the title "Sur la figure de la terre," in *CT* (1822/1820), 284–293. Read on 26 May 1819 (*115*). Cf. *OC*, XIII, 187.

[*1820f*] "Mémoire sur la diminution de la durée du jour par le refroidissement de la terre," in *CT* (1823/1820), 245–257. Incorporated in *Mécanique céleste*, Volume V, Book XI, chapters 1 and 4. *OC*, V, 24–28, 82–88, 91–96. Cf. *OC*, XIII, 213. *RSC* 48.

[*1820g*] "Addition au mémoire précédent) sur la diminution de la durée du jour. . .," in *CT* (1823/1820), 324–327. Incorporated in *Mécanique céleste*, Volume V, Book XI, chapter 4, 88–91. Cf. *OC*, XIII, 214. *RSC* 48.

[*1820h*] "Sur la densité moyenne de la terre," in *CT* (1823/1820), 328–331; *OC*, XIII, 215–220. *RSC* 50.

[*1820i*] "Éclaircissements sur les mémoires précédents, relatifs aux inégalités lunaires dépendantes de la figure de la terre, et au perfectionnement de la théorie et des tables de la lune," in *CT* (1823/1820), 332–337; *OC*, XIII, 221–228.

[*1821a*] "Éclaircissements de la théorie des fluides élastiques," in *AC*, **18** (1821), 273–280; *OC*, XIV, 305–311.

[*1821b*] "Sur les variations des éléments du mouvement elliptique, et sur les inégalités lunaires à longues périodes," in *CT* (1824/1821), 274–307; *OC*, XIII, 229–264.

[*1821c*] "Sur la détermination des orbites des comètes," in *CT* (1824/1821), 314–320; *OC*, XIII, 265–272. Includes an example of the method applied to the comet of 1805 by Bouvard.

[*1821d*] "Sur l'attraction des sphères et sur la répulsion des fluides élastiques," in *CT* (1824/1821), 328–343; *OC*, XIII, 273–290. Read before the Academy on 10 Sept. 1821 (*123*). *RSC* 51.

[*1821e*] *Précis de l'histoire de l'astronomie* (1821). A separate publication of Book V of *Exposition du système du monde*, from the 4th ed. (1813).

[*P1821*] Debate on Article 351 of Code d'Instruction Criminelle, Chambre des Pairs, 30 Mar. 1821, in *AP*, 2nd ser., XXX, 531–532; *OC*, XIV, 379–381. Recommendations on the most equitable modes of reaching a decision in juries.

[*1822a*] "Développement de la théorie des fluides élastiques et application de cette théorie à la vitesse du son," in *CT* (1825/1822), 219–227; *OC*, XIII,

291–301. Submitted to the Bureau des Longitudes on 12 Dec. 1821. *RSC* 55.

[*1822b*] "Addition au mémoire précédent sur le développement de la théorie des fluides élastiques," in *CT* (1825/1822), 302–323; *OC*, XIII, 302.

[*1822c*] "Sur la vitesse du son," in *CT* (1825/1822), 371–372; *OC*, XIII, 303–304. *RSC* 57.

[*1822d*] "Addition au mémoire sur la théorie des fluides élastiques," in *CT* (1825/1822), 386–387; *OC*, XIII, 305.

[*1823a*] "De l'action de la lune sur l'atmosphère," in *CT* (1826/1823), 308–317. Incorporated in *Mécanique céleste*, Book XIII; *OC*, V, 184–188, 262–268. *RSC* 58.

[*1824a*] "Sur les variations de l'obliquité de l'écliptique et de la précession des équinoxes," in *CT* (1827/1824), 234–237; *OC*, XIII, 307–311.

[*P1824*] Debate on conversion of the public debt, Chambre des Pairs, 1 June 1824, in *AP*, 2nd ser., XLI, 125; *OC*, XIV, 382–383. Calculation of the effects of various schemes of amortization on the cost of servicing the debt.

[*1825a*] "Sur le développement en série du radical qui exprime la distance mutuelle de deux planètes, et sur le développement du rayon vecteur elliptique," in *CT* (1828/1825), 311–321; *OC*, XIII, 312.

[*1825b*] "Sur la réduction de la longueur du pendule, au niveau de la mer," in *AC*, **30** (1825), 381–387; *OC*, XIV, 312–317.

[*P1825*] Debate on the conversion of the government bond issue, Chambre des Pairs, 26 Apr. 1825, in *AP*, 2nd ser., XLV, 144–145; *OC*, XIV, 385–387. Favored reduction of the interest rate and an increase of capital.

[*1826a*] "Mémoire sur les deux grandes inégalités de Jupiter et de Saturne," in *CT* (1829/1826), 236–244; *OC*, XIII, 313–322.

[*1826b*] "Mémoire sur divers points de mécanique céleste: I. Sur les mouvements de l'orbite du dernier satellite de Saturne; II. Sur l'inégalité de Mercure à longue période, dont l'argument est le moyen mouvement de Mercure, moins celui de la terre; III. De l'action des étoiles sur le systéme planétaire"; in *CT* (1829/1826), 245–251; *OC*, XIII, 323–330.

[*1826c*] "Mémoire sur un moyen de détruire les effets de la capillarité dans les baromètres," in *CT* (1829/1826), 301–302; *OC*, XIII, 331–333.

[*1827a*] "Mémoire sur le développement de l'anomalie vraie et du rayon vecteur elliptique en séries ordonnées suivant les puissances de l'excentricité," in *MASIF*, **6** (1823/1827), 61–80; *OC*, XII, 549–566.

[*1827b*] "Mémoire sur le flux et reflux lunaire atmosphérique," in *CT* (1830/1827), 3–18; *OC*, XIII, 342–358. Published in Supplement to *Mécanique céleste*, Volume V; *OC*, V, 489–505.

SECONDARY LITERATURE

J. Biographical and General. The only attempt at a comprehensive account is H. Andoyer, *L'oeuvre scientifique de Laplace* (Paris, 1922), a modernized précis that is useful at this level. It is unfair to this unpretentious little (162-page) work to quote here the comment with which the editor of *L'action nationale* prefaced an excerpt from its methodological section (n.s. **18** [Jan.–June 1922], 14–21), but the remark deserves rescue from the oblivion of that journal:

> À l'heure où l'esprit de snobisme et d'aventure, qui semble être la marque de notre temps, croit découvrir dans les théories, peu accessibles mais tapageuses, du mathématicien allemand Einstein, un je ne sais quel nouveau point de départ pour la pensée, une voie insoupçonnée pour la recherche scientifique, il est bon de rappeler l'effort aussi innovateur que parfaitement intelligible de notre Laplace, dont la grande hardiesse s'accompagnait d'une égale modestie.

The most important éloge is Joseph Fourier, "Éloge historique de M. le Marquis de Laplace, prononcé . . . le 15 juin 1829," in *MASIF*, **10** (1831), "Histoire," lxxxi–cii, prefixed to the 1835 printing of the 6th ed. of *Exposition du système du monde* and reprinted in most later eds; there is an English trans. in *Philosophical Magazine*, 2nd ser., **6** (1829), 370–381. J.-B. Biot and S.-D. Poisson delivered eulogies at the funeral, and both texts are prefixed to the 1827 printing of *Exposition du système du monde*. Much later, Biot published a reminiscence of Laplace as scientific father figure that he had delivered before the Académie Française, "Une anecdote relative à M. Laplace," in *Journal des savants* (Feb. 1850), 65–71. François Arago also composed a "Notice," in his *Oeuvres*, III (1859), 459–515; there is an English trans. in *Biographies of Distinguished Scientific Men* (London, 1857), 196–241. In this century innumerable anniversary pieces have appeared, the only ones worthy of record being André Danjon, "Pierre-Simon Marquis de Laplace," in *Notices et discours. Académie des sciences*, **3** (1949–1956) [Paris, 1957], delivered at a 200th anniversary celebration in Caen; René Taton, "Laplace," in *La Nature* (Paris), **77** (1949), 221–223; and E. T. Whittaker, "Laplace," in *Mathematical Gazette*, **33** (1949), 1–12.

It is still useful to consult the articles in certain well-known nineteenth-century biographical encyclopedias; but before doing so, readers should bear in mind that just as entries in these works concerning figures in previous centuries derive from tradition and legend rather than from historical research, so the accounts of near-contemporary persons depend on gossip and reminiscence rather than on scholarship. Hence, the decline of Laplace's personal reputation is the main motif in A. Rabbe, J. Vieilh de Boisjolin, and Sainte-Preuve, *Biographie universelle et portative des contemporains*, 5 vols. (Paris, 1834), III, 151–153; Parisot, in *Biographie universelle, ancienne et moderne*, L. G. Michaud, ed., 85 vols. (Paris, 1811–1862), LXX, 237–260; and E. Merlieux, in *Nouvelle biographie générale*, F. Hoefer, ed., 46 vols. (Paris, 1855–1866), XXIX (1859), cols. 531–548. On Laplace's failure to support the declaration of the Académie Française in favor of the freedom of the press, see also E. Grassier, *Les cinq cents immortels. Histoire de l'Académie française* (Paris, 1906), 148–149.

Notarial records and other official notices are reproduced in B. Boncompagni, "Intorno agli alti di nascità e di morte di Pietro Simone Laplace," in *Bollettino di bibliografia e storia delle scienze matematiche e fisiche*, **15** (1883), 447–465. L. Puisieux, *Notice sur Laplace* (Caen, 1847), may still be consulted. Karl Pearson published extracts from lectures on the life and work of Laplace in *Biometrika*, **21** (1929), 202–216; and appended to it an antiquarian article written at his request by the Abbé G.-A. Simon, "Les origines de Laplace: sa généalogie, ses études," *ibid*., 217–230. G. Bigourdan, "La jeunesse de Laplace," in *Science moderne*, no. 8 (1931), 377–384, takes him up to his election to the Academy. S.-F. Lacroix, *Essais sur l'enseignement en général et sur celui de mathématiques en particulier,* 3rd ed. (1828), is useful for the context of Laplace's teaching in the 1790's. Laplace joined with Legendre in a report read by the latter (11 nivôse an V [31 Dec. 1976], *PVIF*, I, 154–157) on Lacroix, *Traité de calcul différentiel et intégral*, 2 vols. (1797), and printed in full at the end of Vol. I of that work, pp. 520 ff. There is much biographical information in Maurice Crosland, *The Society of Arcueil* (London, 1967). Of more incidental interest are G. Sarton, "Laplace's Religion," in *Isis*, **33** (1941), 309–312; and D. Duveen and R. Hahn, "Laplace's Succession to Bezout's Post of *Examinateur des élèves de l'artillerie*," *ibid*., **48** (1957), 416–427.

There is information on Laplace's relations with Fourier, and further bibliography, in Ivor Grattan-Guinness with J. R. Ravetz, *Joseph Fourier, 1768–1830* (Cambridge, Mass., 1972). Paul Marmottan edited the correspondence of Madame de Laplace with Elisa Bonaparte, *Lettres de Madame de Laplace à Élisa Napoléon, princesse de Lucques et de Piombino* (Paris, 1897). They cover mainly the years 1807 and 1808.

K. Celestial Mechanics. It is worthy of remark that in the nineteenth century interest in Laplace centered on his celestial mechanics; that in the twentieth century it shifted to probability; that very recently attention has begun to be paid to his physics, although mainly for its institutional and political implications; and that surprisingly little scholarship has been addressed expressly to his mathematical work, which has been discussed mostly in sections of large-scale works on the evolution of overall aspects of mathematics.

Of near-contemporary astronomical and physical texts, the most helpful by far is Mary Somerville, *Mechanism of the Heavens* (London, 1831). Her exposition is somewhat more elementary than Bowditch's commentary [F], to which it makes a valuable supplement. Less satisfactory, although still worth mentioning, is Thomas Young, *Elementary Illustrations of the Celestial Mechanics of Laplace* (London, 1821). More valuable historically are two Victorian monuments: Robert Grant, *History of Physical Astronomy From the Earliest Ages to the Middle of the 19th Century* (London, 1852; repr. New York, 1966); and Isaac Todhunter, *A History of the Mathematical Theories of Attraction and the Figure of the Earth From the Time of Newton to That of Laplace*, 2 vols. (London, 1873; repr. New York, 1962). Among more recent works, questions of origin and development bulk much more largely than they did in the writings of Laplace himself. So it is in B. O. Bianco, "Le idee di Lagrange, Laplace, Gauss, e Schiaparelli sull'origine delle comete," in *Memorie della Reale Accademia delle scienze di Torino*, 2nd ser., **63** (1913), 59–110. Jules Vuillemin is both methodological and metaphysical in "Sur la généralisation de l'estimation de la force chez Laplace," in *Thalès*, **9** (1958), 61–76; Jacques Merlau-Ponty is largely metaphysical in "Situation et rôle de l'hypothèse cosmogonique dans la pensée cosmologique de Laplace," in *Revue d'histoire des sciences et de leurs applications*, **29** (1976), 21–49; Stanley L. Jaki is more concrete on the development of Laplace's cosmogonical views, with special reference to the status of the nebular hypothesis in successive editions of the *Système du monde*, "The Five Forms of Laplace's Cosmogony," in *American Journal of Physics*, **44** (1976), 4–11. It has, unfortunately, not been possible to consult the unpublished M.Sc. thesis by Eric J. Aiton, "The Development of the Theory of the Tides in the Seventeenth and Eighteenth Centuries" (University of London, 1953).

L. Probability. The starting place is still Isaac Todhunter, *A History of the Mathematical Theory of Probability From the Time of Pascal to That of Laplace* (London, 1865). Since then the most comprehensive treatment of the subject as a whole is Ivo Schneider, *Die Entwicklung des Wahrscheinlichkeitsbegriffs in der Mathematik von Pascal bis Laplace* (Munich, 1972), a *Habilitationsschrift*. Ian Hacking, *The Emergence of Probability* (Cambridge, 1975), is philosophically motivated and largely nontechnical; in our judgment the resulting account does less than justice to Laplace. Still of interest are the unsigned article by Augustus de Morgan in the form of a review of the *Théorie analytique*, in *Dublin Review*, **2** (1836–1837), 338–354; and **3** (1837), 237–248; and E. Czuber, "Die Entwicklung der Wahrscheinlichkeitstheorie und ihrer Anwendungen," which is *Jahresbericht der Deutschen Mathematikervereinigung*, **7**, no. 2 (1899).

More recent writings are D. van Dantzig, "Laplace probabiliste et statisticien et ses précurseurs," in *Archives internationales d'histoire des sciences*, **8** (1955), 27–37; Churchill Eisenhart, "The Meaning of 'Least' in Least Squares," in *Journal of the Washington Academy of Sciences*, **54** (1964), 24–33; C. C. Gillispie, "Intellectual Factors in the Background of Analysis by Probabilities," in A. C. Crombie, ed., *Scientific Change* (London, 1963), 433–453; and "Probability and Politics: Laplace, Condorcet, and Turgot," in *Proceedings of the American Philosophical Society*, **116**, no. 1 (Feb. 1972), 1–20; H. O. Lancaster, "Forerunners of the Pearson χ^2," in

Australian Journal of Statistics, **8** (1966), 117–126; Wilhelm Lorey, "Die Bedeutung von Laplace für die Statistik," in *Allgemeines statistisches Archiv*, **23** (1934), 398–410; L. E. Maistrov, *Probability Theory; A Historical Sketch*, translated and edited by Samuel Kotz (New York–London, 1974); E. C. Molina, "The Theory of Probability: Some Comments on Laplace's *Théorie analytique*," in *Bulletin of the American Mathematical Society*, **36** (1930), 369–392, largely in defense of generating functions; R. L. Plackett, "The Discovery of the Method of Least Squares," in *Biometrika*, **59** (1972), 239–251; Ivo Schneider, "Rudolph Clausius' Beitrag zur Einführung wahrscheinlichkeitstheoretischer Methoden in die Physik der Gase nach 1856," in *Archive for History of Exact Sciences*, **14** (1975), 237–261, which deals with Laplace's role in the emergence of the idea of treating kinetic theory by methods of the calculus of probabilities; H. L. Seal, "The Historical Development of the Use of Generating Functions in Probability Theory," in *Mitteilungen der Vereinigung schweizerischer Versicherungs-Mathematiker*, **49** (1949), 209–228; an important group of papers by O. B. Sheynin: "Finite Random Sums," in *Archive for History of Exact Sciences*, **9** (1973), 275–305; "P. S. Laplace's Work on Probability," *ibid.*, **16** (1976), 137–187, which is mainly a summary critique of *Théorie analytique des probabilités*; and "Laplace's Theory of Errors," *ibid.*, **17** (1977), 1–61; an excellent series of papers by Stephen M. Stigler: "Laplace, Fisher, and the Discovery of the Concept of Sufficiency," in *Biometrika*, **60** (1973), 439–455; "Cauchy and the Witch of Agnesi," *ibid.*, **61** (1974), 375–380; "Gergonne's 1815 Paper on the Design and Analysis of Polynomial Regression Experiments," in *Historia mathematica*, **1** (1974), 431–447; "Napoleonic Statistics: The Work of Laplace," in *Biometrika*, **62** (1975), 503–517; and a postscript to the article on Laplace, in J. Tanner and W. Kruskal, eds., *Encyclopedia of Statistics in the Social Sciences* (in press); and E. Yamakazi, "D'Alembert et Condorcet–quelques aspects de l'histoire du calcul des probabilités," in *Japanese Studies in the History of Science*, **10** (1971), 59–93.

M. Mathematical Physics. The indispensable sources for Laplace's work in physical science are J. J. Bikerman, "Theories of Capillary Attraction," in *Centaurus*, **19** (1975), 182–206; Bernard S. Finn, "Laplace and the Speed of Sound," in *Isis*, **55** (1964), 7–19; Robert Fox, *The Caloric Theory of Gases From Lavoisier to Regnault* (London, 1971); and "The Rise and Fall of Laplacian Physics," in *Historical Studies in the Physical Sciences,* **4** (1974), 89–136; Eugene Frankel, "The Search for a Corpuscular Theory of Double Refraction: Malus, Laplace and the Prize Competition of 1808," in *Centaurus,* **18** (1974), 223–245; H. H. Frisinger, "Mathematics in the History of Meteorology: The Pressure-Height Problem From Pascal to Laplace," in *Historia mathematica,* **1** (1974), 263–286; Henry Guerlac, "Chemistry as a Branch of Physics: Laplace's Collaboration With Lavoisier," in *Historical Studies in the Physical Sciences,* **7** (1976), 193–276; and T. H. Lodwig and W. A. Smeaton, "The Ice Calorimeter of Lavoisier and Laplace and Some of Its Critics," in *Annals of Science,* **31** (1974), 1–18.

N. Mathematics. Much of the discussion in many of the foregoing titles is of necessity mathematical. In addition, help is available from works primarily concerned with topics in the history of mathematics and rational mechanics, notably H. Burckhardt, "Entwicklungen nach oscillirenden Functionen und Integration der Differentialgleichungen der mathematischen Physik," which is *Jahresbericht der Deutschen Mathematiker-vereinigung,* **10**, no. 2 (1908), an 1,800-page monograph of which Section VI, pp. 398–408, deals especially with Laplace's method for integrating differential equations by means of definite integrals; Florian Cajori, *A History of Mathematical Notations*, 2 vols. (Chicago, 1928–1929), II, *passim*; Theodor Körner, "Der Begriff des materiellen Punktes in der Mechanik des achtzehnten Jahrhunderts," in *Bibliotheca mathematica*, 3rd ser., **5** (1904), 15–62, esp. 52–54; Elaine Koppelmann, "The Calculus of Operations and the Rise of Abstract Algebra," in *Archive for History of Exact Sciences*, **8** (1971), 155–242; E. C. Molina, "An Expansion for Laplacian Integrals," in *Bell System Technical Journal*, **11** (1932), 563–575; Thomas Muir, *The Theory of Determinants in the Historical Order of Development*, 4 vols. (London, 1906–1923); S. S. Petrova, "K istorii metoda kaskadov Laplasa" ("Toward the History of Laplace's Method of Cascades"), in *Istoriko-matematicheskie issledovania,* **19** (1974), 125–131; and "Rannyaya istoria preobrazovania Laplasa" ("The Early History of the Laplace Transform"), *ibid.,* **20** (1975), 246–256; O. B. Sheynin, "O poyavlenii delta-funktsii Diraka v trudakh P. S. Laplasa" ("On the Appearance of Dirac's Delta-Function in the Works of P. S. Laplace"), *ibid.*, 303–308; and Isaac Todhunter, *Elementary Treatise on Laplace's Function, Lamé's Function, and Bessel's Function* (London, 1875).

MARX, KARL (*b.* Trier, Prussian Rhineland, 5 May 1818; *d.* London, England, 14 March 1883), *economics, history, philosophy, political science, sociology, history and sociology of science and technology.*

Karl Marx was the third child and eldest son of Heinrich Marx (born 1782), a lawyer of local distinction and moderate wealth who was appointed magistrate a year after formal conversion to the Evangelical Lutheran church in 1817. The elder Marx combined enlightened Voltairean and deist inclinations with middle-class cultural interests, liberal Prussian patriotism, and a strong paternal affection for Karl. Both Heinrich and his Dutch wife, Henriette Pressburg, came from distinguished rabbinical families, Heinrich's having been of particular prominence since the early fifteenth century in Germany, Italy, and Poland, and Henriette's for a century in Holland, and before that, in Hungary. Although there was no Jewish education or tradition in the upbringing of their children—indeed, the home was deliberately separated from family connections—Jewish self-consciousness was to some extent unavoidable. There were nine children, of whom four survived early childhood.

Marx was educated (1830–1835) at the Friedrich-Wilhelm Gymnasium in Trier, formerly a Jesuit school, where he was influenced chiefly by the headmaster, who was also the history teacher. But greater encouragement came from his father's interest in the poet Gotthold Lessing and the French classics, and from their devoted neighbor, Baron Ludwig von Westphalen, who, with warm-hearted enthusiasm, read Homer, Dante, Cervantes, and Shakespeare, as well as such advanced political thinkers as Saint-Simon, with young Marx. To his mother, Karl was "the best and most beloved"; and he wrote to his father of his "angel of a mother," despite the lack of any mutual intellectual or political sympathy. Heinrich Marx died in 1838, Henriette Marx in 1863.

During 1835–1841, Marx studied at the universities of Bonn and Berlin, reading law at his father's request but turning to philosophy and history. After initial resistance, he studied Hegel thoroughly, in part through the lectures of Eduard Gans but more deeply with an intellectual club of somewhat older philosophers, among them Bruno Bauer and, later, Arnold Ruge. In 1836, Marx became engaged to Jenny von Westphalen, daughter of his beloved older friend; they were married in 1843. Hoping for an academic career, he submitted a dissertation entitled "The Difference Between the Democritean and Epicurean Philosophies of Nature" to the University of Jena in 1841 and was awarded the doctorate. Central to that dissertation was Marx's praise for Epicurus's addition of spontaneity—the famous "swerve"—to the determinism of the Democritean atomic dynamics, and for the Epicurean recognition of an animate level of human will along with the inanimate mechanisms of natural necessity.

Immersed for some time in the history of philosophy, Marx followed Hegel's cultural setting of philosophical thought in an inherently rational and explicable sequence that is the historical as well as the systematic maturation of awareness and self-awareness of the human spirit. The young Marx understood Hegel's work to be also a fundamental advance in logic and methodology of inquiry, one that would enable philosophers to comprehend the movement of ideas in their actuality, their potentialities, their mutual conflicts and inner tensions, and their syntheses. The scope of this outlook was vast, for it was to reach all the achievements of civilization, with every specialty to be understood in its own historical development and in its relations to others: religions and philosophies, but also the arts and literature, fashions and superstitions, wars and revolutions, politics, jurisprudence, technologies, and the sciences of nature and of mankind. Above all, Marx thought that Hegel would make clear the relation of man to his environment, to his fellows, and to himself by a philosophy that was at once an epistemology, a history, and a psychology.

In contrast with the orthodox conservative reading of Hegel (according to which all that exists is to be understood by rational methods, and to be understood and defended as being rational, necessary, and good, the progressive embodiment of reason in history), Marx joined with the Young Hegelians in seeing basic challenge and change to be central for Hegel, with progress the recurring theme of the increasing self-awareness of human consciousness, in the larger society as much as in the philosophical mind. For young Marx the task of philosophic reason was to criticize whatever exists, whether in social institutions, religious doctrine, or the realm of ideas; for what exists is limited, always incompletely rational, and potentially open. Illusions, self-deceptions, group delusions, plain factual errors were to be exposed; the incompletely rational, the spurious, and the idolatrous would be recognized and, partly by being known, righted.

Not unexpectedly, the initial target of these young radical thinkers was religious doctrine, in its logic, its historical evidence, its social roles, and its relation to political interests and to scientific knowledge. Marx's personal hero was Prometheus, "who stole fire from heaven and began to build houses and settle on earth." Philosophy, for Marx, "turns itself against the world that it finds."

If only on ideological grounds, Marx was unable to begin an academic career. His friend Bauer was dismissed from his teaching post at Bonn because of his secular critique of the Christian Gospels, and Marx, seeing his academic hopes disappearing, turned to journalism. He joined the staff of a liberal newspaper in Cologne, the *Rheinische Zeitung*; became editor by October 1842; and resigned early in 1843, just before the paper was closed by the Prussian censor. He met Friedrich Engels briefly in Cologne; and by their second meeting in Paris in 1844, a friendship had flourished that was to last until Marx's death and to be an example of intimate collaboration, personal affection, steadfastness, and mutual respect.

Marx went to Paris in October 1843, already committed to a life that would combine scientific work with political activity. He had begun thorough studies of economics, in particular the writings of Adam Smith and Ricardo, and he was coming to terms with Hegelian and post-Hegelian philosophy. He joined the radical German colony in Paris, and collaborated in a short-lived publication, Arnold Ruge's radical *Deutsch-französischer Jahrbücher*. For the first time Marx met revolutionary members of the urban working class; he knew the French socialist Proudhon, the German poet Heine, and the Russian anarchist Bakunin; he associated himself with a secret communist group, the League of the Just; he became a socialist and a communist. He was in Paris for only three years, but they were the years of his early maturity, of his decisive intellectual, professional, and political transformation. From those years come his incisive and profound notebooks, published a century later (the influential *Economic and Philosophic Manuscripts of 1844*) and his first writings with Engels.

Deported from France in 1845, Marx lived in Brussels until the revolutionary year of 1848, when he returned briefly to Paris at the invitation of the provisional government; he then went to Cologne to organize the *Neue Rheinische Zeitung*. Within six months he had been charged with incitement to rebellion and tried in court. Although acquitted in February 1849, Marx was expelled once more. He stayed briefly in Paris, was again ordered from France, and in July 1849 settled himself and his family permanently in London. Engels came to London in November of that year; and in 1850 he settled in Manchester to work in his father's textile firm, thereby providing Marx's principal financial support.

Aside from some ten years writing political commentary, mainly for the *New York Tribune* (1852–1862), Marx had no regular income. Despite Engels' support, he was often desperately poor and was beset by chronic, and for extended periods very painful, illnesses. In the 1860's he wrote of the family's "humiliations, torments and terrors," yet his three surviving daughters recalled with gratitude his unending storytelling, his games with them, and his entrancing reading aloud from the whole of Homer, the *Niebelungenlied, Don Quixote*, the *Arabian Nights*, and that Bible of the Marx household which was Shakespeare. Only in his last decade, when Engels had retired from his prosperous business to settle in London, was Marx somewhat free from financial trouble.

Marx's political activities were manifold, from his first contacts with working-class people in the early 1840's to his repeated organizational efforts: the German Communist League in Brussels (1847); various workers' and democratic associations in subsequent years; the *Manifesto of the Communist Party*, written with Engels and published in 1848; the International Working Men's Association of 1864, with its several congresses and its national sections (ultimately dissolved in 1876, after a struggle with Bakunin); the uniting of the various German workers' parties in 1875; continuing relations with the Chartists and with other British labor organizations; efforts to assist refugees after the fall of the Paris Commune in 1871; and, throughout his life, a voluminous correspondence with European and American socialists and sympathetic thinkers and activists.

Nevertheless, Marx's principal energies were devoted to his studies of empirical materials and theoretical models relating to the development and functioning of modern European society, the political economy of capitalism. He saw the first volume of his chief work, *Das Kapital*, published in 1867; the second volume (1885) and the third (1894) were edited from Marx's notes and drafts by Engels; further portions (1905–1910) were edited by Karl Kautsky. The important preparatory outlines and studies for *Kapital*, the *Grundrisse* of 1857–1858, were first published at Moscow in 1939–1941 but became widely available only with the Berlin edition of 1953. Aside from these, Marx's works comprise more than a dozen monographs and treatises, and hundreds of shorter articles. Since 1957 the collected *Marx-Engels Werke* have appeared in forty volumes.

Marx and Science. Marx's scientific work was entirely within the social sciences, but on several counts his work related to the natural sciences.

First, he sought to be scientific in his understanding of society. He gave recurrent attention to scientific methodology, at times in the context of comparing a natural science with social science but more often in his appreciative but critical fusion of Hegel's mode of understanding with empirical studies or in his critical studies of the methods of classical political economics. As general methodologist of science, Marx is of historical and systematic interest beyond his great influence upon economics, history, and sociology.

Second, Marx's conception of explanation in social science was entirely historical, with the consequence that he gave particular attention to the nature of historical understanding. Here again his methodological views are of broad interest, to the philosopher of science and to historians of ideas, as well as directly to the historian of science as historiographer, as specialist-investigator, and as the interpreter of science as a component of civilization.

Third, Marx's central conception of natural science as a social phenomenon requires that historians and philosophers of science—and scientists—set their accounts of the cognitive as well as the practical character of science within the framework of an understanding of the societies within which science arises and develops. For Marx himself, as we shall see, this social character of science suggested an agenda of separate issues about the sciences. It required both a coherent Marxist history of science and technology, and the elaboration of a political economy of science, but Marx himself was unable to devote energy to these tasks.

Fourth, Marx's materialist outlook upon mankind as situated within the natural environment, together with his conception of human emancipation through mastery of natural and social forces, brings his theory of nature to a primary position in his social theory as well as in his epistemology and methodology. Here the relations between the Marxian dialectic, the Marxian understanding of materialism, and both of these with Marx's concept of nature, take their place.

Science. The principal contribution of Karl Marx to the understanding of the sciences was his emphasis on their social character. Although he admired the great advances in knowledge that the sciences have provided, especially since the Renaissance—that is, he acknowledged the cognitive successes of the sciences—Marx nevertheless comprehended them as social phenomena. For the sciences to be social meant, to begin with, that they were part of the general social and economic processes of their times, changing with the changes in those historical processes; and if at times they were isolated from social forces, then they were understood as a product of social conflicts and pressures that allow such isolation. To be social meant, further, to respond to socially produced motivations and purposes, and to do so with socially stimulated modes of inquiry and explanation, and criteria of success or failure.

At times a component of leisure-class playtime and the object of curiosity, and often characterized for many scientists by the pleasant fulfillment of creative labor rather than by the imposition of necessity, the sciences were nevertheless not in any full sense promoted by such pleasurable motivation, for in the development of the sciences Marx saw a central contribution to the grim and practical task of mastering nature. By the mid-nineteenth century, mastery had come to a novel and high point in human history, accompanied by the bourgeois revolution and the development of industrial capitalism. Where, Marx wrote, ". . . would natural science be without industry and commerce? Even this 'pure' natural science is provided with an aim, as with its material, only through trade and industry, through the sensuous activity of men" (*The German Ideology* [New York ed.], 36).

As an element in the general historical process, science would be understood only in a completely historical way. Whether Hegelian or not in his historical epistemology, Marx imposed upon himself the task of comprehending science, like other human phenomena, within the political and economic history of mankind. Perhaps it is now evident that engineering, the technologies, and the practical arts must be described and understood in their social context and their historical development, with the external play of economic, military, political, cultural, and other forces upon them, as well as the internal sociology of inventiveness, learning, and genius (these notions, too, would have to be investigated and supplemented, as well as set within historical contexts); but it was surely not so evident when Marx was writing. The noted pioneering works on the development of technology were Johann Beckmann's *Beiträge zur Geschichte der Erfindungen* (5 vols. [Leipzig, 1782–1805]) and J. H. M. Poppe's *Geschichte der Technologie* (3 vols. [Göttingen, 1807–1811]). Both were known to Marx, and neither paid much attention either to the steam engine in particular or to the industrial revolution at large. Even Charles Babbage limited himself to an analysis of individual technological accomplishments, rather than striving for general historical comprehension, in his standard work *On the Economy of Machinery and Manufactures* (London, 1832).

The Marxian analysis is best seen in the detailed studies that constitute chapter 15, "Machinery and Modern Industry," of volume I of *Capital*, particularly section 1, "The Development of Machinery." Marx there sets himself the task of understanding the distinction between the two revolutions in mode of production: that of manufacture with labor power, which uses tools, and that of industrial production, which uses machinery. He sees the historical process to be from handicraftsmen who use tools to manufacturers whose laborers are still craftsmen using tools but are socially linked

MARX

through division of labor, with resulting reduction in labor cost. Then comes the drastically influential entry of machinery on the historical scene. His analysis may be given in several passages:

(1) On the general nature of productive machinery:

> All fully developed machinery consists of three essentially different parts, the motor mechanism, the transmitting mechanism, and finally the tool or working-machine. The motor mechanism is that which puts the whole in motion. It either generates its own motive power, like the steam engine, the caloric engine, the electromagnetic machine, etc., or it receives its impulse from some already existing natural force, like the water-wheel from a head of water, the wind-mill from wind, etc. . . . The tool or working-machine is that part of the machinery with which the industrial revolution of the 18th century started. And to this day it constantly serves as such a starting point, whenever a handicraft, or a manufacture, is turned into an industry carried on by machinery (*Capital*, I, 367).

(2) On machines as distinct from human implements:

> On a closer examination of the working-machine proper, we find in it, as a general rule, though often, no doubt, under very altered forms, the apparatus and tools used by the handicraftsman or manufacturing workman; with this difference, that instead of being human implements, they are the implements of a mechanism, or mechanical implements. . . . The machine proper is therefore a mechanism that, after being set in motion, performs with its tools the same operations that were formerly done by the workman with similar tools. Whether the motive power is derived from man, or from some other machine, makes no difference in this respect. From the moment that the tool proper is taken from man, and fitted into a mechanism, a machine takes the place of a mere implement. The difference strikes one at once, even in those cases where man himself continues to be the prime mover. The number of implements that he himself can use simultaneously is limited by the number of his own natural instruments of production, by the number of his bodily organs. In Germany, they tried at first to make one spinner work two spinning wheels, that is, to work simultaneously with both hands and both feet. This was too difficult. Later, a treadle spinning wheel with two spindles was invented, but adepts in spinning, who could spin two threads at once, were almost as scarce as two-headed men. The [spinning] Jenny, on the other hand, even at its birth, spun with 12–18 spindles, and the stocking-loom knits with many thousand needles at once. The number of tools that a machine can bring into play simultaneously, is from the very first emancipated from the organic limits that hedge in the tools of a handicraftsman . . . (*Capital*, I, 368, 370–371).
>
> . . . apart from the fact that man is a very imperfect instrument for producing uniform continued motion but assuming that he is acting simply as a motor, that a machine has taken the place of his tool, it is evident that he can be replaced by natural forces . . . (*Capital*, I, 370–371).

(3) On the change in scale of power required for industry:

> Modern Industry had . . . itself to take in hand the machine, its characteristic instrument of production, and to construct machines by machines. It was not till it did this that it built up for itself a fitting technical foundation, and stood on its own feet. . . . But it was only during the decade preceding 1866, that the construction of railways and ocean steamers on a stupendous scale called into existence the cyclopean machines [steam engines] now employed in the construction of prime movers . . . capable of exerting any amount of force, and yet under perfect control.
>
> . . .we find the manual implements reappearing, but [also] on a cyclopean scale. The operating part of the boring machine is an immense drill driven by a steam-engine; . . . the tool of the shearing machine, which shears iron as easily as a tailor's scissors cut cloth, is a monster pair of scissors; and the steam hammer works with an ordinary hammer head, but of such a weight that not Thor himself would wield it (*Capital*, I, 373, 380–382).

(4) On the deliberate link of science with industry, and the social implication:

> The implements of labour, in the form of machinery, necessitate the substitution of natural forces for human force, and the conscious application of science, instead of rule of thumb. In Manufacture, the organization of the social labour-process is purely *subjective*; it is a combination of detail labourers; [whereas] in its machinery system, Modern Industry has a productive organism that is purely *objective*, in which the labourer becomes a mere appendage to an already existing material condition of production. In simple co-operation, and even in that founded on division of labour, the suppression of the workman, isolated by the collective, still appears to be more or less accidental. Machinery, with a few exceptions to be mentioned later, operates *only* by means of associated labour, or labour in common. Hence the co-operative character of the labour-process is, in the latter case, a technical necessity dictated by the instrument of labour itself (*Capital*, I, 382).

(5) On the role of science in completing the role of the division of labor:

MARX

(a) . . . Intelligence in production expands in one direction, because it vanishes in many others. What is lost by the detail labourers, is concentrated in the capital that employs them. It is a result of the division of labour in manufactures, that the labourer is brought face to face with the intellectual potencies of the material process of production as the property of another, and as a ruling power. This separation begins in simple co-operation, where the capitalist represents, to the single workman, the oneness and the will of the associated labour. It is developed in manufacture, which cuts down the labourer into a detail labourer. It is completed in modern industry, which makes science a productive force distinct from labour and presses it into the service of capital (*Capital*, I, 355).

(b) The bourgeoisie has stripped of its halo every occupation hitherto honoured and looked up to with reverent awe. It has converted the physician, the lawyer, the priest, the poet, the man of science into its paid wage labourers (*Communist Manifesto, Collected Works*, VI, 487).

(6) On the distinction between science and cooperative labor:

It should be noted that there is a difference between universal labour and co-operative labour. . . . Universal labour is scientific labour, such as discoveries and inventions. This labour is conditioned on the co-operation of living fellow-beings and on the labours of those who have gone before. Co-operative labour, on the other hand, is a direct co-operation of living individuals (*Capital*, III, 124).

(7) On the relations of nature, science, and industry:

. . . historiography pays regard to natural science only occasionally, as a factor of enlightenment, utility, and of some special great discoveries. But natural science has invaded and transformed human life all the more *practically* through the medium of industry; and has prepared human emancipation, although its immediate effect had to be the furthering of the de-humanization of man. *Industry* is the *actual*, historical relationship of nature, and therefore of natural science, to man. . . . In consequence, natural science will lose its abstractly material—or rather, its idealistic—tendency, and will become the basis of *human* science, as it has already become the basis of actual human life, albeit, in an estranged form. *One* basis for life and another basis for *science* is *a priori* a lie. The nature which develops in human history—the genesis of human society—is man's *real* nature; hence nature as it develops through industry, even though in an *estranged* form, is the true nature (*Economic and Philosophic Manuscripts*, 142–143).

Marx early recognized that, like prejudices and religious beliefs, ideas too have their social functions and determinants—and not least scientific ideas, even those of the most confirmed and objectively established sort. Thus, he was an admirer of Charles Darwin's work, which he saw as a penetrating insight and proof of the historical character of biological nature. But he also noted, with amusement, that Darwin's hypothesis saw nature in a social image:

(8) (a) . . . Darwin's book is very important and serves me as a natural-scientific basis for the class struggle in history. One has to put up with the crude English method of development, of course. Despite all deficiencies, not only is the death-blow dealt here for the first time to "teleology" in the natural sciences but its rational meaning is empirically explained . . . (letter to Lassalle, 16 Jan. 1861, *Selected Correspondence*, Moscow ed., 151).

(b) It is remarkable how Darwin recognizes among beasts and plants his English society with its division of labour, competition, opening up of new markets, "inventions", and the Malthusian "struggle for existence". It is Hobbes's *bellum omnium contra omnes*, and one is reminded of Hegel's *Phenomenology*, where civil society is described as a "spiritual animal kingdom", while in Darwin the animal kingdom figures as civil society . . . (letter to Engels, 18 June 1862, *Selected Correspondence*, 156–157).

But Marx also saw Darwin's work as suggestive for human history, and for the instrumental role of the human body, of technology, and of science:

(c) Darwin has interested us in the history of Nature's Technology, i.e., in the formation of the organs of plants and animals, which organs serve as instruments of production for sustaining life. Does not the history of the productive organs of man, of organs that are the material basis of all social organization, deserve equal attention? And would not such a history be easier to compile, since as Vico says, human history differs from natural history in this, that we have made the former, but not the latter? (*Capital*, I, 367).

To Marx, sociological understanding of the origin of scientific ideas was a component of the total appreciation of science. Two further aspects of his thought relate to such a historical sociology of science: the instrumental aspect of science, and of all knowledge, and the flexibility of nature when confronted with humankind. Here Marx consistently treated science under the general heading of labor, and he understood scientific conceptions to be joined with the material basis of human existence, with practical life, and with the social relations among human beings. The previous passage continues:

(d) Technology discloses man's mode of dealing with Nature, the process of production by which he sustains his life, and thereby also lays bare the mode of formation of his social relations, and of the mental conceptions that flow from them. . . . [But the] weak points in the abstract materialism of natural science, a materialism that excludes history and its process, are at once evident from the abstract and ideological conceptions of its spokesmen, whenever they venture beyond the bounds of their own speciality (*Capital*, I, 367).

Marx saw that capital "first creates bourgeois society and [with it] the universal appropriation of nature. . . ." Nature takes an instrumental role in human history.

(9) For the first time, nature becomes purely an object for humankind, purely a matter of utility; ceases to be recognized as a power for itself; and the theoretical discovery of its autonomous laws appears merely as a ruse so as to subjugate it under human needs, whether as an object of consumption or as a means of production (*Grundrisse*, 410).

Such an attitude toward technology and science leads to Marx's notion of freedom, in the now-familiar Marxian theme of reversing the domination of human beings either by the "blind" forces of nature or by the industrial society with its technology. It is technology that is the fundamental, because it is the mediation between man and nature:

(10) The realization of freedom consists in socialized man, the associated producers, rationally regulating their material interchange with nature and bringing it under their common control, instead of allowing it to rule them as a blind force (*Capital*, III, Chicago ed., 954).

This too leads beyond craft technology to science with the impressive modification of human life, which is made possible by the cognitive achievement of science when, and if, it is, in Marx's term, "appropriated":

(11) In this transformation, it is neither the direct human labour he himself performs, nor the time during which he works, but rather the appropriation of his own universal [scientific] productive power, his understanding of nature and his mastery over it by virtue of his presence as a social body—it is, in a word, the development of the social individual which appears as the great foundation-stone of production and of wealth (*Grundrisse*, 705, slightly modified).

To understand the "societal individual" is to understand Marx's theory of society. Here we cannot pursue the main body of Marx's work; but we must indicate his own method, which is also his conception of scientific explanation and scientific inquiry.

Scientific Method. Research into Marx's methods of scientific thought and investigation, both as shown in his works and as deliberately expounded by him, has reached no general scholarly agreement. The principal explicit texts on method in Marx's writings are section 3, "The Method of Political Economy," of the introduction to his *Grundrisse*; *Notes on Adolph Wagner*; the preface to the second edition of *Capital*; section 2 of *The Holy Family*; and the preface to *Critique of Political Economy*.

Engels often praised Marx's method, even above Marx's achievements, which were said to have been due to it. In a letter of 1895 to Werner Sombart, Engels wrote: "Marx's whole manner of conceiving things is not a doctrine, but a method. It offers no finished dogmas, but rather points of reference for further research, and the method of that research. . . ." In the several methodological texts, and from his first writings to the last, Marx consciously worked on methodological problems, explicitly and repeatedly developing his own views by criticizing Hegel for methodological (as well as other) inadequacies; frequently criticizing other economists, historians, philosophers, and political thinkers on grounds of scientific method; and, at the same time, elucidating his own understanding of the conceptual principles of sound scientific thinking. Although much is still disputed among Marxists and by other students of Marx's works, some matters of substance and of conceptual vocabulary seem clear from the relevant texts.

In the preface to the second edition of *Capital*, Marx quoted at length from a Russian article that treated his method in *Capital* in what Marx said was "this striking and generous way":

(12) (a) "The one thing which is of moment to Marx, is to find the law of the phenomena . . . [and] the law of their variation, of their development, i.e., of their transition from one form into another. . . . Marx only troubles himself about one thing; to show, by rigid scientific investigation, the necessity of successive determinate orders of social conditions, and to establish, as impartially as possible, the facts that serve him for fundamental starting points [and] both the necessity of the present order of things, and the necessity of another order into which the first must inevitably pass over. . . . Marx treats the social movement as a

MARX

process of natural history, governed by laws not only independent of human will, consciousness and intelligence, but rather, on the contrary, determining that will, consciousness and intelligence . . . not the idea, but the material phenomenon alone can serve as its starting-point. Such an inquiry will confine itself to the confrontation and the comparison of a fact, not with ideas, but with another fact. For this inquiry, the one thing of importance is both that facts be investigated as accurately as possible, and that they actually form, each with respect to the other, different moments of an evolution; . . . it will be said, the general laws of economic life are one and the same, no matter whether they are applied to the present or the past. This Marx directly denies. According to him, such abstract laws do not exist. On the contrary, in his opinion every historical period has laws of its own. . . . In a word, economic life offers us a phenomenon analogous to the history of evolution in other branches of biology. The old economists misunderstood the nature of economic laws when they likened them to the laws of physics and chemistry . . ." (*Capital*, I, xxvii–xxix).

To this, Marx adds:

(12) (b) . . . what else is he picturing but the dialectic method? Of course the method of presentation must differ in form from that of inquiry. The latter has to appropriate the material in detail, to analyse its different forms of development, to trace out their inner connection. Only after this work is done, can the actual movement be adequately described. If this is done successfully, if the life of the subject-matter is ideally reflected as in a mirror, then it may appear as if we had before us a mere *a priori* construction (*Capital*, I, xxix–xxx).

Marx distinguishes the method of inquiry from the method of exposition. Inquiry (*Forschung*) is factually realistic, beginning with initially uninterpreted data that are subjected to analysis in stages of complexity that demand insightful abstraction, simplification, and subtlety. The factual data (*Tatsache*) are the concrete entities, or wholes; and the results of analysis are abstract principles, analyzed into theoretically formulated "parts," hypothetically guided by theories that have been based upon, and more or less tested by, previous empirical investigations. Inquiry is a complex stage of empiricism and of inductive and hypothetical analysis.

Presentation (*Darstellung*) gives the results their necessary development, which aims to be a conceptual return to the concrete and brings the component parts or qualities of any subject matter together in their "organic" interrelatedness and their evolutionary or historical movements. The return will be mediated by expository as well as theoretical demands so as to clarify the separate qualities and the various relations among them and with their environment. For Marx, the truth will be the whole in its changes; and these in turn relate by historical processes, the Marxian dialectic of contending and negating "forces" within history. Indeed, the negative quality of historical changes links up, for Marx, with his positive notion of liberation of unfulfilled and repressed (alienated) human nature. (We shall see below how this may also comprehend nature.)

Volume I of *Capital* presents a theoretical model of the process of production in capitalism that, like so many models in natural science, isolates the theoretically conceived key qualities by means of simplifying assumptions. For Marx, abstraction was a justified but contrary-to-fact simplification. As he understood the problem of knowledge, scientific thought must be completed by a careful conceptual process of synthesis, by removal of the assumptions stage by stage, and by asymptotic approximation to the concrete complexity of the real world. The abstract model of Marx's volume I was brought closer to the actual economic process of nineteenth-century capitalism, as he hoped, with his series of realistic considerations in volume III.

Abstraction is characteristic of all science but, for Marx, it has a central place in scientific investigation of social phenomena. Furthermore, abstraction is the method of discovering the "inner connections" and "inner movements" of the phenomena; Marx remarked that "all science would be superfluous if the manifest form and the essence of things directly coincided" (*Capital*, III, 797). In his preface to the first edition of *Capital*, Marx wrote:

. . . the body, as an organic whole, is more easy of study than are the cells of that body. In the analysis of economic forms, moreover, neither microscopes nor chemical reagents are of use. The force of abstraction must replace both. But in bourgeois society the commodity-form of the product of labour—or the value-form of the commodity—is the economic cell-form (*Capital*, I, xvi).

Investigation, then, is empirical but also abstract; exposition is dialectical and concrete. Truth in science is concrete. And, as we shall see, Marx was not an inductivist. But while the scientist starts with abstract categories (of thought), he must go from these to the concrete, for the elementary and simple abstraction, although not fictitious, is only one aspect of any object of investigation, and an aspect in relation to man. To go further requires the human side and, hence, the social relations among the categories. Marx's mature methodological reflections

MARX

on this dialectic of abstract and concrete moments of scientific practice were most fully set forth in the 1857 introduction to the *Grundrisse*:

(13) It seems to be correct to begin with the real and the concrete, with the real precondition, thus to begin, in economics, with e.g. the population, which is the foundation and the subject of the entire social act of production. However, on closer examination this proves false. The population is an abstraction if I leave out, for example, the classes of which it is composed. These classes in turn are an empty phrase if I am not familiar with the elements on which they rest, e.g. wage labour, capital, etc. These latter in turn presuppose exchange, division of labour, prices, etc. For example, capital is nothing without wage labour, without value, money, price, etc. Thus, if I were to begin with the population, this would be a chaotic representation [*Vorstellung*] of the whole, and I would then, by means of further determination, move analytically towards ever more simple concepts [*Begriff*], from the imagined concrete towards ever thinner abstractions until I had arrived at the simplest determinations. From there the journey would have to be retraced until I had finally arrived at the population again, but this time not as the chaotic conception of a whole, but as a rich totality of many determinations and relations.

. . . [This] is obviously the scientifically correct method. The concrete is concrete because it is the concentration of many [abstract] determinations, hence a unity of the diverse. It appears in the process of thinking, therefore, as a process of concentration, as a result, not as a point of departure, even though it is the point of departure in reality and hence also the point of departure for observation [*Anschauung*] and conception. Along the first path the full conception was evaporated to yield an abstract determination; along the second, the abstract determinations lead towards a reproduction of the concrete by way of thought.

. . . it may be said that the simpler category can express the dominant relations of a less developed whole, or else those relations subordinate to a more developed whole which already had a historic existence before this whole developed in the direction expressed by a more concrete category. To that extent the path of abstract thought, rising from the simple to the combined, would correspond to the real historical process.

As a rule, the more general abstractions arise only in the midst of the richest possible concrete development, where one thing appears as common to many, to all. Then it ceases to be thinkable in a particular form alone. . . .

This example of labour shows strikingly how even the most abstract categories, despite their validity—precisely because of their abstractness—for all epochs, are nevertheless, in the specific character of this abstraction, themselves likewise a product of historic relations, and possess their full validity only for and within these relations. . . .

It would be unfeasible and wrong to let the economic categories follow one another in the same sequence as that in which they were historically decisive. Their sequence is determined, rather, by their relation to one another in modern bourgeois society, which is precisely the opposite of that which seems to be their natural order or which corresponds to historical development. The point is not the historic position of the economic relations in the succession of different forms of society. Even less is it their sequence "in the idea" (Proudhon) (a muddy notion of historic movement). Rather, their order within modern bourgeois society (*Grundrisse*, 100–108).

To Marx, exposition and articulation, when carefully accomplished, showed the movement of thought, a conceptual dynamic. He was concerned to contrast his understanding of this dialectic with that of Hegel, for whom

(14) . . . the life-process of the human brain, i.e., the process of thinking, under the name of "the Idea", he even transforms into an independent subject, [as] the demiurge of the real world, and the real world is only the external, phenomenal form of "the Idea". With me, on the contrary, the ideal is nothing else than the material world reflected by the human mind, and translated into forms of thought (*Capital*, I, xxx).

For Marx, the concrete-in-thought was real and concrete enough, insofar as thoughts are real, but in no way was it to be considered as the genuine thing, as abstractions that somehow were formed into concrete matters of nature or society. Indeed, Marx focused his methodological criticism of Hegel in 1857 on this point:

(15) In this way Hegel fell into the illusion of conceiving the real as the product of thought concentrating itself [*sich zusammenfassenden Denkens*], probing its own depths, and unfolding itself out of itself, by itself, whereas the method of rising from the abstract to the concrete is only the way in which thought appropriates the concrete, reproduces it as the concrete in the mind. But this is by no means the process by which the concrete itself comes into being. . . . the concrete totality is . . . a product . . . of the working-up [*Verarbeitung*] of observation and conceptual representation into concepts [*Begriffe*]. The totality as it appears in the head, as a totality of thoughts, is a product of a thinking head, which appropriates the world in the only way it can, a way different from the artistic, religious, practical and mental appropriation of this world. The real subject-matter retains its autonomous existence outside the head just as before; namely as long as the head's conduct is merely specu-

MARX

> lative, merely theoretical. Hence, in the theoretical method, too, the subject, society, must always be kept in mind as the presupposition (*Grundrisse*, 101–102).

Marx's empirical side had earlier been pressed against Hegel in 1843 in the *Critique of Hegel's 'Philosophy of Right'*:

> (16) Thus empirical actuality is admitted just as it is and is also said to be rational; but not rational because of its own reason, but because the empirical fact in its empirical existence has a significance [for Hegel] which is other than it itself. The fact, which is the starting point, is not conceived to be such but rather to be the mystical result.
>
> It is evident that the true method is turned upside down. What is most simple is made most complex and vice versa. What should be the point of departure [of the presentation] becomes the mystical result, and what should be the rational result becomes the mystical point of departure (O'Malley ed., 9, 40).

The issue appears thirty years later in the 1873 preface, once more in Marx's well-known image:

> (17) . . . With him [dialectic] is standing on its head. It must be turned right side up again, if you would discover the rational kernel within the mystical shell (*Capital*, I, xxx).

Perhaps the most explicit contrast, in Marx's own estimation, was stated in the unpublished notes for *The German Ideology*:

> (18) First Premises of Materialist Method.
>
> The premises from which we begin are not arbitrary ones, not dogmas, but real premises from which abstraction can only be made in the imagination. They are the real individuals, their activity and the material conditions under which they live, both those which they find already existing and those produced by their activity. These premises can thus be verified in a purely empirical way. . . .
>
> Empirical observation must in each separate instance bring out empirically, and without any mystification and speculation, the connection of the social and political structure with production. . . .
>
> This method of approach is not devoid of premises. . . . Its premises are men, not in any fantastic isolation and rigidity, but in their actual, empirically perceptible process of development under definite conditions. . . .
>
> When reality is depicted, philosophy as an independent branch of knowledge loses its medium of existence. At the best its place can only be taken by a summing-up of the most general results, derived through abstractions which arise from the observation of the historical development of men. Viewed apart from real history, these abstractions have in themselves no value whatsoever. They can only serve to facilitate the arrangement of historical material, to indicate the sequence of its separate strata. But they by no means afford a recipe or schema, as does philosophy, for neatly trimming the epochs of history . . . (*German Ideology*, 42, 46–48).

Marx and Nature. By "nature" Marx meant the natural world of the nonliving and the living, in which the human-social world was situated; but he also meant to include mankind within that natural world, as a species among the mammals, an animal among animals, a living being among all the forms of life, and a material entity of matter and energy, existing in the forms of space and time. In the profound but compressed manuscripts of 1844, Marx set forth his theme of understanding nature as that of understanding the relation between man and nature, between historical man and the external environment. But the environment comprises both the existing situation within which mankind exists, and as such relates to his species history, and also the autonomous world, temporally prior to mankind. What is this human-natural relation? In the long human history of recurring and potential scarcity, man has mainly struggled with nature. Whatever the cognitive forms—whether magical, technological, scientific, or otherwise—human interaction with nature has had domination as its goal. Even when the mode has been one of alliance or harmony with nature, nature has set the conditions and limits; and when the mode is one of successes in conquest, transformation, using and exploiting nature and natural processes, the transformations of nature by human labor (and its allied intelligence) nevertheless must be seen against the inexhaustible properties and impenetrable levels of resistance of matter.

For Marx, man and nature have a history together; man encounters nature in his own species history, each encounter within a specific concrete stage of that history. Without doubt, for Marx, nature had its own history; but that was not so much his own view as one he thought increasingly demonstrated by the natural sciences themselves. For him this was evident from developments in geology, astronomy, and, above all, evolutionary biology. And yet there was also a peculiarly Marxian understanding of nature that had two further aspects.

First, Marx stressed the insight that ideas of nature have their own history, which is a part of general human cultural history, itself a creative product of the material processes of society (and hence Marx's *aperçu* is a principal stimulant to later sociology of knowledge, and of scientific ideas in particular). Such a historical sociology of science reworked the ancient relativism about varying human perceptions of nature from skepticism about knowledge of nature to the (social-scientific) cognitive problem of the history of that knowledge. In the Marxian reconstruction of relativism it remains a difficult research question to locate the sources of success and failure of different approaches to nature, to ascertain the cognitive thread within human practice (and especially among the differing modes of cognitive practice that are revealed by studies in the history of the natural sciences and technologies). In the end, Marx believed practice was always the criterion, but practice is complex. At least, Marx saw, external nature was receptive to human labor, if not ever exactly a simple metaphorical raw and unformed clay to be shaped by the human potter. What was necessary in human development, he also saw, was for man to learn both the facts of natural entities, processes, tendencies, and laws, and the alternative possibilities to which those facts may be understood (with difficulty) as pointing. Here he thought he went beyond the "mere" empiricism of positive science.

In the latter sense, Marx understood the literal role of man within nature as concretely formative; men and women are fully natural beings who seek, choose, and remake the natural world, within the necessary limits. Man is child and maker of nature. Man the maker, for Marx, is even greater than his hero Prometheus, the conqueror of fire and liberator of mankind, because man creates new natural events, materials, qualities—indeed, creates a new nature.

The materialist history of ideas of nature is a history of changing intentional practice, for which implicit as well as explicit ideas have their several functions: cognition, rote aids to learning, conjectures to be tested and often to be generalized. All of this is articulated by means of the developing languages of collaborating scientific workers who are also ideological representatives of class and sectional interests (including interests in the concrete facts, in the truths of those facts and of what they suggest or conceal—or, at any rate, in some partial truths). Ideas of nature, and scientific theories as their modern form, were for Marx a part of the labor process, theoretical practice. To Marx, Hegel had investigated nature only through his logic, vainly seeking a concrete content; orthodox science investigated nature through observation and hypothesis, seeking autonomous laws; Marx investigated nature through man.

Second, at all human times, as nature is encountered historically, it must have its socially conditioned aspects and, increasingly, its socialized transformations. In its transition from the "natural" role of peasants in feudal agriculture to the "commodity" of man and natural processes in capitalist industry, nature changes. Nature has become, and now is, part of human history, which expands human nature so as to make over the external environment, at times, into the larger material body of individual men and of humanity. These metaphors were useful to Marx, to whom the flow of matter and energy between the body and the environment easily suggested that man is more than what his skin encloses, and for whom the social reality equally existed in such a mutual relationship with the natural context. Human bodily processes were natural, and so were social processes; Marx saw his most illuminating natural-science metaphor for social processes in "metabolism" (*Stoffwechsel*).

But the historical situation of nature was not seen by Marx as just metaphorical. Nature as known to concrete human beings is nature as it has been both dominated and understood; for nature to be understood by ideas of nature means nature's being subjected to the specific criteria and requirements of societies the dominant class forces of which have also dominated their forms of rationality. For Marx, while "prior" nature produces the human species in the course of biological, geological, and chemical processes, yet there are historical stages of nature, known to historians of science and technology by periods in the history of the natural sciences; and these, he anticipated, may be linked with the stages of evolution of social-economic formation. It is not too much to say, then, that there is a nature known to feudal society, and a different nature known to capitalist society; different societies raise different questions, work on different problems, use different ideas and methods, labor in different ways, learn differently, generalize differently, and reason differently. (When Marx wrote [see excerpt (5), above] that science is "pressed into the service of capital," he did not refer to applied science alone.)

Marx's early image of man in nature was that man appropriates nature, thereby bringing human purposes into nature. But which human purposes? Marx did not hesitate to link closely human appropriation and exploitation of nature with human exploitation of human beings. If men are treated as things, so will nature be; if human labor becomes the center of exploitation, and then is abstracted into average values for exchange in a commodity society—in a word, commercialized—then commercialized nature will appear (where it had not been); if men are distorted and polluted, then a

MARX

polluted nature will be made. Marx's conception of social tendencies toward the emancipation of mankind from human exploitation was explicit about his grounding of human liberation in a changing relationship with nature. Just as the emancipation of man requires emancipation from necessary labor (or at least minimization, as sketched in *Capital*, III), so it implies an open attitude on Marx's part toward changes in ideas of nature when the relation of man to man is no longer dominated by exploitation.

In bourgeois industrial society, and always in class societies, Marx saw nature as a limiting and resisting material that had increasingly become a productive force; or if nature itself is not literally a productive force, then the social metaphor may be shifted and nature comes to function as abstract matter, to be made, administered, and exploited as men wish, and as abstractly as the labor power of the working men. In the expected future classless society, which Marx in the *Grundrisse* foresaw to be characterized by fully automated and nearly labor-free factory productive processes, human nature may once again see its (new) rationality within nature. That is, if human purpose transcends mere domination, then it may transcend that purpose with respect to external nature too; and nature again would be receptive.

Marx did not pursue the matter of nonexploited nature further, with the singular but crucial exception of the changes in human nature as part of the natural order. Any speculation or development of his suggestions is beyond our concern here, but at least his discussions of the bodily base for aesthetic sensibility may be mentioned. He linked liberation from domination by the social relations of private property to "the complete emancipation of all human senses and qualities" (*Economic and Philosophic Manuscripts of 1844*, 139); indeed, the liberated human being in socialist society, the seemingly quite new man, would be one whose "senses are *other* than those of non-socialized man." For, he argued, ". . . not only the five senses but also the so-called mental senses—the practical senses (will, love, etc.)—in a word, *human* sense—the human nature of the senses—comes to be by virtue of its object, by virtue of *humanized* nature. The *forming* of the five senses is a labor of the entire history of the world down to the present" (*Economic and Philosophic Manuscripts of 1844*, 141).

At any rate, postcapitalist (or, in general terms, postexploitative) nature-for-man would be that part of the universe that is transformed into an environmental context within which the specifically human qualities and faculties will develop and flourish. Marx saw nature, and with it human nature, as flexible, plastic, and, above all, not restricted to a utilitarian function. He went so far as to say that the human senses would "relate themselves to the *thing* for the sake of the thing . . .," but he went on at once to add that "the thing itself is an *objective human* relation to itself and to man and vice versa . . . nature has lost its mere *utility* by use becoming human use" (*Economic and Philosophic Manuscripts of 1844*, 139). These processes of humanization and socialization of natural objects are precise: "The eye has become a *human* eye, just as its *object* has become a social, *human* object—an object made by man for man." And so sensuous human nature, along with all social-historically related external nature, changes as society does; Marx wrote: "The *social* reality of nature, and *human* natural sciences, or the *natural science about man*, are identical terms."

The lesson was completely socialized. A repressive, exploitative society would be expected to produce a dehumanized nature, because the actual known world of science, technology, and their society is a world of things that, in Marx's understanding of political economy, are actually or potentially objectified human labor. Through labor, the primary category of both his philosophy and his social science, we are brought to comprehend Marx's natural science. Man makes himself, following Hegel's famous phrase; but man also makes his natural world, for, as Marx said, nature is man's inorganic body.

> (19) . . . just as the working subject appears naturally as an individual, having a natural existence, so does the first objective condition of his labor appear as nature, as earth, as his inorganic body. The individual himself is not only the organic body of nature but also the Subject of this inorganic nature (*Grundrisse*, 488).

In critical discussion of the destructive use of natural resources, Marx was looking ahead to a nondestructive relationship with nature, which equally would be the work of human labor; praxis, he believed, had the potentiality of treating human beings as human and, at the same time, of accepting both the potentialities and the limitations of nature. Within those potentialities, a fully human home on earth could be designed and constructed, in light of scientific understanding of the fullest range of human potentialities and those of external nature.

Marx took his idea of socialized nature cautious-

MARX

ly. The limitations placed by autonomous nature are genuine, for, as mentioned above, Marx agreed with Giambattista Vico that human beings have made human history, but not natural history. The problem that arises, then, for Marx in his conception of nature can be clarified by his method of investigation: Nature in its autonomy, prior to human history and apart from that history, is, as one commentator remarked, only on the horizon of history. Nature has its own history, and yet it both generates and yields to the human species with its characteristically concrete history. Autonomous nature, then, is—and can be—only abstract for mankind because it has been apprehended neither by ordinary practice nor by any cognition through scientific practice. ". . . nature, taken abstractly, for itself, rigidly separated from man, is *nothing* for man" (*Economic and Philosophic Manuscripts of 1844*, 117, Bottomore ed.).

There must be concrete nature rather than an artificial abstraction; but this shift to the concrete, as we have seen, is what Marx understands to be nature appropriated, exploited—indeed, mediated—by socially organized labor. In 1880, toward the end of his life, Marx wrote: "Only a schoolmaster-professor [could construe] the relations of man to nature as not practical from the outset, that is relations established by action, but as *theoretical* relations . . ." (*Notes on Adolph Wagner*, 190). He went on to clarify: Not first the epistemological relation of scientific practice but, rather, first the socially primary relation of "appropriating certain things of the external world as the means for satisfying their own needs, etc." and by "thus satisfying their needs, therefore they begin with production." Intellectual practice—indeed, all learning from experience and reflecting upon experience in theoretical practice—comes after the fundamental base within material production.

The common-sense Marx prevailed, even while he analyzed socialized nature and speculated upon liberated nature. In *The German Ideology* he wrote:

(20) . . . of course, in all of this, the priority of external nature remains unassailed . . . but this differentiation [between autonomous or presocial, and socially mediated, nature] has meaning only insofar as man is considered to be distinct from nature.

Marx goes on at once, in this comment on Feuerbach:

For that matter, nature, the nature that preceded human history . . . is nature which today no longer exists anywhere (except perhaps on a few Australian coral-islands of recent origin) and which, therefore, does not exist for Feuerbach (*German Ideology*, 63).

And yet, as we know from Marx's sociological comment on Darwin's work, any thought of nature before mankind, or of nature insofar as it is not yet known or appropriated, must, for Marx, be comprehended through the very same socially generated categories as the concretely grasped nature of ordinary labor and scientific practice. And even the autonomous qualities are suspected of being human—with cunning, as Hegel might have said (see excerpt [9] above).

If nature provides the metabolic biochemistry for man in society as in physiology, the metaphor deserves a further caution, since Marx understood that metabolism too has its autonomous properties and laws. Hence, "Man can only proceed in his production in the same way as nature itself, that is he can only alter the *forms of the material*" (*Capital*, 10). But these alterations affect nature too; Marx simply sees man as an agent of nature transforming itself. He speaks of labor power as a "material of nature transferred to a human organism"; and he also sees quickly, in *Capital*, that the very simile of changing the forms of a kind of raw, unformed substance must be otherwise understood: "The object of labour can only become raw material when it has already undergone a change mediated through labour." And yet it is nature that actually participates in such mediation through the emergence of the human species, which brings practical, creative, transformative labor into nature.

(21) Labour is, in the first place, a process in which both man and nature participate, and in which man of his own accord starts, regulates, and controls the material reactions between himself and Nature. He opposes himself to Nature as one of her own forces . . . (*Capital*, I, ch. 7, 156).

BIBLIOGRAPHY

I. Original Works. The current standard ed. of the known published and unpublished writings of Marx and Engels is *Marx-Engels Werke*, 39 vols. plus index (Berlin, 1957–1968), which includes early works from student days, speeches and newspaper articles, and the correspondence with each other and with third parties. Supplementary vols. appeared in 1967 and 1969. There are two Russian-language eds. of the complete works, the *Sochinenia*, 25 vols. (Moscow, 1928–1946), and a 2nd, rev. ed. (Moscow, 1955–); a complete

Oeuvres in French is under way (Paris, 1963–); and the *Collected Works* are in progress in English, 50 vols. plus an index vol. (New York–London–Moscow, 1974–). M. Rubel, *Bibliographie des oeuvres de Karl Marx* (Paris, 1956), is immensely helpful; it includes "Repertoire des oeuvres de Friedrich Engels" as an appendix; a supp. appeared later (Paris, 1960).

Also see M. Klein *et al., Marx-Engels-Verzeichnis: Werke, Schriften, Artikel* (Berlin, 1968). An earlier collected ed., the *Karl Marx-Friedrich Engels: Historisch-kritische Gesamtausgabe*, 11 vols. (Frankfurt–Berlin; Moscow, 1927–1935), commonly referred to as *MEGA*, went only as far as 1848. Despite its limited scope, it was significant and influential as the first publication of major early writings of Marx and Engels and for the bulk of the correspondence between them. A guide to the various collected eds. is G. Hertel, *Inhaltsvergleichregister der Marx-Engels-Gesamtausgaben* (Berlin, 1957). As noted in these various eds. and guides, many of Marx's works were first published decades after his death; the historically influential writings must be seen in that respect.

The principal centers of research in the original materials are the Institut für Marxismus-Leninismus (Berlin), the Institute of Marxism-Leninism (Moscow), and the International Institute for Social History (Amsterdam). A practical introduction to the Amsterdam holdings is the *Alphabetical Catalog of the Books and Pamphlets of the International Institute of Social History*, 12 vols. (Boston, 1970); 2-vol. supp. (Boston, 1975). An annotated variorum scholarly ed. of the complete writings, speeches, notebooks, and correspondence of Marx and Engels is in preparation at the Berlin institute. A preliminary but useful specimen volume is *Karl Marx-Friedrich Engels-Gesamtausgabe (MEGA), Editionsgrundsätze und Probestücke* (Berlin, 1972). Publication of this new *MEGA* began in 1975.

A chronological list of Marx's principal works, as well as those written in collaboration with Engels, includes the following. The date of composition is indicated in parentheses.

"The Difference Between the Democritean and Epicurean Philosophies of Nature," Ph.D. diss. (1841); *Critique of Hegel's 'Philosophy of Right'* (1843); *On the Jewish Question* (1843); *Economic and Philosophic Manuscripts of 1844; The Holy Family* (1844), written with Engels; *Theses on Feuerbach* (1845); *The German Ideology* (1845–1846), written with Engels; *The Poverty of Philosophy* (1847); *Manifesto of the Communist Party* (1848), written with Engels; *The Class Struggles in France, 1848–1850* (1850); *The Eighteenth Brumaire of Louis Bonaparte* (1852); *Grundrisse (Foundations of the Critique of Political Economy—Rough Draft)* (1857–1858); *A Contribution to the Critique of Political Economy* (1858–1859); *Wages, Price and Profit* (1865); *Capital,* written over many years: I was published in 1867; II and III were posthumously edited by Engels and published in 1885 and 1894; IV, *Theories of Surplus Value*, appeared in three parts, 1905–1910, and was edited by K. Kautsky; *The Civil War in France* (1871); *Critique of the Gotha Programme* (1875); and *Notes on Adolph Wagner* (1879–1880), unfinished critique of a textbook on political economy.

II. Secondary Literature. The literature Marx and his work seems endless. Of interest are the biographies, with differing viewpoints, by Franz Mehring (long a standard), Isaiah Berlin, Otto Rühle, Werner Blumenberg, H. Gemkow, David Riazanov, David McLellan, M. Rubel, and the exhaustive joint biographical studies of Marx and Engels by Auguste Cornu (treating only 1818–1846 in 3 vols. thus far). A detailed chronicle of Marx's life, keyed to the current *Werke* ed., is M. Rubel, *Marx-Chronik: Daten zu Leben und Werk* (Munich, 1968), rev. trans. of the French original in *Karl Marx, Oeuvres, Économie*, I (Paris, 1965). A useful detailed chronological study of Marx's life, with full précis of all his works, is given by M. Rubel in *Marx Without Myth* (Oxford–New York, 1975), written with M. Manale.

The topics that might be listed under "Marx and science" range throughout the entire Marx literature, for "science" in his case must include the various social sciences (not excluding historical studies) and their methodologies, along with the natural sciences, mathematics, logic, engineering, and the relevant portions of philosophy (including philosophy of science, metaphysics, and epistemology) as well as their histories. Thus, Marx's methodology in *Capital* has been examined and interpreted; his relationship to Kant, to Spinoza, and to J. S. Mill; aspects of his critique and development of Hegel's thought; his response to Darwin; his sociological and historical understanding of religions; and so on. The following list (see also "Engels" in the *DSB*) includes some works that bear upon Marx's own understanding of nature, natural science, technology, methodology, and epistemology:

L. Althusser, *For Marx* (London–New York, 1969), translated from the French ed. (Paris, 1965); L. Althusser and E. Balibar, *Reading Capital* (London, 1970), translated from the French ed. (Paris, 1968); J. D. Bernal, *Science in History* (London, 1954; Cambridge, Mass., 1971); and *The Freedom of Necessity* (London, 1955); T. Carver, ed. and trans., *Texts on Method of Karl Marx* (Oxford, 1975), annotated texts of the introduction to the *Grundrisse* and the *Notes on Adolph Wagner*; J. Fallot, *Marx et le machinisme* (Paris, 1966); E. V. Ilyenkov, *The Dialectic of Abstract and Concrete in Marx's 'Capital'* (Moscow, 1960), in Russian—the third and central chapter is availabie in German in *Beiträge zur marxistischen Erkenntnistheorie*, A. Schmidt, ed. (Frankfurt, 1969), 87–127; in French in *Recherches internationales* (1968), 98–158; and in a complete Italian ed.; G. Lukacs, *Zur Ontologie des gesellschaftlichen Seins: Die ontologischen Grundprinzipien von Marx* (Frankfurt, 1972), a methodological study; H. Marcuse, *Reason and Revolution* (New York, 1941); S. Moscovici, *Essai sur l'histoire humaine de la nature* (Paris, 1968); B. Ollman, *Alienation* (Cambridge–New York, 1971); and M. Raphael, *The-*

orie des geistigen Schaffens auf marxistischer Grundlage (Frankfurt, 1974), rev. ed. of *Erkenntnistheorie der konkreten Dialektik* (Paris, 1934)—available in English as vol. XLI of *Boston Studies in the Philosophy of Science* (Boston–Dordrecht, 1978).

Also see R. Rosdolsky, *Zur Entstehungsgeschichte des Marxschen 'Kapital'*, 2 vols. (Frankfurt–Vienna, 1968), also in English (London, forthcoming 1978); N. Rosenberg, "Karl Marx on the Economic Role of Science," in *Journal of Political Economy*, **84** (1974), 713–728; "Science, Invention and Economic Growth," *ibid.*, 90–108, both in Rosenberg, *Perspectives on Technology* (Cambridge–New York, 1976); and "Marx as a Student of Technology," in *Monthly Review*, **28** (1976), 56–77; A. Schmidt, *Der Begriff der Natur in der Lehre von Marx* (Vienna, 1962; rev. ed., Frankfurt, 1971); also in English (London, 1971); A. Schmidt, ed., *Beiträge zur marxistischen Erkenntnistheorie* (Frankfurt, 1969), esp. G. Markus, "Über die erkenntnistheoretischen Ansichten des jungen Marx"; J. Zeleny, "Zum Wissenschaftsbegriff des dialektischen Materialismus"; and E. V. Ilyenkov (cited above); P. Thomas, "Marx and Science," in *Political Studies*, **24** (1976), 1–23; R. C. Tucker, *Philosophy and Myth in Karl Marx* (Cambridge, 1961); and J. Zeleny, *Die Wissenschaftslogik bei Marx und 'Das Kapital'* (Berlin, 1968), trans. and rev. from the Czech ed. (Prague, 1962).

See the coordinate article on Friedrich Engels in this volume of the *Dictionary*.

ROBERT S. COHEN

MELLANBY, EDWARD (*b.* West Hartlepool, Durham, England, 8 April 1884; *d.* Mill Hill, near London, England, 30 January 1955), *biochemistry, physiology, biomedical research and administration.*

After pioneering work on dietary deficiency diseases, particularly rickets, Mellanby served as secretary (chief executive) of the Medical Research Council from 1933 to 1949. He was the fourth son and sixth and last child of John Mellanby, an amateur boxing champion from Yorkshire who managed the shipyard of the Furness-Withy Company in West Hartlepool, and his wife, Mary Isabella Lawson of Edinburgh. Reared in religious nonconformity of an evangelical cast, all three sons who survived to maturity became professors in scientific or technical disciplines. The eldest, Alexander Lawson Mellanby (1871–1951), held the chair in civil and mechanical engineering at the Royal Technical College, Glasgow. The second son, John (1878–1939), was a distinguished physiologist who ultimately became professor of physiology at Oxford. Edward taught physiology at King's (later Queen Elizabeth) College for Women from 1913 to 1920 and held the chair in pharmacology at the University of Sheffield from 1920 to 1933.

Like his two older brothers, Mellanby attended Barnard Castle School with support from a bursary. He spent four years there, leaving in 1902 as head boy, having also won the upper school prize and a prize for theoretical and practical physics. He enjoyed perhaps even greater celebrity as an athlete, not only as captain of the cricket and football teams but also in track and field. In the Michaelmas term of 1902, he entered Emmanuel College, Cambridge, where he had won an open exhibition in natural science and where his expenses were further defrayed by a leaving exhibition from Barnard Castle School. At Cambridge, Mellanby took a second class in part I of the natural sciences tripos in 1904 and a first class in part II in 1905, his special subject being physiology.[1] After graduating B.A. in 1905, he remained at Cambridge with a research studentship from Emmanuel College and pursued biochemical research under Frederick Gowland Hopkins, who had been his college tutor.

In 1907, to complete his clinical training and become medically qualified, Mellanby went to St. Thomas's Hospital, London. While in London he continued to receive prizes and degrees from Cambridge—the Walsingham Medal (1907), the Gedge Prize (1908), and the Raymond Horton-Smith Prize (1915), as well as the M.A. (1910), M.B. (1910), and M.D. (1915) degrees. From 1909 to 1911 he was demonstrator in the department of physiology at St. Thomas's Hospital, where from 1910 to 1912 he also held a Beit memorial fellowship for medical research. In 1913 Mellanby published a study of metabolism in lactating women and joined the faculty of King's College for Women, a constituent institution of the University of London "founded to provide an education in household science with a genuinely scientific basis and a full academic status."[2] The following year he married May Tweedy, a fellow student at Cambridge who had taken a second class in both parts of the natural sciences tripos in 1905–1906, and who had since begun her own career in biomedical research at Bedford College, London. Their long and happy partnership, which had no issue in the usual sense, extended into the laboratory.[3] No proper study of Mellanby's career should overlook the important contributions of his wife, who survived him.

Even though Mellanby studied under Hopkins (who in 1929 won a Nobel Prize for his work on vitamins), and even though his own initial research concerned disordered metabolism, his work on rickets did not evolve directly from his earlier interests. His first publications (1907–1908) dealt with the metabolism of creatine and creatinine under embryonic, normal, and pathological conditions. He was particularly struck by the fact that creatine (ordinarily stored in voluntary muscle) appeared in the urine of patients with hepatic cancer, whereas creatinine alone was excreted in the urine of those with healthy livers or with such noncancerous hepatic diseases as cirrhosis. On this basis, Mellanby hoped that urinary creatine could serve as an aid in the diagnosis of cancer of the liver. During the next decade he pursued his interest in creatine metabolism and (partly in collaboration with Frederick Twort) investigated the possible role of intestinal bacteria in the destruction of creatine, in the production of histamine, and in the etiology of infantile diarrhea and cyclical vomiting. He turned to rickets only when invited to do so by the Medical Research Committee (forerunner of the Medical Research Council) in 1914. According to Henry Dale, then a member of the Committee, it was Hopkins who urged that rickets be made an object of special study and that Mellanby be asked to undertake it.[4]

Beginning with the vague mandate to study "experimental rickets and its relations to conditions of oxidation," Mellanby early concluded that nondietary factors played no more than a secondary, contributory role in the etiology of rickets. In his search for the specific dietary factor involved, he conducted hundreds of feeding experiments on puppies kept for the purpose in the Field Laboratories of Cambridge University. The South Kensington laboratories of King's College for Women served as the main site for the associated biochemical, histological, and X-ray work, through which the existence and severity of induced rickets could be determined quite precisely. Between 1918 and 1921, using the crude and time-consuming process of elimination then characteristic of much nutritional research, Mellanby established that while meat and certain vegetable juices had some inhibitory effect on rickets, such animal fats as cod-liver oil, butter, and suet had the most striking preventive action. Of the three "accessory food factors" or vitamins then known, only the "fat-soluble A" had a similar distribution in natural foods. Mellanby therefore suggested that the antirachitic factor was either fat-soluble A itself or a substance closely related to it. At the same time, however, he discussed several possible objections to their absolute identity;[5] and his work formed part of the background for the further separation of fat-soluble A into the carotene derivative now known as vitamin A and the specifically antirachitic vitamin D complex, normally produced through solar irradiation of certain sterols.

By most accounts, Mellanby's work had an immediate and dramatic impact on the prevention and treatment of rickets in England. Indeed, for Henry Dale, Mellanby's achievement represented a major milestone in the emergence of science-based therapy.[6] Nonetheless, many clinicians had already linked rickets empirically with a deficiency of fats; and cod-liver oil had long been prescribed for its antirachitic properties. While acknowledging the work of his clinical predecessors, Mellanby insisted that competing theories and therapies had allowed rickets to remain a prevalent disease and a probable contributor to high infant mortality among the urban populations of Great Britain and the United States. By the early 1930's, presumably as a result of the impact of Mellanby's work, no case of rickets could be found in the London clinics for trials of commercial vitamin D.[7]

Toward the end of World War I, while conducting his work on rickets, Mellanby undertook another "mission-oriented" research project—this time at the request of the Liquor Control Board (established out of governmental concern over the effects of drunkenness on the war effort). Once again with financial support from the Medical Research Committee, he studied the "comparative rates of absorption of alcohol into the blood from different kinds of drinks, and when taken in different relations to food of various kinds."[8] In 1920 Mellanby accepted the newly established chair in pharmacology at the University of Sheffield, attracted partly because of the clinical privileges made available to him through his simultaneous appointment as honorary physician at the Sheffield Royal Infirmary. In fact, his career often reflects his concern to reap practical benefits from basic research and to promote the "interaction of clinical and experimental work," a phrase that served as the subtitle for his book *Nutrition and Disease* (1934).

For some time after he became secretary of the Medical Research Council in 1933, Mellanby spent his weekends at Sheffield, engaged in productive research. Later, both while serving as secretary and after his retirement from the Council in 1949, he worked in its new nutrition building at Mill Hill. Especially during the 1930's Mellanby and his associates directed their efforts partly toward cancer, some of their results being recorded in the annual reports of the British Empire Cancer Campaign for 1934 through 1937. But most of Mellanby's later research had its roots in his work on rickets. During the early 1920's, for example, he reported that dogs on rachitic diets often developed thyroid growths histologically similar to those of patients with exophthalmic goiter. He therefore treated such patients with cod-liver oil, claiming that it had a beneficial effect, especially when combined with the iodine therapy recently recommended by others.

More generally, although he took no direct part in the chemical separation of "fat-soluble A" into vitamins A and D, Mellanby did much to clarify their action and mutual relations. Having early noticed the directly rachitogenic action of certain cereals, even in dogs that seemed to have sufficient dietary vitamin D, he argued that such cereals must contain a toxic substance that actively interfered with the calcifying role of vitamin D. In the late 1930's, after a long search for this "toxamin," Mellanby and his collaborators implicated phytic acid, which in certain cereals blocks calcium absorption by forming highly stable calcium phytate.

A decade later Mellanby excited popular interest by exposing the possible toxicity of wheat flours bleached with nitrogen trichloride as part of the Agene process. In particular, he showed that the ingestion of such flours was regularly associated with "canine hysteria." Although no decisive evidence established the toxicity of agenized flours in human diets, the process was eventually banned in Britain and the United States. Especially because of this work on agenized flours, the toxicity of which his associates traced more specifically to methionine-sulfoximine, Mellanby issued a prophetic warning against "the chemical manipulation of food."

Perhaps the most fundamental and interesting outgrowth of Mellanby's work on rickets was his study of the effects of vitamin A on infections, nerve degeneration, bony malformation, and embryogenesis. Although his search for specifically anti-infective properties in vitamin A was less than conclusive, it reinforced his conviction that improper nutrition plays a significant role in the etiology of infectious diseases. He achieved more decisive results in his investigation of the nerve degeneration and bony malformation associated with vitamin A deficiency. In an elegant series of papers, Mellanby and his collaborators linked the ataxia of vitamin A deficiency with nerve degeneration, especially in the cranial nerves and central tracts of the special senses. These degenerative changes were later traced to the destructive compression of the nerves and their ganglia by improper development of the skull bones and upper vertebrae—and ultimately to distorted events in the osteoblasts and osteoclasts.

Toward the end of his life Mellanby joined Honor Fell of the Strangeways Research Laboratory, Cambridge, in studying the effects of excess vitamin A on embryonic tissues cultivated *in vitro*. Among their exciting results, perhaps the most striking was the induced production of ciliated mucous-secreting epithelium from chicken ectoderm grown in a medium with high vitamin A content, followed by its reversion to normal skin when transferred to a normal medium. These and other metaplastic effects of vitamin A led Mellanby to suggest that it could be considered the "director" of basal cell development, comparable with the "organizers" in embryological growth. To the end of his life—even on the day of his death—Mellanby continued to pursue these studies of the role of vitamin A in embryogenesis and metabolism, having turned at the very end to its effects on sulfate metabolism.

As Henry Dale emphasized, Mellanby's distinguished career in research becomes all the more impressive when one realizes that he simultaneously held, for sixteen years, the most important post in the administration of medical research in Great Britain. To be sure, when he succeeded Sir Walter Morley Fletcher as secretary of the Medical Research Council in 1933, he inherited a stable, efficient, and highly regarded organization. But the immense expansion of funds for medical research, together with the heavy burdens placed upon the Council during World War II, made his task a demanding one. Some appreciation of its full dimensions can be gained from the annual reports of the Council and from A. L. Thomson's two-volume history of the Council.[9] At the time of his selection as secretary, Mellanby was already a member of the Council, with which he had been associated as a major recipient of research funds and as a leading contributor to its work since its origins as the Medical Research Committee. According to Thomson, his selection as secretary "was virtually settled by the scientific members in private meetings";[10] and he accepted with the understanding that every possible provision would be made to enable him to continue his own research with support from the Council. As secretary he was criticized for his brusque manner toward subordinates and outsiders, and for an alleged partiality toward the nutritional and applied fields in which he had made his own greatest contributions.

His leading asset was his ability to distinguish promising avenues of research from other, less fertile competitors for Council funds.

During World War II, Mellanby played a major role in the administration of military medicine and in the setting of dietary standards for British civilians and military personnel. He was a leading international force in efforts to improve nutrition and in the standardization of vitamins, and he undertook advisory missions to South Africa, India, Australia, and New Zealand. Elected fellow of the Royal Society in 1925 and fellow of the Royal College of Physicians in 1928, Mellanby was named K.C.B. in 1937 and G.B.E. in 1948. He received the full panoply of honors one might expect of a man whose career was perhaps unique for the extent to which it combined stature in biomedical administration with eminence in creative research.

NOTES

1. J. R. Tanner, ed., *Historical Register of the University of Cambridge* (Cambridge, 1917), 787, 790. H. H. Dale ("Mellanby," 196) incorrectly states that Mellanby took first-class honors in both parts of the natural sciences tripos.
2. H. H. Dale, "Mellanby," 200. According to Dale (*ibid.*), Mellanby went to King's College for Women as professor of physiology, but B. S. Platt (in *Dictionary of National Biography*) says he went there as lecturer and only later attained the chair.
3. For an early example of their cognate and complementary research, see May Mellanby, "An Experimental Study of the Influence of Diet on Teeth Formation," in *Lancet* (1918), **2**, 767–770, esp. 770: "This work, taken in conjunction with the experiments of E. Mellanby on rickets, puts the close relationship between hypoplastic teeth and rickets on to an experimental basis."
4. Dale, *op. cit.*, 201.
5. See, for instance, Mellanby's "An Experimental Investigation of Rickets," esp. 409–410.
6. Dale, *op. cit.*, 202–203.
7. See *ibid.*; and B. S. Platt, "Prefatory Chapter," 399.
8. Dale, *op. cit.*, 204.
9. A. L. Thomson, *Half A Century of Medical Research.*
10. *Ibid.*, I, 229.

BIBLIOGRAPHY

I. ORIGINAL WORKS. Mellanby published two books: *Nutrition and Disease: The Interaction of Clinical and Experimental Work* (Edinburgh, 1934); and *A Story of Nutritional Research: The Effect of Some Dietary Factors on Bones and the Nervous System* (Baltimore, 1950). In addition, he published almost 100 papers, including several originally delivered as invited lectures. Full bibliographies appear in the articles by Dale and Platt (see below), of which the latter is slightly more valuable. Mellanby's most significant papers on rickets are "The Part Played by an 'Accessory Factor' in the Production of Experimental Rickets," in *Journal of Physiology*, **52** (1918), xi–xii; "A Further Demonstration of the Part Played by Accessory Food Factors in the Aetiology of Rickets," *ibid.*, liii–liv; "An Experimental Investigation of Rickets," in *Lancet* (1919), **1**, 407–412; "Experimental Rickets," *Special Report Series. Medical Research Council*, no. 61 (1921); and "Experimental Rickets: The Effect of Cereals and Their Interaction With Other Factors of Diet and Environment in Producing Rickets," *ibid.*, no. 93 (1925).

II. SECONDARY LITERATURE. Of the available accounts of Mellanby's life and work, the most extensive and valuable is Henry H. Dale, "Edward Mellanby, 1884–1955," in *Biographical Memoirs of Fellows of the Royal Society*, **1** (1955), 193–222, with bibliography. Another fairly lengthy account, which includes autobiographical remarks by Mellanby, is B. S. Platt, "Prefatory Chapter. Sir Edward Mellanby, G.B.E., K.C.B., M.D., F.R.C.P., F.R.S. (1884–1955): The Man, Research Worker and Statesman," in *Annual Review of Biochemistry*, **25** (1956), 1–28, with bibliography, repr. in *The Excitement and Fascination of Science: A Collection of Autobiographical and Philosophical Essays* (Palo Alto, Calif., 1965), 381–405. See also *British Medical Journal* (1955), **1**, 355–358; *Lancet* (1955), **1**, 309–310; and B. S. Platt, "Sir Edward Mellanby, G.B.E., K.C.B., F.R.S.," in *Nature*, **179** (1955), 530–532; and "Sir Edward Mellanby," in *Dictionary of National Biography, 1951–1960*, 731–732. On the work of the Medical Research Council under Mellanby's direction, see Arthur Landsborough Thomson, *Half A Century of Medical Research*, I (London, 1973), *passim*, esp. 229–230.

GERALD L. GEISON

MENELAUS OF ALEXANDRIA. The following additions and corrections are offered to the article in volume IX, pages 296–302.

There is no reason to discount (note 7) the statement in the *Fihrist* that Menelaus composed his work on specific gravities at the commission of Domitian. It would be compatible with the fact that he made observations in the first year of Trajan; and the name of Domitian is associated with Menelaus in several Arabic sources besides the *Fihrist*, notably al-Bīrūnī[1] and al-Khāzinī. Although N. Khanikoff did not translate the whole of the passage by al-Khāzinī relating to Menelaus (see note 28), he excerpted a good deal of the introduction.[2]

It is not made clear in the main article that the work of Menelaus on specific gravities actually exists in Arabic translation; it is extant in the Escorial Arabic MS 960, formerly numbered 955. The Latin title given in note 27 is that of Michael Casiri's eighteenth-century catalog, *Bibliothecae arabico-hispana Escurialensis*, I (Madrid, 1760), 386; but G. J.

Toomer states that a more literal translation would be "The book of Menelaus to the Emperor Domitian [name unrecognizable in the MS] on the mechanism by means of which one can know the amount of each of a number of bodies [or 'substances'] which are mixed together."[3]

It may be taken as certain that Menelaus did not, in fact, draw up a catalog of the fixed stars (page 297 and note 8). A. A. Björnbo's conjecture that he did so was based on his misreading of al-Battānī's misunderstanding of Ptolemy's account in the *Almagest*, as was shown by C. A. Nallino and, after him, by J. L. E. Dreyer and H. Vogt.[4]

NOTES

1. In his work "Maqāla fī 'l-nisabi 'llatī bayna 'l-filizzāt wa 'l-jawāhir" (unpublished), extant only in MS Beirut, Université Saint Joseph, 223(6), which has now disintegrated, but of which photocopies exist, for instance, at the American University of Beirut.
2. *Journal of the American Oriental Society*, **6** (1859), 12–13.
3. The text has never been edited, but there is a German trans. by J. Würschmidt, in *Philologus*, **80** (1925), 377–409. There is an extensive discussion of the content in the dissertation by T. Ibel, *Die Wage im Altertum und Mittelalter* (Erlangen, 1908), 182–187, with a trans. of the intro. to the Escorial MS.
4. C. A. Nallino, *Al-Battānī sive Albatenii Opus astronomicum*, I (Milan, 1903), 124–126, and II (Milan, 1907), 269–270; J. L. E. Dreyer, in *Monthly Notices of the Royal Astronomical Society*, **77** (1917), 534–535; H. Vogt, "Versuch einer Wiederherstellung von Hipparchs Fixsternverzeichnis," in *Astronomische Nachrichten*, **224** (1925), cols. 37–38.

IVOR BULMER-THOMAS

MERRILL, ELMER DREW (*b*. East Auburn, Maine, 15 October 1876; *d*. Forest Hills, Massachusetts, 25 February 1956), *botany*.

For the breadth of his knowledge and his influential international status, Merrill was acknowledged the "American Linnaeus." He and his twin brother Dana were the youngest of five children of Daniel C. Merrill, a sailor, factory worker, and farmer, and Mary Adelaide Noyes Merrill. At Maine State College of Agriculture and the Mechanic Arts (now the University of Maine) he pursued a general science curriculum that included only two botany courses, but his interest in the latter science was stimulated by Francis LeRoy Harvey. He graduated first in his class with the B.S. (1898) and remained as an assistant in the natural sciences.

After three years as an assistant to agrostologist Frank Lamson-Scribner at the United States Department of Agriculture, Merrill went to Manila in 1902 as botanist to the Philippine Bureau of Agriculture. His office consisted of a bare room with only a table and chair: the earlier plant collections and reference books had been either burned or stolen. Characteristically he set about collecting local weeds. When he left Manila in 1923, the Bureau of Agriculture herbarium contained 275,000 Philippine and Malayan specimens. From 1912 to 1918 Merrill devoted eighteen to thirty-six hours a week as "half-time" professor of botany at the University of the Philippines. In 1919 he became director of the Bureau of Science, an appointment that forced him to abandon his hope of writing a detailed flora of the Philippines and led him instead to undertake a catalog that ultimately appeared in four volumes.

Merrill married Mary Augusta Sperry in Manila in 1907; they had four children. Mrs. Merrill and the children remained after his short leave in Washington, D.C., and his return to Manila in 1915, and, except for brief leave in 1921, the family was not reunited until 1924, when he became dean of the College of Agriculture at the University of California, Berkeley. In 1930 Merrill became professor of botany at Columbia University and director of the New York Botanical Garden, remaining there until 1935, when he went to Harvard as Arnold professor, director of the Arnold Arboretum, and administrator of botanical collections. He held those posts in varying combinations until 1948. Merrill was a consultant on tropical botany to the secretary of war; and his survival guide for the armed forces, *Plant Life of the Pacific World* (New York, 1945), a distillation of his years in the tropics, is still widely useful. Merrill's *Botany of Cook's Voyages* (1954) reviewed facts and what he believed to be unfounded theories on the origin and pre-Columbian dissemination of crops and weeds in the Pacific.

Merrill can be considered a builder. Toward a documentary record of tropical floras he amassed a million plant specimens for six institutions; for the ethnobotanist he designed a comprehensive field label for plant vouchers; for editors he advocated one-name journal titles (such as *Hilgardia, Brittonia, Arnoldia*); for botanical nomenclaturists he created a loose-leaf ledger of *Index Kewensis* by interfiling the names from all installments into a master sequence; for the low-budget institution he designed a cardboard herbarium box ("Merrill case") for housing specimens; and he launched the reprinting of hard-to-find botanical classics by photo-offset. Merrill was, in short, an organization man. His friendly support of promising scholars, often behind the scenes, his skill in attracting donors, and his fondness for books, music, and, on occasion, tennis were blended with the frank, critical manner of a man in a hurry.

BIBLIOGRAPHY

I. ORIGINAL WORKS. Merrill published more than 500 books and papers. A chronological list prepared by Lazella Schwarten is appended to W. J. Robbins' memoir (see below); an alphabetical list by title appears in Torrey Botanical Club, *Index to American Botanical Literature, 1886–1966*, III (Boston, 1969), 264–273. Chief among his writings are pioneer accounts of the Philippine, Bornean, and Micronesian floras: *A Flora of Manila* (Manila, 1912); and *An Enumeration of Philippine Flowering Plants*, 4 vols. (Manila, 1922 [1923]–1926); "A Bibliographic Enumeration of Bornean Plants," spec. no. of *Journal of the Straits Branch of the Royal Asiatic Society* (1921); and "An Enumeration of the Plants of Guam," in *Philippine Journal of Science*, sec. C, Botany, **9** (1914), 17–155. Supplements to all of these accounts were issued.

Merrill's *Bibliography of Eastern Asiatic Botany* (Jamaica Plain, Mass., 1938), written with E. H. Walker; "A Botanical Bibliography of the Islands of the Pacific," in *Contributions from the United States National Herbarium*, **30** (1947), 1–322; and *Index Rafinesquianus* (Jamaica Plain, Mass., 1949) are key bibliographies. F. Verdoorn, ed., "Merrilleana," in *Chronica botanica*, **10** (1946), 127–394, is a selection of 23 of his general writings. The first installment—no more published—of an illustrated autobiography appeared in *Asa Gray Bulletin*, n.s. **2** (1953), 335–370. Typical of Merrill's spirited narrative is his account of the first ascent of Mount Halcon, ele. 8,900 ft., in the Philippines, "A Mindoro Fern Adventure," in *American Fern Journal*, **36** (1946), 33–47.

Merrill's library of 2,600 titles was given to the New York Botanical Garden, as was his correspondence from 1902 to 1935, amounting to 2,378 items; 83 letters related to his introduction of Metasequoia alone. His field books have not been located.

II. SECONDARY LITERATURE. There are two complementary accounts of Merrill and his work: R. E. Schultes, in *Taxon*, **6** (1957), 89–101; and W. J. Robbins, in *Biographical Memoirs. National Academy of Sciences*, **32** (1958), 273–333. Shorter sketches, stressing various facets of Merrill's life, include I. H. Burkill, in *Nature*, **177** (1956), 687–688; J. Ewan, in *Journal of the Washington Academy of Sciences*, **46** (1956), 267–268; R. A. Howard, in *Journal of the Arnold Arboretum, Harvard University*, **37** (1956), 197–216; and F. Verdoorn, "Merrilleana," in *Chronica botanica*, **10** (1946), 127–157. Ivan M. Johnston, *A Correspondence Between a Professor at Harvard and the University* (Jamaica Plain, Mass., 1957), which concerns Merrill's final years, was privately printed.

JOSEPH EWAN

MEYERSON, ÉMILE (*b.* Lyublin, Russia [now Lublin, Poland], 12 February 1859; *d.* Paris, France, 4 December 1933), *history and philosophy of science.*

Meyerson was educated in Germany, where he studied from the ages of twelve to twenty-three and passed his *Abitur*. Interested in chemistry, he followed the usual practice of spending time at several universities distinguished for research laboratories: Göttingen, Heidelberg, and Berlin. He also worked in Paul Schützenberger's laboratory at the Collège de France after his arrival at Paris in 1882. His short career as an industrial chemist was blighted by his failure to develop a process for the synthetic manufacture of indigo based on a wrong reaction obtained by Baeyer. Meyerson's excellent command of several languages then led him to become foreign news editor at the Havas News Agency. He joined the M Group—Jean Moréas, Charles Maurras, and Maurice Maindron—which met at the Café Vachette.

In 1898 Meyerson left Havas to work for Edmond de Rothschild's philanthropic organization that sought to settle Jews in Palestine, and became the head of the Jewish Colonization Association for Europe and Asia Minor. He also collaborated on the famous report on the economic situation of Jews in the Russian Empire. Although not a practicing Jew, Meyerson retained an attachment to Zionism. Through the support of Harald Høffding, a longtime friend and correspondent, he was elected to the Royal Danish Academy of Sciences and Letters in 1926; that year he also became a *correspondant étranger* of the Académie des Sciences Morales et Politiques. Although Meyerson was not formally a member of the French academic community, he enjoyed the friendship of the philosophers Dominique Parodi, Léon Brunschvicg, and Lucien Lévy-Bruhl, as well as the scientists Paul Langevin and Louis de Broglie. His intellectual salon, which met weekly to discuss scientific and philosophical topics, included Langevin, de Broglie, Hélène Metzger, Alexandre Koyré, General André Metz, André George, Salomon Reinach, Lévy-Bruhl, Henri Gouhier, and Vladimir Jankélévitch. Meyerson's influence was assured through this informal institution as well as through his articles and books.

Meyerson was greatly influenced by Hermann Kopp's work in the history of chemistry and by Kristian Kroman's *Naturerkenntnis* (German translation, 1883), which argued that the principles of identity and causality are the basic premises of science. An autodidact in philosophy, Meyerson first learned philosophy from the works of Charles Renouvier and was influenced by the neo-Kantian movement. He also was affected by the works and correspondence of Høffding, a close friend of Niels Bohr. The genesis of Meyerson's first and most

MEYERSON

famous work, *Identité et réalité* (1908), was in his Kopp-inspired studies of chemistry before Lavoisier, which he found to be linked to modern chemistry by an ontological common denominator. An extension of his inquiry beyond the development of chemistry led him to master much of the history of the study of the natural sciences from antiquity to his own day. His program was the unfulfilled one of Comte: to discover a posteriori the a priori principles guiding thought in its search for the nature of reality. As George Boas put it: " . . . to discover inductively the *a priori* principles of human thinking. By '*a priori*' M. Meyerson . . . means . . . those principles without which the human mind has not operated to date and which are not discovered by it in experience itself." Although Meyerson's epistemological quest was strongly influenced by German thought, it was an integral part of French philosophy of science in the late nineteenth and early twentieth centuries. Following Descartes, d'Alembert, Ampère, Comte, Cournot, Renouvier, Jules Lachelier, and Émile Boutroux, Meyerson assumed a relationship between philosophy and natural science. He clearly was part of the great French movement in the philosophy of science during the period dominated by Pierre Duhem, Henri Poincaré, Henri Bergson, and Édouard Le Roy.

Meyerson's minute examination of the technical works and stated aims of scientists led him to reject Ernst Mach's phenomenalism and Comte's positivist thesis, which limited science to the functions of action and prediction. As William A. Wallace put it, "Scientists have never been content simply with registering phenomena and summarizing them in laws that make possible the prediction and control of further phenomena." The leitmotiv of Meyerson's epistemology is his distinction between *légalité* and *identité*. The dynamism of science comes from the desire of the scientist to understand the external world. Meyerson, according to Owen Hillman, held that "scientific explanation consists in transforming empirically discovered natural laws into statements of identity in time." Since for Meyerson the principle of causality was only the principle of identity applied to time, *Identity and Reality* is basically the demonstration of the key role of causality in the physical sciences (Abel Rey and Erich Becher arrived at a similar conclusion). Yet the concept of identity is limited to the role of guiding principle in Meyerson's epistemology. And although he held scientific thought to be a continuation of commonsense views—both of them are founded on the concept of a *Ding an sich*—Meyerson did not adopt a realist philosophy based on a fixed philosophical system. (Meyerson valued the chapter on common sense in *Identity and Reality* above the others.)

Few would quarrel with Meyerson's case for science as a progressive rationalization of reality. The difficulty comes in explaining why science never finishes its explanation or, worse still, changes it. Meyerson accepted the Bergsonian idea of a residual irrationality of things expressed by the irreducible specificity of time as compared with space. Not susceptible of reduction to spatial representation, life and conscience escape the mechanistic clutches of science in Bergson's work. Matter itself was given a little irreducibility by Meyerson. The presence of the residual element of irrationality in nature means that there is an element of irrationality in science itself that cannot be totally eliminated by scientific explanation. Reality resists the effort of reason to annihilate the external world; the resistance finds its most general expression in Sadi Carnot's principle, the dissymmetry of which cannot be deduced from mechanism. "In opposition to the illusions of identity to which mechanical theories . . . give rise, Carnot's principle stipulates that the whole universe is modifying itself in time in a constant direction" (Meyerson, *Identity and Reality* [1930], p. 265). (Abel Rey rejected the idea that Carnot's principle shows the resistance of reality to attempts at rationalization; he argued that the principle could conceivably be given a mechanistic interpretation—that is, future kinetics might be so constructed as to integrate it successfully.)

Thus identity can attain only the status of "the eternal framework of our mind"—never a Kantian category—which penetrates but does not make up the totality of science. Since "Carnot's principle is an integral part of science," science has given reality its rightful place, proving that "contrary to what causality postulated, it is not possible to eliminate time." Carnot's principle saves science from the "progressive elimination of reality, which is the consequence of successive identifications." Small wonder that Brunschvicg declared Sadi Carnot to be the real hero of *Identity and Reality*. Abel Rey noted Meyerson's failure to make much of a case for the interpretation of Carnot's principle as an expression of the idea of probability in physics. The reason for this, as George Boas points out, is that if one "remains content with equations of probability, one has abandoned any hope of Meyersonian explanation." The success of this mechanical theory founded on statistical probability partly explains the ultimate failure of Meyerson's philosophy of science to achieve paradigm status among both scientists and philosophers of science.

Variations on Meyerson's thesis can be found in his other works: *De l'explication dans les sciences* (1921), a work that also gives considerable attention to the "global explanation" attempted by such philosophers as Hegel; *La déduction relativiste* (1925), which incorporated relativity into the general evolution of physics, thus showing that Meyersonian explanation applied not only to the relics of past science but also to contemporary scientific thought; and *Du cheminement de la pensée* (1931), which moved beyond scientific explanation to general logic, in order to develop a theory of general knowledge based on the principle of identity. Meyerson thus took his position in a series of European thinkers from Descartes to Claude Lévi-Strauss who have studied the functioning of the mind; his friendship with Lévy-Bruhl reinforced this epistemological interest. The *Cheminement* is also important because it developed Meyerson's philosophy of mathematics. According to Kenneth Bryson, "Meyerson finds in mathematics the ideal instrument for the conversion of reason and reality in plausible propositions. . . . It enables science to retain identities while accounting for differences." Meyerson's philosophy of mathematics is oriented toward representation, the element coming from perception, and thus makes formalism secondary. Meyerson noted that although relativity is a theory of reality, it is also the triumph of the mathematicization of physics over mechanism as the form of modern science. In *Réel et déterminisme dans la physique quantique* (1933), Meyerson admitted that the principle of legality plays a much larger role in science than he had thought and is really derived from the principle of causality. The great revolution—although Meyerson called it an evolution—in scientific understanding was that the precision of the physicist's formulas met an unpassable barrier in the existence of the quantum of action. Yet the physicist of the future would be driven by his desire for a *Weltbild* to look for the physical significance of the concepts born of mathematical reasoning.

Meyerson's philosophy had little influence after the 1930's; its eclipse corresponded to "the decline and fall of causality" in contemporary science, the strength of the Vienna Circle in the philosophy of science, and the shift in interest among philosophers from problems of knowledge to problems of existence. In France there was also the attack by Gaston Bachelard, who exaggerated the continuity and realism in Meyerson. Nevertheless, the continuing literature on Meyerson indicates a fairly strong interest in his work, and his ideas show little sign of becoming extinct. Jacques Maritain's *Degrees of Knowledge* reveals the interest of Thomists in his ideas.

Some of Meyerson's *Essais* (1936) contained early pieces in the history of science, but he did not wish to be known as a historian of science. If we accept the judgment of P. M. Rattansi that "Some of the most exciting work in the history of modern science has come for a generation from intellectual historians like E. A. Burtt, Ernst Cassirer, and Alexander Koyré, who showed how the study of nature is related to larger metaphysical assumptions and is involved in complex ways with other areas of intellectual culture," then we can justly include Meyerson in his pantheon. When Meyerson wrote to Høffding that Koyré would translate Høffding's *Erkenntnistheorie und Lebensauffassung*, he described Koyré as a "young and learned philosopher of true talent." The relation between these two East Europeans who came to France via Germany is not to be ignored, for through the critical mediation and example of Koyré, it is likely that Meyerson has exerted more influence than is generally recognized. As Koyré admitted, this influence is not to be found in fidelity to the subtle dogma of the basic identity of human thought, for most of us follow Koyré in recognizing its different structures in different historical periods. Meyerson's great precept was that we should respect our predecessors who made errors and should seek reasonable explanations of their mistakes as carefully as the explanations of their successes. Like Duhem, Meyerson saw in the history of scientific thought an essential instrument for the study of scientific thought itself: history gives a dimension unattainable through introspection or direct analysis of the processes of science and their development.

BIBLIOGRAPHY

I. Original Works. Meyerson's most famous work is *Identité et réalité* (Paris, 1908); the English trans. (London–New York, 1930) is of the 3rd ed. (Paris, 1926), which is identical with the rev. and enl. ed. (Paris, 1912). Meyerson considered the German trans. (1930), which has a long introduction by the mathematician Léon Lichtenstein, who spread Meyerson's ideas in Germany, better than the English. Meyerson's other works are *De l'explication dans les sciences*, 2 vols. (Paris, 1921; 2nd ed., 1927); *La déduction relativiste* (Paris, 1925); *Du cheminement de la pensée*, 3 vols. (Paris, 1931); *Réel et déterminisme dans la physique quantique* (Paris, 1933); and *Essais* (Paris, 1936), Lucien Lévy-Bruhl, ed. A key work for the study of

Meyerson's ideas is *Correspondance entre Harald Høffding et Émile Meyerson* (Copenhagen, 1939).

II. SECONDARY LITERATURE. George Boas, *A Critical Analysis of the Philosophy of Émile Meyerson* (Baltimore, 1930), has no bibliography; but the critical study by Thomas R. Kelly, *Explanation and Reality in the Philosophy of Émile Meyerson* (Princeton, N.J., 1937), does, The Italian interest in Meyerson is evident in several studies, including Silvestro Marcucci, *Émile Meyerson. Epistemologia e filosofia* (Turin, 1962), which defends Meyerson against Bachelard's criticisms and is critical of Georges Mourélos, *L'épistémologie positive et la critique meyersonienne* (Paris, 1963)—see *Physis*, **5** (1963), 199–205. An early critique is Owen N. Hillman, "Émile Meyerson on Scientific Explanation," in *Philosophy of Science* (1938), 73–80. On general context of the Meyersonian critique, see William A. Wallace, *Causality and Scientific Explanation*, II (Ann Arbor, Mich., 1974). On his metaphysics, see Kenneth A. Bryson, "The Metaphysics of Émile Meyerson: A Key to the Epistemological Paradox," *Thomist*, **37**, no. 1 (1973), 119–132. The 26 Nov. 1960 session of the Société Française de Philosophie was devoted to "Commemoration du centenaire de la naissance de deux épistémologues français. Émile Meyerson et Gaston Milhaud." See *Bulletin de la Société française de philosophie*, **55**, no. 1 (1961), 51–116. P. M. Rattansi's remark is in "Some Evaluations of Reason in Sixteenth and Seventeenth-century Natural Philosophy," in Mibuláš Teich and Robert Young, eds., *Changing Perspectives in the History of Science. Essays in Honour of Joseph Needham* (Boston, 1973).

H. W. PAUL

MICHURIN, IVAN VLADIMIROVICH (*b.* Dolgoye, Ryazan gubernia, Russia, 28 October 1855; *d.* Kozlov [now Michurinsk], U.S.S.R., 7 June 1935), *plant breeding, genetics.*

Born into a family of impoverished gentry who lost their hereditary orchard, Michurin was obliged to leave school during his first year of secondary education. He worked as a railway clerk, signal repairman, and watchmaker until he saved enough money to buy some land and concentrate on breeding varieties of fruit for the harsh climate of north-central Russia. When his nursery proved commercially unsuccessful, he tried, between 1905 and 1908, to have it transformed into a government experiment station. Specialists in the Ministry of Agriculture rejected Michurin's proposal, offering him a couple of medals as consolation prizes, which he accepted, and suggesting grants to aid contract research, which he refused. Somehow he kept his nursery going, although it was in poor condition; published occasional articles in horticultural journals; and developed a deep resentment of the official world of science.

During the land revolution of 1917–1918 Michurin almost lost his thirty acres to the neighboring peasants. Two local agronomists, however, persuaded the new central government to transform the modest nursery into a state institution. The story of Lenin's personal support is unsubstantiated, but Michurin's plant breeding station did get successive increases in government support during the 1920's. Through skillful appeals to journalists and politicians Michurin won a reputation as "the Russian Burbank," glorified for performing miracles without benefit of diplomas and in spite of the disdainful opposition of academic scientists.

Great fame and power came to Michurin in the early 1930's, during Stalin's "revolution from above," which drove peasants into collective farms and pushed agricultural scientists into a centralized system of research and extension work, subject to frantic demands for quick, cheap solutions to overwhelming problems. Michurin was put in charge of an Institute directing fruit breeding for almost the entire country. Criticism of him was sharply rebuked. Indeed, V. L. Simirenko, a leading specialist in horticulture, was denounced as a "wrecker" and was arrested after he had presided over a 1931 conference that refused to endorse Michurin's idiosyncratic methods of breeding.

Michurin died in 1935, when Lysenko was beginning his campaign against genetics. Very quickly Michurin was transformed into the patron saint of that campaign, and "Michurinism" became the official name of Lysenko's doctrine. Some geneticists tried to prove that Michurin's methods and beliefs could be squared with their science; but such efforts were unavailing during the long period of Lysenko's dominance, from the mid-1930's to the mid-1960's. After 1965, when Lysenko lost all political support, official sanction was bestowed on the view that Michurin was a breeder of genius whose unusual methods can be explained by genetics.

Michurin's genuine contributions to horticulture are difficult to determine, in view of the great diversity of claims and the paucity of hard evidence. In 1931, for example, when he was claiming over 300 commercially useful varieties, only one of his creations was officially certified for use in Soviet orchards. Even after political pressure increased that number, few of his hybrids seem to have achieved significant commercial success; but a considerable number may have provided useful breeding stock for further experimentation.

Michurin's methods of breeding were partly commonplace and partly controversial. He insisted on hybridization as the only way to combine such qualities as the hardiness of a northern variety and the lusciousness of a southern one. He favored wide crosses, "to shake up the heredity," followed by intuitive selection of promising seedlings from the resulting mass of wildings. To facilitate wide crosses he urged "vegetative blending" (*sblizhenie*) —that is, grafting different species or genera onto one another to predispose them for cross-pollination. To train hybrid seedlings in desired directions, he recommended the use of "mentors," grafting a seedling onto the variety he wished it to resemble.

In Michurin's writings, justification of these methods is hard to distinguish from the ancient belief in vegetative hybrids—that is, hybridization by grafting, without the mixture of germ plasm. That was how the Lysenkoites read Michurin, and one of the main reasons they named their anti-Mendelian doctrine "Michurinism." (The other main reason was Michurin's reputation as an untutored genius who achieved great things against the opposition of academic scientists.) Soviet geneticists who have tried to separate Michurin from "Michurinism" have argued that physiological changes resulting from grafting may indeed predispose widely separated plants for cross-pollination and that the use of "mentors" may in fact serve the breeder's purpose by affecting the penetrance or dominance of genes in complex situations.

Michurin himself resented efforts to subject his methods and beliefs to rigorous tests in accordance with the basic assumptions of genetics. He was annoyed by the combination of learned criticism and condescending praise that characterized scientific comments on his work down to the 1930's, before the Stalinist regime raised him above criticism. Michurin was separated from modern biological science not only by his confessed ignorance of it but also by his objection, on principle, to some of its most basic aspirations. He was convinced, for example, that heredity "does not yield and in essence cannot conform to any patterns worked out by theoretical science and determined in advance." In plant breeding, he declared, "Not only is it impossible to apply any calculation in accordance with Mendel's law, but it is quite impossible to do any strictly precise work in accordance with a plan worked out in advance." Practical intuition must be the principal guide of the breeder, according to Michurin. At times he revealed the primitive vitalism of many gardeners, ascribing to "every living organism a reasoning [*razumnuyu*] power of adaptability in the struggle for existence." In the 1970's Michurin is largely ignored by Soviet biologists, although he continues to be admired by the ideological establishment.

BIBLIOGRAPHY

I. Original Works. The most revealing collection of Michurin's writings is the first, *Itogi ego deyatelnosti v oblasti gibridizatsii po plodovodstvu* ("The Results of His Activity in the Field of Hybridization of Fruit"; Moscow, 1924), with a preface by N. I. Vavilov and a friendly though critical evaluation by V. V. Pashkevich. Subsequent collections of Michurin's writings dispensed with such evaluations, ultimately substituting Lysenkoite adulation and tendentious editing. For the largest collection, see Michurin, *Sochinenia* ("Works"), 4 vols. (Moscow, 1939–1941; 2nd ed., 4 vols., 1949). The 2nd ed. is the more egregiously tendentious in its editing. The largest selection of Michurin's writings in English translation is his *Selected Works* (Moscow, 1949).

II. Secondary Literature. There is a very large Russian literature on Michurin. For a guide, see E. V. Parkhomenko and F. S. Ginzburg, *Bibliografia trudov I. V. Michurina i literatury o nem* ("Bibliography of I. V. Michurin's Works and of the Literature Concerning Him"; Moscow, 1958). Some of the best scientific evaluations are omitted from this bibliography. See especially N. I. Vavilov, "O mezhdurodovykh gibridakh dyn, arbuzov i tykv" ("Concerning Intergeneric Hybrids of Melons, Watermelons, and Squashes"), in *Trudy po prikladnoi botanike, genetike, i selektsii*, **14**, no. 2 (1924), 3–35; V. V. Pashkevich, "Russky originator-plodovod I. V. Michurin" ("The Russian Fruit Breeder I. V. Michurin"), in Michurin, *Itogi . . . plodovodstvu*, 16–61; P. N. Shteinberg, "O 'chudesakh' i 'charodeiakh' v selskom khoziaistve" ("Concerning 'Miracles' and 'Miracle Makers' in Agriculture"), in *Vestnik znaniya*, no. 117 (1926), 1129–1138; E. A. Aleshin, "I. V. Michurin i nauka" ("I. V. Michurin and Science"), in *Puti selskogo khozyaistva* (1927), no. 6–7, 118–125; G. D. Karpechenko, "Teoria otdalennoy gibridizatsii" ("The Theory of Distant Hybridization"), in N. I. Vavilov, ed., *Teoreticheskie osnovy selektsii rastenii* ("The Theoretical Foundations of Plant Selection"), I (Moscow, 1935), 293–354; and D. D. Romashov, "O metodakh raboty I. V. Michurina" ("Concerning the Methods of Work of I. V. Michurin"), in *Zhurnal obshchei biologii* (1940), no. 2, 177–204.

The most detailed biography is a "Michurinist" work by I. T. Vasilchenko, *I. V. Michurin* (Moscow, 1950; 2nd ed., 1963). The officially approved interpretation since Lysenko's fall is embodied in S. I. Alikhanian, *Teoreticheskie osnovy uchenia Michurina o peredelke rastenii* ("The Theoretical Foundations of Michurin's Teaching on the Transformation of Plants"; Moscow,

MIKLUKHO-MAKLAY

1966); and in N. P. Dubinin, *Teoreticheskie osnovy i metody rabot I. V. Michurina* ("The Theoretical Foundations and the Methods of Work of I. V. Michurin"; Moscow, 1966). See also S. Y. Kraevoi, *Vozmozhna li vegetativnaya gibridizatsia rastenii posredstvom privivok?* ("Is Vegetative Hybridization of Plants Possible by Means of Grafting?"; Moscow, 1967). For an appraisal of Michurin in English, see D. Joravsky, *The Lysenko Affair* (Cambridge, 1970), 40–54, 69–76.

DAVID JORAVSKY

MIKLUKHO-MAKLAY, MIKHAIL NIKOLAEVICH (*b.* Rozhdestvensky, Novgorod gubernia, Russia, 1857; *d.* Leningrad, U.S.S.R., 12 April 1927), *geology*.

Miklukho's father, Nikolay Ivanovich Miklukho, a hereditary nobleman, was an engineer who helped to construct the railroad between St. Petersburg and Moscow. His mother, Ekaterina Semenovna Bekker, was the daughter of an officer. The suffix Maklay was added to the family name by his brother Nikolay Nikolaevich—for reasons that have never been fully clarified.

After graduating from the St. Petersburg Mining Institute in 1886 Miklukho went to Strasbourg University, where he studied under K. H. F. Rosenbusch. He attended lectures for three semesters and determined various minerals in the laboratory using Rosenbusch's physical-optical methods. After returning to Russia, he began work at the museum of the Mining Institute and as curator of the Geological Committee. At the suggestion of the committee's director, Miklukho conducted microscopic research on 180 samples of ore and rock from the second Zavodinsky mine in the Altai. Compiling a detailed map of the excavations, on which he showed the sites from which the samples were taken, he used the results of his analysis to compile a geological cross section of the vein. The deposits, he concluded, represented a thick lode falling almost vertically from northeast to southwest and consisting basically of quartz, hornstone, and porphyry at the edges.

Miklukho began geological fieldwork as a student at the Mining Institute. During the summers of 1885 and 1886 he conducted geological research in the little-studied Novgorod-Volynsk and Zhitomir districts of Volynsk gubernia, especially the northern slope of the watershed of the basin of the Pripet and Southern Bug rivers. Widespread in this region were post-Tertiary granite veins in archean deposits—analogous, in his opinion, to the beds of Saxony. Besides describing the geological structure, he found rich deposits of iron ore, frequently in the form of lumps containing 45 to 50 percent pure iron. He also noted the accumulation of ocher, deposits of clay, and building materials—gneiss, granite, gabbro. Miklukho was especially interested in gneiss, thin sections of which from outcroppings seemed, under the microscope, to be nonhomogeneous in comparison with thin sections of the main rock.

In 1888–1890 Miklukho, on instructions from the Imperial Russian Mineralogical Society, investigated the geological structure of Olonets gubernia. Besides producing an accurate geological map, he was commissioned to make a petrographical comparison of gneisses with massive and ancient sedimentary rock of Olonets and Volynsk gubernias. This could help to determine the origin of slates. In the Olonets district he investigated the basins of the Svir, Olonka, Tuloksa, Vidlitsa, and Tulmozerka rivers, and the shores of Lake Ladoga. Part of the territory, composed of crystalline rock formations, consisted of high cliffs, some with perpendicular and others with mildly sloping sides, with glacial material between them.

Miklukho explored several islands on Lake Ladoga and compiled a geological map that included only those features that he had observed and sampled. His identification of bog iron deposits was particularly important for that region. Later, in 1898, he proposed a method of determining bog iron reserves with the aid of a sand pump used in petroleum production. In the deposits on the shores of Kandalaksha Bay, he noted traces of the lowering of the level of the White Sea, which he believed had occurred at the rate of three or four feet a century.

In 1891 Miklukho conducted microscopic investigation of gneisses and compared them with ancient sedimentary formations and rocks. He concluded that the gneisses of the Kem and Olonets districts were formed by the "breaking up of the rock by mechanical forces at the time of orogenesis and subsequent hydrochemical processes." He proposed to designate these rocks as "dynamo-hydro-metamorphic." In March 1892 Miklukho reported to the Imperial Russian Mineralogical Society on the difference that he had noted in the "glacial deposit" in the Kem and Olonets districts. "The glacial cover that formerly covered the north of Russia," he said, "consisted of separate glaciers of varying sizes." The glacial cover "moved from the borders of Finland to the White Sea" at different velocities. The moraines of different sizes were evidence of the variations in movement.

In the summer of 1892 Miklukho conducted geological research between Archangel and the towns of Onega and the divide between the White Sea and Lake Onega, as well as in the basins of the Ilek, Vama, and Vodla rivers. He concluded that the hypothesis of a former connection between the White and Baltic seas was highly probable and that the wide distribution of stratified formations in Olonets gubernia could serve as confirmation of this theory. Miklukho discovered the origin of the salt springs around Nenoksa, Luda, Uma, and Kyanda and also introduced numerous corrections into the geological map. In the summer of 1893 he continued research on the banks of the White Sea and decided that the shoreline in Onega Bay was receding.

From 1886 to 1897, Miklukho compiled the index to geological literature that was published by S. N. Nikitin as *Russkaya geologicheskaya biblioteka* ("Russian Geological Library," 1886–1900). In 1897 Miklukho traveled from St. Petersburg to Kiev gubernia, then to the Urals, Mangyshlak, and other regions, prospecting for kaolin, manganese, gold, and iron. He did not return to Leningrad until 1923, when he resumed work at the Geological Committee. He also continued to compile bibliographies of geological literature.

BIBLIOGRAPHY

Miklukho's works include "Microskopicheskie issledovania obraztsov rud i porod vtorogo Zavodinskogo rudnika na Altae" ("Microscopic Research on Samples of Ore and Rock From the Second Zavodinsky Mine in the Altai"), in *Izvestiya Geologicheskago komiteta*, **8**, no. 9 (1889), 93; "Geologicheskie issledovania Novogradvolynskogo i Zhitomirskogo uezdov Volynskoy gubernii" ("Geological Research on the Novgorod-Volynsk and Zhitomir Districts of Volynsk Gubernia"), in *Materialy dlya geologii Rossii* ("Materials for a Geology of Russia"), XIV (St. Petersburg, 1890), 3; and "Geologichesky ocherk Olonetskogo uezda i ostrovov Ladozhskogo ozera, raspolozhennykh vokrug Valaama" ("Geological Sketch of the Olonets District and the Islands of Lake Ladoga, Around Valaam"), *ibid.*, XVIII (St. Petersburg, 1897), 173.

There is an obituary in *Geologicheskii vestnik*, **5**, nos. 4–5 (1926–1927), 76. On the origin of the name Miklukho-Maklay, see D. N. Anuchin, "N. N. Miklukho-Maklay, ego zhizn, puteshestvia i sudba ego trudov" (". . . His Life, Travels, and the Fate of His Works"), in *O lyudyakh russkoy nauki i kultury* ("People of Russian Science and Culture"), 2nd ed. (Moscow, 1952), 30.

TATYANA D. ILYINA
VERA N. FEDCHINA

MURALT, JOHANNES VON (*b.* Zürich, Switzerland, 18 February 1645; *d.* Zürich, 12 January 1733), *surgery, medicine, anatomy.*

Muralt was a member of the old noble de Muralto family, which had been driven from its seat in Locarno upon its conversion to Protestantism. The refugees were eventually invested with citizenship in the Reformed Swiss cities of Bern and Zurich, and found new prosperity. Some of Muralt's ancestors were physicians and diplomats; his father, Johann Melchior Muralt, was a merchant.

Muralt was educated at the Zurich Carolinium. When he was twenty he published his *Schola mutorum ac surdorum*, then set out on his academic travels, which took him to Basel, Leiden, London, Oxford, Paris, and Montpellier. He studied anatomy, surgery, and obstetrics with a number of famous teachers, among them Franciscus Sylvius. He returned to Switzerland to take the M.D. at the University of Basel in 1671, with a dissertation "De morbis parturientium et accidentibus, quae partum insequuntur." The following year he settled in Zurich, where he married Regula Escher; they had many children, including the distinguished physician Johann Conrad Muralt.

The Zurich surgeons' guild challenged Muralt's right to practice in that city, and he encountered widespread disapproval for conducting public animal dissections. His success as a physician overcame all opposition, however, and after five years of argument the Zurich Bürgerrat authorized him to dissect the bodies of executed criminals and of hospital patients who had died of rare diseases. Muralt was admitted to the Academia Naturae Curiosorum (with the name "Aretaeus") in 1681; forgetting their old feud, the surgeons also made him an honorary member of their guild.

In 1686 Muralt gave a course of lectures at the surgeons' guildhall, "Zum Schwarzen Garten." His audience was composed of surgeons, their apprentices, medical students, and laymen; the lectures themselves were the first on anatomical subjects to be given in the vernacular. Once a week, for an entire year, Muralt displayed dissected bodies (chiefly animal) and discussed the anatomy, physiology, and pathology of the organs. He expounded the theory of diseases and outlined medical and surgical treatment, including precise directions for the use of medicinal plants and detailed instructions for military surgeons.

In 1691 Muralt was named archiater of Zurich, with duties that comprised devising sanitary measures to protect the city against infectious diseases, advising the municipal marriage court, inspecting apothecaries, supervising the training of midwives, and treating internal diseases in the city's hospital. Ex officio, Muralt also performed all operations for fractures, the stone, and cataracts. In 1691 he was appointed professor of natural sciences at the cathedral school and also became canon of its chapter. He made use of this multitude of offices to transform Zurich into an important center for the study of anatomy and surgery.

Muralt's considerable achievement was largely based upon his surgical skill. He developed new procedures and set them forth systematically in his writings. His work is, however, more notable for the quantity and range of his material than for the depth of his knowledge. His twenty-one titles on anatomy, medicine, and physiology, as well as his thirteen separate publications on mineralogy, zoology, and botany, are marred by repetitiousness. Many of his printed works represent a collection of what are, in effect, his laboratory notes on experiments, natural objects, or the course of a disease (for example, the 174 "Observationes" that he published in *Miscellanea curiosa medico-physica Academiae naturae curiosorum*); others, among them the *Anatomisches Collegium* in 1687, record his lectures more or less verbatim. His principal work on natural history was *Systema physicae experimentalis* . . . (1705–1714); a manuscript regional pharmacopoeia has also been preserved. The last of his writings, *Kurtze und Grundliche Beschreibung der ansteckenden Pest* (1721), remains of interest for its suggestion of the "animal" nature of the plague contagium.

In general, Muralt was a keen observer and a poor critic. He was occasionally prey to superstition, and elements of popular medical beliefs are apparent in his theory of disease. But if some of his therapeutic measures derive from the operations of magic, Muralt was nevertheless an effective physician and a tireless popularizer and communicator of genuinely scientific knowledge.

BIBLIOGRAPHY

I. ORIGINAL WORKS. Muralt's writings include *Vademecum anatomicum sive clavis medicinae* (Zurich, 1677); *Anatomisches Collegium* (Nuremburg, 1687); *Curationes medicae observationibus et experimentis anatomicis mixtae* (Amsterdam, 1688); *Kinder- und Hebammenbüchlein* (Zurich, 1689; 1693); *Chirurgische Schriften* (Basel, 1691); *Hippocrates Helveticus oder der Eydgenössische Stadt- Land- und Hauss-Ärtzt* (Basel, 1692); *Systema physicae experimentalis* . . . (Zurich, 1705–1714), of which the fourth part, *Botanologia seu Helvetiae Paradisus*, was trans. into German as *Eydgenössischer Lust-Garte* (Zurich, 1715); *Schriften von der Wund-Ärtzney* (Basel, 1711); *Kriegs- und Soldaten-Diaet* (Zurich, 1712); and *Sichere Anleitung wider den dissmal grassirenden Rothen Schaden* (Zurich, 1712).

II. SECONDARY LITERATURE. On Muralt and his work, see C. Brunner, *Die Verwundeten in den Kriegen der alten Eidgenossenschaft* (Tübingen, 1903); and *Aus den Briefen hervorragender Schweizer Ärzte des 17. Jahrhunderts* (Basel, 1919), written with W. von Muralt; E. Eidenbenz, "Dr. Johannes von Muralts 'Pharmocopoeia domestica,'" in *Schweizerische Apothekerzeitung*, **60** (1922), 29–31; J. Finsler, *Bemerkungen aus den Leben des Johannes von Muralt* (Zurich, 1833); H. Koller, "Das anatomische Institut der Universität Zürich in seiner geschichtlichen Entwicklung," in *Zürcher medizingeschichtliche Abhandlungen*, **11** (1926); K. Meyer-Ahrens, "Die Ärztfamilie von Muralt, insbesondere Joh. v. Muralt, Ärzt in Zürich," in *Schweizerische Zeitschrift für Heilkunde*, **1** (1862), 268, 423, and **2** (1863), 25–47; O. Obschlager, "Der Zürcher Stadtärzt Joh. von Muralt und der medizinische Aberglaube seiner Zeit," M.D. dissertation, University of Zurich (1926); G. Sticker, *Abhandlungen aus der Seuchengeschichte und Seuchenlehre*, vol. I, *Die Pest* (Giessen, 1910); and G. A. Wehrli, "Die Bader, Barbiere und Wundärzte im alten Zürich," in *Mitteilungen der Antiquarische Gesellschaft Zürich*, **30**, pt. 3 (1927).

J. H. WOLF

MUTIS Y BOSSIO, JOSÉ CELESTINO BRUNO (*b.* Cadiz, Spain, 6 April 1732; *d.* Santa Fe de Bogotá, Nueva Granada [now Bogotá, Colombia], 11 September 1808), *botany, astronomy*.

Mutis received a degree in medicine at Seville in 1755, then continued his education at Madrid in 1757–1760 under the direction of Miguel Barnades the elder (*d.* 1771), studying at the botanic garden of Migas Calientes. In 1760, as a physician to the Marquis de la Vega, viceroy of Nueva Granada, he went to America. There he successfully opposed smallpox inoculation; helped found the Sociedad de Amigos del País (1802); and initiated the study of the exact sciences. In 1762 he introduced the teaching of mathematics in the modern sense; after 1774 he expounded the doctrine of Copernicus; and in 1803 he founded the astronomical observatory at Bogotá.

Mutis was primarily interested in botany. He traveled throughout the viceroyalty, communicating his discoveries to Madrid and corresponding with Linnaeus for several years. In 1783 he was appointed director of a botanical expedition, with Eloy Valenzuela, Diego García, and Salvador Rizo as assistants. He met Humboldt and Bonpland in 1801 and remained friends with them.

Mutis' writings included studies of Bogotá tea (*Symplocos mitonia*), guaco (*Aristolochia anuicida*), Peruvian ipecac (*Psychotria emetica* Mutis), and some species of cinchona; he wrote such monographs on the latter plant as *El arcano de la quina* (1793). His major contribution was *La Flora de la real expedición botánica del Nuevo Reino de Granada*, publication of which began only in 1954.

BIBLIOGRAPHY

Mutis' major work is *La flora de la real expedición botánica del Nuevo Reino de Granada*, 51 vols. (Madrid, 1954–). For the rest of his writings, as well as a bibliography of works on Mutis, see the index to *Índice histórico español* (Barcelona).

An important biography is A. Federico Gredilla, *Biografía de José Celestino Mutis con relación de su viaje y estudios practicados en el Nuevo Reino de Granada* (Madrid, 1911).

J. VERNET

NANSEN, FRIDTJOF (*b.* Store-Fröen, near Oslo, Norway, 10 October 1861; *d.* Oslo, 13 May 1930), *zoology, oceanography.*

Nansen was the son of Baldur Fridtjof Nansen, a lawyer, and Adelaide Johanne Thekla Isidore Wedel-Jarlsberg. An avid outdoorsman from an early age, he excelled at skiing and won several prizes in Norwegian competitions. Although interested in mathematics and physics when he enrolled at the University of Christiania (now Oslo) in 1880, he decided to specialize in zoology, believing that this discipline would afford greater opportunity for outdoor work. In 1882, on the advice of Robert Collett, professor of zoology at the university, Nansen joined the four-and-a-half-month expedition of the sealer *Viking*. The trip was of seminal influence. He was introduced to the harsh conditions of the Arctic as the ship sailed north almost to Svalbard and then west, where it froze fast in the ice and drifted south for several weeks along the eastern shore of Greenland. Nansen's diary of the voyage eloquently describes the region, and many of his drawings and sketches were later used to illustrate his works.

In 1882 Collett obtained for Nansen the post of curator at the Bergen Museum. He first took up the study of myzostomes, a small group of parasitic worms of unusual appearance. His results, published in 1885, are still a basic reference and in 1886 earned Nansen the Joachim Friele Gold Medal of the Bergen Museum. The following year he traveled to Germany and Italy, visiting the newly established marine biological station at Naples and Golgi's laboratory in Pavia, where he observed the technique of silver impregnation of nerve cells. In 1888 Nansen received the Ph.D. at Oslo for a thesis on the central nervous system in which he demonstrated how the nerve fibers, after entering the lower root of the spinal column, divide into T formations. Although this finding was of great importance for the study of nerve fibers, some of his other observations proved to have been based on artificial products that arose from the coloring and fixing processes that he had used.

Nansen's fascination with the unknown interior behind the ice-covered coast of Greenland was sparked by Adolf Erik Nordenskiöld's return in 1883 from an expedition to its west coast. Two Lapps who had accompanied him ascended to the high interior plateau, describing it as an endless snowfield. Many others, however, believed that the interior was ice-free and that temperatures were relatively high; there was even talk of oases. Nordenskiöld himself believed that the Lapps had seen only a snow belt and that the interior was mainly ice, free of snow. Nansen, who had shortly before established himself as a pioneer of high-mountain skiing on a three-day trip from Bergen to Oslo, conceived the idea that it should be possible to ski across Greenland.

On 17 July 1888 a Norwegian sealer carried Nansen and his five-man party as near the east coast of Greenland as the ice permitted, and the group set off in their boats. Driven by adverse currents, they were unable to land until 16 August. After an extremely difficult ascent to the plateau, which reaches a height of almost 9,000 feet, they journeyed across the ice through arduous snow conditions, night temperatures as low as −50°C., snowstorms, and fog, until they reached the village of Godthaab, on the west coast, on 3 October. Having missed the last ship home, the group spent the winter there, making friends with the Eskimos and learning their way of life. The expedition confirmed that Greenland is completely covered with ice, and its meticulous meteorological observations have remained basic to an understanding of the influence of weather conditions in northern Europe and the United States. On his return to Norway in 1889, Nansen married Eva Sars, daughter of the zoologist Michael Sars, and was appointed curator at the University of Christiania.

Nansen subsequently developed a theory that there exists a current from Siberia, across the Arctic Sea, to Greenland. This conjecture was based on his observation that much of the driftwood found in Greenland came from Siberian trees and from the discovery in Greenland of wreckage of the *Jeanette* expedition (1879–1880), which had been trapped north of the Chukchi Sea. In his proposal to the Norwegian Geographic Society and the Royal Geographic Society of London, Nansen planned to drift with the ice current in a specially constructed ship—rather than fighting it, as all polar explorers had previously done. Although his theory was criticized by many, financing was obtained for the *Fram* ("Forward"), the first vessel expressly constructed to withstand the pressure of ice.

Short and broad, the three-master had a powerful engine and a twelve-man crew. On 24 June 1893 it sailed eastward from Oslofjord, skirting the ice packs along the Siberian coast. On 10 September it passed Cape Chelyuskin, the northernmost point on the Siberian coast. Nine days later it set course straight north in open water; but on 22 September the ice closed in, and the *Fram* was made fast at 78°50′ N. lat. After drifting north to 85°55′ N. lat., Nansen and F. Hjalmar Johansen left the *Fram* on 14 March 1895, hoping to make a dash for the North Pole. The way proved more difficult than anticipated, and they were forced to turn back at 86° 14′ N. lat., 268 miles from their goal but the farthest north ever attained by an explorer. A malfunctioning chronometer and an inaccurate map obliged them to halt for the winter. For the next nine months they lived in a small hut that they had built of whalebones, bearskins, and sleds, shooting polar bears and walrus for food before being rescued by an English expedition. In the meantime, on 19 May 1896 the *Fram* started south. It became free of the ice north of Svalbard and reached Norway on 19 June, after almost three years at sea. On his return Nansen received a nonteaching chair in zoology at the University of Christiania. The *Fram* is still preserved in the Fram Museum, Oslo.

The most important contributions of the *Fram* expedition were the discoveries of the great depth of the Arctic Ocean and of the locations of land and water masses; and publication of the results provided an important stimulus to physical oceanography. Nansen also discovered that the ice was drifting not in the direction of the wind but about 45° to the right of it. He explained this as an effect of the rotation of the earth. Lacking the mathematical background to work out the problem, he suggested to Vagn Walfrid Ekman that he investigate it, and the resulting Ekman spiral is a basic tenet of theoretical oceanography. From a study of the temperatures and salinity of the Arctic and Norwegian seas, Nansen surmised the existence of an underwater ridge between Greenland and Svalbard. Later investigated by oceanographic expeditions, it is now called the Nansen ridge.

On the basis of his experiences on the *Fram*, Nansen devised an improved water bottle consisting of an insulated bottle and a pair of specially designed thermometers attached directly to the reversing frame. The so-called Nansen bottle was in general use until the recent development of electronic methods for determining the salinity of seawater. He also constructed an unusually sensitive pendulum-type current meter and a deep-sea bottom sampler.

Nansen explained the dead-water phenomenon on the basis of many samples taken at various depths to measure temperature and salinity. He discovered a sharp dividing line between the top layer of brackish water, which had melted from the ice, and the underlying seawater; and he described the boundary wave between the layers engendered by the motion of a vessel through the upper layer. The mathematical treatment of this work was also turned over to Ekman.

Nansen and his assitant Bjørn Helland-Hansen constructed graphs based on their own determinations of water densities, as well as on Vilhelm Bjerknes' work in the theory of hydrodynamic forces and Knudsen's experiments on seawater. Ekman worked out a set of equations for calculating ocean currents, later called geostrophic currents. A new study of the Norwegian Sea was initiated, and a voyage of the *Michael Sars* was planned and begun by Nansen and continued by Helland-Hansen. Their monograph on the Norwegian Sea became a model for oceanographic studies.

In 1908, reflecting his change of interest, Nansen was appointed to the newly established chair of oceanography at Oslo. He subsequently made various other research voyages and, in 1913, an extensive journey through Siberia. His wife died in 1907, and in 1919 he married Sigrid Sandberg Munthe.

With the outbreak of World War I, Nansen turned to humanitarian work, and in 1920 he became the first Norwegian delegate to the League of Nations, a post he held until his death. In 1921 he was appointed commissioner for refugees for the League and during this period also worked with Herbert Hoover's American Relief Administration. In 1924 he was charged by the League of Nations with finding a solution for "the Armenian question" and recommended resettlement of the Armenians in a new Armenian republic, now a part of the Soviet

 NERNST

Union. In recognition of his humanitarian work Nansen received the Nobel Peace Prize in 1922.

BIBLIOGRAPHY

I. Original Works. For an annotated list of Nansen's works on the polar region, see *Arctic Bibliography*, II (Washington, D.C., 1953), 1780–1788. A more extensive list of his publications related to research and travel can be compiled from Dartmouth College Library, *Dictionary Catalog of the Stefansson Collection on the Polar Regions*, VI (Boston, 1967), 150–160. There is a list of Nansen's articles and lectures for the Norwegian Academy of Sciences in Oslo in Leiv Amundsen, *Det Norske Videnskaps-akademi i Oslo, 1857–1957* (Oslo, 1957), 529–531.

Nansen's publications include *Bidrag til myzostomernes anatomi og histori* (Bergen, 1885), with summary in English; "The Structure and Combination of the Histological Elements of the Central Nervous System," in *Bergens museums årsberetning* (1886), 27–215, his Ph.D. diss.; *Eskimoliv* (Oslo, 1891), also in English (London, 1893); *On the Development and Structure of the Whale. Part I*, in *On the Development of the Dolphin* (Bergen, 1894), written with Gustav A. Guldberg; *The Norwegian North Polar Expedition, 1893–1896*, 6 vols. (Oslo, 1900–1906); *The Norwegian Sea, Its Physical Oceanography Based Upon the Norwegian Researches, 1900–1904* (Oslo, 1909), written with B. Helland-Hansen; "The Sea West of Spitsbergen. The Oceanographic Observations of the Isachsen Spitsbergen Expedition in 1910," in *Skrifter utgitt av det Norske videnskaps-akademi i Oslo*, Math.-naturv. Kl., no. 12 (1912), 1–89, written with B. Helland-Hansen; "Spitsbergen Waters. Oceanographic Observations During the Cruise of the 'Veslemöy' to Spitsbergen in 1912," *ibid.*, no. 2 (1915), 1–132; *Temperature Variations in the North Atlantic Ocean and in the Atmosphere* (Washington, D.C., 1920), written with B. Helland-Hansen; "The Strandflat and Isostasy," in *Skrifter utgitt av det Norske videnskaps-akademi i Oslo*, Math.-naturv. Kl., no. 11 (1921), 1–313; "The Eastern North Atlantic," in *Norske videnskaps-akademi i Oslo. Geofysiske publikasjoner*, **4**, no. 2 (1925), 1–76, written with B. Helland-Hansen; and Steinar Kjaerheim, ed., *Fridtjof Nansen. Brev*, 4 vols. (Oslo, 1961–1963), Nansen's correspondence in Norwegian, English, and German.

II. Secondary Literature. On Nansen and his work, see Per Vogt *et al.*, *Fridtjof Nansen, liv og gjerning* (Oslo, 1961), trans. as *Fridtjof Nansen, Explorer, Scientist, Humanitarian* (Oslo, 1961)—the English ed. includes several additional articles.

Lettie S. Multhauf

NERNST, HERMANN WALTHER (*b.* Briesen, West Prussia [now Wąbrzeżno, Poland], 25 June 1864; *d.* Zibelle manorial estate, near Bad Muskau, Oberlausitz [now German Democratic Republic], 18 November 1941), *chemistry, physics.*

Nernst was a physicist, turned chemist, who was quick to seize on novel ideas no matter what their source. His complete theoretical command of the subject matter of physical chemistry was unparalleled. Above all, he was a superb craftsman with keenly developed technical skills and an imaginative intuitive grasp of what was experimentally feasible. Early in his career the application of the principles of physics to chemical problems became his life's goal. Over a period of four decades his activities in Göttingen and Berlin notably served to extend the boundaries of the traditional domains of both physics and chemistry.

Although Nernst's early worldwide reputation resulted from a broad range of fundamental contributions to the new developments in physical chemistry, especially in electrolytic solution theory, his crowning achievement was in chemical thermodynamics. For this work Nernst received the Nobel Prize for chemistry in 1920. The Nernst heat theorem of 1906, or the third law of thermodynamics, as Nernst preferred to call it, was at first recognized chiefly as a practical means for computing chemical equilibria. The feasibility of directly calculating the entropy constants for gases from quantum theoretical formulations led to a new recognition of Nernst's work prior to World War I. During the 1920's quantum statistical considerations initiated a controversy—even serious reservations in some quarters—over the general validity of Nernst's theorem for solids. By the late 1920's, when Nernst no longer was actively engaged in thermodynamic investigations, several special formulations of the heat theorem, and notably that of Francis Simon, led to the acceptance of Nernst's fundamental idea, in its refined form, as a general law of thermodynamics.

Nernst was the third child of Gustav Nernst, a judge in the Prussian civil service, and Ottilie Nerger. In 1883 he graduated first in his class from the Gymnasium in Graudenz [now Grudziadz, Poland], where his studies had focused on the classics, literature, and the natural sciences. His early ambition was to become a poet. Although that aim faded, he developed a lifelong infatuation for literature and the theater, especially Shakespeare.

From 1883 to 1887 Nernst studied physics at the universities of Zurich, Berlin, Graz, and Würzburg. He attended Heinrich Weber's physics lectures in Zurich and Helmholtz' thermodynamics lectures in Berlin. In Graz, Nernst was deeply impressed by Boltzmann and his emphasis on the atomistic interpretation of natural processes. On Boltzmann's advice Nernst collaborated with his former pupil Albert von Ettingshausen in an investigation of the combined effect of magnetism and the flow of heat on the electric current. They discovered that a magnetic field applied perpendicularly to a temperature gradient gives rise to a potential difference in a metallic conductor. On the basis of this work of 1886 carried out in Graz, Nernst completed his inaugural dissertation the following year under Kohlrausch. Nernst apparently felt that he had taken his thermoelectric investigations as far as he could, for he never returned to the subject. Paul Drude, his future colleague in Göttingen, however, made good use of Nernst's discovery and the corresponding thermomagnetic and galvanomagnetic effects to develop the electron theory of thermal and electrical conductivity.

The circumstances surrounding Nernst's studies in Würzburg evidently provided the stimulus and the opportunities for the shift of his interests toward a career dominated by the application of physics to chemical problems. Emil Fischer, who later became Nernst's colleague in Berlin, and Arrhenius, who had just announced his electrolytic dissociation theory, were both working with Kohlrausch in Würzburg at this time. While Arrhenius and Nernst were visiting Boltzmann in Graz, Arrhenius introduced Nernst to Friedrich Wilhelm Ostwald, with whom he had worked at the Polytechnikum in Riga. With the publication of the volume on *Verwandtschaftslehre* in 1887 in Leipzig, Ostwald's massive *Lehrbuch der allgemeinen Chemie* had been completed. With van't Hoff, who was a proponent of the new chemical theory for weak electrolytes, Ostwald launched in 1887 the *Zeitschrift für physikalische Chemie*. Before the end of the year, Ostwald had accepted a professorship at the University of Leipzig, and Nernst had become his assistant. Although younger by five to ten years than Ostwald, van't Hoff, and Arrhenius (the three dominant figures in the development of physical chemistry), Nernst soon came to be considered one of the founders of the newly created discipline.

In Leipzig theoretical and experimental chemistry were pursued conjointly with impressive zeal. The emphasis fell on the electrolytic theory of ionization; the colligative properties of gases and liquids; and thermodynamics, or "energetics" in the Ostwald sense. Besides Arrhenius, van't Hoff, and Nernst, others who belonged to Ostwald's circle included Tammann (Nernst's successor at Göttingen), Le Blanc (Ostwald's successor at Leipzig), James Walker, Wilhelm Meyerhoffer, Ernst Otto Beckmann, and Julius Wagner—all pioneers in the establishment and exploration of physical chemistry as an academic discipline.

In the context of this group of *Ioner*, as they were called, Nernst soon became totally absorbed with the problems of physical chemistry. Within a year he had published his derivation of the law of diffusion for electrolytes in the simple case when only two kinds of ions are present. Thus for the first time (1888) he was able to calculate the diffusion coefficient for infinitely dilute solutions and to show the relationship between ionic mobility, diffusion coefficient, and electromotive force in concentration cells. This in turn was based upon the idea that, for two solutions of differing concentrations separated by a semipermeable partition, the driving force responsible for the diffusion is given by the difference in osmotic pressures on opposite sides of the partition. The fundamental relationship between electromotive force and ionic concentration was developed more fully in his Leipzig *Habilitationsschrift* of 1889, *Die elektromotorische Wirksamkeit der Ionen*. Applied to ideal solutions, fundamental thermodynamics showed that the electromotive force E, for a galvanic process corresponding to a concentration change from C_1 to C_2, is given by $E = RT/N\mathcal{F} \ln C_1/C_2$, where R is the gas constant; T is the absolute temperature; N represents the gram equivalents that have reacted; and $\mathcal{F}$ is Faraday's constant, so that $N\mathcal{F}$ corresponds to the passage of coulombs. Since the electromotive force of a galvanic cell is directly proportional to the free energy of the cell reaction, in this equation there is a crucial connecting link between thermodynamics and electrochemical solution theory. The Nernst equation, however, does not settle the difficulties associated with the determination of ionic concentrations for strong electrolytes and becomes valid only as the solutions approach infinite dilution.

According to the more general theory, the factors that cause the ions to move can be reduced to the following: (1) forces on the ions, both external (such as electrical) and internal (such as concentration gradients and electrostatic forces due to the presence of the ions themselves); (2) random thermal motions of the ions; and (3) the flow of the solution as a whole. The classical Nernst treatment of ionic diffusion (in terms of mobility and transport number) was based on Arrhenius' theory of electrolytic dissociation and was expressed in terms of the electromotive force of concentration cells and galvanic elements. Nernst had assumed that a metal immersed in an electrolyte acts like a reservoir of ions having properties characteristic of electrolytic solvation pressure. Thus he was able to calculate maximum electric work (electromotive forces) from fundamental principles such as the relation to the gas constant. This work gave Nernst, then in his mid-twenties, an international reputation in electrochemistry.

Nernst's distribution law, which relates the equilibrium concentrations of a solute distributed between immiscible liquid phases, appeared in two papers in 1890 and 1891. The case that Nernst studied theoretically and experimentally was the distribution of benzoic acid between water (phase *w*) and benzene (phase *b*). According to Nernst, benzoic acid in the water phase is present mainly as C_6H_5COOH, but a small fraction α dissociates into $C_6H_5COO^-$ and H^+ ions; whereas, benzoic acid in the benzene phase is present mainly as $(C_6H_5COOH)_2$, but a small fraction β exists as C_6H_5COOH molecules. Using C_w and C_b to represent the concentrations (or, more exactly, the activities) of benzoic acid in water and benzene respectively, Nernst was able to show—by taking quantitative account of the ionization of benzoic acid in water and the incomplete association of benzoic acid in benzene—that at a given temperature the distribution constant K (thereafter called the Nernst distribution constant and, later, partition coefficient) was given by $K = (1-\beta)C_b/(1-\alpha)^2 C_w^2$. Experimentally Nernst was able to confirm this general type of equation in specific cases and to show that since α and β are both small in concentrated solutions, the ratio C_b/C_w^2 is approximately constant. It was seen that Henry's law, according to which the solubility of a gas in a liquid is directly proportional to the pressure of the gas above a liquid at equilibrium, is a special case of the more general Nernst distribution equation (1891). In 1872 Berthelot and E. C. Jungfleisch had carried out experimental investigations of the distribution of a substance between two liquid phases. Nernst's work called attention to the fact that a simple distribution law can be expected to be valid only if, on dissolving, the solute undergoes no changes such as dissociation—that is, only when the concentrations (or activities) of the same molecular species in each phase are considered. The Nernst distribution equation represents an important type of phase equilibrium, and was put to practical use in extraction process calculations and in analyzing the distribution of substances in different parts of a living organism. For example, it was shown that ether tends to concentrate in brain and nerve tissues that are rich in fatty materials rather than in the more aqueous medium of blood.

After a semester at Heidelberg, Nernst returned to Leipzig and in 1891 accepted a post as associate professor in physics at the University of Göttingen. In 1892 he married Emma Lohmeyer, the daughter of Ferdinand Lohmeyer, a distinguished surgeon in Göttingen; they had two sons and three daughters. Nernst's father-in-law was an accomplished musician and played the piano and cello with Brahms and the Joachim string quartet. Although exposed to good music in Göttingen and an appreciative and sympathetic listener, Nernst apparently was not endowed with any noteworthy musical talent, as were some of his closest colleagues—Helmholtz, Planck, Einstein, and Simon.

During the early Göttingen period of his career, Nernst's conception of the goals and significant advances in theoretical chemistry was fashioned and published in his *Theoretische Chemie vom Standpunkte der Avogadroschen Regel und der Thermodynamik* (1893). This textbook was dedicated to Ettingshausen in Graz "*in treuer Erinnerung an . . . Lehr- und Wanderjahre*." As Nernst conceived it, the most important guide to presenting the theoretical treatment of chemical processes was first, to recognize the central importance of Avogadro's hypothesis, which he referred to as "an almost inexhaustible 'horn of plenty' for the molecular theory," and second, to accentuate the law of energy that governs all natural processes. Convinced that theoretical chemistry had begun to attain a certain maturity, Nernst felt that an independent textbook was needed in order to bring together widely different aspects of physics and chemistry. Nernst saw in the development of physical chemistry "not so much the shaping of a new science, as the meeting of two sciences hitherto somewhat independent of each other." As he indicated in the preface, his objective was to present the latest investigation—"all that the physicist must know of chemistry, and all that the chemist must know of physics."

The popularity of *Theoretische Chemie* can be attributed to several factors: it was kept up-to-date with a description of current developments, it was written for students from a not-too-advanced theoretical point of view, it touched upon and interrelated a wide range of phenomena, and it contained an abundance of illustrative materials and descriptions of experiments to facilitate the understanding of the theoretical principles. In Germany until 1926 (the publication date of the fifteenth edition), this work was recognized as the foremost textbook of physical chemistry. Thereafter it was replaced by the texts of Eucken (who followed Tammann at Göttingen) and John Eggert, both of whom were trained under Nernst.

In 1894 Nernst was offered the post of professor of theoretical physics at Munich, left vacant by Boltzmann's move to Vienna. Instead, he bargained for, and was offered, a chair in physical chemistry and a new Institut für Physikalische Chemie und Elektrochemie in Göttingen—the only such post in Germany, apart from Ostwald's in Leipzig. Although Nernst sometimes disagreed with Ostwald on matters of interpretation, he retained a lifelong appreciation of this old master who had turned him in the direction of chemical pursuits. In 1895 Nernst and Schoenflies published their *Einführung in die mathematische Behandlung der Naturwissenschaften*, dedicated to Ostwald; by 1931 it had passed through eleven editions.

From 1891 to 1904 at Göttingen, Nernst managed to assemble an international group of scholars to cooperate in the intensive and comprehensive investigation of experimental and theoretical physicochemical problems. In Nernst's institute it was taken for granted that researchers working on various problems would share their ideas and that all of the work was focused in some way on goals that had been clearly set out by Nernst. For example, the electromotive force theory of 1889 gave rise to the theory of lead accumulators (1900), the study of electrocapillarity (1901), and the theory of polarization (1902, 1908). The study of electrolytic phenomena at the liquid-liquid interface (1901) contributed to the formulation of a theory of electrical nerve stimulus important in nerve physiology (1904, 1908). The electromotive force determinations for metals led to investigations of overvoltage and to the suggestion (1900) that the hydrogen electrode be taken as the normal reference point. From his studies of the influence of the dielectric constant of a solvent on ionic equilibrium (1894) Nernst was led to announce a new method for determining dielectric constants for fluids, using an alternating-current Wheatstone bridge method (1897).

Reminiscent of Helmholtz' enthusiasm for design and refinement of instruments, Nernst and his co-workers became involved in the construction of a microbalance and an ingenious experimental method to measure dielectric constants. They also designed special apparatus for the determination of molecular weights by freezing point depression in dilute solutions (1894) and by vapor density measurement at extremely high temperatures (1903). Nernst's interest in mass action and reaction velocity for gaseous dissociation processes of potential technological significance—as in the case of hydrogen, nitrogen, and ammonia—led to investigations that notably demonstrated his superb mastery of analytical instrumentation. In order to increase the yield in the synthesis of ammonia, he constructed a reaction chamber that could withstand pressures of 75 atmospheres at 1000°C. In 1907 Nernst and Jost achieved a yield of ammonia of about 1 percent at pressures of 50 atmospheres and 685°C., whereas Haber's experiments at pressures of 1 atmosphere and 1000°C. had given a yield of only 0.01 percent. Preoccupied with the heat theorem, Nernst abandoned his research on ammonia, while Haber went on to improve catalytic techniques that made industrial synthesis feasible.

In 1905 Nernst was called to the University of Berlin, upon the retirement of Landolt. It is evident from documents at the archives of Humboldt University that Planck strongly supported Nernst's appointment to the chair of physical chemistry at Berlin. Planck knew that Nernst was the only one in Europe who might be able to lead Berlin out of its chemical doldrums. Besides, Nernst was a chemist who was so interested in physics that he preferred to be recognized as a physicist doing chemistry rather than as a physical chemist. For example, in his introductory Silliman lecture at Yale in 1906 he remarked that the customary separation of physics and chemistry was not altogether advantageous and was "especially embarrassing in exploring the boundary region where physicists and chemists need to work in concert."

The situation in physical chemistry at Berlin at this time can be described only as one of mere tolerance for what the *Ioner* had accomplished. Before the turn of the century the University of Berlin was one of the last strongholds of resistance to the ionic theory of dissociation; thermodynamics also remained almost untouched by chemists. At the university and through his position in the Prussian Academy of Sciences, van't Hoff had given such concerns his blessing, but he was then no longer active in science. As early as 1890/1891 Planck and Helmholtz had lent their support to the ionic theory in general and to Nernst's outstanding contributions to electrochemistry in particular. They were greeted with such cool response that Helmholtz was led to conclude that, while thermodynamics was of great importance in chemistry, the chemists in Berlin obviously were not up to it. While Berlin in 1905 was trying to catch up with what had by then become the old physical chemistry of Ostwald, van't Hoff, and Arrhenius (with its rather exclusive focus on thermodynamics, colligative properties, and ionic theory), Nernst was breaking new ground by defining the limits of applicability of classical thermodynamics for chemical equilibrium, and was simultaneously exploring problems in chemical kinetics. On the whole, Nernst's early investigations fit the pattern of chemistry laid out by Ostwald and the *Ioner*. Until he was about forty years of age, Nernst's efforts were directed predominantly toward the refinement of methods to explore principles already current among chemists. He managed to do this with exquisite finish and expertise. After moving to Berlin, however, he became totally involved, theoretically, in the exploration of new ideas of thermodynamics.

Around Easter 1905 Nernst drove in his open automobile from Göttingen to Berlin to prepare for new duties, which he was to assume that fall at the chemical institute of the University of Berlin in the Bunsenstrasse. On 23 December Nernst was back in Göttingen to present to the Göttingen Academy his now-classic forty-page "heat theorem" paper, "Ueber die Berechnung chemischer Gleichgewichte aus thermischen Messungen." It is apparent that the work and deliberations that led to Nernst's heat theorem and its sequel, the enunciation of the third law of thermodynamics, had been carried out mostly in Göttingen. Undoubtedly that is the reason that led Nernst to present his ideas on this fundamental topic in Göttingen. The paper was published in 1906 in the *Nachrichten von der Gesellschaft der Wissenschaften zu Göttingen*. To comprehend its merits and significance it will be helpful to examine the context in which Nernst's ideas were formulated and to mention briefly the activities of other investigators upon whose work Nernst was able to build.

The twentieth-century advances associated with the thermodynamics of chemical processes and statistical thermodynamics were made largely from work carried out at the beginning of the century by Boltzmann, Planck, Einstein, and Nernst. The third law of thermodynamics, which had its origin in chemistry, was conceived by Nernst in connection with the search for the mathematical criteria of chemical equilibrium and chemical spontaneity. The solution he proposed for predicting the equilibrium conditions for chemical reactions was a novel one, but the problem itself had been of importance to chemists for over a century. It had taken the form of experimental investigations designed to provide an exhaustive catalog of chemical affinity relationships. An old and certainly puzzling question was why certain chemical reactions go while others do not, or more precisely, how far a given reaction will go before it reaches equilibrium. The more general problem may be stated as follows: Given that one knows the energy changes for the transition of a system (chemical or other) from one equilibrium state to another, is it possible—using only the first and second laws of thermodynamics—to calculate theoretically, for that transition, the quantity of maximum useful work (otherwise known as the Helmholtz available work function)?

By 1900 it was known that the thermodynamic calculation of chemical equilibria, using thermal data alone (that is, heats of reaction, specific heats, and the thermal coefficients for both heats of reaction and specific heats), could not be carried through because of what Haber in 1904 called the thermodynamically indeterminate integration constant J that appears in the integrated form of the Gibbs-Helmholtz equation. The theoretically sound point of departure for treating chemical equilibria, indeed, was seen to be the Gibbs-Helmholtz equation, $\Delta F = \Delta H + T(\delta \Delta F/\delta T)_p$, which relates the free energy change ΔF to the heat content or enthalpy ΔH and to the entropy change ΔS, the latter being expressed here in the form of the thermal coefficient of free energy change; since $(\delta \Delta F/\delta T)_p = -\Delta S$.

The Gibbs-Helmholtz equation, derived from considerations of the first and second laws, was expressed in slightly different form by Gibbs (1875–1879) and Helmholtz (1882–1883). The Gibbs free-energy A (Gibbs's function Ψ) relates to isothermal isochoric processes, whereas the Helmholtz free-energy F (Gibbs's function ζ) relates to isothermal isobaric processes. Throughout our discussion we shall use the Helmholtz free energy and represent it as ΔF (Helmholtz' *freie Energie*), because it is more commonly and more conveniently manipulated in dealing with chemical reactions for which ΔH is known. For the sake of uniformity of presentation we shall use the Helmholtz formulation even when Nernst, for example, in some of his papers formulates his heat theorem in the Gibbs form. We recognize that an equivalent expression for the Gibbs free-energy ΔA can readily be written so that the change in internal energy ΔU (Gibbs's ϵ) takes the place of ΔH (Gibbs's X) to give

$$\Delta A = \Delta U + T(\delta \Delta A/\delta T)_v, \text{ where } (\delta \Delta A/\delta T)_v = -\Delta S.$$

From the Gibbs-Helmholtz equation, $\Delta F = \Delta H + T(\delta \Delta F/\delta T)_p$, and the general expression of ΔH as a function of temperature,

$$\Delta H = \Delta H_0 + \alpha T + \beta T^2 + \gamma T^3 + \cdots,$$

and the Kirchhoff law (1858), $\Delta C_p = (\delta \Delta H/\delta T)_p$, where C_p is the heat capacity at constant pressure, it can readily be shown that the free-energy equation takes the integrated form:

$$\Delta F = \Delta H_0 - \alpha T ln T - \beta T^2 - \gamma/2T^3 - \cdots + JT.$$

In this equation, ΔH_0 is the heat of reaction at absolute zero, that is, the integration constant in the integrated form of the Kirchhoff law. It can be evaluated empirically from the knowledge of ΔH at any temperature. Likewise, the heat-capacity coefficients, α, β, γ, $\cdots$, can be calculated from calorimetric data. The only real difficulty encountered in putting this equation into practice is that the integration constant J cannot be evaluated calorimetrically and must be obtained from the knowledge of ΔF at some temperature.

From the mid-1880's until the enunciation of Nernst's theorem in 1906, this integration constant J therefore became the focus of a genuine dilemma in chemical thermodynamics because the integrated form of the Gibbs-Helmholtz equation (given above) merely returned the whole problem of predicting the equilibrium conditions for chemical reactions to the experimentalists. The challenge was straightforward: to invent more ingenious techniques to overcome the almost insurmountable analytical difficulties associated with the experimental determination of ΔF. This problem had plagued all the investigators: Arrhenius, van't Hoff, Ostwald, Le Châtelier, Haber, Richards, Lewis, and, of course, Nernst. The solution that Nernst proposed in 1905 relates to the way in which the J of the integrated Gibbs-Helmholtz equation is interpreted and also to the practice of thermochemistry prior to the 1880's.

During most of the second half of the nineteenth century, chemists used a simple and remarkably serviceable, although theoretically erroneous, energy principle, or rule, to explain and predict the course of chemical reactions. Working independently, H. P. J. Julius Thomsen and Berthelot drew support for the general validity of this energy principle from an enormous number of careful and painstaking calorimetric measurements that were begun in the 1850's and carried out over a thirty-year period. It is fair to say that these important thermochemical studies had very little in common with the thermodynamic discussions of the times.

According to the Thomsen-Berthelot principle, the driving force of a chemical reaction, the free energy ΔF, is equated simply with the heat of reaction ΔH from calorimetric measurements. That is, it was assumed that $\Delta F = \Delta H$. For any given process, say a reaction represented by the chemical equation $Y \rightleftarrows Z$, this principle predicts that the magnitude of ΔH is a direct measure of the driving force of the reaction. For exothermic reactions, where $\Delta H > 0$, the formation of Z from Y is favored. For endothermic reactions, where $\Delta H < 0$, the formation of Y from Z is favored. When $\Delta H = 0$, no reaction takes place. The application of this principle gives almost the right experimental answer most of the time, but not invariably.

Only gradually was the principle seen to be inadequate as experimental data accumulated to show that some exothermic reactions did not proceed spontaneously in the direction of Z, while some endothermic reactions did proceed spontaneously in the direction of Z. On occasion, isenthalpic reactions ($\Delta H = 0$) were seen to go either way. Also, it was learned from a number of galvanic cell studies, that ΔF for the cell reaction, calculated from the electromotive force E (since $\Delta F = -n\mathscr{F}E$, where $\mathscr{F}$ is Faraday's constant), was in agreement with experiment, where ΔH gave poor results. Likewise, in a few classic cases, where the equilibrium constant K could be determined from the equilibrium concentrations of the constituents in the reaction vessel, the driving force could be calculated readily from $\Delta F = -RT ln K$. Here too there was good agreement with experiment, where ΔH gave poor results.

In retrospect it seems that all of this should have been self-evident. The course of spontaneous processes must be governed by both the first and second laws of thermodynamics, and not just the first, as the Thomsen-Berthelot principle implies. Accordingly, values for ΔF, calculated from E or K, but not values of ΔH, should provide reliable information about the driving force or thermodynamic feasibility for chemical spontaneity. Thus, as Nernst and many others recognized clearly at the time, while ΔH can be calculated from ΔF, the reverse is not the case. Values for ΔF could, in general, be calculated from ΔH only by taking into account the entropy term $d\Delta F/dT$ that appears in the Gibbs-Helmholtz equation.

Manifestly, the Thomsen-Berthelot principle owed its long-standing practical successes to the fact that $d\Delta F/dT$ (or the entropy changes) in most chemical reactions is quite small in comparison with ΔF and ΔH. On these premises it became evident that probing into the theoretical significance of the magnitude and sign of $d\Delta F/dT$ might furnish the clue to an explanation of the discrepancies that the thermochemists had tried so diligently, although rather arbitrarily, to bring into line with the Thomsen-Berthelot principle.

Beginning in the early 1890's, Nernst was alert to the central problem of chemical thermodynamics discussed above. From 1905 until the outbreak of World War I, Nernst with singleness of purpose threw himself into the search for an experimentally demonstrable theoretical solution to this one problem in chemical thermodynamics that had been attacked without success, namely, the calculation, from thermal data, of chemical equilibria as related to the search for criteria of chemical spontaneity.

In *Theoretische Chemie* (1893) Nernst had noted that the Thomsen-Berthelot principle was surprisingly accurate for solids. He proceeded to show that for special cases, such as ideal gases and dilute solutions, for which $\Delta H = 0$, ΔF and therefore $d\Delta F/dT$ should also approach zero. At that time he argued that ΔF should have the character of a force function, which, like the gravitational, electrical, and magnetic potential, would be independent of temperature. In 1894 Nernst discussed the Gibbs-Helmholtz equation for the free energy of mixing of concentrated solutions and found a close fit between ΔF and ΔH, plotted as a function of temperature. In Göttingen, between 1894 and 1905, Nernst carried out some experimental work with gaseous reactions in order to test the Gibbs-Helmholtz equation (and therefore the second law) over a wide range of temperature. His analyses of the high-temperature equilibrium conditions for the formation of nitric oxide (1904, 1906), and the work on the synthesis of ammonia (1907) were motivated by the search for methods of nitrogen fixation from the air, in order to provide a potential source of nitrates for fertilizers and explosives. These were specific practical instances of the application of classical thermodynamic principles to the computation of chemical equilibria and the degree of chemical spontaneity for chemical reactions. Their study and the inherent theoretical difficulties connected with the general solution to computing the driving force of chemical reactions, mentioned above, led Nernst to probe the deeper thermodynamic significance of free energies, heats of reaction, and specific heats, especially in the low temperature range.

Other investigators were working along similar lines at about this time. In 1884 in *Comptes rendus* . . . ,. in his comprehensive, 225-page paper on chemical equilibrium, under the heading "Constante d'intégration" Le Châtelier recognized the problem in connection with his own equilibrium studies on cements and blast furnace reactions. In 1888 he explored these problems systematically in a memoir published in *Annales des mines*, in which he stated,

> It is highly probable that the constant of integration, like the [other] coefficients of the differential equation, is a determinate function of certain physical properties of the reacting substances that are present. The determination of the nature of this function would lead to the complete knowledge of the laws of equilibrium. It would permit us to determine *a priori*, independently of any new experimental data, the complete conditions of equilibrium corresponding to a given chemical reaction. It has hitherto been impossible to determine the exact nature of this constant ["Recherches expérimentales et théoriques sur les équilibres chimiques," in *Annales des mines*, 8th ser., **13** (1888), 336].

Nernst later remarked that he had not seen Le Châtelier's work and, in fact, it is unlikely that he would have examined the *Annales des mines* in connection with his thermodynamic interests.

At Harvard, G. N. Lewis (1899) and T. W. Richards (1902) carried out calorimetric and galvanic cell measurements to determine what happens to ΔF and ΔH at low temperatures. For example, Richards found that the thermal coefficient of electromotive force for most galvanic cells dE/dT approaches zero at low temperatures. Representing graphically the relationship of E to ΔH as a function of temperature, he extrapolated to absolute zero. Thus he could conclude that $d\Delta F/dT < 0$ (since $-n\mathscr{F}(\delta E/\delta T)_p = (\delta \Delta F/\delta T)_p$, that $d\Delta H/dT > 0$, and that both become equal to zero as $T \rightarrow 0$.

At Berkeley, Lewis in 1923 and others since then have maintained that the curves presented by Richards very nearly imply the generalizations that were later embodied in the third law of thermodynamics. Richards' paper was a suggestive contribution to physical chemistry at the time and was the subject of van't Hoff's special memoir in the Boltzmann *Festschrift* of 1904. Haber was influenced by the discussions of both Richards and van't Hoff and was particularly impressed that for some reactions of low temperatures on Richards' graph the extrapolated values of ΔF and ΔH practically overlap. Nevertheless, in spite of the stimulus provided by Richards' work, it is doubtful that he saw the real implications for thermodynamics of vanishing ΔF and ΔH values at absolute zero. Nernst later rejected Richards' priority claim to the third law and implied that Richards' work did not reveal even an intimate acquaintance with the second law. Richards' own approach had been to focus on the work of compression of atoms as related to chemical affinity. He had pointed out that since the sign and magnitude of both free energy and heat content were dependent upon the sign and magnitude of the heat capacities during the reaction, then $d\Delta F/dT$ must have some fundamental connection with $d\Delta H/dT$; and this of course is right.

In his 1904 article in the Boltzmann *Festschrift*, van't Hoff examined the probable form of the free-energy curve in the vicinity of absolute zero. In seeking to give a clearer meaning to the work of Lewis and Richards, van't Hoff derived a somewhat different form of the Gibbs-Helmholtz equation. Van't Hoff asssumed that his constant of integration was small enough to be disregarded at low temperatures. Nernst later regarded van't Hoff's enunciation as a somewhat unsatisfactory hypothesis.

In 1905 Haber published a remarkable book, *Thermodynamik technischer Gasreaktionen*. Of considerable critical insight, this work provided an exhaustive critical survey of the thermodynamic data necessary for the calculation of the free-energy changes of the most important gas reactions. Before Haber, gaseous equilibria had been discussed on a mass-action basis. In 1888 Le Châtelier had indicated the significance of specific heats in calculating equilibria over a wide range of temperature, but chemists in general had not recognized the practical importance of his work. On the other hand, Haber was taken seriously because he had treated in detail some well-known and industrially important reactions. Where nineteenth-century chemists had been shy and uneasy about the entropy concept, Haber demonstrated how it could be employed in a practical way.

Most important, Haber attacked head-on the problem that Nernst solved—or thought he had solved—a year later. In 1904 Haber had concluded that if Kopp's law (the additivity of atomic heat capacities) holds for reactions between solids, the integration constant J and therefore the entropy change at absolute zero ΔS must have zero value. Lacking knowledge about specific heats at these low temperatures, Haber felt bound to leave open the possibility that the integration constant might have a small finite value owing to deviation from additivity for atomic heat capacities at low temperature. Thus Haber was led to discuss at some length what he called the "thermodynamically indeterminate constant." Nernst later postulated the validity of Kopp's law near absolute zero.

In his *Gasreaktionen* (1905) Haber took up the problem of the integration constant in relation to gas reactions. As Planck would later do, Haber explained that the nature of the integration constant should be expressible in terms of heat capacity and entropy constants characteristic of the components of the gaseous reaction. In particular he stressed the importance of knowing the variation of these properties with temperature in the vicinity of absolute zero. In the absence of experimental information he proceeded cautiously. Haber could not have formulated, or at least could not have announced in publication, the heat theorem of Nernst without a great deal more experimental support. Haber's contribution therefore seems to stand out above the other physical chemists that we have mentioned. Because of insufficient heat capacity information, he simply left open the question of what happens with gases at absolute zero. For gaseous reactions involving no change in the number of molecules he concluded, however, that the integration constant, if not equal to zero, was probably quite small. His experimental data could support this conclusion, and so he adopted it as a guide in setting up his free-energy equations. By the end of the year Nernst had announced his *Wärme-Theorem*, as he then called it. Haber immediately recognized its immense importance.

During the *Wintersemester* in Berlin, while lecturing on the thermodynamic treatment of chemical processes, Nernst had become more convinced than ever that chemical equilibrium could not be computed from thermal data alone using classical thermodynamics. What was needed, he believed, was a supplementary hypothesis to characterize the change of free energy in the vicinity of the absolute zero of temperature. Nernst perceived more clearly than any of the other investigators mentioned that the work of transition from one state of a system to another could not be calculated theoretically from energy differences using the laws of thermodynamics. What could be inferred thermodynamically was that ΔH and ΔF should be equal at absolute zero. This follows simply from the definition of free energy, namely, that $\Delta F = \Delta H -$

$T\Delta S$ for an isothermal process. There was still no way to determine ΔF from the knowledge of ΔH under any conceivable conditions. A careful study of the available thermochemical data had led Nernst to suspect that the thermal coefficients of heat content ($\delta\Delta H/\delta T$) and free energy ($\delta\Delta F/\delta T$) might approach each other asymptotically in the neighborhood of absolute zero. With as much experimental evidence—scanty at best—as he could muster, Nernst stoutheartedly and with characteristic boldness faced the Göttingen Academy of Sciences to present his *Wärme-Theorem*. He must have been rather confident about its significance, because the proof of the title page, as we learn from Simon's excellent survey of the third law, bore his scribble: "Bitte Revision! Im Ganzen 300 Separata!"

As revealed in the title, the subject of Nernst's paper was a single issue—the calculation of chemical equilibria from thermal measurements. Like the other investigators, Nernst began with the Gibbs-Helmholtz equation, but in the Gibbs form, presenting the following integrated form of the equation in terms of the equilibrium constant K rather than in terms of free energy:

$$lnK = -\Delta H_0/RT + \Sigma\alpha/R\ lnT + \Sigma\beta/2R\ T + \cdots + J.$$

Like Le Châtelier, Lewis, Richards, and Haber, Nernst noted that all of the quantities in this equation can be obtained by thermal measurements, with the exception of the integration constant J. Thus, although the form of Nernst's equation was somewhat different, there was nothing new in it; what is new in this paper is the line of reasoning that Nernst employed to interpret the physical meaning of the equation and to realize the wide chemical implications hidden beneath the constant of integration J. He correctly supposed that within the whole range of possible temperatures, absolute zero should have a very special thermodynamic significance.

The point of departure for Nernst was to recognize from the Gibbs-Helmholtz equation that if the temperature coefficient of total (internal) energy, dU/dT (dH/dT in the Helmholtz form), does not disappear at absolute zero, then the coefficient of free energy, dA/dT (dF/dT in the Helmholtz form), must be infinite; and then nothing can be revealed about the F curve at absolute zero. This implies a thermodynamically indeterminate J. Nernst found his way out of this unacceptable conclusion by shifting from the discussion of gases to that of solids. At this stage of his argument he ignored gaseous reactions, because he knew that for chemical reactions involving gases the sum of the specific heats of the reactants is in general not equal to the sum of the specific heats of the products, since the degrees of freedom may differ. By considering only condensed phases and by assuming that Kopp's law would apply, he could equate the specific heat sums of products and reactants. If so, dU/dT (or dH/dT) would vanish at absolute zero, and dA/dT (or dF/dT) would be finite. Nernst considered this extrapolation acceptable, since classical theory had indicated that with decreasing temperatures the molecular heats of compounds are equal to the sums of their atomic heats.

Accordingly, Nernst reasoned that A and U (or F and H) do not differ markedly at room temperature—nor even at somewhat higher temperatures; so the A (or F) curve is not likely to start at a steep angle at absolute zero. He recognized that the simplest assumption would be for the A and U (or F and H) curves to run together and become tangent to one another as $T \rightarrow 0$. Referring once again to the integrated form of the Gibbs-Helmholtz equation above, we see that

$$(\delta\Delta F/\delta T)_p = -\alpha lnT - \alpha - 2\beta T - 3/2\gamma T^2 - \cdots + J.$$

If we postulate, as Nernst did, that

$$\lim_{T\rightarrow 0} (\delta\Delta F/\delta T)_p = 0,\ \text{then}\ J = 0,\ \text{and}$$

$$\alpha = \lim_{T\rightarrow 0} \Delta Cp = \lim_{T\rightarrow 0} (\delta\Delta H/\delta T)_p = 0.$$

In other words, Nernst's hypothesis reveals that as $T\rightarrow 0$ not only does $\Delta F - \Delta H \rightarrow 0$, but $(\delta\Delta F/\delta T)_p \rightarrow 0$ and $(\delta\Delta H/\delta T)_p \rightarrow 0$. We note that there is no suggestion here that either ΔF or ΔH approaches zero as $T \rightarrow 0$, but only their difference; ΔF and ΔH remain finite quantities, positive or negative approaching each other asymptotically. Nernst recognized that $\lim_{T\rightarrow 0} (\delta\Delta F/\delta T)_p = 0$ was a necessary and sufficient condition for securing a definitive solution to the chemical equilibrium problem, whereas $\lim_{T\rightarrow 0} (\delta\Delta H/\delta T)_p = 0$ was necessary but not sufficient. The important point about Nernst's hypothesis is that since $\Delta F - \Delta H \rightarrow 0$, as $T \rightarrow 0$, the integration constant J is known and there is no longer a need to seek special techniques for determining ΔF, since this can now be done purely calorimetrically. The Nernst theorem simply showed how to compute the value of J as the algebraic sum of other empirically determined "chemical constants" for the constituents of a chemical reaction. The experimentally straightforward but nevertheless time-consuming procedures for carrying out calorimetric measurements at low temperatures later yielded to the perfection of spectroscopic analyses that permits the entropy constants to be calculated with much less effort.

Having postulated his heat theorem for condensed phases in four introductory pages, Nernst turned directly to the question of the determination of the integration constant for gaseous systems. Crediting Le Châtelier and Haber for having recognized the fundamental problem, he proceeded without fanfare or apologies to treat gaseous reactions at sufficiently low temperatures so that all of the constituents could be considered to be in the condensed state. With the help of the van der Waals equation he was able to develop approximation formulas to calculate the free energy for gaseous reactions in the condensed state.

For Nernst this was just the beginning of experimental investigations connected with the theorem, since he sought in various ways to support its theoretical soundness and practical utility. By 1907 his position toward the theorem was essentially fixed—as can be seen from his Silliman lectures. His thesis was that the heat theorem revealed new truths about the relation between "chemical energy and heat," but he suggested that it also would prove useful beyond the solution of problems of chemical equilibrium and spontaneity. He emphasized that a great deal of experimentation would be necessary in order to decide whether the theorem represents only an approximate principle or an exact law of nature similar to the first and second laws. Irrespective of the theoretical status of the theorem, Nernst maintained that approximations based upon the theorem would provide answers to the all-important question about the driving force for chemical reactivity. In this he was right. Still, he suggested that chemists work over the whole field of thermochemistry from the new point of view and, namely, that they undertake an exhaustive experimental analysis of heats of reaction, specific heats, and temperature coefficients over the entire experimentally feasible temperature range.

Thus, Nernst was led to declare that all that thermodynamics can contribute to chemistry is already implied in the Gibbs-Helmholtz equation, provided that the heat theorem be used to furnish an interpretation of the mode of behavior of nature in the vicinity of the absolute zero of temperature. His line of argument was that *F* cannot in general be equal to *H*, because this is contrary to the results of experiment; that the correct relation between *F* and *H* cannot be found from the first and second laws, because the integrated form of the second-law statement contains a constant that was thus far undetermined; and that therefore if a new law of thermodynamics was to be found, it would have to account for the integration constant as the only remaining thermodynamic problem.

The clue to the theoretical treatment of the problem for Nernst lay in the approximate success at ordinary temperatures of the Thomsen-Berthelot principle. The solution that he provided was to assume that this principle was not only approximate at ordinary temperatures but also in the neighborhood of absolute zero, so that *F* and *H*, plotted as a function of temperature, approach each other asymptotically with contrary slopes. Nernst concluded that this theorem would settle the question by making it possible to predict the thermodynamic stability of any stoichiometrically correct chemical equation, based on calculations derived from thermal data alone.

Nernst's conception of the overall significance of the heat theorem can be gathered from his own statement at the end of the 1906 paper:

> If we now bring the result of our considerations together succinctly, we can say that the goal of thermochemistry, namely the calculation of chemical equilibrium from heat of reaction measurements, seems attainable, provided we seek the aid of a new hypothesis, according to which the curves of free energy and the total energy of chemical reactions between only solid or fluid bodies, become tangent to one another at absolute zero. Even if an exact test of the formulae obtained by the help of the above hypothesis about specific heats is not feasible at low temperatures, the approximation formulae developed in this work are seen to be in agreement with experience. The relationships between heat and chemical affinity seem substantially to have been clarified.

Although unaware of it when he announced his heat theorem, Nernst had laid the foundation for the connection between chemical thermodynamics and the quantum theory set out five years earlier by Planck.

For over a decade after the 1906 paper appeared, virtually the entire facilities and personnel of the physical chemistry institute of the University of Berlin were organized into a huge work program to experimentally test Nernst's *Wärme-Theorem*. The immediate consequence of Nernst's new idea was that radically different thermochemical techniques were put into practice to elucidate chemical equilibrium. These involved the determination at very low temperatures (in fact as close to zero as possible) of the specific heats and thermal coefficients of specific heats for the constituents of the chemical reactions under investigation. In a series of seven papers published between 1910 and 1914 Nernst and his co-workers (F. Koref, F. A. Lindemann, and F. Schwers) presented impressive experimental evidence to support Nernst's theorem based on the electrical measurement of electrically induced temperature changes. The rigorous test of the validity of the theorem was soon seen to be an enormously challenging experimental undertaking. To approach this objective the Nernst group constructed ingenious electrical and thermal devices, developed a vacuum calorimeter, and built a small hydrogen liquefier (1911) to achieve temperatures low enough to be able to extrapolate safely to absolute zero. Nernst tackled all these problems imaginatively and successfully, and step by step came increasingly to believe that his hypotheses should be elevated to the rank of a bona fide law of thermodynamics.

Not long after the Nernst heat theorem had become established in its role as a powerful method to predict chemical equilibrium—thus to indicate which reactions were chemically feasible—it was seen that quantum statistical calculations of entropy constants for gases became accessible. As a follow-up to his four classic papers of 1905, Einstein turned to the consideration of radiation theory and treated the specific heat of solids as a problem in quantum mechanics. In Nernst's 1906 paper there had been no mention of Planck's quantum ideas of 1900, nor had he taken seriously Einstein's suggestion in 1907 that quantum theory predicted vanishing heat capacities for solids at absolute zero. During Nernst's visit to Einstein in Zurich in March 1910, the two men discussed the extent of agreement between the Einstein theory of specific heats and the experiments being conducted at Nernst's institute in Berlin. On 13 May 1911 Einstein wrote Michele Besso that his theory of specific heats had celebrated true triumphs, since Nernst had experimentally confirmed virtually everything that his theory predicted. Apparently both Nernst and Einstein gloried in the turn of events: Nernst, realizing that his precious heat theorem was linked with and in agreement with predictions from quantum theory; and Einstein, that his revolutionary quantum conceptions were receiving experimental backing from Nernst's work. Working with Lindemann in 1911, Nernst showed that Einstein's specific-heat equation was in agreement with his data, except for certain systematic deviations. They also showed that a revised Einstein formula was in still closer agreement with the experimental information except at the very lowest temperatures.

Alert to the significance of the heat theorem for the establishment of the quantum theory, Arnold Sommerfeld, in a lecture at the eighty-third *Naturforscherversammlung* in Karlsruhe in 1911, remarked that the work on blackbody radiation carried out during the first decade of the century at the Physikalisch-technische Reichsanstalt constituted one of the pillars of the quantum theory and added: "Perhaps to be estimated as of equal merit is the work of the Nernst Institute which, in the systematic measurement of specific heats, has furnished a second no less powerful pillar to support the quantum theory."

The experimental evidence for the quantum theory was one of the central topics of discussion at the first Solvay Congress in Brussels in late October and early November 1911. Nernst had taken the initiative for setting up and organizing the sessions and seeing to it that the leading physicists would be there; the twenty-two participants included Lorentz, Planck, Rubens, Sommerfeld, Wien, Jeans, Einstein, and Lindemann. Einstein reported on his specific heat theory in relation to the Nernst and Lindemann empirical formula for the thermal energy of solids—just before the problem was solved theoretically by Max Born and Theodore von Kármán (1912). About the same time Peter Debye, who had been Sommerfeld's assistant and was then lecturing on thermodynamics in Zurich, independently presented his theory of specific heat at the Physikalische Gesellschaft in Bern (1912). Born and Kármán had reached their results by a different route than Debye; both had built on the foundations laid by Einstein. For some time these formulations proved satisfactory. Of unique significance was Debye's deviation of the famous T^3 law that gives for the lowest temperatures the proportionality between atomic heats and the third power of the temperature—a relationship that was seen to fit the facts very well.

In 1910 Planck markedly enhanced the usefulness of Nernst's theorem by putting it into the form in which it has most frequently been given ever since. In his paper he promoted the idea that not only the entropy differences during the alterations of a system in the vicinity of absolute zero tend to zero, but that the entropy differences of all of the constituents of the system become zero. That is, given that $\Delta S = S_T - S_0 = \int_0^T C dT/T$, a finite value of S is possible only if the specific heats vanish at absolute zero, since otherwise the integral at the lowest limit becomes infinite. The Planck formulation, however, tells more than that, namely, that the individual entropies at absolute zero not only are finite but that they are zero. Planck stressed that the Nernst theorem was a major extension of the second law because it permits the calculation of absolute entropies. He also pointed out—but in rather cautiously worded statements—that the entropy form of the Nernst theorem necessarily means that the third law, like the second, is intrinsically connected with probability, atomistics, and statistical implications.

Planck's formulation of the heat theorem in terms of vanishing entropies at absolute zero did not appeal to Nernst, who considered it both inappropriate and intuitively too unclear. The paradox of Nernst's position was that his own formulation, which made no reference to entropies, was far more cumbersome and theoretically less elegant. Unquestionably Nernst's peculiarity on this issue was frowned upon by most members of the scientific community. In principle the application of quantum theory to specific heat considerations for solids could have led before 1905 to the conclusion that $\lim_{T \to 0} \Delta S = 0$. Such a formulation might have elicited more conviction from physicists than Nernst's heat theorem, because it follows theoretically from quantum mechanics. By contrast, the Nernst heat theorem could not be deduced from the other laws of thermodynamics, and only experimental evidence could serve to establish its correctness. Nernst and his students eventually amassed sufficient empirical evidence to reverse the attitude of scientists toward one of acceptance of Nernst's heat theorem.

In 1912 Nernst stated his heat theorem in terms of the theoretically decisive principle of the unattainability of absolute zero. According to this principle, it is impossible to build a caloric machine that will allow a substance to be cooled to absolute zero; and from this negative assertion Nernst concluded that the thermal coefficients of all the physical properties of solid bodies would vanish in the approach to absolute zero. The properties of bodies that subsequently were investigated as tests of the third law included direct studies of thermal expansion, surface tension, magnetic and dielectric polarization, and thermoelectric phenomena. More indirect and less unambiguous in an interpretive sense were the studies on fluidity, solutions, mixed crystals, frozen-in phases, crystallographic transformations, and chemical reactions—all designed to demonstrate the disappearance of physical properties as $T \to 0$, that is, to show, in Planck's formulation, that $\lim_{T \to 0} \Delta S = 0$.

Nernst's particular way of enunciating the principle of the unattainability of absolute zero was questioned by physicists, even while the principle itself was seen to be important. Rather than providing a proof, Nernst had demonstrated the consistency of the principle with the impossibility of a *perpetuum mobile*. In order to avoid presenting the idea in terms of entropy-temperature graphs, Nernst presented his proof in the form of a Carnot cycle. He showed that a *perpetuum mobile* of the second type results from taking absolute zero as the lowest temperature of the cycle, thus demonstrating that the attainment of absolute zero is theoretically impossible. The way in which the unattainability of absolute zero was deduced from arguments based on the disappearance of specific heats as $T \to 0$ was rather obscure. In 1913 his proof was challenged by Einstein, who reasoned that *Gedankenexperimente* should be possible in principle even if not in practice and that this was not the case for Nernst's formulation. Nernst stuck to his guns, and so did Einstein.

A long and involved discussion followed about the null-point entropy of frozen-in phases. The best contemporary account of this work was given in 1930 by Franz E. Simon, one of Nernst's most talented students and colleagues. A more precise formulation of Nernst's principle was indicated by postulating that the entropies of chemical reaction between pure condensed phases in internal equilibrium vanish as $T \to 0$. The difficulty that emerged from this alternative formulation was one of defining the criteria for equilibrium. Certain condensed phases, like glasses, as well as mixed solid phases of crystals and solutions, are not at equilibrium. Besides, a system may be at equilibrium with respect to atomic or molecular orientations but not with respect to electron or nuclear orientations. As long as residual questions remain about the deep structure of matter there would be no reason to assume that rock-bottom equilibrium can be reached regardless of how close a system approaches absolute zero.

One of the most serious difficulties confronting Nernst was that classical kinetic gas theory predicts that the heat capacity at constant volume C_v does not tend to zero as $T \to 0$, as the heat theorem demands, but reaches limiting values of $3/2\ R$ for monatomic gases, and $5/2\ R$ for diatomic gases, and so forth. Thus, even if the Nernst theorem had given a fairly acceptable interpretation of the heat capacities of condensed phases at low temperatures, it seemed likely that gases might have to be excluded from the theorem. Of course this would have been very detrimental to Nernst's attempt to elevate the heat theorem to the status of a general law of thermodynamics. Fortunately for Nernst, the developments in quantum mechanics just prior to World War I gave Sackur (1912), Tetrode (1912), and Stern (1913) the means to calculate directly the entropy of a monatomic gas, and later, the entropy of more complex molecules. The calculations agreed tolerably well with the experimental results, showing that theory predicted and experiment confirmed the falling-off of heat capacities at low temperatures.

Nernst interpreted these developments as a step toward confirming his theorem. He proceeded heroically, using some quite primitive arguments and despite the incredulity of many scientists, to postulate (1914) a state of degeneracy (*Entartung*) for gases. Subsequent advances in quantum mechanics showed Nernst to be on the right track. The Bose-Einstein and Fermi-Dirac statistics both confirmed the gas degeneracy idea. The progress of quantum theory notably justified Nernst's idea, if not the reasons for enunciating it, when it was shown that electrons in metals present an example of degeneracy at much higher temperatures and that the same principle suffices to account for the interior physical conditions of stars at very high temperatures and pressures.

With the outbreak of World War I, Nernst's academic pursuits virtually came to a halt as he was drawn into military administration, chemical gas warfare, and service as automobile chauffeur for the German army on the move from Belgium to France. As is evident from the preface to Nernst's 1918 monograph, *Die theoretischen und experimentellen Grundlagen des neuen Wärmesatzes*, which was written during a time of "*Trübsal und Not*," Nernst was able to immerse himself further in the new theoretical physics, namely, quantum mechanics, and to reflect on its meaning and implications for his beloved and not yet controversial *Wärmesatz*. In this volume he presented in a most comprehensive way his mature ideas on chemical thermodynamics.

Until his retirement in 1934 Nernst was again actively involved with the pursuit of physical chemistry at Berlin, but he now took up a number of new topics. With the accumulation of experimental data in the 1920's, Nernst's heat theorem, which had enjoyed the successes of early quantum theory, encountered serious difficulties as experimental anomalies with condensed systems appeared that could not be squared with the general theorem. In testing the validity of the third law these anomalies had to be reckoned with. Most of them were identified and examined in terms of their quantum origins. In time they came to be recognized as the quantum effects resulting from the liberation of energy that takes place during the degeneration of various internal degrees of freedom. At lower temperatures these degenerations were seen to lead to greater order or internal equilibrium with respect to each particular quantum effect. Nernst took relatively little part in the discussion of these problems. He felt that his theorem should apply in a straightforward way to all systems using a thermodynamic mode of reasoning and without appealing to statistical considerations. He was convinced that further experimental and theoretical research would confirm his theorem as a general law.

No person contributed more to providing a theoretically satisfying restatement of the third law than Franz Simon in the 1920's and 1930's. He had steadily gained a wide reputation as an authority in the newly developing field of cryogenics, on the strength of thermodynamic and statistical reasoning that focused on the meaning of the "internal equilibrium" states of systems. Simon had received his doctorate under Nernst in 1921 with a dissertation on the study of specific heats carried out down to the temperature of liquid hydrogen. This early research on undercooled liquids, glasses, and crystalline substances had revealed a λ-type specific heat anomaly for ammonium chloride. In 1929, using liquid helium as a coolant, he discovered the anomalous specific heat of solid orthohydrogen. His single-stroke adiabatic method for liquefying helium was crucial for the development of low-temperature research in general and led to his own pioneering work on specific heats, magnetic cooling, and nuclear cooling and orientation below 1°K. He perfected a great variety of small-scale experiments and demonstrated the effectiveness of working in vessels with "mathematically thin" walls. Toward the end of his career, while at Oxford, Simon worked on nuclear orientation and cooling, utilizing the magnetic moments of atomic nuclei at very low temperatures in a way that is analogous to the use at higher temperatures of paramagnetic moments to achieve cooling by adiabatic demagnetization. Shortly before his death Simon and his colleagues (especially Kurti) attained temperatures of about 2×10^{-5}°K.

From the very start of Simon's interest in the third law in 1920, his central concern was to test the general validity of Nernst's theorem. The desire of some investigators in the 1930's to restrict the heat theorem to pure crystals was unacceptable to Simon because the theorem could then no longer be considered as a general law. Simon, like Nernst, wanted to demonstrate the authenticity of the heat theorem for all processes to which valid thermodynamic reasoning might be applied. The direction that this concern took was to demonstrate from experimental findings on magnetic properties, superconductivity, and the behavior of liquid helium that the apparent anomalies to the third law could be explained either on the basis of the incorrect extrapolation of specific heats to 0°K. or else to the misapplication of thermodynamics to systems that were not in internal equilibrium. Violations of the third law for chemically homogeneous systems, Simon conjectured, could always be attributed to specific heat anomalies at very low temperatures. For example, he felt that there was no valid reason to assume that nuclear spin systems might not lose their entropy at sufficiently low temperatures. It was basically up to experimentalists to devise sufficiently ingenious analytical techniques to clear up such anomalies. Of course Simon realized that it might not be feasible in practice to go reversibly from a given initial state to a final state within a reasonable length of time, especially with certain chemical reactions for which the chemical kinetics and rates are not favorable. Still, his objective was to explain these anomalies.

It was common knowledge in Berlin that Simon's "anomaly consciousness" provided the effective stimulus for the high level of achievement of the many *Doktorands* working in his impressive low-temperature laboratory. Simon's fundamental argument was that the low-temperature specific heat anomalies that showed up for amorphous solids, nonhomogeneous mixed crystals, solid solutions, and glasses could all be accounted for by recognizing that these systems were not in internal equilibrium but were removed more or less from their most probable entropy states by being "frozen-in." Thus, unwarranted thermodynamic reasoning was to be avoided for such nonequilibrium systems. Simon supported this view with his own investigations of glasses, showing with X-ray crystal analyses that the state of disorder persists down to the lowest experimentally achievable temperatures. He was able to explain with a clever argument that in such cases, where experimental data on glasses were cited to show that the third law had been violated, the necessary consequence was that it would be possible to reach absolute zero with such substances. Thus the criticism led to a *reductio ad impossibile*. For, wherever entropy differences exist between two states of a system, a reversible adiabatic process from the lower entropy state should lead to absolute zero, provided that the transition be allowed to proceed at a temperature where the initial entropy state is equal to or smaller than the final entropy state at absolute zero. Thus, for systems genuinely in internal equilibrium, nonvanishing entropies (that is, contradictions to the third law) would provide the means for reaching absolute zero.

In 1927 Simon proposed a new formulation of the third law: "At absolute zero the entropy differences disappear between those states of a system between which reversible transitions are possible at least in principle." In 1930 he expressed similar ideas in another way, namely, that the entropy of all factors within a system that are at equilibrium disappears at absolute zero. In a more acceptable reformulation in 1937 Simon stressed that the entropy contribution of each factor within a system in internal equilibrium becomes zero at absolute zero. This was a rather safe formulation of the third law, seeing that it should be possible in principle to discover, at still lower temperatures and by refinement of analytical techniques, additional hitherto unrecognized nonequilibrium factors (frozen-in states) responsible for apparent anomalies. Simon's formulation was based on the assumption that the lowest energy states of any system are not degenerate. He argued that in that case the apparent contradictions result not from degenerate ground states but rather from a frozen-in disorder that would disappear if the system could somehow be melted out or moved toward greater order catalytically.

The cases of frozen-in disorder that were specially treated by Simon, and that extend beyond the traditional analysis of crystal lattice disorder, concerned foremost disorder in the distribution of isotopes and of magnetic effects at very low temperatures. Examples of the latter were the electron spin disorder of certain paramagnetic salts and nuclear spin systems that would reach internal equilibrium only in the temperature region of 10^{-3} to 10^{-6}°K. Simon argued that there are any number of subsystems that can be studied in connection with the behavior of matter; one subsystem may be in internal equilibrium where another is not. Thus to speak of complete internal equilibrium was to miss the point, because then there would be practically no substance to which thermodynamics would apply. Simon's third-law statement of 1937 took the simple form: The contribution to the entropy of each

NERNST

subsystem that is in internal equilibrium disappears at absolute zero.

In his van der Waals centenary lecture in 1937 Simon summarized his views succinctly:

> We can state that the present experimental evidence indicates the general validity of Nernst's theorem as a law of *thermodynamics*. The possibility that some future experiment may not be in agreement with the theorem obviously cannot be excluded, but unless there is some reason from a theoretical point of view to expect such a result, to anticipate it is mere speculation. So far, no theoretical argument against the theorem exists. On the contrary the assumption, that the state of lowest energy is that of entropy zero, i.e. that of perfect order is theoretically very plausible. I cannot see, therefore, any justification for withholding from Nernst's law the status of a general law of thermodynamics [F. Simon, "On the Third Law of Thermodynamics," in *Physica*, **4**, no. 10 (23 Nov. 1937), 1096].

Over a period of nine years in Berlin, working close to Nernst, Simon published some fifty papers relating to low-temperature studies. Although they contributed to putting Nernst's law on much firmer theoretical footing, Nernst marshaled the sharpest opposition to the new restatements. Nernst wanted no riders attached to his theorem and told Simon that if this should prove unavoidable, he would be prepared to give up his theorem as a general law of nature. Nernst's mind was geared to specific heats and to the expression of the theorem in terms of the capacity factor for isothermally unavailable energy, namely the temperature coefficient for free-energy changes. He wanted to have nothing to do with "entropies," and in his papers he always preferred thermodynamic cycles to entropy diagrams.

It is a paradox that Simon, proceeding from the deep interests in the Nernst theorem that came to dominate his life, subsequently founded a distinguished low-temperature school to demonstrate the theoretical validity of the third law—only to discover that its author did not approve of the conclusions that the investigations led to. In fact, after almost two decades of work, Simon had come to exactly the conclusion that had been Nernst's deep expectation from the start, namely, that the heat theorem be considered valid as a law of thermodynamics. Simon's conception, however, of what constitutes validity was intellectually more appealing than Nernst's had ever been.

The specific problems that claimed Simon's attention, the search for ways to verify the heat theorem with the immense amount of information coming from low-temperature laboratories and the explanation of the numerous anomalies being discovered, are all representative of the general state of dissatisfaction with and confusion about the third law that reigned during the 1920's and 1930's. The severity of the situation is revealed by a statement made in 1932 by Fowler and Sterne:

> We reach therefore the rather ruthless conclusion that *Nernst's Heat Theorem strictly applied may or may not be true, but is always irrelevant and useless—applied to "ideal solid states" at the absolute zero which are physically useful concepts the theorem though often true is sometimes false, and failing in generality must be rejected altogether*. It is no disparagement to Nernst's great idea that it proves ultimately to be of limited generality. The part that it has played in stimulating a deeper understanding of all these constants, and its reaction on the development of the quantum theory itself cannot be overrated. But its usefulness is past and it should now be eliminated [*Review of Modern Physics*, **4** (1932), 707].

By the outbreak of World War II, Simon's views were beginning to find general acceptance. Besides, the need for clarification about various aspects of quantum mechanics and nuclear physics became far more pressing than thermodynamics. Formulated initially to explain gas reactions, by 1940 the third law had outlived most of its intended function, and, in any case, the entropies were then beginning to be treated quantum mechanically and computed largely from spectroscopic data. It may be conjectured, therefore, that all of these matters would have followed, without Nernst's work, from the logical development of quantum mechanics. In his Guthrie lecture of 1956 Simon responded to this by saying:

> Of course, one could say the same of the Second Law, namely that statistical theory would have yielded all the information pronounced first by the empirical statements of Carnot and Clausius. The predictions of the Third Law would eventually have been produced by quantum theory, but I have to remind you only of the fact that we have so far no full quantum statistical explanation of the Third Law to show you that it would have come very much later. Also we must not forget that the Third Law was an extremely strong stimulus to the development of quantum theory and it is perhaps not quite idle to consider how the development would have taken place had Nernst's deliberations started ten years earlier as they might well have done. Perhaps the quantum of action would have been discovered as a consequence of the disappearing specific heats rather than from the ultra-violet catastrophe of the radiation laws.
>
> Things do not always develop in the most direct or logical way; they depend on many chance observa-

NERNST

> tions, on personalities and sometimes even on economic needs. . . . I hope I have shown you how a great mind tackled a very obscure situation, at first only in order to elucidate some relatively narrow field, and how later, as the result of all the interaction and cross-fertilization with other fields, the Third Law emerged in all its generality, as we know it to-day [*Year Book of the Physical Society* (1956), 21].

Irrespective of the status and intrinsic long-range merits of the third law, an impressive number of fringe benefits resulted from the stimuli coming from the Nernst syndrome. Over a period of about three decades a prodigious amount of good low-temperature research had been carried out in highly specialized centers for cryogenic investigation, namely, Berlin, Cracow, Paris, London, Toronto, Leiden, Oxford, and Berkeley. In the process of testing the Nernst theorem, important advances were achieved, especially in the study of the low-temperature properties of matter, the physics and chemistry of solid-state phenomena, gas degeneracy, corresponding states, zero-point energies, λ-point phenomena, magnetic cooling, superconductivity, and superfluidity. More immediately important than the gradual opening of these domains of experimental inquiry was the immediate support that the heat theorem gave to quantum mechanics in the early days of its development, before the results of spectroscopy and the Bohr theory of atomic structure were available.

Over the years Nernst became increasingly possessive about his role in the genesis of the third law. In Berlin, within the shadow of the colossal achievements of Planck and Einstein, Nernst—perhaps understandably—became rather defensive about the merits of his own contributions. Several students of Nernst tell the story—the versions differ somewhat, but not their essential message—that in his lectures Nernst liked to refer to the first law of thermodynamics as having been discovered independently by three investigators (presumably Mayer, Joule, and Helmholtz) and the second law, by two independent investigators (Carnot and Clausius). Concerning the third law of thermodynamics Nernst would say, "Well, this I have just done by myself." According to another version of this anecdote, which also seems to suit his character, Nernst added that it therefore should be obvious that there never could be a fourth law of thermodynamics. In any case, we know that Nernst became rather adamant in insisting that his heat theorem (*mein Wärmesatz*, as he called it) was more than a way of calculating chemical equilibria; it was in fact a law of thermodynamics on a par with the first and second laws. Indeed, after the theoretical low-temperature quantum mechanical interpretation of specific heats by Planck and Einstein, Nernst believed that the collapse of his heat theorem, if possible, would necessarily have to be accompanied by the simultaneous rejection of Planck's and Einstein's views.

Nernst's participation in the extensions, criticisms, and reformulations of his heat theorem was rather that of a spectator; it was his students who were involved in new contributions. Nernst was totally absorbed in other scientific matters. Thus while low-temperature investigations continued to command the attention of experimentalists and theoreticians alike, he explored new leads in photochemistry, chemical kinetics, and chemical astrophysics. This phase of Nernst's career coincides with the Weimar Republic—an incredibly fertile period for the physical sciences. Berlin was an international whirlpool of scientific activity and a hotbed for the germination of radically new perspectives in physics and chemistry. Thus, the more narrowly focused thermodynamic approach to physical chemistry that had marked the approach of Ostwald in Leipzig, and to some extent even the prewar focus in Berlin, was enlarged and upgraded to accommodate new advances in chemical-reaction kinetics, photochemistry, quantum mechanics, spectroscopy, nuclear physics, and radiochemistry.

When Ostwald resigned from his chair in physical chemistry at the University of Leipzig in 1905 to pursue his literary and philosophical predilections, the center for physical chemistry in Germany shifted for a decade to Nernst and his group of collaborators in Berlin. After World War I physical chemistry was pursued in Berlin as vigorously as before, but by then other university centers had also been launched or reinforced. The most consolidated activity in the physical sciences in the 1920's, however, was at the University of Berlin, the Technische Universität, the Kaiser Wilhelm Institut, and the Physikalisch-technische Reichsanstalt—all located within a few miles of each other. It was most fortunate that such government-sponsored institutions had broad-minded and enlightened officials in charge of the funding.

From 1919 to 1933 the most prominent physical chemists included, besides Nernst, whose commanding position was unassailable, four outstanding experimentalists: Fritz Haber, director of the Kaiser Wilhelm Institut for Physical Chemistry and Electrochemistry at Dahlem; Max Bodenstein, a superb experimentalist, who became Nernst's successor in the chair of physical chemistry at the University of Berlin in 1922; Max Volmer, head of the institute for physical chemistry at the Technische Universität; and K. F. Bonhoeffer, Nernst's student, who joined Haber's institute in 1922 and who, more than any of the others mentioned here, took full advantage of the new methods and advances made available in the 1920's. Bonhoeffer later followed Le Blanc in Leipzig.

The whir of excitement generated in seminars, colloquiums, and laboratories by the brilliant constellation of physicists and chemists in Berlin was contagious. The wave of activity was closely related to Nernst's own interests, both in regard to testing the validity and exploring the implications of his heat theorem and in other developments relevant to the new problems in physical chemistry that he had undertaken. Planck and Einstein were working on quantum mechanics and thermodynamics; Laue, on crystal interference; Erwin Schrödinger and Fritz London, on the applications of wave mechanics; Paschen and Ladenburg, on spectroscopy; Otto Warburg, on photochemistry; Otto Hahn, on radiochemistry; Gustav Hertz, on isotype separation by gaseous diffusion; Herbert Freundlich, on capillary and colloid phenomena; and Michael Polanyi, on chemical kinetics. Because of his prominence in the Berlin scientific community, Nernst was at one time or another in contact with all of these investigators.

Like virtually all of his colleagues named here, Nernst had come from a socially prestigious and quite wealthy family. Money was therefore seldom a problem. On the other hand, overwork and high performance expectations did introduce considerable anxiety and discomfort. Among the scientifically elite, solitary retreats from Berlin *zur Erholung* at health resorts and other oases of tranquillity and relaxation became standard fare.

When he returned to his physical chemistry institute in Berlin, Nernst *est revenu à ses premières amours*, as Bodenstein put it. There he was to become entrenched once again in more decisively chemical topics than third-law investigations. In the first instance he worked on photochemistry. According to Einstein's photochemical equivalence law of 1912, a molecule that absorbs one energy quantum of radiation $h\nu$ in a primary photochemical process can initiate secondary chemical reactions no longer dependent on the illumination. This law seemed to hold in a number of instances, but it had been demonstrated that for the formation of HCl from H_2 and Cl_2 (*die boshafte Chlorknallgasreaktion*) at least 10^6 molecules were formed per quantum in place of two as might be expected from the equation $Cl_2 + h\nu = 2Cl$. In 1918 Nernst suggested a simple and ingenious solution to this problem—the idea of a "chain reaction." In this case the suggested process was

$$Cl_2 + h\nu = 2Cl$$
$$Cl + H_2 = HCl + H$$
$$H + Cl_2 = HCl + Cl, \text{ and so forth.}$$

Nernst's theory was fully justified in 1925 by James Franck's calculations of the energy of dissociation of Cl_2 based on absorption-spectrum studies. Nernst's co-workers, mainly John Eggert, Walter Noddack, Friedrich Bonhoeffer, and Max Bodenstein, subsequently set the high standards in photochemical investigations that came to be so essential in the field of chemical kinetics. It was shown that chain reactions of the type suggested by Nernst can be initiated by means other than light, for example, alpha particle bombardment, sparks, and the introduction of sodium vapor. It was also demonstrated that the chain process is terminated only by the removal of active molecules by wall collision or reaction with other molecules.

Nernst left his university post in 1922 to succeed Emil Warburg as president of the Physikalisch-technische Reichsanstalt; but his ambitious plans for major organizational changes were wrecked by severe inflation. After two years of frustration, Nernst returned in 1924 to the university to fill the vacant post of professor of physics that had been created by Heinrich Rubens' death. Until his retirement in 1934, Nernst continued to devote his efforts to physical chemistry and to serve as director of the physical laboratory.

In the late 1920's Nernst turned his attention to cosmological questions. He was inspired in part by discussions of these matters with Einstein, but the principal motivation for this late-in-life escapade into physicochemical astrophysics was a basic uneasiness about the idea that the universe should have a passing existence, its so-called heat-death (*Wärmetod*) being predicted from the second law. The problem essentially was: Why was it that the degradation of energy had not yet reached its maximum? (We should say: Why had the entropy not yet reached a maximum?) Nernst, however, never put it that way; he scrupulously avoided the term "entropy," although, of course, not its mathematical equivalent, namely, the capacity factor for isothermally unavailable energy, dQ_{rev}/T. In an attempt to explain away the *Wärmetod* of the universe, Nernst explored various hypotheses. He argued that cosmological questions about the beginning and the end of the universe were scientifically meaningless. By postulating fluctuations in the zero-point energy in space, or in the ether, as he preferred to put it, he reasoned that a steady-state theory might find scientific support in terms of the balance between energy degradation and energy creation as seen in the appearance of new stars and novae, and in the stages of stellar classifications that range from new to steady-state systems. As a source of energy for these new creations in the universe Nernst postulated what he called the null-point energy of the ether (*die Nullpunktsenergie des Lichtäthers*).

Nernst had talked about such matters as early as 1912, and in 1921 he had written a small volume on the subject, *Das Weltgebäude im Lichte der neueren Forschung*. He continued to explore related ideas in a number of papers on chemical astrophysics (1928), the specific heat of gases at stellar temperatures and pressures (1929), the physics of stellar evolution (1935), and the interstellar radiation temperature (1938). He endeavored—somewhat naively, it appears in retrospect—to provide a thermodynamic synthesis of the observed stellar sequence, the red shift, and so on; but his ideas on a steady-state universe and nuclear processes as the source of stellar energy were later seen to be not completely outlandish.

In all of his work Nernst displayed a cagey, dubious attitude toward the long-range value of abstract, theoretical premises. Committed to "hypotheses" and "theorems" that could be shown to be fruitful in leading to new discoveries, he exhibited little concern for the search—Einstein or Planck fashion—for answers to broad and general philosophical questions. For example, he classed Einstein's theory of Brownian motion above that of relativity because of the former's real physical content. Nernst's dominant inclination was toward experimental investigations and phenomena that could be visualized. Interested in reliable experimental results, he was never fussy about how clumsy or makeshift his apparatus looked. Sometimes he built his own equipment (transformers, pressure and temperature regulators, measuring devices, and even a microbalance); and almost all of the apparatus in Nernst's laboratory was constructed on the premises. The instruments were made as small as possible and were assembled with the minimum waste of materials. In the use of materials or energy Nernst was inordinately frugal, and he looked with contempt upon the misuse of natural resources.

Pure and applied research were identical for Nernst, because theoretical questions were formulated within the context of ongoing experimental investigations. Upon examining his papers, one gets the impression that the areas of pure scientific activity that Nernst investigated grew out of his intense preoccupation with challenging experimental and instrumental situations. It also is singularly evident that Nernst had a deep and sincere interest in the technical applications of physics and chemistry. He was passionately enthusiastic about motor cars, tried out one after another when they came on the market, and investigated various combustible fuels.

On Monte Generoso, Nernst constructed a device for the study of the conversion of atmospheric electricity into useable energy. His experimental investigations (1899–1900) on the electrolytic conduction of solids at very high temperatures were put to use in 1904 in the Nernst lamp, which replaced the older, more fragile, and less efficient carbon filaments by rare-earth oxide wires. His lamp netted him considerable income (he sold the patents outright), but it was relatively short-lived because of the introduction of the tungsten lamp. Several decades later, theoretical interest in similar studies was reinstated in connection with the investigation of the mechanism of ionic conduction in semiconductors.

Nernst's irrepressible, zealous openness to scientific discoveries and their technological application is visible in the way that he followed the new studies of radiation, quantum chemistry, radioactivity, astrophysics, and cosmic rays. On the whole his flair for the fundamentally novel was sound; his suggestions were occasionally based on inadequate comprehension of the situation or overly ambitious speculations. For example, in 1922 Nernst examined the scientifically plausible but musically shallow idea that the concert grand might be replaced with a small piano that was magnetically controlled and furnished with loudspeaker amplification. Nernst called his instrument the *Neo-Bechstein-flügel*. The balance of harmonics was quite unacceptable. Nernst's expression for such extrascientific activities was *physique amusante*.

The story of how Nernst exercised his entrepreneurial talents within the Berlin scientific community of Weimar Germany has been depicted recently with sympathetic deference, vividness of detail, and personal involvement, by Kurt Mendelsohn. Nernst played the conspicuous role of organizer for the first Solvay Conference of 1911. He was the effective promoter in the creation of a post for Einstein at the Berlin Academy of Sciences. He was the prime mover for the establishment of the Kaiser Wilhelm Institut. He was also a founder of the Deutsche elektrochemische Gesellschaft (later called Deutsche Bunsengesellschaft) and for several years edited its *Zeitschrift für Elektrochemie*. With J. A. W. Borchers, beginning in 1895, Nernst edited the *Jahrbuch für Elektrochemie*.

In 1913 Planck and Nernst traveled to Zurich to entice Einstein, then thirty-four years of age, to join them in Berlin. Although his light-quantum theory had not yet found favor with them and the general theory of relativity was still in process of formulation, they were able to offer Einstein a position at the Royal Prussian Academy of Sciences, the directorship of research at the Kaiser Wilhelm Institut, the rank of professor, a special salary, and teaching at his option. The normal requirement for renewal of his German nationality was waived. Einstein accepted the offer.

Some of Nernst's closest colleagues in Berlin, notably Bodenstein, Simon, and Einstein, have told us something about his nonconformist personality. He was short, bald, energetic, impulsive, and candid; his external life was a scenario of innocence, simple charm, and sincerity. It is evident that his personal demeanor did not fit the traditional model of the German professor and *Herr Geheimrat*. His life was quite devoid of academic pedantry, pretense, and deception. In 1942 Einstein commented that Nernst was a personality so original that he had never met anyone who resembled him in any essential way: "So long as his egocentric weakness did not enter the picture, he exhibited an objectivity rarely found, an infallible sense for the essential and a genuine passion for knowledge of the deep interrelations of nature."

Bodenstein, who was closely associated with Nernst and succeeded him in the chair of physical chemistry, remarked in his 1942 address to the German Chemical Society in Berlin that the huge measure of success that Nernst enjoyed in so many areas of science was due to a remarkably clear, prosaic, and mobile mind unhampered by any fear of being wrong. Bodenstein said, "Nernst possessed in a quite extraordinary way a feeling for what is scientifically possible. As a hunter he would forgive me for saying that he had an unusually fine nose for the true and, besides, a blissful phantasy that allowed him to represent graphically difficult matters to himself and to us." Bodenstein made a special point of Nernst's predilection for fertile and lively phantasy and inventions (*geistvolle Apercus*) thrown out freely and spontaneously: "Sometimes [they were] explored more closely, sometimes conducive to further formulation, sometimes lighting up like meteors, and soon thereafter falling into oblivion." His phantasies were easy-come-easy-go, unless they turned out to exhibit great promise.

Self-confident, highly disciplined, and exacting with his students and assistants, Nernst was nevertheless generous and totally impartial, shared ideas freely, and graciously offered help to those able to appreciate his train of thought. When a student disappointed Nernst in regard to some matter, he would say, "For God's sake don't tell anyone that you studied under me!" The impressive list of his students included outstanding personalities such as Bodenstein, Bonhoeffer, Volmer, Eucken, Langmuir, Bjerrum, Warburg, Lindemann (later, Churchill's scientific adviser), and Simon, who, driven from Germany in 1933, became director of the Clarendon Laboratory at Oxford. In 1895 an English student, Miss Moltby, was the first woman to receive the doctorate in experimental physics in Germany, working under Nernst; and Lotte Pusch, who married Volmer, was Nernst's lecture assistant in Berlin.

Nernst's dealings with others frequently were punctuated with a sarcastic sense of humor and spontaneous witty remarks or innuendo concocted to suit the occasion. Behind this facade lay hidden one of the most gifted, versatile, and penetrating minds of the times. If not exactly devoted to him, those with whom he came in contact invariably admired him. To judge from the accounts of his many students, the anecdotes told by and about this funny little man, although undoubtedly apocryphal, seem to be legion. Not notably brilliant as a day-by-day lecturer, Nernst was, nevertheless, long remembered by his students, especially for the personal anecdotes and jokes that he introduced into the classroom and for the salty enthusiasm that he radiated from the podium. This was conspicuous whenever he had a chance to speak about the most recent scientific achievements—and especially his own. In fact, he would then so intimately relate the new scientific developments to his own researches and interpretations that students often came away supposing that he and his assistants had worked out most of physical chemistry. In examinations Nernst was easy on the students; his motto was "Das Wissen ist der Tod der Forschung."

Fond of travel, Nernst visited both North and South America. He was keen on outdoor life, especially on hunting, and took great delight in the role of country gentleman by inviting his guests to a hare-shooting spree at his country estate in Zibelle. He was an enthusiastic carp farmer, arguing on the basis of second-law considerations that fish were a better investment from an energy standpoint than warm-blooded livestock. An avid automobile fan, he owned one of the first automobiles in Göttingen at the end of the nineteenth century and published several investigations on the maximum efficiency of the internal combustion engine (1905–1913). The story has often been told that a special nitrous oxide injector had been installed in his fuel combustion system so that he could call upon this auxiliary energy thrust to go uphill with ease.

Nernst had a deep love for his country, but he was never narrow-mindedly nationalistic. Both of his sons were killed during World War I, and he himself offered his own services to the military when called upon. He was, however, so singularly free from prejudice and so saturated with practical common sense that it was out of the question for him to cover up his almost childlike good nature or to act in an expedient way to protect himself against potential dangers. In August 1920 Nernst joined Arnold Sommerfeld and Otto Rubens to send a public letter to all the major Berlin newspapers in order to protest the position of anti-Semitic organizations that had sought to identify the theory of relativity with Dada. These organizations had even singled out its originator, Einstein, as a plagiarist. After 1933 Nernst did not get along at all with the Nazis and accordingly was no longer welcome in official circles. Like Laue, Planck, Haber, and Von Mises, Nernst refused to cooperate with the anti-Einstein forces in the academy in Berlin in April 1933, or with the subsequent fascist-inspired patronization of *Deutsche Physik*. He warned the faculty at the University of Berlin that the pro-Nazi position of such scientists as Philip Lenard and Johannes Stark would jeopardize the cooperative efforts and free exchange of information among physicists in Berlin.

Two of Nernst's daughters had married men of "non-Aryan" origin and were unable to be with the family during difficult days. Nernst died of a heart attack in 1941 at his estate in Zibelle, now on the German-Polish border about ninety miles southeast of Berlin. Little information about his last years was available in those circles that once had claimed him in their front ranks.

BIBLIOGRAPHY

I. Original Works. The most complete list of Nernst's publications is in Lord Cherwell [F. A. Lindemann] and F. Simon, "Walther Nernst, 1864–1941," in *Obituary Notices of Fellows of the Royal Society of London*, **4** (1942–1944), 101–112. Works published before 1900 include "Ueber das Auftreten elektromotorischer Kräfte an Metallplatten, welche von einem Wärmestrome durchflossen werden und sich im magnetischen Felde befinden," in *Annalen der Physik*, **29** (1886), 343–347, written with A. von Ettingshausen; "Ueber die elektromotorischen Kräfte, welche durch den Magnetismus in von einem Wärmestrome durchflossenen Metallplatten geweckt werden," *ibid.*, **31** (1887), 760–789, his inaugural diss. at Würzburg; "Zur Kinetik der in Lösung befindlichen Körper. I. Theorie der Diffusion," in *Zeitschrift für physikalische Chemie*, **2** (1888), 613–637; "Ueber freie Ionen," *ibid.*, **3** (1889), 120–130, written with W. Ostwald; "Die elektromotorische Wirksamkeit der Ionen," *ibid.*, **4** (1889), 129–181, his *Habilitationsschrift* at Leipzig; "Ueber die Verteilung eines Stoffes zwischen zwei Lösungsmitteln," in *Nachrichten von der Gesellschaft der Wissenschaften zu Göttingen* (1890), 401–416; "Ueber das Henry'sche Gesetz," *ibid.* (1891), 1–14; "Verteilung eines Stoffes zwischen zwei Lösungsmitteln und Dampfraum," in *Zeitschrift für physikalische Chemie*, **8** (1891), 110–139; *Theoretische Chemie vom Standpunkte der Avogadroschen Regel und der Thermodynamik* (Göttingen, 1893), originally written as an introduction to O. Dammer, *Handbuch der anorganischen Chemie* (Stuttgart, 1892), which by 1926 had gone through 15 eds., including translations into English; "Ueber die mit der Vermischung konzentrierter Lösungen verbundene Aenderung der freien Energie," in *Annalen der Physik*, **53** (1894), 57–68; "Methode zur Bestimmung von Dielektrizitätskonstanten," in *Zeitschrift für physikalische Chemie*, **14** (1894), 622–663; "Ueber den Gefrierpunkt verdünnter Lösungen," *ibid.*, **15** (1894), 681–693, written with R. Abbegg; *Einführung in die mathematische Behandlung der Naturwissenschaften—Kurzgefasstes Lehrbuch der Differential- und Integralrechnung mit besonderer Berücksichtigung der Chemie* (Munich–Leipzig, 1895; 11th ed., 1931), written with A. Schönflies; *Die Ziele der physikalischen Chemie. Festrede 1896 zur Einweihung des Instituts für physikalische Chemie und Elektrochemie zu Göttingen* (Göttingen, 1896); "Ueber die Verwendung schneller elektrischer Schwingungen in der Brückenkombination," in *Annalen der Physik*, **60** (1897), 600–624; "Zur Theorie der elektrischen Reizung," in *Nachrichten von der Gesellschaft der Wissenschaften zu Göttingen* (1889), 194–198; and "Ueber die elektrolytische Leitung fester Körper bei sehr hohen Temperaturen," in *Zeitschrift für Elektrochemie*, **6** (1899), 41–43.

Between 1900 and the end of World War I, Nernst published "Ueber die Gaspolarisation im Bleiakkumulator," *ibid.*, **6** (1900), 549–550, written with P. Dolezalek; "Ueber Elektrodenpotentiale," *ibid.*, **7** (1900), 253–255; "Ueber die Leitfähigkeit fester Mischungen bei hohen Temperaturen," in *Nachrichten von der Gesellschaft der Wissenschaften zu Göttingen* (1900), 328–330, written with H. Reynolds; and "Ueber die Bedeutung elektrischer Methoden und Theorien für die Chemie," in *Verhandlungen der Gesellschaft deutscher Naturforscher und Ärzte*, **1** (1901), 83–99. See also "Ueber elektrolytische Erscheinungen an der Grenzfläche zweier Lösungsmittel," in *Nachrichten von der Ge-*

sellschaft der Wissenschaften zu Göttingen (1901), 54–61, written with E. H. Riesenfeld, also in *Annalen der Physik*, **8** (1902), 600–608; "Ueber die Wanderung galvanischer Polarisation durch Platin- und Palladiumplatten," in *Nachrichten von der Gesellschaft der Wissenschaften zu Göttingen* (1902), 146–159, written with A. Lessing; "Ueber Molekulargewichtsbestimmungen bei sehr hohen Temperaturen," *ibid.* (1903), 75–82; "Bildung von Stickoxyd bei hohen Temperaturen," *ibid.* (1904), 261–276; "Theorie der Reaktionsgeschwindigkeit in heterogenen Systemen," in *Zeitschrift für physikalische Chemie*, **47** (1904), 52–55; "Elektrische Nervenreizung durch Wechelströme," in *Zeitschrift für Elektrochemie*, **10** (1904), 664–668, written with J. O. W. Barratt; "Physikalisch-chemische Betrachtungen über den Verbrennungsprozess in den Gasmotoren," in *Zeitschrift des Vereins deutscher Ingenieure*, **49** (1905), 1426–1431; "Ueber die Berechnung chemischer Gleichgewichte aus thermischen Messungen," in *Nachrichten von der Gesellschaft der Wissenschaften zu Göttingen* (1906), 1–40, the classic enunciation of the third law of thermodynamics; "Ueber die Bildung von Stickoxyd bei hohen Temperaturen," in *Zeitschrift für anorganische Chemie*, **49** (1906), 213–228; "Ueber das Ammoniakgleichgewicht," in *Zeitschrift für Elektrochemie*, **13** (1907), 521–524, written with F. Jost; "Die Entwicklung der allgemeinen und physikalischen Chemie in den letzten 40 Jahren," in *Berichte der Deutschen chemischen Gesellschaft*, **40** (1907), 4617–4626, an address in celebration of the fortieth anniversary of the German Chemical Society in Berlin, 11 Nov. 1907, also trans. in *Annual Report of the Board of Regents of the Smithsonian Institution* for 1908 (1909), 245–253; *Experimental and Theoretical Applications of Thermodynamics to Chemistry* (New York–London, 1907), Silliman lectures at Yale University delivered in 1906; "Zur Theorie des elektrischen Reizes," in *Pflüger's Archiv für die gesamte Physiologie des Menschen und der Tiere*, **122** (1908), 275–314; "Zur Theorie der galvanischen Polarisation. Anwendungen zur Berechnung der Reizwirkungen elektrischer Ströme," in *Sitzungsberichte der Preussischen Akademie der Wissenschaften zu Berlin* (1908), 3–13; "Zur Theorie der elektrischen Nervenreizung," in *Zeitschrift für Elektrochemie*, **14** (1908), 545–549; "Untersuchungen über die spezifische Wärme bei tiefen Temperaturen, I–III, and V–VIII," in *Sitzungsberichte der Preussischen Akademie der Wissenschaften zu Berlin* (1910), 247–261, 262–282; (1911), 306–315, 494–501; (1912), 1160–1171, 1172–1176; (1914), 355–370, written with F. Koref, F. A. Lindemann, and F. Schwers; "Ueber neuere Probleme der Wärmetheorie," *ibid.* (1911), 65–90; "Der Energieinhalt fester Stoffe," in *Annalen der Physik*, 4th ser., **36** (1911), 395–437; "Zur Theorie der spezifischen Wärme und über die Anwendung der Lehre von den Energiequanten auf physikalisch-chemische Fragen überhaupt," in *Zeitschrift für Elektrochemie*, **17** (1911), 265–275; "Ueber einen Apparat zur Verflüssigung von Wasserstoff," *ibid.*, **17** (1911), 735–737; "Spezifische Wärme und Quantentheorie," *ibid.*, 817–827, written with F. A. Lindemann; "Application de la théorie des quanta à divers problèmes physico-chimiques," in P. Langevin and de Broglie, eds., *La théorie du rayonnement et les quanta, rapports et discussions de la réunion tenue à Bruxelles, du 30 Octobre au 3 Novembre 1911 sous les auspices de M. E. Solvay* (Paris, 1912); "Thermodynamik und spezifische Wärme," in *Sitzungsberichte der Preussischen Akademie der Wissenschaften zu Berlin* (1912), 134–140; "Zur Thermodynamik kondensierter Systeme," *ibid.* (1913), 971–985; "Ueber den maximalen Nutzeffekt von Verbrennungsmotoren," in *Zeitschrift für Elektrochemie*, **19** (1913), 669–702; "Ueber die Anwendung des neuen Wärmesatzes auf Gase," *ibid.*, **20** (1914), 357–360; "Zur Anwendung des Einsteinschen photochemischen Aequivalentgesetzes," *ibid.*, **24** (1918), 335–336; and *Die theoretischen und experimentellen Grundlagen des neuen Wärmesatzes* (Halle–Salle, 1918), of which the English eds. of 1918 and 1926 were entitled *The New Heat Theorem, Its Foundations in Theory and Experiment.*

Works published during the Weimar Republic include "Zur Theorie photochemischer Vorgänge," in *Physikalische Zeitschrift*, **21** (1920), 602–605, written with W. Noddack; *Das Weltgebäude im Lichte der neueren Forschung* (Berlin, 1921); "Zur Theorie photochemischer Vorgänge," in *Sitzungsberichte der Preussischen Akademie der Wissenschaften zu Berlin* (1923), 110–115, written with W. Noddack; and "Physico-chemical Considerations in Astrophysics," in *Journal of the Franklin Institute*, **206** (1928), 135–142.

Later writings are "Physikalische Betrachtungen zur Entwicklungs theorie der Sterne," in *Zeitschrift für Physik*, **97** (1935), 511–534; "Einige weitere Anwendungen der Physik auf die Sternentwicklung," in *Sitzungsberichte der Preussischen Akademie der Wissenschaften zu Berlin* (1935), 473–479; and "Die Strahlungstemperatur des Universums," in *Annalen der Physik*, 5th ser., **32** (1938), 44–48.

II. SECONDARY LITERATURE. On Nernst and his work, see Kurt Bennewitz and Franz Simon, "Zur Frage der Nullpunktsenergie," in *Zeitschrift für Physik*, **16** (1923), 183–199; Max Bodenstein, "Walther Nernst, 25.6.1864–18.11.1941," in *Berichte der Deutschen chemischen Gesellschaft*, **75** (1942), 79–104; K. F. Bonhoeffer, ed., "Dem Andenken an Walther Nernst," in *Naturwissenschaften*, **31** (1943), 257–275, 305–322, 397–415, which contains articles on current topics related to Nernst's work, by J. J. Hermans, Erich Lange, L. Ebert, Carl Wagner, K. Bennewitz, K. F. Bonhoeffer, G. Damköhler, H. von Wartenberg, Rudolf Edse, A. Eucken, K. Clusius, W. Schottky, G. Wietzel, P. Harteck, and J. Eggert; Max Born and Theodore von Kármán, "Ueber Schwingungen im Raumgitter," in *Physikalische Zeitschrift*, **13** (1912), 297–309; Peter Debye, "Zur Theorie der spezifischen Wärmen," in *Annalen der Physik*, **39** (1912), 789–839; and the following literature by John Eggert: "Das Nernstsche Wärmetheorem und seine Bewährung durch Affinitätsmessungen," in *Natur-*

wissenschaften, **35** (1915), 452–456; "Einführung in die Grundlagen des Nernstschen Wärmetheorems," *ibid.*, **7** (1919), 883–889, 917–921; and "Walther Nernst. Zur hundertsten Wiederkehr seines Geburtstages am 25.Juni 1964," in *Angewandte Chemie*, **76** (1964), 445–455.

See also the following works by Albert Einstein: "Die Planck'sche Theorie der Strahlung und die Theorie der spezifischen Wärme," in *Annalen der Physik*, **22** (1907), 180–190, 800; "Thermodynamische Begründung des photochemischen Aequivalentgesetzes," *ibid.*, **37** (1912), 832–838; **38** (1912), 881–884; "L'état actuel du problème des chaleurs spécifiques," in P. Langevin and de Broglie, eds., *La théorie du rayonnement et les quanta . . .* (Paris, 1912), 407–449; "The Work and Personality of Walther Nernst," in *Scientific Monthly*, **54** (1942), 195–196; and *Albert Einstein-Michele Besso Correspondence, 1903–1955* (Paris, 1972), 19–21.

Other literature includes Arnold Eucken, "Anhang. Die Entwicklung der Quantentheorie vom Herbst 1911 bis Sommer 1913," in *Die Theorie der Strahlung und der Quanten (Abhandlungen der Deutschen Bunsen-Gesellschaft für angewandte physikalische Chemie)*, **7** (1914), 371–405; R. H. Fowler and T. E. Sterne, "Statistical Mechanics With Particular Reference to the Vapor Pressures and Entropies of Crystals," in *Review of Modern Physics*, **4** (1932), 635–722; P. Günther, "Die kosmologischen Betrachtungen von Nernst," in *Zeitschrift für angewandte Chemie und Zentralblatt für technische Chemie*, **37** (1924), 454–457; Werner Haberditzl, "Walther Nernst und die Traditionen der physikalischen Chemie an der Berliner Universität," in *Forschen und Wirken, Festschrift zur 150-Jahr-Feier der Humboldt-Universität zu Berlin*, I (Berlin, 1960), 401–416; Paul Harteck, "Physical Chemists in Berlin, 1919–1933," in *Journal of Chemical Education*, **37** (1960), 462–466; Martin J. Klein, "Einstein, Specific Heats, and the Early Quantum Theory," in *Science*, **148** (1965), 173–180; Hans-Günther Körber, ed., *Aus dem wissenschaftlichen Briefuechsel Wilhelm Ostwalds*, 2 vols. (Berlin, 1961, 1969); N. Kurti, "Franz Eugen Simon, 1893–1956," in *Biographical Memoirs of Fellows of the Royal Society*, **4** (1958), 225–256; Henri Le Châtelier, "Recherches expérimentales et théoriques sur les équilibres chimiques," in *Annales des mines et des carburants*, **13** (1888), 157–382; G. N. Lewis, "The Development and Application of a General Equation for Free Energy and Physico-Chemical Equilibrium," in *Proceedings of the American Academy of Arts and Sciences*, **35** (1899), 3–38; G. N. Lewis and Merle Randall, *Thermodynamics and the Free Energy of Chemical Substances* (New York, 1923), 435–454; Kurt Mendelssohn, *The World of Walther Nernst. The Rise and Fall of German Science, 1864–1941* (Pittsburgh, 1973); *Nobel Lectures in Chemistry, 1901–1921* (Amsterdam, 1966), 347–364; James R. Partington, "The Nernst Memorial Lecture," in *Journal of the Chemical Society*, **3** (1953), 2853–2872; Max Planck, "La loi du rayonnement noir et l'hypothèse des quantités élémentaires d'action," in P. Langevin and de Broglie, eds., *La théorie du rayonnement et les quanta . . .* (Paris, 1912), 93–132; "Ueber neuere thermodynamische Theorien (Nernstsches Wärmetheorem und Quanten-Hypothese)," in *Berichte der Deutschen chemischen Gesellschaft*, **45** (1912), 5–23; and T. W. Richards, "Die Bedeutung der Änderung des Atomvolums," in *Zeitschrift für physikalische Chemie*, **42** (1902), 129–154.

Other works on Nernst and his work are E. H. Riesenfeld, "Walther Nernst zu seinem sechzigsten Geburtstag," in *Zeitschrift für angewandte Chemie und Zentralblatt für technische Chemie*, **37** (1924), 437–439; Otto Sackur, "Die Bedeutung des elementaren Wirkungsquantums für die Gastheorie und die Berechnung der chemischen Konstanten," in *Festschrift W. Nernst* (Halle–Salle, 1912), 405–423; and the following works by Franz Simon: "Die Bestimmung der freien Energie," in *Geiger-Scheel Handbuch der Physik*, **10** (1926), 350–404; "Zur Frage der Entropie amorpher Substanzen," in *Zeitschrift für Physik*, **38** (1926), 227–236, written with F. Lange; "Zum Prinzip von der Unerreichbarkeit des absoluten Nullpunktes," in *Zeitschrift für Physik*, **41** (1927), 806–809; "Fünfundzwanzig Jahre Nernstscher Wärmesatz," in *Ergebnisse der exakten Naturwissenschaften*, **9** (1930), 222–274; "Anomale spezifische Wärmen des festen Wasserstoffs bei Heliumtemperaturen," in *Naturwissenschaften*, **18** (1930), 34–35, written with Kurt Mendelsohn and M. Ruhemann; "On the Third Law of Thermodynamics," in *Physica*, **4** (1937), 1089–1096; and "The Third Law of Thermodynamics. An Historical Survey," in *Yearbook of the Physical Society* (London, 1956), 1–22. See also Arnold Sommerfeld, "Das Plancksche Wirkungsquantum und seine allgemeine Bedeutung für die Molekülphysik," in *Physikalische Zeitschrift*, **12** (1911), 1057–1068; Otto Stern, "Zur kinetischen Theorie des Dampfdrucks einatomiger Stoffe und über die Entropie-Konstante einatomiger Gase," *ibid.*, **14** (1913), 629–632; and H. Tetrode, "Die chemische Konstante der Gase und das elementare Wirkungsquantum," in *Annalen der Physik*, **38** (1912), 434–442; **39** (1912), 255–256.

ERWIN N. HIEBERT

NICOLLE, CHARLES JULES HENRI (*b.* Rouen, France, 21 September 1866; *d.* Tunis, Tunisia, 28 February 1936), *medicine, bacteriology.*

Nicolle grew up in Rouen, where his father, Eugène Nicolle, was a physician at the municipal hospital and professor of natural history at the École des Sciences et des Arts. Following family tradition, he studied medicine although he considered himself more gifted in literature than in science. Nicolle prepared for his medical examinations at Rouen and at Paris, where he passed the competitive examination for a medical residentship in 1889. His older brother Maurice (1862–1932), who had become a noted bacteriologist and pathologist, persuaded him to enroll at the Pasteur Institute. There, under the supervision of Émile Roux and drawing on the teaching of Metchnikoff, Nicolle wrote his doctoral dissertation (1893) on the pathology and etiology of the soft chancre, a venereal disease caused by Ducrey's bacillus.

After obtaining his medical degree, Nicolle returned to Rouen, where he married Alice Avice; they had two children, Marcelle and Pierre, both of whom became well-known physicians. Nicolle was a member of the municipal hospital staff as well as assistant lecturer at the medical school. He also was in charge of a bacteriology laboratory, in which he inoculated monkeys with Ducrey's bacillus and improved techniques for making antidiphtheria serum. Nevertheless, he was disappointed at being unable to create a major center of medical research at Rouen. An enthusiastic man with an unshakable faith in humanistic ideals, and stubborn whenever questions of principle were at stake, Nicolle often found himself in conflict with an inactive and meddlesome bureaucracy. In addition, the deafness from which he had suffered since the age of eighteen created difficulties in his medical practice and in his professional contacts. In 1902, discouraged by the indifferent reception accorded his efforts at Rouen, he accepted the post of director of the Pasteur Institute at Tunis.

When Nicolle arrived in Tunisia, the Pasteur Institute existed only on paper. Through his energetic and devoted work, however, a dilapidated antirabies vaccination unit was transformed into a prestigious institution equipped for large-scale manufacture of vaccines as well as for scientific research. Nicolle remained director of the institute until his death. Built according to his ideas, it became a training ground for bacteriologists and specialists in tropical medicine.

Nicolle's principal discovery, for which he received the Nobel Prize in 1928, was the experimental proof of the role of lice in the transmission of exanthematous typhus. While visiting the native hospital in Tunis, he observed that an outbreak of typhus in the city did not seem to be reflected by any spread of the disease within the wards of the hospital. In his Nobel Prize acceptance speech and in his lectures at the Collège de France he later described how the idea suddenly occurred to him that the factor that ceased to act at the hospital threshold could only be an ectoparasite—in this case, the louse. This initial hypothesis was confirmed by rigorous experiments, the results of which Nicolle reported to the Académie des Sciences in July 1909. He first infected a chimpanzee by injecting it with blood taken from a man ill with typhus. He demonstrated that the disease could be transmitted from this animal to other monkeys only by lice. This discovery had an immense practical significance because of the prophylactic procedures that followed from it. The effectiveness of systematic delousing to combat rickettsiosis, which traditionally plagued armies, was dramatically demonstrated during World War I.

Nicolle's research was not confined to exanthematous typhus. He also examined most of the germs causing infectious diseases in the Mediterranean region. The extraordinary value of his work was due to the combination of imagination and a talent for careful observation. He formulated bold hypotheses, which he scrupulously tested in the light of experimental data. An introvert by nature, Nicolle dreaded social gatherings because his deafness excluded him from the conversation. He divided his time among scientific research, writing, and his family.

Nicolle described African infantile leishmaniasis and differentiated it from kala-azar, generally found in India. He discovered leishmaniasis in the dog, which was thus recognized as the reservoir and vector of this disease; and he developed a culture medium that made it possible to study various types of *Leishmania* (1908). In collaboration with L. Manceaux, he isolated a previously unknown parasite of Tunisian rodents, *Toxoplasma gondii* (1908). Other researchers later established that toxoplasmosis is also a human disease. Nicolle elucidated the mechanisms by which lice spread the infection caused by Obermeier's spirillum and determined the precise role of ticks in epidemics involving certain spirochetes, and of flies in the transmission of trachoma. In addition, he demonstrated the viral nature of influenza. In 1931 he participated in studies of the murine typhus found in Mexico.

Having discovered that injection with serum from a convalescing victim of exanthematous typhus can protect others who have been exposed to the disease, Nicolle sought to apply this finding to other diseases. With the aid of his friend E. Conseil, he won a major victory in the fight against measles; and his work in this area became the point of departure for the use of gamma globulin in preventive medicine.

Nicolle's chief theoretical contribution was the elaboration and utilization of the concept of unapparent infection. On the basis of experiments in which guinea pigs were inoculated with typhus, he and C. Lebailly established that a germ can go through its entire life cycle in an organism that apparently remains healthy. In its broader generalization, this idea of a disease without clinical symptoms has proved very useful in epidemiology.

During his last years, especially after his nomination to the chair of experimental medicine at the Collège de France in 1932, Nicolle began increasingly to consider scientific methodology, the major lines of the historical evolution of diseases, and human destiny. While retaining his post in Tunis, he lectured at the Collège de France every year from 1932 to 1935. These lectures were published in several volumes and were widely read by the French scientific community. Of particular interest are his views on the moral responsibility of scientists and on the biological foundation of creativity. For Nicolle, the birth of an idea is comparable to biological mutation. His statement that scientific creation is essentially similar to poetic inspiration was based on personal experience: he had also distinguished himself as the author of several novels and collections of stories.

BIBLIOGRAPHY

I. Original Works. Nicolle's papers include "Recherches expérimentales sur le typhus exanthématique," in *Annales de l'Institut Pasteur,* **24** (1910), 243–275; **25** (1911), 97–144; and **26** (1912), 250–280, 332–335. His most important books are *Naissance, vie et mort des maladies infectieuses* (Paris, 1930); *Biologie de l'invention* (Paris, 1932); *Destin des maladies infectieuses* (Paris, 1933; new ed., Geneva, 1961); *L'expérimentation en médecine* (Paris, 1934); *Responsabilités de la médecine,* 2 vols. (Paris–Tunis, 1935–1936); and *La destinée humaine* (Paris, 1936).

II. Secondary Literature. See "Hommage à Charles Nicolle," the special commemorative issue of *Archives de l'Institut Pasteur de Tunis,* **25** (1936); and the obituary by F. Mesnil, in *Bulletin de l'Académie de médecine,* **115** (1936), 541–548. Besides the popularized account by G. Lot, *Charles Nicolle et la biologie conquérante* (Paris, 1961), see the pref. to the new ed. of *Destin des maladies infectieuses* (Geneva, 1961), written by Pierre Nicolle's son Charles; and the biography of Nicolle's brother Maurice, written by the latter's son, Jacques Nicolle, *Maurice Nicolle, un homme de la Renaissance à notre époque* (Paris, 1957).

M. D. Grmek

OSTWALD, FRIEDRICH WILHELM (*b.* Riga, Latvia, Russia, 2 September 1853; *d.* Leipzig, Germany, 4 April 1932), *physical chemistry, color science.*

Together with van't Hoff and Arrhenius, Ostwald established physical chemistry as a recognized and independent professional discipline and was its most important spokesman and organizer. His early reputation was based upon investigations into the fundamental principles governing chemical equilibrium and reactivity. A skillful experimentalist, he continued to give chemical affinity a central position in his research on electrolytic dissociation, electrical conductivity, mass action, catalysis, and reaction velocity. Ostwald received the Nobel Prize in chemistry in 1909 for his work in physical chemistry, and especially in recognition of his studies on catalysis. He was also one of the leading twentieth-century researchers in color science, and enriched chromatics through his quantitative theory of colors and his subjective chromatic system. Ostwald was at the same time an inspiring teacher who restored the significance of general chemistry and induced a generation of chemists in Europe and the Americas to adopt a receptive attitude toward theoretical and physical chemistry in their teaching and research. Ostwald was a lucid, imaginative, prolific, yet often controversial writer, synthesizer, expositor, and apostle of scientific ideas. He was a man of notable charm, enormous capacity for work, and many-sided intellectual interests, and was actively involved in the cultural and philosophical debates and humanistic strivings of his time.

Ostwald was the second son of Gottfried Ostwald, a master cooper, and Elisabeth Leuckel, the daughter of a master baker. His parents were descended from German immigrants who came to Livonia from Berlin (on the paternal side) and from Hessen (on the maternal side). Ostwald was educated in Riga, where he attended the Realgymnasium. Even in his student days Ostwald's interests ranged very widely, focusing especially on physics and chemistry as well as literature, music, and painting. He shared his love of music—he played the viola and the piano—with his mother, to whom he felt especially close. Like his father Ostwald developed considerable skill in painting and handicraft. The latter wished his son to become an engineer, but Ostwald's strong inclinations toward chemistry were finally decisive in his choice of profession.

In 1872 Ostwald enrolled at the University of Dorpat (now Tartu), where he studied chemistry under Carl Schmidt and Johann Lemberg and physics under Arthur von Oettingen. Within a short time the highly gifted Ostwald passed all the examinations, and in 1875 he received the candidate's degree. The following year he was awarded the master's degree and became *Privatdozent* at the

University of Dorpat, where he lectured on the theory of chemical affinity and, in 1878, earned the doctorate in chemistry. Ostwald had become an assistant to von Oettingen in 1875; and the modest sum he earned from this work, added to the income he received from a second job teaching physics and chemistry in a Dorpat *Realschule*, enabled him to establish a household. On 24 April 1880 he married Helene von Reyher, the daughter of a Riga surgeon. They had five children, one of whom, Wolfgang, became a colloid chemist.

In 1881 Ostwald was appointed professor of chemistry at the Riga Polytechnic Institute, where he quickly proved to be an outstanding teacher and began two important undertakings that made him widely known. First, in 1885–1887 he wrote the ambitious *Lehrbuch der allgemeinen Chemie* (called "der grosse Ostwald" in order to distinguish it from "der kleine Ostwald" of 1889, *Grundriss der allgemeinen Chemie*). The work was a substantial contribution to the establishment of physical chemistry as a separate branch of the discipline and was published in a partial second edition between 1897 and 1902. It was based on the systematic examination of fifty years of journal literature in physics and chemistry. Second, with van't Hoff, Ostwald began to publish *Zeitschrift für physikalische Chemie*, which became the mouthpiece of the Leipzig school of physical chemistry. This journal quickly established its importance and became the organizational link uniting physical chemists of various countries. In it the chemists associated with Ostwald, Arrhenius, and van't Hoff (who were dubbed "the ionists") broadly disseminated the new ideas of physical chemistry and related fields—such as the existence of ions, which was still disputed by some—in articles that often were aggressive in tone.

Ostwald expanded his activity as a teacher and researcher in September 1887, when he accepted an appointment at Leipzig to the only chair of physical chemistry then existing in Germany; the position had become free when Wiedemann left it to assume the chair of physics at Leipzig. The other candidates, including van't Hoff, withdrew in favor of Ostwald. Ostwald's work in physical chemistry during the next two decades was as fruitful as it was remarkable. He became the head of one of the discipline's most important schools and soon was able to propose his students for professorships at universities throughout the world. His co-workers and assistants included Arrhenius, Nernst, Le Blanc, and R. Luther. Ostwald also introduced new methods in the analytical training of chemists and wrote a textbook on that subject, thereby becoming an important pioneer of new approaches to the education of chemistry students. He laid particular stress on independent work and critical discussion of results obtained by students, thus anticipating modern forms of teamwork.

In 1898 Ostwald celebrated the official dedication of the new physical chemistry institute of the University of Leipzig,[1] which became a training center for generations of physical chemists. Ostwald, however, remained there for barely a decade. As early as 1894 he had wished to be free of teaching and official duties such as elected deanship positions; he wanted to work only as a research professor and continue his literary activities (especially writing his books and editing *Zeitschrift für physikalische Chemie*) and other responsibilities. He retired in 1906, after having been appointed as the first German exchange professor to Harvard University (academic year 1905–1906).[2]

In 1901 Ostwald had purchased an estate in Grossbothen, a village near Grimma. The house on the edge of the woods was baptized "Energie"—as he wrote to Mach in September 1901. Upon his resignation from the University of Leipzig in 1906, he moved his family and tremendous library to "Landhaus Energie" and spent the rest of his life there as an independent scholar and freethinker, devoting his efforts to energetics, scientific methodology, the organizational aspects of science, a world language, internationalism, and pacifism. Not long after he settled at Grossbothen, he enlarged the building and added a laboratory for his color research.[3] He died twenty-six years later, following a short illness.

Ostwald's achievements as a young chemist lay primarily in chemical affinity studies that, along with both general and inorganic chemistry, had been relatively neglected since the work of E. F. Geoffroy (1718), Bergman (1775), C. F. Wenzel (1775), and Berthollet (1803). Ostwald's approach to the subject was to demonstrate that quantitative values for physical properties, such as specific volume and refractive index, could be correlated with the qualitative changes accompanying chemical transformations—and thus provide information about the relative affinities of the constituents of a chemical reaction.

In the mid-1870's, when young Ostwald became interested in the study of chemical affinity (*Verwandtschaftslehre*), the interest of German chemists, both at the universities and in industry, was directed almost exclusively toward organic chemistry. Ostwald's motivation for taking up the study of problems in physical chemistry, therefore, merits some comment. To judge from the style of teaching and research at the University of Dorpat, the intense preoccupation with organic chemistry stopped at the German border. The professor of chemistry, Carl Schmidt (1822–1894), had studied with Heinrich Rose at Berlin, with Liebig at Giessen, and with Wöhler at Göttingen; but at Dorpat he had become totally involved with the study of the mineral content of ground and surface waters as a means of furnishing information concerning the chemical processes of rock formation. In these studies Schmidt followed the direction of the earlier work of C. G. C. Bischof of Bonn—the founder of geological chemistry in Germany. In his *Lebenslinien* Ostwald relates how Schmidt's chief assistant, Johann Lemberg (1842–1902), following some leads given in Bischof's textbooks, early taught him to pay close attention to problems in chemical equilibrium, mass action, and reaction velocity, and to recognize that there are no absolutely insoluble substances in nature. Thus the ground was laid for a way of chemical thinking that eased Ostwald into seriously reckoning with the importance of mass considerations in chemical process. Ostwald remarked that he undoubtedly would have become an organic chemist if he had studied chemistry in Germany.

As a student at Dorpat, while examining the thermochemical investigations of Julius Thomsen, Ostwald conceived the idea of experimentally exploring a problem in chemical affinity that would employ physical methods of investigation but would be more general than the current focus in Dorpat on the weathering of minerals. He chose to study the extent of decomposition of bismuth chloride by water in order to compare his results with Thomsen's calorimetrically determined affinity. Ostwald conjectured—correctly, as it turned out—that, in principle, the magnitude of chemical change in a reaction might be calculated from any experimentally measurable changes in physical property. Accordingly, Ostwald saw Thomsen's method to be but one of many conceivable ways to approach the study of chemical affinity.

Gibbs (1875–1879) and Helmholtz (1882–1883) had demonstrated from considerations of the second law of thermodynamics that the heat of reaction ΔH for a chemical reaction, in general, cannot be relied upon to furnish information about chemical equilibrium and reactivity unless the entropy term ΔS is zero in the equation $\Delta F = \Delta H - T\Delta S$, in which ΔF represents the free energy for the process and T the absolute temperature. Thus Thomsen's thermochemical investigations proved to be more limited than Ostwald's method, since change in many physical properties—such as specific volume, refractive index, viscosity, color, electrical conductivity, and rotation of polarized light—can be correlated with changes in activities (accurately) and changes in concentrations (approximately) to yield the equilibrium constant K, and thus the free energy, since $\Delta F = -RT \ln K$ (R is the gas constant). Ostwald's important contribution was his recognition of the unique advantage of physical methods of investigation in the solution of chemical problems. This is especially important in chemical thermodynamics, since the analysis of the constituents of a reaction by chemical means almost always is rendered impractical by a concurrent shift in the equilibrium during the analysis; physical methods, on the other hand, do not cause chemical changes in the system.

In his *Volumchemische Studien über Affinität* (master's thesis, 1877) Ostwald determined the volume changes that take place during the neutralization of acids by bases in dilute solution. Pycnometers were used to determine the specific volumes before and after the reaction. From the differences in specific volumes (for specific concentrations and constant temperature) Ostwald was able to calculate the chemical action (affinity) for the neutralization reactions. Specifically, he showed that the distribution of a base between two acids can be determined by measuring the specific volume of a solution of each acid and of the base, of the solution formed by mixing each acid separately with the base, and of the solution formed by mixing both acids simultaneously with the base. With minor exceptions Ostwald found that his dilatometric method yielded results in close agreement with the order of "avidities" determined by Thomsen's thermochemical method. Like the latter, Ostwald's method seemed to confirm the Guldberg-Waage law of mass action.

In his *Volumchemische und optisch-chemische Studien* (doctoral dissertation, 1878), Ostwald enlarged his investigations to include the determination of the coefficients of refraction of a large group of acid-base and other double decomposition reactions. Thus he obtained values for chemical reactivity that confirmed those obtained by the specific-volume method, but he felt that the optical method was less trustworthy than the volumetric method. In addition, his chemical affinity studies were extended to incorporate the analysis of both homogeneous and heterogeneous reactions as a function of temperature. In this way Ostwald was able to attach specific numerical values to the term "affinity," which had long been referred to in the literature in a qualitative and often arbitrary way.

In his 1879 review article "Chemical Affinity," M. M. Pattison Muir of Cambridge University stated (p. 182): "The most important contributions made within recent years towards the final solution of the problem of chemical affinity are contained in two papers by Guldberg and Waage [1869 and 1879], and three papers by W. Ostwald [1877 and 1878]." Having given a fairly detailed summary of the results of these papers, Muir concluded (p. 203): "Ostwald furnishes chemistry with a new method for solving some of the most difficult problems; and Guldberg and Waage lead the way in the application of mathematical reasoning to the facts of chemical science."[4] Thus Ostwald's important position in physical chemistry was already recognized when he was but twenty-six years old.

In 1879 Ostwald proposed that the rate at which compounds like zinc sulfide and calcium oxalate are dissolved by different acids be used as a measure of the relative affinities of the acids. He found that the dynamically determined affinity coefficients agreed satisfactorily with those obtained earlier by statistical methods. In his first teaching post at the Polytechnic Institute at Riga, Ostwald pursued the studies on chemical reaction kinetics with resolution. His *Studien zur chemischen Dynamik* (1881) treated the reaction velocity for the acid-catalyzed saponification of acetamides and the hydrolysis of esters. Experiments on the rate of inversion of cane sugar in the presence of various acids (1884–1885) gave Ostwald a second opportunity to evaluate the affinities of acids and to compare his values with those he had measured by other methods. With some exceptions, he found that the various methods gave comparable results. He explained the exceptions on the basis of secondary reactions that influence some chemical processes more than others.

Such was the state of chemical affinity studies prior to the enunciation of the theory of electrolytic dissociation. Ostwald had produced a substantial body of experimental evidence, mostly for acids and bases, that demonstrated that different reactions can be quantitatively characterized by affinity coefficients that depend on the constitution of the acids and their degree of dilution. His general conclusion was that each acid and each base can be assigned an affinity coefficient that quantitatively describes all of its specific reactions and, further, that the relative values of these coefficients, which are independent of the nature of the chemical reaction, can be measured with tolerable accuracy by various physical (static and dynamic) and chemical methods. It is pertinent to add that the thermochemical methods perfected by Thomsen and Berthelot, and the physical methods explored by Ostwald and his students, measured chemical reactivity with fair reliability most of the time, but not invariably. The former were inadequate because of the disregard of the entropy term, and the latter were no more than approximate because of the neglect of nonadditive secondary effects.

In 1884 Ostwald read the memoir on the galvanic conductibility of electrolytes, submitted by Arrhenius as a dissertation to the Swedish Academy of Sciences the previous year.[5] From then on, Ostwald became an enthusiastic proponent and crusader for Arrhenius' new theory of dissociation. In his memoir Arrhenius had demonstrated that the electrical conductivity of acids is proportional to their strength. He met Arrhenius at Stockholm in August 1884, and the two men commenced a close collaboration that led to a lifelong friendship, of which Ostwald gave a detailed account in *Lebenslinien* and that is vividly displayed in the Ostwald-Arrhenius correspondence. Through Ostwald's influence Arrhenius received a five-year traveling scholarship that began in Ostwald's laboratory at Riga (1886) and ended in his laboratory at Leipzig (1889–1891). During the intervening years he came into working contact with Friedrich Kohlrausch at Würzburg (1886), Boltzmann at Graz (1887), and van't Hoff at Amsterdam (1888).

In 1886 Ostwald discovered van't Hoff's *Études de dynamique chimique* (1884), which focused on the application of thermodynamics to chemical problems. Now more than ever Ostwald grasped the deep significance that the concept of electrolytic dissociation would have when applied to questions of affinity, equilibrium, mass action, and chemical thermodynamics in general. From that time on, van't Hoff became the third member of the physical chemistry triumvirate that soon moved its spiritual center to the University of Leipzig. Once there, Ostwald chose Nernst as his chief assistant in physical chemistry, and the laboratory became a mecca for enterprising graduate students from all over the world, especially the United States.[6]

The full-blown theory of electrolytic dissociation that Arrhenius enunciated in *Zeitschrift für physikalische Chemie* in 1887 was, by then, seen to be the logical outcome of a number of important previous studies on solution chemistry. Arrhenius had earlier entertained comparable ideas, but he considered them too bold to state verbally; and his university mentors felt that the twenty-four-year-old student's speculation—identifying electrolytic dissociation with chemical activity—was premature and inadequately supported. Ostwald immediately designed his own experimental program to relate chemical affinity to electrical conductivity. By 1888 Ostwald, Planck, and van't Hoff had independently applied the law of mass action to the equilibrium distribution between the ions and the undissociated portion of an electrolyte.

Hittorf's electrolytic salt solution studies of 1853 and 1859 already had led to the suggestion that the electric current is carried by ions moving at different rates toward the electrodes. In 1874 Kohlrausch had shown that for dilute salt solutions the quotient of electrolytic conductivity and concentration is the same as the sum of both terms. Raoult (1882) deduced a general law for freezing-point depression of solutions containing nondissociating organic solutes. Five years later he announced his general law of the effect of solutes on the vapor pressure of solvents in *Comptes rendus . . . de l'Académie des sciences*. He explained the "anomalous" freezing point depression and vapor pressure of salts—anomalous when compared with organic solutes—by suggesting that the salt molecules dissociate into other molecules.

In his comprehensive statement of the theory of electrolytic dissociation (1887), Arrhenius reasoned convincingly that all of the above-mentioned investigations could be accounted for by postulating, as Clausius had suggested in 1858, the dissociation of molecules into electrically charged ions. Accordingly, compounds that had been thought to be held together by the strongest affinities—such as sodium chloride, hydrogen chloride, and potassium hydroxide—now were seen, in dilute solution, to be largely dissociated.

Arrhenius and Ostwald recognized that an electrically conductive solution—where the conductivity depends on the concentration and the temperature—consists of two different kinds of molecules: active ions and inactive undissociated molecules; moreover, the concentration of the active portion increases with dilution at the expense of the inactive portion and reaches a limiting (maximum) value, presumably when all of the inactive molecules have been transformed into active ions. The important point in terms of chemical affinity studies is that like acids, whose strength increases with electrical conductivity, so too, the chemical activity of electrolytes coincides with their electrical conductivity.

In 1885 Ostwald initiated a comprehensive program to redetermine, by using Arrhenius' electrolytic conductivity method, the affinities of the acids he had studied earlier by other physical methods. He concluded that electrolytic conductivity measurements were far more elegant and less tedious than his own specific volume method. His experiments showed that for strong monobasic acids in aqueous solution, the molecular conductivity gradually increases with dilution and asymptotically approaches a maximum value at infinite dilution. Ostwald's dilution law (*Verdünnungsgesetz*), in its essentially modern form, was theoretically derived in 1888 and was supported with impressive experimental evidence the following year. Ostwald reasoned that since the laws of gas pressure had been shown to be applicable to the osmotic pressure for dilute solutions of nonelectrolytes, the formula for a partly dissociated gas might likewise be applicable to partly dissociated solutions. Ostwald showed that for a binary electrolyte of volume ν, with μ_ν representing the molecular conductivity at volume ν, and μ_∞ representing the limit of conductivity at infinite dilution, it follows that

$$\frac{\mu_\nu^{\,2}}{\mu_\infty(\mu_\infty-\mu_\nu)\cdot\nu}=K_c$$

where Kc is the equilibrium constant for the chemical reaction in question.

This equation shows that the constant has the same value for any given binary electrolyte at all degrees of dilution, and is the product of factors that depend solely on the composition and constitution of the acids or bases the conductivities of which are being investigated. Ostwald confirmed the law for 250 acids, all of which were "weak acids"; the law would not be applicable to strong acids or salts that possess no dissociation constant. Georg Bredig (1868–1944), Ostwald's assistant at Leipzig, studied fifty bases for which the dilution law was seen to be valid. The historical importance of Ostwald's dilution law is that the law of mass action was first applied to dilute solutions of weak organic acids and bases in that form. Most inorganic acids are strong electrolytes in aqueous solution; and salts, even of weak organic acids or bases, are highly ionized and do not obey Ostwald's law. The law was therefore found to hold excellently for all slightly ionized electrolytes, but to fall very wide of the mark for highly ionized electrolytes.

The first issue of the *Zeitschrift für physikalische Chemie* appeared in February 1887. During its first year the journal published articles by van't Hoff, Ostwald, Arrhenius, Lothar Meyer, Raoult, Guldberg, Mendeleev, Julius Thomsen, Le Châtelier, and Planck, among others. Van't Hoff's paper "Die Rolle des osmotischen Druckes in der Analogie zwischen Lösungen und Gasen" provided a crucial link between the thermodynamics of gases and of solutions. In "Ueber die molekulare Konstitution verdünnter Lösungen," Planck argued that harmony between the laws of thermodynamics and freezing point and vapor pressure studies could be maintained only by assuming that the molecules of dissolved solutes are altered in solutions in the way that Arrhenius had suggested. The next-to-the-last article in the *Zeitschrift für physikalische Chemie* of 27 December 1887 was Arrhenius' "Ueber die Dissociation der im Wasser gelösten Stoffe," a forthright and comprehensive statement of the electrolytic theory of ionization. The support for this theory by Arrhenius, Ostwald, van't Hoff, Planck, and Nernst opened a new chapter in the history of physical, experimental, and theoretical chemistry that brought a large and varied range of phenomena under a single point of view. The theory notably served to tie together all of those properties of a system encountered in the study of solutions that depend primarily on the number of molecules involved and not on their nature. Following a suggestion of the philosopher-psychologist Wilhelm Wundt, Ostwald designated such properties as "colligative."

The value of the theory became conspicuously evident by the way in which it reconciled, in a quantitative way, hitherto disparate physical and chemical investigations: the avidity of acids and bases, heats of neutralization, the dissociation of water, the behavior of weak acids and bases, the hydrolysis of salts, the catalytic activity of acids and bases, equilibrium in electrolytic solutions, theory of concentration cells, liquid junction potentials, and many properties of solutions, such as specific volume, refractive index, optical activity, and absorption spectra.

In Ostwald's laboratory at Leipzig concerted efforts were made to show that the many empirically established regularities observed for chemical processes in solution could all be deduced as necessary consequences of the ionic theory of dissociation. The result was a substantial body of experimental evidence, mostly for acids and bases, that clarified the connection between electrolytic dissociation and conduction. It showed that the coefficients of affinity, which measure the degree of dissociation of the acids and bases, were independent of the particular chemical process under investigation but nearly proportional to the conductivities in solution. They depend, in turn, upon the number of free ions present and their velocities. Ostwald's *Grundriss der allgemeinen Chemie* (1889) and the English translation by James Walker (1890) made these ideas of van't Hoff's widely known.

In 1891 Ostwald formulated a theory of acid-base indicators in which he used the principle of ionic equilibrium to account for the ratio of un-ionized weak acid (of one color) to ionized weak acid (of another color). Subsequent investigations, stimulated by Ostwald's contributions, revealed, as Hantzsch and his pupils showed in 1906, that the organic indicators were pseudo acids and pseudo bases—that is, nonelectrolytes susceptible to the formation of metallic derivatives by changing into acidic and basic isomers. The new views did not substantially alter the quantitative formulation of Ostwald's theory. The entire field encompassed by the theory of chemical reaction based on electrolytic dissociation, including indicator theory, was expounded in great detail in Ostwald's *Die wissenschaftlichen Grundlagen der analytischen Chemie* (1894), a work that revolutionized the teaching of analytical chemistry.

Over a period of twenty years at Leipzig, Ostwald championed and, with his students and assistants, furnished experimental support for the electrolytic dissociation theory, demonstrating its remarkable value against the bitter hostility of the chemists. The focus of the opposition to Ostwald and his revolutionary colleagues was concentrated in Berlin, and continued to dominate the chemical scene there until Nernst (who had been with Ostwald at Leipzig from 1887 to 1891) accepted a professorship of physical chemistry at Berlin when Landolt retired in 1905—the year in which Ostwald gave up his teaching career in Leipzig. A common criticism was that the proponents of the theory—called "die Ioner"—had tried too singularly to set the theory in the foreground of chemical research and had attempted to encompass too varied a collection of chemical phenomena. In any case, to explain the formation and stable existence of electrically charged ions in solution, and to account for the role of the solvent in this process, appeared to be too formidable a theoretical undertaking. The reason for ionization and the stability of ions remained a puzzle until the electronic theory of the atom provided clues to the connection between chemical affinity and electricity.

Strong electrolytes had always been an anomaly for the electrolytic dissociation theory, and there was little progress on this matter until the turn of the century. With weak (slightly ionized) electro-

OSTWALD

lytes Ostwald's dilution law was seen to hold quite rigorously, but for strong electrolytes not even approximately. This, of course, brought up the question of the general validity of the theory of ionization in relation to the law of mass action. An important factor in the revision of the theory was introduced with the discovery, by X-ray analysis, that salts and other strong electrolytes in the crystalline state are composed of ionic lattices. This suggested that since the ions preexist in the solid state, these substances are completely ionized at all concentrations. This was about the time that Ostwald lost all desire to continue his chemical studies. Other investigators—Bjerrum (1906), Hantzsch (1906), and Debye, E. Hückel, and Lars Onsager (1923–1927)—approached the problem from a more theoretical and mathematical point of view, drawing on the electronic theory of atomic structure and on the concepts of electrovalence and covalence to extend the original classical theory. In doing so, they elaborated in fine detail the interionic forces, the solvent interaction (solvation), the ionic mobility and ionic atmosphere terms, and the time of relaxation (electrical drag), considerations that led to a far more complex, but also considerably more attractive and useful, theory of ionization.

G. N. Lewis' introduction, in 1907, of "activity" a (or effective concentration) and "activity coefficient" α (or deviation multiplicand)—such that the actual concentration c is given as a/α—manifestly helped to introduce a uniform method of treating the mathematical equations that describe ionization. Accordingly, it became standard practice to adopt the Arrhenius-Ostwald ratio of electrolytic conductivities M_v/M_∞ as the apparent, and not the actual, degree of ionization. This mode of representing the data contributed notably to the elimination of numerous misunderstandings and misinterpretations of the literature.

All things considered, the modern position in chemical solution theory is rather far removed from the classical theory of ionization, notwithstanding its historical significance in the search for a theory of the constitution of electrolytes. From the point of view of order-disorder considerations, it is not surprising to discover that to date our understanding of the structure of pure liquids and solutions has been far less impressive than that of the structure of matter at its extremes in the crystalline and the gaseous forms.

The systematic quantitative investigation of catalysis in Ostwald's laboratory during the last decade of the nineteenth century brought that heterogeneous and untidy subject within the domain of chemical kinetics. In 1835 Berzelius had introduced the term "catalysis" to designate the process by which, through some special kind of force, a relatively small quantity of a substance aroused the slumbering affinities of other substances and hastened a chemical reaction without itself undergoing any change. Many and varied examples of catalysis were reported, classified, and commented upon throughout the century. While in Riga, Ostwald had studied the reaction velocity of chemical processes. In 1890 he reported on the phenomenon of "autocatalysis" and defined it as a process that is "provoked or accelerated through the presence of certain substances with no demonstrable participation of these in the compounds."[7] In an 1894 report Ostwald gave a new rendering of the concept of catalysis that had been introduced by Berzelius: "Catalysis is the acceleration of a slowly proceeding chemical process through the presence of a foreign substance."[8] Ostwald compared catalysts to the action of oil on a machine and of a whip on a lazy horse.

Important experimental work on catalytic processes was carried out in Ostwald's institute. This work is conveniently summarized in his paper "Über Katalyse" (1901). The most important aspects of the experimental contributions on this subject by Ostwald and his co-workers are the ones that treat the process of crystallization from supersaturated solution—for both homogeneous and heterogeneous reactions—and the effect of enzymes. The work on supersaturation and supercooling showed that a system moving from a less stable to a more stable state goes by stages to the one lying closest at hand, and not necessarily to the most stable of all possible states. This is known as Ostwald's law of stages.

Catalytic activity became, for the proponents of the theory of ionization, one of the best ways to measure acid strength. Ostwald took full advantage of such catalytic experiments in his support of Arrhenius' theory. From the time of the investigations of Berthelot and Péan de Saint-Gilles on the equilibrium between acetic acid, ethyl alcohol, ethyl acetate, and water (1862–1863), it had been known that this homogeneous acid-catalyzed esterification reaction is accelerated by the presence of acids and alkalies that remain unchanged in the overall process. Ostwald and Arrhenius recognized that the rate-determining factor in this catalyzed esterification, and similar reactions, was the acid and that the catalytic activity was proportional to the conductivity of the acid (and to the hydrogen ion concentration) but independent of the nature of the anion.

This proved to be true only for reactions catalyzed by weak acids.

In the course of catalytic investigations Ostwald and Oskar Gros developed a photographic contact process that Ostwald called *Katatypie*. Ostwald also applied his knowledge of catalysis to two large-scale industrial chemical projects that, however, did not bring him the success he hoped for. With E. A. Bodenstein and later with Ernst Brauer, who subsequently became his son-in-law, Ostwald developed a process for the synthesis of ammonia from nitrogen and hydrogen gases at high temperature and pressure, using heated bundles of iron wire as the catalyst (1900). After a long series of attempts, a process was developed for the catalytic oxidation of ammonia to nitric acid that would stop short of complete oxidation to free nitrogen. The process was exploited industrially beginning in 1906, but by that time Ostwald and Brauer were no longer associated with the project.[9]

A theoretically significant point, first emphasized by Ostwald and based on a simple argument that excludes a *perpetuum mobile,* was that a catalyst exerts no change in the overall energy relations, and therefore does not alter the thermodynamically stable, reversible equilibrium position of a reaction. Ostwald thus saw that a catalyst must accelerate the forward and the reverse reactions in the same proportion in order for the equilibrium constant in the Guldberg-Waage law to retain its thermodynamic significance. Whether a catalyst necessarily initiates a reaction, and whether it enters into compound formation with the reactants as an intermediary, were questions for which no convincing answers were supplied by Ostwald and his collaborators. Indeed, from the intense investigation into catalysis subsequent to Ostwald's work – including autocatalysis, inhibition phenomena (poisoning), contact and surface effects, enzymes produced by living systems – no single comprehensive and adequate theories resulted, except some that proved to be untestable.

A much-discussed and plausible theory for some cases of catalytic behavior, by Clément and Desormes (1806), was the one according to which the catalyst forms an intermediary, metastable compound with one of the constituents of the reaction and then is regenerated in its original form in the process that leads to the final products of the reaction. Theories for the activity of catalysts and the principles for selecting them in the promotion of specific reactions have been hotly pursued in the twentieth century in connection with the biochemistry of enzymes and the rearrangement of aliphatic and aromatic hydrocarbons.

At the end of the nineteenth century the chemical community had reluctantly accepted Daltonian atomism. For many investigators it apparently was no more than a useful hypothesis for which there was little experimental evidence. Foremost in the vanguard of chemists who were skeptical about the physical reality of atoms and molecules, and who regarded them largely as mental artifices, was Ostwald. His anti-atomistic sentiments were closely connected with an aversion to mechanical doctrines and a strong belief in an energy-based scientific program that he hoped eventually would encompass the natural and social sciences and the humanities in one vast monistic *Weltanschauung*. He began seriously to develop his ideas on energetics around 1890, and thereafter gradually reorganized all of his thoughts and work around energy.

The central position in Ostwald's science of "energetics" was held by the energy concept. For him energy was not, as in physics, a derived quantity but, rather, far more fundamental: "In fact energy is the unique real entity [*das einzige Reale*] in the world and matter is not a bearer [*Träger*] but rather a manifestation [*Erscheinungsform*] of the former."[10] In particular, Ostwald maintained that the principles of energetics would furnish a more tangible basis for chemistry than would the kinetic-molecular theory. He went so far as to declare that the concept of matter was superfluous and that phenomena could be accounted for satisfactorily by analyzing the energy transformations taking place in nature and the laboratory. What is called matter, he argued, is only a complex of energies found together in the same place; differences between substances were reduced to differences between their specific energy content. Ostwald never proved his thesis, and more than one critic pointed out that the experimental facts of stoichiometry could not be deduced at the time from premises that did not already contain them. Unfortunately, Ostwald's premises included the facts.

Despite the obvious pertinence of the atomic conception for stereochemistry by the 1870's, Ostwald apparently could not rid himself of the feeling that atoms were the limping remnants of an irresponsible, speculative *Naturphilosophie*. As early as 1887, in his inaugural lecture at Leipzig, Ostwald spoke on "Die Energie und ihre Umwandlungen." In 1891–1892 he set forth his views on the importance of energy in "Studien zur Energetik." These studies were influenced by Gibbs's thermodynamical writings, which Ostwald translated into German in 1892 and thereafter became required reading for all students in Leipzig. Ostwald mentions in *Lebenslinien* that the English and American students had to read Gibbs in German until Yale University put out a new edition of Gibbs's works after his death. It is appropriate to add here that with his characteristic enthusiasm, Ostwald took laws and conceptions from thermodynamics and generalized them in the form of "physical energetics" that reached beyond the field in which they were known to be valid.

Ostwald, Georg Helm, and Le Châtelier supported and made considerable use of the system of chemistry developed by the philosopher-chemist František Wald of the Czech Technical University of Prague.[11] According to the basic tenets of this system, the composition of chemical compounds depended on the physical circumstances under which they were prepared. The constant composition of substances, and their fixed physical and chemical properties, were explained, much as Berthollet had done earlier, as the end products of the chemists'—and nature's—process of preparation. For example, repeated crystallizations and distillations were said to result in constant composition. The constancy of multiple elemental weight ratios for many compounds, especially in organic chemistry, was explained by Ostwald on the basis of "the law of integral relations."

The science of energetics became a controversial issue as the result of Ostwald's lecture "Die Ueberwindung des wissenschaftlichen Materialismus," delivered at the Lübeck meeting of the German Society of Scientists and Physicians in 1895. The position of Ostwald and Georg Helm, a physical chemist from Dresden—that energy had displaced matter as a concept—was challenged on the floor, and then in the scientific journals, by Boltzmann, Planck, Felix Klein, Victor Meyer, Wislicenus, Nernst, and Arthur von Oettingen.[12] Boltzmann stated bluntly that he saw no reason why energy should not be atomistic. Ostwald was unsuccessful in his attempt to persuade scientists of the value of his energetic views; and within the next few years, as evidence for the particulate nature of matter became more convincing, both physicists and chemists came to hold the view that energetics was an aberration. This lack of acceptance did not deter Ostwald from diligently pursuing and vigorously proclaiming an energy-rooted chemistry as an alternative to atomism. In fact, he extended the scope of his energeticist ideas to the level of a world view that he maintained even after he had adopted the atomic-molecular theory of matter. Ostwald's plan for a systematic philosophical energetics was laid out in his *Vorlesungen über Naturphilosophie* (1902), which was dedicated to Ernst Mach, whom he considered the person who had influenced him most.[13] Mach was of one mind with Ostwald on the rejection of atomism, but had no use for his energetics or that of anyone else. Ostwald's views on energetics were explored in grand style in *Annalen der Naturphilosophie,* which he edited from 1901 to 1914.

Speaking to the Fellows of the Chemical Society in the theater of the Royal Institution, Ostwald said in his Faraday lecture (1904):

> *It is possible, to deduce from the principles of chemical dynamics all the stoichiometrical laws; the law of constant proportions, the law of multiple proportions and the law of combining weights.* You all know that up to the present time it has only been possible to deduce these laws by help of the atomic hypothesis. Chemical dynamics has, therefore, made the atomic hypothesis unnecessary for this purpose and has put the theory of the stoichiometrical laws on more secure ground than that furnished by a mere hypothesis.
>
> I am quite aware that in making this assertion I am stepping on somewhat volcanic ground. I may be permitted to guess that among this audience there are only very few who would not at once answer, that they are quite satisfied with the atoms as they are, and that they do not in the least want to change them for any other conception. Moreover, I know that this very country is the birthplace of the atomic hypothesis in its modern form, and that only a short time ago the celebration of the centenary of the atomic hypothesis has reminded you of the enormous advance which science has made in this field during the last hundred years. Therefore I have to make a great claim on your unbiased scientific receptivity. . . . If my ideas should prove worthless, they will be put on the shelf here more quickly than anywhere else, before they can do harm.[14]

Ostwald proceeded to demonstrate, to his own satisfaction but not to that of more than a handful of his chemistry colleagues, that one could redefine elements, compounds, and solutions without reference to the atomic conception while maintaining consistency with the empirically determined stoichiometric laws of chemical combination. He accomplished this by exploiting Gibbs's thermodynamic work and by classifying the equilibrium of chemical and physical systems in the language of the Gibbs phase rule. At the same time he drew heavily on van't Hoff's law of mobile equilibrium

OSTWALD

and Le Châtelier's principle that a chemical system under stress tends to resist change by moving in the direction that opposes the stress. Ostwald designated a system as a "hylotropic" body when the properties of each of its coexisting phases remain unchanged during the passage from one phase to another. He wrote:

> A substance, or a chemical individual, is a body which can form hylotropic phases within a definite range of temperature and pressure. . . . There are substances which have never been transformed into solutions, or whose sphere of existence covers all accessible states of temperature and pressure. Such substances we call elements. In other words, *elements are substances which never form other than hylotropic phases.*[15]

Drawing on the "emanation" experiments of his close friend Sir William Ramsay, and using the notion that elements conform to regions of low potential energy in matter, Ostwald proposed a stalactite model of element transmutation in which the high energy barriers (long stalactites) represented the lighter elements and the lowest energy barriers (extremely short stalactites) represented the heaviest transmutable elements.

The conversion to atomism, or at least to a granular conception of the structure of matter, by Ostwald and such other diehards as Lord Kelvin, Mach, Wald, and Helm came rather suddenly about 1906, when evidence for the particulate nature of matter became dramatic. Cathode ray experiments had prepared the way for J. J. Thomson's identification of these rays as atoms of electricity—electrons—for which the ratio of charge to mass could be calculated. In their paper "The Cause and Nature of Radioactivity" (1902), Rutherford and Soddy had used their experimental results as a means of "obtaining information of the process occurring within the chemical atom."[16] By 1906 Perrin had drawn attention to experimental observations on Brownian motion that offered a quantitative method for studying the random motion of molecular collisions. Einstein and Smoluchowski had shown theoretically (1905–1906) that the thermal motion of particles could be subjected to accurate statistical analysis by means of the kinetic theory of Brownian motion.

In 1909, in the preface to the fourth edition of his *Grundriss der allgemeinen Chemie*, Ostwald made a straightforward confession about his adoption of the idea of the physical existence of atoms:

> I am now convinced that we have recently become possessed of experimental evidence of the discrete or grained nature of matter, which the atomic hypothesis sought in vain for hundreds and thousands of years. The isolation and counting of gaseous ions, on the one hand, which have crowned with success the long and brilliant researches of J. J. Thomson, and, on the other, the agreement of the Brownian movement with the requirements of the kinetic hypothesis, established by many investigators and most conclusively by J. Perrin, justify the most cautious scientist in now speaking of the experimental proof of the atomic nature of matter. The atomic hypothesis is thus raised to the position of a scientifically well-founded theory, and can claim a place in a text-book intended for use as an introduction to the present state of our knowledge of General Chemistry.

Then he added a comment that seems to say that his rejection of atomism was fully justified before the experimental evidence cited was available: "From the point of view of stoichiometry the atomic theory is merely a convenient mode of representation, for the facts, as is well known, can be equally well, and perhaps better, represented without the aid of the atomic conception as usually advanced."[17]

The paradox connected with the timing of this turnabout by Ostwald is revealed by the fact that during the first decade of the twentieth century, the "billiard ball" model of the kinetic theory of gases and of the Daltonian atom was rapidly being abandoned in favor of an internally structured atom of considerable complexity. To use an appropriate analogy that had been employed thirty years earlier by W. K. Clifford, "an atom must be at least as complex as a grand piano."[18]

By the turn of the century, Ostwald's position on atomism and energetics was relatively obsolete among physicists and most chemists. Scientists and philosophers alike—idealists and materialists—thoroughly opposed his energetics and rejected his suggestion that the concept of matter (conceived by Ostwald as the concept of substance) be subordinated to the concept of energy. Nevertheless, it is fair to add that two elements of his energetics have endured the test of time: his formulation of the second law of thermodynamics as the impossibility of a *perpetuum mobile* of the second kind, and his early insistence on the necessity of employing the free energy functions of Gibbs and Helmholtz, instead of heats of reactions, as a criterion for the feasibility of chemical spontaneity and as a measure of the equilibrium position of chemical reactions.

During the last decades of his life at Grossbothen, Ostwald worked with great determination on the development of color science. Starting from color standardization, which around 1912 was a question of topical interest, he systematically investigated colors, developed a quantitative color theory, and produced color samples and coloring substances in his laboratory. He thereby gave a new impetus to this previously neglected field of applied chemistry, and he considered the problems in color science among the most important that he was able to solve. The way in which Ostwald carried out his research on color is a perfect example of a well-organized and carefully thought-out undertaking. Among the achromatic colors he included the white-gray-black continuum. Following the determination of pure gray, he ascertained the gray scale—the standard achromatic scale required for the measurement of chromatic colors—with the aid of a split-field photometer of his own construction. In this work Ostwald introduced new concepts: saturated colors, unsaturated colors, and completely saturated colors (*Vollfarbe*). For unsaturated chromatic colors he used the relationship $v + w + s = 1$, and for achromatic colors $v + s = 1$, where v represents *Vollfarbe; w, weiss* (white); and *s, schwarz* (black). Postulating that "the colors of bodies are fundamentally mixtures of light,"[19] Ostwald introduced the term *Farbenhalb* for a mixture containing half of all of the wavelengths of the visible spectrum. For measuring the colors of bodies, he employed triangles of equal color tone and color-tone circles. In order to determine color tone, Ostwald developed a polarization color mixer. As indicator numbers for standardizing the bright colors he used the appropriate numbers of the hundred-part color circle and the percentage of the white or black content of the pigments. Thus the "world of color was subordinated to the mastery of measurement and number."[20] In order to achieve the standardization of the color circles and colored bodies (*Farbkreisel*), Ostwald published an atlas of color (1917) and an atlas of color standards (1920).

From color standardization Ostwald turned to color harmony. Because he chose his color scales in accord with the characteristics of human perception—that is, they were logarithmically graded—he was able to construct harmonies, as in music. The gray ladder corresponded to the somber colors; varying the color tone, or the white and black content, corresponded to the bright harmonies. Ostwald enthusiastically set forth his views in *Harmonie der Farben* (1918), *Die Farbe* (1921–1926), and *Harmonie der Formen* (1922). His ideas were criticized and even rejected, especially in artistic circles; but they also found widespread acceptance. Through his color standards and color harmonies Ostwald gave a new and far-reaching impetus to the construction of colors according to a deliberately planned method. In Germany the application of his system was confined mainly to Saxony and to the Bauhaus in Dessau. In Great Britain, on the other hand, it was widely endorsed and was taught in the schools. In the United States his system found important advocates in E. Jacobson and H. Zeishold, through whose influence the system, as presented in *Color Harmony Manual*, became the accepted colors standard.[21]

During the last thirty years of his life, Ostwald spoke and wrote in a grandiloquent style in support of humanistic, educational, and cultural causes. In 1909 he published *Grosse Männer* and classified persons of genius into two broad types according to mental temperament: classicists and romanticists. The laws governing their careers were formulated with reference to mental reaction velocity. Classicists were said to be phlegmatic and melancholic and to have a low reaction velocity, while romanticists, like himself, were sanguine and choleric and had a high reaction velocity.

Ostwald believed that mutual understanding among scholars was indispensable from a humanistic standpoint. In his *Forderung des Tages* (1910), dedicated to Arrhenius, Ostwald, the doyen of the international brotherhood of science, integrated his views on energetics with scientific methodology and systematics, psychology, scientific genius, general cultural problems, public instruction in the sciences, and the introduction of an international language. While at Harvard he had studied Esperanto; and later he created his own artificial language, Ido.

Ostwald had a mania for reform movements. His *Der energetische Imperativ* (1912) was a rousing, prophetic declaration of the urgency for man to adopt internationalism, pacifism, and a systematic plan for the preservation of natural energy resources. The imperative was "Squander no energy. Utilize it." In a similar vein, *Die Philosophie der Werte* (1913) was given over largely to a discussion of the second law of thermodynamics, its history, applications, and prognostic comments.

In various works Ostwald gave mathematical formulas for happiness (G represents *Glück*). In one of these, $G = k(A - W)(A + W)$, A represents the expended energy that is welcomed by the will; W represents the expended energy that corresponds to disagreeable experiences associated with resistance; and k is the factor for transforming the energetic into the psychological process.[22]

Ostwald actively participated in the congresses of the international peace movement (1909–1911) and condemned war as a "squandering of energy of the very worst kind."[23] His support of all scientific efforts led him to join the Deutsche Monistenbund, a civic society that propagated a world view based on science. At the request of Ernst Haeckel, Ostwald was president of this organization from 1910 to 1914. His perception of the individual as a cell in the collective organism of humanity is spelled out with much enthusiasm in his *Monistische Sontagspredigten* (1911–1913). His 1913 lecture delivered at Vienna—*Monism as the Goal of Civilization*—was directed to the organization of "the partisans of Monistic ideas and principles in national Societies in order to formulate one International Organization of all Monists in the whole world."[24] The Monistenbund fell apart after the outbreak of World War I.

Ostwald was active in learned societies and served on commissions. He was a member of the International Commission on Atomic Weights and cofounder and temporary president of the International Association of Chemical Societies. In addition, for many years he was a member of the boards of directors of German chemical societies. He also showed a deep concern with the training of chemists and was an enthusiastic defender and advocate of scientifically oriented secondary school education. This many-sided scientific and organizational activity earned Ostwald, besides the Nobel Prize, many honorary doctorates and honorary memberships in learned societies.

Ostwald strongly advocated the study of the history of science and frequently used historical materials in both his scientific and his more philosophical writings. When *Isis. Revue consacrée a l'histoire de la science* was launched by George Sarton in March 1913, Ostwald's name was among the *comité de patronage*. His most important single work devoted to the history of science was *Elektrochemie. Ihre Geschichte und ihre Lehre* (1896), a book of more than 1,100 pages that exhibits Ostwald's complete command of the scientific literature on electrochemistry and allied areas. It was, however, the only one of Ostwald's major books that was not published in a second edition; neither was it translated. The monumental enterprise known as Ostwalds Klassiker der Exakten Wissenschaften began in 1889 with Helmholtz' 1847 work "Ueber die Erhaltung der Kraft"; 243 volumes had been published by 1938, and 256 volumes by 1977.

In his later years it became clear that Ostwald's deep interest in the history of science was motivated by the belief that man had much to learn from his predecessors about the economical and systematic solution to current problems. He correctly perceived that the most suitable way to realize that goal was through the organization of scientific work; he wished to avoid "energy waste." Ostwald therefore emphasized "organizational activity, which is the great task of the twentieth century." He stated (1926) that "fundamentally, in the present circumstances, I must consider the organizer as more important than the discoverer."[25] In this regard, as in many others, Ostwald was ahead of his time.

In his obituary for Ostwald, Frederik G. Donnan wrote: "It was a rich, full and successful life, in which he endeavoured to make the best use of the abundant energy accorded to him."[26] Many of Ostwald's ideas about the sciences, art, society, and culture are no longer fashionable. His organizational efforts and philosophical generalizations may no longer be appreciated. Nevertheless, he is regarded as an extremely prolific, colorful, and influential early systematizer and spokesman for the new discipline of physical chemistry.

The American physical chemist Wilder Bancroft, who received his doctorate under Ostwald in 1892, and who was one of the most critical of his students, wrote in 1933:

> We can distinguish three groups of scientific men. In the first and very small group we have the men who discover fundamental relations. Among these are van't Hoff, Arrhenius and Nernst. In the second group we have the men who do not make the great discovery but who see the importance and bearing of it, and who preach the gospel to the heathen. Ostwald stands absolutely at the head of this group. The last group contains the rest of us, the men who have to have things explained to us. . . . Ostwald was a great protagonist and an inspiring teacher. He had the gift of saying the right thing in the right way. When we consider the development of chemistry as a whole, Ostwald's name like Abou ben Adhem's leads all the rest. . . . Ostwald was absolutely the right man in the right place. He was loved and followed by more people than any chemist of our time."[27]

NOTES

1. Wilhelm Ostwald, *Das physikalisch-chemische Institut der Universität Leipzig und die Feier seiner Eröffnung am 3. Januar 1898* (Leipzig, 1898).
2. Ostwald had some strong connections with America. The American Chemical Society elected him an honorary member in 1900. Jacques Loeb invited him to Berkeley, where he delivered a lecture: "The Relations of Biology and the Neighboring Sciences," in *University of California Publica-*

OSTWALD

tions in Psychology, **1**, no. 4 (1903); a German version was published in *Abhandlungen und Vorträge auf dem Gebiet der Mathematik, Naturwissenschaften und Technik* (1916), 282–307. In 1904 he was a principal speaker at the International Congress of Arts and Sciences in St. Louis, delivering a lecture before the section on the methodology of science: "On the Theory of Science," in *Congress of Arts and Sciences*, **1** (1905), 333–352. At Harvard his closest associates were William James, Josiah Royce, Hugo Münsterberg, President Charles W. Eliot, and T. W. Richards, who had been his student at Leipzig a decade earlier.

3. Ostwald's estate was given to the German Academy of Sciences at Berlin, and is known as the Wilhelm-Ostwald-Archiv Gross Bothen, Aussenstelle des Archivs der Akademie der Wissenschaften der Deutschen Demokratischen Republik. The research on color science has been continued at Gross Bothen by an industrial research laboratory.
4. M. M. Pattison Muir, "Chemical Affinity," in *Philosophical Magazine*, 5th ser., **8** (1879), 181–203. For an exhaustive report and evaluation of Ostwald's chemical affinity studies, see the chapter "Chemical Change" in Pattison Muir's *A Treatise on the Principles of Chemistry* (Edinburgh, 1889).
5. "Recherches sur la conductibilité galvanique des électrolytes," in *Bihang till K. Svenska vetenskapsakademiens handlingar*, **8** (1884), nos. 13–14, reprinted in Ostwalds Klassiker, no 160 (Leipzig, 1907). For further references see *Aus dem wissenschaftlichen Briefwechsel Wilhelm Ostwalds*, II, 3–14, 357.
6. Ostwald's American students included Wilder D. Bancroft, S. L. Bigelow, Edgar Buckingham, G. W. Coggeshall, Frederick G. Cottrell, Colin G. Fink, H. M. Goodwin, William J. Hall, G. A. Hulett, Harry C. Jones, Louis Kahlenberg, F. B. Kenrick, Arthur B. Lamb, Morris Loeb, J. W. McBain, W. Lash Miller, James L. R. Morgan, Arthur A. Noyes, Theodore W. Richards, G. Victor Sammett, E. C. Sullivan, O. F. Tower, J. E. Trevor, A. J. Wakeman, and Willis R. Whitney.
7. *Zeitschrift für physikalische Chemie*, **8** (1891), 567.
8. *Ibid.*, **15** (1894), 706.
9. See Ostwald's *Lebenslinien*, II, 287–299, and III, 343.
10. *Zeitschrift für physikalische Chemie*, **9** (1892), 771. Also see *Aus dem wissenschaftlichen Briefwechsel. . .*, I, xviii–xxi, 9–23.
11. Michaelis Teich, "Der Energetismus bei Wilhelm Ostwald und František Wald," in *Naturwissenschaften*, supp. entitled *Tradition, Fortschritt-Beiheft zur Zeitschrift für Geschichte der Naturwissenschaften, Technik und Medizin* (1963), 147–153.
12. Erwin Hiebert, "The Energetics Controversy and the New Thermodynamics," in Duane H. D. Roller, ed., *Perspectives in the History of Science and Technology* (Norman, Okla., 1971), 67–86.
13. The circumstances of this dedication and information about Ostwald's plans for the *Annalen der Naturphilosophie* are given in four letters from Ostwald to Mach dated 31 May to 28 Oct. 1901, at the Ernst-Mach-Institut, Freiburg im Breisgau. Also see J. Thiele, " 'Naturphilosophie' und 'Monismus' um 1900. (Briefe von Wilhelm Ostwald, Ernst Mach, Ernst Haeckel und Hans Driesch)," in *Philosophia naturalis*, **10** (1968), 295–315.
14. Wilhelm Ostwald, "Faraday Lecture," in *Journal of the Chemical Society*, **85** (1904), 506–522.
15. *Ibid.*, 515–517.
16. E. Rutherford and F. Soddy, "The Cause and Nature of Radioactivity," in *Philosophical Magazine*, 6th ser., **4** (1902), 370–396; quotation on 396.
17. This work, of which there were 8 eds., was first published in 1889. The quotation is reproduced from the preface to the 3rd English ed., dated Nov. 1908, Gross Bothen, translated by W. W. Taylor, *Outlines of General Chemistry* (1912), vi.
18. Oliver Lodge, *Atoms and Rays* (London, 1924), 74.
19. Ostwald, *Farbenlehre*, II, 118–119.
20. Ostwald, *Lebenslinien*, II, 392.
21. See Ostwald, *The Color Primer* (1969), foreword by Faber Birren, 5–6; and (on Ostwald's color system) E. Jacobson, *Basic Color* (Chicago, 1948). *Color Harmony Manual* is an atlas of colors; it consists of 900 removable color chips in a magnificent portfolio created in 1942 by Egbert Jacobson, who was art director for the Container Corporation of America, Chicago.
22. Ostwald, *Lebenslinien*, III, 3–6.
23. *Ibid.*, 329.
24. Ostwald, *Monism as the Goal of Civilization*, International Committee of Monism, ed. (Hamburg, 1913), 3.
25. Ostwald, *Lebenslinien*, III, 435. Also see H.-G. Körber, "Einige Gedanken Wilhelm Ostwalds zur Organisation der Wissenschaft. Nach einem unveröffentlichten Manuskript ausgewählt," in *Forschungen und Fortschritte*, **31** (1957), 97–103, with references on Ostwald's organization work.
26. Frederik G. Donnan, "Ostwald Memorial Lecture," in *Journal of the Chemical Society* (1933), 332.
27. Wilder D. Bancroft, "Wilhelm Ostwald. The Great Protagonist," in *Journal of Chemical Education*, **10** (1933), 539–542, 609–613; quotation on 612.

BIBLIOGRAPHY

I. ORIGINAL WORKS. Ostwald left an immense literary-scientific work that consists of 45 books, about 500 scientific papers, 5,000 reviews, and the edition of six journals, particularly *Zeitschrift für physikalische Chemie*. His papers on physical chemistry include his master's thesis, *Volumchemische Studien über Affinität* (Dorpat, 1877); and his doctoral dissertation, *Volumchemische und optisch-chemische Studien* (Dorpat, 1878), both repr. in Ostwalds Klassiker der exakten Wissenschaften, no. 250 (Leipzig, 1966), with an introductory essay on Ostwald's work by Gerhard Harig and Irene Strube.

Numerous other papers include "Volumchemische Studien," in *Annalen der Physik und Chemie*, supp. **8** (1876), 154–168, and n.s. **2** (1877), 429–454, 671–672; "Chemische Affinitätsbestimmungen," "Kalorimetrische Studien," "Studien zur chemischen Dynamik," and "Elektrochemische Studien," all in *Journal für praktische Chemie*, **18** (1878)–**33** (1886); "Das Verdünnungsgesetz," *ibid.*, **31** (1885), 433–462; "Über den Einfluss der Zusammensetzung und Konstitution der Säuren auf ihre Leitfähigkeit," *ibid.*, **32** (1885), 300–374; "Über die Dissoziationstheorie der Elektrolyte," in *Zeitschrift für physikalische Chemie*, **2** (1888), 270–283; "Über freie Ionen," *ibid.*, **3** (1889), 120–130, written with W. Nernst; "Über Autokatalyse," in *Berichte über die Verhandlungen der Sächsischen Akademie der Wissenschaften zu Leipzig*, **42** (1891), 190–192; "Studien zur Energetik," *ibid.*, **43** (1891), 272–287, and **44** (1892), 211–237, also in *Zeitschrift für physikalische Chemie*, **9** (1892), 563–578, and **10** (1892), 363–386; "Die Dissoziation des Wassers," *ibid.*, **11** (1893), 521–528; "Über physico-chemische Messmethoden," *ibid.*, **17** (1895), 427–445; "Über Katalyse," in *Verhand-*

 OSTWALD

lungen der Gesellschaft deutscher Naturforscher und Ärzte, **73** (1901), 184–201, also separate ed. (Leipzig, 1902) and in *Physikalische Zeitschrift*, **3** (1902), 313–323, and *Nature*, **65** (1902), 522–526; and "Über Katalyse. Nobelpreisvortrag, gehalten in Stockholm am 12. Dezember 1909," in *Les Prix Nobel en 1909* (Stockholm, 1910), 63–88.

Ostwald's books on physical chemistry include *Lehrbuch der allgemeinen Chemie*, 2 vols. (Leipzig, 1885–1887; 2nd ed., 2 vols. in 3 pts., 1891–1902); *Grundriss der allgemeinen Chemie* (Leipzig, 1889), 1st English ed., *Outlines of General Chemistry*, translated by James Walker (London, 1890); *Hand- und Hilfsbuch zur Ausführung physiko-chemischer Messungen* (Leipzig, 1893), 1st English ed., *Manual of Physico-Chemical Measurements*, translated by James Walker (London, 1894); *Die wissenschaftlichen Grundlagen der analytischen Chemie* (Leipzig, 1894), 1st English ed., *The Scientific Foundations of Analytical Chemistry*, translated by G. McGowan (London–New York, 1895); *Grundlinien der anorganischen Chemie* (Leipzig, 1900), 1st English ed., *The Principles of Inorganic Chemistry*, translated by A. Findlay (London–New York, 1902); *Schule der Chemie*, 2 vols. (Leipzig, 1903–1904), 1st English ed., *Conversations on Chemistry. First Steps in Chemistry*, vol. I translated by E. C. Ramsay (New York–London, 1905), vol. II translated by S. K. Turnbull (New York–London, 1906).

Ostwald's works on other scientific problems include *Vorlesungen über Naturphilosophie* (Leipzig, 1902); *Abhandlungen und Vorträge allgemeinen Inhalts* (Leipzig, 1904); *Grundriss der Naturphilosophie* (Leipzig, 1908); *Die Farbenfibel* (Leipzig, 1916); *Farbatlas* (Leipzig–Gross Bothen, 1918), 1st English ed., *The Ostwald Colour Album* (London, 1932); *Die Farblehre*: I, *Mathematische Farblehre* (Leipzig, 1918); II, *Physikalische Farblehre* (Leipzig, 1919); III, *Chemische Farblehre*, with E. Ristenpart; IV, *Physiologische Farblehre*, by H. Podesta (Leipzig, 1922), for which Ostwald wrote only the introduction; V, "Psychologische Farblehre," is unpublished (some MS chapters are preserved in the Wilhelm-Ostwald-Archives)–English ed., *Colour Science*, translated by J. Scott Taylor, 2 vols. (London, 1931–1933); "Grundsätzliches zur messenden Farbenlehre," in *Sitzungsberichte der Preussischen Akademie der Wissenschaften zu Berlin*, math.-phys. Kl., **22** (1929), 14–26, and **30** (1937), 402–416; "Attribute der Farben," *ibid.*, **30** (1937), 423–436; and *The Color Primer. A Basic Treatise on the Color System of Wilhelm Ostwald*, edited and with a foreword and evaluation by Faber Birren (New York, 1969), with a short bibliography of Ostwald's color papers in English.

Ostwald's books and papers on the history of science include *Elektrochemie. Ihre Geschichte und ihre Lehre* (Leipzig, 1896); *Ältere Geschichte der Lehre von den Kontaktwirkungen. Dekanatsschrift* (Leipzig, 1898); *Leitlinien der Chemie* (Leipzig, 1906), 2nd ed. entitled *Der Werdegang einer Wissenschaft* (Leipzig, 1908); "Psychographische Studien," in *Annalen der Naturphilosophie*, **6–8** (1907–1909); *Grosse Männer. Studien zur Biologie des Genius* (Leipzig, 1909); "Chemische Weltliteratur," in *Zeitschrift für physikalische Chemie*, **76** (1911), 1–20; *Aug. Comte und sein Werk* (Leipzig, 1913); "Geschichtswissenschaft und Wissenschaftsgeschichte," in *Archiv für Geschichte der Mathematik, der Naturwissenschaften und der Technik*, **10** (1927–1928), 1–11; and *Die Pyramide der Wissenschaften* (Stuttgart–Berlin, 1929).

Autobiographic works are "Wilhelm Ostwald," in *Philosophie der Gegenwart in Selbstdarstellungen*, IV (Leipzig, 1924), 127–161; *Lebenslinien*, 3 vols. (Berlin, 1926–1927). His scientific correspondence is in *Aus dem wissenschaftlichen Briefwechsel Wilhelm Ostwalds*, H.-G. Körber, ed., 2 vols. (Berlin, 1961–1969).

Bibliographies (listed chronologically) are P. Walden, "Schriften von Wilhelm Ostwald," in *Zeitschrift für physikalische Chemie* (Jubelband für Wilhelm Ostwald), **46** (1903), xvi–xxvii; G. Ostwald, *Schriften zur Farblehre* (Leipzig, 1936); and *Gesamtregister der Abhandlungen, Sitzungsberichte . . . der Preussischen Akademie der Wissenschaften 1900–1945* (Berlin, 1966), 128–129 (Ostwald's academic papers only). Also consult Poggendorff, III, 991; IV, 1101–1103; V, 929–930; VI, 1928–1929; and VIIa, supp., 476–482, the most comprehensive published secondary bibliography on Ostwald.

II. Secondary Literature. See the following, listed chronologically: J. H. van't Hoff, "Friedrich Wilhelm Ostwald," in *Zeitschrift für physikalische Chemie* (Jubelband für Wilhelm Ostwald), **46** (1903), v–xv; P. Walden, Wilhelm Ostwald (Leipzig, 1904), with bibliography; "Wilhelm Ostwald. Leitlinien aus seinem Leben zu seinem 60. Geburtstag gesammelt," in *Grosse Männer. Studien zur Biologie des Genies*, IV (Leipzig, 1913); *Wilhelm Ostwald. Festschrift aus Anlass seines 60. Geburtstages. 2. September 1913*, Monistenbund in Österreich, ed. (Vienna–Leipzig, 1913), with bibliography; E. Haeckel, "Wilhelm Ostwald. President of the Monistic League," in *Open Court*, **28**, no. 2 (1914), 97–102; H. Freundlich, "Wilhelm Ostwald zum 70. Geburtstag," in *Naturwissenschaften*, **11** (1923), 731–732; M. Le Blanc, "Wilhelm Ostwald," in *Forschungen und Fortschritte*, **8** (1932), 174–175; W. Nernst "Wilhelm Ostwald," in *Zeitschrift für Elektrochemie*, **38** (1932), 337–341; P. Walden, "Wilhelm Ostwald," in *Berichte der Deutschen chemischen Gesellschaft*, Abt. A, **65** (1932), 101–141; R. Luther, "Nachruf auf Wilhelm Ostwald," in *Berichte. Sächsischen Akademie der Wissenschaften*, math.-phys. Kl., **85** (1933), 57–71 (sess. of Nov. 1932); F. G. Donnan, "Ostwald Memorial Lecture," in *Journal of the Chemical Society* (1933), 316–332; Grete Ostwald, *Wilhelm Ostwald. Mein Vater* (Stuttgart, 1953); N. I. Rodnyi and Y. I. Soloviev, *Vilgelm Ostvald* (Moscow, 1969); and Christa Kirsten and Hans-Günther Körber, eds., *Physiker über Physiker, Wahlvorschläge zur Aufnahme von Physikern* *in die Berliner Akademie, 1870–1929* (Berlin, 1975), 167–168.

ERWIN N. HIEBERT
HANS-GÜNTHER KÖRBER

PAGANO, GIUSEPPE (*b.* Palermo, Sicily, Italy, 21 September 1872; *d.* Palermo, 9 August 1959), *physiology.*

Pagano studied medicine at the University of Palermo and, while still a student at the Institute of Physiology, published experimental works on the methods of hypodermic absorption and toxicity of lymph. He graduated in 1895 with honors in medicine and surgery. In his dissertation, "Su di una nuova proprietà del sangue di alcuni animali," he demonstrated the physiological toxicity of blood for certain cellular elements (the agglutination of spermatozoa killed by the blood of the animal from which they came). This work was praised by Richet, who referred to the newly discovered phenomenon as the cytocidal property of blood.

In 1897–1898, on the basis of studies by Paul Heger, Francesco Spallitta, and Michele Consiglio, Pagano began systematic experimental exploration of the sensitivity of the heart and blood vessels. Using a 1 percent potassium cyanide solution as a chemical detector of intravascular sensitive zones, he concluded that the veins of the greater circulation, the endocardium of the right auricle and the right ventricle, and the pulmonary arteries are devoid of zones sensitive to stimulation by potassium cyanide; that injection of this solution into the axillary and femoral arteries leads to a reflex rise in the pressure of the greater circulation; and that injection of the potassium cyanide solution into the common carotid causes immediate and considerable cardiac inhibition, sometimes with long pauses. Such inhibition also can be obtained by injecting defibrinated blood into the common carotid under high pressure. Pagano asserted that the vascular surface that can be stimulated to produce a slowing of the heartbeat—or even cardiac arrest—lies between the origin of the common carotid and its bifurcation. He also stated that the most sensitive region is that closest to the carotid bifurcation.

Pagano's results were published in *Giornale di scienze naturali ed economiche di Palermo* (1899), *Archivio di farmacologia e terapeutica* (1900), and *Archives italiennes de biologie* (1900). Through original investigations he revealed the existence of the carotid reflexes and quickly realized their general importance. He asserted that many of the phenomena previously attributed to direct stimulation of the encephalomedullary centers were of indirect origin, produced through the excitation of sensitive surfaces. Pagano's results were confirmed and amplified in 1900 by Luigi Siciliano's experimental observations, made under Pagano's guidance, of the effects of occlusion and disocclusion of the carotid.

On the basis of a report published by the physiologist François Frank in the *Bulletin de l'Académie des sciences de Paris*, the Paris Académie de Médecine in 1900 awarded Pagano the Bourceret Prize for his studies on the sensitivity of the heart and the blood vessels.

In 1912 P. Kaufmann, having misidentified the site of application of the stimuli, challenged the correctness of the observations of Pagano and Siciliano. In 1923 Bruno Kisch and S. Sakai also rejected their conclusions. That year the first publication by the physiologist H. E. Hering appeared at Cologne. After investigating the mechanism of cardiac and vasomotor reactions elicited by aspecific or specific mechanical stimuli and by electrical stimuli applied to the carotid sinus, Hering made a careful study of the innervation of that sinus, identifying the bundle of sensitive fibers that emanate from it (see *Die Karotissinus reflexe auf Herz und Gefässe* [Leipzig, 1927]). He also demonstrated that the main afferent route of the reflex is the glossopharyngeal nerve, although he was still unable to resolve the controversy concerning the carotid reflexes. Hering's observations were extended by E. Koch and especially by the physiologist Corneille Heymans, who performed an ingenious series of experiments (see *Le sinus carotidien* [London, 1929]) that led in 1929 to the full and irrefutable experimental and theoretical illustration of the highly complex mechanisms involved in the neurohumoral regulation of arterial pressure. He was awarded the Nobel Prize in 1938 for this research.

This formal recognition of Heymans' work left Pagano bitter, for he felt that his achievements had been disregarded. Moreover, several professors at the University of Palermo had nominated him for the Nobel Prize, and another recommendation was made in 1949. Pagano's bitterness was not entirely obliterated the following year, when he was made honorary professor emeritus of human physiology at the University of Palermo. This appointment finally confirmed his achievements and honored his academic career, which had been interrupted some years earlier. In 1900 he had obtained a post as lecturer in physiology, and in 1908 he had become lec-

PAVÓN Y JIMÉNEZ

turer in special medical pathology. For several years he had also taught physiological chemistry—but for various reasons he never became a full professor.

BIBLIOGRAPHY

I. ORIGINAL WORKS. In addition to works mentioned in the text, Pagano's publications include *Sulle localizzazioni funzionali del cervelletto*, (1906), which received the Fossati Prize of the Istituto Lombardo di Scienze e Lettere; and *Sulle funzioni del nucleo caudato* (1913), which won the Lallemand Award of the Académie des Sciences of the Institut de France. He also wrote a *Curriculum vitae* (Palermo, 1950); and an editorial in *Sicilia sanitaria*, **4**, no. 3 (15 Mar. 1951), 1–7.

II. SECONDARY LITERATURE. See Charles H. Best and Norman B. Taylor, *Le basi fisiopatologiche della pratica medica*, Carlo Foà, ed. (Milan, 1955), 311–312, based on the 5th American ed.; L. Condorelli *et al.*, "Sulla sensibilità dell'arteria vertebrale," in *Archivio di scienze biologiche*, **45** (1961), 281–296; Giulio C. Pupilli and Rodolfo Margaria, "Relazione all'Accademia dei Lincei," in *Atti dell'Accademia nazionale dei Lincei. Rendiconti*, classe di scienze fisiche, matematiche e naturali, **16** (1954), 568–571; and Theodore C. Ruch and Harry D. Patton, *La fisiologia e biofisica di J. F. Fulton e W. H. Howell*, **II** (Rome, 1971), 950 and *passim*, a trans. of the 14th American ed. by Vittorio Zagami and Giuseppe La Grutta. On Pagano's investigations of the carotid reflexes, see C. Heymans and E. Neil, *Reflexogenic Areas of the Cardiovascular System* (London, 1958).

BRUNO ZANOBIO
DELFINO LAURI

PAVÓN Y JIMÉNEZ, JOSÉ ANTONIO (*b.* Casatejada, Cáceres, Spain, 22 April 1754; *d.* Madrid, Spain, 1840), *botany*.

The son of Gabriel Pavón and Josefa Jiménez Villanueva, Pavón lived from the age of eleven in Madrid with his namesake uncle, "second pharmacist" of Charles III. This situation no doubt turned his interest toward pharmacy and, hence, to the study of plants. The central event in his life, serving as junior partner to Hipólito Ruiz on the royal botanical expedition to the viceroyalty of Peru (1777–1788), came about through his pharmaceutical training. Pavón worked in the royal pharmacies of Buen Retiro and San Ildefonso from 1773 to 1777, but, unlike Ruiz, apparently never obtained a license to practice. He had studied botany during this time, under Casimiro Gómez Ortega; but he acquired most of his knowledge of plants in Peru and Chile. He married soon after returning from South America (probably 1789) and had at least one son, José Antonio (*b. ca.* 1803), whom he sought to place in a botanical career. Pavón was a member of the Real Academia de Medicina, the Real Sociedad Económica de Amigos del País, and the Real Academia de Ciencias, all in Madrid. He was also elected to societies in Berlin, Lisbon, Montpellier, Bordeaux, and the Moselle, and to the Institut de France. After being rebuffed for five years, he was elected a foreign member of the Linnean Society of London in 1820. No portrait of him is known to exist.

The lack of details on Pavón's life is consistent with his forty years in the shadow of Ruiz, not only during the expedition but also for the many years thereafter in Spain, as they sought to publish their findings. All Pavón's publications but one (of fourteen pages) were as junior author to Ruiz, although he did leave three unpublished manuscripts; and it was Ruiz who wrote the account of the expedition. (One should consult the article on Ruiz for information on the achievements of Ruiz and Pavón.) Following the death of Ruiz in 1816, when Pavón at last took charge, there was lots to do but little was ever done. Three-quarters of the expedition's findings remained unpublished; but the Spanish government apparently did no more than occasionally prod the aging Pavón into brief flurries of action, as each new ministry sought to justify the expense of this project they knew so little about. When Pavón, "touching the threshold of decrepit age," was censured for the final time in 1835, the *Flora peruviana et chilensis* was at last a dead letter, instead of a merely atrophied one.

Ruiz leaves the impression that Pavón was timid in the face of danger; others found him docile, even indolent. The contrast with the irascible Ruiz sometimes turned Ruiz's enemies into friends of Pavón's, and vice versa. Pavón could be obsequious and, in his last years, self-pitying. The most crushing evaluation we have is that of a commission established in 1831 to learn the status of the *Flora peruviana*. The members concluded that the death of Ruiz had "paralyzed in an absolute fashion the scientific works of that Office, there not being in it a person who could continue them."

It is unfortunate that Pavón is so hidden in the records of the expedition, for in fulfilling the essential eighteenth-century role of botanizer, he experienced years of deprivation and hardship in the tropical forests of Peru. And in meeting the needs of a science swamped by new finds, Pavón joined Ruiz in announcing 141 new genera still recognized today. Over 500 species continue to bear names given by Ruiz and Pavón. The Spanish crown had been hard put to find any accomplished botanists when the French Académie des Sciences proposed the expedition. Ruiz and Pavón, when chosen, were only twenty-two years old. Spain sent major botanical expeditions into others of its realms, but only that of Ruiz and Pavón published any of its findings during the participants' lives.

Ironically, Pavón's greatest individual contribution to the world of botany came by chance. The Napoleonic occupation of Spain had halted the botanists' work. In 1814 Pavón found a new source of income: selling herbarium duplicates to the British collector Aylmer Bourke Lambert. Over the next eleven years he disposed of thousands of specimens, including many from the Sessé-Mociño Mexican expedition, of whose finds he had become the guardian. The relationship with Lambert ended on a jarring note in 1825, primarily over financial arrangements. The experience did not prevent Pavón from seeking other customers, among them Philip Barker Webb, to whom he sold 4,500 species in 1826–1827. Webb's collection is today at the Botanical Institute of the University of Florence; Lambert's has been scattered over Europe and, of late, in the United States.

Although the botanical world has suffered confusion from Pavón's mixing of specimens from various Spanish realms, his private merchandising ventures gave scientists easier access to knowledge of Spanish-American plants than they would otherwise have had. To the Spanish government, however, Pavón's actions smacked of treason, for Spain could no longer publish anything not already known to the rest of Europe. Investigators turned up shortages in the property of the *Flora* office; and Pavón found himself, at the age of eighty-one, obliged somehow to repay the government. Whether he ever did is lost to the records—as are many other details of his career.

BIBLIOGRAPHY

I. ORIGINAL WORKS. Aside from the works he wrote with Hipólito Ruiz, Pavón had only one brief publication, "Disertación botánica sobre los géneros *Tovaria, Actinophyllum, Araucaria y Salmia*," in *Memorias de la Real academia médica de Madrid*, **1** (1797), 191–204. He nearly completed "Nueva quinología," which was published by John Eliot Howard as *Illustrations of the Nueva Quinología of Pavón* (London, 1862). Two unpublished MSS are "Laurographia" and an index of the plants of Peru and Chile.

II. SECONDARY LITERATURE. See Agustín J. Barreiro, *Don José Antonio Pavón y Jiménez, 1754–1840*, extracted from the proceedings of the Asociación Española para el Progreso de las Ciencias, 1932 (Madrid, 1933[?]), a seven-page work drawn from documents belonging to Pavón's descendants. Hortense S. Miller, "The Herbarium of Aylmer Bourke Lambert," in *Taxon*, **19** (1970), 489–553, gives details on the disposition of Lambert's herbarium. Arthur R. Steele, *Flowers for the King: The Expedition of Ruiz and Pavón and the Flora of Peru* (Durham, N.C., 1964), from which the quotations in the above article are taken, is a detailed study of the expedition of Ruiz and Pavón and its aftermath.

ARTHUR R. STEELE

PEALE, REMBRANDT (*b.* near Richboro, Pennsylvania, 22 February 1778; *d.* Philadelphia, Pennsylvania, 3 October 1860), *natural history, technology.*

Rembrandt Peale was the son of Rachel Brewer Peale and Charles Willson Peale, a distinguished American artist also noted for the innovative scientific displays at his Philadelphia museum. The family held a respected position in local society; but the museum expenses, an erratic income from Charles Peale's portraiture, and the large number of children created occasional financial strains. Rembrandt Peale received an elementary education in the local private schools and learned to paint and draw under the casual guidance of his older siblings, with more structured tutelage later from his father and other Philadelphia-based artists. He married Eleanora Mary Short in 1798; they had ten children.

About 1801 Peale attended James Woodhouse's chemistry course at the University of Pennsylvania in order to learn to use pigments more effectively. He continued his education abroad, studying with Benjamin West in London during 1802–1803 and with several French artists from 1808 to 1810. These sojourns included informal advice on science from Joseph Banks and Georges Cuvier and on the chemistry of porcelain tints from Alexandre Brongniart. Peale led a nomadic life until he finally settled in Philadelphia in 1834. He lived and painted in Baltimore (1796–1801, 1814–1820), Philadelphia (1803–1808, 1811–1814, 1823–1825), New York (1820–1823, 1825–1828, 1830–1832), Italy (1828–1830), and England (1832–1834). In 1840, four years after his first wife died, he married Harriet Cany.

Although Peale is known to American historians mainly for his portraits, many of which were of scientists, he also contributed to zoology and technology. His interest in inventions began in 1785, when he saw John Fitch's steamboat; and in 1807 he visited Robert Fulton and watched some of his early steam navigation experiments. Peale's reminiscences of these events, written in 1848, constitute a minor historical source on steamboat development. Peale experimented with the gas lighting of Baltimore's streets and buildings in 1816–1820. He also had an abiding interest in technological developments in the fine arts. Besides studying the chemistry of pigments, he was among the first Americans to experiment with lithography in the 1820's; and he is credited with importing from France the encaustic method of painting, in which pigments are fixed in wax rather than in oils.

Peale's fascination with natural history was evident in his earliest published material, some poems celebrating the wonders of science, included in a pamphlet printed by his father (1800). His important scientific work began in 1801, when he accompanied his father to the peat bogs of Orange County, New York, where they excavated the skeletons of mastodons. Peale helped reassemble the bones of one skeleton for display in his father's Philadelphia museum and used the rest of the material to make a second skeleton for a traveling exhibit, which he showed in New York and London in 1802. He wrote three descriptions of the creature to accompany his exhibit, a single sheet published in New York (1802) and two pamphlets printed in London (1802, 1803). In each essay Peale argued with increasing assurance that the teeth were those of a carnivorous animal. He announced in the 1803 pamphlet (pp. 38–39) that an artist "will sooner and with more certainty, establish the character of skeletons, than the most learned anatomist, whose eye has not been accustomed to seize on every peculiarity. . . ." Cuvier's study of mastodons, however, reestablished the scientific hegemony of the anatomist over the artist by convincing naturalists of the herbivorous function of the animals' teeth. A more enduring part of Peale's third essay was the dramatic narrative of how the skeletons were unearthed; the story was a captivating prose analogue to his father's painting "Exhuming the Mastodon" (1806–1808).

John D. Godman, who had married Peale's daughter Angelica in 1821, quoted his father-in-law's narrative of the excavation in his three-volume *American Natural History* (Philadelphia, 1826–1828), a work on mammals written for a popular audience. Godman's text, which was reprinted often through 1862, plus the excerpts from Peale's publications that appeared in English and American journals, made Peale's account known to a diverse readership of several generations. Peale also publicized the mastodon by displaying the skeleton at the museum he founded at Baltimore in 1814; it remained there until the museum's dissolution in 1830. In 1824 Peale published a brief piece on reproduction in the opossum, but it was a flawed study based on an observation of only a few animals. Peale's historical importance to American science remained his interest in technology and his popularizing of the mastodon.

BIBLIOGRAPHY

I. ORIGINAL WORKS. The Charles Willson Peale Papers project at the National Portrait Gallery is preparing a definitive microfilm and selective letterpress ed. of the correspondence and unpublished MSS of Rembrandt Peale, his brothers, and their father. The staff has created biographical and bibliographical files that were useful in writing this article.

Rembrandt Peale's poems about science appeared in Charles W. Peale, *Discourse Introductory to a Course of Lectures on the Science of Nature* (Philadelphia, 1800). His first piece on the mastodon was *A Short Account of the Behemoth or Mammoth* (New York, 1802); a copy of this broadside is at the American Philosophical Society, Philadelphia. The London pamphlets were *Account of the Skeleton of the Mammoth, a Non-Descript Carnivorous Animal of Immense Size, Found in America* (1802); and *An Historical Disquisition on the Mammoth, or, Great American Incognitum, an Extinct, Immense, Carnivorous Animal, Whose Fossil Remains Have Been Found in North America* (1803). For citations to excerpts printed from these works, see Max Meisel, *Bibliography of American Natural History: The Pioneer Century, 1796–1865*, III (Brooklyn, N.Y., 1929), 363–365; and Robert Hazen, *Bibliography of American-Published Geology: 1669–1850* (Boulder, Colo., 1976), entries 10296–10299. Peale's last scientific article was "Interesting Facts Relative to the Opossum," in *Philadelphia Museum*, **1** (1824), 6–8. His recollections of the steamboat appeared in "Letter From Mr. Rembrandt Peale to a Member . . . January 13, 1848," in *Collections of the Historical Society of Pennsylvania*, **1** (1853), 734–736.

II. SECONDARY LITERATURE. The *Dictionary of American Biography* is a convenient source; but for accuracy and detail on any member of the Peale family, Charles C. Sellers, *Charles Willson Peale*, 2 vols. (Philadelphia, 1939–1947), is more dependable. It lists Peale's publications on nonscientific topics and also cites earlier biographical sources, many of which are not reliable. Other useful assessments are Wilbur Harvey Hunter, *The Peale Museum* (Baltimore, 1964); and Detroit Institute of Arts, *The Peale Family* (Detroit, 1967).

MICHELE L. ALDRICH

PENSA, ANTONIO (*b.* Milan, Italy, 15 September 1874; *d.* Pavia, Italy, 17 August 1970), *anatomy, histology, embryology.*

The son of Michele Pensa and Giuseppina Calzini, Pensa graduated in medicine and surgery at Pavia University in 1898, having studied at the institute directed by Golgi. After lecturing in human anatomy at Pavia from 1900 to 1915, he became professor of this subject, first at Sassari (1915–1920), then at Parma (1921–1929), where he was also rector of the university, and finally at Pavia (1930–1948), where he was dean of the faculty of medicine from 1931 to 1945. From 1950 to his death Pensa was director of Pavia's Center for the Study of Neuroanatomy. A distinguished teacher and member of numerous scientific academies, he had many pupils who became professors of anatomy or of other subjects.

Pensa made important contributions to human and comparative morphology, histology, and microscopic anatomy; animal and plant cytology; and embryology. He was the first to demonstrate Golgi's "internal reticular apparatus" in nonnerve cells; the myoid elements of the avian thymus; the ways of development of the *arteria intercostalis suprema*; the structure and development of the avian Hewson-Panizza glands; the morphological changes of the chondriome and Golgi's reticulum in the ossification zone; the presence of certain chondriosome-like and other morphologically comparable formations of the internal reticular apparatus in some plant cells; and the changing aspects of various cytoplasmatic organelles under different conditions. Through extensive research Pensa added to knowledge of the structure of the nervous system, especially to interpretation of the concepts inherent in Golgi's doctrine of the diffused network and in that of the neuron according to Ramón y Cajal. His treatise on general histology (1926) contains many personal contributions.

Endowed with a wide-ranging general and humanist education, and interested in the history of the natural sciences and medicine, Pensa took part in founding the museum of the history of the University of Pavia. He was its director from 1938 until his death.

In 1976 a marble bust of Pensa was placed in the Aula Scarpa in the Anatomical Theater of the University of Pavia.

BIBLIOGRAPHY

I. ORIGINAL WORKS. There is no complete bibliography of Pensa's publications, some of which are listed in Antonio Pensa, *Attività scientifica didattica ed accademica* (Pavia, 1936); and "Notizie e pubblicazioni scientifiche di Antonio Pensa," in *Annuario della Pontifica Accademia delle scienze*, n.s. **1** (1936–1937), 619–627. His most important writings are *Trattato di istologia generale* (Milan, 1926; 5th ed., 1961); *Trattato di anatomia sistematica*, 2 vols. (Turin, 1933–1935), written with Giuseppe Favaro; and *Trattato di embriologia generale* (Milan, 1944).

II. SECONDARY LITERATURE. See Alfonso Giordano, "Antonio Pensa," in *Rendiconti dell'Istituto lombardo di scienze e lettere*, Parte generale e atti ufficiali, **104** (1970), 98–105; Enrica Malcovati, "Ricordo di Antonio Pensa," *ibid.*, 106–109; and Emilio Casasco, "Antonio Pensa," in *Annuario dell'Università degli studi di Pavia per gli anni accademici 1969 . . .* (Pavia, 1975), 1045–1050, with portrait.

BRUNO ZANOBIO

PEURBACH (or **PEUERBACH**), **GEORG** (*b.* Peuerbach, Austria, 30 May 1423; *d.* Vienna, Austria, 8 April 1461), *astronomy, mathematics.*

Georg Peurbach, the son of Ulrich, was born in Upper Austria, about forty kilometers west of Linz. Nothing is known of his early life. He matriculated for the baccalaureate at the University of Vienna in 1446 as Georgius Aunpekh de Pewrbach and received the bachelor's degree in the Arts Faculty on 2 January 1448. Two years later he probably became a licentiate, and on 28 February 1453 he received the master's degree and was enrolled in the Arts Faculty. The last notable astronomer at Vienna, John of Gmunden, had died in 1442, prior to Peurbach's arrival, so it is not clear with whom he studied—if, indeed, he did formally study astronomy at the university. It is possible that he had access to astronomical books and instruments collected by John of Gmunden.

At some time during the period 1448–1453 Peurbach traveled through Germany, France, and Italy. Regiomontanus, in a lecture on the progress and utility of the mathematical sciences delivered at Padua in 1464 (printed with the treatises of al-Farghānī and al-Battānī [Nuremburg, 1537]), says that many in his audience must have heard Peurbach lecture on astronomy in that city. Peurbach also lectured in Ferrara and is said to have been offered positions at Bologna and Padua. It is also said that in Ferrara he made the acquaintance of Giovanni Bianchini, the most noted Italian astronomer of the period, who attempted to persuade him to accept a position at an Italian university. He may have met Nicholas Cusa in Rome at this time and certainly came to know him in later years, since Cusa sent an inscribed copy of his *De quadratura circuli* to Peurbach, who proceeded to point out its errors to Regiomontanus. At any rate, Peurbach seems already to have acquired an international reputation at the time of his Italian sojourn, although, as far as is known, he had written nothing.

After his return to Vienna, Peurbach engaged during the period 1453–1456 in a correspondence (published by Albin Czerny) with Johann Nihil of Bohemia, the court astrologer to Emperor Frederick III in Wiener Neustadt. Ten letters, only two from Peurbach, survive as a result of their inclusion in a collection of specimen letters appended to a treatise on letter writing. On Nihil's advice, Peurbach accepted the position, at a salary of 24 pounds, of court astrologer to King Ladislaus V of Hungary, the young nephew of Frederick. At some later time, perhaps after the death of Ladislaus in 1457, Peurbach became court astrologer to the emperor, since Regiomontanus refers to him as *astronomus caesaris* in the dedication of the *Epitome of the Almagest* and cites his service to Frederick in the lecture given at Padua.

While Peurbach's court appointments were made for his abilities in astronomy and astrology, his responsibilities at the university, to judge by the admittedly scanty evidence, were concerned mostly with humanistic studies. In 1454 and 1460 he lectured on the *Aeneid*, in 1456 on Juvenal, in 1457 possibly on the *Rhetorica ad Herennium*, and in 1458 possibly on Horace. In 1458 he also participated in a disputation, *De arte oratoria sive poetica*, which survives (Munich, Clm 19806, fols. 193–199), as does a treatise from 1458 on letter writing (Clm 18802, fols. 86–97) attributed to Peurbach (written under a pseudonym). A number of undistinguished Latin poems by Peurbach are also known (Vienna, Vin 352, fols. 67a–69b).

Peurbach's student and associate Johannes Müller von Königsberg, known as Regiomontanus, matriculated in the arts faculty at Vienna on 14 April 1450 at the age of thirteen, and received his bachelor's degree on 16 January 1452. His collaboration with Peurbach, to whom he later referred as "my teacher," probably began after Peurbach received his master's degree. Peurbach's own works seem to date from 1454 and later, and a number of them were copied by Regiomontanus in a notebook (Vin 5203) that he kept at Vienna during 1454–1462. The notebook begins with Peurbach's *Theoricae novae planetarum*, completed 30 August 1454, and contains a number of Peurbach's shorter works written during the 1450's. Peurbach says in a letter of 1456 to Nihil (Czerny, 302) that he and Regiomontanus are both calculating ephemerides from Bianchini's tables, checking discrepancies in their calculations by recomputing with the Alphonsine Tables. Both observed Halley's Comet in June 1456; Peurbach mentioned it in a letter to Nihil (Czerny, 298–299) and wrote an astronomical and astrological report on the comet that was not discovered and published until the twentieth century. In June 1457 Peurbach observed another comet; and on 3 September he and Regiomontanus observed a lunar eclipse, finding the observed time of mideclipse to be eight minutes earlier than predicted by the Alphonsine Tables. Evidently Peurbach had not yet completed his own *Eclipse Tables*. In 1460 they observed lunar eclipses on 3 July and 27/28 December, this time comparing the observations with Peurbach's tables, which probably were completed in 1459. These eclipse observations were first published by Johann Schöner (Nuremberg, 1544). Peurbach carried out observations leading to the determination of the latitude 48;22° (correct, 48;13°) for Vienna; and Peurbach and Regiomontanus together found, through some series of observations, an obliquity of the ecliptic of 23;28°.

On 5 May 1460 Johannes Bessarion, archbishop of Nicaea and a cardinal since 1439, arrived in Vienna as legate of Pius II. His mission was to intervene in the continuing dispute between Frederick III and his brother Albert VI of Styria and to seek aid in a planned crusade against the Turks for the recapture of Constantinople. In Vienna he met both Peurbach and Regiomontanus. Bessarion was a figure of considerable importance in the transmission of Greek learning to Italy, and his interests were sufficiently diverse to include the exact sciences. He collected a large number of very fine Greek manuscripts that he later left to the city of Venice, where they form the core of the manuscript collection of the Biblioteca Marciana. One of his plans evidently involved a new translation of the *Almagest* from the Greek to replace Gerard of Cremona's version from the Arabic and to improve upon the inferior translation from the Greek made by George of Trebizond in 1451. He also desired an abridgment of the *Almagest* to use as a textbook. Although Peurbach was unfamiliar with Greek, according to Regiomontanus he knew the *Almagest* almost by heart (*quem ille quasi ad litteram memorie tenebat*) and so took on the task of preparing the abridgment. Further plans were made for Peurbach and Regiomontanus to accompany Bessarion to Italy and there work with him, using Bessarion's Greek manuscripts as the basis of the new translation. Peurbach, however, had completed only the first six books of the abridgment when he died, not yet thirty-eight years old. On his deathbed he made Regiomontanus promise to complete the work, which the latter did in Italy during the next year or two. This account is given by Regiomontanus in his preface to the *Epitome of the Almagest*. The completed work was dedicated to Bessarion by Regiomontanus, probably in 1463, in a very careful and beautifully executed copy (Venice, lat. 328, fols. 1–117).

Peurbach's early death was a serious loss to the progress of astronomy, if for no other reason than that the collaboration with his even more capable and industrious pupil Regiomontanus promised a greater quantity of valuable work than either could accomplish separately. Of their contemporaries, only Bianchini, who was considerably their senior, possessed a comparable proficiency and originality. The equally early death of Regiomontanus in 1476 left the technical development of mathematical astronomy deprived of substantial improvement until the generation of Tycho Brahe.

No systematic effort has been made to collect or enumerate Peurbach's works and the manuscripts containing them, so any catalog is necessarily tentative, in that it probably includes some spurious works and omits some genuine writings that have not yet been located or properly identified. A list was given by Georg Tannstetter Collimitius in the catalog of distinguished mathematicians associated with the University of Vienna that he wrote as an introduction to his 1514 edition of Peurbach's *Tabulae eclipsium* and Regiomontanus' *Tabula primi mobilis*. Tannstetter's list, which was based upon manuscripts collected near the end of the fifteenth century by his teacher Andreas Stiborius Boius (Andreas Stöberl), appears to be generally—and possibly completely—reliable. The works discussed below are listed by Tannstetter, supplemented by some later discoveries. Where manuscripts or printed editions of a given work are known, some, but not all, are listed here. A more extensive list of manuscripts can be found in E. Zinner's *Verzeichnis* (nos. 7691–7761), and some additional manuscripts and printed editions can be found in the text and notes of Zinner's *Regiomontanus*.

Theoricae novae planetarum is an elementary but thorough textbook of planetary theory written by Peurbach to replace the old, and exceedingly careless, so-called *Theorica planetarum Gerardi*, a standard text written probably in the second half of the thirteenth century. The original version of the *Theoricae novae*, completed in 1454 (e.g. Vin 5203, fols. 2a–24a), contained sections on the sun, moon, superior planets, Venus, Mercury, characteristic phenomena and eclipses, theory of latitude, and the motion of the eighth sphere according to the Alphonsine Tables. Peurbach later enlarged the work (e.g. Florence, Magl. XI, 144, fols. 1a–15b) by adding a section on Thābit ibn Qurra's theory of trepidation. Regiomontanus brought out the first printed edition (Nuremberg, *ca.* 1474). Zinner reports no fewer than fifty-six editions through the middle of the seventeenth century; there are also a substantial number of manuscript copies, mostly from the late fifteenth century. A number of printings from the 1480's and 1490's in small quartos (e.g. 1482, 1485, 1488, 1490, 1491), also containing Sacrobosco's *De sphaera* and Regiomontanus' *Disputationes contra Cremonensia in planetarum theoricas deliramenta*, seem to represent the standard school edition and common text, which is generally sound. The colored figures in these editions are copied from Regiomontanus' printing, while contemporary manuscripts contain figures of greater diversity and complexity. The diagrams are of considerable importance, since parts of Peurbach's text would be unintelligible without them.

The *Theoricae novae* contains very careful and detailed descriptions of solid sphere representations of Ptolemaic planetary models that Peurbach based either upon Ibn al-Haytham's description of identical models in his *On the Configuration of the World* (translated into Latin in the late thirteenth century) or upon some later intermediary work. Peurbach's book was of great importance because his models remained the canonical physical description of the structure of the heavens until Tycho disproved the existence of solid spheres. Even Copernicus was to a large extent still under their influence, and the original motivation for his planetary theory was apparently to correct a number of physical impossibilities in Peurbach's models relating to nonuniform rotation of solid spheres.

Since the *Theoricae novae* was intended as an elementary work, much of it is devoted to definitions of technical terms; along with the *Epitome* it helped to establish the technical terminology of astronomy through the early seventeenth century. As the standard textbook of planetary theory, it was the subject of numerous commentaries (see Zinner, *Verzeichnis*, nos. 7700–7714). There were printed commentaries by Albert of Brudzewo

(1495), Joannes Baptista [or Franciscus] Capuan (1495, 1499, 1503, 1508, 1513, 1518), Erasmus Reinhold (1542, 1553), Oswald Schreckenfuchs (1556), Pedro Nuñez Salaciense (1566, 1573), and others. The most interesting are those by Reinhold and Nuñez. The *Theoricae novae* was translated into French, Italian, and Hebrew; there are no modern editions or translations.

Possibly related to the *Theoricae novae* is a short work (Vin 5203, fols. 88a–92a) called *Speculum planetarum*, on the making of manuscript equatoria with revolving disks of paper.

Recognized throughout the sixteenth century as a monument of industry, the *Tabulae eclipsium*, completed probably in 1459, is Peurbach's most impressive work and was still used (although critically) by Tycho near the end of the sixteenth century. There are a substantial number of manuscript copies (especially Venice, lat. 342, and Nuremberg, Cent. V 57, fols. 10a–19b and 108a–153b, both copied by Regiomontanus), and the work was printed very beautifully in a version edited by Tannstetter (Vienna, 1514). The tables are based entirely on the Alphonsine Tables, in that the underlying parameters are exclusively Alphonsine; but Peurbach expanded and rearranged the tables needed for every step in eclipse computation, saving the calculator much time and relieving him of a number of tedious procedures. The tables in the printed version run to fully 100 pages; and earlier manuscripts, which tend to squeeze more on a page, have over ninety pages of closely written digits. Most remarkable, and evidently most laborious to compute, are the forty-eight-page double-entry tables (solar and lunar anomaly) of time between mean and true conjunction or opposition and the twelve-page triple-entry tables (solar longitude, lunar anomaly, time from noon) of the difference of lunar and solar parallax in longitude and latitude for the sixth and seventh climates (latitudes about 45°–49°) that are used to find the time and location of apparent conjunction in solar eclipses.

The tables exist in two forms. Originally they were computed for the meridian of Vienna, and this, with some minor alterations in the instructions, was the version later printed; but a number of manuscripts (such as Vin 5291, fols. 100a–163a) contain a version with the epoch positions shifted 0;30 hours (error for 0;22 hours) to the east to adapt the tables to the meridian of Grosswardein (now Oradea, Hungary). In this version they were dedicated to Johann Vitez, the bishop of Grosswardein, and were known as the *Tabulae Waradienses*.

The instructions for the use of the tables are very clear and are notable for giving two fully worked examples: the solar eclipse of (civil) 18 July 1460 and the lunar eclipse of (civil) 28 December 1460. The latter was observed by Peurbach and Regiomontanus. Comparison of the observation with computation from Peurbach's tables for (astronomical) 27 December is as follows:

	Observation	Computation
Beginning of eclipse	$11;42^h$	$11;32^h$
Beginning of delay (totality)	$12;47^h$	$12;42^h$
End of delay (totality)	$13;55^h$	$13;58^h$

The agreement is good but, as expected, is no better than the comparison with the Alphonsine Tables made using the lunar eclipse of 3 September 1457.

According to Regiomontanus, Peurbach was responsible for the first six books of *Epitoma Almagesti Ptolemaei* (also known by slight variants of this title), the most important and most advanced Renaissance textbook on astronomy, while books VII–XIII were completed by Regiomontanus after Peurbach's death. But this account of the division of labor and credit probably requires some modification. The introduction and first six propositions of book I, giving the general arrangement of the universe, are in part translated and in part paraphrased from the Greek *Almagest* and must be the work of Regiomontanus, possibly with assistance from Bessarion. Further, this section of the work is not in Venice, lat. 329, a manuscript preserving a version of the text with numerous marginal corrections, largely of Greek forms of proper nouns, that are probably in the hand of Bessarion. Venice, lat. 329, is earlier than any other surviving manuscript and contains a preliminary and incomplete version of the text. With one important exception, all other manuscripts descend from a later, complete version. The exception is Venice, lat. 328, which contains a further revision of the text prepared by Regiomontanus for Bessarion, to whom it is dedicated in a note in Regiomontanus' hand. This is in all likelihood the best manuscript of the *Epitome*, although some comparison with others is still necessary to establish the text correctly. The first printing (Venice, 1496) is very careless; later printings were at Basel (1543) and Nuremberg (1550).

Aside from the introductory section, books I through VI are closely based upon the so-called *Almagesti minoris libri VI*, a doubtless unfinished textbook, apparently of the late thirteenth century, that supplements Ptolemy with information and procedures drawn from al-Battānī, Thābit ibn Qurra, Jābir ibn Aflaḥ, az-Zarqāl, and the Toledan Tables. The *Almagestum minor* divides Ptolemy's sometimes lengthy chapters into individual propositions showing the proof of a geometrical theorem, the derivation of a parameter, or the carrying out of a procedure, and there are occasional digressions adding the work of post-Ptolemaic writers. The *Epitome* adopts exactly this arrangement and sometimes follows the *Almagestum minor* nearly word for word, including all of its supplements to Ptolemy. Evidently Peurbach based the *Epitome* upon the earlier work; and, with all due respect to Regiomontanus' account of his teacher's contribution, one may legitimately ask to what extent the present state of the first six books is really the result of Regiomontanus' revision of what Peurbach may have left as little more than a close paraphrase of the *Almagestum minor*.

With the exception of the introductory propositions in book I, the underlying text of the *Epitome* is that of Gerard of Cremona's translation of the *Almagest*. Although the work contains a number of evidently conjectural emendations by Regiomontanus, they seem to have been made without consultation of the Greek text, except possibly for the correction of proper nouns from their Arabic-Latin forms to their Greek forms entered in Venice, lat. 329. These corrections did not extend through the entire work, and hence in all manuscripts except 328 the corrections are only partial; only in 328 are they complete.

However the credit be divided between Peurbach and Regiomontanus, the *Epitome* served as the fundamental treatise on Ptolemaic astronomy until the time of Kepler and Galileo, and remains the best exposition of the subject next to the *Almagest* itself. Although it runs to about half the length of the *Almagest*, the *Epitome* is nevertheless a model of clarity and includes everything essential to a working understanding of mathematical astronomy and even manages to clarify sections in which Ptolemy omits steps or is somewhat obscure. It has not been superseded even by the excellent modern commentaries on the *Almagest*, and the mathematical astronomy of the sixteenth century is in places unintelligible without it. The *Epitome* is the true discovery of ancient mathematical astronomy in the Renaissance because it gave astronomers an understanding of Ptolemy that they had not previously been able to achieve. Copernicus used it constantly, sometimes in preference to the *Almagest*; and its influence can be seen throughout *De revolutionibus*.

None of Peurbach's other works compares in importance with the three previously described. A provisional list of the remaining works is given below.

Iudicium super cometa qui anno Domini 1456to per totum fere mensem Iunii apparuit (St. Pölten Alumnatsbibliothek XIXa, fols. 143a–149b, published in 1960 by Lhotsky and Ferrari d'Occhieppo) is a report on the appearance of Halley's Comet in 1456. It contains observations of its position, an examination of its physical cause and nature, an estimation of its distance and size, and a judgment of its astrological import. Peurbach concludes that the comet must be at least a thousand German miles above the earth, eighty German miles in length (including the tail), and four German miles in thickness. Its significance includes drought, pestilence, and war, especially for Greece, Dalmatia, Italy, and Spain, where the comet reached the zenith, and certain trouble for individuals whose nativities have Taurus in the ascendant.

Compositio tabulae altitudinis solis ad omnes horas consists of tables of solar altitude for latitude 48° and thus is applicable to Vienna. It is in Vin 5203, fols. 54a–58a, and other manuscripts.

Instrumentum pro veris coniunctionibus solis et lunae is a description of an instrument for the rapid determination of the position of true conjunction. It is in Vin 5203, fols. 67a–69a, and other manuscripts.

Canones astrolabii is probably the work in Vin 4782, fols. 225a–270b, and Vin 5176, fols. 156a–162b.

Compositio quadrantis astrolabii is in Vin 5176, fols. 43b–47a, and other manuscripts.

Canones gnomonis (also known as *Quadratum geometricum*) survives in a manuscript in Vin 5292, fols. 86b–93b; the work was printed at Nuremberg in 1516 and was included in J. Schöner's collection (Nuremberg, 1544). Like the *Eclipse Tables*, it was dedicated to Johann Vitez. It consists of a description of an instrument made up of an open square with two graduated sides and a pointer and sight attached to turn on the vertex opposite the graduated sides. The instrument is used for measuring altitudes of heavenly bodies or objects on the earth and, by taking measurements from different positions, for determining the distance of inaccessible objects on the earth. Instructions and tables were provided for each application. Tannstetter mentions *Plura de quadrantibus*, which could refer to this or to other treatises.

There are a number of writings concerned with sundials and time measurement (Zinner, *Verzeichnis*, nos. 7725–7728a). *Instrumentum universale ad inveniendas horas quocunque climate* is in Vin 5203, fols. 80b–86a, and other manuscripts. Georg Tannstetter lists *Extensio organi Ptolemaei pro usu horarium germanicarum ad omnia climata cum demonstratione* and *Modus describendi horas ab occasu in pariete*. The first could be the work in Vin 5203, and the second seems to concern sundials mounted vertically on walls. Tannstetter also mentions a *Compositio novae virgae visoriae cum lineis et tabula nova* and a *Compositio compasti cum regula ad omnia climata* that could have described portable sundials with attached compasses. Other apparently lost works listed by Tannstetter are *Collectio tabularum primi mobilis et quarundam nova compositio cum singulari usu*, which could have been an extensive collection of tables for spherical astronomy on the order of Regiomontanus' *Tabulae directionum*, and a *Tabula nova proportionis parallelorum ad gradus aequinoctialis cum compositione eiusdem*, probably a table giving the fraction of a degree of the equator for a degree of longitude on parallel circles at intervals of one degree of latitude.

Next to the planetary equation tables in book XI of his copy of the *Almagest* (Nuremberg, Cent. V 25, fol. 80a), Regiomontanus mentions that Peurbach had made more accurate equations. Tannstetter lists *Tabulae aequationum motuum planetarum novae, nondum perfectae et ultimum completae*, which he says Johannes Angelus (Engel) (*d.* 1512) attempted to complete. One may guess that these, like the solar and lunar equation tables in the *Eclipse Tables*, were recomputations of the planetary equations at 0;10° intervals using Alphonsine parameters. Such an expansion simplifies interpolation and thus speeds the computation of positions. Tannstetter mentions a *Tabula nova stellarum fixarum*, which could be the Ptolemaic or Alphonsine star catalog corrected for precession to Peurbach's time. There is an *Almanach perpetuum cum canonibus reduxit ad nostra tempora* that appears to be an almanac at intervals of five or ten days running through an integral number of longitudinal and synodic cycles for each planet, as in the almanacs of az-Zarqāl, Abraham Zacuto, and others. Tannstetter also says that Peurbach calculated ephemerides for many years and made *sphaeras solidas* (celestial globes) and many other instruments. A *Computus* by Peurbach is listed by Zinner (*Verzeichnis*, nos. 7750–7757).

Peurbach wrote a short work on the computation of sines and chords, *Tractatus super propositiones Ptolemaei de sinubus et chordis* (Vin 5203, fols. 124a–128a); the work was twice printed (Nuremberg, 1541; Basel, 1561) along with Regiomontanus' *Compositio tabularum sinuum rectorum* and sine tables. He first explains the computation using *kardagas* (arcs of 15°) according, he says, to the method of az-Zarqāl, and then, at somewhat greater length, sets out Ptolemy's derivation from the first book of the *Almagest*. Tannstetter lists a *Nova tabula sinus de decem minutis in decem per multas millenarias partes cum usu, quae plurimum rerum novarum in astronomia occasio fuit*; and such a table of sines at intervals of 0;10° with a *sinus totus* (unit radius) of 600,000 parts survives in Vin 5291, fols. 165a–173b, and Vin 5277, fols. 288a–289b, but without an explanation of its use. A lesser but evidently popular mathematical work was Peurbach's *Algorismus* or *Elementa arithmetices* or *Introductorium in arithmeticam*, a brief elementary textbook on practical computation with integers and fractions that was printed several times in the late fifteenth and early sixteenth centuries (for instance, Hain*13598-601, 1513, 1534).

BIBLIOGRAPHY

Georg Tannstetter Collimitius's list of Peurbach's works in his *Viri mathematici quos inclytum Viennensis gymnasium ordine celebres habuit* is printed in *Tabulae eclipsium magistri Georgii Peurbachii. Tabula primi mobilis Johannis de Monteregio* (Vienna, 1514). The observations of Peurbach and Regiomontanus were first published in *Scripti clarissimi mathematica M. Ioannis Regiomontani . . .*, J. Schöner, ed. (Nuremberg, 1544). The fundamental biography of Peurbach is P. Gassendi, "Georgii Peurbachii et Ioannis Mulleri Regiomontani vita," which appears in Gassendi's *Tychonis Brahei, equitis dani, astronomorum Coryphaei vita . . . accessit Nicolai Copernici, Georgii Peurbachi et Ioannis Regiomontani astronomorum celebrium vita*, 2nd ed. (The Hague, 1655), 335–373, repr. in Gassendi's *Opera*, V (Lyons, 1658), 517–534. The next study of value is F. K. F. A. von Khautz, *Versuch einer Geschichte der österreichischen Gelehrten* (Frankfurt–Leipzig, 1755), 33–57. J. B. J. Delambre, *Histoire de l'astronomie du moyen âge* (Paris, 1819), 262–288, considers principally the precession theory of the *Theoricae novae* and other selected topics, using the commentaries of Capuan, Reinhold, and Nuñez Salaciense. G. H. Schubart, *Peuerbach und Regiomontanus* (Erlangen, 1828); J. Fiedler, *Peuerbach und Regiomontanus* (Leobschütz, Poland, 1870); and J. Aschbach, *Geschichte der Wiener Universität im ersten Jahrhundert ihres Bestehens*, I (Vienna, 1865), 479–493, do not add significantly to earlier sources. Peurbach's correspondence with Nihil is published in A. Czerny, "Aus dem Briefwechsel des grossen Astronomen Georg von Peuerbach," in *Archiv für österreichische Geschichte*, **72** (1888), 281–304.

A distinct advance in research on Peurbach is K. Grossmann, "Die Frühzeit des Humanismus in Wien bis zu Celtis Berufung 1497," in *Jahrbuch für Landeskunde von Niederösterreich*, n.s. **22** (1929), 150–325, esp. 235–254. Grossmann examined numerous MSS, and his study is especially valuable on Peurbach's humanistic activities. The most extensive catalog of MSS containing writings by Peurbach is E. Zinner, *Verzeichnis der astronomischen Handschriften des deutschen Kulturgebietes* (Munich, 1925), 241–243. The most thorough and up-to-date biographical and bibliographical study is E. Zinner, *Leben und Wirken des Joh. Müller von Königsberg genannt Regiomontanus*, 2nd ed. (Osnabrück, 1968). Peurbach is treated separately on 26–49, the notes at the end contain much information on MSS and early printings, and there is a thorough bibliography. Peurbach is considered briefly in most of the standard histories of astronomy and mathematics.

There are no modern eds. or translations of any of Peurbach's major works. The text of the section of the *Theoricae novae* concerning Mercury is given, along with an analysis of the model, in W. Hartner, "The Mercury Horoscope of Marcantonio Michiel of Venice: A Study in the History of Renaissance Astrology and Astronomy," in *Vistas in Astronomy*, **1** (1955), 84–138, repr. in Hartner's *Oriens-Occidens* (Hildesheim, 1968), 440–495, esp. 483–491. Peurbach's report on the comets of 1456 and 1457 is published with extensive analysis in A. Lhotsky and K. Ferrari d'Occhieppo, "Zwei Gutachten von Georgs von Peuerbach über Kometen (1456 and 1457)," in *Mitteilungen des Instituts fur österreichische Geschichtsforschung*, 4th ser., **68** (1960), 266–290; and K. Ferrari d'Occhieppo, "Weitere Dokumente zu Peuerbachs Gutachten über den Kometen von 1456 nebst Bemerkungen über den Chronikbericht zum Sommerkometen 1457," in *Sitzungsberichte der Österreichischen Akademie der Wissenschaften*, Math.-naturw. Kl., Abt. 2, **169** (1961), 149–169.

C. DORIS HELLMAN
NOEL M. SWERDLOW

PIANESE, GIUSEPPE (*b.* Civitanova del Sannio, Campobasso, Italy, 19 March 1864; *d.* Naples, Italy, 22 March 1933), *pathological anatomy and histology*.

After graduating in medicine and surgery at Naples in 1887, Pianese began his scientific career in 1890 as a pupil at the local Anatomical-Pathological Institute, then directed by Otto von Schrön. He became assistant in 1896, supervisor of autopsies in 1897, lecturer in pathological anatomy in 1899 and in pathological histology in 1901, acting professor of the latter subject in 1903, and associate professor of pathological histology in 1904. After qualifying as associate professor of pathological anatomy at Parma and as professor at Turin in 1902, Pianese was appointed associate professor at Cagliari by competitive examination in 1905; from there, a year later, he moved to Naples as professor of pathological histology. He became full professor of pathological anatomy in 1910; and seven years later he succeeded Schrön as head of the Naples Anatomical-Pathological Institute, a post he retained until his death. In 1932 he was appointed to the Accademia d'Italia.

Pianese's scientific work was characterized by scrupulous technique, keen observation, great restraint, and independence of thought, as was shown by his first works on chloralism (1890), on Fede-Riga disease and Cardarelli's cachectic aphthae, on the nerve endings in the pericardium, and on the capsule of the carbuncle bacillus (1892). Although he had received excellent classical anatomical-pathological training, Pianese was above all a histopathologist, with a broad background in microbiology, who was also concerned with general histopathology and experimental biology. An example in this field is his work of 1903 on splenectomy in the guinea pig, in which he notes the manifold effects on bodily development, resistance to infections, various organic reactions, and the morphological constitution of the hematopoietic organs. In some respects he anticipated the successive observations of other writers on the metabolic functions of the spleen, on the possible formation of antibodies in the spleen, on the "splenization" of the liver, and on the revitalization of the lymphoid tissue in the bone marrow after splenectomy. The concern with heredity in this work is also worthy of note.

Pianese's interest in histopathology is also demonstrated by the staining methods he devised, some of which are still used. An area of Pianese's work, important during his lifetime, dealt with minute cytological analysis of malignant tumors. He confirmed the morbidity of the tumor cells and pointed out the paradox that they exhibit great lability and a tendency to degenerative and necrotic processes along with enormous proliferative power.

Pianese's name was particularly linked to infective splenic anemia of infancy, which he first recognized as caused by a protozoan of the genus *Leishmania*. It is distinct from other forms of infantile splenic anemia that have a different and multiple etiological basis; but it is very similar to anemia of the Far East (kala-azar), with which it constitutes a visceral leishmaniasis group, frequently also found in the Mediterranean area.

Other significant studies by Pianese included papers on Sydenham's chorea, the reticuloendothelial system, parasites and protozoans, and histolytic phenomena in tumors.

BIBLIOGRAPHY

Pianese's main published works include *Beitrag zur Histologie und Aetiologie des Carcinoms*, R. Teuscher, trans. (Jena, 1897), which is supp. I of E. Ziegler and C. Nauwerck, *Beiträge zur pathologischen Anatomie und Physiologie; La technica delle autopsie* (Milan, 1911); and *Lezioni di anatomia patologica generale* (Naples, 1927).

On his life and work, see Pietro Rondoni, "Giuseppe Pianese," in *Annuario dell' Accademia d'Italia*, **7–9** (1938), 382–393; and Guglielmo Scala, "Giuseppe Pianese, Accademico d'Italia," in *Rassegna di terapia e patologia clinica*, **4**, no. 3 (Mar. 1932), 172–181. There are obituaries in *Archivio italiano di anatomia e istologia patologica*, **4** (Mar.–Apr. 1933), 145–148; *Folia medica*, **19** (30 Mar. 1933), 343–349; *Riforma medica*, **49** (1 Apr. 1933), 503–504; *Rinascenza medica*, **10** (1 Apr. 1933), ccxv–ccxvii; *Morgagni*, **75** (2 Apr. 1933), 443–444; *Archivio di radiologia*, **9** (May–June 1933), 627–628; and *Pediatria*, **41** (May 1933), 711–712.

BRUNO ZANOBIO

POISSON, SIMÉON-DENIS (*b.* Pithiviers, Loiret, France, 21 June 1781; *d.* Paris, France, 25 April 1840), *mathematical physics.*

Poisson was an example of those scientists whose intellectual activity was intimately linked to a great number of educational or administrative duties and to the authority derived from them. This responsibility and authority earned him more misunderstanding than esteem, and his reputation in the French scientific community was challenged during his lifetime as well as after his death. It was only outside France that certain results of his prodigious activity were best understood and considered worthy of perpetuating his memory. His life and work are thus of special interest for the history and philosophy of science. The same institution that gave him his training, the newly founded École Polytechnique, also assured his success. An exemplary product of a certain type of training and of a particular attitude toward scientific research, he devoted his life to both, exhausting in their service his remarkable capacity for hard work. His activities, which continued unabated through a succession of political regimes, exercised a major influence on French science. Although his ambition to continue Laplace's work by giving a true *summa* of mathematical physics was not to be realized, his numerous efforts toward this goal offer a lesson concerning the application of mathematics to natural phenomena that is still worth examining.

Poisson came from a modest family. His health, like that of several older siblings who died in childhood, was weak; and his mother had to entrust him to a nurse. His father, formerly a soldier, had been discriminated against by the noble officers, and after retiring from military service he purchased a low-ranking administrative post. Apparently it was he who first taught Poisson to read and write. The Revolution, which the elder Poisson welcomed with enthusiasm, enabled him to become president of the district of Pithiviers, which post afforded him access to information useful in choosing a career for his son. The latter had been entrusted to an uncle named Lenfant in Fontainebleau in order to learn surgery, but he lacked the prerequisite manual dexterity and showed little interest in the profession. Having failed as an apprentice, he was guided by his father toward those professions to which access had been eased by measures recently adopted by the republican regime. In 1796 he was enrolled at the École Centrale of Fontainebleau, where he soon displayed a great capacity for learning and was fortunate in having a dedicated teacher. He made rapid progress in mathematics and was encouraged to prepare for the competitive entrance examination at the École Polytechnique, to which he was admitted first in his class in 1798.

On his arrival in Paris, fresh from the provinces, Poisson had to adapt himself in several ways to a radically new life for which he had been little prepared. His easy success in his studies left him time to make this adjustment much more quickly than is usual in such cases. Lagrange, who had just begun his courses on analytic functions, found in Poisson an attentive student always capable of contributing pertinent remarks in class; and Laplace was even more impressed by Poisson's ability to assimilate difficult material. The reputation that Poisson enjoyed among his fellow students is mentioned in an article on him by Arago, whom he preceded by five years at Polytechnique. This comment is certainly an echo of direct testimony, and there is no reason to doubt Arago's assertion that during these years Poisson evinced a lively interest in the theater and in other aspects of cultural life. An openness to every new experience and a passion for learning allowed him to circumvent the difficulties that he might otherwise have encountered on account of his limited early education. However, his teachers were apparently unable to correct his innate clumsiness, for he could never learn to draft acceptable diagrams. This deficiency prevented him from advancing in descriptive geometry, which subject Monge had made a central element of the new school's curriculum and which contributed greatly to its reputation. On the other hand, Poisson possessed undoubted ability in mathematical analysis and displayed this gift in 1799–1800 in a paper on the theory of equations and on Bezout's theorem. At a time when it was difficult to recruit suitably qualified teaching personnel, this asset was sufficient to gain him nomination as *répétiteur* at Polytechnique immediately after his graduation in 1800.

Poisson owed this appointment principally to the backing of Laplace, who unwaveringly supported him throughout a career qualified as "easy" by Victor Cousin. The main reason that "easy" is the right word was that Poisson was enabled to stay in Paris in the milieu of Polytechnique. He was named deputy professor in 1802 and four years later replaced Fourier as titular professor. Thereupon, Poisson had to wait only a short time to obtain supplementary posts outside the school. In 1808 he was appointed astronomer at the Bureau des Longitudes and, in 1809, professor of mechanics at the Faculty of Sciences. It would appear that he never disdained worldly connections or the advantages to be gained through the salons, but it would be completely unjust to assume that he systematically cultivated these means of social advancement. As a participant in the Société Philomathique from 1803 and later in the Société d'Arcueil, he had no intellectual reservations about the idealism that animated Polytechnique; and if his friendships were useful, he did not cultivate them out of self-serving motives.

On 23 March 1812 Poisson was elected to the physics section of the Institute (to the place left vacant by Malus' death), and by 14 April his nomination had received the imperial approbation. This rapid confirmation shows that the authorities had not forgotten the acquiescent attitude that he had adopted in 1804. In that year Poisson prevented the students of the École Polytechnique from publishing a petition against the proclamation of the empire. He had taken this step, however, primarily to avoid a crisis at the institution to which he was devoted. He felt no genuine allegiance to the Napoleonic regime and easily accommodated himself to its overthrow. The restoration of Louis XVIII in 1814 caused no hiatus in his career, and he continued to accumulate official responsibilities. To the responsibilities he already exercised, he added that of examiner at the École Militaire in 1815 and of examiner of graduating students at Polytechnique the following year.

Within this pattern of continuity, however, Poisson's life and career entered upon a new phase at about this time. In 1817 he was married to Nancy de Bardi, an orphan born in England to émigré parents. The marriage constrained his life severely, leaving no time for anything but family, research, and professional obligations. His nomination in 1820 to the Conseil Royal de l'Université introduced him into the national educational system at the highest administrative level—at the very moment when the government's conservative general political stance was issuing in a campaign against the scientific programs and policies adopted during the Revolutionary and Napoleonic periods. Enlisting the aid of colleagues, notably Ampère, he managed to resist this pressure. His efforts in defense of science, which continued until the end of his life, constitute a considerable achievement.

While assuming these weighty pedagogical responsibilities, Poisson was becoming steadily more influential within the Academy of Sciences. Following Laplace's death in 1827, he felt it to be his mission to build upon the latter's scientific legacy; and Cauchy's exile in 1830 contributed still further to casting Poisson in the role of France's leading mathematician. It was primarily against him that Evariste Galois directed his celebrated criticism of French mathematics, and historians have been too eager to accept its validity at face value. Circumstances thrust upon Poisson more responsibility than any one man could have borne. It is all the more remarkable, then, that he published virtually all of his books during the last ten years of his life. To be sure, none of them manifests a profound originality. His many articles and memoirs (of which he himself prepared a list) must therefore be considered in order to arrive at a just assessment of his *oeuvre*. It must also be acknowledged that his books exhibit an uncommon gift for clear exposition and constitute an ambitious project for the instruction of future generations of students. He exhausted himself in the attempt to realize this project and died regretting that he had left it unfinished. Accordingly, any final judgment of his work and influence must give considerable weight to his role as educator.

It was precisely this aspect of Poisson's contribution that was ignored in the unsigned article on him published in Larousse's *Grand dictionnaire universel du XIX siècle* (1874). The author wrote: "The reputation of a mathematician really depends on extraordinary powers of analysis. The experimental scientist, whose physical discoveries the analyst formulates in incomprehensible expressions, is generally incapable of verifying the quality of the help that he is getting. . . . To the layman, what is most striking in the procedures of mathematicians, even though it really has almost no merit, is the art of making transformations. Poisson possessed that skill in a high degree. He amazed people and was taken for a great man. But, in order to be remembered, a scientist needs to have ideas, and Poisson had only those of others. Moreover, when he had to choose, as between two opposing ideas, the one that he would dignify with an application of his analysis, he generally made the wrong choice."

Although this criticism is so hostile as to amount to denigration, it does serve to bring out several of Poisson's characteristic traits. First, it stresses that the tireless manipulation of mathematical equations was his special province. His zeal for extending as far as possible the type of activity for which he had a gift was natural; and the limitations or deficiencies of his results are less interesting than what they reveal about the historical context, which was one of intense scientific activity.

Two authors who were very close to Poisson during the last decade of his life have left accounts that substantially agree as to his position within French science. Their gratitude for his help, moreover, did not lead them to abstain from criticism, and they provide several details that are not available elsewhere.

In 1840 Guillaume Libri, who had not yet become notorious for his bizarre administration of the French national archives, wrote an *éloge* of Poisson in which he affirmed, "Surely no one would dare to say that Poisson lacked inventiveness, but he especially liked unresolved questions that had been treated by others or areas in which there was still work to be done." In a note he added that Poisson, who refused to attend to two matters at once, had a small wallet for papers on which he jotted down information and recorded observations of subjects to be examined later. Libri, who evidently had the document in his possession, stated by way of example that Poisson considered research on algebraic equations and definite integrals to be hopeless. By contrast, he seems to have found it more important to pursue Euler's works on problems of *géométrie dépendant des différences mêlées*, that is, on problems of mathematical physics involving partial differential equations.

Cournot, who owed his university career to Poisson and replaced him in 1839 as chairman of the Jury d'Agrégation in mathematics, recorded in his *Souvenirs* that "the abundance, adaptability, and resourcefulness" that his benefactor displayed to a greater degree than anyone else "in involved calculation" [*dans les hauts calculs*] was combined with an eagerness to examine "all questions and preferably those that are significant for natural philosophy. . . . Despite that, or because of it," Cournot adds with discernment, "he did not enjoy the rare good fortune of developing one of those completely new and striking conceptions that forever establish the fame of their innovator in the history of science. He proceeded steadily along his path, rather than crossing into any new domain."

Although this judgment must be slightly modified, as will be seen below, it epitomizes the essential aspects of Poisson's mathematical work. Poisson was succeeded on the Conseil Royal de l'Université in 1840 by Poinsot, with whose scientific personality Cournot compared Poisson's in a felicitous contrast: "Poinsot took the opposite course to Poisson: he stuck to a few simple, ingenious ideas that were completely his own. He considered them and reconsidered them at his leisure, without worrying about producing a great deal and even (let us speak plainly) without possessing much knowledge." Although Cournot adds only that "it would take too long to explain to someone who is not a member of the profession all the different ways in which mathematics can be cultivated." Cournot obviously preferred Poinsot's originality tempered by laziness. It is also clear that Cournot was more impressed by the "abundance" of Poisson's works than he was concerned to submit them to a genuine historical critique.

Although Poisson's list of his own publications has greatly impressed posterity by its length, it has not aroused adequate critical interest. Historians of science are aware that in citing his memoirs Poisson also listed the extracts derived from them, which appeared mainly in the *Bulletin de la Société philomathique de Paris* and in the *Annales de chimie et de physique*. Accordingly, they have eliminated from the list those works thought to be duplicate or even triplicate entries and have

POISSON

been content to wonder at the remainder, which consists of nearly 300 original titles. But the classification by chronological order and by subject matter obliges the historian, precisely because of the lack of exact bibliographical data, to undertake laborious research in order to reconstruct the actual conditions under which Poisson produced his work during the Empire and the Restoration. The delay in publication of official periodicals such as the *Journal de l'École polytechnique* and the *Mémoires de l'Académie des sciences* deprived them of much of their importance as disseminators of new knowledge. Other private publications, originating in various scientific circles and learned societies, were more effective from this point of view; but they too experienced difficult periods, during which they substituted for each other. One reason for the abundance of Poisson's titles, therefore, lies in the special circumstances of the scientific life of the age.

Before considering Poisson's scientific work, let us try to complete a portrait of its author. At the end of his article Arago delivered a parting shot to the effect that Poisson had the habit of saying "Life is good for only two things: to study mathematics and to teach it." Since we cannot assume that the permanent secretary of the Academy of Sciences wrote this purely and simply for the pleasure of coining a *bon mot*, it is reasonable to believe that it contains an element of truth. Here again, Cournot furnishes the necessary context.

Cournot was told that Poisson wished to receive a visit from him, but the young *docteur ès sciences* postponed the visit for several years, fearing that he would be recruited as a *lycée* professor. Thus, Cournot's image of Poisson was very much like Arago's sketch—that of a narrow mathematical zealot. Cournot's change of opinion after several years of direct contact may therefore be accepted with confidence.

Cournot was eager to elaborate on his first impression of Poisson. Alluding to both his appearance and his style of a straightforward man of the common people, he reports that "With a formal and even elaborate manner Poisson combined great intellectual subtlety and a large store of common sense and forbearance that disposed him toward conservative ideas. Given the distinguished reputation that he enjoyed in the scientific world [in 1815], the royalist party were quite willing to take his conservative outlook for royalism, and he went along with this! . . . When the Revolution of 1830 occurred, Poisson saw no grounds for withholding his support from a government that aimed at being sensible and moderate; and he resolved to tolerate Cousin's philosophical declaration, as he had tolerated others, provided that he was not obligated to subscribe to its tenets personally and that in public life, as in mathematics, things would run their course." This portrait is very probably close to the truth. When in 1837 Poisson accepted elevation to the nobility (he became a baron), even as Laplace had done, he was displaying not so much a political conformism as the desire to neglect no measures likely to be useful in promoting the interests of science.

Reporting further on his contacts with Poisson, Cournot wrote that "At the very end of his life, when it had become painful for him to speak, I saw him almost weep from the chagrin that he had experienced as chairman of a competitive examination [for the *agrégation*], for he had become convinced that our young teachers were concerned solely with obtaining a post and possessed no love for science at all." Perhaps it was when confronted with such an audience and sensing its debased motivation that the aging teacher adopted the habit of bluntly uttering the aphorism cited by Arago. Poisson undoubtedly had forgotten how fully he himself had been protected in his youth from the cares borne by senior colleagues of the scientific establishment. Still, he had always scrupulously fulfilled all his official duties, without ever taking time from his own exhausting program of research. This conscientiousness is proof that Poisson, an unbeliever in religion, had found an ideal to which he had become increasingly willing to sacrifice even his health. Such conduct can evoke only respect.

The number and variety of the subjects that Poisson treated make it impractical to offer an exhaustive account of his scientific work. Accordingly, we shall attempt only to outline its most important aspects. In Poisson's case, a classification by subject matter would be less helpful than one by chronological order, which reveals the shifts in his interests during his long career.

The early stages of Poisson's research can be studied in the eleventh through fourteenth *cahiers* of the *Journal de l'École polytechnique* and in the second series of the *Bulletin de la Société philomathique de Paris*. In each instance the year 1807 is seen to be the *terminus ad quem* of the period in which he completed his mathematical training and was clearly seeking out subjects that readily offered scope for further development. Libri's remarks, cited above, are quite illuminating for this period. They indicate that he was particularly drawn to the integration of differential and of partial differential equations, and to their possible application in the study of the oscillations of a pendulum in a resisting medium and in the theory of sound. Naturally, coming in the wake of the great works of the preceding century, the choice of these topics did not demonstrate any particular originality. Still, this initial selection was decisive for his career, since it continued to guide all his subsequent research.

From 1808 to 1814 Poisson clearly set out to make an original contribution. At the beginning of this period the Academy of Sciences was emerging from an administrative crisis and was preparing to resume its role and assure the publication of its periodicals. This auspicious situation provided Poisson with a favorable opportunity for advancement. Fourier's first memoir on the theory of heat, read to the Institute in December 1807, was not favored with a report; and it was only through Poisson's efforts that it appeared three months later, in abridged form, in the *Bulletin de la Société philomathique de Paris*, of which journal Poisson soon assumed editorial direction. This fact is especially noteworthy. Poisson himself was not impeded by the obstacles encountered in scientific publishing, at their worst around 1810. He submitted three major papers to the Academy, all of which received flattering reports from committees that included Laplace: "Sur les inégalités des moyens mouvements des planètes" (20 June 1808), "Sur le mouvement de rotation de la terre" (20 March 1809), and "Sur la variation des constantes arbitraires dans les questions de mécanique" (15 October 1809). Poisson's career was thus off to a most promising start, and his election to the Academy came just at the moment when Biot was scheduled to report on his fourth major memoir, "Sur la distribution de l'électricité à la surface des corps conducteurs" (9 March 1812). This report was never delivered, since Poisson himself had become one of the judges. During this period he composed two other works, a new edition of Clairaut's *Théorie de la figure de la terre* (1808) and the two-volume *Traité de mécanique* (1811), a textbook. Yet these, taken together with the four memoirs, do not seem to constitute a body of work of sufficient significance to justify such a rapid rise.

Nevertheless, the memoirs are not without a certain importance. In the first, Poisson simply pursued problems raised and mathematically formulated by Laplace and Lagrange regarding the perturbations of planetary motions with respect to Kepler's solution of the problem of motion for two bodies. Even here, however, he improved the demonstration of the stability of the major axes of the orbits and of the mean motions. In finding approximate solutions by means of various series expansions, he also showed that determination of the possibility of secular inequalities required the inclusion of higher-order terms in the calculation. Finally, he simplified the mathematical treatment of these difficult equations by perceptive suggestions regarding the notation and disposition of the terms representing the perturbing function.

Poisson's remarks so impressed Lagrange that he was led to reconsider the bases of his earlier work on the subject. Specifically, he improved the analytic method of integrating the differential equations of motion subject to perturbing forces which appeared in the supplement to Laplace's *Mécanique céleste*, Book VIII (1808), where it is known as the variation of arbitrary constants. Lagrange presented his results to the Institute, and they gave rise in turn to Poisson's memoir of October 1809. In order to grasp fully what was involved in this affair, a certain amount of technical detail is necessary. The terminology and notation in the following account have been somewhat modernized for ease of comprehension.

Consider a material system described by k coordinates q_i satisfying k second-order differential equations (1) of the Lagrange type. From these equations, obtain a second set (2) by adding to the second members of (1) the partial derivatives $\delta\Omega/\delta q_i$ of a function that yields the perturbing force. It is known that every integral of the first set of equations (1), whether or not it can be completely expressed in an analytic form, depends on $2k$ arbitrary constants a_r. The method called variation of constants consists in asking which functions of time can be substituted for a_r in order that the integrals of (1), corresponding to a_r, can satisfy (2).

Lagrange's contribution was to develop fully the implications of Laplace's observation that, since the number of a_r is twice the number of the equations (1) or (2), these constants can be submitted to k well-chosen conditions. Lagrange's choice results in showing that the derivatives of the sought-for functions a_r with respect to time are the solutions of a linear system in which the coefficients of the unknowns are independent of time.

This result, simple in the formulation just cited, is what seemed to Poisson to call for a more direct derivation, which he arrived at thanks to his ingenious idea of modifying the very nature of the general problem. He introduced, besides the parameters q_i, new variables $u_i = \dfrac{\delta T}{\delta q_i'}$, with T being half the live force, a homogeneous quadratic form with respect to the time derivatives q_i. The systems of equations (1) and (2) then take "the simplest form that can be given to them," namely $\frac{du_i}{dt} = \frac{\delta R}{\delta q_i}$, which form subsists on introduction of the perturbation function Ω. One obtains a system of $2k$ first-order differential equations with respect to the $2k$ unknowns q_i, u_i as a function of time. Let (1′) be those that correspond to (1). The a_r previously considered can be identified with the first integrals of (1′): $a_r = f(t; q_i, u_i)$, and Poisson directly expressed the values of da_r/dt needed to satisfy (2′). These values are obviously nothing other than the result of the linear system proposed by Lagrange, the only difference lying in the form of the calculations and of the result.

Accordingly, there is nothing surprising about Lagrange's reaction, as expressed in the second edition of his *Mécanique analytique* (1811; Part Two, Section 8). Poisson, he writes, "has produced a fine memoir," but "perhaps he never would have thought of writing it . . . had he not been assured in advance" of the nature of the results. Lagrange adds, moreover, that the advantage of the new calculations is only apparent. For he contends that the form of the constant coefficients that he himself had given, $[a_r, a_s] = \frac{\delta q_i}{\delta a_r} \cdot \frac{\delta u_i}{\delta a_s} = \frac{\delta q_i}{\delta a_s} \cdot \frac{\delta u_i}{\delta a_r}$, is more practical, since first priority should be given to all the problems in which the integration of (1′) yields expressions of q_i and u_i as a function of a_r and of time.

This was perhaps the first occasion on which Lagrange expressed dissatisfaction with Poisson's work, but his criticism was not totally justified. The property that he called singular, namely the constancy of the coefficients $[a_r, a_s]$ is undoubtedly the same in the case of Poisson's coefficients: $(a_r, a_s) = \frac{\delta a_r}{\delta q_i} \cdot \frac{\delta a_s}{\delta u_i} = \frac{\delta a_r}{\delta u_i} \cdot \frac{\delta a_s}{\delta q_i}$; although one should not be too hasty in minimizing the importance of the changes in form.

Considering a_r, as first integrals, rather than thus designating the constants of integration, opened a new perspective on the problem. For if it is assumed that there are two first integrals α, β of the system (1′), that is to say, that two functions of the q_i, u_i, and t that remain constant because of (1′) can be determined, then the algorithm of Poisson's parenthesis (α, β) also yields a function that remains constant. It may be asked if (α, β) is not a third first integral. As Joseph Bertrand perceptively observed in 1853, (α, β) must not, in being constant, reduce to an identity because of the nature of the functions α, β. This condition limits the scope of "Poisson's theorem"; but to limit is not to destroy, and the theorem retains some interest.

At the end of his memoir Poisson noted that his method of computation immediately clarifies the identity of the mathematical problem involved when passing from the perturbations of the rotation of solids to those of a material point attracted to a center. Lacroix, who was assigned to report on the memoir to the Academy, was much impressed by this fact, but apparently few other members shared his enthusiasm. Hamilton, and then Jacobi, later derived inspiration from Poisson's calculations in creating the mathematical techniques that underlay the great developments in theoretical physics up to the start of the twentieth century. With a bit of patriotic exaggeration Bertrand stated that Jacobi, upon discovering Poisson's "parenthesis" around 1841, found it "prodigious." The truth is more modest. In *Vorlesungen über Dynamik* (1842) Jacobi proclaimed that "Poisson made the most important advance in the transformation of equations of motion since the first version of *Mécanique analytique*." This is a fair judgment. It must simply be added that, in the works preceding his election to the Academy, Poisson displayed something more than skilled calculation. His sense of formalization led him to discover analogies, to unify problems and topics previously considered distinct, and to extend definitively "the domain of the calculus" (*l'emprise du calcul*).

This expression was not peculiar to Poisson, although he used it often. In his efforts to extend the use of mathematics in the treatment of physical problems, he studied many subjects. In some cases he limited himself to brief outlines, one of which, that concerning the potential in the interior of attracting masses (1813), later gave rise to important results in electrostatics. He was more ambitious in the memoirs on electricity and magnetism and on the theory of elastic surfaces. The second of these, read to the Academy on 1 August 1814, represented his attempt to block the progress of Sophie Germain, who he knew was competing—under the guidance of Legendre—for the prize in physics.

Political events once again interrupted French scientific life, and several years elapsed before the delays in publication of specialized journals were reduced to the point where they no longer hindered the exchange of ideas. The period 1814–1827, from the fall of the Empire to the death of Laplace, constitutes the third period of Poisson's career.

He worked closely with Laplace, under whose influence he investigated a number of topics. One of these was the speed of sound in gases; here Laplace in 1816 had simply stated, without demonstration, a correction to Newton's formula. In other studies Poisson considered the propagation of heat, elastic vibration theory, and what was later called potential. He took up ideas that Laplace had proposed and at the same time conducted a vast amount of his own research. It is somewhat difficult to gain a clear view of all this varied activity, but the many debates in which Poisson participated offer considerable insight into his thinking and reveal the beginnings of an ambitious plan.

Poisson's relations with Fourier began to deteriorate in 1815 with Poisson's publication in the *Bulletin de la Société philomathique de Paris* and *Journal de physique* of sketches concerning the theory of heat. Fourier wrote on this occasion: "Poisson has too much talent to apply it to the work of others. To use it to discover what is already known would be to waste it. . . . He adds, it is true, that his method differs from mine and that it is the only valid one. But I do not agree with these two propositions. With regard to the second claim, which is unheard of in mathematics, if Poisson wants others to accept it, he must eliminate from his memoir the part [that I indicated] and take care never to eliminate from below the exponential or trigonometric signs the quantities of which the absolute value is not infinitely small" (manuscript notes written for Laplace, Bibliothèque Nationale, MS Fr. 22525, fol. 91). Fourier's slashing criticism was justified, and Poisson accepted it in three memoirs (1820–1821), which he published in the eighteenth (1820) and nineteenth (1823) *cahiers* of the *Journal de l'École polytechnique* (1820–1821). In a note to one of these papers, moreover, he admitted that he had been to the office of the secretary of the Academy to consult Fourier's prize manuscript of 1812. Nevertheless, in preceding Fourier in publication, he involved himself in an unfortunate enterprise. In dealing with "the manner of expressing functions by series of periodic quantities and the use of this transformation," he proved to be less adroit than in the formalization of mechanics. He scarcely improved either the manner or the use, and his contribution amounted to no more than an emphasis on the necessity of deepening the notion of convergence for the series under consideration.

Furthermore, it was not only with respect to integrals of the differential equations of the propagation of heat that Poisson felt uncomfortable at the kinds of calculation he was encountering. He recurred to a remark that Laplace had made in 1809 to the effect that Fourier's method involved the serious mistake of equating two infinitesimals of different orders. Poisson expended much effort in modifying the infinitesimal derivation of the fundamental equation of the motion of heat, claiming that what he cared about was rigor.

In fact, however, Poisson had misunderstood Fourier's point of view. Where Fourier was considering the flow of heat across a surface, he was thinking—like Laplace before him—of bodies being heated by intercorpuscular action. He held the caloric model of heat to be indispensable to theory. Particles of caloric combine with particles of ponderable matter engendering repulsive forces between the latter at distances too small to be detected. Inasmuch as he substituted this mechanistic model for Fourier's analysis in order to derive the same type of equations as his rival, his reasoning is artificial—not to say byzantine—in its complexity. No more than Fourier did he, at least at this juncture, reckon with the difficulties arising from the nature and physical dimensions of the constants introduced in the course of calculation. Later on, to be sure, he made some headway on that score. We shall come back to that problem.

The physics of heat, which was attracting a growing number of researchers during this period, consists of phenomena other than those that, like radiation, suggest the adoption of a hypothesis concerning events occurring on an infinitesimal, molecular scale. Poisson made worthwhile contributions to the understanding of these other areas, largely through his insistent amassing of theoretical and experimental data.

In "Sur la chaleur des gaz et des vapeurs," published in August 1823 in *Annales de chimie et de physique*, Poisson developed ideas published four months before by Laplace in Book XII of *Mécanique céleste*. Poisson introduced all the precautions needed to render the confused notion of quantity of heat susceptible to mathematical analysis. He called quantity of heat the magnitude that characterizes the transition of a given mass of gas from an arbitrary initial state of temperature and pressure to another state. This definition makes more abstract the quantitative aspect that naturally follows from the concept of heat as a caloric fluid. Poisson could thus deal comfortably with this magnitude, since for him it is simply a function q of p, ρ, and θ (pressure, density, and temperature). The equation of state $p = a\rho(1 + \alpha\theta)$ was already classic, and the growing acceptance of the notions of specific heats, at constant pressure and constant volume, allowed him to write the simple partial differential equation of which q should be the integral. He also showed that independently of any additional hypothesis, and whatever the arbitrary function used in the integration, the adiabatic transformations (the term did not yet exist) correspond to the formulas $p \cdot \rho^{\gamma} =$ constant and $(\theta + 266.67) \cdot \rho^{1-\gamma} =$ constant, γ being the ratio of the specific heats, assumed constant.

Sadi Carnot, who read this article, unfortunately took no interest in these formulas. He devoted his attention solely to the calculations in which Poisson's choice of the arbitrary function of integration was based on a hypothesis taken from Laplace. These calculations led Poisson to propose formulas for the saturated steam pressure H at temperature θ and for the quantity of heat of a given mass of steam. Although Carnot did not pick out from this memoir the results that could have been of greatest help to him, he did discover an interesting numerical fact in another article by Poisson published in the same volume of *Annales de chimie et de physique*.

In this brief note on the relative velocity of sound, Poisson discussed the formula that Laplace had proposed to correct Newton's treatment. The new formulation took account of variations in temperature produced by vibrations of the air. He reported that a comparison of this formula with observations led him to conclude that atmospheric air subjected to a sudden compression of 1/116 is heated 1°C. This "indirect" manner of investigating the physics of gases agrees "very closely" with the interpretation of the experiments of Clément and Desormes. Carnot was their disciple, and mention of them in the article was probably what attracted his interest to it. Considering the place of this numerical value in *Réflexions sur la puissance motrice du feu* (1824), it is important to explain both the origin and the nature of Carnot's concern with this aspect of the subject—especially as the explanation challenges simplistic notions of scientific progress. We must, however, resist the temptation to study all such examples offered by Poisson's activity. These may clearly be seen in the records of his debates with Fresnel on light and with Sophie Germain and Navier on the vibration of elastic surfaces. Poisson continued to depend on the notion of the corpuscle and the forces to which it is subject as his fundamental conceptual tool in studying all areas of physics. Accordingly, it is not surprising that while he became increasingly interested in vibratory models, he never took the final step of admitting transverse vibrations—or, indeed, any oscillation occurring in a direction different from that of the pressures constituting the corpuscular equilibrium. For this reason he was incapable of providing creative solutions to problems arising in this domain.

Poisson was justly accused by his opponents of having failed to cite contemporary scientists in his writings. But this oversight—which, in any case, was never complete—can readily be repaired by internal criticism. For example, Poisson's work reflects Sophie Germain's critical remarks, which were as cutting as Fourier's, concerning his method for mathematizing the treatment of elasticity. Specifically, he added to his molecular hypotheses a coefficient for determining the macroscopic relationship between lengthening or extension of elastic materials and variation in thickness. Similarly, his dispute with Navier over the general laws of elasticity led Poisson to demonstrate the importance of the notion of the normal state. On the same occasion, he showed that in calculating the deformed state it is illusory to conserve coefficients expressed by definite integrals that have a value of zero in the normal state and that remain approximately zero. In sum, it cannot be denied that Poisson showed real tact in "subjecting" physical phenomena to mathematical calculation.

It is in this context that Poisson's contribution to the mathematical treatment of attractive force must be mentioned. His interest in this subject dates from his analysis in 1812–1813 of James Ivory's work; and here again Laplace was the source for his examination of the integral V of the inverse of the distance function, which function was first called potential by George Green. Late in 1813 Poisson pointed out that in addition to the equation for the attraction exerted by a mass on an external point, namely

$$\Delta V = \frac{\partial^2 V}{\partial x^2} + \frac{\partial^2 V}{\partial y^2} + \frac{\partial^2 V}{\partial z^2} = 0,$$

the equation $\Delta V = -4\pi\rho$ must also be considered, where ρ is the density around the point x, y, z, when this point is inside the attracting mass. The demonstration provoked several objections, and Poisson attempted to prove it, notably in two memoirs of 1824 published in the *Mémoires de l'Académie des sciences* (**5** [1821–1822/1826], 247–338, 488–533). In 1826 he stated the triple equation,

$$\Delta V = 0, \quad = -2\pi\rho, \quad = -4\pi\rho,$$

depending on whether the point x, y, z is internal to, on the surface of, or external to the attracting mass.

"Mathematicians have long known of the first of these cases," he added. "An investigation several years ago led me to the third. To it I now add the second, thus completing this equation. Its importance in a great number of questions is well known ("Sur la théorie du magnétisme en mouvement," *ibid.*, **6** [1823/1827], 463).

Actually, it was Green who saw how to exploit Poisson's formulation. Green seized on the importance of the above memoirs in his *Essay on the Application of Mathematical Analysis to the Theories of Electricity and Magnetism* (1828), although he showed little interest in resolving the analytical difficulties still inherent in them. In 1839 Gauss emphasized the need for an improved demonstration of the intermediate equation $\Delta V = -2\pi\rho$, which was not given until 1876 (Riemann). Poisson apparently never knew that he had inspired Green's work. Indeed, up until 1828 the discussions that we have summarized above were continued mainly in relation to an ambitious project for promulgating a charter for applied mathematics.

In the preface to the long "Mémoire sur l'équilibre et le mouvement des corps élastiques" (14 April 1828), the hints yield to explicit declaration. In applying mathematics to physics, Poisson stated, it was necessary at first to employ abstraction and "in this regard, Lagrange has gone as far as possible in replacing physical ties by equations between coordinates." Now, however, "along with this admirable conception," it is necessary to "construct physical mechanics, the principle of which is to reduce everything to molecular actions." In other words, the death of Laplace the previous year enabled Poisson to move boldly ahead with his long-range plans and to present himself as Laplace's successor.

The last period of Poisson's life was dominated by his feeling of being the chosen leader of French science—at a time when he already felt called upon to guide the future of the French university system. Since it was during this last period that he decided to publish books, this sense of a twofold mission determined the form that they took; and these works are therefore treatises, concerned primarily with pedagogical matters. Although ably written and including clear historical accounts of the topics treated, they did little to advance contemporary research. Poisson was always reticent about the contributions of his contemporaries. He even omitted Poinsot's name at the very place in the second edition of *Traité de mécanique* (1833) where he was expounding the new mechanics of solids. Nevertheless, Poisson did have a genuine talent for summarizing the state of knowledge and the theoretical situation in areas of active research. This skill is evident in the report that he presented to the Academy on Jacobi's *Fundamenta nova theoriae functionum ellipticarum*, which was published in 1831.

It would, however, be unfair to reduce Poisson's work during the last period merely to an intelligent reformulation of data commingling his own results with those from other sources. In *Théorie mathématique de la chaleur* (1835), reprinted in 1837 with an important supplement, he offered evidence of his own originality in his treatment of the integration of the auxiliary differential equation

$$\frac{d^2P}{dx^2} - \frac{m}{x^2}P + \alpha^2 P = 0,$$

which is encountered in the problem of heat distribution inside bodies. He showed that the same integral series that is derived from Bessel's work has the sum

$$P = A\,x^{i+1}\int_0^\pi \cos(\alpha x \cos\omega)\sin^{2i+1}\omega\,d\omega,$$

i being such that $m = i(i+1)$.

Although he was concerned only with those cases in which this equation leads to simple results in finite form, the expression for the Bessel function that it gives rise to is often—and justly—called Poisson's integral.

Similarly, as Gaston Bachelard observed, Poisson scored a point in this work by demonstrating how the conductibility of heat in the interior of bodies, far from being contained in the notion of flux as Fourier had held, must be derived from an absorption coefficient that restores a neglected functional dimension. It was in this area that the contribution mentioned in connection with Poisson's mechanical model for conduction of heat was the most fruitful. That conception enabled Poisson to understand on the molecular scale that the complete and correct equation for radiation of heat is

$$C\frac{\partial u}{\partial t} = K\left(\frac{\partial^2 u}{\partial x^2} + \frac{\partial^2 u}{\partial y^2} + \frac{\partial^2 u}{\partial z^2}\right) + \left(\frac{dK}{du}\right)\left[\left(\frac{\partial u}{\partial x}\right)^2 + \left(\frac{\partial u}{\partial y}\right)^2 + \left(\frac{\partial u}{\partial z}\right)^2\right],$$

where u is the temperature at the point x, y, z, as a function of time t; C is the specific heat; and K is the coefficient of conductibility. Fourier's method, on the other hand, led to the omission of the factor dK/du.

Whatever the value of Poisson's argument in this

regard, the admirable program of physical mechanics, as he conceived it, did not provide a model of natural phenomena that could successfully withstand the passage of time.

Poisson believed that he had corrected an error in Laplace's treatment of capillarity by introducing a variation of the liquid density to characterize the action of the container walls, but he was as mistaken in this area as he was in his handling of the radiant heat of the earth and of the atmosphere. Were it not for the titles of these treatises, one would not guess that they were intended as the first draft of a course on mathematical physics.

Toward the end of his life, Poisson turned his attention to other subjects, producing two works of considerable repute. The first, *Recherches sur la probabilité des jugements en matière criminelle et en matière civile* (1837), is significant for the author's participation in an important contemporary debate. The legitimacy of the application of the calculus to areas relating to the moral order, that is to say within the broad area of what is now called the humanistic sciences, was bitterly disputed beginning in 1820 in politically conservative circles as well as by Saint-Simonians and by such philosophers as Auguste Comte. Poisson was bold enough to take pen in hand to defend the universality of the probabilistic thesis and to demonstrate the conformability to the order of nature of the regularities that the calculus of probability, without recourse to hidden causes, reveals when things are subjected to a great number of observations. It is to Poisson that we owe the term "law of large numbers." He improved Laplace's work by relating it explicitly to Jacob Bernoulli's fundamental theorem and by showing that the invariance in the prior probabilities of mutually exclusive events is not a necessary condition for calculating the approximate probabilities. It is also from Poisson that we derive the study of a problem that Laplace had passed over, the case of great asymmetry between opposite events, such that the prior probability of either event is very small. The formula for evaluation that he proposed for this case, which is to be substituted for Laplace's general formula, was not recognized or used until the end of the nineteenth century. The formula states that $P = \sum_{1}^{n} \frac{\omega^n}{n!} e^{-\omega}$ expresses the probability that an event will not occur more than n times in a large number, μ, of trials, when the probability that it will occur in any one trial equals the very small fraction ω/μ. In fact, Poisson's work in this area was not accepted or applied by his contemporaries except in Russia under Chebyshev; it seemed, rather, an echo of Laplace's and was barely accorded the attention appropriate to an excellent popularization. It was many years before the importance of the Poisson distribution was recognized.

The second work in this category, *Recherches sur le mouvement des projectiles dans l'air* (1839), was far better known in its day. It is the first work to deal with the subject by taking into account the rotation of the earth and the complementary acceleration resulting from the motion of the system of reference. A decade after its publication it inspired Foucault's famous experiment demonstrating the earth's rotation. Poisson, who had supervised Coriolis' doctoral research, recognized the importance of his invention of a term to correct for the deviations from the law of motion that arise in a rotating reference system. Unfortunately, Poisson did not consider himself obliged to cite the name of the actual inventor of the term.

The very multiplicity of Poisson's undertakings might be considered a reason for his failure to enjoy the good fortune of which Cournot spoke. He constantly exploited the ideas of other scientists, often in an unscrupulous manner. Still, he was frequently the first to show their full significance, and he did much to disseminate them. In the last analysis, however, what was the scientific value of his own contribution? It is tempting to reply that he merely displayed a singular aptitude for the operations of mathematical analysis. And it is no accident that he rejected the view of Lagrange and Laplace that Fermat was the real creator of the integral and differential calculus. To be sure, he waited until he was free to voice his disagreement; for it was not until 1831, in "Mémoire sur le calcul des variations," that, ignoring French chauvinism, he dared to assert "This [integral and differential] calculus consists in a collection of rules . . . rather than in the use of infinitely small quantities . . . and in this regard its creation does not predate Leibniz, the author of the algorithm and of the notation that has generally prevailed." In discussing this question in its historical context, Poisson clearly revealed that aspect to which he was most sensitive and for which he was most gifted.

Is this all that can be said? We do not think so. Having reproached Poisson for being unable or unwilling to understand him, Galois wrote, perhaps thinking of him: "The analysts try in vain to conceal the fact that they do not deduce; they combine, they compose. . . . When they do arrive at the truth, they stumble over it after groping their way along." But—and this is precisely the point—Poisson was not one of those analysts who attempt to obscure the way in which they really work. He did combine a great deal, compose a great deal, and stumble frequently—and often guess right. He had a tremendous zeal for changing the manner of treating problems, for fashioning and refashioning formulas, and for taking from the experiments of others that material to which his mathematical techniques would be applied. This zeal occasioned criticism and ironic comments, but it was nevertheless rooted in genuine ability, one capable of motivating an experimental approach.

Poisson was certainly not a genius. Yet, just as surely, he was one of those without whom progress in French science in the early nineteenth century would not have occurred.

BIBLIOGRAPHY

I. ORIGINAL WORKS. *Catalogue des ouvrages et mémoires scientifiques*, based on Poisson's MS, groups the titles by the periodicals in which they were published and includes summaries; it was reprinted in *Oeuvres complètes de François Arago*, II (Paris, 1854), 672–689. His books include *Traité de mécanique*, 2 vols. (Paris, 1811; 2nd ed., enl., 1833); *Formules relatives aux effets du tir du canon sur les différentes parties de son affût* (Paris, 1826; 2nd ed., 1838); *Nouvelle théorie de l'action capillaire* (Paris, 1831); *Théorie mathématique de la chaleur* (Paris, 1835), 2nd ed. printed with *Mémoire et notes formant supplément . . .* (Paris, 1837); *Recherches sur la probabilité des jugements en matière criminelle et en matière civile* (Paris, 1837); and *Recherches sur le mouvement des projectiles dans l'air en ayant égard à leur figure et leur rotation, et à l'influence du mouvement diurne de la terre* (Paris, 1839).

II. SECONDARY LITERATURE. On Poisson and his work, see François Arago, *Oeuvres complètes*, II (Paris, 1854), 591–698; Gaston Bachelard, *Étude sur l'évolution d'un problème de physique; la propagation thermique dans les solides* (Paris, 1927; 2nd ed., 1973), 73–88; Joseph Bertrand, "Sur un théorème de Poisson," in J. L. Lagrange, *Mécanique analytique*, 4th ed. (Paris, 1867), 484–491; Henri Pierre Bouasse, *Dynamique générale* (Paris–Toulouse, 1923), 316–317; Antoine Augustin Cournot, *Souvenirs* (Paris, 1913), 160–167; Sophie Germain, *Recherches sur la théorie des surfaces élastiques* (Paris, 1821), preface and 1–12; "Examen des principes qui peuvent conduire à la connaissance des lois de l'équilibre et du mouvement des solides élastiques," in *Annales de chimie et de physique*, **38** (1828), 123–131; and *Oeuvres philosophiques* (Paris, 1879), 344–348, 353; C. G. J. Jacobi, *Vorlesungen über Dynamik* (Berlin, 1866), 6, 51, 67; and Guillaume Libri, "Troisième lettre à un américain sur l'état des sciences en France," in *Revue des deux mondes*, 4th ser., **23** (1 Aug. 1840), 410–437.

PIERRE COSTABEL

RAMSAUER, CARL WILHELM (*b.* Osternburg, Oldenburg, Germany, 6 February 1879; *d.* Berlin, Germany, 24 December 1955), *physics.*

The son of a Lutheran clergyman, Ramsauer graduated from secondary school in 1897. To prepare for a career as a Gymnasium teacher he studied mathematics and physics at the universities of Munich, Tübingen, and Kiel. He taught at a secondary school for a short period but soon decided on a career in science. In 1902 he received his doctorate from Kiel with a dissertation entitled "Über den Ricochetschuss." Its subject, which he chose, was the ricochet of a projectile upon encountering a water surface. After four years at the torpedo laboratory in Kiel, Ramsauer went to Heidelberg in 1907, to study with Philipp Lenard, whom he had met at Kiel. In 1909 he became a researcher at the newly founded radiological institute of the University of Heidelberg. He qualified as a university lecturer that same year with a *Habilitationsschrift* entitled "Experimentelle und theoretische Grundlagen des elastischen und mechanischen Stosses." In 1915 Ramsauer was appointed extraordinary professor, and in 1921 he became full professor and director of the physics institute of the Technische Hochschule in Danzig. He gained international recognition for his discovery, in 1920, of the Ramsauer effect.

In 1927, while in Berlin, Ramsauer accepted an unexpected offer from the Allgemeine Elektrizitäts-Gesellschaft (AEG) to establish and direct a research institute for the company. He held that post until the institute was closed at the end of World War II. He then became the director, with rank of full professor, of the physics institute of the Berlin Technische Hochschule, where he had been honorary professor since 1931. He retired in 1953.

Ramsauer received an honorary doctorate from the Technische Hochschule in Karlsruhe and was a corresponding member of the Academy of Sciences of Göttingen and of the German Academy of Aviation Research. Reacting to the decline and disregard of physics in Germany under the Nazi regime, he lodged vehement protests with the minister of science at considerable personal risk.

From 1940 to 1945 Ramsauer was president of the German Physics Society. His special interest in the teaching of physics in secondary schools is shown in his *Grundversuche der Physik in historischer Darstellung* and his founding of the periodical *Physikalische Blätter*.

In order to pursue research on the ballistics problem treated in his doctoral dissertation, Ramsauer joined the staff of the torpedo laboratory in Kiel. He also continued to deal with ballistics while at Danzig and Berlin. His most important work in this area concerned the explosive pressure impulse resulting from firing shells underwater and the firing of a moving rifle to maximize the pressure impulse.

While at the Heidelberg institute, Ramsauer became acquainted with Lenard's methods and research interests. In 1913 he published an account of what is now called the Ramsauer circular method, which allowed him to ascertain with a high degree of precision the velocity, and thus the energy, of slow electrons, which he had obtained through use of the photoelectric effect. The electrons were forced into circular paths in a uniform magnetic field, separated out by means of suitably mounted diaphragms, and directed toward a collector connected to an electrometer. In this way the number of electrons arriving during each second could be determined exactly. After World War I, Ramsauer used this apparatus to discover the effect named for him.

Initially, Ramsauer studied the absorption in gases of electrons with steadily decreasing velocity. The results at first confirmed expectations; the penetrability of the gas atoms decreased. But at velocities corresponding to one electron volt (EV) the penetrability of argon exhibited a striking increase, attaining a maximum at 0.5 EV—as if only 4 to 6 percent of its kinetic-theory cross section were operative. On this occasion Ramsauer coined the term—and the associated concept—"effective cross section" (*Wirkungsquerschnitt*), which became one of the most common technical terms in atomic and nuclear physics. Analogous findings were made for xenon and krypton; no reasonable explanation of the effect was possible at the time, however, for it is a consequence of the wave nature of the electron and could not be accounted for until the development of wave mechanics. Ramsauer subsequently collaborated with his students on extensive studies of other gases and ions that yielded a large number of findings. Although his work at the AEG left him little time for his own research, he was nevertheless able to attract outstanding coworkers and to consider a host of new problems. The most important project with which he was associated during these years was the development of an electron microscope with electrostatic lenses.

BIBLIOGRAPHY

I. ORIGINAL WORKS. An early work of Ramsauer's is "Über eine direkte magnetische Methode zur Bestimmung der lichtelektrischen Geschwindigkeitsverteilung," in *Annalen der Physik*, 4th ser., **45** (1914), 961–1002. On the Ramsauer effect, see "Über den Wirkungsquerschnitt der Gasmoleküle gegenüber langsamen Elektronen," *ibid.*, **64** (1921), 513–540; and **66** (1921), 546–558; and *Wirkungsquerschnitt der Edelgase gegenüber langsamen Elektronen*, Ostwald's Klassiker der Exacten Wissenschaften, no. 245 (Leipzig, 1954), written with R. Kollath. See also "Negative und positive Strahlen," in H. Geiger and K. Scheel, eds., *Handbuch der Physik*, XXII, pt. 2 (Berlin, 1933), 243–310, written with R. Kollath. Later writings are *Physik, Pädagogik, Technik* (Karlsruhe, 1949); and *Grundversuche der Physik in historischer Darstellung* (Berlin, 1953). A complete bibliography of Ramsauer's works is in Poggendorff, V, 1020; VI, pt. 3, 2116; and VIIa, 668–669.

II. SECONDARY LITERATURE. On Ramsauer and his work, see (listed chronologically) O. Eisenhut, "Zum 60. Geburtstag," in *Zeitschrift für technische Physik*, **20** (1939), 33–36, with portrait; and E. Brüche, "Zum 70. Geburtstag," in *Physikalische Blätter*, **5** (1949), 51–53, with portrait; and "Nekrolog," *ibid.*, **12** (1956), 49–54, with portrait.

F. FRAUNBERGER

RUSCELLI, GIROLAMO (or **Alexis of Piedmont**[?]) (*b.* Viterbo, Italy; *d.* Venice, Italy, *ca.* 1565), *medicine, technological chemistry.*

In the mid-1550's there appeared numerous editions of a book entitled *Secreti* and attributed to an "Alessio Piemontese," otherwise unknown. It was issued first from Italian presses—the earliest edition so far known was published in Venice in 1555—but the Italian text had been translated into French and English by 1558, and by the end of the sixteenth century the work had gone through more than fifty European editions, including Latin and German translations. The author, the presumed Alexis, explained in the preface to his book that in fifty-seven years of travel he had acquired a fund of natural secrets that now, at the age of eighty-four, he had decided to make public for the profit of mankind. The *Secreti* goes on to enumerate a wide range of empirically discovered recipes, including medicinal compounds, cosmetic preparations, and formulas for the chemical technology of pigments and dyes, metallurgy, and jewelry. No trace of Alexis' original version, supposed to have been composed in Latin, has ever been found.

A decade later, in 1567, a collection of similar *Secreti nuovi* was published under the name of Girolamo Ruscelli, a minor literary figure who had died shortly before. Ruscelli's introduction to this work presents quite a different account of the genesis of the *Secreti* of Alexis. Ruscelli explained that some years earlier, living in the Kingdom of Naples, he had belonged to a philosophical academy of perhaps two dozen members. The members had shared a house and laboratory, and had employed assistants (apothecaries, goldsmiths, perfumers) to help them in the experimental study of nature, with the aim of improving the human situation. Not only the *Secreti nuovi*, asserted Ruscelli, but "those earlier ones which I published a few years ago of Donno Alessio Piemontese . . . were in truth all collected in the aforesaid Academy and were tried and found out by our successful company." Which of these two highly circumstantial accounts is the true one is impossible to decide with certainty. But it is worth noting that the *Secreti* of "Alexis" presents one medicinal decoction as having been communicated to the author "in Bologna in 1543 by Signor Girolamo Ruscelli," which makes the claims of the *Secreti nuovi*, ten years later, seem somewhat less gratuitous.

The phenomenal popularity of the original *Secreti* implies a considerable influence. Lynn Thorndike supposed that it was the original stimulus behind the flood of books of secrets that began at this time. The *Secreti nuovi*, however, with its extremely interesting and very early plan for a scientific academy apparently was not widely read and is today a rare book.

BIBLIOGRAPHY

These two works have been studied with some care in John Ferguson, "The Secrets of Alexis. A Sixteenth-Century Collection of Medical and Technical Receipts," in *Proceedings of the Royal Society of Medicine*, **24** (1930), 225–46, which discusses the problem of authorship and gives a bibliography of early editions. Ferguson had seen no editions printed earlier than 1557; J. R. Partington subsequently cited the one of 1555, and claimed once to have seen an earlier edition (Rome, 1540) in a bookseller's catalogue (*History of Chemistry*, II [London–New York, 1961], p. 28). More recently, Stanisław Szpilczyński, "*Tajemnice* Mistrza Aleksego Pedemontana," in *Kwartalnik historii nauki i techniki*, **16** (1971), 27–51, has reviewed the known editions and looked at the work in the context of early experimental science. The contents of the *Secreti* have received little attention: see B. Boni, "Luto e cimatura da fonderia tra i segreti alchemici di Alessio Piemontese," in *Fonderia italiana*, **5** (1956), 3–12; and Jerzy Piaskowski, "Technologia metali w *Tajemnicach* Aleksego Pedemontana," in *Kwartalnik historii nauki i techniki*, **16** (1971), 53–65.

MICHAEL MCVAUGH

RUSSELL, EDWARD JOHN (*b*. Frampton-on-Severn, Gloucestershire, England, 31 October 1872; *d*. Goring-on-Thames, Oxfordshire, England, 12 July 1965), *agricultural chemistry, agronomy*.

Born into a poor family, Russell was encouraged to obtain an education by his father, a schoolmaster and Unitarian minister. He studied at the University College of Wales at Aberystwyth and in 1894–1896 at Owens College (now part of the Victoria University of Manchester), receiving the B.Sc. in chemistry in 1896. For his research work at Owens College on the combustion rates of gases, he received the D.Sc. from London University in 1901. In 1900 he became interested in the chemistry of microorganisms.

The plight of Manchester slum dwellers gave Russell the idea of establishing them on agricultural settlements. In 1901–1907 he taught at the Wye Agricultural College, where, at first, he wished to gain the requisite training to implement his idea. Although obliged to give up the scheme as impractical, Russell obtained a sound knowledge of farm problems and introduced the latest science into his agricultural courses, which formed the basis of his highly significant *Soil Conditions and Plant Growth* (1912). While studying soil oxidations, he found to his surprise that partial sterilization would increase soil productiveness.

In 1901 Russell joined the world-renowned Rothamsted Experimental Station, Hertfordshire, where, on becoming director in 1912, he led teams of experts in applying science to the improvement of crop production. His research disclosed that partial sterilization of soil samples, by destroying protozoa—their presence hitherto unrecognized—would allow a bacterial increase, with consequent increase of nutrients available for plant growth. This discovery represented a new research frontier in soil microbiology, in which Russell was a pioneer, particularly in studying the relationship of nitrate formation to microbial fluctuations, soil gases, and organic debris. In 1917 he was elected a fellow of the Royal Society.

Adding staff and facilities after World War I enabled Russell to expand the research effectiveness of Rothamsted through an emphasis on soil chemistry, physics, microbiology, plant diseases and pests, and the statistical analysis of data. Laboratory results were verified by extensive field tests.

Through his leadership at Rothamsted, Russell attained international recognition as an agricultural expert. Trips abroad extended his influence on agricultural science: the Sudan in 1923, to improve cotton production; Palestine in 1927–1928, to relate increased agriculture to immigration; the Soviet Union on four occasions; India in 1936–1937 and 1951, to examine research projects—as well as journeys to the Continent and lecture tours of North America, New Zealand, and Australia. Russell's travels provided the source material for *World Population and World Food Supplies* (1954), a comprehensive survey of food production capacities and a call for cooperation between the more- and the less-advanced countries to meet increasing nutritional requirements.

In 1931 Russell raised sufficient funds to prevent the loss of Rothamsted to real estate developers. His long directorship, which he resigned in 1943, was characterized by focusing on what research would be most beneficial and on a humane understanding of nutritional problems. Russell's worldwide influence on agricultural policies was attested by his many honors, including knighthood in 1922 and the gold medal of the Royal Agricultural Society in 1954.

BIBLIOGRAPHY

Russell's writings include "Influence of the Nascent State on Dry Carbon Monoxide and Oxygen," in *Transactions of the Chemical Society*, **77** (1900), 361–371; "Oxidation in Soils and Its Connection With Fertility," in *Journal of Agricultural Science*, **1** (1905), 261–279; "The Effect of Partial Sterilization of Soil on the Production of Plant Food," *ibid.*, **3** (1909), 111–144, written with H. B. Hutchinson; *Soil Conditions and Plant Growth* (London, 1912; 9th ed., 1961), *The Fertility of the Soil* (Cambridge, 1913); "The Nature and Amount of the Fluctuations in Nitrate Contents of Arable Soils," in *Journal of Agricultural Science*, **6** (1914), 18–57; *Plant Nutrition and Crop Production* (Berkeley–Los Angeles, 1926); *Man and the Machine* (London, 1931); *English Farming* (London, 1932); *The Farm and the Nation* (London, 1933); "Rothamsted and Its Experiment Station," in *Agricultural History*, **11** (1942), 168–183; "Agriculture in Europe After the War," in *Journal of the Royal Agricultural Society of England*, **104** (1943), 151–164; *World Population and World Food Supplies* (London, 1954; repr. 1956); *Science and Modern Life* (London, 1955); *The Land Called Me* (London, 1956), his autobiography; and *A History of Agricultural Science in Great Britain* (London, 1965).

An obituary is in *Biographical Memoirs of Fellows of the Royal Society*, **12** (1966), 456–477, with complete bibliography.

RICHARD P. AULIE

RUSSELL, HENRY CHAMBERLAINE (*b.* West Maitland, New South Wales, Australia, 17 March 1836; *d.* Sydney, Australia, 22 February 1907), *astronomy, meteorology*.

Russell was the son of the Hon. Bourne Russell. Educated at the West Maitland Grammar School and at Sydney University (B.A. 1858), he went to the Sydney observatory as an assistant in January 1859. In August 1870 he was appointed government astronomer. He immediately began to expand the activities of the observatory and established numerous meteorological stations throughout New South Wales, manned by volunteer observers.

Russell organized and equipped four Australian parties to observe the transit of Venus on 9 December 1874. At the Paris astrophotographic congress of April 1887 he undertook, on behalf of the Sydney observatory, to cooperate in the construction of a photographic chart and catalog of the sky. (He agreed to cover southern declinations between 54° and 62°.) A new objective was purchased, and a new mounting was constructed at the observatory itself, where so much of Russell's equipment was built. When he retired in 1905, the photographic survey was near completion.

Russell's most widely known contributions to astronomy were photographic. He did much useful naked-eye work on nebulae of various sorts. He was the first man to photograph the nebula η Argus, in June 1890, with a portrait lens of six inches aperture. This was not a very satisfactory lens, and Sir David Gill obtained superior results shortly afterward. By April 1891 Russell had taken better photographs and had become aware of a persistent "dark round spot" in η Argus. Some of Russell's claims to have detected a spiral structure in nebulae have not been confirmed. His work on nebulae was badly hampered by his lack of the finest instruments; but his day-to-day astronomical and meteorological recording is still of value, and he played an important part in the establishment of scientific and technical education in Australia.

Russell was elected a fellow of the Royal Society in 1886 and was vice-chancellor of the University of Sydney in 1891. He married Emily Jane Foss in 1861; she, a son, and four daughters survived him.

BIBLIOGRAPHY

I. ORIGINAL WORKS. While at the Sydney observatory Russell published extensively in *Monthly Notices of the Royal Astronomical Society*, and elsewhere. The

IBN SĪNĀ

Royal Society *Catalogue of Scientific Papers*, XI (1896) and XVIII (1923), together list 127 items from before 1900 under Russell's name. His writings include *Photographs of the Milky Way and Nubeculae Taken at Sydney Observatory, 1890* (Sydney, 1891); and *The Climate of New South Wales, Descriptive, Historical and Tabular* (Sydney, 1897).

II. SECONDARY LITERATURE. Obituary notices of Russell are in *Proceedings of the Royal Society*, **A80** (1908), lx–lxiii; and *Monthly Notices of the Royal Astronomical Society*, **68** (1908), 241.

J. D. NORTH

IBN SĪNĀ, ABŪ ᶜALĪ AL-ḤUSAYN IBN ᶜABDALLĀH, also known as **Avicenna** (*b.* Afshana, near Bukhara, central Asia [now Uzbek S.S.R.], 980; *d.* Hamadān, Persia [now Iran], 1037), *philosophy, science, medicine.*

Displaying an extraordinary precocity, Ibn Sīnā rapidly mastered contemporary knowledge of the various sciences and, at the age of sixteen, began to practice medicine. He also was active in the political life of his time. After serving as jurist at Korkanj, teacher of science at Gorgan, and administrator at Rayy and at Hamadān, he was named vizier of Shams al-Dawla. In addition to his government service he found time for an equally demanding scientific career. He died of a mysterious illness, apparently a colic that was badly treated; he may, however, have been poisoned by one of his servants. During the celebration of the nation's millennium, Iran erected an imposing mausoleum over his tomb at Hamadān.

Ibn Sīnā wrote extensively in a number of widely divergent fields; his bibliography comprises nearly 270 titles. Among them is an autobiography that was completed by his disciple al-Jūzjānī.

Ibn Sīnā's major philosophical work is *Al-Shifā*ʾ ("The Cure" [of ignorance]), an immense four-part encyclopedia: logic, corresponding very closely to Aristotle's Organon; physics; mathematics (geometry, arithmetic, music, and astronomy); and metaphysics. He also wrote a compendium of this work called *al-Najāt*, as well as several other general accounts of his philosophy, including *al-Hidāya*, *ʿUyūn al-hikma*, and, in Persian, *Dānish Nāma-i ʿAlāʾi*. One of his last books is *al-Ishārāt waʾl-tanbīhāt*, which has been translated into French by A.-M. Goichon, as *Livre des directives et remarques* (Beirut–Paris, 1951). There is also an English translation by Parviz Morewedge (London, 1973) of the section on metaphysics of the *Dānish Nāma*. His writings on symbolic knowledge include *Ḥayy ibn Yaqẓān*.

Philosophy and Science. Ibn Sīnā was influenced in philosophy and science by three important currents of thought that he brought together in an original synthesis. These were the Koran and its accompanying theological elaborations (concerning theodicy, cosmogony, anthropology, and eschatology); science (geocentricism, Greek astronomy, the circular motion of the heavenly spheres, the hierarchy of the cosmos, and the theory of the four elements); and philosophy, specifically, an Aristotelianism heavily laden with elements of Neoplatonism stemming mainly from Plotinus (via Aristotle's *Pseudo Theology*) and Proclus (*De causis*)—to which were joined some aspects of the Persian tradition.

Metaphysics. Ibn Sīnā's metaphysics is founded on a theory of necessary emanation (*fayḍ*) and progressive descent. The scheme begins with the One or God, necessarily existing per se; in him essence and existence are identical. From him emanate the celestial world, consisting of the bodies, souls, and Intelligences of the spheres—beings that, in themselves, are merely possible but that are made necessary by God—and the sublunar world, containing the mineral, vegetable, and animal kingdoms. In all creatures essence is distinct from existence, and the sublunar world belongs to the realm of the possible.

According to the Plotinian scheme of return to the origin, the entire cosmos is animated by an impulse that leads it toward God through the intermediary of the separate Intelligences. The immediate origin of human souls is the Intelligence of the lunar sphere, the separate Intellect or *dator formarum* (*wāhib al-ṣuwar*), which is the supreme object of happiness for man. Creation in the sense of *fayḍ* is necessary and eternal, and it takes place by means of the separate Intelligences (the angels).

Ibn Sīnā sought to integrate all aspects of science and religion in a grand metaphysical vision. With this vision he attempted to explain the formation of the universe as well as to elucidate the problems of evil, prayer, providence, prophecies, miracles, and marvels. Also within its scope fall problems relating to the organization of the state in accord with religious law and the question of the ultimate destiny of man.

Physics and Cosmology. Ibn Sīnā's conception of science derives from the physics and cosmology of his time and, thus, from Greek science. His understanding and elaboration of this body of

IBN SĪNĀ

knowledge may be outlined as follows (see *al-Najāt*, Cairo edition).

Physics is the study of natural bodies (*jism*) and of movement. On several occasions Ibn Sīnā refutes at length the atomistic conception of body and of reality in general, advocating instead continuity and hylomorphism. In his view, body consists of a material substance that acts as a subject (*maḥall*) and of a form (*ṣūra*) that inheres in the matter. The relationship of matter to form is like that of bronze to a statue.

The common characteristic of all bodies with respect to form is that they have three dimensions. The latter do not exist *in actu* in the body, although they may be assumed to be in it. Therefore, they do not enter into the constitution of matter and are not part of its definition.

Matter, which cannot exist without form, is homogeneous and can receive all possible forms. Matter possesses a first form, the substantial or corporeal form, which is three-dimensionality, the property of having three dimensions. With the substantial form there are others, such as quantity, quality, and place—in fact, the Aristotelian categories. (These are the "accidents," or *aʿrāḍ*.) There is also an external principle that ensures the union of matter and form.

Natural bodies possess two kinds of perfection, primary and secondary. The external principle ensures the secondary perfections through the intervention of powers placed in the body: the primary perfections and certain principles from which the secondary perfections emanate. These secondary perfections include actions.

The powers implanted in natural bodies are of three types. Those of the first type extend throughout bodies and preserve the latter's perfections, forms, natural places, and actions. If the bodies are removed from their natural places or somehow lose their forms or natural modes, the powers return them to their previous condition and maintain them in it. This occurs by imposition (*bi-taskhīr*), without the intervention of knowledge, reflection, or voluntary intention (*qaṣd ikhtiyārī*). These powers are called natural. They are the principle per se of the movements per se of bodies and of their states of rest—and, indeed, of all their perfections. No natural body is without them.

The second type of powers acts on bodies by means of instruments or organs to move them or maintain them at rest, or to preserve their specific essence. Certain implanted powers of the second type act in a permanent fashion involving neither choice nor knowledge—as in the vegetative soul. Others may act or not act, and they can perceive what is agreeable and what is harmful—as in the animal soul. Others comprehend the realities of things by way of reflection and research—as in the human soul.

The third type of powers achieve the same result without instruments by a will oriented in one single way—as in the celestial soul.

Properties associated with natural bodies are movement and rest, time, place, the vacuum, finitude and infinity, contact, continuity, and sequence.

Movement. Movement (*al-ḥaraka*) is an act and a primary perfection of a thing that exists insofar as it is potential. It exists in a time between pure potentiality and pure act. Movement is not something stable and accomplished; thus it can be conceived only as something that can grow or diminish. It is therefore not found in substances, for the generation of substance occurs *in instanti* and not through a movement. Instead, motion is found in quantity, which can exist in greater or lesser degree and which gives rise to increase (*numuww*) and decrease (*dhubūl*), as well as to rarefaction (*al-takhalkhul*) and condensation (*tadākhul*). In the course of these changes the body does not lose its continuity. Movement can likewise be found per se in quality, place (locomotion, for example), and position (rotation, for example).

Rest is the absence of movement in that which has the potential of moving; it is not simply pure negation.

All motion in a body exists only through a cause distinct from the body itself; no movement exists for the body per se. Movement must be attributed exclusively to the cause. This cause is either outside the body, in which case the body is said to move not by itself (*lā bi-dhātihi*), or else it is in the body, in which case the latter is said to move per se (*mutaḥarrik bi-dhātihi*). This cause can sometimes move and sometimes not move; then the body is said to move by choice (*bilikhtiyār*). Or it can move all the time, and the body cannot be devoid of motion; in that case it is said to be "in motion by nature" (*mutaḥarrik bil-ṭabʿ*). The latter movement can be of two kinds: either the body can be moved by imposition (*bil taskhīr*), its cause moving it without volition, whereby it is said to be in motion "by nature" (*bil-ṭabīʿa*); or it is moved by will and intention, whereby it is said to be in motion through the action of the celestial soul.

In general, that which the nature of a thing re-

IBN SĪNĀ

quires per se cannot be separated from the thing; otherwise its nature becomes corrupted. Movement can be separated from a moving body without that body's becoming corrupted. Movement, therefore, is never required by the nature of a moving body. Hence, if a natural body is in motion, it is not in its natural state and is seeking to return to that state and to remain at rest in it. Its removal from its natural place occurs through violence (*bil-qaṣr*). Consequently, all unforced movement (brought about by nature) is a flight "by nature" from a state that is not suitable to the body in question.

This movement is straight if the body is not in its natural place, because it is due to a "natural inclination" (*mayl ṭabīʿī*) that seeks the shortest path. Hence, local circular motion does not derive from nature (*ʿan al-ṭabīʿa*).

Circular motion, accordingly, is never violent. Its principle or origin is a soul—that is, a power moving by choice or will. The circular motion of the stars, too, derives from a soul.

There cannot be an indivisible local movement, as the partisans of atomism claim—either at some maximum or at some minimum velocity.

Movement can be considered in terms of genus, species, or number; and it can be more or less rapid. The opposition between movement and rest is a relationship of privation.

Time, Space, and Infinity. Time is the measure of circular motion with respect to "before" and "after," and not with respect to distance. Place is the inner surface of an enveloping body at the point of contact with the outer surface of the body that it envelops. A vacuum does not exist, nor is there a dimension that is not located in some material substance.

There can be no infinite continuous quantity existing as a whole and having a position. Nor can there be an infinite ordered number (*ʿadad murattab al-dhāt*), just as there can be no power possessing infinite force (*al-shidda*). Further, it is impossible for a power characterized by either infinite duration or infinite number to be divided and shared, even *per accidens*.

The universe is full because a vacuum does not exist. The outer sphere, that of the fixed stars, envelops all that exists. The stars and their spheres move on the inside of this sphere with an eternal circular motion. A knowledge of the relation between the center of the universe (which is also the center of the earth) and the sphere of the fixed stars makes it possible to determine, for every part of the universe, an absolute "up" (toward the sphere of the fixed stars) and an absolute "down" (toward the center of the earth).

Every body is necessarily located in space. By their position (up or down) spaces differentiate bodies with respect to direction. Bodies enveloped by other bodies have places. Every natural body has its own natural place.

Composite bodies are formed by "welding" (*iltiḥām*), a process that does not take place directly between bodies but by the intermediary of sensible qualities. There are four primary sensible qualities that effect such unions: heat, cold, moisture, and dryness. Heat and cold react upon each other in altering bodies and are called the active powers. The other two are passive powers.

The simple bodies that form composite bodies are characterized by a particular combination of these four powers. Every body must contain one active and one passive quality. Accordingly, there are four simple bodies: fire (hot and dry), water (hot and wet), earth (cold and wet), and air (cold and dry).

The natural place of corruptible beings is the sublunar sphere, and that of incorruptible bodies is the supralunar world. The latter are not composed of the four elements, and their spheres are neither light nor heavy.

In Ibn Sīnā's view, the different combinations of the four elements and their qualities, together with the motion of the spheres (which disposes matter to receive forms), are sufficient to explain the formation of the corruptible bodies in the sublunar world: minerals, stones, and metals; plants; animals; and man, who, through his body, belongs to the physical world.

Classification of the Sciences. On the basis of these general physical principles (which he treated extensively in the first book of physics in the *Shifāʾ*) and in accord with his metaphysics of being, Ibn Sīnā elaborated a broad theory of science as wisdom. This approach allowed him to classify organically the spectrum of sciences known in his time. (On this point see his short treatise on the division of the sciences, the essentials of which are outlined below.)

According to the ancient meaning of the term, science is a synonym of wisdom or philosophy; it yields knowledge that is certain by virtue of its insight into causes. It can be either practical or speculative. Practical science acquires knowledge with a view toward acting upon it, while the goal of speculative science is to obtain certain knowledge of beings whose existence does not depend on human action.

IBN SĪNĀ

Practical science, which pertains to human conduct, does not concern us here. Speculative science consists of three parts, established according to the relation of their objects to matter and motion. The first part is physical science (*ʿilm al-ṭabīʿa*); existence and definition of its object are linked with matter and with motion. The second is mathematical science, the object of which is linked with matter only in its concrete existence—matter does not enter into its definition. The object of the third branch, metaphysics, is independent of matter in both its existence and its definition.

Logic, at once an art and a science, is an instrument of science. Ibn Sīnā's view, the subject, with which he deals at length in the *Shifāʾ*, includes the entire Organon of Aristotle.

Natural science or physics consists of eight principal sciences and seven subordinate sciences. The principal sciences are the following:

1. The science of general principles (presented above), which is the subject of the *Kitāb samʿ al-kiyān* (*Physike akroasis*) or *Kitāb al-samāʿ al-ṭabīʿi*.

2. The science of heaven and the world (*al-samāʾ wal-ʿālam*), which studies the celestial and terrestrial bodies of which the universe is composed, as well as the four elements and their movements.

3. The science of generation and corruption (*al-kawn wal-fasād*). This field treats the generation of the primary elements, their interaction, the way in which God links earthly things with those in heaven, and the perpetuation of species despite the disappearance of individuals.

4. The science of meteorology (*al-āthār al-ʿulwiyya*), which investigates the elements before their mixture, the various types of motion, rarefaction, dilatation, and the atmospheric phenomena influenced by the heavens: shooting stars, clouds, rain, thunder.

5. The science of minerals (*Kitāb al-maʿādin*), which is the sequel to meteorology.

6. The science of plants.

7. The science of animals (*ʿilm ṭabāʾiʿal-ḥayawān*).

8. The science of the soul, or psychology (*Kitāb al-nafs*). Ibn Sīnā treated this subject in book VI of the section on Physics in the *Shifāʾ*, which, under the title of *De anima* or *Liber sextus naturalium*, enjoyed an extraordinary success in the Latin Middle Ages.

The subordinate sciences are the following:

1. Medicine (*al-ṭibb*), which seeks to learn the principles of the human body and its condition in health and in sickness. Ibn Sīnā's immense encyclopedia on this subject, the *Canon of Medicine*, became a classic.

2. Astrology (*aḥkām al-nujūm*). For Ibn Sīnā astrology is only a probable science (*takhmīnī*). It attempts, from a knowledge of the configuration of the stars, their reciprocal distances, and their positions in the zodiac, to predict the conditions in the sublunar world—the future of men and nations. Ibn Sīnā wrote an epistle refuting the claims of the astrologers.

3. Physiognomy (*ʿilm al-firāsa*). Ibn Sīnā wrote nothing on this subject; the works attributed to him are apocryphal.

4. Oneiromancy, the science of divination by means of dreams (*ʿilm al-taʿbīr*).

5. The science of talismans (*ʿilm al-ṭilismāt*), "of which the goal is to mix the celestial forces with the forces of certain terrestrial bodies so as to give rise to an extraordinary action in the world here below."

6. Theurgy (*ʿilm al-nīranjāt*), the goal of which is to mix the forces of terrestrial substances in such a way as to produce extraordinary effects. In the last chapters of his *Ishārāt*, Ibn Sīnā attempts to explain in a rational manner "the secrets of prodigies" and extraordinary actions and, more generally, the relationships between the microcosm and the macrocosm.

7. Alchemy (*al-kimyāʾ*). Ibn Sīnā studied the philosophical and scientific foundations of this subject and even undertook alchemical experiments. His conclusion regarding its validity, however, is negative, as can be seen from his epistle on alchemy (now available in French in G. C. Anawati, "Avicenne et l'alchimie," in *Atti dei convegni. Fondazione Alessandro Volta* [Rome, 1971], 285–341).

The mathematical sciences consist of four principal sciences and four subordinate sciences. The principal ones (which Ibn Sīnā discusses in the *Shifāʾ*) are listed below:

1. The science of numbers, or arithmetic (*ʿilm al-ʿadad*).

2. Geometry (*ʿilm al-handasa*), which in Ibn Sīnā's account is based on the theorems of Euclid.

3. Geography and astronomy (*ʿilm al-hayʾa*), based on Ptolemy's *Almagest*.

4. The science of music.

The subordinate mathematical sciences are the following:

1. Hindu computation and algebra (included under arithmetic).

2. Mechanics (*ʿilm al-ḥiyal al-mutaḥarrika*),

IBN SĪNĀ

traction by means of weights (*jarr al-athqāl*), the science of weights and balances, the science of specialized instruments, optics (*ʿilm al'manāẓir wal-mirāya*), and hydraulics (*ʿilm naql al-miyāh*) (all considered part of geometry).

3. The making of astronomical tables and calendars (*ʿilm al-zījāt wal taqāwīm*) (placed under astronomy).

4. The use of foreign instruments, such as the organ (the domain of music).

BIBLIOGRAPHY

The basic bibliographical works on Ibn Sīnā are G. C. Anawati, *Essai de bibliographie avicennienne* (Cairo, 1950), in Arabic, which is based on the manuscripts; and Yahya Mahdavi, *Fihrist-i nuskhahā-yi musannafāt-i Ibn Sina: Bibliographie d'Ibn Sīnā* (Teheran, 1954), in Persian.

Supplementary information is in more recent publications by G. C. Anawati: "Chronique avicennienne 1951–1960," in *Revue Thomiste*, **60** (1960), 613–634; and "Bibliographie de la philosophie médiévale en terre d'Islam pour les années 1959–1969," in *Bulletin de philosophie médiévale*, **10–12** (1968–1970), esp. 343–349. For Ibn Sīnā's more specifically scientific works, see A. C. Crombie, "Avicenna's Influence on the Medieval Scientific Tradition," in G. M. Wickens, ed., *Avicenna: Scientist and Philosopher. A Millenary Symposium* (London, 1952), 84–107; and Seyyed Hossein Nasr, *An Introduction to Islamic Cosmological Doctrines* (Cambridge, Mass., 1964), 177–274.

For Ibn Sīnā's autobiography, see William F. Gohlman, *The Life of Ibn Sina. A Critical Edition and Annotated Translation* (Albany, N.Y., 1974).

G. C. ANAWATI

Medicine. Perhaps it would be appropriate to judge the medical works of Ibn Sīnā in a comparative statement: Among the two *ḥukamāʾ* (physician-philosophers) Ibn Sīnā was the better philosopher and al-Rāzī (Rhazes; *d.* 925 or 935) was the better physician. In compiling his encyclopedia *al-Qānūn* ("Canon"), Ibn Sīnā borrowed extensively from al-Rāzī's *al-Ḥāwī* ("Continens"). In his autobiography, Ibn Sīnā states that he wrote the beginning of *al-Qānūn* at Gorgan; that part of it was later written at Rayy (near modern Teheran), which was al-Rāzī's birthplace (from which his *nisba* is derived); and that it was completed at Hamadān. This book (Ibn Sīnā's major medical work, in about one million words) was very well received by physicians, who favored it over al-Rāzī's *al-Ḥāwī*, over ʿAlī Ibn al-ʿAbbās' (Haly Abbas; *d.* 994) *Kāmil al-ṣināʿa al-ṭibbiyya* ("Complete Art of Medicine"), and even over Galen's works. (Ibn Sīnā and other eminent physicians were honored in their time by the title *Jālīnūs al-Islām*, "Galen of Islam.")

In Córdoba, however, Ibn Zuhr (*d.* 1131), his son, the eminent practitioner Ibn Zuhr (Avenzoar; *d.* 1162), and Ibn Rushd (Averroës; *d.* 1198) criticized *al-Qānūn* ("Principles [or Code] of Laws") bitterly. Some writers suggested that it was complete and sufficient, and could not be improved by additions from other sources. This attitude toward the authority of books kept Arabic medicine (until its decline) and, to a certain degree, early medieval medicine in Europe (which relied in part on Latin translations of Arabic books) in a static condition. Ibn Sīnā, however, was not to blame; physicians who lacked a critical turn of mind and preferred the authority of books and the use of logic to reach conclusions, rather than resort to experiments and observations, were responsible for that status quo. Al-Rāzī's progressive traditions—his refusal to accept statements unverified by experiments, his notion of control experiments, clinical observations, and criticism of Galen and other authorities—were overshadowed by Ibn Sīnā's beautifully presented *al-Qānūn*, which was divided into five books.

Book I, *al-Kulliyyāt* ("Generalities"), the most complicated book, was treated in some ten commentaries, one of which was written by Ibn al-Nafīs (book I of his *Sharḥ al-Qānūn*, in which he described the pulmonary circulation). *Al-Kulliyyāt* contains four *funūn* (treatises). The first treatise is a study of the four elements: fire, air, water, and earth, reactions between which form the temperaments (those particular properties of the four humors: blood, bile, black bile, and phlegm). The four humors intermix in certain proportions, resulting in the homogeneous organs (simple organs), the anatomy of which is included in the first treatise of book I. Ibn Sīnā ends the first treatise with an account of forces: the psychic force, with the brain as its center; the natural force concerned with the preservation of the human being, centered in the liver and the testicles; the animal force (with the heart as its center), which controls the pneuma affecting the senses and locomotion. The second treatise is on etiology and symptoms. The third is on hygiene, the causes of health and sickness, and the inevitability of death. The fourth treatise of book I deals with a classification of the modes of therapy, a general survey of treatment by regimes and diets, principles and rules of the administration of cathar-

IBN SĪNĀ

tic and emetic drugs, and the rules of evacuation, administration of enemas, liniments, and fomentations. An account is also given of the manipulation of cuppings, cautery, bloodletting, and general surgical treatments.

Book II ("Materia Medica") of *al-Qānūn* is divided into two sections: a general account of physical properties of drugs (qualities, virtues, and modes of preservation) and a list of drugs, arranged alphabetically (in which the virtues of each drug are given).

Book III ("Head-to-Toe Diseases") begins with diseases of the brain, followed by those of the nerves, the eye, and the ear, and ends with pains of the joints, sciatica, and diseases of the nails. Anatomical accounts of heterogeneous organs (compound organs) also are given. The anatomy of *al-Qānūn* is divided between book I and book III (Ibn al-Nafīs assembled the anatomy of *al-Qānūn* in his commentaries *Sharḥ tashrīḥ al-Qānūn*, and in book I of his *Sharḥ al-Qānūn*).

Book IV ("Diseases That Are not Specific to Certain Organs") contains details of fevers, their classification, genera, and symptoms, as well as accounts of prognoses, crises, and critical days, and all the principles deemed essential for diagnosis and therapy. This is followed by a study of abscesses, pustules, orthopedics, wounds, poisons, and venomous creatures. The book concludes with diseases of the hair and studies in obesity and emaciation.

Book V ("Compound Drugs") presents accounts of theriacs, troches, electuaries, cathartic drugs, pills, and liniments and their medicinal applications.

The rich information provided in *al-Qānūn* invited numerous physicians (until the nineteenth century) to write commentaries and marginal notes, while others chose to extract epitomes that became very popular among physicians and medical students. In addition to *al-Qānūn*, Ibn Sīnā wrote about forty medical works, most of which are preserved in manuscripts. These works (some are poems) also led renowned physician-philosophers to write their own commentaries.

Al-Qānūn was translated into Latin (Milan, 1473; Padua, 1476, 1479; Venice 1482, 1486) by Gerard of Cremona, and was a textbook at the universities of Montpellier and Louvain until 1650. Gerard also translated *Urjūza fī'al-ṭibb* (*Canticum de medicina seu liber de medicina in compendium reducta*). Andrea Alpago (*d.* 1520) acquired his fame as a translator of Ibn Sīnā's works, including *Aḥkām al-adwiya al-qalbiyya* (*De viribus cordis seu de medicamentis cordialis*), *Dafʿ al-maḍārr al-kulliyya . . .* (*Liber liberationis seu removendis nocumentis, quae accedunt in regimine sanitatis*), and *al-Fuṣūl* (*Aphorismi*).

BIBLIOGRAPHY

I. ORIGINAL WORKS. Works in MSS include *Al-aghdhiya wa'l-adwiya* ("Foods and Drugs"), Aya Sofya (Istanbul), MS 4849 (24), vii; *Aḥkām al-adwiya'l-qalbiyya* ("Rules About Medicines of the Heart"), University of Leiden Library, MSS Or. 958 (46) and Or. 820, iii; Chester Beatty Library (Dublin), MS 3676, i; Dār al-Kutub al-Miṣriyya (Cairo), MS Taymūriyya Ṭibb 314; University of California (Los Angeles), MS Ar. 78, ix; Wellcome Institute for the History of Medicine Library (London), WMS Or. 73; *Asrār al-jimāʿ* ("Secrets of Sexual Intercourse"), Dār al-Kutub al-Miṣriyya, MS Ṭalʿat Ṭibb 549; *al-Bawl* ("Urine"), Glasgow University Library, MS 121, iii; *Dafʿ al-maḍārr al-kulliyya ʿan al-abdān al-insāniyya* ("Repelling General Harm From the Body"), Aḥmad III (Istanbul), MS 3447 (78), x; Āṣafiyya (Hyderabad), MS 41/19; Aya Sofya, MSS 3698 and 3699; Dār al-Kutub al-Miṣriyya, MSS Ṭibb 1867 and Ṭibb 19; Riḍā Library (Rampur), MS 423, v; University of California (Los Angeles), MS Ar. 78, x; Wellcome Library, WMS Or. 47, fols. 4a–29a; and *al-Faṣd* ("Bloodletting"), Aḥmad III, MS 3447 (71), x; Āṣafiyya, MS 41 (27); Aya Sofya, MSS 4829 (3), x and 4849 (32), vii; Kuhdā Bakhsh Oriental Public Library (Patna–Bankipore), MSS 2559, iv and 108, xi; Riḍā Library, MSS 221 (composite vol.) and 76 (composite vol.).

Also in MSS are *al-Fuṣūl* ("Aphorisms"), Aya Sofya, MS 3683; Dār al-Kutub al-Miṣriyya, MS Ṭibb 1867 (composite vol.); Wellcome Library WMS Or. 47, fols. 29b–52b; Khudā Bakhsh, MS 108, xxiii, fols. 295a–322; Riḍā Library, MSS 223 and 76 (composite vol.); *Ḥifẓ al-ṣiḥḥa* ("Hygiene"), Āṣafiyya, MSS 3/730 (1/254) and 41 (25); Aya Sofya, MS 4849 (32), vii; Dār al-Kutub al-Miṣriyya, MS Taymūriyya Ṭibb 378; Khudā Bakhsh, MSS 2559 (5) and 108, xii, fols. 190b–197; Riḍā Library, MS 221 (composite vol.); and *al-Khamr* ("Intoxicating Drinks"), Asʿad Library (Istanbul), MS 3688 (34); Ḥamīdiyya Library (Istanbul), MS 1448.

Additional works in MSS are *al-Qūlanj* ("Colitis"), Āṣafiyya, MS 736/3 (41, 9); Dār al-Kutub al-Miṣriyya, MS Ṭibb, 1867 (composite vol.); Riḍā Library, MSS 480/1 (216) and 712 (76, 14); Riḍawī Library (Meshed), MS 16/19 (57–58); Sohag (Upper Egypt), MS Ṭibb 100 (composite vol.); Vehbi Library (Istanbul), MS 1488 (14); *al-Quwā al-ṭabīʿiyya* ("Natural Forces"), Aḥmad III, MS 3447 (68), x; Aya Sofya, MS 4829 (38), x; Shahīd ʿAlī (Istanbul), MS 2034 (4); Vehbi Library, MS 1488 (20, composite vol.); *al-Sakanjabīn* ("Oxymel"), Aḥmad III, MS 2119 (composite vol.); Āṣafiyya, MS 41/18; Laleli (Istanbul), MS 1647 (3), ix; Riḍā Library, MSS 218 (composite vol.) and 221 (composite vol.); *Shiṭr al-ghibb*

("Semitertian Fever"), Āṣafiyya, MS 3/730 (41/16); Riḍā Library, MSS 1/479 (98) and 712 (76/12); *Urjūza fī asbāb al-ḥummayāt* ("Poem on the Causes of Fevers"), Wellcome Library, WMS Or. 100; *Urjūza fī tadbīr al-ṣiḥḥa fī al-fuṣūl al-arbaʿa* ("Poem on Therapy During the Four Seasons"), Dār al-Kutub al-Miṣriyya, MSS Taymūriyya Ṭibb 440 and Ṭalʿat Ṭibb 561; Wellcome Library, WMSS Or. 17, fols. 49b–52a, and Or. 129, fols. 30a–33a; *Urjūza fī al-tashrīḥ* ("Poem on Anatomy"), Vatican, MS Borg. 87 (g); Wellcome Library, WMS Or. 129, fols. 55a–57a.

Printed eds. include *al-Qānūn* (Rome, 1593), bound with Ibn Sīnā's *al-Najāt; Kulliyyāt qānūn shaykh al-raʾīs,* which is book I of *al-Qānūn* (Teheran, 1867–1868); Būlāq, ed. (Cairo, 1877); also *Ḥummayāt qānūn shaykh al-raʾīs*, Muḥammad Ashraf ʿAlī, ed. (Lucknow, 1878–1879); *al-Kitāb al-mash-hūr bi' l-kulliyyāt min' l-qānūn* (Lucknow, 1880–1881); and *al-Qānūn*, Abū al-Ḥasanāt Quṭb al-Dīn Aḥmad, ed. (Lucknow, 1905), also in an Urdu trans., 3 vols. (Lucknow, 1898–1912), and an English trans. of book I by O. C. Gruner as *A Treatise on the Canon of Medicine of Avicenna* (London, 1930).

P. de Koning published the section of the *Canon* on kidney and bladder stones in his *Traité sur le calcul dans les reins et dans la vessie par Abū Bakr Muḥammed ibn Zakarīyya al-Rāzī* (Leiden, 1896), 228–267; and its section on anatomy in *Trois traités d'anatomie arabes, texte inédit de deux traités* (Leiden, 1903), 432–478. The ophthalmology of the *Canon* was translated into German, with annotations by J. Hirschberg and J. Lippert, as *Die Augenheilkunde des ibn Sīnā, aus dem Arabischen übersetzt und erläutert* (Leipzig, 1902); by E. Michailowsky as *Die Augenheilkunde des Avicenna* (Berlin, 1900); and by T. Bernikow as *Die Augenheilkunde des Avicenna. Nach dem Liber Canonis zum erstenmal ins Deutsche übertragen* (Berlin, 1900). The *Aqrābādhīn* (bk. V) was translated by J. von Sontheimer as *Zusammengesetzte Heilmittel der Araber. Nach dem fünften Buch des Canons von Ibn Sīnā aus dem Arabischen* (Freiburg im Breisgau, 1845); there is also a Russian trans., *Kanon vrachebnoi nauki*, 5 vols. (Tashkent, 1954–1960). For commentaries on *al-Qānūn*, see Ibn al-Nafīs.

An additional published work is *Urjūza fī al-ṭibb* ("Poem on Medicine"), with a French trans. in H. Jahier and A. Noureddine, *Avicenne (370–426 Hégire): Poème de la médecine (Cantica Avicennae I)* (Paris, 1956). Also see their *Anthologie de textes poétiques attribués à Avicenne* (Algiers, 1960); and *Urjūza laṭīfa fī qaḍāyā Buqrāṭ al-khams wa al-ʿishrīn* ("Pleasant Poems on the Twenty-five Doctrines of Hippocrates"), Wellcome Library, WMS Or. 129, fols. 35a–37a.

II. Secondary Literature. See J. Freind, *The History of Physick, From the Time of Galen to the Beginning of the Sixteenth Century*, II (London, 1726), 69–74; F. Adams, *The Seven Books of Paulus Aegineta, Translated From the Greek With a Commentary*, 3 vols. (London, 1844–1847), which includes subject matter from Latin translations of the *Canon* and other works by Ibn Sīnā; L. Leclerc, *Histoire de la médecine arabe*, I (Paris, 1876), 466; Ibn Abī Uṣaybiʿa, *ʿUyūn al-anbāʾ fī ṭabaqāt al-aṭibbāʾ*, A. Müller, ed., II (Cairo–Königsberg, 1884), 2–20; G. J. Fisher, "Abū Ali el-Hosein Ibn Abdallah Ibn Sina, Commonly Called Avicenna, 980–1037," in *Annals of Anatomy and Surgery*, **7** (1883), 23–29, 96–106; J. Eddé, *Avicenna et la médecine arabe* (Paris, 1889); *Ibn al-Qifṭī's taʾrīḫ al-hukamāʾ*, J. Lippert, ed. (Leipzig, 1903), 413–426; E. G. Browne, *Arabian Medicine* (Cambridge, 1921), 57–64; D. Campbell, *Arabian Medicine and Its Influence on the Middle Ages*, I (London, 1926), 77–82; G. Sarton, *Introduction to the History of Science*, I (Baltimore, 1927), 709–713; J. B. Dawson, "Avicenna: The Prince of Physicians," in *Medical Journal of Australia* (1928), **2**, 751–755; and Y. A. Sarkīs, *Muʿjam al-maṭbūʿāt al-ʿarabiyya wa al-muʿarraba . . .*, I (Cairo, 1928), 127–132.

Works from the 1930's are Ibn al-ʿImād, *Shadharāt al-dhahab fī akhbār man dhahab . . .*, III (Cairo, 1932), 234–237; H. P. J. Renaud, "Abulcasis, Avicenne et les grands médecins arabes, ont-ils connu la syphilis?" in *Bulletin de la Société française d'histoire de la médecine*, **28** (1934), 122—also see his "Sur les noms de serpents dans Avicenne," in *Hespéris* (Paris), **24** (1937), 216–220; J. Ruska, "Die Alchemie des Avicenna," in *Isis*, **21** (1934), 14–51—also see his "Zum Avicenna-text des Cod. Vadianus 300," in *Sudhoffs Archiv für Geschichte der Medizin . . .*, **27** (1935), 499–510; A. Bloom, *L'ostéologie d'Abul-Qasim et d'Avicenna, son origine talmudique, suivie d'un chapitre dans le Talmud* (Paris, 1935); A. Soubiran, *Avicenna, prince des médecins: Sa vie et sa doctrine* (Paris, 1935); C. Brockelmann, *Geschichte der arabischen Litteratur*, Supp. I (Leiden, 1937), 812 (1), and I (Leiden, 1943), 589 (1); M. Meyerhof and D. Joannidès, *La gynécologie et l'obstétrique chez Avicenne (Ibn Sina) et leurs rapports avec celles des Grecs* (Cairo, 1938); A. Suhéyl-Ünver, "Four Medical Vignettes From Turkey. 1. Music Therapy for the Insane at Edirne Hospital in the Fifteenth Century. 2. Dietary Habits of the Turks in the Fifteenth Century. 3. Avicenna's 'Journey Into His Own Soul.' 4. The Wisdom of Alexander the Great," in *Indian Medical Record*, **171** (1958), 52–57; and "Aphorisms of Avicenna," in *Journal of the History of Medicine*, **14** (1959), 197–201; and K. Opitz, "Das Lehrgedicht über die Heilkunde (Canticum de medicina)," in *Quellen und Studien zur Geschichte der naturwissenschaften und der Medizin*, **7** (1939), 151–220.

Also see A. A. Khairallah, *Outline of Arabic Contributions to Medicine and the Allied Sciences* (Beirut, 1946), 118–125, 199–200; G. C. Anawati, *Essai de bibliographie avicennienne* (Cairo, 1950); C. Elgood, *A Medical History of Persia and the Eastern Caliphate* (Cambridge, 1951), 184–209; J. O. Leibowitz, "Une pharmacie figurée dans le manuscrit hébreu d'Avicenne," in *Revue d'histoire de la pharmacie*, **42**, no. 141 (1954), 289–292—also see his "Electroshock Therapy in Ibn-Sina's *Canon*," in *Journal of the History of Medicine*, **12** (1957), 71–72 ("Notes and Events");

A. Sanai, "Before Our Time: Avicenna," in *Lancet* (1954), **2**, 329–330; J. Wilczynski, "Contribution oubliée d'Ibn-Sina à la théorie des êtres vivants," in *Archives internationales d'histoire des sciences*, **33** (1954), 35–45; Kh. al-Ziriklī, *al-Aʿlām . . .*, 2nd ed., II (Cairo, 1954), 261–262; *Le livre du millénaire d'Avicenne. Conférence des membres du Congrès d'Avicenne prononcées en persan sur la biographie, l'époque, les opinions et les oeuvres d'Avicenne, 22–27 avril 1954*, II (Teheran, 1955); H. Jahier, "Cantica Avicennae, version latine de l'Urjuzâ fī'l-Tibb d'Avicenne," in *Cahiers de Tunisie*, **3** (1955), 41–48; S. Abdul Latif, "Introduction to 'Heart Drugs'—A Brilliant Work of Research by Avicenna," in V. Courtis, ed., *Avicenna Commemoration Volume* (Calcutta, 1956), 245–254; O. C. Cameron Gruner, "Avicenna's Canon of Medicine and Its Modern Unani Counterpart," in *University of Michigan Medical Bulletin*, **22** (1956), 239–248; D. A. Kronick and A. S. Ehrenkreutz, "Some Highlights of Arabic Medicine (A.D. 750–1400)," *ibid.*, 215–226; W. W. Thoms, "The Story of a Book," *ibid.*, 227–237; and J. Mostafavi, "Avicenna: The 'Prince of Physicians,' Introduced," in *Acta medica iranica*, **1**, no. 1 (1956–1957), 104–120, English trans. by G. Lotf Elahi—also see his "Avicenne, le plus célèbre philosophe et médecin de l'Orient; Le connaissez-vous?" *ibid.*, 121–123.

Further works are M. T. D'Alverny, "Avicenne et les médecins de Venise," in *Medioevo e Rinascimento: Studi in onore di Bruno Nardi*, I (Florence, 1955), 177–198 (see *Journal of the History of Medicine*, **12** [1957], 81); A. J. Arberry, "Avicenna: His Life and Times," in *Hamdard Medical Digest* (Karachi), **1**, no. 6 (1957), 1–5; and no. 7 (1957), 67–72; O. Boucetta, "Ibn Sina (Avicenne). Sa vie, son oeuvre médicale," in *Maroc médical*, **36** (1957), 185–190; R. Levy, "Avicenna—His Life and Times," in *Medical History*, **1** (1957), 249–261; M. Nizamuddin, "A Sketch of Avicenna as a Scientist," in *Indian Journal of the History of Medicine*, **2** (1957), 21–26; O. Temkin, "Avicenna: Poème de la médecine," in *Bulletin of the History of Medicine*, **31** (1957), 380–381, a book review; ʿU. R. Kahhāla, *Muʿjam al-muʾallifīn . . .*, **IV** (Damascus, 1957), 20–23; S. M. Afnan, *Avicenna: His Life and Works* (London, 1958); H. Schipperges, "Das Lehrgedicht des Avicenna," in *Neue Zeitschrift für ärztliche Fortbildung*, **47** (n.s. **1**) (1958), 674–675; Ṣ. el-Munajjed, "Maṣādir jadīda ʿan tārīkh al-ṭibb ʿind al-ʿarab," in *Majallat Maʿhad al-Makhṭūṭāt al-ʿArabiyya*, **5**, no. 2 (1959), 261–264 (nos. 95–120); and P. Rey, "La pensée médicale et philosophique d'Avicenne," in *Praxis* (Bern), **48** (1959), 472–476.

In the 1960's there appeared V. N. Ternovsky, "La doctrine d'Hippocrate dans le 'Canon de la médecine' d'Ibn Sina," in *Actes XVIIe Congrès international d'histoire de la médecine*, **I** (Athens, 1960), 113–116; H. E. Stapleton, R. F. Azo, M. H. Ḥusain, and G. L. Lewis, "Two Alchemical Treatises Attributed to Avicenna," in *Ambix*, **10** (1962), 41–82; H. C. Krueger, *Avicenna's Poem on Medicine* (Springfield, Ill., 1963); M. Nadjmabadi, "Quelques observations et opinions médicales sur Avicenna," in *Revue médicale du Moyen-Orient*, **20** (1963), 530–535; S. H. Nasr, *Three Muslim Sages: Avicenna, Suhrawardi, Ibn ʿArabī* (Cambridge, Mass., 1964), 1–51; J. al-Yasin, "Avicenna's Concept of Physics," in *Bulletin of the College of Arts and Sciences, Baghdad*, **7** (1964), 55–62; N. O. Ameli, "Avicenna and Trigeminal Neuralgia," in *Journal of the Neurological Sciences*, **2** (1965), 105–107; A. M. Aziz and H. H. Siddiqui, "Avicenna's Tract on 'Cardiac Drugs,'" in *Hamdard Medical Digest*, **10** (1966), 1–3—special issue: Greco-Arab concepts on cardiovascular diseases, 41–44; R. D. Clements, "Avicenna the Prince of Physicians," in *Minnesota medicina*, **49** (1966), 187–192; M. H. Shah, *The General Principles of Avicenna's "Canon of Medicine"* (Karachi, 1966); M. D. Grmek, "Influsso di Avicenna sulla medicina occidentale del Medio Evo," in *Salerno*, **1**, no. 4 (1967), 6–21, in Italian and English; E. H. Hoops, "Die sexologischen Kapitel im 'Canon medicinae' des Avicenna, verglichen mit der Schrift 'De coitu' des Maimonides," in *Aesthetische Medizin*, **16** (1967), 305–308; A. Z. Iskandar, *A Catalogue of Arabic Manuscripts on Medicine and Science in the Wellcome Historical Medical Library* (London, 1967), 27–64, 156–166, 207–213; A. Cianconi, "L'aborto e la ritenzione di feto morto nella patologia ostetrica di Ibn-Sina," in *Pagine di storia della medicina*, **21** (1968), 49–59—also see his "Sterilità e difficoltà di concepimento nella ginecopatologia di Avicenna," *ibid.*, 77–87; N. Javadpour, "Avicenna (980–1037)," in *Investigative Urology*, **6** (1968), 334–335; L. Stroppiana, "L'origine delle vene e delle arterie in uno studio morgagnano su Avicenna," in *Medicina nei secoli*, **5**, no. 1 (1968), 3–9; and S. K. Hamarneh, *Index of Manuscripts on Medicine, Pharmacy, and Allied Sciences in the Ẓāhiriyah Library* (Damascus, 1969), 121–127, 262–280.

Later works are M. Ullmann, *Die Medizin im Islam*, Handbuch der Orientalistik, sec. 1, supp. vol. VI (Leiden–Cologne, 1970), 152–156, 333–337; and R. Y. Ebied, *Bibliography of Mediaeval Arabic and Jewish Medicine and Allied Sciences* (London, 1971), 95–101.

ALBERT Z. ISKANDAR

STASZIC, STANISŁAW WAWRZYNIEC (*b.* Piła, Poland, November 1755; *d.* Warsaw, Poland, 20 January 1826), *geology, organization of educational and scientific institutions.*

Staszic was the youngest son of Wawrzyniec Staszic, a miller and mayor of Piła, and Katarzyna Mędlicka, who intended her son to become a priest. After attending secondary school in Poznań, he took minor orders on 2 January 1774 and then studied theology for two years. In July 1778 he was granted a church benefice and was appointed chancellor of the collegiate church in Szamotuły. The following year he published his first works, a Polish translation of Racine's *La réligion* and of Voltaire's *Poème sur le désastre de Lisbonne*.

A legacy from his father enabled Staszic to complete his education abroad. He studied at the universities of Leipzig and Göttingen and then followed courses in the sciences of nature for two years at the Collège Royal de France under Brisson and Daubenton. The latter's influence combined with that of Buffon to arouse in Staszic a strong interest in the sciences of the earth, and in 1786 he published a Polish translation of Buffon's *Époques de la nature*, adding his own comments. Staszic also found occasion in Paris to settle his philosophical outlook. Influenced mainly by the encyclopedists, he adopted the principles of rationalism and utilitarianism, which exerted a considerable effect on his life and activity.

Staszic returned to Poland in 1781 by way of the Alps and the northern Apennines. That year he was employed by the former chancellor, Count Andrzej Zamoyski, as tutor to his two sons. During the next dozen years Staszic's high salary and skill in financial dealings enabled him to earn a considerable fortune, on which he later drew for philanthropic and scientific purposes. In 1782 he obtained his doctorate in canon and civil law from the Academy of Zamość.

Toward the end of 1794, after the suppression of the Kościuszko insurrection, Staszic accompanied Zamoyski's widow and sons to Vienna, where he remained until his young charges had completed their education. He returned to Poland in 1797 and began geologic investigations, traveling to Silesia and Saxony early in 1804 for this purpose. In the summer of 1805 he organized a research expedition to the Tatra Mountains and on 21 August was the first man to climb the Lomnice Peak, the second highest of the range. Staszic was co-founder and organizer of the Towarzystwo Przyjaciół Nauk (Society of the Friends of Science), established in 1800 in Warsaw, becoming chairman in 1808. In 1801 he bought the building that was the society's first home, and between 1820 and 1823 he commissioned and provided most of the funds for the construction of the Staszic Palace and for Bertel Thorvaldsen's statue of Copernicus that stands in front of it.

In 1807, following the creation of the duchy of Warsaw, Staszic published *O statystce Polski krótki rzut wiadomości* ("On the Statistics of Poland"), in which he presented the essential geographic, demographic, and economic data needed by the new administration. Concentrating on the organization of educational institutions and on the development of the mining industry, Staszic participated from 1807 in the work of the chamber of education, and in 1808 he became a member of the supreme examining commission and was elected chairman of the council of the school of law and administration. Beginning in 1809 he supervised the organization of the medical school, and in 1813, during the final period of the duchy of Warsaw, he established the chair of geology and mineralogy at the Jagiellonian University in Cracow.

With the establishment of the kingdom of Poland at the Congress of Vienna in 1815, Staszic was appointed a member of the council of state and of the department of national education; and in 1816 he became head of the directorate of industry and crafts. He was also elected chairman of the general council of the newly established University of Warsaw.

As a high-ranking official in the finance ministry, Staszic initiated a systematic development of the mining and metallurgical industries and of related research. He expanded the industrial center in the Świętokrzyskie Mountains, which had been established by the duchy of Warsaw, and he planned and supervised the construction of a network of mines, metallurgical plants, and factories, supplying them with the best available technical equipment and securing satisfactory working and living conditions and medical assistance for the miners. For the center of this industrial network he chose the town of Kielce, where in 1816 he established the directorate of mining and founded the first mining academy in Poland, staffing them with Polish experts and with specialists from Saxony.

Staszic's investigations of the structure of geological formations was related to his administrative work. Beginning in 1805 he published the results of these studies in the *Rocznik towarzystwa przyjaciół nauk*, compiling them in 1815 as *O ziemiorództwie Karpatów i innych gór i równin Polski* ("On the Geology of the Carpathians and of Other Mountain Ranges and Plains of Poland"). Accompanied by a four-sheet map and numerous plates, the work was the first geologic synthesis of Poland made by a Polish author, and it earned Staszic the sobriquet "the father of Polish geology." His remarkable geologic intuition is evident in his conclusion, while describing some small findings of sulfur southeast of the Świętokrzyskie range, that large deposits of sulfur existed at a considerably greater depth; this hypothesis was fully confirmed after World War II with the discovery of the Tarnobrzeg sulfur basin.

Toward the end of his life Staszic quarreled with

the influential secretary of the treasury Count Ksawery Drucki-Lubecki, and in 1824 he was obliged to relinquish management of the mining industry. His contribution to scientific research and social reform is commemorated in the naming of many schools, the mining academy in Cracow, and several fossil species of animals and plants for him.

BIBLIOGRAPHY

I. ORIGINAL WORKS. In addition to those works cited in the text, Staszic published "Géologie des montagnes de l'ancienne Sarmatie" and "Sur les mélanites trouvés en Pologne," in *Journal de physique, de chimie, d'histoire naturelle et des arts,* **65** (1807). *O. ziemiorodztwie Karpatów* (Warsaw, 1815) was reprinted (Warsaw, 1955) with foreword by Walery Goetel.

II. SECONDARY LITERATURE. *Stanisław Staszic* (Lublin, 1926), published on the centenary of Staszic's death, includes a bibliography of his works, biographical studies, and a discussion of various aspects of his scientific career. On his pedagogic work, see Tadeusz Nowacki, *Materiały do działalności pedagogicznej Stanisława Staszica* (Wrocław, 1957). On his activities until 1795, see Czesław Leśniewski, *Stanisław Staszic. Jego życie i ideologia w dobie Polskie niepodległej, 1755–1795* (Warsaw, 1925). His travel journals were published as *Dziennik podróży*, C. Leśniewski, ed. (Cracow, 1931). A recent biography is Barbara Szacka, *Stanisław Staszic* (Warsaw, 1966).

STANISŁAW CZARNIECKI

THEODORUS OF CYRENE. Since the main article on Theodorus (*DSB*, XIII, 314–319) was written, Wilbur Richard Knorr has produced a novel and attractive theory (see his work cited in the bibliography) to explain how Theodorus proved $\sqrt{3}$ and other numbers to be surds and why he stopped with $\sqrt{17}$. It is based on "Pythagorean triples" of numbers that may be set out as the sides of a right triangle. Given an odd number n, the side of the square of n units in area is constructed as the leg of a right triangle the hypotenuse of which is $(n+1)/2$ units in length; the other leg is $(n-1)/2$ units. Given an even number n, the side of the square of n units in area is constructed as half the leg of a right triangle the hypotenuse of which is $n+1$; the other leg is $n-1$. If the constructed root is commensurable with the unit length, there will be a ratio of integers $a{:}b$ such that the root and the unit are in that same ratio. Knorr shows in the case of $\sqrt{3}$, $\sqrt{5}$, and so on, that the assumption of commensurability leads to the contradiction that at least one of the pair of numbers is both odd and even.

This is an attractive theory in that it is no simple extension of the method used to prove the irrationality of $\sqrt{2}$ (see *DSB*, XIII, 315), but employs the same principle, as anyone investigating the higher irrationals might have been expected to do. It also requires a separate proof for each surd, as the Greek text suggests. Finally, the method encounters difficulties with $\sqrt{17}$. Whereas previous commentators have taken the Greek to mean that Theodorus demonstrated the irrationality of $\sqrt{17}$ and ran into trouble afterward, Knorr's theory requires that he became entangled at $\sqrt{17}$; but this is a possible interpretation of the text.

Contrary to the view taken in the article "Theaetetus," note 9 (*DSB*, XIII, 305), Knorr holds that δύναμις means "power," as in later Greek usage.

For clarity, the stages in proving that the process of finding the greatest common measure of 1 and 17 is endless ("Theodorus," 316) should be presented thus:

(a) $1)\ \sqrt{17}\quad (4$

$$\frac{4}{\sqrt{17}-4}$$

(b) $\sqrt{17}-4)\ 1\quad (8$

$$\frac{8(\sqrt{17}-4)}{1-8(\sqrt{17}-4)}$$

$$=(\sqrt{17}-4)(\sqrt{17}+4)-8(\sqrt{17}-4)$$

$$=(\sqrt{17}-4)(\sqrt{17}+4-8)$$

$$=(\sqrt{17}-4)^2.$$

BIBLIOGRAPHY

See Wilbur Richard Knorr, *The Evolution of the Euclidean Elements* (Dordrecht, 1975), 181–193.

IVOR BULMER-THOMAS

UEXKÜLL, JAKOB JOHANN VON (*b.* Keblas, Estonia [now Estonian S.S.R.], 8 September 1864; *d.* Capri, Italy, 25 July 1944), *biology*.

Uexküll was the third of four children born to Alexander von Uexküll, squire of Heimar and later mayor of Reval (now Tallinn, Estonian S.S.R.), who had traveled in the Ural Mountains as a young geologist and had maintained an interest in natural science. After attending the Domschule in Reval, Uexküll studied zoology at the University of Dorpat (now Tartu, Estonian S.S.R.), where he was influenced by the writings of Karl Ernst von Baer and Johannes Müller. In 1888 he worked with Kühne at

UEXKÜLL

Heidelberg, his research dealing with muscle physiology. After a period of studying marine life at the Zoological Station in Naples, he worked in Paris with Marey, who taught him how to use cinematography to record physiological processes. From 1925 to 1936 he was director of the Institut für Umweltforschung at the University of Hamburg. He received an honorary Ph.D. from the University of Heidelberg in 1907 and from the University of Utrecht in 1936. Uexküll was director of the Hamburg zoological garden and aquarium, and was honorary professor at Hamburg from 1925 to 1944.

Uexküll is known for his *Umweltlehre*, which has stimulated important research in ethology by Konrad Lorenz and Nikolaas Tinbergen, among others. The theory is based on the assumption that, within its own subjective "self-world" (*Umwelt*), a living being perceives only that which its sense organs (receptors) convey to it and deals only with those factors that its locomotive organs (effectors) can affect. Accordingly, by means of its sensory and effector organs, each creature selects a certain portion of the objective (physical-chemical-biological) surroundings suitable to its species; this is its *Umwelt*. This selected subjective world surrounds every creature like a fixed sphere, although it is invisible to the observer.

Nevertheless, through knowledge of the anatomy and physiology of a creature's sensory and effector organs, as well as of its specific needs, the observer can reconstruct its *Umwelt* as a unity composed of a perceptual world and an effector world. He can then analyze its behavior as a sequence of steps in a functional circle that comprises a perceptual sector and an effector sector. For example, in the functional circle model, the sensory organs of a hungry creature endow a previously neutral factor (object) in its external surroundings with definite optical, olfactory, and tactile characteristics. Together these characteristics bestow upon the object the meaning "food"; in effect, they give it a label. This label elicits behavior (an activity of the effector organs) that brings the animal into contact with the object—grasping, biting, chewing, swallowing—and thereby eliminates the characteristics, with the attendant "meaning label." This process occurs either subjectively (through satiation) or objectively (through swallowing). At this point the function circle has been entirely traversed, and the creature's behavior comes to a state of rest.

The concept of the functional circle anticipates many important notions that were subsequently formulated mathematically in cybernetics. The functional circle is, in fact, essentially a control loop,

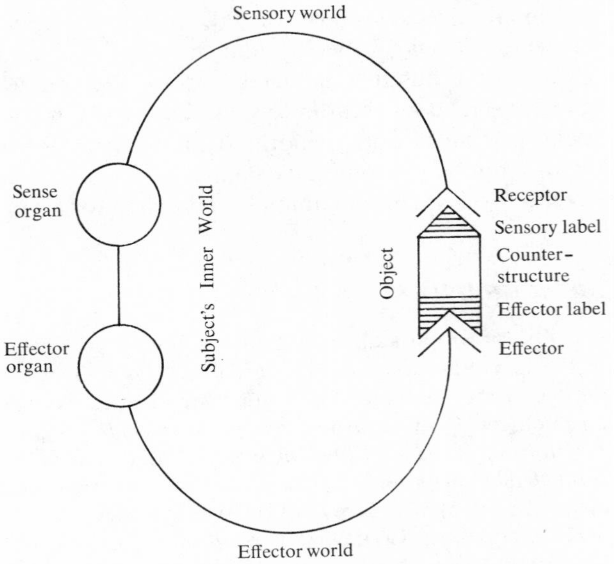

FIGURE 1. Functional circle.

in which the result is retroactive with the beginning and prescribes its direction.

Each *Umwelt* is governed by a particular time, which originates in the subject. The notion of a subjective proper time goes back to von Baer, although he did not introduce a term for the subjective unit of time. Since Uexküll this unit has been known as the "moment." It may be defined as the period within which we experience all stimuli as simultaneous, independent of the objective time sequence. In Uexküll's theory, the basic component of subjective space is called "place." Every designation of a specific location yields a place in the space of the subjective world.

Although ethology has developed only the objective (physiological) aspect of the *Umwelt* theory, sociologists—especially those concerned with the sociology of knowledge—have profited from related insights in their study of subjective phenomena.

BIBLIOGRAPHY

I. Original Works. Uexküll's earlier writings include "Im Kampf um die Tierseele," in *Ergebnisse der Physiologie* (1902); *Leitfaden in das Studium der experimentellen Biologie der Wassertiere* (Wiesbaden, 1905); *Umwelt und Innenwelt der Tiere* (Berlin, 1909); *Bausteiner zu einer biologischen Weltanschauung* (Munich, 1913); "Biologische Briefe an eine Dame," in *Deutsche Rundschau* (1919); *Staatsbiologie* (Berlin, 1920); *Theoretische Biologie* (Berlin, 1920), also in

English, *Theoretical Biology* (London–New York, 1926); "Technische und mechanische Biologie," in *Ergebnisse der Physiologie*, **20** (1922); and "Wie sehen wir Natur und wie sieht sie sich selber?" in *Naturwissenschaften*, **10** (1922).

Subsequent works are "Definition des Lebens und des Organismus," in A. Bethe *et al.*, eds., *Handbuch der normalen und pathologischen Physiologie*, I (Berlin, 1927), 1–26; "Die Einpassung," *ibid.*, 694–701; "Der Wirkraum," in *Pflügers Archiv für die gesamte Physiologie*, **217** (1927), 72–87, written with H. Roesen; *Die Lebenslehre* (Potsdam, 1930); "Die Rolle des Subjekts in der Biologie," in *Naturwissenschaften*, **19** (1931), 385–391; "Biologie oder Physiologie?" in *Nova acta Leopoldina*, **1** (1933), 276–281; and *Streifzüge durch die Umwelten von Tieren und Menschen—Bedeutungslehre* (Berlin, 1934; repr. Hamburg, 1956), written with Georg Kriszat, also in English, *Instinctive Behaviour* (New York, 1957).

II. SECONDARY LITERATURE. On Uexküll and his work, see Rudolf Burckhardt and Hubert Erhard, *Geschichte der Zoologie*, 2nd ed. (Berlin, 1921); *Festschrift für Jakob von Uexküll* (Berlin, 1924), published for his sixtieth birthday; F. Brock, in *Sudhoffs Archiv für Geschichte der Medizin und der Naturwissenschaften*, **27** (1934), with bibliography; and Gudrun Uexküll, *Jakob von Uexküll, seine Welt und seine Umwelt* (Hamburg, 1964).

T. VON UEXKÜLL

VAVILOV, NIKOLAY IVANOVICH (*b.* Moscow, Russia, 25 November 1887; *d.* Saratov, U.S.S.R, 26 January 1943), *botany, agronomy, genetics, phytogeography*.

Vavilov was the oldest son of Ivan Ilich Vavilov, owner of a shoe factory and at one time a member of the Moscow City Duma. His brother Sergey became a well-known physicist and president of the Soviet Academy of Sciences; his sister Lydia, a microbiologist, died of typhus during World War I; his sister Alexandra was a physician. Vavilov was married twice, the second time to his co-worker Elena Ivanovna Barulina; his sons, Oleg and Yuri, both became physicists.

Vavilov's career was shaped by his lifelong dream of mobilizing the sciences in order to transform Russian agriculture. His meteoric rise to prominence following the Russian Revolution began twenty years of prodigious activity involving the creation of a vast research network. The picture that emerges from contemporary accounts and memoirs by some of his many co-workers suggests that his remarkable success as a scientific entrepreneur resulted from his boundless energy and winning personality. During this period Vavilov apparently took no vacations and slept no more than four or five hours a night, often on the train between Moscow and Leningrad or on a sofa in his office, where he lived much of the time. Heterodox and undogmatic, he had an excellent memory for names and faces, and seemed to dazzle almost all those with whom he came into contact. Most regarded charm as his defining characteristic.

In 1906 Vavilov graduated from a commercial high school and entered the Moscow Agricultural Institute (now the Timiryazev Academy of Agriculture). There he organized and led a student science club that conducted botanical and geographical expeditions to the Caucasus, Transcaucasia, and other regions of Russia. As a third-year student he presented a paper on Darwinism and experimental morphology at the institute's Darwin centenary celebration. His student thesis on destructive garden slugs of Moscow Province was awarded the A. P. Bogdanov Prize of the Polytechnical Museum.

After graduating in 1911, Vavilov remained with D. N. Pryanishnikov in the department of special agriculture to prepare for an academic career and conducted research at the institute's Moscow selection station, working with its director, D. L. Rudzinsky. In 1911–1912 he taught agriculture at the Golitsyn Courses for Women. He moved in 1912 to St. Petersburg, where he was a probationary employee at the Bureau of Applied Botany of the Ministry of Agriculture, headed by R. E. Regel, and at its Bureau of Mycology and Phytopathology, headed by A. A. Yachevsky.

In 1913 the department of special agriculture sent Vavilov to England. At Cambridge he attended the lectures of R. C. Punnett and worked in the laboratory of R. D. Biffen. He also worked at the John Innes Horticultural Institution at Merton (London) under William Bateson, whom he always regarded as his teacher. While in England he studied immunity in wheat, using the collection of John Percival, then the world's largest, and conducting experiments on plant immunity simultaneously in Moscow and England.

On his return to Russia, which was hastened by the outbreak of World War I, Vavilov passed the examination for the master's degree and completed his thesis, *Immunitet rasteny k infektsionnym zabolevaniam* ("Plant Immunity to Infectious Diseases"). He continued to teach in the department of special agriculture during the war (he was not drafted because of an eye problem), and in 1916 orga-

VAVILOV

nized and conducted a botanical and geographical expedition to Iran and the Pamir.

In 1917, Vavilov was simultaneously appointed professor of genetics, selection, and special agriculture at the Voronezh Agricultural Institute and professor of agronomy and selection at Saratov University. During the next four years he organized and directed the Saratov branch of the department of applied botany and selection of the Agricultural Scientific Committee of the Commissariat of Agriculture of the Russian Soviet Federated Socialist Republic (formerly the Bureau of Applied Botany of the Ministry of Agriculture).

Vavilov came to the attention of the government in 1920–1921 and rapidly rose in prominence. Following Regel's death in 1920, Vavilov moved to Petrograd to replace him as director of the department of applied botany. In 1924, with Lenin's support, he reorganized the department into the All-Union Institute of Applied Botany and New Cultures, which in 1930 was renamed the All-Union Institute of Plant Breeding (VIR). From 1923, Vavilov was also director of the State Institute of Experimental Agronomy at Leningrad. He reorganized this institute in 1929, forming the All-Union Lenin Academy of Agricultural Sciences (VASKhNIL), with the VIR at its core. He served as its first president until 1935, by which time it had more than 400 experimental stations and a staff of more than 20,000.

During this period Vavilov earned his reputation as "the most widely traveled biologist of our day." In connection with the VIR's collecting activities, he organized more than 180 expeditions, 40 outside the Soviet Union. Collecting expeditions in which he participated include those of 1920 (Astrakhan, Tadzhikistan, and the Pamir), 1921–1922 (United States, Canada, England, France, Germany, Sweden, and Holland), 1924 (Afghanistan), 1925 (Khiva and Uzbekistan), 1926–1927 (Algeria, Tunisia, Morocco, Egypt, Syria, Palestine, Trans-Jordan, Greece, Crete, Capri, Italy, Sardinia, Sicily, Spain, Somaliland, and Ethiopia), 1927 (Germany), 1929 (western China, Taiwan, Japan, and Korea), 1930 (United States, Central America, and Mexico), 1931 (Denmark and Sweden), 1932–1933 (Cuba, Peru, Bolivia, Chile, Brazil, Argentina, Uruguay, Trinidad, and Puerto Rico), 1934 (Caucasus), and 1940 (Belorussia and the western Ukraine). Vavilov was elected president of the U.S.S.R. Geographical Society in 1931 and served through 1940. His own account of his travels is *Pyat kontinentov* ("Five Continents"), published posthumously in 1962.

As a result of such expeditions, by 1940 the VIR had amassed a collection of more than 250,000 specimens, including 36,000 of wheat, 10,022 of maize, 23,636 of legumes, 23,200 of grasses, 17,955 of vegetables, and 12,650 of fruits. These samples were distributed throughout the system of experimental stations for cytological and genetic analysis, breeding and hybridization experiments, and testing under various growing conditions. They formed the basis for the institute's—and Vavilov's own—numerous publications.

Vavilov was elected a corresponding member of the Soviet Academy of Sciences in 1923 and a full member in 1929. Following the death of Y. A. Filipchenko in 1930, he became head of the genetics section of the Commission for the Exploitation of Productive Forces, transforming it into the Soviet Academy's Laboratory of Genetics, which became in 1933 its Institute of Genetics, which he directed until 1940. There he attracted a first-rate group of researchers, including H. J. Muller (1933–1937), and established the leading center of theoretical genetics research in the Soviet Union.

Vavilov won many awards, including the N. M. Przhevalsky Gold Medal of the Geographical Society of the U.S.S.R. (1925), the Lenin Prize (1926), and the Great Gold Medallion of the All-Union Agricultural Exhibition. He was elected a foreign member of many scientific societies, including the International Agrarian Institute of Rome (1924), the Ukrainian Academy of Sciences (1929), the Czechoslovak Academy of Sciences (1936), the Indian National Academy of Sciences (1937), the Royal Society of Edinburgh (1937), the Royal Society of London (1942), and the Linnean Society of London, the New York Geographical Society, the American Botanical Society, the Mexican Agricultural Society, and the Royal Spanish Natural History Society. He was awarded honorary doctorates by the University of Brno (1936) and the University of Sofia. In 1932 Vavilov served as vice-president of the Sixth International Congress of Genetics (Ithaca, New York) and in 1939 was elected president of the Seventh International Congress, although he was unable to attend.

During his lifetime Vavilov published more than 350 scientific books and articles, many of which, according to his own testimony, were written or dictated in the evening, at odd moments, or while traveling. His writings manifest an encyclopedic command of agronomic and biological literature. Vavilov could read, write, and speak English, French, and German; and many of his publications in Russian include texts or lengthy abstracts in

VAVILOV

English. He served as the editor of several journals, most notably *Trudy po prikladnoi botanike, genetike i selektsii* ("Works on Applied Botany, Genetics, and Selection"), and oversaw the translation and publication of works by J. G. Kölreuter, Darwin, Mendel, T. H. Morgan, and H. J. Muller.

Plant Immunity. Vavilov's earliest scientific work concerned the genetic basis of plant immunity. His first studies were undertaken during his school days in Moscow and with Biffen in England (1913–1914). They were published in a series of studies that appeared in 1913–1919, notably "Immunity to Fungous Diseases as a Physiological Test in Genetics and Systematics, Exemplified in Cereals" (1914). In this paper Vavilov focused on taxonomic characters in relation to susceptibility, establishing that in the one variety of *Triticum vulgare* (out of 580) that was immune to mildew, immunity was inherited as a single recessive character. In this and in *Immunitet rasteny . . .* Vavilov developed the concept of the degree of specialization of parasitic fungi, postulating that the more species and genera a parasitic fungus attacks, the less likely it is that resistant varieties will be found in any of the host species.

Vavilov believed that immunity had to be viewed in an evolutionary context: "Immunity is associated with the biological specialization of parasites to genera and species and is caused by divergence of the host and parasite in their evolution." He sought to use reactions to parasites, especially to narrowly specialized parasites, as an aid to systematic botany. One product of Vavilov's research was his discovery of a species of wheat, *Triticum timopheevi*, that is resistant to most of the important diseases of wheat and that has subsequently been used as a source of disease resistance by breeders in the Soviet Union and elsewhere. Like his other works, Vavilov's monographs on plant immunity attempt to synthesize the literature in the field in order to find regularities that would aid the breeder. According to Vavilov, his studies in this field led him naturally into his studies in genetics and plant geography.

Law of Homologous Series. Vavilov first presented his "law of homologous series in hereditary variation" at the Third All-Russian Conference on Selection and Seed-Growing at Saratov (1920). His paper was published in Russian that year, two years later in English, and in its enlarged and revised version in 1935. Punnett later suggested that Vavilov may have gotten the idea for his law in 1913, during one of Punnett's lectures at Cambridge on parallel series of variations in species of rodents and butterflies. This is not certain, however, since the idea was common at the time. As Vavilov himself noted, the phenomenon of "parallel" or "analogous" variations in different races, breeds, species, genera, and families of animals and plants had been commented on by many earlier biologists, including Darwin, Mivart, Naudin, B. D. Walsh, de Vries, J. Duval-Jouve, G. H. T. Eimer, Cope, E. Zederbauer, and P. A. Saccardo.

In his paper Vavilov sought, however, to establish a more general law. By collecting information on morphological variants or mutations in various cereals, he showed that many variants found in a given species also are found in closely related species—the more closely related the species, the greater the number of common variants. Vavilov displayed the data in tabular form and thought he had found a law or regularity (*zakonomernost*) in the range of variability within plant species. He himself drew the parallel between his "homologous series" and the periodic table of the elements, and stated his conclusions in symbolic form reminiscent of chemistry, thus leading the agronomist V. R. Zalensky to comment, "Biology has found its Mendeleev." Vavilov believed that just as the periodic table had served as the basis for predicting the existence of undiscovered elements, his "homologous series" tables could be used to predict the existence of undiscovered plant forms that would fill the "gaps" in his system. He saw his law as evidence that despite the apparently random and unpredictable character of mutation and variation, the variability of species manifests an overall regularity that contributes to the orderliness of the evolutionary process.

A number of other investigators saw in Vavilov's law a challenge to the traditional Darwinian view of evolution as governed by natural selection. L. S. Berg regarded parallel variations as evidence that evolution was an unfolding along predetermined lines. Filipchenko saw the law as evidence that the characteristics that determine species are different from those establishing higher taxonomic categories, thereby agreeing with E. D. Cope and W. L. Johannsen that microevolution and macroevolution are fundamentally different processes with fundamentally different causes. Others understood the law to suggest that mutations can be predicted, or that there are definite limits to species variability.

Responding in 1936 to criticism, Vavilov admitted that in formulating the law in 1920 and 1922, the "striking parallelisms, down to the most minute details" that he had observed in cultivated plants, had led him to be "unnecessarily categorical":

VAVILOV

> We underestimated the variability of the genes themselves. . . . At that time we thought that the genes possessed by close species were identical; now we know that this is far from the case, that even very closely related species which have externally similar traits are characterized by many different genes. By concentrating our attention on the variability itself, we gave insufficient attention to the role of selection . . . ["Puti sovetskoy selektsii," in *Spornye voprosy genetiki i selektsii* (1937)].

In the revised (1935) version of the theory, Vavilov distinguished more carefully between phenotypic and genotypic variation, and between "homologous" variation (caused by the same genes) and "analogous" variation (similar phenotypic traits caused by different genes). He also added information on parallel variations in additional species (such as those in melons, worked out in 1925) and emphasized the importance of selection. Vavilov modified his analogy between genetics and chemistry, commenting:

> In the last decade genetics has moved ahead rapidly and to some degree begins to resemble chemistry, at least the chemistry of complex organic compounds. . . . The regularities in plant polymorphisms . . . may to some measure be compared with the homologous series of organic chemistry, for example hydrocarbons. . . . As with chemical structures, different forms of plants and animals are characterized by a physical structure and bring to mind the systems and classes of the chemistry of crystals [*Zakon gomologicheskikh ryadov v nosledstrennoy izmenchivosti*, 2nd ed. (Moscow–Leningrad, 1935)].

Despite these revisions, however, he maintained the law's validity and importance, and gave several examples of the discovery of forms that had been predicted on the basis of his theory.

The law of homologous series in hereditary variation was the basis of Vavilov's conception of the species, most fully expressed in his 1931 monograph *Linneevsky vid kak sistema* ("The Linnaean Species as a System"). Vavilov regarded the "Linnaean species" (a synonym for Lotsy's term "linneon") as "a complex system of forms, the structure of which conforms to the law of homologous series." For him the species was not an arbitrary division of convenience but a product of a dynamic evolutionary process operating over time and space, a process that both determined its genetic composition and gave it a real existence at any given moment.

Throughout his career Vavilov held what he regarded as a Darwinian view of evolution. As expressed in 1940, it envisioned hereditary variability as the basic material for both natural and artificial selection; the generation of new forms is increased by hybridization; and further evolution proceeds by the dispersion of species, their occupation of new territory, ecological factors, and geographical isolation. For Vavilov "the key factor in evolution, adaptation, and speciation is natural and artificial selection."

Centers of Origin of Cultivated Plants. Vavilov's work on the origin of cultivated plants was largely responsible for his worldwide reputation. His interest in the problem dates from his 1917 article proposing that rye originated as a weed in the southern wheat fields and gradually supplanted wheat by climatic selection in northern and mountainous regions, to which it was better adapted. His first general theoretical statement on the subject was begun in 1924 and published in 1926 as *Tsentry proiskhozhdenia kulturnykh rasteny* ("Centers of Origin of Cultivated Plants"), which in the same year brought Vavilov one of the first Lenin Prizes. His views were elaborated and modified in subsequent monographs (1927, 1931, 1935, 1940).

Vavilov dedicated his book to Alphonse de Candolle, whose *Origine des plantes cultivées* (1882) was the major work in the field prior to Vavilov's. Whereas Candolle had sought to establish centers of origin by using archaeological, historical, and linguistic—as well as botanical—data, Vavilov used these only as "correctives" to his "differential systematic geographical method," which relied heavily on genetic and cytological analysis. As outlined in 1926, this method involved the division of plants into linneons and genetic groups by means of systematic morphology and cytological, genetic, and immunological analysis; the establishment of their geographical ranges; the "detailed determination of the varieties and races of each species (more precisely, the heritable varying traits and the general system of hereditary variability)"; and the geographical mapping of the variant forms of a plant group, as well as those of closely related varieties and wild species, in order to determine "geographical centers of concentrated diversity" (areas that contain the greatest "diversity of endemic varieties"), which Vavilov believed to be "as a rule, the primary center of origin." In 1927 he added, "The primary centers often include a large number of genetically dominant traits."

Using this method, Vavilov found that the centers of origin of many plants coincide with one an-

VAVILOV

other and with areas known to be sites of early civilization. In 1926 he identified five such primary centers: southwestern Asia (from India through Central Asia to the Black Sea); southeastern Asia (China, Korea, and Japan); the Mediterranean coast; Abyssinia and Eritrea; and the mountainous regions of South America and Mexico. Following his expeditions to the Americas and Asia, he increased the number of primary centers to eight (1935), dividing the American center into two (Mexico and Central America, and the Andes) and the southwest Asiatic into three: the Indian (which now included Indochina and the East Indies); the Central Asiatic (northwest India through Uzbekistan); and an area running from eastern Turkey through the Caucasus to Afghanistan. In addition, within the Andean or South American center he acknowledged distinct regions in which different plants had originated (Peru, Bolivia, Ecuador; Chile; Brazil and Paraguay). By 1940 his classification listed thirteen such regions, organized into seven basic centers.

According to Vavilov's final version (1940), these seven centers are the southwest Asiatic (greatest number of species of wheat, barley, rye, and many fruits); the south Asian tropical (sugarcane, rice, many tropical fruits and vegetables); the east Asiatic (millet, soy, many vegetables and fruits); the Mediterranean (the olive, many vegetables, fodder, and certain fruits); the Abyssinian (teff, the Abyssinian banana [ensete], several species of sorghum, and some species of hard wheat); the Central American (corn, upland cotton, many species of beans, pumpkins, tobacco, cocoa, sweet potato, peppers, and tropical fruits [guava, soursop and sweetsop, avocado]); and the Andean (potatoes, okra, the cinchona tree, and the coca shrub). In all, by 1935 Vavilov believed that he and his co-workers had located the center of origin of more than 600 species of cultivated plants.

For Vavilov these primarily mountainous regions, with their wide variety of environmental conditions and natural isolating barriers, provided ideal conditions both for speciation and for human cultivation of new breeds. He never stated why he regarded "centers of diversity" as "centers of origin"; but by 1931 he apparently had recognized the importance of isolation in producing the diversity that he observed in these regions: "Schematically . . . as a result of inbreeding and mutation, advantageously recessive forms separate out and are formed toward the periphery of the ancient range of a species of cultivated plant under isolation (on islands, in mountains)" ("Rol Tsentralnoy Azii . . .," in *Trudy po prikladnoi botanike, genetike i selektsii*, **26**, no. 3 [1931]). Vavilov saw these centers of origin as containing the full variability and "hidden potential" of the various species.

In December 1936 Vavilov acknowledged that in the preceding decade his thinking on the subject had changed.

> At the beginning, we assumed that the majority of genes distinguishing current cultivated plants were really located at these centers, assuming that races and varieties dispersed from definite regions exactly as had whole species. Like geneticists in the first decade of this century, we conceived of genes as being more stable than they later turned out to be. In fact, in these centers we found an enormous number of genes, even more than we supposed at the outset, but at the same time newly formed genes, often remarkably valuable, were found in secondary regions on the periphery, and in the area between the periphery and the centers ["Puti sovetskoy selektsii," in *Spornye voprosy genetiki i selektsii* (1937)].

For Vavilov his studies not only served to clarify the evolutionary process by demonstrating how species are formed but also had a direct "utilitarian goal": by making the full range of hereditary diversity of a species available to the breeder, the theory provided the "key to the mobilization of the plant resources of the globe."

Selection as a Science. Throughout his career Vavilov sought to transform Soviet agriculture by creating new breeds of plants suited to its various environmental conditions. In 1912 he wrote that genetics "gives the foundations for the planned intervention of man in the creation of nature, the rules that govern the creation of forms." In 1925 he held forth "the possibility, in Timiryazev's words, of 'sculpting organic forms at will,' " suggesting that "in the near future man will be able, by means of crossing, to synthesize forms such as are absolutely unknown in nature." At other times he saw his task as "the planned and rational utilization of the plant resources of the terrestrial globe" (1917), or finding "the way to seize the bastions of the fortress" of the plant kingdom (1926). He regarded "the complete control of the organism" as "the final purpose of modern biology" (1934).

Vavilov's approach to this task was synthetic. In 1931 he wrote:

> Biological phenomena connected with speciation are too complex to be reduced to simple physicochemical processes. But even complex biological processes manifest definite rules and laws which the biologist

> must study and which lead the investigator to the mastery of the formative process [*Linneevsky vid kak sistema* (1931)].

He looked upon his law of homologous series and his theory of the centers of formation of cultivated plants as precisely this kind of biological law or rule.

In the mid-1930's Vavilov sought to characterize selection as an independent scientific discipline. In his pamphlet *Selektsia kak nauka* ("Selection as a Science," 1934), he presented what he considered to be its seven basic divisions: phytogeography; studies of hereditary variability (the laws of heredity and the study of mutations); studies of the influence of the environment in creating varieties and affecting their development; hybridization theory; selection theory (centering on different forms of plant reproduction); studies of selection for particular properties (such as immunity, resistance to cold or drought, photoperiodism, and chemical composition); and the study of selection in particular plants.

As conceived by Vavilov, selection was a highly complex science. For controlling heredity it drew on genetics, cytology, and embryology; and in the actual process of selection it drew on physiology, chemistry, technology, phytopathology, and entomology, "transforming and differentiating them in accordance with the final task of creating a variety." He regarded selection as a natural extension of Darwin's theory of evolution and defined selection succinctly as "evolution directed by the will of man."

As director of the VIR, Vavilov put his concept of selection into practice. For him the ultimate purpose of the expeditions he sent or led was to provide Soviet breeders with the full potential of the species, the raw material for sculpting new breeds. Beginning in 1923, he conducted experiments in "geographical sowing" that involved growing varieties at more than 100 sites throughout the Soviet Union in order to ascertain their suitability for planting under various conditions. He set up cyclical interbreeding of geographically close and distant forms of cereal grains, green beans, and flax in both fields and hothouses.

On the basis of these studies, Vavilov developed an agroecological classification of cultivated plants. The data for such a classification were published in brief form in 1940 in *The New Systematics* and fully in 1957 as *Mirovye resursy* . . . ("World Resources . . ."). In this work Vavilov distinguished ninety-five types of ecological areas in relation to climate, soil, and geography, and the varieties of cereal grains, legumes, and flax best suited to them, as well as the role of selection in creating especially adapted varieties.

Vavilov's undertaking was possible only because of the manpower and resources at his disposal, and the latter were made available to him because of his promises to transform Soviet agriculture. In a 1931 newspaper article he wrote that "in practice, agronomy knows no bounds" and suggested that "the cultivation of vegetables, root and tuberous plants, and fodder grasses can go right up to the northernmost boundaries of the Eurasian landmass, not only to the northern boundaries of the taiga . . . but into the tundra." Vavilov apparently believed that the task could be accomplished within a few decades and that the prospects were limitless.

Science and Politics. Throughout most of his career, Vavilov felt that the Soviet Union gave scientists and their work the greatest respect of any country in the world and offered the widest opportunities for serving mankind; therefore the cruelties of the regime had to be overlooked. In 1920 his presentation of his law of homologous series led conference delegates to send telegrams of praise to the commissars of agriculture and education. Sometime during this period, perhaps in connection with the 1921 famine, he apparently came to the attention of Lenin, who had been impressed by K. A. Timiryazev's translation of William Sumner Harwood's *The New Earth: A Recital of the Triumphs of Modern Agriculture in America* (New York, 1906). During the subsequent fifteen years Vavilov enjoyed the strong support of the government, which financed his rapidly expanding activities. He quickly grew in prominence and influence, serving as one of the few non-Communist members of the Central Executive Committee of the government (1926–1935) and, for a time, on the Central Executive Committee of the Russian Soviet Federated Socialist Republic (1927–1929).

From 1928 to 1932, when the Academy of Sciences was "Bolshevized" and greater ideological conformity began to be demanded of scientific specialists, Vavilov started to introduce the language of dialectical materialism into his writings, offering "The Linnaean Species as a System" (1931) as a model for fellow scientists.

Beginning in 1931, Vavilov attracted criticism for failing to produce desired agricultural results. In response, he pointed to what had already been achieved and gave increasing emphasis in his public presentations to techniques that could yield immediate benefit. In so doing, he helped bring to prominence T. D. Lysenko, who had become known in

VAVILOV

the Soviet Union in the late 1920's and early 1930's for his work on "vernalization" (*yarovizatsia*), the treatment of seeds before planting to alter their development in such a way as to permit additional crop yields. In an article in *Izvestia* (6 November 1933), he hailed Lysenko's method as a "revolutionary discovery of Soviet science" that had given "brilliant results" in tests conducted during the preceding year on "tens of thousands" of varieties, and that had made possible the cultivation of wheat from Afghanistan, India, Australia, and South Africa within the Arctic Circle. In 1935 Vavilov termed Lysenko's method a "powerful tool" in the selection of many plants and foresaw a "total revision" of the classification of the world's plant resources on the basis of their reaction to vernalization. He suggested that Lysenko's theory of the "stage development" of plants "also opens exceptional possibilities . . . for the utilization of the world's plant resources."

In 1935 Vavilov was attacked for Soviet agricultural failures by Lysenko and his follower I. I. Prezent, among others. In particular, he was accused of pandering to foreign science, of wasting government money on useless collecting expeditions abroad, and of espousing "idealist" theories of homologous series and centers of origin, which one writer characterized as "fiascoes." The attacks on Vavilov were related to more general attacks on genetics as being a formalistic, "idealistic" science contrary to dialectical materialism and Darwinism, and linked to foreign influences, notably fascism.

Vavilov lost his posts as member of the government's Central Executive Committee and as president of VASKhNIL in 1935. His replacement in the latter post was A. I. Muralov; following Muralov's arrest in July 1937, it was G. K. Meister; and after Meister's arrest in February 1938, Lysenko assumed the post. Vavilov may have been under arrest briefly in 1936. The Seventh International Congress of Genetics, scheduled to be held at Moscow in August 1937, was postponed and was finally held two years later at Edinburgh. As president of VASKhNIL, Lysenko interfered in Vavilov's work at the VIR by appointing one of his followers as Vavilov's assistant, firing members of Vavilov's staff, and generally undermining his direction of the institute.

Until 1939, Vavilov defended genetics and his own work but did not strongly criticize Lysenko, calling instead for cooperation and urging more objective discussions. In that year, however, Vavilov became much more critical of Lysenko. Zhores Medvedev quotes the following line from a speech Vavilov made to a session of the regional bureau of the VIR in March 1939: "We shall go to the pyre, we shall burn, but we shall not retreat from our convictions." The statement proved to be prophetic. On 6 August 1940, while on a collecting expedition in the western Ukraine near Chernovtsy, Vavilov was arrested by agents of the NKVD; shortly thereafter his associate G. D. Karpechenko and others were arrested.

On 9 July 1941 the Military Collegium of the Supreme Court found Vavilov guilty, under Article 58 of the Soviet constitution, of belonging to a rightist organization, spying for England, leading the "Peasant Labor Party," conducting sabotage in agriculture, and maintaining links with émigrés, for which crimes he was sentenced to death. Through the efforts of his brother Sergey and his former teacher Pryanishnikov, his sentence was commuted to ten years' imprisonment. Following his sentencing he was moved to the inner prison of the NKVD in Moscow. In October 1941, during the evacuation of prisoners to the interior of the country, Vavilov was moved to the Saratov prison and was placed for several months in an underground death cell, where he suffered from malnutrition. He died in prison on 26 January 1943 and probably was buried in the Voskresensky Cemetery in Saratov.

After Stalin's death Vavilov's case was reopened on the initiative of his family. On 2 September 1955 he was posthumously rehabilitated by the Soviet Supreme Court, and a week later the Presidium of the Academy of Sciences placed him on its list of deceased members. Since 1957 most of his works have been published, many for the first time. In December 1967 the new All-Union Society of Geneticists and Selectionists was named for Vavilov.

In the West and the Soviet Union, Vavilov has come to be regarded as one of the outstanding geneticists of the twentieth century, a symbol of the best aspects of Soviet science, and a martyr for scientific truth.

BIBLIOGRAPHY

I. Original Works. Vavilov's more than 350 publications include the following early works: *Golye slizni (ulitki), povrezhdayushchie polya i ogorody v Moskovskoy gubernii* ("Destructive Slugs of the Fields and Gardens of Moscow Province"; Moscow, 1910); "Genetika i ee otnoshenie k agronomii" ("Genetics and Its Relation to Agronomy"), in *Otchet Golitsynskikh zhenskikh kursov za 1911 god po khozyaystvennoi i za 1911–1912 uchebnyi god po uchebnoi chasti* ("Report of the Golitsyn Courses for Women for the 1911 Economic Year and the 1911–1912 Academic Year"; Moscow, 1912), 77–87; "Immunity to Fungous Diseases as a Physiological Test in Genetics and Systematics, Exemplified in Cereals," in *Journal of Genetics*, **4**, no. 1 (1914), 49–65; "O proiskhozhdenii kulturnoy rzhi" ("On the Origin of Cultivated Rye"), in *Trudy Byuro po prikladnoi botanike*, **10**, nos. 7–10 (1917), 561–590; "Sovremennye zadachi selskokhozyaystvennogo rastenievodstva" ("Current Tasks of Agricultural Plant Breeding"), in *Selsko-khozyaistvennyi vestnik Yugo-Vostoka*, nos. 19–21 (1917), 3–10; and *Immunitet rasteny k infektsionnym zabolevaniam* ("Plant Immunity to Infectious Diseases"; Moscow, 1919).

Works from the 1920's are *Zakon gomologicheskikh ryadov v nasledstvennoy izmenchivosti* ("The Law of Homologous Series in Hereditary Variation"; Saratov, 1920; 2nd ed., rev. and enl., Moscow–Leningrad, 1935), also in English as "The Law of Homologous Series in Variation," in *Journal of Genetics*, **12**, no. 1 (1922), 47–89; *Polevye kultury yugo-vostoka* ("Field Cultures of the Southeast"; Petrograd, 1922); "O geneticheskoy prirode ozimykh i yarovykh rasteny" ("On the Genetic Nature of Winter and Spring Plants"), in *Izvestiya Saratovskogo selsko-khozyaistvennogo instituta*, **1**, no. 1 (1923), 17–41, written with E. S. Kuznetsova (with English abstract); "K poznaniyu myagkikh pshenits (sistematiko-geografichesky ocherk)" ("Toward an Understanding of Soft Wheats [a Systematic Geographical Essay]"), in *Trudy po prikladnoi botanike, genetike i selektsii*, **13**, no. 1 (1923), 140–257; "K filogenezu pshenits: Gibridologichesky analiz vida *Triticum persicum* Vav. i mezhduvidovaya gibridizatsia u pshenits" ("Toward a Phylogeny of Wheat: An Analysis of the Hybrids of *Triticum persicum* Vav. and Interspecific Hybridization in Wheat"), *ibid.*, **15**, no. 1 (1925), 3–159, written with S. V. Yakushkina (with English abstract); "Ocherednye zadachi selskokhozyaystvennogo rastenievodstva (rastitelnye bogatstva zemli i ikh ispolzovanie)" ("The Next Tasks of Agricultural Plant Breeding [the Plant Riches of the Earth and Their Utilization]"), in *Pravda* (2 Aug. 1925), 5–6; *Tsentry proiskhozhdenia kulturnykh rasteny* ("Centers of Origin of Cultivated Plants"; Leningrad, 1926), with Russian and English texts; "Geograficheskie zakonomernosti v raspredelenii genov kulturnykh rasteny" ("Geographical Regularities in the Distribution of Genes in Cultivated Plants"), in *Trudy po prikladnoi botanike, genetike i selektsii*, **17**, no. 3 (1927), 411–428; "Geographische Genzentren unserer Kulturpflanzen," in *Verhandlungen des V. Internationalen Kongresses für Vererbungswissenschaft, Berlin, 1927*, which is supp. 1 of *Zeitschrift für induktive Abstammungs- und Vererbungslehre* (1928), 342–369; *Zemledelchesky Afganistan* ("Agricultural Afghanistan"; Leningrad, 1929), which is supp. 33 of *Trudy po prikladnoi botanike, genetike i selektsii*, written with D. D. Bukinich; and "Genetika" ("Genetics"), in *Bolshaya sovetskaya entsiklopedia*, XV (1929), 191–201.

Vavilov subsequently published "Gomologicheskie ryady" ("Homologous Series"), in *Bolshaya sovetskaya entsiklopedia*, XVII (1930), 586–587; *Linneevsky vid kak sistema* ("The Linnaean Species as a System"; Moscow–Leningrad, 1931); "Rol Tsentralnoy Azii v proiskhozhdenii kulturnykh rasteny (predvaritelnoe soobshchenie o rezultatakh ekspeditsii v Tsentralnuyu Aziyu v 1929 g.)" ("The Role of Central Asia in the Origin of Cultivated Plants [a Preliminary Communication on the Results of the Expedition to Central Asia in 1929]"), in *Trudy po prikladnoi botanike, genetike i selektsii*, **26**, no. 3 (1931), 3–44; "Meksika i Tsentralnaya Amerika kak osnovnoy tsentr proiskhozhdenia kulturnykh rasteny Novogo Sveta (predvaritelnoe soobshchenie o rezultatakh ekspeditsii v Severenuyu Ameriku v 1930 g.)" ("Mexico and Central America as a Fundamental Center of the Origin of Cultivated Plants of the New World [a Preliminary Communication on the Results of an Expedition to North America in 1930]"), *ibid.*, 135–199, with English and Russian texts; "Sovetskaya agronomia k XVI godovshchine Oktyabrya" ("Soviet Agronomy on the Sixteenth Anniversary of October"), in *Izvestia* (6 Nov. 1933), 3; *Selektsia kak nauka* ("Selection as a Science"; Moscow–Leningrad, 1934); *Nauchnye osnovy selektsii pshenitsy* ("The Scientific Basis of Wheat Selection"; Moscow–Leningrad, 1935); *Uchenie ob immunitete rasteny k infektsionnym zabolevaniam* ("Studies of the Immunity of Plants to Infectious Diseases"; Moscow–Leningrad, 1935); *Botaniko-geograficheskie osnovy selektsii* ("The Phytogeographical Basis of Selection"; Moscow–Leningrad, 1935); and "Puti sovetskoy selektsii" ("The Courses Followed by Soviet Selection"), in *Spornye voprosy genetiki i selektsii* ("Disputed Questions of Genetics and Selection"; Moscow–Leningrad, 1937), 11–38, 462–473; and "The New Systematics of Cultivated Plants," in Julian Huxley, ed., *The New Systematics* (Oxford, 1940), 549–566.

Vavilov's posthumously published writings include *The Origin, Variation, Immunity and Breeding of Cultivated Plants*, which is *Chronica botanica*, **13**, nos. 1–6 (1951); *Mirovye resursy sortov khlebnykh zlakov, zernovykh bobovykh, lna i ikh ispolzovanie v selektsii* ("World Resources of Varieties of Cereal Grains, Legumes, Flax, and Their Use in Selection"; Moscow–Leningrad, 1957); *Izbrannye trudy* ("Selected Works"), 5 vols. (Moscow–Leningrad, 1959–1965); "Zakony estestvennogo immuniteta rasteny k infektsionnym zabolevaniam" ("Laws of the Natural Immunity of Plants to Infectious Diseases"), in *Izvestiia Akademii nauk SSSR*, biol. ser. (1961), no. 1, 117–157; *Pyat kontinentov* ("Five Continents"; Moscow, 1962); *Izbrannye sochinenia: Genetika i selektsia* ("Selected Works: Genetics and Selection"; Moscow, 1966); and *Izbrannye proizvedenia* ("Selected Works"), edited and compiled by F. K. Bakhteev, 2 vols. (Leningrad, 1967), with a biography by P. M. Zhukovsky and a complete bibliography.

II. Secondary Literature. See B. L. Astaurov, "N. I. Vavilov i obshchestvo genetikov i selektsionerov SSSR" ("Vavilov and the Soviet Society of Geneticists and Selectionists"), in *N. I. Vavilov i selskokhozy-*

aystvennaya nauka ("Vavilov and Agricultural Science"; Moscow, 1969), 84–89; F. K. Bakhteev, "Nikolaj Ivanovič Vavilov. Zu seinem 70. Geburtstag," in *T A G; internationale Zeitschrift für theoretische und angewandte Genetik*, **28**, no. 4 (1958), 161–166, with portrait; T. Dobzhansky, "N. I. Vavilov, a Martyr of Genetics," in *Journal of Heredity*, **38**, no. 8 (1947), 226–232, with portrait; N. P. Dubinin, "N. I. Vavilov kak genetik" ("Vavilov as a Geneticist"), in *Genetika*, **4**, no. 3 (1968), 18–27; A. E. Gaissinovitch, "N. I. Vavilov, in Commemoration of the 25th Anniversary of His Death," in *Folia Mendeliana*, no. 3 (1968), 55–58, with portrait; R. Ruggles Gates, "Vavilov and the Soviets," in *Science and Culture*, **12**, no. 3 (1947), 423–427; S. C. Harland and C. D. Darlington, "Prof. N. I. Vavilov, For. Mem. R. S.," in *Nature*, **156** (24 Nov. 1945), 621–622; David Joravsky, "The Vavilov Brothers," in *Slavic Review*, **24**, no. 3 (1965), 381–394; and N. A. Maisurian, "Zhizn i deyatelnost N. I. Vavilova" ("Vavilov's Life and Activity"), in *Genetika*, **4**, no. 3 (1968), 7–17.

See also Paul C. Mangelsdorf, "Nikolai Ivanovich Vavilov," in *Genetics*, **38**, no. 1 (1953), 1–4, with portrait; Zhores A. Medvedev, *The Rise and Fall of T. D. Lysenko*, I. Michael Lerner, trans. (New York–London, 1969), 17–19, 37–41, 52–77; *Nikolay Ivanovich Vavilov* (Moscow, 1967), with intro. by A. A. Fedorov and complete bibliography compiled by R. I. Goryacheva and L. M. Zhukova; M. M. Novikov, "N. I. Vavilov i genetika" ("Vavilov and Genetics"), in *Velikani Rossyskogo estestvoznania* ("Giants of Russian Natural Science"; Frankfurt, 1960), 141–150; A. I. Revenkova, *Nikolay Ivanovich Vavilov* (Moscow, 1962); *Ryadom s N. I. Vavilovym: Sbornik vospominany* ("Side by Side With Vavilov: A Collection of Reminiscences"; Moscow, 1963), compiled by his son, Y. N. Vavilov; and "Uchenye-genetiki o Nikolae Ivanoviche Vavilove" ("Genetic Scientists on Vavilov"), in *Genetika*, **4**, no. 3 (1968), 49–57.

MARK B. ADAMS

VEJDOVSKÝ, FRANTIŠEK (*b.* Kouřim, Bohemia [now Czechoslovakia], 24 October 1849; *d.* Prague, Czechoslovakia, 4 December 1939), *zoology*.

The son of a furrier, Vejdovský graduated from the academic grammar school in Prague and then entered the philosophical faculty of Charles University, from which he received the Ph.D. in 1876. From 1877 to 1907 he was assistant professor of zoology at the Technical University in Prague, as well as associate professor (1884–1892) and full professor (1892–1920) of zoology at Charles University in Prague. He was elected dean of the philosophical faculty in 1895 and rector of Charles University in 1912.

Vejdovský's scientific interest covered all groups of animals except mollusks, insects, and vertebrates (in the last category he dealt only with the evolution of lampreys). His work encompassed zoogeography, comparative anatomy, morphology, embryology, cytology, and taxonomy. Among fauna his best-known discovery was *Bathynella natans*, an ancient species of crustacean that had survived until modern times by being isolated in underground waters; and he contributed to comparative anatomy (particularly of worms) and taxonomy (especially of fungi and *Agaricus maceron*). Vejdovský's most important work consisted of his embryological and cytological studies, conducted mainly on *Rhynchelmis limosella* (Annelida). Although his methods of fixation and of preparation were very imperfect, he achieved remarkable results, particularly concerning the ripening, fertilizing, and grooving of ovules. In 1887, almost simultaneously with Boveri and Edouard van Beneden, Vejdovský showed that nuclear fission in an ovule is preceded by the splitting of the centrosome (which he called the periplast); he also was apparently the first to observe the centriole. In addition, he showed that during the fertilization of the ovule, the centrosome of the ovular nucleus disappears and the male gamete transfers its centrosome into the ovule.

In studying ovular grooving, Vejdovský found that cytological changes occur during the generation of the first blastomeres; he showed that in the four-blastomere stage only the largest blastomere—actually, only one of its parts containing just a few yolk grains—generates mesomeres. He not only managed to disprove Aleksandr Kovalevsky's incorrect opinion concerning the generation of the mesomeres from all four blastomeres, but also was the first to find the presumptive embryonal centers in very young embryos.

Although Vejdovský did not check his results experimentally and at the beginning of the twentieth century cytology began to revert to the study of live, undisturbed cells (as opposed to Vejdovský's tendency to use fixation and staining methods), his cytological and embryological studies belong to the classical beginnings of modern cytology and embryology.

BIBLIOGRAPHY

A complete list of Vejdovský's papers, containing 117 titles, is in *Sborník prací vydaný k 90. narozeninám prof. dr. F. Vejdovského Královskou českou společností nauk a Československou zoologickou společností v Praze* (Prague, 1939).

Unsigned secondary works are "Prof. F. Vejdovský (Note on His 90th Birthday)," in *Nature*, **144** (1939), 276 with portrait; and an obituary, *ibid.*, **156** (1945), 530. See also S. Hrabě, "Prof. Dr. František Vejdovský," in *Zprávy Československé společnosti pro dějiny věd a techniky Praha*, **11** (1969), 23–29.

VĚRA EISNEROVÁ

VITRUVIUS POLLIO (*b.* Italy, early first century B.C.; *d. ca.* 25 B.C.), *architecture, architectural history.*

Life. For the facts of Vitruvius' life we are dependent almost exclusively on the internal evidence of his only known work, the treatise *De architectura*. In the manuscripts of this work and in references to it by other classical writers he is referred to simply by his family name *(nomen)*, Vitruvius. The attempt by Paul Thielscher to show that his full name was Lucius Vitruvius Mamurra, and to identify him with the Mamurra who served as chief engineer under Julius Caesar, is not generally accepted. There does not, on the other hand, seem to be any good reason to question the evidence of the late third-century writer Faventinus (see below) that his last name *(cognomen)* was Pollio.

The known facts of Vitruvius' career are that he worked in some unspecified capacity for Julius Caesar; that he was subsequently entrusted with the maintenance of siege engines and artillery by Caesar's grandnephew and adopted heir, Octavianus, later the Emperor Augustus; and that on retirement from this post he came under the patronage of Augustus' sister, Octavia (I, praef., 2). It is often suggested, on the evidence of Frontinus (*De aquis urbis Romae*, 25), that book VIII of *De architectura* may have been the fruit of personal experience as a hydraulic engineer during Agrippa's construction of the Aqua Julia in 33 B.C.; but Frontinus is in fact quoting Agrippa and Vitruvius as possible alternative sources for his information, and the relevant passages in Vitruvius contain some surprising technical errors. Vitruvius' only excursion into civil architecture was the building of a basilica at Fanum Fortunae, the modern Fano, on the Adriatic Coast (V, 1, 6–10). This commission, coupled with what appears to be a personal knowledge of many of the Roman cities in the Po valley (for instance, I, 4, 11; II, 9, 16; V, 1, 4), suggests that, like many of those prominent in the culture of Augustan Rome, Vitruvius may have been of north Italian origin. It should be noted that in the first century of the Christian era, a freedman of the same family, Lucius Vitruvius Cerdo, is named as architect of the Arch of the Gavii at Verona.

De architectura. Vitruvius' writings belong to the last period of his life (II, praef., 4). The books were all dedicated to his patron, Octavianus, after the latter had achieved undisputed rule of the Roman world by his victory at Actium in 31 B.C. but before the title of Augustus, conferred on him in 27 B.C., had passed into general use. The later title is found only once (V, 1, 7), used in reference to a temple of Augustus (*aedes Augusti*) annexed to the basilica at Fano; otherwise he is addressed throughout as Caesar or *imperator*. Moreover, although Vitruvius makes it clear that his patron was already launched on the great building program that was to change the face of Rome, the buildings specifically cited all belong to that program's earliest years.

De architectura comprises ten books, each with a separate preface. Book I, after a long introductory section defining the nature of architecture and the personality and ideal training of the architect, discusses town planning in very broad terms. Book II covers building materials (brick, sand, lime, stone, timber) and methods. Books III and IV are devoted to religious architecture and to a detailed discussion of the classical orders, and book V to other forms of public architecture, with special emphasis on the theater. Book VI deals with domestic architecture, and book VII with such practical matters as types of flooring, stuccowork, painting, and colors. Book VIII turns to the sources and transport of water, by conduit or aqueduct. After a long excursus on astronomy, book IX describes various forms of clocks and dials; while book X covers mechanics, with particular reference to water engines, a hodometer, and artillery and other forms of military engineering. The illustrations that accompanied the text had already been lost when the earliest surviving manuscripts were transcribed.

To modern readers this may seem a rather curious mixture of subject matter, but antiquity did not recognize the nineteenth-century distinction between architecture and mechanical engineering. The two available sources of architectural training were apprenticeship to an established builder or, as in the case of Vitruvius, service as a military engineer. Thus, the great Roman architect Apollodorus was equally at home building Trajan's Forum in Rome or bridging the Danube for his armies. The scheme of *De architectura* does in fact follow closely the tripartite subdivision of the subject enunciated in the introduction: on building

VITRUVIUS POLLIO

(*aedificatio*) in books I–VII, on the making of timepieces (*gnomonice*) in book IX, and on mechanical devices (*machinatio*) in book X; hydraulics, which included both *aedificatio* and *machinatio*, bridges the transition in book VIII. Whether this classification was derived from some earlier authority, or whether it was Vitruvius' own, designed to embody his special interests, it would not have seemed illogical to a Roman reader.

As defined in book I, Vitruvius' architect is, according to R. Krautheimer, "a strangely ambiguous being . . . both a practitioner and a theoretician, and in the latter capacity a walking encyclopedia: versed not only in draftsmanship, geometry, and arithmetic but also in history, philosophy, and science, with a good smattering of musical theory, painting and sculpture, medicine, jurisprudence, astronomy and astrology." The theme of architecture as one of the liberal arts is ostentatiously picked up and dropped at intervals throughout the work, but at very few points can it be said seriously to illuminate the main subject matter. *De architectura* illustrates the range of scientific knowledge that might be available to a well-read professional man of Vitruvius' time; and it reflects what other, more critical minds held to be the ideal relationship between (to use a modern distinction) science and the arts. But in the context of a treatise on real architectural practice, it is little more than a pretentious literary exercise.

Any appraisal of the historical significance of Vitruvius' treatise has to begin by recognizing that his writings reflect the two distinct aspects of his architectural personality: the practitioner and the theoretician. The former is well represented, for example, in book II (on materials) and in book VII (on the techniques for laying floors and for finishing and decorating walls), both of which contain a great deal of practical information that would have been part of the stock in trade of any competent working builder. Without such knowledge Vitruvius would have been unable to handle the specifications for his basilica at Fano or to supervise the work on it. The mark of personal experience is revealed in his comments on such matters as the qualities of stone available around Rome and how to use them (II, 7); on the relative merits of the concrete building finishes known as *opus incertum* and *opus reticulatum* (II, 8, 1); and, in a section that otherwise relies heavily upon the early Hellenistic writer Theophrastus, his remarks on the qualities of the north Italian larch tree. At the same time, and very characteristically, Vitruvius shows no awareness of the larger significance of the concrete-vaulted architecture of which both *opus incertum* and *opus reticulatum* were manifestations; and from the list of earlier Italian architects whose opinions he would have valued (VII, praef., 17) he omits Lucius Cornelius, the trusted architect of the censor Quintus Lutatius Catulus, whose building of the Tabularium at Rome in 78 B.C. and whose restoration of the Temple of Jupiter Capitolinus, completed in 69 B.C., were among the most important and forward-looking architectural events of their time. Equally characteristic is his sweeping denunciation of contemporary trends in interior decoration (VII, 5, 5–8), as represented in the wall paintings of the Pompeian Second Style and their equivalents in Rome. His familiarity with contemporary building practice did not entail approval of contemporary architectural taste.

That Vitruvius' tastes were strongly conservative is unquestionable. He makes no attempt to conceal his contempt for the innovations introduced by many of his contemporaries. This fact has, however, led to much misunderstanding of the extent of his influence upon the architecture of his own time. It would seem natural to accept Vitruvius as a spokesman for the traditionalist architects of his day. He was living at a time when the forces of traditionalism and of innovation were still very evenly balanced, the former represented by the established formulas of column, architrave, and timber roof inherited from Greece and quintessentially present in the use of the classical orders, and the latter represented by the new, forward-looking, concrete-vaulted architecture of late Republican Latium and Campania. In a great many respects the monumental architecture of the Augustan age was a product of the lively creative dialogue between these two forces; and despite his staunch conservatism, Vitruvius could still have been a significant contributor to the great Augustan building program that in so many respects was to remain the touchstone of architectural excellence for centuries to come.

This view does not stand up to critical examination. Books III and IV, discussing temple architecture and the classical orders, are central to Vitruvius' own interests and to his conception of architecture; yet both in his selection and handling of source material it is evident that he is expressing a highly personal—and on many points a positively antihistoric—point of view. In the preface to book VII he quotes a number of earlier writings, almost exclusively in Greek and consisting largely of accounts of individual buildings written by their builders or treatises on particular aspects of architecture, such as proportions and machinery. He was proba-

VITRUVIUS POLLIO

bly right in claiming that no previous writer had tried systematically to encompass the whole field of architectural theory and practice; his own achievement, he claims, was the first really comprehensive study (*corpus architecturae*: II, 1, 8; see IV, praef., 1, *disciplinae corpus*). But in practice Vitruvius was very selective. His own preferred sources were Pythius, architect of the Mausoleum at Halicarnassus (fourth century B.C.) and, above all, Hermogenes (active *ca.* 200 B.C.); and the models on which he constructed his own system almost exclusively used the Ionic order and were located in Asia Minor. If he had read, for example, Ictinus' account of the Parthenon, he can have had little sympathy with it; and in practice he disregarded it. The great Doric temple architecture of archaic and classical Greece is dismissed (IV, 3, 3) out of hand: "because of this [the difficulty of producing a consistent arrangement of triglyphs and metopes at the outer angles of the frieze] it seems that the ancients avoided the Doric order in their temples." In its place he does, it is true, offer a prescription (IV, 3, 3–10) for laying out a Doric temple in accordance with his own modular principles—how else could he justify his claim to be presenting a conspectus of the whole of architecture?—but the result is patently an exercise in Vitruvian method, not an objective analysis of the work of the great historical masters.

It is this readiness to define perfection in quantitative terms, and to lay down finite laws governing planning and perfection, that constitutes the essence of Vitruvian method. The history of architecture is to be regarded as that of an evolution based on a series of revelatory discoveries leading to certain definitive achievements (*finitiones*) that it was Vitruvius' task to expound. In this view he was following a line of late Hellenistic thinking to which many educated Romans of his day would have subscribed. But whereas, for example, Cicero in *De oratore* could see the possibility of a diversity of manifestations of perfection, Vitruvius' approach lacked any such flexibility. By imposing a system of strict numerical analysis upon his models, he contrived to reduce temple planning to a series of rules based on the "correct" dimensions of each constituent element relative to a constant module. There is no hint of awareness that this modular formulation of the laws governing the proportions of the orders is no more than a convenient device for classifying the infinite variety of real architectural practice. Modular planning was already a familiar concept, but there is nothing in the monuments to suggest that the precise forms propounded by Vitruvius were those actually used by contemporary architects. Many Augustan temples were pycnostyle, in the generalized sense that they had close-set columns (III, 3, 1–2); but none of those preserved was laid out in strict accordance with the Vitruvian formula. Again, many Augustan architects, like Vitruvius, were looking back to Greek models; but many of these models, among them the Erechtheum, were quite different from those preferred by Vitruvius. Even on his chosen ground Vitruvius was not in the mainstream of conservative trends in contemporary architectural thinking.

Vitruvius the theorist left little mark on the official Roman architecture of his time. To us this aspect of his writing is a valuable source of information about current intellectual attitudes toward the arts and sciences, and about many aspects of Hellenistic and late Republican architectural history; but his influence on subsequent Roman architecture seems to have been limited almost entirely to those parts of his work in which Vitruvius the architect and builder was speaking from personal experience. The best evidence for this lies in the works of two late Roman writers, Marcus Cetius Faventinus (*ca.* A.D. 300[?]), who wrote and annotated an abbreviated compendium of parts of *De architectura*, and the somewhat later Palladius, a wealthy landowner who made liberal use of Faventinus' compendium in compiling his own treatise on the management of a typical late Roman estate. Both of these authors were writing manuals for practical use, and both clearly regarded Vitruvius' work as the natural point of departure for their own. Their subject matter tells its own story. Apart from a ritual gesture to culture in Faventinus' introduction, what mattered to them were such things as finding and exploiting a water supply, the siting of domestic buildings, the best use of materials, the techniques of vaulting, and the method of constructing a set square or a simple timepiece. Traditional columnar architecture and the classical orders did not concern them. Such matters were past history.

During the Middle Ages very little of *De architectura* was relevant, but manuscripts of it continued to be copied in monastic scriptoria (the earliest one surviving was produced at Jarrow in the ninth century). In the fifteenth century, classical architecture suddenly became a matter of direct and lively concern to architects and humanists alike. Gian Francesco Poggio Bracciolini's "rediscovery" in 1414 of two manuscripts of *De architectura* was a major event. There was no printed edition until 1486, but there are more than twenty fifteenth-century manuscript copies, made for circulation among humanist scholars, architects, and artists. To the extent that the architecture of the Quattrocento represented a deliberate return to the models of antiquity, *De architectura*, the only surviving ancient treatise on the subject, was bound to become the ultimate authority for true doctrine. When Leone Alberti, between 1452 and 1467, wrote the first great Renaissance treatise on architecture, his debt extended even to the title used, *De re aedificatoria*, and to the work's subdivision into ten books; he wrote in Latin (a self-consciously "purer" Latin than the Hellenized Latin of Vitruvius); he cited Vitruvius frequently and borrowed from him even more frequently. Not that he always agreed with him: there are a great many criticisms, both expressed and implied, of principle and of detail. But for matters of historical fact, for such technical details as the making of bricks or the laying of pavements, for the classical orders, and for the description of a number of classical building types (such as palaestrae, theaters, and forums) about which the Quattrocento had little direct information, he drew heavily on Vitruvius.

Even so, Alberti's debt was often more one of formal presentation and of detail than of real substance. The genuine wish to use Vitruvius as a guide to building in the antique manner came up against formidable difficulties, among them the obscurities of Vitruvius' style, the loss of his illustrations, and the lack of surviving models. On many topics, Krautheimer states, "his book remained sealed, its terminology unintelligible, its references to building types and extant monuments obscure." Moreover, the shifts of intellectual attitude were often too great to be bridged by direct borrowing. However much *De re aedificatoria* may have set out to reshape *De architectura* for contempoary needs, it found itself turning more and more to the monuments of antiquity and to contemporary building practice. To be serviceable, the works of antiquity, monuments and writings alike, had to be interpreted, reconstructed, and, where necessary, improved, in accordance with the Quattrocento vision of antiquity.

In all of this Alberti, the great architectural theorist, was speaking also for the practicing architects of his day. Whatever its ultimate inspiration, Renaissance architecture had to chart a course of its own. Vitruvius continued to be a quarry of detailed information for would-be classical purists, but it was only among scholars that his authority as the source of pure doctrine remained virtually unchallenged. Because of his manifest admiration for Greek architecture, his reputation survived the shock of the subsequent rediscovery of Greece and of the great monuments of Greek classical architecture: indeed, the advent of systematic archaeological research in Italy, which might have supplied a corrective, seemed only to confirm the established opinion that the history of Roman imperial architecture was one of decadence and steady decline from the models of Greek perfection. Where the monuments did not fit the Vitruvian formulas—and few of them did—it was the monuments that were out of step, not Vitruvius. It is only during the twentieth century that a growing appreciation of the true qualities and significance of Roman imperial architecture has enforced a critical reevaluation of his reputation.

BIBLIOGRAPHY

I. Original Works. Critical editions of *De architectura* are V. Rose, ed. (Leipzig, 1899), the Teubner ed.; F. Granger, ed., with English trans., 2 vols. (Cambridge, Mass., 1945–1970), the Loeb Classical Library ed.; and C. Fensterbuch, ed. (Darmstadt, 1964). The Collection Guillaume Budé, with French trans. and full commentary, will run to 10 vols. and be published at Paris. Already issued are vol. VIII, L. Callebat, ed. (1973), and vol. IX, J. Soubiran, ed. (1969). In preparation are vol. II, A. Balland, ed.; vols. III and IV, P. Gros, ed.; vol. VII, P. Liore, ed.; and vol. X, L. Callebat, ed.

II. Secondary Literature. On Vitruvius' life and work in relation to the architecture of classical antiquity, the fundamental bibliographical article is P. Thielscher, in Pauly-Wissowa, *Realenzyklopädie*, 2nd ser., IX, A (1961), 427–489. This article must, however, be used with great caution to the extent that it identifies Vitruvius with Mamurra—a very questionable hypothesis. Other important works on the subject are F. E. Brown, "Vitruvius and the Liberal Art of Architecture," in *Bucknell Review*, **11** (1963), 99 ff.; P. Gros, "Hermodoros et Vitruve," in *Mélanges de l'École française à Rome: l'antiquité*, **85** (1973), 173 ff.; and "Structures et limites de la compilation vitruvienne dans les livres III et IV du *De architectura*," in *Latomus*, **34** (1975), 986–1009; F. Pellati, "La basilica di Fano e la formazione del trattato di Vitruvio," in *Rendiconti della Pontificia Accademia romana di archeologia*, **23–24** (1947–1949), 153–174; H. Plommer, *Vitruvius and Later Roman Building Manuals* (London, 1973); H. Riemann, "Vitruv und der griechische Tempel," in *Archäologischer Anzeiger* (1952), cols. 2–38; and F. W. Schlikker, *Hellenistischer Vorstellungen von der Schönheit des Bauwerks nach Vitruv* (Münster, 1940).

Vitruvius' reputation and influence on postclassical architecture are treated in W. B. Dinsmoor, "The Literary Remains of S. Serlio," in *Art Bulletin*, **24** (1942), 55 ff., esp. 55–61; R. Krautheimer, "Alberti and Vitruvius," in *The Renaissance and Mannerism*, II

(Princeton, 1963), 42–52; H. Millon, "The Architectural Theory of Francesco di Giorgio," in *Art Bulletin*, **40** (1958), 257–261; H. Saalman, "Early Renaissance Theory and Practice in Antonio Filarete's Trattato," *ibid.*, **41** (1959), 89–106; and R. Wittkower, *Architectural Principles in the Age of Humanism* (London, 1949; New York, 1965).

JOHN WARD-PERKINS

Machines and Scientific Instruments. Although *De architectura* is widely cited as one of the very few classical texts that describes interesting machines and scientific instruments in any detail, and although the text also includes some astronomical material, the reader must be warned that the contents are not necessarily typical of the science and technology of the period. These matters are included either incidentally or as addenda to the main treatise on architecture, and constitute embellishments rather than a systematic account. Furthermore, there is some reason to suppose, from the few other texts (such as Hero) and from the artifacts that have survived, that Vitruvius knew only the practitioner arts of his day rather than the more sophisticated mathematics and theoretical astronomy. Even with the water clocks and sundials he may be reporting only a selection of the simpler devices. It should also be noted by historians of science and technology that in the sixteenth and seventeenth centuries Vitruvius was regarded as a living handbook rather than a historical text, and that early editions of the text include pictures of the machines and instruments reconstructed—and thereby popularized in the idiom and technical paraphernalia of the period rather than those of the time of Vitruvius.

At several places throughout the first eight books, Vitruvius draws upon standard Greek physical theory to provide a basis for the properties of materials, the nature of the elements and of climates, and the mathematical proportions governing harmony and pleasing design. The most important passage in these earlier books is, however, an account of the Tower of the Winds constructed in the Roman agora of Athens by Andronicus of Cyrrhus either very shortly before the time of Vitruvius or during his youth. The tower still stands with its frieze of the eight winds, its nine elaborate sundials, and the reservoir and other remains of the astronomical water clock within, the wind vane above, and an element-theory symbolism as architectural design and internal furnishing.[1] Vitruvius introduces a description in the context of a discussion of the winds and their effect in the siting and orientation of buildings, and tells us that the octagonal tower was constructed as an exemplum of the eight-wind theory of Andronicus.

Book IX opens with discursive accounts of the Pythagorean theorem, the anecdotal bathtub discovery by Archimedes, and the Delian problem of the duplication of the cube, all used as illustrations of the cumulative power of ancient authors. In this light, Vitruvius says, he proceeds to an important exposition of gnomonics, the science of sundials. First, however, there is a section on the periods of the planets, the waxing and waning of the moon, and the constellations of the fixed stars.[2] The text then proceeds with the important first discussion of the principle of the analemma, which is used as a basis for much of later mathematical dialing. This geometrical construction has been analyzed and commented on by O. Neugebauer,[3] who also sets it in its context of the development of stereographic projection and its application to the anaphoric clock and then to the astrolabe.

In a much-quoted passage at the beginning of chapter 8, Vitruvius describes the chief varieties of sundials and names their inventors. This list has been matched against known and extant varieties of portable sundials by E. Buchner and by D. de S. Price, although still with some uncertainty.[4] The case for the fixed masonry sundials has become more certain since the publication of the corpus of such surviving dials by S. L. Gibbs.[5] Next follows an account of the water clocks of Ctesibius of Alexandria, in which an inflow pipe of gold or gemstone fills a cylinder with a float that can rise, working *parerga* and/or turning a dial through the action of a rack and pinion. Vitruvius discusses devices for controlling the water flow by wedgelike stopcocks or by using a sort of analemma to raise and lower the level of the output hole. He also considers the anaphoric clock, in which the seasonal variations in the length of day and night are exhibited by using an astrolabic (stereographic) projection of the heavens and the ecliptic that is turned by the water clock at a uniform rate.

The mechanical details of these water clocks have been discussed with admirable competence and thoroughness by A. G. Drachmann,[6] who also supplies the best commentary on the machines discussed in book X. Again, it must be noted that the older reconstructions and diagrams showing these machines and instruments are highly unreliable and should no longer be repeated. The book on machines opens with a discussion of simple and compound pulleys and on cranes using them in building

VITRUVIUS POLLIO

construction, then treats levers in similar fashion. Next discussed are water-raising wheels, mill wheels, the Archimedean-screw water raiser, and the pump of Ctesibius. Chapter 8 describes the water organ, an example of which, somewhat dubiously reconstructed, has survived at Aquincum, near Budapest. Chapter 9 deals with the hodometer devices also described by Hero and considered by most modern commentators to be so unrealistic in their technical descriptions as to be fanciful and "theoretical" rather than accounts of actual working devices. This point is, however, no longer of the essence as evidence for the use of gear trains at this period of classical antiquity, since the evidence of the Antikythera[7] mechanism shows that much more complex gearing systems were known and utilized at this date.

The remaining chapters of *De architectura* deal in good technical detail with such military architectural machinery as catapults, ballistae, battering rams and other siege engines, and a pair of devices called "tortoises" for filling and digging ditches. In all of these the detailed criticism and reconstructions due to Drachmann may be regarded as authoritative, and supersede much of the earlier evaluations. For an account of the most recent studies, see the review by B. S. Hall of the books of E. W. Marsden.[8]

NOTES

1. D. de S. Price, "The Water Clock in the Tower of Winds," in *American Journal of Archaeology*, **72** (1968), 345–355, written with J. V. Noble. For an account of the symbolism, see my chapter, "Clockwork Before the Clock and Timekeepers Before Timekeeping," in J. T. Fraser and N. Lawrence, eds., *The Study of Time, Proceedings of the Second Conference of the International Society for the Study of Time, Japan* (New York, 1975), 367–380; also in *Bulletin of the National Association of Watch and Clock Collectors*, **18**, no. 5 (Oct. 1976), 398–409.
2. The astronomical sections are splendidly analyzed in O. Neugebauer, *A History of Ancient Mathematical Astronomy*, pt. 2 (New York–Heidelberg–Berlin, 1975).
3. *Ibid.*, 843–845.
4. E. Buchner, "Antike Reiseuhren," in *Chiron*, **1** (1971), 457–482; and "Römische Medaillons als Sonnenuhren," *ibid.*, **6** (1976), 329–348; and D. de S. Price, "Portable Sundials in Antiquity, Including an Account of a New Example From Aphrodisias," in *Centaurus*, **14**, no. 1 (1969), 242–266.
5. S. L. Gibbs, *Greek and Roman Sundials* (New Haven–London, 1976).
6. A. G. Drachmann, *The Mechanical Technology of Greek and Roman Antiquity* (Copenhagen–Madison, Wis.–London, 1963); *Ktesibios, Philon and Heron* (Copenhagen, 1948); and "Ktesibios's Waterclock and Heron's Adjustable Siphon," in *Centaurus*, **20**, no. 1 (1976), 1–10.
7. D. de S. Price, "Gears From the Greeks, the Antikythera Mechanism—a Calendar Computer From *ca.* 80 B.C.," *Transactions of the American Philosophical Society*, n.s. **64**, pt. 7 (1974); also published separately (New York, 1975).
8. B. S. Hall, "Crossbows and Crosswords," in *Isis*, **64** (1973), 527–533.

DEREK DE SOLLA PRICE

Sundials. Book IX ostensibly is concerned with the construction of sundials and clocks, which fell within the province of the architect in antiquity. However, in accordance with his claim that the complete architect must be familiar with astronomy (I, 1, 3 and 10), Vitruvius devotes a long section (IX, 1–6) to theoretical astronomy. Although the book would have been useless to a contemporary as a practical guide to the construction of time-measuring instruments (so muddled and incomplete is Vitruvius' account, even when we allow for the corruption of the manuscript tradition), for us it provides much valuable historical information, particularly since no work on the theory of sundials has survived from antiquity. It also throws some light on the obscure area of pre-Ptolemaic astronomy.

After recounting some edifying anecdotes illustrating the importance of mathematical discoveries, Vitruvius attempts to describe the structure of the universe as conceived by astronomers of his time. The chief points of interest in his confused account are a possible reference to the theory that Mercury and Venus revolve about the sun (IX, 1, 6), some fairly accurate figures for the sidereal periods of the outer planets (IX, 1, 10), and a possible example of the notion that the absolute speed of all planets is the same. Vitruvius also retails some curiously primitive notions about the physical reasons for the phases of the moon and the retrogradations of the planets, and provides (IX, 3–5) a detailed but non-numerical description of the relative positions of the chief constellations. A historical notice on the origin and progress of astrology precedes the real matter of the book, gnomonics and timekeeping.

The only practical information Vitruvius provides for the construction of sundials is his description of the analemma (IX, 7), a graphic method of determining the hour lines and day curves in a plane sundial with vertical gnomon. We have a treatise on the analemma by Ptolemy (*ca.* A.D. 150); but the invention of the method is due to Diodorus of Alexandria (first century B.C.), and Vitruvius' description, although sadly incomplete, is a most valuable aid to reconstructing the original form of the theory. In Figure 1, AB represents the gnomon and BC the equinoctial shadow at a given place. The circle with center A and radius AB represents the meridian, DE the celestial equator, and LO the local horizon. BJ

VITRUVIUS POLLIO

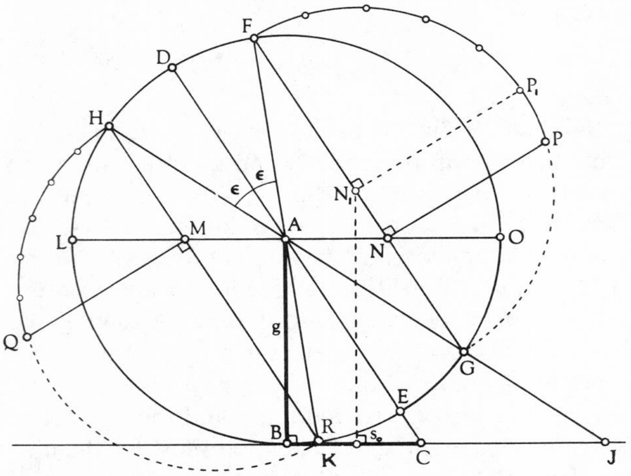

FIGURE 1. Analemma for solstices

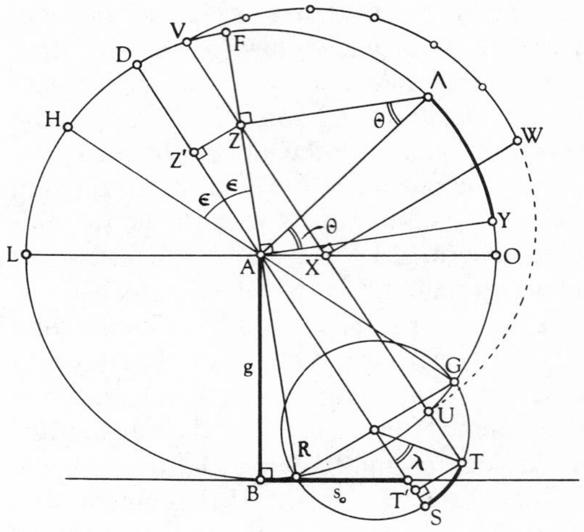

FIGURE 2. Analemma for given date

and *BK*, the lengths of the shadow at summer and winter solstices, are found by marking off the arcs *DF* and *DH* equal to ϵ, the obliquity of the ecliptic (Vitruvius takes ϵ as $^1/_{15}$ of the circle, or 24°), and drawing *HAJ* and *FAK*. If these lines intersect the meridian circle in *G* and *R* respectively, then chords *FG* and *HR* (parallel to the equator *DE*) are the traces of the day circles of the sun at summer and winter solstice respectively. Half of these circles are drawn and rotated into the plane of the meridian, as *FPG* and *HQR*. The perpendiculars from the intersections of day circle and horizon, *NP* and *MQ*, cut off the arcs *FP* and *HQ*, which represent half the length of daylight on the longest and shortest days respectively. The "civil hours" ($^1/_{12}$ the length of daylight) used in antiquity are determined by dividing these arcs into six equal parts. The day circle and civil hour length for any given date are found by means of an auxiliary circle *RSG*, on diameter *RG* (see Figure 2). Vitruvius explains how to construct this circle but not how to use it. If the sun's longitude on the given date is λ, mark off the arc *ST* equal to λ on circle *RST*, and draw *TUXV* parallel to *DA*, intersecting the meridian in *U* and *V* and the horizon in *X*. The semicircle on diameter *UV* represents the sun's path on the day in question, and *XW*, the perpendicular from *X*, cuts off the half-day arc *VW*;[1] the civil hour lengths are again obtained by dividing this arc into six. The hour lines and day curves for a horizontal sundial can then be derived graphically, although Vitruvius omits all explanation of the method.[2]

Vitruvius ends his discussion of sundials with a list of the types of sundial and their inventors. The historical value of this, although considerable, is diminished by his failure to describe any of the sundials named. However, comparison of the meaning of the names with preserved ancient sundials has permitted tentative identification of many types. All the inventors who can be identified belong to the Hellenistic period, with the exception of Eudoxus.

Vitruvius ends the book with an account of what he calls "winter clocks" (*horologia hiberna*), which tell time without the sun, or water clocks. His occasionally obscure description of the water clock of Ctesibius (third century B.C.), with its ingenious methods of ensuring a constant flow of water and of indicating the seasonal hours, is sufficiently detailed to allow modern reconstructions. He also describes the anaphoric water clock, which employed a dial rotating once daily on which the constellations were represented by stereographic projection, and in front of which was fixed a "spider" of wires forming the civil hour curves, also constructed by stereographic projection.[3] His description is useful in reconstructing the two examples of such a clock which have survived, fragmentarily, from antiquity.

NOTES

1. For a proof, see Neugebauer, 844–845.
2. For a modern explanation, see Drecker, 3–4.
3. This is exactly analogous to the astrolabe, except that the roles of "spider" (rete) and dial are reversed in the astrolabe.

Vitruvius provides the earliest unambiguous evidence for the use of stereographic projection, the basic principle of the astrolabe.

BIBLIOGRAPHY

Jean Soubiran, ed., *Vitruve, De l'architecture, Livre IX* (Paris, 1969), provides a text and translation; the extensive commentary is useful but should be read with caution, since it contains some technical and many historical errors. On theoretical astronomy in Vitruvius, see Otto Neugebauer, *A History of Ancient Mathematical Astronomy* (New York–Heidelberg–Berlin, 1975), 782–784, and index; on the analemma, *ibid.*, 839–857, esp. 843–845; and Joseph Drecker, *Die Theorie der Sonnenuhren* (Berlin–Leipzig, 1925), 1–11; and, on Ptolemy's version, P. Luckey, "Das Analemma von Ptolemäus," in *Astronomische Nachrichten*, **230** (1927), 17–46. The only fully informed modern attempt to identify the sundials listed by Vitruvius is Sharon L. Gibbs, *Greek and Roman Sundials* (New Haven–London, 1976), 59–65, with full reference to preserved dials and descriptions of all known ancient sundials.

The best older discussion is Hermann Diels, *Antike Technik*, 3rd ed. (Leipzig–Berlin, 1924), 155–192. For the identification of portable sundials named by Vitruvius, see D. de S. Price, "Portable Sundials in Antiquity," in *Centaurus*, **14**, no. 1 (1969), 242–266. Reconstructions of the water clocks described by Vitruvius are in Diels, 204–219; D. de S. Price, "Precision Instruments to 1500," in Charles Singer, ed., *A History of Technology*, III (London, 1957), 601–605; and esp. A. G. Drachmann, *Ktesibios, Philon and Heron* (Copenhagen, 1948), 16–36. On the anaphoric clock in particular, see also Neugebauer, 869–870, with references (870, n. 5) to the literature on the fragments of clocks found at Salzburg and Grand (Vosges), France.

G. J. TOOMER

WASSERMANN, AUGUST PAUL VON (*b.* Bamberg, Germany, 21 February 1866; *d.* Berlin, Germany, 16 March 1925), *immunology, serology, bacteriology, cancer therapy.*

Wassermann was the second son of Angelo Wassermann, a Bavarian court banker who was elevated to the hereditary nobility in 1910, and Dora Bauer. He attended the Gymnasium in Bamberg and studied medicine at the universities of Erlangen, Vienna, and Munich, receiving the M.D. at Strasbourg in 1888 for a work on the effect of Sulfonal. In 1895 he married Alice von Taussig of Vienna; they had two sons. Attracted by the presence of Robert Koch, Wassermann went to Berlin in 1891 and began work at Koch's hygiene institute as assistant to Bernhard Proskauer. On 1 September 1891 he entered the newly established Institute for Infectious Diseases, headed by Koch, as an unpaid assistant in both the scientific and clinical divisions. In February 1893 he became a temporary assistant assigned to problems related to cholera, and from February 1895 to June 1896 he was inspecting physician at the institute's antitoxin control station for diphtheria, which was transferred in 1896 to the Institute for Serum Research and Testing in the Berlin suburb of Steglitz. Wassermann then returned to the institute itself as an unpaid assistant. During this time he was senior physician in the clinical division, of which he became director on 1 April 1902.

In 1901 Wassermann became *Privatdozent* in internal medicine at the Friedrich Wilhelm University in Berlin, having obtained the title of professor in 1898. He was named extraordinary professor in 1902 and in 1911, honorary professor. In 1906 he became director of the division of experimental therapy and serum research at the Institute for Infectious Diseases, and the following year he was awarded the title of *Geheimer Medizinalrat*. Wassermann left the institute in 1913 to become director of the Institute for Experimental Therapy at the Kaiser Wilhelm Society for the Advancement of Science in Berlin.

During World War I, Wassermann's research was considerably curtailed and was finally suspended completely. As a hygienist and bacteriologist with the rank of brigadier general in the medical corps, he supervised epidemic control on the Eastern Front. He was subsequently appointed director of the Office of Hygiene and Bacteriology of the Prussian Ministry of War.

In the early 1920's Wassermann's institute was expanded and renamed the Kaiser Wilhelm Institute for Experimental Therapy and Biochemistry. Although Wassermann began to suffer in 1924 from Bright's disease, which took his life the following year, he went to the institute whenever possible, remaining director until his death.

Wassermann's first papers, written with Ludwig Brieger, Shibasaburo Kitasato (1892), and R. F. J. Pfeiffer (1893), dealt with cholera immunity; they were followed by his own studies of cholera immunity and diphtheria antitoxin (1894, 1895). In independent investigations Wassermann showed that many people possess diphtheria antitoxin, in varying degrees, in their serum; and he related this finding to differences in the ability to resist diphtheria infection.

Using pyocyaneous bacteria, Wassermann conducted experiments in 1896 on breaking the toxin-antitoxin bonds; this research lent support to Ehrlich's side-chain theory, of which Wassermann became a proponent. His investigations made in 1898 with T. Takaki aroused considerable interest in the theory by furnishing important support for it, for their results indicated that matter from the brain of healthy animals can bond tetanus toxin selectively and thereby detoxify it. Having observed and described the variable rate at which influenza can progress, Wassermann in 1900 attributed its periodic recurrence to a decrease in the immunity that had been acquired in the previous pandemic. He also drew attention in several works to the difference between antibacterial and antitoxic immunity.

Wassermann was the first to point out that a reagent existed in precipitins and that it was far superior in sensitivity to that usually employed in chemistry. On the basis of studies paralleling those of Paul Uhlenhuth, who preceded him in publication but used a different method, Wassermann reported the possibility of differentiating albumen by a serologic procedure, and in 1901 he pointed out its potential practical applications in a joint paper with A. Schütze. Three years earlier he had suggested to Koch that human and animal blood might be differentiated by means of specific antibodies for erythrocytes. Drawing upon his work with Schütze, he also proposed a method for the evaluation of precipitating serums.

Experiments on immunity to swine plague bacteria undertaken with R. Ostertag (1902, 1904) led Wassermann to recognize the necessity of producing polyvalent vaccines and serums, since the antigen pattern of various strains of the same type of bacterium exhibits differences. In work with J. Citron (1905) Wassermann attributed residual post-typhus immunity not to antibodies in the blood but to a specific alteration of the intestinal mucosa and thus to local immunity.

After 1900 Wassermann became increasingly occupied with problems relating to complement, which had not been thoroughly investigated. Considering this omission responsible in certain instances for the failure of serum therapy (1900), he called for the determination of complement in serum (1901), expecting it to yield important clinical benefits. In 1901 Jules Bordet and O. Gengou reported on complement-fixing substances in the serum of patients recovering from cholera and in other serums obtained by specific immunization, for instance, against red blood cells. They noted that the presence of these substances could be demonstrated, in the case of red blood cells, by hemolysis, if these substances could not also react with the complement being in fresh serums.

Working with Carl Bruck, Wassermann conducted experiments on complement fixation, one of which was designed to find antibodies against tuberculin in the vicinity of tubercular foci (1906). A new feature of this research was that the reaction was carried out not only with unchanged bacteria but also with extracts derived by agitation. Employing the latter to effect fixation of the complement, Wassermann determined the strength of the meningococcal antiserum; and with Wilhelm Kolle at the Institute for Infectious Diseases he produced it for practical application. In 1906, with Albert Neisser and C. Bruck, Wassermann published "Eine serodiagnostische Reaktion bei Syphilis," in which the authors examined experimentally induced syphilis in apes and then investigated human patients infected with the disease. Wassermann described the principle guiding their research:

> The so-called fixation of the complement . . . depends upon this principle: that when an antigen is mixed with its homologous immune body a union occurs between the two. If complement—a constituent of every fresh serum—is added at the same time, it becomes anchored through the union of the antigen and antibody. It follows, accordingly, that if the complement is anchored, the conclusion may be drawn that either the homologous antigen or the homologous immune body is present in such a mixture. The determination whether in such an experiment the complement is bound can be made easily and convincingly. For this purpose one needs simply to add simultaneously, or somewhat later, the serum of an animal which has been previously treated with red blood corpuscles, the so-called amboceptor, together with its homologous erythrocytes. If the complement has already become bound as a result of the union between the antigen and immune bodies, then it is no longer available for the haemolytic amboceptor and the red blood corpuscles. Consequently the latter remain undissolved . . . [and] from the appearance or non-appearance of haemolysis, one can draw the conclusion as to whether the sought-for antigen or immune body is present.

Beyond facilitating diagnosis in acute cases of syphilis, the reaction illuminated certain unproved relationships between diseases. In 1906 Wassermann demonstrated with Felix Plaut that in creeping paralysis the Wassermann reaction shows positive when carried out on spinal fluid. The findings

WASSERMANN

of Wassermann and his co-workers not only made it possible to detect syphilis but also established a new basis of therapy.

After a short time, Wassermann had to abandon the theory with which he had begun his research, namely that his experiments demonstrated the presence of specific antibodies against the agent of syphilis or its extracted portions; for the reaction showed positive in persons infected with the illness even when extracts from normal organs were used. In research at his own institute, Wassermann later sought to find a theoretical explanation of the reaction, which was rapidly proving its worth in clinical practice. With this end in view he attempted in 1922 to culture *Spirochaeta pallida*.

Around 1910 Wassermann began to devote his major efforts to discovering a way of treating cancer via the bloodstream. Unfortunately, in treating human patients, he was unable to repeat his success at Koch's institute in 1911, when he had arrested the growth of tumors in mice by administering selenides and tellurium salts. During his last years Wassermann focused his research primarily on obtaining a complement-fixation reaction in tuberculosis (1923).

Although small in stature, Wassermann was not—contrary to assertions—stooped, and he had bright blue (not dark) eyes. He was an impulsive man whose rich intellectual endowment soon became evident in conversation. Gifted with exceptional oratorical ability, he had an outstanding capacity to render complicated theoretical problems comprehensible to the uninitiated. He often lectured at weekly sessions of the Berlin Medical Society and eagerly accepted invitations to speak before local groups and international congresses. He was fond of similes and continually devised new ones in his lectures—comparing specificality, for example, to a light source that has a maximum but also a cone of dispersion. Similarly, he sought to find "railroad tracks" in an organism that might provide a clue for cancer therapy; the particular "car" that circulates on them was of only secondary importance.

With Rudolf Kraus, Wassermann was a cofounder of the Free Association for Microbiology, and he served as president of the Academy for Knowledge of Judaism. Although occasionally sarcastic, he was always helpful and kind, even under difficult conditions. He once characterized himself as a "laboratory worker." His many honors included orders and decorations from Prussia, Belgium, Japan, Romania, Spain, and Turkey. In 1921 Wassermann was the first recipient of the Aronson Foundation Prize.

BIBLIOGRAPHY

I. Original Works. There is no collected edition of Wassermann's more than 150 scientific publications, most of which are listed by K.-E. Gillert and K. Gerber (see below). Wassermann's separately published monographs include *Hämolysine, Cytotoxine und Präzipitine*, Sammlung Klinischer Vorträge, no. 331 (Leipzig, 1902), 2nd ed., rev. by J. Leuchs and M. Wassermann (Leipzig, 1910). His important collaborative effort with W. Kolle resulted in *Handbuch der pathogenen Mikroorganismen*, 6 vols. (Jena, 1903–1909), 2nd ed., enl. (1912–1913), to which he contributed many articles; and he was author of "Allgemeine Lehre der Infektionskrankheiten," in W. Ebstein and J. Schwalbe, eds., *Handbuch der praktischen Medizin*, 2nd ed., 4 vols. (Stuttgart, 1906); and "Schweineseucheserum," in R. Kraus and C. Levaditi, eds., *Handbuch der Technik und Methodik der Immunitätsforschung*, II (Jena, 1909).

Wassermann's memoirs cited in the present article include "Über künstliche Schutzimpfung von Thieren gegen Cholera asiatica," in *Deutsche medizinische Wochenschrift*, **18** (1892), 701, written with L. Brieger; "Untersuchungen über Immunität gegen Cholera asiatica," in *Zeitschrift für Hygiene und Infektionskrankheiten*, **14** (1893), 35–45; "Untersuchungen über das Wesen der Choleraimmunität," *ibid.*, 46–63, written with R. F. J. Pfeiffer; "Über die Gewinnung der Diphtherie-Antitoxine aus Blutserum und Milch immunisierter Thiere," *ibid.*, **18** (1894), 239–250, written with P. Ehrlich; "Über die persönliche Disposition und die Prophylaxe gegenüber Diphtherie," *ibid.*, **19** (1895), 408–426; and "Über tetanusantitoxische Eigenschaften des normalen Centralnervensystems," in *Berliner klinische Wochenschrift*, **35** (1898), 5–6, written with T. Takaki.

Works published after 1900 include "Einige Beiträge zur Pathologie der Influenza," in *Deutsche medizinische Wochenschrift*, **26** (1900), 445–447; "Über eine neue forensische Methode zur Unterscheidung von Menschen- und Thierblut," in *Berliner klinische Wochenschrift*, **38** (1901), 187–190, written with A. Schütze; "Über die Bildungsstätten der Typhusimmunkörper. Ein Beitrag zur Frage der localen Immunität der Gewebe,"in *Zeitschrift für Hygiene und Infektionskrankheiten*, **50** (1905), 331–348, written with J. Citron; "Eine serodiagnostische Reaktion bei Syphilis," in *Deutsche medizinische Wochenschrift*, **32** (1906), 745–746, written with A. Neisser and C. Bruck, in which the test for syphilis was announced; and "Reinkulturen der Spirochaeta pallida in festem und flüssigem Nährboden sowie Übertragung dieser Kulturen auf Tiere," in *Klinische Wochenschrift*, **1** (1922), 1101, written with M. Ficker.

II. Secondary Literature. There is no full-length

biography. Studies of individual aspects of his research and obituaries include G. Bako, "August von Wassermann (1866–1925)," in *Journal of the American Medical Association*, **209** (1969); G. Blumenthal, "August von Wassermann zum 25. Todestage," in *Zeitschrift für Immunitätsforschung und experimentelle Therapie*, **107** (1950), 380–384; T. Brugsch, *Arzt seit fünf Jahrzehnten* (Berlin, 1957), 148–149; "Wassermann: Nachhaltige Reaktion," in *Selecta*, **8** (1966), 606; E. Friedberger, "August von Wassermann†," in *Zeitschrift für Immunitätsforschung und experimentelle Therapie*, **43** (1925), i–xii; K. Gerber, "Bibliographie der Arbeiten aus dem Robert Koch-Institut 1891–1965," in *Zentralblatt für Bakteriologie, Parasitenkunde, Infektionskrankheiten und Hygiene, I. Abteilung, Referate*, **203** (1966), 1–274; K.-E. Gillert, "Bibliographie der Arbeiten August von Wassermann's 1914–1925," *ibid.*, **254** (1977), 289–291; N. Korken, *Jewish Physicians. A Biographical Index* (Jerusalem, 1973), with references to further literature; R. Kraus, "August von Wassermann 1865–1915," in *Seuchenbekämpfung und experimentelle Therapie . . .*, **11** (1925), 104–106; "Wassermann," in *British Medical Journal* (1966), **1**, 436–437; H. Mengel, "August von Wassermann zum 100. Geburtstag," in *Münchner medizinische Wochenschrift*, **108** (1966), 1434–1436; A. Keitner, ed., Menschen und Menschenwerke, I (Vienna, 1924), 669–670; H. Mühsam, *Jüdisches Lexikon*, IV, pt. 2 (Berlin, 1930), cols. 1344–1346; F. Neufeld, "August von Wassermann†," in *Deutsche medizinische Wochenschrift*, **16** (1925), 667–668; I. Mc. I., "August von Wassermann, M.D.," in *British Medical Journal* (1925), **1**, 638; J. Plesch, *János. Ein Arzt erzählt sein Leben* (Munich, 1951), 70, 74–76; H. Reiter, "August von Wassermann†," in *Münchener medizinische Wochenschrift*, **72** (1925), 813–814; H. Sachs, "August von Wassermann†," in *Klinische Wochenschrift*, **4** (1925), 902–903; G. H. Schneider, "August Paul von Wassermann zum 100. Geburtstag," in *Berliner Ärzteblatt*, **79** (1966), 278–279; "Zum 100. Geburtstag von August von Wassermann," in *Deutsches Ärzteblatt*, **63** (1966), 586; and "Persönliche Erinnerungen an August Paul von Wassermann . . .," in *Deutsche Apotherkerzeitung*, **106** (1966), 1908–1909; and E. Witebsky and F. Milgrom, "August von Wassermann (1866–1925). Wassermann Reaction," in *Journal of the American Medical Association*, **204** (1968), 1000–1001.

KARL-ERNST GILLERT

WOLFF, CASPAR FRIEDRICH (*b*. Berlin, Germany, 18 January 1734; *d*. St. Petersburg, Russia [now Leningrad, U.S.S.R.], 22 February 1794), *biology*.

Wolff was the son of Johann Wolff, a tailor who moved to Berlin in the late seventeenth or early eighteenth century, and Anna Sofia Stiebeler. He studied medicine at the Medical-Surgical College in Berlin (1753–1754) and in 1755 enrolled at the University of Halle; his dissertation, *Theoria generationis* (1759), was criticized by Haller and Bonnet. On behalf of the Prussian Academy of Sciences, Euler attempted unsuccessfully in 1760–1761 to obtain for Wolff a post at the St. Petersburg Academy of Sciences. In 1761 Wolff became a field doctor in the Prussian army, which was then at war with Russia; and he also lectured in anatomy at the Breslau Military Hospital. His attempts in 1762 and 1764 to obtain permission to lecture in Berlin were opposed by the professors of the Medical-Surgical College, who had guild privileges to teach medicine.

After returning to Berlin in 1763, Wolff gave private lectures in anatomy, physiology, and medicine. The following year he restated his theory of generation and replied to Haller's and Bonnet's criticism in *Theorie von der Generation*—further decreasing his chances of obtaining a professorship. In 1766 he accepted an invitation from the St. Petersburg Academy of Sciences, extended at Euler's initiative, to join the department of anatomy. He traveled to Russia with his wife in May 1767 and later that year presented to the Academy "De formatione intestinorum praecipue." During the next twenty-seven years he published thirty-one memoirs in the Academy's *Proceedings*, including several that were devoted to anatomical research on the muscles of the heart and on connective tissue. He paid special attention to the study of human monstrosities, which were collected in the Academy's anatomical cabinet (which Wolff directed) of the *Kunstkammer*. Surviving manuscripts indicate that Wolff prepared a major work on the "theory of monsters," in which he attempted to systematize his epigenetic ideas. His sudden death from a brain hemorrhage prevented his completing this project.

Wolff's fundamental achievement was the refutation of the theory of preformation, which considered the development of an organism to be simply the expansion of an invisible, transparent, fully formed embryo. Wolff's detailed studies of plants led him to establish that growth takes place at the apex of any axial organ, in the so-called growing point. In the cabbage and chestnut he observed the gradual formation of the leaf layers and the appearance of veins and petioles. In establishing that the blossom is a modified leaf, Wolff anticipated the theory of metamorphosis, formulated in 1790 by Goethe, according to which all the organs of a plant are the result of transformation of leaves.

In the chick embryo, Wolff followed the development of the heart and blood vessels, and studied the formation of the blood from "blood islets" and the development of the extremities, the mesonephros, and the intestines. He discovered only the embryonic ("primary") kidneys, which become the final ("secondary") metanephros. The primary kidneys which he discovered became known as Wolffian bodies and their ducts, Wolffian ducts. Using the example of the development of the intestine, Wolff established the principles of formation of organs from foliate layers, by means of such processes as proliferation, folding, and wrapping (for example, in tubes and cavities). He thus laid the foundations of the theory of embryonic layers.

Wolff also attempted to give a universal explanation of the developmental process of organs. Because sufficient knowledge of the cellular structure of organisms was lacking during his lifetime, Wolff believed that all growth originates in a liquid substance that is completely lacking in organic structure. In such an "inorganic substance," he asserted, "bubbles" *(vesicula),* or "globules," and vessels are formed. Of primary importance to Wolff was the search for that force which provides for the entry of juices into plants and "nutrient material" into the embryo, and that ensures its subsequent infiltration during the formation of the various parts. Early on Wolff asserted that the presence of an "essential force" (*vis essentialis*) and the "ability to solidify" (*solidescibilitas*) are sufficient to explain nutrition, growth, and development. By means of these two "capacities" he attempted to explain not only individual development but also the obvious differences between organisms—and even the distinction between the plant and animal kingdoms. Wolff later abandoned these ideas, postulating only that "the formation of organic bodies in general is caused by one natural force," which inhabits the animal or plant substance (1768). In his last published treatise (1789), he concluded that the "essential force . . . consists in nothing other than in a certain special and definite kind of attractive and repulsive force." However primitive these views, there is no basis for interpreting them either as vitalistic or animistic; and in his last treatise he spoke out categorically against identifying "essential force" with the *anima* of Stahl. He also vigorously denied Blumenbach's concept of "formative tendency" (*nisus formativus*). In an unpublished treatise Wolff considered the soul to be "an extract of the brain and of the brain matter." Asserting the material nature of the soul, Wolff held that it "is born together with the body, which it inhabits and with which it is connected, but which it does not preexist."

Wolff's works contributed to the development of embryology and especially to the work of Pander and Baer, both of whom repeated, confirmed, and continued his research.

BIBLIOGRAPHY

I. Original Works. Wolff's dissertation, *Theoria generationis* (Halle, 1759), was followed by his polemical restatement, *Theorie von der Generation in zwo Abhandlungen erklärt und bewiesen* (Berlin, 1764); and a rev. and enl. Latin ed., by P. Meckel (Halle, 1774). Modern eds. include that of Paul Samassa, in German, Ostwalds Klassiker der Exacten Wissenschaften nos. 84–85 (Leipzig, 1896); in Russian, *Teoria zarozhdenia*, A. E. Gaissinovitch, ed. (Moscow, 1950), with commentary, unpublished additions by Wolff, and complete bibliography of his writings; and a photo reprint of the 1759 and 1764 eds., in one vol. (Hildesheim, 1966).

Other works are "De formatione intestinorum praecipue . . . Observationes, in ovis incubatis institutae," in *Novi commentarii academiae scientiarum imperialis Petropolitanae*, **12** (1768), 403–507; and **13** (1769), 478–530—a German ed., J. F. Meckel, trans., was published as *Über die Bildung des Darmkanals im bebrüteten Hühnchen* (Halle, 1812); and *Von der eigenthümlichen und wesentlichen Kraft der vegetabilischen, sowohl als auch der animalischen Substanz* (St. Petersburg, 1789). The unfinished *Objecta meditationum pro theoria monstrorum* was published in Latin and Russian by T. A. Lukina (Leningrad, 1973).

A list of Wolff's MSS at the Academy of Sciences in Leningrad was published by L. B. Modzalevsky, in *Vestnik Akademii nauk SSSR*, **3**, no. 3 (1933), 59–66.

II. Secondary Literature. Earlier sources on Wolff's life and work include Alfred Kirchhoff, *Die Idee der Pflanzen-Metamorphose bei Wolff und bei Göthe* (Berlin, 1867); and "Caspar Friedrich Wolff. Sein Leben und seine Bedeutung für die Lehre von der organischen Entwicklung," in *Jenaische Zeitschrift für Naturwissenschaft*, **4** (1868), 193–220; the excellent study by W. M. Wheeler, "Caspar Friedrich Wolff and the *Theoria generationis*," in *Biological Lectures. Marine Biological Laboratory, Woods Hole, Mass.* for 1898 (Boston, 1899), 265–284; W. Waldeyer, "Festrede," in *Sitzungsberichte der Preussischen Akademie der Wissenschaften*, **6** (1904), 209–226; J. Schuster, "Caspar Friedrich Wolff. Leben und Gestalt eines deutschen Biologen," in *Sitzungsberichte der Gesellschaft naturforschender Freunde zu Berlin* (1937), 175–195; "Der Streit um die Erkenntnis des organischen Werdens im Lichte der Briefe C. F. Wolffs an A. von Haller," in *Sudhoffs*

Archiv für Geschichte der Medizin und der Naturwissenschaften, **34** (1941), 196–218; and L. Stieda, *Biographisches Lexikon der hervorragenden Ärzte*, V (Berlin–Vienna, 1934), 983–984, with brief bibliography of secondary literature.

More recent works are L. Y. Blyakher, *Istoria embriologii v Rossii* ("History of Embryology in Russia"; Moscow, 1955), 21–68; A. E. Gaissinovitch, *C. F. Volf i uchenie o razvitii organizmov* ("C. F. Wolff and the Theory of the Development of Organisms"; Moscow, 1961), the most complete account of his life and work, with complete bibliography of his writings; B. E. Raykov, *Ocherki po istorii evolyutsionnoy idei v Rossii do Darvina* ("Sketches in the History of the Idea of Evolution in Russia Before Darwin"), I (Leningrad, 1947), 76–93; *Russkie biologi-evolyutsionisty do Darvina* ("Russian Evolutionary Biologists Before Darwin"), I (Leningrad, 1952), 165–193; and "Caspar Friedrich Wolff," in *Zoologische Jahrbücher*, Systematik, Ökologie und Geographie, **91** (1964), 555–626; and G. Uschmann, *Caspar Friedrich Wolff. Ein Pionier der modernen Embryologie* (Jena, 1955). Of the papers marking the 200th anniversary of the publication of *Theoria generationis*, the best is R. Herrlinger, "C. F. Wolffs *Theoria generationis* (1759). Die Geschichte einer epochemachenden Dissertation," in *Zeitschrift für Anatomie und Entwicklungsgeschichte*, **121** (1959), 245–270.

A. E. GAISSINOVITCH

ZEJSZNER (or **Zeuschner**), **LUDWIK** (*b.* Warsaw, Poland, 1805; *d.* Cracow, Poland, 3 January 1871), *geology, paleontology, mineralogy.*

Zejszner was descended from a family of German chemists who settled in the small town of Skwierzyna, near Poznań, at the beginning of the eighteenth century. There is no definite information on his early years, and even the date of his birth is uncertain. In 1822 he graduated from a *lycée* in Warsaw and then attended the University of Warsaw for two years. From 1824 to 1828 Zejszner studied in Berlin, where he attended the lectures of Hegel, Humboldt, and Ritter, and then at Göttingen. Early in 1829 he obtained a doctorate from the University of Heidelberg with a dissertation on the influence of chemical mixture on crystallization form. While in Germany he met the geologists Bronn, Leonhard, G. Rose, and Heinrich Steffens; and he maintained contact with them after returning to Poland. He also made geological observations, mainly on the occurrence of basalt, the results of which appeared in his first published treatise, *O powstaniu i względnym wieku formacji bazaltowej* ("On the Origin and Relative Age of Basalt Formation"; Warsaw, 1829). In this work Zejszner established the magma origin of these rocks—contrary to the opinions of the neptunists, which still prevailed in Poland at the time. After returning home he used a Polish spelling of his name, and the traditional one, Zeuschner, appeared only in works published in foreign languages.

On 7 January 1830 Zejszner was appointed to the chair of the newly established department of mineralogy at the University of Cracow. He subsequently organized instruction, founded a library and a collection of minerals, and in 1833 published a manual of mineralogy that was not based on Werner's system. He also began geological surveys in the Cracow district and in the Carpathian Mountains.

Despite his achievements, Zejszner was obliged to resign his professorship in 1833, when the repressive measures that followed the collapse of the November Revolution (1830) reached Cracow. Because of his patriotism Zejszner was suspected of having imported and distributed materials published by Polish émigrés in Paris. He was a mining inspector in Cracow until 1837, when he moved to Warsaw and began conducting research privately and publishing the results.

The liberalization that began in Austria at the time of the revolutionary movement in 1848 made it possible for Zejszner to return to Cracow, which since 1846 had been incorporated into Galicia, the Austrian section of partitioned Poland. He built up the mineralogy department, which he again headed, and published *Geologia do łatwego pojęcia zastosowana* ("Comprehensible Geology"; Cracow, 1856), the first Polish lecture on geology. Ignacy Łukasiewicz, who later invented the kerosene lamp (1853) and a method of petroleum distillation, was one of his students.

The increase in German influence at the University of Cracow led Zejszner in 1857 to move to Warsaw, where a Polish university was being established. After lecturing on mineralogy for a year at the Academy of Medicine and Surgery, he entered the civil service, participating in explorations for salt and the compiling of a geological map of the kingdom of Poland. In 1861 he published a new and extensive manual of mineralogy, *Początki mineralogii według układu Gustawa Rose na krystalizaccji i składzie chemicznym opartego* ("Introduction to Mineralogy According to a System by Gustav Rose Based on Crystallization and Chemical Composition"; Warsaw, 1861). Following the collapse of the January Revolution in 1863, repressive measures were taken by the czar, and the remaining elements of Polish self-government were

ZEJSZNER

abolished. Depressed and ill, Zejszner left Warsaw and the civil service. In 1870 he moved to Cracow, where the following year he was murdered during a robbery attempt.

Zejszner was the first Polish geologist in the modern sense of the word. From the time he completed his university training until he died, he conducted field surveys and published the results. He investigated geological formations and the occurrence of useful minerals; he made thousands of barometric measurements at various places; and he studied the temperature of the springs in the Carpathian Mountains. Of his more than 200 works, the great majority are published results of field surveys and synthetic descriptions of Polish regions. Two periods may be distinguished. In the first (1829–1857) Zejszner investigated southern Poland, mainly the Carpathian Mountains. In the second (1858–1870), central Poland, particularly the Świętokrzyskie Mountains, was his area of concentration. Because of their great keenness of observation, his works still retain their validity.

Zejszner's special regional works include geological maps and profiles. He published an unsigned geological map of the Tatras (Berlin, 1844) and prepared a manuscript of a detailed geological map of southern Poland, which has been lost. His paleontological works, which are mainly on the fauna of the Jurassic period, include an important Polish paleontology that could not be completed for lack of funds. After his death his outstanding geological collection was acquired by the Dzieduszycki Museum in Lvov (now the Museum of the Academy of Sciences of the Ukrainian S.S.R.).

Zejszner's participation in international conferences and such organizations as the Deutsche Geologische Gesellschaft and the Society of Naturalists in Moscow enabled him to meet most of the leading European geologists of the period. He traveled widely, and in Poland received many foreign geologists, including Murchison, with whom he went to the Świętokrzyskie and Carpathian mountains. He translated treatises by Davy and the geological part of Humboldt's *Kosmos* into Polish. An active member of the scientific societies of Cracow and Poznań, he also participated in the activities of the scientists and writers who gathered in the editorial office of the periodical *Biblioteka Warszawska*. About a dozen species of animal and plant fossils were named for Zejszner by European paleontologists. In 1960 the Polish Geological Society established a prize in his name, to be awarded annually for the best published work by a young geologist.

BIBLIOGRAPHY

I. Original Works. A list of Zejszner's geological publications is in Regina Fleszarowa, *Retrospektywna bibliografia geologiczna Polski* ("Retrospective Geological Bibliography of Poland"), II (Warsaw, 1966), 416–433; and S. Czarniecki and Z. Martini, *Retrospektywna bibliografia . . . Uzupełnienia* (". . . Supplement"; Warsaw, 1972), 186–187. The Royal Society *Catalogue of Scientific Papers* lists 86 memoirs: VI, 496 (under Zeiszner) and 504–506 (under Zeuschner); VIII, 1300–1301; and XII, 802. See also Poggendorff, III, 1482. S. Czarniecki, *Studia i materiały z dziejów nauki polskiej* ("Studies and Materials for the History of Polish Science"), ser. C, fasc. 4 (Warsaw, 1961), 61–103, presents information on Zejszner's unpublished diary and a reprint of his memoirs on his studies at Berlin in 1824.

II. Secondary Literature. Fleszarowa (see above) cites biographies of Zejszner in Polish and other languages. S. Czarniecki presents a new critical analysis of his life and work in "Ludwik Zejszner," in *Wszechświat*, fasc. 4 (1958), 93–96, 356–357. His geographical work is treated by A. Chłubińska in "Ludwik Zejszner jako geograf" ("Ludwik Zejszner as a Geographer"), in *Kosmos*, ser. A, 53 (1928), 245–286.

Stanisław Czarniecki

TOPICAL ESSAYS

Topical Essays

PREFACE 531

HISTORY OF MATHEMATICAL ASTRONOMY IN INDIA
David Pingree 533

MAN AND NATURE IN MESOPOTAMIAN CIVILIZATION
A. Leo Oppenheim 634

MATHEMATICS AND ASTRONOMY IN MESOPOTAMIA
B. L. van der Waerden 667

THE MATHEMATICS OF ANCIENT EGYPT
R. J. Gillings 681

EGYPTIAN ASTRONOMY, ASTROLOGY, AND CALENDRICAL RECKONING
Richard A. Parker 706

JAPANESE SCIENTIFIC THOUGHT
Shigeru Nakayama 728

MAYA NUMERATION, COMPUTATION, AND CALENDRICAL ASTRONOMY
Floyd G. Lounsbury 759

Preface

The Board of Editors is pleased to conclude the *Dictionary of Scientific Biography* with the essays that follow. The idea of completing the collection with topical articles on the scientific outlook and accomplishments of certain ancient civilizations arose in a conversation between E. S. Kennedy, David Pingree, and the editor in chief during an early stage of the planning for the *Dictionary*. Professor Pingree then undertook to identify, in the regular alphabetical sequence, the more important persons, whether real or mythical, in the Indian astronomical tradition and, in this final volume, to give an account of the various schools in which they figured throughout Indian history up to 1800. He has fulfilled that commitment in the monograph that we now have the privilege of publishing.

The Board also thought that it would be appropriate to treat the sciences of ancient Mesopotamia and Egypt in a similar manner, although the identity of the scribes who recorded the techniques is unknown. We are very grateful to B. L. van der Waerden and to the late A. Leo Oppenheim for essays on Mesopotamia and to R. J. Gillings and Richard A. Parker for their essays on Egypt.

As work on the *Dictionary* progressed, the Board thought an account of early Japanese science would be specially welcome to readers, and we are indebted to Shigeru Nakayama, who has most generously contributed an exposition of the astronomical, medical, and mathematical traditions prior to the westernization that followed the Meiji Restoration in the nineteenth century. It has not seemed feasible to include an essay on traditional science in China. For one thing, a biographical mode of treatment would, in principle, be possible if the scholarship were available. For another, the immense work of Joseph Needham, *Science and Civilisation in China* (Cambridge, 1954–), is still in course of publication. Also *Chinese Science* (Cambridge, Mass., 1973, edited by Shigeru Nakayama and Nathan Sivin) contains scholarly appreciations of that enterprise and of the subject in general. We have, therefore, limited ourselves to a few articles written in the spirit of the *Dictionary* and intended to be representative of major aspects of Chinese thought, namely Shen Kua (XII), Liu Hui (VIII), Ch'in Chiu-shao (III), Li Chih (VIII), Yang Hui (XIV), Chu Shih-chieh (III), Tsu Ch'ung-chih (XIII), and Li Shih-chen (VIII).

Finally, it seemed appropriate to conclude a dictionary published in the western hemisphere with a study of the exact sciences developed by the Maya before the Spanish conquest. The members of the Board feel confident that readers will join them in gratitude to Floyd G. Lounsbury for his precise and lucid treatment of that important and little-known body of material.

HISTORY OF MATHEMATICAL ASTRONOMY IN INDIA

David Pingree

Astronomy[1] shares with other scholarly disciplines in India the characteristic of being repetitive. Indian astronomers did not usually attempt innovations in theory; they wished to preserve their tradition as intact as possible. Most of their energies, therefore, were devoted to devising computational techniques. And they delighted both in simplifications or approximations and in needless complications; each type of change displayed the skill of the master. Much of the history of this science in India, then, must be simply an account of the means by which the traditions were preserved, and a recitation of the often bizarre modifications and elaborations of the basic formulas. In the present essay these basic formulas are presented primarily in section V, and will not be repeated thereafter.

That Indiạn astronomy was not completely static is due almost entirely to the repeated intrusion of new theories from the West. Five times have such intrusions occurred – in the fifth century B.C., from Mesopotamia via Iran; in the second and third centuries A.D., from Mesopotamia via Greece; in the fourth century A.D., directly from Greece; in the tenth to eighteenth centuries, from Iran; and in the nineteenth century, from England. But, although the character of Indian astronomy at each such intrusion was changed, to a greater or lesser extent, these changes were accompanied by the minimum possible alterations of the earlier traditions, none of which ever completely died. The coalescence of these several traditions necessitated transformations and adjustments of both the older, native or assimilated theories and the new, foreign ones that were being adapted. It also meant that some contradictions would always persist. Thus, internal consistency was not expected in any Indian astronomical system, and this tolerance of inconsistency is undoubtedly one important factor that prevented Indian astronomers from making any advances in theory; they were not motivated to examine the logical foundations of mathematical astronomy, but only to tinker with the computational superstructure. The lack of a tradition of observational astronomy also permitted them to iterate models and parameters that did not at all satisfactorily predict the phenomena.

Moreover, Indian astronomers had the misfortune to receive, in the first three transmissions from the West, theories that were either antiquated in their countries of origin or also deviant and second-rate. For instance, despite the enormous influence of Greek astronomy on Indian astronomy in the period contemporary with and immediately subsequent to Ptolemy, the Indians remained ignorant of most aspects of Ptolemaic astronomy until the seventeenth century. For the historian of Hellenistic astronomy, however, this fact and the conservatism of the Indian tradition are of enormous assistance, for much of what we know of Greek astronomy between Hipparchus and the fourth century A.D. is to be found in Sanskrit texts. Admittedly, because of the Indian tendency to modify intellectual imports and also the corrupt nature of the earliest surviving texts, it is often difficult to determine precisely the nature of the Greek texts upon which the Sanskrit are based; but one hopes that careful analyses and comparisons of the sources will gradually achieve a more accurate description of Hellenistic astronomy than is possible from a consideration of the surviving Greek tradition in isolation.[2]

In the preceding paragraph reference was made to the corrupt tradition of the earliest surviving Sanskrit texts on astronomy. The cause of this corruption is usually that the texts had become unintelligible; and this unintelligibility is not unrelated to the style developed by Indian astronomers. The texts proper were composed in verse in order to facilitate memorization, with various conventions for rendering numbers into metrical syllables. The exigencies of the meter often necessitated the omission of important parts of mathematical formulas, or contributed to the imprecision of the technical terminology by forcing the poet to substitute one term for another. This did not matter too much as long as the texts were taught in a school that preserved their interpretation orally, or after a commentary had been published; but the teaching of the earliest texts ceased and no commentaries, if any were ever composed, have survived. Only the texts were copied, without comprehension or care, so that what we have are frequently fragmentary, and certainly incorrect, representatives of the original. Often with such texts one must guess at the problem that is to be solved, invent a solution that seems reasonable within the context of the rest of the work, and then see whether the manuscript readings can somehow be made to coincide with this solution. Success in such endeavors is not of necessity attainable.

Even the existence of a commentary, however, does not always provide the reader with an understanding of the text. The commentators themselves are almost never interested in explaining the derivation of rules; they are more concerned to repeat them in clearer language and to exemplify them. Sometimes they cannot even repeat them correctly. What the commentary does do in all cases, however, is to offer some guarantee of the status of the text in the time of the commentator. Frequently this is a very valuable aid.

In consideration of the character of Indian astronomy as outlined above, this essay is divided into sections, each of which is devoted to a particular tradition. The basic chronological divisions are as follows (the sections in which they are discussed are indicated in parentheses):

1. Vedic, *ca.* 1000 B.C. – 400 B.C. (I).
2. Babylonian, *ca.* 400 B.C. – 200 (II).
3. Greco-Babylonian, *ca.* 200 – 400 (III).
4. Greek, *ca.* 400 – 1600 (V – X).
5. Islamic, *ca.* 1600 – 1800 (XI).

In the Greek period there originated the five pakṣas, or schools of astronomy, that characterized the science in medieval times and to which most of our extant texts belong. The pakṣas, the approximate dates of their inception, and the areas in which they were especially prevalent are

1. Brāhmapakṣa, *ca.* 400, western and northwestern India (V).
2. Āryapakṣa, *ca.* 500, southern India (VI).
3. Ārdharātrikapakṣa, *ca.* 500, Rājasthān, Kashmir, Nepal, and Assam (VII).
4. Saurapakṣa, *ca.* 800, northern, northeastern, and southern India (VIII).
5. Gaṇeśapakṣa, *ca.* 1500, western and northwestern India (X).

The basic features of each pakṣa will be delineated when the initial text is described; later texts within each pakṣa will largely repeat this material, and only their more significant innovations will require description. Furthermore, texts in one pakṣa frequently repeat rules or techniques from another. These repetitions and borrowings are too numerous to be noticed in this essay.

The corpus of texts to which reference is made is restricted because of the lack of editions, good or bad, of the majority of Sanskrit astronomical works. Obviously, the future explorations of the manuscripts will necessitate corrections of many details in this essay. I can only hope that my inadvertent omissions and errors will not require many more.

I. VEDIC ASTRONOMY

Although speculations abound concerning alleged astronomical data in the *Vedas*, in the *Brāhmaṇas*, and in other early Sanskrit texts,[3] there is no convincing evidence that any method of mathematically describing celestial phenomena was known to Indians before the latter half of the first millennium B.C. Rather, the Vedic texts refer almost exclusively, when they do refer unquestionably to astronomical phenomena, to the times for performing sacrifices. In this connection they mention various yugas (periods), saṃvatsaras (years), ayanas (half-years), ṛtus (seasons), māsas (months), adhimāsas (intercalary months), pakṣas (half-months), and specific nights (indicated by ordinal numerals in the feminine, modifying rātri [night] and *not* tithi), as well as nakṣatras (constellations).[4] All of these elements survived into later periods, and profoundly affected the form into which Indian astronomers molded the foreign systems that they adapted to their own use. As characteristic features of Indian astronomy they must now be briefly described, even though they do not in themselves pertain to mathematical astronomy.

In the following discussion, no attempt is made to distinguish sources chronologically since, even where this is possible — as in the case of the priority of the *Ṛgveda* to the *Brāhmaṇas* — it is not always clear that the views expressed in the latter developed historically after the composition of the former. All texts that can reasonably be dated before *ca.* 500 B.C. are here considered to represent essentially a single body of more or less uniform material.

Yugas of two, three, four, five, and six years are mentioned at various places in this vast literature. But, although there was no standard yuga, one of the more frequent variants was the yuga of five years named, respectively: saṃvatsara, parivatsara (or idāvatsara), idāvatsara (or iduvatsara), iduvatsara (or idvatsara or anuvatsara), and vatsara (or udvatsara). This yuga of five years is presumably that which influenced the author of the *Jyotiṣavedāṅga*.

The length of a saṃvatsara is normally taken to be 360 days, or twelve months of thirty days each. Such an "ideal" year is also known in Mesopotamia.[5] Since months were in practice synodic, however, this "ideal" year could not really have been in use.

The year contains two ayanas: in the uttarāyana the sun travels north; in the dakṣiṇāyana, south. Presumably these directions were determined by some such means as observing the point on the horizon at which the sun rises on successive days. There is no compelling reason to believe that the solstices were accurately determined.

There are three ṛtus in the *Ṛgveda*: vasanta (spring), grīṣma (summer), and śarad (autumn). In most texts of the Vedic period, however, there are five: vasanta, grīṣma, vārṣa (rains), śarad, and hemantaśiśira (winter). A final enumeration of ṛtus, which remains standard in India, is vasanta, grīṣma, vārṣa, śarad, hemanta (winter), and śiśira (cold).

The māsa in this period is either the "ideal" month of thirty days, a twelfth of a 360-day year, or a synodic month of twenty-nine or thirty days, which may have been measured either from new moon (amāntamāsa), as in Mesopotamia, or, less probably, from full moon (pūrṇimāntamāsa). There are two systems of naming the months: descriptively and after the nakṣatras. Naturally, many synonyms can be substituted for the names listed in Table I.1.

TABLE I.1

Descriptive Name	Nākṣatra Name	Ṛtu
Madhu	Caitra	vasanta
Mādhava	Vaiśākha	
Śukra	Jyaiṣṭha	grīṣma
Śuci	Āṣāḍha	
Nabha	Śrāvaṇa	vārṣa
Nabhasya	Bhādrapada	
Iṣa	Āśvina	śarad
Ūrja	Kārttika	
Saha	Mārgaśira	hemanta
Sahasya	Pauṣya	
Tapa	Māgha	śiśira
Tapasya	Phālguna	

In some texts Phālguna is the first month of vasanta.

An intercalary month, or adhimāsa, was from time to time added to the normal twelve months of the year, presumably in order to make the beginnings of the sun's ayanas fall in the correct months. Despite the efforts of many scholars to prove the contrary, no systematic intercalation scheme can be attributed to this period.

Pakṣas are the periods of fourteen or fifteen days between new moon and full (pūrvapakṣa; later called śuklapakṣa) and between full moon and new (aparapakṣa; later called kṛṣṇapakṣa).

Since the normal term for a nychthemeron in this period is rātri (night), the nychthemeron presumably began at sunset, as in Mesopotamia.

The nakṣatras are twenty-seven or twenty-eight stars or groups of stars with one of which the moon is supposed to conjoin each night. Which stars were given the names of the nakṣatras preserved from this period cannot be determined; we are on secure ground in making such identifications only with texts written in the fifth century of the Christian era and after. With each nakṣatra is associated a deity, after whom the nakṣatra is sometimes named. All lists of nakṣatras in this period begin with Kṛttikā; a composite list is provided in Table I.2.

TABLE I.2

	Nakṣatra	Deity
1.	Kṛttikā	Agni
2.	Rohiṇī	Prajāpati
3.	Mṛgaśīrṣa (Invakā)	Soma (Marutaḥ)
4.	Ārdrā (Bāhu)	Rudra
5.	Punarvasū	Aditi
6.	Tiṣya (Puṣya)	Bṛhaspati
7.	Āśreṣā (Āśleṣā)	Sarpāḥ
8.	Maghā	Pitaraḥ
9.	Phalgunī (Pūrvā Phalgunī)	Aryaman (Bhaga)
10.	Phalgunī (Uttarā Phalgunī)	Bhaga (Aryaman)
11.	Hasta	Savitṛ
12.	Citrā	Indra (Tvaṣṭṛ)
13.	Svāti (Niṣṭyā)	Vāyu
14.	Viśākhā	Indrāgni
15.	Anurādhā	Mitra
16.	Rohiṇī (Jyeṣṭhā)	Indra (Varuṇa)
17.	Vicṛtau (Mūla)	Pitaraḥ (Nirṛti)
18.	Āṣāḍhā (Pūrvāṣāḍhā)	Āpaḥ
19.	Āṣāḍhā (Uttarāṣāḍhā)	Viśve devāḥ
20.	Abhijit	Brahmā
21.	Śroṇā (Śravaṇa)	Viṣṇu
22.	Śraviṣṭhā	Vasavaḥ
23.	Śatabhiṣaj	Indra (Varuṇa)
24.	Proṣṭhapadā	Aja Ekapād
25.	Proṣṭhapadā	Ahirbudhnya
26.	Revatī	Pūṣan
27.	Aśvayujau	Aśvinau
28.	Apabharaṇī (Bharaṇī)	Yama

When only twenty-seven nakṣatras are listed Abhijit is omitted. We have no information that would allow us to identify the stars to which these names refer.

II. BABYLONIAN ASTRONOMY IN INDIA

Probably late in the fifth century B.C., while the Achaemenids dominated northwestern India, there occurred a massive transmission of Mesopotamian literary material to India. This influence is discernible in the *Gargasaṃhitā*[6] and its successors that deal with omens in essentially the same manner as did the Babylonians. It also can be discovered in the *Jyotiṣavedāṅga* of Lagadha (or Śuci) and in a series of other Sanskrit texts that are analyzed below.[7]

The Ṛk recension of the *Jyotiṣavedāṅga* is evidently earlier than the Yajur recension; the former may be as old as *ca.* 400 B.C., the latter as late as *ca.* A.D. 500. The problem that Lagadha set out to solve in the Ṛk recension is that of establishing an intercalation cycle; this problem had also engaged the attention of Mesopotamian, Greek, Egyptian, and Iranian astronomers in the fifth century B.C. Lagadha's solution involves the use of parameters, mathematical models, time units, and instruments of Mesopotamian origin. There can be no doubt that his system is not indigenous to India.

The basic period relation that he hypothesizes is

$$5 \text{ solar years} = 1{,}830 \text{ sidereal days} = 62 \text{ synodic months} = 1{,}860 \text{ tithis}, \qquad (\text{II}.1)$$

where a tithi (a Mesopotamian time unit) is one-thirtieth of a synodic month and all units of time are treated as mean. The yuga of five years perhaps appealed to Lagadha because of its appearance in the Vedic texts, but its choice ultimately depended on astronomical rather than sentimental considerations. The Egyptian twenty-five-year intercalation cycle, introduced during Achaemenid rule, in the middle of the fourth century B.C., is based on the period relation

$$25 \text{ solar years} = 309 \text{ synodic months}, \qquad (\text{II}.2)$$

whereas five of Lagadha's yugas give the relation

$$25 \text{ solar years} = 310 \text{ synodic months}. \qquad (\text{II}.3)$$

Since for both Lagadha and the Egyptians it holds that

$$1 \text{ solar year} = 366 \text{ sidereal days} = 365 \text{ civil days}, \qquad (\text{II}.4)$$

it is apparent that the Egyptian period relation, in which

$$1 \text{ synodic month} \approx 29;32 \text{ civil days}, \qquad (\text{II}.5)$$

is more accurate than Lagadha's, in which

$$1 \text{ synodic month} \approx 29;26 \text{ civil days}. \qquad (\text{II}.6)$$

It should be noted that a solar year of 365 civil days may have been adopted in Iran as early as the second half of the fifth century B.C.

In order to designate the positions of the sun and the moon on the ecliptic, Lagadha uses the list of twenty-seven nakṣatras to divide the ecliptic into twenty-seven equal arcs, each of which contains 13;20°, in the same way that Babylonian astronomers in the late fifth century B.C. had divided it into twelve equal arcs, each of which was named after a constellation. Since Lagadha places the beginning of the uttarāyana at the beginning of Śraviṣṭhā, the twenty-first nakṣatra in a list of twenty-seven or twenty-second in a list of twenty-eight, it appears that he retained the old nakṣatra list of the Vedic period, beginning with Kṛttikā, and distributed the solstices and equinoxes symmetrically, on the assumption that the vernal equinox occurs when the sun is in Kṛttikā. His adaptation of the earlier lists without any adjustment makes these data useless for determining his own date.

The progress of the sun is stated to be a nakṣatra in $13\frac{5}{9}$ sidereal days, since in five years it traverses 135 nakṣatras, and

$$135 \cdot 13\tfrac{5}{9} = 1{,}830. \qquad (\text{II}.7)$$

Lagadha elsewhere teaches, however, that the sun progresses through a nakṣatra in $13\frac{2}{3}$ days; this parameter yields an impossible 1,845 days in a yuga. He correctly states that the moon travels through a nakṣatra in $1\frac{7}{603}$ sidereal days, since it traverses 1,809 nakṣatras in a yuga and

$$1{,}809 \cdot 1\tfrac{7}{603} = 1{,}830. \qquad (\text{II}.8)$$

In both cases only the mean motion of the luminary is considered.

In one final matter Lagadha demonstrates his dependence on Babylonian astronomy: his determination of the length of daylight. His time-measuring instrument is an outflowing waterclock such as those attested in cuneiform texts from about 700 B.C. on. From approximately the same date the Babylonians used 3:2 as the ratio of the longest to the shortest day in the year; and this is precisely the parameter used by Lagadha, although it is applicable only in part of northwestern India. He further divided the day into thirty equal parts, called muhūrtas, probably in analogy to the division of the synodic month into thirty equal tithis. Then Lagadha established a linear zigzag function, such as the Babylonians had used for several centuries before the fifth century B.C. to describe periodic and regular deviations from a mean, to determine the amount of water to be put into the waterclock so that on each day it will measure the length of daylight. Unfortunately, he does not give the mean value that would measure fifteen muhūrtas at the equinoxes, but says only that one must add a prastha of water each day that the sun is in the uttarāyana, and remove one each day that it is in the dakṣiṇāyana. Lagadha also gives a rule to determine the length of daylight in muhūrtas that is based on a linear zigzag function:

$$d(x) = 12 + 6x/183 = 12 + 2x/61, \qquad \text{(II.9)}$$

where x is the time before or after the winter solstice expressed in days, $d(x)$ is the length of daylight in muhūrtas on that day, 12 muhūrtas is the length of daylight at the winter solstice, 6 muhūrtas is the difference between the longest and shortest days of the year, and 183 is the number of (sidereal) days in an ayana. The parameters, then, are those of Table II.1.

TABLE II.1

M (= maximum)	18 muhūrtas
m (= minimum)	12 muhūrtas
μ (= mean)	15 muhūrtas
d (= difference)	2/61 muhūrtas
Δ (= $M - m$)	6 muhūrtas
p (= period)	366 "days"

Precisely the same five-year yuga as is found in the Ṛk recension (II.1) and the same linear zigzag function for determining the length of daylight (Table II.1) are found in a *Paitāmahasiddhānta*, the epoch of which is 11 January A.D. 80 and that is known to us through a summary in chapter 12 of Varāhamihira's *Pañcasiddhāntikā*. Varāhamihira[8] understood the five-year yuga to contain 1,830 civil rather than sidereal days. This was a common mistake in the sixth century. The *Paitāmaha* also contains equivalents of Lagadha's rules for determining the mean longitudes in nakṣatras of the sun and moon. It adds to the data of the Ṛk recension only the idea of the yoga, which is the period in which the sum of the motions of the sun and moon equals a nakṣatra, or 13;20°. Since in a yuga the sun makes five rotations and the moon sixty-seven, which gives a total of seventy-two, and each combined rotation includes an entire series of the twenty-seven yogas, there are seventy-two series in a yuga of 1,830 sidereal days. Therefore, the number of lapsed yogas at a given ahargaṇa (sum of days) within the yuga is given by the formula

$$\text{yoga} = \frac{72c}{1830} = \frac{12c}{305}, \qquad \text{(II.10)}$$

where c is the ahargaṇa.

The *Paitāmaha* names the seventeenth yoga, Vyatipāta; therefore, it presumably already had a complete list of the twenty-seven, such as is reproduced in Table II.2.

TABLE II.2

1. Viṣkamba	10. Gaṇḍa	19. Parigha
2. Prīti	11. Vṛddhi	20. Śiva
3. Ayuṣmān	12. Dhruva	21. Siddha
4. Saubhāgya	13. Vyāghāta	22. Sādhya
5. Śobhana	14. Harṣaṇa	23. Śubha
6. Atigaṇḍa	15. Vajra	24. Śukla
7. Sukarmān	16. Siddhi	25. Brahman
8. Dhṛti	17. Vyatipāta	26. Indra
9. Śūla	18. Varīyas	27. Vaidhṛti

Further, the method of expressing the rule and the longitude of the conjunction of 11 January A.D. 80 imply that the list of nakṣatras was already considered to start with Aśvinī. This new list of nakṣatras and the ecliptic longitudes of their beginnings is given in Table II.3.

TABLE II.3

1. Aśvinī	0°
2. Bharaṇī	13;20°
3. Kṛttikā	26;40°
4. Rohiṇī	40°
5. Mṛgaśiras	53;20°
6. Ārdrā	66;40°
7. Punarvasu	80°
8. Puṣya	93;20°
9. Āśleṣā	106;40°
10. Maghā	120°
11. Pūrvaphālgunī	133;20°
12. Uttaraphālgunī	146;40°
13. Hasta	160°
14. Citrā	173;20°
15. Svāti	186;40°
16. Viśākhā	200°
17. Anurādhā	213;20°
18. Jyeṣṭhā	226;40°
19. Mūla	240°
20. Pūrvāṣāḍhā	253;20°
21. Uttarāṣāḍhā	266;40°
22. Śravaṇa	280°
23. Dhaniṣṭhā	293;20°
24. Śatabhiṣaj	306;40°
25. Pūrvabhādrapadā	320°
26. Uttarabhādrapadā	333;20°
27. Revatī	346;40°

As we have seen, Varāhamihira in the sixth century understood the *Paitāmaha*'s period relation to be

5 solar years = 62 synodic months
= 1,830 civil days. (II.11)

This leads to the disturbing conclusion that a solar year contains 366 civil days. A similar misunderstanding of Lagadha's Ṛk recension is found in the Yajur recension of the *Jyotiṣavedāṅga*, which otherwise is merely a rearrangement of Lagadha's work with a few additional verses; I would date the Yajur recension to between the third and the fifth centuries of the Christian era. This same misinterpretation is repeated in the verses of Garga cited by Somākara in his bhāṣya on the Yajur recension, and also in the Jaina canonical astronomical works, which we know in a recension of the early sixth century.

Before concluding this section, it should be noted that two texts of the early Christian era—the *Śārdūlakarṇāvadāna*, which is earlier than A.D. 148, and the second book of the *Arthaśāstra* of Kauṭilya, which was given its present form probably in the second century, preserve rules for telling time similar to those of Lagadha, employing the same Babylonian parameters, zigzag functions, and instrument. Another instrument of Mesopotamian origin appears in these texts, however—the gnomon, which the Indians normally divide into twelve digits, although one passage in the *Śārdūlakarṇāvadāna* refers to a gnomon of sixteen digits. Normally the noon shadow at the summer solstice is assumed to be zero digits, which may indicate that Ujjayinī, which lies very close to the Tropic of Cancer, was already a center of astronomical studies; and the length of the noon shadow at the winter solstice is assumed to be twelve digits, with the monthly (or for each zodiacal sign) difference being two digits. The parameters are shown in Table II.4.

TABLE II.4

M	12 digits
m	0 digits
μ	6 digits
d	2 digits
Δ	12 digits
p	12 months

The ratio of the longest to the shortest day is still assumed to be 3:2, which is valid about 11° north of Ujjayinī. Such gross inconsistencies usually did not disturb Indian astronomers, but they often make it difficult to determine the meaning of Sanskrit texts.

III. GRECO-BABYLONIAN ASTRONOMY IN INDIA

In A.D. 149/150, in the realm of the Western Kṣatrapa, Rudradāman I, and probably in Ujjayinī, one Yavaneśvara translated a long Greek astrological treatise into Sanskrit prose. We now possess a large portion of the versification of this made by Sphujidhvaja in 269/270 and entitled *Yavanajātaka*. Chapter 79 of the *Yavanajötaka* is devoted to mathematical astronomy. In part Sphujidhvaja relies on the tradition of the *Jyotiṣavedāṅga*, and in part upon Greek adaptations of the Babylonian astronomy of the Seleucid period; he also mentions Vasiṣṭha (*YJ* 79, 3), who will be discussed below.

Sphujidhvaja's lunisolar yuga, the epoch of which is sunrise of 23 March 144, is based on the relation (*YJ* 79, 3–10)

165 solar years = 1,980 saura months = 2,041 synodic months = 58,231 risings of the moon
= 60,265 civil days = 61,230 tithis. (III.1)

From this relation follow those of Table III.1.

TABLE III.1

1 solar (tropical) year = 6,5;14,32 civil days
1 synodic month = 29;31,38 civil days
1 sidereal month = 27;24,18 civil days
1 saura month = 30;26,11 civil days

Elsewhere in his work, however, Sphujidhvaja states the relations given in Table III.2 (*YJ* 79, 5, 11–13, and 34).

TABLE III.2

1 synodic month = 29;31,54,34 civil days
1 tithi = 63/64 civil days (1 synodic month = 29;31,52 civil days)
1 civil day = 61/60 tithis (1 synodic month = 29;30,28 civil days)
1 ṛtu = 62 tithis (1 yuga = 61,380 tithis)
1 sidereal month = 27;17,10,34 civil days
1 saura month = 30;26,9,52,4 civil days
1 solar year = 6,5;13,58,24,48 ≈ 6,5;14 civil days
1 solar year = 6,5;14,47 (or 6,5;14,48) civil days

Of the items in Table III.2, the first, fifth, sixth, and eighth were intended to be correct, the rest approximate; but none are related to the yuga. The year length in the eighth, if a slight correction is accepted, is that of Hipparchus and Ptolemy (see Table III.14). Sphujidhvaja already employs the four kinds of time measurement that became standard in Indian astronomy: saura (solar), cāndra (lunar), nāksatra (sidereal), and sāvana (civil).

Since each yuga is assumed to be precisely identical with every other yuga, in order to determine the mean longitudes of the planets (in this case, of the sun and moon), it will suffice to determine the number of days lapsed since the beginning of the yuga (the ahargaṇa), and then to apply the "rule of three" (proportion). Indian astronomers traditionally find the ahargaṇa from the lunar calendar date, and Sphujidhvaja is no exception. In the formulations of these rules I use the following symbols:

a: adhimāsa
c: civil day
ϵ: epact
m: synodic month
n: saura month
r: rotation
s: saura day
t: tithi
u: ūnarātra
y: solar year

Capital letters refer to units in a yuga, lowercase letters to lapsed units. Subscript letters refer to the current larger units within which the referent letters have elapsed.

Sphujidhvaja's procedure (III.2) is given below (*YJ* 79, 15–18):

$$y \cdot 12 = n \quad \text{(a)}$$
$$n + m_y = n' \quad \text{(b)}$$
$$30n' + t_m = t \quad \text{(c)}$$
$$t - u = c. \quad \text{(d)}$$

Clearly, (*b*) and (*c*) are not correct, since $n \neq m$ and $30n' = s \neq t$. Instead, we need (III.3)

$$n + a = m \quad \text{(a)}$$
$$30m + t_m = t. \quad \text{(b)}$$

Further, Sphujidhvaja, at least as his text is now preserved, never gives instructions for finding u, such as

$$u = U \cdot \frac{n}{N}. \quad \text{(III.4)}$$

He does, however, give a rule for finding a, which he would need had he given (III.3a) (*YJ* 79, 19):

$$a = \epsilon \cdot \frac{y}{30}, \quad \text{(III.5)}$$

where ϵ = 11;11 tithis. This parameter means that one solar year contains 371;11 tithis; elsewhere Sphujidhvaja has given the approximate value 372 tithis. After (III.5) he states correctly (*YJ* 79, 20) that

$$t = T \cdot \frac{a}{A}. \quad \text{(III.6)}$$

This formula obviously is superior to (III.2a–c).

Once one has found c, one can use (III.16). Sphujidhvaja, however, refrains from taking this simple step, and instead gives crude rules for determining the mean longitudes of the luminaries. These rules assume that the sun's mean progress is 30° in a synodic (instead of a saura!) month, and that the mean moon traverses 30° for every 2;30° traversed by the mean sun. Then Sphujidhavaja describes two linear zigzag functions for obtaining the daily velocities, the parameters of which are given in Table III.3 (*YJ* 79, 23–24).

TABLE III.3

	Solar daily velocity	Lunar daily velocity
M	62′	
m	57′	700′+
μ	(59;30′)	
d	(1′)	12;20′ +
Δ	5′	

Unfortunately, the data for the moon are incomplete in the unique manuscript; and for neither the sun nor the moon does Sphujidhvaja provide enough information so that these functions could actually be used.

The *Yavanajātaka* gives two other linear zigzag functions. The first, for the rising times of the zodiacal signs, is that of Babylonian System A, which also is known from Greek texts.[9] The units are called "muhūrtas," but in fact each equals ten time degrees, as shown below (*YJ* 1, 68; 79, 26).[10]

TABLE III.4

♈ 2 muhūrtas	= 20°	♓
♉ 2 2/5 muhūrtas	= 24°	♒
♊ 2 4/5 muhūrtas	= 28°	♑
♋ 3 1/5 muhūrtas	= 32°	♐
♌ 3 3/5 muhūrtas	= 36°	♏
♍ 4 muhūrtas	= 40°	♎

The second gives the length of daylight in muhūrtas when the sun is in each zodiacal sign; it employs the *Jyotiṣavedāṅga's* ratio 3:2 (Table II.4). This scheme is reproduced in Table III.5 (*YJ* 79, 31).

Table III.5

♑	12 muhūrtas	
♒	13 muhūrtas	♐
♓	14 muhūrtas	♏
♈	15 muhūrtas	♎
♉	16 muhūrtas	♍
♊	17 muhūrtas	♌
	18 muhūrtas	♋

The *Yavanajātaka's* approximate rule for the relation of the time since sunrise, t, in hours, to the length of the shadow of a twelve-digit gnomon, s, in digits (the noon shadow is denoted s_n), is (*YJ* 79, 32)[11]

$$t=\frac{6C}{s-s_n+12}, \tag{III.7}$$

where $12C$ is the length of daylight.

For the planets Sphujidhvaja repeats a Greek adaptation of Babylonian planetary theory of the Seleucid period.[12] First he establishes for each planet a yuga, or number of solar years, in which occur, approximately, a given number (Π) of synodic periods (*YJ* 79, 35–36); these relations are given in Table III.6 (compare Tables III.10–11).

TABLE III.6

Planet	Years	Π	Synodic Period
Saturn	31	30	≈377 days
Jupiter	130	120	≈395 days
Mars	32	15	≈779 days
Venus	115	72	≈583 days
Mercury	1	3	≈122 days

He then gives, with some gaps in the manuscript, the intervals in degrees (and times) between the occurrences of successive Greek-letter phenomena (*YJ* 79, 40–47); these are repeated in Table III.7 (compare Table III.13), where the symbols have the following meanings.

Superior Planets	Inferior Planets
Γ first visibility in the east	Γ first visibility in the east
Φ first stationary point	Φ stationary point in the east
	Σ last visibility in the east
	Ξ first visibility in the west
Ψ second stationary point	Ψ stationary point in the west
Ω last visibility in the west	Ω last visibility in the west

C symbolizes the conjunction of a planet with the sun.

TABLE III.7

	Saturn	Jupiter	Mars
Γ → Φ	8;15° in 112 tithis	16°	162° in 288 tithis
Φ → Ψ	−8° in 100 tithis	−8°	−34°
Ψ → Ω		21°	
Ω → Γ		6; 15°	88; 30°
Γ → Γ	12°		

Venus		Mercury	
		Σ → Ξ	48° in 16 tithis
Ξ → Ψ	258° in 208 tithis	Ξ → Ψ	16° in 8 tithis
Ψ → Φ	−24°in 48 tithis	Ψ → Φ	−x° in 24 tithis
Φ	0° in 5 tithis		
Φ → Φ′	5° in 36 tithis	Φ → Φ′	x° in 16 tithis
Φ′ → Φ″	8° in 16 tithis	Φ′ → Σ′	20° in 32 tithis
Φ″ → Ξ	(at rate of 7° in 6 tithis)	Σ′ → Σ	20° in 12 tithis

Not only is this information incomplete; one is also given no epoch positions, so that the whole is of no practical value. Yet its ultimately Babylonian origin is clear.

One final bit of planetary theory introduced by Sphujidhvaja is the elongation between each planet and the sun necessary for the planet's visibility (*YJ* 79, 50); these arcs are given in Table III.8.

TABLE III.8

Saturn	15°
Jupiter	11°
Mars	15°
Venus	8°
Mercury	15°
Moon	12°

Again the method is crude, but the values are similar to the elongations necessary for Γ found in the *Pauliśasiddhānta* (Table III.17).

In general, the *Yavanajātaka* could not (as we presently have the text, at least) have made possible the computation of the longitudes of the planets necessary for the casting of horoscopes. But the occurrence of this material in such a context gives a strong indication of the motivation of the Indians who adapted Greco-Babylonian, and later purely Greek, planetary theories.

As Vasiṣṭha is cited by Sphujidhvaja, some version of a siddhānta ascribed to him existed before A.D. 269/270. But what we know of—aside from the *Vasiṣṭhasiddhānta* of Viṣṇucandra[13] and a late *Vṛddhavasiṣṭhasiddhānta* that will be discussed in section VIII—is a work similar to *YJ* 79, but preserved only in a summary by Varāhamihira. The epoch of this version of the *Vasiṣṭhasiddhānta* was 3 December 499, although the contents in general probably are derived from a second- or third-century adaptation of a text translated from Greek.

The solar year according to Vasiṣṭha contains 6,5;15 civil days—that is, it is a Julian year. Solar motion is represented by a table of the number of quarter-days that the sun remains in each zodiacal sign (*PS* 2, 1).

Two Babylonian period relations provide the basis of Vasiṣṭha's lunar theory (*PS* 2, 2–6):

$$9 \text{ anomalistic months} = 248 \text{ civil days}$$
$$(1 \text{ anomalistic month} = 27;33,20 \text{ civil days}); \tag{III.8}$$

$$110 \text{ anomalistic months} = 3,031 \text{ civil days}$$
$$(1 \text{ anomalistic month} = 27;33,16,21,49, \ldots \text{ civil days}). \tag{III.9}$$

For each period of 3,031 days (see Table III.14) the moon's longitude increases by 5,37;32° (this implies a mean daily motion of 13;10,34,52,46, . . .°); for each anomalistic month it increases by

3;4,50° (this implies a mean daily motion of 13;10,34,43,3, . . .°). On the epoch date the moon's mean longitude was 69;7,1° and its anomaly 180°; for each period of 3,031 days thereafter one adds the first increment noted above, for each anomalistic month within the current period of 3,031 days, the second increment; the true progress in longitude of the moon during the current anomalistic month is found from the summing of a linear zigzag function of lunar daily velocity based on the parameters of Table III.9 (see Tables III.3 and III.15).

TABLE III.9

M	14;39,8,34,17, . . .°
m	11;42°
μ	13;10,34,17, . . .°
d	$\frac{1;30°}{7} \approx 0;12,51,25,42, . . .°$
Δ	$\approx$ 2;57°
p	27;33,20 civil days

From this it follows that the maximum lunar equation is 5;5°.

In its rules for finding the moon's nakṣatra, the muhūrta (here defined as 1/30 of the time that the mean moon is within a nakṣatra), and the tithi, the *Vasiṣṭhasiddhānta* employs the relations (*PS* 2,7)

$$1 \text{ nakṣatra} = 13;20° = 4/9 \cdot 30°; \qquad \text{(III.10)}$$

$$1 \text{ muhūrta} = \frac{13;20°}{30°} = 4°/9; \qquad \text{(III.11)}$$

and, since a mean tithi occurs when the elongation of the mean moon from the mean sun is 12°, each 30° of mean elongation occurs in 5/2 tithis. Vasiṣṭha (*PS* 2, 8) repeats the linear zigzag function of Sphujidhvaja for finding the length of daylight (Table III.5) and Kauṭilya's linear zigzag function for determining the length of the noon shadow (Table II.4) (*PS* 2, 9–10). What he introduces that is new is a crude formula, related to one of Sphujidhvaja's (III.7), for finding the longitude of the ascendant, λ (*H*), in zodiacal signs (*PS* 2, 11–13).

$$\left.\begin{aligned} \lambda(H) &= \lambda_{☉} + \frac{36}{12+s-s_n} && \text{before noon} \\ \lambda(H) &= \lambda_{☉} + \left(6 - \frac{36}{12+s-s_n}\right) && \text{after noon.} \end{aligned}\right\} \text{(III.12)}$$

This is based on the assumption that, at noon,

$$\lambda(H) = \lambda_{☉} + 3 \text{ signs}. \qquad \text{(III.13)}$$

It seems likely that the first sixty verses of adhyāya 17 of the *Pañcasiddhāntikā* preserve Vasiṣṭha's planetary theory. Like Sphujidhvaja's it is closely related to Babylonian material of the Seleucid period, but probably was also transmitted to India through a Greek intermediary.

Like his lunar theory described above, Vasiṣṭha's planetary theory involves computing lapsed synodic periods since epoch, and then finding the planet's progress within the current synodic period. The epoch dates and longitudes all fall within the year 505, and appear to have been computed (in the case of Venus, wrongly) by Varāhamihira. The synodic periods in days (compare Table III.3) and increments in longitude, $\overline{\Delta\lambda}$, are tabulated in Table III.10.

TABLE III.10

Planet	Synodic Period	$\overline{\Delta\lambda}$
Saturn	378 + 1/10	≈ 12;39°
Jupiter	399 − 1/9	≈ 33;9°
Mars	780 − 0;2,41	≈ 48;43°
Venus	584 − 1/11	215:50°
Mercury	115 + 7/8 + 0;0,15	≈ 114;13°

Only the $\overline{\Delta\lambda}$ of Venus is actually given in the text[14] (the Babylonian value is 215;30°). The rest can be derived from the basic period relations of Vasiṣṭha given in Table III.11; they are all attested in cuneiform texts.[15] These consist of statements that there are Π occurrences of a phenomenon (synodic periods) in *Z* sidereal rotations of the planet; I have added a column of solar years.

TABLE III.11

Planet	Π	*Z*	Years
Saturn	256	9	265
Jupiter	391	36	427
Mars	133	18	284
(Venus	720	431	1151)
Mercury	684	217	217

Vasiṣṭha omits the period relation of Venus.

If the distances between consecutive occurrences of the same phenomenon were always $\overline{\Delta\lambda}$, the longitudes of all the Π points on the ecliptic would be equidistant from each other, since

$$\frac{360°}{\Pi} = \frac{\overline{\Delta\lambda}}{Z}. \qquad \text{(III.14)}$$

However, since more Π points occur in some arcs of the ecliptic than in others, the Babylonians divided the ecliptic into a number of sections of unequal length for each planet, as does Vasiṣṭha (see Table III.12).

TABLE III.12

Saturn	3 sections
Jupiter	3 sections
Mars	6 sections
Venus	1 section
Mercury	8 sections

TABLE III.13

Saturn			Jupiter		
Γ → Γ′	1;20°	16^d	Γ → Γ′	12°	60^d
			Γ′ → Γ″	4°	40^d
Γ′ → Φ	3;52°	56^d	Γ″ → Φ	2°	24^d
Φ → Θ	−3°	55^d	Φ → Θ	−6°	56^d
Θ → Ψ	−4°	60^d	Θ → Ψ	−6°	60^d
Ψ → Ω′	8°	112^d	Ψ → Ω′	12°	80^d
Ω′ → Ω	3°	36^d	Ω′ → Ω	9°	50^d
Ω) → Γ	≈3;30°	≈43^d)	Ω → Γ	7°	29^d
(Γ → Γ	12;42°	378^d	Γ → Γ	34°	399^d)

Mars

		♓ ♈	♉ ♊	♋ ♌	♍ ♎	♏ ♐	♑ ♒
Γ → Φ	186°	267;30^d	267;30^d	267;30^d	267;30^d	267;30^d	267;30^d
Φ → Ψ	−18°	57^d	71^d	72^d	66^d	61^d	51^d
Ψ → Ω	180°	301^{d}305^d	308^{d}311^d	314^{d}311^d	309^{d}306^d	302^{d}299^d	296^d
Ω → C	30°	62^d	69^d	72^d	69^d	63^d	60^d
C → Γ	30°	62^d	69^d	72^d	69^d	63^d	60^d

Unfortunately, Varāhamihira has not understood this procedure, and his summary does not enable us to reconstruct this aspect of Vasiṣṭha's system.

Once the date and longitude of the beginning of the current synodic period have been determined, it is necessary to determine the progress of the planet within the synodic period. This is accomplished by means of schemes similar to those of Sphujidhvaja (Table III.4); they are given in Table III.13.

This sixfold division of the ecliptic is of Babylonian origin.[16] So is the second of two variant schemes for retrogression introduced into the middle of the one given above:

	♑ ♒	♓♈ and ♏♐	♉♊ and ♍♎	♋ ♌
Φ → Θ	−6° 32^d	−6° 42^d	−7° 40^d	−7° 44^d
Θ → Ψ	−9° 39^d	−10° 42^d	−10° 40^d	−11° 40^d
Φ → Ψ	−15° 57^d	−16° 60^d	−17° 63^d	−18° 66^d

The retrograde arcs in this last scheme are identical with the Babylonian System R.[17]

Venus

Ξ	74°	60^d
↓	73°	60^d
	72°	60^d
	20°	27;30^d
Ψ	3°	1;15^d
Ψ	−2°	15^d
	Ω −2° (?)	5^d
Ω → Γ	−4° (?)	10^d
Γ → Φ	−4°	20^d

This section of the text is corrupt and incomplete.

The scheme for Mercury has, in part, a direct parallel in cuneiform:[18]

Mercury

Ξ → Ω Babylonian Arc Time	Ξ → Ω Vasiṣṭha Arc Time	Ω → Γ Vasiṣṭha Arc Time
♈ 36° 36^d	35° 36^d	22° 25^d
♉ 42° 42^d	44° 45^d	17° 23^d
♊ 46° 48^d	48° 45^d	14° 20^d
♋ 42° 44^d	43° 42^d	9° 18^d
♌ 36° 38^d	34° 34^d	9° 16^d
♍ 22° 22^d	27° 26^d	9° 18^d
♎ 14° 15^d	18° 21^d	12° 20^d
♏ 14° 15^d	14° 16^d	18° 25^d
♐ 16° 16^d	15° 16^d	21° 26^d
♑ 20° 22^d	19° 20^d	28° 27^d
♒ 22° 24^d	22° 23^d	25° 26^d
♓ 22° 24^d	24° 24^d	24° 25^d

Γ → Σ Babylonian	Γ → Σ Vasiṣṭha	Σ → Ξ Vasiṣṭha
♈ 12° 14^d	21° 29^d	54° 29^d
♉ 14° 16^d	23° 23^d	69° 49^d
♊ 18° 19^d	27° 26^d	75° 47^d
♋ 22° 24^d	31° 30^d	71° 46^d
♌ 26° 27^d	32° 32^d	70° 45^d
♍ 30° 30^d	35° 33^d	70° 43^d
♎ 34° 36^d	36° 35^d	70° 40^d
♏ 44° 46^d	43° 44^d	64° 38^d
♐ 44° 46^d	43° 42^d	62° 32^d
♑ 42° 44^d	39° 38^d	58° 32^d
♒ 34° 34^d	33° 35^d	60° 35^d
♓ 24° 24^d	24° 29^d	49° 27^d

The *Romakasiddhānta* or *Roman Siddhānta* probably was originally produced in the third or fourth century; of this original version[19] we know that in it the discrepancy between the nakṣatras marking the solstices in the *Jyotiṣavedāṅga* and the zodiacal signs was explained by the theory of the precession of the equinoxes.[20] This implies the use of a tropical year; and, in the summary of Lāṭadeva's version of the *Romaka* presented by Varāhamihira, the length of the solar year is Hipparchus' 6,5;14,48^d (*PS* 1, 9–10), as it also had been at one point in the *Yavanajātaka* (Table III.2). Combining this parameter with the nineteen-year Metonic cycle, the *Romaka* obtains a yuga of 2,850 years, in which there are 35,250 synodic months and 1,040,953 civil days (*PS* 1, 15). The epoch is 21 March 505, sunset at Yavaṇapura (Alexandria), which lies 7⅓ nāḍīs (2;56 hours) west of Ujjayinī (*PS* 1, 8). Lāṭadeva corrects this by 0;43,52, . . . days; 0;45 days would correct it precisely from sunset to midnight.

Although the yuga contains 35,250 · 30 = 1,057,500 tithis and 1,040,953 civil days, in the rule for computing the ahargaṇa Lāṭadeva uses the approximation also found in the *Brāhmasphuṭasiddhānta* (*BSS* 1, 42–43):

$$703 \text{ tithis} = 692 \text{ civil days.} \qquad \text{(III.15)}$$

The only further computation of the *Romaka* about which Varāhamihira informs us is that of solar eclipses (*PS* 8). This involves first the determination of the longitudes of the sun, the moon, the lunar apogee, and the lunar node, for which the *Romaka* uses the well-known rule of proportion

$$r = c \cdot \frac{R}{C}. \qquad \text{(III.16)}$$

The ratios of *R* to *C* are given in Table III.14.

TABLE III.14

Sun	$\frac{1,0,0}{6,5,14,48}$	(Hipparchan: 1 year = 6,5;14,48 days)
Moon	$\frac{38,100}{1,040,953}$	(Metonic: 1 sidereal month = 27;19,17,45,50 . . . days)
Lunar apogee	$\frac{110}{3,031}$	(Babylonian: 110 anomalistic months = 3,031 days)
Lunar node	$\frac{24}{163,111}$	(1 rotation of the node in 6796;17,30 days)

The equations of the center for the sun and moon are found by interpolation in tables, presumably of Greek origin, in which the interval between arguments is 15°. The solar apogee is placed at 75°; the maximum solar equation, 2;23,23° (Ptolemy's is 2;23°), occurs at an anomaly of 90°, although the table is not computed by the "method of sines." The maximum lunar equation, which also occurs at an anomaly of 90°, appears to be 4;46°; one would expect 4;56°.

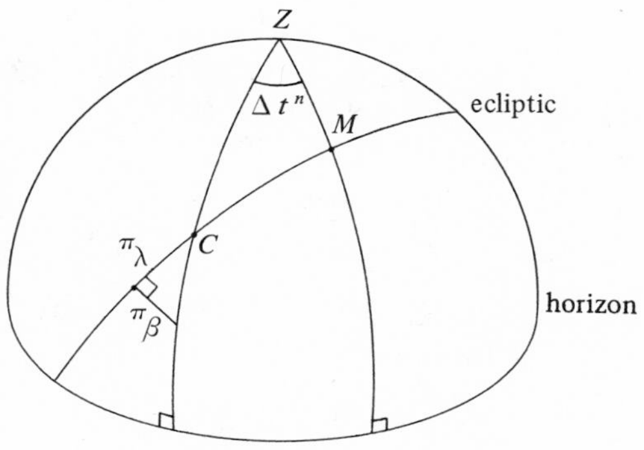

FIGURE 1

The rules for computing parallax (see Figure 1) in the *Romaka* are essentially the same as those in the *Pauliśa*; presumably both have a Greek origin. The horizontal (maximum) parallax, π_0, is assumed to be 4 nāḍīs (0;4 days), which represents about 0;49° of elongation between the sun and the moon or 0;53° of lunar motion. The maximum mean lunar parallax, according to Ptolemy, is 0;53,34° and the maximum solar parallax is 0;2,51°. It is assumed that the longitudinal parallax, π_λ, is 0 on the meridian, and varies sinusoidally between horizon and meridian. Therefore, if the depression of the conjunction from the meridian, measured in nāḍīs, is designated Δt^n and if the Radius, *R*, is taken to be 120 (this value of *R* is not necessarily that of the *Romaka*), the formula is

$$\pi_\lambda = 4 \cdot \frac{\text{Sin}\,(6\Delta t^n)}{R} = \frac{\text{Sin}\,(6\Delta t^n)}{30} \qquad \text{(III.17)}$$

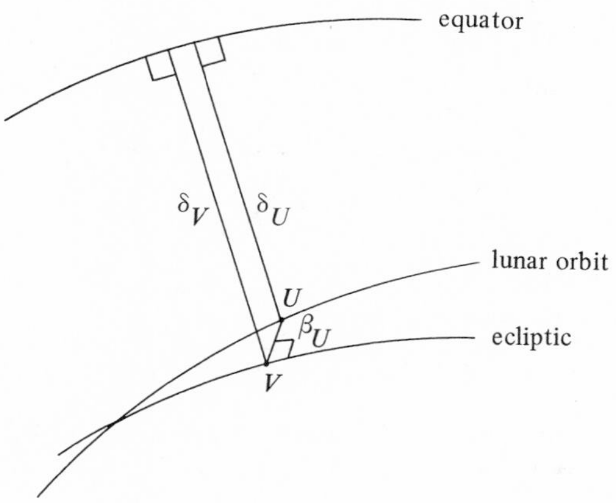

FIGURE 2

To find the latitudinal parallax, π_β, which also is assumed to have a maximum of 0;53° at the horizon, one must first determine the zenith distance of a point on the lunar orbit, U, of which the degree of longitude on the ecliptic coincides with the nonagesimal, V, of the ecliptic (see Figure 2). It is assumed that (III.18)

$$\beta_U = \frac{2}{60} \text{Sin}\, \omega, \qquad \text{(a)}$$

where ω is the distance of λ_V from the moon's node. Correctly, the formula should be

$$\text{Sin}\, \beta = \frac{\text{Sin}\, i\, .\, \text{Sin}\, \omega}{R}, \qquad \text{(b)}$$

where i is the inclination of the lunar orbit; but since β is small, the approximation

$$\text{Sin}\, \beta^\circ \approx 2\beta^\circ \qquad \text{(c)}$$

can be used if $R = 120$. Then it follows that

$$\beta_U^\circ \approx \frac{\text{Sin}\, i \cdot \text{Sin}\, \omega}{2R}, \qquad \text{(d)}$$

and then

$$\frac{\text{Sin}\, i^\circ}{2R} = \frac{2}{60} = \frac{4}{R} \qquad \text{(e)}$$

$$\text{Sin}\, i = 8 \qquad \text{(f)}$$

$$i \approx 4^\circ. \qquad \text{(g)}$$

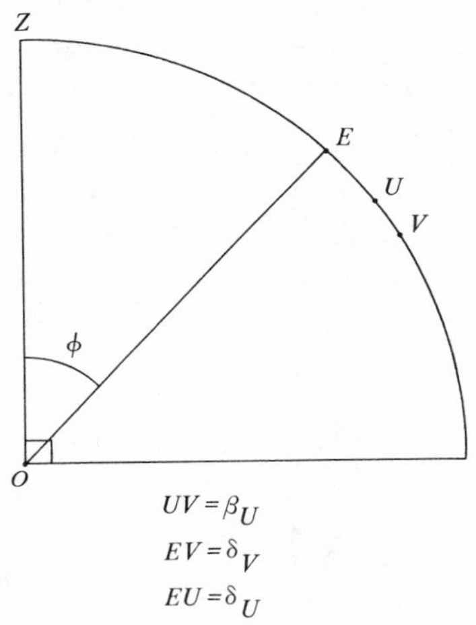

FIGURE 3

Having found β_U, one forms (see Figure 2) (δ symbolizes declination) (III.19)

$$\delta_U \approx \delta_V \mp \beta_U, \qquad \text{(a)}$$

and then (see Figure 3) (z symbolizes zenith distance):

$$z_U \approx \phi \pm \delta_U. \qquad \text{(b)}$$

This is true only if U lies on the meridian; the mistake is often repeated. The latitudinal parallax is found from

$$\pi_\beta = \frac{v_{☾} \cdot \text{Sin}\, z_U}{1800} = \frac{v_{☾} \cdot \text{Sin}\, z_U}{15R}, \qquad \text{(III.20)}$$

where $v_{☾}$ is the daily lunar velocity, $R = 120$, and 1/15 of a day is 4 nāḍīs.

The lunar latitude, $\beta_{☾}$, in minutes is found from (III.21):

$$\beta_{☾} = \frac{21}{9} \text{Sin}\, \omega. \qquad \text{(a)}$$

Again (see III. 18 d),

$$\frac{\beta_{☾}}{60} = \frac{\text{Sin}\, i}{2\text{R}} \cdot \text{Sin}\, \omega. \qquad \text{(b)}$$

$$\frac{60\, \text{Sin}\, i}{2R} = \frac{21}{9} \qquad \text{(c)}$$

$$\text{Sin}\, i = \frac{84}{9} = 9;20 \qquad \text{(d)}$$

$$i \approx 4;40^\circ. \qquad \text{(e)}$$

There is a discrepancy between (III.18g) and (III.21e). The corrected lunar latitude, β_α, is determined by

$$\beta_\alpha = \beta_{☾} \pm \pi_\beta. \qquad \text{(III.22)}$$

The mean diameters, $\bar{d}$, of the sun and moon are defined as 0;30° and 0;34°, respectively; their apparent diameters, d, vary as their velocities according to a Greek principle that was used in determining the distances of the planets:[21]

$$d = \bar{d} \cdot \frac{v}{\bar{v}} \qquad \text{(III.23)}$$

Then the half-duration of a solar eclipse, AB, is given by the formula (see Figure 4)

$$AB = \sqrt{(r_{\odot} + r_{☾})^2 - \beta_\alpha^2}, \qquad \text{(III.24)}$$

where r symbolizes the apparent radius. It is here assumed that the lunar orbit is parallel to the ecliptic. Finally, the magnitude of the eclipse at its middle, CD, is found from

$$CD = r_{\odot} + r_{☾} - \beta_\alpha. \qquad \text{(III.25)}$$

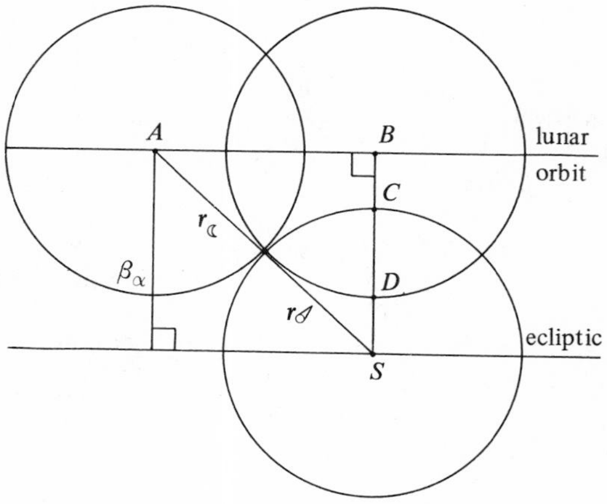

FIGURE 4

Probably in the third or fourth century yet another Greek astronomical text, the *Pauliśasiddhānta*,[22] was translated into Sanskrit; the name Pauliśa may be a transliteration of the Greek Παῦλος. This work, like the *Romaka*, was revised by Lāṭadeva, and is now known to us only through the summary of Lāṭadeva's version included in the *Pañcasiddhāntikā*.

The year in the *Pauliśa* is sidereal, as (*PS* 1, 11–13):

1 solar year = 6,5;15,30 civil days = 6,11;4,14,57 tithis = 12;22,8,29,54 synodic months (1 synodic month ≈ 29;31,48,16,37 civil days). (III.26)

This year length is attributed to the "Egyptians and Babylonians" by al-Battānī;[23] it also is expressed as a yuga in which (*PS* 3,1)

120 solar years = 43,831 civil days. (III.27)

The mean longitude of the sun can be determined from this yuga by means of (III.16). The true longitude is obtained by applying the equation of the center, found by interpolation in a table, presumably Greek in origin, in which the equations are given for mean solar longitudes at the beginnings of the zodiacal signs (*PS* 3, 2–3). The maximum equation is 1;12°; the text fails to give the necessary instruction to double these equations. The solar apogee is at 80°.

Pauliśa gives rules very similar to Vasiṣṭha's (Table III.9) for finding the longitudinal increment of the moon during what has passed of the current anomalistic month. This is in the form of a linear zigzag function for lunar velocity, the parameters of which are recorded in Table III.15 (*PS* 3, 4–9).

TABLE III.15

$M = 14;39°$
$m = 11;42°$
$\mu = 13;10,30°$
$d = \frac{1;30°}{7}$
$\Delta = 2;57°$
$p = 27;33,20$ civil days

This is based on (III.8), but is modified in a very strange and inexplicable fashion; and the maximum lunar equation in it is only 4°. Presumably Pauliśa determined the mean lunar longitude at the beginning of the anomalistic month by means of a procedure similar to Vasiṣṭha's.

The planetary theory incorrectly and incompletely expounded by Varāhamihira in *Pañcasiddhāntikā* 17, 64–80, probably is derived from the *Pauliśa*. His confusion concerning the epoch longitudes and the "ahargaṇa" makes it impossible to utilize the system as he presents it. The structure is apparent, however, and is Greco-Babylonian.

One first operates with the mean synodic arc, $\overline{\Delta\lambda}$, of each planet, expressed in the form of a ratio, $a°{:}b$ synodic periods; these mean synodic arcs, given in Table III.16, are almost identical with attested Babylonian values[24] (see Tables III.7, III.10, III.13).

TABLE III.16

Planet	a/b	$\overline{\Delta\lambda}$	Babylonian $\overline{\Delta\lambda}$
Saturn	1118/3	6,12;40°	6,12;39,22,30°
Jupiter	2752/7	6,33;8,34, . . .°	6,33;8,44,48, . . .°
Mars	3075/4	12,48;45°	12,48;43,18,29, . . .°
Venus	1151/2	9,35;30°	9,35;30°
Mercury	3312/29	1,54;12,24, . . .°	1,54;12, . . .°

Then the mean synodic arcs are divided into sections by the Greek-letter phenomena, and each section is characterized by a certain solar motion and a certain elongation of the planet from the sun. Such elongations for the occurrence of the Greek-letter phenomena are found in a Greek text ascribed to Rhetorius.[25] The elongations according to the *Pauliśa* and to "Rhetorius" are given in Table III.17.

Much of *Pañcasiddhāntikā* 3 is based on the *Pauliśasiddhānta*, although it is clear that some extraneous matter has been introduced. In any case, since all of it seems to belong to the Greco-Babylonian period, it ought to be described here. Some is similar to what we have already encountered, such as the definitions of nakṣatras and tithis (*PS* 3, 16; see III.10) and the scheme for daily solar velocity in which $M = 1;1°$ and $m = 0;57°$ (*PS* 3, 17; see Table III.3)[26]. Others are new.

TABLE III.17

	Saturn			Jupiter			Mars		
	Sun	Pauliśa Elongation Increment	"Rhetorius" Elongation	Sun	Pauliśa Elongation Increment	"Rhetorius" Elongation	Sun	Pauliśa Elongation Increment	"Rhetorius" Elongation
C → Γ	18°	16;30°	10°	16°	12°	10°	36°	15°	10°
Γ	98°	90;30°		54°	44°		188°	60°	
Φ	14°	13°(120°)	120°	70°	64°(120°)	120°	108°	60°(135°)	120°
Φ → Ψ	113°	120°(240°)	240°	109°	120°(240°)	240°	72°	90°(225°)	240°
Ψ	98°	91°		88°	76°		68°	50°	
Ω	13°	12;30°(343;30°)	350°	40°	32°(348°)	350°	240°	70°(345°)	350°
Ω → C	19°	16;30°		16°	12°		56°	15°	
Totals	373°	360°		393°	360°		768°	360°	

	Venus				Mercury		
	Sun	Pauliśa Elongation Increment	"Rhetorius" Elongation		Sun	Pauliśa Elongation Increment	"Rhetorius" Elongation
C → Γ	5°	−9°	−5°	C → Γ	10°	−12°	−3°
Γ → Φ	15°	−21°(−30°)	−40°	Γ → Φ	14°	−5°	
Φ	208°	15°		Φ → Σ	18°	14°(−3°)	−2°
Σ	12°	5°(−10°)	−2°	Σ → Ξ	30°	6°(3°)	3°
Σ → S	48°	10°		Ξ → Ψ	18°	14°	
				Ψ → Ω	16°	−8°(9°)	2°
				Ω → C	8°	−9°	
Totals	288°	0°			114°	0°	

A karaṇa is 1/60 of a synodic month, or half a tithi. Those near conjunction have special names; the other fifty-six form eight series of seven each, as in Table III.18 (*PS* 3, 18–19).

TABLE III.18

1	Kiṃstughna
2,9,16,23,30,37,44,51	Bava
3,10,17,24,31,38,45,52	Bālava
4,11,18,25,32,39,46,53	Kaulava
5,12,19,26,33,40,47,54	Taitila
6,13,20,27,34,41,48,55	Gara
7,14,21,28,35,42,49,56	Vaṇij
8,15,22,29,36,43,50,57	Viṣṭi
58	Śakuni
59	Catuṣpada
60	Nāga

There are two special configurations of the sun and the moon that are called pātas (*PS* 3, 20–22). In this earliest text defining them, they are vaidhṛta, when the sun and the moon are equidistant from and on opposite sides of an equinox, so that

$$\lambda_{\odot} + \lambda_{☾} = 360°; \qquad \text{(III.28)}$$

and vyatipāta, when the sun and the moon are equidistant from and on opposite sides of a solstice, so that

$$\lambda_{\odot} + \lambda_{☾} = 180°. \qquad \text{(III.29)}$$

Varāhamihira introduces into this simple scheme the idea of a trepidation of the solstices and equinoxes—an idea that he evidently derived from a Greek source, and then modified.[27]

The ṣaḍaśītimukhas divide the ecliptic into four equal arcs of 86° each and one remaining arc of 16° (*PS* 3, 23–24). The beginnings of these arcs are at Libra 0°, Sagittarius 26°, Pisces 22°, Gemini 18°, and Virgo 14°. The origin and purpose of the ṣaḍaśītimukhas remain obscure.

The saṅkrānti is the time in nāḍīs, t^n, during which the sun passes a boundary between two zodiacal signs (*PS* 3, 26). It is found from

$$t^n = 60 \cdot \frac{d_{\odot}}{v_{\odot}}, \qquad \text{(III.30)}$$

where $d_{\odot}$ is the diameter of the sun in minutes and $v_{\odot}$ its daily motion in minutes per day.

Previous Indian methods of determining the length of daylight depended on simple linear zigzag functions (Tables II.1 and III.5); the *Pauliśa* gives a series of three coefficients that, when multiplied by the noon equinoctial shadow, s_0, give the differences between the lengths of daylight in vināḍīs for the moments when the sun's longitude is 0°, 30°, 60°, and 90°; symmetry will provide the lengths at other solar longitudes (*PS* 3, 10–11). These coefficients are given in Table III.19.

TABLE III.19

$\gamma_1 = 20$ vināḍīs
$\gamma_2 = 16;30$ vināḍīs
$\gamma_3 = 6;45$ vināḍīs

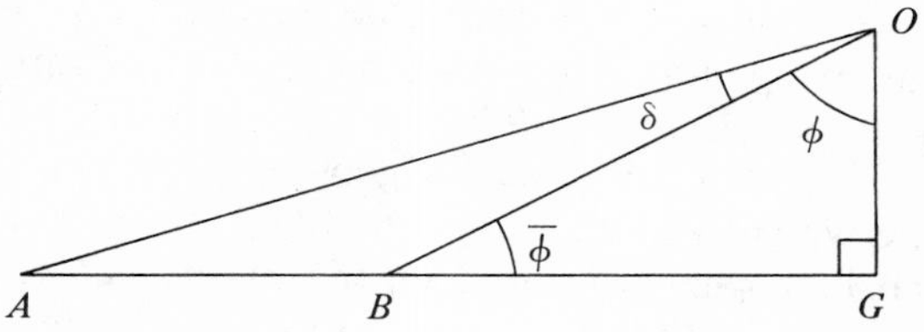

$OG = g$
$BG = s_o$
$AG = s_n$
$OB = h_o$
$AO = h_n$

FIGURE 5

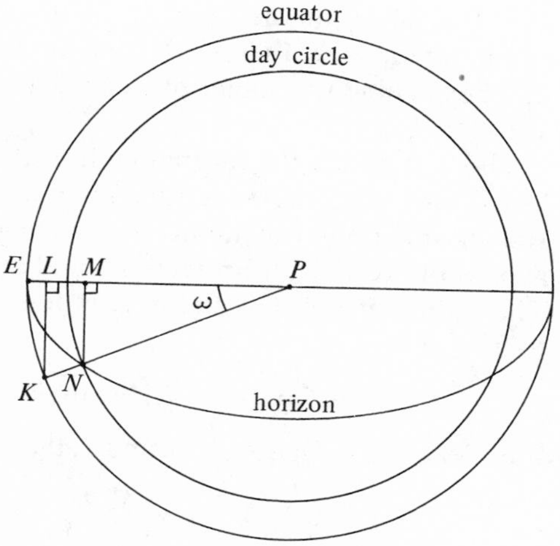

$MN = \text{Sin}_r\omega = e$
$LK = \text{Sin}\,\omega$

(projection in the plane of the equator)

FIGURE 7

$NE = \text{Cos}\,\delta\,(60^\circ)$
$NN_e = EO = \text{Sin}\,\delta\,(60^\circ)$
$N_e S = \text{Vers.}\,\delta\,(60^\circ)$
$OL_e = \text{Sin}\,\omega\,(60^\circ)$
$EL = \text{Sin}_r\,\omega\,(60^\circ) = e$
$ON = OS = R$
$FO = \text{Sin}\,\epsilon$
$FC = \text{Cos}\,\epsilon$

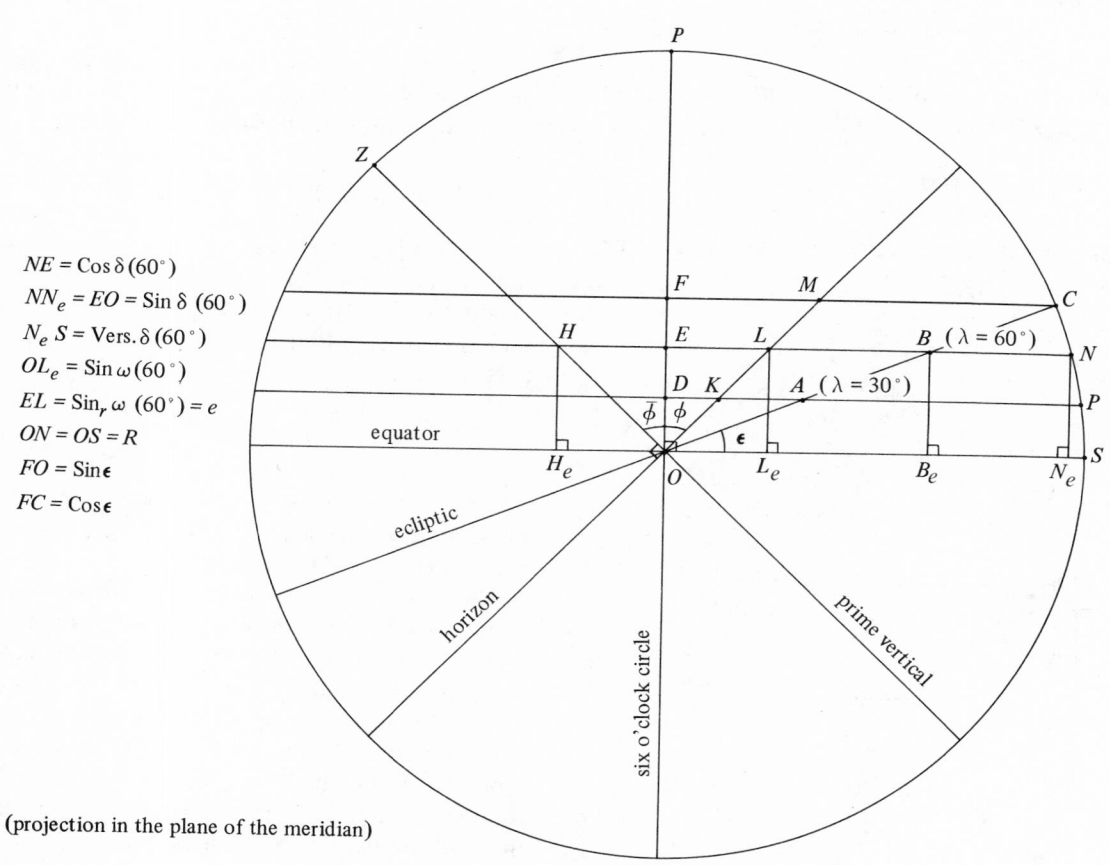

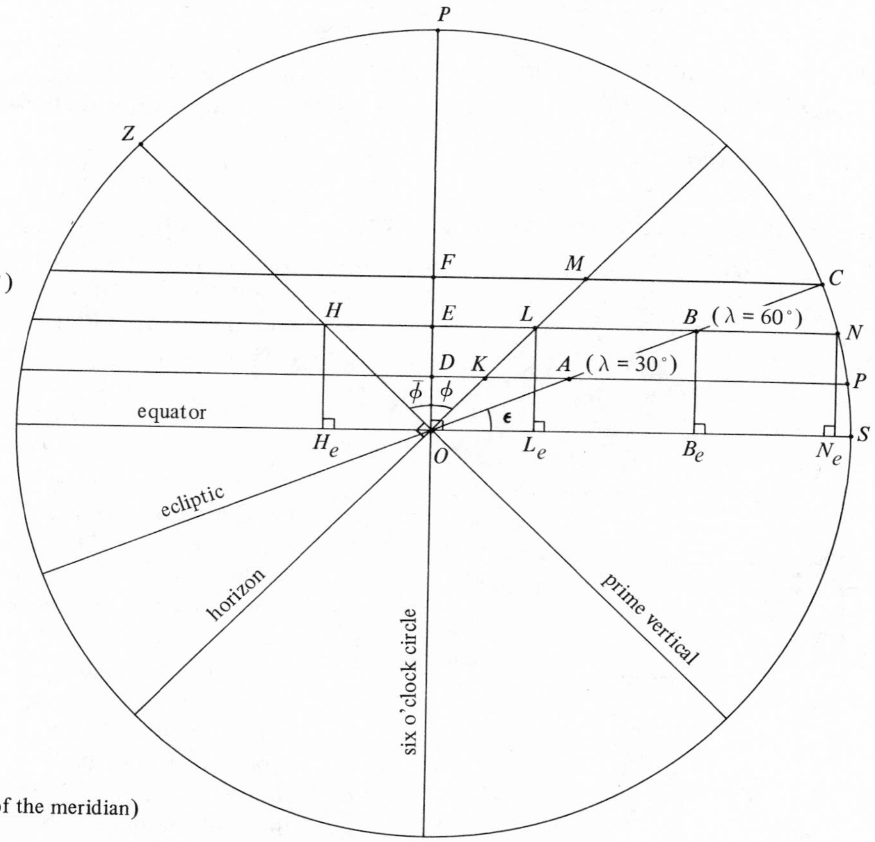

(projection in the plane of the meridian)

FIGURE 6

The use of s_0 implies that ϕ is involved in the computation (see Figure 5), and the computation that lies behind the values of γ can be reconstructed as follows.

The day circle, the diurnal path of the sun parallel to the equator, is a fundamental concept of Indian spherical trigonometry, which operates, if it possibly can, with projections and analemmas.[28] The radius of the day circle, r, is found from (see Figure 6)

$$r = \text{Cos}\,\delta = \sqrt{R^2 - \text{Sin}^2\,\delta}. \qquad \text{(III.31)}$$

It is clear, from Figures 5, 6, and 7, that

$$\text{Sin}_r\,\omega = \text{Sin}\,\delta \cdot \frac{s_0}{g} = e, \qquad \text{(III.32)}$$

where ω is half of the equation of daylight and $\text{Sin}_r\,\omega$ is what is later called the "earth sine," e. From (III.32) it follows that

$$\text{Sin}\,\omega = e \cdot \frac{R}{r}. \qquad \text{(III.33)}$$

From this and (III.18 c) we obtain

$$2\omega^\circ = s_0 \cdot \frac{\text{Sin}\,\delta}{g} \cdot \frac{R}{r}. \qquad \text{(III.34)}$$

And, since $\omega^\circ = 10\,\omega$ vināḍīs, the values of γ should come from this formula

$$\gamma = 10 \cdot \frac{\text{Sin}\,\delta}{g} \cdot \frac{R}{r}; \qquad \text{(III.35)}$$

in fact, this gives results very close to the attested numbers.

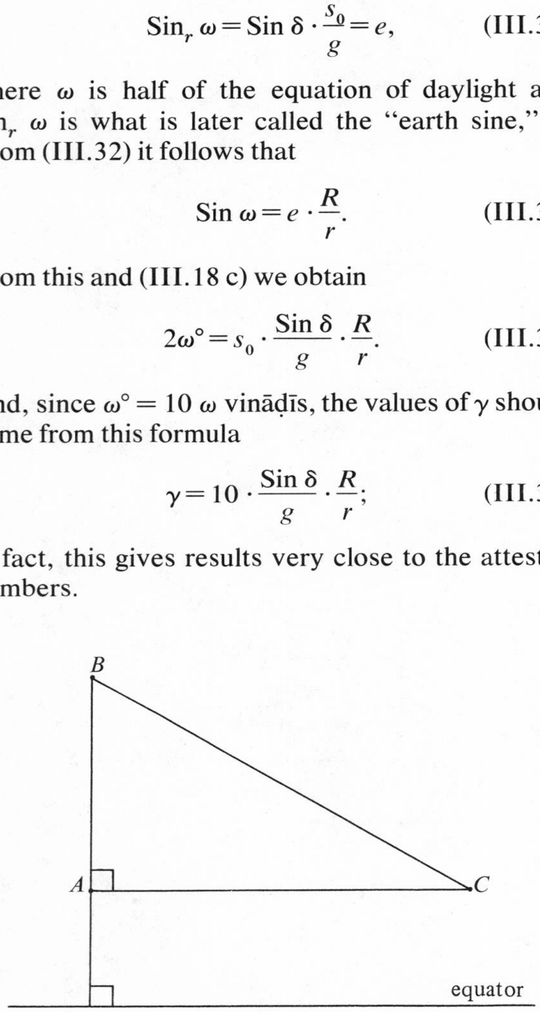

$BC = d$

$BA = \Delta\phi$

$AC = \Delta l$

FIGURE 8

Another, but less successful, attempt to solve a problem in spherical trigonometry occurs in this section of the *Pañcasiddhāntikā* (*PS* 3, 14). The deśāntara is the difference in geographical longitudes between two localities, B and C. If the direct distance in yojanas, d, between the two is known (a virtual impossibility in antiquity) and the terrestrial latitude of each, ϕ_B and ϕ_C, is also known, so that $\Delta\phi = \phi_B - \phi_C$, then the computation is the following (see Figure 8). It is assumed that the circumference of the earth is 3,200 yojanas (as in the Ārdharātrikapakṣa), so that

$$d^\circ = d^y \cdot \frac{360^\circ}{3{,}200} = d^y \cdot \frac{9}{80} \qquad \text{(III.36)}$$

and

$$\Delta l = \sqrt{\left(\frac{9d}{80}\right)^2 - \Delta\phi^2}. \qquad \text{(III.37)}$$

Even though Δl is not an equatorial arc unless C lies on the equator, the text converts it into a time difference in nāḍīs by

$$\Delta t^n = \frac{\Delta l}{6}. \qquad \text{(III.38)}$$

A final section of *Pañcasiddhāntikā* 3, which probably is not connected with Pauliśa, concerns the computation of the moon's latitude, $\beta_{☾}$ (*PS* 3, 30–31). The retrograde motion of the lunar node from its given epoch position on 22 March 505 is stated to be 8° in 151 civil days, to which must be added a bīja (correction) of 0;1° for every revolution; in daily motion this is 0;3,10,44,14,18, . . .°. When the longitude of the node has been found, and thence its elongation from the Moon, ω°, the text states that

$$\beta_{☾} = \omega^\circ \cdot \frac{280}{90^\circ}. \qquad \text{(III.39)}$$

We have seen the maximum lunar latitude of 4;40° previously in the *Romaka* (III.21 e), although the *Romaka* does not use (III.39).

The eclipse theory of the *Pauliśasiddhānta* is summarized in *Pañcasiddhāntikā* 6 (lunar) and 7 (solar). For lunar eclipses, it is first necessary to compute the longitudes of the sun, the moon, and the center of the earth's shadow at sunrise in the usual manner (*PS* 6, 1). Then the distance of the moon from the shadow's center is converted into a time difference in nāḍīs, Δt^n, on the assumption that the elongation (which is equal, of course, to the elongation between the moon and the sun increased by 180°) increases at a rate of 12° per day or 0;12° per nāḍī. The longitudes of the several bodies and of the lunar node are then computed for the new time of opposition. A lunar eclipse will occur if the moon is less than 13;36° from its node, and a darkening (penumbra) if it lies between 13° and 15° from the node (*PS* 6, 2). The given eclipse boundary, 13;36°, implies that

$$i \approx 4°, \qquad \text{(III.40)}$$

that is, the value of (III.18 g).

The *Pauliśa*'s rule for the duration of a lunar eclipse in degrees is similar to (III.24) (*PS* 6, 3):

$$AB = \sqrt{(r_u + r_{☾})^2 - \beta_{☾}^2}\,, \qquad \text{(III.41)}$$

where $r_u = 0;38°$ is the radius of the earth's shadow and $r_{☾} = 0;17°$. This is converted into time, Δt, by

$$\Delta t = \frac{2\,AB}{v_{☾} - v_{☊}}. \qquad \text{(III.42)}$$

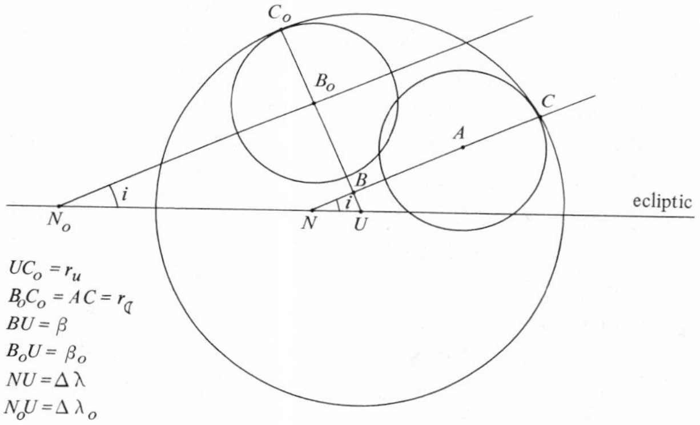

FIGURE 9

The derivation of the formula for the duration of totality of a lunar eclipse, τ, is far more complicated. That formula is (*PS* 6, 5)

$$\tau = \frac{21}{5}\sqrt{4(5 - \Delta\lambda)\,(10 - [5 - \Delta\lambda])}\,, \qquad \text{(III.43)}$$

where $\Delta\lambda$ is the elongation of the node from the center of the shadow, U (see Figure 9). Taking the example where the moon is totally eclipsed for a moment with its center at B_0, and assuming that a small spherical triangle is plane, it is clear that (III.44)

$$\text{Sin}\, i = \beta_0 \cdot \frac{R}{\Delta\lambda_0} \qquad \text{(a)}$$

$$\beta_0 = r_u - r_{☾} = 0;21° \qquad \text{(b)}$$

$$\text{Sin}\, i = 0;21 \cdot \frac{120}{\Delta\lambda_0} \qquad \text{(c)}$$

$$\text{Sin}\, i = \text{Sin}\, 4° \approx 8;22 \qquad \text{(d)}$$

$$\Delta\lambda_0 = 301' \approx 5° \qquad \text{(e)}$$

$$\text{Sin}\, i = \frac{2{,}0 \cdot 0;21}{5} = \frac{2 \cdot 21}{5}. \qquad \text{(f)}$$

Further, combining (III.41) with this, we find that

$$AB = \frac{\text{Sin}\, i}{R}\sqrt{\Delta\lambda_0^2 - \Delta\lambda^2}. \qquad \text{(III.45)}$$

The duration of totality in minutes, τ, is equal to $2AB \cdot 60$; therefore, substituting in (III.45) the values of Sin i and $\Delta\lambda_0$ in (III.44e and f), we have

$$\tau = \frac{2 \cdot 21}{5}\sqrt{5^2 - \Delta\lambda^2}. \qquad \text{(III.46)}$$

This is another way of expressing (III.43).

A characteristic of Indian eclipse computations, which originally may have been motivated by some omens in the Sin section of the Babylonian omen series *Enūma Anu Enlil*,[29] is the projection of the eclipse that permits one to determine the directions of points on the circumferences of the sun or the moon at different phases relative to an "east-west" line that is perpendicular to the great circle passing through the center of the eclipsed body and the north and south points on the local horizon. It is assumed that the angle between the equator and the "east-west" line, the "deflection," γ, has two components. The first, or akṣavalana, depends on ϕ; the second, the ayanavalana, on δ. Pauliśa, as preserved by Varāhamihira, gives a rule only for the akṣavalana, γ_1 (*PS* 6, 8):

$$\gamma_1 = \phi \cdot \frac{t^0}{90}, \qquad \text{(III.47)}$$

where t^0 is the "depression" of the eclipsed body from the meridian—that is, the hour angle in time degrees. For γ_1 one assumes that the ecliptic and equator coincide, so that at midheaven $\gamma_1 = 0$ and at the horizon $\gamma_1 = \phi$.

Another eclipse projection is preserved by Varāhamihira (*PS* 6, 12–15). In this, thirteen lines are drawn parallel to the lunar orbit, their limits being the parallels that pass through the lunar node and through the center of the shadow; they represent the eclipse limit of 13°, and can be used to measure the magnitude of the eclipse. In addition, the direction of the phases with respect to the lunar orbit can be determined by graphic means utilizing three concentric circles with radii r_u, $r_☾$, and $r_u + r_☾$ and, by the use of diameters, reflecting lunar positions onto a quadrant of the innermost circle, which represents the moon. Again, the original motivation was probably to establish omen criteria.

One final section of this chapter on lunar eclipses is also related to omens of a type found in *Enūma Anu Enlil.*[30] This involves criteria for determining the direction of impact and the color of the eclipsed body. The latter in the *Pauliśasiddhānta* involves the altitude of the eclipsed body, its relation to the ascendant or descendant, and its magnitude; the colors are blood-red, reddish-brown, variegated, smoke-colored, and cloud-colored (*PS* 6, 9–10). The later texts that deal with the colors of lunar eclipses make them depend on the phase (see Table V.19).

Pauliśa's computation of solar eclipses is essentially identical with that of the *Romaka*, except that the radii of both the sun and the moon are assumed to be 0;17°; the inclination of the lunar orbit is 4°, as in (III.40). The eclipse limit for a solar eclipse is 8;36° (*PS* 7, 5). Longitudinal parallax is found by (III.17) (*PS* 7, 1). Latitudinal parallax is determined by a formula equivalent to (III.21); this is (*PS* 7, 2)

$$\pi_\beta = 4^n \cdot \frac{5}{23} \cdot \frac{\operatorname{Sin} z_U}{2}, \qquad \text{(III.48)}$$

where $\frac{5}{23} \approx 0;13 \approx v_☾$ and $\frac{4^n}{2} = \frac{1^d}{15\,R}$.

The rule for finding the duration of the eclipse in nāḍīs, t^n, is (*PS* 7, 6)

$$t^n = \frac{3}{4}\sqrt{\Delta\lambda_0^2 - \Delta\lambda^2}, \qquad \text{(III.49)}$$

where $\Delta\lambda_0$ is the appropriate eclipse limit. We have already seen that the duration of totality in minutes, τ, is given by (III.43). To convert minutes of arc into nāḍīs, we divide by 0;12° of elongation per nāḍī, which produces

$$t^n = \frac{2 \cdot 21}{5} \cdot \frac{1}{12}\sqrt{\Delta\lambda_0^2 - \Delta\lambda^0} = \frac{7}{10}\sqrt{\Delta\lambda_0^2 - \Delta\lambda^2}. \qquad \text{(III.50)}$$

Pauliśa replaces 7/10 by 3/4 for indiscernible reasons.

This completes the analysis of all that can be definitely attributed to this phase of the development of Indian astronomy. It is characterized by the use of essentially Babylonian techniques for determining lunar and planetary positions, although they are modified, presumably by Greek intermediaries; there exist, in fact, Greek parallels in the texts ascribed to Rhetorius (Table III.17) and to Heliodorus (Table V.9). Other elements are survivals from the Babylonian period (such as daylight and noon-shadow schemes), and yet others purely Indian developments (such as karaṇas and pātas). Two of the most interesting areas, however, reflect Greek astronomical traditions of which we otherwise are little informed. These are in the development of projections to solve problems in spherical trigonometry (employing the Sine function) and the rough methods of predicting eclipses found in the *Romaka* and *Pauliśa*, which also involve the Sine function and material related to the interpretation of Babylonian omen literature. Because they fit in with these general characteristics, although no authors are named, chapters 4, 5, and 14 of the *Pañcasiddhāntikā* can be regarded as largely derived from the texts produced in the Greco-Babylonian period. The possibility remains, however, that some of the more sophisticated methods in these chapters come from the later Greek phase—for instance, from the *Sūryasiddhānta* of Lāṭadeva.

The Sine table found in chapter 4 (*PS* 4, 6–15) is based on $R = 120$—that is, double the value used in Ptolemy's Chord table. The intervals in the argument are 90°/24 = 3;45°, as became standard in Indian Sine tables; this may be related to the Hellenistic βαϑμοί (15°)[31] halved and halved again. This Sine table is reproduced in Table III.20; there are added, for the sake of comparisons, columns showing $R \sin \vartheta$ and $\operatorname{Sin} \vartheta$ from Ptolemy's Chord table ($\operatorname{Sin}_{120}\vartheta = \operatorname{Chrd}_{60} 2\vartheta$).

The relation of the diameter of a circle, d, to its circumference, c, is given as the crude (*PS* 4, 1; see [V.1])

$$d = \sqrt{\frac{c^2}{10}}. \qquad \text{(III.50)}$$

The following formulas for deriving Sines (*PS* 4, 1–5) are presented (III.51):

$$\operatorname{Sin} 60° = \sqrt{R^2 - \frac{R^2}{4}} \qquad \text{(a)}$$

$$\operatorname{Sin} 45° = \sqrt{\frac{R^2}{2}} \qquad \text{(b)}$$

$$\operatorname{Sin} 30° = \sqrt{\frac{R^2}{4}} \qquad \text{(c)}$$

$$\operatorname{Sin}^2 \vartheta = \left(\frac{R - \operatorname{Sin}(90 - 2\vartheta)}{2}\right)^2 + \left(\frac{\operatorname{Sin} 2\vartheta}{2}\right)^2 \qquad \text{(d)}$$

TABLE III.20

ϑ	Sin ϑ	R sin ϑ	Chrd_{60} 2ϑ (Ptolemy)
3;45°	7;51	7;50,54	7;50,54
7;30°	15;40	15;39,48	15;39,47
11;15°	23;25	23;24,40	23;24,39
15°	31;4	31;3,30	31;3,30
18;45°	38;34	38;34,22	38;34,22
22;30°	45;56	*45;55,20*	*45;55,19*
26;15°	53;5	*53;4,28*	*53;4,29*
30°	60	60	60
33;45°	66;40	66;40,6	66;40,7
37;30°	73;3	73;3,4	73;3,5
41;15°	79;7	79;7,18	79;7,18
45°	84;51	84;51,10	84;51,10
48;45°	90;13	90;13,14	90;13,15
52;30°	95;12	95;12,8	95;12,9
56;15°	99;46	*99;46,34*	*99;46,35*
60°	103;55	103;55,22	103;55,23
63;45°	107;37	107;37,30	107;37,30
67;30°	110;52	110;51,56	110;51,57
71;15°	113;37	*113;37,54*	*113;37,54*
75°	115;55	115;54,40	115;54,40
78;45°	117;42	117;41,40	117;41,40
82;30°	118;59	*118;58,24*	*118;58,25*
86;15°	119;44	*119;44,36*	*119;44,35*
90°	120	120	120

The text gives also an inaccurately computed table of the differences between declinations at intervals of 7;30° (*PS* 4, 16–18); in it $\epsilon = 23;40°$, although elsewhere one finds the rounded and more normal Indian value $\epsilon = 24°$ (*PS* 4, 23–25).

The cardinal directions are determined by describing a circle about a gnomon ($g = 12$), marking the points where the shadow enters and leaves this circle (it is assumed that there is no change in solar declination), and bisecting the line connecting those two points by means of intersecting arcs (*PS* 4, 19). The relations to be derived from observed noon shadows of this gnomon (*PS* 4, 20–23) (see Figures 5 and 10) are (III.52)

$$\text{Sin}\,\phi = s_0 \cdot \frac{R}{\sqrt{s_0^2 + g^2}}; \qquad \text{(a)}$$

$$s_n = \text{Sin}\,(\phi \pm \delta) \cdot \frac{g}{\sqrt{R^2 - \text{Sin}^2\,(\phi \pm \delta)}}, \qquad \text{(b)}$$

where s_n is the noon shadow; and

$$\text{Sin}\,\bar{\phi} = \sqrt{R^2 - \text{Sin}^2\,\phi}. \qquad \text{(c)}$$

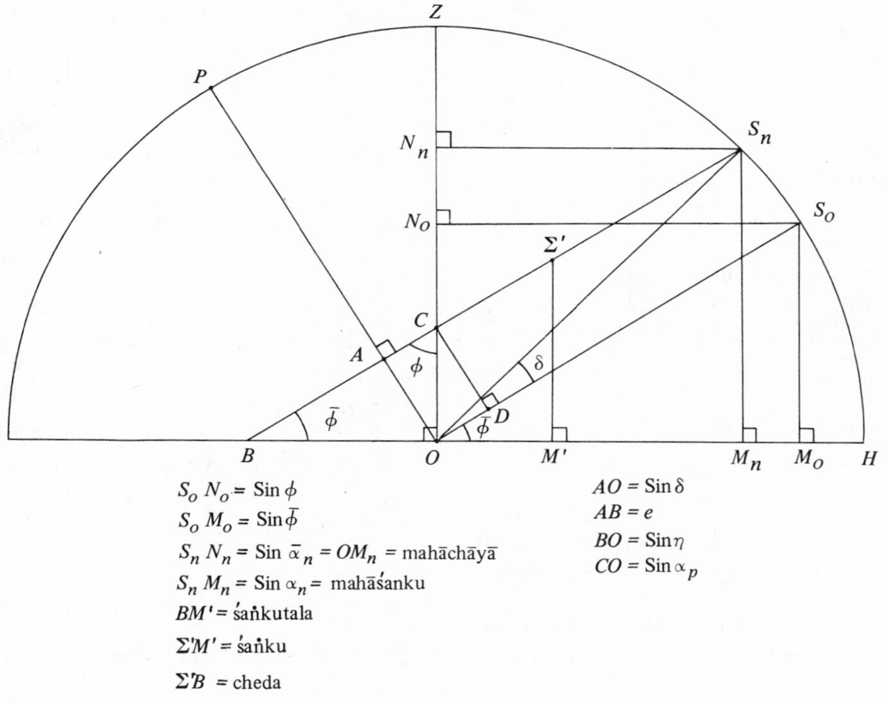

(projection in the plane of the meridian)

FIGURE 10

Other rules are (see Figure 10) (III.53)

$$\text{Sin}\,\bar{\alpha}_n = s_n \cdot \frac{R}{\sqrt{s_n^2 + g^2}} \qquad \text{(a)}$$

where α is the altitude of the sun, $\bar{\alpha}$ its coaltitude; and

$$\phi = \bar{\alpha}_n \pm \delta. \qquad \text{(b)}$$

The altitude of the sun when it is on the prime vertical, α_p, can also be determined from (Figure 10) (*PS* 4, 32–33)

$$\text{Sin}\,\alpha_p = R \cdot \frac{\text{Sin}\,\delta}{\text{Sin}\,\phi}. \qquad \text{(III.54)}$$

And, since (see Figure 6)

$$\text{Sin}\,\delta = \frac{\text{Sin}\,\lambda \cdot \text{Sin}\,\epsilon}{R}, \qquad \text{(III.55)}$$

it follows also that (*PS* 4, 35–36)

$$\text{Sin}\,\alpha_p = \frac{\text{Sin}\,\lambda \cdot \text{Sin}\,\epsilon}{\text{Sin}\,\phi}. \qquad \text{(III.56)}$$

The right ascensions of ecliptic arcs, α, are found by the formula (*PS* 4, 29–30)

$$\text{Sin}\,\alpha = \frac{2R}{d} \cdot \sqrt{\text{Sin}^2\,\lambda - \text{Sin}^2\,\delta}. \qquad \text{(III.57)}$$

This follows from Figure 6, where, for example,

$$\text{Sin}\,\alpha\,(60°) = B_eO = BE \cdot \frac{R}{r}, \qquad \text{(III.58)}$$

which is equivalent to (III.57).

The oblique ascension, ρ, is found from (*PS* 4, 31)

$$\rho = \alpha \mp \omega. \qquad \text{(III.59)}$$

The text tabulates the values of Sin δ, d, and α for λ = 30°, λ = 60°, and λ = 90° (R = 120 and ϵ = 24°) that are given in Table III.21 (compare Table V.17) (*PS* 4, 23–25, 29–30).

TABLE III.21

λ	Sin δ	d	α
30°	24;24	235	278 vināḍīs
60°	42;15	224	299 vināḍīs
90°	48;48	219	323 vināḍīs

We are also given (*PS* 4, 26) a rule for computing ω measured in vināḍīs that is equivalent to (III.33):

$$2\omega^v = 60 \cdot \frac{R \cdot \text{Sin}\,\phi \cdot \text{Sin}\,\delta}{3d \cdot \text{Sin}\,\bar{\phi}}, \qquad \text{(III.60)}$$

and another, equivalent formula where ω is measured in nāḍīs (*PS* 4, 34):

$$2\omega^n = \frac{2R \cdot \text{Sin}\,\phi \cdot \text{Sin}\,\delta}{6d \cdot \text{Sin}\,\bar{\phi}}. \qquad \text{(III.61)}$$

A concept frequently used in Indian astronomy is that of the "earth sine," e (see III.32). From Figure 10 it is clear that (*PS* 4, 27–38)

$$\text{Sin}\,\phi = e \cdot \frac{R}{\sqrt{e^2 + \text{Sin}^2\,\delta}}. \qquad \text{(III.62)}$$

From the same figure it is obvious that the Sine of the rising amplitude of the sun, Sin η, which is BO, can be found from (*PS* 4, 39–40)

$$\text{Sin}\,\eta = R \cdot \frac{\text{Sin}\,\delta}{\text{Sin}\,\bar{\phi}}, \qquad \text{(III.63)}$$

and thence that

$$\text{Sin}\,\bar{\phi} = \text{Sin}\,\delta \cdot \frac{R}{\text{Sin}\,\eta}. \qquad \text{(III.64)}$$

Varāhamihira gives the following new rules involving gnomon shadows (*PS* 4, 41–44):

$$\text{Sin}\,\alpha = \frac{r \cdot \text{Sin}\,\bar{\phi}}{R^2} \cdot (\text{Sin}\,t' \pm \text{Sin}\,\omega), \qquad \text{(III.65)}$$

where $t' = 6t^n \mp \omega°$ and t^n is the nāḍīs elapsed since sunrise or to come until sunset. Thus, in Figure 6, for λ = 60°, when the sun is on the prime vertical, Sin $6t^n$ is H_eL_e and Sin ω^0 is L_eO. Hence (III.66)

$$HE = (\text{Sin}\,t' - \text{Sin}\,\omega) \cdot \frac{r}{R} \qquad \text{(a)}$$

$$HO = \text{Sin}\,\alpha = \text{Sin}\,\bar{\phi} \cdot \frac{HE}{R}, \qquad \text{(b)}$$

which is (III.65). Then, from Figure 11:

$$s = \sqrt{R^2 - \text{Sin}^2\,\alpha} \cdot \frac{g}{\text{Sin}\,\alpha}; \qquad \text{(III.67)}$$

Varāhamihira also presents the inverse of this procedure, which allows one to compute t^n from s (*PS* 4, 45–47).

A more complicated analemma involves the triangle $\Sigma'BM'$ in Figure 10 (*PS* 4, 52–54). In this the śaṅku (the Sine of altitude, Sin α) is $\Sigma'M'$ (which equals ΣM in Figure 11); it is found from

$$\text{Sin}\,\alpha = g \cdot \frac{R}{\sqrt{s^2 + g^2}}. \qquad \text{(III.68)}$$

Then the śaṅkutala, which is BM', is found from

$$\text{śaṅkutala} = \text{Sin}\,\alpha \cdot \frac{\text{Sin}\,\phi}{\text{Sin}\,\bar{\phi}}. \qquad \text{(III.69)}$$

Furthermore, combining (III.55) and (III.63), we have

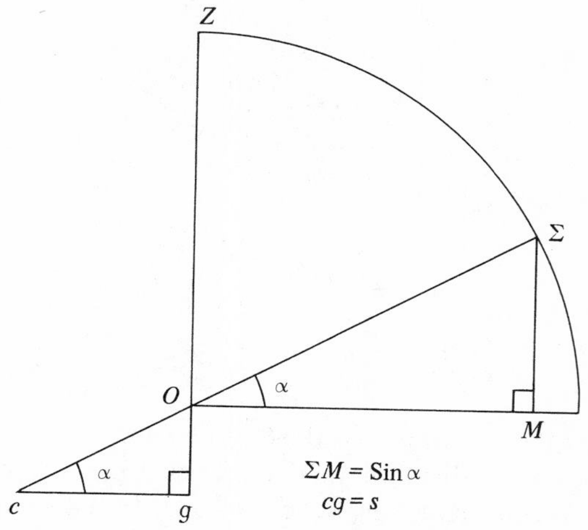

(projection in the plane of the sun's altitude circle)

FIGURE 11

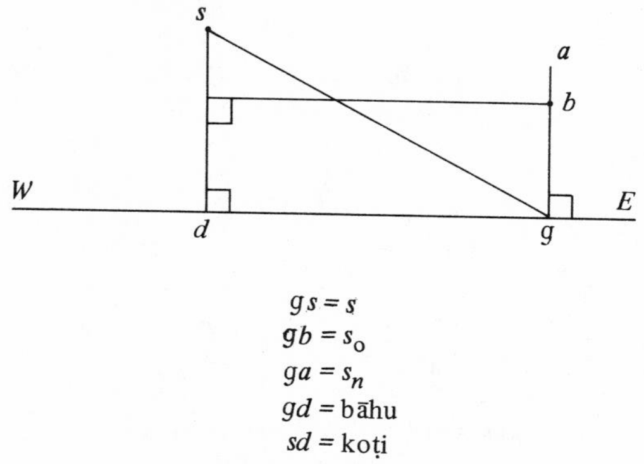

FIGURE 12

$$\text{Sin}\ \eta = \frac{\text{Sin}\ \lambda \cdot \text{Sin}\ \epsilon}{\text{Sin}\ \bar{\phi}}, \qquad \text{(III.70)}$$

and then

$$\text{koṭi} = (\text{śaṅkutala} \pm \text{Sin}\ \eta) \cdot \frac{h_s}{R}, \qquad \text{(III.71)}$$

where the koṭi is the perpendicular distance of the tip of the shadow from the east-west line running through the base of the gnomon (see Figure 12) and

$$h_s = \sqrt{s^2 + g^2}. \qquad \text{(III.72)}$$

Pañcasiddhāntikā 5 is devoted to the problem of lunar visibility, in the solution of which appears the first approximation to what became known as the ayanadṛkkarma, which is the computation of the longitudinal difference, $\lambda\lambda^*$, between the ecliptic longitude and the polar longitude of the moon, a star-planet, or a star (see V.33). If $\Delta\lambda = \lambda_{☾} - \lambda_{\star}$ or $\lambda_{\star} - \lambda_{☾}$ and $\Delta\delta = \delta_{☾} - \delta_{\star}$ or $\delta_{\star} - \delta_{☾}$ (see Figure 13), then (*PS* 5, 1–3)

$$\Delta\alpha = \sqrt{\Delta\lambda^2 - \Delta\delta^2}, \qquad \text{(III.73)}$$

since the small spherical triangle may be considered to be plane. Further, under the same consideration,

$$\lambda\lambda^* = \beta_{☾} \cdot \frac{\Delta\delta}{\Delta\alpha}. \qquad \text{(III.74)}$$

Then, clearly,

$$\Delta\lambda^* = \Delta\lambda \pm \lambda\lambda^*, \qquad \text{(III.75)}$$

and its right ascension is $\Delta\alpha^*$. The moon is visible, according to the text, if $\Delta\alpha^* \geq 2$ nāḍīs $= 12°$.

The sickle of the lunar crescent, σ, is measured in units of which the moon's diameter contains fifteen. Then the number of these units that are illuminated

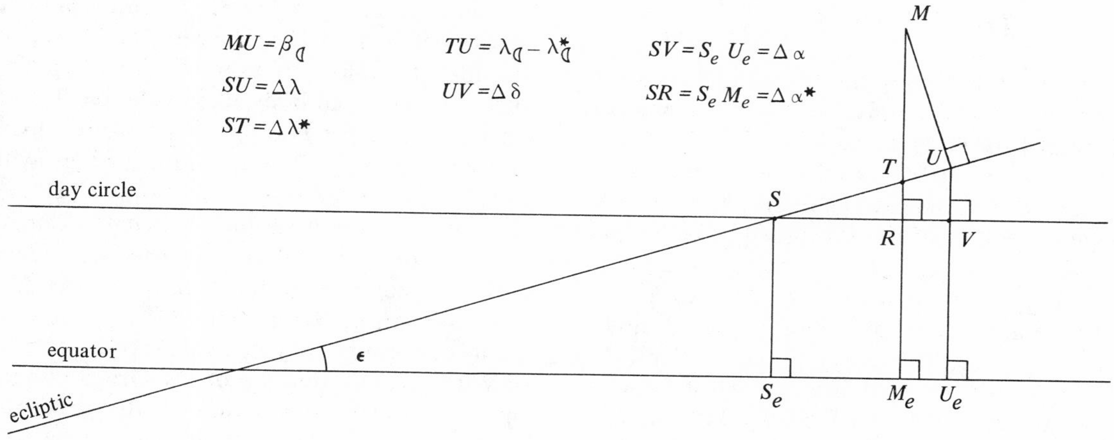

FIGURE 13

TABLE III.22

Nakṣatra	Varāhamihira λ	Varāhamihira β	Identification	Ptolemy λ	Ptolemy β	Magnitude
Kṛttikā	32;40°	+3;9°	(η Tau.	33;40°	+3;20°	5)
Rohiṇī	48;30°	−5;51°	α Tau.	42;40°	−5;10°	1
Punarvasu	88°	+7;12°	β Gem.	86;40°	+6;15°	2
	88°	−7;12°	(λ Gem.	81;40°	−6°	3)
Puṣya	97;20°	+4;3°	(η Canc.	97;40°	+1;15°	4.5)
Āśleṣā	107;40°	+0;54°	γ Canc.	100;20°	+2;40°	4.3
	107;40°	−0;54°	δ Canc.	101;20°	−0;10°	4.3
Maghā	126°	(0°)	α Leo	122;30°	+0;10°	1
Citrā	181;50°	−2;42°	α Virg.	176;40°	−2°	1

along the diameter that, when extended, reaches the sun is made to vary with Δλ, so that (*PS* 5, 4–7)

$$\sigma = 15 \cdot \frac{\Delta\lambda}{180^\circ} = \frac{\Delta\lambda}{12}. \qquad \text{(III.76)}$$

In adhyāya 14, Varāhamihira gives us our oldest information about the coordinates of the yogatārās of at least some of the nakṣatras (*PS* 14, 33–37). These are all of small enough latitudes that they can still plausibly be considered to serve the function of "junction stars" with the moon and planets; some of the yogatārās in the later star catalogs do not meet this requirement. Table III.22 lists the *Pañcasiddhāntikā*'s coordinates along with some possible identifications with stars included in Ptolemy's star catalog.[32]

That Varāhamihira here intends to give ecliptic rather than polar coordinates is proved by the computation, in the following verse, of the distance between the moon at latitude $\beta_{☾}$ and a yogatārā at its latitude (*PS* 14, 38). The chapter concludes with a statement of the rising time in oblique ascension for $\phi = 24°$ of the ecliptic arc between the sun and 90° necessary for the visibility of Agastya—that is, Canopus (α Arg.)(*PS* 14, 39–41).

IV. COSMOLOGY

Before discussing the Indian adaptations of Greek spherical astronomy, it is necessary to describe briefly the early cosmology of the *Purāṇas*, some of the basic concepts of which were taken over by the siddhāntakāras. The text source of the cosmological section of the *Purāṇas* probably was written in the early centuries of the Christian era; some of the concepts it reflects go back to Vedic times, and some show an affinity with Iranian theories.

In the *Purāṇas*[33] the earth is a flat-bottomed, circular disk, in the center of which is a lofty mountain, Meru. Surrounding Meru is the circular continent Jambūdvīpa, which is in turn surrounded by a ring of water known as the Salt Ocean. There follow alternating rings of land and sea until there are seven continents and seven oceans. In the southern quarter of Jambūdvīpa lies India—Bhāratavarṣa.

Above the earth's surface and parallel to its base are a series of wheels the centers of which lie on the vertical axis of Meru, at the tip of which is located the North Polestar, Dhruva.[34] The wheels, bearing the celestial bodies, are rotated by Brahma by means of bonds made of wind. The order of the celestial bodies varies; the earliest seems to be sun, moon, nakṣatras, and Saptarṣis (Ursa Major). Some *Purāṇas* place the grahas (planets) between the moon and the nakṣatras; in others, interpolated verses add Mercury, Venus, Mars, Jupiter, and Saturn (in that order) between the nakṣatras and the Saptarṣis.

Meru provides the explanation for the alternation of day and night. The variation in the sun's rising amplitude is ascribed to the existence of 180 paths for the sun on its wheel; it travels the next, successive path on each day of a half-year. It is not clear how this model could explain some other phenomena, such as the varying periods of visibility of the nakṣatras due to their varying declinations.

The Jaina canonical works,[35] which were codified in the early sixth century A.D., are based on this Purāṇic cosmology, but confuse it by hypothesizing that each celestial body has its double 180° from it on its wheel; the field of vision from a locality in Bhāratavarṣa is only a quadrant, so that, while in a nychthemeron the sun's wheel rotates 180°, half of that nychthemeron is day and half is night. The order of the celestial wheels also is somewhat different: fixed stars, sun, moon, nakṣatras, Mercury, Venus, Jupiter, Mars, Saturn. Also, there are 184 paths of the sun, not just 180.

With the introduction of the Greek concept of a spherical earth surrounded by the spheres of the planets and that of the fixed stars, it became necessary to discard these older ideas; they are assiduously attacked by Varāhamihira, Brahmagupta, and others. But it was also found possible to preserve some of them, although in altered form.

The circumference of the disk of the earth became the equator; at its four quadrants are located Laṅkā at the prime meridian and, proceeding westward, Romakaviṣaya (the Roman territory), Siddhapura (the city of the perfected ones), and Yamakoṭi (the peak of Yama). At the terrestrial North Pole stands Meru; opposite it lies Vaḍavāmukha (the Mare's Mouth). Above Meru still shines Dhruva, and the axis passing through these two and extended to the opposite Pole has wrapped about it bonds of wind that cause it to rotate with the diurnal motion. The individual motions of the planets in their orbits, and the deviations from their mean longitudes—known as their manda and śīghra equations—are also caused by cords of wind; the deviations are due to the pulls on these cords by demons located at the mandoccas and śīghroccas. This reluctance to part with any segment of their tradition that could possibly be saved is one of the most noteworthy characteristics of Indian astronomers, and one that has provided us with information about obsolete systems that in many other cultures would surely have been forgotten.

THE GREEK PERIOD

V. THE BRĀHMAPAKṢA

Probably in the late fourth or early fifth century, during the Gupta domination of western India, a non-Ptolemaic tradition of Greek astronomy, influenced by Aristotelianism, was transmitted to India. The occurrence of several planetary models and sets of parameters in Sanskrit texts indicates that more than one Greek work was translated; it also is clear that the Greek material was modified considerably, both to fit in with the established traditions of Indian astronomy (discussed in the preceding sections) and to conform to a concept of time that was first described in Sanskrit texts of the second century.[36] This concept is one of a system of yugas, of which a Caturyuga or Mahāyuga of 4,320,000 years contains four smaller yugas in the ratio to each other of 4 : 3 : 2 : 1; a Kaliyuga (in one of which we now are) contains 432,000 years (this is a Babylonian period); the Dvāparayuga, 864,000 years; the Tretāyuga, 1,296,000 years; and the Kṛtayuga, 1,728,000 years. A Kalpa contains 1,000 Mahāyugas; and often there are further multiples of the Kalpa forming larger yugas, although these larger yugas were not used for astronomical purposes.

The astronomers who adopted Greek planetary theory expressed the mean motions of the planets as integer numbers of rotations within a yuga, an idea that also lay behind the Hellenistic *magnus annus*.[37] The earlier Indian system, which seems to date from the early fifth century, assumes a true conjunction of the planets, their mandoccas, and their nodes at a sidereally fixed Aries 0° at the beginning and end of a Kalpa (*PS* 3, 20); a later system simplifies the numbers by assuming a mean conjunction of only the seven planets at the beginning and end of a Mahāyuga. Both systems assume a mean conjunction or near conjunction of the seven planets at a sidereally fixed Aries 0° at the beginning of the current Kaliyuga, which is either midnight of 17/18 February or 6 A.M. of 18 February -3101 Julian at Laṅkā and Ujjayinī. Table V.1 shows the mean tropical longitudes of the planets at 6 A.M. of 18 February -3101, as computed by means of Ptolemy's *Almagest*, and their distance from ζ Piscium, whose longitude according to Ptolemy was then 320;37°.

TABLE V.1

Planet	"Tropical" Longitude	Distance from ζ Piscium
Saturn	290;48°	−29;45°
Jupiter	325;4°	+4;27°
Mars	301;55°	−18;42°
Sun	314;38°	−5;59°
Moon	323;2°	+2;25°

If one knew, from a Greek set of tables, the "correct" mean tropical (or sidereal) longitudes of the planets at some time in the fifth century and their period relations, one could derive for each by means of an indeterminate equation an integer number of rotations in a Kalpa such that the conjunction of -3101 and the "correct" mean longitudes in the fifth century would both be accounted for. This, I suggest, is what the Indians did.

The Brāhmapakṣa, it is claimed, was revealed by Brahma (Pitāmaha); it flourished for some 1,500 years, primarily in western and northwestern India. The basic text of this school is the *Paitāmahasiddhānta*,[38] written in the early fifth century; one element from it appeared in a Sassanian work of *ca.* 450,[39] and the work was familiar to Aryabhaṭa I, as will be demonstrated below. It is recognized as the origin of the Brāhmapakṣa by Kamalākara (*STV* 1, 62). Unfortunately, however, it is very imperfectly preserved in a vast compilation of the sixth or seventh century, entitled *Viṣṇudharmottarapurāṇa*.

TABLE V.2

Body	R	$\frac{R \cdot 6,0}{Y}$	$\frac{R \cdot 6,0}{C}$
Saturn	146,567,298	12;12,50,11,21,50,24°	0;2,0,22,51,43,54, . . .°
Manda	41	0;0,0,44,16,48°	
Node	−584	−0;0,0,10,30,43,12°	
Jupiter	364,226,455	30;21,7,56,11,24°	0;4,59,9,8,37,23, . . .°
Manda	855	0;0,0,15,23,24°	
Node	−63	−0;0,0,1,8,2,24°	
Mars	2,296,828,522	3,11;24,8,33,23,45,36°	0;31,26,28,6,47,12, . . .°
Manda	292	0;0,0,5,15,21,36°	
Node	−267	−0;0,0,4,48,21,36°	
Sun	4,320,000,000	6,0°	0;59,8,10,21,33,30, . . .°
Manda	480	0;0,0,8,38,24°	
Venus' śīghra	7,022,389,492	3,45;11,56,50,51,21,36°	1;36,7,44,35,16,45, . . .°
Manda	653	0;0,0,11,45,14,24°	
Node	−893	−0;0,0,16,4,26,24°	
Mercury's śīghra	17,936,998,984	54;44,59,41,42,43,12°	4;5,32,18,27,45,33, . . .°
Manda	332	0;0,0,5,58,33,36°	
Node	−521	−0;0,0,9,22,40,48°	
Moon	57,753,300,000	2,12;46,30°	13;10,34,52,46,30,13, . . .°
Manda	488,105,858	40;40,31,45,26,38,24°	0;6,40,53,56,32,54, . . .°
Node	−232,311,168	−19;21,33,21,1,26,24°	−0;3,10,48,20,6,41, . . .°

In the *Paitāmaha*, the Kalpa contains 1,000 Mahāyugas, grouped into fourteen Manvantaras, each of which consists of seventy-one Mahāyugas or 306,720,000 years; alternating with these fourteen Manvantaras are fifteen Sandhis, each of which is equal to a Kṛtayuga or 1,728,000 years (*Pait.* 3, 4). The revolutions, R, of the celestial bodies in a Kalpa according to the *Paitāmaha* are recorded in Table V.2 (*Pait.* 3, 5), along with the corresponding yearly and daily mean motions. The latter are computed from the number of days, C, in a Kalpa – 1,577,916,450,000 – which implies a year length of 6,5;15,30,22,30 days (*Pait.* 3, 32). The *Paitāmaha*, pretending to be prophetic, gives no indication of the fraction of the Kalpa that has elapsed.

The śīghra motions of the planets are their proper motions on their śīghra epicycles, as counted from a point on a radius of that epicycle parallel to the radius of the deferent that connects the center to Aries 0°; they are, for the inferior planets, the sums of the mean solar motion and the planets' anomalies. The śīghras of the inferior planets are used in Indian tables because of the Greek principle (a variant of which is found in the *Romakasiddhānta*) according to which the velocity of the planet is inversely proportional to its distance from the center of the earth.[40] The mean motions of the inferior planets are, of course, identical with the sun's, and Venus' anomalistic motion is less than the sun's; therefore, the only way to preserve the principle and keep Venus in an orbit below the sun was to use the śīghra motion.

The application of this principle to the problem of the distance of the planets' orbits led the author of the *Paitāmaha* to make the rather arbitrary assumption that each of the 21,600 minutes of the moon's orbit contains 15 yojanas, so that the circumference is 324,000 yojanas.[41] Then it is assumed that each planet travels the same number of yojanas in a Kalpa; that number, O, which is also the circumference of heaven, is 324,000 yojanas · 57,753,300,000 = 18,712,069,200,000,000 yojanas (*Pait.* 3, 6–7). The circumferences of the planetary orbits are given in Table V.3; the circumference of the nakṣatras' is 60 times that of the sun, or 259,889,850 yojanas.

TABLE V.3

	O/R	
Saturn	$127{,}668{,}787\,\frac{8412079}{24427883}$	yojanas
Jupiter	$51{,}374{,}821\,\frac{54182089}{72845291}$	yojanas
Mars	$8{,}146{,}916\,\frac{82430924}{1148414261}$	yojanas
Sun	4,331,497 1/2	yojanas
Venus' śīghra	$2{,}664{,}629\,\frac{1627580383}{1755597373}$	yojanas
Mercury's śīghra	$1{,}043{,}210\,\frac{1561237670}{2242124873}$	yojanas
Moon	324,000	yojanas

The diameters of these orbits are found by using

$$\pi = \sqrt{10}, \qquad \text{(V.1)}$$

the value found in (III.50) (*Pait.* 3, 6); this value of π is traditionally used in the computation of the radii of the planetary orbits in preference to more correct ones.

The numbers of rotations of the sun and the moon yield the units in a Kalpa given in Table V.4.

TABLE V.4

Saura years (Y)	4,320,000,000
Saura months (N)	51,840,000,000
Saura days (S)	1,555,200,000,000
Sidereal months	57,753,300,000
Lunar months (M)	53,433,300,000
Intercalary months (A)	1,593,300,000
Tithis (T)	1,602,999,000,000
Civil days (C)	1,577,916,450,000
Avamas (U)	25,082,550,000

The ahargaṇa, c, or number of lapsed days, is computed from the beginning of the Kalpa (*Pait.* 4, 1), although it is not stated how far back in the past that is (V.2):

$$y = 71 \cdot L \cdot 4{,}320{,}000 + 1{,}728{,}000 \cdot (L+1) + y_L, \qquad \text{(a)}$$

where L is the lapsed Manvantaras and y_L the lapsed years of the current Manvantara;

$$12\,y = n \qquad \text{(b)}$$

$$30\,(n + m_y) + t_m = s \qquad \text{(c)}$$

(it is incorrect to add together n and m_y, and to assume that $t_m = s_n$);

$$a = s \cdot \frac{A}{S} \qquad \text{(d)}$$

$$s + 30a = t \qquad \text{(e)}$$

$$u = t \cdot \frac{U}{T} \qquad \text{(f)}$$

$$c = t - u. \qquad \text{(g)}$$

To compute the remaining fractional part of the current intercalary month in saura days, $-\epsilon$, the author posits (figures between vertical lines are integers; r represents remainders) (*Pait.* 4, 2) that (V.3)

$$a = s \cdot \frac{A}{S}\left(= |a| + \frac{r_a}{S}\right); \qquad \text{(a)}$$

$$-\epsilon = \frac{S}{A} - \frac{r_a}{S} \cdot \frac{S}{A} = \frac{S - r_a}{A}. \qquad \text{(b)}$$

And, to compute the remaining part of the current avama in tithis, $-v$ (*Pait.* 4, 3), one proceeds as below (V.4):

$$u = t \cdot \frac{U}{T} = |u| + \frac{r_u}{T}; \qquad \text{(a)}$$

$$-v = \frac{T}{U} - \frac{r_u}{T} \cdot \frac{T}{U} = \frac{T - r_u}{U}. \qquad \text{(b)}$$

Both $-\epsilon$ and $-v$ could be used in ahargaṇa computations, since ϵ and r_u are used by Brahmagupta (V.52–54). Our version of the *Paitāmahasiddhānta*, however, pursues the subject no further.

With c, one is instructed to find the mean longitudes of the planets by means of (III.16) (*Pait.* 4, 6). These mean longitudes are to be corrected by the planets' mean motions in half the equation of daylight of that day and in the time interval corresponding to the deśāntara, or longitudinal difference, of that locality. For computing the deśāntara it is assumed that the distance in yojanas between Laṅkā and the point where one's local circle of terrestrial longitude crosses the equator has been found by comparing the computed and observed times of an eclipse (*Pait.* 3, 27); the circumference of the earth is stated to be 5,000 yojanas.

The normal Indian model of the planets is derived from a Greek attempt to account for the inequalities of planetary motion while retaining the Aristotelian principle of concentricity; the model seems to have originated in Greek philosophical circles in the second century.[42] It is depicted in Figure 14. The center of the earth and of each planetary orbit is at O. The mean planet rotates on the concentric deferent. The mean position of the planet, $\bar{P}$, is surrounded by epicycles of given circumferences: one (the manda epicycle) about the mean sun and the mean moon, and two (the manda and the śīghra epicycles) about the mean star-planets (the mean longitudes of Venus and Mercury are equal to the mean longitude of the sun). On each of these epicycles is an ucca—M on the manda, S on the śīghra—the longitudes of which, measured from the epicycle radius pointing toward Aries 0°, are equal to those of the planet's manda and śīghra; for all planets, OM_0 is parallel to $\bar{P}M$, and for the superior planets the deferent radius pointing toward $\bar{\lambda}_{☉}$ is parallel to $\bar{P}S$. The points where the lines MO and SO, extended if necessary, cut the deferent are the positions to which each epicycle, if acting alone, would displace the mean planet at $\bar{P}$; their combined effects in the case of the star-planets are determined by various integrations. The sun's true longitude can be

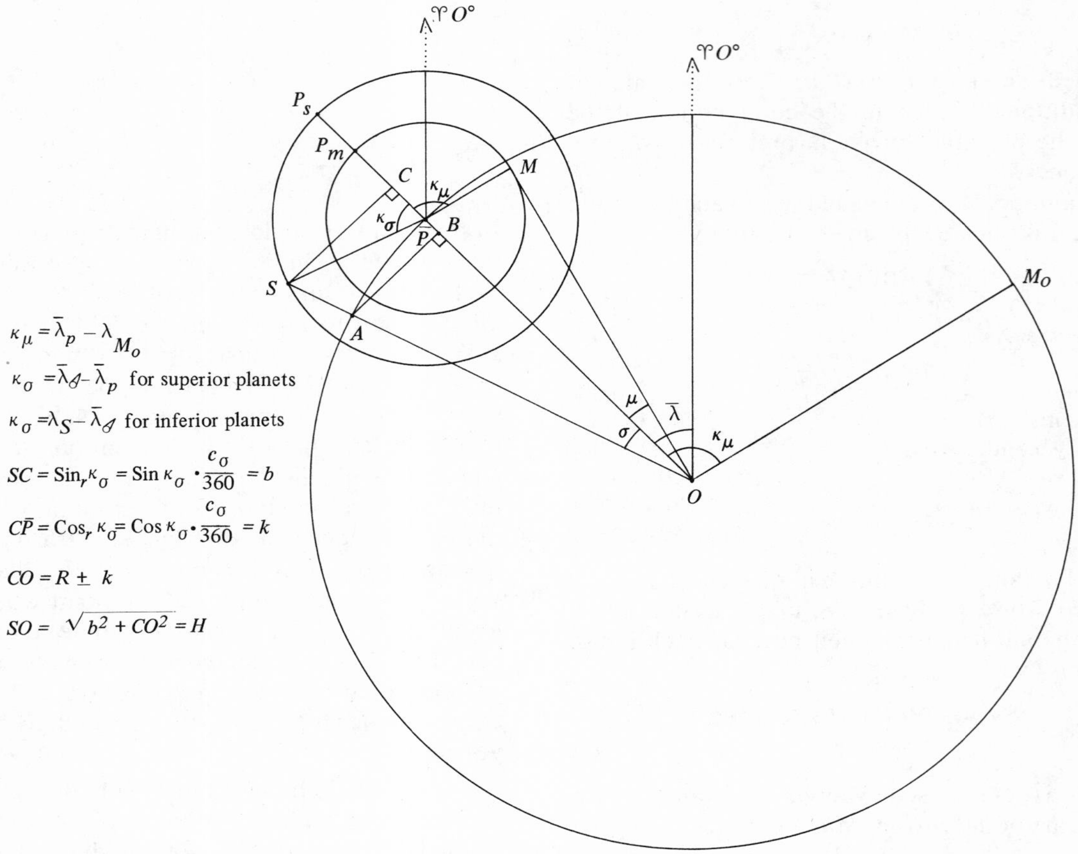

FIGURE 14

found directly by the "method of declinations" (III.28):

$$\text{Sin}\,\lambda_{♂} = \text{Sin}\,\delta_{♂} \cdot \frac{R}{\text{Sin}\,\epsilon}. \qquad (\text{V}.5)$$

The circumferences, c, of the epicycles according to the *Paitāmaha* are listed in Table V.5, along with the corresponding maximum equation, $\mu_{\max}$ and $\sigma_{\max}$, for each. The circumferences are measured in units of which there are 360 in the deferent (*PS* 3, 10–11).

TABLE V.5

Planet	Manda		Śīghra	
	c_μ	$\mu_{\max}$	c_σ	$\sigma_{\max}$
Saturn	30°	4;46,46°	40°	6;22,9°
Jupiter	33°	5;15,33°	68°	10;53,6°
Mars	70°	11;12,28°	243°	42;28,36°
Sun	13;40°	2;10,31°		
Venus	11°	1;45,3°	258°	45;48,4°
Mercury	38°	6;3,31°	132°	21;31,43°
Moon	31;36°	5;2,7°		

In the *Paitāmahasiddhānta* a correct geometric solution is given for both μ and σ (*Pait.* 4, 10). To use σ as an example, in Figure 14 we have (V.6)

$$b = \text{Sin}\,\kappa_\sigma \cdot \frac{c_\sigma}{360}; \qquad (\text{a})$$

$$\text{Sin}\,\sigma = b \cdot \frac{R}{H}, \qquad (\text{b})$$

where $H^2 = \left(R \pm \text{Cos}\,\kappa_\sigma \cdot \frac{c_\sigma}{360}\right)^2 + b^2$.

For the superior planets the method of integration is (V.7)

$$\lambda' = \bar{\lambda} \pm \frac{\mu}{2}; \qquad (\text{a})$$

$$\kappa_\sigma = \bar{\lambda}_{♂} - \lambda'; \qquad (\text{b})$$

$$\lambda'' = \lambda' \pm \frac{\sigma}{2}. \qquad (\text{c})$$

The process is iterated until differences disappear.

For the inferior planets, one first finds σ and then, with the corrected λ', one finds μ.

The preceding computations have utilized the Sine function. A rule for determining the second differences, $\Delta\Delta$, in a Sine table where $R = 3{,}438$ is given by the *Paitāmaha* (*Pait.* 3, 12):

$$\Delta\Delta = \frac{s_n}{s_1}, \qquad \text{(V.8)}$$

where s_1 is the first Sine, that for an argument of 3;45°, and s_n is a given tabular Sine. It has been argued persuasively that 3,438, which is derived from the relation

$$\pi = \frac{c}{d} = \frac{21{,}600'}{6{,}876}, \qquad \text{(V.9)}$$

was also the Radius in Hipparchus' Chord table.[43] The rule (V.8) is not accurate, as can be seen from Table V.6, where I use the Δ's of Āryabhaṭa I, since the *Paitāmaha* gives no numbers.

TABLE V.6

n	ϑ	Sin ϑ	R Sin ϑ	Δ Sin	ΔΔ Sin	$\frac{\text{Sin}_n}{\text{Sin}_1}$
1	3;45°	225	224.8	225		
2	7;30°	449	448.7	224	1	1
3	11;15°	671	670.7	222	2	2
4	15°	890	889.8	219	3	3
5	18;45°	1105	1105.1	215	4	4
6	22;30°	1315	*1315.6*	210	5	5
7	26;15°	1520	*1520.5*	205	5	*6*
8	30°	1719	1719	199	6	*7*
9	33;45°	1910	1910.0	191	8	8
10	37;30°	2093	2092.9	183	8	8
11	41;15°	2267	2266.8	174	9	9
12	45°	2431	2431.0	164	10	10
13	48;45°	2585	2584.8	154	10	*11*
14	52;30°	2728	2727.5	143	11	11
15	56;15°	2859	2858.5	131	12	12
16	60°	2978	*2977.4*	119	12	*13*
17	63;45°	3084	*3083.4*	106	13	13
18	67;30°	3177	*3176.2*	93	13	*14*
19	71;15°	3256	3255.5	79	14	14
20	75°	3321	3320.8	65	14	14
21	78;45°	3372	3371.9	51	14	*15*
22	82;30°	3409	3408.5	37	14	*15*
23	86;15°	3431	3430.6	22	15	15
24	90°	3438	3438	7	15	15

The *Paitāmaha* repeats many of the problems connected with spherical trigonometry that we have previously considered. Here, as later, we shall refrain from specifying these repetitions, and shall indicate only those new formulas that have appeared to us most significant. Some have un-

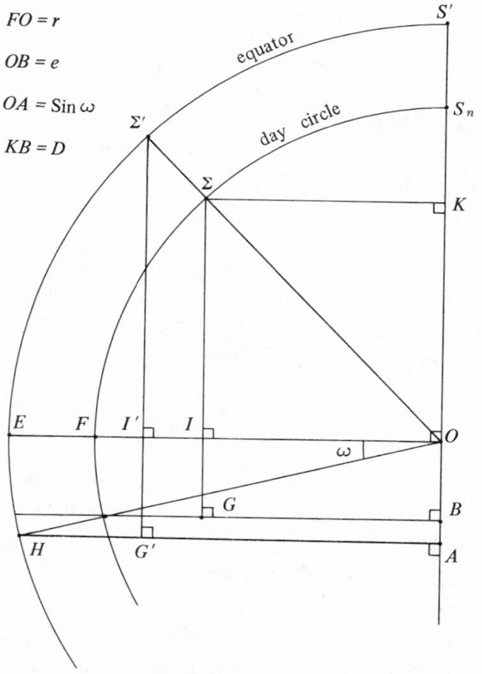

(projection in the plane of the equator)

FIGURE 15

doubtedly escaped our notice that should be mentioned, but brevity has been our general aim.

The main elaboration of the projections of the Greco-Babylonian period that we find in the *Paitāmaha* is the cheda, or divisor, D (*Pait.* 6, 1); this is illustrated in Figure 15, where D is ΣG; it is Σ'B in Figure 10. In Figure 10 the time degrees since sunrise are represented by $\Sigma' H$, and the time degrees since 6 o'clock by $\Sigma' E = \Sigma' H - \omega$. Then (see III.65a)

$$\Sigma I = \Sigma' I' \cdot \frac{r}{R} = \text{Sin}\, \Sigma' E \cdot \frac{r}{R}; \qquad \text{(V.10)}$$

$$D = \Sigma I + e. \qquad \text{(V.11)}$$

With D known, one can find the śaṅku, or Sine of the sun's altitude, Sin α, which is $\Sigma' M'$ in Figure 10 (see III.65b);

$$\text{Sin}\, \alpha = \text{Sin}\, \bar{\phi} \cdot \frac{D}{R}. \qquad \text{(V.12)}$$

Then one can find the shadow with (III.67), which involves also finding $\sqrt{R^2 - \text{Sin}^2\alpha} = \text{Cos}\, \alpha$, the Sine of the zenith distance, called the dṛgjyā in Sanskrit. We have already seen in (III.53b) that $\bar{\alpha}_n = \phi \pm \delta$;

the *Paitāmaha* calls Sin $\overline{\alpha}_n = \text{Cos}\ \alpha_n$ the anaṣṭa, and prescribes (*Pait*. 6, 2)

$$h_n = g \cdot \frac{R}{\text{Cos}\ (\phi \pm \delta)}, \qquad (\text{V}.13)$$

which is illustrated in Figure 16.

Furthermore, in Figure 15, S_nB is the noon cheda, D_n, such that (*Pait*. 6, 3)

$$D_n = r + e. \qquad (\text{V}.14)$$

From this it is clear that h_n and D_n are immediately computable if one knows $\lambda_{☊}$. Then, observing the gnomon's shadow at any given time, one can find h by the Pythagorean theorem; and

$$D \approx D_n \cdot \frac{h_n}{h}, \qquad (\text{V}.15)$$

which is certainly not true. With D one can compute backward to find the time degrees since sunrise; and with those time degrees, the solar longitude, and the local oblique ascensions, one can find the longitude of the ascendant.

The latitude theory of the *Paitāmaha* is different for the moon, the superior planets, and the inferior planets. Each planet has a given orbital inclination, i (*Pait*. 3, 18). Then, for the moon, (see III.18b)

$$\beta = i \cdot \frac{\text{Sin}\ \omega}{R}. \qquad (\text{V}.16)$$

For the superior planets, however,

$$\beta = i \cdot \frac{\text{Sin}\ \omega}{H}, \qquad (\text{V}.17)$$

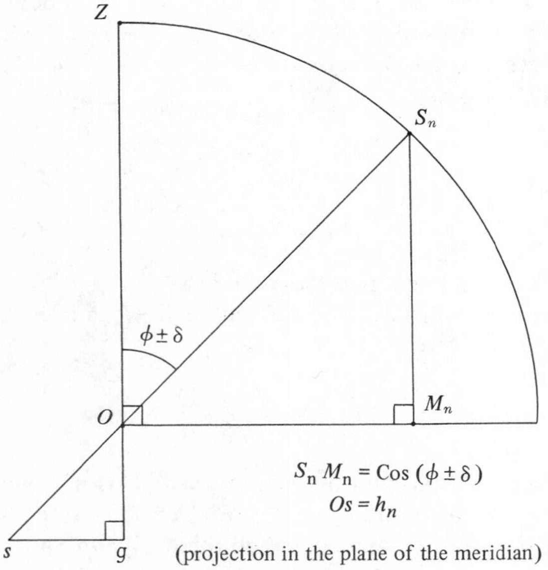

FIGURE 16

where H is the final śīghra hypotenuse. The nodes of the superior planets and the moon lie on their respective orbits, while those of the inferior planets lie on their śīghra epicycles, so that, for them, ω is the difference between the longitudes of the planet's śīghrocca and its node. Thus, the plane of the śīghra epicycle is tilted with respect to the plane of the ecliptic at the angle i so that (see Figure 17) when $\omega = 90°$ and $H = R - r_\sigma$ (r_σ being the radius of the śīghra epicycle), the maximum latitude occurs. For the inferior planets, then, given their special definition of ω, the relation (V.17) is also valid. In Table V.7 are tabulated the values of i and of $\beta_{\text{max}} = i \cdot \dfrac{R}{R - r_\sigma}$.

TABLE V.7

Planet	i	β_{max}
Saturn	2;10°	2;26°
Jupiter	1;16°	1;34°
Mars	1;50°	5;38°
Venus	2;16°	8°
Mercury	2;32°	4°
Moon	4;30°	4;30°

The *Paitāmaha's* fragmentary rules for computing eclipses (*Pait*. 5, 2–4; 9, 5–7) show some development beyond the computations of the Greco-Babylonian period, but the text is both corrupt and incomplete. The diameters of the sun and the moon, as of the other planets (see Table V.8), are given in yojanas (V.18):

$$d^y_{☉} = 6500; \qquad (\text{a})$$

$$d^y_{☾} = 480. \qquad (\text{b})$$

The true radius of the planet's orbit at any given time, measured in yojanas, k^y, is found from (*Pait*. 5, 2)

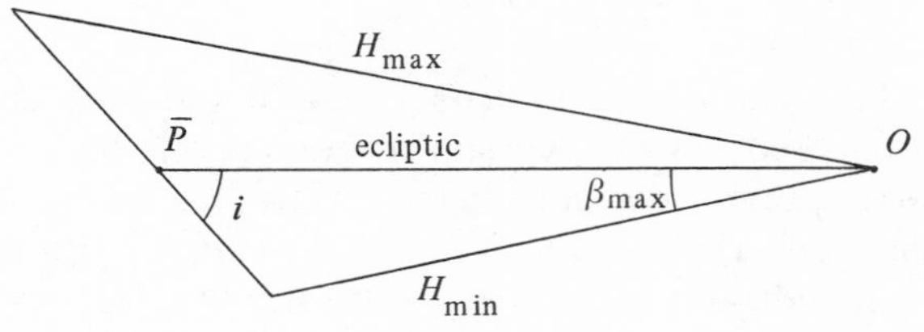

FIGURE 17

$$k^y = \bar{k}^y \cdot \frac{H}{R}, \qquad \text{(V.19)}$$

where H is the manda hypotenuse for the sun and the moon. Since the apparent diameter in minutes, d^m, varies inversely with k^y, it follows that

$$d^m = d^y \cdot \frac{R}{k^y}, \qquad \text{(V.20)}$$

since $R = \frac{21600'}{2\pi}$. From the data given in Table V.3 we know that the mean distances from the earth of the sun and the moon are, respectively, 684,869 and 51,229 yojanas. Therefore, we find that (V.21)

$$\bar{d}^m_{\odot} = 0;32,37,\ldots{}^\circ; \qquad \text{(a)}$$

$$\bar{d}^m_{☾} = 0;32,12,\ldots{}^\circ. \qquad \text{(b)}$$

The formula for finding the distance from the earth of the tip of the earth's shadow, k_u, is given by the *Paitāmaha* as (*Pait*. 5, 3)

$$k_u = (k_{\odot} - r_e) \cdot \frac{d_e}{d_{\odot} - d_e}, \qquad \text{(V.22)}$$

where r_e and d_e are, respectively, the radius and the diameter of the earth, the circumference of which has been given as 5,000 yojanas. As can be seen from Figure 18, one should have

$$k_u = \frac{k_{\odot} \cdot d_e}{d_{\odot} - d_e}. \qquad \text{(V.23)}$$

The distance of the end of the shadow cone from an observer on earth is approximately $k_u - r_e$; this presumably is what the *Paitāmaha* intended to find.

The *Paitāmaha* ascertains the diameter of the earth's shadow at the moon's distance in minutes thus (*Pait*. 5, 4):

$$d^m_u = \frac{k_{☾} \cdot d_e}{k_u} \cdot \frac{R}{k_{☾}}. \qquad \text{(V.24)}$$

This formulation is incomplete, since we have, from (V.20) and the relation $\frac{d_u}{d_e} = \frac{k_u - k_{☾}}{k_u}$, that

$$d^m_u = d^y_u \cdot \frac{R}{k_{☾}} = \frac{(k_u - k_{☾}) \cdot d_e}{k_u} \cdot \frac{R}{k_{☾}}. \qquad \text{(V.25)}$$

The remainder of the computation of a lunar eclipse in the *Paitāmaha* is lost. For the computation of a solar eclipse, all that remains is a corrupt version of rules for obtaining parallax. The longitudinal parallax seems to be found from the right ascension of the difference between the longitudes of the moon and the point where the ecliptic crosses the meridian; one then should employ something like (III.17). To find the latitudinal parallax the *Paitāmaha's* rule is (*Pait*. 9, 6)

$$\pi_\beta = \text{Cos}\,\alpha_n \cdot \frac{v_{☾} - v_{\odot}}{15\,R}, \qquad \text{(V.26)}$$

where Cos α_n, the solar dṛgjyā for noon, also called the anaṣṭa (see **III**.53b), is Sin ($\phi \pm \delta$). The text might have been expected to use the lunar dṛgjyā for noon, which is Sin ($\phi \pm \delta \pm \beta_{☾}$), and $v_{☾}$ rather than the relative velocity, $v_{☾} - v_{\odot}$ (see III.20). The *Paitāmaha* also should have used the zenith distance of the nonagesimal rather than that of a body on the meridian.

In addition, the *Paitāmaha* contains a section on the visibility of the moon, the planets, and the stars

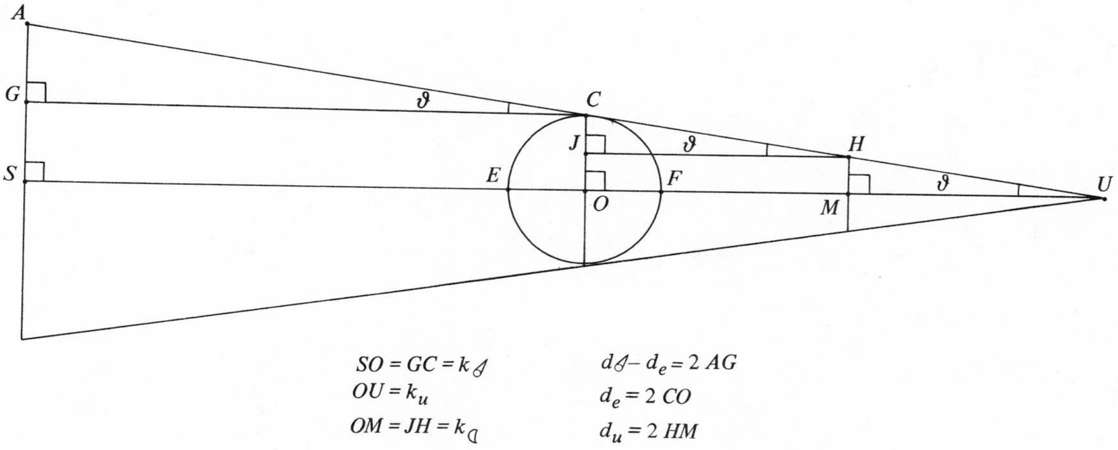

FIGURE 18

(*Pait*. 8, 2–3) that represents an advance over the Greco-Babylonian computation (III.73–75). In this computation, following a convention of Hipparchus,[44] one operates with polar rather than ecliptic coordinates; polar latitude is the angular distance of the body from the ecliptic measured along the declination circle passing through the planet, and polar longitude is the longitude of the point where that declination circle cuts the ecliptic.

The first step is to find the longitudinal difference, $\lambda\lambda^*$, between λ and λ^*; this procedure is called the ayanadṛkkarma, and is similar to the computation of the ayanavalana (see V.94). In Figure 19, since, when $\lambda = 0°$, $< PSP' = \epsilon$, and when $\lambda = 90°$, $< PSP' = 0°$, therefore

$$< PSP' = < \lambda SM = \delta\,(\lambda + 90°). \qquad (V.27)$$

For solving triangle $\lambda S \lambda^*$, which is assumed to be small and virtually plane—$\beta = S\lambda$ is assumed to be very small—the present text of the *Paitāmaha* has (*Pait*. 8,2)

$$\lambda\lambda^* = \frac{\beta}{\epsilon}. \qquad (V.28)$$

This is a corrupt version of what is found in the *Brāhmasphuṭasiddhānta* (*BSS* 6, 3):

$$\lambda\lambda^* \approx \beta \cdot \frac{\operatorname{Sin} \delta\,(\lambda + 90°)}{R}. \qquad (V.29)$$

In fact, Brahmagupta has found λM rather than $\lambda\lambda^*$. More accurate would be

$$\operatorname{Sin} \lambda M = \operatorname{Sin} \beta \cdot \frac{\operatorname{Sin} \delta\,(\lambda + 90°)}{R}. \qquad (V.30)$$

Moreover, one could easily find $\lambda\lambda^*$ with (V.31)

$$\operatorname{Sin}^2 SM = \operatorname{Sin}^2 \beta - \operatorname{Sin}^2 \lambda M; \qquad (a)$$

$$\operatorname{Sin} \lambda\lambda^* = \operatorname{Sin} \lambda M \cdot \frac{\operatorname{Sin} \beta}{\operatorname{Sin} SM}. \qquad (b)$$

Furthermore, one can then find $\beta^* = S\lambda^*$ from

$$\operatorname{Sin}^2 \beta^* = \operatorname{Sin}^2 \beta + \operatorname{Sin}^2 \lambda\lambda^*. \qquad (V.32)$$

The corruption in the *Paitāmaha* may be related to the formulation (see VI. 4)

$$\lambda\lambda^* \approx \beta \cdot \frac{\operatorname{Sin} \epsilon \cdot \operatorname{Sin}\,(\lambda + 90°)}{R}, \qquad (V.33)$$

which follows from (III.55) and (V.29).

When a body with latitude β^* crosses the local horizon, one must find the ecliptic arc between λ^* and the horizon. This computation is called the akṣadṛkkarma, and is related to the computation of the akṣavalana (see III.93). In Figure 19, the required ecliptic arc is $D\lambda^*$; to find it, the present text of the *Paitāmaha* prescribes (*Pait*. 8, 2)

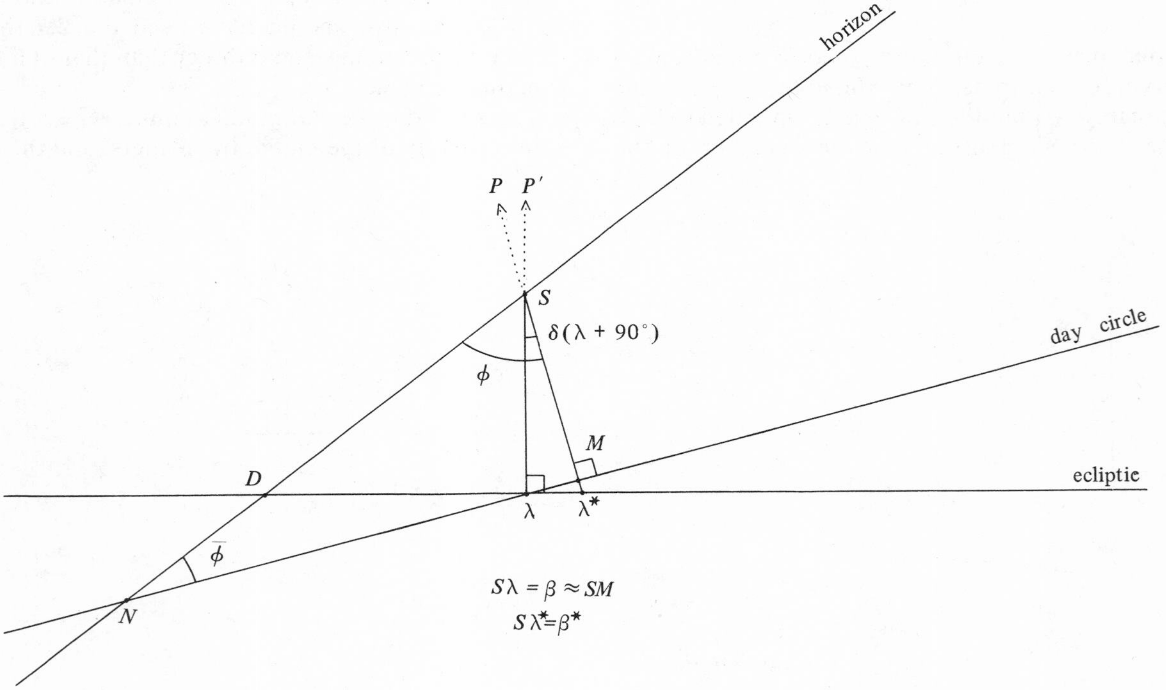

FIGURE 19

$$D\lambda^* = \beta \cdot \frac{\text{Sin}\,\phi}{R}. \qquad \text{(V.34)}$$

Again, this is a corrupt version of Brahmagupta's rule (*BSS* 6, 4)

$$D\lambda^* \approx \beta \cdot \frac{s_0}{12} = \beta \cdot \frac{\text{Sin}\,\phi}{\text{Sin}\,\bar{\phi}}. \qquad \text{(V.35)}$$

Brahmagupta has found NM rather than $D\lambda^*$; and a more accurate formula would be

$$\text{Sin}\,NM = \text{Sin}\,\beta \cdot \frac{\text{Sin}\,\phi}{\text{Sin}\,\bar{\phi}}. \qquad \text{(V.36)}$$

Moreover, one again could have determined $D\lambda^*$ accurately. Since $\angle DS\lambda = \phi - \delta\,(\lambda + 90°)$, it follows that (V.37)

$$\text{Sin}\,D\lambda = \text{Sin}\,(\phi - \delta\,[\lambda + 90°]) \cdot \frac{\text{Sin}\,\beta}{\text{Sin}\,(90° - \phi + \delta\,[\lambda + 90°])}; \qquad \text{(a)}$$

$$D\lambda^* = D\lambda + \lambda\lambda^*. \qquad \text{(b)}$$

The time at which it is computed that D is on the horizon is also the time that the body at S is on the horizon.

The body will be visible if the sun at that time is sufficiently far below the horizon. Indian astronomers convert this distance below the horizon into time; for each planet and for the fixed stars there is an equatorial arc measured in kālāṃśas or time degrees, k (*Pait.* 3, 9), that is the minimum difference in the right ascensions, $\Delta\alpha$, of λ^* of the body and $\lambda_{\mars}$ at which the body can be seen. Time until or since first visibility or first invisibility can be computed by permutations of (*Pait.* 8, 3)

$$t = \frac{\Delta\alpha - k}{v_{\mars} - v_p}. \qquad \text{(V.38)}$$

The planets' kālāṃśas are related to their diameters in yojanas (*Pait.* 3, 8), which must be considered to be their apparent diameters at the moon's distance, where 15 yojanas = 1′. These are tabulated in Table V.8.

The *Paitāmaha* gives another rule for determining the times of the first and last visibilities and first

TABLE V.8

Planet	Kālāṃśas	Diameter Yojanas	Diameter Minutes
Moon	12°	480	32
Venus	9°	240	16
Jupiter	11°	120	8
Mercury	13°	60	4
Saturn	15°	30	2
Mars	17°	15	1

and last stations of the planets, in which the effects of terrestrial and celestial latitude are ignored. This method is based on the śīghra anomalies, κ_0, necessary for the occurrences of these phenomena; it is, thus, closely related to a Greek text computed from Ptolemaic parameters and erroneously attributed to Heliodorus.[45] In Table V.9 are tabulated κ_0 for the first visibilities and first stations; the others are symmetrical (*Pait.* 3, 31). They are, of course, computed from the sizes of the śīghra epicycles.[46]

For a superior planet, the time until phase, t, is found from (*Pait.* 4, 15)

$$t = \frac{\kappa_0 - \kappa_\sigma}{v_{\mars} - v_p}; \qquad \text{(V.39)}$$

for an inferior planet, from

$$t = \frac{\kappa_0 - \kappa_\sigma}{v_p - v_{\mars}}. \qquad \text{(V.40)}$$

To find the actual distance in degrees between the sun and the moon, the *Paitāmaha* establishes a very complicated procedure (*Pait.* 9, 1–2) that later authors such as Brahmagupta (*BSS* 7, 15–16) converted into physical models. For both the sun and the moon one finds Sin α, the śaṅkutala, and Sin η from (III.68–70). Then, for each, the distance of the base of its Sin α from the east-west line, the bāhu (see Figure 20), is found from (see III.71)

$$\text{bāhu} = \text{śaṅkutala} \pm \text{Sin}\,\eta. \qquad \text{(V.41)}$$

Then one finds the dṛgjyā, cos α, of each from (see III.67)

$$\text{Cos}\,\alpha = \sqrt{R^2 - \text{Sin}^2\,\alpha}. \qquad \text{(V.42)}$$

TABLE V.9

Planet	First Visibility *Paitāmaha*		First Visibility "Heliodorus"		First Station *Paitāmaha*	First Station "Heliodorus"
Saturn	17°		17°		113°	112;44°–115;29°
Jupiter	14°		16°		125°	124;5°–127;11°
Mars	28°		22°		164°	157;30°–169;14°
	West	East	West	East		
Venus	24°	183°	12;24°	–	165°	165;52°–168;21°
Mercury	50°	205°	38°	–	146°	144;40°–147;14°

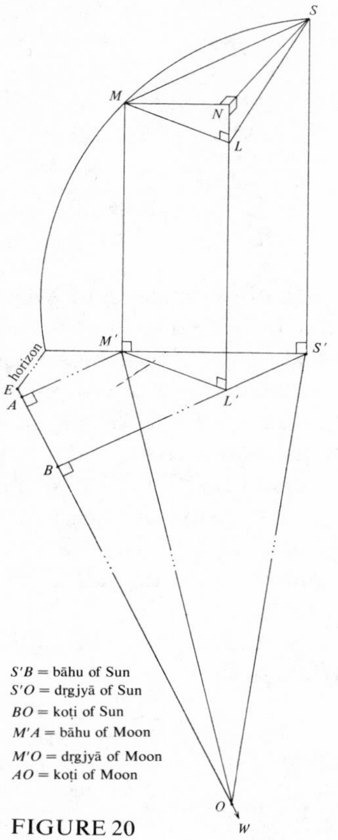

FIGURE 20

By the Pythagorean theorem one finds the koṭis, *AO* and *BO*, of the moon and the sun; their difference is $AB = M'L' = ML$. The text of the *Paitāmaha* becomes corrupt at this point, but the rest of the procedure can be supplied from Brahmagupta (*BSS* 7, 7–10). If $NL' = SS' = \text{Sin}\ \alpha_{♂}$, then the difference of the Sin α's of the sun and the moon is *NL*, and

$$MN^2 = ML^2 + NL^2. \qquad \text{(V.43)}$$

Further, the difference of the bāhus is $S'L' = SN$; therefore

$$MS^2 = MN^2 + SN^2. \qquad \text{(V.44)}$$

MS is the chord of the arc of the great circle that passes through the sun and the moon.

The illuminated portion of the moon's diameter, σ, is, according to the *Paitāmaha* (*Pait.* 9, 3), found by (V.45)

$$\sigma_1 = d_{☾} \cdot \frac{\text{Sin}\ \Delta\lambda}{90} \qquad \text{(a)}$$

in the first and fourth quadrants, and

$$\sigma_2 = d_{☾} - d_{☾} \cdot \frac{\text{Sin}\ \Delta\lambda}{90} \qquad \text{(b)}$$

in the other two quadrants, where $\Delta\lambda$ is the difference between the longitudes of the sun and the moon. Clearly, what is required is the equivalent of (III.76); this is given by Brahmagupta (*BSS* 7, 11–12) (V.46):

$$\sigma_1 = r_{☾} \cdot \frac{\Delta\lambda}{90}; \qquad \text{(a)}$$

$$\sigma_2 = r_{☾} + r_{☾} \cdot \frac{\text{Sin}\ (\Delta\lambda - 90°)}{R}. \qquad \text{(b)}$$

The *Paitāmaha's* text again seems to be corrupt.

To find the sūtra, *s*, which is the radius of the circle the circumference of which lies along the inner edge of the moon's sickle, the *Paitāmaha* (*Pait.* 9, 4), as preserved, gives what appears to be a fragment of Brahmagupta's rule (*BSS* 7, 14):

$$s = \frac{\frac{d_{☾}^2}{4}}{r_{☾} - \sigma} + r_{☾} - \sigma = \frac{r^2}{r_{☾} - \sigma} + r_{☾} - \sigma. \qquad \text{(V.47)}$$

This at least meets the requirement that when $\sigma = 0$, $s = d_{☾}$ and there is no "sickle"; and when $\sigma = r_{☾}$, $s = \infty$ and the inner edge of the "sickle" is a straight line passing through the center of the moon and perpendicular to σ.

We have noted previously that the *Paitāmaha* uses polar coordinates to determine the times of first and last visibility of the planets. In Table V.10 are listed the polar coordinates of the yogatārās ("junction stars") of the naksatras according to the *Paitāmaha* (*Pait.* 3, 30), the ecliptic coordinates computed with adaptations of (V.30) and (V.32), and tentative identifications with stars listed by Ptolemy.[47] The extreme latitudes of some of these stars suggest that they are not members of the Vedic nakṣatras; nor is it conceivable that any astronomer without an armillary sphere could determine the polar coordinates of the stars, and there is no evidence that the Indians had armillary spheres. Therefore, it is likely that the author of the *Paitāmaha* somehow adapted a Greek star catalog to his purposes; however, either his source or he himself created such numerous errors that definite identifications are impossible. The non-Indian origin of these coordinates is also deducible from the discrepancies between Table V.10 and Table III.22.

The *Paitāmaha* redefines the pātas of the sun and the moon given in (III.28–29); they are now said to be the instants when the two luminaries have equal declinations (*Pait.* 5, 9). Vaidhṛta occurs when they are on opposite sides of an equinox, vyatipāta when they are on opposite sides of a solstice.

TABLE V.10

Nakṣatra	*Paitāmaha* λ^*	*Paitāmaha* β^*	*Paitāmaha* λ	*Paitāmaha* β	Identification	Ptolemy λ	Ptolemy β	Magnitude
Aśvinī	8°	+10°	12°	+9;10°	β Ari.	7;40°	+8;20°	3
Bharaṇī	20°	+12°	24;32°	+11;4°	33 Ari. (?)	19;10°	+10;40°	5
Kṛttikā	37;28°	+5°	39;5°	+4;43°	η Tau.	33;40°	+3;20°	5
Rohiṇī	49;28°	−5°	48;9°	−4;49°	α Tau.	42;40°	−5;10°	1
Mṛgaśiras	63°	−10°	61;11°	−9;50°	15 Ori. (?)	50;30°	−8°	4
Ārdrā	67°	−11°	65;16°	−10;51°	λ Ori. (?)	57°	−13;30°	neb.
Punarvasu	93°	+6°	92;52°	+5;59°	β Gem.	86;40°	+6;15°	2
Puṣya	106°	0°	106°	0°	ϵ Canc.	100;20°	+0;20°	neb.
Āśleṣā	108°	−7°	108;53°	−6;56°	β Canc. (?)	97;10°	−7;30°	4.3
Maghā	129°	0°	129°	0°	α Leo	122;30°	+0;10°	1
Pūrvaphālgunī	147°	+12°	142;56°	+11;15°	δ Leo	134;10°	+13;40°	2.3
Uttaraphālgunī	155°	+13°	150;16°	+12;3°	β Leo	144;30°	+11;50°	1.2
Hasta	170°	−11°	174;24°	−10;3°	η Corv.	167°	−11;40°	4
Citrā	183°	−2°	183;49°	−1;50°	α Virg.	176;40°	−2°	1
Svāti	199°	+38°	185;19°	+34;37°	α Boöt.	177;0°	+31;30°	1
Viśākhā	212;5°	−1;30°	212;36°	−1;23°	ι Libr.	204;0°	−1;40°	4
Anurādhā	224;5°	−3°	224;57°	−2;51°	δ Scorp.	215;40°	−1;40°	3
Jyeṣṭhā	229;5°	−4°	230;9°	−3;50°	α Scorp.	222;40°	−4°	2
Mūla	240;4°	−8;30°	241;47°	−8;20°	d Oph.	235;30°	−6;10°	5
Pūrvāṣāḍhā	249°	−5;20°	249;46°	−5;17°	γ Sag.	244;30°	−6;30°	3
Uttarāṣāḍhā	260°	−5°	260;21°	−4;59°	ϕ Sag.	253°	−3;30°	4
Abhijit	265°	+62°	262;48°	+61;53°	α Lyr.	257;20°	+62°	1
Śravaṇa	278°	+30°	279;44°	+29;57°	α Aquil.	273;50°	+29;10°	2.1
Dhaniṣṭhā	290°	+36°	295°	+33;25°	α Delph.	290;10°	+33;50°	3.4
Śatabhiṣaj	320°	0°	320°	0°	λ Aquar.	314;50°	+0;10°	4
Pūrvabhādrapadā	326°	+24°	332;4°	+22;26°	α Peg.	326;40°	+19;40°	2.3
Uttarabhādrapadā	337°	+26°	346;43°	+23;52°	α Andr. (?)	347;50°	+26°	4
Revatī	0°	0°	0°	0°	ζ Pisc.	353°	−0;10°	4
Agastya	87°	−76°	85;25°	−75;43°	α Arg.	77;10°	−75°	1

The second representative of the Brāhmapakṣa, the *Brāhmasphuṭasiddhānta*, was composed by Brahmagupta at Bhillamāla in southern Rājasthān in 628. The first ten adhyāyas, the Daśādhyāyī, are a summary of the *Paitāmaha*; adhyāyas 13–17 contain additions. At some points Brahmagupta displays familiarity with the *Mahābhāskarīya* of Bhāskara I or its source. Adhyāya 25, the *Dhyānagrahopadeśa*, is a karaṇa (that is, it uses a contemporary epoch); adhyāya 21, a description of the celestial spheres. In this essay only these adhyāyas will be discussed, and in some detail; they are fundamental for all later Indian astronomy.

Brahmagupta's division of the Kalpa is identical with the *Paitāmaha's* (*BSS* 1, 7–12), although he adds that the interval between the beginning of the Kalpa and the beginning of the current Kaliyuga (sunrise at Laṅkā on 18 February -3101) is 1,972,944,000 years or 4,567 Kaliyugas (*BSS* 1, 26–28). His values for the rotations of the celestial bodies (*BSS* 1, 14–22) also are identical with the *Paitāmaha's* (see Table V.2), as are, necessarily, the time relations in Table V.4. His computation of the ahargaṇa is very similar to (V.2), but for (V.2 b–e) he substitutes (*BSS* 1, 29–30) (V.48)

$$n = 12y + m_y \quad \text{(a)}$$

$$a = n \cdot \frac{A}{N} \quad \text{(b)}$$

$$m = n + a \quad \text{(c)}$$

$$t = 30m + t_m. \quad \text{(d)}$$

In adhyāya 13 he adds another series of rules for computing the ahargaṇa (*BSS* 13, 11–19). From (V.3b) he derives

$$s = \frac{S \cdot |a| + r_a}{A}, \quad \text{(V.49)}$$

and, from (V.4 a) (V.50),

$$t = \frac{T \cdot |u| + r_u}{U}; \quad \text{(a)}$$

$$a = A \cdot \frac{t}{T}; \quad \text{(b)}$$

$$s = t - 30a. \quad \text{(c)}$$

He then proceeds to develop several more such

rules for finding the ahargaṇa, of which the most important is (V.51)

$$|u| = \frac{c \cdot U - r_u}{C} = t - c; \qquad \text{(a)}$$

therefore

$$t = \frac{c \cdot U - r_u}{C} + c = \frac{c \cdot (U + C) - r_u}{C}$$

$$= \frac{c \cdot T - r_u}{C}, \qquad \text{(b)}$$

substituting T for $U + C$. Moreover,

$$|a| = \frac{t \cdot A - r_a}{T} = \frac{c \cdot T \cdot A - A \cdot r_u}{C \cdot T} - \frac{r_a}{T} =$$

$$\frac{c \cdot A}{C} - \frac{A \cdot r_u + C \cdot r_a}{C \cdot T}. \qquad \text{(c)}$$

Thence,

$$c \cdot A = C \cdot |a| + \frac{A \cdot r_u + C \cdot r_a}{T}. \qquad \text{(d)}$$

This last quantity, $\frac{A \cdot r_u + C \cdot r_a}{T}$, is called the "accurate remainder of the intercalary months"; we will denote it r'_a. Then

$$c = \frac{C \cdot |a| + r'_a}{A}. \qquad \text{(e)}$$

Brahmagupta gives rules for finding the accumulated epact, ϵ—that is, the time between one year's last mean conjunction of the sun and the moon (Caitraśuklapratipad) and the beginning of the next mean solar year (Meṣasaṅkrānti) (*BSS* 1, 39–40). These rules also are not unrelated to (V.3). In the following formulas, a subscript zero indicates the unit in a year (V.52):

$$c_0 = \frac{C}{Y} = \frac{1577916450000}{4320000000} = 360 + 5 + \frac{2481}{9600}; \qquad \text{(a)}$$

$$u_0 = \frac{U}{Y} = \frac{25082550000}{4320000000} = 5 + \frac{7739}{9600}; \qquad \text{(b)}$$

$$s_0 = 360; \qquad \text{(c)}$$

$$c - s = y \cdot \left(5 + \frac{2481}{9600}\right); \qquad \text{(d)}$$

$$u = y \cdot \left(5 + \frac{7739}{9600}\right) = t - c; \qquad \text{(e)}$$

$$c - s + u = t - s = y \cdot \left(10 + \frac{2481}{9600} + \frac{7739}{9600}\right); \qquad \text{(f)}$$

$$\frac{t - s}{30} = |a| + \frac{r}{30}, \qquad \text{(g)}$$

where $r = \epsilon$ in tithis. From (V.52 f) we have

$$\epsilon_0 = 11; 3, 52, 30 \text{ tithis}. \qquad \text{(h)}$$

To compute the civil days, c_m, between Caitraśuklapratipad and Meṣasaṅkrānti from the tithis, t_m, in the same interval (*BSS* 1, 42–43), one must first compute the corresponding avamas, u_m; then (V.53)

$$c_m = t_m - u_m. \qquad \text{(a)}$$

There are two components of u_m: the fractional avama accumulated at Meṣasaṅkrānti—which, from (V.52e) is $\frac{r_u}{9600}$, which Brahmagupta converts into civil days by multiplication by $\frac{C}{T}$ — and any avamas resulting from the difference between the true interval, t_m, and the mean interval, ϵ. Using the approximation in (III.15), Brahmagupta's final formulation is

$$u_m = \frac{U}{T} \cdot (t_m - \epsilon) + \frac{C}{T} \cdot \frac{r_u}{9600} = \frac{11}{703} \cdot (t_m - \epsilon)$$

$$+ \frac{692}{703} \cdot \frac{r_u}{9600}. \qquad \text{(b)}$$

He also gives a rule for finding the civil days, c_n, between mean Caitraśuklapratipad and mean Meṣasaṅkrānti (*BSS* 1, 57–58). This interval in tithis is ϵ; the avamas, u_n, again are both those in ϵ and the fractional avama accumulated at Meṣasaṅkrānti. Brahmagupta's formula, then, is

$$c_n = \epsilon - \left(\frac{11\epsilon}{703} + \frac{r_u}{9600} \cdot \frac{692}{703}\right)$$

$$= \epsilon - \frac{r_u}{9600} \cdot \frac{692}{703} - \frac{11\epsilon}{703}. \qquad \text{(V.54)}$$

The basic formula for finding the mean longitude of a planet in the *Brāhmasphuṭasiddhānta* is that given in (III.16). Brahmagupta, however, adds several variants designed to shorten the computations. In the first, one proceeds as follows. The mean longitude at the last mean Meṣasaṅkrānti is found from (*BSS* 1, 31)

$$r = R \cdot \frac{y}{Y}. \qquad \text{(V.55)}$$

Then the ahargaṇa since mean Meṣasaṅkrānti is multiplied by the mean daily motion of each planet except the moon; these are given in Table V.11 (*BSS* 1, 44–49).

TABLE V.11

Planet	Mean Daily Motion
Saturn	$2' + \frac{2'}{315} = 0;2,0,22,51,25,\ldots°$
Jupiter	$5' - \frac{5'}{354} = 0;4,59,9,9,10,\ldots°$
Mars	$\frac{11°}{21} + \frac{11'}{875} = 0;31,26,28,6,50,\ldots°$
Sun	$1° - \left(\frac{1}{70} + \frac{1}{70 \cdot 129}\right)° = 0;59,8,10,22,44,\ldots°$
Venus' śīghra	$\frac{8°}{5} + \frac{8'}{62} = 1;36,7,44,30,58,\ldots°$
Mercury's śīghra	$4° + \frac{6°}{65} = 4;5,32,18,27,41,\ldots°$
Moon's manda	$\frac{1°}{9} + \frac{1°}{4004} = 0;6,40,53,56,45,\ldots°$
Moon's node	$-\left(\frac{1°}{19} + \frac{1°}{2701}\right) = -0;3,10,48,23,43,\ldots°$

For the moon Brahmagupta gives the special equation (*BSS* 1, 45)

$$\bar{\lambda}_{☾} = \bar{\lambda}_{☉} + 12° \cdot t_c + \frac{703 r_u}{9600} \cdot \frac{3°}{173}, \qquad \text{(V.56)}$$

where t_c represents the integer tithis elapsed since mean Caitraśuklapratipad, and $\frac{703 r_u}{9600} \cdot \frac{3°}{173} = \frac{r_u}{9600} \cdot \frac{703}{692} \cdot 12°$ represents the elongation between the mean sun and the mean moon in the fractional avama—that is, in the time between the end of the mean tithi before mean Meṣasaṅkrānti and the next crossing of the local six-o'clock circle by the mean sun.

In the second method Brahmagupta first gives the kṣepas of the planets for the beginning of Kaliyuga, sunrise at Laṅkā on 18 February -3101 (*BSS* 1, 51–56). The kṣepas are the remainders, ρ, from the application of (V.55), wherein $\frac{y}{Y} = \frac{1972944000}{4320000000} = \frac{4567}{10000}$. I tabulate in Table V.12 the integer rotations, $|r|$; the remainders, ρ; and the corresponding mean longitudes, $\bar{\lambda}$ (compare Table V.1).

Given these kṣepas, one could proceed to apply the first method, with y being the years lapsed since the beginning of the present Kaliyuga.

The published text of the *Brāhmasphuṭasiddhānta* at this point inserts bījas or corrections to the planets' mean longitudes (*BSS* 1, 59–60); the verses describing these bījas, however, seem to be interpolated. (See Table V.22.)

In adhyāya 13 Brahmagupta adds some more rules for finding mean and true longitudes (*BSS* 13, 20–39). First, for the sun and the moon, where $r_u \cdot \frac{A}{C}$ converts the fraction of a civil day between the beginning of the mean tithi and the beginning of the next mean sunrise day, $r_u = c \cdot U - C \cdot |u|$, into a fraction of an intercalary month, and where $r_u \cdot \frac{A}{C} + \frac{r_a}{T}$ is the fraction of an intercalary month that has accumulated at the beginning of that sunrise day, it follows that

$$a = |a| + r_u \cdot \frac{A}{C} + \frac{r_a}{T}. \qquad \text{(V.57)}$$

Then (V.58)

$$r_{☉} = R_{☉} \cdot \frac{a}{A}; \qquad \text{(a)}$$

$$r_{☾} = R_{☾} \cdot \frac{a}{A}. \qquad \text{(b)}$$

The calendar days, c_y, since mean Caitraśuklapratipad are found from (V.59)

$$c_y \approx 30\, m_y - u + t_m, \qquad \text{(a)}$$

where m_y is the lapsed synodic months, u the cor-

TABLE V.12

Planet	$\lvert r \rvert$	ρ	$\bar{\lambda} = \frac{\rho \cdot 6,0°}{Y}$
Saturn	66,937,284	4,305,312,000	5,58;46,33,36°
Jupiter	166,341,221	4,313,520,000	5,59;27,36°
Mars	1,048,961,585	4,308,768,000	5,59;3,50,24°
Sun	1,972,944,000	0	0°
Sun's manda	219	933,120,000	1,17;45,36°
Venus' śīghra	3,207,125,280	4,304,448,000	5,58;42,14,24°
Mercury's śīghra	8,191,827,435	4,288,896,000	5,57;24,28,48°
Moon	26,375,932,110	0	0°
Moon's manda	222,917,925	1,505,952,000	2,5;29,45,36°
Moon's node	−106,096,510	1,838,592,000	−2,33;12,57,36°

responding avamas, and t_m the lapsed tithis of the current month until the beginning of the current tithi, which precedes the current mean sunrise day by $\frac{r_u}{C}$. Therefore,

$$c'_y = c_y + \frac{r_u}{C}, \tag{b}$$

and the mean progress of the moon since mean Caitraśuklapratipad is

$$\overline{\Delta\lambda_{☾}} \approx 13\, c'_y. \tag{c}$$

The lapsed synodic months from mean Caitraśuklapratipad until the beginning of the current mean sunrise day are found by (V.60)

$$m'_y = m_y + \frac{t_m + \frac{r_u}{C}}{30}. \tag{a}$$

Since $n = m - a$, it follows that

$$n_y = m'_y - \frac{r_a}{T}, \tag{b}$$

and the mean progress of the sun is

$$\overline{\Delta\lambda_{☉}} = 30\, n_y. \tag{c}$$

Brahmagupta's remaining rules for finding mean longitudes are simply playful, and need not be explicated here. The mean longitudes are corrected for the deśāntara by a rule similar to (III.37), although the circumference of the earth is taken to be 5,000 yojanas, as in the *Paitāmaha* (*BSS* 1, 33–38):

$$\delta^2 = d^2 - \left(\frac{5000}{360} \cdot \Delta\phi\right)^2, \tag{V.61}$$

where d is the shortest distance between two localities, $\Delta\phi$ the difference of their terrestrial latitudes, and δ their longitudinal difference in yojanas. Clearly one of the two localities must lie on the prime meridian; for the formula to be correct, it should be Laṅkā itself. This criticism was also made by Bhāskara I (*MB* 2, 5–6).

The basic planetary model in the *Brāhmasphuṭasiddhānta* is identical with that in the *Paitāmaha* (Figure 14). The manda epicycles of the sun, the moon, and Venus and the śīghra epicycle of Venus are, however, pulsating (*BSS* 2, 20–21, 34–35). For Venus the maximum manda epicycle, c_M, and minimum śīghra epicycle, c_m, occur at anomalies of 0° and 180°, the minimum manda epicycle and maximum śīghra epicycle at anomalies of 90° and 270°. The corrected epicycle, c', will, then, be (*BSS* 2, 13)

$$c' = c_0 \pm (c_M - c_m) \cdot \frac{\text{Sin}\,\kappa}{R}, \tag{V.62}$$

where c_0 is the mean epicycle. The pulsation of the epicycles of the luminaries depends on the position of the body relative to the local meridian and six-o'clock circles and on the amount of the anomaly. These pulsating epicycles are so arranged that it appears they may be intended to account for horizon refraction, although that is in fact independent of the anomaly; and the magnitude of the effect of Brahmagupta's device, at its maximum effectiveness on the horizon, is only a tenth of that of refraction.

The longitudes of the mandoccas and nodes of the planets in 628, according to Brahmagupta's parameters, are given in Table V.13.

TABLE V.13

Planet	Mandocca	Node
Saturn	260;55°	103;12°
Jupiter	172;32°	82;1°
Mars	128;24°	31;54°
Sun	77;55°	
Venus	81;15°	59;47°
Mercury	224;54°	21;11°

Brahmagupta's values of the circumferences of the epicycles are closely related to, but not identical with, the *Paitāmaha*'s (Table V.5); they are given in Table V.14 (*BSS* 2, 21–22, 34–39), where the *Paitāmaha*'s values of c are in italics.

TABLE V.14

Planet	Manda epicycle		Śīghra epicycle	
	c_μ	μ_{max}	c_σ	σ_{max}
Saturn	*30°*	4;46,47°	35°	5;34,46°
Jupiter	*33°*	5;15,35°	*68°*	10;53,19°
Mars	*70°*	11;12,41°	243;40°	42;37,39°
Sun	c_M 14°	2;13,42°		
	c_0 *13;40°*	2;10,30°		
	c_m 13;20°	2;7,20°		
Venus	c_M *11°*	1;45,3°	263°	46;57,43°
	c_0 10°	1;30,50°	260;30°	46;22,54°
	c_m 9°	1;25,57°	*258°*	45;48,12°
Mercury	*38°*	6;3,33°	*132°*	21;31,30°
Moon	c_M 32;28°	5;10,28°		
	c_0 *31;36°*	5;2,7°		
	c_m 30;44°	4;53,50°		

The differences in the maximum equations between the *Paitāmaha* and Brahmagupta are due to computing with different values of R. Brahmagupta uses the normal Indian "method of sines" to find the manda equation (*BSS* 2, 17).[48]

Brahmagupta's method of integrating the two equations for all the planets save Mars (*BSS* 2, 36) is similar to that of the *Paitāmaha* for the superior planets (V.7); the difference lies in the fact that the corrected anomalies are found by algebraically applying all of the equation rather than just half. For Mars, Brahmagupta (*BSS* 2, 37–40) begins the process by modifying κ_μ by an amount that depends on κ_σ; the maximum correction is 6;40° when κ_σ is 45°, 135°, 225°, or 315°. Such a computation for Mars had already been stated, although obscurely, by the *Paitāmaha* (*Pait.* 4, 8). The integration rule for Mars uses the halving of the equations that Brahmagupta avoids for the other planets.

Brahmagupta gives rules for computing the increase or decrease to the equation, x, resulting from the pulsation of the manda epicycles of the sun and the moon as the angle of depression from the meridian, n, varies (*BSS* 2, 27–28). For the sun this is

$$x = \text{Sin}\,\kappa \cdot \text{Sin}\,n \cdot \frac{191}{R^2}. \qquad \text{(V.63)}$$

The proof of this is as follows. To express μ in minutes, we can form (since $\mu < 3;45°$ for the sun) (V.64)

$$\mu = \text{Sin}\,\kappa \cdot \frac{c_\mu}{360} \cdot \frac{225}{214} \cdot 60. \qquad \text{(a)}$$

From Table V.13 we know that the change in the size of the manda epicycle of the sun is $\frac{1°}{3} \cdot \frac{\text{Sin}\,n}{R}$.

$$x : \mu \approx \frac{\text{Sin}\,n}{3R} : c_\mu \qquad \text{(b)}$$

and, from (a) and (b),

$$x = \text{Sin}\,\kappa \cdot \text{Sin}\,n \cdot \frac{c_\mu \cdot 225 \cdot 60}{360 \cdot 214 \cdot 3R \cdot c_\mu} \cdot \frac{R}{R} =$$

$$\text{Sin}\,\kappa \cdot \text{Sin}\,n \cdot \frac{191\frac{1}{214}}{R^2}, \qquad \text{(c)}$$

which is rounded off to (V.63).

For the moon the change in the size of the manda epicycle is $\frac{13}{15} \cdot \frac{\text{Sin}\,n}{R}$. Therefore (V.65)

$$x : \mu = \frac{13 \cdot \text{Sin}\,n}{15 \cdot R \cdot c_\mu}. \qquad \text{(a)}$$

But, since the effect of the pulsating epicycle for the moon is most noticeable when $\mu > 3;45°$, we use the second sine difference, 213, in

$$\mu = \text{Sin}\,\kappa \cdot \frac{c_\mu}{360} \cdot \frac{225}{213} \cdot 60, \qquad \text{(b)}$$

in which μ is expressed in minutes. Then

$$x = \text{Sin}\,\kappa \cdot \text{Sin}\,n \cdot \frac{13 \cdot 225}{15 \cdot 6 \cdot 213 \cdot R} \cdot \frac{R}{R} =$$

$$\text{Sin}\,\kappa \cdot \text{Sin}\,n \cdot \frac{498\frac{201}{213}}{R^2}. \qquad \text{(c)}$$

Brahmagupta's formula is

$$x = \text{Sin}\,\kappa \cdot \text{Sin}\,n \cdot \frac{499}{R^2}. \qquad \text{(d)}$$

One component of the equation of time is that due to the variability of the daily solar velocity; it is simply $\mu_☉$. The influence of this upon any planet's argument, x_κ, which is called the bhujāntara, is given by the formula (*BSS* 2, 29)

$$x_\kappa = \mu_☉ \cdot \frac{v_p}{21600}, \qquad \text{(V.66)}$$

where v_p is the planet's daily velocity and all terms of the equation are in minutes.

The correction, x_μ, to the mean daily motion, $\bar{v}$, due to the manda equation described by Brahmagupta is an approximation to the cosine rule already expounded in the *Sūryasiddhānta* of Lāṭadeva (*PS* 9, 11–14). The expression in the *Brāhmasphuṭasiddhānta* is (*BSS* 2, 41–44)

$$x_\mu = \Delta\kappa \cdot \frac{\Delta\,\text{Sin}\,\kappa}{225} \cdot \frac{c_\mu}{360}, \qquad \text{(V.67)}$$

where $\Delta\kappa$ is the daily increment in the manda anomaly, and $\frac{\Delta\,\text{Sin}\,\kappa}{225} \approx \cos\kappa$.

The correction, x_σ, due to the śīghra equation, on the other hand, is inexplicable. It is

$$x_\sigma = \sigma \cdot \frac{\text{Sin}\,v_\kappa}{225} \cdot \frac{R}{H}, \qquad \text{(V.68)}$$

where $v_\kappa = \bar{v}_\kappa - \Delta\mu$ ($\bar{v}_\kappa$ is the difference between the mean daily motion of the planet's śīghra and the mean daily motion of the sun) and $\frac{\text{Sin}\,v_\kappa}{225} \approx \frac{\Delta\,\text{Sin}\,\kappa}{225} \approx \cos\kappa$. The presence of σ and of $\frac{R}{H}$ in the formula, however, seems to be incorrect.

Brahmagupta also lists the śīghra anomalies necessary for the occurrences of the Greek-letter phenomena (*BSS* 2, 48–53). This is identical with the *Paitāmaha*'s (Table V.9), except that the first stations of Mars and Mercury occur when κ_σ equals 163° and 145°, respectively.

In adhyāya 14 Brahmagupta briefly explains a

planetary model with an eccentric circle producing the manda equation and an epicycle producing the śīghra equation (*BSS* 14, 10–18). This model, which is surely Hellenistic in origin, was also known to Āryabhaṭa (*A* Kālakriyā 17–19).

Brahmagupta also refers to the portions (bhogas) of the nakṣatras (*BSS* 14, 47–52), 1 bhoga being equal to 790′ (the mean daily motion of the moon). Certain naksatras (Bharaṇī, Ārdrā, Āśleṣā, Svāti, Jyeṣṭhā, and Śatabhiṣaj) have half a bhoga or 395′; others (Punarvasu, Uttaraphālgunī, Viśākhā, Uttarāṣāḍhā, and Pūrvabhādrapadā) have one and a half bhogas or 1185′. The rest of the twenty-seven nakṣatras have one bhoga apiece. Since the sum of these does not equal 21,600′, the remainder—270′—is made the bhoga of Abhijit.

In this chapter he also gives, following Bhāskara I (*MB* 7, 17–19), a remarkable rule for finding $\mathrm{Sin}\,\kappa \cdot \frac{r}{R}$, where *r* is the radius of the epicycle, without first finding Sin κ; it is, in fact, a general formula for finding Sin ϑ. The rule is (*BSS* 14, 23–24)[49]

$$\mathrm{Sin}\,\kappa \cdot \frac{r}{R} = \frac{R \cdot (180-\kappa) \cdot \kappa \cdot r}{10125 - \frac{(180-\kappa)\cdot\kappa}{4}}. \qquad \text{(V.69)}$$

This is equivalent to

$$\sin\kappa = \frac{4R \cdot (180-\kappa)\cdot\kappa}{\kappa^2 - 180\,\kappa + 40500}. \qquad \text{(V.70)}$$

In this formula $R = 1$ since, if $\vartheta = 90°$,

$$\sin 90° = \frac{4R \cdot 90^2}{40500 - 8100} = R. \qquad \text{(V.71)}$$

The computation of a few values of sin ϑ with this formula in Table V.15 and their comparison with the entries in a modern table demonstrate its general approximative value.

TABLE V.15

ϑ	sin ϑ (computed)	sin ϑ (tabular)
1°	.0176	.01745
15°	.2603	.25882
30°	.5	.5
45°	.7058	.70711
60°	.8919	.86603
75°	.9654	.96593

The reverse of (V.69) is also given by Brahmagupta in the form (*BSS* 14, 25–26)

$$\kappa = 90 - \sqrt{90^2 - \frac{10125 \cdot \mathrm{Sin}\,\kappa}{R + \frac{\mathrm{Sin}\,\kappa}{4}}}, \qquad \text{(V.72)}$$

so that

$$\vartheta = 90 - \sqrt{90^2 - \frac{40500 \cdot \mathrm{Sin}\,\vartheta}{4 + \mathrm{Sin}\,\vartheta}}. \qquad \text{(V.73)}$$

The actual Sine table in the *Brāhmasphuṭasiddhānta* uses the odd value $R = 3{,}270$. The table itself, which is very accurately computed, is reproduced in Table V.16 (*BSS* 2, 2–5).

TABLE V.16

ϑ	Sin ϑ	R sin ϑ	ϑ	Sin ϑ	R sin ϑ
3;45°	214	213.8	48;45°	2459	2458.5
7;30°	427	426.8	52;30°	2594	2594.2
11;15°	638	637.9	56;15°	2719	2718.9
15°	846	846.3	60°	2832	2831.9
18;45°	1051	1051.1	63;45°	2933	2932.7
22;30°	1251	1251.3	67;30°	3021	3021.0
26;15°	1446	1446.2	71;15°	3096	3096.4
30°	1635	1635	75°	3159	3158.5
33;45°	1817	1816.7	78;45°	3207	3207.1
37;30°	1991	1990.6	82;30°	3242	3242.0
41;15°	2156	2156.0	86;15°	3263	3263.0
45°	2312	2312.2	90°	3270	3270

Brahmagupta also gives a table of Versines (*BSS* 2, 6–9), as is usual.

Brahmagupta presents many elaborations of the *Paitāmaha*'s formulas in spherical trigonometry, but he also adds some new ones. He analyzes the gnomon shadow (see Figure 21) as follows (*BSS* 3,4):

$$\mathrm{Sin}_h\,\eta = \mathrm{Sin}\,\eta \cdot \frac{h_s}{R}. \qquad \text{(V.74)}$$

Then the bhuja, *bg*, which was called the koṭi in (III.71) (see Figures 10 and 12), is found from

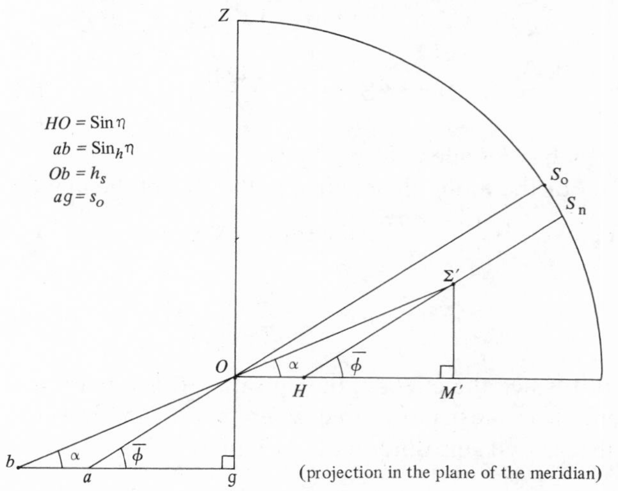

FIGURE 21

$$bg = s_0 \pm \mathrm{Sin}_h\, \eta. \qquad \text{(V.75)}$$

The other side, here called the koṭi, again is found by the Pythagorean theorem.

For finding the right ascension, α, of an arc, λ, Brahmagupta presents a rule (*BSS* 3, 14–17) previously given by Āryabhaṭa (*A* Gola 25):

$$\mathrm{Sin}\, \alpha = \mathrm{Sin}\, \lambda \cdot \frac{\mathrm{Cos}\, \epsilon}{R} \cdot \frac{R}{\mathrm{Cos}\, \delta} = \mathrm{Sin}\, \lambda \cdot \frac{\mathrm{Cos}\, \epsilon}{\mathrm{Cos}\, \delta}, \qquad \text{(V.76)}$$

which is evident from Figure 6.

As well as further exploiting the relation of the cheda triangle to the gnomon triangle (V.10–15), Brahmagupta introduces the iṣṭāntyā, I, which is $\Sigma' G'$ in Figure 15 (*BSS* 3, 29):

$$I = \mathrm{Sin}\, t \pm \mathrm{Sin}\, \omega, \qquad \text{(V.77)}$$

where t is the time degrees elapsed since or remaining before six o'clock. The iṣṭāntyā, therefore, is the Sine of the time degrees that have lapsed since sunrise or remain until sunset. Thus (*BSS* 3, 30–32)

$$D = I \cdot \frac{r}{R}, \qquad \text{(V.78)}$$

and $I \cdot \frac{r}{R}$ can be universally substituted for D. And, as the *Paitāmaha* found the noon antyā, so Brahmagupta finds the noon iṣṭāntyā, I_n, which is $S'A$ in Figure 15 (*BSS* 3, 34):

$$I_n = R + \mathrm{Sin}\, \omega = D_n \cdot \frac{R}{r}. \qquad \text{(V.79)}$$

Finally, along these same lines, he adds (*BSS* 3, 37)

$$I_n - \mathrm{Vers}\, n = I, \qquad \text{(V.80)}$$

where n is the hour angle of the sun from the meridian—$\Sigma' OS'$ in Figure 15. Brahmagupta explores the many ramifications of this relationship (*BSS* 3, 38–46).

But the most extraordinary rule that Brahmagupta gives is the following, for finding the solar altitude (*BSS* 3, 54–56):

$$\mathrm{Sin}\, \alpha = \sqrt{\frac{\left(\frac{R^2}{2} - \mathrm{Sin}^2 \eta\right) \cdot 12^2}{72 + s_0^2} + \left(\frac{12 \cdot s_0 \cdot \mathrm{Sin}\, \eta}{72 + s_0^2}\right)^2} + \frac{12 \cdot s_0 \cdot \mathrm{Sin}\, \eta}{72 + s_0^2}. \qquad \text{(V.81)}$$

This is true only when the sun is in the northern hemisphere and the angle between the shadow and the east-west line is 45°. Clearly there are easier ways to find Sin α for such a moment.

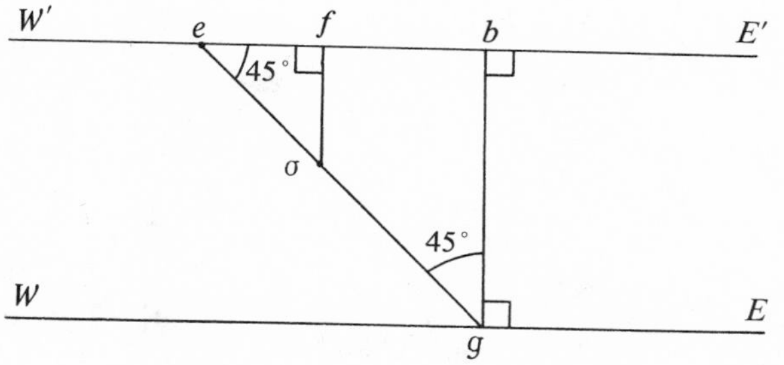

$gb = eb = s_o$
$eg = s$
$\sigma f = ef = \mathrm{Sin}_h\, \eta$

FIGURE 22

From Figure 22 it can be seen that

$$s^2 = 2 \cdot s_0^2 \qquad \text{(V.82)}$$

under the conditions stated above. Further, if $g\sigma$ is the shadow of the sun (at E in Figure 23) on the equator at an hour angle of 45°, then

$$\sigma e^2 = 2 \cdot \mathrm{Sin}_h^2\, \eta. \qquad \text{(V.83)}$$

Now, from Figure 23 it is clear that (V.84)

$$\mathrm{Sin}\, \alpha = SJ + JF. \qquad \text{(a)}$$

To find SJ (using V.83):

$$\frac{SO}{h} = \frac{SE'}{\sigma e} = \frac{SE'}{2 \cdot \mathrm{Sin}_h\, \eta} = \frac{R}{h}; \qquad \text{(b)}$$

$$\frac{R^2}{R^2} = \frac{SE'^2}{2 \cdot \mathrm{Sin}^2\, \eta}; \qquad \text{(c)}$$

$$SE'^2 = 2 \cdot \mathrm{Sin}^2\, \eta; \qquad \text{(d)}$$

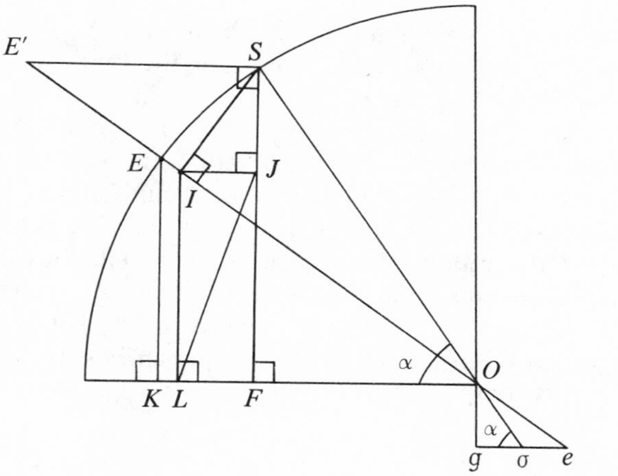

FIGURE 23

$$\frac{SI}{SJ}=\frac{h}{s}; \quad \text{(e)}$$

$$\frac{SI}{SE'}=\frac{g}{h}; \quad \text{(f)}$$

$$\frac{SJ}{SE'}=\frac{s\cdot g}{h^2}. \quad \text{(g)}$$

Then, using (V.82) and (V.84 d):

$$SJ=\frac{SE'\cdot s\cdot g}{g^2+s^2}=\frac{\text{Sin}\,\eta\cdot s_0\cdot 12}{72+s_0^2}. \quad \text{(h)}$$

To find $JF=IL$:

$$\frac{IJ^2}{SI^2}=\frac{EK^2}{EO^2}=\frac{IL^2}{IO^2}=\frac{IJ^2+IL^2}{SI^2+IO^2}=\frac{JL^2}{EO^2}. \quad \text{(i)}$$

Therefore,

$$JL^2=EK^2; \quad \text{(j)}$$

$$IL^2=JL^2-IJ^2=EK^2-IJ^2=EK^2-SI^2+SJ^2. \quad \text{(k)}$$

But

$$\frac{EO^2}{EK^2}=\frac{h^2}{g^2}. \quad \text{(l)}$$

With (V.84 f) and (V.84 l):

$$EK^2-SI^2=\frac{g^2\cdot EO^2-g^2\cdot SE'^2}{g^2+s^2}=$$

$$\frac{\left(\frac{R^2}{2}-\text{Sin}^2\,\eta\right)\cdot 12^2}{72+s_0^2}. \quad \text{(m)}$$

Therefore,

$$IL=\sqrt{\frac{\left(\frac{R^2}{2}-\text{Sin}^2\,\eta\right)\cdot 12^2}{72+s_0^2}+\left(\frac{\text{Sin}\,\eta\cdot s_0\cdot 12}{72+s_0^2}\right)^2}, \quad \text{(n)}$$

and $IL+SJ=\text{Sin}\,\alpha$ has the form of (V.81).

Brahmagupta devotes most of the remainder of this adhyāya to problems involving $\text{Sin}_h\eta$ and $\text{Sin}\,\eta$. The only rule that need be repeated here is the following: To find the length of the string, $e'w'$, connecting the points on the circumference of a circle the center of which is the base of the gnomon and the radius of which is r', at which the sun's shadow falls at sunrise (e') and at sunset (w') (it is assumed that there is no change in the solar declination during daylight), he prescribes (*BSS* 3, 64) (V.85):

$$E'W'=2\sqrt{R^2-\text{Sin}^2\,\eta}; \quad \text{(a)}$$

$$e'w'=E'W'\cdot\frac{r'}{R}. \quad \text{(b)}$$

In adhyāya 15 the principal new relations that Brahmagupta investigates are those involving the time interval between the moments of the sun's being on the prime vertical and its being on the meridian. In Figure 10 the Sine of the corresponding time degrees is DS_0, and the Sine of the time degrees corresponding to the time interval between the moment of sunrise and that when the sun is on the prime vertical is $OD\pm\text{Sin}\,\omega$. Brahmagupta's first formula is (*BSS* 15, 19–20)

$$AC=\text{Sin}\,\delta\cdot\frac{g}{s_0}=OD\cdot\frac{r}{R}. \quad \text{(V.86)}$$

Then, if we multiply all sides of right triangle OAS_n by $\frac{R}{r}$, we have (V.87)

$$AS_n=R; \quad \text{(a)}$$

$$AO=\text{Sin}\,\delta\cdot\frac{R}{r}=OD\cdot\frac{s_0}{g}; \quad \text{(b)}$$

$$OS_n^2=AS_n^2+AO^2=R^2+OD^2\cdot\frac{s_0^2}{g^2}. \quad \text{(c)}$$

Further,

$$\frac{\text{Sin}^2\,\delta}{AO^2}=\frac{R^2}{AS_n^2}, \quad \text{(d)}$$

and therefore

$$\text{Sin}^2\,\delta=\frac{s_0^2\cdot OD^2}{g^2+\frac{s_0^2\cdot OD^2}{R^2}}. \quad \text{(e)}$$

At one point Brahmagupta makes the absurd mistake of substituting $R^2-DS_0^2$ for OD^2, so that he states wrongly (*BSS* 15, 24–25):

$$\text{Sin}^2\,\delta=\frac{s_0^2\,(R^2-DS_0^2)}{g^2+\frac{s_0^2(R^2-DS_0^2)}{R^2}}. \quad \text{(f)}$$

But later, using the relation

$$\text{Sin}\,\delta\cdot\frac{R}{r}=\text{Sin}\,\omega\cdot\frac{g}{s_0}, \quad \text{(g)}$$

he obtains the correct (*BSS* 15, 35–38)

$$\text{Sin}^2\,\delta=\frac{R^2}{\frac{R^2\cdot s_0^2+g^2\cdot\text{Sin}^2\,\omega}{g^2\cdot\text{Sin}^2\,\omega}}. \quad \text{(h)}$$

Then, considering the triangle BOC in Figure 10, he observes that

$$OD \pm \text{Sin}\,\omega = BC \cdot \frac{R}{r} = \text{Sin}\,\eta \cdot \frac{R}{\text{Sin}\,\phi}. \quad (\text{V}.87)$$

Combining this with (III.63) and (III.70), he finds (*BSS* 15, 21–23):

$$\text{Sin}\,\lambda_{\odot} = \frac{BC \cdot \frac{R}{r} \cdot \text{Sin}\,\phi \cdot \text{Sin}\,\bar{\phi}}{R \cdot \text{Sin}\,\epsilon}. \quad (\text{V}.88)$$

Finally, Brahmagupta gives values for the right ascension, α, of the first three zodiacal signs (*BSS* 15, 32–33); in Table V.17 I add Ptolemy's values, the minutes corresponding to Brahmagupta's divided by 6, and the vināḍīs $\approx \frac{\text{minutes}}{6}$, according to the *Pañcasiddhāntikā* (Table III.21).

TABLE V.17

Sign	Brahmagupta	Ptolemy		*Pañcasiddhāntikā*
Aries	27;50°	27;50°	$\frac{1670'}{6} = 278\ 1/3$	278 vin.
Taurus	29;55°	29;54°	$\frac{1795'}{6} = 299\ 1/6$	299 vin.
Gemini	32;15°	32;16°	$\frac{1935'}{6} = 322\ 1/2$	323 vin.

Adhyāya 4 of the *Brāhmasphuṭasiddhānta* is devoted to lunar eclipses, in which five significant phases are distinguished: first contact, beginning of totality, mideclipse, end of totality, and last contact (*BSS* 4, 1–3). Brahmagupta's rule for finding the latitude of the moon is identical with the *Paitāmaha*'s (Table V.7 and formula V.16) (*BSS* 4, 5). But for finding the apparent diameters of the sun, the moon, and the earth's shadow, he develops a procedure different from the *Paitāmaha*'s as described in (V.18–21). If the daily motion of a planet be denoted v, he prescribes (*BSS* 4, 6) (V.89):

$$d_{\odot} = v_{\odot} \cdot \frac{11}{20}; \quad \text{(a)}$$

$$d_{☾} = v_{☾} \cdot \frac{10}{247}; \quad \text{(b)}$$

$$d_u = \frac{v_{☾} \cdot 8 - v_{\odot} \cdot 25}{60}. \quad \text{(c)}$$

From this the mean values given in Table V.18 are derived.

TABLE V.18

Planet	$\bar{d}$
Sun	0;32,31°
Moon	0;32°
Shadow	1;20,49,12°

Brahmagupta's rules for finding the duration of a lunar eclipse, Δt_e, are equivalent to the *Pauliśa*'s (III.41–42); the duration of totality, Δt^{τ}, is found from (*BSS* 4, 8)

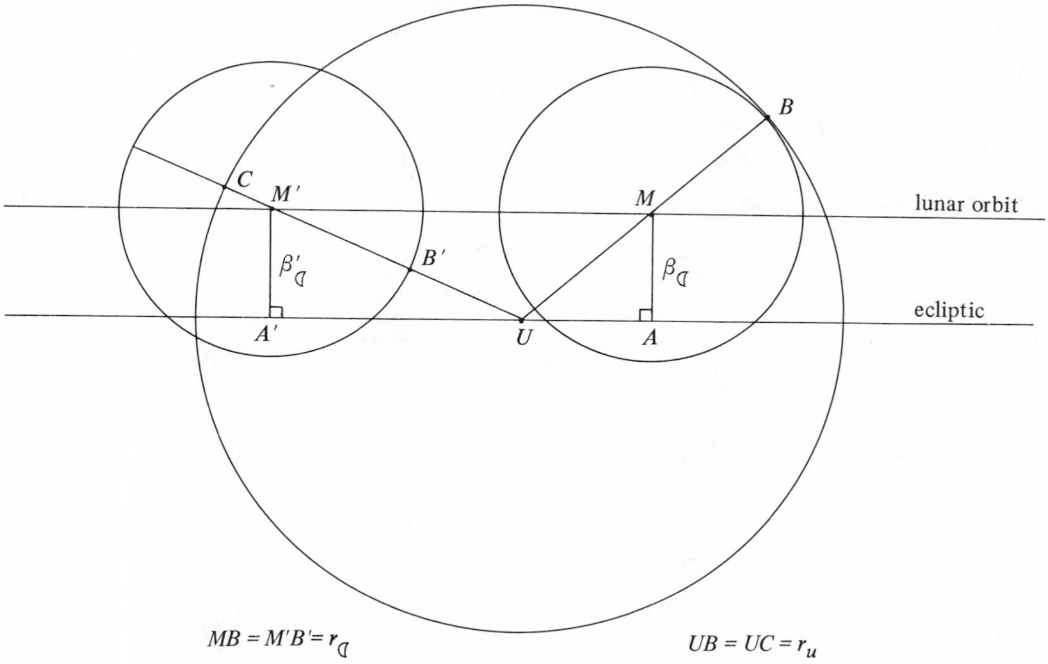

FIGURE 24

$$\Delta t^{\tau} = 2 \cdot \frac{\sqrt{(r_u - r_{☾})^2 - \beta_{☾}^2}}{v_{☾} - v_{☊}}, \qquad (V.90)$$

where the dividend is AU in Figure 24. To find the changes in longitude during the half-durations of the eclipse and of totality, Brahmagupta forms (*BSS* 4, 9)

$$\Delta\lambda = \frac{\Delta t^{\tau}}{2} \cdot \frac{v_{☾}}{60}. \qquad (V.91)$$

To find the magnitude of an eclipse for a given time, t_n, we must first find the elongation of the moon from the shadow center at that time (*BSS* 4, 11–12) (V.92):

$$\Delta t_n = \frac{\Delta t_e}{2} - t_n; \qquad (a)$$

$$\Delta\lambda_n = \Delta t_n \cdot \frac{v_{☾} - v_{☊}}{60}. \qquad (b)$$

Then, in Figure 24, the eclipse magnitude, $B'C$, is (*BSS* 4, 12)

$$B'C = r_{☾} + r_u - M'U. \qquad (c)$$

These rules, with appropriate substitutions, are also considered valid approximations to the durations and magnitudes of solar eclipses, although in adhyāya 5 the changes in lunar latitude are taken into consideration (*BSS* 5, 13–15).

Brahmagupta gives a complete set of rules for computing the two elements of deflection (*BSS* 4, 16–18) like those previously given by Lāṭadeva in his *Sūryasiddhānta* (*PS* 11). The akṣavalana, γ_1, is found from (see III.47)

$$\text{Sin}\,\gamma_1 = \text{Sin}\,t^0 \cdot \frac{\text{Sin}\,\phi}{R}. \qquad (V.93)$$

The ayanavalana, γ_2, which is the angle between the ecliptic and the equator, is found from

$$\gamma_2 = \delta\,(\lambda + 90°). \qquad (V.94)$$

The projection is illustrated in Figure 25.

Brahmagupta repeats Āryabhaṭa's colors (*A* Gola 46) for the different phases of a lunar eclipse; they are noted in Table V.19 (*BSS* 4, 19).

TABLE V.19

Beginning and end	smoky
Partial	black
More than half	black-coppery
Total	tawny

In computing parallax, Brahmagupta makes the

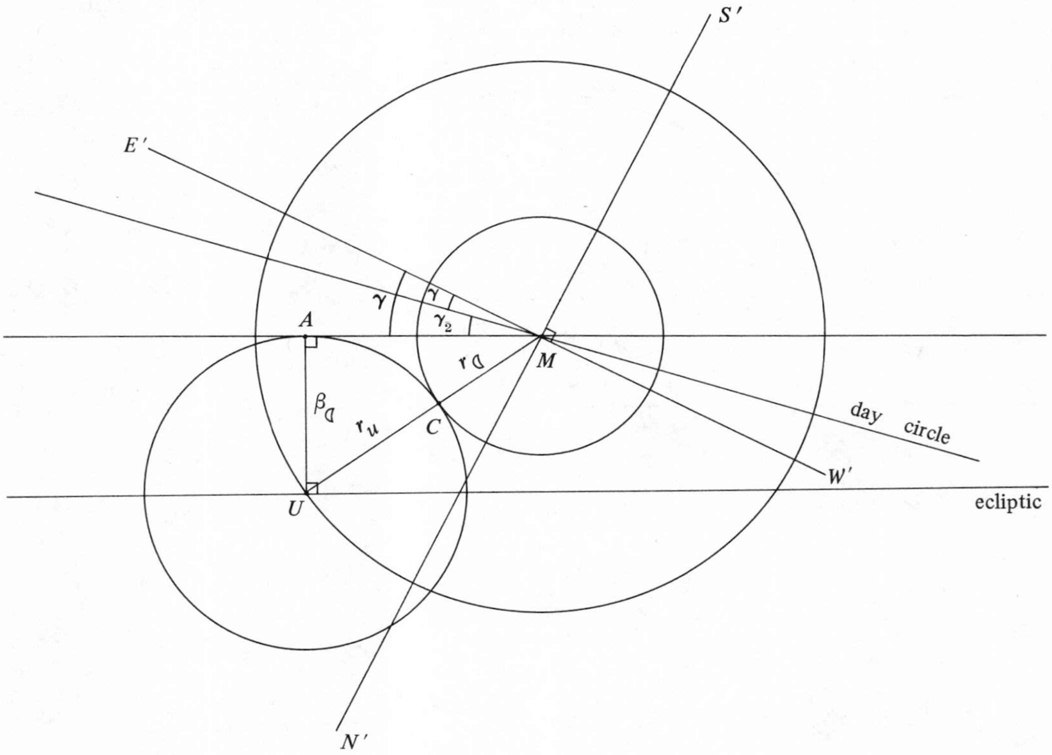

FIGURE 25

same error as does the *Paitāmaha*—that is, he regards the zenith distance of the nonagesimal, ZV, to be the same as the zenith distance of a body on the meridian (see V.26). Thus he states (*BSS* 5, 3)

$$\text{Sin } ZV \approx \text{Sin } \bar{\alpha}_n = \text{Sin } (\phi \pm \delta). \qquad (V.95)$$

If the ecliptic passed through Z (see Figure 1), then it would be true that

$$\pi = \text{Sin } ZS \cdot \frac{\pi_0}{R} = \text{Sin } VS \cdot \frac{\pi_0}{R}, \qquad (V.96)$$

where $ZS = VS$ is the elongation between the sun and the zenith and $\pi_0 = 4$ nāḍīs. When the ecliptic does not pass through Z, it is assumed that the ratio $\frac{\text{Sin } \alpha(V)}{R}$ (where $\alpha[V]$ is the altitude of the nonagesimal) can be used to transform the longitudinal parallax to the ecliptic. Then (*BSS* 5, 4)

$$\pi_\lambda = \frac{\text{Sin } \alpha(V)}{R} \cdot \frac{\text{Sin } VS \cdot 4}{R} \approx \frac{\text{Sin } VS \cdot 48}{R \cdot h_n}, \qquad (V.97)$$

since $\frac{h_n}{g} = \frac{R}{\text{Sin } \alpha_n}$. Moreover, for latitudinal parallax, one must find the approximate zenith distance of the moon, ZM, from (*BSS* 5, 9)

$$\text{Sin } ZM \approx \text{Sin } (\phi \pm \delta \pm \beta_☾). \qquad (V.98)$$

From (V.95) one finds the solar latitudinal parallax (*BSS* 5, 10–11)

$$\pi_\beta = \bar{v}_☉ \cdot \frac{\text{Sin } ZV}{15\, R}, \qquad (V.99)$$

and the lunar latitudinal parallax

$$\pi_\beta = \bar{v}_☾ \cdot \frac{\text{Sin } ZM}{15\, R}, \qquad (V.100)$$

and the combined parallax (*BSS* 5, 22–24)

$$\pi_\beta = (\bar{v}_☾ - \bar{v}_☉) \cdot \frac{\text{Sin } ZM}{15\, R}. \qquad (V.101)$$

In all three formulas one would expect the true rather than the mean daily velocity to be used.

By computing the parallaxes for different phases of the eclipse and iterating the procedure until all differences vanish, the true durations and times between phases can be determined, as can the true magnitudes at different times during the eclipse. The projections are similar to those in a lunar eclipse. If the magnitude of a solar eclipse is less than 1/12 of the sun's apparent diameter, or if the magnitude of a lunar eclipse is less than 1/16 of the moon's apparent diameter, Brahmagupta states that the eclipse is not visible (*BSS* 5, 20).

Adhyāya 16 is devoted to practical rules for constructing projections. Brahmagupta does, however, add the method for finding the longitudinal difference (the deśāntara) from the difference between the calculated and the observed times of eclipses (*BSS* 16, 27–28); he repeats this from the *Paitāmaha* (*Pait.* III, 27). He also gives the rules for "predicting" eclipses from the parvan, which is a period of six months, or 177 civil days. To the mean longitudes at one parvan of the sun, moon, moon's manda, and moon's node are added the cālanas, or increments in mean motion for 177 days, listed in Table V.20 (*BSS* 16, 29–33).

TABLE V.20

Sun	5^s 24;27,6°
Moon	5^s 22;12,53°
Manda	0^s 19;42,56°
Node	0^s 9;22,40°

In adhyāya 6, Brahmagupta deals with the first and last visibilities of the planets. In the course of this he gives the correct formulations of the rules for performing the ayanadṛkkarma (V.29) and the akṣadṛkkarma (V.35), which are corruptly preserved in the *Paitāmaha*. He also repeats the *Paitāmaha*'s kālāṃśas for visibility (Table V.8) (*BSS* 6, 6) and its rule for finding the time until first or last visibility (V.38) (*BSS* 6, 7). Brahmagupta adds only a correction in the kālāṃśas of visibility for Mercury that depends on its śīghra equation, one for Jupiter that depends on its manda equation, and one for Venus that takes into account the fact that its apparent diameter is smaller at superior than it is at inferior conjunction (*BSS* 6, 10–11).

Adhyāya 7 of the *Brāhmasphuṭasiddhānta* is devoted to the determination of the illuminated portion of the moon. One performs the dṛkkarma as in adhyāya 6 in order to find $\beta^*_☾$; then (*BSS* 7, 5)

$$\delta^*_☾ = \delta_☾ \pm \beta^*_☾. \qquad (V.102)$$

With $\delta^*_☾$ one can find the radius of the day circle passing through the moon by means of (III.31), and thereby the time in ghaṭikās that the moon will be above the horizon. To find the arc on the great circle that passes through the sun and the moon, Brahmagupta employs the construction given by the *Paitāmaha* (V.41–44) (*BSS* 7, 7–10). His rules for finding the illuminated portion have already been given (V.46–47). Brahmagupta concludes with instructions on how to transform these calculations into a projection in scale in order to astound one's patrons (*BSS* 7, 15–17). Adhyāya 17 gives further rules for this projection.

In adhyāya 8 Brahmagupta discusses the moon's shadow, $s_{☾}$, which is approximately

$$s_{☾} \approx \text{Sin}\,\bar{\alpha}_{☾} \cdot \frac{g}{\text{Sin}\,\alpha_{☾}}. \quad (\text{V}.103)$$

First one computes the ghaṭikās that the moon is above the horizon, $g_{☾}$, as indicated in adhyāya 7; then one converts the ecliptic arc between the longitude of the ascendant and the polar longitude of the moon, $\lambda^*_{☾}$, into ghaṭikās by means of the local oblique ascensions. If those ghaṭikās equal $\frac{g_{☾}}{2}$, then $\lambda^*_{☾}$ is on the meridian and one can find Sin $\bar{\alpha}_{☾}$ from (III.53 b) (with $\delta^*_{☾}$). Otherwise, with the time in ghaṭikās since moonrise or until moonset, one can find the altitude of the moon from (V.10–12). The shadow, however, is assumed to be cast by the upper rim of the moon, so that one is to form (*BSS* 8, 6–7)

$$\alpha'_{☾} = \alpha_{☾} + r_{☾}, \quad (\text{V}.104)$$

with the radius of the moon expressed in minutes of arc. And this new altitude is affected by parallax to form

$$\alpha''_{☾} = \alpha'_{☾} - \frac{\bar{v}_{☾}}{15}. \quad (\text{V}.105)$$

Thence

$$s_{☾} = \text{Sin}\,\bar{\alpha}''_{☾} \cdot \frac{g}{\text{Sin}\,\alpha''_{☾}}. \quad (\text{V}.106)$$

One can perform a similar operation to find the shadow of the sun.

In adhyāya 9 Brahmagupta tackles the problems of the conjunctions of planets with other planets and with fixed stars. The first step toward any solution is to find the planetary latitudes. He derives the values for the orbital inclinations from the *Paitāmaha* (Table V.7) (*BSS* 9, 1), but "corrects" the *Paitāmaha*'s formula for finding β (V.17) while retaining the separate models for superior and inferior planets. The distance of the mean longitude of an inferior planet from its node on its śīghra epicycle, ω, is corrected by its first manda equation; the distance of the mean longitude of a superior planet from its node on its orbit is corrected by its first manda equation and its first śīghra equation (*BSS* 9, 8–11).

The apparent diameters of the planets depend, according to Brahmagupta, on two factors; their kālāṃśas of visibility and their śīghra hypotenuses, H. The *Paitāmaha* determines their mean diameters in yojanas by a process of halving in which the order of the planets depends on the kālāṃśas of visibility (Table V.8). Brahmagupta's rule to find the mean diameter of a planet in minutes, $\bar{d}$, is (*BSS* 9, 2)

$$\bar{d} = \frac{81}{k}, \quad (\text{V}.107)$$

where k represents the kālāṃśas of visibility. The resulting mean diameters in minutes are presented in Table V.21.

TABLE V.21

Planet	$\bar{d}$
Venus	9′
Jupiter	7 4/11′
Mercury	6 3/13′
Saturn	5 6/15′
Mars	4 13/17′

Brahmagupta's formula (*BSS* 9, 3–4) for finding the true apparent diameter, d, is (V.108)

$$d = \frac{27(R + 3\,\text{Sin}\,\sigma - H)}{\text{Sin}\,\sigma \cdot k}. \quad (\text{a})$$

But, with (V.107), this is equivalent to

$$d = \bar{d} \cdot \frac{3\,\text{Sin}\,\sigma + R - H}{3\,\text{Sin}\,\sigma}. \quad (\text{b})$$

This can be transformed into the following two formulas:

$$\text{if } 90° < \kappa_{\sigma} < 270°,\ d = \bar{d} - \frac{\bar{d}}{3} \cdot \frac{R - H}{\text{Sin}\,\sigma}; \quad (\text{c})$$

$$\text{if } 270° < \kappa_{\sigma} < 90°,\ d = \bar{d} + \frac{\bar{d}}{3} \cdot \frac{H - R}{\text{Sin}\,\sigma}. \quad (\text{d})$$

The factor $\frac{\bar{d} \cdot r}{3}$, which occurs when $R - H$ or $H - R$ equals r, the radius of the śīghra epicycle (one must substitute 1 for Sin σ), is much too large; one needs something on the order of $\bar{d} \cdot \frac{r}{R}$.

With the longitudinal difference between the two planets, $\Delta\lambda$, and the difference of their velocities, Δv, one can find a first approximation to the time until conjunction, Δt (*BSS* 9, 5):

$$\Delta t = \frac{\Delta\lambda}{\Delta v}. \quad (\text{V}.109)$$

For each planet the cālana, c, which is the difference between the longitude of the conjunction and the longitude of the planet at time Δt, is found from (*BSS* 9, 6)

$$c = v \cdot \Delta t. \quad (\text{V}.110)$$

And the longitude of the planet at conjunction, λ_c, is given by (*BSS* 9, 7)

$$\lambda_c = \lambda + c. \tag{V.111}$$

But this is the computation of a conjunction only if both planets (or the planet and the fixed star) are on the ecliptic, for to Brahmagupta conjunction means that both bodies lie on the same great circle passing through the poles; in other words, their polar longitudes, λ^*, must be the same. As was done in adhyāyas 7 and 8, one finds λ^* of each body at its rising and its setting, and the ghaṭikās that it is above the horizon during the night, g. By comparing these elements for the two bodies, one can discover whether or not a conjunction will occur during the night. If one will, one must find the velocities of the planets (measured in minutes per ghaṭikā), v, during the time g from (*BSS* 9, 15)

$$v = \frac{\Delta\lambda^*}{g}, \tag{V.112}$$

where $\Delta\lambda^*$ is the difference between the planet's λ^* at rising, λ^*_U, and that at setting, λ^*_A. Then the time between the rising of the bodies and the conjunction, Δt_U, will be approximately (*BSS* 9, 16–18) (V.113)

$$\Delta t_U = \frac{\Delta\lambda^*_U}{\Delta v}, \tag{a}$$

where $\Delta\lambda^*_U$ is the difference of the λ^*_U's of the two bodies; and the time between the conjunction and their setting is

$$\Delta t_A = \frac{\Delta\lambda^*_A}{\Delta v}. \tag{b}$$

The cālana for the polar longitude, c^*, will be, for each planet:

$$c^* = v \cdot \Delta t_U. \tag{V.114}$$

Then a first approximation to λ_c will be

$$\lambda_c = \lambda^*_U + c^*. \tag{V.115}$$

However, since the latitudes of the planets will have changed in the interval Δt_U, one must recompute the polar longitudes for that time and find the true time and longitude by a process of iteration. By comparing the final values of β^* for the two bodies and their respective radii, one can determine whether an obscuration occurs. A further pair of verses asks that the time of the conjunction be modified by the longitudinal parallax of the planet, that the new β^*'s be computed for that moment, and that they be modified by the latitudinal parallax of the planet (*BSS* 9, 20–21). The planetary parallax, π_p, is found from the proportion

$$\frac{\pi_p}{\pi_{☾}} = \frac{\bar{v}_p}{\bar{v}_{☾}}. \tag{V.116}$$

Finally, at the end of the adhyāya Brahmagupta adds a mistaken rule for checking the computation of the conjunction and correcting the time if an error is found. First he prescribes finding, from the day circles of the two planets at conjunction, the respective ghaṭikās that have risen, n_p and n_q; we have already computed the ghaṭikās that each is above the horizon, g_p and g_q. Then he assumes that when there is a conjunction (*BSS* 9, 22),

$$\frac{n_p}{g_p} = \frac{n_q}{g_q}. \tag{V.117}$$

That this is not true is clear from Figure 26; it is equivalent to saying $\frac{x}{y} = \frac{x-a}{y-a}$. But Brahmagupta proceeds to attempt to determine the time when it will be true (*BSS* 9, 23–24). To accomplish this he finds two time differences, Δt_1 and Δt_2 (V. 118):

$$\Delta t_1 = n_q \cdot \frac{g_p}{g_q} - n_p; \tag{a}$$

$$\Delta t_2 = (n_q - n'_q) - (n_p - n'_p), \tag{b}$$

where n' represents the ghaṭikās that have risen for each planet in a time, Δt_c, that is guessed at as the time between the previously computed time of

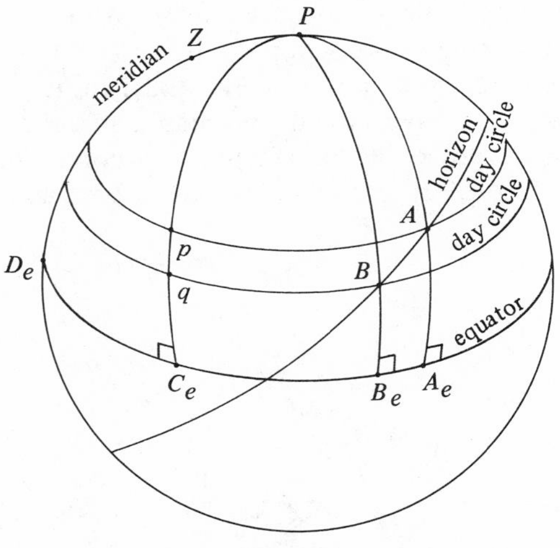

$A_e C_e = n_p$, $B_e C_e = n_q = n_p - A_e B_e$

$A_e D_e = \frac{g_p}{2}$, $B_e D_e = \frac{g_q}{2} = \frac{g_p}{2} - A_e B_e$

FIGURE 26

conjunction, t_c, and the time when (V.117) will hold. From this he forms a new estimate of that interval, $\Delta t'_c$, using

$$\Delta t'_c = \Delta t_c \cdot \frac{\Delta t_1}{\Delta t_1 \pm \Delta t_2}. \qquad \text{(c)}$$

Finally, he asserts that (V.117) will hold at the time t'_c, found from

$$t'_c = t_c \pm \Delta t'_c. \qquad \text{(d)}$$

The reasoning behind this procedure is not apparent. It is checked by computing new values of n_p and n_q for the time t'_c; if (V.117) still does not hold, one must iterate (V.118) until it does.

Adhyāya 10 of the *Brāhmasphuṭasiddhānta* is devoted to the computation of conjunctions of planets with fixed stars. Brahmagupta gives polar longitudes and polar latitudes of the yogatārās of the nakṣatras that are in most cases identical with those of the *Paitāmaha* (Table V.10); his changes are given in Table V.22 (*BSS* 10, 1–9, 35–37, 40). In only one case does Brahmagupta seem to be thinking of a star other than the *Paitāmaha*'s; he substitutes δ Sagittarii for γ Sagittarii as yogatārā of Pūrvāṣāḍhā. Both are third-magnitude stars. Otherwise, his corrections to β are usually in the wrong direction; only for the yogatārās of Kṛttikā, Svāti, Anurādhā, and Mūla (slightly) are his values better than those he corrects. It is tempting to believe that for the yogatārās of Rohiṇī, Citrā, Viśākhā, and Jyeṣṭhā he is merely using the *Paitāmaha*'s β's (usually poorly computed) as his own β*'s. In any case, he does not demonstrate any greater aptitude for observational astronomy than did other Indian astronomers.

Much of the rest of adhyāya 10, then, is parallel to what we have already discussed in connection with the preceding adhyāyas; it involves finding λ* and β* of the planet, δ* of both planet and fixed star, and then, as in adhyāya 9, transforming all differences into time differences—that is, into equatorial arcs. Here, however, Brahmagupta uses a somewhat different procedure (*BSS* 10, 16–23).

This procedure is illustrated in Figure 27 (compare Figure 18). Using (III. 31–33), one can find the equatorial arcs $\omega (= N_e E \approx D_e E)$ and ω^* $(= \lambda^*_e\ E)$; then the sum or difference of ω and ω^*, depending on the relative directions of δ and δ*, will be approximately $D_e \lambda^*_e$, which is the right ascensional arc of the longitudinal difference of the rising point (or setting point) of the ecliptic and the polar longitude of the planet. Thence the longitude of the ascendant (or descendant) when the planet, S, crosses the horizon is easily found. For a fixed star, which has no stated λ, one can simply perform the akṣadṛkkarma, substituting β* for β in (V.35). With these longitudinal differences and a knowledge of the longitude of the current ascendant (or descendant), one can predict the time until the planet and the star rise above (or set below) the horizon.

The rules for constructing a model illustrating the conjunction follow those in adhyāya 7 (*BSS* 10, 24); the rules for the "shadow" follow those in adhyāya 8 (*BSS* 10, 25); and the rules for the time of the conjunction follow those in adhyāya 9 (*BSS* 10, 25–26). Further, comparison of the right ascensions of a planet and the sun allows one to predict the time at which the Greek-letter phenomena will occur in a manner different from the procedure in adhyāya 6 (*BSS* 10, 33–34). Finally, in the *Paitāmaha* the ghaṭikās of visibility for the yogatārās were given as 2, and the corresponding kālāṃśas are 12° (*Pait.* 8, 4); Brahmagupta prescribes 2 1/6 ghaṭikās (13 time degrees) for Sirius and 2 1/3 ghaṭikās (14 time degrees) for the yogatārās (*BSS* 10, 32, 35–38).

In adhyāya 18 Brahmagupta applies the "pulver-

TABLE V.22

Nakṣatra	*Paitāmaha*				Brahmagupta				Identification	Ptolemy		
	λ*	β*	λ	β	λ*	β*	λ	β		λ	β	Magn.
Kṛttikā	37;28°	+5°	39;5°	+4;43°	37;28°	+4;31°	38;55°	+4;13°	η Tau.	33;40°	+3;20°	5
Rohiṇī	49;28°	−5°	48;9°	−4;49°	49;28°	−4;33°	48;17°	−4;21°	α Tau.	42;40°	−5;10°	1
Citrā	183°	−2°	183;49°	−1;50°	183°	−1;45°	183;42°	−1;34°	α Virg.	176;40°	−2°	1
Svāti	199°	+38°	185;19°	+34;37°	199°	+37°	184;48°	+33;20°	α Boöt.	177;0°	+31;30°	1
Viśākhā	212;5°	−1;30°	212;36°	−1;23°	212;5°	−1;23°	212;33°	−1;9°	ι Lib.	204;0°	−1;40°	4
Anurādhā	224;5°	−3°	224;57°	−2;51°	224;5°	−1;44°	224;35°	−1;39°	δ Scorp.	215;40°	−1;40°	3
Jyeṣṭhā	229;5°	−4°	230;9°	−3;50°	229;5°	−3;30°	230;4°	−3;22°	α Scorp.	222;40°	−4°	2
Mūla	240;4°	−8;30°	241;47°	−8;20°	241°	−8;30°	242;40°	−8;10°	*d* Oph.	235;30°	−6;10°	5
Pūrvāṣāḍhā	249°	−5;20°	249;46°	−5;17°	254°	−5;20°	254;37°	−5;18°	δ Sag.	247;40°	−6;30°	3
Śatabhiṣaj	320°	0°	320°	0°	320°	−0;18°	319;54°	−0;16°	λ Aquar.	314;50°	+0;10°	4
Agastya	87°	−76°	85;25°	−75;43°	87°	−77°	85;19°	−76;53°	α Arg.	77;10°	−75°	1
Mṛgahartā					86°	−40°	84;53°	−40°	α Can. Mai.	77;40°	−39;10°	1

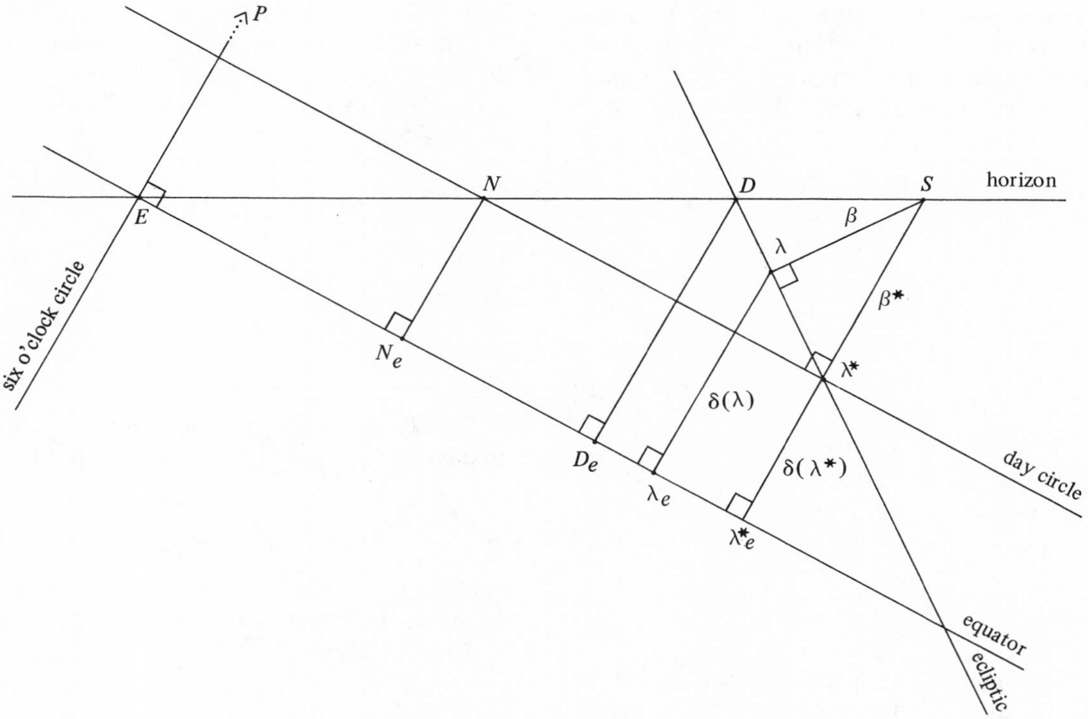

FIGURE 27

izer"—a method for solving indeterminate equations—to problems involving planetary longitudes and the ahargaṇa. In adhyāya 21 he includes the *Paitāmaha*'s rules for determining the circumferences of the orbits of the planets (Table V.3) (*BSS* 21, 11–14) in yojanas, and in this connection he repeats the *Paitāmaha*'s crude value of π, $\sqrt{10}$ (*BSS* 21, 15); this statement by Brahmagupta of the value of π (compare *BSS* 21, 16–23) formed the basis of Śrīpati's Sine table (Table V.33), in which $R = 3415 \approx \frac{1}{2} \cdot \sqrt{21600^2/10}$. Brahmagupta gives new values for the diameters in yojanas of the sun, the moon, and the earth, reported in Table V.23 (compare Table VII.7) (*BSS* 21, 32).

TABLE V.23

Sun	6522 yojanas
Moon	480 yojanas
Earth	1581 yojanas

His formula here for finding the diameter of the earth's shadow, d_u, at the moon's distance is (*BSS* 21, 33)

$$d_u = d_e - (d_☉ - d_e) \cdot \frac{k_☾}{k_☉}. \qquad (V.119)$$

This follows directly from Figure 17.

The most interesting adhyāya is the last, the Dhyānagrahādhyāya, which is an abbreviated karaṇa the epoch of which is 21 March 628. In this adhyāya Brahmagupta introduces a new Sine table in which $R = 150$; this is reproduced in Table V.24 (*BSS* 25, 16).

TABLE V.24

ϑ	Sin ϑ	150 sin ϑ
15°	39	38.8
30°	75	75
45°	106	106.0
60°	130	129.9
75°	145	144.8
90°	150	150

The yearly mean motions of the planets, listed in Table V.25, differ slightly from those of adhyāya 1 of the *Brāhmasphuṭasiddhānta* (Table V.2) (*BSS* 25, 26–31).

TABLE V.25

Planet	Mean yearly motion
Saturn	12; 12, 50, 13°
Jupiter	30; 21, 7, 6°
Mars	3, 11; 24, 8, 33°
Venus' śīghra	3, 45; 11, 52, 4°
Mercury's śīghra	54; 45, 9, 36°
Moon's node	−19; 21, 36, 0°

Brahmagupta also gives the epoch longitudes, and then the crude mean daily motions, which are constants, c, multiplied by the mean daily motion of the sun. The values of c are indicated in Table V.26 (*BSS* 25, 33–36).

TABLE V.26

Planet	c
Saturn	$\frac{2}{59}$
Jupiter	$\frac{7}{83}$
Mars	$\frac{1}{2}+\frac{10}{158}$
Venus' śīghra	$\frac{3}{2}+\frac{1}{12}$
Mercury's śīghra	$4+\frac{28}{184}$
Moon's node	$-\frac{10}{186}$

The longitudes of the mandoccas, given in Table V.27, also differ from those in the earlier part of the work (Table V.13) (*BSS* 25, 37).

TABLE V.27

Planet	Mandocca
Saturn	252°
Jupiter	170°
Mars	127°
Sun	77°
Venus	90°
Mercury	227°

The manda equations are derived by the "method of Sines," with the coefficients listed in Table V.28 (*BSS* 25, 38–39); also given there are the maximum equations, which differ from those in Table V.14.

TABLE V.28

Planet	$\frac{\mu}{\text{Sin}\,\kappa_\mu}(R=150)$	$\mu_{\max}(\kappa_\mu=90°)$
Saturn	$3+\frac{1}{30}$	7;35°
Jupiter	2	5°
Mars	$\frac{32}{7}$	11;25,42, · · ·°
Sun	$\frac{7}{8}$	2;11,15°
Venus	$\frac{2}{3}$	1;40°
Mercury	$\frac{7}{3}$	5;50°
Moon	2	5°

The śīghra equations are tabulated at intervals of 13;20° (a nakṣatra); the maximum equations are listed in Table V.29 (compare Table V.14) (*BSS* 25, 42–57).

TABLE V.29

Planet	$\sigma_{\max}$	κ_σ
Saturn	5;20°	80°–93;20°
Jupiter	10°	93;20°
Mars	40;40°	133;20°
Venus	46;26,40°	133;20°
Mercury	22;40°	106;40°–120°

Finally, Brahmagupta computes the half-equation of daylight in vināḍīs, ω^v, with the formula

$$\omega^v = s_0 \cdot \frac{\gamma}{2}, \qquad (V.120)$$

where the values of $\gamma/2$ are very nearly those in the *Pauliśa* (Table III.19); they are listed in Table V.30 (*BSS* 25, 61).

TABLE V.30

$\frac{\gamma_1}{2}=\frac{159}{16}$ vināḍīs

$\frac{\gamma_2}{2}=\frac{65}{8}$ vināḍīs

$\frac{\gamma_3}{2}=\frac{10}{3}$ vināḍīs

In the late seventh or early eighth century a work based largely on the *Brāhmasphuṭasiddhānta* was written; it apparently was entitled the *Mahāsiddhānta*. This was the basis of the *Zīj al-Sindhind al-kabīr* composed by al-Fazārī in the late eighth century, and thereby became the foundation of the Sindhind tradition in Islamic astronomy, which had a particularly profound impact on western European science through its translation into Latin by Adelard of Bath in 1126 and the translation of associated texts by others in the twelfth century; it also influenced Byzantine science.[50] In most particulars it agreed with *Brāhmasphuṭasiddhānta* 1–5. Al-Fazārī disagrees or has other sources in the following particulars:

1. He gives the rotations of Saturn in a Kalpa as 146,569,284 instead of 146,567,298; this implies a mean daily motion of 0;2,0,22,57,36,16, · · ·°.
2. He knows not only the value of R (3,270) used in the *Brāhmasphuṭasiddhānta*, but also Āryabhaṭa's R = 3,438 and the *Khaṇḍakhādyaka*'s R = 150; this last value, as we have seen, is also found in *Brāhmasphuṭasiddhānta* 25.

3. His models (eccenter and epicycle) and parameters for computing the equations of the planets are adapted from the Sasanian *Zīj al-Shāh*, which was influenced by the Ārdharātrikapakṣa.

4. His eclipse limits are derived from Varāhamihira's *Pañcasiddhāntikā*.

5. Al-Fazārī derived his method for computing planetary latitudes from adhyāya 9 of the *Brāhmasphuṭasiddhānta*, although the parameters come from the *Uttarakhaṇḍakhādyaka*.

The *Siddhāntaśekhara* of Śrīpati is based largely on the *Brāhmasphuṭasiddhānta* of Brahmagupta; Śrīpati is known to have written other works in 1039 and 1056. His description of the Kalpa and its subdivisions is identical with Brahmagupta's (*SSS* 1, 16–19), although he goes further in saying that a nychthemeron of the Creator contains two Kalpas; a year, 360 such nychthemera; and the life of the Creator, 100 such years (*SSS* 1, 20). He repeats Brahmagupta's interval from the beginning of the current Kalpa to the beginning of the current Kaliyuga (*SSS* 1, 23–25). He also iterates the rotations of the planets according to the Brāhmapakṣa (Table V.2), with the exception that he has Saturn's manda rotate fifty-four rather than fourty-one times in a Kalpa (*SSS* 1, 27–31); this means that about A.D. 1000 its longitude was 267° instead of Brahmagupta's 261°.

Śrīpati gives a rule for finding the mean daily motions in rotations, which are simply reductions of *R*/*C* (*SSS* 2, 33), and also gives much cruder ratios for daily motions in zodiacal signs and so on, as recorded in Table V.31 (*SSS* 2, 45–49) (compare Table V.11).

TABLE V.31

Planet	Mean Daily Motion	
Saturn	$\frac{6^s}{5383}$	≈ 0;2,0,22,51, . . .°
Jupiter	$\frac{1^s}{361}-\frac{1'''}{73}$	≈ 0;4,59,9,19, . . .°
Mars	$\frac{4^s}{229}$	≈ 0;31,26,27,46, . . .°
Venus	$\frac{40^s}{749}+\frac{1''}{20}$	≈ 1;36,7,44,24, . . .°
Mercury	$\frac{3^s}{22}+\frac{12'}{143}$	≈ 4;5,32,18,27, . . .°
Moon's manda	$\frac{3^s}{808}-\frac{1''}{11}$	≈ 0;6,40,53,57, . . .°
Moon's node	$-\frac{1^s}{566}$	≈ −0;3,10,48,45, . . .°

Śrīpati lists the mean longitudes of the planets at the beginning of Kaliyuga precisely as in column 4 of Table V.12 (*SSS* 2, 51–54). He gives rounded

TABLE V.32

Planet	Yearly Bīja	Bīja Amplitude	Bīja in A.D. 628
Saturn	$+\frac{4'}{200}=+0;0,1,12°$	+2°	+1;14,34,48°
Jupiter	$-\frac{5'}{200}=-0;0,1,30°$	−2;30°	−1;33,13,30°
Mars	$+\frac{1'}{200}=+0;0,0,18°$	+0;30°	+0;18,38,42°
Sun	$-\frac{3'}{200}=-0;0,0,54°$	−1;30°	−0;55,56,6°
Venus' śīghra	$-\frac{15'}{200}=-0;0,4,30°$	−7;30°	−4;39,40,30°
Mercury's śīghra	$+\frac{52'}{200}=+0;0,15,36°$	+26°	+16;9,32,24°
Moon	$-\frac{5'}{200}=-0;0,1,30°$	−2;30°	−1;33,13,30°
Moon's manda	$-\frac{2'}{200}=-0;0,0,36°$	−1°	−0;37,17,24°
Moon's node	$-\frac{2'}{200}=-0;0,0,36°$	−1°	−0;37,17,24°

For Mercury's bīja Śrīpati gives $+\frac{62'}{200}=+0;0,18,36°$

values for the yojanas in the orbits of the planets listed in Table V.3 (*SSS* 2, 59–65), and he records (*SSS* 2, 91–92) the bījas that are inserted into some manuscripts of the *Brāhmasphuṭasiddhānta*. The period of increase and decrease in these bījas is 12,000 years; there were exactly 164,412 such periods between the beginnings of the Kalpa and of the current Kaliyuga. Table V.32 lists the yearly bījas according to the interpolation in Brahmagupta, the bīja amplitudes, and the bīja in A.D. 628.

Śrīpati's rule for the deśāntara is equivalent to (V.61) (*SSS* 2, 94), although he correctly criticizes it (*SSS* 2, 104), as had Bhāskara I. He lists the following localities as lying on the prime meridian: Laṅkā, Kumārī, Kāñcī, Pānāṭa, Sitādri, Ṣaḍāsya, Vatsagulma, Māhiṣmatī, Ujjayinī, Paṭṭaśiva, Gargarāṭa, Rohita, Sthāṇvīśvara, Śītagiri, and Sumeru (*SSS* 2, 95–96).

Śrīpati's table of Sines uses the unusual value, $R = 3{,}415$, that results from the formula in adhyāya 21 of the *Brāhmasphuṭasiddhānta*; it is reproduced in Table V.33 (*SSS* 3, 3–6).[51]

TABLE V.33

ϑ	Sin ϑ	R sin ϑ	ϑ	Sin ϑ	R sin ϑ
3;45°	223	223.3	48;45°	2568	2567.5
7;30°	445	*445.7*	52;30°	2709	2709.2
11;15°	666	666.2	56;15°	2839	2839.4
15°	884	883.8	60°	2958	*2957.4*
18;45°	1098	1097.7	63;45°	3063	3062.8
22;30°	1307	1306.8	67;30°	3155	3155.0
26;15°	1510	1510.4	71;15°	3234	3233.7
30°	1708	1707.5	75°	3299	3298.6
33;45°	1898	*1897.2*	78;45°	3349	3349.3
37;30°	2079	2078.9	82;30°	3386	3385.7
41;15°	2252	2251.6	86;15°	3408	3407.6
45°	2415	2414.7	90°	3415	3415

Śrīpati's circumferences of the manda and śīghra epicycles (*SSS* 3, 19–38) are identical with Brahmagupta's (Table V.14) except that he makes the minimum of Venus' śīghra epicycle 263° and its maximum 268°.

Śrīpati repeats the cālanas of the parvans from the *Brāhmasphuṭasiddhānta* (Table V.20), although he gives 19;42,26° as the cālana of the moon's manda (*SSS* 7, 3). He adds the cālanas of the pakṣas, which are the mean longitudinal differences between two successive syzygies separated by fifteen civil days. These are recorded in Table V.34 (*SSS* 7, 4).

TABLE V.34

Sun	0^s 14;47,2°
Moon	6^s 17;35,42°
Manda	0^s 1;40,13°
Node	0^s 0;47,45°

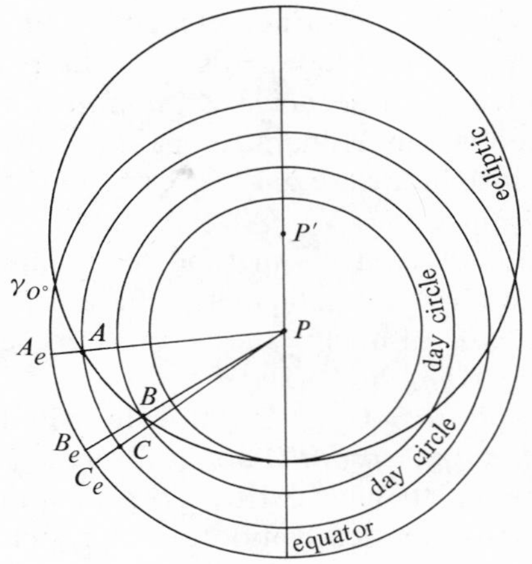

$AB = AC$

$B_e\,C_e$ = equation of time due to obliquity of ecliptic

(projection in the plane of the equator)

FIGURE 28

One component of the equation of time had been given previously by Brahmagupta (V.66) and is repeated by Śrīpati (*SSS* 3, 46). But Śrīpati also recognizes the second component, the udayāntara, which is due to the changing declination of the sun; (see Figure 28). This component, e, he gives correctly as (*SSS* 11, 1)

$$e = \lambda_{\odot} - \alpha, \qquad \text{(V.121)}$$

having found α from

$$\text{Sin}\,\alpha = \text{Sin}\,\lambda_{\odot} \cdot \frac{r_{\min}}{r}, \qquad \text{(V.122)}$$

where $r_{\min} = \text{Cos}\,\epsilon$ and $r = \text{Cos}\,\delta$, so that (V.122) is equivalent to (V.76). He may have derived (V.121) from an Islamic source.

Śrīpati also introduces a rule to compute the evection of the moon, ϑ, here called the (sphuṭa) caraphala; a similar rule had been given by Muñjāla in the tenth century (IX.1). Śrīpati's rule is (*SSS* 11, 2–4)

$$\vartheta = 160 \cdot \frac{\text{Sin}(\lambda_{\odot} - [\lambda_A - 90°])}{R} \cdot \frac{\text{Sin}(\lambda_{☾} - \lambda_{\odot})}{R} \cdot \frac{\text{Vers}(\lambda_{\odot} - \lambda_A)}{H - R} = 160 \cdot \frac{\text{Cos}(\lambda_{\odot} - \lambda_A)}{R} \cdot \frac{\text{Sin}(\lambda_{☾} - \lambda_{\odot})}{R} \cdot \frac{\text{Vers}(\lambda_{\odot} - \lambda_A)}{H - R}, \qquad \text{(V.123)}$$

where ϑ is expressed in minutes and λ_A is the longitude of the moon's apogee. The maximum correc-

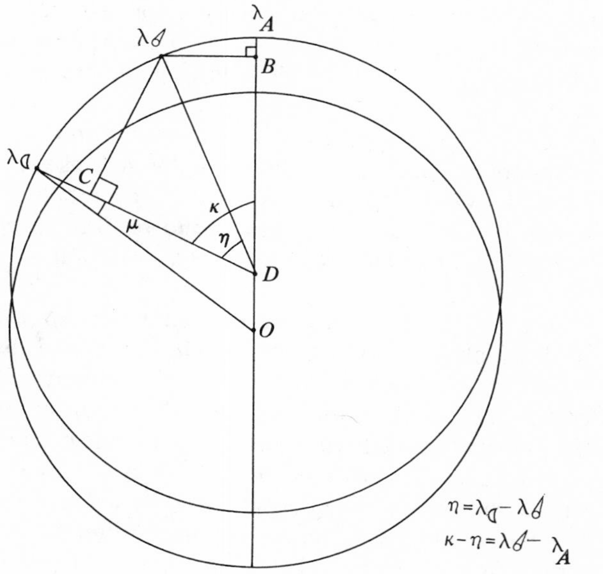

FIGURE 29

tion, then, without the last term, is 2;40° when Cos $(\lambda_{\mathcal{U}} - \lambda_A) = R$ and Sin $(\lambda_{☾} - \lambda_{\mathcal{U}}) = R$ – that is, the moon is at quadrature and its apsidal line points toward the sun (see Figure 29). The value for the maximum is almost that of Ptolemy (2;39°), and Śrīpati's rule must be dependent on a knowledge of the Ptolemaic lunar model.

In the text the last term is expressed only as Vers/$H - R$ (or Vers/$R - H$ when $R > H$). This term will be $0/0 = 1$ at the time of the maximum correction if the arc of which one is to take the Versine is $\lambda_{\mathcal{U}} - \lambda_A$ and if $H = R$ when the anomaly is 90° or 270°. The astronomical significance of this term is not apparent.

Finally, Śrīpati repeats the orbital inclinations of the planets given in Table V.7, except that the value of i for Jupiter is 1;26° (*SSS* 11,8), which gives a β_{max} of 1;46°.

The *Rājamṛgāṅka* is a small karaṇa attributed to the Paramāra rājā of Dhārā, Bhojarāja (*ca.* 1005–1056); its epoch is 21 February 1042. Unfortunately, there exists in print of this important work only a redaction, apparently by one Rāma, of the first two adhikāras, which deal with the determination of the ahargaṇa and the longitudes of the planets.[52]

Bhojarāja changed the planetary mean motion parameters of the Brāhmapakṣa; most later karaṇas and tables of this school follow his lead. These changes depend on the yearly bījas given in Table V.32. The new yearly and daily mean motions as they appear in the tables are given in Table V.35; it is clear that they are not precise. The mean daily motions are the mean yearly motions divided by the year length – 6,5;15,31,17,17 civil days. The difference between this year length and that of the *Paitāmaha* – 6,5;15,30,22,30 – is 0;0,0,54,47 days, or approximately the sun's annual bīja.

Bhojarāja gives rules for finding the ahargaṇa since his epoch (*RM* 1, 3–6), the interval between the beginning of the Kalpa and his epoch (*RM* 1, 8), the mean longitudes of the planets at the epoch (*RM* 1, 11–13), and the mean daily motions of the planets in minutes and seconds (*RM* 1, 14–15). From these elements it is simple to find the mean longitudes for any given day within a reasonable time after the epoch.

Bhojarāja lists the following localities as lying on the prime meridian: Laṅkā, Kāñcī, Vatsagulma, Avanti, Gargarāṭaka, Sthāneśvara, and Sumeru (*RM* 1, 21). This list has obvious affinities to Śrīpati's. Bhojarāja is the first known follower of the Brāhmapakṣa to admit precession; he fixes its rate as 0;1° per year, and the year of coincidence as Śaka 444 = A.D. 522 (*RM* 1, 24).[53] He also gives another method of finding the second component of the equation of time (see V.121–122). His rule is equivalent to that in the *Siddhāntaśiromaṇi* (V.126); e is measured in nāḍīs (*RM* 1, 25):

TABLE V.35

Planet	Old Yearly Motion	Yearly Bīja	New Yearly Motion	Mean Daily Motion
Saturn	12;12,50,11, . . .°	+0;0,1,12°	12;12,51,25,12°	0;2,0,23,3,30,3, . . .°
Jupiter	30;21,7,56, . . .°	−0;0,1,30°	30;21,6,30,42°	0;4,59,8,53,49,56, . . .°
Mars	3,11;24,8,33, . . .°	+0;0,0,18°	3,11;24,9,20,6°	0;31,26,28,9,45,48, . . .°
Sun	6,0°	−0;0,0,54°	6,0°	0;59,8,10,12,41,20, . . .°
Venus' śīghra	3;45;11,56,50, . . .°	−0;0,4,30°	3,45;11,53,48,36°	1;36,7,43,50,57,14, . . .°
Mercury's śīghra	54;44,59,41, . . .°	+0;0,15,36°	54;45,19,2,12°	4;5,32,21,1,33,42, . . .°
Moon	2,12;46,30, . . .°	+0;0,1,30°	2,12;46,40,32°	13;10,34,52,31,44,43, . . .°
Manda	40;40,31,45, . . .°	[0;0,0,36°]	40;40,31,44°	0;6,40,53,55,18,32, . . .°
Node	−19;21,33,21, . . .°	−0;0,0,36°	−19;21,33,59,54°	−0;3,10,48,26,1,14, . . .°

$$e = \frac{40 \text{ Sin}(2\lambda_{♂})}{90R}. \qquad \text{(V.124)}$$

The published text omits division by $90R$. Clearly (V.124) produces the proper curve, for the maximum occurs when λ is 45°, 135°, 225° or 315°, and $e = 0$ when λ is 0°, 90°, 180°, or 270°.

Bhojarāja's formula for the correction to κ due to the first component of the equation of time is the same as (V.6) (*RM* 1, 26). When the two corrections due to the two components of the equation of time are combined with that due to the half-equation of daylight, one has the tryaika, which corrects the mean longitude for mean sunrise into the mean longitude for true sunrise (*RM* 1, 32–33). The half-equation of daylight in vināḍīs, ω^v, is computed by Brahmagupta's formula (V.120), and the values of $\gamma/2$ given in Table V.36 are nearly those in Table V.30 (*RM* 1, 27).

TABLE V.36

$$\frac{\gamma_1}{2} = 10 \text{ vināḍīs}$$

$$\frac{\gamma_2}{2} = 8 \text{ vināḍīs}$$

$$\frac{\gamma_3}{2} = 3;20 \text{ vināḍīs}$$

Bhojarāja, at the beginning of the second adhikāra, lists the longitudes of the mandoccas and nodes of the planets in 1042; these are given in Table V.37 (*RM* 2, 1 and 6–7). Those that differ from Brahmagupta's as listed in Table V.13 are in italics.

TABLE V.37

Planet	Mandocca	Node
Saturn	238;*16,0*°	103;*11,26*°
Jupiter	172;*33,18*°	82;1,20°
Mars	128;24,17°	*38;58,31*°
Sun	77;55,*33*°	
Venus	81;15,*41*°	59;*45,33*°
Mercury	224;53,55°	21;*10,6*°

The equations and latitudes of the planets were tabulated, but these tables are not now available. Bhojarāja also lists the śīghra arguments required for the occurrences of the Greek-letter phenomena, as did Brahmagupta, except that the first visibility of Saturn occurs (as in Lalla) at 20°, the first station of Venus at 165°, the first visibility of Mercury in the west at 51°, and the last visibility of Mercury in the east at 309° (*RM* 2, 33–36).

Daśabala, the only Indian Buddhist known to have written on mathematical astronomy after the author of the *Śārdūlakarṇāvadāna*, composed in Madhyapradeśa two works following the Brāhmapakṣa. The first, the *Cintāmaṇisāraṇikā*, has as epoch 1055 and apparently uses the lunar and solar elements of Table V.28. His second work, a karaṇa entitled *Karaṇakamalamārtaṇḍa*, the epoch of which is 1058, seems to have followed the *Rājamṛgāṅka*; unfortunately little is yet known of it.

Another karaṇa with tables belonging to the Brāhmapakṣa is the *Grahajñāna* composed by Āśādhara in Gujarat;[54] its epoch is Sunday, 20 March 1132. The tables of the mean motions of the planets for one to nine years, one to nine periods of ten years, one to nine periods of one hundred years, and one to nine periods of a thousand years are based on yearly mean motions differing minutely from those in Table V.35; they are listed in Table V.38 along with Āśādhara's bījas.

TABLE V.38

Planet	Yearly Mean Motion	Bīja
Saturn	12;12,51,25,6,40°	+0;6,45°
Jupiter	30;21,6,17,20°	−0;14,15°
Mars	3,11;24,9,20°	+0;6°
Venus' śīghra	3,45;11,53,48,40°	−0;20,15°
Mercury's śīghra	54;45,19,2,13,20°	+0;52,30°
Moon's node	−19;21,34°	

Āśādhara's year length is identical with Bhojarāja's. He gives tables of the true longitudes of the planets of the type I have called "true linear"; in them $\Delta\lambda = 13;20°$ (compare Table V.29), $N = 0$ to 26, and $k = 14$ days.[55]

The only other material of interest to us in the *Grahajñāna* is Āśādhara's list of the periods, in days, of retrogression, visibility, and invisibility of each planet. They are given in Table V.39 (compare Table VI.19) (*GJ* 18–20).

TABLE V.39

Planet	Retrograde	Direct	Synodic Period
Saturn	141 days	237 days	378 days
Jupiter	122 days	277 days	399 days
Mars	74 days	706 days	780 days
Venus	49 days	535 days	584 days
Mercury	23 days	93 days	116 days

Planet	Visible	Invisible	Synodic Period
Saturn	336 days	42 days	378 days
Jupiter	368 days	31 days	399 days
Mars	659 days	121 days	780 days

	Visible Γ→Σ	Invisible Σ→Ξ	Visible Ξ→Ω	Invisible Ω→Γ	
Venus	247 days	78 days	247 days	10 days	582 days
Mercury	33 days	33 days	33 days	16 days	115 days

An expanded version of Āśādhara's *Grahajñāna* is the *Gaṇitacūḍāmaṇi*, composed by Harihara in Saurāṣṭra about 1580.[56]

The last siddhānta of the Brāhmapakṣa that we can consider is the *Siddhāntaśiromaṇi*, written by Bhāskara II in what is now Mysore State in 1150. This work basically follows the *Brāhmasphuṭasiddhānta* of Brahmagupta, although with some reference to later works both within and outside the Brāhmapakṣa and with some additions, of which I will mention those that seem most significant. In the first part of the *Siddhāntaśiromaṇi*, the Gaṇitādhyāya, adhikāra 1 repeats all the parameters of the *Paitāmaha* relating to mean motions. Instead of Table V.12, however, Bhāskara gives the mean longitudes of the planets at the beginning of the Kaliyuga in the form of the seconds by which each longitude is less than a circle; thus, in Table V.40 (*SSB* 1, 1, 3, 19–20):

TABLE V.40

Planet	Dhruvaka	Mean Longitude
Saturn	−4406″	5,58;46,34°
Jupiter	−1944″	5,59;27,36°
Mars	−3370″	5,59;3,50°
Venus' śīghra	−4666″	5,58;42,14°
Mercury's śīghra	−9331″	5,57;24,29°
Sun's manda	−1,016,064″	1,17;45,36°
Moon's manda	−844,214″	2,5;29,46°
Moon's node	−744,422″	2,33;12,58°

Bhāskara's statements of the mean daily motions are crude, but are accompanied in his commentary by more accurate values; compare Brahmagupta's (Table V.11) and Śrīpati's (Table V.31). Both sets of values are given in Table V.41 (*SSB* 1, 1, 5, 15, 17–20).

TABLE V.41

Planet	Mean Daily Motions Text	Commentary
Saturn	$2' + \frac{2''}{5} = 0;2,0,24°$	0;2,0,22,51°
Jupiter	$\frac{1°}{12} - \frac{1'}{71} = 0;4,59,9,18,\ldots°$	0;4,59,9,9°
Mars	$\frac{1°}{2} + \frac{3'}{2} - \frac{1'}{17} = 0;31,26,28,15,\ldots°$	0;31,26,28,7°
Sun	$59' + \frac{3'}{22} = 0;59,8,10,54,\ldots°$	0;59,8,10,21°
Venus' śīghra	$\frac{10°}{6} - \frac{10°}{155} = 1;36,7,44,31,\ldots°$	1;36,7,44,35°
Mercury's śīghra	$4° + \frac{4' \cdot 3}{130} = 4;5,32,18,27,\ldots°$	4;5,32,18,28°
Moon		13;10,34,53,0°
Moon's manda	$\frac{1°}{10} + \frac{1°}{88} = 0;6,40,54,32,\ldots°$	0;6,40,53,56°
Moon's node	$-\frac{30°}{566} = -0;3,10,48,45,\ldots°$	−0;3,10,48,20°

For the moon Bhāskara gives an alternate form of Brahmagupta's rule (V.56) (*SSB* 1, 1, 5, 16).

Bhāskara gives yet another set of crude values for the mean daily motions, presented here in Table V.42 (*SSB* 1, 1, 5, 22–23).

TABLE V.42

Saturn	$\frac{100000°}{2990000} = 0;2,0,24,4,\ldots°$
Jupiter	$\frac{100000°}{1203400} = 0;4,59,9,8,\ldots°$
Mars	$\frac{100000°}{190833} = 0;31,26,27,58,\ldots°$
Sun	$\frac{100000°}{101461} = 0;59,8,9,40,\ldots°$
Venus' śīghra	$\frac{100000°}{62416} = 1;36,7,45,6,\ldots°$
Mercury's śīghra	$\frac{100000°}{24436} = 4;5,32,21,43,\ldots°$
Moon	$\frac{100000°}{151787} \cdot 20 = 13;10,34,53,32,\ldots°$
Moon's manda	$\frac{100000°}{898000} = 0;6,40,53,27,\ldots°$
Moon's node	$-\frac{100000°}{1886800} = -0;3,10,47,57,\ldots°$

The prime meridian is said by Bhāskara to pass over Laṅkā, Ujjayinī, Kurukṣetra, and Meru (*SSB* 1, 1, 7, 2).

Bhāskara's table of Sines is that of the *Paitāmaha* as given in Table V.6, except that Sin 60° = 2,977—a more correct value than 2,978 (*SSB* 1, 2, 3–6). He also gives a Sine table in which $R = 120$ (as in Table III.13) and the interval is 10°; this is recorded in Table V.43 (*SSB* 1, 2, 23).[57]

TABLE V.43

ϑ	Sin ϑ	R sin ϑ
10°	21	20.8
20°	41	41.0
30°	60	60
40°	77	77.1
50°	92	91.9
60°	104	103.9
70°	113	112.7
80°	118	118.1
90°	120	120

Such a Sine table, together with Bhāskara's advanced knowledge of trigonometry,[58] hints at the influence of Islamic astronomy, although this is yet to be substantiated.

Bhāskara's circumferences of the planets' epicycles are identical with the *Paitāmaha*'s (Table V.5), save that Saturn's manda epicycle is 50° and Mars's śīghra epicycle is 243;40° (*SSB* 1, 2, 22–23); the latter value is found in the *Brāhmasphuṭasiddhānta* (Table V.14). He correctly states that the equation of time, E, is found by (*SSB* 1, 2, 61–63; compare 2, 4, 19–20)

$$E = \lambda_{\odot} - \alpha - \mu. \qquad (V.125)$$

In a variant to Bhojarāja's rule in (V.124), he gives the following formula for finding e, the component of E due to the obliquity, in nāḍīs (*SSB* 1, 2, 65):

$$e = \frac{\sin_{120}(2\bar{\lambda}_{\odot})}{270} = \frac{120 \sin(2\bar{\lambda}_{\odot})}{270} = \frac{40 \sin(2\bar{\lambda}_{\odot})}{90}. \qquad (V.126)$$

Bhāskara repeats the theory of the nakṣatrabhogas expounded by Brahmagupta, except that he uses the more accurate mean daily motion of the moon, 790;35′, as one bhoga, so that 1 1/2 bhogas are 1185;52′, 1/2 bhoga is 395;17′, and Abhijit's bhoga is 254;18′ (*SSB* 1, 2, 71–75).

The rest of the Gaṇitādhyāya also is mostly derived from Brahmagupta and from Śrīpati, with some modifications. Of these I note the following:

1. Bhāskara gives the radii of the orbits of the sun and the moon as 689,377 and 51,566 yojanas, respectively. He uses $\pi = 3927/1250 = 3.1416$ instead of the value, $\pi = \sqrt{10}$, used by the *Paitāmaha* and Śrīpati. He does, however, keep Śrīpati's values for the diameters of the sun, moon, and earth given in Table V.26 (*SSB* 1, 5, 3–6).

2. Bhāskara modifies the traditional kālāṃśas of visibility of the planets given in the *Paitāmaha* (Table V.8) so that they are those in Table V.44 (*SSB* 1, 8, 6).

TABLE V.44

Planet	Kālāṃśas
Moon	12°
Venus	10°
Jupiter	11°
Mercury	14°
Saturn	15°
Mars	17°

3. Bhāskara gives values for the mean apparent diameters of the planets in minutes rounded off from Brahmagupta's (Table V.21); these rounded values are listed in Table V.45 (*SSB* 1, 10, 1).

TABLE V.45

Planet	Diameter
Venus	9′
Jupiter	7;20′
Mercury	6;15′
Saturn	5;30′
Mars	4;45′

4. He minutely corrects β^* of Kṛttikā to + 4;30°, of Rohiṇī to − 4;30°, of Viśākhā to − 1;20°, of Anurādhā to − 1;45°, and of Śatabhiṣaj to − 0;20°; otherwise he follows Brahmagupta (Tables V.10 and V.22) (*SSB* 1, 11, 1–7).

The Golādhyāya of the *Siddhāntaśiromaṇi* is a more theoretical treatise than the Gaṇitādhyāya; in it Bhāskara explains the models on which astronomical computations are based. It contains also several parameters not in the Gaṇitādhyāya:

1. The circumference of the earth is 4,967 yojanas and its diameter 1,581 1/24 yojanas; here, then, $\pi = 4967/1581\frac{1}{24} = 3.1413\ldots$ (*SSB* 2, 3, 52).

2. Bhāskara refers erroneously to the theory of trepidation given by the *Sūryasiddhānta* as one of a precession with 30,000 rotations in a Kalpa (only one rotation in 144,000 years or 1° in 400 years!). He also attributes to Muñjāla a theory of precession with 199,699 rotations in a Kalpa, which gives a motion of nearly 1° in sixty years; this theory is not, however, found in Muñjāla's extant works (*SSB* 2, 6, 17–18).[59]

3. Bhāskara gives a rule for finding Sin (ϑ + 1°) if R = 3,438 (*SSB* Jyotpatti 16)[60]:

$$\text{Sin}\,(\vartheta + 1°) = \text{Sin}\,\vartheta \cdot \frac{6568}{6569} + \frac{10\,\text{Cos}\,\vartheta}{573}, \qquad (V.127)$$

which follows from his rule for finding Sin ($\alpha + \beta$) (*SSB* Jyotpatti 21):

$$\text{Sin}\,(\alpha + \beta) = \frac{\text{Sin}\,\alpha \cdot \text{Cos}\,\beta}{R} + \frac{\text{Cos}\,\alpha \cdot \text{Sin}\,\beta}{R}. \qquad (V.128)$$

Bhāskara also wrote a karaṇa, the *Karaṇakutūhala*, the epoch of which is Thursday, 24 February 1183; a set of astronomical tables based on this karaṇa is also extant.[61] In the tables the mean daily motions of the planets are virtually identical with those in Table V.35, and in the text approximations are given (*KK* 1, 7–12). Both are listed in Table V.46.

Bhāskara states that the prime meridian passes over the city of the Rakṣasas (Laṅkā), Devakanyā,

TABLE V.46

Planet	Text	Tables
Saturn	$\frac{1°}{30}+\frac{1°}{9397}=0;2,0,22,59,\ldots°$	0;2,0,23,3,30°
Jupiter	$\frac{1°}{12}-\frac{1°}{4227}=0;4,59,8,54,\ldots°$	0;4,59,8,54,0°
Mars	$\frac{11°}{21}+\frac{11°}{52544}=0;31,26,28,4,\ldots°$	0;31,26,28,9,50°
Sun	$1°-\frac{13°}{903 \text{ days}}-\frac{1'}{64 \text{ years}}=0;59,8,10,13,\ldots°$	0;59,8,10,12,40°
Venus' śīghra	$\frac{16°}{10}+\frac{16°}{7451}=1;36,7,43,49,\ldots°$	1;36,7,43,49,50°
Mercury's śīghra	$4°+\frac{4°}{43}-\frac{1°}{1421}=4;5,32,21,1,\ldots°$	4;5,32,21,1,0°
Moon	$14°-\frac{14°}{17}-\frac{1°}{8600}=13;10,34,52,32,\ldots°$	13;10,34,52,31,50°
Manda	$\frac{1°}{9}+\frac{1°}{4012}=0;6,40,53,50,\ldots°$	0;6,40,53,50,10°
Node	$-\left(\frac{1°}{19}+\frac{1°}{2700}\right)=-0;3,10,48,25,\ldots°$	−0;3,10,48,25,30°

Kāntī, Sitaparvata, Paryalī, Vatsagulma, Ujjayinī, Gargarāṭa, Kurukṣetra, and Meru (*KK* 1, 14). He gives as longitudes of the mandoccas of the planets those listed in Table V.47 (*KK* 2, 1–2).

TABLE V.47

Planet	Mandocca
Saturn	261°
Jupiter	172;30°
Mars	128;30°
Sun	78°
Venus	81°
Mercury	225°

The position of Saturn's mandocca indicates that Bhāskara follows Brahmagupta (Table V.13) rather than Bhojarāja (Table V.37).

The *Karaṇakutūhala's* Sine table is the second one of the *Siddhāntaśiromaṇi* (Table V.43), in which $R = 120$ (*KK* 2, 6). The equations of the center are found by multiplying the Sine of the anomaly by given constants; the equations also are tabulated in the tables. Table V.48 lists the text's formulas and resulting maximum equations, which are also the tables' maximum equations (*KK* 2, 9–10).

The text gives the radii of the śīghra epicycles, called parākhyas, in parts of the radius of the deferent ($R = 120$), and the correct geometrical solution for σ; the tables give the values of σ for every

TABLE V.48

Planet	$\frac{\mu}{\text{Sin } \kappa_\mu}$	$\mu_{\max}$ ($\kappa_\mu = 90°$)
Saturn	$\frac{10}{157}$	7;38,35,...°
Jupiter	$\frac{10}{228}$	5;15,47,...°
Mars	$\frac{10}{107}$	11;12,53,...°
Sun	$\frac{10}{550}$	2;10,54,...°
Venus	$\frac{10}{784}$	1;31,50,...°
Mercury	$\frac{10}{198}$	6;3,38,...°
Moon	$\frac{10}{238}$	5;2,31,...°

degree and the values of the śīghra hypotenuse, H_σ. Table V.49 lists the radii of the epicycles from the text and the maximum equations and corresponding anomalies from the tables (*KK* 2, 2).

The rate of precession adopted by Bhāskara in the *Karaṇakutūhala* is 1° in sixty years (*KK* 2, 17). In other respects the karaṇa offers little that is unusual; but one may note the following:

1. Bhāskara gives a table of lunar latitudes divided by four at intervals of 15° in which $\beta_{\max} = 4;40°$

TABLE V.49

Planet	r_σ	σ_{max}	κ_σ
Saturn	13	6;10,7°	100°
Jupiter	23	10;59,1°	100°
Mars	81	41;17,59°	130°
Venus	97	46;30,28°	140°
Mercury	44	21;36,48°	110°

(*KK* 4, 6) as in the *Romaka* (III.21e; compare III.39).

2. He gives a table of longitudinal parallaxes for every 15° of elongation between the nonagesimal and the sun. The maximum is 4 nāḍīs at an elongation of 90° (*KK* 5, 4).

3. He gives rules for computing the times of the heliacal risings of the planets. These are based on the date of a heliacal rising in the epoch year and the mean synodic periods of the planets (*KK* 6).

The next text belonging to the Brāhmapakṣa is the *Laghukhecarasiddhi* of Śrīdhara;[62] its epoch is 20 March 1227. This work is accompanied by tables, from which the following elements appear.

The mean daily motions of the planets are given for one, ten, and one hundred years of 365 days each (Egyptian years), and for one, ten, and one hundred days. Comparison of Tables V.50 and V.35 will show that they belong to the new Brāhmapakṣa.

TABLE V.50

Planet	Mean Daily Motion
Saturn	0;2,0,23,15,...°
Jupiter	0;4,59,8,54,...°
Mars	0;31,26,28,9,...°
Sun	0;59,8,10,11,...°
Venus' śīghra	1;36,7,43,51,...°
Mercury's śīghra	4;5,32,20,55,...°
Moon	13;10,34,52,26,...°
Manda	0;6,40,53,56,...°
Node	−0;3,10,48,25,...°

The longitudes of the mandoccas for 1227 in Table V.51 are close to Bhojarāja's for 1042

TABLE V.51

Planet	Mandocca	Node
Saturn	238;16,4°	103;10,22°
Jupiter	172;34,34°	82;1,14°
Mars	128;24,48°	21;53,45°
Sun	77;56,16°	
Venus	81;16,40°	59;44,55°
Mercury	224;54,33°	21;9,8°

(Table V.37); the longitudes of the nodes are closer to Brahmagupta's for 628 (Table V.13).

Śrīdhara's manda equations and śīghra equations are rather eclectic (see Table V.52).

TABLE V.52

Planet	μ_{max} ($\kappa_\mu = 90°$)	σ_{max}	κ_σ
Saturn	7;59,0°	5;34,0°	100°
Jupiter	5;16,0°	10;53,0°	100°
Mars	11;12,47°	41;17,16°	130°
Sun	2;10,34°		
Venus	1;26,0°	46;23,0°	140°
Mercury	4;2,0°	21;30,0°	110°
Moon	5;2,17°		

The manda equation of Saturn is greater than it normally is in the Brāhmapakṣa, that of Mercury is less, and none of the others is precisely the same; Saturn's śīghra equation is less than, and Mercury's is equal to, that of the Ārdharātrikapakṣa (Table VII.6).

The *Śīghrasiddhi* of Lakṣmīdhara,[63] the epoch of which is 1278, is the earliest set of tables yet known for determining the tithis, nakṣatras, and yogas. It is a double set based on the parameters of both the Brāhmapakṣa and the Āryapakṣa. This is also the arrangement in the *Pañcāṅgavidyādharī* written by Vidyādhara at Rajkot, Saurāṣṭra;[64] its epoch is 1 April 1643.

The *Mahādevī* of Mahādeva,[65] the epoch of which is 28 March 1316, is a set of tables of the true longitudes of the planets according to the "true linear" type used previously by Āśādhara. In the *Mahādevī*, $\Delta\lambda = 6°$, so that $N = 0$ to 59; and $k = 14$. The mean motion tables are based on the mean yearly motions given in Table V.35; other parameters characteristic of the Brāhmapakṣa are listed in Table V.53.

TABLE V.53

Epact	11;3,53,22,40 tithis per year
Lord of the year	1;15,31,17,17 days per year

Related sets of tables were written by Dinakara at Bārejya in Gujarat; his epoch is 31 March 1578. The *Candrārkī*[66] is concerned with solar and lunar positions, and the *Khetasiddhi*[67] with those of the planets. Dinakara also wrote a *Tithisāraṇī*,[68] which contains tables of the tithis, nakṣatras, and yogas; its epoch is 31 March 1583. A further set of tithi, nakṣatra, and yoga tables based on the parameters of the Brāmapakṣa is the *Tithikalpadruma* of Kalyāṇa;[69] its epoch is 31 March 1605. This Kalyāṇa may also be the author of a set of planetary tables, the *Khecaradīpikā*[70] based on the *Mahādevī*; its epoch is 31 March 1649.

The *Jagadbhūṣaṇa*[71] written by Haridatta II in Mewar, Rājasthān, is another set of astronomical tables based on the parameters of the Brāhmapakṣa; its epoch is 31 March 1638. The tables of true longitudes of the planets in this work, as in the Anonymous of 1704,[72] are of the "cyclic" arrangement, derived from the Babylonian "Goal-year" periods.[73]

The existence of several anonymous sets of tables belonging to the Brāhmapakṣa and datable to the sixteenth through nineteenth centuries attests to the vitality of this tradition in western and northwestern India.[74] The last full-scale work to which we can at present refer, however, is the *Karaṇavaiṣṇava* of Sankara,[75] the epoch of which is 9 April 1766 (Gregorian).

Śaṇkara's yearly mean motions are generally those found in Table V.35, except that the lunar node's is −19;21,33,38°. His list of cities over which the prime meridian passes includes Laṅkā, Devasutāpurī, Kāntī, Karṇātasitācala, Parjali, Vatsagulma, Ujjayinī, Gargarāṭa, Āṭavāśrama, Rohītaka, Kurukṣetra, Himālaya, and Sumeru.

Śaṅkara's longitudes of the mandoccas and nodes are those of Table V.51 rounded to the nearest degree (the longitude of Mars's mandocca is 129°). He uses two Sine tables: in one $R = 24$, and in the other $R = 700$ and the entries are for increments of 5°; the second table is reproduced in Table V.54.

TABLE V.54

ϑ	Sin ϑ	R sin ϑ	ϑ	Sin ϑ	R sin ϑ
5°	61	61.0	50°	536	536.2
10°	121	*121.5*	55°	573	573.4
15°	181	181.1	60°	607	*606.2*
20°	239	239.4	65°	634	634.4
25°	296	295.8	70°	658	657.7
30°	350	350	75°	676	676.1
35°	401	*401.5*	80°	689	689.3
40°	450	449.9	85°	697	697.3
45°	495	494.9	90°	700	700

The first Sine table, in which $R = 24$, is used for finding the manda equation by the method illustrated in Table V.55 (compare Table V.48).

The *Karaṇavaiṣṇava* lists the diameters of the śīghra epicycles of the planets ($R = 24$) as the *Karaṇakutūhala* did their radii (Table V.49;

TABLE V.55

Planet	$\dfrac{\mu}{\text{Sin } \kappa_\mu}$	μ_{max} ($\kappa_\mu = 90°$)
Saturn	$\frac{7}{22}$	7;38,10,54,32,. . .°
Jupiter	$\frac{2}{9}$	5;20°
Mars	$\frac{7}{15}$	11;12°
Sun	$\frac{1}{11}$	2;10,54,32,43,. . .°
Venus	$\frac{1}{16}$	1;30°
Mercury	$\frac{1}{4}$	6°
Moon	$\frac{4}{19}$	5;3,9,28,25,. . .°

$R = 120$); it also gives tables of equations. These are listed in Table V.56.

TABLE V.56

Planet	d_σ	$r_\sigma \cdot \frac{120}{24}$	σ_{max}	k_σ
Saturn	5	12;30	6;20°	90°
Jupiter	9	22;30	11°	90°–105°
Mars	32	80	41;30°	135°
Venus	34	85	46;45°	135°
Mercury	18	45	23°	120°

Śaṅkara's values for the kālāṃśas of visibility and the maximum latitudes of the planets, given in Table V.57, differ somewhat from the standard Brāhmapakṣa values (compare Tables V.7 and V.8).

TABLE V.57

Planet	Kālāṃśas	i	β_{max}
Saturn	15;9°	2;9°	2;24°
Jupiter	11;11°	1;15°	1;32°
Mars	19;30°	1;48°	5;24°
Venus	9;25°	2;15°	7;43°
Mercury	13;36°	2;33°	4;5°
Moon	12;20°	4;30°	4;30°

Thus, for a millennium and a half this school of astronomy survived in India with only one major change—that which occurred in the eleventh century, when the planetary parameters were modified and when some elements apparently derived from Ptolemaic Islamic were introduced.

VI. THE ĀRYAPAKṢA

The Āryapakṣa was founded by Āryabhaṭa I with his *Āryabhaṭīya*, in which he mentions the year 499, when he was twenty-three. He states that the science of astronomy is based on a revelation from Brahma (*A* Gola 50), and it appears that the *Paitāmahasiddhānta* of the *Viṣṇudharmottarapurāṇa* was one of his sources. Although Āryabhaṭa seems to have lived in Pātaliputra (modern Patna in Bihar), and the *Āryabhaṭīya* was still followed by some in northwestern India in the tenth century and its elements were known even later (for instance, by Gaṇeśa), since the seventh century it has been associated primarily with southern India, where it still is influential.

The *Āryabhaṭīya* is organized in a fashion different from that of all other Indian siddhāntas. We will here describe the contents of the Daśagītikā, the Kālakriyāpāda, and the Golapāda, indicating similarities with and differences from the early texts of the Brāhmapakṣa.

The *Āryabhaṭīya* assumes a Kalpa containing 1,008 Mahāyugas or 4,354,560,000 years; these 1,008 Mahāyugas are divided into fourteen Manus, each of which comprises seventy-two Mahāyugas (*A* Daśagītikā 3, Kālakriyāpāda 8). The Mahāyugas are divided into four equal Yugas, each of which contains 1,080,000 years. Since a mean conjunction of the planets at Aries 0° occurs at the beginning of each of these Yugas of 1,080,000 years, Āryabhaṭa is forced to have each planet rotate an integer number of times within each of these Yugas rather than within the much larger Kalpa. Theoretically, this allows the users of the Kalpa to employ better parameters for mean motions than does Āryabhaṭa; that they do not, indicates that they had no means to arrive at better parameters. But Brahmagupta severely criticizes Āryabhaṭa for breaking with tradition, and this criticism caused followers of the Āryapakṣa and Ārdharātrikapakṣa in the eighth century to abandon Āryabhaṭa's equal Yugas.

The fundamental parameters of the mean motions of the planets in the Āryapakṣa are given in Table VI.1; in the Daśagītikā they are expressed as rotations in a Mahāyuga, *R* (*A* Daśagītikā 1–2).

The solar parameter implies a year-length of 6,5;15,31,15 civil days (in excess of the Brāhmapakṣa's parameter by 0;0,0,52,30 days) and 1,577,917,500 civil days in a Mahāyuga. Since Āryabhaṭa has the current Kaliyuga begin at the same moment as does the Brāhmapakṣa—on 18 February –3101 at sunrise at Laṅkā—he places the beginnings of the mean solar years in the fifth and sixth centuries of the Christian era about one day later than does the Brāhmapakṣa. He has, however, so chosen his lunar parameter that mean conjunctions of the sun and the moon will occur on the same day in both pakṣas, for the mean synodic month in the Brāhmapakṣa is 29;31,50,5,43, . . . days (see Table V.4) and that in the Āryapakṣa is 29;31,50,5,40, . . . days. His mean motions of the planets, however, seem unrelated to those of the Brāhmapakṣa (see Table V.2), and apparently are computed from the assumed mean conjunction of the beginning of the Kaliyuga and the mean longitudes of the planets as found from a Greek table for exactly 3,600 years later—that is, for noon of 21 March 499.

The time periods in Āryabhaṭa's Mahāyuga are enumerated in Table VI.2 (compare Table V.4). Āryabhaṭa at times refers to the sidereal days as "rotations of the earth" (*A* Daśagītikā 1, Gola 9), and at times as "rotations of the nakṣatras" (*A* Gola 5). Like other Indian astronomers, he was aware of the mathematical equivalence of the two concepts; unlike others (such as Varāhamihira and Brahmagupta) he did not specifically reject the rotation of the earth on physical grounds.[76]

Āryabhaṭa's rules for computing the orbits of the

TABLE VI.1

Planet	R	$\frac{R \cdot 6,0}{Y}$	$\frac{R \cdot 6,0}{C}$
Saturn	146,564	12;12,49,12°	0;2,0,22,41,41,32, . . .°
Jupiter	364,224	30;21,7,12°	0;4,59,9,0,38,51, . . .°
Mars	2,296,824	3,11;24,7,12°	0;31,26,27,48,54,22, . . .°
Sun	4,320,000	6,0°	0;59,8,10,13,3,31, . . .°
Venus' śīghra	7,022,388	3,45;11,56,24°	1;36,7,44,17,4,45, . . .°
Mercury's śīghra	17,937,020	54;45,6°	4;5,32,18,54,36,24, . . .°
Moon	57,753,336	2,12;46,40,48°	13;10,34,52,39,18,56, . . .°
Manda	488,219	40;41,5,42°	0;6,40,59,30,7,38, . . .°
Node	–232,226	–19;21,8,48°	–0;3,10,44,7,49,44, . . .°

TABLE VI.2

Saura years (Y)	4,320,000
Saura months (N)	51,840,000
Lunar months (M)	53,433,336
Sidereal months	57,753,336
Intercalary months (A)	1,593,336
Sidereal days	1,582,237,500
Civil days (C)	1,577,917,500
Saura days (S)	1,555,200,000
Tithis (T)	1,603,000,080
Avamas (U)	25,082,580

planets in yojanas are based on the same principles as are those in the *Paitāmaha*; but he uses a longer yojana, ten of which equal a minute of arc in the orbit of the moon (*A* Daśagītikā 4). Therefore, the yojanas in the orbits of the planets in the Āryapakṣa are as given in Table VI.3 (compare Table V.3).

TABLE VI.3

Planet	O/R
Heaven (O)	12,474,720,576,000
Nakṣatras	173,260,008
Saturn	$85{,}114{,}493\frac{5987}{36641}$
Jupiter	$34{,}250{,}133\frac{699}{1897}$
Mars	$5{,}431{,}291\frac{44009}{95701}$
Sun	$2{,}887{,}666\frac{4}{5}$
Venus' śīghra	$1{,}776{,}421\frac{255221}{585199}$
Mercury's śīghra	$695{,}473\frac{373277}{896851}$
Moon	216,000

The Daśagītikā gives the diameters of the planetary disks in yojanas as in Table VI.4 (*A* Daśagītikā 5); I add the diameters of the moon and the star-planets in minutes, computed on the assumption that these yojanas represent their apparent diameters at the moon's mean distance.

TABLE VI.4

Planet	Diameter (yojanas)	Diameter (minutes)
Sun	4,410	
Moon	315	31;30′
Venus	63	6;18′
Jupiter	31 1/2	3;9′
Mercury	21	2;6′
Saturn	15 3/4	1;34,30′
Mars	12 3/5	1;15,36′

The order of the planets is identical with that in the *Paitāmaha* (V.22 and Table V.8), although the diameters, when multiplied by 3/2, are not the same. Āryabhaṭa states that the diameter of the earth is 1,050 yojanas, but its circumference 3,375 yojanas (*A* Daśagītikā 9); this yields 3 3/14 as the value of π, although elsewhere he uses $\pi = \sqrt{10}$ and $\pi = 21600/6876 = 3\ 27/191$.

The orbital inclinations according to Āryabhaṭa, as listed in Table VI.5 (*A* Daśagītikā 6), differ from those in the Brāhmapakṣa (Table V.7); neither set is very accurate, Āryabhaṭa's represent values rounded to multiples of 0;30°.

TABLE VI.5

Planet	i	β_{max}
Saturn	2°	2;15°
Jupiter	1°	1;15°
Mars	1;30°	4;26°
Venus	2°	7;37°
Mercury	2°	1;36°
Moon	4;30°	4;30°

Āryabhaṭa accepts the rotation of the mandoccas and nodes demanded by the Kalpa system of the Brāhmapakṣa (*A* Daśagītikā 7; Gola 2), but he assumes that their longitudes since the beginning of the present Kaliyuga can be regarded as fixed. These longitudes are given in Table VI.6 (compare Table V.13).

TABLE VI.6

Planet	Mandocca	Node
Saturn	236°	100°
Jupiter	180°	80°
Mars	118°	40°
Sun	78°	
Venus	90°	60°
Mercury	210°	20°

In the *Āryabhaṭīya* the manda epicycles and śīghra epicycles are pulsating, except for the manda epicycles of the sun and the moon. For the superior planets the maximum manda epicycles occur when $\kappa_\mu = 90°$ or $\kappa_\mu = 270°$ and the maximum śīghra epicycles when $\kappa_\sigma = 0°$ or $\kappa_\sigma = 180°$; for the inferior planets the maximum manda epicycles and śīghra epicycles occur when κ_μ or $k_\sigma = 0°$ or 180°. The effect of the pulsation is minimal, its purpose unfathomable. The dimensions are given in Table VI.7 along with the maximum equations from Haridatta (compare Table V.5) (*A* Daśagītikā 8–9).

In the Daśagītikā, finally, Āryabhaṭa lists the differences in the *Paitāmaha*'s Sine table (Table V.6) (*A* Daśagītikā 10); and in the Gaṇitapāda he it-

TABLE VI.7

Planet	Manda c_{max}	Manda c_{min}	Manda μ_{max} ($\kappa_\mu = 90°$)	Śīghra c_{max}	Śīghra c_{min}	Śīghra σ_{max}	Śīghra κ_σ
Saturn	58;30°	40;30°	9;32°	40;30°	36°	5;44°	97;30°
Jupiter	36°	31;30°	5;43°	72°	67;30°	10;53°	101;15°
Mars	81°	63°	13;7°	238;30°	229;30°	44;53°	131;15°
Sun	13;30°		2;9°				
Venus	18°	9°	1;26°	265;30°	256;30°	53;37°	138;45°
Mercury	31;30°	22;30°	3;35°	139;30°	130;30°	21;57°	112;30°
Moon	31;30°		5;1°				

erates the *Paitāmaha*'s rule for determining the second differences in the Sine table (V.8) (*A* Gaṇitapāda 12).

In the Kālakriyā, Āryabhaṭa defines the vyatipāta as in (III.29) (*A* Kālakriyā 3), and mentions the twelve-year Jovian cycle also referred to in the *Paitāmaha* (*A* Kālakriyā 4). More important, he discusses the equivalence of planetary models with double epicycles to those with eccenter and epicycle (*A* Kālakriyā 17–21); both models are of Greek origin. In integrating the effects of the two equations he prescribes essentially the procedure of the *Paitāmaha* (V.7) (*A* Kālakriyā 22–24).

In the Golapāda, Āryabhaṭa deals with various astronomical problems. In the course of this diffuse chapter he gives the kālāṃśas of visibility of the planets as the *Paitāmaha* had (Table V.8) (*A* Gola 4). He also repeats from the *Paitāmaha* several formulas of spherical trigonometry (*A* Gola 18–32); but his formula for right ascension is that later repeated by Brahmagupta (V.76) (*A* Gola 25).

Āryabhaṭa's rules for computing parallax, however, differ from those of the *Paitāmaha* (V.26) and of the *Brāhmasphuṭasiddhānta* (V.95–101). He mentions only the combined parallax, which he considers to be 0 when the sun is at the nonagesimal at the zenith. His computation (*A* Gola 33–34) is illustrated in Figure 30, where triangle S_nNZ is assumed to be a plane right triangle, with the right angle at N. Then one can find Sin $(S_nZ) = \text{Cos}\ \alpha_n$ from (III.53b) and (VI.1):

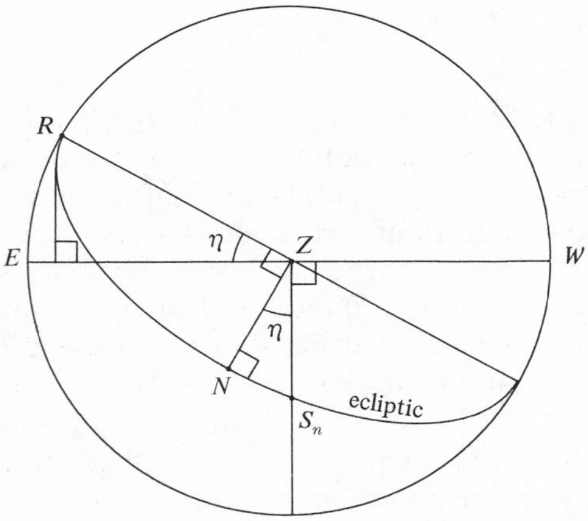

(projection in the plane of the horizon)

FIGURE 30

$$\text{Sin}\ (S_nN) = \text{Sin}\ \eta \cdot \frac{\text{Cos}\ \alpha_n}{R}; \qquad \text{(a)}$$

$$\text{Sin}^2\ (ZN) = \text{Cos}^2\ \alpha_n - \text{Sin}^2\ (S_nN). \qquad \text{(b)}$$

Āryabhaṭa then assumes that the elongation of the sun from the nonagesimal, SN, can be found from a plane right triangle:

$$\text{Sin}^2\ (SN) = \text{Sin}^2\ (ZS) - \text{Sin}^2\ (ZN). \qquad \text{(VI.2)}$$

This is, of course, grossly wrong.

Āryabhaṭa does not proceed to tell one how to obtain the parallax; he only states that the horizontal parallax equals the radius of the earth. Indeed, since the radius of the earth is 525 yojanas and each minute in the orbit of the moon equals ten yojanas, it follows that

$$\pi_0 = 0;52,30°, \qquad \text{(VI.3)}$$

which is essentially correct for lunar parallax.

For finding the distance of the tip of the earth's shadow and the diameter of the shadow at the moon's distance, Āryabhaṭa (*A* Gola 39–40) gives the correct (V.23) and (V.25), where the *Paitāmaha*, as the text is preserved, gives the incorrect (V.22) and (V.24). Āryabhaṭa's rules for finding the durations of eclipses, the durations of totality, and the eclipse magnitudes are the familiar rules in which it is assumed that the lunar orbit is parallel to the ecliptic (*A* Gola 41–44). Moreover, his formulas for the akṣadṛkkarma and the ayanadṛkkarma (*A* Gola 35–36) are those that presumably were originally in the *Paitāmaha* (V.35, V.29), although the latter is replaced by the approximation (compare V.33)

$$\lambda\lambda^* \approx \beta \cdot \frac{\text{Vers}\ (\lambda + 90°) \cdot \text{Sin}\ \epsilon}{R^2}. \qquad \text{(VI.4)}$$

Further, Āryabhaṭa gives the formula for the akṣavalana and refers obliquely to that for the ayanavalana, which are repeated by Brahmagupta (V.93–94) (*A* Gola 45); and he gives the eclipse colors that are found in the *Brāhmasphuṭasiddhānta* (Table V.19) (*A* Gola 46). But he states that a solar eclipse of 1/8 or less than the sun's disk is invisible, whereas Brahmagupta puts the limit at 1/12 (*A* Gola 47).

The ablest commentator on the *Āryabhaṭīya*, and one of India's most competent mathematicians, was Bhāskara I, whose *Bhāṣya* on the *Āryabhaṭīya* was written at Valabhī in Saurāṣṭra in 629, although he is usually associated with Aśmaka in central India. Apparently before composing the *Bhāṣya*, he wrote a summary of the astronomical system of the Āryapakṣa entitled *Mahābhāskarīya*, in which he especially displays his ingenuity in mathematics. The text is in the form of rules based on the parameters of the Āryapakṣa, which parameters are stated by Bhāskara in adhyāya 7.

In adhyāya 1, Bhāskara gives rules for finding the ahargaṇa from Āryabhaṭa's beginning of the current Kaliyuga, and for finding the mean longitudes of the planets. The ahargaṇa rules are identical with, or variations of, Brahmagupta's (V.48) (*MB* 1, 4–7). He also provides formulas to find the mean longitudes of the planets without the ahargaṇa that are similar to those in *Brāhmasphuṭasiddhānta* 13 (*MB* 1, 13–19). Also, similar to (V.52) but based on the parameters of the Āryapakṣa, he states (*MB* 1, 22–28) (VI.5):

$$\epsilon_0 = \frac{1593336 \cdot 30}{4320000} = 11\frac{389}{6000} = 11;3,53,24 \text{ tithis; (a)}$$

$$u_0 = \frac{25082580}{4320000} = 5 + \frac{58043}{72000}$$

$$= 5 + \frac{29}{36} + \frac{43}{72000} \text{ avamas; (b)}$$

$$c_0 = \frac{1577917500}{4320000} = 360 + 5 + \frac{149}{576} \text{ days.} \qquad \text{(c)}$$

The remainder of adhyāya 1 gives rules for finding the mean longitudes of the planets from the grahatanu, which is the number of saura days between the beginning of Kaliyuga and the beginning of the current year, and from the dhruvaka, which is the ahargaṇa in saura days (*MB* 1, 29–39), as well as applications of the kuṭṭaka to problems in mean motions similar to Brahmagupta's in *Brāhmasphuṭasiddhānta* 18 (*MB* 1, 41–52).

Adhyāya 2 is a brief chapter on the correction for the longitudinal difference. The prime meridian is stated to pass over Laṅkā, Kharanagara, Sitorugeha, Pāṇāṭa, Misitapurī, Taparṇī, Sitavara mountain, Vātsyagulma, Vananagarī, Avantī, Sthāneśa, and Meru (*MB* 2, 1–2). For the deśāntara Bhāskara gives Brahmagupta's rule (V.61), but with the circumference of the earth taken to be 3,298 17/25 yojanas (*MB* 2, 3–4); if the diameter is 1,050 yojanas, as Āryabhaṭa states, and $\pi = 3\ 27/191$, the circumference would be 3,298 82/191. Bhāskara correctly criticizes this rule, however, on the grounds that the distance between two cities cannot be accurately known and that a parallel of latitude is not a great circle. To correct this, he proposes using a water clock to measure the time difference between computed moonrise for the prime meridian at one's own latitude and the observed moonrise; from this time difference one can find the true longitudinal difference, and thence the true distance in yojanas of one's locality from the prime meridian (*MB* 2, 5).

In *Mahābhāskarīya* 3, Bhāskara gives many of the usual rules relating to direction, place, and time. Using (III.35) and (III.55) and the value $\text{Sin}\ \epsilon = 1{,}397$ ($R = 3{,}438$, $\epsilon = 24°$), Bhāskara establishes as values for γ in time minutes at solar longitudes of 30°, 60°, and 90° those in Table VI.8 (compare III.19) (*MB* 3, 8).

TABLE VI.8

$\frac{\gamma_1}{2}$	1°
$\frac{\gamma_2}{2}$	0;48°
$\frac{\gamma_3}{2}$	0;20,15°

Bhāskara also gives the values for α_1, α_2, and α_3 found in the *Brāhmasphuṭasiddhānta* (Table V.17) (*MB* 3, 10), and Āryabhaṭa's rule for Sin α (V.76) (*MB* 3, 9). The remaining rules in this adhyāya are repetitions of or elaborations on those occurring in other works. Bhāskara adds descriptions of some observational instruments and a catalog of the ecliptic longitudes and latitudes of the yogatārās; the latter are reproduced in Table VI.9 (*MB* 3, 62–71).

Mahābhāskarīya 4 presents the rules for computing the true longitudes of the planets. Both the epicyclic and the eccentric models are considered, and both are used to solve the concentric with equant model by iteration (*MB* 4, 9–12, 19–20).[77]

Adhyāya 5 is concerned with eclipses. Bhāskara

TABLE VI.9

	Bhāskara			Ptolemy	
Nakṣatra	λ	β	Identification	λ	β
Aśvinī	8°	+10°	ρ Pisc.	0;30°	+9°
Bharaṇī	27°	+12°	35 Arí.	19;40°	+11;10°
Kṛttikā	36°	+5°	16 Tau. (?)	32;10°	+4;30°
Rohiṇī	49°	−5°	α Tau.	42;40°	−5;10°
Mṛgaśiras	62°	−10°	15 Ori. (?)	50;30°	−8°
Ārdrā	70°	−11°	λ Ori. (?)	57°	−13;30°
Punarvasu	90°	+6°	β Gem. (?)	86;40°	+6;15°
Puṣya	105°	0°	ε Canc. (?)	100;20°	+0;20°
Āśleṣā	114°	−7°	α Canc.	106;30°	−5;30°
Maghā	128;30°	0°	α Leo	122;30°	+0;10°
Pūrvaphālgunī	141°	+12°	δ Leo	134;10°	+13;40°
Uttaraphālgunī	154°	+13°	β Leo (?)	144;30°	+11;50°
Hasta	173°	−7°	η Corv. (?)	167°	−11;40°
Citrā	185°	−2°	α Virg.	176;40°	−2°
Svāti	197°	+37°	c Boöt. (?)	188;10°	+41;20°
Viśākhā	212°	−1;30°	ι Lib.	204°	−1;40°
Anurādhā	222°	−3°	δ Scorp.	215;40°	−1;40°
Jyeṣṭhā	228°	−4°	α Scorp.	222;40°	−4°
Mūla	241°	−8;20°	d Oph.	235;30°	−6;10°
Pūrvāṣāḍhā	254°	−7°	δ Sag.	247;40°	−6;30°
Uttarāṣāḍhā	267°	−7;20°	ζ Sag. (?)	256;20°	−6;45°
Śravaṇa	285°	+30°	α Aquil. (?)	273;50°	+29;10°
Dhaniṣṭhā	296°	+36°	α Delph.	290;10°	+33;50°
Śatabhiṣaj	307°	−0;18°	μ Capr.	298;40°	0°
Purvabhādrapadā	328°	+24°	ξ Peg. (?)	320;30°	+19°
Uttarabhādrapadā	345°	+26°	α Andr. (?)	347;50°	+26°
Revatī	0°	0°	ζ Pisc.	353;0°	−0;10°

It is difficult to conceive of how Bhāskara arrived at these longitudes and latitudes.

repeats the diameters of the sun, the moon, and the earth in yojanas given by Āryabhaṭa (Table VI.4) (*MB* 5, 4), but gives as approximate distances in yojanas of the sun and the moon (based on Table VI.3) the figures in Table VI.10 (*MB* 5, 2).

TABLE VI.10

Sun	459,585 yojanas
Moon	34,377 yojanas

For finding the true diameters he gives the rules in (VI.6) (compare V.89) (*MB* 5, 6–7):

$$d_{\odot}=v'_{\odot}\cdot\left(\frac{5'}{9}+\frac{5''}{36}\right); \quad \text{(a)}$$

$$d_{\text{☾}}=v'_{\text{☾}}\cdot\left(\frac{1'}{25}-\frac{1''}{100}\right); \quad \text{(b)}$$

$$d_{u}=v'_{\text{☾}}\cdot\left(\frac{1'}{10}+\frac{1''}{16}\right). \quad \text{(c)}$$

The diameter of the shadow should depend on the difference of the lunar and solar velocities rather than on the lunar velocity alone. From these relationships the mean values in Table VI.11 are (compare Table V.18)

TABLE VI.11

Planet	$\bar{d}$
Sun	0;32,54, . . .°
Moon	0;31,44, . . .°
Shadow	1;19,49, . . .°

Bhāskara finds the moon's latitude by the rule (*MB* 5, 14) (VI.7)

$$\text{Sin}\,\beta_{\text{☾}}=\frac{15}{191}\cdot\text{Sin}\,\omega. \quad \text{(a)}$$

From (III.18 b) we know that

$$\text{Sin}\,i=R\cdot\frac{15}{191}=270, \quad \text{(b)}$$

if $R=3{,}438$. The maximum lunar latitude, then, is 4;30°, as in Table VI.5. Bhāskara's computation of parallax is a completed form of Āryabhaṭa's (VI.1–VI.3) (*MB* 5, 16–32), and his rules for finding the durations and magnitudes of eclipses are elaborations of those in the *Āryabhaṭīya* (*MB* 5, 33–41). For deflection, however, he gives the approximative formulas (compare V.93–94) (*MB* 5, 42–45)

$$\text{Sin}\,\gamma_1 = \text{Vers}\,t^\circ \cdot \frac{\text{Sin}\,\phi}{R}; \qquad \text{(VI.8)}$$

$$\text{Sin}\,\gamma_2 = \text{Vers}\,(\lambda + 90^\circ) \cdot \frac{\text{Sin}\,\epsilon}{R}. \qquad \text{(VI.9)}$$

Compare Āryabhaṭa's ayanadṛkkarma (VI.4) with (VI.9). Bhāskara completes his discussion of solar eclipses with elaborate rules for constructing projections of eclipses (*MB* 5, 47–67). He erroneously prescribes the application of parallax to lunar eclipses as well as to solar (*MB* 5, 68–70).

In *Mahābhāskarīya* 6, Bhāskara discusses the heliacal risings and settings of the planets and their conjunctions. In this he repeats Āryabhaṭa's rules for the akṣadṛkkarma (V.35) (*MB* 6, 1–2) and the ayanadṛkkarma (VI.4) (*MB* 6, 2–3). For the illuminated portion of the moon Bhāskara's formula, when the moon is in the second and third quadrants, is equivalent to Brahmagupta's (V.46b); but when it is in the first and fourth he substitutes for (V.46a) the following (*MB* 6, 5–7):

$$\sigma_1 = d_{☾} \cdot \frac{\text{Vers}\,\Delta\lambda}{2R}. \qquad \text{(VI.10)}$$

Bhāskara devotes many verses to the problems of the elevation of the moon's horns, its projection, and the time of moonrise (*MB* 6, 9–42). He repeats Āryabhaṭa's kālāṃśas of visibility of the planets (Table V.8), but adds that those for Venus when retrograde (at inferior conjunction) are only 4°–4;30° (*MB* 6, 44–45). He computes conjunctions with ecliptic rather than polar coordinates (*MB* 6, 48–55).

In adhyāya 7, besides listing the parameters of the Āryapakṣa (*MB* 7, 1–16), Bhāskara gives the formula for finding Sin ϑ that is also given by Brahmagupta (V.69) (*MB* 7, 17–19). At the end of this adhyāya he describes the characteristics of the Ārdharātrikapakṣa (*MB* 7, 21–35).

The *Laghubhāskarīya* repeats in abbreviated (and sometimes corrected) form the contents of the *Mahābhāskarīya*, but in it Bhāskara follows the order of the Daśādhyāyī. Here one need only note that he has changed some of the ecliptic coordinates of the yogatārās listed in Table VI.9; these changes are displayed in Table VI.12 (*LB* 8, 1–9)

TABLE VI.12

Yogatārā	*Mahābhāskarīya* λ	*Mahābhāskarīya* β	*Laghubhāskarīya* λ	*Laghubhāskarīya* β
Bharaṇī	27°	+12°	26;30°	+12°
Rohiṇī	49°	−5°	50°	−5°
Mūla	241°	−8;20°	241;30°	−8;30°
Pūrvāṣāḍhā	254°	−7°	254;30°	−7°
Uttarāṣāḍhā	267°	−7;20°	266;30°	−7°
Śravaṇa	285°	+30°	284;30°	+30°
Dhaniṣṭhā	296°	+36°	295;30°	+36°

A small karaṇa, the *Grahacāranibandhana*, based on the parameters of the Āryapakṣa, was written by Haridatta, traditionally in 683. It is the fundamental text of the so-called Parahita system of astronomy, prevalent for many centuries in south-

TABLE VI.13

Planet	Period Relation	Mean Daily Motion
Saturn	$\frac{1\text{ rot.}}{10766\text{ days}} - \frac{1''}{2727\text{ days}}$	0;2,0,22,42,58,29, . . .°
Jupiter	$\frac{10\text{ rot.}}{43323\text{ days}} + \frac{1''}{1026\text{ days}}$	0;4,59,8,57,14,20, . . .°
Mars	$\frac{1\text{ rot.}}{687\text{ days}} + \frac{1''}{2804\text{ days}}$	0;31,26,27,47,39,32, . . .°
Sun	$\frac{31\text{ rot.}}{11323\text{ days}} - \frac{5''}{1647\text{ days}}$	0;59,8,10,23,40,23, . . .°
Venus' śīghra	$\frac{500\text{ rot.}}{112349\text{ days}} - \frac{1''}{561\text{ days}}$	1;36,7,44,23,18,37, . . .°
Mercury's śīghra	$\frac{100\text{ rot.}}{8797\text{ days}} + \frac{9''}{884\text{ days}}$	4;5,32,18,19,0,49, . . .°
Moon	$\frac{600\text{ rot.}}{16393\text{ days}} - \frac{1''}{625\text{ days}}$	13;10,34,52,44,54,37, . . .°
Manda	$\frac{1\text{ rot.}}{3232\text{ days}} + \frac{1''}{1230\text{ days}}$	0;6,40,59,27,16,59, . . .°
Node	$-\left(\frac{1\text{ rot.}}{6795\text{ days}} + \frac{3''}{841\text{ days}}\right)$	−0;3,10,43,55,21,30, . . .°

ern India. Haridatta expresses the mean motions of the planets in the period relations of Table VI.13 (compare Table VI.1) (*GCN* 1, 21–29). These expressions take the form of those in Lāṭadeva's *Sūryasiddhānta* (Table VII.2).

The central portion of Haridatta's work is versified tables of the sines of the planetary equations for arguments increasing by 3;45° (*GCN* 2, 1–15 and 3, 35–36); he calls these tables vākyas, whence the name of the Vākya system of astronomy. The maximum equations are those recorded in Table VI.7. He also gives the śīghra anomalies (nonsymmetrical!) necessary for the occurrence of first and last stations, as in Table VI.14 (compare Table V.9) (*GCN* 3, 12–16).

TABLE VI.14

Planet	First Station	Second Station
Saturn	117;50°	243;45°
Jupiter	131;15°	233;45°
Mars	172;46°	196;26°
Venus	163;7°	195;18°
Mercury	153;6°	216;53°

Haridatta gives mean longitudes for the ahargaṇa of 210,389 (= 365;15,31,15 · 576) from the beginning of Kaliyuga (this epoch is 23 February – 2525) (*GCN* 1, 12–18) and for the ahargaṇa of 210,389 · 6 = 1,262,334 (this epoch is 20 March 355) (*GCN* 3, 47–49).

Based on, and repeating much of, Haridatta's work is an anonymous *Grahacāranibandhanasaṅgraha*, the epoch of which is represented by the ahargaṇa 1,472,723 = 210,389 · 7 (25 March 931) (*GCNS* 4). This text preserves bīja corrections to the mean yearly motions since 522 that are ascribed to Haridatta himself by Sundararāja[78] and may be from the former's lost *Mahāmārganibandhana*; these bījas and the resulting mean yearly motions are given in Table VI.15 (compare Tables VI.16 and VI.18) (*GCNS* 17–18).

Immediately following these bījas the *Grahacāranibandhanasaṅgraha* gives another set, also to be applied to the years since 522; they are given in Table VI.16 (*GCNS* 19–22).

One of the earliest Arabic astronomical texts, written at Sind in 742, was the *Zīj al-Harqan*,[79] which evidently utilized parameters of the Āryapakṣa; its existence indicates the survival of this pakṣa in that region of India into the eighth century, although the Arabs primarily encountered texts belonging to the Brāhmapakṣa or the Ārdharātrikapakṣa. Perhaps this *Zīj al-Harqan* is the source of the knowledge of Āryabhaṭa's rotations of the planets in a Mahāyuga (Table VI.1) demonstrated by al-Ahwāzī, who flourished after about 830;[80] al-Ahwāzī changes the rotations of Mars from 2,296,824 to 2,296,828, a figure found also in Vaṭeśvara (Table VI.24). There also are Āryapakṣa parameters in the works of al-Fazārī, who likely derived them from the *Zīj al-Harqan*.

Probably also in the eighth century, although perhaps in the early ninth, Lalla wrote the *Śiṣyadhīvṛddhidatantra*, a work structured like Śrīpati's *Siddhāntaśekhara* and surely antedating it. He may have lived in Mālava, since he mentions

TABLE VI.15

Planet	Mean Yearly Motion	Bīja	Corrected Mean Yearly Motion
Saturn	12;12,49,12°	$+\frac{20'}{235}$	≈ 12;12,54,18,21°
Jupiter	30;21,7,12°	$-\frac{47'}{235}$	30;20,55,12°
Mars	3,11;24,7,12°	$+\frac{45'}{235}$	≈ 3,11;24,18,41,20°
Venus' śīghra	3,45;11,56,24°	$-\frac{153'}{235}$	≈ 3,45;11,17,20°
Mercury's śīghra	54;45,6°	$+\frac{420'}{235}$	≈ 54;46,53,14°
Moon	2,12;46,40,48°	$-\frac{9'}{85}$	≈ 2,12;46,34,27°
Manda	40;41,5,42°	$-\frac{65'}{134}$	≈ 40;40,36,36°
Node	– 19;21,8,48°	$-\frac{13'}{32}$	– 19;21,33,10,30°

TABLE VI.16

Planet	Bīja	Corrected Mean Yearly Motion
Saturn	$+\frac{21'}{235}$	≈ 12;12,54,34°
Jupiter	$-\frac{50'}{235}$	≈ 30;20,53,26°
Mars	$+\frac{50'}{235}$	≈ 3,11;24,19,58°
Venus' śīghra	$-\frac{160'}{235}$	≈ 3,45;11,55,33°
Mercury's śīghra	$+\frac{430'}{235}$	≈ 54;46,55,47°
Moon	$-\frac{25'}{235}$	≈ 2,12;46,34,25°
Manda	$-\frac{144'}{235}$	≈ 40;40,28,56°
Node	$-\frac{96'}{235}$	≈ − 19;21,33,18°

Daśapura (*SDV* 2, 9, 10). Lalla's parameters are those of the Āryapakṣa; but he describes the four yugas in a Mahāyuga in the traditional fashion of the Brāhmapakṣa rather than equal, as Āryabhaṭa has them (*SDV* 1, 1, 14). This, of course, destroys Aryabhaṭa's structure and denies the occurrence of mean conjunctions of the planets at the beginnings and ends of the Mahāyugas. His contemporaries, the authors of the later *Pauliśasiddhānta* and of the *Sūryasiddhānta*, had done essentially the same thing, but had achieved harmony with the Brāhmapakṣa's Kalpa system by delaying the inception of the planets' motions for the requisite number of years to produce a mean conjunction at the beginning of the current Kaliyuga.

Lalla's crude approximations to the mean daily motions of the planets are given in Table VI.17 (*SDV* 1, 1, 39–46).

Lalla names only Laṅkā, Ujjayinī, and Himālaya as lying on the prime meridian (*SDV* 1, 1, 55); he uses Āryabhaṭa's value for the diameter of the earth (1,050 yojanas) but gives its circumference as 3,300 yojanas, which is based on $\pi = 3\frac{1}{7}$ (*SDV* 1, 1, 56–57). But elsewhere he repeats Āryabhaṭa's aberrant value for the circumference of the earth (3,375 yojanas) and gives the diameter as 1,074 yojanas; here $\pi = 3\ 51/358$ (*SDV* 2, 8, 2). His bījas, listed in Table VI.18, are applied to the years following 498; although his period is 250 rather than 235 years, the numerators of the bīja fractions frequently are identical with those in the *Grahacāranibandhanasaṅgraha* (Tables VI.15 and VI.16) (*SDV* 1, 1, 59–60).

Lalla's normal Sine table is that of the *Paitāmaha* and Āryabhaṭa (Table V.6) (*SDV* 1, 2, 1–8), although he also reproduces that in which $R = 150$ (Table V.24) (*SDV* 1, 13, 2). He also retains the same longitudes of the apogees as Āryabhaṭa's (Table VI.6) (*SDV* 1, 2, 9 and 28). His circumferences of the manda epicycles and śīghra epicycles are Āryabhaṭa's for $\kappa = 0°$ or 180° (Table VI.7) (*SDV* 1, 2, 28–29). Among other rules Lalla repeats that of Bhāskara I for computing the concentric with equant model for the sun and the moon

TABLE VI.17

Planet	Mean Daily Motion	
Saturn	$2' + \frac{2'}{365} =$	0;2,0,19,43, . . .°
Jupiter	$5' - \frac{5'}{354} =$	0;4,59,9,9, . . .°
Mars	$\frac{1°}{2} + \frac{1°/2}{21} + \frac{1'}{80} =$	0;31,26,27,51, . . .°
Sun	$1° - \frac{1°}{70} - \frac{2''}{5} =$	0;59,8,10,17, . . .°
Venus' śīghra	$\frac{8°}{5} + \frac{8'}{62} =$	1;36,7,44,30, . . .°
Mercury's śīghra	$4° + \frac{6°}{65} =$	4;5,32,18,27, . . .°
Moon	$13° + \frac{12°}{68} =$	13;10,35,17,38, . . .°
Manda	$\frac{1°}{9} + \frac{1'}{61} =$	0;6,40,59,0, . . .°
Node	$-\left(\frac{1°}{19} + \frac{1'}{28}\right) =$	− 0;3,11,36,59, . . .°

TABLE VI.18

Planet	Bīja	Yearly Mean Motion
Saturn	$+\frac{20'}{250}$	12;12,54°
Jupiter	$-\frac{47'}{250}$	≈ 30;20,55,58,19°
Mars	$+\frac{48'}{250}$	3,11;24,18,43,12°
Venus' śīghra	$-\frac{153'}{250}$	3,45;11,19,40,48°
Mercury's śīghra	$+\frac{420'}{250}$	54;46,46,48°
Moon	$-\frac{25'}{250}$	2,12;46,34,48°
Manda	$-\frac{114'}{250}$	40;40,38,20,40°
Node	$-\frac{96'}{250}$	− 19;21,31,50,24°

by means of an epicycle with varying radius (*SDV* 1, 2, 44). But he lists Brahmagupta's śīghra arguments for the occurrences of the Greek-letter phenomena, except that the first visibility of Saturn is said to occur at 20°, as it is also by Bhojarāja (*SDV* 1, 2, 47–50). Lalla records the retrograde periods, and the periods of visibility and invisibility of the planets in days, as in Table VI.19; his figures disagree with Āśādhara's (Table V.39) (*SDV* 1, 2, 52–53).

In a list of values for $\gamma/2$ Lalla gives slightly different numbers from those of Bhāskara I (Table VI.8); these are noted in Table VI.20 (*SDV* 1, 13, 9)

TABLE VI.20

$\frac{\gamma_1}{2}$	60′
$\frac{\gamma_2}{2}$	49′
$\frac{\gamma_3}{2}$	20′

Most of the rest of Lalla's work is similar to Āryabhaṭa's, to Bhāskara's, or to Brahmagupta's. He does state, however, that the diameter of the moon's disk is 320 yojanas rather than the 315 in Table VI.4 (*SDV* 1, 4, 6). Lalla also is the first to give the cālanas for parvans of 277 and 15 civil days according to the Āryapakṣa. These are enumerated in Table VI.21 (*SDV* 1, 6, 11); they of course differ slightly from the Brāhmapaksa cālanas (Tables V.20 and V.34).

TABLE VI.21

	277 days	15 days
Sun	5ˢ 24;27,6°	0ˢ 14;47,2°
Moon	5ˢ 22;12,53°	6ˢ 17;38,42°
Manda	0° 19;42,53°	0ˢ 1;40,13°
Node	0ˢ 9;22,51°	0ˢ 0;47,45°

He repeats the standard kālāṃśas of visibility of the planets (Table V.8), but adds that when Venus is retrograde, its kālāṃśas are 8°, and when Mercury is retrograde, its kālāṃśas are 12° (*SDV* 1, 7, 5).

In his catalog of coordinates of the yogatārās, Lalla generally repeats those (ecliptic) in the *Mahābhāskarīya* (Table VI.9) (*SDV* 1, 11, 1–9); his changes are recorded in Table VI.22, along with polar coordinates of Brahmagupta (Table V.10 and V.22) that influenced him. This mixture of ecliptic and polar longitudes can be explained only by Lalla's incompetence or, possibly, by the uncertainty of the identifications of the yogatārās among Indian astronomers.

Al-Bīrūnī refers several times to a *Karaṇasāra* composed by Vaṭeśvara, the son of Mahadatta; the epoch of this work was 899. The same Vaṭeśvara wrote a *Vaṭeśvarasiddhānta* at Ānandapura in 904, of which a large part, although not all, survives. It shows a strong influence of the Āryapakṣa and, save for some late tithi, nakṣatra, and yoga tables, is the last major representative of that pakṣa in northern, northwestern, or western India.

Unfortunately, the published text is both inaccurate and incomplete. Therefore, one cannot yet report definitively on its contents. At present it will suffice to say that Vaṭeśvara, like Lalla and the authors of the later *Pauliśasiddhānta* and of the *Sūryasiddhānta*, has made certain changes in an attempt to accommodate some of the system of the

TABLE VI.19

Planet	Retrograde	Invisible	Visible	Synodic Period
Saturn	134 days	36 days	352 days	388 days
Jupiter	112 days	30 days	372 days	402 days
Mars	66 days	120 days	660 days	780 days
Venus	52 days	8 days ($\Omega \rightarrow \Gamma$)	251 days ($\Gamma \rightarrow \Sigma$)	
Mercury	21 days	16 days ($\Omega \rightarrow \Gamma$)	37 days ($\Gamma \rightarrow \Sigma$)	

TABLE VI.22

Nakṣatra	Bhāskara		Lalla		Brahmagupta	
	λ	β	λ	β	λ*	β*
Bharaṇī	27°	+12°	20°	+12°	20°	+12°
Punarvasu	90°	+6°	92°	+6°	(93°	+6°)
Maghā	128;30°	0°	128°	0°		
Pūrvaphālgunī	141°	+12°	139;20°	+12°		
Hasta	173°	−7°	173°	−8°	(170°	−11°)
Citrā	185°	−2°	184;20°	−2°	(183°	−1:45°)
Mūla	241°	−8;20°	241°	−8;30°	241°	−8;30°
Pūrvāṣāḍhā	254°	−7°	254°	−5;20°	254°	−5;20°
Uttarāṣāḍhā	267°	−7;20°	267;20°	−5°	260°	−5°
Śravaṇa	285°	+30°	283;10°	+30°	(278°	+30°)
Dhaniṣṭhā	296°	+36°	296;20°	+36°		
Śatabhiṣaj	307°	−0;18°	313;20°	−0;20°	(320°	−0;18°)
Pūrvabhādrapadā	328°	+24°	327°	+24°	(326°	+24°)
Uttarabhādrapadā	345°	+26°	335;20°	+26°	(337°	+26°)
Revatī	0°	0°	359°	0°		
Agastya			87°	−80°	(87°	−77°)
Mṛgayu (= Mṛgaharttā)			86°	−40°	86°	−40°
Abhijit			267°	+63°	(265°	+62°)

TABLE VI.23

Planet	Apogee		Node	
	R	ρ	R	ρ
Saturn	72,974	236;55,4°	1,542	99;59,0°
Jupiter	13,948	172;48,31°	39,202	89;5,58°
Mars	81,165	128;50,50°	20,684	39;49,48°
Sun	165,801	78;51,37°		
Venus	152,842	80;3,26°	196,127,480,636,835	185;58,4°
Mercury	477,291	226;42,54°	988,271,456,418,719	19;40,6°

TABLE VI.24

Planet	R	$\frac{R \cdot 6,0}{Y}$	$\frac{R \cdot 6,0}{C}$
Saturn	146,568	12;12,50,24°	0;2,0,22,53,30,11, . . .°
Jupiter	364,220	30;21,6°	0;4,59,8,48,46,46, . . .°
Mars	2,296,828	3,11;24,8,24°	0;31,26,28,0,28,30, . . .°
Sun	4,320,000	6,0°	0;59,8,10,12,34,22, . . .°
Venus' śīghra	7,022,376	3,45;11,52,48°	1;36,7,43,40,48,28, . . .°
Mercury's śīghra	17,937,056	54;45,16,48°	4;5,32,20,39,2,6, . . .°
Moon	57,753,336	2,12;46,40,48°	13;10,34,52,32,49,20, . . .°
Manda	488,203	40;41,0,54°	0;6,40,58,42,45,58, . . .°
Node	− 232,238	− 19;21,11,24°	− 0;3,10,44,43,17,4, . . .°
Saptarṣis	1,692	0;8,27,36°	

Brāhmapakṣa. In particular, he has the apogees and nodes of the planets rotate integer numbers of times in 72,000 Kalpas, as in Table VI.23, where R is the number of rotations in 72,000 Kalpas and ρ is the longitudes at the beginning of the current Kaliyuga (compare Table VI.6) (*VS* 1, 1, 15–19 and 1, 4, 56–60).

Vaṭeśvara also gives "new" parameters for the mean motions of the planets, and changes the number of civil days in a Mahāyuga to 1,577,917,560 (see Table VI.2), so that a year is 6,5;15,31,18 days. In Table VI.24 his values for the planets' rotations and the corresponding mean yearly and mean daily motions are recorded (*VS* 1, 1, 11–14). In fact, except in the cases of Mars and Mercury's śīghra, where Vaṭeśvara's numbers are less by 4, the rotations of the planets in a Mahāyuga are identical with those in the Saurapakṣa (Table VIII.1); but the Saurapakṣa has a slightly longer year (6,5;15,31,31,24 days). Vaṭeśvara's figure for the rotations of Mars is identical with al-Ahwāzī's.

Vaṭeśvara's table of Sines uses $R = 3437;44$, but is computed (to a sixtieth of a part) for ninety-six intervals in a quadrant (a fourth of the standard 3;45°); each interval, then, is 0;56,15° (*VS* 2, 1, 2–48). The table in the published text is very corrupt, but has been reconstructed.[81] As presented in the edition, his dimensions of the manda epicycle and śīghra epicycle are unusual and perhaps corrupt; they are reproduced in Table VI.25. (*VS* 2, 1, 49–50).

A description of the remainder of the text must await a better edition.

TABLE VI.25

Planet	c_μ	c_σ
Saturn	46°	32°
Jupiter	33°	65°
Mars	72°	233°
Sun	14°	
Venus	11°	260°
Mercury	22°	138°
Moon	31;30°	

A popular karaṇa belonging to the Āryapakṣa as corrected by Lalla's bījas is the *Karaṇaprakāśa* of Brahmadeva, probably written at Madurai in southern India; its epoch is 11 March 1092. The mean daily motions of the planets are given by Brahmadeva in the form represented in the second column of Table VI.26 (*KP* 1, 4–12). The *Karaṇaprakāśa* gives the rate of precession (or trepidation) as 1° in sixty years, and the year of coincidence as 522 (*KP* 2, 9).

Brahmadeva's Sine table is based on $R = 120$ (compare Tables III.13 and V.43). The entries are recorded in Table VI.27 (*KP* 2, 1).

TABLE VI.27

ϑ	Sin ϑ	120 sin ϑ
15°	31	31.0
30°	60	60.0
45°	85	84.8
60°	104	103.9
75°	116	115.9
90°	120	120

TABLE VI.26

Planet	Mean Daily Motion	Mean Daily Motion
Saturn	$\frac{1^\circ + \frac{1^\circ}{300}}{30} - \frac{1'}{6968} =$	0;2,0,23,29,0,36,...°
Jupiter	$\frac{1^\circ - \frac{1^\circ}{341}}{12} - \frac{1'}{64039} =$	0;4,59,7,9,28,18,...°
Mars	$\frac{10^\circ - \frac{10^\circ}{230}}{19} - \frac{1'}{16080} =$	0;31,26,29,42,26,53,...°
Sun	$1^\circ - \frac{2^\circ}{139} - \frac{1^\circ}{115589} =$	0;59,8,10,13,3,31,...°
Venus' śīghra	$\frac{100^\circ + \frac{100^\circ}{107}}{63} - \frac{1^\circ}{68301} =$	1;36,7,38,15,9,23,...°
Mercury's śīghra	$4^\circ + \frac{11^\circ}{119} - \frac{1^\circ}{19783} =$	4;5,32,35,28,5,7,...°
Moon	$13^\circ + \frac{3^\circ}{17} - \frac{1^\circ}{8315} =$	13;10,34,51,40,11,40,...°
Manda	$\frac{1^\circ + \frac{1^\circ}{440}}{9} + \frac{1'}{7787} =$	0;6,40,55,0,27,56,...°
Node	$-\left(\frac{8^\circ}{151} + \frac{1^\circ}{51348}\right) =$	−0;3,10,47,54,54,42,...°

Keeping Āryabhaṭa's longitudes of the mandoccas (Table VI.6) (*KP* 3, 1), Brahmadeva states the maximum manda equations and Sines of the maximum śīghra equations, as in Table VI.28 (*KP* 2, 5 and 3, 1 and 4).

TABLE VI.28

Planet	$\frac{\mu_{max}}{R}$	μ_{max}	Sin σ_{max}	σ_{max}
Saturn	$\frac{5}{66}$	9;5,27,...°	12	5;48,3,...°
Jupiter	$\frac{5}{108}$	5;33,20°	22	10;38,42,...°
Mars	$\frac{5}{48}$	12;30°	77	40;12°
Sun	$\frac{100}{93 \cdot 60}$	2;9,1,56,...°		
Venus	$\frac{5}{360}$	1;40°	86	45;47,18,...°
Mercury	$\frac{5}{156}$	3;50,46,...°	43	21;12,24,...°
Moon	$\frac{5}{2 \cdot 60}$	5°		

Brahmadeva gives both the śīghra anomalies necessary for the occurrence of the Greek-letter phenomena (*KP* 3, 8 and 10) and the periods of retrogression, visibility, and invisibility (*KP* 3, 9 and 11); these are all presented in Tables VI.29 and VI.30. The synodic periods, which are not given by Brahmadeva, indicate how approximative his periods are.

TABLE VI.29

Planet	First Appearance		First Station
Saturn	20°		113°
Jupiter	18°		125°
Mars	28°		163°
	East	West	
Venus	183°	23°	165°
Mercury	205°	51°	145°

In southern India, probably at the beginning of the second millennium after Christ and professedly based on the Parahita system of Haridatta, the Vākya system of predicting planetary longitudes was developed. This is based on period relations, which are integer numbers of days, and tables of positions for every day within each period. The lunar vākyas, ascribed to Vararuci, give lunar longitudes for each of the 248 days in nine anomalistic months;[82] this is one of the period relations of Vasiṣṭha (III.8). From the ahargaṇa since epoch all periods of 12,372 days or 449 anomalistic months are to be eliminated; since the beginning of the current period of 12,372 days all periods of 3,031 days or 110 anomalistic months are to be eliminated. This last period relation is also found in Vasiṣṭha (III.9), and 12,372 days is simply four periods of 3,031 days plus one period of 248 days. Various epoch dates have been given in different presentations of the vākyas; they range from 1184 to 1756. None, of course, can be shown to be even the approximate date of the inception of the Vākya system. For each period there is an increment in lunar longitude: 4,57;48,10° in 12,372 days; 5,37; 31, 1° in 3,031 days; and 27;44,6° in 248 days.

One of the epoch dates utilized in the tradition is 22 May 1282; this also is the epoch date of the lunar vākyas in the anonymous *Vākyakaraṇa*, which gives vākyas for all the planets. It was composed in about 1300, probably near Kāñcī.

In this text the year length is 365 + 1/4 + 5/576 days (*VK* 1, 2–3), which is the Āryapakṣa's 6,5;15,31,15 days. The true longitude of the sun in degrees is determined approximately as equal to the days elapsed since true Meṣasaṅkrānti; this approximation is corrected by the entries in a vākya table in which the corrections for each ten days up to 370 days are recorded (*VK* 1, 4–5). The longitude of the moon is found as in the *Candravākyas* of Vararuci (*VK* 1, 9–11). It is assumed that the ascending node makes one rotation in 6,792 days (*VK* 1, 17–19).

The longitudes of the planets are computed by means of cycles from an epoch position; one cycle is an approximate synodic period. In each cycle, at stated intervals in days, the planet's true position and a corrective factor are recorded. The numbers of cycles of the planets, the days in each cycle, the

TABLE VI.30

Planet	Retrograde Period	Visibility Period		Invisibility Period		Synodic Period
Saturn	134 days	342 days		36 days		378 days
Jupiter	112 days	372 days		30 days		402 days
Mars	66 days	660 days		120 days		780 days
		Γ→Σ	Ξ→Ω	Σ→Ξ	Ω→Γ	
Venus	52 days	251 days	251 days	71 days	8 days	581 days
Mercury	21 days	37 days	37 days	32 days	16 days	122 days

total numbers of days, and the increments in longitude are given in Table VI.31 (*VK* Appendix III).

TABLE VI.31

Planet	Cycles	Days	Total Days	Longitudinal Increment
Saturn	29	378	10,962	2;16°
Jupiter	11	399	4,389	5,58;3°
Mars	15	780	11,700	5,44;34°
Venus	5	584	2,920	5,9;8°
Mercury	22	116	2,552	5,43;56°

Superimposed on these synodic periods are a number of larger cycles for each planet, which are tabulated in Table VI.32, where the cycles are given in days and the longitudinal increments in degrees (Sundararāja on *VK* 2, 18). These larger cycles, of course, represent mean rather than true motions; the italicized numbers in the last column are the mean synodic periods.

The remainder of the *Vākyakaraṇa* is devoted to the problems related to time, position, and direction, to eclipses, and to the first and last visibilities of the planets. The rate of trepidation accepted by the author is 0;1° in 120/121 years over an arc 24° on either side of sidereal fixed Aries 0° (*VK* 3, 1). His table of sines is based on $R = 43$; it is given in Table VI.33 (*VK* 3, 2–4). Otherwise, the text is unexceptional. The vākyas continued to be popular in Kerala, as is clear from various works of such authors as Mādhava (*ca.* 1340–*ca.*1425),[83] Parameśvara (*ca.* 1380–*ca.* 1460), and Sundararāja (*ca.* 1475).

TABLE VI.33

ϑ	Sin ϑ	R sin ϑ
15°	11	11.12
30°	21.5	21.50
45°	30.5	30.40
60°	37.5	*37.23*
75°	41.5	41.53
90°	43	43

With Parameśvara's institution in 1431 of the dṛggaṇita system, which is really only the Saurapakṣa, the process already visible in Vaṭeśvara's work was accelerated and the Āryapakṣa was replaced by the Saurapakṣa. It lingered only in southern India in the Parahita system and in the Vākya system, although its solar and lunar parameters were used for computing tithis, yogas, and nakṣatras in western India. There is only one known set of tables belonging to this school,[84] although this situation is undoubtedly due to the paucity of southern Indian tables that have hitherto been examined.

VII. THE ĀRDHARĀTRIKAPAKṢA

The Ārdharātrikapakṣa, like the Āryapakṣa, was founded by Āryabhaṭa I in about 500. It is characterized not only by its own parameters but also by the fact that its epoch is midnight rather than dawn,

TABLE VI.32

Planet	Cycles	Increments	Composition of Cycles
Saturn	10,964;32	+6;41°	*378;5,14* · 29
	21,550	+0;43°	(10,964;32 · 2) − 379;4
	182,994;23	−0;13°	(10,964;32 · 17) − (378;4 · 9)
	570,534;8	+0;5°	(182,994;23 · 3) + 21,550
Jupiter	4,387;44	+4;34°	*398;53* · 11
	21,539;48	−10;19°	(4,387;44 · 5) − 398;12
	30,315;17	−1;11°	21,539;48 + (4,387;44 · 2)
	65,018;17	+2;13°	(30,315;17 · 2) + 4,387;44
	125,648;50	−0;9°	(65,018;17 · 2) − 4,387;44
	474,875;27	0°	
Mars	11,699;4	+10;38°	*779;56,16* · 15
	17,158;37	−8;24°	11,699;4 + (779;52,30 · 8)
	28,857;41	+2;13°	11,699;4 + 17,158;37
	132,589;21	+0;27°	(28,857;41 · 4) + 17,158;37
	634,089;9	+0;4°	(132,589;21 · 5) − 28,857;41
Venus	2,919;38	−2;24°	*583;55,36* · 5
	44,962;23	+35;3°	(2,919;38 · 15) + (583;56,30 · 2)
	88,756;53	−0;58°	(44,962;23 · 2) − (583;56,30 · 2)
	174,594;8	+0;29°	(88,756;53 · 2) − 2,919;38
	437,945;9	0°	(174,594;8 · 2) + 88,756;53
Mercury	2,549;15	−7;27°	*115;52,30* · 22
	4,750;53	+2;29°	(2,549;15 · 2) − (115;52,20 · 3)
	16,801;54	−0;1°	(4,750;53 · 3) + 2,549;15

TABLE VII.1

Planet	R	$\frac{R \cdot 6,0}{Y}$	$\frac{R \cdot 6,0}{C}$
Saturn	146,564	12;12,49,12°	0;2,0,22,41,36,36, . . .°
Jupiter	364,220	30;21,6°	0;4,59,8,48,36,56, . . .°
Mars	2,296,824	3,11;24,7,12°	0;31,26,27,47,36,55, . . .°
Sun	4,320,000	6,0°	0;59,8,10,10,37,48, . . .°
Venus' śīghra	7,022,388	3,45;11,56,24°	1;36,7,44,13,7,53, . . .°
Mercury's śīghra	17,937,000	54;45°	4;5,32,17,45,23,13, . . .°
Moon	57,753,336	2,12;46,40,48°	13;10,34,52,6,50,56, . . .°
Manda	488,219	40;41,5,42°	0;6,40,59,29,51,10, . . .°
Node	−232,226	−19;21,7,48°	−0;3,10,44,7,41,54, . . .°

so that the present Kaliyuga begins at midnight of 17/18 February −3101. The work of Āryabhaṭa I from which this pakṣa is derived is lost except for testimonia and fragments; but one of his pupils, Lāṭadeva, is evidently the author of a revision of the original *Sūryasiddhānta* that makes it conform to the Ārdharātrikapakṣa; there is a summary of Lāṭadeva's *Sūryasiddhānta* in Varāhamihira's *Pañcasiddhāntikā*. Lāṭadeva's epoch is midnight 20/21 March 505; the epoch of the original *Sūryasiddhānta* was noon, and its parameters probably were not those of the Ārdharātrikapakṣa.[85]

The relation of the Ārdharātrikapakṣa to the Āryapakṣa is close. The former shares the latter's division of the Mahāyuga into four equal Yugas. The difference in epoch dates of the present Kaliyuga, however, which comes to 0;15 days, amounts to an increment of 0;0,0,15 days per year when distributed over the 3,600 years between that epoch and 499. The Ārdharātrikapakṣa's year length, therefore, is 6,5;15,31,30 days instead of the Āryapakṣa's 6,5;15,31,15; but the rotations of the planets in a Mahāyuga remain essentially the same (see Table VI.1). They are listed along with the mean yearly and mean daily motions in Table VII.1.

Lāṭadeva represents the planets' period relations as indicated in Table VII.2 (*PS* 9, 1–5 and 16, 1–9). Further, Lāṭadeva adds the yearly bījas given in Table VII.3 (*PS* 16, 10–11).

TABLE VII.3

Planet	Yearly Bīja	Yearly Mean Motion
Saturn	+0;0,6,30°	12;12,55,42°
Jupiter	−0;0,10°	30;20,56°
Mars	+0;0,17°	3,11;24,24,12°
Venus' śīghra	−0;0,45°	3,45;11,11,24°
Mercury's śīghra	+0;2°	54;47°

Table VII.2

Planet	Relation	Correction per Rotation	Mean Daily Motion
Saturn	$\frac{1000 \text{ rot.}}{10766066 \text{ days}}$	−5‴	0;2,0,22,41,36,37, . . .°
Jupiter	$\frac{100 \text{ rot.}}{433232 \text{ days}}$	−10‴	0;4,59,8,48,38,42, . . .°
Mars	$\frac{1 \text{ rot.}}{687 \text{ days}}$	+14‴	0;31,26,27,47,35,53, . . .°
Sun	$\frac{800 \text{ rot.}}{292207 \text{ days}}$	0	0;59,8,10,10,37,48, . . .°
Venus' śīghra	$\frac{10 \text{ rot.}}{2247 \text{ days}}$	+10;30″	1;36,7,44,13,8,15, . . .°
Mercury's śīghra	$\frac{100 \text{ rot.}}{8797 \text{ days}}$	+4;30‴	4;5,32,17,45,25,3, . . .°
Moon	$\frac{900000 \text{ rot.}}{24589506 \text{ days}}$	$-\frac{51''}{3120}$	13;10,34,52,6,50,55, . . .°
Manda	$\frac{900 \text{ rot.}}{2908789 \text{ days}}$	$+\frac{10''}{297}$	0;6,40,59,29,51,10, . . .°
Node	$\frac{-2700}{18345822 \text{ days}}$	0	−0;3,10,44,7,54,12, . . .°

From the relations given above, it follows that the time periods in the Ārdharātrikapakṣa are identical with those in the Āryapakṣa (Table VI.2) except for those listed in Table VII.4.

TABLE VII.4

Sidereal days	1,582,237,800
Civil days (C)	1,577,917,800
Avamas (U)	25,082,280

The longitudes of the nodes in the Ārdharātrikapakṣa coincide with those of the Āryapakṣa (Table VI.6), although the longitudes of the mandoccas differ slightly. The latter are given in Table VII.5 (*PS* 16, 13).

TABLE VII.5

Planet	Mandocca
Saturn	240°
Jupiter	160°
Mars	110°
Sun	80°
Venus	80°
Mercury	220°

The circumferences of the manda epicycles and of the śīghra epicycles according to Lāṭadeva are recorded in Table VII.6 (*PS* 9, 7–8 and 16, 12 and 14); the corresponding maximum equations are added (the śīghra equations and anomalies are those of the *Khaṇḍakhādyaka*).

Lāṭadeva's model for the planets is that with two epicycles (*PS* 16, 15–22); his values for the kālāṃśas of visibility are identical with those in the *Paitāmaha* (Table V.8) (*PS* 16, 23).

In *Pañcasiddhāntikā* 9–10, with which 11 seems to belong, Varāhamihira summarizes the eclipse computations found in Lāṭadeva's *Sūryasiddhānta*. First, in order to determine the true longitudes and true velocities, Lāṭadeva proposes the following two rules (*PS* 9, 9):

$$\Delta\mu = \Delta\kappa \cdot \frac{c_\mu}{21600}, \qquad \text{(VII.1)}$$

where $\Delta\mu$ is the daily increment or decrease to the equation and $\Delta\kappa$ is the daily increment in the argument, all measured in minutes. For the sun $\Delta\kappa = \bar{v}_{♂}$, mean daily motion; for the moon $\Delta\kappa = \bar{v}_{☾} - \bar{v}_A$, where $\bar{v}_A$ is the mean daily motion of the moon's manda. The second rule is (V.67), the cosine rule for finding the daily increment or decrease to the mean daily motion (*PS* 9, 13).

Further, Lāṭadeva assumes, with other Indian astronomers, that (*PS* 9, 14)

$$H = R \cdot \frac{\bar{v}}{v}, \qquad \text{(VII.2)}$$

where H is the true distance of the planet in units of R; as presented in the *Pañcasiddhāntikā*, $R = 120$. He measures the actual distance in units of which there are eighteen in the radius of the earth, r_e. He also states that the circumference of the earth is 3,200 yojanas, so that its radius is about 509-1/3 yojanas (*PS* 9, 10); these parameters are slightly smaller than those of the Āryapakṣa.

The distances of the sun and moon are computed by the following formulas (VII.3) (*PS* 9, 15–16):

$$k_{♂} = \frac{5347}{40} \cdot \rho_{♂}; \qquad \text{(a)}$$

$$k_{☾} = 10 \cdot \rho_{☾}. \qquad \text{(b)}$$

Thence the mean values, when $\rho = R = 120$, are (VII.4)

$$\bar{k}_{♂} = 16041 = 891\ 1/6 \cdot r_e \qquad \text{(a)}$$

$$\bar{k}_{☾} = 1200 = 66\ 2/3 \cdot r_e. \qquad \text{(b)}$$

From (VII.4 b) it follows that

$$\pi_0 = r_e \cdot \frac{21600}{c_e \cdot 66\text{-}2/3} = 0;51,34,12°, \qquad \text{(VII.5)}$$

where π_0 is the horizontal parallax. The value of $\bar{k}_{♂}$, 16041, should be derived, if (VII.2) is correct, from

$$\bar{k}_{♂} = \frac{R_{☾} \cdot \bar{k}_{☾}}{R_{♂}}, \qquad \text{(VII.6)}$$

TABLE VII.6

Planet	c_μ	$\mu_{\max}$ ($\kappa_\mu = 90°$)	c_σ	$\sigma_{\max}$	κ_σ
Saturn	60°	9;36,55,...°	40°	6;20°	96°
Jupiter	32°	5;6°	72°	11;30°	108°
Mars	70°	11;13°	234°	40;30°	135°
Sun	14°	2;14°			
Venus	14°	2;14°	260°	46;15°	141°
Mercury	28°	4;28°	132°	21;30°	120°
Moon	31°	4;56°			

where $R_{☾}$ and $R_{☉}$ are the rotations of the moon and the sun in a yuga. But then, with the Ārdharātrikapakṣa's parameters, we would have $\bar{k}_{☉} = 16042;35,36$. One must suspect that the original *Sūryasiddhānta* derived $\bar{k}_{☉}$ in some other fashion.

To find the true diameters, d, of the sun and the moon, in the units in which $k_{☉}$ and $k_{☾}$ are measured, Lāṭadeva uses the relation

$$d=\frac{b}{k}, \qquad \text{(VII.7)}$$

where $b_{☉}$ is given as 517,080 and $b_{☾}$ as 38,640. Using the mean values of k, one finds (VII.8)

$$\bar{d}_{☉}=0;32,14,5,37,\ldots° \qquad \text{(a)}$$

$$\bar{d}_{☾}=0;32,11,57,54°. \qquad \text{(b)}$$

Lāṭadeva's rule for determining the diameter of the earth's shadow at the distance of the moon in units of the radius, R, is related to (V.30), which was used by Āryabhaṭa I; it is (*PS* 10, 1–2)

$$d_u=\left(36-\frac{36\cdot k_{☾}}{\frac{90\cdot k_{☉}}{286}}\right)\cdot\frac{R}{k_{☾}}, \qquad \text{(VII.8)}$$

where the term in parentheses is the numerical expression of (V.123).

In computing parallax, Lāṭadeva follows essentially the incomplete rules of Āryabhaṭa I. He finds Sin SN from (VI.1) and (VI.2), and then (*PS* 9, 19–23)

$$\pi_{\lambda}=\pi_0\cdot\frac{\text{Sin }SN}{R}. \qquad \text{(VII.9)}$$

And, with Sin ZN from (VI.1 b), he forms (*PS* 9, 24–25)

$$\pi_{\beta}=\pi_0\cdot\frac{\text{Sin }ZN}{R}. \qquad \text{(VII.10)}$$

Otherwise, his rules for determining the durations and magnitudes of eclipses are unexceptional. The projection with akṣavalana and ayanavalana described in the *Pañcasiddhāntikā* (*PS* 11, 1–5) is identical with that in the *Brāhmasphuṭasiddhānta* (V.93–94).

A text belonging to the Ārdharātrikapakṣa, known as the *Zīk i Arkand* ("Tables of the Ahargaṇa"), was available in Sasanian Iran in the middle of the sixth century and formed the basis of much of the *Zīj al-Shāh* of Anūshirwān, and of that of Yazdijird III in the 630's.[86]

The next Sanskrit text available to us is the *Mahābhāskarīya*, in which (*MB* 7, 22–35) Bhāskara I repeats the Ārdharātrikapakṣa's parameters. One variation is that the latitudes of the planets are regarded as a combination of the tilting of both the manda epicycle and the śīghra epicycle. Further, the following measures in yojanas are given, as in Table VII.7.

TABLE VII.7

d_e = 1,600 yojanas
$d_{☉}$ = 6,480 yojanas
$d_{☾}$ = 480 yojanas
$\bar{k}_{☉}$ = 689,358 yojanas
$\bar{k}_{☾}$ = 51,566 yojanas

We have seen in Lāṭadeva that 1,600 yojanas is half of the circumference of the earth, not its diameter. And in general these diameters are closer to those in Table V.23 than to Lāṭadeva's. The given values of the mean distances of the sun and the moon correspond to 861-279/400 and 64-183/400 earth radii; these numbers do not agree with Lāṭadeva's either (VII.4). The correctly computed, although rounded, distances of the sun and moon according to the later *Pauliśa* (Table VII.11) are 689,378 yojanas and 51,566 yojanas; the former is 861-289/400 earth radii.

The principal text of the Ārdharātrikapakṣa, however, is the *Khaṇḍakhādyaka* of Brahmagupta;[87] its epoch is 23 March 665. Like his *Brāhmasphuṭasiddhānta*, this work consists of an initial summary of an early text (in this case the Ārdharātrika text of Āryabhaṭa I) followed, in the uttara, by corrections and additions; many of these corrections and additions in the *Khaṇḍakhādyaka* are derived from the *Brāhmasphuṭasiddhānta*.

Brahmagupta gives the mean daily motions of the planets (Table VII.1) in the form of Table VII.8 (*Kh* 1, 1, 8–13 and 1, 2, 1–5); compare Table VII.2, which contains some of the same relations. The mean daily motion of the lunar node is based on −232,218 rotations in a Mahāyuga instead of −232,226. In the uttara section of the *Khaṇḍakhādyaka*, Brahmagupta uses the relation (III.9) to obtain the longitude of the moon's manda (*Kh* 2, 1, 2), and substitutes for the lunar node the approximation (*Kh* 2, 1, 3)

$$-\frac{1\text{ rot.}}{6{,}792\text{ days}}=-0;3,10,48,45,47,42,\ldots^{\circ/d}. \qquad \text{(VII.11)}$$

This parameter is close to that of the Brāhmapakṣa

TABLE VII.8

Planet	Mean Daily Motion	
Saturn	$\frac{1\text{ rot.}}{10{,}766\text{ days}} - \frac{1'}{80{,}450\text{ days}}$	$= 0;2,0,22,41,36,37,\ldots^\circ$
Jupiter	$\frac{1\text{ rot.}}{4{,}332\text{ days}} - \frac{1^\circ}{162{,}621\text{ days}}$	$= 0;4,59,8,48,36,56,\ldots^\circ$
Mars	$\frac{1\text{ rot.}}{687\text{ days}} + \frac{1'}{174{,}259\text{ days}}$	$= 0;31,26,27,47,36,53,\ldots^\circ$
Sun	$\frac{800\text{ rot.}}{292{,}207\text{ days}}$	$= 0;59,8,10,10,37,48,\ldots^\circ$
Venus' śīghra	$\frac{10\text{ rot.}}{2{,}247\text{ days}} + \frac{1^\circ}{77{,}043\text{ days}}$	$= 1;36,7,44,13,7,50,\ldots^\circ$
Mercury's śīghra	$\frac{100\text{ rot.}}{8{,}797\text{ days}} + \frac{1'}{71{,}404\text{ days}}$	$= 4;5,32,17,45,23,13,\ldots^\circ$
Moon	$\frac{600\text{ rot.}}{16{,}393\text{ days}} - \frac{1'}{4{,}929\text{ days}}$	$= 13;10,34,52,6,50,53,\ldots^\circ$
Manda	$\frac{1\text{ rot.}}{3{,}232\text{ days}} + \frac{1'}{39{,}298\text{ days}}$	$= 0;6,40,59,29,51,21,\ldots^\circ$
Node	$-\left(\frac{1\text{ rot.}}{6{,}795\text{ days}} + \frac{1'}{514{,}656\text{ days}}\right)$	$= -0;3,10,43,42,56,6,\ldots^\circ$

(Table V.2). Both parameters also appear in the *Vākyakaraṇa*.

In the first section Brahmagupta repeats the longitudes of the mandoccas recorded in Table VII.5 (*Kh* 1, 1, 13 and 1, 2, 6); in the uttara he changes Jupiter's to 180°, Mars's to 117°, and the sun's to 77° (*Kh* 2, 1, 1 and 2, 2, 1). Brahmagupta tabulates the sun's and the moon's manda equations at intervals of 15°; the manda equations of the star-planets are multiples of the sun's. Their śīghra equations are given for unequal intervals. The maximum śīghra equations have been recorded in Table VII.6; the maximum manda equations are given in Table VII.9 (*Kh* 1, 1, 16–17 and 1, 2, 6–7), and half the synodic periods in Table VII.10 (*Kh* 1, 2, 8–17).

TABLE VII.9

Planet	Multiple of 2;14°	$\mu_{\max}$ ($\kappa_\mu = 90°$)
Saturn	$4\frac{2}{7}$	9;34,17, . . .°
Jupiter	$2\frac{2}{7}$	5;6,17, . . .°
Mars	5	11;10°
Sun	1	2;14°
Venus	1	2;14°
Mercury	2	4;28°
Moon		4;56°

The longitude of the moon is further corrected by 1/27 of the sun's manda equation, so that the maximum correction is 5;1° (*Kh* 1, 1, 18).

In the uttara Brahmagupta prescribes the following modifications (*Kh* 2, 2, 1). The manda equation of Saturn is decreased by 1/5, so that the maximum equation becomes 7;39,25, . . .°. The śīghrocca of Venus is diminished by 74′, so that

$$\kappa'_\sigma = \kappa_\sigma - 1;14°. \qquad \text{(VII.12)}$$

This in effect means that at the beginning of the Mahāyuga the center of Venus' śīghra epicycle was the true rather than the mean sun. Also, the śīghra equation of Mercury is increased by 1/16, so that the maximum equation becomes 22;50,37, . . .°. Further, the manda equation of the sun is decreased by 1/42, so that the maximum becomes 2;11° (*Kh* 2, 1, 5); and the manda equation of the moon is increased by 1/52, so that the maximum becomes 5;1,41, . . .°. These corrections generally bring the equations close to the values in *Brāhmasphuṭasiddhānta* 25 (Tables V.28 and V.29).

In the first part of the *Khaṇḍakhādyaka*, the computation of the longitudinal difference is based on the circumference of the parallel of latitude passing through Ujjayinī being 4,800 yojanas (*Kh* 1, 1, 15). In the uttara the earth's circumference is given as 5,000 yojanas, as in the *Paitāmaha*; and the circumference of the parallel of latitude, c_ϕ, is found by (*Kh* 2, 1, 6)

$$c_\phi = 5000 \cdot \frac{\text{Sin } \phi}{R}. \qquad \text{(VII.13)}$$

TABLE VII.10

	κ_σ	σ	Phase
Saturn	20°	2°	first visibility
	56°	5°	
	76°	6°	
	96°	6;20°	
	116°	6°	first station
	133°	5°	
	155°	3°	
	180°	0°	
Jupiter	14°	2;20°	first visibility
	54°	8;20°	
	90°	11;20°	
	108°	11;30°	
	130°	10°	first station
	144°	8°	
	164°	4°	
	180°	0°	
Mars	28°	11°	first visibility
	60°	23°	
	90°	33°	
	121°	40°	
	135°	40;30°	
	148°	37;30°	
	164°	25;30°	first station
	173°	12;30°	
	180°	0°	
Venus	24°	10°	first visibility in west
	63°	26°	
	96°	38°	
	123°	45°	
	141°	46;15°	
	154°	42°	
	165°	32°	first station
	177°	8°	last visibility in west
	180°	0°	
Mercury	51°	13°	first visibility in west
	89°	20°	
	120°	21;30°	
	146°	16;30°	first station
	155°	13°	last visibility in west
	180°	0°	

Brahmagupta's Sine table in the *Khaṇḍakhādyaka* is identical with that in *Brāhmasphuṭasiddhānta* 25 (Table V.24) (*Kh* 1, 3, 6), as are his values of $\gamma/2$ (Table V.30) (*Kh* 1, 3, 1). The computation of eclipses in the *Khaṇḍakhādyaka* also is essentially identical with that in the *Brāhmasphuṭasiddhānta* (*Kh* 1, 4–5); in the uttara the cālanas for 177 days are also given (Table V.20) (*Kh* 2, 4, 21–22).

The first part of the *Khaṇḍakhādyaka* repeats the familiar kālāṃśas of visibility of the planets (Table V.8) (*Kh* 1, 6, 1), the Āryapakṣa's longitudes of their nodes (Table VI.6) (*Kh* 1, 8, 1), and its values of the orbital inclinations (Table VI.5) (*Kh* 1, 8, 1); in the uttara Mercury's orbital inclination is given as 2;30° instead of 2° (*Kh* 2, 5, 1). Further, the kālāṃśas of visibility of Venus and Mercury are stated to be 10° and 14° at superior conjunction, and 8° and 12° at inferior conjunction (*Kh* 2, 5, 3–4).

In the *Khaṇḍakhādyaka*, Brahmagupta ascribes to the yogatārās of the nakṣatras the same polar longitudes and polar latitudes that he does in the *Brāhmasphuṭasiddhānta* (Table V.10 with Table V.22) (*Kh* 1, 9, 4–12). In the uttara he repeats the theory of the nakṣatrabhogas that he has expounded in *Brāhmasphuṭasiddhānta* 14 (*Kh* 2, 1, 6–9). Both of his works, then, although fundamental for their respective pakṣas, represent the unassimilated conflation of material of diverse origins.

In the early eighth century an Arabic *Zīj al-Arkand*, written in Sind, was dependent on the *Khaṇḍakhādyaka*;[88] its epoch is 735.

At about the same time, in Sthāneśvara, the later *Pauliśasiddhānta* was composed; it is largely based on Ārdharātrika parameters, although the author accepts the traditional division of the Mahāyuga into four unequal yugas that is used in the Brāhmapakṣa. Therefore he dates the first mean conjunction of the Mahāyuga 648,000 years after the beginning of the Kṛtayuga, which is 1,080,000 · 3 years before the beginning of the current Kaliyuga. Therefore, like the roughly contemporary author of the *Sūryasiddhānta*, he must have introduced a period of nonmotion of the planets at the beginning of the Kalpa. The later *Pauliśa* gives precisely the parameters of Table VII.1, and by and large those of Table VII.6, although Saturn's śīghra epicycle is made 39;30° rather than 40°, and Mars's 233° rather than 234°. Following Bhāskara I's summary of the Ārdharātrikapakṣa, the diameter of the earth is given as 1,600 yojanas (Table VII.7), and thence, with π = 3-177/1,250, its circumference as 5,026-14/25 yojanas. The prime meridian passes over Ujjayinī, Rohītaka, Kurukṣetra, the Yamunā (Jumna) River, the Himālayas, and Meru. Because of the above value of π, which is Āryabhaṭa's, the later *Pauliśa* makes $R = 3{,}437\text{-}967/1{,}309$.

Basing the computation on the principles previously enunciated, the later *Pauliśa* gives the circumferences and radii of the heavenly orbits, measured in yojanas, as in Table VII.11.

TABLE VII.11

Body	Circumference	Radius
Heaven	18,712,080,864,000,000	3,001,938,106,524,064
Nakṣatras	259,890,012	41,362,683
Saturn	127,671,739	20,319,541
Jupiter	51,375,764	8,176,688
Mars	8,146,937	1,296,622
Sun	4,331,500	689,378
Venus' śīghra	2,664,632	424,088
Mercury's śīghra	1,043,211	166,031
Moon	324,000	51,566

The later *Pauliśa*'s mean diameters of the star-planets in minutes and of the luminaries in yojanas and minutes are displayed in Table VII.12.

TABLE VII.12

Planet	Diameter in Yojanas	Diameter in Minutes
Sun	6,480	32
Moon	480	32
Venus		16
Jupiter		8
Mercury		4
Saturn		2
Mars		1

The diameters of the luminaries in yojanas are those found in Table VII.7; the order of the star-planets and their diameters are those of their kālāṃśas of visibility in Table V.8.

Al-Bīrūnī, in the late 1020's, was acquainted with both the *Khaṇḍakhādyaka* and the later *Pauliśasiddhānta*, and makes extensive quotations from them. This fact attests to the great popularity of the Ārdharātrikapakṣa in northwestern India in the eleventh century.

A somewhat later Ārdharātrika karaṇa that is more difficult to localize is the *Bhāsvatī* of Śatānanda; its epoch is 1099. Śatānanda claims that his *Bhāsvatī* is based on the *Sūryasiddhānta* summarized by Varāhamihira.

The Ārdharātrikapakṣa continued to flourish, especially on the fringes of India—in Kashmir, Nepal, and Assam. So far very little tabular material has come to light,[89] but this is undoubtedly due to the paucity of manuscripts from the areas mentioned above that have yet been examined.

VIII. THE SAURAPAKṢA

As we have seen, one of the earliest texts of the Ārdharātrikapakṣa was a *Sūryasiddhānta*—that of Lāṭadeva. Another *Sūryasiddhānta* closely allied to the Ārdharātrikapakṣa was composed sometime before Vijayananda of Benares (966), and apparently before Vaṭeśvara of Ānandapura (905) and Govindasvāmin of Kerala (*ca.* 800/850);[90] it seems reasonable to date this text to the late eighth or early ninth century, and to surmise that it was composed in southern India. By the twelfth century commentaries were being written on it in Mysore, by Mallikārjuna (*fl.* 1179[?]), and in Mithilā, by Caṇḍeśvara (1183). In the fifteenth century it was recognized as a strong rival to the Āryapakṣa in southern India, and as the authentic astronomical

TABLE VIII.1

Planet	R	$\frac{R \cdot 6{,}0}{Y}$	$\frac{R \cdot 6{,}0}{C}$
Saturn	146,568	12;12,50,24°	0;2,0,22,53,25,46,...°
Jupiter	364,220	30;21,6°	0;4,59,8,48,35,47,...°
Mars	2,296,832	3,11;24,9,36°	0;31,26,28,11,8,56,...°
Sun	4,320,000	6,0°	0;59,8,10,10,24,12,...°
Venus' śīghra	7,022,376	3,45;11,52,48°	1;36,7,43,37,16,52,...°
Mercury's śīghra	17,937,060	54;45,18°	4;5,32,20,41,51,16,...°
Moon	57,753,336	2,12;46,40,48°	13;10,34,52,3,49,4,...°
Manda	488,203	40;41,0,54°	0;6,40,58,42,31,5,...°
Node	−232,238	−19;21,8,24°	−0;3,10,44,13,35,59,...°

TABLE VIII.2

Planet	R (Ārya) (yr. = 6,5;15,31,15)	R (Ārdharātrika) (yr. = 6,5;15,31,30)	Bīja	R (Saura) (yr. = 6,5;15,31,31,24)	Bīja
Saturn	146,564	146,564	0	146,568	+4
Jupiter	364,224	364,220	−4	364,220	−4
Mars	2,296,824	2,296,824	0	2,296,832	+8
Venus' śīghra	7,022,388	7,022,388	0	7,022,376	−12
Mercury's śīghra	17,937,020	17,937,000	−20	17,937,060	+40
Moon	57,753,336	57,753,336	0	57,753,336	0
Manda	488,219	488,219	0	488,203	−16
Node	−232,226	−232,226	0	−232,238	−12

text in northern and eastern India. The text with which modern scholars are most familiar is the version by Raṅganātha of Benares (1602).[91]

The *Sūryasiddhānta* follows completely the divisions of the Kalpa and Mahāyuga enunciated in the Brāhmapakṣa (*SS* 1, 14–21). The rotations of the planets in a Mahāyuga and their corresponding mean yearly and mean daily motions are displayed in Table VIII.1; the number of civil days in a Mahāyuga is given as 1,577,917,828 (*SS* 1, 29–33, and 37).

Thus, although the equal-Yuga system of Āryabhaṭa I has been abandoned by the author of the *Sūryasiddhānta*, as by Lalla, and by the author of the later *Pauliśa*, the principle remains that his values of R are divisible by 4. In order to produce a mean conjunction at the beginning of the current Kaliyuga, the author of the *Sūryasiddhānta* hypothesizes a period of creation equal to 17,064,000 years at the beginning of the Kalpa (*SS* 1, 24). The Saurapakṣa, like the Ārdharātrikapakṣa, uses midnight epoch.

The year length implied by the number of civil days is 6,5;15,31,31,24 days. It is this year length that generates differences in the mean longitudes of the planets computed according to the Ārya, Ārdharātrika, and Saura pakṣas. The relations of these pakṣas to each other are displayed in Table VIII.2.

The numbers of rotations imply orbits of almost the same sizes in yojanas as those in the later *Pauliśa* (Table VII.11); they are listed in Table VIII.3 (*SS* 12, 80–90).

TABLE VIII.3

Heaven	18,712,080,864,000,000
Nakṣatras	259,890,012
Saturn	127,668,255
Jupiter	51,375,764
Mars	8,146,909
Sun	4,331,500
Venus' śīghra	2,664,637
Mercury's śīghra	1,043,209
Moon	324,000

The changed number of days also alters the following parameters in a Mahāyuga, listed in Table VIII.4 (*SS* 1, 34–40).

TABLE VIII.4

Sidereal days	1,582,237,828
Avamas	25,082,252

Unlike the Ārya and Ārdharātrika pakṣas, the Saura does not just give the current longitudes of the planets' mandoccas and nodes, but records their rotations in a Kalpa, as the Brāhma had. Those are listed in Table VIII.5 (*SS* 1, 41–44); their approximate longitudes in 850 also are given.

TABLE VIII.5

Planet	Mandocca		Node	
	R	λ in 850	R	λ in 850
Saturn	39	236;37°	662	100;24°
Jupiter	900	171;18°	174	79;41°
Mars	204	130;2°	214	40;4°
Sun	387	77;17°		
Venus	535	79;50°	903	59;43°
Mercury	368	220;27°	488	20;43°

The longitudes of the mandocca of the sun (approximately) and of the nodes are from the Āryapakṣa (Table VI.6), those of the inferior planets' mandoccas from the Ārdharātrikapakṣa (Table VII.5); compare Table IX.1.

Like the *Khaṇḍakhādyaka*, the *Sūryasiddhānta* gives the diameter of the earth as 1,600 yojanas (*SS* 1, 59); but, following the *Paitāmaha*, it uses $\pi = \sqrt{10}$ to find the circumference, which then is 5,059+ yojanas. The circumference, c_ϕ, of the parallel of latitude of any locality is found by (*SS* 1, 60)

$$c_\phi = \sqrt{25600000} \cdot \frac{\text{Sin}\,\bar{\phi}}{R}. \qquad \text{(VIII.1)}$$

With this rule, also found in the *Khaṇḍakhādyaka*

TABLE VIII.6

Planet	c_μ $\kappa=0°, 180°$	c_μ $\kappa=90°, 270°$	μ_{max} ($\kappa_\mu=90°$)	c_σ $\kappa_\sigma=0°, 180°$	c_σ $\kappa_\sigma=90°, 270°$	σ_{max}
Saturn	49°	48°	7;40°	39°	40°	6;22°
Jupiter	33°	32°	5;6°	70°	72°	11;31°
Mars	75°	72°	11;32°	235°	232°	40;16°
Sun	14°	13;40°	2;10,32°			
Venus	12°	11°	1;45°	262°	260°	46;24°
Mercury	30°	28°	4;28°	133°	132°	21;31°
Moon	32°	31;40°	5;2,48°			

(VII.13), one computes the longitudinal difference from the prime meridian, which passes over Laṅkā, Rohītaka, Avantī, and Meru (*SS* 1, 60–62). The orbital inclinations of the planets in the *Sūryasiddhānta* are those of the first part of the *Khaṇḍakhādyaka* (Table VI.5) (*SS* 1, 68–70).

The *Sūryasiddhānta* employs the common double-epicycle model for computing the true longitudes of the star-planets. Its Sine table is precisely that of Āryabhaṭa I (Table V.6) (*SS* 2, 17–22). The measures of the circumferences of the epicycles according to the *Sūryasiddhānta* are given in Table VIII.6 (*SS* 2, 34–37), as are the maximum equations recorded in tables belonging to the Saurapakṣa.

In most other respects the *Sūryasiddhānta* simply repeats material previously described in the summaries of earlier texts. For the sake of comparison I reproduce in Table VIII.7 the śīghra anomalies he states are required for the occurrences of the first stations of the planets (*SS* 2, 53).

TABLE VIII.7

Planet	First Station
Saturn	115°
Jupiter	130°
Mars	164°
Venus	163°
Mercury	144°

The *Sūryasiddhānta*, following Maṇindha (Μανέϑων),[92] hypothesizes a trepidation of the equinox over an arc extending 27° east and west of the fixed vernal point, at the rate of 0;0,54° per year. With the longitude of the sun corrected by the amount of trepidation, the text proceeds to review all the gnomon and related problems discussed by earlier authors, including Brahmagupta's rule for finding the sun's altitude when it is in the northern hemisphere and the angle between the shadow and the east-west line is 45° (V.81) (*SS* 3, 28–34).

The diameters of the sun and moon in yojanas adopted by the *Sūryasiddhānta* are those of the *Paitāmaha* (V.22) (SS 4, 1). The remainder of the computation of eclipses is essentially the same as the *Brāhmasphuṭasiddhānta*'s, save that the nonagesimal is not mentioned, but only the meridian point of the ecliptic, in the rules for computing parallax (*SS* 5). Also, like the *Brāhmasphuṭasiddhānta*, the *Sūryasiddhānta* uses polar rather than ecliptic coordinates to compute the conjunction of a planet with another planet or with a fixed star (*SS* 7).

The *Sūryasiddhānta*'s polar coordinates of the yogatārās are identical with the *Paitāmaha*'s (Table V.10), save in the instances noted in Table VIII.8 (*SS* 8, 2–10).

TABLE VIII.8

Nakṣatra	*Paitāmaha* λ*	*Paitāmaha* β*	*Sūryasiddhānta* λ*	*Sūryasiddhānta* β*
Kṛttikā	37;28°	+5°	37;30°	+5°
Rohiṇī	49;28°	−5°	49;30°	−5°
Ārdrā	67°	−11°	67;20°	−9°
Āśleṣā	108°	−7°	109°	−7°
Pūrvaphālgunī	147°	+12°	144°	+12°
Citrā	183°	−2°	180°	−2°
Svāti	199°	+38°	199°	+37°
Viśākhā	212;5°	−1;30°	213°	−1;30°
Anurādhā	224;5°	−3°	224°	−3°
Jyeṣṭhā	229;5°	−4°	229°	−4°
Mūla	240;4°	−8;30°	241°	−9°
Pūrvāṣāḍhā	249°	−5;20°	254°	−5;30°
Abhijit	265°	+62°	266;40°	+60°
Śravaṇa	278°	+30°	280°	+30°
Śatabhiṣaj	320°	0°	320°	−0;30°
Revatī	0°	0°	359;50°	0°
Agastya	87°	−76°	90°	−80°

Most of these changes appear to be simple roundings; some others (for instance, Svāti, Mūla, and Pūrvāṣāḍhā) were anticipated by Brahmagupta (Table V.22). Further, the *Sūryasiddhānta* adds the stars in Table VIII.9 (*SS* 8, 10–12).

TABLE VIII.9

Nakṣatra	λ*	β*	Identification
Mṛgavyādha	80°	−40°	α Can. Mai.
Agni	51°	+8°	β Tau. (?)
Brahmahṛdaya	51°	+30°	α Aur.
Prajāpati	56°	+38°	δ Aur.
Apāṃvatsa	180°	+3°	ϑ Vir. (?)
Apas	180°	+9°	ζ Vir.

TABLE VIII.10

Planet	Relation	Mean Daily Motion
Saturn	$\frac{4 \text{ rot.}}{43{,}063 \text{ days}} - \frac{1''}{3{,}876 \text{ days}}$	= 0;2,0,22,53,25,47, . . .°
Jupiter	$\frac{100 \text{ rot.}}{433{,}232 \text{ days}} - \frac{1''}{22{,}200 \text{ days}}$	= 0;4,59,8,48,35,48, . . .°
Mars	$\frac{1 \text{ rot.}}{687 \text{ days}} + \frac{1'}{8{,}719 \text{ days}}$	= 0;31,26,28,11,8,55, . . .°
Sun	$\frac{800 \text{ rot.}}{292{,}207 \text{ days}} - \frac{1''}{43 \text{ yrs}}$	= 0;59,8,10,10,24,3, . . .°
Venus' śīghra	$\frac{600 \text{ rot.}}{134{,}819 \text{ days}} - \frac{1'}{9{,}980 \text{ days}}$	= 1;36,7,43,37,16,54, . . .°
Mercury's śīghra	$\frac{9{,}000 \text{ rot.}}{791{,}727 \text{ days}} - \frac{1'}{10{,}060 \text{ days}}$	= 4;5,32,20,41,54,2, . . .°
Moon	$\frac{9{,}000 \text{ rot.}}{245{,}895 \text{ days}} - \frac{9{,}000''}{691{,}547 \text{ days}}$	= 13;10,34,52,3,49,8, . . .°
Manda	$\frac{15 \text{ rot.}}{48{,}481 \text{ days}} + \frac{1'}{18{,}633 \text{ days}}$	= 0;6,40,58,42,59,18, . . .°
Node	$-\frac{60 \text{ rot.}}{407{,}671 \text{ days}}$	= − 0;3,10,44,54,40,22, . . .°

The earliest known karaṇa belonging to the Saurapakṣa is the *Karaṇatilaka*, written by Vijayananda of Benares; its epoch is midnight of 23/24 March 966. It is preserved for us, not in Sanskrit, but in an Arabic translation made for al-Bīrūnī about 1030, which has not yet been completely published.

In this karaṇa the computation of time elapsed since the beginning of the Kalpa is done in accordance with the views of the *Sūryasiddhānta* — the mixture of Ārdharātrika and Brāhma ideas. The mean motions of the planets since epoch are derived from the relations in Table VIII.10. These figures are identical with the values in Table VIII.1 except those for the moon's manda and node, which are based, respectively, on 488,211 and −232,234 rotations in a Mahāyuga.

The longitudinal difference is computed on the assumption that the diameter of the earth is 1,600 yojanas and its circumference 5,028 yojanas; in this computation $\pi = 3\text{-}57/400$. Vijayananda uses a variant of (VII.13) to find the circumference of a parallel of given terrestrial latitude.

TABLE VIII.11

Planet	Mandocca	Node
Saturn	240°	100°
Jupiter	171°	80°
Mars	120°	40°
Sun	77;56°	
Venus	80°	60°
Mercury	220°	20°

Vijayananda gives the longitudes of the mandoccas and nodes of the planets as listed in Table VIII.11; compare Table VIII.5.

Vijayananda's ratios for finding the equations are given in Table VIII.12; compare Table VIII.6.

TABLE VIII.12

Planet	Manda Ratio $\frac{\mu_{max}}{R}$ ($R = 200$)	μ_{max}	Śīghra Ratio $\frac{c_\sigma}{360}$	c_σ
Saturn	$\frac{7}{4}$	5;50°	$\frac{1}{9}$	40°
Jupiter	$\frac{61}{40}$	5;5°	$\frac{1}{5}$	72°
Mars	$\frac{22}{7}$	10;28,34, . . .°	$\frac{13}{20}$	234°
Sun	$\frac{49}{75}$	2;10,40°		
Venus	$\frac{7}{12}$	1;56,40°	$\frac{13}{18}$	260°
Mercury	$\frac{27}{20}$	4;30°	$\frac{10}{27}$	133;20°
Moon	$\frac{3}{2}$	5°		

As indicated in Table VIII.12, Vijayananda's value for R is 200; his Sine table is reproduced in Table VIII.13.

The values of the coefficients of the noon equinoctial shadow that produce the equations of daylight according to the *Karaṇatilaka* and the right

TABLE VIII.13

ϑ	Sin ϑ	R sin ϑ
10°	34;42	34;43,48
20°	68;20	68;24,14
30°	100;0	100;0
40°	128;30	128;33,28
50°	153;10	153;12,28
60°	173;14	173;12,21
70°	187;53	187;56,16
80°	196;0	196;57,43
90°	200;0	200;0

ascensions of the signs in palas are given in Table VIII.14.

TABLE VIII.14

Sign	γ	α
Aries	20 vināḍīs	278
Taurus	16 vināḍīs	299
Gemini	7 vināḍīs	323

Finally, in the part of Vijayananda's work that has been published, there is a table of the śīghra anomalies necessary for the occurrences of the first and second stations of the planets; this is close to, but does not agree with, Table VIII.7. I repeat it in Table VIII.15.

TABLE VIII.15

Planet	First Station
Saturn	116°
Jupiter	130°
Mars	163°
Venus	165°
Mercury	146°

The date of composition of the *Somasiddhānta*[93] remains obscure, although there exists a commentary on it composed in South India by Nṛsiṃha in about 1400. This text adheres to the Saurapakṣa, as is indicated by its basic parameters (*Soma* 1, 21–34). The only other indication we have of its origin is that it names Kāñcī, Lohitaka Lake, Avantī, and Vatsagulma as lying on the prime meridian (*Soma* 1, 47); it is not clear whether or not Lohitaka Lake is the source of the Brahmaputra in Tibet, which would be an unusual association for such a list of places and might be relevant to the problem of the provenience of the *Somasiddhānta*.

The dimensions of the manda epicycle and śīghra epicycle according to this text are identical with those for κ_μ, $\kappa_\sigma = 0°$, 180° in Table VIII.6, except that the manda epicycle of Mercury is given as 34° (*Soma* 2, 15–17). Otherwise, the only unusual aspect of the work is adhyāya 10, which contains a long eulogy of himself by Brahmā.

Another undatable text belonging to the Saurapakṣa is a *(Laghu)vasiṣṭhasiddhānta*,[94] which claims to be based on the sixth-century *Vasiṣṭhasiddhānta* of Viṣṇucandra[95] (*LVS* 80). It disguises its parameters of mean motions by replacing them with the yojanas in the orbits of the planetary spheres, as in Table VIII.3 (*LVS* 24–30). It also has the peculiarity of using Śrīpati's Sine table (Table V.33) where $R = 3{,}415$ (*LVS* 38–42); this certainly dates it after the eleventh century. Also, the text as edited omits the dimensions of the manda epicycles, and provides dimensions of the śīghra epicycles that are the means or variants of those in Table VIII.6; I list them in Table VIII.16 (*LVS* 46–47).

TABLE VIII.16

Planet	c_σ
Saturn	39°
Jupiter	71°
Mars	234°
Venus	201° (read 261°)
Mercury	133°

Yet another work ascribed to Vasiṣṭha and belonging to the Saurapakṣa is the *Vṛddhavasiṣṭhasiddhānta*.[96] Although normally an unexceptional text, the *Vṛddhavasiṣṭha* has a few oddities. One is the Sine table with $R = 1000$, given in Table VIII.17 (*VVS* 2, 9–10).

TABLE VIII.17

ϑ	Sin ϑ	*R* sin ϑ
10°	174	173.6
20°	342	342.0
30°	500	500
40°	643	642.7
50°	766	766.0
60°	868	*866.0*
70°	940	939.6
80°	985	984.8
90°	1000	1000

The intervals of 10° hint at a relationship to Vijayananda's Sine table (Table VIII.13). The text is most probably later than the tenth century.

Another non-Saura feature involves the dimensions of the epicycles, which are reproduced in Table VIII.18 (*VVS* 2, 14–15).

TABLE VIII.18

Planet	c_μ	c_σ
Saturn	50°	35°
Jupiter	33°	68°
Mars	70°	243;40°
Sun	13;40°	
Venus	11°	259°
Mercury	28°	132°
Moon	31;36°	

These are to a large extent derived from the Brāhmapakṣa (Table V.5), as are the values of the śīghra anomalies necessary for the occurrence of first station (Table V.9) and the retrograde periods given in Table VIII.19 (*VVS* 2, 30–31).

TABLE VIII.19

Planet	First Station	Retrograde Period
Saturn	115°	133 days
Jupiter	130°	112 days
Mars	164°	72 days
Venus	163°	48 days
Mercury	144°	24 days

The star catalog in the *Vṛddhavasiṣṭha* (*VVS* 8, 2–8; I have made some obvious emendations) repeats that of the *Sūryasiddhānta* (Tables VIII.8, VIII.9), with the alterations noted in Table VIII.20.

TABLE VIII.20

Nakṣatra	*Sūryasiddhānta*		*Vṛddhavasiṣṭha*	
	λ*	β*	λ*	β*
Kṛttikā	37;30°	+5°	37°	+5°
Rohiṇī	49;30°	−5°	49°	−5°
Ārdrā	67;20°	−9°	75°	−9°
Abhijit	266;40°	+62°	267°	+62°
Revatī	359;50°	0°	0°	0°
Lubdhaka	80°	−40°	70°	−40°
Agni	51°	+8°	52°	+8°
Brahmahṛdaya	51°	+30°	52°	+30°
Prajāpati	56°	+38°	57°	+38°
Apāṃvatsa	180°	+3°	180°	+5°
Āpas	180°	+9°	180°	+6°

The changes for Ārdrā (approximately), Agni, Brahmahṛdaya, and Prajāpati are also found in the *Siddhāntatattvaviveka* (Table VIII.32).

A fourth anonymous text belonging to the Saurapakṣa is the *Brahmasiddhānta* of the *Śākalyasaṃhita*;[97] it was known in Benares in the seventeenth century, although how much older it might be cannot at present be determined. It is very interesting for its cosmology and its information on muhūrtas, but need not be discussed further in this essay.

In southern India the chief rival of the Āryapakṣa was the Dṛgganitapakṣa, founded by Parameśvara in his *Dṛgganita*, composed at Ālattūr in Kerala in 1431. This pakṣa is, in fact, simply the Saurapakṣa, as is clear from the mean daily motions given by Parameśvara and reproduced in Table VIII.21 (*DG* 1, 1, 10–22 and 2, 1, 2–10), where the computed values of column 4 do not include the minute corrections of column 3. Clearly, despite his claims to be correcting the Parahita parameters by means of observations, Parameśvara has only computed period relations from the parameters of the Saurapakṣa.

He does, however, make a slight alteration by assuming that there was not a mean conjunction at the beginning of the current Kaliyuga. His mean longitudes for that time are iterated in Table VIII.22 (*DG* 1, 2, 1–5 and 2, 1, 12–15).

TABLE VIII.21

Planet	Period Relation	Correction	Mean Daily Motion
Saturn	$\frac{13\text{ rot.}}{139{,}955\text{ days}} - \frac{1'}{1{,}395{,}953\text{ days}}$	0	0;2,0,22,53,25,26, . . .°
Jupiter	$\frac{53\text{ rot.}}{229{,}613\text{ days}} + \frac{1''}{141{,}478\text{ days}}$	0	0;4,59,8,48,35,46, . . .°
Mars	$\frac{1\text{ rot.}}{687\text{ days}} + \frac{1'}{8{,}719\text{ days}}$	$+\frac{3''}{4{,}320{,}000\text{ years}}$	0;31,26,28,11,8,55, . . .°
Sun	$\frac{116\text{ rot.}}{42{,}370\text{ days}} - \frac{1'}{45{,}486\text{ days}}$	$-\frac{1''}{3{,}000\text{ years}}$	0;59,8,10,10,24,13, . . .°
Venus' śīghra	$\frac{73\text{ rot.}}{16{,}403\text{ days}} + \frac{1'}{37{,}348\text{ days}}$	$+\frac{11;20''}{4{,}320{,}000\text{ years}}$	1;36,7,43,37,16,52, . . .°
Mercury's śīghra	$\frac{33\text{ rot.}}{2{,}903\text{ days}} - \frac{1'}{67{,}452\text{ days}}$	$+\frac{1''}{4{,}320{,}000\text{ years}}$	4;5,32,20,41,51,16, . . .°
Moon	$\frac{143\text{ rot.}}{3{,}907\text{ days}} + \frac{1'}{8{,}310\text{ days}}$	$+\frac{1''}{5{,}791\text{ years}}$	13;10,34,52,3,49,1, . . .°
Manda	$\frac{74\text{ rot.}}{23{,}171\text{ days}} - \frac{1''}{47{,}036}$	0	0;6,40,59,6,10,22, . . .°
Node	$-\left(\frac{29\text{ rot.}}{197{,}041\text{ days}} + \frac{1'}{92{,}920\text{ days}}\right)$	$+\frac{1''}{4{,}320{,}000\text{ years}}$	−0;3,10,44,33,37,34, . . .°

TABLE VIII.22

Planet	Mean Longitude
Saturn	4;8°
Jupiter	357;31°
Mars	0;3°
Sun	0;20,22°
Venus' śīghra	356°
Mercury's śīghra	356;16°
Moon	3;15,2°
Manda	85;34,26°
Node	92;36°

Parameśvara gives to the mandoccas the longitudes listed in Table VIII.23 (*DG* 1, 3, 1–2 and 2, 1, 16–17).

TABLE VIII.23

Planet	Mandocca
Saturn	236°
Jupiter	172°
Mars	127°
Sun	78°
Venus	80°
Mercury	220°

Except for Saturn's, these longitudes are taken from Muñjāla (Table IX.1).

The Sines of the manda equations are tabulated by Parameśvara at intervals of 6°, with the maximum equations at arguments of 90°; he tabulates the Sines of the śīghra equations at intervals of 6° also, but over two quadrants. The maximums are recorded in Table VIII.24 (*DG* 2, 1, 26–43).

TABLE VIII.24

Planet	Sin μ_{max} ($\kappa_\mu = 90°$)	μ_{max}	Sin σ_{max}	σ_{max}	κ_σ
Saturn	420	7;1°	338	5;38°	96°–102°
Jupiter	333	5;33°	649	10;53°	102°
Mars	707	11;52°	2414	44;36°	132°
Sun	129	2;9°			
Venus	86	1;26°	2788	54;13°	132°
Mercury	215	3;35°	1283	21;57°	114°
Moon	301	5;1°			

But, although the Saurapakṣa continued after Parameśvara to maintain its position in southern India, most of what we know of its subsequent history concerns texts and tables composed in northern and eastern India. The first of the northern sets of Saura tables is the *Makaranda*, written by Makaranda at Benares; its epoch is 1478. The *Rāmavinoda* was composed by Rāmacandra, either at Delhi or at Benares;[98] its epoch is 11 March 1590. Rāmacandra's value for the rotations of the lunar node in a Mahāyuga is –232,242, which yields a mean yearly motion of –19;21,12,36°. Mathurānātha Śarman of Bengal was the author of the *Ravisiddhāntamañjarī*; its epoch is 29 March 1609.

The last two major works belonging to the Saurapakṣa were composed by two rival astronomers at Benares in the seventeenth century. The earlier of the two was Munīśvara Viśvarūpa's *Siddhāntasārvabhauma*, completed in 1646. This work follows the Brāhmapakṣa in determining the time elapsed since the beginning of the current Kalpa, as had the *Sūryasiddhānta*; but Munīśvara (*SSBM* 1, 19–29) refers to the *Siddhāntaśiromaṇi* of Bhāskara II as his authority, here as elsewhere. The parameters of mean motion are all Saura (*SSBM* 1, 31–39); but he mentions the bījas of Śrīpati's *Siddhāntaśekhara*, of Bhāskara II's *Siddhāntaśiromaṇi*, of Dāmodara (who was followed by Jñānarāja in his *Siddhāntasundara*, one of Munīśvara's favorite sources), and of the Yavanas (Muslims) (*SSBM* 1, 119–123). He elsewhere refers often to views of the Muslims that he opposes—for instance, with respect to precession (*SSBM* 2, 253–280).

Munīśvara names many localities over which the prime vertical passes: Laṅkā, Kumārī, Kāñcī, Pānāṭa, Sitādri, Ṣaḍāsya, Vatsagulma, Purī, Mahiṣmatī, Ujjayinī, Paṭṭaśikhā, Gargarāṭa, Rohita, Sthāneśvara, Himālaya, and Sumeru (*SSBM* 1, 135–136).

Since $\pi = 600/191 = c/2R$ (*SSBM* 1, 134), Munīśvara makes $R = 191$ and computes the Sine table recorded in Table VIII.25 (*SSBM* 2, 3–18).

TABLE VIII.25

ϑ	Sin ϑ	R sin ϑ	Δ (Sin ϑ – R sin ϑ)
1°	3;20,0,17	3;19,58,37	+0;0,1,40
2°	6;39,56,54	6;39,57,14	–0;0,0,20
3°	9;59,46,12	9;59,48,59	–0;0,2,47
4°	13;19,24,33	13;19,26,58	–0;0,2,25
5°	16;38,48,17	16;38,51,12	–0;0,2,55
6°	19;57,53,46	19;57,54,49	–0;0,1,3
7°	23;16,37,22	23;16,37,48	–0;0,0,26
8°	26;34,55,25	26;34,53,17	+0;0,2,8
9°	29;52,44,20	29;52,41,16	+0;0,3,4
10°	33;10,0,29	33;10,1,44	–0;0,1,15
11°	36;26,40,16	36;26,40,57	–0;0,0,41
12°	39;42,40,5	39;42,38,54	+0;0,1,11
13°	42;57,56,21	42;57,55,37	+0;0,0,44
14°	46;12,25,30	46;12,24,11	+0;0,1,19
15°	49;26,3,58	49;26,4,37	–0;0,0,39
16°	52,38,48,5	52;38,50,3	–0;0,1,58
17°	55;50,34,15	55;50,33,36	+0;0,0,39
18°	59;1,20,6	59;1,22,9	–0;0,2,3
19°	62;11,0,40	62;11,1,55	–0;0,1,15
20°	65;19,33,4	65;19,32,57	+0;0,0,7
21°	68;26,53,48	68;26,55,12	–0;0,1,24
22°	71;32,59,30	71;33,1,50	–0;0,2,20
23°	74;37,14,53	74;37,45,56	*–0;0,31,3*
24°	77;41,12,7	77;41,14,25	–0;0,2,18

TABLE VIII.25 (Continued)

ϑ	Sin ϑ	R sin ϑ	Δ (Sin $\vartheta - R$ sin ϑ)	ϑ	Sin ϑ	R sin ϑ	Δ (Sin $\vartheta - R$ sin ϑ)
25°	80;43,12,19	80;43,13,30	−0;0,1,11	58°	161;58,37,52	161;58,39,10	−0;0,1,18
26°	83;43,44,0	83;43,43,12	+0;0,0,48	59°	163;43,8,14	163;43,10,5	−0;0,1,51
27°	86;42,43,52	86;42,43,31	+0;0,0,21	60°	165;24,58,10	165;24,42,13	*+0;0,15,57*
28°	89;40,8,39	89;40,7,34	+0;0,1,5	61°	167;3,8,31	167;3,8,42	−0;0,0,11
29°	92;35,55,6	92;35,55,21	−0;0,0,15	62°	168;38,34,46	168;38,36,25	−0;0,1,39
30°	95;30,0,0	95;30,0,0	0	63°	170;10,56,5	170;10,58,28	−0;0,2,23
31°	98;22,20,11	98;22,21,30	−0;0,1,19	64°	171;40,10,47	171;40,8,0	+0;0,2,47
32°	101;12,52,29	101;12,52,59	−0;0,0,30	65°	173;6,14,3	173;6,18,45	−0;0,4,42
33°	104;1,33,48	104;1,34,27	−0;0,0,39	66°	174;29,13,51	174;29,16,58	−0;0,3,7
34°	106;48,21,2	106;48,19,2	+0;0,2,0	67°	175;48,59,8	175;48,55,48	+0;0,3,20
35°	109;33,11,9	109;33,13,36	−0;0,2,27	68°	177;5,31,37	177;5,28,58	+0;0,2,39
36°	112;16,1,8	112;16,4,24	−0;0,3,16	69°	178;18,49,54	178;18,49,36	+0;0,0,18
37°	114;56,48,1	114;56,51,25	−0;0,3,24	70°	179;28,53,42	179;28,50,50	+0;0,2,52
38°	117;35,28,50	117;35,27,48	+0;0,1,2	71°	180;35,38,34	180;35,39,33	−0;0,0,59
39°	120;12,0,41	120;12,0,25	+0;0,0,16	72°	181;39,6,28	181;39,8,51	−0;0,2,23
40°	122;46,20,46	122;46,22,24	−0;0,1,38	73°	182;39,15,9	182;39,11,34	+0;0,3,35
41°	125;18,26,11	125;18,26,51	−0;0,0,40	74°	183;36,3,33	183;36,2,22	+0;0,1,11
42°	127;48,14,12	127;48,13,47	+0;0,0,25	75°	184;29,30,35	184;29,33,28	−0;0,2,53
43°	130;15,42,4	130;15,43,12	−0;0,1,8	76°	185;19,35,20	185;19,38,16	−0;0,2,56
44°	132;40,47,6	132;40,48,12	−0;0,1,6	77°	186;6,16,51	186;6,16,48	+0;0,0,3
45°	135;3,26,37	135;3,28,50	−0;0,2,13	78°	186;49,34,17	186;49,35,56	−0;0,1,39
46°	137;23,38,3	137;23,38,11	−0;0,0,8	79°	187;29,26,51	187;29,28,47	−0;0,1,56
47°	139;41,18,48	139;41,16,15	+0;0,2,33	80°	188;5,53,49	188;5,55,21	−0;0,1,32
48°	141;56,26,23	141;56,23,3	+0;0,3,20	81°	188;38,54,30	188;38,55,38	−0;0,1,8
49°	144;8,58,18	144;8,58,35	−0;0,0,17	82°	189;8,28,29	189;8,29,39	−0;0,1,10
50°	146;18,52,10	146;18,49,6	+0;0,3,4	83°	189;34,34,44	189;34,37,22	−0;0,2,38
51°	148;26,5,34	148;26,8,20	−0;0,2,46	84°	189;57,13,15	189;57,11,57	+0;0,1,18
52°	150;30,36,12	150;30,35,40	+0;0,0,32	85°	190;16,23,28	190;16,20,14	+0;0,3,14
53°	152;32,21,47	152;32,24,51	−0;0,3,4	86°	190;32,5,2	190;32,2,15	+0;0,2,47
54°	154;31,20,5	154;31,22,9	−0;0,2,4	87°	190;44,17,40	190;44,17,59	−0;0,0,19
55°	156;27,28,57	156;27,27,32	+0;0,1,25	88°	190;53,1,8	190;53,0,33	+0;0,0,35
56°	158;20,46,14	158;20,47,54	−0;0,1,40	89°	190;58,15,17	190;58,16,51	−0;0,1,34
57°	160;11,7,53	160;11,9,29	−0;0,1,36	90°	191;0,0,0	191;0,0,0	0

Munīśvara follows this table with a long section describing the computation of the Sines of various angles (*SSBM* 2, 23–112). Later he also gives a table of the right ascension of each degree from Aries 1° to Cancer 0° (*SSBM* 2, 289–290).

The circumferences of the epicycles in the *Siddhāntasārvabhauma* are the mean values of the normal, pulsating epicycles of the Saurapakṣa (Table VIII.6); they are listed in Table VIII.26 (*SSBM* 2, 115–116).

TABLE VIII.26

Planet	c_μ	c_σ
Saturn	48;30°	39;30°
Jupiter	32;30°	71°
Mars	73;30°	233;30°
Sun	13;50°	
Venus	11;30°	261°
Mercury	29°	132;30°
Moon	31;50°	

Most of the *Siddhāntasārvabhauma* and its commentary are not yet available. In the future they should prove to be a major source for the study of the reaction of northern Indian astronomers in the sixteenth and seventeenth centuries to Islamic astronomy.

The second important text of this period is the *Siddhāntatattvaviveka* of Kamalākara, which was completed at Benares in 1658. Kamalākara, like Munīśvara, accepts the *Sūryasiddhānta*'s arrangement of the Kalpa (*STV* 1, 36–60) and its rotations of the planets (Table VIII.1), mandoccas, and nodes (Table VIII.5) in a Kalpa (*STV* 1, 94–112). Exceptionally among Indian astronomers—only Pṛthūdakasvāmin is really comparable with him in this, to my knowledge—Kamalākara devotes many verses to the physics of the celestial spheres, referring specifically at many points to the (Aristotelian) views of the Yavanas or Muslims (*STV* 2, 66–126).

Starting from the common belief that the radius of the earth is 800 yojanas, Kamalākara finds its

TABLE VIII.27

Kamalākara			Ulugh Beg[99] or (Astrolabes)[100] or [Geographical Lists][101]			
City	Longitude	Latitude	City	Longitude	Latitude	s_0
Kābula	104;0°	34;40°	Kābulistān	104;40°	35;0°	[8;30]
Amadāvāda	108;20°	23;0°	(Aḥmadābād	108;40°	23;15°)	[5;0, 5;31]
Khambāīta	109;20°	22;20°	(Kanbāyat	109;20°	22;20°)	[4;59]
Burahānapura	111;0°	21;0°	(Burhānbūr	108;0°	20;30°)	[4;31]
Ujjayinī	112;0°	22;31°	(Ujayn	112;0°	22;30°)	[5;0]
Lāhora	109;20°	31;50°	(Lahāwar	109;20°	31;50°)	[7;30]
Indraprastha	114;0°	28;13°	(Dihlī	113;35°	28;19°)	[6;0, 6;3]
Argalāpura	115;0°	26;35°				
Somanātha	106;0°	22;35°	Sawmnāt	106;0°	17;0°	[5;6]
Bijyāpura	118;0°	17;20°	(Bījābūr	105;30°	17;20°)	[3;30, 3;41]
Kāśī	117;20°	26;55°	Bānāras	117;20°	26;15°	
Golakuṇḍa	114;19°	18;4°	(Kūlkundah	114;19°	18;20°)	[3;45]
Lakhanaura	114;13°	26;30°	(Lakhnaw'	114;13°	26;30°)	[5;45]
Ajameru	111;5°	26;5°	(Ajmīr	111;5°	26;0°)	[5;50, 5;52, 6;0]
Devagiri	111;0°	20;30°	(Dawlatābād	111;0°	20;30°)	[4;11, 4;25]
Mulatāna	107;35°	29;40°	Mūltān	107;35°	29;40°	[6;0]
Kanauja	115;0°	26;35°	Qanawj	115;50°	26;35°	[6;0, 6;10]
Māṇḍava	121;0°	27;0°				
Kāśmīra	108;0°	35;0°	Kashmīr	103;0°	35;0°	[7;30, 7;52]
Samarakanda	99;0°	39;40°	Samarqand	99;16°	39;37°	[10;3]

circumference to be 5059;38 yojanas (*STV* 2, 163); thence

$$\pi = \frac{1,24,19;38}{26,40} = 3;9,44,10,30. \quad \text{(VIII.2)}$$

In his discussion of geography, however, he depends on Islamic sources. The prime meridian, he states, runs through a locality on the equator named Khāladātta, which lies 22° west of Romaka (*STV* 2, 172), so that the tūla (from Arabic ṭūl) of Laṅkā is 112°. The edition of the *Siddhāntatattvaviveka* includes a table of the longitudes and latitudes of certain cities (compare Table XI.2) that is reproduced in Table VIII.27 (*STV* 1, 174–175). With this more accurate information, Kamalākara repeats the usual rules for determining the time difference corresponding to the longitudinal difference (*STV* 2, 176–177).

Computing the *Sūryasiddhānta*'s distances of the planetary orbits from the center of the earth (Table VIII.3) (*STV* 2, 231–234), Kamalākara further computes with (V.20) the diameters of the planets in yojanas; these are listed in Table VIII.28 with the corresponding diameters in minutes (*STV* 2, 236–239).

TABLE VIII.28

Planet	Distance	Diameter	
	yojanas	yojanas	minutes (appr.)
Saturn	20,186,123	14,776;25	2;30′
Jupiter	8,123,221	8,324;45	3;30′
Mars	1,288,139	754;20	2′
Sun	684,870	6,500	32;37′
Venus	421,316	493;27	4′
Mercury	164,945	144;54	3′
Moon	51,229	480	32;12′

In the course of his lengthy discussion of the computation of Sines (*STV* 3, 1–183), Kamalākara criticizes Munīśvara (for instance, *STV* 3, 72) and cites Ulugh Beg (*STV* 3, 89). He then produces the Sine table given in Table VIII.29, where $R = 60$; compare Malayendu's table (Table XI.1).

A spot check of some values indicates that this table was computed by a technique similar to Munīśvara's (Table VIII.20); in fact, many are Munīśvara's values divided by 3;11. Again, like Munīśvara, Kamalākara provides a table giving the right ascension of each degree of the ecliptic (*STV* after 4, 91) and a similar table of the oblique ascension of each degree of the ecliptic for places of which the latitude is 26° (*STV* after 4, 214).

Kamalākara preserves the pulsating manda epicycles of the *Sūryasiddhānta* (Table VIII.6), although he differentiates between the variation in size when $\kappa_\mu = 0°$ and when $\kappa_\mu = 180°$. This differentiation apparently is intended to reflect the varying distance from the earth. His values are recorded in Table VIII.30 (*STV* 3, 218–226). The values for $\kappa_\mu = 90°, 270°$ are identical with those in the *Sūryasiddhānta*.

Kamalākara also gives slightly different values of both manda epicycles and śīghra epicycles with the resulting Sines of the maximum equations;

TABLE VIII.29

ϑ	Sin ϑ	ϑ	Sin ϑ
1°	1;2,49,43,15	46°	43;9,37,23,49
2°	2;5,38,17,29	47°	43;52,52,23,58
3°	3;8,24,34,2	48°	44;35,19,16,56
4°	4;11,7,23,54	49°	45;16,57,16,10
5°	5;13,45,38,26	50°	45;57,45,35,59
6°	6;16,18,8,53	51°	46;37,43,31,40
7°	7;18,43,46,41	52°	47;16,50,19,22
8°	8;21,1,23,23	53°	47;55,5,16,13
9°	9;23,9,50,40	54°	48;32,27,40,15
10°	10;25,8,0,23	55°	49;8,56,50,30
11°	11;26,54,44,35	56°	49;44,32,6,56
12°	12;28,28,55,31	57°	50;19,12,50,34
13°	13;29,49,25,40	58°	50;52,58,23,20
14°	14;30,55,7,46	59°	51;25,48,8,13
15°	15;31,44,54,49	60°	51;57,41,59,14
16°	16;32,17,40,8	61°	52;28,37,52,24
17°	17;32,32,7,18	62°	52;58,36,40,48
18°	18;32,27,40,15	63°	53;27,37,24,33
19°	19;32,2,43,17	64°	53;55,39,30,50
20°	20;31,16,21,23	65°	54;22,42,28,55
21°	21;30,7,28,38	66°	54;48,45,49,8
22°	22;28,35,1,27	67°	55;13,49,2,54
23°	23;26,38,55,26	68°	55;37,51,42,45
24°	24;24,15,6,54	69°	56;0,53,22,20
25°	25;21,25,32,40	70°	56;22,53,56,22
26°	26;18,8,10,4	71°	56;43,52,0,44
27°	27;14,21,56,53	72°	57;3,48,12,27
28°	28;10,5,51,27	73°	57;22,41,49,38
29°	29;5,18,52,41	74°	57;40,32,31,35
30°	30;0,0,0,0	75°	57;57,19,58,43
31°	30;54,8,13,27	76°	58;13,3,52,37
32°	31;47,42,33,40	77°	58;27,43,56,2
33°	32;40,42,1,54	78°	58;41,19,52,54
34°	33;33,5,40,2	79°	58;53,51,28,18
35°	34;24,52,30,7	80°	59;5,18,28,39
36°	35;16,1,36,52	81°	59;15,40,40,54
37°	36;6,32,2,42	82°	59;24,57,54,10
38°	36;56,22,52,43	83°	59;33,9,58,8
39°	37;45,33,12,16	84°	59;40,16,43,46
40°	38;34,2,7,25	85°	59;46,18,3,17
41°	39;21,48,45,1	86°	59;51,13,50,5
42°	40;8,52,10,40	87°	59;55,3,58,46,14
43°	40;55,11,38,45	88°	59;57,48,25,7,50
44°	41;40,46,12,29	89°	59;59,27,6,7,45
45°	42;25,35,3,53	90°	60;0,0,0,0,0

TABLE VIII.30

Planet	c_μ ($\kappa_\mu = 0°$)	($\kappa_\mu = 90°,270°$)	($\kappa_\mu = 180°$)
Saturn	42;40°	48°	55;57°
Jupiter	29;29°	32°	35;15°
Mars	61;1°	72°	92;20°
Sun	13;10°	13;40°	14;12°
Venus	10;40°	11°	11;20°
Mercury	26;3°	28°	30;29°
Moon	29;12°	31;40°	34;51°

those are reproduced in Table VIII.31 (*STV* 3, 187–200).

Moreover, despite his familiarity with Islamic zījes, Kamalākara's polar longitudes and latitudes of the yogatārās of the nakṣatras and other stars are identical with the *Sūryasiddhānta*'s (Tables VIII.8 and VIII.9) except in the instances recorded in Table VIII.32 (*STV* 12, 5–16).

TABLE VIII.32

Nakṣatra	*Sūryasiddhānta* λ^*	β^*	Kamalākara λ^*	β^*
Ārdrā	67;20°	−9°	74;50°	−9°
Agni	51°	+8°	52°	+8°
Brahmahṛdaya	51°	+30°	52°	+30°
Prajāpati	56°	+38°	57°	+38°

These changes are anticipated (or followed?) in the *Vṛddhavasiṣṭha* (Table VIII.20).

These brief notes in no way do justice to either Munīśvara or Kamalākara. Their works, and those of their contemporaries, deserve an intensive study to determine the reaction of the astronomers of Benares to their own tradition and to that of the Muslims, and the attempt they evidently made to improve the level of science in India.

At some time in the sixteenth or seventeenth century, probably at Benares, the Saurapakṣa parameters of the mean motions of the planets were adjusted by the application of bījas to their rotations

TABLE VIII.31

Planet	c_μ	Sin μ_{max}	μ_{max}	c_σ	Sin σ_{max}	σ_{max}
Saturn	47;26°	8;5	7;44,...°	40°	6;40	6;17,...°
Jupiter	32;7°	5;22	5;8,...°	72°	12	11;32,...°
Mars	73;29°	12;2	11;35,...°	232°	38;40	40;7,...°
Sun	13;40,34°	2;16,45	2;10,37,...°			
Venus	11°	1;50	1;46,...°	260°	43;20	46;15,...°
Mercury	28;5°	4;40	4;31,...°	132°	22	21;31,...°
Moon	31;47,22°	5;17,53	5;3,56,...°			

TABLE VIII.33

Planet	R (Saura)	bīja	R (Adj. Saura)	$\frac{R \cdot 6{,}0}{Y}$	$\frac{R \cdot 6{,}0}{C}$
Saturn	146,568	+12	146,580	12;12,54°	0;2,0,23,28,54,40, . . .°
Jupiter	364,220	−8	364,212	30;21,3,36°	0;4,59,8,24,56,31, . . .°
Mars	2,296,832	0	2,296,832	3,11;24,9,36°	0;31,26,28,11,8,56, . . .°
Sun	4,320,000	0	4,320,000	6,0°	0;59,8,10,10,24,12, . . .°
Venus' śīghra	7,022,376	−12	7,022,364	3,45;11,49,12°	1;36,7,43,1,47,58, . . .°
Mercury's śīghra	17,937,060	+16	17,937,076	54;45,22,48°	4;5,32,21,29,9,48, . . .°
Moon	57,753,336	0	57,753,336	2,12;46,40,48°	13;10,34,52,3,49,4, . . .°
Manda	448,203	+8	488,211	40;41,3,18°	0;6,40,59,6,10,22, . . .°
Node	−232,238	−8	−232,246	−19;21,13,48°	−0;3,10,45,6,49,10, . . .°

in a Kalpa; the new parameters of the Adjusted Saurapakṣa[102] are presented in Table VIII.33.

The works that are known to belong to this Adjusted Saurapakṣa include the *Grahaprakāśa* composed by Devadatta, probably at Benares (epoch: 18 March 1662);[103] an anonymous *Pañcāṅgānayanasāraṇī* (epoch: 30 March 1718);[104] the *Gaṇitarāja* composed by Kevalarāma Pañcānana at Navadvīpa in Bengal (epoch: 30 March 1728);[105] and the *Pattraprakāśa* composed by Viśrāmaśukla, probably at Benares (epoch: 1777).[106]

IX. ECLECTIC ASTRONOMERS

The *Laghumānasa* is a short and curious karaṇa composed by Muñjāla; its epoch is not known, but the epoch of the same author's lost *Bṛhanmānasa* is noon of 9 March 932. His elements are derived from the Ārya and Ārdharātrika pakṣas (thus bringing the work close to the Saurapakṣa), or are independently arrived at. His method of computing the mean longitudes of the planets is too complicated to be reproduced here. Muñjāla's longitudes of the mandoccas and nodes of the planets in 932 are listed in Table IX.1.[107]

TABLE IX.1

Planet	Mandocca	Node
Saturn	247°	100°
Jupiter	172°	80°
Mars	127°	40°
Sun	78°	
Venus	80°	60°
Mercury	220°	20°

The longitudes of the nodes and of the sun's mandocca are all from the Āryapakṣa (Table VI.6); the longitudes of Venus' and Mercury's mandoccas are from the Ārdharātrikapakṣa (Table VII.5). These longitudes also appear in the Saurapakṣa (Table VIII.5); if Saturn's mandocca were 237°, it also would be a Saura parameter.

Muñjāla gives a very simplified sine table, with $R = 488$; it is reproduced in Table IX.2 (*LM* 12).

TABLE IX.2

ϑ	Sin ϑ	R sin ϑ
30°	244	244.0
60°	427	*422.6*
90°	488	488.0

In sixtieth parts of that Radius, Muñjāla's eccentricities (or the radii of his manda epicycles) are computed in Table IX.3 from the values of the divisors, d_μ, that he himself gives (*LM* 13); the corresponding circumferences were computed by the commentator Yallaya.

TABLE IX.3

Planet	d_μ	$r_\mu = \frac{488}{d_\mu}$	c_μ	μ_{max}
Saturn	63	7;40	48°	7;45°
Jupiter	92	5;14	33°	5;18°
Mars	45	11;11	70°	10;51°
Sun	224	2;10,31	13;40°	2;10°
Venus	320	1;45	11°	1;31°
Mercury	100	4;52,48	38°	4;53°
Moon	97	5;2,45	31;21°	5;2°

The values of c_μ are close to those of the Saurapakṣa for $\kappa_\mu = 90°$, 270° (Table VIII.6).

The similar values for the śīghra epicycle are displayed in Table IX.4 (*LM* 15).

TABLE IX.4

Planet	d_σ	$r_\sigma = \frac{488}{d_\sigma}$	c_σ
Saturn	73-1/2	6;38, . . .	40°
Jupiter	39-1/2	12;21, . . .	74°
Mars	12	40;40	244°
Venus	11	44;21, . . .	266°
Mercury	21	23;14, . . .	139;30°

But the most astonishing formulation in the

TABLE IX.5

Planet	R	$\frac{R \cdot 6,0}{Y}$	$\frac{R \cdot 6,0}{C}$
Saturn	146,571,813	12;12,51,32,38,2,24°	0;2,0,23,4,46,29,...°
Manda	54	0;0,0,0,58,19,12°	
Node	630	0;0,0,11,31,12°	
Jupiter	364,219,954	30;21,5,59,10,19,12°	0;4,59,8,48,38,11,...°
Manda	982	0;0,0,17,40,33,36°	
Node	−190	−0;0,0,3,25,12°	
Mars	2,296,833,037	3,11;24,9,54,39,57,36°	0;31,26,28,15,19,32,...°
Manda	327	0;0,0,5,53,9,45,36°	
Node	−245	−0;0,0,4,24,36°	
Sun	4,320,000,000	6,0°	0;59,8,10,12,29,31,...°
Manda	480	0;0,0,8,38,24°	
Venus' śīghra	7,022,372,148	3,45;11,51,38,39,50,24°	1;36,7,43,39,55,52,...°
Manda	526	0;0,0,9,28,4,48°	
Node	−893	−0;0,0,16,4,26,24°	
Mercury's śīghra	17,937,055,474	54;45,16,38,31,55,12°	4;5,32,20,37,8,37,...°
Manda	356	0;0,0,6,24,28,48°	
Node	−648	−0;0,0,11,39,50,24°	
Moon	57,753,334,515	2,12;46,40,21,16,12°	13;10,34,52,27,20,57,...°
Manda	488,104,634	40;40,31,23,24,43,12°	0;6,40,53,51,54,17,...°
Node	−232,313,235	−19;21,33,58,13,48°	−0;3,10,48,25,44,8,...°

Laghumānasa is that of the evection of the moon, which is virtually equivalent to Śrīpati's (V.123) (*LM* 18–19):

$$\vartheta = (\bar{v}_{☾} - 11°) \cdot \text{Cos}\,(\lambda_{\odot} - \lambda_{A}) \cdot \text{Sin}\,(\lambda_{☾} - \lambda_{\odot}), \qquad \text{(IX.1)}$$

where ϑ is expressed in minutes. If $\bar{v}_{☾} \approx 13;10°$, when each of the last two terms equals R or 8;8, then the maximum value of ϑ is about 2;29°, which is a close approximation to Ptolemy's 2;39° and Śrīpati's 2;40°. The presumption that Muñjāla's source ultimately was Ptolemy, although probably through Islamic intermediaries, is virtually irresistible.

The system of Parāśara, which probably is to be dated to the ninth century, is known to us only from the description by Āryabhaṭa II (*MS* 1, 2). From this it appears that Parāśara accepted the division of the Kalpa of the Brāhmapakṣa; but he makes the number of sāvana days 1,577,917,570,000, so that the length of a year is 6,5;15,31,18,30 days (*MS* 1, 2, 4). His numbers of rotations of the planets in a Kalpa are given in Table IX.5 (*MS* 1, 2, 5–8).

The longitudes of the planets, their mandoccas, and their nodes at the beginning of the current Kaliyuga, when 1,972,944,000 years have elapsed, are given in Table IX.6. It is curious that the rotations of the node of Saturn must be taken to be direct, while those of the other star-planets' nodes are retrograde.

In the late tenth or early eleventh century, Āryabhaṭa II composed a *Mahāsiddhānta*, which seems to be basically a variant of the Brāhmapakṣa. He accepts the Brāhmapakṣa's arrangement of the Kalpa, although he inserts at its beginning a period of creation consisting of 3,024,000 years (*MS* 1, 1,

TABLE IX.6

Planet	Itself	Mandocca	Node
Saturn	358;57,21,36°	238;14,52,48°	103;40,48°
Jupiter	357;2,52,48°	172;35,2,24°	−278;16,48°
Mars	359;14,38,24°	123;7,47,50,24°	−320;56,24°
Sun	0°	77;45,36°	
Venus (śīghra)	356;58,33,36°	80;42,43,12°	−299;54,57,36°
Mercury (śīghra)	351;17,16,48°	210;50,19,12°	−338;58,33,36°
Moon	0;10,48°	125;12,28,48°	−326;32,49,12°

TABLE IX.7

Heavenly Body	R	$\frac{6,0 \cdot R}{Y}$	$\frac{6,0 \cdot R}{C}$
Saturn	146,569,000	12;12,50,42°	0;2,0,22,56,27,53, . . .°
Manda	76	0;0,0,1,22,4,48°	
Node	−620	−0;0,0,11,9,36°	
Jupiter	364,219,682	30;21,5,54,16,33,36°	0;4,59,8,47,51,5, . . .°
Manda	830	0;0,0,14,56,24°	
Node	−96	−0;0,0,1,43,30°	
Mars	2,296,831,000	3,11;24,9,18°	0;31,26,28,9,25,23, . . .°
Manda	299	0;0,0,5,22,55,12°	
Node	−298	−0;0,0,5,21,50,24°	
Sun	4,320,000,000	6,0°	0;59,8,10,12,43,7, . . .°
Manda	461	0;0,0,8,17,52,48°	
Venus' śīghra	7,022,371,432	3,45;11,51,25,46,33,36°	1;36,7,43,27,32,17, . . .°
Manda	654	0;0,0,11,46,19,12°	
Node	−524	−0;0,0,9,25,55,12°	
Mercury's śīghra	17,937,054,671	54;45,16,24,4,40,48°	4;5,32,20,35,42,38, . . .°
Manda	339	0;0,0,6,6,7,12°	
Node	−947	−0;0,0,14,53,9,36°	
Moon	57,753,334,000	2,12;46,40,12°	13;10,34,52,28,51,24, . . .°
Manda	488,108,674	40;40,32,36,7,55,12°	0;6,40,54,3,52,34, . . .°
Node	−232,313,354	−19;21,34,0,22,19,12°	−0;3,10,48,26,5,58, . . .°

TABLE IX.8

Planet	Itself	Mandocca	Node
Saturn	0°	236;9,36°	−260;12°
Jupiter	357;7,12°	172;48°	−279;21,36°
Mars	0°	123;50,24°	−319;40,48°
Sun	0°	77;45,36°	
Venus (śīghra)	357;7,12°	80;38,24°	−339;50,24°!
Mercury (śīghra)	351;21,36°	210;14,24°	−299;31,12°!
Moon	0°	123;50,24°	−302;28,24°

15–19); in this respect he imitates the Saurapakṣa. According to the *Mahāsiddhānta*, a Kalpa contains 1,577,917,542,000 civil days (*MS* 1, 1, 13), so that its year length is 6,5;15,31,17,6 days. The rotations of the planets in a Kalpa according to Āryabhaṭa II, and their mean yearly and daily motions, are displayed in Table IX.7 (*MS* 1, 1, 7–11).

As has been indicated, however, the motions of the planets, mandoccas, and nodes according to Āryabhaṭa II do not begin at the beginning of the Kalpa, but 3,024,000 years later, so that the time from the creation until the beginning of the current Kaliyuga was 1,972,944,000 − 3,024,000 = 1,969,920,000 years. The longitudes of the planets and their mandoccas and nodes at the beginning of the current Kaliyuga are given in Table IX.8 (compare *MS* 1, 1, 37–50).

The Sine table in the *Mahāsiddhānta* (*MS* 1, 3, 4–8) is that of the *Paitāmaha* (Table V.6). The dimensions of the epicycles are presented in Table IX.9 (*MS* 1, 3, 14, and 21–23).

TABLE IX.9

Planet	c_μ	c_σ
Saturn	59;43°	40;40°
Jupiter	28;15°	62;31°
Mars	65;30°	230;59°
Sun	13;41°	
Venus	9;35°	261;30°
Mercury	27;36°	134;30°
Moon	31;34°	

The śīghra anomalies necessary for the occurrence of the first visibilities and first stations of the planets according to Āryabhaṭa II are given in Table IX.10 (*MS* 1, 3, 31–33).

TABLE IX.10

Planet	First Visibility		First Station
Saturn	16;30°		113°
Jupiter	14°		125°
Mars	28°		163°
	East	West	
Venus	182;30°	20°	166°
Mercury	202°	49°	145°

Āryabhaṭa II's inclinations of the orbits of the planets and the corresponding maximum latitudes are given in Table IX.11 (*MS* 1, 3, 39).

TABLE IX.11

Planet	i	β_{max}
Saturn	2;10°	2;26°
Jupiter	1;14°	1;30°
Mars	1;46°	4;56°
Venus	2;10°	7;54°
Mercury	2;18°	3;40°
Moon	4;30°	4;30°

In the *Mahāsiddhānta* the right ascension is tabulated in time minutes for every dṛkāṇa (δεκανός) or 10°, as in Table IX.12 (*MS* 1, 4, 40–41).

TABLE IX.12

λ	α	λ	α
10°	549′		
20°	555′		
30°	565′	Aries	1669′
40°	580′		
50°	598′		
60°	617′	Taurus	1795′
70°	636′		
80°	648′		
90°	652′	Gemini	1936′

The kālāṃśas for the visibility of the planets (*MS* 1, 9, 3) and their mean apparent diameters in minutes (*MS* 1, 11, 1) according to Āryabhaṭa II are recorded in Table IX.13. These values are unique in the history of Indian astronomy.

TABLE IX.13

Planet	Kālāṃśas	Kālāṃśas (retrograde)	Mean Diameters
Venus	8°	7;30°	$\frac{36'}{4} = 9'$
Jupiter	12°		$\frac{29'}{4} = 7;15'$
Moon	12°		
Mercury	13°	12;30°	$\frac{25'}{4} = 6;15'$
Saturn	15°		$\frac{21'}{4} = 5;15'$
Mars	17°		$\frac{19'}{4} = 4;45'$

The star catalog in the *Mahāsiddhānta* also is quite unique; it is reproduced in Table IX.14 (*MS* 1, 12,1–8 and 1, 9, 8).

TABLE IX.14

Nakṣatra	λ*	β*
Aśvinī	12°	+10°
Bharaṇī	24;23°	+12°
Kṛttikā	38;33°	+5°
Rohiṇī	47;33°	−5°
Mṛgaśīrṣa	61;3°	−10°
Ārdrā	68;23°	−11°
Punarvasu	92;53°	+6°
Puṣya	106°	0°
Āśleṣā	111°	−7°
Maghā	126°	0°
Pūrvaphālgunī	140;23°	+12°
Uttaraphālgunī	150;23°	+13°
Hasta	174;3°	−10°
Citrā	182;53°	−2°
Svāti	194°	+37°
Viśākhā	211;33°	−1;30°
Anurādhā	224;53°	−3°
Jyeṣṭhā	230;3°	−4°
Mūla	242;44°	−9°
Pūrvāṣāḍhā	252;33°	−5°
Uttarāṣāḍhā	260;23°	−5°
Abhijit	263°	+63°
Śravaṇa	280;3°	+30°
Dhaniṣṭhā	296;33°	+36°
Śatabhiṣaj	319;53°	−0;20°
Pūrvabhādrapadā	334;53°	+24°
Uttarabhādrapadā	347°	+26°
Revatī	0°	0°
Agastya	85°	−77°

It appears probable that Āryabhaṭa computed the λ's from the Brāhampakṣa's λ*'s and substituted these computed λ's for the latter.

TABLE IX.15

Planet	R	$\frac{R \cdot 6,0}{Y}$	$\frac{R \cdot 6,0}{C}$
Saturn	146,612	12;13,3,36°	0;2,0,25,3,37,9,...°
Jupiter	364,180	30;20,54°	0;4,59,6,50,32,53,...°
Mars	2,296,864	3,11;24,19,12°	0;31,26,29,47,10,42,...°
Sun	4,320,000	6,0°	0;59,8,10,13,3,31,...°
Venus' śīghra	7,022,268	3,45;11,20,24°	1;36,7,38,22,15,43,...°
Mercury's śīghra	17,937,048	54;45,14,24°	4;5,32,49,51,28,57,...°
Moon	57,753,320	2,12;46,36°	13;10,34,51,25,23,43,...°
Manda	488,122	40;40,36,36°	0;6,40,54,43,19,0,...°
Node	−232,300	−19;21,30°	−0;3,10,47,46,37,57,...°

Another aberrant text is the *Tantrasaṅgraha* composed by Nīlakaṇṭha, the pupil of Dāmodara, the son of Parameśvara, in Kerala in 1501. In this work he accepts the Āryapakṣa's number of days in a Mahāyuga, but introduces new numbers of rotations of the planets, as indicated in Table IX.15 (*TS* 1, 15–20).

In his *Golasāra* (*GS* 1, 2–5) Nīlakaṇṭha gives these rotations rounded to the nearest multiple of ten, and then divided by ten, as the planets' rotations in a Kaliyuga.

According to the *Tantrasaṅgraha*, then, there was not a mean conjunction of the planets at the beginning of the present Kaliyuga; the mean longitudes at that time are listed in Table IX.16 (*TS* 1, 35–38).

TABLE IX.16

Planet	Mean Longitude
Saturn	347;20°
Jupiter	12;10°
Mars	347;47°
Venus' śīghra	36;13°
Mercury's śīghra	359;24°
Moon	4;45,46°
Manda	119;17,5°
Node	−202;20°

From these numbers it is clear that Nīlakaṇṭha's Kaliyuga begins after neither 9/10 nor 3/4 of the Mahāyuga has elapsed; it is probable that both these epoch positions and the rotations in a Mahāyuga are specifically designed to satisfy the situation in 1507, when eight of his yugabhogas, of 576 years each, had elapsed.

Nīlakaṇṭha's longitudes of the mandoccas and nodes are strongly influenced by those of the Saurapakṣa (Tables VIII.5 and VIII.11); they are given in Table IX.17 (*TS* 1, 39–40 and 7, 5).

TABLE IX.17

Planet	Mandocca	Node
Saturn	240°	100°
Jupiter	172°	80°
Mars	127°	40°
Sun	78°	
Venus	80°	60°
Mercury	220°	20°

He repeats these parameters in the *Golasāra* (*GS* 1, 5–6).

In another work, the *Siddhāntadarpaṇa*, Nīlakaṇṭha gives the rotations of the planets in a Kalpa as in Table IX.18 (*SD* 2–8; also in *SDP* 1–4).

In computing the mean daily motions I have used the value of C given in the *Siddhāntadarpaṇasiddhaparyayādayaḥ* (*SDP* 4)—1,577,917,839,500—which implies a year length of 6,5;15,31,31,58,30 days.

The author of this same work has computed the

TABLE IX.18

Planet	R	$\frac{R \cdot 6{,}0}{Y}$	$\frac{R \cdot 6{,}0}{C}$
Saturn	146,571,016	12;12,51,18,17,16,48°	0;2,0,23,2,20,40, . . .°
Manda	54	0;0,0,58,19,12°	
Node	−757	−0;0,0,13,37,33,36°	
Jupiter	364,160,611	30;20,48,10,59,52,48°	0;4,59,5,51,55,24, . . .°
Manda	601	0;0,0,10,49,4,48°	
Node	−825	−0;0,0,14,51°	
Mars	2,296,862,137	3,11;24,18,38,27,57,36°	0;31,26,29,40,12,31, . . .°
Manda	754	0;0,0,13,34,19,12°	
Node	−834	−0;0,0,15,0,43,12°	
Sun	4,320,000,000	6,0°	0;59,8,10,10,18,37, . . .°
Manda	353	0;0,0,6,21,14,24°	
Venus' śīghra	7,022,270,552	3,45;11,21,9,56,9,36°	1;36,7,38,25,20,25, . . .°
Manda	272	0;0,0,4,53,45,36°	
Node	−766	−0;0,0,13,47,16,48°	
Mercury's śīghra	17,937,120,175	54;45,36,3,9°	4;5,32,23,39,27,21, . . .°
Manda	494	0;0,0,8,53,31,12°	
Node	−902	−0;0,0,16,14,9,36°	
Moon	57,753,332,321	2,12;46,39,41,46,40,48°	13;10,34,51,51,41,46, . . .°
Manda	488,123,318	40;40,40,35,43,26,24°	0;6,40,55,22,23,6, . . .°
Node	−232,226,745	−19;21,8,1,24,36°	−0;3,10,44,9,53,2, . . .°

mean longitudes of the planets and the moon's mandocca and node at the beginning of the current Kaliyuga (that is, 432,000 · 4,567 years from the beginning of the Kalpa [*SD* 11–12], and of the mandoccas and nodes of the star-planets in 1699 (*SDP* 9–13); these are given in Table IX.19.

TABLE IX.19

Planet	Itself	Mandocca	Node
Saturn	2;36,13,33°	238;16°	99;49°
Jupiter	15;45,39,52°	171;51°	79;46°
Mars	348;37,38,30°	126;57°	40;3°
Sun	0;20,41,48°	77;35°	
Venus (śīghra)	35;59,5,12°	80;10°	60;6°
Mercury (śīghra)	350;46,34,36°	219;44°	20;1°
Moon	4;51,49,27°	119;3,17,57°	−201;2,29,9°

The rules for computing the planetary equations in the *Tantrasaṅgraha* (*TS* 2, 21–80) are based on the dimensions of the epicycles listed in Table IX.20; Nīlakaṇṭha repeats these dimensions in the *Golasāra* (*GS* 1, 7–8) and the *Siddhāntadarpaṇa* (*SD* 9–10).

The dimensions of the śīghra epicycles and of some of the manda epicycles are taken from the Āryapakṣa (Table VI.7). Nīlakaṇṭha also repeats the Āryapakṣa's inclinations of the planetary orbits (Table VI.5) in the *Tantrasaṅgraha* (*TS* 7, 6) and in the *Golasāra* (*GS* 1, 7), although in the *Siddhāntadarpaṇa* (*SD* 8–9; compare *SDP* 5) he gives the values reported in Table IX.21.

TABLE IX.20

Planet	c_μ	c_σ Odd Quadrants	c_σ Even Quadrants
Saturn	$\frac{10}{80}=45°$	$\frac{9}{80}=40;30°$	$\frac{8}{80}=36°$
Jupiter	$\frac{8}{80}=36°$	$\frac{16}{80}=72°$	$\frac{15}{80}=67;30°$
Mars	$\frac{16}{80}=72°$	$\frac{53}{80}=238;30°$	$\frac{51}{80}=229;30°$
Sun	$\frac{3}{80}=13;30°$		
Venus	$\frac{3}{80}=13;30°$	$\frac{59}{80}=265;30°$	$\frac{57}{80}=256;30°$
Mercury	$\frac{14}{80}=63°$	$\frac{31}{80}=139;30°$	$\frac{29}{80}=130;30°$
Moon	$\frac{7}{80}=31;30°$		

TABLE IX.21

Planet	i	β_{max}
Saturn	2°	2;13°
Jupiter	1°	1;14°
Mars	1;30°	4;7°
Venus	2;45°	9;33°
Mercury	5;15°	8;14°
Moon	4;30°	4;30°

TABLE IX.22

Planet	Mean Daily Motion
Saturn	$\frac{1°}{30\text{ days}}+\frac{2;30'}{1\text{ year}}\approx$ 0;2,0,24,38,...°
Jupiter	$\frac{1°}{12\text{ days}}-\frac{1'}{67\text{ days}}\approx$ 0;4,59,6,17,...°
Mars	$\frac{1°}{2\text{ days}}+\frac{30'}{21\text{ days}}+\frac{1'}{77\text{ days}}+\frac{1''}{2\text{ years}}\approx$ 0;31,26,30,24,...°
Sun	$1°-\frac{11°}{764\text{ days}}+\frac{180''}{181\text{ years}}\approx$ 0;59,8,10,13,...°
Venus' śīghra	$1°+\frac{6°}{10\text{ days}}+\frac{1;1'}{8\text{ days}}+\frac{1'}{10\text{ years}}\approx$ 1;36,7,38,29,...°
Mercury's śīghra	$4°+\frac{10°}{108\text{ days}}-\frac{2'}{123\text{ days}}\approx$ 4;5,32,21,28,...°
Moon	$13°+\frac{3°}{17\text{ days}}-\frac{1'}{140\text{ days}}-\frac{1''}{3\text{ years}}\approx$ 13;10,34,51,53,...°
Manda	$\frac{1°}{9\text{ days}}+\frac{1'}{66\text{ days}}+\frac{1''}{1\text{ year}}\approx$ 0;6,40,54,41,...°
Node	$-\left(\frac{1°}{19\text{ days}}+\frac{1'}{46\text{ days}}+\frac{1'}{10\text{ years}}\right)\approx-$0;3,10,47,39,...°

TABLE X.1

Planet	Mallāri	4,000 Days	4,016 Days
Saturn	0;2,0,23,4,37°	0;2,0,23,4,37, . . .°	0;2,0,23,18, . . .°
Jupiter	0;4,59,8,34,17°	0;4,59,8,34,18, . . .°	0;4,59,8,0°
Mars	0;31,26,31,3,36°	0;31,26,31,3,35, . . .°	0;31,26,28,26, . . .°
Sun	0;59,8,10,17,9°	0;59,8,10,16,30, . . .°	0;59,8,10,11, . . .°
Venus' anomaly	0;36,59,40,6,37°	0;36,59,40,6,37, . . .°	0;36,59,29,31, . . .°
Mercury's anomaly	3;6,24,8,7,13°	3;6,24,8,7,13, . . .°	3;6,24,9,30, . . .°
Moon	13;10,34,51,56,0°	13;10,34,51,55,59, . . .°	13;10,34,39,40, . . .°
Manda	0;6,40,51,25,43°	0;6,40,51,25,42, . . .°	0;6,40,55,14, . . .°
Node	−0;3,10,48,25,15°	−0;3,10,48,25,15, . . .°	−0;3,10,47,12, . . .°

Toward the end of the sixteenth century, Acyuta Piṣāraṭi of Kerala wrote a *Karaṇottama*. This contains very crude approximations to the mean daily motions of the planets, which are recorded in Table IX.22 (*K* 1, 3−10).

Acyuta also proposed a method for reducing the mean longitude of the moon in its orbit to an ecliptic longitude in his *Sphuṭanirṇaya*, which he repeats in his *Rāśigolasphuṭānīti* (*R* 47); this correction had previously been made by Yaḥyā ibn abī Manṣūr in the *Zīj al-Mumtaḥan* in the early ninth century,[108] and Acyuta may have heard of it from a Muslim source. Acyuta's formula is

$$\text{Sin } c = \frac{\text{Sin } \omega \cdot \text{Cos } \omega \cdot \text{Vers } i}{\text{Cos } \beta \cdot R}, \qquad \text{(IX.2)}$$

where c is the correction, AB in Figure 31; i is the inclination of the lunar orbit, which Acyuta takes to be 4;48° (Yaḥyā used 4;46°); ω is $NM = NB$; and β is MA.

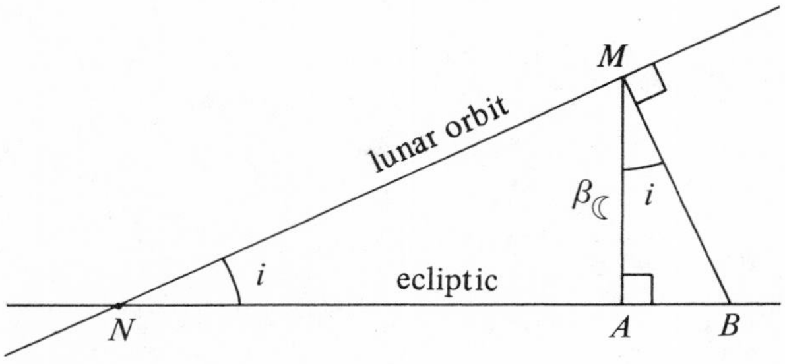

FIGURE 31

X. THE GAṆEŚAPAKṢA

Although allegedly originated by Keśava, the epoch of whose *Grahakautuka* is 1496, this pakṣa is usually named after, and apparently was put into its present form by, his son, Gaṇeśa, the epoch of whose *Grahalāghava* is 18 March 1520. After its creation near the western coast, in Gujarāt, the Gaṇeśapakṣa rapidly spread across northern India and competed there for popularity with the Brāhmapakṣa, from which it appears to be derived. The most southerly adherent of this pakṣa so far identified is Kṛṣṇa, who wrote the *Karaṇakaustubha* near modern Bombay; its epoch is 1653.

The basis of this pakṣa is the division of time into periods of eleven "years," called dhruvāṅkas, each of which contains 4,016 days (*G* 1, 4−5). The mean longitudes for epoch are given (*G* 1, 6−8), as well as the mean motions for a dhruvāṅka and for a day (the latter in *G* 1, 10−14). According to Mallāri (on *G* 1, 10) Gaṇeśa observed the planets and determined that the Saurapakṣa gave correct positions at epoch for the sun and the moon's mandocca; the Saurapakṣa diminished by 0;9° for the moon; the Āryapakṣa for Mars, Jupiter, and the moon's node; the Āryapakṣa increased by 5° for Saturn; the Brāhmapakṣa for Mercury's anomaly; and the mean of the Brāhmapakṣa and the Āryapakṣa for Venus' anomaly. Gaṇeśa's mean daily motions, as recorded by Mallāri, are tabulated in Table X.1 along with those squeezed from the tables for 4,000 and 4,016 days.[109]

Clearly the parameters for the dhruvāṅkas are very crude approximations. The solar mean daily motion according to Mallāri, which is attested elsewhere as the *Grahalāghava*'s,[110] yields a year length of 6,5;15,30,49,46,25 days; that for 4,000 days yields a year-length of 6,5;15,30,53,44,39, . . . days; and that for 4,016 days, 6,5;15,31,27, . . . days.

TABLE X.2

Planet	Mandocca
Saturn	240° (8 · 30)
Jupiter	180° (6 · 30)
Mars	120° (4 · 30)
Sun	78°
Venus	90° (3 · 30)
Mercury	210° (7 · 30)

The longitudes of the mandoccas of the sun (*G* 2, 1) and of the star-planets (*G* 3, 8) are only approximately given by Gaṇeśa; they are recorded in Table X.2.

Gaṇeśa's values of the maximum equations of the planets, as found in the tables belonging to the Gaṇeśapakṣa, are given in Table X.3.

TABLE X.3

Planet	μ_{max}	σ_{max}
Saturn	9;18°	5;42°
Jupiter	5;42°	10;48°
Mars	13;0°	40;0°
Sun	2;10,45°	
Venus	1;30°	46;6°
Mercury	3;36°	21;12°
Moon	5;1,40°	

His śīghra arguments for the occurrences of first stations and first visibilities are listed in Table X.4 (*G* 3, 15–17).

TABLE X.4

Planet	First Visibility		First Station
Saturn	17°		113°
Jupiter	14°		125°
Mars	28°		163°
	East	West	
Venus	183°	24°	167°
Mercury	205°	50°	145°

The last element of the *Grahalāghava* that will be recorded here is its differences from the *Paitāmaha*'s star catalog (Table V.10) in Table X.5 (*G* 11, 1–5).

TABLE X.5

Nakṣatra	Paitāmaha λ*	Paitāmaha β*	Gaṇeśa λ*	Gaṇeśa β*
Kṛttikā	37;28°	+5°	38°	+5°
Rohiṇī	49;28°	−5°	49°	−5°
Mṛgaśiras	63°	−10°	62°	−10°
Ārdrā	67°	−11°	66°	−11°
Punarvasu	93°	+6°	94°	+6°
Āśleṣā	108°	−7°	107°	−7°
Pūrvaphālgunī	147°	+12°	148°	+12°
Svātī	199°	+38°	198°	+37°
Viśākhā	212;5°	−1;30°	212°	−1°
Anurādhā	224;5°	−3°	224°	−2°
Jyeṣṭhā	229;5°	−4°	230°	−3°
Mūla	240;4°	−8;30°	242°	−8°
Pūrvāṣāḍhā	249°	−5;20°	255°	−5°
Uttarāṣāḍhā	260°	−5°	261°	−5°
Abhijit	265°	+62°	258°	+62°
Śravaṇa	278°	+30°	275°	+30°
Dhaniṣṭhā	290°	+36°	286°	+36°
Pūrvabhādrapadā	326°	+24°	324°	+24°
Uttarabhādrapadā	337°	+26°	337°	+27°
Prajāpati			61°	+38°
Brahmahṛd			56°	+30°
Agni			53°	+8°
Agastya			80°	−76°
Apāṃvatsa			183°	+3°
Lubdhaka			91°	−40°

The coordinates of the nakṣatra yogatārās are not seriously different except for Pūrvāsādhā (elsewhere its λ* = 254°) and Abhijit. Some of Gaṇeśa's values of λ* are those of λ according to the Āryapakṣa; this is true for Rohiṇī, Mṛgaśiras, and Viśākhā.

To the Gaṇeśapakṣa belong some anonymous tables[111] and the works of Nṛsiṃha (1588, 1603), the nephew and pupil of Gaṇeśa;[112] of Nāgeśa (1619) and his pupil, Yādava (1663);[113] of Gaṅgādhara (1630);[114] and of the son of Govinda (1773),[115] as well as many others who need not be mentioned here. None seems to have made any startling contributions, although his mode of presentation caused Nṛsiṃha to give variant parameters for the planets' mean yearly motions. These are reproduced in Table X.6. The solar parameter is based on the Ārdharātrikapakṣa's year length: 6,5;15,31,30 days.

XI. ISLAMIC ASTRONOMY IN INDIA

In the course of the preceding discussion we have had occasion to mention some evidence for the introduction of Islamic astronomical concepts into India. In particular, traces of that influence can be discerned in Muñjāla (IX.1) in the tenth century; in Śrīpati (V.121 [?] and V.123) in the eleventh; in Bhāskara II (Table V.43) in the twelfth; in Acyuta

TABLE X.6

Planet	Mean Yearly Motions	Mean Daily Motions
Saturn	12;12,54°	0;2,0,23,28, . . .°
Jupiter	30;21°	0;4,59,7,49, . . .°
Mars	3,11;24,10°	0;31,26,28,15, . . .°
Sun	6,0°	0;59,8,10,10, . . .°
Venus' anomaly	3,45;11,30°	0;36,59,29,41, . . .°
Mercury's anomaly	54;45,20°	3;6,24,10,51, . . .°
Moon's node	−19;21,30°	−0;3,10,47,46, . . .°

(IX.2) in the sixteenth; and in Munīśvara and Kamalākara in the seventeenth. Some knowledge of Arabic astronomical terminology is also shown in manuscripts of tables written in the eighteenth and nineteenth centuries.[116]

At present, however, we know very little about how this transmission took place before the fourteenth century. Virtually the only recognized possibility is too late for Muñjāla, and probably for Śrīpati. About 1030 al-Bīrūnī claims to be translating Euclid's *Elements* and Ptolemy's *Almagest* into Sanskrit, but his ignorance of Sanskrit makes this a very dubious claim.[117] Further documentation is needed for this early period of Islamic influence.

But we do have, in the *Yantrarāja* of Mahendra Sūri with the commentary of Malayendu, a late fourteenth-century Sanskrit treatise on the astrolabe based on Islamic sources. Among the many interesting contents of this work is a Sine table in which $R = 3{,}600$—that is, sixty parts. This is reproduced in Table XI.1 (Malayendu on *Y* 1, 5). I italicize values that do not agree with Kamalākara's (Table VIII.29).

Malayendu also gives a declination table (on *Y* 1, 6) in which he uses an Islamic value for $\epsilon - 23;35°$[118]—and a list of latitudes of cities, including some outside India (compare Table VIII.27); these latitudes are repeated again in Table XI.2.

TABLE XI.1

ϑ	Sin ϑ	ϑ	Sin ϑ	ϑ	Sin ϑ
1°	62;50	31°	1854;8	61°	3148;38
2°	125;38	32°	1907;43	62°	3178;37
3°	188;25	33°	1960;42	63°	3207;37
4°	251;7	34°	2013;6	64°	3235;40
5°	313;45	35°	2064;*52*	65°	3262;42
6°	376;18	36°	2116;*1*	66°	3288;46
7°	438;44	37°	2166;32	67°	3313;49
8°	501;1	38°	2216;23	68°	3337;52
9°	563;10	39°	2265;33	69°	3360;53
10°	625;8	40°	2314;2	70°	3382;54
11°	686;55	41°	2361;49	71°	3403;52
12°	748;29	42°	2408;52	72°	3423;48
13°	809;49	43°	2455;1*1*	73°	3442;42
14°	870;55	44°	2500;46	74°	3460;33
15°	931;45	45°	2545;35	75°	3477;20
16°	992;18	46°	2589;37	76°	3493;4
17°	1052;32	47°	2632;52	77°	3507;44
18°	1112;28	48°	2675;19	78°	3521;20
19°	1172;3	49°	2716;57	79°	3533;51
20°	1231;16	50°	2757;4*5*	80°	3545;1*9*
21°	1290;*8*	51°	2797;4*3*	81°	3555;41
22°	1348;3*6*	52°	2836;50	82°	3564;58
23°	1406;39	53°	2875;5	83°	3573;10
24°	1464;15	54°	2912;2*7*	84°	3580;17
25°	1521;26	55°	2948;5*6*	85°	3586;18
26°	1578;8	56°	2984;32	86°	3591;14
27°	1634;22	57°	3019;13	87°	3595;4
28°	1690;6	58°	3052;58	88°	3597;48
29°	1745;19	59°	3085;48	89°	3599;27
30°	1800;0	60°	3117;4*1*	90°	3600;0

TABLE XI.2

Malayendu		Ulugh Beg or (Astrolabes) or [Geographical Lists]		
City	ϕ	City	ϕ	s_0
Laṅkā	0;0°			
Ādana	11;0°	ʿAdan	11;0°	
Tilaṅga	18;0°	[Tailaṅga		3;55, 4;4]
Anāgūndī	18;10°			
Gaṅgāsāgara	18;20°	[Gaṅgāsāgara		4;56]
Sahābāsa	18;30°			
Devagiri	20;34°	(Dawlatābād	20;30°)	[4;11, 4;25]
Tryambaka	21;0°	{Tryambaka}		
Sañjana	21;0°	{Sañjān}		
Damana	21;0°	[Damaṇa, Damanapura		5;45]
Navasārikā	21;0°	[Navasārī		5;4]
Makkā	21;20°	Makkah	21;40°	
Bhṛgukaccha	22;0°	[Bhṛgukacha, Bhṛgukṣetra		4;48, 4;51]
Nandabhadrava	22;0°	[Khambhādrava		4;51]
Vaṭapatra	22;0°	[Vaṭapatra, Paḍodara		4;52, 4;54]
Stambhatīrtha	22;0°	(Kanbāyat	22;20°)	[4;59]
Dhavalaka	22;0°	[Dholaka		5;2]
Aṃśāvalī	23;0°(?)			
Somanāthapattana	22;15°	Sawmnāt	17;0°	[5;6]
Māṅgalyapura	22;15°			
Raivatakācala	22;31°	[Jūnāgaḍha, Jūnagaḍa		5;0, 5;20]
Dvārakā	22;31°	[Dvārikā, Dvārāvatī		6;5]

TABLE XI.2 (Continued)

Malayendu			Ulugh Beg or (Astrolabes) or [Geographical Lists]	
City	ϕ	City	ϕ	s_0
Navapattana	22;31°			
Ujjayinī	23;30°	(Ujayn	22;30°)	[5;0]
Dhārā	23;30°	{Dhārā}		
Aṇahilapurapattana	24;0°	{Aṇahilapattana}		
Nalapura	25;0°	[Naravara		5;54, 5;55]
Ajameru	26;0°	(Ajmīr	26;0°)	[5;50, 5;52, 6,0]
Nāgapura	26;0°	[Nāgapura		5;0]
Vārāṇasī	26;15°	Bānāras	26;15°	
Lakṣaṇāvatī	26;20°	(Lakhnaw'	26;30°)	[5;45]
Kaḍānagara	26;29°			
Kānyakubja	26;35°	Qanawj	26;35°	[6;0, 6;10]
Māṇikapura	26;49°	(Mānikpūr	26;49°)	
Tīrabhukti	27;0°	{Tirabhukti}		
Jājanagara	27;0°			
Uddīsāsthāna	27;5°			
Ayodhyā	27;22°	(Awadah	27;22°)	[6;7]
Gopālagiri	27;29°	(Gwāliyār	26;29°)	
Bundī	27;32°	[Būndī		5;30]
Gopīmaṇḍala	27;45°			
Kolajalālī	28;4°			
Kapilā	28;10°	[Kapilā, Kampilā		6;15, 6;16]
Śivasthāna	28;15°			
Uccanagara	28;20°	[Ucavāḍhāgaḍha		5;3]
Vaḍanagara	28;28°	[Vadhānagara]		
Dillī	28;39°	(Dihlī	28;19°)	[6;0, 6;3]
Yoginīpura	28;39°	[Yoginīpurapāṭalī		6;24]
Rohitaka	28;45°	{Rohitaka}		
Mehāra	29;20°			
Mūlatāna	29;40°	Mūltān	29;40°	[6;0]
Hāsī	29;45°	(Hānsī	29;15°)	
Hiṃsāra Pirojāvāda	29;48°	[Hiṃsāra, Hisāra		6;4, 6;45]
Sarasvatīpattana	29;50°	[Sarasapattana		6;50]
Sthāneśvara	30;10°	(Tānīsar	30;10°)	[6;30]
Kurukṣetra	30;10°	[Kurukṣetra		6;30, 6;36, 6;55]
Kapisthala	30;30°	{Kāpisthala}		
Jālandhara	30;30°	[Jālandhara		7;0]
Punnāmāsī	30;30°			
Māṇika	30;30°	{Māṇikapura}		
Nepālapura	31;0°	[Naipāla, Naipāl		6;25]
Lāhola	31;50°	Lahāwar	31;50°	[7;30]
Vodhaura	31;50°			
Badaṣasāna	36;22°	Badakhshān	37;10°	
Dārbhaṅgā	36;30°			
Posañja	36;40°			
Balaṣa	36;40°	Balkh	36;41°	[8;7]
Nayasāpura	37;10°	Nīshābūr	36;21°	
Kāśmīra	37;20°	Kashmīr	35;0°	[7;30, 7;52]
Tiramidi	37;35°	{Tirmidh}		
Vāvarada	37;40°			
Cīgāni	37;50°			
Kaluāra	39;0°			
Samarakanda	40;0°	Samarqand	39;37°	[10;3]
Khurāśāna	42;20°	[Khurāsāna		10;3]
Kāsagara	44;0°	Kāshghār	44;0°	
Bulagāra	49;0°	Bulghār	49;30°	

TABLE XI.3

	Mahendra			Ptolemy		
Star	λ	β	Identification	λ	β	Magnitude
(Aśvanābha)	6;43°	+27;0°	α Andr.	347;50°	+26;0°	2.3
Nadyantaka	19;*43*°	−53;30°	ϑ Eri.	0;10°	−53;30°	1
(Matsyodara)	22;43°	+26;20°	β Andr.	3;50°	+26;20°	3
Aśvi	25;*21*°	+ 7;20°	γ Ari.	6;40°	+7;20°	3.4
(Kartitakara)	26;43°	+51;*20*°	β Cass.	7;50°	+51;40°	3
Manuṣyaśīrṣa	48;33°	+23;0°	β Pers.	29;40°	+23;0°	2
(Manuṣyapārśva)	53;43°	+30;0°	α Pers.	34;50°	+30;0°	2
Brāhma	61;33°	−5;10°	α Taur.	42;40°	−5;10°	1
(Mithunapādadakṣiṇa)	68;43°	−31;30°	β Ori.	49;50°	−31;30°	1
Skandhavāma	72;53°	−17;30°	γ Ori.	54;0°	−17;30°	2
Ṣaḍāsya	73;53°	+22;30°	α Aur.	55;0°	+22;30°	1
Hasta	80;53°	−17;0°	α Ori.	62;0°	−17;0°	1.2
Agasti	96;*4*°	−75;0°	α Arg.	77;10°	−75;0°	1
Ārdrā	96;33°	−39;10°	α Can. Mai.	77;40°	−39;10°	1
(Prathamabālaśīrṣa)	102;13°	+9;*40*°	α Gem.	83;20°	+9;30°	2
Vyādhānuja	108;*43*°	−16;10°	α Can. Min.	89;10°	−16;10°	1
Maghā	141;23°	+0;10°	α Leo	122;30°	+0;10°	1
(Uttaraphālgunī)	163;23°	+11;50°	β Leo	144;30°	+11;50°	1.2
(Kākaskandhapakṣa)	*177;33*°	−14;50°	γ Corv.	163;20°	−14;50°	3
Citrā	195;33°	−2;0°	α Virg.	176;40°	−2;0°	1
Svāti	195;53°	+31;30°	α Boöt.	177;0°	+31;30°	1
Viśākhā	213;33°	+44;30°	α Cor. Bor.	194;40°	+44;30°	2.1
Jyeṣṭhā	241;33°	−4;0°	α Scorp.	222;40°	−4;0°	2
Dhanuḥkoṭi	253;43°	+36;0°	α Ophi.	234;50°	+36;0°	3.2
Mūla	260;3°	−13;*10*°	G Scorp.	241;10°	−13;15°	neb.
(Dhanuḥśarāgra)	265;*34*°	+2;50°	$\mu^{1,2}$ Sag.	246;40°	+2;50°	4
Abhijit	276;13°	+62;0°	α Lyr.	257;20°	+62;0°	1
Śruti	292;43°	+29;10°	α Aquil.	273;50°	+29;10°	2.1
(Matsyamukha)	325;53°	−*23;0*°	α Pisc. Austr.	307;0°	−20;20°	1
(Kakudapuccha)	328;3°	+60;0°	α Cyg.	309;10°	+60;0°	2
Hayāṃśa	351;3°	+31;*10*°	β Peg.	332;10°	+31;0°	2.3
(Samudrapakṣī)	353;13°	−9;40°	ι Ceti	334;20°	−9;40°	3.4

Mahendra offers a catalog of astrolabe stars with their longitudes and latitudes for 1370 (the rate of precession is 12° in 800 years or 1° in 66 2/3 years); this is reproduced in Table XI.3[119] (*Y* 1, 22–40). The coordinates are taken from Ptolemy with a precessional constant of 18;53°;[120] Mahendra's errors are italicized.

Malayendu also gives a table of the shadow lengths of twelve- and seven-digit gnomons for solar altitudes of 1° to 90°—that is, a table of Cotangents (on *Y* 1, 70). Later Sanskrit texts on astrolabes and other astronomical instruments, of which many were composed in Gujarāt and Rājasthān in the fifteenth through eighteenth centuries, continued this dependence on Islamic sources. In the seventeenth century this interest in usable instruments spread to Benares, where several important treatises on the subject were composed. This aspect of Indian astronomy, like many others, still awaits a serious and comprehensive investigation.

Another aspect awaiting this kind of exploration comprises the translations of Arabic and Persian astronomical treatises into Sanskrit in the sixteenth, seventeenth, and eighteenth centuries. Among the texts so translated is the *Jīca Ulugbegī* (*Zīj i Ulugh Beg*, composed for and by Ulugh Beg in Samarqand not earlier than 1437/1438); this *zīj* is often cited by the astronomers of Benares in the seventeenth century, and a manuscript of the Sanskrit translation from Surat is now in the Mahārāja of Jaipur Museum in Jaipur (no. 45). Others are the *Samrāt-siddhānta* (Ptolemy's *Almagest* in Naṣīr al-Dīn's recension), translated by Jagannātha for Jayasiṃha in 1732, and the works translated by Nayanasukhopādhyāya in the early eighteenth century, which include the *Ukāra* (*Sphaerics* of Theodosius of Bithynia in Naṣīr al-Dīn's recension) and the *Śarahatajkaraḥ Varjandī* (*Sharḥ al-Tadhkira* of al-Barjandī, which may be a commentary on Naṣīr al-Dīn's *Tadhkira*, although al-Barjandī, or al-Bīrjandī [*d.* 1527/1528], is otherwise known to have written one on Naṣīr al-Dīn's *Taḥrīr al-Majisṭī*, but not on his *Tadhkira*). The name al-Barjandī is probably a mistake for al-Jurjānī (1339–1414), who did comment on the *Tadhkira*. There is a copy of this translation in the Mahārāja of Jaipur Museum in Jaipur (no. 46).

Other Sanskrit translations of Islamic astronomical works may well have been undertaken at Jayasiṃha's court at Jaipur in the early eighteenth century, and undoubtedly many of the Persian astronomical treatises composed in northern India during the Mughal period were based on or incorporated Sanskrit material. But for now there is only space to mention another example of Islamic influence on the Benares astronomers of the eighteenth century; this is the anonymous *Hayata* (Arabic *Hay'a*),[121] in which the year 1764/1765 is used in an example, while the original Persian text refers to the year 1437/1438, which makes certain its identity with the *Zīj i Ulugh Beg* (p. 69). In the *Hayata* the Islamic material is often compared with the opinions expressed in the *Sūryasiddhānta*.

In the early nineteenth century, through the efforts of Lancelot Wilkinson at Sihor and Nīlāmbara Jhā at Alwar, European, and specifically English, astronomical works were translated into Sanskrit. But here again our present lack of specific knowledge prevents any detailed discussion of this phase in the history of Indian astronomy.

XII. CONCLUSIONS

It is hoped that the above essay will at least clarify the historical significance of Indian astronomy as the recipient and remodeler of foreign elements, and as an influential force in the early stages of Islamic mathematical astronomy. It also has established the interrelations of the five pakṣas, their internal histories, and their ultimate derivation from at most two—perhaps only one—Sanskrit version of Greek planetary tables (although a number of Greek texts, describing various other planetary models, also were translated into Sanskrit). And it has been suggested that most elements of Indian mathematical astronomy, as well as the planetary parameters and models, originated in Greece. Among these would be the gnomon problems; the projections; the predictions of eclipses, of the moon's crescent, and of the first and last visibilities and first and last stations of the planets; the projections of eclipses; the star catalog; and probably the Sine function. All of these were modified, and some of them improved or at least expanded in India; and these modifications, improvements, and expansions allow us to characterize this astronomy as Indian rather than Babylonian, Greek, Islamic, or Chinese. But, as Greece was indebted to Mesopotamia, Islam to Greece, Iran, and India, and western Europe to Islam, so India was indebted first to Mesopotamia, then to Greece, then to Islam, and finally to western Europe.

There remain several areas that have not been discussed in this essay for lack of space, time, and ability on its author's part. These include, in particular, astronomical tables (which I have begun to investigate and write about elsewhere), astronomical instruments (I don't believe any serious observations were made in India before the late fourteenth century, so that the contribution of instruments to the development of Indian astronomy prior to that date was minimal), calendaric problems, and mathematics. I particularly regret the exclusion of the last subject, because of its great importance and because of the inadequacies of the books on the subject presently available.[122] I have also, of course, missed many interesting topics in published texts and am ignorant of many more in unpublished texts. This essay can only serve as a preliminary guide to a vast and fascinating field, and perhaps as a spur to its eventual harvesting.

XIII. EPILOGUE

As was the case with other technical subjects in India, astronomy remained an essentially isolated tradition, never to be integrated into a general theory of knowledge. Its function was not to discover the truth about the apparent motions of the heavenly bodies or about the other celestial phenomena, but to train experts who could prepare calendars and astronomical tables for use in determining the proper times for religious observances and in operating the various modes of astrology. It was oriented toward these very practical goals, and normally eschewed all theoretical considerations.

Moreover, like the other Indian śāstras or disciplines, astronomy was preserved and taught within families of practitioners. The members of each such family over many generations remained faithful to a particular pakṣa; in the family library they would have manuscripts of texts belonging to that pakṣa and of works written by members of the family. By and large they would not be interested in educating outsiders or in making innovations in their traditional learning; Gaṇeśa is the most notable exception to this general tendency.

The divine origin of the science was universally proclaimed, and each pakṣa asserted its fidelity to the deity's revelation. Fundamental changes, therefore, were virtually impossible because they usually could attract no adherents; innovation was restricted to computational technique, in which the Indian astronomers excelled, and to the introduction of foreign material. But even in this the normal goal was to achieve the results of the original, revealed tradition; the new mathematical formulas were merely variant methods of arriving, often approximately, at the same conclusions, not the expression of different astronomical concepts. And, although several astronomers claim to obtain better agreement between computation and observation than did their rivals, there were no systematic series of observations (save perhaps those of Parameśvara) carried out as part of a program to improve the basic astronomical models and parameters. So far as we can judge, observations were essentially accidental and haphazard.

Finally, Indian astronomers had minimal concern for the kinematics of celestial motions or for physical astronomy. Nor was their science regarded as occupying a significant place in the intellectual life of India. Therefore, it interacted with no other field of learning in any fundamental way, and was really of little interest to anyone besides the professionals. Other intellectuals learned something of it only through astrology or the necessity of performing certain rituals at astronomically determined times. Thereby it became a tradition largely abstracted from reality, which it dealt with more comfortably through astrological than through astronomical predictions. Its creativity was channeled into developing the mathematics of computation, since it was never forced to grapple with the problems of its relations to physics or to philosophy or with that of the accuracy of its predictions. Nor was any scientific methodology ever developed in India that would have allowed it to conceive of such problems as being real.

The characteristics of Indian science outlined above may help to explain why Indian astronomy did not, and could not, develop in the same way that Western astronomy did. They do not provide a criterion for making a judgment concerning the value of Indian astronomy. Such a criterion resides only in the personal bias of the judge.

ABBREVIATIONS OF TEXTS

A *Āryabhaṭīya* of Āryabhaṭa I
BSS *Brāhmasphuṭasiddhānta* of Brahmagupta
DG *Dṛggaṇita* of Parameśvara
G *Grahalāghava* of Gaṇeśa
GCN *Grahacāranibandhana* of Haridatta I.
GCNS *Grahacāranibandhanasaṅgraha*
GJ *Grahajñāna* of Āśādhara
GL *Grahalāghava* of Gaṇeśa
GS *Golasāra* of Nīlakaṇṭha
K *Karaṇottama* of Acyuta
Kh *Khaṇḍakhādyaka* of Brahmagupta
KK *Karaṇakutūhala* of Bhāskara II
KP *Karaṇaprakāśa* of Brahmadeva
KT *Karaṇatilaka* of Vijayananda
LB *Laghubhāskarīya* of Bhāskara I
LM *Laghumānasa* of Muñjāla
LVS *Laghuvasiṣṭhasiddhānta*
MB *Mahābhāskarīya* of Bhāskara I
MS *Mahāsiddhānta* of Āryabhaṭa II
Pait *Paitāmahasiddhānta* of the *Viṣṇudharmottarapurāṇa*
PS *Pañcasiddhāntikā* of Varāhamihira
R *Rāśigolasphuṭānīti* of Acyuta
RM *Rājamṛgāṅka* of Bhojarāja
SD *Siddhāntadarpaṇa* of Nīlakaṇṭha
SDP *Siddhāntadarpaṇasiddhaparyayādayaḥ*
SDV *Śiṣyadhīvṛddhida* of Lalla
Soma *Somasiddhānta*
SS *Sūryasiddhānta* in Raṅganātha's recension
SSB *Siddhāntaśiromaṇi* of Bhāskara II
SSBM *Siddhāntasārvabhauma* of Munīśvara
SSS *Siddhāntaśekhara* of Śrīpati
STV *Siddhāntatattvaviveka* of Kamalākara
TS *Tantrasaṅgraha* of Nīlakaṇṭha
VK *Vākyakaraṇa*
VS *Vaṭeśvarasiddhānta* of Vaṭeśvara
VVS *Vṛddhavasiṣṭhasiddhānta*
Y *Yantrarāja* of Mahendra
YJ *Yavanajātaka* of Sphujidhvaja

NOTES

1. Biographies and bibliographies of Indian astronomers will be found in D. Pingree, *Census of the Exact Sciences in Sanskrit* (hereafter *CESS*), of which there have so far appeared, of series A, vol. I (Philadelphia, 1970), vol. II (Philadelphia, 1971), and vol. III (Philadelphia, 1976). Of previous attempts to write the history of Indian astronomy, the most notable is the Marāṭhī *Bhāratīya Jyotiḥśāstra* by Śaṅkara Bālakṛṣṇa Dīkṣita (Poona, 1896; repr. Poona, 1931); less informative and reliable are Sudhākara Dvivedin, *Gaṇakataraṅgiṇī* (Benares, 1892; repr. Benares, 1933); and George F. Thibaut, *Astronomie, Astrologie und Mathematik* (Strasbourg, 1899). Of even less value are George R. Kaye, *Hindu Astronomy* (Calcutta, 1924); S. N. Sen, "Astronomy," in *A Concise History of Science in India* (New Delhi, 1971), 58–135; and Roger Billard, *L'astronomie indienne* (Paris, 1971). Other efforts need not be referred to here.

2. An attempt to distinguish the elements in Sanskrit astronomical texts of the second to seventh centuries that are of Greek origin is made in D. Pingree, "The

Recovery of Early Greek Astronomy From India," in *Journal for the History of Astronomy*, **7** (1976), 109–123.

3. See in particular the extraordinary theories expounded by V. H. Vader, R. Shamasastry, and P. C. Sengupta in their articles listed in the bibliographies of *CESS*.

4. A convenient source of references is A. A. Macdonell and A. B. Keith, *Vedic Index of Names and Subjects* (London, 1912; repr. Vārāṇasī [Benares], 1958).

5. For instance, in *MUL. APIN* and in tablet 14 of *Enūma Anu Enlil*.

6. The MSS and contents of the earliest form of the *Gargasaṃhitā* are listed in *CESS* A, II, 116a–117b; I hope to prepare an ed. of this text that will elucidate its relation to *Enūma Anu Enlil, Šumma Ālu*, and other Babylonian omen series.

7. The contents of this section are derived from D. Pingree, "The Mesopotamian Origin of Early Indian Mathematical Astronomy," in *Journal for the History of Astronomy*, **4** (1973), 1–12, where all the exact references will be found.

8. Biographical and bibliographical information concerning many of the individuals mentioned in these pages will be found in the *Dictionary of Scientific Biography* as well as in *CESS*. In general I will avoid referring again to material cited in the articles of this *Dictionary*.

9. O. Neugebauer, *Astronomical Cuneiform Texts* (*ACT*), 3 vols. (London, 1955), I, 47; "The Rising Times in Babylonian Astronomy," in *Journal of Cuneiform Studies*, **7** (1953), 100–102; and "On Some Astronomical Papyri and Related Problems of Ancient Geography," in *Transactions of the American Philosophical Society*, n.s. **32** (1942), 251–263.

10. See Varāhamihira, *Bṛhajjātaka* 1, 19.

11. See *PS* 4, 48; *BSS* 12, 52; al-Fazārī, fr. *Q 2* in D. Pingree, "The Fragments of the Works of al-Fazārī," in *Journal of Near Eastern Studies*, **29** (1970), 103–123.

12. For the Greek knowledge of Babylonian astronomy, see O. Neugebauer, *A History of Ancient Mathematical Astronomy*, 3 vols. (Berlin–Heidelberg–New York, 1975), *passim*.

13. The fragments are given in O. Neugebauer and D. Pingree, *The Pañcasiddhāntikā of Varāhamihira*, 2 vols. (Copenhagen, 1970–1971), I, 10–12.

14. Against the reconstruction of P. Wirth, "Die Venustheorie von Vasiṣṭha in der Pañcasiddhāntikā des Varāha Mihira," in *Centaurus*, **18** (1973), 29–43, see D. Pingree, "Vasiṣṭha's Theory of Venus: The Misinterpretation of an Emendation," *ibid.*, **19** (1975), 36–39. The response by B. L. van der Waerden, "On the Motion of Venus in the Pañcasiddhāntikā," *ibid.*, **20** (1976), 35–43, is unconvincing.

15. *ACT*, II, 283.

16. *Ibid.*, 303.

17. *Ibid.*, 305–306.

18. *Ibid.*, 293–294.

19. For the version by Śrīṣeṇa, see *PS*, I, 12; for later versions, consult D. Pingree, "Astronomy and Astrology in India and Iran," in *Isis*, **54** (1963), 229–246, esp. 237, n. 67.

20. D. Pingree, "Precession and Trepidation in Indian Astronomy Before A.D. 1200," in *Journal for the History of Astronomy*, **3** (1972), 27–35.

21. D. Pingree, "On the Greek Origin of the Indian Planetary Model Employing a Double Epicycle," *ibid.*, **2** (1971), 80–85.

22. The later *Pauliśasiddhānta* is described in section VII.

23. *Zīj al-Ṣābī*ʿ 27.

24. These data follow from the material in *ACT*, II.

25. The text is edited by F. Boll in *Catalogus codicum astrologorum Graecorum*, VII (Brussels, 1908), 213–224; also see D. Pingree, ed., *Albumasaris De revolutionibus nativitatum* (Leipzig, 1968), 245–273.

26. These values do not agree with those given by Paul of Alexandria in Εἰσαγωγικά 28.

27. See the article cited in n. 20.

28. There are many reasons—sufficient to convince the present writer—for believing that the projection method was introduced into India from Greece, although it was further developed in the country of its adoption. This matter and others related to it are discussed by D. Pingree, in the article cited in n. 2. See also O. Neugebauer, "On Some Aspects of Early Greek Astronomy," in *Proceedings of the American Philosophical Society*, **116** (1972), 243–251, esp. 249.

29. Tablets 15 and 20 (21)–21 (22).

30. Tablets 17–18.

31. See O. Neugebauer's article cited in n. 28, 250–251.

32. *Almagest* 7–8.

33. See W. Kirfel, *Die Kosmographie der Inder* (Bonn–Leipzig, 1920; repr. Hildesheim, 1967), 54–177.

34. W. Kirfel, *Das Purāṇa vom Weltgebäude* (Bonn, 1954), 48–55, 196–220.

35. W. Kirfel, *Die Kosmologie*, 208–339.

36. See D. Pingree, "Astronomy and Astrology in India and Iran," in *Isis*, **54** (1963), 238.

37. B. L. van der Waerden, "Das Grosse Jahr und die ewige Wiederkehr," in *Hermes*, **80** (1952), 129–155; many of the statements made in this article are of dubious validity.

38. D. Pingree, "The *Paitāmahasiddhānta* of the *Viṣṇudharmottarapurāṇa*," in *Brahmavidyā*, **31–32** (1967–1968), 472–510.

39. D. Pingree, "The Persian 'Observation' of the Solar Apogee in *ca.* A.D. 450," in *Journal of Near Eastern Studies*, **24** (1965), 334–336.

40. See n. 21 above.

41. This problem is related to the computation of horizontal parallax, which is the radius of the earth seen at the moon's distance. Therefore, an estimate of the earth's circumference will lead directly to an equivalence of yojanas to minutes in the orbit of the moon. The length of a yojana is, however, not fixed; Āryabhaṭa I, for instance, assumes a different number of yojanas in the earth's circumference, and therefore also in the moon's orbit.

42. See n. 21 above.

43. G. J. Toomer, "The Chord Table of Hipparchus and the Early History of Greek Trigonometry," in *Centaurus*, **18** (1973), 6–28. On the rule for obtaining the sines, see R. C. Gupta, "Early Indians on Second Order Sine Differences," in *Indian Journal of History of Science*, **7** (1972), 81–86.

44. So in his commentary on Aratus he measures λ^* on circles parallel to the equator—that is, λ^* is the λ of the point on the ecliptic cut by the declination circle passing through the star. Instead of β^*, however, he gives the polar distance of the star on that declination circle.

45. O. Neugebauer, "On a Fragment of Heliodorus (?) on Planetary Motions," in *Sudhoffs Archiv*, **42** (1958), 237–244; A. Tihon, "Les scolies des Tables faciles de Ptolémée," in *Bulletin de l'Institut historique belge de Rome*, **43** (1973), 51–110, esp. 64–65, 99–102.

46. The sizes of planetary epicycles, of course, were first determined by Apollonius from the retrograde arcs; see O. Neugebauer, "The Equivalence of Eccentric and Epicyclic Motion According to Apollonius," in *Scripta mathematica*, **24** (1959), 5–21.

47. Our differing basic assumptions have led Burgess-Whitney (*Sūryasiddhānta* [New Haven, 1860], 175–213) and me to arrive at different identifications in some instances.

48. E. S. Kennedy and A. Muruwwa, "Bīrūnī on the Solar Equation," in *Journal of Near Eastern Studies*, **17** (1958), 112–121.

49. R. C. Gupta, "Indian Approximations to Sine, Cosine and Versed Sine," in *Mathematics Education*, **6** (1972), sec. B, 59–60; and "Bhāskara I's Approximation to Sine," in *Indian Journal of History of Science*, **2** (1967), 121–136.

50. D. Pingree, "The Indian and Pseudo-Indian Passages in Greek and Latin Astronomical and Astrological Texts," in *Viator*, **7** (1976), 141–195.

51. R. N. Rai, "The Extant *Siddhānta Śekhara*: An Error in One of Its Sine Values, "in *Indian Journal of the History of Science*, **6** (1971), 135–138.

52. K. M. K. Sarma, "The Rājamṛgāṅka of Bhoja," in *Brahmavidyā*, **4** (1940), 95–105; and K. S. Shukla, "A Note on the Rāja-mṛgāṅka of Bhoja Published by the Adyar Library," in *Gaṇita*, **5** (1954), 149–151.

53. See n. 20 above.

54. On Āśādhara see *CESS* A, I, 54b and A, II, 16a; and D. Pingree, *Sanskrit Astronomical Tables in England* (henceforth *SATE*) (Madras, 1972), 69–72. My ed. of the text will appear soon in a collection of astronomical texts to be published in the *Gaekwad Oriental Series*.

55. D. Pingree, "On the Classification of Indian Planetary Tables," in *Journal for the History of Astronomy*, **1** (1970), 95–108.

56. See *SATE*, 30. My ed. of this text will appear soon in the collection mentioned in n. 54.

57. A sine table for every degree from 1° to 90° with $R = 120$ was given by Viśrāma in 1615 in his *Yantraśiromaṇi*, K. K. Raikva, ed. (Bombay, 1936), 85–89.

58. See, for instance, M. G. Inamdar, "A Formula of Bhāskara for the Chord of a Circle Leading to a Formula for Evaluating Sin $\alpha°$," in *Mathematics Student*, **18** (1950), 9–11; and A. A. Krishnaswami Ayyangar, "Remarks on Bhaskara's Approximation to the Sine of an Angle," *ibid.*, 12. An interpretation (not always reliable) of Bhāskara's mathematical astronomy is in D. A. Somayaji, *A Critical Study of the Ancient Hindu Astronomy* (Dharwar, 1971).

59. D. Pingree "Precession and Trepidation in Indian Astronomy Before A.D. 1200," in *Journal for the History of Astronomy*, **3** (1972), 32.

60. For the Indian anticipation of Newton's power series for sine and cosine in the work of Mādhava (late fourteenth century), see C. T. Rajagopal and A. Venkataraman, "The Sine and Cosine Power-Series in Hindu Mathematics," in *Journal of the Royal Asiatic Society of Bengal*, **15** (science) (1949), 1–13. See also R. C. Gupta, "An Indian Form of Third Order Taylor Series Approximation of the Sine," in *Historia mathematica*, **1** (1974), 287–289.

61. D. Pingree, *Sanskrit Astronomical Tables in the United States* (henceforth *SATIUS*) (Philadelphia, 1968), 36a–37a.

62. *SATE*, 73–76.

63. *SATE*, 76–82.

64. *SATIUS*, 60b–61b; *SATE*, 142.

65. *SATE*, 82.

66. *Ibid.*, 101

67. *SATIUS*, 53a–53b; *SATE*, 101–112.

68. *SATE*, 112–114.

69. *Ibid.*, 123–128.

70. *SATIUS*, 61b–62b.

71. *SATE*, 141–142.

72. *SATIUS*, 64b–66a.

73. D. Pingree, "On the Classification of Indian Planetary Tables," in *Journal for the History of Astronomy*, **1** (1970), 104.

74. *SATIUS*, 54a–55b, 62b–63a, 66a–66b, 67a–68b.

75. D. Pingree, "The *Karaṇavaiṣṇava* of Śaṅkara," in *Charudeva Shastri Felicitation Volume* (Delhi, 1974), 588–600.

76. See D. Pingree, "Al-Bīrūnī's Knowledge of Sanskrit Astronomical Texts", in *The Scholar and the Saint*, P. J. Chelkowski, ed. (New York, 1975), 67–81.

77. D. Pingree, "Concentric With Equant," in *Archives internationales d'histoire des sciences*, **24** (1974), 26–28.

78. *Laghuprakāśikā* on *Vākyakaraṇa* 2, 18–20, T. S. Kuppanna Sastri and K. V. Sarma, eds. (Madras, 1962).

79. D. Pingree, "The Greek Influence on Early Islamic Mathematical Astronomy," in *Journal of the American Oriental Society*, **93** (1973), 32–43, esp. 37, no. 37–38.

80. F. I. Haddad, E. S. Kennedy, and D. Pingree, *The Book of the Reasons Behind Astronomical Tables* (in press), sec. 3.

81. R. N. Rai, "Sine Values of the *Vaṭeśvara Siddhānta*," in *Indian Journal of History of Science*, **7** (1972), 1–15.

82. C. Kunhan Raja, *Candravākyas of Vararuci* (Madras, 1948). See also O. Neugebauer, "Tamil Astronomy," in *Osiris*, **10** (1952), 252–276; I. V. M. Krishna Rav, "The Motion of the Moon in Tamil Astronomy," in *Centaurus*, **4** (1956), 198–220; B. L. van der Waerden, "Tamil Astronomy," *ibid.*, 221–234; and G. J. Toomer, "A Note on Tamil Astronomical Tables," *ibid.*, **9** (1963–1964), 11–15; and "A Further Note on Tamil Astronomical Tables," *ibid.*, 254–256.

83. For the important mathematical work of Mādhava and others from Kerala, see K. M. Marar and C. T. Rajagopal, "On the Hindu Quadrature of the Circle," in *Journal of the Bombay Branch of the Royal Asiatic Society*, n.s. **20** (1944), 65–82; and "Gregory's Series in the Mathematical Literature of Kerala," in *Mathematics Student*, **13** (1945), 92–98; C. T. Rajagopal, "A Neglected Chapter of Hindu Mathematics," in *Scripta mathematica*, **15** (1949), 201–209; C. T. Rajagopal and T. V. V. Aiyar, "On the Hindu Proof of Gregory's Series," *ibid.*, **17** (1951), 65–74; and "A Hindu Approximation to Pi," *ibid.*, **18** (1952), 25–30; K. V. Sarma, *A History of the Kerala School of Hindu Astronomy* (Hoshiarpur, 1972), 11–28; and R. C. Gupta, "The Mādhava-Gregory Series," in *Mathematics Education*, **7** (1973), 67–70.

84. *SATE*, 181–182.

85. O. Neugebauer and D. Pingree in *Pañcasiddhāntikā*, I, 13–14.

86. See F. I. Haddad, E. S. Kennedy, and D. Pingree in the work cited in n. 80 above, sec. 7, and 8.

87. I refer here to the excellent ed. by B. Chatterjee, 2 vols. (Calcutta, 1970); see also D. Pingree, "The Beginning of Utpala's Commentary on the *Khandakhādyaka*," in *Journal of the American Oriental Society*, **93** (1973), 469–481.

88. See the work cited in n. 80 above, sec. 4.

89. *SATIUS*, 33a, 39a–39b; *SATE*, 175–176.

90. D. Pingree, "Precession and Trepidation in Indian Astronomy Before A.D. 1200," in *Journal for the History of Astronomy*, **3** (1972), 28–29.

91. F.-E. Hall and B. D. Śāstri, eds. (Calcutta, 1859), translated by E. Burgess (New Haven, 1860). There are several reprints of both of these, as well as other eds. and translations.

92. See n. 90 above.

93. V. P. Dvivedin, ed., in *Jyautiṣasiddhāntasaṅgraha* (Benares, 1912), fasc. 1, 1–36.

94. V. P. Dvivedin, ed. (Benares, 1907).

95. For this work see O. Neugebauer and D. Pingree, *Pañcasiddhāntikā*, I, 10–12.

96. V. P. Dvivedin, ed. in *Jyautiṣasiddhāntasaṅgraha* (Benares, 1912), fasc. 2, 25–78.

97. *Ibid.*, fasc. 1, pt. 2.

98. *SATE*, 114–118.

99. L. P. E. A. Sédillot, ed., *Prolégomènes des tables astronomiques d'Oloug-beg* (Paris, 1853), 257–271.

100. R. Webster, M. Webster, and D. Pingree, *Catalogue of Astrolabes* (in press).

101. *SATIUS*, 73a–75b; *SATE*, 53–55. On an MS of the *Makaranda* copied at Argalāpura, see *SATIUS*, 23a–24a.

102. D. Pingree, "On the Classification of Indian Planetary Tables," in *Journal for the History of Astronomy*, **1** (1970), 99.

103. *SATE*, 142–149.

104. *Ibid.*, 153–158.

105. *Ibid.*, 158–168.

106. *Ibid.*, 170–175.

107. *Bṛhanmānasa*, quoted by Praśastidhara on *LM* 2.

108. See H. Salam and E. S. Kennedy, "Solar and Lunar Tables in Early Islamic Astronomy," in *Journal of the American Oriental Society,* **87** (1967), 492–497, esp. 496.

109. D. Pingree, "On the Classification of Indian Planetary Tables," in *Journal for the History of Astronomy*, **1** (1970), 100.

110. *SATIUS*, 29a, 47a.

111. *Ibid.*, 46b–47b, 50b, 69a–70a; *SATE*, 93–100.

112. *SATE*, 118–123; D. Pingree, "On the Classification of Indian Planetary Tables," in *Journal for the History of Astronomy*, **1** (1970), 101.

113. *SATIUS*, 63a–64b; *SATE*, 149.

114. *SATE*, 134–141.

115. *Ibid.*, 168–175.

116. For instance, *SATIUS*, 45a–45b; *SATE*, 34–38.

117. *India* (Hyderabad, 1958), 106; see n. 76 above.

118. See, for example, E. S. Kennedy, *A Commentary Upon Bīrūnī's Kitāb Taḥdīd al-Amākin* (Beirut, 1973), 33, 39–40, 43, 50.

119. See my ed. of Malayendu's table in the book cited in n. 100 above.

120. This fact was first recognized by G. R. Kaye, *The Astronomical Observatories of Jai Singh* (Calcutta, 1918), 116.

121. V. B. Bhaṭṭācārya, ed. (Vārāṇasī [Benares], 1967).

122. For instance, B. Datta and A. N. Singh, *History of Hindu Mathematics*, 2 vols. (Lahore, 1935–1938); or C. N. Srinivasiengar, *The History of Ancient Indian Mathematics* (Calcutta, 1967).

MAN AND NATURE IN MESOPOTAMIAN CIVILIZATION

A. Leo Oppenheim

I

Mesopotamian man lived in a concrete world that he experienced directly and strove to adapt to his immediate needs and special demands. The data provided to him by his senses were utilized in two essentially different ways by his intellect. On the one hand, he constructed around himself an orderly world in which he could make rational decisions within a frame of predictable events and situations. On the other, after experience had taught him to recognize a pattern in the sequence of certain events and in the predictable features of specific phenomena, he considered any deviations and irregularities to be endowed with meaning—and, more than that, to be meaningful with regard to himself. They were taken to convey a message that referred to impending events, fortunate or unfortunate.

Moreover, Mesopotamian man attempted to construct an integrated whole extending beyond the objects he could touch and see, a whole of which he himself was to be an essential part. Its internal organization and purpose were to provide a setting and a direction for man's role within and beyond the dimensions of observable reality.

In none of these respects were the intellectual achievements of Mesopotamian man outstanding among the cluster of civilizations that had evolved in the ancient Near East and beyond in the last three to four millennia B.C. Still, in a few respects he did succeed in creating unique and characteristic formulations and attitudes, the possible origins and essential consequences of which this essay presents and discusses.

The relationship between man and nature in the ancient Near East is nowhere as pointedly formulated as in Genesis 1:26, where it is said that God gave man "dominion over the fish of the sea, and over the fowl of the air, and over the cattle, and over all the earth, and over every creeping thing that creepeth upon the earth." The parallel version of the Creation story (Genesis 2:19) formulates the same relationship differently, and in a way that is more relevant to the characteristic attitude of those civilizations that relied on writing for the preservation of their intellectual traditions. It says, "God formed every beast of the field, and every fowl of the air; and brought them unto Adam to see what he would call them: and whatsoever Adam called every living creature, that was the name thereof." While it was thus man's privilege as the lord of creation to give names to the animals, the knowledge of all their names and their individual features and behavior was considered the privilege of the sage. This is illustrated by the passage (1 Kings 4:33) that extols the wisdom of Solomon: "And he spake of trees, from the cedar tree that is in Lebanon even unto the hyssop that springeth out of the wall: he spake also of beasts, and of fowl, and of creeping things, and of fishes."

While the last passage clearly focuses on practical wisdom, oriented toward exemplary behavior as illustrated by generally known characteristics of certain animals seen as moral prototypes, a far more encyclopedic view is taken by a small group of Egyptian texts published by Alan Gardiner as *Ancient Egyptian Onomastica*. In these texts, scribes make a conscious effort to organize the entire known world by means of lists of names comprising "everything created by Ptah, recorded by Thoth," as the solemn introductory sentence states. The list includes heaven and what pertains to it; earth and what is on it; all the things upon which the sun god shines; and what grows on earth—specifically, gods and human beings (from kings to foreign tribes), the towns of Egypt, buildings and their parts, cereals and their products, food and drink, cuts of meat and other viands. The purpose of the text—datable to about 1100 B.C., with fragments of similar lists about half a millennium older—is difficult to establish within the framework of the conventional interests of the Egyptian scribes.[1] Its terseness, the restrictions in subject matter, and the exclusive use of nouns in the lists pose problems in view of the grandiose claims of the preamble. Obviously the onomastica are devoid of the "wisdom" connotations of the just-cited verse describing Solomon's knowledge;[2] they should probably be compared to the Genesis passage in which Adam gives names to all the animals. Although the outward expression of the passage is man's dominion over the animal world, its underlying concept may have been that it was man's responsibility not only to give names but also to know the names of all animals (and plants). In a civilization that knew writing, from such an attitude there might have developed the desire to enumerate such names so as to demonstrate both erudition and the power of the human intellect in confronting nature, the world around the scribe.

From Mesopotamia come somewhat similar texts, written on clay; they are far more numerous and more elaborate, and their function is much better known than those from Egypt.[3]

For elementary as well as for advanced training, Mesopotamian scribes used a great variety of sign lists that contained cuneiform signs representing Sumerian words, written one underneath the other, in long and narrow columns. If a sign had several readings—that is, phonetic values—it was repeated as many times as there were values. Because of the application of several principles of arrangement (acrophonic or acrographic sequences, and topical organization) and the accretion of auxiliary columns on both sides of the nuclear list, these syllabaries or "vocabularies" grew into a large body of scholarly literature. From a tool for teaching they turned into a vehicle for philological research. Among the types that evolved, only the topically arranged lists are the concern of this essay.[4] These lists contain nouns referring to man-made objects, trees, plants, minerals, classes of persons, localities, deities, and stars. All nouns are written (syllabically or with word signs) after a class determinative (for instance, wood + plow; silver + ring; green plant + lettuce; stone + lapis lazuli; deity + moon; man + blind), which leads automatically to such topical arrangements. From the middle of the second millennium on, the Sumerian lists were provided with a corresponding column in Akkadian; the column to the left contained the Sumerian noun with its classifier, and the one to the right the translation into Akkadian.[5]

Prominent among the topical vocabularies is the one called ḪAR . r a = *ḫubullu* (hereafter cited as Hh), which contains on twenty-two of its twenty-four tablets (tablets III–XXIV) from eight to ten thousand lexical entries, although about one-third of its content is lost through the fragmentary state of some tablets. The topics treated are trees, wooden objects, reeds and reed objects, earthenware, leather and leather objects, metals and metal objects, domestic animals, wild animals, parts of the human and animal body, precious stones and stone objects, green plants, fish and birds, wool and garments, localities within Babylonia, beer, barley and its products, honey, and other foodstuffs. The series was very much used, and copies of it or of its monolingual Sumerian prototypes are often found outside Mesopotamia wherever the Akkadian language was taught.[6] In the first millennium it was brought up to date, so to speak, with a third column added at the right in order to explain obsolete expressions by means of contemporary words. Mention also should be made of a four-tablet series, likewise extant in Sumerian prototypes, that lists designations of human beings, officials, professions, craftsmen, social classes, and deformed and crippled persons. A similar list exists with the names of the major and minor deities of the Mesopotamian pantheon.[7]

Nearly all Assyriological work on the topical series was, and is, concerned with the reconstruction of the texts on the basis of fragmentary copies, school tablets, and similar sources, as well as with the utilization of this mine of information for the purposes of Sumerian and Akkadian lexicography. The internal history of their composition—that is, their growth by accretion caused by the general Mesopotamian preference for additive elaboration—and the internal logic of their sequences have, however, hardly been touched upon.

In view of the passages from the Old Testament and from the Egyptian onomastica cited above, the Mesopotamian lists appear to be the product of two needs: that for training the scribe and maintaining his bilinguality (considered proof of scholarly status), and that for organizing, classifying, and defining the phenomena of nature in an established order, ranging from the topography of the home country to the stars and embracing animals, plants, and minerals and whatever man was able to produce from them.[8] Unlike the Egyptian onomastica, which are an isolated literary phenomenon, the Sumero-Akkadian word lists are part of an extensive body of similar literature.

Although the sequence of entries in the lists often cannot be explained, and the number of entries assigned to a specific topic (a plant, a stone, an animal, a man-made object) seems to vary according to the practical or the emotional interest it arouses, attempts toward classification are repeatedly in evidence. Such contrasts as male versus female, native versus foreign (often with the names of countries), and domestic versus wild are stressed; colors, and qualitative and quantitative differences, are listed (although sometimes only schematically). At times (see Hh III 37, 216; XVI 151 f.) two different plants, or stones, are listed side by side to fit an atypical entry into existing classifications.

It cannot and should not be claimed, of course, that the word lists containing, for example, the names of plants, animals, or stones constitute the beginnings of botany, zoology, or mineralogy in Mesopotamia. They are not a scientific (not even a prescientific) achievement; rather, they result from a peculiar interaction of a genuine interest in philology (or, at any rate, lexicography) and a traditional Near Eastern concern for giving names to all things surrounding the scribe, thus linking nature to man.

Apparent sequels to the topical lists are the somewhat more elaborate descriptive texts that, likewise in a very standardized way, concern similar subjects. They exhibit a very characteristic pattern in each line: "a stone/plant/snake that has the looks of the . . . stone/plant/snake but has . . . [certain qualifications follow] is called the . . . stone/plant/snake." In these texts the object to be identified is compared, as a whole or through its specific parts (specks, leaves, flowers), with a better-known object by pointing out both similarities and differences; then its name is given. The description of the object often reveals a keen interest on the part of the observer in details of stones, plants, and snakes. Unfortunately, the extant tablets of this type of text are quite fragmentary.[9] Their purpose, however, is clearly stated (in *Materials for the Sumerian Lexicon*, X [1970], p. 68 vi 17 f.): to help the user identify the objects.

Another listlike arrangement of names of plants, minerals, insects, and other medicinal substances should be mentioned here, although the purpose of the composition cannot definitely be established. This is the series Ú u r u . a n . n a = Ú *maštakal*, which seems to enumerate in its two juxtaposed columns the entire range of the Mesopotamian pharmacopoeia. There is no overt indication of the relationship between the two columns, and it has been suggested that the materia medica in the right column might serve as a substitute for those in the left.[10]

Let us turn from the scribe, who, as a means of relating himself to nature, classifies and lists the world around him (thereby fashioning a screen that prevents its immediate perception) to the poet and his relationship to nature. The poetic interest of Mesopotamian literati clearly is attracted more to the impact on man of nature's awe-inspiring destructive forces (fire, storms, flood) than to natural phenomena that appeal to detached observation, be it unselfish interest or admiring curiosity. Man and his works under the impact of the elements are central to the poet's concerns. To a certain extent such preference is created by the tenor of the texts in which poetic imagery based on the observation of nature is used; it is by such comparisons that royal inscriptions, as well as hymns in praise of the major deities, strive to render the *tremendum* of gods and kings alike.[11]

A rare and atypical document such as the letter written by Sargon II of Assyria (721–705 B.C.) to the deities and the citizens of the ancient capital Assur should make us realize, however, that the marvels of nature themselves appealed to the poet under certain conditions. Through the quite unconventionally wide range of interests of the writer of this text, for whom foreign mores are as worthy of description as are the activities of his king during the spectacular raid into Urartu and the sack of the enemy's capital, we learn above all how the wild mountains through which Sargon penetrated impressed the observer.[12] The thickness of the forests (line 16), which even the light of the sun could not penetrate, the thunder of the cascading waterfalls (line 326), the repeated crossings of the rushing mountain streams (line 17), the dizzying chasms (line 21), the snow and ice (lines 100 f.) of the mountaintops towering over each other (line 32), and the sweet smell of the vegetation (line 28) are described by a poet-scribe sensitive to such impressions. His power of observation is directed not only at the grandeur of the scenery, but also at its smallest details. He produces the unique simile that compares the defeated king's fear with that of a partridge fleeing, with palpitating heart, from an eagle (line 149).[13] He reacts to the beauty of a fertile garden, and admires the architectural and agricultural achievements of the enemy (lines 200–212 and KAH 2 141). In short, we are allowed a glimpse into a world of experiences and attitudes that the severe formalism of this genre of literature, and the tyranny of a traditional and restricted imagery, do not as a rule tolerate.

The stance expected of a poet in creative Mesopotamian literature is a distinct disengagement from his oeuvre. This disengagement breaks down so rarely that the few instances we are able to cite as exceptions only serve to underline this rigidity. The famous prayer to the Gods of the Night, with its sophisticated description of the silence of the starred night and its solemn address to the gods, seems to have been well appreciated by the ancient scribes and was repeatedly used by them.[14] The short threnody spoken by a dead wife, touching in its unusual wording, remains, however, unique in Akkadian.[15] In both instances the empathy is on the human level; solely the *topos* of the lonely reed cane or tree, in the isolation of which the poet sees his own loneliness and misery, attests to a link of sympathy between a human being and a part of nature.[16]

Beyond the limits of the urban settlement, the fields and gardens around it, and the far-reaching steppes with their few tracks, and beyond the river or canal that brings the water for irrigation, floods, and traffic, there lies, for Mesopotamian man, the *oikumene*, the totality of the inhabited regions with their many tongues, capitals in which kings rule, and fortresses that guard border points.[17] Those speaking different languages are not considered barbarians,[18] nor are their customs looked down upon,[19] with the exception of the Martu tribes and other groups living in the desert.[20] Not language, therefore, but social practices separate the civilized from the uncivilized. The borders of the *oikumene* are formed by the seas that seem to surround it, although in the direction of the Persian Gulf there was very early contact across the sea with islands and far-off coastal regions.[21]

The need to import into Mesopotamia such essential raw materials as timber for a wider roof span for the abodes of gods and kings, metals for war and peace, and stones for utilitarian and decorative purposes, led to the creation of a series of mythographic terms. The fact that most of the metal, stone, and timber had to come through the passes of the Zagros Mountains gave rise to the assumption that the source of these materials were specific "mountains" in far-off regions. Thus the royal inscriptions and even the geographical lists speak readily of the Gold Mountain,[22] the Cedar Mountain, and the Lapis Lazuli Mountain. These enumerations provide us with a mixture of names of distant mountains in foreign languages, trade names for specific stones, and mythical mountains where certain deities are said to live. A late litany contains the names of mountains yielding gold, silver, copper, and tin (no Iron Mountain is ever mentioned), precious stones and millstones, cedar, boxwood, juniper, oak trees and fruit trees, and even mountains belonging to deities of the Mesopotamian pantheon or inhabited by foreign peoples (Elamites, Gutians, Subarians, Amorites, Lullubi).[23] Thus, the mountain chains that surrounded the Mesopotamians in the northeast-northwest quadrant were thought of as full of treasures and inhabited by hostile tribes, but also as the home of certain of their own gods. The deserts beyond the steppes held more terror for them than the mountain regions. They knew the desert tribes and their strange way of life (tents,[24] camels, the scarcity of water), as we know best from Assurbanipal's record of campaigns against the Arabs.[25] Only rarely do such descriptions speak of imaginary animals encountered, such as the two-headed snakes mentioned in the damaged report of Esarhaddon (680–669 B.C.) that describes in considerable detail a journey to Egypt through the desert.[26] The high seas are avoided by the Mesopotamians. With obvious pride Tiglath-Pileser I (1115–1077 B.C.) reports that he harpooned a killer whale in the Mediterranean Sea;[27] still, there is no evidence of a belief in fabulous sea monsters, as there was in Egypt, Ugarit, and the Old Testament.[28]

After having surveyed what literary texts and lists tell us incidentally about the regions at the border of the *oikumene*, a unique and difficult scholarly tablet must now be adduced. The scribe's name is broken, those of his father and family do not appear in any other document, and the tablet itself is badly damaged.

After a short (eleven-plus lines) introduction, the document shows a not very carefully drawn *mappa mundi* on its obverse. The reverse comments on that map in some twenty-three lines, which are followed by the "title" of that oeuvre: "[(these are) the region]s of the four rims of the entire u[niverse(?)]; the interior of the [. . .] nobody knows."[29]

The map shows a circle surrounding the *oikumene*, which is conceived as circular, surrounded by a body of water that is identified by the words ÍD *marratu* written four times. This designation is probably a double entendre, meaning "salt water river" as well as "circular river."[30] Inside the circle the following conventions can be observed: parallel lines represent watercourses, small circles (sometimes with a dot in the center), cities and countries; a large rectangle must denote Babylon. An oval formed by a curved line at the upper inner rim of the periphery is inscribed *šadû* ("mountain[s]"). It can safely be assumed that the river drawn as flowing out of the mountain region, bisecting the rectangle of Babylon, represents the Euphrates. Downstream from Babylon, the end of the river is marked by its branching into a watercourse of the same size that, to the left, is inscribed with the word *bitqu* ("[small] canal") and, to the right, with the word *apparu* ("swampland"). Before the branching point a horn-shaped arm seems to connect the Euphrates with the swamp. Unfortunately, the signs inscribed in the "horn" are damaged.[31] The lower right quadrant is broken. In the upper right quadrant four circles appear; they are inscribed as follows: City *Ú-ra-*[*x-x*]; Land of Assyria; a point; and City of Dêr. The arrangement fits the identification of the river as the Euphrates, and shows that the top of the circle is oriented toward the northwest. On the left side three cities are indicated: the first is called Habban, the second has a point in the center and no subscript, and the third has the subscript Bīt-Ja'kinu. The position of Bīt-Ja'kinu on the lower Euphrates is correct, but that of Habban poses a problem. A region of that name is known to be located east of the Tigris, toward the foothills of the Zagros.[32] Possibly we have here another of the not-too-rare instances of topographical homonymy in Mesopotamia. In what is left of the lower half of the circle there is the symbol of a city, the name of which is lost in a break.

The occurrence of the geographical names Habban and Bīt-Ja'kinu, and possibly also the representation of Babylon lying on both banks of the Euphrates, suggest that the tablet was written in the first millennium B.C., most likely in Babylon.[33]

But the geographer did not let his map end at the *okeanos*-like circular river. He drew on its outer edge several isosceles triangles; as far as they are preserved, they had written in or near them the words *nagû* ("district, province, region") and, underneath, *n bēru ina birīt* ("*n* double hours [distance] in between"). The triangle at the upper right has an interesting addition: "where the sun is not seen." This, of course, calls to mind the passage in the *Odyssey* (XI, 14 ff.) that refers to the land of the Cimmerians, who live at the rim of the world, where the sun is never seen.[34] The Mesopotamian and Greek traditions reflect travelers' tales about the polar night.

Only five of the triangles are extant, and neither their spatial arrangement nor the distances between them show any regularity. Since the commentary on the reverse of the tablet has eight sections, it seems to follow that there were originally eight such triangles on the map.[35] The purpose of the commentary was to give additional information about these eight mysterious regions beyond the *okeanos*. The above-mentioned addition to the upper-right *nagû* (and whatever additions may be lost in the break) is not repeated or alluded to in the text of the reverse. The commentary follows a definite pattern for each of the regions.[36] Each entry begins: "When you go to the first/second/ . . . region, [the distance(?)] is seven double hours." Unfortunately, none of the descriptions is complete, and nearly all are too damaged to be of any value. Most are short—one line each for the second to the fourth and the sixth to the eighth *nagû*; the description of the fifth is nine lines long, but badly broken. Names of animals are mentioned, and figures and measures are given; but only here and there can a phrase be made out. Thus, it is said of the third region: "[even] the winged bird cannot comple[te the journey(?)]"; and of the seventh: "bull [shaped] [. . .]s that are provided with horns [. . .] they run about and chase [. . .]." The last passage suggests that monstrous creatures were thought to live at the rim of the world, a notion well-known from the Greek world.

It cannot be ascertained to what extent this document (or certain sections of it) represents traditional views or expresses the insight of just one individual scholar. In their damaged state, the introductory lines yield too little information. They speak of "the abolished gods that live on islands in the sea,"[37] and also of the flood hero Utnapištim, and of King Sargon of Akkad and his adversary Nūr-Dagan, all located at or concerned with the very rim of the world,[38] as well as of wild and foreign animals, even of [people(?)] with birds' wings.[39]

It is rather difficult to draw something like a cognitive map of the universe as Mesopotamian man conceived of it beyond the confines of the *oikumene* that was known to him either directly or by hearsay. An important source of information for such an endeavor would have been cosmological and cosmogonic tales. Although a good number of such stories are preserved, they do not, however, shed much light on cosmography, which is my interest in this context. Not only does man occupy the central position in such stories, he is practically their unique concern. Theogony and cosmogony are but prolegomena to human history, meant primarily to establish man's nature and function in the social and moral order of the cosmos.[40] The organizing activities of the creating deity focus on the temple and the city, and their economic requirements, such as water, fields, and workmen. There is hardly a trace here of the universal concern in the Creation story of the first chapter of Genesis that reaches from the luminaries of the sky to all plants and all living beings.

An important but isolated exception is offered in the fifth tablet of the Babylonian Epic of Creation (*Enūma eliš*), which deals in its first sixty-six lines with the organization of the celestial sphere, including atmospheric phenomena.[41] This section contains two parts. In lines 1–46 we learn about the installation of the stars in certain places to determine the length of the year; the creation (*šūpû*) of the moon to fix the length of the month; and—in a very damaged group of eighteen lines—the creation of the sun to establish the boundaries of day and night. In the second part (lines 47–66) Marduk fashions heaven and earth out of the corpse of the slain Tiamat, using her head—out of her two eyes come the rivers Tigris and Euphrates – her body (*kabattu*), her udders (see n. 64), and her tail.[42]

The organization of the stars and constellations, the assignment of certain stars to the twelve months of the year, the role of the polestar, and the paths of the planets along the ecliptic are described in a very succinct way and in technical terms that are difficult to understand.[43] The creation of the moon is couched in a long command (lines 15–26) addressed to the moon god (Nannaru) and concerns shape and timing of the phases of the moon and their relation to the sun. The passage dealing with the sun is almost completely destroyed. It is obvious that all the luminaries of the sky function primarily for the sake of man; they establish the calendar in order to allow him to organize his time – and, more important, his work. In other words, the regularity of agricultural and sacral activities is the sole concern of the gods.

In the verses on the creation of atmospheric phenomena (lines 47–52), Marduk fashions the rain-bearing clouds[44] from the "spittle" of Tiamat and reserves for himself the power to make the wind blow and the rain fall, and to create cool weather and fog.

The badly preserved final lines turn to the completion of the cosmos. They speak of the *durmāḫu*, the Great Band with which Marduk ties together heaven and earth,[45] using for this purpose the tail of Tiamat. Marduk also props up the sky and places it as a roof (*ṣullulu*) over the earth.

This grand vision of a tightly structured and functional cosmos, with its interest in celestial matters, seems to represent the concept of an individual scholar-poet rather than a living tradition. About the latter we learn much more from pertinent passages in religious and literary texts that mention either the marvels of heaven or the secrets of the netherworld only incidentally. As can be expected, they exhibit considerable variation, due partly to secondary elaboration and partly to divergent local traditions. They are mentioned here to give some insight into Mesopotamian cosmographic speculations as one aspect of Mesopotamian man's relation to the world around him.

The opposition heaven-earth is basic in Mesopotamian cosmography, although Sumerian theogonic speculations posit an original sexual union between the two, whereupon the earth gave birth to gods, mankind, and animals.[46] The Babylonian Epic of Creation locates the primeval couple (Apsû and Tiamat) in the lower world as progenitors of the gods, including Anu (the sky god).

Over the solid and stable earth (*erṣetu, ammatu, dannatu*) that extends to the shore of the salty sea, the sky (*šamû, irmeanu*) is spread at unreachable height.[47] It is conceived of as a vault[48] of which the top is called *elât šamê* ("top/crown of heaven"), and its lower part *išid šamê* ("base/root of heaven"), a term that in astronomical contexts refers to the horizon.[49] Heaven and earth come together at a cosmic structure called *šupuk šamê* (Sumerian UL.GAN), probably conceived of as a dikelike structure upon which the base of heaven rests. This zone is important for the astral deities, because through it they enter and leave the sky to manifest themselves to mankind. This is clearly stated in a bilingual incantation (CT 16 19:54 ff.): "Enlil considered the matter and took counsel with Ea and they assigned the *šupuk šamê* to Sin [the moon], Šamaš [the sun] and Ištar [the plant Venus] to organize [it]." It was assumed that these luminaries had to pass through the structure. This is well attested for the sun god, for whom Marduk installed two gates, in the east and in the west (Epic of Creation V 9), but it is not attested for the moon. Of Ištar it is said that she rises heliacally from the *šupuk šamê* after opening the "bars of heaven."[50] Whether the "gate of the great gods" through which the "Gods of the Night" make their appearance refers to this or another entrance to heaven remains uncertain,[51] as does the unique Neo-Assyrian oracle passage in which we read of Aššur: "Leaning down from the gate of heaven I have heard your cry of distress."[52]

The most elaborate description of the sun's gate comes from the ninth tablet (ii 1–8) of the Gilgameš Epic. There the sun is said to enter and leave heaven every day through a mountain called Māšu that reaches up to the *šupuk šamê* and down to the netherworld. This mountain has a gate guarded by two awe-inspiring monsters that combine human and scorpion forms; they are male and female. The use of the same gate for the rising and setting of the sun is difficult to understand (lines 3 and 9), especially because the gate is said to be at the head of a long tunnel – twelve double hours – through which, one has to assume, the sun passes during the night from its setting to its rising point. The iconography of Old Babylonian seal cylinders, often showing the sun god stepping up from between two mountains and brandishing the key to a gate, suggests that the imagination of the poet was influenced more by the idea of a sunrise mountain than by cosmographical considerations.[53] At any rate, there is evidence that the sun was thought to pass the night in the subterranean realm of the *apsû* (É. NUN, Sumerian agrun), where, probably, the stars also stayed during daytime. (See Richard I. Caplice, "É. NUN in Mesopotamian Literature," in *Orientalia*, n.s. **42** [1973], 299–305, esp. 304 f.)

It is often said in Akkadian texts that high mountains touch the sky. In a way this concept is fused with that of mountaintops inhabited by deities, which is well attested for Mesopotamia. Apart from such purely mythological localities as ḫ u r . s a g and k u r in Sumerian texts,[54] we find, for example, in the late litany cited in note 23 of this article, a mountain of Enlil (called Mt. Sâbu) and a mountain called Lil-mun of a storm god of uncertain identity (ᵈIM). Moreover, the letter of Sargon describes a mountain in Urartu as "higher than the mountain, the dwelling place of the goddess Bēlet-ilī" (line 18). We also know that Aššur loved Mt. Ebih (not far from his city), where Tukulti-Ninurta I (1244–1208 B.C.) built a sanctuary for him.[55] Much more evidence for such mountains comes from the West; Mt. Casium, Mt. Sapan, Mt. Hermon, and others are considered the abodes of important deities.[56]

Mountains, moreover, seem to have provided an access to heaven. We learn from the version of the mythical story of Nergal and Ereškigal (STT 28 v 13′ and 42′) that the messenger of the netherworld ascended the *simmilat šamâmi* to arrive at the gate of Anu, Enlil, and Ea, probably the entrance to the heavenly mansion of the great triad. The word *simmiltu* may mean ladder as well as stairs; here, however, stairs would be more appropriate, since the mountains towering over each other appear to be steps of a giant stairway leading up to heaven, the abode of the gods.[57] The fact that diseases are spoken of in an Old Babylonian conjuration (*Journal of Cuneiform Studies*, **9** [1955], 8 A:10) as descending from the "temple tower of heaven" (*ziqqurrat šamê*) confirms this interpretation, as does the poetic expression for mountaintop—*ziqqurrat šadî*—in the Gilgameš Epic (XI 156).[58]

In daytime the surface of the heaven shows—in addition to the sun god moving across the sky with his horse-drawn chariot[59] and his driver—clouds in many colors and shapes.[60] When clouds covered the entire sky, they were called *nalbašu*, as was the fleece of a sheep; therefore the north wind clearing the sky of clouds was, for the poet, the *gallāb šamê* the "shearer of the sky" (Maqlû V 85). The clouds, filled with rain, release their content—despite the claims of Marduk—upon the command of the storm god Adad, who may also withhold rain or use it for destructive purposes when angered.[61] Magic means by which to produce rain are not mentioned,[62] and diviners prognosticated rain only rarely.[63] Rain is also thought, although only in highly literary contexts,[64] to come out of the udders of the primeval monster Tiamat, as do dew and evil spirits. Interestingly, there are traces of a different concept concerning the origin of rain. In a letter, addressed probably to King Sargon II of Assyria (721–705 B.C.), an administrative officer reports that on the fifth day "after the pegs of the rain had been released [? *ú-sa-ri-a*] [it rained] all night and the entire following morning" (ABL 707:5). This suggests the idea that rain was somehow stored in heaven, perhaps in waterskins closed by wooden pegs, and was released by removing these pegs.[65]

At night the sky exhibits the "heavenly writing" (*šiṭir šamê*), that is, the stars in their orderly and lasting arrangement (*riksu*). A late bilingual text speaks of the stars moving side by side (*sunnuqu*), as if in furrows, across the sky.[66] Only late texts speak of the polestar as *markas šamê* ("center of the sky");[67] in earlier texts a "crossing point" (*nēberu*) of uncertain location in the sky is mentioned.[68] The multitude of stars is organized in constellations the names of which appear in early lists;[69] special stars are singled out, and their heliacal risings and settings are used to establish the months of the year.[70] The stars used for dating were called *lumāšu* stars, a term that in the late first millennium was applied to the zodiacal constellations that came into use only then.[71] Apart from these stars and the Milky Way ("river of heaven"),[72] five planets were distinguished; as a category they were called *bibbu* ("wild sheep").[73] The same word also refers to comets, although there is no direct evidence that any were sighted. The description of a comet appears in a text with celestial omens and runs as follows: "the star which has a coma in front and a tail in back."[74] Shooting stars appear quite often in divinatory texts and conjurations.[75] They are also recorded as a prodigy when very numerous.[76]

Although certain ritual acts had to be performed "under, or before, the stars," there is evidence that evil forces (diseases, demons) also were thought to come down from the stars to attack man.[77] Dew was believed to be produced by the stars, probably because it appears during clear and cool nights.[78]

All atmospheric and astral phenomena take place on the visible surface of the heaven (*pan šamê*). Beyond is the inscrutable realm called *qereb šamê* ("interior of heaven"), the abode of the gods. There they dwell or return from stays in their sanctuaries.[79] Only one text, KAR 307, offers a scholar-poet's vision of the entire cosmos.[80] It describes the nature and organization of the hidden heavens in the following way: "The uppermost heaven is of *luludānītu*-stone and belongs to the sky god Anu; the 300 Igigu gods reside in it. The middle heaven is of *saggilmut*-stone and belongs to the [other] Igigu gods;[81] the 'Lord' resides in it on a sublime dais, he resides in it on a dais of lapis lazuli; he made it resplendent like *būṣu*-glass and crystal.[82] The lower heaven is of jasper [?] on which are drawn the divine constellations [reverse lines 30–33]." This tripartite division of the upper world, which corresponds to a similar structuring of the lower world in the same text, suggests careful consideration of the status and function of the gods. The lower heaven shows what is visible to mankind; the middle heaven, the splendor of which the poet obviously stresses, harbors the "Lord" ruling earth and mankind; from the upper heaven Anu, exalted but aloof, as is his wont, presumably rules the entire cosmos.[83]

All three levels are thought of as made of precious stones, a concept for which there is an isolated parallel in the Old Testament: "and I saw the God of Israel and there was under his feet as it were a paved work of sapphire stone, and as it were the body of heaven in his clearness" (Exod. 24:10).[84] That the jasper [?] of the lower level was a greenish or bluish stone, the color of the sky, is also indicated in the series *abnu šikinšu*, which describes that stone as "looking like the clear [variant, "faraway"] sky."[85] The stone of the middle heaven, *saggilmut*, probably has a similar color,[86] while the stone of the upper heaven, *luludānītu*, seems to have had a marblelike texture; the series *abnu šikinšu* describes *luludānītu* as having black, red, and white veins.

The dais of lapis lazuli and other precious material assigned to the "Lord" is reminiscent of the golden chamber (*massuku ša hurāṣi*) in the inner heaven, mentioned in an Assyrian oracle as the place from which the goddess Ištar promises to watch over Esarhaddon (680–669 B.C.).[87] It is probable that all the major deities had abodes of their own in that part of heaven.[88]

For those who live "under the sky" (as it says in the Hattušili Bilingue),[89] the area beneath the surface of the earth consists of several distinct regions, such as the realm of the dead, the reservoirs of the subterranean fresh water that causes the rivers to swell and to inundate the arable land, and a passageway for the sun god to use at nighttime to return to the sunrise point. All was surrounded by the sea. According to a rarely mentioned tradition, the earth, and perhaps the entire world, was tied to a cosmic mooring post (*tarkullu*[90]), a floating but securely anchored structure.

It is very difficult to describe the topography of the netherworld. There the god Nergal and his queen Ereškigal reside in a palace with seven gates; and there is the realm of the dead, about the conditions and needs of which mythological texts, conjurations, and frequent allusions in other literary texts yield a complex picture of several, and at times conflicting, traditions. All this falls outside the concern of this article,[91] as do the traditions about the abode of demons and other nefarious phantoms that plague mankind and are said to originate in the netherworld.

It is to the realm of Ea (Sumerian, Enki), deep in the Apsû, the cosmic and primeval freshwater body,[92] that a number of persistent traditions ascribe the origin of god and man alike. Equally essential is Ea's role in the organization of the social and intellectual life of man. Unlike the other gods, who primarily demand service from man, Ea, on the mythological level (like Prometheus), intervenes for man again and again; on the legendary level, Ea repeatedly instructs man through his fish-shaped emissaries, the *purādu*.[93]

The scholar-poet who described the heavenly mansions also attempted (in KAR 307) to create a well-structured picture of the lower half of the cosmos, likewise seen as divided into three levels. The passage (reverse lines 35–38) is unfortunately quite broken, and my translation is based to a certain extent on reconstructions and guesses. Linguistic difficulties also abound. The poet begins, in my opinion, with the inhabited world: "He [Marduk] placed in secure folds the spirit-endowed [?] mankind[94] [upon the . . .] of the upper level of the 'earth.'" Then follows a reference to the Apsû: "He installed his father Ea in the [. . .] of the middle part of the 'earth.'" The last two lines are quite fragmentary and seem to me to allude to the rebellious gods relegated to the very depth of the abyss.[95] They seem to run as follows: "He does not forget the rebellion of his [. . .]s [and] imprisoned the Anunnaku-gods [?] [in the . . . of] the lowest part of the 'earth.'"[96]

The main motivation for Mesopotamian man's interest in keeping the manifestations of animal and plant life, the movements of the heavenly bodies, and other phenomena under close and constant surveillance was his hope to obtain from them timely warning of impending misfortune or disasters. In a way that is never explicitly stated or even hinted at, Mesopotamian man assumed the existence of an unknown, unnamed, and unapproachable power or will that intentionally provided him with "signs." This is at the base of what Seneca expresses so succinctly: "non quia facta sunt significant, sed quia significatura sunt, fiant"—that is, ominous events happen and ominous features present themselves because they are meant to convey meaning, but they do not convey that meaning because they become manifest.[97] The very necessities of everyday life focus man's attention on the atypical: it shocks and alerts him.

Forms of primitive divination are practically ubiquitous, as are certain elaborations on the folklore level. Good and evil portents are readily attributed by means of subconscious as well as linguistic (paronomastic) associations; and they are often preserved by various mnemonic devices, such as rhymes, alliterations, and numerical sequences. In Mesopotamia, in the first half of the second millennium B.C., such a folklore-level system of divination seems to have undergone expansion and elaboration, probably stimulated by the transfer from an oral to a written tradition. In a civilization where an extensive bureaucracy centered in temple and palace is geared to recording all activities, the desire might easily arise to retain in writing divinatory folklore sayings, individually or in sets.

From small collections united by topical connections, larger ones are bound to develop for practical or prestige purposes. Certain early tablets still show evidence for the growth of such a corpus.[98] This is again an example of the process of additive rather than structural changes that is evidenced in nearly all types of Mesopotamian literary production.[99] Eventually omens—that is, events or features and their interpretation—are more or less freely invented to complement certain sets, such as right-left, above-below, the traditional four-color scheme, and numerical elaboration, in order to increase the usefulness of the collection. The nature of the writing system and the calligraphic interest of the scribe also contribute to both elaboration and standardization, especially in the protasis, the part of the omen that describes what is observed.[100]

Once in the hands of scholar-scribes, these compendiums grew more and more complex and arcane. The preservation of the written text became important to the copyist, and this concern increased the philological difficulties, since a discrepancy developed between the scribe's language and that of the text he copied. Explanatory glosses and commented texts became necessary as divination moved completely into the domain of scholarship.[101]

Even the apodoses, the predictions, changed in the course of this process. Thus they often grant us insight into actual living conditions, into the gamut of expectations and apprehensions shared by everybody, from king to commoner.[102] In the second millennium texts the apodoses were often stated quite specifically and with much detail. In the first millennium they became increasingly standardized, apparently because they came to be viewed solely as either favorable or unfavorable. In this form, moreover, they were more amenable to hermeneutic manipulations, as one can easily see in certain Neo-Assyrian and Neo-Babylonian communications addressed to the Assyrian kings, where the diviner is at work as interpreter of the apodoses.[103]

All this holds true only in a rather general way. Each medium through which the unknown power activated signs (abnormal features inside the body of a slaughtered lamb, the birth of deformed animals and humans, the strange behavior of animals, atypical formations on plants, the movements of the planets) had its own history within Mesopotamian divination, as well as its own circle of practitioners and believers. Some techniques were long-lived, such as extispicy and birth omina; others went out of fashion, such as the observation of oil in water, of the movements of smoke, or dream experiences; others came to the foreground, such as the observation of celestial signs, of the weather, or of the behavior of animals. The king, the wealthy, and the poor made use of different divination methods. We, of course, know only of those that were fixed in writing and were popular enough to be preserved.[104]

Evidently each technique requires a different degree and range of actual observations of natural phenomena. We discuss here only those techniques that evolved a special terminology for exact descriptions. This will make evident the inherent limitations of this "prescientific" attitude toward facts observed.

The most complex and most consistent terminology was created by the diviners who searched the inner parts (*exta*) of a slaughtered sheep for deviations from their normal size, shape, texture, and coloring. They developed special terms for atrophies, hypertrophies, and other abnormalities; they freely used comparisons with well-known objects, and even offered graphic illustrations to identify the deformation with sufficient clarity. They also established a unified terminology to refer to all important parts of the animal's anatomy.[105] And yet they exhibited no curiosity about the role and function of these organs beyond the most rudimentary facts; after all, deviations from the normal state were considered neither congenital nor the result of diseases, but were believed to have materialized inside the slaughtered animal solely to provide a "sign," a message for the person for whom the sheep was inspected, or for his city and state. The prayers that preceded the extispicy in the second millennium asked expressly for propitious signs on the *exta*; in the first millennium the oracle god, Šamaš, was asked in prayers to "write" his decision on the organs as an answer.[106]

With a similar interest in observing nature, the divination-oriented "physician," the *āšipu*, scrutinized his patient to establish the disease affecting him and to predict its outcome.[107] We know of this from a considerable corpus of omens called "If When the *Āšipu* Goes to the House of a Patient."[108] The first two tablets deal exclusively with possible signs and happenings to which the *āšipu* has to pay attention while en route to the patient, because they may portend the outcome of the sickness. The main body of the compendium (tablets 3–40 and a number of unclassifiable fragments) is concerned, in the sequence from head to toe, with the patient's symptoms.[109] Observation is concentrated on the temperature of the skin and its coloring, on the breathing of the patient, on swollen or otherwise abnormal tissues, on the appearance of his excretions, and on a few other symptoms. Subjective complaints also are taken into account. Obviously, we can learn much from this omen collection about anatomic nomenclature and about terms describing normal as well as morbid features and functions of the human body. Blood vessels are carefully examined at several locations for their coloring and blood flow. These observations of the pulse, like all others, are made by the *āšipu* for the purpose of identifying "signs" that portend whether or not the patient will recover and when this will happen.[110] The signs are listed systematically in the compendium. No treatment is ever prescribed, however, in contrast with medical texts, which instruct the physician on how to proceed and what medication to use.

Thus we have the interesting situation that the observation of natural phenomena within a narrowly defined context—the human body—takes place on two distinct levels: on one, the observed features and phenomena are considered "signs" that indicate the nature of the disease; on the other, the same phenomena are observed by the physician, who then applies his treatment (medication, manipulations, even magic means) in order to heal the patient. Still, the two levels are not rigidly separated: the observer of the signs, once the disease is recognized as amenable to a cure, may well hand over the patient to the ministrations of a physician. From the point of view of scribal scholarship, however, the two approaches are carefully kept apart: the form and structure of the omen literature are adopted for the one,[111] and the pattern of technical handbooks for the other.

A similar case for the separation of the two levels on which nature is observed in Mesopotamia can be found in celestial observation. Most of the tablets of the large series Enūma Anu Enlil deal with omens derived from the moon, the sun, and the planets.[112] This compendium has a complex history, the roots of which at times go back to the second millennium. Its internal development is still problematic because of the large number of fragments at hand that come partly from Assyria (Assur and Calah) and partly from Babylonia (from Sippar to Ur).[113] The timing of the phases of the moon, the moon's relationship to the sun, and the eclipses of the moon receive much attention, as do the movements of the planets Venus (heliacal risings and settings), Jupiter, Saturn, Mars, and Mercury. The prognostications concern the well-being of the king and his country, harvests, floods, and so on. The terminology created for the identification of all these phenomena is not nearly as rich and diversified as that evolved by the experts in extispicy and physiology.

At about the same time, and certainly in the same region, the same celestial phenomena came within the ken of the observer's attention on another level. When the Mesopotamian diviners began seriously, perhaps even exclusively,[114] to read the "signs" offered by the heavenly bodies, and when religion-oriented poetry turned its emphasis toward stars and planets, the same phenomena, whose unpredictability made them appear carriers of meaningful messages, were subjected to a fully rational scrutiny. Their timing, especially, was observed and recorded[115] for the purpose of predicting the appearances and disappearances of stars and planets. Thus, on one level, the heliacal rising of a planet was observed and interpreted as presaging certain events; on another, scholars, who related to nature in an entirely different manner, made the same observation in order to establish or to test a pattern in the recurrence of the phenomenon. The mathematical methods used to calculate the parameters of the planets are not our concern, nor is this the place to present the development and achievements of Mesopotamian astronomy. However, it should be stressed that the scholars who computed parameters and predicted eclipses[116] were called Enūma-Anu-Enlil-scribes, as were the diviners who studied the omen compendium Enūma Anu Enlil to decode the messages conveyed by the same phenomena and who reported their interpretations to the king and other clients. As in the case of the observation of a patient, this situation underlines both the separation and the interconnection of the two levels on which nature was observed in Mesopotamia—the mantic and the rational.

The regularity of the celestial bodies was viewed as divinely ordained only insofar as they were related to timekeeping; this is clear from the verses of the fifth tablet of the Babylonian Epic of Creation. The movements of the planets (the "wild sheep"), however, were considered outside such limitations. When, in the middle third of the first millennium B.C., for reasons unknown, a new and different attitude of the observers prevailed and the order in planetary movements was discovered, this knowledge failed to suggest in Mesopotamia the concept of an orderly functioning cosmos (with all its moral consequences), as it did later in Greek thought.

What has been discussed so far makes it rather obvious that one should not speak of Mesopotamian astrology and astrologers, as is often done even by Assyriologists. In Mesopotamia the signs that appear in the sky are in their nature and function as relevant and meaningful as those produced by animals or plants on earth.[117] The Mesopotamian diviners observing celestial signs were no astrologers, if one takes the term astrology—as one obviously must—in its Hellenistic sense. In this sense, astrology presumes that celestial bodies affect this sublunar world directly, instantaneously, and irrevocably, because of their placement in the sky at a given moment and their mutual relationship. From the sky they release their influence, conceived of as a "power" that acts on such events as the conception or birth of a child, or any undertaking begun at the moment this power is released, if it is not modified or counteracted by another *dynamis* created by simultaneous and competing celestial events.[118]

Astrology[119] and Mesopotamian divination based on celestial signs are worlds apart.[120] One of the essential characteristics of Mesopotamian divination based on unprovoked signs (such as the birth of malformed animals, untoward events, or strange behavior of animals) is the belief that all evil portended by such signs can be effectively counteracted by a ritual act. Such ritual acts (*namburbû*[121]) are also used, as we know from contemporary letters, to avert unfavorable celestial signs, although no text of such a conjuration is yet known.[122] This clearly shows that celestial signs were considered on the same level as any other unprovoked sign; they had no special standing.

Another important but only spottily attested development in Mesopotamian divination methods may well have had a lasting influence on the history of astrology. Since the late Old Babylonian period, the concept of intrinsically favorable and unfavorable days has produced a type of text that Assyriologists call hemerologies. These tablets list every day of the month and indicate whether it is propitious or not for certain undertakings, such as a business journey, a marriage, or the performance of a religious act, or whether one should avoid certain or even all actitivies. Later texts dealing with the months of the year (menologies) tell their users about the prospects of undertakings planned during these months.[123] It is therefore not unexpected that the birth date of a child should be taken to indicate its fate. We know about this through a Hittite fragment found in Asia Minor, obviously a translation of an Old Babylonian divination text, that derives predictions from the date of a child's birth.[124] A small and damaged fragment (K 11082) in the library of Assurbanipal (668–627 B.C.) in Nineveh offers further evidence for the existence of this type of divination.[125] After a long gap, a new development of that old idea appears in a group of tablets from Uruk (Erech), datable to the Seleucid period (beginning 311 B.C.). Advice is given concerning the ritual and private activities to be carried out when an eclipse of the moon has occurred in a specific constellation of the zodiac.[126] Among the predictions is one that refers to the fate of a child conceived under such circumstances.[127] In another contemporary text from Uruk (TCL 6 14), studied by Abraham Sachs in *Journal of Cuneiform Studies* (6 [1952], 65–75), predictions are offered for a child born when the moon, the sun, or certain planets were rising above the horizon, or when planets, singly or in conjunction, were in a particular position. At times these predictions refer also or only to the parents or family of the child (reverse lines 7–12). Becaue the zodiac is not mentioned, Sachs insists that these are not "horoscopic omens" in the strict sense.

A group of real horoscopes on clay tablets, however, has been assembled and studied by Sachs.[128] They were written between 410 and 142 B.C. and in general contain the date of birth (sometimes even the hour), the name of the child or of his father, and the report on the positions at that time of moon, sun, and planets in the zodiac; predictions about the child's future usually conclude the text.

What we have here seems to be but a transfer to a new and more sophisticated level of the old Mesopotamian method of predicting the fate of a child on the basis of the date of its birth. In addition to the traditional lunar date, a series of synchronous astral events is now taken into consideration, probably from the information provided by pertinent compendiums of celestial omens. This method, which, according to classical writers,[129] was very popular in Seleucid Babylonia, seems eventually to have moved westward into Hellenistic Egypt and eastward into India and beyond.[130]

The parameters of the planets, established in Babylonia, were brought to Hellenistic Egypt in the course of the extensive exchange of scholarly information that took place during the last centuries of the first millennium B.C. They were used by Greek astronomers in a manner utterly alien to their Mesopotamian counterparts. What the latter regarded as a sequence of points in time (based on observations and projected into the future by computation), Greek thinkers explained by geometry in such a way that a mechanical model could be constructed to produce these "irregularities" automatically. The Greeks posited a universe functioning in time as well as in space, in a continuous and regular circular movement of the planets that, combined with the ingenious invention of secondary circles (epicycles), did what their philosopher is said to have demanded of the astronomer, that is, "to save the phenomena."[131] This resulted in the picture of a nonarbitrary universe, kept in motion by a divine power beyond the religious imagination of Mesopotamia, whether it is called divine love or—gravity.

To investigate the extent of Mesopotamian medical knowledge, it would seem best to establish the relation between diseases and the materia medica prescribed for them. Theoretically, this could shed light on the etiological thinking of the ancient physicians and help us to identify the inventory of their pharmacopoeia.[132] This approach is, however, fraught with difficulties. The botanical identification of the plants and herbs of which the roots, stems, bark, leaves, blossoms, and fruit (seeds) were used is extremely hazardous; and one has to rely too much on etymologies from cognate languages (where the designation is Semitic in origin) or on telltale names.[133] Names of diseases are only rarely transparent—such as the word for dropsy (*aganutillû*, "unending water," or *malia mê*, "full of water"), leprosy[134] (*saharšubbû*, "covered with scales"), itch (*ekketu*), fever (*humṭu*, *ṣurhu*)—or directly descriptive—for example, *miqit irrī* ("fall of the intestines") for prolapse of the rectum or hernia, *kīs libbi* ("stricture of the intestine") for an intestinal disease. There are many nontechnical terms that refer to conditions of the hair, skin, throat, or stomach, or to the abnormal functioning of the intestines or to breathing difficulties.[135] Some of these names make it possible to establish that certain herbs were used as laxatives, diuretics, or cough and stomach remedies. This might well help with their botanical identification by a botanist well versed in the history of pharmacology and at home in the flora of the modern Near East. A caveat should be added here: while at times the relation between remedy and disease seems discernible, at other times such diverse medication is prescribed that a reason for its use other than a knowledge of the specific effectiveness of a given herb for a given disorder must be assumed.[136]

No evidence from the rather numerous letters concerned with the sick and their treatment suggests that rare, and therefore expensive, materials were used for medical purposes.

Apart from herbs—and "herb" was the generic term for "medication"—minerals (such as salt, alum, crushed stones) and animal parts were used.[137] The ingredients were applied either dry (ground and sifted) or wet (soaked, boiled). Taken internally, they were swallowed with beer and other liquids, or were used in enemas and suppositories. Externally, they were applied in lotions and salves, or on bandages. The rarely mentioned medical instruments were scalpels, lancets, spatulas, and metal tubes; no mention is made of syringes for the frequently given enemas. Other instruments may well have been known; but the texts do not refer to them expressly, probably because their use was self-evident.

There was a certain admixture of what we call "magic" practices, even in therapeutic medicine. The preference for magic numbers (three, seven, and their multiples), the requests for both special timing and special persons (a child, a virgin girl, an old woman) for the preparation or application of the medication, as well as the recitation of an occasional conjuration, may seem to us only an intrusion of magic into medicine. The borderline between the two methods for relating man to nature—magic and protoscience—cannot yet be drawn, and any attempt to do it might distort and falsify whatever picture we may eventually obtain from our text material.[138]

In view of the folklore character of Mesopotamian medicine, it is not surprising that surgery plays no role, in contrast with the situation in Egypt, where it is so well developed.[139] The passage in the Codex Hammurapi (secs. 215–220), often cited as evidence of cataract operations, most likely refers to scarifications in the eye region for relief of certain diseases of the eye. Scarification—a common practice in Alexandrian medicine—could have endangered the patient's life through infection. And that is why the law intervened.[140] An Old Babylonian text[141] mentions rather casually a Caesarian section, apparently performed after the death of the mother. We know of such operations as emergency measures not only from primitive tribes but also from classical and Talmudic sources. Neither trepanation, nor excision of teeth, nor circumcision is attested from Mesopotamian sources, either archaeological or documentary. Direct references to castration[142] and midwifery[143] are quite rare in literary and secular texts. Herodotus' low opinion of Babylonian medicine (I,197) is not contradicted by what we know of it today; still, there is no evidence for the Babylonians' bringing the sick to the market to learn about remedies and treatment from passers-by, as he asserts.[144]

Besides the "practical" or "therapeutic" school[145] or tradition, there is a "scientific" one. Just because its lore is couched in the form of omens, we must not fail to realize that the compendium of these prognostic omens represents an achievement of Mesopotamian medicine.

Formally, the medical texts of the practical school represent a fusion of the pattern of omens ("if a person has the following symptoms . . . he suffers of the . . . disease—he will get well" or "he will die") with that of instructional texts, inasmuch as the physician is addressed in the second person in prescribing the treatment. This follows the protasis. The entries are topically arranged in collections dealing with specific diseases or particular symptoms and sometimes are also classified according to the affected parts of the body. Thus, they are useful "handbooks" for the practicing physician. To what extent—if at all—the codification of the medical art affected the development of the discipline can hardly be ascertained. Did the written formulations stifle any interest in observation, any desire to rely on actual experience, not to mention attempts at experimentation—as often happens when traditionalism is rampant? Or did another medical lore develop, independent of the written corpus that had become the domain of scholar-physicians—that is, of the scribes who copied and recopied the texts? Although no answer can be given to the last question, I would like to point out that in the abundant correspondence of the last Assyrian kings with scholars at court and abroad, omen collections are repeatedly cited by title, whereas medical advice is often given or mentioned in these letters without any reference to medical compendiums. This may be accidental; but it may also indicate that, at least at court,[146] medicine was practiced on the basis of experience rather than of textbooks.

A text type (KAR 203 = BAM 1 and many, mostly unpublished, fragments from Nineveh) may be mentioned here because it has the earmarks of a learned composition arranged for systematic, and not for practical, purposes. It has more than 160 lines, each listing first a medicinal herb, then the name of the disease for which it is used, and finally the way it is to be applied. Quite atypically, these latter instructions are couched in infinitives instead of in the second person singular.

In social position, the physician ranked with the diviner—as well as with the baker and innkeeper.[147] This means that when not attached to a court, he lived on payments he received for his services. He is said, at times, to carry a bag, probably containing herbs and certain instruments, a libation jar, and a censer; and he seems to have been attired in a characteristic way.[148] Specialized physicians were very rare; an eye doctor is mentioned once, in a late text.[149] Veterinarians are mentioned from time to time.[150]

Although certain deities with healing powers, such as Marduk and Gula, are called "physicians," no deified physician, such as Imhotep in Egypt and Asclepius in Greece and Rome, is known from Mesopotamia.

II

A systematic and critical inventory of the technological achievements of Mesopotamian civilization is needed to establish the degree to which Mesopotamian man succeeded in mastering nature, an essential aspect of the man-nature relationship in any civilization. This inventory will remain woefully inadequate, however, for two reasons: the paucity of the evidence available and the lack of scholarly interest in material culture.[151]

One would expect that the spade of the archaeologist, by now active in that region for nearly a century, would have yielded ample evidence not only of buildings and works of art, but also of the wide range of manufactured objects produced by the many craftsmen about whom we know from our texts. In contrast with the soil of Egypt, however, the Babylonian alluvium, and even the soil of the higher ground in the piedmont regions, has to a large extent destroyed such wooden objects, fabrics, leather, and bone utensils as the cuneiform tablets list and so often describe in detail. Human habitation in Mesopotamia is determined rather narrowly by such ecological conditions as availability of water for irrigation and transportation, and the need for protection from the often severe flooding. These conditions created a stability of location that, combined with the use of sun-dried bricks as the main building material, produced the phenomenon of "tells"—that is, hills consisting of well-layered debris accumulated during millennia of occupation.[152] Under such conditions the chances for the survival of smaller artifacts are slim, especially since such essential raw materials as metal and stone are scarce. Sanctuaries in particular tended to be bound strictly to their traditional emplacement;[153] their rapidly decaying mud-brick walls repeatedly required complete rebuilding from the foundation up. Thus, their maintenance—a sacred duty of all rulers—destroyed rather than preserved them and their furnishings. The nearly complete absence of prestigious tombs[154] and the paucity of the funerary inventory, typical of Mesopotamian burials, adversely affect the probability of the survival of both artistically important and everyday utensils, personal decorations, and weapons.

Another kind of archaeological evidence bears on the topic at hand—pictorial material. This may supplement what knowledge we are able to derive from actual objects, vestiges of constructions, and descriptions in texts. Sometimes iconographic evidence is nearly all we have for such artifacts as chariots,[155] wagons, thrones and other furniture, weapons, jewelry, and attire; for temples and fortifications, of which normally only the foundations are preserved; and for a number of animals. Indeed, the iconographic repertory retrieved from small seal cylinders and from palace reliefs and murals represents an important source of information. To interrelate representations, physical remains, and linguistic evidence is a task for many scholars.[156] Unfortunately, discrepancies in purpose, style, timing, and subject matter among these three "media" constitute obstacles that grow more serious as the questions that we are trying to answer become more specific.

An aggravating factor is that Near Eastern archaeology has always been, and to a certain degree still remains, a prestige activity of Western as well as Near Eastern nations, and accordingly is directed, consciously or not, toward museum objects. Only as scientific methods became increasingly refined did objects made of clay and nonprecious metals move into the ken of the archaeologist, as their value for dating and coordinating sites became evident. A similar shift in focus is now occurring as one comes to realize the importance of even minute animal and plant remains for the light they shed on a wide range of social and economic areas, such as trade, density of population, and the stages of the domestication of plants and animals. Much valuable evidence of that nature was destroyed or disregarded at a number of famous and crucial sites.[157]

Still, considerable evidence has been gathered from Mesopotamia and neighboring areas. It ranges from imposing mud-brick structures, reliefs, statues and other decorated objects made of stone, and vessels of metal, stone, glass, and clay, to smaller artifacts used as weapons and personal decoration. As a rule, evaluation and technical analysis of individual finds are restricted to their evidentiary value in an immediate context, their artistic excellence, and the provenience of the material. Systematic investigation and appreciation of the techniques used to produce such objects are rather rare in the vast archaeological literature. There are exceptions to this, such as the research done in architecture by the school of the German excavator Robert Koldewey, a number of studies in metallurgy[158] (oriented more toward the origins of the craft than toward the techniques employed), a study on brewing techniques,[159] and a study on glass and glassmaking. Of the other fields of technology, only ceramics and textiles have come under closer scrutiny.[160] Such ample and explicit evidence as human apparel has been left practically untouched,[161] notwithstanding all the statuary, reliefs, and murals that could serve as illustration, together with the abundant evidence provided by texts.

In the time span of nearly three millennia during which we follow Mesopotamian man dealing with his environment, we cannot ignore the question of possible changes in the raw materials (either native or imported) at his disposal and in the techniques and tools employed. Instead of superimposing a developmental scheme upon whatever picture one could piece together, I prefer in this respect to focus on differences created by region, period, and social context. Admittedly the lacunae that beset this type of research must be attributed to the accidents of survival; but even so, two levels of technological advance can be discerned: the level of "subsistence technology," which remains rather stable, evolving only in restricted areas and under special conditions, and that of "prestige technology," subject to more drastic, although sometimes short-lived, changes.[162] The evidence at hand—be it documentary or physical remains—does not illuminate these levels equally; that for the "prestige" level, bearing on the life styles of gods and kings, is far more explicit, coming from the archives of the "Great Organizations," temple and palace; in the same archives the only segments of the "subsistence" level that are recorded are those that pertain to the large-scale agricultural activities (and related manufacturing procedures) that formed the economic basis of the temple and palace "households." Many other segments of the "subsistence" level are rather poorly attested.

The Mesopotamian environment placed at the disposal of its inhabitants a limited number of native raw materials that increased but little prior to the invasion of the Iranian and Greek armies that brought about the end of the political (and cultural) independence of the region.[163] This event affected the "subsistence" level, whereas earlier, outside influences had extended solely to the "prestige" techniques.[164]

The basis of Mesopotamian subsistence was cereal agriculture, made possible in the rain-starved alluvium by irrigation and by the use of cattle for plowing, seeding, harrowing, and threshing. Sheep and goats provided a meat supply and fibers for textiles; the pig had only limited economic importance. Fish were utilized in several ways; ducks and geese were kept and, at least in the first millennium, fattened.[165] Other fowl were known, but their identification is doubtful.[166] The donkey carried loads overland.[167] The dog assisted the shepherd,[168] and the horse and the donkey are attested as draught animals.[169]

Ranking first among the raw materials available locally are clay, reeds, wood, and animal products (wool, hair, skins, bones, horns, and shells), supplemented by easily accessible stones[170] and the local bitumen.[171] Metals,[172] obsidian, amber,[173] stone beads, stones for milling, mineral dyes, and ivory[174] can be assumed always to have been brought into the alluvial plain via petty hand-to-hand trading, as gifts, or as loot. The gamut of techniques required to utilize these raw materials and those yielded by the domesticated plants constituted the basis of both subsistence and prestige production. Over time, the prosperity of the country and the warlike activities or fame of its rulers increased the quantity and the variety of metals, stones, and timber coming from foreign parts. Moreover, finished objects were imported as booty or gifts and made Mesopotamian craftsmen conversant with new techniques of production and decoration.[175]

The techniques invented,[176] adapted, or taken over covered three main areas: food technologies; shelter technologies; and the production of tools, furnishings, and personal attire and decoration.

Most diversified of these are the food technologies. They require special tools and appliances for preparing the soil, for seeding, harvesting, and milling grain, and for pressing oilseeds. They also utilize special methods for cooking, baking, and brewing, as well as containers and buildings for the proper storage of cereals.[177] All this concerns the two main staples, barley and wheat,[178] the latter appearing much more rarely than the former. Millet was equally rare. The identification of the main oilseed plant (sesame) still poses a problem. The most important fruit tree was the date palm, a cultigen in the region;[179] other fruit trees, such as apple, fig, and pomegranate, were relatively rare. Grapes for wine were grown mainly outside of Babylonia proper.[180] Among the vegetables we are able to identify, on linguistic grounds, are those of the genus *Allium* (onion, leek, garlic), chick-peas, lentils, cucumbers, lettuce, mustard, and a number of spice plants, such as coriander and *Ammi*. The proper preparation and preservation of food of animal origin (meat or fish) likewise require special techniques.

The essential materials used in housing technology were clay, wood, and reed; bitumen, stone, (colored) plaster, and even metals were used only on the prestige level. Earth or clay to build walls was treated in several distinct ways (from *terre pisée* to kiln-fired brick). Timber (native or imported),[181] branches, or reed and clay were used to make a watertight roof. Reed (bundled, plaited, or woven) served many purposes, mainly to create temporary protection and shelters for the indigent. Clay containers, baskets, and (rarely) stone vessels provided storage facilities; reed and wood, household furnishings. Metal mountings and inlays of polished stone and shells decorated the furnishings of the rich and of the temples and palaces.

The need for clothing, and for protection and adornment, required techniques that could change wool and hair into fabrics, skins into leather, and wood, metal, and stone into tools for defense and attack. Metal, reed, animal sinews, bones, horns, tusks, mineral dyes, shells, and colorful stones can be put to many decorative uses.

From the vantage point of the archaeologist, the evidence for this extensive complex of techniques is unevenly distributed. Architecture is best represented;[182] of the royal and religious prestige buildings, city gates and walls, and private houses, enough has been left undisturbed to permit us insight into the methods of construction, the functions of certain buildings, and the aspirations of the architect. For weapons, vehicles, furniture, tools, harnessing, clothing and ornaments, and musical instruments we have very few actual remains at hand.[183] Tools connected with agriculture, crafts, and other types of production were made predominantly of wood and have disappeared completely. Some information about these objects can be gleaned, however, from the iconographic repertory preserved on decorated containers, seal cylinders, murals, and reliefs. This repertory often adds another dimension to excavated buildings and shows clothing, weapons, harness, musical instruments, and other utensils in considerable detail, as well as the way in which they were used. Iconographic representations often provide revealing insights into warfare, hunting, and acts of worship about which texts, even when amply available, contain little factual information because of the extremely stylized form in which such events are usually described.

Three types of documents form the main sources of information bearing on technology: administrative texts, lexical lists, and technical manuals. Of course, any tablet—a private letter, a report on historic events, even a poetic text—can speak, either directly or through imagery, of craftsmen and their activities, materials, and tools.[184]

Clay tablets that record the issuing of raw materials to craftsmen and the delivery of the finished work concern, as a rule, those crafts that are of prime importance to the "Great Organizations" (temple and palace). Most of them refer to the textile industry, the brewing of beer, and certain basic agricultural activities. Others deal with the construction of boats, the making of bricks, and the manufacture of precious objects (metalwork, jewelry, furniture). Such an essential craft as that of the potter is hardly mentioned. Still, from texts of this kind we learn little more than the names of materials,[185] of craftsmen, and of finished products, and hardly anything about the technical processes applied. Only exceptionally is a work of art (if wrought of precious materials) described in details that are of interest to both the philologist and the historian of art.[186]

The word lists that enumerate basic materials and the products fashioned from them are quite informative but rarely specific enough to be of much use to students of applied science. One list, dating to the second half of the second millennium B.C., presents nearly one thousand items, many of them names of professions (see note 7), many more than we are able to allocate to specific crafts—a real *embarras de richesse*.

The best source of meaningful information bearing on Mesopotamian man's technical know-how is contained in a special text category that I call "procedural instructions." They are written in the second person singular, thus directly addressing the person who is to carry out the instructions. One set of documents deals with mathematical and astronomical operations or with expiatory and prophylactic ritual acts and medical treatments; another is addressed to what we call craftsmen.

The extant manuals refer to the following crafts: the brewing of beer; the dyeing of wool (to make it resemble purple-dyed wool); the coloring of stones (to look like precious stones); the making of certain alloys (to look like silver or gold). Apart from these isolated texts, there are instructions for the preparation of perfumes (attested in the Middle and Neo-Assyrian periods)[187] and the training of chariot horses (Middle Assyrian and Hittite).[188] The manuals on the making of colored glasses are best attested in a few Middle Babylonian and a large number of Neo-Assyrian copies; apparently there are also a few in Hittite.[189] These instructions, with the exception of those concerning brewing, belong to the category of prestige technologies; instructions for brewing beer are found in an Old Babylonian text from Harmal[190] and in a hymn to the beer goddess.[191] The hymn, which exhorts the beer goddess to make beer and gives her explicit instructions (in the second person singular), actually does not belong to the category of procedural instructions. It might be likened, rather, to an Old Babylonian text (UET 6 414) that addresses the fuller with similarly styled instructions on how to treat a piece of textile.[192] These documents are quite sophisticated literary works that happen to throw considerable light on certain crafts.

In a paper entitled "Mesopotamia in the Early History of Alchemy" (*Revue d'assyriologie et d'archéologie orientale*, **60** [1966], 29–45), I published two fragmentary clay tablets, one of the Middle Babylonian period from Babylon and one from the library of Assurbanipal, both of which are clearly "procedural" in character. The former instructs the craftsman on how to produce imitations of two kinds of precious stones, apparently by applying some kind of glaze to a base carrier; the latter concerns the making of a silver-like alloy from base metal. My reason for treating these two texts together, and under the title chosen, was that the same two techniques are listed side by side in a group of Greek chemical papyri of the second century B.C. found in Egypt.[193] My use of the word "alchemy" in the title, however, was as unwarranted as is the use of "astrology"—criticized earlier in the present article—in connection with Mesopotamian celestial omens.[194] Alchemy, like astrology, is a creation of the Greek genius in Hellenistic Egypt. It is based on the concept of transmutation, the discovery that matter is but a mixture of a restricted number of primary elements, a mixture that can, at least theoretically, be altered. Its roots go back to Aristotle, if not to pre-Socratic atomism. Still, it was correct to relate the two cuneiform tablets to the Greek papyri on the basis of their content. A recently discovered damaged tablet from Babylon dating to the first millennium B.C.[195] contains instructions for dyeing wool in several shades of the coveted purple. The same topic also appears in the Greek papyri mentioned above.

The manuals dealing with the making of perfume and colored glasses, and with the training of fast chariot horses, clearly have their background in the royal court.[196]

In two papers I have stressed the importance of the royal courts of the ancient Near East in the network of exchange of intellectual, artistic, and technological information between civilizations.[197] In fact, these courts provided the only channels through which such contacts could take place. Embassies with gifts were more consequential in acquainting one country with the artistic and technical achievements of another than was tribute or booty. Tribute, normally delivered annually, consisted mainly of raw materials and was therefore less important for the transfer of technical information. This special and ritualized form of commercial relations between countries moved essential metals, timber, domestic animals, and certain chemicals across borders. Even such finished products as were part of the tribute were merely standardized, "mass-produced" items, such as garments and metal bowls. Moreover, most of these objects were simply transferred from one governmental storehouse to another and functioned mainly as a stimulus to the industrial activity of the delivering and the receiving countries—that is, the workshops of the palaces.

Victory in war produced not only booty and tribute but also prisoners of war and deportees. Such persons were always in great demand in Mesopotamia because of the royal policies of forced urbanization and internal colonization. The preference was always for craftsmen, partly to prevent the defeated enemy from rearming or from returning to activities above the mere subsistence level, and partly to alleviate the dire need for skilled labor at home. Thus, the craftsmen who produced the objects normally imported through trade and diplomatic channels[198] could practice their skills on Mesopotamian soil.

The gift "circulation"[199] not only disseminated ideas—methods and forms—but also stimulated new demands. Individual pieces sent from king to king caught the fancy of the ruler and his entourage by their novelty, intrinsic value, or artistic quality. Sumptuous garments, elaborate jewelry, weapons, harnesses, and household furnishings[200] created something like a royal fashion all over the ancient Near East, setting a style of elegant living accepted everywhere. Every king prided himself on having the attire, the jewelry, and the fast horses that proclaimed his status. New things and ideas spread from the realm of personal decoration to that of living standards, leading to new architectural designs that changed the palaces[201] and their furnishings,[202] and may well have influenced religious concepts and literary forms as well.[203] The replacement of cuneiform writing on clay by a technique that used perishable materials (for writing Aramaic), however, has robbed us of this segment of the intellectual history of the ancient Near East.

While it is evident that such skills as the training of chariot horses, and the production of beautiful glass objects and sophisticated perfumes, have their place in royal courts,[204] another social milieu must be assumed for the production of materials that simulate the expensive—that is, that look like certain precious stones, or real purple-dyed wool, or genuine silver. These materials were no doubt produced for patrons who demanded but could not afford the style of living, if not of the court, then of those who were in some way connected with it. There seems to have developed in the early first millennium—this is the date of most of these instructional manuals—a small urban middle class for whom these substitutes for the precious were produced.

I have a final point to make about the socioeconomic background of Mesopotamian technology. Apart from the quasi-industrial crafts of weaving[205] and beer brewing, which were in the domain of the "Great Organizations," there were those that required a less trained personnel: the crafts of the smith, carpenter, leatherworker, potter, and mat and basket weaver. We know, if only spottily, about these craftsmen, from the records of the well-organized shops of the temple and palace. Smaller households probably did weaving and brewing as part of the normal routine. Other crafts in the Babylonia of the second millennium B.C. stayed within certain families, so that an outsider had to be adopted in order to receive instruction and to practice the craft. A change occurred during the next millennium: apprentices were accepted to be trained for a certain number of years, after which they presumably could work on their own;[206] even slaves are attested as apprentices and masters. It is difficult to establish the reason for such a change.[207]

The maintenance of the professional tradition was assured by the unchanging nature of the essential raw materials and tools used. Wood, clay, metals, wool,[208] and colored stones were not replaced by new materials, nor were the techniques used in working them affected by alien influence or the ingenuity of an inventor. The changes that happened after the collapse of Babylonian independence, when the country was suddenly exposed to influences from the West and from Iran (and central Asia), only underline the basic stability—not to say inertia—of Mesopotamian technology.[209] This holds true mainly for subsistence technology,[210] because prestige technology was always—and especially in the middle of the second millennium—ready for changes.

We can draw conclusions about the techniques used by these craftsmen only from their finished work, if available, or from linguistic evidence.

The most frequently applied method is using fire. Although, curiously enough, we still do not know how fire was produced in Mesopotamia, its technical utilization was very diversified. There was variety in the range of temperatures obtained and in the lengths of the processing. The installations are little known apart from domestic hearths and pottery kilns;[211] not much evidence can be gathered from texts about bellows and their use,[212] or the arrangements for controlling the access of air.

In food technology, fire had many uses: for boiling the essential gruellike dishes made of cereals, parching kernels, and baking the yeastless bread, as well as the preparation of vegetables and the rather infrequent meat dishes.[213] The extraction of oil from certain oilseed plants (especially sesame) likewise required the application of heat.[214] Moreover, the brewing of beer required fire for drying the malted kernels and baking the beer "breads." Ashes of certain plants yielded lye, which, combined with oil, served as a detergent.[215] The charcoal made for braziers and censers could, moreover, be used to obtain rather high temperatures in kilns.

The two techniques that rely on high temperatures are metallurgy and glassmaking. We are still ill-informed about metals and their treatment, despite a number of excellent products of Mesopotamian craftsmanship that have survived the ravages of time and men. They bear testimony to the competence with which these people worked gold,[216] silver, copper, bronze, iron, and even antimony, in various distinct techniques.

Goldsmiths, often working with jewelers, provided temple and palace with innumerable serving containers, furnishings, and decorative objects cast of gold,[217] and with wooden objects coated with gold sheets, often embellished with mounted colored stones.[218] The descriptions of royal palaces always use a limited inventory of stereotyped phrases,[219] but we have ample archaeological evidence for the first millennium B.C.; the temples that were built, renewed, or refurbished by the last kings of Babylonia must have been sumptuously decorated with gold and silver, from threshold to roof, according to their eloquent foundation documents, not to mention the paraphernalia of the cult. We may assume the same for earlier periods, even if such evidence is rare;[220] after all, the Babylonia of the last kings was to become the richest satrapy of the Persian empire.[221]

Silver was used in much the same way as gold, but also—and this creates complications—as a means of payment, especially in the first millennium. As such it was often alloyed with base metals,[222] a fact always indicated in contemporary texts in exact figures, although we do not know how this ratio was established or controlled.[223]

Copper and bronze, of course, were used most frequently for weapons,[224] tools, containers, and other items, as the long lists of such objects in lexical texts, administrative documents, and other evidence suggest.[225]

Iron objects, however, are rare, according to archaeological and philological evidence.[226] As to the treatment of iron by the smith—the ironsmith as specialist appears only in first-millennium Assyrian texts—hardly anything is known, for example, about the use of carbonization for improving the quality of certain tools. For the present investigation it is of interest that magnetism was recognized as a quality of certain iron ores.[227]

Tin,[228] lead, and antimony[229] were also used by the Mesopotamian metallurgists.

In general, it would appear that these craftsmen knew much more about the techniques they applied—their timing, effects, and limitations—than we are inclined to assume. This has become clear from the manuals dealing with the production of colored glasses. Although extant metalwork from Mesopotamia is hardly outstanding when compared with the discoveries made elsewhere in the ancient Near East, especially Egypt,[230] we must make allowance for lost evidence.

We have a level of technical information about glassmaking that is unparalleled in the investigation of ancient Near Eastern technology. Our texts not only allow us insight into a new chapter of this art, but do so from an angle not accessible for any other aspect of ancient technical knowledge.[231] We learn about special techniques, tools, and ingredients, denoted by an extensive technical vocabulary that we can trace in other documents and there discover hitherto unsuspected evidence for glass and glassmaking. Thus, the horizon of the investigation expands, leading to the discovery of cross-cultural technological contacts.

The main body of the glassmaking texts from the library of Assurbanipal in Nineveh shows clear traces that they represent at least two earlier collections of prescriptions based on different shop traditions that use distinct sets of technical terms for both materials and procedures. Philological indications suggest that the originals were written down in the last third of the second millennium, a time that saw the great flowering of glassmaking in Mesopotamia. The methods used are basically those of the preindustrial stage of glassmaking: silicate-carrying minerals and ashes of certain plants were crushed and heated together, then allowed to cool; the mixture was repeatedly ground and remelted until a high-quality glass was produced.[232] Sintering and melting were done in special types of kilns. Antimony compounds were added to produce opacity, since the goal of the Mesopotamian glassmaker was to imitate precious stones rather than to make transparent or even translucent glasses. Copper compounds were used as colorants to make the glass red or blue by either preventing or increasing oxidation, according to how the access of air was manipulated. In short, a consistent and well-thought-out procedure was applied to change the base ingredients into a new material of intense coloration and smooth texture that could be fashioned at will when hot and pliable or, when cooled, could be cut, polished, and mounted in precious metal like a stone.[233]

Philological and archaeological indications suggest that the home of this ingenious invention was Upper Syria.[234] At least, it is first attested in that region; there glass technology remained in a state of constant advance into the first millennium of the Christian era. From there it spread as a royal art from court to court, into Egypt as well as into Mesopotamia.[235] It is essential to note here that the glazing of ceramic beads, bowls, and tiles had been practiced in the entire ancient Near East from the fifth millennium B.C. on,[236] so that the change from glaze to glass that occurred in the middle third of the second millennium does not represent a technical breakthrough but an organic, if sudden, development. Still, the rise and spread of the new craft throughout the great civilizations of the region furnishes an impressive example of the degree to which the nature of an invention and the speed of its transmission are related to the socioeconomic stratum in which it materializes.

A final observation concerning the manuals for glassmaking is in order. In three places in the Nineveh collection (one in the introductory section and two in atypical prescriptions [sec. 13 and sec. L]) ritual acts are prescribed; all are quite similar in content and style, as well as in their timing: before the kiln is heated, certain sacrifices are to be made to deities related to fire. In evaluating this practice, one could suggest that the ritual acts performed pertained to the use of fire for technical purposes, and may have been enacted also by metalworkers and potters.[237] Or one could argue that the atypical prescriptions represent an older tradition—the bulk of the prescriptions dispensed with any ritual act—and that an editor relegated all ritual acts to the general introduction, expressing a tendency toward deritualization.

Fire was, of course, also applied to produce ceramic wares destined for storing, preparing, and serving foodstuffs. Most of these uses were on the subsistence level; and our information is quite inadequate, in view of the variety of techniques applied. Of bricks, sun-dried or kiln-fired, and their uses we know much more from excavated constructions, especially from those on the prestige level.[238] Still, Mesopotamian architecture makes use of kiln-fired bricks mainly for utilitarian purposes.[239] The early stamped-earth technique has left its imprint on the way walls are erected, with sun-dried bricks laid in mud mortar and completely covered with mud facing. Height is achieved at the price of thickness of the walls, and the size of wooden beams determines the width of the rooms (and hence the layout of the building).[240] In the constructions of the enormous temple towers, the Mesopotamian architects[241] knew very well how to deal with problems of mass and stress.[242]

Very little is known about the less prestigious crafts and about the work of the artists: the construction of furniture,[243] implements, tools, and weapons;[244] the application of dyes to fabrics; and the making of reed mats and ropes, to mention but a representative selection. There are also those artisans who engrave stamp and cylinder seals; use chemicals to etch designs on carnelian beads; cut and polish precious stones for containers, inlays, weights, and decorative pendants; design and execute statuary and wall reliefs; and paint murals.

The most complex items produced by the carpenter[245] are chariots and wagons,[246] boats,[247] plows,[248] and, on a smaller scale, furniture, certain musical instruments, and looms.[249] The number of words recorded in the Sumero-Akkadian word lists for the parts of all these implements attests to the complexity of their construction; metal, wood, bitumen, ropes, and leather were used ingeniously, as we know from administrative documents that record materials handed out to craftsmen.

There is an exception worth mentioning. We would know nothing about the working methods of the tanner were it not for a few passages in religious texts describing—or, rather, prescribing—certain ritual acts. For unknown reasons, these passages instruct the officiating person on how to prepare certain animal skins to serve specific ritual purposes.[250]

Quite early, many ways were discovered to protect animal skins from decay and to make them into that effective and lasting raw material of many and diverse uses, tanned leather. Pieces of apparel, containers of all sorts, coatings of wooden objects, and straps to reinforce constructions or to secure attachments were made of leather. From administrative texts we know that the Mesopotamian tanner used vegetable substances and certain minerals. The ritual texts tell us that the skins had to be soaked in several liquids, such as water, vinegar, wine, and beer, to which were added various types of flour and, at times, aromatic matters that often cannot be identified. Fats, mainly tallow and oils, were used to make the skins pliable.[251]

One must keep in mind that the nature of these prescriptions was ceremonial rather than practical; thus, certain stages of the tanning process, such as the dehairing of the skins and the scraping and beating of them with stones, are omitted. It is important to note that the chemical used was alum[252] and the vegetable tanning agent was called *ḫuratu*, while in second-millennium administrative texts a mineral called *alluḫaru*[253] and a plant written Ú.HÁB are mentioned in this context.[254] The change in tanning methods, like other developments in technology, seems to have occurred during the crucial centuries of the middle of the second millennium B.C. From then on, alum was imported in large quantities into Mesopotamia, not only for tanning but also as a mordant for dyeing wool, a technique that had become the fashion.

There is another tanning method, applied to goatskins that are tanned and dyed at the same time (to produce cordovan leather). In this process the skins become the color of the stone called *dušû*.[255] This is most likely a shade of red or orange, judging from textual evidence: in an Old Babylonian text a sun disk for a sanctuary is said to be made of *dušû*-stone, and a late medical text mentions *dušû*-colored urine as one of the symptoms of a patient. This type of leather, rarely mentioned in earlier texts, became so common in the first millennium that a new profession arose, that of the cordovan tanner, which is not mentioned in the traditional word lists (*ṣārip dušê*).[256]

We felt singularly fortunate when Leo Oppenheim cordially agreed to contribute the article on natural philosophy and natural history in ancient Mesopotamia. We were correspondingly distressed when word came of his death in 1974. It is fortunate that he had largely completed the article that we are privileged to publish. At Mrs. Oppenheim's request, in which the Board wholeheartedly joined, Dr. Erica Reiner consented to prepare the manuscript for publication. The Board would like to express our very profound gratitude for her generosity.

The Board of Editors

The final draft of this essay had been completed at the time of Leo Oppenheim's death. Although he would have made further changes or additions, I know specifically only of his intention to add a section on scales and weights. Rather than attempt to add anything of my own, I have left the present form of the essay, with its rather abrupt ending, unchanged. Of the additional material that he had accumulated after the final draft was written—mainly bibliographic references—I have made very selective and sparing use.

I am most grateful to my colleagues at the Oriental Institute whom Oppenheim asked to read the draft for their advice, and to Professor Edith Porada, Columbia University, for help with selecting the references that were added. Mrs. Oppenheim read the manuscript and contributed many helpful suggestions. For the editing for style, thanks are due to Mrs. Marjorie Cutler, Palo Alto, California, and to Mrs. Olga A. Titelbaum, Book Editor, Oriental Institute, who also assumed the task of verifying the footnotes and making them consistent.

E.R.
Chicago
April 15, 1975

NOTES AND BIBLIOGRAPHY

ABBREVIATIONS

ABL Robert Francis Harper, *Assyrian and Babylonian Letters*, 14 vols. (London–Chicago, 1892–1914).

ADD Claude Hermann Walter Johns, *Assyrian Deeds and Documents*, 4 vols. (Cambridge, 1898–1923).

Ai Benno Landsberger, *Die Serie* ana ittišu, which is *Materialien zum Sumerischen Lexikon*, I (Rome, 1937).

BAM Franz Köcher, *Die babylonisch-assyrische Medizin in Texten und Untersuchungen* (Berlin, 1963–).

BM Sigla of cuneiform tablets in the British Museum.

CT *Cuneiform Texts from Babylonian Tablets, &c., in the British Museum* (London, 1896–).

GCCI Raymond Philip Dougherty, *Archives from Erech*, Goucher College Cuneiform Inscriptions, 2 vols. (New Haven–London, 1923–1933).

Gilgameš Epic *The Epic of Gilgamish*, text, transliteration, and notes by R. Campbell Thompson (Oxford, 1929).

Hh Lexical series ḪAR . ra = *ḫubullu*, published in *Materials for the Sumerian Lexicon*, V–XI (Rome, 1956–1974).

K Sigla of cuneiform tablets in the Kuyunjik collection of the British Museum.

KAH *Keilschrifttexte aus Assur historischen Inhalts*, Wissenschaftliche Veröffentlichungen der Deutschen Orient-Gesellschaft, 16 and 37 (Leipzig, 1911–1922).

KAR *Keilschrifttexte aus Assur religiösen Inhalts*, Wissenschaftliche Veröffentlichungen der Deutschen Orient-Gesellschaft, 28 and 34 (Leipzig, 1919–1923).

KAV *Keilschrifttexte aus Assur verschiedenen Inhalts*, Wissenschaftliche Veröffentlichungen der Deutschen Orient-Gesellschaft, 35 (Leipzig, 1920).

KUB *Keilschrifturkunden aus Boghazköi* (Berlin, 1921–).

LKA Erich Ebeling, *Literarische Keilschrifttexte aus Assur* (Berlin, 1953).

Maqlû Gerhard Meier, *Die assyrische Beschwörungssammlung Maqlû*, which is *Archiv für Orientforschung*, supp. 2, Ernst F. Weidner, ed. (Berlin, 1937; repr. Osnabrück, 1967).

PBS University Museum, University of Pennsylvania Publications of the Babylonian Section (Philadelphia, 1911–).

R *The Cuneiform Inscriptions of Western Asia*, prepared for publication by Major-General Sir Henry Creswicke Rawlinson, 5 vols. (London, 1861–1884).

STT Oliver Robert Gurney, Jacob Joel Finkelstein, and Peter Hulin, *The Sultantepe Tablets*, 2 vols. (London, 1957–1964).

TCL Musée du Louvre, Département des Antiquités Orientales, *Textes cunéiformes* (Paris, 1910–).

UET *Ur Excavations, Texts,* Publications of the Joint Expedition of the British Museum and of the University Museum, University of Pennsylvania, Philadelphia, to Mesopotamia (London, 1928–).

VAS *Vorderasiatische Schriftdenkmäler der Königlichen Museen zu Berlin* (Leipzig, 1907–).

YOS *Yale Oriental Series, Babylonian Texts* (New Haven–London, 1915–).

NOTES

1. This has to be qualified somewhat. The grammatical texts mentioned by Alan Gardiner in *Ancient Egyptian Onomastica*, I (London, 1947), 4, n. 2, have their counterpart in the Old and Neo-Babylonian grammatical texts published by Richard T. Hallock and Benno Landsberger in *Materialien zum Sumerischen Lexikon*, IV (Rome, 1956), 45–207.
2. The tenor of Solomon's knowledge about trees and animals is indicated by the preceding line (1 Kings 4:32): "and he spake three thousand proverbs and his songs were a thousand and five."
3. See Albrecht Alt, "Die Weisheit Salomos," in *Theologische Literaturzeitung* (1951), 139–144.
4. For the entire text category, see A. Leo Oppenheim, *Ancient Mesopotamia, Portrait of a Dead Civilization* (Chicago, 1964), 244–249; and Wolfram von Soden, "Leistung und Grenze sumerischer und babylonischer Wissenschaft," in *Die Welt als Geschichte*, II (Stuttgart, 1936), 411–464, 509–557, esp. 433–436, of which there is a reprint ed. with additions and corrections, in Benno Landsberger, *Die Eigenbegrifflichkeit der babylonischen Welt* (Darmstadt, 1965)."
5. At times not the entire Sumerian entry, but only its qualifying adjective, is translated into Akkadian (for instance, tree + date palm + early ripening = early).
6. There are such lists with added columns of Ugaritic and Hurrian translations. See Jean Nougayrol, "Vocabulaires polyglottes," in *Ugaritica 5*, which is Mission de Ras Shamra 16 (Paris, 1968), 230–251. From the Hellenistic period come clay tablets with one column or both transcribed in Greek letters; see Edmond Sollberger, "Graeco-Babyloniaca," in *Iraq*, **24** (1962), 63–72.
7. For the series ḪAR . r a = *ḫubullu*, see *Materialien zum Sumerischen Lexikon*, V–XI (1956–1974); for the list of human classes, *ibid.*, XII (1969). There is no definitive edition of the list of deities as yet.
8. My position as outlined here differs in certain ways from that expressed in my *Ancient Mesopotamia*, 248.
9. For the stone text (*abnu šikinšu*), see STT 108 and 109, also K 4751 and BAM 194 and 378; for the plant text (*šammu šikinšu*), which, however, has an explicit pharmacological orientation, see Franz Köcher in BAM 4, p. xxvi ad no. 379; for the snake list (*sēru šikinšu*), see the isolated text CT 14 7 K. 4206+ obv.(!) and Benno Landsberger, *Die Fauna des alten Mesopotamien* (Leipzig, 1934), 52 f. There is an obvious interest in fantastic snakes; see the refences in *Assyrian Dictionary* (Chicago, 1956–) sub *ṣēru* B and the Esarhaddon passage mentioned in no. 26 (below).
10. Lists are also used to describe cities by systematically enumerating divine images, temples and other sacral buildings, fortifications, gates, and other structures. This has been done for Babylon—see Eckhard Unger, *Babylon, die heilige Stadt nach der Beschreibung der Babylonier* (Berlin–Leipzig, 1931)—for Borsippa, *ibid.*, 250 ff.; and for Assur—see Eckhard Unger, *Das Stadtbild von Assur*, which is Der Alte Orient, XXVII, no. 3 (Leipzig, 1929), 12–16.
11. See Albert Schott, *Die Vergleiche in den akkadischen Königsinschriften*, which is Mitteilungen der Vorderasiatisch-Aegyptischen Gesellschaft, XXX, no. 2 (Leipzig, 1926); and Wolfgang Heimpel, *Tierbilder in der sumerischen Literatur* (Rome, 1968).
12. For the purpose and tenor of this text, see A. Leo Oppenheim, "The City of Assur in 714 B.C.," in *Journal of Near Eastern Studies*, **19** (1960), 133–147.

13. A comparison of similar force and subject matter, based again on the observation of wild animals, can be found in the Neo-Assyrian literary work published by Wolfram von Soden as "Die Unterweltsvision eines assyrischen Kronprinzen," in *Zeitschrift für Assyriologie,* **43** (1936), 1–30, ll. 69 f. It is not likely that court poets used imagery of this sort because of its appeal to the king alone, as a hunter, since both texts cited were destined for a wider public. Thus, it would appear that the well-to-do Assyrians to whom these texts were addressed were likewise hunters, not for food but for pleasure, a practice that is not reflected in written documents of any kind.
14. See A. Leo Oppenheim, "A New Prayer to the 'Gods of the Night,' " in *Analecta biblica,* **12** (1959), 282–301.
15. See the text K 890, published by S. Arthur Strong, in "On Some Oracles to Esarhaddon and Ašurbanipal," in *Beiträge zur Assyriologie,* **2** (1894), 634. Its central topos, the ship adrift, reflects a Sumerian prototype. See Claus Wilcke, "Eine Schicksalsentscheidung für den toten Urnammu," in *Actes de la XVIIe Rencontre assyriologique internationale* (Ham-sur-Heure, 1970), 89:214.
16. See *Assyrian Dictionary,* sub *ēdu* (lex. section) and *bīnu* A.
17. Note in this context the first millennium Assur text KAV 92, which lists countries and gives distances spanning the region from the islands of the Mediterranean Sea to Dilmun in the Persian Gulf, and to Makan and Meluhha. For a discussion, see Ernst F. Weidner, "Das Reich Sargons von Akkad," in *Archiv für Orientforschung,* **16** (1952–53), 1–23. Old Babylonian itineraries also giving distances were written for practical purposes: see Albrecht Goetze, "An Old Babylonian Itinerary," in *Journal of Cuneiform Studies,* **7** (1953), 51–72; and William W. Hallo, "The Road to Emar," *ibid.,* **18** (1964), 57–88. For Neo-Assyrian lists of way stations (*mardītu*), see Ernst F. Weidner, "Assyrische Itinerare," in *Archiv für Orientforschung,* **21** (1966), 42–46.
18. Interest in foreign languages is also reflected in lists. See n. 6 (above) and Heinrich Otten and Wolfram von Soden, *Das akkadisch-hethitische Vokabular KBo I 44 + KBo XIII 1* (Wiesbaden, 1968), which is Studien zu den Boğazköy-Texten, 7; for Egyptian and Kassite word lists, see the papers cited in Oppenheim, *Ancient Mesopotamia,* 371, no. 25.
19. Witness the rather detailed description of the coronation ritual of an Urartian king—François Thureau-Dangin, *Une relation de la huitième campagne de Sargon,* which is Musée du Louvre, Textes Cunéiformes, 3 (Paris, 1912), ll. 339–342; see *Journal of Near Eastern Studies,* **19** (1960), 141—and the obvious respect shown in an inscription of Assurbanipal (668–627 B.C.) when describing the temple and sacred grove of the Elamite capital—Maximilian Streck, *Assurbanipal,* II (Leipzig, 1916), 52–54, vi 30–69.
20. For the Sumerian evidence, see Dietz Otto Edzard, *Die "Zweite Zwischenzeit" Babyloniens* (Wiesbaden, 1957), 31–33. It should be stressed that such descriptions are not meant to characterize the living habits of the desert dwellers as barbaric but, rather, as what we would call "primitive." For evidence, see Giovanni Pettinato, *Das altorientalische Menschenbild und die sumerischen und akkadischen Schöpfungsmythen* (Heidelberg, 1971), 20–25.
21. See Ignace J. Gelb, "Makkan and Meluhha in Early Mesopotamian Sources," in *Revue d'assyriologie et d'archéologie orientale,* **64** (1970), 1–8; also John Hansman, "A *Periplus* of Magan and Meluḫḫa," in *Bulletin of the School of Oriental and African Studies,* **36** (1973), 554–587.
22. Gold, however, is somehow associated with the netherworld. See *Assyrian Dictionary,* sub *arallû,* and note the possible connection with the country Harali in Sumerian literary texts. See Géza Komoróczy, "Das mythische Goldland Harali im alten Vorderasien," in *Acta orientalia Academiae scientiarum hungaricae,* **26** (1972), 113–123. Only in an inscription of Tiglath-Pileser III (744–727 B.C.) can one find a reference to gold from a specific known mountain (region), Šikrakki: Paul Rost, *Die Keilschrifttexte Tiglat-Pilesers III* (Leipzig, 1893), 62:32.
23. See Erica Reiner, "*Lipšur* Litanies," in *Journal of Near Eastern Studies,* **15** (1956), 146 f.
24. The sedentary population of Mesopotamia considered the tent to be the main feature of nomadism (see *Assyrian Dictionary,* sub *kuštāru* and *bīt ṣēri*), as did Strabo, referring to the *scenitae* Arabs.
25. Streck, *Assurbanipal,* II, 70–80 viii 79–ix 114; and Rykle Borger, *Die Inschriften Asarhaddons, Königs von Assyrien,* which is *Archiv für Orientforschung,* supp. 9 (Graz, 1956), 112–113, 6–r. 19. The Assyrian army knew well how to survive in the "region of thirst," as the desert was called at times. They dug special deep wells—Borger, *op. cit.,* 112:17—or made water holes in dry riverbeds—Wolfgang Schramm, "Die Annalen des assyrischen Königs Tukulti-Ninurta II," in *Bibliotheca orientalis,* **27** (1970), 147–160, ll. 48, 63. The latter has an interesting parallel in the Old Testament: 2 Kings 3:16.
26. Borger, *op. cit.,* 112.
27. See Ernst F. Weidner, "Die Feldzüge und Bauten Tiglatpilesers I," in *Archiv für Orientforschung,* **18** (1957–1958), 355–356.
28. For Egypt, the Old Testament, and Ugarit, see Otto Kaiser, *Die mythische Bedeutung des Meeres in Ägypten, Ugarit und Israel* (Berlin, 1959), 140 ff.; Godfrey Rolles Driver, "Mythical Monsters in the Old Testament," in *Studi orientalistici in onore di Giorgio Levi della Vida,* I (Rome, 1956), 234–249; Herbert Gordon May, "Some Cosmic Connotations of *mayîm rabbîm,*" in *Journal of Biblical Literature,* **74** (1955), 921.
29. The text is BM 92687, published in Felix E. Peiser, "Eine babylonische Landkarte," in *Zeitschrift für Assyriologie,* **4** (1889), 361–369, and in CT 22 48. Translations of the text are in Ernst F. Weidner, *Der Zug Sargons von Akkad nach Kleinasien,* which is Boghazköi-Studien, 6 (Leipzig, 1922), 86–91; and Unger, *Babylon,* 254–258.
30. See *Assyrian Dictionary,* sub *marratu.*
31. One is tempted to connect the "horn" drawing with the Sumerian name of the city Borsippa: B à d . s i . a . a b . b a ("Fortress-at-the-horn-of-the-sea/lake"). See Eckhard Unger, in *Reallexikon der Assyriologie,* I (Berlin–Leipzig, 1928), 404. Its placement on the map may lend support to this suggestion.
32. See *Reallexikon der Assyriologie,* IV (1972), 71.
33. This date is confirmed by the use of the measure *subbān* on the reverse (see *Assyrian Dictionary,* sub voce), and fits the style of the script.
34. The same is said of the region of the Hyperboreans. For literature see Edward Lipiński, "El's Abode: Mythological Traditions Related to Mount Hermon and to the Mountains of Armenia," in *Orientalia Lovaniensia periodica,* **2** (1971), 13–69, esp. 44, nn. 156, 157.
35. The problem is that if eight triangles were intended, one would expect a symmetrical arrangement, which is not the case. The actual traces and the partly reconstructed copy by Felix E. Peiser, in *Zeitschrift für Assyriologie,* **4** (1889), 361 f., suggest seven triangles. On the other hand, one can find eight sections only if one assumes a dividing line after line 17, copied by Peiser (and assumed by Weidner) but omitted by Thompson.
36. Note that *nagû,* when taken over into Aramaic (*nagwān*), means specifically "island" or "coastal region."
37. The notion of deposed gods is also attested in the expressions *ilāni darsūti*—for which see Rintje Frankena, *Tākultu de sacrale Maaltijd in het assyrische Ritueel* (Leiden, 1954), 13 f.—and *ilāni kamûti*—see *Assyrian Dictionary,* sub *kamû,* adj. usage a. See also n. 95 (below).
38. Utnapištim, the hero of the Babylonian flood story, lives

(as does his Sumerian counterpart Ziusudra) on an island beyond the sea, enjoying eternal life on that prototype of the "Island of the Blessed" known from classical mythology. Sargon of Akkad and his adversary Nūr-Dagan appear in the old legend called "King of Battle" (*šar tamḫāri*), in which a campaign to far-off Asia Minor is the central topic. See Hans Gustav Güterbock, "Die historische Tradition und ihre literarische Gestaltung bei Babyloniern und Hethitern bis 1200," in *Zeitschrift für Assyriologie,* **42** (1934), 86–91; and the Hittite version, *ibid.*, **44** (1938), 45–49; also Wilfred G. Lambert, "A New Fragment of The King of Battle," in *Archiv für Orientforschung*, **20** (1963), 161–162. In Sumerian mythological literature, the island of Dilmun appears as the "pure land," the "land of the living," the "place where the sun rises," and the place where the flood hero Ziusudra was transported, to live there a "life like a god." Still, the reference to three harvests per year (UET 6/1 1 ii 24) points in a different direction, perhaps to a land of Cockaigne.

39. The animals listed correspond only partly to the traditional Mesopotamian enumerations of exotic animals. For Sumerian, see, for example, Adam Falkenstein, "Fluch über Akkade," 1. 21, in *Zeitschrift für Assyriologie,* **57** (1965), 43–124; and the unpublished text 3 NT 385 iv 20. The lion, the wolf, the cat, the ostrich, and various animals always mentioned with the gazelle do not really fit. Only the *kusarikku* (?) is strictly mythological, as is probably the *apsasû* at that period (see *Assyrian Dictionary*, sub voce), while the *lulīmu* and the *pagītu*—the baboon—are not; for baboons (rather than monkeys) in Sumerian literary texts, see Edmund I. Gordon, "Animals as Represented in the Sumerian Proverbs and Fables: A Preliminary Study," in *Drevny mir, sbornik statey akademiku V. V. Struve*, N. V. Pigulevskaya *et al.*, eds. (Moscow, 1962), 228. In Akkadian proverbs the cat from Meluhha and the bear (see *Assyrian Dictionary*, sub *margû*) from Parahše (Persia) are mentioned in the proverb published by Wilfred G. Lambert, *Babylonian Wisdom Literature* (Oxford, 1960), 272:6f.; for Elamite, Persian, and Meluhha dogs, see *Assyrian Dictionary*, sub *kalbu*, meaning 1e. Interest in exotic animals is expressed in repeated demands of the Babylonian king addressed to the pharaoh for representations of them to be sent to Babylon: "The experts who are at your disposal should make lifelike replicas of the strange animals of land and river; the skin [especially] should be like that of a living animal; your messenger should bring [them] here. But, if when my messenger Sindišugab arrives there and there are available some old and finished [replicas], they should load them quickly on wagons and come here with dispatch, and [then] they should make new ones for later." Jörgen Alexander Knudtzon, ed., *Die El-Amarna-Tafeln*, Vorderasiatische Bibliothek, 1 (Leipzig, 1915), no. 10:32–40, letter of Burnaburiaš. See also *ibid.*, letter no. 4:24 and 35 in a letter of Kadašman-Ḫarbe.
40. See Pettinato, *Das altorientalische Menschenbild*, 1–22.
41. See Benno Landsberger and James V. Kinnier Wilson, "The Fifth Tablet of *Enūma eliš*," in *Journal of Near Eastern Studies*, **20** (1961), 154–179; Ephraim Avigdor Speiser, in James Bennett Pritchard, ed., *Ancient Near Eastern Texts Relating to the Old Testament*, 2nd ed. (Princeton, 1955), 67 f.; and Albert Kirk Grayson, *ibid.*, 3rd ed. with supp. (Princeton, 1969), 501 f.
42. Verse 11 of tablet V, the first of the sections dealing with the moon, is out of place because it refers, proleptically, to the creation of the upper half of the cosmos (*elâti*) from the body of Tiamat. The topic has already been taken up in a short passage (IV, 137 f.) which speaks of the victorious Marduk splitting the corpse of Tiamat as one would a fish to prepare it for drying in the sun (also KAR 307 r. 2). He uses one-half to roof the sky.
43. For a parallel to the astronomic section of the Babylonian Epic of Creation, see Ernst F. Weidner, "Die astrologische Serie *Enûma Anu Enlil*," in *Archiv für Orientforschung*, **17** (1954–1956), 89; and Landsberger and Kinnier Wilson, "The Fifth Tablet," 172.
44. For rain clouds pictured with raindrops inside them, see Walter Andrae, *Farbige Keramik aus Assur und ihre Vorstufen in altassyrischen Wandmalereien* (Berlin, 1923), pl. 8.
45. The Great Band is also mentioned in two of the names of Marduk enumerated in *Enūma eliš* VII, 80 and 95.
46. See Pettinato, *Das altorientalische Menschenbild*, 62 f. In the Hittite myth of Kumarbi—see Hans Gustav Güterbock, *Kumarbi*, which is Istanbuler Schriften 16 (Zurich–New York, 1946)—heaven and earth are separated with a cleaver; compare the splitting of the corpse of Tiamat (n. 42). See also Heinrich Otten and J. Siegelová, "Die hethitischen Gulš-Gottheiten und die Erschaffung des Menschen," in *Archiv für Orientforschung*, **23** (1970), 32–38.
47. See the passage 4 R 9:28f. a n . s u d . d a m : *kīma šamê rūqūti* ("like the far-off heaven"); also Lambert, *Babylonian Wisdom Literature*, 148:83. The distance from earth to heaven is vividly and quite naturalistically described in the story of the ascent of Etana on the back of an eagle. The ever-diminishing size of the earth as seen from aloft represents a very rare instance in Mesopotamian literature of the imaginative rendering of a fantastic experience.
48. The comparison in KAR 23:16, "the *apsû* is your [the gods'] *ḫamû*-kettle[?], your censer (*niknakku*) is the heaven of Anu," points to a spherical shape of the cosmos. Note that the word *kippatu* in *kippat šamê/erṣeti* ("totality," literally "circumference of heaven/earth") denotes a hoop, a loop-shaped handle of a container, a loop of a snare, the curl of a tendril of the grapevine.
49. The Sumerian words corresponding, respectively, to *elât* and *išid šamê* are used in certain Assyrian and Neo-Babylonian royal inscriptions (see *Assyrian Dictionary*, sub *elâtu*, meaning 5c-2′) to refer to East and West: "from sunrise [AN.ÚR] to sunset [AN.PA] wherever the sun shines."
50. The text—Friedrich Delitzsch, *Assyrische Lesestücke*, 2nd ed. (Leipzig, 1876), 73:41 f.—speaks of the *šigar šamê*, while an omen tablet—*Archiv für Orientforschung*, **14** (1941–1944), pl. 16, VAT 9436—speaks of the *handūh šamê* (r. 6 and 12) and the *sikkat šamê* as essential parts of the lock of this gate (see *Assyrian Dictionary*, sub *ḫandūḫu*). The Šamaš prayer in *Oxford Editions of Cuneiform Texts*, VI (London, 1924), 45 f.:6 ff., describes in detail the god's leaving the *qereb šamê* by removing locks and bars and opening the doors of heaven.
51. See *Assyrian Dictionary*, sub *abullu*, meaning 2b.
52. James Alexander Craig, *Assyrian and Babylonian Religious Texts*, I (Leipzig, 1895), 22 ii 15; I interpret *attaqallala* as a rare Neo-Assyrian form connected with *šuqallulu*.
53. The connection of the topographical name with *māšu/māštu* ("twin") is also suggested by the Ugaritic passage speaking of the "two hills at the border of the netherworld" (*tlm gśr arṣ*). See Nicholas J. Tromp, *Primitive Conceptions of Death and the Nether World in the Old Testament* (Rome, 1969), 7.
54. For the Sumerian key word k u r , see Samuel Noah Kramer, *The Sumerians, Their History, Culture and Character* (Chicago, 1963), 151–153, 296; for h u r . s a g , see Thorkild Jacobsen, "Sumerian Mythology: A Review Article," in *Journal of Near Eastern Studies*, 5 (1946), 141.
55. See KAH 2 54, edited by Ernst F. Weidner, *Die Inschriften Tukulti-Ninurtas I und seiner Nachfolger*, which is *Archiv für Orientforschung* supp. 12 (Graz, 1959), 36, no. 25; English trans. in Albert Kirk Grayson, *Assyrian Royal Inscriptions*, I (Wiesbaden, 1972), 126, no. 25*.
56. The pertinent material has recently been collected from biblical, Ugaritic, and classical sources by E. Lipiński in "El's Abode." For the Egyptians' quite different concept of chaos and darkness surrounding the edges of the world, see

Hellmut Brunner, "Die Grenzen von Zeit und Raum bei den Ägyptern," in *Archiv für Orientforschung*, **17** (1954–1956), 141 f.; and Rudolf Kilian, "Gen. I 2 und die Urgötter von Heliopolis," in *Vetus testamentum*, **16** (1966), 420–438.

57. There is no evidence from Mesopotamia for pillars supporting the heaven as we have them in the Old Testament (Job 26:11) and in Egypt.
58. For the "ladder" of Jacob, see Gen. 28:12–17; and the discussion by Allan Millard, "The Celestial Ladder and the Gate of Heaven," in *Expository Times*, **78** (1966–1967), 86–87.
59. We know, incidentally, that in Sippar, the city of the sun god, a golden chariot drawn by horses was used at religious ceremonies. See Oppenheim, *Ancient Mesopotamia*, 193 and no. 20.
60. See *Assyrian Dictionary*, sub *erpetu*, usage a.
61. See the passage "[the king . . .] who angered Adad in heaven so that he did not let it rain for three years nor [allow] vegetation to grow for three years," in Erica Reiner, "The Etiological Myth of the 'Seven Sages,' " in *Orientalia*, n.s. **30** (1961), 3:15 ff. No connection, however, is assumed between lightning and rain, as it is in Jer. 51:16.
62. For omens predicting rain, see *Assyrian Dictionary*, sub *zanānu* A, meaning 1a-1′–4′.
63. For traces of rain magic in the Old Testament, see M. Delcor, "Rites pour l'obtention de la pluie à Jérusalem et dans le Proche-Orient," in *Revue de l'histoire des religions*, **178** (1970), 117–132.
64. See *Assyrian Dictionary*, sub *ṣirtu* A. Add there *Revue d'assyriologie et d'archéologie orientale*, 67 (1973), 42:18; *Archiv für Orientforschung*, **19** (1959–1960), 61:9; and the passages sub *ṣerretu* A, meaning 4a. Also see Rykle Borger, in *Journal of Cuneiform Studies*, **18** (1964), 55.
65. Similar concepts exist in the Old Testament: the "bottles of heaven" in Job 38:37; the "windows of heaven" in Gen. 7:11 and 8:2, the opening of which spells abundance for the country (2 Kings 7:2 and 19, Mal. 3:10). See also in Ugaritic the opening of the *bdqt ʿrpt*, Joseph Aistleitner, *Wörterbuch der ugaritischen Sprache* (Berlin, 1967), no. 2290; and David Neiman, "The Supercaelian Sea," in *Journal of Near Eastern Studies*, **28** (1969), 243–249. "Storehouses" for the winds of heaven are mentioned in Jer. 10:13, 51:16; and Ps. 135:7.
66. TCL 6 51 r. 17 f., in *Revue d'assyriologie et d'archéologie orientale*, **11** (1914), 149:29. Note also the passage in *Beiträge zur Assyriologie*, **5** (1906), 653:21, which says of the command of the god Assur "like the stars of heaven it does not miss its *adannu*." The term *adannu*, denoting a normal, expected, computed or prearranged point in time, refers to the punctuality of the stars. See *Assyrian Dictionary*, sub *adannu*, meaning 2a–3′.
67. See the first-millennium hymn in François Thureau-Dangin, *Rituels accadiens* (Paris, 1921), 139:330.
68. For this difficult word, see Landsberger and Kinnier Wilson, "The Fifth Tablet," 172 f.
69. For such lists, see *Materials for the Sumerian Lexicon*, XI (1974), 105 f., 131 f., 134 f., 141 f.
70. See Albert Schott, "Das Werden der babylonisch-assyrischen Positions-Astronomie und einige seiner Bedingungen," in *Zeitschrift der Deutschen morgenländischen Gesellschaft*, **88** (1934), 303–337, esp. 309 ff.; and Ernst F. Weidner, "Ein astrologischer Sammeltext aus der Sargonidenzeit," in *Archiv für Orientforschung*, **19** (1959–1960), 105–113.
71. For the several meanings of *lumāšu*, see *Assyrian Dictionary*, sub voce.
72. See Landsberger and Kinnier Wilson, "The Fifth Tablet," 174.
73. They are Mercury, Venus, Mars, Jupiter, and Saturn, although a late text (Thureau-Dangin, *Rituels accadiens*, 79:33) mentions offerings to seven *bibbu*. For late tablets with pictures of some constellations in which the moon and planets reach their *hypsoma*, see Ernst F. Weidner, "Gestirn-Darstellungen auf babylonischen Tontafeln," in *Sitzungsberichte der Österreichischen Akademie der Wissenschaften*, Phil.-hist. Kl., **254**, no. 2 (1967), 7–11.
74. See A. Leo Oppenheim, "A Babylonian Diviner's Manual," in *Journal of Near Eastern Studies*, **33** (1974), 204.
75. See Erica Reiner, "Fortune-Telling in Mesopotamia," in *Journal of Near Eastern Studies*, **19** (1960), 23–35, esp. 27 f.
76. See CT 29 48:19—"many stars were falling from the sky."
77. Demons were often called "the spawn of heaven" (*reḫût šamê*). See Richard I. Caplice, "É.NUN in Mesopotamian Literature," in *Orientalia*, n.s. **42** (1973), 304 f.; also called "spawn of the stars" in Franz Köcher and A. Leo Oppenheim, "The Old-Babylonian Omen Text VAT 7525," in *Archiv für Orientforschung*, **18** (1957–1958), 63:12 and no. 9; and "spawn of Anu" (CT 16 15 v 3 and *passim*).
78. See PBS 1/2 113:53.
79. There they retreat before the flood that engulfed the earth-Gilgameš Epic XI, 114.
80. A shortened version of the description of the three heavens is given as an insert in a collection of miscellaneous star identifications and "astrological" interpretations studied by Weidner, "Ein astrologischer Sammeltext," 105–113. It reads (iv 20–22, on p. 110): "upper [heaven]: *luludānītu*-stone, for Anu; middle [heaven]: *saggilmut*-stone, for the Igigu; lower heaven: jasper [?], for the stars." See Benno Landsberger, "Über Farben im sumerisch-akkadischen," in *Journal of Cuneiform Studies*, **21** (1967), 154 f.
81. For the designation *igigu* of the heavenly deities, see Wolfram von Soden, "Babylonische Göttergruppen: Igigu und Anunnaku. Zum Bedeutungswandel theologischer Begriffe," in *Compte rendu de l'XIe Rencontre assyriologique internationale* (Leiden, 1964), 102–111; and "Die Igigu-Götter in altbabylonischer Zeit," in *Iraq*, **28** (1966), 140–145; and Burkhart Kienast, "Igigū und Anunnakkū nach den akkadischen Quellen," in *Studies in Honor of Benno Landsberger on His Seventy-Fifth Birthday, April 21, 1965*, which is Assyriological Studies, 16 (Chicago, 1965), 141–158.
82. A. Leo Oppenheim, Robert H. Brill, Dan Barag, and Axel von Saldern, *Glass and Glassmaking in Ancient Mesopotamia. An Edition of the Cuneiform Texts Which Contain Instructions for Glassmakers, With a Catalogue of Surviving Objects* (Corning, N.Y., 1970), 16 and n. 31.
83. For three heavens, see also Gerhard Meier, "Die Zweite Tafel der Serie bīt mēseri," in *Archiv für Orientforschung*, **14** (1941–1944), 142–143. This is reflected in the New Testament (2 Cor. 12:2), while later Jewish folklore and the Koran speak of seven heavens.
84. However, there is also evidence from the Old Testament for a sky conceived of metal (bronze)—"the sky, which is strong, and as a molten looking-glass" (Job 37:18)—which we find also as "brazen heaven" in Homer (*Iliad*, 5.504, 17.425; *Odyssey*, 3.2).
85. See STT 108:76; but the next line likens the color of the stone to that of a storm cloud (*urpat rihṣi*).
86. See *Assyrian Dictionary*, sub *ḫašmānu*.
87. Assuming, against Wolfram von Soden, *Akkadisches Handwörterbuch* (Wiesbaden, 1959–), that *massuku* is Assyrian for *maštaku*.
88. In a telling contrast with such Near Eastern celestial sumptuosity are the heavenly meadows and the great tree (*harikešriaš*) under which the gods assemble, as the Hittite texts tell us. See Maurice Vieyra, "Ciel et enfers hittites," in *Revue d'assyriologie et d'archéologie orientale*, **59** (1965), 127–130, esp. 128 f. See also Jaan Puhvel, " 'Meadow of the Otherworld' in Indo-European Tradition," in *Zeitschrift für vergleichende Sprachforschung*, **83** (1969), 64–69.
89. See Heinrich Otten, "Keilschrifttexte," in *Mitteilungen der*

Deutschen Orient-Gesellschaft zu Berlin, **91** (1958), 83 r. 13′.

90. For *tarkullu*, see Åke W. Sjöberg and E. Bergmann, *The Collection of the Sumerian Temple Hymns* (Locust Valley, N.Y., 1969), 67 n. to line 79.
91. See also Samuel Noah Kramer, "Death and Nether World According to the Sumerian Literary Texts," in *Iraq*, **22** (1960), 59–68; Vieyra, "Ciel et enfers hittites," 127–130; and Anton Jirku, "Die Vorstellungen von Tod und Jenseits in den alphabetischen Texten von Ugarit," in *Ugaritica 6*, which is Mission de Ras Shamra 7 (Paris, 1969), 303–308.
92. See *Assyrian Dictionary*, sub *anzanunzû, asurrakku*; see also *nagbu*, another subterranean source of water, with which compare the "fountains of the deep" in Gen. 7:11, 8:2; and Prov. 8:24, 28; and the "springs in hidden places" 2 Sam. 22:16 and Ps. 18:16, cited in Lipiński, "El's Abode," 36.
93. For details, see Reiner, "The Etiological Myth of the 'Seven Sages,'" 1–11; and Jan J. A. van Dijk, "Die Tontafeln aus dem *rēš*-Heiligtum," in *Vorläufiger Bericht über die von dem Deutschen archäologischen Institut und der Deutschen Orient-Gesellschaft aus Mitteln der Deutschen Forschungsgemeinschaft unternommenen Ausgrabungen in Uruk-Warka*, XVIII (Berlin, 1962), 43–52.
94. See A. Leo Oppenheim, *The Interpretation of Dreams in the Ancient Near East* (Philadelphia, 1956), 235.
95. I connect this line with the passages that speak of banished gods (see nn. 37, 81). The Hittites, too, knew of gods driven out of heaven and relegated to the netherworld—the "dark earth." See Heinrich Otten, "Eine Beschwörung der Unterirdischen aus Boğazköy," in *Zeitschrift für Assyriologie*, **54** (1961), 120:43–48, 132 iii 34 f.; also Vieyra, "Ciel et enfers hittites," 130; and Hans Gustav Güterbock, in Erica Reiner and Hans Gustav Güterbock, "The Great Prayer to Ishtar and Its Two Versions From Boğazköy," in *Journal of Cuneiform Studies*, **21** (1967, published 1969), 265–266. The many mythological names referring to the realm of the dead are not our concern here. For a somewhat outdated survey, see Knut Leonard Tallqvist, *Sumerisch-akkadische Namen der Totenwelt* (Helsinki, 1934). See also Ludwig Wächter, "Unterweltsvorstellung und Unterweltsnamen in Babylonien, Israel und Ugarit," in *Mitteilungen des Instituts für Orientforschung*, **15** (1969), 327–336.
96. The text is as follows: [*ina* KALA]G.GA KI-*tim* AN.[T]A(?) *zi-qi-qu* NAṂ.LÚ.U$_x$.LU *ina* ŠÀ *ú-sar-bi-iṣ* [. . .] KI-*tim* [MURU$_4$]-*tu* ^{d}DIŠ AD-*šú ina* ŠÀ *ú-še-šib* [. . .] KI-*tim* [. . .] *x-si-ḫu ul ú-maš-ši* [. . .] KI-*tim* KI.TA-*tú* 600 ^{d}[*A-nun-na*]-[*ki ina lìb*]-*bi e-sir.*
97. The passage is from *Quaestiones naturales*, 2.32.4: in ea opinione sunt, tamquam non, quia facta sunt, significant, sed quia significatura sunt, fiant.
98. A case in point is furnished by the long (more than 670 short lines) extispicy text YOS 10 31, which deals with signs derived from the shape, size, and other characteristics of the gall bladder of a sheep. The text is obviously compiled from several smaller collections of a similar nature, as shown by the repetition of omens with identical protases and apodoses, and of omens with different apodoses based on identical protases. The former is evidenced in the passage i 5–8, repeated in x 4–7 and xiii 42–45 (all connecting a lizard-shaped gall bladder with Sargon of Akkad); the passage i 47–49, which recurs in x 45–47; the passage iii 20–24, in vi 43–46; the passage ii 42–47, in xiii 46–50; and the passage v 18–24, in vi 15–22. The latter is evidenced, for example, in iii 32–35, as against x 11–14. A right-left pair is treated separately: right in vi 39–42 and left in xiii 46–50. Rhyming omens occur in i 9–11 and v 37–39. Sometimes there are awkward insertions into established sequences that should not occur in a well-integrated collection; for instance, ii 1–12 is inserted into the connected sequence i 47–ii 15.
99. Growth by accretion—that is, by insertions or additions of similar text material prompted by the desire for elaboration, or by repetitions with minor variations—can be observed likewise in epic and hymnic works, as well as in royal inscriptions. Such additions can just as readily be omitted when economy of space or presentation demands. The resulting fluidity in text formation can be controlled only by rigid standardization. Additive instead of structural organization is also characteristic of large-scale Mesopotamian architecture.
100. The slowly evolving practice of writing omens with word signs rather than syllabically contributed to the ever-increasing terseness and formalistic monotony of the predictions.
101. While there exist Elamite and Hittite translations of Akkadian omens, and even Ugaritic versions—see Anson Frank Rainey, *The Scribe at Ugarit, His Position and Influence*, which is *Proceedings of the Israel Academy of Sciences and Humanities*, **3**, no. 4 (1968), 131 f.—divination in Mesopotamian style is attested only rarely in Egypt: for dream omens, see Alan Henderson Gardiner, *Hieratic Papyri in the British Museum*, 3rd ser., Chester Beatty Gift (London, 1935); and Aksel Volten, *Demotische Traumdeutung (Papyrus Carlsberg XIII und XIV verso)*, which is Analecta Aegyptiaca 3 (Copenhagen, 1942); for lunar omens, see Richard Anthony Parker, *A Vienna Demotic Papyrus on Eclipse- and Lunar-Omina* (Providence, R.I., 1959).
102. See A. Leo Oppenheim, "Zur keilschriftlichen Omenliteratur," in *Orientalia*, n.s. **5** (1936), 199–228; and Jean Nougayrol, "Divination et vie quotidienne au début du deuxième millénaire avant J.-C.," in *Acta orientalia neerlandica* (1971), 28–36.
103. See A. Leo Oppenheim, "Divination and Celestial Observation in the Last Assyrian Empire," in *Centaurus*, **14** (1969), 97–135.
104. See the use of arrows for divination mentioned by the prophet Ezekiel (Ezek. 21:21) but not attested in cuneiform sources. For observation of birds (known only from inference), see Oppenheim, *Ancient Mesopotamia*, 209 f.
105. Economic and ritual texts provide us with the names of cuts of meat for culinary and sacral purposes. See also William L. Moran, "Some Akkadian Names of the Stomachs of Ruminants," in *Journal of Cuneiform Studies*, **21** (1967, published 1969), 178–182.
106. Another version of the same query-answer relation is suggested by a passage that has the owner of the sheep whisper his secret request (*tamīt libbišu*) into the ear of the animal before the diviner addresses the sun god and slaughters the sheep. Heinrich Zimmern, *Beiträge zur Kenntnis der babylonischen Religion* (Leipzig, 1901), 98–99:8–9.
107. For the *āšipu*, see Oppenheim, *Ancient Mesopotamia*, 29 ff.; and Edith K. Ritter, "Magical-Expert (=*Āšipu*) and Physician (=*Asû*); Notes on Two Complementary Professions in Babylonian Medicine," in *Studies in Honor of Benno Landsberger*, 299–321.
108. See René Labat, *Traité akkadien de diagnostics et pronostics médicaux* (Paris, 1951); also James V. Kinnier Wilson, "Two Medical Texts From Nimrud," in *Iraq*, **18** (1956), 130–146; and "The Nimrud Catalogue of Medical and Physiognomical Omina," *ibid.*, **24** (1962), 52–62. For earlier texts, see Oppenheim, *Ancient Mesopotamia*, 369, nn. 62, 63.
109. The fact that the text begins with an injunction to the *āšipu* to protect himself with a spell before approaching the patient shows the *āšipu*'s concept of the etiology of diseases.
110. See A. Leo Oppenheim, "On the Observation of the Pulse in Mesopotamian Medicine," in *Orientalia*, n.s. **31** (1962), 27–33; and Kinnier Wilson, "The Nimrud Catalogue," 61. See also Samuel Mendelsohn, "Die Funktion der Puls-

adern und der Kreislauf des Blutes in altrabbinischer Literatur," in *Jenaer medizin-historische Beiträge*, no. 11 (Jena, 1920).

111. For other uses of the formal structure of omen texts in Mesopotamian literature, see Oppenheim, *Ancient Mesopotamia*, 224.
112. It also contains omens derived from atmospheric phenomena (thunder, lightning, rainbows, halos, [miraculous] rain, hail) and earthquakes. See *ibid.*, 368 f., nn. 66, 68.
113. A critical ed. of this series (based on several thousand fragments) is being written by Erica Reiner.
114. For the problem involved, see A. Leo Oppenheim, "The Position of the Intellectual in Mesopotamian Society," in *Daedalus* (spring 1975), 37–46.
115. Greek tradition has it that with Nabonassar (Akkadian, Nabû-nāṣir; 747–734 B.C.) a new era began, and that recorded astronomical observations were available from the time of this king on. See John A. Brinkman, *A Political History of Post-Kassite Babylonia* (Rome, 1968), 226 f. and nn. 1432–1436.
116. In Old Babylonian extispicy texts one rather frequently finds apodoses predicting an eclipse. In the much more numerous omens of the later period these are quite rare, a fact that suggests a conscious process of elimination based on rational considerations. Reports on eclipses are rare, too; apart from the lunar eclipses mentioned by Ptolemy—see John A. Brinkman, "Merodach-Baladan II," in *Studies Presented to A. Leo Oppenheim* (Chicago, 1964), 49 sub 44.3.12—a famous eclipse of the sun is mentioned in an eponym list—see *Reallexikon der Assyriologie*, II (1938), 430 r. 7, for 763 B.C.; and an eclipse of the moon in two letters from Mari. Of course, we have numerous references to eclipses in the letters of the Assyrian and Babylonian experts (see n. 103, above) of the middle third of the first millennium. Observational reports on lunar eclipses, given in detail and arranged in eighteen-year groups ("Saros Canon"), are in *Late Babylonian Astronomical and Related Texts*, copied by Theophilus Goldridge Pinches and Johann Nepomuk Strassmaier, prepared for publication by Abraham Joseph Sachs, with the cooperation of Johann Schaumberger (Providence, R.I., 1955), nos. 1413–1430. The Ugarit text discussed by J. F. A. Sawyer and F. R. Stephenson, "Literary and Astronomical Evidence for a Total Eclipse of the Sun Observed in Ancient Ugarit on 3 May 1375 B.C.," in *Bulletin of the School of Oriental and African Studies*, **23** (1970), 467–489, cannot be considered a report on a solar eclipse.

    The two Mari letters are important inasmuch as both—*Archives royales de Mari*, **10** (1967), no. 124; *Compte rendu de la IIe Rencontre assyriologique internationale* (Paris, 1951), 46 f.—resort to extispicy rather than consulting a collection of celestial omens to establish the nature of the eclipse. Still, Georges Dossin, "Les archives économiques du palais de Mari," in *Syria*, **20** (1939), 101, mentions a tablet from Mari with "des présages tirés d'une éclipse de lune." This text has not yet been published. For an attempt to use for chronological purposes late omen protases mentioning eclipses in connection with names of early kings, see Johann Schaumberger, "Die Mondfinsternisse der dritten Dynastie von Ur," in *Zeitschrift für Assyriologie*, **49** (1950), 50–58.
117. For the special stress placed in the first millennium on the equal value of celestial and terrestrial signs, see A. Leo Oppenheim, "A Babylonian Diviner's Manual," in *Journal of Near Eastern Studies*, **33** (1974), 206.
118. It should be stressed that in the first millennium the "power" of the celestial bodies to give signs was independent of their worship as gods. We have a prayer to Sirius—Eric Burrows, "Hymn to Ninurta as Sirius (K. 128)," in *Journal of the Royal Asiatic Society*, centenary supp. (1924), 33–40; also Erich Ebeling, "Sammlungen von Beschwörungsformeln," in *Archiv orientální*, **21** (1953), 403 f.: 14–19—and, in late texts, to other stars, for example, Thureau-Dangin, *Rituels accadiens*, 119:31; and a list of eight such prayers, *ibid.*, 139:325–332.
119. Literature on the development of astrology in Hellenistic Egypt is listed conveniently in Frederick Henry Cramer, *Astrology in Roman Law and Politics* (Philadelphia, 1954), 1 f. There is some evidence in late Uruk (Erech) texts for a possible application of the *dynamis* concept insofar as it implies a link between stars and precious stones and certain plants. For a presentation of such tablets with eclipse omens, lists of zodiacal constellations (names of stones and plants are added to each), hemerological lists (with similar additions), and other rather obscure text types that all bespeak the originality and vitality of late Mesopotamian scholarship, see Weidner, "Gestirn-Darstellungen."
120. The case of alchemy is somewhat similar.
121. See the series of papers published by Richard I. Caplice in *Orientalia*, n.s. **34–40** (1965–1971), with an index in n.s. **40**, 183.
122. The letters are ABL 23, 46, 337, 470, 629, 647, 895, and others. A fragmentary *namburbû* ritual designed to avert celestial signs is partially preserved in LKA 108:13–23, transliterated by Erich Ebeling, "Beiträge zur Kenntnis der Beschwörungsserie Namburbi," in *Revue d'assyriologie et d'archéologie orientale*, **50** (1956), 26; the pertinent lines are quoted in *Assyrian Dictionary*, sub *anqullu*, usage b-1′.
123. See René Labat, *Hémérologies et ménologies d'Assur* (Paris, 1939); and *Un calendrier babylonien des travaux, des signes, et des mois* (Paris, 1965).
124. For the Hittite text (KUB 8 35), see Bruno Meissner, "Über Genethlialogie bei den Babyloniern," in *Klio*, **19** (1925), 432–434; and Kaspar K. Riemschneider, *Babylonische Geburtsomina in hethitischer Übersetzung*, which is Studien zu den Boğazköy-Texten 9 (Wiesbaden, 1970), 44, n. 39a.
125. See Labat, *Un calendrier babylonien*, 132 f., sec. 64. For an Egyptian parallel, see Abd-al-Muhṣin Bakir, *The Cairo Calendar No. 86637* (Cairo, 1966), esp. 13–50.
126. See Weidner, "Gestirn-Darstellungen," 26f., 32f., 35 ff., 45 f.
127. *Ibid.*, 14.
128. Abraham Joseph Sachs, "Babylonian Horoscopes," in *Journal of Cuneiform Studies*, **6** (1952), 49–75; and "Naissance de l'astrologie horoscopique en Babylonie," in *Archaeologia* (Paris), **15** (1967), 13–19.
129. See the Strabo passage cited in A. Leo Oppenheim, *Letters From Mesopotamia* (Chicago, 1967), 53.
130. See David Pingree, "Astronomy and Astrology in India and Iran," in *Isis*, **54** (1963), 229–246.
131. See A. Leo Oppenheim, "Perspectives on Mesopotamian Divination," in *La divination en Mésopotamie ancienne et dans les régions voisines* (Paris, 1966), 35; for the problem, see Pierre Duhem, *To Save the Phenomena: An Essay on the Idea of Physical Theory From Plato to Galileo* (Chicago, 1969), English trans. of the French original (1908); and Jürgen Mittelstrass, *Die Rettung der Phänomene* (Berlin, 1962).
132. See Dietlinde Goltz, "Mitteilungen über ein assyrisches Apotheken-inventar," in *Archives internationales d'histoire des sciences*, **21** (1968), 95–114.
133. Thus, when a plant is called "herb for sorrows" or "herb for forgetting sorrows." See *Assyrian Dictionary*, sub *azallû*.
134. Although other names are known for this or a similar disease, such as *garābu*, *garāṣu*, and *epqu*, it is difficult to define, especially from the narrow point of view of modern medicine. See also James V. Kinnier Wilson, "Leprosy in Ancient Mesopotamia," in *Revue d'assyriologie et d'archéologie orientale*, **60** (1966), 47–58. The social con-

sequences for those who suffer from such a disease are repeatedly mentioned. Note also that the laws of Hammurapi (1792–1750 B.C.) indicate that a man may divorce a wife suffering from a skin disease called *la'bu* (sec. 149), or, if she prefers, he may maintain her in his home while he takes another wife (sec. 148).

135. For a text group that lists names of diseases, see "List of Diseases," composed by Anne D. Kilmer, edited by Benno Landsberger, in *Materials for the Sumerian Lexicon,* IX (1967), 77–109.

136. For a parallel, see Oppenheim *et al., Glass and Glassmaking,* 80; chemicals, even herbs, that could not have affected the outcome of the process were added to the glass before heating.

137. These are listed in the series Uruanna.

138. The entire domain of magic has been excluded from this presentation, although it admittedly concerns a large segment of the relationship between man and nature in Mesopotamian civilization. I restrict myself to citing here one of the very few revealing passages in magic-oriented texts that speaks of the limitation of the magic approach to reality: "who can work witchcraft against heaven, who can rebel against the netherworld [death]?" Gerhard Meier, *Die assyrische Beschwörungssammlung Maqlû,* which is *Archiv für Orientforschung,* supp. 2 (Berlin, 1937), V 12 and 16.

139. For the problem in Mesopotamia, see René Labat, "À propos de la chirurgie babylonienne," in *Journal asiatique,* **242** (1954), 207–218; Wolfram von Soden, "Der Chirurg im Akkadischen," in *Wiener Zeitschrift für die Kunde des Morgenlandes,* **55** (1959), 53–54; and Helmut Freydank, "Chirurgie im alten Mesopotamien?" in *Altertum,* **18** (1972), 133–137.

140. The knowledge that diseases were contagious (*muštaḫḫizu*) and could be transmitted through contact with the objects used by the sick person is illustrated by the Mari letter in which orders are given that "no one must drink from the cup she [the patient] drinks from, no one must sit in the chair she sits in, no one must lie on the bed she lies on, so that she does not infect other women." See André Finet, "Les médecins au royaume de Mari," in *Annuaire de l'Institut de philologie et d'histoire orientales et slaves,* **14** (1954–1957), 129; the text is published as *Archives royales de Mari,* **10** (1967), no. 129.

141. See A. Leo Oppenheim, "A Caesarian Section in the Second Millennium B.C.," in *Journal of the History of Medicine and Allied Sciences,* **15** (1960), 292–294. See also John Harley Young, *Caesarian Section, the History and Development of the Operation From Earliest Times* (London, 1944).

142. For the practice of castration and the gelding of horses, see A. Leo Oppenheim, "A Note on *ša rēši,*" in *Journal of the Ancient Near Eastern Society of Columbia University,* **5** (1973), 325–334; and a correction by Oppenheim, in *Revue d'assyriologie et d'archéologie orientale,* **68** (1974), 95.

143. The length of pregnancy was set at ten months—see Labat, *Traité akkadien,* 212:7—for which there is a parallel from Virgil, discussed by Otto Neugebauer, "Decem tulerunt fastidia menses," in *American Journal of Philology,* **84** (1963), 64–65.

144. Letters from Nippur in the Kassite period give reports on the health of singers treated in a hospital; see Heinz Waschow, *Babylonische Briefe aus der Kassitenzeit,* which is Mitteilungen der Altorientalischen Gesellschaft, **10**, no. 1 (1936), 25–40.

145. To use an expression coined by Robert D. Biggs, "Medicine in Ancient Mesopotamia," in *History of Science,* **8** (1969), 94–105.

146. For the importance of court physicians, see Oppenheim, *Ancient Mesopotamia,* 304.

147. See Heinrich Zimmern, "Der Schenkenliebeszauber," in *Zeitschrift für Assyriologie,* **32** (1918–1919), 164–184.

148. See Oppenheim, *Ancient Mesopotamia,* 301 f.

149. See VAS 6 242; also note the women physicians mentioned in *Assyrian Dictionary,* sub *asû* A, usage e.

150. In late lists, veterinarians—for early references, see *Assyrian Dictionary,* sub *asû* A, usage e—are called *muna'išu* ("healers")—see *Assyrian Dictionary,* sub voce. Esarhaddon (680–669 B.C.) reports (Borger, *Die Inschriften Asarhaddons,* 114, sec. 80) that he brought one to Assyria among the Egyptian specialists he took as prisoners of war. For prescriptions dealing with veterinary science in Akkadian, see BAM 159 v 33–47; in Ugaritic, Andrée Herdner, *Corpus des tablettes en cunéiformes alphabétiques,* which is Mission de Ras Shamra 10 (Paris, 1963), 245–247: "Textes hippiatriques"; and Maurice Bear Gordon, "The Hippiatric Texts From Ugarit," in *Annals of Medical History,* 3rd ser., **4** (1942), 406–408; in Egyptian: Hildegard von Deines, Hermann Grapow, and Wolfhart Westendorf, "Übersetzung der medizinischen Texte," in *Grundriss der Medizin der alten Ägypter,* IV, pt. 1 (Berlin 1958), 317–319; and Hermann Grapow, *Von den medizinischen Texten* (Berlin, 1959), 88, n. 1. Ilse Fuhr, "Ein sumerischer Tierarzt," in *Archiv orientální,* **34** (1966), 570–573, is not convincing.

151. Much valuable work in the investigation of Mesopotamian material culture has been done by Armas Salonen in an impressive series of books dealing successively (since 1939) with boats and other nautical matters, vehicles of every description, draught animals, furniture and household utensils, doors, footwear, agriculture, fishing, and brickmaking. Salonen's books in many respects constitute pioneering work that of necessity is word-oriented. They do not take into account the specific evidentiary value of the text types in which these words occur and the complexities of the socioeconomic structures that created the documentation. The work of Robert James Forbes, *Studies in Ancient Technology* (Leiden, 1955–   ; 2nd ed., 1964–   ), is much wider in scope, since it incorporates the technologies of the classical world; but it suffers from philological inadequacy in the realm of the ancient Near East, especially Mesopotamia.

152. The names of many late settlements—"Gold Tell," "Brick Tell," "Stone Tell," "Ruin Tell," "*galala*-stone Tell,"—show that the inhabitants knew well what was buried under these mounds. They also seem to have used the rubble of ancient settlements (called *eperu*; see *Assyrian Dictionary,* sub *eperu,* meaning 6) as fertilizer, just as the Egyptians, both ancient and modern, have done. See Ludwig Keimer, "Das 'Sandfahren' der Totenfiguren (wšbtiw) 'um den Sand der Ostseite zur Westseite zu fahren': Die früheste Erwähnung einer künstlichen Düngung im Alten Ägypten," in *Orientalistische Literaturzeitung,* **29** (1926), 98–104.

153. Deviating not a finger's width from the original dimensions, as the Neo-Babylonian kings love to stress in their inscriptions.

154. The best-known exception is furnished by the royal tombs at Ur, discovered by C. Leonard Woolley—*The Royal Cemetery,* which is Ur Excavations 2 (London, 1934). Only a few bricks of Neo-Assyrian sepulchral structures are left (KAH 46 and 47). For a stone sarcophagus and its cover, see *Assyrian Dictionary,* sub *arannu,* usage c. Tomb inscriptions are extremely rare in Akkadian texts.

155. See Yigael Yadin, *The Art of Warfare in Biblical Lands* (New York, 1963); Mary Littauer, "Did the Kassites Influence the Development of the Late-Bronze-Age War and Hunting Chariot?" in *Studi di archeologia e storia dell'arte del vicino oriente,* **1** (1975); and "New Light on the Assyrian Chariot," in *XXIIe Rencontre assyriologique internationale* (Rome, 1974). For a unique example of an extant chariot found in Egypt, see Heinrich Schäfer, *Armenisches Holz in altägyptischen Wagnereien,* which is *Sitzungsberichte der Preussischen Akademie der Wissenschaften zu Berlin,* phil.-hist. Kl., **25** (1931).

156. For an attempt in that direction, see Barthel Hrouda, *Die Kulturgeschichte des assyrischen Flachbildes* (Bonn, 1965), although it lacks the collaboration of a philologist, which would have greatly enhanced the value of the book.
157. For modern methods in archaeology, see, for example, Don R. Brothwell and Eric Higgs, eds., *Science in Archaeology. A Comprehensive Survey of Progress and Research* (New York, 1963); rev., enl. ed., 1969. Note also the periodicals *Archaeometry* and *Archäographie: Archäologie und elektronische Datenverarbeitung*.
158. See now, with previous literature, P. R. S. Moorey, *Catalogue of the Ancient Persian Bronzes in the Ashmolean Museum* (Oxford, 1971); and P. R. S. Moorey and F. Schweizer, "Copper and Copper Alloys in Ancient Iraq, Syria, and Palestine," in *Archaeometry*, **14** (1972), 177–198.
159. See A. Leo Oppenheim and Louis F. Hartman, *On Beer and Brewing Techniques in Ancient Mesopotamia: According to the XXIIIrd Tablet of the Series HAR.ra = ḫubullu, Journal of the American Oriental Society*, supp. no. 10 (1950). See also Miguel Civil, "A Hymn to the Beer Goddess and a Drinking Song," in *Studies Presented to A. Leo Oppenheim*, 67–89; Dietz Otto Edzard, "Brauerei, Bierkonsum und Trinkbräuche im Alten Mesopotamien," in *Jahrbuch der Gesellschaft für Geschichte und Bibliographie des Brauwesens e.V.* (1967), 9–21; Wolfgang Röllig, *Das Bier im Alten Mesopotamien* (Berlin, 1970); and Marten Stol, "Zur altmesopotamischen Bierbereitung," in *Bibliotheca orientalis*, **28** (1971), 167–171.
160. See Marie-Thérèse Barrelet, *Figurines et reliefs en terre cuite de la Mésopotamie antique*, I *Potiers, termes du métier, procédés de fabrication et production* (Paris, 1968); and James Leon Kelso, *The Ceramic Vocabulary of the Old Testament* (New Haven, Conn., 1948). For textiles, see Harmut Waetzoldt, *Untersuchungen zur neusumerischen Textilindustrie* (Rome, 1972).
161. For an early attempt, see Walter Reimpell, *Geschichte der babylonischen und assyrischen Kleidung* (Berlin, 1921).
162. Problems related to the connection between technological changes and social and/or economic developments cannot be discussed as yet in the realm of the ancient Near East. There is not enough evidence even to approach the topic presented by Moses I. Finley, "Technical Innovation and Economic Progress in the Ancient World," in *Economic History Review*, 2nd ser., **18** (1965), 29–45.
163. An interesting exception to this statement is furnished by the probable use of the silk of the "Assyrian silkworm" (*Pachypasa otus Drury*), discussed in A. Leo Oppenheim, "Essay on Overland Trade in the First Millennium B.C.," in *Journal of Cuneiform Studies*, **21** (1967, published 1969), 236–254. This was prior to the importing of real silk (produced by *Bombyx mori*) in the last third of the first millennium B.C.; for problems involved, see *ibid.*, 252 f.
164. On the level of subsistence technology, one might refer to the introduction into the Neart East of the rotary quern instead of the traditional push quern. As for agricultural products, cotton and rice (Strabo, XV.i.18) came into the Near East at that time; for agricultural methods, a passage from Strabo (XVI.4.1) may be quoted: "The vine grows in the marshes, as much earth being thrown on hurdles of reeds as the plant may require; so that the vine is often carried away, and then is pushed back again to its proper place by means of poles." This technique, also attested in the New World (segments of land artificially constructed in lakes or canals, called *chinampas* or *camellones*; see Charles Gibson, *The Aztecs Under Spanish Rule* [Stanford, Calif., 1964], 320 f.), illustrates the intensification of agricultural production in late Mesopotamia, ushered in during the Achaemenid period.
165. See *Assyrian Dictionary*, sub *mušākilu*.
166. A number of domesticated birds, such as the chicken, the turkey (bustard), and the peacock, came into Mesopotamia from the East. They seem to have been considered rare and exotic animals, and had no economic importance. The rooster (not the peacock)—"bird of the god Haya," the watchman of the night—is known from Sumerian literary texts, and the chicken as the "bird that gives birth every day" from an Egyptian source—Georg Steindorff, ed., *Urkunden des ägyptischen Altertums* (Leipzig, 1904–), IV, 700 13–14; in Syriac, the bustard has the name "Akkadian (bird)"; the peacock is not attested.
167. In that role, the donkey is well attested from the Sumerian period on. See Samuel Noah Kramer, *Enmerkar and the Lord of Aratta* (Philadelphia, 1952), ll. 127, 282–284, 331–332; see also Klaas Roelof Veenhof, *Aspects of Old Assyrian Trade and Its Terminology* (Leiden, 1972), 1–45 (with previous literature); and William Foxwell Albright, "Midianite Donkey Caravaneers," in Harry Thomas Frank and W. L. Reed, eds., *Translating and Understanding the Old Testament* (Nashville–New York, 1970), 197–205.
168. This is well attested in *Archives royales de Mari*, **9** (1960), no. 24 ii 23; PBS 1/2 136:34 f. The dog as helper of the hunter is difficult to find in texts, although it does appear in that role in Neo-Assyrian reliefs. Possibly the many dogs mentioned in the administrative texts of the Third Dynasty of Ur were destined for hunting.
169. The problem of the Equidae of Mesopotamia has not yet been treated in all its philological, paleozoological, and iconographic complexities. See also Burchard Brentjes, *Die Haustierwerdung im Orient* (Wittenberg, 1965). For the gelding of horses as a "western" practice, see Oppenheim, "A Note on *ša rēši*."
170. Not only precious stones (strongly colored minerals that took a good polish) but also millstones were imported; see Madeleine Lurton Burke, "Lettres de Numušda-Naḫrâri," in *Syria*, **41** (1964), 75 f., an article based on certain Mari letters—*Archives royales de Mari* (texts in translation and transliteration), **13** (1964), nos. 82 and 90. Lapis lazuli was imported to Mari from Eshnunna, according to *Archives royales de Mari*, **9** (1960), no. 254. For early stone vessels made of imported stones, see Arno Schüller, "Die Rohstoffe der Steingefässe der Sumerer aus der archaischen Siedlung bei Uruk-Warka," in *Vorläufiger Bericht über die . . . Ausgrabungen in Uruk-Warka*, **19** (1961), 56–58. Alabaster vessels were imported but were also imitated locally; see Friedrich Wilhelm von Bissing, "Ägyptische und ägyptisierende Alabastergefässe aus den deutschen Ausgrabungen in Assur," in *Zeitschrift für Assyriologie*, **46** (1940), 149–182; and, from Babylon, "Ägyptische und ägyptisierende Alabastergefässe aus den deutschen Ausgrabungen zu Babylon," *ibid.*, **47** (1942), 27–49. The tracing of cultural contacts, across the Near East and beyond, by the spread of obsidian is shown by Colin Renfrew, J. E. Dixon, and J. R. Cann, "Obsidian and Early Cultural Contacts in the Near East," in *Proceedings of the Prehistoric Society for 1966*, n.s. **32**, 30–72; and by Gary A. Wright, *Obsidian Analyses and Prehistoric Near Eastern Trade: 7500 to 3500 B.C.*., Museum of Anthropology, University of Michigan, Anthropological Papers 37 (Ann Arbor, 1969).
171. Bitumen was found in Hit (near Sippar) and brought from there to the rest of Babylonia. In spite of an abundance of evidence, I know of no adequate presentation of its manifold uses; but see the remarks in Armas Salonen, *Die Ziegeleien im alten Mesopotamien* (Helsinki, 1972), 53–57. For another source of bitumen, see Burke, "Lettres de Numušda-Naḫrâri," 67 f.; and, from Madga, in the region of modern Kirkuk, see Adam Falkenstein, *Die Inschriften Gudeas von Lagaš*, I, which is Analecta Orientalia, 30 (Rome, 1966), 51.
172. The fact that metal (ore) has to be mined is rarely alluded to in Mesopotamian texts; see *Assyrian Dictionary*, sub *ḫurru*, usage c; and, for Sumerian passages, see Gudea Statue

B vi 21–23, Cylinder A xvi 15–17, and l. 86 of the story called Inanna and Epih, as cited in Benno Landsberger, "Tin and Lead: The Adventures of Two Vocables," in *Journal of Near Eastern Studies*, **24** (1965), 291, n. 25, sub A d. For the Old Testament, see Deut. 8:9 and Job 28: 1–2. Smelting operations at the mines seem to be mentioned by Sargon in Arthur Gotfred Lie, *The Inscriptions of Sargon II, King of Assyria* (Paris, 1929), 38:230–232. The situation was different in Egypt, where slaves worked in local and nearby mines; see Alan Henderson Gardiner, *Egypt of the Pharaohs* (Oxford, 1961), 251–255.

173. For amber and the problem of its provenience seen from the point of view of the scientist, see R. C. A. Rottländer, "On the Formation of Amber from *Pinus* Resin," in *Archaeometry*, **12** (1970), 35–52.

174. For ivory, see A. Leo Oppenheim, "The Seafaring Merchants of Ur," in *Journal of the American Oriental Society*, **74** (1954), 6–17, esp. 11 f., telling of the fluctuations in the importing of this raw material and its provenience (India versus Upper Syria). The problem persists in the first millennium B.C., when, for example, ivory is mentioned in the Neo-Babylonian letter ABL 1283 r. 5 as going from Babylon to Nineveh. There is, as yet, no scientific method to establish the provenience of the material used for individual ivory objects. Ivory was sometimes dyed (see *Assyrian Dictionary*, sub *bašlu*, meaning 4) and gilded—see Max Edgar Lucien Mallowan, *Nimrud*, II (London, 1966), 554–556—and in the first millennium, it was used predominantly as inlay for precious furniture. For ivory as writing material, see Donald J. Wiseman, "Assyrian Writing-Boards," in *Iraq*, **17** (1955), 3–13. In addition to elephant tusks, narwhal tusks were used; see E. A. Wallis Budge and Leonard William King, eds., *Annals of the Kings of Assyria* (London, 1902), 373:88.

175. Remains of industrial installations, apart from the ubiquitous metalworker shops and pottery kilns, are very rare. It seems that oil presses from Ugarit can be identified—Claude F. A. Schaeffer, *Ugaritica 4*, which is Mission de Ras Shamra 15 (Paris, 1962), 420–433—as can oil and wine presses and dyeing vats from Palestine—see William Foxwell Albright, *The Excavation of Tell Beit Mirsim*, III (New Haven, 1943), secs. 36–40, 55–57. The city quarters in which certain craftsmen lived and worked together (such as smiths and tanners; see n. 251) have not been located.

176. Pride in invention is very rarely attested in the ancient Near East. In Egypt, Amenemhet claimed to have invented the water clock (see n. 249); see Siegfried Schott, "Voraussetzung und Gegenstand ägyptischer Wissenschaft," in *Jahrbuch. Akademie der Wissenschaften und der Literatur in Mainz* (1951), 290; and, for a trans. of the pertinent text, see Ludwig Borchardt, *Die altägyptische Zeitmessung* (Berlin–Leipzig, 1920), B 60–63. The architect Ineni claimed credit for certain technical innovations, stressing his independence of tradition and his working for future generations; see Helmut Brunner, "Zum Zeitbegriff der Ägypter," in *Studium Generale*, **8** (1955), 589. In Mesopotamia, in one of the few extant inscriptions of provincial governors, we find an elaborate if not fully intelligible description of a locklike arrangement on a canal, apparently invented by Šamaš-rēš-uṣur, the governor of Suhi and Mari, published by Franz Heinrich Weissbach, in *Babylonische Miscellen* (Leipzig, 1903) pl. 2 ff. and no. 4. The same official also stresses that he introduced apiculture in his province. This represents the *topos* of the ruler as inventor of technical innovations, which we encounter also in the inscriptions of Sennacherib (704–681 B.C.). See Daniel David Luckenbill, *The Annals of Sennacherib*, which is Oriental Institute Publications 2 (Chicago, 1924), 141; for collated text, see *Assyrian Dictionary*, sub *bašālu*, meaning 8.

One should also mention the representation of the Sumerian ruler Gudea of Lagash (*ca.* 2140 B.C.) as architect holding the plans of a building; see the statues listed in *Reallexikon der Assyriologie*, III (1957–1971), 682, sub b 1 and b 3. A unique stance is taken by Thutmosis III (1469–1436 B.C.) when proclaiming that he designed (literally, "what his own heart had conceived") precious containers; James H. Breasted, *Ancient Records of Egypt*, II (Chicago, 1906–1907), secs. 164, 545, 775. This is in patent contrast with the concept of the ideal ruler in the Old Testament: Solomon has the temple built by foreign artisans, and even its plan is of divine origin (see 1 Chron. 28: 19; and Num. 8:4). Pride in originality expressed by the artist is very rare in the ancient Near East. I know of examples only from Egypt: see John A. Wilson, "The Artist of the Egyptian Old Kingdom," in *Journal of Near Eastern Studies*, **6** (1947), 231–249, esp. 247 f. and n. 68; see also Hermann Junker, *Die gesellschaftliche Stellung der ägyptischen Künstler im alten Reich* (Vienna, 1959).

177. The domestication of the wild-growing cereals and their adaptation to the ecological realities of the Near Eastern agricultural centers are not the concern of this paper. See Jane Renfrew, *Paleoethnobotany. The Prehistoric Food-plants of the Near East and Europe* (London, 1973). Many vegetables and spice plants have left their seeds in the soil of excavated tells; the matching of these with the equally numerous Sumerian and Akkadian plant names is a task that has not yet been undertaken.

178. In Mesopotamia, all agricultural activities aim at the support of human beings; only in Urartu, as we learn from a text of Sargon II (TCL 3 275; also 180, and see n. 12), was a forage crop grown for horses. This seems to refer to the alfalfa (*Medicago sativa*; in Greek, "median [plant]"), which was grown on the entire Iranian plateau to feed horses; see Berthold Laufer, *Sino-Iranica*, Field Museum of Natural History publication no. 201, Anthropological Series, vol. XV, no. 3 (Chicago, 1919), 208–219. Strangely enough, the Pahlavi *aspast* appears in Mesopotamian texts (see *Assyrian Dictionary*, sub *aspastu* and *asupasāti* and, possibly, *aspatu*), but these words predate the Persian conquest of the region and do not suggest a forage plant. On the other hand, the name of the Persian official *aspastua* (from Achaemenid Nippur) does refer to alfalfa.

179. For the date palm and its history, literature is given in Ingrid Wallert, *Die Palmen im alten Ägypten* (Berlin, 1962); for Mesopotamia, see Benno Landsberger, *The Date Palm and Its By-Products According to the Cuneiform Sources* (Graz, 1967).

180. Nabonidus (555–539 B.C.) states this expressly—Leonard William King, ed., *Babylonian Boundary-Stones and Memorial-Tablets in the British Museum* (London, 1912), no. 37 r. 10: "Wine, the mountain drink, that does not exist in my country." Wine is mentioned much more frequently in Assyrian texts and in those coming from the West than in texts from Babylonia proper, at least up to the prosperous times of the Chaldean empire. When it is mentioned, it is said to be imported from the West. There is evidence from Mari and Assyria that wine was drunk diluted with water, as it was in the classical world.

181. As an illustration of the most coveted imported tree, see Horst Klengel, "Der Libanon und seine Zedern in der Geschichte des Alten Vorderen Orients," in *Altertum*, **13** (1967), 67–76. The locally available gypsum was burned and mixed with sand to produce a fine material for plastering walls and floors. The use of lime, which requires high temperatures, is also attested. Quite early (in Uruk [Erech] and Ur) this material was used to form building "stones."

182. For an example, see David Oates, "The Excavations at Tell al Rimah, 1968," in *Iraq*, **32** (1970), 1–23. For ancient city models and city plans, see Paul Lampl, *Cities and Planning in the Ancient Near East* (London, 1968); the unique detailed plan of Nippur was last published by Inez Bernhardt and Samuel Noah Kramer, "Der Stadtplan von

Nippur, der älteste Stadtplan der Welt," in *Wissenschaftliche Zeitschrift der Friedrich Schiller Universität, Jena,* **19** (1970), 727–730.

183. At times, textiles leave impressions on corroded metal surfaces—see Jacques Jean Marie de Morgan, *La préhistoire orientale*, II (Paris, 1927), 59–61—or on clay—see Robert D. Biggs, *Inscriptions From Tell Abū Ṣalābīkh*, which is Oriental Institute Publications, no. 99 (Chicago, 1974), 22 f. In "Çatal Hüyük—the Textiles and Twined Fabrics," *Anatolian Studies*, **15** (1965), 169–174, Harold B. Burnham presents textiles and twined fabrics of a much earlier period from Çatal Hüyük. Egyptian fifth-millennium fabrics are extant in the Fayum; see Gertrude Caton-Thompson and Elinor Wight Gardner, *The Desert Fayum* (London, 1934), pl. 38. See also Louisa Bellinger, "Textiles From Gordion," in *Bulletin of the Needle and Bobbin Club*, **46** (1962), 5–33.

184. We know of meteoric iron (called "iron fallen from heaven"), for example, only from a religious hymn—col. v, l. 21, of the Papulegarra hymn published by Theophilus G. Pinches in "Hymns to Pap-dul-garra," in *Journal of the Royal Asiatic Society*, centenary suppl. (1924), 63:86—from a litany—see Reiner, "*Lipšur* Litanies," 140:18′ and 34′—and from Sumerian literary texts, which use the term k ù. a n. n a ("metal [fallen] from heaven") for what may be meteoric iron. See the references cited by Landsberger, "Tin and Lead," 290 f., n. 25, esp. 291, sub Cb and Cc.

185. Even the absence of a material can be of importance. See A. Leo Oppenheim, "The Seafaring Merchants of Ur," in *Journal of the American Oriental Society*, **74** (1954), 6–17, on the absence of pearls from Mesopotamia up to the Persian period, despite the proximity of the pearl-rich waters of the Persian Gulf. See also Robert J. Braidwood, "Some Parthian Jewelry," in *Second Preliminary Report Upon the Excavations at Tell Umar, Iraq* (Ann Arbor, 1933), 72 and pl. 42, fig. 1. The stone beads imported from the East into Ur and called "fish eyes" (IGI.KU$_6$) cannot be interpreted as pearls. Rather, they are black and white banded agate made into beads to look like fish eyes. The absence of pearls is the more astonishing since Gilgamesh uses a stone for diving, just as pearl fishers do.

186. For such instances, see the year names of the later kings of the Hammurapi dynasty, which contain, apart from descriptive references to divine and royal statues and sacral furnishings, the characterization of certain copper socles (?) as showing the representations of "mountains and rivers"—*Reallexikon der Assyriologie*, II (1938), 182 ff., the dates no. 153 (Samsuiluna) and 261, 262, 265 (Ammizaduga); the Middle Assyrian treasury inventory, which seems to describe representations of animals, trees, and other objects made of gold inlaid with precious stones and colored glass, and also mentions embroidered hangings—Franz Köcher, "Ein Inventartext aus Kār-Tukulti-Ninurta," in *Archiv für Orientforschung*, **18** (1957–1958), 300–313; and the letter of Sargon II (TCL 3 399–404), which enumerates the booty taken in the capital of Urartu, describing certain metal statues in detail. Some inscriptions of Neo-Babylonian rulers at times describe even buildings in technical terms, such as the bonding of the bricks—Stephen Langdon, ed., *Die neubabylonischen Königsinschriften*, which is Vorderasiatische Bibliothek 4 (Leipzig, 1912), 76 iii 13 f.—or the construction of a double roof plus ceiling in CT 37 8 i 44 f.—parallel *Museum Journal*, **14** (1923), 267 ff. i 48 f.; also Langdon, *op. cit.*, 126 iii 27 ff. and 230 i 21 ff.

187. See Erich Ebeling, *Parfümrezepte und kultische Texte aus Assur* (Rome, 1950), with a Neo-Assyrian fragment from Calah—Donald J. Wiseman and J. V. Kinnier Wilson, "The Nimrud Tablets, 1950," in *Iraq*, **13** (1951), 112 ND 460.

188. See Erich Ebeling, *Bruchstücke einer mittelassyrischen Vorschriftensammlung für die Akklimatisierung und Trainierung von Wagenpferden* (Berlin, 1951); and, for the Hittite material, Annelies Kammenhuber, *Hippologia hethitica* (Wiesbaden, 1961).

189. Oppenheim *et al., Glass and Glassmaking*; also A. Leo Oppenheim, "More Fragments With Instructions for Glassmaking," in *Journal of Near Eastern Studies*, **32** (1973), 188–193; and "Towards a History of Glass in the Ancient Near East," in *Journal of the American Oriental Society*, **93** (1972), 259–266.

190. Published in autograph copy by Jan J. A. van Dijk, "Textes divers du Musée de Baghdad, II," in *Sumer*, **13** (1957), 115, pl. 24.

191. Published by Civil, "A Hymn to the Beer Goddess."

192. For a translation of this difficult text, see Cyril J. Gadd, "Two Sketches From the Life at Ur," in *Iraq*, **25** (1963), 183–188.

193. They are accessible in their trans. by Earle Radcliffe Caley, "The Leyden Papyrus X," in *Journal of Chemical Education*, **3** (1926), 1149–1166; and the text from Stockholm, in "The Stockholm Papyrus," *ibid.*, **4** (1927), 979–1002.

194. This has been pointed out in a critique of my paper by Dietlinde Goltz, "Versuch einer Grenzziehung zwischen 'Chemie' und 'Alchemie,'" in *Sudhoffs Archiv für Geschichte der Medizin . . .*, **52** (1968), 30–47.

195. This is BM 62788, to which Professor A. Kirk Grayson drew my attention.

196. Also belonging here are such isolated texts as Erich Ebeling, "Ein Rezept zum Würzen von Fleisch," in *Orientalia*, n.s. **18** (1949), 171–172 (referring to GCCI 2 394); and Oliver R. Gurney, "An Old Babylonian Treatise on the Tuning of the Harp," in *Iraq*, **30** (1968), 229–233, also discussed by David Wulstan, "The Tuning of the Babylonian Harp," *ibid.*, 215–228.

197. A. Leo Oppenheim, "Towards a History of Glass in the Ancient Near East"; and "The Position of the Intellectual in Mesopotamian Society."

198. In "Towards a History of Glass" I have pointed out just such an instance: prisoners or deported craftsmen from Syria introduced glassmaking into Egypt. In Egypt they first produced containers in forms that were un-Egyptian, but later changed the style of their products to exquisite and typically Egyptian objects. See also Dan Barag, "Mesopotamian Glass Vessels of the Second Millennium B.C.," in *Journal of Glass Studies*, **4** (1962), 23–25. The problem touched upon here has hardly been treated in its implications. Pertinent evidence is not very rare—for example, the list of deported craftsmen and "professionals" taken from Egypt by Esarhaddon (see n. 150, above) and the Bible passages 1 Sam. 13:19 and 2 Kings 24:14 = Jer. 24:1, 29:2. For a related problem, see Jack M. Sasson, "Instances of Mobility Among Mari Artisans," in *Bulletin of the American Schools of Oriental Research*, **190** (1968), 46–54.

199. See Carlo Zaccagnini, *Lo scambio dei doni nel Vicino Oriente durante i secoli XV–XIII* (Rome, 1973).

200. The Amarna letters (Knudtzon, *Die El-Amarna-Tafeln*)—nos. 13 (gifts from Babylon), 14 (from Egypt), and 22 and 25 (from Mitanni)—are the best examples of such lists. See also ADD 810 (= ABL 568).

201. For an example, see A. Leo Oppenheim, "On Royal Gardens in Mesopotamia," in *Journal of Near Eastern Studies*, **24** (1965), 328–333. For the practice of laying out a royal botanical garden to display the king's wealth and his conquest of foreign countries, see Donald J. Wiseman, "A New Stela of Aššur-naṣir-pal II," in *Iraq*, **14** (1952), 33: 40–52. The difference in the style of living at court (and probably also in well-to-do circles) between West and East is a topic worthy of investigation. The use of ice to cool drinks—latest discussions in Stephanie Page, "Ice, Offerings and Deities in the Old Babylonian Texts From Tell-el-Rimah," in *Actes de la XVIIe Rencontre assyriologique internationale* (Ham-sur-Heure, 1970), 181–183; and

Helmut Freydank, "*Bīt šurīpim* in Boğazköy," in *Welt des Orients*, **4** (1968), 316–317—and kings holding flowers or handkerchiefs in their hand illustrate this contrast. The difference in courtly manners also had its moral implications; see the reference from Ugarit to a princess behaving coquettishly in public—Jean Nougayrol, *Le palais royal d'Ugarit 3*, which is Mission de Ras Shamra 6 (Paris, 1955), 43 RS 16.270:25—as well as the case of adultery discussed by William L. Moran, "The Scandal of the 'Great Sin' at Ugarit," in *Journal of Near Eastern Studies*, **18** (1959), 280–281.

202. For the importance of the new style of ivory-inlaid furniture that characterized the palaces of the first half of the first millennium, see Richard D. Barnett, *A Catalogue of the Nimrud Ivories With Other Examples of Ancient Near Eastern Ivories in the British Museum* (London, 1957).

203. The Old Testament offers a revealing exception. The ceaseless fight of the prophets against the strong desire of the kings (of Judah and Israel) to live, to think, and to behave just like all other kings around them illustrates the pervasive force of "royal fashions."

204. Trade in spices and perfumes (that is, their importation for the use of palace and temple) is attested sporadically from the end of the third millennium B.C.—see, for example, John B. Curtis and William W. Hallo, "Money and Merchants in Ur III," in *Hebrew Union College Annual*, **30** (1959), 103–139—to the Chaldean period. Such trading, however, is not comparable in importance with the role of, let us say, the pepper and incense trade of the subsequent two millennia. The camel-borne incense trade of the first millennium bypassed Mesopotamia, with the exception of the encroachments of the Assyrian kings (from Tukulti-Ninurta II to Assurbanipal, incense is mentioned as tribute coming from the Arabs). The incense was drawn out of Arabia into Egypt and Palestine (and eventually into Greece and Rome) by the demands of the cult and of personal uses.

205. For the shifting of the trade pattern in textiles, see A. Leo Oppenheim, "Trade in the Ancient Near East," in *Fifth International Congress of Economic History, Leningrad 1970* (Moscow, 1970), 11.

206. See Oppenheim, *Ancient Mesopotamia*, 283.

207. For a proposal, see Oppenheim, *Letters from Mesopotamia*, 42–54.

208. For certain changes in this respect, see n. 163.

209. A case in point is furnished by the Mesopotamian textile industry. Weaving techniques (in wool and linen) remain quite primitive throughout, producing monochrome fabrics in simple (tabby) weave patterns. Appliqué work and colored surface decorations were used at times to embellish the product; see A. Leo Oppenheim, "The Golden Garments of the Gods," in *Journal of Near Eastern Studies*, **8** (1949), 172–193; and Jeanny Vorys Canby, "Decorated Garments in Ashurnasirpal's Sculpture," in *Iraq*, **33** (1971), 31–53. Certain pavement slabs from Neo-Assyrian palaces show stone imitations of carpets with fringes and characteristic ornamentation. We do not know which craftsmen produced them. The multiheddled (Far Eastern) loom that permits pattern weaving was introduced into Mesopotamia along with silk. For the importation of "Western" textiles from the last third of the second millennium, see A. Leo Oppenheim, "Essay on Overland Trade in the First Millennium B.C.," 246 f.

210. An exception is an alcoholic beverage produced from dates, instead of a beer brewed of malted barley "bread," a change that occurred around the end of the second or the beginning of the first millennium B.C. See Oppenheim, *Letters from Mesopotamia*, 44.

211. See Armas Salonen, "Die Öfen der alten Mesopotamier," in *Baghdader Mitteilungen*, **3** (1964), 100–124; and "Bemerkungen zur sumerisch-akkadischen Brennholz-Terminologie," in *Jaarbericht van het Vooraziatisch-Egyptisch Genootschap Ex Oriente Lux*, **18** (1964), 331–338. For a discussion of the glassmaker's kiln and its parts, see Oppenheim *et al., Glass and Glassmaking*, 69–71.

212. Except that they were activated by both hands and feet and had a leather air bag. For pot bellows preserved in the Iraq Museum, Baghdad, see W. Winton, as quoted by Lamia al-Gailani in "Tell edh-Dhibaʾi," in *Sumer*, **21** (1965), pl. 6–7 and pp. 37–38; N. Avigad in *Bulletin of the American Schools of Oriental Research*, **163** (1961), 18–22; and C. C. Lamberg-Karlovsky, "Archaeology and Metallurgical Technology in Prehistoric Afghanistan, India, and Pakistan," in *American Anthropologist*, **69** (1967), 152. Egyptian foot bellows are discussed by T. G. Crawhill, "Iron Working in the Sudan," in *Man*, **33** (1933), 41–43.

213. Meat was rare food for the common man. This is shown by Nabonidus' (555–539 B.C.) boast that he fed the workmen on his temple-building projects various kinds of bread, fine beer, meat, and wine: *Oxford Editions of Cuneiform Texts* I, p. 36 iii 27. This reflects a similar statement of Sin-iddinam of Larsa (1834–1823 B.C.); see Edmond Sollberger and Jean-Robert Kupper, *Inscriptions royales sumériennes et akkadiennes* (Paris, 1971), 190:50–57, 192 ii 16–35. A detailed bill of fare for a royal banquet given by Aššurnaṣirpal II (883–859 B.C.) on the occasion of the inauguration of his new palace was published in Wiseman, "A New Stela of Aššur-naṣir-pal II," 24–44. See the trans. by Oppenheim in Pritchard, *Ancient Near Eastern Texts*, 3rd ed. with supp. (1969), 560a.

214. Since the seeds of sesame (*Sesamum indicum*) are conspicuously absent from Mesopotamian soil, we cannot establish to what oilseed the Akkadian *šamaššammu* refers. For the problems involved, see Fritz Rudolf Kraus, "Sesam im alten Mesopotamien," in *Journal of the American Oriental Society*, **88** (1968), 112–119. For other oil-producing plants, see Hans Gustav Güterbock, "Oil Plants in Hittite Anatolia," *ibid.*, 66–71.

215. Certain plants—see Reginald Campbell Thompson, *A Dictionary of Assyrian Botany* (London, 1949), 31 ff.; and Immanuel Löw, *Die Flora der Juden*, I (Vienna–Leipzig, 1928), 635 ff.—when burned, yield ashes that contain alkali. When combined with oil, a soaplike liquid is obtained; see Oppenheim *et al., Glass and Glassmaking,* 74 and n. 87. Add to the texts there cited KUB 37 55 iv 26′–29′ and 32′–35′. See also Helmuth T. Bossert, "Zur Geschichte der Seife," in *Forschungen und Fortschritte*, **29** (1955), 208–213; and Eberhard Schmauderer, "Seifenähnliche Produkte im alten Orient," in *Technikgeschichte*, **34** (1967), 300–310. At times, crushed gypsum was added as an abrasive. For the use of alkali in the manufacture of glass, see Robert H. Brill, "The Chemical Interpretation of the Texts," in Oppenheim *et al., Glass and Glassmaking*, 112.

216. For the use of filigree techniques (*Kornfiligran*), see Robert Koldewey, *Das wieder erstehende Babylon*, 4th ed. (Leipzig, 1925), fig. 186. For wire filigree in the Old Testament, see Exod. 39:3. See also Diane Lee Caroll, "Wire Drawing in Antiquity," in *American Journal of Archaeology*, **76** (1972), 321–323.

217. For the use of the *cire perdue* technique for casting metal objects, see *Assyrian Dictionary*, sub *iškuru*, usage b, for second-millennium references; for the first millennium, see Johann Nepomuk Strassmaier, ed., *Babylonische Texte, Inschriften von Nabonidus* (Leipzig, 1889), no. 429:1 f.

218. See K. R. Maxwell-Hyslop, *Western Asiatic Jewellery c. 3000–612 B.C.* (London–New York, 1971); and the important review of this book by Agnès Spycket in *Revue d'assyriologie et d'archéologie orientale*, **67** (1973), 83–90. See also John F. X. McKeon, "Achaemenian Cloisonné-Inlay Jewelry: An Important New Example," in Harry A. Hoffner, Jr., ed., *Orient and Occident, Essays Presented to Cyrus H. Gordon on the Occasion of His Sixty-Fifth Birthday,* which is Alter Orient and Altes Testament 22 (Neukirchen-Vluyn, 1973), 109–117.

219. For the fame of the palace in Mari, see the letter published by Georges Dossin in *Ugaritica I*, which is Mission de Ras Shamra 3 (Paris, 1939), 15 and 16, n. 2; also André Parrot, "Fouilles de Mari," in *Syria*, **18** (1937), 74 ff.
220. In the inscription of the Kassite king Agum-kakrime (*ca.* third quarter of the second millennium) we find a more detailed description of the work on the temple of Marduk, including the doors and doorways, the garments, crowns, and jewelry of the divine images made of gold and decorated with precious stones; for a now somewhat obsolete trans. see Unger, *Babylon*, 276 ff.
221. For the relation between the quality of the artwork and the social position of those who commissioned (or bought) it, see Edith Porada, "Gesellschaftsklassen in Werken altorientalischer Kunst," in *Gesellschaftsklassen im alten Zweistromland und in den angrenzenden Gebieten,* D. O. Edzard, ed., which is *Abhandlungen der Bayerischen Akademie der Wissenschaften*, Phil.-hist. Kl., n.s. no. 75 (1972), 147–157.
222. Since the quality of the silver brought to the temple varied greatly, it was temple practice to melt down all silver offerings. See A. Leo Oppenheim, "A Fiscal Practice of the Ancient Near East," in *Journal of Near Eastern Studies*, **6** (1947), 116–120, with references to papers discussing parallel practices in the temples of Solomon and Zerubbabel, in Persia (according to Herodotus), and in the Egypt of the Abbasid caliphs.
223. Silver used for payment was generally of poor quality. Witness a letter from the authorities of the temple in Uruk to goldsmiths: "If you [again] cast [objects out of] silver provided with the *gin*-mark [currency silver] you will commit a serious crime against the king" (GCCI 2 101:8–10). Another letter (CT 22 40) refers to a royal decree that refined silver, rather than silver with a *gin*-mark, should be given for a specific purpose. Because even small quantities of silver are mentioned as being marked with the sign *gin* ("normal"), it is possible that sheets of silver stamped all over with such a mark may have been cut up and weighed out as payment.
224. A special use of copper comes into the ancient Near East in the second millennium: copper scales sewn on leather serve as mail, for the protection of men as well as of horses. Its foreign designation (in Akkadian, *sa[iriam*; in Hurrian, *šarrijanni*; in Egyptian, *tryn*; in Hebrew, *širjōn*) shows that this protective armor was brought in from the outside. See A. Leo Oppenheim, in *Journal of Cuneiform Studies*, **4** (1950), 192–195; see also Erkki Salonen, *Die Waffen der alten Mesopotamier,* which is Studia Orientalia 33 (Helsinki, 1965), 105–107; for illustrations, see Burchard Brentjes, "Equidengerät, Equiden in der Religion des alten Orients," in *Klio*, **53** (1971), 76–96, esp. 80 ff. A probably similar coat of mail (also used for men and horses) called *gurpisu* (see *Assyrian Dictionary*, sub voce; also Salonen, *op. cit.*, 101–104) appears somewhat earlier in peripheral Babylonian texts (Ishchali and Mari).
225. Copper was also used as means of payment in Assyria until the eighth century B.C.; see Carlo Zaccagnini, "La terminologia accadica del rame e del bronzo nel I millennio," in *Oriens antiquus*, **10** (1971), 129. For the "oxhide" ingots of copper, see Robert Maddin and James D. Muhly, "Some Notes on the Copper Trade in the Ancient Mid-East," in *Journal of Metals*, **26,** no. 5 (May 1974), 1–7. The ratio copper: tin is rarely mentioned; the pertinent evidence is conveniently assembled in James D. Muhly, *Copper and Tin. The Distribution of Mineral Resources and the Nature of the Metals Trade in the Bronze Age* (New Haven, 1973); see also Henri Limet, *Le travail du métal au pays de Sumer au temps de la IIIe dynastie d'Ur* (Paris, 1960), 71, 73.
226. For iron in the second millennium, see Paul Garelli, *Les Assyriens en Cappadoce* (Paris, 1963), 272; see also *Assyrian Dictionary*, sub *amūtu* B and *aši'u*; and Oppenheim, "Essay on Overland Trade," 241. For the iron called *ḫabalkinnu* (*ḫapalki* in Hittite), see Emmanuel Laroche, "Études de vocabulaire VI," in *Revue hittite et asianique*, **15**, fasc. 60 (1957), 9–15. For meteoric iron, see n. 184. For the techniques of cold and hot hammering of copper and iron and their applications, see Oppenheim, *Ancient Mesopotamia*, 322 f. The still unpublished doctoral dissertation of Carlo Zaccagnini, "Il ferro nel vicino Oriente" (Univ. of Rome, 1969), should be mentioned here.
227. The lexical texts list *šadânu*—that is, hematite (for the identification, see Landsberger, "Tin and Lead," 285, n. 1)—beside *šadânu ṣābitu*—that is, "clinging hematite"—which must refer to magnesite, another iron ore similar to hematite but, in contrast, very magnetic.
228. For the problem of the identification of tin, see Landsberger, "Tin and Lead," 284–296, with previous literature. For a new theory about the provenience of tin and a review of the old theories, see Muhly, *Copper and Tin*, 315–338; and James D. Muhly and Theodore A. Wertime, "Evidence for the Sources and Use of Tin During the Bronze Age of the Near East: A Reply to J. E. Dayton," in *World Archaeology*, **5**, no. 1 (1973), 111–122.
229. For the use of antimony for making metal objects, see Oppenheim *et al., Glass and Glassmaking*, 21. For the problem of the use of antimony in glass as an opacifier, see *ibid.*, 20 f., 87 f.; also Brill, *ibid.*, 116 ff.
230. See, for example, Alfred Lucas, *Ancient Egyptian Materials and Industries*, 4th ed. (London, 1962), 214 f.
231. Most of the texts come from the library of Assurbanipal; one was excavated in Babylon, and another (probably likewise from Babylon) was acquired by the British Museum. A few fragments written in Hittite, but very difficult and fragmentary, were excavated in the Hittite capital in central Anatolia. See also Kaspar K. Riemschneider, "Die Glasherstellung in Anatolien nach hethitischen Quellen," in *Anatolian Studies Presented to Hans Gustav Güterbock on the Occasion of His 65th Birthday* (Istanbul, 1974), 263–278.
232. For a modern attempt to use the materials and procedures indicated in the glass texts and its results, see Oppenheim *et al., Glass and Glassmaking* 111 f., "A Laboratory Synthesis of *zukû*," performed by Robert H. Brill at the facilities of the Corning Museum of Glass.
233. For the much earlier use of thin colored glazes on faience-like carriers, see J. F. S. Stone and L. C. Thomas, "The Use and Distribution of Faience in the Ancient East and Prehistoric Europe," in *Proceedings of the Prehistoric Society for 1956,* n.s. **22**, 37–84.
234. For details, see Oppenheim, "Towards a History of Glass," 263 f.
235. This was not an isolated case but part of a general west-east movement of technical innovations; see Oppenheim, "Towards a History of Glass," 264.
236. See Stone and Thomas, "The Use and Distribution of Faience"; see also Klaus Kühne, *Zur Kenntnis silikatischer Werkstoffe und der Technologie ihrer Herstellung im 2. Jahrtausend v.u.Z.*, which is Abhandlungen der Deutschen Akademie der Wissenschaften zu Berlin, Kl. für Chemie, Geologie und Biologie (1969), no. 1.
237. Mircea Eliade, "Symbolisme et rituels métallurgiques babyloniens," in *Studien zur analytischen Psychologie C. G. Jungs*, II (Zurich, 1955), 42–46, represents an unwarranted distortion of the evidence.
238. See Salonen, *Die Ziegeleien im alten Mesopotamien.*
239. A late and probably foreign use of such bricks for decorative purposes can be found in the wall reliefs produced by preformed bricks (temple of Karaindaš in Uruk) that were later given colored glazes (Nebuchadnezzar II in Babylon and the Achaemenid palace in Susa). For earlier preformed mud bricks to form columns, see David Oates, "The Excavations at Tell Al Rimah, 1966," in *Iran*, **29** (1967), 70–96.
240. For plans of buildings, see *Reallexikon der Assyriologie,*

III (1957–1971), 664–668, sub "Grundriss-Zeichnungen"; and Donald J. Wiseman, "A Babylonian Architect?" in *Anatolian Studies,* **22** (1972), 141–147. See also Geoffrey Turner, "The State Apartments of Late Assyrian Palaces," in *Iraq,* **32** (1970), 177–213.

241. The architects (*šitimgallu* and *itinnu*) worked with measuring rods, pegs, and ropes to lay out the foundations. Instructions and dimensions for the construction of a wall appear in the isolated late tablet published in *Proceedings of the Society of Biblical Archaeology*, **33** (1911), 155–157, and pl. 21, by Theophile G. Pinches as "The Gateways of the Shrines of the Gods at Sippar," and reedited and discussed by Wolfgang Röllig in *Wiener Zeitschrift für die Kunde des Morgenlandes*, **62** (1969), 299 f.

242. For Mesopotamian architecture, see Ernst Heinrich, "Sumerisch-akkadische Architektur," in *Propyläen Kunstgeschichte,* XIV (Berlin, 1975) 131–158.

243. See Helmut Kyrieleis, *Throne und Klinen. Studien zur Formgeschichte altorientalischer und griechischer Sitz- und Liegemöbel vorhellenischer Zeit,* which is *Jahrbuch des Deutschen archäologischen Instituts*, supp. 24 (Berlin, 1969).

244. Bows were made by the *sasinnu*, arrows by the *ēpiš qanê*, shields and coats of mail (see n. 224) by the *aškāpu*. See *Assyrian Dictionary*, sub vocis. The composite bow, the most sophisticated weapon used in the ancient Near East, is preserved only in Egypt. See T. Säve-Söderbergh, "The Hyksos Rule in Egypt," in *Journal of Egyptian Archaeology*, **37** (1951), 53–71, pl. 69–70. The earliest representation of the composite bow comes from Mari—see André Parrot, "Les fouilles de Mari. Dix-neuvième campagne (printemps 1971)," in *Syria*, **48** (1971), pl. XIV 4 and p. 269; and Yigael Yadin, "The Earliest Representation of a Siege Scene and a 'Scythian Bow' From Mari," in *Israel Exploration Journal*, **22** (1972), 89–94.

245. The carpenters apparently were quite specialized; some built boats, others made doors and wheels. Other specialized craftsmen were the smiths, who, especially in the Middle and Neo-Assyrian periods, were differentiated by the metals they worked with (gold, silver, copper, and iron); the weavers were differentiated by material (linen) and product (*ēpiš birmi, ēpiš tunši/bašāmi/naḫlapti*), or technique (*kāṣiru*).

246. While wagons retained the disk wheel, chariots changed to spoked wheels. See P. R. S. Moorey, "The Earliest Near Eastern Spoked Wheels and Their Chronology," in *Proceedings of the Prehistoric Society for 1968*, n.s. **34** (1969), 430–432.

247. Not quite as large as boats were the siege engines, consisting of more or less armored, wheeled battering rams. They are already attested in the second millennium and are often shown on Neo-Assyrian palace reliefs. See Heinz Waschow, *4000 Jahre Kampf um die Mauer; der Festungskrieg der Pioniere* (Bottrop, Germany, 1938). We are not well informed about wooden installations for irrigation—apart from primitive weirs, sluices, and runnels. Those of the *shaduf* type are attested iconographically—see Armas Salonen, *Die Hausgeräte der alten Mesopotamier,* I (Helsinki, 1965), 264–272—as well as in certain difficult text passages (such as Ai. IV ii 33–35); irrigation devices using rotary movements are not attested.

248. The plows were provided with a seeding device that appears to have a parallel in the Far East. See Paul Leser, "Westöstliche Landswirtschaft," in *Festschrift publication d'hommage offerte au P. W. Schmidt* (Vienna, 1928), 416–484; and, more recently, Fritz Christiansen-Weniger, "Die anatolischen Sähpflüge und ihre Vorgänger im Zweistromland," in *Archäologischer Anzeiger* (1967), 151–162. For a more primitive tool for making furrows, see Burchard Brentjes, "Der Zugspaten und seine Vorläufer im alten Orient," in *Ethnographisch-archäologische Zeitschrift,* **10** (1969), 535–542.

249. Among the small-scale implements, the apparently complex construction of the royal umbrella may be mentioned here; see A. Leo Oppenheim, "Assyriological Gleanings IV. The Shadow of the King," in *Bulletin of the American Schools of Oriental Research*, **107** (1947), 8. For mechanical devices to replace human teeth, see Don Clawson, "Phoenician Dental Art," in *Berytus*, **1** (1934), 23–29; see also F. Filce Leek, "The Practice of Dentistry in Ancient Egypt," in *Journal of Egyptian Archaeology,* **53** (1967), 51–58. For timekeeping devices, note the clepsydra (in Akkadian, *dibdibbu*), for which see F. Thureau-Dangin, "Clepsydre babylonienne et clepsydre égyptienne," in *Revue d'assyriologie et d'archéologie orientale*, **30** (1933), 51–52; see also Otto Neugebauer, "The Water Clock in Babylonian Astronomy," in *Isis*, **37** (1947), 37–43. For the type called *mašqû*, see A. Leo Oppenheim, "A Babylonian Diviner's Manual," in *Journal of Near Eastern Studies*, **33** (1974), 205, n. 38. Instructions for making a gnomon for astronomical observation are given in nos. 1494 and 1495 in Pinches and Strassmaier, *Late Babylonian Astronomical and Related Texts*, identified by Abraham Joseph Sachs on p. xxxiv.

250. The texts are Thureau-Dangin, *Rituels accadiens*, 3 f. and 14 f. ii 21–25 (paralleled in the Assur text KAR 60 r. 5–8); 4R 28* no. 3 r. 3–5 from the library of Assurbanipal; and (again from Assur) KAR 29 r.(!) 13–15. See also Thureau-Dangin, "Notes assyriologiques," in *Revue d'assyriologie et d'archéologie orientale*, **17** (1920), 29 f.

251. The tanners (both *aškāpu* and *ṣārip dušê* tanners) had to live in special city quarters in the Neo-Babylonian period (mainly according to texts from Uruk). Note in this connection the fuller's field outside the city of Jerusalem in Isa. 7:3. For a city quarter of the *gurgurru* metalworkers in Assur, see *Assyrian Dictionary,* sub *gurgurru,* discussion section.

252. For the importing of alum, see Oppenheim, "Essay on Overland Trade," 243. In Egypt, it was denoted in the New Kingdom by a foreign word; see Wolfgang Helck, *Die Beziehungen Ägyptens zu Vorderasien im 3. und 2. Jahrtausend v. Chr.*, which is Ägyptologische Abhandlungen 5, 2nd ed. (Wiesbaden, 1962), 505, n. 4.

253. See *Assyrian Dictionary*, sub voce.

254. A parallel but not so informative passage comes from the ritual published in Thureau-Dangin, *Rituels accadiens*, 77: 44–47, which indicates which blessings must be said at certain stages of the preparation and the baking of bread destined for the deity's meal.

255. It is noteworthy that the stone itself is mentioned quite rarely in second-millennium texts, and then preponderantly in those from western regions, only to disappear later. Apparently it had been imported, and contact was lost with its provenience.

256. See *Assyrian Dictionary*, sub *dušû* A and *ṣārip dušê*.

MATHEMATICS AND ASTRONOMY IN MESOPOTAMIA

B. L. van der Waerden

MATHEMATICS. The Sources. In 1935–1937, Otto Neugebauer published all mathematical cuneiform texts then known in three volumes entitled *Mathematische Keilschrifttexte* (*MKT*). They contain photographs, transcriptions, and German translations of all texts, with a very good commentary. A later publication by F. Thureau-Dangin contains new transcriptions and French translations of most texts published in Neugebauer's *MKT*, with a new, valuable commentary.

In 1945, Neugebauer and Sachs published another collection, *Mathematical Cuneiform Texts* (*MCT*), supplementing Neugebauer's *MKT*. It contains newly discovered texts and corrections to earlier publications. More recently E. M. Bruins and M. Rutten produced a volume entitled *Textes mathématiques de Suse* (1961), in which cuneiform texts from Susa were published and discussed.

All our knowledge concerning Babylonian mathematics is ultimately derived from these sources. The majority of the texts are either Old Babylonian (dynasty of Hammurapi, probably 1830–1531 B.C.) or Hellenistic (Seleucid era, beginning 311 B.C.).

Babylonian arithmetic, algebra, and geometry were already fully developed under the dynasty of Hammurapi. Mathematical astronomy came into existence much later. Therefore, Babylonian mathematics can be studied without any attention to astronomy.

BIBLIOGRAPHY

E. M. Bruins and M. Rutten, *Textes mathématiques de Suse*, in the series Mémoires de la Mission Archéologique en Iran (Paris, 1961).

O. Neugebauer, *Mathematische Keilschrifttexte*, 3 vols., which is *Quellen und Studien zur Geschichte der Mathematik, Astronomie und Physik,* Abt. A, **3** (Berlin, 1935–1937). Cited as *MKT*.

O. Neugebauer and A. Sachs, *Mathematical Cuneiform Texts*, which is American Oriental Series, 29 (New Haven, 1945). Cited as *MCT*.

F. Thureau-Dangin, *Textes mathématiques babyloniens* (Leiden, 1938).

The Sexagesimal System. Before 2000 B.C. the Sumerians, the inventors of cuneiform script, who lived in southern Mesopotamia, used a positional system with base 60 for writing integers and fractions. "Positional" means that the value of a numeral depends on where it stands. The symbol for 1, a vertical wedge, can denote any power of 60 or any power of 1/60. Integers up to 59 were written in the decimal system. Thus, 21 would be written as 𒌋𒌋𒁹, the symbols for 10 and 1 being repeated as often as necessary. Beyond 59, integers were written as sums of multiples of powers of 60. Thus, $80 = 60 + 20$ would be written as 𒁹𒌋𒌋.

The normal way of writing fractions was to express them as multiples of 1/60, or of $1/60^2$, and so on. The denominator 60^n was not written: it had to be inferred from the context. Thus, 1 1/2 would be written as 1,30; and the reader would have to determine from the context whether 1 1/2 or 90 was meant, or even 90×60^n or $1\ 1/2 \times 60^{-n}$.

After 2000 B.C. the Babylonian and Assyrian scribes inherited this system from the Sumerians, together with cuneiform script, which they adapted to their Semitic languages.

In transcribing Sumerian or Babylonian numerals, I shall follow the method introduced by O. Neugebauer, separating successive sexagesimal places by a comma:

$$1{,}30 = 60 + 30 = 90.$$

If it is known that the first unit has the value 1, integers and fractions will be separated by a semicolon:

$$1;30 = 1 + \frac{30}{60} = 1\frac{1}{2}.$$

On the other hand, if it is known that the first unit has the value 60^2, I shall write

$$1{,}30{,}\ 0 = 60^2 + (30 \times 60) = 5400.$$

BIBLIOGRAPHY

O. Neugebauer, "Zur Entstehung des Sexagesimalsystems," in *Abhandlungen der Akademie der Wissenschaften zu Göttingen*, n.s. **13** (1927), 1–55.

F. Thureau-Dangin, *Esquisse d'une histoire du système sexagésimal* (Paris, 1932).

Methods of Calculation. The Sumerians and Babylonians performed such arithmetical operations as multiplication, division, squaring, and extracting square and cube roots by means of tables. The Sumerians already had multiplication tables, each table containing the multiples of a given "head number." For instance, if the head number is 1,30, the table would contain the items

1,30	times 1
	times 2
	. . . . .
	times 19
	times 20
	times 30
	times 40
	times 50.

All other multiples of 1,30 were obtained by adding items found in the table.

Other tables contain reciprocals. The reciprocal of an integer is a finite sexagesimal fraction, provided the integer contains no prime factors other than 2, 3, and 5. Such integers may be called "regular." A normal table of reciprocals would give the reciprocals of all regular integers from 2 to 1,21 as follows:

1:2=	30
3	20
. . .	. .
1,20	45
1,21	44,26,40.

From the Hellenistic era we have large tables giving the reciprocals of regular numbers up to seven sexagesimal places (see Neugebauer, *MKT*, I, 16).

The usual procedure to divide b by c was to take c^{-1} from a table of reciprocals and to calculate $b \cdot c^{-1}$ by means of a multiplication table. There are multiplication tables headed by numbers that are the reciprocals c^{-1} of simple regular integers. Other tables contain approximate reciprocals of irregular numbers. We also have tables of squares, of cubes, and of square roots and cube roots. Thus, even extensive calculations presented no difficulties for the Sumerian and Babylonian scribes.

BIBLIOGRAPHY

The structure of the system of tables for multiplication, division, and squaring was first described in a series of papers by O. Neugebauer: "Sexagesimalsystem und babylonische Bruchrechnung I–IV," in *Quellen und Studien zur Geschichte der Mathematik, Astronomie und Physik*, Abt. B, **1** and **2** (1930–1932). For further details see *MKT* and *MCT*.

Babylonian Algebra. Babylonian algebra was concerned mainly with the solution of equations. The methods for solving equations were developed in the Old Babylonian period, and the same methods were still used in the Hellenistic age.

Equations and methods of solution were always formulated in words. An unknown x was often called "the side" and its square, "the square." When two unknowns occurred, they were usually called "length" and "breadth," and their product, "area." Three unknowns were sometimes called "length," "breadth," and "height," and their product, "volume."

Despite this geometric terminology, the Babylonians did not hesitate to subtract a side from an area. Thus, the quadratic equation

$$x^2 - x = 14{,}30 \tag{1}$$

was formulated as follows in the text BM (British Museum) 13901: "I have subtracted the side of the square from the area, and the result is 14,30." (See Neugebauer, *MKT*, III, 6, prob. 2.)

The rules for solving equations were always explained as operations on definite numerical coefficients, but in such a way that the general rule is easy to recognize. For example, equation (1) was solved as follows:

> Take 1, the coefficient (of the linear term).
> Take one-half of 1 (result, 0;30).
> Multiply 0;30 by 0;30 (result, 0;15).
> Add 0;15 to 14.30. Result, 14,30;15.
> The square root of this is 29;30.
> Take the 0;30, which you have multiplied by itself, and add it to 29;30.
> Result, 30: this is the [required side of the] square.

Of course, the procedure is equivalent to the modern formula for solving the equation

$$x^2 - ax = b, \tag{2}$$

namely,

$$x = \sqrt{\left(\frac{1}{2}a\right)^2 + b} + \frac{1}{2}a. \tag{3}$$

A complete discussion of all types of equations and all methods of solution occurring in Babylonian texts was given by H. Goetsch. My account is based mainly on his paper.

Six types of equations were solved.

First, there were linear equations in one unknown.

Second, there were systems of linear equations in two or three unknowns. The usual method was to solve one of the equations for one of the unknowns and substitute it into the other equations. In some cases, however, a different method was used. Given the equation

$$x+y=a \tag{4}$$

and another (linear or nonlinear) equation for x and y. If x and y were equal, each of them would be $a/2$. Now suppose x exceeds $a/2$ by an unknown amount, s:

$$x=\frac{a}{2}+s. \tag{5}$$

Then for y we must take

$$y=\frac{a}{2}-s. \tag{6}$$

Substituting these two expressions into equation (4), one obtains an equation for s, which can be solved by Babylonian methods if it is linear or quadratic. Let us call this method the "plus and minus" method. Diophantus often used it if a sum of two unknown numbers $x+y$ was given.

A good example of the "plus and minus" method is found in the Old Babylonian text VAT 8389 (see Neugebauer, *MKT*, I, 323). In this text a pair of equations of the form

$$x+y=a$$
$$bx-cy=d$$

is not solved by eliminating one of the two unknowns, but by the "plus and minus" method. The unknowns x and y are calculated simultaneously according to equations (5) and (6), while s is found as the solution of a linear equation.

In a linear and a quadratic equation, the following standard types occur frequently:

(I)	$x+y=a,$	$xy=b$
(II)	$x-y=a,$	$xy=b$
(III)	$x+y=a,$	$x^2+y^2=b$
(IV)	$x-y=a,$	$x^2+y^2=b.$

All four types could have been solved by the method of substitution, solving x or y from the linear equation and substituting it into the quadratic equation. This method would first have yielded one of the two unknowns, and the other would be found by a formula such as $y=a-x$ or $y=x-a$. No case is known in which such formulas were used. It appears that in cases (I) and (III), in which $x+y$ is known, the "plus and minus" method was used, for x and y are regularly found as $x=a/2+s$ and $y=a/2-s$, while s is calculated as a square root.

In cases (II) and (IV), where $x-y=a$ is given, a similar method can be used. One can write

$$x=s+\frac{a}{2} \tag{7}$$

$$y=s-\frac{a}{2}, \tag{8}$$

where s is a new unknown. In case (II) as well as in case (IV), one finds a pure quadratic equation for s, from which s can be obtained as a square root. The Babylonian expressions for x and y always have the forms of (7) and (8). Diophantus also used this method, as S. Gandz has noted.

Very often, other types of equations were reduced to the standard forms (I, II, III, IV). Thus, in the text AO 8862 (Neugebauer, *MKT*, I, 113) the pair of equations

$$xy+(x-y)=3,3$$
$$x+y\ =27$$

was reduced to the standard form (I) by adding the two equations and writing

$$y=y'-2.$$

(For other examples of nonlinear systems of equations, see the paper by Goetsch, 126–141.)

Among single quadratic equations, two types occur frequently:

$$x^2-ax=b \tag{9}$$

$$x^2+ax=b. \tag{10}$$

The positive root of such an equation is always found by a procedure equivalent to our solution formula, well known since al-Khwārizmī. Most probably the Babylonians found their solution by adding $(a/2)^2$ to both sides of the equation and applying the formulas

$$\left(x-\frac{a}{2}\right)^2=x^2-ax+\left(\frac{a}{2}\right)^2$$

$$\left(x+\frac{a}{2}\right)^2=x^2+ax+\left(\frac{a}{2}\right)^2,$$

with which the Babylonians were just as familiar as the Greeks and Arabs.

The third type used by al-Khwārizmī,

(11) $$x^2 + b = ax,$$

is not found in Babylonian texts.

Pairs of quadratic equations are discussed by Goetsch, who presents two examples (103, 141; for the first example, see also Neugebauer, *MKT*, I, 486; and van der Waerden, *Science Awakening*, I, 70–71). The two equations in this example can be written as

$$0;20\,(x+y) - 0;1\,(x-y)^2 = 15$$
$$xy = 10,0.$$

Cubic equations occur in the following types:

(12) $$x^3 = a$$
(13) $$x^2(x+1) = a$$
(14) $$x(10-x)(x+1) = a.$$

The Babylonians had tables of cube roots, by means of which they could solve equations of type (12). They also had a table giving numbers $n^2(n+1)$ in the first column and their "roots" n (from 1 to 60) in the second column (Neugebauer, *MKT*, I, 76). By means of such a table they could solve equations of type (13). We do not know how they solved (14), but they did give the solution

$$x = 6 \text{ for } a = 2;48$$

(see Goetsch, 149).

Numbers of the form $n^2(n-1)$ with their "roots" n also were tabulated (see Thureau-Dangin, *Textes mathématiques babyloniens*, 123). The neo-Pythagorean Nicomachus of Gerasa (first century) had a special name for the numbers $n^2(n+1)$ and $n^2(n-1)$: *arithmoi paramekepipedoi*. (See O. Becker's paper in *Quellen und Studien zur Geschichte der Mathematik, Astronomie und Physik*, Abt. B, **4** [1938], 181.)

The Babylonians certainly were familiar with such rules of elementary algebra as

(15) $$(a+b)(a-b) = a^2 - b^2$$
(16) $$(a+b)^2 = a^2 + 2ab + b^2$$
(17) $$(a-b)^2 = a^2 - 2ab + b^2.$$

The Greeks and Arabs proved such formulas by drawings of rectangles and squares, and it is possible that the Babylonians also derived these formulas by means of drawings.

BIBLIOGRAPHY

Cuneiform texts dealing with algebraic equations were first discussed by O. Neugebauer, H. Waschow, F. Thureau-Dangin, K. Vogel, and others between 1929 and 1936. Full quotations of these pioneer papers can be found in Neugebauer, *MKT*.

Later books and papers include the following:

S. Gandz, "The Origin and Development of Quadratic Equations in Babylonian, Greek and Early Arab Algebra," in *Osiris*, **3** (1937), 405–543.

H. Goetsch, "Die Algebra der Babylonier," in *Archive for History of Exact Sciences*, **5** (1968), 79–153.

O. Neugebauer, *The Exact Sciences in Antiquity*, 2nd ed. (Providence, R. I., 1957).

O. Neugebauer, *Vorgriechische Mathematik* (Berlin, 1934; repr. Heidelberg, 1969).

Babylonian Arithmetic. From several texts we know that the Babylonians were able to calculate the sum of an arithmetical progression. The problem discussed in these texts is to distribute a sum of money among a number of men according to an arithmetical progression.

In the text AO 6484 (Neugebauer, *MKT*, I, 96–107) we find the summation of a geometrical progression the ratio of which is 2:

$$1 + 2 + 4 + \cdots + 2^9 = 2^9 + (2^9 - 1).$$

In the same text the sum of the squares of the integers from 1 to 10 is computed according to the formula

$$1^2 + 2^2 + \cdots + n^2 = \left(\frac{1}{3} + n \cdot \frac{2}{3}\right)(1 + 2 + \cdots + n).$$

BIBLIOGRAPHY

See the publications mentioned in the bibliography to "The Sources."

Babylonian Geometry. Areas of triangles and trapeziums were calculated by the correct formulas

$$A = \frac{ha}{2} \quad \text{and} \quad A = \frac{h}{2}(a+b),$$

respectively. The area of a circle with radius r was calculated as $3r^2$, and its perimeter as $6r$. The volumes of prisms and cylinders were determined by multiplying the area of the base by the height.

In several Old Babylonian texts, volumes of frustums of cones and pyramids were calculated by the incorrect formula

$$V = \frac{h}{2}(A + B),$$

in which h is the height and A and B are the areas of bottom and top. It is remarkable that at about the same time, the Egyptians used the correct formula for calculating the volume of a pyramid frustum on a square base:

$$V = \frac{h}{3}(a^2 + ab + b^2).$$

If a trapezium is divided into two parts of equal areas by a line parallel to the base, the length x of the dividing line is given by the formula

$$(18) \qquad x^2 = \frac{1}{2}(a^2 + b^2),$$

in which a and b are the two parallel sides. The Babylonians knew this formula (see Neugebauer, *MKT*, I, 131).

In the text VAT 8512 (Neugebauer, *MKT*, I, 342), one of the problems is accompanied by a drawing showing a triangle that is divided into a triangle and a trapezium by a line parallel to the base. The problem solved in the text concerns the areas of the two parts. Neugebauer was able to explain the solution, assuming that the Babylonians knew about the proportionality of the sides of the two similar triangles. Another, more geometrical explanation of the same text, based upon formula (18), was given by P. Huber.

In the texts from Susa published by E. M. Bruins and M. Rutten, one finds approximate calculations of the sides of regular polygons.

BIBLIOGRAPHY

In addition to the publications mentioned in the bibliography to "The Sources," see P. Huber, "Zu einem mathematischen Keilschrifttext (VAT 8512)," in *Isis*, **46** (1955), 104–106.

On the texts from Susa, see E. M. Bruins and M. Rutten (cited under "The Sources") and E. M. Bruins, "Quelques textes mathématiques de la mission de Suse," in *Proceedings K. Nederlandse akademie van wetenschappen*, **53** (1950), 1025–1033.

On the Egyptian pyramid formula, see W. W. Struve, "Mathematischer Papyrus des . . . Museums . . . in Moskau," which is *Quellen und Studien zur Geschichte der Mathematik, Astronomie und Physik*, Abt. A, **1** (1930), 134–145.

The "Theorem of Pythagoras." Several texts, from the Old Babylonian as well as from the Hellenistic period, show clearly that the Babylonian mathematicians knew the formula

$$(19) \qquad d^2 = h^2 + b^2$$

for the sides of a right triangle. They calculated b as the square root of $d^2 - h^2$. (For examples see Neugebauer, *MKT*, II, 53; and III, 22.)

The Old Babylonian text Plimpton 322, published by Neugebauer and Sachs (*MCT*, 38–41), contains an extensive table of "Pythagorean triples"—of rational triples h, b, d satisfying equation (19).

BIBLIOGRAPHY

E. M. Bruins, "Reciprocals and Pythagorean Triads," in *Physis*, **9** (1967), 373–392.

P. Huber, "Bemerkungen über mathematische Keilschrifttexte," in *Enseignement mathématique*, 2nd ser., **3** (1957), 19–27.

D. Price, "The Babylonian 'Pythagorean Triangle' Tablet," in *Centaurus*, **10** (1964), 219–231.

B. L. van der Waerden, *Science Awakening*, I (Groningen, 1954; New York, 1961), 78–80.

Babylonian Influence on Greek Mathematics. We have already seen that the Babylonians had tables of numbers $n^2(n \pm 1)$ and that the neo-Pythagorean Nicomachus had a special name for these numbers. We also have seen that the "plus and minus" method for solving equations in two unknowns x and y, of which the sum $x + y$ or difference $x - y$ is given, was used by the Babylonians, as well as by Diophantus of Alexandria.

The four standard types of pairs of equations in two unknowns recur in Euclid's *Elements*. Theorems 5–6 and 9–10 of book II are the theorems needed for the geometrical solution of the Babylonian standard types. As Neugebauer was the first to see, the single steps in the Greek geometrical solutions are exactly the same as the steps in the algebraic solutions of the Babylonians. Whenever the Babylonians say "Take the square root of A," the Greeks say "Take the side of the square of area equal to A," and so on. It is known that theorems 5–6 and 9–10 of book II of the *Elements* are due to the Pythagoreans.

In Pythagorean geometry a fundamental theorem is the "theorem of Pythagoras." The Babylonians also knew and used this theorem. They had methods to calculate "Pythagorean triples"—triples of integers x, y, z satisfying the equation

$$z^2 = x^2 + y^2.$$

The Greeks also had methods for calculating such triples. The most general method, regularly used by Diophantus, is given by the equations

$$x = p^2 - q^2$$
$$y = 2pq$$
$$z = p^2 + q^2.$$

According to Neugebauer and Sachs (*MCT*, 38–41), it is quite possible that the Babylonians used the same formulas.

All these facts seem to indicate a Babylonian influence in Greek geometry and arithmetic. According to a Greek tradition, Pythagoras went to Babylon and learned the science of numbers, the science of music, and the other sciences from the magi. This may well be true.

BIBLIOGRAPHY

The first to draw attention to the relations between Babylonian and Greek mathematics was Otto Neugebauer in the series of papers "Studien zur Geschichte der antiken Algebra," in *Quellen und Studien zur Geschichte der Mathematik, Astronomie und Physik*, Abt. B, **2** and **3** (1933–1936). See also B. L. van der Waerden, *Science Awakening*, I, 66–101, 116–127.

For the spread of Babylonian algebra and geometry to Egypt, see R. A. Parker, *Demotic Mathematical Papyri*, Brown Egyptological Studies, VII (Providence, R.I.–London, 1972).

ASTRONOMY. Four Stages of Babylonian Astronomy. The history of Babylonian astronomy is much clearer today than it was about 1930. This is due mainly to O. Neugebauer's thorough analysis of Babylonian mathematical astronomy and to the publication and interpretation of observational texts by A. Sachs. We are now in a position to recognize four distinct periods:

The first, or Old Babylonian, period coincides with the reign of the dynasty of Hammurapi (most probably 1830–1531 B.C.).

The second period encompassed the Kassite reign, followed by Assyrian domination (1530–612). It ended with the destruction of Nineveh and its great library in 612 B.C.

The third, or Neo-Babylonian, period spanned the Chaldaean dynasty (611–540).

The fourth period saw the highest development of observational and mathematical astronomy under the Persian, Seleucid, and Arsacid reigns (539 B.C.–A.D. 75).

Each of these four clearly separated periods produced a particular type of astronomy connected with a particular type of astrology.

BIBLIOGRAPHY

The first important publications on Babylonian astronomy were the following:

J. Epping, *Astronomisches aus Babylon* (Freiburg im Breisgau, 1889).

F. X. Kugler, *Babylonische Mondrechnung* (Freiburg im Breisgau, 1900).

F. X. Kugler, *Sternkunde und Sterndienst in Babel*, 2 vols., 3 supps. (Münster, 1907–1935), 3rd supp. written by J. Schaumberger.

Indispensable for all serious students of the subject are three standard publications:

O. Neugebauer, *Astronomical Cuneiform Texts*, 3 vols. (London, 1955).

A. Sachs, *Late Babylonian Astronomical and Related Texts* (Providence, R.I., 1955).

O. Neugebauer, *A History of Ancient Mathematical Astronomy*, 3 vols. (Heidelberg–New York, 1975).

A summary of the entire subject is in B. L. van der Waerden, *Science Awakening*, II, *The Birth of Astronomy* (Leiden–New York, 1974).

Astronomy and Astrology in the Old Babylonian Period. From the Old Babylonian period we have just one astrological and one astronomical (observational) text. The astrological text, published by V. Shileyko in 1927, contains predictions based on the state of the sky on the day when the crescent just becomes visible, at the beginning of a new year. The predictions are of the following kind:

1. If the sky is dark, the year will be bad.
2. If the face of the sky is bright when the new moon appears and [it is greeted] with joy, the year will be good.
3. If the north wind blows across the face of the sky before the new moon, the corn will grow abundantly.

The astronomical text, of which several copies exist, is a list of dates for the first appearance and disappearance of Venus as an evening star or morning star during the twenty-one years of the reign of King Ammizaduga. This reign began just 160 years after that of Hammurapi. The observations were connected with astrological predictions. I shall quote one example from the tenth year: "If on the tenth of Araḫsamna, Venus disappeared in the east, remaining absent two months and six days, and was seen on the sixteenth of Tebitu in the West, the harvest of the land will be successful."

If the astrological predictions (which may or may not be later additions) are ignored, we are left with a sequence of observations of the appearance and disappearance of Venus during twenty-one years. These observations may be used to date the reign of Ammizaduga. Within reasonable limits, four chronological hypotheses are possible: one "long chronology," two "middle chronologies," and one "short chronology." According to P. Huber's careful investigation of all dates, the "short chronology" is much more probable than the others: Hammurapi reigned from 1728 to 1686, Ammizaduga from 1582 to 1562, and the dynasty lasted from 1830 to 1531 B.C.

In Babylonian astronomical texts, most names of stars and constellations are Sumerian. For example:

^{mul}gu · an · na (bull of the sky) = Taurus
^{mul}ur · gu · la (lion or lioness) = Leo
^{mul}gir · tab (scorpion) = Scorpio.

According to Willy Hartner one may conclude from this fact that the Sumerians already regarded these constellations as representing a bull, a lion, and a scorpion respectively.

Hartner also has drawn attention to the numerous pictorial representations of a Lion-Bull combat in ancient Near Eastern art. These pictures show a remarkable constancy from the fourth millennium B.C. (Elamite and Sumerian pictures) through the Achaemenid period up to a Persian miniature from the Mogul period (see the photographs in Hartner's paper). In many of these pictures, stars and other celestial symbols are visible, which seems to indicate that the lion and the bull were meant to represent not just animals, but the constellations Leo and Taurus. Thus, the Sumerian bull attacked by a lion in Hartner's Figure 2 carries a star rosette in his horns.

For the geographical latitude of Persepolis (30°) and for 4000 B.C. Hartner has calculated the dates of heliacal rising or *Morningfirst* of the Pleiades (in Taurus), of Regulus (the main star of Leo), of Antares (the main star of Scorpio), and of β Pegasi (the first rising star of the Pegasus quadrangle). These four easily recognizable stars are nearly 90° apart in the zodiacal circle. Hartner found that these four heliacal rising dates coincide exactly with the spring equinox, the summer solstice, the autumn equinox, and the winter solstice for 4000 B.C. This would explain why the Sumerians regarded Taurus, Leo, and Scorpio as important constellations and why pictorial representations often show two or three of these animals.

Hartner explains the Lion-Bull fight as representing the situation of the sky at nightfall on the date of *Eveninglast* (heliacal setting) of the Pleiades, when the sun is just 20° below the horizon. About 4000 B.C. this date was 10 February (Gregorian), and in the time of the Achaemenids (500 B.C.) it was 28 March. In any case Leo was standing in the zenith and thus was displaying its greatest power, while the Bull was just disappearing below the horizon.

BIBLIOGRAPHY

W. Hartner, "The Earliest History of the Constellations in the Near East and the Motif of the Lion-Bull Combat," in *Journal of Near Eastern Studies*, **24** (1965), 1–16, and pl. I–XVI.

On the astrology of the Old Babylonian period see the following:

T. Bauer, "Eine Sammlung von Himmelsvorzeichem," in *Zeitschrift für Assyriologie*, **43** (1936), 308–314.

V. Shileyko, "Mondlaufprognosen aus der Zeit der ersten babylonischen Dynastie," in *Doklady Akademii nauk SSSR* (1927), ser. B, 125–128.

The Venus tablets of Ammizaduga are discussed in the following:

P. Huber, "Early Cuneiform Evidence for the Planet Venus," paper presented to the annual meeting of the American Association for the Advancement of Science, San Francisco, Feb. 1974.

S. Langdon, J. K. Fotheringham, and C. Schoch, *The Venus Tablets of Ammizaduga* (Oxford, 1928).

B. L. van der Waerden, *Die Anfänge der Astronomie* (Basel, 1968).

J. D. Weir, *The Venus Tablets of Ammizaduga* (Istanbul–Leiden, 1972).

Second Period. We know much more of the second period because many reports and letters from the Assyrian court astrologers to their kings (written 722–612 B.C.) and several other texts from the same period are preserved. Among these texts are lists of fixed stars, circular diagrams of the sky, and many parts of the great omen collection "Enūma Anu Enlil."

BIBLIOGRAPHY

On the letters and reports of the Assyrian court astrologers see the following:

R. F. Harper, *Assyrian and Babylonian Letters*, 14 vols. (Chicago, 1892–1914).

S. Parpola, *Letters From Assyrian Scholars*, I (Kevelaer, Germany, 1970).

R. C. Thompson, *The Reports of the Magicians and Astrologers . . .* (London, 1900).

The Omen Collection "Enūma Anu Enlil." An omen is a statement concerning the astrological signification of a phenomenon observed in the sky. An example of a very old omen is "If Venus appears in the east in the month Airu and the Great and Small Twins surround her, all four of them, and she is dark, then will the king of Elam fall sick and die" (see J. Schaumberger, supp. 3 to Kugler's *Sternkunde*, 344).

The omen collection "Enūma Anu Enlil" consisted of seventy or more tablets, altogether containing some 7,000 omina. It probably was composed before 900 B.C. but was quoted extensively long afterward. If, for example, an Assyrian king wished to know whether the stars were favorable, the royal astronomer would observe the sky and consult the old omen series. He would then report his observation in a letter to the king, quote an appropriate omen, and explain its application to the observed situation.

Omen astrology, which prevailed in the first and second periods, differs in several respects from horoscope astronomy, which came into use in the fourth period. It is concerned mainly with matters of general interest, such as good or bad harvests, peace and war, and the health of the king; whereas the chief aim of horoscope astronomy is to predict the fate of individuals from the aspect of the sky at the moment of their birth or conception. Also, omen astrology does not use the division of the zodiac into twelve signs. In fact, in the numerous extant texts from the second period, zodiacal signs are never mentioned. The phenomena mentioned in the great omen collection are always visible phenomena that everyone can observe in the sky without any knowledge of zodiacal signs.

Between the old and the new astrology there is an intermediate stage, in which the zodiacal signs appear but birth horoscopes are not yet cast. In this "primitive zodiacal astrology" predictions are based either on the zodiacal sign in which Jupiter dwells or on the zodiacal sign in which the moon stays on the day of first visibility of Sirius. These predictions can be found in Greek texts ascribed to Orpheus or to Zoroaster. In *Science Awakening*, II, 180, I have tentatively given some arguments in favor of the hypothesis that this "primitive zodiacal astrology" originated in the third, or Neo-Babylonian, period.

BIBLIOGRAPHY

On the series "Enūma Anu Enlil," see the following:

C. Virolleaud, *L'astrologie chaldéenne* (Paris, 1905–1912).

E. F. Weidner, "Die astrologische Serie Enuma Anu Enlil," in *Archiv für Orientforschung*, **14** (1943), 172–195.

On the astrological texts ascribed to Orpheus and Zoroaster see the following:

J. Bidez and F. Cumont, *Les mages hellénisés*, 2 vols. (Paris, 1938), I, 107–127, and II, O37–O52.

B. L. van der Waerden, *Science Awakening*, II (Leiden–New York, 1974), 177–180.

The "Three Stars Each." The modern, ill-chosen name "astrolabes" denotes a class of texts in which three stars are listed for every month of the year. The Babylonians had a better name for these lists: "the three stars each." The thirty-six stars were divided into three sets of twelve: the "stars of Ea," or northern stars; the "stars of Anu," near the equator; and the "stars of Enlil," or southern stars.

The texts state that the stars become visible only in the months to which they are assigned, but this is not always true. Three of the stars—Venus, Mars, and Jupiter—are actually planets.

There are astrolabes in circular and rectangular form (see Figures 1 and 2). The circular form is probably older (see B. L. van der Waerden, *Science Awakening*, II, 67). The astrolabes were very popular in Babylonia and Assyria. Today we have astrolabes from Assur, Nineveh, Uruk (Erech), and Babylon. The oldest known, from Assur, was written about 1100 B.C.

Numbers are written on some of the astrolabes. In the outer ring of the circular astrolabe the numbers increase from 2, 0 to 4, 0 by equal increments of 20 and then decrease by equal amounts. The numbers in the middle ring are half, and those in the inner ring a quarter, of those in the outer rings. We have here an early example of a "linear zigzag function," a function that increases linearly to a maximum and then decreases linearly to a minimum. In the mathematical astronomy of the fourth period, many astronomical quantities were represented by linear zigzag functions.

The numbers on the astrolabes reach their maximum in the summer and their minimum in the winter. They were supposed to be proportional to the duration of the day. Day and night were divided into three "watches" each, and the duration of the watches was regulated by means of water clocks. A certain amount of water was poured into a vessel, and its becoming empty determined the end of the watch. The numbers in the outer ring of the astrolabe were supposed to determine the weight of the water to be poured into the water clock. If this explanation, given by O. Neugebauer, is accepted, the numbers in the middle ring must refer to half-watches, and those in the inner ring to quarter-watches (twelfth parts of the light day).

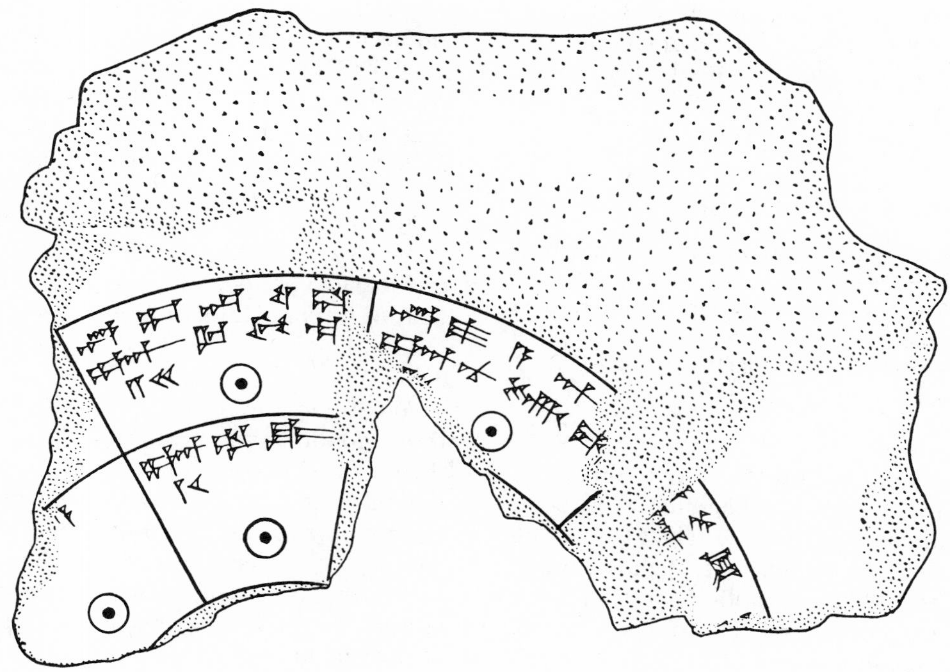

FIGURE 1. Fragment of a circular astrolabe.

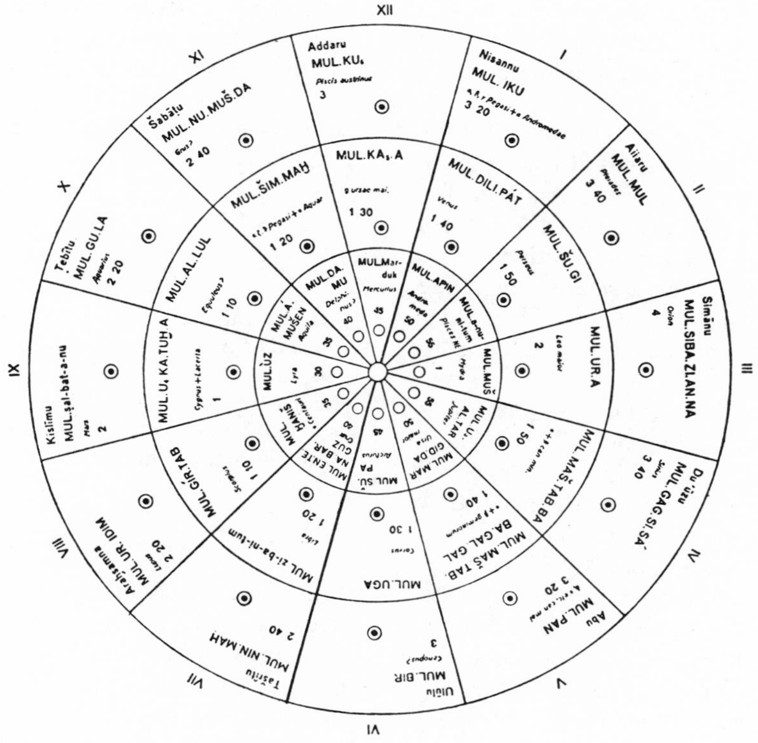

FIGURE 2. Reconstruction of a circular astrolabe according to A. Schott.

This division of the day into twelve hours of varying duration confirms the statement of Herodotus (*Histories*, II, 109): "Polos and Gnomon and the twelve parts of the day did the Greeks learn from the Babylonians."

BIBLIOGRAPHY

O. Neugebauer, "The Water Clock in Babylonian Astronomy," in *Isis*, **37** (1947), 37–43.

A. Schott, "Das Werden der babylonisch-assyrischen Positions-Astronomie," in *Zeitschrift der Deutschen morgenländischen Gesellschaft*, **88** (1934), 302–337.

The Compendium MUL·APIN. About 700 B.C. or earlier, an astronomical compendium was composed, consisting of at least three tablets, of which several copies are preserved. One of these, found in Assur, is dated 687 B.C. Another copy bears the remark "copy from Babylon." The series, which is named after its opening words, MUL·APIN, seems to be a compilation of all or nearly all astronomical knowledge of the period before 700 B.C.

The first tablet begins with a list of thirty-three stars of Enlil, twenty-three stars of Anu, and fifteen stars of Ea, with indications of their relative positions. With few exceptions, all stars of Anu lie in a zone between +17° and −17° declination. The "path of Enlil" is north of this zone, and the "path of Ea" is south of it.

Next comes a list of dates of first visibility of thirty-six fixed stars and constellations. In a separate list the date differences are given. Another list states that certain constellations rise while others set in their daily motion. Still another section names stars that culminate while others rise. Using these lists, C. Bezold, F. X. Kugler, and B. L. van der Waerden have identified many Babylonian constellations.

Another section contains a list of differences between the times of culmination of fixed stars. From a letter written by an astrologer we know that culminations were used to determine the exact time of eclipses. See J. Schaumberger, *Zeitschrift für Assyriologie*, **47** (1941), 127, and **50** (1942), 42.

The division of the zodiacal circle into twelve "zodiacal signs" of 30° each, which is fundamental for all texts of the fourth period, does not occur in MUL·APIN. Yet the author knows that the sun, the moon, and the five planets are always in a zone that he calls "the path of the moon," which coincides with our "zodiacal belt." The text enumerates some eighteen constellations and single stars in the path of the moon. Among these are twelve constellations bearing the same names as the zodiacal signs in later texts. For instance, the constellation Aries appears in MUL·APIN as LU·ḪUN·GA, and the star Spica is called AB·SIN. In texts of the fourth period, LU·ḪUN·GA (often abbreviated to ḪUN) usually denotes the zodiacal sign Aries, a segment of 30° in the ecliptic; and AB·SIN denotes the sign Virgo as well as the star Spica. In a drawing from the Seleucid period, Virgo is represented as a virgin bearing an ear of corn (in Latin, *spica* = ear of corn). On another drawing, UR·GU·LA (Leo) is drawn as a lion.

According to MUL·APIN, the sun dwells for three months in the "path of Enlil" in the summer, then for three months in the "path of Anu," three months in the "path of Ea" in the winter, and again three months in the "path of Anu." Since the three paths are bounded by circles parallel to the equator and since the orbit of the sun was supposed to cross all three paths, it follows that the author was aware of the obliquity of the zodiac.

The text also gives rules for calculating the shadow length of a vertical bar at different times of the year and of the day, and for calculating the times of rising and setting of the moon. Although crude and inaccurate, these rules represent a first attempt toward the calculation of celestial phenomena by means of rising and falling arithmetical sequences.

We find next to nothing in MUL·APIN on the course and periods of the planets. The text merely states that all planets move within the "path of the moon."

BIBLIOGRAPHY

B. L. van der Waerden, "The Thirty-Six Stars," in *Journal of Near Eastern Studies*, **8** (1949), 6–26.

B. L. van der Waerden, *Science Awakening*, II (Leiden–New York, 1974), 70–86.

E. F. Weidner, "Ein babylonisches Kompendium der Himmelskunde," in *American Journal of Semitic Languages and Literatures*, **40** (1924), 186–208.

Diaries and Eclipse Records. According to A. Sachs, cuneiform texts that contain records of daily observations are called "diaries." There are more than 1,200 fragments of astronomical diaries, most of them at the British Museum. Usually a diary contains a record of lunar and planetary observations made during half of a Babylonian year. Not all entries in the diaries are records of actual observations. Some are accompanied by remarks like NU PAP ("not watched for") or DIR NU PAP ("cloudy, not watched for"). Such notations often occur with predicted eclipses.

Most diaries are from the period 400 – 50 B.C., but a few go back to the Neo-Babylonian or even to the Assyrian period. The oldest datable fragment contains observations made in 652 B.C.

Among the planetary phenomena recorded in the diaries are conjunctions of planets with the moon and fixed stars, usually with an indication of the northern or southern distance, and dates of first and last visibility of planets. Among the lunar phenomena are, of course, lunar and solar eclipses. Regular observations also were made of six time intervals that A. Sachs calls the "lunar six." In our texts these intervals are denoted by standard cuneiform signs, which may be read as

na, šu, me, na, mi (or *ge*), *kur*.

The first of these was observed just after the new moon, on the evening of first visibility of the crescent. The next four were observed just before and after the full moon, and *kur* was observed on the day of last visibility of the moon in the morning. The meanings of the terms are given below:

na = time between sunset and moonset
šu = time between moonset and sunrise
me = time between moonrise and sunset
na = time between sunrise and moonset
mi = time between sunset and moonrise
kur = time between moonrise and sunrise.

If bad weather prevented the observation, calculated values were inserted; we do not know how they were made.

Other items are the first and last visibility of Sirius, and equinoxes and solstices. These dates were nearly always calculated, not observed. (On the method of calculation see the paper by O. Neugebauer cited in the bibliography.) All sorts of meteorological events are reported in the diaries: rainbows, halos, thunder, rain, clouds, storms. Also mentioned are the prices of barley, dates, and wool in Babylon.

In other texts, observations of only one planet for several years or long sequences of eclipse records are assembled. These texts probably are compiled from diaries. The oldest texts of this kind contain detailed reports of lunar eclipses ranging over long periods: one text, for example, records eclipses from 731 to 317 B.C.; another goes back to the reign of the Babylonian king Nabonassar (747 – 734 B.C.).

Ptolemy informs us in book III of the *Almagest* that records of eclipse observations beginning with the reign of Nabonassar were still available to him. This statement is confirmed by the cuneiform texts.

The reports of the Assyrian court astrologers contain not only records of observed eclipses but also predictions of eclipses (see R. C. Thompson, *The Reports of the Magicians*). Report no. 273 reads: "On the fourteenth an eclipse of the moon will take place: woe for Elam and Amurru, good for my Lord, the King. Let the heart of my Lord the King rejoice. . . . An eclipse will take place. . . ." Report no. 274F seems to contain confirmation of the prediction: ". . . To my Lord the King I have written: An eclipse will take place. Now it has not gone by, it has taken place. In the occurrence of this eclipse lies happiness for my Lord the King. . . ."

We do not know by what method these predictions were made. (For other examples and for a discussion of possible methods, see B. L. van der Waerden, *Science Awakening*, II, 115 – 122.)

BIBLIOGRAPHY

A. Sachs, *Late Babylonian Astronomical and Related Texts* (Providence, R. I., 1955).

A. Sachs, "Babylonian Observational Astronomy," in F. R. Hodson, ed., *The Place of Astronomy in the Ancient World* (London, 1974), 43 – 50.

R. C. Thompson, *The Reports of the Magicians and Astrologers* (London, 1900).

On the computation of Sirius dates and equinoxes and solstices, see O. Neugebauer, "Solstices and Equinoxes in Babylonian Astronomy," in *Journal of Cuneiform Studies*, **2** (1948), 209 – 222.

The Neo-Babylonian Period. The most important astronomical text from this period is a diary for year 37 of Nebuchadnezzar II (568 B.C.). The text shows that more and more attention was paid to the course of the moon and the planets. Conjunctions of the moon and the planets with fixed stars were regularly noted, as were the dates of first and last visibility of planets and the "lunar six."

For the date 4 July 568 B.C. an "eclipse of the moon that failed to occur" was recorded. The eclipse actually took place during the day, when the moon was invisible. We may conclude that the eclipse was predicted.

We have seen that in the MUL·APIN the zodiacal circle was divided into four parts corresponding to the four seasons and that the sun was supposed to dwell in each for three months. If each part is further divided into three parts corresponding to the months of the schematic year, one obtains a division of the circle into twelve zodiacal signs. Since every month of the schematic year has thirty days, it is natural to divide every zodiacal sign into thirty degrees. Thus, it is only a small step from the zodiacal schema of MUL·APIN to the division of the zodiac into twelve signs of thirty degrees each. There are reasons to suppose that this division was known and used for astrological purposes in the Neo-Babylonian period (see B. L. van der Waerden, *Science Awakening*, II, 180).

BIBLIOGRAPHY

P. V. Neugebauer and E. F. Weidner, "Ein astronomischer Beobachtungstext aus dem 37. Jahre Nebukadnezars II," in *Berichte. Sächsische Gesellschaft der Wissenschaften*, phil.-hist. Kl., **67** (1915), 29–89.

Babylonian Horoscopes. The greatest period of Babylonian astronomy was the Persian reign (539–333 B.C.), during which mathematical theories of the motion of the sun, the moon, and the five classical planets were developed. Historically, these theories are closely connected with horoscope astrology, because astrologers need methods for calculating solar, lunar, and planetary positions. As long as such methods did not exist, astrologers could not cast horoscopes, except in the very rare cases where they could observe the sky immediately after the birth of the child.

The oldest surviving cuneiform horoscope was cast for a date in 410 B.C. At that time lunar and planetary theories were already fully developed. Other Babylonian horoscopes were cast for 263, 258, 235, 230, and 142 B.C.

BIBLIOGRAPHY

A. Sachs, "Babylonian Horoscopes," in *Journal of Cuneiform Studies*, **6** (1952), 49–75.

Chaldaeans in the Persian Era and Later. During and after the Persian reign, Babylonian astronomers and astrologers were called Chaldaeans. According to Herodotus (*Histories*, I, 181), the Chaldaeans were priests in the temple of the highest god, Bel (=Marduk, or, as Herodotus calls him, Zeus-Belos), at Babylon.

When Alexander came to Babylon, two processions of priests met him: first the magi and then the Chaldaeans (Quintus Curtius Rufus 1, *Historia Alexandri Magni*, 22). According to Diodorus, the Chaldaeans made astrological predictions for Alexander and his successors Antigonus and Seleucus.

The geographer Strabo makes a distinction between the Chaldaean people, who lived near the Persian Gulf, and the astronomers, who were called Chaldaeans and lived in a special quarter of Babylon. There were also Chaldaean astronomers at Uruk (Erech) and Borsippa. As far as Babylon and Uruk are concerned, Strabo's statements are confirmed by cuneiform texts, for we have hundreds of astronomical cuneiform texts from Babylon and Uruk. Those from Babylon cover the period from 748 B.C. to A.D. 75, and those from Uruk range from *ca.* 230 to *ca.* 160 B.C.

Strabo also mentions the names of famous Chaldaean astronomers: Kidenas, Naburianus, Sudines, and Seleucus of Seleucia. The first two are also known from cuneiform texts as Kidinnu and Nabu-rimannu. According to B. L. van der Waerden (*Science Awakening*, II, 247–248, 281–283), Nabu-rimannu probably lived under Cambyses (530–522) and Darius I (522–486), and Kidinnu under Artaxerxes I (465–424). In the texts mentioned above, the name Nabu-rimannu is associated with the method of calculation of system A of lunar theory and Kidinnu with system B. (These two systems will be explained below.)

The other two astronomers mentioned by Strabo lived in the Hellenistic age. Sudines was astrologer at the court of Attalus I at Pergamum (241–197), and Seleucus of Seleucia was a follower of the Greek astronomer Aristarchus of Samos (*ca.* 280 B.C.), who first proposed the heliocentric system.

BIBLIOGRAPHY

B. L. van der Waerden, "The Date of Invention of Babylonian Planetary Theory," in *Archive for History of Exact Sciences*, **5** (1968), 70–78.

B. L. van der Waerden, "Die 'Aegypter' und die 'Chaldäer,'" in *Sitzungsberichte der Heidelberger Akademie der Wissenschaften*, math.-naturwiss. Kl. (1972), 197–227.

Systems A and B. For calculating positions of the sun, the moon, and the five planets the Babylonians had two kinds of theories, which Neugebauer calls system A and system B. To explain the characteristic differences between them, let us first consider the motion of the sun.

The fundamental unit of time in all lunar and planetary tables is the mean synodic month (slightly more than 29½ days). According to system A, the motion of the sun in a certain part of the zodiacal circle was 30° per month, and in the remaining part 28°7′30″ per month. In system B, on the other hand, the monthly motion of the sun was supposed to be a function of time, increasing by constant differences up to a maximum *M* and then decreasing in the same manner to a minimum *m*. Such a function is called a linear zigzag function (see Figure 3).

In both systems the velocity of the moon was supposed to be a linear zigzag function of time. For the lunar latitude and the duration of daylight, more complicated piecewise linear functions were adopted, the graphs of which roughly approximate sine waves. Methods were developed that enabled the scribes to calculate, by using purely arithmetical operations, the longitude and latitude of the moon at any time, the time of the new moon and of the full moon, eclipse magnitudes, and times of rising and setting of the sun and the moon. These quantities were regularly calculated from month to month, some of them even from day to day.

In planetary theory the methods were quite similar. Planetary tables that I have called cardinal tables (Neugebauer calls them ephemerides) serve to calculate the longitudes and dates of certain characteristic points in the orbits of the planets called cardinal points. These are of three kinds:

1. The points of first and last visibility of the planets (to be denoted as Morningfirst, Eveninglast, and so on).

2. The stationary points, where the retrograde motion begins or ends.

3. The opposition to the sun (for outer planets).

The first element to be calculated was always the synodic arc, the increase in longitude of any planet from one cardinal point to the next of the same kind (for instance, from one Morningfirst to the next). In system A this synodic arc was supposed to have different values in different parts of the zodiacal circle. For Jupiter the zodiacal circle was divided into two or four parts, and for Mars into six parts. In the case of Mars, each of the six parts consisted of two adjacent zodiacal signs: Pisces and Aries, Taurus and Gemini, and so on. Exactly the same division was used in Egyptian planetary tables of the first two centuries A.D. and in Hindu texts of the sixth century.

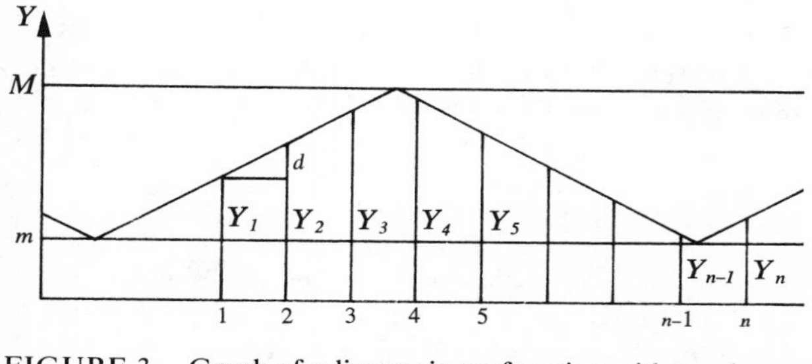

FIGURE 3. Graph of a linear zigzag function with maximum *M*, minimum *m*, and monthly difference *d*.

For computing the dates of cardinal points, the Babylonians applied a principle that I have called the sun-distance principle. It states that at each of the cardinal points, the planet has a fixed elongation from the sun. This is obvious for the oppositions, where the elongation is 180°. The Babylonians also assumed fixed elongations for the other cardinal points. For instance, the first and last visibilities of Mars were supposed to take place at elongations of 15°, and the first stationary point at 120°. (For details see B. L. van der Waerden, *Science Awakening*, II, ch. 6.)

For the motion of planets between cardinal points, the Babylonians, as well as the Egyptians and Hindus, used velocity schemata. A velocity schema for Jupiter on the "fast arc" of the zodiacal circle, drawn from a procedure text for system A, reads as follows:

After Morningfirst 30 days' velocity	15′ per day
3 months to first stationary point	8′ per day
4 months retrograde	5′ per day
3 months after second stationary point	7′40″ per day
30 days to Eveninglast	15′ per day
30 days to Morningfirst	15′ per day

By means of such schemata, the Babylonians were able to calculate dates of entrance of planets into zodiacal signs. These dates were compiled, together with other calculated data, in so-called almanacs (see below) that probably were used for casting horoscopes.

In one text discussed by P. Huber, longitudes of Jupiter between Morningfirst in 165 B.C. and the next Eveninglast are calculated as terms of a third-order arithmetical progression (their third differences are constant).

BIBLIOGRAPHY

The first to explain the methods of calculation of lunar tables were J. Epping and J. N. Strassmaier, followed by F. X. Kugler, who also explained planetary tables. All lunar and planetary tables known in 1955 were assembled and explained in the three-volume *ACT* by O. Neugebauer. (See the bibliography to "Four Stages of Babylonian Astronomy.")

The earliest known lunar texts (system A) were published by A. Aaboe and A. Sachs, "Two Lunar Texts From the Achaemenid Period From Babylon," in *Centaurus*, **14** (1969), 1–22.

Probably the oldest planetary tables are the dateless cardinal tables for Saturn, Jupiter, and Mars published by A. Aaboe and A. Sachs, "Some Dateless Computed Lists of Longitudes of Characteristic Planetary Phenomena . . .," in *Journal of Cuneiform Studies*, **20** (1966), 1–33.

These dateless tables belong to system A. In B. L. van der Waerden's "The Date of Invention of Babylonian Planetary Theory," in *Archive for History of Exact Sciences*, **5** (1968), 70–78, it was shown that the most probable dates for the first lines of the tables for Saturn and Mars are 511 and 499 B.C. Both dates are in the reign of Darius I.

The following papers have contributed to a better understanding of Babylonian planetary theory:

A. Aaboe, "On Period Relations in Babylonian Astronomy," in *Centaurus*, **10** (1964), 213–231.

P. Huber, "Zur täglichen Bewegung des Jupiter," in *Zeitschrift für Assyriologie*, n.s. **18** (1957), 265–303.

B. L. van der Waerden, "Babylonische Planetenrechnung," in *Vierteljahresschrift der Naturforschenden Gesellschaft in Zürich*, **102** (1957), 39–60.

A very good summary of the principles of Babylonian mathematical astronomy is given by A. Aaboe, "Scientific Astronomy in Antiquity," in F. R. Hodson, ed., *The Place of Astronomy in the Ancient World* (London, 1974), 21–42.

For a more extensive treatment, see O. Neugebauer, *A History of Ancient Mathematical Astronomy*, 3 vols. (Heidelberg–New York, 1975).

For the relation of Babylonian theories to Greek, Egyptian, and Hindu astronomy, see the following:

O. Neugebauer, "The Survival of Babylonian Methods in the Exact Sciences of Antiquity and Middle Ages," in *Proceedings of the American Philosophical Society*, **107** (1963), 528–535.

D. Pingree, "The Mesopotamian Origin of Early Indian Astronomy," in *Journal for the History of Astronomy*, **4** (1973), 1–12.

B. L. van der Waerden, *Science Awakening*, II (Leiden–New York, 1974), ch. 8.

Goal-Year Texts and Almanacs. The Babylonian scribes knew that the phenomena of the planet Venus are repeated after eight years—or, more precisely, after ninety-nine lunar months minus four days. Thus, if one has observed a Morninglast of Venus in the year $x - 8$ on the seventh day of a certain month, one may expect a Morninglast in the year x near the third day of a month, ninety-nine months minus 4 days after the month in which the earlier observation was made. Phenomena of Saturn are repeated after fifty-nine years, and so on. These facts were used by the Babylonians to obtain predictions of lunar and planetary phenomena.

In goal-year texts observational data from earlier years are assembled in order to obtain predictions for a certain "goal year." By "observational data" I mean data of the kind found in diaries, without claiming that all these data were actually observed. The data include conjunctions of planets with fixed stars, the "lunar six," and dates of first and last visibility of planets.

In a goal-year text for the year x, we find the following data:

for Jupiter:	years $x-71$ and $x-83$
for Venus:	year $x-8$
for Mercury:	year $x-46$
for Saturn:	year $x-59$
for Mars:	years $x-79$ and $x-47$
for the moon:	$x-18$ and $x-19$.

We have goal-year texts for many years between 240 and 20 B.C. There is also a text in which corrections are indicated, such as the four days to be subtracted from a Venus date in order to obtain the correct date in the goal year.

After having applied the necessary corrections, the Babylonian scribes obtained predictions for the most important lunar and planetary phenomena. These predictions were assembled, together with information from other sources, in an almanac for the goal year. Among the data from other sources are the dates of first and last visibility of Sirius and the equinoxes and solstices, which were calculated from year to year by simple rules.

Two kinds of almanacs are known, which A. Sachs calls normal-star almanacs and almanacs in the narrower sense. In the former, conjunctions of planets with "normal stars" in the zodiac are mentioned, whereas the other almanacs note dates of entrance of planets into zodiacal signs. These dates, which were needed for casting horoscopes, probably were computed by means of velocity schemata.

Apart from these conjunctions and dates of entrance, the two types of almanacs are concerned with the same phenomena: the "lunar six," cardinal points of planets, solstices and equinoxes, appearance and disappearance of Sirius, and predictions of possible eclipses.

BIBLIOGRAPHY

A. Sachs, "A Classification of the Babylonian Astronomical Tablets of the Seleucid Period," in *Journal of Cuneiform Studies*, **2** (1948), 271–290.

THE MATHEMATICS OF ANCIENT EGYPT

R. J. Gillings

When one refers to mathematics in the time of the pharaohs, the significance of the word "mathematics" inevitably comes into consideration. In the long period of ancient Egypt, some three millennia, mathematics meant first arithmetic, then some elementary geometry, then varied problems that by modern standards had an algebraic flavor. The historian would be wise not to judge too hastily the very few genuinely mathematical papyri and ostraca that have come down to us, by making critical comparisons with the more numerous and detailed works of the Greeks, which have been known and studied for over two thousand years. Pharaonic mathematics has been available to the student and the historian for barely a century, although it originated nearly four thousand years ago. One had to wait for a Champollion and a Rosetta Stone before interpretation of hieroglyphs, and the cursive hieratic writing and numbers, became possible. When it appeared that Egyptian arithmetic was based solely upon a complete knowledge of the "two-times" table and an ability to find two-thirds of any number, integral or fractional; that their geometry dealt almost wholly with areas and volumes; and that problems were solved by a kind of literal algebraic reasoning, the question arose whether this can be called mathematics.

It is a matter of semantics; and to help resolve it, one should note what Ernst Mach wrote in 1898: "There is no problem in all mathematics that cannot be solved by direct counting, but with present implements, many operations can be performed in a few minutes, which without mathematical methods, would take a lifetime." And Comte wrote: "There is no inquiry which is not finally reducible to a question of numbers." Acceptance of these statements would perhaps justify saying that Egyptians at the time of the pharaohs did have mathematics, in their own particular way.

Of all the ancient papyri and ostraca that have been recovered from ancient Egypt and are preserved in universities, museums, and other institutions, relatively few are of a mathematical nature. About a dozen of these, often referred to in a study of the history of mathematics, are briefly listed below, not in order of their importance but alphabetically, with their abbreviations that will be used throughout the essay. More detail regarding these papyri will be found following the bibliographical notes.

THE FOUR OPERATIONS

Addition. Modern arithmetic books give special techniques for addition, subtraction, multiplication, and division. Combinations like $3 + 5 = 8$ and $2 + 7 = 9$ are taught first by simple counting, and then they are memorized; at the same time the

Abbreviation	Papyrus	Location
AMP	Akhmim Mathematical Papyrus	Cairo
BP	Berlin Papyrus	Berlin
DMP_1	Demotic Mathematical Papyrus	Cairo
DMP_2	Demotic Mathematical Papyrus	London
EMLR	Egyptian Mathematical Leather Roll	London
KP	Kahun Papyrus	London
Mich P	Michigan Papyrus	Ann Arbor
MMP	Moscow Mathematical Papyrus	Moscow
Ostraca	Various	*Manchester
RP	Reisner Papyrus	Boston
RMP	Rhind Mathematical Papyrus	London

*also, Berlin, Cairo, and London.

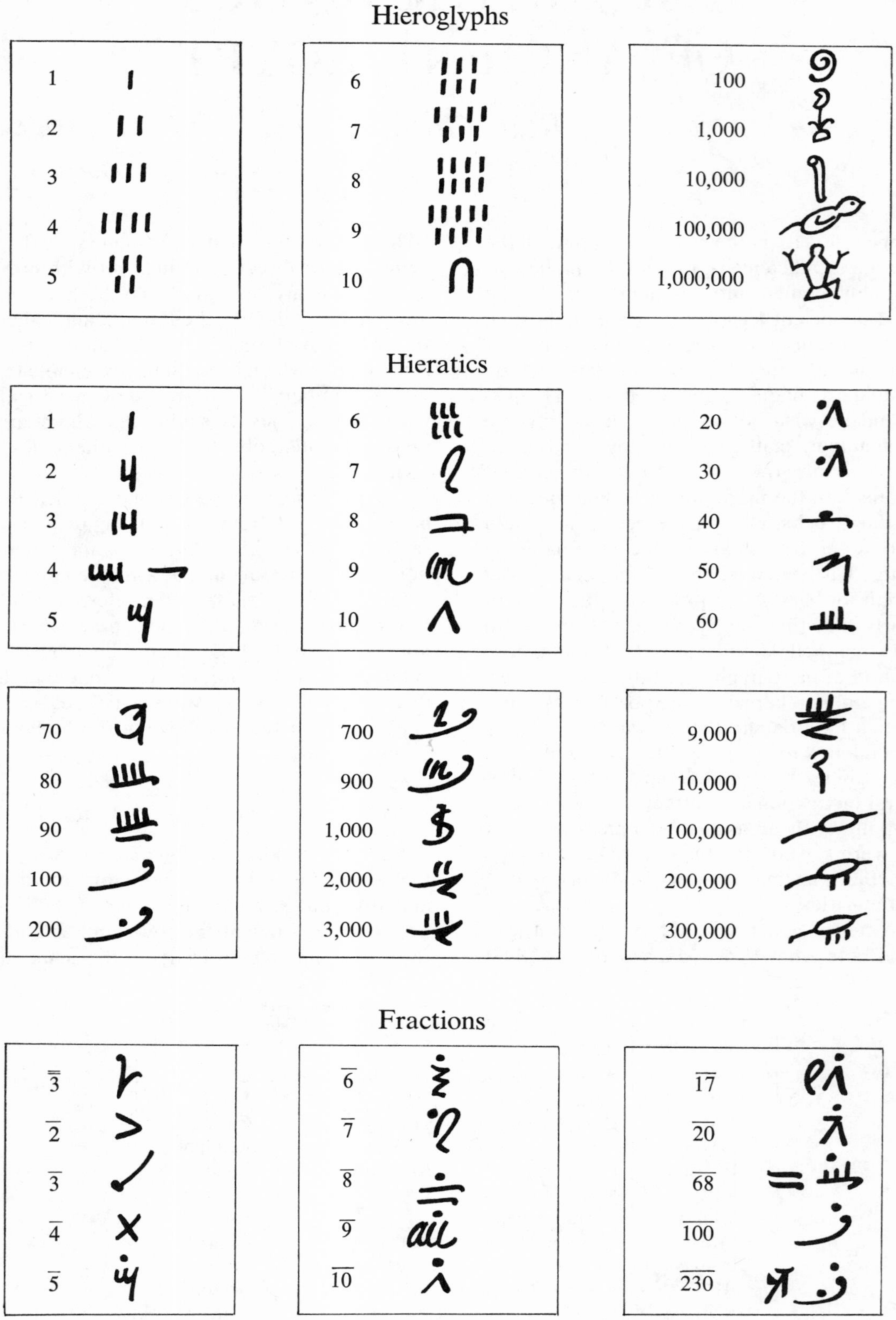

FIGURE 1. Egyptian numbers.

subtractive operations of $8 - 5 = 3$ and $9 - 7 = 2$, being related, are also learned. Subtraction of large numbers is taught in several different ways, however, such as the "borrow and pay back" method or the "equal additions" method. For multiplication and division, the present technique is that all multiplication tables must be learned by heart up to that for twelve—a time-consuming operation for students, many of whom never really learn them.

We are to examine the probable techniques of the Egyptian scribes for the four operations with their hieratic notation, which, although it was based on a decimal system, is still unlike our modern Hindu-Arabic form. Many historians have glossed over considerations of how the ancient scribes performed any of the four operations, with statements like the following: "Addition is simple counting. Subtraction is merely counting backwards. Multiplication is a special form of counting, and division is the reverse of multiplication."[1] How easy it all sounds! "People who could count beyond a million had no difficulty about the addition and subtraction of whole numbers."[2] But they did, and we still do, which explains the invention of the abacus and desk computers. "For ordinary additions and subtractions, nothing needs be said."[3] Oh, but it does!

Keeping in mind that hieratic numbers are written from right to left, note that in the first addition, *A*, the third addend 5 is written in the tens column, and the total is not placed underneath but is to the left of the 5. There is nothing to be "carried." In the second addition, *B*, the addend 192 is slightly out of alignment and the "carrying" is easily enough done, with the total at the bottom. In both *C* and *D* the arrangements of the addends and the totals are closer to what we would expect to see. Note, however, that 320 is not to be included in the addition, because it does not have a check mark (or tick) alongside it. In *E*, the 6 is in the tens' column instead of the units' column, and again the total is not underneath. Finally, in *F*, note that the sign for five thousands is a dual one and, like the third addend, does not have a check mark and thus is not to be included in the total. There are three "carryings" in this sum.

The proper alignment of digits in their correct columns was not so necessary to the Egyptian scribes as it is to us, because the hieratic sign for (say) three units was ///, for three tens it was , and for three hundreds it was —symbols that are quite different. The scribe would therefore know at once not to add units to hundreds merely because they appeared to be in the same column. Nevertheless, all these additions need careful attention, especially the last; and to say that "there are no difficulties about addition and subtraction," and leave it at that, is clearly evading the issue.

Many tables for the addition of fractions have come down to us, notably the EMLR. These will be considered in a separate section.

Subtraction. Tables for the addition of integers at least up to 10, and possibly further, prepared by simple counting, were most surely available, although none has come down to us in the papyri. Such tables would be equally useful for subtraction, for the scribe did not say "Subtract 5 from 8" or "Subtract 7 from 9," but "5, how many to make

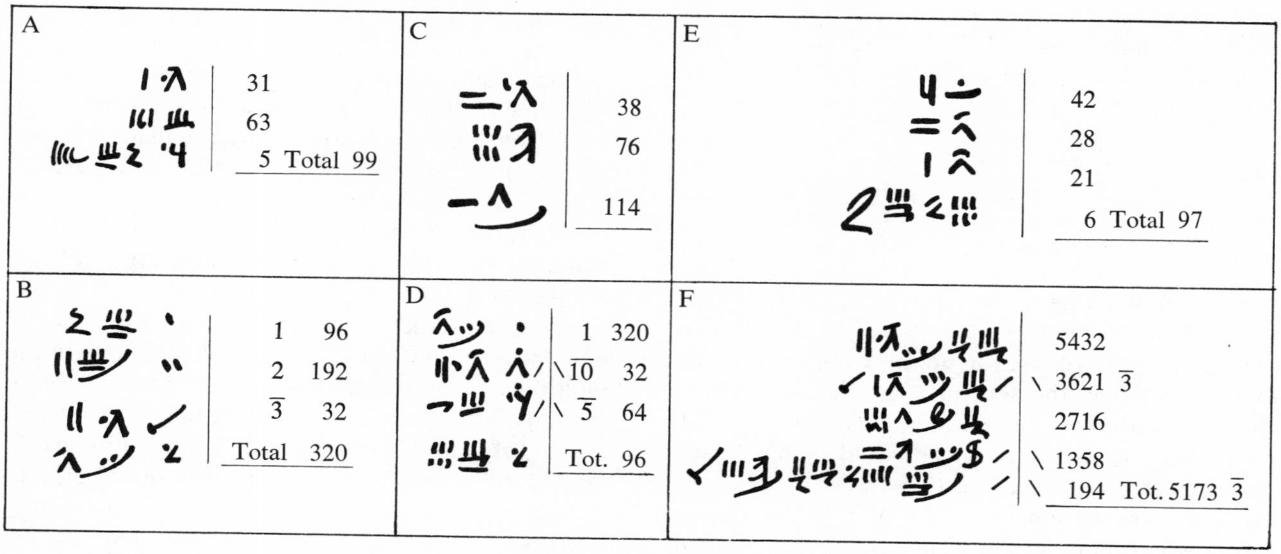

FIGURE 2. Addition sums.

8?" and "7, how many to make 9?" Thus we find in MMP9, "Complete thou the excess of these 10 over these 4. Result 6." And in KP LV, 4, "Make thou the excess of 100 over 45. The result thereof is 55."

Subtraction of unit fractions was done in the same way; and in this case tables of the addition of fractions have certainly come down to us, the best example of which is the EMLR. Examples of such subtractions, taken from the Recto of the RMP, are given below.

$$\begin{aligned} &\text{From } 2 \div 17, \quad 1 - (\bar{4}\ \bar{6}) = \bar{3}\,\bar{4} \\ &\text{From } 2 \div 19, \quad 1 - (\bar{2}\,\overline{12}) = \bar{4}\,\bar{6} \\ &\text{From } 2 \div 23, \quad 1 - (\bar{2}\ \bar{4}) = \overline{12} \\ &\text{From } 2 \div 37, \quad 1 - (\bar{2}\,\overline{24}) = \bar{3}\,\bar{8} \\ &\text{From } 2 \div 41, \quad 1 - (\bar{\bar{3}}\,\overline{24}) = \bar{6}\,\bar{8} \end{aligned}$$

Consider the subtrahend of the last subtraction. To it add $\bar{8}$:

$\bar{\bar{3}}(\bar{8}\,\overline{24}) = \bar{\bar{3}}\,\bar{6}$. EMLR, l. 3: $\bar{4}\,\overline{12} = \bar{3}$[×2].

Now add $\bar{6}$: $= \bar{\bar{3}}(\bar{6}\,\bar{6})$
$= \bar{\bar{3}}\,\bar{3}$ EMLR, l. 5: $\bar{6}\,\bar{6} = \bar{3}$.
$= 1$.

Therefore $1 - (\bar{\bar{3}}\,\overline{24}) = \bar{6}\,\bar{8}$.

In the RMP Recto there are more than twenty such subtractions, some much more complicated than those shown here; but in no case is the working shown. Only the answer is written, and there are no errors.

Multiplication. This process was done by repeated multiplications of 2, and a thorough knowledge of the "two-times" table was all that the scribal arithmetician required. Preferred or favored multipliers were 3, 7, 15, 31, 63, · · ·. Thus, to multiply 29 by 7, the scribe proceeded as shown below.

	1	29
	2	58
	4	116
Totals	7	203

The sums of terms of the series 1, 2, 4, 8, 16, · · · produce, in order, the totals 3, 7, 15, 31, 63, · · ·, which, if they are used as multipliers, means that every number in the second column is included in the final addition. The scribes early found, however, that any multiplier at all could be expressed uniquely as the sum of certain terms of this geometrical progression, a minor difficulty being to locate which terms these were. Thus, suppose that the multiplier of 29 was 21 instead of 7.

	\ 1	29
	2	58
	\ 4	116
	8	232
	\ 16	464
Totals	21	609

The scribe located the numbers in the first column that totaled 21 in one way only – 1 + 4 + 16 = 21 – and placed a check mark alongside them, indicating that the product was the sum of the corresponding numbers in the second column: 29 + 116 + 464 = 609. This particular property of the progression 1, 2, 4, 8, 16, · · · is nowhere referred to in any Egyptian papyri known to me; but there are various other properties of both arithmetical and geometrical progressions that arise in certain problems and will be considered later.

Sometimes other multipliers were used, such as 10, which meant that the hieratic digits of the multiplicand were merely rewritten, the units as tens, the tens as hundreds, and so on, since the notation is of course a decimal one. The multiplier 1/10 was used in a similar manner. The fraction two-thirds (written , but here $\bar{\bar{3}}$) was frequently used, and tables giving $\bar{\bar{3}}$ of numbers were available. The scribal rule for expressing $\bar{\bar{3}}$ of any fraction, in unit fractions, is given in RMP 61B. Two-thirds of quite large numbers occurs often. For example, in RMP 33 we find $\bar{\bar{3}}$ of 5,432 written as 3,621 $\bar{3}$. The rule given for $\bar{\bar{3}}$ of the odd fraction $\bar{5}$ was to add the double of the number to six times the number, giving ($\overline{10}$ $\overline{30}$), and similarly for all odd fractions. The rule holds equally for even fractions, so that $\bar{\bar{3}}$ of $\bar{6}$ would be ($\overline{12}$ $\overline{36}$); it was seldom used, however, because the value $\bar{9}$ was simpler, from the addition to 6 of its half 3. That the reciprocal of 1 $\bar{2}$ was $\bar{\bar{3}}$ was well-known to the scribes; indeed, the hieroglyph for $\bar{\bar{3}}$ was originally 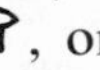, one vertical line being half the length of the other. Or, again, ($\overline{12}$ $\overline{36}$) would be written as $\bar{9}$, as indicated by line 3 of the EMLR, ($\bar{4}$ $\overline{12}$) = $\bar{3}$, which becomes ($\overline{12}$ $\overline{36}$) = $\bar{9}$ on multiplication of the numbers by 3.

For the ordinary multiplication of integers the work was simple enough; but with fractions in the multiplicand, it sometimes became difficult just to multiply by 2. For example, the double of $\overline{20}$ was $\overline{10}$, and the double of $\overline{10}$ was $\bar{5}$; but the double of $\bar{5}$ or any other odd unit fraction was quite another matter. The doubling of the fifty "odd" unit fractions $\bar{3}$, $\bar{5}$, $\bar{7}$, $\bar{9}$, $\overline{11}$, · · ·, $\overline{101}$ is performed by the scribe in the first portion of the RMP, called the

Recto; and it takes up almost one-third of the eighteen feet of the papyrus.

Division. Little further needs to be said about division, because the method of dividing was the same as that of multiplying. To calculate 297 divided by 11, the scribe thought of it as "By what must I multiply 11 to give me 297?" Thus 11 became the multiplicand, and he kept on doubling it until 297 was reached, just like a multiplication sum.

	1	11	/
	2	22	/
	4	44	
	8	88	/
	16	176	/
Totals	27	297	

A further doubling would give 352, which is beyond 297. The problem is now to determine which numbers of the second column will add to 297, or come close to it—a task not always easy. In this case the numbers are 11 + 22 + 88 + 176 = 297, indicated by check marks, so that the required quotient is the sum of the corresponding numbers in the first column: 1 + 2 + 8 + 16 = 27. When the divisor contained one or more odd unit fractions, then trouble brewed, just as it did when the division of two integers did not produce an integral quotient, as 297 ÷ 11 does. To illustrate this, divide 297 by 12.

	1	12
	2	24
	4	48
\	8	96
\	16	192

A study of the right-hand column shows that 192 + 96 = 288 is the closest sum to 297; and these, and/or 16 and 8, are ticked. Also, 288 + 9 = 297. Then $\overline{2}$ of 12 = 6, and $\overline{4}$ of 12 = 3, so that we now have

	1	12
	2	24
	4	48
\	8	96
\	16	192
\	$\overline{2}$	6
\	$\overline{4}$	3
Totals	24 $\overline{2}\,\overline{4}$	297

THE RMP RECTO

The Recto of the RMP gives the ancient Egyptian values chosen and accepted by the scribes for the expression of 2 ÷ (2*n* − 1) as the sum of not more than four unit fractions, where *n* had the fifty values, 2, 3, 4, 5, · · ·, 51. Since the latter part of the nineteenth century, historians of mathematics have discussed and debated how and why the scribes determined the values given in this table, and still no general agreement has been reached. In 1967, with the purpose of examining more closely the various equalities possible, I was able to enlist the services of Prof. C. L. Hamblin of the University of New South Wales to program the computer KDF−9 at Sydney University to produce all the possible answers for these equivalents. The restrictions were such that not more than four unit fractions should be included in any one equality, and that none of the denominators should exceed 1,000.

On this basis KDF−9 produced the grand total of 22,295 equivalents. But for the purpose of proper comparison of these values with the specific value given in each case by A'hmosé, the scribe of the RMP, 22,295 must be reduced by the number of redundancies properly included by KDF−9, according to its instructions, such as

$$2 \div 47 = (\overline{47}\ \overline{47}),\ (\overline{28}\ \overline{376}\ \overline{376}\ \overline{658}),\ \text{and}\ (\overline{30}\ \overline{282}\ \overline{282}\ \overline{470}).$$

because ancient Egyptian scribes would never accept two equal fractions in any *stated* number. Therefore the total of 22,295 must be reduced by 2,024. Further, the program for KDF−9 stated, "No fraction less than 1/1,000 should be included," when it should have been instructed to "include only fractions greater than 1/1,000." This is a fine distinction, but it means that the nine values in which the unit fraction 1/1,000 occurs should be subtracted, so that the total for comparison purposes is reduced to 20,262.

The purpose of this table, giving the numbers of the possible decompositions of the RMP Recto fractions, is to make clear the nature of the problems historians have puzzled over for some years. Some general agreement appears to have been reached that the scribal values for 2 divided by the odd numbers are the very best available, although in some cases there are differences of opinion. On what are these differences based? Some critics appear not to be clear on the purpose of the table, which is primarily to simplify ordinary multiplication and division, and not just an interesting operation in the theory of numbers. Attempts to explain how the scribe arrived at his values have varied considerably; indeed, there is little agreement over what precepts guided the scribe in choosing his spe-

Divisor	No. of Values	Divisor	No. of Values
3	57	53	21
5	201	55	954
7	278	57	531
9	468	59	18
11	332	61	5
13	301	63	1,391
15	1,015	65	758
17	225	67	21
19	253	69	407
21	1,059	71	25
23	260	73	10
25	526	75	882
27	632	77	739
29	183	79	3
31	140	81	308
33	870	83	3
35	1,252	85	503
37	82	87	205
39	762	89	6
41	161	91	439
43	103	93	162
45	1,689	95	323
47	45	97	9
49	338	99	713
51	602	101	1
	Total 11,834		8,437
			11,834
			20,271
		Less	9
		Total	20,262

TABLE I. RMP Recto, $2 \div (2n-1)$ for $n = 2, 3, 4, \cdots, 51$, expressed as the sum of not more than four different unit fractions, none smaller than 1/999.

cific values, although some are of course obvious.

The scribe first states what solution he has selected and then, by ordinary multiplication, proves that the equality is correct. In these proofs there are no scribal errors. Nowhere in any papyrus known to me is there any indication of the scribal technique used by the scribe.[4] Some theories regarding this technique seem to be based on modern mathematics; and attempts to reproduce the scribal values, following these theories but using only the techniques of which the scribe was capable, often are not possible. J. J. Sylvester's treatment is a classic example.[5] I suggest the following as the precepts that guided the scribe in his choice of equalities; and to assist in understanding their significance, I precede them with some of the Recto equalities for reference (bars over the numbers of the unit fractions have been omitted).

Divisor	Unit Fractions			
7	4	28		
9	6	18		
13	8	52	104	
17	12	51	68	
35	30	42		
43	42	86	129	301
79	60	237	316	790

Canon for the RMP Recto.

Precept No. 1: Of the possible equalities, those with the smaller numbers are preferred, but none as large as 1,000.

Precept No. 2: An equality of two terms is preferred to one of three terms, and one of three terms to one of four terms; but an equality of more than four terms is never to be used.

Precept No. 3: The unit fractions are set down in descending order of magnitude—that is, smaller numbers come first—but never the same number twice.

Precept No. 4: The smallness of the first number is the main consideration, but the scribe will accept a slightly larger first number if it will greatly reduce the last number.

Precept No. 5: Even numbers are preferred to odd numbers, even though they might be larger, and even though the number of terms might thereby be increased. (There are more than one hundred even numbers used in the table, but only twenty-four odd numbers.)

There can be no doubt about the general acceptance among authorities of the first three precepts, and I am sure that competent critics will agree with the last two. It is of course possible that they might wish to add further precepts.

Historians' Views on the Recto Equalities.

The following statements represent a cross section of opinions:

O. Becker and Hofmann—"The principle of calculation does not seem to be uniform."

L. Hogben—"They went to extraordinary pains to split up fractions like 2/43 into a sum of unit fractions, a procedure that was as useless as it was ambiguous. The Greeks and Alexandrians continued this extraordinary performance."

A. B. Chace—"Of the discussions which I have seen, the clearest is that by Loria, but no formula or rule has been discovered that will give all the results of the table."

F. O. Hultsch—"Attempts to explain it have hitherto not succeeded."

P. Mansion—"The decompositions are always, from one point of view or another, simpler than any other decompositions."

J. J. Sylvester (1882)—"The very beautiful ancient Egyptian method of expressing all fractions under the form of a sum of the reciprocals of continually increasing integers."

THE FRACTION TWO-THIRDS

In hieratic papyri all fractions had only unity as numerator, although the unit 1 was not written. A number became its own reciprocal by putting the sign 𓂋 (*r*, an open mouth), above it in the hieroglyphs, or a large dot in the cursive hieratic. The solitary exception to this notation was their largest fraction, 2/3, written 𓂌 as a hieroglyph but [hieratic sign] in hieratic; and it was used whenever possible. (There is evidence that a hieroglyph for 3/4 was used.) Tables for finding two-thirds (here written $\bar{\bar{3}}$) of both integers and fractions were available to the scribes, and in RMP 61 there is such a table giving $\bar{\bar{3}}$ of seventeen different fractions. As late as the seventh century of the Christian era, the Greek AMP had an extensive table giving $\bar{\bar{3}}$ of integers up to ten thousand, shortened and condensed in some obvious ways. There are similar tables in Coptic of an even later date (see Crum's *Catalogue*). In the RMP alone there are more than seventy occasions on which the scribe writes $\bar{\bar{3}}$ of both integers and fractions, including seventeen where $\bar{3}$ is written with the intermediary $\bar{\bar{3}}$ value omitted for brevity.

In RMP 61B, the scribe A'hmosé states the rule for odd unit fractions. Chace translates it as follows: "The making of two-thirds of a fraction uneven. If it is said to thee. What is two-thirds of $\bar{5}$, make thou times of it 2, and times of it 6, two-thirds of it this is. Behold does one according to the like, for fraction every uneven, which may occur." Although the rule applies equally well to even unit fractions, it was seldom used for them, because the scribe knew that $\bar{\bar{3}}$ was the reciprocal of 1 $\bar{2}$ (see the hieroglyph for $\bar{\bar{3}}$), and it was easier and simpler merely to increase the even number by its half, like $\bar{\bar{3}}$ of $\bar{6}$ is $\bar{9}$.

Examples of the two-thirds rule are shown below.

RMP 32:

$$\text{Since } \bar{\bar{3}} \text{ of } 1\ \bar{3}\ \bar{4} = 1\ \overline{18}, \text{ then } \bar{3} \text{ of } 1\ \bar{3}\ \bar{4} = \bar{2}\ \overline{36}$$

RMP 33:

$$\bar{\bar{3}} \text{ of } 16\ \overline{56}\ \overline{679}\ \overline{776} = 10\ \bar{\bar{3}}\ \overline{84}\ \overline{1358}\ \overline{4074}\ \overline{1164}$$

RMP 42:

$$\bar{\bar{3}} \text{ of } 8\ \bar{\bar{3}}\ \bar{6}\ \overline{18} = 5\ \bar{\bar{3}}\ \bar{6}\ \overline{18}\ \overline{27}$$

THE EMLR

Because of its brittle condition, the leather roll remained unopened for sixty years. When A. Scott and H. R. Hall finally succeeded, it was found to contain a table of twenty-six unit-fraction equalities, in duplicate. The disappointment of archaeologists over the contents was not shared by historians of mathematics, who have found it of great interest. In common with other scribal tables, no methods of calculation are shown, nor is there even a heading or title that might indicate the use to which the entries could be put. The following selection of entries from the EMLR will interest historians.

4	12	=	3		7	14	28	=	4
5	20	=	4		14	21	42	=	7
9	18	=	6		18	27	54	=	9
10	40	=	8		30	45	90	=	15
12	24	=	8		25	50	150	=	15
15	30	=	10	25	15	75	200	=	8
18	36	=	12	50	30	150	400	=	16

A close study of these equalities shows that the scribe clearly knew that any equality—for example, 9 18 = 6—can be multiplied by 2 or 3 or 4, and so on, to produce other equalities, such as 18 36 = 12; therefore division would do the same, but only if the denominators of the fractions had a common factor, so that division by 3 would produce 3 6 = 2. Other possibilities at once suggest themselves, such as combining 4 12 = 3 and 5 20 = 4, to produce, by simple substitution, 5 12 20 = 3—and so on indefinitely, as Sylvester's treatment suggests. This short table shows how the scribe of the EMLR utilized the two most elementary of his unit fraction equalities, 3 6 = 2 and 2 3 6 = 1, to produce seven of the thirteen equalities shown.

A study of the papyri, particularly the RMP and the MMP, will show how frequently the use of these equalities was necessary. To illustrate this, I choose a rather extreme example from RMP 70, which, on the evidence, the scribe did in his head.

Summation of Sixteen Unit Fractions From RMP 70

= 2 6 12 14 21 21 42 63 84 126 126 168 252 336 504 1,008

= 2 (6 12)(14 21 21) 42 63 (84 126 126) 252 (168 336 504 1,008)

= 2 4 6 42 63 36 252 84

= 2 4 6 (42 63 84 252) 36

= 2 4 (6 18 36)

= (2 4 4)

= 1

There are seven groups of unit fractions within parentheses in this summation. Their respective sums, shown above, are derived from scribal tables like the EMLR and are listed below.

Generators	Equality	Multiplying Factor
(1,2)	$\bar{3}\ \bar{6} = \bar{2}$	2
(2,3,3)	$\overline{14}\ \overline{21}\ \overline{21} = \bar{6}$	1
(2,3,3)	$\overline{14}\ \overline{21}\ \overline{21} = \bar{6}$	6
(1,2,3,6)	$\bar{2}\ \bar{4}\ \bar{6}\ \overline{12} = \bar{1}$	84
(2,3,4,12)	$\overline{14}\ \overline{21}\ \overline{28}\ \overline{84} = \bar{6}$	3
(1,3,6)	$\bar{3}\ \bar{9}\ \overline{18} = \bar{2}$	2
(1,2,2)	$\bar{2}\ \bar{4}\ \bar{4} = \bar{1}$	1

In the above expressions the occurrence of the same fraction twice is an accident of the detail of the calculations involved. Duplication would never occur in a scribal answer to the problem.

An ingenious and competent scribe might conceivably devise a more expeditious method of deriving the answer, unity.[6] Or he could use the "red auxiliaries." When using the red auxiliaries, the scribes invariably chose the largest number of the set of fractions for use as their "common denominator." Each number was then divided into this largest number, and the sum of all the quotients was found as shown below.

Division of 1,008 by	Quotients in Red
2	504
6	168
12	84
14	72
21	48
21	48
42	24
63	16
84	12
126	8
126	8
168	6
252	4
336	3
504	2
1,008	1
	Total 1,008

Quite by chance, 1,008 is here what we call the least common multiple; but usually in such summations this does not happen, and there may be several unit fractions occurring in the quotients. The division of the quotients' total by the "common denominator" chosen is seldom as simple as 1,008 divided by 1,008. Furthermore, the scribes seldom show the actual divisions (sixteen of them in RMP 70), which probably were performed on an odd piece of scribbling papyrus or ostracon; this work might well have been quite voluminous, compared with the apparent brevity and conciseness of the table above.

The detail shown above has been included because this summation is a very small portion of RMP 70, so small that the scribe merely writes the total as 1, with no further comment. But he could never have done this mentally. He is concerned to find the amount of meal in each of 100 loaves made from 7 $\bar{2}\ \bar{4}\ \bar{8}$ *hekats* of meal, as well as their *pesu*—and this is quite a calculation. Just how the scribe performed the summation of these sixteen fractions, the historian can never truly know.

TABLES FOR THE ADDITION OF UNIT FRACTIONS, AND THE G RULE

Of the twenty-six equalities in the EMLR table, ten are examples of, and are derivable from, the simplest of all the additions of unit fractions known to the scribes—$\bar{3}\ \bar{6} = \bar{2}$—which fundamental dual sum I have designated as having the generator (1, 2)—that is, the second term is double the first. A scribe preparing a table of such equalities would naturally begin with it and, by simple multiplication or simple addition, produce the following, which could be extended indefinitely.

Generator (1, 2)

$\bar{3}\ \bar{6} = \bar{2}$
$\bar{6}\ \overline{12} = \bar{4}$
$\bar{9}\ \overline{18} = \bar{6}$
$\overline{12}\ \overline{24} = \bar{8}$
$\overline{15}\ \overline{30} = \overline{10}$ etc.

The equality of EMLR line 3 is $\bar{4}\ \overline{12} = \bar{3}$, which I refer to as of generator (1, 3); and it produces the table below, again by either multiplication or addition, which also could be extended indefinitely.

Generator (1, 3)

$\bar{4}\ \overline{12} = \bar{3}$
$\bar{8}\ \overline{24} = \bar{6}$
$\overline{12}\ \overline{36} = \bar{9}$
$\overline{16}\ \overline{48} = \overline{12}$
$\overline{20}\ \overline{60} = \overline{15}$ etc.

The third table would of course be that resulting from the generator (1, 4), as on lines 2 and 1 of the EMLR.

Generator (1, 4)

$$\begin{aligned} \overline{5}\ \overline{20} &= \overline{4} \\ \overline{10}\ \overline{40} &= \overline{8} \\ \overline{15}\ \overline{60} &= \overline{12} \\ \overline{20}\ \overline{80} &= \overline{16} \\ \overline{25}\ \overline{100} &= \overline{20} \text{ etc.} \end{aligned}$$

The G Rule. An observant scribe looking at the equalities of generator (1, 2) could well have summed up the situation briefly in some such manner as "For adding two fractions, if one is double the other, divide it by three." A little further observation would enable him to change the word "double" to "three times," "four times," and so on, and to divide by four, five, and so on for the generators (1, 3), (1, 4), and so on. This is in essence what I have called the G rule; but using modern terms and introducing some further detail, I express it as follows: "If of two unit fractions, one is K times the other, then their sum is found by dividing the larger number by $(K + 1)$ if, and only if, the quotient is an integer. If it is not an integer, then a unit-fraction sum is not possible." So far as the scribe was concerned, his rule, if he used it as suggested, did not apply to equalities from generators such as (2, 3), (2, 5), · · ·, (3, 4), (3, 5), · · ·, (4, 5), (4, 7), · · ·. Mathematically the G rule still held, although with his notation he could not apply it. It would have been clear to him, however, that pairs of unit fractions like ($\overline{4}\ \overline{8}$), ($\overline{5}\ \overline{10}$), ($\overline{7}\ \overline{14}$), ($\overline{8}\ \overline{16}$), which evolve from the generator (1, 2), cannot add up to another unit fraction.

Let us look at line 1 of the EMLR, which is $\overline{10}\ \overline{40} = \overline{8}$. It is related to line 2, which is $\overline{5}\ \overline{20} = \overline{4}$, while line 4 is $\overline{10}\ \overline{10} = \overline{5}$.

Now the only other dual equalities with $\overline{10}$ as the first term are $\overline{10}\ \overline{90} = \overline{9}$ and $\overline{10}\ \overline{15} = \overline{6}$. The first of these clearly follows the G rule but is not among the twenty-six equalities of the EMLR, although it might well have been. The student of ancient Egyptian mathematics may now ask, "If the scribe knew that $\overline{10}\ \overline{15} = \overline{6}$, how did he find it?"

Three-Term Equalities. The simplest three-term equalities of unit fractions have the generators (1, 2, 4), (1, 3, 6), (2, 3, 6), which produce the following equalities:

Generator (1, 2, 4)

$$\begin{aligned} \overline{7}\ \overline{14}\ \overline{28} &= \overline{4} \\ \overline{14}\ \overline{28}\ \overline{56} &= \overline{8} \\ \overline{21}\ \overline{42}\ \overline{84} &= \overline{12} \text{ etc.} \end{aligned}$$

Generator (1, 3, 6)

$$\begin{aligned} \overline{3}\ \overline{9}\ \overline{18} &= \overline{2} \\ \overline{6}\ \overline{18}\ \overline{36} &= \overline{4} \\ \overline{9}\ \overline{27}\ \overline{54} &= \overline{6} \text{ etc.} \end{aligned}$$

Generator (2, 3, 6)

$$\begin{aligned} \overline{2}\ \overline{3}\ \overline{6} &= \overline{1} \\ \overline{4}\ \overline{6}\ \overline{12} &= \overline{2} \\ \overline{6}\ \overline{9}\ \overline{18} &= \overline{3} \text{ etc.} \end{aligned}$$

Eight of these are included in the EMLR table, five of them from the generator (2, 3, 6) forming lines 14–18:

$$\begin{aligned} \overline{14}\ \overline{21}\ \overline{42} &= \overline{7} \\ \overline{18}\ \overline{27}\ \overline{54} &= \overline{9} \\ \overline{22}\ \overline{33}\ \overline{66} &= \overline{11} \\ \overline{26}\ \overline{39}\ \overline{78} &= \overline{13} \\ \overline{30}\ \overline{45}\ \overline{90} &= \overline{15}. \end{aligned}$$

Line 12 of the EMLR is $\overline{7}\ \overline{14}\ \overline{28} = \overline{4}$, one of the more interesting of all the equalities, which could have been derived in many ways. We look here at one method. The Egyptian table of length was simple enough:

4 digits = 1 palm
7 palms = 1 cubit.

Therefore,

4 digits = $\overline{7}$ cubit
2 digits = $\overline{14}$ cubit
1 digit = $\overline{28}$ cubit.

And by addition,

$$\begin{aligned} \overline{7}\ \overline{14}\ \overline{28} &= 4 + 2 + 1 \text{ digits} \\ &= 7 \text{ digits} \\ &= \overline{4} \text{ cubit.} \end{aligned}$$

Thus

$$\overline{7}\ \overline{14}\ \overline{28} = \overline{4}.$$

Four-Term Equalities. Only two four-term equalities occur in the EMLR:

$$\begin{aligned} \overline{25}\ \overline{15}\ \overline{75}\ \overline{200} &= \overline{8} \\ \overline{50}\ \overline{30}\ \overline{150}\ \overline{400} &= \overline{16}. \end{aligned}$$

One finds it hard to understand why these two particular examples should have been chosen, when so many simpler and more useful ones might have been included:

$$\begin{aligned} \overline{4}\ \overline{8}\ \overline{12}\ \overline{24} &= \overline{2} \\ \overline{7}\ \overline{21}\ \overline{28}\ \overline{42} &= \overline{4} \\ \overline{10}\ \overline{15}\ \overline{20}\ \overline{30} &= \overline{4} \\ \overline{12}\ \overline{18}\ \overline{30}\ \overline{36} &= \overline{5}, \end{aligned}$$

which have generators (1, 2, 3, 6), (1, 3, 4, 6), (2, 3, 4, 6), and (2, 3, 5, 6), compared with generator (5, 3, 15, 40). An explanation why the first two terms are out of normal order would make very interesting reading for the reflective scholar.

AN INTERESTING OSTRACON

Ostracon 153, dating from the early New Kingdom and thus somewhat later than the RMP, contains abbreviated divisions of 2 by 7 and 4 by 7, as in the RMP Recto, but uses the red auxiliaries and a reference number.[7] What the scribe in fact does is to consider the divisions of 6 by 21 and 12 by 42; 6 is partitioned as $(3\,\bar{2}\ 1\,\bar{2}\ 1)$, written in red. When divided by 21, these auxiliaries produce the answer $(\bar{6}\ \overline{14}\ \overline{21})$, which is not as simple or convenient as the Recto value of $(\bar{4}\ \overline{28})$. The scribe proceeded by partitioning 12 as $(10\,\bar{2}\ 1\,\bar{2})$, again in red; when divided by 42, these auxiliaries produce the answer $(\bar{4}\ \overline{28})$.

Of the fifty divisions of 2 by the odd numbers in the Recto, only one, that of $2 \div 35$, discloses anything of the scribe's methods—and it happens to be the same as that of the scribe of Ostracon 153. Thus A'hmosé considers $2 \div 35$ as $12 \div 210$, and then partitions 12 as (7 5), which auxiliaries, on division by 210, produce the answer $(\overline{30}\ \overline{42})$. The only other pair of red auxiliaries he could have chosen is (10 2), which would have given $(\overline{21}\ \overline{105})$—not so acceptable as the first answer.

THE SCRIBAL RULE FOR RECIPROCALS

The following examples show that the scribes were familiar with the operation that in modern notation may be written as "If $a \times K = b$, then $1/b \times K = 1/a$." In RMP 33 it is clearly shown that the reciprocal of $\bar{\bar{3}}$ is $1\ \bar{2}$.

RMP 34. "A quantity, its half, and its quarter, added together, become 10. What is this quantity?" Multiply $1\ \bar{2}\ \bar{4}$ so as to obtain 10.

Line 1	\ 1		1	$\bar{2}$	$\bar{4}$
Line 2	2		3	$\bar{2}$	
Line 3	\ 4		7		
Line 4	\ $\bar{7}$			$\bar{4}$	
Line 5	$\bar{4}$	$\overline{28}$		$\bar{2}$	
Line 6	\ $\bar{2}$	$\overline{14}$	1		
Totals	$5\ \bar{2}\ \bar{7}\ \overline{14}$		10		

Line 3 states that $4 \times 1\ \bar{2}\ \bar{4} = 7$, and therefore line 4 drives that $\bar{7}$ of $1\ \bar{2}\ \bar{4} = \bar{4}$. Line 5 is line 4 multiplied through by 2, so that from the Recto the scribe finds that 2×7, which is $2 \div 7$, is $\bar{4}\ \overline{28}$. Again, multiplying by 2, line 6 shows that $\bar{2}\ \overline{14}$ of $1\ \bar{2}\ \bar{4} = 1$. The answer is $5\ \bar{2}\ \bar{7}\ \overline{14}$.

RMP 70.

The scribe finds that $8 \times 7\ \bar{2}\ \bar{4}\ \bar{8} = 63$.

He then writes that $\overline{63}$ of $7\ \bar{2}\ \bar{4}\ \bar{8} = \bar{8}$.

RMP 32.

The scribe finds that $144 \times 1\ \bar{3}\ \bar{4} = 228$.

He then writes that $\overline{228}$ of $1\ \bar{3}\ \bar{4} = \overline{144}$.

RMP 38.

The scribe finds that $22 \times \bar{6}\ \overline{11}\ \overline{22}\ \overline{66} = 7$.

He then writes that $\bar{7}$ of $\bar{6}\ \overline{11}\ \overline{22}\ \overline{66} = \overline{22}$.

RMP 33.

The scribe finds that $\bar{\bar{3}}$ of $42 = 28$.

He then writes that $\overline{28}$ of $42 = 1\ \bar{2}$.

DEMOTIC MATHEMATICAL PAPYRI

For the translations of the DMP that follow, I am indebted to R. A. Parker of Brown University. (See the bibliography and "Mathematical Papyri and Ostraca.")

Cairo Papyrus (JE 89127). Discovered at Tuna el-Gebel in 1938 and first examined by Parker in 1962, this papyrus contains forty problems of a mathematical nature. Of these, eight are obscure or fragmentary, and fourteen have counterparts in the much earlier hieratic papyri. The Cairo papyrus dates from 300 B.C. or possibly earlier, and it is interesting to examine what advances or developments are to be found in its mathematical techniques.

First we note that there are nine problems dealing solely with Pythagoras' theorem—for instance, "A ladder of ten cubits has its foot six cubits from a wall; to what height will it reach?" With one exception, the numbers used are those of the triads (3, 4, 5), (5, 12, 13), and (20, 21, 29); and no such problems have been noted in the earlier papyri. The one exception requires the calculation of an approximate square root.

Two problems deal with rectangles having areas of sixty square cubits and diagonals of thirteen and fifteen cubits, respectively. It is required to find the sides. The directions for their solution are identical with the steps of the following algebra, but of course they are entirely descriptive.

$$\begin{aligned} x^2 + y^2 &= 169 \\ xy &= 60 \\ \hline 2xy &= 120 \\ (x+y)^2 &= 289 \\ (x-y)^2 &= 49 \\ \hline x+y &= 17 \\ x-y &= 7 \\ \hline 2y &= 10 \\ y &= 5 \\ x &= 17-5=12 \end{aligned}$$

The second problem is the same except for the numbers, and so the square roots of 345 and 105 are to be found. The formula used is the approximation usually attributed to Archimedes or Hero, $\sqrt{a^2+b} = a + \frac{b}{2a}$. Thus we find the square roots to be $18\ \bar{2}\ \overline{12}$ and $10\ \bar{4}$ from

$$\sqrt{345} = \sqrt{18^2+21} = 18 + \frac{21}{36} = 18\ \bar{2}\ \overline{12} \qquad \sqrt{105} = \sqrt{10^2+5} = 10 + \frac{5}{20} = 10\ \bar{4}$$

correct to .05 percent — correct to .03 percent

There are two problems on the areas of circles in which the equivalent of π is 3, the Old Testament value, as compared with the RMP value of more than a millennium earlier, 256/81, or 3.16. The problems are to find the diameters of circles with areas of 100 square cubits and 10 square cubits. The answers given are $11\ \bar{2}\ \overline{20}$ and $3\ \bar{\bar{3}}$, as further approximations from the formula of Hero given above.

Finally, there are three problems concerning the areas of circles that circumscribe two equilateral triangles and a square. Rather surprisingly, the scribe finds the areas of the triangles and the square separately, and then the areas of the segments of the circles. The arithmetic for the areas of the segments is equivalent to using the formula $A = 1/2\ s(s + c)$, where s is the height (*sagitta*) and c is the chord length of the segment. This is unexpected, and the earliest reference I can find to it is the *Chui-chang suan-shu* from China (*ca.* 300 B.C.).[8]

In the Cairo papyri the old notation for unit fractions is clearly retained, but there are signs of slight changes developing. In the forty problems there are demotic signs for 5/6 and 2/3. In addition, the fractions 6/47, 39/47, 17/53, 35/53, and at least a dozen others are used, written with both numerator and denominator on the same line, but the numerator underlined to distinguish it. These are the first nonunit fractions.

P. Dem. Heidelberg 663. Parker's translation of the fragmentary demotic papyrus P. Dem. Heidelberg 663 (*Journal of Egyptian Archaeology,* **61** [1975], 189–196) deals with four problems concerning the dimensions of isosceles trapezoidal fields. Of great interest is the division of one such field into two equal areas, by a straight line parallel to the top and bottom sides, and the determination of its length. Parker remarks that the details of the papyrus are "so scanty that it is rarely possible to do more than suggest a connected translation, but with the help of the four figures, it is possible to reconstruct the aim of each problem." He then offers a careful translation of the papyrus, which dates from the Ptolemaic period of the first or second century B.C.

The second problem deals with the length of the parallel line x that divides the field into two equal areas.

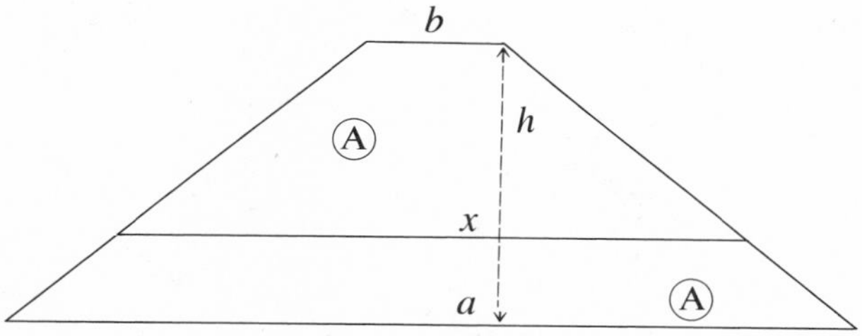

The detail of this calculation is not clear in the papyrus, but Parker suggests that the formula $x = \sqrt{a^2 - \frac{(a-b)}{h} \times 2A}$ or its equivalent in demotic terms could have been used and that the square root, if not integral, could be found from Hero's approximation,

$$\text{if } k = \sqrt{p^2+q},$$

$$\text{then } k = p + \frac{q}{2p} \text{ approximately.}$$

For problem 2 the formula gives

$$x = \sqrt{18^2 - \frac{(18-2)}{6} \times 60}$$

$$x = \sqrt{324 - 160}$$

$$x = \sqrt{164}.$$

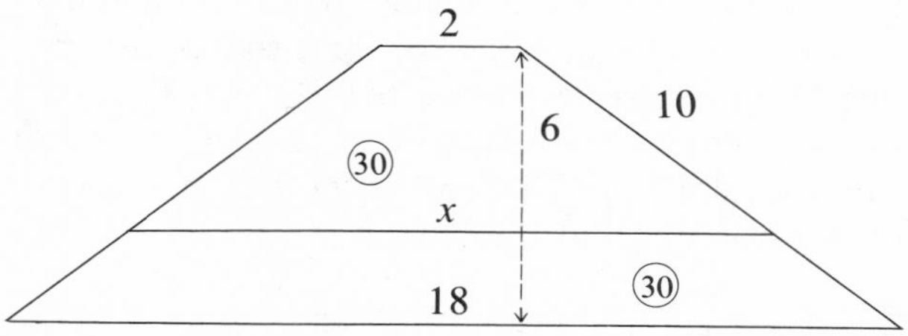

Because 164 is not a perfect square, its square root from Heron's formula would be $\sqrt{144 + 20} = 12 + \frac{20}{2 \times 12} = 12\ 5/6$ approx. Because of the demotic units of measure, the value of x had to be expressed 12 1/2 1/4 1/16, a reasonably close approximation. One naturally wonders whether the dimensions of the trapezoid could have been chosen so that the bisector x would turn out to be a whole number, and it so happens that this can be done.

Yale Babylonian Collection clay tablet 4675, dating some centuries earlier than Heidelberg 663, contains a similar problem in which the parallel sides of the trapezoid are 7 and 17. From Mathematical Cuneiform Texts, American Oriental Society, 1975, by Neugebauer and Sachs. The translation of this tablet indicates in the cuneiform sexagesimal number system that the method for calculating x using only

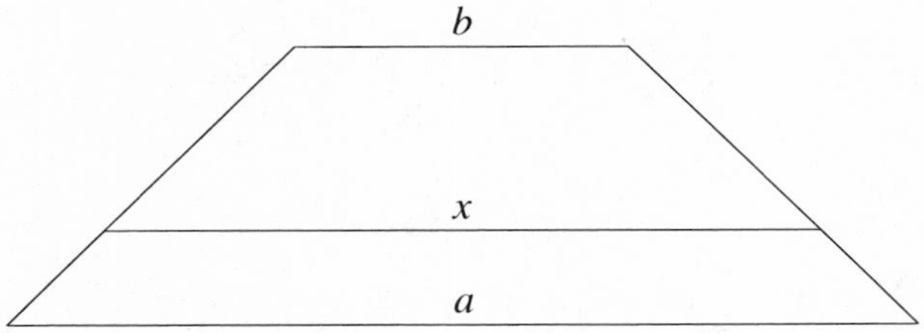

the values of a and b would have followed in modern terms the steps indicated by the formula

$$x=\sqrt{a^2-\tfrac{1}{2}(a+b)(a-b)},$$

in which there is no reference to perpendicular heights or to sides or areas. With the sides a and b given as 17 and 7, we have

$$x=\sqrt{17^2-\tfrac{1}{2}(17+7)(17-7)}$$
$$x=\sqrt{289-120}$$
$$x=\sqrt{169}$$
$$x=13.$$

We ask ourselves if the Babylonian scribes had some method of their own in this type of problem for avoiding awkward square roots, even though they may have had some sort of a square root rule like Hero's many centuries before he had made it known. Indeed, we find another related problem in the clay tablet VAT 7535[3]. at the Berlin Museum, dating from ancient Babylonia.

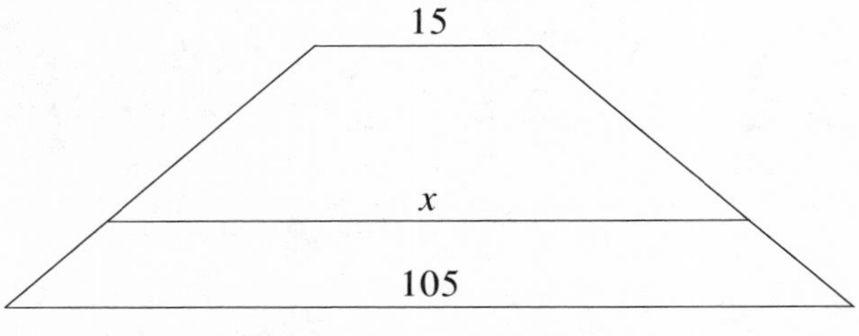

Using the same formula we have

$$x=\sqrt{a^2-\tfrac{1}{2}(a+b)(a-b)}$$

so that

$$x=\sqrt{105^2-\tfrac{1}{2}(105+15)(105-15)}$$
$$x=\sqrt{11025-5400}$$
$$x=\sqrt{5625}$$
$$x=75.$$

There are many ways of finding the length of the area-bisecting parallel of a trapezoid, isosceles or otherwise; and methods or formulas may be devised to include sides, perpendicular heights, and areas. The simplest formula in modern terms is

$$x=\sqrt{\tfrac{1}{2}(a^2+b^2)}.$$

A further problem that presented itself to the Babylonian and Egyptian scribes was, what values should be given to a and b to ensure that the value of x should be an integer or an exact number?

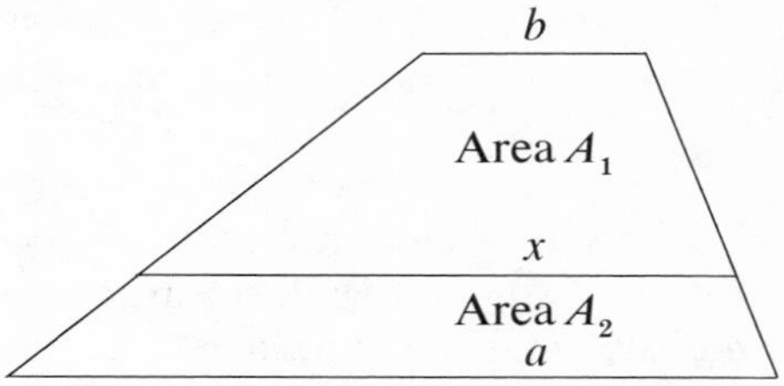

The apparently simple problem suggested by the ancients is thus conveniently shown in Figure 5, in which it is required to devise integral values of a, b, and x, so that $A_1 = A_2$.

A solution of the problem treated purely as theory of numbers was given by A. Guibert in 1862; reference to it is made in L. E. Dickson's *History of the Theory of Numbers.* There the general solution in positive relatively prime integers of $a^2 + c^2 = 2b^2$ is stated to be

$$a=\pm(p^2-q^2-2pq),$$
$$b=(p^2+q^2),$$
$$c=(p^2-q^2+2pq),$$

where p and q are relatively prime and one even.

Thus solutions for values of a, b, and x all less than 100 can be found for the trapezoid.

b	x	a	b	x	a
1	5	7	31	41	49
7	13	17	17	53	73
7	17	23	49	61	71
17	25	31	23	65	89
1	29	41	47	65	79
23	37	47	71	85	97

We note that the scribe of VAT 7535 utilized the first trio (1, 5, 7) multiplied through by 15, while the scribe of YBC 4675 used the second trio (7, 13, 17). It would be of great interest to the historian of mathematics if it were ever determined exactly how these values were determined.

British Museum Papyrus (10399, 10520). These papyri date from the period of the Ptolemies and the early Romans, say 300 B.C. They contain twenty-seven problems, of which about half are similar to those found in the hieratic papyri. Thus, six deal with the four operations, mostly operations with unit fractions; two with the areas of rectangles; three are multiplication tables—64 times the numbers 1 to 16, $\overline{90}$ times the numbers 1 to 10, and $\overline{150}$ times the numbers 1 to 10; and two are not sufficiently preserved to be translated. But the other half are of interest to the historian.

There are four similar problems on frustums of cones and the calculations of their volumes. This is done by averaging the top and bottom diameters; finding the area of the circle with this diameter, using the formula $A = 3/4\ d^2$; and then multiplying by the height. This is a better approximation than that of averaging the areas of the top and bottom circles. Then there are six consecutive very closely allied problems, the first of which is worded "The fraction $\overline{5}$ is added to 1. Determine what fraction of 1 $\overline{5}$ must be subtracted from it to give 1 again." The succeeding questions are similarly worded, with the fractions to be added to unity changed to $\overline{6}$, $\overline{7}$, $\overline{8}$, $\overline{9}$, and 5/6. The answers given are $\overline{6}$, $\overline{7}$, $\overline{8}$, $\overline{9}$, $\overline{10}$, and 5/11. There are two questions requiring the square roots of 10 and $\overline{2}$. The controversial (only regarding its origin) formula used is $\sqrt{a^2 + b} \div a + b/2a$, and $\sqrt{10}$ is found as 3 $\overline{6}$; but $\sqrt{\overline{2}}$ requires preliminary treatment. The scribe replaces $\overline{2}$ by 18/36 and thus finds $\sqrt{\overline{2}} = \overline{\overline{3}}\ \overline{24}$. The division of 2 by 35, almost exactly as A'hmosé indicated it should be done in RMP Recto, is calculated to be $\overline{30}\ \overline{42}$, as in the Recto. Parker, like Peet, expected to find $\overline{21}\ \overline{105}$.[9]

The last remaining problem is of particular interest. It deals with the sum of a number of terms of certain progressions, the first of which is the arithmetical progression formed by the natural numbers. In order to understand clearly what series the scribe is summing to ten terms, I write them first, because the scribe did not so write them but merely described them in his own way. He plans to find the sum of the first ten terms of the following series, but actually does so for only the first two, which are 1 "filled once up" to 10, and 1 "filled twice up" to 10. The third (and fourth, and so on), which would have been 1 "filled thrice (and so on) up" to 10, are not included.

(0)	1	2	3	4	5	6	7	8	9	10
Once (A)	1	3	6	10	15	21	28	36	45	55
Twice (B)	1	4	10	20	35	56	84	120	165	220
Thrice (C)	1	5	15	35	70	126	210	330	495	715

The scribe writes:

You shall say, 1 up to 10 amounts to 55.
You shall reckon 10, 10 times. Result 100.
You shall add 10 to 100. Result 110.
You shall take the half. Result 55.
You shall say, 1 up to 10 amounts to 55.
You shall add 2 to 10. Result 12.
You shall take the third of 12. Result 4.
You shall reckon 4, 55 times. Result 220.
You shall say, 1 is "filled twice up" to 10. Result 220.

The scribe's directions hold, of course, for any number in the natural number series, merely by replacing the 10 that he chose for his example. Let us first rewrite the above as follows.

First,

$$10^2 = 100$$

$$10^2 + 10 = 110$$

1 is "filled once up" to 10 $\quad \tfrac{1}{2}(10^2 + 10) = 55$

Then,

$$10 + 2 = 12$$

$$\tfrac{1}{3}(10 + 2) = 4$$

1 is "filled twice up" to 10 $\quad 4 \times 55 = 220$

Now we replace the 10 by n, representing any term of the natural number series.

1 is "filled once up" to n, $S_n = \tfrac{1}{2}(n^2 + n)$

1 is "filled twice up" to n, $S_n = \tfrac{1}{3}(n + 2)\ \tfrac{1}{2}(n^2 + n)$.

And if he proceeded to the next series, he would have written

1 is "filled thrice up" to n, $S_n =$

$$\tfrac{1}{4}(n + 3)\ \tfrac{1}{3}(n + 2)\ \tfrac{1}{2}(n^2 + n).$$

We may now say that in our modern notation, we have the formulas for the sum to n terms of the series, as follows.

Series (O) $\quad S_n = \tfrac{1}{2} n(n + 1)$

for instance, $S_5 = \tfrac{1}{2}\, 5 \times 6 = 15$

Series (A) $\quad S_n = \dfrac{1}{3!} n(n + 1)(n + 2)$

for instance, $S_5 = \frac{1}{6}\, 5 \times 6 \times 7 = 35$

Series (B) $S_n = \frac{1}{4!}\, n(n+1)(n+2)(n+3)$

for instance, $S_5 = \frac{1}{24}\, 5 \times 6 \times 7 \times 8 = 70$

AN INTERPRETATION OF RMP 79

RMP 79 consists of a house inventory.

	1	2,801	Houses	7
	2	5,602	Cats	49
	4	11,204	Mice	343
Totals	7	19,607	Spelt	2,401
			Hekats	16,807
			Total	19,607

In the second column the scribe starts with 7, multiplies it by 7, then multiplies this product by 7, repeats these multiplications by 7 a certain number of times, then adds all the products for a grand total. This operation is self-explanatory. In order to explain the multiplication of 2,801 by 7 (I use the index notation for brevity), we take the foregoing step by step.

1 term No additions	$7 = 7$	$= 7$	$= 7$
2 terms One addition	$7 + 7^2 = 7[1 + (7)]$	$= 7(1+7)$	$= 56$
3 terms Two additions	$7 + 7^2 + 7^3 = 7[1 + (7 + 7^2)]$	$= 7(1+56)$	$= 399$
4 terms Three additions	$7 + 7^2 + 7^3 + 7^4 = 7[1 + (7 + 7^2 + 7^3)]$	$= 7(1+399)$	$= 2,800$
5 terms Four additions	$7 + 7^2 + 7^3 + 7^4 + 7^5 = 7[1 + (7 + 7^2 + 7^3 + 7^4)]$	$= 7(1+2,800)$	$= 19,607$

If the scribe wished to go one step further, he would need only to add unity to the total 19,607, and then multiply by 7. In the scribe's setting for RMP 79, the shorter method was shown first—that of multiplying 2,801 by 7—and then its correctness proved by the longer, detailed calculation. It would appear that in this problem, the scribe is concerned only with a property of certain geometrical progressions, an exercise in the field of numbers. The title "A House Inventory" (or "The Contents of a House"; there is some doubt in the translation) and the names given to the successive totals—Houses, Cats (in the houses), Mice (that the cats would destroy), Spelt (ears of grain the mice would eat), and *Hekats* (the number of particles of grain that would be saved)—are all put in to give the problem an air of verisimilitude. But if this was the scribe's plan, it failed, because the grand total he is seeking cannot have any sensible meaning in regard to the contents of a house.

If, as is elsewhere claimed, this problem is the origin of the Mother Goose nursery rhyme "As I was going to St. Ives, I met a man with seven wives . . . how many were coming from St. Ives?" (not "going to St. Ives," which is the storyteller's trick), then the change to seven wives, seven sacks, seven cats, seven kits removes this disability and makes the problem sensible. Not only that, the multiplications are reduced and the problem slightly simplified to the multiplication of 400 by 7.

PESU PROBLEMS

Pesu is a number indicating the strength of bread or of beer made from grain. If one *hekat* (about 1/8 bushel) of grain were used to produce one loaf of bread or one *des* jug of beer, then its *pesu* was one. If one *hekat* of grain produced two loaves or two jugs of beer, their *pesu* was two, and so on, so that the higher the *pesu* number, the weaker the bread or beer. The relation was expressed as

$$\text{Pesu} = \frac{\text{Number of Loaves or Jugs}}{\text{Number of Hekats of Grain}}.$$

There are twenty problems concerning *pesus* in the RMP and MMP. Generally speaking, their arithmetic is simple, as the examples below indicate.

RMP 69. If 3 $\bar{2}$ *hekats* of meal are made into 80 loaves, find the amount of meal in each loaf, and the *pesu*.

RMP 73. 100 loaves of *pesu* 10 are exchanged for loaves of *pesu* 15. How many of the latter will there be?

RMP 78. 100 loaves of *pesu* 10 are exchanged for *des* jugs of beer of *pesu* 2. How many *des* jugs of beer will there be? (A *des* jug of beer is roughly 7/8 pint.)

Two problems, RMP 74 and 76, merit closer

attention than the others, even though at first glance they appear simple enough and, indeed, very similar.

RMP 74. 1,000 loaves of *pesu* 5 are exchanged, half of them for loaves of *pesu* 20 and half for loaves of *pesu* 10. How many of each will there be?

RMP 76. 1,000 loaves of *pesu* 10 are exchanged for a number of loaves of *pesu* 20 and the same number of loaves of *pesu* 30. How many of each will there be?

The scribe's solutions are shown below.

RMP 74. 1,000 loaves of *pesu* 5 required 200 *hekats*, and half of 200 is 100 *hekats*. Multiply 100 by 10; it makes 1,000, the number of loaves of *pesu* 10. Multiply 100 by 20; it makes 2,000, the number of loaves of *pesu* 20. The answer is 1,000 and 2,000 loaves.

RMP 76. For loaves of *pesu* 20, $\overline{20}$ *hekat* produces one loaf. For loaves of *pesu* 30, $\overline{30}$ *hekat* produces one loaf. Then, $\overline{20}\ \overline{30} = \overline{12}$ *hekat* produces two loaves, one of each kind. The number of *hekats* in 1,000 loaves of *pesu* 10 is 100. Multiply 100 by 12; the result is 1,200, the number of loaves of each kind for the exchange.

We now restate the scribe's processes for these two problems in modern terms, for our own information.

RMP 74. 1,000 loaves of *pesu* 5 required 200 *hekats* of grain. The arithmetic mean of the two new *pesus* is 1/2 (20 + 10) = 15. Then the number of loaves received in exchange is greater, in the ratio of 15 to 5—that is, 3 to 1. Therefore the number of loaves received is 1,000 × 3, or 3,000 total. The new *pesus* are in the ratio of 20 to 10, or 2 to 1. Hence there must be 2,000 loaves of *pesu* 20 and 1,000 loaves of *pesu* 10.

RMP 76. 1,000 loaves of *pesu* 10 required 100 *hekats* of grain. The harmonic mean of the two new *pesus* is 2(20 × 30) ÷ (20 + 30) = 24. Then the number of loaves received in exchange is greater in the ratio of 24 to 10, that is, 12 to 5. Therefore the number of loaves received is 1,000 × 12 ÷ 5 = 2,400 total. Hence there must be 1,200 loaves of *pesu* 20 and 1,200 loaves of *pesu* 30.

Now consider the problems RMP 73 and 72, in that order.

RMP 73. 100 loaves of *pesu* 10 are exchanged for loaves of *pesu* 15. How many of the latter are there?

RMP 72. 100 loaves of *pesu* 10 are exchanged for loaves of *pesu* 45. How many of the latter are there?

RMP 73 Solution. The number of *hekats* in 100 loaves of *pesu* 10 is 100 ÷ 10 = 10. Multiply 10 by 15; it makes 150, the number of loaves exchanged.

RMP 72 Solution. Instead of a simple solution like the above, of 10 × 45 = 450 loaves, the scribe has the following:

The excess of 45 over 10 is 35

$45 - 10 = 35 \qquad Q-P$

Divide this 35 by 10; it makes 3 $\bar{2}$

$35 \div 10 = 3\tfrac{1}{2} \qquad \dfrac{Q-P}{P}$

Multiply 3 $\bar{2}$ by 100; it makes 350

$3\tfrac{1}{2} \times 100 = 350 \qquad \dfrac{Q-P}{P}L$

Add 100 to 350; it makes 450

$350 + 100 = 450 \qquad \dfrac{Q-P}{P}L + L$

The historian desiring to examine the reasoning in this "round-about way," as A. B. Chace put it, of solving RMP 72 would write L for the original number of loaves, P for the original *pesu*, N for the number of new loaves, and Q for the new *pesu*. The only data available to the scribe is the relation

$$\text{Pesu} = \frac{\text{Loaves}}{\text{Hekats}};$$

and since the number of *hekats* is constant, he would have

$$\text{Hekats} = \frac{\text{Loaves}}{\text{Pesu}}.$$

Therefore, $\dfrac{N}{Q} = \dfrac{L}{P}$

or (alternando), $\dfrac{N}{L} = \dfrac{Q}{P}$

and (dividendo), $\dfrac{N-L}{L} = \dfrac{Q-P}{P}$

so that $N - L = \left(\dfrac{Q}{P} - 1\right) \times L.$

Hence, $N - L = \dfrac{Q}{P}L - L$

and, finally, $N = \dfrac{Q}{P} \times L,$ adding L to both sides, which is the formula of the method the scribe adopted for RMP 73 and all the other *pesu* problems of a like nature. The incidence of the step that we today term "dividendo" is unavoidable, because of the instruction to find the excess of the final *pesu* of 45 over the original *pesu* of 10.

Although the *pesu* problems of the RMP and the MMP are in general similar, there are one or two (as we have seen) that call for special mention. The scribe of the MMP was not as careful, nor as good a writer of hieratic as was A'hmosé; and MMP 21 is one such problem, which may resemble RMP 76. It is included here for interest and inquiry.

MMP 21. This problem involves the method of calculating the mixing of sacrificial bread.

> If one names 20 measured as $\bar{8}$ of a *hekat*,
> And 40 measured as $\overline{16}$ of a *hekat*,
> Compute $\bar{8}$ of 20. Result 2 $\bar{2}$.
> Compute $\overline{16}$ of 40. Result 2 $\bar{2}$.
> The total of both these halves is 5.
> Compute the sum of both halves. Result 60.
> Divide thou 5 by 60.
> Result $\overline{12}$. Lo! The mixture is $\overline{12}$.
> You have correctly found it.

EGYPTIAN MEASURES

Listed below are some of the Egyptian measures.

Arura	A unit of area equal to that of a square with a side of 100 royal cubits, thus a square *hayt* or *khet*.
Cubit	A royal cubit was 20.6 inches and a short cubit 17.7 inches; hence the phrase "a cubit and a hand's breadth."
Deben	A weight of about 3.2 ounces or 90 grams, used for metals.
Des Jug	A unit of volume used generally for beer, approximately half a liter or 7/8 of a pint.
Digit	Or finger. A quarter of a palm, and thus 1/28 cubit.
Double Remen	The length of the diagonal of a square with a side of one cubit, thus 29.13 inches.
Finger	The same as the digit.
Hayt	100 royal cubits in length. The same as a *khet*.
Hekat	A dry measure for barley, flour, spelt, corn, and grain in general; about 1/8 of a bushel.
Hinu	A smaller unit for grain, 1/10 of a *hekat*.
Khar	Two-thirds of a cubic cubit, a measure more commonly used for the contents of granaries.
Khet	100 royal cubits.
Palm	Or hand's breadth, 1/7 of a cubit.
Pesu	A number giving the relation between the number of loaves of bread (or *des* jugs of beer) and the number of *hekats* of grain required to produce them.
Remen	The half of a double remen, 14.56 inches; that is, half the diagonal of a square with a side of one cubit.
Ro	The smallest named unit for grain, 1/320 part of a *hekat*.
Seked	The inclination of the face of a right pyramid, measured as so many horizontal palms or cubits per vertical palm or cubit.
Setat	One square *khet*, or 10,000 square cubits.

Areas and Volumes. The areas of triangles and rectangles were found in the usual way, half the base times the height, and length times breadth; but the areas of circles were something quite different. RMP 50 states the rule as "Take away one-ninth of the diameter, and square the remainder." In modern terms we would express this as

$$A=\left(\frac{8}{9}d\right)^2$$

$$=\left(\frac{16}{9}r\right)^2$$

$$=\frac{256}{81}r^2.$$

We may therefore say that the ancient Egyptian equivalent of π was $^{256}/_{81}$ or 3.1605, an unexpectedly close value, in excess of the true value by 0.6 percent.

Those problems concerned with finding the contents of cylindrical granaries, expressed in *khars*, where 1.5 *khars* equals a cubic cubit, were solved by multiplying the area of the circular base by the height in cubits, and then multiplying by 1.5. The area of the circular base was found by the formula above.

There are two problems, KP IV, 3 and RMP 43, where this calculation was done differently. The directions read: "Add to the diameter its one-third part, square the total, and then multiply by two-thirds of the height. This is the contents in *khar*." How the scribes thought out this variation is not explained in the papyri; but it worked correctly, as we can check by setting down both methods in modern form.

METHOD 1

$$\text{Volume in } khars=\left(\frac{8}{9}d\right)^2 h\times\frac{3}{2}$$

$$=\left(\frac{16r}{9}\right)^2 h\times\frac{3}{2}$$

$$=\frac{256}{81}\times\frac{3}{2}\times r^2h$$

$$=\frac{128}{27}r^2h$$

METHOD 2

$$\text{Volume in } khars = \left(\frac{4}{3}d\right)^2 \times \frac{2}{3}h$$
$$= \left(\frac{8r}{3}\right)^2 h \times \frac{2}{3}$$
$$= \frac{64}{9} \times \frac{2}{3} \times r^2 h$$
$$= \frac{128}{27} r^2 h$$

A challenge remains for the historian of mathematics: to explain the scribal thought process of transforming the first method into the second, without using any symbolic algebra.

Pyramids. The problems that concern pyramids show that the scribes were able to calculate the slope or inclination of the sides of a pyramid, the volume of a pyramid, and, perhaps surprisingly, the volume of a truncated pyramid or the frustum of a pyramid. The slope or *seked* of the side of a pyramid was stated as the relation between the vertical height and half the base, and was given as so many palms horizontally for each cubit measured vertically–usually 5 palms, or 5 palms, 1 finger, or 5 $\overline{25}$ palms, possibly because of the *seked* of Cheops was 5 $\bar{2}$ palms.

Struve's translation of MMP 14 is as follows:

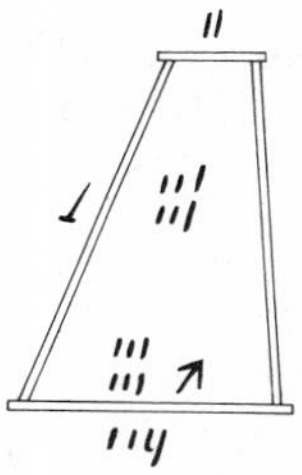

Method of calculating a truncated pyramid.

If it is said to thee, a truncated pyramid of 6 cubits in height,
Of 4 cubits of the base by 2 cubits of the top,
Reckon thou with this 4, its square is 16.
Multiply this 4 by 2. Result 8.
Reckon thou with this 2, its square is 4.
Add together this 16, with this 8, and with this 4. Result 28.
Calculate thou $\bar{3}$ of 6. Result 2.
Calculate thou with 28 twice. Result 56.
Lo! It is 56! You have correctly found it.

If we replace the base of 4 cubits by a, the top of 2 cubits by b, and the height of 6 cubits by h, we find that the scribe has calculated the volume of the frustum, according to the modern formula $V = \frac{1}{3}h\,(a^2 + ab + b^2)$. It is generally accepted that the Egyptians knew that the volume of a right square pyramid was one-third that of a right square prism of equal base and height, although this is nowhere specifically attested (to my knowledge) and although there are several simple ways in which the scribes may have found the equivalent of the modern formula $V = \frac{1}{3}a^2h$.[10] Even with the powerful help of such a formula or its equivalent, however, it is still not easy to establish just how the scribes arrived at this relatively erudite method for finding the frustum of a pyramid, an achievement that, in the words of B. G. Gunn and T. E. Peet, "has not been improved upon in 4,000 years."

Area of the Surface of a Hemisphere. Struve's translation of MMP 10 is as follows.

Method of calculating a basket.

If it is said to thee, a basket with an opening,
Of 4 $\bar{2}$ in its containing, Oh!
Let me know its surface.
Calculate thou $\bar{9}$ of 9, because the basket
Is the half of an egg. There results 1.
Calculate thou the remainder as 8.
Calculate thou $\bar{9}$ of 8.
There results $\bar{\bar{3}}$ $\bar{6}$ $\overline{18}$.
Calculate thou the remainder of these 8 left
After taking away these $\bar{\bar{3}}$ $\bar{6}$ $\overline{18}$. There results 7 $\bar{9}$.
Reckon thou with 7 $\bar{9}$, 4 $\bar{2}$ times.
There results 32. Lo! This is its area.
You have correctly found it.

In modern terms, we would express the operations above in the form $A = 2(\frac{8}{9}d)^2$, which is the equivalent of $A = 2\pi r^2$.

W. W. Struve (*d.* 1965), the original translator of the MMP, was thoroughly convinced that the above is the correct interpretation of MMP 10. T. E. Peet (*d.* 1934) thought that a semicylinder may have been concerned, but had to introduce two new terms, "diameter and heights," into the translation. B. L. van der Waerden and R. J. Gillings agree with Struve; and T. G. H. James, while acknowledging the translation difficulties raised by Peet, inclines to concur with Struve's conclusion.[11] O. Neugebauer still has some doubts that the paleography is definite enough to convince him beyond question.[12] Thus far, no authorities have expressed any views or opinions, in scientific or historical journals, contrary to those expressed here; and MMP 10 and MMP 14 remain the outstanding mathematical achievements of the ancient Egyptians. A detailed discussion of the views of the above-mentioned writers is given by Gillings.[13]

EQUATIONS

In this section we consider problems requiring for their solution what we might call algebraic reasoning, but which are treated by the scribes quite literally and perhaps termed "rhetorical algebra." For clarity and brevity, I will not give complete translations from the hieratic writing of the papyri; and so that we may examine the thought processes involved, I use the standard *x*, *y*, *z* for unknowns, and *a*, *b*, *c* for knowns. For obvious reasons, however, I retain the standard notation for the ancient Egyptian unit fractions. Thus, in a limited space we may cover a considerable field of study of the mathematical techniques of the scribes.

First-Degree Equations, One Unknown. Below are several first-degree equations, with one unknown.

RMP 24	$x+\bar{7}x=19$	Answer $16\ \bar{2}\ \bar{8}$
RMP 25	$x+\bar{2}x=16$	Answer $10\ \bar{\bar{3}}$
RMP 26	$x+\bar{4}x=15$	Answer 12
RMP 27	$x+\bar{5}x=21$	Answer $17\ \bar{2}$

These are straightforward enough, largely arithmetical operations with unit fractions. In each case the scribe gives the proof that his answer is correct.

MMP 19	$x+\bar{2}x+4\ =10$	Answer 4
KP LV, 3	$x-\bar{2}x-\bar{4}x=\ 5$	Answer 20
RMP 30	$\bar{\bar{3}}x+\overline{10}x\ =10$	Answer $13\ \overline{23}$
RMP 34	$x+\bar{2}x+\bar{4}x=10$	Answer $5\ \bar{2}\ \bar{7}\ \overline{14}$

This second group of equations represents a slight advance on the first four, each of which was solved by a method that today is called "false position" or "false assumption." Methods for the second four vary.

RMP 31	$x+\bar{\bar{3}}x+\bar{2}x+\bar{7}x=33$
RMP 32	$x+\bar{3}x+\bar{4}x\ =2$
RMP 33	$x+\bar{\bar{3}}x+\bar{2}x+\bar{7}x=37$

Answers

RMP 31	$14\ \bar{4}\ \overline{56}\ \overline{97}\ \overline{194}\ \overline{388}\ \overline{679}\ \overline{776}$
RMP 32	$1\ \bar{6}\ \overline{12}\ \overline{114}\ \overline{228}$
RMP 33	$16\ \overline{56}\ \overline{679}\ \overline{776}$

If the scribe A'hmosé faltered, he did so here. His method of solution was that of division; and if he was teaching a technique, his choice of fractions and of integers, or both, was unfortunate, for the answers, all of which are correct, show that the problems were not of a practical nature. So here he was, lost in a maze of four-digit unit fractions, in one instance (RMP 33) finding it necessary to add sixteen unit fractions, the last twelve of which have the denominators

84	776	1552
112	1164	4074
392	1358	4753
679	1358	5432,

which he found by using red auxiliaries. This could all have been neatly avoided had he chosen 97 instead of 37, and his answer would have been 42. And in RMP 31, had he written $\bar{8}x$ instead of $\bar{7}x$, he would have simplified it so greatly that the answer would have been $14\ \bar{3}\ \overline{15}$. Finally, the answer to RMP 32 could have been simplified from $1\ \bar{6}\ \overline{12}\ \overline{114}\ \overline{228}$ to $1\ \bar{4}\ \overline{76}$ in the scribal working, as the G rule shows, since

$$\bar{6}\ \overline{12}=\bar{4}$$

and

$$\overline{114}\ \overline{228}=\overline{76}.$$

Second-Degree Equations, Two Unknowns. In the discussion of the demotic papyri, the solution of simultaneous second-degree equations is shown, so they are not repeated here. DMP 34 and 35 were

$$x^2+y^2=169$$

$$xy=\ 60,$$

$$x^2+y^2=225$$

$$xy=\ 60.$$

DMP 34, in exactly the form shown above and with the same numbers, occurs in W. E. Paterson's *School Algebra* ([London, 1916], 250), where the textual work on solutions is the same as that of the scribe. Two problems from BP 6619, restored and translated by H. Schack-Schackenburg, are

$$x^2+y^2=100$$

$$4x-3y=\ 0,$$

$$x^2+y^2=400$$

$$4x-3y=\ 0.$$

F. L. Griffith, in his translation of a problem in KP IV, 4 (1897), was not able to understand certain incomplete lines, which have since been restored with reasonable surety and produce another problem of simultaneous equations, details of which may be studied in *Mathematics in the Time of the Pharaohs* (162–165):

$$xy=12$$

$$x-\tfrac{3}{4}y=\ 0.$$

Progressions. RMP 40 speaks of an arithmetic progression of five terms, in which the sum of the three largest terms is seven times the sum of the two smallest terms. Chace's translation reads: "100 loaves for 5 men, one-seventh of the total of the largest 3 shares shall equal the total of the smallest 2 shares. What is the common difference of the shares?"[14]

KP IV, 3, is solely a column of eleven decreasing numbers, and next to it is a short multiplication. There are no words! I have judged the calculations shown to be an answer to the problem "The sum of 12 terms of an arithmetic progression is 110, and the common difference is 5/6. What is this series?"[15]

RMP 64 gives the sum of ten terms of an arithmetic progression as 10, and the common difference as one-eighth. What are the terms of this series? Since the scribal statement of the problem speaks of dividing ten *hekats* of barley among ten men, the fractions used are all Horus-eye fractions, which were used for grain: $\bar{2}$, $\bar{4}$, $\bar{8}$, $\overline{16}$, $\overline{32}$, $\overline{64}$, of a *hekat*, written in a special way. The scribe's method of solving RMP 64 is quite unexpected; and if one follows each step of the solution, and replaces each word and number with the modern notation used in algebra texts – such as S for the sum, d for the difference, a and l for the first and last terms, n for the number of terms – one has, at the conclusion,

$$S_n=\frac{n}{2}\left[2l-(n-1)d\right].$$[16]

The problem stated in RMP 79 has been treated elsewhere in some detail. Here we can state it as "Find the sum of five terms of a geometrical progression of which the first term is 7, and of which the common ratio is also 7."

Think of a Number. RMP 28 reads in Chace's translation:

> Two-thirds is to be added, then one-third is to be subtracted.
> [The foregoing is in red, all else is in black.]
> There remains 10.
> Make one-tenth of this, there becomes 1. The remainder is 9.
> Two-thirds of it, namely 6, is to be added. The total is 15.
> One-third of this is 5. Lo! 5 is that which goes out, and the remainder is 10.
> *The doing as it occurs.*[17]

Chace concludes, "The solution does not seem to be complete. The words '*The doing as it occurs*,' are usually put at the beginning, and in no other problem are they at the end. Peet has suggested that in copying, the scribe let his eye pass to the same words in the next question." There are no such words in the next question!

I draw attention to the title page of the RMP, where the first sentence (again in red) is "Accurate reckoning of entering into the knowledge of all existing things, and all mysteries and secrets." RMP 28 is the first problem in which the scribe discloses one of his secrets and shows how his magic works. Having done this, he writes, "*The doing as it occurs*!" It is in its proper place. I restate the problem in modern terms. "Think of a number. Add to it its two-thirds part. From this number take away its third part. What is your answer? Suppose you are told the answer is 10. Then you (mentally) subtract its tenth part, and say that the number first thought of was nine! That is how you do it. That is your magic!"

RMP 29 is a second example of "think of a number" problems given by A'hmosé, and they are probably the earliest in recorded history: "Think of a number. Add to it its two-thirds part. To this number add its third part. Find one-third of this number. What is your answer? Suppose you are told the answer is 10. Then you (mentally) add a quarter and a tenth part to this 10, and say that the number you first thought of was $13\frac{1}{2}$! That is how you do it. That is your magic!"

In the section on DMP, the summation of the terms of the natural number series is treated, as are certain other series derived therefrom.

ADDENDA

J. J. Sylvester's Formula. This formula is quoted in the additional bibliographical notes. From his formula the mathematician can derive unit fraction equalities very similar to those found in the EMLR, when it was successfully unrolled. I have found the following:

$\bar{2}\ \bar{2}=1$	$\bar{2}\ \bar{3}\ \bar{6}=1$
$\bar{3}\ \bar{6}=\bar{2}$	$\bar{3}\ \bar{7}\ \overline{42}=\bar{2}$
$\bar{4}\ \overline{12}=\bar{3}$	$\bar{4}\ \overline{13}\ \overline{156}=\bar{3}$
$\bar{5}\ \overline{20}=\bar{4}$	$\bar{5}\ \overline{21}\ \overline{420}=\bar{4}$
$\bar{6}\ \overline{30}=\bar{5}$	$\bar{6}\ \overline{31}\ \overline{930}=\bar{5}$
$\bar{7}\ \overline{42}=\bar{6}$	$\bar{7}\ \overline{43}\ \overline{1{,}806}=\bar{6}$
etc.	etc.

Other sets are possible. A careful scrutiny will show that they are interrelated.

Two-Term, Three-Term, and Four-Term Equalities for Unit Fractions. Note that the overbars have been omitted.

Hekat Measures for Grain. One *hekat* of grain was approximately one-eighth of a bushel. For stating the contents of larger containers, the unit used was sometimes a double *hekat*, or even a quadruple

2	2	=	1	10	15	=	6	21	28	=	12	36	45	=	20	55	66	=	30
3	6	=	2	12	24	=	8	24	40	=	15	40	60	=	24	60	84	=	35
4	12	=	3	14	35	=	10	27	54	=	18	44	77	=	28	65	104	=	40
5	20	=	4	16	48	=	12	30	70	=	21	48	96	=	32	70	126	=	45
6	30	=	5	18	63	=	14	33	88	=	24	52	117	=	36	75	150	=	50

2	3	6	=	1	11	22	33	=	6
3	9	18	=	2	18	24	36	=	8
5	10	30	=	3	21	35	42	=	10
7	14	28	=	4	19	57	76	=	12
10	15	30	=	5	26	65	78	=	15

2	4	6	12	=	1
10	15	20	30	=	4
7	21	28	42	=	4
12	18	30	36	=	5
17	51	85	102	=	10

hekat. For granaries, an even larger unit was used, "100 quadruple *hekats*." A cubic cubit of grain was equal to thirty *hekats*. For smaller quantities, fractions of a *hekat* were used–but only 1/2, 1/4, 1/8, 1/16, 1/32, 1/64, which were called Horus-eye fractions. They were not written like ordinary unit fractions, but as portions of the eye of the god Horus, which he lost in battle. The very smallest measure used was the *ro*, which was 1/320 of a *hekat*–about a tablespoonful. Some other non-Horus-eye units were also sometimes used: the *hinu*, which was one-tenth of a *hekat*, and the *khar*, which was twenty *hekats*.

The RMP Recto. In the section dealing with the RMP Recto, the views of some historians are given on how they thought the scribes calculated the fifty values of $2 \div (2n - 1)$, expressed in unit fractions. My own theory has been published in *Archive for the History of Exact Sciences*.

I do not consider it proper, however, to discuss that article in detail until it has been read and studied or reviewed by competent judges, and its merits evaluated. It is therefore sufficient to state that I considered the one equality of the Recto that was entirely unique in its context (as verified by the computer KDF–9): the very last, $2 \div 101 = \overline{101}\ \overline{202}\ \overline{303}\ \overline{606}$, which is the only possible four-term unit fraction value for this division. From this equality, expressed generally as $2 \div n = \overline{n}\ \overline{2n}\ \overline{3n}\ \overline{6n}$, I derive every other equality of the Recto, using those methods and techniques accepted as being attributable to the scribes. I direct the attention of those interested in the problem to my article.

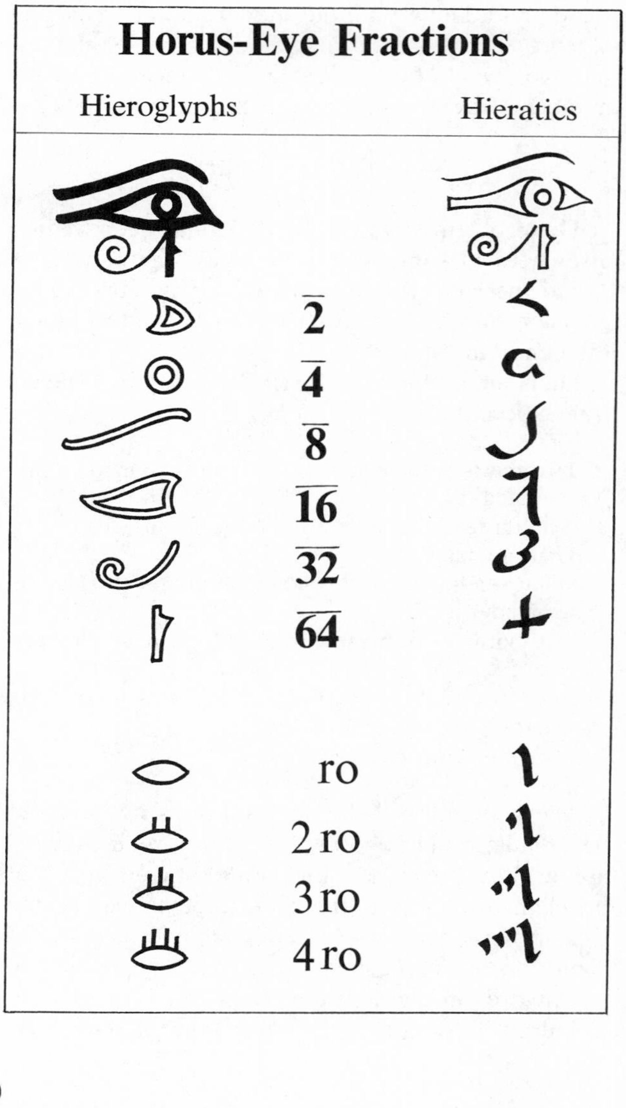

NOTES

1. R. W. Sloley, "Science," 166.
2. E. T. Peet, "Mathematics in Ancient Egypt," 412.
3. O. Neugebauer, *The Exact Sciences in Antiquity*, 73.
4. In the papyrus KP IV, 2, a portion of the Recto is included: 2 divided by 3, 5, 7, · · ·, 21. Proofs are identical but briefer.
5. Following Archibald: If $U_{x+1} - U_{x^2} + U_x - 1 = 0$, then $\sum \frac{1}{U_x} = \frac{1}{U_0 - 1} - \frac{1}{U_{x+1} - 1}$. Thus, for example, $1/5 + 1/21 = 1/4 - 1/420$.
6. See, for example, R. J. Gillings, "The Addition of Egyptian Unit Fractions," in *Journal of Egyptian Archaeology*, **51** (1956), 95.
7. See W. C. Hayes, *Ostraca From the Tomb of Sen-Mut*.
8. Further references are in *Heronis Alexandrini Opera quae supersunt omnia*, 5 vols. (Leipzig, 1899–1914), III, 73; IV, 357; V, 187.
9. Compare the *Canon for the RMP Recto*. See R. J. Gillings, *Mathematics in the Time of the Pharaohs*, 99.
10. *Ibid.*, ch. 17, "Pyramids and Truncated Pyramids," 185–193.
11. B. L. van der Waerden, letter (1967); R. J. Gillings, "The Area of the Curved Surface of a Hemisphere in Ancient Egypt," in *Australian Journal of Science*, **30**, no. 4 (1967), 113–116; T. G. H. James, letter (1970).
12. O. Neugebauer, letter (1967).
13. Gillings, *Mathematics in the Time of the Pharaohs*, ch. 18.
14. A. B. Chace, L. Bull, H. P. Manning, and R. C. Archibald, *The Rhind Mathematical Papyrus*, **II** (Oberlin, Ohio, 1929), prob. 40, pl. 62.
15. Gillings, *Mathematics in the Time of the Pharaohs*, 176–180.
16. *Ibid.*, 173–175.
17. A. B. Chace *et al.*, *The Rhind Mathematical Papyrus*, II, pl. 51.

BIBLIOGRAPHY

This bibliography, which should not be regarded as complete, may suggest further sources of information to students of the history of ancient Egyptian mathematics.

R. C. Archibald, *Bibliography of Egyptian Mathematics*, I and supp. (Oberlin, Ohio, 1927–1929).

O. Becker and J. E. Hofmann, *Geschichte der Mathematik* (Bonn, 1951).

V. V. Bobynin, *Bibliotheca mathematica*, 2nd ser., **3** (1889), 104–108; **4** (1890), 109–112; **8** (1894), 55–60; **10** (1896), 97–101.

L. Borchardt, *Gegen die Zahlenmystik an der grossen Pyramide bei Gise* (Berlin, 1922).

C. B. Boyer, *A History of Mathematics* (New York, 1968).

J. H. Breasted, *A History of Egypt* (London, 1946).

E. M. Bruins, "Ancient Egyptian Arithmetic: 2/*N*," in *Indagationes mathematicae*, **14** (1952), 81–91.

E. A. W. Budge, *The Rosetta Stone* (London, 1922; rev. eds., 1950, 1957).

E. A. W. Budge, "The Rosetta Stone," ch. 3, pt. 3, of M. Wheeler, *A Second Book of Archaeology* (London, 1959).

A. B. Chace, L. S. Bull, H. P. Manning, and R. C. Archibald, *The Rhind Mathematical Papyrus*, 2 vols. (Oberlin, Ohio, 1927–1929).

J. F. Champollion, *Précis du système hiéroglyphique* (Paris, 1824).

L. Cottrell, *Life Under the Pharaohs* (London, 1957).

W. E. Crum, *Coptic Ostraca From the Collections of the Egypt Exploration Fund, Cairo Museum* (London, 1902).

W. E. Crum, *Catalogue of Coptic Manuscripts in the British Museum* (London, 1905).

G. Daressy, "Calculs égyptiens du moyen-empire," in *Recueil de travaux relatifs à l'archéologie* (Paris, 1906).

D. Davidson and H. Aldersmith, *The Great Pyramid, Its Divine Message* (London, 1924).

L. E. Dickson, *History of the Theory of Numbers*, 3 vols. (Washington, D.C., 1919–1923; repr. New York, 1934), 437–438.

J. Edgar and M. Edgar, *The Great Pyramid Passages and Chambers* (London, 1923).

I. E. S. Edwards, *The Pyramids of Egypt* (London, 1952; rev. ed., 1961).

A. Eisenlohr, *Ein mathematisches Handbuch der alten Ägypter, RMP* (Leipzig, 1877).

R. Engelbach, "The Volume of a Truncated Pyramid," in *Journal of Egyptian Archaeology*, **14** (1927).

H. Eves, *An Introduction to the History of Mathematics* (New York, 1964).

A. H. Gardiner, *Egyptian Grammar*, 3rd ed., rev. (London, 1969).

O. Gillain, *La science égyptienne, L'arithmétique au moyen empire* (Paris, 1927).

R. J. Gillings, *Mathematics in the Time of the Pharaohs* (Cambridge, Mass., 1972).

R. J. Gillings, "The Recto of the Rhind Mathematical Papyrus. How Did the Ancient Egyptian Scribe Prepare It?" in *Archive for the History of Exact Sciences*, **12**, no. 4 (1974), 291–298.

R. J. Gillings, "What Is the Relation Between the EMLR and the RMP Recto?" *ibid.*, **14**, no. 3 (1975), 159–167.

S. R. K. Glanville, "The Mathematical Leather Roll in the British Museum," in *Journal of Egyptian Archaeology*, **13** (1927), 232–239.

S. R. K. Glanville, ed., *The Legacy of Egypt* (Oxford, 1942; repr., 1963; 2nd ed., 1971).

F. L. Griffith, *Hieratic Papyri From Kahun and Gurob* (London, 1898).

W. C. Hayes, *Ostracon 153. Ostraca and Name Stones From the Tomb of Sen-Mut at Thebes*, Metropolitan Museum of Fine Art Publication 15 (New York, 1942).

L. Hogben, *Mathematics for the Million* (London, 1945), 261, 297, 555.

F. Hultsch, *Die Elemente der ägyptischen Theilungsrechnung* (Leipzig, 1895).

P. E. B. Jourdain, "The Nature of Mathematics," I, ch. 1, of J. R. Newman, ed., *The World of Mathematics* (New York, 1956).

L. C. Karpinski, "Algebraical Developments Among the Egyptians and the Babylonians," in *American Mathematical Monthly*, **24** (1917), 257–265.

M. Kline, *Mathematics, a Cultural Approach* (Reading, Mass., 1962).

G. Loria, "Studi intorno alla logistica greco-egiziana," in *Giornale di matematiche di Battaglini*, **32** (1894), 28–57, decompositions of $2 \div (2n-1)$.

P. Luckey, "Anschauliche Summierung der Quadratzahlen und Berechnung des Pyramideninhalts," in *Zeitschrift für mathematischen und naturwissenschaftlichen Unterricht,* **61** (1930), 145–158.

P. Mansion, "Sur une table du papyrus Rhind," in *Annales de la Société scientifique de Bruxelles*, **12** (1888), 44–46.

E. K. Milliken, *Cradles of Western Civilisation* (London, 1955), 86–98.

O. Neubert, *The Valley of the Kings* (London, 1957).

O. Neugebauer, *The Exact Sciences in Antiquity* (Copenhagen, 1951; New York, 1962), 91–96.

J. R. Newman, "The Rhind Papyrus," I, ch. 2, of J. R. Newman, ed., *The World of Mathematics* (New York, 1956).

C. F. Nims, "The Bread and Beer Problems of the Moscow Mathematical Papyrus," in *Journal of Egyptian Archaeology*, **44** (1958), 56–65.

R. A. Parker, "A Demotic Mathematical Papyrus Fragment," in *Journal of Near Eastern Studies*, **18**, no. 4 (1959), 275–279.

R. A. Parker, *Demotic Mathematical Papyri* (Providence, R.I.–London, 1972), prob. 56 (pl. 20: C8–16), pp. 65–66.

T. E. Peet, *The Rhind Mathematical Papyrus* (London, 1923).

T. E. Peet, "Mathematics in Ancient Egypt," in *Bulletin of the John Rylands Library*, **15**, no. 2 (1931), 409–441, esp. 412.

W. M. F. Petrie, *The Pyramids and Temples of Gizeh* (London, 1883).

G. Posener, *Dictionary of Egyptian Civilization* (New York, 1959).

V. Sanford, *A Short History of Mathematics* (London, 1930), 225.

G. Sarton, *The Study of the History of Mathematics* (Cambridge, Mass., 1936; New York, 1957).

H. Schack-Schackenburg, "Der Berliner Papyrus 6619," in *Zeitschrift für ägyptische Sprache*, **38** (1900), 135–140.

A. Scott and H. R. Hall, "Egyptian Leather Roll of the 17th Century B.C.," in *British Museum Quarterly*, **2** (1927), 56–57.

K. H. Sethe, "Von Zahlen und Zahlworten bei den alten Ägyptern," in *Schriften der Wissenschaftlichen Gesellschaft in Strassburg*, no. 25 (1916), 85–119.

J. W. S. Sewell, "The Calendars and Chronology," ch. 1 of S. R. K. Glanville, ed., *The Legacy of Egypt* (London, 1942; repr. 1963; 2nd ed., 1971).

W. K. Simpson, *The Papyrus Reisner*, 3 vols. (Boston, 1963–1969).

R. W. Sloley, "Science," ch. 6 of S. R. K. Glanville, ed., *The Legacy of Egypt* (London, 1942; repr. 1963; 2nd ed., 1971).

Piazzi Smyth, *Our Inheritance in the Great Pyramid* (London, 1864 and repr. eds.)

D. J. Struik, *A Concise History of Mathematics* (New York, 1948), 19–23.

W. W. Struve, *Mathematischer Papyrus des Staatlichen Museums der schönen Künste in Moskau*, which is Quellen und Studien zur Geschichte der Mathematik, Abt. A, I (Berlin, 1930), esp. 98 ff.

D. P. Tsinzerling, "Geometria v drevnikh egiptyan" ("Geometry in Ancient Egypt"), in *Izvestiya Rossyskoy Akademii nauk SSSR*, 6th ser., **19** (1925), 541–568.

B. A. Turaev, "The Volume of the Truncated Pyramid in Egyptian Mathematics," in *Ancient Egypt* (1917), 100–102.

B. L. van der Waerden, *Science Awakening*, translated by Arnold Dresden (Groningen, 1954), 15–36.

K. Vogel, "Erweitert die Lederolle unsere Kenntniss ägyptischer Mathematik?" in *Archiv für Geschichte der Mathematik* (1929), 386 ff.

K. Vogel, "The Truncated Pyramid in Egyptian Mathematics," in *Journal of Egyptian Archaeology*, **16** (1930), 242–249.

K. Vogel, *Vorgriechische Mathematik*, I, *Vorgeschichte und Ägypten* (Hanover, 1958).

N. F. Wheeler, "Pyramids and Their Purpose," in *Antiquity*, **9** (1935), 5–21.

J. A. Wilson, *Signs and Wonders Upon Pharaoh. A History of American Egyptology* (Chicago, 1964).

ADDITIONAL BIBLIOGRAPHICAL NOTES

Names included here may be referred to in general reading on the history of ancient Egypt. Such references may concern articles on specific topics, notes on archaeological work or Egyptology generally, or a published text from which lines may have been quoted. However brief the notes, the names included are of interest to the historian.

Abd el-Rasoul. Dealer in Egyptian antiquities who sold a papyrus to Golenischev in 1893, "pour une somme assez modique," that was early referred to as the Golenischev Papyrus.

J. Baillet, *The Akhmim Papyrus* (Paris, 1892). A Greek mathematical papyrus of A.D. 750, that still used Egyptian unit fractions, as in RMP Recto.

G. B. Belzoni. A six-foot, six-inch strong man who discovered the tomb of Seti I at Thebes and the entrance to the Cephren pyramid at Giza in 1818.

L. Borchardt. Archaeologist who excavated in Egypt. Found the now famous head of Queen Nefertiti at Tell-el-Amarna in 1907.

J. H. Breasted. First professor of Egyptology in the United States, 1895. Author of many books. Original director of the Oriental Institute at the University of Chicago, 1919. Died 1935.

H. K. Brugsch. Wrote a dictionary of demotic, 7 vols., and *Aegyptische Studien* (Leipzig, 1855). Articles on RMP. Professor of Egyptology, Göttingen, 1868. Died 1894.

L. S. Bull. One of the authors of vol. II of RMP: photographs, transcriptions, transliterations, and literal translations of the papyrus.

H. Carter. Associated with Lord George Carnarvon from 1907 in Egyptian excavations. Responsible for the discovery and clearance of the tomb of Tutankhamen, Valley of the Kings, in 1922. Died 1939.

J. Černý. Wrote on the religion of the ancient Egyptians in *Annales du Service des antiquités de l'Égypte,* **43** (1943), 179 ff.

E. Collignon. Discussed the decomposition of fractions of the form $2 \div (2n - 1)$, as in the RMP. Recto (Paris, 1881).

W. R. Dawson, *Who Was Who in Egyptology* (London, 1951).

G. M. Ebers. Professor of Egyptology, Leipzig. Obtained a medical papyrus at Luxor (1872) that bears his name. Died 1898.

J. P. A. Erman. Professor of Egyptology at Berlin. Produced Egyptian dictionary (1926). A major figure. Died 1937.

M. Eyth. A pyramid mystic. Wrote *Der Kampf um die Cheops Pyramide*, 2 vols. (Heidelberg, 1902). Quotes value of π to forty decimal places.

A. Fakhry. Details the history of Cheops in *The Pyramids* (Chicago, 1961).

A. Favaro. An early disputant on Eisenlohr's translation of the RMP Recto and unit fraction values of $2 \div (2n - 1)$. See "Sulla interpretazione matematica del papiro Rhind pubblicato ed illustrato dal Prof. Augusto Eisenlohr," in *Memorie della R. Accademia di scienze, lettere ed arti in Modena*, 2nd ser., **19** (1879), 89–143.

V. S. Golenischev. Member of the tsarist nobility who became professor of Egyptology at Cairo University. His collection of antiquities, loaned to the Moscow Museum in 1912 "contre une rente viagère," included the famous Golenischev Papyrus. After the 1917 Revolution, payments ceased; the collection became government property; and the roll became known as the Moscow Mathematical Papyrus. Died 1947.

H. Grapow. Edited an Egyptian dictionary in association with Erman.

B. G. Gunn. Worked for the Service of Antiquities and for the Cairo Museum. Professor of Egyptology at Oxford in 1934. Wrote article on finger numbering for *Zeitschrift für ägyptische Sprache*, **57** (1922), 71–72, dealing with Horus-eye fractions. Died 1950.

H. R. Hall. Keeper of Egyptian antiquities in the British Museum, 1924. His *Ancient History of the Near East* (London, 1913) went through many editions. With Dr. A. Scott he unrolled the Egyptian Mathematical Leather Roll, which had remained rolled in the museum for sixty years, owing to its brittle condition. The EMLR came to the trustees, with the RMP, from A. H. Rhind in 1864.

Herodotus. Greek historian (*ca.* 500–*ca.* 424 B.C.) who visited Egypt and recorded stories of the pyramids 2,000 years after they were built. H. W. Turnbull refers to one of his "obscure" passages, "which implies that the area of the triangular face of Cheops equals the square of the vertical height." This led to the *Sectio aurea*, or the golden section, in geometry.

E. Iversen. In his book *The Myth of Egypt and Its Hieroglyphs in European Tradition* (Copenhagen, 1961), Iversen discusses some of the misconceptions regarding the deciphering of the hieroglyphic writings.

T. G. H. James. Editor of *Journal of Egyptian Archaeology*, published annually by the Egypt Exploration Society, Manchester Square, London.

A. Jarolimek, *Die Rätsel der Cheops Pyramide* (Berlin, 1910), supports the pyramid mysticism of Taylor and Smyth, particularly with regard to π and the golden section, and surveys the works of other mystics: Nairz, Neikes, and Eyth.

K. R. Lepsius. Curator of Egyptian antiquities at Berlin, 1865. Produced *Denkmäler aus Aegypten und Aethiopien*, 12 vols. (plus) (Berlin, 1859). An authority on the Egyptian language and monuments. Died 1884.

A. F. F. Mariette. A famous Egyptologist. Wrote *Les papyrus egyptiens du Musée de Boulag* (Paris, 1872). Discoverer of the Serapeum at Sakkarah. Became curator of Egyptian monuments at Cairo. Died 1881.

G. C. C. Maspero. Professor of Egyptology at Paris, 1869. Publications include *Dawn of Civilization* (London, 1894); *Manual of Egyptian Archaeology* (London, 1895); *Guide to Cairo Museum* (Cairo, 1895); *Struggle of the Nations* (London, 1896); *Passing of the Empires* (London, 1900); *Art in Egypt* (New York, 1912). Ranks with Erman and Petrie. Received a British knighthood in 1909. Died 1916.

G. Möller. In "Die Zeichen für die Bruchteile des Hohlmasses und das Uzatauge," in *Zeitschrift für ägyptische Sprache*, **48** (1911), 99–106, Möller discusses the unit fractions used as portions of the *hekat*, the measure of capacity, referred to as Horus-eye fractions.

G. A. Reisner. Professor of Egyptology at Harvard, 1914. Excavator of renown who died at Giza in 1942. Many publications, including *History of the Giza Necropolis*, 2 vols. (Cambridge, Mass., 1942).

A. H. Rhind. A Scotsman who went to Thebes for health reasons and became interested in excavating. In 1858 he purchased the RMP and the EMLR, which after his death (1863) came to the British Museum. He wrote *Thebes, Its Tombs and Their Tenants* (London, 1862).

M. Simon. Wrote at least four articles on the mathematics of the RMP–1904, 1905, 1907, 1909–dealing with π and with the unit fractions of the Recto.

E. Smith. A dealer in Egyptian antiquities and hieratic papyri, including the Ebers Papyrus and the Edwin Smith Papyrus. Died 1906.

J. J. Sylvester. Considered the $2 \div (2n - 1)$ of the RMP Recto as a problem in number theory, to which he gave the name "fractional sorites." In "On the Theory of Vulgar Fractions," in *American Journal of Mathematics*, **3** (1880), 332–385, 388–389; and *Educational Times*, **37** (1882), 42–43, 80; he wrote: "It was their [the ancient Egyptians] curious custom to resolve every fraction into a sum of unit fractions according to a certain traditional method."

J. Taylor. One of the earliest pyramid mystics, perhaps the first. His theories were expressed in *The Great Pyramid, Why It Was Built, and Who Built It?* (London, 1859; 2nd ed., 1864). They were supported and further developed by the astronomer royal for Scotland, Piazzi Smyth (see Bibliography).

H. W. Turnbull. In *The Great Mathematicians* (London, 1951), 2–5, Turnbull refers to the geometry of the Great Pyramid, to the golden section, and to some of the theories of the pyramid mystics.

Q. Vetter. Discussed the methods of Egyptian division, especially with regard to $2 \div (2n - 1)$ of the RMP Recto, in *Egyptiské deleni* (Prague, 1921).

H. E. Winlock. One of the more successful excavators in Egypt. Discovered statues and relics of Queen Hatshepsut at Deir-el-Bahri in 1927. Became director of the New York Metropolitan Museum of Art in 1932. Highly regarded in the field of Oriental scholarship. Died 1950.

T. Young. British physicist, originator of the wave theory of light. An early decipherer of certain hieroglyphs, in 1814 he recognized that the cartouches contained the names of pharaohs and queens, but did not pursue these studies. Died 1829.

Mathematical Papyri and Ostraca:
Abbreviations Used, Approximate Dates, Other Details

ABBREV.	TITLE	APPROX. DATE	LOCATION	OTHER DETAILS
AMP	Akhmim Mathematical Papyrus	A.D. 750	CAIRO Cat. no. 10,758	Greek papyrus containing tables of unit fractions and 50 problems, as in the RMP, 2,400 years earlier. See Baillet in additional bibliographical notes.
BP	Berlin Papyrus	1850 B.C.	BERLIN Cat. no. 6619. Acquired by the Berlin Museum in 1887.	Hieratic. Contains problems on simultaneous equations. Square root referred to. See Schack-Schackenburg in the bibliography.
DMP	Demotic Mathematical Papyrus (Cairo Mus.)	300 B.C.	CAIRO Cat. no. 89127–89137. Discovered 1938 at Tuna el-Gebel.	Demotic. Contains 40 problems on equations, series, volumes of pyramids, Pythagorean theorem. Uses fractions like 5/6 and 6/47.
DMP	Demotic Mathematical Papyrus (Br. Mus.)	Ptolemaic Period	LONDON British Museum. Acquired 1868. Provenience unknown. Cat. nos. 10399, 10520, 10794	Demotic. Contains 27 problems on summation of series, volume of truncated cones, square roots, and several multiplication tables. See R. A. Parker in the bibliography.
EMLR	Egyptian Mathematical Leather Roll	1650 B.C.	LONDON British Museum. Bought at Luxor, Egypt, by A. H. Rhind, 1858. See additional bibliographical notes.	Hieratic. Contains, in duplicate, a collection of the sums of Egyptian unit fractions, such as $\overline{18}\ \overline{27}\ \overline{54} = \overline{9}$.
KP	Kahun Papyrus	1850 B.C.	LONDON British Museum. Found by W. M. F. Petrie at Kahun, Egypt, in 1889.	Hieratic. Contains portion of RMP Recto, 10 entries. Volume of a cylindrical granary, and solution of equations. See F. L. Griffith, in bibliography

Mathematical Papyri and Ostraca:
Abbreviations Used, Approximate Dates, Other Details (*continued*)

ABBREV.	TITLE	APPROX. DATE	LOCATION	OTHER DETAILS
Mich P	Michigan Papyrus	A.D. 350	ANN ARBOR University of Michigan. From the Fayum, Egypt. Cat. no. 621, also 145.	Greek. Contains several multiplication tables, all in unit fractions, and some problems. Since 6,000 drachmas made a talent, many fractions of 6,000 for tax offices.
MMP	Moscow Mathematical Papyrus	1850 B.C.	MOSCOW Bought by Golenischev in 1893, loaned to Moscow Museum in 1912. Cat. no. 4576, Pushkin Museum of Fine Arts.	Hieratic. Contains 25 problems, 11 on *pesus* of beer and bread, 6 on triangle areas, one on volume of truncated pyramid, one on area of hemisphere. See Abd el-Rasoul and Golenischev in additional bibliographical notes.
Ostraca	Various	Various From 2000 B.C. to A.D. 600	CAIRO [Thebes tablets (25367) 2000 B.C.] Hieratic. BERLIN [Elephantine Ostraca. A.D. 250] Greek. LONDON [Dendarah (480) A.D. 550] Coptic. MANCHESTER [Numbers tables (6221) A.D. 600] Coptic.	Hieratic, Demotic, Greek, and Coptic. These contain mostly tables of numbers, divisions of unit fractions, and varied problems. See W. E. Crum, G. Daressy, and W. C. Hayes in the bibliography.
RP	Reisner Papyrus	1880 B.C.	BOSTON Museum of Fine Arts. Found by Dr. G. Reisner at Giza in 1904	Hieratic. Concerns calculations of blocks of stone, and payment of workmen, for building a temple in the reign of Sesostris I. See W. K. Simpson in the bibliography.
RMP	Rhind Mathematical Papyrus	1650 B.C. Copy of a papyrus 200 years older	LONDON British Museum. Cat. no. 10,057–8. Bought with the EMLR by A. H. Rhind at Luxor, Egypt, in 1858.	Hieratic. The Recto contains divisions of 2 by the odd numbers 3 to 101 in unit fractions, and the numbers 1 to 9, by 10. The Verso has 87 problems on the 4 operations, solution of equations, progressions, volumes of granaries, *sekeds* of pyramids, *pesus*, the two-thirds rule, tables of Horus-eye fractions. See A. B. Chace, J. R. Newman, E. T. Peet, A. H. Rhind in bibliography and additional bibliographical notes.

EGYPTIAN ASTRONOMY, ASTROLOGY, AND CALENDRICAL RECKONING

Richard A. Parker

INTRODUCTION

More than any other ancient people, the Egyptians seem to have occupied themselves with the reckoning of time. They were the first to come to an approximation of the true length of the natural year and to devise a calendar based on it. They were the first to divide the night and the day into twelve hours each, and they were the first to make these hours equal. The story of these developments is at the same time the story of the Egyptians' astronomical knowledge, because to a very high degree Egyptian astronomy was the severely practical servant of Egyptian time-reckoning. This was true over the long span of Egyptian written history, the three millennia before the Christian era.

Although during this time there were many references to the sun, moon, and stars in texts, and the planets were known and named, there is nothing approaching an astronomical treatise, such as those known from Babylonia, where the movements of the celestial bodies were studied and recorded. Before the Ptolemaic period, in fact, there is but one text, the Cosmology of Seti I and Ramses IV, that exhibits some thinking about astronomical matters; but these too are concerned mainly with time-reckoning. It was only with the Persian conquest in the middle of the first millennium B.C. and, two centuries later, with the advent of the Ptolemies, that Egypt became a fertile field for the transplant of foreign speculation and the study of the cosmos and its activities. Theoretical astronomical treatises appeared then, as did the zodiac and astrology. But this was long after the early splendid achievements of the Egyptians in calendar development and time-reckoning, and it is to these that we must first direct our attention.

THE FIRST CALENDAR

The Egyptians, no more than any other primitive people, did not suddenly, one day, invent a calendar. It must have been the result of a long, slow process of accretion and experiment. Before the desiccation of the Sahara plain of North Africa drove primitive man into the valley of the Nile, he was a hunter and food-gatherer. As such he would have come to some terms with the rhythm of the seasons. From the simple concept of the presence of the sun in the sky as day and its absence as night, he would surely have begun to notice the other large heavenly body, the moon, which had a regularity in its change from its reappearance as a thin crescent to its growth as full moon and then its gradual change back to crescent and final disappearance from the sky for two or three days. When he began to count, he would discover that from one disappearance to the next, or from one new crescent to the next, or from one full moon to the next, twenty-nine or thirty days would elapse. This concept of month might then be carried to the realization that usually twelve or thirteen of these larger units would cover the time from one natural event, such as the reappearance of a certain edible fruit, to its next occurrence after a period of time; this would be a year.

When the lack of rainfall in North Africa forced primitive man to abandon his nomadic life as a hunter and settle down near water, he had perforce to become mainly an agriculturist. In eastern North Africa the Nile valley proved to be an admirable haven. Some hunting was still possible in the marshes along the river, and the valley itself proved most fertile. Very quickly the primitive Egyptian discovered that the river was the dominant force in his new activity as a food producer. This mighty waterway had a life of its own. At a certain time it would begin to rise until it overran its banks and covered all the valley between the eastern and western cliffs except the higher ground of habitation. Slowly the water would retreat, the river would return to its bed, and planting could be done. As the river dropped, growth would take place and finally crops could be harvested. After a period of low water, the cycle would repeat itself. Through long experience the valley dweller came to divide his years into three seasons of four months each, although he noticed that at times one season would appear to be five months long.

Through the years he came to notice that a remarkable celestial event took place just about the time that the Nile, after its long period of low water, began to rise again. The brightest star in the sky, Sirius (named Sothis by the Egyptians), would reappear on the eastern horizon just before sunrise, after a lengthy period of invisibility. This event is now called its heliacal rising, and in the Julian calendar (the predecessor of our Gregorian calendar) it fell around July 17/19 throughout Egyptian history.

The heliacal rising of Sirius-Sothis was thus recognized as heralding the coming inundation, and the first calendarial genius in Egypt used it as a peg on which to construct and keep in place a formal calendar year. The first season, *Akhet* (since the Egyptians wrote only consonants, the vocalizations are makeshifts), "Flood" or "Inundation," coincided with the rise and overflow of the river; the second was *Peret*, "Emergence"; and the third was *Shomu*, "Low Water" or "Harvest." Each season had four months, and each month was named after a special festival that took place in it. Thus, for example, the first month was named from the *Tekhy* feast; the third, from the feast of the goddess Hathor; the eighth (the fourth month of the second season), from the festival of the goddess of the coming harvest, Ernutet; and so on for all except the fourth month of the third season. Whereas all the other month names came from festivals that took place on days determined by the moon (such as the first day, the day of the first quarter or last quarter, the day of full moon), the twelfth month was named *Wep-renpet*, "Opener of the Year." This was taken from the name the Egyptians gave to the heliacal rising of Sirius-Sothis and was, of course, a stellar event. To keep this festival always within its proper month was a problem, since twelve months of varying length, either twenty-nine or thirty days, would average but 354 days, shorter than the natural or solar year by some eleven days.

The problem was solved by applying a simple rule. Whenever Sirius-Sothis rose heliacally (*Wep-renpet*) in the last eleven days of the twelfth month, an intercalary month was added to the year, lest in the next year the festival fall out of its month. Fittingly, the intercalary month was named *Thoth*, from a god associated with the moon. To illustrate this rule, if Sirius-Sothis rose heliacally on the third day of the month *Wep-renpet* in one year, in the following year it would rise on the fourteenth day, and in the year after that on the twenty-fifth day. The intercalary month of *Thoth* would then be added to keep the feast of *Wep-renpet* from falling in *Tekhy*, the first month of the next year. Such a procedure would need to be utilized every three or, more rarely, two years.

Parenthetically, it should be noted that the Egyptians began a lunar month not with the reappearance of the crescent in the western sky, as did most ancient peoples using a lunar calendar, but with the day on which the old crescent was no longer visible in the eastern sky just before sunrise. Their calendar day, then, ran from sunrise to sunrise and not from sunset to sunset nor, like ours, from midnight to midnight.

How early this lunistellar calendar was systematized and adopted, we do not know. Written proof of it is lacking before the Fourth Dynasty, the middle of the third millennium B.C.; but there can be little doubt that it was employed for centuries before.[1]

LATER CALENDARS

While other countries, such as Babylonia, that employed a lunar calendar with an intercalary month to keep it in harmony with the natural year, continued using such a calendar year exclusively throughout their history, Egypt took a different course. During the Predynastic Period and early into the Dynastic Period (the First Dynasty began about 3110 B.C.), the oscillation of the lunar year about the heliacal rising of Sirius-Sothis and its irregular length—now twelve, now thirteen months—were relatively minor inconveniences. With the advent of the dynasties, however, Egypt became a highly organized and efficient state. Such a year of irregular months, with the months themselves of irregular length, must have come to be regarded by officialdom as a great nuisance.

At this point the second unknown calendarial genius in Egypt made his contribution. He determined the length of the natural year to be 365 days. There are two ways he could have done this. He might have averaged the lengths of several lunar calendar years in succession (the average of eleven years would come very close to 365) or, more probably, he counted the days from one heliacal rising of Sirius to the next, since this event was the lunar calendar's control. Only very few years would be necessary to establish the figure of 365 days from one *Wep-renpet* to the next. With this figure firmly established, he organized the new calendar as follows: The three seasons remained, but the four months in each were given a constant length of thirty days. Unlike the lunar month, which divides more or less evenly into four "weeks" based on "first day" to "first quarter" (seven days), "first quarter" to "full moon" (eight days), "full moon" to "last quarter" (seven days), and "last quarter" to "last day" (seven or eight days); the new thirty-day month was given ten-day "weeks" termed "first," "middle," and "last." In the entire year, then, there were thirty-six such ten-day weeks, or decades. The five remaining days were taken as a small intercalary month and were termed "the days upon the year" (later the Greeks called them the "epagomenal days"). This calendar with its constant length of 365 days has been acclaimed as "the only intelligent calendar which ever existed in human history."[2] The fact that it was constant made it the ideal calendar year for astronomy. Hellenistic astronomers and their successors until Copernicus used this Egyptian calendar for their planetary and lunar tables.

The introduction of the new "civil" year, as we may term it, did not mean the abandonment of the lunar year. Quite the contrary. Almost all important festivals continued to be determined by the moon—in this respect much like Easter Sunday, still set by the moon. The new civil year was an artificial creation, useful for accounting and administrative purposes but devoid of religious significance.

We do not know exactly when the new calendar was introduced, but we can set probable limits for this act. On the plausible assumption that when its use was inaugurated, the first days of both the lunar and the civil calendars coincided, this event must have fallen between *ca.* 2937 and *ca.* 2821 B.C. If the first day of the lunar year was only twelve days after *Wep-renpet*, the year would be *ca.* 2821 B.C.; if as much as forty-one days after (eleven plus an intercalary month of thirty days), the year would be *ca.* 2937 B.C.

The reason for this latitude of over a century lies in the fact that the true length of the natural year is, of course, not 365 days but almost exactly 365.25 days. The practical result of this was that the beginning of the new civil year, since its calendar did not have a leap year every four years, as ours has, began imperceptibly to move forward in the natural year. The civil year was, to be sure, planned to run concurrently with the lunar year. But since the latter was so variable, it would be a long time before the progress forward of the civil year became apparent. After fifty years—an ordinary lifetime—the two years would still seem to be in good general agreement. After two centuries, however, there could be no doubt that the two calendars were in difficulty. Never, now, would the first month of the civil year and the first month of the lunar year coincide for even one day.

Superficially, it would have been easy to adjust this situation simply by adding some fifty days onto one civil year and thus bringing it back to the pattern of concurrency at its inauguration. For whatever reason—we may suppose that Egyptian bureaucracy would not have it—this solution was rejected. Instead, since this was apparently regarded as a religious problem, a new lunar year was created, one designed specifically to run in partnership with the civil year and so give substance to the latter's artificiality. Like the original lunar year, the new lunar calendar had an intercalary month when it was necessary to keep the lunar calendar in place. A very simple rule of intercalation was used. Whenever the first day of the lunar year would fall before the first day of the civil year, an intercalary month was added.

From this time on (about 2500 B.C.), Egypt had three calendar years, all of which remained in use until the end of pagan Egypt. As the centuries passed, certain festivals were given fixed dates in the civil year, while the new lunar calendar determined the lunar festivals that were not dependent on the natural year, such as the monthly full-moon feasts. The original lunar year, kept in place by the heliacal rising of Sirius-Sothis in its twelfth month, maintained its general agreement with the natural year and gave dates to all agricultural and seasonal events.

The civil calendar is the one most familiar to students, and the most important for historians, because all records were dated in it. A typical date would give the regnal year of the king who happened to be on the throne, followed by the month (I to IIII) of *Akhet*, *Peret*, or *Shomu* and the day of the month (1 to 30). Throughout most of Egyptian history, regnal and calendar years coincided, with the last fractional year of one king and the first fractional year of his successor together making one civil year. For a while, however, beginning in the Eighteenth Dynasty, a pharaoh began his reign with a full first regnal year that began on the day of his accession to the throne and ran for 365 days, thus encompassing parts of two civil years and at times creating dating problems for historians.

The civil year of course continued its slow progression through the natural year, until after some 1,460 years (4 × 365) it regained its original place vis-à-vis the natural year. This period is known as the Sothic cycle, since the important chronological event of the heliacal rising of Sirius-Sothis (besides *Wep-renpet*, now also called *Peret-Sepdet*, "the Coming Forth of Sothis") would, generally speaking, fall for four years on I *Akhet* 1, for example, then for four years on I *Akhet* 2, and so on through the whole civil year. Occasional dates of *Peret-Sepdet* are of signal importance for chronology, since it is firmly established on the authority of Censorinus that there was a coincidence of *Peret-Sepdet* with I *Akhet* 1 in A.D. 139. Reckoning backward, we would expect a similar coincidence in 1322 B.C. and, before that, in 2782 B.C. In fact, since the star Sirius has a movement of its own, the coincidences must be computed from astronomical tables, and come out as more likely in *ca.* 1317 and *ca.* 2773 B.C. For example, there is a temple record dating one occasion of *Peret-Sepdet* to IIII *Peret* 16 of the seventh year of Sesostris III of the Twelfth Dynasty. Other historical data place this dynasty in the early second millennium B.C. The Sothic date narrows the seventh year of Sesostris III to 1870 B.C. ± about six years. Some lunar dates of the same king enable us to pin the year down to 1872 B.C.; and since we know how long the kings of this dynasty reigned, we can fix them all chronologically.[3]

By the New Kingdom, and from then on, the months of the civil year were at times referred to by names, some of which are the same as those of the original lunar calendar months, while others witness the introduction of new and important feasts. As written later by the Greeks, the names are shown below.

FIRST SEASON	SECOND SEASON	THIRD SEASON
Akhet	*Peret*	*Shomu*
1. Thoth	5. Tybi	9. Pachons
2. Phaophi	6. Mechir	10. Payni
3. Athyr	7. Phamenoth	11. Epiphi
4. Choiak	8. Pharmuthi	12. Mesore

In these names it is possible to recognize the goddess Hathor in the name of the third month and, in the eighth month (with slight change), the harvest goddess Ernutet (*Pha* or *Pa* means "The One of").

There is no evidence whatever to suggest that after the civil calendar was installed, it was adjusted to arrest its forward movement in the natural year prior to 239 B.C., when Ptolemy III Euergetes issued the Decree of Canopus, which made every fourth year a leap year with a sixth epagomenal day. No attention, however, was paid to this decree and the civil year continued exactly as before.[4] It was not until the time of Augustus, over two centuries later, in 30 or possibly 26 B.C., that a leap year was firmly established. From that time on, the reformed civil calendar was known as the Alexandrian calendar.

Nevertheless, the old civil calendar remained in use—and not only by astronomers. One of these, however, whose name is still unknown, wrote a papyrus (Papyrus Carlsberg 9) that represents the one astronomical and mathematical text we now have from the last centuries B.C. that is certainly Egyptian in origin. Although it was written no earlier than A.D. 144, it surely goes back for its material to the fourth century B.C.[5] The text itself is a cyclical scheme over twenty-five years for beginning the months of the later lunar year without the necessity of observing the moon. While the papyrus gives only dates in alternate months of the civil calendar, it is possible to deduce, from other sources, the rules governing the dates in the intervening months and thus to reconstruct the entire cycle.[6] Table 1 gives the completed cycle.

Underlying the cycle are the astronomical facts that twenty-five years have 9,125 days and 309 lunar months have 9,124.95231 days. Over twenty-five civil years, nine intercalary months are necessary to prevent the first day of the lunar year from falling before the first day of the civil year; they occur in cycle years 1, 3, 6, 9, 12, 14, 17, 20, and 23. As we should expect, the first lunar month begins on I *Akhet* 1. For 357 B.C., when it is possible that the cycle was installed (plus or minus some fifty years) and the morning of crescent invisibility remained the first day of the new month, a test by calculation for the twenty-five dates in cycle years 1 and 2 shows that eighteen of them are in agreement with observation. It is likely that the figure of 72 percent correctness would hold for the entire cycle; and given the simplicity of the scheme, this is no mean achievement. In time, however, since the lunar months were .04769 of a day short over twenty-five years, this deficiency would accumulate. In five hundred years the cycle would be about one day off and would no longer agree with the mornings of crescent invisibility but, rather, with the evenings of first crescent visibility, normally one day later and the basis for the month's beginning in the Macedonian, Hebrew, and Babylonian calendars. We know that this result was anticipated by the Ptolemies using the Macedonian calendar, since they took over the

	AKHET				PERET				SHOMU				
Months	I	II	III	IIII	I	II	III	IIII	I	II	III	IIII	Epag.
Year 1	1	1	1–30	30	29	29	29	28	27	27	27	26	–
2	20	20	19	19	18	18	18	17	16	16	16	15	–
3	9	9	8	8	7	7	7	6	5	5	5	4	4
4	28	28	27	27	26	26	26	25	24	24	24	23	–
5	18	18	17	17	16	16	16	15	14	14	14	13	–
6	7	7	6	6	5	5	5	4	3	3	3	2	2
7	26	26	25	25	24	24	24	23	22	22	22	21	–
8	15	15	14	14	13	13	13	12	11	11	11	10	–
9	4	4	3	3	2	2	2	1	1–30	30	30	29	–
10	24	24	23	23	22	22	22	21	20	20	20	19	–
11	13	13	12	12	11	11	11	10	9	9	9	8	–
12	2	2	1	1	1–30	30	30	29	28	28	28	27	–
13	21	21	20	20	19	19	19	18	17	17	17	16	–
14	10	10	9	9	8	8	8	7	6	6	6	5	5
15	30	30	29	29	28	28	28	27	26	26	26	25	–
16	19	19	18	18	17	17	17	16	15	15	15	14	–
17	8	8	7	7	6	6	6	5	4	4	4	3	3
18	27	27	26	26	25	25	25	24	23	23	23	22	–
19	16	16	15	15	14	14	14	13	12	12	12	11	–
20	6	6	5	5	4	4	4	3	2	2	2	1	1
21	25	25	24	24	23	23	23	22	21	21	21	20	–
22	14	14	13	13	12	12	12	11	10	10	10	9	–
23	3	3	2	2	1	1	1–30	30	29	29	29	28	–
24	22	22	21	21	20	20	20	19	18	18	18	17	–
25	12	12	11	11	10	10	10	9	8	8	8	7	–

TABLE 1. The Lunar Calendar Cycle (R. A. Parker, *The Calendars of Ancient Egypt*, 25).

Egyptian twenty-five-year cycle in the third century B.C. and added one day to every date in it, thus effectively converting the cycle to suit their own requirements.[7]

THE HOURS OF THE NIGHT AND DAY

Thus far we have been concerned with time measurement as determined by natural events—the day, the month, and the year—and have seen that no smaller unit goes evenly into the larger one. The lunar month has no exact number of days, and the natural or solar year has neither an exact number of days nor an exact number of lunar months. These elements of time measurement have to be fitted to one another by some calendrical scheme that, while arbitrary, attempts to preserve the essential relationship of the units. Thus our own calendar still keeps to twelve months in the year and twenty-eight to thirty-one days to a month, although the result is an awkward instrument.

There are, however, no natural subdivisions of the day beyond light and darkness. A minute of sixty seconds, an hour of sixty minutes, and a day of twenty-four hours have completely arbitrary subdivisions. Nothing in nature requires that the day should be divided into twenty-four parts. That it is so divided is a by-product of the established civil calendar and the ancient Egyptians' desire to break the night into smaller units.

Why the Egyptians desired to do this is a matter of inference, since there is nothing in the written records that have come down to us to explain their action. From late temple calendars of feasts, we do know that some feasts were celebrated at night; and occasionally we are told that certain rites must be performed at a specific hour of the night. There can be little doubt that what was the late custom also obtained in much earlier times. Over and above such individual feasts, however, was the obsession of the Egyptians with the passage of the sun god Re through the Other World, or Duat, from sunset until sunrise of the next day. There are a number of religious treatises, such as the "Book of Him Who Is in the Underworld," the "Book of

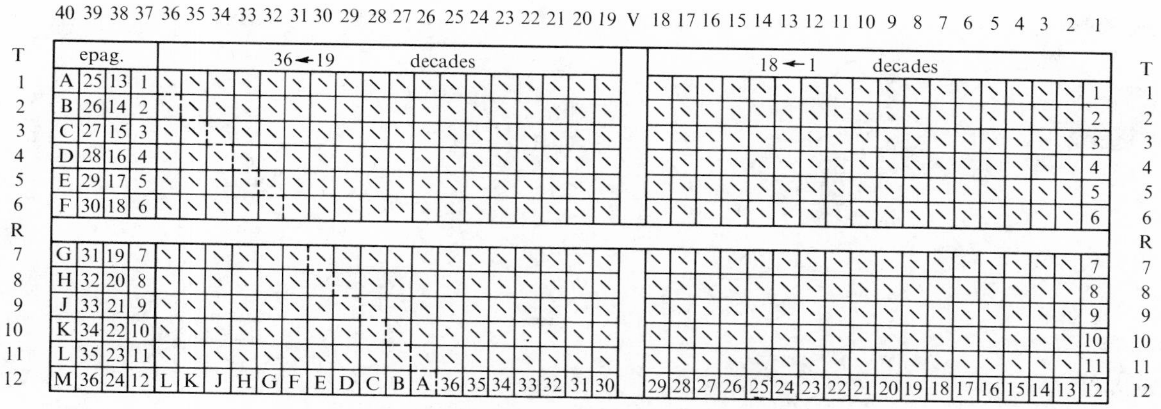

FIGURE 1. Schematic star clock.

Gates," and the "Book of Day and Night," that trace his passage in his night bark, with his crew of gods and goddesses, through dangers and difficulties until once again he rises triumphantly in his day bark and illuminates the Two Lands, Upper and Lower Egypt. In all of these the Other World is divided into twelve regions, and in each of these regions the sun god spends one hour of the night. We do not have such books written earlier than the New Kingdom (*ca.* 1500 B.C.); but in one of the pyramid texts of the last king of the Fifth Dynasty, Unas (twenty-fourth century B.C.), he "clears the night and dispatches the hours." In Egyptian the word for "hour" is written with a star.

At some time before Unas, then, but after the civil calendar was adopted, another unknown Egyptian astronomer devised a scheme to divide the night into parts by using the apparent movement of the stars. We know that this apparent movement is due to the rotation of the earth on its axis and its travel about the sun. He did not. For him the stars did rise, traverse the sky, and set, like the other heavenly bodies. He had studied the stars and had grouped many of them into constellations to which he gave names such as *Nakht* ("Giant"), *Reret* ("[female] Hippopotamus"), Sah (our Orion), *Meskhetiu* ("Foreleg" or "Adze," our Big Dipper). He knew that certain stars, which we call "circumpolar," never set. These he called the "indestructible stars." He had studied Sirius-Sothis, the most important star; and he knew that when it disappeared from the sky, it was some seventy days before it reappeared in its heliacal rising. He knew that other stars followed the pattern of Sirius, with a similar period of invisibility and then a heliacal rising. His concentration was on the eastern horizon, more than the western. It was there that Sirius reappeared to announce the coming inundation. It was there that the last crescent was no longer to be seen and the new lunar month began. It was there that the morning star appeared, to which there are so many references in the earliest religious texts in the pyramids and with which the king sought identification. These texts never mention an evening star.

Our unknown astronomer also knew that a star rising heliacally, which might then for a time be called the morning star, did not stay on the horizon but every day was a little higher in the sky at sunrise. After some days another bright star might rise heliacally, and this in turn might be named the morning star. His brilliant scheme to divide the night was simply to go through one year of the civil calendar and pick a star or group of stars, which we now conventionally term "decans," that rose heliacally on the first day of each of the thirty-six decades or ten-day weeks of the year. During the night, then, he called the interval between the rising on the eastern horizon of one such decanal star and the rising of its immediate follower an "hour."

It is not, however, until the middle of the twenty-second century B.C. that we know with certainty that these "hours" totaled only twelve. The proof comes from diagrams on the underside of the lids of some coffins, buried during the Eleventh Dynasty, that were first known as "diagonal calendars" but are now correctly termed "star clocks." Figure 1 is a schematic version of such a clock, and all the examples we have follow it to some degree of completeness. It is to be read from right to left. The upper horizontal line T is the date line, beginning with the first decade of the first month of *Akhet* in the first column and running to the thirty-sixth column with the last decade of the last month

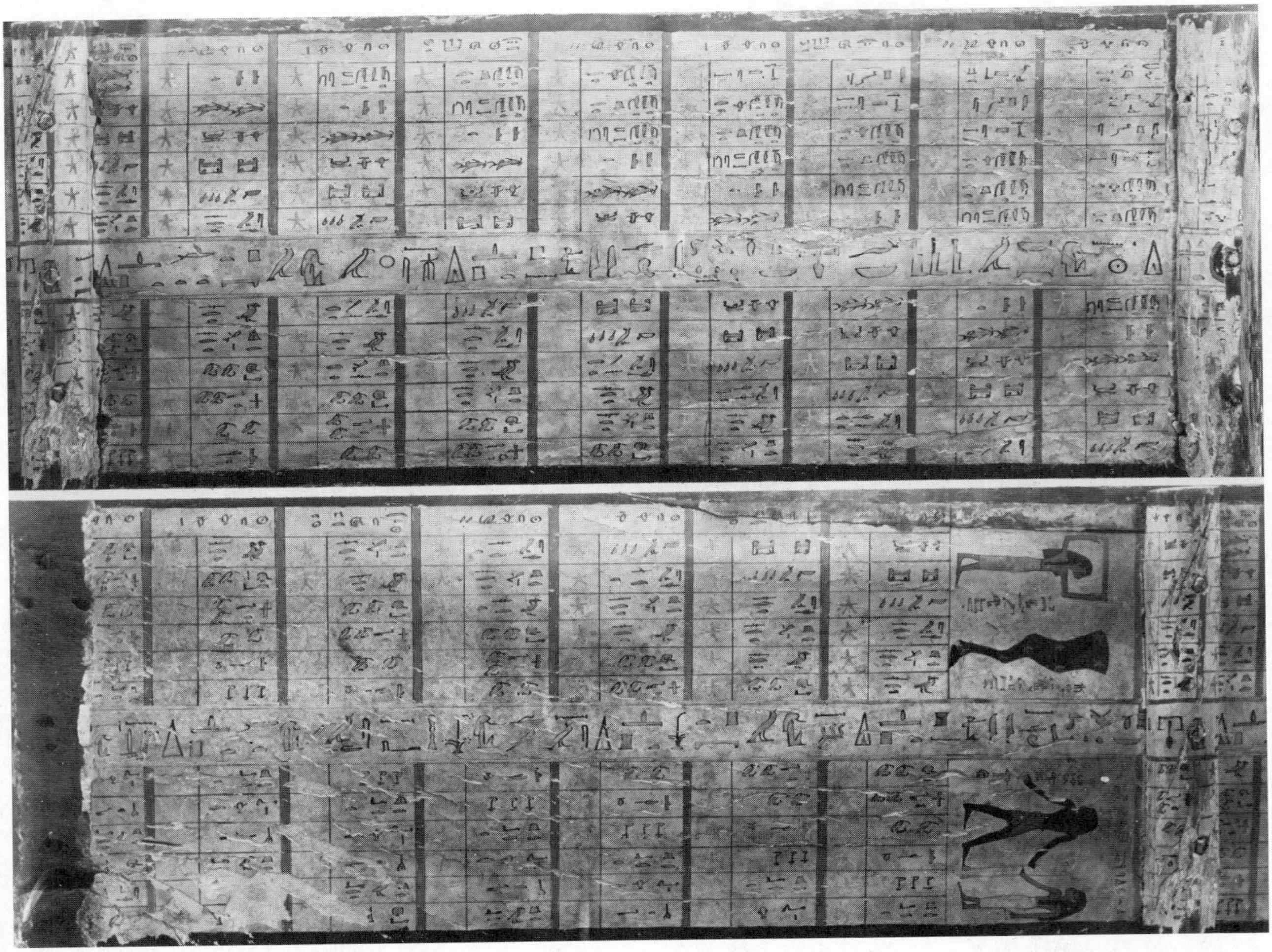

PLATE 1. Star clock of Idy (Sammlung des Ägyptologischen Instituts der Universität Tübingen, no. 6).

of *Shomu*. In the decade numbered 26 there begins a triangle of alternate decans to tell the hours of the five epagomenal days. Intruding before the epagomenal days column (40) are three columns that merely list the thirty-six decans of the preceding columns. Each column lists twelve hours. Between the sixth and seventh hours is a horizontal inscription (R) that is a funerary prayer that the sun god Re and other celestial deities will provide the dead person with offerings. Separating the eighteenth and nineteenth columns is a space (V) wherein are depicted Nut, the goddess of the sky, the Foreleg of an Ox (the Big Dipper), the constellation *Sah* (Orion) as a god, and Sothis as a goddess.

The actual star clock for a man named Idy is shown on Plate 1. It is not complete, since it has only eighteen decade columns; but it well illustrates the features just discussed and provides graphic evidence of the progress of a decanal star from the twelfth hour to the first hour, after which it drops from the clock. The probability is that the rising of a decanal star showed the end, not the beginning, of its hour. From still later texts in the cenotaph of Seti I at Abydos,[8] we learn that there was a conscious effort by the early astronomer to choose decanal stars invisible for seventy days, just like Sirius. Such stars would all fall in a band south of and parallel to the ecliptic. Figure 2 shows Sirius and Orion, a constellation that provided several decanal stars, in the decanal band.[9]

It is obvious that hours determined in this fashion were not all of equal length. As the night grew longer, the last hour before dawn would be longer; and this process would go on until the night began to shorten, at which time dawn would begin to come earlier and the last hour before it would shorten. Such hours of varying length may be called "sea-

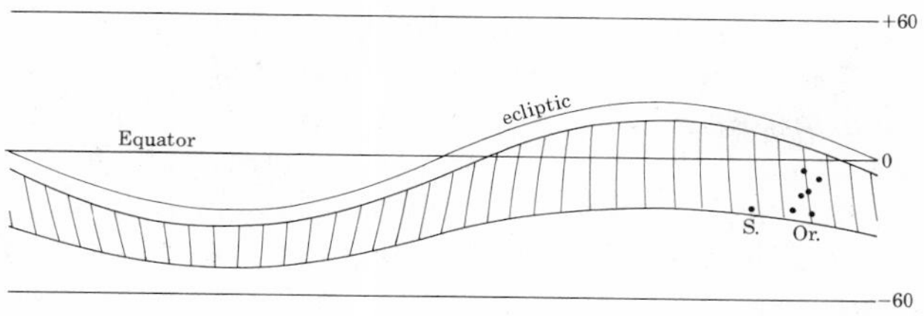

FIGURE 2. The decanal band, with Orion and Sirius-Sothis as hour stars.

sonal" hours, since twelve winter night hours are individually much longer than twelve summer night hours, although, strictly speaking, "seasonal" hours are one-twelfth the time from sunrise to sunset or from sunset to sunrise.

That there were but twelve hours reckoned, when at any one time there were always eighteen decanal stars visible from horizon to horizon, is explained by the necessity to take into account morning and evening twilight. The length of time it takes for the brightest stars to appear is termed "civil" twilight, and it averages about half an hour. From sunset until all stars are visible ("astronomical" twilight) takes longer, and may be well over an hour. The decans measured only the time of total darkness. Moreover, in any one night only the interior hours (two to eleven) would be of the same length throughout a decade. One may well assume that the first hour would begin with darkness and run until the rising of a certain decanal star. This first hour would be longest at the beginning of a decade, since on each night thereafter the star ending the hour would rise a little earlier. The twelfth hour, however, would gradually lengthen if, as is probable, it would be taken as ending with morning twilight, even though the decanal star that rose heliacally on the first day of the decade would be visible a little earlier in darkness on each succeeding day of the decade.

It is important to observe that the twelve-hour division of total darkness, with two to three "hours" of twilight after sunset and before dawn, is the direct result of the division of the civil calendar into thirty-six decades. We have only to assume that the year had been divided not into ten-day but five-day "weeks." There would then have been thirty-six pentades of stars from horizon to horizon, and the division of total darkness would have been into some twenty-six "hours."

There appears to be no other explanation for the hours of the day being taken as twelve except analogy with those of the night.[10] Presumably the earliest information we have on the day hours is from a

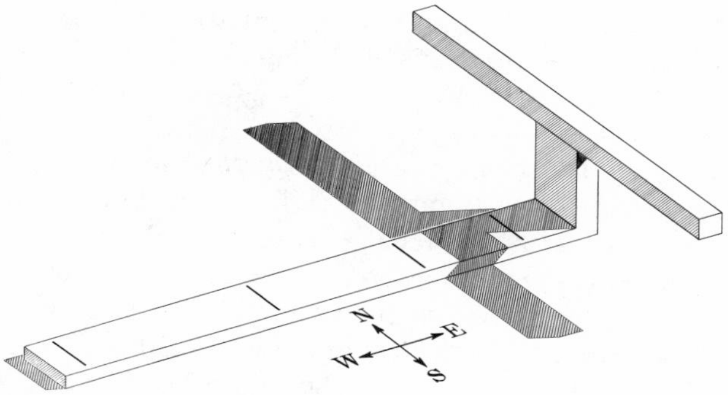

FIGURE 3. Shadow clock.

funerary text—and so a text of some antiquity—in the cenotaph of Seti I (1303–1290 B.C.) that gives directions for constructing a shadow clock. This has a base with four divisions marked on it and with a raised crossbar at its head (Figure 3). The text informs us that, with the head to the east, four hours are indicated by shadows that grow shorter and shorter. The instrument is then reversed with head to the west, and four more hours are marked off. More importantly, the text ends with the statement that two hours pass before the shadow clock tells the first hour and two more pass after the clock itself has finished. It is a reasonable conclusion that these two hours, before and after the eight hours determined by the shadow clock, are divided by the phenomena of sunrise and sunset; and the first and last hour of the whole period of daylight were morning and evening twilight, respectively.

From the division of total darkness into twelve hours and of daylight, including the two twilights flanking it, into twelve hours, the next step for which we have evidence was the division of daylight, from sunrise to sunset, into twelve hours. A shadow clock from the time of Thutmose III (1490–1436 B.C.) has five divisions on its base, so that the first and twelfth hours must have been those before and after the sun shone on it. The corresponding division of the period from sunset to sunrise into twelve parts could not have been effected by the stars and must have been done by a water clock. The earliest water clock we have, however, dating from Amenhotep III (1397–1360 B.C.) but reflecting the calendar situation about 1540 B.C., still divided only total darkness and was used, as a later text informs us, only when the stars could not be seen.

With the water clock, however, came the means for developing the concept of "equal" as opposed to "seasonal" hours. This is hinted at in a sadly mutilated inscription in the tomb of one Amenemhet (time of Amenhotep I, 1545–1525 B.C.), who is thus the first astronomer we can name in ancient Egypt.[11] The change in length of the nights is discussed with a presumable ratio of the longest to the shortest as fourteen to twelve. That this is a very poor approximation is obvious, but it paved the way for a text on papyrus that clearly proves the existence of equal hours. Again, although the papyrus itself is from the Ramesside period (twelfth century B.C.), it reflects an earlier calendrical situation, about 1300 B.C., since the shortest night falls in the last month of the third season, the last month of the year. For every month the text gives the hours of day and night, with the extremes of eighteen to the day and six to the night, and vice versa six months later. Such figures can be explained only on the basis that the six night hours were those of total darkness on the shortest night of the year, and then the astronomical scribe simply employed a linear increase and decrease of two hours to each month.[12] The correct ratio for Egypt of longest to shortest day or night is more correctly fourteen to ten.

Egypt's contribution to our time-reckoning did not go beyond the concept of twenty-four equal hours to the day. It is to the Babylonians, with their sexagesimal system, that we owe the division of the hour into sixty minutes and the minute into sixty seconds. We have not, however, exhausted the Egyptian efforts to make easier the telling of the hours of the night.

THE DECANS IN TRANSIT

We have seen that the star clocks of the Eleventh Dynasty determined the hours by the risings of the decanal stars. With adjustments from time to time, necessitated by the forward movement of the civil calendar in the natural year, such clocks continued in use into the Twelfth Dynasty (1991–1786 B.C.). At that time, however, it would appear that decanal risings were abandoned in favor of the transits of the decans. We learn this from later texts that have already been mentioned, the Cosmology of Seti I and Ramses IV.[13] Although there is much mythology in the content of the texts that surround a ceiling depiction of the goddess of the sky, Nut, bending over the earth, Geb, and supported by Shu, god of the air, there are a few texts that embody the first astronomical thinking of the ancient Egyptians so far known to us.

Most important is the concept that the stars and sun are linked as heavenly bodies that disappear and reappear in the sky. At sunset the sun goes to the Duat, where he spends the twelve night hours journeying from west to east. The decanal stars also disappear, for the lengthy period of seventy days that they spend in the Duat. When a star leaves the Duat, it is born again at its heliacal rising. It then spends eighty days in the eastern sky before embarking upon 120 days of work in telling the hours (ten days for each hour) with its transit of the meridian. When the star finishes work, having indicated the first hour of the night, it spends ninety days in the western sky and then dies again. One text says:

> As for (what is) between the star which makes the first hour and the star which will be enclosed by the Duat, it is 9 (stars). Now as for (what is) between the star of birth to the star which makes the first hour, it is 20 stars; which gives 29, being those which live and work in the sky. One dies and another lives every decade (of days). Now as for (what is) between the star of birth and the star which will be enclosed by the Duat, it is 29 through that breadth of the sky as stars.[14]

[In this, as in the following extracts, the symbols ( ) indicate additions by the translator, [ ] restorations, and < > emendations.] Thus at any one time there were seven stars in the Duat, eight in the east, twelve in the middle of the sky working, and nine in the west. It is clear that we have here a simplified scheme of stellar activity, based on a year of only 360 days. As such, corrective factors would need to be continuously applied, but there is no evidence for them.

A further consequence of a shift from rising to transiting decans was an extensive rearrangement of the decans themselves. While successive rising stars on the horizon might mark off an hour nicely, the same stars, in transit, might be too close or too far from one another. In the list of thirty-six rising decans of the star clock nearest in time to the transit list, only twenty-three remain in the same relative positions. The remaining thirteen stars are either newly selected ones or have changed their places.

After the Twelfth Dynasty there is no contemporary evidence for any decanal star clocks. One from the time of Merneptah (1223–1211 B.C.), while the latest we have, is purely funerary in purpose; and its antiquity is such that its arrangement of stars in the civil year could suit only a time some six centuries earlier.

We do have many lists of decans from later monuments extending well into the Roman period, but these are never in the form of a star clock. One can surmise, of course, that any observer had only to memorize a list of thirty-six decans, rising or transiting, watch to see which one was rising or transiting, just after evening twilight, and then use the risings or transits of the next twelve decans to mark the night hours. This could have been done, but again there is no evidence for it.

THE RAMESSIDE STAR CLOCKS

Following the discovery of the water clock, alluded to above, a new type of star clock was invented no later than the middle of the second millennium B.C.[15] Three centuries later the ceilings of three royal Ramesside tombs were adorned with the twenty-four tables (two to a month) of a new clock (Plate 2). The basis for telling the hours remained transits, but these were not always of the meridian. Short distances before or after the meridian were indicated by such phrases as "on the right eye" or "on the left eye," "on the right ear" or "on the left ear," "on the right shoulder" or "on the left shoulder," while the meridian itself was known as "opposite the heart." "Right" and "left" are from the viewpoint of a seated observer facing a seated man, the target figure, to his exact south. In Plate 2 the target figure is shown below a chart of seven inner vertical lines crossed by thirteen lines of text and with thirteen stars entered on it. The first line gives the position of the star that begins the night; and the following twelve lines detail the stars that end the twelve hours, with their positions. In the example shown, six stars are "opposite the heart" and so on the meridian, the central vertical. The first of these begins the night. Two stars are "on the right eye" and five "on the right shoulder," although three of these are to one side of the vertical.

Such relatively fine distinctions in transit clearly imply that a water clock was utilized to construct the tables. There are, as well, other differences from the decanal star clocks based on risings. In the latter the list of stars remains constant from one decade to the next, as do the lengths of the hours. In the new clock, stars change position from table to table and at times drop out entirely, replaced by new stars, with the lengths of hours varying accordingly. Very few decanal stars appear in the new clocks, but these include Sirius-Sothis and Orion, so it is a safe conclusion that the new set of hour stars is in the southern sky and in a band that parallels and probably slightly overlaps the decanal belt. This would be a convenient arrangement for an observer sitting due north of and facing the target figure, perhaps on a temple roof, and using a plumb bob (in Egyptian, a *merkhet*) to determine the hours.

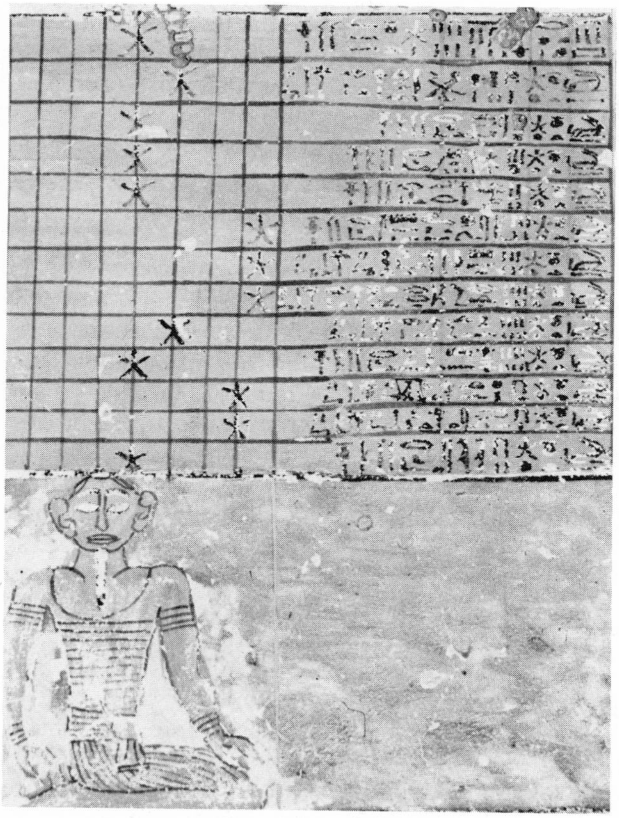

PLATE 2. Hour table, tomb of Ramses VII.

ASTRONOMICAL MONUMENTS

In our discussion of the development of the hours, we have become acquainted with some of the sources for our knowledge of the astronomy of the ancient Egyptians. The earliest monuments we have are the star clocks on the insides of coffin lids. These number twelve, ranging from the Ninth to the Twelfth dynasties (between *ca.* 2150 and *ca.* 1800 B.C.). The Cosmology of Seti I and Ramses IV, together with the Ramesside star clocks (*ca.* 1300–1100 B.C.), introduced us to astronomical monuments on the ceilings of royal tombs. In all, more than eighty monuments are now known, apart from the early coffin lids, that are concerned to some degree with astronomy. Most of these are on ceilings of royal or private tombs or of temples. The insides of some coffin lids from the Greco-Roman period bear zodiacs. Very few of the monuments are nonfunerary; these are several water clocks that have astronomical depictions on the outside.

Many of the monuments incorporate lists of decans, although they are not arranged so as to serve as star clocks. When all these lists are compared with one another, the variants in them permit a grouping into five families, three of rising decans, one of decans in transit, and one that cannot be assigned with certainty to either. Named from the first example of each, we have the Senmut, Seti I A, and Seti I C families of rising decans. The Seti I B family, made up of the decans associated with the cosmologies we have already discussed, is one of transiting decans. Tanis names the family of uncertain application. It is late in origin, and seems to be of artificial construction and unusable as a true star clock.

Many other elements can enter into the composition of an astronomical monument. The decans are given individual figures, and various deities are associated with them. The planets are shown in characteristic images. Various constellations, particularly a group called "northern," appear. In the Greco-Roman period zodiacs are incorporated. A monument also may have cosmic deities, such as those of the sky, air, and earth; calendar elements; figures of the hours of day and night; the goddesses of the cardinal points; and mythological figures of the four winds, although these are all less frequent.

As an illustration of an astronomical ceiling we may analyze the earliest one presently known. It is from the unfinished tomb of Senmut, an official of Queen Hatshepsut (*ca.* 1473 B.C.). Plate 3 is oriented so that the top is south and the bottom north. Correctly, then, since we have seen that the decanal belt is south of the ecliptic, the list of rising decans is above. It begins on the right, or east, and the first six columns list eleven decans, after which a horizontal line separates single decans from their associated deities below the line. What this means is that the decanal arrangement of the Senmut ceiling is based on the first column and the twelfth hour line of a diagonal star clock (see Figure 1). While some decans are reversed and there is confusion in the Orion decans, the model is perfectly clear. The last decan is that of Sirius-Sothis, in the first large column, portrayed as the goddess Isis in a bark. Her position as decan 36 means that the star clock that served as the model for this arrangement must be dated four centuries earlier, at the end of the Twelfth Dynasty—another example of funerary antiquarianism. Before Isis-Sothis is the figure of Orion, also in a bark, with body facing the goddess but with head turned away. Among the other figures shown before Orion are those of a bark and a sheep (not to be confused with the zodiacal Aries), both of which have several decans associated with them.

After the thirty-sixth decan of a star clock we expect to find the twelve "triangle" decans for the epagomenal days. Six of these do appear after Isis-Sothis (one is shown as two turtles), but the others have been replaced by four of the five planets. Jupiter and Saturn, both in the image of the sky god Horus in barks, precede the decans. Mercury and Venus follow them, with Venus shown as a heron in the last column. Mars is omitted, whether intentionally, because it was invisible in the sky when this arrangement was first drawn, or through error. On other, later ceilings Mars follows Saturn and is usually depicted as a form of Horus, like Saturn and Jupiter, standing in a bark.

Below a central section of five lines of texts, the largest of which lists the titles of Senmut, is a very interesting combination of astronomical and calendrical elements. Roughly in the center is the group of constellations called "northern" because they are always found in that part of an astronomical monument that exhibits orientation. These divide twelve circles with twenty-four segments each (that presumably represent the twenty-four hours) into three groups of four. From the names above the circles we know that they represent the months of the lunar calendar, beginning with the first month of *Tekhy* in the upper right, running across, and returning to the twelfth month of *Wep-renpet*, below *Tekhy*. The three groups are thus the three seasons. Flanking the northern constellations and facing them are two rows of deities, whom we know from other sources to be those of days of the lunar month,[16] except for Isis, who heads the right-hand file. It became conventional on later ceilings to retain the lunar day deities with the northern constellations while omitting the lunar month circles, thus establishing an apparent association of deities and constellations that had no basis in fact.

THE NORTHERN CONSTELLATIONS

On the Senmut ceiling, at the top of the constellations and, by comparison with other monuments,

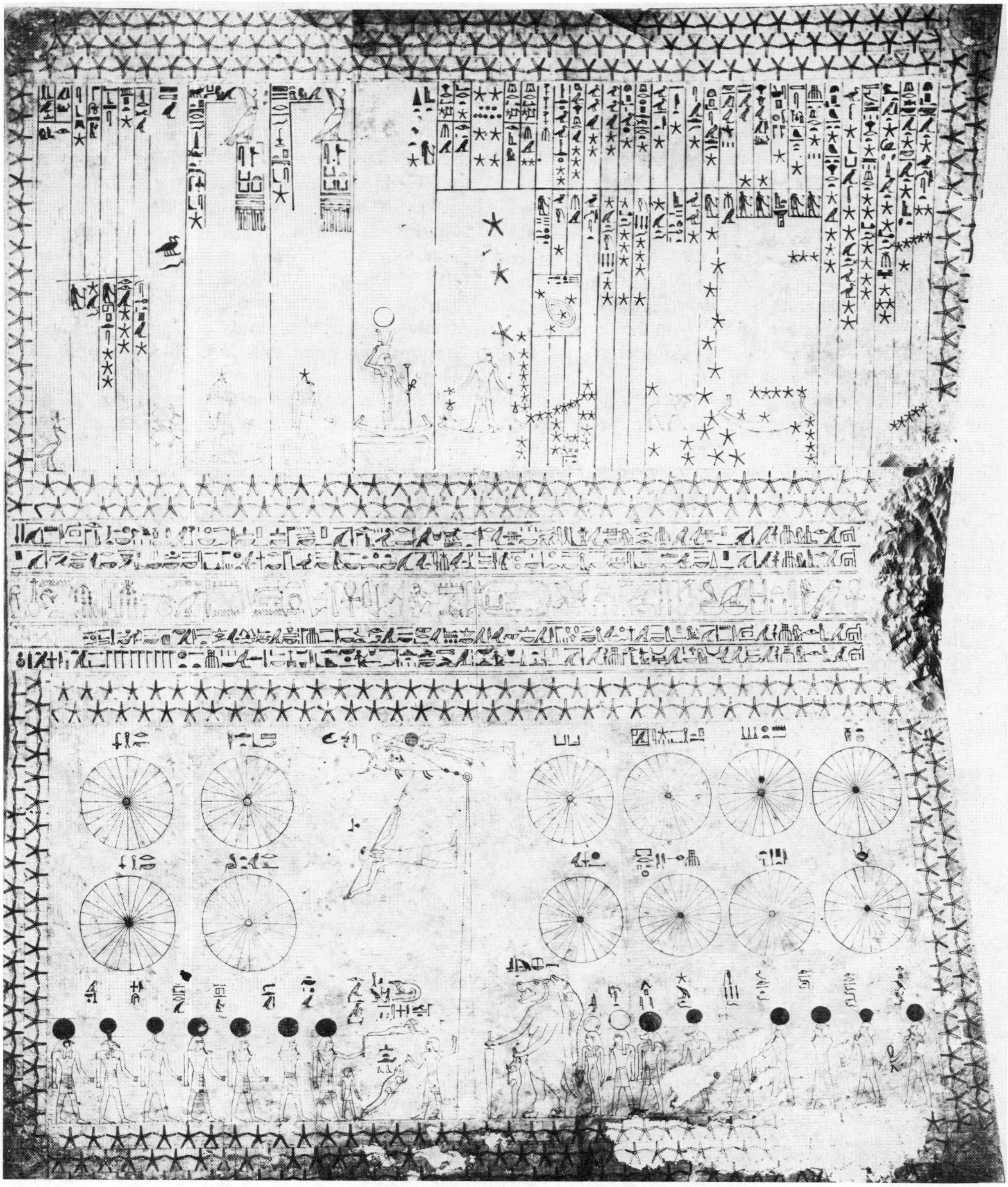

PLATE 3. Ceiling, tomb of Senmut.

rather out of place through the necessity of separating the month circles into seasons, is a group including the figure of a bull with tiny legs on an oval body. A ceiling in the tomb of Seti I (Plate 4) has a much more realistic bull. Both are portrayals of the constellation we call the Big Dipper—the one certain identification that can be made in all the northern group. Called *Meskhetiu* by the Egyptians, it may be depicted merely as the foreleg of a bull, such as on the coffin lid of Idy (Plate 1) and on certain later monuments. To the bull's right stands a female hippopotamus with a crocodile on her back and her hands on a mooring post (Seti I) or on a mooring post and a smaller crocodile (Senmut). This is the other constellation always shown with the Foreleg or Bull, from Senmut's time on. The other elements may or may not be present, but these two were essential. Their relationship is mentioned in a number of mythological texts, such as this from the Book of Day and Night (time of Ramses VI): "As to this Foreleg of Seth, it is in the northern sky, tied to two mooring-posts of flint by a chain of gold. It is entrusted to Isis as a Hippopotamus guarding it."[17] Isis, as the wife of Osiris, thus protects him from his inimical brother Seth. The chain is visible on Seti I's ceiling but not on that of Senmut.

Like the decan lists, the monumental depictions of the northern constellations fall into groups with variations in details although the main elements are commonly present. Thus on Senmut a falcon-headed god An is apparently spearing the Bull, while on Seti I he supports it. The constellation of the goddess Serket is above and behind the Bull on Senmut, while on Seti I she is to its left. On Seti I the constellation Lion has a Falcon over its head that is replaced in Senmut by a small Crocodile (*Sak*). In both an unnamed Man is apparently spearing a second crocodile constellation under Lion, although the spear is lacking. Lion's name in full is "Divine Lion, Who Is Between Them"—that is, between the two Crocodiles. On Seti I another unnamed Man holds the chains between the Hippopotamus and the Bull, but on Senmut he is not present.

Moreover, on Senmut there are traces of an earlier and different arrangement of Lion and Crocodile, with the spearing Man omitted. It is just such variations as these, in inclusion or omission of constellations or in their positions relative to one another, that make any attempt at identification of the constellations other than the Big Dipper a hazardous one. We can be reasonably sure that they all belong in the northern sky in opposition to the decanal belt, but we cannot be sure that they were all circumpolar and never-setting (and thus called in Egypt the "indestructible stars"), although no doubt some of them were.

PLATE 4. Northern constellations, tomb of Seti I.

THE PLANETS

All the planets were surely recognized and named long before we have textual evidence to support such a statement. Anyone using a star clock could not fail to notice the five bright stars changing their positions among the fixed ones. But it is not until the ceiling of Senmut that we have the planets depicted, with only Mars absent. Mars is present with the others, less than two centuries later, on the tomb ceilings of Seti I and Ramses II; and they are all frequently found on later ceilings. The usual order on the earlier monuments is Jupiter, Saturn, Mars, Mercury, and Venus—but this order, rather surprisingly, has no demonstrable factual basis. When the planets are separated, as by the triangle decans, the first three, the outer planets, are usually grouped together, with the second group made up of the inner planets, Mercury and Venus. Jupiter, Saturn, and Mars were all considered to be aspects of Horus, the falcon god of the sky. Jupiter was "Horus Who Bounds the Two Lands" or "Horus Who Illuminates the Two Lands" at first, and later "Horus Who Illuminates the Land" or "Horus Who Opens Mystery." Saturn was "Horus Bull of the Sky" or "Horus the Bull," with no later change. Mars was "Horus of the Horizon" or "Horus the Red." When depicted, the planets were usually shown as falcon-headed gods with human bodies, a star above the head, standing in barks. On the latest monuments Jupiter may be human-headed or be only a falcon; Saturn may be bull-headed with human or falcon body; and Mars may be human-headed or a falcon with a serpent tail. Still other variations were possible.[18]

Mercury had the name *Sebeg(u)*, but its meaning is not known. The god associated with Mercury was Seth, an enemy of Horus and his father, Osiris. Shown with human body and characteristic animal head, Seth's figure was frequently mutilated or replaced on early monuments by reason of his hostile nature. Venus was early pictured as a heron, with the name "Crosser" or "Star Which Crosses." Later she was known as "Morning Star" and was frequently given human form, either falcon-headed or, occasionally, two-headed or two-faced.

The names of the planets and the bits of text with them tell us little about the outer planets except that "Horus the Red" can only be Mars. We are better off with the inner ones. Venus' name "Crosser," together with two faces or two heads, should mean that it was recognized as both morning and evening star as early as the Senmut ceiling, where it bears the name although it is shown as a heron. The other inner planet, Mercury, was certainly known to be both evening and morning star at an early date, since we have a text from the tomb of Ramses VI (1148–1138 B.C.) stating that Mercury is "Seth in the evening twilight, a god in the morning twilight." The suggestion is that as evening star and Seth, Mercury was malevolent, while as morning star he may have been quite different. It seems probable, then, that by the middle of the second millennium B.C. the Egyptians were aware that the morning and evening stars were but one star, crossing the sun, and that this was either one of the two inner planets.

Identification of the planets from their names only was speculated about by early scholars; but it was not until the Stobart Tablets, written in Demotic of the Roman period, were successfully read by Heinrich Brugsch as giving the dates of entry of the planets into the zodiacal signs for years in the reigns of Vespasian, Trajan, and Hadrian that certain identification was made.[19] The identifications have all been amply confirmed by other late texts, in particular the horoscopes of the Greco-Roman period.

With the introduction of the zodiac to Egypt, the planets are frequently found on monuments in their special astrological signs, such as their exaltations (among others, the Esna temple ceiling and the Dendera B ceiling). On the Dendera E ceiling they are found in their day and night houses.

THE DECANS IN THE ZODIAC

Zodiacs on astronomical monuments do not occur in Egypt until the Greek period. The first one we know of, now entirely destroyed but fortunately copied by the French expedition under Napoleon, was a ceiling in the temple of Esna (Ptolemy III–V, 246–180 B.C.). The latest, from the Roman period, is from a private tomb at al-Salāmūni (about A.D. 150). In all there are twenty-five known and published zodiacs, all but two either on ceilings in temples or tombs or on the inside of coffin lids. Unpublished, and possibly as late as the fourth century A.D., are three zodiacs in tombs found recently by Ahmed Fakhry at Qaret al-Muzawwaqa in the oasis of al-Dakhla.

The concept of the zodiac was not Greek in origin but Babylonian, as is proved by the forms of the signs. For example, the images of the goat-fish (Capricorn) and the two-headed archer on a winged, scorpion-tailed horse (Sagittarius) are found on much earlier Babylonian boundary stones. Egypt enthusiastically received this new concept of ordering the sky and the path of the sun, as transmitted to it by the pervading Hellenistic culture of the time, and proceeded to incorporate into its depiction the traditional Egyptian elements of decans, planets, constellations, sun, and moon. It was the decans that were most affected. Since they were all in a band just south of the ecliptic, and it was the ecliptic that the zodiac divided into twelve signs, the names of three presumably adjacent decans were taken over by each sign and were used to divide that sign into three ten-degree subdivisions.

More than one decanal family was thus introduced into the zodiac. On one ceiling alone, that of the destroyed Esna temple, the decans of the Seti I B family were in a strip above the strip of zodiacal signs, while the decans of the Tanis family were in another strip just below. A comparison of the two

	SETI I B	HEPHAESTION		TANIS
Cancer	*spdt*	σωθις		*knm(t)*
	št(w)	σιτ		*ḫry (ḫpd) knm(t)*
	knm(t)	χνουμις		*ḥꜣt ḏꜣt*
Leo	*ḫry ḫpd knm(t)*	χαρχνουμις		*ḏꜣt*
	ḥꜣt ḏꜣt	ηπη		*pḥwy ḏꜣt*
	pḥwy ḏꜣt	φουπη		*tm(ꜣt)*
Virgo	*tm(ꜣt)*	τωμ		*wšꜣt(i)*
	wšꜣt(i) bkꜣt(i)	ουεοτεβκωτ		*bkꜣt(i)*
	ipsd	αφοσο	αφοσο	*ipsd*
Libra	*sbḫs*	σουχωε	σουχωε	*sbḫs*
	tpy-ʿ ḫnt	πτηχουτ	πτηχουτ	*tpy- ʿḫnt*
	ḫnt ḥr(t)	χονταρε		*ḥry-ib wiꜣ*
Scorpio	*ḫnt ḫr(t)*		στωχνηνε	*s(ꜣ)pt(i) ḫnwy*
	tms (n) ḫnt		σεσμε	*sšm(w)*
	spt(y) ḫnwy		σισιεμε	*sꜣ sšm(w)*
Sagittarius	*ḥry-ib wiꜣ*	ρηουω		*knm(w)*
	sšmw	σεσμε		*tpy-ʿ smd*
	knm(w)	κομμε		*pꜣ sbꜣ wʿty*
Capricorn	*tpy-ʿ smd*		σματ	*smd*
	smd		σρω	*srt*
	srt		ισρω	*sꜣ srt*
Aquarius	*sꜣ srt*		πτιαυ	*tpy-ʿ ꜣḫw(y)*
	ḫry ḫpd srt		αευ	*ꜣḫw(y)*
	tpy-ʿ ꜣḫw(y)		πτηβυου	*tpy-ʿ bꜣw(y)*
Pisces	*ꜣḫw(y)*		βιου	*bꜣw(y)*
	tpy-ʿ bꜣw(y)		χονταρε	*hnt(w) ḥr(w)*
	bꜣw(y)	πτιβιου		*ḫnt(w) ḫr(w)*
Aries	*ḫnt(w) ḥr(w)*	χονταρε		*ḳd*
	ḫnt(w) ḫrw	χονταχρε		*sꜣ ḳd*
	sꜣ ḳd	σικετ		*ḫꜣw*
Taurus	*ḫꜣw*	χωου		*ʿrt*
	rt	ερω		*rmn ḥry*
	rmn ḥry	ρομβρομαρε		*ṯsʿrk*
Gemini	*tsʿrk*	θοσολκ		*rmn ḫry*
	wʿrt	ουαρε		*wʿr(t)*
	tpy-ʿ spdt		φουορι	*pḥwy ḥry*

TABLE II. Lists of decans from Seti I B, Hephaestion, and Tanis (O. Neugebauer and R. A. Parker, *Egyptian Astronomical Texts*, III, 170–171).

decanal lists reveals that only twelve common decans are in the same position in both lists, sixteen are common to both lists but differ in position, and eight variants are found. This underlines the artificial character of the whole arrangement. It is obvious that such decans can hardly function as hour stars in clocks any longer and are merely lending their names to divisions of the zodiac.

This conclusion can be forcefully substantiated in yet another way. In Hellenistic astrology the Egyptian decanal names were rendered in Greek. So far as is now known, the first complete list of decans in the zodiac is that of Hephaestion of Thebes (fourth century A.D.). A comparison of Hephaestion's list with that of Seti I B or Tanis shows only partial agreement with either. It will be found very instructive, however, to set the three lists together, the Seti I list beginning as usual with *Sepdet* (Sothis) and Tanis with *Kenme(t)*. Hephaestion, in the center of Table 2, shows agreement with Seti I B in the left column of Greek names and with Tanis in the right column.

This tabular pairing shows twenty-four agreements between Hephaestion and Seti I B and fifteen between Hephaestion and Tanis, with but three in common. Such a result shows convincingly that the list compiled or taken over by Hephaestion was a quite arbitrary and eclectic one. The Tanis list, to be sure, is of suspicious origin; but the list of Seti I B was originally based on transits and must have been established through direct observation. Were there any real desire to have decanal names in the zodiac bear a direct relationship to the location of the decans in the decanal belt, we should certainly have had a much higher agreement between Seti I B and Hephaestion. But many of the decans are off by ten or twenty degrees and one, Sesme, is repeated thirty degrees away.

THE ZODIACS OF DENDERA

The most famous monument, the circular zodiac (Plate 5), now in the Louvre, was once part of the ceiling of the East Osiris Chapel on the roof of the temple of Hathor. To be dated to the years before 30 B.C., it appears to be the most successful of all the known astronomical monuments in Egypt in representing the heavens with some approach to accuracy. Not shown on the plate are the four goddesses of the cardinal points, who are correctly oriented, and four pairs of falcon-headed deities, all of whom support the circular sky. About the edge of the circle are the thirty-six figures of the decans of the Tanis family (no. 1 is *Kenmet*). In the center are the Hippopotamus (*A*) and the Foreleg (*C*), the sole representatives of the northern constellations. The pole, unmarked, must be between them. In a circle about it, but a circle that, like the ecliptic, is properly askew with respect to the pole, are the twelve zodiacal figures. Between the zodiac-ecliptic circle and the pole are depicted a number of constellations (*A – M*) that must all be considered north of the ecliptic. The planets are placed with the zodiacal signs in which they are considered to be most influential – in astrological terms, in exaltations. Venus is in Pisces, Mercury in Virgo, Mars in Capricorn, Saturn in Libra, and Jupiter in Cancer. The zodiacal belt touches the decans at the top of the plate, but below there is a crescent-shaped area between the two belts that is occupied by other constellations (*N – Y*) that must be regarded as south of the ecliptic. Some of them are surely in the decanal band, since Orion (*P*) and Sothis (*S*) are readily identifiable, the latter as a cow (with a star between the horns) recumbent in a bark. Two goddesses, Satis (*T*) and Anukis (*U*), may not represent constellations, since they commonly accompany Sirius-Sothis. The other and unidentifiable constellations south of the ecliptic may also be in the decanal belt or may be somewhat south of it.

Another ceiling in the Dendera temple, of the time of Tiberius (A.D. 14 – 37), and so about a century later than the circular zodiac, has many of the same figures as the latter; but this time they are arranged in long strips on the ceiling of the outer hypostyle hall (Dendera E). The constellations are located in particular signs; and this information, combined with the location north or south of the ecliptic in the circular zodiac (Dendera B), permits the drawing up of a catalog that can at least give a general idea of where a constellation may be sought. Briefly, north of the ecliptic are the following:[20]

- *A* (hippopotamus). Dendera E places *A* and *C* with the god An between Sagittarius and Capricorn. Dendera B has *A* above and between them and the pole.
- *B* (jackal on hoe). Dendera E, in Scorpio.
- *C* (foreleg). Dendera E (bull-headed leg), see *A*. Dendera B, between Gemini and the pole.
- *D* (small lion on *C*). Omitted by Dendera E.
- *E* (god). Dendera E, in Gemini. Dendera B, near Gemini.
- *F* (small seated god). Only in Dendera B, above Leo.

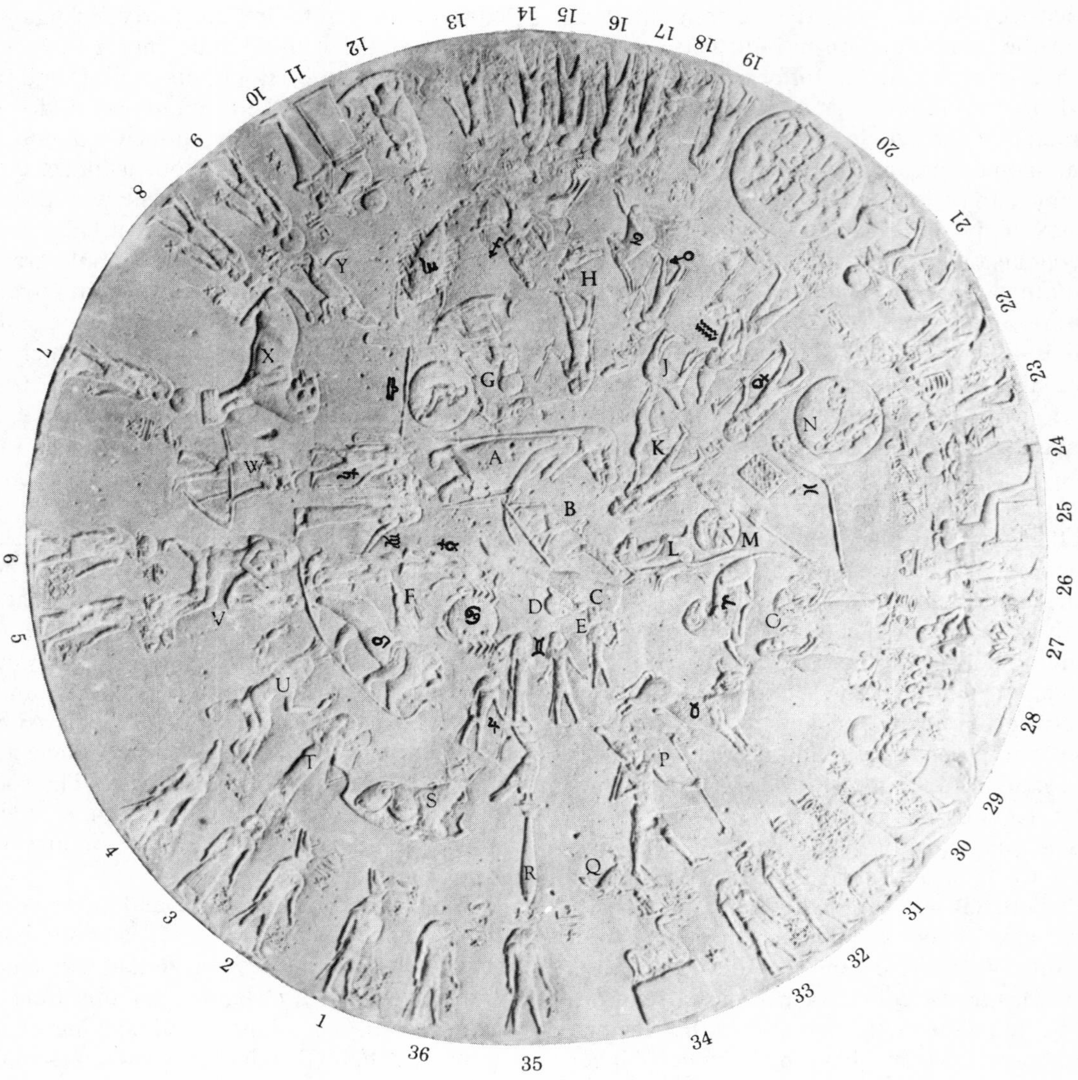

PLATE 5. Zodiac ceiling, Dendera.

G (group of two gods, one falcon-headed, and jackal). Only in Dendera B, near Libra and Scorpio.

H (group of god and goose). In Dendera B, near Scorpio and Sagittarius. Dendera E has them in Aquarius.

J (headless body). Near Aquarius in Dendera B and in Aquarius in Dendera E.

K (group of god and oryx). In Dendera B, near Aquarius and Pisces. In Aquarius between *H* and *J*, in Dendera E.

L (group of oryx and baboon with falcon on head). Nearest to Aries in Dendera B. In Taurus in Dendera E.

M (eye in disk). Only in Dendera B, between Pisces and Aries.

For the constellations south of the ecliptic we have the following:

N (god in *B*, goddess in *E*, and pig in disk). Near Pisces in Dendera B. In Pisces in Dendera E.

O (group of two goddesses, one lion-headed, in B; two gods, one lion-headed, in E). Near Aries and Taurus in Dendera B. In Taurus in Dendera E.

P (Orion). Near Taurus and Gemini in Dendera B. After Gemini and before Cancer in Dendera E.

Q (bird). Only in Dendera B, and near Gemini.
R (falcon on column). Under Gemini in Dendera B. After Orion in Dendera E.
S (Sirius-Sothis as cow in bark). Under Cancer in Dendera B. After *R* in Dendera E.
T (Satis). Goddess-companion of Sothis in both Dendera B and E.
U (Anukis). Goddess-companion of Sothis in both Dendera B and E.
V (seated woman holding child). Near Leo in Dendera B. In Leo in Dendera E.
W (bull-headed god, with hoe). Under Virgo in Dendera B. In Virgo in Dendera E.
X (lion with forefeet in water). Only in Dendera B, near Virgo and Libra.
Y (god with human upper body and hippopotamus lower body). In Dendera B near Libra and Scorpio. In Scorpio in Dendera E.

Dendera E has two constellations that do not appear in Dendera B (one in Libra and one in Leo), and these might be either north or south of the ecliptic. On both ceilings there are variants in the forms of the constellations; and, as has already been remarked, their locations are at best only approximate.

EGYPTIAN ASTROLOGY

After the conquest of Egypt by the Persians in 525 B.C. and during their control of it for over a century, there was a constant cultural interchange among all the countries of that empire. Egyptian priests are known to have been in Persia; indeed, we have the personal account of one Udjeharresnet, who was commanded to return to Egypt by Darius I (521–486 B.C.) and there reform the Houses of Life, the centers in the precincts of the temples where medical and religious books were written.[21] It may well have been through him that the first astrological literature reached Egypt from Babylonia. Although the two texts we have[22] are copies made in the Roman period (second or third century), the concordance of Text A between the Egyptian and the Babylonian lunar calendars is such that it must date from six or seven centuries earlier. Moreover, it has no mention of the zodiac, and this places it surely before the fourth century B.C. Text B may be assumed to be of the same antiquity.

Text A is exceedingly fragmentary, with perhaps only one-fifth of it remaining; but it has nevertheless been possible to establish a fairly comprehensive understanding of what it treated. It turns out to be an example of the earliest type of astrological literature, that of judicial astrology. The concern is with the fate of countries and their rulers (not with that of the individual, whose destiny was the focus of the later, horoscopic astrology), and the text is a compendium of predictions to be drawn from eclipses of the sun and moon. It was important to know in what month of the year (as opposed to what zodiacal sign) the eclipse took place, at what hour of the day or night, and where in the sky. From this information it was possible to tell what country would be affected and what that effect might be. The countries, besides Egypt, are Crete, Syria, Amor, and the land of the Hebrews, those to the north and east of Egypt and those with which Egypt was then more directly involved. It is evident that the text as we have it had been edited and adapted to reflect the special concerns of Egypt.

There were two systems employed to assign the months to countries. In System I, for the sun, IIII *Akhet* to III *Peret* were said to belong to a country of which the name is now missing, IIII *Peret* to III Shomu belonged to Hebrew (land), and IIII *Shomu* to III *Akhet* belonged to Egypt. It is the beginning with IIII *Akhet* that proves a Babylonian origin, since the text states (II, 18–20): "[Nisan (is) the lunar month] IIII *Akhet*; Iyyar [is] the lunar month [I *Peret*; Siwan is the lunar month II *Peret*; Tammuz] is the lunar month III *Peret*; the[se 4 months belong to . . .]." Nisan to Tammuz are the first four months of the Babylonian year, and it is that year that obviously controls the concordance. For the moon, in System I, the year was divided into four three-month groups, assigned in turn to Hebrew (land), Amor, Egypt, and Syria. Again the concordance began with Nisan being IIII *Akhet*.

In System II the assignments are the same for both sun and moon; but instead of the months being grouped, they were alternated in the order Crete, Amor, Egypt, and Syria, so that to Crete, for example, IIII *Akhet*, IIII *Peret*, and IIII *Shomu* were assigned. Nowhere in either system is there a mention of an intercalary month, although one would have occurred every two or three years.

The hours of the day, like System I for the sun, were divided into three groups of four, to which were assigned Egypt (1–4), Crete (5–8), and Amor (9–12). Again in agreement with System I for the moon, the hours of the night were divided into four groups, assigned to Egypt (1–3), Hebrew (land) (4–6), Amor (7–9), and Syria (10–12).

For solar eclipses the sky was divided into three

assigned areas, the northern to Hebrew (land), the middle probably to Crete, and the southern to Egypt. In the moon section of Text A there is an unfortunately broken passage that speaks of "the four places of the sky." It may be that for lunar eclipses the sky, in analogy to the four-part divisions of the months and the hours, was also divided into four; but this remains uncertain.

From such assignments it would appear that the effect of an eclipse could vary considerably, being restricted to one country or spread over several. As an example, a solar eclipse in II *Akhet*, in hours one to four of the day and in the southern sky, presumably would concern only Egypt, while one in IIII *Peret*, in hours nine to twelve of the day and in the middle of the sky would involve Crete, Amor, and Hebrew (land) as well as Egypt.

The predictions that are preserved in Text A are only those for the months according to System II. Thus, for example, the most complete passage reads (II, 30–31): "If the sun be eclipsed in [I *Peret*], (since) the month belongs to the Amorite, it means: The chief [of the Amorite shall . . .] occur (in) the entire land." One example of a lunar eclipse reads (IV, 26–27): "If [the moon be eclipsed in II *Shomu*, (since) the month belongs to <Egypt>], it [means]: The chief of the land named shall be captured. The army shall fall to [battle]-weapo[ns]." What may have happened to the other countries involved, determined by the hours and the sky, is nowhere stated. Nor is there any statement about what may happen when the month assignments of Systems I and II conflict.

That such a treatise as Text A was a source of inspiration to Egyptian scholars from the Persian period on may be readily assumed from the second-century B.C. writings of Nechepso and Petosiris, who were the "ancient Egyptians" upon whom Hephaestion of Thebes drew for Chapter 21 of his Book I, in which he discusses eclipses and their omens. It would seem that throughout these later centuries the Egyptians observed eclipses and carefully noted what happened. Nechepso and Petosiris were still concerned with judicial astrology alone, but the sphere of interest was widened to include up to thirty-nine countries, although Egypt predominated. The zodiac was employed and the hour pattern of day and night was by threes, although not consistently. In all, we seem to have in their work a highly complex and developed system of predictions built upon the framework of Text A.

Text B is quite different. It is concerned with happenings in and about the disk of the full moon, such as stars or disks inside or outside, or changes in coloration. For each case a colored vignette illustrates the text and is accompanied by the specific predictions. These are still judicial, but concern Egypt almost exclusively. If the original text was Babylonian, as seems most likely, it has been thoroughly revised to suit Egypt. Or it may have been that only the idea of such a study of the moon came to Egypt—after which, as with Text A, data were accumulated through the centuries.

It is worth stating here that science as we have it had its origins in just such patient study and accumulation of data. We know that *post hoc, ergo propter hoc* is usually fallacious; but this was by no means so apparent to the peoples of the ancient world, and the hypothesis that celestial events foreshadow earthly events was to them a hypothesis that had been proved correct. Just so have some more recent hypotheses been taken as proved, although later discarded. The activity described was commented upon by Herodotus after he had visited Egypt in 460 B.C.:

> I pass to other inventions of the Egyptians. They assign each month and each day to some god; they can tell what fortune and what end and what disposition a man shall have according to the day of his birth. This has given material to Greeks who deal in poetry. They have made themselves more omens than all other nations together; when an ominous thing happens they take note of the outcome and write it down; and if something of a like kind happen again they think it will have a like result.[23]

In passing it should be remarked that predictions from "the day of his birth" are not from the positions of the planets but, rather, according to calendars of lucky and unlucky days, one of which goes back as far as the New Kingdom in the middle of the second millennium B.C.[24]

About twenty vignettes with their predictions are preserved in Text B. For example, the upper half of col. IX has in the vignette a dark yellow disk with plain center and black disks to right and left. The text reads (1–13):

> Another. If you see the disk colored completely, its scent(?) red downward in it, there being one black disk on its right and another black disk on its left, [you are to] say about it: Enmity shall happen (in) [the entire land]—another version: the land of Egypt—and king shall approach king [. . ..] Great fighting shall happen (in) Egypt. Barley and emmer ⟨shall⟩ be plentiful and every harvest likewise with respect to every plowing and every field (in) the entire land. Good things and satisfaction shall occur everywhere, so that they quarrel, they drink . . . and they eat the knife.

Judicial astrology continued in Egypt well into the Roman period, as is witnessed not only by the texts just discussed but also by another of the relatively few astrological texts written in the Egyptian language, *Papyrus Cairo* 31222.[25] This text too is badly preserved, but its structure is clear; it has to do with predictions associated with the heliacal rising of Sirius-Sothis, when the moon and the various planets are in certain zodiacal signs. Thus, in lines 5–8:

> If it (Sothis) rises when Jupiter is in Sagittarius: The king of Egypt will rule over his country. An enemy will be [his and] he will escape from them again. Many men will rebel against the king. An inundation which is proper is that which comes to Egypt. Seed (and) grain(?) will be high as to price (in) money, which is . . . . The burial of a god will occur in Egypt. . . . [will come] up to Egypt and they will go away again.

We have already remarked that horoscopic or personal astrology developed later than judicial astrology, and the suggestion has been made that the original impetus for it came in Babylonia as a further development from the older celestial omens.[26] In fact, the first horoscope we now have is dated to 410 B.C. and is Babylonian.[27] Horoscopic astrology probably came to Egypt with the zodiac around the turn from the fourth to the third century B.C. as a developing Hellenistic "science" to which Egypt made little or no contribution that can be specifically identified, except for the decanal divisions of the zodiacal signs.

The earliest Egyptian-language horoscope we now have is on an ostracon and is datable to 39 B.C.[28] The next oldest horoscopes are from 10 and 4 B.C. and are in Greek, although written in Egypt on papyrus.[29] All others, both Egyptian[30] and Greek, fall into the Roman period, with the vast majority of them in Greek. Less than a dozen horoscopes in Egyptian are presently known. The latest of these are the one of Heter, on the inside of his coffin lid (A.D. 93 but written A.D. 125), and the horoscopes of two brothers (*A*, Ib-pmeny and *B*, Pa-mehit), on the ceiling of their tomb at Athribis (A.D. 141 and 148).[31]

The last two are pictorial rather than written out, and one of them (*A*) may serve to illustrate the zodiac of the time and the individualized figures of the planets (Plate 6). The upper row of the zodiac begins with Scorpio on the left and runs to Aries, with the lower row beginning with Taurus on the right and running to Libra. Between and below the rows are the planets. Jupiter is a falcon with outstretched wings and the hieroglyphic sign (horns) for "open" on its head, in Aquarius. Saturn is a bull-headed falcon with outstretched wings, in Gemini. Mars is a falcon with outstretched wings, three serpents for a head, and a serpent's tail, in Leo. Mercury is a falcon with the head of Seth and a serpent's tail, in Capricorn with the sun. Venus is a standing god, two-faced (human and lion?), in Pisces. The moon, shown both crescent and full, is in Sagittarius. Below the zodiac are Sirius-Sothis, as a cow, and Orion in barks, as well as the *ba* or "soul" birds of the two brothers and the barks of the sun and the moon. About the zodiac are mythological figures and texts.

All these Egyptian horoscopes are concerned primarily with the date of birth and the planetary positions and other data of that day. There are few or no predictions given. These no doubt had to be derived from the abundant literature of the Hellenistic period. In this literature there were a few astronomical treatises written in Egyptian Demotic. Two of these, *Papyrus Carlsberg* 1 and *Papyrus Carlsberg* 9, were surely native in origin, since the first of them is a commentary on the Cosmology of Seti I and Ramses IV and the second provides the twenty-five-year lunar cycle. All the others might as well have been written in Greek as Egyptian, since they are purely Hellenistic in content.[32] The more important are *Papyrus Berlin* 8279 and the Stobart Tablets, which are tables for some of the years between 16 B.C. and A.D. 133, giving the dates of entry of the planets and the moon into the zodiacal signs and thus useful for the casting of horoscopes.

SUMMARY

An appropriate summation of the position of Egypt in astronomy, astrology, and calendrical reckoning may be found in the words that an astronomer of the third century B.C., Harkhebi by name, had inscribed on his statue. He describes himself thus:

> Hereditary prince and count, sole companion, wise in the sacred writings, who observes everything observable in heaven and earth, clear-eyed in observing the stars, among which there is no erring; who announces rising and setting at their times, with the gods who foretell the future, for which he purified himself in their days when Akh [decan] rose heliacally beside Benu [Venus] from earth and he contented the lands with his utterances; who observes the culmination of every star in the sky, who knows the he-

PLATE 6. Zodiac A, Athribis (W. M. F. Petrie, *Athribis*, pl. XXXVII).

liacal risings of every . . . in a good year, and who foretells the heliacal rising of Sothis at the beginning of the year. He observes her [Sothis] on the day of her first festival, knowledgeable in her course at the times of designating therein, observing what she does daily, all she has foretold is in his charge; knowing the northing and southing of the sun, announcing all its wonders [omina?] and appointing for them a time[?], he declares when they have occurred, coming at their times; who divides the hours for the two times [day and night] without going into error at night . . .; knowledgeable in everything which is seen in the sky, for which he has waited, skilled with respect to their conjunction[s] and their regular movement[s]; who does not disclose [anything] at all concerning his report after judgment, discreet with all he has seen.[33]

Harkhebi's text continues with a recital of his skill as a charmer of snakes and scorpions, but with this we need not concern ourselves. We see that as an astronomer he was involved with the regulation of calendar years through his interest in Sothis, with the measurement of the hours of day and night, with the movements of the stars, the sun, and the planets, with predictions to be derived from any sort of celestial omens, and with at least the beginnings of horoscopic astrology in the exaltation of Venus.

Egypt's greatest achievement was its civil calendar. This, with the observations of the decanal stars, led to our twenty-four-hour day. Otherwise Egyptian astronomy remained at a very elementary level. In the Hellenistic world Egypt was certainly an enthusiastic student of Greek and Babylonian astronomy and astrology, and no doubt there were peculiarly Egyptian contributions to this pervasive learning besides the decans in the zodiac; but at least for the present these remain unknown to us.

NOTES

1. R. A. Parker, *The Calendars of Ancient Egypt*, chs. 1, 3.
2. O. Neugebauer, *The Exact Sciences in Antiquity*, 81.
3. Parker, *op. cit.*, excursus C.
4. R. A. Parker, "Sothic Dates and Calendar 'Adjustment.'" On 239 B.C. as the correct date for the decree, see R. A. Parker's forthcoming article in *Studies in Honor of George R. Hughes*.
5. Parker, *The Calendars of Ancient Egypt*, secs. 49–107.
6. *Ibid.*, secs. 49–119; O. Neugebauer and R. A. Parker, *Egyptian Astronomical Texts*, III, 220–225.
7. A. T. Samuel, *Ptolemaic Chronology*, 54–61.
8. Neugebauer and Parker, *op. cit.*, I, ch. 2.
9. *Ibid.*, 100.
10. *Ibid.*, 116–128.
11. L. Borchardt, *Altägyptische Zeitmessung*, 60–63.
12. Neugebauer and Parker, *op. cit.*, I, 119–120.
13. *Ibid.*, ch. 2.
14. *Ibid.*, 58.
15. *Ibid.*, II.
16. Parker, *The Calendars of Ancient Egypt*, sec. 222.
17. Neugebauer and Parker, *op. cit.*, III, 190.
18. *Ibid.*, 177–181.
19. H. Brugsch, *Nouvelles recherches sur la division de l'année des anciens Égyptiens* . . ., 19–61; Neugebauer and Parker, *op. cit.*, III, 175.
20. Neugebauer and Parker, *op. cit.*, III, 200–202.
21. A. H. Gardiner, "The House of Life."
22. R. A. Parker, *A Vienna Demotic Papyrus on Eclipse- and Lunar-Omina*.
23. A. D. Godley, *Herodotus*, bk. II, 82.
24. Neugebauer, *op. cit.*, 188.
25. G. R. Hughes, "A Demotic Astrological Text."
26. Neugebauer, *op. cit.*, 171.
27. A. Sachs, "Babylonian Horoscopes."
28. O. Neugebauer and R. A. Parker, "Two Demotic Horoscopes."
29. O. Neugebauer and H. B. Van Hoesen, *Greek Horoscopes*, 16–17.
30. O. Neugebauer, "Demotic Horoscopes."
31. Neugebauer and Parker, *Egyptian Astronomical Texts*, III, 93–98.
32. *Ibid.*, 217–255.
33. *Ibid.*, 214–215.

BIBLIOGRAPHY

See L. Borchardt, *Altägyptische Zeitmessung* (Berlin, 1920); H. Brugsch, *Nouvelles recherches sur la division de l'année des anciens Égyptiens, suivies d'un mémoire sur des observations planétaires consignées dans quatre tablettes égyptiennes en écriture démotique* (Berlin, 1856); A. H. Gardiner, "The House of Life," in *Journal of Egyptian Archaeology*, **24** (1938), 157–179; A. D. Godley, *Herodotus*, in Loeb Classical Library (London, 1931); G. R. Hughes, "A Demotic Astrological Text," in *Journal of Near Eastern Studies*, **10** (1951), 256–264; O. Neugebauer, "Demotic Horoscopes," in *Journal of the American Oriental Society*, **63** (1943), 115–126; and *The Exact Sciences in Antiquity* (Providence, R.I., 1957); O. Neugebauer and R. A. Parker, *Egyptian Astronomical Texts*, 3 vols. (Providence, R.I., 1960–1969): I, *The Early Decans*; II, *The Ramesside Star Clocks* (1964); III, *Decans, Planets, Constellations and Zodiacs*; and "Two Demotic Horoscopes," in *Journal of Egyptian Archaeology*, **54** (1968), 231–235; O. Neugebauer and H. B. Van Hoesen, *Greek Horoscopes* (Philadelphia, 1959); R. A. Parker, *The Calendars of Ancient Egypt* (Chicago, 1950); "Sothic Dates and Calendar 'Adjustment,'" in *Revue d'égyptologie*, **9** (1952), 101–108; and *A Vienna Demotic Papyrus on Eclipse- and Lunar-Omina* (Providence, R.I., 1959); W. M. F. Petrie, *Athribis* (London, 1908); A. Sachs, "Babylonian Horoscopes," in *Journal of Cuneiform Studies*, **6** (1952), 49–75; and A. T. Samuel, *Ptolemaic Chronology* (Munich, 1962).

JAPANESE SCIENTIFIC THOUGHT

Shigeru Nakayama

Professional Views of Nature.

In seventeenth-century Europe the goals and approaches of modern science were established by the scientific revolution. The professionalization of science in the nineteenth century, sometimes called the second scientific revolution, was no less important. When science became a full-time vocational activity, even the perception of nature was reorganized.

The second scientific revolution has not attracted the interest of many intellectual historians because it has seemed to be a revolution only of the social system of science. How intellectual its motivations were is a moot point; it has important intellectual implications. As the disciplines separated out of the ancient unity of science, each professional learned to view nature from the standpoint of his own field. The physicist's nature overlapped little with that of the botanist; the same could be said of the mechanical engineer and the professional philosopher (who in the nineteenth century often, especially in Germany, thought of himself as a *Wissenschaftler*). Although the assumption survived that one nature underlay the work of all scientists, there was no longer a consensus among professionals as to what it was and how it was articulated. The unity of nature was abandoned to the layman, but the technical perspectives of scientists moved so quickly and divergently that there remained no standpoint from which an overview was possible. The fragmented conceptions of and assumptions about nature, centered in academic specialties and heavily colored by their prejudices, will be referred to here as professional views of nature.[1] People within the specialties tend to see them as universal. The metallurgist or biochemist, when he sets out his opinion on broad questions of natural philosophy, often is unable to transcend the ideology of his own scientific community.

The views of nature imported into Japan in the Meiji modernization period (late nineteenth century) were of the kind described above. When the theory of organic evolution was introduced, the fact that Japan was not a Christian country was largely beside the point in determining its reception. By and large the Japanese sedulously assimilated views that had become international in character. The confrontations of values that one might expect did not occur. Furthermore, since science was professionalized by fiat of the central government rather than by the piecemeal effort of scientists, there was no natural continuity between traditional Japanese outlooks on nature and the new disciplinary views. The new Japanese scientist could not draw directly upon his native heritage in establishing a career; all that mattered was to conform to the conventional view from abroad.

In order to find views of nature that are in some sense genuinely Japanese, it will be necessary to examine the state of affairs before the Meiji reforms cut off the confrontation of Eastern and Western science. This is not to say that there was a single characteristic Japanese view of nature that can be reconstructed by reading the works of individual thinkers and examining anecdotes about them. The view of nature varied from one period to the next and depended greatly upon social background. It is, in fact, worth pursuing the hypothesis that even in traditional Japan, views of nature were rooted in and bound to professions—although the definition of a profession and the character of membership were very different from those of modern times.

During the Tokugawa shogunate (1603–1867) class hierarchy was tightly maintained, with the hereditary warrior class, the samurai (about 5 percent of the population), at the top of a rigid structure of farmers, artisans, and merchants. The major professions, independent of the four classes, were those of the Confucian scholar, the physician, and the Buddhist priest. Vocation was hereditary in feudal Japan, and even professionals were bound by their inherited callings to a partial view of nature. Given the lack of social mobility, collective, static views of reality are more prominent than the individualistic activity that certainly existed at the same time.

In seeking to identify views of nature, we shall

pay particular attention to the prefaces and postscripts of scientific writings. The texts generally were concerned with stylized technical subjects, and there was no place for direct and outspoken expression of the author's assumptions. In the front matter and end matter, on the other hand, basic philosophical matters often were argued. Frequently, these discussions of fundamentals were merely ornamental, adapted from Chinese arguments using the yin-yang principle and the Five Phases doctrine. Nevertheless, a comparison of Chinese and Japanese prefaces reveals differences in views of nature.

Also useful are prefaces contributed by writers other than the authors. Such prefaces, which tend to be complimentary rather than critical, provide excellent sources for the criteria of evaluation in each profession. Even today these vary; good philosophers are profound, scholars are erudite, and mathematicians bring elegance to their proofs. In old Japan, astronomers, mathematicians, medical doctors, natural historians, linguists, and Confucian scholars differed in their excellences. The differences, I argue, reflect different views of nature.

Astronomy.

Pseudoscience and Science. Technical professions began in Japan with the immigration of Korean and Chinese experts in the sixth to eighth centuries. As their Chinese view of nature began to take root in Japan, it was institutionalized on continental models in the College of Confucian Studies and the Board of Yin-yang (divination) Art. The Yin-yang Board had three departments: observational astrology, calendar making (mathematical astronomy), and yin-yang divination. In principle it was a miniature reproduction of the Astronomical Bureau at the Chinese court. A closer look discloses significant remodeling to meet local requirements. In China the Divination Bureau was administratively separate from, and much inferior in status to, the organizations that computed the ephemerides and observed celestial phenomena for astrological purposes. In Japan all of these activities were subsumed under the single Yin-yang Board, the name of which indicates a clear priority for divination.[2]

The Japanese yin-yang art was a complex of magical divination techniques. These techniques had little in common with the portent astrology practiced in the Chinese and Japanese courts, which was based on a belief that the natural and political orders influenced each other in such a way that changes in the former could be taken as warnings about inadequacies in the latter. Throughout the sphere of Chinese culture, calendar making was the paradigmatic exact science, used for computing solar, lunar, and planetary positions, forecasting eclipses, and composing a complete lunisolar ephemerides.[3]

Although both the yin-yang art and portent astrology were ways of forecasting changes in human affairs, the latter depended upon unpredictable omens, such as comets and irregular eclipses, as indexes of mundane crises. The yin-yang art, because it is less passive, was more important in the everyday life of the court—it determined the dates of court ceremonies, fixed propitious directions in which to begin journeys, and so on.

Much of this divination (in particular, the kind called hemerology) was based on cyclic notations of the year, month, and day, and therefore was an outgrowth of calendar making. In general the goals of mathematical astronomy are universal; local differences in the motions of the sun and moon are trivial. Given a Chinese manual from which to determine basic parameters and computational procedures, there was little that local talent or preference could add, at least to the routine work of making the yearly ephemerides. On the other hand, when unforeseen and ominous celestial phenomena were observed, they had to be interpreted without delay. No Chinese book could cover every contingency, and there was no time to consult with the astrologers of the Chinese court. In astrology the Japanese were thrown upon their own resources. In Japan, as elsewhere, the practical applications of imported knowledge were valued over basic theory.

Theoretical elements from Chinese natural philosophy played an important part in the interpretations of the yin-yang art and of astrology. There is an old belief, for instance, that the northeast (*Kimon,* the gate of the demon) was a channel of unlucky influences. The yin-yang art explained it as "the direction from which the god of Yang enters and the *ch'i* [energy] of Yin goes out." But this notion did not imply a strictly deterministic causal principle. It was merely a warning, so that countermeasures might be devised; otherwise the art would have no practical use. For instance, people avoided building houses at places where the configuration of the land opened toward the northeast. Nor was astrology thoroughly deterministic. Before a predicted solar eclipse appeared to cast doubt upon the emperor's virtue, he could defend himself by calling in Buddhist monks to exorcise it.

Although astrology and alchemy often are called

pseudosciences, they are neither misconceived sciences nor forerunners of modern science. Their goals are in no sense those of science, which may be defined as a pursuit of regularities that underlie natural processes. Astrology assumes that the future may be predicted and that defensive measures may avert undesirable futures; alchemy of the Chinese kind assumes that eternal life is possible. It is true that they assume certain regularities in nature, but these are means rather than goals.

Although the astronomers who computed the calendar were considered to be mere functionaries, the master of astrology and the practitioner of alchemy were regarded by laymen as possessing more than ordinary wisdom. As in early Europe, astrology was the higher art. Only in the Tokugawa period, when Confucian rationalism became intellectually predominant, could the official astronomers of the shogunate (the military government) attain high status by monopolizing the scientific aspects of calendar reform. Even so, traditional prerogatives kept them formally subordinate to the Abe family, hereditary masters of the yin-yang art.

Western Orientation Toward the Regular and Japanese Orientation Toward the Extraordinary. As we have just seen, it was not the regularities or eternal truths of mathematical astronomy, but the unforeseeable omens of the astrologers, that attracted attention in Japan. Only after exposure to Western influence did academic disciplines (*gakumon*; Chinese, *hsueh-wen*) come to be considered predominantly as parts of a converging search for eternal laws and for enduring realities.

We may juxtapose these two tendencies as orientation toward the regular and orientation toward the extraordinary. Exaggerating the difference for heuristic reasons, we may say that the former assumes that there are eternal and universal truths, and seeks to formulate underlying laws. The latter denies that such truths are attainable and therefore is not disposed to debate their existence. Those who relentlessly pursue regularity overlook the individual and the accidental. Those who value the extraordinary pay little attention to persons or events that conform without deviation to stereotyped patterns. If the former are unresponsive to change because of their preoccupation with order and system, the latter reject change reflexively because they lack set principles against which to measure it. Philosophers, especially natural philosophers, strive to discover underlying laws; while historians, including students of natural history, are attracted to discontinuities.

These two tendencies were rivals in the formative period of philosophy in both Europe and the Far East. Over time the emphases became divergent in the two cultural spheres. The main current of the Western academic tradition remained centered upon philosophical and logical inquiries in the Platonic and Aristotelian traditions. Eastern scholarship definitely inclined to history, with the *Shih-chi* (*ca.* 100 B.C.) of Ssu-ma Ch'ien as the prototype.[4]

Various conjectures have attempted to explain this bifurcation. Joseph Needham argues that the absence of an anthropomorphic lawgiver in their religion left the Chinese with little motivation to conceive laws of nature.[5] While intellectual centers shifted from state to state in the West, in China there was one polity and one elite culture; historical records were accumulated in a single language. The record of the past was conceived as that of a single people. The early appearance of a true bureaucracy encouraged the development of chronology and the systematic compilation and classification of administrative precedents. The early availability of paper and the currency of printing by A.D. 1000 made historical literature more subject to ideological control and thus more central to political and social concerns.[6]

Historical analogy rather than tightly constructed chains of logical reasoning became predominant in Eastern learning. This was true even in natural science, so that although there were considerable overlaps of subject matter in East and West, there were enormous differences of style. Abstraction and involved theoretical argument were by no means rare in Chinese science but, as already noted, they were vastly less important in Japan.

It is well known that in ancient China historical scholarship grew out of the recording of astrological portents to provide an empirical foundation for future prognostications. From the time of Ssu-ma Ch'ien, whose duties included astrology and history, such omens were an important component of the imperial chronicles. The positivistic view of history predicts that the horror of celestial omens, such as eclipses, should evaporate with the development of rationality. This was not the case in China, because such foreboding is a social rather than a psychological phenomenon. The astrologer-historians also were mathematical astronomers and strove to remove phenomena from the realm of the ominous by making them predictable. Once that happened, such events lost their significance in astrology. What could be predicted no longer had news value and no longer needed to be individually recorded in the annals.

The Platonic conviction that eternal patterns underlie the flux of nature is so central to the Western tradition that it might seem no science is possible without it. Nevertheless, although Chinese science assumed that regularities were there for the finding, they believed that the ultimate texture of reality was too subtle to be fully measured or comprehended by empirical investigation. The Japanese paid less attention to the general but showed a keener curiosity about the particular and the evanescent. In the early West, in keeping with the orientation toward regularity, phenomena that could not be explained by contemporary theory, such as comets and novae, were classified as anomalous and received scant attention. In the history-oriented East, extraordinary phenomena were keenly observed and carefully recorded. The incomparable mass of carefully dated astrological records that thus accumulated has proved invaluable to astronomers today.

In the classical Western tradition there is an urge to fit every phenomenon into a single box; those unassimilable to the pattern thus formed are rejected. In the Eastern tradition, in addition to the box in which all the regular pieces are assembled, there are a great number of others in which irregularities can be classified. Sorting exceptional phenomena into proper boxes was as satisfying for the Japanese as fitting together the puzzle in his single box was for the Platonist (the Chinese preference was intermediate). If science is defined, as Europeans conventionally do, as the pursuit of natural regularities, the Far Eastern tradition is bound to appear weak because it lacked analytical rigor. Judged less ethnocentrically, there is some merit in its relatively catholic and unprejudiced interest in natural phenomena.[7]

When a court astronomer in Peking or Kyoto found that the position of the moon was radically different from what he computed, one would expect him to consider his theory to be compromised. Such crises often occurred in China, but there was an alternative that can be seen with some frequency both there and, quite often, in Japan. The phenomenon could simply be labeled "irregular." It was not the astronomer's fault that the moon moved erratically.

This attitude may be seen in the career of Shibukawa Harumi, the first official astronomer to the Japanese shogun, in a form more distinct than that which existed among his Chinese contemporaries. In the preface to his early treatise *Shunju jutsureki* ("Discussions on the Calendar Reflected in the *Spring and Autumn Annals*," the oldest Chinese chronicle), he stated:

> Astronomers have rigidly maintained that when Confucius dated the events in his *Annals of the Spring and Autumn Era* he made conventional use of the current calendar with little care for its astronomical meaning, so that the dates are not very reliable. This error is due to their commitment to mathematical astronomy, so that they do not admit that extraordinary events happened in the sky. . . . Extraordinary phenomena do in fact take place in the heavens. We should therefore not doubt the authenticity of [Confucius'] sacred writing-brush.

In his own work on mathematical astronomy, Shibukawa remained thoroughly positivistic; but he also left a somewhat problematic astrological treatise, *Tenmon keito* ("Treasury of Astrology," 1698). Careful examination of these eight volumes of astrological formulas and interpretations of recorded portents discloses that a large portion was inattentively copied from a famous Chinese handbook, Huang Ting's *T'ien-wen ta-ch'eng kuan-k'uei chi-yao* ("Essentials of Astrology," 1653).[8] In this work he repeatedly expressed the skepticism toward astrological interpretations that might be expected of a practical astronomer, and would often write "We do not know the basis [of this interpretation]. . . . Is this unreliable?"

Shibukawa believed that a professional astronomer must be thoroughly competent in both major branches of celestial studies: portent astrology and calendrical science. His Jokyo calendrical reform (*ca.* 1684) provided a "box" for regularities. It was no less important to furnish the means by which astrological portents might be classified. He was convinced that the heavens could not be fully comprehended through mathematical regularity. The sky was a unity of such depth that the tools of no single discipline could plumb it. Although he found astrological interpretations to be often equivocal, the vast historical accumulation of omen records suggested that portent astrology had to be taken seriously. There must have been, he thought, justified passion and reason behind that tireless activity of the ancients.

Once admitting, as Shibukawa did, that regular motion was too limited an assumption, one could easily conceive such notions as that astronomical parameters could vary from century to century. In the official Chinese calendar of the thirteenth century and earlier, the discrepancy between ancient records and recent observations was explained by a secular variation in the length of the tropical year.[9] Shibukawa revived this variation in the Japanese calendar, and Asada Gōryū extended it[10] to other basic parameters to account for both Western and Eastern observations then available to him.

The variation terms used in Chinese and Japanese astronomy were too large to survive empirical testing and eventually were discarded. Wherever the Aristotelian notion of an unalterable universe was followed rigorously, irregular motions in the sky were inconceivable. Even the mathematically justified variation in the precession of the equinoxes, which had a brief acceptance in Europe, came from Islam. After Newton, variations in parameters were acceptable to the extent that they could be given mechanical explanations. In the West, the first systematic study of variations in basic parameters was delayed until Laplace, in the late eighteenth century. It is significant for the history of ideas that in China and Japan there was no reason to resist such variations.

In the Far East, not only were irregular motions of the celestial bodies admissible, but the algebraic approach to mathematical astronomy made it unnecessary to take a stand on the spatial relations of the sun, moon, and planets. The earliest astronomical schemes in China (first century B.C.) depended heavily upon a cyclical view of nature. These numerical models explained all of the calendrical phenomena by a vast construction of interlocking constant periods. The cycles of the sun and moon, the synodic periods of the planets, and cycles of recurrence for such phenomena as eclipses were tied together by larger cycles determined by their least common multiples. By the end of the Han period, however, improved observational precision and recording accuracy made it clear that the heavenly courses were too complex to fit such simple assumptions. Eventually the metaphysical commitment to cyclical recurrence was abandoned.[11] Periods of recurrence became no more than algebraic constants to be used alongside a great variety of other numerical devices. Neither celestial morphology nor cosmic ontology was of further professional interest to astronomers.

The Chinese tradition of astronomy, including its offshoots in Japan, Korea, and Vietnam, thus did not depend for its direction of development upon a dialectical relation between metaphysics and observation. Computational schemes neither challenged nor strengthened philosophers' conceptions of cosmic design.

Differences in Chinese and Japanese Views of Science. Ogiu Sorai (1666–1728), the most influential Japanese Confucian philosopher, had some interest in astronomy. He commented on the variation of astronomical parameters (*Gakusoku furoku*): "Sky and earth, sun and moon are living bodies. According to the Chinese calendrical technique, the length of the tropical year was greater in the past and will decrease in the future. As for me, I cannot comprehend events a million years ahead." Since the heavens were imbued with vital force, the length of the year could change freely, and constancy was not to be expected in the sky. Indeed, only a dead universe could be governed by law and regularity. The study of such a world would be of no interest to the natural philosopher. Since it was precisely the vital aspects of nature that interested Ogiu, he remained an agnostic in physical cosmology.

Indifference toward the search for regularities in nature prevailed in the School of Ancient Learning (*kogaku*), which emerged in the late seventeenth century led by Ogiu. Its anarchistic and dynamic cosmology was bathed in historicism. "All scholarship should finally converge in historical studies," said Ogiu.[12] Because he was a Confucian philosopher, "history" meant humanistic history. Whenever the philosophers of the Ancient Learning School looked at nature, they saw it in the light of social and ethical concerns.

This moralistic, anthropocentric, and often anthropomorphic view of nature was common among Confucian thinkers throughout the Far East. Many of them were unable to imagine that mathematical astronomy could make any greater contribution than to provide an accurate calendar.[13] Nevertheless, there were some notable differences between the views of Chinese and Japanese Confucians on the search for regularities in nature, especially with reference to calendrical science. Although these views were not imposed upon astronomers as corresponding ideas were in the West, the importance of philosophy in education makes them worth examining.

In China, computational astronomy was an integral part of the imperial bureaucracy. The head of the Astronomical Bureau, unlike his subordinates, was not a technical expert but a civil service generalist on his way up the career ladder. Many Confucian scholars wrote competently on astronomy, and books on the subject often were ornamented with prefaces and colophons by high officials.

In feudal Japan occupations were hereditary. The post of official astronomer to the shogun was created to recognize the personal achievement of Shibukawa Harumi and was passed down to his descendants. It had no significance beyond the technical acumen and thus was of no interest to the generalist. Technical posts of this kind were from their origin separated from the general samurai bureaucracy. When the official astronomer and his subordinates were compiling the ephemerides, Confucian scholars were not consulted. Even the Tsuchimikado family, for many centuries astrologers to the imperial court in Kyoto, was accorded the courtesy of an invitation to contribute a preface to the ephemerides.

A popular Chinese book of negligible depth, the *T'ien-ching huo-wen* ("Queries on the Astronomical Classics," 1675) by Yu I, exerted considerable influence on Japanese cosmological notions. Among the many Japanese editions, the only preface by a Confucian scholar was that of Irie Shukei. Irie states in his preface that he was motivated to write a commentary on the simple work because most astronomical writings were so full of mathematics and technical terms that, although they might be "useful for the narrow calculations of small men engaged in the divinatory and computational arts, they are of no use for the greater mathematical concerns of gentlemen and scholars." It was no doubt commonly believed in China as well that calendrical astronomy, which Irie looked down upon, had lost its ideological implications and had become nothing but a collection of techniques. Nevertheless, the Chinese, particularly from the mid-seventeenth century on, continued to think of astronomy as part of the Confucian system of learning. As Juan Yüan (1764–1849), a high official and patron of learning, put it, mathematics and astronomy were "a proper study for those scholars who search out the facts to get at the truth, and not a tool for technicians scraping up a living." In China many Confucian scholars contributed prefaces to *T'ien-ching huo-wen*, not merely for ornament but often to discuss fundamental technical matters.

What accounts for this difference? Almost without exception the computational schemes and theories used in eastern Asia were discovered by the Chinese. To the Chinese they were integral parts of native culture; to the Japanese they were importations. In China the lingering excitement of discovery clung to knowledge of regularities in nature. In Japan these foreign regularities constituted one more routine skill prerequisite to established occupations.

This was true not only of science but also of Confucianism itself. In China, Confucianism was more than a philosophy; it was the basis of political legitimacy. The government's use of it as a political ideology demanded that great care be given to defining what orthodox Confucianism should be—just as the imperial monopoly of the calendar made it necessary to have one official system of astronomical computation. Official philosophy and official astronomy were exported to maintain China's cultural suzerainty over her satellites and neighbors. Confucian philosophy in its contemporary interpretations endorsed and justified these concerns. The commitment of the Chinese elite to civil service channeled a great deal of intellectual energy in this direction. What interpretations should be orthodox and what sorts of learning should be propagated were central subjects of philosophic inquiry. Not all thinkers shared the official view at any given time; but because it determined the content of the civil service examinations, about which much of early education was organized, the official view was enormously influential.

In Japan there was no social or political reason for philosophic orthodoxy to be an important issue. Although nominally based on the centralized Chinese model, Japanese government until the mid-nineteenth century was imposed upon a feudal society and thus remained multifocal. Although dynastic legitimacy could not be taken from the imperial court in Kyoto, real political power lay entirely in the hands of the military dictator, the shogun, in what is now Tokyo. He was able to retain that power only by leaving local authority in the fiefs (*han*). Certain prerogatives in astronomy belonged by tradition to the Tsuchimikado family, the imperial court astrologers; and others were divided between the shogunate astronomers and those of the fiefs. Satsuma, one of the larger fiefs, issued its own calendar. There was no occasion to establish a single orthodoxy, either political or intellectual.

Just as political and astronomical orthodoxy were related in China, so their absence was related in Japan. This contrast is apparent even in the art of divination. The great Chinese treatise *Wu-hsing ta-i* ("Fundamental Principles of the Five Phases," *ca.* 600) set out a coherent synthesis of contemporary knowledge and belief. The early Japanese treatise *Hoki naiden* ("Ritual Implement," undated), equally influential upon later practice, was an undigested juxtaposition of hemerological practices from Shinto, Buddhism, and perhaps Taoism. In Japan freedom to choose between several paradigms seems to have been as desirable as the search for a unitary principle was in China.

When the Japanese did originate something, there was no expectation that it would be universally accepted or that it would have influence outside Japan. Although the Chinese did occasionally acknowledge Japanese originality in connection with one development or another, the Japanese did not believe, prior to the twentieth century, that they could contribute to universal systems of knowledge.

From the seventeenth century on, when Western knowledge began to exert claims of its own in the background of Chinese learning, Japanese thinkers were critically attentive. Once convinced that European technical knowledge was superior, they promptly switched to the new paradigm. This was not the first time that the Japanese had modified their attitudes smoothly and quickly to conform to desirable goals presented from abroad. For the Chinese the encounter with European ideas was traumatic; to accept them was to reject traditional values, and to reject them would leave no defense against dismemberment by the Western powers.

The Academicism of Shibukawa Kagesuke. When Shibukawa Harumi founded the shogunate astronomical office, he was in the rare position of initiating a tradition. The older astronomical institutions of Japan were devoted to inherited responsibilities from which the incumbent could not freely deviate. Shibukawa defined his responsibilities in a way that involved considerable political activity. He enlisted the support of powerful figures for a calendar reform not based upon a borrowing from China. He had to overcome the opposition of Confucian scholars, who saw no merit in native independence. The problem that Shibukawa set and the solution that he envisioned constituted a paradigm (in Thomas Kuhn's sense of the word)—a paradigm of purely local applicability.

As an exact science, calendrical science could be given the solid foundations that enabled normal science to proceed. Shibukawa left to his descendants and disciples the responsibility to work toward greater precision and to improve agreement between observation and theory. For some time there was no need to question the validity of his conception of astronomy. Historical questions that also interested Shibukawa, such as the gradual variation in astronomical parameters, did not retain interest for later generations. The historical orientation of such problems made it impossible to adapt them to the concern of "normal science" with the improvement of precision.

The success of Shibukawa's program depended on the quality of his successors. Since the family stipends of samurai, even those in professional posts, were inherited only by eldest sons, there was a conflict between the rigid ideal of social structure in feudal Japan and the need for technical talent. There was more than one astronomical institution because new ones had been created at various times to make room for gifted scientists. Certain established families took advantage of another means to maintain their intellectual standards: adopting as successor to the head an intelligent second or third son of a samurai family. The Shibukawa family maintained its tradition in this way. Adopted sons probably contributed more to the cumulative achievement of the family than did eldest natural sons.

Shibukawa Kagesuke (1787–1856), an adopted heir, was perhaps the greatest professional, as well as the last important figure, of traditional astronomy. In his youth he suffered bitterly when his talented brother, Takahashi Kagesuke, was executed after being involved in an attempt by a foreigner, F. von Sieboldt, to smuggle forbidden materials out of Japan. Shibukawa never deviated from the behavior expected of a professional astronomer and maintained unblemished authority in his discipline.

Shibukawa's passion for rectitude made him particularly apprehensive about criticisms of the official calendar. In order to safeguard the prestige of his office, it was essential that he have first access to newly imported astronomical literature. This was a time when foreign threat, social change, and natural calamity jeopardized the authority of the shogun, who attempted to minimize unrest and maintain a useful monopoly by prohibiting the diffusion of Western knowledge. The translation of Western literature for official use was confined to the Astronomical Bureau. Shibukawa took full advantage of his privileged position. He left voluminous notes on his wide reading in books forbidden to others—Chinese and European, ancient and contemporary.

Shibukawa's motivations emerge clearly from the story of his conflict with Koide Shuki (1797–1865), a scholar of astronomy who did not hold a position in any of the government astronomical institutions.[14]

The Kansei calendar, then in official use, was almost completely based on *Li-hsiang k'ao-ch'eng* ("Compendium of Observational and Computational Astronomy," draft completed 1722, printed 1724), in which the outdated European astronomy introduced to China by the Jesuits was adapted to the needs of a traditional calendar reform. In addition, the compilers of the Kansei computational system had incorporated Asada Gōryū's cyclical variation in length of the tropical year. The official calendrical treatise was not published, for laymen had no business using it. On the other hand, nonreligious writings of the Jesuits in Chinese had recently been exempted from the ban on importation, so the *Li-hsiang k'ao-ch'eng* was available for study by private scholars. If the Kansei calendar had been based entirely upon it, anyone could have checked the validity of the official calendar. In the late 1820's, Koide had an opportunity to calculate an ephemerides on the basis of the *Li-hsiang k'ao-ch'eng* and compare it with the official calendar. He found considerable disagreement. He suspected that Shibukawa Kagesuke's father, a disciple of Asada, had introduced Asada's variation in the last calendar reform, in order to conceal the system of computation under a cloak of security and thus avoid criticism from amateurs. He was unable to prove this suspicion, since the value of Asada's variation was unavailable to him; it belonged to the private teaching of Asada's successors.

Determined to obtain the formula, Koide enrolled in the academy of the Tsuchimikado clan, the hereditary imperial astrologers, where one of Asada's four major disciples had taught. In 1834, after eight years of discipleship, he was permitted access to the details of Asada's variation. His heart's desire having been fulfilled, as he wrote, he immediately calculated the current value for the length of the tropical year. Koide next made precise observations from which to determine the winter solstice, and found a discrepancy of a quarter of an hour between his computations and those given by the official calendrical system. When he ignored Asada's variation and calculated on the basis of the *Li-hsiang k'ao-ch'eng* alone, his results coincided closely with observation. Thus he determined that Asada's formula was a "fake supported by blind belief."[15] In 1835 he visited Edo (now Tokyo), became a disciple of Shibukawa Kagesuke, and questioned him about his view of Asada's variation. Shibukawa would not give him a clear answer.

Shibukawa had already read Lalande's *Astronomie* and had a good understanding of the European astronomy of the previous century. He knew perfectly well that Asada had misled his successors. He did not, however, dare to voice his doubts publicly. In 1835, in fact, Shibukawa drafted a manuscript entitled "Saishu shochō kō" ("On the Variation of the Length of the Tropical Year"),[16] in which he analyzed the origin of, and tried to give a rationale for, Asada's variation.

Why did he write this essay in that particular year? He must have been influenced by Koide's criticisms. Although Koide had no official standing, his connection with the Tsuchimikado (an older institution and, to some extent, a rival for power) gave him some authority.

Koide was not prepared to do more than point out numerical discrepancies. Shibukawa had access to vastly greater information and technical skills, and was able to form an analytical overview of the variation question. Since his bureau had come to be responsible for the actual calculation of ephemerides, he felt that a frank answer to Koide's inquiries might compromise his own official responsibilities. But he was now aware that Koide had found the weak point in the official calendar and would be driven by his remarkable determination to press an attack that was bound to be successful. Although Shibukawa Kagesuke was not responsible for the failure of the Kansei calendar, his office would suffer for it.

Shibukawa had two lines of defense. The only permanent defense was to carry out a calendar reform as soon as possible, discarding Asada's variation in the interest of accuracy. But calendar reforms in Japan were so infrequent that they could not be proposed and carried out overnight. If a crisis arose too soon, Shibukawa wanted to be prepared with an official interpretation of Asada's variation to interpose against attack. That was apparently the point of his treatise.

Shibukawa's fears were soon realized. Koide's prediction of a solar eclipse in 1839 (based entirely on the *Li-hsiang k'ao-ch'eng*) proved to be more accurate than that given in the official ephemerides. Koide submitted to the prime minister (*rōjū*) a proposal for a calendar reform based on the Chinese source.

The official astronomers had no choice but to hurry their own reform. The new Tenpo system of computation became official in 1842. Shibukawa, unlike Koide, did not have to depend upon a century-old treatise. He had the authority to mobilize the best translators in Japan so that his system could be based upon contemporary European data.

Through the years of mounting conflict between Shibukawa and Koide, the goal of both was complete agreement between computation and observation. Nevertheless, they differed in their conceptions of the astronomer's work and thus of astronomy. From Shibukawa's viewpoint, Koide's painstaking efforts had no significance whatever for the advancement of astronomy. Koide was simply duplicating outmoded results. The only positive service he could perform was to prevent the corruption of the astronomical functionaries who monopolized research facilities and tools under government protection. Koide began with a simple puzzle: why the official calendar failed to agree with observation. Since he had only limited access to scientific and political information, all he could do was deal with each step as it unfolded from the last. What began as a technical exercise in the testing of computational theory was reduced to the unraveling of bureaucratic prerogatives.

Here the conflict between Koide and Shibukawa finally lay. Although both accepted the commitment of the astronomer to accurate prediction, Shibukawa's dedication to his profession gave the administrative order precedence over the celestial order. This Koide could not accept; but he was bound to be drawn into intrigue, for Shibukawa's professional standing reserved to him alone the initiative to define the rules of the game.

Western Cosmology and the Traditional Calendrical Science. Elaboration and precision were the criteria by which calendrical treatises were evaluated in prefaces written by astronomers. The holders of astronomical sinecures needed to fear no crisis so long as these criteria were satisfied. In view of the moderate level of precision needed for the traditional ephemerides, why should Shibukawa have wanted to involve the Astronomical Bureau in the active dissemination of European natural philosophy? Such a major departure from his inherited duties would seem to carry as much risk as gain. Nevertheless, Shibukawa wrote an account of Copernican theory and Newtonian mechanics for government use (not for publication) in his "Shinpo rekisho zokuhen" ("Sequel to the Astronomical Treatises According to the New Methods" [to the series of works compiled by the Jesuits in China before 1635]).[17] Again his motive seems to have been bureaucratic caution. It was essential to formulate an official view of European cosmology as awareness of it gradually became more general in Japan; otherwise a query by a high official might result in grievous embarrassment.

Heliocentric theory had been previously introduced by Motoki Ryoei (1735–1794), an official interpreter; Shizuki Tadao (1760–1806), an independent intellectual; and Shiba Kokan (1739–1819), a free-lance popularizer. Partly because cosmology was not traditionally an important topic in Japan, and partly because they understood Copernicanism only superficially, these writers and those who read their works were not shaken by heliocentricism as Europe had been. Motoki considered it merely an exotic European curiosity. Shiba adhered to it for the sake of sensation. Shizuki treated it incidentally to his main interest, the philosophical implications of Newtonianism, which he tried to assimilate to Chinese natural philosophy.

Shibukawa disdained these amateur writings. For his account of Copernicanism he depended heavily upon an explanation written in Chinese by the French Jesuit missionary Michel Benoist (1760). It was little known in Japan and did not so much as mention Newton. This work allowed Shibukawa to deal with the subject while completely ignoring the Japanese literature (based upon much later secular sources). Although not enthusiastic about the sun-centered system, Shibukawa did accept it with critical reservations. He agreed with Benoist that the difference between heliocentricism and geocentricism was merely a matter of transforming coordinate systems. He contributed the notion that theories of a moving earth were not original in the modern West but had existed long before in China and India—a theory first advanced by Shizuki Tadao.[18] Shibukawa's discussion of the technical aspects of Newtonianism was superior to that of Shizuki, but the latter's philosophic depth was totally missing. For Shibukawa, Newton's work was not the foundation of a world view but a peripheral issue on the margins of calendrical science. Shibukawa was successfully guided by his motto, "Let us melt down the mathematical principles of the West and recast them in the mold of our tradition," a cliché in earlier Chinese writing. Newtonianism was so well assimilated in Shibukawa's presentation that it did not perceptibly challenge the traditional definition of mathematical astronomy. Like his ancestors who had headed the Astronomical Bureau before him, Shibukawa devoted his erudition and energy to the perfection of routine, not to the development of new fields of investigation.

In the scientific revolution of seventeenth-century Europe, astronomy played a great role because of its implications for cosmology and world view. Japanese astronomy had no such implications, and its social matrix gave it no scope for free inquiry into nature. During the seventeenth and eighteenth centuries it remained the vanguard of disciplines oriented upon mathematical regularity. Its paradigm was so well insulated from challenge by professionalism that even the impact of Western exact science did not throw it into doubt. The most important function of European astronomy was to furnish numerical data and computational techniques, by use of which traditional calendrical goals could be better met.

The Tenpo calendar reform brought the accuracy of prediction to as high a level as the traditional calendrical art envisioned (solar eclipses to the nearest quarter-hour, for instance). It was no coincidence that the esteem of intellectuals for astronomy was practically lost in the process. At a time when the revolutionary implications of foreign science were gradually coming to be understood, the astronomical profession was seen as too routinized and unimaginative to make any important contribution to change. It was finally the physicians, who lacked both the advantages and the disadvantages of a tightly organized profession, and whose proficiency in the exact sciences was inferior to that of the astronomers, who were able to introduce the true core of Western scientific thought. The structure of the scientific revolution they brought about in Japan was bound to be different from the one led by astronomy in seventeenth-century Europe.

Medicine.

The Chest as the Seat of the Mind. Even as late as the middle of the nineteenth century, the Japanese did not believe that thought takes place in the head. As Shibukawa Kagesuke put it, "Mathematical principles all originate in the breast of the mathematical astronomer. . . . [Thoughts] are stored in the chest."[19] He was typical in situating both memory and arithmetical thinking in the chest.

To the Japanese such expressions as "a dull head" or "a clear head" have a modern and exotically occidental flavor; they were never used until the Tokugawa period. The cognitive and imaginative functions of the brain were unknown, and their anatomical substrate undemonstrated in Japan, until the beginning of Westernization.

Traditional Chinese medicine, upon which the learned tradition of Japan depended, was concerned primarily with function and only secondarily with tissues and organs. The sites of function to which most attention was paid were two groups in the thorax, a set of six *fu* that fermented food, separated energy from waste, and excreted the latter, and five *tsang* that stored the refined energy. These spheres of function were identified with the familiar viscera; but the physiological nature of the latter was of such minor importance that little was known about it, and it played only a small abiding role in medical discourse. Occasional drawings of the body ignore both the interior of the head and the nerve tissues, neither of which was assigned specific functions, at least in the Chinese medical writing that was influential in Japan.

To my knowledge, the brain's function was first discussed in China in *Wu-li hsiao-chih* ("Notes on the Principles of the Phenomena," 1664), by the idiosyncratic Fang I-chih, who was acquainted with the writings of the Jesuit missionaries. His knowledge originated in Western medicine.

In Japan thought was first located in the brain in an Amakusa edition (1593) of Aesop's *Fables*, in which it was said, "if we have intelligence in our heads. . . ." Again it is clear that this idea was imported into Japan along with Catholicism. The notion did not spread until the study of the Dutch language (and consequently of secular sources) became widespread in Japanese society; and in the early period there was not the slightest influence upon developing knowledge of anatomy, physiology, or pathology.

It is well to remember that the interrelation between the brain and mental processes could not be proved before the development of cerebral physiology. The idea has a long history in Europe, but it is the history of a belief rather than of a fact.[20] Plato and Aristotle held quite different views on the location of mental processes. Plato believed that the immortal and holy rational soul is located in the brain. Aristotle placed the center of sensation and perception in the heart, and did not believe that it was related to the brain or spinal cord in any way.

These were not isolated opinions but were integral with coherent views of the body and its functions. Plato and others who placed mental functions in the head have tended to think of them as quite separate from the physical body; schemata that consider the heart and mind as identical have tended to think of mind and body as integral. Traditional Chinese and Japanese views must be classed with the organismic and naturalistic group to which Aristotle belongs rather than with Plato's idealists. The experimental work of Galen settled the matter in the West, providing an authoritative basis for the doctrine that the brain is the center of perception and of all other mental processes. The introduction of this idea into the Far East had implications at least as revolutionary as those of Copernican astronomy. It challenged the doctrine of bodily functions and the rather negligible notions about internal organs related to them, and posed a range of questions about the physical basis of sensation that had not been previously considered.

Ōgawa Teizo has located[21] the first appearance in Japanese of terms that correspond to "nerve" and to "consciousness" as associated with the brain in the *Kaitai shinsho* ("New Book of Anatomy," 1774), by Sugita Genpaku and others. Otsuki Gentaku, in his *Chotei kaitai shinsho* ("New Book of Anatomy, Revised" [compiled 1798, published 1826], enthusiastically described the significance of the new study of the cerebral and nervous system in this way:

> We have not come across anyone in the long Chinese and Japanese traditions who discussed the active functions of these organs of sentience. They were taken up superficially and in an elementary way only at the close of the Ming dynasty [early seventeenth century]. Most regrettably, in the two hundred years since that time no one has taken up the problem in closer detail. It is a great pleasure that now we are able to explore it more deeply. This is not particularly due to my personal endeavor, since we are all influenced by the trends of the times.

The learned treatises of the Chinese and Japanese medical traditions lacked terminology not only for brain functions but also for mental processes. Conventional Chinese discourse was not much concerned with what we would consider epistemology. Some late Confucian schools were somewhat concerned with how knowledge becomes certain but tended to connect this problem with that of attaining enlightenment. The vocabulary for mental operations remained rudimentary and to a considerable extent was borrowed from Indian Buddhism.

Although the spheres of function within the body were thought to process nutriment and to store the energy refined from it—and Japanese terms that predate Chinese influence, such as *kusowatafukuro* and *yuharifukuro*, are literally types of containers—knowledge was never thought of as localized and stored. There was no reason to investigate the physiological basis of cognition.

In short, the need to explore the relations between mind and brain did not exist in China because the Chinese assumed neither the mind-matter dualism nor the dualism between the self and the outer world. They saw all of nature as united in a single pattern of function in which the patterns of function of individual things (*li*) participate. The dualistic terminology used today in Japan, except for a few terms borrowed from fundamentally religious Buddhist dualism, was for the most part devised by Nishi Amane and others at the beginning of the Meiji period (1870–1890).

Anatomy and Energetics. In the East, the apparently rudimentary association of physical functions with internal organs does not indicate a low state of medical theory. Although the Chinese lacked the sophistication of Galen's anatomy, attempts to study rigorously the Chinese language of function in its own terms, a very recent development, suggest an artfulness that was obscured by the imposition of modern viewpoints. From the historical point of view, the fundamental question is not whether, before modern times, the Chinese or the European tradition incorporated the greater number of correct facts, but how their theoretical paradigms, and the views of nature on which they depended, differed.

In Western medicine the rivalry before modern times between solidists and humoralists is well known. The aim of the former was to locate the seat of a disorder in a solid part of the body, such as the stomach or brain. The motivation to pursue anatomical research is obvious. The humoralists, on the other hand, thought of health primarily as a balance of the various humors that circulate in the body. Anatomy had a great deal less to contribute to their holistic diagnoses.

Traditional Chinese theories of bodily function and of pathology are closer to the humoralist tradition than to that of the solidists. Health was related to the balance of *ch'i,* which is the basis of material organization and function not only in the human body but throughout the physical universe.

Ch'i was not a ponderable fluid, as the humors were. It originated very early as the word for air—not as an inorganic gas but as an enveloping substance that maintains vital functions. Its closest analogue in the West was the Stoic pneuma. In medical theory its vital or energetic aspect—in a purely qualitative sense—became preponderant in discussions of etiology. *Ch'i* was not only inspired air but also the energy refined from food that circulated throughout the body and was responsible for all vital functions. The concept of *ch'i* as a material pneuma was to some extent reconciled with this energetic approach and was never abandoned; for instance, certain tumors and internal swellings were thought to be stagnated or congealed *ch'i.* Indeed, the seventeenth-century Japanese physician Goto Gonzan attempted to explain the cause of all medical disorders by stagnation of this kind. Since *ch'i* was involved in processes in physical nature and in the body, it assumed different qualities or characteristics in different phases of such processes. If the whole process was analyzed into two phases, the two different types of *ch'i* were characterized as yin and yang; if a fivefold analysis was used, the five types of *ch'i* were described by the language of the Five Phases theory. A dynamic balance between the two or five types of *ch'i* defined health; ethical disorders were always identified with an imbalance.

The language of yin-yang and the Five Phases theory were used to establish sets of correspondences that governed bodily function. For instance, the Five Phases corresponded to the five spheres of function (loosely identified with the heart, lungs, spleen, liver, and kidneys). But discourse about health and pathology was never anatomical. The system identified with the spleen amounted to the ensemble of functions that would be ascribed today to the urogenital system and was thought of in functional terms.

Internal disorders were never local in Chinese medicine. Although they might be concentrated in a particular sphere of function, the connection of the spheres by the energetic circulation system meant that the whole body was affected and that as the pathological process developed, its seat would move. There was no value in local treatment, for the site of treatment often was far removed from the center where the disorder was concentrated for the moment. Abstract correspondences often were used in discussions of pathology and therapy—for instance, Five Phases correspondences between the heart and the ears and between the liver and the eyes. These were not so much statements about physical connections (although such connections were claimed to make the model plausible pneumatically) as about similarities and analogies of function.

Early Far Eastern anatomical charts were extremely simple and crude. As Lu Gwei-djen and Joseph Needham have said,[22] they incorporate a much more rudimentary level of knowledge than the texts that they accompany. Why were Chinese physicians satisfied with them? Their purpose obviously was different from that of Western anatomical diagrams. They were simply meant to depict the broad outlines of the general system of physical function. One might think of them as half anatomical diagram and half flow chart.

According to this view of the theoretical entities of Chinese medicine, reconstructed largely through the painstaking work of Manfred Porkert,[23] it is possible to conclude that it was closer to the European humoralist point of view, although it was pneumatic in a sense that does not fit the European theory. It had no use for exact anatomy. For the latter to be accepted in the Far East, its utility would have to be proved; and it could be proved only by an appeal to a different conception of nature and of the human body.

Anatomy in Japan. When anatomical inquiry began in mid-eighteenth-century Japan, did its demand for an analytical approach to the human body have revolutionary consequences?

Even before the serious study of anatomy began in Japan, Yamawaki Toyo questioned traditional Chinese anatomical charts on the basis of his own anatomical findings (1759), and his criticism entitles him to be considered the forerunner of anatomical students. Yamawaki's interest in anatomy must have been stimulated by access to a Western anatomical chart. Although he could not read the legends, his experience must have convinced him that the Western schema was a great deal more accurate than the Chinese. What led him to evaluate both schemas as anatomical rather than as functional?

First of all, when Western knowledge began filtering into Japan during the Tokugawa period, it was natural that it should have been compared with the official Chinese academic knowledge, since the latter had become firmly entrenched not very long before. It was also natural to ask which set of ideas was better—the situation was different from the case in China, where traditional ideas were so strongly rooted that such a question could only be radical. In mathematical astronomy, criteria of predictive accuracy were so obvious that Western superiority was quickly recognized. This was equally true in China, since the criteria for that recognition could be traditional ones. In medicine, however, there is good reason to doubt that there was any difference in therapeutic efficacy between the two systems before the late nineteenth century. It is above all in the comparison of anatomical charts that the strength of Western medicine would be apparent. But if the foregoing argument is correct, the difference between Chinese and European ideas about the interior of the body would be anatomically significant only after acceptance of the idea of anatomy and of the more general medical and philosophical ideas on which it was based.

Yamawaki Toyo was one of the leaders of a new group called the Koihō ("Back to Ancient Medicine School"), who rejected the theoretical entities of Chinese medicine and undertook an empirical approach to clinical treatment. Their utilitarian goals made the very elaborate conceptual superstructure of the Chinese tradition seem an impediment. Because they wished to confront as directly as possible the ills of the body, its role as a microcosm of physical nature could be rejected. As Yoshimasu Todo (1701–1773), the foremost figure of this school, declared, "Yin and yang are the *ch'i* of the universe, and thus have nothing to do with medicine."[24] This group was prepared, then, to take a position much closer to that of the solidists than had previously been possible in Japan. Functional analysis lost its importance, and the physical organs could be studied for their own sake. From this point of view the traditional anatomical charts were recognized as crude and inaccurate representations of material organs. This was nothing less than a change of gestalt.

Nevertheless, this conceptual radicalism was circumscribed. Yamawaki's accomplishment was not to do away with the old scheme of six processing spheres and five storage spheres but, rather, to alter them so that they made sense in anatomical terms. He had no reason to be curious about the contents of the skull. It was only later figures with considerable knowledge of Western anatomy, such as Sugita Genpaku, who could abandon the Chinese tradition entirely and display as much anatomical interest in the brain as in the viscera. At that point the confrontation between champions of the two systems becomes interesting.

Lately there has been a tendency to emphasize the value of organismic and synthetic thought, of the sort that Joseph Needham has found predominant in Chinese science, to the detriment of the early modern habits of physical reductionism and remorseless analysis. Although the pneumatic *ch'i* doctrine, and the yin-yang and Five Phases theories that qualified it, are not precisely reductionist, they are not modern either. All of these concepts, although originally taken from everyday phenomena, were too abstract to have fixed empirical significances. They remained satisfactory only because, as N. Sivin has shown,[25] the goal of Chinese science was not complete understanding of the natural world but limited knowledge for practical purposes. The body was clearly not the cosmos, but the correspondence between the two set limits upon what could be asserted about the body.

Because of the special character of Chinese medical thought as it was received in Japan, we have examined in some detail the reasons that traditional doctors would find anatomy, and thus dissection, irrelevant to the improvement of medical therapy. There were other objections as well. The idea was deeply ingrained in Confucian ethics that keeping intact the body one has received from one's parents is a major obligation of filial piety. This prohibition against mutilating the body effectively ruled out dissection in China. In Japan, however, it had practically no effect on medical specialists.

A second objection to dissection originated in traditional physiology. In his *Hi zoshi*[26] ("A Refutation of the Anatomical Charts," 1760) Sano Antei said, "What the *tsang* [the word for the spheres of function and their associated viscera] truly signify is not a matter of morphology; they are containers in which vital energy with various functions is stored. Lacking that energy, the *tsang* become no more than emptied containers." In other words, the internal organs were characterized not by their morphology but by the differences in their functions, which were defined by the energy they stored. Nothing could be learned by dissecting a cadaver, since it lacked this vital energy. The anatomical charts that captured the imagination of Yamawaki, since they were based on dissection, could cast no light on the dynamic functions of the body. The same point emerges in another criticism that Sano made. He noted that Yamawaki's anatomical charts did not distinguish the large and small intestines. He did not believe, in fact, that those organs were morphologically or physiologically dissimilar. What made them utterly different was that the large intestine was responsible for absorbing and excreting solid wastes, while the small intestine performed those functions for fluid wastes. He emphasized that this crucial difference would be undetectable in a dead body. Figure and appearance could be significant only to the extent that they were related to function. Sano, unlike the Koihō radicals, had no use for pure empiricism. "The observation of two obvious facts is of much less value than groping speculation . . . even a child is as good an observer as an adult." A scholar who refrained from speculatively tracing the connections between form and function was no better than a child.

Beginnings of a Solidist Approach. After accurate European anatomical charts were introduced into China, even the traditional schools of medicine that admitted anatomy as the basis of surgery (an undeveloped art in the Chinese tradition) still adhered to an energetic and functional view as the basis for internal medicine. But a shift to a solidist approach had at least begun.

It would be a mistake to see this shift purely in terms of the increasing accuracy of anatomical description. The Koihō school, like Western empiricism, could not dispense with metaphysical entities, but depended upon them without acknowledging them. Its physiological and pathological ideas were not only less explicit than those of earlier speculative Chinese medicine but also a great deal less sophisticated. The move toward solidism was not a rejection of models, but the construction of a new model.

Yoshimasu Todo, for instance, rejected the elaborate Chinese theories but was unable to translate his solidist thinking into diagnosis without the aid of a theory that Chinese doctors would have considered primitive: he saw all disease as the action of one fundamental poison on the various organs and tissues of the body. This was not really a pharmacological theory about the effect of poison, but merely a rationale for locating the part of the body on which treatment should be concentrated. He also rejected the traditional pulse diagnosis, which had served as a way of reading functional characteristics of the *ch'i* circulation. Thus faced with the problem of how to determine the condition of the internal organs without dissection, he did not so much eliminate pulse reading as substitute abdominal palpation for it. This technique had been used to a very limited extent in traditional medicine, chiefly to determine whether existing abdominal pain increased or decreased when the belly was pressed. Yoshimasu enormously increased its importance as the most direct way of learning about the conditions of the internal organs and thus founded a Japanese diagnostic tradition that still flourishes among traditional doctors.

The solidist tradition begun by the Koihō school eased the way for Western anatomy. In the second half of the eighteenth century, Sugita Genpaku took up the study of anatomy because it seemed the most tangible, and therefore the most comprehensible, part of Dutch medicine. Following the solidist breakthrough, the successors of Sugita in medicine studied physics and chemistry, thus opening up the world of modern science. The Copernican influence was minor by comparison, because the Japanese cosmos had not been defined by religious authority. The impact of anatomy challenged the energetic and functional commitments not only of medicine but also of natural philosophy. Its effect was bound to be revolutionary.

Medicine and Science After "Kaitai Shinsho." Publication of *Kaitai shinsho* (1774), the first Japanese anatomical treatise based directly on Western materials, not only led to recognition that Western knowledge of the interior of the body was superior to that of China but also provided a new paradigm for Japanese science. Once the Japanese were prepared to compare accounts of the interior of the body from a purely morphological point of view, the superiority of the West became obvious. Chinese-style conservatives could dismiss European anatomical charts as superficial, but they could not convince others.

The first Japanese to realize the power of Western anatomical knowlege naturally assumed that the European system also was therapeutically more effective, although there is no reason to believe that this proved to be the case. Indeed, on therapeutic grounds there is very little to choose between the systems of internal medicine evolved in the various high civilizations before the end of the nineteenth century. Moreover, it is unlikely that the relatively frequent resort to surgery in the European tradition led to consistently higher recovery rates before the introduction of anesthesia and asepsis. Some scholars actually give the edge to Chinese internal medicine because it tended to use milder and less drastic drugs than were prevalent in Europe. It is ironic that one of Yoshimasu Todo's innovations was the frequent use of poisonous drugs to "fight poison with poison."[27]

Among the great diversity of schools in Japan were eclectic groups that prescribed both Chinese and Western drugs for a single symptom, but that was about as far as eclecticism could go. The views that underlay Chinese and Western medicine, or even Koihō medicine and practice of a more traditional kind, were irreconcilable.

It was quite possible to introduce European data into traditional calendrical astronomy without challenging the paradigm on which the latter was based. An analogous accommodation was impossible in internal medicine, for there was little overlap of the conceptions of relevance. Acceptance of the European view of the body came only with the publication of *Kaitai shinsho.* The Koihō school can be considered a vanguard in this scientific revolution. Such a transition did not take place in China because the Chinese maintained their traditional medical world view much more rigidly than did the Japanese.

An important characteristic of early modern science was mechanical reductionism, in which every phenomenon was believed to be ultimately explainable in terms of matter and motion. This reductionism gave birth to the positivists' hierarchical arrangement of the sciences. Comte ranked the abstract sciences in the order in which he believed they would be entirely quantified, beginning with mathematics, then astronomy, then physics, with sociology at the end of the list. At about the same time Japanese physicians were constructing an analogous but very different schema.

After the publication of *Kaitai shinsho*, some medical practitioners, exploring the newly available writings on European physical science, recognized and responded to its reductionism. In the prefaces to Aochi Rinso's *Kikai kanran* ("Contemplating the Waves in the Ocean of *Ch'i*," 1825) and Kawamoto Komin's sequel, *Kikai kanran kogi* (1851), the authors claimed that physics must be the basis of medicine and the other practical sciences. Kawamoto described a hierarchical order from physics to physiology to pathology, eventually encompassing practical therapy. It is not clear how seriously his fellow practitioners took him. It is likely that they saw no clear role for physics in medicine, except perhaps for embellishing prefaces, as disquisitions on yin-yang had done in traditional books of therapy.[28] Physics and chemistry were introduced into Japan by European medical men only for their limited direct value to clinical medicine, just as anatomy, physiology, and pathology were subordinated to the same use. Hoashi Banri, a natural philosopher whose background was in medicine, was disappointed and disillusioned with Western science when he examined books on microscopy and chemistry and found them of no help in the understanding of drug therapy. The disciplines such books represented underlay techniques of measurement in materia medica and of extracting the active essences from herbs. Those applications were the basis for their initial study by physicians. Their value for a new philosophy unfolded only gradually.

Social Status of Medical Practitioners. In Japan mathematical astronomers were minor bureaucrats, responsible for preparing the national ephemerides. Although they were the earliest to recognize superior aspects of Western science, they overlooked its basic paradigms and remained within the traditional mold. Their academic style, as we have seen, tended to be greatly shaped by their proximity to sources of power.

Medical practitioners, who first took up the challenge of the Western sciences, constituted the largest scientific profession during the Tokugawa period. Medicine, unlike astronomy, was a private concern and thus free of one kind of constraint upon the response to new ideas. Because there was no public health program at the time, medical practice was essentially a relationship with individual patients. There usually was a private physician in each community. The samurai class had its government doctors and fief doctors, and townsmen and peasants had their local practitioners; but the profession was not tied together or controlled by the central government. Although Edo, as the seat of the shogunate, was a center of professional activity, the important schools of medicine were scattered as far as Nagasaki. This decentralization made medicine one of the few geographically mobile professions in Japan.

Toward the close of the Tokugawa period, in the first half of the nineteenth century, it became conventional for medical students to visit the various centers of instruction and to be initiated into the different schools of clinical medicine. Moreover, practitioners who distinguished themselves often were called to serve the fief governments or the shogunate. Although their stipends were small, the prestige they gained raised their fees in subsequent private practice. The competitive market for medical practices in Japan was most untypical of the society as a whole. Western-style physicians took advantage of it as they moved into spheres of health care that previously had been monopolized by traditional practitioners. There is a loose analogy between this situation and that of the nineteenth-century German academic market described by Joseph Ben-David.[29] In Japan, furthermore, there was no guild organization of physicians to limit or control competition.

Unlike the medical profession in Europe, which was well integrated into society and could reproduce itself in the universities, Japanese doctors were socially marginal. Their mobility was anomalous in a society where status was supposed to be hereditary and where the only elite was supposed to be the hereditary warrior class, the samurai.

The Taki family, hereditary physicians to the shogunate, once tried to centralize medical standards by founding an official medical school that all sons of doctors were to attend and at which they were to be examined for a license to practice medicine. This attempt failed, in contrast with the ease with which central authority was established in other fields. The main impediment to uncontrolled competition in medicine was not guild or government organization but the hereditary system on which the Tokugawa social order was based.

The samurai, the military elite, inherited ranks and stipends that depended upon the contributions of their ancestors to the foundation of the Tokugawa shogunate in the early seventeenth century.

Merchants, artisans, and others did not depend upon fixed stipends as the samurai did, but their social class was fixed through inheritance. This system could not easily find a place for intellectual professions, which could flourish only in situations where advancement was based on talent rather than birth. In the Tokugawa period there were three such professions: Confucian philosopher, medical doctor, and mathematical astronomer. In all of them people of outstanding ability often remained subordinate to incompetent samurai and, if they worked for the government, received lower stipends.

Attempts were continually made to subordinate these professions on the hereditary principle. It was expected, for instance, that the son of a doctor would eventually be registered as a doctor, regardless of how little intelligence or motivation he might have. At the same time, the shogunate and the fief governments needed talented professionals. The conflict often was resolved by the governmental authorities, who would advise a professional family to adopt a gifted youngster.

Government employment was only one possible source of income for a physician. Osaka, for instance, was famous for a medical center patronized mainly by merchants. The clinical experience of the therapist mattered a great deal more than his formal education. The son of a village doctor would begin by working with his father, then spend many years as an apprentice to more distinguished doctors, and finally return to his native village to take over his father's practice. From generation to generation the number of patients would gradually increase until such a medical family was expected to provide doctors for the whole village. Because such hereditary traditions were quite independent of the government hierarchy, doctors were among those most responsive to liberal thought in the period shortly before the modernization of Japan. Mathematical astronomers were not independent in the same way.

Although Confucian scholars formed a professional group, they lacked the social mobility and economic security of physicians. Their official social status was a good deal higher, but the revenue that they might earn by private teaching could not compare with the fees of the doctors. In essays of the Tokugawa period the social commitments of the two often were compared, to the detriment of physicians. Confucian scholars were concerned with society as a whole, and physicians only with individuals; Confucian scholars dealt with the mind, and physicians only with the body; Confucian scholars were generally poor, while physicians gouged fees and lived in luxury. It is clear from such remarks that establishment values favored the scholars, and found the practical skills of the doctor a little too close to those of the artisan.

This difference should not be overstressed, since Chinese-style medicine emphasized that practice must be based upon Confucian ethics. Young men who chafed under this devaluation of medicine as a pursuit in its own right were especially attracted to Western medicine, which seemed free of philosophic and moral constraints.

Those attracted to intellectual pursuits found them most attainable if they left clinical medicine and became teachers or public figures. Western-style medical doctors gradually distinguished themselves into two groups, one that concentrated upon medical practice and one that mainly taught foreign languages and Western science. In the difficult international situation following the Opium War (1839–1842), it was from medical schools of the latter group, such as that of Ogata Koan, that there appeared political activists such as Hashimoto Sanai who renounced their inherited professions to pursue political careers.

Of the three intellectual professions in the Tokugawa period, only physicians were able to achieve an independent stance from which to view the world in a new light. It was naturally they who brought modern universal science to Japan. But their independence was bought at the cost of alienation from the true sources of power in Japan.

The Science of the Physician and the Science of the Samurai. The pattern of response to Western science changed fundamentally when Japan was opened to the free flow of foreign ideas in the 1860's. The initiative passed from the physicians to the samurai. Among graduates in the class of 1890 at Tokyo University, the percentage of those who came from samurai families were as follows:[30]

Medicine	– 40.8
Agriculture	– 55.9
Law	– 68.3
Literature	– 75.0
Science	– 80.0
Engineering	– 85.7
(Total population)	– 5 (approximately).

Why was the proportion of students from the elite class so high in science and engineering, and so much lower in medicine and agriculture? As we have seen, it was customary for the sons of doctors to become doctors. They did not belong to the samurai class, unless they were employed by the government. Similarly, many agriculture students were the sons of wealthy farmers. The downfall of the samurai regime (which made the foundation of the European-style university possible) had little effect on the livelihoods of either doctors or farmers.

On the other hand, that transformation was catastrophic for the hereditary military elite, whose traditional occupations in the bureaucracy of the old regime, as well as their hereditary stipends, were lost. The fields of medicine and agriculture were largely occupied. The law was not considered a dignified occupation; and until the Civil Service Examination Act of 1890, the law schools did not provide a route to upper-level civil service careers.

The only promising professions left for the samurai were science and engineering. In nineteenth-century Europe, the upper class tended to occupy the legal and medical professions, and science and engineering were largely shaped by the rising middle class. In Meiji Japan, on the contrary, scientists and engineers were drawn largely from the top 5 percent of the population.

In the wealthier European countries the scientific and engineering professions drew on the ideology of the middle class rather than on that of the old aristocracies, whereas in Meiji Japan they were entirely subservient to the social and international aims of the imperial government. Scientists were indispensable to the policy of modernizing and Westernizing, and engineers played key roles in building the physical structure of a modern state. Much of the increased demand for engineers was for survey work and telegraph network construction. From the time these professions began to form on the Western model, those who entered them were government officials. The descendants of samurai, who valued public over private occupations, were thus attracted to engineering. The result was a group of public-spirited engineering professionals oriented toward civil service, in contrast with English engineers in Britain, for example, who came mostly from the class of skilled mechanical people and served private interests.

Before the Meiji period, Western-style physicians took up the physical sciences as intellectual pursuits; those who continued to study them did so largely out of an interest in natural philosophy. Although they legitimized their study, in the Meiji period the formation of scientific and engineering professions became the concern of the samurai and the doctors largely returned to their family practices. Samurai entered science and technology because of their contributions to the state. Their response was not so much intellectual as institutional.

Mathematics.

The Royal Society and Tokugawa Mathematicians. Many would consider the appearance in 1660 of such a disinterested group as the Royal Society of London to be quite unthinkable outside the sphere of Western civilization, but Japanese mathematicians of the Tokugawa period (*wasanka*) were similar in many ways. Members of the Royal Society secured a charter from the king for reasons of prestige and frequently studied subjects of no economic importance. Leisured gentlemen constituted the entire membership and dues financed most of the activities. Similarly, groups of mathematicians in Japan were purely private in nature, consisting of samurai, rich merchants, and affluent peasants; they gathered solely for leisure activities.[31]

In its early period the Royal Society tried to realize certain Baconian ideals, but the activities of the amateurs declined in prominence after the first generation.[32] Learned societies in the British provinces also tended to become philosophical societies after the model of the Royal Society. That is, salaried officials—journal editors, directors of libraries and museums—became central in both while the socially more prominent members were relegated to support functions and retreated from front-line research.[33] In Japan, however, interest in *wasan* activities grew over time, spread outside urban centers, and expanded until the nineteenth century. Participants were increasingly recruited from lower strata of society. Because mathematical activities lacked a significant occupational base, distinctions between professionals and amateurs did not arise. Mathematics was enjoyed by leisured groups in the same way as *waka*, *haiku*, or the tea ceremony. In fact, modern historians of Japanese mathematics have commonly observed that *wasan* was more of an art form than a field of scholarly inquiry.[34] Here, however, I should like to raise the question of how art and scholarship differ from each other and to consider *wasan* in that context.

Scholarship and Art. While *wasan* may be considered an art form, it is by no means easy to distinguish art forms (*gei*) from scholarship (*gaku*). Activities that individuals consider scholarly in nature are not necessarily so regarded by society; such evaluations depend on the views of certain social groups in specific locations during a particular period. They may also depend on the value standards of intellectuals and be subject to influence by the presence or absence of official authorization as well as by popular impressions.

I shall not attempt conceptually rigorous definitions of scholarship and art here, but merely note that scholarship is usually thought to have some public function, while art forms are often regarded as private indulgences that may or may not have significant social value. This distinction was consciously employed by many writers during the Tokugawa period. For example, Seki Takakazu (Seki Kowa), often called the *sansei* or "sacred mathematician" for creating the dominant *wasan* paradigm, wrote on a student's diploma in 1704, as a way of conferring legitimacy on his field: "Mathematics, after all, is more than an art form."[35] He thus refused to define mathematics as an art and tried, instead, to establish it as a dignified, prestigious form of scholarship. He even referred to mathematicians as "scholars."[36] Among themselves mathematicians acted as if they were pursuing an art form, but in the presence of nonmathematicians they tried to present their work in such a way as to give it prestige. Some kind of legitimation was necessary for mathematicians to succeed in this effort. While a leisure activity does not require public legitimation, scholarship does; and there had to be some basis on which to differentiate the one from the other. The evidence of this concern for social legitimation is best seen in Seki's introductions.

Introductions to mathematical treatises usually did not reflect the authors' personal views, since they were written by Confucian scholars according to a fixed, decorative formula in order to partake of the prestige of Confucianism. Confucian scholars were asked to insert hackneyed phrases into mathematical texts even when they confessed ignorance of the subject. One of these phrases declared that mathematics had been mentioned in the *Chou Li* as one of the six classical arts. Another alluded to its association with divination and numerology (esoteric doctrines about the nature of the universe). Mathematicians who wrote the introductions themselves said the same things in other words. These introductions, however, had no connection whatever with the highly technical matters discussed in the main text and were essentially empty, formalistic passages.

The basis for the contention that contemporary people should respect mathematics was its high status in antiquity, stemming from Confucius' esteem for it, and from its designation by the sages as one of the six classical arts. This Confucian form of legitimation derived from the conceptions of classical scholars. Tokugawa mathematicians, however, generally were socially marginal curiosity-seekers and thus did not care whether mathematics had been one of the six classical arts. During the more than 2,000 years since the time of Confucius, Confucianism had become securely established; but astrology, mathematical astronomy, medicine, and arithmetic had come to be considered lesser crafts and were consigned to a peripheral, low status in the hierarchy of disciplines. Arithmetic, which was associated with such mundane matters as surveying and tax collecting, was assigned a status well below that of astrology and even further below that of mathematical astronomy, which described the principles governing the heavens and earth. Even so, unlike chemistry or other sciences, arithmetic had a guaranteed position in the Chinese bureaucratic system; and in Japan's prefeudal period there were doctors of mathematics and official arithmeticians.

During the Tokugawa period, however, mathematics was not formally recognized in the governmental structure. Astronomy had its hereditary doctors in the Tsuchimikado family of Kyoto; the more competent astronomers in the shogun's government were officially, if nominally, subordinate to them. The Tokugawa mathematical tradition, however, existed entirely in the private sector and had no link to the prefeudal tradition of the mathematical doctors. In fact, Tokugawa mathematicians had no interest in the tradition of court mathematics.[37] From the introductions to their writings one perceives instead a recognition of Seki Kowa as forebear or perhaps of Mori Shigeyoshi or Yoshida Koyu. Historical awareness of founding fathers, in other words, did not antedate the Tokugawa period. Consequently, Japanese mathematicians, unlike the school of such Ch'ing mathematicians as Mei Wen-ting, did not try to use the ancient designation that made arithmetic one of the six classical arts as a basis for defining their own identity.

Belief of the Pythagorean type that numbers permeate all objects in space, or constitute the basic principle of the cosmos, was certainly part of the Chinese mathematical tradition; it was specifically the creed of Chinese diviners and specialists in yin-yang cosmology. Even Kawakita Chorin, a mathematician at the end of the Tokugawa period, wrote: "Numbers constitute the elements of the heavens, the earth and all of nature. Everything that happens is a result of their presence."[38] This kind of belief was often used to justify the activities of mathematicians; but such an argument was more commonly espoused by cosmologists, astronomers, and specialists in calendrical science than by mathematicians themselves. An attempt like the Pythagorean to explain the cosmos by numerical cycles and the cyclical world view as such existed in the Chinese tradition into the second century A.D.; but as astronomical observation became more precise, cycles came to be described algebraically in more complicated ways, and this cosmological view collapsed.[39] Experts in calendrical science in the T'ang era (seventh through ninth centuries) who became specialists in exact, empirical science did not consider expatiating on the cosmos to be a legitimate part of their work and they rejected cosmologizing altogether.[40] In the eleventh and twelfth centuries, cosmology once again became a subject of discussion among Confucian scholars; but the calendar makers were inclined to think that Chu Hsi "talked nonsense because he did not understand mathematics" and generally refused to consider the problem.

During the earlier period Japanese mathematicians, especially their founding father, Seki Kowa, frequently studied problems in calendrical science. Seki's investigations, however, were confined to the technical aspects of the Shou-shih calendar. As scientific mathematical activity of a kind that might be called "normal science" continued, the classical Pythagorean view of nature as based on mathematical principles disappeared from the *problematique* of Seki's followers. This was at a time when the essayist Nishimura Tosato was claiming that *sugaku* (the scholarly study of numbers) was an important subject while *san* (arithmetic) was a minor practical art.[41] Propounding a mathematical view of nature, he designated *sugaku* as a learned discipline that "investigates basic principles, constitutes a major element of divination, and was expounded by the Sages." Nishimura also criticized the activities of the *wasan* mathematicians as "arithmetic done by people of little consequence." This mathematical view of nature had to conform to the Confucian values of a Confucian society. Consequently, mathematical studies were not considered scholarly unless they made a contribution to "self-cultivation, husbandry and the pacification of society." They were devalued if the practitioner admitted a "desire to investigate mathematical principles merely for amusement."[42]

It is said that mathematicians such as Wada Yasushi made a living by practicing divination during the Tokugawa period, but this report may be based on a popular misconception deriving from the fact that diviners and mathematicians both used *sangi* (computing rods). In fact, mathematical calculations and divination based on calculating rods were entirely different.

Mathematics during the prefeudal period was completely practical and thus was at least socially legitimate. Even in the Tokugawa period appeals to practicality appear in introductions to mathematical works designed for such purposes as surveying, and these appeals apparently were accepted at face value. Practical mathematics, however, did not interest most mathematicians. *Wasan* lost even its practical character and became explicitly nonutilitarian, a situation that readily allowed Nishimura Tosato to dismiss it as simply an art form.

Let us compare this experience with that of the West. According to Alexander Koyré, Platonism had an important influence on early modern science because of its program for the mathematicizing of nature.[43] Although Platonism dominated intellectual discourse during the Renaissance, its indispensability for the Galilean school may well be doubted. A close reading of Plato's *Timaeus* shows that it had almost nothing in common with Koyré's notion of Platonism.[44] Galileo did, however, use popular Platonist ideas to legitimize his mathematical methods, an effort that had precisely the social effect intended. Thus Galileo presented Plato as the defender of geometry against the Aristotelians, who emphasized logic; and thus Platonism had an important role justifying the inclusion of mathematics in school curricula, and eventually in the emergence of modern science.

Unfortunately, Japanese mathematicians had no charismatic figure to invoke in opposing the Confucian tradition. If Mo-tzu had not been forgotten for two millenia, this might have been possible; but Mo-tzu and the yin-yang natural philosophers had almost totally disappeared from intellectual discourse in Tokugawa Japan. The mathematicians would have needed something in their own tradition approximating Galileo's Platonism to have become full members of intellectual society. No reiteration of references to divination or the six classical arts in the introductions to their books could have raised their status significantly.

Social Position of Tokugawa Mathematicians. Whenever scholars demand legitimacy from society, they display a sense of mission that reinforces their commitment. This sense of mission is associated with the rise of professions that in Western society are intellectually based associations not explicitly connected with worldly gain. Theology, law, and medicine were recognized as professions in medieval universities. In the early modern period, scientific researchers and technicians were also recognized as professional men. The largest profession in Tokugawa Japan was probably the class of Confucian scholars. One might also consider physicians, astronomers, and specialists in calendar making as professionals. But whether the mathematicians could be called professional is a difficult question.

A sense of mission includes a desire to achieve a lofty objective beyond immediate personal interest. Scholarship or science for their own sakes seemingly represent an early modern form of consciousness that developed after the emergence of scholarly elitism, especially that of the universities. Before this modern attitude became established, a scholarly discipline could form only when learning proved to be useful to the sacred or secular establishment that monopolized universes of meaning.

But as society became more complicated, intellectual groups managed to secure autonomy as third parties between various powerful agencies and were expected to ignore matters of direct economic interest. Since the Middle Ages, universities have proclaimed their independence as professional bodies, meanwhile maneuvering between religious and secular authorities. Even in Tokugawa Japan, with its very strict regulations, officially sponsored Confucian schools, and even heterodox Confucian schools were generally recognized as authentic disciplines by the establishment. Nor was there any problem with medicine, even for doctors in the private sector, because practical utility assured recognition. Fields of knowledge related to commercial production or manufacturing were invariably supported on their own terms. It was difficult, however, to discover in either Confucianism or in Baconian pragmatism any grounds on which to legitimate the activities of the *wasan* mathematicians.

The popular image of mathematics was that of the abacus. The social position of mathematicians was probably based on demand for their services in teaching people how to use this device. But daily use of the abacus did not require anything like the elaborate technique of the mathematicians. In merchant families, abacus manipulation was considered a form of spare-time study. Apprentices were introduced to it by the head of the business. There was a saying to the effect that "While use of the abacus is one of the most important things a merchant must learn, he should not take it too seriously. Excessive study will hurt business."[45] Studying more advanced mathematics than was required in business generally was forbidden, being considered a form of dissipation. Some mathematicians managed to make a living by opening schools. The majority of such schools were run by masterless samurai. The mathematical training they gave usually stopped with simple arithmetic and the calculation of interest rates. Thus, according to Professor Oya Shin'ichi, the *enri* calculus, which included the most sophisticated problems studied by the mathematicians, was not generally taught in these schools.

From the government's point of view, mathematics was closely associated with simple calculation and land surveying. But in the *Ryochi shinan* ("Introduction to Surveying") one finds such statements as "People who study mensuration say that not all mathematics is intended to be used by surveyors. According to them, there is nothing about mathematicians' theories that is contradictory to mensuration; but if you look at their work, it seems too much involved in mathematical theory and divorced from practice. And in general mathematicians' talk about surveying is all of this sort."[46] Or "Mathematicians' techniques are a distraction, with no utility whatever."[47] The traditions of the academic mathematician and of the practical surveyor were quite distinct. Because of an attack on the *Sampo jikata taisei* ("Manual of Practical Mathematics," 1837), written by Akita Giichi and edited by Hasegawa Kan'ei (both mathematicians of the Seki school), surveyors subject to feudal authority were reluctant to publish significant mathematical works for fear of their lords' reactions. Even Seki Kowa, founder of the *wasan* mathematical tradition, was warned about this. He disregarded the admonition, however, and later wrote a text describing approximate or simple methods for solving problems in the style of Yoshida Koyu's *Jinkoki.* Seki's disciples, however, considered that text a disgrace to their school.

Astronomy offered mathematicians far more sophisticated problems than did surveying. Examples from traditional Chinese astronomy include indeterminate procedures for calculating multiple conjunctions, the problem in spherical trigonometry of transforming equatorial coordinates into ecliptic coordinates on the sphere, and interpolation procedures for handling the equation of the center. The trigonometry and algebra that came to Japan with the Jesuits' later transmission of European astronomy may have opened up new mathematical vistas. There also were problems in navigational astronomy that the mathematicians Honda Toshiaki and Sakabe Hironao investigated. These topics were, however, considered astronomical problems and, as such, were separated from the mainstream of mathematical activity.

In the *han* schools, which were for the training of the sons of samurai, mathematics apparently became part of the curriculum about the 1780's.[48] Textbooks published by various domains seem to have included problems in applied mathematics taken from surveying, calendrical science, and navigational astronomy. Some mathematicians served as Bakufu astronomers, and others worked as accountants or surveyors in the domains. They saw their public duties as separate from their research in mathematics, however, considering the latter a private activity. They apparently honored this distinction to the point of not publicizing their research.[49] Astronomers and astrologers employed by the domains all used the algorithms of *wasan* mathematicians when they calculated but, aside from the writings of Nishimura Tosato, one finds no significant public comment on the *wasan* tradition in works by astronomical specialists. This may have been due to their bureaucratic consciousness. They gave no thought to the work of mathematicians, who belonged to the private sector. The result was that mathematics had no real place either in the government or in private life.

Nor did mathematical research have any base in the occupational system. Mathematicians' researches were separate from their occupations—which ran the gamut from warrior to farmer, artisan, or merchant.[50] The orthodox Seki school was typical in this respect.

In general, the mathematicians' greatest achievements appeared in their early years, although Mikami Yoshio's research showed that the *wasan* mathematicians were most active in their later lives.[51] According to him, this was due to the shallowness of the roots of *wasan* mathematics in the educational system, the inadequacy of the available textbooks, and the extraordinary amount of time required to become proficient. All of these factors surely had an important impact, but an even more influential factor may have been that the time and money required to indulge in such a leisurely activity came only in later life. Anyone who gave up his regular occupation for mathematics encountered financial problems, including difficulties in paying publication costs. In particular, samurai employed in the government had scruples about participating in such activities and, for the most part, published mathematical works only after retirement.

Unlike astronomers or physicians, mathematicians did not have to deal with occupational inheritance. Because of the special ability required to produce original achievements in mathematics, the occupation could not be passed on—indeed, in the economic sense it was not worth passing on. Mathematics thus did not establish itself as a specialty of certain families. The status of *wasan* teacher had no economic implications (such as guild protection), despite the licensing system established by Seki Kowa. Nor, in consequence, was there any basis for giving academic autonomy to mathematics, such as the modern university has provided in the West.

The prestige of any field of scholarship is bound up with the social status of those engaged in it. From that point of view, there is little reason to think that mathematicians were particularly respected by society in general. In China, Chu Shih-chieh and Ch'eng Ta-wei, authors, respectively, of the *Suan-hsüeh ch'i-meng* and the *Suan-fa t'ung-tsung*, from which *wasan* mathematics derived, were itinerant teachers. In Japan there also were mathematicians who traveled from place to place and were supported by wealthy patrons.[52] In point of livelihood, these intellectual salesmen were no different from traveling artists.

The gentlemen members of the Royal Society were not professionals in the sense of earning their livings by research but, rather, in the sense that royal patronage gave them a degree of institutional independence. In a formative period in either scholarship or the arts, when institutional protection is lacking, a leader will assemble a supporting constituency and create an occupational base for his activity. When scholarship becomes systematized, institutionalization and professionalization occur simultaneously. The Royal Society, which was officially recognized yet lacked an occupational base, was exceptional; but its recognition made its dignity and status as a scholarly organization secure. *Wasan* mathematicians had no such recognition. There was no system for training or recruitment. The occupational base consisted solely of the inadequate patronage of a few wealthy individuals. As a result, there was no agency to press the claims of mathematics upon society. Because their organizations lacked status, *wasan* mathematicians could not defend the scholarly aspects of their work effectively. They received little or no social recognition. Nevertheless, in certain respects, they were more active than the Royal Society.

The artistic character of *wasan* mathematics helps to explain this fact. In publishers' catalogs of the Tokugawa period, books on *wasan* were classified with materials on the tea ceremony and on flower arranging, which suggests that *wasan* was mainly viewed as a popular art. Art, however, involves the pursuit of aesthetic pleasure, while purely intellectual matters are regarded as scholarly. During the Tokugawa period there were several pursuits that were not scholarly occupations and had no academic prestige—for example, Japanese chess (*go* and *shogi*). It was probably the aesthetic and playful aspect of mathematics which provided scope for the development of *wasan* and its diffusion down to the commoners.

In his book *Les jeux et les hommes* (published in English as *Man, Play, and Games*), Roger Cailloit defines recreation as activity that is free, isolated, indeterminate, unproductive, rule-bound, and unrealistic. The activities of *wasan* patrons fit these six conditions perfectly.

Internal Logic of Wasan Development. What distinguished *wasan* from poetry, haiku, and the arts in general? In what sense was it a scholarly rather than an artistic form? While it would not be quite accurate to say that the methodology of *wasan* was that of a modern scientific discipline, it definitely did come closer than any other field of inquiry existing during the Tokugawa period. It had, for example, a way of asking questions that was very similar to that of modern science. Thomas Kuhn states that all scientific traditions begin with a paradigm or model for raising and answering questions in terms of which scientific progress will necessarily occur.[53] *Wasan* conformed to this pattern rather well. It did not command a very extensive body of knowlege, but it stated its problems and questions in a precise way. Among the disciplines of the Tokugawa period, *wasan* and mathematical astronomy were methodologically closer to modern exact sciences than were the schools of moral philosophy or of clinical medicine. Since the questions raised in *wasan* were not limited by traditional structures (save those it evolved itself), it was free to move off in new directions.

In its formative period *wasan* mathematics developed the unique custom of *idai* (bequeathed problems). A mathematician would pose scores of problems of several kinds at the end of a book. Another mathematician would publish answers to these problems and present his own in the same manner. According to convention a third mathematician might propose answers to the second set of problems and issue his own, in relay fashion. This interest in mathematical puzzles greatly stimulated the formation of *wasan* groups. They were influenced by Chinese experience, but there does not appear to have been anything like the custom of posing *idai* in China. The tradition began with twelve problems from the *Shimpen jinkoki* (1641). A succession of mathematical lineages soon developed and reached a peak during the lifetime of Seki Kowa. Certain themes continued to appear in these problems, which passed through a number of phases. Practically all of the important problems in the history of *wasan* date from the period of these *idai*.[54]

Mathematics has a strong puzzle-solving character. Problems need not be constrained by physical or social reality. Mathematicians can freely create new intellectual worlds which may violate the logical forms of daily language.

During the early period there were two types of *idai*—problems used in daily computation and those relating to geometry. In the Chinese mathematical tradition that influenced *wasan*, practical problems were more common, but later more whimsical problems were added. The *Jinkoki*, which defined the popular image of *wasan*, also emphasized practical problems. As *idai* passed through many generations, a trend toward purely intellectual or recreational problems developed among the heirs to the tradition, enthusiastic puzzle-solvers uninhibited by utilitarian constraints. The puzzlelike character of *wasan* mathematics made this development entirely predictable. Given the sense of *problematique*, practitioners were not inclined to select problems with practical applications. Moreover, other factors reinforced the trend. Once a problem had been abstracted in the form of a diagram, people did not concern themselves with its utility and could enlarge or develop it freely. The purer the mathematical character of a problem and the greater its detachment from practicality, the greater was the enthusiasm with which it was received. We shall consider below a typical example, the *yojutsu,* which involved fitting various large and small circles into a triangle.

It also seems significant that the impracticality of *wasan* precluded links to mechanics or optics like those of mathematics in Western science. The seventeenth-century *idai* were probably responsible for the leisure-oriented character of *wasan*.[55]

Commerce, surveying, and calendar making represented practical applications of mathematics.

The first two, however, did not offer very sophisticated problems. The degree of precision required in their calculations was much less than that demanded in the theory of errors or in higher-degree equations. Calendar making, of course, was an exact science; but the major problems presented by the Shou-shih calendar had already been solved by such eminent mathematicians as Seki Kowa and Takebe Katahiro. In theory, the planetary motions should have generated problems fully as interesting as the theory of epicycles in Western astronomy. The Japanese art did not emphasize planetary movements, so they were not investigated very thoroughly. Ultimately the absence of kinematic and dynamic problems in Japan's scientific tradition handicapped and retarded *wasan*'s approach to analysis and proved decisive in its race with the Western tradition.

When *idai* were popular, from about 1650 to the early eighteenth century, the natural sciences did not develop to any extent. Science from the West had not yet been imported. It was during this period and from these *idai* that *wasan* mathematics created and established its significant problems, although in certain respects prematurely. The topics investigated were taken from such concrete problems as the volume of a rice bag or measuring cup. They were often focused on diagrams of circles or cones that had to be solved for a numerical value. As this trend progressed, the enthusiasts of pure mathematics gave little thought to the relationship between observation and its practical meaning; they simply invented fictitious problems whenever they wished. Once they discovered a strategic or paradigmatic problem, practical problems of astronomy, trigonometry, or Western logarithms were no longer considered legitimate. The eccentric genius Kurushima Yoshihiro wrote: "In mathematics it is more difficult to raise a problem than to give the answer. Only mathematicians who cannot invent problems borrow them from calendrical science."[56] In short, the idea that looking for subject matter in society or nature was undesirable had already developed by the eighteenth century, when the notion of mathematics for its own sake emerged.

In the practical problems of applied mathematics, it is important to obtain an answer stated as a numerical value. Given the determination of topics in this field by social or natural conditions, it is entirely appropriate that obtaining numerical solutions should be considered more important than inventing problems. On the other hand, the asking of questions is essentially unlimited in pure mathematics, as Kurushima implied, and is necessarily considered supremely important. As with crossword puzzles, inventing the question is more difficult than supplying the answer. That one person paves the way by inventing a problem, while many others follow in trying to solve it, is part of the normal science tradition and further underscores the process by which normal science is conducted.

This was not, however, the way in which *idai* were developed and passed on in the early period (before Seki Kowa's time). The form of the problems at that time was not fixed; and when people invented problems, they did not simply adhere to those devised by predecessors. During that time there was a shift from practical to pure mathematical problems. Numerical solutions converged toward diagrammatic problems using the *tengen* algebra (based on the use of computing rods). *Wasan* did not yet have a characteristic type of problem.

As time passed, the problems became more intricate and multifaceted. Inventors did not simply present problems that they themselves had already solved, since other mathematicians considered that too simple—even foolish. There was an emphasis on solving problems by some unusual means or in presenting problems for which it was not known whether a satisfactory solution existed. People would take up a mathematical challenge and expend considerable energy trying to solve it. The way in which *idai* were presented gave an enormous stimulus to the competitive spirit of later mathematicians. Difficult problems constituted an enduring challenge. There have been, of course, similar examples in the history of Western mathematics—for instance, the unsolved problem of trisecting an angle, which has continued to engage the interest of mathematicians.

As problems became more complex, however, impossible problems appeared, and mathematicians began expending excessive energy to little effect. In the *Sampo kokon tsuran* the following comment appears: "The shallowness of present technique and reasoning, together with the enormously complicated effort expended, lead one to think that the really interesting problems have all been exhausted." Such problems could not constitute paradigms, could not lay the groundwork for normal science, and could do nothing but create confusion. Indeed, the confusion suggests that Japanese mathematics was at a preparadigm stage. Consequently, what brought Seki Kowa his enormous reputation was his creation of the basic paradigm for posing and answering questions in an intellectual setting where almost total confusion had prevailed earlier.

After the importation of the *tengen* technique, changes occurred in the way questions were posed through the *idai*. In the earlier period, *idai* were highly diverse and multifaceted; now they emphasized higher-degree equations solved through the use of the computing rods. The confusion of the period was compounded by the popularity of *idai* deliberately designed to solve increasingly complicated problems by transforming systems of simultaneous equations into a single equation of increasingly higher degree. These problems were known as *handai* (troublesome problems). The tendency may have been an aberration, but it prompted Seki Kowa to introduce an important innovation—the *tenzan* algebra, a system for expressing unknowns in symbols similar to *A, B, C,* in order to use simultaneous equations freely.[57]

Seki also developed a theory of equations based on the existence of negative and imaginary roots, which he treated in his *Daijutsu bengi* ("Discussion of Problem Specification") and *Byodai meichi* ("Clarification of Impossible Problems").[58] From China's pragmatic mathematical tradition *wasan* had inherited the idea that an equation can have only one root.[59] Seki, however, introduced discussion of negative and imaginary roots, and tried to interpret their meaning. Taking an approach characteristic of *wasan*, he transformed and "corrected" such problems to provide real positive solutions, rejecting the implications of the original problem setting. Thus his theory of equations ruled out further development toward imaginary and complex number theory in *wasan*.[60]

Seki's writings on the theory of equations perpetuated *wasan*'s orthodox way of asking questions. These writings were included in a seven-volume work transmitted esoterically by the Seki school. They were studied and passed on by pupils who were eminent enough to devise new problems themselves. Since Seki had previously laid down guidelines for solutions in the *tenzan* algebra, one may say that he fully paved the way for *wasan*'s later development.

Seki Kowa was not the only outstanding mathematician of the period. He certainly had the intuition of a genius; but if he had been an isolated figure too far ahead of his time, the paradigm he laid down would not have paved the way for normal science and there would have been no Seki school. During Seki's lifetime his school was extended by such eminent disciples as Takebe Katahiro and Matsunaga Ryohitsu. Seki was placed on a pedestal within the tradition. One may suppose that the special treatment accorded him was instrumental in establishing the diploma system. In the introductions to *wasan* books, the origin of the discipline is always attributed to Seki, not only in his own school but also in the Saijo school of Aida Yasuaki. Certainly the tendency of the *wasan* mathematicians to revere Seki limited their field of vision, yet it would seem from comparing such works as the *Suan hsüeh ch'i-meng* and the *T'ien-wen ta-ch'ang kuan-k'uei chi-yao* with his own that Seki himself was more influenced and informed by Chinese writings than his later disciples ever imagined. Everything in the *wasan* tradition was, however, referred back to Seki, and the original Chinese mathematical works were ignored.[61] This was very different from the attitudes of Chinese mathematicians at the Ch'ing period, who turned their attention toward ancient Chinese mathematics as they mastered the Jesuits' astronomy.

Seki Kowa left answers to a large number of *idai* during the period of their greatest popularity. In fact, the theory of equations seems to have been born from the many different kinds of *idai* that he considered. There is no indication, however, that Seki himself left behind any *idai*; and after his time the practice of issuing them declined considerably. Their frequent appearance corresponded to a preparadigm stage in the delineation of mathematical problems. *Idai* were scrutinized and restated according to the principles of equation theory, and the significant ones were passed on. The paradigm emerged when the techniques for solving these problems by means of *tenzan* algebra were given; and at that point a way of asking and answering mathematical questions emerged that was quite different from what had existed in the preparadigm period. Creation of the paradigm was not the achievement of Seki Kowa alone; but circumstances led to his being given credit for it and he was, in fact, at the center of the *wasan* tradition. Since the procedures for raising and answering questions were fixed, it was meaningless to set forth a large number of *idai*. With the issues about structure clarified, subsequent mathematicians readily resolved particular problems. *Enri* (circle theory) calculus was one of the areas that attracted attention. This technique developed from mensuration of the circle and led to the development of linear progressions and analysis. It probably began with either Seki or his leading disciple, Takebe Katahiro.[62] In any event, its inclination to the analytical approach of Ajima Naonobu is apparent. *Enri* calculus was applied not only to problems involving the circle, but also to curves and curved surfaces in general. Wada Yasushi helped to develop mathematical analysis by compiling tables of definite integrals and applying them to the mathematically infinite and infinitesimal. The Takuma school of Osaka developed a calculus that in some respects was superior to that of the Seki school.

Mathematical problems are not limited to those posed in nature or by society. But a pattern of development in mathematics will change according to the kinds of problems taken up; different choices of problems may create different mathematical worlds. The *enri* calculus, which developed from problems concerning the arc, an important topic in astronomy, coincided in its results with the Western-style calculus. The course it followed, however, was completely different from that of Western mathematics, which began with problems in dynamics. We should say, on the other hand, that dynamical problems offered mathematics greater scope for development than those concerned with arcs and circles, simply because the element of time was involved. There were greater limits to the problem development possible in *enri*.

Another stimulus to the development of *wasan* was the *ema sangaku* (pictured mathematical tablet) form that came after the *idai* tradition. On these wooden plaques were written both problems and answers; they were offered at shrines and displayed there. The best mathematicians made their accomplishments known through books, but it was largely the custom of *sangaku* that supported the activities of the *wasan* enthusiasts. That *wasan* was a hobby costing money to pursue is best shown by the elegant diagrams that embellished such work. In fact, the offering of *sangaku* had a strong attraction for local gentlemen over other art forms—drama, music, poetry—as a way of making their work known; and their frequent indulgence in it shows a desire to keep themselves more or less before the public at all times.

The most important kind of problem taken up in *sangaku* was called *yojutsu* (packing problems), which involved the attempt to inscribe the largest possible number of small circles in a larger circle. *Sangaku* also treated problems involving solar eclipse predictions in the Shou-shih calendar; but table of eclipses, with their arrangements of letters and numbers, could not attract attention even if displayed as framed pictures. Popular interest was limited to *yojutsu* decorated with designs of circles and squares. The pictures were intended, as far as possible, to produce the impression that the designer had obtained the solution to a difficult problem through a complex diagram. Thus, while *yojutsu* seemed to concern itself with very difficult problems, its emergence was not very significant mathematically.

In the early development of *yojutsu*, there was a tendency for different techniques to compete on the same problem. Later, however, the method of solving them became fixed. There was a tendency to devise new problems to which the standard technique could be applied. Problems became increasingly complicated while the technique scarcely developed.

Enri calculus and *yojutsu* both developed as normal sciences but in somewhat different ways. The *enri* calculus developed step by step: when one problem was solved, its result was used to solve a problem requiring deeper investigation; and in the process there appeared what might be called subparadigms. In principle *yojutsu* was also a problem-solving technique, but its pattern of development amounted to merely a series of transformations and variations. Moreover, its technique was not based on a demonstrational logic in the Euclidean style, and did not aspire to general principles of problem solving. During its 200-year history, *yojutsu* yielded several useful by-products, but its reliance on casual inspiration ranks it well below *enri* calculus in scholarly value.[63] One could have referred to *yojutsu* as a Japanese form of geometry only if it had gone beyond mere puzzle solving and had achieved a general methodology. In fact, it did not move very far toward analytical geometry. Because it was purely a puzzle-solving technique, it did become an acceptable recreation for amateurs, who were inclined to consider logical rigor in bad taste.

Wasan mathematicians apparently did not fully realize the importance of logical foundations; rather they valued insignificant, complicated, and overly elaborate problems.[64] When Euclid's *Elements* first appeared in Japan, people said the simplistic, poorly developed, and inferior character of European mathematics compared with *wasan* could be determined just from looking at its pictures—a reaction suggesting that *wasan* was not the kind of discipline to raise basic questions. One might even call it an art form that had as its major goal the refining of trivialities. On rare occasions someone like Takebe Katahiro, who respected precision, would come along; but because the *wasan* mathematicians did not generally raise basic questions, they made no revolutionary breakthroughs and confined themselves to refinements and improvements on the paradigm set forth in Seki Takakazu's time.

Mathematics, especially a pure form like *wasan*, differs from the natural and social sciences in the absence of checks imposed on it by the objects it investigates. It does not follow the same development as physics, in which the interpretation of a phenomenon can change completely during a scientific revolution. Non-Euclidean geometries can coexist with Euclidean geometry, and the replacement of the latter by the former is far from inevitable. Non-Euclidean geometries are not so much replacements as variations on Euclidean geometry. But unless one counts such trivialities as different assortments of diagrams in *yojutsu* problems, one would have to say that *wasan* had very few basic variations and, what is more, that almost none of its basic notions were conceptually deep.

Perhaps the conceptual poverty of *wasan* can be explained by the derivative character of the culture in which it developed. Most of Japan's basic cultural patterns were of foreign origin. Highly refined art forms evolved from these patterns, but the fact that they were borrowed precluded critical examination of them. By contrast, the Chinese were forced to reconsider their mathematical heritage during the intellectual crisis precipitated when the Jesuits introduced European exact sciences. The Chinese approach was characteristic of a society with strong autochthonous values; scholars used the new learning to resuscitate forgotten mathematical conceptions from China's past, and thus preserve the identity of traditional science. In Japan, *wasan* mathematicians raised neither the question of historical origin nor any other involving the theoretical foundations of mathematics.

Given that Japanese astronomers and physicians constantly compared China and Europe and took from either what they judged to be good, why did the mathematicians remain in their own world? In the first place, there was a difference in disciplinary structure between medicine or the natural sciences and mathematics. Practitioners of the former could readily determine what was better and what was worse by having both Eastern and Western examples before them. In natural science the search for a single truth could more or less readily lead to replacement of inferior Chinese conceptions by Western ones. In mathematics the belief that *wasan* and Western mathematics differed only in style allowed the two to exist side by side. This difference was rather like that between Japanese *shogi* and European chess. In this sense *wasan* and other forms of pure mathematics are closer to art forms than scholarship. The *wasan* mathematicians did not feel threatened by the importation of Western mathematics and remained in their own artistic world. Comparing the applied mathematics of China, which emphasized astronomical orientation, with the pure mathematical tradition of *wasan*, one perceives a clear difference in the extent to which practitioners of each considered the Western impact threatening.

According to the *Tokyo-fu kaigaku meisai sho* ("Survey of Schools in Tokyo"), a report on private academies in Tokyo prepared for the inauguration of the early Meiji school system, Western-style mathematicians trained at the Survey Office or Nagasaki's Naval Training Center, and the more numerous *wasan* mathematicians of the Seki tradition belonged to sharply differentiated groups. Contact or movement between the two groups seems to have been quite difficult. Modern Japanese mathematics really began when *wasan* mathematicians were superseded by those who had adopted Western styles. It would seem that differences in the ability of astronomers, physicians, and mathematicians to restructure their disciplines and respond to foreign knowledge is explained primarily by the fact that the first two groups were professions and the last had no recognized occupational base. Occupational concerns of the first two groups made some awareness of Dutch studies inevitable.

It seems significant that *wasan* mathematics persisted longest in the northeast, where contacts with Europe and general cultural development lagged furthest.[65] During the Meiji period, Western mathematics came to dominate in the cities, following the establishment of the modern school system, and *wasan* was essentially banished to the countryside. It became, so to speak, an exotic flower blooming by the roadside of civilization.

Japanese Mathematics and the Pure Mathematics of the West. Could early modern mathematics in the West also be described as an art form? After all, people no longer accept the Pythagorean notion that cosmic mysteries can be discovered in the nature of numbers, nor do they believe that divine attributes can be deduced from the transcendental axioms of Euclidean plane geometry. J. W. N. Sullivan, author of *A History of Mathematics in Europe* (1925), asks why mathematicians enjoy greater social esteem than chess players, given that they no longer claim to be pursuing a single absolute truth.[66] The artistic spirit of *wasan*, moreover, seems to embody the essential ethos of mathematics; and one suspects that Platonism gave mathematics an excessively authoritarian aura that may not be essential to its nature.

The European public apparently viewed mathematics as the handmaiden of science. This handmaiden proved to be more competent in the West than did its Chinese or Japanese counterpart. It helped bring about the seventeenth-century scientific revolution and constituted one of the keystones of the mechanical view of nature. The common image of the mathematical practitioner in that period was apparently that of a man who makes a living by producing and selling calendars or maps, or by conducting land surveys. The names of such people seldom appear in the histories of mathematics—in contrast with those who enjoyed royal patronage. Thomas Wright, a notable figure in the history of cosmological theory, was such a man. He came from a tradition separate from academic mathematics, and is thus excluded from histories of scientific astronomy.

Academic mathematics, however, retained its links with science well into the eighteenth century. *Wasan*'s weaning from science, by contrast, occurred quite early. In the eighteenth century, applied mathematicians sought numerical solutions. The more academic applied mathematics became, however, the greater was its tendency to seek elegant solutions of differential equations in preference to crude numerical calculations. This trend seems to have been somewhat similar to the dominant mentality in *yojutsu.*

Since applied mathematics dealt with issues posed in nature or society, numerical solutions were essential. In contrast, because *wasan* problems were recreational in character, practical issues and numerical values were beside the point. Nevertheless it remained characteristic of *yojutsu* that solutions were expressed concretely.

During the nineteenth century, pure mathematics in Europe tried to detach itself from various practical applications. Thus, even if numerical solutions were obtained, they remained nonessential and had little meaning. To this extent mathematics inevitably became more abstract. By contrast, *wasan* could be described as a form of pure but not abstract mathematics. The kind of numerical answers sought by *wasan* had no practical significance. Numerical solutions were sought only to determine the winner of what amounted to a sporting competition.

After the seventeenth century, the abandonment of Archimedean mathematical rigor in the West opened the way for the development of a normal scientific tradition in the application of calculus to mechanics.[67] In the nineteenth century, through a search for precision and logical rigor, mathematics became independent of science for the first time. Mathematicians of that era were very proud of this development and sought the origins of their discipline not in the Renaissance or the seventeenth century but in classical Greece. They even styled themselves the Greeks' successors.[68] Thus Greece was revived as the fount of "modern" mathematics.

In this context the differences between the Euclidean tradition and that which developed from the *Jinkoki* would seem to be clear. Western mathematics was moving toward greater reliance on generalization and rigor, while *wasan* emphasized the solution of puzzles by finely honed intuition. Although *wasan* won its independence before Western mathematics arrived, it did so in a rather problematical manner. *Wasan*'s independence was merely separation from science and practical application. Separate though it was, computational mathematics did not extricate itself from the form of mathematics that seeks numerical solutions to problems. So long as the major emphasis was placed on quantitative calculation, questions about the quality of the underlying theory were difficult to raise.

Nor was Western mathematics prompted to raise basic questions through a concern with such quantitative problems as how to obtain an approximate decimal fraction. Potentially revolutionary developments were hidden in problems of quality. A new paradigm finally did separate itself from, and became independent of, science by returning to qualitative fundamental questions. From the fact of its independence or separation it developed the possibility of a new intellectual universe.

In trying to explain how this separation occurred, one must necessarily look to the institutional background. In the Western academic tradition, Platonism supported the recognition of mathematics as a legitimate and established subject. Thus, in the early modern period, calculus and analytic geometry, unlike other new disciplines, could be recognized immediately because there were posts for mathematicians in the university system.

In the history of science, the nineteenth century was a period during which the German university system occupied center stage. Kant's *Der Streit der Fakultät* tells how the universities reconstituted, under the impact of modern science, an organizational structure that had existed since the Middle Ages. The newly constituted faculty of philosophy in particular acquired equal status with the traditional faculties of theology, law, and medicine. As such, it was able to offer more than preparatory training for the three higher faculties. Elementary algebra and Euclidean plane geometry were introduced into the Gymnasium, while differential and integral calculus and analytic geometry were taught in the university. It is conceivable that mathematics might have had to serve as the exclusive handmaiden of such physical sciences as mechanics, physics, and astronomy. In the nineteenth century there were disciplinary wars among the various specialized subjects, particularly over the constitution of chairs in the faculty of philosophy. Were mathematics to hold a subordinate position to physics, it might well be absorbed into the dominant disciplines. In the eyes of the public the value of a new field was recognized to the extent that it contributed to society; but in the universities, disciplines that raised fundamental questions had higher status than pragmatic ones, and were more appealing to students capable of responding to their professors' enthusiasms. Each specialty and every chair justified itself in this academic system.

For mathematics to become independent in the universities it first had to sever its ties to physics. Rigorously questioning the logical basis of mathematical assertions used by physicists elevated mathematics to be the queen of the sciences. A typical example is found in the career of Karl Weierstrass (1815–1877), preeminent figure in the German academic community and founder of the mathematics seminar at the University of Berlin. We do not know how far Weierstrass played academic politics as he developed his discipline. We may assume, however, that students who decided to prepare at the university for careers in mathematics were glad that he had enhanced the prestige of the mathematical profession.

Fundamental paradigms (such as that of Galois) by their nature emerge outside the established line of development, and often outside the universities. They are eventually accepted there, developed there into disciplines, and follow the course of a normal science. All disciplines represented by chairs in the German universities were considered *Wissenschaften.* Even if a discipline had no stronger link to a more comprehensive system of values than that of the advancement of learning, its status, once it became part of the university system, was secure.

Since no tradition of university scholarship existed in Tokugawa Japan, it is unlikely that any of *wasan*'s supporters would have rejoiced at (or even comprehended) the advent of a Weierstrass. There was no university environment to tolerate or even encourage questioning of fundamentals. As for the local devotees of *yojutsu* who patronized the Japanese mathematicians, they viewed the study as an embellishment, and were uninterested in close scrutiny of fundamentals.

As we have seen, Western mathematics, by precise and rigorous questioning of fundamentals, overcame its position as the handmaiden of science, and came to be widely considered the most basic and independent of disciplines. But Tokugawa mathematics merely separated itself from practical problem-solving and indulged in aesthetic pursuits. It would appear that the difference between European and Japanese mathematics is related to a difference in patronage. Through its recognition in a university system, the former shared state patronage redistributed according to academic criteria. The latter had only the direct patronage of individuals enthusiastic about an art form, with their own ideas about how their money was to be spent.

Conclusion: The Japanese View of the Laws of Nature.

Primarily through discussing Tokugawa practitioners, I have tried to describe the dominant view of science and of the laws of nature that existed during that period. What I have tried to argue in part is that aside from notions of law, the Japanese at that time had a conception of nature different from what we have today.

Our present conception of law in nature, stated in value-free terms, was produced in the nineteenth-century university. Earlier intellectual activity pursued much wider goals than the rigor of modern science allows. In that early intellectual world, modern concepts of the laws of nature were applied only in limited situations. Similarly, in Japan one readily discovers, from the tone of the introductions to books on calendrical science, medicine, and mathematics, that the academic notion of science for its own sake did not exist in Tokugawa society.

In earlier times, science in the West was pursued on the assumption that its investigations would demonstrate the glory of God; in Japan the ideology of Tokugawa science derived from Confucian emphasis on individual moral cultivation and social harmony. Morality was the basis of law; laws of nature conformed to and were necessarily subordinate to it. Thus, astronomy and medicine in Japan ultimately had to subordinate themselves to an essentially Confucian set of priorities in order to guarantee respect for their status as disciplines. Pursuits like *wasan*, which diverged from moral values or had nothing to do with them, not only became isolated from the Confucian framework but remained a recreational art.

In its unconcern for moral values *wasan* was the discipline closest to modern science in the Tokugawa period. Takebe Katahiro consciously used the term "law" in his methodological writings, saying: "The establishment of laws provides the basis for technique; thus mathematics needs to have laws."[69] Modern science's objectives consist in trying to establish or prove certain laws, not in defining the metaphysical realities that preoccupied scientists for centuries. Neither does modern science have such grandiose and far-reaching objectives as alchemy, which tried to prolong life indefinitely, or astrology, which sought to predict the course of nature and human affairs. It tries, rather, to achieve immediately foreseeable objectives. As Japan accepted modern science, astronomers, unlike Shibukawa Harumi, maintained a mechanistic framework, and avoided using historical changes in the celestial movements to explain changes in the heavens. In coping with disease, early modern physicians emphasized solidistic explanations rather than supposing that the human body is governed by temporal vicissitudes as doctors in the Chinese tradition had done. Attempts to prolong life indefinitely and to interpret celestial phenomena as portents were believed to impede technical progress, and thus were abandoned.

As science has restricted itself to universal and objective features of physical phenomena, the idea that values inhere in the physical world has been excluded; as the center of activity shifted from one European state to another national tendencies have dropped out. These tendencies toward objectivity and universality would have been uncongenial to Tokugawa thinkers, to whom the ideals of modern Japanese scientists would make little sense.

Unlike Joseph Needham, I do not believe that the Chinese and Japanese scientific traditions were converging toward the same ends as early Western science. Perhaps because of Needham's great esteem for Chinese science, he stresses China's priorities in discovery and invention.[70] Some of these claims of priority are unquestionable, but when Needham compares and evaluates the contributions of China and the West, he uses proximity to today's standard of knowledge and technique as his criterion of value.

Today's standard, however indispensable to the working scientist, is too parochial to be useful in evaluating the past. It gives unwarranted normative value to knowledge much of which will shortly become outdated, and it accepts the research emphases of fields shaped and dominated by the educational, career, and fiscal patterns of North American, Western European, and Soviet instititutions.

Treating today's understanding as more than transitory may seem to Needham unavoidable if he is to make China's great contributions intelligible to Westerners. Nevertheless there is ample room for doubt that, as Needham believes, Chinese and Western science had the same objectives.[71] Would Chinese science or that of Tokugawa Japan have developed in the same direction as that of Europe in the absence of influence from the latter? Needham can offer no convincing support, beyond a profession of faith, for his affirmative opinion. It seems to me more likely that East Asian and European science were diverging. The goals that were explicitly stated in China and Japan differed fundamentally from those articulated in the West, and the intellectual frameworks too were so different that the implicit goals we deduce from them also do not greatly resemble their Occidental counterparts. Many discoveries which seem to have been made on both sides of the world lose much of their similarity (especially similarity of significance) once they are examined carefully in context.

Japanese science developed within a framework of Confucian values. When Chinese culture was first accepted by Japan, science was included as part of the court culture and bureaucratic system. Japan was earlier practically a *tabula rasa* so far as the intellectual aspect of science was concerned. But by the time Western science entered Japan during the Tokugawa period, the ideological aspects of traditional Japanese science had been fortified by the recently developed Japanese Confucian system. To put it as simply as possible, the naturalistic and metaphysical aspects of Japanese science had been deemphasized, while its moral dimensions were stressed. The collision of Western science with traditional values was headed off by transforming the new science according to the dictum of Sakuma Shozan about "Eastern morality, Western art forms." The outcome in the early Meiji period was a highly pragmatic science, utilitarian and materialistic, quite lacking in moralistic connotations. Whether strictly speaking it was an art form was debatable, but its impact on the intellectual and spiritual transformation of Japan was delayed and minimized.

NOTES

1. At the end of the nineteenth century, Theodore Merz, in *A History of European Thought in the Nineteenth Century*, 4 vols. (London, 1903–1914), classified views of nature as astronomical, atomistic, mechanical, physical, morphological, genetic, vitalistic, psychophysical, and statistical.
2. Shigeru Nakayama, *A History of Japanese Astronomy, Chinese Background and Western Impact* (Cambridge, Mass., 1969), 19.
3. *Ibid.*, ch. 6.
4. Kawakatsu Yoshio, "Shigaku ronshu" ("Treatise on Historiography"), in Asahi shinbun (*Chugoku bunmei sen* 12, 1973), 24–25.
5. Joseph Needham, *Science and Civilisation in China*, II (Cambridge, 1956), sec. 18.
6. Shigeru Nakayama, "Educational Institutions and the Development of Scientific Thought in China and the West," in *Japanese Studies in the History of Science*, no. 5 (1966), 172–179.
7. Shigeru Nakayama, *Rekishi toshite no gakumon* ("Academic Traditions"; Tokyo, 1974), 74–78.
8. I showed a table of contrasts between these two treatises in Nihon shisoshi taikei, *Kinsei kagaku shiso, ge* ("Japanese Thought [series], Modern Scientific Thought, II"; Tokyo, 1971).
9. Shigeru Nakayama, "Accuracy of Pre-modern Determinations of Tropical Year Length," in *Japanese Studies in the History of Science*, no. 2 (1963), 101–118.
10. Shigeru Nakayama, "Cyclic Variation of Astronomical Parameters and the Revival of Trepidation in Japan," *ibid.*, no. 3 (1964), 68–80.
11. Nathan Sivin, "Cosmos and Computation in Early Chinese Mathematical Astronomy," in *T'oung Pao*, **55** (1969), 1–73.
12. *Sorai sensei tomonsho* ("Queries and Answers of Master Sorai"; 1727).
13. Shigeru Nakayama, "Edo jidai niokeru jusha no kagakukan" ("Confucian Views of Science During the Tokugawa Period"), in *Kagakusi kenkyu*, no. 72 (1964), 157–168.
14. The main source of information on this conflict is Koide Shuki, "Rarande yakureki zenbun" ("Preface to the Translation of Lalande"), preserved in the Japan Academy.
15. *Ibid.*
16. Preserved in Kunaisho, Zushoryo.
17. Preserved in Naikaku Bunko.
18. Shigeru Nakayama, "Diffusion of Copernicanism in Japan," in *Studia Copernicana*, **5** (1972), 153–188.
19. "Tengaku zatsuroku" ("Miscellaneous Records on Astronomy"; n.d.), preserved in Naikaku Bunko.
20. Gregory Zilboorg, *A History of Medical Psychology* (New York, 1941).
21. Ogawa Teizo, "Meiji zen Nihon kaibo gakushi" ("History of Anatomy in pre-Meiji Japan"), in *Meiji zen nihon igakushi* ("History of Medicine in pre-Meiji Japan"), I (Tokyo, 1955), 159–166. Also see his "Kindai igaku no senku" ("Forerunners of Modern Medicine"), in *Nihon shiso taikei, yogaku, ge* ("Japanese Thought [series], Western Learning, II"; Iwanami, 1972), 506–509.
22. Joseph Needham, "Science and China's Influence on the World," in Raymond Dawson, ed., *The Legacy of China* (Oxford, 1964), p. 239.
23. Manfred Porkert, *The Theoretical Foundations of Chinese Medicine* (Cambridge, Mass., 1974).
24. Tsuruoki Genitsu, *Idan* ("Medical Critique"; 1795).
25. "Shen Kua," in *Dictionary of Scientific Biography*, XII, 369–393.
26. Ogawa Teizo, *op. cit.*, 92 ff.; and Uchiyama Koichi, "Nihon seiri gakushi" ("History of Physiology in Japan"), in *Meiji zen nihon igakushi*, II, 122 ff.
27. *Yakucho* ("Pharmacological Effects"), vol. II; repr. in Nihon shiso taikei, *Kinsei kagaku shiso, ge*, 256. Also see Otsuka Keisetsu, "Kinsei zenki no igaku" ("Medicine in the Early Tokugawa Period"), *ibid.*
28. Shigeru Nakayama, "Kindai kagaku to yogaku" ("Modern Science and Western Learning"), *ibid.*, 460–461.
29. "Scientific Productivity and Academic Organization in Nineteenth-Century Medicine," in Bernard Barber and Walter Hirsch, eds., *The Sociology of Science* (New York, 1962), 305–343.
30. Amano Ikuo, "Kindai Nihon ni okeru koto kyoiku to shakai ido" ("Higher Education and Social Mobility in Modern Japan"), in *Kyoiku shakaigaku kenkyu*, no. 24 (1969), 84.
31. Maruyama Kiyoyasu, "Chiho ni okeru wasanka no shiso to seikatsu" ("Thoughts and Lives of Local Traditional Mathematicians"), in *Shiso*, no. 356 (Feb. 1954); repr. in *Kyodo shiso no bunkenshu*, no. 2 (1966), 13–31.
32. Robert G. Frank, "Institutional Structure and Scientific Activity in the Early Royal Society," in *XIVth International Congress of the History of Science, Proceedings* (1974), IV, 82–99.
33. Arnold Thackray, "Natural Knowledge in Cultural Context: The Manchester Model," in *American Historical Review*, **79**, no. 3 (June 1974), 672–707.
34. For instance, Mikami Yoshio, *Bunkashijo yori mitaru Nihon no sugaku* ("Japanese Mathematics From the Viewpoint of Cultural History"; Tokyo, 1947); and Ogura Kinnosuke, *Nihon no sugaku* ("Japanese Mathematics"; Iwanami, 1940).
35. Hirayama Akira, *Seki Takakazu* (Koseisha, Tokyo, 1959), 178–179.
36. For instance, *Daijutsu bengi* ("Discussions on Problems and Solutions"), reprinted in *Sekiryu sanpō shichibusho* ("The Seven Books of the Seki School"; Tokyo, 1907).
37. There is not much citation of traditional doctors of mathematics, but Aida Yasuaki wrote a few words on the Miyoshi and Kotsuki families in his preface to *Seiyo sanpō*.
38. Kawakita Chorin, *Ken'o sanpo* (Tokyo, 1863), preface.
39. Sivin, *op. cit.*
40. *Hsin T'ang shu* ("New History of the T'ang"), Li chih ("Treatise on Mathematical Astronomy"). Also see Yabuuchi Kiyoshi, *Zuitō rekihoshi no kenkyū* ("Researches in the History of Calendrical Science During the Sui and T'ang Periods"; Tokyo, 1944).
41. "Sugaku, sangaku no hanashi" ("Topics on Mathematics and Arithmetic"), in *Sudo shodan*, I (1773).
42. *Ibid.*
43. Koyré's assertion that "the allusions to Plato so numerous in the works of Galileo . . . are not superficial ornament born from his desire to conform to the literary mode of the Renaissance . . . nor to cloak himself against Aristotle in the authority of Plato. Quite the contrary, they are perfectly serious and must be taken at their face value" is groundless. Alexandre Koyré, "Galileo and Plato," in Philip Wiener and Aaron Noland, eds., *Roots of Scientific Thought* (New York, 1957), 174.
44. Shigeru Nakayama, "Galileo and Newton's Problem of World-Formation," in *Japanese Studies in the History of Science*, no. 1 (1962), 76–82.
45. Okumura Tsuneo, "Kinsei shōnin no sanyo ishiki" ("Mathemetical Concern of Modern Merchants"), in *Shuzankai*, **19** (Sept. 1969), 5.
46. Murai Masahiro, *Ryochi shinan, kohen* ("Introduction to Surveying"; II, 1754), preface.
47. *Ibid.*, 4.
48. Kasai Sukeharu, *Kinsei hanko niokeru gakuto gakuha no kenkyu* ("Researches on Academic Tradition and Schools in Modern Fief Schools"), I (Tokyo, 1969), 9, 75.
49. Akabane Chizuru, "Shomin no wasanka to hanshi no wasanka" ("Commoner Mathematicians vs. Samurai

Mathematicians"), in *Kagakusi kenkyu*, no. 34 (1955), 25.
50. Hirayama Akira, "Wasanka no shokugyo ("Occupations of Japanese Traditional Mathematicians"), in *Kyodo sugaku no bunkenshu*, no. 1 (1965), 192–198.
51. Mikami Yoshio, "Sugakushijo yori mitaru nihonjin no dokuso noryoku" ("Japanese Creativity in the History of Mathematics"), in *Wasan kenkyu*, no. 11 (Oct. 1961), 7–8. Originally written in 1927.
52. Mikami Yoshio, "Yureki sanka no jiseki ("Facts About Wandering Mathematicians"), in *Kyodo sugaku no bunkenshu,* no. 1 (1965), 166–184.
53. Thomas Kuhn, *Structure of Scientific Revolutions* (Chicago, 1962).
54. Hosoi So, *Wasan shisō no tokushitsu* ("Characteristics of Mathematical Thought in the *Wasan* Tradition"; Tokyo, 1941), 44.
55. For instance, Ogura Kinnosuke, *Nihon no sugaku* ("Japanese Mathematics"; Tokyo, 1940).
56. Ajima Naonobu, *Seiyo sanpō* (1779), postscript.
57. Hosoi, *op. cit.*, 48–49.
58. Both repr. in *Sekiryu sanpō shichibusho* (see note 36).
59. Fujiwara Matsusaburo, *Nihon sugakushi yo* ("Epitome of Japanese Mathematics"; Tokyo, 1952), 127.
60. Hosoi, *op. cit.*, 69–78.
61. Yabuuchi Kiyoshi, *Chugoku no sugaku* ("Chinese Mathematics"; Tokyo, 1974), criticizes the narrowness of Japanese mathematicians.
62. Mikami Yoshio, "Enri no hatsumei ni kansuru ronsho ("Arguments on the Discovery of *Enri*"), in *Sugakushi kenkyu*, no. 47 (1970), 1–43, and no. 48 (1971), 1–42. Originally written in 1930.
63. Hosoi, *Wasan shiso*, 298–300.
64. *Yamaji Kunju sensei chawa* ("Table Talks of Master Yamaji Kunju"), repr. in *Rekizan shiryo*, I (1933).
65. Mikami, *Bunkashi . . .* , 77–80.
66. J. W. N. Sullivan, *A History of Mathematics in Europe* (London, 1925), intro.
67. Dirk Struik, *A Concise History of Mathematics* (New York, 1948).
68. Mori Tsuyoshi, *Sugaku no rekishi* ("A History of Mathematics"; Tokyo, 1970), 126.
69. Takebe Katahiro, "Tetsujutsu" (1722), in Saegusa Hiroto, ed., *Nihon tetsugaku zensho*, VIII (Tokyo, 1936), 375.
70. Joseph Needham and Shigeru Nakayama, "Chugoku no kagakushi o megutte" ("On the History of Chinese Science"), in *Gekkan Economist* (Oct. 1974), 86–87.
71. Joseph Needham, "The Historian of Science as Ecumenical Man," in S. Nakayama and N. Sivin, eds., *Chinese Science* (Cambridge, Mass., 1973), 1–8. Sivin has outlined an approach to Chinese science less bound to such assumptions in his preface to *Science and Technology in East Asia* (New York, 1977).

MAYA NUMERATION, COMPUTATION, AND CALENDRICAL ASTRONOMY

Floyd G. Lounsbury

The following article is divided into seven sections: I. Introduction; II. Numeration and Notation; III. Calendar, Chronology, and Computation; IV. The Venus Calendar; V. Eclipse Reckoning; VI. Numerology; VII. A Time Perspective.

I. INTRODUCTION

The civilization known to archaeologists as Classic Maya flourished in the lowlands of southern Mexico, Guatemala, Belize, and western Honduras from the late years of the third century to the end of the ninth century of our era, according to most estimates. This area is dotted today with the architectural remains—in ruins except where recently reconstructed—of the numerous administrative and ceremonial centers of that civilization. The societies that erected and occupied these centers may be described as primitive kingdoms and incipient city-states. Their rulers took on the classic characteristics of "divine kings." They caused monuments to be erected in their commemoration and for the affirmation of their pedigree. A related priestly class attended to their needs for mythological charter and supernatural sanction. Among those in priestly occupations were the sky-watchers and the experts in numeration and the calendar. Scribes and artists, some of them with great sensitivity for line and composition, were in courtly service. It is difficult to see how the temples, pyramids, and palaces of these centers could have been erected without some form of coerced labor. There is evidence of wars of conquest and there are records of the taking of high-ranking prisoners; a warrior class and a military organization are to be presumed. There is some evidence of alliances by marriage between ruling families of different centers, or between major centers and satellites. The fortunes of the centers varied as first one and then another rose to prominence and dominating influence. The autochthonous development of this civilization came to an end during the ninth century when, one after another, the Classic Maya kingdoms collapsed. The abandonment of their centers was concomitant with influences—political, cultural, and military—emanating from the Mexican highlands to the west. The ensuing "Mexican" period, or "early postclassic" of Toltec domination, lasted until about the end of the twelfth century, by which time its foreign Nahua-speaking elite appear to have become culturally and linguistically assimilated to the Mayan subject populations. During the "late postclassic," from the thirteenth to the sixteenth centuries, concentrations of political power and "high" culture were absent from the central lowlands, where the monuments of the earlier civilization had fallen into ruin and their sites were reclaimed by the forest. Meanwhile to the north in the peninsula of Yucatan there arose a state or federation with centralized control, with lingering but faded Mexican impress, and lacking most of the excellence that had characterized the culture of the classic period in the central lowlands. This lasted for something over two centuries, until the overthrow of its capital city of Mayapan. Sixty years later the Spaniards arrived, conquered, and consolidated their rule at least in the temporal domain, while somewhat more slowly and less securely in the spiritual. Today in these areas there live some two and a half million speakers of about thirty different Mayan languages, assimilated in varying degrees to Spanish Catholic culture, but yet conserving much that is anciently Mayan.

Our interest here is in the attainments of the Maya astronomers, numerators, and calendar priests of the classic and postclassic periods. Information on that subject comes from various sources: from inscribed stone monuments of the classic period, from hieroglyphic codices of the postclassic period, from Spanish chronicles and native ethnohistorical manuscripts of the postconquest period, and from modern ethnography. The primary sources, however, are the stone monuments and the presumably twelfth-century Codex Dresdensis (named for the library which has had it since 1739). Other sources either furnished the keys for the initial unlocking of the content of the primary sources or have offered supplementary and comparative evidence for the interpretation of that content.[1]

The Maya left no treatises on mathematical or astronomical methods or theories. There is no posing of a problem, proof of a theorem, or statement of an algorithm—none of the usual kinds of source material for the history of a science. Their writing system, if not actually prohibitive of such disquisitions, was at least conducive to brevity in the extreme. What they left are the various end products of the application of their methods. It is up to the students of these remains to decipher what the problems were and what may have been the methods employed in their solution. This essay therefore, in some of its parts, must take the form of an analysis of the remains, and an argument, with the aim of deducing from them what knowledge of facts and of methods may have lain behind them. It can hardly take the form of a history of Maya mathematics, calendrics, and astronomy. Although the names of the rulers of the Classic Maya principalities are known, or at least the hieroglyphic forms of their names, the contributors to the Maya "sciences" remain totally anonymous. That is not to say, however, that some of the developments in this domain cannot be placed at least approximately in a historical time scale.

By way of an introduction to the content of the remaining pages of this essay, the following brief summary is presented.

1. Maya numeration was vigesimal. (It still is.) In the enumeration of days it was modified to accommodate a 360-day "chronological year."

2. Chronology was by means of a continuous day count, reckoned from a hypothetic zero day some three millennia B.C. (Precise correlation of the Maya with the Julian day count remains uncertain.)

3. Basic calendrical cycles were the "trecena" of 13 days, the "veintena" of 20 days, the "sacred almanac" of 260 days (the product of the trecena and the veintena), the "calendar year" of 365 days, and the "calendar round" of 52 calendar years or 18,980 days (the lowest common multiple of the sacred almanac and the calendar year). Others were of 9 days, 819 days, and 4 × 819.

4. A concurrent lunar calendar characterized days according to current moon-age, moon-number in lunar half-years, and moon duration of 29 or 30 days.

5. The principal lunar cycle, for warning of solar eclipse possibilities, was of 405 lunations (11,960 days = 46 almanacs), in three divisions of 135 lunations each, with further subdivisions into nine series of 6-month and 5-month eclipse half-years. The saros was a station in this cycle (the end of the fifth series) but was not recognized as the basic cycle.

6. Venus cycles were the mean synodic Venus year of 584 days, an intermediate cycle of 2,920 days (the lowest common multiple of the calendar year and the Venus year, equal to 8 of the former and 5 of the latter), and a "great cycle" of 37,960 days (the lowest common multiple of the sacred almanac, the calendar year, and the Venus year, equal to 104 calendar years or 2 calendar rounds). Reckonings with the periods of other planets are more difficult to establish.

7. The calendar year was allowed to drift through the tropical year, the complete circuit requiring 29 calendar rounds (1,508 calendar years, 1,507 tropical years). Corrective devices were applied in using the Venus and the eclipse calendars, to compensate for long-term accumulations of error owing to small discrepancies between canonical and true mean values of the respective periods. For numerology, canonical values were accepted at face value.

II. NUMERATION AND NOTATION

The history, or prehistory, of mathematics can be said to begin with the discrimination of *one*, *two*, and *many*, and with the naming of these concepts. Although admitted as true in a sense, it may hardly seem worth the saying; for we assume that people in all societies must enumerate things and that all languages must have number vocabularies capable of reaching at least into the hundreds and thousands. But when one has encountered a human language lacking this capability, one no longer takes numbers for granted. Such languages do exist; and the counting and computation done by speakers of those languages—whether as effect or as cause—is correspondingly limited. The Mayan languages are not among them, however. The names of the Maya numerical units up to 20^6 (or 64,000,000) are known, and in the enumeration of days, written numerals with higher-order units up to 360×20^{12} are recorded in inscriptions. The Maya probably had the highest development of numeration in indigenous America. But since it is exceptional, it is appropriate to view it as such, and to see it against the background of the mathematical resources of certain other aboriginal societies of the western hemisphere—for which purpose we shall make a brief detour into some comparative data. If the Maya were at the top of the scale in regard to the development of numerals, then near the bottom of that scale are various tribal groups along the tributaries of the Amazon and on the central plateau of Brazil. Frequently reported are numeral systems consisting of the expressions "one," "two," "two and one," "two and two," "two and two and one," and "many"; for example, that of the Cachuianã, a people speaking a language of the Carib family and dwelling along the Rio Cachorro (a tributary of the Trombetas, one of the northern tributaries of the Amazon): *tuinerô* (1), *ahsakô* (2), *ahsakô tuinerô* (3), *ahsakô ahsakô* (4), *ahsakô ahsakô tuinerô* (5), *chicarauhô* ("many," "a lot"). A similar system has been reported by the Salesian Fathers Colbacchini, Venturelli, and Albisetti, and by General Rondon, from the Bororo Indians along the Araguaia and the São Lourenço in the uplands of Mato Grosso. In Rondon's version, with minor substitutions in orthography, it is as follows: *mito* (1), *pobe* (2), *augere pobe ma awo metuya bokware* (3), *augere pobe augere pobe* (4), *augere pobe augere pobe awo metuya bokware* (5), *augere pobe augere pobe augere pobe* (6), and so forth, up to 2 + 2 + 2 + 2 + 2 (10), but noting also an alternative expression for five, *awo kera upodure* ("this my hand all together"). Colbacchini and Albisetti reported the same or similar expressions, with the remark that ordinarily they simply show the fingers of the hand, or of the two hands, saying *inno* or *ainna* ("thus") or *ainó-tuǰé* ("only this many"). They also noted that to indicate five they show the left hand open, saying *ikera aubodure* ("my hand complete"), that for ten they show both hands saying *ikera pudugidu* ("my hands together"), and that going beyond ten they employ the toes of one foot, and beyond fifteen the toes of the other foot; and that when the objects counted are more than a number that they can express easily, they say *makaguraga* ("many") or *makaaguraga* ("very many"). In the summer of 1950 the present writer made a brief visit to the Bororo village of Pobore on the Rio Vermelho and was able to coax one man (one of the few who had enough knowledge of Portuguese for the undertaking) as far as "thirty" in the elicitation of numerals. Following is a sample of the results: (1) *ure mitótuǰe* ("only one"); (2) *ure póbe* ("two of them," or perhaps better, "a pair of them"); (3) *ure póbe ma ǰéw metúya bokwáre* ("a pair of them and that one whose partner is lacking"); (4) *ure póbe púibìǰi* ("pairs of them together, in reciprocity, or in sequence"); (5) *ure ikéra aobodúre* ("as many of them as my hand complete"); (6) *botúre ikéra aobowúto* (precise significance uncertain, apparently something about changing to the other hand); (7) *ikéra metúya pogédu* ("my hand and another with a partner"); (8) *ikerakó boeyadadáw* ("my middle finger," that is, of the second hand); (9) *ikerakó boeyadadáw mekíw* ("the one to the side of my middle finger," again omitting to specify that it is of the second hand); (10) *ikerakó boeǰéke* ("my fingers all together in front"; . . . (13) *ičare butúre ivúre boeyadadawúto pugéǰe* ("now the one on my foot that is in the middle again"); . . . (15) *ičare ivúre iyádo* ("now my foot is finished"); . . . (20) *avúre ičare maka réma avúre* ("your feet, now it is as many as there are with your feet"); (21) *otúre turegodáǰe pugéǰe* ("starting them over again"); (22) *ure póbe turegodáǰe pugéǰe* ("two of them, starting over again"). These were given as if they were names for numerals, without accompanying gestures of finger showing or pointing to toes.

The Bororo case has been cited because it shows vividly the basis for vigesimal systems (it is clear what counters are employed and named in the tally). It also enables one to appreciate the nature and length of the path that lies between a primitive one like the Bororo, with its long descriptive phrases, and an advanced one like the Mayan, with its brief unanalyzable roots for the numerals up to "ten," "eleven," or "twelve," rather than only up to "two" as in Bororo. A long history of ellipses, of accumulation of phonetic changes with resulting loss of etymological transparency, and of repeated abbreviations accompanying increased frequency of use, must lie between these two stages in the development of a number vocabulary.

An intermediate stage can be glimpsed in the numerals of the Iroquois Indians of the region of New York State, lower Ontario, and southern Quebec. These numerals are conveniently short words (in a language of the polysynthetic variety which is otherwise given to long and morphologically complex words), their etymologies do not intrude into the thoughts of the speakers who use them (only linguists notice their apparent origins), and they serve as excellently as do the numerals of any language in all of those uses to which spoken number words are put. But the etymologies of at least seven of the first ten numerals show through and reveal their original significations. Enumerations in amounts of "one" and "two," whenever the underlying stem will permit of it, are accomplished by inflection of the stem with appropriate prefixes and suffixes. The suffixes, however, are only apparently such; formally they are stative verb roots which incorporate the noun stems. The independent words for "one" and "two" are composed of such inflectional elements, but without an incorporated noun stem; formally they are petrified stative verb forms. They are used only with expressions whose morphology will not accept the affixes. The word for "three" is homonymous with the verb root meaning "(to be) in the middle of it." The word for "four," in its unabbreviated form, is homonymous with the inflected word meaning "it is complete," probably owing its origin to a widespread ritual significance attached to the number four in Amerindian cultures. The word for "five" is unanalyzable in Iroquoian, but there is a word for "hand" which is of this form (and related forms) in some of the Caddoan languages, these being a very distantly related family. The word for "six" is homonymous with the bare verb stem for "to cross over," in the form which it has when no incorporated object noun stem is present. An original reference to crossing from one hand to the other seems to be implied. The words for "seven," "eight," and "nine" are opaque and at best are only weakly suggestive of possible origins. Two of the words for "ten" are transparent, or relatively so, while another is not. One of them is derived from a participial inflection of the same verb root as is contained in the word for "four" and must owe its origin to the fact that with "ten" a tally on the fingers is "completed." The Iroquois system is decimal.[2]

The Mayan numerals, in contrast to these, are opaque through "eleven" or "twelve." It is perhaps possible, although not proven, that words for "ten" may be derived from a root meaning "end" and "to finish" or "complete." Whether or not this is historically correct, there is nonetheless a decimal stratum within the otherwise vigesimal system of numeration, as can be seen in thirteen through nineteen of the following set from Yucatec Maya:

1. hun	11. buluc
2. caa, ca, c-	12. lahca
3. ox	13. oxlahun
4. can	14. canlahun
5. hoo, ho	15. hoolahun, hoolhun
6. uac	16. uaclahun
7. uuc	17. uuclahun
8. uaxac	18. uaxaclahun
9. bolon	19. bolonlahun
10. lahun	20. hun kal

[NOTE: Following tradition, the colonial Yucatec orthography is used in citations of words from that language. The letter *x* is for *š*; prevocalic *u* is for *w*; *c* is for *k*, regardless of whether it is before a back vowel or a front vowel; the letter *k* is for the midvelar glottalized stop *k'*. Thus *kal* in the above list is for *k'al*. The same word in highland Guatemalan languages, with the same initial consonant, was written *4al* in colonial orthographies and is mostly *c'al* in modern orthographies in that country, where the letter *k* is used to represent the postvelar stop in the indigenous languages, corresponding to *q* of standard international phonetic transcriptions. The glottalized postvelar was written *ℰ*. In Yucatec the Mayan postvelars were anciently merged with the midvelars. Doubled vowels in Yucatec orthography (except with syllable-initial *u* for *w*) are for vowels interrupted or checked by a glottal stop. Use of the convention, however, was inconsistent in Spanish colonial sources.]

Words for twenty, or score, in the Mayan languages are *kal* (*c'al*, and so forth; see above), *may*, and *uinic*, or forms cognate to one or another of these. The last of these is the word for "man" or "human being," implying in this context a reference to the totality of his digits. The others appear to be related to words for tying and bundling and may reflect practices of counting and packaging in ancient commerce and rendering of tribute.

Multiples of twenty in Yucatec are *hun kal* (20), *caa kal* (40), and so forth, up to *buluclahun kal* (380), after which is *bak* (400). Highland Mayan languages have *q'o* or *oq'ob* as the equivalent of *bak*. Some of the languages have separate words for certain of the intermediate multiples of twenty, such as *tu:c* (40), *much'* (80), and *lah* (200). Thus in Kekchi 200 is expressed alternatively as *laheb c'a:l* ("ten score"), *ho tu:c* ("five forties"), or *hun lah* ("one 200"); and 400 is *laheb tu:c* ("ten forties"), *ca'ib lah* ("two 200s"), or *hun oq'ob* ("one 400"), and in Cakchiquel, also *woo' much'* ("five eighties").

The powers of twenty, as far as their names are known in Yucatec, are *kal* (20), *bak* (20^2), *pic* (20^3), *calab* (20^4), *kinchil* (20^5), and *alau* (20^6). The Cakchiquel equivalent of Yucatec *pic* (8,000) is *chuwi*, which is also a word for "sack." Its use as a numeral is said to derive from the custom of packing cacao beans—an important commodity and also a medium of exchange—in quantities of 8,000 to the bag. Multiples of the higher powers of twenty are enumerated in the same way as are those of the first power.

There are two different methods of naming numbers that intervene between the multiples of any power of twenty. In Yucatec as it was spoken at the time of the Spanish conquest, and as it continued until about the end of the nineteenth century, the predominant method was to name the intervening quantity and to place it in the ordinal-numbered score or other power of twenty. Thus, for example, forty-one was "one in the third score" (*hun tu yox kal*), and 379 was "nineteen in the nineteenth score" (*bolonlahun tu bolonlahun kal*). This sytem still prevails today in most, although not all, of the Mayan languages. Ordinal numerals are formed by preposing the third-person possessive pronoun to the cardinal number. The preposition contracts with the pronoun; *tu* in the above examples is such a contraction. By analogy with these examples, for 399 one might expect "nineteen in the twentieth score"; but "twenty score" was regularly replaced by "one 400," so that 399 was expressed as "nineteen in the first 400" (*bolonlahun tu hun bak*), an expression that would be wholly misunderstood if interpreted either literally or in accord with the elliptical patterns about to be mentioned. For proper interpretation, *u hun bak* must first be understood as if *u kal kal*, that is, "twentieth score" rather than "first 400." This was the case only with the first multiple of *bak*, not with higher multiples. Thus, "nineteen in the third 400" (*bolonlahun tu yox bak*) was not to be understood as "nineteen in the sixtieth score," that is, 1,199. Neither, however, was it to be taken literally, in which case it would have had the value 819. Rather, its value was 1,180, for a reason that follows.

There were several curious but systematic ellipses. Between higher powers of twenty, simple numerals stood for multiples of the next lower power rather than for multiples of unity as if taken literally. Thus, for example, "five in the third 400" (*ho tu yox bak*) stood for "five SCORE in the third 400" (as if it were *ho kal tu yox bak*). This is the reason for the seemingly odd value given to the last example in the paragraph above. Another systematic ellipsis involved the word for two in references to the second score, the second 400, and so forth. The word was omitted (except in cases that called for yet a third and complementary type of ellipsis). Thus, whereas forty-five was "five in the third score" (*ho tu yox kal*), twenty-five was simply "five in its score" (*ho tu kal*), with the word for two suppressed. Similarly, whereas nine hundred was "five in the third 400" (*ho tu yox bak*), meaning "five SCORE in the third 400," five hundred was simply "five in its 400" (*ho tu bak*), meaning "five SCORE in the SECOND 400." This last expression is doubly elliptic. The third type of ellipsis (in a scale from the seemingly well-motivated to the seemingly whimsical) involved the syllable *tu*, which is the contraction of the preposition *ti* and the ordinal-forming pronoun *u*. It was something less than obligatory but was at least usual, if not wholly regular, in the particular context that permitted it. That context was a preceding word for ten (*lahun*) or for fifteen (*hoolahun* or *hoolhun*). It was not just a matter of the presence of the morpheme for ten, because none of the other teens precipitated this ellipsis. Nor was it a matter of multiples of five, because the first multiple of five also did not precipitate it. Thus thirty was "ten two score" (*lahun ca kal*), meaning "ten IN ITS two (that is, in the second) score." Similarly, six hundred was "ten two 400" (*lahun ca bak*), meaning "ten SCORE IN ITS two (that is, in the second) 400." This last expression also is doubly elliptic, although in a partially different manner from the expression for five hundred (*ho tu bak*). The last two types of ellipsis were complementary: a single expression could not combine both; but either of them could be combined with an ellipsis of the first type.

The second method of expressing compound numerals was to do it as we do: with a conjunction, either expressed (*catac* in Yucatec) or implied by juxtaposition of two orders of components, and proceeding from the higher-order to the lower-order components. Thus, for example, fifty-one could be either "two score and eleven" (*ca kal catac buluc*) or "eleven in the third score" (*buluc tu yox kal*) in the manner described above. Higher complex numerals could be expressed entirely in one of these ways, or entirely in the other, or by employing combinations of the two methods.[3]

In all of the preceding, attention has focused on numerals in spoken language. But for record keeping, and for computations other than the most simple, notational devices are required to give the numerals a permanence and retrievability lacking in the spoken word and in the unaided memory of the word. Iconic notation of low-order numerals is a ready adaptation of the tally or of the finger count, and this, together with symbols for extending the device to higher-order units, is one of the first systematic components of a primitive writing system. Thus, for example, in the Aztec system, numerals from one through nineteen were represented by aggregations of dots (in linear, broken-linear, or L-shaped array, and with various subgroupings, not necessarily by fives); scores were represented by flags, four-hundreds by treelike figures, and eight-thousands by ceremonial pouches such as shown being carried by divinities and by priests for the containment of copal incense.

The Maya notation employed dots for units up to four, bars for fives up to fifteen, and combinations of bars and dots for the intervening numbers. Thus the system of written numerals, unlike that of the spoken numerals, was quinary below and between the vigesimal values. Twenties were represented by the unit symbols positioned in second place (the base position, above or preceding the units position) in a system of place notation. Similarly, four-hundreds, eight-thousands, and so forth, were represented by the same symbols in higher-order places. (The apparent exception in the enumeration of days is only superficially such; see "Calendar, Chronology, and Computation," below.) Place notation requires a zero symbol for otherwise unoccupied places, and for this purpose a "shell" sign and several other hieroglyphs were employed.

There were also other symbols for twenty. By itself or in numerals up to thirty-nine, in contexts where place notation was not employed, that number was represented by the hieroglyphic sign that in other contexts refers to the moon or to the lunar month (a circumstance to be explained in the section "Calendar, Chronology, and Computation"). It is quite certain that the sign had different readings in these two functions; in Yucatec it would have been *kal* when interpreted as "twenty," and *u* (and perhaps sometimes *uen*) when interpreted as "moon" or as "lunar month." There were in turn two different methods of employing this sign in the value "twenty, which reflect the two different methods of expressing such numbers in the spoken language. In that which is most frequently attested, the sign for twenty is placed first and is followed by the numeral representing the excess over twenty. This, like the place numeration, reflects the conjunctive or "*catac*" method of expression in the spoken numerals. The other manner, attested only three times, reflects the more colloquial expression, including its idiomatic ellipsis; e.g., "nine in the SECOND score" (*bolon tu kal*), "sixteen in the SECOND score" (*uaclahun tu kal*), and so on. In these, the hieroglyphic sign having the phonetic value *tu* is prefixed to the *kal* sign.

III. CALENDAR, CHRONOLOGY, AND COMPUTATION

The system of numerical notation was at least seven or eight centuries old by the beginning of the "classic" period, late third century A.D. It is probable that the Maya received it ready-made, along with their chronological system and some of the components of their calendar, from other peoples in Oaxaca, Tabasco, and Vera Cruz. No doubt from its inception it was used—as it was when first observed by Europeans—in records of trade, levies of tribute, mensuration, census, and other functions of government and religion. Of special importance among the latter were chronology and the regulation of the calendar. It is almost exclusively in these that records of its use have survived for modern study. Although documents exhibiting other uses have survived from the Aztec and other peoples to the west, for example, listings of items of trade and tribute (from late postclassic and postconquest times), none such has survived from the Maya. There are several instances, however, of enumeration of objects for offerings: nodules of copal, rolls of rubber, beans of cacao, "precious things" (whatever they were), and others yet undeciphered. Presumably then, as today, the numbers of things offered were as important as their kinds; for numbers had esoteric meanings and efficient powers in their own right. Particular combinations of numbers and kinds were prescribed—suited to the occasion, the recipient, or the day in the calendar. But outside of these, the only records of Maya numerical notation that we have are in its calendrical and chronological applications.[4]

The basic calendar of the Maya, as of other Mesoamerican peoples, was the 260-day almanac. It was the product of two component cycles: one of day numbers, from "one" to "thirteen," and one of day names, twenty in all. Any day in the compound cycle of 260 can thus be specified by a pair of coordinates: for example, *1 Imix, 2 Ik*, and so forth, to *13 Ahau*; or translated into numerical equivalents: (1, 1), (2, 2), · · · , (13, 20), expressing the position of the day in two simultaneous cycles of different moduli. The Maya name for the almanac is not known for certain; but in Mayanist literature it is commonly called the *tzolkin* ("the sequence of days," "the naming-in-order of the days"), on the grounds that it is an apt term, that an analogous one is documented for the sequence of calendar months (*u tzol uinal*—a syntactic phrase rather than a compound noun), and that even if this is not what it was actually called, the Maya could well have understood the term in that sense. In the present account, the term "almanac," or "sacred almanac," will be used. The component lesser cycles will be called the "trecena" and the "veintena," employing convenient Spanish terms. The names of the days of the veintena, in Yucatec (in a conventional but inadequate transcription), are *Imix* (1), *Ik* (2), *Akbal* (3), *Kan* (4), *Chicchan*

(5), *Cimi* (6), *Manik* (7), *Lamat* (8), *Muluc* (9), *Oc* (10), *Chuen* (11), *Eb* (12), *Ben* (13), *Ix* (14), *Men* (15), *Cib* (16), *Caban* (17), *Etznab* (18), *Cauac* (19), and *Ahau* (20). They have other names in other Mayan languages, some cognate and others not; but the hieroglyphs were the same, including variants, throughout the territory in which they were used, regardless of the local language. In non-Mayan language areas of Mesoamerica, not only the names, but also the hieroglyphs, were different.

The 365-day year was the other principal component in the calendar. Its name in Yucatec was *haab*, with cognate forms in other Mayan languages. The apparent derivation of the term implies a reference to the seasonal year (the "rains" being for the count of years in the tropics what "winters" have been in the temperate zones); but there is evidence for its nonspecificity of use, being interchangeable even with the word for the 360-day chronological year, with a corresponding interchangeability of the hieroglyphs, but with context serving to clarify the reference. In Mayanist literature the uncorrected 365-day year is commonly called the "vague year," or sometimes the "*haab*." Here it will be called the "calendar year," or simply the "year" (when the reference is clear from context). Any other kind of year—such as the "tropical year" of 365.2422 days, the "chronological year" of 360 days, or the "computing year" of 364 days—will be specially designated as such.

The calendar year was divided into eighteen named "months" of twenty days each, and a residual period of five days; and within each of these the constituent days were numbered or otherwise specified as to position. The numbers employed for days in the months were 1 to 19, and 1 to 4 in the residual period. In place of 20 in the months, and in place of 5 in the residue, the last day of any period was sometimes designated with a glyph whose reading was *tun*, a word with several senses (homonymous), one of which was "end" or "final." But far more frequently it was designated as the "seating" or "installation" day of the next month, employing a glyph of that meaning (also employed for the seating or installation of rulers) whose reading is assumed to have been *cul*, but for which there are also other possibilities. Thus, for example, the twentieth day of the month *Mol* was designated either as "end of *Mol*" or as "seating of *Chen*," the latter method being predominant. In transcriptions of glyphic texts, the seating day of a month is conventionally numbered "0," with the name of the month seated, since it is in effect the zero day of the month about to begin.

The term "month" is customarily employed for these periods in spite of the incongruity between the name and their duration. The reasons are (1) that some of the Maya are known to have done the same, and (2) that the hieroglyph for "twenty" is the "moon" sign. In regard to the first point, it has been reported in an early twentieth-century ethnographic study that specialists in the calendar among the Jacaltec Maya used the same word for the twenty-day month as did laymen in the community for the thirty-day period, which is also their word for "moon."[5] In regard to the second point, the glyphic usage is reasonably although speculatively explained on the assumption that a similar polysemy existed in ancient times also, with ultimate derivation of the special sense from the more general sense (possibly with the institution of an arithmetically motivated calendar, supplanting a lunar one in the function of subdividing the vague seasonal year). The Yucatec Maya names of the months are *Pop* (1), *Uo* (2), *Zip* (3), *Zotz* (4), *Tzec* (5), *Xul* (6), *Yaxkin* (7), *Mol* (8), *Chen* (9), *Yax* (10), *Zac* (11), *Ceh* (12), *Mac* (13), *Kankin* (14), *Muan* (15), *Pax* (16), *Kayab* (17), *Cumhu* (18); and the residue, *Uayeb*.[6] The names of the months are highly variable from one Mayan language to another; yet the hieroglyphs are universal, including variants, in the area in which hieroglyphic inscriptions are found. In about half of the cases, the forms of the hieroglyphs are explainable only by reference to non-Yucatec names for the months, particularly those of Kekchi, Pokom, and Jacaltec. The Yucatec readings, however, are standard in transcriptions of glyphic texts.

The days, named for their places in the calendar year—*1 Pop, 2 Pop*, and so forth, through all of the months, to the end of the *Uayeb* (or "seating of *Pop*")—constituted the third principal cycle in the calendar. Compounded with the trecena and veintena of the almanac, it yielded a greater cycle, the "calendar round." Because 260 and 365 share a common factor of 5, the length of the calendar round is 52 calendar years, or 73 almanacs, or 18,980 days. A day in the calendar round was specified by its positions in the component cycles; for example, *12 Lamat 1 Muan*, which may be translated into numerical coordinates as (12, 8, 281), or vigesimally (12, 8, 14.1), representing a triple classification of the day according to its places in three cyclical schemes with moduli 13, 20, and 365 (18.5) respectively. Also because of the common factor of 5 in the veintena and in the calendar year and its subdivisions, any day in the veintena was restricted as to the positions it could occupy in a month of the year, or in the *Uayeb*. Thus, the 5th, 10th, 15th, and 20th days of the veintena (*Chicchan*, *Oc*, *Men*, and *Ahau*) could fall on any of the days *3, 8, 13*, or *18* of a month, and on day *3* of the *Uayeb* period; with the others falling in place accordingly. And because the days of the year were eighteen score *and five*, the positions of the year days shifted five places in the veintena from one year to the next. Similarly, because they were one more than a multiple of thirteen, each year they fell one place later in the trecena.

The calendar round, with its three component cycles, provided a unique characterization for every day in a 52-year span of time. It was the most heavily employed system for the dating of events. Employed as it was with an occasional anchor to the day count, its specifications became unique in their reference, not just within a 52-year span, but for all time.

The Maya "day count" was the ultimate resort for absolute chronology. It was, like the Scaliger or "Julian" day count, a method of assigning a distinct number to each and every day in sequence, conceivably forever, or at least for the duration of an era, starting from a conventional zero point in time. That zero point for the Maya count—the theoretical start of their current chronological and cosmological era—was a day some three millennia before the beginning of the Christian era. Other eras, however, had preceded this one, and some dates (of mythological events or retrospective astronomical projections) were given in the chronology of the last preceding era. The duration of that one was treated as having been of thirteen *baktuns* (about 51¼ centuries). The zero day of the current era was thus also the final day of the preceding era, and was designated as the "end of thirteen *baktuns*," or 13.0.0.0.0 in the day count of that era. Its place in the calendar round was *4 Ahau 8 Cumhu*, or in numerical translation (4, 20, 348), or with the year day expressed vigesimally and the veintena day by its modular equivalent (4, 0, 17.8). There is no evidence as to whether a similar duration was envisioned for the current era (if it was, its end will soon be upon us) or for other preceding eras. In other words, grounds are lacking that would warrant our considering the day count also as cyclical.

Fundamental to the day count is the "chronological year" of 360 days. The hieroglyph designating this period is known from several kinds of evidence to have been read *tun*, and this was the Maya term for the chronological year. The glyph is the same as that used for the homonymous *tun* meaning "end" or "final," used at times for the designation of the last day of a calendar month. The chronological years, or *tuns*, were counted vigesimally. The next higher unit was the *katun*, equal to twenty chronological years. The name for the unit is well attested from postconquest documentary sources; it is quite surely from *kal* ("twenty" or "score") plus *tun*, with loss of syllable-final *l* (under phonological conditions in which such loss has ample precedent). The unit beyond the *katun* was that of 400 chronological years, presumably called the *baktun*, although the term is unattested in postconquest sources, and its hieroglyph does not in itself compel that reading. In Maya historical chronology no further higher-order units are needed; but in the numbers employed for long-distance projections into the future and into the past, hieroglyphs are found whose values, by powers of 20, range up to 64 million chronological years or *tuns*. It is supposed that the names for these units were most likely compounded from the corresponding numerals and the word *tun*, and on this assumption they are commonly called the "*pictun*," "*calabtun*," "*kinchiltun*," and "*alautun*"; but these compounds are not attested in postconquest sources, although their constituent morphemes are.

The count was not just one of chronological years, but of days. Units of order lower than the *tun* were the *uinal*, amounting to twenty days, and the *kin*, single days. In its adaptation to the count of days, the numerical system was thus vigesimal in all of its places but one, namely in the third. That is, every higher-order unit is equal to twenty units of the next lower order, except the third; that one, the *tun*, is equal to eighteen of the next lower order—a detail that must not be forgotten when doing additions and subtractions or other arithmetic manipulations with Maya chronological numerals. Although *kal* is the general term in Yucatec Maya for aggregates of twenty, and was employed in some contexts also for spans of twenty days, in the day count the term *uinal* and its particular glyph were employed. The term is documented in postconquest sources both in this sense, that is, of twenty-day units in counting, and also in the special sense of the twenty-day named months of the calendar year.

In Mayanist literature the count of days from *4 Ahau 8 Cumhu*, 13.0.0.0.0, is commonly called the "long count" (abbreviated LC), although some have extended that term also to long reckonings from special bases antedating the normal zero, and have extended the term "initial series" (abbreviated IS, see below) to designate such reckonings as are made from the *4 Ahau 8 Cumhu* zero date and are placed at least in the almanac, whether they are "initial" to any text or not. Here it will be called the "day count," or "Maya day count" when a contrast with the Julian is required. A date is fixed by its "day number."

In hieroglyphic inscriptions on stone monuments, dates spelled out in full in the day count were usually confined to the initial passage of a text and were given the most artful elaboration and prominent display in the inscription. The initial passage opens with a standard "introducing glyph," often double or quadruple the size of the other glyphs, which may have had a conventional reading. Infixed into it is a variable element, varying according to the calendar month which appears further on in the elaboration of the date, and was thus in some way symbolic of that month (patron deity?, hieroglyph of an ancient name?). Following the introducing glyph, the numbers of *baktuns, katuns, tuns, uinals*, and *kins* that constitute the day number are given, sometimes simply with bars and dots prefixed to standard forms of the signs for the unit periods, but often with ornate personified forms symbolic of the numerals, and zoomorphic forms for the periods, these being portrayed either as heads or more elaborately as full figures. Following this numerical specification of the place of the date in the day count there are given its positions also in the almanac and in the calendar year, as well as in other supplementary cycles. One of these is a cycle of nine days, the position in this being given by naming the one of the nine lords of the night under whose regency the given date falls. Another is the lunar calendar, the position in this being specified by giving the age of the current moon on this date, the numerical position of the moon in its lunar half-year, and the prescribed duration of this lunar month (or the preceding?) whether of 29 or of 30 days in the lunar calendar. Placement in these cycles—almanac, calendar year, lords of the night, and lunar calendar—was standard procedure for the initial date of a major inscription, in addition to its placement in the day count. There was yet another cycle, of 819 days and somehow involving the rain god, which attained importance at some sites, and in which the date might also be placed by specifying how many days it was past the last station (or zero day) in that cycle; and since the stations rotated with the cardinal directions and colors, which also were specified, this became in effect a 4 × 819 or 3,276-day cycle. The order of all of these characterizations of a date was fixed, as just enumerated, except for some variability as to where placement in the calendar year was inserted. Sometimes, although relatively infrequently, this followed immediately after the placement in the almanac. Sometimes, but even less frequently, it came just after the phrase pertaining to the lord of the night. Most frequently it followed the lunar data; and in at least one inscription it followed the placement in the 819-day cycle.

In the alphabetic code designations employed in Mayanist literature—a code imposed before the meanings of the glyphs were known—the so-called "supplementary series" of glyphs is as follows, and in this order: (*G*) the place in the nine-day cycle, presumed lord of the night; (*F*) a constant glyph that follows glyph *G*, or into which glyph *G* is infixed, which perhaps names the standard event of which the particular lord of *G* is protagonist—or the constant predicate of which the variable *G* is subject; (*E*) the excess over 20 days, if any, in the age of the current moon on the date specified; (*D*) the age of the moon up to and including 20 days, supplemented by the preceding glyph *E* when over 20 days; (*C*) the position of the month in the lunar half-year; (*X*) a variable glyph, in a cycle of the same magnitude as that of glyph *C*, and partially constrained by the value of *C*, the meaning of which is not known; (*B*) a constant glyph—except for apparently free alternation among supposedly equivalent signs in two of its component parts—whose meaning is not known; and (*A*) the calendrical duration of the moon in question, always 29 or 30 days. Placement of the date in the almanac precedes this series; placement in the calendar year may precede it, or may intervene (although but rarely) between *F* and *E* or *D*, or may follow *A* (its most usual position). Placement in the 819-day cycle, if present, is usually last.[7]

Only after such characterization of a date in all of its important aspects, which constitutes a compound temporal-adverbial phrase or clause in the sentence, only then are the event and its protagonist stated. The events are of a suitable importance: the births of gods, the births of rulers, and the rulers' observances with the proper religious rites of momentous turns in the day count. The initial date and its event furnish the anchor, both thematically and chronologically, for the events and dates enumerated in succeeding passages. These dates were usually given only in the calendar round, that is, by naming the pertinent almanac day and year day. The rest of the attributes of such a date, including its number in the day count, could be determined from its calendar-round position in relation to that of the initial date. The information for anchoring these "secondary" dates in the chronology was sometimes made explicit in the form of "distance numbers" that linked the dates of successive passages to those of preceding ones. Sometimes, however, the function of distance numbers was to connect two dates of a pair in a single passage to each other, rather than to connect either of them with the date of a preceding passage. But their relation to a preceding date was clear nonetheless, it being understood that the indicated calendar-round day was the *next* such after the calendar-round day of the preceding passage, or after the later one of a linked pair in that passage, *unless otherwise specified.* And there were several ways of making it clear if that was not to be the case. Thus the day number corresponding to such a date could be supplied, provided only that one knew how to compute the interval between any two calendar-round days.

One of the ways of anchoring secondary dates was to insert into the sequence the calendar-round day corresponding to a round-numbered period-ending: for example, "*8 Ahau 13 Ceh*, completion of nine *baktuns*," meaning 9.0.0.0.0; or "the period-ending" of *5 Ahau 3 Chen*, therefore 9.8.0.0.0, it being understood that "period-ending" meant "*katun*-ending" unless otherwise specified, in which case this date is the only possibility within 18,980 chronological years (18,720 calendar years); or "the period-ending of *5 Ahau 18 Tzec*, *oxlahuntun*," therefore 9.8.13.0.0, this being a case where the "otherwise" is specified. Such chronological anchors, employed when necessary or pertinent, could be linked by distance numbers to other dates in a concatenation; or in unconcatenated texts they could be paired with other dates when necessary to remove potential ambiguities. Or in sequences without distance numbers, they could be inserted as chronological guideposts. In one known case the number of *katun*-endings intervening between two dates is given, thus assuring that the later one is not the next but the second-next occurrence of that calendar-round day. It is apparent that care was taken to eliminate ambiguities in such sequences. There are remarkably few; and some dates once thought to be unanchorable turn out only to be anchored by means then undetected.

The texts accompanying these secondary dates have rather limited and stereotyped subject matter: births, heir designations, and accessions of rulers; in some cases also marriage and the seating of a wife as queen mother and nominal coruler; a few records of the taking of high-ranking captives; performances of rites for the gods; some records of deaths or interments of rulers; lines of lineal descent; mythological ancestry; events involving the gods in ancient times, possibly with astronomical or mythico-cosmological references; a single probable record of a historical eclipse and possibly one or two others; and other matters not yet understood.[8] The syntactic patterns are formulaic, in the majority of cases simple but in some quite complex; and some of the latter preserve a couplet structure—parallel statements of same or similar import, the second rephrasing or amplifying the first with different but synonymous or partially synonymous expressions. It is a pattern that appears in certain postconquest pieces of Mayan traditional literature (recorded by native writers in romanized alphabets) and of which some beautiful specimens have been encountered in the oral literature also of other Amerindian peoples, of both North and South America. The fact that a few applications of this structural formula made their way into some of the hieroglyphic texts, however, should not be construed so as to suppose that they might be read as poetry; but they do give a glimpse of the linguistic and poetic resources that must once have existed for Mayan oral literature, and of the kinds of oral forms that may have been associated with the glyphic texts as prior versions, expansions, or traditional commentaries on them.

In addition to the other technical and artistic skills that went into the compilation and execution of the hieroglyphically inscribed monuments, a considerable skill and labor must have gone into the calendrical work contained in them. There are occasional arithmetical errors: a few rare ones that display real incompetence and apparent bluffing, a few that suggest copying errors (although one wonders why preliminary sketches on stone were not proofread before carving), and others that reflect the all too human kinds of mistakes in arithmetic or in table consultation. But in proportion to the amount of computation that must lie behind the Maya inscriptions and codices, the frequency of error is indeed low. Certain kinds of problems were faced over and over again. There were, of course, simple additions and subtractions of numerals in the modified vigesimal system of the day count, involving "carrying" and "borrowing." Whether the steps in these were carried out mentally, or with the aid of counters of different varieties or in different places, or with scorings on the ground or on sand tables or on other flat surfaces, is a minor matter. They were done, and for the most part they were done accurately. It is doubtful that the Maya did multiplications, other than by two and by twenty, except by repeated additions. There are no records of elementary multiplication tables, although in the codices there are tables of multiples of important longer periods (a few of which will be seen in following sections). Multiplication by twenty was relatively simple. Any day-count number that had a zero in second place (the *uinals* place) could be multiplied by twenty simply by adding a zero to the terminal end of the number. If the second-place number in the multiplicand was not zero, then the first step was to proceed as before and the second step was to double the second-place number of the multiplicand (now the third-place number in the intermediate product) and contribute it to the second place of the product. Of division there was surely less need in day-count arithmetic than in ordinary arithmetic, say as employed in commerce; but in ordinary arithmetic all processes were simpler because in that the base factor was uniform in all places of a number.

Beyond the elementary processes, as modified for day-count arithmetic, there were two important types of reckoning in applying the count to the calendar round. The two most typical problems were that of adding an increment (or decrement) to a given position in the calendar round, and that of determining the interval between any two given positions in the calendar round. We may pose these according to our fashion, not knowing what phrasings the Maya may have given to them. The first is the simpler one.

I. Proceeding from a given initial day (t_0, v_0, y_0) in the calendar round, what will be the day (t, v, y) reached in that round after the lapse of n_5 *baktuns*, n_4 *katuns*, n_3 *tuns*, n_2 *uinals*, and n_1 single days?

Our approach would be first to write three formulas in which the positions of the terminal day in the trecena (t), in the veintena (v), and in the calendar year (y) are expressed as functions respectively of t_0, v_0, and y_0, and the relevant n_i. For historical time in the day count, i ranges to 5; so for the time being we may restrict it to that maximum. The formulas may then take the following forms:

(1) $t = t_0 - n_5 - 2n_4 - 4n_3 + 7n_2 + n_1$, mod 13;
(2) $v = v_0 + n_1$, mod 20;
(3) $y = y_0 + 190n_5 - 100n_4 - 5n_3 + 20n_2 + n_1$, mod 365;

or with the last one in equivalent vigesimal form; where each coefficient in higher positions is derived from the next lower one (the one to its right) by multiplying it by the pertinent base factor (18 for n_3, 20 for all others) and reducing the result by the appropriate modulus (13 for t, 20 for v, and 365 or vigesimal 18.5 for y). In the case of the expression for v, all higher coefficients vanish because of the equal magnitudes of the first base factor and the modulus. The choice of a positive or a negative coefficient is obviously open in each case (for example, the expression for y could just as well be written as $y_0 - 175n + \cdots$, mod 365), such choices being made as will best facilitate mental calculation.

How did the Maya calendar priests do it? Direct evidence is lacking. It is possible to imagine all sorts of ways. But it is hard to imagine that there was not among them a method that was essentially equivalent to the one above—minus, of course, the algebraic format. First, it was an elementary matter to add single days (compare coefficients of unity for n_1). Second, it was surely common knowledge—as it is wherever the almanac survives today—that each successive recurrence of any day of the veintena had a place in the trecena that is seven places ahead of its previous one; thus, for example, the next *Ahau* after *1 Ahau* is *8 Ahau*, and so forth (compare coefficient of 7 with n_2 in the above expression for t). And it must have been equally well known that the second-next occurrence of any day of the veintena was just one place further along in the trecena; for example, that the second-next *Ahau* after *1 Ahau* is *2 Ahau*. These forty-day or two-uinal intervals have been remarked upon for this property by Jacaltec Maya, who call them *u yoc habil*, the "feet," or perhaps "footsteps," of the year, because they measure off one-day advances in the trecena associated with a constant place in the veintena.[9] Third, no one can have escaped knowing that every major chronological period ended on a day *Ahau*, and that these—the *tuns*, the *katuns*, and the *baktuns*—were named for their terminal almanac days, always *Ahau*, but varying their positions in the trecena. These names appear to have been as much common parlance as our numerical names for the years, decades, and centuries; so it is probable that the cycles of their positions in the trecena were well known. But if not, at least it was common knowledge for the specialists. The Paris Codex outlines the cycle of *tuns: 7 Ahau, 3 Ahau, 12 Ahau, 8 Ahau*, and so forth, always regressing four places in the trecena (compare the coefficient of −4 with n_3 in the above expression for t); and it does the same for the *katuns: 4 Ahau, 2 Ahau, 13 Ahau, 11 Ahau*, and so forth, always regressing two places (compare the coefficient of −2 with n_4 in the same formula). And circular arrangements of the *katuns* in postconquest sources, the so-called "*katun* wheels," present the same information in another format well designed for rapid calculation.[10] Finally, the order of the *baktuns*, with regressions of just one place in the trecena (compare the coefficient of −1 with n_5 in the expression for t), was so simple as hardly to require a visual aid; although wheels, or linear arrays understood to repeat, could well have been employed to facilitate each step in a computation. Determination of the place attained in the trecena after the lapse of any amount of time, whether a few days or thousands of years, could thus have been a fairly simple task. The place in the veintena was even easier to ascertain, since it depended only on the lowest-order digit in the distance number.

Reckoning of the place in the calendar year might have been done by analogous procedures, not different in principle, but more cumbersome and thus inviting the invention of other sorts of aids. For this we have even less evidence; we know that it was done, but we can only guess at the means. In any case, the combination of a vigesimal system of numeration and a vigesimally subdivided calendar year certainly facilitated the procedure. There would have been an advantage in reckoning *baktuns* in pairs, up to the last one if their number was odd. A tabulation such as the following makes the task quite simple:

Day-Count Increments	Calendar-Year Moves	
2 *baktuns*		+ 15 days
1 *baktun*	+ 9 *uinals*	+ 10 days
1 *katun*	− 5 *uinals*	
1 *tun*		− 5 days
1 *uinal*	+ 1 *uinal*	
1 day		+ 1 day

A convenient visual aid would have been a simple list of the hieroglyphs of the eighteen 20-day months, together with the *Uayeb*, for tallying and place keeping. Crossing the *Uayeb*, of course, requires a five-day adjustment in the month day (subtraction if going forward over the *Uayeb*; addition if regressing in the sequence) and sometimes a concomitant change in the month.

Finding the calendar-round day corresponding to a date given in the day count is the same kind of problem. For this, *4 Ahau 8 Cumhu* is the initial day, and the Maya day number is the increment.[11]

This subject requires no further elaboration and can be left by posing a problem for the reader: *Determine the calendar-round day corresponding to the day number 9.9.2.4.8.* (The numerical equivalents of the veintena day names and of the month names have been given above.) The answer is *5 Lamat 1 Mol.* It was the accession day of the ruler known as Pacal, or Sun Shield, whose remains lie in the sarcophagus within the crypt under the Pyramid of the Temple of the Inscriptions at Palenque, in present-day Chiapas, Mexico.

Should this have given no difficulties, a second may be in order, which will require expansion of the above formulas (according to the stated rule) so as to accommodate a seven-place distance number: *Find the calendar-round day that is 7.18.2.9.2.12.1 after a day 1 Manik 10 Tzec.* The answer is the same: *5 Lamat 1 Mol.* In the inscription of this monument the date of accession of the ruler is related first to his birth date: "12.9.8 from the birth of SHIELD-Pacal on *8 Ahau 13 Pop* to the accession of 'Great-Sun' SHIELD-Pacal on *5 Lamat 1 Mol*" (he was then a boy of twelve, his mother having acted as regent until he was old enough to be named ruler). Then the accession date is anchored in the day count: ". . . which was 2.4.8. after the *3 Ahau 3 Zotz* period-ending" (that is, after, 9.9.0.0.0). And then it is related to the "enthronement" of a mythological ruler or ancient deity who is said to have ascended to power about one and a quarter million years earlier: "7.18.2.9.2.12.1 from the enthronement of XXX (name of deity) on *1 Manik 10 Tzec* to the enthronement of 'Great-Sun' SHIELD-Pacal of the Royal Line of Palenque."

The other major type of calendrical problem may be posed as follows:

II. Given any two days in the calendar round, (t_1, v_1, y_1) and (t_2, v_2, y_2) respectively, what is the magnitude of the interval from the first to the second? ("Interval" here is normally to be understood to mean "minimal interval," that is, less than the duration of one full calendar round; longer intervals with the same initial and final days may be derived from the minimal interval by adding to it appropriate integral multiples of the calendar round, 2.12.13.0, or 18,980 days.)

This was possibly the most frequent type of problem that the Maya calendar mathematicians had to solve; for dates were most recorded, and presumably most remembered and talked about, in terms of their calendar-round specifications. These must normally have been the "givens"; distance numbers and placements in the day count probably required computation in most cases. Again we have ample evidence that the Maya solved these problems successfully, but no direct evidence as to how they did it. Our natural approach, of course, is to derive such formulas as are necessary and use them. We can only guess at theirs. Because of the constraints of the problem, however, we can anticipate that there must necessarily be some underlying formal equivalences between their methods and ours, however much these may be masked by differences in the phraseology and devices employed.

The solution to the problem requires first the separate determinations of the minimal intervals in each of the three component cycles, that is, in the trecena (Δt), in the veintena (Δv), and in the calendar year (Δy); then, from the first two of these, also the minimal interval in the almanac (Δa); then, from the almanac interval and the year interval, the number of whole calendar years (n_Y) that are contained in the calendar-round interval; and then, finally, from that integral number of years together with the fractional year interval, the number of days in the calendar-round interval (Δcr) is determined. The formulas may take the following forms:

(4) $\Delta t = t_2 - t_1$, mod 13. $\Delta v = v_2 - v_1$, mod 20.
$\Delta y = y_2 - y_1$, mod 365.
(5) $\Delta a = 40(\Delta t) - 39(\Delta v)$, mod 260.
(6) $n_Y = \Delta a - \Delta y$, mod 52.
(7) $\Delta cr = 365\ (n_Y) + \Delta y$.

The derivations of formulas (5) and (6), a trifle lengthy, need not be presented here. The others are mere expressions of definitions.

Again we should ask: How might the Maya have done it? Only the tasks corresponding to formulas (5) and (6) require consideration. Equation (5) may be rewritten as

(5*a*) $\Delta a = \Delta v + 40\,(\Delta t - \Delta v)$, mod 260,

which expresses the almanac interval as the sum of a different pair of two parts: the first consisting of a series of one-day steps that move the veintena day from its initial position to one that agrees with its final position; and a second consisting of a series of forty-day steps (*u yoc habil!*) that move the trecena day by an amount equal to the difference between its initial position and its final position, less the amount already accounted for by the moves it has made in accompanying the veintena day in its shift. It will be remembered that the special property of the forty-day "footsteps" was that they advanced their position in the trecena by one day per step while remaining stationary in the veintena. Since the Maya were clearly aware of this property, at least some of them having given the period a name that expressed the property, it is reasonable to assume that they may once have put it to some use; perhaps *this* use. If so, they would have reckoned almanac intervals more or less in the following fashion. Let us pose a particular problem: *What is the interval from 8 Ahau 13 Pop to 5 Lamat 1 Mol?* A Maya computer might have proceeded thus, tabulating his positions or counting them off with the aid of some device:

8 Ahau
9 Imix
10 Ik
11 Akbal
12 Kan
13 Chicchan
1 Cimi
2 Manik
3 Lamat – *4 Lamat* – *5 Lamat.*

Proceeding down the column at the left takes eight moves of one day each; and proceeding to the right in the last row takes two more moves amounting to forty days each. Therefore the interval within the almanac from *8 Ahau* to *5 Lamat* is 88 days, or in Maya notation 4.8, that is, four score and eight days.

Next we would wish the interval within the calendar year, from *13 Pop* to *1 Mol.* The latter is the first day of the eighth calendar month, numerically "one in the eighth score" (*hun tu uaxac kal*), or in the alternative manner of expression "seven score and one" (*uuc kal catac hun*), whence the notation 7.1. Similarly, the year day of the earlier date, *13 Pop*, is 0.13. Subtracting the earlier from the later, we find the interval between them, Δy, to be 6.8 or 128 days. The computation of intervals within the calendar year was the simplest sort of problem for the Maya, even when crossing the *Uayeb* was involved, or when application of the modulus 18.5 was required. (The task is more difficult for us in our calendar, with months of variable lengths and unrelated to our arithmetic base.)

For the next step numerous alternatives are open. Proceeding as we would, according to formulas (6) and (7), they would have found the number of whole years contained within the desired calendar-round interval to be (88 – 128) + 52, or 12 calendar years. Twelve of these are 12.3.0 in day-count notation. To this sum would be added the fractional year just determined above, which was 6.8, resulting in 12.9.8 as the calendar-round interval between *8 Ahau 13 Pop* and *5 Lamat 1 Mol*, that is, twelve chronological years (*tuns*) plus nine score and eight days.

It is not known whether this was the Maya way. It probably was not. It is more likely that one or two other methods were in use, and there is circumstantial evidence favoring one of them. At any rate, they must have been aware that the difference between what we have called Δa and Δy, the interval in the almanac and the interval in the year, is always divisible by five (if it is not, someone has made a mistake).

Consider that the probable procedure for reckoning intervals in the almanac was that sketched in the fourth preceding paragraph above, the strategy of which was to move toward the goal in the veintena by one-day steps and then, holding the veintena day constant, to move toward the goal in the trecena by forty-day steps. This precedent suggests analogous procedures for reckoning intervals also in the calendar round. For example, one may move toward the goal in the almanac by the method just mentioned, and then, holding the almanac day constant, move toward the goal in the calendar year by steps of some magnitude that is a multiple of the almanac and affords a convenient measure. Or alternatively, one may move first toward the goal in the calendar year by the simple arithmetic method shown above, and then, holding the year day constant, move toward the goal in the almanac by steps of some magnitude that is a multiple of the length of the calendar year. In either case, whether the minimal interval in the calendar round was desired, or an interval of some larger approximate magnitude (say greater than some known number of *katuns* or *baktuns*, or passing over some specified number of *katun*-endings), a table of multiples of the calendar round would have been a necessary adjunct to reduce or augment the result (sometimes negative) to the desired proportion. Although such a table (2.12.13.0; 5.5.8.0; 7.13.3.0; and so forth) has not survived – except for the preface to the Venus table, which has the even-numbered multiples as far as to eight calendar rounds, 1.1.1.14.0 – tables of multiples of other intervals have survived in the codices. Therefore since the construction of such a table was well within Maya competence, and since the need existed, we may assume that the tables did also. (Mayanists have constructed them and use them constantly.)

From equations (6) and (7) the following facts are readily apparent:

A. If the year day and the veintena day are held constant and increments are applied to the trecena day, then each one-day increment in the trecena reflects a 40-year increment, or a 12-year decrement, in the calendar round.

B. If the almanac day is held constant (that is, in both the trecena and the veintena) and increments are applied to the year day (they can be applied only in multiples of five, because of the constraints mentioned earlier), then each five-day increment in the year reflects a 5×364-day decrement in the calendar round, and conversely, each five-day decrement in the year reflects a 5×364-day increment in the calendar round. Note that 5×364 is equal to 7×260, which is 5.1.0 in Maya numeration, or 1,820 days.

Other such facts, reflecting other manipulations of the variables in the specification of calendar-round days, can also be brought out; although for present purposes these will suffice. But in such manipulations it must be remembered that if the year day is held constant, the value of the veintena day cannot be altered except in amounts that are multiples of five; and that if the almanac day, or the veintena day by itself, is held constant, then the year day cannot be altered except by similar quinary amounts. The trecena day, however, can be varied freely in computation problems.

Two additional alternatives are thus open for the second step in our illustrative problem of determining the calendar-round interval between *8 Ahau 13 Pop* and *5 Lamat 1 Mol*. Having determined the interval between the almanac days, we may exploit *B* above; or having determined the interval between the year days, we may exploit *A*.

Reckoning by means of *A* would go as follows. From *8 Ahau 13 Pop*, which is on year day 0.13, to the next *1 Mol*, year day 7.1, is an interval of 6.8 (by subtraction of the former year day from the latter). But the next *1 Mol* would be calendar-round day *6 Lamat 1 Mol* (because an increment of 6.8 added to almanac day *8 Ahau* reaches almanac day *6 Lamat*; compare formulas [1] and [2]). Further, from *6 Lamat 1 Mol* to *5 Lamat 1 Mol* is an interval of 12 calendar years, or 12.3.0 in Maya day-count numeration (because there is a decrement of 1 in the trecena value while all else is constant, and according to *A* this must result from an increment of 12 years in the calendar round). Adding now to this the fractional part of a year as first determined, we have 12.3.0 + 6.8 = 12.9.8.

Reckoning by means of *B* would go as follows. From *8 Ahau 13 Pop* to the next *5 Lamat* is an interval of 4.8 (as determined by the method for calculating intervals within the almanac; compare formulas [4] and [5]). But this next *5 Lamat* would be the calendar-round day *5 Lamat 1 Xul* (because an increment of 4.8 added to *13 Pop*, year day 0.13, reaches year day 5.1 or *1 Xul*). Further, from *5 Lamat 1 Xul* to *5 Lamat 1 Mol* is a decremental interval of 40×364 days, $(2.0) \times (1.0.4)$

in Maya day-count notations, or −2.0.8.0; which, modulo 2.12.13.0, is equivalent to an increment of 12.5.0 (because in this pair of calendar-round days the almanac portions are constant and the interval from year day *1 Xul* (5.1) to year day *1 Mol* (7.1) is 2.0 or 40 days, and because, according to *B*, this must result from a decrement in the amount of that number of 364-day intervals, or from the complementary increment). Adding now to this increment the fractional part of an almanac as first determined, we have 12.5.0 + 4.8 = 12.9.8.

There are thus three (that is, at least three) alternative procedures for determining intervals between calendar-round days, given the interval in the almanac and/or that in the year. Analogously there are three alternatives for determining intervals between almanac days, proceeding from the interval in the trecena and/or that in the veintena. Of these latter, only two have been described. The third is like the second, except that it moves first toward the goal in the trecena by one-day steps, and then toward the goal in the veintena by 39-day steps, each effecting a one-day regression in the veintena; instead of, as before, moving first toward the goal in the veintena by one-day steps, and then toward the goal in the trecena by 40-day steps, each effecting a one-day advancement in the trecena. The point to be made is that the two nonalgebraic methods for computing intervals in the almanac, and the two similar methods for computing intervals in the calendar round, employ a single uniform strategy for dealing with two-variable problems of a sort inherent in a calendrical system employing simultaneous cycles of differing lengths. It is a strategy whose discovery and execution are entirely compatible with the stage of mathematical development that had been attained by the Maya at the very beginning of the classic period, or by their precursors even earlier. One ventures to suppose that it may have been their method. Circumstantial evidence possibly supportive of this hypothesis are (*a*) the modern ethnographic relic of the recognition and naming of the 40-day steps that serve in one of the alternatives for reckoning almanac intervals; and (*b*) the preservation in the Dresden Codex of several tables of multiples of 5.1.0, that is, of 1,820 days, which is equal to 5 × 364, and to 7 × 260. The tables are subdivided in various ways, passing either through 364 or through 260 in their buildup to 1,820. Two of them reach 400 × 364, or Maya 1.0.4.8.0. Because of some of its other useful properties in relation to the almanac and the calendar year, the 364-day period has been called the "computing year." Its potential role in the computation of intervals within the calendar round is comparable to that of the 40-day "footsteps" in the computation of intervals within the almanac. One thing is certain: the Maya calendar specialists regularly had to make such calculations, and most of the time they did them correctly. Still needed, however, is a thorough analytical study of the errors that they made, both in the inscriptions and in the codices. Such a study, not yet done, might furnish further and more substantial evidence on the computational methods of the Maya.

It should be added that the problem of placing a calendar-round date in the day count is simply a special case of the problem just discussed. In this, the initial calendar-round position is *4 Ahau 8 Cumhu*, and the final one is whatever day it is that is to be placed in the day count. The minimal interval between the two positions is determined by one of the methods above. To this, then, is added whatever multiple of the calendar round will bring the date into the anticipated range in the day count, or will allow it to have a certain moon age, or will be compatible with any other piece of information or circumstantial evidence about some aspect of the date or of the personages named in the inscription. If the range or the constraints are poorly defined because of paucity of context, plausible alternatives may at least be enumerated. Usually, however, they can be fixed with fair certainty.

This discussion of computation problems has focused on the almanac, the year, the calendar round, and the day count. There are others, inasmuch as the complete specification of a date in the initial passage of a monument normally included also its position in the nine-day cycle of glyphs *G*, the lunar data, and at least optionally its place in the 819-day cycle. The first of these is easily determined from the lowest two places in the day number, once it is known that the one occurring with all *tun* endings (multiples of 360 in the day count) is conventionally called *G*9. Thus:

(8) $$G = 2n_2 + n_1, \bmod 9.$$

The numerical *G* values refer to hieroglyphs whose identities are well known but whose precise significations are not.

The place in the 819-day cycle requires, and probably required for the Maya, a table of multiples of 819, Maya 2.4.19, up to the twentieth, which is 2.5.9.0, and then higher multiples of 2.5.9.0. This last is one of the numbers that had numerological importance, being the lowest common multiple of the almanac and the 819-day period, as well as of the 364-day computing year. (Factored, it is $2^2 \cdot 3^2 \cdot 5 \cdot 7 \cdot 13$.) It was employed, for example, in fixing two important mythological dates at Palenque.

The position in the trecena of all 819-day stations was the same, since the number is divisible by 13. The trecena position was 1. All positions in the veintena were occupied, with one-day regressions in that cycle from one 819-day station to the next. Three days before the zero of the day count is a convenient station to use as a reference point in reckoning; although it is not known whether the Maya had a traditional zero point for their reckoning. The earliest one recorded was at a distance of 6.15.0 before the beginning of the current era. Directions and directional colors were assigned to the stations. Those on days 1, 5, 9, 13, and 17 of the veintena (*Imix, Chicchan, Muluc, Ben,* and *Caban*) were of the east. The others followed from these, the assignments being in agreement with the normal counterclockwise rotation when proceeding forward through the days of the veintena; but since successive 819-day stations regress in the veintena, the sequence here is clockwise. Colors associated with directions were red with east, white with north, black with west, and yellow with south.

In the lunar series there were three items of data: the age of the current moon, the place of the moon in the current lunar half-year, and the duration of the current moon (or the immediately past one?) in the lunar calendar. There were variations from site to site and from time to time in the evaluations of moon ages. One-day variations might be attributable to variability of conditions, both meteorological and topographic, affecting visibility of the new moon, but there are also other factors that may enter into this when the recorded ages are results of computations rather than of observation. Three-day variations must be attributed instead to different conventions regarding the zero point for the count of days or nights in the age of the moon; and occasional four-day variations may be attributed to combinations of factors. It is concluded that, at many of the Maya sites, moon age was reckoned from the first visibility of the new crescent, an age of one day not being accorded until one full day had elapsed since the "birth" of the moon; but that at other sites an attempt was made to reckon from an estimated or computed conjunction time. (An alternative hypothesis would have the former sites attempting to reckon from conjunction, and the latter from the last visibility of the waning crescent, the "death" of the moon. This hypothesis has disadvantages which, in the opinion of the writer, outweigh its advantages: it would require the rejection of the best potential eclipse dates—observed or predicted—of which there are records; and it would necessitate discounting also the meaning or the relevance of the "birth" sign that appears as the principal constituent of the hieroglyph glossing moon ages at some sites belonging to the first of these categories.)

It has been assumed that some of the recorded moon ages were actual records of observation; but many of them may have been—and some of them must have been—the results of computation. In monuments in which the date of the initial passage is the most recent, and in which further content pertains to prior events, that initial date can probably be regarded as approximately contemporary with the erection of the monument and its moon age could perhaps be the permanent record of a lunar observation. In others, in which the date of the initial passage is one of many years prior to the last date of the inscription, or to the date of erection of the monument, unless there was available a log of observations reaching considerable distances back into the past (of which no specimens have survived in any form) the recorded moon ages may have been the products of computation. And in the cases of mythological dates that reach thousands of years back into past time, the recorded moon ages must of necessity be the results of computation. The subject of these computations is discussed below in "Eclipse Reckoning"; for the eclipse-reckoning tables were surely employed also as a general reckoning device for moon ages. And there is good evidence that the cycle on which eclipse reckoning was premised was a discovery that long antedated the table of the Dresden Codex.

The place of the moon in the lunar half-year is another datum in which there was variability from site to site and from time to time. Prior to about 9.12.15.0.0 (which would be A.D. 687 according to a chronological correlation accepted by some) there was variability among the sites. From this time, however, until about 9.16.5.0.0 (A.D. 756 in this correlation) there was uniformity. After that time there was variability again. A grouping of moons into half-years would seem to have only two possible motivations deriving from natural phenomena: either to relate the moons to the recurrence of eclipse seasons, or to relate them approximately to the halves of the tropical year as determined by the equinoxes or by the solstices. In the former case, half-years are usually of six months but sometimes of five; in the latter they are usually of six but sometimes of seven. Thus, the latter is excluded for the Maya groupings, because their moon numbers never exceed six. If the motivation was the former, uncertainty and variability are well understandable (see "Eclipse Reckoning"). Curiously, during the period of uniformity a standard six-month lunar half-year was adopted and spread the length and breadth of the area of Classic Maya civilization, with uniform numbering of the moons. This can have related to no natural phenomenon and must have been either an arbitrary convention or one born of an erroneous theory that took some seventy years to discredit. The date of the beginning of the gradual abandonment of this method is of interest. According to the Dresden Codex, 9.16.4.10.8 appears to be the initial epoch for the system there presented, although that particular recension and arrangement must pertain to an epoch some 300 years later. The close proximity of this date to the first abandonment of the uniform system of moon numbers in inscriptions is perhaps not an accident. Yet, the knowledge of that eclipse cycle, of 405 lunations to 11,960 days, must be considerably older. It was definitely used at Palenque just prior to 9.14.10.4.2 (*circa* A.D. 725) in an interesting piece of numerology (see "Numerology," below), and the same ratio was used also at Palenque about 9.13.0.0.0 (*circa* A.D. 692) to compute moon ages attributed to mythological dates of 1.18.5.4.0, 1.18.5.3.6, and −6.14.0 (the first two *circa* 2360 B.C., and the third *circa* 3120 B.C.).

Thus it appears that a proper theory may already have been in existence—perhaps even from the beginning of the recording of moon numbers, which was at least as early as 8.16.0.0.0 (A.D. 357)—and that something more persuasive than convenience or theory may have been behind the rapid spread of the seemingly ill-conceived uniform system. It has been supposed that it may reflect an episode in the political history of the area—a hypothesis that will require new testing, now that more potential evidence is coming to light both from archaeology and from advances in the understanding of the inscriptions.

The remaining item of lunar data associated with the initial date of a monument is that of the moon's duration (glyph *A*). This was a calendrical rather than an observational matter. The problem is not a trivial one for a lunar calendar, and so should be stated. A simple alternation of 30-day and 29-day months must be broken from time to time, so as to allow for more of 30 days than of 29, since the mean length of the lunar month is greater than 29½ days, namely, 29.530588 days. The optimum arrangement is one that provides for two 17-month groups followed by one 15-month group [2(17) + 1(15)], each group beginning and ending with a 30-day month, with the alternation thus being broken at the junctures of the groups where two 30-day months come in immediate succession. The resulting ratio is one of 49 lunar months to 1,447 days [(26 × 30) + (23 × 29)], which is equivalent to a mean lunation of 29.530612 days. There are other arrangements with other ratios that are workable, but none as good as this for relatively short-term determinations. It is not known how many Maya solutions there were to this problem. One, which was employed at Copán, in what is now Honduras, involved a ratio of 149 moons to 4,400 days [(79 × 30) + (70 × 29)], equivalent to a mean lunar month of 29.530201 days—good enough for relatively short spans of time, but not one of the best (it is eighth-best among such possibilities), and not good enough to be accurate for the long-range use to which it was applied. It has not been determined how the astronomers of Copán arrived at this ratio, or on what theory it was based. Its arrangement of moons however, as required for the problem of glyph *A*, requires seven 17-month groups and two 15-month groups, each group beginning and ending with a 30-day month, possibly in the sequence [4(17) + 1(15) + 3(17) + 1(15)], but with other compromises and *ad hoc* arrangements possible also.

Another ratio, employed at Palenque, can be expressed as 81 moons to 2,392 days, equivalent to a mean lunar month of 29.530864 days. Neither is this among the best possibilities (it is sixth-best, erring in the opposite direction from that of Copán), and it too was not good enough to provide accuracy for the long-range projections in which it was used, up to almost four millennia. (Of course the Palenque astronomers could not have known this, as neither could those of Copán have known of the deficiency in theirs; but had they known, it is quite apparent that it would have been a matter of concern.) The theory behind the Palenque ratio, unlike that of Copán, is known; it is derived from the eclipse cycle of 405 lunations to 11,960 days, which is the same ratio. The arrangement of the sequence of 30-day and 29-day months in this scheme is governed in part by the structure of the eclipse cycle, which introduces considerations extraneous to the simple problem of finding a suitable arrangement for a lunar calendar. (These matters are treated in "Eclipse Reckoning.")

The fact that the juxtaposition of two 30-day months in immediate succession was governed by a theoretical scheme of some sort is illustrated by the occurrence of such a sequence in a projection far back into mythological antiquity at Palenque, in a pair of dates just fourteen days apart. To the earlier of the pair, 1.18.5.3.6, the initial date of the Temple of the Sun, is ascribed a moon number of 4, a moon age of 26 days, and a lunar month duration of 30 days. To the later one, 1.18.5.4.0, the initial date of the Temple of the Foliated Cross, is ascribed the moon number of 5, a moon age of 10 days, and month duration again of 30 days. Since these were backward projections of a little over three millennia at the time they were made, they are necessarily the products of computation according to some theory; and at this juncture in the theoretical scheme—perhaps codified as a table of arrangements within a cycle—two successive 30-day months were prescribed.

Questions not entirely settled are whether the moon number (glyph *C*) and the moon duration (glyph *A*) pertain to the current lunar month or to the preceding one. For example, in regard to moon number, is "moon 4, age 26" in the Temple of the Sun intended to mean that on that date 26 days have been completed in the fourth moon of the current half-year, or that 4 moons and 26 days have been completed? Is it 3 lunar months and 26 days, or 4 lunar months and 26 days that have passed? Crucial cases for a decisive test are elusive. The same is true for the question of lunar month duration. Were it not for the circumstance that *both* of the above-mentioned inscriptions specify 30-day durations, they would have provided a crucial test. The interval between the dates is 14 days. If 14 days are added to moon age 26 with moon number 4, which are the data of the earlier date, then a moon age of either 10 or 11 days with moon number 5 is reached for the second date, depending on whether the moon just completed was of 30 or of 29 days. But the recorded moon age for the second date is 10 days, so the moon just completed was of 30 days. If the data with the earlier date had prescribed a duration of 29 days and that with the later had prescribed 30, then it would be known that the prescriptions referred to the last completed month. Had the reverse been the case, it would be known that they referred to the current month. But since both are 30, the question remains unanswered. (This is the pair of dates that enabled John Teeple [about 1924] to decipher the meanings of the hieroglyphs pertaining to the lunar calendar.) Related to these uncertainties is the question of whether any recorded moon ages at all were actually the products of current observation, or whether they too were prescriptions of a lunar calendar. The role of observations may have been to develop a theory and to attempt to perfect the lunar calendar, while the recording of moon ages could well have been by calendrical prescription in all cases. These, and the theories and practices that must be seen to lie behind them, varied from site to site, except during the "period of uniformity."

The moon's behavior appears to have been a subject of ongoing research and perhaps of fundamental disagreements among the Classic Maya astronomers. To most of us it must seem that they did remarkably well (see also "Eclipse Reckoning"). Why, for example, should we care what is beyond the third decimal point in the figures for the mean duration of the lunar month which are equivalent to the Maya ratios, especially when they lacked decimal (or vigesimal) fractions? A comparison of their achievements can be expressed in another way. Employment of the Copán ratio accumulates an error of one full day (negative) in a little less than 209 years. Employment of the Palenque ratio accumulates an error of one full day (positive) in slightly less than 293 years. They had not discovered the optimum short-range ratio mentioned first in this discussion, employment of which accumulates an error of one full day (positive) only after nearly 3,369 years. This would have been suitable to the time spans over which they attempted predictions. Somewhere and at some time, perhaps wherever the Dresden Codex came from, and perhaps in the early postclassic period, it appears that a means was discovered for making periodic corrective adjustments to the lunar calendar based on the eclipse cycle that accomplished a double purpose, namely, to preserve an important relationship between that cycle and the almanac, and to make possible far more accurate long-range reckonings of the moon. If we can believe that the apparent applications of this device in the relations between certain dates in the Dresden Codex were not merely fortuitous, then they attest to a method having a precision such that the accumulation of an error of one full day (negative) would require nearly 4,492 years. (This topic is touched on further, although briefly, in "Eclipse Reckoning.")

IV. THE VENUS CALENDAR

Pages 46–50 of the Dresden Codex contain the table shown here in Figure 1. Its division into five periods of 584 days each identify it as a table of the synodic years of Venus. The further subdivisions of these into intervals of 236, 90, 250, and 8 days, can be assumed to represent the canonical values ascribed by the Maya to the periods of Morning Star, Superior Conjunction, Evening Star, and Inferior Conjunction, respectively, of that planet.[12]

The 584-day figure is the average of the approximate lengths of the synodic periods of Venus in a cycle of five, which, to the nearest whole day, are of 580, 587, 583, 583, and 587 days, respectively. It approximates closely the true mean value of 583.92 days. The eight days ascribed to inferior conjunction are a fair approximation to a mean value for a period of invisibility that can vary from a couple of days to a couple of weeks. As for the other subdivisions, they are of appropriate orders of magnitude, although more nearly equal intervals might have been anticipated for the two periods of visibility. A reason for the slight disproportion is not readily apparent, although considerations of lunar reckoning (236 days are eight lunar months, 250 are about eight and a half), or of canonical days chartered by myth, have been suggested as possibilities.

Since the Maya calendar year was of 365 days (without leap year corrections), and since 365 and 584 share a common factor of 73, a cycle of five Venus years coincides with a cycle of eight Maya calendar years. The length of this combined cycle is 2,920 days, or 8.2.0 in Maya calendrical numerals. This is the accumulation of days after one procession through the five pages of the table. It is recorded as such on page 50, in the black number of the fourth column in the middle section of that page.

The 260-day sacred almanac is not contained without remainder in this five-Venus-year/eight-calendar-year cycle. Its veintena component, 20, is contained in 2,920, but its trecena component, 13, is not. Hence it requires thirteen processions through the five pages of the table to return to the same day in the sacred almanac. The almanac days of the stations in this greater cycle are given in the thirteen lines of the top sections of the five pages of the table. The veintena days (represented by their hieroglyphic signs) repeat line after line, but the trecena days (the numerical coefficients) vary. The length of the greater cycle is 13 times 2,920, or 37,960 days, which is equal to two calendar rounds, or 65 approximate synodic years of Venus, or 104 Maya calendar years.

The calendrical and numerical data are found in the first four columns (the left-hand half) of each of the five pages of the table. Certain other items are also included. From top to bottom, the left-hand halves of the pages display the following information.

Upper Section: thirteen lines of positions in the 260-day almanac, twenty per line (four on each page), naming the almanac days reached by adding the cumulative totals (black numbers, bottom of middle section) to a base, or starting point, falling on almanac day *1 Ahau*.

Middle Section: (1) a line of positions in the 365-day year, naming the year days reached by adding the same cumulative totals to a base falling on year day *13 Mac*; (2) a line of repetitions of a constant "event" glyph, repeated in each of the twenty columns; (3) a line of direction glyphs, north, west, south, and east, repeated in this order on each of the five pages; (4) a line of name glyphs of "gods," variable, with only a single repetition in the line of twenty; (5) a line of "Venus" glyphs, in two variant but equivalent forms—so-called half and full—twenty in all; and (6) a series of black two-place and three-place numerals, giving the cumulative totals of the intervals specified by numbers in red at the bottom of each page (see below).

Lower Section: (1) a second line of positions in the 365-day year, naming the year days reached by adding the cumulative totals to a base falling on year day *18 Kayab*; (2) a line of repetitions of a second constant "event" glyph—in line 3 on page 46, omitted on page 47, and in line 2 on pages 48 to 50—different from the "event" glyph of middle section line 2; (3) a line of variable name glyphs of "gods," the same ones (a few with variant but equivalent glyph forms) and in the same order as those of middle section line 4, but offset one column to the right—in line 2 on pages 46 and 47, in line 3 on pages 48–50; (4) a second line of constant "Venus" glyphs, only in the "full" variant—omitted on page 48; (5) a line of direction glyphs, in the same order as in line 3 of the middle section, but offset one column to the right, thus reading east, north, west, south; (6) a third line of positions in the 365-day year, naming the year days reached by adding the cumulative totals to a base falling on year day *3 Xul*; and (7) a series of two-place numerals, in red, naming the intervals 11.16, 4.10, 12.10, and 0.8, which are equal to 236, 90, 250, and 8 respectively, specifying the canonical durations of the four subdivisions of the Venus year.

The right-hand sides of the five pages are given over to astrological interpretations, pictorially illustrated, and to prognostications associated with each of the five calendrical varieties, or celestial regions, of heliacal risings of Venus. The upper illustrations are of deities seated on celestial thrones consisting of bands of planetary and astral signs. They are the deities named in the five "east" posi-

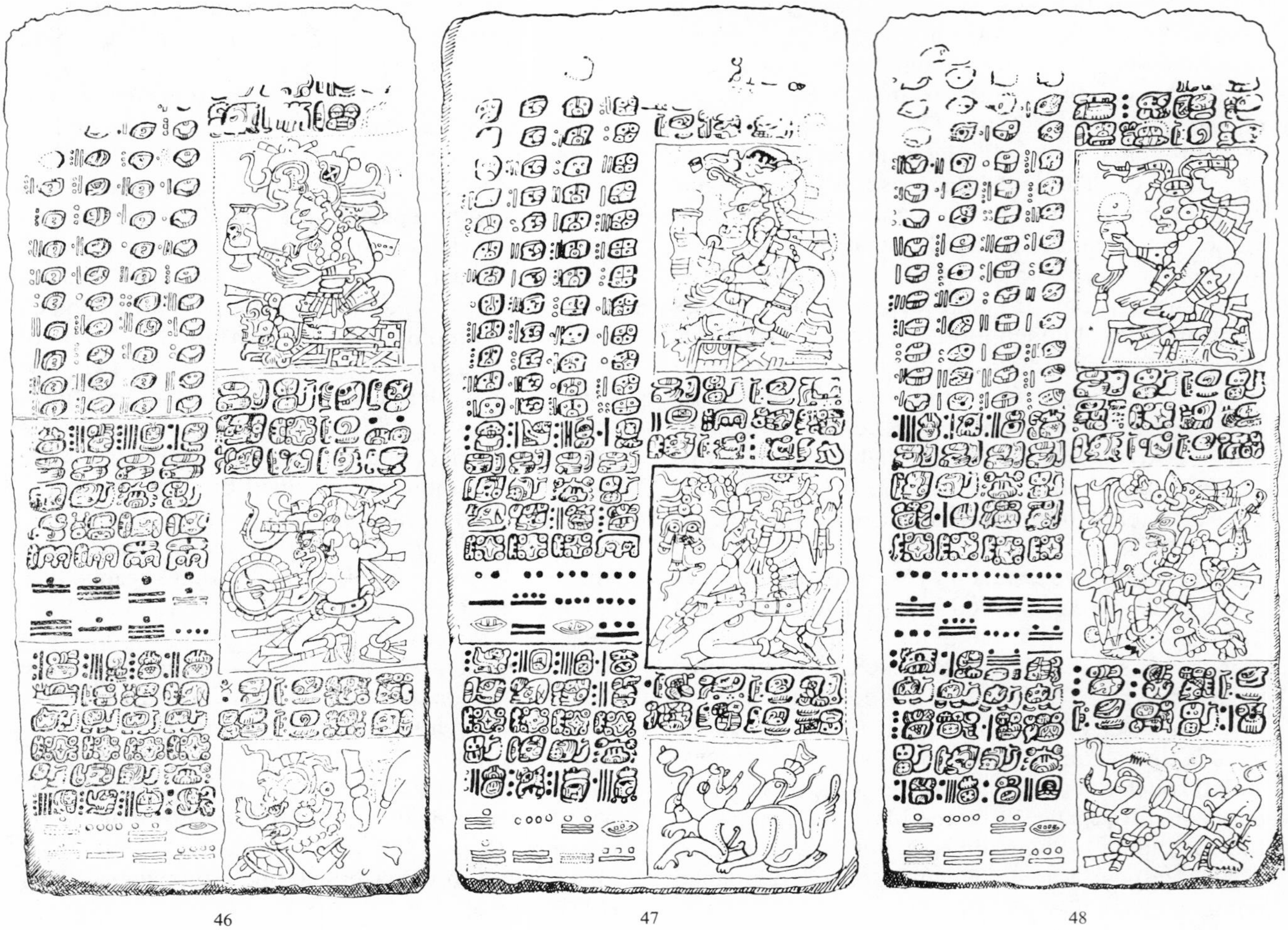

FIGURE 1. Pages 46–50 of the Dresden Codex.

tions in the two lines of god names on the left-hand sides of the pages. (The name of the deity pictured on any given page is found in the first column of the lower section of that same page, and in the fourth column of the middle section of the preceding page.) The iconography seems to warrant their being regarded as presiding figures in the different celestial regions in which heliacal risings of the planet take place. The middle illustrations are of deities equipped with instruments of warfare (shield, spears, spear thrower). They are assumed, on the basis of analogy with information contained in Mexican ethnohistorical sources, to represent the guises or manifestations of Venus at each of its canonical heliacal risings. The shafts emanating from the Morning Star on these occasions were said to have "speared," each time, a different order of victims. The lower illustrations are of deities or other symbolic figures that represent the primary victims, each being shown pierced, or about to be pierced, with a spear. The illustrations are accompanied with hieroglyphic annotations. These name the various protagonists in the drama, the "spearing" event, the prognostications of ill, and the categories of other things, personages, and deities susceptible to misfortune on these occasions. The six glyphs in the first pair of columns over the middle picture of each page state the following kinds of information:

Event glyph ("appear"?)	"East" (*lakin*)
Name of spearer (variable)	"Venus" (*chac ek*)
Name of victim (variable)	"his spearing" (*u hul*).

Thus, on page 46 this sequence can be interpreted,

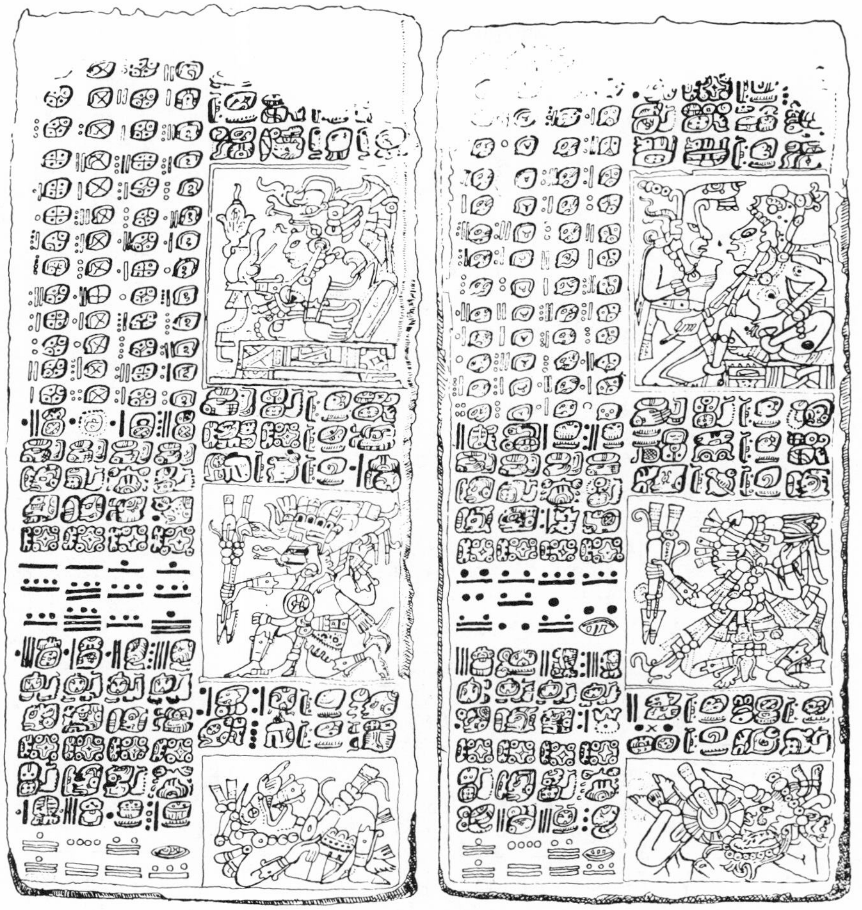

49 50

quite in accord with Maya grammar, as follows: "Appears in the east, god *L* as Venus; god *K* is the object (the victim) of his spearing." Similarly, but with other protagonists, on the other pages.

The calendrical and numerical content of the left-hand portions of the pages, excluding all other matters, are transcribed in Table 1. The first thirteen lines present the trecena positions (the coefficients of the day signs) from the first thirteen lines of the Dresden Codex table. Effaced portions at the tops of the pages of the codex are restored in the transcription. The veintena days, which are constant for each column and are repeated redundantly in the codex, are indicated only once in the transcription, in line 14, by means of numerals according to the scheme of the table of "Days of the Veintena" (p. 781). Year days, from the three alternative sets provided in the codex table, are given in the transcription by their numerical equivalents, Maya fashion, in lines 15, 16, and 17. Thus, for example, the number "6" standing over the number "4" in the first entry in line 15 (corresponding to column *A* on page 46 of the codex) represents the Maya numeral otherwise transcribed as 6.4, meaning "six score and four," that is, the 124th day of the year. This is the day *4 Yaxkin*, as recorded in this position in the first line of year days in the Maya table. Direct conversion from the Maya name of any day of the year to its numerical equivalent, or vice versa, can be made by filling in its day-of-the-month number in the appropriate blank in the table of "Months of the Year" (p. 781); conversion to Arabic follows then from reading the first of the two digits as scores

Table 1. Calendrical and Numerical Data From the Venus Table of the Dresden Codex.

Line no.	Category of information	Page 46 *A*	*B*	*C*	*D*	Page 47 *A*	*B*	*C*	*D*	Page 48 *A*	*B*	*C*	*D*	Page 49 *A*	*B*	*C*	*D*	Page 50 *A*	*B*	*C*	*D*
1	Trecena day	3	2	5	13	2	1	4	12	1	13	3	11	13	12	2	10	12	11	1	9
2	" "	11	10	13	8	10	9	12	7	9	8	11	6	8	7	10	5	7	6	9	4
3	" "	6	5	8	3	5	4	7	2	4	3	6	1	3	2	5	13	2	1	4	12
4	" "	1	13	3	11	13	12	2	10	12	11	1	9	11	10	13	8	10	9	12	7
5	" "	9	8	11	6	8	7	10	5	7	6	9	4	6	5	8	3	5	4	7	2
6	" "	4	3	6	1	3	2	5	13	2	1	4	12	1	13	3	11	13	12	2	10
7	" "	12	11	1	9	11	10	13	8	10	9	12	7	9	8	11	6	8	7	10	5
8	" "	7	6	9	4	6	5	8	3	5	4	7	2	4	3	6	1	3	2	5	13
9	" "	2	1	4	12	1	13	3	11	13	12	2	10	12	11	1	9	11	10	13	8
10	" "	10	9	12	7	9	8	11	6	8	7	10	5	7	6	9	4	6	5	8	3
11	" "	5	4	7	2	4	3	6	1	3	2	5	13	2	1	4	12	1	13	3	11
12	" "	13	12	2	10	12	11	1	9	11	10	13	8	10	9	12	7	9	8	11	6
13	" "	8	7	10	5	7	6	9	4	6	5	8	3	5	4	7	2	4	3	6	1
14	Veintena day	16	6	16	4	20	10	20	8	4	14	4	12	8	18	8	16	12	2	12	20
15	Year day (*a*)	6	10	4	5	17	3	15	16	9	14	8	9	2	7	1	1	13	18	12	12
		4	14	19	7	3	8	18	6	17	7	12	0	11	1	6	14	10	0	5	13
16	Year day (*b*)	10	14	9	9	3	7	1	2	14	0	12	13	6	11	5	5	17	4	16	16
		*9	*19	4	12	3	13	18	6	2	7	17	5	16	6	11	19	15	0	10	18
17	Year day (*c*)	16	3	15	16	9	14	8	8	2	6	1	1	13	17	12	12	*6	10	4	5
		19	4	14	2	13	3	8	16	7	17	2	10	6	16	1	9	0	10	15	3
18	Cumulative totals			1	1	2	2	3	3	3	4	4	4	5	5	6	6	7	7	8	8
		11	16	10	11	5	9	4	4	16	2	15	15	9	13	8	8	2	7	1	2
		16	6	16	4	0	10	0	8	*4	14	4	12	8	18	8	16	12	2	12	0
19	Increments	11	4	12	0	11	4	12	0	11	4	12	0	11	4	12	0	11	4	12	0
		16	10	10	8	16	10	10	8	16	10	10	8	16	10	10	8	16	10	10	8

NOTE: The four numbers with prefixed asterisks are corrections of scribal copying errors.

and the second as units, and interpreting the result. Finally, the cumulative totals in the progression through the five-Venus-year cycle, and the increments contributing to those totals—which are the black numbers and the red numbers respectively of the codex table—are transcribed directly in lines 18 and 19.

The three alternative bases, in the order of their listing in the table, are the calendar-round days (*a*) *1 Ahau 13 Mac*, (*b*) *1 Ahau 18 Kayab*, and (*c*) *1 Ahau 3 Xul*—respectively (1, 20, 12.13), (1, 20, 16.18), and (1, 20, 5.3) in Maya numerical equivalents, or (1, 20, 253), (1, 20, 338), and (1, 20, 103) in Arabic. Their common almanac position, *1 Ahau*, is the last one in the Maya table, at the end of the thirteenth line of these on page 50, column *D*. Their three alternative positions in the Maya year are at the ends of the three lines of year-day data, also on page 50, column *D*. Since it is a reentering table, these end points are also its beginning points or "bases."

Since the table has three alternative bases and three corresponding alternative series of year days marking the phenomena of Venus, it is in effect three tables, each self-contained and complete. The implication must be that they were designed to replace one another. The motive for replacement can have come only from the obsolescence, or anticipated obsolescence, of a single table. Obsolescence is the inevitable result of the small discrepancy previously noted between the length of the canonical Venus year of 584 days and the true mean value of the period over longer spans of time, namely, 583.92 days. The error, of 0.08 days per Venus year, accumulates to one of 5.2 days in the 65 Venus years (104 calendar years) covered by the

Days of the Veintena, With Numerical Equivalents
1 = Imix
2 = Ik
3 = Akbal
4 = Kan
5 = Chicchan
6 = Cimi
7 = Manik
8 = Lamat
9 = Muluc
10 = Oc
11 = Chuen
12 = Eb
13 = Ben
14 = Ix
15 = Men
16 = Cib
17 = Caban
18 = Etznab
19 = Cauac
20 = Ahau

table. This, however, is still within the range of variability and ambiguity in the beginnings and endings of the periods of visibility of the planet. For the accumulation of error to pose an acute problem might have required more than one run through the table. But toward the end of a second run it would surely have been perceived as critical. Heliacal risings, on an average, would then be coming about ten days ahead of the prescribed calendrical positions. Correction would be in order. To make a correction of five days, for example, at the end of one run of the table, or of ten days at the end of two, would require a complete rewriting of the table; for the base position in the almanac would no longer be at *1 Ahau*, but at a day somewhat prior to that, say *9 Men* for a five-day correction, or *4 Oc* for one of ten days. Yet the base day provided by the Maya table remains at *1 Ahau* in each of the three recorded alternatives. This suggests that the day *1 Ahau* was somehow sacrosanct as the day for a Venus epoch—which is known, in fact, to be true: "One-Ahau" was the calendrical name of the mythical hero who "became" Venus. It also implies that corrections were accomplished not by making small foreshortenings of the table in amounts equal to the magnitude of the perceived error, but rather by making foreshortenings of whatever magnitude might be required to locate another almanac day *1 Ahau*, somewhere in the cycle, that would come satisfactorily close to coincidence with a heliacal rising. (The alternative, that error be allowed to accumulate until it amounts to the length of the almanac, 260 days, is excluded not only by the impracticality of carrying along an error of such magnitude, but also by the length of time that it would take for it to accrue—fifty runs of the table, or a little over five millennia. It is also excluded by the particular choices of year days specified for the *1 Ahau* bases in the codex.)

The choice of year days assigned to the three alternative bases must provide a clue as to the manner of introducing corrections into the table, as well as to the order in which the alternative bases, and the table premised on them, were to succeed one another. With three such, there are three possible pairings and six permutations. The corresponding minimal intervals between members of ordered pairs are as follows (employing the letters *a*, *b*, and *c*, as in the second paragraph above, for the alternative bases *1 Ahau 13 Mac*, *1 Ahau 18 Kayab*, and *1 Ahau 3 Xul*, in their order of listing in column *D*, codex page 50; and letting *ab* designate the interval from *a* to *b*; *ac* the interval from *a* to *c*; *bc* that from *b* to *c*; and so forth):

"Months" of the Year, With Conversion of Year Days to Numerical Equivalents

Number of prior months completed ↓	Day in current month (supply) ↓	Name of current month ↓	(ordinal number)
0.	___	Pop	(first month)
1.	___	Uo	(second)
2.	___	Zip	(third)
3.	___	Zotz	(fourth)
4.	___	Tzec	(fifth)
5.	___	Xul	(sixth)
6.	___	Yaxkin	(seventh)
7.	___	Mol	(eighth)
8.	___	Chen	(ninth)
9.	___	Yax	(tenth)
10.	___	Zac	(eleventh)
11.	___	Ceh	(twelfth)
12.	___	Mac	(thirteenth)
13.	___	Kankin	(fourteenth)
14.	___	Muan	(fifteenth)
15.	___	Pax	(sixteenth)
16.	___	Kayab	(seventeenth)
17.	___	Cumhu	(eighteenth)
18.	___	Uayeb	(remainder)

Reading Down

(1) *ab*: 19.9.0, or 7,020 days
(2) *ac*: 2.6.4.0, or 16,640 days
(3) *bc*: 1.6.13.0, or 9,620 days

Reading Up

(4) *ba*: 1.13.4.0, or 11,960 days
(5) *ca*: 6.9.0, or 2,340 days
(6) *cb*: 1.6.0.0, or 9,360 days

The second three are the calendar-round complements of the first three. That is to say, the sum of the first and the fourth is equal to 2.12.13.0, or 18,980 days, the length of the calendar round; and so also is the sum of the second and the fifth; and that of the third and the sixth. This follows from the definition of *minimal* intervals. (For the method of computation of intervals between calendar-round days, see "Calendar, Chronology, and Computation.")

Three of these intervals share the property of being an integral multiple of 6.9.0, or 2,340. The fifth (*ca*) is equal to that quantity, the first (*ab*) is its third multiple, and the sixth (*cb*) is its fourth multiple. The other intervals do not contain this factor. Thus the fifth, the first, and the sixth intervals form a proper set, while the second, the fourth, and the third form a complementary set. These hold more promise of being significant sets than do the ones tabulated above, which were premised on spatial arrangement of the bases on page 50 of the codex. The common-factor criterion realigns them as follows:

Set I, Multiples of 2,340

(5) *ca*: 6.9.0, or 2,340 days
(1) *ab*: 19.9.0, or 7,020 days
(6) *cb*: 1.6.0.0, or 9,360 days

Set II, Not Multiples of 2,340

(2) *ac*: 2.6.4.0, or 16,640 days
(4) *ba*: 1.13.14.0, or 11.960 days
(3) *bc*: 1.6.13.0, or 9,620 days

In set I both occurrences of *c* are initial, both of *b* are final, while one instance of *a* is final and the other initial. Therefore the sequence of bases in set I is necessarily *c, a, b*, and the interval *cb* is the sum of the intervals *ca* and *ab*. By a similar argument the proper sequence for the set II ordering is shown to be *b, a, c*, and the interval *bc* is the sum of *ba* and *ac* (modulo 2.12.13.0, or 18,980). Thus the order of replacement of the bases, whichever of these two orders it be, is not the order of their vertical alignment in the codex.

It should now be noted that 2,340 is equal to 4 × 585, that is, to four synodic years of Venus plus four days. It is also equal to 9 × 260, which means that any increment or decrement in this amount, or in any multiple of this amount, will preserve a given position in the almanac. These two facts point to the significance of the intervals of set I. Any one of these quantities, if subtracted from the *1 Ahau* terminal day of the table, will locate another *1 Ahau* in the table, that much prior to its termination, which will be respectively 4, 12, or 16 days earlier in relation to a mean time of heliacal rising of Venus than would be the terminal *1 Ahau*. Thus, it locates a potential new base, for a new Venus epoch, which will effect a 4-day, 12-day, or 16-day correction of accumulated error.

The intervals determined above were the minimum intervals for the six possible permutations of the calendar-round days recorded as alternative bases. Now it can be seen that only for three of these permutations, those of set I, are minimum intervals the relevant ones. Those of set II cannot be relevant, because they lead respectively to days *1 Ahau* which are 296, 304, and 308 days prior to expected heliacal rising, rather than 4, 12, and 16 days. This is because the length of the complete Venus table is equal to two calendar rounds, not one. Thus, the intervals for the permutations of set II should be derived by subtracting those of set I from two calendar rounds (not from just one), or from some multiple of two calendar rounds.

If the interval *ca*, 6.9.0 or 2,340 days, the smallest of set I, is subtracted from two calendar rounds, that is, from 5.5.8.0 or 37,960 days, the complementary interval *ac*—that is, from *1 Ahau 13 Mac* to *1 Ahau 3 Xul*—will be 4.18.17.0, or 35,620 days. This is equal to 61 times 584, less 4 days. In 61 Venus years, reckoned at 584 days each, the accumulated error reaches 4.88 days. The almanac day reached is *5 Kan* (page 46 of the codex, line 13, column *D*). By stopping four days short of *5 Kan*, the almanac day *1 Ahau* is reached, and the accumulated error is reduced from 4.88 to 0.88 days. Thus, it is appropriate that the interval 6.9.0, the shortest of set I, be subtracted from two calendar rounds, that is, from the length of the full Venus table. This locates a more suitable day *1 Ahau* for a new Venus epoch than would be the one at the end of the table. But it has a different year day, that is, *3 Xul* instead of *13 Mac*.

*n*CR	Numerical equivalent	Complement of 1.6.0.0	Same in Arabic	Same in V, less days	Error in days	Correc- tion	Residual error
2CR	5. 5. 8.0	3.19. 8.0	28,600	49V - 16d	3.92	−16	−12.08
4CR	10.10.16.0	9. 4.16.0	66,560	114V - 16d	9.12	−16	− 6.88
6CR	15.16. 6.0	14.10. 6.0	104,520	179V - 16d	14.32	−16	− 1.68
8CR	1.1.1.14.0	19.15.14.0	142,480	244V - 16d	19.52	−16	+ 3.52

If now the interval *cb*, 1.6.0.0 or 9,360 days, the largest of set I, also is subtracted from two calendar rounds, the complementary interval *bc*—that is, from *1 Ahau 18 Kayab* to *1 Ahau 3 Xul*—will be 3.19.8.0 or 28,600 days. This is equal to 49 times 584, less 16 days. In 49 canonical Venus years the accumulated error reaches 3.92 days. Thus it is not appropriate that 1.6.0.0 be subtracted from two calendar rounds, for it would overcorrect, substituting an error of −12.08 days for one of +3.92 days. The appropriate complement to 1.6.0.0, then, should be that obtained by subtracting it from some larger even-numbered multiple of the calendar round. The possibilities are enumerated in the tabulation at the head of this page.[13] The optimum is obviously the six-calendar-round complement. The interval *bc* then—that is, from *1 Ahau 18 Kayab* to *1 Ahau 3 Xul*—will be 14.10.6.0, or 104,520 days, which is equal to 179 times 584, less 16 days. In 179 canonical Venus years the accumulated error reaches 14.32 days. The almanac day reached is *4 Cib* (page 49 of the codex, line 10, column 4). By stopping sixteen days short of *4 Cib*, a day *1 Ahau* is reached, and the accumulated error is reduced from 14.32 days to −1.68 days. But this *1 Ahau* falls on year day *3 Xul* rather than on *18 Kayab*.

The above determination of the magnitudes of intervals *ac* and *bc* of set II results also in the determination of *ba*, for it has been shown that the proper sequence for set II is *b, a, c*, and that *bc* is the sum of *ba* and *ac*. The interval *ba*, therefore, is equal to the difference between *bc* and *ac*. This interval, the apparently intended length of time from *1 Ahau 18 Kayab* to *1 Ahau 13 Mac*, must then be 9.11.7.0, or 68,900 days, which is equal to 118 times 584, less 12 days. In 118 canonical Venus years the accumulated error reaches 9.44 days. The almanac day reached is *13 Eb* (page 48 of the codex, line 11, column 4). By stopping twelve days short of *13 Eb*, a day *1 Ahau* is reached, and the accumulated error is reduced from 9.44 days to −2.56 days. Such overcorrections, one supposes, might have been balanced in the long run by successive applications of the 6.9.0 foreshortening, which undercorrects in the amount of 0.88 days. Three of these would be the appropriate number.

Since the 16-day correction inherent in the greatest of these intervals is equal to the sum of the 12-day and 4-day corrections inherent in the two lesser ones, and since the intervals themselves stand in a similar relationship, the question arises whether the 12-day correction might not also be resolved into the sum of an 8-day and a 4-day correction, these being made separately and sequentially. Inasmuch as the interval that effects a 12-day correction is equal to four calendar rounds less 19.9.0, the two separate intervals effecting 8-day and 4-day corrections would be, respectively, two calendar rounds less 13.0.0, and two calendar rounds less 6.9.0. These are the intervals 4.12.8.0

1 Ahau 18 Kayab	to	1 Ahau 8 Yax:	4.18.17.0	or	35,620	(=2CR − 6.9.0)
+ 1 Ahau 8 Yax	to	1 Ahau 13 Mac:	4.12. 8.0	or	33,280	(=2CR − 13.0.0)
= 1 Ahau 18 Kayab	to	1 Ahau 13 Mac:	9.11. 7.0	or	68,900	(=4CR − 19.9.0)
+ 1 Ahau 13 Mac	to	1 Ahau 3 Xul:	4.18.17.0	or	35,620	(=2CR − 6.9.0)
= 1 Ahau 18 Kayab	to	1 Ahau 3 Xul:	14.10. 6.0	or	104,520	(=6CR − 1.6.0.0)
or:						
1 Ahau 18 Kayab	to	1 Ahau 18 Uo:	4.12. 8.0	or	33,280	(=2CR − 13.0.0)
+ 1 Ahau 18 Uo	to	1 Ahau 13 Mac:	4.18.17.0	or	35,620	(=2CR − 6.9.0)
= 1 Ahau 18 Kayab	to	1 Ahau 13 Mac:	9.11. 7.0	or	68,900	(=4CR − 19.9.0)
+ 1 Ahau 13 Mac	to	1 Ahau 3 Xul:	4.18.17.0	or	35,620	(=2CR − 6.9.0)
= 1 Ahau 18 Kayab	to	1 Ahau 3 Xul:	14.10. 6.0	or	104,520	(=6CR − 1.6.0.0)

(33,280 days) and 4.18.17.0 (35,620 days). No further resolution is possible, since each of these is a foreshortening of just two calendar rounds, that is, of one run through the Venus table. Depending on which of these might precede the other, one or the other of the two sequences tabulated at the bottom of page 783 is anticipated.

The last three lines of either of these sequences present the inferred intervals (the optimal ones) between the bases recorded on page 50 of the codex. The potential intermediate bases presented in the top pairs of lines in the two alternative sequences are not recorded on that page. They are hypothetic possibilities, either *1 Ahau 8 Yax* or *1 Ahau 18 Uo*.

Pages 46–50 of the Dresden Codex would have been known as pages 25–29 if the screenfold manuscript had not become unhinged and separated into two parts, or if it had been realized when pagination was assigned that the two parts were once joined. The five pages of the Venus table were immediately preceded by another (before the separation), this being page 24, which contains related and prefatory material. This page, and a transcription of it, are reproduced here as Figure 2.

The page contains a table of multiples of 8.2.0 (2,920) on the right, with other numerals intrusive in the fourth of the five tiers; and on the left is a chronological matter, with hieroglyphic annotations. As is usual in tables of multiples in the codex, the lowest number, of which the others are multiples, is in the lower right-hand corner. Reading is from right to left in any tier, and from bottom to top in passing from tier to tier. This is contrary to the left-to-right and top-to-bottom reading, in pairs of columns, which is otherwise standard.

The number 8.2.0, or 2,920 (equal to 5 × 584) is that which is recorded also on page 50, column *D*, at the end of the line of cumulative totals (the black numbers of the codex, or line 18 of the transcription in Table 1). In its occurrence on page 24, it stands over the sign of the almanac day *9 Ahau*. The base or zero day for the table, then, is *1 Ahau*, for the addition of 8.2.0 to *1 Ahau* yields *9 Ahau*. The second number is 16.4.0, the second multiple of 8.2.0, and the day recorded below it is *4 Ahau*. Thus, the first through the fourth multiples are recorded in the bottom tier, the fifth through the eighth in the second tier, and the ninth through the twelfth in the third tier. The trecena coefficients of the *Ahau* days reached by these twelve intervals are 9, 4, 12, 7, 2, 10, 5, 13, 8, 3, 11, and 6, and are so recorded. This is the same series of trecena coefficients that is found prefixed to the first twelve *Ahau* signs of column *D* of page 50 of the codex. The series terminates there with *1 Ahau* on the thirteenth line. Thus, the thirteenth multiple of 8.2.0, namely, 5.5.8.0, standing over a *1 Ahau*, would be anticipated to follow also in this series on page 24. The *1 Ahau* is there, but 5.5.8.0 is not. The numerals of the fourth tier are not multiples of 8.2.0. But the expected number is found in the fifth tier. Although partially obliterated, sufficient remains of the numerals in this topmost tier to identify them as 5.5.8.0, 10.10.16.0, 15.16.6.0, and 1.1.1.14.0, which are 13th, 26th, 39th, and 52nd multiples of 8.2.0. These are 2, 4, 6, and 8 calendar rounds, respectively, or the first four multiples of the number of days covered by the entire table of pages 46–50.

Two of the numerals in the intrusive fourth tier are already familiar. One of these, 4.12.8.0, is the doubly foreshortened cycle (2CR−13.0.0) that effects an eight-day correction. It is equal to 57 Venus years less 8 days. Proceeding from the *1 Ahau* base of the table of pages 46–50, an increment of 57 Venus years leads to the day *9 Lamat* (page 47, line 12, column *D*). By stopping eight days short of *9 Lamat*, a day *1 Ahau* is reached for a new Venus epoch, effecting a correction of eight days against the accumulated lag of the table behind the planet. If the application is to the base *1 Ahau 18 Kayab*, it leads to *1 Ahau 18 Uo*. It will be noted that a day *1 Ahau 18 Uo* is recorded on page 24 of the codex, at the bottom of the third column on the left-hand side. This was one of the two alternative hypothetic possibilities previously noted for the intermediate base.

The next numeral in the fourth tier, 9.11.7.0, is also familiar. It is the sum of one doubly foreshortened cycle (2CR−13.0.0) and one singly foreshortened cycle (2CR−6.9.0), which together make up the triply foreshortened double cycle (4CR−19.9.0) that provides for a twelve-day correction and that, as already seen, best mediates between *1 Ahau 18 Kayab* and *1 Ahau 13 Mac*. Assuming that these numbers are indeed to be applied to *1 Ahau 18 Kayab*, which is the earliest of the alternative bases recorded in the table, they constitute supporting evidence for the second of the alternative sequences posited above. An additional singly foreshortened cycle then leads to *1 Ahau 3 Xul*.

The remaining two numerals in this tier are new, in that neither of them has been implied in any way by the content of pages 46–50 of the codex. The first of these, 1.5.5.0, is equal to 9,100 days. If it is applied, as were the others, to *1 Ahau 18 Kayab*, it

	A	B	C
1	1 Ahau	18 Kayab	4 Ahau
2		Event 'B'	8 Cumhu
3		VENUS	Event 'B'
4	East	Event 'A'	Spearer p. 46
5	VENUS	Regent p. 48	VENUS
6	Venus	Regent p. 49	Spearer p. 47
7	VENUS	Regent p. 50	Venus
8	Venus	Regent p. 46	Victim p. 46
9	VENUS	Regent p. 47	Victim p. 47
10	Ill Omen		Victim p. 48
11	”		Victim p. 49
12	”		Victim p. 50
13		2340V =72 CR	LC Day Number
14		9	9
15		9	9
16	RN 6 2 0	16 0 0	9 16 0
	4 Ahau	1 Ahau	1 Ahau
	8 Cumhu	18 Kayab	18 Uo

260V =8CR	195V =6CR	130V =4CR	65V =2CR
1 1 1 14 0	15 16 6 0	10 10 16 0	5 5 8 0
1 Ahau	1 Ahau	1 Ahau	1 Ahau
10 CR −8(V+1)	4 CR −12(V+1)	2 CR −8(V+1)	16(V+1) −13.0
1 5 14 4 0	9 11 7 0	4 12 8 0	1 5 5 0
1 Ahau	1 Ahau	1 Ahau	1 Ahau
60V	55V	50V	45V
4 17 6 0	4 9 4 0	4 1 2 0	3 13 0 0
6 Ahau	11 Ahau	3 Ahau	8 Ahau
40V	35V	30V	25V
3 4 16 0	2 16 14 0	2 8 12 0	2 0 10 0
13 Ahau	5 Ahau	10 Ahau	2 Ahau
20V	15V	10V	5V
1 12 *8 0	1 4 6 0	16 4 0	8 2 0
7 Ahau	12 Ahau	4 Ahau	9 Ahau

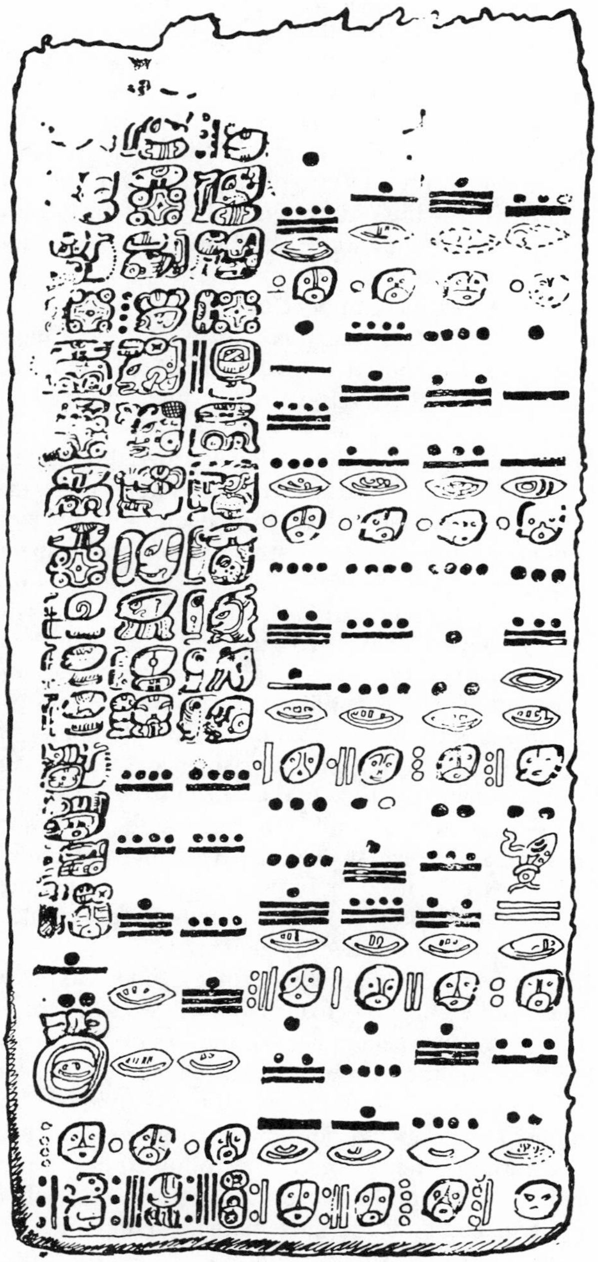

FIGURE 2. Page 24 of the Dresden Codex with transcription.

leads to a day *1 Ahau 13 Pax*; but, unlike those, it falls nowhere near a heliacal rising of Venus. This number is equal to 15 Venus years plus 340 days. It leads to a day that is 14 days after the *13 Cimi* of page 46, line 4, column *B*, and is thus 14 days after the expected or scheduled beginning of the Evening Star period of that Venus year. If the number that was intended here is that which is written, and if it is to be applied to the same base as are all of the others, then its purpose must be something different from the purpose of the others, which was to approximate as best as possible the *Ahau* heliacal risings of Venus and the optimal *1 Ahau* positions for new Venus epochs and new starts through the table. Nothing is recorded in the glyphic annotations on the left side of the page that offers any clue as to what phenomenon this date may have reference to. Neither is a year day *13 Pax* recorded in any of these pages. There is no generally accepted interpretation of this number or of the day to which it apparently leads. An accompanying oral tradition may once have supplied the missing information.

The last of the numbers of this tier, 1.5.14.4.0, if applied to *1 Ahau 18 Kayab*, leads again to a day *1 Ahau 18 Uo*; but it is one that is eight calendar rounds later (1.1.1.14.0) than that attained by 4.12.8.0. The accumulation of error in eight calendar rounds, or four unabridged runs through the entire table, is 20.8 days. It is not likely, then, that two such days, separated from each other by eight calendar rounds, were both intended to be designated as Venus epochs. (The interval 1.5.14.4.0 is 17.36 days longer than 317 mean Venus periods of 583.92 days.) These two numbers, the second and the fourth in this tier (counting from the right), were either to be applied to two different bases *1 Ahau 18 Kayab*, eight calendar rounds apart, not both reckoned as heliacal risings of Venus, and were to lead to the same *1 Ahau 18 Uo*; or else they were to be applied to the same *1 Ahau 18 Kayab* and were to lead to two different days *1 Ahau 18 Uo*, not both of which were to be counted as days of heliacal rising of Venus. The first alternative is the one generally accepted.

The left side of this page of the codex is given over to the chronological placement of an event, or pair of events, seemingly having to do with the institution of the corrective mechanism for the Venus calendar. The data are arrayed in three columns, to be labeled here as *A, B*, and *C*. In these the normal order of reading is followed in the hieroglyphic portion of the text, which is from left to right in pairs of columns, with any unpaired column being read singly. Columns *A* and *B* are thus paired, and column *C* follows. Considering the calendrical and chronological matters first, attention is directed to the numbers and days recorded in the lower left.

At the bottom of column *A* is *4 Ahau 8 Cumhu*, which is the calendrical position of day zero of the day count (something over three millennia B.C.). Over it stands the number 6.2.0 with a "ring" – a band of cloth, looped and knotted at the top – encircling its last digit. A number so marked is a negative base, at that distance prior to the beginning of the current era. Historical dates in the codex (presumably, but not assuredly, of some astronomical import) are most often reckoned from such negative bases rather than from chronological zero. The items determining such a date consist of the following: (1) a "ring number" standing over *4 Ahau 8 Cumhu*, giving the chronological position of the selected pre-zero base, these being at the bottom of the column if the whole is arrayed in one column, or at the bottom of the left-hand column if it is in two; (2) the calendrical position of the pre-zero base, this standing at the top of the column, or of the pair of columns; (3) a long reckoning or distance number, similar to a day number in the normal day count but applied to the special pre-zero base rather than to the normal base; (4) the calendrical position of the terminal date reached by applying the distance number to the pre-zero base, this following immediately the distance number and standing over the ring number if they are in one column, but at the bottom of the second column if they are in two; and (5), in some cases, glyphic annotation, inserted between (2) and (3) if in a single column, or between (2) and (1 and 3) if in two columns. In only one case, namely the present one, is there recorded also (6) the normal day number of the terminal date. This is near the bottom of column *C*, with the calendrical position of its base – the normal one, *4 Ahau 8 Cumhu* – standing at the top of that column, and with further glyphic annotations in between. The data recorded in the present example are tabulated below.

The special pre-zero base and the terminal date here have the same calendar day, which is frequently but not always the case with such reckonings. The standard place for the calendar day of the base is at the top of the column or pair of columns. These glyphs are obliterated on this page, but they can be restored with confidence on the basis of precedent and have been included in the transcription of the page. The calendar day of the terminal date also appears in its normal position, immediately below

(1*b*) Normal base, day zero of the day count:	. . . . *4 Ahau 8 Cumhu*
(1*a* and 2) Ring number and pre-zero base:	– 6. 2.0, *1 Ahau 18 Kayab*
(3) Distance number applied to pre-zero base:	9.9.16. 0.0 (add)
(6 and 4) Terminal date, normal day number:	9.9. 9.16.0, *1 Ahau 18 Kayab*.

the distance number at the bottom of column *B*. It is not repeated at the bottom of column *C*, possibly because of its redundancy, or more likely because some place had to be found for the *1 Ahau 18 Uo*, which is required by two of the intervals already reviewed from the fourth tier on the right side of the page. In any case, the calendar day at the bottom of column *C* belongs with those intervals, and not with the day number above it.

Something of the motive for choosing negative ring number bases from which to reckon dates, rather than employing the normal base, can be detected in the nature of the distance numbers that are applied to them. These are contrived numbers. They are multiples of the values of various periods or cycles such as are in some way relevant to the commemorated date and event. They are, moreover, the smallest such multiples that will just exceed the normal day number. The date of a ring number base, then, is a like-in-kind to the historical date that is counted from it, inasmuch as it occupies the same position in one or several relevant calendrical cycles as does the historical date. And it is the last possible date that bears such a likeness in kind before the beginning of the current chronological era, that is, before *4 Ahau 8 Cumhu*, day zero of the current era, or the close of the preceding era. In the present case the date of the ring-number base and the historical date occupy the same positions in the 260-day almanac, the 365-day year, the 584-day canonical Venus year, a 2,340-day cycle that unites that of the almanac with that of the nine lords of the night and with a 117-day cycle pertaining to the rain god (besides being also the corrective foreshortening of the Venus great cycle), and various higher-order cycles deriving from these, such as the calendar round (18,980 days), the unabridged great cycle of Venus (37,960 days), and so forth. Thus:

$$\begin{aligned} 9.9.16.0.0 = 1{,}366{,}560 &= 5{,}256\ (260) \\ &= 3{,}744\ (365) \\ &= 2{,}340\ (584) \\ &= 584\ (2{,}340) \\ &= 468\ (2{,}920) \\ &= 72\ (18{,}980) \\ &= 36\ (37{,}960). \end{aligned}$$

The motive for commemorating the date 9.9.9.16.0, *1 Ahau 18 Kayab*, is less clear. It is the day of this same name coming eight calendar rounds later that is the best candidate for being the *1 Ahau 18 Kayab* heliacal rising of Venus in column *D*, page 50 of the codex. On this assumption, the chronology of these pages is as follows:

[13. 0. 0. 0. 0],	4 Ahau 8 Cumhu [day no., old era]
−6. 2. 0	(ring number)
[12.19.13.16. 0],	1 Ahau 18 Kayab [day no., old era]
+9. 9.16. 0. 0	(72 CR) [add; subtract 13 baktuns]
9. 9. 9.16. 0,	1 Ahau 18 Kayab [day no., current era]
+1. 1. 1.14. 0	(8 CR)
10.10.11.12. 0,	1 Ahau 18 Kayab
+4.12. 8. 0	(2 CR − 13.0.0)
10.15. 4. 2, 0,	1 Ahau 18 Uo
+4.18.17. 0	(2 CR − 6.9.0)
11. 0. 3. 1. 0,	1 Ahau 13 Mac
+4.18.17. 0	(2 CR − 6.9.0)
11. 5. 2. 0. 0,	1 Ahau 3 Xul.

The pivotal date is the third *1 Ahau 18 Kayab*, at 10.10.11.12.0. This and the following dates can be assumed to be deliberate approximations to heliacal risings of Venus, chosen as epoch days for new starts through the Venus calendar. The two earlier *1 Ahau 18 Kayab* dates appear to be numerological projections backward in time, as if the canonical length of the Venus year, 584 days, were to be assumed as accurate and valid before the pivotal date. After the pivotal date, however, it is no longer quite valid, and the corrective mechanism is instituted. That mechanism cuts short the great cycles of Venus by either of two amounts, 13.0.0 (equal to 4,680 days) or 6.9.0 (equal to 2,340 days), in order to locate days *1 Ahau* at or suitably close to heliacal rising for beginning new cycles. Only these two amounts are deducted from a single complete great cycle (two calendar rounds). Larger amounts, such as 19.9.0 and 1.6.0.0, result as sums of successive applications of the lesser corrections, in succeeding great cycles, and are consequently deductible from some higher even-numbered multiple of calendar rounds.

As a result of this corrective mechanism, the *1 Ahau* starting point of any new great cycle will be found close to the end of either the 57th or the 61st Venus year of the preceding cycle, rather than at the end of the 65th which formally completes the table. For the four-day correction this is four days before the *5 Kan* of line 13, column *D*, of codex page 46; and for the eight-day correction it is eight days before the *9 Lamat* of line 12, column *D*, of page 47. The next 236 days then lead from the one or the other of these directly to the *3 Cib* at the top of column *A* of page 46, bypassing the remainder of the table in line 13, or in lines 12 and 13.

The hieroglyphic commentary in column *C* of page 24 reflects this. In glyph-block *C*3 (column *C*, line 3) is an event glyph of common occurrence whose meaning and reading are still uncertain. This is followed in *C*4 and *C*5 by the name glyph of the

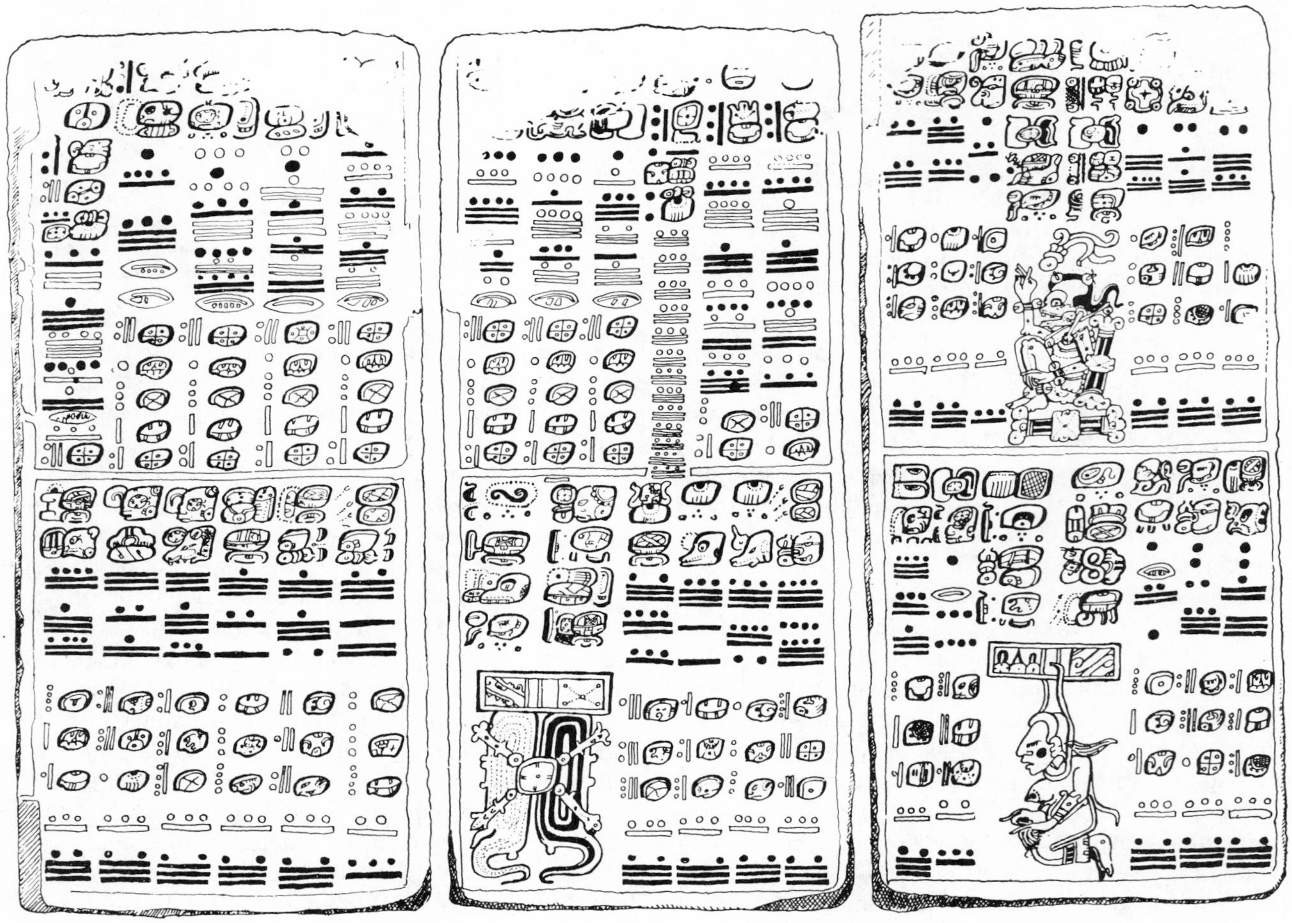

FIGURE 3. Pages 51–58 of the Dresden Codex.

Venus deity—"The Spearer"—of the heliacal risings of page 46, qualified, as also on that page, by the Venus glyph that identifies his role. These are followed in turn, in *C*6 and *C*7, by the name glyph of the spearer of page 47, again with the Venus glyph. These two are the only spearing deities named here, but their appellatives are followed, in *C*8 to *C*12, by the names of all five of the victims of pages 46–50. This sequence, by both its context and its content, appears intended to apply to the 4.12.8.0 abridgment of the great cycle. This is the one that began on *1 Ahau 18 Kayab* and was terminated on *1 Ahau 18 Uo*, eight days prior to the *9 Lamat 6 Zip* of line 12, column *D*, of page 47 (the year day, *6 Zip*, is in the first line of the lower section, same column). The implication is that in this terminal short run only two of the spearing deities were to play their roles, but that none of the victims would escape. In columns *A* and *B* of page 24 there is also an oblique reflection of the termination of the cycle with page 47. In glyph-blocks *AB*5–9 the five presiding deities or "Venus regents" are named, but those of pages 46 and 47 are named last.

The *18 Kayab* base was probably obsolete and *18 Uo* already installed at the time of this recension of the table. The *13 Mac* and *3 Xul* bases would have been prescriptions for the future. There is no record of application or prescription of corrections beyond *3 Xul*. The device could of course have been reapplied indefinitely. The year day of each successive revised base for a single correction (four days) is 215 days later in the year calendar, or 150 days earlier, than the one before. For a double correction (eight days) one of these is skipped. The optimum mix of double and single corrections is

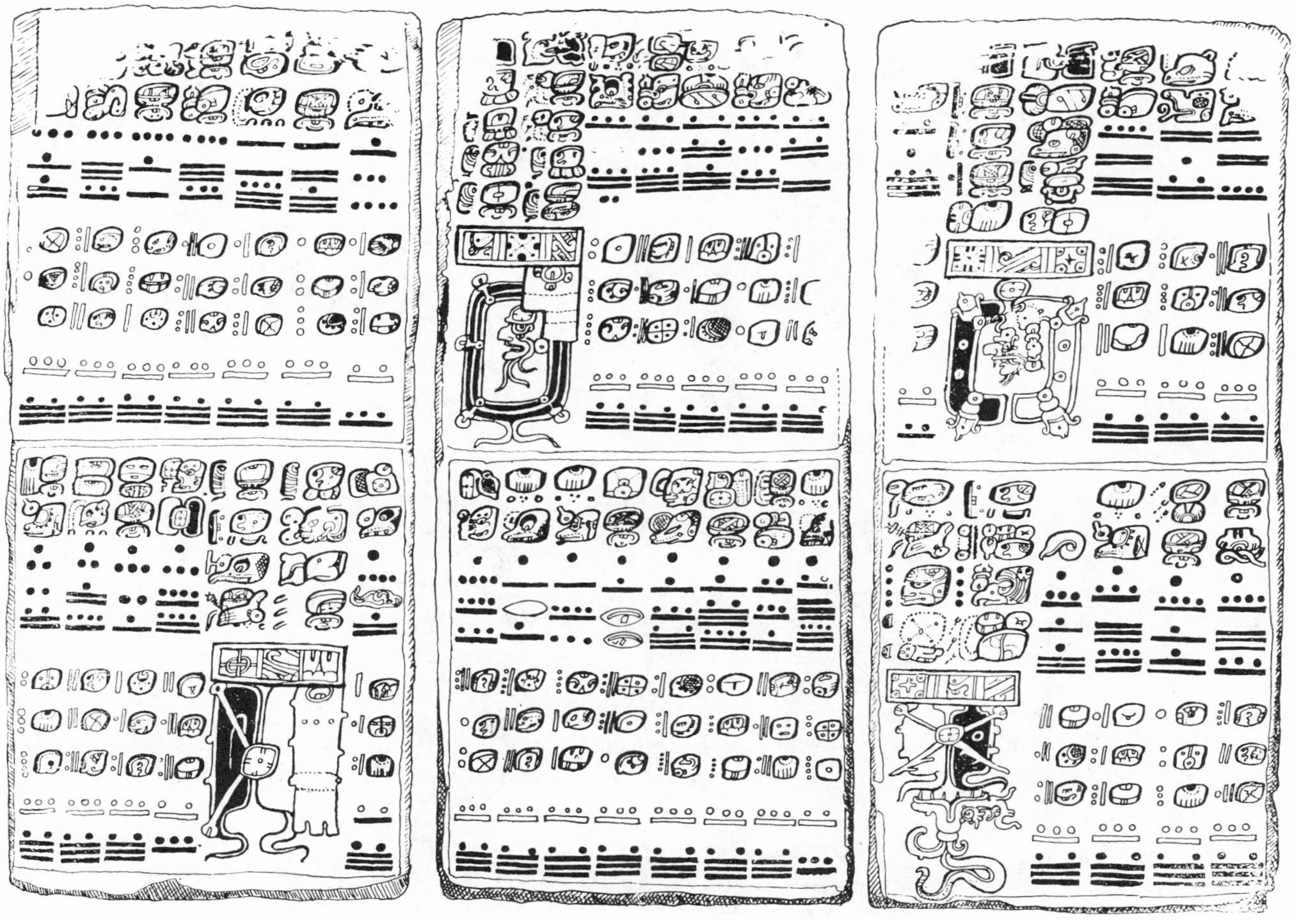

54 55 56

(continued on next page)

one to four. This ratio results in the assignment of 301 Venus years to (301 × 584) −24 days. The result is accurate to within 0.08 part of a day for this span of time, which is over 481 years. It yields a mean Venus year of 583.92026 days. The Maya calendar priests and star watchers were doubtless aware that they had a good scheme worked out for this planet, but one wonders whether they could have known how good it really is. (It should not be forgotten that they lacked fractional arithmetic.) Its long-term accuracy is of course greater than its short-term accuracy. But even for short terms the accumulated error never exceeds the magnitude of the deviation of the planet's actual periods from their mean value. The table was probably used as a warning table, geared to anticipate the phenomena, thus favoring negative errors over positive ones, or early predictions over late ones.

V. ECLIPSE RECKONING

Pages 51–58 of the Dresden Codex contain the table pictured here in Figure 3. Its division into intervals of 177 days (occasionally 178), and of 148 days, identifies it as a lunar table treating of groupings of six and of five lunar months. The number and distribution of the five-month periods among those of six months mark it moreover as having to do with the prediction of solar eclipse possibilities. There are nine of the five-month periods distributed among sixty of the six-month periods, these being in three major divisions of 23 groups each (20 of six months and 3 of five) for a total of 135 lunar months per division, or 405 months in all.[14] Data extracted from the Maya table are aligned in Table 2.

The structure of the eclipse table of the Dresden

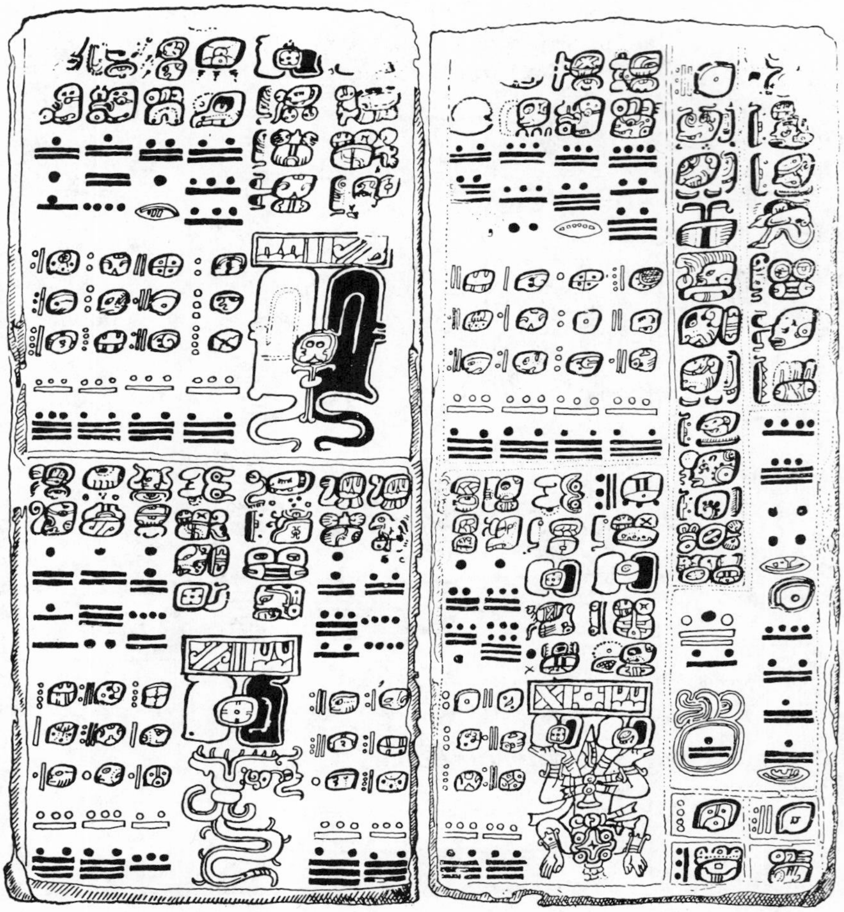

57 58

Codex, and its implications, can be appreciated most readily if we set ourselves the task of constructing our own "primitive" solar-eclipse-prediction table. For this we require three items of information that can be derived from naked-eye observations and record keeping with a calendar, or inferred from an accumulation of such records, namely, the average length of the lunar month, the average length of the interval between lunar nodes, and the ecliptic limit for solar eclipses. Today we would use 29.530588 days for the first of these, 173.30906 for the second, and from 15 to 18 days for the last. The Maya, or at least some of them, used the ratio of 81 moons to 2,392 days for the first of these, three "nodes" to 520 days for the second (apparently with the understanding that these ratios, equivalent to 29.530864 and 173.33333 respectively, entailed small errors requiring correction as they accumulated), and it may be inferred that they used ±18 days for the last—although other and perhaps more plausible inferences may also be drawn from the numbers and their arrangement in the table of the Dresden Codex. While there is no need to quibble about differences between the Maya and our notions of "lunar month," it should perhaps be noted that the Maya concepts corresponding to "node" and "ecliptic limit" can hardly have entailed even the minimum understanding of celestial mechanics that is implied in our use of these terms; nor would they have needed to have such understanding. It was sufficient for them merely to have observed over time that eclipses, whether solar or lunar, never occurred except within three circumscribed sectors of their sacred 260-day almanac, and that these "eclipse-possible" periods came three times within two rounds of that almanac, that is, three times in 520 days, their midpoints being 173 or 174 days apart.

For our experiment, however, let us use the modern figures and avail ourselves of the convenience of decimal fractions. But let us choose a slightly narrower hypothetic ecliptic limit, say 14.765 days, or one-half of the mean duration of a lunar month. This will be at little if any cost to the predictive capacity of the table in relation to its intended use, namely, as a predictor of "eclipse-possible" times (a warning table) for use at a tropical latitude. What is lost through this restriction of the ecliptic limit is the prediction of those eclipse possibilities that come at maximum distance from the nodes. In instances where it is possible for two solar eclipses to occur within the same eclipse season, one month apart, the one at the greater distance from the node will thus be eliminated from the table. The loss is of minimal consequence, since the probability of the excluded eclipse being visible in the tropics is virtually nil. (Such an eclipse is a partial eclipse of minimal magnitude, observable in either the far northern or the far southern latitudes, but rarely if ever within the tropics.)

The results of the experiment are tabulated in the first three columns of Table 3. Starting from a hypothetical lunar node passage, column 1 gives the times, in days and hundredths, of the next 69 nodes. The ordinal numbers of the nodes are adjacent, in square brackets. In column 3, starting from a lunar-solar conjunction hypothetically coinciding with the node passage, are given the times of those successive conjunctions that fall within ±14.765 days of the enumerated node times of column 1. These are the "eclipse-possible" times, of which it is the function of the table to forewarn. The ordinal numbers of the lunations are adjacent, in square brackets. Column 2 gives the differences $(l - n)$ between the lunation times and the corresponding node times. Those eclipse times for which $(l - n)$ is less than about eleven days may be expected to correspond to central eclipses (total or annular) visible as such somewhere on the face of the earth. There is of course nothing in such a simple table to give an indication of the ranges of longitude or of latitude through which any given eclipse may be visible, except to the extent that the fractional parts of the numbers in column 3 may be employed as a rough guide to relative longitude for short-range predictions, and to the extent that low numbers in column 2 may offer a somewhat better-than-chance guide to probable visibility in the tropics.

The experimental table has been constructed for the purpose of offering a reference scheme against which the Maya scheme may be compared, in order better to understand the latter. Data from the Maya eclipse table of the Dresden Codex are translated and presented in Table 2. Some items of the data have also been transferred to Table 3, columns 6 and 7, with elaboration in columns 4 and 5, in order to facilitate comparison with the reference scheme that was derived from our experiment (columns 1–3). Before proceeding to the comparison, however, a word must be said about some problems encountered in the translation of the numerals from the Maya table.

The table of the Dresden Codex is not without imperfections. There are indications that it is not an original version, but a copy with revisions of an earlier version. There are a few rather gross errors that surely must be laid to faulty copywork. For instance, the very first number of the series of cumulative totals (the first Maya number in the upper series beginning on page 53*a* of the codex: see Figure 3) is written as 7.17. But obviously it should be 8.17, since the first cumulative total of the series must necessarily be the same as the first contribution to that total, which is written below as it should be, as 8.17, in the series of eclipse half-year intervals (same column, lower series of numbers, below the three lines of day signs). It is clear that one dot was omitted in the *uinal* or "scores" position by the copyist, which makes for an error of 20 days.

Another copying error on the same page of the codex is in the fourth cumulative total, written as 1.15.14; but this should be 1.15.19. It is obvious that one bar was omitted in the *kin* or "units" position by the copyist, resulting in an error of 5 days. And there are others, all of which are easily corrected.

There are two internal sources of evidence for every such correction, since the table is two ways redundant. The table could be reconstructed in its entirety from any one of three separate kinds of data contained in it, if the data of that kind were without error. As it is, the three kinds of data serve as checks on one another. They are (*a*) the series of intervals, mostly 8.17 or 7.8, equal to 177 and 148 days respectively, and corresponding to six-month and five-month eclipse half-years; (*b*) the series of cumulative totals, giving the day numbers, within the cycle, of the designated eclipse possibilities; and (*c*) the three series of day signs with numerical coefficients, naming the days in the 260-day almanac on which the respective eclipse possibilities

TABLE 2. Calendrical and Numerical Data from the Eclipse Table of the Dresden Codex.

1. Dresden Codex page number	2. Column number (=node number)	3. Cumulative totals as written in the Codex	4. Corrections required by indicated almanac days	5. Cumulative totals required by indicated almanac days, in Maya and in Arabic numerals [with cumulative month totals]	6. Day, from lower line of almanac days (with its number in the double almanac)	7. Increment
-----	(0)	---------------		0 0 [0]	*13 Muluc (169)	-----
53a	1	7.17	1.0	8.17 177 [6]	8 Cimi (346)	177
	2	17.13	1	17.14 354 [12]	3 Akbal (3)	177
	3	1. 7. 2		1. 7. 2 502 [17]	8 Chuen (151)	148
	PICTURE					
	4	1.15.14	5	1.15.19 679 [23]	3 Lamat (328)	177
	5	2. 6.16		2. 6.16 856 [29]	*11 Chicchan (505)	177
	6	2.15.13		2.15.13 1,033 [35]	6 Ik (162)	177
54a	7	3. 6.11		3. 6.11 1,211 [41]	2 Ahau (340)	178†
	8	3.15. 8		3.15. 8 1,388 [47]	10 Caban (517)	177
	9	4. 6. 5		4. 6. 5 1,565 [53]	5 Ix (174)	177
	10	4.15. 8	−6	4.15. 2 1,742 [59]	13 Chuen (351)	177
	11	5. 5.19		5. 5.19 1,919 [65]	* 8 Lamat (8)	177
	12	5.10.16	4.0	5.14.16 2,096 [71]	3 Chicchan (185)	177
	13	6. 3. 4	1.0	6. 4. 4 2,244 [76]	8 Ben (333)	148
55a	PICTURE					
	14	8.13. 2	−2.0.0	6.13. 2 2,422 [82]	4 Chuen (511)	178†
	15	7. 3.18	1	7. 3.19 2,599 [88]	12 Lamat (168)	177
	16	7.12.16		7.12.16 2,776 [94]	7 Chicchan (345)	177
	17	8. 3.13		8. 3.13 2,953 [100]	* 2 Ik (2)	177
	18	*8.12.10		8.12.10 3,130 [106]	10 Cauac (179)	177
56a	19	*9. 1.18		9. 1.18 3,278 [111]	2 Manik (327)	148
	PICTURE					
	20	9.10.15		9.10.15 3,455 [117]	10 Kan (504)	177
	21	10. 1.12		10. 1.12 3,632 [123]	5 Imix (161)	177
	22	10.10. 9		10.10. 9 3,809 [129]	13 Extnab (338)	177
57a	23	11. 1. 6	1	11. 1. 7 3,987 [135]	9 Cib (516)	178†
	24	11.10. 4		11.10. 4 4,164 [141]	4 Ben (173)	177
	25	12. 1. 0	1	12. 1. 1 4,341 [147]	12 Oc (350)	177
	26	12. 8. 8	1	12. 8. 9 4,489 [152]	4 Etznab (498)	148
	PICTURE					
58a	27	12.17.*5	1	12.17. 6 4,666 [158]	12 Men (155)	177
	28	13. 8. 2	1	13. 8. 3 4,843 [164]	7 Eb (332)	177
	29	13.17. 0	1	13.17. 1 5,021 [170]	3 Oc (510)	178†
	30	14. 7.17	1	14. 7.18 5,198 [176]	11 Manik (167)	177
51b	31	14.16.14	1	14.16.15 5,375 [182]	6 Kan (344)	177
	32	15. 7.11	1	15. 7.12 5,552 [188]	1 Imix (1)	177
	33	15.16. 8	1	15.16. 9 5,729 [194]	9 Etznab (178)	177
	34	16. 7. 5	1	16. 7. 6 5,906 [200]	4 Men (355)	177
	35	16.16. 2	1	16.16. 3 6,083 [206]	12 Eb (12)	177
	36	17. 5.10	1	17. 5.11 6,231 [211]	4 *Ahau (160)	148
52b	PICTURE					
	37	17.14. 8	1	17.14. 9 6,409 [217]	13 Etznab (338)	178†
	38	18. 5. 5	1	18. 5. 6 6,586 [223]	8 Men (515)	177
	39	18.14. 2	1	18.14. 3 6,763 [229]	3 Eb (172)	177
	40	19. 4.19	1	19. 5. 0 6,940 [235]	11 Muluc (349)	177
53b	41	19.13.16	1	19.13.17 7,117 [241]	6 Cimi (6)	177
	42	1. 0. 3. 4	1	1. 0. 3. 5 7,265 [246]	11 Ix (154)	148
	PICTURE					
	43	1. 0.12. 1	1	1. 0.12. 1 7,442 [252]	6 Chuen (331)	177
	44	1. 1. 2.18	1	1. 1. 2.19 7,619 [258]	1 Lamat (508)	177
	45	1. 1.11.15	1	1. 1.11.16 7,796 [264]	* 9 Chicchan (165)	177
54b	46	1. 2. 2.12	1	1. 2. 2.13 7,973 [270]	4 Ik (342)	177
	47	1. 2.11. 9	1	1. 2.11.10 8,150 [276]	12 Cauac (519)	177
	48	1. 3. 2. 6	1	1. 3. 2. 7 8,327 [282]	7 Cib (176)	177
	49	1. 3. 9.14	1	1. 3. 9.15 8,475 [287]	12 Kan (324)	148

(continued on next page)

TABLE 2. Calendrical and Numerical Data from the Eclipse Table of the Dresden Codex *(continued)*.

1. Dresden Codex page number	2. Column number (=node number)	3. Cumulative totals as written in the Codex	4. Corrections required by indicated almanac days	5. Cumulative totals required by indicated almanac days, in Maya and in Arabic numerals [with cumulative month totals]		6. Day, from lower line of almanac days (with its number in the double almanac)		7. In-cre-ment
	PICTURE							
	50	1. 4. 0.11	1	1. 4. 0.12	8,652 [293]	7 Imix	(501)	177
55b	51	1. 4. 9. 8	1	1. 4. 9. 9	8,829 [299]	2 Etznab	(158)	177
	52	1. 5. 0. 6	1	1. 5. 0. 7	9,007 [305]	11 Cib	(336)	178†
	53	1. 5. 9. 3	1	1. 5. 9. 4	9,184 [311]	6 Ben	(513)	177
	54	1. 6. 0. 0	1	1. 6. 0. 1	9,361 [317]	1 Oc	(170)	177
	55	1. 6. 8.17	1	1. 6. 8.18	9,538 [323]	9 Manik	(347)	177
	56	1. 6.17.14	1	1. 6.17.15	9,715 [329]	4 Kan	(4)	177
	57	1. 7. 8.11	1	1. 7. 8.12	9,892 [335]	12 Imix	(181)	177
	58	1. 7.15.19	1	1. 7.16. 0	10,040 [340]	4 Muluc	(329)	148
56b	PICTURE							
	59	1. 8. 6.16	1	1. 8. 6.17	10,217 [346]	12 Cimi	(506)	177
	60	1. 8.15.14	1	1. 8.15.15	10,395 [352]	8 Kan	(164)	178†
	61	1. 9. 6.11	1	1. 9. 6.12	10,572 [358]	3 Imix	(341)	177
	62	1. 9.15. 8	1	1. 9.15. 9	10,749 [364]	11 Etznab	(518)	177
57b	63	1.10. 6. 5	1	1.10. 6. 6	10,926 [370]	6 Men	(175)	177
	64	1.10.15. 2	1	1.10.15. 3	11,103 [376]	1 Eb	(352)	177
	65	1.11. 4.10	1	1.11. 4.11	11,251 [381]	6 Ahau	(500)	148
	PICTURE							
	66	1.11.13. 7	1	1.11.13. 8	11,428 [387]	1 Caban	(157)	177
	67	1.12. 4. 4	1	1.12. 4. 5	11,605 [393]	9 Ix	(334)	177
58b	68	1.12.13. 1	1	1.12.13. 2	11,782 [309]	4 Chuen	(511)	177
	69	1.13. 3.18	1	1.13. 3.19	11,959 [405]	12 Lamat	(168)	177

NOTE: Prefixed asterisks in column 6 mark reconstructions of missing, illegible, or obviously erroneous data where adjacent items supply the necessary information.

may fall. No one of these is wholly free from errors; but the triple series of almanac days has the fewest and, but for a single case, offers the surest guide to the intended intervals and the proper cumulative totals.

Column 3 of Table 2 lists the cumulative totals in Maya numbers as written in the codex. Column 4 gives the corrections that are required to bring these into accord with the indicated almanac days. Column 5 gives the corrected totals, in both Maya and Arabic numerals, together with the count of lunations. Column 6 lists the almanac days (from the third of the three lines of these in the codex) and, in parentheses, their numerical equivalents in the 520-day double almanac. Column 7 gives the proper increments, as determined by the almanac days of column 6.

Points of interest in a comparison of the Maya table with the experimental one will be the following: (1) the structures of the two tables, as seen in the groupings of moons into eclipse half-years and of half-years into divisions of the cycle; (2) the magnitude of the ecliptic limits and the location of the nodes in the Maya table, as compared with those of the experiment; (3) the relevance of the pictures to their places of insertion in the Maya table; and (4) the degree of correspondence between the Maya computations and our own. Other matters to be considered are (5) the distribution of 30-day and 29-day months in the lunar calendar; (6) the preface to the Maya table; and (7) correction devices to compensate for small accumulations of error.

1. The Structures. In the experimental table (columns 2–3 of Table 3) a pattern can be observed in the groupings of moons into eclipse half-years, and of half-years into divisions of the cycle. Beginning after the first five-month half-year (after 23 lunations, column 3), which is after the first drop in the abnodal-distance function (that is, *after* the first negative number in column 2), the grouping of moons is as follows:

seven groups of six, and one of five;
seven groups of six, and one of five;
six groups of six, and one of five;

for a total of 135 moons, in 23 groups, corresponding to that many nodes, and spanning a period of 3,986 and approximately 2/3 days. On repetition of the 135-month scheme, the arrangement found in the third line of the above ("six groups of six, and

TABLE 3. Tabulations of eclipse syzygies, restricted to one per node, (*a*) as determined by hypothetic ±14.765-day ecliptic limit, and (*b*) as selected in Dresden Codex; with dividing lines after each drop in abnodal-distance function, corresponding to picture dividers in Codex.

(*a*) Selection for ±14.765-day Ecliptic Limit					(*b*) Selection Corresponding to Dresden Codex				
1. Nodes: Days	[No.]	2. Abnodal Distances	3. Lunations: Days	[No.]	4. Abnodal Distances	5. Lunations: Days	[No.]	6. $a_n - a_0$	7. Δa
0.00	[0]	0.00	0.00	[0]	0.00	0.00	[0]	-----	-----
173.31	[1]	+ 3.87	177.18	[6]	+ 3.87	177.18	[6]	177	177
346.62	[2]	+ 7.75	354.37	[12]	+ 7.75	354.37	[12]	354	177
519.93	[3]	+11.62	531.55	[18]	−17.91	502.02	[17]	502	148
693.24	[4]	−14.04	679.20	[23]	−14.04	679.20	[23]	679	177
866.55	[5]	−10.16	856.39	[29]	−10.16	856.39	[29]	856	177
1,039.86	[6]	− 6.29	1,033.57	[35]	− 6.29	1,033.57	[35]	1,033	177
1,213.17	[7]	− 2.42	1,210.75	[41]	− 2.42	1,210.75	[41]	1,211	178†
1,386.48	[8]	+ 1.46	1,387.94	[47]	+ 1.46	1,387.94	[47]	1,388	177
1,559.79	[9]	+ 5.33	1,565.12	[53]	+ 5.33	1,565.12	[53]	1,565	177
1,733.10	[10]	+ 9.20	1,742.30	[59]	+ 9.20	1,742.30	[59]	1,742	177
1,906.41	[11]	+13.08	1,919.49	[65]	+13.08	1,919.49	[65]	1,919	177
2,079.72	[12]	−12.58	2,067.14	[70]	+16.95	2,096.67	[71]	2,096	177
2,253.03	[13]	− 8.71	2,244.32	[76]	− 8.71	2,244.32	[76]	2,244	148
2,426.34	[14]	− 4.83	2,421.51	[82]	− 4.83	2,421.51	[82]	2,422	178†
2,599.65	[15]	− 0.96	2,598.69	[88]	− 0.96	2,598.69	[88]	2,599	177
2,772.96	[16]	+ 2.92	2,775.88	[94]	+ 2.92	2,775.88	[94]	2,776	177
2,946.27	[17]	+ 6.79	2,953.06	[100]	+ 6.79	2,953.06	[100]	2,953	177
3,119.58	[18]	+10.66	3,130.24	[106]	+10.66	3,130.24	[106]	3,130	177
3,292.89	[19]	+14.54	3,307.43	[112]	−14.99	3,277.90	[111]	3,278	148
3,466.20	[20]	−11.12	3,455.08	[117]	−11.12	3,455.08	[117]	3,455	177
3,639.51	[21]	− 7.25	3,632.26	[123]	− 7.25	3,632.26	[123]	3,632	177
3,812.82	[22]	− 3.37	3,809.45	[129]	− 3.37	3,809.45	[129]	3,809	177
3,986.13	[23]	+ 0.50	3,986.63	[135]	+ 0.50	3,986.63	[135]	3,987	178†
4,159.44	[24]	+ 4.37	4,163.81	[141]	+ 4.37	4,163.81	[141]	4,164	177
4,332.75	[25]	+ 8.25	4,341.00	[147]	+ 8.25	4,341.00	[147]	4,341	177
4,506.06	[26]	+12.12	4,518.18	[153]	−17.41	4,488.65	[152]	4.489	148
4,679.37	[27]	−13.54	4,665.83	[158]	−13.54	4,665.83	[158]	4,666	177
4,852.68	[28]	− 9.66	4,843.02	[164]	− 9.66	4,843.02	[164]	4,843	177
5,025.99	[29]	− 5.79	5,020.20	[170]	− 5.79	5,020.20	[170]	5,021	178†
5,199.30	[30]	− 1.92	5,197.38	[176]	− 1.92	5,197.38	[176]	5,198	177
5,372.61	[31]	+ 1.96	5,374.57	[182]	+ 1.96	5,374.57	[182]	5,375	177
5,545.92	[32]	+ 5.83	5,551.75	[188]	+ 5.83	5,551.75	[188]	5,552	177
5,719.23	[33]	+ 9.70	5,728.93	[194]	+ 9.70	5,728.93	[194]	5,729	177
5.892.54	[34]	+13.58	5,906.12	[200]	+13.58	5,906.12	[200]	5,906	177
6,065.85	[35]	−12.08	6,053.77	[205]	+17.45	6,083.30	[206]	6,083	177
6,239.16	[36]	− 8.21	6,230.95	[211]	− 8.21	6,230.95	[211]	6,231	148
6,412.47	[37]	− 4.33	6,408.14	[217]	− 4.33	6,408.14	[217]	6,409	178†
6,585.78	[38]	− 0.46	6,585.32	[223]	− 0.46	6,585.32	[223]	6,586	177
6,759.09	[39]	+ 3.41	6,762.50	[229]	+ 3.41	6,762.50	[229]	6,763	177
6,932.40	[40]	+ 7.29	6,939.69	[235]	+ 7.29	6,939.69	[235]	6,940	177
7,105.71	[41]	+11.16	7,116.87	[241]	+11.16	7,116.87	[241]	7,117	177
7,279.02	[42]	−14.50	7,264.52	[246]	−14.50	7,264.52	[246]	7,265	148
7,452.33	[43]	−10.62	7,441.71	[252]	−10.62	7,441.71	[252]	7,442	177
7,625.64	[44]	− 6.75	7,618.89	[258]	− 6.75	7,618.89	[258]	7,619	177
7,798.95	[45]	− 2.87	7,796.08	[264]	− 2.87	7,796.08	[264]	7,796	177
7,972.26	[46]	+ 1.00	7,973.26	[270]	+ 1.00	7,973.26	[270]	7,973	177
8,145.57	[47]	+ 4.87	8,150.44	[276]	+ 4.87	8,150.44	[276]	8,150	177
8,318.88	[48]	+ 8.75	8,327.63	[282]	+ 8.75	8,327.63	[282]	8,327	177
8,492.19	[49]	+12.62	8,504.81	[288]	−16.91	8,475.28	[287]	8,475	148
8,665.50	[50]	−13.04	8,652.46	[293]	−13.04	8,652.46	[293]	8,652	177
8,838.81	[51]	− 9.16	8,829.65	[299]	− 9.16	8,829.65	[299]	8,829	177
9,012.12	[52]	− 5.29	9,006.83	[305]	− 5.29	9,006.83	[305]	9,007	178†
9,185.43	[53]	− 1.42	9,184.01	[311]	− 1.42	9,184.01	[311]	9,184	177
9,358.74	[54]	+ 2.46	9,361.20	[317]	+ 2.46	9,361.20	[317]	9,361	177

(continued on next page)

TABLE 3. Tabulations of eclipse syzygies, restricted to one per node, (*a*) as determined by hypothetic ±14.765-day ecliptic limit, and (*b*) as selected in Dresden Codex; with dividing lines after each drop in abnodal-distance function, corresponding to picture dividers in Codex *(continued)*.

(*a*) Selection for ±14.765-day Ecliptic Limit					(*b*) Selection Corresponding to Dresden Codex				
1. Nodes: Days	[No.]	2. Abnodal Distances	3. Lunations: Days	[No.]	4. Abnodal Distances	5. Lunations: Days	[No.]	6. $a_n - a_0$	7. Δa
9,532.05	[55]	+ 6.33	9,538.38	[323]	+ 6.33	9,538.38	[323]	9,538	177
9,705.36	[56]	+10.20	9,715.56	[329]	+10.20	9,715.56	[329]	9,715	177
9,878.67	[57]	+14.08	9,892.75	[335]	+14.08	9,892.75	[335]	9,892	177
10,051.98	[58]	−11.58	10,040.40	[340]	−11.58	10,040.40	[340]	10,040	148
10,225.29	[59]	− 7.71	10,217.58	[346]	− 7.71	10,217.58	[346]	10,217	177
10,398.60	[60]	− 3.83	10,394.77	[352]	− 3.83	10,394.77	[352]	10,395	178†
10,571.91	[61]	+ 0.04	10,571.95	[358]	+ 0.04	10,571.95	[358]	10,572	177
10,745.22	[62]	+ 3.91	10,749.13	[364]	+ 3.91	10,749.13	[364]	10,749	177
10,918.53	[63]	+ 7.79	10,926.32	[370]	+ 7.79	10,926.32	[370]	10,926	177
11,091.84	[64]	+11.66	11,103.50	[376]	+11.66	11,103.50	[376]	11,103	177
11,265.15	[65]	−14.00	11,251.15	[381]	−14.00	11,251.15	[381]	11,251	148
11,438.46	[66]	−10.12	11,428.34	[387]	−10.12	11,428.34	[387]	11,428	177
11,611.77	[67]	− 6.25	11,605.52	[393]	− 6.25	11,605.52	[393]	11,605	177
11,785.08	[68]	− 2.36	11,782.70	[399]	− 2.38	11,782.70	[399]	11,782	177
11,958.39	[69]	+ 1.50	11,959.89	[405]	+ 1.50	11,959.89	[405]	11,959	177

one of five") works its way forward. For three repetitions the pattern is:

7 (6) + 1 (5); 7 (6) + 1 (5); 6 (6) + 1 (5);
7 (6) + 1 (5); 6 (6) + 1 (5); 7 (6) + 1 (5);
7 (6) + 1 (5); 6 (6) + 1 (5); 7 (6) + 1 (5);

for a total of 405 moons, in 69 groups, corresponding to that many nodes, and spanning a period of approximately 11,960 days. The internal arrangement of this scheme is a familiar one, known from Babylonian astronomy of the Seleucid period and earlier. The first five subdivisions of the above nine constitute the saros.[15]

Attention should be given to the positions of the five-month periods in Table 3*a* (immediately above the horizontal dividing lines of columns 2 and 3). Note that they terminate only at those conjunction times that are closest to the *prenodal* ecliptic limit. This is important for our understanding of the significance of the location of the five-month periods in the Maya table, and of the pictures that follow them.

The Maya table deviates somewhat from the above pattern of month groupings. It is of the same length, and with the same major divisions (3 × 135 moons = 3 × 23 groups), but the internal arrangement of the divisions is a bit different. These may be seen either in Figure 3 or in Table 2, in the placement of the "picture" breaks, or in columns 4–7 of Table 3, in the placement of the horizontal dividing lines across those columns. Beginning after the first break (after the picture of the "death god" on page 53*a* of the codex) the scheme is as follows:

9 (6) + 1 (5); 5 (6) + 1 (5); 6 (6) + 1 (5);
9 (6) + 1 (5); 5 (6) + 1 (5); 6 (6) + 1 (5);
8 (6) + 1 (5); 6 (6) + 1 (5); 6 (6) + 1 (5).

This also totals 405 moons, in 69 groups, for approximately 11,960 days. (For explanation of the apparent cumulative total 1.13.3.18, or 11,958, see below.) Although the pattern deviates from what might have been anticipated on the basis of our experiment and on the basis of Babylonian precedents, it shares with these the features (*a*) that each of the three major divisions consists of 20 six-month and 3 five-month groups, and (*b*) that the breaks *following* the five-month periods, which are marked by insertion of the pictures into the table, come (mostly) at times that are at or fairly near a *prenodal* ecliptic limit. The deviations from the experimental pattern must necessarily have correlates in differing durations, and in variability of duration, of the eclipse seasons implicit in the Maya table.

2. Ecliptic Limits and Nodes. In our experimental table (Table 3, columns 1–3) the hypothetical ecliptic limit employed was one-half of a lunar month, or ±14.765 days, and the zero line for the table assumed a precise coincidence of a conjunction and a node passage for the starting point of the table. This, of course, is a somewhat rare happening, and we can have no *a priori* grounds for assuming that the Maya astronomers made a similar assumption for the construction of their table. Inspection of the table, however, may reveal whatever assumption underlies it in this regard.

Column 6 of Table 2 gives the names of the almanac days from the lowest of the three lines of such in the Dresden Codex table, and in parentheses in the same column are their translations into ordinal positions within the double almanac of 520 days (= about 3 nodes). If these are retabulated into three columns, corresponding to the three eclipse seasons that are found within a 520-day period, it will then be seen that the outer limits of the first eclipse season are days 151 (*8 Chuen*) and 185 (*3 Chicchan*), those for the second are 327 (*2 Manik*) and 355 (*4 Men*), and those for the third are 498 (*4 Etznab*) and 12 [= 532 mod 520] (*12 Eb*). From these the durations of the apparent eclipse seasons may be determined (35 days, 29 days, and 35 days respectively) as well as their midpoints and the values of the ecliptic limits that they imply. These are: day 168 (*12 Lamat*) ± 17 days for the first season; day 341 (*3 Imix*) ± 14 days for the second; and day 515 (*8 Men*) ± 17 days for the third. The midpoint days, then, are the implied node days; and they are indeed properly spaced to be such: from *12 Lamat* to *3 Imix* is 173 days; from *3 Imix* to *8 Men* is 174 days; and from *8 Men* to *12Lamat* is again 173 days. (See "Calendar, Chronology, and Computation" for a method of computing the interval between any two almanac days.)

The end-day of the table is *12 Lamat* (day 168), one of the implied node days just mentioned. The table thus appears to end, as our experimental one began, with a hypothetic coincidence of a node day with a lunar-solar conjunction—in other words, with the optimum condition for a central solar eclipse. This is marked in the codex by the tenth or "extra" picture that is inserted into the table (the only one that is not immediately preceded by a five-month half-year). The table, begins, however, not with *12 Lamat* (day 168), but with *13 Muluc* (day 169). Although the starting day is not written at the beginning of the table (there being no "zero" column), it can be derived from the day names and numbers in the columns that follow. For example, *8 Cimi* (in column 1), less Maya 8.17, is *13 Muluc*; similarly *8 Chuen* (from column 3), less 1.7.2, is also *13 Muluc*; again, *10 Kan* (column 20), less 9.10.15, is *13 Muluc*; and so forth. Thus the table, as represented by any one of the three lines of day names, spans 11,959 days rather than the expected 11,960—that is, 1.13.3.19 rather than 1.13.4.0, the latter of which is equal to 23 double almanacs, or 46 × 260. That the recognized length of the Maya cycle was 11,960 days, however, is apparent from the table of multiples of 1.13.4.0 that precedes the eclipse table proper on pages 51–52*a* of the codex. The one-day foreshortening of the cycle in the table suggests that it may have represented the occasion of a shift from a previously effective *13 Muluc* base to a new one on *12 Lamat*, one day earlier in the almanac. Since 405 mean lunar months are about eleven one-hundredths of a day short of 11,960 days (see last line of Table 3), such a foreshortening of the cycle would be a necessity approximately every ninth time that the cycle is employed. Such a shift is also suggested by the relationship between two of the dates that figure importantly in these pages. The earlier of the two is 9.12.10.16.9, *13 Muluc 2 Zip* (defined, on page 58 of the codex, by the interval 9.12.11.11.0 reckoned from a *13 Muluc* "ring" base of minus 12.11); the other is 9.16.4.10.8, *12 Lamat 1 Muan*, recorded as such on page 52*a*, as well as by means of a 9.16.4.10.0 increment to a *12 Lamat* reckoning base of plus 8, on page 51a). The interval between these two dates is 3.13.11.19, or 26,519 days. It is an eclipse interval, equal to two full eclipse cycles plus the lesser eclipse interval of 2,599 days (Maya 7.3.19, see column 5 of Table 2); it is about half a day in excess of 898 mean lunations, and about two and a half days in excess of 153 mean internodal intervals. Whether either of these two dates was the occasion of an eclipse observed by the Maya is not known. They could have been; one or both may have been; but the possibility that one or both of them may have been the products of computation—perhaps for use as proximate reckoning bases—cannot be excluded.

The three rows of almanac days, with one-day intervals between successive rows, are also compatible with the hypothesis of a periodic one-day regressive shift of the base of the cycle. A *13 Muluc* line appears to have been superseded by a *12 Lamat* line; and future regressive shifts to *11 Manik* and *10 Cimi* appear to have been anticipated in the upper two lines.

In summary, 405 mean lunar months fall about one-ninth of a day short of 11,960 days, accumulating very close to a one-day shortfall in nine repetitions of the cycle, which amounts to nearly 295 years or, more exactly, to 299 chronological years (14.19.0.0). Compensation is attainable by advancing the base of the table by one day after the lapse of this span of time. Three items of circumstantial evidence suggest that the Maya astronomers may—at least once—have employed this stratagem, whether or not they were aware of the precise length of time that would necessitate a repeat application. The table appears to represent the institutionalization of a new base, occasioned by the slow drift of the "zero" conjunction day.

The accumulation of shortfall in node positions proceeds at a more rapid pace than it does in lunations. Sixty-nine mean internodal intervals fall about 1.61 days short of 11,960 (again see last line of Table 3). In nine repetitions of the cycle the nodes therefore regress about fourteen and a half days, or close to half the effective length of an eclipse season for a tropical latitude. After this many repetitions, a locus in the table which at first predicted a maximally prenodal eclipse possibility, will now, on the ninth or tenth run of the table, predict one very close to a node; and by the eighteenth or nineteenth run of the table it will predict an eclipse possibility that is nearly maximally postnodal. Such loci in the table—which are the ones at the ends of the five-month intervals (just before the pictures) when the table is new—will be good as theoretically "eclipse-possible" times for about eighteen runs of the table. But those that mark nodal or near-nodal eclipse possibilities in the original table—especially such as those at the beginning of the table and after 88, 135, 223, and 358 months—will be good as predictors only for about nine runs of the table. And those that mark maximally postnodal eclipse possibilities—such as those just before the five-month intervals—will become obsolete after a single run of the table. Thus one envisions the possibility that the table may have been under more or less continual revision, perhaps every 33 years or so (the length of the table is about 32¾ years). If this were done systematically, and on the basis of adequate theory, then three or four of the five-month half-years would have to be moved one position to the left each time through. But if it were done haphazardly and pragmatically (as when an actually observed eclipse showed the table to be wrong), then some earlier "eclipse-possible" positions might survive their obsolescence for some number of runs through the cycle, thus distorting the structure of the table. An "adequate" theory, such as might have allowed the Maya astronomers to move the five-month periods systematically and with proper precision, would have required only that they had known the rate of regression of the node days, or of the eclipse seasons, through their sacred almanac. There is no clear evidence to indicate that they knew this, although the possibility cannot be excluded.

3. The Pictures. The "picture" breaks into the sequence of eclipse half-years, marking divisions of the cycle, have invited speculation as to their possible import. An early hypothesis, tested and rejected, was that they symbolized a series of solar eclipses that were actually observed by the Maya, and that they marked the dates of such eclipses.[16] Considering the structure of the table, however, this would seem, *a priori*, not a good hypothesis. The pictures follow positions in the cycle that are from middling to maximally prenodal. Their implied prenodal distances can be read from column 4 of Table 3, immediately above each of the nine horizontal dividing lines. These values, of course, are predicated on placement of the hypothetic nodes at the midpoints of the three eclipse seasons determined by the recorded almanac days, as previously explained. If the table is regarded as synchronic, then there is no room at all for juggling of the hypothetic node positions, for two of the three indicated eclipse seasons are represented as of about maximum possible duration, namely 35 days.[17] But if the table is assumed to contain anachronisms, including some already obsolete far-postnodal positions as a result of a "pragmatic" approach such as suggested above, then there is the possibility of moving the hypothetic nodes to positions a few days earlier than those determined above. But even with this provision, there is no way that the nodes can be placed sufficiently earlier in the table so that some of the "picture" positions will be postnodal. And if they were, the five-month periods would then be wrongly placed, and the table would lose up to 50 percent of its predictive potential. The pictures therefore cannot plausibly be assumed to have marked a series of actually observed eclipses. Such a series would have consisted solely of prenodal eclipses, most of them quite far or very far prenodal, which is most improbable as a sequence of empirical events. Yet the pictures are full of eclipse symbolism; and their placement exactly after the drops to negative maxima in the abnodal-distance function requires explanation.

It should be noted that to the extent that the pictures are consistently placed in relation to the ideal structure of such a table, as several of them are, they mark places where a partial eclipse of the sun might occur either on the indicated date or on a date one month later, or on both dates (somewhere on earth), with optimum conditions for a total eclipse of the moon at opposition time halfway between them. The Maya moon goddess hanging by her neck from a celestial band in the picture on codex page 53b is suggestive of such a reference. So also are some of the hieroglyphs in the legend over that picture, the first three of which make reference to the "death" and the "darkening" of the "moon." And in the legend over the preceding column, which marks the end of a five-month eclipse half-year, there is a "solar eclipse" sign followed by the head of the sun god.

In a time-span of 32¾ years—the length of the table—about fifty lunar eclipses occur, and from any one location on earth about half of these are visible, namely those that happen during local nighttime. Of these, approximately a third are total eclipses. An eclipse table of this length constructed for lunar eclipses would have the same major divisions and subdivisions, but could not allow for as many lunar eclipse possibilities as there are solar. Not every half-year permits of one, nor does a half-year ever permit of more than one (there cannot be two, a month apart, as there can solar). They are variably spaced, most frequently at intervals of six months, but sometimes with intervals of eleven, twelve, seventeen, or twenty-three months. Yet the partitioning of the series into subdivisions of 47 and 41 months and the arrangement of these in the familiar pattern (47 + 47 + 41 for the 135-month division, another 47 + 41 completing the saros, and then repeating) are as for solar eclipses.

It may be helpful to be reminded of some of the differences between the conditions for a solar eclipse and those for a lunar eclipse. A partial solar eclipse occurs when the sun enters the moon's penumbra, and a central eclipse (total or annular) when it enters the umbra or the inverted projection of the umbra beyond its vertex. Because the ratios of the diameters of the sun and the moon to their respective distances from the earth are approximately equal (varying in relation to each other slightly due to orbital eccentricities) the vertex of the moon's umbra comes close to the earth's surface during an eclipse syzygy, falling just short of it during an annular eclipse, and just reaching it or penetrating it (geometrically speaking) to a slight depth during a total eclipse, thus describing in either case a narrow path of centrality on a portion of the face of the earth, but never engulfing the earth. A visible lunar eclipse on the other hand, whether total or partial, requires entry of the moon, totally or partially, into the earth's umbral cone. Entry into the penumbra does not suffice, because the unshadowed portion of the sun (in what would be seen by an observer on the moon as a partial terrestrial eclipse of the sun) is sufficient—because of the sun's intensity—for full illumination of the moon while it is in the earth's penumbra. Thus, although the earth casts a larger shadow into space than does the moon, the ecliptic limits for a visible lunar eclipse are narrower than those for a solar eclipse, and lunar eclipses are correspondingly rarer (as a phenomenon for the earth as a whole, although of course not for a single point of observation). The pertinent fact for present purposes is that lunar eclipses therefore, because of their shorter eclipse seasons but yet more frequent opportunities for verification at a single location, offer more clues and a somewhat better guide to the construction of a primitive solar-eclipse table than do the solar eclipses themselves. (A stationary observer has a 50 percent chance of verifying a lunar eclipse, whereas with solar eclipses it is closer to about 8 percent.)

Referring now to Table 3, it is seen that the horizontal dividing lines (which correspond to the picture dividers in the Maya table) immediately follow the greatest negative values in the abnodal distance function, these being where they are because of the shortening of the immediately preceding eclipse half-years from six to five months. By adding half a lunar month to these, one finds minimal abnodal distances for oppositions, and hence the most favorable times for total lunar eclipses. It is almost necessary to conclude that these facts played some role in the placement of the five-month periods in the layout of the Maya table, and that the pictures relate to this. One is led to suspect for a given recension of the table, when some or all of the five-month periods required relocation, that their placement may have been governed to some extent by the record of recent lunar observations together with a rule that a total eclipse of the moon, or the first in a single series of such, should be preceded by a five-month eclipse half-year. In this case, the circumstances of observability of total lunar eclipses—whether nighttime or daytime locally—would result in departures from the "ideal" pattern of lengths of the subdivisions of the eclipse cycle. Missing the first two of such a series (consisting usually of three or four at six-month intervals) would transform the normal pattern of Table 3*a* into the Maya pattern as in Tables 2 and 3*b* and Figure 3. A pattern emerging from such a series of observations might well have been made canonical and turned into a formula for prediction into the future, or for warning of possibilities that have precedent. There is reason to believe that the glyphic annotations at the heads of the 69 columns of the Maya table, and over the pictures, may be a log of precedents, or a conflation of several such. As a final point in this argument it may be noted that at several of the divisions in the table, in the pictures or in the legends over them, there are references to both an eclipse of the sun and an eclipse of the moon. In five cases there are paired eclipse symbols, the first solar and the second lunar. A sixth case, represented differently, has already been noted. These appear to be an explicit acknowledgment of the fact that a lunar eclipse close to a node may be preceded by a solar eclipse (partial) half a month earlier, which then is more or less maximally prenodal and entails shortening the preceding half-year to five months.

There is still the matter of the tenth or "extra" picture to dispose of. A table of this length can have only nine natural subdivisions (it is one 41-month subdivision short of two saroi). The first nine of the picture breaks initiate eclipse seasons, coming after five-month half-years, and at positions that are more or less maximally prenodal. The tenth picture, however, as noted earlier, comes in the middle of an eclipse season, after a six-month half-year, and at or very close to a node. Yet it has a prominent display of the paired eclipse symbols, solar and lunar like the others, both in the picture and in the glyphic legend above it. This shows it to be an anachronism, for there cannot be a lunar eclipse either at this time (it is conjunction) or half a month later as in the other cases (in this case it would be several days past the lunar ecliptic limit and is in one of the longer periods when a lunar eclipse is excluded). The tenth picture, then, must be seen as a retention from an earlier version of the table, one in which a *12 Lamat* conjunction was approximately fifteen days prenodal rather than at a node as in the present version. Its current function must be that of a historical marker, memorializing the date and circumstance of the institution of *12 Lamat* as a canonical base for eclipse cycles.

4. The Maya Computations. The Maya numerals and their Arabic equivalents in Table 2, column 5, are the distances from the *13 Muluc* base of the codex table to the successive almanac days named in the lowest of the three lines of these in the same table. To be compared with these values are the numbers with decimal fractions in Table 3, column 5, which are the intervals from a hypothetic zero point (positing coincidence of a lunisolar conjunction with a node passage) to the ends of successive six-month or five-month eclipse half-years grouped as these are in the table of the codex (Figure 3). The numbers of the first of these sets are whole integers, the results of computations by Maya primitive astronomers, employing whatever methods they had at their disposal for reckoning lunations and conjunction days, and without benefit—so far as we know—of fractional arithmetic. The second set consists of multiples of the number 29.530588 computed on a modern handy calculator and rounded off to two decimal places. Had they been rounded off to the nearest integer in each case, they would have been in near-perfect agreement with the Maya values. Of the 69 Maya values, 54 are exactly those that would be obtained from our experimental table by rounding off the values in this way, following the usual conventions. Of the other fifteen Maya numbers, eleven differ from the rounded-off experimental figures by −1, and four by +1; but the actual fractional differences are mostly small, averaging only 0.17 of a day in their deviation from the dividing line between fractions dropped and fractions counted as wholes in the rounding-off process. It is worth noting that the four instances of +1 deviations in the Maya integers would vanish if we took as our primary evidence the cumulative totals as written in the codex rather than the almanac days recorded there. However, had that policy been followed throughout, it would have introduced 35 additional −1 deviations and increased nine others from −1 to −2. The string of −1 errors in the cumulative totals beginning on codex page 57a and continuing to the end of the table on 58b can be attributed to a single error in that amount that went uncorrected, and so was carried forward to the end of the table. An alternative hypothesis, sometimes entertained, is that the "error" was deliberate, signaling an intended shift from the lower line of almanac days to the middle line, thus shortening this run of the cycle to 11,958 days as written, for a correction in the amount of two days. Yet another current hypothesis is that the cumulative totals were not intended to be in complete agreement with any single line of almanac days, the one series being empirical and the other theoretical—but with three choices each time to allow for plus or minus deviations in the amount of one day. The description given above, however, has followed the hypothesis that a single line was the intended referent throughout, that the one-day shortening of the cycle embodies a rule for its use every ninth time, or when perceived as necessary, and that these are consequently the occasions for shifting from one line of almanac days to the line above. It is the aforementioned relation between the *13 Muluc* date 9.12.10.16.9 and the *12 Lamat* date 9.16.4.10.8, as well as the fact that each line of almanac days in the table exhibits the one-day foreshortening, that has prompted this choice from among the available hypotheses. There is evidence, however, as will be seen later, of the employment of another and more drastic foreshortening device—somewhat similar in principle to that employed with the Venus table—that would make possible the holding of epochal conjunction days to any given day in the almanac, say *12 Lamat*. But whether held, or allowed to slip a day every nine cycles, there is no way of containing such an epochal almanac day, or series of adjacent ones, within eclipse seasons for longer than eighteen or twenty cycles. To this end, the slipping procedure does slightly better. In either case, the span of time is in the neighborhood of six centuries.

5. The Distribution of 30-Day and 29-Day Months. A simple alternation of 30-day and 29-day months in a lunar calendar yields a mean lunar month of 29.5 days, causing the pace of the calendar to exceed that of its object. Since the Maya eclipse table spans 405 moons, such an unmodified alternation would cause it to fall twelve and a half days short of its proper length: 11,960 − (405 × 29.5) = 12.5. Additional days are required to compensate for the shortfall. This in effect means that the ratio of 30-day to 29-day months, and the pattern of their alternation, must be modified. Permitting two consecutive 30-day months, and shifting the ordinal positions of each of the two varieties of months from odd to even and even to odd, effects a half-day compensatory increment. If this method were employed throughout the table, twenty-five modifications would be required. An alternative, substituting a 30-day month for a single 29-day month, without changing the ordinal positions of the others, and thus permitting a sequence of three consecutive 30-day months, effects a whole-day compensatory increment. The method employed in the Maya eclipse table is transparent. There are nine 5-month eclipse half-years, of 148 days each, distributed among sixty of the 6-month half-years. Each of the 5-month periods has three months of 30 days to two of 29 days. By its change of the ratio, each such period automatically contributes a half-day compensatory increment, for a total of 4½ days. This leaves eight to be provided by other means in a full run of the cycle, or seven in this special 11,959-day truncated run. Among the sixty 6-month periods, fifty-three are of 177 days each [(3 × 30) + (3 × 29)], but seven are of 178 days [(4 × 30) + (2 × 29)]. (These are marked † in column 7 of Table 2. Only one of them is so indicated in the series of intervals in the codex table, Figure 3, bottom pair of lines. Their determination is from the series of almanac days.) Each of these 178-day periods provides one whole-day compensation, which could be made either at once (for example, 30 + 29 + 30 + 30 + 30 + 29) or in two half-day increments (for example, 30 + 30 + 29 + 30 + 30 + 29). An error graph, which can be constructed from the differences between the integral values and the fractional values of columns 6 and 5 of Table 3, confirms that there are nine half-day and seven whole-day compensations. Had the eighth whole-day intercalation been included, the second-last half-year would be of 178 days, and the normal length of the cycle, 11,960 days, would be attained. As has been seen, the end result of the Maya computations is remarkably close to what they would have had if we had done the figuring for them.

6. The Preface to the Maya Eclipse Table. Pages 51a–52a of the table contain prefatory matters that raise more questions than can be answered. The content may be summarized as follows (labeling the five columns on page 51a as *A* to *E*, and the six on page 52a as *F* to *K*).

Column A (· · ·, *4 Ahau, 8 Cumhu, 12 Lamat*, 8 days · · ·, 9.16.4.10.0 [black number], 10.19.6.1.8 [red number], *12 Lamat*). The glyph at the top of the column is obliterated. *4 Ahau 8 Cumhu* refers to the zero day of the Maya day count. *12 Lamat,* as indicated by the following "8 days · · ·" notation, refers to the eighth day after the zero of the day count, which is at that position in the almanac. This serves as a special base from which to reckon the following interval or intervals (see "Numerology" for analogy with "ring number" bases). The black distance number is written in the codex as 8.16.4.10.0, which does not accommodate the terminal day *12 Lamat* written below it. The initial "8" can only be a copyist's error for 9, and is so corrected. The number 9.16.4.10.0, an integral multiple of the length of the almanac (5,434 × 260), added to the special base, reaches the date 9.16.4.10.8, which is on a day *12 Lamat* (*1 Muan*). This date recurs in column *K*. The interscribed red number 10.19.6.1.8, applied to *4 Ahau 8 Cumhu*, specifies the date with that number in the day count, which is also a day *12 Lamat* (*6 Cumhu*). Its distance from the special base is again a multiple of the almanac (6,073 × 260). The interval between these two *12 Lamat* dates is 1.3.1.9.0, or 166,140 days, or 14 eclipse cycles less five almanacs (1,300 days). It is equal to 5,626 lunations (of 29.530588 days) plus the 0.91 part of a day, or to 5,626 Maya moons (of 29.530864 days) minus the 0.64 part of a day. It is not a multiple of the internodal interval however, or close to one (it is about 64 days short of such a multiple), so that if one of the two dates was within ecliptic limits, the other was not.

Columns B–H. Below the glyphic annotations at the top, which are mostly obliterated, these seven columns contain thirteen numbers, ten of which are multiples of the length of the eclipse cycle (1.13.4.0, or 11,960 days) and the other three of which are equal to multiples of that number with increments that one is perhaps obliged to suppose intentional and somehow significant, but some of which may nonetheless have been due to errors in compilation. It will suffice to note that the list contains the following values, although not in this order:

1.18. 5.0, = 1 EC + 5.1.0 (1,820 days,
3. 6. 8.0, = 2 EC or 7 almanacs)
4.19.12.0, = 3 EC (entered twice)
6.12.16.0, = 4 EC (entered twice)
8. 6. 2.0, = 5 EC
9.19.12.0, = 6 EC + 6.0 (120 days)
1. 6.11.10.0, = 16 EC
1. 8. 4.14.0, = 17 EC
1. 9.18. 0.0, = 18 EC
2.11.10.11.0, = 31 EC + 13.0 (260 days,
3. 4.15.12.0, = 39 EC or one almanac)

Below the numerals are the almanac days *12 Lamat, 1 Akbal, 3 Etznab, 5 Eb*, and *7 Lamat*. These are separated from each other in the almanac by 15-day intervals, increasing in the order of listing, for a total of 60 days. They are repeated, identically, in each of the seven columns.

Column I (· · ·, 8 days, 1 *uinal* and 5 *tuns*, 2 · · ·, thirteen 13's). The uppermost glyph is mostly obliterated, but a surviving detail (detectable only in the photographically based editions of the codex) appears to narrow down the choice to either the moon sign or the death sign. The "8 days" recalls the similar entry in column *A*, and is assumed thus to refer to the special *12 Lamat* base, eight days after the start of the day count. The "1 *uinal* and 5 *tuns*" are the period 5.1.0, or 1,820 days. This period, equal to seven times the 260-day almanac and to five times the 364-day computing year, was apparently one of the much used Maya reckoning units. It is the subject of four different multiplication tables in the Dresden Codex and of one in the Paris Codex, and it is a factor in five of the numerologically contrived long-reckoning numbers that are applied to "ring" bases in the Dresden Codex. It recalls the discrepancy between what is expected (1.13.4.0) and what is written (1.18.5.0) in the table of multiples in columns *B–H*. But the next glyph and the thirteen 13's following the "1 *uinal* and 5 *tuns*" suggest yet another possible significance. It can be seen in Figure 3 that this glyph (column *I*, fourth glyph), except for its numerical prefix "2," is the same as that following the "8 days" in the fifth position of column *A*. Whatever its inherent meaning may be, its function there could be seen as that of marking a quantity to be added to what precedes the statement of that quantity. The prefix "2" which accompanies it in column *I* might then be interpretable as an indicator of a double addition, which, if taken together with the thirteen 13's that follow, produces the number 10.10.9 (= 3,809 days). This is one of the good solar eclipse intervals, as can be seen in the last column of codex page 56a, or in Tables 2 and 3. There are thus two alternative possibilities of relevance here for the "1 *uinal* and 5 *tuns*," but as yet no other suspected significance for the thirteen 13's.

Columns J–K. Four dates are recorded here, reckoned from the normal base of the day count, *4 Ahau 8 Cumhu*, which is recorded at the head of each column. There are two copying errors, an "8" for 18, and a "10" for 11. These are corrected (with prefixed asterisks) in the following transcription. From earliest to latest they are:

K-black: 9.16.4.10.8, *12 Lamat* (*1 Muan*)
K-red: 9.16.4.11.3, *1 Akbal* (*16 Muan*)
J-black: 9.16.4.*11.18, *3 Etznab* (*11 Pax*)
J-red: 9.19.*18.7.8, *7 Lamat* (*16 Zac*)

The first one is the date determined also in column *A*. The first three are at 15-day intervals, as are the corresponding almanac days in columns *B–H*. That series proceeds by two further 15-day intervals, to *7 Lamat*. If the series of dates here did also, it would reach 9.16.4.13.8, *7 Lamat* (*1 Cumhu*). Instead, a date 26,520 days later is indicated. This interval, less one day, is an approximate eclipse interval (898 moons plus about half a day, and 153 mean internodal intervals plus about 2½ days) and is also the interval between the *13 Muluc* date of previous reference and the *12 Lamat* date, 9.16.4.10.8. However, if either or both of the *13 Muluc* and *12 Lamat* dates were eclipse dates, then neither of the *7 Lamat* dates can be; although all four are possible conjunction dates.

These dates have invited attempts to place the eclipse table in Maya chronology. It has been proposed that the first three, which are at 15-day intervals, may be the dates of two solar eclipses a month apart and of a lunar eclipse between them; that a node must therefore have been close to the *1 Akbal* date, and that the *12 Lamat* date, 9.16.4.10.8, must then have been approximately 15 days prenodal; and that since the *12 Lamat* at the end of the eclipse table is obviously at or very close to nodal position, it cannot possibly be the one of the date 9.16.4.10.8, but must be one of about nine or ten eclipse cycles later, say 10.11.3.10.8 or 10.12.16.14.8, by which time the node would have migrated to a position about 15 or 16 days earlier in the cycle and in the almanac (9 × 1.61 = 14.49 days; 10 × 1.61 = 16.1 days).[18] The conclusion follows correctly, and is therefore at least as good as the premise. In regard to this, however, it needs to be remembered that two solar eclipses a month apart cannot have been visible from the same region of the earth, so it is not possible for the two hypothetic solar eclipse dates to have been based on Maya observations. At least one of them has to have been a prediction, in line with their knowledge about the possibilities either side of a total lunar eclipse. The safest part of the premise, then, is the supposition of a lunar eclipse, perhaps total, on the date 9.16.4.11.3, *1 Akbal* (*16 Muan*). Embarrassing for the hypothesis, and for the conclusion that follows from it, is the presence of the seven times repeated sequence of these same almanac days *and two more—also at 15-day intervals*—in the adjacent columns *B* to *H*. (Three solar eclipses at 30-day intervals are, of course, impossible, as are two lunar eclipses a month apart.) Favoring the hypothesis, however, is at least one very plausible eclipse date from an inscription, and two other quite plausible ones, that are so distributed as to be optimally placed in relation to nodes if the *1 Akbal* date of this passage in the codex is also at a node. These would make the *12 Lamat* date the more likely candidate for the predicted but unobserved one of the pair of supposed solar eclipse dates, which would be in conformity with the hypothesis about how lunar eclipses were employed as guides to the placement of the five-month periods in the eclipse-warning table. All of this is also concordant with the earlier conclusion concerning the anachronism of the tenth picture in the table. It must be of the epoch of 9.16.4.10.8, while the others are of the current epoch, some nine or ten eclipse cycles later.

The purpose of the series of five almanac days, repeated under each column of multiples of the eclipse cycle, remains open to conjecture. There is the possibility of a double relevance for the 15-day intervals. On the one hand they are approximations to the intervals from conjunction to opposition and again to conjunction, so that any three of the almanac days could represent a set of hypothetic eclipse possibilities, solar and lunar; and, on the other hand, they approximate the distance that the nodes migrate in the length of time that it takes for conjunctions to migrate one full day in the same cycle, namely in about nine repetitions. Any subset of three out of the five almanac days, then, taken in order from top to bottom, could represent the epochal eclipse possibilities for one given run of the cycle, while successive such sets, moving from bottom to top, could represent the node migrations corresponding to the lengths of time that require one-day corrections (or equivalent adjustments, see below) in the calendrical reckoning of lunations. If this second potentiality was a factor in its determination, then the significance of the recorded *7 Lamat* date can only have been commemorative or something extraneous to the reckoning of eclipses.

7. The One-Day Corrections. As noted earlier, the discrepancy between 405 lunations and 46 almanacs (the length of the Maya eclipse cycle) is about $^{11}/_{100}$ of a day, so that approximately every ninth run through the cycle would call for a one-day adjustment in the locations of conjunction days in the almanac. In one interpretation, this is the function of the three lines of almanac days in the table proper, a shift from a lower to the next higher line accomplishing a one-day compensation for the accumulated shortfall of lunation times in nine or so runs of the table. If this was their intended function, and if the length of the eclipse seasons was properly appraised, then there should be just three such lines, allowing for two shifts. During this time a given series of adjacent epochal days—such as the three in the sixty-ninth column of the table—would have passed from an initial maximum prenodal position, to a nodal position, to a maximum postnodal position, after which its relevance is past. Then a new epochal series would need to be located and instituted, which would require a more drastic abridgment of the final cycle of a series. There is circumstantial evidence suggesting that such a procedure was indeed in effect, and that, employed more often, it provided a way even to dispense with the one-day shifts in the almanac positions of the epochal day.

This possibility rests on the felicitous circumstance that ten lengths of the almanac amount to a little over a day more than a good eclipse interval. If this amount is subtracted from an eclipse cycle, then the remainder is a little over one day *less* than another good eclipse interval. Such a foreshortened cycle can therefore be employed periodically, in place of a full cycle, to effect a compensation for accumulated shortfall in lunation times, while still preserving the same almanac day—say *12 Lamat*—for the beginning and the end of the cycle. The optimum ratio is ten full cycles of 1.13.4.0 or 11,960 days each, to one short cycle of 1.6.0.0 or 9,360 days. It has a remarkable accuracy, both in holding epochal lunation days to a fixed place in the almanac, and in reckoning lunar phenomena over long periods of time. It allots 4,367 lunations to a period of 17.18.4.0 or 128,960 days, which is equivalent to a mean lunar month of 29.530570 days. This is an improvement over the Maya "first approximation," as in the single eclipse cycle, which is equivalent to a mean lunar month of 29.530864 days; and it may be compared with our modern figure of 29.530588 days.

The conclusion that the Maya astronomers and calendar priests had discovered the possibility of this manner of correction, and even the optimum ratio of full to abridged cycles, rests perhaps a bit precariously on a few pieces of circumstantial evidence. One such piece consists in a pair of dates, recorded in the "ring number" and "serpent number" manners, that are separated by exactly twice this interval, as if there had been two successive applications of the formula.

10. 7. 4. 3. 5, *3 Chicchan 13 Yaxkin*
− 8.11. 7.13. 5, *3 Chicchan 8 Kankin*
= 1.15.16. 8. 0, = 2 × (17.18.4.0),
= 2 × (128,960 days).

Both of the dates have the same position in the almanac, *3 Chicchan*. Both have a moon age, reckoned from conjunction, of between three and four days, which is an amount by which recorded moon ages in inscriptions at many Maya sites fall short of true moon ages, because of their being reckoned from the first appearance—or "birth"—of the new crescent moon, rather than from estimated conjunction days (as at certain other sites). A part of the significance of the dates therefore can be ascribed to their being dates of the rebirth of the moon. In the case of the later one, this appears to be borne out by the "celestial birth" symbolism of the open jaws of the serpent, by the rabbit form assumed by the emerged deity (the rabbit is a moon symbol in Mexican iconography and in postclassic Mayan), and by the presence of a "birth" hieroglyph in the accompanying brief text. Also in the case of the later date it can be shown (1) that eighteen days earlier was the date of a lunar eclipse (on the assumption that the *1 Akbal* date of the preface to the eclipse table was a node day, or very close to one), and (2) that three days earlier was potentially a date for a solar eclipse (predicted, surely, if not actually observed); so that the *3 Chicchan* date was that of a rebirth of the moon under somewhat special circumstances. Whatever else of astronomical phenomena may have lent special significance to either or both of these dates is unknown.

The *3 Chicchan* dates can have only one place in the eclipse cycle: some three days after the completion of 335 lunations, or four days after the *12 Imix* in the Maya table (Table 2, following node 57; Table 3, line 57; Figure 3, page 55b, second-last column; note that the theoretical figure, 9,892.75, would have called for one more day here, 9,893 rather than 9,892, or *13 Ik* rather than *12 Imix*). This makes it possible to specify the corresponding epochal dates, which of course have the same interval between them. It is of interest that the later of these is separated from the presumed date of the institution of *12 Lamat*, 9.16.4.10.8, by an interval that is equal to five full eclipse cycles plus one short one. This, although apparently involving a different ratio, is more likely an instance of the same. It is what would be expected if the corrective short cycle were placed in the middle of the ten full cycles rather than at the end; that is, if the error were corrected when it reached a magnitude of a little over half a day, substituting smaller errors, positive and negative, for an otherwise larger range of errors, only positive.

10. 7. 4. 3. 5, *3 Chicchan 13 Yaxkin* (*b.* ☽)
−4 interval, conj. to cresc.
10. 7. 4. 3. 1, *12 Imix 9 Yaxkin* (conj.)
−1. 7. 8.13 = 335 lunations
10. 5.16.12. 8, *12 Lamat 11 Tzec* (epoch)
− 9.16. 4.10. 8, *12 Lamat 1 Muan* (epoch)
9.12. 2. 0 interval between epochs
−8. 6. 2. 0 = 5 full eclipse cycles
1. 6. 0. 0 = 1 short cycle,
= 9,360 days, or 36 almanacs.

The difference between the length of the full cycle and that of the short cycle, namely 10 almanacs, is equivalent approximately to 88 lunations and to 15 internodal intervals; 10 and 88 are even numbers, 15 is odd. Therefore, in reckonings concerned with lunations but where eclipse syzygies are not at issue, the corrective foreshortening can be halved and applied more frequently, permitting a yet closer adherence of the timing of ordinary lunar phenomena to a schedule of almanac days.[19] There are a few pairs of dates in the Dresden Codex that seem to imply such an application. Without intermediate dates it is not possible to distinguish between two applications of a 5-almanac foreshortening and one application of a 10-almanac foreshortening. Such an intermediate date is that of 9.17.8.8.5, *3 Chicchan 18 Xul*. It has the same almanac position and the same moon age (to within

$^1/_{100}$ of a day) as does the later one of the two dates discussed in the preceding paragraph; and like that one, this too is given in the "serpent number" manner. It may be assumed that there was yet some other astronomical significance, besides its moon age, for this date; for over the jaws of the serpent in this case is the Maya rain god (celestial identification unknown). The interval between these two dates is equal to 6 eclipse cycles less 5 almanacs, or 5 eclipse cycles plus 10,660 days.

10. 7. 4. 3. 5, *3 Chicchan 13 Yaxkin*
− 9.17. 8. 8. 5, *3 Chicchan 18 Xul*
9.15.13. 0, = 5 × (1.13.4.0) + 1.9.11.0,
= 5 × (11,960) + 10,660 days.

The problem for the Maya astronomers was not simply that of determining the natural cycles of celestial phenomena but, equally important, that of integrating these with the cycle of days in the sacred almanac. This second desideratum required an ideal cycle—for lunar reckoning as for Venus reckoning—that was a multiple of the almanac. The almanac was peculiarly well suited for the definition of eclipse seasons and for the isolation of their nodal midpoints, as it was also for the delimitation of the 135-lunation divisions. And in its fifth and tenth multiples the almanac also offered a peculiarly fortunate opportunity, apparently grasped by the Maya, for refining this reckoning instrument to one of great precision. But the primacy of the almanac seems to have placed an impediment in the way of discovery of the special significance of the saros. The saros is there, as it has to be, but only as one out of sixty-nine specified eclipse possibilities. Its nonrecognition makes for some eventual awkwardness in the major groupings of lunar half-years: the sequence of divisions, of 135+135+135, and so forth, is forced; the more natural one would have been 135+88+135+88, and so forth, where the 135's are optimally 47+47+41, and the 88's are 47+41. But to have integrated this arrangement with the almanac into a single repeating scheme would not have been possible, except in a scheme of undue length. Similarly, the Metonic cycle appears to have attracted no particular attention, It too is there, as it has to be (at 235 lunations), but again only as one of many eclipse possibilities. Any concern for the tropical year was subordinate to that for the almanac.

VI. NUMEROLOGY

The prime use of Maya astronomy was to learn the habits of the celestial powers so as to make predictable the hazards of living under their influence. Keeping track of the schedules of the heavenly bodies, and predicting their appearances, disappearances, and crises, depended on the auxiliary calendrical and numerical sciences. In its discovery of regularities in phenomena, and in its descriptive formulations of these, Maya calendrical astronomy was akin to science (in a kind of ancestral relationship); but in its interpretative edifice, and in its applications, it pertained rather to the domains of astrology, demonology, and divination. The predictions that it made possible must at times have seemed awesome. It appears that rulers and their priesthood found uses for this highest form of divination. One of them was to forewarn against dangers emanating from the celestial scene, so that preventive measures might be taken—rituals, offerings, and sacrifices directed to the appropriate deities. Another was to gain the most auspicious times and circumstances for projected undertakings. Both of these were in application to matters of the more or less immediate future. But another was to the fixing of events in very distant times, past and future, providing mythological and numerological charter for the positions of rulers, and seemingly forecasting their continued existence and influence in afterlife.

In the section "Calendar, Chronology, and Computations," above, there was brief reference to a passage from an inscribed text at the site of Palenque that documents the accession, to at least titular rulership, by the boy Pacal at an age of approximately $12^1/_3$ years. Although introduced there only as a computational problem, it also illustrated one of the seemingly fanciful uses to which the labors of the court arithmeticians and calendar keepers were put. There are several of these, from several sites—Tikal, Quirigua, Copan, Yaxchilan, Palenque, and others. They relate events in the lives of rulers to similar events in the careers of mythological ancestors and gods of the far distant past, or to anniversaries, as it were, of these events in the distant future when these rulers too shall presumably have become gods. This was one type of royal numerology. There was a second, which operated with numbers of much more modest magnitudes, covering spans of time that were either contained within the current chronological era or that extended only minor intervals beyond its beginning. In one respect these are the more interesting, because it is possible to discern in some of them a motive—other than the attainment of sheer magnitude—for their being the particular numbers that they are. A few examples of each category will be cited here, all from Palenque, together with relevant portions of their immediate contexts, which illustrate something of the quality of the Mayan concern with numbers and days.

The accession of Pacal is documented in at least eight locations in the hieroglyphic texts of Palenque, so that its date and the age of the boy on that occasion are well secured. One of them, from the east panel of the Temple of the Inscriptions, states that the accession was 17.13.12 before the period-ending *1 Ahau 8 Kayab* (which is 9.10.0.0.0), this chronological anchor then being made further explicit, redundantly so, by stating that it was the end of the tenth *katun* and the midpoint of the current *baktun*. The indicated arithmetic fixes the accession date at *5 Lamat 1 Mol* (9.9.2.4.8). A passage from the west panel in the same temple, to which reference was made earlier, has it that it was 12.9.8 from the day of his birth, *8 Ahau 13 Pop*, to the day of his accession, *5 Lamat 1 Mol*. Then it explains further that this latter day was 2.4.8 after the *3 Ahau 3 Zotz* period-ending (the *katun*-ending 9.9.0.0.0). This doubly confirms the information derived from the passage in the east panel, and further, and redundantly, fixes his birth date of *8 Ahau 13 Pop* at 9.8.9.13.0—which fact is known also from other texts, including an initial series that displays the same day number with head-variant numerals. Thus far the text is sober history. But it continues from there with the statement that it was 7.18.2.9.2.12.1 from the assumption of rulership by an ancient deity or celestial forebear, on a day *1 Manik 10 Tzec*, to the identical event on the part of the young Lord Pacal of Palenque on *5 Lamat 1 Mol* (9.9.2.4.8). And it goes on further to state that it will be 10.11.10.5.8 from the birth of Pacal (which was on *8 Ahau 13 Pop*, 9.8.9.13.0) to yet another *5 Lamat 1 Mol*, and that this *5 Lamat 1 Mol* will be just eight days after the completion of one *pictun* (1.0.0.0.0.0) in the current day count.

The passage is of interest for its projections into the past and into the future, and also for its structural parallelisms. The appeal to the past event, of identical kind, not only equates the status assumed by the young heir to the throne, on the historical date 9.9.2.4.8, with that assumed by this deity in a far past era, but appears to go beyond that, to draw a special parallel, or to assert a special likeness, between these two accessions, the secret of which is perhaps locked in the meaning of the great distance number. The reference to the future event is of a different sort. Here the parallelism is between two pairs of dates, the earlier one in each pair being the same, namely, *8 Ahau 13 Pop* (9.8.9.13.0), the day of Pacal's birth, and the later one in each pair having the same position in the calendar round, *5 Lamat 1 Mol*, the calendar day of Pacal's accession, but the one of the second pair being the eightieth calendar-round anniversary of the one in the first pair (4,160 years later), and being moreover the first *Lamat* (just eight days) after the end of one *pictun*, that is, after 8,000 chronological years counting from the beginning of the current era. It must be supposed that, to have a point made of it, something auspicious must have been seen in that coincidence.

Another great distance number, 5.18.4.7.8.13.18, mediates between a *9 Ik 10 Mol* date and a contemporary date of *9 Ahau 3 Kankin* (9.11.1.2.0) in the inscription of Temple XIV at Palenque, but the hieroglyphs naming the events for these two dates are not yet understood. The human protagonist of the contemporary date is Chan-Bahlum, the son and successor of Pacal. The others are gods.

The motivations for the particular values of these great distance numbers are not clear. The one in the Temple of the Inscriptions is equal to 455,393,761 days, or 1,247,654 calendar years plus a remainder of 51 days. That of Temple XIV is 340,469,558 days, or 932,793 calendar years and 113 days. Projections that far into the past, pending any better hypotheses, are assumed to have been intended to reach dates of cosmic importance and to represent the results of calculations into the past, perhaps involving several astronomical variables. The events ascribed to such dates, where their glyphs are understandable, are "accessions" of deities who rose to power in eras past. The current era was about halfway into its tenth *baktun* at that time (about 3,800 years), and the preceding era had had only thirteen *baktuns* all told (about 5,125 years); so the ages of the past must have been envisioned either as having spanned a great many more eras of comparable lengths, or else a lesser number of eras of much longer durations.

As numbers, more instructive are those of the more modest projections that go back merely to the beginnings of things pertaining to the present era. These allow us a glimpse into numerology at work, and in some cases, of myth in the making. Mention has already been made (in the section "The Venus Calendar") of those five-place numerals in the Dresden Codex that on the surface appear like any other numerals of comparable magnitudes fixing dates in the day count of the current era, but which, unlike ordinary Maya day numbers, are not reckoned from *4 Ahau 8 Cumhu* (the start of the era and normal zero of the count) but are reckoned instead from various special bases or epochal days at relatively short intervals prior to the beginning of the era. The intervals from day zero back to these negative bases—each base unique to the historical date that is reckoned from it—are given by the so-called ring numbers. Those leading forward from the pre-zero bases to the respective historical dates will here be called "ring-based day numbers," in order to distinguish them from the normally based variety. One of these was encountered in the prefatory page to the Venus tables (Dresden Codex, page 24). In that example the ring-based day number was 9.9.16.0.0, and the special ring-number base was at −6.2.0, which is equivalent to the date of 12.19.13.16.0 in the preceding chronological era. If we subtract 6.2.0 from 9.9.16.0.0, or if we add 9.9.16.0.0 to 12.19.13.16.0 and cast off 13 *baktuns*, we arrive at the date in the current era that is determined by this device, 9.9.9.16.0 in the normally based day count. (In verifying this, it should not be forgotten that a unit in third place of a day-count numeral is equal to eighteen units in the second place, whereas in all other places the ratio is one to twenty.)

Why the roundabout way of specifying a date in the current era? The answer has already been anticipated above in the discussion of the Venus example. The numbers that tell the days from the pre-zero "ring" bases to the related historical dates are contrived numbers—contrived in such a way as to provide, for each historical date, a base which is its like-in-kind in respect to its position in one or more pertinent calendrical or astronomical cycles. The day of the special base thus has significant attributes in common with the historical day whose chronological position is reckoned from it; and it is, moreover, the *last possible one* to do so before the beginning of the current era. The difference in quality between the two categories of day numbers, the ring-based and the normally based, can usually be seen when those which are associated with a single date are reduced to their prime factors. As a category, the ring-based ones are composite to a degree and in a manner that would not be expected by chance, whereas the category of normally based day numbers are in respect to their composition as if chosen at random from within the same general range of magnitudes (roughly from 1,200,000 to 1,600,000). Those of the one set are contrived; those of the other just "happened." There are fifteen ring-based day numbers in the Dresden Codex (eighteen examples, but with three repetitions). Eleven of them are multiples of the almanac, and four of these are also multiples of the triple almanac (one suspects an interest in Mars, with mean synodic period very nearly 780 days). Other periods which are contained integrally in one or more of these numbers are the computing year of 364 days (in five of them), the five-Venus-year or eight-calendar-year cycle of 2,920 days (in the example in the section "The Venus Calendar"), the 81-moon period of 2,392 days corresponding to the Palenque moon ratio, and in two instances a factor of 9 that may perhaps relate to the Lords of the Night and the glyph *G* cycle. A few of the ring-based day numbers are integral multiples of only one of these basic periods (the almanac is the most frequent), but others are multiples of two, three, or four of them. The ring-number bases, except for one of them, are at intervals ranging from 17 to 2,200 days prior to the beginning of the era. The exception is at 51,419 days (7.2.14.19) before the era; and the ring-based day numbers (two of them in this instance, interscribed red and black) that are reckoned from this anomalous base are exceptional in that they do not contain the more usual clusters of factors. One from this pair contains a prime factor of 59,167. Unless a component of such magnitude were of interest in the computation, there would be no apparent reason for having such a large ring number.

The events that are ascribed to the various historical dates reached by these contrived counts in the Dresden Codex are not known, because the meanings of crucial hieroglyphs are still elusive. But there are analogs to these, and kindred manners of reckoning, in at least a few of the monumental inscriptions of the classic period; and although the historical dates in these cases are of events in human history—in contrast to the presumably astronomical events of the Dresden Codex—similar ends are served by the numerological contrivances. They establish likeness of kind: between the dates, between the events, and here apparently also between the protagonists.[20]

The initial date of the Temple of the Cross at Palenque is 12.19.13.4.0, *8 Ahau 18 Tzec*. This is a date in the previous chronological era, at a distance of 6.14.0 before its end and before the beginning of the current era; it is thus equivalent to a ring-number date of −6.14.0. Considering that in the majority of instances of reckonings from such bases (that is in eleven out of the fifteen in the Dresden Codex) the base date and the historical date share the same position in the almanac, one would test for that possibility here. Since the base date is an *8 Ahau*, another *8 Ahau* date is sought that would be in historical time at Palenque. There is one, but in this instance it is not found in the same inscription. It is the *8 Ahau 13 Pop* of Pacal's birth, 9.8.9.13.0, recorded in several other locations, two of which have been cited above. The interval from a day 6.14.0 before zero to one that is 9.8.9.13.0 after zero is equal to 9.8.16.9.0, or 1,359,540 days. By way of a test, the number may be decomposed into its prime factors. It is found then to be equal to $2^2 \cdot 3^2 \cdot 5 \cdot 7 \cdot 13 \cdot 83$. This cluster of basic Maya primes marks it indubitably as a contrived interval. It is an integral multiple of the almanac ($2^2 \cdot 5 \cdot 13$), the glyph *G* cycle (3^2), the computing year ($2^2 \cdot 7 \cdot 13$), the triple almanac ($2^2 \cdot 3 \cdot 5 \cdot 13$), the 819-day cycle ($3^2 \cdot 7 \cdot 13$), and some of the compound cycles that required attention earlier, such as the 1,820-day period, the 2,340-day period, and the 3,276-day circuit. These, of course, are not all independent contributors to the interval, some being by-products of combinations of others. The almanac and the 819-day cycle alone would suffice to determine all but the last factor. Or the almanac, the computing year, and the cycle of the nine lords of the night would suffice for the same. The lowest common multiple of this collection of cycles is 16,380 days, equal to the day-count numeral 2.5.9.0, or in factor notation $2^2 \cdot 3^2 \cdot 5 \cdot 7 \cdot 13$. It was an important Maya number, equal to 20 × 819, of which a table of multiples must have been employed, together with one also of the first twenty multiples of 819 leading up to it, for the calculation of stations in the 819-day cycle. The remaining factor of 83 in the long interval is not basic to any of the Maya cycles, but was determined by the other principal requirement of a ring-number base, namely, that it be the last possible day in the old era, before the beginning of the new, that would satisfy the pertinent cyclical desiderata for the case in hand. These, in the present case, appear to have been that it have the same position in the almanac (*8 Ahau*), that it be under the same lord of the night (*G8*), and that it have the same location in the 4 × 819-day cycle (20 days after a "south" station in that cycle). These are three of the four principal attributes of a day, other than the lunar data, that accompany its day number in an initial-series passage at Palenque. But such a date has ascribed to it also an event. In the Temple of the Cross the old-era date of 12.19.13.4.0 (equal to −6.14.0 in respect to the current era) is declared to be the date of the birth of a female deity, an apparent mother-goddess and the implied ancestress of the royal line of Palenque. The likeness in kind between this date and the date of Pacal's birth has obvious implications. One's destiny was intimately related to the day of one's birth (this is amply documented throughout Mesoamerica from immediate postconquest times virtually to the present, and in some parts of the area both gods and men carried the name of the almanac day on which they were born as a principal component in their own names). The common attributes of the birth dates thus imply common attributes in the personages. The initial date of the temple appears thus to provide calendrical and numerological charter attesting to the legitimacy of the position of the ruler and of the dynasty that he founded.

This is not the only piece of numerology based on the number 2.5.9.0 at Palenque. The mother-goddess had a consort who was born shortly before her, at −8.5.0. Of their progeny, the second generation of gods and more or less triplets, one was born at 1.18.5.3.2 (as recorded in a further passage in the inscription of the Temple of the Cross), one at 1.18.5.3.6 (recorded in the first two passages of the inscription of the Temple of the Sun), and one at 1.18.5.4.0 (in the first two passages in the Temple of the Foliated Cross). The firstborn of these was the namesake of their sire, but the lastborn—a snake-footed deity—is the one who is calendrically and numerologically identified with him. (A scepter in the image of this deity became one of the principal symbols of authority at Palenque and at several other sites.) The interval from −8.5.0, which is the birth date of the sire, to 1.18.5.4.0, the birth date of the lastborn, is equal to 1.18.13.9.0, or 278,460 days, or $2^2 \cdot 3^2 \cdot 5 \cdot 7 \cdot 13 \cdot 17$. Here again are the factors of the Maya number 2.5.9.0, of which this interval is its seventeenth multiple.

The remaining birth dates of the second generation appear to be numerologically contrived also, in different ways and according to different desiderata, but subject to a further common condition that requires all three of them to be very near to each other, and, according to an early but still credible hypothesis,[21] to be within the calendar year in which the equinoxes (or solstices) are for the first time reversed in comparison with their calendrical positions in the year of the beginning of the era. These pieces of numerology will be passed over here—the one because of less than full certainty about it (chance cannot be entirely ruled out), and the other because of the complexity of the hieroglyphic evidence on which the demonstration depends. It will suffice to note that the latter establishes a special relationship of similarity between Pacal's son and immediate successor and the secondborn of this trio of gods. Instead of this, a simpler case from another inscription will provide the final example.

It will have been observed that the mother-goddess was some 760 years old at the time of giving birth to this brood. This happened before her "accession." A following passage in the inscription of the Temple of the Cross says that it was over two *baktuns*—some 800 years—between the date of her birth and the date of her "accession to rulership" on a day *9 Ik 0 Zac*. (The glyphic phrase for accession in this case is the same one that is employed in similar "birth to accession" statements, in later passages of the inscription, for historical rulers who acceded to power at normal ages for human beings. It should be noted also that female occupancy of royal title and office is not an anomaly at Palenque.) The *9 Ik 0 Zac* date is brought up again in an inscription from Temple XVIII, which records events in the life of a later ruler. The accession passage in this inscription states that it was 2.3.16.14 (a little over 43 years) from this ruler's birth to his accession, and that the latter event was 7.14.9.12.0 (something over 3,045 years) after the accession of the mother-goddess on *9 Ik 0 Zac*. The ruler's birth date is recorded in the initial series of the inscription as 9.12.6.5.8, *3 Lamat 6 Zac*. His accession date was 9.14.10.4.2, *9 Ik 5 Kayab*. That of the mother-goddess was 2.0.0.10.2, *9 Ik 0 Zac*. The Maya numeral that intervenes between these two accessions, 7.14.9.12.0, makes the interval equal to 1,112,280 days. Factored, this is $2^3 \cdot 3 \cdot 5 \cdot 13 \cdot 23 \cdot 31$, which contains the components of the almanac, the triple almanac (of possible relevance to Mars), and the Maya eclipse cycle (46 almanacs) as known from the much later Dresden Codex. It was not an eclipse date, however. The age of the moon is given in the initial passage as 19 days, that is, counted from appearance of the crescent (from conjunction it was approximately 22 days). Neither was it close to an eclipse season. The only apparent reason for an integral multiple of the 11,960-day eclipse cycle then is numerological. Something of the historical context can be sketched. An aging predecessor had either died or failed to rule to the end of his days (if dead, he was not yet buried). An also aging brother of his, five years his junior, was installed as successor a little over a year and a half before the date in question. Of this one there is no further record; he may not have lasted long, having been 71 years old when he replaced his brother. For whatever reason, quite possibly the death of the elderly recent successor, the time was at hand for the installation of someone of the next generation. Apparently an auspicious day was sought for the occasion, perhaps with some urgency in the face of a rival claim. None of the other good numbers when applied to an auspicious day of the past would produce one now, but this one did. It may have been seen as a safe place in the eclipse cycle and a good day in the almanac; but in any case, in this combination it had auspicious precedent in the accession of the ancient one. However good it may have looked, it was not good enough. This ruler's death is recorded as of a year later, and he was succeeded by another—six years his senior—in less than a year and a half after his own accession. In his successor's tablet he is treated rather curiously, as if an interloper.

VII. A TIME PERSPECTIVE

The earliest recorded dates that are indubitably Maya, and that are contemporaneous with the objects on which they are engraved, are at 8.12.14.8.5 and 8.14.3.1.12 in the day count. There are yet earlier ones recorded in the same count, also contemporaneous with their objects, going back as far as 7.16.3.2.13; but these are not "Maya" in any accepted sense of the term. The distinction hinges on several archaeological criteria and on the character of the associated hieroglyphs, whether these be in the tradition of Maya writing or in a different tradition, as well as on the question of whether there are reasons for assuming that the peoples who produced the objects were ethnically and linguistically ancestral to Maya peoples.

In order to put these dates and others into a more familiar frame of reference for history, it is necessary to adhere—at least with consistency if not yet entirely with conviction—to some one of the several hypotheses that have been advanced for a correlation between the Maya and the Christian chronologies. At an earlier point in this review, in section III, the need arose for approximate translations of Maya dates, and use was made of a correlation characterized there as "accepted by some." That was the one commonly known as the Goodman-Martinez-Thompson correlation, which is not just one, but is a family of successive revisions within a six-day range, positing from 584,280 to 584,285 as the number to be added to a Maya day number to convert it to the equivalent Julian day number. It began as a historically based correlation (rather than as an astronomical one), resting on interpretations of evidence in documents of the postconquest period. Goodman's original conversion figure was 584,280. At the present point in the understanding of the problem, the favored alternatives within this set are 584,283 (Thompson's third and last revision) and 584,285 (his first).

The greater of these, preferably increased by yet another day, is the only one of this set that would be consistent with the interpretation of moon ages and of dates in the eclipse table that has been presented here. Employment of this correlation will be continued in what follows, but it will not be carried to the point of specifying the precise day that is implied for each Maya date. These equations with Christian years should be regarded as tentative. If untrue, perhaps they are not so grossly untrue as seriously to distort the temporal relationships between Mesoamerican and Old World history. Most archaeologists at the present time are satisfied that the Goodman-Martinez-Thompson correlation provides dates that are suitably within the ranges established by their methods. Some others who have concerned themselves with the correlation problem, however, will wish to see a greater antiquity in the day-count dates than is given to them here.[22]

By this correlation the two early Maya dates cited above are from the years A.D. 292 and A.D. 320. If Maya "history" is to be counted from its oldest chronicles of events, then it begins with the accessions of two rulers in these years, the earlier from the ancient city of Tikal (in the Peten of present-day Guatemala), and the other from some location in the same sphere of cultural influence. The earlier record is on Stela 29 of Tikal; the second is on a small celt-shaped jadite pendant known as the Leiden Plaque (after the city in whose ethnographic museum it is preserved). The inscription of the Leiden Plaque has an "initial-series introducing glyph" that anticipates the classic form, having three of its four classic constituents, including the variable part appropriate to the month named further on in the series. It has bar-and-dot numerals prefixed to head-form period signs for the five orders of units in the day number (*baktuns*, *katuns*, *tuns*, *uinals*, and *kins*). And it has the proper almanac day, glyph *G*, and year day. But it has no lunar data. The full specification of the date is "8.14.3.1.12, *1 Eb*, G 5, *0 Yaxkin*." The date is followed by an accession statement, with the "seating" glyph and the name of the ruler who was seated on that date and who is depicted in symbolic regalia on the front side of the plaque. It is worthy of note that already in this second-earliest surviving Maya inscription the twentieth day of a month is designated as the day of the seating of the following month (transcribed as *0 Yaxkin*), and that the seating sign appears twice, with appropriately different suffixes, in two of the uses which it has in later inscriptions: for the seating of months and for the seating of rulers. (In the latter function it is one of five different but more or less synonymous glyphic expressions used in Classic Maya inscriptions for the installation of rulers.) Stela 29 of Tikal, despite its difference in size, is similar in composition as far as it is preserved, including the depiction of a ruler in his symbolic vestments on the front side; but the bottom portion of the stela is missing, being broken off just beyond the day number and through the trecena number of the almanac day. It is probable that this inscription was of a length and form comparable to that of the Leiden Plaque. The hieroglyphs in both inscriptions are of Maya form.

The day count, the 260-day almanac, and the 365-day year divided into twenty-day "months," all have a history that antedates their Maya use.[23] There are at least four pieces—one from Chiapas, two from Vera Cruz, and one from the southern highlands of Guatemala—that have earlier dates in the day count. The earliest (Stela 2 of Chiapa de Corso, Chiapas) is at 7.16.3.2.13, which would be in the year 36 B.C. by the correlation employed here. The next (Stela *C* of Tres Zapotes, Vera Cruz) has the date 7.16.6.16.18, corresponding to 31 B.C. Another (Stela 1 of El Baul, Guatemala) is at 7.19.15.7.12, which is in the year A.D. 36. The fourth (the Tuxtla Statuette, from San Andrés Tuxtla, Vera Cruz) has 8.6.2.4.17, in the year A.D. 162. Thus the day count was already in use for at least 328 years before the earliest surviving record of it that can be identified as Maya, and had quite a wide geographic distribution. All four of these earliest surviving instances employ bar-and-dot numerals. None of them employ signs for the periods, depending simply on place notation. In this they differ from the Classic Maya usage, but conform to that which appears again in the postclassic Dresden Codex. Thus the use of place values in numerals appears to precede the more explicit graphic expressions from which the convention might otherwise be assumed to have been derived. All four of the inscriptions specify the almanac days, and in these, all of them employ signs for the days in the veintena that are quite outside of the Maya tradition of day signs. In three of them, the almanac day follows the day number, but in one (El Baul) it precedes it. None of the four, so far as can be known, specified the year days; although because of breaks in two of the pieces the negative evidence is lacking in these. Two pieces (Stela *C* of Tres Zapotes and the Tuxtla Statuette) exhibit early forms of the "initial-series introducing glyph," even though neither of them has the chronological series that is so introduced in a position which is initial to the inscription. (It might thus better be called a "day-number introducing glyph.") Both introducing glyphs have a tripartite superfix similar (one very much so, and one somewhat less so) to that which is a standard part of this glyph in Classic Maya inscriptions. They also have instances of the "variable element" in them, one definitely and the other possibly corresponding to the appropriate calendrical twenty-day "months." Thus the introducing glyph, in at least two of its standard components, is pre-Maya. The inclusion of a variable element, especially the identifiable one in Stela *C* of Tres Zapotes, implies that the vigesimally based subdivisions of the calendar year were already instituted, even though they are not named in separate year-day specifications as they are in Maya inscriptions.

The 260-day almanac and the 365-day year, as well as bar-and-dot numerals, have a still older history. Inscriptions from the Valley of Oaxaca give evidence of these as far back as about the middle of the first millennium B.C. In these inscriptions, bar-and-dot numerals for days in the trecena accompany signs for days of the veintena; but they follow them, being written immediately below the day signs, rather than preceding them as prefixes or as superfixes as in all later usages. These inscriptions give evidence of the 365-day year by naming years for their "year bearers," the almanac days on which they begin. These, restricted to four days of the veintena, but combining with all positions in the trecena, name the 52 years of the calendar round and fix dates within this cycle. This is a practice for which there is no evidence in Classic Maya inscriptions, but for which there is good evidence in all three Maya codices (Dresden, Paris, and Madrid), as well as in Central Mexican sources, and which was very much in evidence in Yucatan and elsewhere at the time of the conquest and until recently. It is an element of Mesoamerican calendrical practice which has had a history of approximately two and a half millennia. It is uncertain whether the Oaxacan inscriptions give evidence for vigesimally based subdivisions of the year. Numerals higher than thirteen in certain contexts have suggested that possibility, but they are open also to other possible interpretations.

The early history of the Mesoamerican numerical and calendrical usages may be outlined roughly as follows, where the dates refer to the *earliest* reliable attestations of each usage. The first two groups of items are pre-Maya; after A.D. 292 they are Maya.

ca. 500 B.C.: Bar-and-dot numerals
260-day almanac
365-day year
Year-bearer names for years

36/31 B.C.: Day-count chronology
Place notation in numerals
Day-count introducing glyph
20-day months

A.D. 292/320: Period signs
Year days (month and day)
9-day cycle of Glyph *G*

A.D. 357: Lunar data

A.D. 668: 819-day cycle

The earliest secure evidence for the recording of lunar data is in an inscription from Uaxactun (north of Tikal), bearing a date of 8.16.0.0.0, which would place it in the year A.D. 357. It records a moon age (25 days), a position in the half-year (no. 1), and a duration (29 days). The lunar calendar, including a concept of lunar half-years (presumably eclipse half-years), and the classic pattern for recording lunar data were thus established before this date. The moon age was apparently reckoned from appearance of the crescent, for an attempt to reckon from conjunction would have yielded an age of 27 days for this date. It is possible that the lunar calendar and the concept of half-years were considerably older than this date. There is an early polychrome vase, also from Uaxactun, which has a recorded date with seven *baktuns* and which also records a moon age, moon number in the half-year, and a moon duration; but the date is inconsistent both with the lunar data and with the calendar-round position ascribed to it. If it really was from before the completion of eight *baktuns*, then the origin of the lunar calendar, with its half-years and the pattern for recording lunar data, would have to be attributed to a time prior to A.D. 42.

The 819-day cycle apparently did not enter the picture until about the middle of the classic period. The earliest evidence for it is from Palenque (Chiapas) in a stucco glyph panel commemorating an event in the life of the ruler Pacal in 9.11.15.15.0, which would be in A.D. 668. The panel may have been placed there later, but probably no later than the next *katun*-ending, 9.12.0.0.0, or A.D. 672. The 819-day station was 9.11.15.11.11. An earlier one, also from Palenque (the Palace Tablet), is at 9.10.10.11.2 for an initial date of 9.10.11.17.0; but the panel records history up to 9.14.4.8.15 and was probably erected or dedicated about half a year later, at 9.14.5.0.0, in A.D. 716. There are earlier stations recorded, going all the way back to mythological antiquity (the earliest being 6.15.0 before the current era), but they are on monuments erected no earlier than 9.13.0.0.0. The last one known is at Quirigua (Stela *K*) with a station at 9.18.14.7.10 for an initial as well as dedication date of 9.18.15.0.0, in A.D. 815. The interest in this cycle, late to become manifest, appears thus to have lasted for only a century and a half. Only fourteen examples of it are known to the writer: eight from Palenque, three from Yaxchilan, two from Copan, and one from Quirigua.

The 405-lunation eclipse cycle, of 11,960 days, is known principally from the Dresden Codex, which is estimated to be from the late twelfth or early thirteenth century. The codex, however, is a copy or new recension of an earlier work. The prototype of that eclipse table may be assumed to belong to the epochal date 9.16.4.10.8, in A.D. 755, when its final *12 Lamat* was about fifteen days prenodal. The version appearing in the codex, however, is assumed to be from nine or ten eclipse cycles later, 10.11.3.10.8 or 10.12.16.14.18, in A.D. 1050 or 1083, when the then final *12 Lamat* was approximately at a node or at most a day postnodal. The earlier of these two possibilities, A.D. 1050, appears more likely at present. The bottom line of almanac days with a *13 Muluc* base, and the eclipse interval of 26,519 days from the *13 Muluc* date of 9.12.10.16.9 (represented as a ring-based day number, −12.11 + 9.12.11.11.0, on page 58 of the codex) to the *12 Lamat* date of 9.16.4.10.8, suggest knowledge of the eclipse cycle going back at least to 9.12.10.16.9, or A.D. 683. The eclipse cycle was employed at Palenque in a piece of numerology (in Temple XVIII, described here in section VI) to determine the accession day of a ruler. That accession day was 9.14.10.4.2, at the beginning of the year A.D. 722. The cycle was employed in another piece of numerology also at Palenque (in the Temple of the Sun, to which reference was made in section VI but which was not described there); and the so-called Palenque moon ratio of 81 moons to 2,392 days, which was used to predict moon ages for the mythological dates of 1.18.5.4.0, 1.18.5.3.6, and −6.14.0, is nothing more than the ratio in the eclipse cycle, of 405 moons to 11,960 days. The dedication date of the temples in which these appear was 9.13.0.0.0, or A.D. 692. It is thus clear that the eclipse cycle was fully formulated by the last decades of the seventh century A.D.

There is a question as to how much earlier it may have been known. The numbering of moons in half-years that are never greater than six months in length suggests that knowledge of eclipse seasons may be as old as that practice, which would put it back to before A.D. 357. But knowledge of seasons should not be equated with knowledge of the cycle. The earliest definite evidence for the latter is from A.D. 683 or 692.

The so-called period of uniformity in regard to moon numbering, with its inflexible six-month half-years from 9.12.15.0.0 to the beginning of its demise about 9.16.5.0.0, or roughly from A.D. 687 to 756, is a curious interlude in a period when the length of the cycle and the nature of its subdivisions must already have been well known. It was pervasive while it lasted; but its dominance was broken within seventy years. A reformulation of the arrangement of divisions, subdivisions, and half-years in the eclipse cycle appears to have been under way before 9.16.4.10.8, the Dresden Codex epochal date. At Yaxchilan a stela with an initial date of 9.16.1.0.0 (A.D. 752) has its date and its lunar data recorded twice, with the latter according to two different interpretations. On the side of the stela it is given with a moon number agreeing with what that date should have according to the uniform system (moon age 12 days, moon number 5, duration 30 days), but on the front it appears in the new interpretation (moon age 12 days, moon number 4, duration 29 days). The fact that the moon duration also differs in the new interpretation gives additional evidence that a new lunar calendar was being inaugurated. The moon number "4" in this interpretation is possible only if the last division of the eclipse cycle just prior to 9.16.4.10.8 was one of 47 months (or more), that is, seven (or more) half-years of six months and one of five months, and further, only if glyph *C*, the moon number, is the ordinal number of the current moon. There is no possibility that the moon number here can refer to the number of completed moons in the half-year. Whether this was a new interpretation of moon number in relation to that of the preuniform period cannot yet be known; but it was new, at least in one respect, in relation to the concept of moon number held during the period of uniformity. As for moon age, both recordings of this inscription give witness to an attempt to reckon it from a hypothetic conjunction (overshooting it, apparently, by something over half a day). It cannot have been reckoned from visibility of the crescent.

The discovery of the calendrical method for more refined long-range lunar computations, with compensation for the small error (about one-ninth of a day in the length of the eclipse cycle) as it is multiplied by repetition through long periods of time, cannot be dated with any certainty, because it is not known whether the late dates involved in its application were contemporary, already in the past, or were predictions into the future. This last, however, is least likely. Exclusion of this hypothetic possibility, if it be a cause of error, will result in error only on the conservative side, in not risking ascription of dates that may be too early. If the latest of the *3 Chicchan* dates discussed in section V is taken as a guide (10.7.4.3.5), the discovery might be placed at, or after, A.D. 972. If the latest of the set of dates in the codex that contains this be chosen (10.11.5.14.5), then it would be A.D. 1052 or thereabouts, which is just two years after the most probable epochal date of the eclipse table in the form which it has in the Dresden Codex, with its *12 Lamat* base at a nodal position in the cycle. If the latest date recorded in the Dresden Codex, another *12 Lamat* (10.19.6.1.8), which also may represent an application of the compensation method (it is five almanacs less than five eclipse cycles after the probable epochal date of the codical version of the table), then the date would be A.D. 1210, or not long thereafter, that is, about the presumed time of the codex itself.

If 10.11.3.10.8, in A.D. 1050, be the epochal date for the current version of the eclipse table, then it is quite in line for 10.10.11.12.0, in A.D. 1038, to have been the epochal date for the current Venus table, as was posited in section IV and as is most generally accepted. That was the date *after which* corrections to the Venus cycle were applied by periodic foreshortenings of the cycle in order to find a new base at the foreordained place in the almanac.[24] There is a notable similarity between the Venus corrective device, instituted as of 10.10.11.12,0, with its 13.0.0 and 6.9.0 foreshortenings (eighteen and nine almanacs respectively) preserving its *1 Ahau* base, and the corrective device for the eclipse and lunar-reckoning table, with periodic foreshortenings in amounts of 7.4.0 or 3.11.0 (ten and five almanacs respectively) preserving its *12 Lamat* base. They employ a common strategy, and may well have been products of the same school of calendrical astronomy, and of about the same time, in which case the invention of the latter would most plausibly be ascribed to the epochal date in the year A.D. 1050. The two current epochal dates, of the Venus table and of the eclipse and lunar-reckoning table, are within twelve years of each other. Before that time, a shifting base, allowing one-day foreshortenings of the cycle as needed, must be assumed for the eclipse table. This is in agreement with the interpretation given in section V to the three successive lines of almanac days in that table.

The Mesoamerican day count, which was in use as early as 36 B.C., disappeared also early from the non-Maya and from the highland Maya areas, and its continuation was principally in the hands of the lowland Maya peoples. It did not survive, however, to the coming of Europeans. (If it had, there would be no "correlation problem.") The last stone monuments erected with dates in the day count were 10.3.0.0.0, in a year that we are supposing to be A.D. 889. The last surviving use of the count engraved on stone was on a jade piece with the date of 10.4.0.0.0, in A.D. 909. But the count was still employed in other media, at least until the time of the compilation or copying of the Dresden Codex, whose last recorded date—presumably more or less contemporary—was 10.19.6.1.8, in A.D. 1210. There is no further trace of it, although employment of its numerical system persisted in other contexts. The day count had by this time been superseded by a method that grew out of one that had coexisted with it almost from the beginning of the Maya use of the count. This was the one exhibited in the anchoring of secondary dates, by relating them to period-endings—principally *katun*-endings—as described in section III. So long as the complete calendar-round specification of a period-ending was employed, including its positions in both the almanac and the calendar year, the resulting specification was unique and unambiguous within a span of 18,720 calendar years, or 18,980 chronological years. But the naming of *katuns* by their ending days came to be abbreviated—even sporadically in inscriptions of the classic period—to include just the name of the almanac day. Suppression of the year day from the *katun* name became general at sites in the peninsula of Yucatan by the late decades of the ninth century A.D. The *baktun*, *katun*, *uinal*, and *kin* numberings were dropped. Of the five orders of units enumerated in a classic day number, only the *tuns* (the chronological years) continued to be numbered within the *katuns*. The *katuns* were known by their terminal *Ahaus* with trecena coefficients (diminishing by successive decrements of 2 as described in section III). And the days were specified by their positions in the calendar round, for which the year-day positions as well as the almanac days were still used. This system of characterizations apparently served adequately, according to the colloquial manner that had developed in the spoken language, for chronological orientation and for the record of historical events; and it is possible to translate such dates back into the day count. For example, a date in this manner, repeated three times in inscriptions at Chichen Itza, is expressed as: "Nine Lamat the day, on the eleventh of Yax, in the course of the thirteenth tun of One Ahau (*katun*)," which is equivalent to 10.2.12.1.8, *9 Lamat 11 Yax*, in the classic manner of statement.[25] The year is A.D. 881. Dates expressed in this late manner have the same degree of specificity as had those related by distance numbers to fully expressed period-endings, that is, they are unique within a cycle of 18,720 calendar years, or 18,980 chronological years, which is equivalent to 360 calendar rounds, or 949 *katuns*, or 2.7.9.0.0.0. The earliest date in this manner was from A.D. 743. The rest are from between A.D. 867 and 916.

By the end of the classic period the day count was on its way to abandonment, at least for civil use and commemorative records and monuments. In the hands of the astronomers, however, it survived for some time. Its last recorded date in the Dresden Codex, from A.D. 1210 (we are presuming), has already been mentioned. It may have been understood, and possibly even used, by copyists and possessors of codices into later times, but no record survives to give evidence. The Paris Codex, believed to be of only slightly later date than the Dresden, has interscribed numbers in various colors, several of them five positions in length. It cannot be shown that these are dates, but they are quite certainly intervals, some of them apparently with astronomical import. The Madrid Codex, estimated to be a product of the mid-fifteenth century, has a single instance of what may be a five-place numeral (with ten *baktuns*), but again there is no assurance that it is a date. One can only wonder how much of the chronological and astronomical content of the Dresden Codex was understood by its last Maya possessor. What survived of the chronological system after that, in the postconquest conflations of history and prophecy known as *Books of Chilam Balam* (written in romanized Maya), is the naming and ordering of the *katuns* by their numerically qualified *Ahaus*, with occasional additional precision lent by placements of dates either in the almanac or in the European calendar and Christian years.

The 260-day almanac survives today in some parts of Mesoamerica, not only in certain Maya communities, but also in a few of other linguistic areas. In no place is it common knowledge, but is the more or less guarded knowledge of religious specialists, curers, and diviners. The 365-day year calendar, with its twenty-day months and five-day residue period, also survives in some communities both within and outside of the Maya area. But in the years immediately following the conquest this became anchored to the European calendar, so that its days no longer drift through the seasons. Its anchoring is not by conscious intercalation of a day in leap years, but rather by equating a particular day of their calendar year with a local patron saint's day in the Roman church calendar, the fixing of which is out of their control. These anchorings may have taken place at different times in the early postconquest years, since there is variability in amounts up to a week or more in the equations of the native calendars with the European calendar from one community to another. There is also variability as to where the five-day residue is inserted into the sequence of the eighteen months. Knowledge of the year calendar, where it survives, is more public and less guarded than knowledge of the almanac is in its places of survival. There are few if any communities where both are still known. Where the almanac survives, it represents an almost unbroken continuity from the ancient past. Their agreements are very close, to within a day or two when not to the precise same day, even in widely separated areas. The implication is that there has been very little of losses or slippage during the more than two millennia of its existence.

The outline of early history of Mesoamerican numerical and calendrical usages tabulated earlier in this section, extending from *ca.* 500 B.C. to A.D.

668, can be supplemented now with the following outline of later developments:

A.D. 683/692: Early evidence for knowledge of eclipse cycle

687: Beginning of period of "uniform" moon numbering, in disregard of eclipse seasons

752: First evidence for reformulation of lunar calendar

755: Epochal date of eclipse table with 15-day-prenodal *12 Lamat* base (9.16.4.10.8)

756: Beginning of abandonment of "uniform" system of moon numbering

815: Last recognition of 819-day cycle

867–916: Dates in the "Yucatecan" method (an isolated early example in A.D. 743)

889: Last stone monuments with dates in the day count (10.3.0.0.0)

909: Last smaller object (jade) with inscribed date in the day count (10.4.0.0.0)

1038: Epochal date for Venus table as appearing in Dresden Codex (10.10.11.12.0)

1050: Probable epochal date of eclipse table as appearing in Dresden Codex, with approximately nodal *12 Lamat* base (10.11.3.10.8)

1129: First application of corrective device for Venus calendar (10.15.4.2.0)

1210: Latest day-count date recorded in Dresden Codex (10.19.6.1.8)

1536: First anchoring of Maya calendar year to European year

Today: Sporadic survivals of 260-day almanac and of Maya 365-day year with 18 twenty-day "months" and five-day residual period

It is perhaps superfluous to add that this should be taken as no more than one person's attempt at a chronology of developments in Maya calendrics and astronomy, that it is incomplete in many respects, and that the dates rashly ventured here are held subject to revision.

NOTES

1. The primary historical source from the early colonial period is Diego de Landa's *Relación de las cosas de Yucatán* (*ca.* 1566), Tozzer, translator and annotator (1941).

   Sources for Maya inscriptions are too numerous to be listed here. The principal collection has been Maudslay's *Archaeology* (1889–1902). The compilation of a *Corpus of Maya Inscriptions* has only recently been inaugurated, and four installments, by Graham and Von Euw (1975, 1977), have appeared. For other sources, see the list and the references at the end of Thompson's *Catalog of Maya Hieroglyphs* (1962), 404–424.

   The codex that is our principal source for Maya astronomical knowledge is that of Dresden. For its history, and for a description of the codex and its various published editions, see the essays by Deckert and Anders that accompany *Codex Dresdensis . . .* (Graz, 1975). See also the essay by Thompson (1972), 3–27, accompanying his edition and commentary. The first publication of the Dresden Codex was by Lord Kingsborough (1831), after drawings by Agostino Aglio, printed and hand-colored. The editions by Förstemann (1880, 1892; reissued 1964) are the only photographically based color reproductions of the codex in its condition prior to water damage suffered during World War II. The edition by Villacorta and Villacorta (1930; reissued 1976), which includes also the Paris and the Madrid Codices, is from black-and-white hand-drawn copies of photographs that are remarkably true to the originals. The Venus and the lunar tables reproduced as figures in the present essay are from that edition. The edition by Gates (1932), in color, is no attempt at facsimile rendition but is a complete recasting in Gates's own personal but very legible style of drawing. Thompson's edition (1972) is from recolored black-and-white photographs of Förstemann's 1892 edition. The "Graz" edition, *Codex Dresdensis . . .* (1975), in color, is from photographs of the codex in its present damaged state. A slightly retouched reprint of the Villacorta drawings accompanies it, offering a comparison and compensating for details now lost. The essays by Deckert and Anders, cited above, are in the volume that contains this reprint.

   The Paris Codex, known also as Codex Perez or Codex Peresianus, was published by de Rosny (1887, 1888), Gates (1909), Villacorta and Villacorta (1930), and by the Akademische Druck- und Verlagsanstalt of Graz (1968). There are editions of the Madrid Codex, known also as the Codex Tro-Cortesianus, by Brasseur de Bourbourg (the Troano portion, 1869–1870), de Rosny (the Cortesiano portion, 1883), de la Rada y Delgado (the Cortesiano portion, in color, 1892), Villacorta and Villacorta (1930), Gates (black-and-white photographic, 1933), and the Akademische Druck- und Verlagsanstalt (color photographic, 1967). Fragments of eleven pages from a fourth codex, containing a portion of a Venus table, have been published by Coe (1973), 150–154.

2. The Cachuianã numerals are from Vinhaes (1944), 197–202. The Bororo examples are from Colbacchini and Albisetti (1942), 282–283, Rondon and Faria (1948), 25, Albisetti and Venturelli (1962), I, 800–801, 866–870, and from the author's field notes. Iroquois information derives from the author's field studies. Two of the important Yucatec Maya sources are Beltrán de Santa Rosa (1746, 1859) and Pío Pérez (1866–1877). For the numerous other sources on Yucatec, as well as on the other languages of the Mayan family, see the *Bibliography of Mayan Lan-*

guages and Linguistics compiled by Lyle Campbell and collaborators (in press). An early collection of comparative data on Mayan and other Mesoamerican numeral systems, both verbal and graphic, is by Thomas (1900).

3. In the citation of Yucatec Maya forms it has been necessary to choose between fidelity to the sources quoted and fidelity to the known character of the language. The latter would have required *hoo* and *caa* in several of the cited examples, or retranscription of them as *ho'* or *ho'o* (depending on context) and similarly *ca'* or *ca'a*.
4. Of the numerous introductions to Maya calendrics the following should be mentioned: Bowditch (1910), Teeple (1930), 33–64, and Thompson (1950), 66–262.
5. LaFarge and Byers (1931), 157.
6. The spellings employed here for the Yucatec month names are those from Landa's *Relación de las cosas de Yucatan* (*ca.* 1566; see Tozzer [1941]). Although they are inadequate in many respects as transcriptions of Maya, they are traditional and serve well enough as labels for calendrical entities and for the hieroglyphs named by them. The writer sees no point in applying halfway corrective measures to their spellings, as some have done, when the resulting forms are still faulty anyway. A linguistic treatment of the terms is beyond the scope of this essay.
7. The determination of the values of the glyphs of the "supplementary series" is an interesting chapter in the history of decipherment. The lunar character of a portion of the series was anticipated by Morley in 1916 and the significance of glyphs *A* and *C* was recognized. The meanings of glyphs *D* and *E* were determined by Teeple (1925). The significance of glyphs *G* was apprehended by Thompson in 1929, when he determined that they were essentially a set of nine (although with variants) and discerned in them a Maya analogue to the nine Aztec lords of the night. The 819-day cycle was discovered in 1943, also by Thompson. Additional attributes of the cycle were determined by Berlin and Kelley (1961), and further examples of its representation in inscriptions have come to light since then. (See references in Bibliography.)
8. The first demonstration of historical content in Maya inscriptions was by Proskouriakoff (1960); see also (1961), (1963), and (1964). Prior to her discoveries a prevailing view among Mayanists was that the inscriptions focused on the "journey of time" as an object of religious veneration; cf. Thompson (1950), 64–65, 155. A wholly erroneous view prevailed concerning the function of the "secondary series," that is, of passages containing dates and distance numbers within a text; see Morley (1946), 289–291. This derived from one of the few serious errors in Teeple's remarkable work, his theory of "determinants" (1930), 70–85; see also Thompson (1936) and (1950), 317–320. In spite of the now demonstrable and predominant historical content in the inscriptions, it must still be granted that the Maya concern with time was in some considerable part a religious matter. Even in modern times the twenty named days of the veintena, or the 260 of the almanac, have been regarded as deities and addressed in prayers of supplication; see La Farge and Byers (1931), 153–173.
9. La Farge and Byers (1931), 158.
10. See Bowditch (1910), 324–331.
11. Cf. Bowditch (1901), Zimmermann (1935), Lizardi Ramos (1939), and Thompson, *Maya Arithmetic* (1941).
12. Important previous studies of the Venus tables are by Teeple (1930), 94–98, and Thompson (1950), 217–229. It was Teeple who discovered the significance of the alternative bases and the nature of the corrective procedures. Barthel (1952) has dealt with the iconography, hieroglyphs, attributes of the deities named in the table, and the structural relations symbolized by them.
13. Unavoidably the "period" mark is employed in two different senses in the punctuation of numerals, depending on whether it appears in a Maya or an Arabic numeral. In the columns headed "Error in Days" and "Residual Error" it is the decimal point, whereas in the second and third columns it is a place divider for Maya numerals. Elsewhere the context of the numeral or the magnitudes of its parts suffice to make clear the nature of the numeral and the interpretation of the period.
14. Some previous studies and commentaries on this table are those of Förstemann (1906), 200–215, Meinshausen (1913), Guthe (1921), Willson (1924), Teeple (1930), 86–93, Makemson (1943), 187–208, Satterthwaite (1947), 142–147, and Thompson (1950), 232–236; (1972), 71–77; (1975), 88–90. A brief review of contributions prior to 1950 is given in Thompson (1950). Teeple's study is of particular importance. It was he who first isolated the nodes of the table and who posited the method by which the Maya might have defined the eclipse seasons. Meinshausen's paper is important for its demonstration of the role that observations of lunar eclipses could have had in the construction of the table.
15. See Aaboe (1972) for apparent instances of employment of a 135-lunation cycle in the prediction of eclipse possibilities in ancient Old World astronomy.
16. See Willson (1924) for an account of the testing of this hypothesis.
17. The ecliptic limit for solar eclipses varies between a minimum value of 15.4 degrees and a maximum value of 18.5 degrees; that for lunar eclipses varies between 9.5 and 12.1 degrees; see Smart (1965), 398. For present purposes, "degrees" (of the earth's angular distance in its orbit from a node) may be translated approximately into "days" by multiplying these figures by 0.96 days per degree.
18. See Makemson (1943). A somewhat distorted version of Makemson's hypothesis is attributed to her by Thompson (1950), 234; and (1972), 73.
19. The possibility of this method of correction was anticipated by Thompson in 1950 and was developed in 1972 and 1974, but only in respect to the five-almanac foreshortening which is not applicable to eclipse reckoning. See Thompson (1950), 235–236; (1972), 74–75; (1974), 89–90.
20. For fuller treatment of ring numbers and ring-based day numbers in the Dresden Codex, and of two of the Palenque cases described here, see Lounsbury (1976). Other instances of Palenque numerology are not yet in print.
21. Bowditch (1906).
22. For reviews of the correlation problem see the following, listed chronologically: Teeple (1930), 104–109, 115; Thompson (1935); Andrews (1940); Thompson (1950), 306–310; Satterthwaite and Ralph (1960); Satterthwaite (1965), 625–631; and Kelley (1976), 30–33. Further bibliography pertinent to the numerous proposed correlations will be found in these.

   It will be observed that the interpretations of the eclipse and Venus tables put forward in the present essay, as these affect the correlation problem, are more in accord with Thompson's views than with any of those that are seriously at variance with his. For example, see criteria nos. 6, 7, and 8 in the review by Kelley. In the matter of reckoning moon ages, however, the interpretation here concurs better with Thompson's earlier views (1935) than with his later opinion (1950), 310, or with Satterthwaite's (1965), 629–630. The historical problem that motivated their shift to a 584,283 correlation constant, and to a reevaluation of the meaning of moon ages, appears now to have another resolution and is less compelling.

   The suggested one-day increment to Thompson's first constant, giving 584,286 rather than 584,285, has the advantage that it permits acceptance of the best apparent candidate for the recording of a solar-eclipse observation on a stone monument. The item came to Teeple's attention only in time to be noted in a one-page addendum to his *Maya Astronomy* (1930). The event, recorded on Stela 3 of Santa Elena Poco Uinic, Chiapas, Mexico, is dated

9.17.19.13.16, which is day number 1,425,516 in the Maya count, and would be JD 2,009,802 or July 16 (Julian) A.D. 790 by the 584,286 hypothesis. This is the date of eclipse no. 4768 in Oppolzer, a total eclipse whose path of centrality swept across this precise location and whose conjunction time was approximately 48 minutes after 12:00 noon, local time at 91.8 degrees west longitude. The record is on a katun-ending monument with initial date 9.18.0.0.0, which records events of the preceding period concerning the local ruler. These are badly eroded, so that neither their precise dates nor their nature can be established; but the last event, the presumed eclipse, with both its glyph and its distance number connecting it to the initial date, is well preserved due to having been below ground. It has still to be proven, however, that the glyph designating this event is indeed a "solar eclipse" glyph. It has the "sun" sign flanked by a pair of over-arching appendages, the arrangement being formally similar to that of the eclipse glyphs of the codices, and the flanking appendages having the form of the principal element of glyph *B* of the lunar series that accompanies initial dates of monumental inscriptions.

A disadvantage, not only of the 584,286 hypothesis but of the entire Goodman-Martinez-Thompson set of alternatives, is that it requires all of the apparent "eclipse" dates of the Dresden Codex to be products of computation rather than records of observation. Neither the *12 Lamat* date 9.16.4.10.8 nor the *13 Muluc* date 9.12.10.16.9 would have been eclipse dates by this correlation, although both would have been half a month prior to dates of total lunar eclipses, and one month prior to partial solar eclipses. Thus, of the three dates on page 52*a* of the Dresden Codex that are separated by fifteen-day intervals, the first would not have been an eclipse date; the second would have been within about half a day of a lunar eclipse, but one visible only in the eastern hemisphere; and the third would have been within a day of a partial solar eclipse, but one not visible in Central America. As attempted computations of eclipse-possible times, however, all of them are reasonable.

The great strength of the Thompson correlation lies in its accommodation of data from postconquest historical sources as well as twentieth-century survivals of the almanac, while still accommodating computational if not observational astronomy.

23. For more detailed reviews of the early dated monuments see Coe (1957) and (1976), and Marcus (1976). Acceptance of the evidence for the pre-Mayan dates was slow in coming; compare the earlier views of Morley (1946), 41, and of Thompson, *Dating of Certain Inscriptions* (1941), in regard to these monuments, and in particular in regard to the reconstruction by Stirling (1940) of the 7-baktun date of Stela C of Tres Zapotes. Any lingering doubts were laid to rest in 1973 with the discovery of the missing portion of that stela, with seven baktuns, as had been predicted from the trecena number together with the lower four places in the day number, thus proving the correctness both of the date and of Stirling's assumption that it belonged to the same day count as do Maya dates.
24. The first such application was toward the end of that Venus great cycle, on 10.15.4.2.0, in A.D. 1129.
25. The decipherment of dates expressed in this manner was by Thompson (1937); see also (1950), 197–203.

BIBLIOGRAPHY

Asger Aaboe, "Remarks on the Theoretical Treatment of Eclipses in Antiquity," in *Journal for the History of Astronomy*, **3** (1972), 105–118; César Albisetti and Ângelo Jayme Venturelli, *Enciclopédia Bororo*, I: *Vocabulários e etnografia* (Campo Grande, Brazil, 1962); E. Wyllys Andrews IV, "Chronology and Astronomy in the Maya Area," in *The Maya and Their Neighbors* (New York, 1940; repr. Salt Lake City, 1962), 150–161; Thomas Barthel, "Der Morgensternkult in den Darstellungen der Dresdner Mayahandschrift," in *Ethnos*, **17** (1952), 73–112; Pedro Beltrán de Santa Rosa, *Arte de el idioma Maya* (Mexico City, 1746; 2nd ed., Mérida, 1859); Heinrich Berlin and David H. Kelley, "The 819-Day Count and Color-Direction Symbolism Among the Classic Maya," in *Middle American Research Publication* **26** (New Orleans, 1961), 9–20; Charles P. Bowditch, *A Method Which May Have Been Used by the Mayas in Calculating Time* (Cambridge, Mass., 1901); *The Temples of the Cross, of the Foliated Cross and of the Sun at Palenque* (Cambridge, Mass., 1906); and *The Numeration, Calendar Systems and Astronomical Knowledge of the Mayas* (Cambridge, Mass., 1910); Charles Étienne Brasseur de Bourbourg, *Manuscrit troano. Études sur le système graphique et la langue des Mayas*, 2 vols. (Paris, 1869–1870).

Lyle Campbell *et al.*, *Bibliography of Mayan Languages and Linguistics* (Albany, N.Y., in press); Michael D. Coe, "Cycle 7 Monuments in Middle America: A Reconsideration," in *American Anthropologist*, **59** (1957), 597–611; *The Maya Scribe and His World* (New York, 1973); and "Early Steps in the Evolution of Maya Writing," in Henry B. Nicholson, ed., *Origins of Religious Art and Iconography in Preclassic Mesoamerica* (Los Angeles, 1976), 107–122; Antonio Colbacchini and César Albisetti, *Os Boróros Orientais Orarimogodógue do planalto oriental de Mato Grosso* (São Paulo, 1942); Ernst W. Förstemann, *Commentary on the Maya Manuscript in the Royal Public Library of Dresden*, Papers of the Peabody Museum of American Archaeology and Ethnography, vol. IV, no. 2 (Cambridge, Mass., 1906); and *Die Mayahandschrift der königlichen öffentlichen Bibliothek zu Dresden* (Leipzig, 1880; 2nd ed., Dresden, 1892; reissued, 1964); William E. Gates, *The Dresden Codex, Reproduced From Tracings of the Original, Colorings Finished by Hand* (Baltimore, 1932); and *The Madrid Maya Codex* (Baltimore, 1933), a black-and-white photographic ed.; Ian Graham and Eric Von Euw, *Corpus of Maya Hieroglyphic Inscriptions*, 2 vols. (Cambridge, Mass., 1975–1977); Carl E. Guthe, *A Possible Solution of the Number Series on Pages 51 to 58 of the Dresden Codex*, Papers of the Peabody Museum of American Archaeology and Ethnography, vol. VI, no. 2 (Cambridge, Mass., 1921); David H. Kelley, *Deciphering the Maya Script* (Austin, Tex., 1976); Edward K. Kingsborough, "Facsimile of an Original Mexican Painting [Dresden Codex] Preserved in the Royal Library at Dresden, 74 Pages," in *Antiquities of Mexico*, III (London, 1831); Oliver La Farge II and Douglas Byers, *The Year Bearer's People* (New Orleans, 1931); César Lizardi Ramos, "Computo de fechas Mayas," in *Actas, XXVII Congreso internacional de americanistas* (Mexico City, 1939; repr. Vaduz, 1976), I, 356–359; Floyd G. Lounsbury, "A Rationale for the Initial Date of the

Temple of the Cross at Palenque," in Merle Green Robertson, ed., *The Art, Iconography, and Dynastic History of Palenque*, pt. III of *Proceedings of the Segunda Mesa Redonda de Palenque 1974* (Pebble Beach, Calif., 1976), 211–224; Maud W. Makemson, *The Astronomical Tables of the Maya*, Carnegie Institution of Washington, Contributions to American Anthropology and History no. 42 (Washington, D.C., 1943), 183–221; Joyce Marcus, "The Origins of Mesoamerican Writing," in *Annual Review of Anthropology*, **5** (1976), 35–67; Alfred P. Maudslay, *Archaeology*, 6 vols. (London, 1889–1902); M. Meinshausen, "Über Sonnen- und Mondfinsternisse in der Dresdner Mayahandschrift," in *Zeitschrift für Ethnologie*, **45** (1913), 221–227; Sylvanus G. Morley, *The Ancient Maya* (Stanford, Calif., 1946); and "The Supplementary Series in Maya Inscriptions," in *Holmes Anniversary Volume* (Washington, D.C., 1916), 366–396.

Theodor von Oppolzer, *Canon der Finsternisse* (Vienna, 1887), translated by Owen Gingerich as *Canon of Eclipses* (New York, 1962); Juan Pío Pérez, *Diccionario de la lengua Maya* (Mérida, 1866–1877); Tatiana Proskouriakoff, "Historical Implications of a Pattern of Dates at Piedras Negras, Guatemala," in *American Antiquity*, **25** (1960), 454–475; "The Lords of the Maya Realm," in *Expedition*, **4** (1961), 14–21; and "Historical Data in the Inscriptions of Yaxchilan," in *Estudios de cultura Maya*, **3** (1963), 149–167, and **4** (1964), 177–201; Candido Mariano da Silva Rondon and João Barbosa de Faria, *Esbôço gramatical e vocabulário da língua dos índios Borôro* (Rio de Janeiro, 1948); Linton Satterthwaite, *Concepts and Structures of Maya Calendrical Arithmetics* (Philadelphia, 1947); and "Calendrics of the Maya Lowlands," in Robert Wauchope, ed., *Handbook of Middle American Indians*, III (Austin, Tex., 1965), 603–631; Linton Satterthwaite and Elizabeth K. Ralph, "Radiocarbon Dates and the Maya Correlation Problem," in *American Antiquity*, **26** (1960), 165–184; William M. Smart, *Text-Book on Spherical Astronomy*, 5th ed. (Cambridge, 1965); Matthew W. Stirling, *An Initial Series From Tres Zapotes, Vera Cruz, Mexico* (Washington, D.C., 1940).

John E. Teeple, "Maya Inscriptions (I): Glyphs C, D, and E of the Supplementary Series," in *American Anthropologist*, **27** (1925), 108–115; "Maya Inscriptions (II): Further Notes on the Supplementary Series," *ibid.*, 544–549; *Maya Astronomy*, Carnegie Institution of Washington. Contributions to American Archaeology, vol. I, no. 2 (Washington, D.C., 1930), 29–116; Cyrus Thomas, *Numeral Systems of Mexico and Central America*, Smithsonian Institution, Bureau of American Ethnology, Annual Report, pt. 2 (Washington, D.C., 1900), 853–955.

J. Eric S. Thompson, "Maya Chronology: Glyph G of the Lunar Series," in *American Anthropologist*, **31** (1929), 223–231; *Maya Chronology: The Correlation Question*. Carnegie Institution of Washington, Contributions to American Archaeology, vol. III, no. 14 (Washington, D.C., 1935), 51–104; "The Dates of the Temple of the Cross, Palenque," in *Maya Research*, **3** (1936), 287–293; *A New Method of Deciphering Yucatecan Dates With Special Reference to Chichen Itza*, Carnegie Institution of Washington. Contributions to American Archaeology, vol. IV, no. 22 (Washington, D.C., 1937), 177–197; *Dating of Certain Inscriptions of Non-Maya Origin*, Carnegie Institution of Washington, Division of Historical Research, Theoretical Approaches to Problems, vol. I (Cambridge, Mass., 1941); *Maya Arithmetic*, Carnegie Institution of Washington, Contributions to American Anthropology and History no. 36 (Washington, D.C., 1941), 37–62; *Maya Epigraphy: A Cycle of 819 Days*, Carnegie Institution of Washington, Notes on Middle American Archaeology and Ethnology, vol. I (1943), 137–151; *Maya Hieroglyphic Writing: An Introduction* (Washington, D.C., 1950; 2nd ed., Norman, Okla., 1960; 3rd ed., 1971); *A Catalog of Maya Hieroglyphs* (Norman, Okla., 1962); "A Commentary on the Dresden Codex," which is *Memoirs of the American Philosophical Society*, **93** (1972); and "Maya Astronomy," in *Philosophical Transactions of the Royal Society*, **A276** (1974), 83–98.

Alfred M. Tozzer, *Landa's Relación de las cosas de Yucatan*, which is Papers of the Peabody Museum of American Archaeology and Ethnography, vol. XVIII (Cambridge, Mass., 1941); J. Antonio Villacorta and Carlos A. Villacorta, *Códices mayas, reproducidos y desarrollados . . .* (Guatemala, 1930; reissued, 1976); Ernesto Vinhaes, "Vocabulário Cachuianã," in *Aventuras de um repórter na Amazônia* (Pôrto Alegre, Brazil, 1944), 197–202; Robert W. Willson, *Astronomical Notes on the Maya Codices*, Papers of the Peabody Museum of American Archaeology and Ethnography, vol. VI, no. 3 (Cambridge, Mass., 1924); Günter Zimmermann, "Einige Erleichterungen beim Berechnen von Maya-Daten," in *Anthropos*, **30** (1935), 707–715.

See also *Codex Peresianus: Manuscrit hiératique des anciens Indiens de l'Amérique Centrale . . ., publié en couleurs, avec une introduction, par Léon de Rosny* (Paris, 1887; 2nd ed., 1888)—cover has title *Codex Peresianus: Manuscrit yucatèque; Codex Perez, Maya-Tzental. Redrawn and Slightly Restored, and With the Coloring As It Originally Stood. . . . Accompanied by a Reproduction of the 1864 Photographs; . . . Drawn and Edited by William E. Gates* (Point Loma, Calif., 1909); *Codex Peresianus (Codex Paris)* (Graz, 1968), with introduction and summary by Ferdinand Anders.

Codex Cortesianus: Manuscrit hiératique des anciens Indiens de l'Amérique Centrale, conservé au Musée Archéologique de Madrid, photographié et publié pour la première fois, avec une introduction et un vocabulaire de l'écriture yucatèque par Léon de Rosny (Paris, 1883), in black and white; *Codice Maya denominado Cortesiano que se conserva en el Museo Arqueológico Nacional (Madrid). Reproducción fotocromolitográfica ordenada en la misma forma que el original. Hecha y publicada bajo la dirección de D. Juan de Dios de la Rada y Del-*

gado y D. Jerónimo López de Ayala y del Hierro, Vizconde de Palazuelos (Madrid, 1892); *Codex Tro-Cortesianus (Codex Madrid)* (Graz, 1967), facsimile reproduction with introduction and summary by Ferdinand Anders.

Codex Dresdensis: Maya Handschrift der Sächsischen Landesbibliothek Dresden (Berlin, 1962); *Codex Dresdensis . . ., vollständige Faksimilieausgabe des Codex im Originalformat* (Graz, 1975), with commentary by Helmut Deckert and Ferdinand Anders and reproduction of the drawings of the Dresden Codex from Villacorta and Villacorta (Graz, 1975).

DICTIONARY
OF
SCIENTIFIC BIOGRAPHY

PUBLISHED UNDER THE AUSPICES OF
THE AMERICAN COUNCIL OF LEARNED SOCIETIES

The American Council of Learned Societies, organized in 1919 for the purpose of advancing the study of the humanities and of the humanistic aspects of the social sciences, is a nonprofit federation comprising forty-five national scholarly groups. The Council represents the humanities in the United States in the International Union of Academies, provides fellowships and grants-in-aid, supports research-and-planning conferences and symposia, and sponsors special projects and scholarly publications.

MEMBER ORGANIZATIONS

AMERICAN PHILOSOPHICAL SOCIETY, 1743
AMERICAN ACADEMY OF ARTS AND SCIENCES, 1780
AMERICAN ANTIQUARIAN SOCIETY, 1812
AMERICAN ORIENTAL SOCIETY, 1842
AMERICAN NUMISMATIC SOCIETY, 1858
AMERICAN PHILOLOGICAL ASSOCIATION, 1869
ARCHAEOLOGICAL INSTITUTE OF AMERICA, 1879
SOCIETY OF BIBLICAL LITERATURE, 1880
MODERN LANGUAGE ASSOCIATION OF AMERICA, 1883
AMERICAN HISTORICAL ASSOCIATION, 1884
AMERICAN ECONOMIC ASSOCIATION, 1885
AMERICAN FOLKLORE SOCIETY, 1888
AMERICAN DIALECT SOCIETY, 1889
AMERICAN PSYCHOLOGICAL ASSOCIATION, 1892
ASSOCIATION OF AMERICAN LAW SCHOOLS, 1900
AMERICAN PHILOSOPHICAL ASSOCIATION, 1901
AMERICAN ANTHROPOLOGICAL ASSOCIATION, 1902
AMERICAN POLITICAL SCIENCE ASSOCIATION, 1903
BIBLIOGRAPHICAL SOCIETY OF AMERICA, 1904
ASSOCIATION OF AMERICAN GEOGRAPHERS, 1904
HISPANIC SOCIETY OF AMERICA, 1904
AMERICAN SOCIOLOGICAL ASSOCIATION, 1905
AMERICAN SOCIETY OF INTERNATIONAL LAW, 1906
ORGANIZATION OF AMERICAN HISTORIANS, 1907
AMERICAN ACADEMY OF RELIGION, 1909
COLLEGE ART ASSOCIATION OF AMERICA, 1912
HISTORY OF SCIENCE SOCIETY, 1924
LINGUISTIC SOCIETY OF AMERICA, 1924
MEDIAEVAL ACADEMY OF AMERICA, 1925
AMERICAN MUSICOLOGICAL SOCIETY, 1934
SOCIETY OF ARCHITECTURAL HISTORIANS, 1940
ECONOMIC HISTORY ASSOCIATION, 1940
ASSOCIATION FOR ASIAN STUDIES, 1941
AMERICAN SOCIETY FOR AESTHETICS, 1942
AMERICAN ASSOCIATION FOR THE ADVANCEMENT OF SLAVIC STUDIES, 1948
METAPHYSICAL SOCIETY OF AMERICA, 1950
AMERICAN STUDIES ASSOCIATION, 1950
RENAISSANCE SOCIETY OF AMERICA, 1954
SOCIETY FOR ETHNOMUSICOLOGY, 1955
AMERICAN SOCIETY FOR LEGAL HISTORY, 1956
AMERICAN SOCIETY FOR THEATRE RESEARCH, 1956
SOCIETY FOR THE HISTORY OF TECHNOLOGY, 1958
AMERICAN COMPARATIVE LITERATURE ASSOCIATION, 1960
AMERICAN SOCIETY FOR EIGHTEENTH-CENTURY STUDIES, 1969
ASSOCIATION FOR JEWISH STUDIES, 1969

DICTIONARY
OF
SCIENTIFIC BIOGRAPHY

CHARLES COULSTON GILLISPIE

EDITOR IN CHIEF

Volume 16

INDEX

CHARLES SCRIBNER'S SONS · NEW YORK

Editorial Staff

Panel of Consultants

Index Staff

JULIA A. McVAUGH, *Director*

Assistants: ELIZABETH G. BROWNRIGG

FRANCINE M. HUNNICUTT

JANE R. LUDINGTON

JAMES F. MAURER

EVA C. METZGER

JUDITH M. SAMS

JO ANNE G. WEST

Contents

A NOTE ON THE INDEX	xiii
INDEX	1
CONTRIBUTORS	441
SOCIETIES INDEXED	471
PERIODICALS INDEXED	475
LISTS OF SCIENTISTS BY FIELD	479

A Note on the Index

In compiling the Index the following criteria have been adopted:

People. In the case of passing references to persons who are subjects of articles, all but the most trivial are included, in the belief that someone looking in the *Dictionary* for complete information on a scientist may want to follow all personal encounters and influences. References to other persons are indexed unless they are isolated or insignificant. References to living historians of science are omitted, unless there is significant discussion of their work.

Periodicals. References to these are indexed when information on the founding, editorship, policies, or influence of the periodical is given. A separate listing of these periodicals will be found in this volume.

Societies. Only significant mentions are indexed. Foreign societies given English names in the text are indexed in the English form. A separate listing of these societies appears in this volume.

Universities, museums, medals, lectureships, prizes. These are indexed only when significant information is given. For example, individual awards of the Nobel Prize are not indexed, since complete lists are readily available elsewhere; but discussions of Nobel committees, disputes, and policies are included.

Countries. Scientific topics are listed under individual countries only when the specifically national characteristic of a science is discussed.

Bibliographies. Two types of references are indexed from the bibliographical sections of the biographies:

(1) Bibliographical information on the subject of another article that does not appear in that subject's main article.

(2) Historical information on scientific topics not mentioned in the biographical section of the article.

FORM OF REFERENCES

General. Page references are in the form "**3**:271a"; the volume number is in boldface, followed by a colon and the page number; "a" indicates the left-hand column and "b" the right-hand column. An asterisk following a page number indicates a reference to the Bibliography section; "(n)" indicates a reference to the Notes section.

Subjects of articles. The name of the subject of an article, dates of birth and death, and the location of the biography appear in boldface; when a further biographical note exists in the supplement volume, this reference is also in boldface.

ALPHABETIZATION

Entries are arranged alphabetically following the word-by-word method (up to a comma); apostrophes and hyphens are ignored, as are diacritical marks. Abbreviations—DNA, for example—are treated as single words. When a spelling is both common noun and proper noun, the order is: persons, places, things.

Plate, Ludwig Hermann	Heat
Plate of heavens (astron. instrum.)	Heat, animal
Plate tectonics	Heat, innate
Platearius, Matthaeus	Heat engine

Names with prefixes are treated as single words:

Defoe, Daniel
De Forest, Lee
Deformation, geological

Names beginning with *Mc, St., or Ste.* are listed as though they were spelled out:

MacDonald, Donald Laurie
McDonald Observatory
McDougall, Alexander
Macellama
Macewen, William

A Note on the Index

Medieval and royal personages. Given names are generally listed in the English form: Henry rather than Heinrich, William rather than Guillaume. These names have been alphabetized by the next major element, ignoring "of" and "the." Their listing precedes that of the same name used as a surname, headed by saints and monarchs:

Henry VIII of England	Peter I (the Great) of Russia
Henry (Aristippus)	Peter of Alexandria
Henry Bate of Malines	Peter Bonus
Henry, Aimé	Peter of Dacia
	Peter Deacon

Arabic names. "Ibn" and "al-" have been ignored in alphabetizing, whether at the beginning of the name or between major elements. The diacritical marks ʿ and ʾ have also been ignored:

Abū Saʿīd Shādhān
Abu'l-Ṣalt Umayya
Ibn Abu'l-Shukr
Abū Sulaimān

Chinese names. The family name is treated as a separate word:

Li Yen-Wen
Liapunov

German names. The umlaut has been ignored in alphabetizing:

Schönherr, Carl Johan
Schonland, Basil Ferdinand Jamieson
Schönlein, Johann Lucas

Russian names. The transliteration of Russian names is generally according to the Library of Congress system, omitting diacritical marks. Cross-references are given when necessitated by the text. Names of subjects of articles, however, are indexed in the form in which they appear in the text.

Subheadings. Alphabetization is by principal words, ignoring prepositions:

Hydrogen
  absorption
  active
  in air
  atomic volume

The listing of subheadings deviates from strict alphabetical order in two situations:

(1) When a science has chronological subheadings, these are listed first; other subtopics follow in alphabetical order.

(2) When the subject of an article has written many books, the listing of these in alphabetical order follows the alphabetical listing of other subtopics.

ORGANIZATIONAL METHODS

Subheadings. Chronological divisions are delimited as follows:

ancient Greek	300 B.C.–A.D. 100
ancient Roman	100 B.C.–A.D. 400
Byzantine	A.D. 300–A.D. 600
medieval (Arabic, Hebrew, Latin)	500–1450
Renaissance	1450–1600
early/late seventeenth century	1600–1650/1650–1700
early/late eighteenth century	1700–1750/1750–1800
early/mid-/late nineteenth century	1800–1830/1830–1870/1870–1900
early/mid-twentieth century	1900–1930/1930–1970

It must be emphasized that "ancient Greek," "ancient Roman," and "Byzantine" are used primarily as chronological rather than cultural divisions.

Topical subheadings. These have been dictated more often by the text than by a predetermined scheme. The reader is advised to scan all subheadings under a given topic or person to ensure finding the particular information sought.

Biological nomenclature. Plants and animals are referred to in the text both in Latin and in English, and they are indexed in the language used in the text. The reader should look under both names when searching for particular biological species.

Scientific congresses and meetings. These are not indexed under their individual titles, but are grouped by scientific field under the heading "Congresses, scientific."

Expositions and exhibitions. These will be found under the cities in which they took place.

Broad categories. Entries such as "Biology" and "Mathematics" have been avoided as far as possible. The reader is encouraged to search for information on a more specific level—for example, "Embryology" or "Geometry," or, even more specifically, "Gastrulation" or "Triangle." Individual chemical compounds, biological species, atomic particles, etc., are listed when significant information is given, and cross-references are provided to lead the reader from general topics to more specific ones.

Errata. Readers are advised to consult the Errata section to determine whether there are entries that affect the subjects in which they are interested.

My appreciation is due to the members of the Editorial Board for their advice and encouragement; to the staff of the Louis Round Wilson Library, University of North Carolina at Chapel Hill, for many investigations cheerfully undertaken; to the authors and other scholars both here and abroad who have responded to requests for information; and especially to Michael McVaugh for services and support at all stages of this enterprise.

JULIA A. MCVAUGH

DICTIONARY
OF
SCIENTIFIC BIOGRAPHY

DICTIONARY OF SCIENTIFIC BIOGRAPHY

INDEX

A

A.A.A.S.
see American Association for the Advancement of Science
Abacus, 2:233b; 3:214a; 4:605a,b; 5:365a, 365b–366a; 6:302a; 7:418b; 12:94a; 14:542a; 15:10a–b, 747a–b
Abailard, Pierre (1079–1142), 1:1–4; 1:276a,b
philosophy, 7:330b
theology, 14:153a
universals, 6:377a
Abano, Pietro d' (1257–*ca.* 1315), 1:4–5; 5:448a; 13:349b
astronomy, 15:35b
Conciliator, 1:620b; 8:54a
on Dioscorides, 4:121b
translations by, 5:233b; 6:274b*
Abastumani Astrophysical Observatory, 10:159a
ᶜAbbās ibn Firnās (*d.* A.D. 887), 1:5
al-ᶜAbbās ibn Saᶜīd al-Jawharī
see al-Jawharī
Abbasid culture, 1:32b; 2:148b; 4:438b–439a, 555b; 7:64a, 187b, 222b, 223a; 9:36a, 160a; 13:229b, 230b, 538a
Abbe, Cleveland (1838–1916), 1:6; 4:593a*; 5:139a*; 8:1a, 2b; 9:108a*; 11:163a; 12:210a*
Abbe, Ernst (1840–1905), 1:6–9
acquaintances of, 7:401b; 12:211b, 423a; 14:633b; 15:124b
correspondence of, 10:134b*
influences on, 5:144a; 7:404a; 12:255b
instrumentation, 2:575b; 7:422b, 448a; 11:208a
Abbot, Charles Greeley, 6:106a
Abbot, John, 12:133b; 14:417a,b
Abbott, Alexander Crever, 3:447a
Abbott, Leonard, 2:459b
ᶜAbd al- or ᶜAbdallah
see last element of name
Abdera, 8:269a
Abderhalden, Emil, 2:585a,b, 586a; 4:90b; 5:208b; 11:129a,b
Abdomen
anatomy of, 1:619a, 620b; 3:99b
bacterial flora, 3:518a
dilation of veins, 3:490a
disorders of, 11:99b
muscles of, 10:598b
tumors of, 2:464a
Ibn ᶜAbdūn of Seville, 15:175a
Abegg, Richard Wilhelm Heinrich, 8:290a; 13:40b
Handbuch der anorganischen Chemie, 2:429a; 3:334a–b; 11:551a
Abeille, Louis-Paul, 15:334a
Abel, Frederick Augustus, 4:78b; 6:461b
Abel, John Jacob (1857–1938), 1:9–12; 1:443a; 10:205b; 14:249b, 496a; 15:99b, 104b*
Abel, Niels Henrik (1802–1829), 1:12–17; 3:466b; 5:96b
Abelian equations, 1:16b
Abelian functions
see Functions (math.), Abelian
Abelian integrals, 2:85a; 3:315a; 5:260a, 263b, 471b, 482a; 6:214b; 7:398b, 477b, 478b, 479a; 10:25b*, 244b, 595a*; 11:54b, 548a–b; 14:223b
Abelian transcendentals, 4:341a
Abel-Ruffini theorem, 1:15a; 11:598b
Abel-Tauber theorem, 6:113b; 13:263b
and Académie des Sciences, 3:134b
acquaintances of, 5:94b; 7:550b; 13:14b, 128a
algebra, 1:207a; 2:468a; 3:169a; 5:151a, 304b; 7:103a; 8:139a; 13:571b; 14:630b
analysis, 2:276a; 3:137a, 228b; 4:205b; 5:309b, 572b; 6:92b; 7:52a,b, 478b; 8:383b; 10:139b; 11:449a; 13:238a; 15:386a
on Cauchy, 3:133a, 135b
correspondence of, 11:599a
editions of works, 6:474a; 8:324a; 13:216a
and Galois, 3:168a; 5:260a–263b *passim*
and Holmboe, 1:13a, 14a, 15a, 16a; 6:474a
influence of, 2:105a; 3:228a; 7:109b; 10:249b; 14:220a
and Jacobi, 1:16b–17a, 17b; 7:51a, 52a,b, 53a, b
and Legendre, 1:15b, 16b–17a; 8:138a, 141a
theorems of, 3:315a, 467b, 560a; 6:307a, 547b; 7:53a; 8:141a; 12:150a; 13:81b
generalization of, 3:141b; 14:202a–b
translations of works, 8:458b
Abel, Othenio (1875–1946), 1:17–18; 4:148a*; 7:61b*, 480a; 14:203b*
Abel, Rudolf, 8:450a
Abelard, Peter
see Abailard, Pierre
Abenare
see Ibn Ezra
Abenguefit
see Ibn Wāfid
Abenmoat
see al-Jayyānī
Aberdeen University
formation of, 1:404a
Abernethy, John, 2:483a; 6:567b; 8:97a,b, 98a; 10:260b; 14:128a
Aberration
chromatic, 1:9a, 135a,b; 3:284a; 4:128a, 482a; 6:603b; 7:110b; 10:60a; 11:583b*; 12:413b
of electron-optical systems, 15:8b
theories of, 9:118b
constant of, 2:99a, 469a*, 616b; 8:369b; 9:514b; 12:449a; 13:112a*, 139b*
spherical, 1:7b; 3:98a; 5:143a; 6:195a,b, 603b; 7:214b, 404a; 12:250b
laws of, 8:460a
stellar, 1:201b; 2:100b; 3:283b–284a; 4:168a, 322b; 5:363a, 466a, 491b; 6:347a, 512b; 8:493a; 11:379a; 15:338b
constants of, 3:374b
discovery of, 2:387b; 5:25b; 9:464a; 11:526a
naming of, 9:77b
tables of, 2:388b
and theories of light, 2:329a; 8:493a; 9:214a–b, 215a, 220b, 267a, 372b; 13:76b, 104a
Abetti, Antonio (1846–1928), 1:19; 4:162a; 8:504b*; 13:232b
Abhandlungen aus dem mathematischen Seminar der Hamburgischen Universität
founding of, 2:191b
al-Abharī, Athīr al-Dīn, 4:440b; 12:91a; 13:510b, 514b
Abhayadeva Sūri, 1:309b; 12:597b
Abhomeron
see Ibn Zuhr, Abū Marwān ᶜAbd al-Malik ibn Abi'l-ᶜAlāᵓ
Ibn Abī Uṣaybiᶜa al-Saᶜdī al-Khazrajī, 1:538b; 4:541b; 6:189b, 190a, 199b; 7:223b; 9:602b; 11:238a–b; 12:1b, 90a,b; 14:584a; 15:230b, 234b, 238b, 261b
ᶜUyūn al-anbāᵓ, 15:230a, 231a,b
Abich, Otto Hermann Wilhelm (1806–1886), 1:19–21
Abies, 9:390a
Abiogenesis
see Spontaneous generation
Ablation, 5:44b, 208a; 6:441a, 518b; 8:536a, 552b; 9:20b, 177a; 12:355b, 473b–475a; 15:109a
Abney, William de Wiveleslie (1843–1920), 1:21–22; 9:201b
photography, 12:248b
Åbo
observatory at, 1:241a–b
Åbo, University of
chemistry at, 5:215b
Aboaly
see Ibn Sīnā
Abortion
contagious, 12:482b
opposition to, 9:68b
Abraham
see also Ibrāhīm
Abraham bar Ḥiyya ha-Nasi (*fl.* before 1136), 1:22–23; 1:31b, 512b; 3:150a; 8:280b
Calculation of the Courses of the Stars, 13:400b

Abraham bar Ḥiyya ha-Nasi (cont.)
Ḥibbur ha-meshīḥah we-ha-tishboret, **1**: 227a
translations of, **4**:609a; **11**:31b–32b
translations by, **11**:31b–32a
Abraham ibn Dā᾽ūd, **5**:592a
Abraham ben Jacob
see Ibrāhīm ibn Ya^cqūb
Abraham ben Meir ibn Ezra
see Ibn Ezra
Abraham Hebraeus, **6**:197b
Abraham Judaeus
see Abraham bar Ḥiyya ha-Nasi *and* Ibn Ezra
Abraham, Edward Penley, **5**:43a
Abraham, Karl, **5**:178a, 179a; **7**:190b
Abraham, Max (1875–1922), 1:23–25; **5**:64a; **6**:348a; **8**:10b; **11**:475b
electron theory, **2**:559b; **4**:319b; **5**:597b; **7**:263b–264a; **8**:496b, 497a,b; **14**:218a
equations of, **11**:13b
gravitation theory, **7**:26b
Theorie der Elektrizität, **1**:24a
Abreu, Aleixo de (1568–1630), 1:25–26
Abrin, **4**:297a–b, 298a
Abscesses, **3**:218b
drainage of, **8**:405b
indolent, **8**:403b
Absolute temperature, **6**:470a; **13**:378a
concept of, **5**:461b–462a
and radiation theory, **11**:150b
Absolute zero, **8**:2a; **15**:94b, 95b
attainment of, **3**:620a
concept of, **1**:138b
definition of, **6**:292a–b
entropy, **15**:439a–448a
measurement of, **3**:316a; **8**:1a
Absorptiometry, **2**:588a; **12**:271a
Abstraction, logical, **10**:485a; **15**:410b–411a, 412a–b
extensive, **14**:307a–b
Absynthium, **1**:526a
Abu'l-ᶜAbbās Aḥmad ibn Muḥammad ibn ᶜUthmān al-Azdī
see Ibn al-Bannā᾽ al-Marrākushī
Abu'l-ᶜAbbās al-Ma᾽mūn
see al-Ma᾽mūn, Abu'l-ᶜAbbās
Abū ᶜAbd Allāh ibn Makhlūf al-Sijilmāsī, **1**:437a
Abū ᶜAbdallāh, Muḥammad ibn Muᶜadh
De crepusculis
translations of, **15**:180a–b
Abū ᶜAbdallāh Muḥammad ibn ᶜAbd al-Raḥīm
see Abū Ḥāmid al-Gharnāṭī
Abū ᶜAbd Allāh Muḥammad al-Tijnarī
Zahr al-bustān wa-nuzhat al-adhhān, **1**:350b
Abū Aflaḥ ha-Saraqosṭī, **7**:38a
Abū Aḥmad al-Nahrajūrī, **15**:250a
Abu'l-ᶜAla᾽ Zuhr, **14**:637b, 638b*
Abū ᶜAlī ibn Abi'l-Ḥusayn al-Ṣūfī, **13**:150a
Abū ᶜAli al-Ḥasan ibn al-Jīlī, **9**:84a
Abū ᶜAlī al-Ḥasan al-Milyānī, **1**:437b
Abū Bakr al-Ḥasan ibn al-Khaṣīb, **11**:32a
Abū Bakr Muḥammad ibn al-Ḥasan al-Karajī, al-Ḥāsib
see al-Karajī
Abū Bakr Muḥammad ibn ᶜUmar ibn al-Farrukhān, **13**:538a
Abū Bakr al-Qallūsī (al-Fār), **1**:437a
Abū Bakr al-Rāzī
see al-Rāzī, Abū Bakr Muḥammad ibn Zakariyyā (Rhazes)
Abū Bakr al-Ṭurṭūshī, **1**:29b
Abu'l-Barakāt al-Baghdādi, Hibat Allāh (*ca.* 1080–*ca.* 1164/1165), 1:26–28; **12**: 91b
dynamics, **1**:410b
Kitāb al-muᶜtabar, **13**:510a
philosophy, **11**:326a
Abu'l-Barakāt al-Balāfiqī **1**:437b
Abū Bishr Mattā, **1**:275a,b; **15**:261b
Abu'l-Faraj Yūḥannā ibn al-ᶜIbrī al-Malaṭī (Bar Hebraeus), **1**:539a; **4**:442a, 542a; **9**:555a
Candelabrum Sanctorum, **10**:111b
Abu'l-Fidā᾽ Ismāᶜīl ibn ᶜAlī ibn Maḥmūd ibn . . . Ayyūb, ᶜImād al-Dīn (1273–1331), 1:28–29, **11**:248a
Taqwīm al-Buldān, **9**:36b
Abū Ḥāmid al-Gharnāṭī (1080–1169), 1:29–30; **11**:231a
Abū Ḥanīfa
see al-Dīnawarī, Abū Ḥanīfa
Abu'l-Ḥasan ᶜAlī, **14**:593a,b
Abu'l-Ḥasan ᶜAlī ibn Abi'l-Rijāl, **1**:513b
Abu'l-Ḥasan ᶜAlī of Granada, **1**:408b
Abū'l-Ḥasan ᶜAlī ibn Hārūn al-Zanjānī, **15**:250a
Abu'l-Ḥasan Thābit ibn Ibrāhīm al-Ḥarrāni, **2**:619a
Abū Jaᶜfar ibn Ḥabash, **5**:612a
Abū Jaᶜfar Hārūn al-Tajālī, **12**:1b
Abū Jaᶜfar al-Khāzin
see al-Khāzin
Abū Jaᶜfar Yūsuf ibn Ḥasday
see Yūsuf ibn Ḥasday, Abū Jaᶜfar
Abu'l-Jūd
see Ibn Layth, Abu'l-Jūd Muḥammad
Abū Kāmil Shujāᶜ ibn Aslam ibn Muḥammad ibn Shujāᶜ (*ca.* 850–*ca.* 930), 1:30–32; **4**:541b
algebra, **4**:611b, 612a; **7**:241a, 244a; **15**: 256a
Kitāb fī al-jabr wa'l-muqabala, **12**:91b
translations of, **15**:187b
Abu'l-Kasim
see Abu'l-Qāsim
Abū al-Khayr Khayr Allāh Khān, **7**:81a
Abu'l-Khayr al-Shajjār, **1**:350b
Abū Māhir Mūsā ibn Sayyār, **9**:40b; **13**:229a
Abū Manṣūr al-Baghdādī, **11**:229b–230a
Abū Maᶜshar al-Balkhī, Jaᶜfar ibn Muḥammad (787–886), 1:32–39; **14**:537b, 575b
acquaintances of, **15**:261a
astrology, **2**:155a; **6**:272b–273a; **7**:82b, 362b; **9**:160a
astronomy, **9**:168a
commentaries on, **12**:431b
on contemporary scientists, **13**:538a
Great Introduction to the Science of Astrology, **15**:173b
translations of, **15**:175b
al-Hazārāt, **7**:338a
influence of, **1**:91b, 382a; **5**:361a, 402a; **7**:187a; **9**:363b; **11**:322a; **14**:388a
influences on, **15**:125a
Kitāb al-ulūf, **7**:222b
translations of works, **1**:62a, 165b; **6**: 272b, 273a
Zīj
critics of, **7**:334b
Abū Muḥammad ibn Abi'l-Baghl, **15**:250a
Abū Naṣr al-Fārābī
see al-Fārābī
Abū Naṣr Manṣūr
see Manṣūr ibn ᶜAlī ibn ᶜIrāq, Abū Naṣr
Abu'l-Qāsim ᶜAlī ibn Aflaḥ, **7**:38a
Abu'l-Qāsim al-Anṭāqī, **4**:438a
Abu'l-Qāsim Khalaf ibn ᶜAbbas al-Zahrāwī
see al-Zahrāwī
Abu'l-Rayḥān Muḥammad ibn Aḥmad al-Bīrūnī
see al-Bīrūnī
Abū Saᶜd al-ᶜAlā ibn Sahl, **11**:240a
Abū Sahl al-Faḍl ibn Nawbakht, **1**:33a
Abū Sahl al-Masīḥī, **9**:84a
Abū Saᶜīd Shādhān
see Shādhān, Abū Saᶜīd
Abu'l-Ṣalt Umayya, **1**:290a; **4**:439b
Ibn Abu'l-Shukr
see Muḥyī al-Dīn al-Maghribī
Abū Sulāimān al-Manṭiqī al-Sijistānī, **7**:39b
Abū Sulaymān ibn Maᶜshar al-Bustī al-Maqdisī, **15**:250a
Abū ᶜUbayd al-Bakrī
see al-Bakrī
Abū ᶜUmar ibn Ḥajjāj
al-Muqniᶜ, **1**:350b
Abu'l-ᶜUqūl, **14**:578a, 579a
Abū ᶜUthmān al-Dimashqī, **1**:39a, 275b; **4**:438a, 439b, 445b; **10**:300a; **15**: 243a(n)
Abu'l-Wafā᾽ al-Būzjānī, Muḥammad ibn Muḥammad ibn Yaḥyā ibn Ismāᶜīl ibn al-ᶜAbbās (940–997/998), 1:-39–43
arithmetic, **7**:532a,b; **13**:546a; **15**:9b
astronomy, **2**:149a; **5**:613a; **11**:240a
commentaries of, **4**:117a, 438a; **7**:241a
geometry, **4**:596b; **9**:157a
translations into Arabic, **1**:31a
trigonometry, **7**:38a,b, 352b, 532a; **9**:84b–85a, 555b; **11**:229a; **13**:510b
Abū Yaᶜqūb Yūsuf, **12**:1b, 2a–b, 8a
Abū Yūsuf al-Rāzī, **4**:438b
Abū Yūsuf Yaᶜqūb al-Qaṣrānī, **1**:33a
Abū Zayd ᶜAbd al-Raḥmān . . . al-Lajā᾽ī, **1**:437b
Abulcasis
see al-Zahrāwī
Abunazar
see al-Fārābī
Abuteus (Andreas) Levita, **9**:361a, 362a
Abyssal plains, **10**:564a
Abysses, maritime
fauna in, **1**:72a
Academia Caesarea Leopoldina Carolina Naturae Curiosorum, **10**:12a–b
Academia Leopoldina Naturae Curiosorum, **5**:366b
Academia Naturae Curiosorum Sacra, **10**: 567b
Academia Operosorum (Ljubljana), **5**:366b
Academia Parisiensis, **2**:353b; **4**:46a; **6**:112b; **9**:316b, 599a; **10**:330b, 342a; **12**:585a; **14**:368a,b
Academia Secretorum Naturae
see Accademia dei Segreti
Academic Assistance Council, **13**:227b
Académie Bourdelot
see Bourdelot, Pierre Michon, academy of
Académie de Médecine (Paris)
in nineteenth century, **10**:385b, 388a, 389b, 391a,b, 404b, 405a; **11**:100a
Académie des sciences, Paris
in seventeenth century, **2**:201a, 312a, 349a, 592b; **3**:63b; **4**:221b; **5**:259a–b, 525a; **6**:599a; **8**:162a; **9**:114a–b, 116b; **10**:145a; **11**:423a,b; **12**:585b
in early eighteenth century, **3**:281a; **4**: 214b; **5**:352a; **9**:33b; **10**:320a; **11**: 334a–b; **12**:117b, 128a–b
in late eighteenth century, **1**:58b, 76b; **2**:74a; **3**:5a–b, 72a, 108b, 114b, 440a,b; **4**:23b; **5**:92a; **7**:581b; **8**:74b, 84a; **9**:471b–472a; **11**:523a; **12**:63b; **15**:276a–b, 302b, 312a–b, 317a, 341b–342a
in early nineteenth century, **1**:15b, 17a,b, 45a; **3**:134b, 440b, 545b; **4**:16a; **5**: 94a,b, 251b, 260b–261a, 411a; **15**: 361a, 484a

in mid-nineteenth century, **3**:287a; **4**:67b; **10**:196a
in late nineteenth century, **3**:135a; **4**:66a; **10**:404b; **13**:411a–b, 551b
in early twentieth century, **7**:549a
aerostatics, **8**:79a; **9**:343a
anatomy, **4**:267a; **5**:327b–328a
blood circulation experiments, **4**:37b; **5**:327b; **10**:477a
astronomy, **10**:615b; **14**:589b; **15**:310a
botany, **4**:135b; **9**:92b, 114a–b
chemistry, **2**:353a; **6**:236b, 477b–478a; **8**:55b; **12**:64a–65a
competitions of
early eighteenth-century, **2**:42b–43a, 343a; **3**:39a, 215b, 282b, 444a, 460a, 486a; **4**:468b; **8**:179a, 611a; **14**:84a
late eighteenth-century, **2**:43a–43b, 328b, 335a; **3**:71a, 440a; **4**:14b; **5**: 410b; **7**:562a–564b *passim;* **8**:113b; **9**:250a; **15**:331b
early nineteenth-century, **2**:600a; **3**: 132a, 258b; **4**:240a, 241a; **5**:95a, 168b, 375b–376a; **10**:545b; **11**:548a; **13**:311b, 402b, 403a; **15**:361b, 363a, 376b
mid nineteenth-century, **3**:331a, 612b; **4**:258a; **8**:62b, 254a,b, 389b; **9**:431a; **10**:367b; **13**:395a, 555b–556a; **15**:86b
late nineteenth-century, **4**:514b; **7**:478a; **9**:411b; **14**:38b
early twentieth-century, **11**:477b
electricity, **14**:79b
forerunners of, **9**:498b
founding of, **2**:201a, 349a, 592b; **3**:63b; **4**:221b; **6**:598b; **9**:92b; **10**:519b, 546b, 595b; **11**:486b; **13**:335a–b
geodesy, **3**:107b; **4**:603a; **7**:183a; **9**:554b; **13**:530b; **15**:270a–271b
and industry, **2**:201b; **3**:454b; **4**:87a, 224b*; **7**:581a; **8**:621b; **9**:62a; **11**: 328a–b
and Lavoisier, **3**:7a; **5**:138a; **8**:69b–71a, 74b, 84a
mechanics, **10**:443b
metrology, **3**:106a; **4**:14b–16b; **5**:491b; **7**:569a; **8**:85a; **9**:472a; **15**:333b–336a
natural history, **5**:356a
observatory
see Paris, observatory
Pascal-Newton controversy, **3**:214a
physics, **9**:119a
population studies, **15**:304b, 309a–b
prehistory, **15**:51a,b
publications of
Comptes rendus hebdomadaires des séances de l'Académie des sciences, **3**:132b; **4**:410b; **5**:44a
éloges, **5**:61a–b, 88a
histories, **5**:60b–61a, 434b–435a
Mémoires . . . , **3**:281b, 282b, 283a; **5**:435a, 510b; **15**:483a
Procès-verbaux, **15**:389b*–390a*
see also *Connaissance des temps, Descriptions des arts et métiers, Mémoires pour servir à l'histoire naturelle des animaux, Mémoires pour servir á l'histoire des plantes*
reorganizations of, **3**:199b, 440b–441a; **4**:16a; **5**:60b, 91b, 92a, 94a, 259a, 411a, 567b, 603a; **9**:473a
and the Revolution, **3**:440b–441a; **5**:91b; **15**:333a–336b, 389b*
spontaneous generation, **10**:372a, 382b; **11**:110b
technology, **9**:118a, 342b
Académie des Sciences, Belles-Lettres et Arts (Rouen), **5**:58b
Académie des Sciences et Beaux-Arts des États-Unis, **11**:564b–565a
Académie Française
in seventeenth century, **9**:497a–b
in eighteenth century, **1**:117a
in nineteenth century, **5**:94b; **15**:376b
competitions, **12**:469a; **13**:127a–b
Académie Internationale d'Histoire des Sciences, **9**:341b–342a
Académie Montmor
see Montmor, Henri Louis Habert de, Académie Montmor
Académie Royale (Paris), **4**:221a
Académie Royale de l'Architecture (Paris), **2**:201a,b
Académie Royale de Peinture et de Sculpture (Paris), **4**:49a
Académie Royale des sciences, des lettres et des beaux-arts de Belgique, **3**:213a; **10**:10a
Académie royale des sciences, Paris
see Académie des sciences, Paris
Académie Royale des Sciences et des Belles Lettres de Berlin
see Akademie der Wissenschaften (Berlin)
Academies, scientific
see listing under Societies, scientific
Academy (Plato's)
see Athens, Academy
Academy of Sciences, Turin, **2**:291b–292a
Academy of Sciences (USSR)
see Akademiia nauk USSR
Acanthaceae, **10**:13b
Acanthocephala, **1**:602b; **8**:270b
Acapnia, **9**:547a
Acariasis, **3**:259b, 332b
Acarida, **3**:180b
Acarina, **10**:149a
Acarus, **5**:566a
Accademia Cosmografica degli Argonauti, **3**:421a
Accademia degli Inquieti, **9**:77a, 485a, 511a; **13**:566b
Accademia degli Investiganti, **2**:306b, 309a, 311a
Accademia degli Oziosi, **11**:96b
Accademia dei Fisiocritici, **12**:517a
Accademia dei Lincei
see Lincei, Accademia dei
Accademia dei Percossi, **13**:434a, 438a
Accademia dei Segreti, **11**:96a–b
Accademia del Cimento, **2**:306b, 309a–310a, 311a, 592a,b, 593a;, **3**:101a; **4**:506a; **5**:55b; **9**:62b, 121a(n), 485a, 595b; **11**:341b, 411a; **13**:30b
founding of, **9**:3a
Saggi, **11**:404a
Accademia dell' Esperienza, **2**:312a
Accademia della Crusca, **2**:311b; **9**:3b; **11**: 342b; **13**:438a
Accademia della Fucina, **2**:307b, 308a, 311b
Accademia della Traccia, **9**:485a
Accademia delle Scienze dell'Istituto di Bologna, **9**:485a
founding of, **9**:135b
Accademia d'Italia (Rome), **4**:579a
Accademia Fisica-matematica, **2**:312a
Accademia Reale (Rome), **2**:312a
Accademia Scientifica-Litteraria (Milan)
founding of, **2**:470b
Acceleration, **1**:253a–b, 254b;, **3**:282b
complementary, **3**:418a
composite motions, **3**:418a
of falling bodies
medieval theories, **1**:28a, 101a; **2**:606a–b; **4**:157b; **5**:402b; **8**:229a; **10**:225b–226a, 421a
Renaissance theories, **1**:495b, 606a–b, 607b; **2**:590b; **8**:229a, 231b–233b
seventeenth-century theories, **1**:424a, 567a–b; **5**:239a, 240a–241a, 242b, 246a, 247a; **6**:610b; **8**:267b–268b
and earth's rotation, **6**:606b
gravitational
medieval study, **1**:94b; **2**:606a–607a
Renaissance study, **12**:136a
seventeenth-century study, **5**:158b; **6**: 607b
eighteenth-century study, **1**:326b
nineteenth-century study, **2**:102a; **4**: 379a, 380a; **5**:78b–79a; **7**:160a, 263a
twentieth-century study, **4**:577a
equivalence principle, **4**:324b–328b *passim*
variation of, **4**:380a
radial, **8**:159a
relative, **3**:418a
uniform, **5**:360b; **7**:487a
see also Falling bodies, Gravitation, Motion
Accelerators, particle, **1**:366a; **3**:329b–330a; **4**:582b; **5**:272a; **7**:155a, 527a; **8**:96a; **12**:33b; **13**:227a, 569b–570b; **14**:340b
circular
see Cyclotron
linear, **6**:104a–105a
Stanford, **3**:369b
tandem principle, **13**:570a,b
theory of, **13**:600b–601a
Van de Graaff, **2**:338b; **11**:460a; **13**:570a
Accident
ancient Greek discussion, **1**:250b–251a, 251b–252a, 277b, 278a
medieval discussion, **2**:232a
Accum, Friedrich Christian (1769–1838), 1: 43–44; **2**:336a, 420a
Accumulator, electrical, **11**:474a
theory of, **15**:435a
Acenaphthene, **2**:67b
Acetaldehyde, **10**:22b
condensation of, **7**:282b
Acetals, **8**:335b
formation of
reversibility of, **4**:20b
Acetic acid, **1**:65a
composition of, **5**:125b, 325a, 381a; **8**: 337b; **14**:530b
compounds of, **11**:87a
derivatives of, **7**:280a; **8**:520b
substitution, **3**:10b; **5**:373b
fermentation of, **10**:530b
formula for, **3**:289b, 449a; **8**:338b
heat effects, **2**:65b
isolation of, **8**:520b
neutralization, heat of, **11**:577a
synthesis of, **1**:286b; **2**:65b; **4**:134a; **7**: 452a; **8**:336a
vapor density, **3**:10b
Acetoacetic ester
derivatives of, **14**:455a
structure of, **14**:455a–b
synthesis of, **5**:381a–b
Acetobacter, **7**:406a
Acetone
early reference to, **1**:571b
preparation of, **1**:286b; **3**:232b; **4**:41b; **7**:224a; **8**:312a
microbiological, **14**:247b
properties of, **4**:41b
ring compound, **7**:224a
synthesis of, **5**:420b
Acetonuria
experimental, **8**:556b
Acetous acid, **1**:65a
Acetyl radical, **8**:337b–338a

Acetylcholine, **8**:454a–455a
physiological properties, **15**:106b–107a
Acetylene, **3**:299a
combustion of, **8**:118a
as disinfectant, **1**:575a
formula for, **4**:400a
for illumination, **2**:70a
isolation of, **2**:67b
manufacture of, **9**:451b
naming of, **2**:67b
polymers of, **3**:86b; **10**:122a
reactions of, **2**:67b; **10**:121b
synthesis of, **2**:67a–b; **14**:478a
see also Hydrocarbons, acetylenic
Achard, Franz Karl (1753–1821), 1:44–45; **4**:41b; **7**:394b; **9**:105b
Acharius, Erik (1757–1819), 1:45–46; **5**:191a
Acheulean (stratig. stage), **5**:163a; **8**:105a
Achilles Tatius, **6**:617b; **10**:180b, 590a,b; **15**:216a–b
Achillini, Alessandro (1463–1512), 1:46–47; **1**:617b; **4**:520b
In quaestione de motuum proportionibus, **2**:590b
Achorion, **5**:566a; **12**:203a
Achroödextrin, **2**:531b
Achúcarro, Nicolás, **11**:465b
Acid-alkali hypothesis, **2**:381a; **8**:344b; **13**:234b–235a
Acidity
definitions of, **2**:77a; **10**:81a
and electricity, **1**:348a
hydrogen theory of, **5**:372a; **8**:341a–b
measurement of, **3**:290b; **6**:109a; **13**:228a–b; **14**:196b; **15**:461b
oxygen theory of, **1**:142b–143a, 347b; **5**:323b; **8**:77a–b, 82b; **13**:312a
disproof of, **3**:602a–603a
scale of, **1**:347b–348a; **13**:228b
theories of, **2**:170b; **6**:108b, 109a, 395a; **8**:58a,b, 76a–77b, 293a, 338a–342b; **9**:244a, 245b; **10**:184b; **12**:64b; **14**:125a
see also pH
Acidosis, **13**:574b; **15**:73b
theories of, **6**:265a
Acids
affinity coefficients, **15**:458a–b, 459b
aminosulfon ("Cleve's acids"), **3**:321b
basicity of, **5**:372a–b
biliary, **3**:341b; **11**:129a
catalytic activity, **15**:461b–462a
characterization of, **2**:381a
composition of, **11**:146a
conversion to amines, **3**:511b
definition of, **2**:499a; **4**:239a; **5**:372a
gases produced, **6**:41a, 42a
heteropoly, **2**:199b; **11**:551a,b
hydracid, **5**:323b, 324b–325a, 372a
hydroxy, **4**:400a
inorganic
complex, **5**:394a; **12**:465a–b
isohydric solutions of, **1**:300a
isopoly, **11**:551b
Lewis acids, **8**:293a
manufacture of, **3**:201a; **8**:553b; **13**:559a
metallic, **5**:595b
mineral, **2**:381b; **3**:201a, 419a; **6**:236b; **11**:146a
distillation of, **8**:312a
synthesis of, **5**:420b
nature of, **8**:173b
normal acid, **5**:325b
oils, reaction with, **3**:419b
organic, **2**:92b; **3**:242a; **5**:488b; **7**:208a; **10**:499a,b; **11**:278a; **12**:146a, 148a–b; **13**:389a
affinities, **14**:124b
anhydrous, **14**:396a
composition of, **5**:125b; **7**:452a–b; **8**:60a, 338b–343b; **9**:425a
formulas of, **5**:371a,b
metabolism of, **13**:393b
molecular weights, **4**:76b
neutralization of, **11**:436b
in plants, **9**:461a
pyrogenic, **8**:339a, 340b, 341a
structure of, **4**:75b; **7**:280a, 281a; **8**:58a; **11**:92b
synthesis of, **4**:553a; **8**:76a, 520b; **9**:355a–b; **10**:516b; **13**:177a, 309b
oxaminated, **5**:100a
oxygenated, **13**:312a–b
polyacetic, **3**:347b
polybasic, **1**:168a; **2**:66a, 170b; **3**:289b; **13**:359a
concept of, **5**:493b
synthesis of, **13**:388b
polythionic, **14**:112a
pseudo acids, **6**:108b, 109a
reactions of
quantitative studies, **11**:435b–437a
reactivity of, **9**:305a
reduction of, **1**:390a
saturation capacity, **8**:341a–b
solutions of, **13**:547b
starch, effects on, **7**:378b
strength of, **1**:298b–299a; **2**:170b, 381a; **6**:109a, 478a; **7**:388b
tetronic, **1**:168b
theories of composition, **2**:76b–77a; **3**:289b; **4**:531b; **6**:283b*; **7**:603a; **8**:76b–77b; **9**:110a
unsaturated, **5**:13a; **7**:281a; **10**:578b
see also Acidity, Amino acids, Fatty acids, Nucleic acid, specific acids
Acidum pingue, **2**:177a; **8**:76b; **9**:346b; **11**:523a; **12**:65a; **14**:224b–225a, 333b*
Ackeret, Jakob, **8**:597b
Ackerknecht, Erwin Heinz, **2**:448b; **12**:241b–242a
Ackermann, Dankwart, **15**:105b
Acland, Henry, **1**:539b
Acne, **5**:30b*
Aconite, **15**:252b
Acosta, Cristóbal (*ca.* 1525–*ca.* 1594), 1:47–48
Tractado de las drogas y medicinas de las Indias orientales
translations of, **8**:120b
Acosta, José de (1539–1600), 1:48
Natural and Moral History of the Indies
translations of, **9**:123b
Acoustics
medieval, **5**:550b; **14**:155b
Renaissance, **3**:99b; **5**:249b
seventeenth-century, **1**:483b; **3**:153a; **4**:51b; **7**:376b; **9**:318a–319b, 486a; **10**:76a, 520b; **11**:41a
eighteenth-century, **3**:258a–b; **4**:88b, 468b; **5**:511a; **7**:561b, 565a, 599a; **9**:477a; **12**:127a–128b; **14**:565a; **15**:306b
early nineteenth-century, **3**:8b, 9a, 258b–259a; **4**:534b; **10**:183a; **14**:204a, 206a, 289b; **15**:356b–358a, 484a
mid nineteenth-century, **4**:222a,b; **5**:305a; **6**:242b, 245a–246a, 278b, 553b; **7**:445a–b; **8**:122a, 398a–399b; **9**:166b; **11**:453b, 454a; **12**:450b; **13**:522a
late nineteenth-century, **3**:420a; **4**:588b; **6**:432a*; **7**:267b, 526a; **8**:597a–b; **9**:230b; **12**:54a–b; **13**:11b, 101b, 102b, 103b
twentieth-century, **2**:565b; **4**:305b, 306a–b; **5**:202a; **7**:84b, 470a; **8**:11b, 473b; **9**:386a–b; **11**:264b–265b; **12**:53b, 54b; **13**:104b
architectural, **9**:386b; **12**:53b, 54a–b
linguistic, **7**:445b; **14**:290a
loudness measurement, **1**:455b
mathematical, **5**:249b; **13**:76a–b; **15**:200a, 486a, 487a
molecular, **13**:103b
musical, **7**:84b, 445a–b; **9**:386b; **11**:264b–265b; **12**:129b; **13**:11b, 76b; **14**:289b
naming of, **12**:127b
optical methods, **8**:398a–399b
polyphonic, **4**:588b
reverberation time, **12**:53b, 54b
see also Electroacoustics, Hearing, Sound
Acquired characteristics, **4**:12b, 402a; **5**:447a; **8**:586a; **11**:19a, 137a; **12**:298a; **14**:297a
inheritance of, **2**:10b, 580b; **3**:429a, 493a, 574a, 579b; **6**:8a, 9a, 613b; **7**:114b, 590a,b, 591b; **8**:125a, 596a; **11**:518b–519a; **12**:571a; **14**:233a–237a *passim,* 313a; **15**:93a
critics of, **7**:166b, 498b; **8**:27a
Acrania, **5**:453b
Acridine, **2**:59b; **5**:489a
Acroerythrosis, **1**:580a
Acroleins
synthesis of, **5**:420b
Acromegaly, **9**:109a
Acrose, **5**:3a
Acrylic acid
synthesis of, **4**:553b
Acta arithmetica
editors of, **12**:426b
Acta astronomica
founding of, **1**:430a
Acta crystallographica, **4**:522a
Acta eruditorum, **3**:183b; **5**:499b
Acta litterarum ac scientiarum Regiae Universitatis hungarica . . .
editors of, **7**:312a
Acta Mathematica, **3**:53b
editors of, **8**:366b
founding of, **9**:426b–427a
Acta medica et philosophica hafniensa
founding of, **1**:483a
Acta physicochimica URSS, **1**:361b
Acta scientiarum mathematicarum
founding of, **5**:608a; **11**:459a
Actedius, Petrus
see Artedi, Peter
Acteon (ship), **7**:20a
ACTH
see Adrenocorticotrophic hormone
Actiniae, **3**:550a
Actinium, **2**:259a–b, 260a; **5**:394b; **6**:15b
decay series, **6**:16b
derivatives of, **3**:502b
discovery of, **2**:258b; **3**:499b, 507b
radioactivity of, **3**:499b; **7**:159b*; **9**:261a–b; **12**:29a
Actinium X, **5**:394b
Actinoelectric series, **4**:355a
Actinoelectricity, **6**:96b; **7**:526b; **13**:80b
Actinology, **1**:304a
Actinometer, **1**:556a; **3**:204a; **4**:182b–183a; **6**:325a, 326b; **14**:38b, 39a, 174b
Actinometry, **4**:283a*; **10**:320b; **12**:248a; **13**:11a
Actinomycosis, **10**:19a
Actinomyxidia, **3**:148b
Actinon, **5**:394b; **11**:280b
Actinouranium, **6**:16b
Action
contiguous, **8**:494a; **9**:205a–b
at a distance, **1**:67a, 424b; **2**:166b, 324b, 593a; **3**:431a, b, 444b; **4**:481b, 536a–b; **5**:48a, 136b; **6**:251a, 279b, 343b, 344a, 345a; **7**:228a, 380b, 381a; **8**:490b, 491a, 492a,b; **9**:115a, 204b,

548a, 549a; **10**:28a, 59b, 134b, 146b; **12**:250a; **13**:375a–b; **14**:32b, 33a, 71b, 305b, 353a; **15**:358b, 360a
rejection of, **6**:346a; **13**:278b
at an instant, **1**:253b
quantity of, **10**:3b
and reaction, **4**:481a; **5**:239a; **13**:200b
heat theory, **9**:132b–133a
quantification of, **1**:409b; **13**:195b–197a
reciprocal, **5**:193b
see also Least action, principle of
Activity, chemical
concept of, **8**:290a–b
Acton, Edward Hamilton, **3**:582a
Actualism, **1**:234b; **2**:556a; **5**:205a; **6**:456b–457b; **8**:241a–243a, 251b–252a, 463b; **9**:533b; **12**:38b, 272a, 514b
vs. uniformitarianism, **6**:457a–b
see also Uniformitarianism
Actuarial science
seventeenth-century, **6**:538a; **14**:466b–467b
eighteenth-century, **4**:470b; **8**:611b; **9**:454a–b
nineteenth-century, **2**:318b, 319a; **5**:307a, 466a–b, 480a; **8**:61b; **13**:339a; **15**:31a
twentieth-century, **2**:616a; **8**:458b; **10**:443b
see also Insurance
Actus sorboniens, **5**:115a
Acupressure, **8**:404b
Acupuncture, **7**:205a; **13**:282b
Acyl phosphate, **14**:173b
Acyuta, **12**:115b, 599a
Acyuta Piṣāraṭi (*ca.* 1550–1621), 1:48–49; **15**:625b
Horāsāroccaya, **12**:599a
Karaṇottama, **15**:624a
Rāśigolasphuṭānīti, **15**:624a
Sphuṭanirṇaya, **15**:624a
Adalbero, **5**:365a,b
Adam of Bodenstein (1528–1577), 1:49–50; **4**:169b; **10**:306b
Adam, Jehan, **3**:273a
Adam, Melchior, **1**:49b; **2**:218a
Ibn al-Adamī, **7**:223a–b; **14**:575b
Adami, John George, **5**:80b; **8**:447a; **15**:100a
Adami, Tobia, **1**:447b
Adamkiewicz, Albert
Adamkiewicz reaction, **6**:500b
Adams, Edward, **12**:73a
Adams, Frank Dawson (1859–1942), 1:50–53; **3**:609b*; **6**:565b*; **8**:364a
Adams, George, **9**:595a; **14**:452b
Adams, John Couch (1819–1892), 1:53–54; **6**:402b
celestial mechanics, **10**:41b; **11**:207a; **15**:332b
on contemporary scientists, **8**:279a*
editions of works, **12**:96a*
influences on, **12**:524b
lunar theory, **3**:169b–170a
and Neptune discovery, **1**:86b; **2**:359b; **3**:186b–187a; **5**:257b; **8**:277b, 299b; **10**:479b
on Newton, **10**:61a, 91b(n), 94a*
Adams, Leason Heberling, **5**:186a
Adams, Robert
Stokes-Adams syndrome, **4**:398a; **5**:366b
Adams, Roger (1889–1971), 15:1–3; **3**:86a
Adams catalyst, **15**:2b
Adams, Walter Sydney (1876–1956), 1:54–58; **7**:238b
acquaintances of, **10**:473b; **12**:20a, 348b; **13**:98a
Adams-Russell effect, **12**:21b
astronomy, **4**:279b, 280b; **13**:98b
on contemporary scientists, **6**:29a, 32b, 533a*; **12**:73b*
spectroscopy, **5**:199b*; **6**:29b; **12**:21b–22a, 73a
Adanson, Michel (1727–1806), 1:58–59; **7**:199a; **8**:377a; **9**:254a; **12**:536a,b; **13**:443a
Adaptation, biological, **3**:493a, 575a–b, 576a; **5**:438b; **6**:319b, 613b; **7**:64b, 252b, 409b; **8**:125a; **9**:24b, 44b, 375a; **10**:427b, 434a; **14**:134b, 434a–b
embryonic, **10**:523a
functional, **11**:571a–b
Adder
poison gland, **3**:342b
Addison, Joseph, **2**:16b
Addison, Thomas (*ca.* 1793–1860), 1:59–61; **2**:464a,b
Addison's anemia
see Anemia, pernicious
Addison's disease, **1**:60a–b; **10**:205a
treatment of, **13**:53b; **15**:258a,b
Addisonia, **2**:477a
Addition, **10**:299b
algebraic, **7**:241b
Egyptian techniques, **15**:681b, 683a–b
geometric, **9**:431a
logical, **2**:296a, 297b; **8**:154b
theorems of, **2**:53a; **4**:478a, 516a
Adductor pollicis
oblique head of, **3**:40b
Adeimantus, **11**:23b, 24a
Adel, Arthur, **12**:455a
Adelaide, University of
geology, **9**:197b–198a
Adelaide System (geol.), **9**:198a
Adelard of Bath (*fl.* 1116–1142), 1:61–64; **1**:381b; **11**:322a, 504a; **14**:35a
Natural Questions, **1**:380a
translations by, **1**:36a, 61a; **3**:24b; **4**:443b–448a *passim,* 611b; **7**:361a, 363a; **15**:175a,b, 580b
Adelea, **12**:423b, 424a
Adenine
discovery of, **7**:466b
synthesis of, **4**:604a; **5**:2b
Adenoma
basophile, **3**:518b
Adenosine monophosphate
metabolism of, **10**:326b
Adenosine triphosphate (ATP), **3**:416a
role of, **9**:359b
Adenyl phosphoric acid, **4**:360b
Adenylic acid, **4**:486a
Adequality, **4**:569b–572a
Adet, Pierre-Auguste (1763–1834), 1:64–65; **6**:164b; **7**:16a*; **8**:81a
Adhesion (med.), **6**:566b
Adhesion (phys.), **8**:373b, 468b, 469a; **13**:11a
molecular, **4**:211b; **15**:359b
work of, **6**:118b
Adiabatic invariants, **8**:284b
Adiabatic principle, **4**:293b–294a
Adiabatic processes, **3**:80b–83a *passim,* 304a, 305a, 620a; **4**:230b, 410b; **5**:462a, 610b; **6**:97b–98a, 545b; **7**:493a; **8**:482b; **13**:381a; **15**:356b, 357b, 377b, 378b
Adiantum, **7**:320a
Adipic acid, **2**:515a; **14**:455a
Adirondack Mountains
naming of, **4**:363b
Adjunction (math.), **3**:145a
Adler, Alfred, **5**:179a; **7**:190b
Adler, August, **9**:156b
Adler, Saul, **3**:566a
Adonis (asteroid), **4**:26b
Adrain, Robert (1775–1843), 1:65–66; **3**:399b*; **5**:414a
Adrastus, **3**:14b; **13**:325b, 326a
Adrenalin
see Epinephrine
Adrenocorticotrophic hormone (ACTH), **3**:352b–353a
Adriaanson, Adriaan
see Metius, Adriaan
Adrian, Edgar Douglas, **2**:2a; **5**:65a,b; **8**:533b, 534a,b; **12**:397a, 400a
Adsorption, **1**:415a, 537b; **2**:300b; **4**:80a, 203a; **6**:118a,b; **8**:23b, 427b; **12**:191b, 404b–405a; **13**:419b, 420b–421b, 487a; **14**:412a; **15**:159b
concept of, **8**:24a
hydrolytic, **12**:404b
on oxides, **6**:118b
theories of, **8**:520a
ᶜAḍud al-Dawla, **2**:619a; **7**:260a; **9**:40b; **11**:240a; **12**:431b, 447b; **13**:149a
Adventure (ship), **3**:397a; **5**:16a
Aecidiospores, **3**:613a
Aedes aegypti, **4**:620a
Aegean civilization, **10**:550b
Aegidius
see Giles
Aegilotriticum, **13**:478b
Aelbert (archbishop), **1**:104b
Aelian, **1**:120a,b; **2**:256b; **4**:463b; **5**:379b; **10**:180a
Aenesidemus, **12**:340b
Aeolipile, **9**:491b
Aepinus, Franz Ulrich Theodosius (1724–1802), 1:66–68; **6**:178b; **14**:352b, 419b*
electricity, **3**:51b; **5**:133a, 136a–b; **10**:147a; **11**:433a; **14**:71b, 72b, 209a, 352b, 353a,b
gravitation, **11**:173b
magnetism, **3**:444a, 445b
pyroelectricity, **2**:5a
Aerial Experimental Association, **1**:583a
Aerobacter, **2**:86b; **6**:298a, 405a; **15**:14a
Aerodynamics
eighteenth-century, **4**:479a
nineteenth-century, **8**:19b; **13**:483a–b; **14**:404a, 621a
early twentieth-century, **3**:196a; **7**:246b–248b, 613b; **8**:301a, 597a–b; **9**:419a–b; **11**:123b–124b, 306a–b; **12**:496b, 613a; **13**:492a; **14**:437a, 520a–b
mid twentieth-century, **4**:376a; **7**:437a; **10**:20a–b, 514a; **13**:492b
high-speed, **14**:621b
Aerohydrodynamics, **3**:196a
Aerojet Engineering Corporation, **7**:247b
Aerology
late nineteenth-century, **6**:98b; **12**:187b
early twentieth-century, **5**:187b–188a
Aeromechanics, **3**:195b
Aerometer, **7**:341a, 346a, 376a; **11**:434b
Aeronautics
eighteenth-century, **3**:78b*; **9**:492b; **10**:609a–b
nineteenth-century, **2**:439a; **3**:12a; **7**:76b; **9**:291a; **13**:483a–b; **14**:621a
early twentieth-century, **5**:424a, 433a; **7**:247a–248a, 613b; **8**:532b; **9**:419a–b; **13**:431b, 483b–484a, 492a; **14**:348b, 437a, 520a–521a, 621a–b
mid twentieth-century, **6**:104b; **13**:492a–b
see also Aviation
Aerophotometry, **12**:352b
Aerosols, **14**:319a
Aerostatics
eighteenth-century, **2**:474b; **3**:153b, 207b–208a, 326b; **4**:548b; **5**:602a–b; **7**:494a; **9**:342b–343b; **11**:166b
nineteenth-century, **3**:12a; **12**:512a; **13**:483a

Aerostatics (cont.)
early twentieth-century, **4**:361a; **12**:249b
see also Balloon ascensions, Balloons
Aerotonometer, **5**:150a
Aeschylus, **6**:416b
Aesthenosphere, **1**:470b
Aesthetics
Renaissance, **4**:260b–261a; **5**:106b–107a; **8**:234b; **12**:135a
eighteenth-century, **4**:88b
mid nineteenth-century, **4**:558a; **10**:185b; **14**:128b; **15**:414a–b
late nineteenth-century, **8**:7b; **14**:343a
twentieth-century, **2**:146a; **3**:278a, 456a; **11**:546a
experimental, **4**:558a
see also Art
Aestivation, botanical, **2**:519b
Aether
see Ether
Aëtius of Amida (*fl. ca.* A.D. 540), 1:68–69
astronomy, **10**:590b
historiography, **5**:233a; **6**:381b, 382a; **10**:179b, 180a, 181a, 590a; **11**:602b; **13**:303b
influences on, **10**:230b; **12**:542a(n)
Placita philosophorum, **4**:526a; **13**:333a
Afanasev, Mikhail Ivanovich, **10**:431a
Affect (psychol.), **5**:176a
Affinity, chemical
early eighteenth-century study, **1**:451b; **5**:353b–354a; **11**:109b
late eighteenth-century study, **1**:527b; **2**:6b–7a, 74b, 76b, 78a–82a, 180b–181a; **3**:539b; **5**:601a; **6**:165a*, 386a; **7**:388a–b; **8**:81b, 298a, 469b, 620b–621a
early nineteenth-century study, **1**:141b, 145a, 348a–b, 349a; **3**:601b; **5**:324b, 431a; **7**:254a; **8**:373b
mid nineteenth-century study, **1**:579a; **4**:554b; **5**:412b, 587a; **6**:354a–b; **9**:446a*; **14**:109a
late nineteenth-century study, **4**:229a; **6**:244b; **13**:578b–579a; **15**:436b, 456b–461a
twentieth-century study, **2**:499a; **3**:334a; **15**:439a, 441b
affinity numbers, **1**:348b; **6**:386a
in atmosphere, **3**:540b–542b
and atomic weight, **4**:134b
constants, **2**:499a
definitions of, **8**:621a
distance effects, **2**:80b–81a
elective, **1**:141b, 145a; **2**:7a–b, 180b–181a; **4**:530a
and electrochemistry, **1**:348b; **3**:601b; **4**:536a; **5**:7a, 292a; **10**:182b; **13**:359a; **14**:124b; **15**:459a–461a
measurement of, **2**:499a; **5**:601a; **7**:388a–b
and pressure, **12**:543a
and refractive index, **1**:348b; **15**:361a–b, 457b
and specific heat, **1**:348a–b
and specific volume, **15**:457b
tables of, **2**:6b, 7a, 180b; **3**:419a, 495a; **5**:353b–354a; **8**:620b; **10**:446b*
criticism of, **2**:78a–b
temperature effects, **2**:78b, 80b–81a
theories of
eighteenth-century, **2**:6b, 7a, 79a–82b, 94b–95a; **5**:601a; **8**:260a; **11**:168b; **15**:315b
nineteenth-century, **9**:287a; **11**:169a; **14**:265b–272a; **15**:358b, 361a,b
twentieth-century, **4**:553b
and thermochemistry, **13**:358b–359a; **15**:441b
variability of, **2**:6b–7a, 78b–81a; **4**:238b
wet vs. dry reactions, **1**:527b; **2**:7a
work of, **13**:578b
Affinity, mathematical, **4**:259b–260a
Affinity, theory of, **2**:135a; **15**:361b
Afflacius, Johannes, **3**:393a,b, 394a
Afhandlingar i fysik, kemi och mineralogi, **2**:93a
Ibn Aflaḥ
see Jābir ibn Aflaḥ
Africa
botany, **9**:34b–35a; **13**:342b; **15**:148a
circumnavigation of, **9**:36b
ethnology, **2**:505a; **7**:31a
and Europe, geodesic union of, **7**:2a
exploration of, **5**:265b; **8**:274b, 613b
geography, **8**:190a–b; **10**:617b; **13**:84b
geology, **3**:533a; **4**:262b; **6**:168b; **9**:104a; **11**:336a; **14**:402a
medieval visitors to, **1**:516a, 517a
natural history, **1**:58b–59a; **8**:190a
ornithology, **8**:274b–275a
paleontology, **2**:110a; **11**:336a–b
and South America, similarities between, **4**:262b, 310a; **14**:215a, 216a
volcanoes, **11**:336b
water problems, **3**:617a
see also East Africa; South Africa; specific cities, countries, regions
African Association, **11**:376b
African coast fever, **7**:428b–429a
Africanus, Sextus Julius, **5**:232b
Afzelius, Johan, **2**:90b; **3**:473b; **4**:343b; **5**:215b
Agache, Roger, **8**:30b
Agamennone, Giovanni, **9**:444b
Agapius, **10**:6a
Agar, **12**:125a
Agardh, Carl Adolph (1785–1859), 1:69–70; **5**:190b; **7**:610b; **8**:518b; **13**:178b*
Agardh, Jacob Georg (1813–1901), 1:70–71
Agardh, John Mortimer, **9**:460a
Agardh Herbarium, Lund, **7**:535b
Agassiz, Alexander (1835–1910), 1:71–72; **1**:74b; **2**:501b; **7**:447a; **9**:112a, 589b; **10**:522b; **11**:113a, 114a*; **14**:314a
Challenger expedition, **9**:589a
zoology, **11**:113b
Agassiz, Jean Louis Rodolphe (1807–1873), 1:72–74; **2**:282b; **6**:553a; **8**:103b; **9**:540b; **12**:479a; **13**:467a
acquaintances of, **1**:364b; **2**:108a, 425b, 497b; **3**:535b; **5**:569a,b, 533a–b; **6**:517a; **8**:260b, 264a; **9**:99b, 100a, 388b; **12**:167a; **13**:65b; **14**:57b, 533a; **15**:130b, 227b
and Baird, **1**:405a
biographies of, **9**:100a
correspondence of, **2**:375a; **6**:57b; **15**:126b
and Desor, **4**:73a–b
editions of works, **7**:480a
and Emmons, **4**:364b
Études sur les glaciers, **3**:211a
and evolution, **1**:71b; **3**:553a, 574a; **5**:513a
glacial theory, **2**:498a, 556b, 568a, 570b; **3**:211a; **5**:337b; **6**:438a; **7**:45a; **10**:240a; **12**:167b
critics of, **9**:173a, 585a
and Guyot, **5**:599a–b
influence of, **9**:101a
oceanography,**11**:113b
paleontology,**6**:594b; **7**:252a; **8**:104a; **9**:389b
Principles of Zoology, **5**:478a; **11**:113a
students of, **2**:141b, 501b; **6**:613a; **7**:67a, 169b, 370a; **8**:123a; **9**:416a, 536b; **10**:272b, 482b; **11**:113a, 219a; **12**:264a, 343b; **14**:1a, 314a
Système glaciaire, **5**:599b
zoology, **12**:578b; **13**:552b*
Agate
artificial, **8**:213b
polarization of light, **2**:451b
Agathias, **1**:121a; **12**:440b
Agathinus, Claudius (*fl. ca.* A.D. 50), 1:74–75; **1**:212a, 324a
Agathodemon, **1**:33a,b; **10**:208b; **14**:632a
Agave, **13**:456b
Age
see Aging
Agglutination, **6**:432b; **12**:482a
of blood, **7**:623a–b; **10**:387b
from cold, **7**:623b
discovery of, **5**:564a–b
Vibrio comma, **2**:300b
Agglutinins, **7**:624b
Aggregata, **4**:132b, 207a; **12**:424a
Aggregate (math.)
see Sets
Aghānīs, **5**:346b; **7**:79a; **10**:6a,b
Agilon, Walter, **1**:290a
Aging, **2**:58b; **8**:283a; **9**:416a; **10**:611a; **11**:135a; **15**:74a
causes of, **6**:337a; **13**:227b
Agnesi, Maria Gaetana (1718–1799), 1:75–77; **5**:499a
Analytical Institutions, **9**:159a
"Witch of Agnesi"
see Versiera
Agnesi, Pietro, **1**:75b
Agnosticism
concept of, **6**:596b
Agorius Praetextatus, Vettius, **1**:269b; **2**:229b
d'Agoty, Fabien Gautier
Histoire naturelle, **11**:523a
Agouti, **3**:569b, 570a
Agraphia
motor, **3**:203b
Agricola, Georgius (1494–1555), 1:77–79; **2**:143a; **8**:208b
drawings of, **4**:87a
geology, **8**:147a; **9**:532b
influence of, **5**:420a; **7**:355a; **8**:312a; **11**:288b; **14**:507b; **15**:94a
De re metallica, **4**:393b; **8**:70a, 214a, 464b; **11**:607a, 608a
Agricola, Rudolf, **11**:286b, 287a
Agriculture
ancient Greek, **2**:256b
ancient Roman, **13**:588a–b
Byzantine, **11**:186a
medieval Arabic, **6**:79b; **14**:112b, 118a–b; **15**:232a
Renaissance, **10**:280b; **11**:97a; **12**:316b–317a
seventeenth-century, **6**:142a*; **11**:315b
early eighteenth-century, **1**:486a; **2**:390a; **4**:224a; **12**:116a
late eighteenth-century, **1**:233b, 554b; **2**:107a, 321b, 509b; **3**:7b, 172b–173a, 200b, 579b–580a; **4**:71a; **6**:219b, 578b, 584b–585b; **7**:89a, 201b; **10**:283b; **12**:116a–b; **13**:495a–b, 497a; **14**:145a
early nineteenth-century, **1**:492b, 586a; **3**:7b; **10**:118b; **12**:116b–117a; **13**:168a, 596a
mid nineteenth-century, **1**:587a; **3**:605a*; **5**:350a; **9**:277b–278a, 279a; **10**:150a–b
late nineteenth-century, **3**:479b, 608b; **4**:144a, 145a; **5**:64a; **6**:237b; **13**:417a, 459b–460a; **14**:98b
twentieth-century, **5**:523b; **9**:44b; **13**:477b–479a; **15**:492b–493a
biological competition, **2**:322b
and chemistry

see Chemistry, agricultural
and climate, 4:144a
entomology and, 1:329a; 2:221b; 5:12a, 70a,b; 6:524b, 525a
experimental
Renaissance, 9:609b; 12:317a
seventeenth-century, 5:419b–420a
eighteenth-century, 8:84a–b; 13:603a–b
nineteenth-century, 1:325a–b, 357a; 8:92b–93b, 346b–348a; 9:11a, 287b; 11:131b–132a, 155a
early twentieth-century, 4:270b–271b; 14:552b–553b
Mesopotamian, 15:648a, 649b
and meteorology, 6:281a*
as a science, 1:554b; 9:287b; 10:150a–b; 12:233b
statistics in, 9:279a; 14:552b–553b, 574a
technology, 5:64a; 13:614a
tropical, 14:255a
see also Agronomy, Crops, Horticulture
Agriculture, U.S. Department of
biological survey, 9:313b
Bureau of Chemistry, 14:357a–b
entomology, 6:524b
Experiment Stations, Office of, 1:325b
Agriculture Society of Edinburgh, 14:132b
Agrimensores, 1:227a; 3:59a; 4:443a–b, 609a, 611a
Agrippa, Heinrich Cornelius (1486–1535), 1:79–81; 10:304b, 310a; 14:375b(n)
alchemy, 5:580a
De incertitudine et vanitate omnium scientiarum et artium, 5:119b
De occulta philosophia, 2:540a; 6:397a; 7:40b
Agronomy
medieval, 1:350b–351a
eighteenth-century, 2:257a–b, 509b; 4:223b–224a; 8:303a; 9:53b–54b; 13:411b
nineteenth-century, 2:570b; 3:44b, 45a; 4:144a
twentieth-century, 14:551b; 15:505a–511b
see also Agriculture, Horticulture
Agrostology, 15:224a–b
Ague, 8:375a
Aguilon, François d' (1546–1617), 1:81; 12:74b
Ahargaṇa (period), 15:537b
calculation of, 15:543a, 557a–b, 565a–566a, 579a, 583b, 593a, 605a
definition of, 15:539a
Aharonov, Yakir, 9:217b
Ahlmann, Hans Wilhelmsson, 13:167a
Aḥmad ibn Ibrāhīm al-Uqlidīsī
see al-Uqlidīsī
Aḥmad Kamāl, 9:604a
Aḥmad ibn Muḥammad ibn ʿAbd al-Jalīl al-Sijzī
see al-Sijzī
Aḥmad ibn Muḥammad ibn al-Bannāʾ
see Ibn al-Bannāʾ
Aḥmad ibn Mūsā ibn Shākir
see Banū Mūsā
Aḥmad the Persian
Introduction to Astrology, 1:39a
Aḥmad ibn al-Ṭayyib al-Sarakhsī, 1:33b
Aḥmad ibn Yūsuf (*fl. ca.* 900–905), 1:82–83; 4:541b
De arcubus similibus
translations of, 15:177a
De proportione et proportionalitate
translations of, 15:177b
Aḥmad ibn Yūsuf al-Tīfāshī
see al-Tīfāshī
A'hmosé (Egyptian scribe), 15:685b–686b, 687a, 690a, 693a, 698a–b, 699b
Ahran ibn Aʿyan, 7:187a
al-Aḥwal, Abu'l-ʿAbbās Muḥammad ibn al-Ḥasan ibn Dīnār, 15:234b
al-Ahwāzī, 15:596b, 600a
Aida Ammei
see Aida Yasuaki
Aida Yasuaki (1747–1817), 1:83–84; 15:751b
Aikin, Arthur, 2:336a
Ailly, Pierre d' (1350–1420), 1:84; 3:280b; 10:419b
astrology, 6:273a
on calendar, 1:383a
geography, 15:88a
Imago mundi, 1:381b
Ainu people, 1:175a; 5:294a
Air
analysis of, 3:292b; 5:315b, 318b, 367a; 6:23b–24a, 550b, 551a; 7:620b–621a; 8:75b; 9:140a; 10:558a; 12:145b, 147a, 478a–b; 13:311a
microscopic methods, 7:501b
carbonic acid in, 9:155a
in chemical reactions, 7:188b
color of, 9:118a
compressibility of, 3:195b, 7:182a
condensation of, 1:152a
condensation coefficient, 8:509b
as a conductor, 1:547a, 4:355a
density of
Renaissance determination, 3:66b
eighteenth-century determination, 12:123b
and altitude, 5:575a
dephlogisticated, 3:157b; 8:75b, 77a, 531a; 10:496b; 14:75a
disinfection of, 5:603b
elasticity of, 2:377b, 379b; 4:44b; 5:575a; 6:41b, 42b–43a; 7:494b; 8:72a–b, 73a, 469a; 9:117a–b
electrical conductivity of
and altitude, 14:422b
and radioactivity, 3:498b–499a, 507a
as element, 1:549b; 6:43b; 8:72a–b, 131a; 10:181a, 309b
opposition to, 7:538b
evaporating capacity, 8:425b
expansion of, 7:182a
and health, 5:137b; 6:44b, 421b; 10:558a,b; 12:561a; 14:638a
as the infinite, 1:152a
inflammable, 3:157b, 465b; 8:76b, 77b–79b; 14:74b–75b
specific gravity of, 3:156b
ionization of, 1:559a; 4:356a–b
and life, 6:40b, 41a–b, 43b; 7:538a,b; 12:557b
light scattering, 11:265b–266a
liquid, 7:220b–221a; 8:365b; 10:206a; 13:454a
manufacture of, 3:299a; 4:79b; 6:94a; 10:206a; 13:454b; 14:522b
microbes in, 1:497a; 3:189a, 359b; 10:362b, 367b, 368b–372a, 383a–b; 11:37b, 110a–b; 12:555a
sterilization of, 13:522a
molecules
size of, 8:508b–510a
motive force, 9:114b
nature of, 8:72a–73b, 585a–b; 9:15a; 12:64b, 307b–308a
opalescence, 12:497b
oxygen content, 7:160a–b
phlogisticated, 3:157b; 8:531a; 12:605b
pressure of
see Pressure, atmospheric
rare gases in
separation of, 4:80a
salubrity of, 5:57a; 7:14b*, 620b; 14:74b
solubility in water, 9:117b
specific heat of, 3:156b; 9:237a; 15:86a, 358a
temperature of
and vibrations, 15:487a
weight of, 1:424a; 4:52b; 5:245b–246a, 288a; 7:612a; 9:117a; 10:333a,b, 334a; 11:506b, 507a; 12:119b, 123b; 13:438a
work capacity of, 5:575a
see also Atmosphere; Elements, Aristotelian; Factitious air; Fixed air; Pumps, air; Ventilation
Air bladder, 7:501b, 519a
Air Defence Research Development Establishment (G.B.), 3:330a
Air Force, U.S., 7:248a
Air masses
analysis of, 11:558a; 12:366b
concept of, 7:389a
Air pump
see Pumps, air
Aircraft
see Aeronautics, Airplane, Aviation, Glider plane, Helicopter
Airplane, 1:304a; 7:465a
bombing from, 14:621b
control, 14:520a
de-icing, 8:24b
design of, 8:301a; 9:578a; 13:483a, 492a–b; 14:520a–b, 621b
jet, 13:483b, 492b
fuel, 13:484a
lift
theory of, 14:621a–b
models, 8:19b
propulsion of, 5:433a
stability of, 2:466a; 8:532b; 14:621b
stabilizer, 12:575a
streamlining of, 11:124b
tracking of, 5:103a
see also Aviation, Wing
Airscrew, 8:213b; 9:343b
Airy, George Biddell (1801–1892), 1:84–87; 2:134a; 3:260b*; 4:354a; 5:25b; 6:86a; 9:566a
acquaintances of, 1:136a; 4:35a; 5:413a; 6:402b; 8:277a; 10:483a; 14:55b, 293a
Airy transit circle, 5:405a
celestial mechanics, 15:354b
correspondence of, 3:94a*; 13:110a
horology, 9:163b
influence of, 3:261a; 7:481b
instruments, 11:552b
isostasy, 2:373a; 6:188a
lunar theory, 6:103b
magnetism, 12:50a
and Neptune discovery, 1:53b, 86b; 3:186b–187a; 8:277b
observations of, 1:403a; 2:99a
optics, 13:78a; 15:200b
statics, 9:226b
tides, 4:591a–b; 14:295a*, 569a
time signals, 3:288b
transit studies, 11:163a
Aisne
geology, 1:210a
Aither
see Ether
Aitken, John, 14:253b
Aitken, Robert Grant (1864–1951), 1:87–88; 3:509b*; 9:504b*, 10:321a*; 14:521b
Aiton, William (1731–1793), 1:88–89
Aiton, William Townsend (1766–1849), 1:89–90
Hortus Kewensis, 1:89a–b; 2:519b; 12:516b
Aiuta, Philippus, 13:210a
ʿAjāʾib literature, 9:171a
Ibn al-Ajdābī, 11:246b

Ajima Chokuyen
see Ajima Naonobu
Ajima Naonobu (*ca.* 1732–1798), 1:90–91; **1**:83b; **15**:751b
Ajtai, Miklós, **13**:226b
Akademi Internasional de Lingu Universal, **10**:443b
Akademie der Wissenschaften (Berlin), **4**: 200b, 290a; **8**:539a; **9**:105a, 187a–b, 188b; **12**:615b; **15**:449b, 451a
Akademische Sternkarten
see *Berliner Akademische Sternkarten*
astronomical observatory, **2**:56b, 220b
Astronomisches Jahrbuch
see *Berliner astronomisches Jahrbuch*
chemistry, **9**:424b
competitions of, **2**:161b; **3**:71a; **4**:471b; **5**:490b; **6**:341b, 344b; **8**:135a, 305b
Egyptian expedition, **4**:288b–289b
and Euler, **4**:470a–b, 472a
and fluid mechanics, **1**:113b–114a
founding of, **8**:150b
mathematics, **7**:506b
Akademie der Wissenschaften (Munich)
competitions of, **7**:201b
Akademiia nauk USSR, **4**:599a; **8**:251b
Astronomical Council, **5**:364a
Biochemical Institute, **1**:361b
competitions of, **6**:218b
geology at, **3**:532a
mineralogy at, **4**:598a; **5**:211b
oceanography, **12**:415a
Åkerman, Richard, **3**:234a
Akhbar (usages of the Prophet), **14**:547a–b
Akhiezer, Naum Ilich, **4**:562a; **9**:127a; **15**: 22b
Akkadian language, **15**:635a,b
Akṣadṛkkarma (computation), **15**:562b, 575b, 592b, 595a
al-
see next element of name
Alabama
botany **6**:122b
Geological Survey, **6**:122b
geology, **3**:392a
Alabaster, **11**:523b*
Alain de Lille (*d.* A.D. 1203), 1:91–92
Alakaluf tribe, **9**:144a
Alalakh (ruins), **14**:504b–505a
Alamogordo, **10**:216a
Alanus de Insulis
see Alain de Lille
Alaska, **2**:14b
birds, **5**:545a
botany, **7**:285a
ethnography, **6**:527b–528a
existence posited, **2**:551b
exploration of, **2**:499b–501a; **3**:536a; **4**: 361b; **5**:396a; **9**:313b; **10**:480a
geography, **2**:500a; **13**:28b
geology, **2**:499b–501a
historiography of, **2**:501a
natural history, **3**:536b; **13**:28a–b
Alaska Railroad, **2**:500b
Albany, N.Y.
Albany Academy, **6**:277a–b
Institute of History and Art, **6**:277b
Albategni
see al-Battānī
Albatenius
see al-Battānī
Albedo, **2**:284b
Albert of Bollstädt
see Albertus Magnus
Albert the Great
see Albertus Magnus
Albert of Lauingen
see Albertus Magnus
Albert I of Monaco (1848–1922), 1:92–93; **2**:347a, 557b; **3**:204a; **11**:101b, 429b
archaeology, **11**:484a
Albert of Saxe-Coburg-Gotha, Prince consort of England, **4**:11a; **6**:461b; **8**: 572a; **11**:36b; **12**:276a
Albert of Saxony (*ca.* 1316–1390), 1:93–95
acquaintances of, **2**:607a
influence of, **3**:420b; **4**:228a; **8**:241a, 242a; **9**:132b; **10**:420b; **12**:136a; **13**:349b
natural philosophy, **1**:199b
physics, **2**:590b, 607b; **8**:223b, 228a; **10**: 421a; **13**:185b
Questio de quadratura circuli, **1**:226b; **2**:360b
Questiones de caelo, **4**:157b
Tractatus proportionum, **4**:237b
Alberta (province)
geology, **14**:120b
Alberti, Friedrich August von (1795–1878), 1:95–96
Alberti, Gasparo
see Berti, Gasparo
Alberti, Leone Battista (1404–1472), 1: 96–98; **4**:259a; **5**:113a; **10**:269b; **11**: 406a; **14**:636a; **15**:517a
architecture, **13**:568b
Ludi matematici, **11**:406a
perspective, **2**:419a, 535b; **8**:234b; **10**: 103b
De re aedificatoria, **15**:517a
Alberti, Michael, **7**:188a
Alberti, Salomon (1540–1600), 1:98; **4**:508b; **10**:598b
Albertini, Ippolito Francesco, **9**:511a; **13**: 566b
Albertis, Antonius de, **1**:229b
Albertsson, Per-Åke, **13**:421b
Albertus Magnus, Saint (*ca.* 1200–1280), 1: 99–103; **1**:276b, 377b; **4**:94b; **9**:435a; **10**:21b
acquaintances of, **14**:388a
alchemy, **1**:199a; **2**:317b; **5**:580a; **6**:306a; **10**:555a
De animalibus, **1**:266a
astronomy, **13**:12a; **15**:34b
De caelo, **15**:35a
chemical theory, **12**:312a
commentaries by, **4**:448a; **9**:362a, 364a
critics of, **12**:429a
editions of works, **4**:237b; **7**:66b, 418b
influence of, **1**:512b; **2**:22b; **3**:343a; **4**:92b; **5**:361a, 402a, 448a; **13**:534b, 535a; **14**:35a,b; **15**:68b
influences on, **5**:553a; **8**:242a; **9**:362b; **10**:21b; **11**:31b, 244b; **12**:428b; **15**: 263b
optics, **4**:93a
De quindecim problematibus, **5**:401b
De sententiis antiquorum et de materia metallorum, **7**:187b
Speculum astronomiae, **7**:38a; **15**:178b–179a, 179b–180a
students of, **2**:20b; **6**:272b; **13**:347b, 534a
translations by, **1**:196b
Albian (stratig. stage), **6**:228b
Albin, Eleazer, **3**:129b
Albinism, **3**:120a*; **6**:567b; **9**:187a; **11**:137a
masking nature of, **3**:493a
Albinus (logic), **1**:269b; **2**:228b, 229b
Albinus, Bernard (1653–1721), 15:3–4
Albinus, Bernard Siegfried (1697–1770), 15:4–5; **2**:226a; **3**:37b; **6**:61b, 62a, 569a; **7**:11a; **8**:327b; **9**:483a; **14**:320a, 450b*
Albinus, Christiaan Bernard (*ca.* 1698–1752), 15:5
Albinus, Frederik Bernard (1715–1778), 15: 5–6
Albinus family, **15**:3a
Albinus Flaccus
see Alcuin of York
Albion (astron. instrum.), **7**:119b; **11**:415a–b
Albitskii, Vladimir Aleksandrovich, **12**:368a
Albohazen
see Abu'l-Ḥasan
Albrecht, Carl Theodor (1843–1915), 1:103
Albrecht, Walter, **1**:105b
Albritton, Claude Carrol, Jr., **2**:549a
Albucasis
see Abu'l-Qāsim
Album der Natuur
founding of, **6**:138a
Albumasar
see Abū Maʿshar
Albumin, **8**:342b
composition of, **8**:348b
crystallization of, **12**:546b
differentiation of, **15**:522a
in electric field, **15**:201b
formula of, **9**:558a
hydrolysis of, **11**:129a
lysozyme in, **5**:29b
in plants, **11**:179a
precipitation of, **10**:307b
purification of, **14**:529b
serum, **3**:336a; **5**:149a; **13**:420b
structure of, **8**:608a
sugar in, **1**:471a
in urine, **2**:464a,b
coagulation of, **3**:437b
Albuminoids, **15**:11b
Albuminuria, **4**:398a; **14**:253b
Albuzale
see Abu'l-Ṣalt Umayya
Alcabitius
see al-Qabīṣī
Alces, **10**:16b
Alchemy
ancient Greek, **2**:256b; **6**:305b; **7**:41a
ancient Roman, **1**:378a, 150a–b; **10**:301a; **14**:631a–632b
Byzantine, **10**:208a–b; **11**:183b; **13**:37b, 225b
medieval Arabic, **6**:79b–80a; **7**:39b–42b, 223b; **9**:39b–40a; **11**:324a; **13**:511b; **14**:118a; **15**:497b
medieval Latin, **1**:62a, 101b, 199a–b, 291a, 305b, 378a, 380b; **2**:22a–23a, 393a; **6**:306a; **9**:363b, 364b; **10**:554b–555b; **11**:322b; **13**:348a; **14**:35b, 155a–156a
Renaissance, **1**:49a–b; **2**:143a, 293a; **3**: 66a, 471b; **4**:6a, 169b–170a, 209a–b; **5**:447b–448a, 579b; **10**:308a–310a, 515a; **11**:606a–b; **12**:335a–b; **13**: 397a–b, 534b–535b
early seventeenth-century, **1**:317a,b, 448b; **4**:75a, 95b, 183b, 184a, 550b; **5**:47b–48a, 419b–423a; **6**:145b, 257a; **7**:355b–356a; **8**:310a–311b; **9**:23b–24a, 336a; **12**:307a–308a; **13**:559a–b; **14**:451a–b
late seventeenth-century, **1**:451a, 549a–550b; **2**:317a–b, 354a; **5**:419b–423a; **6**:478a; **7**:376b, 524b–525a; **10**:81a–83b; **11**:40b; **12**:616a–b
eighteenth-century, **2**:227a; **4**:129a; **5**: 353a–b; **9**:106b; **14**:332a
and biology, **7**:440b, 441b
Chinese, **15**:730a
and cosmology, **4**:209a–b
critics of
Byzantine, **11**:183a
medieval, **4**:525a; **7**:64b, 322a–b; **8**:550a

Renaissance, **2**:143a; **4**:393b; **10**:512b
seventeenth-century, **5**:579b–580a; **7**:284a–b; **11**:511b
eighteenth-century, **14**:333a
Jābirian, **1**:33b
Mesopotamian, **15**:649b–650a
philosophical, **4**:170a
psychoanalysis of, **1**:366a; **7**:192b
and religion, **4**:75a, 170a
subdivisions of, **8**:311a
see also Historiography, of alchemy; Illustration, alchemical; Transmutation of metals
Alchimech (star), **7**:124b
Alciopida, **6**:299b
Alcmaeon of Crotona (*b. ca.* 535 B.C.), 1:103–104; **4**:385b; **6**:317a; **11**:29b
Alcock, Nathaniel, **8**:452b
Alcohol(s)
aliphatic
synthesis of, **2**:66b; **6**:462b
amino, **5**:100a
amyl, **1**:417a
optical properties, **10**:358b, 361b
analysis of, **5**:325a; **8**:82b
anhydride, **8**:520b
aromatic, **3**:47a
benzyl, **14**:476b
synthesis of, **2**:66b
cetyl, **4**:245a
composition of, **4**:244a–245b; **8**:333b–338a
concentration, **10**:308a
definition of, **2**:66a
dihydroxy, **14**:531a
distillation of, **3**:201a; **8**:312a; **13**:534b–535a
effects on man, **2**:586a; **3**:257a; **5**:366b; **6**:43b–44a, 519b*; **7**:414a; **10**:445a; **12**:270a
absorption rates, **15**:418b
electrical conductivity of, **7**:203b
Erlenmeyer rule, **4**:400b
ethyl
dehydration of, **14**:561a
and potato oil, **3**:10a–b
synthesis of, **2**:66b
formulas of, **1**:345b; **3**:289b; **5**:371a; **8**:508a
heat effects on, **2**:65b
isomers of, **9**:304b
isopropyl
synthesis of, **2**:66b
manufacture of, **3**:202a; **7**:22a
from beetroot, **10**:361b
from potatoes, **3**:316b; **4**:250a
theory of, **7**:379a
medical use, **13**:535a
methyl, **3**:10b
synthesis of, **2**:66b, 324a
naming of, **10**:310b
nature of, **3**:193b–194a; **7**:452b; **14**:394b–395a
octyl, **7**:65a
oxidation of, **12**:148b
polyatomic, **2**:66a
polysulfides of, **3**:10b
as preservative of animal tissues, **3**:482b
propyl
discovery of, **3**:193b
reactions of, **5**:92b; **8**:333a, 335a, 336a, 337b; **9**:304b–305a; **12**:47a, 619b; **13**:310a; **14**:395a–b
secondary
synthesis of, **13**:549b
solutions of, **9**:287b
as solvent, **8**:618b
strength of, **1**:528a; **6**:386a
structure of, **9**:304b; **11**:92b
synthesis of, **2**:66a–b; **4**:553b; **5**:540a,b; **11**:438b*
thermocatalytic decomposition of, **7**:21a
in thermometry, **4**:518a
triatomic, **7**:203b*
vaporization of, **3**:9a
vinylization of, **4**:553a
in wines, **2**:420b; **5**:340b
see also Esterification
Alcoholism, **5**:74a; **10**:493b; **12**:458a
heredity and, **10**:464a
neurological effects, **9**:95a
statistical studies, **11**:123a
treatment of
late nineteenth-century, **5**:73b
Alcuin of York (*ca.* 735–804), 1:104–105; **1**:565b; **10**:271a
Alcyonarians, **9**:112b
Alcyone (star), **9**:2b
Aldebaran (star), **3**:406a; **7**:140a; **8**:613b
position of, **3**:318b–319a
Aldehydes, **4**:603b, 604a; **5**:1b; **8**:336a
composition of, **8**:335b, 337b
condensation of, **1**:389b; **2**:316b; **7**:282b, 283b; **12**:164a; **14**:454b
derivatives of, **4**:20b; **13**:406b
fuchsine test, **12**:164a
and ketones, **7**:452b
polymerization of, **7**:282b
synthesis of, **3**:247a*; **4**:553a; **7**:623a; **12**:47a; **13**:406b
Alder, Kurt (1902–1958), 1:105–106; **4**:91a–b
Alderotti, Taddeo (1223–*ca.* 1295), 1:107; **9**:467b
Aldine press, **1**:77b, 269a; **4**:431b; **8**:360b
Aldini, Giovanni (1762–1834), 1:107–108; **4**:201b; **5**:268b; **14**:77b–78a, 80a
De animale electricitate, **1**:107b
Essai théorique et experimentale sur le galvanisme, **1**:107b
Aldohexoses
structures of, **5**:3b
Aldol, **2**:316b
synthesis of, **14**:531b
Aldonolactones, **6**:538b; **10**:15a
Aldrich, Thomas Bell, **10**:205b
Aldrovandi, Ulisse (1522–1605), 1:108–110; **4**:509b
acquaintances of, **1**:167a; **9**:309a; **11**:527b
botany, **5**:383b
on contemporary scientists, **12**:90a
influence of, **1**:293b; **7**:164b; **9**:179a, 440b; **13**:451a; **14**:413b, 507b
influences on, **1**:596b
students of, **3**:342a, 343a
taxonomy, **11**:331b
Aleander, Johann Abraham, **3**:320b
Aleaume, Jacques, **4**:48a
Aleksandrov, Pavel Sergeevich, **3**:170b; **6**:177a, 496a; **8**:558a; **9**:249a; **13**:549a
Alekseev, Vissarion Grigorevich, **5**:472b
Alembert, Jean le Rond d' (1717–1783), 1:110–117; **1**:240a; **2**:330a, 334b; **4**:478a; **9**:47a
acquaintances of, **1**:479a; **3**:6a, 459b; **4**:85a; **6**:468b; **7**:596b; **9**:342b, 470b, 500b; **13**:496a, 565b; **15**:276a, 302a, 317a
d'Alembert's principle, **1**:112a, 600b; **2**:41a; **3**:74b, 454a; **5**:305a; **8**:494a, 602b; **15**:85a
d'Alembert's theorem, **1**:116a
analysis, **1**:207a; **2**:275a; **7**:566b, 568a; **8**:139b; **9**:51a, 471a; **10**:327b; **15**:278a, 294a
biographies of, **2**:88b
Cause des vents, **7**:567a
celestial mechanics, **7**:562a, 580a; **15**:287b, 289a
lunar theory, **3**:283a; **8**:179a
characteristics, method of, **4**:478b
and Clairaut, **1**:111a; **3**:282b, 283b; **7**:563a
commentaries on, **5**:195b
and Condorcet, **3**:383b–386a *passim*
on contemporary scientists, **14**:180a
correspondence of, **3**:14a*, 286a*; **15**:276b
critics of, **7**:619a
Diderot on, **4**:88b, 89a–b
disputes, **3**:561b, 562a; **4**:471a, 472a, 475a, 477a; **5**:55a
Encyclopédie, **1**:114b–115a; **2**:335a; **4**:86a,b; **5**:567b; **7**:581a
and Euler, **1**:113b, 114a, 117a; **4**:475a, 481a
geometry, **3**:166b; **7**:565a
hydrodynamics, **3**:137a; **8**:561b
influence of, **2**:342b; **7**:490a; **15**:349b, 350a, 423a
influences on, **11**:392b
and Lagrange, **7**:560b–567a *passim,* 570b
mechanics, **7**:55a; **8**:159b; **11**:62a; **13**:586a
philosophy, **8**:81b
probability, **8**:321a; **15**:280a, 285a
telescope, **3**:284a
tides, **15**:296b
Traité de dynamique, **1**:111a–113a
on vibrating strings, **2**:41a–b; **7**:561b
Alephs (math.), **3**:56a; **10**:484a–b
Aleppo, **11**:312b
Aleurone grains
discovery of, **6**:136b
Aleutian Islands, **2**:14b; **7**:63a
exploration of, **3**:536a
natural history, **3**:536b
Alexander VII (Pope), **3**:23a, 61a, 101a; **4**:128a, 506a
Alexander of Aetolia, **13**:326a
Alexander of Aphrodisias (*fl.* 2nd–3rd cent. A.D.), 1:117–120; **1**:268a–b
astronomy, **15**:33b
De augmento
translations of, **15**:180b–181a
commentaries by, **1**:269b, 270b, 272a, 275a, 276a; **2**:230a, 590b; **4**:461b, 462a,b; **6**:411b–412b, 416b; **7**:25a, 66a; **9**:437a; **11**:244b, 245a; **13**:304b; **14**:611a
critics of, **5**:105a; **12**:441a
dynamics, **1**:409b
geometry, **12**:441b
influence of, **5**:346b; **7**:42b; **9**:28b, 30a; **11**:71b, 444b; **14**:582a
De intellectu
translations of, **7**:25a; **15**:180b–181a
philosophy, **9**:111b
Problemata
translations of, **9**:132b
De sensu
translations of, **15**:180b–181a
De tempore
translations of, **15**:180b–181a
De unitate
translations of, **15**:187a
works translated, **9**:437b
Alexander of Ephesus, **13**:326a
Alexander the Great, **1**:250a; **4**:414b, 415a; **9**:269a; **13**:331a,b; **15**:678a–b
Alexander of Hales, **1**:377b; **14**:388a
Alexander of Myndos (*fl. ca.* A.D. 25–50), 1:120–121
Alexander of Tralles (*fl. ca.* A.D. 550), 1:121; **1**:212b; **5**:233a
works translated, **5**:586a

Alexander of Villa Dei, **6**:377a
Alexander, Andreas, **11**:458a
Alexander, Ernst, **6**:366a
Alexander, James Waddell, **2**:513a; **6**:496a; **13**:600a; **14**:311a
Alexander, Jerome, **5**:186b
Alexandria, **11**:187b
anatomy, **6**:316b
Aristotelian school, **1**:148a
astronomy, **1**:248a
Euclidean school, **1**:213a
library, **4**:30b, 388a–b
mathematics, **1**:213a, 223b; **4**:414b, 415b
medicine, **5**:228b; **10**:230b, 417a; **15**:235a
museum, **4**:383a, 415a; **13**:91a, 321b
philosophy, **6**:616a; **7**:134a
politics, **6**:616a
technology, **10**:587a
Alexandrov
see Aleksandrov
Alexia, **1**:496a
Alexin, **2**:300b; **4**:299a
Alexis of Piedmont
see Ruscelli, Girolamo
Alfakini, son of Abizarch, **11**:32a
Alfalfa
vitamins in, **10**:243b
Alfarabius
see al-Fārābī
Alfonso VII (King of Leon and Castile), **14**:592b
Alfonso X of Castile, el Sabio (1221–1284), 1:122; **1**:414a, 512a; **2**:405a; **6**:197b; **9**:39b; **13**:150a*; **14**:594a,b
tables of, **1**:122a, 382b, 383a; **2**:401b–402a; **4**:185b; **6**:274a; **7**:120a, 122b, 123a, 123b–124b, 125b, 128b, 129b–130a, 139a–140b, 142a; **11**:322a, 415a; **13**:324a; **14**:390b; **15**:154a, 156a, 187b, 474b, 476a
editions of, **7**:123b
Alfonso el Sabio
see Alfonso X of Castile, el Sabio
Alfraganus
see al-Farghānī
Alfred of Sareshel, **1**:271b; **10**:111b; **11**:244b, 504a; **14**:35b
Alfurs, **5**:277a
Algae, **2**:19a; **5**:605a
and altitude, **2**:430b–431a
anatomy
reproductive organs, **7**:610a; **12**:125a
blue, **13**:395a, 405a
brown, **7**:535a; **12**:125a–126a; **13**:395a
calcified, **7**:610b
chemical composition, **5**:297b*; **7**:535a
pigments, **7**:534a, 535a
crustaceous, **12**:328a
culture of, **3**:259b
ecology, **13**:165a
evolution of, **7**:252b–253a
fossil, **7**:252b–253a; **9**:579a; **14**:298b
freshwater, **2**:426b; **3**:43b, 321b, 337b; **7**:533b; **10**:13a; **11**:153b; **13**:506b; **14**:496a
genetics of, **3**:259b
geographic distribution of, **7**:610b
green, **7**:534a; **13**:395a; **15**:14a
in lichens, **3**:613b; **10**:166a
marine, **12**:328a; **13**:165a, 394b–395b
pelagic, **9**:587b
physiology, **3**:259b; **7**:395b, 535a
ionic exchanges, **8**:29b
photosynthesis, **4**:362b–363a
red, **3**:338a; **7**:535a–b, 610a; **8**:512b
enzymes in, **4**:220a
life cycle, **13**:165a–b
reproduction, **13**:395a–b
starch granules, **7**:534a
reproduction, **12**:125a–126a; **13**:165a–b, 394b–396a
conjugation, **13**:595b–596a
developmental history, **3**:337b; **7**:535a–b
germination, **1**:70b
polymorphism, **3**:259b
sexuality, **3**:337b; **5**:217a; **7**:152a; **11**:152a, 153a–154b; **13**:595b–596a
spores, **7**:13b; **11**:153a–b; **13**:596a
symbiotic, **14**:203a
taxonomy of, **1**:69a–b, 70b, 71a; **2**:185a; **7**:395b, 535a–b, 610a–b; **13**:506b
unicellular, **3**:337a,b; **9**:587b
see also Algology, Cryptogams
Algarotti, Francesco, **9**:77b, 187a; **12**:554a; **14**:83b
Il newtonianismo per le dame, **3**:215b
Algebra
ancient Greek, **1**:31b, 183b–184b, 188a–189a; **13**:399b–400a
ancient Roman, **4**:111a–117a; **15**:119a–122a
medieval Arabic, **1**:22b, 30b–31b, 437b; **4**:117a–b; **6**:208b*; **7**:240b–245a, 256b–259a, 324a–b, 327b–329a, 358b–360a; **9**:21b; **11**:229b–230a, 240b; **12**:92a–94a, 363b*; **13**:514b–516a; **14**:472a
medieval Latin, **4**:605a–612b *passim;* **7**:129a, 175b–176a, 362b
Renaissance, **2**:280a–281b, 618b; **3**:64b–65b, 272a–277b, 312a; **4**:586b–588a, 595a–596b; **8**:198a, 234b; **9**:191b, 610a; **10**:270b–271b, 494a; **11**:288a, 339a–b, 350a, 457a–458a, 533a–b, 589b–591a; **13**:48a–b, 59a–60b, 259a–b, 260a; **14**:19b–24b, 325a–b
"règle des premiers", **3**:273a, 274a, 275b–276a
early seventeenth-century, **3**:116b, 128b; **4**:549b, 550a, 566b–569b; **5**:382a, 408b–409a, 409b–410a; **6**:124b, 125a, 128a, 271b, 449b–450a; **8**:267a–b; **10**:254a–255a; **11**:488a; **12**:205a, 206a, 610b
late seventeenth-century, **1**:474b; **2**:47a; **3**:621b–622a; **5**:159a–b; **6**:537a–538a; **8**:161b–162a, 163a; **10**:51b, 495b; **11**:404a, 512a–513a; **13**:479b–480b; **14**:147b, 150a–b
early eighteenth-century, **1**:76a–b; **3**:284b–285a, 460b–461a; **4**:475a–b, 515b; **5**:520a; **6**:130a*; **9**:17a; **12**:283a
late eighteenth-century, **2**:112a–114a, 467b–468a; **7**:564b–567a, 570a; **8**:306a–b; **9**:158b–159a; **12**:283b; **13**:571a–b; **14**:180a, 279b–280b
early nineteenth-century, **3**:142b–143a, 555a; **5**:111b, 299a–300a; **6**:510a–511a; **8**:139a; **11**:61b, 598b–599a; **13**:129a–b; **15**:226a
mid nineteenth-century, **1**:591a–b; **2**:105a–b, 288a–b, 295b–298b, 299a, 351a, 469b–470a; **3**:163a–170a; **5**:260a–264b, 490b; **6**:90a–92b, 95a–b, 307a–308a, 357a–b; **7**:55a, 103a–b, 164a, 506b; **8**:384b–385a, 429a, 432b; **9**:410a; **10**:249b, 437b, 480a–b; **12**:86b–87a, 171a–b, 534a; **13**:217b–221a; **15**:194a–b, 195a–198b
late nineteenth-century, **2**:465a; **3**:263a–b, 265a, 560a; **4**:84a; **5**:472a–473a; **6**:388b–390b; **7**:167b–168b, 408a, 435a–436b, 444a–b, 508a, 573b, 621b; **8**:61b, 250a–b, 617a; **9**:175a, 358a; **10**:24a–b, 442a, 443a, 483b–484a, 485a, 544a; **11**:55b–56a, 407b, 409a, 543b, 589a, 611a–b; **12**:217a; **13**:250a; **14**:202b, 303a–b, 630a–b; **15**:65b
twentieth-century, **2**:351b; **3**:221b–222a, 346a; **5**:508b, 609a; **6**:176b, 286b, 297b; **8**:326b, 458b, 585a; **9**:71a–72a; **10**:137b–139a; **11**:260b, 470b–471a; **12**:237a–b, 451b; **13**:23a–b, 125b; **14**:211b–212a, 317a
abstract, **1**:307a–b; **5**:198b; **6**:389a; **10**:138a–139a, 437b; **11**:598b–599a; **13**:23a–b
analysis in, **7**:175b–176a, 243b–245a
applied, **2**:481b
arithmetization of, **4**:83a,b; **7**:241a–245a
associative, **1**:307b; **3**:166a; **4**:83a; **6**:90b, 91a; **10**:480b, 483b–484a; **14**:211b
in astronomy, **1**:314a
Boolean, **2**:295b–298b; **4**:36b; **7**:105a; **10**:485a; **11**:94b
Cartesian, **2**:358b; **4**:55b–58b
Cayley division, **4**:83a
Chinese, **3**:254a–256a, 265b–270b; **6**:511a; **7**:257a, 325b–326a; **8**:313b–320a, 420a–423a; **12**:290b; **14**:540a–541b, 543a–544b, 545b
classification of problems, **12**:93b
cohomological, **6**:497a
commutative, **14**:90b
complementary, **3**:331a
cossist, **4**:55b–56a, 549b, 612a; **8**:160b; **11**:339a, 457a–458a, 589b–591a; **13**:59a–60b; **14**:325a
definition of, **7**:328b
differential, **11**:471a
Diophantine, **1**:65b; **4**:111a–117a; **7**:444b
division, **4**:83a; **6**:90b, 91a; **14**:212a
double, **4**:35b
Egyptian, **1**:31b; **4**:606a,b; **15**:690b–699a
enveloping, **11**:55b–56a
etymology of, **7**:359b
formal, **7**:405a
foundations of, **2**:269a; **5**:108a, 260a, 523a; **6**:90a–91a, 95b; **8**:267a–b; **12**:364b; **14**:303b–304a
fundamental theorem of, **1**:239a, 239b–240a; **2**:444a; **3**:136b, 137b, 138b; **4**:58a; **5**:299b, 307a, 409b; **8**:139a, 385a; **14**:280b
geometrical, **2**:280a; **4**:608a–b; **5**:410a; **6**:414b–415a; **7**:242a, 328a–329a, 335a; **13**:289a–291b; **14**:544a–b; **15**:670a, 671b
see also Areas, application of
and geometry, **3**:451a; **5**:382a–b; **9**:475b; **14**:21a, 24a–b, 150a
Grassmann, **15**:196b–197a
historiography of, **14**:150a
Indian, **1**:30b, 31a, 309b; **2**:115b, 116b–117b, 417a; **5**:159a; **9**:613b; **12**:597b–598a
invariant, **3**:167b
Japanese, **8**:314b; **12**:290b–291b; **15**:751a
Lie, **3**:95a–b; **8**:326a,b; **9**:71b; **11**:55b–56a; **13**:219b
representation of, **14**:284a–b
linear, **3**:166a, 275a; **4**:83a; **7**:530b, 573b; **9**:72a; **10**:139a; **15**:195b, 197b, 198a
critics of, **11**:362b
Lipschitz, **8**:389a
local, **14**:630a–b
Maltsev, **9**:71b
Mesopotamian, **1**:30b, 31b; **4**:115a; **15**:668b–670b
modern, **3**:314a–b; **15**:198a
noncommutative, **2**:248a; **5**:309a; **6**:91a–92b; **10**:139a; **11**:55b; **14**:90b–91a

notation, **2**:281a; **3**:460b–461a; **4**:576a(n), 607a–b, 609a, 611b, 612a; **14**:19b, 20b–21a, 24b
number system of, **4**:618b
philosophy of, **3**:120a*; 7:242b, 243b; **12**:325b; **14**:19b–20b
of polynomials, 7:241a–245a; **12**:92a–93a
proportional, **3**:128b*
of quaternions
see Quaternions
rhetorical, **15**:698a–699a
semisimple, **1**:307b; **14**:211b–212a, 284b
simple, **14**:212a
symbolic, **3**:179a; **4**:56a, 112b; **6**:389a; **10**:437b; **11**:229b, 230a; **12**:92b, 325b; **13**:60a; **14**:19b
theory of, **9**:457b–458a
and time, **6**:90b–91a
topological, **8**:326b; **9**:71b
two-valued, **2**:297b–298b
von Neumann, **14**:90b–91b
see also Equations, Invariant theory, Number theory
Algebra i logika. Seminar
founding of, **9**:72a
Algeria
botany, **9**:24b, 35a
natural history, **4**:67b
Algidity
choleraic, **12**:96b
Algol (star)
double nature of, **14**:56a–b
observations of, **9**:486a; **12**:153a
orbit of, **3**:510a
variability of, **5**:468a–b; **7**:373b
Algology
eighteenth-century, **4**:178a; **7**:13b; **9**:368b
nineteenth-century, **6**:162b–163a; **7**: 533b–534b, 610a–b; **9**:587b; **11**:502a; **13**:154a, 394b–396a; **14**:93a
twentieth-century, **3**:259b; **7**:535a–b; **12**:125a–126b, 328a; **13**:165a–b
see also Algae
Algoman disturbance (geol.), **8**:99b
Algorism, **1**:22b–23a, 30b; **3**:272b; **4**:605a; 7:363a, 418b, 532b; **15**:10a–b, 256a
theory of, **9**:72a
Alhazen
see Ibn al-Haytham
ʿAlī
see last element of name
Aliboron
see al-Bīrūnī
Alidade, **10**:299a
Alikhanian, Artemii Isaakovich, **15**:8a
Alikhanov, Abram Isaakovich, **15**:8a, 251b
Alimentary canal
morphogenesis of, **5**:283a–b
Aliphatic acids, **3**:246b
Aliśiṅgarāja
commentaries of, **2**:116a
Alizarin
see Dyestuffs
Alkahest, **5**:422a; **13**:234b
Alkali(s)
seventeenth-century study of, **13**:234b
eighteenth-century study of, **2**:173b, 175a–177b, 180b–181a; **5**:353a–b; **9**:106a
nineteenth-century study of, **14**:600b; **15**:7b
"animal," **2**:381a
characterization of, **2**:381a; **8**:82a,b; **12**: 320b
compounds of, **11**:109a
electrolysis of, **3**:602a
identification of, **2**:381a
medical use of, **14**:320b
nature of, **8**:173b
in plant nutrition, **6**:381a; **8**:344b, 346a
production of, **7**:277b; **12**:478b; **14**:197b
adulterants, **6**:385b
from vegetable sources, **4**:250a
salts of, **6**:255a
spectra, **11**:478b–479b
strength of, **2**:381a
"vegetable", **2**:381a; **5**:422a; **9**:106a; **12**: 320b; **14**:332b
Alkali metals
electrochemical properties of, **6**:372a
isolation of, **2**:92b
oxides of, **1**:579b
radiation studies, **5**:459a
solutions of, **7**:497b
spectra, **3**:124b; **4**:253b; **7**:267b–268a; **10**:422b; **11**:580b; **12**:20b
thin films of, **4**:253a–b
photoelectric effect, **7**:36b
Alkaline earth metals
compounds of, **11**:109a
isolation of, **2**:92b, 588b; **3**:602a
neutralization of, **11**:436a–b
salts of
separation of, **5**:165a
solutions of, **7**:497b–498a
spectra, **3**:124b; **7**:268a; **10**:422b; **12**:20b–21a
Alkalinity
and electricity, **1**:348a
theory of, **3**:602a
Alkaloids, **1**:156b; **3**:159b–160a, 300a; **5**: 100a; **6**:293b; **9**:9b; **13**:601b; **15**:10b
action of, **8**:454b; **9**:296a, 557b; **14**:256a
analysis of, **4**:243a; **8**:331b–332a; **9**:178a; **10**:498a
discovery of, **10**:497b, 498a
formulas for, **4**:243a; **14**:411b
localization of, **4**:401b
metabolic occurrence of, **5**:315b
nature of, **3**:279b; **6**:462a–b; **12**:320b
oxidation of, **11**:278b
physiological effects, **9**:19b; **11**:615b; **15**:103b
production of, **3**:246a; **5**:394a, 421a; **8**: 329b
quinine, **12**:452a–b
structure of, **5**:589a; **10**:517b; **14**:334b; **15**:252b, 257b
synthesis of, **7**:552a; **15**:2b
tropine, **14**:411b
Alkalosis, respiratory, **6**:265a
Alkanes, **14**:602a
synthesis of, **14**:531a
Alkarsine, **2**:587b
Alkindus
see al-Kindī
Alkylation, **3**:246b
catalytic, **7**:21b
Allamand, Jean Nicolas Sebastian, **5**:509b, 510a; **9**:596a,b; **14**:193b
Allan, Thomas, **6**:18b
Allantoin, **8**:343a–b
metabolism of **9**:284b
Alleaume, Jacques, **6**:271b; **14**:23a
Allee, Warder Clyde, **7**:99a
Allegheny Mountains
geology, **8**:615b
Allegory
ancient Greek, **10**:208a
medieval, **1**:91b; **6**:396b–397b
eighteenth-century, **3**:579a
Allen, Bennet Mills, **12**:473b
Allen, Edgar (1892–1943), 1:123–124
Allen, Frederick Madison, **1**:443a
Allen, John (medic.), **8**:37b; **12**:523b
Allen, John Frank, **8**:478a
Allen, Thomas, **2**:435b; **4**:95b; **6**:127b
Allen, Willard, **2**:345a
Allenes, **4**:553a
stereoisomers, **9**:403a
Allergy, **9**:545b; **14**:624a
concept of, **7**:427a
hemorrhagic, **12**:96b
Allgemeine berg- und hüttenmännische Zeitung
founding of, **6**:143a
Allgemeine Elektrizitäts-Gesellschaft (AEG), **15**:490b, 491a
Allgemeine nordische Annalen der Chemie, **5**:558b
Allgemeines Journal der Chemie, **3**:465b
editors of, **7**:254a; **15**:171b
Allgemeines Magazin, **13**:425b
Allgemeines Repertorium der Mineralogie
editors of, **8**:245b
Alliaco, Petrus de
see Ailly, Pierre d'
Allibone, Thomas Edward, **3**:329b; **12**:33a
Alligator
nervous system, **8**:122a
Alling, Harold Lattimore, **3**:313b
Allison, Anthony Clifford, **3**:572b
Allison, Samuel King, **4**:197b*, 582b; **8**:94b
Allison, Victor Douglas, **5**:29b
Allman, George Johnston, **6**:410b, 411b; **9**:269a, 272b, 273b, 274a
Allochthony, **8**:544a
Allophane, **2**:587b
Allose, **8**:275b
Allotropy, **2**:70b; **6**:177b, 438b; **12**:495b; **13**:246b, 468a
concept of, **2**:95b
of minerals, **8**:119a
see also Polymorphism; specific elements
Alloxan, **8**:343a,b
Alloxantin, **8**:343b
Alloxic acid, **8**:343b
Alloys, **2**:143a; **15**:649b
aggregation, **9**:60b–61a
analysis of, **7**:325b, 339a–343a *passim*, 347a–348a, 529a–b; **8**:119a, 312a; **9**:300b; **11**:485b; **13**:245b–246a, 310b
cooling, **11**:550a
cooperative phenomena, **5**:103b
corrosion of, **7**:57a
cracking of, **3**:239a
hardening in, **9**:312b; **11**:101a
phase rule, **11**:534b
phases
theory of, **6**:560b–561b
physical properties
electrical conductivity, **5**:576b; **7**:529a–b; **8**:119a; **11**:446b
expansibility, **5**:582b–583a
flow pressure, **7**:529b
hardness, **7**:529b
thermal conductivity, **11**:446b
solidification of, **11**:485b
spectroscopy of, **1**:166a
structure of, **8**:119a; **11**:550a; **12**:544a; **13**:245b–246a
systems, **6**:561a
testing, **13**:331a
theories of, **8**:119a; **11**:486a
thermal treatments, **5**:583b
types
antifriction, **8**:210a
coinage, **11**:563a
eutectic, **3**:212a; **11**:486a; **13**:8a
fusible, **9**:106b
magnetic, **1**:293a
minimum-melting-point, **13**:327a
nonferrous, **3**:87a

Alloys (cont.)
steam-resistant, **3**:239a
superconducting, **8**:480a
see also Metallurgy; specific metals and alloys; Steel
Allsopp and Sons (brewers), **5**:536b
All-Union Association of Scientists and Technicians
founding of, **1**:361b
All-Union Institute of Plant Breeding (VIR), **15**:506a–b, 510a
All-Union Lenin Academy of Agricultural Sciences (VASKhNIL), **15**:506a
Alluviation, **2**:549a; **8**:241b
Allyl, **14**:622b
Almanacs
ancient Greek, **4**:172a, 460a
medieval Arabic, **1**:437b; **13**:401a; **14**: 593a–b
medieval Latin, **7**:124b; **14**:390a–b
Renaissance, **4**:97a; **14**:583b; **15**:154b, 478a
seventeenth-century, **4**:214a, 551b; **7**: 373a–b; **9**:485a; **14**:285b–286a, 446b
eighteenth-century, **2**:220b; **5**:434b; **6**: 512b; **7**:374a; **11**:471b
nineteenth-century, **4**:36b
twentieth-century, **5**:405a; **7**:6a–b
Chinese, **12**:375a
Mayan, **15**:760a, 771a–773b, 777a–b, 782b, 783a–784a, 790b, 802b–804a, 807a, 810a, 813b
medical, **11**:301a
Mesopotamian, **15**:679b, 680a–b
nautical, **4**:481b; **5**:404b; **7**:6b; **10**:614b; **13**:184a; **15**:353b
perpetual, **7**:124b
mechanical, **9**:529b
Almansori
see Yaḥyā ibn Abī Manṣūr
Almenar, Juan, **14**:28a
Almonds
analysis of, **11**:495a; **14**:476b, 477a
see also Hydrocyanic acid
Almoravids, **1**:408b, 414a; **14**:637b
Almucantar (astron. instrum.), **3**:194b
Aloes, **9**:152b
Alpago, Andrea, **1**:539a; **9**:603b–604a; **12**: 8a; **15**:499a–b
Alpago, Luigi, **1**:109a
Alpetragius
see al-Biṭrūjī al-Ishbīlī
Alpha particles
absorption of, **2**:398a; **3**:564a
binding energy, **9**:38a
bombardment by, **5**:272a; **6**:117b–118a; **7**:153a–b, 154a–b; **11**:281a; **12**:32b–33a, 33b
charge-to-mass ratio, **12**:28b
counting of, **5**:330b–331b; **12**:30a–b
deflection of, **9**:260a; **12**:28b
detection of, **5**:394b; **14**:422a
emission of, **3**:329b, 480a; **5**:272a; **12**:33b
energy loss, **2**:241b–242a
Geiger-Nuttall law, **12**:32a
and helium, **12**:504b
impact of, **2**:250b–251a
ionization density, **9**:261b
nature of, **5**:331a; **12**:30a,b; **13**:107b
and nuclear transmutation, **2**:338a–b; **12**:28b
photographic quantum sensitivity, **10**: 136b
ranges of, **2**:398a; **5**:331a; **7**:158a
resonance of, **5**:272a
scattering of, **2**:240b, 337b–338a; **5**:331a; **9**:543b; **12**:30b–31b; **13**:369a; **15**:42b
scintillation effects, **4**:356a; **11**:347b–348a; **12**:30b
sources of, **2**:338b; **7**:152b
spectrum of, **3**:502a–b
total charge, **4**:194b
Alphabets
universal, **1**:573a; **14**:68b–69a
Alphanus, **10**:21b
Alpharabius
see al-Fārābī
Alpher, Ralph Asher, **5**:273a
Alphonso
see Alfonso
Alpini, Alpino, **1**:124b, 125a
Alpini, Prospero (1553–1616), 1:124–125; **2**:226a
botany, **14**:12b
Alps, **4**:27b
Allgau, **6**:229a
Apuan, **3**:328a; **11**:575b
Atesine, **1**:234a
Bavarian, **2**:489a; **12**:167b
biogeography of, **5**:66b; **13**:25b
botany, **1**:525b, 526a; **3**:318a; **5**:379a, 380a,b; **6**:221a, 533b; **8**:121a; **10**: 575a; **12**:169a; **14**:116b, 117a
eastern, **13**:284b, 285a
French, **1**:208a; **2**:191a; **6**:168b
genesis of, **1**:236b; **5**:221a,b; **13**:610a
geography, **1**:208a; **2**:191a; **6**:550a
geology
Renaissance, **8**:241b, 242b
eighteenth-century, **1**:234a–b; **2**:554a; **3**:210b–211a; **4**:402b–403a; **12**:119b–122b, 159a
early nineteenth-century, **2**:482a, 495b, 555b, 569a; **4**:275b*; **10**:8b
mid nineteenth-century, **3**:328a, 434a; **4**:349b; **9**:583a; **10**:212a; **11**:375a, 439a; **12**:167b, 276b; **13**:123b–124a, 143b, 145b–146b
late nineteenth-century, **2**:89b; **3**:553b; **6**:168b, 227a–228a; **7**:44a; **9**:367b; **10**:29a; **12**:139b–140b
early twentieth-century, **1**:235b–237a; **2**:489a; **5**:562b; **6**:228b–229b; **8**: 543a–544b; **11**:575b–576a; **13**:283a–285b
mid twentieth-century, **1**:394a
geomorphology, **9**:150a–b; **10**:505a–b, 508a; **11**:277a
glaciers, **4**:73a; **5**:69a, 337b, 339a, 599b; **8**:544a; **9**:241a; **10**:502a, 504a, 505b; **11**:272a
Glarus, **13**:146a
High Calcareous, **11**:375a
high-altitude research, **2**:62b
mineral springs, **5**:380b
paleontology, **11**:386a
Pennine, **1**:236a; **8**:543b, 544a
Prealps, **3**:533a; **5**:221a–b; **8**:543b, 544a,b; **11**:375a; **12**:139b–140b
Rhaetian, **1**:525b
Savoy, **2**:495b, 569a; **8**:543a, 544a
Swiss, **4**:402b–403a; **5**:221b; **6**:221a, 227a–228a, 228b–229b; **12**:159a; **13**:123b–124a, 285a
tunneling, **9**:267b
western, **1**:236b; **6**:168b
Alsace
geology, **14**:88a
Alsted, Johann Heinrich (1588–1638), 1: 125–127; **2**:163a–b; **7**:194b
Encyclopaedia, **3**:359b
Alston, Charles, **2**:173b, 175a–b, 176b; **9**: 482b
Altai Mountains
geology, **3**:434a; **6**:238b; **10**:168a–169a
Altamira (cave), **2**:450b; **9**:539a; **10**:607a
Altazimuth instrument, **3**:93b; **7**:259b; **11**: 285a; **14**:485a
Altenstein, Karl von Stein zum, **10**:11b, 12b
Alternation of generations, **2**:160a, 371b, 492a, 520a; **3**:30a,b; **5**:539b; **6**:465b, 466a, 590b; **9**:332a; **11**:152a; **12**:107a, 125a,b, 183a; **13**:9b, 88a, 89b, 165a–b; **15**:78a, 82b
theories of, **8**:5a–b; **11**:154a–b
"antithetic" theory, **2**:371b
Althoff, Friedrich Theodor, **1**:576b; **4**:297b, 298a, 299b; **7**:425b, 426b; **8**:181a; **10**:18a; **11**:613a; **12**:527a; **13**:243b; **14**:171b
Altimetry
see Altitude, determination of
Altitude
adaptation to, **6**:319b
and atmospheric pressure, **2**:40a; **3**:64a; **5**:428b, 528b; **6**:512b; **8**:267b; **9**:117b; **10**:333a,b; **13**:267a; **15**:355a–b
and blood count, **9**:380b
and boiling point, **3**:16b; **9**:443a
and climate, **6**:550a
determination of
Renaissance, **9**:254b
medieval Latin, **6**:544b; **14**:31a
eighteenth-century, **5**:380a
nineteenth-century, **8**:101a, 187a; **13**: 111a–b, 120b
twentieth-century, **13**:114a
algal growth, **2**:430b–431a
by barometer, **2**:343b; **4**:28b; **8**:68b; **9**:471b; **13**:445a
by boiling point of water, **3**:16b
trigonometric, **6**:240b
and fertility, **6**:551b
and health, **4**:171a–b
and life, **6**:221a
and magnetism, **2**:135a
optical effects, **1**:519b
physiological effects, **1**:454b; **2**:61b–62b; **6**:24b; **9**:546b, 547a; **12**:159b
erythrocyte number, **2**:62b
oxygen dissociation curve, **1**:454a
and plants, **3**:16b, 45a; **6**:221a, 550a; **11**:272a
and solar radiation, **4**:171a–b
see also Hypsometry
Altitude and azimuth instrument, **2**:471b; **4**:175b; **13**:471a*
Altmann, Richard, **1**:613b
Altomare (Donato Antonio), academy of, **11**:96a
Alton, Johann Samuel Eduard d', **4**:146a
Alton, Joseph Wilhelm Eduard d', **10**:287a,b
Altona
observatory, **10**:542b; **12**:234b
Altrose, **8**:275b
Altshuler, Semen Aleksandrovich, **13**:240b
Alum, **4**:250a; **15**:653a
analysis of, **3**:201a–b, 316a; **9**:106a; **10**: 177b
crystallography, **5**:186b; **9**:106a
isomorphism, **5**:324a
manufacture of, **3**:201a, 202a, 316a; **4**: 74a; **6**:381a; **8**:147b
medical use of, **11**:177b
mining of, **10**:555b
natural formation of, **5**:72b
Alumina, **9**:106a
as catalyst, **7**:21a; **12**:47a
in cements, **8**:117b
colloidal, **3**:488b
Aluminic acid, **6**:371a
Aluminosilicates
structure of, **13**:618a
Aluminum
alloys, **3**:87a; **5**:576b, 583b; **7**:57a, 92b; **11**:486a, 550a

alpha bombardment of, **5**:272a; **7**:154a–b
cold-rolled, **3**:87a–b
compounds of, **11**:437a–b; **14**:475b
electrochemical properties of, **6**:370b–371a, 371b–372a
isolation of, **2**:588b; **14**:475a–476a
light alloys of, **3**:239b
metallography, **3**:87b
in minerals, **13**:618a
precipitation of, **3**:193a
production of, **3**:87a–b; **4**:77b; **6**:50b–51a, 319a–b; **8**:116a
separation from its sulfide, **2**:559a
toxicity, **6**:443a
X-ray absorption, **14**:218a
Aluminum Company of America
founding of, **6**:50b
Alvarez, Luis Walter, **3**:370b; **8**:96a; **13**:570a
Alvarus Thomas
see Thomaz, Alvaro
Alveolar theory, **2**:627a
Alves, Francisco de Paula Rodrigues, **15**:96b
d'Alviano, Bartolomeo, **5**:104b–105a
Alzate y Ramírez, José Antonio (1738–1799), 1:127
Alzheimer, Alois, **10**:130b–131a
Amagat, Émile (1841–1915), 1:128–129; 2:458a; **13**:244b
Amaldi, Edoardo, **4**:578b, 579b, 580a
Amalgamation (metall.)
Renaissance techniques, **10**:515a
eighteenth-century techniques, **2**:315b; **4**:345b, 346b
nineteenth-century techniques, **3**:481a*
Amalgams, **2**:92b; **11**:550a
properties of, **11**:418a
Amarāja
commentaries of, **2**:417b
Amarantite, **5**:162a
Amareśa, **2**:416b
Amaryllidaceae, **6**:296a
Amatus Lusitanus
see Lusitanus, Amatus
Amazon (river)
botany, **12**:594a–b
exploration of, **6**:551a; **12**:578b; **14**:133b–134b
Amber, **5**:442a; **11**:39b, 225b; **15**:648a
artificial, **8**:213b
deposits of, **13**:525b
distillation of, **10**:498a
medical use of, **10**:237a
Amber effect, **5**:397b–398a
Ambergris, **7**:205a
isolation of ambrein from, **3**:159b
Amblystoma, **4**:397b
Amboceptor, **4**:299a
Ambrein
isolation of, **3**:159b; **10**:497b
Ambronn, Hermann, **9**:601a; **12**:423a
Ambronn, Leopold, **1**:352a; **5**:554b; **13**:474b
Ambrose, bishop of Milan, **1**:333b; **7**:27b
Ambystoma, **3**:332a; **15**:126b
Ameba
see Amoeba
Ameghino, Carlos, **1**:131a
Ameghino, Florentino (1854–1911), 1:129–132
Ameghino, Juan, **1**:131a
Amenemhet, **15**:662a(n), 714a
America
botany, **15**:267b
discovery of, **1**:381b; **10**:617b; **15**:89b
naming of, **12**:199b, 323b; **14**:127b
natural history, **13**:530b
pharmacy, **9**:466a–b
settling of, **6**:528a
see also Canada, Latin America, New England, North America, South America, United States
American Academy of Arts and Sciences, **12**:350a; **13**:350b
American Air Almanac, **15**:129a
American Almanac and Repository of Useful Knowledge, **10**:479b
American Association for Cancer Research
founding of, **4**:499b
American Association for the Advancement of Science, **3**:130b, 293b–294a, 392a; **9**:552b; **10**:12b; **11**:219a–b; **12**:350a
founding of, **9**:197a; **11**:340b–341a, 504b; **12**:434a
American Association of Anatomists, **9**:57a–b, 58a, 416b
American Association of Economic Entomologists, **12**:98b
American Association of Geologists
founding of, **6**:437b
American Association of Neurological Surgeons, **3**:517b
American Association of Pathologists and Bacteriologists
founding of, **3**:447b
American Association of Physical Anthropologists
founding of, **6**:527b
American Association of University Professors, **8**:517a
founding of, **3**:130b
American Association of University Women, **15**:124a
American Association of Variable Star Observers
founding of, **10**:601a
American Astronomical Society, **6**:529b
founding of, **6**:29a
American Breeders Association
see American Genetic Association
American Breeders' Magazine, **3**:123a
American Cancer Society
founding of, **4**:499b
American Chemical Journal
founding of, **11**:371a
American Chemical Society, **4**:182a; **9**:394b
formation of, **3**:293b–294a; **6**:564b
Petroleum Research Fund, **15**:2b
American Chemical Society Scientific Monographs
editors of, **10**:158a
American Commission to Negotiate Peace, **2**:500b
American Crystallographic Association, **4**:522a
American Cyanamid Company
Lederle Division, **4**:220b
American Ephemeris and Nautical Almanac, **2**:332b; **6**:399a–b; **10**:34a, 479b; **11**:552b, 553a
American Epidemiological Society
founding of, **7**:170b–171a
American Ethnological Society
founding of, **12**:204b
American Expeditionary Force
Geologic Section, **2**:500b
American Genetic Association
founding of, **3**:123a
American Geographical Society
South American studies, **2**:374a
American Geological Society
founding of, **12**:434a
American Geologist
editors of, **13**:531b; **14**:443a
founding of, **14**:439b
American Institute of Physics, **9**:256b
American Journal of Anatomy
founding of, **9**:57a
American Journal of Mathematics
editors of, **14**:454a
founding of, **13**:218a
American Journal of Physical Anthropology
founding of, **6**:527b
American Journal of Physiology, **2**:367b
American Journal of Psychology
founding of, **6**:53b
American Journal of Science, **3**:553a
editors of, **10**:42a; **13**:473b; **14**:1a
founding of, **12**:433b
American Journal of Science and Arts
critics of, **9**:100b
editors of, **12**:434b
American Mathematical Monthly
editors of, **9**:388a
American Mathematical Society
Bulletin, **3**:345b–346a; **14**:559a
founding of, **4**:618b; **9**:502a
American Medical and Philosophical Register, **6**:521b
American Medical Association, **15**:72b
American Men of Science
founding of, **3**:130b
American Meteorological Society, **11**:558a
American Metrological Society
founding of, **10**:42a
American Midland Naturalist
founding of, **10**:122a
American Museum of Natural History
in late nineteenth century, **10**:241a–b; **14**:312b–313a
anthropology, **11**:219a
expeditions, **5**:488a*, 500b
ichthyology, **3**:610b
paleontology, **5**:500a–b
zoology, **14**:291b, 329a
American Naturalist, **2**:103a; **15**:93a
founding of, **9**:536b; **10**:272b–273a
American Ornithologists' Union
founding of, **3**:438b; **9**:313b
American Philosophical Society
in early eighteenth century, **3**:345a
in late eighteenth century, **7**:89b; **11**:472a,b
in early nineteenth century, **1**:364a
founding of, **1**:487b; **5**:129a–b
women in, **3**:50b
American Photographic Society, **4**:179a
American Physical Society
and Compton, **3**:368b–369a
founding of, **1**:490b; **11**:579a
American Physiological Society, **15**:71b, 72b
founders of, **2**:367b; **6**:526a; **9**:142b, 143a
American Phytopathological Society
founding of, **4**:220b
American Psychoanalytic Association, **5**:179b, 180a
American Psychological Association, **14**:550b
American Public Health Association
Standard Methods of Water Analysis, **12**:481a
American Revolution, **1**:435b, 436a; **10**:438a; **11**:471b, 616b
American Society for Horticultural Science
founding of, **1**:396b
American Society for Pharmacology and Experimental Therapeutics
founding of, **1**:10b
American Society of Agronomy
founding of, **4**:220b
American Society of Biological Chemists
founding of, **1**:10b; **8**:556a
American Society of Mammalogists
founding of, **9**:313b
American Society of Mechanical Engineers
founding of, **13**:399a

American Society of Naturalists
founding of, 6:613a
American Society of Zoologists
founding of, 14:314b
American-Soviet Medical Society, 15:75b
American Telephone and Telegraph Company, 7:107b
American Union Academy of Literature, Science and Art, 4:182a
Amerindians
archaeology, 11:177a; 14:443a
ethnology, 6:527b–528a; 9:540b; 10:285a; 12:204a–b
languages, 2:212b; 11:119a–b; 14:328b; 15:761b–762a
literature, 14:456b
North America, 1:489a–b; 3:344b, 351a, 608b; 6:567b; 7:210b; 14:328b
classification of, 11:119b
Mexico, 1:48a
Northwest Coast, 2:207b, 209a, 212a–b
numeration, 15:761b–762a
South America
Argentina, 1:130a
Brazil, 14:328b
Peru, 1:48a
Ames, Joseph Sweetman (1864–1943), 1: 132–133
Ames, Louise Bates, 5:378a
Ames, William (1576–1633), 1:133–135
Amethyst
medical use of, 6:397a
Ametus
see Aḥmad ibn Yūsuf
Amici, Giovan Battista (1786–1868), 1:135–137; 2:492a; 4:161b; 6:325a; 10:267a; 12:36b, 174b, 268a
Amici objective, 1:19a
fertilization theory, 6:266a
Ibn al-ʿAmīd, 4:438b; 7:343a
Amide radical, 4:244b
Amides, 9:183a
Amination, 3:246b
Amines, 5:373a
discovery of, 14:530b
isolation of, 5:4b; 6:462b; 15:10b–11a
isomerism, 11:88a
sympathomimetic, 15:11a
synthesis of, 4:20b; 5:215a; 6:462a
Amino acids, 2:170b; 3:336a; 5:149b, 273a; 15:15b
aromatic, 5:6a
derivatives of, 3:511b; 15:10b
determination of, 8:275b; 10:242b–243a
discovery of, 13:575a; 14:492a
measurement of, 12:546b
metabolism of, 4:360a; 5:53b; 8:555a
nutritional role, 5:451b, 452a; 6:501a; 9:284b, 285a
photochemistry of, 14:247b
physical chemistry of, 13:421a
polymers of, 12:164a
sequences of, 7:467a
structures of, 15:257a
synthesis of, 5:4b, 5a, 215a; 12:546a–b; 14:602a–b, 603a
see also specific amino acids by name
Aminoacetic acid
synthesis of, 10:516b
p-Aminobenzoic acid, 4:358a
Ammeter, 6:342a
Ammines
cobalt, 5:349b–350a
Ammocoetes, 6:167b
Ammonia
as chemical type, 6:462a–b
composition of
eighteenth-century analysis, 2:77a; 6: 395b; 8:82a
nineteenth-century analysis, 6:285a; 11:437b
compounds of, 11:541a; 14:267a–271b
with metals, 3:301b, 321b; 7:224a, 497b–498a
condensation of, 9:152b
as coolant, 8:365b
decomposition of, 7:518b; 13:313a
derivatives of
organic, 14:530b
discovery of, 7:42b; 11:145a
liquefaction of, 5:603a
liquid
as solvent, 7:497b–498a
in metabolism, 10:326b; 11:179b
metallic nature of, 3:602b
in rainwater, 2:356b, 357a; 8:344b, 345b
reactions of, 5:320a; 8:334b; 14:445b
salts of, 5:90b; 8:171b; 13:310b
strength of, 5:340b
synthesis of, 5:459a; 8:118b; 12:146a; 15:462a
industrial, 2:323b–324a; 3:202a, 239a, 299a–b, 436b–437a; 4:165b; 5:621b–622a; 8:118b; 9:427a,b, 467a; 12: 520b; 15:435b, 438a
urinary, 13:575a
Ammonites, 1:285a; 2:110a, 490a; 8:242b; 10:428a; 13:145b
animal nature of, 7:197b
evolution of, 6:168b, 613a,b, 614a
geometrical regularity of, 7:252a
and stratigraphy, 1:285b; 10:128a, 212a
suture lines, 2:556a
taxonomy of, 1:285b; 2:109b; 7:252a; 9: 579a; 10:29b, 429b
Ammonium, 7:498a
salts
in fertilizers, 8:346b, 347a–b; 11:179b
optical activity, 11:85b–86a
Ammonium chloride
manufacture of, 3:347a*
vapor density, 13:299a
Ammonium compounds, 3:347a; 11:318b
metal, 8:109a; 12:164a
stereochemistry, 9:403a
Ammonium nitrate, 6:395b
as explosive, 9:59b
Ammonius, son of Hermias (*d.ca.* A.D. 517–526), 1:137; 1:272a; 2:228b, 230a; 4:488a,b; 7:134a; 12:440b
Almanac, 14:593a–b
commentaries by, 7:25a; 9:437a, 438b
philosophy, 12:441a
Ammonius Saccas, 2:229a
Ammonoidea, 15:53b–54a
Ammophila, 4:504b
Amnion
formation of, 8:357b
Amoebas
intestinal, 8:296b
motions of, 7:520a; 9:167a
nonpathogenic, 2:534a
Amoebiasis, 1:26a; 12:482a
Amoebocyte, 8:447b
Amontons, Guillaume (1663–1705), 1:138–139; 7:111a; 8:211a
friction, 3:441b, 442b
Amor (country), 15:723b
Amor (asteroid), 4:26b
Ampère, André-Marie (1775–1836), 1:139–147; 2:106a; 3:79a, 379a; 4:241b(n)
acquaintances of, 1:557b; 3:545b; 5:167a; 8:37b; 13:127a,b, 128b; 15:481b
and Avogadro's hypothesis, 3:47a
correspondence, 8:36a
critics of, 13:11b
electrochemistry, 8:36a–b
electrodynamics, 1:143b–146a, 202b; 5: 171a; 7:380b; 8:510b; 9:166b; 10: 134b; 15:193a
electromagnetism, 1:555b–556a; 4:532b, 533b; 5:407b; 6:277b, 278a; 9:204b–205a, 206b; 10:27b–28a; 13:183b
equations of, 10:28a
formula of, 1:129a; 6:251b–252a; 14:206b, 207a
geophysics, 6:503a
influence of, 9:548a; 10:135b; 13:128a; 15:423a
magnetism, 3:446a; 8:38a–b, 490b; 14: 208a
molecular theory, 1:346a, 518a–b; 3:48a; 4:243b; 8:10b
optics, 5:169b
scientific explanation, 1:141a–142a
solenoids, 1:203a
students of, 5:294b, 295a; 8:384b; 11:76b; 12:197a; 13:127b
Ampere (elect. unit), 14:205b
Amphibians, 11:503a; 12:216a
anatomy, 13:402b
cardiovascular system, 3:604b
classification of, 2:493b; 9:572b; 13:25b
embryogeny, 6:132a–133b; 10:39b; 11: 255b, 571b–573b, 597a–b
asymmetry, 10:104b–105a
endocrinology, 12:473b–474a
paleontology, 2:489a,b; 3:608b; 5:296a; 6:594b; 14:410a
Amphiboles, 14:60a
Amphioxus, 2:61a; 6:167b; 9:573a
anatomy of, 8:9a
embryogeny of, 3:390a; 7:475a; 8:586a
excretory organs, 2:364b
histology of, 11:234b
morphology of, 11:380a
Amphipoda, 13:8b–9a
Amphitrite (asteroid), 11:114b
Amplifiers, electronic, 4:7a; 5:652; 7:107b–108a
crossed-field, 6:546b
Amputation, 1:212b; 2:546b; 3:67b; 6:509a; 8:402a; 9:12a; 10:620b; 12:137a
Amsler (later Amsler-Laffon), Jakob (1823–1912), 1:147–148
Amsterdam
Athenaeum, 10:253b
Amsterdam, University of
astronomy at, 10:290a
founding of, 14:96a
physics at, 14:597b, 598b
Amulets, 1:212b; 11:244b
Amundsen, Roald, 9:443a; 13:166b
Amwald, Georgius, 8:309b–310a
Amy, Alain, 7:613a*
Amygdalin
discovery of, 8:333a; 11:495a
reactions of, 8:342b; 14:477a
Amyl nitrite, 2:547b; 11:419a
Amyl series, 3:10b
Amylases, 7:517b
Amylene
as anesthetic, 12:502b
Amyloid, 10:175b
extracellular origin of, 4:276b–277a
Amylopectin
structure of, 9:354b
Amyotrophy
Charcot-Marie, 3:205b
Anabaptists, 5:119a, 121a
Anabiosis, 3:588b; 5:56b; 7:273b
Anabolism, 5:282a–b; 6:300a
oxidative, 2:184b
Anaconda Copper Mining Company, 14: 441b
Anaerobiosis, 7:184a

Anaerobism, **10**:381b–382a
concept of, **10**:365a
obligate, **10**:389a
Anagram, as priority device, **6**:604a
Analemma, **14**:577a–b, 578a; **15**:210b–211a, 518b, 519b–520b, 548a, 552b
Anales de ciencias naturales
editors of, **11**:171b*
Anales de historia natural, **3**:154b
Analgesics, **14**:256a
Analogy
in astronomy, **6**:334a
in biology, **1**:262b–265b *passim*; **5**:192a; **6**:38a, 39b, 40b, 161a, 590a–b; **7**: 607a; **8**:27a, 127a, 128a–b, 415b; **9**: 65b–66b, 354b; **11**:217a, 570b–571a; **13**:402b; **14**:351b
concept of, **10**:263a
in chemistry, **4**:245a; **5**:370b–371a, 373b; **8**:56a–57a, 60a, 298a; **11**:389b; **12**: 197b, 198a
concept of, **1**:277b
in geology, **2**:567b, 569b; **8**:566b, 568a, 569b, 575a; **10**:32b
in mathematics, **11**:52b; **14**:347a
in medicine, **3**:577b; **6**:428a; **11**:617b
in optics, **11**:97b
in philosophy, **12**:156a
in physics, **3**:32b–34b; **5**:511a; **6**:89b, 277b; **7**:213b, 216a; **8**:255b–256b; **9**:117b, 205b, 206a, 210a–211a, 212a, 217a–b, 220b; **10**:56b, 60a; **11**:474a–b; **13**:76b, 365b, 377a–b; **14**:565a, 568a, 569a; **15**:287a,b, 312b, 315a, 343b
in psychology, **5**:253b
Analysis (math.)
ancient Greek, **6**:411a; **8**:192b
ancient Roman, **10**:300a
medieval Arabic, **7**:3a–b
seventeenth-century, **2**:47a–48b, 50b, 131a; **3**:615b–616a; **4**:55b–58a, 567a; **8**:161b–166b, 304a–305a; **9**:49b–51a, 303a–304a, 447a; **10**:45a–53a, 65a–68b, 120a–b; **11**:392a; **13**:586a; **14**: 148a–149a, 150a–b
early eighteenth-century, **1**:76a–b; **2**:54a, 330a; **3**:460a–461b, 485b; **4**:473b–480b; **5**:449b–450a; **7**:490b; **8**:610b–611b; **11**:392a, 399b–401a; **12**:443b–444b; **13**:67b–69b, 265a–267a
late eighteenth-century, **3**:75a–78a, 385a–b; **4**:473b–480b, 515a; **5**:375a; **7**:549b, 560a–571a *passim,* 619a–b; **8**:135a–140a, 299b; **9**:55a, 471a, 476a–b; **10**:327a–b, 572b–573a; **11**:164a, 401b–402b; **14**:180a; **15**:276b–283a, 292b–296a, 302b–311b, 316b–333a, 339a, 352a, 382b–384a
early nineteenth-century, **2**:274a, 274b–277a; **3**:132a; **4**:501b–502a; **5**:94b–98b, 302a, 304a,b, 306a; **6**:324a; **7**: 37a–b; **8**:137b–142a; **10**:111a, 437b, 573b–574b; **11**:6b; **15**:225a–226a, 199b–200b, 481a–490a
mid nineteenth-century, **2**:88a–b, 294a–295a, 350a, 351a; **3**:53a, 213b, 223a, 224a, 227a, 314a, 416b, 451a; **4**: 125b–126b, 205b–206b; **5**:261a, 471b; **6**:87b–90a, 95b, 230b, 307a–309a; **7**:50a–55a, 506b; **8**:62b–63b, 382a–385b, 432a–b; **9**:307a–308a, 410a; **10**:247b–250b, 437b; **11**:206b–207a, 447b–455b, 548a–b; **12**:289b–290a, 318b–319a, 469b, 524a, 534a; **13**:78a, 128a–b, 129b–130a, 237a, 238a; **14**:219b–224a; **15**:66b, 198b
late nineteenth-century, **1**:194b; **2**:356b, 470a; **3**:53b, 178a, 559b, 560a; **4**: 103a, 407a–b; **5**:203b–204b, 213a–b, 472a–473a; **6**:3b–4a, 451a–b, 472a–473a, 522a, 545b; **7**:167b, 397b–399a, 444a, 446a, 459b–460a, 477b–479b, 507b–508b, 530b, 573b–575b, 621a–b; **8**:61b–62a, 389a–b; **9**:124b–128a, 175a, 232a–b, 426a–427a; **10**:25a, 244b–245a, 441a–442a, 443a, 482a, 593b–594a, 610a–b; **11**:52a–57b, 407a–409b, 452a–b, 498b; **12**:212b–213a, 512b–513a, 535a; **13**:55a–58a, 81b, 216a, 249b–250b, 263b–264a; **14**:86a, 158a–b, 202a–b, 343a, 630a; **15**:65b
early twentieth-century, **1**:406b, 407a; **2**:144a–145b, 198a–b, 238a–239a, 320a, 358a–b; **3**:95a–96b, 220b; **4**: 177a–b, 547b–548a, 561b–562a; **5**: 103a, 201a–b, 482a, 608a–609a; **6**: 113a–114a, 176b, 235b, 286a–b, 391b–392b, 570b–572b; **7**:101b–102a, 408a, 453b, 514a–b, 515a, 616a–b; **8**:110a–112a, 253a–b, 284a–b, 366b–367a, 558a–559a, 562a–b; **9**:410b, 419b, 502b; **10**:117b, 513b; **11**:106b, 149a, 156b–157a, 260b–261a, 268a–269a, 327a–b, 459a–b, 470b–471a; **12**:187b–190a; **13**:26b–27b, 35b–36a, 138b, 428a–b, 548b, 561b–562b; **14**:13b–14a, 87a, 188a–189a, 281a–285a, 317a–b, 345a–346a, 572b–573a; **15**:22b–23b
mid twentieth-century, **4**:173b–174a; **5**: 342a–b, 508b; **6**:147a–148a, 548b; **8**:457a–b; **10**:514a, 553a–554b; **11**: 259b–260a; **12**:78a; **13**:424a–b; **14**: 89a–92a, 453b–454a; **15**:49b
abstract, **2**:304a
arithmetization of, **7**:507b–508b; **9**:307a
classical, **2**:303b; **3**:53b; **5**:351a
complex, **1**:194b; **4**:562a; **5**:307b; **7**:478b–479a; **11**:449a–450a; **12**:78a
Diophantine, **1**:367b, 584b; **3**:275a; **4**: 111b, 113a–117a, 474b; **5**:528a; **7**: 566a; **8**:306b; **9**:50a; **11**:512a
finite, **3**:77a; **15**:366a
foundations of, **2**:269a, 274a, 274b–277a; **3**:278b; **4**:205b–206a; **5**:97b, 351a; **11**:598a, 599a; **12**:364b, 446b; **13**: 250b, 599b
Fourier, **3**:370a; **4**:579a; **6**:245b; **9**:175a; **10**:3a, 4a; **15**:380a, 386a,b
functional, **1**:427a–428a; **3**:171a; **6**:5a; **7**:436b; **8**:384a; **9**:414a; **10**:610a–b; **11**:459a–b; **12**:78a, 187b–190a; **14**: 282a
general, **1**:427b; **2**:144a; **9**:502b
geometric, **15**:193a, 195b–198b
global, **2**:145b
harmonic, **4**:561b; **6**:98b; **8**:367a; **10**:585a; **13**:127b; **14**:282a, 345b–346b; **15**: 298a–300b, 385b
of electrical current, **3**:329a
mechanical, **12**:526a
spherical, **9**:225a–b
indeterminate, **2**:131a; **4**:117b, 389a, 605a, 606a,b, 607a, 610a–b; **7**:566a, 567b; **8**:135b, 138a, 161b
Chinese methods, **3**:250a–254a; **14**: 545b
Diophantine, **4**:111b, 113a–117a; **7**: 244a–245a
technical terms, **3**:253b
infinitesimal, **1**:591b; **3**:77a, 151a, 152a, 77a; **4**:125a, 476b–479b; **7**:567a–568a; **8**:164b–165a; **9**:50b–51a; **11**: 409a, 599a; **12**:75a–76a; **13**:586a; **14**:46b; **15**:366a,b
opposition to, **11**:512b
Japanese, **15**:751b
and molecular physics, **1**:143a
multivariate, **5**:10b; **14**:382a–383a
nature of, **2**:358b; **4**:567a
numerical, **6**:147a–148a, 510b
p-adic, **6**:286b
real variable, **5**:97b
spectral theory, **14**:281b–282a, 345b, 346a–347a
and synthesis, **7**:3b
see also Calculus; Combinatorial analysis; Geometry, analytic
Analysis, philosophical
ancient Greek, **1**:188a; **4**:416a–b, 425b; **11**:28a
medieval Arabic, **6**:203a–b
early seventeenth-century, **6**:445b, 447a; **7**:269a
late seventeenth-century, **10**:85b
early eighteenth-century, **14**:483b
late eighteenth-century, **3**:381a, 382b; **8**:81b
early nineteenth-century, **3**:1b, 2b
late nineteenth-century, **5**:211b–212a
Analysis, structural, **3**:117b–118b, 496a–b; **9**:268a
Analysis situs
see Topology
Analyst, **1**:65b
Analytical engine, **1**:356a
Analytical Society (Cambridge University), **6**:324a; **10**:437b
Anamorphosis, **10**:103b–104a
Anania Shirakatsi
see Shirakatsi, Anania
Anaphrodisia, **4**:67a*
Anaphylaxis, **1**:92b; **5**:316b, 317a; **6**:432b; **7**:97a; **9**:266a, 545b; **10**:155b, 243a; **14**:624a; **15**:105b–106a
discovery of, **9**:10b; **11**:101b, 430a
theory of, **11**:430b–431a
Anarchism, **11**:337b*
Anaritius
see al-Nayrīzī
Anas, **10**:16b
Anatexis, **4**:409b; **10**:126b; **12**:273a
Anaṭoli, Jacob, **4**:542a; **8**:280b
Anatolius of Alexandria (*fl. ca.* A.D. 269), 1:148–149; 4:111a
Anatomical Society of Great Britain and Ireland, **13**:504a
Anatomy
ancient Greek, **4**:106a; **11**:128a
Byzantine, **10**:20b–21b
medieval, **1**:4b, 102a; **6**:276a–b; **9**:41a, 603b–604a; **12**:80b–83a
Renaissance, **2**:336b–337a; **5**:580b–581a; **8**:194a, 554a; **11**:390b–391a, 604b, 605a; **12**:106a; **15**:80b
early seventeenth-century, **5**:288b, 425b–427a; **6**:151a–157a; **12**:333a–b
late seventeenth-century, **2**:237b, 309a, 311a; **5**:484b; **6**:183b, 484a; **9**:581b–582a; **10**:477a–478a, 564b, 567b–568a; **11**:586b–587b, 588a; **13**:30b–33b; **14**:509b
eighteenth-century, **3**:15b–16a; **4**:352a–b; **6**:395b, 566a–568a; **8**:146b; **12**:136b–138a; **15**:524b
early nineteenth-century, **2**:442b–443a
mid nineteenth-century, **5**:469a–470b; **6**:268a–270b; **7**:23b–24a; **9**:560b; **12**:231a, 354a
late nineteenth-century, **4**:12a; **10**:175b; **11**:381b–383a, 570a–571b; **13**:503b–504a; **14**:126a–b
early twentieth-century, **12**:473b–475b
anomalies, **9**:66a
applied, **6**:618b

Anderson, Thomas (1819–1874), 1:155–156; 8:403a, 11:278b
Anderson College, Glasgow, 10:511a
Anderson's Institution (Glasgow), 5:276b; 13:547a–b
Andersson, Johan Gunnar, 2:171b; 5:500b
Andes Mountains, 2:374a
exploration of, 10:240a,b
geology, 3:567b, 568a,b; 8:570b; 10:240a, 506b–507a
Andesite, 4:141a, 564a
Andoyer, Henri (1862–1929), 1:156–157; 6:548b; 13:423b*
historiography, 15:294b
Andrachne, 4:295b*
Andrada e Silva, José Bonifácio de, 15:7b
Andrade, Edward Neville da Costa, 12:32a
André, Charles Louis François (1842–1912), 1:157–158; 11:320b*; 14:480a*
André, Gustave, 2:70b
Andrea, Francesco d', 2:309a
Andrea, Paolo d', 9:146b
Andreae, Johann Valentin (1586–1654), 1: 158–160; 3:362b; 9:169b*; 14:226a
Chymische Hochzeit des Christiani Rosencreutz, 4:550b
Andreev, Nikolai Nikolaevich, 7:621b
Andréossy, Antoine-François, 1:327a,b
Andrewes, Christopher Howard, 12:492a,b
Andrews, Ethan Allen, 14:426a
Andrews, Roy Chapman, 5:500b; 13:275a
Andrews, Thomas (1813–1885), 1:160–161; 5:462a; 13:81a, 236b
physical chemistry, 1:128a; 3:9a, 11b; 5:389a; 13:244b; 14:109b
Andrey Pervozvanny (ship), 7:410b
Androgens
isolation of, 8:359a
Andromachos, 1:235a; 11:602b
Andromeda (constellation)
galaxies, 1:353a; 4:282a; 9:330a
gamma Andromedae
third component of, 3:605b
nebulae, 3:509a; 4:506a; 6:322b; 9:247a; 11:470a, 485a; 12:269b, 349a
distances of, 6:529a–b; 12:347b
photography of, 11:484b
radial velocity, 12:455a–b
supernova, 6:150a
Andromedid meteors, 2:126a
Andronia (chem.), 10:183a, 184a
Andronicus of Cyrrhus, 15:518a,b
Andronicus of Rhodes, 1:268a; 4:460b; 13: 329a
Andronov, Aleksandr Aleksandrovich (1901–1952), 15:6–7; 9:77a
Andrusov, Nikolai Ivanovich (1861–1924), 1:161–163
Andry, Nicolas
Traité de la génération des vers dans le corps de l'homme, 8:171b
Anemia
copper, 6:135b, 136a
diagnosis of, 4:296b
effects of, 13:53b
experimental, 9:417a
hemolitic, 9:95a
hypochromic progressive, 1:439a
treatment of, 9:417a
iron, 6:136a
milk-induced, 4:357b
miners', 10:527a–b
pernicious, 1:60a; 3:335b; 9:417a
splenic, 15:480a
Anemometer, 2:354b; 3:358a, 559a*; 8:198a; 14:292b, 495a
recording, 6:522b
Anencephaly, 11:99b
Anes, Pêro, 8:390a
Anesthesia
chloroform, 1:168b; 2:367a; 5:45a, 566a–b; 6:110a; 8:335a, 402a; 10:620a; 12:165a, 502b
colloidal phenomena, 1:431b
colonic, 13:158a
coniine, 14:256a
endotracheal, 3:518a
ether, 2:27b, 31a; 5:478b, 566a–b; 8:401a; 10:620a; 12:165a, 502a–b; 13:158a
opposition to, 9:10b
hypnosis, 9:242a
intravenous, 10:620a
local, 3:518a; 15:2b
neuroregional, 6:77b
rectal administration, 10:620a
soporific sponges, 2:314b
from spinal cord section, 2:525a
stages of, 12:502b
stovaine, 5:99b
Anesthesiology
medieval, 2:314b
nineteenth-century, 2:27b, 31a, 62b; 5: 478b, 566a–b; 6:110a; 7:45b; 10:620a; 11:99a, 462b; 12:165a, 502b
twentieth-century, 3:518a; 9:266a; 11: 482a; 13:262b
accidents of, 14:496a
Aneurysms, 2:478a; 4:173a; 6:78a, 509a*, 521a, 566b, 567a; 7:614a; 12:137a; 14:6a
Angeli, Stefano degli (1623–1697), 1:164–165; 2:310b, 311a; 3:183a; 5:524b, 526a, 527a,b; 11:399b
Angelica, 6:123b
Angelology, 5:549b; 11:231b
Angelus, Jacobus, 11:202b, 349a–b
Angelus (Engel), Johannes (*ca.* 1453–1512), 1:165–166; 15:478a
Angermannus, Petrus
see Artedi, Peter
Angina
epidemic, 5:384b
pectoris, 2:547b; 8:7b, 295b
Vincent's, 5:317a
Angiography
cerebral, 4:286b
Angiology, 14:26a
Angiosperms, 6:163a; 7:198a; 8:354a
as distinct from gymnosperms, 2:517a, 520a; 10:1b
embryogeny, 5:581b
fertilization in, 5:582a; 10:1b
growth of, 9:600b
morphology of, 3:187b
origin of, 8:354b
reproduction, 5:539b; 8:128a
taxonomy, 2:103a–b; 10:1b
Anglada, Joseph, 1:416a,b
Angle(s), 4:462b
about a point, total of, 7:80b
of contact, 2:618b; 3:311b; 10:494a; 14: 21b
of contingence, 3:515b
curvilinear, 8:239a
dihedral, 7:29b
division of, 3:183b
Euler, 4:479b, 480a
horn, 2:395b; 4:447a, 451b; 12:75a–b; 14:21b
Laplace's, 15:319a
measurement of, 8:137a; 15:336a
centesimal units, 4:15b–16a
repeating principle, 9:234a
multisection of, 2:48b; 14:24a
reentrant, 4:609b
right, 4:417a; 10:299a–b
signed, 3:469a
solid, 3:615a
trisection of, 1:41b, 224b, 227a,b, 245a, 445b; 2:280b; 3:25a, 61a; 4:103b, 104a; 6:406b, 407a–b, 410b; 7:87b*, 175a; 10:114b–115a, 161b, 295a–b, 300a, 342b, 544a; 11:240b; 12:75a, 431b, 432a; 13:291a; 14:21b, 49b
equation for, 7:258b–259a, 328a; 14:21a
Anglesea, 6:288b
Anglicanism, 2:568b–569a, 570a; 5:415a; 14:157a–b, 484b–485a
Anglo-Saxon language, 11:339b
Ango, Pierre, 10:315a
Angostura
bark, 4:560a
Angot, Charles Alfred, 6:98b
Ångström, Anders Jonas (1814–1874), 1: 166–167; 4:251a
spectroscopy, 1:425b; 14:55b
Ångström, Johan, 10:165a
Anguillae, 12:557b
Anguillara, Luigi (*ca.* 1512–1570), 1:167; 1:108b; 5:383b; 14:335b
Anguillidae
metamorphism of, 4:12a
Anguillula, 5:502b
reproduction, 5:56b
Angular momentum
see Momentum, angular
Angyal, Stephen John, 5:6b
Anhydrides, 3:10b
preparation of, 5:373b
Anhydrite
and gypsum, 2:159a
Anhydrobiosis, 5:385b
Aniline
ammonium compounds, 14:444a
and phenol, 6:462a
salts
as catalysts, 7:282a
see also Dyestuffs, aniline
Animal earth, 5:223a
Animal gravitation, 9:326a
Animal heat
see Heat, animal
Animal husbandry, 6:585a
Animal Industry, Bureau of (U.S.), 12:480b, 481a
Animal magnetism, 1:401b; 3:208b; 5:531a; 6:226b; 8:84a; 9:325b–327b, 568a
and electricity, 13:183b
opposition to, 12:592a
see also Hypnosis, Mesmerism
Animal oeconomy, 3:244a; 6:37a, 566a, 567a; 9:237b, 306b*; 13:122a
Animal spirits, 1:620a; 5:56a, 286b; 6:38a, 459a–b; 7:274b, 459a; 9:243b, 244b; 10:21a, 54b; 12:601b; 13:170a; 14:9b, 10b, 406a–b, 408a
Animalcules, 4:289b, 290a
Animalculism
opposition to, 2:579b; 8:580a
Animals
acclimatization of, 5:359b
anesthesia of, 2:27b
automatism, 4:61b–62a, 64b
breeding of, 3:570b, 574a
chemical composition of, 13:619b
colonies, 4:266b; 10:523a
coloration
changes in, 1:102a
inheritance of, 13:134a
survival value, 9:550a–b; 14:136a
theories of, 11:137a
decapitated, 13:543b
domestic, 1:175a; 3:219b, 521a, 570b, 574a; 15:648a

Animals (cont.)
artificial insemination, 7:31b–33a
fossil, 10:16b
domestication of, 11:567b
and coloration, 11:137a
physical effects, 10:16b, 209b
ecology, 1:405b; 2:221b; 3:550b; 4:285b
experimental use of
eighteenth-century, 6:36a,b, 37b–38a
nineteenth-century, 8:133b; 10:580b; 11:98b, 419a, 593a; 12:354b; 14:187b, 495b, 496a, 558b
twentieth-century, 5:40b; 6:519a; 8:591a; 11:443a; 15:100b
extinction
prevention of, 7:33a
geographic distribution of
see Zoogeography
iatromechanics of, 2:309a, 312a
individuality of, 6:590b
instinct, 3:521a; 5:45a*; 11:567b
intelligence, 3:553a; 5:45a*; 11:518a–b
killing of, 11:325b
in Koran, 3:548b, 549a
language of, 7:64b
as machines, 13:543b
medical use of, 2:422b; 3:548b; 7:459a
mummified, 3:526b; 5:356b; 10:16b
nutrition
see Nutrition, agricultural
plants, comparisons with
ancient Greek discussion, 13:332a
early seventeenth-century discussion, 7:537b–538a
late seventeenth-century discussion, 5:535b; 8:127a, 128a–b; 9:114b; 15:67b
early eighteenth-century discussion, 6:38a, 39b, 40b; 7:197b*, 607a
late eighteenth-century discussion, 3:37b, 426a; 6:549b; 7:586a; 9:151b
early nineteenth-century discussion, 3:579b–580a; 4:264a–b; 6:551b
mid nineteenth-century discussion, 3:337a–b, 612a–b; 9:345a; 10:364a; 11:217a; 12:231a–232a, 243a; 13:462b, 556a,b
late nineteenth-century discussion, 13:88b
chemical relationships, 10:22b
psychology of, 1:265b; 2:581a; 3:521a–b; 6:229b; 8:45b, 529a; 9:432a, 513a–b; 14:549b–551a
type theory, 5:443b; 15:167a, 168a,b
see also Dissection, zoological; Ecology; Ethology; individual species; Taxonomy, zoological; Veterinary science; Vivisection, animal; Zoology
Animism
Renaissance, 1:373b; 2:540a, 541a, 543a–b
seventeenth-century, 5:422a, 426b–427a; 6:145b; 7:301b; 15:69b
eighteenth-century, 12:601a–604a; 13:231b–232a, 543a; 14:321a–b
opposition to, 2:3a, 226b
universal, 2:540a
see also Naturalism, Vitalism
Anions
large, 7:498a
Aniruddha, 12:115b
Anisic acid, 3:10b
Anisole, 3:10b
Anisotropy, 9:601a
Anixiopsis, 6:100a
Ankyropteris, 12:260a
Anlagen (biol.), 14:234b, 235b
Annalen der Chemie, 11:359b
editors of, 7:464a; 14:335a
Annalen der Naturphilosophie
editors of, 15:463b
Annalen der Pharmacie und Chemie
editors of, 8:344a
Annalen der Physik, 4:189b
Beiblätter zu den Annalen der Physik, 2:266a
editors of, 15:172a
founding of, 5:532a
Annalen der Physik und Chemie
editors of, 11:50a,b; 14:329b, 330a
Annalen für Meteorologie und Erdmagnetismus
founding of, 7:608b
Annales de chimie (et de physique), 1:346a,b; 3:465b; 8:387b; 13:615a; 14:530a
editors of, 4:242b; 5:325a, 602b; 10:497a*; 12:304a
founding of, 1:64b; 5:90b; 8:81a; 9:472a
Annales de Gergonne
see Gergonne, *Annales de mathématiques . . .*
Annales de la médecine physiologique, 2:508a
Annales de l'agriculture française, 2:322a
Annales de l'Institute Océanographique
founding of, 1:92b
Annales de l'Institut Pasteur, 4:211a, 212a; 10:407a
Annales de mathématiques pures et appliquées, 1:237b, 240a; 2:214a,b; 3:461b; 5:367b; 8:382b
editors of, 8:306b
Annales de parasitologie humaine et comparée
founding of, 2:534a
Annales des mines
editors of, 12:304a
Annales des sciences naturelles
founding of, 1:328b; 2:492a; 4:242b
Annales d'hygiène publique et de médecine légale
editors of, 8:272a–b
Annales scientifiques de l'École normale supérieure
founding of, 10:355a
Annali di matematica pura ed applicata, 5:201a
editors of, 12:285a
founding of, 2:469b
Annals and Magazine of Natural History, including Zoology, Botany, and Geology, 2:19a
Annals and Magazine of Zoology and Botany
see *Annals and Magazine of Natural History, including Zoology, Botany, and Geology*
Annals of Anatomy and Physiology, 5:470a
Annals of Botany
founding of, 1:422b
Annals of Electricity, 13:126b
Annals of Eugenics
founding of, 10:464b
Annals of Mathematical Statistics
editors of, 14:383b
Annals of Mathematics
editors of, 14:211b
Annals of Medicine and Surgery
editors of, 11:172a
Annals of Philosophical Discovery and Monthly Reporter of the Progress of Science and Art, 13:126b
Annals of Philosophy, 10:108b; 13:372b, 373a, 414a
Annals of Philosophy, Natural History, Chemistry, Literature, Agriculture, and the Mechanical and Fine Arts, 5:276b
Annals of the New York Academy of Sciences
editors of, 11:84a
Anne, Queen of England, 3:295a,b; 6:36a, 68a; 7:276a; 8:416a; 10:83b; 12:458a
Annealing, 6:368b
Annecy
museum, 9:538a
Année biologique
founding of, 4:12b
Année psychologique
founding of, 2:131b
Annelida, 1:600b; 2:322a; 5:109a–b; 8:609a; 9:112b, 328b, 575b
anatomy, 6:269a; 9:560b
cytology, 12:423b; 15:513b
embryogeny, 3:148b, 389b; 14:426a–b
life cycles of, 12:107a
phylogeny, 8:3b; 11:234a–b
regeneration of, 7:33b
Annihilators (math.), 3:169b; 13:219b–220a
Annuaire astronomique et météorologique, 5:21b
Annuaire de chimie, 9:401b
Annual Record of Science and Industry, 1:405a
Annuities, 8:530a; 9:454a–b; 14:466b–467b; 15:280a, 373b, 374a
Anopheles, 5:503a–b; 9:82b; 11:556a
Anorthite
discovery of, 11:539b
Anorthosites, 1:51a
Anosov, Pavel Petrovich, 1:581a; 3:233a
Anoura, 3:588b
Anoxemia, 2:61b
Ansa
Haller's, 3:489b
Anschütz, Richard (1852–1937), 1:168–169; 3:286b, 448a,b; 7:282b, 283a; 10:578a, 579a
Anselm, Saint, 7:483a,b, 484a–b
Anselme of Laon, 1:1b
Ant, 12:348a
colonies, 8:528b
hearing, 8:528b
intelligence, 8:529a
life-span, 8:528b
social instinct, 5:73b; 13:134b; 14:291b
taxonomy, 5:73b
vision, 8:529a
The Antarctic Manual, 9:587b
Antarctic Ocean
exploration of, 1:594b–595b; 3:203b–204a
ice in
classification of, 1:595b
South American, 3:204a
Antarctica
botany, 6:489a
discovery of
by exploration, 1:594b–595a
by lunar eclipse, 1:311b
exploration of, 3:204a, 549b, 557b, 591a, 616b; 6:327a, 489a; 9:124a, 198a; 11:554b; 13:167a
search for, 3:396b–397a
structure of, 9:589b
zoology, 13:552a
Antares (star)
position of, 3:318b–319a
Antelme, Pierre, 1:169a
Antelope
fossil, 2:494b
Antennae (biol.)
regeneration of, 3:492b
Antennas (elect.)
dipole, 9:98b
directional, 2:427b; 6:104b; 9:99a
resonant circuit, 8:444b
sparkless circuit, 2:427b
Anthelme, Voituret (*ca.* 1618–1683), 1:169
Anthelmintics, 11:37b

Anthemius of Tralles (*fl. ca.* A.D. 550), 1: 169–170; 1:121a, 223b; 4:488a,b; 7: 28b–29a
critics of, 15:264a,b
optics, 6:195a, 204a(n)
On Remarkable Mechanical Devices, 10:298a
Antheridium, 3:30a,b; 6:219a, 465b; 9:368b
Antherozoids, 9:601a; 13:542b
Anthoceros, 5:539b; 6:466a
Anthocerotaceae, 3:30a–31a
Anthocyanins, 3:359a; 14:412a; 15:257a
Anthonisz (Metius), Adriaen (*ca.* 1543–1620), 9:335–337; 5:409a
Anthoniszoon, Cornelis, 10:239a
Anthozoa
paleontology, 11:502a
Anthracene, 1:156b, 168b; 9:376b
compounds of, 8:58b
derivatives of, 5:488b–489a; 8:328b–329a
discovery of, 8:56a
formula of, 5:488b
solid-state transformation, 8:485b
structure, 8:329a
vapor density, 9:356a
Anthracite, 8:463a; 9:591b
Anthracosaurus, 6:594b
Anthranilic acid, 5:197b
Anthraquinone, 12:236b
synthesis of, 7:283a
vapor density, 9:356a
Anthrax, 1:575a; 5:219a, 269b; 7:422b; 10: 19a; 11:429a; 12:88a
bacillus, 3:340a, 587b–588a; 5:128b; 7: 421b–422a, 423a; 10:386a–b, 387b, 393b–394a, 397a,b, 493a
discovery of, 11:69a–70b
etiology of, 3:189a, 587b–588a; 4:276b; 7:421b–422a, 424a–b; 8:407b; 10: 384b, 386a–388b; 11:69a–70b, 569a
mechanism of, 10:155a
transmission of, 3:588a; 10:388a, 397b, 70a
treatment of, 10:387b–388a
vaccination, 3:189a; 4:211a; 5:270b; 7: 286b, 424b; 10:392b–398b, 407b
Anthropithecus, 11:270a
Anthropogeography, 3:533a; 7:513a
Anthropological Review
founding of, 6:563a–b
Anthropological Society of London, 6:563b
founding of, 6:563a
Anthropologie, 6:527b
L'anthropologie
founding of, 2:347a
Anthropologischer Anzeiger
founding of, 9:145a
Anthropology
medieval, 1:102a
Renaissance, 5:119b, 120a
eighteenth-century, 2:203b–204b, 581a–b; 3:2a, 38a, 198a; 5:75b, 77a; 6:469a; 12:509b
early nineteenth-century, 7:547b; 11: 136b–138a, 243a–b; 14:44b
mid nineteenth-century, 1:387b–388a; 2:161b–162a, 320b, 321a, 478a–b, 617b; 5:505a–506a; 6:8b, 138a, 280b, 563a–b; 8:44b, 528a–b; 9:540b–541a; 11:177b–178a, 234a, 235a, 399a, 514b; 12:175a; 14:57b, 113a, 128b
late nineteenth-century, 2:207a–209b; 3: 536b; 5:109b–110a, 196a–197a; 6:9b; 7:421b; 8:9a; 9:86a, 143b–144a; 11: 219a, 483b–484a; 12:298a; 13:504a; 14:42a–b, 126b; 15:133b
twentieth-century, 2:171b–172a, 505a; 4:353b; 5:144b–146a, 294a; 7:278b–279a; 8:104b–105b; 12:520a–b; 14:3a
comparative, 5:77a, 120a, 145b; 9:144a; 13:288a
criminal, 3:99b
historical approach to, 2:209a–213a
philosophical, 3:112a; 5:505a–506a; 8: 514b; 9:143b
physical
early eighteenth-century, 9:480a
early nineteenth-century, 5:277a; 11: 383a
mid nineteenth-century, 1:388a; 5:255a; 6:436a, 454a, 593b; 11:380b–381a; 13:396b, 404a
late nineteenth-century, 1:130a,b, 131a–132a, 173a, 562a–563a; 5: 46a–b, 109b–110a, 196a; 9:332b
twentieth-century, 2:211a–b, 212b; 3: 555b–556a; 6:527b–528a; 8:28b–29a; 9:144a–145a
"scientific" approach, 2:208a–213a; 9: 144b
serological, 6:433a
see also Archaeology; Classification, anthropological; Ethnology; Man; Paleontology, human; Prehistory; Somatology; specific countries
Anthropometry
eighteenth-century, 14:297a
mid nineteenth-century, 2:367a–b; 11: 514b
late nineteenth-century, 2:207a; 3:130a; 5:266a; 11:383a
early twentieth-century, 2:211b; 3:113a–b, 555b–556a, 590b; 9:144a,b; 10: 470b*
Anthropomorphism
critics of, 8:125a
Antibacterial agents, 5:29b–30a
see also Chemotherapy, Disinfectants, Immunization, specific agents, Vaccination
Antibiosis, 11:37b
Antibiotics, 4:220b; 5:29b–30a, 42a–43a; 8:273b–274a; 15:150b
concept of, 10:362b
effects of, 8:608a
interference phenomena, 8:274a
pharmacological action
and molecular structure, 9:296a
Antibodies
in blood, 1:575b; 2:300b; 5:317a; 7:624a; 12:483a, 492b; 13:420b; 15:522a
and fever, 5:270b
formation of, 15:479b
and heredity, 5:599a
placental transmission of, 5:599a
site of production, 12:48b
Anticlericalism, 2:315b; 6:446b, 448b, 468b; 7:157b, 607a
Anticoagulants, 6:526b
Anticosti Island, Quebec, 13:518b
Anticyclones, 5:266a; 9:108a
theories of, 6:98a
Antidotarium Nicolai, 6:123a
Antidotes
for arsenic, 2:72b, 587a–b
Antifreeze, 5:622a
Antigen-antibody reaction, 2:300b; 8:556b; 12:48b; 14:624a; 15:522b
Antigens, 2:300b; 4:297a; 5:317a; 8:273b
constituents of, 7:624b
heat, influence of, 14:624a
preservation of, 6:233a
stimulation of, 11:271b
Antigone of Carystus
see Diocles of Carystus
Antihistamines
derivatives of, 5:100a
Anti-hormones, 3:353a
Antilia, mythical island of, 1:574a
Antilles
botany, 4:67a*; 11:47b
geography, 10:617a
Antimony, 8:174a; 13:141a; 15:651b
allotropy, 13:70a
alloys, 13:310b
atomic weight, 3:398a
calcination of, 8:131a
colloidal, 9:295b–296a
compounds of, 3:201b, 617b; 12:572b; 13:70a, 559a
oxides, 5:157b; 11:169b; 12:306a; 13: 310b
sulfides, 11:169b
electrical conduction in, 10:500a
electrodeposition of, 5:474a
medical use of, 4:209a; 5:419b, 421a, 439b; 7:612b; 8:311b; 11:48b; 13: 490b, 559a–b
separation of, 6:165a*
Antinomies (math.), 2:593b; 3:278b*, 456a; 8:262b; 11:285b, 413b; 12:13a–b
in set theory, 3:56a, 57a–b
Antiperistasis, 9:133a; 13:330b
Antiphon (*fl. ca.* 430 B.C.), 1:170–172; 1:119b
squaring of circle, 1:119b; 6:412a,b, 416a; 8:236b–237a
Antiproton
discovery of, 8:96a
Antiquarianism, 1:317a, 317b–318a, 411b; 10:239b; 11:339b; 12:394a; 14:506a
Antiquaries, Society of, 3:350b
Antirrhinum, 8:513a
Anti-Semitism
medieval, 3:219a; 14:388a
Renaissance, 8:554a–b; 10:237a, 237b–238a
eighteenth-century, 5:466a
nineteenth-century, 3:336b; 4:402a; 11: 368a; 12:165a; 13:556a
twentieth-century, 8:180b, 182a–b, 284a; 11:347b; 12:530b–531a, 615a,b; 14: 121a–b; 15:451a
see also Nazism
Antisepsis
antibiotic, 5:30a
chemical, 1:493b, 574b; 2:130a; 3:185a, 189a; 11:37b; 12:294a, 295a–296a
intravenous, 5:29b
oral, 8:403a
electrical, 1:304b
hormonal, 1:303b
theory of, 8:403a–409a
see also Disinfection, Sterilization
Antitoxins, 4:211b, 404b; 7:427a
assay, 2:584a; 4:298a
discovery of, 11:569b
formation of, 12:88b
manufacture of, 4:297b–298a; 8:409a
standardization of, 12:481b–482a; 15: 107a, 521b
theory of, 1:301b, 575a,b, 576b; 7:392a; 15:522a
see also Serums
Antivenins, 10:141b
Antivitamins, 12:208a
Antoniadi, Eugène M. (1870–1944), 1:172; 12:161b–162a
Antonoff, Georges, 12:505b
Anuchin, Dmitrii Nikolaevich (1843–1923), 1:173–175; 7:512b, 513a; 9:594b*; 11:182a*; 15:428a*
Anūshirwān
Zīj al-Shāh, 7:360b; 15:581a, 605a
Anville, Jean-Baptiste Bourguignon d' (1697–1782), 1:175–176
Anwā᾽ (astron.), 11:246b; 12:447b

Anyscom, François-Xavier, **1**:341a
Aochi Rinso
Kikai kanran, **15**:742a
Aorai, Mount, **3**:551a, 552b
Aorta, **1**:267a, 619b
diseases of, **14**:26a
ligation of, **10**:620a
sinuses, **8**:204a
valve cusps, **8**:204a
Apáthy, Stephan (1863–1922), 1:176–177
Apatite, **4**:598b
deposits of, **5**:212b
Apeiron, **1**:151a, 152a
Apelt, Ernst Friedrich (1812–1859), 1:178
Apennines, **13**:257b
botany, **1**:525b
geology, **2**:496a–b; **5**:226b
natural gas, **12**:562a
Aperture, numerical, **1**:8b, 135a, 135b–136a
Apes
adaptability, **5**:359a
anatomy of
comparative, **12**:224a–b
anthropoid, **2**:162a; **7**:278b; **8**:43b
anatomy of, **10**:261b
brain, **5**:359a
classification of, **5**:359a
fossil, **8**:105a
and man, **3**:574a–b; **5**:356b, 359a–b; **6**: 592b–594a; **7**:278b, 606b; **8**:378b; **11**:514a, 592b; **13**:527b–528a
psychology of, **14**:550a–551a
Aphacia, **5**:590b
Aphasia, **2**:478a; **3**:203b, 205b; **5**:255a; **14**: 277b
motor type, **1**:439a
theories of, **5**:173a–b; **9**:109a
Wernicke's, **1**:496a; **14**:277a
Aphid
parthenogenesis of, **2**:286a; **3**:180b; **8**: 580a; **11**:333a–b
reproduction of, **1**:417b
Aphidophorus fly
see Ichneumon fly
Aphtha
cachectic, **15**:479b
vaccination, **11**:271b
Apian, Peter (1495–1552), 1:178–179
acquaintances of, **11**:395b
Arithmetic, **8**:161a
Astronomicum Caesareum, **14**:358b–359a; **15**:154a
astronomy, **2**:403a; **4**:168b; **7**:39a*
cartography, **14**:275b; **15**:156a
Cosmographia, **2**:401b
editions of, **5**:349a
influence of, **4**:97a; **7**:269b; **13**:60b
influences on, **11**:415b; **13**:150a*; **14**:275b
Instrumentum . . . primi mobilis, **15**:156a
Introductio geographica, **14**:275b
Liber Iordana Nemorarii . . ., **7**:174b
mathematics, **10**:338b(n)
Apian, Philipp, **9**:167b, 168a
Apiculture, **2**:625a; **8**:518a–b; **9**:279a; **15**: 662a(n)
Apiole, **3**:279b
Apion, **6**:492b
Apocrenic acid, **9**:558a–b
Apocynaceae, **2**:519a; **3**:43a
Apoda, **12**:336b
Apogamy, **3**:613b; **8**:5a; **9**:282a; **11**:154a
Apoidea, **8**:518a
Apolant, Hugo, **4**:298b, 299b
Apollodorus, **1**:151b; **10**:179b, 324a
Apollonius of Byzantium, **15**:253b
Apollonius of Myndos, **1**:191a
on comets, **3**:100b
Apollonius of Perga (*fl. ca.* 200 B.C.), 1:179–193; 3:391a
acquaintances of, **4**:414b
algebra, **10**:300a
arithmetic, **10**:294b
conics, **4**:104a, 421a, 427a–b; **7**:576b–577a; **8**:252b; **9**:270a,b, 272b, 598b; **10**:295a; **14**:619a
cosmology, **1**:247b; **7**:143a; **11**:188a, 195a, 196b; **15**:211a, 218b, 632a(n)
critics of, **4**:415b
editions of works, **4**:392b, 416a; **12**:500a; **13**:509b; **14**:472a
geometry, **4**:105a,b; **10**:114b; **13**:302b; **14**:274b
Pi, value of, **1**:189a
influence of, **1**:22b, 82b; **3**:149a, 494a; **4**:47a; **6**:616a–b; **11**:288a; **13**:433b; **14**:458b
lost works, **5**:382a; **10**:295b, 296b–297a
method of, **7**:3b
optics, **6**:195a
problems of, **4**:489b; **5**:209b, 368a, 382a; **8**:175b; **11**:533b; **13**:14b, 125a
theorem of, **12**:228a
Conics, **1**:170a, 245a,b, 444a, 446a, 474a, 541b; **2**:310a; **4**:108b, 428a,b, 568a,b; **6**:201a; **7**:133b; **8**:191b; **10**:331a; **14**:465a
commentaries on, **3**:364b; **4**:488a, 489a, 491a–b; **6**:615b; **10**:295b; **12**:314a
construction theorems, **4**:57b
editions of, **2**:308b; **4**:491a; **9**:191a, 555b; **14**:49a
translations of, **6**:71a; **13**:289a; **14**:49a; **15**:187a–b
Determinate Section, **4**:569a; **10**:295b
Formula for Quick Delivery, **4**:491a
Inclinationum libri duo, **5**:382a; **10**:295b
Plane Loci, **4**:567a, 567b–568a; **10**:295b; **12**:206b
commentaries on, **12**:446b
Sectio rationis, **6**:71a; **10**:295b
Sectio spatii, **6**:71a; **10**:295b
Tangencies, **10**:295b; **14**:22b
Apollonius of Tyana, **11**:231a
The Secret of Creation, **7**:41a; **13**:407b
Apologetics, **5**:415a–417a; **7**:483b; **8**:548a–550a; **10**:335b; **14**:157a
Apomixis, **1**:521a
Apoplexy, **4**:386a; **11**:560a; **14**:256a, 408a
cerebral, **1**:391b
Apospory, **2**:371a; **8**:5a–b; **11**:154a–b
Apotelesmata astrologiae christianae, **3**:280b
Apothecaries
seventeenth-century, **2**:353a
Apothecaries, Society of (London), **9**:390a,b
Appalachian Mountains
geology, **2**:500a; **3**:553b, 593b, 594a, 595b; **4**:364a–b, 493a; **6**:229b; **7**: 144b–145a; **8**:260b; **11**:505a; **13**:532a; **14**:298a, 301a, 402b
stratigraphy, **11**:321a
surveying of, **5**:600a
Apparatus
alchemical, **14**:632a
chemical
Renaissance, **8**:213b; **13**:535a
seventeenth-century, **5**:420a–b; **7**:354b; **8**:131a
eighteenth-century, **1**:527a; **3**:201a; **6**: 42a–43a, 176a, 383b; **8**:77b–78a, 79a; **9**:5b; **11**:145a; **14**:213b, 508a
nineteenth-century, **2**:69b, 70a, 93a–b, 258a, 587a; **4**:41b, 240a; **5**:127a*, 319a, 320b, 325b, 431b; **6**:54b–55a, 115a; **8**:332a; **9**:353a, 356a, 424b, 425b, 445b; **11**:417a
early twentieth-century, **1**:320b; **7**: 497b; **13**:70b, 71a, 574a; **15**:442a, 449a
electrical, **5**:496a–b
microbiological
nineteenth-century, **3**:188b; **8**:406a
mineralogical
eighteenth-century, **2**:5a–b; **5**:602a
oceanographic, **14**:2a
physics
seventeenth-century, **9**:115b; **11**:489a
eighteenth-century, **5**:130a,b; **8**:21b; **10**:145b; **14**:70a, 71a–b, 73a–b, 78a–b
nineteenth-century, **3**:477a; **4**:559a; **5**: 292a–b; **7**:445a–b; **11**:603b–604a; **13**:126a–b
twentieth-century, **2**:458a–459a; **4**:522a
physiological, **6**:24a; **12**:59b–60a
pneumatic, **6**:311b
surgical
twentieth-century, **3**:91b
see also Instrumentation, scientific
Appel, Otto, **7**:417b; **12**:58b
Appell, Paul (-Émile) (1855–1930), 1:193–195; 1:407a; **2**:88b, 215b, 303a; **3**: 428b, 498a, 501b; **8**:563b*
Appell functions, **1**:194b
commentaries on, **7**:7a
theorem of, **5**:201b
Appellatio, **1**:3a
Appendicitis, **2**:464a; **4**:586a; **10**:321b
surgical treatment of, **1**:438b
Appendicularia, **6**:590b
Appendix, vermiform, **1**:619a; **8**:203b; **9**: 165b
Apperception, **14**:528b
Appert, Nicolas, **10**:366b, 367a, 370a; **12**: 241a, 563b
Apples
respiratory analysis of, **2**:184b–185a
Appleton, Edward Victor (1892–1965), 1: 195–196; 12:33b
Appleyard, Edgar Thomas Snowden, **8**:480b
Applied Mechanics Reviews
editors of, **14**:590b
Approximation, method of
ancient Greek, **1**:171b; **2**:417b; **6**:314b, 315a; **13**:315a
medieval Arabic, **11**:230a; **13**:536b, 540a
medieval Latin, **4**:607b–608a, 610a,b
Renaissance, **3**:274a; **14**:23a
seventeenth-century, **3**:61a; **6**:599b–600a; **9**:598b; **10**:72b; **12**:75a, 292a; **14**:161b
eighteenth-century, **3**:388a(n); **4**:475b, 478b; **9**:156b
nineteenth-century, **3**:228b–229a, 555a; **4**:561b, 562a; **5**:98b, 573b; **6**:309a, 510b; **9**:127a, 409a
twentieth-century, **7**:514b; **13**:562a
Cauchy-Lipschitz, **3**:143a, 196a; **5**:204a
in celestial mechanics, **3**:283b; **7**:580a; **11**:57b
Chebyshev, **3**:224a, 226b; **5**:608b
Chinese, **14**:540b
Diophantine, **2**:197b, 303b; **3**:228a; **4**: 114a, 115b; **5**:342b; **6**:113b; **14**:282b
theory of, **7**:351b
EIH (Einstein-Infeld-Hoffmann) method, **7**:10a
iterative method, **7**:258b–259a
in philosophy, **3**:515a,b
in physics, **2**:356a; **3**:195b; **4**:618b; **5**:97a; **6**:147b; **7**:215a; **8**:561a, 562a–b; **9**:38a
Poisson, **9**:454a
in probability theory, **9**:452b–454a
successive approximations, **8**:384a; **10**: 249a, 594b; **11**:229b, 477b; **13**:80a, **15**:338a, 366a–b, 371b, 383a–b

theory of, 8:389a
topological, 2:513a; 11:459b; 12:246b
see also Integration (math.), approximate
Apriorism, 5:193a–b
Apūcha Śarman (Jhā)
commentaries of, 7:315a,b
Apuleius, 2:21a; 4:463b; 5:592a; 10:112b
Apurinic acid, 4:603b
Aqāṭun, 12:447b
Ibn ᶜAqnīn, Joseph, 7:38a
al-Āqṣarāᵓī, Jamāl al-Dīn Muḥammad ibn Muḥammad, 9:604a
Aqua fortis, 2:421b
preparation of, 10:308a
Aqua regia, 1:518b; 5:323a
Aquapendente
see Fabrici, Girolamo
Aquaria, 9:431b; 15:123b–124a
marine, 3:338a
Aquarius (constellation), 9:36a
Aqueducts, 11:5a
Aquifers
perennial yield, 9:257b–258a
salt-water encroachment, 9:258a
Aquila (naturalist), 14:35b
Aquila (constellation)
alpha Aquilae
parallax of, 2:468b; 13:110b, 120b
eta Aquilae
variability, 10:608a
novae in, 1:598a; 4:498b*
Aquinas, Saint Thomas (*ca.* 1225–1274), 1: 196–200
acquaintances of, 5:402b
alchemy, 5:580a
astrology, 10:512b
commentaries by, 1:276a; 9:346b–347a
critics of, 4:92b, 255b; 5:403a; 12:429a
dynamics, 1:410b
editions of works, 6:562b*
influence of, 2:23a, 539a, 540a, 590b; 3: 56b, 280b, 322b; 5:592a; 13:349b; 14:35a, 318a; 15:68b
influences on, 1:99a–102b *passim*; 7:66b; 9:31b; 10:21b; 14:118b, 388a
and Moerbeke, 9:435a
observation, use of, 1:278b
philosophy, 10:173b; 11:73b
students of, 2:20b; 5:401b
Summa contra gentiles, 8:548a
De unitate intellectus, 12:428b
Aquitanian (stratig. stage), 9:240b
Ibn ᶜArabī, 11:230b, 248a, 250a; 12:2b–3a
Arabia
flora, 5:74b
geography, 9:88b; 10:117a
Arabic language
grammar, 15:232a
teaching of in Europe, 3:262a
Arabic science
see specific sciences, *e.g.,* Algebra, medieval Arabic
Araceae, 13:405a; 15:147a
Arachnids, 6:267a; 12:216a
classification of, 8:48b–49a
Arachnoidea
phylogeny, 8:27a
Arachnology, 2:322b; 4:504a
see also Spider
Arago, Dominique François Jean (1786–1853), 1:200–203; 3:302a; 4:350a
at Académie des Sciences, 1:143b; 2:72a; 4:410b; 5:44a; 8:265b; 11:63a
acquaintances of, 3:132b, 545b; 4:238b; 5:166b, 271b; 6:325a, 552a; 8:37b, 288b; 9:474b; 10:546a; 11:77a, 78b; 12:519a, 523a; 13:127a,b, 128a,b; 15:347b
biographical works by, 3:388a*; 9:74b*
on contemporary scientists, 4:14b, 242a*, 256b*, 349b; 5:99a*, 171a*;6:3a*, 336a*; 11:165b*; 14:571b*; 15: 401b*, 480b–481a, 483a
correspondence of, 5:20b; 8:36a, 63a; 9: 177a; 12:523b; 13:238b
critics of, 15:379a
electromagnetism, 6:341b
éloges académiques, 5:582b
geodesy, 2:135b
geophysics, 12:49b
influence of, 15:199b
influences on, 15:361a–b, 376b
and Le Verrier, 5:257b; 8:277a, 278b
magnetism, 3:260a; 4:534b; 7:608b; 12: 50a–b
on meteorites, 1:615b
optics, 5:167b; 9:74a; 14:290a
polarization, 1:201b, 202a–b; 2:137a; 5:169b, 170a; 14:567a
refraction, 1:348b; 2:135a; 9:215a, 267a; 15:361a–b
speed of light, 2:439a; 5:19a, 85b
steam pressure, 4:241b; 5:78b
students of, 3:79a; 4:555a; 5:18b; 11:498a; 12:469a
Tract on Comets, 4:546b
translations of works, 6:97a*; 12:53a*
Arago, Jacques, 5:224b
Aragon
mapping of, 8:65b
Aragonite
and calcite, 2:135a–b; 5:124a; 13:311a
morphology, 2:355b
refractivity of, 2:135a
Aral Sea, 1:622a
level of, 13:412b
Aran, François Amilcar, 3:490a
Aranzio, Giulio Cesare (*ca.* 1529–1589), 1: 204; 1:522b; 3:342a; 4:520a; 13:587a
embryology, 4:510b–511a
Aratus of Soli (*ca.* 310-*ca.* 240), 1:204–205; 1:283b, 338b; 4:172a, 391b
Phaenomena, 1:419a; 3:14a; 4:14a, 467a; 5:345b; 9:79b
commentaries on, 13:323b; 15:208a, 210b, 216b–217a, 217b, 220b
editions of, 10:615b; 11:18a; 12:580a
Arbacia, 8:357b
Arbelos, 10:294b–295a
Arber, Agnes Robertson (1879–1960), 1: 205–206; 5:446b*, 535a; 7:541a*
Arber, Edward Alexander Newell, 8:354b; 13:345a
Arbogast, Louis François Antoine (1759–1803), 1:206–208; 5:110b; 6:510b; 7:494b, 566b; 8:140b
Calcul des dérivations, 12:325b
library of, 5:111b
Arboriculture, 4:144a, 495a–b; 9:54a
Arbos, Philippe (1882–1956), 1:208
Arbuthnot, John (1667–1735), 1:208–209; 1:531a; 3:244a; 7:276b*; 13:67b, 68a; 14:503a*
Arbutin, 9:95b
Arc (geom.)
bisection of, 9:156b
concept of, 7:72a
elliptic, 7:619a
hyperbolic, 7:619a
trisection of, 5:410a
Arc, electric, 5:597b; 10:236a, 552b; 11:559b
and chemical reactions, 4:79a
Arc lamp, 2:200b, 439a; 5:87a; 6:223a; 12: 574b; 13:286b, 361b
Arcesilaus, 1:338b; 4:388a, 415a; 13:328b
Arcet, Jean d'
see D'Arcet, Jean
Arcetri Astronomical Observatory (Italy), 1:19a, 136a; 4:162a
Archaeologia, 3:350b
Archaeology
medieval, 4:164a; 11:468b
Renaissance, 1:525a; 9:309a
seventeenth-century, 2:523a; 3:153a, 209a; 6:68a–b; 11:40b, 41a; 14:505b
eighteenth-century, 2:326b; 3:350b; 4: 156b; 5:162b–163a; 6:84a; 7:197b; 10:117a; 11:65b; 13:232a*; 14:518b; 15:52a
early nineteenth-century, 4:288b; 5:94a; 7:615a; 13:357b–358a
mid nineteenth-century, 1:20b; 2:341a, 570b; 4:73b; 5:72b; 6:509b–510a; 8: 44a, 528a, 573b–574a; 9:538a–b; 11:131a, 514b; 13:9b; 14:398a, 404a, 506a–b, 533b; 15:50b–52a
late nineteenth-century, 2:346b; 3:532a; 5:109b, 144b; 6:215b; 8:442b–443a; 9:489a–490a, 539a, 592b; 10:285a, 549b–550b, 606b–607b; 11:5b–6b, 177a, 219a; 12:179b–181b, 545a; 13:288a; 14:42a–b, 183a, 443a
early twentieth-century, 2:212b, 450b; 3:162a, 606b, 622b–623a; 4:353b; 6:217a; 8:539b; 9:176a, 547a; 10: 550a–b; 11:210a–b; 12:520a–b; 14: 504a–b
mid twentieth-century, 2:549b; 4:262b; 8:104b–105b; 13:505b
and astronomy, 8:442b–443a
and botany, 13:542b
chemical, 2:64b; 5:190a; 6:110b
Chinese, 12:382a, 382b–383a
geology and, 2:549b; 14:183a
Mesopotamian, 14:504b; 15:646b–647b, 649a, 651b
methods, 12:181a
cross-dating, 10:550b
sequence-dating, 10:550a
of Roman Gaul, 3:622b
and stratigraphy, 12:181b; 14:506a
see also specific countries and sites
Archaeopteryx, 10:262a, 428b
Archebiosis, 1:497a
Archegoniatae, 2:160a
Archegonium, 3:30a,b; 6:219a, 465b; 9:368b, 601a; 13:88b
Archeion
editors of, 9:378a,b
Archencephala, 6:593a; 10:261b, 262b–263a
Archer, Robert Samuel, 7:92b; 9:312b
Arches (arch.), 3:118b, 441b; 9:578a
nineteenth-century theories of, 1:326b; 5:530b
rampant, 2:201b
stability of, 7:602a
thrust of, 11:163b–164a
vaulted, 3:442a
Archetype, biological, 5:443a–444a; 6:590b–591a; 13:506b
Archetype, psychological, 7:191b–192a, 193a
Archeus, 5:426b; 6:255b, 256b; 7:539a; 10: 309a, 310b
Archiac, Étienne-Jules-Adolphe Desmier de Saint-Simon, Vicomte d' (1802–1868), 1:209–212; 2:347b; 4:66a, 67b, 218b*
Archiannelida, 5:109a–b
Archibald, Raymond Clare, 4:426a; 12:345a; 13:615b

Archigenes (*b. ca.* A.D. **54,** *fl.* **98–117), 1: 212–213; 1**:75a, 324a; **12**:541a(n)
Archil, 7:224a
Archimedes (*ca.* **287–212), 1:213–231; 1**: 267b; **4**:53a, 414b, 424b, 622a; **5**: 310a; **6**:310a; 7:81a
Archimedes' principle, **1**:223a, 228b, 605b, 607b; **2**:153a; **5**:238b–239a; 7:344a, 345a; **8**:226a; **9**:119b; **13**:49a; **15**:518b
derivation of, **6**:602a
axiom of, **4**:419b, 422a; **6**:207a*; **10**:344a; **12**:56a
balance, **9**:300b
ballistics, **1**:214a
biographies of, **8**:505a
Cattle problem, **4**:117a, 389a
commentaries on, **9**:264a, 437b; **14**:274b
and Conon, **3**:391a; **4**:171b–172a
cosmology, **1**:247a–b, 190b; **3**:403a
critics of, **11**:288a, 533b; **12**:580a
editions of works, **2**:311b; **4**:451b; **8**:191a; **9**:191a,b; **13**:509b
see also specific works below
on Eudoxus, **1**:220a, 221b; **4**:466b; **6**: 416a
geometry, **1**:41b, 82b; **2**:349a; **3**:149a, 149b–150a, 272b; **4**:108b, 414a, 609a; **5**:60b; **6**:208a*, 314b; **7**:129a, 327b, 335a; **8**:240b; **9**:430a; **10**:51b, 294a, 295a–b, 298b; **11**:240b, 487a; **12**: 432a; **13**:515b; **15**:691a
conics, **1**:179b–181b, 186a–b; **4**:428b; 7:133b; **9**:270b
exhaustion, method of, **4**:422a, 422b–423a, 424a
pi, value of, **1**:189a, 222a, 223b, 224a, 227a, 445a; **6**:314b; **7**:258b; **9**:556a
quadrature, **1**:171a; **2**:618a; **3**:128b*; **4**:571a–b; **8**:237a,b; **10**:295a; **13**:435a
hydrostatics, **1**:97a, 165a; **4**:431a; **5**:241b, 382a; **7**:129a, 340b, 341a,b, 342b, 343a
influence of, **2**:590b; **3**:66a; **4**:567a; **5**: 238a, 246b, 299a, 375a,b; **7**:42b, 583a; **9**:487b; **11**:288a, 508b; **13**: 48a,b, 433b; **14**:29b, 458b, 603b
Lemma of, **1**:215a–b; **6**:415b–416a; **13**: 623b
method of, **1**:220a–222a, 252a, 444b, 474a; **3**:515b; **4**:416a; **6**:599b, 600b; **7**:487a; **8**:164b, 602a; **10**:336a,b; **13**:235b, 434b, 435a, 560b, 561a
numeration, **1**:189a, 222a
optics, **6**:195a; **11**:201a
burning glass, **3**:153a; **7**:376a
problems of, **9**:21b
spurious works
De ponderibus Archimenidis, **8**:215b, 216a, 225b, 226a
statics, **6**:310b, 312a,b; **7**:174b; **8**:225b
translations of works, **1**:605b; **4**:105b, 611b; **9**:436b, 437b, 438b; **13**:259b–260a, 289a, 615b; **14**:49b
trigonometry, **15**:209a,b
Conoids and Spheroids, **3**:364a, 391a; **4**: 172a, 429a, 570b
De dimensionibus
see Archimedes, *Measurement of the Circle*
Equilibrium of Planes, **1**:256b, 419b; **7**: 558a; **8**:223a–224b; **9**:488a
commentaries on, **4**:488b, 491a
Floating Bodies, **6**:275b; **8**:225b, 226a
commentaries on, **3**:364a–b
De insidentibus aquae, **9**:438b
Lemmata, **2**:309a
commentaries on, **11**:240b
editions of, **9**:615a
Measurement of the Circle, **1**:381a, 468a; **3**:181a, 364a; **4**:104b; **5**:526a; **6**:314b
commentaries on, **4**:488b–491a *passim;* **6**:412a; **7**:29a
editions of, **11**:533b
translations of, **4**:8b; **5**:360a; **7**:29b, 128b; **11**:32a; **15**:175a, 177a
Method, **4**:389a; **5**:360b; **6**:315a, 416a; **12**:75b
commentaries on, **13**:319b, 320a
Quadrature of the Parabola, **3**:364a; **4**: 172a; **6**:415b–416a; **13**:290a
The Sandreckoner, **1**:246b, 247a, 248b; **3**:364a, 403a
Sphere and Cylinder, **1**:474a; **3**:391a; **4**: 109a–b, 172a, 414a; **5**:409b; **6**:206b*, 207a*; **11**:244b; **15**:116a,b
commentaries on, **4**:488a–490b *passim;* **7**:29a; **9**:270a, 369b; **15**:115b
editions of, **11**:240a
translations of, **7**:25a, 29b
Spirals, **3**:364a, 391a; **4**:172a; **14**:21b
On Triangles, **12**:447b
On Uprights, **6**:310b
Archimeter, **11**:406a
Archipendulum, **4**:609b
Architecture
ancient Greek, **13**:320a; **15**:515b–516a
ancient Roman, **15**:514a–518a, 519b–520b
Byzantine, **7**:28b–29a
medieval Latin, **14**:30a–31a
Renaissance, **1**:97b; **2**:534b–535a; **4**:260a–b; **5**:113a; **9**:146b; **10**:271a; **13**:233b; **14**:29a–b; **15**:516b–517b
seventeenth-century, **2**:201a–b, 419a–b; **4**:49a, 496a; **6**:483b; **7**:578a; **10**:520b–521a; **11**:587b; **14**:510a
eighteenth-century, **2**:326b; **7**:576a; **11**: 65b
nineteenth-century, **7**:615a
twentieth-century, **8**:100a; **14**:468a
civil
eighteenth-century, **1**:581b–582a
nineteenth-century, **2**:104a–b
Chinese, **12**:383a
earthquake-proofing, **7**:9b; **12**:342a
Gothic, **6**:56a; **14**:404a
hydraulic, **1**:581b–582a
Mesopotamian, **15**:648b, 650b, 652b, 658b(n)
military
Renaissance, **9**:147b; **15**:519a
seventeenth-century, **11**:508b; **13**:592a–b
eighteenth-century, **1**:581b
modular, **12**:383a
naval
seventeenth-century, **4**:214a; **5**:344a
eighteenth-century, **2**:343b; **4**:224b*; **7**:183b; **11**:496a
nineteenth-century, **2**:84b; **5**:200a–b; **9**:292b; **11**:293b; **15**:25a
twentieth-century, **7**:513b–514a; **12**: 341b
see also Engineering; Historiography, of architecture; Illustration, architectural; Strength of materials
Archiv der Mathematik und Physik, **5**:471b
Archiv der Pharmazie
editors of, **14**:112a
Archiv der reinen und angewandten Mathematik, **6**:404a
Archiv för mathematik og naturvidenskab
founding of, **8**:324a
Archiv für Anatomie, Physiologie und wissenschaftliche Medizin, **4**:201a; **9**:253a; **11**:360b
editors of, **6**:269a; **9**:253a
Archiv für Anatomie und Physiologie, **4**: 372b*
Archiv für Anthropologie
founding of, **1**:388a
Archiv für Bergbau und Hüttenwesen
editors of, **7**:254b
Archiv für die gesamte Physiologie des Menschen und der Tiere
editors of, **10**:580a
Archiv für die Physiologie, **9**:252b
editors of, **9**:253a
founding of, **11**:364a
Archiv für Entwicklungsmechanik der Organismen
founding of, **11**:574a
Archiv für Geschichte der Medizin und der Naturwissenschaften, **13**:142a
Archiv für Hygiene
founding of, **10**:559b
Archiv für Hygiene und Bakteriologie
editors of, **11**:585b
Archiv für Mikrobiologie, **12**:207b
Archiv für mikroskopische Anatomie und Entwicklungsmechanik
founding of, **12**:232a
Archiv für Mineralogie, Geognosie, Bergbau- und Hüttenkunde, **3**:623b
editors of, **7**:254b
Archiv für Naturgeschichte, **9**:345b
Archiv für pathologische Anatomie und Physiologie, und für die klinische Medizin, **14**:40b
Archiv für praktische Geologie
founding of, **11**:103a
Archiv für Psychiatrie und Nervenkrankheiten, **5**:570b
Archiv für Rassen- und Gesellschaftsbiologie
founding of, **11**:19b
Archiv für wissenschaftliche Kunde von Russland, **4**:401a
Archiv für Zellforschung
founding of, **5**:453b
Archives d'anatomie microscopique
founding of, **1**:418a; **11**:295b
Archives de Genève, **5**:598a
Archives de zoologie expérimentale et générale
founding of, **7**:546a
Archives des sciences physiques et naturelles. Supplément à la Bibliothèque universelle de Genève, **8**:36a
editors of, **9**:109b
Archives internationales d'histoire des sciences, **9**:378b
Archives of Pathology
editors of, **6**:233a
Archives of Universal Science, **14**:128a
Archivio di storia della scienza
founding of, **9**:378a
Archivio italiano di anatomia e di embriologia
founding of, **3**:245b
Archivio italiano di biologia
founding of, **9**:546b
Archivio per l'anthropologia e la etnografia
founding of, **9**:86a
Archivio per le scienze mediche
founding of, **2**:164b
Archivos de histología normal y patólogica
founding of, **11**:465b
Archytas of Tarentum (*fl. ca.* **375** B.C.), **1: 231–233**
acquaintances of, **1**:282a–b; **8**:192a,b; **10**:589b; **11**:22b, 24b, 25a

critics of, **4**:466b; **11**:28a
geometry, **1**:227a, 445b; **4**:105a, 489b, 609a; **6**:407a; **9**:270a, 272b–273a, 274a, 300a; **11**:221b
music, **11**:223a
students of, **4**:466a, 467b; **11**:27b
Arcolani, Giovanni, **9**:133a
Arconati, Luigi Maria, **3**:116a; **8**:194b
Arctaedius, Petrus
see Artedi, Peter
Arctic Regions
ethnography, **1**:29b
exploration of, **3**:204a; **7**:92a; **9**:617b; **10**:148a–b; **12**:49a, 412a; **14**:50a–b, 491b; **15**:430a, 431a
ice, **9**:43b; **14**:50b
laws of, **14**:635a
surveys of, **14**:635a
oceanography, **12**:415a; **14**:634b–635a; **15**:430b–431a
paleobotany, **6**:221a–b; **9**:617b
zoology, **10**:509b; **12**:141b; **13**:552a
Arcturus (star), **9**:36a
parallax of, **2**:468b
proper motion of, **6**:71a, 512a
spectrum, **14**:56a
Arcueil, Société d', **2**:133b–137a *passim;* **3**:199b, 346b, 545b; **4**:238b, 239b; **5**:94b, 320a, 322a; **13**:309a; **15**:343a, 347b, 361a,b, 376b
founding of, **2**:74a
Ardennes
geology, **5**:437a, 475b–476a; **8**:463b
natural history, **5**:150a
Ārdharātrikapakṣa (school of astronomy), **1**:308a; **2**:114b, 417b; **4**:556a; **7**:582b; **14**:546b; **15**:534a, 548b, 581a, 588b, 590b, 595b, 596b, 602b–608b, 609a–b, 618a–b, 625b
founding of, **15**:602b
Arduino, Giovanni (1714–1795), 1:233–234; **9**:534a; **11**:304a
Arduino, Pietro, **1**:233b
Area
application of areas, **1**:183b–184b, 188a–189a; **4**:417b–421a *passim,* 424a, 425b, 428b, 463a; **9**:274a–b
Cavalieri principle, **3**:150b–152a
of circle, **1**:90b, 216a–217a, 219b, 220a, 227a, 444b; **3**:128b*; **4**:104b, 422b–423a, 424a, 466b, 609b, 611b–612a; **5**:526b; **6**:314b, 415b–416a; **10**:298b; **14**:22a; **15**:670b, 691a, 696b
see also Quadrature, of circle
of circular segment, **3**:255a; **6**:315a; **8**:419b
composition of, **11**:61b
of cone, **1**:217a–b, 227a,b
of curved figures, **1**:90b, 214b, 215a
of curved surface, **12**:247a
of cylinder, **1**:227b
division of, **13**:259b
of an ellipse, **3**:152a; **12**:292a–b; **13**:291a
of hyperbola, **5**:526b
of lemniscate, **4**:516b
minimal, **12**:246b
of parabolic segment, **1**:218a–b; **6**:315a
and perimeter, **13**:19b
of polygons, **1**:591a; **4**:609b
projective method, **9**:234a
proof of existence, **9**:303b–304a
of quadrangles, **1**:40a; **3**:254b–255a
of quadrilaterals, **5**:409a; **6**:314b; **12**:500a
of rectangle, **4**:609b
of rhomboids, **4**:609b
signed, **3**:469a
of sphere, **1**:219b, 220a, 221b, 227a,b, 445a; **12**:247a
of square, **1**:31a; **4**:609b; **6**:414b
theory of, **3**:152a; **8**:112a; **12**:364a–b
of trapezium, **15**:670b, 671a
of trapezoids, **4**:609b
of triangle, **1**:226a, 227a, 445a; **3**:254b, 277a; **4**:609b; **5**:365b; **6**:314b; **7**:175a; **12**:314a; **15**:670b, 671a
spherical, **5**:409a; **6**:125a
weighted, **3**:469a
see also Quadrature
Areism, **9**:150b
Areometry, **1**:527a
Aretaeus of Cappadocia (*fl. ca.* A.D. 50), 1: 234–235
editions of works, **2**:226a
Aretino, Rinucio, **1**:274b
Arfvedson, Johann August (1792–1841), 15: 7–8; **2**:94b
Argand, Émile (1879–1940), 1:235–237; **8**:544a; **12**:140b; **13**:146a
geology, **13**:285b
Argand, François-Pierre Ami (Aimé), **1**:237b; **2**:589a; **9**:492b, 493a
Argand lamp, **7**:89a; **9**:343a
Argand, Jean Robert (1768–1822), 1:237–240; **12**:325b
algebra, **8**:139a; **14**:279b, 280b
diagram of, **1**:237b; **6**:90a
geometry, **15**:196a
imaginary numbers, **5**:111b
Argelander, Friedrich Wilhelm August (1799–1875); 1:240–243, **11**:545a
acquaintances of, **2**:98a; **6**:553a; **12**:192a, 200b
influence of, **3**:92b; **10**:601a
star catalogs, **2**:100a; **4**:30a, 369b
students of, **7**:505b; **14**:472b(n)
Argentina
anthropology, **1**:130a, 131a–b
botany, **5**:546a
geology, **1**:129b, 131a,b; **10**:506b–507a; **11**:575b, 576a; **14**:402a
National Observatory, **10**:526b
founding of, **5**:480a
Pampas
paleontology, **1**:130a, 131a
prehistory, **1**:130a, 131a–b
science, **15**:228b
Argenville, Antoine-Joseph Dezallier d' (1680–1765), 1:243–244; **1**:58b
Arginine, **7**:467a
nutritional role, **6**:501a
Argoli, Andrea (*ca.* 1570–1657), 1:244–245
Argon, **2**:429a
atomic weight of, **2**:267a
discovery of, **9**:292b; **11**:279a–b; **13**:102b–103b
liquefaction of, **10**:206b; **11**:279b; **13**:454a
penetrability, **15**:491a
properties of, **13**:453b–454a
spectrum of, **13**:473a
Argonne National Laboratory, **3**:371b
Argulus, **9**:575b
Arguments (math.)
logarithms of, **13**:69a–b
Arguments (philos.)
classification of, **2**:232b
Argus (constellation)
nebulae, **15**:493b
9 Argi, **13**:118a
Argyrodite, **14**:447a
Argyropulos, John, **1**:274a
Arianism, **2**:540b; **3**:294b, 295b; **10**:81b; **11**:140a; **12**:322b; **14**:295b
ᶜArīb ibn Saᶜd al-Kātib al-Qurṭubī, **15**:186b–187a
Aricine, **10**:498a
Aridity
index of, **9**:150b
and settlement policies, **11**:119a, 119b–120a
Ariel (satellite), **8**:46a
Aries (constellation)
alpha Arietis, **2**:411a
magnitudes of stars, **3**:174a
Arima Yoriyuki, **1**:83b, 90a
Arisi, Francisco, **15**:173b
Aristaeus (*fl. ca.* 350–330 B.C.), 1:245–246; **1**:180a, 215a; **4**:392b, 416a; **6**:616b; **13**:306b(n)
conics, **4**:427b; **9**:270b, 272b
Conics, **4**:428b
De locis solidis secunda divinatio geometrica, **14**:49a
Aristarchus of Samos (*ca.* 310–230), 1:246–250; **1**:190b, 283b; **7**:81a; **10**:295b, 591a; **13**:95a; **15**:678b
cosmology, **1**:222a, 468a; **3**:402b–403a; **11**:188a, 224b, 489a; **12**:162b
editions of works, **13**:509b
observations of, **15**:218b
Sizes and Distances of the Sun and Moon, **3**:402b
commentaries on, **3**:364b
translations of works, **11**:245a, 489a
trigonometry, **15**:209b
Aristides Quintilianus, **5**:228a; **9**:141a
Aristippus
see Henry (Aristippus)
Aristocles, **1**:117b
Aristolochia, **3**:422a
Aristophanes, **4**:389a
The Birds, **9**:337a, 339a
on squaring of circle, **1**:171a
Aristotelian Society for the Systematic Study of Philosophy (London), **3**:90a
Aristotelianism
ancient Greek, **1**:117b–119b, 137a–b, 268a–b; **2**:229a–232b *passim;* **4**:415a, 424b; **5**:246b; **11**:103b; **13**:91b–94b; **14**:607b
opposition to, **4**:381b
ancient Roman, **1**:269b–270a, 333b; **3**:15a; **5**:230b, 232a; **11**:201b; **12**:310a
opposition to, **1**:279b
Byzantine, **7**:66a; **13**:307b–308b
medieval Arabic, **1**:33a,b, 408b–410a; **2**:619b; **4**:523b–525b; **6**:79b, 190a, 191a; **7**:330b; **11**:230b, 231a, 444b, 445a; **12**:1b, 5b–7Bb; **13**:229b, 513a; **15**:33a–35a, 262a,b, 494b
opposition to, **6**:80a; **11**:323b–325a
medieval Hebrew, **9**:27b–31b
medieval Latin, **1**:99a–102a *passim,* 197a, 197b–199b, 268b, 270a–273b; **2**:23a, 193a, 391b–392b, 394a, 603b–607b, 609a–610b; **3**:27b, 218b, 470a; **4**:92b–94b; **5**:115b–116a, 549a–553a; **7**:65b–66b, 116a–b, 174b; **9**:136a–137b; **10**:228b, 420b–421a, 555b; **11**:469b; **12**:429a; **13**:185b, 187b–188a, 189a, 196b, 197a–b, 340a
opposition to, **1**:279b–280a; **2**:391b; **5**:402b, 403a; **10**:474a
Renaissance, **1**:274a–b, 495a, 587b; **2**:143a, 401a, 540a, 541a, 590a–b; **3**:66b, 172a, 262a; **4**:285a; **8**:227b, 309b; **9**:132b–133b; **10**:512a–b; **11**:71a–74b, 391a, 528a; **12**:135a, 151a, 335b, 548a; **13**:278a–279b, 441b; **14**:47a, 581a–582a, 586b; **15**:45a–b, 68b, 80b
opposition to, **1**:607a; **2**:293a; **4**:213a; **5**:238b–239a; **10**:416b; **11**:286a–287b, 339b; **12**:97b, 106a; **13**:278a

Aristotelianism (cont.)
early seventeenth-century, **1**:480b; **2**:527a, 601b; **3**:100b; **4**:96a, 510b; **6**:152a, 158a–159a, 448b; **7**:269a; **9**:15a, 98a, 316b, 527b; **10**:137a; **12**:311a–312b; **13**:583b; **14**:286a, 301b–302a
late seventeenth-century, **7**:219b; **8**:130b, 149a, 157a; **10**:43b, 60a, 315a; **11**:506b–507a
seventeenth-century opposition to, **1**:278a, 280a; **2**:13a, 293a, 378a–379a; **5**:47a, 241b, 285b, 415b; **7**:539a–b; **9**:16a, 25a–b; **12**:333a–b, 431a; **13**:32a, 169b; **14**:364b; **15**:69a
eighteenth-century, **4**:129b; **8**:378b, 379b; **12**:603a–b, 604a; **14**:590a
opposition to, **13**:39a
nineteenth-century, **6**:80b; **9**:252b; **10**:580b; **12**:217a
Catholic condemnation of, **1**:99b, 280a; **5**:402b, 403a; **10**:173a
Indian, **15**:555a
in Japan, **15**:730b, 732a, 746b
Aristotherus, **1**:338b
Aristotle (384–322), 1:250–281; 2:549b; **3**:516a; **4**:367a, 368a, 433a(n), 463a, 467b; **6**:406a; **8**:269a
acquaintances of, **9**:269a; **11**:27b, 28a; **14**:534b
astronomy, **2**:24a, 407a; **4**:213a; **6**:416b; **11**:188a, 223b; **12**:3b
biographies of, **7**:187a
biology, **1**:358b–266a; **3**:548b; **6**:420a
anatomy, **1**:4b, 264a–265a, 266a–267a, 316a; **4**:519b, 520a; **5**:356b; **12**:81b
botany, **15**:81a
embryology, **1**:109b, 262a, 265a, 266a–b; **3**:343a; **4**:509a, 510a, 511a; **6**:159b, 160b, 161a
entomology, **1**:263a–265b *passim*
ornithology, **1**:263a, 264a; **13**:502a
physiology, **1**:261b, 262a, 264b, 266a–267a, 324a–b; **4**:384a, 385b; **5**:234a; **6**:158b; **10**:598b; **11**:127b; **13**:92b; **15**:737b
spontaneous generation, **1**:261b, 262a, 263b, 265a; **2**:591a–b; **8**:128a
taxonomy, **1**:251a,b, 259b, 260b–261a, 262a–b; **5**:443a; **11**:316b, 317a
on Callippus, **3**:21b
commentaries on, **1**:270a–276b *passim*
ancient Greek, **7**:1a; **10**:111b, 112a
ancient Roman, **2**:230a; **5**:232a
Byzantine, **10**:208b; **11**:184a–b; **12**:440b–443a; **13**:37b
medieval Arabic, **3**:470a; **4**:525a; **7**:41a; **12**:1b–7Bb *passim;* **13**:292b; **15**:262a
medieval Latin, **5**114b*, 116b*; **9**:136a–b; **10**:172b; **12**:61a
Renaissance, **10**:122b–123a; **12**:151a; **14**:581b–582a
seventeenth-century, **12**:152b*
see also specific works below
cosmology, **1**:199a, 248a, 254a,b, 256b–259b *passim*; **3**:410a; **4**:466b; **5**:115b; **8**:537b; **10**:590a,b; **13**:326b
critics of, **3**:319b; **7**:134b–135b; **13**:93a
see also Cosmology, Aristotelian
critics of, **7**:41a; **11**:323b; **13**:292b, 564b; **14**:388b; **15**:262b
on Democritus, **4**:30b, 31a
editions of works, **8**:360b
errors of, **1**:100a–b; **2**:619b
and Eudemus, **4**:460b–461b
geography, **1**:381b; **15**:88a
geology, **13**:85a, 330b, 407b
on Hippocrates of Chios, **6**:410a–b
influence of, **7**:42b; **9**:364a; **11**:312a; **13**:565a; **14**:112b, 457b, 586a, 606a
logic, **1**:2b; **2**:295a, 296a–b, 609a–b; **4**:462a–b; **8**:546b; **12**:6a, 55b; **14**:607a, 613a
dialectic, **1**:251a–258a *passim*
mathematics, **1**:251a–b, 256a–258a, 279a; **4**:58a, 424b; **7**:327a, 329b; **13**:304a–b, 315a, 334a
quadrature, **1**:171a; **2**:550a; **3**:515b
meteorology, **2**:283a; **9**:363b; **11**:104b; **12**:310a
method, **1**:251a–258a, 263b–265a, 277a–279b, 589a–b; **4**:52b; **9**:317a; **10**:344a
analogy, **1**:262b–265b *passim*
classification, **1**:252a, 256a–b, 259b, 262a–b, 277b–278a, 375a
natural philosophy, **4**:416b; **11**:288a–b; **13**:91b–94b *passim*
"accident," **1**:250b–251a, 251b–252a, 277b, 278a
atomism, **2**:378a; **4**:31b–32a, 32b, 382a; **15**:650a
causation, **1**:198a–b, 255b, 259a–261a; **2**:502a; **5**:106b; **13**:206b, 340a
change, **1**:118b, 251b, 253a, 254b–255a, 258a; **7**:137a
elements, **1**:254a, 255a, 261a–b; **4**:368b
infinite, **7**:138a
location, **1**:27b–28a, 252b–253a, 256b, 257a
matter-form doctrine, **1**:255a–b, 278a
"natural places" doctrine, **1**:248a, 254a; **3**:404a; **12**:429a; **15**:495a
phenomena, **1**:252b, 277b, 278b
potentiality, **1**:258b, 277a–b; **7**:1a
time, **14**:608a–b, 609a
nutrition, **6**:499a
paraphrases of works, **13**:308a
philosophy, **2**:544b, 545a; **4**:43b; **6**:290a,b; **7**:23a; **12**:575b, 576a; **13**:291b; **14**:375a(n)
physics, **1**:34a, 126b, 223a, 229b; **4**:452a–b; **5**:115b–116a; **10**:548a; **11**:161a
critics of, **3**:470a; **7**:135b–138b
dynamics, **1**:28a, 253a–256a, 409b; **2**:605b; **4**:431a; **6**:378a; **13**:292a
kinematics, **1**:605b, 607a; **8**:231b
mathematics and, **1**:256a–258a
motion, **1**:100b–101a, 253a–256b *passim,* 258a, 260a,b, 279b; **3**:403b–404a; **4**:382a; **7**:1a; **9**:133a; **10**:161a; **13**:93a–b; **14**:608b–609b
optics, **4**:158b; **7**:136a; **10**:475a; **11**:526a
theory of, **1**:251b, 252b, 254b, 589a–b
on Plato, **11**:23a, 25b, 28b; **12**:441a; **13**:91b
on Pre-Socratics, **1**:151a, 152a; **10**:324b; **13**:295a,b, 296b–297a; **14**:536b
psychology, **1**:26b–27a, 118a–b, 260a–b, 265b; **4**:585b; **7**:1a, 137a–b; **9**:572a; **11**:71b
on Pythagoreans, **10**:591a
questiones on, **1**:275b–276b; **9**:33a
students of, **1**:282a, 283a; **4**:81b, 105b, 106a, 107a; **5**:249b; **6**:421a; **15**:202a
terminology, **1**:253a, 277a–279b
theology, **9**:29a
and Theophrastus, **13**:328b–333b
works translated, **1**:267b–276b; **3**:534a; **7**:25a, 65b–66b; **9**:361a, 362a–363a, 436a–438b; **10**:223a; **15**:31a, 176a, 230b, 231b, 261b, 263a
revisions of, **9**:436a
see also specific works below
Against Alcmaeon, **1**:103b–104a
Analytica Posteriora, **1**:94a, 251a,b, 252a, 256a, 258a, 269b–279a *passim,* 377b; **2**:229b, 230b, **2**:231a; 4:93a, 523b; **11**:287b
commentaries on, **5**:549a–b, 551a, 553a; **12**:6b; **13**:340a; **15**:176b
translations of, **7**:65b, 66a; **15**:175a, 176b
Analytica Priora, **1**:117b, 269b–275b *passim*; **2**:229b, 230b, 232a
commentaries on, **4**:237b; **15**:176b
De anima, **1**:84a, 199b, 260b, 270b–277a *passim*; **2**:604a, 609b; **7**:25a; **11**:244b
commentaries on, **5**:402b
translations of, **7**:66a
De animalibus, **1**:268a, 272a–277a *passim,* 409a
De bona fortuna (spurious), **1**:273a
De caelo et mundo, **1**:33b, 250b, 252a–258a *passim,* 271a–276a *passim;* **2**:391b; **4**:237a; **5**:238a; **6**:275b; **11**:469a; **14**:606a–b; **15**:34b
commentaries on, **1**:93b, 94a, 197b, 198b, 279a; **2**:590a, 604a, 605b, 607a,b, 609b; **3**:21b, 172a, 407a; **4**:157a–b; **12**:2a; **13**:206a; **15**:231b
translations of, **3**:406b; **5**:549b; **15**:180b
Categories, **1**:2a, 94a, 269b–278a *passim;* **2**:229b, 230a,b, 231b
De causis proprietatum elementorum (spurious)
translations of, **15**:180b
De coloribus (spurious), **1**:272a
Corpus recentius, **1**:273a
Corpus vetustius, **1**:273a
Dialectics, **2**:401a
Ethica ad Eudemum, **1**:273a, 274a; **4**:461a–b
Ethica Nicomachea, **1**:271a–275b *passim;* **4**:461a
translation of, **5**:549b
Ethics, **1**:94a; **2**:231a, 604a, 609b
De generatione animalium, **1**:259b, 260b, 261b, 264a, 265a
De generatione et corruptione, **1**:33b, 252b, 254a, 258a, 260b, 271a–275a *passim;* **11**:245a; **13**:279a
commentaries on, **1**:94a, 197b, 409a; **2**:590a, 609b; **3**:172a; **5**:403a
translations of, **15**:180b
Historia animalium, **1**:250b, 259a–267a *passim;* **5**:148a; **7**:64b; **9**:573a; **12**:576a
translations of, **13**:353a*
De incessu animalium, **1**:259a
De insomniis, **1**:271a, 272a, 273b
De interpretatione, **1**:2a, 94a, 269b–277a *passim;* **2**:229b, 230b, 231b; **4**:237a, 462a
Liber de causis (spurious), **1**:271b, 275b, 377b; **9**:438b
translations of, **15**:175a, 180b
Liber lapidum (spurious), **15**:187a
De lineis insecabilibus (spurious), **1**:272a
Lumen luminum (spurious), **13**:348a
Magna moralia (spurious?), **1**:272a, 273a, 274a; **4**:461a
Mechanica (spurious), **1**:250b, 256b, 419b, 607a; **5**:148a, 239b, 246b; **6**:312a; **7**:172b–173a, 342b; **8**:227b
De memoria et reminiscentia, **1**:199b, 272a, 273b
Metaphysics, **1**:117b, 251a, 256b, 257b, 260b, 270b–277a *passim,* 377b; **2**:231a, 604a,b, 606a, 609b; **3**:21b; **4**:461b; **13**:333b; **15**:262b
commentaries on, **12**:2a, 3a, 5b–6a, 7a; **15**:181a–182a
translations of, **7**:66a
Meteorologica, **1**:33b, 84a, 94a, 117b,

197b, 251b, 262a, 270b–276b *passim,* 409a; **2**:609b; **4**:157a, 237a; **5**:400a; **6**:275b; **13**:91b, 330b
commentaries on, **7**:134b
translations of, **5**:402a; **15**:180b
Mirabilia, **1**:265b
De mirabilibus auscultationibus (spurious), **1**:272a; **13**:91a
De motu animalium, **1**:256b, 279a; **2**: 609b
De mundo (spurious), **1**:250b, 272a, 274b, 275a,b
De naturali auditu
translations of, **15**:180b
De Nilo, **1**:272a
Oeconomica, **1**:94a, 272b, 274a
Organon, **1**:409a; **2**:230a; **4**:462b; **15**:497a
De partibus animalium, **1**:251a, 259a,b; **12**:2b
Parva naturalia, **1**:259a, 270b, 272a,b, 273a; **2**:604a, 609b
translations of, **7**:66a
Physics, **1**:33b, 250b–259b *passim,* 268a–279b *passim,* 377b; **2**:193a, 391b, 394a,b, 401b, 610a; **3**:420b; **4**:237a,b, 461b
commentaries on, **1**:93b, 94a, 198b, 409a,b; **2**:590a, 604a,b, 605b, 607b, 609b; **3**:172a; **4**:463a; **5**:113b–114a, 402b, 549b, 550b–551b, 553a; **6**: 411b, 412b; **7**:134a; **10**:123a; **12**:548a; **13**:91a; **15**:181a
Questiones, **13**:210a–b
translations of, **7**:66a; **11**:244b–245a; **15**:175b
Physionomia (spurious), **1**:272a; **15**:231b
De plantis (spurious), **1**:250b, 268a, 271b, 377b; **10**:111b, 112a; **11**:504a
commentaries on, **13**:292b
translations of, **11**:18a
Poetica, **1**:268b, 272a,b, 274a, 275a,b, 277a
Politica, **1**:268b, 272a, 273b, 274a; **2**:604a, 609b; **3**:378a; **4**:461a; **6**:419a
translations of, **9**:438a–b
De principiis (spurious), **1**:272a
Problemata (spurious), **1**:250b, 272a,b, 305b; **2**:609b; **6**:196a; **13**:91a
Rhetorica, **1**:271a–274b *passim*
Rhetorica ad Alexandrum (spurious ?), **1**:272b; **9**:438a
Secretum secretorum (spurious), **1**:380b
De sensu, **1**:94a, 117b, 199b, 271a, 272a, 273b
De somno et vigilia, **1**:271a, 272a, 273b
Sophistici elenchi, **1**:269b–276b *passim,* 377b; **2**:230b; **6**:376b, 412b; **12**:579b
commentaries on, **7**:66a; **13**:340a
On the Soul, **1**:275a
De spiritu (spurious), **13**:94b
Topics, **1**:117b, 250b, 251a, 258a, 269a–278a *passim;* **2**:230b; **13**:304a
Aristoxenus (*b. ca.* 375–360 B.C.), 1:281–283; **4**:82a; **11**:220a; **13**:326a
music theory, **4**:430b; **5**:249b; **10**:112b–113a; **11**:201a
Aristyllus (*fl. ca.* **270 B.C.), 1:283;** **3**:402b
astronomy, **11**:188a; **15**:217a,b
Arithmetic
ancient Greek, **4**:392a–b, 421b; **11**:221a–b; **13**:303a, 323b
ancient Roman, **2**:230b; **3**:109b; **4**:111a; **9**:140b–141a; **10**:113a, 294b
Byzantine, **4**:159b; **12**:406a
medieval Arabic, **4**:502a–b; **6**:208b*; **7**: 256b–257b, 325b–327b, 359a–b, 360b, 531b–533a; **9**:614a–615a; **11**: 227a, 229b–230a, 244b; **12**:31a, 61a, 363b*; **13**:510a, 539a–540b, 544b–546a; **15**:9b–10b
medieval Latin, **1**:62b–63a; **2**:395a, 435a; **4**:604b–606b, 607b, 611a–612b; **5**: 365a; **6**:302a; **7**:129a; **10**:540a; **11**: 18a; **12**:62b–63a; **15**:163a
Renaissance, **2**:618b; **3**:65a, 126a; **4**:260a; **7**:418b–419a; **8**:41a–b; **9**:191b, 610a; **10**:238a–b, 269b–271b; **11**:288a, 338b–339a, 457b–458a, 533b–534a, 590a, 591a; **13**:48a, 59a–60b, 259b, 491b; **14**:19b, 274b, 325a–b; **15**:155a, 478b
seventeenth-century, **3**:183b, 621b; **4**: 550a; **5**:59a, 409a; **7**:194a; **10**:254a–255a, 336a; **11**:508b; **13**:235b; **14**: 148a
eighteenth-century, **3**:462a*; **5**:520a; **7**: 565a–566a
nineteenth-century, **2**:274b–275a; **3**:177b; **5**:298b–299a; **7**:509a; **8**:62a
twentieth-century, **12**:411a
bibliography of, **4**:36b
business
medieval Arabic, **1**:22b, 30b, 40a–41b; **6**:205b*; **9**:39b; **15**:10a
medieval Latin, **4**:606a
Renaissance, **2**:279b–280a; **3**:272a,b, 280a, 311b; **8**:41a–b; **10**:238b, 270b, 494a
seventeenth-century, **3**:348a
eighteenth-century, **10**:319b
Chinese, **7**:325b–326a; **8**:419b; **14**:539a–b, 541b–543a
Egyptian, **1**:41a; **4**:111a, 612a; **6**:315a; **15**:681a–690b
formal, **5**:154b
foundations of, **2**:269a; **3**:278b; **4**:3b–4b, 101b; **5**:152b–153b, 351a; **7**:326b; **8**:263a; **10**:606a; **12**:217a, 229a; **14**: 89b
geometry and, **1**:474b; **8**:267a–b
Indian, **1**:309b; **7**:325b–326a, 531b–533a; **9**:22a–b, 613b, 614a–615a; **11**:18a; **12**:597b–598a; **13**:544b–546a; **15**:9b–10b
infinite, **3**:178a
infinitesimal processes, **3**:126a
Japanese, **15**:745b, 746a
laws of, **11**:543b
Mayan, **15**:769a
Mesopotamian, **3**:126a; **13**:540a; **15**:668a, 670b
Neoplatonic, **1**:148b
notation, **5**:414a
political, **10**:565b–566a; **11**:236b
theory of, **2**:233a–b; **7**:176a–b, 507b, 621b; **8**:100b
see also Calculation, Number theory, Numbers
Arithmetica universalis, **3**:514b
Arizona
geology, **4**:265b; **11**:118b, 209b
Arkadiev, Vladimir Konstantinovich (1884–1953), 1:283–284; **8**:107b
Arkansas
geology, **10**:258b
Arkell, William Joscelyn (1904–1958), 1: 284–286
Arlandes, François Laurent, Marquis d', **8**:78b; **9**:492b; **10**:609a
Arloing, Fernand, **1**:498a; **3**:219b
Arm
muscles of, **3**:40a
Armadillo, **3**:185b
fossil, **3**:569b, 570a
polyembryony, **3**:493a
Armagh
observatory, **5**:562a
Armengaud of Montpellier, **10**:540b; **12**: 7Bb; **13**:400b
Armenian science, **1**:274b; **12**:405a–406a
Armillary sphere, **1**:5b, 179a, 507b, 509a; **2**:118a, 133a, 406a, 411a; **3**:402a, 558b; **4**:260a, 556a; **5**:365a; **6**:144a; **7**:125b–126a, 259b, 353b; **12**:377a–b; **13**:510b–511a; **14**:273b, 575a–b; **15**: 564b
Armor
medieval, **3**:610b–611a
twentieth-century, **3**:611a
Armsby, Henry Prentiss, **1**:325b
Armstrong, Edward Frankland (1878–1945), 1:286–287; **1**:289a
Armstrong, Edwin Howard (1890–1954), 1: 287–288; **11**:213a
Armstrong, Henry Edward (1848–1937), 1: 288–289; **5**:127b*; **6**:463b*
acquaintances of, **9**:379b
chemistry, **5**:126b; **11**:85a
on contemporary scientists, **12**:28b
correspondence of, **3**:481b*
students of, **1**:461b; **8**:31b; **11**:84b
water analysis, **14**:169b
Armstrong College (Newcastle), **3**:238b
Army, U.S.
and statistics, **14**:553b
Veterinary Corps, **7**:286b
Army Medical College, **11**:346a
Arnald of Villanova (*ca.* 1240–1311), 1:289–291; **2**:23a, 317b, 393a; **5**:448a; **13**: 535a; **15**:266b(n)
alchemy, **1**:49b; **5**:580a; **10**:555a
Rosarium, **1**:49b
translations by, **11**:244b; **14**:584b
Arnauld, Antoine (1612–1694), 1:291–292; **4**:496b*; **8**:149b, 151a, 152b; **9**:47a, 48a, 49a; **10**:335b
Arndt, Johann
True Christianity, **1**:159b
Arneth, Arthur
Geschichte der reinen Mathematik . . . , **3**:59a
Arnold of Saxony, **4**:121b
Arnold, Christoph, **7**:373a
Arnold, Friedrich, **2**:127a, 160b; **5**:469b; **7**:437b; **15**:166a
Arnold, Harold de Forest (1883–1933), 1: 292–293; **4**:7a; **7**:108a
Arnold, John, **9**:163b
Arnold, John Oliver, **10**:246b
Arnold, Joseph, **2**:443a; **11**:261b, 262a
Arnold, Julius, **2**:165a; **6**:132a; **11**:176a
Arnold, Vladimir Igorevich, **6**:394a
Arnold, William, **4**:363a
Arnozan, Charles Louis Xavier, **1**:440b
Aromatari, Giuseppe degli (1587–1660), 1: 293–294
Aromatari, Renier, **1**:293a
Aromatization, **1**:415b
Aron, Max, **2**:345b
Aronhold, Siegfried Heinrich (1819–1884), 1:294–295; **3**:164a; **6**:356b
algebra, **6**:357b; **13**:219b
Aronhold's equation, **1**:294b
invariant theory, **3**:164b, 314b; **5**:472a
process of, **1**:294b
Arons, Leo Martin, **11**:582a
Aronsohn, Edward Hermann, **1**:450a
Arouet, François-Marie
see Voltaire
Arp, Halton Christian, **1**:353a
Arragona, Giovanni d', **7**:16b
Arran, Isle of, **11**:276b

Arrest, Heinrich Louis d' (1822–1875), 1: 295–296; 5:257a; 9:429b; 14:54a–b
comet of, 1:295a,b, 464a; 12:161a
Arrhenius, Svante August (1859–1927), 1: 296–302
acquaintances of, 4:485a; 7:203b; 12: 108b; 13:243a, 579b; 15:433a, 456a
on contemporary scientists, 4:299a
critics of, 1:288b
electrolytic dissociation theory, 2:170a; 3:176a, 619b–620a; 4:554b; 6:354b, 366a, 439b; 7:203b, 449b; 11:279a, 298b–299a; 13:244a, 364a, 579b; 15:433b, 458b–461a
critics of, 7:208b; 9:286b, 291b
equation of, 1:415a
historiography, 13:180b*
influence of, 2:498b; 11:298a
influences on, 2:186b; 11:299b
and Ostwald, 15:458b, 461b–462a, 465b
students of, 7:162a; 11:482b
Arrogant (steamship), 5:16b
Ars medicinae
see *Articella*
Arsanilic acid, 4:300a
Arsberättelser över vetenskapernas framsteg, 2:96b
Arsenic
alloys, 1:44b
compounds of, 2:421b; 3:617b; 5:100a; 8:618b; 13:70a; 15:172b
organic, 7:280b; 15:257a
oxides, 11:169b
salts, 2:572b*, 587a–b
sulfides, 13:310b
therapeutic use of, 4:300a–301a; 5:315b
detection of, 4:160b; 5:7b, 165a, 315b; 6:293b; 10:557a
isolation of, 1:101b
medical use of, 10:307a
metallic nature of, 2:421b
molecular weight of, 4:243b
oxidation of, 2:75a
physiological effects, 9:19b
poisoning by, 1:529b; 3:160a; 5:165a; 6: 17b; 15:172b
antidote for, 2:72b, 587a–b
as pollutant, 6:117a
sublimation of, 2:314a
in tin, 1:529b; 9:106a
vapor density of, 3:47b
Arsenic acid, 8:618b; 10:496a; 12:146b
Arsonval, Arsène d' (1851–1940), 1:302–305; 2:27a, 525a; 3:299a; 4:40a; 5:496a
calorimetry, 11:427b
galvanometer, 4:333b
d'Arsonvalization, 1:304b
Arsphenamine
invention of, 4:300b–301b
Art
medieval, 13:327a; 14:30b
Renaissance, 2:534b–535b; 4:258b–261a; 5:113a; 8:193b–194a, 201b, 241a–b, 242b
eighteenth-century, 12:537a
nineteenth-century, 10:372b
"scientific" impressionism, 3:241b–242a
anatomy and, 3:38a, 245b; 4:266b; 5: 196b; 7:416a*; 9:153b
in anthropology, 2:210b–211a
artists as scientists
medieval Latin, 11:468b; 14:30a–31a
Renaissance, 1:96b; 2:333b–334a; 4: 258b–261a; 8:192b–243b; 9:146b–147b, 309b
seventeenth-century, 5:439a; 7:576b–577a, 578a
eighteenth-century, 1:243b; 4:295a; 7: 411b–412b; 10:438a–439a; 11:502b–503a; 14:418b–419b
nineteenth-century, 1:235b, 330a–331b, 583b; 11:532a; 12:552b; 13:506a–b; 15:471b
twentieth-century, 5:467a
canons of, 4:260b, 261a
cave, 11:484a
mathematics and, 4:258b–261a
Mesopotamian, 15:649a, 651b
Paleolithic, 2:450b; 9:539a; 10:607a; 11: 484a
prehistoric, 8:44a; 14:3a; 15:51a
and science, 7:402b; 15:515a, 744b–745a, 749a, 753a
scientific analysis of, 2:531b; 3:115a*
theory of, 2:531b; 4:258b–261a; 8:195b, 234b; 12:88b, 155b
see also Aesthetics; Historiography, of art; Illustration; Paintings; Perspective
Artedi, Peter (1705–1735), 1:305–306
Artemisia absinthium, 5:439a–b
Artephius, 5:420a, 448a
Arterial branches, 2:3a
Arteries, 12:231a
ancient ideas on, 4:384b; 6:317b, 318a
cerebral, 1:569b–570a
elasticity of, 2:2b; 9:9a, 10b; 14:200a
function of, 6:153a–b, 154b
histology, 8:145b
ligature, 14:584b
presence of blood in, 5:234b–235a
structure of, 11:216b
suturing of, 3:91a
theories of, 10:520b
valves in, 9:41a
see also Aorta, Blood, Capillaries, Cardiovascular system, Carotid arteries, Pulmonary artery
Arteriometer, 10:205a
Arteriosclerosis, 10:493a
Arthritis
chronic, 1:400a
treatment of
steroid, 3:353a; 15:258b
Arthrodires, 3:610b
Arthropathy
neurogenic, 3:205b
Arthropods, 2:502a
microanatomy, 4:207a
morphology, 12:131a
neurology, 3:88a
parasites of, 8:146a
phylogeny, 15:123a
and vertebrate evolution hypothesis, 5: 283a–b
Arthur, Joseph Charles, 1:396a
Articella (Ars medicinae), 9:469a; 12:82a; 15:183a
Articuleae, 8:354b
Artifacts
classification of, 9:489b
stone, 14:505b
typological sequences, 11:6a
Artificial insemination, 2:287a; 7:31b–33a; 9:86a; 12:559a
Artificial respiration, 2:367a; 9:266a; 10: 267b; 11:102a; 12:356a; 14:187b
gases for, 6:264b–265a
see also Resuscitation
Artillery
Renaissance, 9:610a
seventeenth-century, 13:592b
eighteenth-century, 3:39b, 119b, 459b, 562a
nineteenth-century, 3:224b, 234a, 366a; 4:39b–40a; 7:602a; 9:60b; 15:24b
twentieth-century, 3:195b
historiography of, 4:217b*
see also Ballistics, Missiles, Rockets
Artin, Emil (1898–1962), 1:306–308
acquaintances of, 2:191b
algebra, 11:55b
on contemporary scientists, 14:212b*
influence of, 14:90b
influences on, 13:23b
number theory, 6:214b, 297a,b, 389b, 394a,b
law of reciprocity, 3:222a; 4:474b; 6: 393b
Artinsk Stage, 7:252a, 253a
Artiodactyls, 12:38a
Arts and trades, 7:578a; 10:496b–497a; 11:43b, 328a
Encyclopédie, 4:86b–87b
Artsimovich, Lev Andreevich (1909–1973), 15:8–9
Arum
heat production in, 5:278a
tuber of, 9:18a
Arvicolidae, 10:16b
Āryabhaṭa I (*b.* A.D. 476), 1:308–309; 1:33a; 7:582b; 8:46b; 9:580b, 613b; 10:419b, 426a; 15:602b, 607b, 610a
algebra, 1:30b
Ārdharātrikapakṣa, 12:115b
Āryabhaṭīya, 2:416a; 4:556a; 6:115b; 7: 363b(n); 10:313b; 15:590a–602a *passim*
commentaries on, 2:114b; 10:129b
astronomy, 15:574b, 580b, 605a,b, 631b(n)
critics of, 15:590a–b
equal Yuga system, 15:590a–b, 603a, 609a
influences on, 15:555b, 570a
trigonometry, 5:612a,b; 15:559a, 571a
Āryabhaṭa II (*fl.* between *ca.* 950 and 1100), 1:309–310; 12:597b
Mahāsiddhānta, 15:619b–621b
sine table, 15:620a
Aryans
origin of, 11:210b
Āryapakṣa (school of astronomy), 2:114b; 7:582b; 13:590b; 15:534a, 588b, 590b–602b, 604a,b, 607b, 609a–b, 613b–614a, 618a, 622a, 623a, 624b, 625a
founding of, 15:590a
Arzachel
see al-Zarqālī
Der Arzt, 13:543a
ʿaṣabiyya (psychol.), 7:322a,b
Asada Gōryū (1734–1799), 1:310–314; 7:18b
calendric reform, 15:732a, 735a
Asada Yasuaki
see Asada Gōryū
Āśādhara, 15:588b, 598a
Grahajñāna, 15:584b–585a
Asahina, Yasuhiko, 15:150a
Asbestos, 1:30a
Ascariasis
control of, 12:453a
Ascaris, 5:453b; 7:557a
cytology, 12:423b
development of, 3:588a; 8:270b
embryology of, 1:601b
mitosis in, 1:601b; 2:362b–364b
Ascension Island, 7:544b
geology, 3:547b
Aschaffenburg, Gustav, 7:190a
Aschkinass, Emil, 11:582a
Aschoff, Hermann, 9:424b
Aschoff, Ludwig, 8:9a
Asci, 5:192b

Ascidiacea, **3**:148a
Ascidians, **4**:12a; **5**:385a; **8**:414b; **12**:131a
anatomy, **6**:590b
budding, **6**:442a
cell-lineage, **3**:120b
embryogeny, **1**:601a; **3**:184b; **4**:187a; **7**: 475a; **9**:524a
regeneration of, **4**:187a
Ascites, **4**:106a
Asclepiadaceae, **2**:518b, 519a,b; **5**:424b
Asclepiades (*ca.* 130–*ca.* 40), 1:314–315; **1**:567a; **4**:119b; **7**:187a
Asclepiads, **6**:418b, 420b
Asclepius, **2**:21a; **4**:461b; **5**:232b; **12**:88b, 440b
translations of works, **11**:245a
Asclepius, **2**:539b, 540a; **6**:306a
Ascocarp, **6**:121b
Ascoli, Cecco d', **2**:540b
Ascomycetes, **2**:19a, 437a,b; **4**:401b; **5**:584b; **6**:100a, 121b; **14**:93b
cytology, **9**:34b
Ascorbic acid
see Vitamin(s), C
Ascus, **5**:584b
Aselli, Gaspare (1581–1625), 1:315–316; 1: 482a; **11**:507b
anatomy, **5**:226b; **13**:31a
De lactibus, **10**:490b
lymphatics, **6**:317b; **10**:477a–b
Asepsis, **1**:493b; **3**:91a; **8**:408a–b
al-Asfizārī, al-Muẓaffar ibn Ismāᶜīl, 7:336b–345b *passim*
Ash, volcanic, **2**:481b
al-Ashᶜath, Aḥmad ibn Abī, **9**:40b
Asher, Leon, **7**:519b; **8**:454a
Ashmole, Elias (1617–1692), 1:316–318; 2: 523a; **13**:96a; **14**:285b, 286a, 287a
and Tradescant, **13**:451a–b
Ashmolean Museum
see Oxford, Ashmolean Museum
Asia
botany, **1**:422b
climatology, **11**:181b
exploration of, **10**:167b; **11**:180b–182a, 210a–b; **12**:300b–301b
frontier with Europe, **5**:428b
geography, **2**:191a; **6**:215a–217a; **11**: 108a–b
geology, **4**:598b; **9**:104a, 592a–594b; **10**: 167b; **11**:209b, 539b; **13**:275a–b
glaciers, **14**:517a
mountain systems, **11**:181b
paleobiogeography, **2**:171b
volcanoes, **11**:336b
see also specific countries and regions
Askesian Society, **13**:280b
Āsmakatantra, **2**:114b
Asophus
see al Sūfī, ᶜAbd al-Raḥmān ibn ᶜUmar
Asparagine
derivatives of
optical properties, **10**:358a–b
isolation of, **11**:494b
metabolic role, **10**:575b, 576b; **11**:179b; **14**:66b
Asparagus
analysis of, **11**:494b
Aspartic acid, **8**:340b
polymers of, **12**:164a
preparation of, **10**:359a–b
Aspergillus, **2**:87a; **4**:402a; **11**:310b
Asphalt
mining, **7**:44a
occurence of, **6**:229a
Asphyxia
mechanism of, **2**:61b
treatment of, **2**:74b, 445a; **6**:265a; **12**:502a
see also Artificial respiration
Aspirin, **3**:246a
Asratian, Ezras Asratovich, **10**:435a
Ibn al-ᶜAssāl (Abu'l-Faraj Hibatullāh), **15**: 232b
Assalti, Pietro (1680–1728), 1:318–319
Assaying
Renaissance, **4**:393a–394a
eighteenth-century, **3**:41b; **6**:236b; **8**: 621a; **12**:63b; **13**:411b, 597a; **15**:94a
nineteenth-century, **11**:485b; **12**:329b
Assembly line, **13**:614a
Assimilation, geological, **2**:370a
Association, of ideas, **2**:132a; **4**:85b; **6**:139a–140a, 556a, 557a
Association, statistical
coefficient of, **14**:574a
Association Française pour l'Avancement des Sciences, **14**:530a
founding of, **11**:235a
Association Internationale de Botanistes
founding of, **8**:512b
Association Internationale de Géodésie, **10**:524a
Association of American Geographers
founding of, **3**:592b
Association of American Geologists
founding of, **3**:392a; **4**:364a; **11**:504b; **12**:434a
Association of German Mathematicians
founding of, **3**:53a
Association Polytechnique, **3**:375a
Associationism, **11**:518b
Associative law, **2**:295b, 298a; **3**:166a
Associativity (math.)
Jacobi, **8**:326a,b
Assurbanipal, **15**:637a–b
Assy, Geoffroy d', **4**:14a
Assyria
see Mesopotamia
Astarte, **11**:500a
Astasia, **8**:536b
Astat (Eustathius), **1**:275b; **10**:230b
Astbury, William Thomas (1898–1961), 1: 319–320; 6:133b; **8**:485a
Asten, Friedrich Emil von, **1**:371b
Astereae, **10**:13b
Asteriidae, **3**:492b
Asterina
embryology, **8**:586a
Asteroids
nineteenth-century study of, **1**:295b; **2**: 532b; **3**:92b; **5**:256b, 257b, 258a–b; **6**:49a
early twentieth-century study of, **2**:126b; **3**:472b; **4**:26b; **8**:504a
and comets, **7**:386a
dimensions of, **1**:466b; **6**:334b
discovery of
late eighteenth-century, **13**:425a; **14**: 583a
early nineteenth-century, **6**:112a, 334b; **10**:198b–199a, 591b–592a
mid nineteenth-century, **4**:369b; **6**: 402b; **8**:557b; **10**:543b; **12**:160a
late nineteenth-century, **5**:258b; **6**:282a; **10**:279b, 528b; **13**:36b; **14**:481b, 482a
early twentieth-century, **1**:352b; **3**: 357b; **4**:219a; **10**:30b
mid twentieth-century, **1**:352b
distances from sun
Kirkwood's gaps, **2**:511a; **7**:385a–b
ephemerides of, **4**:514b; **5**:256b; **10**:159a
families of, **2**:511a; **6**:431a; **7**:385b
Hilda group, **7**:385b
identification of, **4**:514b
mean motions of
distribution of, **6**:431a
naming of, **6**:334b; **10**:592a–b
orbits of, **3**:14a; **4**:250b, 359a, 514b; **6**: 103b; **9**:514a; **10**:33b–34a, 159a, 543b; **13**:422b; **14**:242b
calculations of, **10**:212b, 219a; **11**: 114b–115a
origin of, **13**:270b
perturbation of, **1**:157a; **2**:442a; **4**:369b; **6**:539b; **9**:460b; **10**:528b; **11**:552a; **12**:17b; **13**:422b–423a
photography of, **2**:196b
positions of, **10**:279b; **14**:482a
and solar parallax, **1**:157b; **4**:351a; **5**:258a, 404b, 405a
spectra, **12**:269a
Trojan group, **2**:516b
Asteroma, **3**:45a
Asteroxylon, **8**:5b
Asterozoans
paleontology, **9**:346a
Asthenia, **8**:536b
Asthma, **1**:98b; **3**:64b; **7**:605b; **11**:445a; **14**: 407b
bronchial
theories of, **9**:266a
cardiac, **11**:559b, 560a
causes of, **6**:255b–256a
colloidal phenomena, **1**:431b
Astian (stratig. stage), **9**:241a
Astigmatism
see Vision
Aston, Francis William (1877–1945), 1:320–322; 5:459a,b*; **6**:16b, 117a,b; **8**:369a; **9**:110a; **12**:33b, 505a, 507a(n); **13**: 369b
Astra, **10**:308b; **12**:335b
Astrakanite, **11**:534b
Astrarium, **4**:164b
Astrobotany, **13**:408b–409a
Astrochemistry, **8**:47a
Astrogeology, **5**:212b
Astrograph, **4**:26b; **8**:364a; **9**:46b; **10**:30b; **11**:553a; **13**:45b, 408a, 481b
Astrographic Catalogue, **4**:359a
Astrography
eighteenth-century, **7**:543b
nineteenth-century, **5**:405a–b; **6**:282a, 326b, 332b, 522b; **9**:551b; **12**:153a; **15**:493b
twentieth-century, **1**:429a; **5**:183a; **11**:49a
see also Atlases, stellar; Globe, celestial
Astrolabe (astron. instrum.)
ancient Greek discussion, **13**:323a–b
ancient Roman discussion, **11**:198a
medieval discussion, **1**:62b, 82a, 438a; **2**:149a, 154a, 435a; **3**:217b, 219a*; **4**:502b; **5**:365a,b; **6**:544b; **7**:2b, 119a, 140b, 338b, 362a–b; **9**:435a, 615a*; **13**:149b, 509b; **14**:400a; **15**:163a
Renaissance discussion, **1**:165b; **14**:273b; **15**:154a, 477b
eighteenth-century discussion, **2**:133a
astrological, **6**:273a–b
construction of
ancient Roman, **10**:299a
Byzantine, **6**:616a; **13**:225b
medieval, **2**:152b; **4**:556a; **6**:301b; **7**: 334b, 353b, 531b; **9**:39b, 85a, 160b, 555b; **10**:534a; **11**:240b, 321b; **12**: 361b, 432a; **13**:514b; **15**:162b–163a
Renaissance, **3**:423b, 558b; **5**:348a; **6**: 144a; **10**:161a; **13**:621b–622a
eighteenth-century, **9**:234a
geometry of, **3**:27a; **4**:543a; **11**:198a; **15**: 154b–155a
impersonal, **3**:298b
Indian discussion, **9**:22b; **15**:628a
instruments similar to, **7**:125b, 126a, 260a; **13**:400b; **15**:520b(n)
invention of, **13**:323a–b

Astrolabe (cont.)
linear, **13**:514b
mechanized, **12**:362a
plane, **15**:219b
prismatic, **3**:298b
shadow functions, **2**:153b; **6**:301b
spherical, **1**:122a, 382a; **10**:6b
tables for, **2**:152b; **7**:334a–b
users of, **3**:402a; **4**:584b; **7**:81a,b, 419a; **9**:21b; **14**:575a, 583b; **15**:219a–b
see also Saphea
Astrolabe (ship), **5**:224b; **11**:243a
Astrolabes (texts), **15**:674b–676a
Astrology
ancient Greek, **1**:248a; **2**:256b; **5**:345b; **6**:305b, 306a; **10**:547b–548b; **11**: 105a; **13**:320a; **14**:549a–b, 606b; **15**:125a–b, 220a
ancient Roman, **3**:175b; **4**:621b–622b; **9**:79b; **10**:301a, 419a–b; **11**:42a, 198a–b, 201b; **12**:310a–b
Byzantine, **8**:191a–b; **11**:161a, 185a; **13**: 37b
medieval Arabic, **1**:23a, 33a–39b, 62a, 509a, 513a–b; **2**:149a, 153b–154a, 155a; **4**:502b, 542b–543a, 555b; **5**: 614a; **6**:80a, 206b*, 381a–b; **7**:42a, 79a, 82b, 222b–223b, 324b–325a, 335a, 336b, 358b, 362b, 531b; **8**:191b, 547b; **9**:159b–162a, 555b; **11**:226a–b, 231b, 444a; **12**:431b–432a; **13**:149b, 230b, 509a, 510a, 538a–539a; **14**: 118a, 537b, 574a, 578a, 594b; **15**: 265b, 497b
medieval Hebrew, **8**:281a–b
medieval Latin, **1**:4b, 84a, 199a, 290b, 378a, 382a–b; **2**:193a, 393b; **3**:27a, 219a*; **4**:157a; **5**:361a, 402a, 552b; **6**:272a–274a; **7**:112a–b, 125b, 131a, 140b; **8**:550a; **9**:361b, 362a, 363a–b; **10**:421a; **11**:322a, 468b, 504a; **13**: 348a; **14**:399b–400a
Renaissance, **1**:165b, 495b, 605b, 607b; **2**:401b, 403a,b, 407b, 410a–b; **3**:65a; **4**:6a, 584b, 619a,b; **5**:401a; **6**:144a, 562b; **9**:12b; **10**:152b–153a, 153b, 245b, 306b, 308b–309a, 310b, 311a; **11**:395b, 396b–397a; **12**:199b, 439a–440a; **13**:440a–b; **14**:273b, 583b; **15**:154b, 474a–b, 477b
early seventeenth-century, **1**:244b; **2**: 526a; **3**:59b–60a, 152b, 153a; **7**:289a, 289b–290a, 297b, 299b, 301a–b, 302b, 306b, 537b; **9**:527b; **12**:342a–b; **14**:285b–286a; **15**:68b–69b
late seventeenth-century, **1**:317a–b; **2**: 349a; **3**:100a, 248a; **4**:168b; **9**:310b; **14**:446b
eighteenth-century, **4**:23a; **7**:374a
autobiographical, **6**:272a–b, 273a
Babylonian, **10**:547b–548b
catarchic, **7**:315b, 583a; **12**:598b; **15**:125b
Chinese, **15**:729b, 730b
contact-action theory, **1**:495b
continuous, **15**:125b
cosmology and
heliocentric theory, **3**:248a
critics of
ancient Greek, **1**:335a–b
medieval Arabic, **2**:156a; **4**:525a; **7**: 322a–b; **12**:94a
medieval Latin, **5**:361a; **6**:275b; **7**:118a–b; **10**:224b, 225a; **14**:388b
Renaissance, **2**:537b; **4**:213a, 387a, 585a; **10**:512b; **11**:97b; **12**:97b
seventeenth-century, **2**:131a, 463a; **3**: 61a; **7**:459a; **9**:486b; **12**:103a
and divination, **15**:644a–b
Egyptian, **10**:547b–548b; **15**:706a–727b
"experimental," **3**:248a
Indian, **2**:154b; **5**:275b; **6**:115b; **7**:222b–223b, 315a,b, 583a; **10**:314a*, 548a; **11**:264a; **12**:576b, 598a–599a; **13**: 581b–583a; **14**:549a–b
Japanese, **15**:729a,b, 730a, 731b
judicial, **2**:410b; **4**:6a; **5**:361a; **9**:12b; **10**: 512b; **11**:504a; **12**:324a; **13**:440b; **15**:723b, 724a, 725a
critics of, **9**:485a; **14**:388b
justifications for, **1**:33a–34a; **12**:439b
Mayan, **15**:777b, 804b
and medicine, **1**:49b, 244b, 382a–b, 399b; **2**:337a; **4**:585a; **5**:48a, 547a; **6**:123b; **7**:142b, 459a; **8**:192a*; **9**:74b; **10**: 153b; **12**:324a; **14**:399b; **15**:154b
Mesopotamian, **15**:635b, 644a, 672a,b, 673b–674a, 677b, 678a–b
military, **13**:582b–583a
political, **8**:191b
and religion, **4**:213a
see also Divination; Education, astrological; Genethlialogy; Historiography, astrological; Horoscope; Meteorology, astrological
Astrometer, **6**:326b; **9**:563b
Astrometry
eighteenth-century, **2**:99a
nineteenth-century, **2**:196a; **4**:369b; **5**: 256b, 480a; **10**:528b
twentieth-century, **2**:333a; **5**:363b, 554b; **6**:149b; **7**:6b; **12**:18b, 176a–177a; **13**:139a
photographical, **7**:472a–b; **11**:320b
Astronautics **5**:433a–b
Astronomical Journal, **4**:590b; **5**:480b; **6**:49b
editors of, **9**:514a
founding of, **5**:479b
Astronomical Newsletter
editors of, **13**:119a
Astronomical Papers Prepared for the Use of the American Ephemeris and Nautical Almanac
founding of, **10**:35a
Astronomical Society of Berlin
see Astronomische Gesellschaft
Astronomical Society of London, **3**:93b; **4**: 185b; **9**:184b; **12**:524b, 551b
criticism of, **12**:552a
expeditions sponsored by, **5**:404b
founding of, **1**:354b, 402b; **6**:325b
women in, **3**:50b
Astronomical Society of the Atlantic, **7**:238b
Astronomical Society of the Pacific
founding of, **6**:470b, 471b
Astronomical unit, **1**:466b; **3**:409a; **11**:163a, 424a
Astronomichesky ezhegodnik, **7**:7a
Astronomische Gesellschaft, **7**:481a; **14**:54b, 449a
founding of, **1**:339b
observational program, **1**:54a, 243a, 340a
star catalog
see *Berliner Akademische Sternkarten*
Astronomische Nachrichten, **5**:257b; **6**:112a; **13**:98b
editors of, **10**:542a,b
founding of, **6**:103a; **12**:234b
Astronomisches Jahrbuch, Berlin
see *Berliner astronomisches Jahrbuch*
Astronomisches Rechen-Institut, Heidelberg, **4**:27a
Astronomy
ancient Greek, **1**:189b–191a, 205a; **3**:21a–22a, 318b–320a, 391a–b; **4**:391b, 429b, 459b–460a, 466b–467a; **5**: 344b–347a; **6**:291a, 410b, 416b, 616b–617b; **9**:337a–339a; **10**:179b–180b; **11**:28b, 187b–188b, 223a–224b; **13**:295a–296a, 319b, 320a; **15**:208a–221a, 538a–625b *passim*, 645a
ancient Roman, **1**:148b, 335b; **3**:14b–15a, 176a; **9**:141a, 296b–297a; **10**:295b; **11**:186b–198a; **12**:547a, 580a; **13**: 325b, 326a–b; **15**:518a–b, 519b–520b
Byzantine, **1**:137a–b; **11**:161a; **12**:405b–406a; **13**:37b, 225b
medieval Arabic, **1**:23a*, 122a–b, 437b–438a, 507b–513b; **2**:148b–155a *passim*; **4**:542a–543b, 555b–556b; **5**:612a–620a; **6**:79b, 197a–199b; **7**:2b, 25a, 38b, 79a, 82a–b, 255a–256a, 259a–260a, 324b–325a, 334a–335a, 336a–338b, 358b, 362a, 532a–b; **9**: 21b, 36a, 39a–b, 83b–85a, 555a–b; **10**:5b; **11**:202a–b, 231b, 240a,b, 246b–247b, 248b–249a, 249b, 250b–251b; **12**:2a, 3a–5b, 91a, 162a, 357b–362b, 431b, 447b–448a; **13**:149a–150a, 291b–292a, 400b–401a, 509a–511a, 513b, 514a–b, 535b–537a, 538a–539a; **14**:546b, 574b–579b, 592b–594b; **15**:33a–35b, 259a–260b, 265b, 497b, 596b, 605a, 607b, 624a, 625b–629a
medieval Hebrew, **8**:280a–281b
medieval Latin, **1**:91b, 101a, 199a; **2**:20b, 24a–b, 435a–b, 550b–551a; **3**:25a–27b, 217b–218a; **4**:94b, 164a–b; **5**: 361a, 365a–b, 402a, 552b–553a; **6**: 273a–274a, 275b, 301a–302a; **7**: 112a–b, 117a–120b, 122b–126a, 128b–131a, 141b–142b; **9**:363a–b; **10**:540a–541b; **11**:321b–322b, 415a–b, 504a, 468b–469a; **12**:61a–62b; **13**:38b, 534b; **14**:35b, 389b–390b, 400a–401a; **15**:162b–163b
Renaissance, **2**:401b–414a; **3**:401b–410b, 585a; **4**:97b, 213a, 619b; **5**:106a, 349a, 400a–b; **7**:419a; **9**:167b–168b, 191b–192a, 335b; **10**:153b–154a, 515a; **11**:339a, 348b–352a, 402b–403a, 561b; **12**:199b; **13**:49a–b, 440a–b; **14**:18b; **15**:154a–156a, 473b–478b
early seventeenth-century, **2**:83a–b; **3**: 60a, 115b, 116a, 262a, 457b, 528a–b; **4**:221a, 551b; **5**:240b–244b *passim,* 382b, 542b–543a; **7**:193b, 194a, 269b, 289a–308a; **8**:271b; **10**:176a–b, 489a–490a; **11**:411a–b; **12**:151b–152a, 163a, 332a–b, 342a–343b, 500b; **13**:29a, 256b; **14**:254a–b, 285b–286a, 637a
late seventeenth-century, **1**:481a; **2**:592b–593a; **3**:534b; **4**:221b–222a*, 506a; **5**:22b–25b, 58b, 59a–b, 520b, 525a–526a, 529a; **6**:67b–69a, 148b, 603b–604a; **7**:373a–b, 584a; **9**:310a–311a, 465a, 485a–486b, 554b; **10**:546b; **11**:423a–424b, 525b–526b; **12**:85b, 431a; **13**:96a–b, 335a; **14**:177b–178a, 446b, 509a
early eighteenth-century, **2**:220b–221a, 325a, 387a–389b, 468a–b; **3**:344b, 430b–431a; **4**:40b, 41a, 166a, 481a; **5**:87b–88a; **6**:69b–71a, 129b; **7**:373b–374a, 576a; **9**:77a–78a; **13**:179a; **14**: 178a–179a, 518b
late eighteenth-century, **2**:56b, 216b, 327a–329a, 599b; **3**:16b, 158a; **4**: 131b–132a, 565b; **5**:195a, 210a, 467a–469a; **6**:511b–512a; **7**:494a–b, 579b–582a, 598a–b; **9**:162b–163b, 164a–165a, 370a–371a; **11**:67a, 472a,

497b*, 609b; **12**:554a; **13**:424b–425a; **14**:414b–415a; **15**:343a–345b
early nineteenth-century, **1**:615b, 616a; **2**:368a–b, 421a, 468b; **3**:135b, 555a; **5**:299b–302b, 309b, 466a, 518a; **6**: 85b, 107a; **7**:18b–20a; **8**:369a–b; **9**: 547b; **11**:236a–b; **12**:234a–b; **13**:4a; **14**:489b, 500a
mid nineteenth-century, **2**:430b, 431a, 441b; **3**:186b–187b, 238b, 302a–b; **4**:178b–180b, 182b, 555a–b, 590b; **5**:86a–87b, 256b–258b; **6**:453a; **10**: 207a–b; **12**:160a–b, 169b–170a, 266a–269b, 289b, 498b–499a; **13**: 22b; **14**:480b–481a
late nineteenth-century, **1**:371b–372a, 449a–b; **2**:432b–434b, 616a; **3**:374a–b; **4**:178b–180b; **5**:33b–34a; **7**:270a–271a; **8**:250b; **9**:189b, 194a–195a; **12**:160b–162a; **13**:45a–b, 82a, 422b, 472a–b, 481a–482a, 500a–b; **14**:56b, 139b, 242b, 479b
early twentieth-century, **2**:196a–b, 429a; **3**:50a–b; **4**:269b–270a, 272b–273a; **5**:181b–182b, 199a–b; **6**:105b–106a; **8**:539a–b; **9**:552a–b; **10**:289b–290a; **11**:18b–19a, 542a–543a, 552a–553a; **12**:95b–96a, 176a–177a, 280a, 345b–349b, 448a–449b; **13**:12a, 95a–b, 98a–b, 98b–99b, 138b–139a, 474b–475a; **14**:521a–b
mid twentieth-century, **5**:183a–184a; **8**: 364a–365a; **10**:317a–318a
Alfonsine, **3**:217b
algebraic, **1**:314a
and archaeology, **8**:442b–443a
Chinese, **1**:313b; **3**:249a, 269b; **7**:374a; **12**:372a, 376a–379b; **13**:485b; **14**: 159a–165b; **15**:729b, 732a–b, 733a,b, 747b
Egyptian, **3**:391a; **12**:547a; **15**:706a–727b
epistemology of, **1**:589a–b
experimental, **1**:56a
fundamental constants, **2**:511b; **4**:176a; **9**:514b; **10**:35a; **12**:449a–b
and geodesy, **6**:240b
geometry and, **1**:248b–249b; **2**:348b; **13**: 54b–55a; **14**:160b; **15**:208b
Greek-letter phenomena, **15**:540a, 545b, 569b, 578b, 584a, 598a, 601a
handbooks
Greek, **3**:319a
Indian, **2**:154b
Islamic, **2**:154b; **10**:5b; **14**:574b–579a; **15**:580b
Indian
see India, astronomy
Japanese, **1**:310a–314b; **7**:18b–20a; **12**: 403a–404a; **15**:729a–737a, 745b
mathematical
ancient Greek, **1**:231b; **15**:209b–220b
ancient Roman, **11**:187b–198a
Renaissance, **1**:589a–b; **4**:213a; **11**: 365b–366b; **15**:475a–477b
seventeenth-century, **6**:515a–516a
eighteenth-century, **7**:542a–b, 564b; **10**:614b–615b; **11**:472a; **13**:417b; **15**:286a–295a, 302b–303b, 350b–356a
early nineteenth-century, **2**:359b–360a, 600a; **11**:544b–545b
mid nineteenth-century, **1**:449b; **2**: 284b, 616b; **4**:369a–370a; **5**:256b–258a; **6**:103a–104a, 263a; **7**:51b, 481b–482a; **8**:530a; **10**:41a–b, 479b; **11**:114a–115a, 206b, 207b
late nineteenth-century, **2**:198a; **3**:18b–19a; **5**:223b–224a, 414a; **6**:94b, 398b–400a; **8**:504a; **10**:34a–b; **13**:339a
twentieth-century, **4**:359a; **7**:6a–b, 369b; **9**:410b–411a; **11**:49a–b, 83b–84a; **12**:573b–574b
Chinese, **3**:269b; **12**:374b–375a; **15**: 730b
Egyptian, **15**:692a–693a
Indian, **15**:533a–633b
Japanese, **15**:729a,b, 730a, 731b, 732a,b, 736b, 739b, 743a, 745b, 747b, 749a, 753b
Mesopotamian, **15**:672a, 676b, 679b
Mayan, **15**:759a–818b
measurement in
astronomical unit, **1**:466b; **3**:409a; **11**: 163a, 424a
distance, **1**:56a–b, 157b, 353a, 398a, 466b; **4**:169a; **6**:352a–b, 529b, 530a–531a, 532b
micrometric, **1**:466a–b
Mesopotamian, **6**:616b; **10**:290a, 426a; **11**:188a,b, 191b, 195b, 223a–224a; **12**:547a, 576b; **13**:295b–296a, 582a; **14**:593b; **15**:209b–212b *passim*, 216a, 218b, 219a, 220b, 331a, 640b, 643b–644a, 672a–680b
and metaphysics, **12**:2a
nautical, **3**:158a; **10**:572a
nomenclature, **1**:242a, 530a; **9**:23a; **13**: 149b; **15**:673a
observational
ancient Greek, **1**:283b; **15**:220b
medieval Arabic, **1**:42a
medieval Latin, **7**:130a
Renaissance, **1**:530a–b; **2**:401b–414a; **5**:547a; **14**:358b–359b
early seventeenth-century, **1**:566b; **4**: 128a; **5**:287b; **6**:124b, 126a–b, 514a–b, 520a–b; **7**:374b–375a, 376a; **8**:25b; **9**:122a–b, 247a–b, 527b–528a; **11**: 395a
late seventeenth-century, **1**:169a; **2**: 310a–b, 349a; **3**:23a, 100a–104a; **4**: 168b–169a; **6**:360b–364a, 538a; **7**: 577b; **8**:132a; **9**:115b, 508a*; **10**: 595b–596b; **11**:532a–b; **13**:445a; **14**: 451b
early eighteenth-century, **2**:328a; **3**: 104b–105b, 108a, 173b–174a; **4**:22b–24a, 602b–603a; **5**:434b–435b, 515b–516a; **6**:513a–b; **8**:143b–144b, 459b, 469a, 611b; **9**:90a–b, 463b–464a; **10**:255b–256a, 572a; **11**:66a*; **14**: 589a–b; **15**:270b–271a
late eighteenth-century, **1**:310a–311b, 400b–401a; **2**:4b–5a, 141a; **3**:52a, 106a, 197b–198a; **4**:14a–b; **6**:233b–234b, 322b–323a, 328a–335b, 512a–b; **7**:86b–87a, 542a–545a, 596b; **8**: 67b, 178b–179b, 299a–b, 321b; **9**: 231a–232a, 233a–234b, 250a–b, 329b–330b; **10**:197b–198b, 607b–608b; **12**:226a–b, 413b–414a, 463a; **13**:39b; **14**:452b–453a, 485a–b, 589b–590a
early nineteenth-century, **1**:202b, 241a–243b, 402b–403a, 568b–569a; **2**:97b–101b, 125b–126a, 451a; **6**:112a, 323b, 324b–326a; **9**:429a–430a, 459b; **10**: 110b–111a, 199a–b; **11**:82a–83a, 545a–b; **12**:518b, 551b–552a; **13**: 108b–109b
mid nineteenth-century, **1**:86a–87a, 241a–243b, 295a–b; **2**:147a–b, 236a, 284b–285a, 285a–b, 471b–472a; **3**: 288a, 605b–606a; **4**:19a, 354a, 369b–370a; **5**:479a–480b; **6**:48b–49a, 326a–327b, 402b–403a, 540b–541b, 553b; **7**:145b–146a, 481a, 607b–608b; **8**: 46a, 487b, 557b, 577b, 613b; **9**:2a–b, 195b, 421b, 615b–616a; **10**:33b, 218b–219a, 328a–329a, 480a, 542b, 543a–b, 584a–b; **11**:379a; **12**:192a, 200b, 239b; **13**:109b–111b, 120a–121a; **14**:448b–449a
late nineteenth-century, **1**:19a, 54a, 157b–158a, 397b–398a, 463b–467b, 597a; **2**:146b, 433a, 502b–503a, 532b–533a, 614b–615a; **3**:35a–36a, 194a–b; **4**:185a–b, 250b–251a, 351a–b, 514a–b; **5**:21a–22a, 403b–406a; **6**:49a–b, 119a–b, 239b, 282a–b, 470b–471b, 522b–523a; **7**:76b–77a, 400a–b, 461a–b; **8**:465a–b, 503b–504a; **9**:183b–184b, 460a–b, 551b; **10**:34b, 119b, 279a–b, 483a, 542a, 602a; **11**:156a, 162b–163a, 297a, 484b–485a; **12**:153a, 579a–b; **13**:36b–37a, 113b–114a, 115a; **14**:54b, 481b, 557a–b; **15**:493b
early twentieth-century, **1**:87b, 172a–b, 398a, 429b–430a; **2**:126b; **3**:35b–37a, 357b, 428a,b, 472a–b, 509a–b, 510a; **4**:26b–27a, 218b–219a, 249a–b, 278a, 407b; **5**:489b–490a, 554a–b; **6**:146b–147a, 149b–150a, 507b–508a, 528b–533a, 574b–575a; **7**:17b–18a, 239b, 471b–472b; **8**:50a, 105b–106b, 520b–523a, 582a–583a; **9**:503b–504a; **10**: 30a–31a, 158b–159a, 231a–232a, 233b–234b, 320b–321a, 473a–b, 526a–b, 585a; **11**:470a–b; **12**:265a–b, 279b, 367a; **13**:113a–b, 115b–116b; **14**:482a–b
mid twentieth-century, **1**:352a–353b; **7**:265a–266a; **8**:581a–b; **10**:107a–b; **12**:352b–353a, 453b–454b
Chinese, **12**:378a–b; **15**:729a, 746a
Egyptian, **15**:709b
Japanese, **12**:403a; **15**:731a, 736a
Mayan, **15**:774b, 776a, 790a–b, 791a, 797b, 802a
Mesopotamian, **15**:676a–678a
systematic measurement, **1**:401a; **14**: 583a
techniques, **6**:324b
and philosophy, **14**:165a–b
philosophy of
ancient Greek, **5**:346b
medieval Arabic, **6**:197b–198a, 199a–b; **11**:250b; **12**:4a–5a
medieval Latin, **5**:550b
Renaissance, **11**:288b
seventeenth-century, **14**:162b, 509a
eighteenth-century, **10**:278b
nineteenth-century, **3**:377a–b
and physics, **1**:256b; **7**:292a–b, 303b–304a
positional
seventeenth-century, **5**:23a–25b, 278b; **6**:69b
eighteenth-century, **4**:14b; **15**:322b–333a
early nineteenth-century, **3**:13b–14a; **10**:591b–593a
mid nineteenth-century, **3**:14a; **4**:29b–30a; **6**:263a–b; **7**:209a–b
late nineteenth-century, **1**:19a; **2**:332b–333a; **3**:261a–b, 298a–b; **4**:16b, 175b–176a; **7**:235b–236b; **8**:539a
early twentieth-century, **3**:373b, 374b; **4**:407b, 547b; **9**:514a–b
practical, **2**:330a, 471b, 614b; **3**:92b–93b; **11**:94b
probability in, **15**:280b, 282a–283a, 285b, 286a, 290b–292b, 325b, 344b–345b, 364a, 370b, 372a

Astronomy (cont.)
spherical
ancient Greek, **1**:338b–339a; **4**:429b, 466b–467a; **13**:323a–b; **15**:215b
medieval, **1**:509a; **2**:154b; **5**:613a–614b; **7**:259b; **11**:415a; **12**:358a, 360b–362a; **13**:291b; **14**:577a–578a, 578b–579a; **15**:259a–260b
Renaissance, **14**:273b–274a
seventeenth-century, 7:303a; **9**:310b
nineteenth-century, **10**:542b
twentieth-century, **3**:373b; **5**:489b; **10**:35a
statistics and, **1**:242b; **5**:363b; **6**:74b, 332b, 352b–353a, 431a, 530b, 532a, 539b–540a; **7**:236b–239b; **8**:364b, 539b; **9**:410b; **10**:41b–42a, 317a; **12**:461b; **13**:98a–b, 475b*; **14**:178b
theoretical
seventeenth-century, **6**:514b–516a
eighteenth-century, **4**:107b; **6**:512b; **8**:143b; **9**:90a
nineteenth-century, **2**:442a, 532b; **3**:583a; **6**:103b–104a; **7**:384b–386a; **9**:460a–b; **10**:42a, 212b; **12**:282a–b
twentieth-century, **1**:428b–429b; **4**:278a–282a; **7**:6b–7a; **8**:520b–523a; **10**:232a
see also Computers; Congresses, scientific; Cosmology; Expeditions, astronomical; Historiography, of astronomy; Illustration, astronomical; Instrumentation, scientific, astronomical; Mechanics, celestial; Photography, in astronomy; Photometry; Reduction, of astronomical observations; Spectroscopy; Sun, Stars, etc.; Tables, astronomical; Terminology, astronomical
Astronomy and Astro-Physics
founding of, **6**:28b
Astrophotometer
invention of, **14**:628b–629a
Astrophysical Journal, **5**:199b
editors of, **7**:271a; **13**:119a
founding of, **6**:28b–29a
Astrophysics
seventeenth-century, **2**:348b; **5**:527b
mid nineteenth-century, **4**:161b–162a, 167b–168a; **6**:541b–543a; **7**:74a–77a; **8**:1b–2a; **11**:559b; **12**:36b–37b, 266b–269b; **13**:121a, 383a; **14**:185b, 628b–629b
late nineteenth-century, **1**:340a, 597a–598b; **2**:433a–434a; **7**:270b–271a, 461a; **8**:20a–b, 440a–442b, 465a–b, 503b; **9**:563b–564a; **10**:528b, 599b–601b; **11**:412a–b; **12**:95b, 153a, 187b; **13**:232a–233a, 344a–b; **14**:38b–39a, 54b–57a, 414a–b
early twentieth-century, **1**:352a, 429b–430a; **2**:424a; **3**:327a–b; **4**:69a–b, 279a–280b, 497b–498a, 513b–514a; **5**:101b–102b, 363a–364a, 489b–490a; **6**:27a–32b, 74b, 105b, 350a–353b, 539b–540a; **7**:85a–86a, 186b, 268a–b; **8**:500b; **9**:414b–415a, 564a; **10**:31b, 106a–b, 233b–234b, 519a; **11**:542b; **12**:18a–23b, 70b–71a, 72b–73a, 248a–251b, 454b–456b; **13**:12a, 408a–409a; **15**:448b–449a
mid twentieth-century, **4**:360b–361a; **7**:617a; **8**:364a–b; **9**:415a; **10**:107b, 290a; **12**:367b–368b; **13**:116b–119b
and nuclear physics, **5**:272b
theoretical, **5**:103a, 272a; **9**:238b–239a, 404a–406a
and thermodynamics, **12**:70b–71a
Astruc, Jean (1684–1766), 1:322–324; 4:589b; **5**:355a; **8**:506a
physiology, **13**:46b
Astudillo, Diego de, **13**:210b, 350a
al-Asṭurlābī, ʿAlī ibn ʿIsā, **14**:538a
Asuntos varios sobre ciencias y artes
founding of, **1**:127b
Aswan Dam, **4**:353b
Asylums
late eighteenth-century, **10**:612a, 613b
nineteenth-century, **4**:252a; **5**:570a–b; **6**:440a
regulation of, **10**:321b
Asymmetry
and life, **10**:360b–361b
Asymptotes, **5**:567a
Asymptotics, **3**:146b; **5**:608b
Asystole, **2**:35b
Atavism, **3**:574a
experimental production of, **3**:493a
explanation of, **3**:493a
Ataxia, **1**:580a; **5**:207b
cerebellar, **8**:536b
Ataxophragmium, **11**:386a
Atchana (city), **14**:504b–505a
Aten, Adrian Hendrik Willem, **11**:535a
ATG (alloy), **3**:239a
Atheism, **6**:468b, 469a; **7**:151b, 157b, 605a; **9**:148b
Athenaeus of Attalia (*fl. ca.* 25 B.C.), 1:324–325; 1:75a, 235a; **6**:406a
Deipnosophistae, **1**:120b
historiography, **1**:213b; **3**:491a; **4**:388b
influences on, **11**:105b(n); **12**:541a(n)
translations of works, **3**:534a
Athenaeus the Mechanic, **3**:491b
Athénée Médical
founding of, **7**:556b–557a
Athens, **2**:229b–230a
Academy, **1**:250a, 251a, 258a, 259a; **3**:21b; **4**:105a, 415a, 489b; **8**:192b; **9**:111a; **11**:22b–23a, 26b–28b, 160a; **12**:575b; **13**:91a, 334a; **14**:534b–535a, 606a; **15**:202a
medicine, **11**:29b
Lyceum, **1**:246a, 250a, 282a; **3**:21b; **4**:383a–b; **13**:328b
mathematics, **6**:410a,b
Atherosclerosis, **5**:43a
Atherton, Edward (pseudonym)
see Holden, Edward Singleton
Ibn al-Athīr
al-Kāmil fi' l-taʾrīkh, **1**:28b
Athrepsia, **4**:299b
Athyrium, **2**:371a
Atichiales, **9**:78b
Atkins, William Ringrose Gelston, **4**:131b*
Atkinson, Robert, **5**:272a
Atlantic Ocean
basin, **5**:596b
exploration of, **3**:204a
maps of, **13**:440b–441a
salinity, **8**:188b
thermometry, **5**:137a
Atlantis (mythical island), **2**:321a; **3**:15a; **11**:587b
Atlases
Renaissance, **9**:309b; **14**:127b
early seventeenth-century, **9**:13a
first use of word, **9**:309b
oceanographic, **1**:92b; **2**:185b; **4**:214a; **12**:115a
paleontological, **15**:54b
sociocultural, **10**:505b
stellar, **1**:242a–243a, 295b; **2**:220b–221a, 441b, 533a; **3**:428b; **5**:490a; **6**:112a, 282a; **10**:207b, 279b, 543b; **12**:384b; **14**:485b
photographic, **10**:600b
world
Renaissance, **10**:239a
seventeenth-century, **2**:185b; **3**:421a
twentieth-century, **11**:524b
Russian, **1**:173b
see also Cartography, Maps
Atlasov, Vladimir Vasilevich, **7**:495b
Atmometer, **8**:425b
Atmosphere
adiabatic expansion, **4**:410b
air currents, **12**:366b
available potential energy, **9**:107b–108a
circulation in, **2**:168a; **4**:591b–592b; **6**:98a–b; **9**:108a; **11**:181b, 558a–b; **13**:166a–b; **14**:50b, 53a
and earth's rotation, **12**:595a
general, **9**:196b
prediction of, **11**:558b
composition of, **3**:541b; **5**:318a–b; **6**:40b–41a; **10**:520b; **12**:123b
and altitude, **10**:289a
carbonic acid, **8**:345a
trace elements, **11**:558b
variations in, **3**:190b
convection in, **3**:593a; **4**:592a–b; **5**:137a
currents in, **6**:508a
density of, **3**:432a
dust in, **6**:355b
electrical conductivity of, **6**:356a*
equilibrium of, **9**:222b
extent of, **14**:489a–b
friction with earth, **9**:443a
frontal surfaces, **9**:108a
gases of, **4**:600a; **5**:602b; **8**:73b; **11**:279a–280a; **13**:103a
carbon dioxide, **13**:389a
origin of, **13**:619b
oxygen, **9**:140a, 530b–531a
ozone, **3**:328a*; **4**:514a, 600a; **5**:102a; **13**:107b
height of, **9**:117b
ionization in, **6**:355a–b; **8**:12b*
layers of, **14**:327b
asymmetry of, **1**:202b
liquid particles in, **8**:24b
mass of, **9**:155a
motion of, **9**:443a
noncondensed matter in, **1**:528a
oscillation in, **9**:107b
pressure field, **9**:107b
pulsation in, **13**:336a–b
radioactivity in, **4**:356b; **5**:341b; **12**:29a, 218a
and solar radiation, **5**:68b; **13**:522a; **14**:38b–39a, 53a
selective absorption, **8**:20b
spectrum, **12**:499a
temperature in, **9**:291a; **12**:512a
and altitude, **5**:187b; **8**:262a; **9**:266b, 444b; **15**:355a–b
inversions, **9**:108a
thermochemical effects in, **1**:528a
thermodynamics of, **2**:168a; **4**:361a; **9**:107b–108a; **11**:558a; **12**:187a–b, 366b; **14**:216a
transparency of, **12**:352b
turbulence in, **11**:558a
vertical fluxes in, **5**:188a
vortices in, **5**:188a; **11**:558a,b
water in, **3**:303b, 540b–541a; **13**:76a
waves in, **2**:147a
see also Air; Chemistry, atmospheric; Electricity, atmospheric; Ionosphere; Meteorology; Pressure, atmospheric; Refraction, atmospheric; Tides, atmospheric
Atmungsferment, **14**:173b, 175a
Atolls, **3**:591a
theories of, **3**:550b–552a, 568b

glacial control, **3**:547b, 569a
subsidence, **1**:72a; **3**:551a; **8**:570b
Atom(s)
absorption edges of, **2**:488a
active powers of, **1**:373b
binding energy, **1**:321b; **2**:241b, 242b; **9**:38a
Cannizzaro's law of, **3**:48a–b
chemical vs. physical, **4**:244a
chemical vs. radioactive properties, **2**: 240b–241a
collisions of, **2**:242a, 245a; **5**:117b; **7**:617a
combining power, **14**:531a–b
composite nature of, **9**:349b
compressibility theory, **11**:417b–418a
concept of, **1**:346b; **2**:621b; **3**:543a–544b; **4**:244a; **8**:510a
dissociation of, **8**:441a–b
division of, **1**:348b
doubly ionized, **10**:348a
and earth's structure, **4**:599b–600a
electric charge, **13**:82b
energy levels, **12**:529a
excited, **3**:373a
decay periods, **14**:338b
excitation potential, **5**:117b
metastable levels, **5**:117b
existence established, **2**:267b
form factor, **3**:619b
frequencies of motion, **6**:164a
indivisibility of, **4**:382a; **14**:67b
magnetic moment of, **9**:38a; **13**:42a–b; **14**:243b, 244b–246a
models of
see Atomic structure
molecules, distinct from, **1**:344a, 347a; **2**:242a; **3**:47a–49a; **4**:243b–244a; **14**:531a
mutual influence of, **2**:622b; **9**:291a, 606b
"packing" in, **1**:321b; **4**:281b; **6**:117b; **9**:110a
packing of, in crystals, **1**:460b–461a; **12**:520a
radiating, **2**:170a
resonance frequencies, **2**:241b
rotational energy of, **2**:170a
scattering power of, **15**:62b–63a
size of, **1**:462a; **5**:457a, 609a; **8**:510a
stable trajectories, **2**:466a
stationary states, **2**:243b, 244a, 245b, 246b, 248a; **9**:38a; **12**:43b; **15**:42a
submolecular grouping of, **1**:518b
vibration frequencies, **4**:316a–b; **11**:10b
vortex theory, **6**:250a
see also Atomic number; Atomic structure; Atomic weight, etc.; Electrons; Nucleus, atomic
Atomic beams, **4**:253b; **13**:42a–b
Atomic bomb, **2**:252a–b; **3**:197b, 371a; **4**: 313a, 581b–582a; **5**:118a–b, 273a; **6**:17a; **7**:527b; **8**:95b–96a, 479b, 482a; **9**:263a; **12**:438a; **13**:227b
Bikini test, **5**:273a
development of, **10**:215a–216a
fallout, **13**:138a
testing of, **10**:216a
Atomic energy, **3**:330a, 371a, 565a*; **4**:582a, 600a; **6**:104b, 185a; **7**:159a, 406b, 527b
harnessing, **3**:330b; **7**:155b, 493b, 527b; **10**:215a–216a; **15**:26a
as heat, **3**:508a
implications of, **1**:321b; **3**:508a; **6**:17a; **10**:216a–b, 217a; **12**:11a, 505b; **13**: 619a; **15**:44a
regulation of, **8**:24b; **10**:216b, 217a; **13**: 227b; **15**:107a
Atomic Energy Authority (G.B.), **3**:330a; **8**:368b–369a
Atomic Energy Commission (U.S.), **10**:216b, 217b, 218a; **14**:89a
Atomic Energy Research Establishment (G.B.), **3**:330a
Atomic mass, **4**:314b; **13**:369b
and electromagnetism, **5**:16a
Atomic number
concept of, **7**:469a; **9**:543a–544b; **12**:44a
determination of
by X-ray scattering, **1**:457a
and nuclear charge, **2**:241a
and nuclear stability, **6**:117b
and spectral properties, **6**:365b
vs. magnetic moment, **3**:4a
Atomic structure
mid nineteenth-century study, **6**:463b
late nineteenth-century study, **5**:294b–295a; **9**:256b; **13**:368a–369a
early twentieth-century study, **1**:321b, 371a; **3**:125a; **4**:578b; **9**:231a, 606b; **10**:106b, 347b, 422b–423a; **11**:475b–476a, 478b; **12**:43b–44a; **15**:464b
of crystals, **2**:431b; **5**:295a, 557b; **9**:183a–b; **15**:40b–41a, 62b–63b
cubic, **8**:289b–290a, 291a–292a
electron model, **7**:498b; **11**:281a, 291b–293a; **14**:208b
nuclear model, **11**:281a; **13**:369a; **14**:497b
Bohr-Rutherford, **2**:240b–242b, 246a–b, 247a; **3**:124a, 564a; **4**:195b, 196b; **5**:160b; **6**:365a; **7**:469a–b, 553b; **8**: 11a, 23b, 181a; **9**:262b, 543b; **10**: 106b; **11**:476b; **12**:30b–32a, 43b, 528b–529a, 614b; **15**:63a
Saturnian, **9**:606b
plum pudding model, **12**:31a
quantum theory, **1**:500a; **5**:609a–b; **9**: 256a; **12**:43b; **15**:41b–43a
spectroscopic techniques, **12**:20b–21b, 43b–44a, 251b
wave theory, **12**:219a–b
X-ray techniques, **4**:195b
Atomic theory
in seventeenth century, **10**:565b
in early nineteenth century, **2**:93a; **3**:47a, 539b–540a, 541b–544b; **4**:240a–241a, 243b–244a; **5**:320a–b, 431a; **6**:285a,b; **9**:578b; **10**:545b–546a; **11**: 170a–171a, 173b; **13**:373a; **14**:488a–490a
in mid-nineteenth century, **1**:518a; **2**: 484b, 485b, 515b, 621b; **3**:47a–49a, 377b, 586a; **5**:203a; **6**:278a, 354a; **8**:508a–510b, 596a; **11**:541b; **14**: 395b, 529b–531b
in late nineteenth century, **2**:169b, 261a–267b; **9**:291b; **11**:538a; **13**:359a, 367a–368a; **14**:67b; **15**:462b–463a
in early twentieth century, **2**:600b; **5**: 160b, 213a; **7**:268a; **8**:120a; **10**:525a; **11**:583a; **13**:42a; **15**:40b–41a, 464a–b
in mid-twentieth century, **10**:424b
critics of, **3**:602b, 604a; **4**:228a, 232a; **8**:374a, 598a–599a, 601a; **12**:198b, 607b; **13**:313b; **14**:123b; **15**:462b–464a
foreshadowing of, **6**:383b–384a, 386a
mathematical problems of, **2**:266a–b
proof of, **10**:525b; **14**:489a–b
testing of, **4**:314b–315a; **10**:185b; **12**:496b
see also Atomism, Avogadro's hypothesis
Atomic volume, **9**:290a; **10**:125a; **11**:36b
and atomic weight, **9**:351a
and electronegativity, **1**:349a
of hydrogen, **1**:129a
of oxygen, **1**:129a
of solids, **4**:78b
Atomic weight, **3**:289a–b; **7**:451b; **8**:507b, 510a
and atomic volume, **9**:351a
changes in, **2**:624b
determination of
early nineteenth-century, **2**:93b; **3**: 542a–543a; **4**:241a, 243b–244a; **9**: 424a–b; **13**:373a–b, 499a–b
mid nineteenth-century, **1**:344b, 347a–b; **3**:47b; **9**:110b; **10**:510b; **12**:619b–620a
late nineteenth-century, **2**:428b–429a; **3**:398a, 476b; **7**:552b; **9**:531a; **11**: 417a–b; **12**:465a
early twentieth-century, **1**:321a–b; **5**: 595b, 597a; **6**:15a, 481a–b; **10**:158a; **13**:389b, 546a–b; **14**:318b–319a
hydrogen:oxygen ratio, **10**:158a
hydrogen standard, **3**:48a; **9**:91b, 351b–352a; **11**:173a
isotopes, **2**:241a
numerical relations between, **3**:398a; **4**: 134b, 247a–b; **9**:348b–352a; **10**:37b–38a, 557b; **12**:619b–620a; **13**:373a
oxygen standard, **2**:429a; **6**:463b; **9**:110a, 351b–352a
prediction of, **6**:117b
and properties of elements, **1**:528b; **5**: 412a; **9**:288a–291a, 351a–b; **12**:42a
revision of, **3**:294a*; **4**:244a, 395a; **5**:125b; **8**:60b–61a; **9**:91b, 288b, 289b, 290a–b, 351b; **10**:178a; **11**:417b
silver standard, **3**:480b(n)
and specific heat, **1**:348b; **2**:588b; **3**:48b; **4**:240b–241a; **13**:411a
and spectra, **7**:268a
and vapor density, **3**:47b, 48b
variability of, **3**:479a
whole-number rule, **1**:321b; **11**:173a–b; **12**:619b
see also specific elements
Atomicity, **1**:345a–b; **4**:281a–b
Atomism
ancient Greek, **1**:255b, 315a; **2**:378a, 379a; **4**:30b, 31b, 32a, 368b, 381b–382b; **5**:285b; **8**:269a–b; **11**:29b
opposition to, **1**:255a; **4**:382a
ancient Roman, **8**:537a–538b
opposition to, **12**:310a
medieval, **1**:101a–b; **4**:525a; **5**:402b; **6**: 207b*; **11**:324b; **15**:265a
opposition to, **2**:394a–395a; **15**:495a
Renaissance, **1**:495a–b; **2**:293a, 543a; **3**: 403a–b
early seventeenth-century, **1**:373a–b, 567a, 568a; **2**:13a–14a, 378b; **4**:96a; **5**:285b–288b *passim;* **6**:127b; **9**:14b–15b; **12**:311a–312b
late seventeenth-century, **1**:474a; **3**: 209a,b; **4**:495b; **6**:608b–609a; **8**:152a, 157b, 159a, 173b; **9**:64a
seventeenth-century opposition to, **5**:543a
early eighteenth-century, **3**:296a; **4**:89a; **7**:227b; **8**:468a–469a, 471b; **10**:320a; **12**:604b; **13**:39a
late eighteenth-century, **2**:326a; **4**:529b; **6**:335b; **9**:6a; **11**:496b–497a; **12**:407b
eighteenth-century opposition to, **7**:227b
early nineteenth-century, **3**:602b; **4**:530a, 531a, 532b–533a; **6**:395b; **8**:262a; **12**:155b
mid nineteenth-century, **4**:537b; **6**:86b–87a, 90a, 279a; **7**:181a, 382b; **9**:287b; **11**:344a; **12**:74a, 243a; **14**:205a
late nineteenth-century, **14**:343b
nineteenth-century opposition to, **5**:391a–b; **6**:564b; **10**:184a
and biology, **1**:259a; **5**:286b
Democritean, **2**:378a; **4**:31a–33b, 381b; **5**:107a, 286b, 287b; **9**:14b–15b; **13**: 93a

Atomism (cont.)
elasticity and, 1:568a; 2:266b
geometrical refutation of, 2:394b–395a
and geometry, 1:381a; 2:394b–395a
in India, 4:31a
in Japan, 12:407b
logical, 8:159a, 160a; 14:468b
logical criticism of, 12:607b–608b, 610a
mathematical, 2:394b–395a; 6:86b–87a
and medicine, 12:538b
pre-Socratic, 4:30b–31a
qualitative, 2:13b–14a
quantitative, 2:13b; 5:156b
and theology, 3:209a,b; 9:316b
see also Atomic theory
Atony, 8:536b
Atoxyl, 4:300a
ATP
see Adenosine triphosphate
Atrophy, 3:204b*; 15:642b
muscular, 9:109a
myelopathic, 3:205b
neuropathic, 3:205b
progressive, 11:374a
treatment of, 5:56a; 8:17a
Atropine, 2:111a; 3:246a; 7:552a; 15:103a
Atropisomerism, 7:517b
Attalus of Rhodes, 15:216b
Attenuation
of fungi, 2:35a
see also Bacteria, attenuation of; Viruses, attenuation of
Attica
paleontology of, 5:295b–296a
Attraction, gravitational (Newtonian, universal), 2:577b; 3:183b, 215b, 432a; 5:521a; 6:305a; 8:384b; 9:341b
in biology, 2:579b; 6:40a; 7:274b, 367b, 409a–b; 9:9a; 13:231b–232a; 15:525a
in capillarity, 6:171b–173a; 8:200b; 12:430a
in chemistry, 9:6a; 15:315b, 358b
constant of, 2:102a; 3:420a; 5:273a; 7:458a; 15:59b–60a
cosmic, 4:281a
variation in, 14:305b–306a
critics of, 2:166b; 3:103b, 105b; 5:62a; 6:609b, 611a
field equations, 4:324a–331b; 9:377a–b
nonlinear, 4:326b
singularities of, 4:329b, 330b, 331a; 7:10a
in flat space-time, 2:145b
galactic, 10:317b
geometrization of, 4:325b, 327a–331b
induction effect, 4:329b
inverse-square law, 3:431b; 5:521a; 6:486a; 8:259a–b; 10:61a–64b *passim,* 69a, 74a, 77b, 82b; 12:288a; 15:286b–287a, 359a
laws of, 7:276a
mathematical treatment, 1:14a; 2:47a–b; 4:281a; 5:555a; 7:109b; 8:611a; 15:350b, 487b–488a
of ellipsoids, 1:147a; 2:134b; 7:37a–b, 565a
and matter theory, 7:228a–b; 12:288a, 288b–289a; 14:488a; 15:287a–b
in chemistry, 4:529b; 10:80b–81a
molecular, 14:111a–b; 15:315a–b, 359a, 360a–b
in outer space, 6:330b; 7:580a
of a point, 15:487b–488a
relativistic theory, 4:324a–331b; 8:284b; 14:305a–306b
speed of propagation, 3:221a
in rotating liquid mass, 3:225a; 15:295b–296a, 317b, 351b
of spheroids, 5:158a; 7:37b; 8:135a, 136a; 10:61a–b, 63b, 68b, 74a; 15:296a–b, 316a–322a, 337b, 351a,b, 356b, 380a
and theology, 3:244b; 7:275b
theories of
ancient Greek, 1:101a, 198a–b, 253b
medieval, 1:101a, 198a–b, 410a; 2:193b–194a, 606a–b; 4:94b; 12:429a
Renaissance, 3:410a
early seventeenth-century, 1:541b; 5:288a; 6:515a; 7:307b; 10:342b; 11:489a,b
late seventeenth-century, 2:46b, 311a, 593a; 3:183b; 6:485b, 487b; 8:152a–b; 9:52a–53a; 10:54b, 61a–65a, 71a–78b
early eighteenth-century, 2:38b, 42b; 3:114b, 216a, 282b, 296a, 344b, 431a–b; 5:62a, 511a; 6:40a, 171b–173a; 9:34a; 11:580a; 12:430a; 14:83b
late eighteenth-century, 2:180a; 3:158a, 206b; 5:195a; 6:69a; 7:228a–b; 8:259a–260a; 12:407b–408b; 14:415a; 15:286a–288b, 343b–344a, 353b, 354b, 358a–b
early nineteenth-century, 1:141b; 3:213b; 6:291b, 330b; 9:204b; 11:173b; 14:500a; 15:377a
mid nineteenth-century, 9:236b; 13:374b
late nineteenth-century, 2:350b; 4:378b–380a; 12:282b; 14:209a
twentieth-century, 3:204a; 4:281a, 329b, 331b; 7:26b; 10:212b; 12:222b; 14:305a–306b
modifications of, 2:38b; 3:216a, 283a; 4:324a–331b, 481a–b; 6:68b, 291b, 344a; 15:286a–288b
universality of, 9:163b
velocity of, 15:288a–289a, 356a
waves of, 4:329b; 12:280b
see also Acceleration, gravitational; Falling bodies; Force; Gravity; Mechanics, celestial; Newton, Isaac
ATV (alloy), 3:239a
Atwater, Wilbur Olin (1844–1907), 1:325–326
physiology, 1:610a; 8:555b
Atwood, George (1745–1807), 1:326–327; 11:497b*
Atwood's machine, 1:326b; 9:5b
Aubert, Jacques, 4:209a
Aubert Dupetit-Thouars, L. M.
see Dupetit-Thouars, L. M. Aubert
Aubertin, Simon Alesandre Ernest, 4:594a; 6:440b
Aubrey, John, 2:523a; 3:209b*; 5:278b (n); 6:127b, 151a, 444b, 445b, 450a, 487b; 9:310a; 10:254b, 495b, 564b; 14:363b, 380a(n)
Aubriet, Claude, 13:442b, 443b
Aubuisson de Voisins, Jean-François d' (1769–1841), 1:327–328; 2:553a; 14:257a
Audayaka system, 2:114b
Audiometry, 3:353a
Auditory system
see Ear, Hearing
Audouin, Jean Victor (1797–1841), 1:328–329; 1:493b; 4:242b; 5:566a; 8:48b; 9:408a; 12:130b, 202b
Audubon, John James (1785–1851), 1:329–332; 1:405a; 2:282b; 6:120a; 10:164b; 14:187b, 417a–b
critics of, 14:187a
taxonomy, 1:405b
Audubon, John Woodhouse, 1:331a
Audubon, Victor, 1:331a
Auenbrugger, Joseph Leopold (1722–1809), 1:332–333; 3:427a; 9:512a
Auer, John, 9:266a
Auer von Welsbach, Carl, 9:544a
Auerbach, Heinrich Stromer von, 1:77b
Auerbach, Leopold, 6:337b
Aufrère, Léon, 9:150b
Auger, Pierre
Auger effect, 3:430a; 9:261b
Augite, 4:141a
Augitite, 7:249a
Augmentation and diminution, 13:195a
Augsburg
Collegium Medicum, 14:639b
Augury, 1:264a
Augustine of Hippo, Saint (354–430), 1:333–338; 4:259a; 5:592a
on Biblical interpretation, 1:588a,b
Categoriae decem, 1:269b; 2:229b
editions of works, 6:562b*
epistemology, 1:334a, 588b
historiography, 4:383a; 6:305b; 11:105a
influence of, 4:396a; 7:27b; 14:388a
optics, 10:475a
rule of, 3:149a
scientific explanation, 1:588a–b; 5:107a
Augustinianism
medieval Latin, 1:197a; 4:92b, 93a, 255b; 8:547b; 10:474a
Renaissance, 4:237b; 5:119a
seventeenth-century, 6:257b; 7:611a; 9:47b, 48b; 10:337b
Augustinians, 9:5a, 277b; 10:419b–420a; 12:60b; 13:58b
Auk, 9:406b
Aurelia (anat.), 12:355b
Aurelian Society, 6:184a
Aurelius Augustinus
see Augustine of Hippo
Aureomycin
discovery of, 4:220b
Auria, Giuseppe d', 9:191a
Auriga (constellation)
alpha Aurigae, 13:120b
beta Aurigae, 9:194b; 14:56b
novae in, 1:598a; 3:36a
variable stars, 8:539b
Aurignac, 8:44a
Aurignacian age, 12:520a
Aurillac, Gerbert d'
see Gerbert
Aurochs, 7:33a
Aurora borealis, 1:300b; 2:327a, 328a, 431a; 3:173b, 436a, 462a*; 4:23b, 162a; 5:520a; 6:278b, 471a; 8:469a; 9:33a, 78a; 10:9a; 11:561b; 13:232b; 14:589b, 597b
distribution of, 8:487b
experimental studies, 13:83a; 14:32a
height of, 2:4b; 3:158a; 8:67b
photography of, 13:83a
spectrum, 1:166b; 5:183b; 12:456a, 499a; 14:55a
and terrestrial magnetism, 3:260a; 5:258b; 6:277b, 553b; 8:487b
theory of, 1:528a; 3:540a; 5:434b; 6:71b, 235a*; 8:36b; 11:446b; 14:32a
Auscultation, 4:197b; 12:202b, 450a–b, 451a
direct, 1:332b–333a
mediate method, 7:557a; 10:613a
Austen, Robert A. C.
see Godwin-Austen, Robert A. C.
Austenite, 10:246b
Austin, Louis Winslow (1867–1932), 1:338
Austin-Cohen formula, 1:338a
Austin, William, 6:284b, 385a
Australia, 10:518a
botany, 1:520b; 2:517b–519b; 12:472a*
algae, 7:610b
colonization of, 1:436a
Council for Scientific and Industrial Research, 11:482b–483a

discovery of, **3**:397a
education, **15**:493b
existence predicted, **3**:116a
geology, **2**:569a; **3**:297a–298a, 591a–b; **4**:140b–141a, 262a; **5**:14b; **7**:185b; **9**:198a; **12**:293a
glaciation, **3**:591a
hydrography, **4**:492a
meteorology, **3**:297b
natural history, **1**:434b
and South America, possible former connection, **6**:594b
zoology, **3**:192b; **5**:481a; **12**:298a
Australian Academy of Science
founding of, **9**:198a
Australian National University, **5**:41b–42a
Australopithecus, **2**:505a; **6**:9b; **8**:105a; **13**:276a
Austria
botany, **7**:58b*
chemistry, **7**:58a
education, **14**:72a
geodesy, **8**:350b–351a; **10**:219a
Geological Survey, **4**:140b; **6**:19b
geology, **6**:168b; **13**:143b, 145b
meteorology, **6**:97a–b
physics, **12**:464a
zoology, **12**:216a
Austria-Hungary
botany, **8**:121a
Austrian Academy of Sciences, **12**:29b
Autenrieth, Johann Heinrich Ferdinand von, **2**:595a; **5**:356b–357a, 429b; **7**:368a
Authority
medieval use of, **1**:378b
and science, **11**:544a
Autochthony, **8**:544a
Autoclave, **7**:21a
development of, **3**:188b; **8**:406a
Autocollineations, **7**:185a
Autolycus of Pitane (*fl. ca.* 300 B.C.), 1:338–339; **1**:22b; **4**:416a; **7**:81a
astronomy, **4**:429b; **10**:295b
editions of works, **3**:585a; **9**:190b; **13**:509b
On the Moving Sphere, **4**:415a; **13**:319b; **15**:180a
editions of, **9**:191a
translations of works, **7**:25a; **11**:244a, 245a; **13**:400b
Automata, **6**:312b–313a; **8**:207b; **14**:31a, 366a, 404a
self-reproducing, **14**:92b
theory of, **14**:92a–b
Automation, **8**:214a; **12**:575a
Automobile, **1**:304a; **7**:316b, 613a–b; **15**:449b, 451a
electric, **13**:24b
electric starter, **7**:316a
Automorphisms
maximal order of, **6**:572a
Autoplasty, **11**:99a
Autopsies
see Post-mortems
Autotomy, **5**:149b, 385b
Autotypy, **4**:283a
Autoxidation, **5**:621b; **12**:198a; **13**:453a
Autumn mineral, **12**:383b–384a
Auvergne, William of
see William of Auvergne
Auvergne (province)
geology, **2**:346b; **8**:566a; **12**:261b, 262a
Auvergne Mts.
volcanoes of, **2**:554b; **3**:586a; **4**:71a–72a, 152a; **5**:578a
Auwers, Arthur Julius Georg Friedrich von (1838–1915), 1:339–340; **2**:533b*; **5**:405b; **9**:563b; **14**:55b
star catalog **2**:99a
Auwers, Karl Friedrich von (1863–1939), 1:340–341; **9**:356b; **14**:265a–b
Auxanography, **15**:14a
Auxanometer, **12**:59b
Auxins, **14**:255b
Auzout, Adrien (1622–1691), 1:341–342
acquaintances of, **3**:102a; **6**:598b; **9**:498a,b; **10**:477a, 595b; **13**:334b, 335a
comets, **12**:85b
experiments of, **11**:506b
hydrostatics, **10**:334b
instrumentation, **5**:278b; **13**:336a
longitude, **3**:63b
physiology, **4**:37b
Avalanches
prevention of, **10**:124b
Avdeev, Ivan, **9**:288b
Avebury, Baron
see Lubbock, John
Avempace
see Ibn Bājja
Avenare
see Ibn Ezra
Avenarius, Richard, **8**:183b, 603b; **12**:218a; **15**:38b
Avendauth
see Abraham ibn Dāᵓūd
Avenzoar
see Ibn Zuhr, Abū Marwān ᶜAbd al-Malik ibn Abi'l-ᶜAlāᵓ
Averroës
see Ibn Rushd
Averroism, **2**:590b; **7**:116a–b; **10**:419b–420b; **12**:135a, 136a, 312b, 548b
condemnation of, **1**:108b, 197a; **5**:402b
Latin Averroism, **1**:5a, 197a; **2**:193a; **9**:136b
Paduan, **4**:212b
unity of intellect, **1**:5a
Avery, Oswald T. (1877–1955), 1:342–343
Aviation
medieval, **1**:5b
Renaissance, **3**:558a; **8**:213a–b
seventeenth-century, **6**:482a
eighteenth-century, **13**:178b
nineteenth-century, **1**:490a, 583a; **3**:388b
early twentieth-century, **3**:195b; **4**:398b; **5**:187b–188a; **7**:247a, 465a; **8**:19b–20a, 368b; **10**:275a; **11**:431a; **12**:17b–18a
blind landings, **4**:398b; **6**:104b
and medicine, **5**:208a
supersonic, **7**:248a
weather service, **11**:557b
see also Aeronautics, Airplane, Flight
Avicenna
see Ibn Sīnā
Avogadro, Amedeo (1776–1856), 1:343–350; **4**:243b; **5**:320a–b; **14**:73a
atomic volume, **1**:349a
Avogadro's hypothesis, **1**:344a–347b, 518b; **3**:47a–49b; **5**:371b, 372a; **8**:509b; **9**:347b; **11**:173b, 370b; **13**:579a–b; **14**:531a; **15**:434b
confirmation of, **1**:141b, 143a; **4**:78a; **5**:294b; **7**:509b
physical argument for, **3**:307a
Avogadro's number, **1**:519b; **3**:370a; **8**:510b
determination of, **4**:314a–315a; **8**:501b; **10**:525a; **12**:30b
"Avogadro's reversal," **1**:347b
and Berzelius, **5**:371b
voltmeter multiplier, **1**:347b
Avogadro, Felice, **1**:343b
al-ᶜAwfī, **15**:250a
Ibn al-ᶜAwwām Abū Zakariyyā Yaḥyā ibn Muḥammad (*fl. ca.* A.D. 1170), 1:350–351
Axes, prehistoric, **9**:539a; **15**:50b–51a
Axiomatization
in mathematics
ancient Greek, **1**:252a, 258a; **4**:415a, 416b, 417b, 466a
medieval, **4**:447a
late nineteenth-century, **4**:229b, 373b; **6**:391a–b; **10**:343b–344b; **11**:60b–61a
twentieth-century, **2**:352a–b; **3**:63a; **4**:9a; **5**:108a–b, 152b, 154b; **6**:570a; **11**:413a; **12**:364a–b; **13**:599b; **14**:614b–615a; **15**:23b
philosophy of, **1**:251a, 374b–376b; **4**:100b–102a; **6**:249a; **9**:317a
in physics
ancient Greek, **4**:431a
medieval Latin, **2**:394a–b; **13**:198a, 201b
seventeenth-century, **8**:151b
nineteenth-century, **14**:67b
twentieth-century, **4**:228a–b, 229b; **6**:393b; **9**:377a; **14**:304b
rejection of, **2**:512b; **14**:67b
Axolotl, **7**:101a; **15**:126b
Axonometry, **12**:246a
Axons, **5**:56b
function of, **5**:460b
myelination, **5**:27a
regeneration, **8**:283a
Axoplasm
fluidity of, **5**:56b
Ayabe Yasuaki
see Asada Gōryū
Ayana (half-year), **15**:535a,b
Ayanadṛkkarma (computation), **15**:562a, 575b, 592b, 595a
Ayanavalana (angle), **15**:593a
definition of, **15**:574b
Aye-aye (animal), **12**:537a
Aylesbury, Thomas, **6**:127b, 128a
Ayyūb al-Ruhāwī, **15**:236b, 237b, 238b
Azafea
see Saphea
Azara, Félix de (1742–1821), 1:351–352
Azarquiel
see al-Zarqālī
Azeotropes, **4**:229b; **9**:131b
Azides
conversion to isocyanate, **3**:511b
Azilian, **10**:607a
Azimuth
determination of, **7**:500b; **14**:577b
Azimuth instruments, **3**:102b; **7**:81b; **10**:536b, 537a, 572a; **14**:400a
al-ᶜAzīz (Nizār Abū Manṣūr) (caliph), **14**:574b, 575a
Azoester, **1**:105b–106a
Azores
charting of, **14**:514a–b
geography, **10**:617b
Azoth, **8**:311a
Azotobacter, **14**:37b; **15**:14a
Azotometry, **2**:316b
Azov, Sea of, **7**:410b
Azoxyanisole, **9**:183a
Azra, Jean-Pierre, **5**:262a
al-Azraq, **15**:234b

B

Baade, Wilhelm Heinrich Walter (1893–1960), 1:352–354; **4**:282a; **6**:532b; **12**:347a–b; **13**:475a
Baba Masatoda, **1**:90a
Babbage, Charles (1792–1871), 1:354–356; **3**:545b, 605a; **6**:325a; **11**:384b; **12**:524a; **13**:238a

Babbage, Charles (cont.)
calculating machine, **6**:324a; **9**:267b; **13**:431b–432a
The Decline of Science, **3**:545a
and education, **7**:551a; **10**:437b; **14**:292b–293a
geomagnetism, **1**:203a; **12**:49a,b
historiography, **15**:406b
and Sabine, **12**:49a, 51b–52a
Babbage, Henry Prevost, **1**:356a
Babcock, Harold Delos, **6**:30a; **12**:73a
Babcock, Horace Welcome, **6**:30a
Babcock, Stephen Moulton (1843–1931), 1: 356–357; 6:135b
Babesiosis, **12**:315a
Babinet, Jacques (1794–1872), 1:357–358; 4:350a, 410b; **11**:498b*; **12**:519a
Babinet's rule, **6**:19a
compensator of, **1**:358a
principle of, **1**:358a
Babington, Benjamin Guy, **3**:531a
Babington, Charles Cardale (1808–1895), 1: 358–359; 5:80a; **6**:289a
Flora of Cambridgeshire, **11**:314a
Babington, William (1756–1833), 1:359–360
Babinski, Joseph François Félix, **2**:35b; **3**: 205a; **4**:286a
Babirusa, **11**:243a
Babo, Lambert Heinrich Clemens Anton, **9**:566a–b; **11**:299a
Babylonia
see Mesopotamia
Baccelli, Guido (1830–1916), 1:360
Bach, Aleksei Nikolaevich (1857–1946), 1: 360–363
Bach-Engler theory, **1**:362a
Bacharach, Isaak, **3**:167a
Bache, Alexander Dallas (1806–1867), 1: 363–365; 4:411a*; **5**:479b; **6**:165b; **9**:197a; **10**:479b, 480a; **11**:113a,b
Bache Fund, **1**:365a
Bachelard, Gaston (1884–1962), 1:365–366; 2:546a; **6**:427b, 429a; **15**:424a, 488b
Formation de l'esprit scientifique, **6**:426a
Bachelier, Louis (1870–1946), 1:366–367; 9:129b
Bachet de Méziriac, Claude-Gaspar (1581–1638), 1:367–368; 2:131a; **3**:275a; **8**:271b; **10**:264a; **11**:512a
on Diophantus, **4**:117b
number theory, **4**:612b; **5**:409a,b; **7**:565b
Problèmes plaisans et délectables, **3**:276b; **5**:159b
Bachman, John, **1**:331a,b, 405b; **10**:164b
Bachmann, Andreas, **1**:368b
Bachmann, Augustus Quirinus (1652–1723), 1:368–370; 8:379a; **9**:17b
taxonomy, **6**:401a
critics of, **4**:98b–99a; **11**:315a
Bachmann, Friedrich, **11**:362b
Bachmann, Paul Gustav Heinrich (1837–1920), 1:370; 4:5a; **5**:314a*; **7**:522b; **14**:222a
Niedere Zahlentheorie, **15**:66b
Bacilli, **11**:428b
abortus, **12**:482b
actinoides, **12**:482b
amylobacter, **13**:405b
anthrax, **3**:340a, 587b–588a; **5**:128b; **7**: 421b–422a, 423a; **10**:386a–b, 387b, 393b–394a, 397a,b, 493a
cholera, **5**:219b, 503a; **7**:391b, 425a, 518a; **10**:400a; **12**:96b
coli, **12**:481a, 483a
diphtheria
see Bacilli, Klebs-Loeffler
Ducrey's, **15**:454a
dysenteriae, **4**:404b; **5**:40a; **6**:298a, 432b; **7**:392b; **12**:96b; **13**:520b
edema, **7**:100a
Flexner bacillus, **5**:40a
glanders, **7**:425b; **8**:448b
Hansen's, **6**:101b
hog cholera, **12**:480b, 482a
icteroides, **11**:346b
Klebs-Loeffler, **4**:404b, 276b; **7**:425b; **8**: 448b–449b; **11**:569a; **14**:551b
Koch-Weeks, **4**:618a; **7**:425a
life cycle of, **3**:339a–b; **7**:421b–422a
Metchnikoff, **5**:269b
pathogenic, **3**:339b–340a; **4**:276a–b
discovery of, **3**:587b–588a; **6**:101b
pigmented, **9**:176a
plague, **5**:40a; **7**:392a, 427b
Schaudinn's, **12**:141b
soil, **5**:42b
subtilis, **2**:437a; **3**:338b; **10**:364a, 397b
thermoresistant endospores, **3**:339a–b
swine erysipelas, **7**:425b
syphilis, **11**:515b
tetanus, **7**:391b
tubercle, **3**:22b; **5**:220a*, 270b; **7**:423b, 424a; **8**:273a, 275b; **12**:481a,b, 482b; **14**:551b
acid-fast nature of, **4**:296b
immunochemistry of, **8**:608a
respiration in, **10**:155b
typhoid, **4**:276a,b; **5**:128b, 219a–b; **7**: 425b; **8**:450b
Back, Ernst E(mil) A(lexander) (1881–1959), 1:370–371; 2:245a; **10**:348a
Paschen-Back effect, **1**:370b; **10**:347b, 423a
Back
anatomy of, **3**:99b
Backgammon, **3**:549a*
Backhouse, William, **1**:317b
Backlund, Jöns Oskar (1846–1916), 1:371–372; 7:386a
Backstaff, **6**:126a
Backwater, **4**:208a
Bacon, Francis (1561–1626), 1:372–377; 5: 397a; **7**:488b
acquaintances of, **6**:151a; **13**:573a
and Baranzano, **1**:447b
and Cartesianism, **1**:376a
on contemporary scientists, **2**:543b; **5**: 399b; **12**:335b
critics of, **14**:367b
editions of works, **14**:288a
Historia experimentalis, **1**:567b
and Hobbes, **6**:445a–b
influence of, **2**:16a, 309a, 377b; **3**:362b; **5**:426b; **6**:141b; **9**:501a; **10**:495a; **11**:40a, 41a; **12**:581b; **13**:214b, 564a; **14**:83a, 209b, 288b, 510b
influences on, **6**:127b; **11**:289a; **12**:136a; **14**:377b(n)
on language, **14**:372a
method of, **1**:374b–376b; **4**:54b; **7**:106a, 539b; **9**:317a
natural philosophy, **1**:567b; **5**:135a; **8**: 152a; **14**:364b
atomism, **1**:373a–b; **2**:378b
heat theory, **2**:180a
New Atlantis, **1**:159b
Novum organum, **1**:134b; **2**:378b; **3**:248b
and Shakespeare, **3**:58a*
Sylva sylvarum, **1**:567b
translations of works, **6**:445b
tree of knowledge, **1**:114b–115a
on winds, **5**:401a
Bacon, Roger (*ca.* 1219–*ca.* 1292), 1:377–385; 7:134a*
acquaintances of, **14**:388a
alchemy, **5**:580a; **6**:306a; **10**:555a
astrology, **1**:378a, 382a
biographies of, **12**:440a; **14**:157b
Communia mathematica, **10**:534a
Communia naturalium, **15**:35b
on contemporary scientists, **1**:99b; **3**:24a; **4**:446a; **5**:549a; **9**:362a,b; **14**:391b
cosmology, **7**:142a
critics of, **1**:198b; **4**:92b
experientialism, **1**:278b, 380a–381a
influence of, **2**:163b; **4**:157b; **5**:488a; **10**: 474a; **14**:157b
influences on, **1**:63b, 226a; **5**:553a; **7**:66b; **8**:550b*; **11**:203a, 244b; **15**:264a
mathematics, **1**:379a, 380b–381b
medicine, **15**:265b
De multiplicatione specierum, **14**:458b
natural philosophy, **1**:100a,b; **14**:459a
optics, **1**:81a; **10**:475a; **14**:459b, 460a
Opus maius, **14**:458b; **15**:35b
Opus tertium, **10**:533a–534a
Perspectiva, **6**:197a
Baconianism, **2**:451b; **5**:416b; **10**:564b, 565b; **11**:141b, 146b; **13**:335b; **15**:744b, 747a
Bacq, Zénon Marcel, **15**:74a
Bacteria, **8**:127b
acetic-acid, **6**:100b
agglutination of, **5**:564a–b
anaerobic, **10**:155a
anaphylactic reactions, **11**:430a
animal vs. vegetable nature of, **3**:337b; **4**:236b
attenuation of, **7**:424b; **10**:391a, 391b–392a, 393b–394b, 398a,b, 399b, 400a–b
carcinogenic, **7**:101a
cellulose-decomposing, **14**:37b
chemosynthesis, **14**:37a–b
chemotaxis, **4**:372a; **10**:576a
classification of, **3**:338b–339b; **5**:564a
cold resistance, **3**:338b
cytology of, **5**:30a, 584b
definition of, **3**:338b
destruction of, **1**:438b; **5**:269b
developmental history of, **3**:337b; **14**:37a
DNA in, **1**:342b–343a
enzymes in, **6**:405a
evolution of, **13**:520a,b
fecal, **8**:450b
and fermentation, **1**:497a; **5**:128b; **6**:100b, 110b, 501b; **14**:45b
and fungi, **3**:338b
fusiform, **12**:96b
in gastric juice, **11**:428b
genetics of, **13**:137a
inherited variation, **1**:343a
mutation, **5**:317a; **6**:405a; **13**:227b, 520b
gram-positive, **5**:30a
growth of, **6**:404b–405a
heat-resistant, **1**:496b; **3**:338b, 339a
identification of, **5**:564b; **6**:110b–111a; **7**:100b
immunological specificity of, **1**:342b–343a
intestinal, **4**:358a, 403b; **8**:169b; **9**:335a
iron, **14**:37a
isolation of, **5**:30a; **15**:14a
light, effects of, **4**:621a
metabolism, **5**:6a; **7**:406b; **14**:37a–b
morphology, **7**:422a
nitrogen-fixing, **2**:70b; **14**:37a–b; **15**:14a
nutrition of, **3**:338b; **13**:520b
pathogenic, **3**:463b, 588a; **7**:421b–430a
in plants, **12**:468a
pathogenic vs. putrefactive, **3**:338b
phosphorescent, **3**:271b; **15**:14a
pressure, effects of, **8**:608a
purple, **14**:181b

pyocyaneous, **15**:521b
radiation, effects of, **4**:220b
reproduction, **9**:86a
specificity of, **5**:219a; **7**:422a,b; **15**:14a, 15a
spirillum, **12**:96b
spore formation, **4**:133a*; **7**:421b–422a
structure of, **13**:3a
sulfur, **14**:37a
toxins, **8**:512b; **10**:155a
variability of, **5**:564a; **7**:171a; **12**:480b; **14**:37a
in water, **8**:556b
Bacteriology
mid nineteenth-century, **3**:337b–340a; **8**:405a–406a; **13**:154a–b
late nineteenth-century, **1**:438b–439a; **2**: 528a–b, 583b–584b, 598b; **3**:94a–b, 184b, 188b–189a, 219b, 613b–614a; **4**:276a–277b, 310b–311a, 403b–405a; **5**:52a, 128a–b, 219a–220a, 269a–270b, 495a–b, 564a–b, 582a; **6**:11b–12b, 101b–102a; **7**:100a–b, 391a–392b, 420a–430b; **8**:125a, 407b, 448a–450b, 556b–557a; **9**:333b–334b; **10**:17b–18a, 154b–155b, 493a, 527a–b; **11**:176b, 345b–347a, 569a–b; **12**:87b–88b, 480a–481b; **13**:456a, 522a; **14**:218b–219a, 249a–250a, 278a–b, 511b, 551a–b; **15**:13b–14a, 454a
early twentieth-century, **1**:342a–343a; **2**:300a–301a; **3**:22b, 204a, 271b, 290b; **4**:133a*; **5**:29a–30a, 40a–b, 316b–317a; **6**:298a–b, 432b–433a; **7**:170b–171a, 286b–287a; **8**:273a–274a; **10**:141b–143a; **11**:595b; **12**: 96b, 190b, 468a, 481b–483b; **13**: 228b–229a, 519b–520b; **14**:551b, 623a–624b; **15**:454a–455a, 521b
mid twentieth-century, **7**:624b; **13**:227b
chemical, **6**:110b–111a
medical, **12**:88a
opponents of, **1**:496b–497b
see also Microbiology
Bacteriolysis, **2**:300a–b; **5**:29b–30a; **7**:427a,b
Bacteriophages, **2**:300b; **5**:317a
discovery of, **5**:270b; **6**:298a; **13**:520a–b
prophylactic use of, **6**:298a–b
Bacteriotherapy, **10**:387b
Bacterium coli
see *Escherichia coli*
Bacterium icteroides, **12**:96b
Bacterium termo, **3**:337b, 338b, 339b
Baden (Germany)
observatory, **14**:481b
al-Badī al-Asṭurlābī al-Baghdadī al-Isfahanī, **7**:353b
Badīᶜ al-Zamān al-Jazarī
see al-Jazarī
Badische Anilin- und Sodafabrik, **2**:4a, 59a, 323b; **3**:85a; **5**:621b–622a; **9**:354a, 427a–b
Badovere, Jacques (Giacomo Badoer), **5**: 241a
Baekeland, Leo Hendrik (1863–1944), 1:385
Baentsch, Otto, **7**:597b
Baer, Erich, **5**:6a
Baer, Hans Helmut, **5**:6b
Baer, Karl Ernst von (1792–1876), 1:385–389; 11:217b*; **14**:114b*
acquaintances of, **2**:422b, 595b, 596a; **4**:146a; **9**:374b; **10**:287a; **11**:307b; **12**:420b–421a
anatomy, **13**:170b–171a; **15**:168b
Baer's law, **12**:337b–338a
biographies of, **7**:517a
correspondence of, **15**:123b
embryology, **5**:485a; **8**:128a, 254a; **11**: 133a, 380b; **14**:113a
influence of, **1**:476b; **2**:161a; **6**:592a; **7**: 475a; **12**:38b, 316a, 570b; **15**:503b
influences on, **11**:215b, 593a; **12**:155b, 156b; **15**:525b
students of, **11**:360b
translations of works, **2**:443a; **6**:591a
Ueber die Entwicklungsgeschichte der Thiere, **2**:596b
Baeyer, Adolf Johann Friedrich Wilhelm von (1835–1917), 1:389–391; 5:12b; **9**:354a
acquaintances of, **1**:426a; **3**:286b, 511a; **5**:464a, 488b; **6**:440a; **7**:283a; **8**:328b; **9**:347b, 355a; **10**:158a, 597b; **13**:337b; **14**:124b
benzene, **3**:300a,b; **9**:356a; **10**:517b
condensation reactions, **1**:385a
dye chemistry, **3**:85a; **10**:516b; **15**:422b
phytochemistry, **1**:362a; **3**:511b; **5**:2a
students of, **2**:560a,b; **5**:1a; **7**:372a; **9**: 393b; **10**:14b, 22a, 517a; **11**:15b(n), 585a; **12**:493b; **14**:411b
structural chemistry, **1**:288b; **2**:587b; **7**: 280b; **13**:338a, 388b; **14**:476a
Baeyer, Johann Jacob, **1**:389a; **2**:102a; **6**: 239b, 553a; **7**:608b; **10**:483a
Baeyer, Otto von, **6**:16a; **11**:582b–583a
Baffin Island
geography of, **2**:207b
Baghdad, **4**:555b; **11**:312b
Dār al-Ḥikma (House of Wisdom), **1**: 444a; **7**:358b; **15**:261a
founding of, **9**:160a; **13**:538a
library, **6**:189a
observatory, **14**:537b–538a
translation movement, **7**:24b
al-Baghdādī, Abū Manṣur ᶜAbd al-Qāhir ibn Ṭāhir ibn Muḥammad ibn ᶜAbdallah, al-Tamīmī, al-Shafiᶜī (*d.* 1037), 15: 9–10
translations of works, **15**:178a
Baglivi, Georgius (1668–1707), 1:391–392; 1:593b; **6**:45b(n), 64b, 183b; **12**: 100b*, 104a*
medical theory, **1**:594a; **2**:226a; **6**:37a; **10**:568a; **12**:103b
Opera omnia
editions of, **10**:612a
Bahāʾ al-Dawla
Khulāṣat al-Tajārib, **1**:399b
Bahama Islands
natural history, **3**:129b
Bahr, Johann Friedrich, **9**:290b
Baiandurov, Boris Ivanovich, **10**:435a
Baier, Johann Jacob (1677–1735), 1:392–393
Baikal, Lake
geology of, **3**:529b, 532a,b
origin of, **10**:167a–b
Bailak al-Qabajaqī
see Baylak al-Qibjāqī
Bailey, Edward Battersby (1881–1965), 1: 393–395; 1:472a; **5**:337a; **6**:228b*, 588a*,b*; **8**:576a*
Bailey, Jacob Whitman, **1**:397a,b; **9**:196a, 588b; **11**:113a
Bailey, Liberty Hyde, Jr. (1858–1954), 1: 395–397; 12:537b*; **13**:456b*
Bailey, Loring Woart (1839–1925), 1:397; 4:291b
Bailey, Percival, **3**:518b
Bailey, Solon Irving (1854–1931), 1:397–398; 8:106a; **10**:483a, 601b*; **12**:346b
Bailey Hortorium
founding of, **1**:396b
Baileya, **1**:396b
Baille, Jean Baptiste Alexandre, **3**:420a
Baillet, Adrien, **4**:46a, 49a, 51b, 52b
Baillie, Matthew (1761–1823), 1:398–399; 3:323b, 487b; **6**:569b; **9**:512a
Baillou, Guillaume de (*ca.* 1538–1616), 1: 399–400
Bailly, Jean-Sylvain (1736–1793), 1:400–402; 3:6a, 7a; **5**:138a; **9**:327a; **11**: 521a; **14**:14b; **15**:302b
astronomy, **5**:88a; **8**:145a*; **9**:330a
Traité de l'astronomie indienne et orientale, **15**:329b
Baily, Francis (1774–1844), 1:402–403; 5: 79b*; **7**:146b*, 543b; **13**:76a, 537b*
"Baily's beads," **1**:402b, 403a, 429b; **6**: 523a
Baily, Walter, **4**:589a
Bain, Alexander (1818–1903), 1:403–404; 3:90a, 290a*; **4**:593a; **5**:253b; **6**:140a; **9**:385b*
Baines, Henry
Flora of Yorkshire, **1**:412b
Baird, Spencer Fullerton (1823–1887), 1: 404–406; 1:331a–b; **3**:438b, 536a; **4**:265b; **5**:214a, 406b; **6**:280b; **7**:169b; **8**:2b, 169b; **9**:313b; **11**:443a; **13**:25a; **14**:1b, 424b
Baire, René Louis (1874–1932), 1:406–408; 2:304a; **6**:451b; **8**:558b
analysis, **14**:572b
Baire's function, **1**:408a; **8**:110a, 111a
category method, **9**:249a–b
theorem of, **1**:428a
al-Bairūnī
see al-Bīrūnī
Ibn al-Baiṭār
see Ibn al-Bayṭār
Baja California
geography, **5**:214b
Ibn Bājja, Abū Bakr Muḥammad ibn Yaḥyā ibn al-Ṣāʾigh (*d.* 1139), 1:408–410; 1:198a; **9**:28b
mechanics, **13**:186b
philosophy, **9**:30b; **12**:7Aa–b
Bakelite resins, **1**:385b
Baker, Alan, **6**:393b
Baker, Clement John, **2**:529a
Baker, Edmund Gilbert, **1**:412b
Baker, Henry (1698–1774), 1:410–412; 3: 350b; **14**:508b
Bakerian lectures, **1**:411b; **14**:508b
translations of works, **10**:572a*
Baker, Henry Frederick, **4**:373b; **13**:221a
Baker, Howard Bigelow, **14**:216a
Baker, John Gilbert (1834–1920), 1:412–413; 14:190b*
Baker, Joseph, **3**:399b
Bakewell, Robert (agric.), **6**:585b
Bakewell, Robert (1768–1843), 1:413; 5:533a
Introduction to Geology
editions of, **12**:433a–b
Bakh, Aleksei Nikolaevich
see Bach, Aleksei Nikolaevich
Baking powder, **6**:517b
al-Bakrī, Abū ᶜUbayd ᶜAbdallāh ibn ᶜAbd al-ᶜAzīz ibn Muḥammad (*ca.* 1010–1094), 1:413–414; 6:79b; **7**:4a
Balabhadra, **12**:115b
Balaenoptera
anatomy of, **4**:12a
Bâlagovinda
commentaries of, **5**:274b
Balance
see Equilibrium, physical; Lever, theory of
Balances
medieval Arabic, **7**:325b, 336b, 339a–348a
Renaissance, **8**:198a
aperiodic, **3**:506b

Balances (cont.)
assay, **4**:393b; **9**:300b; **12**:132a
Brillouin, **2**:466a
chronometric, **7**:342b, 343a
electromagnetic, **1**:557b
electrostatic, **1**:555b
exchanging, **7**:342a–b, 343a
gas buoyancy, **13**:71a
glass, **5**:340b
hydrostatic, **5**:238b; **7**:336b, 339a–348a
construction of, **7**:346a–348a
invention of, **7**:340b
Mohr, **9**:445b
Oertling, **3**:476b
osmotic, **13**:163a
philosophical, **7**:339b–340a
piezoelectric, **3**:504b
precision, **5**:78a–b; **7**:160a; **11**:129b; **12**:132a
Roberval, **11**:488b
spring, **7**:160a
torsion, **3**:158a, 440a, 443a, 444a,b, 505b; **4**:8a; **10**:159a; **11**:122b; **14**:207a
discovery of, **9**:370b
Eötvös, **2**:466a; **4**:379a–380a
see also Microbalance
Balandin, Aleksei Aleksandrovich (1898–1967), 1:414–415; 14:602b*
Balanophoraceae, **5**:441b; **6**:492a; **8**:512b
Balanus, **2**:591b
Balard, Antoine Jérome (1802–1876), 1:416–417; 2:63b; **4**:210b–211a; **8**:55b, 109b; **10**:357a, 369b
Balbi, Paolo, **3**:15b
Balbiani, Édouard-Gérard (1823–1899), 1:417–418; 11:295b
vesicle of, **1**:418a
Balboa, **10**:619a
Balbus (Balbus Mensor?) (*fl. ca.* A.D. 100), 1:418–419
Baldi, Bernardino (1553–1617), 1:419–420; 3:365a*; **5**:249b; **9**:254b, 487b; **10**:272a*; **11**:468b*
Baldinger, Ernst Gottfried, **3**:465a; **10**:197b; **14**:332b
Baldness, **1**:315a
Baldwin, James Mark, **3**:130a
Baldwin, William, **1**:485a; **3**:563a
Balfour, Francis Maitland (1851–1882), 1:420–422; 5:80b; **6**:535b; **7**:402a; **10**:241a, 273b; **13**:352b; **14**:251a
cytology, **14**:235b
embryology, **5**:83a; **11**:573b
Balfour, Isaac Bayley (1853–1922), 1:422–423; 3:614a*
Balfour, John Hutton (1808–1884), 1:423; 1:422a,b; **2**:598a
Baliani, Giovanni Battista (1582–1666), 1:424–425; 3:3a; **5**:245b; **8**:268a; **9**:369a; **10**:103b; **13**:438b
Balīnās or Balīnūs
see Apollonius of Tyana
al-Balkhī, Abū Zayd Aḥmad ibn Sahl, **9**:88b
al-Balkhī, Jaᶜfar ibn Muhammad, Abu Maᶜshar
see Abū Maᶜshar
Ball, John, **6**:492a
Ballestilla
see Cross staff
Ballistae, **8**:213b
Ballistics
ancient Greek, **1**:214a
Renaissance, **3**:66b; **4**:97a,b; **8**:209b, 213b–214a; **13**:260a
seventeenth-century, **2**:201b–202a; **5**:246a–b; **6**:71a, 127a, 255a; **9**:486b; **13**:436a–437a
eighteenth-century, **1**:581b; **2**:54a; **4**:470b; **5**:449a; **6**:577a; **8**:135a; **11**:493b–494a; **13**:350b–351a; **15**:98b
nineteenth-century, **1**:616a; **3**:224b; **4**:39b–40a; **5**:111a; **6**:35a, 545a; **7**:522a; **8**:516a; **9**:504b; **10**:248a, 250b; **14**:85b–86a, 620b
twentieth-century, **2**:198a; **4**:407b; **5**:103a, 202a; **6**:147b, 532b; **7**:555a; **8**:11b, 500b; **9**:404a, 404a, 552b; **12**:251b; **14**:14a; **15**:490b–491a
equations, **4**:177b
Ballistocardiography, **2**:601a
Ballonius
see Baillou, Guillaume de
Balloon ascensions
eighteenth-century, **3**:7a, 208a; **7**:89a, 582a; **8**:78b–79a; **10**:609a; **11**:166b; **15**:355b
nineteenth-century, **2**:134b; **5**:21b, 318a–b, 413a; **13**:311a
twentieth-century, **6**:355a; **10**:597b
for astronomical observation, **6**:105b; **7**:76b–77a; **9**:291a
for meteorological observation, **14**:422b
see also Aerostatics
Balloons, aerostatic
design of, **3**:207b–208a; **8**:78b–79a; **9**:343b, 471b; **10**:609b
dirigible, **9**:343a,b; **13**:431b, 483a
equilibrium, **9**:343b
gases for, **5**:602a–b; **14**:86b
high-altitude, **11**:348a
military use of, **5**:603a; **6**:1a; **9**:472b; **14**:86b
rupture, conditions of, **8**:124a
sounding, **3**:12a; **12**:366b
tracking of, **14**:216a
vertical motion, **9**:324b
Ballot, Christoph Buys
see Buys Ballot, Christoph
Balloting
see Votes
Ballou, Clinton Edward, **5**:6b
Balmer, Johann Jakob (1825–1898), 1:425–426; 1:133a; **7**:267b
Balmer lines, **7**:553b; **13**:95b
equation of, **5**:609b
spectral formulas, **6**:164a; **7**:267b, 268a; **11**:476a, 478a–479a, 611b; **12**:42b–43a, 235b; **13**:104a, 364b
Balmes, Abraham de, **6**:197b
Balneology
medieval, **4**:164a; **14**:112b
Renaissance, **3**:343a; **4**:387a, 521a; **5**:586b; **10**:7b, 306b; **13**:502b
eighteenth-century, **11**:387a–b
nineteenth-century, **11**:363b
see also Baths, Hydrotherapy
Balsam, **1**:124b–125a; **14**:349b
Balsamo-Crivelli, Giuseppe, **1**:493b
Balthazard, Victor, **3**:508a
Baltic Geodetic Commission, **1**:429b
Baltic Sea, **7**:410b; **14**:634a
algae, **13**:165a
chemistry, **14**:233b
falling level of, **3**:174a
geography of, **13**:111b
geology, **15**:428a
hydrology, **12**:412b
surveying, **12**:114b
Baltimore (Maryland)
public health, **14**:250a
Baltzer, Richard, **2**:271a; **6**:522a
Baluchitherium, **5**:500b
Baly, Edward, **11**:279b
Bamberg
observatory, **6**:149b
Bamberger, Eugen (1857–1932), 1:426–427; 14:265b
Bamberger, Ludwig, **6**:108a–b; **7**:623a
Ban, Ichitarō, **8**:577a
Banach, Stefan (1892–1945), 1:427–428; 13:428b; **15**:226b*
Banach space, **1**:428a; **2**:513a; **11**:459a
Banach-Steinhaus theorem, **1**:428a
Hahn-Banach theorem, **1**:428a
Banachiewicz, Thaddeus (1882–1954), 1:428–430
Banana plant, **1**:124b
Banatite, **3**:434a
Bancroft, Joseph Austen, **1**:52a; **3**:484a
Bancroft, Nellie, **5**:605a
Bancroft, Thomas Lane, **9**:82a
Bancroft, Wilder Dwight (1867–1953), 1:430–431; 9:395a*; **11**:535a; **15**:466b, 467a(n)
Bandaging, **12**:540a
Bandini, Sallustio, **12**:517a
Banester, John, **5**:348b
Bang, Bernhard Laurits Frederik, **2**:528a; **7**:100a,b; **12**:88a
Bang's disease, **12**:482b
Bang, Ivar, **4**:299a–b
Banister, John (1650–1692), 1:431–432
Banks, John, **3**:537b
Banks, Joseph (1743–1820), 1:433–437; 3:324a, 365b*, 397a; **5**:217a; **6**:479b*; **12**:561b
acquaintances of, **1**:520b–521a; **2**:509b, 510a; **4**:295a; **5**:218a; **6**:493a, 521a; **8**:371a; **11**:261b; **12**:471a, 487b, 515b–516b, 536a; **14**:562b
botany, **2**:390a; **7**:147b, 205b*
and Brown, **1**:434b; **2**:517b, 518a–b
collections of, **5**:216b; **7**:95b, 198b; **10**:323b; **11**:218b
on contemporary scientists, **14**:187a, 487b
correspondence of, **4**:540a,b; **5**:76b; **7**:69b, 408b; **8**:303b; **9**:152b; **11**:262a; **14**:78a, 79a,b
geology, **4**:549a
herbarium of, **1**:433b; **2**:517b, 518a–b; **13**:177b
influence of, **12**:535b
Kew Gardens, **1**:88a–b, 89a; **6**:494a
philology, **10**:324a
and Royal Society, **2**:186a; **3**:68a, 603b; **5**:518b; **6**:577a; **10**:608a
students of, **15**:471b
Ibn al-Bannāᵓ al-Marrākushī (1256–1321), 1:437–438; 14:593a
arithmetic, **13**:539b, 545b, 546a
commentaries on, **11**:229b
Banqueri, José Antonio, **1**:351a*
Bansha (Association of Foreign Learning), **12**:116b
Banti, Guido (1852–1925), 1:438–440
Banti's disease, **1**:439a
Banting, Frederick Grant (1891–1941), 1:440–443; 3:352a, 353a, 354a*; **5**:6a
insulin, **1**:613b; **8**:614b
Bantu tribe, **7**:414b
Banu Amājūr, **14**:575b, 577a
Banū Mūsā (*fl.* A.D. 850), 1:443–446; 4:541b; **6**:199b, 205a*; **7**:358b; **9**:300a; **13**:288b; **14**:537b, 538a, 575b; **15**:236a,b, 237a, 238a,b, 242a, 253b, 261a
astronomical measurements, **1**:510b
Geometria
translations of, **15**:177b
geometry, **1**:224a
proposition of, **6**:208a*
translations of works, **15**:255a
Verba filiorum, **1**:227a–b; **7**:175a, 335a
translations of, **4**:609b, 612a
Banyuls
Laboratory Arago, **7**:546a

Baobab tree, **1**:124b
Baoussé-Roussé, **11**:483b–484a
Bâpû
  see Vyeñkaṭa
Bär, Nicolai Reymers
  see Ursus (Bär)
Bárány, Robert (1876–1936), 1:446–447
Baranzano, Giovanni Antonio (1590–1622), 1:447–448
Barba, Alvaro Alonso (1569–*ca.* 1640), 1: 448–449
  El arte de los metales, **4**:345b
Barbaro, Ermolao, **1**:274a; **8**:248a, 249a; **13**:209a; **14**:636a
Barberini, Maffeo
  see Urban VIII
Barberry, **7**:441a
  and wheat rust, **3**:613a–b
Barber-Surgeons, Company of, **6**:568b
Barbier, Joseph-Émile (1839–1889), 1:449
  theorem of, **13**:221a
Barbier, Philippe Antoine, **5**:540a,b, 541a
Barbital
  synthesis of, **5**:2b
Barbiturates, **1**:389b
Barbituric acid derivatives
  synthesis of, **5**:2b
Barbour, Henry Gray (1885–1943), 1:449–450
Barbour, Thomas, **13**:25b
Barcelona
  latitude of, **9**:250b–251b
Barchusen, Johann Conrad (1666–1723), 1: 450–452
Barclay, John (1758–1826), 1:452; **7**:414a,b; **10**:260b; **14**:128a
Barcroft, Henry, **1**:452b
Barcroft, Joseph (1872–1947), 1:452–455; **5**:81a; **6**:24a; **7**:502b; **8**:14b, 535a(n); **10**:220b
Bardeen, John, **7**:221b; **8**:477b, 480b
Bardeleben, Heinrich von
  Handbuch der Anatomie des Menschen, **6**:224a
Bardi, Giovanni, **5**:238a
Barents Sea, **7**:410b; **9**:331a
  ecology, **4**:42a–b
  hydrology, **4**:42b; **12**:414b
  ice, **14**:634b
  surveying of, **8**:417b–418a; **14**:634a
Barez, Stephan Friedrich, **14**:40b
Barfuss, Friedrich, **1**:7a
Barger, George (1878–1939), 15:10–11; **15**:105b
Bargmann, Valentine, **4**:331a
Barhebraeus
  see Abu'l-Faraj Yūḥannā ibn al-ᶜIbrī al-Malaṭī
Barigazzi, Faustino, **1**:617b
Barigazzi, Giacomo
  see Berengario da Carpi, Giacomo
Barite
  decomposition of, **7**:378a
  refining of, **7**:378b
Barium, **5**:323a
  compounds of
    in cements, **8**:117a–b
  as fission product, **6**:16b–17a; **9**:262b–263a
  isolation of, **2**:588b; **3**:602a
  isotopes of, **7**:158b
  oxide of, **2**:565a
  salts of, **3**:316a; **14**:463b
    solubility of, **4**:238b–239a
al-Barjandī (or Bīrjandī), **15**:628b–629a
  Sharḥ al-Tadhkira
    translations of, **15**:628b
Barker, John, **2**:184b
Barker, Thomas Vapond, **1**:461a; **9**:379b
Barkhausen, Heinrich Georg (1881–1956), 1:455–456; **9**:462a
  Barkhausen effect, **1**:455b
  Barkhausen-Kurz oscillator, **1**:455b
Barkhausen, Johann Conrad
  see Barchusen, Johann Conrad
Barkla, Charles Glover (1877–1944), 1:456–459; **3**:564a
  and Bragg, **2**:398b, 399b; **9**:543a
  X rays, **15**:61b
Barks
  medicinal, **4**:560a
Barlaamo, **3**:585a; **4**:437b, 449a
Barletta, Mariano Santo da, **14**:28a
Barletti, Carlo, **14**:72a,b, 79a, 80b
Barley, **5**:524a*; **15**:648b
Barlow, Alfred Ernest, **1**:51b
Barlow, Guy, **11**:122b
Barlow, Peter (1776–1862), 1:459–460
  electricity, **10**:188a–b
  tables, **3**:374a*
Barlow, William (1845–1934), 1:460–463; **9**:379b
  Barlow lens, **1**:460a
  crystallography, **2**:431b; **11**:86b; **12**:512a
Barlow, William
  Magnetical Advertisements
    editions of, **13**:126b*
  The Navigator's Supply, **14**:514a
Barnabites, **1**:447b; **5**:195a; **13**:613b
Barnacle(s), **9**:508a*; **13**:355b
  classification of, **3**:573b
  complemental males, **3**:573b
  fossil, **3**:573b
Barnard, Edward Emerson (1857–1923), 1: 463–467; **5**:199b; **11**:553a; **13**:119a; **14**:481b
Barnaud, Nicolas, **2**:293a
Barnes, Howard Turner, **3**:19b, 20a
Barnett, Miles, **1**:195b
Barnett, Samuel Jackson
  Barnett effect, **9**:211b
Barney, Ida, **12**:176b
Barocius, Franciscus (1537–1604), 1:468; **3**:312a*; **9**:487b
  commentaries by, **12**:61b
  Geometricum problema tredecim modis demonstratum, **7**:133b
Barograph, **9**:189b
Barometer, **2**:141a, 444b; **3**:534b, 540a; **5**: 17b, 491a; **7**:578a; **11**:66a*; **15**:312b
  seventeenth-century development of, **1**: 138a,b, 341a; **2**:83b–84a; **6**:255a, 487a
  in altitude measurement, **2**:343b; **4**:28b; **5**:380a, 428b; **9**:486a
  balance, **9**:529b; **11**:432b
  changeable-scale, **11**:579b
  diagonal, **9**:529b
  differential, **7**:463b
  luminosity, **2**:55a; **6**:170a
  mercury, **9**:607b
    adjustable, **5**:78b
    self-registering, **6**:522b
  metal, **2**:439a
  phosphorescence in, **4**:214b; **10**:596b
  standard, **11**:174a
  water, **3**:557a
  wheel, **6**:487a
Barometric elevation, **14**:635b
Barometry
  seventeenth-century, **1**:483b; **5**:288b, 575a; **7**:577b; **8**:267b; **9**:117a–b, 311a, 486a; **10**:332b–334a; **13**:438a–439b, 445a
  eighteenth-century, **2**:474b; **4**:28b–29a, 518a; **13**:267a; **15**:355a
  nineteenth-century, **2**:147a; **11**:272b; **12**: 512; **15**:372b, 380b–381b, 382a, 527a
Baron, George
  Mathematical Correspondent, **1**:65b
Baron, Hyacinthe-Théodore, **5**:352b; **8**:619a
Baros (alloy), **3**:239b
Barozzi, Francesco
  see Barocius, Franciscus
Barrande, Joachim (1799–1883), 1:468–469; **4**:364b; **6**:57b; **9**:101a*, 584b; **13**: 143b
Barrandian area, **1**:469a
Barré de Saint Venant
  see Saint Venant, A. J. C. Barré de
Barrel
  volume of, **3**:152b; **7**:300b, 418b; **10**:572a; **13**:435b
Barrelier, Jacques
  Plantae per Galliam, Hispaniam et Italiam observatae, **7**:197b
Barrell, Joseph (1869–1919), 1:469–471; **5**:396a*; **10**:507a; **13**:518b
Barren Lands, **13**:525a
Barreswil, Charles-Louis (1817–1870), 1: 471; **2**:25b
Barrett, William Fletcher, **6**:5b; **10**:246a
Barro Colorado Island, **15**:224b
Barrois, Charles (1851–1939), 1:471–473; **3**:160b; **4**:66b*; **5**:475b, 476b*; **8**: 31b*; **14**:627b*
Barrois, Jules, **1**:471b; **5**:52a
Barron, Moses, **1**:440b
Barrow, Isaac (1630–1677), 1:473–476
  acquaintances of, **14**:370b
  analysis, **9**:50a; **10**:48a; **13**:436b
  Archimedes, **1**:229b; **3**:348b
  on contemporary scientists, **2**:463a; **3**: 348a–b
  edition of Euclid, **4**:451b–452a; **10**:45b
  geometry, **3**:458b; **4**:419b, 441a; **7**:83b
  influence of, **1**:475a; **2**:47a, 358b; **5**:524b; **14**:619a
  influences on, **5**:288a
  Lectiones geometricae, **2**:49b; **5**:529a
    publication of, **3**:348b
  Lectiones opticae et geometricae, **8**:161a, 164b
  manuscripts of, **3**:348b
  and Newton, **1**:475a–b; **10**:43b, 44a, 45b, 48a, 49b, 55a
  optics, **6**:125b; **9**:159a
  philosophy, **14**:375a(n)
Barrows (archaeol.), **14**:506a
Barry, Martin (1802–1855), 1:476–478; **2**: 161b
Bartels, Johann Martin Christian, **5**:298b, 302b; **7**:206b; **8**:428b
Bartels, Julius (1899–1964), 1:478
Barth, Jeremias, **1**:571a
Barth, Joseph, **11**:158b
Barth, Ludwig, **7**:283a
Barth, Thomas Fredrik Wieby, **4**:409a; **5**: 458a
Barthélemy, Jean Jacques, **3**:523b
Barthélemy, Toussaint, **1**:558b
Barthez, Paul-Joseph (1734–1806), 1:478–479; **1**:617a; **2**:122b; **3**:379a
Bartholin, Caspar (1585–1629), 1:479–481; **1**:481a; **2**:13a; **3**:98b; **4**:507b
  Institutiones anatomicae, **1**:482a,b; **13**: 32b, 222b
Bartholin, Caspar II (son of Thomas), **1**: 482a; **13**:33b; **14**:449b
  Bartholin's duct, **1**:482b
  Bartholin's glands, **1**:482a; **4**:268b
Bartholin, Erasmus (1625–1698), 1:481–482; **6**:520b; **11**:525a–b, 527a; **12**:205b
  crystallography, **9**:73a; **11**:523a
  editions by, **3**:614b, 615a
  geodesy, **11**:525b

Bartholin, Thomas (1616–1680), 1:482–483; 1:481a; 2:224b, 317a; 4:65a; 6:509a; 11:525a; 12:333a; 13:30a, 32a; 14:13a
anatomy, 1:480b; 5:226b; 10:477a,b, 478a; 11:467a–b; 13:31b
lymphatic system, 11:587a, 588a
Bartholomaeus of Salerno, 12:82a–b, 83a
Bartholomeus Anglicus, 2:293b; 4:121b
Bartholomew of Messina, 1:271b, 272a
Bartlett, Maurice Stevenson, 14:383a
Bartol Research Foundation, 13:176a
Bartoli, Adolfo, 2:266a
Bartoli, Daniello (1608–1685), 1:483–484
Bartolomeo da Varignana, 1:107a
Bartolotti, Gian Giacomo (*ca.* 1470–*ca.* 1530), 1:484
Bartolotti, Pellegrino, 1:484a
Barton, Benjamin Smith (1766–1815), 1: 484–486; 3:562b; 4:21b; 10:163b, 509b; 11:218a; 14:558b
Elements of Botany, 1:489b
Bartonella, 10:142b
Bartonian (stratig. stage), 9:240b
Bartram, John (1699–1777), 1:486–488; 1: 488b–489a; 3:345a, 350a, 351a, 563a; 4:492b, 493a; 7:210b; 8:460a; 9:537b; 12:515b
Bartram, Moses, 12:133b
Bartram, William (1739–1823), 1:488–490; 1:488a; 9:366a, 537b; 11:263a; 12: 132b; 14:416b–417a
illustrations, 1:485b*
Travels, 11:263b*
Bartsch, Jacob, 7:306b, 307a
Barus, Carl (1856–1935), 1:490–491; 7:5b*, 370b
Bary, Heinrich Anton de
see De Bary, (Heinrich) Anton
Barysphere, 13:148a
Baryta, 2:8a
discovery of, 12:146b
purification of, 11:495a
Basal metabolic rate, 1:610b; 8:555a
Basalt, 4:149b, 262b; 5:518b, 562b; 12:120a
assimilation, 4:564a
composition of, 2:494a; 3:293b
crystallization of, 2:370a; 7:63a
fused, 4:141a
inclined, 11:269b
magnetization of, 9:181b
massive, 11:303b
melts of, 6:54b
origin of
eighteenth-century discussion, 1:234b; 2:439b; 4:71a–72a, 150b–151a, 548b–549a; 6:84a–b; 7:277b; 11:303b–304a, 387b; 12:329a; 14:60b, 132b, 261a–262a, 312a
nineteenth-century discussion, 1:327a; 2:481a, 494a, 554b–555a; 3:313b, 412a; 6:456b; 7:69a, b, 70b; 8:246a, 616a; 15:526a–b
twentieth-century discussion, 1:154a
prismatic (columnar), 4:71a–72a, 150b; 5:578a; 6:84b; 8:564a; 9:61a; 11: 303b–304a
types of, 7:548b
Basch, Samuel Siegfried Carl von, 7:504b
Baseilhac, Jean, 8:114b
Basel
municipal hospital, 5:418b–419a
Basel, Council of, 3:513a
Basel, University of
in eighteenth century, 4:468b–469a
Basements (geol.), 12:272a–274b
fracture patterns, 12:273b
reactivation of, 12:272b
Bases (chem.)
definition of, 2:499a
organic, 5:104a; 6:462a
partition of, 12:46b
pseudo bases, 6:108b
strength of, 1:298b–299a, 300a; 2:170b
Basicity
measurement of, 14:124b
theory of, 1:579a; 2:170b; 6:108b; 8:293a; 14:125a, 396a
Basidia, 5:192b
Basidiomycetes, 2:19a, 437a,b; 6:121b; 14: 93b
cytology, 9:34b
Basidiospores, 9:34b; 13:489b
Basil Valentine
see Valentine, Basil
Basilides of Tyre, 6:616a–b
Basilius de Varna (anagram)
see Libavius (or Libau), Andreas
Basilosaurus, 6:120b
Basins, stratigraphic
concept of, 5:476a
Basins, submarine, 2:551b
Basis (math.)
integral, 2:90a
Basophilism
pituitary, 3:518b
Basophils, 4:296a
Ibn Baṣṣal, 1:350b; 14:112b
Bassani, Francesco (1853–1916), 1:491–492
Bassi, Agostino Maria (1773–1856), 1:492–494; 1:329a; 10:372b–373a; 12:202b–203a
Bassi, Laura, 12:553b, 555a, b
Bassler, Raymond Smith (1878–1961), 1: 494–495; 5:50b*; 13:531b–532a, 534a*
Basso, Sebastian (*fl.* 1550–1600), 1:495
Bastian, Adolf, 2:207a–b
Bastian, Henry Charlton (1837–1915), 1: 495–498; 4:594a; 8:406b; 13:523a
Bastian's law, 1:496b
and Pasteur, 10:372b, 382a–383a
on spontaneous generation, 3:188b, 339a
Bastianelli, Giuseppe, 5:503a, b
Bastnäs' tungsten, 3:473b
Bat, 14:299b
anatomy of, 8:204b, 205a
flight of, 12:562b
sound generation, 10:605a
Bataillon, Jean Eugène (1864–1953), 1:498–499; 5:584a
Batavian Club, 13:469a
Bate, George, 5:425b
Bate, Henry
see Henry Bate of Malines
Bate, John, 10:43a
Bateman, Harry (1882–1946), 1:499–500; 12:29b; 13:266a
Bateman's expansion, 1:500a
Bateman's function, 1:500a
Bateon
see Buteo, Johannes
Bates, Henry Walter (1825–1892), 1:500–504
Batesian mimicry, 1:501b, 503a–504a; 9:560b; 14:136a
and Wallace, 14:133b, 134a, b, 136b
Bateson, William (1861–1926), 1:505–506
acquaintances of, 8:513b; 11:212b; 14: 424a
on contemporary scientists, 9:516a
correspondence of, 11:211b–212a
embryology, 8:586a
evolution, 14:251b
genetics, 9:520a; 13:158a; 14:100b
influence of, 3:121a; 6:22b
Mendelism, 3:123b, 493a, 590b; 13:157a
and Pearson, 10:462b, 463a–b
students of, 15:505b
Bath Philosophical Society, 6:328b
Bather, Francis Arthur (1863–1934), 1:506–507
Batholiths, 1:470a; 8:42a, 99a–b
Bathonian formations, 1:285b
Baths
England, 3:13b; 5:415a
Slavic, 7:4a–b
see also Balneology, Hydrotherapy
Bathurst, Ralph, 4:40a; 6:129b; 9:243b; 12:308a, 580b; 14:369a, 405a, 407b
Bathybius Haeckeli, 4:291b; 14:146a
Bathymetry, 1:92b, 595a; 3:265a; 8:532a; 9:196a; 13:552a
Bathynella, 15:513b
Bathyrheometer, 4:12b
Bathyscaphe, 9:26b; 10:597b–598a
Batrachians, 2:493b
development of
light, role of, 4:285b
embryogeny, 11:133a
fossil, 5:296a
freezing of, 15:126b
parthenogenesis of, 1:498a–499a
Rusconian orifice, 4:266b–267a
al-Baṭrīq, Abū Yaḥyā, 13:538a
al-Battānī, Abū ʿAbd Allāh Muḥammad ibn Jābir ibn Sinān al-Raqqī al-Ḥarrānī al-Ṣābiʾ (*fl. ca.* A.D. 880), 1:507–516; 1:382b; 2:405a; 7:38a; 9:297a
astronomical tables, 14:593a
revision of, 9:39b
astronomy, 3:405b–406a; 7:363a; 9:84b; 12:62a; 15:545a
on calendar, 5:553a
cosmology, 1:247b; 7:142a; 12:3a
influence of, 2:24a; 3:26b; 7:338a; 8:280b; 9:36a; 11:31b; 13:323b; 14:388a; 15:421a, 477a
De motu stellarum, 1:23a
observations of, 15:349b
translations of works, 11:349b; 12:162a
trigonometry, 7:532a; 14:471a(n)
al-Zīj al-Ṣābiʾ, 7:532a; 11:202a; 14:574b, 575b, 576b
translations of, 11:32a,b
Battery (elect.), 4:248b; 5:132a
Bunsen, 2:587a, 588b
calomel cell, 3:334a
carbon arc, 3:601a
chemical theory of, 4:503b
Daniell cell, 2:439a; 3:334a, 557a; 4:19a; 5:292a, 559b
thermochemistry of, 11:298a
dry pile, 4:74a; 11:474a
fuel cell, 5:559b–560a, 621a
galvanic, 2:588b; 3:521a; 4:29a; 5:7a; 6: 243a, 252b
electrode potentials, 8:188b
electromotive force, 8:503a; 10:481b
energetics of, 6:244b
at low temperatures, 11:418a
quantitative measurement of, 4:556b
structure of, 10:552b
theories of, 15:433b–434a
Grove cell, 5:559a–b
Leclanché, 2:439a
non-polarized, 1:557b
photoelectric, 11:460b
standard cells, 3:19b, 288b, 334a
storage, 4:195a, 284a; 13:473b
voltaic cell, 1:557a–b; 5:7a
heat effects, 7:65b
theory of, 5:292a
Zamboni, 5:292a

Ibn Baṭṭūṭa (1304–*ca.* 1368), 1:516–517
Bauch, Martin, **12**:143b, 144a
Baudin, Nicholas, **2**:320a; **8**:266a; **9**:366b
Baudrimont, Alexandre Édouard (1806–1880), 1:517–519; 4:246a
chemical theory, **4**:246b; **8**:56b
Baudrimont, Marie Victor Ernest, **1**:518a
Bauer, Edmond (1880–1963), 1:519–520; 2:456a; **8**:10a; **14**:246b*
Bauer, Ernst, **8**:473a
Bauer, Erwin, **3**:422b
Bauer, Ferdinand Lucas (1760–1826), 1:520; 2:517b–518a
Bauer, Franz Andreas (1758–1840), 1:520–521; 5:539a
Bauer, Georg
see Agricola, Georgius
Bauer, Hans, **13**:136a
Bauer, Louis Agricola (1865–1932), 1:521–522; 13:156a
Bauer, Oswald, **6**:368b
Bauhin, Gaspard (1560–1624), 1:522–525; 1:480a; **3**:534a; **8**:309a; **9**:528b; **11**:527b; **14**:413b
anatomy, **1**:46b, 98b; **14**:9a
Catalogus plantarum circa Basileam sponte nascentium, **11**:313b
Pinax theatri botanici, **1**:526b; **3**:129a; **8**:120b; **11**:587b; **13**:443a
revisions of, **4**:99a; **12**:394a
Theatrum anatomicum, **6**:151a
Bauhin, Jean (1541–1613), 1:525–527; 1:522a,b; **7**:60b(n); **9**:528b; **11**:311b, 313b; **13**:502b; **14**:413b
Historia plantarum universalis . . . , **11**:312a, 314a
Bauhin, Jean Gaspard, **1**:523a
Baum, William Alvin, **1**:353a
Baumann, Eugen, **4**:201a; **6**:506a; **9**:284a,b
Baumbach, Otto, **12**:30b
Baumé, Antoine (1728–1804), 1:527; 3:419a; **5**:600b; **8**:80a, 621b; **9**:347a; **11**:616b; **12**:63b
affinity tables, **2**:7a
chemistry, **3**:5a,b, 562a; **7**:254a; **8**:73a; **12**:64a, 65a
and Macquer, **8**:619a–b
Baumès, Pierre Prosper François, **3**:531a
Baumgarten, Paul Clemens von, **3**:490a; **7**:423b
Baumgartner, Andreas, **9**:278a
Naturlehre, **10**:192b
Baumhauer, Edouard Henri von (1820–1885), 1:527–528
Baumhauer, Heinrich Adolf (1848–1926), 1:528–529; 9:290a
Bauxite
deposits of, **7**:530a
discovery of, **2**:72a
origin of, **1**:623a
Bavaria
geology, **14**:627a
Bawden, Frederick Charles, **13**:152b
Baxter, Gregory Paul, **6**:481b; **11**:417b, 418a
Bayen, Pierre (1725–1798), 1:529–530; 7:603a; **8**:74b, 75a, 247b*; **9**:106a; **10**:496a; **11**:389b, 563a; **13**:602b, 603a
Bayer, Johann (1572–1625), 1:530–531; 7:412a
Bayerische Akademie der Wissenschaften, **4**:269a; **9**:149a; **15**:172a
Bayes, Thomas (1702–1761), 1:531–532; 7:106b, 317b; **8**:306a
Bayesian analysis, **15**:366a, 375a
theorem of, **1**:531b; **3**:386a, 386b–388a; **10**:456a, 457a; **11**:599b; **15**:281a–b, 284b, 304a
al-Bayhaqī, Abu'l-Ḥasan, **7**:324a, 330b, 335b, 336a,b, 338a,b, 531b
Baylak al-Qibjāqī (*fl. ca.* 1250), 1:532–533
Bayle, Gaspard Laurent, **2**:444b, 464a; **7**:556b; **12**:450a
Bayle, Pierre, **1**:606a; **2**:542b; **3**:484b; **5**:58b, 60a; **6**:555b; **8**:151b; **9**:25b; **10**:492a; **11**:1b, 493a
critics of, **3**:485a
Dictionnaire historique et critique, **14**:152b–153a
Bayliss, Leonard Ernest (1900–1964), 1:533–535
Bayliss, William Maddock (1860–1924), 1:535–538; 2:599a; **8**:535b*; **15**:73b
Principles of General Physiology, **1**:533b–534a; **5**:485a
and Starling, **1**:536b–537a; **12**:617b, 618a
students of, **15**:104b
Ibn al-Bayṭār al-Mālaqī, Ḍiyāʾ al-Dīn Abū Muḥammad ʿAbdallāh ibn Aḥmad (*ca.* 1190–1248), 1:538–539; 4:121b
pharmacology, **1**:414a
Bazaine, Pierre Dominique, **7**:601a
Bazzani, Matteo, **5**:226a
Bazzoni, Charles Blizard, **11**:420b
BCG vaccine, **3**:22b
Bdellostoma, **3**:610b
Beadle, George Wells, **9**:525b; **13**:134b, 135b, 137b, 269a
Beagle (ship), **3**:565b–569b *passim,* 573a; **5**:16a–17a, 481a; **6**:289a; **8**:528a
Beal, William James, **1**:396a; **2**:102b
Beale, John, **6**:41a; **7**:194b
Beale, Lionel John, **1**:539b
Beale, Lionel Smith (1828–1906), 1:539–541; 10:291b
Beale's cells, **1**:540b
Beale's solution, **1**:540b
translations of works, **15**:78b
Beale, Peyton Todd Bowman, **1**:539b
Beams (arch.)
breaking strength, **9**:119b
buckling, **9**:578a
cantilever, **3**:442a; **9**:578a; **10**:470b*
continuous, **9**:446b
equal resistance, **2**:201a,b
flexure of, **3**:441b, 442a; **5**:530b; **6**:452a–b; **8**:212b; **10**:468a*; **11**:123b, 496b
neutral surface of, **3**:442a
partial forces in, **8**:222a–223a
rupture of, **3**:441a
stress of, **3**:496b; **10**:319b
neutral point, **11**:496b
vibrations of, **6**:545b
weight distribution, **8**:199a
Bearden, Joyce Alvin, **3**:370a
Bearings (mech.)
ball, **8**:210a, 210b (fig.), 212a (fig.)
bushings, **8**:210a
Beatley, Ralph, **2**:145b
Beats, acoustic, **12**:127b–128a
Beattie, James, **1**:404a
Beaufort, Sir Francis, **5**:17a; **14**:184b
Beaugrand, Jean (*ca.* 1595–*ca.* 1640), 1:541–542; 6:299a; **9**:322a(n), 528a; **10**:103b; **14**:24a
and Desargues, **4**:46b, 47a, 48a
Geostatice, **4**:46b; **7**:537a
mathematics, **5**:527a; **6**:112b
Beaulieu, J. de (pseudonym), **8**:132a–b
Beaumont, Élie de
see Élie de Beaumont, Jean B.
Beaumont, John, **2**:613b; **8**:416a
Beaumont, William (1785–1853), 1:542–545; 4:252a
experiments, **4**:399a; **11**:426a
Beaune, Florimond
see Debeaune, Florimond
Beaunis, Henri, **2**:131b
Beauty, classical ideal of, **2**:531b
Beauvais, Vincent of
see Vincent of Beauvais
Beaver
social instinct of, **3**:521b
Bebb, Michael Schuck, **1**:396a
Beccari, Jacopo (Giacomo) Bartolomeo, **1**:547b; **5**:226a, 267a; **13**:566b; **15**:54a
Beccari, Nello (1883–1957), 1:545–546; 3:113a
Beccaria, Giambatista (1716–1781), 1:546–548; 1:68a; **5**:267b; **7**:560a; **9**:189a; **14**:70a, 419b*
Dell'elettricismo lettere . . . al chiarissimo Signor Giacomo Bartolommeo Beccari, **7**:371b
electricity, **3**:51b, 153b; **13**:224b–225a; **14**:70b, 71b, 73a
Elettricismo artificiale, **14**:71a
and Franklin, **1**:548a; **10**:147a
magnetism, **13**:183b
Bech, Philippus, **1**:78b
Béchamp, Pierre Jacques Antoine (1816–1908), 15:11–12; 5:315a
Becher, Johann Joachim (1635–1682), 1:548–551; 1:451a; **7**:188a, 284a, 524b; **14**:145a
chemistry, **12**:604b
Bechert, Karl, **3**:125a
Bechhold, Heinrich Jakob, **4**:299b
Beck, Albert van
see Beek, Albert van
Beck, Heinrich, **8**:571a
Beck, Lewis Caleb, **4**:364a; **6**:277b
Beck, R. & J., Ltd., **8**:414a,b
Beck, Theodoric Romeyn, **6**:277a,b
Becke, Friedrich Johann Karl (1855–1931), 1:551–552; 5:457a, 562b; **12**:271b; **13**:148b, 285a, 477a*
Becke line effect, **1**:154a, 551b
Becke volume rule, **1**:551b
Becker, Ernst, **12**:247b
Tafeln zur Berechnung der Precession, **4**:176a
Becker, George Ferdinand (1847–1919), 1:552–553; 4:366b, 367a*, 395b*; **7**:5a; **8**:370a
Becker, Heinrich, **3**:530a
Becker, Herbert, **2**:338b; **7**:153a
Becker, Richard, **8**:476a, 477b
Electromagnetic Fields and Interactions, **1**:24a
Beckmann, Ernest Otto (1853–1923), 1:553–554; 5:622a; **14**:455b*; **15**:433b
Beckmann thermometer, **1**:553b
Beckmann transformation, **1**:553b
Beckmann, Friedhelm, **4**:420a
Beckmann, Hermann, **11**:582b
Beckmann, Johann (1739–1811), 1:554–555; 7:202b*; **8**:379b; **14**:224b; **15**:206a,b, 406b
Becquerel, Alexandre-Edmond (1820–1891), 1:555–556; 1:557a, 558b, 561a; **4**:183a; **8**:36a,b; **9**:177a,b; **11**:111b*
Becquerel, Antoine-César (1788–1878), 1:557–558; 1:555b, 558a; **3**:497a; **4**:74b*
battery, **5**:559b
electricity, **10**:188a–b
Becquerel, [Antoine-] Henri (1852–1908), 1:558–561; 1:557a, 561a; **4**:217a*; **9**:292b
and Curies, **1**:559b, 560a–b; **3**:500a
influences on, **11**:58b

Becquerel, [Antoine-] Henri (cont.)
radioactivity, **2**:257b, 258b; **3**:498b, 507a; **12**:26b, 27b, 191b
physiological effects, **3**:508a
on uranium X, **3**:480a
Becquerel, Jean, **1**:558b; **2**:203a; **5**:610b; **7**: 492b, 494a*
Becquerel, Paul (1879–1955), 1:561–562
Becquey, Louis, **4**:348a,b, 349a
Beddard, Frank Evers, **1**:504a
Beddoe, John (1826–1911), 1:562–563; 6: 563a
Beddoes, Thomas (1760–1808), 1:563–564; 6:385a, 386a; **12**:564a*; **14**:198b*
Chemical Experiments and Opinions . . . , **9**:245b
and Davy, **3**:599a
Pneumatic Institution, **3**:598b, 600a–b
Bede, the Venerable (672–735), 1:564–566; 1:104b
critics of, **6**:302a
De temporum ratione, **3**:26b
Bee, **3**:149a*, 180a; **5**:502b; **9**:440b; **12**:309a, 588b; **13**:29b, 347b
communication, **11**:332b
dance language, **1**:265b
genetics, **9**:282b
identification of, **8**:518a
leaf-cutting, **14**:413a
natural history of, **11**:332b
parthenogenesis, **12**:421b
queen, **11**:332b
reproduction, **1**:264a, 265b
sexual characters, **2**:364b
vision, **8**:529a
see also Apiculture
Beebe, William, **10**:598a
Beech tree, **3**:45a; **6**:100a
Beecher, Charles Emerson, **6**:57b; **12**:229b, 230a
Beeckman, Isaac (1588–1637), 1:566–568; 5:285a, 288a; **6**:520a; **7**:73b, 269b; **9**:316b, 318a; **11**:289a(n)
and Descartes, **1**:567a, 568a; **4**:51b, 55b, 56a
hydrodynamics, **9**:14a
optics, **6**:197a
Beek, Albert van, **1**:145a; **9**:459b
Beeldsnijder, François, **1**:135a
Beer, August, **6**:392a
Beer, Wilhelm (1797–1850), 1:568–569
lunar map, **2**:147b; **9**:2a
Beer, **9**:558b
alcohol concentration, **14**:109a
brewing of, **10**:354a–b, 377a, 380a; **15**: 647b, 649a,b, 651a
concentrates of, **5**:421a
microbiology of, **6**:100a–101a
preservation of, **10**:380a
testing of, **5**:203a
Beet sugar, **1**:44b–45b; **3**:7b, 202a, 316b, 481a*; **4**:41b; **5**:203a; **6**:237b; **9**: 105a–b; **10**:436a; **11**:359a; **13**: 309b
Beetle, **10**:437a; **12**:201a
diseases of, **9**:333a
Beevor, Charles Edward (1854–1908), 1: 569–570; 1:496b; **5**:26b
Béghin, Henri (1876–1969), 1:570–571
Begoniaceae, **3**:43a
Beguin, Jean (1550–1620), 1:571–572; 3: 596b; **6**:46a(n); **12**:311b; **13**:507b
Tyrocinium chymicum, **6**:146a*
Béguyer de Chancourtois, Alexandre-Émile (1820–1886), 1:572–573; 2:89a; **9**: 349a
Behaim, Martin (1459–1507), 1:573–574; 5:377a; **8**:281b; **11**:351a
Behavior
animal, **10**:434b
classification of, **10**:434b
functional circle model, **15**:504a–b
innate vs. learned, **14**:314b–315a
mechanism of, **12**:270a–b
objective analysis of, **8**:446b
origins of, **3**:332a; **7**:604b
theory of, **10**:433a, 434b–435a
see also Ethology, Psychology
Behaviorism, **3**:89a; **8**:446b; **10**:433a; **11**:518b
Behnisch, Robert, **4**:155b
Behrend, Anton Friedrich Robert (1856–1926), 15:12–13
Behrends, Johann Bernard, **12**:138a
Behring, Emil von (1854–1917), 1:574–578; 4:305a*; **7**:425b
antitoxins, **4**:297b; **8**:449b; **11**:569b
immunology, **3**:22a; **7**:391b, 427a; **9**:334b
Behring Institute, **1**:576b
Behringer, Johann Barthel Adam
see Beringer, Johann Bartholomaeus Adam
Beiblätter zu den Annalen der Physik, **14**: 330a
Beighton, Henry, **4**:44a
Beijerinck, Martinus Willem (1851–1931), 15:13–15; 7:35a
Beilby, George, **11**:549a
Beilstein, Konrad Friedrich (1838–1906), 1: 578–579; 2:624a; **5**:12b; **9**:347b
Handbuch der organischen Chemie, **5**: 431a
Being
infinite vs. finite, **4**:255a
univocal, **4**:255a
see also Essence; Ontology
Beiträge zur Biologie der Pflanzen, **3**:340a
founding of, **3**:338a
Beiträge zur Krystallographie und Mineralogie
founding of, **5**:456b*
Beiträge zur Morphologie und Physiologie der Pilze
editors of, **14**:93a
Beiträge zur Physiologie der Sinnesorgane
editors of, **7**:458a
Beketov, Nikolai Nikolaevich (1827–1911), 1:579; 7:34b, 496a; **13**:416b
Bekhterev, Vladimir Mikhailovich (1857–1927), 1:579–581; 5:26b; **12**:270b
Bekhterev cells, **1**:580a
Bekhterev's disease, **1**:580a
Bekhterev Institute
founding of, **1**:580a
Bekkerhinn, Karl, **7**:494a
Beklemishev, Vladimir Nikolaevich
Osnovy sravnitelnoi anatomii bezpozvonochnykh, **7**:34a
Belaiew, Nicholas Timothy (1878–1955), 1: 581
Beldamandi, Prosdocimo de, **3**:513a; **10**: 270b; **12**:61a, 62a; **15**:10a–b
Belemnites, **9**:87a; **10**:429b; **15**:53b
Belenkii, Semen Zakharovich, **13**:241a
Belga, Dominicus Gnosicus, **4**:170b*
Belgica (asteroid), **4**:26b
Belgium
archaeology, **5**:109b
geology, **5**:475b–476b; **8**:463a–b
National Committee on Astronomy, **4**: 27a
paleontology, **5**:109b; **7**:460b–461a; **12**: 184a
Royal Observatory (Uccle), **4**:26b–27a
Service de la Carte Géologique, **5**:475b
stratigraphy, **2**:108b
Beliaev, Nikolai Timofeevich
see Belaiew, Nicholas Timothy
Bélidor, Bernard Forest de (1697/1698–1761), 1:581–582; 4:44a; **5**:101a; **8**: 213a; **13**:594a(n)
Architecture hydraulique, **11**:164a
Le bombardier françois . . . , **2**:202a
editions of works, **10**:3a,b
Science des ingénieurs, **3**:441b
Belknap, Jeremy, **9**:537b
Bell, Alexander Graham (1847–1922), 1: 582–583; 9:372a
telephone, **1**:303b; **2**:439a; **4**:248a–b
Bell, Alexander Melville, **1**:582b
Bell, Charles (1774–1842), 1:583–584; 6:59b; **9**:241b, 242a; **15**:110a
Bell-Magendie law, **1**:584a; **8**:17a; **9**:9b; **14**:128b
confirmation of, **9**:570b
Idea of a New Anatomy of the Brain, **9**:241b
Bell, Eric Temple (1883–1960), 1:584–585; 5:314a*; **9**:22b*
geometry, **14**:356b
Bell, Florence, **1**:319b
Bell, John, **1**:452a, 583b
Bell, Julia, **6**:22a
Bell, Robert (1841–1917), 1:585; 8:461b
Bell metal
casting of, **3**:560b
Bell Telephone Laboratories, **1**:292b–293a; **3**:597a; **7**:108a
Bella Coola Indians, **2**:207b, 212a
Belladonna, **4**:252a; **11**:215b
Bellani, Angelo (1776–1852), 1:585–587
Bellardi, Luigi (1818–1889), 1:587
Bellarmine, Robert (1542–1621), 1:587–590; 3:311b; **5**:244b, 382b; **12**:105b
Bellaso, Giovanni Battista, **1**:605b
Bellavitis, Giusto (1803–1880), 1:590–592; 1:239b; **13**:623a
analysis, **10**:443a; **11**:512b
geometry, **15**:196a
number theory, **6**:522a
Belleval, Pierre Richer de (1564–1632), 1: 592
Belli, Onorio, **1**:125a
Belling, John, **13**:137a
Bellingshausen, Fabian von
see Bellinsgauzen, Faddei
Bellini, Lorenzo (1643–1704), 1:592–594; 1:391a,b; **2**:226a, 309a, 311a; **4**:520b; **11**:2b; **14**:586b
De urinis et pulsibus et missione sanguinis, **11**:2a
Bellinsgauzen, Faddei F. (1779–1852), 1: 594–595; 8:101a
works translated, **3**:617a*
Belloni du Chaillu, Paul, **14**:533b
Bells
vibrations of, **9**:174b
Beloglazov, Konstantin Fedorovich, **7**:530a
Belon, Pierre (1517–1564), 1:595–596; 4: 412b; **13**:451a; **14**:507b
works translated, **8**:121a
Belopolsky, Aristarkh Apollonovich (1854–1934), 1:597–599; 2:434b; **4**:168a, 288a*; **5**:363a; **6**:105b; **9**:203b; **10**: 30a; **13**:482a
Belt, Thomas
The Naturalist in Nicaragua, **1**:502a
Beltrami, Eugenio (1835–1899), 1:599–600; 2:470a; **3**:331b*; **6**:534b; **7**:506b; **8**: 504b, 505a; **12**:55a, 173a*; **14**:86b
on contemporary scientists, **11**:409a
geometry, **1**:591a; **2**:271a; **6**:522a; **7**:397a, 600a; **8**:433a,b; **9**:410a
surfaces, **4**:102b–103a
Belts (geol.), **12**:273b, 274a
Bembo, Pietro, **5**:104b, 105b; **9**:192a

Bémont, Gustave, **3**:499a, 507b; **4**:356a; **12**:26b
Ben Nevis (mtn.), **3**:569a
Bence-Jones, Henry, **4**:201b; **5**:340b
Benches (geol.)
origin of, **8**:30b
Benda, Ludwig, **4**:299b
Beneden, Edouard van (1846–1910), 1:600–602; **2**:383b, 384a,b; **5**:109a,b; **11**:254b, 256a*; **13**:157b
cytology, **14**:235b
embryology, **2**:362b, 363a; **6**:338a; **15**:513b
Beneden, Pierre-Joseph van (1809–1894), 1:602–603; **1**:600b
Benedetti, Alessandro (1450–1512), 1:603–604; **1**:47a, 229b; **5**:104b; **8**:248b, 249a
Anatomice
editions of, **2**:537b*
mechanics, **7**:486b; **9**:488a
Benedetti, Giovanni Battista (1530–1590), 1:604–609; **5**:246b–247a; **13**:260b; **14**:461a
commentaries of, **4**:450b
mechanics, **1**:419b, 424b; **5**:239a
music theory, **5**:249b
Benedetto, Giovanni, of Tirna, **1**:608b*
Benedicks, Carl Axel Fredrik (1875–1958), 1:609
Benedict XIV (Pope), **2**:324b; **3**:23a; **4**:515b; **9**:5b; **11**:65b
Benedict, Francis Gano (1870–1957), 1:609–611; **1**:325b; **8**:555b
Benedict, Ruth, **2**:210a
Benedicta, **6**:123b
Benedictines, **3**:97a, 115b, 393a, 394a; **6**:396b; **11**:253b, 414b; **13**:185a, 278a; **14**:155b; **15**:275b
Benedikt, Moritz
Benedikt's syndrome, **3**:203b
Bengal Chemical & Pharmaceutical Works, Ltd., **11**:319a
Benioff, Hugo, **5**:596b
Benivieni, Antonio (1443–1502), 1:611–612
Bennet, Abraham, **14**:73b, 565a–b
Bennett, Alfred William, **13**:341b
Bennett, John Hughes, **2**:598a; **8**:405a
Bennett, John Joseph, **2**:518b, 520b
Bennett, Willard, **13**:570a
Bennettitales, **8**:354a,b; **12**:260a
Bennewitz, Peter
see Apian, Peter
Bénoist, Louis, **14**:218a
Benoist, Michel, **15**:736b
Benoit, Justin, **1**:612b
Benoit, Justin-Mirande René (1844–1922), 1:612–613; **9**:155a
Bensley, Arthur Benjamin, **1**:613b
Bensley, Robert, **1**:613b–614a
Bensley, Robert Russell (1867–1956), 1:613–614
Bensley, Sylvia Holton, **1**:613b
Bentham, George (1800–1884), 1:614–615; **4**:36a; **5**:539a
Genera plantarum, **2**:18b
Handbook of British Flora, **6**:491b
taxonomy, **6**:491b
Bentham, Jeremy, **1**:614b; **9**:189a, 383a; **11**:140b; **12**:571a
Bentley, Richard, **3**:296b(n), 430b, 431a; **6**:35b–36a; **10**:45a; **14**:27a, 296b
critics of, **7**:275b
Bentley, Robert, **2**:382a; **6**:331b; **14**:496a
Benzaldehyde, **3**:47a; **11**:495a
reactions of, **14**:476b–477a
Benzenberg, Johann Friedrich (1777–1846), 1:615–616; **2**:421a; **12**:192a
Benzene, **14**:602a
bromination of, **3**:448a,b
carboxylic acids of, **1**:390a
composition of, **9**:425a; **13**:359a
compounds of, **5**:488b; **11**:92b
chlorination of, **1**:578b
oxidation of, **11**:371a
terminology, **5**:489a
derivatives of, **7**:552a; **9**:425a; **14**:269b
constitution of, **9**:355b
formulas, **8**:508a
substitution, **7**:281b–282a
discovery of, **4**:531b
hydrogenation, **12**:47a
oxidation of, **10**:157b
ring, **8**:475b
dimensions, **8**:485a
substitution in, **2**:515a; **3**:46b; **9**:352b
structure of, **1**:288b, 390a; **2**:68a; **3**:300a–b, 448b, 449a; **4**:78b, 400a; **7**:552a; **8**:61a; **9**:352a–b; **13**:337b–338a
ring theory, **1**:390a–b; **7**:281b–283b; **8**:508a
synthesis of, **2**:65b; **5**:420b; **7**:282b; **9**:425a
tests for, **9**:356a–b
and thiophene, **6**:107b–108a
Benzi, Ugo, **9**:75a
Benzidam, **5**:197b
Benzidine, **14**:622a
structure of, **7**:517b
Benzoates
in foods, **14**:357b
Benzoic acid, **1**:578a; **8**:331b, 333a, 339a; **14**:476a
composition of, **9**:425a; **14**:476b
derivatives of, **9**:355b; **13**:406b
esterification, **9**:357a, 619b
hydrogenation of, **14**:156b
metabolism of, **7**:520a
solutions of, **15**:434a
Benzoin, **8**:32a; **14**:622a
distillation products, **3**:10b
Benzoline
isolation of, **5**:104a
Benzopyrone, **3**:347b
Benzoquinone
structure of, **5**:13a
Benzoyl series, **8**:59a
Benzyl, **14**:622a
Bérard, Étienne, **1**:616a
Bérard, Jacques Étienne (1789–1869), 1:616–617; **1**:416a; **5**:315a; **15**:347b
specific heat, **1**:348a; **3**:84a(n); **15**:95b, 358a, 363a, 378b
Bérard, Joseph Frédéric (1789–1828), 1:617; **11**:560a
Béraud, Père Laurent, **2**:334b
Béraut, Laurent, **3**:206a
Berberich, A., **7**:160a
Berberis
see Barberry
Bercé, Henri de, **3**:97b
Berek, Max, **5**:212b; **12**:195a, 304b
Berengario da Carpi, Giacomo (1460–1530), 1:617–621; **1**:47a; **11**:600a*
anatomy, **3**:40b; **5**:585b; **9**:165b, 166a, 469a; **14**:10a,b
Commentaria, **14**:8b
Isagogae, **2**:337a
psychology, **10**:21b
Beresovite, **7**:249a
Berg, Ernst Julius, **13**:24a
Berg, Lev Simonovich (1876–1950), 1:621–624; **1**:174b; **5**:467a; **7**:495b*, 512a*; **11**:108b*, 182b*; **12**:302a*; **13**:249b*, 413b*
evolution, **15**:507b
Berg, Otto, **10**:136b
Berg- und hüttenmännische Zeitung, mit besonderer Berücksichtigung der Mineralogie und Geologie
founding of, **6**:143a
Berga, Antonio, **1**:605a, 606b
Bergasse, Nicolas, **9**:327a
Bergedorf observatory, **12**:186a–b
Bergen, University of
geophysical institute, **2**:168b
Berger, Claude, **5**:352b; **8**:171a
Berger, Hans (1873–1941), 2:1–2
Berger, Johann Gottfried (1659–1736), 2:2–3
Bergerac, Cyrano de, **5**:59b
Bergeron, Tor, **11**:558a
Bergius, Friedrich (1884–1949), 2:3–4; **2**:324a; **9**:354a
Bergman, Torbern Olof (1735–1784), 2:4–8
acquaintances of, **5**:222b–223a
chemistry, **5**:90b; **8**:76b; **9**:534a; **14**:332b
affinity tables, **2**:7a
nomenclature, **5**:601b; **10**:446b*
on contemporary scientists, **5**:329a
correspondence of, **7**:388a; **8**:621b, 622a(n); **10**:147a; **11**:142a; **12**:64b
critics of, **11**:522b
crystallography, **11**:521b
influence of, **2**:92a; **5**:77a; **6**:181a; **7**:394b; **12**:67b(n)
influences on, **2**:421b
mineralogy, **2**:291b, 588a; **9**:5b, 559b
De niccolo, **3**:473b
Opuscules physiques et chymiques
translations of, **5**:601b
and Scheele, **12**:143b–149a *passim*
Sciagraphia regni mineralis
editions of, **7**:604a
translations of, **14**:464a
students of, **4**:346b; **5**:215b
translations of works, **1**:563b
Bergmann, Arno, **10**:347a; **11**:478b–479b
Bergmann, Carl Georg, **8**:270b; **11**:428a
Bergmann, Ernst von, **7**:425a, 426a,b; **8**:408a; **12**:142b
Bergmann, Margot, **4**:522b
Bergmann, Max (1886–1944), 15:15–16
Bergschrund, **7**:149a
Bergson, Henri Louis (1859–1941), 2:8–12; **1**:27a; **2**:544b; **3**:456a; **7**:71b, 483a, 485b; **8**:257a, 489b; **12**:609b; **15**:423a
critics of, **7**:115a; **15**:80a
Energie Spirituelle, **3**:90b
L'évolution créatrice, **13**:274b–275a
philosophy, **15**:423b
vital impulse, **5**:439a
Bergstrand, Osten, **8**:364a; **11**:542b
Beriberi, **5**:208b; **9**:81b; **13**:490b
cause of, **4**:310b–311b; **5**:541b; **10**:493a
and rice, **4**:311a–b; **5**:209a; **14**:392b–393a
treatment of, **5**:209a; **6**:111b
Berichte der Deutschen chemischen Gesellschaft
editors of, **13**:406b
Berichte über die Leistungen der niederen Thiere
editors of, **8**:271a
Bérigard, Claude Guillermet de (*ca.* 1578–1664), 2:12–14; **6**:448b
Bering, Vitus (1681–1741), 2:14–15
expeditions, **5**:428a,b; **13**:28a–b
Bering Sea, **4**:42b; **8**:418a
Bering Strait, **3**:397a
Beringer, Johann Bartholomaeus Adam (*ca.* 1667–1738), 2:15–16; **8**:4a
Beringer, Johann Ludwig, **2**:15a
Berkeley, George (1685–1753), 2:16–18; **3**:296a, 381b, 382a, 598b; **6**:86b, 555b, 556b, 557b; **7**:227a; **9**:463b, 465b
Analyst, **1**:531a; **2**:274b; **8**:610b
critics of, **11**:494a, 579b

Berkeley, George (cont.)
Essay Towards a New Theory of Vision, **9**:465a
mechanics, **10**:70a
on tar water, **3**:344b
works translated, **5**:567b
Berkeley, Miles Joseph (1803–1889), 2:18–19; **1**:493b; **3**:612a; **5**:192b; **6**:289a, 491a, 596a; **13**:489b
Berkhan, Gustav Waldemar, **9**:358b
Berlin
Anatomisch-zootomische Museum, **9**: 572a–b
Bauakademie
founding of, **4**:501b
Berlin Enlightenment, **6**:549a
Botanical Garden, **14**:386b; **15**:148a
Charité Hospital, **4**:296b, 352b
Ethnological Museum, **14**:42b
Health Department, **7**:423a, 425a–b
Institut für Infektionskrankheiten, **5**:220a
Institute for Plant Physiology, **5**:623b
Museum für Naturkunde
see Berlin, University of, Zoological Museum
Naturforscherversammlung, **5**:304b–305a
physics institutes, **4**:190a; **6**:341b–341a; **11**:8b
Physikalisch-technische Reichsanstalt, **6**:470a; **7**:528a–b; **10**:348b; **11**:8b; **14**:172a
founding of, **6**:243a–b
Preussische Akademie, **7**:597a
competitions of, **7**:597b
public health, **14**:41b
Royal Observatory, **4**:369b; **5**:181b–182a, 256b; **7**:598b
founding of, **7**:373b–374a
Berlin, University of
in nineteenth century, **9**:569a; **11**:592a
anatomy at, **15**:170a
chemistry at, **6**:15b, 461b; **15**:435b–436a, 441b
Faculty of Medicine, **9**:568b
founding of, **6**:549a; **14**:239b–240a; **15**: 206b
mathematics at, **4**:2a, 342b, 560b; **6**:472a; **7**:522a; **10**:24a; **11**:447b; **14**:221a–222a; **15**:755a
medicine at, **11**:363b, 593a; **12**:203a; **14**:126a–b
Museum für Meereskunde, **11**:439b
oceanography, **10**:503a
Pathological Institute, **5**:196a; **14**:41a
physics at, **5**:117a; **6**:243a; **9**:260b; **12**: 221b; **14**:171b–172a
physiology at, **4**:201a; **7**:504b
Zoological Museum, **4**:289a, 291a; **9**:432a
paleontology, **2**:110a
Berlin Academy
see Akademie der Wissenschaften (Berlin)
Berlin Geographical Society, **4**:193b
Berlin Medical Society, **12**:142b
Berliner Akademische Sternkarten, **1**:295b, 340a; **2**:100a, 236a, 332b, 441b; **4**: 251a, 369b; **5**:257a–b; **6**:112a; **10**: 207b, 599b; **13**:22b
Berliner astronomisches Jahrbuch, **2**:220b, 441b; **4**:370a; **5**:256b; **8**:557b; **9**:563b; **11**:114b, 115a; **12**:234b
founding of, **7**:598b
Berman, Harry, **8**:42a
Bermuda
natural history, **3**:129a; **14**:2a
surveying, **10**:151a
Bern, University of, **13**:556a
anatomy at, **13**:557b–558a
biochemistry at, **10**:22a
chemistry at, **7**:471a
physiology at, **7**:504b
Bernal, John Desmond (1901–1971), 15:16–20; **4**:521b, 522a
Bernal chart, **15**:17b
Bernard of Chartres (*d. ca.* 1130), 2:19–20; **1**:91b; **13**:339b
Bernard of Clairvaux, **1**:2a; **6**:396a; **13**:327a
Bernard of Le Treille (Trilia) (*ca.* 1240–1292), 2:20–21; **12**:62a
Bernard Silvestre (Bernardus Silvestris) (*fl.* 1150); 2:21–22; **1**:91b; **2**:19b
Cosmographia, **13**:340a
Bernard of Tours (Bernardus Turonensis)
see Bernard Silvestre
Bernard of Trevisan (*fl. ca.* 1378), 2:22–23
Bernard of Verdun (*fl. ca.* 1280), 2:23–24; **2**:20b; **14**:401a; **15**:35b
Bernard, Claude (1813–1878), 2:24–34; **1**: 479b; **2**:525a; **5**:83b, 566a; **8**:30a
acquaintances of, **1**:303a, 417b; **3**:588b; **4**:163a, 201b; **9**:546b; **11**:373b, 491b; **12**:270a
biochemistry, **1**:471a; **3**:416a; **9**:11a; **10**: 499b; **14**:430a
fermentation, **10**:379a–b, 381a
glucose metabolism, **3**:220a; **9**:506a; **15**:109b
glycogen, **6**:287a; **15**:109b
respiration, **6**:505a; **14**:175a
biographies of, **10**:372b
La chaleur animale, **2**:484a
critics of, **7**:546a; **10**:379a–b
curare, **14**:187b
diabetes, **1**:441b; **7**:519b; **14**:492b
experiments of, **15**:108a
influence of, **9**:101b, 505b; **11**:295b, 425b, 426a, 565a; **12**:470a; **15**:109a, 228b
influences on, **12**:137b
Introduction à l'étude de la médecine expérimentale, **10**:196a
forerunners of, **12**:309a
Leçons sur le diabète, **15**:109b–110a
and Magendie, **2**:25a–b; **9**:7b–8a, 8b, 9b, 10a
neurology, **2**:525b; **8**:16a; **9**:241b, 505b; **14**:143b
Horner-Bernard ocular syndrome, **2**: 30b
Phénomènes de la vie commune aux animaux et aux végétaux, **1**:537b
philosophy, **2**:122b–123a; **3**:376b, 379a; **5**:484b, 485a
on seeds, **1**:561a
students of, **2**:60a,b, 366a, 598a; **6**:434b, 476a; **7**:520a; **9**:422a, 505a; **11**:135b, 295a; **13**:416b; **15**:107b
see also *Milieu intérieur*
Bernard, Noël (1874–1911), 2:34–35; **14**: 552a*
Bernardston Series, **4**:361b
Bernardus Carnotensis
see Bernard of Chartres
Bernardus Sylvanus, **11**:203a; **14**:275b
Bernelinus, **5**:365a
Bernhardi, A., **10**:502a
Bernhardt, Adolph, **11**:215b
Bernheim, Hippolyte (1840–1919), 2:35–36; **5**:174a
Bernheim syndrome, **2**:35b
Bernier, François, **5**:289a(n); **8**:173b; **9**:497b
Bernoulli, Daniel (1700–1782), 2:36–46; **1**: 114a, 209a; **3**:459b; **5**:209b, 380a, 449a,b, 450a; **6**:241b; **7**:490a, 561b
algebra, **11**:611a
analysis, **4**:125b; **9**:471a
and Bernoulli (Johann I), **2**:37a
Bernoulli's principle, **2**:40a
critics of, **15**:288a, 300b
electricity, **3**:444a; **11**:142a
and Euler, **2**:36b, 38b–39a, 41a; **4**:468b–481a *passim*
Exercitationes quaedam mathematicae, **11**:400b
Hydrodynamica, **3**:73a; **6**:292a; **9**:119b
hydrodynamics, **1**:113b; **2**:54b
method of, **8**:602a
on oscillations, **2**:101b
probability, **1**:116b; **5**:98b; **8**:321a; **15**: 282b, 285a, 374a
students of, **7**:442a
tidal theory, **4**:591a; **8**:611a; **15**:288a, 296b
Bernoulli, Jakob (Jacques) I (1654–1705), 2:46–51; **2**:51b–53b *passim,* 101b; **12**:206a
analysis, **4**:516a; **5**:449b; **7**:52a, 567b; **8**: 163a, 164b, 166a, 304a
Ars conjectandi, **1**:531a; **2**:57a; **5**:299a, 449b; **9**:453a, 499b, 500a; **15**:280a
on Barrow, **1**:475a
and Bernoulli (Johann I), **2**:48a–b, 53a–b
Bernoulli equation, **2**:40a, 48a, 52b
Bernoulli numbers, **1**:54b; **2**:50a; **3**:302b; **4**:475b; **6**:572a; **11**:268a; **12**:291b
computation of, **1**:356b*
Bernoullian inequality, **2**:49b
Bernoullian polynomials, **12**:535a
Bernoulli's problem, **12**:118a
correspondence of, **8**:164a; **13**:480b
critics of, **6**:148b
dynamics, **8**:163b, 212b
editions of works, **3**:460a, 461b
geometry, **9**:55b; **14**:49b(n)
Opera omnia, **2**:57a
optics, **1**:475a
probability, **2**:319a; **3**:231a, 451b; **8**:306a; **9**:127b, 159a; **11**:599b; **15**:282a, 304b, 489a
students of, **2**:56b; **6**:304b
theorem of, **3**:230b
generalization of, **9**:249b*
Bernoulli, Jakob (Jacques) II (1759–1789), 2:51
Bernoulli, Johann (Jean) I (1667–1748), 2: 51–55; **5**:61a; **10**:67b
acquaintances of, **3**:281b, 459b
analysis, **2**:47a–48b *passim;* **4**:476b; **5**: 54b, 449b; **7**:52a, 560a; **8**:163a, 166a, 304b, 384a; **9**:50a; **11**:400a; **13**:266b
and Bernoulli (Daniel), **2**:37a
and Bernoulli (Jakob I), **2**:48a–b, 53a–b
and Bernoulli (Nikolaus II), **2**:57b
Bernoulli series, **2**:52b
and Mme. du Châtelet, **3**:215b, 216a
on contemporary scientists, **12**:118b
correspondence of, **1**:206b; **3**:119b, 460a; **5**:61b; **8**:164a, 305a; **9**:452b; **13**:480b, 585a,b; **14**:481a
and Crousaz, **3**:485b, 486a
editions of works, **3**:460a, 461b
and Euler, **4**:468a–470a *passim,* 477a, 479b, 483a*
geometry, **3**:182a, 282a, 616a(n); **11**:402b, 513a
problems, **6**:600b; **8**:304a; **10**:52b
Histoire des ouvrages des savants, **2**:48a
Hydraulica, **2**:37a, 40b
hydrodynamics, **2**:40a
influence of, **3**:62b, 281a; **11**:392a, 401b; **12**:118a
influences on, **13**:267a
and Leibniz, **2**:52a–54a *passim*
mechanics, **2**:49a; **3**:454a; **8**:163b
Opera omnia, **2**:56a
optics, **10**:89b(n)

phosphor, **6**:170a
students of, **3**:282b; **5**:380a; **6**:61b, 305a; **7**:442a
and Taylor, **2**:48b, 57b; **13**:265b, 266b, 267a
Bernoulli, Johann (Jean) II (1710–1790), 2:56; **3**:444a; **9**:188b
Bernoulli, Johann (Jean) III (1744–1807), 2:56; **2**:43b; **6**:403b; **7**:563a, 565b
Bernoulli, Nikolaus I (1687–1759), 2:56–57; **2**:49b–50a; **4**:516a; **5**:448b, 449b; **6**: 304b; **10**:67b; **11**:65b; **13**:67b
correspondence of, 3461b; **4**:475a, 483a*; **9**:499b, 500a
geometry, **13**:68b
theorem of, **4**:477b
Bernoulli, Nikolaus II (1695–1726), 2:57–58; **2**:36b; **4**:468b; **5**:449a
analysis, **11**:400b
and Bernoulli (Johann I), **2**:57b
Bernoulli family, **1**:111a, 113a; **2**:330a, 418a; **3**:282a, 458b; **4**:475b; **5**:96a; **6**:403b, 600b; **9**:186b
correspondence of, **11**:65b
critics of, **11**:494a
genealogical table, **2**:56a
mathematical estate of, **2**:56b
Bernstein, Benjamin Abram, **7**:623b
Bernstein, Felix (1878–1956), 2:58–59; **3**: 57a; **15**:21a
theorem of, **5**:201a
Bernstein, Julius (1839–1917), 15:20–22; **2**:58a
Bernstein, Sergey Natanovich (1880–1968), 15:22–24; **3**:230a,b, 231a; **9**:127a, 129b
Bernthsen, Heinrich August (1855–1931), 2:59
Berr, Henri, **11**:388b, 389a
Berriasian (stratig. stage), **6**:229a
Bert, Paul (1833–1886), 2:59–63; **1**:303a; **2**:27a, 366b; **5**:149a, 506b*; **7**:505a; **10**:386b–387b, 389a; **11**:425b
Bertagnini, Cesare, **3**:46a, 47a
Bertheim, Alfred, **4**:298b, 300a
Berthelot, Claude François, **8**:211a
Berthelot, Jacques Martin, **2**:63b; **4**:131a
Berthelot, Pierre Eugène Marcellin (1827–1907), 2:63–72; **9**:409a*, 619a*
acquaintances of, **1**:417a; **4**:20b; **6**:434b; **10**:440b; **11**:425b, 426a; **14**:265b
correspondence of, **11**:319a; **13**:454a
critics of, **4**:225b; **5**:621a; **14**:529b, 531a
and Duhem, **4**:225b–226a
first principle, **2**:69a
historiography, **5**:236b(n); **7**:39b, 40a–b, 41a, 464a,b; **8**:86b(n); **10**:208a; **11**: 319a; **14**:631b
influence of, **2**:624b
organic chemistry, **2**:621b; **3**:242b–243a; **4**:75b; **10**:440b; **11**:426a; **12**:47a; **13**:468a
physical chemistry, **7**:620a; **15**:434a, 461b
physiological chemistry, **2**:31a–b, 561b; **10**:379a,b
second principle, **2**:69a
students of, **12**:46b; **13**:416b
thermochemistry, **6**:244b; **8**:289b, 545b; **13**:359a; **15**:437a–438a, 441b
third principle (law of maximum work), **2**:69a; **6**:354b
Berthier, Pierre (1782–1861), 2:72; **4**:347b
Berthier de Sauvigny, Louis-Bénigne-François de, **2**:509b, 510a
Berthold of Reichenau, **6**:301a, 302a
Berthold, Arnold Adolphe (1803–1861), 2: 72–73
Berthollet, Amédée Barthélemy, **6**:285a
Berthollet, Claude Louis (1748–1822), 2:73–82; **1**:141b, 346a; **3**:240a, 347a*, 561a*; **4**:239b; **5**:601b; **6**:285a; **7**: 569b; **8**:79a, 87b*
acids, **5**:323b; **13**:309b
acquaintances of, **2**:91b, 133b, 186b; **3**:5a, 131b, 199b, 200a, 346a,b, 545b; **5**: 90a, 202b; **8**:612b; **9**:72b, 343a; **10**: 602b; **11**:540b; **12**:317b; **13**:571a; **14**:239b; **15**:346a, 361a
affinity, theory of, **2**:7b; **5**:431a; **8**:373b; **11**:169a, 541a; **13**:312b
ammonia, **8**:82a,b
analytical chemistry, **13**:310b
and Avogadro, **1**:347b, 348a,b
on bleaching, **3**:202a; **14**:197a
and Clément, **3**:316a
correspondence of, **10**:118b; **14**:197a
critics of, **13**:310b
dyeing, **3**:489a; **8**:621b
and education, **15**:342b, 347a
Élémens de l'art de la teinture, **3**:201a–b
Essai de statique chimique, **11**:169a, 437a; **15**:347b, 358b
on gases, **3**:541a
and Gay-Lussac, **5**:317b–324a *passim*
and Haüy, **6**:181b
influence of, **10**:546a; **13**:309a, 311b; **15**:136a, 358b, 361b
and Lavoisier, **8**:79b, 80b, 81a
method, **3**:200b
and Monge, **9**:472a,b, 473a, 477b
nomenclature, **6**:164b
organic chemistry, **7**:393a
physical chemistry, **15**:377a
and Proust, **2**:81b–82b, 93a; **11**:168a–171a *passim*
Recherches sur les lois de l'affinité
translations of, **11**:437a
students of, **1**:616a,b; **2**:72a; **4**:238b, 247a; **5**:558a
translations of works, **13**:548a*
Berthollide compounds, **7**:529b
Bertholon, Pierre (1741–1800), 2:82–83; **13**:183b
Berthoud, Ferdinand, **3**:198a; **10**:615a
Berti, Gasparo (*ca.* 1600–1643), 2:83–84; **9**:13b, 25a; **10**:332b; **13**:438b
Bertin, Henri-Léonard-Jean-Baptiste, **4**: 348a; **5**:578b; **9**:478b; **12**:65b, 66a
Bertin, Louis-Émile (1840–1924), 2:84–85; **2**:471a; **8**:387b
Bertini, Eugenio (1846–1933), 2:85; **8**:504b
Bertini transformations, **3**:468b
geometry, **12**:330b
Bertot, Henri, **9**:446b
Bertrand, Alexandre, **1**:194a; **2**:87b; **6**:307a
Bertrand, Charles-Eugène (1851–1917), 2: 85–86; **2**:500a; **8**:354a
Bertrand, Claude, **5**:602b
Bertrand, Gabriel (1867–1962), 2:86–87; **4**:485a; **8**:607b; **9**:328b; **12**:47b*
Bertrand, Joseph Louis François (1822–1900), 2:87–89; **2**:89a; **7**:572a*
acquaintances of, **1**:194a; **3**:59a, 559b; **4**:222b; **6**:307a,b; **8**:367b, 382b; **9**: 268a
Bertrand's paradox, **2**:88b; **7**:481b
calculus, **12**:319a
on contemporary scientists, **1**:449a; **3**: 132b, 133a,b, 163a, 214b*, 286a*; **4**:257b*; **5**:87b*, 265a*; **7**:602a; **8**: 64a*, 279a*, 385a; **11**:62b*; **12**: 306b8; **13**:424a*; **15**:31a, 485a–b
correspondence of, **5**:569b
critics of, **6**:250a
geometry, **8**:141b
influences on, **2**:288a
number theory, **3**:227b
postulate of, **3**:142b
students of, **4**:102b; **8**:305a,b
Traité d'arithmétique, **14**:85b
Bertrand, Marcel-Alexandre (1847–1907), 2:89–90; **1**:235b; **3**:160b, 161a; **8**: 288b, 463b; **12**:140a; **13**:146a
on contemporary scientists, **13**:148a–b
geology, **1**:472a; **6**:168b, 228a; **8**:543b; **13**:285a
and Termier, **13**:283b–284a, 284b
Bertrand, Paul, **1**:472b; **2**:86a
Bertrucius (Bertruccio), Nicolaus, **3**:218a; **9**:468a
Bérulle, Pierre (Cardinal) de, **4**:52a
al-Bērūnī
see al-Bīrūnī
Berwerth, Fritz, **5**:562b
Berwick, William Edward Hodgson (1888–1944), 2:90
Beryl
analysis of, **11**:437b
crystallization of, **6**:178a
Beryllia, **13**:597a
Beryllium
atomic weight of, **2**:428b; **9**:288b, 290b
bombardment of, **2**:338b; **7**:153a
discovery of, **7**:394b
isolation of, **14**:476a
production of, **3**:401a; **13**:71a
valence of, **9**:290a
wave mechanics of, **6**:615a
Beryllium oxide, **3**:401a
Berzelius, Jöns Jacob (1779–1848), 2:90–97; **2**:421b; **3**:243a; **4**:246b; **5**:216a, 411b
acquaintances of, **4**:242a*; **5**:223a, 430b; **7**:450b; **8**:519b; **9**:109b, 541b; **10**: 436a; **11**:270b, 539a, 540b; **14**:474b, 486b
alloys, **13**:245b
analytical chemistry, **3**:301a; **7**:394b; **8**: 331b; **11**:169a, 387b; **13**:310b; **14**: 475a–b, 478a; **15**:7b
elements, discovery of, **3**:473b; **6**:436b
water, composition of, **3**:601b; **4**:239b
atomic weights, **3**:47b; **4**:241a, 244a; **9**: 91b, 110a, 424a–b; **11**:173b; **13**:499b; **14**:489a
and Avogadro, **5**:371b
biochemical analysis, **1**:542b; **2**:385b
on catalysis, **2**:561a; **4**:241b; **8**:394b; **9**: 425a; **12**:47a; **15**:461b
chemical notation, **1**:345b–346a; **3**:49a, 289a; **5**:373a
chemical theories, **1**:346b, 348a; **3**:47a–b, 48a, 49a; **5**:320b; **8**:344a; **14**:395a
acid theory, **5**:125b, 493b; **8**:338a–340a, 342a
definite proportions, **1**:346a, 347a; **8**: 38a; **11**:170b–171a
dualistic theory, **1**:348a; **2**:199b; **3**:47b, 601b; **5**:125b, 370a–b, 381a; **6**:462a; **7**:451b, 452a; **8**:332b
organic structure, **1**:345b–346a; **2**:587b; **8**:333a–342b *passim;* **9**:558a,b; **14**: 477a
on contemporary scientists, **11**:437a; **13**:373a–b, 403b
correspondence of, **11**:172a, 593a; **14**: 474b–477a *passim*
critics of, **1**:70a; **4**:395a; **5**:104a; **8**:342a; **9**:286b; **10**:177b; **11**:465a, 537b–538a
on Cronstedt, **3**:474a
fermentation, **10**:361b; **13**:312a
geology, **2**:556b
influence of, **9**:558a; **10**:546a; **11**:172b, 379b, 541b; **12**:163b
isomerism, **8**:331a, 332a–b; **14**:474b

Berzelius, Jöns Jacob (cont.)
Jahresberichte
see *Jahresbericht über die Fortschritte der Chemie*
and Laurent, 8:55a, 57b
Lehrbuch der Chemie, 14:474b
and Liebig, 8:336a, 338a–344b *passim;* 14:478a
mineralogy, 6:18b; 9:620b
and Mitscherlich, 9:423b–425b
physiology, 14:113b
students of, 1:155b; 6:534a; 7:450b; 9:18b; 14:447a; 15:7b
translations of works, 9:558a; 11:270b, 465a*; 14:474b, 477b
Berzelius Museum, 13:243b
Besicovitch, Abram Samoilovitch, 2:239a
Besold, Christoph, 1:158b–159a
Bessarion, Iohannes, 1:274a; 4:117b; 11:348b–351b *passim;* 13:233b; 15:474b–475a, 476b
Bessel, Friedrich Wilhelm (1784–1846), 2:97–102; 6:103a; 11:378b*
acquaintances of, 2:616b; 4:400b; 5:301a, 309a; 7:50b, 51a, 52a; 10:25a, 199b; 11:377b, 544b; 12:226a,b; 13:109b
astronomy, 7:209b
instruments, 2:616a; 10:542a
observations, 1:242b; 2:468b; 3:92b
parallax determination, 3:404b; 5:144a; 6:263b
Bessel equations, 4:479a; 13:78a
Bessel functions, 2:41a, 101b; 4:103a*, 478a, 479a; 5:96b; 6:95b, 230b, 309a, 572a; 7:6b, 492b; 8:389a; 9:173b; 12:172b–173a; 14:158b, 317a; 15:385a, 488b
theory of, 14:188b
celestial mechanics, 1:429a
comets, 1:372a
on contemporary scientists, 13:110a
correspondence of, 5:303b, 308a; 7:494a,b; 10:199a; 12:234b
Fundamenta astronomiae, 8:369b
geodesy, 1:389a; 12:234b
hydrodynamics, 13:76a
influence of, 1:340a; 2:330a, 441b; 10:27a; 12:170a
mathematics, 7:494b
statistics, 11:237a
students of, 1:241a, 294a; 2:299a; 3:302a,b; 7:108b; 10:207a, 542b; 12:289b; 13:22b
Bessemer, Henry (1813–1898), 15:24–25; 4:532a
Bessemer process, 3:234a; 4:265b; 9:286b, 591b; 13:8a, 346a–b; 15:24b–25a
Bessey, Charles Edwin (1845–1915), 2:102–104
influence of, 15:224b
phylogeny, 3:318a
students of, 3:317b; 15:224a
Bessey, Ernst Athearn, 2:103b
Besson, Jacques, 8:211b, 212b; 10:281a
Best, Charles Herbert, 1:441b–442a, 442b, 443a; 3:352a; 8:614b; 15:106a
Bestelmeyer, Adolf, 11:13b
Bestiaries
Arabic, 3:548b–549a
Beta decay
discovery of, 7:158a
theory of, 10:424b–425a
Beta particles, 12:29a
absorption, 9:260b
biological effects, 9:550a
charge conjugation, 11:117b
counting of, 5:331b
emission of, 3:480a; 9:542b
energy of, 7:152b
magnetic deflection of, 1:559b; 6:16a
nature of, 9:260b
scattering of, 2:337b–338a; 12:31b; 13:368b–369a
spectrum of, 5:331b; 6:16a
theory of, 13:240b
total charge, 4:194b
Betancourt y Molina, Augustin de (1758–1824), 2:104; 6:2b*
Betatron, 3:330b*
radiation loss, 15:8b
Betelgeuse (star)
see Orion (constellation), alpha Orionis
Beth, Evert Willem, 5:351a
Bethe, Albrecht Theodor Julius, 11:275a
Bethe, Hans Albrecht, 2:242a; 4:579a; 5:273a, 463a; 6:615a; 12:529b, 530a
carbon cycle, 4:280a; 5:272b
Betti, Enrico (1823–1892), 2:104–106; 5:264a; 14:49b, 86b
acquaintances of, 1:599b; 7:506b; 11:448a
analysis, 4:103a
Betti numbers, 2:105b; 3:95b; 11:57a, 60a
invariance of, 2:513a
electrodynamics, 10:27b
engineering, 3:118a
influence of, 2:470a
students of, 2:121a; 4:102b, 373b; 10:610a; 11:407a; 14:86a
theorem of, 2:105b
topology, 11:449b
Betula
see Birch tree
Betulaceae
flower clusters, 2:219a
Beudant, François-Sulpice (1787–1850), 2:106; 6:179a, 182a; 8:288b; 12:304a; 15:114b
law of, 2:106b
Bevatron, 8:96a
Beverley, Robert
History and Present State of Virginia, 1:432a
Bevis, John, 14:194a
Bewick, Thomas, 1:331a; 2:375a; 6:577a; 13:463a
Bewley, George, 3:537b, 538a
Bexon, Gabriel-Léopold-Charles-Amé (1747–1784), 2:106–108
Histoire des oiseaux, 2:474b
Bexon, Scipion, 2:107a
Beyer, Johann Hartmann, 4:511b
Beyer, Max Philipp Johannes, 5:490a
Beyrich, Heinrich Ernst (1815–1896), 2:108–110; 8:511b; 9:431b; 11:539b
on contemporary scientists, 11:501b
Beyschlag, Franz Heinrich August, 14:59a
Bezoar stone, 1:523b; 8:484a
Bezold, Albert von (1836–1868), 2:110–111; 4:371b; 7:519b; 15:170a
Bezold-Jarisch reflex, 2:111a–b
Bezold's ganglia, 2:111a
Bezold, Johannes Friedrich Wilhelm von, 11:524b; 12:595a
Bezout, Étienne (1739–1783), 2:111–114; 2:335a; 3:383b; 4:107b; 6:178a; 7:567a; 9:471b; 13:571a; 15:302a, 310a, 317a
algebra, 3:284b
Bezoutiant, 2:112b–113a
theorem of, 2:113a, 444a; 12:228a,b, 331a; 15:481a
works translated, 4:546b
Bhabha, Homi Jehangir (1909–1966), 15:25–28; 4:579a
Bhabha scattering, 15:26b
Bhāskara I (*fl.* 629), 2:114–115; 15:568a, 570a, 582a, 597b, 598a, 607b
Bhāṣya, 15:593a
commentaries of, 1:308a
commentaries on, 10:313b*
Laghubhāskarīya, 15:595b
Mahābhāskarīya, 15:565a, 593a–596a, 598b, 605b
Bhāskara II (*b.* 1115), 2:115–120; 1:309b; 5:159a; 15:625b
Brahmatulya, 4:100a
commentaries of, 7:582b
commentaries on, 9:580a
Karaṇakutūhala, 15:586b–587b
Līlāvatī, 5:275b; 9:613b; 11:290a
Siddhāntaśiromaṇi, 15:583b, 585a–587a, 614b
sine table, 15:585b, 587a
Bhaveśa
commentaries of, 2:116a
Bhoga (portion), 15:570a
Bhojarāja, 3:584b; 15:586a, 598a
Rājamṛgāṅka, 15:583b–584a, 584b
Bhūtiviṣṇu
commentaries of, 1:308b
Biaggio Pelicani
see Blasius of Parma
Białobrzeski, Czesław (1878–1953), 2:120–121
Bianchi, Giovanni Battista, 1:318b; 3:489b
Bianchi, Leonardo, 5:27b
Bianchi, Luigi (1856–1928), 2:121; 2:105b, 191b; 3:323a; 4:103a,b*, 336a, 373b; 5:200b, 201a; 10:545a; 11:408b
Lezioni di geometria differenziale, 4:336b
surfaces of, 4:337b*
Bianchini, Francesco, 9:78a, 135a, 486b; 12:162a
Bianchini, Giovanni, 4:111b; 11:349b; 15:155a, 473b, 474b, 475a
Bible
and Aristotelianism, 5:549b
astronomical records in, 9:184b
commentaries on, 5:549a,b; 13:340a, 410b; 14:29a, 226b
contradictions in, 5:119b, 120b
dating of, 1:323a–b
editions of, 12:323a
and geology, 3:526a–b, 527b; 4:28a; 5:32a; 6:580a,b, 584b; 7:389a; 12:262b, 278a, 433b; 14:296a–b
interpretation of
ancient Roman, 3:109b
Renaissance, 1:587b–589b; 12:323a
seventeenth-century, 2:613a–b; 4:550a–551b; 5:242a,b, 286b, 287a
eighteenth-century, 10:81b–82a; 13:180a
allegorical, 2:613a–b; 5:549b
literalist, 1:333b, 336a–b
numerological, 4:550a–551b; 13:58b–59a, 13:60b
theory of, 6:448a
as literature, 6:596a,b
and science, 4:129b; 5:47b, 115b; 6:72a, 580b; 11:317b; 15:53b, 634a–b
hexamera, 1:565a
translations of, 5:606b
Bibliographia genetica, 8:513a
Bibliography
in Renaissance, 5:379a
alchemical, 10:82a–b
astronomical, 6:471a; 7:580b
botanical, 6:66a, 122b; 8:375a; 13:342b
chemical, 6:375a, 525b; 11:551b
critical, 12:112b–113a
geological, 2:341b; 8:370b; 9:103b, 104a; 10:128b; 15:428a

historiography of science, **9**:378a
mathematical, **4**:36b; **6**:76b*; **7**:317a; **8**: 505a; **12**:532b; **15**:256b
medical, **2**:535b; **6**:66b, 233a*; **15**:237a
optical, **11**:21b
ornithological, **3**:439a; **5**:545b*
paleontological, **1**:494b; **15**:54b
physiological, **11**:139a
scientific, **4**:559b; **6**:66a–b; **11**:50b; **12**: 264a; **15**:206a
zoological, **15**:78b
Biblioteka Matematyczno-Fizyczna, **4**:84a
Bibliotheca sacra, **14**:517a
Bibliothèque britannique
editors of, **8**:38a
founding of, **10**:603a
Bibliothèque italique, **15**:53a
Bibliothèque universelle de Genève
editors of, **8**:35b–36a, 38a
founding of, **10**:603a
Bibracte (city), **3**:622b
Bibron, Gabriel, **15**:126b
Bichat, Ernest, **2**:203a
Bichat, Jean-Baptiste, **2**:122a
Bichat, Marie-François-Xavier (1771–1802), 2:122–123
acquaintances of, **7**:556b; **15**:126a
anatomy, **5**:514b; **6**:269b; **9**:64b
critics of, **7**:590a; **9**:7a–b
editions of works, **9**:7b
influence of, **9**:505b; **14**:456b
influences on, **1**:479b; **10**:612b
neurology, **9**:506a
pathology, **2**:445a; **8**:352b; **9**:512a
physiology, **2**:187a; **5**:253a; **6**:65a; **9**:8a; **11**:592b
Bickerton, Alexander William (1842–1929), 2:123; **12**:25b, 26a
Bicquilley, Charles François de, **7**:549b
Bidder, Friedrich Heinrich (1810–1894), 2: 123–125; **13**:556b
Bidder's ganglia, **2**:124b
Bidder's organ, **1**:545b; **2**:125a
physiology, **2**:162a; **10**:568a; **14**:63b, 113b; **15**:109a
Biddulphia, **9**:587b
Bidloo, Govard (1649–1713), 15:28–30; **6**: 231a; **15**:4a
critics of, **12**:41a
Bieberbach, Ludwig, **6**:394a
Bieberbach hypothesis, **8**:457a
Biedermann, Benedikt, **14**:226a
Biedl, Artur
Innere Sekretion, **2**:73a
The Internal Secretory Organs, **1**:441b
Biela, Wilhelm von (1782–1856), 2:125–126
comet of, **2**:126a; **5**:258a; **6**:263b; **10**:42a; **12**:160a, 161a
Bielid (or Belid) meteors, **2**:126a; **10**:42a
Bienaymé, Irénée-Jules (1796–1878), 15:30–33; **3**:142a, 230b; **9**:126a, 127b
Bienaymé-Chebyshev inequality, **15**:32a
Bienewitz, Peter
see Apian, Peter
Biermer, Anton, **1**:60b; **10**:17b
Biesterfeld, J. H.
see Bisterfeld, Johann Heinrich
Big Bone Lick (Kentucky), **1**:487b; **14**:456b
Big Dipper (constellation), **15**:718a,b
Bigelow, Frank Hagar, **6**:29b; **9**:107b
Biggs, Charles Henry Walker, **6**:211b
Biggs, Hermann, **7**:429b
Bignami, Amico, **5**:503a,b
Bignon, Abbé Jean-Paul, **4**:214b; **5**:58b, 60b; **6**:477b; **7**:558b; **10**:568b, 569a; **11**: 331b; **13**:442b
Bigollo, Leonardo
see Fibonacci, Leonardo
Bigourdan, Camille Guillaume (1851–1932), 2:126–127; **4**:15a, 25b*; **7**:77b*, 78a*, 126a(n); **9**:251b*; **10**:615b; **13**:37b*; **14**:254a, 480b*; **15**:402a*
critics of, **15**:276b
Bīja (correction), **15**:567b, 581a–582a, 583a–b, 584b, 596b–603b *passim,* 609a–b, 614b, 617b, 618a
Bijl, A., **8**:481b
Bijvoet, Johannes Martin, **3**:430a
Bikini (atoll), **3**:568b
Bildungstrieb, **2**:204a
Bile, **5**:234a
acids of, **3**:341b; **11**:129a
structure of, **14**:334b–335a, 444b
analysis of, **13**:309b, 403b
black, **5**:230b; **9**:468b
composition of, **15**:108b, 110a
detection of, **10**:557a
and fat metabolism, **2**:124a–b; **13**:403b
pigments of
tests for, **5**:431b
proteolytic properties of, **2**:28a
role of, **6**:255a–b; **12**:244a; **13**:122a
as a secretion, **2**:124a, 237b, 335b; **15**: 108b
Bile salts, **13**:263a
Bilfinger, George Bernhard, **2**:42b; **5**:428a, 449a
Bilharz, Theodor (1825–1862), 2:127–128; **4**:198a; **12**:421b
Bilharzia, **2**:127a,b, 534a; **5**:502b
Bilin spa, **11**:387a–b
Bilirubin, **15**:157a,b
in urine, **4**:296b–297a
Billet, Félix
Traité d'optique, **3**:419b
Billiards
mathematical theory of, **3**:418b*
Billich, Anton Günther, **12**:78b, 312b
Billing, Johan Samuel, **11**:380a
Billings, Elkanah (1820–1876), 2:128–129; **4**:364b; **8**:461b
Billings, John Shaw, **1**:610a; **9**:422b; **11**:346a; **14**:249b
Billings, Joseph, **12**:114b
Billingsley, Sir Henry, **4**:5b, 450a
Billion, **3**:273a; **5**:409a
Billroth, Christian Albert Theodor (1829–1894), 2:129–131; **2**:445b; **6**:434b; **9**:259a; **12**:203a; **14**:113b
on art, **2**:531b
bacteriology, **3**:338b–339a; **5**:564a; **7**: 422b
critics of, **12**:87b
Billwiller, John, **2**:3b
Billwiller, Robert, Jr., **9**:189b
Billy, Edouard-Louis-Daniel de, **4**:349b
Billy, Jacques de (1602–1679), 2:131
Diophantus, **5**:528a
Doctrinae analyticae inventum novum, **4**:574b
editions of works, **13**:253b
Bils, Louis de, **6**:508b; **12**:40a; **13**:31b
Binder, Nicolaus, **7**:534a
Binding energy, atomic, **1**:321b; **9**:38a
and frequency, **2**:241b, 242b
Binet, Alfred (1857–1911), 2:131–132; **7**:67b
tests of, **5**:377b
Binet, Jacques Philippe Marie, **3**:142b, 418b*; **8**:382a; **10**:247b; **11**:206b; **12**:172a
Binney, Amos, **5**:478a; **6**:56b; **9**:536b
Binney, William Greene, **5**:478b; **13**:65b
Binomial theorem, **1**:15a; **2**:50a, 577a; **5**: 298b, 521a; **7**:162b–163a, 326a, 560a; **9**:453a–b; **10**:48a; **14**:149a; **15**:99a
discovery of, **10**:45b–47a
Binomials
development of, **7**:243a
Binswanger, Otto, **2**:1a
Binz, Carl, **1**:575a
Bioassay, **1**:11b; **5**:10b
quantal-response, **14**:437b
Biochemica et biophysica acta, **14**:524a
Biochemical Journal, **6**:111b
Biochemistry
early eighteenth-century, **2**:227a; **5**:226a–b; **9**:302b
late eighteenth-century, **6**:283a–b; **8**:83b–84a; **12**:145b
early nineteenth-century, **2**:92b; **5**:431b; **11**:172a–173a; **12**:124a; **13**:311b–312a, 403b; **14**:199b–200a, 477a, 492a–b
mid nineteenth-century, **1**:471a; **2**:65a, 67a, 531b, 587a–b; **4**:76a, 162b, 554b, 616a; **5**:126b, 432a–b; **6**:269b; **7**:421a, 519a–520b; **8**:272a, 342b–345b, 348a–349a; **9**:18b, 91b, 259b, 347b–348b, 456b–457a, 558a–b; **10**:557a–b; **11**:51b, 133a, 492a, 541b, 565a–b; **12**:59a, 270a, 319b; **13**:452a–453a, 557a; **14**:63a–66b, 233a–b, 454b; **15**:11b, 109b–110a
late nineteenth-century, **1**:543a–b; **2**:86b–87a, 316b, 366a,b, 585a; **3**:256b–257a, 341a–b; **4**:210b–212a, 296b–301b; **5**:148b–149b, 315b; **6**: 505a–506b; **7**:113b, 466b–467b; **8**: 28b; **9**:380b–381b, 506a; **10**:22a–23a, 281b–282b, 493a; **11**:129a, 135b; **12**:95a, 452a; **13**:243a; **15**:13a
early twentieth-century, **1**:9b–11b, 301b, 361a–363a, 453a–454b; **2**:339a, 560b–562b; **3**:257a–b, 271b, 279b, 290b, 347b, 351b–353a, 359a–b; **4**: 359b–360b, 485a–486a, 603a–604a, 618a; **5**:6a–b, 37b–38a, 53a–b, 208b–209a, 438b, 451b–452a; **6**:21a, 110b–111b, 135b–136a, 260a–262a, 498a–502a; **7**:405b–407a; **8**:275a–276b, 446a, 583b–584a, 614a–615a; **9**:95b–96a, 183a, 284a–285b, 359a–b, 394a, 461a–b; **10**:243a–b, 326b; **11**:102a, 179a–180a, 576b–577b; **12**:546a–b; **13**:152a–b, 228a–b, 262a–263a, 393a–394a, 486a–488a, 574a–575a; **14**:45b–46a, 173a–176a, 247b, 255a–b, 334b–335a, 444a–446a, 523b–524a; **15**:157a–158a, 201b, 252b, 257a–b, 258a, 417a–418b
mid twentieth-century, **3**:335b–336a, 415b–416b; **4**:357b–358a; **5**:42a–b, 118a, 273a, 589a–b; **6**:22a, 366b; **7**: 273a–274a, 517b–518b, 519a; **8**: 607b–608a; **12**:207a–208a; **13**:152b, 227b, 418b–421b; **14**:174a–177a, 412a–b, 524a–b; **15**:15b, 257b, 258b, 418b–420a
bacterial, **6**:500b–501a, 501b
emergence of, **1**:536b; **3**:257b
evolutionary, **10**:22b
industrial, **1**:361a, 362a, 363a
philosophy of, **6**:498a
physical, **7**:274a; **13**:418b–421b; **15**:434a–b
techniques, **13**:487a–488a; **14**:523b
Bioclimatology, **4**:171b, 285b
Biocoenosis, **5**:70b; **9**:432a
Bioenergetics, **11**:585a–586a
Bioethics, **14**:42a
Biogenesis, law of, **1**:601a; **3**:434b; **5**:447a; **6**:8b, 9a, 435b–436a; **10**:523a; **12**: 337b
Biogeochemistry, **13**:619b

Biogeography
eighteenth-century, **10**:283b–284a
nineteenth-century, **3**:552b; **5**:66b–68a; **8**:569a, 574b; **9**:374b–375a; **11**:386a; **12**:38b; **13**:148a; **15**:83a
twentieth-century, **1**:623b; **11**:309b
and geology, **5**:67a–b
theories of, **5**:67b; **12**:328b
see also Ecology, Phytogeography, Zoogeography
Biographie universelle (Michaud), **3**:525b
Biography
see Historiography, biographical
Bioherms, **1**:163a
Biokhimiia, **1**:361b
Biologia Centrali-Americana, **1**:502a; **6**:507a
Biological Abstracts, **4**:220b
founding of, **7**:447b; **8**:589b
Biological control, **5**:70b; **6**:524b, 525a; **8**: 450a; **9**:333a
Biological field, theory of, **5**:594b
Biological Society of Washington (D.C.), **3**:536a
Biologicheskii zhurnal
founding of, **7**:456a
Biologische Untersuchungen, **11**:382a
Biology
ancient Greek, **1**:251a, 258b–267a; **4**:34a
ancient Roman, **3**:175b
medieval Arabic, **11**:232b
medieval Hebrew, **2**:221b
medieval Latin, **1**:199b
seventeenth-century, **2**:591a–b; **9**:485b; **13**:169a–174b
eighteenth-century, **9**:187a, 188a–b; **10**: 9b–10b; **15**:52b, 524b–525b
nineteenth-century, **3**:120a–b, 192a–b; **4**:12a–b; **5**:385a–386a; **8**:124b–125a; **11**:109b–110b
twentieth-century, **3**:389a–390b; **6**:319b–320a
applied, **4**:545b; **7**:447b, 456a; **10**:352a
and atomism, **1**:259a; **5**:286b
chemical theory of, **4**:296b, 297a, 301b
compensation, principle of, **5**:443a–b
complementarity in, **2**:253b
and cosmology, **1**:259a–b
experimental approach, **2**:286a–287a; **3**: 390a; **5**:594b; **7**:454b; **9**:485b, 516b–517a; **12**:103b–104a; **14**:423a, 434b; **15**:479b
and geometry, **1**:256b; **5**:470b
laws of, **15**:509b–510a
marine
early eighteenth-century, **9**:135b; **10**: 568b–569b
late eighteenth-century, **9**:575b; **11**:67a; **12**:560b–561a
mid nineteenth-century, **1**:74a,b; **2**: 61a–b; **12**:193a; **14**:57b
late nineteenth-century, **1**:405a; **7**:401b; **9**:78b, 536b, 587b, 588b–589b; **11**: 176a; **12**:45b, 545a; **13**:360a–b; **14**: 291b
twentieth-century, **1**:92b, 534a; **3**:17a; **6**:442a; **7**:410b, 447a–b, 535a; **12**: 190b–191a; **13**:281b–282a
and mathematics, **13**:498a–b
and medicine, **3**:515a; **10**:195b
molecular, **1**:319a–b; **2**:400a; **5**:140a–141b, 594b; **6**:133b; **10**:326b; **12**: 223a; **13**:3a–b; **15**:63b
origins of, **1**:342b–343a; **15**:18a–b
naming of, **2**:123a; **7**:589a; **11**:491b
periodicity in, **5**:172b
philosophy of
seventeenth-century, **6**:256b; **7**:612a–b; **8**:127a; **9**:52b
eighteenth-century, **2**:286b–287a; **5**: 443a–444a; **6**:469a–b; **7**:232b–233a, 367b–368a; **8**:579b–580b; **9**:458a, 576a; **11**:363a
early nineteenth-century, **4**:264a–265a; **7**:585b–593a *passim;* **9**:7b; **10**:194b–196a; **13**:460b–461b
mid nineteenth-century, **3**:377b–378a, 453a–b; **4**:235b–236a; **5**:505a; **8**: 514a–515b; **10**:210a, 364b; **11**:491b–492a, 567b; **12**:242a–243a
late nineteenth-century, **5**:385b; **6**:224b; **8**:125a, 445a–446a; **9**:516b, 518a; **10**:431b; **11**:574a; **12**:567b; **13**:417a; **14**:425a–426a, 434b–435a
twentieth-century, **3**:389b, 390a; **4**: 165b; **5**:222a*; **6**:224a, 260b–262a; **7**:94b, 99a–b; **8**:446b; **9**:494b–495a, 525a; **12**:77a, 470a; **13**:241b, 269a, 353a; **14**:139b; **15**:509b–510a
physicochemical methods, **2**:184b; **7**: 455a–456a; **14**:434b–435a
and religion, **7**:99b; **10**:9b–10a, 11b, 210a, 371b
social implications of, **4**:271b–272b
structure vs. function, **4**:235b; **14**:596b
survey techniques, **7**:447a
see also Anatomy; Botany; Historiography, of biology; Life; Microbiology; Morphology; specific fields and organisms; Physiology; Taxonomy; Zoology
Bioluminescence
see Luminescence
Biomechanics, **4**:12a
Biometer, **13**:262b
Biometrics
eighteenth-century, **6**:65b
early nineteenth-century, **3**:44b
mid nineteenth-century, **5**:266a; **9**:281a
late nineteenth-century, **3**:590a–b; **10**: 449a–454b, 461a–463a; **14**:251a–252a
early twentieth-century, **3**:113a–b; **5**:7b, 8a–11a, 70b; **7**:239b; **8**:512a; **10**: 444b–445a, 462b–465a; **12**:190b–191a; **14**:87a, 574a
mid twentieth-century, **6**:22a, 22b–23a
and genetics, **3**:259b; **5**:10b; **8**:3b
naming of, **10**:462b
opposition to, **1**:506a; **14**:252a
see also Demography; Statistics, in biology
Biometrika
founding of, **10**:462b–463a, 464a; **14**:251b
Bion, Nicolas (*ca.* 1652–1733), 2:132–133; **8**:21b
translations of works, **4**:166a–b
Biondini, Leonardo, **10**:598b
Bionomics, **2**:583a; **5**:486b
Biophysics
mid nineteenth-century, **4**:203a, 264b, 614b–616b; **13**:557a
late nineteenth-century, **1**:10b, 303a–304b; **4**:401b–402a; **11**:585a–586a
early twentieth-century, **2**:325b; **4**:195a; **6**:24b; **7**:501b–503b; **8**:102a–b; **9**: 295b
mid twentieth-century, **2**:601a; **3**:430a; **7**:406b–407a; **8**:96a; **10**:236b
molecular, **4**:615a
see also Instrumentation, scientific, biophysical
Biopoesis, **15**:18a
Biopsy, **6**:101b
Biorhythms, **14**:44b
Biosphere, **4**:145b; **13**:275a, 618b, 619b
concept of, **13**:145b
Biostatistics
see Biometrics
Biostratigraphy, **3**:520a–b; **4**:68a–b; **10**: 211b–212a, 222a–b; **11**:386a; **12**: 182a–b
see also Stratigraphy
Biot, Jean-Baptiste (1774–1862), 2:133–140; **2**:287b; **3**:446b*; **5**:97b, 111a; **7**: 572a*; **10**:177a; **15**:484a
acoustics, **3**:258b; **15**:356b
speed of sound, **15**:357a–b
acquaintances of, **1**:201a, 557b; **2**:88a, 473a; **3**:521a; **4**:65b; **6**:18b, 325a; **7**: 615a; **8**:382a; **11**:540b; **12**:288a, 523a; **13**:310b–311a, 375a; **15**:126a
Biot-Savart law, **1**:145b, 146a; **2**:137a; **6**:250a; **12**:129a
on contemporary scientists, **3**:132b, 133a,b, 327a*, 447a*; **7**:572a*; **10**: 546a*; **14**:80b; **15**:347a–b, 384a, 401b*
crystallography, **10**:357b
geophysics, **1**:200b; **5**:78b, 318a
heat theory, **5**:95b, 96b
historiography, **5**:110b; **9**:74b*
influence of, **3**:445b–446a; **6**:241b; **10**: 186b; **12**:289a; **14**:72b
influences on, **14**:491a; **15**:361a–b
instrumentation, **15**:392b*
law of rotatory dispersion, **2**:137b–138b, 139b
law of solar radiation, **2**:136b
mathematics, **15**:346a, 347a
optics, **1**:201b, 348b; **5**:167a, 295a; **9**:74a; **12**:281b, 304b; **14**:567a; **15**:361a–b, 377a
organic chemistry, **10**:357a,b, 358a, 359a, 532a
and Pasteur, **10**:352b, 357a, 361a,b, 366b
students of, **11**:111a
Traité de physique expérimentale et mathématique, **3**:83b(n); **15**:377a
translations of works, **4**:546b; **6**:277b
Biotin, **4**:358a; **15**:257b
Biotite, **2**:139a; **7**:161a
Biphenyl
discovery of, **5**:13a
Bipinnaria, **12**:107a
Birch, Thomas, **5**:535a; **14**:193a
History of the Royal Society, **12**:583b
Birch tree, **6**:221b; **10**:1a,b; **14**:413a
Birckenbach, Lothar, **5**:595b
Bird, John (1709–1776), 2:140–141; **2**:388b; **4**:131b; **5**:491b, 492a; **6**:511b, 512a; **8**:178b; **9**:156b, 231b, 234a
Bird of paradise, **8**:265b
Birds, **5**:358b
anatomy, **3**:342b–343a; **8**:204b, 205a; **12**:224b, 298a; **13**:402b
cerebral, **9**:253a
ear, **5**:267a; **13**:461b
eyes, **3**:207a
kidneys, **5**:267a
nervous system, **13**:461b
respiratory system, **11**:380a; **12**:224a
skeletal system, **2**:383b
stomach, **14**:331a
banding of, **1**:331b
bathing habits, **1**:265b
body temperature of, **4**:285b
bones of
air spaces in, **3**:38a
breeding of, **1**:405b
digestion, **11**:334a
ecology of, **1**:405b; **5**:70b
embryogeny, **6**:573b; **11**:307b
endocrinology, **10**:570a–b; **15**:473a
flight of, **7**:503b

flightless, **2**:321a; **10**:262a
fossil, **5**:611a–b; **10**:16b, 40b, 262a
geographical distribution of, **8**:519a; **14**: 138b
hybridization, **5**:359b
identification of
artificial key, **3**:438b
migration of, **1**:331b, 405b; **3**:129b, 350a; **7**:97b; **9**:441a; **12**:562a
oil spills, death from, **11**:102a
parasites of, **5**:503a,b
phylogeny, **4**:266b; **9**:134b
psychology of, **6**:229b
reproduction, **4**:509a–510a
spermatogenesis, **5**:598b
sanctuaries, **14**:187b
sexuality of, **14**:595b
songbirds
taxonomy of, **9**:573a
sounds, **6**:229b; **10**:164a
stomach contents of, **5**:69b–70a
taxonomy of, **1**:331b; **2**:474a; **3**:438b; **5**: 406b, 480b; **6**:594b; **11**:316b, 443b
toothed, **9**:134b
see also Flight; Ornithology; specific names
Birefringence
see Refraction, double
Birge, Edward Asahel (1851–1950), 2:141–142; **7**:184a
Biringuccio, Vannoccio (1480-*ca.* 1539), 2: 142–143; **5**:420a; **8**:208b
metallurgy, **1**:448b
Pirotechnia, **10**:515a
Birkeland, Olaf Kristian, **3**:479b; **12**:451b; **13**:83a; **14**:32a
Birkhoff, George David (1884–1944), 2:143–146; **2**:198b; **4**:618b; **7**:351b; **9**:502a; **10**:479a; **14**:90b
algebra, **11**:56a, 60b
dynamics, **13**:36a
topology, **12**:411a
Birmingham, John (*ca.* 1816–1884), 2:146–147
Birmingham, University of
founding of, **8**:444b
Birnie, Frederick, **3**:487b
Birt, William Radcliff (1804–1881), 2:147
al-Bīrūnī, Abū Rayḥān Muḥammad ibn Aḥmad (*b.* 973, *d.* after 1050), 2:147–158; **1**:309b; **10**:425b; **12**:447b; **13**: 150a*; **15**:611a
and Abū Naṣr Manṣūr, **9**:83b–85a
algebra, **7**:258b
on Archimedes, **1**:226a
astronomy, **1**:510b, 511a; **5**:613a; **7**:337a, 338a, 353b; **9**:84a–b; **13**:538b–539a, 590b
on al-Battānī, **1**:512b
On the Calculation of Chords in Circles, **4**:543a–b
Chronology, **1**:35a, 39a; **7**:222b; **12**:431b
on contemporary scientists, **1**:42b; **11**: 240a,b
correspondence of, **12**:431b
errors of, **1**:510a; **10**:419b
geography, **1**:29a; **9**:36b
geometry, **13**:289b
historiography, **15**:420b
India, **7**:363b(n); **13**:582b
influence of, **11**:231a, 248b, 251b
influences on, **1**:308a; **2**:418a*; **9**:580b; **10**:5b, 6b; **15**:598b, 608a
on al-Khāzin, **7**:334b, 335a, 338b
Kitāb fī istīʿāb, **12**:432a
Kitāb al-jamāhir fī maʿrifat al-jawāhir, **7**:342a; **13**:511a–b
optics, **6**:195b
on proportions, **7**:342b
al-Qānūn al-Masʿūdī, **7**:256a
Rasāʾil, **9**:160b
on al-Rāzī, **11**:326a
On Shadows, **14**:546b
statics, **7**:336b, 341a–b, 342b, 343a, 346a
Taʾamē lūḥōt al-Chowārezmī, **4**:502b
translations by, **14**:28b; **15**:626a
trigonometry, **13**:510b; **14**:577a
works translated, **4**:502b
Biscay, Bay of, **7**:92a
Bischof, Carl Gustav Christoph (1792–1870), 2:158–159; **6**:395b*; **14**:625a; **15**:457a
Lehrbuch der chemischen und physikalischen Geologie, **2**:440b; **10**:171a
Bischof, Georg, **11**:278a
Bischoff, Ernst Christian Heinrich, **2**:160b
Bischoff, Gottlieb Wilhelm (1797–1854), 2: 159–160; **6**:337b; **8**:349a; **11**:437a
Bischoff, Johann, **2**:430a
Bischoff, Theodor Ludwig Wilhelm (1807–1882), 2:160–162; **8**:270a; **12**:421b; **13**:404a,b*; **14**:63a
embryology, **1**:477a
physiology, **3**:245b; **14**:63b, 64b
Bishop, George, **6**:402b
Astronomical Observations at South Villa, **3**:605b
Bishop, George Holman, **4**:398b
Bismarck, (Prince) Otto Eduard Leopold von, **5**:254b; **10**:28b, 559b; **14**:41b
Bismarck brown (dyestuff), **3**:85a
Bismuth, **5**:595b
alloys, **3**:560b; **8**:321a; **9**:106b; **11**:534b
analysis of, **5**:100a
in animal body, **6**:366a
compounds of, **9**:557a; **11**:109a
diamagnetism, **14**:208a
electrical conduction in, **10**:500a
isotopes, **10**:288b
medical use of, **8**:273b
oxides, **11**:169b
physical properties, **11**:460b
spectrum of, **1**:371a
Bison, **7**:33a; **10**:16b
Bisterfeld, Johann Heinrich (*ca.* 1605–1655), 2:163–164
Biston, **6**:22a
Bistrzycki, Augustyn, **7**:471a; **14**:247a
Bitangents, **3**:167b
Ibn al-Biṭrīq
see Yaḥyā ibn al-Biṭrīq
al-Biṭrūjī al-Ishbīlī, Abū Isḥāq (Alpetragius) (*fl. ca.* 1190), 15:33–36; **1**:380b; **2**: 24a; **14**:593b
cosmology, **1**:101a; **12**:3a
critics of, **8**:280b
influence of, **5**:402a; **13**:534b; **14**:35b, 388a,b
influences on, **7**:38b; **13**:488b–489a; **14**: 593a
De motibus caelorum, **12**:2a
translations of works, **9**:361a, 362a–b; **13**:401b
Bitter almonds
see Almonds
Bitumen, **7**:44a; **15**:648a
distillation of, **10**:498a
medical use of, **1**:596a–b
Biuret reaction, **12**:164a
Bixin, **7**:518a
Bizzozero, Giulio Cesare (1846–1901), 2: 164–166; **4**:277a; **5**:459b; **12**:319a; **13**:589a
Bjerknes, Carl Anton (1825–1903), 2:166–167; **2**:167a–b; **8**:323b; **11**:448a
hydromagnetic theory, **2**:169a
Bjerknes, Jacob Aall Bonnevie, **2**:168b; **9**: 107b, 108a
Bjerknes, Vilhelm Frimann Koren (1862–1951), 2:167–169; **2**:166a, 167a; **4**: 344a; **6**:614b, 615b; **11**:10b; **13**:166a–b, 391a*
meteorology, **11**:557b, 558b
ocean currents, **14**:634b
oceanography, **15**:431b
Bjerrum, Jannik Nielsen, **2**:169b
Bjerrum, Niels Janniksen (1879–1958), 2: 169–171; **1**:300a; **10**:345b–346a; **11**: 583a; **14**:478b(n); **15**:450b, 461a
Bjoernbo, Axel Anthon, **2**:430a, 550b; **6**: 412b; **9**:297b, 298b; **15**:188b, 421a
Björnsson, Stephán, **8**:299b
Blaauw, Anton Hendrik, **14**:255a
Black, Davidson (1884–1934), 2:171–172
Black, James (1787–1867), 2:172–173
Black, Joseph (1728–1799), 2:173–183; **2**: 524a; **7**:277a; **14**:322b
acquaintances of, **3**:494b; **4**:250a; **6**: 579a,b; **9**:482b, 484a; **11**:48b, 304b, 499a,b; **14**:196b, 197a
causticity, theory of, **2**:173b, 175a–177b, 180b–181a; **5**:493b; **6**:42a; **7**:58a; **8**: 76b; **9**:347a
chemical nomenclature, **8**:80b
critics of, **12**:65a
diagrams of chemical reactions, **2**:7b, 180b–181a; **3**:495b*
on fixed air, **7**:388b; **13**:602b
influence of, **3**:464b; **5**:276a; **8**:73b–74a; **14**:197b
influences on, **14**:320b
lectures of, **14**:562a
pneumatic chemistry, **11**:144a,b
specific and latent heats, **2**:177b–180a, 483b; **3**:156a; **4**:28b; **5**:137b; **8**:73a; **9**:5b; **14**:197b, 354a; **15**:94b, 95b
students of, **6**:54a, 495b; **8**:261b, 530b; **10**:445b, 582a; **11**:496a, 616a,b; **12**: 24b; **13**:280b, 372a
translations of works, **8**:85b(n)
and Watt, **2**:179a–b
Black Forest
geomorphology, **10**:507b, 508a
Black Hills, **14**:312b
geology, **14**:443a
Black holes, **12**:251b; **15**:345b
Black Sea, **1**:163a; **4**:308b; **7**:410b
altitude of, **13**:111a
density of, **9**:43a
oceanography, **1**:162a,b; **12**:412a–b
Blackband ironstone, **9**:591a
Blackburn, Ashton, **10**:509b
Blackett, Patrick Maynard Stuart, **3**:370b; **6**:117a,b; **12**:33a; **14**:422a; **15**:19a, 20a, 26b, 43b
Blackleg, **7**:391b
Blackloism, **14**:301b
Blacklow
see White, Thomas
Blackman, Frederick Frost (1866–1947), 2: 183–185; **3**:582b*; **5**:37b
Blackman reaction, **2**:184b
theory of limiting factors, **14**:255a
Blacktongue, **4**:358a
Blackwater fever, **7**:428a
Blackwelder, Eliot, **4**:562b
Bladder
anatomy of, **4**:520b
histology of, **6**:269a–b
semipermeability of, **5**:7a
tumors of, **1**:322b
see also Calculi, urinary; Cystitis
Bläes (Blasius), Gerhard, **13**:31a, 169a; **15**: 28b

Blaeu, Willem Janszoon (1571–1638), 2: 185–186; 2:406a; 13:583a
instrumentation, 12:500b
Novus atlas, 8:25b; 10:239b
works translated, 6:520b*
Blagden, Charles (1748–1820), 2:186; 8:78a, 79b; 11:298a; 12:516b; 15:388a*
Blagden's law, 2:186b; 14:191a
Blainville, Henri Marie Ducrotay de (1777–1850), 2:186–188; 1:417b; 5:66a, 504b, 505a, 506a; 7:603a, 604b*; 11:242b, 243b; 13:554b
taxonomy, 12:550b
zoology, 11:243a; 12:551b*
Blair, Edgar Allan, 4:399a
Blair, Patrick (*d.* 1728), 2:188–189
Botanick Essays, 6:38b
Blaise
see Blasius of Parma
Blaise, Armengaud
see Armengaud of Montpellier
Blake, Carter, 5:206b
Blake, Francis, 12:444a,b
Blake, William (poet), 8:213a
Blake, William Phipps, 9:100b
Blakeman, John, 10:461a
Blakeslee, Albert Francis, 14:101b
Blanc, Alberto-Carlo (1906–1960), 2:189–190
Blanc, Gian-Alberto, 2:189a
Blanchard, (Charles) Émile, 1:329a; 9:409a*; 11:234a
Blanchard, Jean-Pierre, 10:609b
Blanchard, Raoul (1877–1965), 2:190–191; 1:208a
Blanchard, Raphaël Anatole Émile, 2:533b; 11:302a
Blanchet, Pierre Henry, 3:144b, 145a
Blanco, [Father] José María, 1:131b
Bland, Theodorick, 1:432a
Blane, Gilbert, 8:363a; 9:483b
Blankaart, Stephan, 6:183b
Blaschke, Josef, 2:191a
Blaschke, Wilhelm Johann Eugen (1885–1962), 2:191–192
editions of works, 11:362b
geometry, 7:397b; 14:283a
Blasius of Parma (*ca.* 1345–1416), 2:192–195; 9:136b; 14:461a
commentaries by, 2:393b; 12:62a
optics, 10:475b
Tractatus de ponderibus, 8:215b, 216a, 226a
Blasius, Gerard, *see* Bläes
Blasius, Johann Heinrich, 7:319b
Blast furnace, 9:591a; 11:463b
"bears," 13:8a
reactions in, 8:118a,b
Blastema, 11:491b; 12:174b, 242a, 243a
Blasting gelatin, 10:132b–133a
Blastoderm, 4:266b
naming of, 10:287a
structure of, 10:287a
Blastoidea, 11:500b
Blastomeres, 8:355a; 10:105a; 15:513b
separation of, 4:187a
Blastomycosis, 10:19a; 11:442b
Blastopore, 2:384a; 8:27a
Blastozooid, 3:148b
Blastula, 4:187a; 14:426b
Blatchley, Willis Stanley, 6:507a; 12:264b
Blauw, Willem
see Blaeu, Willem
Blavatsky, Elena Petrovna, 3:476a
Blazhko, Sergei Nikolaevich (1870–1956), 2:195–197; 13:46a*, 482b*
Blazhko effect, 2:196a
Bleaching agents
alkali-steam method, 3:202a
bleaching powder, 3:289a; 6:385b
formula of, 10:178b*
chlorine, 1:416b; 2:77b; 3:202a, 399b; 6:58b, 283a; 7:389b, 391a; 14:197a
photochemical, 9:60b
salts, 14:394a
Bleeding
reflex, 3:492b
Bleeding, therapeutic
see Phlebotomy
Blemmydes, Nicophorus, 11:18a
Blepharoceridae, 7:286a
Bleuler, Eugen, 5:179a; 7:190a; 12:298b
Blichfeldt, Hans Frederick (1873–1945), 2: 197; 9:388a*
Blignières, Célestin de, 13:250b
Blindness
of cave animals, 4:310a; 10:273b; 12:216a
of fishes, 14:534a
hysterical, 10:307a
reading aids
raised type, 4:252a
Bliss, Eleanor Albert, 4:154b
Bliss, Gilbert Ames (1876–1961), 2:198–199; 4:618b; 9:503b*
Bliss, Nathaniel, 9:164a
Bliss, Sidney William, 15:74a
Blisterfeld, J. H.
see Bisterfeld, Johann Heinrich
Bloch, Felix, 2:242a; 4:579a; 6:104b–105a; 8:477b
Blodgett, Katherine, 8:25a
Bloemfontein, South Africa
observatory, 6:574b
Blomstrand, Christian Wilhelm (1826–1897), 2:199–200; 6:166b; 9:290a
chain theory, 7:179a–b; 14:267a–b
Blondel, André Eugène (1863–1938), 2:200
Blondel, Nicolas-François (*ca.* 1618–1686), 2:200–202; 7:578a
Blondlot, Nicolas, 1:543a, 544a; 2:202b
Blondlot, René-Prosper (1849–1930), 2:202–203; 10:518b; 15:108a
Blood
acid-base exchange, 3:351b; 6:265a; 13: 574b
aeration of, 3:356b
agglutination of, 7:623a–b; 10:387b
analysis of, 2:572b; 6:505a; 8:272a; 10: 204b; 11:51b, 133a; 13:574b–575a; 14:26a, 523b
microchemical, 5:53a
and animal heat, 14:407b
antibodies in, 1:575b; 2:300b; 5:317a; 7: 624a; 12:483a, 492b; 13:420b; 15: 522a
buffers in, 1:454b; 13:574b
carbon dioxide in, 2:161a; 3:600a; 5:149a; 6:21b, 24a, 261b; 9:18b; 11:138b; 12:319b
carbon monoxide in, 9:347b, 348b
carbonic acid in, 11:428a
chemistry of, 2:366b; 5:431b, 432b; 6:22a; 11:139a, 541b
comparative, 3:351b
compounds, 13:274a
circulation of
ancient Roman discussion, 5:233b
medieval Arabic discussion, 9:603b–604a
Renaissance discussion, 1:98a; 2:543a–b; 3:41a; 4:209b, 508b–509a; 11:605a; 15:80b
early seventeenth-century discussion, 1:244b, 567a; 4:54b, 62b; 5:47b; 6: 508b; 10:477a, 490b; 11:467a, 511a; 13:222b, 474a; 14:146b
late seventeenth-century discussion, 1:483a, 593b; 2:312a; 6:386b–387a; 7:612b; 8:128b; 9:323a–b; 11:341b, 587a; 13:33b, 170a; 14:405a, 509b
eighteenth-century discussion, 2:2b–3a; 6:37a–38a, 63b–64a, 458b, 459b; 8: 327b; 10:80a; 11:158b–159a; 12:556a, 556b–557a
nineteenth-century discussion, 2:596b; 5:47b, 184b; 8:133a, 541b; 9:101b, 546b; 10:431a–432b; 11:63a, 133a; 12:224a; 13:53b; 14:200a–b
twentieth-century discussion, 2:597b; 3:113b*; 6:264b, 265a; 12:618a
ammonia, effects of, 2:172b
confirmation of, 2:3a; 9:63a
discovery of, 4:505b; 6:155a–158a; 11:390b–391b; 12:106a, 324b
disorders of, 8:295a,b
gravity, effects of, 12:556b
growth, effects of, 12:556b
hydrodynamics of, 14:200a
intracranial, 8:614a
"local," 2:26b, 30b
mathematical analysis, 7:494a
peripheral resistance, 6:38a; 14:321b
pharmacology of, 15:100a–101b
pressure phenomena, 1:537a; 4:615b; 10:205a
regulation of, 5:426a; 12:618a; 15:106a
and respiratory movements, 2:599a; 6:64a
tidal theory, 4:368b
velocity of, 7:274b; 8:526a; 11:159a
coagulation of, 1:106b; 2:300b, 531b; 5: 149a; 6:367b, 432b, 526a–b; 8:401b; 9:417a; 10:493a; 12:224a; 14:511b
chemical studies, 11:463a
electrical, 1:304b
theories of, 9:64b–65a
and toxins, 15:228b
color of, 11:68b; 13:46b; 14:253a
color differences
theories of, 8:525a–526b; 9:244b
cultures of, 6:233a
cytocidal property, 15:469a
detection of, 13:274a
fate of, 4:146b–147a; 8:203b
formation of
ancient theories, 1:266b; 4:62b, 384a, 385a; 5:234a
Renaissance theories, 14:9b
seventeenth-century theories, 7:612b; 13:31b
nineteenth-century theories, 11:173a
twentieth-century theories, 12:48a
in embryo, 7:438a
fractionation of, 15:76b
gases in, 1:453a; 5:149a; 6:24a; 8:540b; 9:347b, 348b; 11:135b
analysis of, 13:574a
and atmospheric pressure, 2:61b–62b
heat of, 5:137b; 6:63b; 15:95a
individuality of, 7:623b
iron in, 5:226b; 9:302b
microscopy of, 13:174a
and muscle contraction, 3:483a
nitrogen in, 2:62a
osmotic processes, 4:311b; 5:149a–b; 7: 503a
oxygen in, 2:161a; 4:615b; 6:165a*, 261b; 9:18b
atmospheric pressure, 2:61b–62b
dissociation curves, 1:453b–454a, 454b
saturation of, 1:453a
tension of, 1:453a; 2:62a; 7:502a

paleochemistry, **8**:584a
parasites in, **8**:66a
pathology of, **2**:597b; **4**:499a–b; **9**:296a
pH of, **1**:534a
phagocytic assay, **5**:29b
as physicochemical system, **6**:261b–262a; **13**:574b
plasma
alkali reserve, **3**:351b
composition of, **8**:584a
fractionation of, **3**:336a
lipids in, **8**:608a
preservation of, **3**:353a; **7**:623b; **10**:371b; **15**:76b
protein constituents of, **1**:11a; **5**:149a; **12**:319b; **13**:419b, 420b
putrefaction, **12**:87b
replacement of
saline, **1**:537b; **7**:504b
Rh factor, **5**:11a; **6**:433a; **7**:624b
rheology of, **1**:534a–b
role of, **1**:593a–b; **5**:234a; **6**:152a–b, 153b, 157b–158a, 160b, 255b; **8**:128b; **11**:146a; **12**:335b; **14**:407b–408a
salinity of, **5**:149a–b
serum
acidity, **13**:228b
anaphylactic phenomena, **11**:430a
antitoxins in, **1**:575a–577a
buffer capacity, **13**:228b
lysozyme in, **5**:29b
osmotic pressure of, **3**:351b
specific gravity, **7**:346a
spectrum of, **12**:544a
sugar in, **8**:614b; **14**:492b, 523b
transfusion of, **3**:101a; **8**:526a; **9**:417a, 547a*; **10**:202a, 477a; **11**:133a
in animals, **4**:37b; **5**:327b; **8**:524b–525a; **9**:485b; **11**:429a; **12**:137a
in man, **4**:37b–38a; **6**:233a; **7**:612a, 623b, 624b; **8**:525a; **9**:498b; **11**:431a; **15**:38a–b
typing, **7**:624b
viscosity of, **1**:534a–b
volume of, **2**:3a, 161a; **5**:291a
determination of, **6**:225a; **13**:556b–557a
and salts, **9**:285b
and spleen, **1**:454b–455a
see also Cardiovascular system; Erythrocytes; Hematology; Hemoglobin; Hemolysis; Injection, intravenous; Leukocytes; Plasmaphaeresis; Serology; Vividifusion
Blood banks, **1**:11a; **6**:233a; **7**:623b
Blood count
and altitude, **9**:380b
Blood groups, **6**:432b
chemical properties of, **7**:455b
classification systems, **9**:545a–b
determination of, **7**:623a–b
discovery of, **7**:623a–b
evolution of, **6**:433a
frequencies of, **2**:58a
genetics of, **2**:58a–b; **5**:11a; **6**:433a; **7**:623b; **12**:76b; **15**:21a
naming of, **6**:433a
Blood poisoning
treatment of, **7**:208a
Blood pressure
arterial
regulation of, **15**:229a, 469b
capillary, **1**:537a; **12**:618a
diastolic, **4**:398b
and emotional stress, **9**:547a
and heart beat, **2**:366b; **9**:101b
hypertension, **2**:547b
and kidneys, **12**:470b
therapy, **11**:482a
intracardiac, **3**:219b
measurement of, **1**:536b–537a; **3**:518b; **4**:398a; **6**:36a,b, 37b; **7**:504b; **9**:19b; **11**:63a, 482a
sounds of Korotkoff, **4**:398b
oscillations of
categories of, **5**:150a
respiratory, **6**:300a
pulse vs. arterial pressure, **4**:398a
recording of, **2**:366b
regulation of, **10**:432b
and respiration, **11**:63a
and spinal pressure, **3**:518a
venous, **1**:537a
Bloodletting
see Phlebotomy
Bloodstains
detection of, **6**:293b
Blowpipe
in chemical analysis, **2**:5a–b, 93b; **3**:474a; **5**:455b; **6**:293b; **9**:105b; **11**:33b, 463b; **15**:94a
oxyhydrogen, **3**:291b, 617b; **6**:115a; **12**:433a
Blue, Rupert, **5**:451b
Blumenbach, Johann Friedrich (1752–1840), 2:203–205; 13:402b; **14**:113a
acquaintances of, **5**:75b
anatomy, **3**:522a; **5**:442b; **14**:14b
anthropology, **2**:321a
correspondence of, **4**:29a*
critics of, **11**:137a, 593a; **15**:525a
influence of, **11**:137b, 380b
influences on, **5**:77a
Short History of Comparative Anatomy, **8**:97b
students of, **5**:518b; **6**:455b; **7**:366b; **8**:245b; **9**:252b; **10**:197b; **12**:182a, 420b, 509b; **14**:328a
Blumenthal, Otto von, **10**:594a
Blundevile, Thomas, **14**:514a
Blunt, Edmund, **11**:340b
Blunt, George William
American Coast Pilot, **11**:340b
Blunt, Thomas, **9**:607b
Blushing, **14**:322a
Blyth, Edward (1810–1873), 2:205–207; 8:98a; **14**:135a
and Darwin, **2**:206a
B.M.R.
see Basal metabolic rate
Boar
fossil, **12**:184a
Boas, Franz (1858–1942), 2:207–213; 11:219a
Boat(s)
electric-powered, **7**:56a
propulsion, **8**:287b
see also Ships
Bobart, Jacob, the younger, **9**:528b; **12**:394a,b
Bobillier, Étienne (1798–1840), 2:213–216; 3:468a; **5**:367b
Bobillier construction, **2**:215b
Boccius, Father Norbert, **1**:520b
Hortus botanicus, **1**:520a
Bochart de Saron, Jean-Baptiste-Gaspard (1730–1794), 2:216–217; 9:330b; **15**:316a
Bôcher, Maxime (1867–1918), 2:217–218; 2:144a; **3**:345b; **5**:313a*; **7**:602a; **8**:386a; **10**:244b; **13**:130a, 132a*
Bochner, Salomon, **2**:238b; **14**:90a
Bock, Jerome (1498–1554), 2:218–220; 2:536b; **4**:138b; **8**:484a; **13**:463a
Bode, Johann Elert (1747–1826), 2:220–221; 6:329a; **7**:569b; **10**:572b, 592a; **12**:518a
Bode's law, **2**:221a; **5**:298b; **10**:198b; **13**:425a
Bodenheimer, Fritz Simon (1897–1959), 2:221–222
Bodenstein, Adam of
see Adam of Bodenstein
Bodenstein, Ernst August Max, **2**:3b; **15**:462a
Bodenstein, Max (1871–1942), 15:36–38; 6:481a; **9**:427b; **15**:447b–448a, 448b, 450a–b
Bodenstein number, **15**:37a
Bodmer, Karl, **14**:328b
Body
atomistic view, **12**:538b
fluid/solid ratio, **7**:274b–275a
heat of
see Body temperature; Heat, animal
and mind
see Dualism, philosophical
and soul, **3**:245a, 321a*; **6**:59b, 573b; **7**:137a–b; **8**:115a, 514b–515a; **12**:224b; **13**:179b; **15**:262a
Body cavities
embryogeny of, **6**:435a
Body fluids, **2**:336a, 597b
acidity of, **13**:228b
analysis of, **13**:574a–b
nature of, **13**:31b
osmotic pressure, **2**:339b
paleochemistry, **8**:584a
pathology of, **5**:565b
water-shifting mechanisms
hypothalamic control, **1**:450b
see also specific fluids
Body snatching, **7**:415a–b; **9**:480a; **14**:7b
Body temperature, **2**:619b
and air temperature, **2**:186a–b; **9**:133a; **10**:579b–580a
anatomical variations in, **1**:610b
and health, **12**:103a
lowering of, **2**:30b–31a
of newborns, **4**:285b
regulation of, **1**:610b; **4**:163a; **5**:149b; **11**:427a–428b; **12**:165a
brain and, **1**:450a–b; **2**:483b–484a; **10**:252b; **11**:427b–428b
chemical, **15**:126b
vasomotor reactions, **1**:450a
in thermometry, **4**:517b
see also Heat, animal
Boë, Franz de la
see Sylvius, Franciscus
Boeck, Carl Wilhelm, **6**:101b
Boegehold, Hans, **5**:591a; **7**:448a
Boehme, Jacob (1575–1624), 2:222–224; 3:362b; **5**:120a; **7**:483a,b, 485a; **12**:156b; **14**:226a, 365b, 370a, 372a
critics of, **14**:367a
Boehmer, Georg Rudolf, **7**:412a
Boer, Jan de, **8**:481b
Boer War, **13**:357a; **14**:511b
Boerhaave, Hermann (1668–1738), 2:224–228; 4:268a; **5**:420a, 484b; **12**:103b; **13**:553b
acquaintances of, **3**:37a; **4**:517a,b; **8**:375a; **9**:479b; **12**:41a, 394b
anatomy, **9**:153b; **14**:287a; **15**:4a
Aphorisms
commentaries on, **3**:427a; **13**:181b
biographies of, **3**:334a; **5**:61a
chemical theories, **2**:179b; **5**:137b, 511a; **8**:620a
on contemporary scientists, **15**:4a*
correspondence of, **11**:526b; **13**:449a
critics of, **2**:122b; **14**:354a
Elementa chemiae, **2**:177b, 178b; **11**:140a
influence of, **1**:451b; **2**:16b; **5**:130a, 509b;

Boerhaave, Hermann (cont.)
11:616b, 617a; **12**:365b; **13**:541a; **15**:3b
influences on, **10**:311a; **11**:2b
Institutiones medicae, **12**:185a; **13**:122b; **14**:321a
editions of, **6**:66a
lectures of, **13**:181a,b
mechanistic philosophy, **1**:479a
medicine, **1**:594a; **3**:15b; **5**:92b; **7**:605b; **8**:363b*; **11**:48b
Methodus studii medici, **14**:562a
editions of, **6**:66a
New Method of Chemistry, **6**:46b(n)
physiology, **6**:61b–64b *passim;* **8**:506a, 526b; **9**:571a
students of, **2**:421a; **3**:245a, 494a; **5**:380a; **6**:231a; **7**:11a, 605a; **8**:327a–b; **9**:595a; **10**:25b, 500b; **11**:48a; **13**:181a; **15**:4a, 6a
and Swammerdam's works, **13**:169b, 171a, 172a
translations of works, **7**:605b; **12**:365b; **14**:193a, 333b*
Boethius, Anicius Manlius Severinus (*ca.* 480–525), 2:228–236; **1**:137a,b; **4**:462a; **9**:79a
arithmetic, **1**:62b; **2**:395a; **3**:214a; **4**:446a; **8**:100b; **13**:545a
Arithmetica, **1**:381a; **2**:435a; **4**:443b; **7**:176a
editions of, **3**:280b
on chance, **1**:27b
commentaries on, **13**:340b*
De consolatione philosophiae, **1**:27b, 91b; **4**:396a
translations of, **3**:217b
De differentiis topicis, **1**:2a
Geometria, **3**:513b; **5**:365b
geometry, **1**:418b; **4**:443b, 444a
De hebdomadibus (= *Quomodo substantiae),* **1**:91a
influence of, **1**:267b, 381b, 565b; **2**:393a; **3**:272b; **5**:365a; **9**:364a; **10**:238b, 270b; **13**:60b, 62a(n), 349b; **14**:35b, 225b, 388a
De institutione musica, **1**:231b; **4**:444b
music, **4**:259a; **5**:550b; **7**:128b; **11**:203a
number theory, **6**:302a; **9**:191b
philosophy, **2**:21a; **5**:592a; **6**:544a; **7**:27b; **15**:182a
translations by, **1**:63a, 226a, 269a–270b *passim*, 272a, 273b, 276a; **4**:443a–444a; **7**:66a; **10**:113b*
revisions of, **9**:436a, 437a
Boethus of Sidon, **1**:268a
Boetius de Boodt, Anselmus
see Boodt, Anselm de
Boettger, Rudolph Christian von
see Böttger, Rudolf Christian von
Bogdan, Martin, **10**:477a; **11**:587a
Bogdanov (Malinovsky), Aleksandr Aleksandrovich (1873–1928), 15:38–39
Bogdanovich, Karl Ivanovich, **9**:593a, 594a
Bogenhausen
observatory, **12**:518b
Boggio, Tommaso, **2**:594a
Bogoliubov, Nikolai Nikolaevich, **8**:477b
Bogotá
observatory at, **15**:429b
Bogs, **5**:537a; **13**:9b
Boguslavsky, Heinrich Georg, **2**:236a
Boguslavsky, Palm Heinrich Ludwig von (1789–1851), 2:236; **5**:256b, 257b
Böheim, Martin
see Behaim, Martin
Bohemia
geology, **11**:385b–386a, 387a–b
science, **13**:44a
stratigraphy, **1**:468b–469a
Bohemian Brethren, **3**:359b–360a, 360b, 362b
Bohemian Scientific Society (Prague), **2**:315b
Bohl, Piers (1865–1921), 2:236–237; **2**:58b*
number theory, **14**:282b
Bohlmann, Ferdinand, **4**:553b
Bohm, David Joseph, **9**:217b
Böhm, Johann, **6**:371a
Bohn, Johannes (1640–1718), 2:237–238; **1**:593b; **2**:2b
Circulus anatomico-physiologicus, **2**:2b
Bohnenberger, Johann Gottlieb Friedrich von, **8**:369b
Bohr, Christian, **1**:453a–b; **2**:238a, 239a, 239b–240a; **5**:150a; **7**:501b, 502a,b, 617a
Bohr, Harald (1887–1951), 2:238–239; **2**:239b, 240a; **7**:616b; **10**:118a*
analysis, **2**:236b; **13**:36a
Bohr-Fourier series, **2**:238b
Bohr-Landau theorem, **2**:238b
Bohr, Niels Henrik David (1885–1962), 2:239–254; **4**:580b; **7**:417a*; **10**:350b*, 425a; **12**:29b, 35b*, 44b*
acquaintances of, **3**:429b; **4**:294b, 561b; **5**:117b, 272a; **6**:365a, 366a; **7**:492a, 494a*; **8**:94b; **9**:262b, 543a; **10**:423b; **11**:422a; **15**:25b, 422b
astrophysics, **7**:268a; **10**:601a
atomic theory, **2**:600b; **3**:597a; **5**:331b; **6**:366a; **10**:422b–423a, 424b; **11**:583a; **12**:528b–529a; **13**:42a, 359a; **14**:588b
tests of, **11**:420b
Bohr atomic model, **2**:242a–b, 246a–b, 247a; **3**:124a; **4**:195b, 196b; **5**:160b; **7**:469a, 553b; **8**:11a, 23b; **9**:262b, 543b; **10**:106b; **12**:31b, 43b; **14**:497b
critics of, **8**:291b–292a; **12**:614b–615a
Bohr-Kramers-Slater theory, **4**:196a
complementarity principle, **2**:249b–250b, 253a–b; **4**:318b; **8**:322b(n)
on contemporary scientists, **13**:370a(n); **14**:67a
critics of, **12**:219a; **13**:364b
and Einstein, **2**:250a–b
influence of, **3**:564a; **7**:518b; **9**:256a
influences on, **15**:83b
method of, **2**:239b, 245b, 249a–250b
quantum mechanics, **4**:318b; **12**:219b
quantum theory, **2**:338a, 466a; **3**:369a–b; **4**:317a–b; **5**:117a–b, 609a; **7**:468b, 491b; **8**:497b–498a, 578b; **9**:544b; **11**:13a; **12**:220a, 221b, 528b; **15**:41b, 43a, 48b
and Rutherford, **2**:240b, 241a, 244a,b; **12**:31b
spectra, theory of, **5**:102a–b; **9**:397b; **10**:347b–348a; **11**:446a; **12**:21a,b
critics of, **13**:104a
Bohsted, Gustav, **6**:136a
Boilers, steam, **4**:241b; **13**:398b
explosion of, **1**:203a, 364a
multitubular, **12**:287b
Boiling, **9**:19a; **14**:191a–b
effects of, **9**:60b
Boiling point, **3**:292b
and atmospheric pressure, **4**:28b, 518a
and chemical constitution, **12**:209a
of organic substances, **5**:372b; **7**:463b
of solutions, **4**:517b; **11**:298b
and temperature, **14**:561a
and vapor pressure, **5**:324a; **12**:495b
Boils
etiology of, **10**:389b
Boisbaudran, Paul Èmile Lecoq de (called François) (1838–1912), 2:254–255; **2**:428b; **3**:321b; **9**:110b, 291a; **14**:530a
spectroscopy, **3**:478a; **5**:497b; **9**:290b
Boisduval, Jean Alphonse, **1**:502b
Boislaurent, François Budan de
see Budan de Boislaurent, François
Boissier de Sauvages, François, **8**:379a
Boissière, Claude de
Rythmomachia, **1**:468a
Boissonade, Jean François, **4**:159b, 160a
Bol, Gerrit, **2**:192a
Bolívar, Simón, **6**:551a
Bolk, Lodewijk (Louis) (1866–1930), 2:255–256; **5**:359a
Boll, Franz Christian, **7**:519b, 520a
Boll weevil, **6**:524b
Bologna
Academy of Sciences, **9**:77a
meridian of, **5**:542b–543a
observatory, **6**:507b; **9**:77b, 78a
Bologna, University of
in Middle Ages, **12**:62b
in Renaissance, **1**:618a
in seventeenth century, **3**:100b–102a *passim*; **9**:485a
anatomy at, **1**:204a; **14**:5a–b
mathematics at, **9**:12b; **10**:610a
observatory, **14**:589a–b
Bologna stone, **4**:215a; **9**:106a
Bolognetti, Pompeo, **4**:595b
Bolometer, **2**:600b; **7**:528a; **8**:20a,b; **10**:346a; **11**:581b–582a; **14**:38b, 339a
surface, **8**:551b
Bolometry, **5**:597b; **10**:345b–347b
Bolos of Mendes (Bolos the Democritean) (*fl. ca.* 200 B.C.), 2:256–257
Bolotov, Andrei Timofeevich (1738–1833), 2:257
Bolshaia meditsinskaia entsiklopediia, **1**:361b
Bolshaia sovetskaia entsiklopediia, **1**:361b
Bolthead (chem. vessel), **6**:42a
Boltwood, Bertram Borden (1870–1927), 2:257–260
radioactivity, **12**:28b–29a, 30b
radiochemistry, **6**:14b, 15b; **12**:504b
Boltzmann, Ludwig (1844–1906), 2:260–268; **5**:390b; **6**:343a; **12**:527b
acquaintances of, **1**:297b; **7**:380a; **8**:489b, 511a; **11**:8b, 10a, 12a; **15**:458b
atomic theory, **7**:220b; **14**:67b; **15**:463a
Boltzmann factor, **2**:261a
Boltzmann's constant, **2**:262a; **8**:10b; **9**:222a
physical significance of, **4**:314a
electrodynamics, **9**:228a(n)
gas theory, **5**:391b; **9**:221b, 222b, 223a–224a; **10**:235b; **12**:496b; **14**:614a; **15**:49a
H-theorem, **2**:263b–264a; **4**:293b
influence of, **9**:606b; **10**:525a; **12**:217b–218a, 526b; **13**:40b
mathematics, **11**:12b, 57b
Maxwell-Boltzmann equations, **4**:375b–376b; **9**:222b
mechanics, **4**:293b, 314a; **6**:347b
mechanistic philosophy, **8**:599a
optics, **8**:492b–493a; **9**:212b
probability, **3**:310a
Stefan-Boltzmann law of radiation, **2**:266a; **11**:150a,b; **13**:11a; **14**:339b, 340a
students of, **4**:293a; **5**:122a; **6**:163b; **9**:260a, 606a; **10**:212b, 604b; **13**:115a; **15**:433a
thermodynamics, **3**:309a, 310a; **5**:462a; **8**:11a, 493a, 510b, 598b; **9**:227a–b; **11**:11a, 420a; **12**:497a

Vorlesungen über Gastheorie, **9**:219a
translations of, **2**:466a
Bolyai, Farkas (Wolfgang) (1775–1856), 2: 268–269; **2**:270a; **4**:417b; **5**:301a, 302b, 308a; **7**:206b; **12**:255a
editions of works, **12**:599b
Bolyai, János (Johann) (1802–1860), 2:269–271; **2**:269a; **5**:301a
editions of works, **8**:433a; **12**:599b
geometry, **2**:274b; **5**:302b; **6**:522a; **7**:397a; **8**:141a, 432b
works translated, **6**:76b
Bolza, Oskar (1857–1942), 2:271–273; **9**: 502a; **10**:97b*; **14**:222a
problem of, **2**:198a, 271b–272a
Bolzano, Bernard (1781–1848), 2:273–279; **4**:167b, 168a*; **7**:485b; **10**:484a; **11**: 214a
analysis, **3**:75b; **4**:4a; **5**:154b; **6**:95b; **13**: 81b
Bolzano function, **2**:276a; **3**:136a
Bolzano-Weierstrass theorem, **2**:275b
Paradoxien des Unendlichen, **4**:4b
Bombay
Plague Research Laboratory, **6**:12a–b
Bombelli, Rafael (1526–1572), 2:279–281; **3**:128a, 275a, 277b; **4**:515b; **5**:409b; **10**:338b(n); **14**:20b
Algebra, **1**:367b; **2**:320a; **4**:117b, 588a, 595b, 596a,b; **10**:270b
Bombinator, **5**:447a
Bon jardinier
editors of, **14**:33b
Bonacosa, **12**:7Bb
Bonamico, Francisco, **7**:486b
Bonanni, Filippo
see Buonanni, Filippo
Bonaparte, Lucien Jules Laurent, called Charles Lucien (1803–1857), 2:281–283; **1**:331a; **12**:257b
American Ornithology, **7**:70a; **10**:439b
Bonaventura, Federigo (1555–1602), 2:283
Bonaventure, Saint, **1**:197a, 198b, 378a; **5**: 107a; **12**:428b
Boncompagni, Baldassare (1821–1894), 2: 283–284
historiography, **15**:173b, 177b, 183b, 187a, 188a, 189a
of astronomy, **7**:141b*; **10**:154b*
of mathematics, **3**:135a, 312a*; **4**: 605a,b, 609a, 612b; **6**:299b*; **8**:281b*; **9**:191b; **10**:272a*; **11**:32b*, 405a*; **13**:261a*; **15**:402a*
of natural philosophy, **10**:538b*
influences on, **14**:472a–b
Bond, George Phillips (1825–1865), 2:284–285; **6**:48b, 470b; **10**:543b*; **14**:449a*
Saturn, **3**:605b; **10**:479b
Bond, Henry, **5**:343b
Bond, William Cranch (1789–1859), 2:285; **1**:364a; **2**:284b; **3**:605b; **6**:350b; **8**:46a; **14**:449a*
Bonding, chemical
nineteenth-century view of, **2**:96b; **5**:374a; **7**:220a; **13**:363a, 365b
twentieth-century view of, **2**:242a–b; **7**: 469a–b; **8**:47a, 290a, 291a–292a, 473a–475b; **13**:388b
carbon, **1**:390a; **5**:371a; **7**:281a, 552a
and catalysis, **1**:414b–415a
concept of, **2**:621b–622a
covalent, **1**:461a; **7**:469a–b; **12**:419a–b
crystallographic study, **5**:456a
dative, **12**:419a
double, **3**:86a; **4**:400a; **7**:281a; **8**:508a; **13**:337b–338a
hydrogen, **1**:519b; **5**:589b; **7**:622b; **8**:47a; **9**:38a; **12**:418b–419a
ionic, **1**:461a; **12**:419a
nitrogen, **1**:553b; **6**:108a
nonpolar, **8**:291a
shared electron pair, **8**:291b–292a, 293a
triple, **4**:400a; **8**:508a; **13**:337b–338a
see also Compounds, Molecules, Valency
Bones, **3**:355a; **6**:509a
absorption of, **7**:439b
analysis of, **5**:92b, 218a; **6**:166b
cavities in, **4**:413a; **6**:183a–b
chemical composition, **12**:146a
coloring of, **5**:226a
diseases of, **3**:174b; **4**:268b*; **5**:220b; **11**: 596a
formation of, **6**:65b; **9**:146a; **11**:383a, 492a; **12**:315b
nourishment, **8**:171b–172a
trajectorial theory, **3**:496b
fossil, **2**:569a–b; **3**:324a, 525b; **5**:156a, 162b, 206b–207a; **11**:483b; **12**:183b–184a
dating of, **12**:318a
gelatin from, **3**:560b
growth of, **5**:45a; **6**:40b, 183b, 566a; **9**: 546b
malformation, **15**:419a–b
phosphorus in, **5**:223a
physiology of, **12**:165a
regeneration of, **7**:414b; **8**:34a–b; **13**:465a
softening of, **5**:220b
strength of, **9**:480a
structure of, **3**:496b; **4**:522b; **5**:220b; **6**: 183a–b, 566a; **9**:145b–146a, 226b; **10**:267b; **11**:216b
see also Fractures, Marrow, Orthopedics, Osteology, Skeletal system
Bonet, Théophile
see Bonnet, Théophile
Bonhoeffer, Karl Friedrich, **15**:448a,b, 450b
Bonie, Johannes de, **15**:156a
Boniface VIII (Pope), **3**:24a; **4**:254b; **5**:402b; **10**:540a; **15**:70b
Bonin Islands
gravimetry, **9**:181a
Bonite (ship), **12**:550a–b
Bonn, University of
in nineteenth century, **9**:568a
anatomy at, **12**:231a
botanical garden, **10**:11b
botany at, **13**:88a, 89b–90a
Chemical Institute, **7**:282b, 283a
chemistry at, **6**:461b
observatory, **1**:242a–243b
Physical Institute, **6**:343a; **7**:268a
physiology at, **10**:579b
Bonnemaisoniaceae, **12**:125a
Bonner Durchmusterung, **1**:87b, 242b–243a, 340a; **2**:100a, 146b; **4**:273a; **7**:236a; **12**:200b
Bonnet, Charles (1720–1793), 2:286–287
acquaintances of, **12**:119b, 308b
biographies of, **15**:110a
chain of being, **3**:522b; **13**:564a
Considérations sur les corps organisés, **12**:555a
Contemplation de la nature
translations of, **12**:556a; **13**:425a
correspondence of, **2**:204a; **3**:461b; **10**: 10b*; **13**:183b
critics of, **5**:577b; **11**:159a
embryogeny, **1**:293b; **8**:160a; **15**:524b
influence of, **11**:493a; **15**:50b
influences on, **8**:579b
physiology, **5**:253a
plant physiology, **5**:277b
on regeneration, **12**:555b
on reproduction, **8**:580a; **11**:333b; **12**: 559a,b
and Spallanzani, **12**:555a–563a *passim*
Bonnet, Pierre-Ossian (1819–1892), 2:287–288; **2**:350a, 351a; **3**:559b; **7**:109b; **8**:382a,b; **9**:409b; **12**:319a
Gauss-Bonnet theorem, **2**:288a
mathematics, **3**:331a; **9**:409b; **10**:544b; **12**:319a
Bonnet, Théophile, **3**:294b; **7**:164b; **11**:508a
Sepulchretum, **9**:511b
Bonney, Thomas George (1833–1923), 2: 288–290; **5**:437b*; **9**:88a*, 393b*; **11**:131b*; **12**:264a*, 519b; **13**:272b, 273a
Bonnier, Gaston (1853–1922), 2:290–291; **3**:359a; **8**:145b; **13**:406a*; **14**:600a*
botany, **1**:561a; **3**:318a; **9**:78b
Bonnier, Jules, **5**:385b
Bonomo, Giovan Cosimo (1666–1696), 2: 291; **3**:180b; **11**:342b
Bonpland, Aimé, **3**:16b; **6**:550b–551a; **14**: 386b; **15**:267a,b, 430a
Bontekoe, Cornelis, **8**:127a
Bontius, Jacobus, **10**:621b
Bonvicino (also known as Bonvoisin or Buonvicino), Costanzo Benedetto (1739–1812), 2:291–292
Boodt, Anselmus Boetius de (*ca.* 1550–1632), 2:292–293
Book of Changes, **3**:251b, 253b; **12**:385b, 386b, 390b, 408b; **14**:162b
Book of Nature, **1**:159b; **10**:308b, 310a; **15**:69b
Bookkeeping
double-entry, **3**:170a*; **10**:270b
Books of Chilam Balam, **15**:813b
Boole, George (1815–1864), 2:293–298; **1**: 207b; **5**:523a
algebra, **13**:219b; **14**:303b
critics of, **7**:105a; **13**:612a
editions of works, **13**:427b*
historiography, **15**:281b
influence of, **2**:484b; **12**:16a, 217a; **15**: 387a(n)
influences on, **12**:326a
invariant theory, **3**:164a, 169b; **5**:472a
Laws of Thought, **1**:532a; **4**:138a; **8**:590b; **13**:612a,b
logic, **4**:36a; **5**:152b; **6**:82a,b; **7**:105a–b; **8**:159b; **10**:442b; **11**:94b; **12**:12a
Mathematical Analysis of Logic, **4**:36a
probability, **7**:106b; **9**:219a; **13**:611b
Booth, Edward, **5**:491a
Booth, Felix, **11**:554b
Boothia Peninsula, **11**:554b
Bopyridae, **5**:385b
Boracite, **5**:497a
Boranes, **13**:70b, 71a
Borates
in foods, **14**:357b
Borax, **3**:5a; **5**:351b
compounds of, **11**:109a
determination of, **5**:325b
manufacture of, **10**:436a
refining of, **13**:566a
Borax beads, **12**:544b; **15**:94a
Borch, Oluf
see Borrichius, Olaus
Borchardt, Carl Wilhelm (1817–1880), 2: 298–299; **4**:2b; **6**:76b, 230a, 286a; **7**:51a,b, 55b*, 109b, 110a*, 506a–b; **8**:383b; **12**:170b
editions of works, **7**:507a,b
Borda, Jean-Charles (1733–1799), 2:299–300; **2**:475b*; **3**:387a; **4**:208a
and Académie des sciences, **15**:333a, 336a
acquaintances of, **3**:439b, 440b, 441a
biographies of, **7**:550b

Borda, Jean-Charles (cont.)
and French Revolution, 6:178b; 7:569a
hydrodynamics, 2:39b
influence of, 1:201b; 3:446a; 6:107a
instrumentation, 9:234b, 250b; 15:334b
and metric system, 4:14b, 15a,b, 16a; 15:334a, 335b
Tables trigonométriques décimales, 4:16a
Bordeaux
observatory, 11:320b
Bordet, Jules (1870–1961), 2:300–301; 5:316b, 317a; 9:334b
immunology, 4:299a, 305*; 11:430b; 15:522a–b
Studies in Immunology
translation of, 5:316b
Bordet, Paul, 2:300a, 301a
Bordeu, Théophile de (1722–1776), 2:301–302; 4:88a, 89b; 9:306a
Bordini, Antonio, 2:469b
Bordone della Scala, Giulio
see Scaliger, Julius Caesar
Bordonius
see Scaliger, Julius Caesar
Boreel, Willem, 7:73a,b
Borel, Émile (Félix-Édouard-Justin) (1871–1956), 2:302–305; 1:408a; 2:58b*; 4:177b; 6:548a*; 10:276a*
acquaintances of, 1:194a; 10:513b; 14:243b
analysis, 3:63a; 10:594a; 11:54a; 14:572b
Collection Borel, 1:407a
on contemporary scientists, 13:251a*; 14:32b*
game theory, 14:92b
Heine-Borel theorem, 2:303b; 6:230b; 13:599b
influence of, 6:451b
influences on, 13:250a
measure theory, 8:110b
probability, 7:351b
series, 3:178a
set theory, 6:572a; 8:558b
Borel, Pierre (*ca.* 1620–1671), 2:305–306; 2:13a; 7:73b; 10:492b*; 13:30b
Discours nouveau prouvant la pluralité des mondes, 5:59b
Borelli, Filippo, 2:307a–b
Borelli, Giovanni Alfonso (1608–1679), 2:306–314; 9:369b; 11:240b; 14:18b
and Academia del Cimento, 9:3a
acquaintances of, 3:115b; 9:14a
anatomy, 2:2b
celestial mechanics, 7:487b–488a
correspondence of, 3:348b; 9:63a; 10:48b
cosmology, 1:164b–165a
critics of, 12:55b
historiography, 14:49a
hydraulic laws, 1:593b
influence of, 2:37a, 52a, 226a, 237b; 4:589a; 6:37a; 9:62b, 191a; 13:231b, 566a; 14:587b
influences on, 13:170b
mechanics, 10:62b
mechanistic philosophy of, 1:479a
De motu animalium, 1:593a; 3:483a
physics, 2:592b; 8:209b
physiology, 2:3a; 4:65a; 13:32b, 46b
students of, 1:592b; 9:3b
Borg, Erik, 3:320b
Borgia, Cesare, 8:193b, 206b, 213a, 214a, 242a
Borgnis, Guiseppe Antonio, 9:147a; 11:80a, 384b
Borgognoni of Lucca, Theodoric (*ca.* 1205–1298), 2:314–315; 6:276b; 10:307a
Boric acid, 12:145b
decomposition of, 5:321b
esters of, 12:164a
manufacture of, 5:351b
medical use of, 7:208a; 10:385b–386a, 390a
Borlase, Bingham, 3:598b, 599a
Bormann, Elisabeth, 15:41a
Born, Gustav, 2:383b; 6:132a–b; 11:571b
Born, Ignaz Edler von (1742–1791), 2:315–316; 1:234b; 4:345b, 346b; 11:304a
translations of works, 9:479a*; 11:304b
Born, Max (1882–1970), 15:39–44; 4:357a
acquaintances of, 2:337b; 3:597b; 4:577b; 5:117a–118a; 8:51b; 10:423b; 12:222a,b; 13:41a,b
Born approximation method, 15:42b–43a
Born-Infeld theory, 7:10a; 15:43b
Born-Kármán theory, 11:266b
chemistry, 11:88b
on contemporary scientists, 4:319a; 8:182a, 495a; 12:253b*; 14:317b
correspondence of, 10:214a
Dynamik der Kristallgitter, 6:614b
influence of, 2:245b; 12:529b
influences on, 9:377b; 12:250b
mathematics, 2:248a; 12:221a
matrix mechanics, 2:248b; 4:318a–b
quantum theory, 8:475a; 12:221b
specific heat, 7:246b; 15:442b
students of, 6:303b, 614a–b; 11:355a; 15:48a
thermodynamics, 4:229b
Bornet, [Jean-Baptiste] Édouard, 10:166a; 12:125a; 13:394b, 395b; 14:93a
Bornetia, 9:78b
Börnstein, Richard, 7:620a
Physikalisch-chemische Tabellen, 14:448a
Borodin, Aleksandr Porfirevich (1833–1887), 2:316–317; 8:275a; 10:1a; 14:106a, 622a,b*
Boron, 13:454b
bombardment of, 3:329b–330a; 7:153a, 154a,b; 10:289a
compounds of, 13:70b, 71a, 468a
discovery of, 5:321b
isolation of, 3:602a; 4:77b; 13:311b; 14:478a
molecular weight, 4:243b
transmutation to carbon, 2:338b
valence, 8:502b
wave mechanics of, 6:615a
Boron trifluoride, 5:320a, 321b
Borough, William, 5:343b, 593a,b
Borrel, Amédée, 9:328b; 14:551b
Borrel, Jean
see Buteo, Johannes
Borrelly, Alphonse, 13:36b
Borri
see Borro, Girolamo
Borrichius (or Borch), Olaus (1626–1690), 2:317–318; 4:75a
Borries, August von, 12:527a
Borries, Bodo von (1905–1956), 2:318
Borro, Girolamo (*or* Borri *or* Hieronymus Borrius) (1512–1592), 15:44–46; 5:238a; 10:512b
Bortkiewicz (or Bortkewitsch), Ladislaus (or Vladislav) Josephowitsch (1868–1931), 2:318–319; 15:31b
Bortolotti, Ettore (1866–1947), 2:319–320; 2:280b; 3:127b; 4:595b, 613b*; 9:55b*, 304b*; 10:610b*; 11:600b*; 13:261b*, 440a*
Boruttau, Heinrich, 2:339a; 4:201b; 9:259b
Bory de Saint-Vincent, Jean Baptiste Georges Marie (1778–1846), 2:320–321; 4:256b*; 7:610a
Dictionnaire classique d'histoire naturelle, 7:610a
Bos, 10:16b
Bosanquet, Bernard, 3:90a; 6:26b
Bosanquet, Claude Henry, 15:63a
Bosc, Louis Augustin Guillaume (Dantic) (1759–1828), 2:321–323; 3:172b; 9:366b, 418b; 13:390a
Bosch, Carl (1874–1940), 2:323–324; 2:4a; 9:427b
ammonia synthesis, 2:323b–324a, 436b; 5:621b–622a; 8:118b; 9:427a, b
Bose, Ernst Gottlob, 6:218a
Bose, Georg Matthias (1710–1761), 2:324–325; 10:147a
electricity, 7:403a; 9:596b; 10:146a; 11:433a; 14:193b
Bose, Jagadis Chunder (or Jagadischandra) (1858–1937), 2:325; 15:46–47; 15:47b
Bose, Satyendranath (1894–1974), 15:47–50; 11:265a
Bose-Einstein statistics, 4:317b–318a, 361a, 578a; 5:473a; 8:478a–479a; 12:219a; 15:48b–49b
Bose Research Institute (Calcutta), 15:46b
Bošković, Rudjer Josip (1711–1787), 2:326–332; 3:156a, 539b; 6:335b; 8:159b, 350b, 351a; 13:39b, 235b, 375a; 14:70a; 15:352b
atomism, 6:86b–87a, 90a, 279a, 586a; 8:262a; 11:496b–497a
on comets, 2:216b; 15:309b–310a
critics of, 12:608b; 15:310a
matter theory, 11:142b, 143a, 147a; 14:70b
natural philosophy, 1:462a; 4:529b–530a, 532b
statistics, 15:376a
Boson (particle), 15:48b
Bosporus
currents in, 9:42b–43a, 135a
Boss, Benjamin, 2:333a
Boss, Lewis (1846–1912), 2:332–333; 13:98b
Bosse, Abraham (1602–1676), 2:333–334; 7:537a, 577a
and Desargues, 4:46a–49a *passim*
Manière universelle de M. Desargues pour pratiquer la perspective . . . , 4:47b
Bossu, Nicholas, 5:76b
Bossuet, Jacques-Bénigne, 12:117b; 14:449b
Bossut, Charles (1730–1814), 2:334–335
and Académie des sciences, 7:563b; 13:412a*; 15:310a, 333a
acquaintances of, 3:439b, 440a, b; 9:470a, 471a, b; 13:496a
celestial mechanics, 15:289a
correspondence of, 13:417b
historiography, 7:572b*; 12:61a; 13:585a, 586a
hydraulics, 1:116b; 3:387b(n); 4:208a; 11:63b
mathematics, 2:38a; 3:78b*, 383b
Bostock, John (1773–1846), 2:335–336
Boston (Mass.)
Huntington Hospital, 4:195a, 196a
Symphony Hall, 12:54b
Boston Nutrition Laboratory, 1:610a
Boston Society of Natural History, 6:56b
Boston University
chemistry at, 11:577b
Boswell, William, 14:368a–b
Botallo, Leonardo (*ca.* 1519–*ca.* 1587), 2:336–337; 1:204a; 5:448a
Botallo's duct, 2:336b
Botallo's foramen, 2:336b
De curandis vulneribus sclopetorum, 9:12a
works edited, 6:509a
Botanical Exchange Club, 1:412b

Botanical gardens
medieval Arabic, **14**:112b
Renaissance, **3**:413a; **5**:383b, 448a; **9**:308b; **14**:336a
seventeenth-century, **7**:536a–537a; **9**:92b, 122b, 528b; **10**:491b; **11**:511a, 587b; **13**:450a–b
eighteenth-century, **3**:172b, 365a, 578b; **6**:539a; **7**:199a–b; **8**:177b, 375a,b; **9**:253b, 366a, 390a,b, 434a; **11**:66b; **12**:327a, 394a
nineteenth-century; **6**:184b, 493b, 494a, 494a–b, 521a; **9**:149a, 340a, 618a; **11**:218b, 262a; **13**:44a, 89b, 154b, 395b, 459a–460a, 462b; **14**:386b
twentieth-century, **7**:496a; **8**:513a; **12**:77a; **13**:456b; **15**:148a
Mesopotamian, **15**:663b(n)
Botanical Magazine
editors of, **6**:491b
Botanical Miscellany
founding of, **6**:495a*
Botanical Register
editors of, **8**:372b
Botanical Society of America
founding of, **1**:396b; **15**:224a
Botanical Society of Edinburgh
founding of, **1**:423a
Botanische Jahrbücher für Systematik, Pflanzengeschichte und Pflanzengeographie
founding of, **15**:147b
Botanische Zeitung, **3**:611b
founding of, **9**:442a
Botanisches Zentralblatt, **3**:188a; **8**:513a
Botany
ancient Greek, **2**:256b; **10**:111b–112a; **13**:331b–333a
ancient Roman, **4**:119b–121b
medieval Arabic, **1**:517a
medieval Latin, **1**:101b; **11**:601b; **14**:35b
Renaissance, **1**:109a, 124b–125a, 167a–b, 522b, 525a–526b, 592a–b, 596a; **2**:218a–219b, 535b–537b; **3**:413a, 413b–414b, 533b–534a; **4**:138b–139b; **5**:361b–362b, 379a–b, 383a–384a, 448a; **8**:205b–206b, 435a–436a, 483b–484a, 554a; **9**:75b, 178b–180a, 308b; **10**:116b, 236b–238a; **11**:311b–312b, 595a; **12**:135a; **13**:502a–503a; **14**:335b–336a, 586a–b; **15**:80b–81a
early seventeenth-century, **1**:524a; **3**:179a–180a; **4**:135b; **6**:387a; **7**:147a–148a, 536a–539b; **9**:122a–b; **10**:426a–b; **11**:511a–b; **12**:577a; **13**:450a–b; **14**:12b, 13a
late seventeenth-century, **1**:432a; **2**:317a–b; **4**:96a; **5**:418b; **7**:204b–205a, 459a; **8**:307b; **9**:17b–18a, 52b, 92a–93a, 313a, 400b, 528b–529a; **10**:520a–b; **11**:47a–b, 313a–316a, 587b–588a; **12**:40b, 394a, 457a–b; **13**:282b, 442b–443b, 449a–b, 526a; **14**:413a; **15**:67a–68a
early eighteenth-century; **1**:486a–b; **2**:188b, 225a, 227a–b, 390a, 576b; **3**:129a–b, 344a, 349b–351a, 535a; **4**:98b–99b, 216a–b, 223b–224a, 295a, 603a; **5**:351b, 428b, 429a, 577b; **6**:61b, 65b–66a, 231a, 533b; **7**:197a–b, 199a–b, 210a–b, 495b; **8**:177a–178a, 374a–380a, 459b–460a; **9**:368a–b; **11**:64a–b, 83a–b; **12**:394a–b, 459a; **13**:553a–554a; **14**:193a
late eighteenth-century, **1**:45b–46a, 58b–59a, 127b, 433a–434b, 487a–488a, 520a–b; **2**:257a–b, 322b, 342b, 510a–b, 516b, 16b, 37b, 154b, 172b–173a, 177a, 365a–b, 426a, 538a, 578a–b; **4**:156b–157a, 178a; **5**:74a–b, 75a–b, 77a, 216b–217a, 380a–b, 424b, 443a,b, 444a; **6**:218a–220a, 400b–401a, 521a, 539a; **7**:58b, 198a–b, 390b–391a, 440b–442a, 584b–586b; **8**:251b, 303b, 373b; **9**:53b–54a, 140a, 253b–254a, 365b–366b, 432b–433a, 434a–b, 575b, 576a; **11**:272a–b, 605b; **12**:119b, 185a,b, 326b–327a, 471a–b, 587b–588b; **13**:177b–178b, 390a–b, 391b–392b, 497a, 506a, 601a–b; **14**:132a–133a, 386a–387b, 463b; **15**:113b, 429b–430a, 470a–471a
early nineteenth-century, **1**:69a–71a, 136a, 484b–485a, 485b, 520b; **2**:320b, 341a, 516b–521b; **3**:43b–45b, 291b, 562b–563a; **4**:21a–22a, 67a, 234a, 273b; **5**:190b–192b, 218a–b, 518a; **6**:184b, 492b–495a, 549a,b, 550b, 551a,b; **7**:196b, 408b–409b, 609b–610b; **8**:265a, 371a–372b, 564a; **9**:148b–149a, 418a–b; **10**:11b–13b, 163b–164b, 531a–b; **11**:218a–b, 262b–263b, 592a–b; **12**:215a, 591b; **13**:44a, 354a–b, 506a–b, 595b–596a, 601b; **14**:116a–117b; **15**:81b–83a, 267a–b
mid nineteenth-century, **1**:358b–359a, 423a–b, 614b–615a; **2**:18b–19a, 159b–160b, 425b–426b, 430b–431a; **3**:42b–43a, 196b–197a, 301a, 336b–338a, 586a, 575a–576a, 611a–612b; **4**:288b–289a, 306b–307a, 518b, 560a; **5**:440b–442a, 512a–514a, 539a–b, 546a–b, 611a; **6**:6b–7a, 72b–73a, 162b–163a, 220b–222b, 265b–266b, 288b–289a, 295b–296a, 326b, 464a–467a, 488b–492a; **7**:165a–166b, 533b–534b; **9**:60b, 340a–b, 344b–345b, 417b, 441b–442b, 561a; **10**:165a, 165b–166a, 575a–b; **11**:130a, 151b–154b, 502a; **12**:58a, 167a–168a, 169a, 173a, 173b–175b, 594a–b; **13**:168a, 394b–396a, 432a–433a, 463a, 463b–464b, 489a–490a, 542a–b; **14**:95b, 350a–b, 398a–b, 496a; **15**:130a–131a
late nineteenth-century, **1**:358b–359a, 396a, 412a–b, 422a–b; **2**:86a, 102b–103b, 183b–185a, 290a–b, 370b–371a, 436b–438a, 475b–476b, 477a; **3**:29a–30b, 187b, 259a–b, 340a, 421b–422a, 575a–576a, 581a–582a, 613a–614a; **4**:130b–131a, 401b–402a, 545b–546a; **5**:50a, 437b–439a, 581b–582a, 604b–605a, 623b–624a; **6**:99b–100a; **7**:94b, 285a, 417a–b; **8**:354a–b, 529a; **10**:1a–2a, 285a–b, 575b–577b; **11**:108b; **12**:99a–100a, 124b–125a, 166a–b, 259a–b; **13**:87b–90a, 154a–b, 341a–343a, 405a–406a, 456a, 459a–460a; **14**:181a–182a, 192a–b; **15**:13b–15a, 147a–148b, 268a–269a
early twentieth-century, **1**:205b–206a, 422b, 561a–b; **2**:34b–35a, 371a–372a, 582a–583b; **3**:30b–31a, 188a, 204a, 317b–318a, 429a–b; **4**:131a–b, 583b–584a; **5**:293a–294a, 584a–585a; **6**:121a–b, 122a–b; **7**:90b–91a, 102a–b, 535a–b; **8**:5a–6a, 512b–513b, 518a–b, 523a; **9**:24b–25a, 34a–35a; **10**:121b–122a; **12**:190b, 207a, 260a, 328a–b; **13**:477b–479a; **15**:60b, 148b–151a, 421a–b
mid twentieth-century, **9**:403b
applied, **3**:259b, 359a; **4**:546a; **6**:494b; **9**:455b; **10**:130a; **13**:459b; **14**:349b
chemistry and, **1**:70a; **8**:345b
culture methods, **4**:131a
developmental, **7**:395b–396a
economic
Renaissance, **2**:219a, b; **8**:554b
eighteenth-century, **7**:210b; **10**:324a
nineteenth-century, **6**:491a–b, 493b, 494b; **9**:123a–b, 618a–b; **13**:342a–b, 354b, 459a–b, 464a, 601b; **14**:96a, 349b
twentieth-century, **6**:122b
experimental
late seventeenth-century, **12**:357a–b
eighteenth-century, **3**:44a
nineteenth-century, **2**:160a; **5**:437b–438b; **9**:277a–283a, 618a–b
handbooks, **2**:160a, 290b; **4**:273b, 274a; **6**:467a; **7**:102b
historical, **4**:560a; **14**:387a–b
international exchange, **4**:216b
medical
ancient Greek, **4**:106b–107a
medieval Latin, **11**:601b
Renaissance, **2**:536a–b; **3**:40b; **4**:139a; **5**:361b, 383a, 384a; **8**:120b–121a; **10**:237a–b; **13**:502b; **15**:161a
seventeenth-century, **7**:536a–539b; **10**:622a; **11**:47a
eighteenth-century, **3**:44a, 177a, 578a–b; **4**:603a*; **5**:352b; **6**:400b, 401a, 539a
nineteenth-century, **2**:422b; **3**:45a; **10**:253b; **11**:263b*; **12**:166a; **13**:464a
twentieth-century, **8**:427b
Chinese, **12**:384a
Mesopotamian, **15**:645a
philosophy of, **6**:220a; **8**:378b; **13**:332a, 406a; **14**:586a–b
physiological approach, **13**:341b, 343a
and religion, **1**:423b; **14**:350b
separation from materia medica, **1**:524a; **13**:443b; **15**:81a
tropical, **3**:30a; **5**:624a; **8**:120b–121a; **13**:282b; **14**:181a; **15**:147a, 148a, 421b
see also Agriculture; Agronomy; Algology; Chemistry, vegetable; Cytology, botanical; Education, botanical; Herbaria; Historiography, botany; Horticulture; Illustration, botanical; individual species; Mycology; Nomenclature, botanical; Paleobotany; Physiology, botanical; Phytogeography; Plants; specific countries; Taxonomy, botanical; Terminology, botanical
Boteo
see Buteo, Johannes
Botfly, **7**:273a
Bothe, Walther Wilhelm Georg (1891–1957), 2:337–339; 3:369b; **5**:331b, 333a*; **7**:153a; **11**:348b*
Bothkamp Observatory, **14**:54b–55a
Botrychium, **8**:5b
Botryllus, **6**:442a
Botryopteris, **12**:260a
Botrytis, **1**:493b; **5**:503a
Botta, Paul-Émile, **2**:341b
Bottazzi, Filippo (1867–1941), 2:339–340; 12:564a*
Böttger, Johann Friedrich, **11**:109a, 330a; **13**:480a
Böttger, Rudolph Christian (1806–1881), 2:340; 12:198b
Böttger, Wilhelm, **6**:374a; **13**:226a
Böttiger, Carl Conrad, **5**:537a
Bottini, Enrico, **8**:406b

Botulin
detection of, **7**:286b
Bouchard, Charles Jacques, **3**:205b, 508a; **5**:269b; **8**:273a
Boucher, Étienne Jean, **3**:454b
Boucher de Crèvecoeur de Perthes, Jacques (1788–1868), 2:341; 15:50–52; 4: 350a; **5**:163a; **8**:43b–44a, 528a, 573b; **11**:131a, 235a
Boudinage, **8**:463b
Boudouard, Octave, **8**:118b
Boué, Ami (Amédée) (1794–1881), 2:341–342; 1:210b; **4**:349b; **7**:70b; **15**:51a
Bouelles, Charles de
see Bouvelles, Charles
Bougainville, Louis Antoine de (1729–1811), 2:342–343; 4:88a
expedition of, **3**:365a–b
Voyage autour du monde
editions of, **5**:75a, 76b
Bougainville Island, **2**:342b
Bougainville Strait, **2**:342b
Bougainvillea, **2**:342b
Bouguer, Jean, **2**:343a
Bouguer, Pierre (1698–1758), 2:343–344; 1:326b; **2**:43b; **4**:468b
editions of works, **7**:544b
geodesy, **3**:39a; **3**:109*; **5**:435b; **7**:183a; **7**:200a; **15**:270a–271a, 271b, 287b
law of, **2**:343b; **7**:599a; **9**:371a
photometer, **1**:81b
Bouillaud, Ismael
see Boulliau, Ismael
Bouillaud, Jean-Baptiste, **4**:594a; **5**:196b; **6**:440b
Bouillaud syndrome, **2**:35b
Bouilles, Charles de
see Bouvelles, Charles
Bouin, Pol André (1870–1962), 2:344–346; 1:152b–153a; **10**:571a*
Boulder clay, **10**:502a
Boule, Marcellin (1861–1942), 2:346–347; 5:296a, 297a*; **11**:270a*; **13**:274b, 275a
Boulenger, Julius Caesar, **6**:299a
Boulger, George Edward Simonds, **2**:476a
Boullanger, Nicolas-Antoine (1722–1759), 2:347–348
Boullay, Polydore, **4**:244a; **5**:325a; **8**:332b–334b *passim*, **8**:338a
Boulliau, Ismael (1605–1694), 2:348–349; 2:13a; **4**:166a; **5**:23b; **8**:26a; **9**:369a, 527b; **12**:342b
acquaintances of, **6**:360b, 361b, 538a, 598b; **9**:498a; **12**:84b; **13**:335a
Astronomia Philolaica, **7**:308a; **14**:177b
astronomy, **13**:96a
Boulliau's construction, **4**:17a–b
cosmology, **9**:310b, 486a; **14**:446b
photometric law, **3**:116a
Boulonnais massif, **5**:476a
Boulton, Matthew, **3**:578a, 579a; **6**:511b, 579a; **7**:277a; **11**:141a, 304b, 354a, 499b; **14**:196b–197a, 463a
Bouncing, **2**:606a
Bound, greatest lower, **2**:275a–b
Boundary layer theory, **11**:123b, 124b
Boundary-value problems, **2**:144a–b, 198a, 217b; **3**:145a; **4**:126b; **5**:95b–96b, 305a; **6**:391b, 392a, 394a; **7**:7b*, 411b*, 514a; **8**:562b; **9**:125b–127b; **10**:25a, 248b, 553b; **11**:59b, 452a, 477b–478a; **12**:246a,b, 526a, 527a; **13**:26b–27a
Bounty (ship), **1**:435b; **12**:516a
Bouquet, Jean-Claude (1819–1885), 2:349–350; 2:471a; **3**:141a–b, 178a; **5**:204a, 481b; **9**:479b; **11**:56b, 206b
Théorie des fonctions doublement périodiques, **8**:383b
Bour, Edmond (1832–1866), 2:350–351; 8: 382b
surfaces, theory of, **2**:288a; **3**:331a
Bourbaki, Charles-Denis-Sauter, **2**:351b; **4**: 266b
Bourbaki, Nicolas (*n.d.*), 2:351–353; 1:195a, 307b, 407b; **8**:384a; **11**:79b
Bourdelin, Claude (1621–1699), 2:353; 8: 173a; **9**:62a
Bourdelin, Louis-Claude, **3**:5a; **8**:619b; **11**: 562a
Bourdelot, Pierre Michon (1610–1685), 2: 353–354; 4:267a; **7**:111a; **8**:172b; **11**:342b*
academy of, **2**:353b–354a
Bourdin, Pierre, **2**:13a
Bourdon, Amé, **7**:274b
Bourdon, Eugène (1808–1884), 2:354–355
Applications de l'algèbre à la géométrie, **3**:560a*
Bourdon tube, **2**:355a
Bourges, University of, **12**:62b
Bourgne, Robert, **5**:262a
Bourguet, David
Chemisches Handwörterbuch nach den neuesten Entdeckungen entworfen, **11**:437b
Bourguet, Louis (1678–1742), 15:52–59; 3: 210b
Bourignon, Antoinette, **13**:169a
Bournon, Jacques-Louis, Comte de (1751–1825), 2:355; 11:523a
collection of, **2**:106a
Bournonite, **2**:355b; **4**:190b; **14**:625a
Bourseul, Charles, **4**:248a
Boussac, Jean, **6**:168a; **13**:284a
Boussinesq, Joseph Valentin (1842–1929), 2:355–356; 11:498b*; **12**:74b*
Boussingault, Jean Baptiste Joseph Dieudonné (1802–1887), 2:356–357; 6: 553a; **8**:93b, 349a; **10**:436a; **11**:352b; **13**:416b; **14**:33b, 63b
Bouton d'Orient, **12**:315a
Boutron-Charlard, Antoine François, **8**: 333a; **11**:495a; **14**:477a
Boutroux, Émile, **2**:357b, 544b, 545a; **9**: 382b; **11**:388a; **13**:249b, 252a; **15**: 423a
Boutroux, Pierre Léon (1880–1922), 2:357–359; 2:544b; **3**:220b
Bouty, Edmond, **3**:456a, 498a; **8**:10a; **11**: 388a
Bouvard, Alexis (1767–1843), 2:359–360; 1:202b; **8**:277a; **12**:523a; **15**:382b, 388a*
celestial mechanics, **15**:348b–349a, 353a,b, 354a,b, 356a, 376a, 381b, 382a
Bouvard, Charles, **7**:536b
Bouvelles, Charles (*ca.* 1470–*ca.* 1553), 2: 360–361
Boveri, Theodor (1862–1915), 2:361–365; 15:124a
acquaintances of, **5**:453a; **10**:427b; **14**: 424b, 427a–b
on contemporary scientists, **6**:9b
cytology, **1**:601b; **4**:187a; **5**:52a,b; **6**:338b; **8**:283a, 588b; **11**:255b, 573b; **14**:236a, 427a–b, 428b, 430a
embryology, **13**:281b–282a; **15**:513b
genetics, **13**:158a
students of, **10**:276a; **12**:567b
Bovet, Daniel, **4**:154b; **5**:100a
Bovillus
see Bouvelles, Charles
Bowditch, Henry Pickering (1840–1911), 2: 365–368; 6:52b; **7**:504b; **9**:142b, 416a; **13**:53b; **15**:71b, 72b
Bowditch clock, **2**:366b
physiology, **8**:533a; **11**:125b
Bowditch, Nathaniel (1773–1838), 2:368–369; 1:66a; **2**:365b; **4**:546b; **10**:479a; **15**:351b, 354b
Bowen, Ira Sprague, **4**:280b; **6**:542a; **7**:270b; **9**:397a–b; **14**:521b
Bowen, Norman Levi (1887–1956), 2:369–370; 4:409a, 564a–b; **6**:475b; **8**:42a; **10**:124b, 125b; **12**:274b
Bower, Frederick Orpen (1855–1948), 2: 370–372; 6:468a*; **8**:6a
acquaintances of, **3**:30a; **5**:604b; **12**:259a; **13**:342a
on contemporary scientists, **5**:605b*; **13**:343b*
influence of, **2**:503b; **8**:5a
influences on, **5**:80b
phylogeny, **3**:29b, 30b
reproduction, **7**:314b*; **11**:154b
Bowie, William (1872–1940), 2:372–373; 6:188a, 241a
Bowles, George, **7**:147a,b
Bowman, Isaiah (1878–1950), 2:373–374; 10:509a*
The New World
translations of, **2**:539a*
Bowman, John Eddowes, **2**:375a
Bowman, William (1816–1892), 2:375–377; 1:539b; **4**:163a; **5**:565b
Bowman's capsule, **2**:376a; **6**:226a
Bowman's membrane, **2**:376b
Bowman's muscle, **2**:376b
Bowman's operation, **2**:376b
Bowman's probes, **2**:376b
Bowman's tubes, **2**:376b
on urinary secretion, **6**:226a; **7**:23b; **15**: 101b
Boyce Thompson Institute for Plant Research, **14**:552a–553a
Boyd, George (geol.), **6**:57a
Boyd, George Edward (chem.), **6**:118b
Boyd, Sprott, **11**:216b
Boyd, Thomas Alvin, **7**:316a, 498a
Boyer, Alexis, **9**:7a
Boyle, Robert (1627–1691), 2:377–382; 3: 209a, 537b; **5**:288a, 416b, 420a; **6**: 141a; **7**:525b; **8**:70a, 437b
acquaintances of, **2**:46b, 317a; **4**:221b; **6**:141b, 161b(n), 387a, 458b; **8**:149b, 161a, 437a; **9**:105b, 313a; **10**:200b, 201a, 202b, 292b; **11**:532a; **12**:357a,b, 457a,b, 584b; **13**:214a,b; **14**:369a, 405a
alchemy, **10**:81a; **12**:311b
alcohol, **3**:482b
anatomy, **14**:407a
on atmosphere, **5**:288b; **6**:40b–41a, 183a
biographies of, **14**:157b
Boyle's law, **1**:128b, 138b, 143a, 203a; **2**:377b; **3**:32b; **4**:241b, 243b; **5**:78b; **6**:482a–b; **8**:469a; **9**:117b, 119b, 291a; **10**:75a, 286b; **11**:353b; **13**:445a, 579a; **15**:377b
calcination, **3**:206a
combustion, **6**:484b
on contemporary scientists, **10**:254b; **14**:363b
correspondence of, **3**:214a; **4**:37b, 95b, 495b; **7**:194a; **8**:524a,b; **9**:465a; **10**: 44b, 80a, 83a; **12**:210b; **14**:406a, 407a
critics of, **1**:451a; **7**:459a; **9**:509b, 510a
editions of works, **12**:365b
electricity, **6**:170b
Essay on Nitre, **5**:422a
experiments, **6**:36a, 170a; **12**:210b

"fixed air," 6:41b
heat theory, 2:179b
and Hobbes, 2:377b; 6:450a
and Hooke, 6:482a–b, 487b; 9:245a
influence of, 1:549b; 2:16a, 226b, 237b, 309a; 3:495a; 4:26a; 5:130a; 6:130a, 598b; 8:468a; 10:43b, 55b, 80b; 11:40b; 13:618b; 14:27a, 405b
influences on, 10:311a; 14:374b, 377a(n)
lectures, 2:382a; 3:294b; 5:130b
natural history, 5:415a
New Experiments Physico-Mechanical, 13:445a
"Origine of Form and Qualities," 2:106b
philosophy, 1:134a; 6:43b; 8:174b; 9:15a
phosphorus, 7:525a
physiology, 14:321a, 407b
pneumatics, 3:482b; 4:184b; 5:575a; 9:117a, 243a, 244a, 244b, 245b; 10:334a; 13:169b
and Royal Society, 14:510b
The Sceptical Chymist, 1:550a; 4:209b
students of, 6:477b
translations of works, 10:202b*
on water as element, 1:549b; 8:71b; 10:80a
Boym, Michael, 8:394b
Boyneburg, Johann Christian von, 8:149a,b, 152b
Boys, Charles Vernon (1855–1944), 15:59–61; 5:78a; 11:122b
Boysen-Jensen, Peter, 3:581b
Boze, Claude de, 1:323a
Bozzini, Philipp, 3:531a
Bozzolo, Camillo, 2:164b; 4:198a
Braccio (unit of length), 8:198a
Bracciolini, Gian Francesco Poggio, 8:538a; 15:516b
Brace, De Witt Bristol (1859–1905), 2:382–383; 8:496b
Brachet, Albert (1869–1930), 2:383–385
Brachet, Jean, 2:384a
Brachinus, 6:500b
Brachiopods, 3:536a; 7:460b; 9:536b; 13:143b
paleontology, 3:237a; 5:486b; 9:579a; 10:222a; 12:229b–230b; 13:145b; 14:120b, 312b
Brachistochrone, 2:48a, 53a–b; 3:459a; 6:538a; 8:163b, 304a, 611a; 10:52b; 12:118a
Braconnot, Henri (1780–1855), 2:385–386; 11:171a
Bradbury, John, 10:163b
Bradford, Sir John Rose, 1:536b, 537a; 2:528b
Bradford, Joshua Taylor (1818–1871), 2:386–387
Bradley, Albert James, 6:561a
Bradley, Francis Herbert, 3:90a; 6:26b
Bradley, James (1693–1762), 2:387–389; 5:24b; 6:511b; 14:485a
aberration of light, 1:201b; 3:283b–284a; 5:491b; 6:512b; 9:77b, 214a, 463b–464a; 11:526a; 15:338b
acquaintances of, 9:162b, 164a
Astronomical Observations, 6:512a
astronomical tables, 14:178b
on contemporary scientists, 5:492a*
correspondence of, 8:179a; 9:267a
observations of, 1:340a; 2:98b, 99a; 8:179a; 9:164b; 13:110a
reduction of, 2:97b–98a, 616b; 3:194b; 5:25b; 9:563b
at Royal Observatory, 2:141a; 6:6a
Bradley, James Chester, 6:507a
Bradley, Richard (*d.* 1732), 2:390; 2:188b
Philosophical Account of the Works of Nature, 2:188b
Bradwardine, Thomas (*ca.* 1290–1349), 2:390–397; 1:82b, 94b, 226a; 4:237b; 5:360b; 6:276a*; 9:32b, 132b
Arithmetica speculativa
editions of, 3:280a
Bradwardine's law, 1:290b; 2:194b, 392a–394a; 7:116a; 10:224a; 13:186a, 194a, 197a–207a *passim;* 14:156a; 15:265a
critics of, 9:133a–b
Geometria speculativa, 1:381a; 2:360b; 3:513b, 515a
editions of, 3:280a,b
mechanics, 13:186a
Tractatus de proportionibus velocitatum in motibus, 1:94a; 6:378b; 13:197a, 349b
Brady, Henry Bowman, 3:89a, 519b
Bradycardia, 3:518a
Bradypus, 12:224b
Bradyseisms
concept of, 7:30b–31a
Bragelogne, Christophe de, 3:460a
Bragg, William Henry (1862–1942), 2:397–400; 1:195b, 319a; 3:369a; 5:457a; 6:303a; 8:51b, 484b; 12:532a*; 13:524b*; 14:440a, 525b; 15:17a
and Barkla, 2:398b
Bragg's law, 2:399a; 4:196b
X rays, 1:457b–458a; 2:487b; 9:542b; 12:31b–32a; 15:61a–63a
Bragg, Sir William Lawrence (1890–1971), 15:61–64; 3:564b; 4:196b, 521b; 5:103b, 457a; 6:303a; 8:51b; 9:543a; 14:440a; 15:17a, 20b*
Bragg angle, 15:62a
Bragg's law, 15:62a
crystallography, 1:461a; 14:525b
X rays, 2:399a–b, 487b; 3:370a, 565a*; 9:543a; 12:32a
Brahe, Tycho (1546–1601), 2:401–416
acquaintances of, 2:185b; 9:247a, 611a; 11:561b; 12:334b, 499b; 14:359a, 470a–471a
Astronomiae instauratae mechanica, 13:622a
biographies of, 2:453a
on comets, 4:213a; 7:302b
on contemporary scientists, 14:274a, 359b
correspondence of, 1:447b; 4:619b; 11:561b; 14:359a
cosmology, 3:409b; 8:281a
critics of, 3:61a, 262b; 5:243b; 6:520b
editions of works, 4:185b
geodesy, 12:500b
influence of, 1:311a; 3:60a; 5:415b; 6:514a; 7:269b; 12:431a; 13:433b, 621b
influences on, 1:513a; 11:366a–b; 14:274b, 461a; 15:476a
instruments, 6:362b
and Kepler, 2:413b; 7:293a, 293b–295a, 298a, 302b
manuscripts of, 1:481a; 11:525b
observations of, 1:244b; 4:168b; 6:515b; 7:294a, 295a, 297a, 306b; 10:596a; 12:170a, 500b
observatory, 10:490a
Progymnasmata, 9:167b
and Severin, 12:332a–b
star catalog, 1:403b*; 3:528a; 5:23a, 24b; 6:364a; 7:305b; 11:202b
students of, 3:59b; 9:336a
translations of works, 13:99b
trigonometry, 8:28a; 14:470b–471a
see also Cosmology, Tychonic
Brahma, 15:554b, 555b, 590a, 612a
Brahmadeva (*fl. ca.* 1092), 2:416
Karaṇaprakāśa, 2:119b; 15:600a–601a
sine table, 15:600b
Brahmagupta (598–665), 2:416–418; 5:159a; 8:422b; 15:557b, 590a,b, 592a, 593a,b
Brāhmasphuṭasiddhānta, 2:154a; 4:555b–556a; 7:360b, 361a; 8:46b; 13:582a, 590b; 15:543a, 562a–582b *passim,* 585a,b, 586a,b, 592a, 593a,b, 605a,b, 606b, 607a–b, 610b
cosmology, 15:555a
critics of, 13:590b
Dhyānagrahopadeśa, 15:565a
geometry, 12:500a; 15:564a
influence of, 15:584a, 586a, 587a, 598a,b
Khaṇḍakhādyaka, 5:613b; 15:604a, 605b–607b, 608a, 609b–610a
commentaries on, 7:582b–583a
sine tables, 15:579b, 607a
Brahmah, Joseph, 11:603b
Brāhmaṇas, 15:534b
Brahmapakṣa (school of astronomy), 3:584b; 4:555b, 556a; 7:315a; 9:21a, 580a; 15:534a, 555a–590b, 591b, 596b, 597a, 598b, 599a, 605b, 607b, 609a, 613a, 619b, 621b, 624b
Brahmaputra (river)
source of, 6:216a
Brahmaśarman
commentaries of, 2:416b
Brahmasiddhānta, 15:613b
Braid, James, 9:327b
Braids (math.), 1:307b
Braidwood, Thomas, 5:468a
Braikenridge (Brakenridge), William (*ca.* 1700–1762), 2:418–419
Braikenridge-Maclaurin theorem, 2:418b; 13:17b
Brain
ablation studies, 5:44b, 208a; 6:441a, 518b; 8:536a, 552b; 9:20b, 177a; 12:355b, 473b–475a; 15:109a
activity
and psychic phenomena, 2:1a
anatomy of
ancient Greek, 4:385a–b; 6:317a
Byzantine, 10:21a
Renaissance, 1:46b, 204a, 620a; 4:199a–b; 7:17a; 8:203a; 9:166a; 13:587a–b; 14:10b
seventeenth-century, 4:268a; 5:419a; 9:400a; 13:33a, 223a; 14:25b–26a, 256a, 406a–407a
eighteenth-century, 4:275a; 9:483a; 10:266b; 12:509b, 510a; 14:14b–15a
early nineteenth-century, 1:386b, 583b–584a; 2:595a–596a; 5:251a–b; 9:10a; 11:364b, 510b; 12:596b–597a
mid nineteenth-century, 2:478a; 4:593b; 5:571a; 6:573b; 12:224a–b; 15:110a
late nineteenth-century, 1:580a; 4:353a–b, 593b; 5:73b, 172a; 8:15a; 11:383a; 14:277a–b
early twentieth-century, 1:545a, 545b–546a, 580a; 6:321a; 8:552b; 12:48a
comparative, 5:46b, 196a, 505a; 6:593a–b; 10:262b; 11:383a, 514a; 12:315b; 14:229a
dissection techniques, 5:251b; 13:587b
membranous envelopes, 1:391b; 10:266b
myelogenesis technique, 5:27b
pathological, 8:352b
and body temperature control, 1:450a–b; 2:483b–484a; 10:252b; 11:427b–428b
chemistry of, 5:432a; 10:511b; 13:597a
circulation in, 1:569b–570a; 2:1a, 2a; 9:547a; 14:407a
concussion of, 2:305b

and Freud, **2**:447b–449a; **5**:172b, 173a,b, 174a, 176a, 179a
Breuil, Henri Édouard Prosper (1877–1961), 2:450–451; **2**:172a, 189a,b; **3**:622b; **10**:607b*; **11**:484a; **12**:520a; **13**:275a, 276a, 277a*
Breunnerite, **6**:18b
Brevium, **6**:16b; **12**:505b
Brewer, Thomas Mayo, **1**:405b
Brewer, William Henry, **10**:150b; **12**:436a; **14**:192b, 315b, 423b
Brewing
see Beer, brewing of; Fermentation
Brewing industry, **4**:211a; **5**:476b–477a, 536b
Brewster, David (1781–1868), 2:451–454; **14**:198a*
acquaintances of, **1**:201a; **5**:68b; **6**:18b; **11**:496b; **12**:523a, 524a; **13**:238a, 499a
color theory, **9**:200a–b, 201b–202a; **13**:77a
correspondence of, **12**:288a
critics of, **2**:134a; **6**:246a–b
crystallography, **12**:543a; **14**:241b(n)
Edinburgh Encyclopaedia, **2**:336a
historiography, **3**:261a*; **10**:45b, 86b(n)
influence of, **6**:19a
instrumentation, **2**:390a; **14**:290a
light, theories of, **6**:88b
periodicals, **7**:69b
polarization, **1**:201b; **9**:74a; **12**:281b; **14**:567a
reflection, **9**:74a
refraction, **9**:199b, 221a
spectroscopy, **7**:74b
Breyn, Jacob, **9**:17b
Exoticarum plantarum centuria prima, **13**:282b
Brianchon, Charles-Julien (1783–1864), 2:454–455; **4**:602a; **5**:368a; **6**:1b; **9**:476a
theorem of, **2**:215b, 455a; **3**:555a; **5**:368a
Briand, Joseph, **5**:297b
Bridges, Calvin Blackman (1889–1938), 2:455–457; **9**:515b–516a, 564b; **13**:134a; **14**:432a
genetics, **3**:122a; **9**:521b, 522a–b; **10**:276b; **13**:136a–b; **14**:101a, 431b
Bridges
Burr arch-truss, **3**:496a
construction of
Renaissance, **13**:233b
eighteenth-century, **10**:528a; **11**:163b–164a
nineteenth-century, **2**:104a; **5**:63a–b; **9**:138b; **11**:398a; **14**:473b
design of, **11**:496b
equilibrium of, **6**:577a
metal, **3**:287a, 495b; **7**:602a; **9**:577b; **10**:3a; **11**:355a, 496b; **13**:78b
stone, **10**:528a
suspension, **10**:4a; **11**:398a; **12**:287b; **13**:614a
Town lattice-truss, **3**:496a
tubular, **6**:452b
vibration of, **12**:527a
wind effects, **7**:247a
wooden
United States, **3**:495b–496a
see also Drawbridge
Bridgewater treatises, **2**:570a, 583b; **3**:238b; **7**:366a; **11**:172b, 173b
Bridgman, Laura, **4**:160b
Bridgman, Percy Williams (1882–1961), 2:457–461; **4**:102a, 197b*; **6**:52a*; **8**:579a*; **10**:213b; **12**:609b–610a; **13**:245a
Brieger, Ludwig, **1**:575b; **4**:297b; **7**:426b; **15**:521b
Brien, Victor, **8**:463b
Briggs, George Edward, **6**:22a
Briggs, Henry (1561–1630), 2:461–463; **4**:450b; **5**:593b; **6**:127b, 271b; **9**:127a; **10**:151b; **13**:501a; **14**:51b, 365a, 377b(n), 513a,b, 514b
Arithmetica logarithmica, **9**:610a, 613a; **14**:51a
Logarithmorum chilias prima, **9**:613a
and Napier, **9**:610a, 612b–613a
tables of, **8**:142b*; **9**:554b
translations of works, **14**:51a
Trigonometria Britannica, **5**:343b; **14**:51b
Briggs, Isaac, **7**:89b
Bright, Charles, **3**:288b
Bright, Richard (1789–1858), 2:463–465; **1**:60a; **14**:253b
Bright's disease, **2**:464a
Brill, Abraham Arden, **5**:180a
Brill, Alexander Wilhelm von (1842–1935), 2:465; **2**:430a; **3**:135b, 167b, 315a; **8**:542b*; **11**:407a; **14**:344a*
analysis, **11**:451b
Chasles-Cayley-Brill theorem, **3**:167b
Die Entwicklung der Theorie der algebraischen Functionen . . . , **11**:448b
geometry, **12**:330b
transformations, **12**:284b
Brill, Nathan Edwin
Brill's disease (Brill-Zinsser's disease), **5**:451b; **14**:623b
Brill, Rudolf, **4**:522a; **6**:304a
Brillouin, Léon, **2**:466a; **11**:266b
Brillouin, Marcel Louis (1854–1948), 2:465–466; **3**:620a; **8**:9b; **10**:525a
Mandelshtam-Brillouin effect, **9**:76b
Brinell, Johan August (1849–1925), 2:467
Bring, Ebbe Samuel, **2**:468a
Bring, Erland Samuel (1736–1798), 2:467–468; **7**:103a,b
Brinkley, John (1763–1835), 2:468–469; **3**:432b; **6**:85b
Brinton, William, **1**:544a
Brinvilliers, Marie Madeleine, Marquise de, **5**:417b
Brioschi, Carlo, **10**:593a
Brioschi, Francesco (1824–1897), 2:469–470; **1**:599a; **2**:85a, 105b; **3**:467a,b; **7**:506b; **12**:171a–b; **14**:49b
algebra, **3**:164a; **6**:309a; **7**:398a; **9**:55a
Briot, Charles Auguste (1817–1882), 2:470–471; **2**:349b, 350a; **3**:141a–b, 178a; **5**:204a, 481b; **11**:206b
analysis, **11**:56b
Théorie des fonctions doublement périodiques, **8**:383b
Brisbane, Thomas Makdougall (1773–1860), 2:471–472; **2**:468b
Brissaud, Edouard, **9**:109a
Brisson, Barnabé (1777–1828), 2:473; **9**:474a; **14**:18a
Brisson, Mathurin-Jacques (1723–1806), 2:473–475; **5**:355b; **8**:623b*; **10**:147a; **15**:502a
and Académie des sciences, **3**:5b; **7**:569a; **8**:79a; **9**:554b; **15**:333b, 336a
chemistry, **12**:64a
Ornithologie, **12**:536b
Brissot, Jacques-Pierre, **15**:302a
Brissot, Pierre, **5**:233b; **9**:264a; **14**:4b
Bristol, University of
physics at, **11**:117b
Bristow, Henry William, **5**:68a
Bristow, Jack Randolph, **8**:480b
Brit, Walter
see Brytte, Walter
British Association (for the Advancement of Science)
in mid-nineteenth century, **2**:567a; **3**:586a; **5**:66a; **6**:86a, 110a; **7**:51a; **9**:583b; **11**:419a
in late nineteenth century, **3**:479b; **5**:413b–414a; **6**:5b; **13**:104b, 368a
in early twentieth century, **3**:597b; **8**:443a
anthropology, **6**:563b
botany, **5**:604b; **13**:343a
Committee on Electrical Standards, **5**:423b; **7**:93b; **13**:102a
and Compton effect, **3**:369a
Dalton and, **3**:545b
and Darwinism, **3**:574a; **5**:17a; **6**:592b–593b
founding of, **1**:354b; **2**:172b, 451b; **5**:68b; **6**:289a, 325b; **10**:12b, 583b
geomagnetism, **8**:426b; **12**:49b–52a *passim*
optics, **6**:88b
seismology, **9**:406b, 407a
star catalog, **1**:403b*
British Astronomical Association, **3**:472b
and eclipses, **4**:176b*
founding of, **4**:175b, 176a; **9**:184b
British Columbia
geology, **14**:120b
British East India Company, **11**:494a
British Encyclopaedia, **10**:109a
British Institute of Preventive Medicine, **6**:110b; **8**:409a
British Magazine
editors of, **6**:401a
British Medical Association
reorganization, **6**:519a
British Medical Journal, **8**:407b
British Meteorological Society
see Royal Meteorological Society
British Mineralogical Society
founding of, **1**:359b
British Museum, **12**:488a
archaeology, **14**:504a
botany, **2**:476a, 517a, 518a–b; **9**:587a–b; **14**:194b
Dawson Collection, **3**:606a–b
founding of, **12**:459a
mineralogy, **12**:572b–573a
Natural History Departments, **5**:45b, 46a; **10**:260b–261a; **12**:515b, 516b
British Ornithologists' Club, **12**:258b
British Oxygen Company, **6**:94a
British Physiological Society
founding of, **5**:83a
British School of Archaeology, Egypt
founding of, **10**:549b–550a
British Science Guild, **8**:443a
British Tar Company, **4**:250a
Brittany
geography, **9**:149b
geology, **1**:472a–b; **4**:218a
Britte, Walter
see Brytte, Walter
Britten, James (1846–1924), 2:475–476; **1**:89b; **8**:304a*; **9**:588a*
Brittleness
of steel
measurement of, **3**:212a
Britton, Elizabeth Gertrude Knight, **2**:477a
Britton, Nathaniel Lord (1859–1934), 2:476–477
Brittonia, **2**:477a
Brizard, Léopold, **2**:487b
Broc, Pierre-Paul, **2**:442b
Broca, Pierre Paul (1824–1880), 2:477–478; **1**:562b; **4**:266b; **5**:504b; **11**:178b*; **14**:17a*
acquaintances of, **13**:288a
Broca's convolution, **2**:478a
cerebral localization, **4**:594a; **5**:196b, 255a; **6**:440b

Broca, Pierre Paul (cont.)
critics of, **9**:109a; **11**:178a
influence of, **2**:366a; **3**:245b; **7**:48a
Brocard, Pierre René Jean-Baptiste Henri (1845–1922), 2:478–480; **8**:387a*
Brocard configuration, **2**:479b–480a
Brocchi, Giovanni Battista (1772–1826), 2: 480–482
Brochant de Villiers, André-Jean-François-Marie (1772–1840), 2:482; **4**:347b; **8**:288b; **14**:239b, 241b*
geological map, **4**:218a, 348b–349b
Brochu, Michel, **8**:30b
Brockmann, Hans, **14**:445a
Brödel, Max (1870–1941), 15:64–65
Brödel's line, **15**:65a
Brödel's suture, **15**:65a
Broderip, William John, **2**:566a–b, 569b
Brodhun, Eugen, **8**:551b
Brodie, Benjamin Collins (1783–1862), 2: 482–484; **3**:324a; **6**:479a; **14**:187b
Brodie, Benjamin Collins, Jr. (1817–1880), 2:484–486; **2**:70a, 515a; **5**:374a; **6**: 109b, 565a; **10**:177a,b; **13**:341b; **14**: 396a
chemistry, **1**:161a; **3**:476b, 478b
Brøgger, Waldemar Christopher (1851–1940), 2:486–487; **5**:457a; **12**:271b; **14**:58a,b, 59a
Broglie, Louis-César-Victoir-Maurice de (1875–1960), 2:487–489; **8**:11b; **12**: 35b*; **13**:447b*; **15**:48a, 422b
de Broglie waves, **4**:318a
electron theory, **3**:597b, 598a; **11**:421b
quantum theory, **12**:218b, 219a
wave mechanics, **6**:89b
Broglie, Louis Victor de, **2**:248b, 466a, 488a; **4**:514a*; **8**:14a*, 52a; **13**:42b; **15**:48a, 49b
Broili, Ferdinand (1874–1946), 2:489–490
Bromelius
see Bromell, Magnus von
Bromell, Magnus von (1679–1731), 2:490
Bromides, **9**:422b
Bromination, **9**:619b
Bromine
atomic weight of, **3**:398a
compounds of, **2**:199b, 316a, 366a; **12**: 321b
discovery of, **1**:416a–b
and hydrogen, photochemical reaction of, **3**:197b
medical use of, **9**:9b
reactions of, **7**:203b
in sea water, **1**:416a–b; **3**:437a*; **13**:67a*
Bromoform, **12**:321b; **14**:478a
formula of, **4**:245b
Bromwich, Thomas John I'Anson (1875–1929), 2:490–491
Bronchitis
chronic, **14**:408a
Bronchopneumonia, **12**:482b
etiology of, **9**:185b
Brondgeest, Paul Quirin, **9**:20a
Brongniart, Adolphe-Théodore (1801–1876), 2:491–493; **1**:328b; **4**:242b, 265a*, 560a,b*; **7**:199a*; **10**:32a; **11**:373a
acquaintances of, **7**:196b; **12**:104b; **15**: 152a
paleobotany, **13**:44a; **14**:88a
Brongniart, Alexandre (1770–1847), 2:493–497; **4**:218b; **6**:179a
acquaintances of, **1**:328b; **2**:108a; **4**:242b; **8**:288b, 564b; **12**:317b; **15**: 50b, 126a, 152a
critics of, **11**:133b; **14**:600a
and Cuvier, **2**:493b, 494a; **3**:525a, 526a
Essai sur la géographie minéralogique des environs de Paris, **2**:481a
influence of, **4**:365a; **5**:578b
mineralogy, **1**:557a; **3**:313a–b
paleontology, **7**:589a
on species, **12**:318a
stratigraphy, **1**:210a; **2**:568b; **5**:533a; **10**:209b; **14**:211a
students of, **4**:65b; **8**:54b; **13**:489a, 581a; **15**:471b
Brongniart, Antoine-Louis, **2**:493b; **8**:619b; **13**:597a
Brongniart, Mathilde, **1**:329a
Bronn, Heinrich Georg (1800–1862), 2:497–498; **7**:475a; **14**:626a; **15**:78a, 526a
Lethaea geognostica, **11**:500b
Brønsted, Johannes Nicolaus (1879–1947), 2:498–499; **6**:481a
chemical kinetics, **2**:170a–b
Brontotheres, **9**:134b
Bronze, **5**:583b
early use of, **1**:388a; **9**:539a; **15**:651b
specific gravity, **7**:346a
Bronze Age, **3**:622b; **14**:398a, 506a,b
lake dwellers, **4**:73b
subdivision of, **9**:489a–b, 539a
Brooke, Henry James, **4**:66a; **6**:181a
Brooke, John Mercer, **4**:291b
Brooklyn Institute of Arts and Sciences
Biological Laboratory, Cold Spring Harbor, **3**:590a
Brooklyn Polytechnic Institute, **4**:521b–522a
Brooks, Alfred Hulse (1871–1924), 2:499–501
Brooks, Thomas Benton, **2**:500a
Brooks, William Keith (1848–1908), 2:501–502; **6**:614a*; **9**:142b, 519a; **14**:434a
students of, **1**:505a; **3**:389b; **6**:131b; **9**: 515b, 516a; **13**:281b; **14**:424a, 425a–b, 426b
Brooks, William Robert (1844–1921), 2: 502–503; **1**:465a
Broom, Robert (1866–1951), 2:503–506
Broome, Christopher Edmund, **2**:19a
Broscius, Joannes
see Brożek, Jan
Brougham, Henry, **1**:433a; **2**:174a, 175a, 451a,b; **12**:522b, 524a; **14**:565a,b
Broun, John Allan, **2**:472a
Brouncker, William (1620–1684), 2:506–507; **3**:128a; **4**:95b, 576a(n); **6**:482b; **9**:508a*; **10**:495b; **12**:584b
analysis, **13**:69a; **14**:148b
number theory, **5**:159b; **10**:48a; **14**:149a,b
Broussais, François Joseph Victor (1772–1838), 2:507–509; **2**:574b; **5**:255b*; **10**:612b, 613a; **11**:560a
Broussonet, Pierre-Auguste-Marie (1761–1807), 2:509–511; **2**:321b; **3**:172b; **4**:21b; **13**:497b*
Brouwer, Dirk (1902–1966), 2:511–512; **10**:36a*; **12**:177a*, 449b; **15**:129a
Brouwer, Luitzen Egbertus Jan (1881–1966), 2:512–514
acquaintances of, **13**:548b
on contemporary scientists, **7**:466a*
influence of, **14**:283a, 285a
influences on, **12**:196a
intuitionism, **6**:393a, 473a; **11**:60b; **12**:15a
topology, **2**:236b, 265a; **3**:55a; **6**:496a,b; **7**:312b; **8**:542b; **11**:53a
Brown, Alexander Crum (1838–1922), 2: 514–516; **2**:446b; **4**:78a, 400a; **5**:45a; **9**:246b*; **10**:177b; **12**:25*; **13**:352b; **14**:495b
formulas, **5**:126a; **8**:508a
notation, **4**:78b
physiology, **11**:215b
Brown, Ernest William (1866–1938), 2:516; **1**:156b; **3**:583a; **6**:400a*, 431a; **10**: 36a*
celestial mechanics, **2**:511a; **6**:399a; **8**: 523b*; **10**:34a,b; **10**:602b*; **15**:129a
Brown, Fay Cluff, **11**:420a
Brown, Frank Emerson, **6**:118a
Brown, George Lindor, **15**:106b
Brown, George Matthews, **5**:38a
Brown, John (medicine), **5**:276a, 411a–b; **6**:458a; **12**:155a
see also Brunonianism
Brown, John Stafford, **9**:258a
Brown, Robert (1773–1858), 2:516–522
acquaintances of, **1**:521a; **6**:493a; **8**:371a; **12**:173b
and Banks, **1**:434b; **2**:517b, 518a–b
botany, **1**:520b; **11**:262a
Brownian movement, **2**:170b, 267b, 517a, 520a–b; **4**:314b–315a, 316b; **5**:483b; **7**:104a, 493a; **9**:129b; **10**:525a–b; **15**:464a,b
in colloids, **13**:159b
discovery of, **7**:13b
Einstein-Wiener theory, **1**:366b; **8**:11a
mathematical study of, **14**:345a–b
measurement of, **2**:487b
theories of, **9**:578b, 616b; **10**:461a; **11**:278b; **12**:496b–497b; **13**:159b; **14**:343b; **15**:449a
cytology, **9**:600b; **12**:174a, 241b
and Darwin, **2**:517b, 521a
editions of works, **10**:12b–13a
influence of, **6**:465a
pollination, **12**:588a
taxonomy, **1**:89b, 167b; **7**:148a, 198b; **8**:372a
Brown, Thomas, **6**:81a
Brown Institution, **2**:599a
Browne, Benjamin Chapman, **13**:608a,b
Browne, Edward, **2**:522b, 523a
Browne, John Symonds Lyon, **3**:352b
Browne, Thomas (1605–1682), 2:522–523; **4**:495a; **6**:45b(n); **11**:121b, 316b
Brownian movement
see Brown, Robert
Brownrigg, William (1711–1800), 2:523–524; **6**:43a; **11**:144a
Brown-Séquard, Charles-Édouard (1817–1894), 2:524–526
acquaintances of, **1**:303a; **6**:434b; **9**:546b; **11**:491b
Brown-Séquard syndrome, **2**:525a
on contemporary scientists, **7**:24a
influence of, **7**:47b
neurology, **2**:30a; **11**:275a; **14**:143b
physiology, **10**:205a, 353b
Brożek, Jan (1585–1652), 2:526–527
Epistolae, **11**:533a
Brożek, Nicolas, **2**:527a
Bruce, Alexander, **9**:507b
Bruce, David (1855–1931), 2:527–530; **4**: 132b
Bruce, James (1730–1794), 2:530
Bruce, Thomas, 7th earl of Elgin, **7**:97a
Brucella, **2**:528a
Brucellosis, **2**:528a
Bruch, Philipp, **12**:169a
Brüche, Ernst, **2**:318a
Bruchia, **15**:149a
Brucine, **5**:421a, 589a; **9**:9b
discovery of, **10**:497b
isolation of, **3**:159b
Bruck, Carl, **10**:18b; **15**:522b
Brücke, Ernst Wilhelm von (1819–1892), 2: 530–532
acquaintances of, **4**:200a; **6**:242a, 434a, 618b; **7**:519b; **8**:540b; **12**:295b
critics of, **11**:566a
and Freud, **5**:171b, 172b
influence of, **4**:615a
influences on, **11**:215b

optics, **5**:445b
philosophy, **8**:515b; **12**:243b
Physiological Institute, **5**:172a, 174b, 175a
physiology, **2**:32b; **4**:200b, 203b, 614b; **6**:246a
students of, **5**:184b; **6**:476a; **11**:255a; **12**:231a, 319a; **14**:349a
Brückner, Eduard, **5**:339a; **10**:502a, 503b, 504a
Bruckner, Keith, **4**:583a
Bruguières, Jean-Guillaume, **3**:172b; **7**:588b
Brühl, Julius Wilhelm, **7**:620a; **10**:516b
Bruhns, Karl Christian (1830–1881), 2:532–533; **4**:369b, 370a*; **6**:239b, 553a; **12**:282a; **14**:54b
Brumpt, Émile (1877–1951), 2:533–534
Brunauer-Emmett-Teller (BET) method, **6**:118b
Brunck, Heinrich von, **3**:85a; **5**:489a
Brunel, Isambard Kingdom, **5**:200a
Brunelleschi, Filippo (1377–1446), 2:534–535; **1**:96a, 97a; **8**:208a; **13**:233b, 440b
Brunfels, Otto (*ca.* 1489–1534), 2:535–538; **2**:218a,b; **4**:138b; **13**:463a
Herbarum vivae eicones, **15**:161a
Brunhes, Jean (1869–1930), 2:538–539; **1**: 208a; **9**:151a*; **11**:310a*, 338a*
Bruni, Leonardo, **1**:274a; **13**:209a
Brunner, Johann Conrad, **10**:567b–568a; **14**:256a
Brünnich, Morten Thrane, **9**:5b
Brünnow, Franz Friedrich Ernst, **1**:6a, 616a; **2**:469a*; **4**:369b; **6**:48b; **8**:557b
Astronomie, **6**:49a
Bruno, Giordano (1548–1600), 2:539–544; **5**:411a; **8**:538b, 550a
cosmology, **5**:288a, 400b, 575b
critics of, **9**:316b, 317a
philosophy, **3**:359b, 470a; **9**:572a
on vacuum, **13**:438a
Brunonianism, **2**:595a; **12**:155b, 592a; **14**: 351a
Bruns, (Ernst) Heinrich, **6**:92b, 146b; **14**: 222a
Bruns, Paul, **8**:408a
Brunschvicg, Léon (1869–1944), 2:544–546; **2**:359a*; **7**:483a; **9**:341b; **11**:389a; **12**:609b; **15**:422b, 423b
Brunschwig, Hieronymus (*ca.*1450–*ca.* 1512), 2:546–547; **2**:220a*; **10**:308a; **13**:535b*
Brunswick (Germany)
Polytechnikum, **4**:2b
Brunswick, House of, **8**:150a–b
Brunswyck
see Brunschwig, Hieronymus
Brunton, Thomas Lauder (1844–1916), 2: 547–548; **2**:366b, 531a; **7**:519b; **14**: 495b
Brunyer, Abel, **9**:92b
Brush, Charles Francis, **9**:531a; **13**:454a
Brussels
Institut Pasteur du Brabant, **2**:300b
Royal Observatory, **11**:236b
Brussels, Gerard of
see Gerard of Brussels
Brute, Walter
see Brytte, Walter
Bryan, George Hartley, **11**:393b
Bryan, Kirk (1888–1950), 2:548–549; **3**: 596a*
Brydges, Kempe, **3**:399b
Bryologist
editors of, **15**:150b
Bryology, **2**:477a; **6**:218a–219a; **8**:264a; **10**: 575b; **12**:169a; **13**:553b
Bryonia, **3**:422b
Bryophyta
growth of, **9**:600b
taxonomy, **15**:149a–150a
Bryozoa, **6**:613a; **12**:229b
paleontology, **1**:494b; **10**:222a; **11**:502a; **13**:531b, 532a
reproduction, **13**:458a–b
taxonomy, **2**:617b; **13**:355b–356a
Bryson of Heraclea (*n.d.*), 2:549–550; **1**: 119b, 171a; **2**:618a; **6**:416a
Brytte, Walter (*fl.* 1350–1400), 2:550–551; **2**:435a
Buache, Philippe (1700–1773), 2:551–552; **4**:70b
Bubbles, **4**:80b
Bubo
axillary, **3**:218b
Buch, [Christian] Leopold von (1774–1853), 2:552–557
acquaintances of, **3**:70a, 623b; **4**:350a; **5**:155a–b; **6**:456a, 550b; **7**:319b; **14**: 116b, 239b, 257a
botany, **14**:117b
correspondence of, **8**:245b
critics of, **12**:263a
dolomite, **4**:309a
erratic blocks, **3**:210b
influence of, **1**:20a; **2**:108a
mineralogy, **14**:240a
students of, **1**:19b; **11**:235b
tectonics, **4**:402b; **8**:564a; **13**:145b–146a
Travels Through Norway and Lapland, **7**:70a
on volcanism, **3**:433b, 552a; **4**:349b; **8**: 573a; **11**:134a
Buchan, Alexander, **1**:434a
Buchanan, John Young (1844–1925), 2:557–558; **9**:588b; **14**:146a
Bucher, Walter Herman (1889–1965), 2: 558–559; **4**:563a*; **7**:145a*; **10**:169a
Bucherer, Alfred Heinrich (1863–1927), 2: 559–560
electron theory, **7**:264a
equations of, **11**:13b
Buchler & Co., **5**:394a
Buchner, Eduard (1860–1917), 2:560–563; **7**:504b
fermentation, **4**:485a,b; **6**:111a; **10**:379b
Buchner, Ernst (medic.), **2**:560a
Büchner, Ernst Hendrik (math.), **11**:535a
Büchner, Friedrich Karl Christian Ludwig (1824–1899), 2:563–564
Büchner, Georg, **10**:195b
Buchner, Hans, **2**:300b, 561b; **5**:564a; **7**:424a
Buchner, Johann Andreas, **15**:172b
Bucholz, Christian Friedrich (1770–1818), 2:564–565; **11**:437b; **15**:172b
Buchwald, Johannes de, **14**:449b
Buckingham, Edgar (1867–1940), 2:565–566; **15**:467a(n)
Buckingham, Richard Arthur, **8**:186a
Buckland, William (1784–1856), 2:566–572; **10**:584a; **14**:398a
acquaintances of, **3**:264a, 395a, 585b; **4**: 27b; **7**:365b; **8**:104a, 461a; **12**:318a, 524a; **14**:397b; **15**:151b, 152a
archaeology, **12**:520a
on contemporary scientists, **13**:44a
critics of, **3**:567a; **5**:32a,b; **12**:262a
influence of, **9**:583a,b; **12**:276b, 525a; **15**:51a
influences on, **5**:337b
paleontology, **3**:264b; **9**:87a,b; **12**:184a
Reliquiae diluvianae, **8**:242b
stratigraphy, **3**:395b
students of, **5**:436b; **8**:564a
Buckle, Henry Thomas, **4**:183a; **9**:331b–332a; **11**:237b; **13**:611b; **14**:128b
Buckley, Harold Eugene, **9**:380a
Bucknill, John Charles, **4**:593b
Bucquet, Jean-Baptiste Michel (1746–1780), 2:572–573; **2**:73b; **3**:199a; **5**: 90a; **8**:74b, 80b
Introduction à l'étude des corps naturels tirés du règne mineral, **8**:74a
and Lavoisier, **2**:572b; **8**:76a, 77b
Bucretius, Danieles (pseudonym)
see Rindfleisch, Daniel
Budan de Boislaurent, Ferdinand François Désiré (*fl. ca.* 1800–1853), 2:573–574; **5**:98a
Budapest
observatory, **7**:461b
Budapest, University of
chemistry at, **13**:298b; **14**:447b–448a
Budd, William (1811–1880), 2:574–576; **10**:559a; **12**:503a
Budde, Emil, **3**:310a
Buddhism, **12**:385b, 386a, 387b; **14**:163a, 164a; **15**:738a
Budge, Julius, **14**:142b, 143a
Buechner, William Weber, **13**:570a
Buée, Abbé Adrien-Quentin, **1**:239b; **5**:111b
Buenos Aires, University of
physiology at, **15**:228a,b
Buerg, Johann Tobias
see Bürg, Johann Tobias
Buff, Heinrich, **5**:163b; **9**:566a; **14**:478a
Buffer solutions, **13**:228b; **14**:182b
Buffers, chemical
effects of, **12**:546b
Henderson-Hasselbalch equation, **6**: 260b–261a
theory of, **2**:170b; **13**:574b
Buffon, Georges-Louis Leclerc, Comte de (1707–1788), 2:576–582; **3**:527b; **6**: 47b*, 587a; **8**:72b; **10**:87b(n), 97a*; **12**:68a(n)
acquaintances of, **1**:486b; **2**:530b; **3**:459b, 535a; **4**:548b; **7**:199a, 585a, 586b; **11**:99b; **12**:120a; **13**:565b; **15**:113b, 271b, 272a
biographies of, **3**:384b; **5**:45a*; **15**:110a
biology, **12**:554b
chemistry, **5**:601a
and Clairaut, **2**:577b; **3**:283a
on contemporary scientists, **12**:549b
correspondence of, **3**:461b; **10**:10a; **15**:54b
cosmogony, **15**:345a
critics of, **3**:283a, 382a; **5**:577b; **7**:88b; **8**:580a; **10**:284a; **11**:137a,b, 521a; **12**:555a, 556b, 557a–b; **13**:257b, 494b; **15**:112b
crystallography, **11**:521b
Les époques de la nature, **2**:348a*; **3**:526a; **15**:502a
evolution theory, **2**:320b, 342b; **3**:525b, 579b; **5**:356a,b, 359a,b, 443a; **9**:27b
geology, **15**:55b, 56a
Histoire des oiseaux, **2**:107a, 474b; **12**: 536b
Histoire naturelle générale, **1**:59a, 139a; **2**:107a–b, 474a; **3**:537b; **10**:10a; **15**: 111b–112b
continuation of, **7**:547a–b
critics of, **9**:54a
translations of, **8**:252a
influence of, **2**:107a–b, 321a; **3**:522a; **4**: 88a; **5**:77a; **7**:589a, 592b; **11**:493a; **14**:259b, 296b; **15**:50b, 358b, 502a
at Jardin Royale des Plantes (Paris), **1**: 58b; **2**:321b; **4**:216b; **8**:619b; **11**:562a; **15**:111a
medicine, **14**:15b
physics, **5**:133b, 134b; **10**:146b, 147a; **15**:359a

Buffon, Georges-Louis Leclerc (cont.)
probability, **15**:372b–373a
stratigraphy, **12**:489b
students of, **13**:390a
Théorie de la terre, **8**:68b
Bufo
cellular sexuality, **1**:545b
germ cells, **1**:545b
Bufonides
Bidder's organ, **2**:125a
Bugaev, Nicolai Vasilievich (1837–1903), 15:65–66
Buhl, Ludwig, **10**:559a
Buisson, Ferdinand, **2**:131b; **4**:513b
Buitenzorg
botanical gardens, **13**:459a–460a
Bukhara
natural history, **10**:287b
al-Bukhārī, Aḥmad (ibn) al-Faḍl al-Massāḥ, 7:343a
Bukhtīshūᶜ ibn Jibrāᵓīl, **15**:231a, 236a, 237a, 238b
Bulbs (bot.)
physiology of, **2**:290b
Bulgaria (country)
economics, **1**:155a
Bulgaria (bot.), **9**:176a
Bull
fossils, **5**:443a
Bullein, William, **5**:348b
Bullen, Keith, **5**:596b
Buller, Arthur Henry Reginald (1874–1944), 2:582–583; **2**:438a
Buller phenomenon, **2**:583a–b
Bulletin astronomique
founding of, **9**:551b; **13**:423b*
Bulletin biologique de la France et de la Belgique, **3**:148b
Bulletin . . . de Férussac
see *Férussac's Bulletin*
Bulletin de la Société Géologique de France, **2**:341b
Bulletin de la Société philomathique de Paris
editors of, **15**:484a
Bulletin de l'Institut Océanographique
founding of, **1**:92b
Bulletin de l'Institut Pasteur
founding of, **9**:328b
Bulletin de pharmacie
founding of, **3**:6a
Bulletin des sciences et de l'industrie, **1**:15b, 346b; **3**:450b; **8**:382b
editors of, **13**:128a
Bulletin des sciences mathématiques
editors of, **6**:522a; **13**:249b–250a
founding of, **3**:560a*
Bulletin des sciences médicales, **9**:7a
Bulletin des sciences naturelles et de géologie
editors of, **15**:115a*
Bulletin du jardin botanique de Buitenzorg, **13**:460a
Bulletin du Musée Océanographique
founding of, **1**:92b
Bulletin hydrografique
editors of, **7**:416b
Bulletin of the American Mathematical Society
see American Mathematical Society, *Bulletin*
Bulletin of the Astronomical Institutes of the Netherlands, **12**:449b
Bulletin of the New York Mathematical Society, **5**:414b
Bulletin of the Torrey Botanical Club
editors of, **15**:150b
Bulletin scientifique de la France et de la Belgique, **5**:386a
Bulletin scientifique du département du Nord, **3**:148b; **5**:386a
Bulletin statistique
editors of, **13**:352b
Bullettino Boncompagni, **2**:284a
Bullettino di bibliografia e di storia delle scienze matematiche e fisiche
see *Bullettino Boncompagni,*
Bullettino di paletnologia italiana, **7**:31a
Bullialdus
see Boulliau, Ismaël
Bulliot, Gabriel, **3**:622b
Bulloch, William (1868–1941), 2:583–585; **3**:463b; **4**:132b; **12**:565b*; **13**:519b
on contempory scientists, **3**:338a, 464a*; **7**:433b*; **8**:412a; **10**:365a, 400b; **12**: 486a*
History of Bacteriology, **5**:128b
Bülow, Friedrich Gustav von, **14**:54b, 55a
Bumm, Ernst, **14**:278b
Bunge, Alexander von, **2**:585a
Bunge, Gustav von (1844–1920), 2:585–586; **8**:447a; **9**:380b; **13**:243a
criticism of, **4**:204a
Buniakovskii, Vikter Iakovlevich
see Bunyakovsky, Viktor Yakovlevich
Bunsen, Robert Wilhelm Eberhard (1811–1899), 2:586–590; **5**:431b
acquaintances of, **2**:72b; **4**:65b; **6**:357a; **7**:280a, 380a, 460a; **8**:540b; **9**:287a; **11**:36b; **14**:54a
actinometer, **4**:182b–183a
atomic weight, **9**:290b
Bunsen battery, **2**:587a, 588b; **5**:559b
Bunsen burner, **2**:587a, 589a; **7**:381b, 620a
flame of, **12**:493b
cacodyl compounds, **3**:5a; **5**:125b, 372b
correspondence of, **11**:538a
on explosions, **4**:130a
gas analysis, **5**:125a
influence of, **8**:541a; **11**:537a
and periodic law, **9**:290a
photochemical law, **1**:21b; **11**:537b; **12**: 248b
spectroscopy, **3**:476a; **5**:20a, 143b; **7**:379b, 381b; **8**:440b; **11**:47a; **12**:37a
students of, **1**:389b, 426a, 578a; **2**:59a, 428a, 436b; **3**:85a, 333a, 498a, 511a; **4**:127b, 140a, 378a; **5**:214b–215a, 349b, 488b; **6**:13b; **7**:220a, 450b, 451a, 551b, 619b; **8**:180a, 328b, 553a; **9**:178a, 347b, 354b–355a, 360a, 467a; **11**:241b, 536b, 547a, 551a; **12**:208b, 216b, 493b; **13**:298b, 389a, 416b; **14**:108b, 169a
translations of works, **11**:538a
Bunter sandstone
see Sandstone, variegated
Bunyakovsky, Viktor Yakovlevich (1804–1889), 15:66–67; **3**:223b, 224b, 225b
Buonamici, Francesco (*d.* 1603), 2:590–591; **1**:608a; **2**:607b; **5**:238a; **15**:45a
Buonanni, Filippo (1638–1725), 2:591–592
Buono, Paolo del (1625–1659), 2:592; **9**:484b
Buonvicino, Costanzo Benedetto
see Bonvicino, Costanzo Benedetto
Buot, Jacques (*d. ca.* 1675), 2:592–593
ballistics, **2**:202a
Burali-Forti, Cesare (1861–1931), 2:593–594; **10**:442b, 444a, 605b, 606a
al-Burāq, **3**:549a
Burat, Amédée, **1**:327b
Burbank, Luther, **13**:478a
Burbury, Samuel Hawksley, **7**:84b
Burci, Carlo, **1**:611b
Burck, William, **13**:154b, 459b
Burdach, Karl Friedrich (1776–1847), 2: 594–597; **1**:386a,b; **10**:287a; **12**:155b
Die Physiologie als Erfahrungswissenschaft, **1**:387a
Bürde, Friedrich
Abbildungen merkwürdiger Säugethiere, **2**:422b
Burdenko, Nicolai Nilovich (1876–1964), 2: 597–598; **10**:621a*
Burdick, Charles Lalor, **4**:82b
Burdin, Claude, **5**:101a
Burdon-Sanderson, John Scott (1828–1905), 2:598–599; **1**:535b; **5**:83a; **6**:23b; **8**: 542a*; **11**:520a*; **15**:47a
acquaintances of, **5**:80b; **7**:422a; **12**:354b, 355a
germ theory, **8**:406a; **10**:382b, 383b
Handbook for the Physiological Laboratory, **2**:547b
students of, **11**:517a
Bureau Central Météorologique, **8**:278b; **9**: 154b
Bureau de Consultation Pour les Arts et Métiers, **9**:343a
Bureau des Longitudes, **2**:133b–135b *passim,* 439a; **3**:132b, 298a–b; **5**:271b; **7**:569a; **15**:336a, 359a, 361a
Annuaire, **2**:360a
founding of, **4**:16b
observatory, **9**:551b
Bureau International de l'Heure, **2**:126b
Bureau International des Poids et Mesures, **1**:612b
Burette, **9**:140a, 445b
naming of, **5**:325b
Bürg, Johann Tobias (1766–1834), 2:599–600; **15**:353b–354a
Burger, Eduard, **2**:601a
Burger, Herman Carel (1893–1965), 2:600–601
Burgers, Johannes Martinue, **4**:294a
Burgersdijk (or Burgersdicius), Frank (1590–1635), 2:601–602
Burghley, William Cecil, **3**:12b; **5**:361b
Bürgi, Joost (1552–1632), 2:602–603; **2**: 419a; **11**:561b
astronomy, **14**:359a
instrumentation, **2**:406a; **14**:470b
logarithms, **4**:551a; **9**:610b–611a, 613a
observations of, **12**:500b
trigonometry, **14**:470b, 471a
Burgundio of Pisa, **1**:107a; **10**:21b
Burial customs, **5**:145a
urn burials, **2**:523a
Buridan, Jean (*ca.* 1295–*ca.* 1358), 2:603–608
Aristotelian commentaries, **4**:237b; **5**: 588a
Buridan's ass, **2**:603b
dynamics, **1**:94b; **8**:228a,b; **10**:223b, 224a
impetus, **1**:198b; **5**:113b; **9**:136b, 137a; **10**:226a, 421a
earth's rotation, **3**:407a
influence of, **1**:93b, 94a; **3**:420b; **4**:157b, 228a; **12**:136a
influences on, **5**:403a
logic, **1**:94b
philosophy, **1**:199b
scientific method, **3**:410a,b
students of, **10**:223a
Burke, William, **7**:415a
Burkhardt, Heinrich, **3**:135b; **7**:71b; **9**:358b
Burkholder, Paul Rufus, **15**:150b
Burley, Walter (*ca.* 1275–*ca.* 1345), 2:608–612; **1**:94a,b, 198a; **2**:590b; **4**:237b
natural philosophy, **1**:100b; **10**:421a; **13**:187b, 208a

Burma
ethnology, 7:31a
Burmeister, Karl, 1:129b
Burn, Joshua Harold, 15:105b
Burnet, Thomas (*ca.* 1635–1715), 2:612–614; 11:43a; 14:296b
critics of, 7:275b; 9:532a
Telluris theoria sacra, 13:179a; 14:296a
Burnham, Sherburne Wesley (1838–1921), 2:614–615; 1:87b, 465a, 466a; 6:523a
Burnouf, Émile, 12:180b
Burnside, William (1852–1927), 2:615–616; 3:264b; 10:447a
Burrau, Carl Jensen (1867–1944), 2:616; 13:339a
Bursae mucosae, 9:483b
Burser, Joachim, 1:480b; 11:587b
Burstrøm, Hans, 7:503b
Burton, Decimus, 6:494a
Burton, William, 2:226b
Bury, Richard de, 2:390b
Busch, August Ludwig (1804–1855), 2:616
Busch, Hans, 1:7b
Büsching, Anton Friedrich, 5:75a
Bush, Katharine Jeanette, 14:1b
Bushman, 2:505a
Bushnell, David, 9:343b
Busk, George (1807–1886), 2:616–618; 5:207a*
Bussey Institution
see Harvard University
Bussy, Antoine Alexandre Brutus, 1:471a; 3:160a; 13:597a
Butadiene
synthesis of, 1:415b; 2:316b; 4:553a
Butanes, 2:622b; 7:452a
Butenandt, Adolf, 7:516b, 517b; 14:445b
Buteo, Johannes (*ca.* 1492–*ca.* 1564–1572), 2:618; 8:41a
on Euclid, 4:449b
geometry, 10:494a,b
De quadratura circuli, 2:360b; 15:155b
Ibn Buṭlān, Abu'l-Ḥasan al-Mukhtār ibn ʿAbdūn ibn Saʿdūn (*ca.* 1000–1068), 2:619–620; 11:445a
medicine, 15:246b(n)–247a(n)
Tacuini sanitatis, 15:185a
Butlerov, Aleksandr Mikhailovich (1828–1886), 2:620–625
acquaintances of, 11:92b; 14:530a, 622a
on contemporary scientists, 3:449b; 5:198a*; 9:292a; 14:622b*
influences on, 3:301b*
polymerization reactions, 7:21b, 452b, 457a
structural theory, 1:579a; 2:68a; 3:49b; 9:131a
students of, 4:553a, 597a; 9:130b; 13:549b
Bütschli, Otto (1848–1920), 2:625–628; 4:133b*; 8:271a*
cytology, 12:192b
protozoa, 1:417b
students of, 5:453a, 502b; 7:454a,b; 12:297b
Butter
nutritional role, 9:285a
Butterflies, 2:620b; 3:320a; 9:440b–441a; 11:102a
dimorphism, 14:234a
distribution of, 14:135b
genetics, 7:516b
identification of, 4:494a
Indian, 4:493b–494a
life cycle, 11:333a
mimicry, 1:503a–504a; 3:572b; 11:212a; 14:136a
pigments of, 6:500b
South American, 1:501a
taxonomy, 1:504a; 4:493b–494b
see also Lepidoptera
Büttner, Christian Wilhelm, 2:203b
Butyl alcohol, 2:622b
formula for, 3:449a
preparation of, 14:247b
Butyn, 15:2b
Butyral
discovery of, 3:193b
Butyramide
discovery of, 3:193b
Butyric acid, 9:619b
Butyrin
synthesis of, 10:499b
Butyrone, 3:193a–b
Butyronitrile
discovery of, 3:193b
Buwayhid dynasty, 2:148b; 6:381a,b; 7:222b; 13:149a; 15:250b
Buxbaumia, 6:492b
Buxton, Dudley, 11:462b
Buys Ballot, Christoph Hendrik Diederik (1817–1890), 2:628; 5:86b; 6:137b; 13:576b, 577b
Doppler effect, 4:167b
on gases, 3:307b; 9:218a
law of, 2:628b
meteorology, 2:533a; 4:175a; 9:443a; 12:595b
Buzareingnes, Girou de, 5:357a
Buzzard, Edward Farquhar, 4:594a
Bylica, Martin, 11:350b
Byzance, Louis, 11:392a
Byzantium
mathematics, 4:612a

C

Cabala, 1:80a, 126a; 2:541b; 8:547b; 13:58b
criticism of, 3:280b
Cabanis, Pierre-Jean-Georges (1757–1808), 3:1–3; 2:123a; 3:383a; 5:253a; 7:591b, 592b; 10:611b, 612a; 14:44b
Cabeo, Niccolo (1586–1650), 3:3; 2:13a
electricity, 1:547b; 5:516a; 6:170b; 10:146a
Philosophia magnetica, 5:588b
Cabinet du Roi, 1:59a; 13:553b
Cable car, 13:432a
Cables, electric
high speed, 7:108a
insulation of, 12:425a
submarine, 7:93a–b; 12:425b
theory of, 8:502a–b
see also Telegraphy, electric, Atlantic cable
Cabrera, Blas (1878–1945), 3:3–4
Cacao, 13:342a,b
Cacciatore, Niccolò, 10:592b
Cachexia
marshy, 5:502b
miners', 4:198a
Cacodyl, 5:125b, 372b
compounds of, 2:587b; 3:5a
Cactaceae, 3:44b; 13:154b; 15:130b
classification of, 15:130b
Cadet (*or* Cadet de Gassicourt *or* Cadet-Gassicourt), Louis-Claude (1731–1799), 3:4–6; 3:6a, 7a; 4:41a; 8:623a*,b*; 11:563a; 12:63b
chemistry, 1:527b; 2:587b; 8:69b, 75a; 12:64a, 65a
Cadet de Gassicourt (*or* Cadet), Charles-Louis (1769–1821), 3:6
Cadet de Vaux (*or* Cadet-Devaux *or* Cadet Le Jeune), Antoine-Alexis-François (1743–1828), 3:6–8; 3:173a*; 10:325b, 326a; 11:562b, 563a
Cadet family, 3:4b
genealogical table, 3:5a
Cádiz
Friendly Literary Society, 7:183b
Naval Observatory, 2:141a; 7:183a
Cadmium
alloys, 11:534b
discovery of, 2:586b
optical activity, 7:518b
spectrum, 6:94b
Caecilius of Calacte, 1:170b
Caelius Aurelianus, 1:315a; 5:586a; 7:27b; 12:538b, 539b, 540a
Caenogenesis, 15:169b
Caesar, Augustus, 13:83b, 84a, 588a; 15:514a,b
Caesar, Julius, 3:296a; 4:388a; 6:68a; 12:547a; 15:514a
Caesarean section, 1:323a, 523b; 10:418b; 15:646a
Il caffè, 5:195b
Caffeic acid
structure of, 5:6a
synthesis of, 13:406b
Caffeine, 5:464a; 9:558b
discovery of, 3:159b; 10:497b; 11:494b
industrial production of, 3:246a; 5:2b
structure of, 5:2a–b
synthesis of, 5:2a
Cagniard de la Tour, Charles (1777–1859), 3:8–10; 3:531a
critics of, 12:243b
engine, 3:78b*, 82b
fermentation, 2:561a; 8:402b; 9:425a; 10:361b; 12:241b
yeast, 13:506b
Cagniardelle, 3:8b, 9a
Cahinca acid
isolation of, 3:160a; 10:498a
Cahours, Auguste André Thomas (1813–1891), 3:10–11; 2:587b; 3:46a; 6:462b; 10:499b
Cail, Jean-François, 4:41b
Cailletet, Louis Paul (1832–1913), 3:11–12; 4:79a; 6:177b; 7:220b; 10:604a; 13:244b; 14:522b
law of, 14:561a
Caiman, 4:67a
Cain, Arthur James, 3:572b
Caïnca, 3:45a
Cairo (Egypt)
latitude of, 14:578b
library, 6:189a
Caisse Nationale des Sciences, 2:345b
Caisson disease, 2:62a; 6:24a; 8:614a
prevention of, 13:361b
Caius, John (1510–1573), 3:12–13; 3:40a; 9:440a
Cakracūḍāmaṇi
commentaries of, 2:118a
Calamines, 12:494b
Calamites, 12:259b
Calamostachys, 12:259b
Cālana (increment in motion), 15:575b, 576b, 582a, 598b
Calandrelli, Giuseppe (1749–1827), 3:13–14; 3:14a
Calandrelli, Ignazio (1792–1866), 3:14; 11:379a
Calandrini, Giovanni Ludovico, 3:459a–b
Calcar, Jan Stephen of, 14:4b, 8a
Calcareous earth
heat, effects of, 3:560b
see also Calcium carbonate

Calceolaria, **3**:422a
Calcidius (*fl.* fourth or fifth century A.D.), 3:14–15; **1**:104a; **15**:204a
commentaries by, **15**:203b
Calcification
pathological, **14**:252b
Calcination, **6**:165a*, 478a
in closed containers, **8**:470a
increase in weight, **2**:75a; **3**:206a–b, 495a; **5**:600b–601a; **6**:381a; **7**:388b; **8**:74a, 131a, 311b–312a; **9**:24a, 105b, 245b, 346b; **12**:135b; **13**:47a*, 494b
methods of, **8**:131a
and phosphorescence, **4**:215a
theories of
Renaissance, **12**:135b
seventeenth-century, **1**:550a; **8**:131a; **11**:389a–b
eighteenth-century, **1**:529b; **2**:175b–177a; **3**:206a–b; **5**:156b; **8**:74a–75b, 76b, 77b, 531a, 620b; **14**:75b
nineteenth-century, **7**:587a
phlogistic, **2**:6a; **7**:188b, 603b
Calcite, **2**:552b
and aragonite, **2**:135a–b; **5**:124a; **13**:311a
crystal shape, **6**:178a, 179b, 181a, 182a
Iceland spar, **4**:65b
double refraction, **1**:481b; **2**:451b; **6**:609b, 610a; **9**:73a; **14**:491b
piezoelectricity, **1**:557a
polarization, **2**:137a, 139a; **6**:610a; **10**:109b–110a; **12**:281b
structure of, **9**:472a
morphology, **2**:6a–b, 355b
optical properties, **2**:135a; **12**:304a
Calcium
in blood, **3**:352b; **10**:493a
compounds of, **2**:175a–177a *passim*
in cements, **8**:117a–b
interstellar, **11**:18b; **14**:56b
isolation of, **2**:588b; **3**:602a; **4**:243a
metabolism of, **4**:357b; **6**:136a; **15**:419a
hormonal control of, **3**:352a–b
physiological effects, **8**:453a; **11**:462b–463a
preparation of, **9**:178a
in sea water, **5**:72a
separation from magnesium, **2**:565a; **4**:134b
in solar spectrum, **12**:72b; **13**:232b
spectrum, **12**:21a
Calcium carbonate
dissociation of, **3**:617b
Calcium chlorate, **6**:478a
Calcium phosphide
discovery of, **10**:446a
Calculation
ancient Greek, **1**:222a, 231b
medieval Arabic, **13**:510a
medieval Hebrew, **8**:279b–280a
medieval Latin, **5**:365a, 365b–366a
seventeenth-century, **9**:554b, 610b
eighteenth-century, **7**:559a
nineteenth-century, **6**:85b; **12**:87a, 344b–345a
twentieth-century, **11**:613a
analogue method, **3**:217b–218a; **4**:591b
Chinese, **12**:374b, 375b
finger reckoning, **7**:532a–b
as fundamental science, **1**:231b
graphical, **10**:170a
Indian, **10**:129a
testing methods, **7**:509a
Calculatores, **3**:172a, 420b; **4**:237b; **6**:376a–379b; **9**:133b; **12**:548a; **13**:185a–211a, 349a–350a
Calculators
see Computers
Calculi, **4**:352b; **9**:469a
analysis of, **2**:420a; **5**:92b; **6**:386b*; **8**:485b; **11**:494b; **14**:492a
biliary, **1**:612a; **3**:159b, 91a; **5**:226b, 432b; **9**:4b; **10**:497b; **13**:292b
explanations of, **3**:208b
renal, **5**:226b; **10**:418a
theories of, **10**:418a
treatment of, **7**:15a*
chemical, **6**:44a–b; **12**:365b; **14**:320a–b
surgical, **6**:479a*; **12**:103a, 137a
urinary, **1**:612a; **2**:175a–b; **3**:38a; **5**:92b–93a, 226b; **6**:139a; **8**:114b, 623b*; **9**:309a; **12**:365b; **14**:320a–b, 492a
analysis of, **12**:148b
see also Lithotomy
Calculus
seventeenth-century, **1**:474b–475a; **2**:47a–48b; **4**:572b; **5**:527a, 528a, 529a; **6**:304b, 600b; **7**:300b; **9**:50a–b; **10**:47a–51a, 84a–b, 335b–337a
early eighteenth-century, **1**:76a–b, 531a; **2**:43b, 330a, 342b; **3**:244b, 458a–b; **5**:500a; **9**:454b
late eighteenth-century, **1**:115b–116a; **2**:43b; **7**:560a; **8**:306a,b; **9**:156b; **12**:430a; **15**:98b–99a
early nineteenth-century, **3**:135b–141b; **4**:546b; **5**:110b–111a; **6**:474a; **7**:550a
mid nineteenth-century, **4**:35b; **5**:523a; **11**:454b
late nineteenth-century, **8**:488b
twentieth-century, **1**:429a–b; **2**:351b; **5**:160a; **8**:516a; **13**:357a; **14**:437a
absolute differential, **4**:280b; **5**:201b; **8**:284a–b, 389b; **11**:407a–409b, 455a; **14**:46b
barycentric, **1**:590b–591a; **5**:309a; **6**:95b; **9**:430a–431a; **15**:196a
of concepts, **8**:153b–154a
cross-product, **7**:473a
of derivations, **1**:207a; **5**:110b–111a
differential, **1**:207a; **2**:47b–48a, 52a–54a, 342b, 350a, 577a; **3**:615b; **4**:3b, 469b, 470b, 477b–478a, 571a; **5**:295b; **7**:494b; **8**:62a, 163a–166b, 304a–305a, 384a, 385a, 617a; **9**:392b; **10**:68a, 437b; **11**:379a, 392a, 488a; **12**:319a; **13**:436a, 561b, 586a; **14**:573a; **15**:99a, 489b
covariant, **4**:336b
criticism of, **2**:17a; **6**:304b
foundations of, **2**:275a; **4**:57b; **7**:570b; **8**:304a–b; **12**:325b–326a
enumerative, **6**:394a
of errors, **1**:365b
exponential, **2**:53a; **15**:366a
of extension, **14**:303b
of exterior differential forms, **3**:95b–96a
extreme-value problems, **2**:53b
of finite differences, **1**:355a; **2**:320a; **3**:223a; **4**:469b, 475b, 477b; **5**:160a, 343a*; **8**:140b, 161a; **9**:127a; **13**:68b–69b, 267a; **15**:278a–279a, 285b
fluxional, **3**:458b
foundations of, **3**:71a, 75a–76a; **7**:619a; **10**:120a, 501a; **14**:500a; **15**:366a
functional, **2**:473a–b; **6**:5a; **8**:384a
fundamental theorem of, **1**:474b; **5**:527a; **8**:111a; **13**:436b
geometric, **1**:590b; **5**:527b; **10**:443a,b, 444a; **15**:192b
graphical, **3**:496a
historiography of, **5**:490b
infinitary, **4**:206a; **9**:303b
infinitesimal, **1**:229b, 355b, 449b; **2**:42b, 47a–48b, 52a, 320a, 330a, 358b; **3**:71a, 75a–76a, 114a, 115b, 117b, 178b, 183a, 281a, 454a*, 485b; **4**:205b–206a; **5**:60b, 111a, 158b; **7**:109a; **9**:303b, 470b; **10**:66a–68b, 441b–442a; **11**:61b; **12**:117b, 118a; **13**:435a; **15**:277b
critics of, **7**:577b, 578b
foundations of, **15**:99a
theory of, **5**:61b
integral
ancient Greek, **4**:466b
early seventeenth-century, **3**:615b; **11**:487a; **13**:436a
late seventeenth-century, **7**:583a; **8**:163b; **10**:68a, 335b–337a; **13**:586a; **14**:49b(n)
early eighteenth-century, **3**:282b; **4**:478a, 516a; **8**:304b; **10**:120b; **11**:392a
late eighteenth-century, **2**:342b, 577a; **3**:383b, 385a–b; **5**:54b; **7**:561b, 567b–568a; **11**:402a; **15**:277a–279a, 292b–293a, 294a
early nineteenth-century, **3**:136a
mid nineteenth-century, **2**:471a; **3**:223a; **5**:572b; **10**:247b, 437b; **15**:489b
late nineteenth-century, **1**:449b; **2**:350a; **3**:228a–b; **8**:62a; **13**:561b
early twentieth-century, **2**:357b; **4**:177a–b; **8**:253b
invention of, **7**:163a, 276a
of limits, **3**:136a, 137a, 140a, 143a,b
logical, **8**:154a–155a, 159b–160a, 263a
matrix, **2**:248a
notation, **1**:354b; **3**:136a, 458b; **4**:546b; **5**:54b, 96a, 97a; **8**:250b; **14**:500a
numerical, **9**:410b
of observations, **5**:304a; **14**:318a
operational, **1**:207b; **2**:294b–295a, 491a; **5**:96b, 523a; **6**:211b; **8**:253b; **12**:325b; **15**:303a, 306b, 386b
in chemistry, **2**:484b–485b, 515a–b
of points, **15**:196a, 197a, 198a
of principal relations, **6**:89b
of probabilities, **1**:449b, 591b; **2**:356a; **3**:117a, 222b, 383b, 384b, 386a–b, 451b; **6**:103b; **7**:317a–b, 494b, 561a, 568b; **8**:62a*; **9**:128a; **10**:334b–335a; **11**:599b; **15**:366a,b, 367a
propositional, **8**:546a–b, 590b; **11**:107a
of relations, **12**:13b–14a
of residues, **3**:139b–140a, 144a; **5**:96b; **8**:367a
of sequents, **5**:350b
of series, **7**:570b
of sets, **2**:297a–b
spinor, **14**:317a
symbolic, **6**:548b; **8**:140b, 162b; **10**:514a
Clebsch-Aronhold, **3**:314b; **5**:472a,b
tensor, **3**:263b; **4**:337a; **8**:284b; **11**:455a; **14**:283b, 317a
of variations
seventeenth-century, **2**:48b, 53b; **10**:53a
early eighteenth-century, **3**:282a; **4**:479a–b; **5**:54b
late eighteenth-century, **2**:299b; **3**:441b, 442a; **7**:549b, 560b–561a, 561b, 562a; **8**:135b; **9**:470b–471a
early nineteenth-century, **3**:139a, 466a; **5**:490b; **7**:494b, 570b; **14**:500a
mid nineteenth-century, **3**:314a; **4**:19b; **5**:305a; **6**:87b, 89b; **8**:62b; **10**:249a–250a; **13**:427b; **15**:489b
late nineteenth-century, **2**:272a; **4**:230a; **6**:473a; **7**:530b; **8**:389a; **9**:232b; **12**:246b; **14**:223a, 614a
twentieth-century, **2**:192a, 198a–b, 271b; **3**:62b–63a; **4**:288a; **5**:78a*, 122b, 201b, 608b; **6**:391b, 394a, 614b; **7**:408a, 621b; **8**:112a; **10**:244b; **11**:

256a, 260a,b, 261a, 327a, 477b–478a; **12**:411a
discontinuous solutions, **11**:327a–b
and geometrical optics, **3**:63a
inverse problem, **4**:174a
in physics, **4**:480b
of zeroes, **4**:478a
see also Analysis
Calcutta, University of, **15**:47b–48a
physics at, **12**:71a–b
Caldani, Floriano, **3**:16b
Caldani, Leopoldo Marcantonio (1725–1813), 3:15–16; **5**:55b, 267b; **11**:66b
and Haller, **3**:15b
Caldas, Francisco José de (1768–1816), 3:16; **6**:551a
Historia de los árboles de quina, **13**:464a–b
Calder Ironworks, **9**:591a
Calderoni, Mario, **13**:550b
Calders, **11**:336b
Caledonian disturbance, **1**:394a; **2**:89b
Caledonian mountains, **1**:394a
Caledonian zone, **14**:59b, 60a
Calendar
ancient Greek, **1**:205a; **2**:154b; **3**:391a; **4**:172a, 459b–460a, 466a; **5**:345a,b, 347a; **7**:142a; **10**:179b–180b; **11**:197b; **15**:218b
Byzantine, **9**:36a; **12**:406a
medieval Latin, **1**:565a–b; **3**:26b–27a; **7**:118b–119a, 123b, 124a, 129b, 142a; **9**:363a; **10**:540a–b, 541a–b; **12**:62b; **14**:389b–390a
Renaissance, **9**:191b; **10**:8a; **15**:154b
eighteenth-century, **5**:112a*; **10**:92b(n)
twentieth-century, **6**:512b
Alexandrian, **15**:709b
Arabic
medieval, **1**:23a*, 512a; **2**:152a–b, 154a,b; **4**:542b; **7**:82a–b, 142a, 259b, 324b; **9**:36a; **10**:6b; **11**:231b, 247a; **14**:576a–b
pre-Islamic, **1**:33a; **2**:152a
historiography of, **7**:325a
astrological, **2**:403a; **7**:289b–290a
Callippic period, **3**:21b–22a; **9**:338a; **15**:218b
Chinese, **1**:310a,b, 312b, 313a; **3**:250b–251a, 269b; **7**:259b; **12**:377b–379a, 403a–b; **13**:485b; **14**:545b; **15**:729b, 732a,b, 733a,b, 735a, 741b, 746a
Feng-yuan, **12**:377b–379a
Grand Cycle, **3**:250b–251a
Hsuan-ming, **12**:403a,b
Huang Chi, **3**:269b
Lin Te, **3**:269b
Shih-hsien, **12**:403b
Shou-shih, **1**:310a,b, 312b, 313a; **12**:378b, 403a–b
Ta-ming, **13**:485b
Ta-t'ung, **12**:403b
T'ung-t'ien, **1**:313a
Coptic, **14**:576a–b
ecclesiastical, **9**:488b*, 508a*; **14**:390a
golden number, **1**:383a; **7**:131a
Egyptian, **5**:345a; **9**:338a,b; **11**:190a; **15**:536a–b, 706a–727b
Great Year, period of, **10**:179b–180b
Greek, **9**:337a–339a
Greek-Syrian, **7**:259b
Gregorian, **1**:383a, 565b; **2**:406b; **3**:14a, 312a, 558b; **7**:324b, 373b; **10**:248a; **11**:411b*; **14**:178b; **15**:335b, 343a, 707a
opposition to, **5**:588a; **9**:168a; **14**:23a–b, 153a
historiography of, **7**:7a
Indian, **2**:154a,b; **3**:584b; **4**:100a; **5**:274a; **10**:426a; **12**:343a; **15**:534a–630a *passim*
intercalation, **2**:152a; **4**:172a, 459b–460a, 502b; **5**:345b; **7**:130b–131a, 362a; **9**:337a–b, 339a; **15**:218b, 535a–b, 536a, 557a, 566a, 567b, 707a–b, 708a,b, 709b, 723b
Japanese, **1**:310a–313b; **15**:729a, 731b, 732a
Hōryaku, **1**:310b
Jōkyō, **1**:312b; **12**:403b
Kansei, **1**:313b; **7**:18b; **15**:734b, 735a–b
Shou-shih, **15**:746a, 750a, 752a
Tenpo, **1**:313b; **15**:735b, 737a, 746a
Jewish, **2**:152b; **7**:362a
Julian, **1**:382b; **3**:312a, 558b; **5**:345a; **7**:131a; **11**:231b; **12**:62b, 547a; **14**:23a, 286a; **15**:707a
lunar, **14**:592b; **15**:709a, 716b, 723a, 760a, 765b, 774a, 775a,b, 776a,b, 793b, 800a, 810b, 814a
lunisolar, **9**:337a–339a; **12**:378b–379a
Macedonian, **15**:709b
Mayan, **15**:759a–814b
Mesopotamian, **9**:337b, 338b–339a; **15**:535a, 536a–554a, 589a
Metonic cycle, **1**:205a, 565a–b; **3**:21b; **4**:459b–460a; **5**:553a; **9**:337a–339a; **10**:180a; **12**:406a; **14**:390a
oktaeteris (eight-year cycle), **4**:466a; **10**:179b–180a
perpetual, **8**:25b
Persian, **7**:142a, 259b; **11**:231b; **14**:576a–b
pocket, **9**:529b
reform of
ancient Roman, **12**:547a
Byzantine, **12**:405b
medieval, **1**:84a, 382b–383a; **5**:553a; **7**:130b–131a, 259b, 324b; **12**:62b
Renaissance, **1**:165b, 606b; **3**:262b, 312a, 558a–b; **4**:6a; **14**:23a–b
seventeenth-century, **9**:310a,b; **14**:160a, 286a
eighteenth-century, **11**:65b; **15**:335b–336a
twentieth-century, **1**:407a; **12**:71b
China, **12**:372a, 377b–379b
England, **1**:383a
Japan, **1**:310b–313b *passim*; **12**:403a–404a; **15**:730a, 731b, 734a–b, 735a,b
solar, **12**:379a
Syrian, **14**:576a–b
Venus calendar, **15**:776b–789a, 812a–b
see also Almanacs; Computus; Easter; Eras, calendric; Tables, calendric; Year
Calhoun, John Caldwell, **12**:204a
Calibration of instruments
eighteenth-century, **1**:527a
nineteenth-century, **5**:340b
twentieth-century, **14**:553b–554a
Calico, **3**:400b*, 481a*, 488b; **4**:250a
California
botany, **7**:102a–b, 285a; **14**:192b
earthquakes, **8**:99b; **11**:362a; **14**:402b
Geological Survey, **5**:214a; **6**:57b; **7**:148b, 370b; **12**:436a; **14**:315b–316a
geology, **8**:578a; **9**:100a; **11**:439a
gold, **12**:436b
oil, **12**:436a–b
paleontology, **13**:72a
zoology, **5**:545b
California, University of
in nineteenth century, **8**:122a, 123a
biochemistry at, **5**:6a–b
California, University of (Berkeley)
bacteriology at, **5**:316b
Biology Library, **7**:447a
chemistry at, **8**:290b–291a
founding of, **14**:315b
geology at, **8**:99b
Museum of Vertebrate Zoology, **5**:545a–b
physics at, **8**:94b–96b
Radiation Laboratory, **8**:95b, 96a
California Academy of Sciences
founding of, **7**:285a; **14**:315b
California Botanical Society
founding of, **7**:102b
California Institute of Technology
in early twentieth century, **4**:82a–b; **9**:396b; **10**:156b, 157a
aeronautics at, **7**:247b
astronomy at, **6**:32a–b
biology at, **9**:525a–b; **13**:134a–b
founding of, **6**:31b–32a
seismology at, **5**:596a
Calipers, **2**:133a
beam, **5**:491b–492a
Calkins, Gary Nathan (1869–1943), 3:16–17; **2**:626b; **7**:99a; **14**:424b, 432b
Callan, Nicholas (1799–1864), 3:17–18
Callandreau, Pierre Jean Octave (1852–1904), 3:18–19
Callaway, Charles, **5**:337a; **6**:57b; **8**:33a
Callendar, Hugh Longbourne (1863–1930), 3:19–20; **12**:27a
on Carnot, **3**:83b(n)
Calley, John, **10**:36a
Callinicos of Heliopolis (*fl. ca.* 673 A.D.), 3:20–21
Callippus (*b. ca.* 370 B.C.), 3:21–22; **1**:338b
calendar, **3**:21b–22a; **9**:338a; **12**:547a; **15**:211b, 218b
cosmology, **1**:199a, 252b, 257b–258a; **4**:466b; **9**:269a,b, 338b; **10**:590b; **12**:3a; **13**:326b
Callis, Conral Cleo, **7**:498a
Callisen, Heinrich
System der Wundarzneikunst, **1**:333a
Callistus, Andronicus, **1**:274b
Callose
discovery of, **9**:79a
Callus
formation of, **3**:489b
Calmette, Albert (1863–1933), 3:22; **3**:189a; **8**:66a; **14**:551b
Calomel, **10**:310b
atomic structure, **9**:183b
medical use of, **4**:209a; **11**:48b
oral absorption of, **3**:487a
purification of, **14**:639a
Caloric theory
see Heat, theories of, caloric
Calorie
calculation of, **2**:69b; **11**:585b
naming of, **4**:554b
Calorific rays
see Radiation, infrared
Calorimeter, **9**:493a*; **11**:586a; **13**:351b
adiabatic, **11**:418a, 577a
bomb, **2**:69b
electrical, **3**:19b–20a
ice, **2**:179b, 587a, 588b; **6**:354a; **8**:79b; **14**:213b; **15**:312b–313a, 314a
iron paddle-wheel, **7**:182a
mercury, **2**:69b; **4**:554b
siphon, **11**:427b
steam, **2**:589b
invention of, **7**:161a
vacuum, **15**:442a
vapor, **2**:587a
water, **2**:69b, 179b; **4**:241a
Calorimetry
eighteenth-century, **2**:178a–179b; **3**:156a–b; **7**:596b; **8**:79b–80a, 83b, 620b;

Calorimetry (cont.)
11:432a–b; **14**:353b–354b; **15**:95a, 313b–316a
nineteenth-century, **1**:303a, 325b, 616b; **2**:69b; **3**:19b–20a; **4**:241a, 554b; **5**: 126b; **6**:271a, 354a; **7**:182a–b; **8**:545b; **11**:427b–428b, 485b; **13**:358b
twentieth-century, **1**:610a–b; **4**:80b; **6**: 118b, 470a; **8**:555a–b; **11**:418a
low-temperature, **4**:80b; **8**:47a
Calorimotor, **6**:115a
Calotype, **13**:239a
Calver, George, **3**:366a
Calvert, Mary, **1**:467a
Calvin, John, **5**:121a; **12**:324b; **14**:375a(n)
Calvin, Melvin, **8**:293b
Calvinism
Renaissance, **3**:262a; **4**:212b; **11**:286b
seventeenth-century, **1**:126a, 133b, 134a–b, 159a; **14**:153a
eighteenth-century, **13**:448a
nineteenth-century, **9**:537a; **14**:516a
and natural philosophy, **1**:134a–b
opposition to, **1**:447b; **2**:540b, 541b; **6**: 254b
Calvisius, Sethus, **5**:588a
Calvus, Fabius, **1**:418b
Calyx, **9**:18a
Camaldolese Order, **5**:498b–499a; **9**:189b–190a
Cambium, **8**:206a
composition of, **9**:418b
Cambrian period, **7**:45a
naming of, **9**:584a
paleontology, **1**:397a; **3**:572b, 608b; **8**: 594a; **9**:617b; **10**:204a; **14**:120a–b
stratigraphy, **8**:33a,b, 463b; **9**:584a–585a; **12**:277a–b; **13**:519a, 532a, 620b–621a; **14**:59b
Cambridge Antiquarian Society
founding of, **1**:358b
Cambridge Mathematical Journal
founding of, **5**:522b; **8**:592a
Cambridge Mathematics, **4**:546b
Cambridge Natural Philosophy, **4**:546b
Cambridge Philosophical Society
founding of, **3**:291b, 497a; **6**:288b; **10**:44b; **12**:275b
Cambridge Scientific Instrument Company, **2**:184a; **3**:566b
Cambridge University
in late seventeenth century, **10**:43a–44b; **11**:313b
in nineteenth century, **8**:565b
in twentieth century, **11**:84b
agriculture at, **5**:83a
astronomy at, **5**:516a; **10**:31b
biochemistry at, **5**:83a; **6**:498b–500b
biology at, **1**:420b; **5**:80a–81b; **11**:211b
Botanical Garden, **6**:289a
botany at, **2**:183b–184a; **6**:288b–289a; **12**:339a–b
Cavendish Laboratory, **3**:329a–330a, 367b–368a; **5**:423b; **9**:199a, 227b; **12**:26a–27a, 32b, 33b–34b; **13**:105a, 365a–b, 366a–b, 368b–370a; **14**:420b, 421a
chemistry at, **14**:26b–27a
computing laboratory, **6**:148a
crystallography at, **15**:17b
engineering at, **4**:500b
genetics at, **11**:212a
geography at, **3**:617a
geology at, **2**:289a; **12**:275b–276a; **14**: 502b
herbarium, **1**:359a
Lucasian professorship, **13**:74b
mathematics at, **5**:77b–78a; **6**:451a,b; **9**: 199a; **10**:437b, 447a; **11**:568b; **13**: 362a–b; **14**:500a
observatory, **1**:85b; **5**:257b
philosophy at, **9**:98a
physics at, **5**:81b, 103a–b; **13**:78a–b, 102a, 374b; **14**:420b
physiology at, **5**:80b–81b, 82b–83a; **8**:14b, 15b, 17b, 532b; **9**:142b; **15**:104b
reform, **12**:276a
science at, **6**:35b–36a; **7**:273b–274a; **14**: 293a
see also individual colleges
Cambridgeshire
botany, **11**:313b–314a
Cambro-Silurian period
stratigraphy, **2**:486b
Camden, William, **10**:239b, 489a
Britannia, **8**:307b, 308a; **10**:259b
Camel
wild, **11**:181b
Camellia, **8**:395a
Camera
automatic, **3**:12a
equatorial, **2**:196a
rotating-lens, **15**:60a
Camera lucida, **6**:326b; **13**:238b; **14**:492a
Camera obscura, **1**:97a; **2**:535a–b; **3**:116a; **4**:44b; **6**:195b–196b; **7**:212b, 298a; **9**:480a; **11**:97b; **14**:492a
in astronomy, **8**:280b–281a
theories of, **8**:281a
Camerarius, Joachim, **3**:342a,b; **8**:310b; **11**:312a
Camerarius (Camerer), Rudolph Jakob (1665–1721), 15:67–68; **2**:188b, 257a; **7**:412b; **13**:463a
botany, **1**:293b; **4**:98b; **7**:441a; **11**:83a
Cameron, Alexander Thomas, **11**:280b, 281a; **12**:30a, 507a(n)
Cameron, Hector, **8**:404b, 406a
Cameron, Robert Horton, **14**:347a
Camien, Merrill Nelson, **5**:149b
Camis, Mario, **1**:453b
Camouflage, **7**:314b; **8**:102b; **9**:550a–b
Campanella, Tommaso (1568–1639), 15:68–70; **1**:159a, 447b; **2**:163b, 419a; **3**: 362b; **11**:96a; **12**:333a
Apologia pro Galileo, **14**:364b, 376a(n)
and Borelli, **2**:307a–b
critics of, **9**:316b
imprisonment, **2**:307a
Campani, Giuseppe (1635–1715), 3:23; **3**: 101b, 102b; **4**:128a; **12**:85a
telescope, **2**:310b
Campani degli Alimeni, Matteo, **3**:23a
Campanulaceae, **3**:43a
Campanularia, **8**:519a
Campanus of Novara (*ca.* 1220–1296), 3:23–29; 15:70; **1**:381b; **2**:590b; **3**:272b; **4**:157b; **8**:100b; **9**:435a; **12**:61a, 62a; **13**:60b, 349b; **14**:400b
astrology, **15**:154b
astronomical tables, **10**:539a*
cosmology, **7**:120a
criticism of Aḥmad ibn Yūsuf, **1**:82b
edition of Euclid, **1**:63b; **4**:446a–b, 448a–451a *passim;* **10**:271a
instruments, **6**:273b; **7**:119b, 125a; **10**: 541a; **15**:163a
mathematics, **3**:275a; **11**:415a
Theorica planetarum, **2**:24a; **4**:164b
Campbell, Archibald, **6**:489b
Campbell, Colin, **3**:458a; **5**:529a
Campbell, Douglas Houghton (1859–1953), 3:29–31; **13**:89b
Campbell, Lewis, **9**:199a, 219a; **11**:23a; **13**:236b
Campbell, Marius Robinson, **4**:562b
Campbell, Norman Robert (1880–1949), 3: 31–35; **11**:55b; **13**:108a*
Campbell, Walter Ruggles, **1**:443a; **3**:352a
Campbell, William Wallace (1862–1938), 3: 35–37; **1**:55b; **3**:509a; **4**:249a; **6**:471b; **7**:271b*; **9**:503b; **10**:36a*, 602b*; **13**:474b; **14**:521b
Camper, Adriaan, **9**:152b
Camper, Peter (1722–1789), 3:37–38; **2**: 204a; **5**:442b; **9**:151b, 483a; **12**:509b; **15**:5a*, 112a
Camper's fascia, **3**:38a
Camper's line, **3**:38a
Campestrius, **10**:547b
Camphene, **2**:65b
rearrangement of, **9**:608b
Camphor, **2**:65b; **8**:31b
analysis of, **8**:174b
chemistry of, **9**:608b
composition of, **8**:332b–333a
compounds of, **11**:85a–b
derivatives of, **7**:372b; **10**:158a
intramolecular change, **8**:32a
medical use of, **1**:332b; **10**:620b; **11**:301a
preparation of, **10**:499a
reactions of, **1**:288b; **8**:58b
refining of, **13**:566a
structure of, **8**:32a; **10**:517b; **13**:388b
Camphor tree, **13**:282b
Camphoric acid, **2**:565a; **10**:158a
composition of, **8**:332b–333a
Campi, Carlo Giuseppe, **14**:74b
Campolongo, Emilio, **1**:522b
Campomanes, Pedro, **1**:351a
Camus, Charles-Étienne-Louis (1699–1768), 3:38–40; **2**:216b, 334b; **3**:281b, 439b; **6**:237a*; **8**:22a; **9**:186b
Canada
astronomy, **11**:18b
botany, **7**:210b; **11**:218b
boundaries, **15**:152b
chemistry, **9**:393b, 394b
education, **8**:584a
exploration of, **1**:585b; **12**:293a–b; **13**: 525a–b
Geological Survey, **1**:585a; **2**:128b; **3**: 608a; **5**:49b; **6**:564a; **8**:461a–462a; **12**:293a–b; **13**:525a–b
Organic Remains Series, **2**:128b
geology, **1**:585b; **3**:607b–608b; **5**:49b; **12**:230a
glaciation, **3**:608b; **13**:525b
mining, **13**:525b–526a
Organic Remains Series, **2**:128b
paleontology, **2**:128a–b; **6**:58b*
petrography, **1**:51a–52a
stratigraphy, **1**:397a; **13**:518b
medical research, organization of, **3**:353a
natural history, **1**:585a–b
science, **13**:7a–b
Canada balsam, **10**:109b
Canadian International Boundary Commission, **3**:547b
Canadian Naturalist & Geologist
founding of, **2**:128a
Canadian System (geol.), **13**:532a,b
Canal rays, **5**:332a; **7**:36b; **10**:347a; **12**:191b, 614b; **14**:217b, 218a
charge/mass ratio, **14**:32a
discovery of, **5**:458b, 459a
Doppler effect, **12**:613b–614a
nature of, **14**:31b–32a
see also Positive rays
Canals (anat.)
lacrimonasal, **3**:489b
Canals (eng.), **2**:473a; **3**:440b; **4**:402b; **5**: 195b; **6**:578b; **8**:207a, 212b, 213a; **12**:487a; **13**:496a

cleaning of, **8**:26a; **13**:50a
see also Locks
Canano, Antonio Maria, **3**:40a; **9**:75a
Canano, Giovan Battista (1515–1579), 3:40–41; **4**:487b, 508b, 519b; **8**:554a
Canary
genetics, **3**:590b
Canary Islands, **4**:603a
biology, **8**:572a–b, 573a
flora, **2**:510a
geology, **2**:555b; **8**:572a–b
Cancer, **2**:575b*; **3**:17a, 463b; **6**:479a*; **10**:611a
biochemistry of, **4**:486a; **5**:209a; **14**:176a–b; **15**:418a,b
breast, **3**:38a, 463a, 490a; **6**:77b; **8**:402a
cell growth in, **1**:123b; **5**:599a
cytology, **11**:465b
diagnosis of, **3**:490a; **14**:126a; **15**:418a
cytologic, **10**:291a–292a
experimental studies, **8**:447b; **9**:549b–550a; **15**:95b*
hereditary, **3**:493a; **9**:586a; **14**:252b
histology of, **1**:438b, 439a
immunological studies, **4**:299b
public awareness of, **9**:586b
research organization, **4**:499a; **9**:586a–b
serology, **11**:429b
skin, **2**:386b
statistical studies, **8**:122a*
stomach, **1**:612a
theories of, **7**:157a*; **14**:176b
humoral, **10**:418a
viral, **5**:270a
transmission of, **9**:586a
transplantation of, **7**:101a
treatment of, **3**:345a*
ancient, **1**:212b
chemotherapy, **4**:155b; **5**:599a; **7**:208b; **15**:523a
dietary, **14**:176b
radiation, **4**:195a, 195b–196a; **9**:549b–550a
surgery, **6**:77b; **14**:278b–279a
uterine, **10**:291a–292a
see also Leukemia, Oncology, Tumors
Cancer (constellation)
nebulae in, **3**:528b
Cancer Research
editors of, **9**:586b
Cancrin (Cancrinus), Franz Ludwig von (1738–1812 or 1816), 3:41–42
Candalla, Franciscus Flussatus (François de Foix, Comte de Candale), **4**:449a–b
Caṇḍeśvara
commentaries of, **15**:608b
Caṇḍīdāsa
commentaries of, **2**:119a
Candle
chemical history of, **4**:539b
Candle (photometric unit), **2**:200a
concept of, **13**:351b
Candolle, Alphonse de (1806–1893), 3:42–43; **2**:491b; **8**:307a*; **9**:600a; **10**:9b*; **12**:255b
Flore française, **1**:614b
Géographie botanique raisonnée, **6**:490b
Origine des plantes cultivées, **15**:508b
phylogeny, **2**:492a; **3**:571a,b
Prodromus, **1**:615a
taxonomy, **1**:615a; **2**:103b; **15**:268b
Candolle, Augustin-Pyramus de (1778–1841), 3:43–45; **4**:21b; **8**:35b, 38a
acquaintances of, **5**:44a; **6**:493a; **9**:432b; **11**:132b; **12**:215a, 309a, 523a; **13**:595b; **15**:126a
anthropology, **11**:137b
Astragologia, **3**:172b
botany, **7**:585a; **8**:445a, 567b; **12**:124a; **13**:390a, 601b
on contemporary scientists, **11**:135b*
Histoire naturelle des plantes grasses, **3**:172b
influence of, **3**:192b; **10**:9a
Prodromus, **3**:42b; **10**:13b
students of, **4**:242a
taxonomy, **5**:512b; **6**:491b; **7**:198b; **8**:372a; **13**:432b, 596a
Candolle, Casimir de, **3**:42b
Candragupta II, **10**:425b
Candraśekhara Paṭanāyaka
commentaries of, **2**:116a
Canella (tree), **7**:200b
Canis Major (constellation)
alpha Canis Majoris
see Sirius
Canis Minor (constellation)
alpha Canis Minoris
see Procyon
Canivet, **8**:22a, 351a
Cannabis, **15**:67b, 68a
alkaloids, **15**:2b
analysis of, **10**:530b
medical use of, **14**:496a
Canning, George, **3**:579a
Cannizzaro, Stanislao (1826–1910), 3:45–49; **2**:66b; **3**:279a; **7**:281a; **9**:378a
acquaintances of, **9**:287a; **12**:163b
atomic weights, **6**:463b; **9**:348b, 349a; **10**:37b, 178a; **11**:370b
on Avogadro, **1**:346a, 347a–b
Cannizzaro reaction, **3**:47a
editions of works, **9**:347b–348a
law of atoms, **3**:48a–b
Cannon, Annie Jump (1863–1941), 3:49–50; **5**:34a,a*; **6**:350b; **8**:106a,b; **9**:194b; **10**:600a–b; **12**:348b
Cannon, Walter Bradford (1871–1945), 15:71–77; **1**:544a; **3**:518a; **5**:283a; **6**:226b, 262a*; **8**:454a; **11**:545b; **12**:470a, 474a
homeostasis, **1**:450a; **2**:33a
neurophysiology, **8**:18a(n); **11**:545b–546a
Cannon, William Austin, **13**:157b
Cannon
design of, **3**:562a; **8**:26a
historiography of, **9**:491b–492a
manufacture of, **4**:346b; **5**:603a; **8**:213b; **9**:472b
rifled, **11**:354a, 472a
steam, **8**:211b–212a
De Canonio, **1**:228b–229a; **8**:215b, 216a, 222b
Canopus (star), **9**:36a
Cantal region
geology, **11**:269b–270a
Canterbury (N.Z.), University of
founding of, **5**:611a
Canterbury Philosophical and Literary Institution, **10**:39a
Cantharides, **1**:329a; **11**:494b
Canton, John (1718–1772), 3:51–52; **5**:138b*; **11**:140b, 143b
Canton's phosphor, **3**:52a
electricity, **14**:73a, 352b, 353a
pyroelectricity, **2**:5a
Cantor, Georg (1845–1918), 3:52–58; **2**:269a; **5**:108a; **10**:344b
acquaintances of, **2**:303b; **6**:230b; **7**:522b; **9**:427a; **14**:222b
analysis, **5**:97b
Cantor set, **3**:55a
critics of, **7**:507b, 508b
and Dedekind, **3**:54a–57a *passim*
geometry, **13**:16b
historiography, **4**:117b
influence of, **3**:456a; **4**:4a,b; **7**:485b; **9**:502b; **14**:304a
number theory, **2**:277a; **8**:383b; **9**:307b; **10**:484a; **12**:13a
series, **2**:513a; **3**:54a
set theory, **1**:407b; **2**:265a, 277b, 593b; **6**:388b, 392b; **7**:444b; **11**:60b; **12**:216b–217a; **14**:614a,b
students of, **2**:58a
theorems of, **12**:15a
Cantor, Moritz Benedikt (1829–1920), 3:58–59
acquaintances of, **13**:252b
on contemporary scientists, **3**:512b; **4**:343a*; **5**:312b*; **6**:95b, 358a*; **7**:110a*; **9**:431b*; **13**:22a*
critics of, **14**:336b
historiography, **1**:41a; **2**:52a, 430a; **3**:461b, 466b*, 622a; **4**:602a, 605b; **6**:408a; **9**:191b, 274b; **14**:603b
Canyons
submarine, **3**:548a; **7**:144b
Cão, Diogo, **1**:573b
Capacitance
measurement of, **7**:93b
naming of, **6**:212a
Cape Dezhnev, **2**:14b
Cape of Good Hope, **4**:131b
Cape Philosophical Society, **6**:326b
Cape Photographic Durchmusterung, **5**:405a; **7**:236a
revisions of, **7**:18a
Cape Town, South Africa
observatory at, **5**:404b–405b; **6**:326a; **8**:613b
Cape Verde Islands, **10**:615a
fauna, **3**:569b–570a
geology, **3**:567b; **4**:140b
Cape Zone Catalogue, **3**:50a
Capella, Martianus
see Martianus Capella
Capillaries
circulation in, **2**:172b; **14**:321b
contraction of, **10**:582b; **11**:566a
diameter of, **8**:552b
discovery of, **1**:483a; **8**:128b; **9**:63a; **11**:238b
neoproduction of, **2**:165a
perforation of, **14**:143b
permeability of, **12**:617b
regulation of, **7**:502b–503a
see also Arteries, Cardiovascular system, Veins
Capillarity
seventeenth-century study, **6**:483a; **9**:485b; **11**:507b; **13**:335a, 336a, 445a
eighteenth-century study, **6**:173b; **9**:472a; **15**:312b, 358b
nineteenth-century study, **2**:88b, 356a; **4**:203a, 205b, 228b; **6**:278a; **8**:427b; **13**:102b; **14**:185b, 360a
constant of, **4**:378b
law of Eötvös, **4**:378b
measurement of, **11**:241b
and temperature, **14**:479a
theories of
Renaissance, **8**:200b
seventeenth-century, **1**:165a; **4**:506b
eighteenth-century, **6**:40a, 171b–173a; **12**:430a; **13**:265b; **15**:359a
nineteenth-century, **3**:591b; **5**:305a; **8**:262a, 596a; **9**:174b, 223a, 549a*; **10**:546a; **15**:358a, 359b–360a, 379a, 489a
Capitaine, Louis, **3**:108b
Capitaine, Pierre-Charles, **3**:108b
Capital gains
calculation of, **2**:42a

Capitalism, **15**:133a, 135a, 136a,b, 139a, 405b, 406b, 409a, 410b
Cappeler, Moritz Anton, **3**:210b; **11**:521b
Capra, Aurelio, **3**:59b
Capra, Baldassar or Baldesar (*ca.* 1580–1626), 3:59–61; **5**:240b
  and Galileo, **12**:105b
Capreolus, Joannes, **1**:198b; **13**:349b
Capron, Horace, **8**:577a
Caprylene, **14**:156b
Capsala, **2**:322a
Capsian culture, **11**:336b
Capuchins, **9**:4a–b, 15b
Caqué, Joseph, **5**:204a
Carabidae, **6**:507a
Caramuel y Lobkowitz, Juan (1606–1682), 3:61
Carangeot, Arnould (1742–1806), 3:61–62; **11**:521b, 523a
Carathéodory, Constantin (1873–1950), 3:62–63; **4**:229a,b; **5**:350b; **6**:395a*; **7**:400a*; **8**:368a*; **11**:256a; **12**:247a*
  analysis, **4**:548a, 562a; **7**:573a*; **11**:327a
  topology, **11**:452a
Carbamates, **8**:614b
Carbazole, **5**:489a
Carbide, **6**:480b–481a
Carbohydrates, **5**:6a–b, 324b; **6**:185a
  classification of, **2**:66a
  disaccharides, **1**:286b
  fermentation of, **14**:247b
  glycosides, **1**:286b
  metabolism of, **2**:28b–29b; **4**:360a–b; **7**: 517b, 519a; **8**:452a,b, 614b–615a; **10**:580a; **14**:454b
    hormones and, **3**:415b–416a
    insulin and, **3**:352a; **15**:228b–229a
    in muscle, **2**:29b; **6**:501b
    and phosphorus, **10**:326b
    and pituitary, **15**:228b–229a
  oxidation of, **5**:126b
  reabsorption of, **2**:531b
  structures of, **5**:2b–4b; **15**:13a
  see also Sugar, etc.
Carbolic acid
  see Phenol
Carbon
  absorbent powers of, **5**:57a; **12**:404b–405a
  acids, effects of, **3**:8b
  allotropy of, **5**:187a; **8**:119b
  assimilation of by plants, **1**:361a, 362a, 389b; **7**:13a; **8**:345a–b
    illumination effects on, **2**:184a–b
  atomic weight, **2**:135a; **4**:244a, 395a; **8**: 332b; **9**:91b; **12**:619b
  bombardment of, **3**:329b–330a
  bonding of, **1**:390a; **5**:371a; **7**:281a, 552a
  burning of, **2**:75a–b
  carbon-13, **12**:368a
    production of, **8**:482a
  carbon-14
    dating techniques, **3**:430a
  classes of, **5**:140a
  compounds of
    see Compounds, organic
  crystallization of, **3**:8b
  determination of, **8**:331b–332a
    microchemical, **11**:129b
  electrochemical nature of, **4**:246b
  elemental nature of, **6**:284b
  equivalent weight, **5**:104a
  fusion of, **12**:433a
  graphitization, **5**:139b–140a, 141b
  heat of combustion, **4**:554b
  hydrogenation of, **2**:4a, 324a
  and iron, **3**:212a; **5**:601b; **6**:368b; **8**:119a; **9**:591b; **13**:246a–b
  oxidation of, **5**:621b
  oxides of, **5**:603a
  in plants
    source of, **8**:344b, 345a
  structure of, **5**:139b–140a, 203a
    tetrahedral models, **2**:622a,b; **7**:282a,b; **8**:109b; **10**:359a; **11**:87a, 88a; **13**: 576b–577b; **14**:265a, 455a
  valence of, **3**:448b
    tetravalence, **1**:390b; **3**:448b; **7**:280b; **8**:508a; **10**:178a
    trivalence, **3**:246b
  wave mechanics of, **6**:615a
  white, **5**:187a
Carbon cycle, nuclear, **4**:280a; **5**:272b
Carbon dioxide, **6**:257a; **13**:468b
  acid nature of, **2**:572b
  in atmosphere, **13**:389a
    and earth temperature, **1**:300b
  in blood, **2**:161a; **3**:600a; **5**:149a; **6**:21b, 24a, 261b; **9**:18b; **11**:138b; **12**:319b
    tension of, **7**:502a
  compressibility, **1**:128a
  critical temperature, **1**:161a
  detection of, **13**:262b
  discovery of, **6**:254b
  electric discharge through, **13**:468b–469a
  liquefaction of, **10**:206a
  measurement of, **2**:184a; **9**:95b; **10**:558a
  in metabolism, **5**:37b–38a; **10**:22b; **11**: 102a
  osmotic pressure, **13**:40b
  and plant nutrition, **7**:13a
  poisoning by, **2**:61b
  in respiration, **8**:83b–84a
  and respiration mechanism, **6**:24a, 264b–265a
  in sea water, **7**:503b
  spectrum, **10**:345b
  transport of, in body, **5**:149a
  see also Chokedamp, "Fixed air"
Carbon disulfide, **11**:495a
  composition of, **3**:316a
Carbon monoxide, **3**:600b
  in blood, **9**:347b, 348b
  composition of, **3**:316a
  discovery of, **6**:254b; **9**:152b; **11**:146a
  dissociation of, **11**:281a
  liquefaction of, **10**:206a; **14**:522b
  liquid
    as coolant, **3**:299a
  oxidation of, **5**:621b
  poisoning by, **2**:26a, 31a; **6**:460a
    treatment for, **6**:265a
  reactions of, **8**:118a,b
  and respiration, **14**:175a
  toxicity of, **6**:24a, 505a
Carbon suboxide, **4**:90b
Carbonates
  biodeposition of, **9**:589b
  composition of, **14**:488a
  crystallography, **14**:490b
Carbonatite, **2**:486b
Carbonic acid
  in air, **9**:155a
  assimilation of, **1**:362a
  composition of, **12**:619b
  in natural waters, **7**:502b
  in plants, **12**:124a
  in respiration, **10**:282a; **14**:64b–65a
  salts of, **8**:520b
Carbonic anhydrase, **7**:273b
Carboniferous period, **8**:511b; **10**:258a
  deformation, **3**:70b; **7**:251b
  division of, **2**:109a
  glaciation, **4**:262a–b
  paleobotany, **8**:5a–b, 264a; **11**:373a; **12**: 259b, 277a; **13**:44a; **14**:298a, 398b, 462b
  paleogeography, **7**:250a–b, 320a
  paleontology, **1**:472b; **2**:109b, 110a, 491b–492a; **5**:109b, 441b, 497b, 538a; **6**: 58a*, 221b; **7**:252a, 460b–461a; **8**: 572a; **9**:346a; **11**:501b–502a; **14**:495b
  stratigraphy, **1**:210a; **3**:395b, 530a; **4**: 262b, 363b, 540b; **5**:475b, 476a, 538a; **7**:253a; **8**:486a–b; **9**:583a–584a; **10**: 584a; **11**:35a, 567a; **13**:620b–621a; **14**:311b
  volcanic activity, **5**:336a
Carbonization
  artificial, **2**:3b
Carbonyls, **5**:541a
  discovery of, **2**:70b; **13**:299a
Carboxyl, **7**:452a
Carboxylase, **6**:111b
Carboxylic acid, **1**:390a
  structure of, **6**:109a; **14**:169a
  synthesis of, **5**:540b
  trimethylene, **2**:562b
Carboxytartronic acid, **7**:283a–b
Carbuncle, **3**:218b
  bacillus, **15**:479b
  diagnosis of, **2**:300b
  intestinal, **12**:96b
Carbylamine, **3**:272a; **5**:315a; **6**:462b
Carcano-Leone, Giambattista, **1**:315b; **3**: 355b
Carcavi, Pierre de (*ca.* 1600–1684), 3:63–64; **4**:46a; **9**:118a, 498a; **10**:335b; **13**:335b
  correspondence of, **1**:541a; **4**:574a; **5**: 159a; **8**:149b; **10**:339b(n); **13**:434a
Carchemish (Turkey), **14**:504a
Carcinogens, **9**:586a,b
Carcinoma, **4**:299b
  see also Cancer
Card games, **3**:66a
Card tricks, **9**:471a
Cardano, Girolamo (1501–1576), 3:64–67; **2**:590b, 591b; **3**:249a; **7**:39a; **8**:212b; **10**:512b; **14**:3b
  algebra, **2**:280b; **4**:586b–587b, 595b; **5**: 409b, 410a; **7**:103a; **12**:92b; **13**: 61b(n); **14**:24a, 147b
  Ars magna, **2**:280a; **4**:587a,b, 595b; **10**: 245b; **13**:259a
  Artis arithmeticae tractatus de integris, **4**:612b
  Astronomical Aphorisms, **9**:168b(n)
  Cardano's formula, **4**:596a; **6**:472b; **8**: 162a
  Cardano's suspension, **3**:66a
  correspondence of, **10**:245b
  critics of, **3**:311b; **9**:488a; **12**:135b–136a
  and Ferrari, **4**:586a–587b
  geometry, **9**:157a
  influence of, **2**:320a; **4**:596b; **5**:246b; **8**: 271b; **10**:494a; **11**:457b; **13**:62a(n); **14**:20b
  influences on, **8**:242b; **13**:210b
  and Leonardo da Vinci, **3**:66a
  Practica arithmetice, **2**:280a; **4**:586b; **10**:270b
  rule of, **3**:65b, 120a*
  De subtilitate, **10**:281a; **12**:135b
  and Tartaglia, **3**:64b–65b; **4**:586b–587b; **13**:259a
Cardiac output, **4**:615b; **6**:25a, 37b–38a; **13**:53b
  nitrous oxide method, **7**:502b
  venous return, effects of, **6**:264b; **7**:502a
Cardinalis (master at Montpellier), **1**:290a
Cardiograph, **11**:296b
Cardiology
  medieval Arabic, **15**:499b*
  eighteenth-century, **5**:366b; **12**:302b–303a; **14**:26a

early nineteenth-century, **3**:427a; **9**:252b
late nineteenth-century, **2**:35b; **13**:53b
early twentieth-century, **5**:281b; **7**:278b; **8**:294b–295a, 295b; **9**:95a; **13**:300b; **15**:100a–101b
exercise-tolerance methods, **8**:295a
see also Heart
Cardiometer, **15**:100a
Cardiovascular system
ancient discussion of, **1**:262a, 264b, 266b–267a, 316a–b; **4**:106a, 384a–385a; **5**:233b–235b; **6**:317b–318a; **11**:127b–128b
medieval Arabic discussion of, **9**:41a, 603b–604a; **11**:238b
medieval Latin discussion of, **11**:601a; **12**:81b, 82b
Renaissance discussion of, **4**:487a, 488a, 520a–b, 586a; **8**:201a; **11**:390b–391a; **12**:324a–b; **13**:569a–b; **14**:9a; **15**:80b
seventeenth-century discussion of, **6**:152b–158a; **8**:128b; **11**:586b–587a
eighteenth-century discussion of, **2**:226a–b; **6**:62b, 63b–64a
nineteenth-century discussion of, **12**:45a
anatomy of, **5**:504b; **8**:327b, 328a
arteriovenous anastomoses, **12**:556b; **13**:590a
in cold-blooded vertebrates, **4**:268a–b
collateral circulation, **11**:98b–99a
cytology of, **5**:43a
diseases of, **2**:547b; **8**:447b
aortic insufficiency, **14**:26a
embryogeny of, **11**:570b–571a; **15**:525a
fetal, **1**:204a; **4**:510b; **6**:65b; **9**:323a–b; **11**:133a
histology of, **9**:57a
mechanics of, **14**:200a
neurophysiology of, **2**:367a; **6**:225b; **10**:432b, 582b
innervation of vessels, **1**:537a; **6**:434b
pulmonary circuit, **4**:199a, 520b; **6**:153b, 155a, 318a; **13**:569b
critics of, **10**:599a
discovery of, **3**:355b–357a; **12**:324a–b
temperatures within, **3**:604b
sensitivity studies, **15**:469a–b
synanastomoses, **5**:234b, 235a; **15**:80b
tumors of, **3**:518b
venous system
liver as center, **1**:316a–b
see also Aorta; Arteries; Blood; Capillaries; Heart; Hemodynamics; Pulmonary artery; Pulmonary vein; Vasomotor system; Veins; specific disorders, e.g. Thrombosis, etc.
Carellus, Giovanni Battista
Ephemerides, **2**:401b
Carex, **1**:396a,b, 412a; **5**:191a
Caribbean
botany, **11**:47a–b
exploration of, **15**:89a
gravimetry, **14**:515b
Caridroit, Fernand, **10**:570a
Caries, **5**:317a
Carina (constellation)
eta Carinae (nebula), **6**:326b
Carius, Ludwig, **7**:474b, 551b
Carleman, Tage Gillis Torsten, **4**:561b; **9**:126b; **14**:90a, 282a
Carlini, Francesco, **11**:7a; **12**:160a
Carlisle, Anthony (1768–1840), 3:67–68; **1**:43b, 44a; **3**:600b; **6**:567b; **10**:108b; **14**:129b(n)
Carlsberg breweries
laboratory, **6**:100a,b; **7**:393a; **10**:380a
Carlsberg Foundation, **7**:393a; **8**:501a; **12**:190b
Carlsberg Laboratory, **7**:113b, 393a; **10**:380a; **12**:546a
Carlson, Anton Julius (1875–1956), 3:68–70; **9**:553b*; **15**:75b
Carlson, Chester Floyd, **8**:322a
Carlson, John Franklin, **10**:214b
Carmichael, Dugald, **2**:19a
Carmichael, Robert Daniel, **2**:144b; **8**:617b
Carmine, **3**:159b
Carnall, Rudolf von (1804–1874), 3:70
Carnap, Rudolf, **7**:151a; **11**:355b, 356a; **12**:609b
Carnegie, Andrew, **4**:194b; **7**:429b; **14**:121a; **15**:100a
Carnegie Corporation, **6**:31b; **14**:393a
Carnegie Foundation, **5**:433b
nutrition laboratory, **1**:610a
Carnegie Institute of Technology
physics, **13**:43a
Carnegie Institution (Washington), **3**:130b; **9**:284b; **10**:601a; **14**:503b
astronomy, **1**:55b; **2**:333a; **6**:29a, 30a
botany, **3**:317b
cosmic ray studies, **3**:370b
Department of Embryology, Johns Hopkins Medical School, **13**:97a
Eugenics Record Office, **3**:590a
founding of, **14**:121a
Geophysical Laboratory, **7**:5a
geophysics, **1**:521b, 552b; **3**:484a; **4**:484a, 563b; **5**:596a
history of science, **12**:110b–112b
seismology, **8**:99b
Station for Experimental Evolution, Cold Spring Harbor, **3**:590a
Carnelian
medical use of, **6**:397a
Carnelley, Thomas, **14**:560b
Carnivores
body temperature of, **4**:285b
classification of, **5**:46a
Carnot, Hippolyte, **3**:72a, 79b–80a, 80b
Carnot, Lazare-Nicolas-Marguerite (1753–1823), 3:70–79; **1**:582a; **2**:299b; **3**:81a, 132a, 213a, 417a, 554b; **5**:97b; **7**:569a,b; **9**:418a, 472b, 493b*; **10**:118b; **11**:164a; **12**:234a; **15**:346a
Géométrie de position, **7**:500a; **12**:325a
geometry, **1**:591a; **2**:418b
on heat engine, **3**:8b
historiography, **13**:592a
mechanics, **3**:82b; **10**:3b
theorem of, **1**:570b
Carnot, Nicolas Léonard Sadi (1796–1832), 3:79–84; **3**:72a,b, 73b, 74b; **5**:97b; **6**:243b; **15**:487a
Carnot cycle, **2**:499a; **3**:75a, 81b, 83a, 287a, 304b; **5**:387b; **6**:244b; **11**:292b; **14**:197b; **15**:443b
Carnot engine, **3**:81b, 83a
Carnot function, **3**:82a, 83a, 306a; **11**:292b
Carnot's principle, **15**:423b
and Clément, **3**:316a–b
Réflexions sur la puissance motrice du feu, **3**:287a, 316a
theorems of, **3**:81b–82a, 83a, 287b; **5**:387b; **13**:130b
thermodynamics, **3**:305a; **13**:375a, 377b–382a *passim;* **15**:85b
Carny, Jean-Antoine Allouard, **5**:603a
Caro, Heinrich (1834–1910), 3:84–85; **5**:489a; **8**:553a; **10**:516b
Caro's acid, **3**:85a–b; **5**:189b
Caroline of Anspach, princess of Wales, **3**:295b, 296a; **8**:151b
Caroline Islands, **3**:568b; **4**:256a; **8**:418a
gravimetry, **9**:181a
Caroline University (Germany), **3**:522a,b, 527a
Caroline War
historiography of, **1**:603b
Caronic acid
structure of, **13**:388b
Carotene, **7**:35a; **14**:601a
discovery of, **12**:544a; **14**:112a
structure of, **7**:518a; **15**:257b
Carotenoids, **4**:362b; **15**:257b
naming of, **13**:487b
synthesis of, **15**:257b
Carothers, Wallace Hume (1896–1937), 3:85–86; **13**:3a
Carotid artery
reflexes, **15**:469a–b
Carp
circulatory system, **4**:268a
distribution of, **1**:623b
Carpathian Mountains
botany, **14**:116b, 117a
geology, **8**:544a; **10**:29a; **11**:439a; **15**:502b, 526b
glaciation, **9**:149b
stratigraphy, **2**:108b
Carpenter, Geoffrey Douglas Hale, **12**:417a
Carpenter, Henry Cort Harold (1875–1940), 3:87; **6**:560b; **13**:8a*
Carpenter, James, **9**:616a
Carpenter, Philip Herbert, **3**:87b, 89a
Carpenter, Philip Pearsall, **3**:87b
Carpenter, William Benjamin (1813–1885), 3:87–89; **3**:87a, 477a; **7**:92b*; **11**:113b, 537a
on Crookes, **3**:475a–b
and Darwin, **3**:88a,b
The Microscope and Its Revelations, **8**:328a
ocean currents, **3**:471a
oceanography, **13**:360a–b
Carpentier, Jules Adrien Marie Louis, **4**:40a
Carpi, Berengario da
see Berengario da Carpi, Jacopo
Carpi, Girolamo da, **3**:40a
Carpoidea, **7**:61a
Carpology, **5**:216b–217a, 217b, 218a
Carpospores, **13**:395b
Carr, Ezra Slocum, **4**:274b; **6**:57a
Carr, Francis Howard, **15**:10b
Carr, Herbert Wildon (1857–1931), 3:89–90; **3**:383a*; **8**:168a*
Carré, Ferdinand, **8**:262a
Carré, Louis, **9**:50b
Carrel, Alexis (1873–1944), 3:90–92; **5**:40a
Carrick, John, **3**:494a
Carrier (med.), **9**:546a*; **11**:346a
healthy, **8**:449b
Carrington, Richard Christopher (1826–1875), 3:92–94; **12**:579b
Carro, Jean de, **7**:97a; **12**:57b
Carroll, Benjamin, **4**:522b
Carroll, James (1854–1907), 3:94–95; **4**:620a; **11**:346b
Carroll, Lewis (pseudonym)
see Dodgson, Charles Lutwidge
Carronades, **11**:499a
Carson, John Renshaw, **15**:386b
Carst, Agathe, **7**:554a
Cartailhac, Émile, **1**:562b; **2**:346b–347a, 450b
Cartan, Élie (1869–1951), 3:95–96; **2**:192a; **3**:559b; **4**:83a; **10**:524b*; **14**:14a*, 243b, 283b; **15**:198b
algebra, **8**:326b; **9**:457b; **11**:55b
analysis, **10**:573b; **11**:53b; **14**:13b
Cartan's problem, **9**:71b
Cartan's theorem, **9**:71b
geometry, **4**:336b, 337a; **11**:455a; **12**:172b; **14**:311a

Cartan, Élie (cont.)
group theory, **14**:90a, 284a–b
Cartan, Henri, **2**:351b; **11**:54b
Carte du Ciel, **3**:428b; **4**:251a; **5**:316a, 405a–b, 561a; **6**:282a–b, 351b; **9**:551b; **11**:320b, 412b, 484b; **12**:153a; **13**:423a, 500b
Carte Géologique Internationale d'Europe, **11**:375b
Carter, Henry Vandyke, **5**:515a
Cartesian devil (or diver), **9**:14a; **13**:438b
Cartesianism
seventeenth-century, **1**:474a, 481a; **2**:46b, 354a, 601b; **3**:103b, 320b–321a; **4**:65a, 96a, 505b; **5**:59a–b, 415b, 520b; **6**:380a; **7**:459a, 611a; **8**:127a, 152a, 173a–b, 437a–438b; **9**:47a–53a, 244a, 245a, 497b–499a, 499b, 509b–510a; **10**:120a, 314b–315a; **11**:506a–508b, 587b; **13**:31a, 169a,b, 170a, 445a, 479b–480a; **14**:26a, 482b
eighteenth-century, **1**:111a,b; **3**:104b, 105b, 107a–b, 114a–115a, 283a, 431a, 485b, 486a; **4**:214b–215a, 471b; **5**:61a–62a, 354a; **6**:36a, 459a; **7**:606a; **8**:115a, 259b; **9**:33b–34a, 326a, 500a; **10**:145b, 146a; **11**:157b–158a, 432a; **13**:179b; **14**:296b
nineteenth-century, **2**:274a; **6**:596a–b; **12**:240a
twentieth-century, **13**:254a; **15**:423a, 424a
earth, figure of, **3**:103b, 104b, 107a
and electricity, **1**:547a
mathematics, **1**:115b; **2**:349a
and Newtonianism, **11**:544a
opposition to, **3**:295a, 380b, 381a,b, 622a; **5**:511a; **7**:111a, 275b, 276a; **8**:151a, 152b–153a, 156a, 163b, 171b; **9**:509b–510a; **10**:61b, 319a; **11**:317a, 508a–b; **13**:33a
Cartesius, Renatus
see Descartes, René
Carthusians, **1**:290b; **9**:90a
Cartier, Jacques, **3**:608b; **6**:20a
Cartier, Pierre, **15**:53b, 54b
Cartilages, **3**:355a; **6**:505a
arytenoid, **1**:619b
cytology of, **11**:383a
structure of, **11**:216b; **12**:242a
Cartography
ancient Greek, **4**:389b–390a; **13**:84b–85a
ancient Roman, **11**:198b–200a
medieval Arabic, **7**:8a–9a; **9**:89a–b
medieval Latin, **1**:382a; **9**:190a
Renaissance, **1**:179a, 574a; **3**:423b–424a; **5**:347b, 377a; **8**:214a; **9**:309b, 335b, 581a; **10**:197a, 239a–240a, 616b–619a; **11**:202b–203a, 396a, 403a; **13**:440b–441a; **14**:127b, 275a–b; **15**:156a
seventeenth-century, **2**:185b, 201a; **3**:421a–b; **4**:497a*; **8**:25b, 65a–b; **9**:13a
early eighteenth-century, **1**:176a; **2**:348a; **3**:105b, 107b–108b, 344b; **4**:22a–b, 23a; **9**:233a; **10**:255b–256a, 319b
late eighteenth-century, **1**:176a; **2**:551b–552a; **3**:39a, 106a; **4**:70b, 477b, 492b–493a; **7**:86b, 87a; **8**:351a; **10**:117a; **12**:549a; **14**:279b
nineteenth-century, **1**:573a, 594b–595a; **2**:109a, 288a; **6**:456b; **7**:19b–20a; **11**:337b; **14**:116b
twentieth-century, **3**:616b–617a; **7**:265a–b; **9**:104a; **10**:505b–506a; **11**:524b
Chinese, **12**:380a
contour lines, first use of
for ocean depths, **2**:551b; **4**:70b
geological, **8**:577a
International Millionth Map, **2**:374a; **10**:506a
mathematical approach, **3**:226a–b; **6**:125a; **7**:265a–b
Mesopotamian, **15**:637b–638b
and metric system, **1**:573a
naval, **4**:214a; **14**:513b–514a
topographic, **6**:216b–217a
1:25,000 scale, **2**:109a
route mapping, **6**:216a–b
see also Atlases; Globe; Historiography; Hydrography; Maps; Projections; specific countries and areas
Carus, Julius Victor (1823–1903), 15:77–79; 1:418a; **6**:299b; **10**:263a; **15**:167a
Carus, Paul (1852–1919), 15:79–80; 2:533a; **10**:486a
Carus Monographs, **2**:198b
Carvallo, Emmanuel, **2**:594a
Carver, Jonathan, **9**:500b, 537a
Cary, William, **14**:485a
Casa Biochemica (research institute), **5**:209a
Casa di S. Pantaleo, **2**:312a
Casal Julian, Gaspar Roque Francisco Narciso (1680–1759), 3:97
Casali, Giovanni, **8**:232a
Casamicciola, John, **1**:289a
Casaubon, Isaac, **6**:306a
Casca bark, **2**:547b
Cascades, method of, **11**:512b
Cascarilla
bark, **4**:560a
Case, Edwin Martin, **6**:22b
Casein, **13**:160b, 161a
composition of, **8**:348b
formula of, **9**:558a
Casey, John, **8**:175b; **12**:284b
Casimir, Hendrik Brugt Gerhard, **8**:475b; **10**:424a; **13**:240a
Casing, theory of, **15**:164b
Casiri, Michael, **1**:437a, 513b; **14**:112b; **15**:420b
Casorati, Felice, **2**:85a, 470a; **3**:135b; **8**:504b; **12**:513a
Caspary, Ferdinand, **15**:195b
Caspary, Robert, **2**:425b; **9**:340a
Caspersson, Tobjörn, **7**:448b
Caspian depression, **12**:300b
Caspian Sea, **4**:289b
altitude of, **5**:428b; **13**:111a,b
geography, **13**:85a
geology, **4**:309a
hydrology, **7**:410b; **14**:53b
level of, **8**:187a
natural history, **4**:308a–b
oceanography, **12**:414b
Cass, Lewis, **12**:204a,b
Cassegrain (*fl.* 1672), 3:97–98
Casseri (or Casserio), Giulio (*ca.* 1552–1616), 3:98–100; 1:480a; **12**:577a
anatomy, **1**:480b; **4**:507b; **14**:7a
and Fabrizio, **3**:98a–b
illustrations, **1**:482b; **12**:577b
Cassin, John, **12**:257b
Mammalogy and Ornithology, **10**:439b–440a
Cassini, Alexandre Henri Gabriel, **3**:106a
Cassini, Gian Domenico (Jean-Dominique) (Cassini I) (1625–1712), 3:100–104; 2:202a, 306b, 308b, 312a; **3**:23a, 105b; **4**:22a,b; **5**:352a; **9**:89b–90a, 310b, 485a,b; **10**:596a; **14**:637a
astronomical tables, **3**:430b; **9**:77b; **14**:178b
astronomy, **1**:311a; **7**:376a; **9**:115b; **10**:546b; **11**:203b(n), 525b–526a; **12**:162a
geodesy, **5**:543a; **11**:525b
meteorology, **9**:117b
Ovals of Cassini, **3**:103b; **5**:521b
Recueil d'observations faites en plusieurs voyages, **12**:585b
sundial, **14**:589b
tides, **15**:339b–340a
Cassini, Jacques (Cassini II) (1677–1756), 3:104–106; 3:103b, 173b, 197b, 486a; **4**:602b; **5**:226a; **7**:542a; **8**:143b; **9**:90a–b, 532a; **10**:255b
astronomical tables, **4**:22b, 23a; **14**:178b
astronomy, **4**:23b; **14**:415a
De la grandeur et de la figure de la terre, **1**:581b
geodesy, **3**:107a–b, 281a; **7**:576a
tides, **15**:340a
Cassini, Jean-Dominique (Cassini IV) (1748–1845), 3:106–107; 3:108a; **4**:505b; **5**:61a; **7**:563b; **8**:136b
astronomy, **3**:198a; **11**:425a*
Carte de France, **2**:216b; **3**:108b; **4**:107b
Cassini's curve, **9**:55a
geodesy, **4**:15a; **8**:135b; **9**:250b
Cassini de Thury, César-François (Cassini III) (1714–1784), 3:107–109; 3:104b, 198a; **8**:143b, 177a; **9**:90a–b; **11**:65b
cartography, **4**:349b; **7**:86b, 87a; **8**:22a
geodesy, **3**:39a, 105a; **7**:542b; **9**:231b; **10**:255b
Cassini family, **3**:100a; **4**:15a
earth, shape of, **3**:282b; **9**:34a
Cassinoid, **5**:521b
Cassiodorus Senator, Flavius Magnus Aurelius (*ca.* 480-*ca.* 575), 3:109–110; 1:226a, 269b, 564b; **2**:228a, 229a; **3**:175b; **4**:121a; **5**:365a; **7**:27b; **12**:81a
Institutiones, **2**:229b, 233b; **4**:443a,b
Cassiopeia (constellation)
alpha Cassiopeiae, **2**:403a
variability of, **2**:147a
eta Cassiopeiae, **13**:110a, 120b
mu Cassiopeiae, **13**:120b
nova, **2**:402b–404a; **9**:192a
Cassirer, Ernst Alfred (1874–1945), 3:110–112; 3:457a*; **7**:235a; **8**:167b*; **9**:189a; **11**:73b, 355a; **12**:112a; **15**:424b
Die Platonische Renaissance in England und die Schule von Cambridge, **14**:378b(n)
Cassius Dionysius, **3**:174b
Cassius Felix, **5**:233b; **7**:27b
Castaldi, Luigi (1890–1945), 3:112–114; 10:268a*; **11**:598a*; **13**:565a*, 589b*
Castel, Louis-Bertrand (1688–1757), 3:114–115; 5:54b, 510a; **15**:58b*, 270a
critics of, **15**:57a*
Castellani, Aldo, **2**:528b–529a
Castelli, Benedetto (1578–1643), 3:115–117; 1:424a, 541a; **2**:83b, 306b, 307a,b, 308a; **3**:149a; **4**:128a; **5**:239b, 242a, 244a, 246b, 247b; **9**:13b, 369a; **11**:404a; **13**:433b–434a
Castelli, Pietro, **2**:308b, 309a; **10**:307b
Castelnuovo, Guido (1865–1952), 3:117; 4:373a,b, 374a,b*, 522b; **10**:139b, 606b*; **12**:286a*; **13**:623b; **14**:87b*
geometry, **5**:569a; **8**:542b; **11**:55a, 536a; **12**:330b, 331a
Kronecker-Castelnuovo theorem, **3**:117a
Castigliano, (Carlo) Alberto (1847–1884), 3:117–119; 9:578a
and Menabrea, **3**:118a–b; **9**:268a,b
theorems of, **3**:118a
Castiglione, Giovanni
see Castillon, Johann
Castiglioni, A. (pseudonym), **11**:159a
Castilloa elastica, **3**:177a*

Castillon, Johann (Giovanni Francesco Melchiore Salvemini) (1704–1791), 3: 119–120; 7:563a; 8:160a; 10:87b(n)
Castillon problem, 3:461b
correspondence of, 3:286a*
Casting
medieval, 13:327a; 15:253b
Renaissance, 2:143a; 11:486a; 15:255a
eighteenth-century, 3:560b
nineteenth-century, 9:60b
Mesopotamian, 15:651b
Casting out nines, 4:606a, 611a; 13:540a, 544b, 545b
Castle, William Ernest (1867–1962), 3:120–124; 4:272a; 9:520b; 10:610b
Castor (star), 6:512a
satellite of, 13:110a
Castor bean
proteins of, 9:284b
Castoreum, 1:212b
Castration, 15:646a
experimental, 7:64b
parasitic, 5:385b
physiological effects, 10:570a
Castro, João de, 10:160b
Casuarinaceae, 5:441b
fertilization in, 13:460a
Cat, 3:521a
parasites of, 5:502b
Catabolism, 5:282a–b; 6:300a
Catalan, Eugène Charles, 3:177b; 6:307a; 8:250a; 9:81a; 10:25a
Catalán, Miguel Antonio (1894–1957), 3: 124–125; 1:371a; 8:151a
Catalan (language)
as literary medium, 8:547b–548a
Catalase, 2:562a; 7:273b; 13:152b
Cataldi, Pietro Antonio (1552–1626), 3:125–129; 4:451a; 9:12b, 303a
continued fractions, 1:591b; 2:320a, 507a
Catalepsy, 15:3b
Catalysis
seventeenth-century study, 5:420b
nineteenth-century study, 2:589a; 3:259b; 9:401b; 12:198b; 13:313a; 15:461a–462a
twentieth-century study, 4:357b, 553b; 12:404b; 15:2b
analytical use of, 13:226b
autocatalysis, 7:462a–b
biological, 1:361b, 362b; 2:561a–b; 6:505b; 7:273a–b, 406a
compound, 9:427a–b
concept of, 2:95b; 9:352b, 425a
contact, 1:431b; 4:414a; 9:427a–b; 11:34a; 13:312a, 313a
enhancement of activity, 1:415a
heterogeneous, 1:286b; 7:21a–b, 462a; 15:36b
homogeneous, 4:545a
industrial, 2:323b; 7:378b–379a
irreversible, 14:602a
kinetics of, 1:414b–415b; 6:404b; 15:36b
by metals, 3:197b; 4:20b; 7:457b; 12:47a–b; 13:313a
platinum, 1:586a; 3:86a; 4:134a, 241b; 5:595b; 6:285a; 13:313a; 14:602a
prediction of, 1:415a
by minerals, 4:134b; 6:177b, 178a
negative, 2:67b
organic, 14:601b–602a
oxide, 14:602a
poisoning of, 1:415a; 12:47b; 14:602a
selective, 7:21b
theories of, 2:499a; 7:21b, 462a–b; 12:47a–b; 14:395b; 15:462a
multiplet theory, 1:414b–415a
Catapult, 3:491b; 6:312b, 313a; 10:586b, 587a–588b
Cataract (ophth.), 3:116b, 219a*; 9:75a, 464a*
black, 4:285b
cause of, 2:305b; 6:231a
location of, 11:511a
surgery, 11:112a, 158b; 12:137a
treatment of, 8:114b; 10:418b
Catarrh, 9:469a
theories of, 8:526b
Catarrhini, 2:110a; 6:8b
Catastrophism
seventeenth-century, 6:486b
eighteenth-century, 4:28a, 72a–b, 151a–b; 6:584b; 10:283b, 284a
early nineteenth-century, 1:413b; 2:495a, 496a, 568a; 3:395a,b, 526a; 4:402b; 5:32a,b; 6:456b–457a; 10:323a; 11:593a; 12:262a; 15:380a
mid nineteenth-century, 2:570b; 3:552b–553a, 566b–567a; 4:309a; 5:519a; 6:10a, 437b, 454b; 8:260b, 575a; 9:583a, 585a; 10:212a, 222a; 11:505b; 12:262b
late nineteenth-century, 1:130a; 7:370b–371a
opposition to, 11:235b; 12:182b; 13:460b
Categories, 1:381a; 4:255a; 5:193a; 12:6b–7a; 13:510a; 15:495a
Aristotelian, 1:100b, 277b, 278a; 2:232b, 610b; 6:378a–b; 7:136a; 8:549a
Augustinian, 1:334a
Catelan, Abbé de, 9:50a, 51a–b
Catena, Pietro, 1:108b
Catenary, 1:65b; 2:47b, 48b, 49b, 52a, 215b; 5:521b; 6:601a–602a; 7:194a; 8:163b
Catenata, 4:142b
Catenoid, 9:343b
Caterpillars
anatomy, 8:580a; 13:171b–172a
physiology, 8:445b
Catesby, Mark (1683–1749), 3:129–130; 1:331a, 487b; 3:349b, 350b; 9:537b
Natural History, 9:366a
Catfish, 1:265b
Catgut, surgical, 2:584b
Catharsis theory, 2:447b–448b; 5:174a
Cathedrals
architecture, 14:404a
Catherine II, Empress of Russia (Catherine the Great), 2:576b; 3:41b, 115a, 198a, 285a; 4:88a, 472a; 6:383a; 7:309b*; 8:147b; 9:231b; 10:283b; 12:522b; 15:4b
Catheterization
cardiac, 2:25a, 30b; 3:219b, 220a
Catheters, 5:137b
flexible, 13:465a
Cathetometer, 5:271b
Cathode
back bombardment of, 6:546b
dark space, 1:320b; 3:477b; 14:217b
Edison effect, 4:284a
oxide, 14:218a
Cathode ray(s), 1:320b; 3:498b; 5:341a, 458b; 6:343a; 11:420a, 530b; 12:191b, 201b; 14:217b, 338b
absorption of, 8:181a; 13:367b
charge-mass ratio, 12:238b
chemical effects, 14:31b
energy of
and X-ray frequency, 4:195a
magnetic deflection of, 7:263b; 14:32a
nature of, 3:477b–478a; 5:458b; 6:342a, 439b; 7:264a; 8:180b–181a, 496a–b; 10:525a; 11:446b; 13:365b, 366a–367b; 14:31b–32a, 340a; 15:464a
secondary, 7:468b
velocity of, 13:366a, 366b–367a; 14:340a
Cathode tubes, 2:427b; 4:355b; 6:439b; 8:180b, 181a
Catholicism
Renaissance, 1:587b–590a; 2:539a–542b; 3:13b
seventeenth-century, 2:526b; 4:95b; 9:98a; 14:301b
opposition to, 9:23b
eighteenth-century, 1:139b; 3:15b
nineteenth-century, 2:187a, 273b; 3:52b, 132b–133a; 4:225a; 9:81a, 428b; 15:167b–168a
and Aristotelianism, 1:99b, 280a; 5:402b, 403a; 10:173a
and evolution, 13:276a–b
see also Heresy; *Index Librorum Prohibitorum;* Inquisition; specific religious orders
Cations
large, 7:498a
Caton, Richard, 2:2a
Catoptrics
ancient Greek, 1:189b; 4:430a–b; 6:313a; 13:322b–323a
ancient Roman, 11:200b
medieval Arabic, 6:194a; 7:215a–b
medieval Latin, 14:460a–b
Renaissance, 9:192a
seventeenth-century, 1:475a; 5:521a, 525b, 526a; 7:375a; 8:164a, 166a; 9:13a
eighteenth-century, 3:454b*
see also Mirrors, Optics, Reflection
Catskill Mountains
geology, 9:173a
surveying of, 5:600a
Cattell, James McKeen (1860–1944), 3:130–131; 6:53a; 14:528a
Cattle, 3:200b; 15:648a
diseases of, 7:100b, 428a,b; 8:450a–b; 12:480a, 480b–481a, 482b
distemper, 10:563a
feeding of, 1:357a; 8:591a
hybridization, 7:33a
twins, 8:358a–b
Cattle plague, 3:479b; 7:614a
etiology of, 2:575b, 598b; 3:332b; 7:427b
vaccination, 7:286b
Cattle problem
see Archimedes, Cattle problem
Caucasian oil, 3:246b
Caucasus, 4:308b
geology, 1:19b–20b, 163a; 9:592b, 593b
Cauchoix, Robert Aglaé, 2:137b
Cauchy, Augustin-Louis (1789–1857), 3: 131–148; 1:344a; 2:350b*; 3:287a, 417a; 4:225a, 342b; 6:230b; 8:386b(n); 10:2b; 15:481b
acquaintances of, 6:307b; 10:247b; 11:61b, 498a; 13:375a
algebra, 6:403b; 13:129a,b, 571b; 15:194a
elimination theory, 2:114a; 8:384b–385a
substitution, 5:262b
Analyse algébrique, 11:599a
analysis, 2:350a, 473a–b; 3:75b, 164a; 4:205b, 477b; 6:307a, 308b; 7:550a; 8:63b, 384a; 9:175a; 11:56b, 449a; 13:81b; 14:223b; 15:386a
continuity, 1:207a; 2:276a,b
differential equations, 2:358a; 4:478b; 5:204a
integration, 3:151a; 10:594b; 11:454b
residues, 5:96b; 8:367a
series, 2:275b; 7:566b, 570b

Cauchy, Augustin-Louis (cont.)
biographies of, **2**:88b, 134a
Cauchy coefficients, **3**:146b
Cauchy principle, **3**:136a; **8**:62b–63a
Cauchy problem, **3**:143a; **6**:4a,b, 5a; **10**: 554a
Cauchy sequences, **3**:53b–54a; **12**:189b
Cauchy-Goursat integral theorem, **3**: 138a,b, 139a, 140a,b; **5**:482a; **8**:162b; **11**:156b–157a
extension of, **9**:502b
proofs of, **11**:149a
Cauchy-Kovalevsky theorem, **3**:143b; **7**:478a–b
Cauchy-Lipschitz method of approximation, **3**:143a; **5**:204a
Cauchy-Riemann equations, **3**:137a,b, 138a, 139a; **4**:477b; **11**:449a
celestial mechanics, **11**:57b
"mixed method," **3**:146b
on contemporary scientists, **1**:15b; **2**: 431b; **5**:111a; **8**:62b–63a; **11**:77b
correspondence of, **11**:598b
Cours d'analyse, **2**:275a
critics of, **4**:125b; **8**:63a; **11**:77a; **15**:31a, 32a
elasticity theory, **3**:259a; **9**:199b; **14**:61b
Exercices de mathématiques, **3**:134a, 135a
and Galois, **5**:260b, 261b
geometry, **11**:62a, 77a
group theory, **3**:168a; **5**:262a, 263a; **13**: 216a
hydrodynamics, **13**:75b
influence of, **2**:304a; **3**:163a, 416b; **4**:4a, 228b; **6**:1b; **11**:207a, 344a; **13**:78b; **15**:199b
influences on, **11**:599b; **12**:130a
"Mémoire sur les quantités géométriques," **1**:239b
number theory, **1**:238b; **2**:271a; **8**:138b; **13**:219a
optics, **2**:452b; **8**:490b, 510a, 592a,b; **9**: 74a, 154b, 174a; **10**:27b; **11**:115b; **13**:77b; **15**:200b
students of, **11**:206b; **13**:127b
theorems of, **2**:266a; **13**:129b
Cauchy, Louis-François, **3**:131a–b; **6**:307a
Caulerpa, **9**:601b
Caullery, Maurice (1868–1958), 3:148–149; 3:185a*, 212b*; **4**:514a*; **5**:386b*, 584a; **7**:272b; **8**:146a*; **9**:186a*, 328a–b; **11**:334b(n); **12**:423b
Cauloid theory, **8**:354b
Caus, Salomon de, **14**:376b(n)
Causalgia, **9**:422a
Causation, **3**:470a; **4**:374a, 382a
Aristotelian, **1**:255b; **5**:106b; **13**:340a
in biology, **1**:259a–261a
motion, **1**:198a–b, 260a; **2**:605b–606b; **13**:206b
types of, **1**:260a–b
biological, **2**:570b; **4**:102a; **5**:447a; **6**:425a
concept of, **1**:277b, 278a; **3**:111b
Cournot, **3**:452b
efficient cause, **8**:156a
external, **1**:141a
final cause, **1**:255b, 260a; **3**:375b; **5**:286a; **8**:155b, 156a,b; **12**:602b
formal cause, **5**:397b–398a
geometrical theory of, **1**:379a
illumination and, **1**:337b
material cause, **5**:397b–398a
in medicine, **15**:240b–242a
middle cause, **5**:106b
natural, **1**:62a; **4**:383b
occasional cause, **9**:48a, 48b–49b
in physics, **2**:545b, 607b; **9**:235b–237b; **14**:91b–92a
statistical, **2**:247b, 250a; **15**:280b–284b, 303b–304b, 344b–345b, 365a, 366a, 369a, 371a, 375b
and symmetry, **3**:505b
theories of
ancient Greek, **11**:103b, 105a; **13**:329b–330b
medieval, **9**:137a; **12**:7Ba; **15**:261b–262a
seventeenth-century, **5**:59b, 415b–416a; **8**:155b–156a; **9**:48b–49b; **14**:302a
eighteenth-century, **2**:17b; **3**:344b–345a; **6**:558a–559b; **7**:232b–234a
nineteenth-century, **6**:247b–248a; **8**: 159b; **14**:343b
twentieth-century, **5**:122b; **8**:11b; **11**: 14b–15a, 356b; **15**:423a
Causis, Liber de, **1**:271b
Caussin de Perceval, Armand-Pierre, **2**: 360a; **14**:575a
Causticity
theories of, **2**:75a, 173b, 175a–177b, 180b–181a; **5**:493b; **6**:42a; **7**:58a; **8**: 74a, 76b; **9**:346b–347a
Caustics (math.), **2**:47b; **6**:85b; **9**:192a; **13**: 127b–128a, 130b
Cautery, **2**:314b; **10**:315b, 316a, 418b
Cautley, Proby, **4**:518b, 519a
Cavalieri, Bonaventura (*ca.* 1598–1647), 3: 149–153; 2:308b; **3**:100b; **5**:499b
acquaintances of, **1**:541a; **10**:103b; **11**: 404b
analysis, **8**:162a
Exercitationes geometricae sex, **1**:164a; **5**:588b–589a
and Galileo, **3**:149a,b
geometry, **5**:409a; **9**:303b; **10**:495b; **11**: 487a; **13**:434a, 434b–435b; **14**:49b, 148a
Geometry by Indivisibles, **5**:245b, 588b; **10**:103b, 336a
influence of, **4**:505b; **5**:195b; **10**:336b; **12**:205a, 459b; **13**:436b
influences on, **1**:229b; **2**:306b; **3**:115b; **5**:247b; **9**:13a; **13**:561a
principle of indivisibles (Cavalieri principle), **3**:126a, 150b–152a, 182b, 183a; **4**:506a; **6**:599b; **8**:151b; **12**:75b; **13**: 434b, 561a
students of, **1**:164a; **2**:308a, 309a; **9**:303a
theorem of, **3**:152b
Cavallo, Tiberius (1749–1809), 3:153–154; 1:411b; **10**:108a; **14**:73a, 81a(n)
Cavanilles, Antonio José (1745–1804), 3: 154; 2:510a; **11**:605b
Cavendish, Charles (*fl. ca.* 1630), **6**:445a,b; **7**:194a
Cavendish, Charles (*d.* 1783), **3**:155a
Cavendish, Henry (1731–1810), 3:155–159; 1:68a; **2**:177a; **3**:545a; **7**:277a; **9**:140a; **12**:494b; **15**:199b
chemistry, **2**:524a; **14**:488b
critics of, **14**:419a
editions of works, **9**:199a
electricity, **3**:445a; **6**:278b; **11**:142a; **14**: 72b
experiments of, **5**:32b–33a
geodesy, **9**:164b
geometry, **10**:495b
gravitational constant, **3**:420a; **9**:227b
optics, **14**:570b(n)
pneumatic chemistry, **6**:43a; **11**:144a,b, 145a, 279a; **14**:74b
fixed air, **8**:74a
nitrogen, **12**:25a; **13**:103a
torsion balance, **3**:446a(n); **4**:379a
water controversy, **2**:186a; **8**:78a; **9**:343b; **11**:146a
Cavendish, Thomas, **4**:214a
Cavendish, William, Duke of Newcastle, **6**:445a, 446b, 448b; **9**:598b
Cavendish family, **6**:444b–446a, 449a, 450a
Caventou, Joseph-Bienaimé (1795–1877), 3: 159–160; 5:297a; **10**:497b
chemistry, **5**:91a; **10**:498a; **11**:494b
Caves
art, **11**:484a
blindness of animals, **4**:310a; **10**:273b; **12**:216a
fossils in, **3**:264a–b, 532a; **7**:31a; **11**:483b–484a; **12**:184a, 318a
origin of, **3**:595a
Paleolithic, **2**:450b; **5**:109b
see also Speleology
Cavia, **3**:113a; **10**:16b
Cavitation, **11**:393b
Caxton, William
The Mirrour of the World, **13**:491b
Cayenne Island, **11**:423b–424b
Richer's expedition to, **3**:103a; **6**:606b
Cayeux, Lucien (1864–1944), 3:160–162
phtanites, **1**:472b
Cayley, Arthur (1821–1895), 3:162–170; 5: 78a; **7**:570a; **13**:79a*
acquaintances of, **5**:413b, 414a; **7**:383b; **13**:217b
algebra, **6**:451a, 572b; **8**:617a; **12**:171a
determinants, **3**:323a; **6**:357b
invariants, **1**:294a; **3**:314b; **5**:472a; **6**: 309a, 388b, 389b
Cayley transformation, **14**:90a
Cayley-Bacharach theorem, **3**:167a
Cayley-Plücker equations, **3**:168a
Cayley-Salmon theorem, **3**:167b
Cayley-Zeuthen equations, **3**:168a
Chasles-Cayley-Brill theorem, **3**:167b
correspondence of, **12**:86b, 171a
critics of, **12**:284b
dynamics, **6**:93b*; **9**:211a
geometry, **2**:350a; **3**:315b; **4**:419b; **6**:91b; **7**:163b, 397a, 574a; **8**:433a,b; **10**: 139b; **11**:46a; **12**:172b, 533a; **13**:20b; **14**:618b; **15**:198a
influence of, **10**:486a
students of, **5**:77b; **10**:447a
and Sylvester, **3**:163a, 164a, 167a, 170a; **13**:217a, 218a–b, 219b, 221a
and Tait, **3**:163b, 166a
Caytoniales, **13**:345b
Cazré, Pierre, **8**:267b–268a
Cebes, **11**:25a
Table (Pinax), **1**:484b
Čech, Eduard (1893–1960), 3:170–171, 4: 336a; **5**:202a
Cedar tree, **6**:492a
Cedrenus, Georgius, **3**:20b
Čelakovský, Ladislav Josef, **2**:371b; **3**:30b; **11**:154b
Celastraceae, **2**:519b
Celaya, Juan de (*ca.* 1490–1558), 3:171–172; 3:420b; **4**:237a, 238a; **8**:100b; **12**: 547b; **13**:210a–b, 350a
Celebes (island), **14**:203b
Celery
oil of, **3**:279b
Celestial elements, method of, **8**:314b–320a
Celestial navigation
see Navigation, celestial
Celiectomy, **8**:556b
Cell lineage, **1**:601a; **3**:389b–390a; **8**:355a, 357a; **9**:600b; **14**:426a–427a
Cell theory, **1**:476b; **2**:32b–33a, 426b, 520a, 626a; **3**:29b, 337a; **5**:447a, 469b–470a; **6**:132a, 269a, 591a; **7**:438b, 440a; **9**:287b, 442b; **11**:152b–153a, 361a; **12**:141b, 174a–b, 241b–243b, 244b; **13**:269a, 460b; **14**:236a, 596b; **15**:166b

in botany, **11**:592a–b; **13**:506b
critics of, **9**:494b–495a; **11**:234b, 491b–492a
mechanistic vs. biological, **2**:532a; **14**:41b
precursors of, **2**:287a; **4**:264b; **10**:164a, 195a; **11**:217a, 300b; **12**:241b; **13**:556a–b
reform of, **12**:230b–232a
Cellari, Bartolomeo, **1**:125a
Celli, Angelo, **5**:461a, 503a; **9**:95a
Cellini, Benvenuto, **1**:621a; **5**:580b
Cellobiose
synthesis of, **6**:538b; **14**:603a
Cells (biol.), **5**:535b
active transport, **10**:257a
apical, **3**:30a
biochemistry of, **6**:499a
chemical interactions, **14**:430a–b
definition of, **3**:337b; **11**:300b; **12**:231b, 242a
division of, **2**:626a; **3**:30a, 259b; **4**:546a; **5**:34b–36a, 447a; **6**:465b; **9**:442a, 600a–b; **11**:153a, 255a–b, 369a–b; **12**:192b, 193a–b; **13**:88a–89a, 458b; **14**:235a
amitotic, **5**:36a; **6**:435a; **11**:368b
and cytoplasmic volume, **6**:337a
radiation, effects of, **9**:549b–550a
and surface tension, **2**:627a
theory of, **6**:266a
electrical nature of, **1**:303a
energy of, **15**:257b
equilibrium among, **14**:228b–229a
formation of, **6**:270a; **9**:442a, 600a–b; **11**:152b–153a; **12**:174a–b, 242a–243a, 592a; **13**:542a
endogenesis theory, **2**:165b
micellar theory, **9**:601a–b
fusion of, **9**:442a
genetic equivalence of, **6**:338b
growth of, **9**:394a
by apposition, **3**:422a
intussusception theory, **3**:422a
immunity of, **2**:300b
in vitro cultivation of, **3**:91a–b; **5**:624a; **6**:132a; **8**:282b–283a, 447b
inheritance in, **1**:343a
interstitial, **8**:302a–b
ionic equilibria, **4**:165b
membrane
electric potential, **15**:21b
migration of, **5**:43a
movements within, **9**:345a, 442a
naming of, **6**:484a; **13**:556b
as physicochemical systems, **2**:626b–627a; **6**:499a, 500a; **7**:455a; **12**:242b; **14**:41b, 433b
polarities of, **11**:255b
prospective potency, **4**:187a
protoplasmic connections between, **13**:89a
secretory, **8**:302a
specialization of, **9**:494b–495a
stretching of, **14**:96a, 97b–98a
structure of, **7**:455a; **9**:442a
and age, **9**:416a
skeleton, **7**:455a
swarm cells, **3**:337b
vesicular, **8**:302b
walls of, **2**:426b; **4**:401b–402a; **9**:455b
chemical nature of, **6**:138a
formation of, **7**:395b
growth of, **3**:422a; **11**:152b; **13**:89a; **14**:97b, 350a
imbibition, **6**:467a
lysis of, **5**:30a
permeability, **10**:257a–b
structure of, **9**:601a,b; **14**:98b
see also Cytology, Cytoplasm, Meiosis, Mitochondria, Mitosis, Nucleus
Cellulose, **5**:325a; **9**:79a, 354b
decomposition of, **7**:406a
fermentation of, **6**:505b; **7**:503a
manufacture of, **8**:553b; **9**:426a
sulfite waste liquor, **3**:341b
in mushrooms, **2**:385b
naming of, **10**:436b
physical chemistry of, **13**:163a–b
structure of, **13**:1b; **15**:257a
technology of, **5**:344b
transformation into coal, **5**:157b–158a
Cellulose nitrate, **3**:207a
Celoria, Giovanni, **13**:440b
Cels, François, **3**:173a
Cels, Jacques-Philippe-Martin (1740–1806), 3:172–173
Celsius, Anders (1701–1744), 3:173–174; **3**:105a, 281b, 540a; **7**:374a; **8**:519a; **14**:178a
Celsius, Olof, **8**:374b
Celsus, Aulus Cornelius (*fl. ca.* A.D. 25), 3:174–175; **1**:315a; **4**:105b, 383a; **6**:317a, 318b, 422b; **8**:249b; **12**:81b
De medicina libri octo, **1**:612a
Celtium, **9**:544a; **13**:546b
Celts
archaeological study, **3**:622b; **9**:539a
Cement, **4**:548b; **6**:383b; **7**:44a
hydraulic, **5**:157b; **10**:557b
mechanism of, **8**:117a–b
theory of, **5**:203a
Roman, **3**:316b*
Cementation, **13**:327a
of iron, **3**:212b*
Cementite, **6**:480a
Cenozoic era, **1**:286a; **3**:161a; **5**:476a; **10**:168a; **13**:275b
paleontology, **1**:491b, 587a; **3**:536a; **9**:255b
stratigraphy, **9**:579a
see also Quaternary period, Tertiary period
Censorinus (*fl. ca.* A.D. 230), 3:175–176; **4**:172a, 388b, 443a; **10**:180a; **15**:709a
De die natali, **15**:222a(n)
Census, **10**:565a; **11**:236b; **12**:301a
Centaurea, **10**:426b*
Centaurus (constellation)
alpha Centauri
parallax of, **2**:101b; **4**:351a; **6**:263b; **8**:613b
omega Centauri, **6**:69b
Centnerszwer, Mieczyslaw (1871–1944), 3:176–177
Central America
botany, **5**:546a
ethnography, **6**:527b
geography, **10**:617b
limnology, **7**:184a
zoology, **12**:258a; **13**:552a
Centre International de Synthèse, **9**:341b, 342a, 378a; **11**:388b
Centre National de la Recherche Scientifique
founding of, **9**:411a; **10**:525b
Centrifuge, **12**:60a
optical, **13**:160a–b
Centriole, **15**:513b
Centroid, **3**:469a; **4**:506a
of algebraic curves, **6**:537a
concept of, **9**:430a
of triangle, **4**:516a
Centrosome, **13**:89a; **15**:513b
discovery of, **1**:601b
origin of, **14**:429a
role of, **2**:363a; **5**:51a–52b, 582a
Centrosphere, **4**:546a
Century Dictionary, **3**:438b; **10**:486b
Cephalic index, **11**:381a
Cephalization, **3**:553a
Cephalograph, **6**:138a
Cephalopods, **14**:57b
embryology, **7**:437b
eye, **6**:287b
paleontology, **4**:147b; **5**:50b; **6**:168b; **7**:460b; **9**:346a, 579a; **11**:594a
taxonomy, **10**:221b, 261a; **13**:9b; **14**:1b
and vertebrates, analogy between, **8**:49a
Cephalosporin, **5**:43a
Cepheus (constellation)
delta Cephei
variability, **10**:608a
variable stars, **5**:468b; **8**:539b
Ceramics
ancient Roman, **3**:622b
Renaissance, **10**:280a
seventeenth-century, **11**:41a
eighteenth-century, **3**:326a; **4**:250a; **11**:330a–b; **14**:213a–b
nineteenth-century, **2**:493b; **7**:379a; **8**:55a, 117b
crystalline, **11**:330b
Mesopotamian, **15**:647b, 652a–b
physical chemistry of, **14**:182b
production of, **15**:651a
see also Porcelain
Cerasky, Vitold Karlovich
see Tserasky, Vitold Karlovich
Ceratitids, **3**:530a
Ceratium, **14**:93b
Cerceris, **4**:504a,b
Cereal crops
diseases, **15**:507a
hybridization, **13**:478a–b
mutations in, **15**:507b
nutrition of, **11**:416a
polyploidization, **12**:76b
smut diseases, **2**:437b, 438a
toxicity, **15**:419a
vernalization, **5**:523b
Cerebellum, **5**:44b
ablation studies
and recovery, **15**:109a
anatomy of, **2**:255b; **4**:385a–b; **5**:251a–b; **8**:552b; **9**:93b; **14**:26a
compression of, **8**:506a
and equilibrium, **9**:10a, 20a–b
extirpation, **9**:93b
functioning of
tonic nature of, **7**:48b
involuntary nervous functions, **1**:584a; **14**:406b–407a
lesions of, **5**:208a; **6**:518b
localization of function, **7**:48b
neurohistology, **11**:274b
physiology of, **8**:536a; **9**:422a; **10**:220b
removal of, **8**:536a–b
role of, **11**:510b
Cerebral cortex
anatomy of, **5**:251a,b; **14**:15a
descending degeneration, **9**:93b
equipotentiality of, **8**:45b
excitation of, **8**:536a
gyri of
and heredity, **3**:245b
internal organ regulation, **1**:580a
and learning, **5**:175a–b
localization of function, **1**:569b; **2**:478a; **3**:205b; **4**:594a–b; **5**:27b, 208a, 461a, 463b; **6**:139b, 140a, 518b; **8**:15a, 45b, 536a; **10**:131b*; **12**:355b, 596b; **14**:277a
opposition to theory, **2**:525b
origins of theory, **5**:252a–255b; **7**:48a
proof of, **5**:196b; **6**:440b–441a; **11**:274b
motor center, **6**:441a, 518b; **7**:48a,b

Cerebral cortex (cont.)
compensation for cerebellar lesions, **5**:208a
lesions of, **5**:208a
mapping of, **1**:569b; **4**:594a–b; **8**:552b
stimulation of, **2**:1b; **4**:594a; **13**:529a–b
myelination, **5**:27b
neurohistology, **11**:274b
right and left dominance, **14**:277a
taste center, **1**:580a
visual area, **4**:594b; **6**:441a; **8**:445a
Cerebrospinal fluid, **2**:25a; **3**:437b; **9**:10a
sugar in, **2**:29a
Cerebrum, **4**:385a
ablation studies, **15**:109a
activity of, **10**:434a–b
function
reciprocal interchangeability, **5**:463b
gyri of, **2**:596a
removal of, **5**:463b
role of, **5**:44b
and evolution, **5**:208a
voluntary nervous functions, **1**:584a; **14**:406b
Ceres (asteroid), **6**:334b
discovery of, **10**:198b–199a, 591b–592a; **13**:425a; **15**:353b
orbit of, **5**:299b–300a, 301a, 309a
Ceresins, **9**:608b
Cerite, **3**:473b
Cerium, **9**:110b
analysis of, **9**:541b
atomic weight of, **2**:428b; **9**:289b, 290a
discovery of, **2**:92b; **3**:473b; **6**:436b; **7**:394b
isolation of, **2**:588b
Čermak, Johann Nepomuk
see Czermak, Johann Nepomuk
CERN, **2**:252b; **3**:330a; **11**:117b
Cerotic acid, **2**:484b
Cerotin, **2**:484b
Cerruti, Valentino (also Vincenzo), **2**:320a; **3**:117a, 118a
Certainty, logical, **9**:382a
Cervantes, Julian, **3**:177a
Cervantes, Vicente (1755–1829), 3:177; 12:327a
Cesalpino, Andrea (Andreas Caesalpinus) (1519–1603), 15:80–81; **5**:238a, 383b; **6**:152a; **8**:311b; **9**:308b–309a; **10**:512b
botany, **7**:194b; **11**:314b; **15**:67b
critics of, **11**:315b
taxonomy, **8**:375b, 377b, 378b; **9**:528b
Cesarini, Virginio, **5**:243a
Cesàro, Ernesto (1859–1906), 3:177–179; **5**:210a; **6**:472b
curves of, **3**:178b
summability, **2**:238b
Cesàro, Giuseppe, **3**:177b
Cesena, Michael of, **10**:172a
Cesi, Federico (1585–1630), 3:179–180; 2:284a; **5**:241b, 244a; **6**:309b; **11**:96a–b; **13**:29a
biology, **13**:29b
taxonomy, **13**:29b–30a
Cesium
discovery of, **2**:589a; **3**:476a; **7**:381b
in electronics, **8**:24a
spectrum of, **11**:580b
Cesspools, **10**:326a
cleaning of, **3**:7a
disinfection of, **3**:7a
Cestodes, **2**:322a; **5**:109a; **12**:421b
life cycle
discovery of, **1**:602b
Cestoni, Giacinto (1637–1718), 3:180–181; **2**:291a; **11**:342b
Cetacea, **1**:18a, 603a; **2**:423a; **3**:521a; **7**:547a; **13**:504b*; **15**:432a*
anatomy, **2**:443a; **11**:528b
classification of, **2**:474a; **5**:46a; **11**:316b, 317a
embryology, **3**:184b
spouting of, **11**:101b
teeth, **3**:184b
Cetus (constellation)
Mira, **2**:349a; **7**:194a, 373b, 472b; **8**:539b; **9**:486a; **12**:368a
Ceulen, Ludolph van (1540–1610), 3:181; **4**:8b; **9**:336a; **11**:533a,b; **12**:206b, 499b
Arithmetische en geometrische Fondamenten, **12**:205a
pi, value of, **7**:258b; **8**:28a; **12**:500b
translations of works, **12**:500a
Ceva, Giovanni (*ca.* 1647–1734), 3:181–183; **12**:55a
theorem of, **3**:182a
Ceva, Tomasso (1648–1737), 3:183–184; 2:182a; **12**:55a
De natura gravium, **12**:55a
Ceylon
botany, **6**:490a; **13**:342a
geology, **1**:52a
rubber industry, **6**:491a–b
C.G.S. system, **5**:407b
Chabaneau, François, **4**:345a,b
Chabrey, Dominic, **1**:526a
Chabry, Laurent (1855–1893), 3:184–185; **1**:498a
embryology, **14**:429b
Chadwick, James, **2**:251a, 338b; **3**:329b; **5**:331b; **6**:118a; **7**:153b; **8**:95a; **9**:38a, 261a; **12**:32b–33a, 34a,b; **15**:27a
Chaetoceros, **9**:587b
Chaetognatha
embryogeny, **7**:475a
Chaetopterus, **8**:357a
Chagas, Carlos Ribeiro Justiniano (1879–1934), 3:185–186; **2**:533b; **15**:97b
Chagas' disease, **2**:533b; **3**:185b–186a, 566a–b
Chaillou, Auguste, **11**:569b
Chain, Ernst Boris, **5**:30a, 42a–b, 43b
Chain of Being
see *Scala naturae*
Chain reaction
see Fission, nuclear, chain reactions; Reactions, chemical, chain reactions
Chain rule, **4**:611a,b
Chais, Charles, **7**:11b
Chalazogamy, **10**:1b; **13**:460a
Chalcedony
artificial, **8**:213b
Chalcidius
see Calcidius
Chalcography, **4**:496a
Chalcolite (torbernite), **3**:499a, 507a
Chaldaean astronomers, **15**:678a–b
Chaldaean Oracles, **1**:33b; **11**:184b
Chaldeanism, **11**:161b
Chales, Claude Milliet de
see Dechales, Claude François Milliet
Chalk, **2**:176a,b; **5**:411a; **14**:398a
composition of, **12**:65a
Chalk formation (geol.), **2**:494a–495b, 568b; **3**:161a,b, 162a; **4**:540b; **5**:14a–b, 476a
fossils in, **4**:549a; **9**:87a; **12**:543a
marine, **6**:229a; **8**:566a
Challenger (ship), **2**:557b; **3**:89a, 161b; **4**:127b, 492a; **6**:9a, 535b; **7**:92a; **8**:188b, 609a; **9**:588a–590a; **11**:113b, 372a; **12**:107b, 258a, 520a; **13**:8b, 237a, 360b–361a, 504b*
Challis, James (1803–1882), 3:186–187; 1:53b, 54a; **3**:92a, 93b, 583a; **9**:202b; **13**:76a
acoustics, **13**:76a
hydrodynamics, **13**:75a,b
Chalmers, Thomas, **2**:451a, 568b; **5**:32a; **11**:116a
Chamaerops, **15**:68a
Chamberlain, Alexander Francis, **2**:208a
Chamberlain, Charles Joseph (1863–1943), 3:187–188; **13**:89b, 90b*
Chamberlain, Owen, **4**:582b; **8**:96a
Chamberland, Charles Edouard (1851–1908), 3:188–189
anthrax, **3**:588a; **10**:410b(n); **11**:569a
and Pasteur, **10**:382b, 383b, 385b, 392b–399b *passim*
septicemia, **10**:407b
Chamberlin, Rollin Thomas, **3**:189b, 190a
Chamberlin, Thomas Chrowder (1843–1928), 3:189–191; **4**:562b; **9**:180b; **12**:84a; **13**:269b, 270a,b, 573a*; **14**:517b
cosmogony, **7**:85b; **8**:522a; **9**:552b
geology, **3**:594b; **6**:188b; **12**:83b; **13**:532a
Chambers, Ephraim, **2**:227b
Cyclopaedia, **4**:86a; **5**:567b; **6**:401a
Chambers, Robert (1802–1871), 3:191–193; **5**:252a, 255a, 333b; **14**:133b
critics of, **12**:278a
on evolution, **6**:592a
Vestiges of the Natural History of Creation, **1**:502a; **3**:573b; **5**:67b; **6**:592a; **10**:262b; **11**:116a; **14**:134b, 138a
critics of, **9**:389a; **12**:278a, 570b; **14**:190a
Chambers, Robert (*fl. ca.* 1920), **13**:268b
Chambers, William, **1**:88a
Chameleon, **1**:264a; **9**:441a
color changes, **2**:532a; **6**:455a*; **10**:520a
Chamié, Catherine, **3**:502b
Chamisso, Adelbert von (1781–1838), 15:81–83; **4**:288b, 407a; **13**:9b
Chamisso, Louis Charles Adélaïde
see Chamisso, Adelbert von
Chamousset, Claude Humbert Piarron de, **1**:529a; **3**:5a, 7a
Champeaux, William of
see William of Champeaux
Champier, Symphorien, **9**:75b; **11**:253b; **12**:323b
Champlain Group (geol.), **4**:364a
Champollion, Jean François, **14**:569b; **15**:681a
Champy, Christian, **2**:344b
Chance
implications of, **10**:335a
theories of, **1**:27a–b, 251b; **15**:366a,b
see also Game theory, Probability
Chancel, Gustav Charles Bonaventure (1822–1890), 3:193–194; **14**:395a
and Gerhardt, **3**:193b
and Laurent, **3**:193b
Chancel, Jean, **7**:22a
Chancourtois, A. E. Béguyer de
see Béguyer de Chancourtois, A. E.
Chancre
soft, **15**:454a
Chandler, Seth Carlo (1846–1913), 3:194; **3**:510a; **5**:480b*; **9**:226a
latitude, **7**:369b; **10**:35a, 232a
Chandoux, Sieur de, **4**:51b
Chandrasekhar, Subrahmanyan, **9**:405a; **13**:116b; **14**:92a
Chang Ch'ao, **14**:544b
Chang Heng, **3**:254b; **8**:419b; **13**:484b
Chang, Min Chueh, **10**:611a
Chang Ting, **8**:396a*

Chang Ts'ang, **8**:419a
Chang Ts'un-hui, **8**:391a, 396b*–397a*
Chang Tun-jen, **3**:250a; **8**:419b
Chang Ch'iu-chien suan ching, **3**:250b; **14**: 545b
Change, **4**:367b
Aristotelian, **1**:118b, 251b, 253a, 254b–255a, 258a; **7**:137a
first and last moments, **2**:610a–b
limits for
intrinsic vs. extrinsic, **2**:610a–b
as ontological principle, **6**:290b–291a
opposition to, **3**:335a
quantitative description, **6**:378a–b
Chanut, Pierre, **3**:320b; **10**:332b
Chao Ch'i-mei, **3**:250a
Chao Yu-ch'in, **8**:420a
Chao ch'a (finite differences), **3**:269b–270b
Chaos, **7**:538b
Chapelain, Jean, **5**:448a; **9**:497b, 498a,b; **10**:491a; **12**:84b
Chaplygin, Sergei Alekseevich (1869–1942), 3:194–196; 5:188b; **8**:559a; **10**:20a
Chaplygin-Zhukovsky postulate, **3**:196a; **14**:621b
formulas of, **3**:196a; **8**:301a
mechanics, **14**:620b
and Zhukovsky, **3**:194b–195a, 196a
Chapman, Alvan Wentworth (1809–1899), 3:196–197
Chapman, David Leonard (1869–1958), 3: 197
equation of, **3**:197a
Chapman, Ernest Theophron, **14**:169b
Chapman, Sydney, **2**:263a; **4**:375b, 376a, 561b; **8**:185b; **9**:222b, 224b, 225b, 404a
Chappe d'Auteroche, Jean-Baptiste (1728–1769), 3:197–198
Venus, **2**:5a
Voyage en Californie, **3**:106a
Chappuis, James, **6**:177b
Chaptal, Jean Antoine (1756–1832), 3:198–203; 1:616a; **2**:134b, 475a, 509a, 510a,b; **3**:317a*; **4**:178a; **7**:588a; **10**:326a, 446b, 611b; **12**:65b; **13**:309a, 310a
translations of works, **10**:108b; **15**:207b*
Chaptalization, **3**:200b–201a
Chapter Coffee House Society, **10**:108a
Char (fish), **8**:146a
Chara, **11**:154a
Characeae, **2**:160a; **13**:394b
Characidae, **4**:310a
Characteristic function, **8**:563a
discovery of, **6**:90a; **9**:128a–b
in dynamics, **6**:88b–89a
in optics, **6**:85b, 87a,b, 88a,b, 92b
in thermodynamics, **4**:228b
Characteristic triangle, **10**:337a
Characteristica universalis, **8**:153b–154a, 155a, 159a,b
Characteristics (math.)
method of, **3**:132a, 143a; **4**:205b, 478b; **5**:608b; **10**:544b
theory of, **3**:213b; **6**:545b; **7**:562a; **9**:476b; **10**:544b; **12**:227b–228b
trajectory of, **9**:476b
Characters, genetic
continuously varying, **4**:271a
coupling of, **3**:422b
degenerative, **4**:310a
fundamental, **13**:136a
quantitative, **10**:130a
see also Genes
Charas, Moyse, **2**:334a
Charcoal, **15**:651a
activated, **14**:601b
adsorptive power, **4**:80a; **8**:520a; **12**:146b, 404b
animal, **4**:41b; **10**:436a
combustion of, **13**:280b
conductivity of, **11**:142a
manufacture of, **11**:359a
Charcot, Jean-Baptiste (1867–1936), 3:203–205; 9:78b
Charcot, Jean-Martin (1825–1893), 3:205; 1:580a; **2**:35b; **5**:174a; **7**:67b; **8**:27a; **9**:327b; **11**:425b, 559b
Charcot-Marie amyotrophy, **3**:205b; **9**: 109a
Charcot's disease, **3**:205b
students of, **2**:131b, 366a; **3**:203b; **5**:172a; **9**:108b, 109a
Chardenon, Jean Pierre (1714–1769), 3:206–207; 5:600b, 601a
Chardonnet, Louis-Marie-Hilaire Bernigaud, Comte de (1839–1924), 3:207
Charge, electric
absolute unit of, **13**:445b–446a
capacity for, **5**:135b
conservation of, **5**:131a–133b
distribution of, **2**:137a; **3**:445a–b; **5**:132b, 133b, 135b; **8**:384b, 385b; **13**:375b–376a
intra-atomic, **6**:147b, 164a
electrostatic vs. electromagnetic, **7**:450a–b; **13**:365a; **14**:205a, 207a
elementary, **11**:347b–348a
fundamental value of, **12**:30b
interaction of, **7**:380b
leakage of, **3**:445a; **4**:44b–45a; **10**:190b
longevity of, **5**:517a
motion of, **6**:212a
nature of, **8**:491b, 494b
point, **15**:43b
residual, **6**:504b
Charlemagne, **1**:104b; **5**:365a
Charles I, King of England, **4**:13a,b; **6**:150b–151a, 387a; **7**:147a; **9**:507a; **12**:581b; **13**:450a, 501a, 507b; **14**:147a, 370b
Charles II, King of England, **4**:495b; **6**:68a, 448b; **8**:131a; **9**:400b, 507a, 529b; **10**:255a
patronage and appointments, **1**:317a; **2**: 522b; **3**:209a; **4**:38a, 496a; **5**:24b, 507a; **6**:449a–b; **8**:130a; **9**:528b; **10**: 565a; **12**:584b; **14**:147a, 150b, 361a, 405b, 510a
Charles V, Holy Roman Emperor, **3**:420b, 423b; **4**:164b, 586b; **9**:190b; **10**:122b; **11**:96a; **12**:322b; **14**:3b, 5b, 6b
Charles VIII, King of France, **12**:439b
Charles le Géomètre, **3**:208a
Charles, Jacques-Alexandre-César (1746–1823), 3:207–208; 3:346a; **5**:318a, 602a; **8**:78b–79a; **9**:492b; **11**:166b; **12**:66a; **14**:79b
Charles's law, **2**:262a; **3**:32b, 208a, 316a, 541a; **4**:243b; **5**:318a; **13**:579a
Charles F. Kettering Foundation, **7**:316b
Charles Louis, *Ger.* Karl Ludwig, elector palatine, **5**:606b; **14**:364a, 366b, 377b(n)
Charles University, Prague, **2**:428a–b; **14**: 586a
Charlesworth, Edward, **8**:570b–571a
Charleton, Walter (1620–1707), 3:208–210; 4:377b, 495a; **6**:258a*; **10**:43b, 60b; **11**:316b; **13**:508a*; **14**:364a
Physiologia Epicuro-Gassendo-Charletoniana, **2**:378a
Charlier, Carl Wilhelm Ludwig, **1**:429a; **10**:231a; **13**:98a
Charnockite, **4**:262a, 410a*
Charophytes, **7**:252b–253a
Charpentier, Johann (Jean) de (1786–1855), 3:210–211; 2:556b; **4**:347b; **5**:445a; **13**:604b–605a
Charpentier, Johann Friedrich Wilhelm de, **3**:210a; **4**:152a, 346b
Charpentier, Toussaint, **3**:210a
Charpy, Augustin Georges Albert (1865–1945), 3:211–212
Chartists, **15**:405b
Chartres
medieval school of, **2**:19b
Chase, Agnes, **15**:224a–b
Chase, Charles, **2**:459b
Chase, Frederick Lincoln, **4**:351b
Chase, Merrill Wallace, **7**:624b
Chaslavskii, Vasilii Ivanovich, **4**:143b
Chasles, Michel (1793–1880), 3:212–215; 2:214b; **5**:368a; **6**:1b; **9**:476a; **11**:61b
acquaintances of, **2**:299a; **3**:59a, 80a; **4**: 350a–b; **8**:367b
Aperçu historique . . . , **3**:182a; **8**:253a; **9**:301a(n); **14**:343a
translations of, **12**:511a–b
biographies of, **2**:88b
correspondence principle, **6**:571b; **12**: 228b
geometry, **1**:192a; **4**:426b; **8**:250b, 385a; **9**:598b; **11**:79b, 80a; **12**:172a
influence of, **1**:194b; **3**:560a; **4**:429a; **7**: 163b; **9**:80a
problem of, **6**:75a
students of, **2**:287b; **3**:178a, 559b; **10**:41a; **11**:414a; **14**:618b
theorem of, **3**:167b; **6**:570b
Chassaignac, Pierre Marie Édouard, **3**:490a; **8**:405b
Chastelain, Jacques, **1**:322b
Chateaubriand, François René de, **1**:489b
Châtelet, Gabrielle-Émilie Le Tonnelier de Breteuil, Marquise du (1706–1749), 3:215–217; 3:281b, 485b; **4**:216b; **7**: 442b; **14**:83a–b, 84a
Institutions de physique, **7**:602b
and Maupertuis, **3**:215a,b; **9**:186b, 187a, 188b
Chatton, Walter, **2**:394a
Chaucer, Geoffrey (*ca.* 1343–1400), 3:217–218; 2:435b; **9**:160b; **10**:541a
Chaudé, Joseph Victor Ernest, **5**:297b
Chaudesaigues, **10**:525b
Chauliac, Guy de (*ca.* 1290–*ca.*1367), 3:218–219; 2:314b, 546b; **5**:348a; **6**:276a
Chirurgia, **9**:468a; **15**:185a
Chaulmoogra oil, **4**:155a; **11**:121a; **15**:2b
Chautard, Jules Maria Augustin, **4**:76a
Chauveau, Jean-Baptiste Auguste (1827–1917), 3:219–220; 2:598b; **8**:541a; **9**:103a*, 505b, 506a; **10**:392b, 393b; **11**:429a
Chavanneau, François, **13**:530b
Chayes, Felix, **1**:394b
Chazelles, Jean Matthieu de, **9**:90a
Chazy, Jean François (1882–1955), 3:220–221; 2:359a*; **4**:408b*, 548a*
Chazy Limestone, **4**:363b
Chebotaryov, Nikolai Grigorievich (1894–1947), 3:221–222; 1:307a; **5**:508b, 509a*; **10**:102b*; **12**:365a*; **14**:631a*
Chebyshev, Pafnuty Lvovich (1821–1894), 3:222–232; 2:425a; **3**:142b; **4**:483a*; **5**:508b, 608b; **12**:512b; **15**:31a
analysis, **4**:475a; **9**:126a; **10**:249b; **15**: 23a–b
Bienaymé-Chebyshev inequality, **15**:32a
Chebyshev's inequalities, **9**:126b, 127b, 128a
editions of works, **12**:535a; **15**:22b
formulas of, **3**:178b

Chebyshev, Pafnuty Lvovich (cont.)
hydrodynamics, **8**:559b, 560a, 561b–562a
polynomials, **3**:229b; **9**:127a; **13**:26b
probability, **8**:562b; **9**:127b–128b; **15**:489a
school of mathematics
see Petersburg mathematical school
students of, **4**:474a,b; **7**:513b; **8**:559b; **9**: 125a,b, 324a; **12**:364a; **14**:630a
translations of works, **12**:535a; **15**:31a
Chebyshev, Vladimir Lvovich, **3**:222a
Cheese
chemistry of, **6**:135b
cold curing of, **1**:357a
formation of, **4**:211b–212a
manufacture of, **9**:287b
Cheirostrobus, **12**:259b
Chelates
in analytic chemistry, **3**:272a
Chelius, Maximilian Joseph von, **5**:430a
Chelsea Physic Garden, **9**:390a,b; **12**:458b
Chemical Abstracts
editors of, **10**:158a
founding of, **10**:157a
Chemical equivalents, **5**:371b, 372b
Chemical Institute of Canada
founding of, **9**:394b
Chemical mass, **2**:79b–80a, 81b; **11**:436a–b
Chemical News, **3**:475a, 476b, 481a*
Chemical Reviews
editors of, **10**:158a
Chemical Society, London, **3**:488b; **4**:19a; **10**:177a; **14**:394b
founding of, **3**:557b; **5**:292a, 492b; **6**: 293a–b; **9**:392a
Chemical warfare
see Warfare, chemical
Chemical Warfare Service (U.S.), **5**:465b
Chemical-Physical Society (Vienna), **8**:511a
Chemicals
identification of, **2**:381a
manufacture of
Renaissance, **13**:397a
seventeenth-century, **13**:235a; **14**:451a
eighteenth-century, **1**:527a; **4**:250a; **5**: 602a; **6**:383b; **11**:499a–b
nineteenth-century, **7**:452b; **8**:122a, 553a–b; **9**:467a; **13**:597b; **14**:447a
twentieth-century, **14**:247b, 248a; **15**:1b
contact process, **14**:447a
regulation of, **12**:478b
purity of, **2**:381a
see also Classification, chemical; Drugs
Chemische Annalen für die Freunde der Naturlehre, Arzneygelahrtheit, Haushaltungskunst und Manufacturen, **3**:465a,b
Chemisches Journal für die Freunde der Naturlehre, Arzneygelahrtheit, Haushaltungskunst und Manufacturen, **3**:465a
Chemisorption, **12**:47a
Chemistry
Byzantine, **3**:20b–21a
medieval, **14**:35b
Renaissance, **2**:143a, 402b, 412b; **4**:387b, 393b–394a; **8**:213b; **10**:306b, 308a; **13**:141a–b
early seventeenth-century, **1**:571a–b; **2**: 305b; **3**:596a–b; **6**:127a, 254a–255a, 257a; **7**:354b–355a, 538a–539b; **8**: 310b–311a; **11**:389a–b, 511a–b; **12**: 78b–80a; **13**:559a; **14**:209a–210a
late seventeenth-century, **1**:318a, 451a–b, 549b–550b; **2**:311a, 317a–b, 377a–b, 380a–381b; **5**:417b–418a; **7**:150a–b, 284a–b, 459a, 524b–525b; **9**:507b; **11**:40b; **12**:600a; **14**:26b–27a, 639a
early eighteenth-century, **2**:16b–17a, 225b, 227a,b, 421a–b, 490a; **4**:129a, 214b, 223b; **5**:156b; **6**:37a, 259a–b, 477a–478a; **9**:62a; **10**:25b–26a; **11**: 562a–563b; **12**:365b
late eighteenth-century, **1**:527a–b, 529a–b; **2**:5a–8a, 74b–82b, 173b–181a, 523b–524a, 572a–b; **3**:5a–b, 7a–b, 200a, 206a–b, 326a–b, 382b, 419a–b, 464b–465b, 494a–495a, 540b–541a; **4**:343b–344a, 345b, 346b; **5**:57a, 90a–93a, 215b–216a, 276a, 411a–b, 532a–b, 600b–603b; **6**:18a, 54a, 283a–b, 382b–384a, 384b–386a, 468b, 495b, 578a–579b; **7**:58a, 202b*, 277a–b, 387b–389b, 586b–587b; **8**:247a, 519b–520b, 618a–621b; **9**:5a–6a, 148a–b, 152a–b, 471b–472a, 477a–b; **10**:325b–326a, 445b–446b, 496a–497a; **11**:565a, 616a–b; **12**:63b–65a; **13**:47a*, 468b–469a, 601b; **14**:191a–b, 197a–b, 213b–214a, 224b–225a, 332a–333a, 508a–b; **15**:205b–206b
early nineteenth-century, **1**:142b–143b, 344a–349a, 586a; **2**:81a–82b, 90b–96b, 420a–b; **3**:6a, 291b, 315b–316b, 400a, 541a–544b, 556b–557b, 598b–604a, 604b; **4**:74a, 233b–234a, 528a–241b, 274a, 528b–532a; **5**:7a–b, 202b–203a, 317b–326a, 429b–431b, 518a; **6**:114b–115a, 277a,b, 284a–285b, 324a, 325a, 436b; **8**:37b, 373a–b; **12**:253b–254b, 432b–433a; **13**: 372a–374a; **15**:171b–172b
mid nineteenth-century, **1**:416a–417a, 518a–b; **2**:436b, 484b–485b; **3**:289a–b, 586a; **4**:19a, 242a–247b, 394b–395a; **5**:197b, 297a–b; **6**:109b–110a, 564a–565a; **7**:224a, 460a, 509a–b; **8**:277a; **9**:18b–19a, 178a, 391a–392a, 401b; **12**:478a–b, 619b–620b; **13**: 467b–468a; **14**:529a–531b
late nineteenth-century, **3**:176a; **4**:130a–b; **5**:563b; **12**:493b; **13**:389a–b
twentieth-century, **3**:290a–b; **5**:189b–190a; **6**:525a–b; **8**:108b–109a
as academic discipline, **1**:451a; **15**:1b–2a
agricultural
late eighteenth-century, **4**:250a–b; **14**: 145a
early nineteenth-century, **3**:601a
mid nineteenth-century, **1**:156a, 518a; **2**:356b–357a; **8**:92b–93b, 344a–348a; **9**:18b, 558a–b; **10**:150a–b; **12**:233a,b
late nineteenth-century, **1**:325a–b, 357a; **2**:70b, 356b–357a; **3**:321b; **6**: 237a–238a; **10**:242a–b; **13**:459b–460a; **14**:357b–358a
early twentieth-century, **1**:288b–289a, 357a; **6**:135b, 443a; **7**:530a; **8**:591a; **10**:243a–b; **11**:179a–180a; **14**:45b–46a; **15**:492b
mid twentieth-century, **2**:170b; **4**:357b; **13**:66b; **15**:2a
analytical
seventeenth-century, **2**:353a, 380b–381b; **6**:380b–381a; **7**:150b; **8**:173b–174a, 312a; **12**:79a–b
early eighteenth-century, **5**:352b–354a; **6**:236b, 460a, 477b–478a; **8**:171b; **9**:104b–106b; **11**:109a–b, 562a; **15**: 94a
late eighteenth-century, **2**:5a–6a, 291b–292a, 564b–565a; **3**:5a–b; **5**:90b–91a, 216a; **7**:390b–391a, 394a–b; **8**:82b, 298a, 520a–b; **9**:106a–b, 478a–b, 559a–b; **11**:145a, 166a–169a, 434b–437b, 564a–b; **12**:287a, 494b; **13**: 280b–281a, 411a–b, 602a–603b; **14**: 332b, 463b–464a
early nineteenth-century, **2**:93a–94b, 385a–386a, 565a; **3**:159a–160a, 232b, 242a–243b, 316a, 600b; **4**:41a–b, 531b; **5**:318b, 319b–326a; **6**:166a–b, 284b–285a, 354a; **7**:379a; **9**:423b–424b, 527a; **10**:497b–498a; **11**:169a–171a, 172b–173a, 437b, 494b–495a, 540b–541b; **13**:310a–311a, 499a–500a, 547a–b, 597a; **14**:474a–b, 475a–476a, 486b–488a; **15**:7b
mid nineteenth-century, **1**:471a, 528a; **2**:65a, 67a, 587a–b, 589a–b; **3**:84b, 193a, 237b–238a, 488b–489a; **4**: 554a–b; **5**:72a–b, 73a, 125a–126b, 163b–165b, 349b–350a, 394a, 475a; **6**:293a–294a, 517a; **8**:54b–55a, 330a–333b; **9**:91b, 425b, 445b, 541a–b; **10**:37a, 440b, 499a–b, 510b–511a, 530a–b, 532b; **11**:51a–b, 359a–b, 615b; **12**:233a; **14**:156a–b, 169a–b, 477b–478a
late nineteenth-century, **2**:258a–b; **3**: 11b, 292a–293b, 321b, 507a–b; **4**: 127b; **5**:315b, 464a; **6**:498b; **7**:393a–b; **9**:95b, 305a, 530b–531b; **10**:157b–158a, 242a–b; **13**:7b–8a, 298b–299a, 517b; **14**:123a, 357a–b, 447a–b, 447b–448a; **15**:84a
early twentieth-century, **2**:170b; **3**: 235b, 272a; **5**:53a–b, 595b; **6**:365a–366a, 372a–b, 374a–b, 480b–481b; **7**:529a–530a; **10**:136a–b, 158a, 243a–b, 288b–289a; **11**:129a–b, 551a–b; **13**:546a–b, 574a–b; **14**:37b, 357b–358a, 448a, 587b–588b
mid twentieth-century, **4**:545b; **6**:366a–b; **13**:226a–b
catalytic methods, **13**:226b
combustion methods, **1**:451a; **2**:95b; **5**:126b, 353a; **8**:82b, 173b, 330b, 331b–332a; **9**:425b; **13**:311b
electrolytic method, **2**:258b; **6**:372a–b, 374a–b; **15**:84a
flame tests, **2**:589a; **9**:105b
immediate principle, concept of, **3**:243a
indirect analysis, **5**:164b
microchemical, **1**:10b, 613b; **8**:584a; **11**:129a–b; **14**:174a; **15**:57b
microscopic, **8**:520a; **9**:105a; **15**:157b
optical methods, **2**:453a; **4**:191b; **5**: 412b; **13**:238b
organic reagents in, **3**:272a; **4**:20b
oxidizing agents, **5**:324b
physicochemical methods, **7**:529a–530a
precipitation analysis, **2**:5b–6a; **5**:164a, 165a
radioactive techniques, **6**:366a–b
solvent extraction, **2**:5b; **13**:602a
thermal analysis, **5**:164b; **7**:529a; **14**: 182b
vapor density techniques, **2**:587b; **4**: 243b
volumetric, **3**:600b; **5**:216a, 321a, 325a–326a; **13**:547b
wet analysis, **2**:5b; **5**:90b, 163b–165b; **9**:105a
see also Chemistry, qualitative analysis; Chemistry, quantitative analysis
animal, **2**:92b, 95b, 385a, 386a, 483b; **3**:201b, 242a, 324a; **5**:90b–91a, 92b–93a, 324b; **8**:171b, 348a; **11**:172a, 593a
antiphlogistic, **1**:64b, 563b; **2**:90b, 92a, 180a–b; **3**:465b; **4**:343b; **5**:90a–b, 411b, 532b, 602b; **6**:386a, 395a, 549b;

7:388b–389a, 395a; **8**:73b, 373b; **11**:438a; **14**:600b
applied
medieval Latin, **13**:327a–b
Renaissance, **15**:491b–492a
seventeenth-century, **6**:381a; **14**:451a–b
eighteenth-century, **1**:44b–45a; **2**:5a; **3**:200a–202a, 494b; **6**:383b, 385b–386a, 395b; **8**:297a–298b, 469b–470b, 621a–b; **9**:62a, 105a, 343a; **13**:596b–597a; **14**:144b–145a, 197a–b, 225a, 332b
early nineteenth-century, **1**:43b–44a; **3**:79b; **4**:134a–135a, 531b–532a; **6**:115a; **7**:378a–379a; **13**:309a–b, 310a–b; **15**:172a–b
mid nineteenth-century, **1**:528a; **2**:340a–b, 620b; **4**:242b–243a; **5**:125b–126b; **6**:517a–b; **8**:330a; **9**:401b; **10**:557b; **13**:358b, 548a; **14**:394b
late nineteenth-century, **1**:553a; **6**:110a; **8**:116a–119a; **9**:451b; **11**:538a–b; **14**:123a, 447a–b, 633a; **15**:449b
early twentieth-century, **1**:553a–b; **5**:189b–190a; **8**:22b–24b; **13**:71a, 243b–247b, 454b; **14**:247a–248a; **15**:462a, 465a
mid twentieth-century, **7**:518a–b
see also Chemistry, technological; Industrial research, chemistry
archaeological, **2**:64b; **5**:190a; **6**:110b
aromatic, **3**:85a
atmospheric, **3**:204a; **11**:558b
atomic physics, effects of, **2**:247a
Baltic school, **3**:176b
biological
see Biochemistry
and botany, **1**:70a; **8**:345b
catalytic, **1**:414b–415b; **9**:427a–b
clinical, **13**:574a–b
colloid, **1**:431a–b; **2**:171a, 627b; **4**:142a, 165b, 545a; **5**:492b, 494a–b, 595b; **6**:118b; **8**:427b; **10**:251b; **14**:588a, 632b–633b; **15**:159b–160a, 201b
biological applications, **15**:160a
crystallographic, **5**:213a, 577a–b; **9**:183b; **10**:125b, 578b; **11**:270b–271a
definition of, **8**:310b
descriptive, **12**:418b, 420a
of explosives, **1**:390a; **2**:70a, 455a; **4**:239a–b; **5**:315a
food chemistry, **1**:471a; **5**:165a; **9**:314b; **10**:325b–326a
geological
see Geochemistry
handbooks, **5**:418a; **10**:321b; **12**:144a
industrial
see Chemistry, technological; Industrial research, chemistry
inorganic
late eighteenth-century, **8**:113b–114a, 469b–470b; **12**:145a–148a; **15**:315a–316a
early nineteenth-century, **1**:616a–b; **3**:455a–b; **4**:134a–b; **11**:495a; **13**:310b–313b; **14**:475a–476a
mid nineteenth-century, **1**:579a–b; **2**:66a, 586b–589b; **3**:301a–b, 617b; **4**:77b–78a; **5**:394a, 412a, 492b–494b; **9**:109b–110b, 424b–425a; **11**:537b–538b; **12**:227a; **14**:478a
late nineteenth-century, **2**:65a, 69a, 199a–b, 254b–255a, 428a–b, 557b; **3**:197a, 321b, 397b–398b, 478a–479b; **5**:381a; **7**:529a; **8**:116a–119a, 553a; **9**:450b–451b, 557a; **11**:318b; **12**:46b, 465a–b, 546a; **13**:389b; **14**:265b–272a
twentieth-century, **2**:429a; **3**:197b, 235a–b; **4**:20b–21a; **5**:595a–b; **10**:329b, 578a–b; **11**:551a–552a; **13**:70a–71b
systematization of, **1**:347b
judicial, **3**:237b; **4**:531b; **5**:7a, 164b, 297b; **10**:511a
marine
see Oceanography, chemical
mathematical, **2**:484b–485b, 515a–b; **4**:411b–412a; **5**:472b; **8**:468a–b; **9**:127a; **11**:172b, 434b–437a
see also Stoichiometry
medical
seventeenth-century, **3**:596b; **5**:418a; **7**:536b
eighteenth-century, **11**:48a–b
nineteenth-century, **1**:518a; **2**:336a; **5**:315b; **6**:506a; **10**:557a–b; **15**:11b
twentieth-century, **11**:576b–577a
and medicine, **1**:10a, 563b, 571a; **2**:317b; **4**:170a, 208b–209b; **5**:90a,b; **6**:293b; **9**:62a; **11**:51b; **12**:311a–312b, 604a; **13**:507a–508a, 535a–b
metallurgical, **6**:259b, 560b–561b
and meteorology, **1**:300b
mineral, **2**:572b; **11**:109a–b, 270b–271a; **12**:146a–b, 147b–148a, 329a,b; **13**:546a–b; **14**:463b–464a
and mineralogy, **2**:552b; **5**:203a; **14**:257b–258a
and natural history, **2**:572a; **5**:90a
notation
eighteenth-century, **1**:64b; **2**:6b, 7b
Black's diagrams, **2**:180b–181a
Cullen's diagrams, **2**:180b
nuclear, **6**:16b–17a; **8**:47b
optical analysis, **2**:453a; **4**:191b; **5**:412b; **13**:238b
organic
early eighteenth-century, **11**:562a
late eighteenth-century, **8**:82b–83a; **11**:564a–b; **12**:148a–b
early nineteenth-century, **1**:345b–346a; **2**:95b–96b, 135a; **3**:240a–243b; **5**:323b–324a, 324b–325a, 431a–b; **11**:171a; **12**:320b, 321b; **13**:309b–310a, 311b–312a; **14**:474b–475a, 600b
mid nineteenth-century, **1**:155b–156b, 347a, 389b–390b, 518b, 578a–b; **2**:63b–70b, 316a–b, 587b, 620a–624b; **3**:10a–b, 46a–49b, 193a–194a, 448a–449b; **4**:75b–76a, 77b, 78b, 243a–247a, 395a, 399b–400b; **5**:103b–104a, 125b–126a, 369a–374a, 474b–475a, 531a, 536b–537a; **6**:461a–463b; **7**:279b–283b, 450b–453a, 463b; **8**:54b–61a, 329b–350a; **9**:91a–b, 109b, 287a, 425a; **10**:357a–360b, 499a–b, 532a, 575b; **11**:36b, 300b, 353a, 541b; **12**:163b–164a, 236a–b; **14**:168b–169b, 394a–396a, 454b, 476a–478a, 529b–531b, 600b–601a, 622a–b; **15**:11b
late nineteenth-century, **1**:426a–b, 553b, 579a; **2**:560a; **3**:87a, 279a, 286a–b, 299b–300b, 321b, 341b, 510b–511b; **5**:1a–5a, 12b–13a, 128a–b, 214b–215a, 315a–b, 381a–b, 394a, 488b–489a; **6**:107b–108b; **7**:65a, 471a–b, 551b–552b, 623a; **8**:109b, 328b–329b; **9**:130b–131b, 354b–357a, 619a–b; **10**:14b–15a, 154b, 156a; **11**:277b–278b, 370b–371a; **12**:208b–209a, 236b, 452a–b, 593a; **13**:337b–338b, 406b, 410a–411a, 549b–550a; **14**:124b, 141b–142a, 265a–266a, 411b, 454b–455b, 560b–561a, 601b
early twentieth-century, **1**:286b–287a, 288a–289a; **2**:3b–4a; **3**:86a–b, 246a–b, 271b–272a, 286b, 347a–b; **4**:20b, 90a–92a, 553a–554a; **5**:5b–6b, 100a, 464a–465b, 540a–541a, 589a, 595a; **6**:108b–109a, 185a–b, 480b; **7**:372a–b; **8**:31b–32a, 108a–b; **9**:95b–96b, 183a, 354a–b, 402a–403b, 608a–b; **10**:121b–122a, 578b; **11**:84b–88b; **12**:46b–47b; **13**:1a–3a, 66b, 177a–b, 388a–b; **14**:125a, 247a–b, 392b–393a, 411b–412b, 601b–602b, 603a–b; **15**:1a–2a, 10b–11a, 13a, 15a–b
mid twentieth-century, **6**:398a–b; **7**:323a, 517a–518b; **13**:3a–b, 66b; **14**:248a; **15**:2a–3a
analytic, **2**:95b; **3**:488b; **5**:324b–325a, 412b; **11**:494b–495a; **13**:77b
definition of, **5**:372b; **6**:564b
handbooks, **1**:578b; **3**:341b
structural, **1**:106a, 168a–b, 340b–341a, 389b–390b; **3**:299b–300b; **4**:78b, 400a–b; **5**:126a, 488b–489a; **6**:108a–109a, 185b; **8**:60a–61a, 484b–485b; **9**:352a–353a, 355b–357a; **10**:517a–b; **11**:92b; **13**:549b–550a; **14**:334a–335a, 444a–446a; **15**:252b, 257a–b
synthetic, **1**:105b–106b, 390a; **2**:65b–69a, 316a–b, 562b, 621a, 622b, 624a; **3**:86a–b, 246b, 271b, 286b, 300a, 510b–511b; **4**:91a–b, 400a, 553a–554a; **5**:126a, 464a–465b, 540a–541a; **6**:107b–108a; **7**:21a–b, 457a–b; **8**:56a; **9**:355a–b, 360a; **10**:14b–15a, 515b–516b; **13**:66b; **14**:169a, 455a, 474b–475a, 477b, 531b; **15**:1b, 2b, 13a
patents, *see* Patents, chemistry
pathological, **2**:584a; **6**:293b; **11**:364b
pharmaceutical
Renaissance, **3**:471b–472a; **4**:209a
seventeenth-century, **6**:145b–146a; **8**:130a–b
eighteenth-century, **2**:291b, 564b; **3**:5b, 419b; **5**:351b; **6**:460a; **10**:26a
nineteenth-century, **2**:560b; **3**:546a; **4**:243a; **10**:575a; **13**:466a
twentieth-century, **3**:246a–b; **5**:99b–100b; **15**:252b, 258b
philosophy of
seventeenth-century, **8**:130b, 131a; **12**:312b
eighteenth-century, **2**:179b; **8**:81b–82a; **13**:603a
nineteenth-century, **3**:377b; **8**:336b; **9**:286b–287a; **13**:499b–500a
twentieth-century, **8**:119b–120a; **12**:494a; **14**:123b–124a
physical
eighteenth-century, **2**:186a–b, 227a; **8**:470a
early nineteenth-century, **1**:348a–349a; **4**:240a–241b; **10**:552b–553a
mid nineteenth-century, **1**:389b, 579a–b; **2**:68a–70b, 628a; **4**:78a; **5**:492b–494b, 587a–b; **7**:463a–b; **9**:110a, 287a; **10**:440b; **11**:372b; **12**:198a–b; **13**:236b, 358a–359b; **14**:108b–109a, 185b, 329a–330b, 360a–b
late nineteenth-century, **1**:296b–300b, 431a, 537b; **2**:258a, 559a; **3**:333a–b, 619b; **4**:76b, 78b–81a, 228b–230a, 485a; **5**:389b–390b, 597a–b; **6**:108b, 177b–178a, 244b, 519b–520a; **7**:203b, 461b–462b, 619b–620a; **8**:427b; **9**:107a–b, 304b–305a, 352b–353a, 616a–b, 619b; **10**:156a, 206a–b, 481b, 516b, 577a–b; **11**:8b, 278b–280a, 297b–299b, 446a, 485b, 534a–

Chemistry (cont.)
535a; **12**:191a–b, 271a, 593a–594a; **13**:156a, 242b–243b, 299a, 389b, 453b–454b, 576a, 577b–580a; **14**: 31b, 98a–b, 124b, 560a–561a; **15**: 432b–435b, 455b–463a
early twentieth-century, **1**:320b–321b; **2**:169b–171a, 498b–499b; **3**:176b, 333b–334a, 335b–336a, 436b–437a, 620a–b; **4**:82b, 141a–142a, 165a–b, 194b, 413b–414a; **5**:595b, 621a–623a; **6**:117a–118b, 365b, 538b; **7**:59a–b, 151b–153a, 161a–162a, 208a–b, 497b–498a, 529a–530a; **8**:22b–24b, 32a, 289a–293b, 446a, 502b–503a; **9**:378a, 393b–394b, 427a–b; **10**:156a–b, 251b, 525a–b; **11**:417b–418b, 482b; **12**:191b, 404b, 418b–419a, 495b, 546b; **13**:158b–162a, 177a, 243b–247b, 429a–b, 505a; **14**:125a, 182a–b, 318b–319a; **15**:36b–37b, 159b–160a, 201b, 435b–451a
mid twentieth-century, **1**:519b; **3**:290b; **4**:545a–b; **5**:103b, 118a–b, 139b–141b; **6**:118b–119a, 404a–405a; **7**: 387a–b, 518b–519a; **8**:46b–47b; **12**: 419b–420a, 437b–439a; **13**:7a, 162a–164a
and biology, **2**:184b
emergence of, **1**:536b; **15**:433b
and geology, **2**:369b–370b; **4**:408b; **7**: 5a–b; **10**:125b, 126b; **14**:58a–59a
and immunology, **1**:301b; **14**:624a
mineralogical, **4**:141a–142a
and pharmacy, **8**:427b
physiological
see Biochemistry
pneumatic
late seventeenth-century, **9**:244a–b, 245b–246a
early eighteenth-century, **2**:227a; **6**:35b, 40b–43b
late eighteenth-century, **2**:77a, 524a; **3**:156b, 157b, 561a; **5**:601a; **6**:165a*, 550b; **7**:13a; **8**:73b–79b, 247a, 585a–b; **9**:140a, 347a; **11**:143b–147a; **12**: 24b–25a, 145b–147b, 427b, 562b; **13**:468b–469a
early nineteenth-century, **6**:551a
see also Pneumatics
preeminence of, **3**:600a
professionalization of, **5**:394a*
qualitative analysis
eighteenth-century, **2**:5a–b
nineteenth-century, **5**:163b–165b; **8**: 331a; **10**:156a
twentieth-century, **2**:429a; **10**:156b
quantitative analysis
seventeenth-century, **12**:79b
eighteenth-century, **2**:5b–6a, 176a–181a, 227a, 292a; **5**:602a; **7**:58a, 394b; **8**:71b–72a; **11**:167a–169a
nineteenth-century, **2**:258b, 385b; **4**: 239b; **5**:164b, 203a; **11**:416b–417a
twentieth-century, **5**:497a–b; **11**:417a–b
and quantum theory, **5**:621b
and religion, **3**:398b; **5**:104a
sociology of, **12**:209a
structural
nineteenth-century, **2**:515a; **7**:179a–180a, 280b–283b; **8**:507b–508a; **9**: 130b–131b; **14**:265a–272a
physico-chemical methods, **1**:168a
see also Chemistry, organic, structural
of surfaces, **8**:24a
technological
Renaissance, **10**:23a–b
eighteenth-century, **10**:497a; **11**:109a–b, 499a–b
early nineteenth-century, **11**:171a; **15**: 206b–207a
mid nineteenth-century, **2**:65a; **7**:22a–b; **9**:287b; **10**:436a–b, 511a, 532b; **11**: 34a; **12**:520b–521a
late nineteenth-century, **7**:462b; **12**: 211b
early twentieth-century, **1**:288b, 301b; **5**:595a; **11**:319a; **12**:546b; **15**:435b
see also Chemistry, applied; Industrial research, chemistry
theoretical
seventeenth-century, **3**:596b; **7**:194b–195a, 524b–525a, 538b–539a; **8**: 130b–131a, 173a–174b; **9**:15a; **12**: 79b, 311a–312b; **13**:234b–235a
early eighteenth-century, **5**:351b, 353a–354a; **6**:477b–478a; **7**:188a–b; **8**: 468a–469a; **10**:26a, 80b–81a; **11**: 562a–b; **12**:303a, 604a–605b
late eighteenth-century, **3**:201b, 465a–b; **5**:532a, 600b–601a; **6**:383b–384a, 385a, 386a, 395a, 585b–586b; **7**:395a, 587a–b, 590a, 603a,b; **8**:66b–83a, 298a–b, 321a, 530b–531a, 612b–613a, 620a–621a; **9**:6a, 346b–347a, 477a–b; **10**:108a–b; **11**:146a–147a, 167b–169a, 437b–438a, 523a; **12**: 64b–65a, 143b–149a, 430a; **13**:465b–466a, 494b
early nineteenth-century, **3**:602a–603a; **4**:29a, 529b–531b; **5**:431a, 558a–559a; **6**:284b; **7**:254a; **10**:183a–184b; **11**:169a–171a, 173a–b; **12**:320b; **13**: 313b, 499b–500a; **14**:488a–490a
mid nineteenth-century, **3**:46b, 301b; **4**:399b–400a; **5**:294b–295a, 370a–374a; **6**:354a, 461b–462a, 463b, 564b; **7**:254b, 280a–283b, 451a–453a; **8**: 54b–61a; **9**:286b–291a, 347b–351b; **10**:37a–38b, 177a–178a; **12**:196b–198b; **14**:394b–396a, 530a–531b
late nineteenth-century, **3**:300a–b; **5**: 374a, 381a–b; **6**:519b; **9**:291a–b, 292b, 351b–353a, 360a–b; **12**:593a, 607b; **13**:363a–b; **14**:123a–b; **15**: 434b–435a, 462a–463a
twentieth-century, **8**:23b, 186a–b, 473a–479a; **13**:546b; **14**:123b–124a; **15**:463b–464b
theories of
acid-alkali hypothesis, **2**:381a; **8**:344b; **13**:234b–235a
Avogadro's hypothesis, **1**:344a–347b, 518b; **3**:47a–49b
azoammonium theory, **2**:199b
chain theory, **2**:199b; **7**:179a–b
contact theory, **14**:395a
coordination theory, **2**:199b; **3**:272a, 301b; **4**:20b–21a; **5**:350a*; **7**:179b; **11**:551b; **13**:546b; **14**:265b–272a
electrical, **6**:252b
mechanistic, **5**:156b
nucleus theory, **8**:60a
oxygen theory, **3**:200b, 201b; **13**:313b
resonance theory, **6**:108b
structural, **2**:621a–623b, 624a–b; **3**:49b, 448a–449b; **4**:400a; **5**:13a; **7**:451b; **9**:287a
unitary theory, **2**:199b; **4**:245b–246b; **5**:374a; **9**:286b
see also Dualistic theory; Phlogiston, theory of; Radical theory; Substitution theory; Type theory
vegetable, **2**:70b, 385a–386a, 572b; **3**: 242a; **5**:91a, 92a–b, 324b, 351b, 352b–353a; **6**:166b, 477b; **8**:55a, 171b, 173b–174a; **13**:602a
extractive principle, theory of, **2**:385b
see also Alchemy; Apparatus, chemical; Atomism; Classification, chemical; Compounds; Congresses, scientific; Crystallography; Distillation; Dyeing; Education, chemical; Electrochemistry; Elements, chemical; Engineering, chemical; Historiography, of chemistry; Iatrochemistry; Industrial research, chemical; Kinetics, chemical; Magnetochemistry; Matter; Mineralogy; Nomenclature, chemical; Phlogiston; Photochemistry; Photography, chemistry of; Phytochemistry; Precipitation; Radiochemistry; Reactions, chemical; Terminology, chemical; Thermochemistry; Titration
Chemnitz, **1**:78a
Chemostat, **13**:227b
Chemotaxis, **4**:372a; **10**:576a
Chemotherapy
medieval Arabic, **9**:41a
eighteenth-century, **1**:392a; **11**:37b
late nineteenth-century, **4**:296b–297a
early twentieth-century, **1**:11a; **4**:298b, 299b–301b; **5**:99b–100b, 317a; **8**: 273b–274a; **9**:328b; **15**:252b
mid twentieth-century, **2**:534a; **4**:154a–156a; **5**:344b; **9**:296a
Ch'en Ch'eng, **8**:396b*
Ch'en Chen-sun, **3**:249b
Ch'en Tsai-hsin, **3**:266b
Ch'en Ts'ang-ch'i, **8**:392b
Chenevix, Richard (1774–1830), 3:232–233
chemistry, **14**:487a
Ch'eng Ta-wei, **8**:419a; **14**:540b, 544b
Suan-fa t'ung-tsung, **15**:748b
Ch'eng lü (multipliers), **3**:252a, 253a
Cherenkov, Pavel Aleksandrovich, **13**:241a, 599a
Cherler, Jean-Henri, **1**:525b
Chernov, Dmitri Konstantinovich (1839–1921), 3:233–235; **1**:581a; **7**:529a
metallurgy, **12**:544a
Chernozem, **4**:143b; **14**:52a
Chernyaev, Ilia Ilich (1893–1966), 3:235–236; **3**:271a, 272a*
Chernyshev, Feodosii Nikolaevich (1856–1914), 3:236–237; **7**:250a; **10**:128b*
Cherokee Indians, **7**:23b
Cherskii, Ivan Dementevich, **10**:167a, 168a,b, 169a; **12**:301a
Cherts, **3**:161b; **13**:518b
plant-bearing, **8**:5b
radiolarian, **11**:594a
Chervets, **7**:379a
Cherwell, Lord
see Lindemann, Frederick Alexander
Chesapeake Zoological Laboratory, **2**:501b
Chéseaux, Jean Philippe Loys de, **6**:70b
paradox of, **8**:617b–618a
Cheselden, William, **4**:172b; **6**:566a; **9**:479b, 480a; **14**:320a
Chess
medieval Arabic discussion, **2**:152b, 154a; **3**:549a*
medieval Latin discussion, **4**:606b
Renaissance discussion, **10**:270a, 271a
eighteenth-century discussion, **9**:471a; **13**:571b
nineteenth-century discussion, **11**:536a; **12**:229a
computer mechanism, **13**:431b
mathematics of
knight's tour, **13**:265b

Chest
  diseases of, **7**:557a
  percussion of, **1**:332b–333a; **3**:427a
  see also Auscultation
Chestnut tree
  diseases of, **9**:79a
Chestnuts
  analysis of, **10**:325b
Chetognates, **5**:502b
Chetverikov, Sergei Sergeevich, **7**:455b; **9**: 282b
Chevalier, Auguste, **1**:135b; **5**:261b; **13**:250b
Chevalier, Charles, **11**:603b
Chevalley, Claude, **2**:351b; **8**:326b
  theorem of, **4**:83a
Chevallier, Jean-Baptiste-Alphonse (1793–1879), 3:237–238
Chevallier, Temple (1794–1873), 3:238; **3**: 92a
Chevenard, Pierre Antoine Jean Sylvestre (1888–1960), 3:238–240; **5**:583a
Chevreul, Michel Eugène (1786–1889), 3: 240–244; **2**:65b, 66b, 86b, 385b; **3**: 10a, 46a; **4**:243a; **5**:157b; **9**:342a*; **10**:436a
  chemistry, **4**:244b, 245a; **6**:398a; **10**:369b; **13**:310a
Cheyne, George (1671–1743), 3:244–245; **1**:594a; **5**:366b; **8**:506b; **9**:454b; **11**:2b
  Fluxionum methoda inversa, **10**:56a
  Methodus, **8**:165a
Cheyne, John, **7**:69a
  Cheyne-Stokes breathing, **2**:35b
Cheyne, Watson, **8**:407a, 408a
Chézy, Antoine de, **3**:442b; **5**:410b; **10**:528a; **11**:164b; **14**:494b
Chi Tsai-Yü, **13**:50b
Ch'i, **12**:374a, 379a, 381a, 407b, 408b, 409a; **15**:738b–739a, 740a, 741a
Ch'i shu (residues), **3**:252a, 253a
Chi square criterion, **2**:319a; **5**:8a; **10**:457a, 459a–460b; **15**:32a
Chia Hsien, **3**:265b, 268a; **14**:544a
  Shih-so suan-shu, **14**:540b
Chiang Chou, **3**:266b; **8**:319a
Chiao ts'ao to, **3**:270a
Chiaramonti, Scipione, **7**:302b
Chiarugi, Giulio (1859–1944), 3:245–246; **1**:545a; **3**:112b; **8**:282a,b; **10**:268a*
Chiatura manganese deposit, **1**:20b
Chibnall, Albert Charles, **10**:243b
Chicago
  Academy of Sciences, **13**:65b–66a
  Century of Progress Exposition, **9**:552b
  Columbian Exposition, **1**:130a; **2**:208a; **6**:30b; **11**:219a; **12**:229b; **13**:287a; **14**:38a
  Cook County Hospital, **6**:232b, 233a
  Dearborn Observatory, **6**:522b
  Electrical Exposition, **4**:589a
  Field Museum, **2**:208a
  hygiene, **7**:171a
  as medical center, **6**:232b–233a
Chicago, University of
  astronomy at, **13**:116a–b
  botanical garden, **3**:188a
  botany at, **3**:187b
  geology at, **3**:190a; **12**:83b–84a
  Institute for Nuclear Studies, **3**:371a; **4**: 582a–b
  mathematics at, **9**:502a
  meteorology at, **11**:557b, 558b
  zoology at, **8**:355a–356a
Chicago World's Fair, **4**:561b
Chichibabin, Alexei Yevgenievich (1871–1945), 3:246–247; **9**:130b
  Chichibabin reaction, **3**:246b
Chicken
  embryo, **5**:357a; **6**:159b–161a, 573b; **8**: 355b, 357b; **10**:287a; **11**:133a; **13**: 555b; **14**:12b; **15**:525a
  genetics, **3**:590b
  see also Egg(s), chick
Chicken cholera, **4**:211a; **10**:406a–b; **11**: 428b; **14**:278a
  vaccination, **7**:424b; **10**:390b–392b, 393a
Chicken pox, **5**:581a
Ch'ien Pao-tsung, **3**:249a, 256a; **8**:418b, 419a,b, 423a; **13**:485a
Chigoe, **10**:622a
Chih-nan suan-fa, **14**:545b
Chikhachev, Pëtr Aleksandrovich, **1**:210b; **10**:168b
Child, Charles Manning (1869–1954), 3: 247–248; **13**:151b*
Child, Clement Dexter
  Child-Langmuir space charge equation, **8**:23b–24a
Childe, Gordon, **9**:490a
Children
  child care, **5**:378b
  handicapped, **5**:377b–378a
  psychology of, **2**:131b–132a; **5**:377b–378b
    observation methods, **5**:378a
    questionnaire method, **6**:53a
Childrey, Joshua (1623–1670), 3:248
Chile
  exploration of, **4**:603a
  flora, **4**:603a
  geology, **10**:170b
  natural history, **9**:458a
Chiliasm, **1**:159a; **2**:163b; **3**:359b–362b *passim;* **5**:120a
Chilinder, **6**:301b–302a
Chilomonas, **9**:167a
Chilopods, **4**:207a
Chilovsky, Constantin, **3**:505a; **8**:11b
Chimpanzee, **12**:224b
  anatomy, **13**:490b, 527b–528a
  neurology, **5**:207b
Ch'in Chiu-shao (*ca.* 1202–*ca.* 1261), 3:249–256; **3**:265b, 266b, 268a; **8**:315a,b, 316a, 317a, 320a
  Shu-shu chiu-chang, **14**:538b
China
  alchemy, **15**:730a
  archaeology, **6**:215b, 217a; **12**:382a, 382b–383a
  astrology, **15**:729b, 730b
  astronomy, **1**:313b; **3**:249a, 269b; **7**:374a; **12**:372a, 374b–375a, 376a–379b; **13**:485b; **14**:159a–165b; **15**:729a–733b *passim,* 746a, 747b
  botany, **3**:45b
  calendar, **1**:310a,b, 312b, 313a; **3**:250b–251a, 269b; **7**:259b; **12**:377b–379a, 403a–b; **13**:485b; **14**:545b; **15**:729b, 732a,b, 733a,b, 735a, 741b, 746a
  economics, **12**:372b, 381a
  education, **8**:13b*; **12**:375a, 388a–b; **14**: 163a
  exploration of, **3**:190a–b; **6**:215b–217a; **11**:180b
  geography, **6**:215b–217a; **11**:403a, 439b; **12**:373b
    cartography, **12**:380a
    geodesy, **3**:250a
  Geological Survey, **5**:486a, 500b; **6**:228b; **13**:275a–b
  geology, **10**:167b; **11**:209b, 439a–440a; **12**:380a–b; **14**:402a
    mining, **6**:13b
    paleontology, **2**:171b–172a; **5**:486b, 487b, 500b; **13**:275a
  and Islam, **13**:513b
  and Japan, **15**:729a, 732b–734a, 737a–739b, 756b
  mathematics, **3**:249a–256a, 265b–270b; **4**:117a, 606a,b, 612a; **5**:159b; **6**:511a; **7**:257a, 325b–326a; **8**:313b–320a, 418b–424b; **12**:290b, 374b–376a, 378a–b, 385b; **13**:484b–485a; **14**: 161a–b, 538b–546a; **15**:32b, 745b–746a
  medicine, **1**:323a; **8**:390b; **12**:374a, 383a–384a, 384b; **15**:737a–739b, 740b, 743b
    pharmacy, **8**:390b–397b
  meteorology, **3**:256a; **12**:380b–381a
  missions, **11**:402b–403a; **14**:159b–165b *passim;* **15**:193b
  music, **12**:376a
  natural history, **8**:391a; **12**:536b
  nutrition, **14**:524a–b
  philosophy, **12**:385a–391a; **14**:163a–165b; **15**:733b
  physics
    magnetism, **12**:379b–380a
    optics, **12**:380b–382a
  politics, **12**:369b–371b, 372a–373b; **14**: 159a–b; **15**:733b
  technology, **12**:382a–383a, 384b; **15**:730b
    paper, **14**:349b
    porcelain, **11**:330a
Chinaroot, **4**:138b; **14**:6a
Chinch bug, **4**:220a
Ch'ing dynasty, **14**:159b, 163b–165a
Chinook Indians, **2**:212a
Chiococca, **3**:160a
Chioniades, Gregory, **11**:18a
Chippewa, **12**:204a
Chirac, Pierre, **1**:322b; **4**:216b; **12**:457b; **14**:26a
Chirikov, Aleksei Ilich, **5**:428a
Chiromancy, **6**:562b
Chironomus
  sexual cell formation, **1**:417b
Chisholm, Alexander, **8**:297b
Chitin, **9**:354b
Chitosamine, **8**:276a
Chittenden, Russell Henry (1856–1943), 3: 256–258; **4**:160b; **6**:264a; **7**:519b; **8**: 556a*; **9**:143a*, 266b*, 284a
Ch'iu Hsü-chü, **14**:539a,b, 544b
Ch'iu i shu (method of searching for unity), **3**:251b
Chiu-chang suan-ching, **3**:254b
Chiu-chang suan-shu, **3**:268b; **8**:418b–423a
Chladni, Ernst Florenz Friedrich (1756–1827), 3:258–259; **4**:241a; **14**:199a–b, 204a, 205b
  Chladni figures, **4**:534b; **5**:375b; **8**:399a; **11**:478a; **12**:128b(n), 129b; **14**:289b
  on meteorites, **1**:615b
Chlamydospores, **14**:93b
Chloral, **8**:335a, 336a
  formula of, **4**:245b; **8**:333a
Chloral hydrate, **9**:422b; **10**:530b
Chloralism, **15**:479b
Chlorates
  tests for, **2**:340b
Chlorella, **14**:174b
Chloric acid, **5**:323a
Chlorides, **12**:46b
  determination of, **9**:445b
  medical use of, **5**:421a
Chlorination, **9**:352b, 619b
  of benzene compounds, **1**:578b
Chlorine
  eighteenth-century study of, **2**:76a–b; **10**:496b
  as analytical reagent, **8**:331a–b

Chlorine (cont.)
atomic weight, **3**:398a; **7**:159a*; **9**:110a, 290b; **10**:510b; **11**:173b, 417b; **13**:499b
as bleaching agent, **1**:416b; **2**:77b; **3**:202a, 399b; **6**:58b, 283a; **7**:389b, 391a; **14**:197a
and combustion, **4**:531b
compounds of
metallic, **3**:604b
organic, **4**:531b; **11**:353a, 541a; **12**:321b
discovery of, **1**:143a; **12**:146b
as disinfectant, **3**:189a, 202a, 238a; **5**:603b
electrochemical nature of, **4**:246a,b
elementary nature, recognition of, **2**:96a; **3**:604b; **5**:321b–322b; **13**:311b
gas, **6**:254b
in warfare, **5**:622a
and hydrogen, photochemical reaction of, **3**:197b; **5**:322a; **13**:311b; **15**:37a–b
isotopes of, **2**:499a; **6**:117b; **7**:519a
manufacture of, **8**:553b; **9**:467a
medical use of, **10**:620b
molecules of, **8**:61a
naming of, **3**:603a
and organic compounds, **4**:245a–b, 246a; **5**:373a; **8**:333a; **11**:353a
preparation of, **2**:76a; **3**:399b; **5**:603b
reactions of, **5**:319b, 324a, 325a; **10**:496b
photochemical, **11**:537b; **15**:448a–b
in sea water, **5**:72a
see also Oxymuriatic acid
Chlorites, **5**:186a
crystallography, **9**:183b
Chloritoids, **10**:124a
Chloroargentic acid
discovery of, **4**:345b
Chloroform, **2**:367a; **8**:335a, 402a
anesthetic use of, **5**:45a, 566a–b; **6**:110a; **8**:402a; **10**:620a; **12**:165a, 502b
formula of, **4**:243a, 245b; **8**:335a
measurement of, **6**:110a
synthesis of, **1**:168b
toxicity of, **12**:165a
Chlorogenic acid
structure of, **5**:6a
Chlorophyceae
naming of, **7**:534a
Chlorophyll, **4**:362b, 363a; **9**:558b; **14**:174b
in animals, **12**:231a
chemistry of, **6**:505b; **9**:96a; **13**:486a–487b; **14**:412a
composition of, **9**:96a; **15**:157b
degradation products, **10**:22b
derivatives of, **9**:95b
fluorescence of, **14**:499b*
naming of, **3**:159b; **10**:497b
role of, **11**:154b–155a; **13**:416b; **14**:412a
spectrum of, **6**:138a, 294a; **7**:35a; **12**:236b, 544a; **13**:416b
synthesis of, **15**:157b–158a
Chlorophyllase, **14**:412a
Chlorophyta, **6**:465a
Chloroplasts, **7**:35a; **12**:59a; **13**:88b; **14**:98b
Chloroprene, **3**:86b; **10**:122a
Chlorosis
Egyptian, **4**:198a
tropical, **4**:198a
Chlorosulfonic acid, **14**:396a
Chlortetracycline
discovery of, **4**:220b
Chocolate, **12**:458b
analysis of, **10**:325b
Chodat, Robert (1865–1934), 3:259; **1**:363a; **12**:207a
Choice
see Free will
Choice, axiom of, **3**:57a; **5**:108a; **10**:442a; **14**:615a
Choiseul Island, **2**:342b
Chokedamp, **2**:524a
Cholanic acid, **14**:444b
Cholera
early nineteenth-century study, **2**:63b, 508a; **5**:224b–225a; **8**:272b; **13**:354a–b
mid nineteenth-century study, **2**:524b; **7**:44b
late nineteenth-century study, **2**:300a–b; **5**:219a, 220a; **11**:516a*
mid twentieth-century study, **8**:313a*
Asiatic, **6**:120a, 285a; **10**:237b; **12**:450a
bacillus, **5**:219b, 503a; **7**:391b, 425a, 518a; **10**:400a; **12**:96b
discovery of, **10**:267b
cholera morbus, **5**:277a
control measures, **7**:425b, 427a–b
diagnosis of, **5**:564b*; **7**:425b, 427b
epidemics, **5**:316b; **7**:421a; **8**:254a; **11**:177b; **12**:233a, 245a*, 502a; **14**:42a
etiology of, **1**:493b; **2**:617a; **5**:269b, 270a; **6**:73a; **7**:424b–425a; **8**:296b, 556b; **10**:267b, 558b–560a; **11**:176b; **12**:96b, 321a; **14**:249b
immunity, **15**:521b
infantum, **4**:403b
phagotherapy, **6**:298b
transmission of, **12**:502b–503a
treatment of, **9**:10b, 153a; **10**:267b; **12**:503a
vaccination, **2**:575a; **4**:211a; **5**:270a; **6**:11b; **10**:198a
see also Chicken cholera, Hog cholera
Cholesterol, **5**:91a; **6**:505b
derivatives of, **14**:445b
structure of, **4**:90b–91a; **14**:334b–335a, 444a–b
and vitamin activity, **14**:444b–445a
Cholic acid, **3**:341b
reactions of, **11**:129a
structure of, **14**:334b–335a
Choline, **5**:431b; **9**:20a, 284b
Cholinergics, **13**:262b
Cholinesterase, **8**:454b–455a
Chondriosomes, **8**:282b; **14**:433a; **15**:473a
see also Mitochondria
Chondrocranium, **8**:282b
Chondrosamine, **8**:276a
Chordariales, **7**:535a
Chordates
embryogeny of, **3**:120b, 389b–390a; **8**:586a
Chords (math.)
ancient Greek discussion of, **9**:297a–b
ancient Roman discussion of, **11**:188b–189a
medieval discussion of, **2**:153b
ideal, **11**:77a, 79b
tables of, **15**:208b–209b
Chorea, **4**:405a; **5**:27a; **7**:48b; **10**:307a; **15**:480a
electric, **4**:198a
posthemiplegic, **9**:422a
Sydenham's, **13**:215a
Chorine, Vitaly, **4**:155b
Chorography, **1**:179a; **11**:440b
Chorology, **6**:8a, 10a; **11**:440b
Chosroes, King of Persia, **12**:442a
Chou Shu-hsüeh, **14**:540b
Chou Su, **1**:498a, 499a
Chou-k'ou-tien, **13**:275a
Chou-pei suan-ching, **3**:268b
Chremastatistics, **8**:61b
Chrétien, Henry, **9**:46a; **11**:470a
Christ Church College, Oxford
museum, **7**:366a
Christian V, King of Denmark, **2**:317a; **9**:446b; **11**:526b; **14**:505a
Christian Catholicism, **10**:11b
Christianity
in antiquity, **2**:231a; **4**:621b
medieval, **1**:379a; **2**:156a
condemnations of 1277, **4**:255a
and Islam, **2**:619a
reform of, **1**:289b
Renaissance, **11**:72b, 73a
early seventeenth-century, **5**:47a–b
late seventeenth-century, **5**:60a; **8**:439a
and atomism, **3**:209a,b
and Epicureanism, **5**:285b–286b
evolution of, **10**:448b
and materialism, **6**:7a; **13**:523a
and medicine, **1**:235a; **10**:20b
and natural philosophy, **8**:538a
Neoplatonism and, **1**:334a, 336a; **4**:396b; **11**:161a
and Platonism, **1**:137a; **10**:20b
prophecy of, **2**:539b
psychology of, **6**:53b; **7**:192b
and science, **1**:334b; **2**:605a–b; **8**:31a
Christians and Turks, math. problem of, **1**:368a
Christiansen, Christian (1843–1917), 15:83–84; **7**:416b
Christiansen, Jens Anton, **7**:493b
Christie, Samuel Hunter (1784–1865), 3:259–261
electricity, **14**:290a
experiments of, **5**:79a
Christie, William Henry Mahoney (1845–1922), 3:261; **3**:260a; **9**:184a
Christina, Queen of Sweden, **2**:306b, 311b–312a, 354a; **3**:101a,b, 320b; **4**:52a; **10**:332a; **11**:312b, 587a
Christison, Robert, **7**:70a; **11**:494b, 495a; **13**:499a
Christmann, Jacob (1554–1613), 3:262
Christmas Island
mining, **9**:590a
Christmas-Dirckinck-Holmfeld, John de, **12**:88a
Christoffel, Elwin Bruno (1829–1900), 3:263–264; **4**:2b; **7**:522b; **11**:407a; **13**:155a
analysis, **11**:407a,b, 408a, 409a
geometry, **5**:555a; **8**:389b
index symbols, **3**:263a–b; **4**:336b; **8**:389b
Riemann-Christoffel symbols, **3**:263b
Christol (Cristol), Jules de (1802–1861), 3:264; **15**:51a
Christopher of Paris, **2**:23a
Christ's Hospital, London
mathematics at, **12**:445b–446a
Christy, Henry, **8**:44a
Chromatids, **5**:35b
Chromatin, **2**:362b; **5**:35a–b, 51b; **6**:339b; **14**:429b, 430a
composition of, **7**:467a
diminution of, **2**:364a–b
Chromatography, **13**:419b, 421a–b; **15**:257b
development of, **13**:487a–488a
gas-phase, **5**:139b
paper, **11**:615b; **12**:198b
Chromic acid
electrochemical regeneration, **8**:113a
and organic substances, **4**:282b
Chromidia, **6**:337a
Chromium
alloys of, **3**:239a; **5**:189b
compounds of, **1**:471a; **2**:170b–171a; **5**:157b; **9**:450b; **13**:389b, 452a
coordination compounds of, **7**:179a; **10**:578a–b
deposits of, **3**:609b
discovery of, **7**:394b; **11**:464b; **13**:597a
isolation of, **2**:588b; **8**:520a; **9**:451b
spectrum of, **3**:125a

Chromosome cycle, **3**:30b
Chromosome theory, **1**:505a,b; **2**:361b, 362b–364b, 456a; **3**:122a, 123b, 422b; **8**:586b, 587a–588b; **9**:496b, 519a, 519b–525a, 564b; **13**:157a–158a; **14**:101a–b, 139b, 427b–434a
  confirmation of, **10**:276b
Chromosomes
  nineteenth-century study, **1**:601b; **2**: 362b–364a, 626b; **3**:29b; **5**:35a–b, 51b, 581b–582a; **6**:465b; **12**:193a–b; **14**:236a,b
  twentieth-century study, **2**:456a–b; **3**: 318a; **5**:293b–294a; **8**:586b–589b; **9**: 496a–497a, 518b–519a
  accessory, **8**:586b, 587b–588b
  bivalents, **4**:131a; **5**:35b
  chiasmata, **13**:158a
  complexes, **14**:101a,b
  continuity of, **11**:255b
  crossing-over, **9**:521b, 564b; **14**:432b–433a
    double, **3**:122a; **13**:135a
    frequency of, **13**:135a
    induction of, **12**:417a
    interference, **9**:522a, 523a
    suppressors, **13**:135b
    unequal, **13**:136a–137a
  differential value of, **2**:363b–364a; **13**: 281b–282a
  dimensions of, **4**:546a
  and disease, **10**:277a
  division of, **5**:35b
  elimination of, **13**:281b
  heterochromosomes, **9**:496a
  individuality of, **1**:601b; **2**:362b–363a; **6**:338b; **8**:587b–589a; **9**:496a–b; **11**: 255b; **13**:158a; **14**:236a, 429b, 430a
  inversion, **13**:135b–136a, 137b
  mapping of, **3**:122a–b; **6**:22a; **9**:521b–522a, 523a, 564b; **10**:276b; **11**:212b; **13**:135a–137b *passim*; **14**:432b
  microscopy of, **13**:136a
  naming of, **14**:126b
  number of, **2**:362b; **5**:293b, 598b; **8**:587b; **10**:277a; **13**:88b
    additive, **13**:478b
    and malignancy, **2**:364b
  pairing of, **9**:496a–b; **14**:431a, 432b
  photography of, **7**:448b
  radiation, effects of, **9**:549b–550a
  reduction of, **4**:546a; **10**:256b; **14**:235b, 236b, 431a
  role of, **7**:95a; **12**:192b; **13**:157a–158a; **14**:427a–b, 429b–434a
  structure of, **7**:455b; **10**:276b–277a
  X, **6**:22a, 267a–b; **8**:586b, 587b–588b; **9**:496b, 518b–519a, 521a; **14**:431a–432a, 595b
    nondisjunction, **9**:522b; **14**:431b
  Y, **14**:431b
  see also Segregation (biol.)
Chromosphere
  see Sun, chromosphere
Chronaxie, **5**:66a*; **8**:29a–b
  and fiber diameter, **8**:29b
Chronodeik (astron. instrum.), **10**:279b
Chronology
  ancient Greek study, **4**:389a, 392b; **6**: 213a; **9**:337a–339a
  Byzantine study, **11**:183b, 185b–186a
  medieval Arabic study, **2**:152a–b; **4**:502b; **6**:186a; **7**:259a–b; **9**:555b, 615a*; **11**:246b–247a; **14**:538a
  medieval Latin study, **5**:401b; **6**:302a
  Renaissance study, **3**:262b; **9**:168a
  seventeenth-century study, **7**:306a
  eighteenth-century study, **10**:82b, 615a
  archaeological, **9**:489b–490a; **13**:357b; **14**:506a–b
  astrological, **7**:82b
  biblical, **10**:10a; **14**:295b
  Mesopotamian, **15**:672b–673a
  see also Geology, time scale
Chronometers, **2**:285a,b, 438b; **3**:106a, 198a, 397a; **4**:354a, 481b; **5**:17a, 491b; **6**: 70a, 130b, 482b; **8**:162a; **9**:310a, 311a, 421b; **13**:111b, 120b; **14**:357a
  factors affecting, **10**:542a
  pocket, **10**:133b
  recording, **6**:523a
  testing of, **9**:163b
  see also Clock; Horology
Chrysalides
  dissection of, **13**:171b–172b
Chrysarobin, **8**:329a
Chrysene, **5**:489a
  discovery of, **8**:58b
Chrysippus the Stoic, **4**:33b–34a; **11**:104b–105a, 128b; **14**:606a, 607a
  commentaries on, **5**:232a; **9**:275b(n); **11**:105b
  on *pneuma,* **1**:324a,b
Chrysippus the Younger, **4**:383a
Chrysoberyl, **15**:7b
Chrysococces, George, **7**:337a
Chrysogonis, Federicus de
  see Grisogono, Federico
Chrysoidine, **3**:85a
Chrysoloras, Manuel, **1**:274a
Chrysophanic acid, **8**:329a
Chrystal, George (1851–1911), 3:264–265; **3**:169b; **8**:381b; **9**:227b; **14**:211b
  Algebra, **6**:451b
Chu Hsi, **12**:380b, 381a, 409a; **14**:163a, 164a; **15**:746a
Chu Hsiao, **8**:397a*
Chu Shih-chieh (*fl.* 1280–1303), 3:265–271; **8**:314b; **10**:338b(n); **12**:375b
  Ssu-yüan yü-chien, **8**:319a; **14**:538b, 540b
  Suan-hsüeh ch'i-mêng, **12**:290b; **15**:748b
Ch'ü Tseng-fa, **8**:419a,b
Ch'u River, **12**:300b
Chugaev, Lev Aleksandrovich (1873–1922), 3:271–272; **3**:235a; **8**:108b; **9**:294b*; **11**:283a*
Chui-chang suan-shu, **15**:691a
Chukchi Sea, **4**:42b
Chuprov, Aleksandr Aleksandrovich
  see Tschuprow, Alexander Alexandrovitch
Chuquet, Nicolas (*fl. ca.* 1480), 3:272–278; **2**:618b
  algebra, **12**:92b
  arithmetic, **5**:409a; **8**:41a; **14**:471b(n)
Church, Alonzo, **11**:107a; **13**:479b, 498a, 600a
Church, Arthur Herbert, **10**:515b; **13**:341b
Churchill, Winston, **8**:368b–369a
Churchill College, Cambridge, **8**:369a
Chusovaya (river), **8**:466a
Chwistek, Leon (1884–1944), 3:278–279
  logic, **12**:15a
Chydenius, Johan Jakob, **9**:290b
Chyle
  formation of, **2**:301b
  transfer of, **2**:531b; **6**:508b–509a; **8**:401a
  transformation of, **8**:526a; **9**:63a–b
Chylous vessels
  see Lymphatic system, lacteals
Chyluria, **8**:296b
Chymotrypsin, **4**:522a
Ciamician, Giacomo Luigi (1857–1922), 3: 279–280
Ciampini, Giovanni Giustino, **2**:312a
Cibo da Roccacontrada, Gherardo, **5**:383b
Cicada, **3**:327b*
Cicatrization, **11**:295b
Cicero
  acquaintances of, **1**:314b; **11**:103b
  Cato major, **8**:460a
  commentaries on, **2**:224b; **9**:1a
  cosmology, **9**:1b; **13**:92a
  historiography, **1**:213a–b, 214a; **4**:30b, 32a, 443a; **6**:381b, 382a; **11**:23a, 25a,b, 30a(n)
  influence of, **5**:365a; **7**:27b; **11**:39a
  influences on, **4**:82a; **8**:536b; **11**:105b(n)
  philosophy, **4**:382b; **11**:324a; **15**:516a
  Topica, **1**:269a; **2**:229b, 230b, 232a,b
Cichlidae, **4**:310a
Cicindelidae, **5**:477b
Cider, **4**:495b
Ciel et terre, **4**:27a
  founding of, **10**:119b
Cigna, Giovanni Francesco, **7**:561a; **11**:510a
  electricity, **1**:547b, 548a; **13**:224b–225a; **14**:70a, 71a,b
Cilianic acid, **3**:341b
Ciliary motion, **12**:354a
Ciliates, **4**:133a*; **8**:145b
  classification of, **7**:447a
  conjugation of, **2**:626a,b; **3**:17b
    discovery of, **9**:185a
  cytology, **6**:337a
  encystment, **13**:268b
  food vacuoles, **5**:424b–425a
  life cycle, **2**:626b
  neuromotor apparatus, **13**:268a
  reproduction, **8**:127b
  senescence, **9**:185a
  sexuality of, **2**:627a
  unicellular nature of, **2**:626b
Cimento, Accademia del
  see Accademia del Cimento
Cimone, Monte, **9**:135b
Cinchona, **4**:156b; **7**:200a; **11**:47b, 605b; **15**:268b, 270b, 271a
  bark, **1**:481a; **3**:159b; **4**:560a; **5**:226b; **6**: 294a*; **10**:567b; **12**:457b; **13**:215a
    analysis of, **5**:91a; **10**:498a
  cultivation of, **6**:494b; **9**:123a; **12**:594a–b
  physiology, **8**:512b
  taxonomy, **3**:16b; **13**:464b; **15**:430a
  see also Quinine
Cinchonine, **5**:91a
  discovery of, **10**:497b
  isolation of, **3**:159b
Cincinnati, University of
  geology at, **4**:562b
Cinematography, **9**:102a
  forerunners of, **7**:77a
  in physiology, **9**:102a; **15**:504a
  in psychological research, **5**:378a
  see also Microcinematography; Motion pictures
Cinnabar, **12**:514a
  in animal body, **8**:9a
  atomic structure, **9**:183b
  manufacture of, **7**:378a–b
Cinnamic acid, **8**:329b
  synthesis of, **10**:516b
Cinnamon, **4**:156b; **7**:200b; **10**:13a
Ciona, **3**:120b
C.I.P.W. system, **3**:484a; **7**:5b; **14**:183b
  critics of, **14**:58b
Circa instans, **11**:601b
Circia, **9**:600a
Circle, **3**:399a; **4**:619a
  area of, **1**:90b, 216a–217a, 219b, 220a, 227a, 444b; **3**:128b*; **4**:104b, 422b–423a, 424a, 466b, 609b, 611b–612a; **5**:526b; **6**:314b, 415b–416a; **10**:298b; **14**:22a; **15**:670b, 691a, 696b
  see also Quadrature, of circle

Circle (cont.)
chords of, **11**:533b
circumscribed, **3**:277a–b; **4**:610a; **6**:415a; **8**:300a
division of, **4**:390b, 426a; **7**:54b; **8**:234b–235a, 239b–240a; **9**:488b; **10**:133b–134a; **11**:95b; **13**:470b; **14**:162b
fundamental theorem, **13**:17a
inscribed, **3**:515b; **4**:610a; **8**:300a, 316a; **10**:294b–295a; **13**:259b
intersections of, **7**:206b; **13**:14b
Lemoine, **8**:176b
nine-point theorem, **2**:455a
osculatory, **2**:330a
rectification of, **1**:219b, 228a; **4**:104b; **12**:291b–292a
sector of
center of gravity, **7**:558a
segments of, **6**:412b, 415a,b, 416a–b; **14**:22a
area of, **3**:255a; **6**:315a; **8**:419b
series of, **13**:15b
systems of, **11**:398b
tangent, **5**:209b, 368a, 587a; **8**:175b; **10**:297a; **11**:533b; **12**:228a; **13**:14b, 15a–16a; **14**:22b
see also Pi; Quadrature, of circle
Circolo Matematico di Palermo
founding of, **5**:569a–b
publications of, **5**:569b
see also *Rendiconti del . . .*
Circuits, electric
analysis of, **13**:24a
design of, **2**:298b
multiple-wire, **6**:504b
mutual magnetic effects, **13**:11b
mutual potential, **10**:28a
oscillations of, **4**:559a
resonant, **4**:559b; **9**:215b
self-induction, **9**:166b
thermionic
"noise" in, **3**:31b
vibrations in, **9**:77a
Circulaire de l'Observatoire de Cracovie
founding of, **1**:430a
Circular motion
see Motion, circular
Circulation, concept of, **6**:156a, 157b; **11**:391a–b
in nature, **1**:550a; **4**:368a; **5**:47b, 194a–b, 422a
Platonic, **2**:543a–b; **7**:485b
Circulation of blood
see Blood, circulation of
Cirey, château of, **3**:215a–b
Cirques (geol.), **2**:289b; **7**:149a
Cirrhosis, **15**:103b
cardiac, **11**:100a
histogenesis, **4**:618a
Cirripedia, **8**:169b; **13**:355b
Ciruelo, Pedro (Pedro Sánchez Ciruelo) (1470–1554), 3:280; **12**:62a, 547b
Cissoid, **4**:105a; **5**:499a, 521a, 527b; **6**:598a; **12**:460a; **14**:148b, 149b; **15**:118a
Cistercian Order, **3**:61a; **14**:30b
Cisterna chyli
see Lymphatic system
Citraconic acid, **11**:495a
Citrenes, **13**:410b
Citric acid, **8**:342a
formula, **8**:338b–341a *passim*
isolation of, **12**:148b
manufacture of, **8**:93b
polybasicity of, **8**:545b
Citron, Julius Bernhard, **15**:522a
Citrus, **2**:221b
Cittadini, Antonio, **1**:484b
City Philosophical Society (London), **4**:528b, 529a
City planning, **11**:607b; **13**:592b; **15**:662b(n)
Ciucci, Antonio Filippo, **4**:511b
Civil War (U.S.), **1**:364b; **2**:386b; **3**:196b, 438b; **8**:122a, 123a, 571b; **11**:118a; **15**:109a–b
chemistry, **6**:517b
Civil War, Great Britain, **6**:150b–151a, 387a; **13**:501a; **14**:285b–286a
and science, **1**:317a
Der Civil-Ingenieur
founding of, **14**:617b
Cladobotryum, **9**:176a
Cladocera, **7**:516b
Cladonia, **15**:149b, 150a–b
Cladoselache, **3**:610b
Claesson, Stig Melker, **13**:421a
Claij, Jacob
see Clay, Jacob
Clairaut, Alexis-Claude (1713–1765), 3:281–286; **3**:114b
acquaintances of, **3**:459b, 561b; **5**:54b; **7**:442a; **8**:178b; **10**:145a; **14**:83b
and d'Alembert, **1**:111a; **3**:282b, 283b; **7**:563a
analysis, **7**:568a
and Buffon, **2**:577b; **3**:283a
celestial mechanics, **1**:113a–114b *passim;* **4**:471a; **7**:580a; **8**:179a; **15**:349a
lunar theory, **1**:400b; **4**:481a
and Mme. du Châtelet, **3**:215a, 216a–b
Clairaut's equation, **3**:282b; **8**:562a
commentaries on, **7**:7a
correspondence of, **3**:461b; **13**:68a
editions of works, **7**:551a*; **15**:484a
formula of, **3**:282b; **6**:240a; **13**:76a
geodesy, **3**:39a; **8**:611a–b; **9**:186b; **13**:69b; **15**:270a
geometry, **4**:476a, 479b; **9**:475a
influence of, **3**:454a; **7**:442b, 490a, 550b; **12**:444a; **15**:358b
influences on, **2**:334b
natural philosophy, **15**:358b, 359a
Clairaut, Jean-Baptiste, **3**:281a, 286a*, 561b
Clairvaux, Bernard of
see Bernard of Clairvaux
Claisen, Ludwig (1851–1930), 3:286; **1**:168a, 286b
Claisen condensation, **3**:286b
Clapeyron, Benoit-Pierre-Émile (1799–1864), 3:286–287; **3**:82a; **5**:85a, 388b; **6**:243b; **7**:601a
on Carnot, **3**;83a
Clausius-Clapeyron equation, **3**:304a; **6**:520a; **13**:578b
equation of, **10**:518b
mechanics, **3**:118a; **9**:446b; **14**:473b
steam engine, **10**:286b
thermodynamics, **3**:305a; **6**:244a–b; **13**:375a, 377b, 378a
Clarendon Press
botanical translations, **1**:422b
Clark, Alvan (1804–1887), 3:288
instrumentation, **13**:121a
Clark, Alvan Graham (1832–1897), 3:288; **3**:288a
Clark, George Bassett (1827–1891), 3:288; **3**:288a; **10**:600a
Clark, George Lindenberg, **4**:196a
Clark, John Goodrich, **14**:278b
Clark, Jonas Gilman, **6**:52b
Clark, Josiah Latimer (1822–1898), 3:288–289; **7**:93b, 288a
Clark cell, **3**:19b, 288b, 334a
Clark, Thomas (1801–1867), 3:289–290
Clark, William, **7**:89b
Clark, William Mansfield (1884–1964), 3:290
Clark (Alvan) and Sons, **2**:614b; **3**:288a; **6**:28b; **10**:35a, 600a,b
Clark University (Worcester, Mass.), **5**:179b
anatomy at, **9**:56a–b
founding of, **6**:52b
Physical Institute, **6**:343a
Clarke, Edward Daniel (1769–1822), 3:290–292; **3**:297a; **9**:67a–b; **12**:261b
Clarke, Frank Wigglesworth (1847–1931), 3:292–294; **4**:598a; **5**:394a*; **9**:531b*
Clarke, Geoffrey R., **8**:482b
Clarke, Hyde, **6**:563b
Clarke, John, **3**:294b, 296b(n)
Clarke, John Mason, **4**:364b; **6**:57b, 58a; **11**:593b
Clarke, Joseph Ryle, **9**:550a
Clarke, Robert Henry, **6**:518b
Clarke, Samuel (1675–1729), 3:294–297; **2**:16b, 382a; **3**:487b; **5**:130b, 509b; **6**:36a, 555b; **10**:79a; **11**:68a, 508a; **14**:296b
Being and Attributes of God, **10**:277b
and Leibniz, **3**:90a, 295b–296a, 450a; **7**:225b; **8**:151b, 152b, 158b, 159a; **10**:84a, 85a; **14**:83b
Clarke, Timothy, **4**:37b; **8**:524b
Clarke, William Branwhite (1798–1878), 3:297–298
Clarke (geochemistry), **4**:598a
Clasmatocyte, **5**:317a
Class field theory, **1**:306b–307b; **6**:214a–b, 297b, 390a–b
Class number, **6**:571b
Class theory, **3**:278a; **6**:308b, 309a; **12**:12b–13b, 15b–16a, 16b
and types, **3**:278b
Classen, Alexander (1843–1934), 15:84
Classici della Scienza e della Filosofia, **9**:378a
Classification
as aim of science, **3**:382b; **4**:226b, 227b; **7**:41b
anthropological, **1**:388a; **2**:208b
chemical
medieval Arabic, **7**:41b
Renaissance, **10**:308a
seventeenth-century, **2**:380b–381a
eighteenth-century, **2**:7a–8a; **3**:495a
carbohydrates, **2**:66a
catalysis, **1**:415a
ethnological, **1**:388a
medical, **12**:202b
natural history, **3**:248b
observational criteria, **5**:548a
see also Aristotle, method, classification; Drugs, classification of; Elements, chemical, classification; Minerals, classification of; Sciences, classification of; Taxonomy
Claude, François Auguste (1858–1938), 3:298–299
Claude, Georges (1870–1960), 3:299; **1**:303b; **3**:239a; **6**:94a
Claudian
In Rufinum, **1**:91b
Claudication
intermittent, **8**:295b
Claus, Adolf Carl Ludwig (1838–1900), 3:299–301; **7**:283a; **13**:577b
Claus, Carl Ernst (1796–1864), 3:301–302; **2**:620a,b; **9**:333b
Clausen, Thomas (1801–1885), 3:302–303; **6**:417a(n)
Staudt-Clausen theorem, **3**:302b; **13**:6a–b
theorem of, **3**:302b

Clausius, Rudolf (1822–1888), 3:303–311; **2**:261a; **6**:343a
acquaintances of, **10**:577a,b; **12**:425a; **14**:360a
on Carnot, **3**:83a
chemistry, **1**:362b
Clausius-Clapeyron equation, **6**:520a; **13**:578b
Clausius-Mossotti equation, **3**:618b
on contemporary scientists, **3**:147a(n); **6**:250b
correspondence of, **1**:297a; **9**:239b
dielectrics, **8**:106b
electrochemistry, **6**:439a–b; **15**:459a
electrodynamic equation, **3**:309b–310a
electrodynamics, **8**:491a, 494b; **15**:193a
engine cycle, **3**:305b
entropy equations, **6**:244b, 245a
gas theory, **2**:262a; **8**:508b, 509a, 510a; **9**:217b–220a *passim*
heat theory, **9**:220a; **11**:292b
influence of, **1**:298b; **5**:391b; **13**:40b, 237b, 614b; **14**:109b
and Joule-Mayer dispute, **3**:306a
and Maxwell, **3**:307b–308a
Mechanische Wärmetheorie, **11**:8a
method of, **11**:10a
students of, **3**:286a; **8**:365a; **11**:208a; **12**:593a
and Tait, **3**:306a–b; **5**:388a
thermodynamics, **3**:287b; **5**:387b–388a; **9**:462a; **11**:9a,b, 149b, 293a; **13**:381b, 382a
Wärmetod theory, **1**:301a
Clausthal
Mining School, **11**:502a
Clavasio, Dominicus de
see Dominicus de Clavasio
Clave, Étienne de, **5**:288b(n); **9**:527b
Clavenna, Jacobus Antonius, **3**:534a
Claviceps, **13**:489b
Clavicylinder, **3**:258a
Clavius, Christoph (1537–1612), 3:311–312; **13**:60b, 320b*
acquaintances of, **10**:160b; **11**:532b; **12**: 74b; **14**:29b
on calendar, **1**:383a; **5**:588a; **9**:169b*; **14**:23a–b
commentaries by, **12**:62b
correspondence of, **12**:97b
critics of, **3**:262b; **14**:23b
The Elements of Euclid, **4**:450b, 451a–b; **14**:29b
and Galileo, **1**:588a; **3**:311b; **5**:238b; **13**: 12a
geodesy, **2**:83b
Geometria practica, **13**:622a
geometry, **2**:618b; **4**:619b; **10**:494a,b
influence of, **4**:55b; **5**:382a,b; **6**:271b; **7**: 558a; **13**:30b, 621b; **14**:375b(n)
influences on, **12**:61b; **13**:401a
instrumentation, **2**:412b; **10**:161a; **13**: 622a
In Sphaeram Ioannis de Sacro Bosco commentarius, **9**:192a
students of, **11**:402b; **13**:560b
tables of, **11**:533b
translations of works, **11**:403a
trigonometry, **8**:28a; **11**:4a; **14**:471a
Clavus, Claudius, **10**:197a
Clay (Claij), Jacob (1882–1955), 3:312–313; **3**:370b; **9**:399a
Clay, **8**:567a
composition of, **2**:494a
experimental study, **8**:621b
fossils in, **2**:495b, 568b; **5**:14a–b
medical use of, **9**:12a
plastic, **8**:463a
writing on, **15**:635a, 649a, 650b
Clayfield, William, **6**:575b
Clayton, John, **1**:431b; **3**:350b
Cleanthes, **1**:248a; **10**:181a; **14**:606a
Cleavage (geol.)
cause of, **3**:567b–568a; **12**:542b–543a
and folding, **3**:325b; **11**:505b
plane of, **3**:567b
slaty, **6**:116a; **13**:521b
Cleaveland, Parker (1780–1858), 3:313
Clebsch, Rudolf Friedrich Alfred (1833–1872), 3:313–315; **3**:164a, 167a; **7**: 54b; **8**:388b; **11**:46b
acquaintances of, **6**:356b; **7**:506b, 522b; **14**:343a
algebra, **3**:461b; **6**:357b
analysis, **6**:388b, 389a; **7**:444a; **11**:451b
Clebsch-Lueroth method, **8**:542b
on contemporary scientists, **15**:195a
geometry, **6**:547b; **9**:55b; **10**:139b
and Gordan, **3**:315a; **5**:472a
hydrodynamics, **3**:195a
influence of, **12**:86b; **14**:202a
mechanics, **13**:26b
students of, **2**:465a; **8**:367b, 542b; **11**: 445b; **13**:81a
transformations, **2**:198a; **12**:284b
translations of works, **9**:81a
Cleland, Ralph Erskine, **14**:101b
Clelia (curve), **5**:499b
Clemence, Gerald Maurice, **2**:511b; **3**:221a; **15**:129a
Clement IV, Pope, **1**:378a
Clement VI, Pope, **3**:218a; **5**:115a; **7**:128a, 131a; **8**:279b
Clement VII, Pope, **1**:619a, 621a; **3**:363b; **6**:275a
Clement VIII, Pope, **2**:542a; **9**:308b; **10**: 416a; **13**:560b; **14**:29b; **15**:80b
Clement XI, Pope, **1**:3118b, 392a; **4**:487b; **7**:614a; **11**:65b
Clement of Alexandria, **6**:305b, 406a
Clément, Nicholas (*ca.* 1778–1841), 3:315–317; **3**:79b, 80a, 82b; **4**:74a; **9**:494a*
and Carnot, Sadi, **3**:316a–b
catalysis, **15**:462a
gases, **15**:487a
iodine, **3**:455a–b
law of saturated vapors, **3**:80b, 83a
on vapors, **5**:603a; **15**:357b
Clément-Desormes
see Clément, Nicholas
Clément-Mullet, Jean-Jacques, **1**:533a; **7**: 343b; **13**:407a–b
Clemente, Simon de Rojas, **3**:154b
Clements, Edith Schwartz, **3**:317b, 318a
Clements, Frederic Edward (1874–1945), 3: 317–318
Cleomedes (*n.d.*), 3:318–320; **1**:247a
influences on, **11**:105b(n)
De motu circulari corporum caelestium, **4**:390a,b; **5**:345b, 347a
Cleonides, **4**:430b
Cleps, **11**:95b
Clepsine, **14**:314a
Clepsydra, **1**:380a; **3**:491a; **4**:368b; **5**:402b; **6**:318a; **10**:588b; **13**:614a; **15**:666b(n)
calibration of, **12**:376b–377a
and longitude determination, **1**:138a
see also Clock, water
Clerck, Carl Alexander (1709–1765), 3:320
Clerk, George, **6**:578b, 579a
Clerselier, Claude (1614–1684), 3:320–321; **9**:47b, 498a; **11**:506a, 508b; **13**:479b
and Descartes' works, **1**:475a; **3**:614b; **4**:64b, 575a,b; **5**:527a; **12**:85a
Cleve, Clifton
Hints on Domestic Sanitation, **6**:293b
Cleve, Per Teodor (1840–1905), 3:321–322; **1**:297a; **4**:485b; **9**:111a*, 290b
helium, **11**:279b
Cleveite, **3**:479b
Clifford, George, **1**:305b; **8**:375a
Clifford, William Kingdon (1845–1879), 3: 322–323; **8**:599a; **10**:483a; **12**:469b
Clifford surfaces, **3**:323a
The Common Sense of the Exact Sciences, **10**:448b
geometry, **14**:141a
parallelism, **5**:200b, 201b
philosophy of science, **8**:159b; **12**:609b; **15**:464b
Clift, William (1775–1849), 3:323–325; **2**: 569b; **6**:478b, 479b*; **10**:260b
Climate
and agriculture, **4**:144a
altitude, effects of, **6**:550a
and atmosphere, **1**:300b
and biological change, **2**:492a; **7**:585b, 587b
classification of, **6**:98b–99a
geomorphological effects of, **2**:549a–b; **8**:568a; **10**:505b
and health, **2**:619b; **4**:171a–b; **8**:554b; **9**:306b; **12**:289b; **13**:541a–b; **14**:464a
and metabolism, **4**:311b
and oceans, **10**:564a
periglacial, **2**:549b
and race, **11**:243a–b
variation of
and sedimentation, **1**:470a
and vegetation, **5**:546b; **6**:551b; **14**:117a
see also Ecology; Environment; Plants, and environment; Weather
Climatology
ancient Greek, **3**:319b
medieval, **5**:551a; **11**:445a; **15**:265b
seventeenth-century, **1**:482a*
eighteenth-century, **4**:492b
early nineteenth-century, **6**:551b; **8**:568a; **12**:215a
mid nineteenth-century, **3**:470b–471a; **4**:174b–175a; **5**:413a; **12**:210a
late nineteenth-century, **3**:536b; **5**:21b, 258b; **6**:98b–99a; **9**:113b*; **11**:181b; **12**:187b; **14**:52b–53b
twentieth-century, **1**:622b, 623a; **4**:273a; **5**:211b; **9**:150b, 189b; **10**:585a; **11**: 524b–525a; **14**:215a
comparative, **7**:390a*
historical-statistical study, **4**:273a
statistical, **6**:98b–99a
see also Meteorology, Paleoclimatology
Clinical Science, **8**:296a
Clinometer, **7**:578a
Clinostat, **12**:60a
Clinton, De Witt, **4**:273b; **5**:11b; **6**:57a
Clinton Engineer Works, **3**:371b
Clock
anaphoric, **15**:518b, 520b
astronomical
medieval, **4**:164b; **11**:415a,b; **15**:162b, 163a,b
Renaissance, **3**:585a; **14**:359a; **15**:155a
eighteenth-century, **11**:471b; **14**:485a
synchronization of, **14**:479b
atomic, **7**:36b*
balance
spiral spring, **4**:40b
candle, **14**:580a*
cuckoo, **3**:491b
electric, **2**:438b; **4**:40a, 248b
transmitting, **2**:439a
escapement, **11**:415b; **12**:384b

Clock (cont.)
deadbeat, **5**:491a
frictionless, **6**:130a
heliotropic, **7**:376a
marine, **6**:130b
metaphor of, **10**:223b, 224a
parastatic, **3**:491b
parerga, **3**:491b
pendulum, **5**:246b; **6**:604b–605a; **8**:210b; **9**:485b, 507b; **11**:411b, 423a; **12**:118a
compensating, **5**:491a; **11**:471b; **12**:131b
conical, **5**:86a; **9**:267b
errors in, **2**:42b
free-pendulum, **12**:96a
gridiron, **6**:130a; **8**:350b
invention of, **6**:598b; **11**:423b
sidereal seconds, **2**:389a
perpetual motion, **4**:183b, 184b
polar, **14**:290a
shadow, **15**:713b
silent, **3**:23a
spring, **6**:482b, 605a
star, **15**:711b–712b, 714a,b, 715b, 716a, 719a
synchronization of, **4**:321b–322a; **14**:479b
synchronous, **4**:270a
talking, **4**:407b
traveler's, **15**:163b
water, **3**:515b; **6**:208a*; **7**:342b; **8**:198a; **10**:521a; **12**:384b; **14**:592b–593a; **15**:253b–255a, 518a,b, 520b, 536b, 537a, 593b, 662a(n), 674b, 713b, 715a
invention of, **3**:491a–b
see also Chronometer, Clepsydra, Horology, Time, Watch
Cloëtta, Max, **4**:615a
Cloëz, Stanislaus, **3**:46a
Cloos, Hans (1885–1951), 3:325–326; 3:587a; **13**:609a
Cloquet, Hippolyte, **14**:14b
Closedness, theory of, **13**:27a–b
Clostridium
culturing of, **7**:391b
isolation of, **7**:391b; **10**:155b; **14**:37b, 247b, 249b
Closure (math.), **1**:428a
Clothing
and health, **10**:558b
Cloud chamber, **15**:26b
Barus, **1**:490b
Wilson, **1**:366a, 490b; **2**:398b; **6**:117b; **7**:152b, 153b, 158a; **9**:261b; **11**:117a; **12**:33a; **13**:367b; **14**:421a–422a
Clouds
altitudes of, **5**:542b
blanket effect, **3**:327b*
cirrus, **7**:400b
electric charges in, **3**:51b; **4**:355a; **5**:133b–135b, 161b; **14**:422a–b
negative vs. positive, **5**:135b
formation of, **1**:547b; **13**:76a; **14**:421a
latent heat, role of, **4**:410b
luminous, **13**:482a
measurement of, **12**:595b
movement of, **9**:444b
nacreous, **13**:83a
noctilucent, **12**:353a; **13**:83a
optical phenomena, **14**:216a
seeding of, **8**:24b
theories of
Renaissance, **1**:607b; **5**:400b–401a
nineteenth-century, **10**:500b
Clouet, Jean-Baptiste-Paul-Antoine, **3**:326b–327a*
Clouet, Jean-François (1751–1801), 3:326–327; 3:561a*; **6**:1a; **9**:471b, 477b
Clouet, Pierre-Romain (Abbé), **3**:326b*; **12**:66a, 67a
Clough, Charles Thomas, **1**:153b, 394a
Clover, **14**:96a
Cloves
cultivation of, **11**:64b
oil of, **3**:279b
Clubroot, **14**:93b
Clusius, Carolus
see L'Écluse, Charles de
Clusius, Klaus Paul Alfred, **6**:481a
Clyde ironworks, **9**:590b–591a
Cnidian Maxims, **6**:420a,b
Cnidosporidia, **8**:145b
Cnidus, school of, **6**:420a–428b *passim*
Cnogler, Quirino, **1**:316a
Coal
analysis of, **11**:48b
brown, **10**:167a
composition of, **2**:86a; **7**:390a*
deposits of, **3**:236b–237a, 395b, 434a, 530a, 607b; **4**:10b, 73b, 262a; **5**:39a, 187a, 498a, 537a,b, 611a; **6**:238b; **7**:253a–b, 370b; **8**:33b, 260b, 461a, 463b, 577a; **9**:173a, 585a; **10**:128a, 170b, 204a; **11**:131a, 209b, 439b; **12**:487b; **13**:86b–87a; **14**:301a
age of, **3**:297b; **5**:476a; **10**:32b
prediction of, **5**:437a; **9**:583b
theories of, **5**:498b
distillation of, **5**:420b
economics of, **7**:104b
Franco-Belgian basin, **1**:472b
as fuel, **12**:381a; **13**:603b
gasification of, **13**:454b
microscopy of, **1**:472b; **2**:86a; **10**:40b
mining of, **1**:472b; **2**:292a, 523b; **3**:358a, 411b–412b; **4**:348a; **5**:441b, 602a; **6**:24a, 576a; **7**:78a, 389a; **9**:291b; **12**:487a
origin of, **2**:579a; **7**:91a; **8**:461a; **10**:171a, 322a–b; **11**:373a; **13**:405b; **14**:298a
paleobotany of, **2**:86a; **3**:608a; **4**:262a; **5**:441b, 497b–498b; **10**:32a,b; **13**:345a; **14**:398b, 600a
physical properties of, **5**:139b
production of
artificial, **2**:3b; **5**:157b–158a
natural, **2**:172b; **4**:309b
schistification, **5**:498b
seams, **2**:375a; **12**:491a; **13**:87a
origin of, **5**:441b
search for, **14**:298a
structure of, **5**:139b–140a
types of, **2**:86a; **4**:309a–b; **6**:579a
Coal balls, **14**:398b
Coal gas
calorific value of, **15**:60a
Coal Measures, **11**:562b; **12**:276b; **13**:87a, 525a; **15**:152a
paleobotany, **13**:44a
Coal tar, **4**:250a; **6**:461b
constituents of, **2**:67b; **5**:197b; **8**:349a; **11**:615b
derivatives, **5**:13a; **6**:462a–463a; **8**:553b
distillation of, **8**:56a; **12**:209a
production of, **7**:254a
Coal tar bases, **4**:79a
Coast and Geodetic Survey, U.S., **1**:6a, 364a–b; **2**:500a; **3**:536a; **4**:591a,b; **5**:479b; **6**:165b; **10**:479b–480a, 482b–483b; **11**:113a–b, 552b; **12**:132a, 209b–210a, 479a; **13**:66a; **14**:448b, 449a
Division of Geodesy, **2**:372b
formation of, **7**:89b
magnetic work, **1**:521b
Coastline
surveying of, **5**:17a
Coates, Benjamin Hornor, **6**:119b
Cobalt
atomic weight, **11**:417a
compounds of, **13**:310a–b; **14**:268a–271b
coordination compounds of, **7**:179a
discovery of, **2**:421b; **8**:147a
mining of, **8**:54b
ores of, **8**:147a
physical properties of, **3**:400b
pigment, **8**:146b
processing of, **14**:447a
salts of, **3**:271b, 400b; **12**:546a
ammines, **5**:349b–350a
separation from nickel, **2**:565a; **5**:7a
Coblentz, William Weber (1873–1962), 3:327–328
Cobra, **10**:237b
Cocaine
action of, **4**:160b; **8**:453a
addiction to, **5**:172a; **6**:77b
as anesthetic, **6**:77b
derivatives of, **10**:154b
production of, **3**:246a; **8**:329b
structure of, **14**:411b
synthesis of, **5**:394a
therapeutic use of, **5**:172a
toxic effects, **4**:297a
Cocchi, Igino (1827–1913), 3:328
Coccidia, **4**:133a*; **8**:145b, 270b; **9**:328b; **12**:141b, 142a
life cycle, **12**:423b–424a
Coccidiomorphs, **4**:207a
Coccidiosis, **12**:482a
Coccobacteria, **3**:338b
Coccosphaera, **9**:587b
Coccus aquaticus, **2**:4b
Cochin China, **11**:64b
Cochineal (dyestuff), **3**:159b; **4**:184b–185a; **7**:471a; **8**:329a, 621b; **11**:47b*
analysis of, **10**:497b
Cochineal (insect), **2**:321b
generation of, **5**:226b
Cochlea (math.), **1**:164b; **9**:488a
Cochon de Lapparent
see Lapparent, Albert
Cochrane, Archibald
see Dundonald, Archibald Cochrane, Earl of
Cockburn, William, **11**:2b
Cockchafer, **6**:221a
Cockcroft
see Cockroft
Cockerell, Theodore Dru Allison, **8**:518a
Cocker's Arithmetick, **4**:36b; **12**:443b
Cockroach
anatomy, **10**:40b
cytology, **4**:546a
embryology, **14**:291a
Cockroft, John Douglas (1897–1967), 3:328–331; 5:272a; **8**:95a; **12**:33b, 35a*, 201b; **15**:26b, 28a*
Coconut oil
analysis of, **5**:475a
Coction, **6**:422a
Cod (fish), **6**:442a
Codazzi, Delfino (1824–1873), 3:331; 2:288a, 351a
equations, **10**:544b
Gauss-Codazzi relations, **2**:288a
Codeine, **3**:246a; **9**:9b
discovery of, **11**:494b
synthesis of, **5**:531a
Cod-liver oil, **15**:418b, 419a
Coefficients
Bessel, **2**:101b
binomial, **7**:243a, 326a; **13**:59b

Cauchy, **3**:146b
Fourier, **1**:194b; **4**:476a; **6**:176b, 472b; **12**:188b
Gaussian, **4**:125a; **7**:168b
irrational, **1**:30b–31a
Laplace, **10**:34b; **13**:139b
law of, **12**:93a
linear, **1**:294b
partial differential, **4**:103a
periodic, **8**:561a–b
substitution, **1**:294b
undetermined, **2**:50a, 56b; **8**:162b, 164a, 165a, 306b
Coelenterates, **4**:236a; **11**:503a
anatomy, **9**:560b
contact toxin, **11**:101b
digestion
intracellular, **9**:333a
ecology of, **3**:550b
embryogeny, **7**:402a, 475a; **14**:425b
freshwater, **1**:411a; **7**:401b, 402a; **11**:520b
freshwater vs. marine forms, **2**:106a
life cycles, **2**:502a; **12**:107a
morphology, **6**:336b
physiology, **3**:550b
regeneration of, **5**:577b; **9**:408b
reproduction, **12**:421a
taxonomy, **3**:550a–b; **6**:590a–b; **7**:610b; **8**:270a; **14**:1b
Coelho de Amaral, Nicolas, **10**:160b
Coelom
embryogeny of, **8**:586a
taxonomic significance, **6**:336b
Coelom theory, **6**:339a
Coelomoducts, **5**:467a–b
Coeloptychium, **14**:626b
Coelostat, **6**:29a; **8**:387b–388a
Coelum
founding of, **6**:508a
Coenzymes, **2**:562a; **4**:358a; **14**:174a; **15**:257b
Coesite
artificial production, **2**:459b
Coffee, **3**:7b
as beverage, **11**:312b; **13**:351b
chemistry of, **5**:6a
Coffee plant, **14**:12b
cultivation of, **7**:197a
description of, **1**:124b; **4**:173a*; **7**:197a
Coffin, James Henry, **2**:628b; **14**:52b
Coggia, Geronimo Eugenio, **13**:36b
Coghill, George Ellett (1872–1941), 3:331–332; **6**:320b, 321a,b
Cogrossi, Carlo Francesco (1682–1769), 3:332–333; **13**:564a
Cohen, Ernst Julius (1869–1944), 3:333–334; **4**:518b*; **7**:14a(n), 222b*; **10**:577b*; **13**:579b
Cohen, Hermann, **3**:110b
Cohen, Jacques, **3**:333a
Cohen, Louis
Austin-Cohen formula, **1**:338a
Cohen, Morris Raphael (1880–1947), 3:335; **10**:485b; **11**:549b
Cohen, Paul Joseph, **3**:57b; **5**:108a; **6**:393b; **14**:614b
Cohesion
seventeenth-century study of, **4**:506b; **5**:245b; **6**:609a
eighteenth-century study of, **2**:330b; **3**:441b, 443b–444a; **5**:601a; **8**:260a; **11**:432a
nineteenth-century study of, **5**:124a; **9**:548a; **14**:360b
and heat, **14**:360b
work of, **6**:118b
Cohn, Edwin Joseph (1892–1953), 3:335–336; **8**:553a; **12**:546b*; **15**:76b
Cohn, Emil, **4**:305b; 306a
Cohn, Ferdinand Julius (1828–1898), 3:336–341; **2**:436b
acquaintances of, **9**:332a; **10**:17b
bacteriology, **5**:564a; **10**:364a, 372a, 383a, 397b
botany, **11**:153b, 154a
on contemporary scientists, **3**:614b*; **5**:442a*; **6**:73a; **11**:155b*
cytology, **1**:540a; **4**:236a; **12**:231b
influence of, **4**:296a; **10**:382a; **12**:87b; **15**:147b
influences on, **10**:13a
and Koch, **3**:340a; **4**:296a; **7**:422a,b
students of, **14**:228a
Cohn, Lassar (1858–1922), 3:341–342
Cohnheim, Julius Friedrich, **4**:296a; **7**:422a, 423a,b, 519b; **12**:87b; **14**:41a
histology, **9**:333b–334a; **14**:228b
students of, **1**:569a; **3**:447a; **8**:8a,b; **10**:17b, 493a; **14**:228a, 249a
Coi, Zuanne da, **4**:586b, 587b
Coignet, Michel, **5**:240a, 382a; **14**:23a
Coinage
seventeenth-century, **7**:558b
nineteenth-century, **4**:36b
Coincidences, method of, **2**:338a,b
Coincidentia oppositorum
see Opposites, coincidence of
Coiter, Volcher (1534–1576), 3:342–343; **1**:109b, 293b; **4**:509b, 519b; **8**:54a; **11**:527b
Coke, Thomas, **12**:487b
Coke
in metallurgy, **7**:78a,b
Coke oven, **6**:370a
Colbert, Jean Baptiste, **3**:102a; **5**:259a; **8**:149b; **10**:521a,b; **13**:594a(n)
and Académie des sciences, **4**:87a, 135b; **6**:598b, 599a; **7**:578a; **9**:498b; **10**:519b, 546b; **11**:328a; **13**:335a
critics of, **4**:566b
and Paris observatory, **10**:520b; **11**:525b
patronage and assignments, **2**:201a; **3**:63b, 103b; **6**:477b; **9**:92b, 310b; **11**:512a
Colburn, Zerah, **6**:85b
Colchicene
structure of, **14**:446a
Cold, **13**:571a
theories of, **13**:279a
as vital quality, **1**:261b–262a
Cold Spring Harbor, N.Y., **3**:590a, 610b
Cold waves (meteorol.)
theories of, **9**:530b–531a
Colden, Cadwallader (1688–1776), 3:343–345; **1**:487a; **4**:492b; **7**:210b; **8**:459a
and Collinson, **3**:351a
History of the Five Indian Nations, **3**:351a
and Linnaeus, **3**:344a, 351a
Principles of Action in Matter, **3**:351a
Colden, David, **3**:344a
Colden, Jane, **3**:344a
Colding, Ludvig August (1815–1888), 15:84–87
Cole, Frank Nelson (1861–1926), 3:345–346; **9**:387a; **10**:244b
Cole, Sydney William, **6**:500b
Colebrook, Leonard, **4**:154b, 155b, 512a
Coleochaete, **11**:153b, 154a
Coleoptera, **1**:502a, 504a; **2**:321b; **5**:605b; **6**:507a
fossil, **12**:264b
Coleridge, Samuel Taylor, **1**:489b; **3**:600a; **4**:530b, 531a; **6**:86b–87a; **9**:69b; **11**:137a, 140a, 391b*
Coles, William, **8**:416a
Colet, John, **1**:79b; **8**:360b
Coley, Henry, **7**:308a
Colic, **5**:418b; **6**:146a*
Colin, Jean Jacques, **5**:297a; **11**:495a
Collado, Luis, **4**:487a; **7**:17a
Colladon, Daniel, **13**:127a–b, 128a,b, 130a–b
Colladon, Jean Antoine, **4**:286a; **12**:309a
Colladon, Theodore
Adversaria, **8**:54a
Collagen diseases
treatment of
steroid, **3**:353a
Collection of Czechoslovak Chemical Communications, **2**:429a
founding of, **6**:375a
Collège de France, **5**:567b; **7**:603a
anatomy at, **11**:295b
biophysics at, **1**:303a
history of science at, **2**:357b–358a; **13**:254b
mathematics at, **6**:3b; **7**:577b–578a; **11**:468a
medicine at, **5**:580b, 581a; **9**:10b
philosophy at, **13**:253b
physics at, **7**:155a; **8**:10a–b, 11b
Collège de la Trinité (Lyons), **4**:505a,b
Collège des Quatre Nations, **8**:67a–b
mathematics at, **13**:585a
College Entrance Examination Board, **14**:383a
Collège Mazarin
see Collège des Quatre Nations
Collège Royal
see Collège de France
Collegio Romano
observatory, **12**:266b, 267a
Collegium Curiosum sive Experimentale, **4**:166a,b
Collegium Medicum Carolinianum (Kassel), **9**:434a
Collembola, **5**:328b; **8**:528b
Collenuccio, Pandolfo, **8**:249a,b
Collet, Léon William, **1**:394b
Collet-Descotils, Hippolyte-Victor (1773–1815), 3:346–347; **15**:347b
chemistry, **5**:92a; **11**:464b; **13**:281a
Collette, Bastiaan Jacob, **13**:611a
Colley, Richard, 1st Marquis Wellesley, **7**:97a
Collie, John Norman (1859–1942), 3:347–348; **1**:390b; **11**:280b, 283b*
Collimator, **3**:299a*
Collineation, **9**:430b
Collins, Anthony, **3**:295a–b, 296a
Collins, Frank Shipley, **12**:328a
Collins, John (1625–1683), 3:348–349; **6**:129b
acquaintances of, **8**:149b, 163a; **9**:310a; **13**:479b
correspondence of, **2**:507b; **5**:22b, 23b; **7**:584a*; **10**:44b, 48a,b, 66b, 86b(n); **13**:265b, 480a
editions of works, **7**:276b*
and the Gregorys, **5**:521a, 524b–529a *passim*
influence of, **6**:130a; **9**:311a
influences on, **9**:303a
manuscripts of, **7**:163a
mathematics, **1**:474b; **2**:507b; **9**:447a
Collins, Samuel
A Systeme of Anatomy, **13**:527b
Collinson, Peter (*ca.* 1693–1768), 3:349–351; **1**:434a; **3**:129b; **5**:216b; **6**:45a(n); **7**:412b; **12**:515b
and J. Bartram, **1**:486b, 487a, 488a
and W. Bartram, **1**:488b, 489a
correspondence of, **3**:444a; **8**:459b; **9**:596b

Collinson, Peter (cont.)
and Franklin, **3**:350a, 535a; **5**:130b–134b *passim*
Collip, James Bertram (1892–1965), 3:351–354
insulin, **1**:442b, 443a; **8**:614b–615a
Collision
seventeenth-century discussion, **1**:567a; **6**:127a–b; **8**:152a, 153a; **9**:51a–52a, 97a–b, 115b–117a; **10**:60b; **11**:532a
eighteenth-century discussion, **2**:330b; **3**:74a–75a; **6**:171b
nineteenth-century discussion, **9**:236a; **14**:568b–569a
elastic
laws of, **9**:115b–116a
inelastic
laws of, **9**:115b, 119b
laws of, **11**:507a; **12**:462b
oblique, **9**:116a
in presence of friction, **3**:418b
see also Impact, Percussion
Collodion, **2**:171a; **4**:283a; **13**:598a
discovery of, **12**:198a–b
Colloids, **8**:446a; **11**:485b; **12**:497b; **13**:158b–163a
Brownian movement, **4**:314b–315a; **10**:525a–b
definition of, **10**:251b
detection of, **12**:593b
flocculation, **15**:160a
flow behavior, **10**:251b
fluctuations in, **12**:497a
metallic, **5**:595b; **13**:160a–b
micelles, **3**:620b; **6**:118b; **9**:601a–b
microscopy of, **14**:633a–b
naming of, **5**:494b
optics of, **10**:251b
osmotic pressure of, **1**:537b
particle sizes, **13**:160a–b
pharmacology, **9**:295b–296a
preparation of, **5**:494b; **13**:160a
protective, **5**:595b
sedimentation, **13**:160b–161a
structure analysis of, **2**:627a–b; **3**:619b
thermodynamics of, **4**:314b–315a
thixotropy, **15**:160a
see also Chemistry, colloid
Colman, James, **5**:215a
Cologarithms, **3**:61a
Colombe, Lodovico delle, **3**:116b*; **5**:240b, 241b, 242a
Colombia
botany, **13**:463b–464b
discovery of, **15**:89b
Colombo, Matteo
see Colombo, Realdo
Colombo, Realdo (*ca.* 1510–1559), 3:354–357; **3**:99a; **4**:519b; **10**:599a; **12**:97b; **13**:569a; **14**:11a
anatomy, **2**:336b; **4**:510b, 511a, 520b; **6**:153b; **14**:12a
and Falloppio, **3**:355b
on heartbeat, **6**:154a
De re anatomica, **4**:519b; **12**:324b
and Vesalius, **3**:354b–355a
Colon, **5**:586a
Colonialism
Renaissance, **6**:124a
seventeenth-century, **1**:549a
eighteenth-century, **2**:342a
Colonna, Egidio
see Giles of Rome
Colonna, Francesco Maria Pompeo, **9**:533a
Color(s)
chromatic circle, **3**:241a–b; **5**:445b–446a; **10**:56b; **15**:465a
color keys, **11**:443b
complementary, **4**:557a; **6**:246b
effects on heat, **1**:364a; **3**:116b
fading of, **8**:102a
harmony, **15**:465a
matching of, **3**:31b
mixing of, **5**:445b; **9**:200b–201a; **15**:194a
additive vs. subtractive, **6**:246b
mathematical theory, **9**:234a
and musical tones, **3**:115a; **10**:56b
nomenclature, **9**:202a
number of colors in spectrum, **2**:453a; **4**:94b; **9**:200a; **14**:491b
of plates, **14**:565a, 566a,b
from polarized light, **2**:137a–139a
dependence on thickness, **2**:137a–b
primary, **3**:489a; **6**:246b–247a; **9**:200a,b
psychology of, **5**:445b–446a; **13**:351b
saturated, **15**:465a
simultaneous contrast of, **3**:241a–b
standardization of, **15**:464b–465b
theories of
medieval, **4**:94a–b; **5**:552a; **6**:191b, 275b; **7**:216a–217b; **13**:511b
seventeenth-century, **1**:475a; **2**:379b; **4**:60a; **5**:543a–b, 544a; **6**:125b, 484b; **9**:52b, 53a, 97b, 118a–119a; **10**:53a–60a *passim,* 315a; **11**:507a
eighteenth-century, **2**:328b; **3**:345a, 454b; **7**:587a; **8**:470a; **9**:267a; **11**:68a
nineteenth-century, **3**:241a–242a, 489a, 599b; **5**:302a, 445b–446a, 558b; **6**:246a–247a, 300a–b; **7**:458a; **9**:463b; **11**:532a; **12**:281a; **13**:76a; **14**:565b–567a
twentieth-century, **12**:218b
quantitative, **15**:465a
wavelengths of, **10**:59a, 59b–60a
Color organ, **3**:115a
Color vision
see Vision
Colorado
Geological Survey, **7**:148b
geology, **3**:483b–484a; **4**:562b, 563a; **6**:187a; **11**:118b
mining, **4**:366b
oil shale, **3**:610a
Colorado River
exploration of, **11**:118b
Colorado Scientific Society, **4**:366a
Coloration
see Animals, coloration of; Pigmentation; Variegation
Colorimeter, **11**:208a
stellar, **5**:490a
wedge, **13**:409a
Colorimetry, **5**:53a,b; **7**:36a–b; **9**:200a, 201a–b; **12**:593b
in astronomy, **12**:352b–353a; **13**:408a,b
tristimulus, **7**:36b
Colostrum
antibodies in, **12**:483a
Colpoda, **13**:268b
Colson, John, **1**:76b; **2**:576b; **9**:159a; **10**:48b
Columbia University
in nineteenth century, **10**:241a–b; **13**:433a
astronomy at, **6**:400a; **12**:37b; **15**:128b
biology at
"fly room," **9**:522b–523a; **13**:134a; **14**:432a–b
College of Physicians and Surgeons, **14**:248b, 249a
physiology at, **15**:108a
Computing Laboratory, **15**:129a
nuclear physics at, **4**:580b–581b
pathology at, **11**:176b
School of Mines, **10**:32a
Columbite, **11**:541a
Columbium
discovery of, **6**:166b
Columbus, Bartolomeo, **15**:87b, 89a,b
Columbus, Christopher (1451–1506), 15:87–91; **1**:84a, 381b; **5**:377a; **10**:153b; **11**:104a, 351a; **13**:441a
cartography, **10**:617a–b
magnetic declination, **3**:424a; **10**:536a
Columbus, Diego, **15**:87b, 89b
Columella, L. Junius Moderatus, **2**:256b; **3**:174b, 343a
De re rustica, **1**:350b
Columns (arch.), **2**:201b
strength of, **4**:480b
theory of, **11**:496b
Colvius, Andreas, **1**:566b, 568a; **14**:254b
Coma, **14**:408a
Coma Berenices (constellation), **3**:391a,b
nebulae in, **9**:330a; **10**:607b
Comas Solá, José (1868–1937), 3:357–358
Comatulae, **3**:89a
Combarelles, Les (cave), **2**:450b
Combe, George, **5**:255a; **14**:189a,b
Combériac, Gustave, **2**:594a
Combes, Charles-Pierre-Mathieu (1801–1872), 3:358; **3**:80b; **4**:348a, 350a
Combes, Raoul (1883–1964), 3:359
Combinants, **13**:220b
Combinations (logical), **2**:163b–164a
Combinations (math.), **2**:233b; **4**:502b; **5**:159b; **9**:159a; **15**:369b–370a
notation, **1**:240a; **6**:403b
theory of, **2**:50a
Combinatorial analysis
ancient Greek, **15**:220a
medieval, **8**:279b, 548b–550a
Renaissance, **13**:259b
seventeenth-century, **8**:153b–154a, 160b–161a; **10**:334b–335a
eighteenth-century, **3**:386b; **6**:403b; **7**:598a; **9**:454a, 471a, 499b–500a
nineteenth-century, **2**:262a; **7**:384a, 494b; **9**:223a, 277b, 278a, 280a–b; **12**:216b; **13**:13a, 20a; **15**:32a
twentieth-century, **8**:617a; **10**:24b; **11**:269a, 362b; **12**:451b; **14**:89a
Chinese, **12**:375b–376a
Combining volumes, law of, **1**:141a–b, 344a–346b *passim,* 616b; **2**:135a; **3**:47b; **4**:243b; **5**:294b, 318b–323b *passim*
and combining weights, **11**:173a
and organic chemistry, **8**:57b
Combustion
medieval Latin study, **12**:429a
eighteenth-century study, **2**:75a–b; **6**:42b; **7**:13a; **8**:74b–75a, 78a, 79b–80b; **11**:499b*; **12**:64a
nineteenth-century study, **1**:586a; **2**:588a; **4**:130a–b; **5**:126a–b, 128a, 558b; **8**:118a; **9**:59a–b; **10**:552b; **12**:493b
twentieth-century study, **7**:617b
and atmospheric pressure, **5**:126a–b
chlorine and, **4**:531b
heat of, **2**:69a,b; **4**:241a, 554b; **6**:243b; **7**:462b; **8**:545b; **13**:351b; **15**:316a
increase in weight, **2**:180a
mathematical analysis of, **9**:462b
and respiration, **6**:484b, 485a; **15**:95a
spontaneous, **10**:446a
theories of
seventeenth-century, **1**:550a; **6**:484b–485a; **7**:195a; **9**:244a; **12**:312a
early eighteenth-century, **6**:43a–b; **12**:604b–605b
late eighteenth-century, **2**:75a–b, 180a; **5**:215b–216a, 601a; **6**:395a, 585b–586b; **7**:58a; **8**:531a, 612b–613a,

620b; 9:6a; 11:523a; 12:144b, 145a–b; 13:469a; 15:95a
nineteenth-century, 10:184b
see also Phlogiston
see also Chemistry, analytic, combustion methods
Côme, Frère
see Baseilhac, Jean
Comenic acid, 8:341a,b
Comenius, John Amos (1592–1670), 3:359–363; 1:126a,b, 160a; 2:163a–b; 6:141a; 9:322a(n); 10:495a; 11:258a; 12:584a; 14:379b(n)
Pansophiae prodromus, 5:607a
and Royal Society, 3:362b; 5:606b
Comets
ancient Greek study, 11:104b
ancient Roman study, 12:310a
medieval Arabic study, 7:338b
medieval Latin study, 1:101a, 199a
Renaissance study, 1:179a, 611b; 3:585b*; 4:213a; 9:168a; 11:561b; 12:199b; 13:440b; 14:273b
early seventeenth-century study, 6:126b; 7:376a; 12:74b
late seventeenth-century study, 2:131a, 592b; 4:168b; 6:363b, 538a; 7:373a; 8:25b; 9:486b, 508a*; 11:411b*, 532a; 12:343a
eighteenth-century study, 3:105a, 198a; 8:321b, 469a; 9:370b; 11:472a
nineteenth-century study, 3:92b, 187a, 428a; 6:49a; 8:613b; 10:110b; 11:379a; 12:192a, 498b, 523b, 579a
twentieth-century study, 2:126b; 4:26a, 249a, 408a*; 8:504a; 9:415a; 10:233b–234b
of 1066, 13:12a
of 1264, 5:361a, 402a
of 1577, 2:406b–409b, 410b, 411b; 4:213a; 7:289a; 9:168a; 12:97b
of 1580, 12:170a
of 1618–1619, 3:528a–b; 4:551b; 8:271b*; 10:176a, 489b; 12:163a, 500b
of 1664–1665, 1:169a, 481a; 2:310a–b; 12:85b
of 1680, 1:169a; 4:168b; 5:58b; 6:68b, 69a; 7:373b; 10:64a, 64b–65a
of 1770, 3:302b; 8:299b; 15:355a
of 1807, 10:608a
of 1811, 1:241a; 2:285a; 6:335b, 437a
of 1843, 2:285b; 4:555a; 10:479b
of 1862, 12:160a–b
d'Arrest's, 1:295a,b, 464a; 12:161a
and asteroids, 7:386a; 10:234a
astrological use of, 6:275b; 10:547b, 548a; 12:135b, 500b; 15:474b, 477b
Biela's, 2:126a; 5:258a; 6:263b; 10:42a; 12:160a, 161a
fragments of, 12:266b
Bode's, 10:197b
brightness of, 11:542a
capture of by planets, 10:42a
catalogs of, 3:472b; 5:257b; 10:615b
Chinese study, 15:729b
classification of, 10:234a
Coggia-Winnecke's, 3:472b; 8:442a
Crommelin's, 3:472b
density of, 2:578b
discovery of, 1:295a,b, 352a–b, 430a, 464a, 464b–466a; 2:360a, 502b–503a; 3:357b; 4:26b–27a, 514b; 5:256b; 6:112a, 149b, 323a, 402b; 7:373b, 374a; 9:250a, 329b, 421b; 10:30b, 107a, 197b–199a, 607b, 608a; 11:82a–b; 12:139a; 13:36b; 14:589a
Science Observer Code, 3:194a
distance from the earth, 2:407b; 4:169a; 6:362a; 7:598b; 12:500b; 15:311a, 477b
Donati's, 2:284b, 433a, 503a; 4:161b
drawings of, 7:209b
Dutoit-Neujmin-Delporte, 4:27a
and earth, possibility of collision, 4:107b; 7:581a
electrical forces, influence of, 2:434a
electromagnetic properties, 11:393a
Encke's (Encke-Backlund), 1:371b, 372a; 2:472a; 4:369b; 5:256b; 6:263b, 326b; 9:250a, 460a; 11:82b, 114b; 12:161a
acceleration of, 7:386a
ephemerides of, 5:257a; 10:30b; 11:545a
Faye's, 9:460a–b; 13:112a*
Forbes', 3:472b
Giacobini-Zinner, 12:161a
groups of, 7:386a; 8:522b
discovery of, 6:453a
Halley's, 10:65a
in 1305, 6:69a
in 1380, 6:69a
in 1456, 6:69a; 13:440b; 15:474b, 477b
in 1531, 1:179a; 6:68b–69a; 7:87a*
in 1607, 2:97b; 6:68b–69a, 126a; 7:87a*, 302b
in 1682, 6:68b–69a; 7:87a*
in 1758/9, 1:400b; 3:283a–b; 4:23b; 6:69a; 7:87b*, 544b, 580a; 9:329b; 11:610a*; 14:452b, 589b
in 1835, 5:256b; 6:326b; 7:209a; 8:487b; 11:114b
in 1910, 3:457a, 472b; 10:233b; 11:412b; 12:251a; 13:12a
historiography of, 3:472a
orbit of, 2:97b; 7:209a; 8:487b; 9:430a; 10:199b; 11:545b
Herschel's, 2:216b, 221a; 6:328b–329a; 8:299b; 13:425a; 15:311b–312a, 354b
historiography of, 5:402a
and insects, 5:439b
light of, 10:78b, 234a
Mesopotamian study, 15:640b
meteors and, 4:555a; 5:258a; 7:386a; 10:41b–42a; 12:160a–161a
Morehouse's, 10:233b
motion of, 2:310b; 3:528a–b; 9:324b; 11:49a
acceleration of, 1:371b, 372a
orbits of, 1:481a; 2:216b, 236a, 237b, 333a, 360a, 388b, 407a,b, 408a–b; 3:14a, 35a, 108a, 101b, 472b, 509a, 528b, 555a*; 6:103b, 539b; 7:37b, 308b, 598a; 8:137b, 299b; 9:442b, 514a; 10:30b, 78b, 479b; 11:84a, 544b; 12:139a, 160a, 170a, 325b; 13:4a, 139b; 14:242b
calculation of, 2:97b; 3:302b, 431a; 4:27a, 369b, 555a; 5:402a; 7:373b, 544b, 564b, 580a; 9:250a, 460a–b; 10:197b, 198b, 199a, 219a, 212b, 615b; 11:114b; 15:309b–312a, 316b, 354b–355a
cometary zodiac, 3:103a
as conics, 3:431b; 6:68b
distribution of, 10:41b–42a; 15:290b–292b, 364a–365a
elliptical elements, 2:97b, 328b; 10:64b–65a
hyperbolic, 4:514b
parabolic, 2:328a–b; 4:168b–169a; 6:362b; 10:64a,b, 65a; 11:545a; 13:99a; 14:589a; 15:311a, 351a, 352a
perturbations in, 2:433a; 3:283a–b; 4:369b, 514b; 5:210a; 7:480b, 564a–b, 581a; 10:30b, 42a; 13:422b–423a; 15:354b–355a
vectorial elements, 1:429a
periodic, 1:295a; 2:126a, 552a*; 3:472b; 4:369b, 555a; 5:25b; 6:68b–69a; 8:277a, 522b; 9:460a–b; 10:30b, 65a, 199a, 234a; 11:82b
short-period, 2:472a; 7:386a; 8:299b
photography of, 1:466a; 5:405a; 10:234b; 12:251a
and planets, 15:289b–290a, 323b
Pons', 3:472b; 4:369b; 9:504a; 11:544b; 13:4a
positions of, 1:467a, 597a; 4:27a; 6:149b; 9:168a
shape of, 11:498a
size of, 2:408b; 3:528a
mass, 8:299b; 10:234a; 15:355a
spectra of, 2:433b; 4:161b, 180a; 5:102a, 199b; 6:542a; 7:268b, 461b*; 8:442a; 9:184a, 504a; 10:234a,b; 11:319b, 542a; 12:268b, 456a; 14:55a
structure of, 2:433b; 3:258b; 10:234a; 12:266b, 268b
and sun, 1:301a; 2:433a–b; 3:93a; 4:555a; 10:234a–b
tailless, 2:409a; 11:82b
tails of, 1:202a, 466a,b; 9:449b*
classification of, 2:433b; 12:160a
radiation in, 12:251a
spectra, 5:102a; 10:234b
and sun, 2:403a, 407a, 407b–408a; 4:168b; 9:212b; 10:234a
theories of, 1:352b; 2:421a; 4:555a; 5:527b; 6:416b; 8:107a; 10:83a, 231b; 12:160a, 249a
theories of
ancient Greek, 2:407a; 6:416b; 15:202b
ancient Roman, 12:310a
Byzantine, 4:159b
medieval, 5:402a, 551a
Renaissance, 2:406b–409b; 4:213a, 387a; 5:400b
early seventeenth-century, 3:528a–b; 5:242b–243a; 6:515a; 7:269b, 302b; 12:500b
late seventeenth-century, 1:169a; 2:46b; 3:100b; 5:58b; 6:362a–b, 485b; 10:315a, 546b; 12:431a
eighteenth-century, 1:400b; 2:328a; 3:105a, 431a; 4:481a; 8:179a; 10:615b; 15:345a–b
nineteenth-century, 2:433a–434a; 4:514b; 6:335b, 453a; 7:385b–386a; 10:199a; 14:629b
twentieth-century, 1:301a; 9:449a; 10:234a–b; 11:412b; 13:99a
mechanical, 2:433a–434a; 10:233b–234a
see also Parallax, of comets
Commandino, Federico (1509–1575), 3:363–365; 3:585a; 4:612b; 8:224b; 9:487b; 12:205a; 13:561a; 14:29b
and Baldi, 1:419a
translations by, 1:191b, 229b, 468a; 3:149b; 4:448a–451a *passim;* 5:588b; 12:75a
Commensurability (math.), 3:127b; 10:224b; 15:503a
test of, 13:304b
Commercium epistolicum
see Newton, Isaac, *Commercium epistolicum*
Commerson, Philibert (1727–1773), 3:365–366; 2:342b; 12:536a
collections of, 7:198a
Commission Géologique Internationale, 11:375a–b
Common, Andrew Ainslie (1841–1903), 3:366; 1:466a
Commonwealth, 6:447b, 448b

Comptes rendus mensuels des travaux chimiques, 5:370a; 8:55b
Compton, Arthur Holly (1892–1962), 3:366–372; 2:247b; 8:94b; 10:216a; 11:420a, 422a; 12:464b
Compton effect, 1:459a; 2:338a, 488a; 3:367b–370a; 4:196b, 317b; 5:331b; 9:261b; 12:528a
and conservation laws, 15:8a
inverse of, 3:369b
opposition to, 4:197a
quantitative theory of, 3:619b; 5:473a
cosmic radiation, 8:618a; 9:398b
and Thomson, J. J., 3:368a
Compton, Henry (bishop), 1:432a,b
Compton, Karl Taylor (1887–1954), 3:372–373; 3:366b, 368b; 11:420a; 13:362a*, 569b, 570a
Computation
see Calculation
Computers
medieval Latin, 10:540b, 541a
seventeenth-century, 8:149b, 161a, 162a; 9:336b*, 529b, 610b; 10:331b–332a; 12:163a
eighteenth-century, 8:166a; 11:66a*, 579b
nineteenth-century, 1:354b–356a; 3:225a,b; 4:591b; 7:105b; 8:531b; 9:267b
twentieth-century, 1:429a; 2:511b; 3:373b; 6:147b, 614b; 8:186b; 9:552b; 13:497b, 498a; 14:92a–b; 15:129a
analogue, 11:579b
astronomical, 2:511b; 10:540b, 541a; 12:163a; 13:153a; 15:128b–129a
binary, 2:298b; 8:531b; 9:610b
differential analyzer, 6:147b
electromechanical, 13:431b–432a
in mathematics, 3:373b
mechanical
algebraic, 13:431b
NORC, 15:129a
in physics, 6:614b; 8:186b
programming of, 14:92a
SSEC, 15:129a
tide-predicting, 4:591b
Computing rod, 15:746b, 751a
Computus, 1:105a, 564b–565a; 3:26b–27a, 262b*; 6:302a; 7:140b–141a; 9:191b; 11:504a
critics of, 7:130b–131a
Comrie, Leslie John (1893–1950), 3:373–374; 15:128b
Comstock, George Cary (1855–1934), 3:374; 5:480b*; 12:248a
Comte, Isidore Auguste Marie François Xavier (1798–1857), 3:375–380; 2:357b, 358b; 15:489a
acquaintances of, 2:187b; 10:2b; 11:491b
chemistry, 14:395b
on contemporary scientists, 5:376b; 14:488b
critics of, 2:31b; 4:374a; 12:607b
editions of works, 6:3a*
influence of, 1:404a; 2:26b, 88b; 11:389a; 12:108a,b, 109a; 13:252a; 15:423a
influences on, 3:383a; 5:255a
law of three states, 3:375b–376b, 378b
philosophy of science, 1:178a; 2:123a; 3:452b; 5:506a; 8:440b; 15:423a, 681b, 741b–742a
positivism, 4:183a; 5:97a
students of, 14:394a
Conant, James Bryant, 11:418a,b; 12:112b
Conatus (physics), 6:606b–607a, 609b, 610a–b; 8:151b–152a, 159a; 10:61b
Concentration, chemical
and activity, 8:290a–b
Concentrobiepicyclic, 3:409b
Concept
Aristotelian, 3:111a
definition of, 3:32a; 5:154a
extension of, 5:154b
function of, 3:111a
theory of, 8:155a, 157a–b
Conceptacles, 2:371a
Conceptualism, 2:545a; 7:330b
Conch
generation of, 2:591b
Conchoid, 4:103b, 259b, 489b; 5:499a, 521a, 527b, 528b; 7:577b; 10:51b, 300a, 342b
discovery of, 10:114b–115b
generalization of, 10:115b
Conchology
seventeenth-century, 8:415b–416a
eighteenth-century, 1:244a; 5:568a; 7:588b
nineteenth-century, 2:109b, 566b, 567b; 3:87b, 536a; 4:67b, 68a; 5:477b–478b; 9:579a; 12:133a
see also Malacology, Shells
Conciliarism, 3:513a–b
Concrete
ancient Roman, 15:515a–b
reinforced, 6:534b
see also Cement
Condensation, chemical, 4:553a–b; 13:3a, 388b
additive, 9:360a
of aldehydes, 1:389b; 2:316b; 7:282b, 283b; 12:164a; 14:454b
Claisen condensation, 3:286b
of formaldehyde, 1:385a
of phenols, 1:385a, 389b
resin production
Bakelite, 1:385a–b
Condensation, physical
medieval discussion, 2:394a
eighteenth-century discussion, 12:408b; 14:225a
twentieth-century discussion, 11:117a
of air, 1:152a
atmospheric, 4:28b
film, 10:162b
of gases, 4:537b; 7:267b; 9:152b
retrograde, 7:515b
and ionization, 1:490b; 14:421a–422a
of metals, 5:160b
of water
and ionization, 1:490b
Condensator (elect.), 14:71a, 73a, 74a, 78b, 79b
Condensers, electrical, 1:67a; 3:506b; 15:312b
air, 14:353a
capacity of, 1:347b; 8:480a
discovery of, 2:324b; 7:403a
flat plate, 5:267b
parallel-plate, 1:547b; 5:132a; 13:224b
plate, 3:498b; 14:353b
see also Leyden jar
Condillac, Étienne Bonnot, Abbé de (1714–1780), 3:380–383; 3:2b, 388a*, 461b; 4:85a; 7:550b; 10:613b; 14:44b
Logique, 8:81a–b
sensationalism, 1:111a, 114b, 140b; 3:1b; 4:85b; 5:253a, 286b; 7:591b; 10:612a
Condon, Edward Uhler, 5:272a; 8:474b
Franck-Condon principle, 5:118a
Condorcet, Marie-Jean-Antoine-Nicolas Caritat, Marquis de (1743–1794), 3:383–388; 4:107b; 7:549b; 8:84b
and Académie des sciences, 15:309b, 333a,b, 334a–b
acquaintances of, 1:401a; 3:1a, 2b, 6a, 7b; 9:343a, 470b, 471a; 10:609a, 612a; 13:495b–496a
analysis, 7:566b, 568a; 13:266a; 15:278a, 294b
biographies of, 7:550b
on contemporary scientists, 4:224b; 5:567a; 8:619b; 13:494b, 495a; 15:276a, 293b, 302a, 304a–b
correspondence of, 7:550a; 9:476a; 12:64a; 13:494b; 15:388a*
éloges, 4:473a; 5:55a*, 88a; 8:622a(n)
historiography, 1:209a; 2:593a; 3:562a; 15:344a
hydrodynamics, 1:116b; 2:335a
influence of, 3:375b, 379a, 383a, 450b, 451b, 561b, 598b; 8:86a(n); 9:68a; 12:109a
and Lagrange, 3:384b; 7:563b–567b *passim*
Mémoires sur l'instruction publique, 8:136b–137a
metric system, 4:14b
on probability, 15:280a, 280b–281a, 281b
on progress, 3:378b
Conductance, 7:527a
measurement of, 14:182b
naming of, 6:212a
Conduction, electrical, 1:547a; 4:215a–b; 9:155a; 11:142a; 14:171a–b, 205b
and absorption of light, 4:191a; 8:491b
in air, 1:547a; 4:355a
and altitude, 14:422b
in alloys, 5:576b; 7:529a–b; 8:119a; 11:446b
and composition, 7:529a–b
anomalous, 7:203b
boundary effects, 3:329a
in cables, 10:594b
in compounds, 6:561a; 14:265a, 268b–269b
in crystals, 2:427b; 12:305b; 14:171b; 15:251b
definition of, 10:190a–b
discovery of, 5:516b
electrolytes, 1:297a–300b
of fused salts, 8:503a
infinite, 7:381a
laws of, 3:260b; 8:177a; 10:189b–191b
in liquids, 5:86b; 8:290b
compression effects, 13:130b
measurement of, 1:298a; 8:502a; 11:433a
in metals, 6:51b; 8:475b–479a, 480a–481a, 491b, 502a; 9:178a; 10:188a–b, 500a; 11:142a, 420b, 446b–447a, 474a; 13:246a, 365a, 429a–b; 14:33a, 330a
and optical absorption, 9:212b
pointed bodies, 4:129a; 5:131a, 134a–135b
of solutions, 3:334a; 6:439a,b; 7:203b, 449a–b, 462a
dilution effects, 7:449b
temperature effects, 7:449b
superconductivity, 1:284b; 2:252a; 7:221b; 8:52a, 475b–479a, 479b–481a
alternating-current effects, 8:475b–476a, 480a,b
electrodynamics of, 8:476a–478a, 480a
intermediate state, 7:617b
microscopic theory, 8:477b, 479a, 480b
in organic molecules, 8:479a
penetration depth, 8:476a,b, 480a,b
and superfluidity, 8:478b–479a, 482a
surface energy, 8:480a
thermodynamics of, 8:477b
two-fluid model, 8:475b–476a, 478b
and temperature, 4:192a–b; 6:271a; 8:502a; 11:446b–447a; 14:522b
theories of, 3:445a–b; 8:261b–262a; 9:174b; 10:28b; 13:374b–377b; 14:208b; 15:251b, 433a

Conduction, electrical (cont.)
through gases, **1**:320b; **6**:439b; **11**:461a; **12**:29a, 238b, 496a, 613b; **13**:364a; **14**:421a
unidirectional, **2**:427b
in vapors, **12**:191b
Conduction, thermal, **1**:166a; **2**:356a; **5**:68b; **9**:260a; **10**:481b; **13**:127b; **15**:83b
in anisotropic media, **1**:365b
in crystals, **12**:305b
electron theory, **4**:189b, 192a–b; **11**:446b; **14**:208b; **15**:433a
in gases, **2**:263a; **7**:526a; **9**:222a–b; **13**:11a
in liquids, **10**:249a; **13**:11a
mathematical study, **5**:94b–97b; **7**:408a; **10**:248b–249b; **12**:526a; **13**:27a, 374b, 376a–b
by metals, **2**:136a; **5**:137a–b; **7**:13b; **8**:502a; **14**:208b, 218b, 330a
physical models of, **5**:95b–96a; **15**:488b
and temperature, **8**:502a; **11**:446b–447a; **12**:438a
theory of, **1**:147a; **3**:308a
Conductors, electrical
capacity of, **11**:433a; **14**:72a–b, 73a
distribution of electricity, **7**:380b–381a; **15**:484a
leakage, **11**:433a
spiral, **6**:278b
see also Semiconductors
Condylarths, **5**:501a*
Cone, Lee Holt, **5**:465a
Cone
area of, **1**:217a–b, 227a,b
horizontal slicing of, **4**:33b–34a
quadrature of, **11**:487a
second-degree, **13**:17b
volume of, **1**:220a, 227a,b; **4**:424a, 466b; **8**:421b; **15**:670b, 693a
see also Conics
Conestoga wagon, **8**:459a
Confervae, **4**:178a
fertilization of, **3**:43b
Confidence limits, **5**:10a
Configliachi, Pietro, **14**:80a
Configuratio, **10**:226a–228b
Confucianism, **15**:733b, 743a–b, 747a
and science, **12**:385b, 387a–391a, 407a–409a; **15**:730a, 732b, 733a, 734a, 740b, 745a, 746a,b, 755b, 756b
Confucius, **15**:731b
Congestin, **11**:430a,b
Conglomerates (geol.), **5**:445a; **8**:463a; **12**:272b
intraformational, **8**:463b
Conglutination, **4**:277a
Congo
geology, **8**:464a
Congo red
osmotic pressure of, **1**:537b
Congresses, scientific
anatomy, **2**:255b
anthropology, **9**:538b, 539a; **11**:235a, 399a
arts and sciences, **4**:404b; **7**:238a
astronomy, **4**:176a; **9**:551b; **10**:600b; **15**:493b
botany, **1**:396b; **4**:307a; **5**:192b; **15**:224a, 268b, 269a
chemistry, **3**:47a; **4**:75b
electricity, **2**:200b; **9**:155b; **14**:39a
general, **1**:136a; **3**:46a; **10**:12b, 195b; **11**:265b
geography, **10**:504b, 506a
geology, **1**:50b, 163b, 236b–237a; **4**:365a(n), 366a; **5**:330a; **6**:564b; **7**:249a,b; **8**:285b–286a, 544a; **11**:375a–b; **13**:284b, 617b
history of science, **9**:341b, 378b
horticulture, **2**:35a
hygiene, **1**:575b; **5**:52a; **8**:408b; **10**:397b
mathematics, **2**:217b; **3**:53a, 584a; **4**:9a, 173b, 617b; **5**:340a; **9**:427a; **10**:610a
medicine
nineteenth-century, **1**:360a; **4**:594b; **8**:7b, 406b; **10**:396a–b; **11**:516a
twentieth-century, **4**:298b, 300b, 301b
meteorology, **4**:162a; **9**:196a–b
natural science, **4**:144a
navigation, **12**:412a
neurology, **4**:286b
philosophy, **4**:373a; **10**:442b; **11**:258a
physics, **4**:578a; **5**:272a, 583a; **8**:489b; **15**:8b
physiology, **4**:603b; **5**:83b; **7**:504b; **9**:102b; **15**:74b
prehistory, **8**:105a
psychoanalysis, **7**:190b
seismology, **10**:231b
statistics, **12**:301a
time, **2**:126b
weights and measures, **2**:200b
zoology, **1**:130a
see also Karlsruhe Congress, Eranos Conferences, Solvay Congresses
Congreve, William (1772–1828), 3:388–389; **9**:491b
Congruences, **4**:336a; **5**:202a*; **6**:249a, 571b; **11**:398b, 407b, 408a–b, 509b; **12**:213b, 469a
linear, **12**:228a
of polyhedra, **4**:9a
quadratic, **11**:390a
of straight lines, **3**:171a, 428a; **4**:257b; **6**:85b
theory of, **1**:591b; **2**:395a; **3**:223b; **4**:573b; **5**:263b; **10**:343b–344a
Congruum, **4**:610b–611a
Conics
ancient Greek, **1**:179b–188b; **3**:391a; **4**:33b–34a, 110a, 421a, 427a–429a; **9**:270a–275a; **11**:28a; **14**:619a; **15**:115b–118a
ancient Roman, **10**:295a–296b, 298a; **12**:314a
Byzantine, **1**:170a
medieval Arabic, **1**:445b–446a; **12**:432a; **13**:289b–290a, 291b
medieval Latin, **7**:133b
Renaissance, **3**:364a–b; **4**:260a; **9**:191b; **14**:274a–b
early seventeenth-century, **3**:149b, 152b, 153a; **4**:47a,b, 568a; **5**:588b; **9**:598a–b; **10**:330b–331b; **12**:74b–75a, 75b
late seventeenth-century, **3**:182b, 621b; **5**:521a; **7**:577a–b; **8**:304b; **10**:51b–52b, 72b; **14**:148a, 465a–466b
eighteenth-century, **2**:326b; **3**:119b; **7**:599b; **8**:609b–610a; **12**:446b
nineteenth-century, **2**:214b, 215a, 444a; **3**:169a, 213b, 468a, 554b–555a; **11**:390a; **13**:17a–19a
twentieth-century, **3**:399a; **13**:125a
conjugation, **1**:184b, 186a–b
construction of, **6**:206a*; **7**:577a; **10**:331a; **12**:42a
definition of, **10**:330b; **13**:5b
diameter-and-ordinate property, **10**:295a
discovery of, **9**:270a–275a
evolute, **1**:187b
focus-directrix property, **1**:187a, 245a; **4**:427b, 429a; **9**:272a, 274b; **10**:295a; **15**:116a–b
proof of, **10**:298a
generation of curves, **1**:180b–184b, 186a, 187a; **4**:427b; **8**:252b–253a, 610a; **9**:272a, 273a–b; **12**:225b
kinematic, **12**:206a,b
mechanical, **9**:272b, 274a–275a
higher-order, **3**:183b
historiography of, **3**:152b
inscribed, **3**:467b; **13**:18b–19a
intersection of, **15**:116b
inversion of, **3**:555a
as loci, **4**:427b–428b; **9**:272b; **14**:465a–466a
and *neusis*, **1**:219a, 225b, 227b, 245a
normals to, **1**:187b; **7**:109b
and optics, **3**:152b
osculatory, **11**:45a
parameters of, **9**:598b
poles, **13**:128a
projective, **4**:47a; **7**:577a; **8**:252a–253a; **9**:173b; **10**:52b, 330b–331b; **11**:76b–77a, 79b; **13**:17a–19a; **14**:465b
quadrature of, **5**:526a–b
systems of, **3**:213b; **6**:75a; **14**:618b
terminology, **1**:180b, 183b–184a, 185a, 215a, 245a; **4**:421a; **9**:270b
theory of, **9**:475b
see also Ellipse, Hyperbola, Parabola
Conifers, **8**:484a
anatomy, **5**:441a
fossils, **5**:441b
life history, **6**:466a
reproduction, **6**:266a; **13**:88a–b
Coniine, **7**:552a; **14**:256a
Conimbricenses, **1**:495a; **2**:601b
Coniothecium, **6**:73a
Conjugation, chemical, **6**:108b
Conjunctiva
cytology, **13**:556b
Conjunctivitis
granular, **10**:142b
Conklin, Edwin Grant (1863–1952), 3:389–391; **2**:502a; **3**:184b; **4**:161b*; **8**:359b*; **9**:496a; **12**:486a*; **14**:239a*
embryology, **8**:355a
Connaissance des temps, **5**:224a, 435a; **6**:234a; **8**:132a; **10**:596b
editors of, **7**:87a, 580a–b; **9**:90b, 250b; **11**:207b
Connectedness, **14**:591a
Connecticut
Agricultural Experiment Station, **4**:271b
botany, **15**:149a, 150a
geology, **3**:593a
Connecticut Valley
geology, **4**:361b
Connection (math.), **3**:96a; **14**:283b; **15**:193a
Connectives, logical, **3**:387a
Connex, **3**:314b
Connor, William Stokes, **14**:553b
Cono, John, **10**:21b
Conodonts, **1**:494b; **13**:532a
Conoids, **1**:219b; **9**:476a
Conolly, John, **5**:570a
Conon of Samos (*fl.* 245 B.C.), 3:391; **1**:180a, 213a; **4**:171b–172a; **14**:604a; **15**:116a
Conrad, Timothy Abbott (1803–1877), 3:391–393; **4**:364a; **6**:57a; **13**:581b
critics of, **8**:104a
Conradi, Heinrich, **7**:428b
Conring, Hermann, **2**:317b; **7**:284a, 412b
Conscience, **5**:178a; **8**:125a
Consciousness, **5**:175a; **7**:68b; **8**:157b; **15**:738a
levels of, **2**:448a
loss of
from pressure on neck, **3**:530b
stream of, **7**:67b
unity of, **14**:527a
Conseil, Ernest Alfred, **15**:454b
Consequentiae, **1**:3a; **2**:232a–b, 609b
Conservation, of natural resources
seventeenth-century, **13**:593a
mid nineteenth-century, **9**:421a

late nineteenth-century, **3**:438b; **7**:169b; **9**:313b; **12**:424b; **13**:572b
early twentieth-century, **5**:545b; **6**:228b; **9**:314a–b; **15**:465b
mid twentieth-century, **4**:600a; **8**:24b; **13**:519a
Conservation principle, in physical sciences, **13**:437a
general law, **8**:469b, 470a; **9**:235b–240a; **10**:424b
see also Charge, electric, conservation of; Energy, conservation principle; Heat, conservation of; Matter, conservation of; Momentum, conservation of; Motion, conservation of; *Vis viva,* conservation of
Conservatoire des Arts et Métiers (Paris), **3**:200a; **11**:111a
Consiglio, Michele, **15**:469a
Consilia, **1**:107a; **9**:469a
Consistency (math.), **5**:350b, 351a; **11**:107a
Consolidation, geological
theories of, **6**:581a, 583b–584a
Constable, Frederick Hurn, **1**:415a
Constance, Council of (1414), **1**:84a
Constance, Lake, **15**:158b
Constancy, principle of, **2**:448a
Constant proportions law
see Compounds, definite proportions theory
Constantin, Robert, **3**:533b
Constantine IX Monomachus, **11**:183a
Constantine the African (*fl.* 1065–1085), 3: 393–395; **6**:123a; **11**:603a; **15**:188a
De gradibus liber, **6**:123b
Pantegni, **9**:41b*; **12**:81a, 82a
spurious works, **11**:244b
translations by, **1**:107a; **4**:121b; **7**:23a; **12**:81a,b; **13**:38a,b
Constantinople
Aristotelian study at, **1**:268b–271a *passim*
council of, **3**:513b
education, **11**:183a
Hagia Sophia, **4**:488b; **7**:28b–29a
Constantinople, University of, **4**:524a
Constants
absolute, **11**:9a, 11a,b, 14b
astronomical, **1**:311a; **13**:120a
variation in, **1**:313a
physical, **11**:393a; **12**:43a
see also specific constants, e.g., Planck's
Constellations, **1**:242a, 530a; **3**:391a; **4**:391b, 467a; **5**:345b; **11**:247a; **13**:149b, 150a; **15**:519b, 640b
limits of, **4**:27a; **6**:327a
naming of, **6**:364a; **7**:543b–544a; **15**:216b, 673a, 711a
northern, **15**:716b, 718a–b, 721b
risings and settings of, **15**:216b, 676a
southern, **1**:530b; **7**:543b–544a
visibilities, **15**:676a
see also Nakṣatras; specific names
Constructions, geometrical, **3**:178b, 277b; **4**:259b–260a; **6**:407a; **7**:599b; **10**: 544a; **12**:325b; **13**:18a–b
analemma, **14**:577a–b, 578a; **15**:210b–211a, 518b, 519b–520b, 548a, 552b
bisection, **9**:157b
center of curvature, **2**:215b
compass alone, **9**:156b–157b, 447a–b
conics, **6**:206a*; **7**:577a; **10**:331a; **12**:42a
fixed-compass, **1**:41b–42a, 605a; **4**:596b; **9**:447b; **13**:259b, 260b
insertion, **1**:219a; **5**:382a,b
limitation of instruments, **10**:181a
with measuring standard, **1**:307b
normal (perpendicular), **4**:57b; **10**:180b
and number representation, **1**:237b–240a
parallel lines, **1**:468a
in perspective, **13**:267a
pointwise, **3**:153a
projective pencils, **3**:153a
theory of, **7**:530b; **8**:175b–176a
verging, **1**:218b
see also Cube, duplication of; *Neusis*
Constructions, mathematical
axiomatic, **4**:177b
transfinite, **1**:428a
see also Constructions, geometrical
Constructivism, **3**:57a
Consumption (med.)
see Tuberculosis
Contact element (geom.), **8**:324b–325a
Contact-sensitivity, **2**:325a
Contagion
see Disease(s), infectious
Contagium vivum theory, **1**:492b, 493a; **2**: 575b; **3**:332a–b; **7**:35a; **12**:57b; **13**: 564a; **15**:14b
Contarini, Marcantonio, **14**:5a
Conti, Andrea, **3**:13b, 14a
Conti, Giovan Stefano, **2**:329a
Conti, Niccolò de', **9**:190a
Continental drift, **1**:236b, 394a, 623b; **2**: 558b; **4**:261b, 263a; **5**:221b, 487a; **6**:475a–b; **12**:190b; **13**:270b–271a, 610b; **14**:215a–217a
opposition to, **13**:285b
Continental Gas Company, **3**:556b
Continents
origin of, **2**:568a–b, 578a; **3**:548b*; **13**: 64a,b, 147a, 270b–271a; **14**:402b
permanency of, **3**:552b
polygonal fields, **3**:325b
previous connections between, **4**:262b, 310a; **6**:489a; **14**:215a
Contingency, philosophical, **4**:255a–b
Contingency, statistical
tables of, **5**:10b; **14**:383a, 574a
theory of, **10**:459a–460b
Contini, Francesco, **2**:83a
Continuation (math.)
analytic, **8**:367a
nonanalytic, **2**:303b
Continuity, mathematical
ancient Greek discussion, **1**:215b, 257a–b, 277b
seventeenth-century discussion, **1**:164a; **5**:245b; **12**:56a; **14**:148b
eighteenth-century discussion, **1**:114a, 115b–116a
early nineteenth-century study, **11**:77a, 79b
mid nineteenth-century study, **3**:136a, 137a; **9**:554a*
late nineteenth-century study, **1**:406b; **3**:456a; **6**:472a; **10**:484a–b; **11**:53a, 59b; **12**:228a–b; **14**:343a
twentieth-century study, **1**:407b–408a; **6**:4b
absolute, **1**:427b; **8**:111a
complete, **12**:188b
of complex functions, **3**:137b; **8**:61b
definition of, **3**:137a; **4**:3b, 126a; **8**:432b; **11**:459b
and differentiability, **2**:276a
and monotonicity, **2**:276a
and probability, **1**:366b
right vs. left, **2**:275a
semicontinuity, **1**:407b
uniform, **2**:275b–276a; **3**:137a; **6**:230b
see also Functions, continuous
Continuity, physical
ancient discussion, **2**:231b; **14**:608a–611a
medieval discussion, **1**:61b–62a, 198a; **2**:391a–b, 394a–395a; **4**:447a; **5**:115b
seventeenth-century discussion, **2**:48b; **8**:157a, 268a–b
nineteenth-century discussion, **13**:75b
twentieth-century discussion, **4**:288a; **14**:308b
in machine motion, **3**:73a
of nature, **4**:89a–b
Continuum (math.), **4**:4a; **6**:377b–378a, 402a*
Cantor's hypothesis, **3**:55a, 57a,b
cardinality of, **6**:393b
composition of, **3**:150b
connectedness, **9**:248b
cut points, **14**:591a
definition of, **3**:55a
indecomposability of, **2**:513a,b; **9**:249a
locally-connected, **14**:591a
one-dimensional, **13**:549a
oscillation of, **9**:248b
quadratic, **12**:172a–b
reality of, **10**:485a
Contraception
medieval discussion of, **9**:41a–b
nineteenth-century discussion of, **9**:383a, 385a
opposition to, **9**:68b
Contraceptives
oral, **10**:611a
Contractility (biol.), **3**:337a, 338a; **5**:44b, 56a; **6**:158b
mechanism of, **7**:455a
Contraction
Lorentz-FitzGerald hypothesis
see Lorentz, Hendrik Antoon, Lorentz-FitzGerald contraction
thermal, **6**:477b; **9**:14a
Contraction hypothesis (geol.)
see Orogenesis
Contradiction, principle of, **8**:156a–b; **14**: 483b
proof by, **10**:8a
Contraries, rule of, **1**:444b, 445a
Contravariants, **1**:294b; **3**:164b
Contributions, Cushman Laboratory for Foraminiferal Research, **3**:520a
Control, automatic
see Regulators, mechanical
Convection, **8**:502a; **11**:174a; **13**:351a,b
atmospheric, **3**:593a; **4**:592a–b; **5**:137a
solar, **8**:1b, 2a
coefficient of, **6**:453a
mathematical analysis, **10**:162b–163a
Convergence (math.), **1**:207a, 427a; **2**:50a, 57a, 507a; **3**:227b, 265a; **5**:96a, 342b; **7**:515a, 559a; **9**:127a, 128a
absolute, **8**:558b
Cauchy's criterion, **2**:275b; **3**:136a; **7**: 566b
circle of, **4**:547b, 548a; **5**:204b
conditional, **2**:275b
continua of, **14**:591a
definition of, **3**:136a
general theorem of, **6**:451b
of infinite algorithms, **2**:320a
nonuniform, **12**:289b
radius of, **3**:136a
speed of, **7**:514a
tests of, **4**:206b, 476a; **13**:562b
uniform, **3**:137a; **4**:287b
weak, **1**:428a
Conversations on Geology, **2**:569b
Convoluta
neurology, **4**:12a
Convolution (math.), **14**:285a
Convolvulus, **10**:622a
Convulsions
in hysteria, **3**:205b
Conway, Henry Seymore, **6**:383a
Conybeare, John Josias, **2**:566a; **3**:585b

Conybeare, William Daniel (1787–1857), 3: 395–396; **2**:566a,b, 570a,b; **3**:585b; **12**:276a, 490b
geology, **3**:567a; **5**:32a; **8**:564b; **9**:583a
Outlines of the Geology of England and Wales . . . ,**5**:519a; **9**:87a; **10**:323a, 586a
paleontology, **4**:549a; **9**:87a
Cook, Allen Fairchild, **9**:203b
Cook, Florence, **3**:475b
Cook, James (captain) (1728–1779), 3:396–397; **2**:509b; **5**:76a, 216b; **7**:198b; **8**:379b; **12**:516a; **13**:495b
Endeavour voyage, **1**:434a–b; **12**:414a, 515b–516a
Franklin and, **1**:436a
Resolution voyage, **5**:75a,b, 76b–77a
scurvy, **6**:499a; **8**:362b–363a, 585b
Cook County Hospital (Chicago), **6**:232b, 233a
Cooke, Hereward Lester, **3**:367a; **11**:420a
Cooke, Josiah Parsons, Jr. (1827–1894), 3: 397–399; **9**:349a, 393b*; **11**:416b
Cooke, William Fothergill, **14**:290a
Cooking
technology of, **13**:351b
Cookson, Isabel Clifton, **8**:6a
Coolants
ammonia, **8**:365b
carbon monoxide, **3**:299a
ethylene, **14**:522b
nontoxic, **7**:316b
see also Refrigerants
Coolidge, Albert Sprague, **8**:474b
Coolidge, Julian Lowell (1873–1958), 3:399; **11**:79b; **12**:286a*
Coolidge, William David
Coolidge tube, **4**:195a,b
Cooling
medieval Arabic study, **11**:326a
seventeenth-century study, **9**:508a*
nineteenth-century study, **2**:136a–b; **4**: 240a; **14**:503b
by adiabatic demagnetization, **7**:493a; **12**:438a; **15**:444b
biological, **4**:171b
curve
and chemical composition, **13**:245a
by evaporation, **1**:44b; **3**:495a; **8**:262a
by expansion, **3**:11b, 580b*; **7**:182b; **8**:1a
laws of, **2**:136a–b, 178a; **4**:240a; **10**:545b; **11**:6b, 432b; **13**:10b–11a
by liquid flow, **2**:356a
in a vacuum, **4**:240a
see also Cryogenics, Low-temperature studies, Refrigeration, Supercooling
Cooper, Anthony Ashley (Earl of Shaftesbury), **4**:85a; **5**:426b; **8**:437a–b
Cooper, Astley Paston, **3**:490a; **6**:120b; **14**: 397b
Cooper, John Niessink, **7**:221b; **8**:480b
Cooper, Leon N., **8**:477b
Cooper, Thomas (1759–1839), 3:399–400; **11**:147a*
and Priestley, **3**:400a
Coordinates, geographical
and geometry, **6**:240a
measurement of, **1**:19a; **5**:490a
see also Latitude; Longitude; Tables, geographic
Coordinates, geometric
avoidance of, **2**:594b
barycentric, **9**:430a–b; **11**:45a; **15**:196a, 197a
bipolar, **10**:52a
Cartesian, **1**:189a; **4**:57a; **5**:96a
circular, **3**:169a
curvilinear, **2**:145b; **3**:560a*; **4**:287b; **7**: 601a–b, 602a; **9**:287b
Gaussian, **4**:479b
Grassmann, **15**:197b
homogeneous, **3**:315a; **4**:602a; **5**:201b–202a; **7**:396b; **11**:45a
discovery of, **9**:430b
isothermic, **2**:288a
line, **3**:167a; **11**:44b, 46a
minimal, **3**:169a
negative, **10**:52b
nonhomogeneous, **11**:44b, 45a
oblique, **5**:111a; **7**:109b
orthogonal, **2**:281a; **4**:569a
Plücker's coordinates, **3**:167a; **9**:430b; **11**:44b, 46a
polar, **1**:147b; **10**:52a; **15**:562a
reference point of, **3**:282a
space, **10**:319b
spherical, **5**:96b
systems of, **3**:59a, 183a; **5**:111a
accelerated, **4**:325a
transformation equations, **4**:321a–322b, 325a, 327a
tangential, **2**:288a
transformation of, **6**:2a; **9**:475b; **10**:52a
triangular, **11**:45a
Weierstrass, **8**:433b
Coordinates, space-time, **4**:321a–322b, 323b, 325b
and gravitation, **4**:327a–331b
Coordination, neurological, **12**:397a, 398b, 400b
concept of, **5**:44b
Coordination complexes, **5**:595b
Copaifera, **5**:582a
Copal, **7**:394a
Copaux, Hippolyte Eugène (1872–1934), 3: 400–401
Cope, Edward Drinker (1840–1897), 15:91–93; **6**:187a; **7**:480a; **10**:241a, 273b; **12**:39a*; **15**:507b
evolution theory, **2**:504a; **6**:613b
paleontology, **6**:595a; **9**:134a,b; **10**:240b*; **14**:409b
Tertiary Vertebrata, **8**:169b
Copenhagen
Institute for Theoretical Physics, **2**:244b–245a
National Museum, **14**:506a
observatory, **11**:526b
Polytechnic Institute, **15**:84b–85a
Copenhagen, University of
physics at, **15**:83b
Copenhagen Academy
competitions of, **5**:303b
Copenhagen Telephone Company, **7**:101b
Copernicanism
see Cosmology, Copernicanism
Copernicus, Nicholas (1473–1543), 3:401–411; **2**:405a, 406a, 526a; **5**:287b
acquaintances of, **5**:104b; **7**:418b; **13**:140b
and Aristarchus' theory, **1**:248a–b
and al-Battānī, **1**:512b–513a
biographies of, **3**:512a; **8**:321b
commentaries on, **7**:7a
Commentariolus, **2**:404a; **3**:402a–b, 403a, 405a, 406a,b, 409b
on contemporary scientists, **14**:274a
correspondence of, **2**:527a
cosmology, **3**:402a–410b; **7**:485b–486a; **9**:141a; **10**:153b–154a; **15**:35b
see also Cosmology, Copernicanism
critics of, **1**:468a; **3**:262b, 311b; **9**:191b; **11**:288b; **14**:22b
and decree of 1616, **1**:588b, 589b
geometry, **6**:144a
Hermeticism, **2**:540a
De hypothesibus motuum coelestium a se constitutis commentariolus
see *Commentariolus*
influence of, **1**:607b; **2**:308b, 403b; **5**:426b; **11**:288b; **12**:431a; **14**:286a
influences on, **1**:382b; **2**:607b; **10**:591a; **11**:202a–b, 349a; **13**:401a, 511a; **15**:34b, 189a, 215a, 475b, 477b, 708a
De revolutionibus orbium coelestium, **1**: 248a; **3**:262a, 402b–410b *passim;* **7**: 269b, 289b; **9**:167b; **15**:204b
commentaries on, **11**:366a–b
editions of, **7**:485b; **10**:245b
printing of, **3**:410b; **11**:395b–396b
translation of, **4**:97b
and Rheticus, **11**:395b–396b
and trepidation, **1**:312b; **13**:324a
trigonometry, **7**:39a; **9**:191a; **11**:396a–b
Copho of Salerno, **6**:123a; **9**:468b
Anatomia porci, **12**:81b–82b
Copleston, Edward, **2**:569b; **14**:287b, 288a
Copper, **15**:651b
alloys, **3**:87a; **6**:561a; **7**:57a; **9**:60b, 591a; **11**:485b, 550a
atomic weight, **11**:173b, 417a
as catalyst, **12**:47a
colloidal, **9**:295b
copper-zinc couple, **5**:412b
deposits of, **3**:462b; **4**:262a; **5**:155b, 444b; **7**:267a; **11**:608a; **12**:195a; **14**:441b–442a
origins of, **11**:209b
determination of, **5**:394a
electrical properties, **14**:522b
low-grade ores, **2**:158b
in metabolism, **4**:357b; **14**:174a–b
metallurgy of, **3**:87b, 473a; **4**:394a; **5**: 457b; **7**:78a,b; **12**:382b
refining, **13**:179a
mining of, **2**:315b; **5**:155b, 156a; **8**:146b; **14**:59a
nutritional role, **6**:135b, 136a
oxides, **11**:169b
precipitation of, **12**:79b; **13**:559a
explanations of, **7**:194b; **12**:312b
salts of, **3**:202a, 232b, 271b; **4**:554a; **12**: 46b
derivatives of, **5**:189b
in foods, **14**:357b
separation from bell metal, **3**:560b; **10**: 496b
separation from silver, **2**:565a
separation from tin, **5**:91a–b
smelting of, **3**:41b, 316b*
by-products, **5**:222b
specific gravity, **7**:346a
tests for, **2**:381b
toxicity, **4**:352b; **6**:443a
Copper peroxide, **3**:488b
Coppet, Louis de, **2**:186b; **11**:298a
Coprinus, **6**:100a
Coprolites, **2**:569b–570a; **11**:386a
phosphates in, **2**:570b
Copromonas, **4**:133a*
Copulatio, **1**:3a
Copying machine
eighteenth-century, **3**:578b; **8**:322a
Coquebert, Romain, **2**:492a
Coquebert de Montbret, Charles Étienne, **4**:348b; **10**:209a
Coquille (ship), **4**:256a; **5**:277a; **8**:265a–b
Coral, **2**:490a; **14**:1b
animal nature of, **9**:135b; **12**:560b; **14**: 193b
chemical analysis of, **12**:437a*
deep-sea, **11**:113b
fossil, **5**:486b; **8**:486a–b; **12**:257a
geographical distribution of, **3**:551b

islands, formation of, **1**:595b; **3**:550a–552b, 595a
morphology of, **7**:545b
nature of, **9**:18a; **10**:568b–569a
physiology of, **3**:550a–b; **4**:289a, 290a
reefs, **2**:289b, 321a; **3**:190a, 547b, 550b–552a, 568b–569a, 591a, 595a; **5**:77a; **8**:570b; **12**:328a; **13**:519a
tetra, **1**:494b
Coral Rag (limestone), **12**:491a
Corallian formations, **1**:285a,b; **5**:14b
Corbino, Orso Mario, **4**:577b–580b *passim*
Cordials, **8**:131a
Cordier, Pierre-Louis-Antoine (1777–1861), 3:411–412; 5:295b; **10**:32a, 221a
Cordierite, **3**:412b
Cordite
invention of, **4**:78b
Cordus, Euricius (1486–1535), 3:412–413
Botanologicon, **3**:414a
Cordus, Valerius (1515–1544), 3:413–415; 1:596a
Corethra, **7**:501b
Corey, Elias James, **5**:589b
Cori, Carl Ferdinand, **3**:415b–416a; **15**:228b
Cori, Gerty Theresa Radnitz (1896–1957), 3:415–416; 15:228b
Coriolis, Gaspard Gustave de (1792–1843), 3:416–419; 1:582a; **4**:222a; **8**:382b; **10**:4b*; **13**:128a
Coriolis force, **3**:418b; **4**:344b; **9**:43a, 203a; **12**:415b
Coriolis law, **9**:443a
mechanics, **3**:72b, 73a; **10**:3b; **15**:489b
Cork
formation of, **12**:99b
structure of, **6**:484a
Cork borer, **9**:445b
Cork oak
diseases of, **9**:24b
Cormorant, **13**:25b
anatomy of, **3**:342b
Corn
anatomy of, **9**:455b
cultivation of, **10**:326a
Indian
see Maize
Cornaeus, Melchior, **5**:575a
Cornbrash, **1**:285a
Cornea, **4**:371b
astigmatism, **5**:590b
histology of, **2**:376b; **4**:276a; **6**:434a,b; **7**:439a; **13**:556b
pricking of, **14**:587a
Corneille, Pierre, **5**:57b, 59b
Corneille, Thomas, **5**:57b, 58b
Dictionnaire des arts et des sciences, **5**:61b
Cornelio, Tommaso, **12**:333a
Progymnasmata physica, **2**:84a, 309a
Cornell University
engineering at, **13**:399a
horticulture at, **1**:396a
Loomis Laboratory, **4**:499a
Corner, George Washington, **2**:345a; **5**:485a–b; **12**:82a, 473b–474a
Cornets de Groot, Jan
see De Groot, Cornets
Cornette, Claude-Melchior (1744–1794), 3:419
Cornil, Victor André, **11**:295a, 373b
Cornu, Marie Alfred (1841–1902), 3:419–420; 2:200a; **4**:68b; **5**:21a*; **9**:372a; **11**:379a; **13**:344b*; **14**:525a
Cornu spiral, **3**:420a
physics, **5**:19b, 20b, 407b
Cornus, **8**:178a, 303b
Cornwall
geology, **4**:10b, 11a; **5**:39a; **12**:276b
Corona Borealis (constellation)
novae in, **2**:146b; **6**:542a
R Coronae Borealis
variability, **10**:608a
variable stars, **8**:539b
Coronagraph, **5**:364a; **6**:106a; **8**:581a–b
Coronel, Antonio, **3**:420b; **9**:32b; **12**:547b
Coronel, Luis Nuñez (*d.* 1531), 3:420–421; 2:395b; **4**:238a; **9**:32b; **12**:547b
Physice perscrutationes, **13**:210b
Coronella, **10**:16b
Coronelli, Vincenzo Maria (1650–1718), 3: 421
Coronium, **11**:280b; **12**:44a
Corps du Génie, **3**:70b–71b, 439b–440a
Corpus Hermeticum
see Hermeticism
Corpus luteum, **1**:123a–b; **2**:345a; **5**:226b; **15**:107b
glandular nature of, **5**:485a
stigma, **5**:485a
Corpuscularian philosophy
see Atomism; Matter, structure of, corpuscular theory
Corrasion, **7**:144b
Correlation, geological
see Strata, correlation of
Correlation, geometrical, **3**:77a–78a
Correlation, statistical, **2**:211b; **5**:266a; **10**:449a–464a
correlation coefficient, **5**:8a; **10**:451b, 456b, 458a–b, 461b; **14**:251a
partial, **10**:453a
tetrachoric, **10**:453a–b
multivariate, **10**:451b, 453a
partial correlation, **10**:453a; **14**:574a
rank-order, **10**:453b
ratio of, **10**:452a, 457a–b
skew, **10**:450a, 452a–b
Correns, Carl Franz Joseph Erich (1864–1933), 3:421–423; 5:453a; **9**:281b; **10**:576b; **11**:574b
genetics, **7**:623b; **9**:277a, 601b; **13**:478a,b; **14**:96b, 99b, 431a
Correspondance astronomique, géographique, hydrographique et statistique, **14**:583a
Correspondance mathématique et physique
see Quetelet, Lambert-Adolphe-Jacques
Correspondence (math.), **10**:484a; **13**:133a
theory of, **3**:167b–168a, 171a; **6**:571b; **8**:504b
Correspondence principle (physics), **2**:246b–247b, 248a–b, 250b; **7**:491b, 493b, 553b; **11**:13a; **13**:240a; **15**:41b
relativity theory, **8**:495b–497a
Correspondents, scientific
see Transmission of science, via correspondence
Corrigan, Dominic John, **14**:26a
Corrin, Myron Lee, **6**:118b
Corrosion
geological, **7**:513a
of metals, **3**:176b; **5**:189b, 190a; **7**:57a; **9**:60b
Corrosive sublimate, **8**:174b
manufacture of, **3**:202a
Corsali, Andrea, **1**:530b
Corsets
effects of, **12**:510a; **14**:450a
Cort, Henry, **2**:174b; **3**:87a
Cortés de Albacar, Martín (*d.* 1582), 3:423–424
Breve compendio de la esfera y de la arte de navegar
translations of, **5**:593a
Cortese, Giovanni Battista, **9**:4a
Corti, Alfonso Giacomo Gaspare (1822–1876), 3:424–425; 6:618a; **13**:557b
Corti, Bonaventura (1729–1813), 3:425–426; 9:345a
Corti, Matteo, **14**:4b
Corticoid hormones, **3**:352b–353a; **15**:258b
Cortisone, **15**:258b
Corundum, **3**:232b
Corvisart, Jean-Nicolas (1755–1821), 3:426–428; 1:479b; **2**:122a, 444b; **3**:1b, 2b; **4**:197b; **7**:556b
Essai sur les maladies . . . du coeur, **1**:333a
percussion method, **1**:333a; **10**:613a; **12**:450a
Corycium, **12**:272b
Corydaline
discovery of, **14**:112a
Corysanthine, **5**:100a
Cos, school of, **6**:419b–429a
Coscinodiscus, **9**:587b
Cosine, **5**:414a
Arabic use of, **1**:511b
law of, **7**:118a; **8**:27b–28a; **11**:350a; **14**:470b
theorem of, **11**:4a; **13**:291b; **14**:274a, 577b
Cosmas Indicopleustes, **1**:121a
Cosmimetry, **6**:544b
Cosmogony
ancient Greek, **1**:149b, 151a, 152a; **4**:31b; **8**:269b
ancient Roman, **8**:538a
medieval, **2**:21a–b; **5**:550b–551a; **11**:324b–325a; **13**:230b
seventeenth-century, **2**:223a; **4**:52a; **7**:275b
early eighteenth-century, **9**:26a–b; **13**:179b; **14**:295b–296b
late eighteenth-century, **6**:332a, 333a; **7**:231b–232b; **12**:409a; **15**:344a–345b
early nineteenth-century, **12**:155b
mid nineteenth-century, **3**:192a; **5**:21b; **11**:498a; **13**:120b
late nineteenth-century, **2**:123b; **3**:583b, 470b–471a; **4**:555a–b; **7**:385b–386a; **8**:441b–442b; **9**:594a; **12**:247b; **13**:270b
early twentieth-century, **1**:56b–57a, 470b; **4**:69b; **7**:85b; **8**:522a–b, 561b, 562b, 617b–618a; **9**:552b; **10**:234a–b; **12**:280a–b; **13**:99a
mid twentieth-century, **4**:409b; **9**:449a; **12**:368b
condensation theory, **7**:232a
historiography of, **14**:479b–480a
impact hypothesis, **2**:578a–b; **14**:296b
mathematical, **3**:583b
Mesopotamian, **15**:638b–639a
meteoritic hypothesis, **8**:441b–442b
nebular hypothesis, **3**:190b, 191a, 192a; **7**:385a; **9**:552b, 594a; **11**:21b; **15**:343a, 344a–345a
critics of, **7**:385b
planetesimal hypothesis, **3**:191a; **9**:552b
tidal theory, **7**:85b
vortex theory, **12**:409a
see also Creation
Cosmography
ancient Roman, **9**:1a–b
medieval Arabic, **1**:29b–30a; **11**:230b–232b
medieval Latin, **15**:163a
Renaissance, **1**:468a, 179a; **3**:558a–b; **4**:138b; **5**:349a, 547a; **9**:191a, 254b, 581a; **10**:160b–161b; **14**:18b, 127b; **15**:154b
seventeenth-century, **2**:133a; **3**:421a; **6**:272a*; **8**:65a–b

Cosmography (cont.)
eighteenth-century, **2**:5a; **5**:195b
nineteenth-century, **12**:213b
Indian, **2**:117b; **14**:548b
mathematical, **1**:247a
Mesopotamian, **15**:638b, 639a–b
Cosmology
ancient Greek, **1**:171b, 254a,b, 256b–259b *passim,* 338b; **3**:21b, 319b; **4**:32b, 368a, 463b, 466b–467a; **6**:290b–291a, 381b–382a; **9**:269b; **10**:548a, 590a–591a; **11**:29b, 104a, 187b–188b, 223a–224b; **13**:91b–94a, 296b–297a; **14**:606a–607a, 608a; **15**:202b–204b, 211a–212b, 218b–219a, 645a
ancient Roman, **3**:14b–15a; **8**:537b–538b; **9**:1a–b; **11**:188b–197b; **12**:310a; **13**:326a–b; **15**:519b
Byzantine, **7**:134b–135b; **12**:405b–406a
medieval Arabic, **2**:154a, 154b–155a; **6**:198b–199b; **7**:41b–42a, 64b, 334b–335a; **9**:30a; **11**:250b–251b; **12**:2a, 3a–5b, 6b, 358a–360b; **13**:488b–489a, 511a; **14**:538a, 593b–594a; **15**:33a–35b, 494b–496b
medieval Hebrew, **8**:280b–281a
medieval Latin, **1**:101a, 198b–199a, 565a; **2**:551a, 606b, 607a–b; **3**:14b; **5**:114a, 115b, 403a, 550b; **6**:396a–397a; **7**:27b–28a, 112a–b, 120a, 124b, 142a, 143a; **9**:136b–137b; **10**:223b–225b, 474b; **11**:468b–469a; **12**:61b; **14**:388b
Renaissance, **1**:468a; **2**:403a–b; **3**:514a–b; **4**:209a–b, 213a; **5**:106a, 239a, 399b–400b; **8**:201a–b; **9**:167b–168b; **10**:8a, 161b, 309b–311a, 515a; **11**:72a, 533a, 606b; **12**:135b, 136a; **13**:279a–b; **14**:22b, 226b; **15**:475a–476a
early seventeenth-century, **1**:244b, 373a, 424a, 588a–b; **4**:52a,b, 58b, 60a; **5**:47a–48a, 240b; **7**:269b, 376a; **9**:317a–318a; **11**:489a; **12**:74b, 151b, 152a; **13**:236a; **14**:160a–b, 161b–162b, 286a, 302a, 363a, 364b–365a
late seventeenth-century, **2**:310b; **3**:100b, 101a–b, 103b; **4**:168b; **5**:59a–b, 526a, 575b–576a; **6**:610a–611b; **9**:310b–311a, 486a; **10**:76b–78b; **11**:507a; **13**:96a–b; **14**:177b–178a, 446b
early eighteenth-century, **4**:41a; **5**:511a, 517a; **9**:77b; **11**:44a, 158a; **12**:118a–b; **13**:179a; **14**:83b, 84a, 483b–484a, 518b–519b
late eighteenth-century, **1**:311a–b, 312b, 314a–b; **6**:331b–333a, 469a; **7**:231b–232b, 490b, 580a, 598b; **9**:326a; **12**:430a; **15**:290b–291a, 343b–345b
nineteenth-century, **3**:470b–471a; **6**:9b; **7**:20b*; **8**:599b–600a; **12**:139b, 282b; **14**:343b; **15**:144b–145b
early twentieth-century, **3**:190b–191a; **4**:279a, 281a–282a; **5**:188a–b, 271b; **6**:188b, 529a–533a; **11**:393b; **12**:248b–249a, 250b, 280b, 449a; **13**:275a; **14**:304b; **15**:448b–449a
mid twentieth-century, **4**:340a*; **5**:273a; **7**:10b; **9**:405b–406a; **12**:222b; **13**:429b
alchemy and, **4**:209a–b
algebraic, **1**:312b, 314a
Aristotelian, **1**:198b–199a, 588a, 589a; **2**:410a; **5**:243b, 552b; **9**:362a–b
see also Aristotle
biology and, **1**:259a–b
Cartesian, **2**:329b; **6**:604a, 610a–b; **9**:26a
see also Descartes, René du Perron
Chinese, **12**:376a–b, 378a; **15**:746a
Copernicanism
Renaissance, **1**:468a; **2**:401b, 403b, 409b–410a, 411a, 414a, 540a, 541a; **4**:97b; **5**:400b, 401a; **9**:168b; **10**:245a–b; **11**:339a, 351b–352a, 395b–396b, 366a–b, 561b; **12**:106b; **13**:49a–b; **14**:359a
early seventeenth-century, **1**:126b, 447a, 567a, 588a–b; **2**:542b; **3**:115b; **4**:52a; **5**:239b–245a *passim,* 285a–288a *passim*; **6**:520a–b; **7**:289a–308a *passim;* **8**:28a; **9**:317a, 336a; **11**:411a; **12**:103a; **13**:433b; **14**:254a, 364b, 365a
late seventeenth-century, **1**:164b–165a; **2**:310b, 348b; **3**:103b, 596b; **5**:574b, 575b; **6**:604a, 611a–b; **14**:446b
eighteenth-century, **3**:105b; **4**:41a, 166a; **12**:408b–409a; **15**:429b
nineteenth-century, **1**:178b
forerunners of, **11**:250b–251b; **12**:358a
Galileo and, **1**:336b, 588a–590a; **2**:306b, 307b, 348b, 542b; **3**:115b, 116b; **4**:52a, 62a; **5**:242a–245a, 287b–288a; **9**:168b, 317b; **10**:491b; **13**:560b; **15**:69b
in Japan, **12**:408b–409a; **15**:736a–b, 741a
observational proof of, **2**:388a; **5**:241b; **10**:77a
religion and, **1**:336b, 588a–590a; **2**:540a; **3**:179b; **5**:242a–245a *passim,* 248a, 286b–287a; **7**:291b; **9**:317a; **14**:254a, 365a
see also Copernicus, Nicholas
geocentric theory, **1**:199a, 244b, 247b–248a, 258a, 588b; **3**:100b, 423b; **9**:77b; **10**:8a, 555b; **13**:326a; **15**:736b
and astrology, **3**:248a
see also Cosmology, Aristotelian; Cosmology, Ptolemaic
and geodesy, **2**:329b
geoheliocentric theory, **2**:408a–411b; **3**:15a; **9**:1b; **14**:538a; **15**:202b–204b
see also Cosmology, Tychonic
heliocentric theory
ancient Greek, **1**:222a, 246a–248b; **11**:188a, 224a,b; **12**:162b; **15**:204a
ancient Roman, **1**:336b; **9**:141a
Renaissance, **2**:543b; **5**:400b
seventeenth-century, **1**:590a; **2**:348b–349a; **12**:152a
eighteenth-century, **1**:311a; **15**:736a–b
and astrology, **3**:248a
observational proof of, **2**:388a
opposition to, **1**:126b, 468a; **2**:410a
precursors of, **10**:591a
see also Cosmology, Copernicanism; Cosmology, Keplerian; Cosmology, Newtonian
Indian, **1**:308a; **7**:338a, 582b; **10**:129a; **13**:582a; **15**:554a–555a
island universe theory, **3**:509a–b; **6**:529b; **12**:347b; **15**:144b–145a
Keplerian, **1**:311a; **2**:310b; **3**:457b; **5**:285a, 287b; **6**:514b; **7**:487b–488a; **9**:320b(n); **14**:364b, 446b
kinematic, **9**:405b
mechanistic, **1**:314a–b; **3**:583b; **4**:60a; **5**:114a
Mesopotamian, **14**:549a; **15**:638b–641b, 643b, 645a
Neoplatonic, **2**:234a
Newtonian, **2**:329b; **3**:281b; **4**:45b; **14**:296a
see also Newton, Isaac, cosmology
observational, **6**:332a
Ptolemaic
ancient Roman, **10**:298b–299a; **11**:188b–197b
Byzantine, **11**:161a
medieval Arabic, **1**:34b, 509a–510b; **2**:154b; **5**:616a–618b; **6**:197b–199b; **9**:160a; **12**:3a, 357b–360b, 362a; **14**:576b
medieval Latin, **1**:101a, 122a, 199a; **2**:20b, 24a; **3**:25a–26a, 218a; **5**:552b; **14**:35b; **15**:35a–b
Renaissance, **1**:179a, 589a; **2**:410a; **3**:311b, 423b; **9**:12b; **11**:366a; **12**:548a; **13**:49b; **14**:18b; **15**:154a, 475b
seventeenth-century, **1**:164b–165a; **2**:310b; **12**:500b; **14**:162a, 446b
opposition to, **1**:409a; **7**:293a; **8**:280b; **11**:251a; **12**:4a–b; **13**:511a; **15**:33a–35b
see also Ptolemy
relativistic, **7**:10b
religion and, **1**:248a; **4**:506a; **5**:242b, 244b, 286b–287a; **9**:406a; **13**:236a; **14**:518b–519b; **15**:638b–641b, 643b
Tychonic, **1**:246b, 312b, 424a; **2**:310b, 407a–411b, 413b; **3**:101a–b; **4**:166a; **5**:285a, 287b; **9**:336a; **14**:160a, 162a–b, 286a; **15**:203b–204a
see also Brahe, Tycho
see also Astronomy; Cosmogony; Planets; Universe
Cosmolysis
see Evolution, theories of
Cosmos, **12**:289a
Cossali, Pietro, **4**:612b
Cosserat, Eugène Maurice Pierre (1866–1931), 3:428–429; **4**:230b, 232a; **7**:446b*
Cosserat, François, **3**:428a; **4**:230b, 232a
Cossists
see Algebra, cossist
Cossus, **8**:580a
Costa ibn Luca
see Qusṭā ibn Lūqā
Costa, Christovão da
see Acosta, Cristóbal
Costa, Oronzo Gabriele, **11**:380a
Costa Rica
discovery of, **15**:89b
geography, **5**:214b
geology, **5**:214a
Costantin, Julien Noël (1857–1936), 3:429; **9**:176b*; **13**:406a*
Costantini, Giuseppe, **9**:534a
Coste, Jacques-Marie-Cyprien-Victor, **7**:31b; **14**:113a
Coste, Pierre, **3**:380b
Coster, Dirk (1889–1950), 3:429–430; **2**:247a; **6**:366a; **7**:469a; **9**:262a; **12**:464a; **13**:546b
Costus, **7**:58b
Cosyns, Max, **10**:597b
Cotangents
Arabic use of, **1**:511b–512a
tables of, **13**:536b; **14**:577a
Cotelle, Sonia, **3**:502b
Cotes, Roger (1682–1716), 3:430–433; **5**:516a; **6**:35b, 36a, 41a,b; **12**:477a
analysis, **4**:477a; **13**:68b
Canonotechnia, **2**:462b
Harmonia mensurarum, **3**:432a
and Newton, **3**:430b, 431a; **10**:67b, 85a,b
Newton-Cotes formula, **3**:432b
optics, **9**:225b; **12**:477b(n)
theorem of, **3**:432b; **6**:324a
Cotta, Carl Bernhard von (1808–1879), 3:433–435; **2**:563b; **5**:162a, 548b; **6**:13b, 457a; **9**:620b*; **10**:168b; **11**:209a

Cotte, Louis (1740–1815), 3:435–436
Cottereau-Duclos, Samuel, **2**:353a
Cotterill, James Henry, **3**:118a
Cotton, Aimé Auguste, **3**:4a; **11**:477a; **14**: 243b
Cotton's phenomenon, **9**:616b
Cotton, **13**:354b
diseases of, **4**:220a
growing of, **5**:523b
insects, **13**:456a
oil of, **11**:121a
Cottonseed oil, **15**:2b
Cottrell, Frederick Gardner (1877–1948), 3: 436–437; 15:467a(n)
patents, **14**:121a, 393a
Cotugno, Domenico Felice Antonio (1736–1822), 3:437–438; 12:137b
Cotyledons, **2**:492a; **3**:422a; **4**:511a; **5**:217a; **6**:9a
Couder, André, **9**:46a
Coué, Émile, **2**:35b
Couerbe, J. P., **4**:246a; **10**:498a
Coues, Elliott (1842–1899), 3:438–439; 1: 331b, 485b
Coughing
convulsive, **14**:407b
reflex nature of, **6**:60a; **9**:571a
Coulomb, Charles Augustin (1736–1806), 3: 439–447; 1:582a; **8**:209b
and Académie des sciences, **3**:71a; **4**:15b; **7**:569a; **8**:79a; **13**:183b; **15**:336a
acquaintances of, **5**:271b; **10**:145b
on contemporary scientists, **1**:68a
Coulomb's law, **2**:43b, 87b, 166b; **8**:510b; **9**:205b; **14**:74a
electricity, **4**:537a; **10**:184b; **13**:375a–376a, 377a
hydraulics, **11**:164b
influence of, **2**:300a; **4**:538b; **6**:178b; **10**: 190b; **13**:76a; **15**:360b
magnetism, **1**:129a, 146a; **10**:184b
mechanics, **3**:145b, 147b; **4**:222b; **6**:452a; **10**:3b; **11**:496b
torsion balance, **4**:379a; **9**:370b
Coulon, Louis (Abbé), **4**:348a
Coulson, John, **7**:163a
Coulter, John Merle, **3**:30b; **14**:192b
Coumarin, **3**:347b
synthesis of, **10**:516b
Councilman, William Thomas (1854–1933), 3:447–448; 9:142b; **14**:249b
Counter-Reformation, **7**:293b, 294b, 300a, 305b
Couper, Archibald Scott (1831–1892), 3: 448–450; 1:168a; **14**:530a
chemistry, **2**:515a, 622a, 623a; **5**:374a
critics of, **14**:531b
Couple (physics), **11**:62a
Couplet, Pierre, **9**:90a
Counting board, **3**:254a, 256a, 267a; **8**:315a; **14**:540b–541b, 542a, 543a–b
Courant, Richard, **3**:135b; **5**:350b; **14**:282a; **15**:42a
Methoden der mathematischen Physik, **12**:220a
Cournot, Antoine-Augustin (1801–1877), 3: 450–454; **3**:379a; **7**:105a; **10**:486a; **13**:128a; **15**:31a, 423a, 482b, 483a–b*
Cournot's point, **3**:451a
Traité de la théorie des fonctions, **2**:350a*
Courrier, Robert, **2**:345b
Courtivron, Gaspard le Compasseur de Créquy-Montfort, Marquis de (1715–1785), 3:454–455
Traité d'optique, **9**:267a
Courtois, Bernard (1777–1838), 3:455; 1: 143a; **5**:322b, 602a
Courtois, Jean-Baptiste, **5**:602a
Courtship, animal, **8**:415b
Cousin, Victor, **2**:545a; **6**:81a; **13**:254a; **15**: 481a, 483b
Coutchiching Series (geol.), **8**:99a
Couthouy, Joseph Pitty, **3**:549b, 550a; **5**: 478a
Couturat, Louis (1868–1914), 3:455–457; 8:156b; **11**:95a; **14**:379a(n)
Covariance
analysis of, **5**:9a
Covariants, **1**:294b; **3**:164a, 164b–165a, 169b; **4**:342b; **8**:542b
theory of, **12**:86b
Cow
ovary, **5**:484b
Cowdry, Edmund Vincent, **4**:155a; **15**:74a
Cowell, Philip Herbert (1870–1949), 3:457; 2:516b; **3**:472a
Cowell's method, **3**:457a; **7**:17b
Cowley, Abraham, **12**:580b–581a, 584a
Cowling, Thomas George, **9**:405a; **13**:117a
Cowper, William, **15**:5b
The Anatomy of Humane Bodies . . . , **15**:29a–b
Cowper's glands, **9**:323b
Myotomia reformata, **10**:501a; **13**:527b
Cowpox, **3**:463b; **5**:317a; **6**:73a; **12**:57a–b
and smallpox, **7**:96b–97a
transmission of, **7**:96b; **11**:175a
virus, **7**:96b; **10**:390a; **12**:492b
Cox, John, **12**:27a
Coxe, William, **11**:272a
Coxwell, Henry, **5**:143a
Cozymase, **6**:111a; **9**:359b
chemical nature of, **4**:486a
role of, **4**:485b; **14**:45b
CP conservation, **7**:617b
Crab
autotomy, **5**:149b
biometrics, **14**:251b
reflexes in, **7**:516b
reproduction, **13**:355a–b
Crabtree, William (1610–1644), 3:457–458; 5:279a(n); **6**:514a,b, 515b; **12**:342b
manuscripts of, **13**:445a
Crace-Calvert, Frederick, **11**:536b; **12**:478b
Cracking, chemical, **7**:21b; **9**:608b
theories of, **4**:553b
thermal, **2**:4a
Cracovians (math.), **1**:429a–b
Cracow, University of
geology, **15**:502b
mineralogy, **15**:526b
Cracow Academy of Sciences, **9**:388a
Crafts, James Mason, **3**:292a; **7**:552a
Craig, John (*fl. ca.* 1590), **9**:611a
Craig, John (*d.* 1731), 3:458–459; 1:475a; **5**:520b; **10**:87b(n)
interpolation lemma, **5**:351a
Craig, Lyman, **15**:252b
Craigie, John Hubert, **2**:583b; **6**:298b
Cramer, Carl, **9**:600a
Cramer, Gabriel (1704–1752), 3:459–462; 1:591a; **2**:576a, 577a; **3**:167a; **5**:567a; **8**:259a
algebra, **13**:571b
correspondence of, **3**:216a, 282a, 283a, 286a*
Cramer's paradox, **3**:460a–461b *passim*; **11**:45b
Cramer's rule, **3**:460a, 460b–461a; **5**: 151b; **7**:566b
problem of, **3**:39b*; **5**:209b
Cramer, Johann Andreas (1710–1777), 15: 93–94; 3:473b
Cramp, **5**:157b*
Cranach, Lucas, **3**:414a
Cranch, John, **13**:355a
Crane, Charles Richard, **8**:356a
Crane (machine), **6**:312a, 313a; **8**:211a; **15**:518b
Crane (zool.)
anatomy of, **3**:342b
breeding, **9**:441a
Craniology
early nineteenth-century, **12**:578b
mid nineteenth-century, **1**:388a; **2**:478a–b, 617b; **5**:571a; **6**:138a, 454a; **9**:540b; **11**:178a, 381a, 514b; **12**:87b; **13**:396b, 404a
late nineteenth-century, **5**:46b; **8**:125b; **10**:16b; **11**:235a, 383a; **13**:504a; **14**: 42a
twentieth-century, **2**:211b; **6**:528a*
Craniosclerosis, **6**:574a*
Cranioscopy, **5**:254a,b
Craniotomy
in Renaissance, **1**:618b
Cranston, John Arnold, **12**:505b
Crantor of Soli, **14**:535a–b
Crassulaceae, **3**:44b
Crataegomespilus, **5**:624b
Craters of elevation, **1**:20a; **2**:554b; **3**:552a; **4**:349b–350a; **8**:573a–b; **10**:209b; **11**:336b; **12**:263a
Crates of Mallos, **9**:1b
Crateuas, **4**:120b, 121a
Cratylus, **6**:290b; **11**:25b
Crawford, Adair (1748–1795), 15:94–96; 2: 483b; **3**:541a; **5**:215b; **6**:495b; **8**:37b; **9**:5b; **14**:75b; **15**:363a
Crawford, Albert Cornelius, **1**:10b
Crayfish, **11**:503a
neurophysiology, **8**:534a
Creatine, **4**:400a; **9**:284b
metabolism of, **15**:418a
Creatinine, **4**:400a
discovery of, **10**:557a
metabolism of, **15**:418a
Creation
ancient Roman views of, **1**:334a, 336a–337a
Byzantine views of, **12**:441b
medieval views of, **1**:197a, 565a; **4**:255a–b; **5**:549b; **9**:29b, 30a; **11**:324b–325a; **13**:340a–b; **14**:388b
eighteenth-century views of, **8**:257a–b
nineteenth-century views of, **10**:322b; **11**:308b; **12**:244b
continuous, **5**:296a; **7**:604a; **10**:322b
of earth, **1**:393a; **2**:556a, 568b, 613a–b; **4**:28a; **6**:581b, 587a; **9**:48a
of man, **2**:21a, 321a, 568b; **3**:392b; **6**: 454a–b; **9**:48b; **10**:195b; **14**:114a, 296a
Mesopotamian views of, **15**:638b
separate (successive) creations, **1**:73b; **2**:172b, 498a; **3**:526a, 553a, 567a; **4**:309a; **5**:32a, 356b, 357b; **6**:454b; **10**:222a; **12**:318a–b
of space and time, **2**:13b–14a
of species, **3**:552b–553a; **5**:67b, 68a; **6**: 266a; **14**:138a
of universe, **1**:549b; **2**:21a; **3**:14b, 527b; **6**:306a; **7**:134b, 135a–b, 231b–232a; **12**:441b; **14**:388b; **15**:249a
continual, **3**:479a; **8**:617b–618a
see also Cosmogony
Credner, (Karl) Hermann Georg (1841–1913), 3:462; **7**:473b; **10**:502a; **12**: 182b*; **13**:64a; **14**:183a
Creep (phys.), **4**:229b
Creeth, James Michael, **5**:589b
Cregut, Friedrich Christian, **2**:3a
Creighton, Charles (1847–1927), 3:463–464; 7:97b*

Creighton, Harriet, 14:432b
Crell, Lorenz Florenz Friedrich von (1745–1816), 3:464–466; 2:174a; 5:215b, 532a; 7:395a*; 12:149b*; 13:281a
Crelle, August Leopold (1780–1855), 3:466–467; 3:302a; 4:341a, 501b; 5:304b, 471b; 13:15a
and Abel, 1:14b–17b *passim*
Crelle's Journal, 1:14b–17b *passim;* 2:299a; 3:54a, 466a–b; 4:125b, 341b, 342a; 5:203b, 367b; 7:51a; 13:15a
editors of, 6:76b, 286b; 7:506a–b, 507a
Cremation, 9:456b
Crémieu, Victor, 4:306a
Cremona, Antonio Luigi Gaudenzio Giuseppe (1830–1903), 3:467–469; 3:117a, 496b; 12:469b*; 13:132b
acquaintances of, 2:469b, 470a; 3:178a; 7:506b
biographies of, 8:505a
Castigliano-Menabrea controversy, 3:118b
correspondence of, 12:173a*; 13:623a
Cremona transformation, 2:85b; 3:323a, 467b, 468b; 4:523a; 5:569a; 7:164a; 8:505a; 11:536a; 12:330b; 13:133a
critics of, 12:284b
geometry, 2:85b; 3:164b, 167b; 7:163b; 13:20a,b
influence of, 10:139b, 140a
students of, 2:85a; 5:568b
Cremona, Gerard of
see Gerard of Cremona
Cremonini, Cesare, 5:239b; 6:150b, 152a
Crenic acid, 9:558a–b
Creosote, 11:359a
Crepidula, 3:389b, 390a
Crepitations, 7:557a
Crescas, Hasdai (*ca.* 1340–1412), 3:469–470
Cretaceous period, 3:392a; 4:401a; 5:519a; 10:204a
deformation, 5:476a; 7:251b; 10:507a
paleobotany, 12:104b, 260a, 339b
paleogeography, 6:216b; 7:145a, 250b; 13:147b
paleontology, 1:471b–472a, 491b; 2:110a, 491b–492a; 3:608a; 4:147b; 5:214a, 437a, 500b; 6:221a; 7:43b; 9:87a, 100b, 134a,b, 540b; 10:29b, 32b, 40b, 221b, 222a; 11:386a, 500a–b, 501b–502a; 13:525a; 14:313a, 409b; 15:91b, 92a, 151b, 152a
stratigraphy, 1:131a, 210a; 2:89b, 108b, 341b; 3:433b; 4:262b, 540b; 5:14a–b, 155b, 335a; 6:186b, 187a, 228b–229a; 9:87a, 240b, 579a; 10:128a, 428a, 428b–429a, 584a; 13:87a, 581b; 14:210b–211a
Crete, 15:723b
flora, 1:125a
Cretinism, 3:113a; 6:518b; 14:115a
and goiter, 10:307a
Crèvecoeur, Michel Guillaume St. Jean de (called J. Hector), 1:488a; 13:497a
Crichton, Alexander, 7:378b
Crichton-Browne, James, 4:593b, 594a
Crick, Francis, 5:140b–141a, 273a, 589b; 12:223a; 15:18a
Cricket, 14:300a
chirps of, 5:199a
Crimea
geography, 10:283b
geology, 1:161b–163b; 13:620b
paleontology, 4:68a
Crimean Astrophysical Observatory, 12:368b
Crimean War, 10:620b; 12:425a; 15:24b
Criminality
inheritability of, 3:590b
Crinoidea, 3:89a; 11:113b; 12:107a; 13:355a
anatomy of, 2:109b
paleontology of, 2:109b, 128b; 6:58a*; 9:346a; 13:467a–b; 14:312b
Cristobalite, 4:563b
Cristol
see Christol
Critical days, medical doctrine of, 5:231a–b; 10:123b*
Critical state
see Matter, states of, critical states
Criticism
aesthetic, 11:140a
biblical, 6:448a
historical
medieval Latin, 1:1b
literary
ancient Greek, 4:392b
textual, 1:2b; 5:402a; 11:140a, 141a
Croatia
meteorology, 9:443b
seismology, 9:443b–444b
Crocetin, 7:518a
Crocodile, 4:147b; 5:357b
physiology of, 5:281b
Croconic acid, 5:431b
Croft, William Noble, 8:6a
Crofton, Morgan, 13:221a
Croll, James (1821–1890), 3:470–471; 7:186a, 386b*
Croll, Max, 11:440a
Crollius, Oswald (*ca.* 1560–1609), 3:471–472; 1:126b; 4:388a; 6:146a; 8:310a; 10:310b; 12:307a
Cro-Magnon, 8:44b; 11:483b–484a
Crommelin, Andrew Claude de la Cherois (1865–1939), 3:472–473; 3:457a; 4:176b*; 11:19a*
comet of, 3:472b
Cromwell, Oliver, 9:310a, 529b; 10:200b, 471a*, 565a; 12:580b; 13:213b; 14:153a, 361a, 370a
Cronstedt, Axel Fredrik (1722–1765), 3:473–474; 2:490a; 14:132a
chemistry, 2:421b; 7:389a
Försök til Mineralogie
translations of, 9:5a,b, 478b*; 14:258b
mineral analysis, 2:5a–b
Crooke, Helkiah
Microcosmographia, 1:523b; 8:54a
Crookes, William (1832–1919), 3:474–482; 6:343b, 461b, 541a; 9:287b; 10:517a*; 13:356b; 14:629b
cathode rays, 3:599b; 5:458b; 6:439b; 8:180b; 13:365b
chemistry, 6:565a; 9:110a, 290b; 11:279a–b
Crookes dark space, 1:320b; 3:477b
Crookes tube, 3:477b, 498b; 11:530b–531b
radioactivity, 4:80b, 355b, 356a; 12:27b–28a
radiometer, 9:212b, 224a
spiritualism, 6:541a
uranium X, 1:560a
Croom, Hardy Bryan, 3:196b
Croone, William (1633–1684), 3:482–483; 13:170b
De ratione motus musculorum, 13:32b
Crops
intermediate, 6:237b
rotation of, 2:356b; 3:45a; 8:346a
Cross, Charles Whitman (1854–1949), 3:483–484; 4:367a*, 564b; 7:5a–b; 8:41b–42a; 14:183b
Cross ratio, 3:213a,b, 214a; 13:5a, 6a
Cross staff (astron. instrum.), 1:573b; 2:133a; 3:262a, 423b; 4:168b; 5:344a, 593b; 7:419a; 8:280b, 281b; 14:273b
Brahe's, 2:402a
Crossbow, 6:312b; 8:213b, 214a; 12:383a
Crosthwaite, Peter, 3:538a, 540a
Crotalaria
alkaloids, 15:2b
Crotonaldehyde
formation of, 7:282b
Crotonic acid
separation from croton oil, 3:159b; 10:497b
stereochemistry of, 1:340b
Crotti, Francesco, 3:118a,b
Croup, 2:445a; 4:404a
Crousaz, Jean-Pierre de (1663–1750), 3:484–486; 7:442b; 10:527b
critics of, 12:118a; 15:57a*
Crowfoot, Dorothy, 4:522a
Crown gall disease, 12:468a
Crucible
Gooch crucible, 3:292b
Cruciferae, 2:519b; 3:44b; 8:354a; 9:253b; 13:154b
Cruickshank, William (chem.), 5:603a; 15:84a
Cruikshank, William Cumberland (1745–1800), 3:486–488; 3:67b; 6:569a,b
museum of, 3:488a
Crum, Walter (1796–1867), 3:488–489; 2:514b; 13:374a*
Crum Brown, Alexander
see Brown, Alexander Crum
Cruquius, Nicolas, 2:551b
Crusades, 5:147a; 7:131a; 10:533a
Crusca, Accademia della
see Accademia della Crusca
Crusius, Christian August, 7:597b; 8:160a
Crustacea, 2:491b; 7:184a; 9:328b; 12:560b; 13:65b; 15:513b
anatomy, 1:264b, 329a; 9:408a–b; 12:479a–b
microscopic, 4:207a
circulation in, 4:12a
daily vertical migration of, 15:158b
development of, 9:560b; 15:123a
distribution of, 12:479b
embryology, 14:234a
hearing of, 8:529a
histology, 8:302b
larval stages, 2:502a
metamorphism, 13:355a–b
morphology, 3:148b; 12:131a
paleontology, 2:496a; 9:346a
parasitic, 5:385b; 9:112b
physiology, 1:329a
regeneration in, 11:333b
sessile, 8:3b
taxonomy, 3:552b–553a; 8:48b–49a; 9:575b; 10:273b; 11:67a; 12:479a–b; 13:8b–9a; 14:1b
Cruveilhier, Edouard, 3:490a
Cruveilhier, Jean (1791–1874), 3:489–491; 2:574b
Cruz, Oswaldo Gonçalves (1872–1917), 15:96–98; 3:185a,b
Cryogenics, 1:304a; 4:79a–80b; 8:365b–366a, 482a–483a; 13:454b
see also Low-temperature studies
Cryohydrates, 5:587b
Cryolite, 13:358b
Cryopedology, 2:549b
Cryoscopy, 4:131a; 5:149b; 7:65b
Cryptanalysis, 6:402a; 8:579b; 14:150b–151a
Cryptococcus, 9:176a
Cryptogams, 2:160a, 371a–b, 425b; 3:337a; 4:560a; 5:218a, 441a; 6:492b–493b;

7:442a*; **9**:78b, 92b; **13**:449a, 554a; **14**:463b
analogies with phanerogams, **3**:29b; **6**: 219a
anatomy, **5**:539b
developmental history, **11**:152b
fossil, **14**:600a
growth of, **9**:600b
life cycle, **6**:219a
morphology, **12**:185a
reproduction, **5**:582a; **6**:218b–219b, 465b–466a; **8**:5a–b; **11**:152a, 153a–154b; **13**:463a
sexual organs, **4**:98b
sexuality in, **3**:337b; **5**:217a, 539b
spores of, **3**:180a; **5**:217a
subterranean, **6**:549b
taxonomy, **4**:98b–99b; **6**:66a; **11**:154a; **13**:178a; **15**:148a
tropical, **3**:30a
vascular, **1**:412b; **2**:371a–b
see also individual species and groups of cryptogams
Cryptography, **6**:402a; **10**:589a; **11**:97a; **14**:18a, 290a–b, 365b, 368b
Cryptology, **14**:365b
Cryptorchism, **2**:344b
Crysanilic acid, **5**:197b
Crystallization, **3**:556b
cause of
eighteenth-century theories, **2**:6b, 81a; **9**:533a; **15**:54a
nineteenth-century theories, **8**:374a
twentieth-century theories, **5**:557b
and cell formation, **12**:242b
dynamical theory of, **3**:232b
experimental study of, **4**:141b; **8**:113b, 373b, 520a; **9**:379b; **10**:496b
force of, **4**:141b
forced vs. spontaneous, **8**:520a
fractional, **5**:394b, 475a; **13**:546a
as fundamental process, **7**:603b, 604a
laws of, **13**:245a; **14**:58b
and organic growth, **5**:194a–195a; **15**: 53b–54b
metamorphic, **1**:551b
as purifying technique, **3**:317a*
of steel, **3**:234a
water of, **5**:186a; **8**:69b, 73b; **13**:420a
see also Crystals, growth of; Recrystallization
Crystallography
Renaissance, **5**:379b
seventeenth-century, **6**:486b; **13**:34a
early eighteenth-century, **1**:411a; **8**:379a; **15**:53b–54b
late eighteenth-century, **2**:6a–b, 81a, 555a; **3**:61b; **7**:494a; **8**:520a; **11**:520b–523a; **14**:258a
early nineteenth-century, **1**:141b; **2**:106b, 135a–b, 355a–b, 452b, 482a, 494a; **3**:291b; **4**:234a; **6**:18b, 178a–182b, 358a–b; **9**:423a–424b, 448a–b, 620a; **10**:27a, 109b–110a, 586a; **14**:240a–241b, 490a–491a
mid nineteenth-century, **1**:518b; **2**:431a–b, 441a, 471a, 587a; **5**:124a–b, 203a, 305a; **7**:463a, 464a; **8**:59a, 507b; **9**: 286b, 392b–393a; **10**:357a–361a; **11**:235b, 539a–b; **12**:304a–306b, 543a; **14**:108b; **15**:114b–115a
late nineteenth-century, **1**:460b–462a, 528b–529a; **2**:254b; **3**:504b–505b; **5**:211a–213b, 455a–456a, 556b–557b; **8**:148b; **9**:58b–59b, 379a–b; **12**:195b–196a, 511b–512a, 520a; **13**:616b–617b; **14**:61b–62a, 525a–b
early twentieth-century, **1**:288b, 552a; **3**:400b; **4**:82b, 597b; **5**:186a–187a, 457a–b, 610b; **7**:59a–b; **9**:182b–183b, 341a, 379b–380a; **10**:124b–125b; **11**:85a–86b; **13**:517a–518a; **14**:525b–526a; **15**:17a–b, 251b
mid twentieth-century, **4**:521b–522b; **6**: 303a–304a; **11**:266b–267a
as analytic tool, **9**:423b
Beudant's law, **2**:106b
chemical, **5**:213a, 557a–b; **9**:183b; **10**: 125b, 578b; **11**:270b–271a
definition of purpose, **5**:455b–456a
discontinuity principle, **6**:181a
geometry and, **1**:591a; **5**:212a; **11**:235b; **12**:195b–196a
law of decrement, **6**:179b–181a
law of rational indices, **6**:180a
low-temperature techniques, **4**:522b
mathematical, **14**:293a
of metals, **1**:581a; **3**:87a, 619b; **4**:522b; **11**:486a, 539b, 549a; **13**:8a, 246a; **15**:63b
optical
early nineteenth-century, **6**:324b; **9**: 424b
mid nineteenth-century, **1**:358a; **6**:19a; **9**:566a; **12**:304a–305b, 519a; **13**:238b
late nineteenth-century, **1**:558b; **4**: 66a–b, 189b–190b; **9**:58a–59a; **13**: 517b
early twentieth-century, **2**:466b; **5**: 186a–187a; **8**:41b–42b; **9**:183a
mid twentieth-century, **6**:614a; **14**:515b
Becke line effect, **1**:154a, 551b
universal instrument, **5**:212a–b, 557a
terminology, **1**:463a*
X-ray, **1**:319a–b, 461a; **2**:399a–b; **3**:370a, 430a, 619b; **4**:82b, 196a, 521b–522a; **5**:139b–141b, 186a, 186b–187a, 212a, 457a, 557b; **6**:546a, 560b; **7**:274a, 469b; **8**:484b–485b; **9**:183a–b, 226b, 376b–377a; **13**:227a; **14**:525b; **15**: 17b, 49b, 61a–64a, 461a
discovery of, **8**:51a–b
divergent-beam, **8**:485b
see also Goniometry; Historiography, of crystallography; Morphology, crystallographic; Notation, crystallographic; Symmetry, crystallographic; Terminology, crystallographic
Crystalloids
naming of, **5**:494b
Crystals
classification of, **5**:455b; **6**:358a; **8**:56b, 288b; **11**:521b; **14**:240a–241a
cleavage of, **1**:461a; **6**:179b; **11**:522a; **15**:114b
corrosion of, **5**:187a
dissolving of, **5**:390b, 456a
electrical properties, **6**:96a–b
actinoelectricity, **6**:96b; **7**:526b; **13**:80b
conductivity, **2**:427b; **12**:305b; **14**:71b; **15**:251b
electromagnetism, **2**:466a–b
piezoelectricity, **1**:557a; **3**:504b–505b; **4**:231a; **6**:96b, 178b, 546a; **7**:526b; **10**:605a; **11**:530a,b
pyroelectricity, **1**:461a; **3**:504a–b, 505a; **4**:228b, 231a; **5**:497a; **6**:178b; **11**: 530a,b
thermoelectricity, **6**:96b; **7**:526b
electron bombardment of, **3**:597b
etch figures, **1**:528b; **14**:491a
excitation in, **5**:160b
fluid cavities, **10**:110a
growth of, **1**:461a; **3**:505a; **7**:469b
overgrowth, **5**:124a, 186b, 456a
seeding, **8**:520a
theory of, **5**:187a; **9**:379b, 380a; **13**: 245a; **14**:525b
see also Crystallization
heat, effects of, **4**:66b; **8**:485b
liquid inclusions in, **12**:543a
melting point, **8**:368a
migration in, **5**:161a
optical properties, **9**:379b; **13**:217a; **15**: 83b
absorption, **1**:558b; **5**:160b; **6**:19a
binormal angle, **12**:304b–305a
birefringence, **1**:461a; **6**:178b, 324b
and composition, **14**:440b, 588b
conical refraction, **6**:88a–b; **8**:426a
diffraction, **4**:196b
fluorescence, **1**:559a; **4**:356a
optical activity, **2**:137b–139b; **5**:170a; **6**:303b; **10**:357a–358b; **12**:304a–b; **14**:525a
phosphorescence, **1**:558b
vs. physical properties, **4**:190a–b
polarization, **4**:66a; **6**:293b, 324b; **9**:59a
scattering, **7**:622a; **9**:76b
spectra, **11**:266b
total reflection, **12**:305a
variability of constants, **4**:190b
X-ray effects, **2**:398b, 399a; **3**:564a, 619b; **6**:303a; **9**:543a; **12**:527b; **15**: 251b
physical properties, **9**:59a; **14**:599a
cohesion, **5**:124a
and composition, **14**:440b
directional dependency of, **12**:304a
elastic constants, **8**:485b; **14**:61b–62a
energy ranges, **3**:430a
forces within, **5**:455b–456a
hardness, **5**:124a, 457b
magnetic, **8**:485a–b; **11**:46b, 47a; **13**: 521b
oscillations, **15**:40b
strength, **12**:464b
and structure, **12**:305b
surface energy, **14**:525b
thermal, **9**:59b; **11**:530a,b
powders, **13**:247a
structure analysis of, **3**:619b
quantum mechanics of, **13**:240b
slipping of, **13**:246a,b
structure of, **5**:457a–b
atomic, **2**:431b; **4**:82b; **5**:295a, 557b; **9**:183a–b; **15**:40b–41a, 62b–63b
axiality, **2**:139a; **14**:240a–241a
and chemical composition, **5**:556b, 557b; **9**:286b, 423b–424b; **11**:86b; **13**:517b, 617b; **15**:526a
classification of, **6**:303b
constancy of angles, **13**:34a
cubic, **1**:460b–461a; **14**:490b–491a
curved faces, **5**:187a
Debye-Scherrer method, **3**:619b
deformation of, **5**:151a, 161a, 556b, 557a
dimorphism, **2**:135b; **9**:424a; **10**:357b; **12**:306a; **13**:311a
dissymmetry, **2**:471a; **10**:357a–358b
and electrical properties, **6**:96b
electron distribution, **6**:304a
and energetics, **13**:617b
gliding, **13**:617a
hemihedrism, **2**:431a; **3**:504b; **5**:187a; **10**:357b–358b; **14**:240b, 241a; **15**: 114b–115a
hemimorphism, **6**:180a; **9**:620a
holohedral, **5**:187a
interatomic distances, **4**:196a
interfacial angles, **6**:486b; **11**:521b, 523a

Crystals (cont.)
structure of (cont.)
isodimorphism, **5**:124a
isomorphism
see Isomorphism (cryst.)
macromolecular, **13**:2b–3a
mapping of, **5**:455b; **9**:393a
models of, **5**:161a, 455b; **8**:520a
molecular orientation, **8**:485b
monoclinic, **9**:620a
morphotropy, **5**:556b
orthorhombic, **5**:124a
packing of atoms, **1**:460b–461a; **12**: 520a
plesiomorphism, **15**:115a
point system, **14**:343b
polymorphism, **1**:461a; **6**:177b–178a, 179a, 182a
primary forms, **6**:179a
real vs. ideal, **10**:125a–b
rhombic, **5**:124a
right- or left-hand configurations, **3**: 430a
secondary, **11**:521b
superlattices, **13**:245b
syngony, **1**:461a
tetartohedrism, **9**:620a
tetragonal, **9**:620a
theory of, **5**:211a,b; **6**:178a–181b, 303a–b; **9**:379b; **11**:86b, 521a–522b; **12**: 464b; **14**:293a, 488a, 490a–491a, 525a,b; **15**:61b–63b, 114b–115a
variation of angles, **14**:241a
zones, **5**:455b; **9**:393a; **10**:27a; **14**:241a
systems of, **2**:441a; **9**:448a–b, 620a; **14**: 240a–241a
twinning, **5**:186a, 456a; **9**:58b
theory of, **1**:461a; **5**:161a, 186b; **9**:59a
types of
anisotropic, **9**:59a
biological, **4**:522a
liquid, **2**:600b; **4**:230b; **5**:186b; **8**:148a–b; **9**:182b, 183a; **10**:236a; **14**:525b
metallic, **2**:459b; **5**:610b; **7**:470a
mixed, **5**:556b; **6**:179a, 181b–182a; **11**:534b; **13**:245a,b; **14**:58b, 588a–b
monocrystals, **3**:505a
negative, **5**:187a
organic, **14**:440b
two-element, **1**:460b
vibration in, **10**:125b
see also Goniometry; Lattices; Morphology, crystallographic; specific chemicals and minerals; Symmetry, crystallographic
Ctenophores, **5**:51a; **6**:167b
embryogeny, **4**:187a; **7**:475a; **9**:517a
Ctesias, **1**:259a, 263a,b, 265b; **6**:419b, 420b
Ctesibius (Ktesibios) (*fl.* 270 B.C.), 3:491–492; 6:310b
catapult, **10**:586b, 587a, 588b
influences on, **13**:93a, 95a
pneumatics, **10**:588b
pump, **15**:519a
water clock, **15**:518b, 520b
Cuba
botany, **12**:326b
discovery of, **15**:88b
geography, **6**:551b; **10**:617a
Cubature, **4**:506a, 571a–b; **5**:526b–527a; **6**: 602a; **10**:103b; **11**:487a–b; **12**:75b, 460a; **13**:418a
of sphere, **12**:292a
Cube
duplication of
ancient Greek discussion, **1**:232a–b; **4**:105a, 389a, 489a,b, 609a–b; **6**:407a, 410b–411b, 415a; **8**:192b; **9**:270a–275a, 300a–b; **10**:114b–115a, 587a–588a; **11**:28a; **15**:116b, 118a
ancient Roman discussion, **12**:579b–580a; **15**:518b
medieval Arabic discussion, **1**:41b; **9**: 556b
Renaissance discussion, **2**:618a; **4**:260a; **8**:238a–239a; **10**:161b, 494b; **14**: 21a,b, 274b
early seventeenth-century discussion, **6**:112b, 449b; **7**:269b; **12**:460a; **14**: 49b, 152a
mechanical solution, **4**:391b–392a, 466b; **9**:274a–275a
volume of, **3**:277a
Cube law, **3**:269b
Cubics
twisted, **3**:468a
see also Curves, cubic; Equations, cubic
Cubit, **14**:29b; **15**:221a(n)
Cubitt, William, **12**:131b
Cuckoo, **1**:263b; **14**:299b
breeding habits, **7**:96a
Cucurbitaceae, **1**:396a; **3**:44b
Cudworth, Ralph (1617–1688), 3:492; 2: 612b; **9**:98a; **11**:318a; **14**:378b(n), 379a(n)
philosophy, **5**:427a; **7**:539a; **15**:54a
Cuénot, Lucien (1866–1951), 3:492–494; 1: 505b; **3**:121a; **9**:151b*
Cuff, John, **8**:328a
Culex, **4**:620a; **9**:82a
Cullen, Glenn, **13**:574b
Cullen, Thomas Stephen, **15**:64b, 65a
Cullen, William (1710–1790), 3:494–495; 14:322b
acquaintances of, **6**:568b; **7**:11b; **9**:484a; **11**:48b, 499a; **14**:320b
and Black, **2**:173b–181a *passim*
on contemporary scientists, **8**:297b
correspondence of, **6**:569a,b
diagrams of chemical reactions, **2**:7b, 180b
First Lines of the Practice of Physic
translations of, **10**:612a
influence of, **3**:464b, 579b; **8**:73a; **11**:616b, 617a; **12**:354a; **14**:132a
influences on, **6**:458a; **13**:215a
Materia medica
translations of, **6**:18a
medicine, **10**:612b
students of, **10**:582a; **13**:448a
Culmann, Karl (1821–1881), 3:495–497; 5: 502a; **9**:446b; **11**:390a
Culpeper, Nicholas, **7**:164b
Culture
classification of, **2**:208b
complementarity in, **2**:253a,b
concept of, **2**:209b
diffusion of, **2**:209a–b, 210b; **4**:353b; **9**: 490a; **11**:309a; **14**:506b
and environment, **11**:309a–b
evolution of, **4**:34a; **8**:528a–b; **11**:6a; **14**:506b
and human nature, **6**:447b
and language, **2**:208b, 210a; **4**:34a; **15**: 635a
philosophy of, **3**:112a; **9**:385a
and race, **2**:208b, 211a–b; **11**:137b–138a
Cuming, Hugh, **11**:130a
Cumming, James (1777–1861), 3:497
Cumulenes, **4**:553b
Cunaeus, Andreas, **9**:596a
Cuneiform, **10**:117a; **15**:635a, 650b, 667a,b
Cunha, José Anastácio da (1744–1787), 15: 98–99
Cuningham, James, **10**:559a
Cunitz, Martin, **7**:308a
Cunningham, David Douglas, **8**:296b
Cunningham, Ebenezer, **2**:144b, 560a*
Cunninghamella, **9**:176a
Cunoniaceae, **2**:519b
Cupellation, **4**:393b; **13**:327a, 411b
Cupola
Viviani's problem, **5**:499b, 521b; **7**:185b*; **9**:300a; **12**:55a; **14**:49b
volume of, **13**:290a
Cupping, **1**:315a; **10**:418b
Cupping glass, **5**:402b
Curabelle, Jacques, **4**:48b
Curare, **2**:31a, 111a; **5**:56b; **10**:499b; **11**:135b; **14**:187a–b
analysis of, **10**:498a
neurological effects, **8**:7a, 16b; **13**:558a
Curcuma, **7**:471a
Curcumin, **7**:471a
Curie, Eugène, **7**:157b
Curie, Eve, **3**:502a
Curie, Irène
see Joliot-Curie, Irène
Curie, Jacques, **3**:503b, 504a–505a
Curie, Marie (Maria Sklodowska) (1867–1934), 3:497–503; 4:194b
acquaintances of, **7**:152a,b; **8**:10a; **9**:292b; **14**:32a
atomic weight, **6**:15a, 481b
and Becquerel, **1**:559b, 560a–b
influence of, **4**:599a; **9**:606b
letters of, **4**:197b*
radiation, **4**:356a
radioactivity, **3**:503b, 504b, 506b–507b; **7**:159b*; **12**:26b, 28a
theory of, **4**:355b
thorium, **1**:559b; **2**:258b; **12**:191b
students of, **2**:259b; **4**:253a; **7**:157b–158a
Curie, Pierre (1859–1906), 3:503–508; 3: 498a–502a *passim;* **4**:194b; **7**:263b; **8**:9b, 10a; **9**:182b, 292b, 606b; **12**: 70a*; **13**:546a; **14**:32a
and Becquerel, **1**:599b, 560a–b; **2**:203a*
crystallography, **14**:525b
Curie law, **14**:244b
Curie point, **3**:506b; **7**:493a; **14**:243b, 245b
Curie-Weiss law, **3**:4a; **14**:245b
Curie-Wulff principle, **14**:525b
magnetism, **8**:11a; **14**:244b
radioactivity, **1**:559b; **4**:80b; **12**:26b, 28a, 34a
Curie (phys. unit), **3**:501a
Curie Foundation, **3**:502a
Curietherapy, **3**:502a
Currency, **9**:485b, 486b; **12**:500a; **15**:665a(n)
bimetallic parity, **7**:319a
conversion tables, **10**:238b
decimal, **13**:359b; **15**:334b–335a
exchanging, **7**:342a–b
floating exchange rate, **7**:318a
gold standard, **7**:317b, 318a
supply of, **12**:372b
unforgeable, **3**:388b; **13**:413b
velocity of, **10**:566a
Currents, electrical
alternating, **5**:597b; **6**:504a–b; **13**:11b, 24a
effects of, **1**:304a; **14**:32b–33a
high frequency, **1**:304a; **2**:427b; **13**: 287a, 473a
intensity of, **2**:200b
measurement of, **14**:206b
penetration depth, **8**:476a
self-induction, **14**:32b–33a
attenuation of, **8**:502a–b
branching law of, **8**:188b
conduction currents, **4**:305b
convection currents, **4**:305b
definition of, **10**:191a
diaphragm, **11**:241b

displacement currents, 4:305b–306a; 7:381a; 9:209a, 212a
distribution of, 7:380b–381a
effects of
cold, 8:188a
heat, 1:304a–b, 555b; 7:181a–182a; 8:188a; 11:581b; 14:32b–33a
physical deformation, 1:303b–304a
electrodynamic unit, 14:206b
harmonic analysis of, 3:329a
higher-order, 6:278b
induced, 9:19a
intensity of, 11:581b
interaction of, 1:143b–146a; 7:380b; 14:207a–208a; 15:193a
laws of, 10:188b
in magnetic fields, 6:51b; 15:433a
mathematics of, 7:288b
measurement of, 14:290a
open, 6:251a–252a, 344a–b
oscillating
velocity of, 7:381a
saturation current, 13:80b
stationary, 7:380b
theory of, 7:380b–381a
thermoelectric, 9:19a
vs. hydroelectric, 10:134b–135a
transversal waves, 7:381a
variable, 7:380b
see also Conduction, electrical; Electricity, as power source; Electrodynamics; Galvanism; Induction, electrical
Curtin, John, 5:41b
Curtis, Heber Doust (1872–1942), 3:508–509; 3:510a; 6:30b; 12:455b
and Shapley, 6:529b; 8:582b; 12:347b–348a
Curtis, Moses Ashley, 5:512b
Curtis, William, 1:88b; 6:521a
Curtis's Botanical Magazine, 6:494b
editors of, 13:342b
Curtis's Flora Londinensis, 6:494b
Curtiss, Glenn, 8:20a
Curtiss, Ralph Hamilton (1880–1929), 3:510
Curtius, Jacobus, 13:622a
Curtius, Theodor (1857–1928), 3:510–512; 2:560a,b; 5:99b
chemistry, 2:562b; 5:4b
Curtius reactions, 3:510b–511a, 511b
Curtz, Albert
Historia coelestis, 1:481a
Curtze, E. L. W. Maximilian (1837–1903), 3:512
Grundzüge der allgemeinen Theorie der Oberflächen in synthetischer Behandlung, 3:467b
historiography, 2:430a; 3:59a; 4:158a*, 163b*; 6:545a*; 7:120b(n), 127a*; 8:281b*; 10:154a(n), 154b*, 541b(n); 11:352a(n); 14:276b*, 458a; 15:177a,b, 178a
editions of works, 4:431a,b; 7:177a*,b*; 12:63a*
Curvature, 3:460a,b; 4:336a; 7:573b; 10:52a
Bernoulli's problem, 3:459a
center of, 2:215b; 8:164a, 301a; 13:266b
Gaussian, 5:376a,b
geodesic, 2:288a; 8:385b
concept of, 9:409b–410a
integral, 6:496b
lines of, 4:257b; 9:471a; 14:231b
concept of, 9:476a
mean
concept of, 5:376a–b
minimal, 9:127a
principal, 6:497b
radius of, 2:47b, 48a, 52a; 5:209b; 6:600b; 9:504b; 13:266b–267a
relative, 8:385b
of rods, 15:164b
of surfaces, 4:102b, 479b; 5:376a–b; 11:455a
of thin plates, 15:164b
total, 8:385b
vector, 8:389b
Curved figures
area of, 1:90b, 214b, 215a
Curves
algebraic, 3:428b*, 460a–461a, 560a; 5:521a; 6:537a, 547a–b; 7:108b; 8:166a, 250a–b, 505a; 9:358a; 10:553b; 11:54b–55a, 509b, 512b–513a; 12:213b; 13:20a–b, 480b
classification of, 3:460b; 6:75b; 11:600a*; 12:86b
foci, 7:573b; 11:45b
general formula, 6:75b
generation of, 2:47b; 15:193a, 195b, 198a–b
intersection of, 4:475b; 10:140a–141a; 11:45b
multiple points, 11:450b
rank of, 11:55a–b
representation of, 11:53a,b, 54a
singularities of, 2:465a; 3:315a, 468b; 5:567a, 569a; 6:75b, 547b; 11:45b
theory of, 1:245b; 5:567a; 7:163b–164a; 10:139b–141a; 11:45a–b, 390a; 14:336b
three-dimensional, 2:465a; 6:75a
transformation of, 10:275a; 12:284b
anallagmatic, 7:573b
associated, 7:446b
beams of, 7:163b–164a
of Bertrand, 2:88a
cardioid, 3:119b
Cassini's, 9:55a
Cesàro's, 3:178b
characteristic, 9:476b
characterization of, 12:196a
classification of, 1:591a; 4:568b; 5:368a; 13:319a(n)
concept of, 7:72a
of constant descent
see Isochrone
contact, 13:418a
with corners, 3:62b
cubic, 1:76a–b; 3:555a; 8:610a; 13:67b
class equation, 12:171a
classification of, 2:49b; 3:167a; 5:567a; 10:52a; 13:68b
construction of, 8:250b; 11:45a
generation of, 3:282a; 12:225b
quadrature of, 9:599b
theory of, 3:461a; 4:479b; 6:357a; 13:219b
cyclic, 7:185a
definition of, 11:394b
derivation of, 2:40b
of double curvature, 2:88a; 3:281a, 282a, 467b; 5:158a, 569a; 6:547b; 7:615b*; 9:300a, 470b; 12:74a; 13:418a
elastic, 2:47b, 48a, 50a; 4:480b
elementary, 7:185a
elliptic, 3:315a; 6:571b
equations of, 11:400b
evolutes of, 2:47b; 7:615a; 9:475b
families of, 13:437a
fields of, 3:62b, 63a
flower-shaped, 5:499b; 11:402a
fourth-order
see Curves, quartic
frequency
division of, 2:616a
gauche
see Curves, of double curvature
generation of, 1:474b; 9:186a
higher-order, 3:183b
hyperelliptic, 11:548b
"imaginary," 4:56a
integral, 9:476b; 11:56a–57a
intersection of, 3:461a; 5:368b; 7:577a; 10:52b
irrational, 3:460b
irregular, 3:460b
isoperimetric, 2:48a
isotomous, 1:65b
isotropic, 13:125a
Jordan, 2:274b, 513a; 4:174a, 562a; 7:72a; 13:562a, 599b
lengths of, 3:432a, 459a; 8:111b; 10:443a; 11:487b; 12:247a
definition of, 7:167b
logarithmic, 3:184a, 459a; 5:499b, 528a; 8:162a, 163b
rectification of, 2:49b; 8:505a
mechanical, 3:460b
multiple points of, 8:611b
multivalued, 12:118a
pairs of
generation of, 2:53a
paradoxical, 9:300a–b
pedal, 13:19a
plane, 4:107b, 412a*
algebraic, 3:315a, 399a; 8:585a; 11:45b
cubic, 1:294b, 591a; 11:45a; 12:87a
points of inflection, 6:357b
quartic, 1:294b; 7:163b; 11:45b
systems of, 5:569a
theory of, 3:167a, 468a; 7:163b
transformations of, 3:467b, 468a–b
universal, 9:249b
polar, 5:368b
projection of, 7:164a
quadratic, 8:592a
theory of, 2:418b; 4:479b; 13:18b–19a
quartic, 3:103b, 167b, 314b, 315a, 460a; 4:516a; 7:185a, 573b; 8:250b, 610a; 10:115b
of quickest descent
see Brachistochrone
rational
theories of, 9:358a
rectifying surface of, 7:615a
regular, 3:460b
resolution, 7:408a
rolling, 9:199b
singularities of, 5:569a
sixth-degree, 11:510a
source book of, 2:479b
space, 6:2a; 9:475b, 476a; 10:244b; 11:22a; 12:246b; 13:6a, 418a
cubic, 12:225b, 228a
quartic, 12:225b
theory of, 4:479b; 5:158a; 7:408a, 615a; 8:385b; 12:319a
transformation of, 3:468a
space-filling, 10:442a
spiric, 10:529a–530a
squarable, 6:305a
Steiner, 13:20a
surface, 10:544b
theory of, 2:88b; 4:47a; 5:567a; 7:185a, 621a–b; 8:61b; 11:45a–b
four-vertex theorem, 7:408a
topological characterization of, 9:249a
transcendental, 3:460b; 4:57b; 5:567a; 8:162a, 505a; 10:42b*; 12:150a
transformations of, 3:460b; 5:527a
uniformization, 6:394a
uni-rational, 8:542b
W, 7:397a
see also Brachistochrone, Caustics, Conchoids, Conics, Elastica, Isoperimetry, Lunules, Quadratics, Spiral, *Versiera*

Cusa, Nicholas (*ca.* **1401–1464), 3:512–516; 1**:80a, 158b; **3**:362b; **4**:170a; **6**:46a(n), 256b; **8**:157a, 550a; **13**:62a(n), 440a,b; **14**:390b; **15**:473b–474a
De circuli quadratura, **2**:360b; **15**:474a
cosmology, **7**:488a
influences on, **11**:162a
instruments of, **1**:97a; **6**:147a*
De mathematicis complementis, **1**:229a
Opera omnia, **6**:274a
Perfectio mathematica, **12**:501a
Reparatio calendarii, **1**:383a
De staticis experimentis, **6**:37b
Cusanus, Nicolaus
see Cusa, Nicholas
Cuscuta, **15**:130b
Cushing, Harvey Williams (1869–1939), 3: 516–519; 3:447b, 490a; **5**:207b, 208a, 463b; **9**:586a; **11**:482a; **14**:250b*
Cushing's disease, **3**:518b
Cushing's solution, **3**:518a
endocrinology, **12**:474a
historiography, **5**:269a*, 348b*; **14**:12a*
Cushing, Henry Kirke, **3**:516b
Cushman, Joseph Augustine (1881–1949), 3:519–520
Cushny, Arthur Robertson (1866–1926), 15: 99–104; 5:281b; **8**:452b; **14**:495b
Cusp (math.), **3**:167a
Custer, George Armstrong, **14**:443a
Cut (math.), **5**:350b
Cuthbertson, John, **9**:152a; **10**:446a
Cuticula, **14**:350a
Cutler, Sir John, **6**:483a,b
Cutleriaceae, **12**:125a
Cutlery industry, **4**:86b
Cuts and fills (geometry), **1**:194b; **4**:257b; **9**:471a, 475b–476a
Cuvier, Frédéric (1773–1838), 3:520–521
Cuvier, Georges (1769–1832), 3:521–528; 1:328b; **3**:520b, 521a,b; **7**:603a; **8**: 265b; **10**:518b*; **11**:243b; **12**:321b
acquaintances of, **2**:566b–570b *passim;* **3**:43b, 324a, 545b; **4**:234b, 235a; **5**: 218a; **6**:18b, 268b, 454a; **7**:366b, 367a, 368a, 603b; **8**:43a, 564b; **9**: 252b, 570b; **10**:260b; **11**:242b, 243a; **12**:130b, 523a; **13**:554a; **15**:50b, 126a, 127a, 152a
The Animal Kingdom, **2**:205a; **4**:68a
anthropology, **2**:341b; **3**:264a,b; **5**:46b
on Banks, **1**:436b
and Brongniart, **2**:493b, 494a; **3**:525a, 526a
catastrophism, **1**:413b; **2**:495a, 496a; **5**: 32b, 519a; **6**:10a, 454b, 456b
on contemporary scientists, **2**:323a*; **3**: 173a*, 561b*; **4**:28a, 73a*, 241b(n); **5**:225b*, 251b, 379b; **7**:547a, 604b*; **8**:88a*, 177b; **10**:195b, 284b, 510a*; **11**:273a*; **12**:122a; **13**:597b*; **15**:112a
correspondence of, **12**:536b
critics of, **9**:346a; **10**:209b; **11**:133b, 134a, 234b, 593a; **12**:167b, 262a; **13**:460b; **15**:126b
Discours préliminaire, **2**:567b, 568a
Essai sur la géographie minéralogique des environs de Paris, **2**:481a
Essay on the Theory of the Earth
translations of, **7**:70a
and Geoffroy Saint-Hilaire, **3**:523a–b, 526a–527a; **5**:356a–359a *passim;* **6**: 10a; **8**:49a, 534b
geology, **1**:413b; **2**:568a; **4**:365a; **5**:533a, 578b; **8**:565a; **14**:211a
Histoire des poissons, **11**:243a; **13**:554b
historiography, **14**:35b
influence of, **2**:502a; **3**:395b; **5**:371a; **8**: 594b; **9**:101a, 408a, 573b; **10**:9a, 263a, 323a; **11**:133b, 379b; **12**:216a, 316a; **15**:51a, 379b
and Lamarck, **3**:522b, 523a, 526a–527a; **7**:592b, 593a
Leçons d'anatomie comparée, **15**:126a
translations of, **9**:252b
paleontology, **7**:60a; **9**:87a, 152b; **10**:322b; **15**:51a, 472a
phylogeny, **1**:351b, 387a; **2**:187a, 282a, 492a,b, 498a; **3**:378a; **4**:290b, 291a; **5**:32a; **6**:120a, 593a; **7**:461a, 475a, 588b; **8**:49a, 254a, 270a; **9**:27b; **10**: 261a; **13**:355b; **15**:168b
Recherches sur les ossements fossiles, **4**: 549a; **5**:357b
Le règne animal, **11**:234a
students of, **1**:72b, 73a; **2**:356b; **5**:44a,b; **7**:59b, 414b; **10**:209a; **12**:317b; **13**: 402a; **14**:113a, 351a; **15**:226b, 471b
on voltaic pile, **2**:136b
Cyanamide, **3**:46a
discovery of, **12**:321b
Cyanates, **5**:315a; **14**:474a–475a, 476a
and fulminates, **8**:331a
Cyanic acid, **8**:331a, 332b; **13**:468a
formula, **14**:476a
Cyanides, **5**:125b
double, **2**:587b
poisoning by
treatment of, **9**:296a
preparation of, **5**:104a, 431b; **10**:499a
Cyanin, **14**:412a
Cyanines
heat and light, effects of, **13**:598a
Cyanogen, **3**:176b; **6**:177b; **13**:468a
compounds of, **5**:315a, 324a
derivatives of, **14**:530b
discovery of, **5**:323b–324a
Cyanohydrin, **8**:32a
Cyanometer, **1**:202a; **13**:409a
Cyanophyceae, **5**:584b
Cyanuric acid, **8**:331a, 332b, 340b
discovery of, **12**:321b
formula for, **3**:449a; **8**:339a,b
Cybernetics, **2**:33a, 325b; **5**:65b; **9**:227a; **11**:546a; **12**:575a; **13**:227a; **14**:347a; **15**:504a
precursors of, **15**:39a
Cycadaceae, **3**:187b; **9**:417b
Cycadofilicales, **12**:260a; **14**:600a
Cycads, **2**:492b; **3**:187b; **12**:260a; **13**:343a
phylogeny, **3**:188a
Cycas, **5**:539b
Cyclides, **3**:560a; **4**:257a,b; **12**:284b
Cyclobutane
derivatives of, **10**:517b
Cycloheptane compounds, **2**:562b
Cyclohexane, **14**:601b–602a
Cycloid, **1**:425a; **2**:48a, 53b, 361a; **3**:64a–b, 183b; **5**:527a; **6**:598a; **7**:577a, 583b; **8**:304a; **9**:320b(n); **10**:103b, 336a–337a, 342a; **11**:404a; **12**:460a; **13**: 435a; **14**:148b, 149b
length of, **9**:599b
normal to, **4**:58a
quadrature of, **2**:48b, 54a; **4**:58a, 506a; **8**:161b; **11**:487a; **13**:434b
rectification of, **10**:337a; **11**:487b; **14**:509b
solid, **9**:14a
tangents to, **13**:436b
tautochronism of, **6**:605a–b, 607a, 608a
Cyclones, **2**:147a; **4**:555a; **8**:278b; **10**:176b; **11**:390a
theories of, **6**:98a–b; **8**:487a; **9**:107b, 108a, 155a; **11**:557b; **14**:338a, 353b
thermal structure of, **6**:98a
Cyclopentane, **14**:601b–602a
Cyclopropane
derivatives of, **10**:517b
Cyclops, **9**:575b
Cyclostomata, **11**:380a; **12**:337a; **15**:126b
anatomy, **9**:572b, 573a
Cyclotomy, **4**:341b, 342a; **7**:506a, 508a
Cyclotron, **2**:338b; **3**:329b, 330a; **4**:582a, 583a; **7**:155a, 527a; **9**:262b; **12**:33b; **13**:163b
invention of, **8**:94b–95a
Cygnus (constellation)
alpha Cygni, **10**:30a
parallax of, **2**:468b
novae in, **1**:598a; **8**:465a; **13**:114b; **14**:55b
61 Cygni, **6**:49b
orbit of, **10**:542a
parallax of, **2**:101a; **6**:263b; **13**:120b
variable stars, **11**:542a
Y Cygni, **4**:251a
Cylinder
area of, **1**:227b
quadrature of, **11**:487a
sections of, **13**:291a
and sphere, **1**:213a–b, 219b, 220b–221b
transformation of, **14**:274b
volume of, **1**:227b; **8**:421b; **15**:670b
Cylindrite, **5**:162a
Cymograph, **14**:404a
Cynareae, **10**:575b
Cynthia
embryogeny, **3**:389b–390a
Cyon, Élie de, **2**:27a, 531a; **10**:431a
Cyperaceae, **10**:13b
Cyprian, Abraham, **10**:25b
Cyprinidae, **3**:588b
Cyriaque de Mangin, Clément
see Henrion, Denis
Cyrtandreae, **2**:521a
Cysat, Johann Baptist (*ca.* 1586–1657), 3: 528–529
Cyst(s), **3**:490a
hydatid, **14**:584b
ovarian, **1**:323a
Cysticerci, **1**:602b
Cysticercosis, **10**:527a
Cystids, **2**:128b
Cystine, **4**:79a
discovery of, **14**:492a
sulfur in, **1**:518b
Cystitis, **4**:404b
Cystoidea, **1**:494b
Cystoliths, **9**:79a
Cystopus, **3**:612b
Cystoseira, **12**:125b
Cystotomy
supra-pubic, **8**:402a
see also Lithotomy
Cytoblast, **12**:174a
Cytoblastem, **6**:270a
Cytochemistry
nineteenth-century, **9**:418b, 442a; **11**: 300b
early twentieth-century, **4**:604a; **6**:339a–b
Cytochrome
role of, **7**:273a–b
Cytogenetics, **3**:318a; **9**:565a; **10**:276a–277a
Cytologia
editors of, **8**:589b
Cytology
late eighteenth-century, **5**:56b
early nineteenth-century, **5**:195a; **13**:506b
mid nineteenth-century, **3**:337a–b; **5**: 469b–470a; **6**:266a, 434b–435a; **7**: 438b, 440a; **8**:302a–b; **9**:441b–442a; **11**:300b–301a, 367b–369b; **12**:230b–232b, 241b–243b; **13**:542a, 556a–b; **14**:113a, 350a
late nineteenth-century, **1**:421b; **2**:361b, 362a–364a, 625b–627b; **3**:29b, 422a;

5:34b–36a, 51a–52b, 461a, 581b–582a; **6**:267a, 337b–338b; **11**:255a–b, 382b, 383a; **12**:192b–193b; **14**:126b, 234a, 235a–b, 236a, 425a, 427a–430b; **15**:513b
twentieth-century, **1**:545a–b; **2**:345a, 364a–b, 456a–b; **3**:259b; **4**:132b, 207a; **5**:43a, 293b–294a, 453b, 594a–b, 598b; **6**:337a; **7**:455a–456a, 516b; **8**:282b, 586b–589b; **9**:176a, 495b–497a, 518b–519a, 521b, 522a, 524a; **11**:465b; **12**:423b–424a; **13**:136a, 156b–158a, 281b–282a; **14**:430b–434a; **15**:473a
bacterial, **5**:30a, 584b
botanical
nineteenth-century, **3**:187b; **4**:131a, 545b–546a; **6**:136b, 465b, 467a; **9**:600a–601b; **10**:1b, 256b; **11**:152b–153a; **12**:174a–b, 592a; **13**:88a–89b, 459a, 595b
twentieth-century, **5**:584b–585a; **6**:121b; **7**:91a; **9**:24b–25a, 34b; **15**:473a, 508b
chemical
see Cytochemistry
comparative, **10**:1b
and embryology, **2**:161b; **8**:357a
exfoliative, **10**:291a–292a
and heredity, **14**:431a–433a
philosophy of, **14**:429a–b
techniques, **9**:497a; **13**:88b
see also Cells (biol.), Microbiology
Cytophysiology, **7**:519b, 520a
nuclear variations, method of, **9**:24b–25a
Cytoplasm
and differentiation, **14**:428b–429a
and heredity, **2**:363b, 364a; **3**:422b; **14**:433a
inclusions in, **5**:584b–585a; **14**:433a
Balbiani vesicle, **1**:418a
naming of, **13**:88b
and nuclear control, **2**:364a–b; **14**:430b, 433a
nucleic acids in, **4**:603b–604a
organelles, **15**:473a
as physical-chemical system, **2**:627a
reserves in, **3**:259b
volume of, **6**:337a
Cytosine, **4**:604a
discovery of, **7**:466b
Cyzicus
mathematical school, **9**:269a
Czajewicz, Aleksander, **4**:84a
Czapski, Siegfried, **1**:8a; **5**:212b
Czechoslovakia
Academy of Sciences, **6**:371a
chemistry, **6**:371a, 374b–375b
German occupation of, **6**:371a
science, **11**:217a
Czekanowski, Aleksander Piotr (1833–1876), 3:529–530; 3:531b, 532a
Czermak, Johann Nepomuk (1828–1873), 3:530–531; 11:135b, 215b, 595b; **13**:493b
Czermak lines, **3**:530b
Czermak spaces, **3**:530b
Czerski, Jan (1845–1892), 3:531–532; 3:530a

D

Dadoxylon, **12**:260a
Daedalus hyperboreus, **13**:179a
Daffodil
chemical analysis of, **3**:159a
Dagelet, Joseph, **5**:480a
D'Agostino, Oscar, **4**:580a
Daguerre, Louis Jacques Mandé, **1**:203a; **5**:18b; **6**:120b; **10**:119a; **14**:492a
Daguerreotypy, **4**:178b, 182a, 283a*; **5**:18b, 84b; **6**:517a; **12**:436b; **13**:23a, 238b; **14**:348b
in astronomy, **2**:285a, 285b, 616b; **4**:179a; **5**:86a; **12**:266b
in physics, **12**:132a
D. A. H. (deranged action of the heart), **8**:295a
Dainelli, Giotto (1878–1968), 3:532–533
Dairy science, **1**:357a; **5**:225b–226a; **6**:135b; **7**:100b; **14**:45b
see also Cheese, Milk
al-Dakhwār, Muhadhdhab al-Dīn ʿAbd al-Raḥīm ibn ʿAlī, **9**:602b
Dakin, Henry, **3**:91b
Daldy, Thomas Mee, **1**:60b
Dale, Henry Hallett (1875–1968), 15:104–107; 4:298b; **5**:81a, 291a; **8**:451a,b; **12**:492a; **15**:73b
on contemporary scientists, **4**:302a*, 305a*; **5**:84b*; **8**:18a(n), 453a,b, 454a,b; **9**:21a*; **12**:35b*; **15**:73b, 418a,b, 419b
pharmacology, **4**:133a*; **15**:11a
Dale, Samuel, **3**:129a
Dale, Thomas Pelham, **5**:412b; **7**:620a
Daléchamps, Jacques (Jacobus Dalechampius) (1513–1588), 3:533–534; 11:527b
Historia plantarum generalis, **1**:525b, 526b; **11**:312a
D'Alembert, Jean le Rond
see Alembert, Jean le Rond d'
Dalencé, Joachim (*ca.* 1640–1707), 3:534–535; 10:596b
Dalgarno, George, **8**:153b; **14**:379b(n)
Dalibard, Thomas François (1703–1779), 3:535; 4:129a; **5**:133b, 134b; **10**:146b
Dall, William Healey (1845–1927), 3:535–537; 1:406a; **3**:392b*; **5**:214b*, 407a*, 478b; **8**:104b*; **12**:134a*; **13**:66a*
Dalton, John (1766–1844), 3:537–547; 5:216a, 492b
acquaintances of, **5**:71b, 518a; **6**:283b, 285a; **9**:424b
atomic theory, **1**:344a,b, 345a, 346a; **2**:93a; **3**:47a, 539b–540a, 541b–544b; **5**:320a; **6**:384a, 386a; **8**:38a; **9**:531a; **11**:170a–171a, 172b; **13**:373a; **14**:488a, 489a
biographies of, **6**:285b; **11**:538a; **12**:478b
and Charles' law, **3**:208a; **5**:318a
chemical combination, **4**:243a–b; **5**:323b; **12**:619b
evaporation, **8**:256a
heat, **13**:10b; **15**:95b, 363a
influence of, **7**:181a; **11**:344a
law of partial pressures, **2**:588a; **3**:540b–542b; **4**:28b; **6**:285a; **9**:222a; **10**:108b; **15**:377b
library of, **3**:546b
meteorology, **3**:540a–541a; **4**:410a; **6**:585b
natural philosophy, **3**:537a–540a
A New System of Chemical Philosophy, **11**:170b
notebooks of, **6**:111b
organic compounds, **2**:420b; **6**:284b
students of, **6**:452a; **7**:180a
table of vapor pressures, **3**:80b
vision, **9**:200a; **14**:564b
water, **6**:495b
Dalton, John Call (1825–1889), 15:107–110; 2:27a; **6**:77a,b
Dalton, Jonathan, **3**:537a, 538a, 539b, 546b
Daltonide compounds, **7**:529b
Dal Vesco, Ezio, **10**:126b
Daly, Reginald Aldworth (1871–1957), 3:547–548; 2:369b; **3**:595b*; **4**:261b; **7**:62b; **12**:127a*
on atolls, **3**:569a
Damas, Désiré, **6**:442a
Damascius, **1**:137a, 279b; **4**:159b, 462b; **10**:6a; **12**:440b, 442a
Damascus
latitude of, **15**:259b
observatory, **14**:537b
Damascus steel
see Steel, Damascus
Damianos
see Heliodorus of Larissa
Damigeron, **4**:121b
al-Damīrī, Muḥammad ibn Mūsā (1341–1405), 3:548–549
Damocrates, **11**:602a,b
Dāmodara, **15**:614b
commentaries of, **2**:116a, 416b; **12**:599a
students of, **1**:48b; **10**:129a; **15**:622a
Damour, Augustin-Alexis, **4**:66b
Damps
see Mines, damps from
Dams
overflow, **3**:116a
siting of, **14**:621a
stability of, **10**:469a*
stress in, **15**:165a
Dana, James Dwight (1813–1895), 3:549–554; 1:364b; **14**:392b
acquaintances of, **3**:297b, 462b; **6**:57b; **12**:433b; **14**:516b, 517b
collections of, **13**:65b
on contemporary scientists, **5**:600a; **12**:435a
critics of, **9**:100b
crystallography, **14**:491a
and Darwin, **3**:551a–b, 553a
and Emmons, **4**:364b
geology, **6**:58a, 565a; **7**:185b; **13**:146b
influence of, **10**:504b; **11**:505b
Manual of Geology, **4**:364b
mineralogy, **5**:556b; **12**:306a
students of, **1**:405a; **6**:13b; **7**:370a
System of Mineralogy, **2**:440b; **9**:379b
Dandelin, Germinal Pierre (1794–1847), 3:554–555; 8:432b; **11**:236a
theorem of, **3**:554b
transformations, **3**:468a
Dandy, James Edgar, **2**:476a
Dandy, Walter Edward, **4**:286b
Danforth, Charles Haskell (1883–1969), 3:555–556; 13:269b*
Dangeard, Pierre Augustin, **5**:584a; **9**:34b
Dangschat, Gerda, **5**:6a
Daniel of Morley, **11**:504a; **15**:173b, 174a, 175b
Daniell, John Frederic (1790–1845), 3:556–558; 4:410a; **9**:391a
crystallography, **14**:491a
Daniell cell, **2**:439a; **3**:334a, 557a; **4**:19a; **5**:292a, 559b
on Deluc, **4**:28b
electrolysis, **6**:439a; **9**:391b
Meteorological Essays and Observations, **4**:28b
Daniels, Farrington, **11**:418a; **13**:429b
Danielssen, Daniel Cornelius, **6**:101b
Danilevskii, Nikolai, **12**:300a
Danilov, Vladimir Vladimirovich, **13**:111b
Danjon, André, **3**:298b
Dankov, Pavel Davydovich, **13**:241a
Danne, Jacques, **3**:508a

Dante Alighieri, **1**:107a; **4**:542b, 605a; **5**: 238b; **6**:327a; **9**:81a, 140b; **10**:529b; **15**:35a
Divina Commedia, **9**:549a*; **11**:202b; **12**:428b
Danti, Egnatio (Pellegrino Rainaldi) (1536–1586), 3:558–559; 3:100b; **13**:440b
Danti, Giovanni Battista, **3**:558a
Danti, Pier Vincenzio, **3**:558a
Danti, Vincenzio, **3**:558a
Dantic
see Bosc, Louis Augustin Guillaume
Danube (river), **9**:135b; **10**:505b
control of, **13**:144b
Danzig
botanical garden, **7**:401a
Gymnasium, **7**:268b–269a
Institute for Experimental Physics, **7**: 469b, 470a
Daphnia, **9**:575b
reproduction, **8**:528a; **9**:575b
vision, **8**:529a
Darbishire, Arthur Dukinfield, **3**:121a; **4**: 546a; **6**:21b
Darboux, Jean-Gaston (1842–1917), 3:559–560; 1:407a; **6**:394b; **7**:446b*, 572a*; **8**:382b, 383a
acquaintances of, **8**:323b; **13**:249b
analysis, **4**:174a; **9**:410a
on Cayley, **3**:163a
on contemporary scientists, **5**:482a; **6**: 307a; **11**:61a*; **14**:231b
geometry, **3**:178b; **4**:336a; **12**:284b
influence of, **2**:302b; **3**:96a, 428a; **4**:335b; **5**:481b; **7**:446a
influences on, **2**:350a; **3**:428b; **10**:545a
Leçons sur la théorie générale des surfaces, **2**:88b; **12**:285a
Leçons sur les systèmes orthogonaux et les coordonnées curvilignes, **4**:287b
students of, **1**:560b; **3**:178a; **9**:382a
surfaces of, **4**:338a*
D'Arcet, Jean (1725–1801), 3:560–561; 9: 471a–b; **10**:496a; **11**:563a; **12**:64b; **13**:565b, 597a
chemistry, **11**:564b; **12**:65a
D'Arcet, Jean Pierre Joseph, **3**:202a; **8**:113b
Darcy, Henri
Darcy's law, **9**:257b
D'Arcy (or D'Arci), Patrick (1725–1779), 3: 561–562; 15:310a
chemistry, **12**:64a
principle of areas, **15**:289b, 301b, 339a, 343b
Dareste, Camille, **8**:254b
Darlington, William (1782–1863), 3:562–563; 1:485a; **11**:218b
Darlu, Alphonse, **2**:544b
Darmois, Georges, **3**:221a
Darmstädter, Ludwig, **4**:298b
Dart, Raymond, **2**:505a
Dartmouth College
mathematics at, **14**:559a
Darton, Nelson Horatio, **7**:62b
Dartos, **5**:586a
Darwin, Charles (*d.* 1778), **3**:578a; **14**:492b
Darwin, Charles Galton (1887–1962), 3:563–565; 5:103a; **6**:148a*; **12**:29b; **13**: 119a; **15**:43b
crystallography, **8**:51b; **9**:542b–543a; **15**:63a
translations of works, **12**:298a
Darwin, Charles Robert (1809–1882), 3:565–577; 2:288b, 564a; **3**:581a; **5**:265b; **8**:169b, 378a; **10**:278a, 287b; **14**:234a
acquaintances of, **1**:562b; **7**:498b; **8**:27a, 528a; **10**:241a, 261b; **11**:517a; **12**: 258a, 277a; **13**:352b; **14**:96b
and Bates, **1**:501b, 503b
Beagle voyage, **5**:16a–17a, 77a; **10**:261b
biographies of, **7**:279a, 402b; **9**:86b*
botany, **2**:599a; **3**:575a–576a; **5**:513b; **6**: 6b, 489a, 491b; **7**:409b, 417b, 442a; **12**:588a–b
and Brown, **2**:517b, 521a
and Carpenter, **3**:88a,b
on contemporary scientists, **11**:571a; **12**:570a
correspondence of, **2**:376b; **9**:560b; **11**: 518a; **12**:105a*; **13**:10a, 504a; **14**:97b, 190a–b, 238b, 534a; **15**:78a,b, 123b
critics of, **1**:71a; **2**:426b; **3**:422a; **4**:504b; **6**:563b; **7**:93b; **9**:589b; **12**:59b; **14**: 96b, 434a
and Dana, **3**:551a–b, 553a
The Descent of Man, **8**:395a
The Different Forms of Flowers on Plants of the Same Species, **5**:513b
ecology, **5**:70a, 71a; **6**:594b
editions of works, **7**:480a
The Effects of Cross and Self Fertilization in the Vegetable Kingdom, **9**:561a; **13**:477b
evolution theory, **1**:71b, 130a, 173a; **3**: 42b, 527b, 569b–575a; **6**:8a; **7**:99a, 475a, 498b, 593a; **8**:105a, 572b–575a *passim;* **10**:433a; **11**:20a; **12**:338a; **14**:138a, 236b, 433b–434b; **15**:507b
corroboration of, **1**:501b, 502a–b, 503b–504a; **2**:617b; **6**:490a–b; **10**:29b
influences on, **2**:205b–206a; **9**:69a; **10**: 284b; **14**:253b, 299b
Expression of Emotions in Man and Animals, **1**:491b
and Fitzroy, **5**:16a–17a
forerunners of, **1**:351b; **2**:257a; **3**:192b, 434b; **5**:359b; **6**:466b; **7**:320a; **8**:98a; **11**:137a, 262b; **12**:607a
genetics, **5**:218b, 266b; **6**:9a, 23a
geology, **1**:73b; **3**:210b, 550b–551b, 566b–569b, 591a; **5**:335b; **7**:185b, 186a; **8**:570b; **14**:398a
and Gray, **5**:513a–b
and Henslow, **6**:289a
and Herbert, **6**:296a–b
and Hooker, **6**:489b, 490a–b
and Huxley, **6**:592b
ill health, **3**:566a–b
influence of, **3**:120b, 322b; **4**:266b; **5**:290b; **7**:320a; **10**:240a; **12**:52a; **14**:98b, 99a, 133b; **15**:139b, 144a, 408b, 415b
influences on, **12**:588a–b; **14**:190a
and Jenkin, **7**:93b
letters of, **3**:582a
and Lyell, **8**:570b, 572b–575a *passim*
The Movements and Habits of Climbing Plants, **14**:97b
On the Origin of Species, **2**:161b, 206a, 289a; **3**:298a, 553a, 573b–574a, 579b; **4**:368a, 504a; **5**:68a, 192a, 296a, 357b; **6**:222b, 280b, 489a, 510a; **9**: 282b–283a; **12**:588a; **13**:416a; **14**:95b, 138b, 173b, 516b, 534a; **15**:123a
defenders of, **6**:7b, 490b, 592a–594a; **14**:233a
opposition to, **1**:387a; **5**:17a, 46b; **10**: 262b
reviews of, **1**:211a; **3**:88b; **6**:592b
translations of, **2**:498a
sociology, **5**:178b
translations of works, **15**:78a–b, 507a
Variation of Animals and Plants Under Domestication, **8**:395a
and Wallace, **5**:513a; **14**:135a–136a, 137a
see also Evolution, theories of
Darwin, Erasmus (1731–1802), 3:577–581; 12:569b
acquaintances of, **7**:277a; **11**:141a; **14**: 463a
correspondence of, **6**:579a
critics of, **5**:411b; **10**:278b
on evolution, **3**:573b; **6**:140a; **8**:378a
influence of, **6**:120b
physiology, **11**:215a; **14**:44b
and Withering, **3**:578a
Zoonomia, **6**:140a; **10**:612b
Darwin, Francis (1848–1925), 3:581–582; 2:183b; **3**:566b; **5**:82b; **6**:44b; **12**:58b; **15**:46b
Darwin, George Howard (1845–1912), 3: 582–584; 2:516a; **3**:471a, 553b, 566b, 582a; **7**:85a,b; **13**:96a*, 365a
astronomy, **8**:562b; **12**:19a
geophysics, **6**:503a
Darwin, Horace, **2**:184a; **3**:566b
Darwin, Leonard, **3**:566b, 572b
Darwin family
genealogy, **3**:563a–b
Darwinism
mid nineteenth-century, **3**:88b; **4**:309a; **5**:444a; **6**:7b–10b, 138a, 490b–491a; **7**:498b; **8**:528a; **9**:332a, 560b; **11**: 254b, 308b; **12**:422a, 570b; **13**:87b–88a, 383b; **14**:58a, 534a; **15**:167a–170a
late nineteenth-century, **1**:173b, 396a; **2**:502a; **3**:338a, 553b; **4**:183a, 307a; **5**:385b, 438a, 447a; **6**:53a, 102a–b, 167a, 535b; **7**:67b, 68a, 90b–91a, 114b, 169b, 474b, 475a, 480a–b, 593a; **8**:183b, 529a; **10**:29a–b, 128a, 429b; **11**:6a, 517a, 518a–519a; **12**: 259a; **13**:9a, 343a, 417a; **14**:136a, 234a, 236b, 409b, 626b; **15**:147a, 148a
twentieth-century, **2**:583a; **3**:390a; **4**: 148a; **5**:467b; **6**:22a; **7**:115a, 253a, 447a; **9**:519a, 536a; **11**:19a–20a; **12**: 417b; **14**:434a–b; **15**:505b, 508b
Christian, **14**:516b
inorganic, **9**:287b
and Mendelism, **5**:10b; **9**:282b–283a
opposition to, **1**:506a, 540a; **2**:10b–11a, 505b; **3**:392b, 608b; **5**:17a, 45a; **6**: 339a; **9**:100a, 428b, 585a; **10**:209b–210a, 427b; **11**:235a, 361a; **12**:232a, 277b–278b; **13**:10a; **14**:42a, 350b
prehistory of, **6**:9b–10a
and religion, **3**:88b, 553b, 576a–b; **5**:513b; **7**:498b; **9**:428b; **15**:158a
social, **6**:339b*; **10**:448a; **11**:120a; **15**: 139a–140a
support of
American, **1**:405b; **5**:513a–b
see also Evolution, Natural selection, Neo-Darwinism
Daśabala (*fl.* A.D. 1050), 3:584; 15:584a–b
Cintāmaṇisāraṇikā, **15**:584b
Karaṇakamalamārtaṇḍa, **15**:584b
Daseki zyutu, **12**:291b
Dasht-i-Kavir (desert), **6**:216a
Dastin, John, **2**:23a
Dastre, Albert, **2**:27a; **9**:505a,b; **11**:101b
Dasypodius, Cunradus (*ca.* 1530–1600), 3: 585; 4:449a, 451a, 619a; **7**:194b
Date palm, **7**:204b; **15**:648b
alcoholic beverage from, **15**:664a(n)
Datholite, **5**:497a
Datiscin, **9**:95b
Datura, **9**:618b; **10**:237b
Daubenton, Edmé, **15**:113b
Daubenton, Louis-Jean-Marie (1716–1800), 15:111–114; 2:474a; **7**:198b, 603a

acquaintances of, **2**:107b, 509b; **3**:523a,b; **11**:522b; **12**:130b; **13**:554a, 565b
influence of, **7**:547b, 589a
medicine, **14**:15b
students of, **5**:355b; **6**:178a; **14**:14a; **15**: 502a
zoology, **3**:524b; **14**:299b
Daubeny, Charles Giles Bridle (1795–1867), 3:585–586; **2**:566a; **11**:174a*; **14**:489b
Daubrée, Gabriel-Auguste (1814–1896), 3: 586–587; **2**:556b; **4**:218b*; **8**:116b; **9**:58b; **11**:608b*; **13**:124b*, 621a*
D'Aubuisson de Voisins, Jean-François
see Aubuisson de Voisins, Jean-François d'
Dā᾽ūd ibn Ḥunayn, 7:24b; **15**:231a, 236a
Dā᾽ūd ibn Isḥāq, 7:24b
Daudin, Jean, **2**:322a
Daussy, Pierre, **8**:64b
Dauvillier, Alexandre, **2**:488a
Davaine, Casimir Joseph (1812–1882), 3: 587–589
anthrax, **4**:276b; **7**:421b; **10**:384b, 386a, 387a, 390b; **11**:70a,b
Davenport, Charles Benedict (1866–1944), 3:589–591; **3**:120a; 7:98b; **10**:462b; **14**:315a*; **15**:71a–b
Davenport, Gertrude Crotty, **3**:590b
Daventry Academy, **11**:139b–140a
David, Jacques Louis, **1**:330a; **8**:71a
David, Jean-Pierre, **8**:115a
David, Tannatt William Edgeworth (1858–1934), 3:591; **9**:197b
Davidov, August Yulevich (1823–1885), 3: 591–592; **10**:544b
Davidson, Charles Rundle, **13**:95b
Davidson, William
see Davison, William
Davie, James, **6**:578a
Daviel, Jacques, **8**:114b
Davies, Gordon Leslie, **7**:70b; **13**:442a
Davies, Mansel Morris, **1**:319b
Davies, Peter, **6**:21b
Daviet de Foncenex, François, **7**:567a
Da Vigevano
see Guy Da Vigevano
Davis, Alva Raymond, **6**:443b
Davis, Bradley Moore, **14**:100b–101a
Davis, Charles Henry, **10**:479b
Davis, Henry Jameson, **9**:98b
Davis, William Morris (1850–1934), 3:592–596
acquaintances of, **10**:502b; **11**:210a, 576a; **14**:516b
on contemporary scientists, **4**:593a*; **5**: 396a*; **8**:261a*; **11**:120b*
geology, **9**:149b; **10**:504a
geomorphology, **5**:395b; **9**:150b; **10**:505a–b, 508b
influence of, **4**:562b; **8**:31a; **10**:503a,b; **11**:524b
influences on, **7**:144b
meteorology, **4**:411a*
and Penck, **10**:507b
students of, **2**:373b, 500a; **5**:49b; **7**:143b
works edited, **7**:144a
Davison, William (1593–*ca.*1669), 3:596–597; **8**:130a
Commentariorum in . . . Petri Severini Dani Ideam medicinae philosophicae . . . prodromus, **12**:335b–336a
Davisson, Clinton Joseph (1881–1958), 3: 597–598; **11**:422a; **13**:447a
Davy, Edmund, **2**:67a
Davy, Humphry (1778–1829), 3:598–604; **1**:411b; **2**:515a; **3**:604b; **5**:216a; **6**: 285b*; **7**:409b; **14**:486b, 493a, 508a
acid theory, **8**:58b, 341b
acquaintances of, **1**:43b, 436b, 563b; **2**: 91a–b, 420a; **3**:324a; **5**:71b; **6**:384a, 386a; **8**:37b; **9**:424b, 583a; **10**:322b; **11**:514a; **12**:288a, 523b; **13**:352a; **14**:562b
biographies of, **13**:389b*
biology, **2**:483b
chemical elements
isolation of, **1**:142b–143a; **5**:321b; **6**: 495b; **9**:451a; **13**:311b; **15**:7b
nature of, **2**:485b; **3**:455b; **4**:531b; **5**: 322b; **6**:284b; **10**:184b; **13**:280b
combustion, **5**:126a
Consolations in Travel, **5**:518a
correspondence of, **3**:291b
crystallography, **12**:543a
and Dalton, **3**:541a, 545a
Davy lamp, **3**:603a; **5**:558b; **7**:509b
editions of works, **3**:605a
electricity, **3**:19b; **10**:188a
electrochemistry, **1**:348a; **2**:92b; **3**:68a; **5**:320b–321a; **8**:38a; **10**:108b; **14**:80a, 475a–b
Elements of Agricultural Chemistry
forerunners of, **4**:250a–b
and Faraday, **3**:603a,b, 605a; **4**:529a–531b
gases, **6**:285a
and Gay-Lussac, **5**:320b–322b *passim*
and Herapath, **6**:292b
influence of, **1**:345b, 557b; **3**:83b; **6**:19a, 277b; **11**:172b; **13**:311a; **15**:134a, 136a
matter, theory of, **3**:542b, 544a; **13**:373a; **14**:488b
organic chemistry, **12**:321b
photochemistry, **4**:182a
and Royal Society, **12**:52a
translations of works, **15**:527b
on volcanism, **3**:586a
Davy, John (1790–1868), 3:604–605; **4**: 535a–b; **11**:514a
Dawes, William Rutter (1799–1868), 3:605–606; **6**:402b, 541b
Dawson, Charles (1864–1916), 3:606–607
Dawson, George Mercer, **12**:293b; **13**:525a
Dawson, John (*fl. ca.* 1760), **5**:276a; **13**:55a
Dawson, John William (1820–1899), 3:607–609; **1**:50b; **3**:31a; **8**:462a, 571b, 572a
Dawson, Paul, **4**:45a
Day, Arthur Louis, **2**:369b; **6**:470a; **7**:59a; **13**:243b; **14**:338b
Day, David Talbot (1859–1925), 3:609–610
Day
civil, **5**:106a; **15**:566b, 609a
divisions of, **15**:209b–210b, 674b, 676a, 710a–b, 713b
epagomenal, **15**:708a, 709b, 712a, 716b
length of, **6**:616b–617a; **9**:141a, 555a*; **10**:161b; **12**:62b; **14**:28b, 390a; **15**: 380a
determination of, **3**:319b; **5**:613b–614a; **10**:540b, 541a–b; **11**:189b, 396a; **12**:361a,b; **15**:520a, 536b–537a, 539b–540b, 546b–548a, 580b, 584a, 611b–612a, 679a
and geographical latitude, **5**:345b; **12**: 61b; **15**:210a–b
Mayan
naming of, **15**:764b–765a, 765b, 796a
sidereal, **3**:319b; **5**:106a; **15**:590b
solar, **3**:319b; **5**:106a; **11**:191a; **15**:707b
Ibn al-Dāya
see Aḥmad ibn Yusuf
Dayton Engineering Laboratories Company (Delco), **7**:316a; **9**:375b
DDT
discovery of, **9**:576b–577b
Deacon, Henry, **12**:520b
Dead Sea, **12**:64a
salt deposition sequence, **2**:159a
Deafness, **5**:137b
education of deaf
seventeenth-century, **6**:253b; **9**:322a(n); **14**:152a
eighteenth-century, **1**:410b–411a; **5**: 468a
nineteenth-century, **1**:582b
vs. hardness of hearing, **1**:98b
hereditary, **1**:582b
Dean, Bashford (1867–1928), 3:610–611; **5**:284a(n); **13**:151a; **14**:623a
Death
by asphyxia, **2**:61b
atomistic explanation of, **4**:32b
definition of, **5**:447a
forecasting of, **5**:233a
from high temperature, **2**:31a
muscle effects
acidity and alkalinity, **2**:31a
signs of, **1**:266a
sudden
detection of, **2**:29a; **7**:614a
see also Mortality statistics, Rigor mortis
De Bary, (Heinrich) Anton (1831–1888), 3: 611–614; **10**:166a; **12**:58b, 169b*
acquaintances of, **2**:436b; **3**:581a; **14**:93a
botany, **1**:493b; **5**:624a; **6**:73a; **7**:395b; **11**:153b
on contemporary scientists, **9**:422b
critics of, **2**:437b
influence of, **10**:382a
students of, **2**:426b, 437a; **5**:437b; **7**:466b; **9**:587a; **12**:166a; **14**:36b
translations of works, **2**:371a
Vergleichende Anatomie der Vegetationsorgane der Phanerogamen und Farne, **12**:100a
Debeaune, Florimond (1601–1652), 3:614–616; **9**:599a; **10**:43b, 45b; **12**:206a
algebra, **12**:206b
curve of, **3**:616a
geometry, **1**:481b*; **11**:402b
problem of, **2**:52b; **4**:58a–b; **8**:162b, 163b
De Beers Consolidated Mines, **4**:262b
Debenham, Frank (1883–1965), 3:616–617
Debierne, André, **1**:560a; **2**:258b, 259a,b; **3**:499b, 500b, 507b; **5**:394b
Debray, Henri Jules (1827–1888), 3:617; **6**:177b; **9**:451a; **13**:468a
Debye, Peter Joseph William (1884–1966), 3:617–621; **10**:235b, 423b
acquaintances of, **4**:579b; **9**:262a; **11**:422a; **12**:527b, 528b, 529b; **14**:244a
on contemporary scientists, **11**:613b
Debye temperature, **3**:619a
Debye-Hückel theory, **1**:300a; **2**:170a, 499a; **7**:208b; **10**:156b; **13**:156a
influence of, **12**:528a
optics, **8**:501b; **11**:266b
physical chemistry, **8**:475a; **15**:461a
specific heat, **4**:414a; **7**:246b; **15**:40b, 442b
Debye, Peter Paul Ruprecht, **3**:618a
Debye (phys. unit), **3**:619a
Decagon
sides of, **4**:610a
Decaisne, Joseph, **9**:601a, 618a,b, 619a*; **13**:395a
Decans (stars), **15**:711b–721a *passim,* 727a
families of, **15**:716a
Decapoda
spermatozoa, **7**:455a
Decastello, Alfred von, **7**:623a
Decerebration, **9**:10a, 20a–b; **12**:400b
decerebrate rigidity, **12**:396b, 397b

Dechales, Claude François Milliet (1621–1678), 3:621–622; 1:81b; 4:452a, 505b; 13:320b*
Cursus seu mundus mathematicus, 2:202a
Dechambre, Amédée
Dictionnaire encyclopédique des sciences médicales, 11:515b
Déchelette, Joseph (1862–1914), 3:622–623; 9:489b
Dechen, Heinrich von (1800–1889), 3:623–624; 8:511b
Decidua, 12:354a
Decimal system
medieval Arabic use of, 4:502a–b; 7:326a, 532b–533a; 9:614a–b; 13:545b
Renaissance use of, 3:273a; 13:48a
Amerindian, 15:762a
Chinese use of, 14:543a
in measurement, 1:206b; 5:593b; 7:89a; 8:137a; 10:248a
notation, 2:602b; 3:273a; 10:255a; 14:19b
popularization of, 9:610b
theory of
medieval, 1:23a; 2:149a; 7:360b
Decision theory, 10:335a; 14:122a
Declination, magnetic, 9:254b; 10:546b; 12:379b; 13:49b
charting of, 6:71b, 107a
measurement of, 6:144a, 470a; 7:375b, 609a; 10:103b; 14:354b
observation of at sea, 2:343a; 5:397a–b, 435a; 8:65a, 390a
theory of, 5:398b–399a
variation of, 5:593b; 8:390a; 15:88b
secular, 3:424a; 5:343b; 11:320b*
Decomposition, chemical, 11:541a; 13:429b, 578a
double, 11:435a–b; 13:177a; 14:395a–b; 15:457b
free-radical, 6:404b
in gas phase
pressure effects, 6:404b
heats of, 9:619b
kinetics of, 13:70a–b
of solid mixtures, 6:404b
thermal, 15:36b
Decomposition, mathematics
see Number(s), decomposition of
Decompression, respiratory, 2:62a; 5:208a; 6:24a
De Decker, Ezechiel, 14:51a
Dedekind, (Julius Wilhelm) Richard (1831–1916), 4:1–5
acquaintances of, 1:370a; 3:303a; 5:307a; 7:506b; 11:448a; 14:202a
algebra, 6:388a; 14:202b
analysis, 7:398b; 11:451b
and Cantor, 3:54a–57a *passim*
on contemporary scientists, 11:448b
critics of, 4:4b; 7:507a, 508b
Dedekind cut, 4:3b–4a, 419b
and Dirichlet, 2:238a; 4:2a, 3a–b, 125b, 126b
and Gauss, 4:1b–2a, 3a
influence of, 10:138a; 11:409b
number theory, 4:125a, 466a; 5:153b; 6:214a; 7:523b; 9:307b; 10:344b, 442b, 484a; 11:256b; 12:12b–13a, 15a
philosophy of mathematics, 7:508a
and Riemann, 4:1b, 2b, 3a–b
set theory, 2:277a,b; 3:54a–55a
zeta function, 6:214b
Deduction, logical, 1:252a, 279a, 375a; 2:232b–233a; 4:423b; 8:263a*; 9:288b, 289b; 10:192a, 344b
criticism of, 14:288b–289a
indirect method of, 3:515b
natural, 5:350b
Dee, Arthur, 1:317b
Fasciculus chemicus, 1:318a*
Dee, John (1527–1608), 4:5–6; 1:126b, 383a; 2:435b; 4:171a*, 450a; 5:553b, 593a; 14:377b(n)
acquaintances of, 3:364b; 4:97b, 170a; 6:127b; 10:239b
Deeds, Edward Andrew, 7:316a
Deen, Isaac van, 4:162b
Deer
reproduction, 2:161b; 6:159b, 161a
Deering, George, 4:98b
Deferent (astron.), 1:509b; 2:412a, 551a; 3:406b, 409b–410a; 10:298b; 11:190b; 12:4b
Deficiency diseases
beriberi, 4:310b–311b; 5:208b, 209a, 541b; 9:81b; 13:490b
concept of, 5:541b
pellagra, 1:360a; 3:97a,b; 4:358a; 5:209a, 451b–452b; 13:448b
rickets, 5:37a, 209a, 425b; 6:136a, 183b, 232a*; 9:243a; 10:243b; 15:418a–b
scurvy, 1:26a, 98b, 434b; 2:617a; 3:397a; 5:209a; 10:426b*; 14:407b
Deficiency states, 9:9a
amino acids, 5:451b
folic acid, 1:60a
iodine, 3:113a
trace elements, 2:87a; 5:209a
Vitamin B, 3:185a
Vitamin B_{12}, 1:60a
Definite proportions, law of
see Compounds, definite proportions theory
Definition, 1:315b; 2:601b
by abstraction, 2:277a,b
coordinating, 11:356b
in mathematics, 1:22b; 2:274b; 3:111a; 5:154b, 346b; 9:311a; 12:12b
recursive, 4:4a
ostensive, 7:151a
theory of, 10:274a; 12:55b
Deflagrator, 6:115a; 12:433a
Defoe, Daniel, 1:411a; 13:488b
De Forest, Lee (1873–1961), 4:6–7
triode, 1:287a,b, 292b; 7:108a; 9:99a
Deformation, geological, 1:394a–b, 472a, 552b; 2:89b, 558b; 3:325b, 568a; 5:395b; 13:64a
geosynclinal, 6:228a
isobase of, 13:270b
"Pennic" type, 1:236a
theory of, 10:507b–508b
see also Diastrophism; Faulting; Folding; Inversion, stratigraphic; Tectonics
Deformation, geometric, 9:598b
Deformation, physical, 11:123b
critical state, 6:534b
plastic, 11:549a–b
Defries, Robert Davies, 1:442b
Degen, Carl Ferdinand, 1:13a–b, 14a; 8:138b
Degen, Jakob
see Schegk, Jakob
Degeneration (biol.)
morphophysiological, 11:234b, 307b–308a; 12:338b, 551a; 14:237a,b
of species, 7:585b, 587b, 588b; 11:234b; 12:551a
De Giovanni, Achille, 3:113b
Deglutition, 7:504b; 15:71b
mechanism of, 9:9a
DeGolyer, Everette Lee (1886–1956), 4:7–8
Degradation (geol.), 4:151b
aqueous, 4:72a
De Groot, Jan Cornets (or Janus Grotius) (1554–1640), 4:8; 3:181a
De Haas, Wander J.
see Haas, Wander J. de
Dehérain, Pierre-Paul, 5:158a; 9:450b
Dehn, Max (1878–1952), 4:9; 5:530a*; 6:235b, 394b*; 7:312a; 10:344a; 12:196b*; 14:612a*
Dehn's lemma, 4:9a; 14:311b
theorem of, 6:393b; 7:207b
Dehydracetic acid
structure of, 3:347b
Dehydration, 14:561a
Dehydrocholic acid, 3:341b
Dehydrocyclization, 1:415b; 7:21b
Dehydrogenases, 6:501b; 7:273a–b; 10:282a; 13:394a
Dehydrogenation, 14:601b–602a
catalytic, 7:21b
of cyclic hydrocarbons, 1:414b
Deiman, Johann Rudolf, 7:12a; 10:446a; 13:468b, 469a
Deinostratus
see Dinostratus
Deism, 2:542b; 3:295a; 4:85a; 5:355b; 6:579b, 586b; 7:135a–b; 14:68b, 361a,b, 483a,b
De Koninck, Laurent-Guillaume, 5:109a,b
De la Beche, Henry Thomas (1796–1855), 4:9–11; 9:585a
acquaintances of, 2:566b, 567a; 5:67a; 10:511b, 583b; 12:489a
critics of, 8:486a; 9:583b, 584a
Geological Manual
translation of, 3:623b; 4:349b
geology, 2:567b; 8:461a; 12:277a
influence of, 2:570b; 5:436b
paleontology, 3:396a*
Report on the Geology of Cornwall, Devon, and West Somerset, 10:584a
Delafield, Francis, 4:498b; 11:176a,b; 14:248b
Delafond, Frédéric, 4:39a
Delafond, Onésime, 5:566a; 11:70a
Delafontaine, Marc, 9:290b
Delafosse, Gabriel (1796–1878), 15:114–115; 2:106b, 431a; 4:66a; 6:179a, 180b; 10:357a,b
Delage, Yves (1854–1920), 4:11–13; 4:207a; 5:386a*
Delamain, Richard (*fl. ca.* 1630), 4:13; 10:255a
Delambre, Jean-Baptiste Joseph (1749–1822), 4:14–18; 9:329b, 493b*; 10:572b*
and Académie des Sciences, 2:216b; 7:569a; 15:336a
acquaintances of, 5:88a; 13:184a
astronomy, 12:239b; 15:331b, 355a
on contemporary scientists, 7:550b, 567a; 8:145a*; 9:251a; 12:518b; 15:365b–366a
correspondence of, 1:436b; 15:388a*
éloges by, 3:447a*; 7:560a
geodesy, 7:500a,b; 9:234b; 10:596a; 12:518a; 15:376a
Histoire de l'astronomie du moyen âge, 1:508a, 513a; 9:191a
historiography, 3:103b, 104a, 105a, 410b, 446b*; 7:302b, 307b, 545a, 577b; 9:90b; 11:32b, 189b, 195a, 201b; 13:323b; 14:575a, 593a; 15:219a, 220a, 344a
and Lalande, 4:14a; 7:580b
parallax determination, 6:263b
Rapport historique sur les progrès des sciences mathématiques depuis 1789, et sur leur état actuel, 10:327b
standard meter, 2:135b; 9:250b, 251a; 15:335a–b
tables of, 15:331b, 337b, 338a,b, 354b
Delaroche, François, 8:261b
on cooling, 2:136a–b; 13:10b

specific heats, **1**:348a, 616b; **3**:84a(n); **15**:95b, 358a, 363a, 378b
De la Rue, Warren (1815–1889), 4:18–19; **6**:350b; **8**:329a; **11**:156a
observations, **12**:268a
Delaunay, Charles-Eugène (1816–1872), 4: 19–20; **5**:360a*; **6**:282a; **8**:62b
celestial mechanics, **1**:86b; **9**:239b, 449b; **13**:422b; **15**:332b
critics of, **13**:422b
Delaunay method, **4**:19b
geometry, **13**:130a
and Leverrier, **4**:20a; **8**:278b
Théorie du mouvement de la lune, **6**:399b
Delaware
geography, **4**:492b
Delbrück, Max, **9**:525b; **12**:222b–223a; **15**: 43a
Delépine, Stéphane-Marcel (1871–1965), 4: 20–21; **5**:100b*, 295a*; **8**:110a*; **10**:498b*; **13**:177b*
Delépine reaction, **4**:20b
Delesseriaceae, **7**:535a
Delian problem
see Cube, duplication of
Delile (or Raffeneau-Delile), Alire (1778–1850), 4:21–22; **9**:8a; **11**:343b*
Delirium, **10**:613b; **14**:408a
Delirium cordis, **15**:101a
Delisle, Claude, **4**:22a
Delisle, Guillaume (1675–1726), 4:22; **2**: 551b
letters of, **3**:174a*
maps of, **2**:14b; **4**:24a
Delisle, Joseph-Nicolas (1688–1768), 4:22–25; **4**:469a,b; **5**:87b, 434b; **6**:364b*; **7**:579b; **8**:143b; **9**:329b
astronomy, **6**:69b; **7**:580b; **10**:615a
Delisle de la Croyère, Louis, **4**:23a; **5**:428a
Delius, Christoph Traugott, **9**:534a
Della Porta, Giambattista
see Porta, Giambattista della
Dell'Aquila, Sebastiano, **1**:484b
Della Torre, Giovanni Maria (1713–1782), 4:25–26; **5**:56b
Della Torre, Marcantonio, **5**:104b; **8**:194a
Dell'Uso e dell'attivitá dell'arco conduttore, **1**:107b–108a
Del Monte, Guidobaldi
see Monte, Guido Ubaldo del
Delos, island of, **3**:162a
Del Papa, Giuseppe, **1**:483b
Delphinidae
classification of, **5**:46a
Delphinium, **5**:429a
Delpino, Giacomo Giuseppe Federico, **3**: 422a
Delporte, Eugène Joseph (1882–1955), 4: 26–27
Del Rio, Andrés Manuel
see Rio, Andrés Manuel Del
Delta deposits, **1**:470a
Deluc, Guillaume Antoine, **4**:27b
Deluc, Jean André (1727–1817), 4:27–29; **2**:568b; **3**:540a; **6**:585b; **8**:321a; **11**: 134a,b; **13**:183b; **14**:76a; **15**:95b
erratic blocks, **2**:556b; **3**:210b
and Saussure, **4**:28b–29a
Deluc, Jean André (nephew of above), **4**:27b
Deluge
see Diluvialism, Flood
Demagnetization
cooling by, **7**:493a
shape effects, **1**:284b
Demarçay, Eugène, **3**:499a
Dembowski, Ercole (1812–1881), 4:29–30
Dementia paralytica
see Paresis
Démeste, Jean, **11**:521b, 523a; **12**:66a
Demetrius of Alexandria, **5**:229a; **9**:300a
Demetrius of Phalerum, **4**:415a–b
Demiurge, **1**:259b, 337a; **7**:41b
Democracy, **9**:385a
biological basis of, **3**:390a
Democrates, **4**:34a
Democritus (*fl.* late fifth century B.C.), 4: 30–35; **1**:263b; **4**:375a*; **10**:181b(n)
acquaintances of, **6**:419b
astronomy, **6**:416b; **15**:33b
atomism, **1**:101b; **4**:381b; **12**:333b
biographies of, **8**:269a
botany, **13**:331b
Causes Concerning Animals, **1**:259a
Francis Bacon on, **1**:373a
geometry, **1**:221b; **3**:150a; **4**:424b, 463a, 466b; **6**:416a; **9**:270a
influence of, **1**:38a, 350b, 495a; **2**:256a,b, 317b; **3**:114b; **6**:422b; **7**:187a, 612a; **9**:509a; **12**:333a; **14**:632a
influences on, **3**:492a
motion, **4**:382a
natural philosophy, **1**:259b; **8**:538a,b; **15**:404a
Demography
medieval Arabic, **7**:8b
seventeenth-century, **1**:208b–209a; **5**: 507a–b; **10**:565a, 565b–566a
eighteenth-century, **2**:42b, 418b; **3**: 387b(n); **4**:107b–108a, 470b; **5**:129a; **9**:454a–b; **13**:448a; **14**:179a–b; **15**: 308b–309b
nineteenth-century, **7**:2a; **9**:69a; **15**:31a–b, 66b, 502a
twentieth-century, **1**:155a; **6**:122b; **7**: 513a; **8**:512a; **10**:445a, 503a–b; **14**: 87a
see also Biometrics, Population, Statistics
De Moivre, Abraham
see Moivre, Abraham de
Demokedes of Crotona, **1**:103b
Demonology, **11**:183b, 185a; **15**:804b
Demonstratio per impossible, **4**:416b
Demonstration, scientific, **1**:279a; **9**:382b
De Morgan, Augustus (1806–1871), 4:35–37; **3**:433a*
acquaintances of, **6**:91a; **7**:383b; **13**:216b, 217a,b
algebra, **14**:303b
on contemporary scientists, **5**:466b
correspondence of, **14**:438a
critics of, **13**:612a
Formal Logic, **13**:612a
and Hamilton, **6**:82b–83b
historiography, **3**:52b*, 621b; **4**:421b; **6**: 449b; **10**:90b(n), 100b*; **14**:500b*; **15**:281b
influence of, **1**:382b; **6**:510b; **11**:568a; **12**:217a
logic, **2**:295a, 298a; **8**:159b; **10**:484a; **12**:13b
philosophy, **14**:293b
probability, **13**:611b
proportions, **4**:419b
students of, **7**:106b; **13**:426b
Dempster, Arthur Jeffrey, **1**:321a; **7**:36b
Dendrerpeton, **8**:572a
Dendrite (mineral.), **7**:412a
Dendrites (zool.)
function of, **5**:460b
Dendrochronology, **8**:206a
Dendrology, **7**:285a
Dene holes, **3**:607a*
Dengue, **5**:451a
Denis, Jean-Baptiste (*ca.* 1640–1704), 4:37–38; **8**:525a; **9**:498b
Denjoy, Arnaud, **1**:407a; **4**:206b; **8**:111a; **11**:54a, 56b
Denmark
archaeology, **14**:505b, 506a–b
botany, **9**:575b
education, **12**:88b
geography, **10**:239a
geology, **5**:71b–72b
mathematics, **10**:544a
medicine, **1**:482b, 483a; **6**:123b
organization of, **1**:483a
meteorology, **15**:87a
monetary system, **13**:359b
natural history, **13**:10a
pharmacy, **14**:600b
zoology, **9**:575b–576a
Density
of air
see Air, density of
critical, **14**:561a
fluctuations of, **9**:76b
of gases, **1**:344b, 345b, 616b; **2**:135a; **4**: 376b; **5**:319a, 320a; **13**:102b–103a; **14**:319a
fluctuations of, **12**:497a–b
and index of refraction, **7**:620a; **8**:493a; **11**:201a; **15**:355a,b, 359b
matrix of, **7**:617a
measurement of, **1**:321a; **2**:474b; **4**:518a; **5**:382a; **7**:340b, 341b; **9**:578b
of metals, **7**:340b, 341b
of solid vs. liquid forms, **5**:400b
of solutions, **3**:211b; **13**:547b
and temperature, **7**:346a
of water, **3**:66b, 482b; **4**:577a; **6**:495b; **7**:341b
see also Rarity and density
Dentin, **3**:530b
Dentistry
medieval Arabic, **14**:584b
eighteenth-century, **6**:460b*
Dentition
see Teeth
Denudation, geological, **4**:540b–541a; **7**: 185b; **8**:575a; **9**:594a; **10**:504b; **11**: 276b–277a; **12**:263b
chronology, **7**:144b–145a
Denumerant, **13**:219a
Denys the Carthusian, **13**:534b
Deontology
ancient Greek, **6**:425a
medieval, **9**:41a; **11**:444a–b; **13**:229b; **14**:584a
Renaissance, **2**:337a
eighteenth-century, **3**:2a; **5**:221a
nineteenth-century, **3**:489b; **7**:520b; **11**: 559b
Deparcieux, Antoine (1703–1768), 4:38–39; **14**:179a; **15**:31a
Dependency (psychol.), **5**:177b
Depéret, Charles (1854–1929), 4:39
Depolarizers, **1**:298a
Depression (psychol.)
theories of, **8**:7b, 506b
Deprez, Marcel (1843–1918), 4:39–40; **4**: 588b, 589a; **5**:496a
galvanometer, **1**:303b; **4**:333b
Depsides, **5**:6a
Depth finder, **8**:12a
De Puydt, Marcel, **5**:109b
Derby Philosophical Society, **3**:578b; **12**: 569b
Derbyshire
geology, **4**:540a–541a; **14**:311b–312a
Derham, William (1657–1735), 4:40–41; **2**: 382a; **3**:458a*; **5**:130b, 279a(n), 516a; **6**:130a; **10**:95b*; **11**:43a, 313b, 316b, 527a*
Physico-Theology, **10**:120b
Deriugin
see Deryugin

Derivative (cont.)
574a; 6:510b
of arbitrary index, 8:384a
concept of, 12:461a; 13:23b, 436a
covariant, 11:407b, 408a
definition of, 3:164a; 11:454b
maximum value, 9:127a
mean-value theorem, 10:442a
partial, 3:145b, 385b, 592a; 4:103a; 10:47b
Dermatology
Renaissance, 9:75a
eighteenth-century, 8:506a–b
nineteenth-century, 5:566a; 10:17b–19a
twentieth-century, 10:18b–19a
see also Skin
Derosne, (Louis-) Charles (1780–1846), 4:41–42; 3:159b
Derosne, François, 4:41a
Derosne, Jean-François, 4:41a–b
Deryugin, Konstantin Mikhailovich (1878–1938), 4:42–43
Desaguliers, John Theophilus (1683–1744), 4:43–46; 3:105a; 4:352b; 5:130a, 510b; 6:37a, 46a(n), 173b, 176a; 9:141b; 10:92a(n); 13:68a
acquaintances of, 5:509b, 516a; 9:595a; 10:145b
physics, 3:444a; 8:255b
translations by, 5:510a; 9:119b; 10:264b*; 11:3*
translations of works, 10:572a*
Deśantara (longitudinal difference), 15:568a, 575b, 582a, 593b
definition of, 15:548b, 557b
Desargues, Girard (1591–1661), 4:46–51; 1:191b; 3:622a; 6:112b
acquaintances of, 1:541a; 6:598b; 9:498a; 10:342a; 13:335a
and Beaugrand, 4:46b, 47a, 48a
Brouillon projet, 1:541b; 7:577a; 10:330b, 331b
critics of, 4:48a,b
and Descartes, 4:46a,b, 47a,b, 49a
geometry, 10:331a
influence of, 7:576b, 577a, 578a; 11:79b
perspective, 2:333b
rediscovery of, 1:541b
theorem of, 2:333b; 4:48b; 13:5a, 128a; 14:212a
Desbois de Rochefort, Louis, 3:427a
Descartes, René du Perron (1596–1650), 4:51–65; 1:191b, 495a; 3:622a; 7:55a
acquaintances of, 6:597b; 9:336a, 497b, 528a, 598a; 10:137a, 333a; 12:205b; 13:222b
astronomy, 4:17b
and Beeckman, 1:567a, 568a; 4:51b, 55b, 56a
and Carcavi, 3:64a
on contemporary scientists, 3:321a, 360b; 12:610b; 14:254b
correspondence of, 4:95b, 505b; 5:158b; 6:520a; 7:537a; 9:382b, 509b; 10:495b; 13:254a, 435b; 14:254b, 368a, 376b(n), 379b(n)
cosmology, 4:52a,b, 58b, 60a; 5:288a
vortex theory, 1:113a; 2:38b; 6:604a; 7:228a, 490b; 10:91b(n); 12:118a–b
see also Cosmology, Cartesian
critics of, 1:541a–b; 7:611b–612a; 9:25b, 527b; 10:296b, 314b, 337b; 11:488b; 12:501b; 14:84b, 565a
crystals, 2:431a
and Debeaune, 3:614b, 615b
and Desargues, 4:46a,b, 47a,b, 49a
and Fermat, 4:567b, 571a, 574b–575b; 5:527a; 6:112b; 9:598a
and Galileo, 5:247b; 7:486b, 487a
and Gassendi, 3:321a; 4:52a; 5:285a–288a *passim*
geology, 2:348a; 12:554a
heat theory, 6:292a
and Hobbes, 4:52a; 6:446a,b
and Huygens, 4:52b, 59b; 6:598a, 599b, 602a–603a, 608b, 610a
hydrostatics, 10:333b
influence of, 1:305b, 322b; 2:13a, 226b, 309a, 357b, 377b, 545b; 3:110b, 111b, 208b; 5:426b; 6:445b; 9:16a, 382b; 10:43b, 55b, 60b, 520b; 12:431a, 581b, 585a, 600b; 14:619a
influences on, 11:289a(n); 14:461a
and Leibniz, 4:52b, 64b; 8:151a, 152b–153b
manuscripts of, 13:479b
mathematics, 4:55b–58b; 5:158b*; 7:490b, 550b; 9:599a
algebra, 5:409b; 6:537a–b, 538a; 7:328a, 329a; 8:162a; 11:288a, 512a; 13:129a; 14:147b, 150a
analysis, 1:164b; 3:460a,b; 4:479b; 9:49b–50a
geometry, 4:550b, 551a; 5:382b, 410a; 6:112b, 599b; 7:577a; 9:199b; 10:52a, 103b, 331a; 11:402b; 12:206b, 460a, 610b; 13:434b; 14:148a, 465b; 15:118a
method of tangents, 1:474b–475a; 3:615b, 616a(n)
notation, 3:274a; 10:254b; 15:368a
rule of signs, 2:573b; 5:97b–98a, 567b; 7:573b; 12:283a
terminology, 1:238a
matter, theories of, 2:2b, 378b, 379a; 5:574b
and Mersenne, 4:51b, 53a, 58a, 62a; 9:316b, 318a–b, 320a, 598a; 13:436a
method of, 3:114b; 4:51b–55a; 12:582a; 13:33a
philosophy, 2:31b; 3:295a–b, 381a, 382a; 5:59a, 415b; 7:483b–484a, 505b, 539a; 8:436b; 9:120a, 318b; 13:278b
causation, 6:558a
physics, 3:114b; 4:52a–54a; 14:376b(n)
barometry, 10:388a(n)
dynamics, 2:606b; 7:228a, 487a–b; 10:71b–72a
impact, 6:602a–603a
motion, 3:295b; 8:268b
optics, 1:475a; 3:454b; 4:58b–60a, 575a; 6:194b, 197a, 446a; 7:215b, 298a–b; 9:116a, 118b, 159a, 199b; 10:56b–57a, 104a; 11:526a
physiology, 4:61b–65a; 9:571a; 12:101b; 13:170a, 222b; 14:321a, 405b, 406b
and Roberval, 3:64a; 4:571a
theology, 7:483b–484a, 488a
writings
editions of, 13:253a, 253b–254a
manuscripts, 1:206b–207a
publication of works, 3:320b–321a
Compendium musicae, 1:567a; 4:51b
translation of, 2:507b
La description du corps humain, 4:54b
La dioptrique, 4:52a, 54a, 58b–60a, 64a, 574b; 5:525b; 8:327b; 9:336a; 12:501b
critics of, 9:598a
Discours de la méthode, 4:46b, 51b–54b *passim,* 61b, 64a,b; 5:61a; 7:194a, 484a
Explication des engins . . ., 4:60b
Géométrie, 1:245b, 474a, 541b; 2:348a, 545a; 3:64a; 4:46b, 52a, 55b–58b, 567b, 568a,b, 570a; 5:567a; 6:536b; 10:43b, 115b, 296a–b; 11:488a; 12:205b; 13:436a
commentaries on, 3:348a
editions of, 12:206a–b; 14:465a
influence of, 10:45b, 47b
Latin edition of, 2:47a; 3:614b; 6:359a
Notes brèves, 3:614b, 615b
translations of, 8:161b; 12:206a
Lettres
editions of, 3:321a
Méditations, 1:292a; 4:52a, 58b; 7:484a
editions of, 3:320b–321a
publication of, 4:47b
translations of, 9:465a
Météores, 4:52a, 54a, 60a
Le monde, ou Traité de la lumière, 4:52a, 54a, 63b
editions of, 3:321a
Olympica, 4:51a–b
Les passions de l'âme, 4:52a, 61b, 64a
Principia philosophiae, 2:378b; 4:52a–54a *passim,* 58b, 61b; 10:61b; 11:507b
criticism of, 10:319b–320a
Regulae ad directionem ingenii, 1:292a; 4:52a,b, 55b–56a, 61b–62a, 63b
De solidorum elementis, 4:55b
Traité de la formation du foetus, 3:321a; 4:64b
Traité de l'homme, 4:52a, 54a, 62a–64a; 9:47b; 13:33a
editions of, 3:321a
see also Cartesianism
Des Cloizeaux, Alfred-Louis-Olivier Legrand (1817–1897), 4:65–67; 2:588a; 7:548b; 8:288b, 289a; 15:115b*
Manuel de minéralogie, 9:393a
Descloizite, 4:66b
Descotils
see Collet-Descotils, Hippolyte-Victor
Descourtilz, Jean Théodore, 4:67a*
Descourtilz, Michel Étienne (1775–1836), 4:67
Description de l'Égypte, 9:474a
Description des arts et métiers, 2:201b; 3:454b; 4:224b*; 7:581a; 8:621b; 9:62a; 11:328a–b
Descroizilles, François, 3:201b
Descroizilles, Henri, 5:325a,b
Descurain, François, 5:577a–b
Deserts
origin of, 14:627a
Desfontaines, René Louiche, 2:321b, 492a; 4:21b; 5:218a; 9:418b; 13:390a; 15:267a
Deshayes, Gerard Paul (1797–1875), 4:67–68
paleontology, 8:567b, 568b, 569b, 571a
on species, 12:318a
Desiccation
by vacuum, 9:607b
Deslandes, François, 12:128a
Deslandres, Henri (1853–1948), 4:68–70; 1:172b; 2:487b; 5:315b*; 6:28a; 7:77b*, 78a*; 11:481a(n)
Desmarest, Anselme-Gaétan, 2:496a; 4:71b; 12:120a
Mammalogie, 6:120a
Desmarest, Nicolas (1725–1815), 4:70–73; 3:387b(n), 436b*; 5:578b; 6:175a*, 579a; 9:534a; 11:269b; 13:412a*, 494b
basalt, 1:327a; 4:149b, 548b; 6:84a–b; 11:303b
influences on, 11:563a
Desmids, 7:533b; 14:398a
chloroplasts of, 2:372a

Desmier de Saint-Simon, Étienne-Jules-Adolphe
see Archiac, Étienne-Jules-Adolphe Desmier de Saint-Simon, Vicomte d'
Desmobacteria, **3**:338b
Desnoyers, Jules, **2**:341b; **4**:67b; **8**:567b; **12**:318a
Des Noyers, Pierre, **9**:16a
Desoille, Henri, **4**:399a
Desor, Pierre Jean Édouard (1811–1882), 4: 73–74; **5**:533b; **6**:57b; **7**:43b; **9**:538a; **12**:139b, 169b*
Desormes, Charles-Bernard (1777–1862), 4: 74; **3**:315b–316a; **5**:603a; **9**:494a*
catalysis, **15**:462a
gases, **15**:357b, 487a
iodine, **3**:455a–b
Desoxyribose, **8**:276a
Despagnet, Étienne, **1**:541a; **4**:75b–75a
Despagnet, Jean (*fl. ca.* 1625), 4:74–75; **1**: 317b
Arcanum hermeticae philosophiae, **1**: 318a*
Despretz, César Mansuète, **2**:586b; **6**:243b, 307a; **12**:197a, 433a
Dessaignes, Victor (1800–1885), 4:75–76; **3**:511a; **10**:359a–b
Dessau, Bernardo, **11**:461a
Destiny
see Fate
Detectors (radio)
see Radio, detectors
Detergent action, **12**:593b–594a
Determinants, **2**:113a, 299a, 465a, 469b; **3**:135b, 142b, 164a, 165b–166a, 166b, 314b; **4**:137a; **5**:573b*; **6**:392a, 403b; **7**:109a, 530b, 565a; **8**:62a, 165b; **9**: 388a*; **10**:24b, 451b; **12**:291b; **13**: 220a; **15**:295a
Cauchy, **3**:136b
circulants, **8**:250a
equivalence classes, **9**:412a
expansion by minors, **7**:435b
functional, **7**:55a
Hessian, **6**:357a–b
infinite, **6**:399b; **7**:515a; **11**:59b
Jacobian, **3**:136b, 142a; **7**:55a; **10**:442a
multiplication of, **3**:323a; **7**:564b
symmetric, **6**:95a
theory of, **7**:506b; **8**:250a; **9**:471a; **13**:218b
use in geometry, **3**:165b; **7**:109b
Vandermonde, **3**:142b; **13**:571a–b
Wronskian, **10**:249b, 442a; **15**:226a
Determinism, **4**:482a; **10**:4b*
astrological, **1**:93b, 378a, 382a; **2**:193a; **9**:137a; **11**:74a; **15**:729b
biological
eighteenth-century, **6**:140a; **7**:605a, 606b, 607a
nineteenth-century, **1**:303a; **2**:32b
experimental, **2**:26b
geological, **2**:191a
opponents of, **2**:9a–b, 538b; **3**:111b, 485b; **10**:173b; **11**:599a
philosophical, **2**:356a, 395b; **4**:255b, 374a; **6**:449a; **7**:68a; **9**:382b; **11**:72b–73b, 143a; **14**:483a–484a; **15**:144a
Detonation, **3**:20a
Detonation wave, **4**:130b
Detonators, **10**:132b
Detoxification, **4**:76a; **9**:9a; **10**:307a, 308a
Detrital rock, **5**:445a
Dettonville, A. (pseudonym), **10**:336b
Deuterium, **3**:330a
bombardment of, **12**:34a
as catalyst, **4**:545a–b
discovery of, **8**:95a
isolation of, **8**:292b–293a
nuclear reactions of, **8**:293a; **15**:8b
Deuteron(s), **13**:241a
acceleration of, **3**:329b–330a
bombardment by, **12**:34a
magnetic moment, **13**:42b
Deutsche Botanische Gesellschaft
founding of, **12**:256b
Deutsche Chemische Gesellschaft, **1**:578b; **5**:431b; **8**:328b; **9**:351b
founding of, **6**:461b; **9**:18b
Deutsche elektrochemische Gesellschaft, **15**:449b
Deutsche Geologische Gesellschaft, Berlin, **3**:70a
founding of, **2**:108a; **3**:434a; **11**:539b, 560b
Deutsche Gesellschaft für Metallkunde
founding of, **6**:368b
Deutsche Gesellschaft zur Bekämpfung der Geschlechtskrankheiten, **10**:18b
Deutsche Mathematiker-Vereinigung, **14**: 202a
Deutsche Physikalische Gesellschaft, Berlin, **2**:400a; **6**:342a; **8**:53a; **12**:615a–b; **14**:360
founding of, **4**:200b; **7**:509a
Deutsche Wochenschrift für Staat, Kirche und Volksleben, **15**:193b
Deutsches Museum (Munich), **4**:269a
Devadatta
Grahaprakāśa, **15**:618a
De Valera, Eamon, **12**:222a,b
Devaux, Henri (1862–1956), 4:76–77; **4**:401b
Development, geometric, **4**:260a
Developmentalism, **1**:335a, 336b–337a
Deviation, standard
see Standard deviation
Devik, Olav Martin, **2**:168b
Deville, Charles, **4**:77b; **8**:116a, 118a
Deville, Henri Étienne Sainte-Claire (1818–1881), 4:77–78; **3**:11b, 617a–b; **4**:79a; **5**:315a; **6**:51a, 177b, 319a, 565a; **9**: 451a; **10**:498a; **11**:111b*, 270b; **13**: 467b–468a; **14**:522a
chemistry, **3**:49a; **6**:50b; **14**:478a
mineralogy, **5**:89a
Devīsahāya
commentaries of, **2**:116a
Devonian period, **5**:476a; **8**:511b
deformation, **1**:394a–b; **7**:251b
paleobotany, **8**:5b
paleogeography, **3**:236b, 237a; **7**:250a, 251b
paleontology, **1**:210b, 472b; **2**:109a, 110a, 128b, 371b; **3**:30b–31a, 236b, 392a, 608a, 610b; **5**:109b, 486b; **6**:594b; **7**:61b, 252b, 460b; **8**:463a; **11**:501b–502a; **14**:392a
stratigraphy, **4**:262b, 361b; **5**:436b, 475b; **7**:250b; **8**:486b; **9**:584a; **12**:277a; **13**:620b; **14**:392a
volcanic activity, **5**:336a
Devonian System
see Devonian period, stratigraphy
Devonshire, Albert Frederick, **8**:186b
Devonshire (county)
geology, **4**:10b, 11a; **5**:436b; **8**:486a–b; **12**:276b, 277a
Dew, **4**:216a
formation of, **14**:253b; **15**:640b
Dew point, **3**:541a, 556b
Dewar, James (1842–1923), 4:78–81; **3**:477a; **6**:476b
Dewar flask, **4**:80a
low-temperature studies, **6**:5b, 94a; **9**: 451a; **10**:206b
organic chemistry, **1**:156a; **3**:300a
radiometer, **9**:224a; **13**:237a
spectra, **11**:611b
Dewey, Henry, **5**:39a
De Winton, Dorothea, **6**:22b
De Witt, Lydia, **1**:441a
Dew-Smith, Albert George, **5**:82a,b
Dextrine, **2**:139b; **10**:532a
Dextrose
differential solubility of, **1**:551b
Deyeux, Nicolas, **10**:325b, 326a
Dezallier D'Argenville, Antoine-Joseph
see Argenville, Antoine-Joseph d'
Dhanarāja of Padmāvatī, **9**:21a
Dhaneśvara
commentaries of, **2**:116a, 118a
Dharmeśvara, **7**:315a
Dhruvaka (period), **15**:585a
definition of, **15**:593a
Dhruvāṅka (period)
definition of, **15**:624b
Diabase, **7**:370b
Diabetes, **4**:360a; **10**:611a
biochemistry of, **2**:29a; **5**:209a; **8**:452b, 555a, 614b; **13**:452a, 574b; **14**:492b
clinical detection of, **1**:471a
and diet, **1**:443a; **10**:558b
experimental, **3**:69a; **7**:519a; **8**:452a; **10**: 568a
puncture, **1**:441b; **2**:29b; **7**:519b; **9**:10a; **12**:165a
and insulin, **1**:440b–443a; **3**:352a
mellitus, **1**:441a; **3**:352a; **13**:452a; **14**:408a
and metabolic rate, **1**:610b; **14**:65b
and pituitary, **15**:229a
polyneuritis, **14**:408a
and pregnancy, **3**:69a
theories of, **10**:580a; **11**:565a–b; **13**:589a
treatment of, **1**:610b; **2**:547b–548a
Diabolism, **2**:542b
Diacaustic(s), **2**:47b
of spherical interface, **1**:475a
Diacetic acid, **3**:347b
Diacetylacetone, **3**:347b
Diacrisis, **12**:312b
Diadkovskii
see Dyadkovsky
Diagenesis, **3**:161b, 162a; **10**:126b; **11**:372a; **12**:277a
Diagnostics
ancient Greek, **10**:581b
ancient Roman, **5**:231a; **12**:538b
seventeenth-century, **12**:102a
eighteenth-century, **1**:332b–333a
nineteenth-century, **6**:58b; **9**:101b; **12**: 450b
twentieth-century, **6**:233a
differential, **6**:433a; **12**:102a
Diagonal process (math.), **3**:54b, 55a–b
Diakinesis, **2**:626b
Diakonov, Ivan Aleksandrovich, **4**:553b
Diakonow, C., **6**:505b
Dialectic
ancient Greek, **1**:251a–258a, *passim;* **4**: 423b
ancient Roman, **2**:232b; **3**:109b
medieval, **1**:1a,b; **4**:94b, 525a; **12**:91a
nineteenth-century, **15**:410a–412a
twentieth-century, **8**:184a–b; **14**:469a
of nature, **15**:135b, 141b–145b
and technology, **1**:366a
Dialuric acid, **8**:343b
Dialysis, **7**:208a
Dialyzer, **5**:494a–b, 595b
Diamagnetism, **1**:557b; **3**:565a*; **6**:251b; **9**: 155a; **13**:521b
of aromatic compounds, **8**:475b
atomic nature of, **3**:3b–4a, 506a–b
discovery of, **4**:538b

Diamagnetism (cont.)
Landau diamagnetism, **7**:617a
measurement of, **11**:241b
and temperature, **3**:506a
theories of, **1**:555b–556a; **2**:266a; **8**:10b–11a; **10**:185b; **14**:208a
Diamond, **2**:289b, 293a; **14**:258a
artificial, **2**:459b; **9**:451a–b
combustibility, **8**:620a; **11**:564b; **13**:280b
crystallography of, **4**:597b; **5**:187a, 456a
deposits of, **4**:261b; **6**:552a
destructibility of, **3**:560b
electron distribution, **6**:304a
false equilibrium, **4**:230a
heat, effects of, **3**:5b, 560b
luminescence of, **4**:215a; **11**:348a
mining of, **3**:548a*
optical properties, **11**:266–267a
structure of, **5**:203a; **8**:485b; **11**:266b; **15**:63a
Dianthus, **7**:441b
Diapedesis, **11**:374a
Diaphragm
anatomy of, **6**:64a
embryogeny of, **2**:383b
role of, **13**:32b
Diaphthoresis, **1**:552a; **13**:148b
Diaphysectomy, **8**:114b
Diaries (astron.), **15**:676b–678a, 680b
Diario literario de México
founding of, **1**:127b
Diarrhea
infantile, **4**:404b
etiology of, **15**:418a
infectious, **7**:100b
Dias, Bartholomeu, **10**:513a; **15**:88b
Diastases, **4**:211b; **8**:342b; **10**:436a, 532a; **13**:557a
see also Enzymes
Diastole, **3**:220a; **4**:586a; **5**:234b, 235a; **8**:536a; **12**:165a
Diastrophism, **2**:558b; **3**:191a, 594b; **5**:395b; **7**:371a; **8**:99a; **10**:508a,b; **13**:532a; **14**:298b
Diathermancy, **8**:261b; **9**:566b
Diathermometer, **13**:11a
Diathermy, **1**:304b
Diathesis
rheumatic, **1**:400a
Diatom ooze, **9**:588b; **11**:372a
Diatoms, **1**:397b; **3**:321b–322a; **5**:72a, 531a; **14**:398a
as index fossils, **3**:322a
reproduction, **9**:587b
staining, **8**:512b
walls of, **7**:533b, 534a
Díaz de Solis, Juan, **10**:617a
Diazo reaction, **4**:296b–297a; **5**:536b
Diazoacetic ester, **2**:562b
discovery of, **3**:511a
Diazobenzene, **7**:282a
Dicaearchus of Messina (*fl.* 310 B.C.), 4:81–82; **4**:389b; **5**:345a; **11**:225b
Di Capoa, Leonardo, **2**:309a
Dice games, **3**:66a; **6**:600a; **15**:283b, 370a–b
Dichloroacetic acid
medical use of, **7**:208b
Dichogamy, **2**:257a; **12**:588a
Dichroscope, **6**:19a
Dicke, Robert Henry, **5**:273a; **14**:306b
Dickinson, Roscoe Gilkey (1894–1945), 4:82
Dickinson, William Lee, **8**:15b, 17a
Dickinson College, **11**:617b
Dickson, James Douglas Hamilton, **10**:451a
Dickson, Leonard Eugene (1874–1954), 4:82–83; **2**:197b; **5**:375b; **9**:502a, 503b*; **14**:211b
historiography, **6**:77a*; **8**:386a; **9**:191b, 388a*; **11**:61b; **14**:438b
History of the Theory of Numbers, **15**:66b, 692b
Dickstein, Samuel (1851–1939), 4:83–84; **7**:72b*; **8**:307a*; **12**:513a*; **15**:226a*
Dicotyledons, **2**:491b, 492a; **8**:513a; **15**:148a
polystely, **5**:604b–605a
Dictionaries, language, **15**:232a, 635a–636a
Dictionaries, scientific
medieval Arabic, **2**:155a; **7**:357b–358a
medieval Latin, **7**:27a–28a; **14**:35b
Renaissance, **2**:537a–b; **4**:170a
seventeenth-century, **11**:342b
eighteenth-century, **4**:85a; **11**:463b*
biological, **4**:273b; **11**:574b
chemical, **3**:6a; **7**:150b; **8**:619b–620a; **13**:547b
geographical, **14**:547b
mathematical, **7**:405a
medical, **4**:252a; **6**:269a; **11**:492a
mineralogical, **12**:329b
natural history, **8**:619b
Dictionary of National Biography, **3**:463b
Dictionnaire classique de l'histoire naturelle, **2**:320b
Dictionnaire de l'industrie commerciale et agricole, **1**:518a
Dictionnaire des arts et métiers, **1**:527b
Dictionnaire des sciences médicales, **1**:617a; **7**:557a
Dictionnaire des sciences naturelles, **3**:521a; **7**:198b; **15**:127a
Dictionnaire portatif des arts et métiers, **8**:619b
Dictyosiphonales, **12**:125b
Dicyemida, **1**:601a; **14**:314a
Dicynodon, **2**:504a
Dicynodonts, **10**:40b
Didactics
see Education
Diderot, Denis (1713–1784), 4:84–90; **3**:200a, 284b, 285a, 380b; **4**:203b; **6**:469a; **7**:592b; **8**:68a; **15**:127b
on contemporary scientists, **3**:381b; **13**:603a
Encyclopédie, **1**:58b, 114b, 115a–b, 117a; **2**:301b, 335a; **4**:85a, 86a–87b; **5**:567b; **8**:177b; **13**:465a
Interprétation de la nature, **8**:67a
Lettre sur les aveugles, **3**:381b
probability, **15**:280a
Supplément au voyage de M. de Bougainville . . ., **3**:365b
Didymium
composition of, **2**:254b; **3**:478a
discovery of, **9**:541b
isolation of, **2**:588b
non-elemental nature of, **3**:321b
Didymus (the musician), **1**:282b; **11**:201a–b
Didymus Chalcenterus, **1**:170b
Diebl, Franz, **9**:278a, 279b
Dielectric constant, **8**:47a; **9**:376b
Clausius-Mosotti equation, **3**:618b
complex, **1**:284a
Debye equation, **3**:619a
frequency dependence, **5**:391a
and index of refraction, **2**:265b–266a; **9**:212b
and ionic equilibrium, **3**:620a; **15**:435a
measurement of, **8**:106b; **15**:435a
and resistance, **1**:298a
temperature dependence, **3**:618b–619a; **6**:163b
Dielectric dispersion, **3**:619a–b
Dielectrics, **2**:120b; **10**:518b; **14**:573b
characteristics of, **5**:461b; **13**:376b; **15**:251b
currents produced
in nonuniform electric field, **4**:306a
"Debye behavior," **3**:619b
displacement current
magnetic field of, **4**:305b–306a; **13**:364b
electric breakdown in, **5**:161a; **7**:527a
inductive powers, **5**:483b; **6**:345a; **7**:527a
magnetic effects in, **11**:530a–b
residual charge, **6**:342a
resistance of, **11**:531b
theory of, **3**:445a,b; **8**:106b; **9**:548a–549a
Diels, Ludwig, **4**:90a
Diels, Otto Paul Hermann (1876–1954), 4:90–92; **15**:1a
Diels-Alder reaction, **1**:105b–106b
Diels's acid, **4**:90b
Dienes
in biosynthesis, **1**:106b
polymerization, **7**:457a–b
synthesis of, **1**:105b–106a; **4**:91a–b, 553a; **7**:457a
Diet
see Dietetics, Nutrition
Dieterich, Helwig, **13**:234b
Dieterici, Conrad, **7**:458a, 469b; **10**:345b; **14**:561a
Dieterici, Friedrich Heinrich, **15**:250a
Dietetics
ancient Greek, **6**:318b, 421b, 422a–b, 426a, 428a–b
ancient Roman, **3**:174b; **5**:230b–231a
medieval Arabic, **12**:90b; **14**:112b, 585a, 637b
medieval Latin, **9**:364b
Renaissance, **4**:412b; **8**:554b; **9**:441a; **12**:323b
seventeenth-century, **7**:354b–355a; **9**:369b
eighteenth-century, **8**:362a–363a, 506b*, 585b; **11**:363b; **12**:458b
nineteenth-century, **1**:543b; **5**:126b; **11**:429b; **13**:541a
twentieth-century, **5**:315b*; **9**:335a
quantitative, **5**:126b
see also Nutrition
Dietrich, Baron de
see Holbach, Paul Henri Thiry, Baron d'
Dietrich von Freiberg (*ca.* 1250–*ca.* 1310), 4:92–95; **1**:337b
De iride et radialibus impressionibus, **7**:218a–b
rainbow theory, **1**:101a; **4**:158b
Dieudonné, Jean, **2**:351b
Difference (philos.)
concept of, **1**:278a
Difference engine, **1**:355b, 356b*
Differences (math.), **2**:294a–b; **3**:255b, 432b; **6**:324a; **9**:311a
method of, **1**:355a–b; **2**:473a; **3**:269b–270b, 385b; **5**:466b*; **8**:137a; **9**:554b; **14**:161b; **15**:383b
Differential (math.), **3**:432b; **6**:403b
Abelian, **7**:398b
algebraic, **6**:547b; **11**:207a
binomial
integration of, **3**:228a; **9**:304a
contravariant, **5**:201b
definition of, **4**:222b(n); **8**:304a–b, 384a; **15**:99a
elliptic, **3**:228b; **14**:630b
nature of
eighteenth-century discussion, **1**:115b; **7**:567b
partial, **5**:54b
semiexact, **12**:331a–b
transformation of, **7**:567b
Differential equation
see Equation, differential

Differentiation, geological, **4**:141b
 see also Magma, differentiation of
Differentiation, mathematical, **3**:282b; **4**:477b; **10**:46b, 47a–b, 48b–49a; **12**:535a
 cogredient, **8**:389b
 continuous, **8**:165b
 covariant, **3**:263b
 definition of, **11**:402a
 differentiability, **13**:562a
 generalized, **10**:47a–b; **11**:106b
 partial, **8**:166a
 scales of, **5**:111a
 theory of, **12**:78a; **13**:23b
Diffraction, **3**:528b
 discovery of, **5**:543a–544b
 at edge of screen, **5**:483b; **14**:339a
 of electrons, **3**:369b, 598a; **7**:470a; **11**:422b
 intensity of, **1**:319a
 of light
  phase difference, **14**:617a
 mathematical analysis of, **9**:376b; **12**:526a,b
 in microscopy, **1**:8a–b
 size effects, **1**:284b
 theories of, **8**:501b; **9**:118b; **10**:54b, 59b, 79a; **15**:376b–377a
  dynamical, **13**:76b–77a
  and quantum theory, **4**:196b
  and wave theory of light, **1**:201b, 357b–358a; **5**:143b, 166a, 167b–169b
 of X rays, **1**:319a; **2**:398b, 431b; **3**:367a, 564a; **7**:272a, 468b; **8**:51a–b; **9**:542b–543a; **12**:464b, 527b; **14**:617a; **15**:61b–63a
  magnetization, effects of, **3**:367b
  mathematical analysis of, **12**:527a
Diffraction bench, **12**:519a
Diffraction grating, **1**:357b–358a; **2**:424a; **3**:420a, 430a; **4**:183a, 283a; **5**:143b; **7**:267b, 270b; **9**:373b; **10**:134a; **11**:578b–579a; **12**:37a, 132a, 499a; **14**:338a
 displacement law, **11**:472a
 grating equation, **5**:143b
 manufacture of, **7**:107b; **14**:497a,b
  photographic, **13**:101b
 reflecting, **5**:143b
 resolving power, **13**:101b
 sound waves, **3**:620a
 X-ray spectra, **3**:370a
Diffraction images
 see Images, diffraction
Diffractometer
 X-ray, **15**:61b–63a
Diffusion, **4**:203a, 264a–b
 of electrolytes, **15**:433b–434a
 of fluids, **13**:11a, 351a
 of gases, **2**:263a, 588a; **3**:307b; **4**:375b; **7**:416b, 509b; **8**:510b; **9**:222a–b; **11**:146a
  Graham's law, **5**:493a
  and temperature, **13**:11a
 of liquids, **4**:615a–b; **5**:494a–b
 models of, **9**:129b
 theory of, **5**:161a; **9**:222b
 thermal, **4**:375b; **14**:503b
  mathematics of, **9**:174b; **15**:384a–385a
 through membranes, **8**:541b
 velocity of, **3**:334a
Digby, Kenelm (1603–1665), 4:95–96; 2:13a; **4**:576a(n); **5**:159a, 415b; **10**:43b, 60b; **14**:301b
Digestion
 medieval study, **5**:147a
 Renaissance study, **11**:528a
 seventeenth-century study, **2**:237b, 308a–b; **4**:62b; **5**:427a; **6**:183b, 255a–b, 257a
 eighteenth-century study, **11**:334a, 616a–b; **12**:558a–b, 560a; **13**:46b
 early nineteenth-century study, **1**:542b–544a; **10**:582b; **11**:172b–173a; **13**:403a–b; **14**:558b
 mid nineteenth-century study, **1**:471a; **2**:25a–b, 28a–29a, 124a–b, 531b; **4**:252a; **7**:520a; **11**:216b; **12**:165a, 244a; **15**:108a–b
 late nineteenth-century study, **2**:547b; **3**:257a; **5**:149a; **10**:431b, 433a; **11**:426a–b
 artificial, **1**:543b
 chemical aspects of, **1**:543a–b; **5**:431b; **13**:403a
 enzymes in, **4**:211b; **10**:433a
 gastric, **1**:542b–544a; **2**:28a–b, 525a; **4**:211b; **10**:493a; **12**:558a–b
  theories of, **10**:307a–b
 as a gradual process, **3**:257a
 and hibernation, **6**:566b
 infantile, **4**:403b
 intracellular, **9**:333a–334b
 microbial, **4**:211b
 pancreatic, **2**:28a; **4**:211b
 peptic, **2**:531b
 psychic influences upon, **1**:543b
 and stimulants, dietetic, **1**:543b
 theories of
  ancient Greek, **4**:384a; **11**:127b
  ancient Roman, **5**:233b–234a
  Renaissance, **8**:203b, 205b
  seventeenth-century, **6**:183b; **13**:171a, 223b; **14**:405a
  eighteenth-century, **12**:558b; **13**:46b
  nineteenth-century, **10**:433a; **11**:133a, 173a; **14**:558b
 see also Dyspepsia
Digestive system
 anatomy of
  ancient Greek study, **1**:265a; **6**:317b
  Renaissance study, **1**:619a; **8**:230b; **9**:165b; **14**:9b
  seventeenth-century study, **5**:426a, 427a; **10**:520a, 567b–568a
  eighteenth-century study, **5**:226b–227a; **8**:327b
  comparative, **5**:535a
  dye studies, **4**:290b
  models, **8**:327b
 diseases of, **4**:252a; **10**:621b–622a
 duodenum
  diseases of, **2**:464a
  naming of, **6**:317b
  papillae, **12**:100b
  retrograde passage, **1**:543b
  yellow staining of, **1**:619a
 embryogeny of, **9**:57a, 253a; **10**:194b
 evolution of, **9**:333b
 ileocecal valve, **1**:46b, 98b; **13**:490b
 intestines, **15**:740b
  absorption processes, **6**:226a
  blockage of, **3**:206b*
  chyle vessels, **13**:490b
  dissection of, **4**:198a
  embryogeny of, **6**:573a
  erosion, **14**:638a
  flora, **4**:358a, 403b; **8**:169b; **9**:335a
  histology of, **9**:56a
  hormonal regulation, **9**:20a
  innervation of, **1**:537a; **9**:259b
  movement of, **1**:537a; **10**:567b
  perforation of, **1**:612a
  resection of, **4**:398a; **14**:587a
  stimulus conduction, **9**:19b
  surgery of, **3**:518a
  villi, **2**:531b; **5**:226b; **6**:269a; **8**:327b; **11**:592b
 mechanism of, **15**:71b–72a, 72b–73a
 Peyer's patches, **2**:531b
 physiology of, **5**:426a, 427a; **10**:582b
 pylorus
  rhythmic movement of, **6**:255a
 see also Peristalsis, Stomach
Digges, Leonard (*ca.* 1520–*ca.* 1559), 4:97; 1:383a; **10**:151a
 works printed, **5**:347b–348a
Digges, Thomas (*ca.* 1546–1595), 4:97–98; 5:397b
Dighton Rock, **5**:520a
Digitalin, **4**:160b
Digitalis (drug)
 chemistry of, **14**:444a
 effects of, **2**:547b; **8**:453a; **11**:215b; **12**:612a; **14**:464a–b; **15**:100a–101b
 pharmaceutical use of, **3**:578a–b; **15**:100a–101b
Digitalis (plant), **13**:154b; **15**:161a
Digitonin, **14**:444b
Dihedron, **7**:398a
Diisopropenyl
 polymerization of, **7**:457a–b
Dijon Academy, **3**:419b; **5**:602a–b
 chemistry, **5**:600b–601a
 competitions of, **3**:71a
Dikes (geol.), **1**:153b; **5**:39b*; **8**:565b; **11**:547b; **12**:272b–273b; **13**:273a
 ring, **1**:394b
 swarms of, **1**:394a–b; **10**:9a
Diketopiperazines, **14**:602b
Dilatancy, **11**:393b
Dilatometer, **8**:119a; **11**:208a
 recording, **11**:550a
Dilatometry, **1**:612b; **2**:343a–b; **3**:239a–b; **15**:457b
 of minerals, **3**:333b; **8**:119a
 of steel, **3**:212b*
Dillenius, Johann Jacob (1687–1747), 4:98–100; 3:351a; **6**:220a; **8**:375a; **12**:394a,b; **13**:554a
 Historia muscorum, **6**:218a
Diller, Joseph Silas, **7**:267a
Dilogarithm, **7**:619b
Dilution
 heat of, **4**:228b; **11**:418a, 577a
 Ostwald's law of, **1**:300a
Diluvial period
 see Quaternary period
Diluvialism
 late seventeenth-century, **2**:613a–b; **6**:129b; **8**:416a; **12**:257a; **14**:501b
 early eighteenth-century, **1**:393a; **2**:348a; **3**:129b, 210b; **5**:519b; **9**:534a, 539b; **13**:179a; **15**:55a–56a
 late eighteenth-century, **4**:28a; **6**:580a; **7**:412a; **10**:284b
 early nineteenth-century, **2**:496b, 556b, 568a–570b; **3**:395b, 567a; **6**:456b
 mid nineteenth-century, **5**:337b; **15**:51a
 critics of, **5**:32a; **8**:4a, 308a; **9**:532a–b; **10**:280b; **11**:133b; **12**:262a
Diluvium, **5**:337b, 519a
al-Dimashqī
 see Abū ᶜUthmān al-Dimashqī
Dimension
 theory of, **1**:407b; **2**:513a, 565b–566a; **3**:54b, 171a; **5**:97a–b; **6**:176b; **8**:112a; **13**:190a, 549a–b
  see also Measure theory
Dimensional analysis, **2**:460a, 565b; **3**:73b; **10**:162a
Dimethylgloxime, **3**:272a

Dimorphism (cryst.), **2**:135b; **9**:424a; **10**: 357b; **12**:306a; **13**:311a
Dimsdale, Thomas, **7**:11b
Dinakara (*b. ca.* 1550), 4:100
Candrārkī, **15**:588b
commentaries of, **5**:275b; **12**:598a
Kheṭasiddhi, **15**:588b
Tithisāraṇī, **15**:588b
Dinant, basin of, **5**:476a
Dinantian (stratig. stage), **8**:463b
al-Dīnawarī, Abū Ḥanīfa, **7**:532b; **9**:614b; **11**:246b, 247a
Kitāb al-nabāt, **1**:350b
Dines, William Henry, **12**:366b
Dingler, Hugo Albert Emil Hermann (1881–1954), 4:100–102; **8**:159b
Dingler, Johann Gottfried, **12**:196b–197a
Dinglers polytechnisches Journal, **12**:197a
Dini, Ulisse (1845–1918), 4:102–103; **2**:105b; **14**:46b
analysis, **4**:206a; **8**:111a; **15**:386b
students of, **2**:121a; **4**:373b; **5**:200b; **9**: 378a; **10**:610a; **11**:407a; **14**:86a
Dinichtys, **8**:463a
Dinitrogen tetroxide
color changes, **4**:239b
Dinoflagellata, **4**:142b; **7**:447a
Dinornis, **10**:262a
Dinosaur, **3**:606b; **4**:147b; **5**:500b; **6**:594b–595a; **11**:336a; **13**:525a; **14**:409b; **15**:92a
armored, **9**:87b
classification of, **9**:134b
discovery of, **9**:87a–b
tracks, **6**:438a
Dinostratus (*fl.* fourth century B.C.), 4:103–105; **6**:407b; **9**:269a; **13**:334a
Dinotherium, **2**:423a
Diocles (*fl. ca.* 190 B.C.), 4:105; 15:115–118
On Burning Mirrors, **1**:186b, 187a; **4**: 489b; **9**:272a; **10**:298a
translations of, **14**:604a; **15**:178a
geometry, **4**:490b; **14**:274b
Diocles of Carystus (*fl.* late fourth century B.C.), 4:105–107; **4**:383a–385b *passim;* **6**:419b; **11**:128b
botany, **13**:331b
physiology, **6**:317a; **11**:127b
psychology, **13**:94b
Diode, **5**:33a; **6**:546b
Diodorus of Alexandria, **15**:519b
Diodorus Siculus, **1**:213b; **9**:337b; **10**:179b, 181a, 300a; **15**:678b
Bibliotheca historica, **4**:34a
Dioecia, **13**:392b
Diogenes of Apollonia, **10**:181a
Diogenes Laërtius, **1**:103b, 120a, 150b, 338b; **2**:378a, 550a; **4**:31b, 382a–b, 388b; **5**:285a; **6**:289b, 381b, 382a, 407a, 411a; **8**:192b; **11**:23a, 105b(n); **12**: 575b; **13**:91b, 295a, 314b, 328b; **14**: 535a, 608a
Dion, **11**:25a, 29a; **12**:575b
Dionaea, **2**:599a
Dionis du Séjour, Achille-Pierre (1734–1794), 4:107–108; **2**:216b; **3**:387b(n); **5**:567a; **15**:291a, 292b, 310b
and Académie des sciences, **15**:309b, 310a, 333b
Dionysius the Areopagite, **3**:514b; **11**:161b; **14**:225b
Dionysius Exiguus, **1**:565b
Dionysius the Younger, **6**:406a; **11**:29a
Dionysodorus (*fl.* third-second centuries B.C.), 4:108–110
geometry, **4**:490b; **6**:315a; **14**:274b; **15**: 116b
Dionysodorus of Amisene, **4**:108b
Dionysodorus of Caunus, **4**:108b
Dionysodorus of Melos, **4**:108b
Diophantine analysis
see Analysis (math.), Diophantine
Diophantus of Alexandria (*fl.* A.D. 250), 4: 110–119; 15:118–122; **1**:40a; **3**:622a; **4**:424b, 569b; **15**:10a
algebra, **1**:30b, 31a; **7**:360a, 362b; **15**: 669a,b, 671b
Arithmetica, **1**:367b–368a; **2**:280b, 281a; **4**:55b, 567a,b, 573a, 574a,b, 606b, 612a,b; **7**:51b; **11**:185b; **14**:19b, 20a,b
commentaries on, **6**:615b
editions of, **11**:18a
translations of, **4**:212b; **7**:241a, 244a–b
commentaries on, **11**:244b
Diophantine equations, **2**:507a; **4**:83a, 124a; **7**:565b; **11**:55b, 61b, 512a; **12**:451b
Diophantine problem, **11**:55a–b
editions of works, **5**:409a; **13**:252b, 253a,b
manuscripts of, **11**:349b
number theory, **4**:610b, 611a; **11**:221b; **13**:540a
De polygonis numeris, **6**:617b
translations of works, **5**:409a; **11**:244b
Diopside
viscosity of, **4**:141b
Dioptra (astron. instrum.), **6**:311a, 312b; **7**:338b; **15**:213b, 219a,b
Dioptrica, **6**:538a
Dioptrics
ancient Roman, **11**:201a
medieval Arabic, **6**:194a–b, 195b; **7**:212b–215b
medieval Latin, **5**:552a–b; **10**:475b; **14**: 460b–461a
Renaissance, **5**:106a; **9**:192a; **11**:96b, 97b
seventeenth-century, **1**:475a; **3**:615b; **5**: 408b, 521a, 525b, 526a; **6**:148b; **9**: 118a–119a; **10**:104a; **11**:507a
eighteenth-century, **2**:328b; **3**:454b*; **4**: 482a; **5**:209b; **7**:404a; **13**:267a
nineteenth-century, **5**:306b; **9**:430b; **12**: 289b
see also Optics, Refraction
Diorismos, **4**:416a; **8**:189a,b; **13**:334a
Dioscorea, **3**:422a
Dioscorides, also known as Pedanius Dioscorides of Anazarbus (*fl.* A.D. 50–70), 4:119–123; **1**:235a, 538b; **2**:219b, 536b; **4**:139a; **7**:539b; **8**:483b; **11**: 601b
botany, **1**:167a; **13**:343a
Codex vindobonensis, **1**:520a
commentaries on, **1**:539a; **5**:384a
editions of works, **2**:537a; **3**:534a
influence of, **6**:79b; **7**:27b, 187a; **11**:231a, 311b, 312a, 444b; **14**:112b
De materia medica, **1**:150b; **3**:414a; **4**: 138b; **10**:418b
commentaries on, **3**:414b; **8**:554a; **9**: 179b–180a
translations of, **7**:187b; **9**:179b–180a; **11**:595a; **15**:238b
pharmacology, **9**:41a
translations of works, **9**:75b; **13**:502b; **15**:230b
Dioxane, **4**:554a
Dip sector (instrum.), **14**:491b
Diphenyls
stereochemistry of, **7**:517b–518a
vapor density, **9**:356a
Diphenylurea
discovery of, **3**:193b
Diphosphopyridine nucleotide (DPN), **4**: 486a; **6**:111a
Diphtheria, **2**:575b, 584b; **3**:344b, 447b; **4**: 404a; **5**:451b; **14**:249b
antitoxin, **4**:211b, 297b, 298a; **7**:427b; **8**:409a; **12**:88b, 481b; **15**:521b
avian, **2**:300b
bacillus, **4**:276b, 404b; **7**:425b; **11**:569a; **14**:551b
discovery of, **8**:448b–449b
description of, **1**:399b; **5**:384b–385a
false membrane, **2**:445a; **5**:384b; **8**:449a
individualization of, **2**:445a
serotherapy, **1**:575a, 576a; **4**:404b; **12**:88a
skin, **10**:19a
toxin, **11**:569a–b
discovery of, **8**:449b
toxin "spectra," **2**:584a
treatment of, **11**:569b
vaccination, **1**:577a; **9**:545b; **11**:271b; **12**:88b
Diphyllobothrium
development of, **8**:270b
Diplococcus
pneumonia, **4**:276b
Diplograptus, **11**:594a
Diploidization, **2**:583a–b
Diplomacy, scientists and
ancient Greek, **6**:406a
ancient Roman, **10**:111a
Byzantine, **13**:225b
medieval, **5**:115a–b; **9**:136a
Renaissance, **2**:283a; **4**:208b, 212b; **5**: 447b; **9**:308b; **10**:616a–b; **13**:613a
seventeenth-century, **2**:163a, 201a; **4**: 221b; **5**:382b–383a, 574a–b, 606b; **8**:149b–150b; **9**:3b, 15b, 97a, 529b; **10**:200b; **12**:307a
eighteenth-century, **2**:326b, 327a; **5**:93b, 138a; **6**:68a, 550a; **7**:89a, 183a; **9**:26a
nineteenth-century, **6**:215a, 456a, 551b, 552a; **9**:267b–268a; **15**:152b–153a
Dipnoi, **7**:314a; **15**:126b
Dipodium, **2**:322a
Dipole, magnetic
concept of, **11**:472a
Dipole moment, **3**:619a
Diprioden valensis, **3**:606b
Diptera, **1**:587a; **12**:612a
balancers of, **1**:432a
embryology, **14**:233b, 234a
life cycles, **7**:273a
taxonomy, **13**:134b, 138a; **14**:410b
Dipterocarpaceae, **13**:459b
Dirac, Paul Adrien Maurice, **2**:248a,b; **3**: 369b, 564a; **4**:281b, 578a, 579a; **7**: 153b, 154a; **10**:213b, 214a,b, 423a; **12**:222a,b; **13**:240a; **15**:26b, 48b
Dirac equation, **5**:473b; **8**:498b; **9**:38b, 398a–b
Dirac function, **14**:91b
Dirac statistics, **4**:577a; **5**:103a; **8**:478b–479a
Direction finder, **7**:265b
Directorium (astron. instrum.), **7**:125b, 126a; **14**:389b
Dirichlet, Gustav Peter Lejeune (1805–1859), 4:123–127; **2**:192a; **4**:205b, 206a, 342a,b; **5**:96a, 97b, 308a; **6**: 553a; **7**:50b, 55b*
acquaintances of, **1**:15b; **3**:450b, 466b; **5**:94b, 301a, 304b, 307a; **7**:51a–b, 52a, 521b; **8**:138b, 382b; **11**:447b, 448a
analysis, **8**:432b; **13**:26b
correspondence of, **13**:250b
and Dedekind, **2**:238a; **4**:2a, 3a–b, 125b, 126b
Dirichlet integrals, **5**:201b; **11**:451b

Dirichlet series, **1**:306b; **2**:238b; **4**:103a, 125a; **6**:114b*, 214b, 571b; **7**:408a; **11**:256b; **12**:451b
Dirichlet's principle, **3**:315a; **4**:125a; **5**: 202a*, 608b; **11**:53b, 54a, 449b–450a, 451b, 452a; **12**:246a; **14**:222b
proof of, **6**:391b
Dirichlet's problem, **4**:126b, 561b; **6**:4b; **8**:562b; **10**:553b; **11**:58b–59a, 59b, 261a, 449b, 459b, 478a; **12**:246a; **14**:95a, 345b; **15**:23a
generalization of, :54a
editions of works, **4**:4b
hydrodynamics, **2**:166a–b; **4**:5a
influence of, **7**:109a, 522a, 523a,b; **11**: 453b; **12**:469a
and Jacobi, **4**:124b, 126a
number theory, **9**:173b, 412a
philosophy of mathematics, **7**:508a
prime number theorem, **3**:221b; **4**:124b–125a, 475a; **6**:214b, 308b; **8**:139a
students of, **1**:294a, 370a, 389b; **2**:299a; **3**:59a, 263a; **4**:340b; **5**:256b; **6**:230a; **7**:108b, 505b, 506a; **8**:388b; **12**:170b, 225b, 289b; **14**:231b, 329b, 480b
Vorlesungen über Zahlentheorie, **2**:238a; **5**:299b
editions of, **4**:4b
publication of, **4**:125b
Disaccharides
structures of, **6**:185b; **14**:603a
Discharge, electric
afterglow, **13**:107b
arc, **4**:79a; **5**:597b; **10**:552b
temperatures, **11**:559b
colors seen, **6**:395b
corona, **5**:117a
disruptive, **14**:218a
electrostatic, **10**:553a
electrochemical effect of, **4**:535b–536a
magnetic effect of, **4**:535b; **13**:183b
in gases, **1**:559b; **3**:31b; **4**:19a; **5**:117b, 292a–b, 340b–341a, 483b; **7**:84b, 554a–b; **8**:24a–b, 148b; **10**:345a; **11**:46b, 47a, 145a, 580b; **12**:238b; **13**:80b, 236b, 365b–366b, 468b, 473a; **14**:171b, 172a, 330a
X-ray effects, **12**:26a–b
luminous, **6**:439b; **10**:596b
striation, **5**:292a–b
point, **14**:171b, 172a
and pressure, **5**:292b
and radiation, **4**:356a
and shape, **5**:517a
silent
use in chemistry, **2**:70a–b
spark, **4**:559a,b; **9**:166b; **11**:142a; **13**:446b; **14**:171b, 172a
from battery, **5**:292a
duration of, **11**:531b
in gases, **9**:472a
mechanical effects, **8**:597a
spectrum of, **14**:290a
in a vacuum, **9**:166b
Discharge tubes, **1**:320b; **10**:348a, 604b; **12**:33b
analysis of, **14**:217b–218a
see also Cathode tubes, Electron tubes
Dischidia, **12**:259a
Discontinuity, **1**:207a; **2**:466a; **6**:240a; **10**: 554a; **12**:496a–b; **13**:427b; **14**:346a
Discovery (ship), **3**:397a; **9**:124a, 587b
Discovery, scientific, **10**:486a; **11**:60b
imagination, role of, **3**:476a
Discriminants, **2**:294b; **3**:164a; **5**:409b; **7**: 444b; **9**:412b; **11**:450b
concept of, **12**:291a–b
Disease(s)
acute vs. chronic, **12**:538b
from animal bites, **1**:526a
and body temperature, **10**:387b–388a
chemical tests for, **3**:160a
chronic, **13**:215a
classification of, **3**:579a–b; **8**:352b, 379a–b; **9**:75a; **10**:612b–613a; **11**:33a; **12**: 538b; **15**:128a
by electrical reaction, **2**:83a
congenital
concept of, **13**:97a
contagious, **11**:462a
elimination of, **3**:332b
pathogenesis of, **1**:492b–493b, 540b; **3**:332b; **10**:374b–376a; **12**:203a; **13**: 564a
prevention of, **1**:493a–b; **15**:660a(n)
theories of, **2**:444b–445a, 575a–b, 589b; **5**:106a, 107a; **6**:269b, 285a–b; **7**: 420b–421a; **11**:37a–38a, 148a
crisis, **6**:422a
and environment, **10**:376a, 389a
enzymatic, **3**:416a; **14**:252b
fungal, **9**:176a
hereditary, **2**:584b; **3**:577b, 590b; **5**:426a; **10**:373a–376a; **14**:231a
infectious, **2**:164b; **5**:219a; **9**:333a; **10**:19a; **11**:33b, 175a
bacterial, **4**:154a–156a; **10**:155a
concept of, **3**:489b; **6**:521a; **7**:16b; **9**:10b
diagnosis of, **5**:317a
evolution of, **15**:455a
incubation period, **11**:37b
and nutrition, **15**:419a
pathogenesis of, **1**:438b; **4**:498b
serology, **2**:300b–301a
theories of, **10**:155a
transmission of, **7**:171a; **12**:503a
treatment of, **1**:574b–577a; **3**:220a; **4**: 154a–156a
viral, **2**:300b; **5**:317a, 451a–b; **8**:273b
insects as cause of, **1**:526a; **5**:451b; **12**: 315a, 468a, 481a
metabolic studies, **14**:65b
neurological, **2**:464a; **6**:60a
diagnosis of, **1**:569b, 580a; **2**:1b
treatment of, **9**:422b
relapse, **6**:422a
respiratory, **4**:276b; **7**:13a, 505a
and soil, **10**:258b
specificity of, **2**:444b–445a; **13**:215a
theories of
ancient Greek, **6**:421b–422a, 423a–b, 427a–428a; **10**:581b
ancient Roman, **12**:538b, 539b
medieval Latin, **6**:396b
Renaissance, **4**:387b; **10**:307b–308a; **12**:323b
seventeenth-century, **1**:593a–b; **4**:388a; **5**:425b; **6**:255b; **9**:582a; **11**:2a; **12**: 311b
eighteenth-century, **3**:579a; **8**:114b, 379a; **9**:511b; **11**:364b, 617a; **13**:232a
nineteenth-century, **2**:33a; **3**:463b; **6**: 269b; **10**:383b–384b, 386a–408a *passim;* **12**:57b; **15**:11b–12a
Japanese, **15**:741a, 756a
Mesopotamian, **15**:640a,b, 658b(n)
ontological, **2**:508b; **6**:255b; **10**:307b, 384a; **12**:296a
sociological, **14**:41b–42a
waterborne, **7**:171a
see also Chemotherapy; Deficiency diseases; Epidemiology; Fevers; Germ theory; Humors; Immunology; Infection; Neuropathology; Nosology; Occupational diseases; Ophthalmology; Parasitology; Pathology; specific diseases; Venereology
Disgregation, **5**:388a
concept of, **3**:308a
mechanical analogue, **3**:308b–309a
Disinfection, **1**:493a–b; **3**:6b, 7a,b; **7**:424a; **8**:448b
of air, **5**:603b
chemical, **3**:185b; **14**:15b
chlorine, **3**:189a, 202a, 238a; **5**:603b
iodoform, **1**:574b–575a
mercuric chloride, **7**:423a; **8**:408a
phenol, **3**:479b; **4**:299b; **7**:423a–b; **8**: 403a–b, 408a; **12**:478b
by heat, **6**:285b
internal, **1**:574b–575a
of milk, **1**:576b; **8**:450b; **11**:177a
with steam, **5**:219a; **7**:423a; **8**:408a
of water, **8**:450b
see also Antisepsis, Sterilization
Dislocations, **2**:314b
treatment of, **2**:546b; **6**:420b; **10**:418b
Dispensatory of the United States of America, **14**:496a
Dispersion phenomena, **3**:146b; **4**:506b
anomalous, **4**:191b; **7**:526a, 553a–554b; **9**:376b; **11**:580b; **13**:107a*, 409a; **14**:498b*; **15**:83b
dielectric, **1**:519b
mathematical theory of, **5**:170b
measurement of, **2**:328b
molecular theory of, **2**:337b
negative, **7**:553b, 554a–b
optical, **5**:143a,b, 391a; **7**:518b; **8**:493a, 496b; **9**:155a, 174a, 606b; **11**:115b
and quantum theory, **2**:247a–b, 248a; **7**:492a
theory of, **7**:553b–554b
X-ray, **3**:430a
Displacement, geological, **5**:395b
Displacement, mathematical, **3**:74b
parallel, **4**:281a
Displacement, psychological, **5**:177b
Dissection, human
ancient Greek, **1**:104a, 264b; **6**:316b
ancient Roman, **5**:230b; **11**:602b; **12**:80b–81a
medieval Latin, **3**:218b; **9**:468a, 469a; **12**:83a
Renaissance, **1**:522b, 523a–b, 603b–604a, 618b; **2**:546b; **3**:13a, 40a,b, 342a,b; **4**:199a, 285a, 413a, 486b, 487b, 519a,b; **5**:585b; **8**:203b; **10**:315b; **14**:4a–5b
seventeenth-century, **1**:482b; **7**:612a; **9**: 322b, 581b; **10**:490b; **11**:466b, 511a; **14**:25b, 287a, 405a
eighteenth-century, **3**:37b, 437b; **6**:231a, 568b; **9**:479b, 511a; **11**:112a; **12**:509b; **14**:180b
nineteenth-century, **2**:127b; **7**:415a–b; **10**:620b
cadavers, source of, **7**:415a–b; **9**:480a; **14**:7b
Chinese, **12**:384a; **15**:740b
pattern of, **4**:487b; **9**:468a; **14**:8b
public
ancient Greek, **4**:383a
Renaissance, **1**:619b; **3**:40a, 98a–b, 354b; **11**:33a, 253b
seventeenth-century, **2**:224b; **5**:419a; **12**:40a–b, 577a; **13**:490a–b; **14**:449b
eighteenth-century, **3**:15b
Dissection, zoological
ancient Greek, **1**:262b, 263b, 264a, 266b; **4**:106a; **6**:317a
ancient Roman, **5**:234a

Dissection, zoological (cont.)
medieval, **12**:81a–b, 82b, 83a
seventeenth-century, **4**:52b, 267b; **5**: 327b–328a; **6**:159b, 161a; **9**:62b, 581b; **10**:519b; **14**:287a
eighteenth-century, **3**:38a, 522a; **12**:509b
nineteenth-century, **3**:219b
pattern of, **12**:81b
see also Vivisection
Dissenters, **3**:537a, 538b, 539a, 556b; **6**:283a; **11**:139a–141b; **12**:466a, 471b
see also Daventry Academy, Manchester Academy, Warrington Academy
Dissociation (chem.), **3**:617b; **4**:79a, 228b; **5**:156b; **8**:118a; **11**:278b, 281a; **13**: 468a
in alcohol-water mixtures, **3**:333b
constant of, **13**:228a
of electrolytes
see Electrolytes, dissociation theory
of gases, **3**:11b, 373a; **4**:229a; **5**:560a; **8**: 23a,b; **9**:619b; **15**:36b–37a, 435b
heat of, **1**:300a; **6**:519b–520a; **8**:23b
temperature effects, **1**:300a; **3**:49a; **4**:78a; **9**:356a
thermal, **13**:299a
Distance
definition of, **3**:165b
measurement of, **4**:463a; **7**:19a
by lunar eclipse observation, **6**:312b
optical, **10**:5b, 6b; **11**:95a; **15**:478a
by radio, **1**:196a
see also Astronomy, measurement in, distance; Length
Distance of travel
calculation of, **2**:590b; **15**:593b
Distillation
in Renaissance, **2**:143a, 546a,b; **11**:97b
in seventeenth century, **5**:420b; **8**:312a
in eighteenth century, **13**:494b
of alcohol, **3**:201a; **8**:312a; **13**:534b–535a
analytical
of plants, **2**:353a; **5**:352b–353a; **8**:171b, 427b
apparatus for, **1**:527a; **2**:546b; **4**:41b; **6**: 41b–43a; **7**:354b; **8**:213b; **13**:535a; **14**:508a
continuous, **4**:41b
destructive, **1**:156a; **3**:556b; **9**:105a; **10**: 498a; **14**:156b
dry, **5**:602a; **9**:425a; **14**:601a
fractional, **1**:321a; **5**:4b; **6**:463a; **8**:56a, 482a; **12**:435b; **13**:454a; **14**:156b, 561a
low-temperature, **8**:427b, 482a
of solutions, **7**:462a
theories of, **5**:156b; **12**:312b
vacuum methods, **1**:168a; **12**:46b
of water, **8**:71b–72a
of wine, **1**:527a
Distomum haematobium, **2**:127b
Distribution, statistical, **2**:303b; **15**:385b
binomial, **3**:386b; **9**:452b–454a; **11**:237a
discontinuous, **5**:9b
exponential, **9**:219a, 222b
laws of, **1**:155b; **7**:352a; **9**:219a, 223a–b; **15**:365a, 372a
logarithmic, **5**:10b
multinomial, **14**:383a
normal, **3**:231a; **5**:8a, 10a, 477a; **9**:218b, 453b; **11**:237a–b; **15**:363b
bivariate, **10**:451a, 453a, 453b–454a
multivariate, **10**:451b
sampling, **5**:8a; **10**:454b, 457a–b, 459a; **14**:382b–383b
skew, **10**:458b
Distributive law, **3**:166a
Dithiocarbamates, **4**:20b
Dithionic acid, **5**:323a
Dithiourethanes, **4**:20b
Dittmar, William (1833–1892), 4:127–128; 11:536b; **12**:208b
Ditto, Raymond Clyde, **11**:420a
Ditton, Humphry, **6**:175b; **12**:446a
Diuresis, **1**:534a; **2**:227a; **8**:452b; **10**:307a
mechanism of, **15**:100b, 102b–103a
Divâkara, **7**:219a,b; **9**:580a
commentaries of, **5**:275b; **7**:315a,b; **9**:42a
Divâkara of Golagrâma, **5**:274a,b; **7**:219a
Divariants, **1**:294b
Diverticulosis
colic, **3**:490a
Dividing engine, **2**:141a; **10**:133b–134a; **11**:284b, 354b, 377b; **12**:132a
Divination, **3**:251b; **11**:105a, 161b; **12**:310a, 374a, 386a–b; **13**:582a,b; **14**:606b; **15**:642a–643b, 733b, 745a, 746a,b, 804b
and astrology, **15**:644a–b
magical, **15**:729a–b
Diving, submarine, **6**:68b
Diving bell, **10**:151a
Divini, Eustachio (1610–1685), 4:128; 1: 341b; **3**:101b, 102b; **9**:485b
Divining rod, **2**:600b; **3**:240b
Divinyl
polymerization of, **8**:108a–b
production of, **8**:108a
Diviš, Prokop (1698–1765), 4:128–130
Divisibility (math.), **4**:573a–b
infinite, **7**:327a; **14**:612b
theory of, **4**:474a–b; **14**:630a
Divisibility (philos.)
infinite, **1**:257a–b
Division (math.), **3**:181a; **7**:533a; **12**:94a; **14**:539a,b, 542a–543a
algebraic, **7**:241b; **12**:93a
Egyptian techniques, **15**:685a–686b, 690a
logical, **2**:296b–297a, 297b, 601b; **7**:105a
synthetic, **6**:511a
tables of, **14**:542b
by zero, **2**:274b–275a
Divisor, **5**:414a
common, number of, **3**:178a
elementary, **7**:396b
greatest common, **2**:444a; **3**:127b; **6**:537b; **7**:602a; **11**:512a
sum of, **8**:385b
Diwisch, Procopius
see Diviš, Prokop
Dix, John Adams, **6**:437b
Dixon, Harold Baily (1852–1930), 4:130; 3:197a; **6**:110b*; **10**:179a*; **11**:538b*
Dixon, Henry Horatio (1869–1953), 4:130–131
Dixon, Jeremiah (1733–1779), 4:131–132; 9:164a; **15**:339b
Mason and Dixon line, **4**:131b–132a
Dixon, Walter Ernest, **15**:106b
Dizé, Michel Jean Jérôme, **8**:113b; **10**:497a
DNA, **6**:505b
and bacterial variation, **1**:342b–343a
depolymerization, **4**:604a
distribution of, **4**:603b–604a
extraction of, **4**:603a–b
formation of, **6**:366b
polymerization of, **1**:343a
structure of, **1**:319b, 342b–343a; **4**:603a–604a; **5**:140a–141b, 273a, 589b; **15**: 18a
Dnieper (river), **8**:466b
Dniester (river), **8**:465b, 466a,b
Doan, Richard Lloyd, **3**:370a, 371a
Dobbie, James, **11**:278b
Dobell, Cecil Clifford (1886–1949), 4:132–133; 7:112a*; **8**:130a*, 297a; **12**:492a
Döbereiner, Johann Wolfgang (1780–1849), 4:133–135; 5:203a; **8**:335b
catalysis, **4**:241b; **6**:285a; **13**:313a
chemical elements, **5**:412a, 431a; **6**:495b; **9**:348b
editions of works, **9**:348b
Dobzhansky, Theodosius, **3**:572b; **9**:523b; **13**:134b, 137b, 276a
Dochez, Alphonse Raymond, **1**:342a–b
Dochmius, **5**:502b
Docimasy, **4**:347b
Doctor
see Physician
Dodart, Claude-Jean-Baptiste, **4**:136a
Dodart, Denis (1634–1707), 4:135–136; 4: 590a; **6**:477b; **9**:92b
Mémoires pour servir à l'histoire des plantes, **2**:334a
Dodder (plant), **9**:461a–b
Doddridge, Philip, **11**:139b–140a
Dodecahedron, **6**:92b; **11**:222a,b
inscribed, **6**:616b
stellated, **7**:301a
Dodgson, Charles Lutwidge (1832–1898), 4: 136–138
Symbolic Logic, **13**:612b
Dodo, **13**:354b–355a, 450a
Dodoens (Dodonaeus), Rembert (1516–1585), 4:138–140; 8:121a; **11**:312a; **13**:463a; **14**:586b
Cruydeboeck
translations of, **8**:120b
Stirpium historiae pemptades sex
translations of, **5**:362a–b
Dodwell, Henry, **3**:295a
Doelter (Cisterich y de la Torre), Cornelio August Severinus (1850–1930), 4: 140–142
Handbuch der Mineralchemie, **11**:271a
Doering, Adolph, **1**:130a,b
Doering, Oskar, **1**:130a
Dog, **3**:13a, 521a; **15**:648a
anatomy of, **4**:511a
decerebrate, **5**:463b
and disease, **15**:454b
fertilization in, **2**:161a–b
kidney of, **4**:520b
neurophysiology of, **5**:463b; **8**:15a
physiology of, **5**:279b; **14**:63b–65b
rapid breathing of, **11**:427b
Dogel, Valentin Alexandrovich (1882–1955), 4:142–143; 7:476b*
Dogfish
placentoid structure, **1**:265a
Dogmatism, **5**:285b
criticism of, **5**:415a–416a; **8**:257a; **14**: 84a–b
Dohrn, Carl August, **15**:122b–123a, 123b
Dohrn, Felix Anton (1840–1909), 15:122–125; 6:10a, 535a; **7**:401b, 402a, 475b, 517a; **14**:424a
comparative embryology, **1**:420b
students of, **1**:176b; **8**:3b, 586a; **12**:423b; **14**:3a, 314a
Dohrn, Reinhard, **15**:124b
Doisy, Edward Adelbert, **1**:123b
Dokuchaev, Vasilii Vasilevich (1846–1903), 4:143–146; 7:34b, 496a; **8**:285a; **9**: 535a; **13**:249a, 616b–617a; **14**:106b–107a
Dolde, Leonhard, **8**:309a,b, 310b
Dolerite, **4**:262a,b
cracking of, **1**:394b
Döllinger, Ignaz (1770–1841), 4:146–147; 1:73a, 386a; **10**:287a; **12**:202a, 420b, 511a*; **14**:351a
Dollo, Louis Antoine Marie Joseph (1857–1931), 4:147–148; 1:18a
Dollond, John (1706–1761), 4:148–149; 4: 472a, 482a; **7**:404a–b; **11**:284b; **12**: 413b; **14**:419b

instrumentation, **2**:216b; **3**:284a, 426a; **14**:589a
Dollond, Peter, **4**:148b, 149a; **5**:143a; **6**:112a, 512a; **9**:231b; **14**:452b, 485a
Dolomieu, Dieudonné (called Déodat) de Gratet de (1750–1801), 4:149–153; **3**:411b; **6**:178b; **7**:603b
critics of, **11**:169a
imprisonment of, **1**:436b
mineralogy, **6**:182a; **11**:167b
Dolomite (mineral), **2**:552b; **4**:140b, 150b, 152b–153a
Dolomites (mountains)
geology, **6**:168b
Dolomites (rocks), **2**:159a; **3**:161b; **8**:463a
origin of, **4**:309a
Dolomitization, **3**:161b
theories of, **6**:19a
Dolomys, **10**:16b
Dolphin
anatomy, **13**:526b, 527a
neuroanatomy, **12**:612a
Dolz del Castellar, Juan, **3**:171b
Domagk, Gerhard (1895–1964), 4:153–156; **5**:30a, 344b
Domain (math.)
n-dimensional, **2**:513a
Dombey, Joseph (1742–1794), 4:156–157; **3**:172b; **8**:303b; **11**:605b; **13**:495b
Dome (arch.), **2**:535a; **7**:28b–29a
Domestication
see Animals, domestication
Dominance, genetic, **3**:121a; **5**:10b, 218b; **7**:409a; **9**:188a
concept of, **9**:280a
modification of, **5**:454a
Domingo de Soto
see Soto, Domingo de
Dominic Gundissalinus
see Gundissalinus, Dominicus
Dominicans
in Middle Ages, **1**:99a, 289b; **2**:20a–b, 314a; **4**:92a–b; **5**:361a, 401b; **12**:547b; **13**:534a; **14**:34b
in Renaissance, **3**:558a; **5**:119a
in seventeenth century, **15**:68b
critics of, **1**:378b
philosophy, **2**:539a, 540a; **10**:172a
theology, **8**:548a
transmission of science, **1**:99b, 196b–197b, 227b; **9**:434b
Dominicus de Clavasio, ***also known as*** **Dominicus de Clavagio, Dominicus Parisiensis,** ***or*** **Dominic de Chivasso (*fl.* mid-fourteenth century), 4:157–158;** **7**:120a
optics, **10**:475b
Dominion Astrophysical Observatory (Victoria, British Columbia), **11**:18b
Dominion Observatory (Ottawa), **11**:18b
Dominis, Marko Antonije (1560–1626), 4: 158–159
Domite, **2**:554b
Domninus of Larissa (*fl.* fifth century A.D.), 4:159–160
Don (river), **8**:466b
Donaldson, Henry Herbert (1857–1938), 4: 160–161; **5**:26b; **9**:284b, 416b*
Donaldson, John Calvert, **4**:161a
Donath, Julius, **7**:623b
Donath-Landsteiner test, **7**:624a
Donath, Willem Frederik, **14**:393a, 446a
Donati, Giovan Battista (1826–1873), 4:161–162; **1**:19a
comet of
see Comets, Donati's
spectroscopy, **12**:269a
Donati, Luigi, **3**:118a,b
Doncaster, Leonard, **1**:505b; **3**:121a
Donders, Franciscus Cornelis (1818–1889), 4:162–164; **2**:376b; **4**:371b; **6**:137b, 535a; **8**:7b; **9**:456b; **13**:579b
Donders' law, **6**:248a
Dondi, Giovanni (1318–1389), 4:164–165
Dondi dall'Orologio, Jacopo (Giacomo) de', **4**:164a
Donets Basin, **3**:326b; **6**:238b
Doni, Giovanni Battista, **9**:316b, 319b
Dönitz, Wilhelm, **4**:297b, 298b; **7**:425b
Donkey, **15**:648a
Donnan, Frederick George (1870–1956), 4: 165; **3**:334b*; **6**:370b; **9**:246b*; **11**: 281a; **13**:581a*; **15**:160a*, 466b
chemistry, **1**:537b; **6**:371b, 372a; **8**:446a
historiography, **5**:393b*; **9**:467b*; **11**: 283b*
Donnay, Gabrielle Hamburger, **5**:187a
Donnay, Joseph Desire Hubert, **5**:186a
Donné, Alfred, **5**:84b
Donne, John, **7**:306a; **14**:157a
Donny, François, **4**:131a
Doody, Samuel, **1**:432a; **4**:98b
Doolittle, Eric, **1**:87b
Dootson, Frederick William, **4**:375b
Dopes, **3**:20a
Doppelmayr, Johann Gabriel (1671[?]–1750), 4:166–167
Neu-entdeckte Phaenomena, **11**:433a
Doppler, Johann Christian (1803–1853), 4: 167–168
Doppler effect, **3**:368b; **4**:322b; **5**:20a–b; **14**:61b
in astronomy, **1**:56a, 598a; **3**:420a, 510a; **4**:69a, 167b–168a, 251a; **5**: 20a–b; **6**:347a, 531a, 542a–b; **12**: 268b; **13**:344a–b; **14**:56a–b, 629a
critics of, **9**:2b
light, **4**:513b; **5**:458b, 459a; **7**:36b; **12**: 69a
in physics, **12**:613b–614a; **14**:339b
sound, **4**:167b; **5**:20a; **8**:596a
verification of, **1**:598b; **4**:167b, 168a, 513b
students of, **9**:278a
Doppler-Fizeau principle
see Doppler, Johann Christian, Doppler effect, in astronomy
Dordogne
caves of, **11**:484a
Dörffel, Georg Samuel (1643–1688), 4:168–169
Dorfman, Iakov Grigorevich, **5**:160b
Dormancy, **2**:221b
breaking of, **7**:113b–114a
Dorminal, **5**:2b
Dorn, Ernst, **12**:27a
Dorn, Gerard (*fl.* 1566–1584), 4:169–171; **1**:49b; **5**:448a; **10**:306b
Dorna, Alessandro, **3**:118a; **9**:268a; **13**:232b
Dorno, Carl W. M. (1865–1942), 4:171
Dorodnov, Anatolii Vasilevich, **3**:222a
Dorotheus of Sidon (*fl.* first century A.D.), 15:125; **1**:33a, 35a; **6**:79b, 381b; **9**: 161b, 162a; **11**:226a; **12**:432a
translations of works, **9**:160a; **13**:538a–b
Dorpat, University of, **1**:386a; **3**:529a–b; **9**:2a
chemistry at, **15**:457a
mathematics at, **9**:409a–b
observatory, **13**:108b–109b
Dörpfeld, Wilhelm, **12**:180b, 181a,b
Dorset
geology, **5**:14a
Dositheus (*fl. ca.* 230 B.C.), 4:171–172; **1**: 187a, 213a; **3**:391a; **15**:116a
Dositheus of Pelusion, **4**:172a
Dossie, Robert, **8**:298a; **11**:167a–b
Doubleday, Edward, **1**:501a
Doubleday, Thomas, **4**:250a
Doubt
methodical, **1**:2b, 278b; **4**:51b
Douet de La Boullaye, Gabriel-Isaac, **12**:66a
Douglas, Claude Gordon, **6**:21b, 24b
Douglas, James (1675–1742), 4:172–173; **6**:568b; **15**:4b*
pouch of, **4**:172b
semilunar fold, **4**:172b
Douglas, Jesse (1897–1965), 4:173–174; **11**:22a, 260a, 454b
Douglas, John, **4**:173a
Douglas, Stewart Ranken, **9**:334b; **12**:492a
Dounce, Alexander Latham, **13**:152b
Dourine
treatment of, **4**:300a,b
Dove, Heinrich Wilhelm (1803–1879), 4: 174–175; **2**:628b
acquaintances of, **4**:200a,b; **6**:553a; **7**:50b; **10**:27a
influence of, **4**:592a; **6**:504a*
meteorology, **6**:97b
students of, **1**:294a; **4**:340b; **5**:256b; **7**: 505b; **12**:579a; **14**:329b, 628a
Dovewood, **1**:521a
D'Ovidio, Enrico, **3**:117a; **8**:504b; **10**:441a, 442a; **12**:284a
Dowlat, Amir, **1**:539a
Downing, Arthur Matthew Weld (1850–1917), 4:175–177
Dowsing, **3**:240b
Doxography, **1**:151a, 152a; **4**:30b, 34b; **13**: 333a–b
Doyère, Louis-Michel-François, **2**:375b, 376a; **11**:566b
Doyon, Maurice, **9**:506a
Draba, **4**:584a; **7**:166a
Drabbe, Jean, **4**:237b
Dracaena, **12**:259a
Drach, Jules Joseph (1871–1941), 4:177; **13**:250a
analysis, **14**:13b
Draco (constellation)
gamma Draconis
deviation of, **2**:388a
parallax of, **2**:100b, 387b–388a; **9**:463b
position of, **3**:319a
Drag, aerodynamic, **11**:124b
Dragonfly, **13**:171b
Drainage (geog.), **9**:150b
Drake, Francis, **8**:121a
Drake, James, **6**:183b
Draparnaud, Jacques-Philippe-Raymond (1772–1804), 4:177–178
Draper, Henry (1837–1882), 4:178–181; **4**: 182b; **6**:471a, 542b; **9**:194b
Draper, John William (1811–1882), 4:181–183; **2**:589a; **4**:178b; **5**:21a(n); **12**: 132a
Draper's law, **4**:182b
Draper Catalogue of Stellar Spectra, **9**:194b
Drawbridge, **2**:49a; **11**:76b
Dreams, **4**:12b
interpretation of, **3**:548b; **5**:176b, 177a–b; **7**:190b–191a, 191b; **10**:300b
origin of, **13**:94a
prediction of, **1**:38a
as wish-fulfillment, **5**:176a–b, 177a
see also Oneiromancy
Drebbel, Cornelius (1572–1633), 4:183–185; **14**:377a(n)
Drechsel, Edmund, **1**:9b, 10b; **8**:275b
Dreissenoids, **1**:162a
Dresden
as electronics center, **1**:456a
Dreyer, Georges, **12**:88b
Dreyer, Johann Louis Emil (1852–1926), 4: 185–186; **5**:26a*; **12**:170b*
historiography, **7**:124a; 293a; **15**:421a

Dṛgganitapakṣa (school of astronomy), **1**: 48b; **10**:129a, 313b; **15**:602b, 613b–614a
Driencourt, Joseph Fernand Ludovic, **3**: 298b
Driesch, Hans Adolf Eduard (1867–1941), 4:186–198; **6**:10b, 224a; **7**:402a; **11**: 575b*; **14**:424b
critics of, **4**:204a; **7**:94b, 99b
Driesch's law, **8**:282b
embryology, **3**:390a; **6**:132b, 133a; **9**:517a; **13**:281b–282a; **14**:427b–428b, 429b
Driffield, Vero Charles, **1**:21b
Drigalski, Karl Wilhelm von, **7**:428b
Dṛk system, **2**:119b
Dropsy, **1**:315a; **3**:437b; **6**:256a
treatment of, **3**:578a; **10**:307a, 309a; **14**: 464a
see also Edema
Drosera, **3**:576a
sensitivity in, **3**:422a
Drosophila
cytology of, **2**:456a–b; **14**:431b
experimental use of, **3**:121a; **9**:520b
genetics of, **2**:455b–456b; **5**:454a; **7**:455b; **9**:520b–525b, 565a; **12**:76b, 417a; **13**:134a–138a; **14**:101a, 432a–b
geographical races of, **3**:572b
mutant types, **2**:456a; **5**:293b
podoptera effect, **5**:454a
sex determination, **2**:456a
Droste, Gottfried von, **9**:262a
Droste-Huelshoff, Albert (Freiherr) von, **11**:85b
Drowning
physiology of, **12**:302b
Druce, George Claridge, **7**:94b
Drude, Paul Karl Ludwig (1863–1906), 4: 189–193; **2**:138b; **5**:117a; **6**:97a*, 350a*; **10**:347a; **11**:347b, 580b, 581b; **12**:304b
electricity, **4**:306a; **11**:421a; **15**:433a
electron theory, **11**:447a; **14**:208b
Drug addiction, **5**:172a; **6**:62a, 77b; **9**:41a
colloidal phenomena, **1**:431b
Drugman, Julien, **5**:186b
Drugs
absorption
mechanism of, **9**:9a
theories of, **9**:8a–b
action of, **9**:8a
adulteration of, **3**:159a, 238a, 414b
chemical vs. physical properties, **15**: 103a–b
classification of, **4**:120a; **8**:352b, 391b–392b, 395a; **14**:495b
compound, **2**:392b, 435b; **4**:120a; **15**: 265a–b, 499a
detoxification, **10**:307a
evaluation of, **10**:613a
local action principle, **9**:8a
longevity of, **4**:120a
manufacture of
medieval Arabic, **14**:585a
eighteenth-century, **1**:527a
early twentieth-century, **5**:2b, 99b
regulation of, **7**:378a
names of
ancient synonymy, **4**:120a
medieval synonymy, **2**:155a–b; **4**:121b; **15**:232a
Renaissance synonomy, **10**:237a–b
Chinese synonomy, **8**:392b
natural history of, **2**:291b; **6**:396a
origin of, **3**:414b
poisonous, **15**:741b
simple, **14**:112b
standardization of, **15**:107a
storage of, **4**:120a–b
substitutions for, **4**:121b
see also Industrial research, pharmaceutical; Materia medica; Pharmacopeia
Drummond, Adam, **9**:479b
Drummond, George, **9**:480b, 481a
Drummond, Jack Cecil, **6**:499b
Drummond, Thomas
Drummond light, **6**:115a
Druses, **11**:177b
Dryander, Jonas, **1**:88b, 89a; **2**:517b, 518a
Drygalski, Erich von (1865–1949), 4:193–194; **5**:574a*; **11**:441a*
Dualism, philosophical
seventeenth-century discussion, **2**:224b; **9**:98a
eighteenth-century discussion, **12**:601a, 602a–b
nineteenth-century discussion, **2**:9b–10a; **5**:252b–253a; **8**:514a,b; **11**:364b; **13**: 461a; **14**:3a, 114a
twentieth-century discussion, **4**:188a; **8**: 517a; **11**:546a; **12**:401a–b
Cartesian, **4**:52a, 53b–54a, 62a–64b; **6**: 459a
rejection of, **3**:492a
Japanese, **15**:738a–b
opposition to, **14**:535b
Dualistic theory, chemical, **1**:578a; **2**:92b, 94b–95a, 199b; **3**:47b, 232b, 601b; **4**:243b, 244a–246b; **5**:370a–b, 372b, 381a
critics of, **14**:530a–531a
Duality, mathematical principle of, **2**:214b, 297b; **3**:170b, 171a, 213a; **5**:608b; **9**:430b; **11**:45a, 60a, 80a; **13**:5a,b, 16b
discovery of, **5**:368a; **11**:77b
Duane, William (1872–1935), 4:194–197; **3**:501a; **12**:464b; **13**:473a
and Compton effect, **3**:368b, 369a
Duane-Hunt law, **4**:195b
Dubiago, Dmitri Ivanovich, **7**:481a
Dubini, Angelo (1813–1902), 4:197–198; **10**:527a
Dublin, Ireland
Institute for Advanced Studies, **12**:222a–b
water supply, **9**:60b
Dublin Journal of Medical and Chemical Science
founding of, **7**:224a
Dublin Philosophical Society, **9**:465a
Dubois, Eugène, **8**:29a; **13**:504a
DuBois, Eugene Floyd, **8**:555b
Du Bois, Henri, **6**:480a; **11**:582a, 583a
Dubois, Jacques (1478–1555), 4:198–199; **4**:412b; **12**:323b; **14**:4a
anatomy, **13**:223a; **14**:6a, 9a
Vaesani cujusdam calumniarum in Hippocratis Galenique rem anatomicam depulsio, **14**:11b–12a
Du Bois-Reymond, Emil Heinrich (1818–1896), 4:200–205; **4**:205a, 371b; **6**: 253a*, 288a*, 553a; **7**:526b*; **9**:259b, 570a, 574b*; **11**:217a; **15**:22a*, 83a*
acquaintances of, **1**:580a; **2**:531a; **5**:149a; **6**:242a, 342a; **8**:540b; **11**:8b, 611b; **12**:270a, 425a; **14**:360a; **15**:170a
on contemporary scientists, **12**:241a; **13**:557a
influence of, **9**:101b; **13**:81b
influences on, **12**:156b
philosophy, **4**:614b; **7**:235a; **8**:515b; **12**: 243b
physiology, **1**:303b; **2**:532a; **4**:265a, 372a; **5**:268b; **6**:245a; **7**:520a,b; **9**:177a,b; **11**:475a*; **12**:241a
students and assistants, **2**:110b; **5**:196a; **6**:225a, 440a, 476a; **7**:466b, 504a,b, 519b; **8**:8a; **10**:447a, 579a, 580b; **13**:53a; **14**:2b, 527a; **15**:20b
Untersuchungen über thierische Electricität, **9**:177b
Du Bois-Reymond, Estelle, **4**:201a
Du Bois-Reymond, Felix Henri, **4**:200a
Du Bois-Reymond, Paul David Gustav (1831–1889), 4:205–206; **6**:309b*, 472a; **7**:477b, 522b; **9**:503a*; **14**: 224a*; **15**:22a*
analysis, **5**:608b; **8**:111b; **13**:35b
Théorie générale des fonctions
translations of, **9**:382a
Du Bois-Reymond, René, **4**:203a
Dubos, René, **5**:42b; **10**:352a, 376a, 404a, 406b, 408a
Duboscq, Jules, **12**:519a,b
Duboscq, Octave (1868–1943), 4:206–207; **8**:146a; **12**:424a
Dubourguet, J. B. E., **2**:444a
Dubreuil, Jean
Perspective pratique, **4**:48a
Du Buat, Pierre-Louis-Georges (1734–1809), 4:207–208; **11**:63b; **14**:494b
paradox of, **4**:208a; **14**:621a
Duchâteau, Ghislaine, **5**:149b
Duchenne, Guillaume Benjamin Amand, **3**:205b, 490a; **4**:266b
Duchesne, Joseph (also known as Josephus Quercetanus) (*ca.* 1544–1609), 4: 208–210; **1**:105a, 571a; **6**:255a; **7**: 540b(n); **8**:309b, 310a; **10**:307b
critics of, **11**:466b
medicine, **13**:507b
Duclaux, Émile (1840–1904), 4:210–212; **1**:194a; **2**:86b, 584a; **4**:621a; **11**:179b, 569a,b
and Pasteur, **4**:210b, 211a; **10**:355b, 373b, 380a, 385b, 389b
on Pasteur, **10**:362a, 363a, 369b, 372a, 374a,b, 378b, 379b, 392a–b, 406a
Duclos, Charles, **1**:323a
Ducrotay de Blainville, Henri Marie
see Blainville, Henri Marie Ducrotay de
Ducts (anat.)
biliary, **1**:619a
ligation of, **14**:587a
sphincter of, **10**:175b–176a
lacrymal, **2**:376b
lacteal, **9**:480a
parotid, **13**:31a
thoracic, **4**:487a; **6**:508b
Wirsüng's, **10**:175b
Dudith (Duditius), Andreas (1533–1589), 4: 212–213; **13**:308b
Dudley, Harold Ward, **15**:106a,b
Dudley, Paul, **12**:133b
Dudley, Robert (1573–1649), 4:213–214
Dudley Observatory (Albany, N.Y.), **2**:333a; **5**:479b; **10**:480a
Dufay (Du Fay), Charles-François de Cisternai (1698–1739), 4:214–217; **2**:324b, 576b; **4**:223b; **5**:61a; **6**:169a; **10**: 145a,b
electricity, **4**:45a, 166b; **5**:517a; **6**:175a*; **7**:371b; **10**:146a; **11**:433a; **13**:224b; **14**:352b
Dufour, Alexandre, **12**:69b
Dufour, Charles, **15**:158b, 159a
Dufour, Guillaume-Henri (1787–1875), 4: 217; **13**:124a
Dufourt, E., **9**:506a
Dufrénoy, Ours-Pierre-Armand (1792–1857), 4:217–218; **4**:65b; **15**:114b
geological map, **2**:482b; **4**:348b–349b
volcanism, **4**:349b–350a

Voyage métallurgique en Angleterre, **3**: 586b
Dugan, Raymond Smith (1878–1940), 4: 218–219; **6**:523b*; **10**:321a*; **12**:19b, 23a; **14**:482a, 558a
Dugdale, Sir William, **1**:317a; **2**:523a
Duggar, Benjamin Minge (1872–1956), 4: 219–221
Dugong, **11**:262a
du Halde, Jean Baptiste, **8**:394b–395a
Du Hamel, Jean-Baptiste (1623–1706), 4: 221–222; **4**:55a(n); **5**:60b, 259a; **13**: 585a
on contemporary scientists, **9**:120b; **11**: 486b
Duhamel, Jean-Marie-Constant (1797–1872), 4:222–223; **2**:87b, 88a; **6**:307b; **8**:399a; **13**:128a
geometry, **11**:487b
heat theory, **10**:248b
mathematical physics, **13**:128b
principle of, **4**:222a,b; **13**:75a
Duhamel, Jean Pierre François Guillot, **12**:65b, 66a, 67a
Duhamel du Monceau, Henri-Louis (1700–1782), 4:223–225; **7**:78a, 199a; **8**: 170b; **9**:54b, 253b–254a; **13**:497a
biology, **11**:343b*; **13**:412a*, 465a
chemistry, **6**:237a*; **9**:106a
Traité des arbres fruitiers
editions of, **13**:506a–b
Duhem, Pierre-Maurice-Marie (1861–1916), 4:225–233; **1**:279b, 565b; **5**: 116b(n); **7**:129b, 132a(n); **8**:599a; **9**:175b*; **13**:251a*
and Berthelot, **4**:225b–226a
chemistry, **7**:462a
correspondence of, **8**:563b*
critics of, **7**:178b*–179a*; **14**:390a,b, 401a, 636a
Duhem-Margules equation, **4**:228b; **9**: 107b
Gibbs-Duhem equation, **4**:229a
historiography, **1**:419b; **2**:20b; **3**:66a; **4**: 94b, 431a,b; **6**:274a; **7**:139a, 140b, 141a(n), 142a, 611a; **8**:218b, 231b, 242b; **10**:281a, 420b, 421a; **11**:321b, 322a; **13**:184b–185a, 210b–211a, 534b, 585b; **14**:29b; **15**:424b
influence of, **3**:32a; **4**:101a; **12**:108a; **15**: 423a
philosophy of science, **3**:376b, 377a, 379a; **5**:391a
physics, **2**:266b, 358b; **8**:289b
Système du monde, **1**:589a; **7**: 126a(n),b(n); **15**:179a
theorem of, **4**:229b
Dujardin, Félix (1801–1860), 4:233–237
and Ehrenberg, **4**:235a–b
zoology, **1**:417b, 602b; **3**:337a, 612b; **10**:221b; **12**:231a,b; **14**:113a
Dullaert of Ghent, Jean (*ca.* 1470–1513), 4: 237–238; **3**:171b, 420b; **8**:100b; **9**: 32b; **13**:210a; **14**:47a
Dulong, Pierre Louis (1785–1838), 4:238–242; **2**:91b, 134b; **3**:545b; **5**:271b; **6**:243b; **8**:330a; **11**:111a; **13**:127b
acid theory, **8**:58b, 338b; **10**:496b; **14**: 529b
catalysis, **13**:313a
heat of combustion, **4**:554b
heat theory, **3**:82a,b; **9**:237a
and Petit, law of, **1**:348b; **2**:93b; **3**:48b; **4**:239b–241a, 316a; **9**:347b; **10**:545b–546a; **13**:10b–11a; **14**:38b, 39a; **15**: 377a
approximate nature of, **11**:353a,b
pressure studies, **1**:203a; **5**:78b
Dumas, Jean-Baptiste-André (1800–1884), 4:242–248; **3**:48a, 49a, 379a; **4**:210b–211a, 554a; **5**:360a*; **8**:37a*, 55a, 87a*, 279a*; **9**:409a*; **11**:82a*
acids, **8**:339b–340a, 342a–b
acquaintances of, **1**:328b, 416a; **3**:243a, 448a; **6**:177b; **8**:54b; **9**:287a; **10**:532a; **11**:353a,b; **13**:127a, 375a; **14**:156a, 394a, 529b
alcohol, **3**:10b
alkaloids, **3**:160a; **8**:331b, 332a
analytical chemistry, **14**:156b
atomic weight, **4**:395a; **9**:349a; **10**:37b; **11**:173a,b
and Avogadro's hypothesis, **1**:346a; **3**: 47a–b, 307a
biology, **8**:349a; **11**:132b, 133a
critics of, **2**:91b, 96b; **5**:370a; **14**:478a
hydrocarbons, **8**:56a–b, 57a, 332b–334b *passim*
influence of, **5**:371a; **9**:351b; **13**:468a
Leçons sur la philosophie chimique
translations of, **11**:270b
and Liebig, **8**:335a–342b *passim,* 349a; **14**:601a
nitrogen determination, **3**:341b; **7**:393a; **9**:91b
and Pasteur, **10**:352b, 354a, 356b, 366b, 372b, 373b, 377a, 400a
and Pelletier, **10**:498a
and Pelouze, **8**:340a, 342a, 344a
and Stas, **12**:619b
students of, **1**:160b; **3**:397b; **5**:369b; **9**: 109b
substitution principle, **2**:96a; **5**:325a; **8**: 56b, 58a, 332b, 335a, 336a, 342a, 344a; **14**:156b, 530a
type theory, **1**:518b; **2**:621b; **5**:373a; **14**: 156b
Dumbarton Oaks Conference, **2**:374b
Dumbleton
see John of Dumbleton
Duméril, André-Marie-Constant (1774–1860), 15:126–127; **3**:525a; **7**:547b; **13**:554a
Duméril, Auguste, **15**:126b
Du Moncel, Théodose Achille Louis (1821–1884), 4:248–249
Dumont, André Hubert, **8**:463b
Dumont d'Urville, Jules-Sébastien-César, **4**:256a; **5**:224b; **8**:265a,b; **11**:243a
Dun Echt observatory, **5**:404a
Dunamis, **1**:253a, 277b; **15**:503b, 644a
Dunbar, William Philip, **7**:427a
Duncan, Andrew, **9**:482a, 484a
Edinburgh New Dispensatory, **8**:298a
Duncan, James, **10**:37a
Duncan, John Charles (1882–1967), 4:249
Duncan, John Shute, **2**:566a
Duncan, Philip Bury, **2**:566a
Duncan, William
Elements of Logick, **14**:288a
Dundonald, Archibald Cochrane, Earl of (1749–1831), 4:249–250
Dunér, Nils Christofer (1839–1914), 4:250–251; **13**:344b
Dung beetle, **4**:504b
Dungern, Emil Adolf Wilhelm Joseph von, **4**:298b; **6**:432b, 433a; **7**:623b
Dunglison, Richard James, **4**:252b*
Dunglison, Robley (1788–1869), 4:251–253; **4**:619b
gastric juice, **1**:542b
Dunham, Kingsley Charles, **1**:394b
Dunham, Theodore, Jr., **1**:57a; **6**:30a
Dunker, Wilhelm, **9**:346a
Dunkin, Robert, **3**:599a
Dunlop, James, **2**:472a
Dunn, Leslie Clarence, **3**:122a,b, 123a; **4**: 272a; **9**:523b; **14**:99b
Dunn, Samuel, **2**:5a
Dunn Foundation (Sir William), **5**:37a
Dunoyer de Segonzac, Louis Dominique Joseph Armand (1880–1963), 4:253–254; **13**:42a
Duns Scotus, John (*ca.* 1266–1308), 4:254–256; **1**:199b; **2**:590b; **5**:115a,b, 553a; **8**:151b; **9**:32b, 138a*; **11**:73b; **13**: 349b
Commentary on the Sentences, **2**:394a
critics of, **10**:172a, 173b–174a
dynamics, **1**:410b
natural philosophy, **12**:548a
Dunsink Observatory (Ire.), **6**:86a; **11**:49b
Dunthorne, Richard, **5**:402a
Duodenum
see Digestive system
Duparc, André, **1**:472b
Duperrey, Louis-Isidore (1786–1865), 4: 256–257; **8**:265a
Voyage autour du monde . . ., **5**:277a
Dupin, Pierre-Charles-François (1784–1873), 4:257–258; **2**:88b; **3**:83b(n), 591b; **4**:18b*; **5**:368a; **6**:3a*; **11**:82a*, 165b*
acquaintances of, **8**:288b
geometry, **5**:368b*; **9**:476a; **12**:172a
indicatrix of, **1**:449b; **4**:257b
influence of, **3**:559b, 560a
optics, **6**:87b
theorem of, **4**:257b
Duplication of cube
see Cube, duplication of
Dupont, Édouard-François, **5**:109b
Du Pont de Nemours, Pierre Samuel, **1**: 436b; **7**:581b; **8**:84b
Du Pont de Nemours and Company, **3**: 86a–b; **10**:122a; **13**:66b
uranium project, **3**:371a–b
Dupotet de Sennevoy, Jules, **9**:242a
Duppa, Baldwin Francis, **5**:126a, 381b; **10**:516b
Dupré, Athanase Louis Victoire (1808–1869), 4:258
Dupuis-Torcy, Pierre-Louis, **2**:473a
Dupuytren, Guillaume, **2**:122a, 443a; **3**: 489b; **7**:556b; **9**:8b; **15**:126a
Dura mater, **1**:391b, 392a; **4**:385b; **10**:266b
Duralumin, **7**:57a; **9**:312a–b
Durand of St. Pourcain, **2**:611a
Durand, Élie, **11**:263b
Durande, Jean François, **5**:601a; **14**:225a
Durante, Francesco, **1**:360a
Dürer, Albrecht (1471–1528), 4:258–261; **3**:128b*; **4**:49b; **5**:119a; **6**:144a; **7**: 269b, 412a,b; **13**:62a(n)
acquaintances of, **14**:273a
"Dürer's leaf," **4**:259b
geometry, **9**:157a, 474b; **13**:48a
illustrations, **15**:4b
paintings of, **12**:459a
perspective, **2**:419a; **10**:103b
Duret, Henri, **1**:569b
Durham, Herbert Edward, **5**:564a–b
Durham, University of
geology at, **6**:474b
observatory, **3**:238b
physics at, **3**:238b
Durkheim, Émile, **2**:12a; **7**:71b; **12**:108b
Durocher, Joseph Marie Élisabeth, **14**:59a
Durret (Duret), Noël, **7**:308a
Duruy, Victor, **5**:296b; **7**:75b; **8**:44b
Dury, John, **1**:160a; **2**:163a; **3**:360a; **5**:606b, 607a; **6**:141a,b; **10**:200b, 495a; **14**: 368a
Dusch, Theodor von, **10**:367a

Du Séjour, Achille-Pierre
  see Dionis du Séjour, Achille-Pierre
Dushman, Saul, **11**:421b
Düsseldorf
  electron-microscopy institute, **2**:318a–b
  observatory, **1**:616a
Dusser de Barenne, Joannes Gregorius, **5**:208a
Dust board, **7**:257a, 532b, 533a; **13**:509b, 544b–546a
Dust Bowl, **3**:317b
Dušû (stone), **15**:653b
Dutch East India Company, **2**:185b; **13**:282a,b
Dutch East Indies
  zoology, **14**:203a–b
Dutch language, **13**:48a; **15**:737b
Dutch oil, **8**:335b–336a, 337b; **13**:469a
Dutch science
  in Orient, **1**:311b–312a
Du Toit, Alexander Logie (1878–1948), 4:261–263; 14:216a
Dutour, François, **3**:444a
Dutrochet, René-Joachim-Henri (1776–1847), 4:263–265; 4:182a, 234a; **6**:268b; **7**:409b
  anatomical theory, **12**:241b
  Dutrochet's membrane, **4**:264a
Dutton, Clarence Edward (1841–1912), 4:265–266; 2:373a; **3**:593b; **5**:395b; **6**:188a; **11**:118b
Dutton, Joseph Everett, **2**:529a; **7**:429a
Duval, Mathias Marie (1844–1907), 4:266–267; 9:144a
Duverney, Joseph-Guichard (1648–1730), 4:267–268; 3:437b; **8**:171b; **9**:145b
  anatomy, **3**:524b; **9**:323a; **13**:567a
  students of, **7**:274a; **8**:114b; **10**:567b; **12**:457b; **14**:449b–450a
Duvernoy, Georges Louis, **3**:525a; **5**:185b; **7**:368a; **8**:254a; **10**:195b
Duvernoy, Johann Georg, **5**:428a; **6**:61b
Duvillard de Durand, Emmanuel Étienne, **15**:31a–b
Dux, Walter, **15**:37a
Dwarfs, **6**:567b
Dwarfs and giants, comparison of, **2**:19b
Dwyka stage, **4**:262a
Dyadics, **8**:161b, 163a, 165b; **15**:195b
Dyadkovsky, Iustin Evdokimovich (1784–1841), 15:127–128
Dyar, Harrison Gray, **6**:524b, 525a
Dybowski, Benedict von, **3**:529b, 532a
Dyck, Walther Franz Anton von (1856–1934), 4:268–269
Dye industry
  seventeenth-century, **4**:184b–185a
  eighteenth-century, **2**:292a; **8**:621b
  nineteenth-century, **1**:390a; **2**:59b; **3**:84b–85b, 240b–241a; **5**:488b–489a; **6**:463a; **7**:283a; **8**:328b, 349a–b; **14**:11b
  twentieth-century, **1**:288b; **8**:113a; **9**:354a; **13**:66b; **15**:2b
  and pharmacy, **4**:154a
  regulation of, **4**:216a
Dyeing, **3**:202b*, 400b*, 481a*, 560b; **15**:649b, 650a
  chemistry of, **3**:241a; **4**:243a; **5**:13a; **6**:236a–b, 283a, 385b; **8**:618b–619a; **10**:578b
  mordants, **4**:250a; **6**:283a; **15**:653a
  technology of, **3**:241a; **4**:184b–185a
  theories of, **1**:431a; **8**:329b
    chemical theory, **2**:77b; **3**:489a; **8**:621b
    mechanical theory, **3**:489a; **6**:236b; **8**:621b
  waters used in, **2**:292a
Dyer, William Turner Thiselton
  see Thiselton-Dyer, William Turner
Dyestuffs, **4**:216a; **5**:1b, 620b–621a; **13**:388b*
  acridine group, **2**:59b; **5**:489a
  adsorption of, **13**:471b
  alizarin, **4**:296b, 504a; **5**:488b–489a; **7**:471a; **10**:516a; **14**:247b
    synthesis, **3**:85a; **8**:328b
  aniline, **6**:463a; **11**:615b; **14**:622b
    composition of, **3**:84b–85a; **5**:2a; **8**:349b
    discovery of, **5**:197b; **8**:349a–b; **10**:516a
    in histology, **4**:296a
    manufacture of, **15**:11b
    naming of, **5**:197b
  aniline black, **14**:411b–412a
  azine group, **2**:59b
  azo, **3**:85a; **5**:344b, 536b; **10**:515b
    antibacterial action, **4**:154a–b
  Bismarck brown, **3**:85a
  brazilin, **10**:517b
  chervets, **7**:379a
  chrysoidine, **3**:85a
  classification of, **7**:471a
  cochineal, **3**:159b; **4**:184b–185a; **7**:471a; **8**:329a, 621b; **10**:497b; **11**:47b*
  Congo red, **1**:537b
  curcumin, **7**:471a
  cyanine
    heat and light, effects of, **13**:598a
  eosin, **3**:85a
    in photography, **4**:283a
  euxanthic, **4**:395a; **5**:489a; **7**:471a
  fast red, **3**:85a
  flavone, **7**:471a
  fluorescein, **4**:296b
  fluorescent, **3**:85a
  fuchsine, **10**:516a; **15**:83b
  galloflavin, **5**:489a
  hematoxylin, **4**:395a; **8**:329b; **10**:517b
  henna, **3**:347a*
  Hofmann's violets, **6**:463a–b
  indigo
    see Indigo
  indophenol blue, **4**:296b
  induline, **3**:85a
  intermediates, **14**:248a
  iodine, **5**:488b
  Martius yellow, **3**:85a; **5**:489a
  methylene blue, **2**:59b; **3**:85a; **8**:293a
    in histology, **4**:296b
    in metabolism, **13**:393b–394a
    therapeutic use of, **4**:296b
  naphthol yellow, **3**:85a
  natural, **3**:241a; **12**:236a–b
  orange, **3**:85a
  oxidation-reduction potentials, **3**:290b
  Paris green, **1**:553a
  Perkin's mauve, **3**:84b; **10**:515b–516a
  in photography, **4**:283a
  phthalein, **1**:389b
  Prussian blue, **3**:326b*; **5**:216a, 324a; **6**:165a*; **8**:618b–619a; **9**:105b; **12**:145b, 148b; **14**:394a
  purpurin, **8**:329a
  quinones, **3**:359a; **5**:488b–489a; **14**:247a–b, 411b, 477b
  rhodamine, **2**:59b
  rosaniline, **3**:85a; **5**:2a; **6**:463a; **7**:283a; **8**:329b
  rosolic acid, **5**:489a
  safranine, **2**:59b
  from shellfish, **4**:223b; **11**:331a–b
  structures of, **7**:471a–b
  substantive, **5**:537a
  synthesis of, **7**:471a
  thiazinic, **9**:296a
  tolyl red, **2**:59b
  trypan red, **4**:299b–300a
  vanadium, **11**:538b
  vegetable, **3**:240b; **7**:471a; **13**:327b; **14**:349b
  Wurster dyes, **14**:411b
  xanthorhamnin, **8**:329b
Dygograph, **4**:253a
Dykes
  see Dikes
Dynamics
  ancient Greek, **1**:28a, 253a–256a, 409b; **2**:605b; **4**:431a; **6**:378a; **13**:292a
  Byzantine, **7**:135b–138b
  medieval Arabic, **1**:28a, 409a–410b
  medieval Latin, **1**:94b, 198a–b, 290b; **2**:194a, 391b–394a, 605b–607b; **5**:113b, 116a; **8**:228a,b; **10**:223b–224a
  Renaissance, **8**:227b–231b; **10**:512b; **12**:135b–136a
  seventeenth-century, **1**:567a–b; **5**:239b, 525a, 528b–529a; **6**:485a–486a; **8**:151a–153a, 156b, 163b; **10**:60a–78b
  eighteenth-century, **3**:156a, 282b; **5**:55a; **8**:159a–b; **12**:408a–b
  nineteenth-century, **2**:470a; **3**:417b, 560a; **5**:193b, 303b, 306b; **6**:395a–b; **12**:325b; **13**:386a
  twentieth-century, **2**:168b; **8**:496a–498b
  analytic, **2**:145a–b; **3**:96a; **5**:98b; **8**:163b; **9**:211a–212a; **14**:317b
  inverse problems, **9**:324b
  laws of
    in astronomy, **15**:350a–b
  mathematical, **2**:245b–246a, 491a; **3**:169b; **6**:87a, 88b–90b; **7**:442b–443a, 487a, 619a; **15**:752a
  punctual, **8**:384b
    variable mass, **9**:324a–325a
  rotational terms, **8**:592b
  symbolic, **2**:145b
  topological, **2**:145b; **4**:337a
  as ultimate science, **2**:515a
  see also Aerodynamics, Electrodynamics, Force, Gravitation, Hydrodynamics, Mechanics, Motion, Thermodynamics
Dynamics, geological
  see Geodynamics
Dynamide, **8**:181a
Dynamism
  seventeenth-century, **5**:286b
  eighteenth-century, **7**:367a–368a
  nineteenth-century
    critics of, **12**:608b
Dynamite
  invention of, **10**:132b
Dynamo, **4**:40a; **14**:206b, 290a
  alternating-current, **13**:286b
  drum armature, **6**:223a
  Edison-Hopkinson, **6**:504b
  Gramme, **2**:439a; **5**:496a–b; **13**:286a
  invention of, **4**:534b–535a
  Pacinotti, **4**:248b; **5**:496a
  self-excitation principle, **3**:18a; **10**:268b; **12**:425b
Dynamometer, **4**:616a; **6**:223a; **8**:209b; **11**:80a; **14**:207a
Dynamometry, **3**:8b; **4**:40a
Dyne centimeter
  see Erg
Dysentery, **5**:316b; **7**:605b; **10**:237b, 518a*, 621b–622a; **11**:148a
  amoebic, **3**:447b; **4**:132b; **7**:425a; **12**:142a
  bacillary, **4**:404b; **5**:40a; **6**:298a, 432b; **7**:392b; **12**:96b; **13**:520b
  experimental, **5**:40a

phagotherapy, **6**:298b
treatment of, **14**:322a
Dyson, Frank Watson (1868–1939), 4:269–270; 3:158b*, 261b*; **5**:406a*; **6**:34a*, 282b*; **7**:238b; **12**:573b
Dyson, Freeman John, **10**:217a; **13**:241a
Dyspepsia, **10**:582a,b
Dysprosium
analysis of, **6**:366b
discovery of, **2**:254b

E

e (math.), **15**:65b
first use of, **4**:474a
irrationality of, **4**:475a; **7**:599b
transcendence of, **5**:472b; **6**:308a, 309a; **15**:368b
value of, **12**:345a
Eagle
fossil, **5**:611a–b
Ear, **13**:566b–567a; **14**:26a
of birds, **5**:267a; **13**:461b
cochlea, **1**:98b; **3**:437b; **4**:520a; **6**:287b, 618a; **13**:567a
microanatomy of, **3**:425a; **6**:246a
comparative anatomy of, **2**:443a; **5**:355a; **6**:566b, 618a; **10**:520b; **11**:383a; **12**: 137a–b; **14**:199b
embryogeny of, **5**:514b; **6**:133b, 573b; **13**:96b
equilibrium mechanism
see Ear, semicircular canals, function of
Eustachian tube, **4**:487a; **13**:567a
fetal, **4**:268a
innervation of, **3**:99b; **4**:268a; **12**:137b
labyrinth, **12**:137a–b; **13**:567a
function of, **13**:567a
measurements of, **13**:567a
ossicles, **1**:46b, 619b; **3**:355b; **4**:268a, 520a; **9**:166a; **13**:569a
discovery of, **4**:487a; **7**:17a
evolution of, **2**:504a–b; **14**:199b
function of, **6**:246a; **13**:567a
naming of, **14**:7b–8a
otolith system, **13**:53b
function of, **2**:447a; **9**:20b
outer muscles, **13**:567a
semicircular canals, **3**:99b, 437b; **4**:520a; **13**:567a
function of, **2**:446b–447a, 515b; **4**:12a; **5**:44b–45a, 463a; **9**:20b; **12**:137b
tympanic membrane, **3**:99b
vascularization of, **3**:99b
vestibule, **3**:437b; **12**:137b
see also Hearing, Otology
Ear trumpet, **9**:486a
Earlobe
shape of
and criminal tendencies, **3**:99b
Earnshaw, Thomas, **9**:163b
Earth
absolute motion of, **4**:320a–b; **8**:495b, 496b; **9**:372a–b, 386b
age of
Renaissance discussion, **10**:515a
seventeenth-century discussion, **6**:71b–72a
eighteenth-century discussion, **2**:348a, 579a; **3**:579b; **4**:72a; **8**:379a; **12**:550a*
nineteenth-century discussion, **3**:191a, 568a; **5**:97b; **7**:93b, 370b; **8**:1b, 575a–b; **13**:383a–b; **14**:503b
twentieth-century discussion, **1**:470a–b; **2**:260a; **4**:563b; **6**:475a; **7**:161a; **12**: 29a; **13**:140a*
axis
motion of, **8**:530a; **9**:174b; **10**:443b; **15**:348b
center of gravity, **2**:607b; **5**:588a; **10**:225b; **15**:320b
chemical history of, **4**:599b–600a; **8**:47a
as conductor, **13**:287a
contraction hypothesis, **1**:154a; **2**:158b, 558b; **7**:251a–b; **9**:61b, 239a–b, 594a; **10**:429b–430a; **12**:168a
cooling of, **13**:382b–383b, 610b; **15**:380a
core, **9**:444a; **10**:203b
depth of, **14**:327b
creation of, **1**:393a; **2**:556a, 568b, 613a–b; **4**:28a; **6**:581b, 587a; **9**:48a
crust, **5**:395b
age of, **7**:370b; **12**:20a
average composition of, **3**:293a–b; **12**: 273b
composition of, **4**:598a–b, 599b–600a; **10**:136b; **13**:148a, 617b–619a
deformations of, **1**:623b; **2**:558b; **3**: 325b; **6**:502b; **8**:463b–464a; **13**:609b–610a
evolution of, **12**:274a
geosynclinal belts, **13**:64a–b
layers of, **1**:470b; **11**:562b; **12**:341b; **13**:148a, 246b, 618b–619a
movements of, **3**:552a; **6**:229b, 475a; **7**:30b–31a, 250a, 251a, 251a–b, 595a; **8**:566a; **10**:265b–266a, 503b; **11**:439a; **13**:271a, 285b
origin of, **14**:259a–263a
pentagonal grid, theory of, **4**:350a
statics of, **2**:348a, 373a; **6**:241a; **8**:33b
strength of, **1**:470b; **3**:548a
structure of, **14**:327b
thickness of, **6**:503a; **8**:516b; **9**:444a; **11**:127a; **12**:342a
density of, **1**:403a; **2**:329b; **3**:158a; **6**:577a; **7**:160a; **8**:593b; **9**:163b, 164b; **11**: 122a,b, 498a; **14**:293a
elasticity of, **10**:203b
electric field, **4**:355a; **8**:426b; **13**:83a; **14**:422a–b
evolution of, **2**:321a, 570b; **5**:599b; **7**: 368a; **8**:379a; **12**:280a
figure of, **3**:281b, 514a; **4**:44a, 82a, 158b, 555b; **5**:304a, 435a–b, 491b; **6**:165b; **7**:442b, 542b, 544a; **8**:611a–b; **9**:34a, 186b–187a; **10**:73b, 76b, 78b, 256a; **12**:187a, 209b, 405b; **13**:583b; **15**: 295b, 338b–339b
area method, **6**:188a
ellipsoid theory, **2**:136a; **3**:282b; **6**: 240a–b, 241a; **13**:609a; **15**:287b, 339a–b, 351b–352b
ellipticity measurements, **1**:66a, 403a; **5**:79a; **7**:263a; **15**:320a, 352b
elongation of, **3**:103b, 104b, 105a, 107a–b
and fluid mechanics, **1**:113a
oblateness, **3**:103a, 105a, 107a–b, 173b; **5**:435b; **6**:241a; **7**:183a; **8**:321a; **11**: 126b–127a; **13**:69b; **15**:270a, 271b, 318a, 320b–321b, 348b, 353b, 354a
spheroid, **7**:497a; **10**:483a; **11**:424a–b
tidal deformations of, **10**:231b
heat of, **3**:294a–b*, 586b; **4**:600a; **12**:139b
concept of, **6**:581a, 583b–584a
origin of, **2**:578b; **3**:412a, 436a; **7**:161a; **9**:33b, 594a; **11**:521a; **12**:262b; **13**: 619a
thermal history, **6**:475a; **8**:575a–b; **10**:209b; **15**:380a
transport of, **13**:148a
see also Geothermometry
history of, **3**:434b; **8**:147a; **9**:490b, 533b; **10**:284b; **12**:139b; **13**:34a, 147a–b, 441b; **14**:387a, 439b; **15**:380a
interior, **3**:548a; **6**:503a; **7**:497a
cooling process in, **13**:246b
light of, **13**:409a
magnetic field, **13**:166b
charged particles in, **13**:83a
dissymmetry of, **9**:399a
variation of, **9**:181b; **13**:52b
mantle, **3**:567b; **9**:444a
convection currents, **6**:475b; **13**:610b–611a; **14**:216b
physical chemistry of, **13**:610b–611a
viscosity of, **13**:610a
and moon
mutual attraction of, **5**:400a, 401a; **6**: 291b; **10**:61a; **15**:325b–326a, 332a–b, 348a
mutual illumination, **3**:116a
motion of
ancient Greek theories, **10**:590a–591a; **11**:224a–b
Renaissance theories, **3**:402b–405a, 514a; **5**:400b; **7**:219b; **11**:351b–352a
early seventeenth-century theories, **1**: 568a; **5**:240a, 244a, 288a, 588a–b; **7**:303a; **8**:28a; **9**:317b; **14**:364b–365a
late seventeenth-century theories, **4**: 506a; **5**:527b; **10**:62b
eighteenth-century theories, **12**:409a; **15**:353b
nineteenth-century theories, **9**:154b
twentieth-century theories, **8**:10b
denial of, **5**:115b; **9**:527b; **13**:236a, 279a–b; **14**:286a
inertial, **4**:592a
mean, **10**:207b
precession, **8**:40b
proofs of, **1**:589b; **6**:512b; **11**:122a*
secular inequalities, **15**:290a, 356a
translation, **10**:225b
see also Earth, absolute motion of; Earth, revolution of; Earth, rotation of
orbit of, **1**:246b–248b; **5**:405b; **7**:294a, 295b
aphelion, **3**:406a
eccentricity of, **3**:169b, 470b; **5**:88a; **6**:512b, 520b; **15**:332a, 349a
radius of, **2**:100a; **6**:515b
semimajor axis, **12**:574a
origin of
seventeenth-century theories, **2**:613a
eighteenth-century theories, **14**:296a
nineteenth-century theories, **3**:190b, 412a; **5**:32a; **7**:385a,b; **13**:448b
twentieth-century theories, **1**:623b; **3**: 190a, 191a, 192a; **7**:385a; **9**:552b
radioactivity of, **4**:356b; **12**:29a; **13**:619a
revolution of, **3**:404b–407a
evidence of, **1**:616a; **3**:404b; **9**:77b
rigidity of, **1**:552b; **6**:503a; **7**:595a; **9**:373b
rotation of
ancient discussion, **1**:246b–247a; **6**: 381b–382a; **13**:320a; **15**:203a–204b
medieval discussion, **2**:607a–b; **10**:225b
Renaissance discussion, **3**:404a–b, 407a; **5**:239a, 399b, 400b
seventeenth-century discussion, **1**:568a; **7**:290a; **12**:332b
eighteenth-century discussion, **5**:195a; **13**:179a; **15**:355b
nineteenth-century discussion, **15**:484a
twentieth-century discussion, **9**:373b; **13**:12a
deflecting force of, **5**:86b
geological changes and, **3**:583a

Earth (cont.)
Indian discussion, 15:590b
measurement of, 3:367a
and motion at surface, 4:591b–592a
and ocean currents, 4:344b; 15:431a
and orogeny, 15:56a
pressure from, 2:425a
and projectile motion, 15:355b, 489b
proofs of, 3:404b; 5:86a–b; 7:220a, 230b
and tidal friction, 4:20a, 591a–b; 9:239a–b
variability of, 2:511a, 516b; 4:270b*; 6:229b; 7:18a; 8:40b; 9:411a*; 10:34a; 12:449a, 574a
and wind, 2:628b; 4:591b–592a; 15:289a
shadow of, 7:219b; 8:1145a*; 10:299a; 15:549a, 561a–b, 573b, 579a, 592b, 605a, 798a
shape of
see Earth, figure of
size, 3:25b; 7:400a; 9:191b; 10:299a
circumference, 3:319a, 320a; 4:82a, 390a–b, 542b, 556a; 5:347a; 6:301b; 9:1b; 11:104a; 15:220a, 548b, 557b, 568a, 586b, 591b, 593b, 597a, 604b–611a *passim*, 616a
diameter, 1:249a–b; 2:150b; 4:542b; 6:301b; 15:579a, 591b, 597a, 607b, 609b, 611a
mass, 9:164b
radius, 1:381b–382a; 4:108b–109a; 10:61a–b, 595b; 11:411b; 15:321a–b, 592b, 604b, 605b, 615b
volume, 1:248b
structure of, 1:552b; 2:551b; 8:516a–b; 9:594a; 10:73b; 11:498b
Mohorovičić discontinuity, 9:444a,b
seismic studies, 5:596b
tetrahedral theory, 2:89b; 4:362a*
surface of
apparent movements of, 11:207a
fluid masses, 3:418b
structure types, 7:251a
thermal gradients, 2:158a–b
water, proportion of, 1:381b; 11:411b
temperature of
see Geothermometry
see also Atmosphere; Geodesy; Geology; Geophysics, etc.; Magnetism, terrestrial; Orogeny; Poles, terrestrial; Tectonics; Tides; Zones, geographical
Earth, elemental, 5:353a–b, 400a; 7:538b; 8:131a, 174a, 620a; 10:309b
three types of, 1:549b; 12:604b–605a
Earth acids, 2:199b
Earth pressure, 3:442a
Earthquakes, 2:83a, 552a*; 3:436a, 567a–b, 568a; 4:151b, 406b; 6:47b*, 84a, 471b, 486b; 7:495a; 8:31b*; 9:234b*; 10:509b; 11:37b, 67a; 12:309b, 310a
aftershocks, 10:210b
air pressure theory, 1:138b
catalogs of, 9:61a, 407a, 593a; 10:204a, 211a
classification of, 9:444a
divination by, 8:191b
effects of, 7:149b; 9:444b; 14:402b
elastic rebound theory, 11:361b–362a
epicenters, 5:596b
determination of, 9:407a, 444b
focus of, 10:211a
depth of, 3:357b
geography of, 5:596b
intensity scales, 10:211a
and isostasy, 9:181a
mechanism of, 12:342a
origin times, 13:500b
precautions, 7:9b; 8:100a; 9:444b
predicting of, 7:9a–b; 10:210b
theories of
ancient Greek, 4:32b; 11:104b; 13:93b, 297a
eighteenth-century, 8:174a–b; 9:370b; 11:302b–303a
nineteenth-century, 2:158b; 9:61a, 593a, 620b; 12:262b, 263a
twentieth-century, 8:99b; 13:146a–b
wave propagation
see Wave(s), seismic
zoning, 10:210b
see also Seismology
Earthshine, 9:167b
Earthworm, 14:426a,b
anatomy of, 3:40b; 14:408a
and disease, 10:388a–b
effect on soil, 3:576a
generation of, 9:253a
East, Edward Murray (1879–1938), 4:270–272; 3:121b–122a, 123a
East Africa
prehistory, 8:104b–105b
East India Company, 2:472a; 6:20b; 7:146a; 9:67b; 10:204a; 11:261b; 13:342a; 14:513b
East Indies
natural history, 1:47b
Easter, dating of
ancient Roman, 1:148b; 3:26b
medieval, 1:105a, 565a–b; 5:553a; 14:576b
eighteenth-century, 3:462a*
nineteenth-century, 4:176b*
twentieth-century, 10:585a
astronomical calculation, 1:383a
perpetual cycle of, 1:565b
see also Computus
Easter Island, 3:397a
Eastman Kodak Company, 11:553a
Easton, Cornelis (1864–1929), 4:272–273; 7:240a*
Eaton, Amos (1776–1842), 4:273–275; 4:363b, 364a; 5:11b, 12a; 6:56b, 57a, 437a, 517a, 524a; 12:433b; 13:432a, 433a
critics of, 15:152a, 153a
Geological and Agricultural Survey of the District Adjoining the Erie Canal, 6:437b
geology, 15:153a
Eaton, Daniel Cady, 4:274b; 11:176a; 12:328a; 14:192a; 15:149a
Eaton, Nathaniel, 1:133b
Ebbinghaus, Hermann, 7:457b, 458a; 9:562a
Ebel, Johann Gottfried (1764–1830), 4:275; 2:569a
Ebelmen, Jacques Joseph, 3:11b; 6:2a
Eberhard, (Paul Alexander Julius) Gustav, 8:539a,b
Ebermaier, Johann Christoph, 3:159a
Eberth, Carl Joseph (1835–1926), 4:275–277; 5:219a
Eccentreccentric, 3:409b
Eccentrepicyclic, 3:409b
Eccentrics (astron.), 1:189b–190a, 247b–248a; 2:20b, 24a, 403b; 3:25b, 406b; 4:542b; 5:403a; 11:190a, 196a; 12:4b; 13:326a; 14:162b; 15:35a, 211a,b, 213a
movable, 14:22b
Ecchellensis, Abraham, 2:308b; 14:49a
Eccles, John Carew, 12:399a,b
Eccrinids
development of, 4:207a
Ecgonine, 14:411b
Echeverría, Athanasio, 9:432b
Echinococcosis, 10:17b
Echinococcus, 5:454a*; 10:527a
Echinoderms, 4:236a, 289a; 6:294a*; 9:112b
anatomy, 3:492b; 8:519b; 12:298a; 13:402b
cytology, 12:423b
development of, 9:573a–b
embryogeny, 1:71b; 4:187a; 5:35b; 6:7b; 7:475a; 8:586a; 13:281b–282a; 14:429a
morphology, 1:507a; 2:128b; 6:7b; 10:522b
nervous capacity, 4:290b
paleontology, 1:210a, 494b; 2:128b; 4:73b; 7:61a; 9:568b; 10:222a; 11:500b; 12:257a
taxonomy, 1:507a; 7:401a; 8:270a; 14:1b
Echinoids, 9:314a
Echo, 5:259b*; 7:376b; 9:508a*
Eck, Johannes, 3:179b; 13:29a
Eckart, Carl, 12:530a
Eckert, Wallace John (1902–1971), 15:128–130; 2:511b
Eckhardt, Arnold, 8:163a
Eckhart, Johann Georg von, 2:15b
Eckmann, Beno, 6:497a
Eclecticism, 1:212a, 268a; 3:175a; 4:389a; 5:232a, 285b; 7:539a; 8:427a,b
Eclipses
ancient Greek discussion, 13:321b–322a
medieval Arabic discussion, 1:508a; 4:543a; 14:576b
medieval Latin discussion, 2:24a, 435a; 14:400a
Renaissance discussion, 3:402a
seventeenth-century discussion, 2:83a; 7:376a; 9:508a*; 13:445b*
eighteenth-century discussion, 2:324b, 325a, 327a; 3:52a; 4:23a, 565b; 5:88a, 435a, 515b; 9:330b; 10:256a; 11:66a*; 13:39b
nineteenth-century discussion, 1:364a; 4:176b*; 6:471a; 10:543b; 13:500b
twentieth-century discussion, 2:462a; 3:35a–b, 509a; 4:278a; 8:364b; 10:526a; 12:249b–250a
astrological use of, 1:513b; 10:547b; 14:286a
Babylonian discussion, 11:188b, 223a; 15:212a–b
catalogs of, 7:118b, 130b, 139b, 140b, 259b, 543a; 10:219a–b; 15:212a–b, 216a
Chinese discussion, 15:729b
chronological use of, 10:615a
computation of
ancient Greek, 11:188a; 13:295a, 295b–296a, 323b; 15:216a
ancient Roman, 10:299a; 11:192b, 194b
Byzantine, 12:406a
medieval Arabic, 1:509a; 2:155a; 7:82b, 338a, 361a; 9:21b
medieval Latin, 7:119b, 124a; 10:540b, 541a
Renaissance, 2:411b; 15:476a–b
eighteenth-century, 4:24b*; 7:494a; 10:197b, 614b, 615a; 13:39a
nineteenth-century, 4:590b; 10:207a, 219a–b; 11:114b
Chinese, 12:375a, 378a; 14:159b; 15:729b
Indian, 1:48b, 49a; 2:118a, 417b; 10:313b; 14:28b; 15:548b–550a, 592b–593a, 594a–595a, 604a–605b, 610a–b
Japanese, 1:310a, 311a; 12:403b
Mesopotamian, 15:643b, 676a, 677a,b, 680b
cycles of, 11:191b; 13:295b–296a; 15:

212a, 216a, 775a, 791b–799a *passim,* 802a,b, 803a, 808a, 811a,b, 812a, 814a
effects of
on terrestrial magnetism, **1**:521b
and Einstein's predictions, **3**:36a–b
and geodesy, **1**:429b
Indian discussion, **1**:308a, 309b; **2**:114b, 117b, 119a; **10**:129a–b, 313a, 426a
lunar
ancient Greek discussion, **1**:251b; **5**: 345b, 346a; **6**:310a; **15**:212a–213a
medieval discussion, **2**:149a,b, 150b; **7**:259a; **14**:576a
Renaissance discussion, **2**:403a, 410b, 411b, 412a,b; **15**:474b
seventeenth-century discussion, **3**:528b; **9**:486b; **11**:532b; **14**:254a
eighteenth-century discussion, **5**:435b; **9**:78a; **14**:589b
nineteenth-century discussion, **6**:326b; **12**:37b*
twentieth-century discussion, **3**:457a
calculation of, **3**:51a; **4**:169a; **5**:274b; **9**:21b; **15**:548b–550a, 560b–561b, 573b–575a, 774b, 802a–b
colors of, **1**:607b; **15**:550a, 574b, 593a
definition of, **15**:798a–b
and distance measurement, **6**:312b
duration of, **15**:549a, 573b–574a
Egyptian study, **15**:723b, 724a
Indian study, **2**:416a,b, 417b; **3**:584b; **5**:274a; **7**:219b, 512b, 582b; **10**:5b; **12**:598a
and longitude determinations, **2**:149a, 153a; **3**:198a; **5**:343b–344a, 435a; **9**:233b; **10**:490a; **14**:275a; **15**:220a
magnitude of, **1**:511a; **15**:549b, 574a–b, 575a, 679a
Mesopotamian discussion, **15**:643b, 644b, 677a
and moon's distance, **1**:249a,b; **5**:435a
"paradoxical," **3**:319b
phases of, **15**:573a, 574b
photography of, **14**:479b
tables of, **15**:474b, 476a–b
mechanism of, **3**:319b; **10**:474b; **12**:61b, 404a, 406a
photography of, **3**:261a; **10**:30a; **11**:95b
prediction of
see Eclipses, computation of
solar
ancient Greek discussion, **4**:368a, 467a; **5**:345b; **15**:213b
ancient Roman discussion, **3**:391a; **4**: 463b; **10**:294a
Byzantine discussion, **8**:191b
medieval Arabic discussion, **2**:150a; **7**:82a; **9**:21b
medieval Latin discussion, **7**:130a; **11**:469a
Renaissance discussion, **2**:401b, 411b, 412b, 413a
early seventeenth-century discussion, **6**:360b, 361b, 520b; **7**:294b, 298a; **9**:122b, 247a; **10**:490a, 595b
late seventeenth-century discussion, **3**:101b; **5**:25b, 290a*, 522b*; **7**:219b, 512b; **9**:486b
early eighteenth-century discussion, **3**:431a; **4**:23b, 166b*; **7**:374a; **8**:459b; **9**:77b, 78a
late eighteenth-century discussion, **3**: 108a, 198a, 396b; **6**:234a; **9**:233b; **10**:591b, 607b; **13**:531a; **14**:589b
early nineteenth-century discussion, **2**:236a, 285a
mid nineteenth-century discussion, **3**: 14a, 93a, 238b; **4**:161b; **6**:49a, 402b, 523a; **10**:207b, 480a, 482b; **11**:320a, 498a; **12**:579a; **13**:37a*, 111b, 120b, 422b
late nineteenth-century discussion, **2**: 434a; **4**:498a; **7**:270b; **9**:184a, 291a, 564a; **10**:119b; **11**:297a; **13**:232b, 472b; **14**:242b, 449a
early twentieth-century discussion, **3**: 510a; **4**:270a, 359a; **5**:102a, 182a,b, 363b–364a; **7**:6a; **10**:231a, 321a; **11**:320b, 412b, 542b; **12**:449a; **13**:95b
mid twentieth-century discussion, **12**: 353a, 368a; **13**:153b, 408b
annular, **1**:511a; **3**:319b; **4**:24a*
Baily's beads, **1**:402b, 403a, 429b; **6**: 523a
Chinese study, **12**:170a
computation of, **4**:463b; **7**:19a, 481b; **9**:234a; **10**:232a; **15**:215b, 543a, 544b, 550a, 561b, 752a, 760a–b, 789b–797a, 802a–b
duration of, **15**:550a
Egyptian study, **15**:723b–724a
geometry of, **13**:339a
Indian study, **2**:416a,b, 417b; **3**:584b; **5**:274a; **7**:582b; **12**:598a; **15**:593a
magnitude of, **15**:575a, 679a
mapping of, **3**:101a; **11**:194b
Mesopotamian discussion, **15**:677a
photography of, **1**:597a; **2**:616b; **3**:472a; **6**:105b, 508a; **9**:230b, 552b; **12**:139a, 268a
and planetary distances, **1**:338b
spectroscopy of, **4**:180a; **5**:199a; **8**:503b; **14**:557b
tables of, **15**:752a, 774b, 789b–799a, 800a–802b, 811a,b, 812a, 814a
timing of, **1**:429b
Ecliptic, **1**:247a, 260a, 339a; **3**:25b, 218a; **5**:619b
astrography of, **6**:282a
division of, **6**:616b; **15**:536b, 541b–542b, 546b, 720a–b
obliquity of, **1**:509a; **2**:99a, 153a, 468b; **3**:101a, 405a,b; **4**:419a; **5**:24b, 195a, 613a; **8**:143b, 144b; **11**:321b; **14**: 538b; **15**:520a, 582b
determination of, **3**:402b; **4**:391a, 463b, 542b, 543a; **7**:81a, 337b, 338a, 353a–354a, 361b; **9**:84a–b, 555a; **10**:179b; **11**:189a–b, 202a, 223a; **12**:358a; **13**:536a; **14**:575a, 576a, 578b; **15**: 219b, 259b, 474b
discovery of, **4**:463b; **10**:179b
variability of, **1**:510a–b; **5**:24a, 435a; **7**:304a, 335a, 564a; **8**:144a; **11**:207a; **14**:254a, 594a; **15**:290a, 338b–339a
Eclogite, **2**:289b; **4**:141a, 408b, 409a–b; **6**: 475b
École des Beaux-Arts (Paris), **10**:372b
École Centrale des Arts et Manufactures (Paris), **3**:417a; **5**:583b
founding of, **3**:416b–417a; **4**:242b
Écoles Centrales (Paris), **3**:1a
École Militaire (Paris)
observatory, **7**:86b–87a
École des Mines (Paris), **1**:327a; **2**:72a; **3**: 200a, 346b, 358a; **4**:217b–218b, 347b, 348b; **5**:185b; **14**:88b
founding of, **12**:65a–67a
geological collections, **1**:573a
library, **4**:67b
paleontology, **14**:599b
École Municipale de Physique et Chimie (Paris), **3**:503b, 504a
École Normale de l'An III (Paris), **1**:206b; **9**:473a; **15**:341b
École Normale Supérieure (Paris)
in nineteenth century, **10**:355a–b, 362b–363a, 385a
chemistry at, **4**:77b
École de Physique et de Chimie Industrielles (Paris), **3**:400b
École Polytechnique (Paris)
in late eighteenth century, **1**:206b; **2**:74a, 133b; **3**:200a; **5**:92a, 93b, 375a; **15**: 302b, 341b, 342a–b, 361a, 480b–481a
in early nineteenth century, **2**:213b; **3**:79a, 416b; **4**:222a–b; **5**:320b; **9**:473a, 474a,b; **10**:2a–b, 545b, 546a; **11**:61b; **13**:311a; **15**:481a–b
in mid-nineteenth century, **2**:87b; **3**:417a; **6**:307a; **11**:79a
in late nineteenth century, **5**:317b–318a
founding of, **5**:603a; **6**:1a–b
publications of, **6**:1b
reorganization of, **6**:1b; **15**:346b–347a
Société Amicale des Anciens Élèves de l'École, **9**:80a
École des Ponts et Chaussées (Paris), **10**: 527b–528a; **11**:164b; **12**:65a–b
École Pratique des Hautes Études
marine research, **3**:203b
École Supérieure d'Électricité
founding of, **1**:304a
Ecology
in antiquity, **2**:256b
Renaissance, **2**:219b
eighteenth-century, **2**:204a, 257b; **8**:177a, 378a; **13**:258a
early nineteenth-century, **3**:44b, 301a; **4**:285b; **12**:124a; **13**:541a; **15**:83a
mid nineteenth-century, **1**:405b, 412b; **3**:44b, 550b, 571b; **5**:358b, 533b, 546b; **6**:221b; **9**:374b–375a, 408a
late nineteenth-century, **3**:259a, 492b; **5**:69b–71a, 441a; **6**:10a; **9**:432a, 496a; **12**:299b; **14**:1b
twentieth-century, **2**:221b; **3**:317b, 318a; **5**:545a–b; **6**:122a–b; **7**:102b; **8**:425b, 512a; **9**:328a, 535a–536a, 550b; **13**: 345a; **14**:291b, 551b–552a
competitive exclusion principle, **5**:70b
definition of, **6**:8a
"dynamic," **3**:318a
ecologic niche, **5**:545b
experimental, **9**:44a
Malthusian, **2**:570b
marine, **4**:42a–b; **6**:267b, 442b; **7**:410b; **12**:125b; **13**:165a; **15**:124b
microbiological, **14**:37b
naming of, **14**:181b
physiological, **8**:425b; **12**:299b
predator-prey relation, **5**:70a–b
quantitative, **3**:576a
statistical, **5**:70b
see also Altitude, Climate, Environment, Ethology, Paleoecology, Phytoecology
Econometrics, **1**:155a
Economic Geology, **14**:443b
founding of, **8**:370b; **14**:442a
Economic Journal
editors of, **7**:319a
Economics
medieval Latin, **10**:229b*
Renaissance, **1**:77b, 549a
seventeenth-century, **10**:566a; **13**:593a–b; **14**:466b
early eighteenth-century, **2**:42a; **3**:380a,b
late eighteenth-century, **1**:554a–b; **3**:200b, 386b, 461b; **4**:503a; **5**:567b–568a; **6**:369b; **7**:201b–202a, 568b; **8**:84b; **11**:135a; **13**:448a, 494a, 495b, 571a
early nineteenth-century, **1**:615b; **3**:44a,

Economics (cont.)
  45b, 79b, 202b, 450b, 451a; 4:257a; 6:551b; 7:615a; 9:67a–70a; 12:116b–117a, 286b
  mid nineteenth-century, 1:70a; 7:104a–105a; 9:384b–385a; 12:261a–b; 15:132b–134b, 405a,b, 406b–415b
  late nineteenth-century, 1:9a; 10:35a, 486a; 12:301a; 13:145a
  twentieth-century, 7:317b–319a; 12:274b, 461a–b, 504a–b; 14:92b
  business cycle theory, 12:461a
  Chinese, 12:372b, 381a
  "macrostatics," 7:318b
  mathematical, 3:450b–451a; 7:104b–105a; 10:486a; 11:285b–286a; 14:122b, 466b
  and medicine, 5:452b
  monopolistic competition, 3:451a
  national income statistics, 7:318b
  and statistics, 1:155a; 2:319a; 7:104b–105a, 318b; 8:61b, 62a; 10:566a
  value theory, 2:42a; 12:461b
  see also Finance
Economist
  editors of, 12:570a
Ecphantus, 1:248b; 3:402b; 11:224b
Ecthesis, 7:194a
Ectocarpaceae, 12:125a
Ectoderm, 4:187b; 11:369a,b
  sexual significance of, 1:601a
Ectosperma, 13:595b–596a
Ecumenicalism
  Renaissance, 3:513a–b, 515b–516a
Eddington, Arthur Stanley (1882–1944), 4:277–282; 4:270b*; 5:406a; 6:26b, 34a*; 7:240a*; 8:41a*; 12:214a, 253b*, 573b; 14:309b*
  affine geometry, 4:331a
  astronomy, 1:56b–57a; 2:266a; 3:36b; 5:182b; 6:74b, 352b; 7:86a
  astrophysics, 12:22a–b, 250a; 13:474b
  cosmology, 12:449a
  critics of, 9:405a
  E-number theory, 4:281b
  influences on, 12:18a, 449a, 455a
  and Jeans, 12:23a
  The Nature of the Physical World, 14:305a
  physics, 14:306a
  and Shapley, 12:346a, 346b–347a
Eddy, Henry Turner, 2:266a; 9:531a
Eddystone lighthouse, 12:462a
Edema, 2:464a, 597b; 6:256a; 7:503a; 14:253a–b
  bacilli, 7:100a
  bronchial, 5:384b
  causes of, 8:608a
  pulmonary, 5:384b
    pathogenesis of, 14:249a
  see also Dropsy
Eden, Richard, 1:606a; 5:348b, 593a
Edentates
  classification of, 5:46a
Eder, Josef Maria (1855–1944), 4:282–283; 1:385a
Edestidae, 7:252a–b
Edestin, 10:242b
  preparation of, 9:284a
Edgeworth, Francis Ysidro, 10:451a,b, 458a
Edguardus, David
  see Edwardes, David
Edict of Nantes
  revocation of, 9:452a
Edinburgh
  hospitals, 9:480b, 481a
  Medical Essays, 8:297a
  observatory, 5:404a; 12:498b
  Oyster (club), 2:174b–175a
  Philosophical Society, 8:610b
  Royal College of Physicians, 11:1b
    founding of, 11:1a
Edinburgh, University of, 4:500b; 6:81a; 13:503b
  anatomy at, 5:470a
  botany at, 1:422b
  geology at, 5:334a
  mathematics at, 3:264b–265a; 13:54a–b
  medicine at, 2:173b, 225b; 3:494b; 9:479b–481b, 483a, 484a; 11:1a, 48a; 12:354a; 14:320a,b
  museum, 5:470a; 7:69a, 70a
  natural history at, 7:69a,b
  observatory, 14:140b
  pharmacology at, 15:100a
  physics at, 15:43b
Edinburgh Botanical Club
  founding of, 1:423a
Edinburgh Encyclopaedia, 2:336a; 6:325b
Edinburgh Journal of Science
  editors of, 13:499a
Edinburgh Mathematical Society
  founding of, 7:413b
Edinburgh Pharmacopoeia, 8:297b
Edinburgh Philosophical Journal, 5:68b
  founding of, 7:69b–70a
Edinburgh Review, 6:81a
Edinburgh School of Arts
  founding of, 2:451b
Edison, Thomas Alva (1847–1931), 4:283–284; 4:180a, 601a; 6:345a; 7:288a; 12:574b; 13:286b
  critics of, 13:361b
  Edison effect, 4:284a; 5:33a
  electricity, 13:356b
  phonograph, 4:500a; 7:445b
Edlefsen, Niels Edlef, 8:94b
Edlén, Bengt, 6:615a
Edlund, Erik, 1:296b–297a, 297b, 298a; 8:491a
Edmondson, George, 5:125a; 13:521a–b
Edmunds, Charles Wallis, 15:101a
Edsall, John Tileston, 3:336a
Education
  ancient Greek, 6:405b, 409a(n); 11:25a, 28a; 13:321b
  ancient Roman, 3:109b–110a
  Byzantine, 8:191a
  medieval Arabic, 15:584b
  medieval Latin, 1:1a, 2a, 104b–105a, 377b, 378b, 564b–565a, 565b; 2:19b; 4:396a–b; 5:115a, 365a, 550b; 6:274a; 8:547a–b, 548a; 9:140b; 12:62b; 13:588b; 14:35a; 15:189a
  Renaissance, 2:401a–b; 5:238a–b, 239b, 448a; 11:608a,b
  seventeenth-century, 1:474a; 7:610b–611a
  eighteenth-century, 2:326b; 3:485a, 538b–539a; 5:567b; 10:609a; 11:139b–140a; 13:457a
  nineteenth-century, 2:133b; 3:89a; 4:84a, 203b; 5:530b; 6:326b, 509b; 7:169a; 8:571b; 9:7a, 348a; 11:610a; 13:144a, 145a
  twentieth-century, 6:26a; 9:553a; 12:618b–619a
  academic freedom, 3:130a
  aeronautical, 14:620a
  agricultural
    nineteenth-century, 5:83a; 10:150b; 11:179a–b; 13:390a
  alchemical, 13:397b
  anatomical
    medieval, 6:276a; 9:468b–469a
    Renaissance, 1:523a–b, 604a; 5:585b; 14:7a–b
    seventeenth-century, 9:581b–582a; 14:13a
    eighteenth-century, 2:122a, 204a; 3:487a,b; 6:231a, 567a–b, 568b–569b; 7:367a; 8:114b; 9:479b–482a, 483a; 11:99b
    nineteenth-century, 1:452a; 5:470a, 515a; 6:436a, 618a,b; 7:414a, 415a,b; 9:56b, 416b; 13:288a; 14:126b, 331a, 456a–b; 15:166a
    twentieth-century, 3:113a; 9:108b
  anthropological
    early twentieth-century, 5:145a; 9:144b
  architectural
    seventeenth-century, 2:201a–b
    eighteenth-century, 3:39a
  astrological
    medieval Arabic, 1:35a
  astronomical
    medieval Arabic, 7:81a
    medieval Latin, 7:119b–120a
    nineteenth-century, 9:463a; 14:557b
    twentieth-century, 6:353a; 12:23a; 14:558a
  bacteriological, 12:481b
  biochemical, 6:499b–500b
  biological
    nineteenth-century, 2:501b; 5:80a–83a; 6:595b; 10:11b; 12:479b; 14:238a, 424b, 425b–426a, 533a
    twentieth-century, 2:345a; 7:314a, 456a; 8:27a*, 355b, 587a, 589b; 9:565b; 12:416b, 417b–418a; 13:269a
    evolutionary approach, 5:81a
    laboratory method, 5:81a
  botanical
    Renaissance, 3:414a; 5:383a–384a
    eighteenth-century, 14:386b
    nineteenth-century, 1:423b; 2:103a, 290b; 3:611b; 4:401b; 5:513b; 6:288b–289a; 12:174b–175a; 13:89b–90a, 341a,b, 456a
    twentieth-century, 2:185a; 9:45a
    laboratory method, 1:396a; 2:103a
  chemical
    seventeenth-century, 1:450b–451a; 3:596b; 8:130b, 172b
    early eighteenth-century, 6:259b; 11:48a–b, 562a, 563a–b; 14:144a–b
    late eighteenth-century, 1:527a–b; 2:5a, 173b–174b; 3:494a–b; 5:215b, 601a–b; 6:382b–383a, 395a; 7:58a; 8:469b–470a, 613a, 619a–b; 11:564a–b; 13:603a; 14:225a
    early nineteenth-century, 5:430a–b; 6:115a; 9:527a; 13:313a–b, 372b, 373b, 499a
    mid nineteenth-century, 2:624a; 3:47b–48a, 397b–398a, 557b; 4:162b; 5:164b, 203a, 393b; 6:461a–b, 463b; 7:45a, 280b; 8:330b–331a; 9:425b–426a; 10:499b; 11:270b–271a, 504b, 540b; 12:433b, 435a; 14:394b
    late nineteenth-century, 9:130b, 394b, 557a–b; 10:14b; 11:278a, 371a, 537a–b, 538a; 15:434b, 456a–b
    twentieth-century, 5:621b; 8:291a; 10:156a–157b; 12:505b; 13:71a; 14:125a; 15:1b–2a
  Chinese, 8:13b*; 12:375a, 388a–b; 14:163a
  of deaf
    seventeenth-century, 6:253b; 9:322a(n); 14:152a
    eighteenth-century, 1:410b–411a; 5:468a

nineteenth-century, **1**:582b
dialogue method
medieval Latin, **1**:104b
engineering
eighteenth-century, **3**:39a; **10**:528a; **11**:164a–b
early nineteenth-century, **4**:347b, 501b–502a; **10**:2b–4a; **11**:77b–78a, 164b
mid nineteenth-century, **4**:274b; **11**:294a, 344a–b
late nineteenth-century, **4**:500b; **6**:188a; **7**:287b; **8**:365b; **13**:398b–399a
twentieth-century, **7**:246b, 247b; **8**:500b; **10**:593b–594a, 605a; **13**:24b
geographical
Renaissance, **6**:20a
nineteenth-century, **3**:593a; **5**:599b; **7**:496b; **11**:439b
twentieth-century, **1**:208a–b; **9**:150a
geological
nineteenth-century, **4**:218b, 309b; **5**:339a; **7**:603a; **8**:31a, 463a; **10**:428b; **11**:103a; **12**:140b; **14**:132b
twentieth-century, **3**:595b; **5**:187a; **10**:124b
histological
nineteenth-century, **10**:267b
history of science
twentieth-century, **12**:110a
horticultural
nineteenth-century, **1**:396a
laicization of, **2**:60a, 64a, 65a; **11**:608a
logic, **1**:107a; **11**:287b
mathematical
ancient Greek, **6**:315a–b
medieval Latin, **7**:117b–118a, 119b–120a
Renaissance, **3**:585a; **11**:288a, 338b
seventeenth-century, **1**:292a; **2**:310a, 526b; **3**:284b; **4**:550a; **5**:520b; **6**:299a; **9**:51a; **13**:29a, 236a; **14**:147b–148a
eighteenth-century, **2**:111b–112a, 334b–335a; **3**:284a–285a, 459b; **5**:210a, 520a; **8**:136b; **9**:473a; **11**:392a–b; **12**:446a; **14**:452b; **15**:342b, 429b
early nineteenth-century, **2**:112a; **4**:546b; **6**:1a–b, 86a, 213b; **10**:187a, 437b; **11**:414a–b; **13**:12b–13b; **14**:500a
mid nineteenth-century, **2**:471a; **3**:260a, 466a; **7**:51a, 109a, 522a; **9**:409a–b; **10**:248a; **11**:568b; **12**:86b; **13**:427a; **14**:221a–222a
late nineteenth-century, **2**:465a, 469b; **3**:264b–265a, 450b; **5**:339b; **6**:3b; **8**:367b, 389a, 584b; **9**:125a, 324a, 502a; **10**:24a, 544a; **12**:599b; **13**:81b; **14**:158b, 202b, 343a
early twentieth-century, **2**:146a, 217b; **4**:373b, 560b–561a; **5**:188b; **6**:113a–b; **7**:399b; **10**:245a, 444a, 464a–b; **11**:157a, 613b; **12**:229a, 532b; **13**:470a; **14**:337a, 437a, 559b
mid twentieth-century, **2**:352a; **3**:171a; **13**:428b; **14**:384a–b, 590b
Japan, **15**:748a, 753b
medical
ancient Greek, **6**:316b
ancient Roman, **5**:227b–229a
Byzantine, **15**:235a
medieval Arabic, **11**:444b–445a; **12**:7Bb, 8a; **13**:415a; **14**:584b; **15**:230a, 239a–b
medieval Latin, **1**:107a, 290a; **3**:394a
Renaissance, **3**:13a, 98a–b; **5**:585b; **6**:360a; **10**:305a; **11**:311a–b; **13**:141a; **14**:4a–b, 581a; **15**:161a
early seventeenth-century, **11**:466a–b; **12**:102a
late seventeenth-century, **5**:419a; **6**:458b; **9**:581b–582a; **13**:223a; **15**:28b–29a
early eighteenth-century, **1**:322b; **2**:225a–b, 227a; **4**:172b; **5**:355a; **6**:231a–b; **11**:48a; **12**:100b; **14**:320a
late eighteenth-century, **3**:2a–b, 323b, 427a; **5**:137b; **9**:483a; **13**:541a; **14**:15b, 320b–321a
early nineteenth-century, **3**:323b–324a, 565b; **4**:252a; **7**:366a; **11**:364b, 379b, 593a; **12**:202a; **13**:541b
mid nineteenth-century, **2**:375b, 464a–b; **8**:400b; **9**:10b, 241b; **10**:267a, 558b, 560b; **11**:345b; **14**:529b
late nineteenth-century, **2**:164b, 367b; **6**:78a; **8**:555b; **9**:83a; **10**:155a; **11**:462a–b; **13**:53b, 63a, 503b–504a; **14**:228a–b, 248b–250a
early twentieth-century, **5**:598b; **6**:232b, 526b; **9**:57b–58a, 506b*; **11**:126a; **15**:72a
mid twentieth-century, **5**:180b; **8**:295b
case system, **15**:72a
China, **8**:390b
clinical, **2**:464a–b; **3**:2a–b; **5**:419a, 495b; **8**:295b; **9**:57b; **11**:364b, 462a–b, 559b; **13**:223a
Japan, **15**:742b–743a
paramedical, **11**:364b
metallurgical, **6**:561b; **10**:511b
meteorological
nineteenth-century, **3**:593a
mineralogical
eighteenth-century, **11**:464a–b; **12**:65b–67a; **14**:257a
nineteenth-century, **3**:291a–b; **5**:203a
twentieth-century, **14**:440b
mining, **6**:369b; **12**:65a–67a
monastic, **1**:564b–565a
natural history
nineteenth-century, **5**:69b; **10**:195a
naval
early twentieth-century, **4**:500b
ophthalmological
nineteenth-century, **4**:163a
pharmaceutical
seventeenth-century, **6**:145b
eighteenth-century, **11**:563b; **14**:332b
nineteenth-century, **11**:120b; **14**:112a
Ph.D. degree, **4**:182a
philosophical, **1**:408b; **4**:525a; **11**:356a; **13**:308a; **14**:468a
philosophy of
ancient Greek, **11**:24b, 26a–27a
Renaissance, **11**:287a–288b; **14**:47b
seventeenth-century, **10**:564b
eighteenth-century, **7**:550b–551a; **11**:140a; **15**:272a
nineteenth-century, **2**:103a; **5**:81a; **6**:595b–596a; **10**:187a, 486a
twentieth-century, **8**:120a
physics
Renaissance, **11**:288a–b
early eighteenth-century, **3**:431a; **5**:509b–510a, 519b–520a; **8**:469a; **9**:595b–596a; **10**:145b; **11**:67b, 579b–580a
late eighteenth-century, **2**:335a, 474b; **6**:165a*; **11**:141b, 142a–b; **14**:72a, 452b
early nineteenth-century, **6**:178b; **14**:293a
mid nineteenth-century, **4**:174b; **8**:596a; **9**:566a,b; **10**:28b; **11**:78b; **12**:435a; **13**:374a–b, 377b, 614b–615a
late nineteenth-century, **6**:51b; **7**:446a; **8**:597b; **9**:324a; **10**:599b; **11**:242a; **13**:80a, 102a, 473a
twentieth-century, **5**:188b; **6**:615a; **7**:622b; **9**:395a–b, 462a; **11**:441b; **12**:527b–528a; **14**:68a, 420b, 498a; **15**:490b
physiological
nineteenth-century, **2**:27a–b, 366b–367a; **3**:531a; **5**:80a–83a; **8**:541a; **9**:8b–9a, 142a–b; **10**:576b; **11**:216a, 217a; **14**:113b; **15**:21a, 108a
psychology of, **3**:89a; **6**:53a–b, 294b–295a; **12**:597a; **14**:382b
public
Byzantine empire, **8**:191a
seventeenth-century, **9**:369a
eighteenth-century, **11**:135a
nineteenth-century, **8**:529a–b; **9**:69b
France, **1**:206b; **2**:60a–b; **3**:384a,b, 441a, 450b; **4**:17b; **5**:92a, 93a; **7**:550a; **15**:341b–342b
Germany, **2**:15a
Great Britain, **6**:596a
Hungary, **4**:380a–b
Italy, **1**:344a, 360a
Russia, **5**:210a
Switzerland, **5**:339b; **8**:35b
United States, **1**:363b–364a; **7**:89b
reform of
medieval Latin, **1**:378b
Renaissance, **11**:286b–289a
seventeenth-century, **1**:549a; **3**:360a–362a; **5**:47a, 48a; **6**:141a; **7**:193b, 269a; **14**:177b, 209b
eighteenth-century, **1**:68a; **2**:15a; **5**:600b; **8**:84b; **9**:472b–473a; **12**:120a–b; **13**:39b
nineteenth-century, **2**:60a–b; **3**:523b–524a, 527a; **5**:69a; **8**:35b, 37b–38a, 565b, 572a; **11**:116a–b, 373b; **12**:511b; **13**:427a
twentieth-century, **8**:12a, 489a; **10**:428b; **12**:493b; **13**:45b, 250a, 551a
of retarded children, **2**:131b; **5**:377b
scientific
medieval Arabic, **15**:188a–b
medieval Latin, **1**:104b–105a; **7**:27a
Renaissance, **1**:81a
seventeenth-century, **7**:269a, 376b; **14**:363a
eighteenth-century, **6**:1a; **9**:472a–473a; **13**:68a, 495a; **15**:342a–b, 346b–347a
nineteenth-century, **2**:370b, 451b; **3**:224b, 586a; **4**:274a–b, 546b; **5**:11b, 125b, 393b–394a; **6**:56b, 595b–596a; **7**:365b–366a; **8**:601a; **11**:504b; **12**:278b; **13**:309b, 427a, 521b; **14**:292b–293a; **15**:481b
twentieth-century, **1**:288b; **8**:12a, 584a; **13**:35b, 369b; **15**:48a, 466a
seminars, **10**:28b
surgical
medieval, **12**:83a
seventeenth-century, **12**:40a–b
eighteenth-century, **2**:122a; **6**:231a–b, 567a–b; **8**:114b
nineteenth-century, **6**:77b, 78a; **8**:401b, 402a; **11**:364b; **14**:128a–b
twentieth-century, **3**:518a
technical
seventeenth-century, **1**:549a
eighteenth-century, **2**:292a; **6**:1a, 234a; **8**:258b; **9**:472b–473a; **11**:464a
early nineteenth-century, **3**:199b–200a; **4**:274b; **6**:213b; **9**:492a; **15**:206b–207a
mid nineteenth-century, **7**:160a

Education (cont.)
late nineteenth-century, **2**:470b; **8**:553b; **11**:537a
early twentieth-century, **6**:368b; **7**: 222a; **11**:123b
mid twentieth-century, **8**:369a
theological
medieval, **6**:544a
therapy, **6**:295a
zoological
nineteenth-century, 7:546a; **11**:514b
see also specific countries; specific schools and universities; Women, education of
Edwardes (or Edguardus), David (1502–*ca.* 1542), 4:284–285
Edwards, George, **3**:129b
Gleanings of Natural History, **1**:488b
Edwards, William Frédéric (1776–1842), 4: 285–286
Edwards, William Henry
Voyage up the River Amazon . . . , **1**:501a; **14**:133b
EEG
see Electroencephalography
Eel
electric, **13**:556b
life cycle, **12**:190b
reproduction, **5**:502b; **8**:128a; **12**:562a
Eel-spy, **8**:128b
Effenterre, André van, 7:57a
Effervescence
theories of, **8**:72b–73a, 173b; **13**:602b
Efficiency, **3**:442b
industrial, **13**:272a
Effluvium, **6**:71b
electric, **5**:398a, 516a–517a; **6**:171a–b; **10**:80b; **12**:407b; **14**:193b
Effort syndrome, **8**:295a
Egas Moniz, Antonio Caetano de Abreu Freire (1874–1955), 4:286–287
Egg(s)
amphibian
organizer theory, **2**:364b
Ascaris, **2**:362b–364b; **14**:595a
canine, **2**:161a–b
centrifuging, effects of, **9**:524a
chalaza, **4**:509b
chemistry of, **5**:432a
chick, **1**:109b; **5**:514b; **6**:65a, 160a–161a, 387a, 427a
incubation, **9**:485b
periodic observation of, **1**:109b, 262a, 265a, 266a; **3**:343a, 482b; **4**:510a
cleavage, **5**:51b; **9**:524a; **11**:571b–572a; **14**:425b, 429b
cell lineage, **1**:601a; **3**:389b–390a; **8**: 355a, 357a; **9**:600b; **14**:426a–427a
mechanisms of, **8**:357a
plane of, **10**:39b
cytology of, **8**:357a
cytolysis, **1**:499a
dispermic, **2**:363b–364a, 384b
doe, **2**:161b
experimental constriction of, **12**:568a–b
in fallopian tube, **2**:161a
germinal disc, **4**:509b; **6**:160b; **11**:215b
guinea pig, **2**:161b
hermaphrodite nature of, **1**:601a
holoblastic vs. meroblastic, **11**:369b
leech, **2**:4b
mammalian, **10**:105a, 610b
discovery of, **1**:386b–387a; **5**:485a; **8**: 128a; **11**:380b
division of, **2**:161b
nucleolus, **14**:113a
penetration by spermatozoon, **1**:476b, 477a; **2**:161b
segmentation of, **1**:476b; **2**:161b
membranes of, **14**:595a
thickening of, **1**:499a
nucleus, **5**:51b; **6**:337b–338b
in ovary, **4**:264a; **10**:579b
penetration by spermatozoon, **10**:39b
polar bodies, **2**:626a; **6**:338b; **14**:235b
discovery of, **8**:519a
polarity of, **2**:364b
rabbit, **2**:161b; **3**:487a
rhythmic changes, **14**:429a
sea urchin, **1**:264b; **2**:363a–b, 364b; **5**:51b; **6**:337a, 338a; **8**:446a; **9**:517a
segmentation of, **1**:601a; **4**:266b; **7**:438b; **11**:133a
chemically-induced, **8**:446a
sex of
and cell dimensions, **4**:504a
shapes of, **13**:352b–353a
symmetry of, **7**:95a
transplantation of, **10**:611a
unsheathing of, **1**:498b
vitelline membrane, **6**:338a
yolk quantity
and germ-layer formation, **1**:421a
see also Embryo, Embryology, Ovogenesis
Eggert, John, **12**:70b; **15**:435a, 448b
Eggshell
chemical composition, **13**:554b
Egidius
see Giles
Ego, **5**:175b, 177b, 178a–b
and soul, **1**:27a
type theory, **7**:191a–b
Egophony, **7**:557a
Egorov, Dimitrii Fedorovich (1869–1931), 4:287–288; **8**:558a; **10**:545a, 553b; **11**:156b; **13**:35b, 548b; **15**:65b
theorem of, **4**:287b
Egypt
agriculture, **5**:410b
archaeology, **4**:353b; **5**:94a; **6**:509b–510a; **10**:549b–550b
astronomical monuments, **5**:99a*; **15**: 721a–723a
astrology, **10**:547b–548b; **15**:706a–727b
astronomy, **3**:391a; **12**:547a; **15**:706a–727b
calendar, **5**:345a; **9**:338a,b; **11**:190a; **15**: 536a–b, 706a–727b
encyclopedias, **15**:634b–635a
ethnology, **5**:197a; **6**:528a*; **9**:540b; **11**: 178a
geography, **14**:68b
paleogeography, **13**:85b
geology, **1**:286a; **2**:77b–78a; **13**:274b
paleontology, **2**:110a
German expedition of 1820, **4**:288b–289b
government of, **5**:99a*
historiography, **7**:322b; **9**:26a
mathematics, **2**:320a; **15**:681a–705b
algebra, **1**:31b; **4**:606a,b; **15**:690b–699a
arithmetic, **1**:41a; **4**:111a, 612a; **6**:315a; **15**:681a–690b
geometry, **1**:31b; **13**:296b; **15**:670b–671a, 681a, 690b–693a, 696b–697b
number theory, **15**:692b
medicine, **1**:125a; **11**:177b–178a
mythology, **11**:137b
Napoleonic expedition to, **2**:74a; **3**:346a–b, 411b, 526b; **4**:21b, 150a; **5**:93b, 110b, 356a–b, 410b; **7**:615a; **9**:72b, 473b, 474a; **12**:130b
natural history, **1**:124b–125a; **2**:127b; **4**: 21b, 288b–289b; **5**:74b, 356a–b; **14**: 12b
pyramids, **6**:327a; **12**:499a; **14**:366b; **15**: 697a–b
religion, **2**:539b, 540a; **4**:467b; **6**:305b, 306a
technology, **2**:480b; **4**:541b
Egypt Exploration Society
founding of, **10**:549b
Egyptian Astronomy, Astrology, and Calendrical Reckoning, 15:706–727
Egyptianism, **2**:539b–540a, 542b; **7**:1b
Egyptology, **7**:375a; **11**:596a
Ehrenberg, Christian Gottfried (1795–1876), 4:288–292; **4**:133b*; **11**:360b*
acquaintances of, **1**:476b; **6**:552a; **11**:539a, 541a
and Dujardin, **4**:235a–b
influences on, **7**:533b; **9**:196a
microbiology, **1**:417b; **4**:235a–b; **5**:425a
students of, **2**:124a; **3**:336b, 611b; **8**:519a; **9**:431b; **11**:367b
zoology, **12**:231a
Ehrenfest, Paul (1880–1933), 4:293–294; **7**:491b; **8**:499b*; **10**:218a(n)
acquaintances of, **3**:429b; **4**:312b, 577b; **7**:186b, 617a; **10**:214a; **11**:476b, 477a; **13**:41a
correspondence of, **11**:9a
Ehrenfest's theorem, **4**:294a
electromagnetism, **1**:499b
influence of, **2**:263b, 265a; **8**:498b
quantum theory, **2**:245b; **4**:196b
statistical mechanics, **4**:293b; **9**:129b
students of, **2**:247b; **7**:491a
thermodynamics, **8**:491b
Ehrenfest, Tatiana Alekseevna Afanaseva, **4**:293a,b; **9**:129b
Ehret, Georg Dionysius (1708–1770), 4:295
Ehretia, **4**:295a
Ehrlich, Paul (1854–1915), 4:295–305; **7**: 430a, 433b*; **15**:523b*
acquaintances of, **1**:301b; **4**:359b; **7**:422a, 427a, 616a; **10**:17b, 18b; **11**:174b; **12**:87b, 481b; **14**:228a, 512a; **15**: 257a
chemotherapy
Salvarsan, **5**:29b; **15**:252b
on contemporary scientists, **14**:228b
correspondence of, **13**:228b–229a
critics of, **4**:299a–b
Ehrlich finger test, **7**:623b–624a
histology, **5**:495b; **7**:423b; **11**:382b
immunology, **7**:426a; **11**:430b; **15**:522a
antibodies, **1**:575b; **2**:300b; **5**:564b; **7**: 392b
antitoxins, **2**:584a
serum standardization, **1**:576a
influence of, **4**:154a; **9**:334b; **12**:480b
students of, **8**:273a; **13**:300a; **15**:104b
techniques, **11**:274b
Ehrlich (Paul) Institute, **1**:576a; **4**:298a–301b *passim*
Eichenwald, Aleksandr Aleksandrovich (1864–1944), 4:305–306
Eichenwald experiment, **4**:306a
Eichler, August Wilhelm (1839–1887), 4: 306–307; 2:426b; **9**:149b*; **15**:148a
Flora Brasiliensis, **15**:147a
Eichwald, Karl Eduard Ivanovich (1795–1876), 4:307–309
Eickemeyer, Rudolf, **13**:24a
Eifel plateau, **5**:476a
Eiffel Tower
structural design, **3**:496a
Eigenfunctions, **4**:577b; **7**:408a; **10**:248b; **12**:188a–189a, 219b, 221a; **13**:26b–27a, 424a–b; **14**:282a
Eigenmann, Carl H. (1863–1927), 4:309–310; **7**:184a

Eigenvalues, **5**:96b, 97a, 152a; **6**:392a; **9**:38b; **10**:248b; **11**:59a–b, 476a; **12**:188a–189a, 246b; **13**:27a; **14**:92a, 282a
Eijkman, Christiaan (1858–1930), 4:310–312; **5**:541b; **6**:500a; **10**:493a
Eikonal, method of, **6**:92b
Eilenberg, Samuel, **6**:497a
Eimer, Gustav Heinrich Theodor, **5**:386a; **15**:507b
Eimmart, George-Christophe, **6**:364a
Einarsson, Oddur
see Islandus, Otto
Einstein, Albert (1879–1955), 4:312–333; **2**:398b; **3**:617b; **4**:339a*; **5**:117a; **8**:11b, 389b; **12**:218b
acquaintances of, **4**:294b; **5**:122b; **7**:186b; **9**:260b; **11**:475b; **12**:71a, 528a, 530b; **13**:40b–41a, 73a, 227a; **14**:244a, 283b; **15**:37a, 41a, 251b, 442a, 448a,b, 450a
and Bohr, **2**:250a–b
Bose-Einstein statistics, **4**:317b–318a, 361a, 578a; **5**:473a; **8**:478a–479a; **12**:219a; **15**:48b–49b
on Compton effect, **3**:369a
on contemporary scientists, **4**:281a, 293a; **7**:555a; **8**:10b, 51b, 499b*; **12**:253b*, 498b*, 529b; **14**:171a; **15**:48a
correspondence of, **8**:292b; **10**:422b; **12**:218b, 219a, 251b; **14**:173a; **15**:44a
cosmology, **5**:364a; **6**:531a
critics of, **8**:182a; **11**:84a; **13**:364b; **15**:451a
editions of works, **15**:48b
EIH (Einstein-Infeld-Hoffmann) method, **7**:10a
Einstein effect, **1**:56b–57a
Einstein-de Haas effect, **5**:610a; **14**:172a
Einstein-Smoluchowski theory, **11**:265a; **14**:345a; **15**:464a
Einstein-Wiener theory, **1**:366b; **8**:11a
electrodynamics, **1**:24b–25a; **4**:320a–323b; **8**:498a; **9**:211b
electromagnetic equations, **1**:500a; **4**:323a–b
equivalence principle, **6**:164a; **11**:357a
The Evolution of Physics, **7**:10b
field theory, **4**:316b–317a, 319a,b, 325b–331b; **5**:188b; **12**:251b, 449a
and Freundlich, **5**:181b–182b
gravitation theory, **7**:26b; **8**:599b; **11**:409a; **12**:222b
and Grossmann, **5**:555a
and Infeld, **7**:10a–b
influence of, **3**:620b; **8**:184b; **15**:40a
influences on, **11**:455b
isothermal coordinates, **3**:221a
and Kaluza, **7**:211a–b
and Lenard, **8**:181b, 182a
and Lorentz, **4**:312b; **8**:489a, 490a, 498a–499a
on Mach, **8**:598a, 599a, 599b–600a, 605b
mass-energy conversion equation, **6**:117b; **8**:95a, 290b; **9**:262b
confirmation of, **12**:33b
optics, **12**:497b, 528a
photochemical law, **10**:136a; **14**:175a, 176a; **15**:37a, 448a
confirmation of, **14**:172a
photoelectric effect, **3**:372b; **9**:396a–b; **11**:420b
on photon, **3**:368b
predictions of, **3**:36a–b; **4**:328b–329a; **5**:182a
quantum theory, **2**:244b; **4**:315a–318b, 326a–b; **6**:74a; **8**:52a–b, 181b, 497b; **11**:13a; **12**:218b–219a; **15**:49a
radiation law, **6**:164a; **7**:553a; **11**:11b
relativity, **1**:24b–25a; **2**:121b, 145b, 249b, 383a, 559b; **3**:111b; **4**:281a, 319a–331b; **5**:181b–182b, 597b; **6**:26b; **7**:36b, 264a; **8**:284a, 290b, 498a–499a, 599b–600a; **9**:414b; **11**:113b, 479b; **12**:177b; **13**:474b; **14**:317b
specific heats, **2**:170a; **3**:619a; **4**:316a–b, 414a; **6**:164a; **11**:583a; **13**:41b; **15**:40b, 442a,b, 443b, 447b
statistical mechanics, **4**:314a–318b, 325b–326a; **9**:129b; **10**:235b
Brownian movement, **10**:525a–b; **12**:496b; **13**:159b; **15**:449a
students of, **7**:26b; **11**:355b
thermodynamics, **4**:314a–317b, 320b; **8**:503a; **13**:41a
on Time, **2**:11a–b, 460b; **4**:321b–322a
translations of works, **7**:26b
see also Relativity, theory of
Einthoven, E., **12**:497a
Einthoven, Willem (1860–1927), 4:333–335; **7**:186b; **12**:95a
Eisenberger, John Percyval, **11**:125b
Eisenhardt, Louise, **3**:518b
Eisenhart, Luther Pfahler (1876–1965), 4:335–340; **3**:560a*; **4**:618b
theorem of, **4**:337a
Eisenhower, Dwight David, **8**:96a
Eisenschitz, Robert Karl, **8**:186a, 475a
Eisenstein, Ferdinand Gotthold Max (1823–1852), 4:340–343; **3**:164a, 466b; **5**:307a; **6**:553a; **7**:109a, 506a, 521b; **12**:469a
Eisenstein series, **4**:342b; **6**:214b, 571b
number theory, **9**:173b, 411b; **11**:55a
Eitingon, Max, **5**:179a
Ejaculation
reflex nature of, **9**:571a
Ekaaluminum
see Gallium
Ekaboron
see Scandium
Ekanātha
commentaries of, **2**:119a
Ekasilicon
see Germanium
Ekazirconium, **9**:288b
Ekeberg, Anders Gustaf (1767–1813), 4:343–344
Ekholm, Nils, **1**:300b
Ekman, Vagn Walfrid (1874–1954), 4:344–345; **12**:415a; **15**:431a,b
Ekman layer, **4**:344b
Ekpyrosis, **6**:290b
Elaeometer, **5**:432b
Elam, Constance Fligg, **3**:87b
Élan vital
see Vital force
Elasmobranchs, **7**:61a–b
anatomy, **11**:379b
biochemistry, **5**:149b
embryology, **1**:420a, 420b–421b
excretory system
development of, **1**:421a–b
morphology, **11**:212b
physiology, **12**:470a
Elasmotherium, **2**:423a
Elastic shock, **1**:424b
Elastica, **2**:47b, 50a; **4**:480b
rectification of, **2**:48a
Elasticity, **7**:449a
of air, **2**:337b, 379b; **4**:44b; **5**:575a; **6**:41b, 42b–43a; **7**:494b; **8**:72a–b, 73a, 469a; **9**:117a–b
of anisotropic media, **3**:146a; **9**:174b
and atomism, **1**:568a; **2**:266b
axes of, **12**:129b–130a
coefficients of, **9**:61a
elastic constant, **11**:292a
elastic potential, **3**:146a
of fibers, **14**:206b
finite, **4**:230a,b
of gases, **1**:128b; **8**:468b; **9**:221a, 291a
intramolecular, **3**:443b
of isotropic media, **3**:146a
of lamellae, **3**:146a
limits of, **3**:443b
mathematical analysis of
seventeenth-century, **2**:47b, 49a
eighteenth-century, **2**:40b–42a; **4**:480b
early nineteenth-century, **3**:132a, 135b, 142b–146a *passim;* **5**:376b; **10**:3a; **15**:363a–b, 376b, 485b, 487b
mid nineteenth-century, **2**:299a; **3**:314a; **6**:452a; **7**:601b; **8**:63b; **9**:199b–200a, 226b, 268a; **10**:248a, 249a; **11**:293b–294a; **12**:73b; **13**:75b, 377a; **14**:329b, 473a–b
late nineteenth-century, **1**:128b, 600a; **2**:105b, 356b; **3**:117b–118b, 178b, 428a; **4**:177b; **5**:501b; **7**:595a; **8**:287b, 502a, 516a–b; **9**:174b; **10**:447b, 448b; **14**:61b–62a
early twentieth-century, **4**:230a, 230b–231a; **5**:608b; **7**:453b; **8**:301a,b; **9**:419b; **10**:594b; **11**:409a, 476a, 477b–478a; **14**:87a, 282a; **15**:40a, 164a–165a
mid twentieth-century, **7**:618a
modulus of, **7**:458a
physical theories of
seventeenth-century, **1**:567a, 568a; **6**:485a; **9**:117a; **10**:520b, 565b
eighteenth-century, **1**:113a; **3**:440a, 443b–444a, 486a; **4**:44b
nineteenth-century, **2**:266b; **3**:146a; **5**:169b; **9**:220a–221a; **11**:292a; **13**:384b–386a; **14**:206b, 569a
twentieth-century, **5**:161a; **6**:534b; **11**:476a
molecular theory, **3**:146a
of plane plates, **3**:146a; **15**:164b
reciprocity law, **2**:105b
of rectangular beam, **3**:146a
of rubbery substances, **5**:161a
of solids, **1**:128b; **3**:303b; **6**:342a; **7**:601b; **9**:166b
temperature effects on, **11**:8a, 10a
and viscosity, **9**:220b
of water, **9**:119b
Eleatics, **4**:31a,b
Elections
see Votes
"Electric column," **4**:29a
Electric discharge studies
chemistry, **2**:67a–b, 69b, 70a; **5**:459a
physics, **5**:458b–459a, 483b
see also Discharge, electric
Electric light
see Lighting, artificial
Electric wind, **6**:170b–171a; **11**:142a
Electrical images, **13**:375a
The Electrician, **6**:211b
Electricity
seventeenth-century study, **2**:523a
early eighteenth-century study, **2**:324b; **3**:350a, 486a; **4**:44b–45a, 166b, 215a–216a; **5**:130b–136b, 511a, 516a–517a; **6**:169a, 170a–171b; **7**:403a–b; **8**:177a–b, 179b, 469a; **10**:145a–147a; **11**:433a–b; **14**:193b–194b
late eighteenth-century study, **1**:67a–b, 107b–108a, 547a–548a; **2**:5a, 83a; **3**:51b–52a, 153b, 156b–157b, 444a–445b, 535a; **4**:129a–b; **7**:13a, 371b, 499b, 547a; **8**:258b; **9**:152a, 607b;

Electricity (cont.)
11:135a, 141b–142b; **12**:119b–120a, 427b; **13**:224a–225a; **14**:69b, 76b–80b, 352b–353b, 418b–419b
early nineteenth-century study, **1**:347b; **2**:91a, 136b; **3**:557a, 603b; **4**:528a–b, 532a–538a; **5**:518a; **6**:2b; **8**:38a; **10**:187b–193a, 552b–553a; **11**:473a–474b; **12**:68b*; **15**:485b
mid nineteenth-century study, **1**:555b; **3**:18a, 260a–b, 288b; **4**:248a–b, 556b; **5**:292a–b, 305b, 306a; **7**:380b–381a, 450a–b; **8**:1a; **9**:166b; **10**:500a; **11**:603b–604a; **13**:126a–b; **14**:289b–290a
late nineteenth-century study, **2**:203a, 427b; **3**:265a; **5**:32b–33a; **6**:341a–347a; **8**:387b; **9**:155a; **10**:518b; **11**:122b, 241b, 393a, 460a–461a; **13**:11b, 24a, 287a, 356b, 473a–b; **14**:573b
early twentieth-century study, **4**:305b–306a; **5**:202a; **7**:527a
animal, **1**:107b–108a, 303b; **4**:200a–b, 201b; **5**:267a–268b; **9**:483b; **14**:76b–80a
theories of, **5**:268a–b; **9**:177a–b
see also Electrophysiology
applications of
see Electricity, as power source; Electronics; Electrotechnology; Lighting; Telegraphy, etc.
atmospheric
early eighteenth-century study, **5**:133b–135b; **8**:177b
late eighteenth-century study, **1**:547b, 548a; **2**:83a, 474b; **3**:51b, 153b; **4**:129a; **8**:470a
early nineteenth-century study, **5**:318a
mid nineteenth-century study, **2**:147a; **7**:608b; **11**:111b
late nineteenth-century study, **4**:355a; **5**:21b; **9**:155a; **11**:446a
early twentieth-century study, **1**:521b; **3**:204a; **4**:356b; **5**:341b; **6**:356a*; **13**:176a; **14**:243a*, 422a–b
mid twentieth-century study, **3**:312b; **12**:201b
biological effects of, **5**:267b
detection of, **14**:73a–74a
harnessing of, **15**:449b
measurement of, **6**:96b; **14**:327b
theories of, **1**:300b; **3**:436a; **5**:161b; **10**:500a–b; **12**:49b; **13**:52b; **14**:74a, 422b; **15**:312b
capillary, **8**:503a
conduction of
see Conduction, electrical
contact effects, **11**:446b
equilibrium of, **13**:80a
in flames, **4**:355a; **6**:96b
generation of
see Battery, elect.; Generators, elect.
and heat, **2**:123b; **3**:238b; **9**:205b; **10**:135a, 192a; **13**:374b–377a; **14**:75b, 354b; **15**:312b, 433a
and hydrodynamics, **9**:206a–b
identity of, from different sources, **1**:144b; **3**:260b; **4**:535a–b; **5**:292a–b; **9**:152a; **11**:474a; **14**:78a–80a
inertia of, **6**:341a, 344b
and light, **8**:36b; **10**:146a; **11**:530a
and magnetism
eighteenth-century discussion, **1**:66b–67a; **3**:535a–b; **13**:183b
nineteenth-century discussion, **1**:143b–146a, 202b–203a; **4**:588b; **5**:292b; **6**:279a–b; **8**:490b; **9**:206b; **10**:184b–185b; **11**:474a–b, 530a–b; **13**:377a–b; **15**:433a
twentieth-century discussion, **4**:320a, 322b; **5**:407b
and matter, **3**:499b, 501a; **8**:497a; **9**:207a; **10**:146a; **11**:421a
measurement of, **3**:498b, 506b; **4**:40a, 535b; **5**:320b; **6**:342b; **8**:188a
quantity vs. intensity, **2**:95a; **10**:190a
mechanical effects, **9**:211b
medical use of
eighteenth-century, **3**:153b; **4**:129b; **10**:146b; **13**:469a; **14**:194a–b, 323b*
nineteenth-century, **4**:67a*; **10**:579a
twentieth-century, **1**:582b; **6**:94a
see also Electrotherapy
naming of, **2**:523a; **5**:397b–398a
and natural philosophy, **5**:194b
network theory, **11**:213a
and optics, **4**:189b–191b, 588b; **6**:344a, 345b–346a; **9**:212b–213a
physiological effects, **1**:304a–b; **10**:582b
and plant growth, **2**:83a; **7**:13a
as power source, **3**:18a; **8**:184b; **11**:306a; **12**:424b, 425b; **13**:286b–287a
distribution networks, **5**:407a
in industry, **1**:304a; **4**:40a
production methods, **3**:299a; **7**:527b
switching processes, **11**:588b
transmission of, **1**:304a; **4**:40a, 588b, 589a; **5**:496a–b; **13**:24b, 570b
see also Hydroelectricity
speed of, **1**:108a, 202a; **2**:203a; **3**:288b; **5**:19a, 20a; **7**:450b; **8**:490b–491a, 494b; **12**:131b–132a; **14**:289b–290a
standard symbols, **3**:288b
standard units, **3**:288b; **4**:588b, 589a; **5**:407b, 423b–424a; **6**:212a; **7**:93b, 182b, 288b; **9**:155b, 178a, 209b; **11**:579a; **13**:102a; **14**:205b
theories of
early eighteenth-century, **4**:166b, 215b–216a, 482a; **5**:131a–136b, 516a–517a; **7**:403a–b; **10**:134b–136a, 146a–147a; **11**:433a; **14**:193b–194a, 418b
late eighteenth-century, **1**:547a–b; **3**:156b–157a, 444a–b, 599b; **4**:129b, 503b; **5**:195a; **6**:178b; **7**:13a, 371b, 587a; **8**:258b, 470a; **11**:66b–67a, 146b; **12**:407b; **13**:224b–225a; **14**:70a–74a, 352b–353b, 419a
early nineteenth-century, **1**:144b–145b, 202b; **2**:94b; **4**:528a–b, 533a; **9**:548a; **10**:184b–185a, 188b–192b; **11**:474a; **12**:254a; **14**:130b*
mid nineteenth-century, **2**:87b; **3**:322b–323a; **4**:535a–538a; **6**:96b, 278b, 279a; **8**:382a; **9**:205a–206b, 208b–209b, 216a–217a, 548a–549a; **13**:374b–377b, 384a–b; **14**:207a–208a, 330a
late nineteenth-century, **2**:88b; **4**:189b–191b; **5**:496b; **6**:252b, 343b–344a, 346b, 347a; **8**:490b–497a; **13**:364a–368a, 473a
twentieth-century, **9**:606b; **13**:241b
action-at-a-distance, **1**:67a; **2**:324b; **3**:444b; **4**:536a–b; **6**:251a, 279b, 343b, 344a, 345a; **7**:380b, 381a; **8**:490b, 491a, 492a,b; **9**:204b, 548a, 549a; **10**:28a, 134b, 146b; **13**:375a–b; **14**:32b, 33a, 71b, 305b, 353a
atomistic, **6**:343b–344a
caloric, **10**:135a–b
contact, **10**:190b, 192a; **14**:70b, 77b–79b
effluence and affluence, **10**:146a–147a
electrical atmospheres, **1**:67a, 547b–548a; **3**:51b; **4**;215b; **5**:132b–133a, 136a–b; **7**:371b; **14**:70b–71b, 72b–73a, 353a
electrical plenum, **6**:279a–b
mathematical, **1**:499b; **2**:87b, 466a; **3**:263a; **7**:84a; **8**:470a; **10**:248a, 594b; **13**:374b–377b; **15**:199b
particulate, **11**:446a–b; **12**:289a
single-fluid, **1**:67a; **3**:156b–157a, 376b, 444b; **4**:528a,b; **5**:131a–136b, 138a; **6**:178b; **8**:36a–b, 258b, 491a; **9**:152a, 548a–549a; **11**:142a; **13**:374b–377b; **14**:70a–74a, 76b
strain theory, **4**:535a–538a *passim*
two electricities, **4**:215b–216a; **7**:371b; **10**:188b; **13**:224b–225a
two-fluid, **1**:144b–145b, 547b; **2**:5a, 94b; **3**:444b; **4**:528a,b; **8**:490b; **14**:208a–b, 353a
vibrational, **4**:528a–b, 533a
see also Electron theory
types of
"aerial" and "proper," **3**:51b
"artificial" and "natural," **1**:547a,b
dynamic and static, **8**:36a
positive and negative, **5**:131a–135b *passim*
resinous and vitreous, **4**:45a, 166b; **5**:135b; **14**:352b
vindicating, **1**:547b–548a; **14**:71a–b
in vacuum, **1**:547b; **5**:292b
see also Apparatus, electrical; Charge, electrical; Conduction, electrical; Conductivity; Congresses, scientific; Currents, electrical; Discharge, electrical; Electrostatics; Engineering, electrical; Force, electrical; Galvanism; Hydroelectricity; Induction, electrical; Piezoelectricity; Thermoelectricity
Electroacoustics, **1**:292b–293a; **7**:470a; **10**:604b–605a
Electrocapillarity, **5**:483b; **6**:371a–372a, 374a; **8**:387b; **15**:435a
Electrocardiography, **2**:601a; **8**:294b–295a, 295b; **12**:95a; **15**:101a
development of, **4**:333b–334a
Electrocautery, **3**:518b
Electrochemical series, **6**:96b
Avogadro's, **1**:347b, 348b, 349a
Berzelius', **2**:95a
Volta's, **1**:347b
Electrochemistry
late eighteenth-century, **13**:468b–469a; **14**:78a–80a
early nineteenth-century, **1**:347a,b, 557a–b; **2**:91a, 92a–b, 94b–95a, 136b–137a; **3**:68a, 544a, 600b–602a; **4**:74a; **5**:7a, 329b–321a, 558a–559a; **6**:436b; **10**:108b, 182b, 184a–b, 552b–553a; **11**:473a–474b; **12**:254a; **13**:311a; **14**:80a
mid nineteenth-century, **1**:555b; **2**:588b; **3**:557a; **4**:535b–537a; **5**:474a–b, 559a–560b; **6**:438b–439b; **7**:451a; **8**:36a–b, 188b; **9**:60b, 178a, 391a–b; **10**:500a; **11**:297b–298a; **12**:198a; **13**:359a
late nineteenth-century, **3**:333a,b, 334a; **4**:141a; **5**:597a; **6**:50b; **7**:65a–b, 449a–b; **8**:112b–113a; **9**:155a, 394a; **10**:345a–b; **12**:465a; **13**:228a; **14**:124b; **15**:84a, 433b–434a, 458b–461a
early twentieth-century, **1**:385a, 431a; **4**:141b, 413b; **5**:595b, 621a–622b; **6**:370b–373a; **7**:498a; **8**:289a–293b, 503a; **10**:156b; **11**:418a; **12**:546b; **13**:446b; **14**:125a; **15**:435a
mid twentieth-century, **3**:336a; **6**:373a–375b; **7**:56b
critics of, **9**:286b

decomposition of salts, **2**:92a; **4**:536a–537a
of fused salts, **8**:503a
historiography of, **15**:466a
laws of, **4**:536a; **8**:36b
terminology, **4**:537a
theories of, **5**:372a; **12**:198a
chemical theory, **8**:36a–b
contact theory, **8**:36a,b
dualistic theory, **2**:92b, 94b–95a; **4**:244b, 531b
strain theory, **4**:536a–537b
see also Electrochemical series, Electrolysis, Electroplating
Electro-coagulation, **1**:304b
Electrode potential, **5**:621a
Electrodeposition, **1**:431a; **2**:340b; **5**:474a; **7**:56b; **13**:23a
see also Electroplating
Electrodes, **5**:65b
aluminum, **6**:372a
amalgam, **6**:372a
carbon, **5**:474a
electromotive forces of polarization, **8**:188b
gas, **7**:208a
glass, **5**:621a
hydrogen, **3**:290b; **8**:113a; **13**:228a; **15**:435a
hydrogen overvoltage, **6**:373a–b
Langmuir probe, **8**:24b
magnetite, **13**:24a
mercury, dropping, **6**:371b–374b
negative
mechanical effects on, **5**:292b
mercury, **2**:92b
platinum-iridium, **5**:474a
quinhydrone, **12**:546b
reversible potential, **6**:373a–b
screen, **6**:546b
silver chloride, **1**:303b
space charge, **8**:23b–24a
unpolarizable, **2**:367a; **4**:202b
Electrodynamics
eighteenth-century, **3**:157b
early nineteenth-century, **1**:143b–146a, 202b; **5**:171a
mid nineteenth-century, **8**:384b; **9**:204b–217b; **10**:27b–28b; **11**:453b; **14**:205a–b, 206b–208a; **15**:193a
late nineteenth-century, **2**:167b; **3**:309b–310a; **5**:462a; **6**:250a–252b, 343b–347a, 348a–b; **8**:40b, 490b–496b; **9**:107b, 174b; **11**:9a, 10a–12a, 407a; **13**:364b–368a
early twentieth-century, **1**:24a–25a; **4**:320a–323b; **5**:161a; **8**:496b–499a; **9**:376b–377b; **11**:12b–13b, 421b; **12**:69b, 250a, 528a–b; **13**:239b–240a; **15**:49a
mid twentieth-century, **7**:10a, 617b; **8**:476a–478a
atomic physics, conflict with, **2**:239b, 241b–250b
classical, **3**:368b–369a
conservation principle, **6**:244b, 250a
and hydrodynamics, **2**:166b–167a
logic of, **2**:460a; **11**:479b–480a
mathematics of, **11**:409a
and mechanics, **6**:96b
of moving bodies, **6**:346b–347a
non-linear, **7**:10a; **9**:377b; **15**:43b
and relativity, **1**:24b–25a; **4**:322b; **11**:479b–480a; **12**:528a, 529b; **13**:176a, 239b
theories of
nineteenth-century, **6**:243a, 245a, 250a–252b
see also Electromagnetism, Quantum electrodynamics
Electrodynamometer, **14**:206b
Electroencephalography, **5**:65b
discovery of, **2**:1b–2a
Electrogravimetry, **15**:84a
Electroluminescence, **6**:170a–171b
Electrolysis, **2**:258b; **5**:621b; **6**:284b, 319a; **9**:19a; **10**:108b
energetics of, **6**:244b
fundamental constant of, **13**:446a
secondary
crystal production, **1**:557b
theories of, **2**:136b–137a; **3**:601b; **5**:558a–b, 559a; **6**:439a–b; **8**:112b–113a
see also Electrochemistry
Electrolytes, **3**:3b; **14**:523b
activity coefficient, **1**:298b, 299b
in blood, **13**:574b
concentration
change during electrolysis, **6**:439a; **9**:391b
conduction in, **1**:297a–300b
decomposition voltage, **8**:112b–113a
diffusion of, **15**:433b–434a
dilution of, **1**:298a
dissociation theory, **1**:297a–300b; **2**:170a; **3**:176a, 619b–620a; **4**:554b; **6**:354b, 366a, 439b; **7**:65b, 203b, 208b, 449b; **9**:286b, 291b; **11**:279a, 298b–299a, 577a; **13**:244a, 364a, 579b; **15**:433b, 458b–461a
impurities, **14**:171b
ionic theory of, **14**:395b
isotonic coefficient, **13**:579b
resistance of
dilution effects, **1**:298a
solid, **12**:191b
solutions of, **1**:288b
nature of, **7**:498a
ordering in, **3**:620a
thermodynamics of, **3**:620a
solvoelectrolytes, **7**:462a
specific heats of, **11**:577a
strong, **1**:300a; **2**:170a
dissociation of, **13**:156a
theory of, **7**:493a
theory of, **14**:125a
transference numbers, **14**:182b
see also Electrochemistry
Electrolytic polishing, **7**:56b–57a
Electromagnetism
eighteenth-century study, **3**:535a
early nineteenth-century study, **1**:202b–203a, 460a, 557b; **3**:497a; **4**:533a–535a; **12**:129b, 254a–b; **13**:126a–b
mid nineteenth-century study, **2**:137a; **6**:277b–279b; **7**:380a; **8**:187a–188b, 510b; **14**:206b; **15**:85a
late nineteenth-century study, **2**:202b, 265b–266a; **5**:63b; **6**:343a; **8**:39b–40a; **11**:460b–461a, 578a–b; **12**:26a; **13**:102b, 473a
twentieth-century study, **1**:283b–284b; **4**:194b; **5**:433a; **10**:214b
applications of, **4**:194b, 248b
in crystals, **2**:466a–b
discovery of, **10**:184b–185b; **12**:281b
effect of motion of bodies on, **2**:559b
field equations, **4**:315a, 327b; **8**:495a; **9**:211b–212a, 377b
mechanical models for, **2**:266a
geometrization of, **4**:325b, 330a–331a
and gravitational field, **4**:325a–b, 326b
and light, **7**:313b
longitudinal fluxes, **4**:231b
mathematical study, **1**:499b, 500a; **2**:491a; **7**:595a; **9**:211a–212a
and mechanics, **4**:315a, 321a, 323a, 330b
and relativity, **13**:176a
in solids, **5**:433a
as source of power, **3**:18a; **7**:180b–181a
theories of, **4**:231a–232a, 313b, 319a–b, 323a–b; **6**:279a–b; **8**:284b; **9**:206b–217b; **11**:578a–b; **12**:254b; **13**:11b, 384a–b
ether pulse theory, **1**:457a–458a
and thermodynamics, **2**:266a; **8**:387b
in vacuum, **6**:279a
X rays, **1**:457a–458b
see also Electrodynamics; Fields (phys.), electromagnetic; Force, electromagnetic; Radiation, electromagnetic; Telegraphy; Waves, electromagnetic
Electromagnets
attractive force of
and magnetization current, **8**:188a
design of, **3**:329a; **6**:278a; **13**:126a–b
lifting capacities, **9**:459b
mathematical study of, **4**:248b
in motor design, **4**:248a
power of, **7**:56a
superconducting, **8**:480a, 482a–483a
Electromerism, **8**:293a–b
Electrometer, **4**:356b, 535b; **5**:331a; **6**:96b; **7**:450a; **8**:258b; **9**:259b; **11**:433a, 496b; **12**:120a
bottle, **14**:73a–b
capillary, **1**:536b; **2**:1b; **4**:333b; **8**:387b, 533a
Curie's, **3**:498b, 506b, 507a
flame, **14**:73b
floating, **3**:562a
induction, **11**:460a
Kelvin's, **3**:504b
Lindemann, **8**:369a
quadrant, **6**:73b–74a
straw, **14**:78a,b
see also Electroscope
Electromicrophotometer, **11**:542b
Electromotive series, **11**:50a
Electron(s), **2**:622b
absorption of, **15**:491a
acceleration of, **3**:329b; **6**:104a–105a; **12**:33a; **13**:570a; **15**:8b
annihilation of, **4**:280a
charge of
determination of, **3**:370a; **9**:395b–396a
inconstancy of, **5**:273a
charge/mass ratio, **4**:281a; **7**:211b, 263b; **8**:496a–b; **11**:13b; **13**:367a–368a; **14**:218a
collisions of
energy transfer, **5**:117a–118a
in ionized gases, **3**:372b
Compton wavelength, **3**:367b, 368b
detection of, **9**:261a
diamagnetism of, **3**:565a*
diffraction of, **3**:369b, 598a; **7**:470a; **11**:422b
discovery of, **3**:475a; **13**:367a–b
ejection of, **7**:153b
electric moments, **7**:518b
emission of, **4**:284a, 579b; **9**:260b–261a
chemical, **11**:422a
cold, **9**:397a
thermal, **3**:372b; **8**:23b–24a; **11**:420a–421b
energy of, **9**:261a; **10**:423b; **12**:528a; **15**:491a
entropy, **8**:47a
exclusion principle, **4**:578a
and field, **8**:498b
finite size, hypothesis of, **3**:367b, 368a,b
free, **5**:160b–161a
heat capacity of, **5**:160b
high-speed
and matter, **15**:8a–b

Electron(s) (cont.)
interference of, **3**:598a; **7**:469b–470a; **8**:52a
itinerant, **5**:160b
magnetic moment, **13**:43a
magnetic orientation, **3**:367b; **5**:160b
mass of
electromagnetic, **1**:24a–b; **7**:263b–264b; **14**:218a
and motion, **2**:559b; **8**:496a–497a; **14**:340b
relativistic increase, **12**:529a
specific, **2**:599b
transverse, **2**:599b
variation of, **5**:597a–b
and velocity, **7**:263b–264b; **8**:496a,b, 497a; **11**:13b
mean free path, **8**:480b; **13**:446b
in metals, **4**:376a; **5**:160b; **7**:617b; **12**:529b, 530a; **15**:444a
momentum, **8**:477a; **9**:217a–b
motions of, **2**:241b–242b
in electromagnetic field, **4**:322b–323a, 326a; **6**:546b; **8**:494a
naming of, **13**:82a–b
nature of, **3**:499b
negative energy states, **13**:240a
number of, **4**:281b; **7**:468b–469b; **13**:368b, 369a
orbits of, **8**:11a, 485a–b
eccentricities of, **12**:529a
pairing of, **8**:477b, 480b
quantum mechanical theory of, **3**:564a; **7**:491b; **8**:477b; **9**:544b; **13**:240a–b
radiation of, **13**:241a
relativistic theory of, **3**:369b, 564a; **6**:615a; **7**:26b; **15**:40a
ring configurations, **2**:241b, 242a; **7**:468b–469b; **13**:368b
scattering of, **2**:338a,b; **3**:597b; **8**:181a
secondary, **3**:430a; **8**:10a; **9**:260b; **14**:218a; **15**:26b
distribution-in-angle, **3**:597b
shell structure, **2**:246a–b, 488a
systematics of, **2**:247b
showers of, **7**:617b
slow, **9**:261b
spectra, **9**:261a
spin of, **1**:371a; **2**:248a; **3**:125a; **4**:281b; **5**:473b; **9**:397b; **10**:423a; **11**:420b; **12**:219b, 438a, 529b; **13**:240a
symmetrical theory of, **9**:38b
temperature, concept of, **8**:24b
velocity, **13**:446b
wave theory of, **3**:430a; **11**:421b; **13**:447a
confirmation of, **3**:597b–598a
Electron compounds, **6**:561a
Electron gas, **11**:421b
Electron microscope
see Microscope, electron
Electron theory, **1**:24a–25a, 457a–b; **2**:383a, 559b; **4**:189b, 191a, 315a–326a *passim,* 355b; **5**:161a, 391a, 473b, 597b–598a; **7**:263b–264b; **8**:10b–11a, 39b–40b, 488b, 493b–499a; **9**:217a, 395b; **11**:58b, 420a–421b, 476b, 478b; **12**:527b; **13**:156a, 367a, 368a; **14**:61b, 208b–209a, 218a; **15**:39b, 40a, 433a, 464a
practical application of, **4**:191b–192b
and quantum theory, **8**:497b–498a
Electron tubes, **6**:546a–b; **7**:36b, 468b; **8**:24a, 51b; **12**:34a; **14**:217b–218a
carbonic acid, **5**:292b
coefficients for, **1**:455b
microwave
velocity modulation, **1**:455b; **6**:104b
noise in, **6**:546b
thermionic, **5**:291a; **11**:420a–421b
Electronastism, **11**:474a
Electronics
nineteenth-century, **2**:325a, 427b–428a; **4**:559a–b
twentieth-century, **1**:292b–293a, 455b–456a; **3**:373a; **4**:6b–7b; **5**:33a, 65a, 433a; **6**:546b; **7**:108a, 468b; **11**:420a–422a; **14**:217b–218a
microwave, **6**:104b–105a
see also Circuits, electric; Electron tubes; Radio, etc.
Electrophoresis, **12**:496b; **13**:418b–421b
Electrophorus (instrum.), **1**:548a; **7**:13a; **8**:322a; **10**:147a; **13**:225a, 468b; **14**:71a–b, 72b
invention of, **14**:353b
Electrophysiology
late eighteenth-century, **5**:267b–268b; **14**:76b–79a
early nineteenth-century, **11**:473a–474b; **13**:547b
mid nineteenth-century, **4**:200a–203a; **9**:177a–b, 259b, 570a; **10**:579a; **12**:270a
late nineteenth-century, **1**:9b, 536b; **2**:599a; **5**:196b; **6**:440b–441a; **13**:53a–b; **14**:3a; **15**:21a–b
early twentieth-century, **2**:1b; **12**:95a–b
mid twentieth-century, **14**:201a
sources of error, **4**:203a
theories of, **9**:259b
see also Instrumentation, electrophysiological
Electroplating, **3**:557a; **5**:7a, 474a; **12**:424b, 425a; **15**:24b
see also Electrodeposition
Electroscope
condensing, **14**:78b
"electric column," **4**:29a
gold-leaf, **14**:78a
pith-ball, **3**:51b, 153b
straw, **14**:73b–74a, 79b
thread, **6**:171a
see also Electrometer
Electrostatic balance, **1**:555b
Electrostatics
Renaissance, **5**:397b–398a
seventeenth-century, **5**:575b; **6**:477b; **7**:578a
eighteenth-century, **1**:59a, 67b; **2**:43b; **6**:170a–171b, 173b; **9**:477a, 596a–b; **14**:70a–74a; **15**:312b
nineteenth-century, **2**:137a, 397b; **4**:535b; **9**:155a, 205b, 215b; **10**:553a; **12**:42a; **13**:80a; **15**:83b, 360b–361a, 485b
conservation principle, **6**:244b
theories of, **3**:51b, 445b; **4**:317a, 537a–b; **5**:397b–398a; **6**:171a–b, 173b, 279b; **9**:216b
Electrostenolysis, **5**:559a
Electrotechnology
mid nineteenth-century, **2**:439a; **4**:248a–b, 559a; **7**:56a; **12**:424b, 425a–426a; **13**:386b–387a
late nineteenth-century, **2**:200a; **4**:40a, 283b–284b; **8**:502a–b; **10**:518b–519a; **11**:93a–94a; **13**:80b, 356b–357a
early twentieth-century, **4**:514a; **5**:33a, 407a; **7**:621b; **8**:23a–24b; **10**:604b–605a
see also Engineering, electrical; Radio; Telegraphy
Electrotherapy, **1**:304a–b; **4**:202b; **5**:137b–138a; **11**:369b
see also Electricity, medical use of
Electrotonus, **4**:202a–b, 616a
Electrotyping, **4**:19a; **9**:60b
discovery of, **7**:56a
Elements, Aristotelian
ancient Greek discussion, **1**:149a, 252b, 254a,b, 255a, 261a–b; **4**:31a, 367a, 367b–368b; **6**:427b; **11**:104a; **13**:93b, 330a; **14**:606b
ancient Roman discussion, **4**:622a
Byzantine discussion, **12**:405b
medieval Arabic discussion, **1**:33b, 34a; **7**:23a; **11**:231b–232a, 324b; **13**:511b; **15**:240a, 496b, 498b
medieval Latin discussion, **1**:197a, 199a–b; **4**:94b; **5**:550b–551a; **7**:27b–28a; **8**:548b, 550a; **13**:187b–188a, 206a, 340a–b; **14**:155b
Renaissance discussion, **1**:495a; **2**:410a; **3**:66a, 514b; **4**:208b, 209b, 387b; **8**:199a–b; **12**:335a–b
seventeenth-century discussion, **1**:549b, 571b; **2**:293a, 380a–b; **5**:47b, 426a; **7**:165a, 269a, 538b; **8**:131a; **9**:15a; **11**:389b; **12**:312a–b; **13**:234b–235a
eighteenth-century discussion, **3**:129b; **4**:352b–353a; **5**:353a–b; **6**:40b, 43b; **7**:587a–b, 603b; **8**:72a–73b, 82b, 619a, 620a; **11**:562b; **12**:604a
nineteenth-century discussion, **2**:179b
in alchemy, **2**:22b; **12**:311b
in astrology, **7**:297b
continuity of, **1**:61b–62a
denial of, **5**:397b, 400a; **8**:72a
distribution of, **7**:28a
forms of, **8**:242a
motion of, **3**:403b–404a
in physiology, **4**:106a
relative volumes of, **1**:606b
transmutation of, **1**:495a; **4**:184a; **5**:551a; **8**:71b–72a
see also Air, Earth, Fire, Water
Elements, chemical
artificial, **10**:136b
classification of, **3**:398a; **10**:178a; **11**:173b
by atomic weights, **4**:247a–b; **5**:412a; **9**:348b–352a
genetic system, **11**:136a
geochemical, **4**:599b–600a; **5**:457b; **13**:618b–619a
by law of octaves, **9**:349a; **10**:38a
by reactions, **1**:143a–b; **5**:164a
triads, **4**:134b; **5**:431a; **9**:348b; **10**:37b
see also Elements, chemical, periodic system
complexity of, **3**:476b; **9**:289b
compound nature, possibility of, **2**:485a–b; **3**:602b; **9**:351b
concept of, **3**:543a, 544a; **6**:384a, 478a, 543a; **7**:188b, 194b–195a; **8**:80b, 81b–82b, 613a; **11**:562b; **15**:464a
dissociation of, **4**:79a
distribution of, **4**:599b–600a; **5**:457a–458a; **10**:136b
in body, **6**:366b
frequency distribution, **4**:598a–b; **5**:273a
evolution of, **1**:321b; **3**:292a, 478a, 478b–479b; **9**:110a; **10**:106b; **11**:173b
Group 5B, **2**:199b
migration of, **1**:623b; **4**:598b; **13**:618b–619a
mineralogical associations of, **1**:572b; **13**:617b–618a
paragenesis, **13**:618a
periodic system, **10**:37b–38b; **12**:42a
completeness of, **9**:543b–544a
critics of, **9**:290a–b
discovery of, **9**:288a–291a

displacement law, **2**:241a; **6**:16a, 481a; **12**:32a, 505a–506b
electron configurations and, **2**:241a, 246a; **6**:366a; **12**:32a, 44a; **13**:367b, 368b
and magnetic coefficient, **6**:480a
Mendeleev's table
see Mendeleev, periodic table
and physical properties, **11**:486a
predictions from, **2**:254b; **3**:321b; **9**: 288b, 289a,b, 290b, 292b, 375b; **11**:279b; **13**:359a; **14**:447a
rare earth elements and, **2**:248b; **3**: 478b–479b
spiral table, **1**:528b, 572a–b; **9**:290a; **12**:44a
telluric screw, **1**:572a–b; **9**:349a
and valence, **9**:349b–351a
see also Elements, chemical, classification of
spectra of
and atomic spectra, **7**:268a
formulas for, **1**:425b; **7**:268a
regularities in, **7**:267b–268a
theories of, **2**:254b–255a; **10**:106b
X-ray, **1**:456b, 458b; **9**:543a–544b
transformation of, **1**:560a; **2**:338a–b; **4**: 580a; **10**:289a; **11**:173a, 281a; **15**: 464a
opposition to, **9**:292b
transition elements, **3**:333b
transuranic, **6**:16b–17a; **9**:261b–263a
see also Isotopes; Radioelements; specific elements and groups of elements
Elements, Empedoclean
see Elements, Aristotelian
Elephant, **15**:662a(n)
anatomy of, **1**:263a; **2**:188a; **3**:38a; **4**:267b; **12**:224a
fossil, **2**:494b; **5**:296a; **11**:41a; **12**:182a; **13**:257b–258a
histology of, **3**:425a
Elephantiasis, **8**:296b; **11**:445a; **13**:274a
surgical treatment, **9**:81b
Elevation, geological, **3**:567a–568a; **4**:265b, 349b; **5**:329b
of islands, **3**:551b, 567b, 569a; **11**:303a
recent, **3**:567a, 568a; **8**:567a, 570b
theories of, **2**:568a, 569a, 570a; **3**:552a, 594b; **6**:55b, 502b, 581a, 583b, 584a–b; **8**:568a; **11**:134a; **13**:85a–b
El Goresy, Ahmed, **5**:187a
Elhuyar (or Elhuyart), Fausto d' (1755–1833), 4:345–346; 4:346b
Elhuyar (or Elhuyart), Juan José d' (1754–1796), 4:346–347; 4:345a–b
Eli Lilly Company
insulin, **1**:443a
Elias (philos.), **1**:137a
Elias of Nisibis, **15**:232a
d'Eliçagaray, Bernard Renau, **2**:54b
Élie de Beaumont, Jean-Baptiste-Armand-Louis-Léonce (1798–1874), 4:347–350
acquaintances of, **1**:572a; **2**:108a, 482b; **13**:145a
biographies of, **2**:88b
on contemporary scientists, **8**:143a*
correspondence of, **7**:45b
critics of, **2**:341b; **6**:504a*; **11**:134a; **12**: 262b, 263a
erratic blocks, **3**:210b
geology, **1**:210a; **4**:218a; **8**:463b, 573a; **12**:431a
influence of, **4**:73a; **6**:178a
influences on, **2**:158b; **4**:67b
orogenesis, **3**:395b; **12**:276b; **13**:145b–146a
students of, **7**:44b; **8**:30b; **13**:620a; **14**: 315b
Voyage métallurgique en Angleterre, **3**: 586b; **4**:218a
Elimination, algebraic, **3**:64a; **4**:475b; **6**: 537b; **9**:410a
theory of, **2**:112b–114a, 299a, 444a, 465a; **7**:444b; **8**:61b, 385a; **12**:171a–b
Elinvar (alloy), **5**:583a
Eliot, Charles William, **2**:366a; **5**:393b–394a; **10**:599a–b; **12**:54b, 481b, 482a; **15**: 72b, 467a(n)
Eliot, Jared, **3**:350b
Elixir of life, **5**:422b; **7**:41a–b
Elizabeth I, Queen of England, **1**:383a; **2**: 540b; **3**:13a; **5**:348a, 361b, 396b; **6**: 20a,b
Elk, **2**:423a
Elkin, William Lewis (1855–1933), 4:350–352; 5:405b; **10**:42b*
Ell (unit of length), **6**:236b
Ellagic acid, **2**:385b
Eller von Brockhausen, Johann Theodor (1689–1760), 4:352–353; 8:71b, 72a–b, 76b
critics of, **14**:191a
Ellerman, Ferdinand, **4**:251a; **6**:28b; **10**:320b
Elles, Gertrude, **8**:33a
Ellicott, Andrew, **6**:166a
Ellicott, John, **3**:51a; **4**:149a; **14**:452b
Elliot, Robert, **9**:479b
Elliot Smith, Grafton (1871–1937), 4:353–354; 2:171b, 172a; **8**:282b; **11**:595b, 596a
Elliotson, John, **11**:172a
Elliott, Stephen (bishop), **8**:103b; **14**:417a
Elliott, Thomas Renton, **5**:81a; **8**:18a(n), 451b, 452b, 453a, 454a; **15**:73b, 106b
Ellipse, **8**:252b
area of, **3**:152a; **12**:292a–b; **13**:291a
construction of, **1**:170a, 446a; **7**:599b; **14**:465a; **15**:60b
definition of, **9**:598b; **10**:228a
equation of, **9**:311b
inscribed, **10**:573b; **13**:18b
meaning of word, **4**:421a
normals to, **7**:109b
rectification of, **4**:516a; **8**:139b; **11**:487b, 609a
sector of, **7**:558a
spherical, **5**:209b
see also Conics
Ellipsoid, **1**:600a
attraction of, **1**:147a; **2**:134b; **3**:169b, 213b; **4**:126b; **7**:37a, 54b; **8**:611a; **15**:200a, 317b, 351a
equilibrium of, **7**:37b
functional equation of, **3**:428a
geodesics, **8**:385a–b
geometry of, **6**:240a
of revolution, **7**:55a; **8**:561b–562a; **11**:58a
rotation of, **11**:454a
surface area of, **3**:432a
thermodynamics of, **7**:601b
Ellis, Charles Drummond, **12**:34a
Ellis, George, **3**:579a
Ellis, James Hawes, **4**:82b
Ellis, John, **1**:434a; **3**:350b, 351a; **6**:44b, 566b; **12**:515b, 516b
Ellis, Robert Leslie, **5**:522b; **7**:106b; **9**:419b
Ellis, William (1828–1916), 4:354; 9:261a,b
Ellison, John, **6**:383b
Elsasser, Walter, **3**:597b; **15**:43a
Elsberg, Charles Albert, **3**:517b
Elster, Johann Philipp Ludwig Julius (1854–1920), 4:354–357; 12:506b(n)
Elster-Geitel activation number, **4**:356b
and Geitel, **3**:480b(n); **4**:354b–357a; **5**: 341b, 394b; **13**:80b, 367b
radioactivity, **12**:28a
Eltzbacher, Fritz, **1**:168b
Elution methods, **13**:421a
Elvehjem, Conrad Arnold (1901–1962), 4: 357–359; 5:209a; **6**:136a
Elvey, Christian Thomas, **13**:116b, 117b, 118a
Elwell, Royston William Dunlap, **1**:394b
Emanation, philosophy of, **1**:33b, 334a; **7**: 23a; **15**:249a–250b, 494b
Emanium, **5**:394b; **6**:15b
Emanuelli, Pio (1888–1946), 4:359
Embalming, **6**:121a*; **12**:40b–41a
Embden, Gustav (1874–1933), 4:359–360; 5:6a
Embden ester, **4**:360b
Embden, Meyerhof, Parnas (EMP) scheme, **10**:326b
Emboîtement, **11**:333b; **13**:174a
Embolism, **3**:489b; **14**:40a, 249b
pulmonary, **14**:408a
Embryo
blood formation, **7**:438a
chick, **10**:105a; **11**:368b
death
from radiation, **9**:565a
delamination in, **1**:420b
differentiation in, **2**:383b–384b; **4**:147a; **6**:435a; **9**:517a–518a; **11**:572a–573b; **14**:428a–b, 429b
progressive, **10**:105a
fusion of, **4**:187a
grafting, **6**:132a–133b
human, **4**:106a; **5**:356b; **11**:361b*; **12**: 510a
collections of, **9**:57a
measurement of, **13**:97a
models of, **9**:56b
pathology of, **3**:245b
injury experiments, **11**:572a–573a
layers, **14**:596b
mammalian, **10**:105a
in vitro cultivation, **2**:384b
morula, **1**:476b
membranes of, **1**:386b
models of, **2**:383b
nutrition of, **4**:509b
preservation of, **3**:482b
primitive streak of, **1**:421a; **6**:287b
regeneration in, **8**:357b; **11**:573a
tubulation in, **1**:386b, 387a
see also Egg(s), Embryogeny, Embryology, specific animals
Embryo sac, **5**:539b, 581b
Embryogeny
air, effects of, **12**:240b
botanical, **9**:418a–b; **13**:405b–406a
causal analysis, **6**:435b; **11**:572a–574b
cellular movement, effects of, **1**:421a; **11**:597a
chemical studies, **7**:95a, 467a; **11**:173b
chromosomal influences, **13**:281b–282a
environmental influences, **3**:245b; **9**:517b; **13**:282a; **14**:428a
temperature effects, **5**:523b
and evolution, **1**:71b–72a, 421a; **5**:447a; **6**:65a–b; **7**:475a; **12**:337b–338b; **14**: 426b; **15**:169b–170a
growth control, **6**:132b–133a
and heredity, **9**:523b; **11**:573b
induction, **4**:187b; **10**:105a; **12**:568a–569a
laws of, **11**:133a
nucleus, role of, **11**:572b–573b
and regeneration, similarity of, **9**:517b–518a

Embryogeny (cont.)
theories of
ancient Greek, 1:260b–261a
medieval Arabic, 11:238b–239a
Renaissance, 4:509b–511a
early seventeenth-century, 1:293a–b; 6:152b, 159b–161b
late seventeenth-century, 3:482b; 4:64b, 135b; 5:427a; 6:149a; 8:157b; 9:65a–b; 10:520a,b
early eighteenth-century, 2:204a; 8:580a; 9:187a; 10:10a–b; 11:332b–334a; 13:564*; 15:54a
late eighteenth-century, 2:287a; 6:65a; 7:441a; 8:160a; 9:576a; 12:559b; 15:524b–525a
early nineteenth-century, 1:387a; 9:253a
mid nineteenth-century, 4:291a; 8:254a–b; 10:263a; 11:307b–308a, 361a, 368b–369b
late nineteenth-century, 3:247b; 7:439b; 8:3b; 9:517a–518a; 11:255a–b, 572b–574a; 14:236b, 427b–430b
early twentieth-century, 2:384b; 3:611a*; 6:132a; 11:597a–b; 12:568a–569b; 13:97a; 14:314b
chemical affinity, 9:188a
coelom theory, 6:339a
concrescence, 6:435a–b
embryogenic acceleration, 10:523a; 15:92b
gastraea theory, 1:421a, 601a; 5:447a; 6:9a, 339a; 7:475a; 11:361a
germ-layer, 1:387a, 421a; 15:525a
gradient, 3:247b
mosaic, 3:389b–390a; 4:187a; 11:573a–b; 14:427b–428b
organizer concept, 4:187a; 14:428b
parablast, 6:435a
regulative development, 3:389b–390a
substitution theory, 7:402b
see also Epigenesis; Pangenesis; Preformation, theory of
and vitamin A, 15:419b
see also Morphogenesis
Embryology
ancient Greek, 1:262a, 265a, 266a–b; 4:106a; 6:427a
medieval Arabic, 9:41a; 13:230b, 292b
medieval Latin, 1:102a; 5:147b, 403a
Renaissance, 1:109b, 596a; 3:343a
early seventeenth-century, 1:293a–b; 4:509a–511b; 6:159b–161b, 387a; 12:577b; 14:12b
late seventeenth-century, 3:482b; 4:135b; 6:149a; 9:52b, 65a–b; 11:317a; 13:33b, 173b
early eighteenth-century, 5:355a; 6:65a–b; 8:115a
late eighteenth-century, 15:524b–525a
early nineteenth-century, 1:385b–387b; 2:443a, 595a, 596b; 4:147a, 264a–b; 5:356b–357a; 9:253a; 10:287a; 11:215b; 12:315b–316a; 13:402b
mid nineteenth-century, 1:417b–418a, 476b–477a, 518b; 2:161a–b; 4:289a, 504a; 5:469b, 514b; 6:7b–9b, 573a–b, 590a–592a; 7:437b–440a, 546a; 8:254a–b, 519a; 9:332a–b, 456b, 570a–b; 10:39b, 579b; 11:307b–308a, 360b–361a, 368b–369b; 12:240b; 13:555a–556a; 14:57b, 233b–234a; 15:108b, 123a–b, 169b
late nineteenth-century, 1:71b–72a, 420a–422a, 601a–b; 2:362a–364a, 383b–384a, 502a; 3:120b, 184b, 245b, 247b, 389b–390a, 493a; 4:12a, 266b, 397b; 5:35b, 51a–52b, 385b, 447a, 581b–582a; 6:267a, 287b, 337b–339a, 435a–436a; 7:402a–b, 454a, 474b–475b, 535b; 8:3b, 27a; 9:56a,b, 333b; 11:254b–255b, 382b, 383a; 12:298a; 14:126a–b, 251a, 291a, 314a–b, 331a, 423b–430b; 15:513b
twentieth-century, 1:455a, 545b; 2:364a–b, 384a–b, 504a; 3:148b, 245b, 331b–332a, 610b; 5:453b, 467a–b, 594b; 7:33b–34a, 278b, 314a, 516b; 8:282b, 283a, 355a–359a; 9:57a, 331a; 11:175a, 597a–b; 12:48a–b, 409b–410b, 473a–474a; 13:96b–97b, 151a, 268b, 281b–282a; 14:255a, 595a–b, 596b; 15:473a
botanical
nineteenth-century, 6:465a–b; 10:1b; 13:459a, 460a, 462a
comparative
seventeenth-century, 4:510a
early nineteenth-century, 5:536b–537a; 14:199b
mid nineteenth-century, 8:254a–b; 9:332a; 11:360b–361a; 15:123b
late nineteenth-century, 1:420b–421b; 2:383b–384a; 5:52a; 6:535b–536a; 7:402a, 475a; 9:416a, 516a–b; 10:273b; 11:254b–255b; 14:426b
early twentieth-century, 5:467a–b; 8:586a
mid twentieth-century, 7:33b–34a
and comparative anatomy, 5:505a
and cytology, 2:161b; 8:357a
experimental
early nineteenth-century, 11:133a
late nineteenth-century, 1:421b, 498a–499a; 2:383b–384b; 3:184b, 245b; 6:337a; 8:446a; 9:517a–518a; 11:570a–574b; 14:427a–430b; 15:124b
early twentieth-century, 1:153a; 4:186a–188b; 6:132a–133a; 7:94a–95a; 8:357b, 358b–359a; 12:567b–569b
mid twentieth-century, 7:455b; 9:524a; 10:104b–105a; 12:410a
founding of, 5:357a–b
methods, 6:436a
physiological approach, 7:402a; 11:574a
and taxonomy, 1:603a; 7:475a; 11:234b
see also Egg(s); Embryo; Embryogeny; Historiography, of embryology; Illustration, embryological; specific animals and groups of animals
Embryotectonics, 1:236b
Emden, Robert (1862–1940), 4:360–361; 4:279a; 8:1b; 12:249b
equation of, 5:103a
Emerald
crystallization of, 6:178a
heat, effects of, 3:560b
medical use of, 6:397a
specific gravity, 7:346a
Emerald Table
see *Tabula Smaragdina*
Emerson, Benjamin Kendall (1843–1932), 4:361–362; 8:260b
Emerson, Ralph Waldo, 2:258a; 6:53a; 7:44b; 12:607a; 13:521a
Emerson, Robert (1903–1959), 4:362–363
Emerson, Sterling, 13:134b, 137a; 14:101b
Emery, Josiah, 9:163b
Emetine
discovery of, 9:8a
isolation of, 10:497b
Emission
thermionic, 3:372b; 8:23b–24a
see also Alpha particles, Beta particles, Cathode rays, Electrons, Gamma rays, Neutrinos, Neutrons, Positive rays, Positrons, Protons, Radiation, X rays
Emmanuel College, Cambridge, 15:16b–17a
Emmanuele, Pietro, 2:308a
Emmenology, 5:157a
Emmerez, Paul, 4:37b
Emmerich, Rudolf, 7:427a; 10:560a
Emmerling, Adolph, 1:390a
Emmons, Ebenezer (1799–1863), 4:363–365; 4:274b; 6:56b, 57a; 13:581b
Taconic system, 1:469b; 9:100b
Emmons, Ebenezer, Jr., 6:57a–b
Emmons, Samuel Franklin (1841–1911), 4:365–367; 3:434a; 6:13b; 7:62b
Emotions
in animals, 3:575a
in man, 3:575a; 6:573b; 8:7b; 13:170a, 543a; 14:322a
outward signs of, 8:322a
physiological bases of, 7:67b–68a
physiological effects of, 9:547a; 12:602a–b; 14:322a; 15:73a
theories of, 11:104b–105a
Empedocles of Acragas (*ca.* 492–*ca.* 432 B.C.), 4:367–369; 1:82b, 263b, 270b, 495a; 2:13b; 6:419b; 11:25a, 29b; 12:340b
biology, 1:262a; 4:106a, 385b; 7:538a; 11:127b; 13:331b
elements, 1:149a, 261a
natural philosophy, 1:259b
Emphysema, 7:557a; 14:408a
treatment of, 6:367a
Empiricism
ancient Greek, 1:252b; 6:423b–424a
ancient Roman, 12:340a
Byzantine, 1:121b
medieval Latin, 1:100a, 290a, 291a, 381b; 2:604b–605a; 4:92b–94b; 10:173a, 173b–174b
Renaissance, 10:310b, 311a; 11:528a; 12:97b, 135a, 335a; 14:47b, 581a, 582a
seventeenth-century, 1:374a–b; 7:539a–b; 8:127a, 153b, 438b–439a; 9:120a–b; 15:69b–70a
eighteenth-century, 3:2a, 381b, 382a, 419b, 495a, 522b; 5:510b–511a; 6:37a, 140a, 469a, 477b–478a, 549b; 7:367a; 8:147b, 378a–b, 580a; 9:52b, 531b–532a; 10:120b–121a; 11:168b, 522b
nineteenth-century, 2:31b; 3:527a; 4:204a; 5:298b, 301b; 6:243a, 248a–249a; 7:546a; 8:20b, 487a, 602a–604b; 9:7b–9a, 252a, 253a, 384a; 11:50b, 474a; 12:276b; 13:499b; 15:412a
twentieth-century, 1:506a; 2:358b, 545b; 4:227a; 11:356a–358a; 12:417b
Chinese, 12:383b
critics of, 3:376b; 4:100b–102a, 616b; 6:558a–559a; 15:33b, 127b, 142a
Japanese, 1:310a; 15:740b
philosophical, 3:381b, 382a, 383a; 6:7b
radical, 7:68b
Empiriomonism, 15:38b
Empyema
diagnosis of, 1:333a
treatment of
drainage, 14:6a
surgical, 7:16b; 8:554b; 12:450b
Emulsin, 5:4a; 8:342b
Emulsions, 1:431a; 6:118b
in photography, 4:282b–283a; 12:235a–b
in physics, 11:117a–b
Enamels, decorative, 3:326a

Enantiomorphism, **1**:461b; **2**:562b; **5**:557b
Enantiotropy, **3**:333b–334a; **12**:197a–b
Encephalitis, **5**:317a
herpes, **14**:624a
lethargic, **4**:198a; **8**:273b
Encephalomalacia, **11**:559b, 560a
Encephalomyelitis
equine
transmission of, **7**:286b–287a
Encephalon
movements of, **8**:506a
Encephalopathy, **14**:277b
Encke, Johann Franz (1791–1865), 4:369–370; 6:103b; **8**:369b
acquaintances of, **6**:325a; **9**:2a; **10**:133b
astronomy, **6**:234b; **10**:34b
cometary theory, **9**:460a–b
at Berlin Observatory, **1**:295b; **2**:98a
biographies of, **2**:533a
comet of
see Comets, Encke's (Encke-Backlund)
correspondence of, **5**:304b; **10**:199a
and Galle, **5**:256b–257b
mathematics, **7**:506a
students of, **2**:532b; **6**:230a; **7**:505b; **8**:557b; **10**:543a; **12**:289b, 579a; **14**:480b
Encyclopädisches Wörterbuch der medicinischen Wissenschaften, **6**:269a
Encyclopaedia Britannica, **3**:265a, 463b; **5**:144b–145a, 276a, 335b; **6**:327b, 452a; **9**:199a; **11**:496a–b; **13**:372a, 504b*, 523b; **14**:563a, 567b, 569b
Encyclopaedia metropolitana, **6**:325b
Encyclopedias
alphabetical, **3**:421a
botanical
nineteenth-century, **15**:147b
twentieth-century, **1**:396b
chemical
nineteenth-century, **5**:158a
Chinese, **3**:249b
Egyptian, **15**:634b–635a
general
ancient Roman, **3**:174a–b, 176a; **13**:588b
medieval Arabic, **7**:357b–358a; **15**:494a
medieval Latin, **1**:99a; **6**:274a; **7**:27a–28a; **8**:548a; **14**:34b–36a
Renaissance, **3**:66a
seventeenth-century, **1**:126b; **3**:359b, 361b–362a
eighteenth-century, **1**:58b; **2**:321b; **3**:383b, 384a, 421a; **6**:129b–130a
nineteenth-century, **1**:405a; **2**:321b–322a; **6**:325b
mathematical, **9**:358a–b
medical
ancient Roman, **3**:174b–175a; **5**:233a
Byzantine, **1**:68b–69a; **10**:417a
medieval, **13**:230b; **14**:584a–585a; **15**:498a–499a
twentieth-century, **1**:361b
Russian, **1**:361b
scientific
ancient Roman, **11**:39a–40a
Byzantine, **11**:185a–b
medieval Latin, **13**:347b–348b
Renaissance, **14**:507b
eighteenth-century, **4**:26a
technological, **9**:287b; **11**:328a–b
Encyclopédie de mécanique appliquée, **8**:124a
Encyclopédie de Panckoucke, **1**:58b; **2**:321b–322a; **3**:383b; **4**:87a; **13**:465a
Encyclopédie méthodique, **2**:335a; **3**:106b*, 208a, 383b; **4**:150a; **7**:588b; **8**:256a, 258b; **14**:16a; **15**:111a
astronomical section, **7**:581a
chemical section, **5**:90a–b, 601b, 602b
Dictionnaire de botanique, **1**:59a; **7**:585a
Dictionnaire de physique, **9**:477a
Encyclopédie ou Dictionnaire raisonné des sciences, des arts, et des métiers, **1**:58b, 59a, 110b, 115b, 139a,b, 479a; **2**:201b, 301b, 335a, 348a; **3**:383b; **4**:26a, 71a,b; **5**:567b, 601b; **6**:468b, 469a; **8**:73a, 177b; **9**:62a, 306a; **11**:388b–389a; **13**:494b, 602b; **15**:111a, 280a
Discours préliminaire, **1**:114b
license suspended, **1**:115a; **4**:86a
translations of, **7**:202b*
Encyclopédie populaire, **11**:22a
Encyclopedists, **5**:62a; **7**:202a; **10**:611b; **14**:15b
Encyklopädie der mathematischen Wissenschaften, **10**:244b, 422a, 441b, 442a; **12**:285a
editors of, **10**:212b; **12**:527a
founding of, **4**:269a; **7**:399b; **9**:358a–b
Encyklopädisches Lexikon, **2**:422b
Endan zyutu (method), **12**:291a
Endeavour (ship), **1**:434a–b; **3**:396b–397a; **7**:95b; **10**:323a–324a; **12**:515b–516a
Endemism, **7**:102b
Endless Mountains
see Appalachian Mountains
Endlicher, Stephan Friedrich Ladislaus, **7**:534a; **13**:542a
Prodromus florae Norfolkicae, **1**:520b
Endoblast, **1**:601a
Endocarditis, **4**:586a
bacterial, **8**:295b
ulcerous, **8**:7a
Endocrinology
seventeenth-century, **13**:527b; **14**:286b–287a
eighteenth-century, **2**:301b; **8**:246b
nineteenth-century, **1**:303a–b, 537a; **2**:73a, 525b; **5**:469b–470a; **6**:518b; **9**:86a, 109a; **10**:205a–b; **12**:165a, 355b–356a; **13**:589a–b
early twentieth-century, **1**:10b–11b, 123a–b, 152b–153a, 440b–443a; **2**:255b; **3**:69a, 113a, 351b–353a, 518b, 555b; **7**:101a; **8**:358b–359a; **10**:570a–b; **11**:576b–577a; **12**:472b–475a, 617b; **13**:53b; **15**:228a, 258a–b
mid twentieth-century, **10**:610b–611a; **14**:595b; **15**:228b–229a
all-or-none law, **10**:570a
beginnings of, **1**:60b
comparative, **2**:345b
sexual, **1**:152b–153a; **2**:344b–345b
see also Glands, Hormones
Endocrinology
editors of, **11**:577a
Endoderm, **6**:339a; **11**:368b, 369b
movements in, **10**:105a
sexual significance of, **1**:601a
Endogenesis, **2**:165b
Endolimax, **4**:133b*
Endolymph, **2**:446b
source of, **3**:425a
Endomycetaceae, **5**:584b
Endosmosis, **4**:615b; **5**:7a; **11**:297b; **12**:496b; **13**:453a; **14**:330a
Endosperm, **3**:422a; **5**:217a; **8**:128a; **10**:1b; **13**:405b–406a
Endothelium, **2**:165a; **5**:43a; **6**:435a; **9**:57a
Endotryptase, **2**:562a
Enemas
ancient Greek use of, **1**:315a
seventeenth-century use of, **10**:426b*
Energeticism, **4**:225b, 227a, 231b; **8**:598b
critics of, **8**:184a
Energetics
early nineteenth-century, **3**:72b, 75a, 118a–b; **5**:171a
mid nineteenth-century, **6**:243b–245a; **9**:205b–206a, 210b; **11**:293a
late nineteenth-century, **2**:266b; **4**:228a; **5**:391b; **13**:399b*; **15**:462b–464b
early twentieth-century, **4**:229b, 231a, 313b, 600a; **13**:417a, 617b
mid twentieth-century, **2**:499b*; **7**:211b
opposition to, **15**:463a, 464b
Energy
of activation, **1**:300b, 362a, 414b–415b; **7**:517b; **13**:429b, 578a
chemical, **7**:519a; **15**:441a
complementary, principle of, **3**:118b
concept of, **1**:277a; **3**:111a; **8**:287b, 598b; **14**:568b–569a; **15**:85b–86b, 462b, 464b
conservation principle
in ancient Greece, **4**:369a
in Renaissance, **8**:209b
in seventeenth century, **6**:608a, 611b
in eighteenth century, **2**:54b
in early nineteenth century, **3**:73a, 74b; **14**:569a
in mid-nineteenth century, **3**:83a; **4**:163a, 554b, 557a; **5**:560a,b; **6**:242a, 354b; **8**:187b, 188a, 491a; **9**:205b, 217b, 237a, 287b, 445a; **11**:292b, 353b; **12**:288b; **13**:76a, 380b, 386a; **14**:205b, 207b; **15**:85b–86b
in late nineteenth century, **2**:356a; **3**:309b; **4**:204a; **6**:250a–b, 545b; **8**:597b, 598b; **11**:7b, 8b, 10a–b; **13**:416b; **14**:123b
in twentieth century, **3**:369a–b, 479b; **4**:317b, 323a, 327a,b; **5**:331b; **8**:618a; **9**:377a; **15**:8a
biological, **4**:616a; **11**:585a–586a
general covariance, **4**:329a
philosophy of, **6**:243b–244a
and quantum theory, **2**:247b, 338a; **7**:492a; **10**:424b
and radioactivity, **3**:500a; **12**:28a
conversion of, **4**:533b; **7**:519a; **8**:500a; **9**:237a, 360a; **14**:175b–176a
dissipation of, **9**:616b; **13**:382b
of dissociation, **5**:118a; **15**:448b
of distortion, **6**:534b
distribution of, **4**:314a; **5**:392a–b; **9**:219b–220a, 223a–b; **14**:338b, 339b–340a
in quantum systems, **5**:103a; **15**:49a
electric, **9**:210a, 213b–214a
electrochemical, **7**:65a–b
electrodynamic, **4**:231a; **9**:205b, 207a
electromagnetic, **8**:474b
equipartition of, **4**:583a; **5**:392b; **7**:84b; **9**:616b; **12**:464a; **13**:104a
exchange of, **9**:218a
flow of, **11**:122b
fluctuation of, **4**:314a–318a
of formation, **8**:290a
free, **6**:244b; **8**:289b, 290a–b, 291a, 292b; **9**:394a; **15**:436b–441b, 457a–b, 464b
and heat, **14**:360b
inertia of, **8**:13a*
intrinsic, **3**:305a; **5**:387b, 388b, 389b–390b, 392b; **12**:218b
kinetic, **3**:296a, 306b–309a *passim,* 417b, 454a; **4**:229b; **7**:443a; **9**:107b, 108a, 214a; **11**:409a; **13**:363b; **14**:569a
of light, **4**:322b; **14**:175b–176a
localization of, **14**:338a
magnetic, **9**:210a, 213b–214a

Energy (cont.)
  and mass, **4**:323a–b, 324a–325a; **6**:164a; **7**:37a*; **8**:10b, 95a, 290b
  and matter, **8**:618a
  measurement of, **5**:407b; **8**:209b
  motion of, **4**:306a
  nuclear, **2**:252a–b, 338b; **5**:610b; **7**:93a
  from the ocean, **3**:299b; **4**:600a
  potential, **3**:308b–309a; **5**:117b; **8**:211a; **9**:107b–108a, 209a; **11**:292b, 293a; **13**:363b; **14**:205b
  propagation of, **13**:438a
  psychic *(P-Energie),* **2**:1a–b
  radiant, **5**:461b, 462a; **10**:345b
    distribution of, **7**:85a
    measurement of, **8**:107b
    and temperature, **4**:183a; **11**:150a
  rotational
    of atoms, **2**:170a
  solar, **8**:1b–2a; **13**:482a
    density of, **9**:214a
    distribution of, **9**:404b
    measurement of, **2**:589a; **6**:325a
    origin of, **10**:525b
    utilization of, **3**:279b; **13**:24b
  sources of, **8**:184b; **13**:614a
  stellar
    distribution of, **9**:404b
    source of, **4**:280a; **6**:117b; **7**:86a; **8**:618a; **12**:22b; **15**:449a
  strain, **3**:118a; **6**:534b; **9**:578a
  surface, **13**:41b
  theories of, **11**:292b–293a
  thermal, **3**:597b
  and time, **13**:241b
  transfer of, **5**:117a–118a; **9**:219b; **13**:103b; **14**:184b
  transmission of, **11**:393b
  units of, **5**:407b
  zero-point, **13**:41a
  see also Atomic energy, Free energy, Power, Work
Engel, Friedrich (1861–1941), 4:370–371; **2**:191b; **8**:324a,b, 327a*, 434b*; **10**: 344b*; **13**:125b*; **14**:222a
  analysis, **8**:325b, 326a
  linguistics, **15**:198b
  Theorie der Parallellinien von Euklid bis auf Gauss, **7**:600a; **12**:255a; **13**:264a–b
Engel, Johann
  see Angelus, Johannes
Engelhart, Erich, **8**:454b
Engelmann, George (1809–1884), 15:130–131; **5**:512b; **13**:456a–b
Engelmann, Theodor Wilhelm (1843–1909), 4:371–373; **5**:196a, 280b; **11**:585b; **14**:127a*; **15**:78b, 166a
  critics of, **15**:22a*
Engels, Friedrich (1820–1895), 15:131–147; **8**:183b; **12**:209a
  and Marx, **9**:70a; **15**:404b–405b, 409b
Engeström, Gustav von, **9**:5b; **14**:225a
Engineering
  seventeenth-century, **5**:245a–b
  nineteenth-century, **3**:19a–20a, 79b–81a, 82b; **4**:500a–b; **5**:63a–64a; **12**:131b; **15**:744a
  twentieth-century, **5**:433a–b; **7**:513b–514a; **8**:561b
  aeronautical
    see Aeronautics
  chemical
    eighteenth-century, **14**:197a–b
    nineteenth-century, **1**:288b; **3**:316b; **6**: 94a; **8**:553a–b; **9**:467a
    twentieth-century, **2**:323b–324a; **3**: 299a–b; **6**:147b; **12**:575a
    high-pressure, **2**:324a
  civil
    medieval Arabic, **7**:240b
    Renaissance, **1**:97b, 605a; **2**:534b–535a; **8**:242a; **9**:146b; **13**:47b, 50a, 233a–234a
    seventeenth-century, **2**:201a–b; **3**:421a; **5**:574a; **8**:26a, 150a; **9**:486b
    eighteenth-century, **1**:581b–582a; **3**: 440b; **4**:501a–b; **6**:370a; **7**:615a; **10**: 109a, 527b–528a; **11**:5a, 163b–164b; **12**:461b–462b, 487a–b
    early nineteenth-century, **1**:326b, 459b, 460a; **2**:104a–b, 473a; **3**:131b, 411b, 416b, 466a; **5**:166a–b, 530b; **7**:601a, 602a; **10**:2b–4b; **11**:165a, 355a; **12**: 287b–288a; **13**:604b
    mid nineteenth-century, **3**:287a–b, 495b–496a; **4**:242b; **6**:292b, 293a; **10**:286a–b; **11**:95a, 291a–b; **13**:78b, 144b; **14**:184a–b; **15**:84b–85a
    late nineteenth-century, **4**:305b; **5**:63b; **6**:340b–341a; **8**:287a–b; **9**:446a–b; **11**:392b–393b, 398a
    early twentieth-century, **8**:99b; **12**: 612b–613a
    Chinese, **12**:371b, 372a, 382b
  electrical
    mid nineteenth-century, **5**:496a–b; **7**: 93a–94a; **10**:268b–269a; **12**:425b
    late nineteenth-century, **4**:40a, 588a–589a; **5**:63b, 597b; **6**:73b–74a, 211b–212a, 223a–b, 504a–b; **7**:101b, 287b; **12**:574b–575a; **13**:286a–287b, 361b
    early twentieth-century, **3**:329a; **5**:407a; **7**:107b–108a, 288a–b; **10**:236a–b; **11**:213a, 588b–589a; **12**:575a; **13**: 24a–b, 361b
    mid twentieth-century, **2**:318a–b; **6**: 104a–105a; **13**:569b–570b, 600b
    communications, **1**:455b; **4**:601a–b; **9**: 98b, 99a
  handbooks
    eighteenth-century, **1**:582a
    nineteenth-century, **3**:118b; **4**:501b; **9**: 578a
  hydraulic
    see Hydraulics
  industrial, **8**:119b
  Japanese, **15**:744a
  management, **13**:271b–272a
  and mathematics, **12**:526b, 527a–b
  mechanical
    ancient Roman, **15**:514b
    medieval Arabic, **12**:362a; **15**:253a–255a
    Renaissance, **8**:207b–214b; **13**:50a–b, 233a–234a
    seventeenth-century, **2**:49a; **8**:212a; **10**:521a; **12**:211a; **13**:613b–614a
    eighteenth-century, **2**:104a; **10**:36a–b; **11**:496a; **12**:463a; **13**:178b–179a; **14**:196b–197a
    early nineteenth-century, **3**:8a–9a, 75a, 316a; **5**:530b; **6**:213b; **11**:340b, 354b–355a
    mid nineteenth-century, **3**:222b, 224a, 227a; **5**:101a, 387a, 501b; **6**:431b–432a; **9**:60a–b, 226b–227a, 615b; **11**:343b–344a; **12**:424b; **14**:403b–404a, 473a, 617b–618a; **15**:24b, 225b
    late nineteenth-century, **4**:39b; **5**:502a; **7**:287b; **8**:365a–b, 500a–b; **10**:551a; **11**:383b–385a; **13**:72b–73b, 271b–272a, 398a–399a; **14**:106a–b
    twentieth-century, **1**:570b; **7**:316a–b, 613a–b; **8**:124a, 301a–b; **9**:462a–b; **10**:162a–163a, 597b–598a; **11**:305b–306b; **12**:527a; **13**:431a–432a; **14**: 468a; **15**:7a, 451a
    model building, **14**:403b
  military
    ancient Greek, **10**:588b–589a
    ancient Roman, **15**:514a,b
    Byzantine, **7**:29a
    Renaissance, **2**:535a; **4**:260a–b; **8**:206b–207a, 213b–214a; **9**:146b–147b, 335b–336a; **11**:97b, 405b–406a; **13**: 50a, 260a
    seventeenth-century, **2**:201a, 419a; **3**: 621b; **4**:549b, 551a–b; **5**:408b; **8**:26a; **9**:485b; **11**:423b; **12**:127b; **13**:590b–594a, 621b, 622a–b
    eighteenth-century, **1**:581b; **3**:16b, 71a–b, 439b–440a, 459b; **4**:207b; **6**: 305a; **8**:611b; **9**:342b, 470a; **11**:354a, 493b–494a
    nineteenth-century, **3**:79b, 554b; **4**: 217a–b; **5**:110b, 111b; **6**:471a, 545a; **7**:499b; **8**:64b; **9**:268a; **11**:33b, 76b, 80a
  mining
    eighteenth-century, **5**:222b
    nineteenth-century, **3**:411b–412a; **4**: 347b–349a; **8**:124a–b; **9**:58b, 59a–b; **10**:265a; **14**:232a–b
  naval
    seventeenth-century, **4**:213b, 214a
    eighteenth-century, **2**:576b; **6**:68a
    nineteenth-century, **2**:84a–b; **4**:257a; **6**:34b–35a; **7**:383a; **11**:345a
    twentieth-century, **9**:458b; **12**:341b
  petroleum, **8**:99b, 301b
  philosophy of, **8**:500b
  and physics, **6**:105a
  structural
    Renaissance, **2**:534b
    eighteenth-century, **3**:441a–442b; **6**: 577a; **11**:496a–b
    nineteenth-century, **3**:117b–118b; **5**: 63a–b; **7**:383a–b; **9**:60b, 138b–139a, 267b–268b, 577b–578a
    early twentieth-century, **15**:164a–165a
    see also Statics, Strength of materials
  see also Radio, Railroads, Steam engine, Technology, Telegraphy, Television
Engines
  air-cooled, **7**:316b
  diesel, **7**:316b, 613b
  electric, **5**:496a
  electromagnetic, **13**:378b
    efficiency of, **7**:180b–181a
  gasoline
    knock in, **3**:20a; **7**:316a, 498a
  high-compression, **7**:498a
  hydraulic, **3**:82b; **11**:80a, 354b–355a
  ideal, **3**:81b, 82b, 305a,b
  internal combustion, **3**:20a, 31b, 239a; **6**:599a; **10**:118b
    efficiency of, **15**:451a
    knock in, **9**:375b–376a
  pneumatic, **14**:509a
  reaction, **5**:433a
  supercharged, **11**:306b
  see also Heat engine; Machines; Newcomen, Thomas, Newcomen engine; Steam engine; Turbines
Engis (cave), **12**:184a
England, Joseph Loveday, **14**:515b
England
  agriculture, **6**:578b
  archaeology, **5**:162b–163a; **11**:40b
  biology
    genetics, **14**:251b–252a
  colonization by, **6**:20a–b
  economics, **10**:566a
  education, **5**:81a; **8**:565b, 571b, 572a; **13**:102a; **14**:209b
  geology, **5**:518b–519a; **6**:116a, 502b,

509b; **8**:564a–566a; **10**:583b, 586a; **11**:130b–131a; **12**:488b
and France, hypothetical isthmus, **4**: 70b
paleobotany, **12**:339b
paleontology, **2**:495a; **8**:570b–571a; **10**:584a
stratigraphy, **4**:540a–541a; **5**:14a, 436b–437a; **12**:276a–277b, 487a–491b; **14**:210b–211a, 311b–312a; **15**:152a
horticulture, **1**:88a–89a, 89a–b, 422b, 423a; **3**:350a, 351a, 556b–557a; **5**: 361b
logic, **14**:288a
mathematics, **7**:37b, 551a; **13**:491b; **14**: 180a
medicine, **3**:12b–13a; **8**:361a–b
epidemiology, **14**:405b
pharmacy, **13**:507b
rabies, **10**:405b
natural history, **9**:539b; **11**:40b–41b
navigation, **14**:514b
politics, **6**:25b–26a
population of, **5**:507b; **11**:236b
sociology, **15**:133a
technology
industry, **9**:467a; **13**:548a
metallurgy, **4**:218a
mining, **4**:218a, 349a; **13**:87a
railroads, **6**:292b
scientific instruments, **9**:5a,b
theology, **14**:361a, 370b, 371b–372a
see also Great Britain; specific cities, counties, etc.
Engler, Heinrich Gustav Adolf (1844–1930), 15:147–148; **6**:221a; **14**:117b*
Engler, Karl, **1**:362a
Engler's Jahrbücher
see *Botanische Jahrbücher für Systematik, Pflanzengeschichte und Pflanzengeographie*
English (language)
dialects, **2**:476a
grammar, **11**:140a; **14**:151b–152a
English Channel, **9**:43a
geological history, **5**:437a
mapping of, **4**:70b; **6**:71b
tunnel, **8**:30b–31a; **11**:131a
The English Cyclopaedia, **13**:79a*
Engramelle, Jacques-Louis-Florentin, **3**:62a
Engraving, **9**:343a
copper plate, **4**:48b; **7**:412a–b
Eniwetok (atoll), **3**:568b
Enlightenment, **2**:273b, 326a, 552b; **3**:7b, 114b, 119b, 378b, 380b, 382b, 383b–384a, 454b; **4**:70b, 85a, 86a, 129b; **5**:62a, 138a, 247b; **6**:294b; **7**:588a, 604a; **8**:67a, 81b; **9**:47a, 148a, 383a, 540b; **10**:145b, 326a, 573a; **11**:493b, 617b; **12**:536a, 591b; **13**:480b, 601b; **15**:98b, 341b, 738a
rejection of, **11**:43a; **12**:156a
Enoch-Idrīs, **1**:34b
Enri calculus, **12**:291b–292a; **15**:747b, 751b, 752a,b
Enrichment, mineral
secondary, **4**:366b
Enriques, Federigo (1871–1946), 4:373–375; **3**:117a; **5**:569a; **8**:505a; **9**:342a; **10**: 139b; **13**:255b*, 551a*
geometry, **11**:55a; **12**:330b, 331a
Enseignement mathématique, **11**:413a
Ensemble theory, **2**:265a
Enskog, David (1884–1947), 4:375–376; **2**: 263a; **9**:222b
Ent, George (1604–1689), 4:376–377; **1**: 566b, 567a; **5**:426a; **6**:151a; **7**: 540b(n); **14**:287a
Entamoeba, **12**:142a
Entelechy, **4**:188a; **6**:256b
Entelodon, **7**:480a
Enteritis, **6**:298a
Enterocele, **1**:601a
Enterokinase, **10**:433a
Enteropneusts, **9**:112b, 328b
Enthalpy, **7**:182b; **8**:289b; **13**:73b, 358b; **15**:436b
concept of, **9**:462a–b; **13**:245a
Enthymeme, **3**:60a
Entomological Society (Cambridge, Eng.)
founding of, **1**:358b
Entomological Society (of London), **6**:184a
Entomology
ancient Greek, **1**:263a–265b *passim*
medieval Latin, **1**:102a
Renaissance, **9**:440a–441a; **14**:507b–508a
early seventeenth-century, **3**:180a; **5**: 439a–440a; **12**:616a; **13**:29a
late seventeenth-century, **1**:431b–432a; **2**:591b; **3**:101a, 180b; **11**:317a, 341b–342b; **13**:169a–b, 171a–174b; **14**:413a
early eighteenth-century, **2**:286a–b; **3**: 320a–b; **5**:226b; **7**:111b*; **8**:378b–379a, 579b–580b; **11**:331b–333b, 503a
late eighteenth-century, **2**:4b, 321b; **3**:62a; **4**:512a–513a; **5**:76b, 90a, 328a–329a, 355a, 424b; **9**:560b–561a, 575a–b; **10**:437a; **14**:300a
early nineteenth-century, **1**:328b–329a; **2**:444b; **3**:426b*; **4**:407a*; **5**:605b; **6**:184a–b, 492b–493a; **7**:615a; **8**:48b–49a; **11**:510b; **12**:132b–133b, 201a, 216a, 318a; **15**:127a
mid nineteenth-century, **1**:388a, 500b–504a, 587a; **2**:620b; **4**:504a–b; **5**:11b–12a, 477b; **6**:136b, 220b–221a; **10**: 39a–b, 165b; **12**:479a, 612a; **14**: 233a,b, 496a; **15**:123a, 127a
late nineteenth-century, **5**:69b–71a, 73b–74a, 502b, 555b; **6**:167a, 267a, 507a; **8**:528b–529a; **10**:272b–273b, 427a; **12**:264a–b, 300a, 479b; **14**:2a, 234a, 410a–b
twentieth-century, **2**:221b; **4**:493b–494b; **7**:273a, 285b–286a; **8**:518a–b; **12**: 473a; **13**:124b, 134b; **14**:291a–292a
agricultural, **1**:329a; **2**:221b; **5**:12a, 70a,b; **6**:524b, 525a
apparatus, **3**:320a
applied, **1**:329a; **2**:221b; **5**:502b–504a; **6**:524b–525a; **11**:332a
economic, **5**:12a; **10**:272b, 273a–b; **11**: 332a; **12**:98b–99a, 133b; **13**:456a
experimental, **2**:286a–b; **5**:439b
medical, **6**:524b
quantitative, **11**:332b
and statistics, **12**:98b–99a
see also Insects; specific insects
Entomophthorales, **2**:437b
Entomostraca, **9**:575a,b
Entrecolles, François-Xavier, **11**:330a
Entrochites, **15**:54a
Entropy, **2**:499b; **5**:389b, 392b; **6**:245a; **8**: 481a; **11**:10a,b, 293a; **13**:363b–364a
at absolute zero, **15**:439a–448a
in chemical reactions, **6**:244b; **15**:436b–444a
concept of, **3**:305b; **5**:387b–388b, 389a; **8**:289b; **11**:8a
constant of, **13**:41b
critics of, **12**:571a
definition of, **4**:229b; **11**:11a
of electrons, **8**:47a
of gas, **4**:578a; **8**:482b; **13**:41b; **15**:432b, 439a–448a
H-function, **2**:263b–264a
and information, **13**:227a
interpretation of, **3**:306b
irreversible increase of, **3**:309a, 310a
mathematical expression of, **3**:308b; **8**: 289b
and molecular distribution, **2**:262b; **9**: 227a–b
in optics, **8**:50b
physical interpretation of, **4**:314a
of radiation, **14**:339a
statistical interpretation of, **2**:264b; **4**: 294a; **8**:598b–599a
tables of, **8**:291a
and universe, **1**:301a; **2**:264a–b; **15**:448b–449a
of vapors, **9**:461b; **13**:73b
Entsiklopedicheskii slovar, **9**:293a
Entwicklungsmechanik, **9**:517a; **11**:570a, 573b–574b
Enūma Anu Enlil, **15**:549b, 550a, 643b, 673b, 674a
Envelopes (math.), **2**:215b; **3**:171a, 213b; **4**:257b; **8**:164a, 166a; **9**:475b; **11**:44b, 398b; **13**:437a
Environment
and culture, **11**:309a–b
and disease, **10**:376a, 389a
and evolution, **1**:163a–b; **2**:321a, 347a, 498a; **4**:512b; **5**:356a, 357b; **7**:170a, 590b, 591a,b; **10**:209b–210a; **11**: 567a–b; **12**:571a, 588a; **13**:276a, 460b–461a; **14**:313a; **15**:93a
geological, **12**:514a
and health, **5**:452b; **6**:421a, 422a, 427a–b, 459b
and heredity, **2**:211a–b, 580b; **3**:390a; **7**:166a–b; **12**:417a
physiological effects, **7**:395b–396a; **9**: 171b; **11**:178a; **12**:190b; **15**:127b–128a
transplant studies, **2**:61a
and plant nutrition, **11**:416a
and plant structure, **2**:290b; **3**:359a, 429a; **6**:296a
response of organism to, **3**:1b; **10**:433b
temporal change, **11**:469a–b
see also Climate; Ecology; Plants, and environment
Environmentalism, **4**:183a
Enzymes, **1**:286b; **2**:531b; **5**:209a
analysis of, **2**:86b
assay procedures, **14**:412a
bacteriolytic, **5**:29b–30a, 42a
chemical effects of, **2**:124a; **6**:501b
cofactors, **4**:485b; **6**:111a, 502a
concept of, **8**:342b–343a
crystallization of, **14**:174a–b
depolymerizing, **4**:604a
and development, **4**:187b, 188a; **14**:432a; **15**:14a
in digestion, **4**:211b; **10**:361b
and disease, **3**:416a; **14**:252b
in fermentation, **2**:66a; **4**:485b–486a; **5**: 4a–b; **6**:111a; **14**:45b
glycolytic, **9**:359a–b
hydrolytic, **1**:362a; **7**:379a
inhibition of, **4**:485b; **8**:454b
isolation of, **8**:342b
in leukocytes, **4**:618a
mechanism of reaction, **1**:288b, 537b; **5**:4b; **6**:22a, 538b; **7**:273b, 517b; **13**:452b; **14**:412b
mitochondrial, **8**:608a
oxidizing, **1**:362a, 362b–363a; **2**:86b; **6**: 501b; **13**:393b–394a
in plants, **4**:363a; **5**:582a; **14**:45b, 255a
precursors of, **7**:520a; **10**:493a
pressure, effects of, **8**:608a

Enzymes (cont.)
protein nature of, **1**:11b; **13**:152a–b; **14**:412b
proteolytic, **2**:562a; **3**:257a; **7**:520a; **15**:15b
purification of, **14**:412a
receptor-destroying, **7**:518a
respiratory, **6**:501b; **10**:282a; **14**:173b, 175a–b, 176b
soluble enzymes vs. ferments, **2**:561a–b
specificity of, **5**:4a–b; **7**:517b; **9**:354b; **13**:452b; **15**:15b
structure of, **13**:3a
sugar-forming, **7**:393b
and temperature, **13**:243a
and trace elements, **2**:86b–87a
yellow, **7**:518a; **14**:174a
see also Coenzymes; specific enzymes
Enzymologia, **12**:207b
Enzymology, **2**:561a–562b; **3**:416a; **4**:211b–212a; **6**:111a,b; **7**:273a–b; **9**:354b; **13**:152a–b
Eoanthropus
see Piltdown man
Eocene epoch, **6**:229a
deformation, **7**:251b
naming of, **8**:569b
paleobotany, **12**:104b
paleontology, **1**:131b, 491b; **4**:67b; **5**:501a*; **9**:134b; **12**:38b; **13**:72a, 274b; **15**:91b
stratigraphy, **2**:108b, 497b; **4**:68a; **13**:274b
Eohippus, **6**:595b
Eosin, **3**:85a
in photography, **4**:283a
Eosinophils, **4**:296a
Eötvös, Roland, Baron von (1848–1919), 4:377–381; **2**:466a; **9**:180b; **11**:278b
Eozoön, **3**:89b, 608b; **8**:462a; **9**:432a
Epact
definition of, **15**:566a
Eparchean peneplain, **8**:99b
Epeirogenesis, **5**:395b
Ephedrines, **5**:100a
Ephemerides
see Tables, astronomical, ephemerides
Ephemerides of Eclipsing Binaries
founding of, **1**:430a
Ephestia, **7**:516b
Ephesus
medical school, **12**:538a, 540b
Epicaridea, **5**:385b
Epictesis, **8**:29b
Epictetus
Enchiridion, **12**:442a
Epicureanism, **1**:315a; **2**:13a, 378a; **3**:209a; **4**:495b; **5**:285a–286b, 347a; **7**:612a; **8**:537a–538b; **14**:612b
and Christianity, **5**:285b–286b
critics of, **3**:319a; **12**:310a
Epicureans, **4**:108b
atomism, **1**:255b
cosmology, **13**:92a
Epicurus (341–270 B.C.), 4:381–382; **3**:114b; **12**:538b; **13**:91a, 307b
commentaries on, **2**:224b; **5**:232a
critics of, **3**:492a; **14**:606a
Letter to Herodotus, **8**:537a–b
natural philosophy, **4**:30b; **8**:269a, 537a–b; **14**:612b; **15**:404a
Epicyclepicyclet, **3**:409b
Epicycles (astron.), **1**:189b–190a, 246b, 247b–248a, 290b, 312b, 313b, 409a, 509b; **2**:20b, 24a, 403b, 412a, 551a; **3**:15a, 25b, 406b, 408a, 409b–410a; **4**:542b; **5**:403a, 435a, 616a, 617a–b; **7**:143a, 295b, 297a; **9**:430a; **10**:298b; **11**:190b, 195a, 224a; **12**:4b; **13**:326a; **14**:446b, 593b; **15**:34b, 154a,b, 203a, 204a, 211a,b, 212b, 213a, 218b, 568a–569b, 645a
change in obliquities, **6**:198a–b
circumference, **15**:558a, 568b, 582a, 586a, 587a–588a, 597b, 604a, 610a, 612b, 615a, 618b
crank mechanism, **11**:193a–194a
manda, **15**:556a–b, 557b, 558a, 568a,b, 569a–b, 576a, 581a, 586a–592b *passim,* 597b, 600a, 601a, 604a–614a *passim,* 616b, 618b, 623a–b, 624a–b
pulsation, **15**:568a–b, 569a–b, 591b, 616b
secondary, **12**:359a–360b
śīghra, **15**:556a–560b *passim,* 568a,b, 576a, 580b, 586a–592b *passim,* 597b–606b *passim,* 612a,b, 616b, 618b, 623a
size of, **11**:192a–b
Epicycloids, **2**:47b; **7**:577b, 578a
Epidemics
and social history, **3**:463b
Epidemiological Society (England)
founding of, **12**:503a
Epidemiology
Renaissance, **1**:399b; **3**:13a; **5**:105a, 107a–b; **7**:16b; **8**:554b
seventeenth-century, **2**:308a–b; **13**:215a; **14**:405b
early eighteenth-century, **3**:97a, 332a–b; **4**:589b; **5**:220b–221a, 366b–367a, 384b; **7**:614a
late eighteenth-century, **2**:523b; **3**:344b; **13**:495b, 541a; **14**:15b–16a, 194b
early nineteenth-century, **2**:445a; **3**:6b; **5**:224b–225a; **12**:321a
mid nineteenth-century, **2**:574b–575b; **6**:120a; **7**:519b; **8**:254a; **9**:10b; **10**:267b, 558b–559b; **11**:177b; **12**:245a*, 450a, 451a, 502b–503a; **14**:42a
late nineteenth-century, **3**:463b–464a; **4**:620a–b; **5**:219b–220a; **6**:11b–12b, 101b; **7**:424b–425b, 427a–429b; **8**:65b–66a, 313a, 556b; **10**:394a, 559b–560a; **11**:346a–347a; **14**:551b
early twentieth-century, **2**:528a–529a, 583a; **3**:185b–186a; **5**:40a–b, 270a–271a, 451a–452b; **7**:171a; **8**:273b; **9**:95a; **11**:175a; **12**:315a–b, 453a; **13**:63a, 229a; **14**:623a–b; **15**:96b–97b, 454b–455a
mid twentieth-century, **6**:433a
experimental, **4**:620a
fungi, **2**:583a
historiography of, **12**:88b; **13**:142b
and statistics, **14**:574a
Epidermis
Malpighian layer, **2**:165a; **8**:8b, 9a; **9**:63b
Epidiascope, **13**:71a
Epidosite, **7**:249b
Epidote, **14**:588a
Epigenesis, **1**:387a; **2**:161b, 204a, 384b, 492a, 579b; **3**:247b; **4**:187a; **6**:65a, 160b–161b; **8**:580a, 589a; **9**:187a, 518b; **10**:10b; **11**:159a; **12**:315b, 335b; **15**:524b
in lower animals, **13**:171b–174b
Epiglottis, **4**:384a
Epilepsy, **1**:212b; **3**:203b; **5**:27a; **6**:256a; **9**:23a; **11**:99b; **12**:225a*
artificial production of, **2**:525b
chorenic, **1**:580a
diagnosis of, **6**:287b
focal, **7**:47b–48a
Jacksonian, **3**:205b; **7**:48a
theories of, **4**:387a; **6**:60a, 423b, 427b; **7**:47b–48a; **8**:536a; **9**:97b–98a; **11**:127b; **14**:407b
uncinate, **7**:48a
Epilobium, **4**:401b
Epimagma, **7**:63a
Epinasty, **2**:286b; **14**:97a
Epinephrine, **15**:74b
and carbohydrate metabolism, **3**:415b–416a
effects of, **8**:16b, 454b; **9**:266a
isolation of, **1**:10b–11a; **10**:205b
reversal, **15**:105b, 106a–b
site of production, **13**:589b
tests for, **8**:453a
Epiphyses, **2**:504b; **13**:33a
Epiphytes, **12**:166a
Epistasis, **1**:505b
Epistemology
ancient Greek, **4**:33a–b, 82a; **6**:290b; **10**:324b–325a; **11**:22b, 25b–26a, 28b, 29b; **13**:329b–330a; **14**:536b, 612b–613a
ancient Roman, **1**:118a–b, 588b; **12**:340b–341a
medieval Arabic, **6**:194a, 203b; **9**:29a–31a; **12**:6a–7Ab; **13**:488b; **15**:261b
medieval Latin, **1**:1b, 197b, 290b, 334a, 335b, 378b; **7**:116a; **10**:173b–174b
Renaissance, **3**:513b–514b; **5**:106b, 119b–121a; **10**:122b–123a, 308b; **11**:287b; **12**:105b–106a; **14**:47b, 225a–226b
early seventeenth-century, **3**:361b–362a; **4**:52b–54b; **5**:286a–287b; **9**:316b–318b; **11**:489a; **14**:362a–363b, 367a–b
late seventeenth-century, **5**:415b–416b; **8**:153b–154a, 437b–439a; **9**:49a–b, 118a, 120a; **13**:214b; **14**:372a–373b
early eighteenth-century, **2**:16b–18a; **3**:215b, 216b, 380a–383a; **5**:511a; **6**:139a–140a, 557a–559b; **8**:151b; **11**:43a
late eighteenth-century, **1**:111a–b; **2**:328a; **7**:226a–227a, 233b–235a, 597b–598a; **14**:180a
early nineteenth-century, **3**:375a–379a; **5**:193b, 252b–253a; **6**:294b; **12**:155a, 156a; **15**:127b–128a
mid nineteenth-century, **2**:31b; **3**:322b, 452a–453b; **5**:262a; **6**:81a–b, 242b, 243a, 247b–249a; **9**:210a, 383b–384b, 569b; **12**:607b; **15**:137b, 138b–139a, 142a, 143b, 167a–b, 410b–412b
late nineteenth-century, **2**:624b; **4**:616b; **6**:349a*, **8**:600a–605b; **9**:382b; **10**:484a–486a; **12**:607b–610a; **14**:67b; **15**:79b
early twentieth-century, **2**:9a–12a, 239a–240a, 248b–250b; **3**:32a–34b, 111a–112a, 335a–b; **4**:225a, 226b–227b; **6**:26a–b; **8**:257a, 473a, 517a–b; **9**:341a–342a; **11**:355b–358a, 388b–389a; **12**:177b–178b; **14**:305a–b, 306b–307a, 308a; **15**:38b–39a, 423a–424a
mid twentieth-century, **2**:252b–253b; **4**:281b–282a; **5**:122b; **11**:546a
analytic, **1**:126b
of astronomy, **1**:589a–b
Chinese, **12**:385a–387a
genetic analysis of, **3**:380b–381a
historical, **3**:375b
medical, **13**:214b
Neoplatonic, **2**:234a
particularity vs. universality, **3**:335a–b
Platonic, **3**:471b
and probability, **15**:304a, 308b, 369a
revelational, **1**:32b, 34b, 335b; **2**:222b; **3**:361b–362a
scientific knowledge, role of, **1**:178a, 365b–366a
sensationalist
ancient Greek, **1**:315a; **4**:33a
medieval Latin, **2**:193b; **4**:255a

Renaissance, **11**:71b; **13**:278b–280a
seventeenth-century, **5**:286b; **6**:446b–447a; **8**:437b–438b
eighteenth-century, **1**:111a, 113a, 114b; **3**:1b, 381a–382a; **4**:85b; **7**:603b; **8**:81b; **10**:611b
nineteenth-century, **1**:140b; **5**:252b–253a; **7**:591b; **9**:383b
see also Historiography, of epistemology
Episynthetic school of medicine, **1**:75a
Epitaxy, **5**:124a, 186b, 456a
Epitedeiotes, **7**:137a
Epithelium, **7**:438b
ciliary, **3**:424b; **4**:276a
ciliary motion
discovery of, **11**:216b
physiology of, **13**:555b
histology of, **2**:164b–165a; **6**:269a–b; **11**:374a; **13**:556b
sensory, **5**:34b
vascular, **3**:425a
Epizootics, **2**:528a–b, 575b; **3**:332b; **9**:550b; **10**:406b, 568a; **12**:480b; **14**:14b, 15b–16a
Epochs, astronomical
Arabic use of, **1**:512a; **2**:152a; **7**:338a
Epochs, geological
see specific names
Epsilontics, **3**:137a
Epsom salts, **3**:201b; **5**:601b
Epstein, Paul Sophus, **4**:196b; **7**:468b; **12**:529a,b
Eptatretus, **3**:610b
Equalitarianism, **11**:323a
Equant, **1**:509b; **2**:403b; **3**:25b, 406b; **5**:617a; **7**:143a, 294a,b, 295a; **11**:196a; **13**:96a
objections to, **6**:198b–199a; **11**:202a; **12**:4b
Equations
Abelian, **1**:16b
affine, **11**:512a
algebraic, **1**:13a, 66b; **2**:217b; **5**:472a; **7**:102a, 408a; **12**:237b
arrangement of terms, **3**:460a,b; **5**:409b
Chinese representation of, **3**:254a–b
classification of, **7**:359a
Graeffe's method, **3**:555a
reduction of, **6**:472b, 537a–b; **7**:103a–b
roots of, **2**:37b–38a, 602b; **3**:229b, 555a; **4**:103a, 475a; **5**:450a, 567a–b; **8**:383b, 385a; **11**:401b, 512b, 599a; **12**:86b, 283a; **13**:129a–b
solution of, **1**:591b; **2**:105a, 112a–114a; **3**:314a, 455a*; **4**:57b; **5**:198a, 260a–263b *passim,* 472b, 490b; **6**:390a; **11**:611a; **13**:59b, 571b; **15**:226a
theory of, **1**:13a; **2**:444a; **3**:65b; **7**:506b; **8**:385a; **9**:173b; **10**:245b, 594a; **11**:598b–599a; **12**:291a–b, 534a
Aronhold, **1**:294b
Bernoulli, **2**:40a, 48a, 52b
Bessel, **4**:479a; **13**:78a
binomial, **3**:274b; **5**:263a; **13**:571b
solvability of, **5**:263b
canonical, **10**:250a
Cauchy-Riemann, **3**:137a,b, 138a, 139a; **4**:477b; **11**:449a
Cayley-Plücker, **3**:168a
Cayley-Zeuthen, **3**:168a
chemical
concept of, **8**:86b(n)
Clairaut, **3**:282b; **8**:562a
classification of, **7**:328a,b
cubic, **2**:281b; **3**:119b; **4**:108b, 611b; **6**:577b*; **7**:258b; **9**:21b, 159a; **11**:240b, 590b; **13**:291a
classification of, **7**:328a,b; **13**:515a
distribution of roots, **7**:328b–329a
solution of, **2**:280a,b; **3**:65a,b, 267b; **4**:58a, 490a–b, 568b, 586b–588a, 595a–596a, 610a; **5**:409b–410a, 528a; **6**:472b, 537b; **7**:257b, 327b–329a, 335a; **8**:162a; **9**:555b; **10**:270a; **11**:401b; **12**:460a, 610b; **13**:259a, 260a, 515a–516a, 536b; **14**:21a,b, 23b, 24a, 147b, 540b–541b; **15**:116b, 670a
cyclotomic, **1**:16a; **2**:54a; **5**:299a,b
decomposition of, **5**:55a
definition of, **14**:20a
difference, **2**:144a,b, 217b, 294b; **3**:136b, 143a; **5**:523a; **6**:572a; **9**:127a, 454a, 471a, 472a, 476b; **14**:188a; **15**:383b
expansion of, **15**:366a
solution of, **15**:277a–279a, 281a, 306b, 364b, 366a, 382b, 385b
differential, **1**:194b; **2**:88b, 198a, 294a–b, 357b, 473a; **3**:135b, 136a,b, 220b–221a; **4**:177a, 370b, 618b; **5**:78a*, 97a, 203b–204b, 471b, 521b; **6**:309a, 394a; **7**:574a–575b; **8**:164a,b, 389a; **11**:60a, 452b; **12**:246b; **13**:424a, 561b, 562a; **14**:13b, 188a; **15**:366a
algebraic, **3**:220b; **5**:523a; **6**:472b; **8**:458b
approximate solution of, **8**:559a
asymptotic solutions of, **2**:356b; **15**:386b
canonical system, **2**:144b
in celestial mechanics, **2**:144a, 145a; **9**:449b
critical points, **11**:56b–57a
in dynamics, **2**:351a; **4**:480a–b, 481a, 547b; **8**:384b; **10**:250a; **13**:73a, 78b; **15**:289b, 483b–484b
homogeneous, **11**:400a,b
hyperbolic, **4**:230a; **11**:454a
integral curves of, **4**:58b, 269a; **10**:594a,b
integration of, **1**:110b, 207a; **2**:52b, 350a,b; **3**:385a, 457a; **4**:103a, 478b; **5**:209b, 449b; **7**:444a, 515a, 567b, 568a; **8**:136a, 139b; **9**:125a; **11**:56a, 399b–400b, 401b, 613a; **15**:164b, 484b–485b, 488b
last multiplier, principle of, **7**:54b
linear, **4**:523a; **5**:201a, 204a,b; **6**:74b, 75a,b; **7**:168b, 408a, 562a; **8**:458b; **9**:124b; **10**:594a–b; **11**:53a, 55b, 407a; **12**:599b; **13**:249b, 250a, 264a; **14**:86a, 281b; **15**:327b
nonlinear, **2**:299a; **5**:97a
*n*th-order, **5**:54b, 204a
in optics, **4**:190b–191a; **12**:220a
p discriminant, **3**:169b
in physics, **2**:144b; **6**:473a
qualitative theory of, **13**:36a
self-adjoint, **14**:281b
singularities of, **2**:358a; **3**:169b; **4**:478b; **5**:203b, 204b; **10**:275a–b; **13**:266b; **15**:293b
solution of, **1**:500a; **2**:53a, 271b; **3**:136a, 143a–145a, 196a, 457a, 615b; **5**:96b, 103a, 204a–b; **7**:10b, 514b; **8**:61b, 162b, 299b, 384a; **9**:410a, 449b; **10**:48b, 244b, 327b, 442a; **15**:277a, 294a–295a, 302b–303b, 382b, 383a, 384a
systems of, **2**:236b; **3**:385a–b; **7**:460a; **8**:560b–561a; **10**:554a
theory of, **2**:217b, 470a; **3**:559b; **4**:205b, 478a–479a; **5**:482a; **6**:4b; **7**:54b, 435b, 460a, 573b; **8**:323b, 324a–326b, 366b, 383b–384a, 560b–561b, 617b; **9**:127a, 232b; **10**:249a–b; **11**:56a–57b, 471a; **13**:55b, 128a, 129b–130a; **14**:202a, 317a; **15**:7a, 23a, 65b, 226a
transformation of, **7**:530b
variation of parameters, **4**:478b
dimension of, **4**:569a
Diophantine, **2**:507a; **4**:83a, 124a; **7**:565b; **11**:55b, 61b, 512a; **12**:451b
solvability of, **6**:393b–394a
double, **4**:574b
Euler, **4**:479a, 480b; **8**:504a; **11**:327a
expansion of, **13**:250a; **15**:348b
Fermat, **1**:76b; **4**:124a,b; **5**:528a; **7**:565b; **8**:138b
forty-fifth degree, **3**:181b; **11**:533a–b; **14**:22a–b
Fredholm's, **5**:151a–152a; **7**:436a,b
Fuchs, **2**:144b; **7**:435b
functional, **1**:307a; **3**:428a; **6**:214b; **7**:561b; **9**:410b
general
solvability of, **5**:263b; **8**:383b
general algebraic, **3**:460b
Hamiltonian, **2**:145a; **7**:54b; **10**:250a
Hill, **6**:399b
hyperbolic, **6**:4b
theory of, **10**:554a
hypergeometric, **14**:317a
indeterminate, **1**:30b, 31a; **4**:57b, 116a, 474b, 568a,b, 573b, 608b, 610a, 611a; **5**:187b; **7**:565a,b; **8**:135a, 422b; **14**:630a
solution of, **15**:579a
solvability of, **8**:135b
systems of, **12**:469a
integral, **1**:14a, 427a,b, 500a; **4**:103a; **5**:150b, 151a–152a, 482a; **6**:235b, 391b–392b; **7**:408a; **9**:502b; **10**:514a, 594b; **11**:156b, 459a,b; **12**:237b; **13**:424a, 548b; **14**:87a
nonlinear, **5**:201b; **8**:561b; **10**:20b
theory of, **2**:238b; **4**:288a; **8**:112a, 384a; **10**:20a; **11**:476a; **12**:187b–190a, 513a; **13**:27b, 428a
Klein-Gordon, **15**:49b
Kummer's, **5**:482a; **8**:366b
Laguerre's, **7**:574a–575b
Lamé, **6**:308a, 309a; **9**:127a
Laplace, **2**:105b; **3**:559b; **4**:126b; **5**:96a; **6**:4b; **7**:601a–b; **10**:553b; **11**:449b; **12**:285a; **13**:26b; **14**:317a
integration of, **2**:105b
linear, **15**:372a
Hesse's normal form, **6**:358a
infinite, **12**:189a–b
second-order, **4**:478a–b
solution of, **3**:166b; **4**:103a; **15**:669a, 698a–b
systems of, **7**:244a; **8**:165b
theory of, **1**:429a; **4**:478a–b
Volterra, **13**:549b*
literal, **7**:566a–b, 567a
Mathieu's, **9**:175a
modular, **5**:263b; **7**:53b; **11**:268b
ninth-degree
solution of, **6**:389b
*n*th-degree, **14**:472a
numerical, **4**:618b; **10**:443a; **14**:22b–23a, 540b
roots of, **13**:218b–219a
partial differential, **3**:62b–63a, 143a–145a; **4**:370b, 478b; **5**:152a, 201a; **6**:4a–b; **7**:477b, 549b; **9**:376b, 553b; **11**:260a; **12**:187b; **15**:23a
application of, **4**:336a; **15**:318a, 366b, 371b
Cauchy problem, **3**:143a; **6**:4a,b, 5a; **10**:554a
characteristics method, **3**:132a, 143a; **4**:205b, 478b; **5**:608b; **10**:544b
elliptic type, **2**:217b; **5**:151a; **15**:23a
generalized solutions, **9**:174b
integration of, **1**:66b, 142b, 407b; **3**:

Equations (cont.)
385a; **5**:110b, 111a, 482b*, 509a*; **7**:568a; **8**:136a; **9**:232b; **10**:573b–574b; **14**:87a
level surfaces of, **3**:213b
linear, **1**:294b, 600b; **3**:169b; **10**:248b; **15**:293a, 306b
in physics, **1**:113b; **3**:145b, 146a; **4**:222a, 316b–317a, 319a, 480b–481a; **5**:95b–96a,150b–151a, 376a, 608b; **6**:89a–b, 147b; **9**:174a; **10**:3a, 4a; **11**:58b–59b, 448a, 453b, 476a; **12**:221a; **13**:130b; **14**:13b, 202b; **15**:293a, 482b, 483b–484b, 487a
second-order, **9**:476b; **12**:285a; **13**:36b
solution of, **1**:116a, 499b, 500a; **4**:577b; **12**:188a; **14**:232a; **15**:200a, 364b, 368b, 383b, 385a, 386a
systems of, **8**:458a*; **10**:554a
theory of, **2**:473a; **3**:314a; **5**:110b; **7**:478a–b; **8**:323b–324a, 325a–b, 383b–384a; **9**:471a, 472a, 476a–b; **10**:554b; **13**:55b
Pellian, **4**:117a, 125a; **10**:238b
Pfaffian, **3**:143a; **4**:370b; **8**:324b; **10**:574a; **12**:214a
integration of, **15**:198b
planar
Hesse's normal form, **6**:358a
Poisson, **6**:472a
quadratic, **4**:111b, 112b, 117a, 612a; **15**:668b, 698b
canonical forms, **4**:568a–b; **14**:466a–b
construction of, **7**:328b
roots of, **3**:312a; **4**:422a; **8**:423a; **11**:339a–b; **14**:279b
solution of, **1**:22b, 30b–31a; **4**:608a–609a, 609b, 610b; **7**:328b; **8**:319a–b; **9**:159a; **11**:32b; **13**:289a, 291a; **14**:543a–544b; **15**:669a–670a
quartic, **1**:30b; **3**:267b, 284b; **5**:410a; **7**:329a; **9**:50a; **14**:147b
roots of, **4**:422a
solution of, **2**:280a,b; **3**:65b; **4**:58a, 568b, 587a, 587b–588a; **5**:409b–410a, 528a; **7**:257b; **14**:21b, 23b–24a, 541b, 544b
quintic, **5**:472b; **7**:50a; **9**:50a; **11**:61b
reduction of, **2**:467b; **9**:55a
roots of, **7**:103a–b; **13**:219a
solution of, **1**:13a–b, 14a, 15a; **2**:90a, 467b–468a, 470a; **5**:260b, 263a, 304b, 528a; **6**:92b, 307a, 308a, 309a; **7**:398a
theory of, **6**:309a
resolvent, **5**:263a; **9**:55a
resultant, **2**:112b–114a, 470a; **12**:171a
Bezoutiant, **2**:112b–113a
degree of, **2**:113a–b
Riccati, 2:37b, 57a; **4**:478b; **5**:414a, 449b; **7**:561b; **8**:384a; **11**:400b
roots of, **6**:510b; **9**:50a; **13**:48b; **15**:751a
distribution of, **3**:221b
geometric construction of, **1**:474b; **7**:328a–329a; **9**:471a; **11**:512b–513a; **12**:283b
root-coefficient relations, **2**:112a; **3**:65b; **14**:23b
squaring method, **8**:432b
Schrödinger, **4**:339b*; **7**:492a–b, 574b; **8**:284b, 473b, 475a, 498b
seventh-degree, **5**:472b; **6**:394a
simultaneous, **8**:385a
linear, **2**:618b; **3**:250b, 256a; **6**:208b*; **13**:219a; **14**:545b; **15**:295a
solution of, **2**:112a–114a; **3**:460b–461a; **4**:610b; **8**:422b; **12**:291b; **13**:220b, 399b–400a; **15**:119b–120a, 690b–691a, 698b
sixth-degree, **3**:345b; **5**:472b
solution of, **2**:470a
solution of, **1**:481b*; **2**:281a, 573a–574a; **3**:65b; **4**:107b, 341a; **15**:668b
by balancing, **4**:115a
by completion, **4**:115a
geometrical, **12**:460a; **15**:671b
by graphs, **3**:460b; **4**:568b; **11**:579b
Horner's method, **3**:254b, 267b; **4**:610a; **6**:510a–511a; **7**:257a; **8**:315b, 423a; **14**:540b, 541b, 544a,b
iteration method, **3**:222b; **6**:511a
Newton-Raphson method, **3**:222b; **5**:98b; **6**:511a
numerical, **8**:139a
plus and minus method, **15**:669a–b, 671b
by radicals, **1**:15a, 16a,b; **4**:475b; **5**:261b–262a, 263a–b
solvability, **3**:168a; **5**:204a; **8**:165b
Sturm-Liouville, **7**:408a
systems of, **9**:174b; **13**:6a
theory of
Renaissance, **3**:274b–275a
seventeenth-century, **1**:481a; **3**:348a; **4**:57b–58a, 566b–569b *passim*, 572a–b; **5**:528a; **6**:124b, 125a, 537a–538a; **9**:50a; **11**:488a
eighteenth-century, **2**:112a–114a; **5**:55a, 449b–450a; **7**:566a–567a
nineteenth-century, **1**:13a, 16b; **2**:444a, 573a–574a; **3**:169a, 428b*; **5**:96b, 97b–98b, 260b–263b; **6**:510b; **7**:103a–b; **8**:61b; **13**:218b; **15**:481a
twentieth-century, **2**:90a, 358a; **4**:83b*; **6**:235b
Chinese, **14**:543a
rule of Budan and Fourier, **2**:573a
transcendental, **5**:96b; **8**:383b
transformation of, **1**:590b; **2**:573a–b; **3**:65b, 221b, 267a–268a; **6**:357b; **12**:75a, 228a; **14**:23b–24a; **15**:278b–279a, 293a, 364a, 484b–485b
trinomial, **7**:566b
in two unknowns
Renaissance use of, **3**:275b–276b
variational, **11**:57b
Equator, **1**:574a
ellipticity of, **6**:241a
measurement of, **5**:435a–b
Equator, celestial, **3**:404b
Equatorial (astron. instrum.), **2**:471b, 532b; **3**:92b, 102b; **5**:271b, 316a; **6**:103b; **7**:81b; **10**:329a; **11**:377b
Equatorium (astron. instrum.), **3**:25b–26a, 27b; **4**:164b; **6**:273a,b; **7**:119a, 125a–b, 126a, 142b–143a, 260a; **10**:540b, 541a; **11**:415a–b; **12**:199b; **15**:154a, 163a
paper, **15**:476a
theory of, **7**:119b
Equidae, **3**:264
Equilibrium, biological, **2**:446b–447a; **5**:44b–45a; **11**:215b
and brain centers, **1**:446b; **9**:10a, 20a–b
caloric nystagmus, **1**:446b
and nervous system, **1**:447a
Equilibrium, chemical
nineteenth-century study, **2**:68b–69a; **3**:212a*, 310a; **5**:324a–b, 389b–390b, 412a–b, 587a; **6**:177b; **8**:118a–b; **9**:619b; **12**:593b; **13**:578a–b, 579b; **14**:109a; **15**:459b
twentieth-century study, **5**:465a; **7**:529a–530a; **13**:244b–245a; **15**:432b, 436a–442a, 463b–464a
false, **4**:232a
thermodynamics of, **4**:229b–230a
heterogeneous, **11**:482b; **13**:243b, 244b–245a
phase rule, **1**:431a; **5**:390a; **6**:561a; **7**:204a*, 529a; **11**:534a–535b; **15**:463b
homogeneous systems, **5**:412a–b
Le Châtelier principle, **8**:118a–b
mobile, **13**:578b
in nature, **3**:293a
stability of, **4**:228b
temperature effects, **8**:118a–b
in tertiary systems, **1**:431a
theories of, **8**:373b
see also Kinetics, chemical; Reactions, chemical
Equilibrium, physical
ancient Greek study, **6**:312a
medieval Arabic study, **1**:446a; **7**:41b, 42a, 341a, 345b; **13**:292a
Renaissance study, **3**:66b; **8**:216a–b, 223a–225b; **9**:487b–488a
eighteenth-century study, **1**:111b, 112a; **3**:441b, 454a
nineteenth-century study, **2**:242b; **3**:145b, 591b; **10**:275b; **11**:61b; **15**:359b, 377a–b
of elastic plane, **3**:145a–b
of ellipsoid, **7**:37b
of floating body, **3**:591b; **6**:602a
hydrostatic, **7**:37b; **8**:226a–227a
of nebulae, **5**:363b
pear-shaped figure, **3**:583b–584a
of planes, **1**:222a–223a
of rotating liquid mass, **3**:225a, 583b; **7**:85a–b; **8**:559b–562b; **11**:498a
of spinning tops, **1**:419b
in structural analysis, **3**:468b, 496a
of surfaces, **8**:124a
see also Isostasy; Lever, law of; Statics
Equilibrium, thermal, **8**:63b; **13**:52a, 374b
Equilibrium, thermodynamic, **2**:261a–265b; **4**:314a, 317b; **15**:377b
general theory of, **5**:389b–390b; **15**:443b–446a
stability of, **5**:388b, 390b
Equimultiples, **4**:420a
Equinoctial ring, **7**:259b–260a
Equinoxes, **4**:460a; **13**:320a; **15**:673a
dates of, **11**:189b; **15**:680b
determination of, **15**:219b, 677a
observation of, **2**:150b; **7**:129b; **14**:576a; **15**:208a, 218a
position of, **1**:383a; **2**:411a; **3**:101a; **5**:24b; **7**:6b; **9**:514b; **11**:191a; **15**:35a
precession of
see Precession of equinoxes
standard, **3**:373b
vernal, **3**:558b; **5**:553a; **15**:536b
position of, **2**:99a–b; **13**:536a
Equipartition theorem, **4**:314a,b, 316a; **9**:220a, 223a,b; **12**:426a, 464a; **14**:185a
proof of, **13**:237a
Equipollences, method of, **1**:590b–591a; **10**:443a; **15**:196a
Equisetales, **2**:160a
Equisetene, **7**:208b
Equisetum, **5**:539b, 605a; **12**:99b; **13**:596a; **15**:67b
Equivalence, chemical, **14**:488b–489a, 489b
Equivalency (math.), **1**:294b; **2**:277a–b; **3**:263a; **5**:299b; **12**:331a
Cantor-Bernstein theorem, **3**:57a
Equivalent volumes, law of, **9**:220a
Equivalent weights, **3**:47b; **5**:104a, 431a; **9**:110a, 348b–349a; **10**:510b; **11**:435b, 436b–437a
Eranos conferences, **7**:192a–b
Erard, Jean, **13**:592a

Eras, calendric
of Alexander, **2**:152a, 154b
of Antoninus, **2**:152a
Arabic use of, **1**:509a, 512a; **2**:152a, 154b; **14**:538a
of Augustus, **2**:152a
Christian, **1**:565b; **3**:26b
conversion of, **7**:123b
Coptic, **1**:512a
of Creation, **2**:152a
of Diocletian, **1**:512a; **2**:152a
of the Flood, **2**:152a
of Hegira, **2**:152a, 154b; **3**:26b
Malikī, **7**:324b
Muᶜtaḍid, **2**:152a
of Nabunassar, **1**:512a; **2**:152a
of Philippus Arrhidaeus, **1**:512a; **2**:152a
of Yazdigird, **2**:152a, 154b
Eras, geological
see specific names
Erasistratus (*b. ca.* 304 B.C.), 4:382–386; **4**:105b; **5**:232a; **12**:102b; **13**:95a, 328b
anatomy, **1**:316a; **3**:482b; **5**:230b; **6**:317b, 318b; **13**:490b
physiology, **5**:234a–235b
d'Erasmo, Geremia, **1**:491b
Erasmus, Desiderius, **1**:77b; **3**:420b; **5**: 119a,b, 121a; **8**:248a, 360b, 361a; **9**:75b; **10**:305a; **12**:135a; **13**:491b; **14**:47b
Erastianism, **4**:387a; **6**:448a
Erastus (Lieber), Thomas (1523–1583), 4: 386–388; **2**:293a; **3**:342b; **8**:309a
on comets, **4**:213a
on Paracelsus, **4**:170a, 208b; **12**:335b; **13**:141a
Eratosthenes (*ca.* 276-*ca.* 195 B.C.), 4:388–393; **4**:414b; **10**:179b, 181b(n)
and Archimedes, **1**:213a, 221b, 252a; **3**: 150a
astronomy, **4**:419a; **11**:189b; **13**:326a
critics of, **10**:114b; **15**:219b–220a
geography, **4**:82a; **6**:212b, 213a; **11**:225b; **13**:84b
earth's circumference, **3**:320a; **6**:301b; **9**:1b
zones, **5**:345b, 347a
harmonics, **1**:282b; **4**:430b
influence of, **11**:104a; **13**:84a
mathematics, **9**:272b
cube duplication, **4**:105a, 466b, 489b; **6**:407a, 411a; **9**:270a
sieve method, **4**:392a–b, 573b; **8**:162a; **10**:113b(n)
Platonikos, **13**:326a
spurious works, **9**:270a, 274a,b, 275a
Erb, Wilhelm Heinrich, **1**:569a
Erbia, **3**:478a; **9**:110b
Erbium, **3**:321b
discovery of, **9**:541b
Ercker (also Erckner or Erckel), Lazarus (*ca.* 1530–1594), 4:393–394; **2**:143a; **5**:420a; **8**:312a; **15**:94a
Beschreibung allerfürnemisten mineralischen Ertzt und Berckwercksarten . . . , **8**:464b
Erdelyi, Arthur, **1**:500a
Erdmann, Axel, **9**:541b
Erdmann, Otto Linné (1804–1869), 4:394–396; **3**:62b; **5**:197b, 369b; **9**:91a,b
Erdnähten, **3**:325b
Erdös, Paul, **4**:562a
Erebus (Mtn.), **3**:591a; **11**:554b
Erebus (ship), **6**:489a
Erectility
mechanism of, **11**:565b
Erfurt, University of, **12**:62b
Erg, **5**:407b
Ergodic hypothesis, **2**:265a–b; **4**:293b, 577b; **9**:129b, 223b; **11**:56b
Ergodic theorem, **2**:145b; **7**:351b; **11**:459a; **13**:36a; **14**:90b, 91b, 346b
Ergograph, **9**:547a
Ergometer
bicycle, **7**:502b
Ergonomics, **3**:441a, 442b
Ergosterol
isolation of, **4**:91b
and vitamin D, **14**:445a
Ergot, **8**:177a; **13**:489b; **15**:105a,b, 106a
alkaloids, **15**:10b
Ergotherapy, **10**:613b
Ergotism, **4**:135b; **14**:93b
Ergotocine, **7**:323b
Eridu (city), **14**:504a–b
Erie (lake), **8**:571a
Erie Canal, **4**:273b–274a; **5**:11b; **9**:134a
Erinaceus, **6**:535b
Eriometer, **14**:570b(n)
Eriugena, Johannes Scottus (*fl.* ninth century A.D.), 4:396–397; **1**:91b; **14**:225b
De divisione naturae, **8**:547b; **9**:364b
Erkenntnis
founding of, **11**:355b
Erlangen, University of, **10**:137b
mathematics at, **13**:4b
Erlanger, Joseph (1874–1965), 4:397–399; **5**:290b, 291a; **8**:29b; **15**:75b
Erlanger Programm, **2**:513a; **4**:522b; **7**:397a–b; **8**:325b, 434a; **9**:358b; **11**:362b; **12**:214b; **13**:599b
Erlenmeyer, Richard August Carl Emil (1825–1909), 4:399–400; **2**:621a; **4**: 100b, 396a*; **5**:12b, 164b, 341a*, 488b; **7**:551b; **9**:355a
chemistry, **1**:362a; **9**:130b
Erlenmeyer flask, **4**:400a
Erlenmeyer rule, **4**:400b
Ermak (ship), **9**:42b, 43a–b, 292b
Erman, Georg Adolph (1806–1877), 4:400–401; **6**:553a; **10**:191b
Ernst II, Duke of Gotha, **6**:456a; **11**:473a
Ernst August, Elector of Hannover, **5**:306a; **8**:150a,b
Eros (asteroid), **3**:261b*; **4**:269b; **5**:258b; **12**:17b; **13**:113b, 114a
photography of, **7**:472b
and solar parallax, **4**:251a; **6**:147a; **10**: 526a, 529a*; **12**:574a
Erosion, **3**:568a; **4**:402b; **5**:577b; **6**:227a, 228a, 229a, 580a; **9**:594a; **11**:41a; **12**:380b; **15**:55b
cycle of, **3**:593b–595b; **5**:395b; **6**:580b–582a
differential, **5**:338a
glacial, **6**:116a; **7**:149a–b; **9**:150a; **11**:277a
marine, **7**:185b–186a
Neptunist denial of, **2**:554a–557a *passim*
prevention of, **10**:280b
regressive, **13**:257b
shore, **3**:548a
control measures, **14**:494a–b
stream, **3**:551a, 552b; **6**:584b; **7**:144b, 185b–186a; **8**:198b; **10**:32b; **12**:38b, 262a, 263b; **13**:257b
subaerial, **5**:338a, 395b; **6**:502b; **7**:185b
submarine, **2**:578a; **6**:229a
theories of, **2**:549a, 569a 570a; **3**:568b; **7**:511a, 513a; **10**:504b; **11**:576a
wind, **14**:627a
see also Weathering
Erotian, **4**:119a, 486b; **6**:419b
Erratic blocks (geol.), **5**:599b; **10**:502a
charting of, **11**:500b
theories of, **2**:496b, 556b, 568a; **3**:210b–211a; **4**:28a, 347b, 402b–403a; **5**: 445a; **6**:55b, 502b–503a, 582b; **7**: 511b; **8**:570a, 575a, 616a; **9**:585a; **12**:122a; **13**:410a
Errera, Léo-Abram (1858–1905), 4:401–402; **12**:175b, 620b*
Errors
calculus of, **1**:365b
compensation of, **3**:75b–76a
estimation of, **5**:98b
instrumental, **10**:461a
in astronomy, **2**:98b, 100a, 332b; **5**: 304a, 554b; **10**:159a
mean, **15**:365a, 367a
normal (exponential) law of, **1**:65b
observational, **10**:450a, 460b
in astronomy, **4**:185a; **6**:364a, 512b; **8**:50a; **10**:479b–480a; **14**:479b; **15**: 352a,b, 355b, 364b, 365a, 372a, 381b
probable, **10**:458b, 465b(n)
standard, **5**:9b; **10**:451b, 457b–458b
statistical prediction of, **15**:372a, 373a
systematic, **7**:239b; **8**:582b, 583a; **9**:386b; **14**:554a
in astronomy, **10**:207b; **14**:414a
theory of
early eighteenth-century, **2**:42b; **3**:432b
late eighteenth-century, **2**:329b–330a; **15**:282a–283a, 320b–321a, 325a, 329b–331a, 339a–b, 351b–352b
early nineteenth-century, **3**:141b–142a; **15**:363b–365b, 372a–b, 373b, 376a–b, 381b
mid nineteenth-century, **1**:591b; **2**:88b; **8**:432b; **10**:479b–480a; **11**:237a
late nineteenth-century, **5**:414a; **6**:240a; **7**:107a; **13**:339a
twentieth-century, **3**:31b; **5**:477a; **8**: 367a; **11**:49a; **14**:437a
combination of observations, **2**:330a; **15**:364b
indeterminate corrections, method of, **8**:136a
Ertel, Hans, **6**:553b
Ertel, Traugott Lebrecht, **3**:14a; **5**:302a; **13**:109b
Erycinidae, **1**:504a
Erycinus, **10**:294b
Erysipelas, **4**:154b; **11**:373b; **12**:296a
swine, **7**:100b, 425b; **8**:449b; **10**:399b–400b
Erysiphaceae, **13**:489b
Erythea
editors of, **7**:103a*
Erythema, **5**:566a
Erythralgia, **8**:295b
Erythroblasts, **9**:296a
Erythrocytes
altitude effects on, **2**:62b
composition of, **4**:486a; **5**:226b
discovery of, **8**:128b; **9**:65a
formation of, **1**:477a–b; **6**:366b; **7**:438a; **11**:368b, 369b
disruption of, **9**:295b–296a
in marrow, **2**:165a
malarial infection of, **1**:360a
metamorphosis of, **2**:598a
permeability of, **13**:53b
shape of, **5**:56b; **6**:63b–64a, 367b; **8**:414b
sickle cell, **3**:572b
size of, **5**:495a; **8**:128b, 414b
structure of, **14**:113a, 596b
Erythrodextrin, **2**:531b
Erythromelalgia, **9**:422a
Erythrophloeum, **9**:296a
Erzgebirge, **3**:433b
triangulation of, **6**:239a

Eschatology
ancient Greek, **4**:367a
medieval Latin, **1**:289b
Escher von der Linth, Arnold, **4**:402b; **6**: 227a–b; **13**:123b–124a, 145a, 146a
Escher von der Linth, Hans Conrad (1767–1823), 4:402–403; **3**:210b; **8**:543b
Escherich, Theodor (1857–1911), 4:403–406
Escherichia coli, **4**:404b
varieties of, **6**:110b
Escholt, Mikkel Pedersön (*ca.* 1610–1669), 4:406
Eschscholtz, Johann Friedrich (1793–1831), 4:406–407; **15**:83a
Esclangon, Ernest Benjamin (1876–1954), 4:407–408; **2**:236b; **11**:321a*
Esenbeck, Nees von
see Nees von Esenbeck, Christian Gottfried (Daniel)
Esenbeckia, **10**:13a
Eserine
see Physostigmine
Eskers, **2**:496b; **14**:516b
Eskimos, **2**:207b; **6**:527b, 567b; **12**:51b; **15**: 430b
physiology, **7**:502a
Eskola, Pentti Elias (1883–1964), 4:408–410; **12**:275b*
Esmark, Jens, **3**:210b
Esophagotomy, **10**:433a
Esophagus
anatomy of, **12**:384a; **13**:32b
resection of, **2**:130a
rupture of, **2**:226a
Esper, Eugen Johann Christoph, **12**:184a
Esperanto (language), **3**:456b; **9**:415b
Esperienza, Accademia dell'
see Accademia dell' Esperienza
Espin, Thomas Espinell, **2**:146b
Espy, James Pollard (1785–1860), 4:410–411; **4**:592b; **8**:1b, 487a; **11**:340b
Esquirol, Jean-Étienne Dominique, **8**: 272a,b; **10**:612a, 613b
Essais et recherches de mathématiques et de physique, **10**:319b
Essence (philosophical), **1**:250b–251a, 277b; **9**:30b; **15**:181a–182b, 261b
and existence, **2**:233a; **10**:173b; **14**:388a; **15**:182a–b, 494b
and motion, **13**:292b
see also Being
Esson, William (*ca.* 1838–1916), 4:411–412; **6**:110a; **13**:218a
Este, Meliadus d', **1**:97a
Este family, **2**:142b; **3**:426a; **8**:150a; **9**:4b
Esterification reactions, **1**:579a; **2**:66a, 68b–69a; **4**:360a–b; **5**:5a; **10**:440b; **13**:309b
acid-catalyzed, **15**:461b
kinetics of, **9**:304b–305a
reversibility of, **9**:304b
steric hindrance, **9**:357a
see also Esters
Estermann, Immanuel, **3**:329a
Esters, **3**:242b–243a; **10**:499a
composition of, **4**:244a–b, 246b
heat of combustion of, **2**:69a
naming of, **5**:431b
reactions of, **13**:310a
synthesis of, **3**:286b; **5**:381a–b; **13**:309b–310a; **14**:531b
see also Esterification reactions
Esthesiology, **3**:99b; **8**:7b
Esthesiometry, **2**:132a; **5**:184b–185a
Estienne (Stephanus), Charles (*ca.* 1505–1564), 4:412–413
Estimation, statistical, **5**:9a, 477a
interval estimation, **5**:9b–10a
Estivo-autumnal fever, **7**:428a; **9**:95a
Estonia
geology, **6**:238b
Estragon oil, **8**:59b
Estriol
isolation of, **3**:352b
Estrogen
discovery of, **1**:123a–b
and malignancy, **1**:123b
oral administration, **3**:352b
Estrous cycle
see Reproduction, biological
Estuaries
formation of, **11**:393b
Eternity
Byzantine discussion, **7**:134b, 138b
medieval discussion, **2**:395b; **3**:470a
Ethane, **7**:452a; **12**:47a, 209a
Ether, Aristotelian
ancient Greek theories, **1**:254a, 261b
medieval Arabic theories, **11**:324b
medieval Latin theories, **1**:80a
Renaissance theories, **1**:495a–b
nineteenth-century theories, **3**:57a
criticism of, **7**:134b–135a
Ether, chemical, **2**:367a, 572b; **6**:255a; **10**: 308a
acetic, **13**:310a
allophanic, **14**:476a
anesthetic use of, **2**:27b, 31a; **5**:478b, 566a–b; **8**:401a; **10**:620a; **12**:165a, 502a–b; **13**:158a
discovery of, **7**:45b
opposition to, **9**:10b
composition of, **4**:244a–b; **5**:325a; **6**:236b; **8**:333b–338a
cyanohydric, **5**:315a
diethyl, **5**:325a; **13**:309b
discovery of, **3**:414b; **5**:478b
formula for, **1**:345b
halogens, reactions with, **3**:10b; **8**:58b–59a
hydrodynamics of, **11**:63a–b
medical use of, **10**:307a
methyl-ethyl, **3**:193b
mixed, **5**:373a; **14**:395a
muriatic, **13**:309b–310a
nature of, **3**:193b–194a; **14**:394b–395a
nitric, **13**:309b
preparation of, **1**:527b; **3**:193b, 202a, 414b; **8**:312a
reactions of, **8**:335a
rearrangement of, **3**:286b
as solvent, **11**:299a
sulfuric, **8**:520b; **13**:309b
vaporization of, **3**:9a
vinyl, **4**:553a
see also Etherification
Ether, physical
seventeenth-century theories, **2**:49a, 378b; **6**:609a, 609b–610a; **8**:152a; **10**:54b, 80a
eighteenth-century theories, **4**:44b, 481b–482a; **5**:195a; **6**:173b; **7**:228b–229a; **10**:79a–80b; **12**:118b; **14**:193b–194a; **15**:289a
early nineteenth-century theories, **2**:452a, 453b; **3**:145b; **9**:548a; **14**:490a, 565b–566a, 567a, 568a–b
mid nineteenth-century theories, **2**:471a; **3**:377b; **6**:96b, 279a–b; **8**:36b, 508b; **9**:207b–215b, 264b; **10**:27b; **13**:76b–77a, 384a–386b; **14**:208b–209a
late nineteenth-century theories, **5**:391a; **6**:346b–347a; **7**:211a; **8**:40a–b, 444a, 490b–497a *passim;* **9**:291a, 372a–373a; **13**:52b, 104a; **14**:61b, 67a, 343b
early twentieth-century theories, **2**:169b, 356a; **4**:319a–321a; **8**:181b, 497a–498b; **9**:292b, 377a; **11**:393b; **12**:69b–70a; **14**:348a
conduction of, **4**:306a
convection of, **3**:310a
density of, **4**:190a; **9**:214a,b; **10**:27b
dielectric nature of, **6**:251a–b, 252a
dragged, **8**:493b, 494a, 495a; **9**:214b, 372b; **13**:76b
elasticity of, **4**:190a; **8**:40a; **9**:209a, 213a–b; **13**:75b, 76b–77a
electric force of, **8**:495a–b
and electrodynamics, **1**:145a,b; **6**:251a–b
as electromagnetic medium, **1**:24a, 25a; **6**:346a–b; **8**:40a, 491a–498b; **13**:365a
Fresnel drag coefficient, **8**:51a, 495a; **9**: 214b, 372b; **14**:598b–599a
and gravitation, **6**:291b; **12**:118b
labile, **3**:146b
magnetic properties of, **1**:556a
and matter, **4**:319a; **6**:347a; **8**:40a–b, 493a,b, 494a–b, 497a, 508b; **9**:213b, 548a–b
and motion, **5**:20a–b; **6**:346b–347a, 453a; **8**:40a–b, 444a, 493b; **9**:215a–b
negative compressibility, **3**:146b
null-point energy, **15**:449a
opposition to, **2**:166b; **4**:319b; **8**:321b; **10**:75b, 80b; **11**:292a, 479b–480a, 580a; **12**:289a
and planetary motion, **4**:481a
resistance of, **2**:335a
stationary, **8**:494a–b; **12**:69b; **13**:104a
vibrations in, **5**:170a; **10**:54b
wave motion in, **5**:168b
Ether drift, **4**:320a–b; **8**:495b–496a; **9**:214a–215b, 372a, 373a; **13**:471b
see also Michelson, Albert Abraham, Michelson-Morley experiment
Ether pulse theory, **1**:457a–458a
Etherification, **3**:194a*; **11**:541b
continuous, **14**:395a–b
of phenol, **3**:10b
theory of, **5**:92b; **8**:333a–b; **10**:496b; **12**: 320b; **14**:394b–395b
Etherin, **8**:333b–336b, 337b–338a; **14**:601a
Etherion, **13**:454a
Ethics
ancient Greek, **4**:34a–b; **11**:24b, 29b, 104b–105b; **14**:606a, 607a
ancient Roman, **12**:310b; **14**:607b
Byzantine, **13**:308a
medieval Arabic, **11**:249b, 325a–b; **13**: 292b, 509a, 510a, 513a; **15**:262a
medieval Hebrew, **9**:31a
medieval Latin, **6**:396a; **14**:35a
Renaissance, **11**:73b; **13**:279b–280a; **14**: 47b
seventeenth-century, **1**:549a; **2**:601b; **6**: 447a,b; **8**:439a–b; **12**:357b
eighteenth-century, **3**:1b–2a; **6**:469b, 556a,b
nineteenth-century, **1**:404a; **6**:596b; **9**: 385a; **12**:570b; **13**:426b
twentieth-century, **2**:545b–546a; **5**:222a; **11**:357b–358a; **12**:177b, 178b; **13**: 272a
science and, **5**:474b*; **7**:606b
and theology
"practical divinity," **1**:134a
see also Deontology, Moral sense
Ethionic acid, **8**:335b
discovery of, **8**:334a; **9**:18b
Ethiops mineral, **11**:48b
Ethnography
ancient Greek, **6**:213a, 316a
medieval Arabic, **1**:29b, 414a; **7**:4a–b
Renaissance, **1**:48a; **6**:124a; **10**:616b

eighteenth-century, 1:489a,b; 5:77a, 428b; 7:210b, 495a,b, 544a; 8:251a–b; 10:256a; 13:392b
early nineteenth-century, 1:387b; 5:225a*; 11:137b–138a
mid nineteenth-century, 1:388a; 9:344b; 11:178a; 12:299a, 301a; 13:358a
late nineteenth-century, 2:207b–212a *passim;* 3:532a; 4:141a; 11:108a–b, 308b–309b; 12:298a; 14:203a
twentieth-century, 10:502b–503a; 12:279b; 13:166b
Arctic, 1:29b
prehistoric, 2:450a
see also specific countries
Ethnology
ancient Greek, 10:111b; 11:105a
ancient Roman, 11:39a
medieval Latin, 9:363b
Renaissance, 10:197a–b, 237b
eighteenth-century, 2:204a; 3:344b, 397a; 4:275b; 6:469a; 7:88b, 490b; 12:537a; 15:271a
early nineteenth-century, 2:585b; 6:551a; 14:328b–329a; 15:83a
mid nineteenth-century, 2:478a; 4:286a, 400b–401a; 6:280b, 593b; 9:149b*, 540b–541a; 10:209b; 12:204a–b; 14:136b, 533b
late nineteenth-century, 2:207a–209b; 3:532a; 7:31a, 149a; 10:285a; 11:119a–120a; 15:268a, 430b
twentieth-century, 2:189b–190b, 210a–213a, 505a; 3:328b; 5:196b–197a; 7:191b–192a; 9:313b; 11:210a–b; 14:291b
in psychology, 7:191b–192a
see also Anthropology; Classification, ethnological; Culture; Man; Race; specific countries
Ethnolysis
see Evolution, theories of
Ethology
eighteenth-century, 11:331b–332b
nineteenth-century, 3:492b, 521a–b, 575a; 4:401b; 5:385a
twentieth-century, 9:328a; 14:314b–315a; 15:504a–b
comparative, 14:315a
of insects, 1:329a; 4:504a–b; 5:73b; 8:528b–529a; 11:331b–332b
paleontological, 4:147b–148a
Ethyl acetoacetate, 9:354a
Ethyl radical, 8:334b–335a, 337b; 9:91b
Ethyl sulfuric acid
constitution of, 9:91b
Ethylene, 2:621a; 4:531b
composition of, 4:244a–b, 400a; 6:284b
compounds of, 2:66b; 9:18b; 10:578b
monomeric sulfides, 4:20b
oxides, 5:100a
as coolant, 14:522b
polymerization, 7:21b
Ethylene glycol
preparation of, 3:448b; 14:531a
Ethylene series
polymerization, 2:624a
hydro-dehydro, 9:608b
Etiolation, 2:286b; 10:281b; 12:59a
Etna (volcano), 1:270b; 2:312a; 10:543a; 11:263a, 336b; 12:120a, 261b, 562a
eruptions of, 2:311a; 5:89a; 7:376a, 549a; 9:192a; 11:412b
structure of, 4:350a; 8:567a, 568a, 573a,b; 11:303b
Etna Observatory, 11:412a; 13:232b
Etruscan language, 15:56a
Etten, Hendrik van (pseudonym), 8:271b*
Ettingshausen, Albert von, 15:433a, 434b
Ettingshausen, Andreas von, 6:51b; 8:596a; 9:278a; 11:343b; 14:349a, 480b
Etymology
ancient Roman, 12:540a–b
medieval Arabic, 2:155a; 7:358a
seventeenth-century, 11:317b
eighteenth-century, 11:609b
Euanthostrobiles, 8:354b
Eucalyptus, 9:618b
Eucken, Arnold Thomas (1884–1950), 4:413–414; 12:437b; 15:435a, 450b, 453a*
Euclid (*fl. ca.* 295 B.C.), 4:414–459
acquaintances of, 11:24a
algebra, 1:31a, 183b, 189a; 4:114a, 596a; 11:512a
continued fractions, 3:127b
arithmetic, 3:127b; 7:176a
astronomy, 1:339a; 4:429b; 7:81a; 9:301a(n); 10:295b
biographies of, 7:187a; 9:269a
commentaries on, 1:40a; 4:221b*; 7:300b
critics of, 11:288a
editions of works, 3:585a; 4:392b; 5:521a–b; 14:21b, 472a
epigram, 4:431b
geometry, 1:171b; 2:619b; 8:240b; 9:297b; 10:270b, 271a; 13:304b; 15:155a
conics, 1:245b; 12:432a
constructions, 1:41b; 3:277b; 9:156b; 10:181a
criticism of, 3:284b; 14:612b
influence of
in Roman antiquity, 4:159b
in Middle Ages, 1:31b, 226b; 2:233b; 3:272b, 275a, 515a; 5:365a; 9:364a; 10:474a; 14:457b
in Renaissance, 2:618a; 4:8b; 9:611b; 13:48a
in seventeenth century, 1:381b; 2:527a; 3:149a; 5:238a, 239b; 13:235b
in nineteenth century, 2:105b
influences on, 1:267a; 11:24a
mechanics, 4:431a–b; 7:341a, 342b, 344a, 345b
music, 4:430b–431a; 10:113a
number theory, 4:160a, 573b; 10:113a; 11:221a,b; 15:10a
optics, 1:81a, 379a, 475a; 4:430a–b; 5:552b; 6:190b, 193a; 10:475a; 11:201a; 15:263b, 264a–b
parallel postulate, 2:268b–269a, 270a, 274a, 310a, 331a; 3:120a*, 128a–b; 4:417a–b, 424b; 5:346b; 6:522a; 7:404b–405a; 8:141b, 429b; 12:55a–56b; 13:510a–b
critics of, 13:414b
proofs of, 3:311b; 4:441b, 451a; 5:302b; 7:79a–80a, 206b, 329b–330a, 335a, 600a; 8:432a; 10:6a–b; 11:201b; 13:290b–291a; 15:66b
porisms of, 3:78a; 5:408b
postulates, 1:251a; 4:416b–417b; 6:249a, 570a
criticism of, 5:346b
on ratios and proportions, 1:247a, 381a; 4:4a; 7:83a–b, 326b–327b; 8:100b; 9:304a
scientific method, 4:104b; 10:344a–b
translations of works, 2:231a; 3:585a; 4:5b, 546b; 13:60b, 349b; 15:261b
trigonometry, 7:405a
On the Balance, 1:224b; 4:431a–b, 442b
De canonio, 4:431a, 438b
Catoptrica (spurious), 4:430a–b, 442b, 444b; 13:322b–323a; 14:458b
On Composition, 4:438b
Conics, 1:185b, 186a; 4:427a–429a
Data, 1:474a; 4:425a–b, 438b, 442b–443a, 444b, 568b, 575b(n); 6:202b–203a, 203b(n); 10:45b, 295b; 14:274b
commentaries on, 10:300a
editions of, 6:112b; 12:446b; 13:322b
translations of, 6:271b; 7:25a; 11:458a(n); 13:400b; 15:178a
Divisions of Figures, 1:22b; 4:425b–426a, 438b, 442b, 609b; 6:315a
translations of, 3:364b
Elements, 1:215a,b, 218a, 437a; 3:176a, 281a; 4:137b, 619a; 6:275b, 412b, 416a; 7:81a, 118a, 120a, 176b; 8:189b; 10:43a; 11:222a,b; 13:315a, 316a, 319b; 15:671b
ancient Greek commentaries on, 4:415b, 437a–438b; 10:180b; 13:302b
ancient Roman commentaries on, 4:437a–b, 439b; 10:293b, 299a–300a
Byzantine commentaries on, 4:437b, 449a, 463a; 11:160b; 12:442b
medieval Arabic commentaries on, 1:82b; 3:512a; 4:437a–442a, 445a–b, 448a, 525a; 6:201b–202b; 7:79a–80b, 326b–327b, 329a, 335a; 9:21b–22a; 10:5b–6b; 11:244b; 13:289a–b, 290a–291a
medieval Latin commentaries on, 4:448a, 605a–611b *passim;* 15:177b
medieval Hebrew commentaries on, 8:280a
Renaissance commentaries on, 1:607a; 3:311b, 364b; 4:450b–451a; 9:170a*, 191b, 488b; 10:494a; 11:288a
seventeenth-century commentaries on, 4:451a; 9:13a; 14:49b
eighteenth-century commentaries on, 4:516a
nineteenth-century commentaries on, 2:274b
authorship of, 1:231b–232a; 4:423b–424b, 438b; 13:303a–304a
condensations of, 4:439b, 446a, 451b–452a; 6:299b*
constructions, 9:447a
editions of, 1:245b; 2:310a, 462b, 470b; 3:23b–24a, 24b–25a, 27b, 128b*, 621b, 622a; 4:416a, 437b–441a *passim,* 446a–452a *passim;* 6:615b; 9:311a, 555b; 11:18a; 12:446b; 13:322a–b, 509b; 14:295b; 15:154a
filiation chart, 4:458–459
handbook tradition, 4:451b–452a
influence of, 1:292a, 444b, 445a, 474a, 604b; 2:394a, 395a, 401b; 3:277a; 4:259a, 424b–425a; 5:409b; 7:360a; 8:191a, 235a–236a, 609b; 9:140b–141a, 615a; 10:8a, 45b; 11:34b; 12:11a, 91b, 93b, 445b; 13:427a; 14:19b; 15:497b, 752b
interpretations of, 3:494a; 7:242a–b; 8:267b
precursors of, 4:466a–b; 6:414b–415b
problems in, 1:605a
proofs of, 2:618b; 9:299b
scholia, 5:346b; 10:300a
spurious books, 6:616a–b; 7:29b
summary of contents, 4:416a–423b
translations of, 1:62b, 63a–b, 444a, 468a; 3:364b; 4:373b, 438a–439b, 442a–452a, 611b; 6:271b; 7:25a, 62a, 81a; 8:235b, 236a; 10:5b, 270a, 271a, 300a; 11:249a–b, 339a, 403a, 508b; 13:259b, 400b, 401b; 14:49b, 164b, 274b; 15:175a, 176b–177a, 626a
transmission of, 4:117a, 437a–452a; 6:189b, 411b; 13:544b

Euclid (cont.)
Elements of Music, **4**:430b, 438b
Heavy and the Light (spurious?), **4**:431a, 438b, 442b; **8**:215b
Introduction to Harmony, **4**:430b
Optics, **4**:415b, 430a, 438b, 442b–443a, 444b; **14**:458a, 461a(n)
commentaries on, **9**:264a; **10**:295b
editions of, **13**:322b–323a
translations of, **7**:25a
Perspective
translations of, **3**:559a*
Phaenomena, **1**:338b; **4**:415a,b, 427b, 429b, 430a, 438b, 442b–443a, 444b; **13**:319b
editions of, **9**:191a
translations of, **9**:190b
De ponderibus, **4**:431b
Porisms, **3**:214a; **4**:426a–427a, 438b, 567a; **9**:299b; **10**:295b, 296b; **12**: 446a–b
commentaries on, **10**:297a–b
Pseudaria, **4**:429a–b
On Resolution, **4**:438b
Sectio canonis (Division of the Scale), **1**:231b; **4**:430b, 444b; **7**:326a
De speculis, **4**:444b
Surface Loci, **1**:187a, 245a; **4**:427b, 429a, 568b–569a; **10**:298a
Euclid of Megara, **2**:550a; **4**:415a
in Plato, **13**:301b–302a
Euclideans, **1**:213a
Eucnemidae, **6**:507a
Euctemon (*fl.* fifth century B.C.), 4:459–460; **9**:339a,b(n); **10**:180a; **11**:189b, 197b
Eudemus of Cyprus, **4**:460b
Eudemus of Pergamum, **1**:179b; **4**:108b
Eudemus of Rhodes (*fl.* fourth century B.C.), 4:460–465; **1**:22b, 252b, 268a; **6**:408a
astronomy, **11**:223b, 224a
Eudemian summary, **4**:463a; **6**:410b, 411b, 414a–b
historiography, **4**:103b, 415b, 489b; **8**: 192b(n); **9**:268b; **10**:180b; **11**:30a(n); **13**:296a–b, 302a,b, 314b, 326b, 334a
History of Astronomy, **4**:463b; **10**:179b
History of Geometry, **4**:463a,b; **6**:411b, 412a, 412b–413b
Eudiometer, **5**:57a; **7**:160a, 620b–621a; **9**:5b; **12**:561a; **14**:75a
Eudiometry
eighteenth-century, **3**:153b, 157b; **6**:550b; **7**:12b; **11**:145a; **14**:74b–75a
nineteenth-century, **1**:586a; **2**:588a; **5**: 318b; **9**:530a
Eudoxus of Cnidus (*ca.* 400–*ca.* 347 B.C.), 4:465–467; **4**:391a; **8**:189a
at Academy, **11**:27b–28b
acquaintances of, **11**:22b
Archimedes on, **1**:220a, 221b; **6**:416a
astronomy, **1**:214a; **11**:188a
calendar, **4**:172a, 460a; **5**:345a; **12**:547a
cosmology, **1**:199a, 252b, 257b–258a, 338b; **3**:21b; **4**:463b; **5**:106a; **9**:269b; **10**:529b, 590b; **11**:28a, 29b, 30a(n); **12**:3a; **13**:326b; **15**:35a
axiom of, **1**:215a; **7**:79a,b
critics of, **11**:28a; **12**:5a
editions of works, **3**:21b
and Euclid, **4**:414b, 415a, 423b–424a, 429b, 433a(n); **13**:303a
geography, **4**:82a
influence of, **11**:225a; **13**:84a, 326a
mathematics
axiomatization, **10**:344a,b
cube duplication, **4**:105a, 489a–b
exhaustion, method of, **1**:215b, 257a, 444b; **4**:104b
geometry, **6**:314b; **9**:270a, 274a, 300b; **13**:319b
proportionality, **4**:4a, 441a, 446b; **7**: 326b; **13**:304a; **14**:611b
Phaenomena, **1**:205a
commentaries on, **15**:216b
students of, **9**:269a
Euganean Hills, **1**:234b
Eugenics
nineteenth-century, **2**:375a; **5**:266a–b
twentieth-century, **3**:91b, 122b, 564b, 590b; **4**:271b–272b; **5**:7b, 11a, 294a, 598b; **9**:513b*, 565a,b; **10**: 464a–b
Eugenius of Sicily, **11**:200b, 203a
Eugenol, **3**:279b
Euglena, **3**:337a
Euler, Christoph, **4**:472a,b
Euler, Johann Albrecht, **4**:472a,b, 473a, 483a*; **7**:563b; **8**:299a
Euler, Leonhard (1707–1783), 4:467–484; **2**:328b; **3**:215b, 384b, 386a; **4**:485a; **6**:305a; **13**:140b*; **14**:84a
acoustics, **15**:306b
acquaintances of, **1**:67a; **3**:459b; **5**:209b; **7**:579b, 597a; **8**:299a; **9**:187b; **15**:524b
and d'Alembert, **1**:113b, 114a, 117a; **4**: 475a, 481a
algebra, **4**:475a–b, 612b; **5**:204a; **7**:508a; **10**:495b; **13**:220b
continued fractions, **3**:128a; **4**:474b, 475a, 476a, 478b; **7**:565a
Fermat's last theorem, **5**:375b
series, **2**:38a, 50a, 57a; **4**:475a–478b *passim;* **9**:500a
summation, **2**:49b, 53a; **4**:475b–476a, 476b
theory of equations, **2**:112b
analysis, **2**:271b; **4**:473b–480b, 516a; **7**: 52a–b, 514b, 550a, 566b, 568a, 619b; **8**:139b, 140a, 384a; **9**:127a, 410a, 471a; **10**:574a; **11**:400b, 599a; **13**: 238a; **14**:149a; **15**:293a,b, 368b
calculus, **1**:113b; **2**:48b; **3**:228a, 282b, 385b, 432a; **4**:469b, 470b; **5**:54b; **15**:278a, 383a
theory of special functions, **1**:207a; **3**: 137a,b; **4**:229a, 477b; **5**:572b; **7**:567b
arithmetic, **3**:223b
ballistics, **4**:470b; **11**:494a
and Bernoulli (Daniel), **2**:36b, 38b–39a, 41a; **4**:468b–481a *passim*
biographies of, **8**:559a
celestial mechanics, **2**:578b; **4**:471a, 480a–b, 481a–b; **7**:564b, 580a; **9**:226a; **15**:289b
lunar theory, **3**:283a; **4**:478b, 481a–b; **6**:399a; **7**:563b; **8**:179a; **13**:424b
combinatorial analysis, **6**:403b; **8**:617a
commentaries on, **2**:467b
on contemporary scientists, **14**:180a
correspondence of, **3**:163b, 216a, 282a; **4**:129a; **8**:469a; **11**:65b, 609b; **12**: 283a; **13**:39b, 68a, 183b
critics of, **3**:345a; **7**:619a; **8**:306a; **11**:496b; **15**:290a
editions of works, **7**:55a; **8**:560a; **10**:573a; **11**:589a–b; **12**:599b; **14**:158b
equations of, **4**:479a, 480b; **8**:504a; **11**: 327a
expansion of, **13**:250a; **15**:348b
Euler angles, **4**:479b, 480a
Euler-Lambert theorem, **7**:481b; **13**:139a
Euler-Maclaurin formula, **4**:475b–476a; **10**:249b; **12**:535a
Euler-Poincaré formula, **11**:60a
Euler substitution, **4**:478a
Eulerian integrals, **1**:15b, 591b; **4**:2a, 478a; **8**:137b, 142a
Eulerian transformations, **2**:50a; **4**:479b
Euler's constant, **1**:54b; **4**:476a; **9**:156a
value of, **12**:345a
Euler's formula, **4**:477a, 515b
Euler's function, **4**:474a; **5**:414a
Euler's theorem, **3**:169a
geometry, **1**:591a; **2**:288a; **4**:479b, 480a; **5**:508b, 567a; **8**:306a; **9**:343a, 475a,b; **12**:171b, 283b; **15**:482b
and Goldbach, **5**:449a
hydromechanics, **2**:40b, 43b; **3**:591b; **4**: 469b, 477a, 480b–581a; **6**:249a–b; **12**:283a
influence of, **1**:13a, 200b, 239a, 240a, 326b; **2**:111b; **3**:227a; **4**:340b; **5**:299a, 375a; **6**:307a, 452a; **8**:137b, 299a–b, 469b; **10**:573a–b; **11**:344a; **12**:283b; **15**:365b, 383a
Koenig dispute, **4**:471b; **7**:443a
and Lagrange, **7**:560a–567b *passim,* 571a
latitude variation, **3**:194b; **7**:369b
logic, **13**:612b
magnetism, **9**:207b
mechanics, **2**:38b–39a; **3**:226b; **4**:228b, 479b–480b; **5**:96a; **11**:62a, 496b; **12**:283b
elasticity, **2**:40b; **4**:480b; **15**:164b
number theory, **3**:223b, 228a, 302b; **4**: 124a, 474a–475a, 515b, 550a, 573a; **5**:342a, 342b–343a, 450a,b; **6**:214b; **7**:565b, 566a; **8**:138b; **11**:256b; **13**: 218a, 219a; **14**:438a
optics, **4**:469b, 482a; **5**:167a,b; **7**:404a; **14**:565a,b
resonance theory, **1**:166a
students of, **11**:609a–b
telescope, **3**:284a; **4**:148b
tidal theory, **2**:43a; **4**:591a; **8**:611a
translations of works, **11**:610a*
trigonometry, **7**:405a
on vibration, **2**:41a–b, 43b, 101b; **3**:145b; **4**:88b, 125b, 476b–477a, 478b; **7**: 561b
vortex theory, **3**:444a
zeta function, **3**:227b; **4**:475a; **11**:453a
Algebra
see Euler, Leonhard, *Volständige Anleitung zur Algebra*
Dioptrica, **4**:472b, 473a, 482a
Institutiones calculi differentialis cum eius usu in analysi finitorum ac doctrina serierum, **2**:50a; **4**:471a, 477a,b; **10**:327b
Institutiones calculi integralis, **4**:472b, 478a, 479b; **15**:383a
Introductio in analysin infinitorum, **3**: 460a; **4**:471a, 476b, 477a, 479b; **7**: 50a; **10**:572b
editions, **3**:119b
errors in, **11**:45a
Lettres à une princesse d'Allemagne sur divers sujets de physique et de philosophie, **2**:166b; **4**:471a–b, 482a
editions of, **9**:566b
Mechanica sive motus scientia analytice exposita, **4**:479b–480a
Methodus inveniendi lineas curvas maximi minimive proprietate gaudentes, **4**:479a, 480b; **7**:560b; **10**:3a; **15**:276b
Nova theoria lucis et colorum, **4**:482a
Scientia navalis—Théorie complette de la construction et de la manoeuvre des vaisseaux, **4**:472b, 480b
Solutio facilis problematum, **4**:602a
Theoria motus corporum solidorum seu

rigidorum ex primis nostrae cognitionis principiis stabilita, **4**:480a
Theoria motuum lunae, nova methodo pertractata, **4**:472b, 481b; **8**:299a
Vollständige Anleitung zur Algebra, **3**: 276b, 284b–285a, 461a–b; **4**:472b, 474b, 475b; **5**:210a
editions of, **7**:565b, 566b
Euler, Ulf Svante von, **4**:485b; **15**:74b
Euler-Chelpin, Hans Karl August Simon von (1873–1964), 4:485–486; **2**:561b; **4**: 358a; **6**:111a,b; **14**:45b
Eunapius, **10**:208a, 230a,b
Eupatorium, **10**:557a
Euphonium, **3**:258a
Euphorbiaceae, **7**:196b; **8**:512b
Euphrates (river), **15**:637b–638a, 639a
Euplotes, **13**:268a,b
Eureka, Nevada, **6**:14a; **7**:5b*
Europäische Gradmessung, **5**:258a; **6**:239b
Europe
and Africa, geodesic union of, **7**:2a
botany, **5**:546a
ethnology, **1**:562b–563a
frontier with Asia, **5**:428b
geography, **7**:4a; **9**:581a
geology, **1**:551b; **13**:621a
prehistory, **9**:490a
triangulation of, **7**:209b
Europium
discovery of, **2**:254b
Eurotium, **6**:73a
Eurypterids, **11**:594a
Eusebius of Caesarea, **4**:391b; **14**:631b
Chronicle, **1**:565a
Ecclesiastical History, **1**:148a–b
Eusporangiateae, **3**:31a
Eustachi, Bartolomeo (*ca.* 1500–1574), 4: 486–488; **3**:41a, 342a,b; **6**:509a; **14**: 11a
anatomical tables, **7**:614a
anatomy, **1**:316a, 480b; **4**:508b, 520b; **10**:477b; **12**:577b
editions of works, **2**:226a; **15**:4a
Eustachian tube, **4**:487a
Eustachian valve, **4**:487a; **9**:323b
influence of, **11**:33a; **13**:566b
Tabulae anatomicae
editions of, **10**:266b
Eustacy, **4**:39a
Eustathius
see Astat
Eustratius of Nicaea, **1**:270a
Euthanasia, **10**:613b
Euthymia, **4**:34b
Eutocius of Ascalon (*b. ca.* A.D. 480), 4: 488–491; **3**:272b, 277b; **4**:609a; **5**: 346b; **10**:299a; **14**:458b
commentaries by, **1**:170a, 189a, 191b, 223b–229b *passim;* **3**:364a,b; **4**: 105a,b, 108b, 109a; **6**:412a; **7**:29a; **9**:270a–274b *passim,* **10**:586b, 587b–588a; **12**:314b*, 579b; **14**:274b; **15**: 115b, 116b, 118a
commentaries on, **9**:437b
geometry, **1**:227b; **4**:391b, 392a, 463b; **6**:407a, 411a; **10**:115b, 588a
Introduction à l'Almageste, **15**:178b
translations by, **15**:261b
Euxanthic acid, **4**:395a
Euxanthone, **5**:489a; **7**:471a
Evadne, **8**:519a
Evangelicalism, **2**:569a
Evans, Alexander William (1868–1959), 15: 148–151
Evans, Alice, **6**:135b
Evans, Frederick John Owen (1815–1885), 4:491–492
Evans, Griffith, **4**:548a
Evans, Herbert McLean, **9**:56b; **12**:473b–474a
Evans, Lewis (1700–1756), 4:492–493; **1**: 486b; **5**:434b; **9**:537b
Evans, Ralph Merrill, **9**:202a
Evans, William Harry (1876–1956), 4:493–494
Evaporation, **1**:495a; **6**:342a; **8**:72b; **11**:278b; **13**:11a
atmospheric, **4**:28b
definition of, **13**:494b
electricity produced, **13**:237a; **14**:70b, 74a
law of, **11**:432b
of odorous substances, **4**:76b
and surface tension, **13**:11a–b
theories of
Renaissance, **1**:495a
eighteenth-century, **8**:255b–256b; **14**: 75b–76a
nineteenth-century, **3**:306b, 540a–b
twentieth-century, **8**:301a
of water, **1**:495a; **2**:178b; **3**:304b; **8**:72b, 256a
see also Cooling, Vaporization
Evaporites, **10**:171a
Evapotranspiration, **9**:150b
Evdokimov, Nikolai Nikolaevich, **13**:114a
Evection (astron.), **2**:349a; **6**:516a; **15**:349a, 353b, 354a
Evelyn, John (1620–1706), 4:494–497; **7**: 485b, 537a; **10**:202b*, 491a; **12**:584b; **14**:369b, 375b(n)
on contemporary scientists, **6**:141b; **8**: 131a; **10**:566b*; **12**:586a; **14**:379a(n)
and Royal Society, **2**:523a; **14**:510b
Eveninglast (astron.), **15**:673a, 679a
Everett, Ebenezer, **13**:369b
Evergreens
chemical composition, **8**:344b
Evermann, Barton Warren, **7**:170a
Evershed, John (1864–1956), 4:497–498; **11**:297a; **12**:72b; **13**:96a*
Evershed effect, **4**:498a
Evershed, Sydney, **4**:497b
Evesham, Walter of
see Walter of Odington
Evidence, theory of, **2**:604b–605b, 607b
Evolutes
theory of, **6**:600b, 605a
Evolution, **3**:44b, 148a–b, 237a, 390a, 398b; **7**:286a
adaptive vs. inadaptive, **7**:480a
convergent, **5**:385b
creative, **2**:10b–11b; **8**:257a–b
degenerative, **4**:310a
and ecology, **5**:70a–71a
and embryogeny, **1**:71b–72a, 421a; **5**: 447a; **6**:65a–b; **7**:475a; **12**:337b–338b; **14**:426b; **15**:169b–170a
environmental effects, **1**:163a–b; **2**:321a, 347a, 498a; **3**:317b, 318a; **4**:512b; **5**:356a, 357b; **7**:170a, 590b, 591a,b; **10**:209b–210a; **11**:567a–b; **12**:571a, 588a; **13**:276a, 460b–461a; **14**:313a; **15**:93a
experimental study, **8**:513a
of fishes, **3**:610b; **7**:169b–170a; **12**:337a
genealogical reconstruction, **1**:505b
and geology, **5**:357b; **6**:120b; **14**:215a
and glaciation, **1**:73b; **5**:513a
and heredity, **1**:505a–506a; **3**:121a–122a, 590a; **5**:10b–11a, 385b–386a; **7**:114b; **9**:519b–520b, 523b, 618b; **12**:417a; **14**:99a, 238a, 432b, 433b–434a
histological, **14**:596b
of horse, **3**:573a; **5**:46b; **6**:595a–b; **7**:480a; **9**:134b
and hybridization, **8**:513a–b; **9**:618b
and idealism, **1**:73a–b
and immunity, **15**:507a
inorganic, **6**:565a; **8**:441b–442b, 464a, 522a–b, 616a; **11**:135b–136a
physical geography, **1**:490b; **3**:434b, 594a
of invertebrates, **10**:29b
irreversibility of, **4**:148a
isolation, role of, **11**:519a
logical demonstration of, **4**:101b–102a
of mammals, **2**:504a–505b; **5**:46b; **6**:536a; **7**:480a; **12**:38a–b
of man, **2**:347a; **3**:390a, 553b, 556a, 573b–574b; **5**:221b–222a, 354a*; **6**:9a–b, 469a, 527b–528a; **7**:279a, 590b; **8**: 125b, 574a–b; **10**:290b; **13**:275a–276b, 504a; **14**:136b–138a, 439b; **15**:93a, 140a, 140b–142a, 144a–b
meaning of word, **5**:357a; **6**:591a
and mutations, **5**:453b
neurological, **4**:353b; **5**:172a, 208a; **7**:48b; **8**:534b
and paleontology, **1**:473a, 491b–492a, 587a; **3**:569b–570a, 572b–573a; **4**: 147b–148a; **5**:295b–296b, 357b; **7**: 411b–412b, 480a–b, 588b–589b; **8**: 486b; **9**:134b; **12**:38a–b; **14**:409b, 626b
parallel, **3**:30b; **5**:359a,b; **15**:148a
physiological, **12**:470a
of pig, **7**:480a
of plants
medieval discussion, **1**:101b–102a
nineteenth-century discussion, **2**:158a; **3**:29b; **5**:513a–b; **6**:296a; **7**:90b–91a; **8**:513a, 354b; **9**:112a, 277b–278a, 278b, 281a; **11**:154a; **12**:104b, 256b, 259b–260a; **13**:89b, 154b, 405a; **15**: 147a–148a, 268b
twentieth-century discussion, **2**:371b–372a; **3**:317b, 318a; **5**:498a; **9**:24b; **13**:165b; **15**:149b
prehistory of, **2**:576a–581b
of protozoa, **4**:142b
psychological, **12**:571a–b; **13**:9a
rate of, **3**:573a; **7**:99a
religion and, **3**:573b–574a; **5**:16b–17a; **8**:98a, 123a–b; **9**:143a, 389a–b, 533b–534a; **12**:45b, 570a–b; **13**:274b–276b, 542a; **14**:137a–138a, 439b
and sexual reproduction, **4**:271a
and spontaneous generation, **1**:496b, 497a
statistical studies, **3**:590b; **10**:450a, 461a–462a; **14**:251a–252a, 574a
stellar, **1**:353a; **2**:221a; **3**:470b; **4**:280a; **5**:272b; **6**:30a; **7**:86a; **8**:2a, 442a–b; **12**:18a,b, 19a–b, 20a, 22b, 280a, 367b; **13**:118b, 120b, 475a; **15**:449a
meteoritic hypothesis, **6**:543a
study methods, **1**:506a; **15**:169a–170a
theories of
ancient Greek, **1**:151a
medieval Latin, **1**:199b
eighteenth-century, **2**:204b; **3**:579b; **4**: 512b; **6**:585a–b; **7**:607a; **9**:458a, 533b–534a; **10**:284a; **14**:133a, 297a
early nineteenth-century, **2**:321a; **3**: 565b; **5**:192a–b, 356a, 356b–357b; **6**:120b; **7**:585b–593a *passim;* **10**: 287b; **11**:262b; **12**:156b; **13**:460b–461a; **14**:253b; **15**:127b
mid nineteenth-century, **1**:501b, 502a–504a; **2**:205b–206a, 498a; **3**:192a–b, 569b–575a; **5**:296a–b, 359b; **6**:7b–10b, 138a, 222a–b, 238b, 490a–491a, 592a–594a; **7**:93b, 320a, 384a; **8**: 169b, 574a–b, 596a; **9**:69a, 280b–

Evolution (cont.)
281a, 282b–283a, 332a, 559b, 560b–561a, 601b, 618b; **10**:209b–210a, 212a, 262b–263a, 381b; **11**:270a, 567a–b; **12**:167b, 318a–b, 570a–571b; **13**:542a; **14**:134b–139a, 189b–190a; **15**:166a–170a
late nineteenth-century, **1**:131b; **3**:470b, 553a–b; **4**:266b; **5**:385b–386a, 599b; **6**:140a, 613a–614a; **7**:161a, 166b, 169b–170a, 371a, 480a, 498b, 511b; **8**:3b, 27a, 123a, 354b, 604a–b; **9**: 432a; **10**:241b, 273b, 427b, 463a, 485a, 522b–523a; **11**:308b, 517a–519a, 571a–b; **13**:8b–9a, 504a; **14**: 181b, 234a–238a, 251b–252a, 350b; **15**:92b–93a
early twentieth-century, **1**:470b; **2**: 505b, 10b–11b; **5**:213a, 467b; **6**: 53a–b; **7**:91a, 99a, 114b–115a; **8**: 513a–b; **9**:513a–b, 519a–520b, 523b; **11**:19b–20a, 257b, 535b; **12**:337a–338b; **13**:165b, 274b–275a; **14**:291b, 314b, 434a–b; **15**:507b–509b
mid twentieth-century, **5**:467b; **11**: 321a; **12**:410a–b; **13**:275b–276b
cosmolysis, **2**:190a–b
and developmentalism, **1**:336b–337a
emergent evolution, **6**:260b; **9**:513a–b
ethnolysis, **2**:189b–190a
foreshadowings of, **5**:424b; **6**:466b; **9**: 93a; **12**:514b; **13**:564a
gastraea, **1**:421a
genetic, **4**:271a; **13**:136b, 137b
and geology, **3**:471a; **6**:238b
historiography of, **8**:517b
idealistic, **1**:70b–71a, 73a–b
morphobiological, **12**:337b
mutation-selection, **2**:190a
opposition to, **1**:74a, 211a, 255a; **2**: 213a; **4**:291a, 504b; **5**:32a, 67b; **6**: 454b; **8**:49a–b; **10**:322b; **13**:154b
orthogenesis theory, **14**:314b
phrenology and, **5**:255a
physiological, **11**:571a
preadaptation, **3**:493a
recapitulationism, **1**:387b; **3**:519b; **5**: 356b–357a; **6**:8b, 338a, 613b–614a; **7**:367b–368a; **9**:252b; **12**:316a, 337b, 338a; **15**:92b
teleological, **13**:275b–276b
transformism, **1**:130a,b, 387a–b; **4**: 309a; **5**:385b
uniformitarianism, **1**:130a, 413b; **2**: 556a, 570b; **3**:162a, 392b, 434b, 552b, 594a; **4**:72a; **5**:31b–32b, 205a; **6**:187a
see also Neo-Darwinism
vertebrate, **2**:505b; **5**:283a–b, 467b; **6**: 535b–536a; **7**:61a–b; **9**:346a; **10**:241b; **11**:361a; **12**:337a, 409b; **15**:123a, 513b
see also Darwin, Charles Robert; Darwinism; Lamarckism; Natural selection; Neoteny; Phylogeny; *Scala naturae;* Variation, biological
Ewald, Julius Richard, **1**:446b; **5**:463a; **9**:20b
Ewald, Paul Peter, **3**:564a; **4**:522a; **6**:303a; **7**:468b; **8**:51a,b, 473a; **10**:125a; **12**: 529b
Ewing, Alfred Cyril, **7**:234a
Ewing, James (1866–1943), 4:498–500
Ewing's sarcoma, **4**:499b
Ewing, James Alfred (1855–1935), 4:500–501; **7**:94a*; **9**:407a; **11**:549a
magnetism, **6**:480a; **14**:244a–b
metals, **13**:246a
Ewing, John, **9**:165a
Excerpta medica, **12**:207b
Exchange integrals, **8**:474a,b
Excitability (physiol.), **4**:264a; **5**:44b; **14**: 105b
central, **12**:399b
membrane theory, **15**:21b
theories of, **8**:102a–b, 534a–b
time course of, **15**:21a–b
Exciton theory, **5**:118a, 160b–161a
Excluded middle, law of, **5**:350b; **8**:546a
Excluded third, principle of, **2**:512b, 513b–514a; **12**:364b
Exclusion, method of (math.), **5**:159b
Exclusion principle (physics), **2**:247b, 248a; **4**:578a; **10**:422b–423a; **15**:49b
Exeligmus, **5**:346a
Exercices de mathématiques, **3**:134a, 135a
Exercise, physical, **9**:41a
Exhaustion, method of, **1**:164a, 171b, 215a, 215b–218b, 227a,b, 444b, 445b; **2**: 550a; **3**:150a, 151b, 152a; **4**:104b, 422a, 422b–423a, 424a, 466a,b; **5**: 527a; **7**:116b; **13**:290a, 292a; **14**: 150a–b
approximation method, **1**:215b, 217b–218b
compression method
"difference" form, **1**:215b–217a
"ratio" form, **1**:215b–216a, 217a–b
forerunners of, **12**:375b
Exhibitions
see under specific cities
Exhumations, **10**:326a
Existence (math.), **2**:512b; **4**:4a; **5**:204b; **6**:3b; **11**:60a, 408b; **12**:246b; **14**:91b
theorems of, **2**:198a, 271b; **3**:57a, 331a; **5**:306b; **7**:398a; **11**:452a; **12**:599b
Cauchy-Kovalevsky, **3**:143b; **7**:478a–b
Existentialism, **3**:112a; **12**:156b
opposition to, **11**:363a
Exner, Felix Maria, **6**:97b; **9**:108a
Exner, Franz Seraphin, **6**:163b, 355a; **8**:511a; **12**:218a, 496a
Exner, Sigmund, **2**:446a, 531a; **5**:174b
Exobasidium, **14**:93b
Exogamy, **5**:145a
Exothermic reactions, **2**:69a
Exotic blocks (geol.), **12**:140a
Exotoxins
discovery of, **3**:189a
Expansion
coefficients of, **14**:360b
of gases, **3**:304a, 305a,b; **5**:78b, 318a; **6**: 244a–b; **7**:160a, 182a,b; **9**:19a; **11**: 111b, 353b; **13**:381a; **15**:357b, 377b
of liquids, **2**:136a; **4**:240a
thermal, **1**:612b; **3**:599b; **4**:241a; **5**:318a; **6**:477b; **9**:14a; **11**:472a; **15**:312b, 315a
biological, **4**:615a, 616a
of water, **13**:351a
Expansion and contraction, doctrine of, **5**: 47b
Expansions (math.), **3**:223a, 229b, 230a; **4**: 474b–475a, 476a; **5**:299a, 521a; **9**: 311a; **11**:207a; **15**:370a
asymptotic, **2**:144a–b; **6**:90a; **14**:188a
binomial, **2**:56b; **5**:528a–b; **15**:371b
Fourier, **2**:144b; **4**:407b, 476a; **5**:96a; **6**: 90a; **10**:573a
Laplace, **7**:435b; **15**:295a
orthogonal, **4**:561b
periodicity of, **3**:228b
Taylor's, **2**:275a, 276b; **5**:521a; **8**:140b, 366b; **10**:573b; **13**:265b–266b; **15**: 198b, 384b
trigonometric, **5**:96a,b, 97b; **13**:422b
Weierstrass, **2**:304a
see also Series, expansion in
Expectation, mathematical, **2**:319a; **3**:230b, 231a, 461b; **14**:466b–467a; **15**:280b, 285a, 374a
and moral value, **2**:42a, 50a
Expeditions, scientific
late nineteenth-century, **5**:83b
anthropological, **2**:212b; **9**:144a
archaeological, **2**:549b
astronomical
seventeenth-century, **3**:103a; **11**:423a–424a
eighteenth-century, **2**:237a; **3**:106b*, 198a, 396b; **6**:69b, 234a; **7**:579b; **9**: 164a,b; **10**:615a–b; **11**:609b; **14**:452b
mid nineteenth-century, **2**:533a; **10**: 480a, 482b; **12**:498b–499a, 579a; **13**:111b, 120b, 422b
late nineteenth-century, **2**:333a; **3**:35a, 261a; **4**:179b, 180a, 498a; **5**:196a, 404b; **6**:49a, 105b, 119a, 149b, 471a, 523a; **7**:270b; **9**:184a, 564a; **10**:34b, 119b, 219a, 543b; **11**:297a; **12**:282a; **13**:115a, 232b, 481b, 500b; **14**:242b, 449a, 557b; **15**:493b
early twentieth-century, **3**:36a, 509a, 510a; **4**:219a, 270a, 278a, 280b, 498a; **5**:102a, 182a,b, 363b–364a; **6**:508a; **9**:504a; **10**:31b, 321a, 526a; **11**:412b, 542b; **12**:449a; **13**:95b
biochemical, **1**:454a,b
biological
nineteenth-century, **3**:17a
botanical
seventeenth-century, **13**:442b
eighteenth-century, **1**:486b–488a; **3**: 16b, 177a; **7**:210a–b; **9**:432b–433a; **12**:326b, 327a, 515b; **15**:429b, 470a–471a
nineteenth-century, **2**:320b; **6**:6b, 492a; **15**:130b
twentieth-century, **3**:188a, 318a; **12**: 328a; **15**:506a
circumnavigational, **4**:256a; **5**:224b; **8**: 101a, 187a, 418a; **9**:42b, 43a, 588b; **11**:242b–243a; **12**:550a–b
epidemiological
nineteenth-century, **5**:219b
geodetic
seventeenth-century, **3**:103b
eighteenth-century, **2**:343a–b; **3**:39a, 105a, 107b, 281b; **5**:435a–b; **7**:183a; **9**:186b; **10**:256a; **13**:69b; **15**:270a–271b
nineteenth-century, **1**:200b, 201b; **2**: 135b–136a; **5**:329b; **13**:111a
geographical
Renaissance, **1**:573b
eighteenth-century, **2**:14b; **5**:428a–429a; **13**:28a–b
nineteenth-century, **1**:173a; **4**:400b; **7**: 510b; **11**:108a–b; **12**:301a
twentieth-century, **2**:374a
geological
nineteenth-century, **2**:106a, 341b, 588a; **3**:236b, 529b–530a, 532a–b; **4**:65b; **5**:211a; **6**:78b; **7**:319b; **10**:240a; **11**: 133b
twentieth-century, **4**:563b; **6**:229b; **10**: 128a–b
geophysical
nineteenth-century, **2**:430b–431a; **5**: 78b–79a; **6**:327a
hydrogeological, **1**:162a
natural history
eighteenth-century, **1**:434a–b; **5**:74b, 428a–429a
early nineteenth-century, **2**:320b; **4**: 288b–289b, 406b–407a; **5**:276b–277a; **8**:266a; **10**:439b

mid nineteenth-century, **1**:387b–388a, 501a–b; **5**:481a; **10**:221a; **11**:539a–b, 541a
late nineteenth-century, **4**:144a; **7**:314a
oceanographic
mid nineteenth-century, **4**:291a–b; **7**: 545b; **13**:360a–b
late nineteenth-century, **1**:162a; **2**:557b; **3**:89a; **5**:52b; **9**:443a; **10**:522b; **11**: 113b; **12**:166b; **13**:360b–361a, 552a
early twentieth-century, **2**:373b; **3**: 203b–204a; **4**:42b; **10**:503a; **12**:190b, 411b–412a; **13**:166b–167a
mid twentieth-century, **10**:564a; **12**: 414b; **13**:167a; **14**:634b–635a
paleontological
nineteenth-century, **9**:134a
twentieth-century, **2**:489b
physiological, **6**:24b
polar, **9**:155a
purposes of, **1**:203a
zoological
nineteenth-century, **2**:423a; **5**:406b; **14**:203a
twentieth-century, **11**:19b
see also Egypt, Napoleonic expedition to; Exploration; Lewis and Clark expedition
Experientialism
ancient Roman, **1**:334a
medieval Latin, **1**:380a–381a
Renaissance, **4**:209b; **5**:106a–b; **14**:581a; **15**:45b, 68b
seventeenth-century, **15**:69a
nineteenth-century, **11**:573b–574b; **14**: 434b
twentieth-century, **12**:417b
Experiment
ancient Greek use of, **4**:383b; **6**:426b–427a; **13**:94b
ancient Roman use of, **11**:201a
in Middle Ages, **1**:380b; **5**:552b; **6**:190b–191a, 195a; **7**:215b–216a; **14**:388b
in Renaissance, **2**:143b; **5**:249b–250a; **8**: 198b–199a
in seventeenth century, **1**:317b, 373a, 375a,b, 376a, 567b; **2**:309a–b, 377b–381b; **3**:116b; **4**:53b, 54a–b; **5**:247a; **6**:450a; **7**:524b; **9**:25a–b, 318a, 369b, 485b; **12**:103b–104a
in eighteenth century, **2**:286a, 287a; **8**: 81b; **9**:596a; **12**:309a; **13**:563a
critics of, **3**:114a–115a; **9**:306a
in nineteenth century, **2**:26b–27a, 31b–32a; **3**:581a; **6**:226b; **14**:423a, 434b; **15**:479b
in twentieth century, **3**:390a; **4**:101a; **5**: 8b–9a, 594b; **7**:454b; **9**:516b–517a, 524a–525b; **10**:125b
design of, **8**:133b; **14**:553a–554b
multifactor experiments, **5**:8b
historical repetition, **2**:572a,b; **5**:32b–33a, 246a; **9**:202a; **10**:55b
see also Thought experiments
Experimentalism
see Experientialism
Experimentum crucis, **6**:292a; **9**:52b; **10**:53b, 54a, 55b, 56b, 315a
Explanation, scientific
ancient Greek, **1**:251b, 254b, 589a–b; **3**: 21b; **13**:297b
ancient Roman, **11**:42a
medieval Latin, **10**:173b
early seventeenth-century, **1**:100b, 588a–589b; **5**:243a, 247a
late seventeenth-century, **11**:506b, 507b
late eighteenth-century, **3**:382b; **11**:146b–147a
early nineteenth-century, **1**:141a–142a
mid nineteenth-century, **3**:452a; **7**:234b; **13**:375b–377b, 384b
late nineteenth-century, **14**:434b
early twentieth-century, **2**:208b; **3**:32b–34b; **4**:374a; **15**:423a–424a
Chinese, **12**:385a–b
equivalence method, **13**:375b–377b
see also Method, scientific
Exploration
ancient Greek, **11**:225a–b
Renaissance, **3**:423b; **6**:20a–b; **10**:617a–619a; **14**:127b; **15**:87b–90a
early eighteenth-century, **4**:602b–603a; **5**:428a–429a; **7**:200a–b
late eighteenth-century, **1**:434a–b; **2**: 342b–343a, 530b; **3**:365a–b, 396b–397a; **5**:75a–b, 76b–77a; **8**:147b, 274b; **10**:117a, 283a–b; **12**:516a
early nineteenth-century, **5**:224b; **7**:89b; **8**:265a–b; **9**:149a; **12**:49a, 133a, 578a–b; **14**:187a; **15**:82a
mid nineteenth-century, **1**:572a, 573a; **2**:430b; **3**:549b, 557b, 565b–566a; **4**:491b–492a; **5**:225a, 265b, 470a, 478a, 512a,b, 611a; **6**:186b, 326b, 489a, 524a, 552a; **10**:240a–b, 439b; **11**:118b; **12**:299a; **13**:65b; **14**:133b–134b; **15**:130a
late nineteenth-century, **1**:585b; **2**:207b; **3**:439a, 532a–b, 536a; **4**:361b; **5**:396a; **6**:215a–b; **9**:587b, 617b; **10**:148a–b, 167b; **11**:108b, 180b–182a; **13**:525a–b; **14**:551b; **15**:91b, 430a, 430b–431a
twentieth-century, **3**:203b–204a, 533a, 591a, 616b; **4**:193b; **9**:589b; **11**:210a–b
geological, **2**:500a–501a; **9**:198a
polar, **1**:573a; **2**:207b; **3**:203b–204a, 617a; **5**:470a; **9**:123b–124a, 198a, 589b; **10**:148a–b; **11**:554a–b; **12**:279b; **14**: 50a–b
see also North Pole, exploration of; South Pole, exploration of
see also specific countries, oceans, regions
Exploring Expedition, U.S. (1838–1842), **3**:549b; **5**:478a, 512a,b
Explosions
emission spectra, **4**:79a
gaseous, **4**:130a–b; **8**:118a
kinetics of, **2**:70a; **3**:197a; **4**:130a–b; **5**: 272b–273a; **6**:404b
theory of, **7**:617b
Explosives, **2**:64a, 455a; **4**:78b; **5**:128a, 622a; **8**:118a; **13**:429a
ammonium nitrate, **9**:59b
chemistry of, **1**:390a; **2**:70a, 455a; **4**:239a–b; **5**:315a
liquid oxygen in, **3**:299a
production of, **5**:465b; **8**:553b; **10**:132b–133a
see also specific chemicals
Exponents
medieval use of, **2**:392b, 393b; **7**:241a
Renaissance use of, **3**:273a; **13**:60a
seventeenth-century use of, **7**:194a
fractional, **10**:224a–b
laws of, **4**:112a; **15**:99a
notation, **1**:90b; **5**:410a*; **7**:194a; **14**:180b
see also Powers (math.)
Expositions
concept of, **15**:50b
see also under specific cities
Extension, analytic, **4**:547b; **14**:303b
Extension of bodies, **1**:256b; **7**:135a; **10**:61b; **11**:324a; **13**:190a,b, 196b–197a, 200a
Extensometer, **3**:118b
Extinction (biol.)
see Species, extinction of
Extinction (optics)
molecular theory of, **2**:337b
Extispicy, **15**:642b–643a, 643b, 658a(n), 659a(n)
Extrapolation, **10**:159a
Extrema, **3**:428a; **4**:477b, 479a; **5**:111a; **7**: 530b; **11**:487a, 487b–488a, 512b; **12**:118a, 206b; **13**:18b, 259b; **14**:92a
Exudates, **13**:566b
Eydoux, Fortuné, **12**:550b
Eye(s)
accommodation mechanism, **5**:590b; **6**: 247b, 287b; **10**:198a, 500b; **11**:215a, 565b–566a; **14**:564a–b
anatomy of
ancient Greek study, **6**:317b
medieval study, **1**:380a; **6**:192a–b; **10**: 475a; **14**:460a
Renaissance study, **3**:355a
seventeenth-century study, **10**:491a
eighteenth-century study, **2**:37a; **7**: 110b; **11**:112a; **12**:41a
nineteenth-century study, **2**:124b, 531a; **6**:268b, 287b, 326b; **8**:169a; **11**:380a; **12**:510b
comparative, **7**:414b; **8**:204b; **10**:520b
of birds, **3**:207a
blind spot, **4**:615b
discovery of, **9**:114a, 115a
color of, **2**:255b
chemistry of, **5**:430a
embryogeny of, **8**:115a
genetics of, **3**:590b; **7**:516b; **9**:521a
compound, **9**:569b
description of, **1**:81b
dioptrics of, **6**:247a–b
diseases of, **12**:137a
electrical condition of, **4**:79a; **7**:520b
embryogeny of, **6**:132b–133a, 573b; **7**: 439b; **11**:255b
examination of, **11**:215a
of fish, **5**:469b
heat absorption, **7**:74a
histology of, **2**:376b; **5**:56b, 514a; **6**:434b; **9**:93b; **11**:565b–566a; **13**:556b; **14**: 113b
humors of
reconstitution of, **14**:587a
streaming of, **4**:296b
induced defects, **5**:598b
of insects, **2**:481a; **6**:484a; **10**:176b; **12**: 318a
interior stimuli, **9**:570a
larval, **8**:169b
lens
embryogeny, **12**:568a
mathematics of, **5**:590b
nature of, **14**:564a–b
regeneration of, **12**:567b
suture of, **6**:508a
luminescence in, **2**:531a
movement of, **6**:248a
nerves of, **5**:571a; **12**:509b
neurophysiology, **9**:506a; **11**:112b; **14**: 143a–b
pressure within, **4**:615b
reflex images, **11**:215a
reflex response, **5**:56a
resolving power, **8**:414b
surgery, **2**:376b; **8**:97b; **10**:418b; **11**:112a; **14**:584b; **15**:645b
see also Cataract (ophth.); Cornea; Iris; Ophthalmology; Optics, physiological; Retina; Vision
Eyeglasses
seventeenth-century, **6**:603b; **7**:299a
eighteenth-century, **2**:133a
nineteenth-century, **14**:491b–492a
bifocal, **5**:130b

Eyeglasses (cont.)
colored, **2**:452b
invention of, **11**:342b
periscopic, **14**:491b
see also Lens(es), ophthalmological
Eyk, Cornelis van, **3**:333b
Eyring, Carl Ferdinand, **9**:397a
Eysenhardt, Carl Wilhelm, **2**:596a
Eytelwein, Johann Albert C. (1764–1848), 4:501–502
Eyzies-de-Tayac, Les, **8**:44a
Ibn Ezra, Abraham ben Meir, also known as Abū Isḥāq Ibrāhim al-Mājid ibn Ezra (*ca.* 1090– *ca.* 1164), 4:502–503; **1**:23a, 512a; **6**:272b–273b; **7**:223b; **8**:280b; **9**:161b; **14**:593a

F

Faà di Bruno, Francesco, **10**:441a, 442a
Fabbroni, Giovanni Valentino Mattia (1752–1822), 4:503; **9**:5b; **15**:388a*
Faber, Knud, **11**:569b
Fabre, Antoine-François-Hippolyte, **2**:442b
Fabre, Jean Henri (1823–1915), 4:503–505
Fabri, Honoré (or Honoratus Fabrius) (1607–1688), 4:505–507; **5**:544b; **8**:267b
Tractatus physicus de motu locali, **8**:268a
Fabri, Nicolas de Peiresc
see Peiresc, Nicolas Fabri de
Fabrici, Girolamo (or Fabricius ab Aquapendente Geronimo Fabrizio) (*ca.* 1533–1619), 4:507–512; **3**:99a
acquaintances of, **12**:101b, 106a
anatomy, **10**:598b; **14**:7a, 12a
venous valves, **1**:98a–b, 523b; **3**:41a; **6**:153a
and Casseri, **3**:98a–b
editions of works, **15**:4a
embryology, **3**:343a; **6**:159b, 160a
influences on, **1**:110a; **6**:152a; **10**:418b
optics, **10**:475b
students of, **1**:293a, 480a, 522b; **6**:150b; **12**:577a
Fabricius ab Aquapendente
see Fabrici, Girolamo
Fabricius Hildanus, **8**:125b
Fabricius, David, **2**:412b, 413a; **7**:295b, 296a; **9**:486a
Fabricius, Johann (astron.), **7**:193b; **12**:151b
Fabricius, Johann Christian (1745–1808), 4:512–513; **2**:322a; **8**:48b, 379b; **10**:323b; **12**:516a
Fabris, Pietro, **6**:84a
Fabrizio, Geronimo Fabrizio
see Fabrici, Girolamo
Fabroni, Giovanni Valentino Mattia
see Fabbroni, Giovanni Valentino Mattia
Fabry, Charles (1867–1945), 4:513–514; **5**:171a*, 497b*
interferometry, **8**:551b; **10**:518b, 519a
on standard meter, **1**:612b
Fabry, Louis (1862–1939), 4:514–515
Face
see Head
Fachgemeinschaft der deutschen Hochschullehrer der Physik, **12**:615a
Fachs, Modestin, **8**:312a
Facies (geological), **7**:320a
concept of, **5**:533a–b; **14**:88a
horizontal variability of, **2**:495b; **6**:168b
isopens, **6**:229a
sedimentary, **2**:87b; **6**:229a
lateral, **5**:486b
synchronism of, **5**:476a
Facies, metamorphic, **4**:361b
high-pressure, **4**:409b
mineral, **4**:408b–409b
classification of, **4**:409a
Factitious air, **3**:156b; **6**:41a
medical use of, **3**:154a*, 600a–b
Factorials, **13**:571a
approximation to, **9**:453a–b
concept of, **7**:494b
conversion of, **13**:68b–69a
geometrical, **3**:141b
Factories
ventilation in, **6**:576a; **9**:153a
Factorization method, **7**:10b
Factors (math.)
discontinuous, **3**:136b
integrating, **3**:282b; **8**:299b
invariant, **2**:491a
prime, **3**:302b; **4**:421b; **7**:599b; **11**:269a
ideal, **7**:523b
al-Faḍl ibn Hātim al-Nayrīzī
see al-Nayrīzī
al-Faḍl ibn Sahl al-Sarakhsī, **13**:538a; **14**:537b; **15**:186b
Faeroe Islands
exploration of, **3**:203b; **5**:225a
geology, **5**:72a
Faget, Guy Henry, **4**:155a
Fagnano dei Toschi, Giovanni Francesco (1715–1797), 4:515
Fagnano dei Toschi, Giulio Carlo (1682–1766), 4:515–516; **7**:560a,b, 567a,b
analysis, **7**:52a, 619a; **8**:139b; **13**:238a
ellipse of, **4**:516a
Fagon, Guy Crescent, **4**:216b; **5**:352b; **8**:171a; **9**:17b; **11**:47b; **13**:442b, 553b
Fahlberg, Constantine, **11**:371a
Fåhraeus, Robin, **1**:534a; **13**:2a–b, 160b
Fahrenheit, Daniel Gabriel (1686–1736), 4:516–518; **3**:334a; **4**:26a; **11**:526b
heat studies, **2**:177b, 178b
instrumentation, **2**:227a; **8**:327b; **9**:595b
Fainberg, Vladimir Iakovlevich, **13**:241a,b
Fairbairn, William, **6**:452b, 503a
Fairbank, William Martin, **8**:482b
Fairchild, Thomas, **3**:129a
Faith and reason, **1**:2a–b; **4**:255b; **10**:173a
Faivre, Joseph, **3**:220a
Fajans, Kasimir, **2**:241a, 259b; **6**:14b, 16a,b; **12**:30b, 32a, 505a,b
Fajardo, Francisco, **3**:185a
Fakhr al-Dīn al-Rāzī, **1**:28a; **4**:442a; **7**:343b; **11**:249b, 251b; **12**:90a; **13**:509a, 513a
Falck, Richard, **2**:436b, 437b–438a
Falconar, Magnus, **6**:367b
Falconer, Hugh (1808–1865), 4:518–519; **15**:51b
Falconer, John, **5**:383b, 384a
Falconnet, Camille, **1**:478b; **5**:62a
Falconry, **1**:62a; **2**:314a; **5**:147b–148a; **14**:35b
Falkenberg, Paul, **12**:125a
Falkland Islands, **2**:342a,b; **5**:277a
botany, **6**:489a
geology, **4**:262b
Fallacies
medieval discussion, **2**:604b; **6**:376b
Falling bodies
Byzantine theories of, **7**:135b
medieval Latin theories of, **1**:198a–b; **10**:225b–226a; **13**:198a–199b
Renaissance theories of, **8**:200b, 231b–233b; **9**:488a; **12**:548a; **13**:349b–350a
seventeenth-century theories of, **3**:183b; **7**:583b; **9**:116a; **10**:62b; **13**:586a
nineteenth-century theories of, **9**:236b
displacement toward east, **1**:616a; **2**:311a
experimental study
ancient Greek, **13**:93b
Renaissance, **4**:8b; **8**:200b; **12**:106b; **15**:45b
seventeenth-century, **3**:3a–b; **5**:158b, 288a, 542b; **6**:126b–127a, 379a; **7**:578a; **9**:97b, 116b, 317b; **11**:411a–b
eighteenth-century, **6**:172a–b, 173b
nineteenth-century, **1**:615b–616a
laws of
Renaissance, **4**:237b; **6**:379a; **9**:488b
early seventeenth-century, **1**:424a; **2**:348b; **4**:58a; **5**:240a–241a, 243b, 287b; **6**:607a; **8**:267b–268b
late seventeenth-century, **5**:247b, 527b, 528b; **6**:610b; **7**:486b–487a, 488a; **10**:69b
eighteenth-century, **3**:432b
mean-speed theorem, **13**:350a
on moving ship, **5**:288a
resistance of medium, **1**:606a; **3**:12a; **5**:246a; **6**:607b
theories of
buoyancy theory, **1**:605b–606b, 607a–b; **5**:238b–239a
see also Acceleration, gravitational; Gravity; Kinematics
Falloppio, Gabriele (*ca.* 1523–1562), 4:519–521; **1**:621a; **3**:41a, 99a; **4**:487a; **9**:179a
acquaintances of, **3**:40a; **14**:336a
anatomy, **1**:98b, 316a; **10**:477b; **14**:7a, 12a
critics of, **4**:508b; **8**:54a
and Colombo, **3**:355b
embryology, **4**:510b
fallopian tubes, **4**:521a; **13**:563b
influence of, **2**:224b; **5**:580b; **6**:154b; **11**:33a; **12**:97b
influences on, **14**:11a
lectures of, **3**:342a,b
Observationes anatomicae, **3**:355b; **7**:17a; **14**:6b
students of, **1**:315b; **2**:336b; **4**:507a
Fallows, Fearon, **6**:263a; **7**:146a
False position, method of, **4**:115a, 608b; **8**:421b–422a
double, **3**:273b, 275b–276a; **4**:607a, 611b
simple, **3**:273b; **4**:607a,b
Falta, Wilhelm, **9**:284b
Famennian (stratig. stage), **8**:463b
Family, taxonomic
concept of, **2**:580a–b; **7**:197a; **8**:435b; **9**:18a; **13**:443a
Fan fa (transformation), **3**:267a–268a
Fan theorem, **2**:513b, 514a
al-Fanārī, Mullā Shams al-Dīn Muḥammad, **11**:227a
Fang Chung-t'ung, **14**:544b
Fang I-chih, **14**:166a(n)
Wu-li hsiao-chih, **15**:737b
Fankuchen, Isidor (1905–1964), 4:521–522; **15**:18a
Fano, Gino (1871–1952), 4:522–523; **6**:391a; **10**:344b
Fan-yung-shu (reduced use numbers), **3**:253a
al-Fārābī, Abū Naṣr Muḥammad ibn Muḥammad ibn Ṭarkhān ibn Awzalagh (*ca.* 870–950), 4:523–526; **1**:267b, 276a; **9**:28b
commentaries by, **4**:441a,b; **15**:176b, 181a
critics of, **12**:5b
influence of, **1**:381b; **11**:444b; **12**:4a; **13**:534b; **14**:35b, 388a
influences on, **15**:265b
Kitab mabādi'
translations of, **13**:401b
music, **11**:249b
De ortu scientiarum, **5**:592a
translations of, **5**:592a

philosophy, **1**:408b; **9**:111b; **15**:249b, 250b
De scientiis, **7**:42b
translations of, **15**:175b, 181b–182b
De tempore (spurious), **15**:181a
Faraday, Michael (1791–1867), 4:527–540; 1:411b; **2**:589a; **5**:127b; **9**:177a, 202a; **10**:177a; **13**:105a, 413b
acoustics, **3**:258b; **4**:534b
acquaintances of, **1**:136a; **3**:474b, 557b; **5**:518a; **6**:87a, 120b; **9**:424b; **12**:131b, 288a, 524a; **13**:352a, 356b; **14**:289b
biographies of, **5**:413a*; **13**:357a
and Brande, **2**:420a
chemistry, **4**:528b–532a; **14**:475b, 489a, 633a
liquefaction of gases, **4**:79b
correspondence of, **8**:3a*, 36a; **11**:21a, 46b; **12**:198b; **14**:292b
critics of, **12**:197a
and Davy, **3**:603a,b, 605a; **4**:529a–531b
Electrical Researches, **11**:578a
electricity, **4**:528a–b, 532a–538a, 559a; **9**:216b, 548a, 549a; **11**:604a; **13**: 375a–377a
electrochemistry, **1**:347b; **3**:601b; **4**:535b–537a; **5**:292a, 559b; **6**:439a; **7**:181a; **8**:36b; **9**:391b
electromagnetism, **1**:146a, 202b, 203a; **3**:288b; **4**:533a–535a; **5**:305b; **6**:341b, 344a; **9**:204b, 205a–b; **10**:28a
Experimental Researches, **4**:283b–284a
Faraday effect
see Optical activity, magnetic
Faraday-Mossotti hypothesis, **4**:231a–b
Faraday's law, **1**:555b; **3**:334a; **11**:418a; **14**:33a
demonstrations of, **8**:503a
field concept, **2**:326a, 330b; **4**:325b, 482a; **6**:279a
and (Joseph) Henry, **6**:277b, 278a, 278b–279a
influence of, **3**:31b; **9**:548a, 558a; **11**:241b; **13**:521b; **14**:32b, 598a
influences on, **14**:73a
magnetism, **1**:556a, 557b; **3**:506a; **4**:538a–539a; **6**:251b; **9**:196b, 206a–208a *passim;* **10**:185b; **12**:49b–50a, 52a; **13**:521b; **14**:208a
matter, theory of, **3**:477b, 599b; **4**:599b–600a
metallurgy, **6**:5b
students of, **10**:510b
Faraj ben Salīm, **15**:183b
Farbius, Antimus
see Fabri, Honoré
Farbwerk Meister, Lucius and Co., **5**:488b
Farbwerke Cassella & Co., **4**:299b
Farey, John (1766–1826), 4:540–541; 5: 519a; **12**:463a*, 490b
Farey, John, Jr., **4**:540a
Farfán, Agustín, **6**:309b–310a
al-Farghānī, Abu'l-ᶜAbbās Aḥmad ibn Muḥammad ibn Kathīr (*d.* after A.D. 861), 4:541–545
astronomy, **7**:362a; **12**:62a; **13**:538b
commentaries on, **11**:226b
cosmology, **12**:3a
Elements of Astronomy, **6**:197b
translations of, **11**:349b
geography, **1**:381b–382a
influence of, **3**:25b; **8**:280b; **9**:363b, 364a; **11**:349b; **14**:388a
Jawāmiᶜ, **14**:472b(n)
Liber de aggregationibus stellarum, **11**: 469a
Liber continens capitula
translations of, **15**:175a,b
and Ptolemy's *Almagest,* **1**:508b, 509a, 512b; **11**:202b
Rudimenta, **5**:361a
translations of works, **3**:262a; **15**:178a–b
Farīd al-Dīn al-Dāmād, **13**:509a
al-Fārisī
see Kamāl al-Dīn Abu'l-Ḥasan Muḥammad ibn al-Ḥasan al-Fārisī
Farkas, Laszlo (Ladislaus) (1904–1948), 4: 545
Farkov, Gelii Frolovich
see Zharkov, Gelii Frolovich
Farlow, William Gilson, **1**:422b; **2**:103a; **4**:220a; **11**:154a; **12**:328a; **13**:299b, 395b, 456a
Farmer, John Bretland (1865–1944), 4:545–546; 13:344a*
Farmer, Moses, **12**:425b
Farner, Christopher, **5**:419b
Farnese family, **2**:142b
Faro, game of, **2**:37b
Farrar, John (1779–1853), 4:546–547; 2: 112a
Ibn al-Farrukhān
see ᶜUmar ibn al-Farrukhān
Fáry, István, **7**:312b
Fasciola, **15**:29b
Fasciolopsiasis, **2**:617a
Fascism, **4**:580b; **5**:201a; **8**:283a; **14**:86b–87a
Fasting
physiology of, **8**:536a
Fat(s)
analysis of, **2**:385b
chemistry of, **2**:63b, 65b–66a; **3**:242a–243a; **5**:432a–b; **6**:398a–b; **9**:285b
circulation of, **9**:146a
as esters, **3**:242b–243a
evolutionary significance, **6**:398b
formation of, **3**:347b; **4**:243a; **9**:146a; **14**:65a, 66a
hydrogenation of, **6**:398a
metabolism of, **2**:28a, 124a–b; **4**:243a; **8**:555a; **9**:285b; **13**:403b; **14**:454b
hormonal control, **15**:229a
oxidation of, **5**:126b
reabsorption of, **2**:531b
splitting of
by sulfuric acid, **5**:157b
Fat cells, **5**:34b
Fate, **1**:119a; **12**:386a–b; **14**:606b–607a
Fathom (unit of length), **8**:351a
Fatigue, **3**:442b
bending tests, **14**:473b
in dynamics, **1**:409b–410a
electrically induced, **15**:47a
industrial, **12**:399a
Ibn Fāṭima, **1**:29a
Fāṭimid caliphate, **15**:250a–b
Fatio de Duillier, Jean-Christophe, **2**:50a, 52a; **3**:282b; **6**:599b
Fatio de Duillier, Nicolas, **2**:51a, 52a; **8**: 164b; **10**:45a, 84a, 87b(n); **13**:480b
Lineae brevissimi descensus investigatio, **8**:165a
Fatou, Pierre Joseph Louis (1878–1929), 4: 547–548; 8:112a; **11**:156b
Fatty acids, **2**:316a; **5**:432b; **6**:398a–b; **8**:58b, 276a
discovery of, **3**:242a–b; **5**:475a
isomerism, **9**:131a
metabolism of, **4**:360a; **10**:22b
structure of, **11**:121a
Faujas de Saint-Fond, Barthélemy (1741–1819), 4:548–549; 3:291b; **4**:150b; **7**:589a; **8**:90a*; **9**:493b*; **10**:281a*, 610a*; **13**:44a, 390a; **15**:113b
Recherches sur les volcans éteints du Vivarais et du Velay, **5**:578a
Voyage en Angleterre, en Écosse et aux Iles Hébrides, **5**:335b
Faulhaber, Johann (1580–1635), 4:549–553; 4:55b
Faulhorn (mountain), **2**:431a
Faulting (geol.), **3**:70b, 325b; **5**:395b; **13**:87a; **14**:402b
block diagrams, **4**:540b
dynamics of, **1**:153b; **10**:429b–430a
horizontal, **5**:476a
inclined, **5**:476a
stepped, **5**:476a; **8**:30b
submarine, **1**:394b
theories of, **1**:472b; **4**:309a; **5**:476a–b; **6**: 502b; **10**:167a, 168a,b
thrust, **5**:337a–b; **8**:462a; **11**:505b
Faunal succession, principle of, **2**:481a, 494a–497a; **5**:533a; **9**:27a; **11**:505a
Faust, Otto, **13**:246a
Faventinus, Marcus Cetius, **15**:514a, 516b
Favorsky, Alexei Yevgrafovich (1860–1945), 4:553–554; 2:623b; **7**:21a; **8**:108a
Favre, Léopold
Dictionnaire du vieux François, **2**:305b
Favre, Pierre Antoine (1813–1880), 4:554–555; 2:69a
Favus, **1**:493b; **5**:566a; **10**:384a; **12**:202b–203a
Faye, Hervé (1814–1902), 4:555; 1:202b; **8**:387a*
Faye's comet, **9**:460a–b; **13**:112a*
Fayet, Gaston, **4**:514b
al-Fazārī, Muḥammad ibn Ibrāhīm (*fl. ca.* 760–790), 4:555–556; 13:538a; **15**: 596b
astronomy, **14**:546a
Zīj al-Sindhind, **1**:34b; **4**:556a; **6**:79b; **7**:223a, 360b; **15**:580b–581a
Feather, Norman, **12**:34a
Feathers
regeneration of, **8**:359a
structure of, **10**:520a
Featherstonhaugh, George William (1780–1866), 15:151–153; 9:172b
February Revolution of 1848, **3**:132b; **4**:68a
Febvre, Lucien, **11**:388b
Fecampia, **9**:328b
Feces
fossil
see Coprolites
pharmaceutical use of, **5**:421a; **6**:425b
prognostic use of, **2**:435b
surgical use of, **9**:12a
Fechner, Gustav Theodor (1801–1887), 4: 556–559; 2:1b, 446a; **6**:299b, 342b; **8**:514a, 605a; **9**:561b; **11**:359b, 360b*; **12**:254a; **14**:205a
electrodynamics, **14**:205a
Lehrbuch des Galvanismus und der Elektrochemie, **10**:192b
Massbestimmungen über die galvanische Kette, **10**:192b
physiology, **6**:52b, 247b
Professor Schleiden and the Moon, **12**: 175b
Revision der Hauptpunkte der Psychophysik, **9**:561b
Weber-Fechner law, **4**:557b–558a; **7**: 458a; **9**:561b; **14**:201a
Feddersen, Berend Wilhelm (1832–1918), 4: 559
Federal Communications Commission, **1**: 287b
Fede-Riga disease, **15**:479b
Fedorov, Evgraf Stepanovich
see Fyodorov, Evgraf Stepanovich
Fée, Antoine-Laurent-Apollinaire (1789–1874), 4:560; 9:340b; **10**:531b*

Feedback, **14**:347a
physiological, **14**:595b
Fehling, Hermann, **5**:125b
Fehling's solution, **1**:471a; **4**:603b
Feigl, Georg (1890–1945), 4:560–561
Feistmantel, Ottaker, **3**:297b
Feit, Walter, **2**:615b
Fejér, Lipót (1880–1959), 4:561–562; **2**:239a; **3**:178a; **5**:608a; **13**:562a
Feldberg, Wilhelm, **15**:106b
Feldspar, **1**:20a; **3**:313b; **4**:598b; **7**:6a*
association laws, **7**:249b
composition of, **5**:212b; **6**:358b; **9**:367a; **13**:476a
crystallization of, **6**:178a
iridescent, **11**:267a
naming of, **13**:410a
plagioclase, **2**:369b; **5**:89b; **6**:358b
weathering of, **5**:72b
Feldspathization, **13**:283b, 284b
Feletti, Raimondo, **5**:503a
Félibien, André, **2**:201b; **10**:521b
Felix, Arthur, **6**:298b
Fell, Honor, **15**:419b
Feller, William, **8**:321b; **9**:129b
Felsite, **3**:434a
Fénéon, Jean Filibert, **4**:349b
Fenger, Christian, **6**:232b
Fenneman, Nevin Melancthon (1865–1945), 4:562–563; **13**:532b(n)
Fenner, Clarence Norman (1870–1949), 4: 563–565; **6**:528a*; **14**:183b*
Férat, Jean-Baptiste-Pierre François, **3**:61b
Ferber, Johann Jacob, **1**:234b; **2**:315a; **11**: 304a; **14**:259b
Ferchault, René Antoine
see Réaumur, René Antoine Ferchault de
Ferdinand I, king of Naples, **9**:146b; **10**: 592b–593a
Ferdinand II, Grand Duke of Tuscany, **2**: 309b; **3**:23a; **13**:30b, 434a, 437a–b, 439a
Ferdinand II, Holy Roman Emperor, **2**: 602a; **4**:214a; **6**:150b; **7**:305b, 375a; **12**:307b; **14**:636b
Ferdinand III, Holy Roman Emperor, **2**: 592a; **3**:61a; **7**:306a, 307a; **9**:96b
Ferdinand VI of Spain, **13**:530b
Ferdinand, Carl Wilhelm, Duke of Brunswick, **5**:298b, 300b, 301b
Ferenczi, Sándor, **5**:179a; **7**:190b
Ferguson, Adam, **2**:174b, 175a; **6**:579a
Ferguson, Donald Fraser, **12**:345a
Ferguson, James (1710–1776), 4:565–566
Ferguson, R. Munro, **13**:525b
Fermat, Pierre de (1601–1665), 4:566–576; **3**:615a, 622a; **5**:382b
acquaintances of, **1**:541a
algebra, **4**:566b–569b
analysis, **6**:599b; **10**:336a; **14**:148a
and Carcavi, **3**:63b–64a
on contemporary scientists, **1**:368a; **4**: 46b, 47a
correspondence of, **2**:507a; **3**:66a; **4**:95b; **7**:583a,b; **9**:316b; **10**:334b–335a, 342b; **13**:435b; **14**:49b
critics of, **14**:152a
and Descartes, **4**:567b, 571a, 574b–575b; **5**:527a; **6**:112b; **9**:598a
and Despagnet, **4**:75a
editions of works, **1**:206b–207a; **13**:252b, 253a,b, 254a
equation of, **1**:76b; **4**:124a,b; **5**:528a; **7**: 565b; **8**:138b
Fermat's principle, **2**:53b; **3**:454b; **4**:575a; **6**:87b; **8**:156a; **9**:616b
Fermat's theorem, **2**:527a; **7**:523b; **8**: 13b*, 138a,b; **14**:438a; **15**:391b*
game theory, **3**:66a; **4**:574b; **9**:599b
geometry, **3**:152a, 226a; **5**:499a; **6**:112b, 359b; **10**:115b, 336b; **12**:206a, 460a; **13**:434b, 435b; **14**:49b, 509b; **15**:118a
problems, **5**:158b–159b; **13**:436a
quadrature, **4**:571a–573a; **5**:527a
tangent algorithm, **1**:474b–475a; **4**: 570a–571a, 575a
and Greek mathematics, **1**:187b, 191b, 245b; **4**:117b; **12**:446a
and Huygens, **3**:64a; **4**:571a, 572b, 573a
infinite descent, **4**:474b, 573a, 574a–b; **5**:159a; **7**:566a; **8**:138a–b; **10**:484a; **11**:55b
influence of, **10**:45b, 103b; **15**:489b
last theorem of, **3**:134a, 143a; **4**:474b, 574a; **7**:601b
general proofs of, **5**:375b; **8**:531b
manuscripts of, **12**:205b
maxima and minima, **4**:57b, 58a, 569a–571a, 575b; **6**:537b; **12**:460a; **13**:436a
mechanics, **1**:541b; **8**:267b
number theory, **2**:131a; **3**:302b; **4**:83a, 474a,b, 567b, 573a–574b, 612b; **5**: 450a,b; **7**:54a; **9**:599b; **14**:149a–b
optics, **6**:197a
probability, **3**:451b; **10**:334b–335a
Analysis ad refractiones, **4**:575a–b
Isagoge ad locos ad superficiem, **4**:568b–569a
Isagoge ad locos planos et solidos, **3**:64a; **4**:567b–568b, 569a
Loca plana restituta, **4**:75a
De maximis et minimis, **10**:342b
Methodus ad disquirendam maximam et minimam, **4**:569b
Novus secundarum et ulterioris ordinis radicum in analyticis usus, **3**:64a; **4**:569a
Observations sur Diophante, **4**:567b
De solutione problematum geometricorum per curvas simplicissimas . . . , **4**:568b
Varia opera mathematica, **3**:64a; **4**:572b
Fermat, (Clément-) Samuel de, **3**:64a; **4**: 118b*, 566a; **5**:528a
Fermentation
seventeenth-century study, **2**:51b, 311a, 378a; **5**:427a; **7**:612a; **14**:26a
eighteenth-century study, **3**:199b; **5**:156b; **14**:332b
nineteenth-century study, **4**:211b; **9**:445b; **10**:499a
twentieth-century study, **3**:22b, 259b; **14**:247b
acetous, **2**:562b; **5**:367a; **10**:365a–366a, 376b, 378a
alchemical view of, **1**:550a
alcoholic
eighteenth-century study, **2**:177a; **3**: 200b–201a; **5**:367a; **8**:82b–83a
early nineteenth-century study, **2**:420b; **3**:316b*; **4**:134b–135a; **13**:311b–312a
mid nineteenth-century study, **3**:9a–b; **4**:211a; **8**:553a; **9**:401b, 425a–b; **10**: 363a–b, 365b; **12**:124a, 241a–b, 242b, 243b–244a
late nineteenth-century study, **2**:31a–b, 560b–561b; **6**:100a–101a; **7**:34b; **10**: 376b, 378a–379b
twentieth-century study, **2**:561b–562b; **4**:485a; **5**:6a; **6**:111a, 501b; **7**:406a; **14**:412b
intermediate products, **2**:562a–b
mechanism of, **14**:174a–b
ammoniacal, **13**:405a
bacteria in, **1**:497a; **5**:128b; **6**:100b, 110b, 501b; **14**:45b
butyrous, **2**:562b; **3**:338b, 339b; **10**:363b–364a, 384b; **13**:405b
cell-free, **2**:561a, 561b–562a
chemistry of, **1**:320a; **2**:560b–562b; **5**: 128b; **6**:505b; **12**:124a; **14**:173b, 174b
citrous, **2**:562b
diastatic, **10**:377b
enzymes in, **2**:66a; **4**:485b–486a; **5**:4a–b; **6**:111a; **14**:45b
gaseous products, **3**:156b; **6**:42a
industrial, **1**:361a; **10**:365a–366a
kinetics of
and multiplet theory, **1**:415b
lactic, **8**:407b; **10**:362a–b, 368a–b; **11**: 426a, 431a; **14**:45b
and optical activity, **10**:360b, 361b
origin of term, **6**:46b(n)
oxygen and, **10**:364a–365a, 376b–377b
physiological nature of, **7**:534b
propionic, **14**:45b
salts in, **6**:111b
theories of
seventeenth-century, **6**:256b; **9**:244a; **10**:83a; **14**:405a–b
eighteenth-century, **3**:345a; **5**:367a; **10**:80a
nineteenth-century, **2**:70b; **8**:349b, 402b, 405b, 407b; **9**:425a; **10**:361b–366a, 367b–368b, 370a, 376b–382b; **12**:241a; **13**:312a; **14**:350a–b; **15**:12a
chemical, **10**:378a–379b
mechanistic, **2**:561a
vitalistic, **2**:561a–b; **10**:361b–366a, 367b–368b, 376b–382b
see also Yeasts
Fermentation tube, **12**:481a
Ferments
see Enzymes
Fermi, Enrico (1901–1954), 4:576–583; **9**: 38a,b
acquaintances of, **4**:294b; **13**:227a; **15**: 25b, 43a
atomic bomb, **10**:216a
Fermi constant *(G),* **4**:579b–580a
Fermi liquid, **7**:617b
Fermi statistics, **4**:578a,b; **5**:103a; **8**:478b–479a; **15**:49b
nuclear physics, **2**:250b; **3**:371a; **7**:154b, 158b; **9**:261b
beta decay, **10**:425a; **13**:240b
students of, **9**:37b
Fermions, **4**:579b
Fernald, Merritt Lyndon (1873–1950), 4: 583–584; **6**:122a; **15**:149b, 224b*
Fernbach, Ernest, **3**:189a
Fernel, Jean François (*ca.* 1497–1558), 4: 584–586; **3**:533b; **5**:448a; **12**:401a
chemical theory, **12**:311b, 312a
commentaries on, **11**:466a
medicine, **1**:399b–400a
"Physiologia," **10**:598b
physiology, **2**:2b
students of, **12**:323b
Fernico (alloy), **6**:546b
Ferns, **1**:412b; **2**:219a; **6**:494b; **9**:418a
anatomy of, **5**:605a; **9**:601a
apogamy in, **3**:613b; **8**:5a; **11**:154a
classification of, **2**:371b; **4**:560a; **9**:340b; **11**:130a; **13**:178a
development of, **2**:160a
eusporangiate, **3**:30a,b
fossil, **5**:441b, 498a; **7**:197b; **11**:373a
fructification, **14**:600a
gametophytes, **3**:29a; **6**:219b
leptosporangiate, **3**:30b
and liverworts, **3**:30a–31a
reproduction, **6**:219a–b, 266a,b; **14**:350a; **15**:81a

Fibromas, **3**:490a
Fichot, Lazare-Eugène (1867–1939), 4:614
Fichte, Johann Gottlieb, **2**:545a; **5**:193a; **6**: 241b, 294b; **9**:578b; **11**:214a; **12**: 153b–156a *passim;* **14**:241b
Wissenschaftslehre, **12**:154a, 155a,b
Fichtelgebirge range, **2**:158a
Ficino, Marsilio, **1**:611a,b; **2**:539b, 541a, 543a; **5**:121a, 448a; **6**:306a; **8**:195a, 234b, 538b; **10**:310a; **11**:42a; **15**:69a–b
Fick, Adolf Eugen (1829–1901), 4:614–617; **5**:196a; **6**:435b; **8**:445a; **9**:101b, 566b
Fick's law, **4**:615b; **6**:444a
physiology, **5**:126b; **14**:65b, 454b
Fick, Ludwig, **4**:614b; **8**:540a
Fiducialism, **5**:10a–b
Field, Richard Montgomery, **1**:394b
Field theory, **8**:599b–600a; **9**:377a–b; **12**: 214a, 222b; **14**:304b
ancient anticipation of, **14**:606b
cosmological term, **8**:600a
quantum, **4**:326a–b; **7**:617b; **8**:498b; **9**: 377a
symmetries in, **10**:425a
unified, **14**:306a; **15**:49b
Fields, John Charles (1863–1932), 4:617
Fields medal, **4**:617b
Fields (math.)
Abelian, **1**:307a; **5**:206a; **6**:394a; **14**:202b
theory of, **6**:390a
abstract, **7**:530b
algebraic, **1**:15a; **3**:221b; **6**:393b; **13**:23a–b
algebraic function, **6**:286b; **7**:398b, 399a
algebraic number, **1**:306b; **4**:4b, 125a–b; **5**:263a, 342a–b; **6**:214b, 297b, 308b, 572a; **7**:508a; **10**:138b; **11**:55a; **12**: 237b; **14**:630a
arithmetical theory, **6**:286b
arbitrary, **6**:394a; **10**:140a
commutative, **14**:212a
complex, **2**:358a; **10**:140a; **13**:6a
concept of, **5**:263a; **6**:391b; **13**:23a
cubic, **14**:94b
extension, **13**:23b
finite, **1**:306b; **4**:83a; **9**:502b
formal-real, **1**:307a–b
function, **1**:306b; **6**:389b
Galois, **1**:307a; **9**:502b; **14**:212a
ideal classes of, **4**:5a
perfect, **7**:530b
quadratic number, **1**:306b; **2**:90a; **4**:124b, 125a; **5**:206a; **6**:214a–b, 390a,b
real-closed, **1**:307a–b
recurrent, **4**:337a
skew, **6**:391b; **14**:212a
of solution curves, **3**:62b, 63a
tensor, **3**:143a
theory of, **1**:255b; **2**:326a, 330b; **4**:2b, 316b–317a, 319a,b, 325b–331b, 539a–b; **7**:10b
arithmetical, **7**:530b
cosmological term, **4**:330a; **5**:188b
valuated, **6**:286b
vector, **3**:143a; **6**:496b, 497a; **11**:408b
Fields (phys.)
displacement, **4**:281a
electrical
effects of, **2**:203a; **14**:497b
field current, **9**:397a
and light, **7**:313b
from motion of electrically charged body, **4**:305b
symmetry of, **3**:505b
theories of, **6**:147b; **7**:493a; **13**:365b
electromagnetic, **2**:397b; **4**:579a; **5**:15b–16a; **6**:344a, 347a; **7**:211a–b; **8**:491b, 492b, 494b, 498b; **9**:209b–217b *passim,* 377a; **13**:365a
and charged particles, **5**:597b–598a
and electrical charges, **4**:191a
electrons in, **4**:322b–323a, 326a; **11**: 446b
energy flow, **11**:122a,b
equations, **8**:494a
high-frequency effects, **1**:283b–284b
mechanical theory of, **4**:315a
motion of, **4**:324a, 328a–b
propagation of, **13**:471b
and quantum theory, **2**:250b; **7**:26b
and relativity theory, **7**:26b
representation of, **11**:122b
rotating, **4**:588b–589a
transformation of, **4**:322b; **8**:40b
electrostatic, **11**:241b
states in, **5**:473b
wave functions, **5**:473b
five-dimensional theory, **7**:26b, 211b
force relationships, **3**:443a; **7**:492b; **8**: 186b
gravitational, **4**:228b, 324a–331b; **14**:306a
curvature of, **4**:379a
deflection of light by, **4**:270a, 278a, 280b
fluctuating, **14**:92a
of point mass, **12**:251b
and quantum theory, **7**:26b
transformation of, **4**:328b
see also Attraction, gravitational
hydrodynamic, **2**:168a
magnetic, **2**:167a; **3**:329a; **4**:228b, 492a, 539a; **5**:483b; **9**:181b, 207b, 211b, 459b; **11**:241b, 461a, 578a
of convection currents, **4**:305b–306a; **13**:364b
of displacement current, **4**:305b–306a
effects on electric current, **6**:51b; **15**: 433a
effects on radiation, **1**:559b; **14**:340b, 497b
extraterrestrial, **6**:29b
galactic, **12**:368b
measurement of, **4**:305b–306a; **8**:478a; **12**:129a,b
molecular, **14**:245a
right-hand rule, **5**:33a
rotation of light
see Optical activity, magnetic
and spectra, **11**:478a–b
see also Zeeman, Pieter, Zeeman effect
strong, **7**:221b
structural, **14**:244b–245a
symmetry of, **3**:505b
and matter, **4**:315a, 325b–326a; **8**:492b, 494b, 498b; **9**:377a
molecular, **3**:4a
self-consistent, **6**:147b
Fiessinger, Noël (1881–1946), 4:617–618
Fiessinger-Leroy-Retter disease, **4**:618a
Fifteen Schoolgirls Problem, **7**:384a
Fig tree, **9**:18a
Figueredo, Manuel de, **10**:160b
al-Fihrist
see Ibn al-Nadīm, *al-Fihrist*
Fiji Islands, **3**:551a
Filaria, **2**:534a; **5**:502b; **8**:296b
see also F.S.H. *(Filaria sanguinis hominis)*
Filariasis, **9**:75a
Filhol, Henri, **11**:269b
Filicales, **2**:371b, 519a
Filipchenko, Yurii Aleksandrovich, **7**:33a; **15**:507b
Filippi, Filippo de, **3**:533a; **6**:8b; **9**:546b
Films, physical
adsorption of, **6**:118b
alkali metals, **4**:253a–b
distortions in
from air jet, **4**:81a
from sound, **4**:81a
electrical conductivity, **8**:480b–481a
liquid, **4**:80b–81a; **8**:481b; **11**:21b–22a
molecular, **15**:201b
mono-molecular, **6**:118a–b
and molecular weight, **4**:76b
optics of, **6**:484a–b
solid, **4**:80b–81a
Filon, Louis Napoleon George, **10**:458a–b
Filter pump, **2**:587a, 589b
Filtering (math.), **14**:346b–347a
Filters, optical, **13**:408a
monochromatic, **8**:581a
Filtration
biological, **5**:128b
bacterial filter, **10**:353a, 382b
porcelain filter, **3**:188b–189a
chemical, **3**:398b*
electric, **3**:437a*
theory of, **8**:301b
Finalism
ancient Greek, **1**:260a,b
seventeenth-century, **5**:286a, 287a
opposition to, **7**:612a
twentieth-century, **3**:493a; **5**:222a; **13**: 276a
opposition to, **2**:11a
see also Causation, final cause
Finance
eighteenth-century, **7**:558b–559a; **8**:84b
nineteenth-century, **9**:292a
see also Economics
Finch
speciation, **3**:569b–570a
Fine, Henry Burchard (1858–1928), 4:618–619
Fine, Oronce (1494–1555), 15:153–157; 2: 618a; **4**:449a,b; **7**:133b; **9**:160b; **10**: 161b; **11**:226b, 339a, 533b
cartography, **14**:275b
De quadratura circuli, **2**:360b
Fingal's Cave, **8**:564a
Finger, Ernest, **7**:624a
Finger reckoning, **15**:9b–10b
Fingernails
formation of, **2**:72b
structure of, **9**:66a
Fingerprinting, **5**:266b; **6**:323b
Fingerprints
classification of, **11**:215a
Finitism, **2**:304a
Fink (Fincke), Thomas (1561–1656), 4:619; **1**:480a
Geometria rotundi, **8**:27b; **11**:4a
Finland
botany, **10**:165a
entomology, **10**:165b
ethnology, **3**:328b
Geological Commission, **12**:271b–272a
geology, **4**:408b–409b; **12**:271b–274b; **13**:409b
politics, **10**:165a
Finlay, Carlos Juan (1833–1915), 4:619–620; **3**:94b; **11**:346b
Finlay-Freundlich, Erwin
see Freundlich, Erwin Finlay
Finsbury Technical College, London, **13**: 356b–357a
Finsen, Niels Ryberg (1860–1904), 4:620–621
Finsler, Paul
Finsler spaces, **3**:96b*; **4**:336b
Fior, Antonio Maria, **4**:586b, 587b, 595b; **13**:259a

Fir tree
diseases of, **9**:79a
Fire
chemical analysis by
see Chemistry, analytical, combustion methods
color of, **7**:135a
definition of, **10**:59b
as element, **5**:353a–b; **7**:587a,b; **8**:131a; **9**:5b; **10**:181a, 309b, 591a
criticism of, **13**:330a
luminosity
cause of, **3**:603a
mechanical power of, **3**:316a
as ontological principle, **6**:290b
propagation of, **3**:486a
protection against, **7**:405b*; **13**:23a
see also Fireproofing
subterranean, **8**:174a
and technology, **15**:651a, 652b
theories of
ancient Greek, **13**:330a–b
Renaissance, **1**:495b
eighteenth-century, **3**:215b–216b; **4**:352b; **5**:411a, 511a; **8**:171b; **14**:84a
atomistic, **4**:32b
see also Combustion; Elements, Aristotelian; Flame(s)
Fire alarm, **6**:223a
Fire engine, **9**:153a
Firearms
see Guns
Firedamp, **2**:523b; **6**:236b; **7**:451a
analysis of, **3**:603a
Firefly, **12**:560b
physical study of, **3**:327b*
Fireplace
design of, **13**:352a
Fireproofing, **5**:203a; **7**:379a; **8**:100a
Fireworks
see Pyrotechnics
Firmicus Maternus (*fl.* A.D. 330–354), 4:621–622; 10:419a, 548a,b; **15**:125a
Firmin de Belleval, **7**:131a; **8**:281b; **12**:440a
First aid, **11**:100a
Firth College, **12**:544b
Fischel, Alfred, **11**:574b
Fischer, Albert, **9**:381a
Fischer, Alfred, **1**:613b
Fischer, Bernhard, **7**:423a, 424b, 425a
Fischer, Emil Hermann (1852–1919), 5:1–5; 1:390b; **5**:6b*; **6**:107b; **9**:260b; **12**:164a
acquaintances of, **3**:279a; **5**:182a, 215a; **6**:15b; **13**:70a; **14**:603a; **15**:15a–b, 433a
analytical methods, **10**:243a
on carbohydrates, **2**:621a; **4**:603b; **13**:407a*
on contemporary scientists, **5**:165b*, 537a*; **6**:463b*; **11**:551a
critics of, **6**:339a; **9**:95b
indole synthesis, **5**:1b
structural theory, **3**:511a–b; **4**:299a; **14**:124b; **15**:13a
students of, **1**:286b; **3**:286b; **4**:90a,b, 485a; **5**:5b–6a, 99b; **7**:59a, 623a; **8**:275b; **11**:129a; **14**:173a, 252b, 444a; **15**:157a, 252b
Fischer, Ernst Gottfried, **2**:133b; **4**:546b; **11**:437a
Fischer, Ernst Sigismund, **5**:97b; **8**:112a; **10**:137b
Riesz-Fischer theorem, **11**:459a
Fischer, Friedrich, **12**:197a–b
Fischer, Hans (1881–1945), 15:157–158; 14:173b, 175b
chlorophyll, **14**:412a
Fischer, Hermann Otto Laurenz (1888–1960), 5:5–7
Fischer, Nicolaus Wolfgang (1782–1850), 5:7
Fischer's salt, **5**:7a
Fischer, Otto, **1**:390b; **5**:2a, 595a; **6**:110b
Fischer, Paul Henri, **1**:210b; **9**:579a
Fischer, Theobald, **2**:207a
Fischer-Petersen, J., **13**:99a
Fish Commission, U.S., **1**:406a; **3**:519b, 610b; **7**:169b; **12**:479a; **14**:1a,b
establishment of, **1**:405a
Fisher, Ronald Aylmer (1890–1962), 5:7–11; 2:211b; **3**:572a,b; **9**:282b; **10**:465a,b(n); **14**:553a
genetics, **6**:22a; **10**:464a; **14**:574a
and Pearson, **10**:454b–457b
Statistical Methods, **14**:552b–553a
Statistical Tables, **14**:553a
statistics, **5**:477a; **10**:458a, 460b; **14**:382a, 383a; **15**:31b, 376a
Fisheries, Bureau of (U.S.), **13**:151a
Fishes
age determination, **6**:442a; **9**:550b
anatomy, **5**:502b; **6**:535b; **9**:511a; **11**:379b
bones, **14**:398b
brain, **5**:196a
circulatory system, **4**:268a–b
comparative, **1**:306a; **12**:298a
eyes, **5**:469b
fins, **6**:131b
heart, **13**:402b
scales, **5**:467b; **11**:331a
skeleton, **6**:453b
swim bladders, **2**:136a; **6**:137b; **11**:316b
teeth, **14**:398b
blind, **14**:534a
cartilaginous, **11**:379b
chimaeroid, **3**:610b
cytology, **4**:546a
development of, **14**:398b
digestion in, **11**:426a
distribution of, **1**:622b, 623b; **4**:310a; **5**:70b, 507b
electric, **2**:127b; **3**:157b; **4**:200a, 201b, 202b, 289a; **5**:196a,b, 268b; **6**:269a; **9**:177a, 259a; **10**:267b; **11**:491b, 566b; **12**:232a; **14**:76b, 78b–79a
embryogeny, **10**:105a; **13**:151a; **14**:57b
evolution of, **3**:610b; **7**:169b–170a; **12**:337a
fertilization of
ancient Greek speculation on, **1**:264a
artificial insemination, **7**:31b
fossils
see Paleoichthyology
freshwater, **4**:310a; **14**:157b
ganoid, **3**:606a, 610b
gills, role of, **4**:268a–b
hearing in, **3**:38a; **5**:355a; **6**:287b, 566b
heart motion, **3**:324a*
life histories, **5**:406b; **8**:609a; **12**:190b
marine
death in fresh water, **2**:61a
Mediterranean, **3**:365a
migration, **1**:622b; **6**:267b; **12**:562a
milieu intérieur
external influences, **11**:102a
neurology, **3**:430a; **6**:321a
pigmentation, **13**:151b
poisonous, **12**:536b
reproduction, **1**:622b
respiration, **7**:503a; **11**:528a; **12**:333b
roe
chemical analysis of, **5**:92b
schools of, **9**:331b
stomach contents, **5**:69b–70a
swimming of, **3**:184b
taxonomy of, **1**:306a; **2**:282a, 509b; **3**:525a; **4**:42a, 310a; **5**:406b, 467b; **7**:169b–170a; **9**:573a; **11**:316b; **13**:554b
tropical, **12**:536b
see also Elasmobranchs; Ichthyology; Paleoichthyology; Pisciculture; specific fishes
Fishing industry, **1**:405a; **4**:225a*; **6**:442a–b; **8**:609a; **9**:113b*; **13**:552a
conservation measures, **7**:169b
curing procedures, **3**:348b*
international control, **6**:442a–b
mathematics of, **6**:267b, 287b
oceanography and, **3**:204a; **4**:42b; **9**:331a–b
processing, **14**:109a
Fiske, Thomas Scott, **5**:414b
Fission, nuclear, **5**:332a; **7**:554b
applications of, **3**:330a
chain reactions, **2**:252a; **3**:371a; **4**:580b–582a; **7**:155a, 527a; **13**:227a
concept of, **9**:262a
discovery of, **4**:580a,b; **6**:16b–17a; **9**:262b–263a
theories of, **2**:251b–252a; **5**:161b, 272a; **7**:493b
Fissure eruptions, **5**:336b
Fistula(s), **13**:403b
biliary, **2**:124a–b; **12**:244a
duodenal, **15**:108b
of Eck, **10**:433a
gastric, **15**:108a
and digestion study, **1**:542b–544a; **11**:426a
lachrimal, **8**:114b
pancreatic, **2**:28a
rectogenital, **4**:507b
Fit, goodness of (math.), **5**:8a
Fitch, Asa (1809–1879), 5:11–12; 4:274b
Fitch, John, **15**:472a
Fitch, Walter Hood, **6**:489a,b, 493b
Fitness, concept of, **6**:261a, 262a; **7**:137a–b
Fittig, Rudolph (1835–1910), 5:12–13; 1:168a, 578a; **11**:277b–278a, 370b; **13**:577b; **14**:560b
organic chemistry, **7**:281b; **13**:338a
Wurtz-Fittig reaction, **5**:13a
Fitting, Hans, **3**:581b; **10**:139a
Fitton, William Henry (1780–1861), 5:13–15; 6:588a*; **8**:564b; **9**:87a; **12**:491a
Fitz, Reginald Heber, **13**:300a
FitzGerald, George Francis (1851–1901), 5:15–16; 6:350a*; **8**:40b; **9**:204a; **13**:364b; **15**:39b
acquaintances of, **6**:343b, 345a; **13**:356b, 365a, 471a
electrodynamics, **9**:217a
Lorentz-FitzGerald contraction
see Lorentz, Hendrik Antoon, Lorentz-FitzGerald contraction
optics, **7**:313b; **8**:40a, 592b; **9**:213b, 372b
Fitzgerald, John Gerald, **1**:442b
Fitzroy, Robert (1805–1865), 5:16–18; 3:565b; **8**:278b
Fitz-Smith, Richard, **3**:248a
Five Phases (Chinese science), **8**:391b; **12**:374a, 376a, 390b; **15**:729a, 733b, 739a, 740a
see also *Wu-hsing ta-i*
Fixatives, histological, **5**:36a
"Fixed air" (carbon dioxide), **2**:173b, 176a–177b, 180b, 181a, 329a, 378a, 572a–b; **6**:41b, 43b; **8**:72b–73a; **11**:564b; **12**:146a
as antiscorbutic, **11**:144a
decomposition of, **10**:446a
formation of, **7**:388b, 389a
medical uses of, **8**:74a

"Fixed air" (carbon dioxide) (cont.)
and plants, **6**:283a–b
reactions of, **8**:74a–76b
and respiration, **8**:83a; **12**:24b–25a, 145b
specific gravity of, **3**:156b
tests for, **8**:585b
Fixed Nitrogen Research Laboratory, **3**: 436b
Fixed-point theorem, **2**:513a
Fizeau, Armand-Hippolyte-Louis (1819–1896), 5:18–21; 3:420a*; **4**:378a; **9**: 154b
astronomy, **13**:36b
Doppler-Fizeau effect
see Doppler, Johann Christian, Doppler effect, in astronomy
electricity, **11**:604a
and Foucault, **5**:18b–19b, 85b–86b
optics, **2**:439a; **4**:167b; **6**:542b; **14**:290a
speed of light, **3**:420a; **6**:453a; **8**:50b–51a, 495a, 496b; **9**:214b, 372a,b; **14**:598b
Fizès, Antoine, **4**:589b
Fjords
stratification of water, **4**:344b
Flacherie, **10**:374b–376a, 397b
Flacius, Matthias, **13**:59b
Flack, Adrian
see Vlacq, Adriaan
Flack, Martin, **5**:281b; **7**:278b
Flagellata, **4**:133a*; **8**:145b; **9**:328b
chemotaxis, **10**:576a
cytology, **11**:174b
embryogeny, **11**:175a
metabolism, **9**:167a
sexual reproduction, **4**:207a
Flame(s)
definition of, **10**:59b
electrification of, **4**:355a; **6**:96b
illuminating power of, **5**:126b
nature of, **12**:605a
sensitive, **8**:122a
singing, **8**:399a
spectroscopy of, **4**:79a; **5**:143a, 483b; **9**: 267a, 391b
structure of, **12**:493b
see also Combustion, Fire
Flamethrower, **9**:491b
Flamingo, **4**:173a
Flammarion, Camille (1842–1925), 5:21–22; 1:172a; **4**:359a; **7**:78a*; **9**:331a*
La planète Mars, **3**:357b
Flamsteed, John (1646–1719), 5:22–26; 5: 279a(n), 491a
acquaintances of, **3**:104b; **5**:520b, 525a; **8**:459b; **11**:284b, 526b; **13**:178b, 445a
astronomical tables, **3**:430b; **4**:169b*; **6**: 516a; **7**:163b*; **8**:67b, 179a, 459b
Atlas céleste
editions of, **5**:78b
catalog of, **4**:16b; **6**:329b
publication of, **5**:25a
revision of, **6**:323a
on contemporary scientists, **5**:521b; **14**: 446b
correspondence of, **9**:465a
and Halley, **1**:403a; **5**:24b, 25a,b; **6**:69b, 72a; **14**:511a
Historia coelestis Britannica, **1**:403a; **5**: 520b–521a; **6**:364a; **7**:81a
publication of, **14**:511a
influence of, **2**:220b; **5**:516a
influences on, **13**:96a
lunar theory, **8**:178b
and Newton, **3**:431a; **5**:25a,b, 516a; **10**: 44b, 64a,b, 78b, 84a
observations of, **1**:86a; **6**:70a; **10**:84a; **15**:330b
students of, **5**:515b
Flanders
geography, **2**:190b
geology, **5**:476a–b
Flare, electrical, **7**:403a
Flash ranging, **3**:372b
Flavin, **14**:174a
Flavone, **7**:471a
Flavonoids, **15**:257b
Flax, **1**:436a
Flea
anatomy, **6**:484a
and disease, **5**:270b, 503a; **14**:623b
metamorphic cycle, **3**:180b
microanatomy, **8**:128a
reproduction, **8**:128a
Flechsig, Paul Emil (1847–1929), 5:26–28; 1:580a; **7**:504b
Flechsig's tract, **5**:27a
Fleck, Alexander, **12**:505a
Fleischer, Johannes (1539–1593), 5:28
Fleming, Alexander (1881–1955), 5:28–31; 4:154b, 156a; **5**:42a, 43b; **14**:511b
Fleming, John (1785–1857), 5:31–32; 3:395b; **5**:333b
Fleming, John Ambrose (1849–1945), 5:32–33; 4:6b, 80b, 284a; **9**:99a; **11**:420b
Fleming, Williamina Paton (1857–1911), 5: 33–34; 8:106a; **9**:194b; **10**:600a
Flemløs, Peter Jacobsøn, **2**:406a, 410a,b
Flemming, Walther (1843–1905), 5:34–36; 7:454a; **8**:283a; **9**:381a; **11**:256a*, 573a
cytology, **2**:626a; **3**:29b; **5**:582a; **13**:89a; **14**:235a
Flemming's fluid, **5**:36a
Flemming's stain, **5**:36a
Flerov, Georgii Nikolaevich, **7**:527a
Fletcher, Andrew Almon, **1**:443a; **3**:352a
Fletcher, Horace, **3**:257a
Fletcher, John, **14**:513a
Fletcher, Lazarus, **9**:379b
Fletcher, Walter Morley (1873–1933), 5:36–38; 5:81a; **8**:18a*, 453a, 532a, 535b*; **15**:73a–b, 419b
physiology, **6**:501a–b; **8**:452b
"Fletcherism," **3**:257a
Flett, John Smith (1869–1947), 5:38–39; 5:337a, 339b*; **6**:56b*, 116b*
Flexner, Abraham, **9**:58a
Flexner, Simon (1863–1946), 5:39–41; 5: 316b; **6**:232b; **8**:275b, 447a; **10**:141b–142a, 144b*; **11**:177a*; **12**:482b, 483b
Flexner bacillus, **5**:40a
at Rockefeller Institute, **15**:252b
Flexure
theory of
see Beams (arch.), flexure of
Fliess, Wilhelm, **5**:172a–174b *passim,* 176b–177a, 179a
Flight, **6**:229b
mechanical reproduction of
ancient Greek, **1**:233a
Renaissance, **8**:199a
seventeenth-century, **14**:366a
nineteenth-century, **6**:167a
physiology of, **7**:503b; **9**:102a
theory of, **2**:88b, 366b; **5**:148a; **8**:199b, 200b, 205a, 213a; **10**:520a
see also Aviation
Flinders, Matthew, **1**:520b; **2**:517b–518a
A Voyage to Terra Australis, **2**:519b
Flint, **12**:318a
knapping, **8**:105a
liquor of flints, **5**:420b, 422a
prehistoric, **2**:346b; **5**:162b–163a; **7**:197b; **11**:270a; **12**:520a
Floating bodies, **3**:115b, 331a; **4**:257b
equilibrium of, **3**:591b; **6**:602a
rule of metacenters, **4**:230b
see also Hydrostatics
Flocculation reaction, **11**:271b
Flodin, Per, **13**:421b
Floerke, Heinrich Gustav, **1**:46a
Floetz formations, **2**:553a,b, 554a
Floirac
observatory, **11**:320a–b
Flood
biblical
geological evidence for, **2**:568b–569b; **11**:469a
scientific explanation of, **6**:72a; **14**:296a, 400a
control measures, **3**:421a; **11**:401a–b
see also Diluvialism
Flora (asteroid), **5**:258a; **6**:402b
Flora, **5**:439a*
Florence
Academy of, **2**:590b
botanical garden, **5**:383b
medical school, **10**:267a
Museum of the History of Science, **5**:55b
Museum of Physics and Natural History, **5**:55b
Florence, University of
anatomy at, **9**:153b
Florentine enigma
see Viviani, Vincenzo, Viviani's problem
Flores (island), **14**:203a
Florey, Howard Walter (1898–1968), 5:41–44; 5:30a
Florida
botany, **6**:122b, 163a
Geological Survey, **6**:122b
Florideophycidae, **7**:535b
Florkin, Marcel, **5**:149b
Flotow, Julius von, **9**:344b; **10**:13b
Flour
agenized
toxicity of, **15**:419a
analysis of, **3**:7b; **10**:325b; **11**:51b
enrichment of, **14**:393b
preservation of, **3**:7a; **10**:326a
Flourens, Marie-Jean-Pierre (1794–1867), 5:44–45; 2:27a; **3**:521b; **9**:101b; **12**: 315b
acquaintances of, **5**:566a; **13**:557a; **14**: 142b; **15**:126b
on contemporary scientists, **3**:203a*, 528a*; **5**:251b, 253a; **7**:199a*; **9**:7a, 8b
experiments of, **2**:124b, 446b; **6**:440b
historiography, **10**:569b*
influence of, **2**:478a; **5**:463b; **6**:441a
Des manuscrits de Buffon, **2**:107b*
neurophysiology, **1**:446b; **5**:505b; **8**:506a; **11**:215b; **12**:400b; **15**:109a
Flow
discontinuous, **6**:545b
inhomogeneous, **7**:493a
in pipes, **11**:123b, 393a; **13**:10b
supersonic, **13**:73b
theory of, **11**:124a
velocity of, **11**:5a–b
measurement of, **3**:182b; **8**:200b; **11**: 164b; **14**:494b–495a
volume of
measurement of, **3**:182b
see also Hydraulics, Hydrodynamics, Outflow, Overflow, Turbulence (hydrodyn.), Underflow
Flower, William Henry (1831–1899), 5:45–47; 3:574a; **6**:593b; **10**:263a,b*; **11**: 515a*

Flowering
chemistry of, **12**:124a
physiology of, **3**:359a
Flowers
color of
genetics of, **6**:22b
and insects, **3**:575a; **8**:518a
heterostylism, **3**:422a, 575a
opening and closing, **10**:577a
spectra of, **11**:267a
structure of, **1**:521a; **2**:219a; **4**:307a; **5**: 424b; **11**:314b; **12**:588a; **15**:67b, 267b
symmetry of, **4**:306b
see also Pistil, Stamen, etc.; Plants, morphology, floral; Wild flowers
Flöz, **5**:444b–445a
Fluctometer, **9**:43a
Fluctuation phenomena, **4**:314a–318a; **12**: 497a–b
Fludd, Robert (1574–1637), 5:47–49; 6:151a, 397b; **9**:327a; **14**:372a
cosmology, **7**:302a
critics of, **9**:316b, 320b(n); **14**:365b, 367a
editions of works, **9**:23b
Tractatus apologeticus, **14**:209b
Fluents, **3**:432a–b, 458b; **5**:54b
Flügel, Johann Gottfried, **6**:553a
Flügge, Carl, **7**:425b; **9**:259b; **10**:559b; **14**: 249a
Fluid mechanics
see Hydrodynamics, Outflow
Fluidity
nature of, **2**:311a, 378b; **4**:352b
Fluids
compressible, **10**:75a
motions of, **5**:188a
density of, **10**:75a
dielectric properties, **7**:387a
diffusion of, **13**:11a, 351a
frictionless, **2**:166b, 167a, 168a
heat transfer, **11**:393a
interior pressure, **1**:128b
internal friction, **13**:75a–76a, 78a
molecular distributions, **7**:387a
Navier-Stokes, **4**:230a,b
opalescence of, **4**:315a; **10**:235b
percussion of, **7**:565a
resistance in, **2**:299b, 335a, 356a; **5**:200a–b; **11**:494a
rotating, **11**:58a
specific heat, **1**:128b–129a; **8**:545b
as thermodynamic systems, **2**:168a; **5**: 390b
viscous, **4**:230a–b; **10**:514a
rotation of, **9**:107b
vortical motion in, **5**:188a
work done by, **3**:417b
see also Gases, Hydrodynamics, Liquids
Fluke (trematode), **2**:617a
Fluorene, **2**:67b; **5**:13a
Fluorescein, **4**:296b
Fluorescence
see Luminescence
Fluorine
bombardment of, **3**:329b; **4**:580a
chemistry of, **9**:451a
cold, effects of, **4**:80b
compounds of
organic, **2**:316a; **13**:177a–b
isolation of, **5**:474a; **9**:451a
liquefaction of, **4**:80b; **10**:206b
in organic substances, **5**:315b
preparation of, **2**:429a
solidification of, **4**:80b
in teeth, **9**:527a
toxicity of, **6**:136a
in water, **9**:258a
Fluorite, **12**:146a
crystal form, **6**:179a
optical properties, **11**:582a–b
photoelectricity, **6**:96b
spectrum, **10**:345b
Fluoroscopy, **11**:213a; **15**:72b
Fluorspar
cleavage of, **14**:490b
volatility of, **9**:106b
Flüter, Balthasar, **1**:49b
Fluvialism, **2**:570a; **11**:277a; **12**:38b, 263b
Flux
luminous, **2**:200a–b
magnetic
quantization of, **8**:477a–b, 478a
Flux, universal, **6**:290b–291a
Fluxions, **1**:531a; **2**:54a; **3**:344b, 432a–b; **5**: 54b, 195b*, 466a,b, 520a; **7**:162b, 163a; **8**:610b–611a; **9**:452b; **10**:47b–51a, 53a, 66a, 67b; **11**:579b; **12**:444a; **13**:265b, 266a; **14**:569a; **15**:99a
definition of, **12**:444b
and infinite series, **10**:51a
inverse method of, **10**:68a
Fluxus (math.), **4**:506a
Fluxus formae, **1**:100b; **2**:610b–611a; **5**:116a
Fly
development of, **13**:173a
and disease, **5**:503a; **15**:454b
eye of, **6**:484a; **10**:176b
generation of, **5**:226b
larvae, **7**:437b
life cycle, **7**:272b–273a
see also *Drosophila,* Housefly
Fly (ship), **7**:185b
Flysch, **13**:284a, 285a
exotic boulders, **6**:229a; **12**:140a
Flywheels, **8**:209a
Foà, Pio, **2**:164b; **10**:205b
Foams, **11**:242a
Fock, Vladimir Aleksandrovich, **6**:147b
Focke, Wilhelm Olbers
Die Pflanzenmischlinge, **3**:422a; **9**:283a
Focus, geometrical
definition of, **4**:49b
theory of, **7**:573b
Fodder
preservation of, **14**:45b–46a
Foerste, August Frederick (1862–1936), 5: 49–51
Foerster
see Förster
Foetus
see Fetus
Fog chamber
see Cloud chamber
Fogel, Martin, **7**:375a–b, 376b
Föhn (wind), **6**:97b
Foix, François de, **14**:21b
Fok, Vladimir Aleksandrovich, **11**:580b, 581a; **15**:43a
Fokker, Adriaan Daniël, **8**:498b; **11**:13a
Fol, Hermann (1845–1892), 5:51–53; 6:338a
critics of, **14**:429a
cytology, **12**:192b; **14**:429a
Folding (geol.), **1**:286a; **3**:325b; **4**:364a; **5**: 437a, 476a; **9**:367b; **11**:35b
anticlinal, **8**:463b
basement folds, **1**:237a
complex, **1**:235b–237a; **12**:277a
cover folds, **12**:140b
epochs of, **13**:147a
experimental, **3**:586b; **14**:402b
explanations of, **2**:158b; **3**:553b; **4**:309a, 350a; **6**:55a–b; **9**:27a–b; **11**:505a–b; **12**:276b
isoclinal, **8**:32b
and oceans, **6**:168a
overturning, **2**:89b; **6**:227a, 228a; **8**:462a, 543b
posthumous flections, **1**:472a
recumbent folds, **1**:394a; **8**:543b, 544a; **12**:140b
vestigial ranges, **7**:250b–251a
see also Nappes, Tectonics
Folger, Timothy, **5**:136b
Foliation (geol.)
cause of, **3**:567b–568a
plane of, **3**:567b
Folic acid, **4**:358a
Folie, François, **8**:250b
Foligno, Gentile da, **1**:619a
Folin, Otto (1867–1934), 5:53; 13:152a; **14**:523b
Folkes, Martin (1690–1754), 5:53–54; 1: 411a; **4**:565b; **8**:609b; **12**:444a; **13**: 68a; **14**:418b
Folklore
Arabic, **3**:549a
Jewish, **15**:657b(n)
Mesopotamian, **15**:642a
of Northwest Coast, **2**:210a–b, 212a–b
Renaissance, **5**:119b
Folliculin, **2**:345b
Font de Gaume (cave), **2**:450b
Fontaine (Fontaine des Bertins), Alexis (1704–1771), 5:54–55; 3:386a; **7**: 560b; **13**:571a
Fontaine, Hippolyte, **5**:496b
Fontana, Felice (1730–1805), 5:55–57; 3: 560b; **4**:503a; **5**:267b; **11**:510a; **14**: 74b, 637a
Fontana's canal, **5**:56b
De irritabilitatis legibus, **12**:138a
on respiration, **7**:12b
Fontana, Francesco, **13**:437b
Fontana, Gregorio, **11**:598b; **12**:561b
Fontenelle, Bernard le Bouyer (or Bovier) de (1657–1757), 5:57–63; 3:114a; **9**:33b
at Academy of Sciences, **2**:286a; **4**:221a
acquaintances of, **3**:459b; **8**:115a, 149a; **9**:62a; **11**:68a, 512b
on contemporary scientists, **4**:136a*, 217a*, 221b, 473a, 515b; **5**:352b, 354b*, 418a; **7**:576b; **8**:172b, 305a; **10**:319a; **11**:158a; **12**:127b; **13**:593b
correspondence of, **3**:486a
cosmology, **9**:26a
critics of, **8**:306a
on Descartes, **4**:51a
electricity, **10**:146a
Élémens de la géométrie de l'infini, **5**:510a
Éloge de Newton, **4**:51a
influence of, **1**:323a; **2**:545a; **6**:129b
Food
additives, **3**:257a; **14**:357b
adulteration of, **1**:44a; **3**:6b, 238a; **11**: 318b; **14**:357b
detection of, **10**:326a
caloric values of, **1**:610b; **5**:126b; **11**: 585a–b
classification of, **4**:412b; **11**:173a
digestibility of, **1**:543b
infection of, **10**:527a
inspection of, **1**:528a; **8**:270b
manufacture of, **11**:171a
hazards of, **15**:419a
nitrogen in, **6**:517b*
poisons in, **1**:529b; **6**:443a; **7**:171a
and population, **4**:272a; **9**:67b–70a; **10**: 444b
preservation of, **3**:7a,b; **10**:366b, 588b; **11**:371a
canning, **4**:554b
vacuum, **10**:292b
rations of, **6**:517b

Food (cont.)
supplies of, **8**:29b
technology of, **15**:648b, 651a
vitamin-enrichment, **6**:499b
wild plants, **8**:397a*
world production, **4**:271b
see also Chemistry, food chemistry; Dietetics; Digestion; Nutrition; specific foods
Food and Drug Act (U.S.), **14**:357b
Food and Drug Administration (U.S.), **3**:69a; **10**:444b
Food industry, **9**:285b
Foot, Nathan Chandler, **13**:263a
Foot (unit of length)
standardization of, **9**:554b
Foot-and-mouth disease, **7**:100b; **10**:568a
etiology of, **3**:94b; **8**:450a–b
Foothill Electronics Museum (Los Altos Hills, Calif.), **4**:7a–b
Föppl, August (1854–1924), 5:63–64; 8:599a; **9**:446a; **11**:123b, 124a; **12**:527a
Einführung in die Maxwellsche Theorie der Elektrizität, **1**:24a
Foraminifera, **9**:432a, 573b; **12**:141b, 231a
fossil, **1**:210b; **9**:579a; **11**:386a, 501b, 502a
geological role, **14**:398a
Holocene, **3**:519b–520a
morphology, **3**:89a
dimorphism, **12**:183a
planktonic, **3**:520a
as stratigraphic tools, **3**:520a
taxonomy, **3**:89a, 519b–520b; **4**:234b–235a; **10**:221b; **11**:386a–b
Forberger, George, **1**:49b; **10**:306b
Forbes, Alexander (1882–1965), 5:64–66; 2:73a
Forbes, David, **6**:565a
Forbes, Edward, Jr. (1815–1854), 5:66–68; 6:266a, 591b; **7**:70a
biogeography, **5**:599b; **6**:490b
biographies of, **5**:335b
evolution, **14**:135a
geology, **14**:211a*
marine biology, **5**:437a, 469b; **7**:91b; **14**:145b
paleontology, **10**:204a
zoology, **5**:470a,b
Forbes, George Shannon, **11**:418a
Forbes, J. M., **14**:533b
Forbes, James David (1809–1868), 5:68–69; 5:334a, 335b; **9**:199b; **10**:9b*; **13**:52a, 237a
critics of, **13**:523a
glaciology, **3**:211b; **9**:220b; **13**:522a
oceanography, **13**:360a
physics, **9**:200b; **11**:115b; **13**:238b; **14**:185a
statistics, **9**:218b–219a
students of, **4**:78a; **9**:199a, 207b; **11**:291a; **13**:51b
Forbes, Stephen Alfred (1844–1930), 5:69–71
Forbush, Scott Ellsworth, **3**:370b
Forcadel, Pierre, **1**:229b; **4**:449b; **13**:540a
Force
aerodynamic, **3**:196a; **7**:437a
central, **2**:54b, 57a; **4**:529b, 532b, 533b–534a; **5**:511a; **7**:466a, 560b; **8**:491a; **10**:72a; **11**:399b; **14**:207a
centrifugal, **10**:72a; **12**:608b–609a
law of, **7**:276a; **9**:52a; **10**:62b, 75a–b
theories of, **4**:60b; **6**:606b–607a, 609b, 610b–611a; **8**:152a–b; **10**:60b–61a; **12**:118a
centripetal, **6**:610b; **10**:69a–75a *passim,* 77b
concept of, **10**:70a
circular, **4**:532b, 533b–534a
composition of, **11**:488a,b
concept of
ancient Greek, **1**:253a–b; **13**:92a–b
medieval Arabic, **1**:409a–410a
medieval Latin, **2**:606a; **13**:195b–196b, 197a–198a
Renaissance, **8**:199b, 228a
seventeenth-century, **4**:60a–b; **6**:607b–608b; **8**:151a, 152a–153a, 156b–159b *passim;* **10**:62a, 70b; **14**:162b
eighteenth-century, **1**:111b, 112a–b; **2**:54b, 330a–b; **7**:227b–228b; **11**:580a; **12**:462b; **14**:84a; **15**:343b
nineteenth-century, **2**:166b, 563b; **3**:73b; **5**:193b; **9**:235b–240a; **10**:183b–185b; **12**:74a
conservation of, **6**:608a; **9**:235b–240a; **15**:85a–86b
conversion of, **4**:530b–531a, 534b–535a, 536a, 538a; **5**:560a; **9**:236b, 237b–238a; **10**:184b–185b
correlation of, **5**:560a; **9**:204b, 205a
electrical, **1**:67a–b; **8**:491a, 495a–b; **9**:548a; **15**:356b
inverse-square law, **3**:157a, 444b; **5**:136a; **11**:142a, 496b
lines of, **9**:205b
measurement of, **3**:444b; **11**:433a
electrodynamic, **9**:217b
inductive vs. ponderomotive, **6**:251b–252a
law of, **1**:146a; **6**:250b–251a; **14**:207a–208a
measurement of, **1**:145a–b
electromagnetic, **12**:408a
effects of, **14**:32b–33a
mathematical law of, **10**:187b–191b
measurement of, **9**:209b; **13**:364b
theories of, **2**:166b; **4**:532b; **12**:407b
electromotive, **3**:19b, 333a,b, 334a; **5**:390b; **8**:36a, 188a, 503a; **9**:207a, 217a,b, 394a; **10**:135a, 192a, 481b, 553a; **11**:297b, 298a; **13**:359a, 578b–579a; **14**:77b, 78a; **15**:435a, 437b
and chemical equilibrium constant, **13**:579b
and ionic concentration, **15**:433b
kinetic derivation, **13**:429a–b
magnetic fields, effect of, **2**:559a; **14**:32b–33a
measurement of, **1**:555b; **7**:450a; **11**:50a; **15**:434a
thermal coefficient, **15**:438b
electroscopic, **10**:190a–192b
electrostatic, **7**:469a,b; **13**:377a; **14**:70a–71a
measurement of, **3**:562a; **9**:209b
and fatigue, **1**:409b–410a
frictional, **4**:222b
hydrodynamic, **2**:167a, 167b–168a
lifting, **7**:437a
impressed, **5**:113b–114a; **9**:136b–137b; **10**:70b, 71a
interatomic, **8**:473b, 474b
laws of, **8**:185b–186a
intermolecular, **1**:141b; **2**:6b; **3**:305a, 619b, 620b; **4**:314a, 534b, 536b–537b; **5**:15b; **8**:475a; **9**:215b, 222b–223a; **10**:4a; **13**:177b; **14**:125a
internal, **3**:417b, 496a; **5**:391a
interparticulate, **15**:315b, 359a–360a, 377a–b, 378a
laws of
criteria for, **6**:250b
lines of, **9**:205a–210a *passim*
concept of, **4**:532b–539a *passim*
magnetic, **2**:137a; **9**:206b; **11**:22a; **13**:377a
components of, **5**:306a
conversion to electric force, **4**:534b
inverse-square law, **3**:444b; **9**:234b
measurement of, **3**:443a, 444b, 505b–506a; **4**:215a
and matter
see Matter, and force
measurement of, **10**:71a, 72a
molecular, **8**:496a, 497a, 498a; **11**:21b–22a; **13**:522a; **15**:358b–363a *passim,* 379a
and motion, **9**:51a
motion-dependent, **8**:491a
normal, **3**:441b, 442b
nuclear, **9**:37b, 38a; **13**:240b–241a; **15**:27a
parallelogram of, **2**:38b; **3**:182b, 496a; **7**:611a; **8**:200b; **13**:584b, 585b–586a
attempts to prove, **2**:38b
patterns of, **4**:529b–530a, 532b, 536b–539a *passim*
resolution of, **3**:441b, 442a; **8**:200b
transmission of, **3**:73b
unity of, **4**:530a–531a, 535a; **9**:445a; **12**:289a
universal, **15**:86a
Bošković's law, **2**:330a–b; **4**:529b; **14**:488a
weak interaction, **4**:579b–580a
see also Action; Attraction, gravitational; Dynamics; Energy; Impetus; Inertia; Matter, and force; Momentum; Power; *Vis viva*
Force-displacement, **3**:417b
Forchhammer, Johan Georg (1794–1865), 5:71–72; 4:127b
Ford, Edmund Brisco, **3**:572b
Ford, Lester Randolph, **14**:317a
Ford, Peter John, **8**:483a
Fordos, Mathurin-Joseph (1816–1878), 5:72–73
Fordyce, George, **3**:67b, 487b; **6**:567b
Forel, Auguste-Henri (1848–1931), 5:73–74; 1:174b; **4**:160b; **6**:440a
heredity, **12**:298b
neurohistology, **5**:571a; **11**:274a; **14**:143a
translations of works, **12**:298a
Forel, François Alphonse (1841–1912), 15:158–159; 2:142b; **8**:544a
Foresight, **3**:90a
Forestry
eighteenth-century, **2**:576b; **4**:224a
nineteenth-century, **3**:7b; **6**:136b–137a; **9**:366b; **13**:596a
twentieth-century, **2**:345b; **3**:317b; **6**:122b; **9**:534b–536a; **10**:130a; **14**:106b–107a
and ecology, **14**:107a
plantation type, concept of, **9**:536a
see also Arboriculture, Trees
Forgeries, literary
nineteenth-century, **3**:214a
see also Hoaxes, scientific
Forges, **3**:454b
Forir, Henri, **8**:463b
Forlanini, Carlo, **11**:482a
"Form"
ancient Greek discussion, **1**:251a; **4**:466a; **11**:25b–26a, 29b
Byzantine discussion, **7**:135a
medieval discussion, **12**:7a–7Aa, 429a
Renaissance discussion, **1**:261a; **2**:611a; **8**:240b
seventeenth-century discussion, **7**:611b; **9**:15a
eighteenth-century discussion, **12**:603a–b
nineteenth-century discussion, **15**:79b
twentieth-century discussion, **7**:486a
in art, **4**:259b, 260b

continuity of, **12**:311b–312a
intension and remission of, **2**:193a, 194a, 609a, 610b–611a; **3**:420b; **4**:237b; **6**:378b; **10**:62a; **13**:196b, 209b
latitudes of, **7**:116a, 118a, 120a
magnetic, **5**:398a–b
and matter, **1**:254b–255b, 277b, 278a; **9**:252b; **11**:244b; **12**:7a–b, 59b; **15**:495a
multitude of, **13**:195b, 196b, 197a, 200a; **14**:459a
as object of science, **1**:259b–261a, 374b–375b; **3**:452a
in optics, **6**:191a–b, 192a–193b; **14**:459a–460a
rejection of, **13**:279a
and soul, **1**:260a–b; **6**:152a, 459a
unity of, **5**:401b, 402b
Forma fluens, **1**:100b; **2**:610b–611a; **5**:116a
Formaldehyde
in foods, **14**:357b
polymerization of, **2**:621a
synthesis of, **2**:324a; **6**:462b
Formalism, **2**:304b, 513b; **3**:57a,b; **4**:323b, 331b–332a; **5**:438b; **6**:393a; **12**:15a
critics of, **4**:206a; **5**:154a,b; **6**:473b
Formamide
synthesis of, **9**:354a
Formations, geological
classification of, **11**:562b–563a
concept of, **5**:205a; **8**:285b
correlation of, **8**:99b
definition of, **5**:205a
primary, **5**:475b
sedimentary, **1**:209b–210a
classification of, **7**:253a
see also Stratigraphy
Formic acid, **6**:381a; **8**:60a; **9**:105a; **10**:499a; **15**:172a
synthesis of, **2**:66b; **5**:431b; **10**:499a
Formol titration, **12**:546b
Forms, mathematical, **2**:470a; **3**:164b, 314b; **4**:342a,b; **11**:55a,b
algebraic, **3**:263a; **5**:472a–b; **7**:444b; **8**:250a; **10**:442a; **11**:407b; **12**:86b
bilinear, **2**:491a; **3**:166a; **6**:497a; **12**:188b, 189a, 284a; **13**:428a
normal, **13**:428a
binary, **3**:164a; **5**:472a
canonical, **13**:220a
contravariant, **13**:219b
covariant, **4**:342b; **5**:201b
cubic, **4**:341a,b, 342b; **13**:250a
decomposable, **4**:125a
differential, **11**:60a, 471a; **15**:198b
extreme, **10**:573b
integrability of, **5**:54b
quadratic, **3**:263a,b; **11**:407b; **13**:600a
expression of, **11**:536a
exterior differential, **3**:95b–96a
Hermitean, **6**:308a
homogeneous, **5**:151a; **6**:357b; **7**:573b
inertia form, **6**:572a
infinite, **13**:428a
linear, **4**:125a; **13**:428a
modular, **4**:342b
*n*th-order, **3**:164a; **8**:389a–b
Pfaffian, **10**:573b–574b
quadratic, **4**:124b, 125a, 341b, 474b, 573b; **5**:471b; **6**:214b, 235b, 308b, 394a; **7**:515a; **8**:138b; **11**:407b, 408a; **12**:170b, 284a; **13**:218a, 428a
binary, **3**:164a; **6**:571b; **9**:125a; **13**:561b
classification of, **2**:491a; **11**:408b
composition of, **5**:198a, 299b; **6**:572b; **12**:172b
generalization of, **6**:308a
indefinite, **9**:126a
inertia of, **12**:172a; **13**:220a
minima, **11**:58a; **14**:630b
n-ary, **9**:411b–412a
quadratic, **3**:263a,b; **11**:407b; **13**:600a
reduction of, **6**:308a, 571b; **9**:412a
ternary, **9**:412a; **11**:54a
theory of, **6**:286b; **7**:54b, 168b, 565b; **8**:385b, 386a; **9**:173b; **12**:469a–b; **14**:94b
transformation of, **3**:166a; **6**:357a, 392a; **13**:250a
representation of, **6**:394a
semi-invariant, **13**:219b, 220a
skew, **8**:324b–325a
systems of, **5**:472a–b; **11**:471a
ternary, **3**:164a; **4**:125b; **6**:571b; **8**:542b
Formulas, chemical
see Compounds, inorganic; Compounds, organic; Notation, chemical
Formulas, dietary, **4**:405a
Fornax (constellation), **12**:349a
Forrest, George, **1**:422b
Forschbach, Joseph, **1**:441b
Forsskl (also Forsskhl or Forskl), Peter (1732–1763), 5:74; 1:124b; **8**:379b; **9**:456a
Forssman, John, **4**:299b
Förster, A., **4**:154a
Forster, (Johann) Georg Adam (1754–1794), 5:75–76; 5:76b–77a; **6**:549b; **9**:456a
Forster, Johann Reinhold (1729–1798), 5:76–77; 5:75a; 7:604b*; **12**:516a
Förster, Wilhelm Julius, **4**:369b; **5**:258b; **8**:539a; **9**:2b, 563b; **11**:114b; **14**:55a,b, 629b
Forster, William, **4**:13b
Forsyth, Andrew Russell (1858–1942), 5:77–78; 3:170a*; **5**:414a; **6**:451b; **10**:97b*; **12**:10a
Theory of Functions, **14**:317b
Fortieth Parallel
geologic survey of, **4**:365b, 366a; **6**:13b; **7**:370a,b; **14**:192a–b, 312b
Fortification
see Architecture, military; Engineering, military
Fortin, Jean Nicolas (1750–1831), 5:78
Fortin barometer, **5**:78b
Fortschritte der Mathematik, **9**:358b
editors of, **14**:158b
Die Fortschritte der Physik
editors of, **7**:509a
Foscarini, Paolo Antonio, **1**:588a, 589b; **5**:242b
Fossils
medieval discussion, **1**:30a, 101b; **11**:469a
Renaissance discussion, **1**:97a; **10**:280b
seventeenth-century discussion, **4**:222a*; **5**:515b; **8**:307b–308a; **11**:41a, 317b; **12**:257a; **13**:29b, 33b–34a; **14**:500b–501a, 501b
eighteenth-century discussion, **1**:393a; **3**:189b; **5**:205a; **6**:580a, 581b; **8**:251b–252a; **9**:531b–534a, 539b; **13**:179a, 410a; **14**:119b; **15**:52b–54a, 56a
bibliographic indexes of, **1**:494b; **4**:541a
castings of, **11**:594a
Chinese discussion, **12**:380b
chronological sequence of, **2**:497b, **7**:60a, 460b
collecting of
Renaissance, **1**:526a
seventeenth-century, **8**:307b; **14**:502a
eighteenth-century, **1**:487b; **7**:412a; **8**:4a–b; **12**:159a, 459a; **14**:119b, 133a; **15**:52b
early nineteenth-century, **6**:120a; **8**:564b, 566b; **12**:202a, 488a, 553a*; **15**:152a
mid nineteenth-century, **2**:497b; **3**:297b; **4**:10b; **5**:519b, 533a–b, 538a; **6**:57a,b; **7**:460b; **8**:43b; **9**:86b, 240b; **10**:211b, 212a; **11**:501b
late nineteenth-century, **3**:606a; **8**:264b; **9**:134a; **10**:241b; **12**:339b; **14**:312b–313a; **15**:92a
colonies of, **1**:469a
comparative study, **5**:481a
dating of, **2**:204b; **3**:607a; **5**:109b; **8**:105a
descriptions of, **8**:4a
guide, **1**:472b; **2**:556a; **11**:502a; **13**:532a
index, **1**:472b; **3**:322a; **5**:205b, 488a*; **13**:532a
intercontinental correlation, **3**:391b
marine
inland occurrence, **3**:66b; **8**:4a; **9**:532b–533a; **10**:240a; **11**:272b–273a; **13**:564a; **14**:296a
see also Micropaleontology, marine
medical use of, **1**:482a; **5**:500b
new species of, **1**:131a, 469a, 472b; **2**:128b; **4**:308b; **5**:295b–296a; **9**:87a
origin of, **8**:4a, 415b–416a
phylogeny, **5**:486b
reconstruction of animal forms, **2**:567b; **3**:525b
serial sections, **12**:520a
stratigraphic significance of, **2**:109a–110a, 481a, 486b, 494a–497b; **5**:14a–b, 475b, 476a; **7**:460b; **8**:32b–33b, 486b, 594a; **9**:532a, 534a, 583a; **10**:322b, 586a; **11**:134a; **12**:182a–b, 487a–491b, 549a; **13**:87a; **14**:133a, 392a
trace, **2**:109b; **10**:204a
type specimens, **3**:520a
utility of, **7**:253a
and zones, **1**:285b; **8**:33a
see also Bones; Geology, time scale; Micropaleontology; Paleobotany; Paleontology; Shells; specific countries, formations, etc.; specific geological time divisions; specific organisms, families, classes, etc.; Teeth
Fossombronia, **6**:219a
Foster, Edward Waddington, **11**:422b
Foster, George Carey, **3**:470b; **7**:93b; **8**:444a; **10**:38a; **13**:156a
Foster, Henry (1797–1831), 5:78–79
Foster, James T., **4**:364b
Foster, John Wells, **4**:73b; **7**:45a
Foster, Michael (1836–1907), 5:79–84; 2:598b; **6**:597a*,b*; **9**:143a*
acquaintances of, **6**:498b, 595b; **12**:354b; **14**:424a
Elements of Embryology
translations of, **7**:402a
on heartbeat, **5**:280a,b, 282a
influence of, **2**:599a; **5**:279a,b; **9**:516b
physiology, **11**:518a
students of, **1**:420a,b; **2**:370b, 501b; **5**:36b; **8**:14a–b; **9**:142a; **11**:517a; **12**:395b; **13**:352b; **15**:104b
Textbook of Physiology, **12**:397a
Treatise on Physiology, **2**:339b
Fothergill, John, **1**:489a; **3**:537b; **4**:295a; **5**:133b; **8**:459b, 460a; **9**:482a; **11**:48a; **12**:516b; **14**:194b
Foucault, Jean Bernard Léon (1819–1868), 5:84–87; 2:88b; **15**:489b
astronomy, **11**:62a, 114b; **13**:36b
and Fizeau, **5**:18b–19b, 85b–86b
Foucault's pendulum, **3**:404b; **4**:592a; **5**:86a–b, 307b

Foucault, Jean Bernard Léon (cont.)
optics, **1**:202a; **5**:20b; **8**:399a; **9**:371b–372a; **10**:34b; **14**:290a
Fouchy, Jean-Paul Grandjean de (1707–1788), 5:87–88
and Académie des sciences, **8**:70a
acquaintances of, **4**:23a; **7**:542a; **8**:178b; **10**:145a
astronomy, **1**:401a; **9**:528a*
on contemporary scientists, **3**:198b*, 486a; **4**:25b*, 590a*; **5**:63a*, 352a*, 436a*; **6**:48a*; **7**:78b*
and Société des Arts, **3**:281a
Foulis, Andrew, **13**:413b
Found purse, mathematical problem of, **4**:606b, 607a, 612b
Fountains, **9**:119b
Fouqué, Ferdinand André (1828–1904), 5:88–89; **1**:472a; **2**:500a; **7**:548b; **12**:180a; **13**:617a
geology, **3**:161a; **7**:549a; **9**:367a
Four elements, method of
see *Ssu-Yüan shu*
Four-color problem, **9**:430b–431a; **10**:483b
Fourcroy, Antoine François de (1755–1809), 5:89–93; **2**:74a; **3**:561a*; **5**:318a, 602b; **10**:609a
acquaintances of, **3**:6a; **5**:202b; **11**:494b; **14**:14b, 79b
chemistry, **9**:245b
analytical, **2**:5b; **3**:346b; **12**:287a; **13**:281a; **14**:492a
nomenclature, **1**:345a; **6**:164b
theoretical, **8**:82b
on contemporary scientists, **3**:7b, 201a; **8**:88a*; **13**:603b
critics of, **2**:385b; **12**:68a(n); **15**:172a
and education, **9**:473a; **10**:612a
Élémens d'histoire naturelle et de chimie
editions of, **8**:81a
translations of, **10**:108a
influence of, **4**:233b; **13**:309b, 313b; **14**:16a
and Lavoisier, **8**:80b, 81a
medicine, **8**:352a
metallurgy, **10**:496b
Philosophie chimique, **11**:164a
students of, **2**:385a; **3**:455a; **5**:558a; **10**:209a; **13**:309a
translations of works, **10**:108b
and Vauquelin, **5**:91a, 92a–93a; **8**:331b; **13**:596b, 597a–b
Fourcy, François (de), **2**:6b
Fourier, Jean Baptiste Joseph (1768–1830), 5:93–99; **5**:261a; **15**:342b, 481a
acquaintances of, **6**:325a; **7**:51a; **10**:2b, 247b; **11**:6b; **13**:127a
algebra, **13**:129a
analysis, **3**:229a; **10**:248b
on contemporary scientists, **3**:208b*; **4**:16b; **15**:382b
critics of, **15**:384a
equations, theory of, **2**:573b, 574a
Fourier analysis, **3**:370a; **4**:579a; **6**:245b; **9**:175a; **10**:3a, 4a; **15**:380a, 386a,b
Fourier coefficients, **1**:194b; **4**:476a; **6**:176b, 472b; **12**:188b
Fourier expansion, **2**:144b; **4**:407b, 476a; **5**:96a; **6**:90a; **10**:573a
Fourier integrals, **2**:238b; **3**:143b; **4**:578a; **5**:95a, 96b, 97a; **13**:424a; **14**:221b, 282a; **15**:385a,b, 386a
Fourier inversion formula, **3**:140a
Fourier series
see Series, Fourier
Fourier system, **5**:608a
Fourier transformations
see Transformations (math.), Fourier
geometry, **7**:615a
influence of, **1**:365b; **3**:379a; **4**:124a; **10**:186b, 188a, 192a; **11**:236b; **13**:78b, 130b, 374b; **15**:199b, 384a–385a
influences on, **9**:476a; **11**:134b
and Laplace, **15**:377a, 380a, 384a–385a
mathematical physics, **4**:478b
heat, **2**:136a; **3**:33a–b, 377b; **4**:126a, 222b; **10**:248a, 249a; **13**:27a, 127b, 376b, 378b, 382b–383a; **15**:360b, 363a, 380a, 384a–385a, 484a, 486a–b, 488b
on metric system, **4**:16a
Oeuvres, **3**:560a
and Poisson, **15**:484a, 486a
statistics, **11**:236b–237a
Théorie analytique de la chaleur, **4**:125b; **7**:435b; **15**:377a
translations of works, **8**:458b
Fourneau, Ernest (1872–1949), 5:99–100; **4**:154b; **8**:273b
Fournet, Joseph Jean Baptiste Xavier, **2**:431a
Fourneyron, Benoît (1802–1867), 5:101
Fournier, Louis, **8**:273b
Fourreau, Jules, **7**:166b
Fowl cholera
see Chicken cholera
Fowl typhoid, **6**:298a
Fowler, Alfred (1868–1940), 5:101–102; **3**:124a; **4**:251b*, 270a; **12**:71a
and Bohr, **2**:244a
optics, **13**:107b
Fowler, Ralph Howard (1889–1944), 5:102–103; **1**:320b; **3**:563b; **6**:148a; **8**:185b; **9**:405a; **10**:214a; **12**:34a
electrodynamics, **9**:397a; **11**:421b; **15**:446b
gas physics, **4**:280a
Fowls
polyneuritis, **4**:311a
Fownes, George (1815–1849), 5:103–104
Fox, Cyril Fred, **9**:489b
Fox, Hermann Herbert, **4**:155b
Foxglove
see Digitalis
Foy-Breguet instrument, **2**:439a
Fracassati, Carlo, **2**:309a; **9**:62b, 63a
Fracastoro, Girolamo (*ca.* 1478–1553), 5:104–107; **13**:430a, 564a
astrology, **1**:495b
medicine, **2**:575b; **7**:614a
Fractionation, chemical, **3**:478a–b
Fractions
ancient Greek use of, **6**:315a
ancient Roman use of, **1**:418b; **4**:111a, 112b
medieval Arabic use of, **1**:40a–41a, 437b, 508b; **7**:257a–b; **9**:614b–615a; **15**:9b
medieval Latin use of, **4**:605b–606a; **7**:176b; **12**:61a
Renaissance use of, **3**:273a–b, 311b–312a; **9**:132b
classification of, **12**:94a
common denominator, **15**:688a–b
continued, **1**:30b, 437b; **2**:507a; **3**:126a, 222b, 227a, 229b, 230a; **4**:474b, 475a, 478b; **5**:260b, 573b; **6**:236a*, 309a, 572a; **7**:565a–b, 568a, 573b, 599b; **8**:135a, 138b, 139a, 306b; **9**:21b, 430b; **11**:599a; **12**:289b, 469b; **13**:266b; **14**:94b, 148b
analytic theory of, **13**:55a–58a
anticipation of, **5**:409a
convergent, **7**:575a,b; **9**:126b
discovery of, **1**:591b; **2**:320a
limited vs. unlimited, **3**:127b
method of, **3**:127a–128a, 224b; **4**:606a; **9**:125a–126b, 128a
periodic, **3**:128a, 228b; **5**:471b; **7**:566b
theory of, **1**:84a; **2**:38a; **7**:351b; **8**:383b; **9**:410a, 412b
transformation of, **4**:476a
decimal, **7**:129a, 257a–b, 259a, 533a; **11**:457b, 590b; **14**:543a
dual, **8**:165b
Egyptian use of, **15**:682a–689b, 693a, 698a–b, 699a,b, 700a–b
expansion of, **1**:40b–41a
factoring of, **4**:606a
Farey, **6**:571b
Horus-eye, **15**:699a, 700a–b
Indian use of, **9**:22b, 615a
interpolation of, **3**:273b–274a
irrational denominators, **4**:596a–b
Lambert's, **7**:599b
lowest common multiple, **3**:311b; **8**:420b; **15**:688a
notation, **3**:311b–312a; **4**:112b, 605b–606a; **7**:257a; **8**:191a; **13**:539b; **15**:667b, 691a
sexagesimal, **1**:508b; **7**:118a, 126a, 129a, 532a; **8**:280a; **15**:155a, 163b
subtraction of, **9**:614b
transformations of, **7**:566a
unit factors, **4**:606a, 612a; **6**:315a; **8**:420b; **15**:688b–689b, 693a, 698a–b, 699b, 700a–b
Fracture, theory of, **5**:245b; **12**:464b
Fractures (anat.), **2**:314b
extraosseous tissues, importance of, **3**:489b
healing, **9**:480a
treatment of, **2**:122a, 546b; **6**:429a; **8**:402a, 403a–404a; **10**:418b; **11**:254a
plaster casts, **10**:620b; **14**:584b
Fraenkel, Adolf Abraham (1891–1965), 5:107–109; **8**:459a*
set theory, **12**:13b; **14**:615a
Zermelo-Fraenkel system, **3**:57b
Fraenkel-Conrat, Heinz, **5**:141a
Fragmentum mathematicum Bobiense, **4**:105b
Fraipont, Julien (1857–1910), 5:109–110; **7**:460b, 461a*; **8**:464a
Fram (ship), **4**:344a–b; **9**:443a; **15**:431a
Français, François (Joseph) (1768–1810), 5:110–112; **1**:206b, 207b
Français, Jacques Fréderic (1775–1833), 5:110–112; **1**:207b, 237b–238a; **12**:325b
Français (ship), **3**:203b
France
agriculture, **9**:618a–b; **12**:317a
archaeology, **9**:538b–539a
art, **12**:537a
astronomy, **7**:75b
Atomic Energy Commission, **7**:155b, 159a
botany, **1**:525b; **4**:234a; **7**:165b, 585a; **8**:177a
canals, **13**:496a
chemistry, **4**:75b; **8**:173a; **10**:499b
colonies, **12**:535b–536a
Corps du Génie, **3**:70b–71b, 439b–440a
economics, **8**:84b
education, **8**:12a, 84b; **9**:7a, 472a–473a; **11**:373b, 414a; **13**:250a; **15**:481b
and England, hypothetical isthmus, **4**:70b
Ferme Générale, **8**:70b–71a, 85a
fortifications, **13**:591a–592b
Freemasonry, **7**:581b
geography, **1**:208a–b; **3**:105b, 106a, 107b–108b; **9**:150a; **10**:255b; **13**:84b

cadastral survey, 11:164a; 15:335b
geodesy, 10:523b–524a
mapping of, 7:87a, 577b; 9:90b; 10:596a
geological surveys
eighteenth-century, 4:71b, 348a–b; 5:577b–578b; 8:68b; 9:478b
nineteenth-century, 1:572a; 2:482a,b; 3:587a; 4:218a, 348b–349b, 350a; 5:475b; 13:283b
geology, 5:89b, 338a, 475b–476a; 6:168b; 8:566a–b; 10:209a; 11:134a; 12:263a, 304a, 549a
mineralogy, 4:348a–349a; 7:548b; 8:71a
paleontology, 2:346a–347a; 3:264a–b; 5:296a, 497b–498a
petrology, 3:161b; 4:71a–72a
stratigraphy, 1:209b–211a; 2:347b–348a; 4:39a; 5:89b; 14:88a
volcanoes, 2:554b; 4:548b; 5:88b–89b; 11:269b; 12:261b
mathematics, 6:271b; 11:61b, 392a
medicine, 7:605a
Naval Health Service, 11:242b
public health, 3:2a, 6b, 7a–b
natural history, 1:243b–244a; 2:107a, 282a, 305b, 321b
Office de la Recherche Scientifique Coloniale, 3:359a
physics, 13:614b–615a
population of, 4:107b–108a; 11:236b; 15:304b–305b, 309a–b, 333b, 373a, 394b*
Régie des Poudres, 3:562a
technology, 11:165a; 12:286b; 15:333b
industry, 7:78a–b; 8:119b, 621b; 11:328b–331a
metallurgy, 8:119a–b
mining, 3:411b–412b; 4:348a–349a; 9:478b
scientific instrumentation, 8:21a–22b, 170b
veterinary science, 14:15a–16a
see also French Revolution; Paris
Francesca, Piero della (Piero dei Franceschi), also known as Petrus Borgensis (*ca.* 1410–1492), 5:112–113; 10:269a,b
geometry, 10:271a–b
translations of works, 10:271a
Francesco di Siena
see Martini, Francesco di Giorgio
Franchimont, Nicolas, 7:282b, 283a
Francis of Marchia (*fl.* first half of the fourteenth century), 5:113–115; 1:198b; 2:605b, 607b; 10:421a
Francis of Meyronnes (*ca.* 1285–*ca.* 1330), 5:115–117; 4:228a; 5:114b*
Francis, John Wakefield, 6:521b
Franciscans, 1:289b, 378a,b, 381b; 2:23b; 4:254b; 5:113b, 115a, 549a; 9:15b; 10:171b, 172a, 269b, 270a, 474a; 11:253b; 13:397a, 601a
Franck, James (1882–1964), 5:117–118; 5:623a; 6:16a; 8:53a*; 11:584b*; 13:41b, 227a,b; 14:171a,b, 499b*; 15:41a
atomic physics, 2:245a; 3:597b
experiments of, 15:42b
Franck Report, 5:118b
Franck-Condon principle, 5:118a
physical chemistry, 15:448b
Franck, Sebastian (1499–1542), 5:119–122; 1:80a; 7:485a; 10:310b; 14:225b
Franco of Liège
De quadratura circuli, 2:360b; 5:365a
François, André, 3:160a; 10:498a
François, Jean, 10:522a
Francon of Poland, 2:24a; 7:119b; 14:401a
Franco-Prussian War, 2:60a; 4:20a, 266b; 8:323b; 10:354a–b, 385a; 11:353a; 12:169a; 14:41b, 529b
Comité Scientifique pour la Défense de Paris, 2:64a, 70a
Frank, François, 15:469b
Frank, Frederick Charles, 9:380a
Frank, Ilia Mikhailovich, 13:241a, 599a
Frank, Johann Peter, 1:333a; 2:595a; 4:146a; 9:252b
Frank, Nathaniel Herman, 12:530a
Frank, Otto, 12:618b; 14:66a
Frank, Philipp (1884–1966), 5:122–123; 8:457b*
Fränkel, Carl, 1:575b
Frankenheim, Moritz Ludwig (1801–1869), 5:124
Frankfurt
Frankfurt Electrical Exposition, 4:589a
Senckenberg Foundation, 14:228a–b
Frankfurt, University of, 8:503a
Frankfurt an der Oder
Physiological Institute, 4:359b
Frankland, Edward (1825–1899), 5:124–127; 11:536b; 13:389a, 524a*
acquaintances of, 7:551b; 13:521b
chemistry, 1:345b; 2:515a, 587b, 588b; 5:372b, 381b; 14:530b
critics of, 10:177b
influence of, 5:540b; 14:395b
and Kolbe, 7:451a, 452a
students of, 1:288a
and Wanklyn, 14:168b–169b
water analysis, 5:128a; 14:169b
Frankland, Percy Faraday (1858–1946), 5:127–129; 1:320a,b; 5:189b
Franklin, Benjamin (1706–1790), 5:129–139; 1:363b; 4:194a; 6:43b; 8:78b; 9:327a; 11:472b
acquaintances of, 1:486b, 489a; 3:7b, 345a; 4:492b, 503a; 5:434a; 6:367a, 383a; 7:11b, 12a, 210b; 8:78a, 84a, 460a; 9:607b; 10:611b; 11:140b; 12:120a, 413b, 516b; 14:194b, 452b, 453a
and Beccaria, 1:548a
and Capt. Cook, 1:436a
and Collinson, 3:350a; 5:130b–134b *passim*
on contemporary scientists, 12:549b
correspondence of, 3:7a; 9:54a, 596b; 11:303a
critics of, 14:419a
electrical theories, 1:67a, 547a–b; 2:5a, 83a; 3:51b, 52a; 4:129a; 6:169a, 178b, 278b; 7:13a, 371b, 608b; 8:177b, 258b; 9:152a, 548b; 11:142a; 13:224b; 14:70a, 71a, 73a, 193b, 194a, 352b, 353a
evaporation, 8:256a
experiments of, 2:327a
duplicators of, 3:535a; 5:134b
Experiments and Observations on Electricity, 11:433a
publication of, 3:350a
translation of, 3:535a; 5:133b, 136a
influence of, 3:207b; 7:588a; 14:74b
inventions, 5:130b
lightning rod, 1:435b; 5:134a–135b; 7:13a
on magnetism, 3:535a–b
and Nollet, 1:547a; 2:474b; 4:216a; 10:146b–147a
as statesman, 7:88b, 89a
stove, 3:7a; 5:130a,b; 11:471b; 13:412a*
translations of works, 14:352b
Franklin, Charles Samuel, 9:99a
Franklin, Edward Curtis, 7:497b
Franklin, Fabean, 13:219b, 220a
Franklin, Fred Aldrich, 9:203b
Franklin, John, 5:470a; 11:554b
Franklin, Rosalind Elsie (1920–1958), 5:139–142; 5:589b
Franklin Institute (Phila.), 1:364a
Franklinia altamaha, 1:488a
Franks, William Sadler, 11:484b
Fransden, Hans
see Ripensis, Johannes Franciscus
Franz, Johann Michael, 9:233b
Franz, Rudolph, 8:502a; 9:566b
Wiedemann-Franz ratio, 8:502a; 11:446b–447a; 14:330a
Franz Josef Land, 7:511a; 14:634b
Franzen, Hartwig, 3:511b
Fraser, Helen Charlotte Isabelle, 5:604b
Fraternitas Medicorum, 9:265b
Fraud
see Forgeries, literary; Hoaxes, scientific
Fraunhofer, Joseph (1787–1826), 5:142–144; 3:302a; 11:355a
Fraunhofer lines, 1:166a,b, 556a; 2:453a; 4:183a; 5:86b, 143a–144a; 7:381b, 528a; 9:566a–b; 12:269a; 14:491b
instrumentation, 1:357b; 3:288a; 5:271b, 302a; 10:198a; 11:354b; 13:109b
Frazer, James George (1854–1941), 5:144–146
Golden Bough, 5:178b
Frazier, Charles Harrison, 3:517b
Fréchet, René Maurice, 1:428a; 2:304a; 6:176b; 9:129b; 11:459a
Frederick I, King of Prussia, 6:458b; 7:373b; 10:25b; 12:600a; 15:3b
Frederick II (Frederick the Great), 2:4b, 56b, 325a; 3:71b, 119b; 4:352b, 470a, 471b, 472a; 6:370a; 7:201b, 563a, 568b, 597a, 605a; 8:129b; 9:105a, 187a, 188b; 11:109a,b, 134b; 12:563a
Koenig dispute, 4:471b; 7:443a
Frederick II of Hohenstaufen (1194–1250), 5:146–148; 4:605a, 610a; 7:133b; 9:361b, 363a,b; 11:600b; 12:83a
De arte venandi, 9:362b
Frederick VI of Denmark, 12:234b
Frederick of Württemberg, 12:307a
Fredericq, Léon (1851–1935), 5:148–150; 4:402a*; 11:138b, 427a; 12:245b*
Frederiks, Vsevolod Konstantinovich, 5:188b
Fredholm, (Erik) Ivar (1866–1927), 5:150–152; 3:428a; 9:426b
analysis, 1:427b; 5:482a; 6:392a; 7:408a; 10:594b; 12:188a,b
Fredholm integrals, 14:13b
Fredholm's equation, 5:151a–152a; 7:436a,b
influences on, 11:59b, 476a
Free association, 5:174a–b, 180a
Free Association for Microbiology, 15:523a
Free energy
see Energy, free
Free will, 1:118b–119a; 2:9a–b, 356a, 395b, 410a; 4:213a; 5:120b, 510a, 550a; 6:449a; 7:67b; 8:445a; 10:173b; 11:73b, 599a–b; 12:7Ba, 23b, 97b; 14:84b, 124a, 343b, 607a; 15:404a, 409a
and science, 3:295a
see also Determinism, Will
Freeman, Walter Jackson, 4:287a(n)
Freemartins, 8:358a–b; 12:137a
Freemasons, 2:315b, 542b; 3:465a; 4:43b, 394b; 7:581a–b; 11:497a
Freeze-drying technique, 1:613b
Freezing
ancient Greek study, 6:426b–427a
seventeenth-century study, 1:483b; 2:311a, 379b, 593a; 7:269b; 9:115a

Freezing (cont.)
eighteenth-century study, 2:177b, 186b; 3:158a; 6:171b; 8:470a; 14:191a
twentieth-century study, 11:348a
Freezing index, 14:634b, 635a
Freezing point
late nineteenth-century study, 4:228b, 229a; 11:298a–299b; 13:579b
in alloys, 8:119a
and electrolytic dissociation, 1:299b
and pressure, 13:380b–381a
of solutions, 11:298a–299b; 15:459a
Frege, Friedrich Ludwig Gottlob (1848–1925), 5:152–155; 2:269a; 8:262b; 14:304a, 468a
class theory, 12:12b, 15b, 16b
on contemporary scientists, 4:4b; 12:229a
logic, 5:350b; 6:82b; 8:159b; 10:344b; 11:356a
Freiberg, 11:607b–608a
Bergakademie
in eighteenth century, 2:552b; 4:346b; 5:155a; 6:369b; 11:464a; 14:256b–257a, 261a
in nineteenth century, 2:440a; 3:210a, 433b–434a; 5:161b–162a; 6:213b; 7:70b; 14:232a–b, 447a
founding of, 6:259b; 8:147b
mineral collection, 3:434a
Freiberg, Dietrich of
see Dietrich von Freiberg
Freiburg, University of
physics at, 9:566a
zoology at, 14:233b, 238a
Freiesleben, Johann Karl (1774–1846), 5:155–156; 6:550a; 14:259a
Freind, John (1675–1728), 5:156–157; 12:303a
Chymical Lectures, 6:41b
physiology, 13:231b
Frémy, Edmond (1814–1894), 5:157–158; 2:86b; 9:450b, 451a; 13:546a, 554b
on fermentation, 10:376b, 377a,b, 380b
French, John, 14:287a
French Astronomical Society
founding of, 5:21b–22a
French East India Company, 11:64b
French language, 10:494a
in science, 8:267b; 10:494a
French Revolution, 2:74a; 3:173a, 326b, 383b; 4:108a; 6:178a–b; 7:547a, 586b; 14:68b; 15:333a–336b, 389b*
calendar, 15:335b–336a, 343a
Committee of Public Instruction, 3:384a; 5:91b–92a; 8:136b
Committee of Public Safety, 2:74a; 3:71b–72a; 4:15b; 5:92a, 603a; 6:1a, 164b; 12:66b–67a, 286b
and education, 3:1a, 2a, 199b, 284b, 441a; 7:198b; 9:472b–473a; 12:427a–b; 15:341b–342b, 346b–347a
historiography of, 2:134a; 3:2a
and medicine, 3:1a–2b; 10:612a; 14:16a
and metric system, 2:475a; 3:106a; 4:15b; 5:91b; 7:569a–b; 9:250b; 15:333b–336a
opponents of, 2:342a; 3:6a, 106a–b, 522b; 7:581b, 603b; 13:418a
Society of 1789, 8:84b
supporters of, 1:563b; 2:321b, 510a; 3:1a–2b, 7a, 70b–72a, 199b–200a, 400a; 4:149b, 150a, 178a; 5:55b, 91a–92a, 93b, 355b–356a, 602b–603a; 6:1a, 164b, 468b, 549b; 7:277b; 8:84b–85a; 9:472a–473a; 11:141b; 12:153b; 13:571a, 596b
victims of, 1:479a–b, 529b; 2:320b, 475a, 509b–510a; 3:106b, 172b, 419a, 440b–441a; 4:149b, 150a, 178a; 5:93b; 8:303b; 9:418a, 500b; 11:272b; 12:121a, 535b; 13:565b; 14:14b; 15:82a
guillotined, 1:401b; 2:216b; 9:53b
Frend, William, 9:67a, 159a
Frenet, Jean-Frédéric (1816–1900), 5:158; 2:288a; 12:319a
formulas, 5:201b
Frenet-Serret formulas, 5:158a
Frenguelli, Joaquín, 1:131b
Frenicle de Bessy, Bernard (*ca.* 1605–1675), 5:158–160; 3:384b; 4:573a; 9:498a; 13:335a
editions of works, 7:577b, 578b
number theory, 9:599b; 14:149a
Frenkel, Yakov Ilyich (1894–1954), 5:160–161; 13:240a; 15:43a, 252a*
Frenkel's defects, 5:161a
Frenkel's exciton, 5:160b–161a
theory of, 3:329a
Frenzel, Friedrich August (1842–1902), 5:161–162
Frenzy, 14:408a
Freon, 7:316b; 13:177a
Frequency modulation (FM), 1:287b–288a
Frere, Hookham, 3:579a
Frere, John (1740–1807), 5:162–163; 15:51a
Frerichs, Friedrich von, 4:296a–b, 297a; 5:196a; 8:8a
Fresenius, Carl Remigius (1818–1897), 5:163–165; 1:553a; 5:349b; 8:331a
Fresenius, Georg, 3:611b
Fresenius Training and Research Institute, 5:164a–b
Fresnel, Augustin Jean (1788–1827), 5:165–171; 2:88b; 6:1b; 12:129a; 15:60b
acquaintances of, 5:271b; 9:424b; 12:519a
critics of, 11:292a
on ether, 8:493a, 494a; 9:214b, 215a, 372a
Fresnel drag coefficient, 8:51a, 495a; 9:214b, 372b; 14:598b–599a
Fresnel's integrals, 5:20b, 169a; 9:213a
influence of, 8:488a; 10:186b; 12:304a; 13:374b
magnetic theory, 1:144b–145a
optical theory, 1:201b–202a, 357b; 3:83a–b, 145b–146a, 146b; 4:190a; 5:19a, 86b, 143b; 6:88a,b, 453a; 8:426a, 490b, 492b, 592a; 9:74a, 264b, 372b; 10:27b; 12:519a; 13:76b, 77b, 217a, 386a,b; 14:566b, 567b–568a, 568b, 598b; 15:376b, 487a
Freud, Anna, 5:178b
Freud, Sigmund (1856–1939), 5:171–181; 2:446a, 531a–b; 14:114b
on Anna O., 2:447b
and Bernheim, 2:35b–36a
and Breuer, 2:447b–449a; 5:172b, 173a,b, 174a, 176a, 179a
and Brücke, 5:171b, 172b
critics of, 5:146a; 15:80a
on dreams, 5:176a–177b
Eros-Destructiveness, 4:368b
and Fliess, 5:172b–174b *passim,* 176b–177a, 179a
influences on, 3:205b; 8:322a; 9:327b
and Jung, 7:190a–b
psychoanalytic method, 1:365b; 7:191b
psychological life, 5:172b
self-analysis, 5:172b, 176b
on sexuality, 5:176a–180b *passim*
Totem and Taboo, 5:145b
Freudenberg, Karl, 7:518b; 13:1b
Freudenthal, Hans, 6:496b–497a; 11:455b
Freundlich, Erwin Finlay (1885–1964), 5:181–184
Freundlich, Herbert Max Finlay (1880–1941), 15:159–160; 15:448a, 468b*
Frey, Hans, 4:258b
Frey, Maximilian Ruppert Franz von (1852–1932), 5:184–185; 4:617a*; 7:504b; 10:256b
Freycinet, Louis Claude de, 4:256a; 5:224b; 8:266a; 11:242b; 12:550b
Frey-Wyssling, Alfred, 9:601a
Frézier, Amédée François, 3:441b; 4:603a; 9:458a
Fribergius, Kalbius
see Rülein von Calw, Ulrich
Fricke, Karl Emanuel Robert, 7:398b
Friction
eighteenth-century study, 2:39a; 3:440a, 441b, 442b, 444a
nineteenth-century study, 4:501a*; 10:275b
twentieth-century study, 15:201b
boundary theory, 12:613a
coefficient of, 3:441b; 8:200b, 209b–210a
dynamic, 3:442b
in fluids, 13:75a–76a, 78a
force of, 4:222b
in gases, 7:526a; 14:170b, 171a
temperature effects, 3:83b
heat of, 1:424a; 3:83a; 10:59b; 15:86a
laws of, 1:138b; 4:40a; 8:209b; 10:319a, 551a–b
in liquids, 4:591b; 6:249a, 250a; 10:551a–b; 12:595a, 613a
temperature effects, 3:83b
in machines, 1:138b; 3:71a, 358b*; 6:431b; 8:209a, 209b–210b; 10:319b, 521a, 551a–b
and normal force, 3:442b
in pipes, 9:119b
static, 3:442b
thermodynamics of, 4:229b
tidal, 3:583b; 4:20a, 591a–b; 12:449a; 14:569a; 15:301b
Friedel, Charles, 1:578a; 2:431b; 3:358b, 503b, 504a, 505a; 4:75b, 141a; 5:185b; 7:452b, 552a; 13:546a; 14:530a; 15:11b
Friedel, Edmond, 5:185b, 186b
Friedel, Georges (1865–1933), 5:185–187; 5:497a; 8:148b; 9:59a
Friedel's law, 5:186b
Friedenthal, Hans, 13:228b
Friedländer, Carl, 4:277a; 5:495a; 10:291b
Friedmann, Aleksandr Aleksandrovich (1888–1925), 5:187–189; 4:330a; 5:271b; 6:531a; 13:26a
Friedrich, Max, 14:527b
Friedrich, Walter, 4:196a; 6:303a; 7:468b; 8:51a; 11:531a; 12:527b; 15:61b
Friedrich Wilhelm IV, King of Prussia, 7:51a,b; 11:368b, 397a–b; 12:154a–b
Friedrich Wilhelms Institut, 1:574b; 14:40a
Friend, John Albert Newton (1881–1966), 5:189–190
Friends, Society of, 2:223b; 3:349a–b, 391b, 537a–b; 4:282a; 14:226a, 456a
Fries, Elias Magnus (1794–1878), 5:190–192; 9:35a; 10:531a
Systema mycologicum, 10:531b
Fries, Jakob Friedrich (1773–1843), 5:192–195; 9:359a, 429b; 11:437a
Neue Kritik der Vernunft, 1:178a
Frigorimeter, 4:171a
Frisch, Karl von, 8:529a
Frisch, Otto Robert, 2:251b; 6:17a; 7:155a; 9:262b–263a; 13:43a
Frisi, Paolo (1728–1784), 5:195
Frisius, Gemma
see Gemma Frisius, Reiner

Fritillaria, 10:1b
Fritsch, Gustav Theodor (1838–1927), 5: 195–197; 1:569b; 4:201a, 202b, 204b; 7:422a
neurophysiology, 4:594a,b; 5:255a, 463b; 6:440b, 518b; 7:48a
Fritschi, Jakob, 13:2a
Fritz, Kurt von, 10:181a; 13:317b(n)
Fritzsche, Carl Julius (1808–1871), 5:197–198
Froben (Frobenius), Johann, 4:223b; 10: 304b–305a
Frobenius, Georg Ferdinand (1849–1917), 5:198; 4:2b; 6:91a, 571a
acquaintances of, 14:222a
algebra, 8:324b
on contemporary scientists, 7:506a, 508b
group theory, 2:615b; 9:457b, 458a; 11: 55b; 14:283b, 284a, 285a, 630b
influence of, 1:307a; 14:211b
influences on, 4:617a
problem of, 3:221b
students of, 7:616a; 12:237a; 14:613b
Frobenius, Wilhelm, 4:403b
Frog, 2:322a; 11:503a
anatomy, 9:63a
circulatory system, 4:268a
croaking of, 3:38a
egg, 13:174b
electrophysiology, 4:201b; 5:267b–268b; 14:76b–77b
embryogeny, 6:132b, 573b; 7:454a; 9: 517b; 11:361a; 13:173b
heart, 5:82a, 280a; 8:451a, 453a–455a; 12:612a; 15:100a
larvae
digestive organs, 3:424b
neurophysiology, 5:463a; 9:570b; 10: 579a; 14:249a
physiology, 5:279b, 281b
respiration, 7:501b; 10:579b
reproduction, 10:39b; 12:558b–559a
spinal, 13:122b–123a
tongue, 14:143a
see also specific names
Fröhlich, Alfred, 8:453a
Fröhlich, Herbert, 8:477b; 12:530a
Froidmond, Libert, 8:28a; 14:365a
Frolov, Iurii Petrovich, 10:435a
Frolova, S. L., 12:76b
Fromondus, Libertus
see Froidmond
Fronde, 2:354a
Frontier, concept of, 2:374a
Frontius, Sextus Julius, 1:418a,b
De aqueductibus urbis Romae, 3:115b
Froriep, Robert, 14:40a,b
Frosch, Paul, 7:428a; 8:450a–b
Frost, Edwin Brant (1866–1935), 5:199; 1: 55a,b, 467a; 3:167a; 6:528b; 10: 321a*, 543b*; 12:153b*; 13:113a*, 116a; 14:558a*
Frost forms
as geologic indicators, 2:549b
Froude, Robert Edmund, 5:200a
Froude, William (1810–1879), 5:199–200
Froude's law of similarity, 5:200a–b; 11:345a
Fructose
synthesis of, 5:3a
Fruit
alcohol production, 10:377a–b, 380b–381a
analysis of, 5:165a; 12:124a
breeding, 15:425a–426a
citrus
and scurvy, 8:363a
classification of, 5:217a
preservation of, 3:359b
ripening of, 1:616b; 12:124a
rot, 3:588b
Fruitfly
see *Drosophila*
Frullani, Giuliano, 11:599a
Frustum
volume of, 8:421b; 15:670b
Fruton, Joseph Stewart, 15:15b
Fry, Thornton Carl, 8:321b
F.S.H. *(Filaria sanguinis hominis),* 8:296b; 9:82a
Fubini, Guido (1879–1943), 5:200–202
analysis, 8:111b
geometry, 3:170b; 12:285a; 13:125a
Fucaceae, 2:371a
Fuchs, Christian Joseph, 11:70a
Fuchs, Immanuel Lazarus (1833–1902), 5: 203–204; 14:224a*
acquaintances of, 4:617a; 7:506b, 508a, 522b
analysis, 7:168b; 11:53a, 56a, 407a, 452b
Fuchsian functions, 6:547b; 9:410b; 11: 52b, 53a
Fuchsian groups, 11:53b, 54a
Fuchsian theory, 5:204b; 7:460a
Fuchs's equation, 2:144b; 7:435b
influence of, 13:249b
students of, 8:253a; 14:613b
theorems of, 13:250a
Fuchs, Johann Nepomuk von (1774–1856), 5:202–203; 5:124b; 10:557a,b
Fuchs, Leonhart (1501–1566), 15:160–162; 1:604a; 2:218a,b; 3:342a; 8:248b; 9: 75b; 12:323b
botany, 2:536b; 4:138b, 139a; 13:463a
De historia stirpium, 13:502b
Füchsel, Georg Christian (1722–1773), 5: 205–206; 14:259b
Fuchsin, 10:516a; 15:83b
Fucina, Accademia della
see Accademia della Fucina
Fucus, 7:610a
reproduction, 3:337b; 5:582a; 9:601a; 11:153b; 12:125b; 13:395a
Fuel
combustible, 15:449b
mineral, 13:331a
nuclear, 3:235b
see also specific fuels
Fuel technology, 13:454b
Fuess, Rudolf, 12:595a
Fueter, Karl Rudolf (1880–1950), 5:206; 11:480b(n), 481b*, 589b*; 13:21b*
Fugacity, 8:289b, 290a
Fuhlrott, Johann Karl (1804–1877), 5:206–207
Fujita Sadasuke, 1:83a
Seiyo sampo, 1:83b, 90a
Fukuzawa, Yukichi, 7:392a
Fuller, Myron Leslie, 14:298a
Fuller's earth, 4:598b
Fulminates, 4:185a; 5:315a; 7:280b; 8:330a–b; 9:148b; 14:474b
and cyanates, 8:331a
Fulminic acid, 14:334a
Fulton, John Farquhar (1899–1960), 5:207–208
acquaintances of, 3:517b
on contemporary scientists, 12:400a, 402b*
historiography, 5:107b*, 269a*; 7:45b*; 8:90b*, 527b*; 9:21a*; 12:324b*, 434a*; 13:123a; 14:464b*
neurophysiology, 4:286b
Fulton, Robert, 6:521b; 9:491b; 15:472a
Fumariaceae, 8:354a
Fumaric acid, 8:339a; 10:516b
discovery of, 7:281a
structure of, 14:455a
Fumaroles, 4:563b–564a; 5:89a
Fumigation, 5:270a
Funafuti (atoll), 3:568b, 591a
Function spaces, 2:145b; 9:249b; 14:345a–b
Functionals
linear, 1:428a; 6:5a
theory of, 14:86a, 87a
Volterra, 14:346b–347a
Functions (math.)
Abelian, 3:315a; 4:341b; 6:308b, 547b; 7:53a; 12:213a; 14:202a
theory of, 2:471a; 5:472a; 7:478b–479a; 11:54a,b, 448a, 449a; 12:284b; 14: 219b–220a, 220b–221a, 223a
transformation of, 2:470a; 6:309a
aggregate, 6:402a
algebraic, 1:194b; 2:198b; 3:166a; 4:476a, 617a; 7:408a; 10:139b; 11:470b
branching, 11:449a
classification of, 8:383b
differentiation, 10:47b, 48b
integration of, 3:222b, 223a, 224a, 228a; 8:162b, 163b; 10:249b; 11:449a, 450b–451a
theory of, 1:15b; 2:465a,b; 3:315a; 6: 286b, 357a; 7:621a–b; 9:410a; 10: 249b; 11:448b, 452a; 14:202a,b; 15: 65b
almost periodic, 2:236b, 238b–239a; 4: 407b; 13:36a; 14:90a
analytic, 2:304a; 3:136b; 5:181b; 6:3b–4a, 214b, 472a–b; 11:451b
boundary properties, 8:558b
limits of, 4:547b; 6:472b
metric distribution theory, 2:303b
quasi analyticity, 6:4a
representation of, 9:426b
singularities of, 2:358a; 4:476b; 6:472b; 8:366b; 9:249b; 11:54a, 156b
theory of, 1:207a; 4:476a, 477a–b; 7: 570a–b, 616a; 8:366b; 10:594a; 11: 54a–b, 156b–157a; 12:512b–513a, 599b; 13:424b; 14:46b, 223b
unicity of, 11:157a
Appell polynomial, 1:194b
arbitrary, 1:207a; 4:476b; 5:96a,b, 97a,b; 7:352a, 408a; 9:471a, 476a; 15:306b
expansion of, 13:27a–b
asymptotic, 14:95a
automorphic, 3:96b; 5:201b; 6:214b, 572a; 7:398b–399a; 10:594a; 11:452b; 12: 212b
discovery of, 11:52a–54a
theory of, 2:144b; 8:433b, 617b; 12: 247a; 14:317a
Baire's, 1:408a
Bessel's, 4:103a*, 478a; 5:96b; 6:95b, 230b, 309a, 572a; 7:6b, 492b; 12: 172b–173a; 14:158b, 317a; 15:385a, 488b
applications of, 2:41a, 101b; 4:479a; 9:173b
expansion of, 6:90a
theory of, 8:389a; 14:188b
best approximation of, 3:224a,b, 228b–230b; 14:630b; 15:23a–b
beta, 3:136b; 4:478a
Bolzano, 2:276a
bounded, 7:167b; 8:110b–111a
circular, 2:299a; 4:126a; 5:527b; 11:401b, 402a
classification of, 1:408a; 3:385a; 4:476b; 6:95b
complex, 2:121b; 3:135b, 227a; 4:103a, 342b, 477b, 547b; 5:103a, 201b, 608b;

Functions (math.) (cont.)
complex (cont.)
6:308b; 7:453b, 570b; 11:22a, 149a; 13:561b
continuity of, 3:137b; 8:61b
theory of, 1:240a; 2:304a; 3:132a, 137a–141b, 228a; 5:78a*, 342b; 6:3b, 95a, 572a; 7:411b; 8:110b; 9:308a; 11:156b, 206b, 207a, 447b, 449a–450a, 452a; 12:512b, 526a; 14:317b, 591b–592a, 620b
composition of, 10:514a
concept of, 2:358a; 5:108a; 9:307b
construction of, 5:342b; 8:389a
continuous, 2:356a; 5:151a, 608b; 6:230b, 392a, 394a; 8:111a, 142a; 9:248b–249b; 10:244b, 442a; 11:449b; 14:285a, 345a
absolute, 14:46b
approximation to, 13:562a
construction of, 4:206a
definition of, 2:275a; 3:136a, 137a; 4:126a
differentiability, 2:276a; 6:95b; 8:111b
expansion of, 13:27a, 549a
limit functions of, 1:407b–408a; 2:275b–276b
periodic, 9:249a
convex, 2:58a; 6:472b
inequalities, 7:102a
cylindrical, 10:117b
theory of, 12:535a
definition of, 3:385a; 4:126a, 476b, 477a, 479a; 5:153a, 154a; 8:432b
derivable, 14:86a
determinate, 6:545b
development into series, 2:350a; 5:528b, 572b, 573a, 608b; 7:408a; 12:325b; 15:226a
Dirac, 14:91b
discontinuous, 1:207b, 407a; 2:304a; 8:110a, 112a; 11:454b; 13:58a; 15:306b
discriminant, 5:8a
elementary, 4:477a, 478a; 11:470b–471a
elliptic, 1:16b, 584b; 2:121b, 271b, 350a; 3:223a, 224a, 592a; 4:342b, 516a; 5:206a, 304b, 309b, 471b, 572b–573a; 6:308a, 390a, 548b*; 7:408a, 459b, 601b; 8:61b, 253b; 9:175a, 553b; 11:55a–b; 12:213a; 13:55b, 216a
applications of, 6:76a; 8:384b; 9:55b; 10:483b; 15:65b
arithmetization of, 8:531b
construction of, 8:140b
generalization of, 12:331b
modular, 2:468a, 470a; 5:414b; 6:214b, 309a; 7:103b
periods of, 1:194b; 3:141a; 5:342b; 6:308b; 7:53b, 54a; 11:449a; 13:250a
representation of, 7:53b–54a
theory of, 1:13b; 2:105a–b, 465a, 471a; 3:169a; 4:103a, 341a, 342a, 617b*; 5:260a–263b *passim;* 6:75b–76a, 307a–b; 7:51a, 52a–54b, 444a, 478b, 506b, 508a, 567b, 573b; 8:62a, 135b–141a *passim,* 383b; 9:426b; 11:269a, 401b, 548a–b; 12:225b, 469b, 513a, 534b; 14:219b–220a, 223a
transformation of, 1:16a; 2:470a; 6:307b; 7:53a–b, 619b; 9:457b; 11:53b
entire, 4:562a; 7:102a; 8:366b; 10:594a; 11:54a; 13:424a
Euler's, 4:474a; 5:414a
theory of, 8:383b
expansion of, 3:222b; 6:545b; 7:570b; 8:140b, 366b, 384a, 611a; 13:130a; 15:23b, 366b
exponential, 1:406b; 2:53a; 5:342b, 414a, 449b; 6:378b; 8:164a
differentiation of, 8:164b
theory of, 15:98b–99a
families of, 14:149a
fluctuating, 6:90a
fractional parts of, 2:236b; 5:342b
Fuchsian, 6:547b; 9:410b; 11:52b, 53a
fundamental, 13:26b
gamma, 4:478a; 6:472b; 7:102a; 8:137b, 140b, 142a, 253b, 367a; 10:117b; 12:172b; 13:55b, 69a, 264a; 14:149a
generating, 8:617a; 9:454a; 15:303b, 306a–307a, 317a, 318b, 354b, 364b, 366a,b, 367a, 368a, 369a, 376b, 385b, 386a
Green's, 3:145a; 4:174a; 6:392a; 8:458b*; 9:215b; 11:454a
growth of, 13:35b
Hamiltonian *H,* 4:579a,b; 6:89a,b
Hankel, 6:95b
harmonic, 8:301a, 558b; 12:213a
integral representation of, 14:317a
theory of, 5:201a,b; 11:157a
holomorphic, 1:207b; 3:178a; 4:548a*; 6:4a; 7:102a, 398b; 8:558b
homogeneous, 1:294b; 4:477b; 5:54b
hyperbolic, 2:53a; 5:573b*; 7:599b; 11:401b–402a
hyperelliptic, 2:470a; 3:141a, 166b, 167b; 4:617b*; 7:53a; 11:509b
transformation of, 2:271b; 6:307a–b, 547b
hyperfuchsian, 10:594a
hypergeometric, 1:194b; 2:101b; 5:302a; 9:175a; 10:594a; 11:269a, 407a, 452b
implicit, 2:198a, 271b; 3:140a; 10:442a
infinitely differentiable, 6:4a
infinity of, 2:320a
integrability of, 10:442a
integral, 3:139b; 5:342a,b; 6:4a; 10:111a
relative, 6:394a
inverse, 3:222b; 7:52b, 53b, 55a, 601b; 13:266b
irrational, 3:224a; 8:299b
integration of, 8:166a
jump, 3:141a
Kleinean, 11:52b
Lagrangian, 6:89b; 11:13b, 568b; 12:599b; 13:363b, 364a
Laguerre, 7:574b
lambda-definable, 13:497b–498a
Lamé, 6:230b; 7:601b; 12:513a; 14:317a
theory of, 7:492b; 14:158b
Laplacian, 7:561b, 601b; 15:319b
Legendre, 6:309a; 14:317a
lemniscatic, 6:572a
linear zigzag, 15:537a, 539b, 541a, 545a, 674b, 679a
Lobachevskian, 8:430a
logarithmic, 3:615b; 4:477a; 5:8a; 10:249b
Mathieu, 9:175a; 14:317a
measurable, 4:288a*; 14:89b, 345b
C property, 8:558b
meromorphic, 1:307a; 3:139b; 5:152a; 8:253b; 9:426b, 427a; 11:52b–53a, 54a, 59a, 471a
Möbius, 9:430b
modular, 6:214a–b, 570b; 11:53b, 256b, 269a, 452b; 12:173a; 14:219b
elliptic, 7:398b
theory of, 4:342b; 6:309a, 571b
monogenic, 2:304a; 11:54a
monotonic, 5:201b; 8:111a,b, 457a–b
multiform, 2:358a
multivalued, 2:240a; 3:136a, 140b–141a; 11:449a, 452a
nonalgebraic, 3:385a
nonanalytic, 10:554a
non-differentiable, 2:276a
nonuniform, 11:207a
number of zeros, 3:139b
numerical, 1:584b
omega, 12:469b
orthogonal, 11:459a
systems of, 14:46b
orthonormal, 12:188b
of *p*-adic variables, 5:342b
periodic, 1:194b; 2:238b; 3:141a; 5:471b, 483b, 572b; 8:561a; 11:470b, 548b
doubly, 3:140b, 169a; 6:308b; 7:478b; 8:141a, 383b; 11:449a; 14:592a
multiply, 1:584b
permutable, 10:513b, 514a
pluriharmonic, 11:54a
primitive, 8:558b
quasi-analytical, 15:23b
quasi-periodic, 2:236b; 4:407b
of quaternion variable
theory of, 5:206a
random
theory of, 12:461a
rational, 1:306b; 3:274b, 468b; 4:547b; 6:389b, 537b, 572a; 11:470b
decomposition of, 8:385a
integration of, 2:54a; 8:166a; 10:249b, 442a
of real variables, 2:288b
series development of, 4:103a
theory of, 1:407a–408a; 2:275b, 303b; 3:63a; 4:205b–206a, 287b; 8:558a,b; 10:553b; 11:149a, 256a, 260b; 12:78a, 216b, 426b; 13:35b–36a; 14:46b, 221b, 572b–573a
recursive, 2:514a; 9:72a; 11:107b, 260a; 13:498a
representability of, 8:558b
Riemann, 8:367a; 11:454a–b
sawtooth, 5:96a
of several variables, 3:63a; 5:111a; 10:442a
sigma, 14:189a
special, 10:117b
theory of, 4:478a; 5:572b–573a; 6:451b; 12:512b–513a, 535a
spherical, 4:103a*, 126a; 5:306a; 6:230b; 8:562b; 9:215b; 14:158a,b
Sturm's, 2:299a; 13:129b, 217a, 218b
subharmonic, 11:157a, 459b
symmetric, 2:299a; 7:566b; 8:161b, 617a; 13:250a
theory of
early nineteenth-century, 1:13b–17a, 354b; 5:572b–573a
mid nineteenth-century, 3:224b, 226b, 454a*; 5:472a; 6:451a–b; 7:521b, 523a; 12:172b–173a; 14:472a
late nineteenth-century, 2:105a–b, 275b–277a, 615a; 3:560a; 4:206a–b, 269a; 5:77b–78a, 203b; 6:3b–4a, 95a,b; 7:397b–399a; 8:62a, 366b–367a, 389a; 9:125b–127a, 502b; 10:244b–245a; 11:52a–54b, 452a–b, 611a; 12:599b; 13:250a,b, 263b, 469b; 14:222b, 223b
early twentieth-century, 1:407a–408a; 2:303a, 304a, 513b; 3:345b; 4:562a; 6:114a, 571a; 8:253b, 558a–559a; 9:248b–249b; 11:269a, 459b; 12:237b, 364b; 13:139a
mid twentieth-century, 5:342a; 13:424b
constructive, 15:23a–b
geometric, 7:398a
metric, 7:351b
theta, 3:166b, 169a, 263a, 315a; 4:475a; 5:471b; 6:308b, 309a; 7:54a, 621b;

11:54b, 450b, 451a,b, 548b; **12**:213a, 469b, 513a; **13**:221a
double, **5**:77b
mock, **14**:189a
transcendental, **1**:500a; **2**:358a, 465a, 468a; **3**:315a; **4**:177a, 478a; **6**:308b, 572a; **7**:398a, 494b; **8**:139b, 162a; **9**:502b; **11**:149a; **12**:518b*
Abelian, **4**:341b; **14**:223a
classification of, **1**:15b; **8**:383b
construction of, **3**:385a
elliptic, **1**:13a–b; **4**:341b; **8**:385a
transitive, **9**:174a, 175a
triangular, **7**:398a
trigonometric, **1**:508b; **2**:53a; **4**:477a; **6**:451b; **11**:397a, 561b
concept of, **7**:405a
trisection of, **8**:140b
uniform, **4**:103a
Weierstrass, **6**:214b; **14**:343a
Whittaker, **14**:317a
zeta, **1**:306b; **2**:238b; **3**:227b; **4**:475a; **6**:4a, 113b, 214b, 571b; **11**:453a–b; **13**:424a, 498a, 561b, 562a
zeros of, **7**:102a
see also Eigenfunctions
Fundamenta mathematicae, **3**:170b
editors of, **9**:248b
founding of, **7**:71b, 72a; **9**:248a; **12**:426b
Fungi, **2**:371a
anatomy, **9**:368b
attenuated culture, **2**:35a
bacteria and, **3**:338b
biochemistry of
vitamins, **12**:207b–208a
cytology, **6**:121b; **9**:34b
developmental history, **3**:337b, 612a–613b; **13**:489b
edible
nutritive value, **9**:284b
entomogenous, **13**:299b
enzyme formation, **14**:255a
isolation of, **2**:34b
in lichens, **3**:613b
morphology, **3**:612a, 613b; **6**:100a
in orchids, **2**:34b–35a
parasitic, **1**:493a–b; **2**:437b; **3**:612a–613b; **5**:56b, 566a; **6**:73a; **10**:1a–b, 253b; **11**:132a, 491b; **12**:202b–203a; **13**:299b, 489b
degree of specialization, **15**:507a
pathogenic, **2**:19a; **3**:340b*, 612a–613b; **4**:220a; **5**:566a; **7**:323b; **9**:176a; **12**:202b–203a
phosphorescent, **4**:505a*
phototropism, **6**:100a
phylogeny, **2**:583a
physiology, **7**:395b; **14**:36b
respiration, **9**:44a
pleomorphism, **3**:612a; **9**:176a
reproduction, **9**:176a; **13**:405a; **14**:93b
salts, effects of, **9**:458b
sexuality of, **2**:437a, 583a; **3**:612b; **4**:288b; **6**:121b; **11**:153b–154b; **12**:207a
social organization in, **2**:583a
soil, **12**:468a
taxonomy, **2**:19a, 583a; **5**:191a–b, 192b; **6**:100a; **7**:197b, 395b; **10**:1a, 13a–b, 531a–b; **13**:178a; **14**:93b; **15**:513b
see also Cryptogams, Mushrooms, Mycology, Spores
Fungine, **2**:385b
Funicular diagram, **3**:468b, 496a
Funk, Casimir (1884–1967), 5:208–209; 5:451b; **7**:471b
Funk, Christlieb Benedict, **6**:403b
Furan, **4**:603b
synthesis of, **3**:246b
Fürbringer, Max, **12**:299a; **14**:203a; **15**:166a,b
Furfural, **1**:553b
color bases, **12**:164a
isolation of, **5**:104a
Furnaces
Renaissance, **2**:143a; **4**:393b
seventeenth-century, **5**:419b, 420a
eighteenth-century, **3**:454b; **4**:87a; **7**:78a
nineteenth-century, **3**:557a; **4**:218a
twentieth-century, **2**:324a
analysis of gases from, **3**:11b; **11**:36b
cupola, **3**:234a
electric, **3**:87a, 212a; **6**:115a, 319b; **9**:451a–b
fuels for, **5**:601b
gas
regenerative, **12**:424b
gradient, **11**:549b
high-temperature, **13**:454b
open-hearth, **8**:116a
reverberatory, **1**:448b
see also Blast furnace
Furtwaengler, Philip, **1**:307a; **4**:5a; **6**:241a, 571a
Furunculosis, **9**:550b
Fusaria, **9**:176a
Fusibility, **6**:478a
Fusion
fusion mixture, **9**:425b
heat of, **2**:178a–b
Fusoris, Jean (*ca.* 1365–1436), 15:162–164; 15:155b
Fuss, Nicolaus (or Nikolai Ivanovich Fus) (1755–1826), 5:209–210; 4:470b, 472b, 473a, 483a*; **7**:564a; **8**:300a, 306a
Fuss, Paul Heinrich von, **3**:224b, 228a; **7**:55a
Fyodorov, Evgraf Evgrafovich, **5**:211b
Fyodorov, Evgraf Stepanovich (1853–1919), 5:210–214; 14:525a
correspondence of, **12**:196a
crystallography, **1**:460b, 461a,b; **2**:431b; **5**:455a; **6**:180b, 303a; **12**:195b–196a, 512a; **14**:94b, 525b
Fyodorov-Groth nomenclature, **5**:212a

G

Gabb, William More (1839–1878), 5:214; 14:312b, 315b
Gabbro, **2**:496a; **13**:273a
Ibn Gabirol, **5**:592a; **14**:459a
Gabriel, Siegmund (1851–1924), 5:214–215; 11:551a
Gabriel synthesis, **5**:215a
Gadd, Pehr Adrian, **5**:215b
Gaddum, John Henry, **15**:106b
Gadfly, **11**:342a
Gadolin, Axel Wilhelm, **1**:461b
Gadolin, Johan (1760–1852), 5:215–216; 3:291b; **4**:343b
Gadolinite, **3**:479a; **5**:216a
Gadolinium, **3**:478a,b
discovery of, **2**:255a; **5**:216a; **9**:110b
Gadow, Hans, **5**:282a; **13**:168a; **15**:166a
Gaede, Wolfgang, **4**:253b; **14**:340b
Gaertner, Joseph (1732–1791), 5:216–217; 5:217b, 218a; **12**:516a
Gaertner, Karl Friedrich von (1772–1850), 5:217–219; 5:217a; **7**:368a
hybridization, **9**:280a,b; **15**:68a
Versuche und Beobachtungen über die Bastardzeugung im Pflanzenreich, **9**:278b
Gaetano da Thiene, **9**:132b; **13**:208b, 209a,b
Gaffarel, Jacques
Curiositez inouyes, **9**:15a
Gaffky, Georg Theodor August (1850–1918), 5:219–220; 4:276b; **7**:433b*; **8**:450b, 451a*
and Koch, **7**:423a–429a *passim;* **8**:448a
Gagliardi, Domenico (1660–*ca.*1725), 5:220–221
Anatomia ossium, **6**:183b
Gagnebin, Élie (1891–1949), 5:221–222; 8:544b
Gahn, Johan Gottlieb (1745–1818), 5:222–223; 2:6a; **12**:67b(n), 143b, 146a,b; **13**:280b, 281a, 310a; **14**:447a, 486b
Gaillot, Aimable Jean-Baptiste (1834–1921), 5:223–224; 8:278b–279a
Gaimard, Joseph Paul (1796–1858), 5:224–225; 11:242b–243a
Gaines, Walter Lee (1881–1950), 5:225–226
Galactose, **11**:577a
Galambos, Robert, **12**:562b
Galápagos Islands
fauna, **3**:536b*, 569b–570a; **8**:572b
Galaxies, **1**:352b
Andromedan, **1**:353a; **4**:282a; **9**:330a
catalogs of, **4**:185b; **10**:30b
classification of, **6**:530a
clusters of, **6**:530b, 531a; **9**:330a; **13**:36b
contents of, **6**:530a
distances of, **1**:353a; **6**:530b–531a; **8**:106a; **9**:405b
distribution of, **6**:530a, 531b–532a; **12**:349a
dwarf, **12**:349a
photographs of, **1**:353a; **4**:249a,b
radial velocities, **6**:530b–531b; **13**:36b
rotation of, **1**:57a; **3**:36b
size of, **3**:509b
spectra of, **6**:531a
spiral arms
rotation of, **6**:532a–b
structure of, **6**:530a
spiral hypothesis, **4**:273a; **8**:364b; **12**:265b
see also Milky Way
Galaxy, the
center of, **5**:363b; **6**:352b; **9**:411a; **11**:18b–19a; **13**:98b
dynamics of, **5**:363b; **7**:238a, 238b–239a
interstellar substratum, **5**:363b; **6**:74b
magnetic fields in, **12**:368b
photography of, **14**:521b
rotation of, **6**:540a; **7**:481b; **10**:317b; **11**:18b–19a; **13**:98b, 114a
differential, **8**:364b
size of, **8**:106a; **13**:475a
structure of, **1**:56b, 398a; **5**:363b; **6**:531b; **7**:235b–240a; **10**:290a, 317a–318a, 483a; **11**:163a; **12**:250b
subsystems of, **10**:317b
see also Milky Way, Solar system
Gale, Henry Gordon, **1**:56a; **6**:29b; **9**:373b, 395b
Galeati
see Galeazzi, Domenico Gusmano
Galeazzi, Domenico Gusmano (1686–1775), 5:226–227; 5:267a; **9**:135b, 302b
Galen (*ca.* 129-*ca.* 200), 5:227–237; 1:212a, 315a; **2**:619b; **6**:421a; **11**:105b
anatomy, **1**:315a, 316a, 592b; **3**:40b, 482b; **4**:199a–b, 521a; **5**:228a, 229a, 230b, 426a; **6**:159b, 192a, 318b; **12**:80b; **13**:223a; **15**:737b, 738b
cardiovascular system, **1**:316a–b; **2**:336b–337a; **3**:356a–b; **5**:233b–235b; **6**:153a; **8**:53b
critics of, **9**:604b

Galen (cont.)
Aristotelianism, **1**:267b, 268a
botany, **13**:343a
chemical theory, **12**:312a,b
commentaries of, **3**:393a; **4**:462a; **5**:230b, 231a, 232a
commentaries on, **1**:107a; **4**:199a–b; **7**:17a; **9**:469a; **11**:445a; **12**:102b, 103a; **13**:37b
on contemporary scientists, **12**:538b, 539a,b, 541a(n)
critics of, **1**:121b, 619a–620b *passim;* **4**:387a, 525a; **5**:233b; **6**:151b, 155a; **9**:31b; **11**:323b, 325b–326a; **12**:335a, 447b; **14**:7a, 8b–11a
on earlier authors, **1**:324b; **4**:105b, 106a, 119a,b, 383a–386a *passim;* **5**:230b, 232a; **6**:316b, 317b, 419b, 420a,b; **11**:127b, 128a
editions of works, **1**:77b; **3**:12b; **6**:509a; **10**:230b; **12**:7Bb; **15**:160b
embryology, **1**:109b; **4**:510a,b, 511a; **6**:159b–160a, 161a
errors of, **1**:47a; **4**:199a–b; **5**:233b; **14**:5a–b, 8a, 11a
influence of, **1**:69a, 199b, 538b; **2**:23a, 314b, 401b, 546b; **3**:218b; **4**:212b; **5**:233a–b; **6**:123a; **7**:27b, 42b, 187a; **9**:40b; **10**:21a; **11**:312a, 466b; **12**:82a, 311a–312b; **14**:112b, 457b, 458b; **15**:498b
influences on, **11**:105b(n); **14**:607b
medicine, **1**:75a; **4**:367a; **5**:227b–232a; **6**:276b; **12**:102a, 540b
methodology, **7**:194a; **10**:8a; **11**:112a
natural philosophy, **11**:324b
optics, **8**:280b
pharmacology, **2**:536b; **5**:231b; **9**:41a
philosophy, **13**:278b
physiology, **10**:21a, 598b
spurious writings, **5**:232a, 232b–233a; **12**:540a
translations of works, **1**:621a; **2**:537a; **3**:534a; **5**:233b, 586a; **7**:25a; **8**:248b, 360b–361a; **11**:245a, 445a; **12**:592a*; **15**:160b, 230b, 231b–232a, 234b–238b, 241a
De alimentorum facultatibus
translations of, **9**:435a, 438a
De anatomicis administrationibus, **3**:12b; **5**:229a, 230b
translations of, **1**:621a; **5**:586a; **8**:248b; **15**:237a–238a
Ars medica (Ars parva), **8**:249a; **11**:253b
commentaries on, **9**:75b*; **11**:445a; **12**:102a
De complexionibus
translations of, **15**:183a
De compositione medicamentorum
translations of, **15**:242a
De creticis diebus
translations of, **15**:183a
De crisibus
editions of, **1**:620b
translations of, **15**:183a
De demonstratione, **15**:230b
De differentiis morborum, **13**:100a
De dissectione venarum et arteriarum, **1**:621a
De elementis
translations of, **15**:182b
De expositione libri Ypocratis in pronosticatione
translations of, **15**:183a
Expositiones super librum Ypocratis de regimine acutarum egritudinum
translations of, **15**:182b–183a
De Hippocratis et Platonis decretis, **3**:12b; **5**:228b, 235b
De hirudinibus, revulsione etc., **1**:621a
De humoribus
commentaries on, **10**:598a
De ingenio sanitatis
see Galen, *Methodus medendi*
De libris propriis, **6**:190a
De malicia complexionis diverse
translations of, **15**:183a
Methodus medendi, **6**:190a
translations of, **3**:393a; **15**:235b–236a
De motu musculorum, **1**:621a; **3**:12b
De naturalibus facultatibus
translations of, **15**:235b
De rigore, **1**:290a
De secretis (spurious), **15**:183a
De simplicium medicamentorum facultatibus, **5**:383a
translations of, **15**:183a
Tegni, **15**:188a
translations of, **15**:173b, 174a, 185b, 187b–188a
De usu partium, **3**:356a; **5**:230b, 232b, 235b
translations of, **9**:468b
see also Medicine, Galenic tradition
Galeopithecus, **6**:535b
Galeotti, Gino, **8**:556b, 557a
Galerkin, Boris Grigorievich (1871–1945), 15:164–165
Galerkin's method, **15**:164b
Galeus, **4**:377b
Galilei, Galileo (1564–1642), 5:237–249; 1:165a, 513a; **9**:596a; **10**:486b; **13**:564a
acoustics, **9**:319a,b
acquaintances of, **1**:541a; **4**:508a; **6**:445b; **9**:369a; **10**:488b; **11**:96b; **12**:101b, 105b, 106a; **13**:256b, 434a, 438a; **14**:49a
on Archimedes, **1**:229b
astronomy, **2**:410a; **6**:330b, 332a, 333b, 361b; **10**:176b; **11**:525b; **12**:151b, 152a; **14**:364b
and Baliani, **1**:424a,b; **5**:245b
and Bellarmine, **1**:588a–590a; **5**:244b
and Benedetti, **1**:607b–608a; **5**:246b–247a
on Biblical interpretation, **1**:336b, 588a,b, 589b; **5**:242a, 286b, 287a
biographies of, **2**:453a; **5**:195b; **7**:486a–487b; **11**:404b; **13**:26b; **14**:49a
blindness, **13**:474a
and Buonamici, **2**:590a
and Capra, **3**:59b–60b; **5**:240b
and Castelli, **3**:115b–116b; **5**:242a, 244a
and Cavalieri, **3**:149a,b
chemistry, **7**:195a
Clavius and, **1**:588a; **3**:311b; **5**:238b; **13**:12a
commentaries on, **7**:7a
and Copernican controversy, **1**:336b, 588a–590a; **2**:306b, 307b, 348b, 542b; **3**:115b, 116b; **4**:52a, 62a; **5**:242a–245a, 287b–288a; **9**:168b, 317b; **10**:491b; **13**:560b; **15**:69b
correspondence of, **3**:214a; **6**:520a; **9**:13b; **12**:102b, 151b; **13**:29a, 256b, 433b, 560b
cosmology, **2**:403b; **5**:239a–245a *passim;* **9**:317b
critics of, **2**:13a; **3**:59b–60b *passim;* **4**:53b, 60a, 96a; **7**:302b; **9**:12b, 527b; **11**:411a; **12**:152a
and Descartes, **5**:247b; **7**:486b, 487a
editions of works, **5**:499a; **9**:369b
family of, **5**:238a, 240a, 246b
and Guidobaldo, **9**:487b, 488a
influence in seventeenth century, **1**:315b, 484a; **2**:291a, 306a–309b *passim,* 380b, 419a; **3**:182b; **5**:285b, 289a(n), 415b; **6**:126a; **9**:15a, 16a, 62b, 63b, 484b; **10**:60b, 103b; **11**:404a; **12**:103b, 431a; **13**:30b, 436b, 566a,b; **14**:587b
influences on, **11**:405b–406a; **14**:461a, 581b; **15**:45a–b
instruments
collections of, **5**:55b
compasses, proportional, **2**:602a
microscope, **3**:113b*
telescope, **4**:51a; **5**:241a–b, 278b; **9**:498a; **10**:176a; **11**:97b; **12**:74b, 105b
thermoscope, **12**:103a; **13**:437a–b
and Kepler, **5**:239b, 241b; **7**:293a, 299a, 302b, 307b–308a
longitude determination, **3**:101b; **5**:242b; **6**:520b
and Lorenzini, **3**:60a
magnetism, **10**:536b
mathematics
geometry, **6**:601a; **7**:194a
indivisibles, **1**:164b; **3**:150b; **5**:245b; **10**:336a; **13**:434b
and Mayr, **9**:247a–b
mechanics, **4**:26a, 227b; **5**:250a; **7**:486b–487b; **8**:209a; **9**:488a–b; **13**:186b; **14**:376b(n)
ballistics, **2**:202a; **5**:246a–b; **13**:436a–b
dynamics, **2**:606b, 607b; **8**:153a; **10**:72a
hydrostatics, **5**:238b, 241b, 242a; **10**:332b, 334a
kinematics, **1**:254b, 606a–b, 607a–b; **3**:73a; **4**:8b; **5**:238b–242b *passim,* 245a, 246a–247a, 288a; **6**:379a–b, 603a, 607a, 608b, 610b; **8**:267b–268b; **9**:115b, 116a, 318a; **10**:69b, 228b
statics, **5**:238b, 239a; **7**:174b; **8**:212b; **9**:119b
methodology, **1**:233b; **3**:111a; **4**:93a; **5**:59a, 97b, 115b, 238b, 247a; **6**:448b; **9**:135a
optics, **11**:526a
personality, **5**:247b–248a
and Platonism, **15**:746b
scientific explanation, **1**:100b, 588a–589b; **5**:243a, 247a
students of, **9**:13b
tides, theories of, **1**:568a; **5**:239b–240a, 244a
translations of works, **5**:247b; **9**:317b
on vacuum, **13**:438a–b
Assayer, **9**:247b
La bilancetta, **5**:238a–b; **11**:406a
Il compasso, **3**:60b
Dialogue Concerning the Two Chief World Systems, **1**:244b, 254a, 589a; **2**:377a, 542b; **3**:116b, 179b; **5**:240a, 243a–245a, 246a, 247b; **6**:515a; **7**:293a, 303a; **10**:43b, 61b; **13**:279b, 433b; **14**:364b
commentaries on, **5**:158b
criticism of, **2**:13a
licensing of, **1**:589b; **5**:244a
Discorsi e dimostrazioni matematiche intorno a due nuove scienze, **2**:83b, 201a,b; **3**:3b, 150b; **4**:96a, 451a; **5**:245a–246a, 247b, 528b; **9**:97a,b, 114b, 488b; **10**:228b; **13**:433b
Istoria e dimostrazioni intorno alle macchie solari
publication of, **13**:29b
Juvenilia, **2**:590a; **6**:379a; **12**:548b
Letter to the Grand Duchess Christina, **3**:115b; **5**:242a, 247b, 286b, 287a; **14**:362a, 376a(n)

Lettere sulle macchie solari, **3**:179b; **5**: 242a
Le meccaniche, **5**:239b; **11**:488a
De motu, **1**:605b, 607b; **2**:590a; **5**:238b–239a; **7**:487a; **15**:45a,b
Le operazioni del compasso geometrico e militare, **3**:59b
Il saggiatore, **3**:179b; **5**:243a; **9**:63b; **10**: 176a
publication of, **13**:29b
Sidereus nuncius, **3**:115b, 311b; **5**:241a–b, 382b, 525b; **7**:299a; **10**:176a, 489a; **14**:364b
Two New Sciences
see Galilei, Galileo, *Discorsi . . .*
Galilei, Vincenzio (*ca.* 1520–1591), 5:249–250; 1:606b; **5**:239a
Dialogo della musica antica e moderna, **7**:301a
music theory, **5**:238a,b; **13**:50b
Galilei, Vincenzio (the younger), **3**:115b; **5**:240a; **6**:604b
Galite (Jezīret Jālita), **7**:30b
Galitzin, B. B.
see Golitsyn, B. B.
Gall, Franz Joseph (1758–1828), 5:250–256; 10:500a; **11**:243a
histology, **14**:15a
phrenology, **2**:508a
critics of, **5**:44b, 45a; **8**:272b; **11**:593a
and Spurzheim, **12**:596a–597a
students of, **13**:402a
Gall, botanical
see Plants, pathology, galls
Gall bladder
diseases of
treatment of, **5**:366b
histology of, **6**:269a
pathology of, **8**:352b
theories of, **1**:265a; **8**:54a
Gall insects, **11**:317a, 342a; **13**:173b, 563b; **15**:14a
see also Plants, pathology, galls
Gallatin, Albert, **11**:119a
Galle, Andreas, **5**:258b
Galle, Johann Gottfried (1812–1910), 5:256–259; 6:553a; **8**:557b
astronomy, **1**:53b, 157b; **8**:277b; **10**:479b; **12**:160b–161a
catalogs of, **3**:472b
and Encke, **1**:295b; **4**:369b; **5**:256b–257b
Gallic acid, **8**:339a, 340b, 341a
in photography, **13**:238b–239a
Gallieni, Joseph, **7**:548b
Galliformes, **6**:594b
Gallium
discovery of, **2**:254b, 428b; **9**:290b
prediction of, **9**:288b, 289b
Galloflavin, **5**:489a
Gallois, Jean (1632–1707), 5:259; 2:354a; **3**:63b; **5**:522b*; **11**:512b; **12**:84b; **13**:335b
correspondence of, **12**:85b
Gallon, **14**:486b
Gallstones
see Calculi, biliary
Galois, Evariste (1811–1832), 5:259–265; 5:94b, 368a; **7**:10b; **8**:386a(n)
acquaintances of, **6**:307a; **11**:414b; **13**: 128a
algebra, **11**:598b; **13**:571b
analysis, **8**:383b
on contemporary scientists, **4**:341a; **5**: 309b; **15**:481b, 489b
editions of works, **8**:383a, 385a
Galois imaginaries, **5**:261a, 263b
Galois theory, **2**:105a; **3**:132a, 167b, 168a,b, 221b, 222a; **4**:2b, 177a; **6**: 472b; **7**:167b–168a; **8**:165b, 326a; **9**:173b; **10**:594a; **12**:364b
group theory, **4**:83a; **7**:567a; **8**:434a
manuscripts, **13**:250b
Galton, Francis (1822–1911), 5:265–267; 2:375a; **3**:130a, 163b; **8**:529a; **14**:251b
biographies of, **3**:582a; **10**:464b
on contemporary scientists, **12**:570a
eugenics, **3**:590b; **10**:464a
Galton's law, **5**:266b; **14**:100a
heredity, **7**:114b; **10**:463b; **14**:234b
on intelligence, **3**:566b; **7**:190a
Natural Inheritance, **10**:449a–b, 461b; **14**:251a
and Pearson, **10**:447b, 461b–462b, 464a
Record of Family Faculties, **10**:461b
statistics, **3**:113a; **9**:129b; **10**:451a, 461a, 465b(n); **14**:99a; **15**:32a
Galvani, Luigi (1737–1798), 5:267–269; 3: 114a*; **12**:155a; **14**:79a
biographies of, **15**:110a
electrophysiology, **1**:107b–108a; **7**:620b; **9**:177a; **14**:76b, 77a, 253a
De viribus electricitatis in motu musculari, **1**:107b
and Volta, **14**:77b–78a
Galvanic Society, **2**:92a
Galvanism, **1**:44b, 107b–108a; **2**:92a, 136b, 336a; **3**:68a; **5**:194b, 320b, 431b, 518a; **6**:549b, 550a,b; **14**:76b–80a
and chemical affinity, **1**:348b; **3**:601b; **4**:536a; **5**:7a, 292a; **10**:182b; **13**:359a; **14**:124b; **15**:459a–461a
conservation principle, **6**:244b
discovery of, **5**:267b–268b; **8**:133a
as fundamental force, **7**:603b, 604a
and medicine, **13**:468b
theories of, **4**:503b; **10**:188b–192b; **12**: 254a; **14**:77a–80a, 130b*
see also Electricity
Galvanometer, **1**:303b; **2**:600b; **4**:535b; **10**: 345b, 346a, 500a; **12**:254a–b; **14**:330b
ballistic, **14**:207a
cosine, **13**:473a
differential, **1**:557b
Edelmann, **2**:1b
Einthoven, **2**:2a; **8**:294b
invention of, **3**:497a
mirror, **4**:333b; **9**:259b; **13**:387a
moving-coil, **14**:617a
shielded, **11**:582a
sine, **11**:50b, 111b
string, **2**:2a; **4**:333b, 334a; **5**:65a; **8**:533a; **12**:95a
tangent, **7**:450a; **11**:111b; **14**:206b
Thomson, **1**:536b
universal, **12**:425b
Galvanoplastic phenomenon
see Electroplating
Galy-Cazalat, Antoine, **1**:128b
Gama, Vasco da, **9**:36b–37a; **14**:583b; **15**:89b
Gamaland, **2**:14b
Gamaleya, Nikolay Fyodorovich (1859–1949), 5:269–271
Gambey, Henri-Prudence (1787–1847), 5: 271; 1:202b
Gambier Islands, **3**:550b–551a
Gambusia, **5**:503b
Game theory
seventeenth-century, **3**:66a; **4**:574b; **6**: 537a
eighteenth-century, **9**:499b–500a; **15**: 279a–285a *passim*
nineteenth-century, **8**:62a*; **15**:366b, 374a
twentieth-century, **2**:304a–b; **14**:92b
asymmetries, **15**:283b–284b, 369a, 373a–b, 489a
division of stakes, **10**:334b–335a; **15**:282a, 369b
Games
see Recreations, mathematical; specific games
Gametes
purity of, **3**:121a–122a
Gametogenesis, **6**:267a
biochemistry of, **7**:467a
Gametophytes, **3**:29a,b, 30a,b; **6**:219a, 465b
Gamgee, Arthur, **2**:547b; **13**:352b
Gamma globulin, **3**:336a; **15**:454b
Gamma rays, **2**:338b; **7**:159a*; **12**:29a
absorption of, **15**:26b
rate of, **6**:355a
attenuation of, **9**:261b
bombardment by, **7**:153b–154a; **15**:26b
discovery of, **14**:32a
nature of, **7**:153a; **9**:260b, 261b
neutral-pair theory, **2**:398a–b
relativistic deflection, **5**:183b
from rocks, **6**:355b
scattering of, **3**:368a; **9**:261b; **13**:369a
secondary, **15**:26b
wavelengths, **12**:32a
Gammaridae, **13**:9a
Gamow, George (1904–1968), 5:271–273; 3:329b; **5**:331b; **9**:261a; **12**:33a, 34a; **15**:60a
Gamow-Teller selection rule, **5**:272b
Gaṇapati Bhaṭṭa, **12**:115b
Gandavo, Joannes de
see Dullaert of Ghent, Jean
Gandolfo, Giovanni, **1**:108b
Gaṇeśa (*b.* 1507), 5:274–276; 9:580a; **15**: 629b
commentaries by, **2**:116a; **7**:315a
commentaries on, **7**:219a,b
Grahalāghava, **7**:512b; **15**:624a, 624b–625a
commentaries on, **9**:579b
influences on, **15**:590a
Gaṇeśapakṣa (school of astronomy), **9**:607a; **15**:534a, 624a–625b
founding of, **15**:624a
Gaṅgādhara
commentaries of, **2**:115b; **5**:274b; **12**:115b
Gaṅgādhara Varman, **5**:274b
Ganges (river), **11**:376b
Ganglia, **9**:505b–506a; **11**:380a, 492a; **12**: 137a
autonomic, **2**:124b
Bezold's, **2**:111a
Bidder's, **2**:124b
in cardiac tissue, **5**:82a
peripheral, **8**:15b–16a
resection of, **8**:92a
spinal cord, **3**:343a; **8**:282b; **11**:296a; **12**:137b
sympathetic, **12**:137b
theory of, **11**:518a
Gangliosides, **7**:518a
Gangrene
exudate from, **13**:566b
gas gangrene
bacillus, **5**:42b; **7**:100a; **10**:155b; **14**: 249b
hospital, **8**:400b
Gans, David Manus, **6**:118a
Garbasso, Antonio, **11**:582b
Garbett, Samuel, **11**:499a
García, Celso-Ramón, **10**:611a
García, Manuel, **3**:531a; **13**:493a–b
Gard basin, **5**:498a
Garda, Lake, **9**:135b
Garden, Alexander, **1**:487a,b; **3**:345a, 350b; **10**:509b; **12**:515b; **14**:253a

Gardiner, James Henry, **3**:475a, 478a, 481a*
Gardner, Nathaniel Lyon, **12**:328b
Garga, **13**:582a; **15**:538a
  Gargasaṃhitā, **15**:536a
Garibaldi, Giuseppe, **3**:46a,b; **9**:268a; **10**:37a, 620b
Gariopontus, **12**:81a
Garland, George Minot, **2**:367a
Garnet
  crystal shape, **2**:6b; **6**:178a, 181a
Garnett, Thomas (1766–1802), 5:276; 3:601a
Garnier, Jean Guillaume, **11**:236a
Garnier, René, **3**:220b; **4**:174a
Garnjobst, Laura, **13**:268b
Garnot, Prosper (1794–1838), 5:276–277; 4:256a; **8**:265a,b
Garonne (river), **8**:466b
Garreau, Lazare (1812–1892), 5:277–278
Garrigou, Félix, **11**:269b
Garrod, Archibald, **6**:498b
Gärtner, Karl Ludwig, **8**:246a
Gas dynamics, **3**:195a–b; **7**:84a,b, 436b, 555a; **8**:301a, 597a–b; **10**:20a; **11**: 390a
Gas mask, **6**:21b, 265a; **12**:404b; **14**:601b; **15**:159b
Gascoigne, William (*ca.* 1612–1644), 5:278–279; 1:341b; **3**:457b; **12**:342b
  manuscripts of, **13**:445a
Gases
  absorption of, **2**:588a; **14**:448a, 522b*
  activation of, **7**:208a
  adsorption, **6**:118a,b; **12**:404b
  analysis of, **2**:70b, 588a; **3**:11b; **5**:125a; **6**:24a; **7**:451a, 621a; **9**:78b; **11**:36b, 279a–280a; **12**:233a; **14**:447a
    microscopic, **7**:502a,b
  asphyxiating, **8**:256b
  atmospheric, **4**:600a; **5**:602b; **8**:73b; **11**: 279a–280a; **13**:103a
  atomic weights, **4**:243b–244a; **14**:318b
  in blood, **1**:453a; **5**:149a; **6**:24a; **8**:540b; **9**:347b, 348b; **11**:135b
  chemistry of, **2**:572a; **3**:465b; **5**:319a–320b; **7**:277a; **13**:7a; **15**:439a–447a
  cleaning of, **7**:555b*
  combining volumes law
    see Combining volumes, law of
  combustibility, **4**:130a; **7**:13a; **8**:118a; **15**:492b
    ignition temperatures, **4**:130b
  compound nature of, **1**:345a
  compressibility, **1**:128a–b; **5**:610a; **6**:470a; **9**:291a; **10**:185b; **11**:111b; **14**:319a
    non-continuous flow, **3**:195b
    temperature effects, **1**:128a
  compression of, **2**:136a; **3**:11b
    thermal effects, **7**:509b; **15**:377b–378a, 379a
  condensation of, **4**:537b; **7**:267b; **9**:152b
    coefficient of, **8**:509b
  corresponding states, law of, **1**:129a; **7**: 220b–221b; **14**:109b–111a
  critical temperature, **1**:161a; **3**:11b; **5**: 389a; **9**:287a; **14**:109b–111a
  degeneracy, **4**:578a; **5**:103a; **15**:444a
  density, **1**:344b, 345b, 616b; **2**:135a; **4**: 376b; **5**:319a, 320a; **13**:102b–103a; **14**:319a
    fluctuations of, **12**:497a–b
    microcanonical, **9**:223b–224a
    and temperature, **8**:262a
  diffusion of, **2**:263a, 588a; **3**:307b; **4**:375b; **7**:416b, 509b; **8**:510b; **9**:222a–b; **11**: 146a
    Graham's law, **5**:493a
    through a septum, **2**:184a
    and temperature, **13**:11a
  discovery of, **6**:255a, 256a–b, 257a, 384b*
  dissociation of, **3**:373a; **4**:229a; **5**:560a; **8**:23a,b; **15**:36b–37a, 435b
  drying of, **9**:531a
  elasticity, **1**:128b; **8**:468b; **9**:221a, 291a
  electric discharge through, **1**:559b; **3**:31b; **4**:19a; **5**:117b, 483b; **7**:84b; **8**:24a–b, 148b; **10**:345a; **11**:46b, 47a, 145a, 580b; **13**:80b, 236b, 468b, 473a; **14**: 171b, 172a, 330a
    Lichtenberg figures, **7**:470a
    spectra, **6**:439a
    striae, **5**:560a
    theories of, **5**:292a–b, 340b–341a; **7**: 554a–b; **12**:238b; **13**:365b–366b
    X-ray effects, **12**:26a–b
  electrical conduction in, **1**:320b; **6**:439b; **11**:461a; **12**:29a, 238b, 496a, 613b; **13**:364a; **14**:421a
  energy states, **13**:103b
  entropy, **4**:578a; **13**:41b; **15**:432b, 439a–448a
  equation of state, **2**:40a; **3**:308b, 309a–b; **4**:376b; **5**:587b; **8**:186b; **9**:291a; **15**: 48b
  expansion of, **5**:78b, 318a; **6**:244a–b; **7**: 182a,b; **9**:19a; **11**:111b; **13**:381a; **15**:377b
    adiabatic, **3**:304a, 305a; **15**:357b
    coefficient of, **3**:305b; **7**:160a; **11**:353b
  flow of
    through capillary tubes, **5**:493a
    density effects, **1**:327b–328a; **5**:493a
    high-speed, **7**:555a
  friction in, **3**:83b; **7**:526a; **14**:170b, 171a
  identification of, **5**:341a
  imperfect
    equation for, **3**:20a
  inductive capacity
    and refraction, **8**:492b
  inert, **2**:429a; **7**:469a
    discovery of, **3**:479b; **6**:94a; **9**:292b; **11**:279a–280a; **13**:453b–454b
    prediction of, **13**:359a
    separation of, **3**:299a
    spectroscopy of, **4**:79a; **14**:599a
  inflammable, **2**:420b
  internal energy, **5**:319a
  interstellar, **1**:57a; **12**:455a; **13**:117b
  ionization of, **2**:397b; **3**:31b, 372b, 373a; **4**:355a,b; **12**:26b, 70b–71a; **13**:366a
    equilibrium states, **5**:103a
  ionized, **8**:10a, 24a–b; **13**:446a–447a
  juvenile, **13**:148a
  kinetic theory of, **2**:40a, 169b–170a, 261a–265b, 266b–267a, 466a, 520b; **3**:306b–308a; **4**:321a, 513b; **5**:559a; **6**:214b, 292a–b, 392b; **7**:382b, 493a, 617a; **8**:185b, 510a,b, 596a; **9**:217b–225b, 616a–b; **12**:218b; **13**:11a, 237a; **14**:170b–171a, 184a–186a, 395b, 614a; **15**:36b
    applications of, **4**:376a; **13**:82a
    foundations of, **3**:32b–33b; **7**:84b, 509a,b
    generalization of, **4**:375b–376b
    opposition to, **2**:266b–267a; **12**:610a
    and radiometer, **3**:477a
    and thermodynamics, **3**:307a; **5**:391b; **12**:497a–b; **15**:443b–444a
    verification of, **4**:253a–254a; **7**:416b; **8**:102a; **12**:496b
  light, pressure of, **8**:107b
  liquefaction of, **1**:129a, 161a, 304a; **2**:65a; **3**:603b; **4**:79a–80b; **6**:94a; **7**:220b–221a, 272a–b; **8**:47a; **9**:291a, 477b; **10**:206a–b; **14**:109b, 522a–b
    cascade process, **10**:604a
  magnetic properties, **8**:11a; **11**:46b, 241b; **13**:241a–b
  mechanics of, **6**:545a–b
  medical use of, **1**:563b; **3**:600a–b; **6**:283a
  mixtures of, **4**:228b; **7**:272a; **11**:534b
    behavior of, **8**:118b
    condensation of, **7**:515b
    explosive, **8**:118a; **9**:356a
    separation of, **4**:80a; **5**:493a; **9**:224a
  molecular volume, **13**:299a
  molecules of, **5**:294b; **8**:508b–510a; **11**: 173b
    diameter of, **11**:291a
    distribution, **2**:261a–262a; **9**:220a
    free, **15**:41a
    velocity, **4**:375b–376a; **7**:416b; **13**:42a
  monatomic, **13**:446b
    isothermal lines, **6**:470a
  motive power of, **3**:82a
  naming of, **6**:254b; **8**:620b
  optical properties, **14**:497b
    dispersion, **7**:553a–554b; **9**:155a
    luminosity, **7**:619b–620a; **12**:493b; **13**: 107b
    and molecular weights, **4**:241a; **7**:620a
    opalescence, **12**:497a–b
    optical activity, **7**:526a; **11**:530a
    refraction, **10**:546a; **15**:361a–b
    refractive indices, **1**:348b; **2**:135a; **4**: 241a; **7**:620a; **8**:492b; **11**:142b
    scattering, **11**:266a
  partial pressures, law of, **3**:540b–542b; **6**:285a; **9**:222a; **14**:75b–76a
  penetrability
    effective cross section, **15**:491a
  phase transitions, **12**:438b–439a
  physical properties, **3**:82a; **9**:287a, 291a
    laws of, **11**:353a–b
  pressure
    measurement of, **7**:514a
    temperature effects, **1**:138b; **15**:357b
  quantum mechanics of, **8**:478a–b
  rare
    see Gases, inert
  rarefied, **7**:416b
    dynamics of, **9**:224a–225a; **15**:378a
  relaxation time, **9**:220a–221a, 224b
  respiratory, **10**:579b
    partial pressures, **2**:62a–b
  solidification of, **4**:79b
  solubility, **2**:170a; **12**:271a
    Henry's law, **2**:588a; **3**:541b; **5**:493a; **6**:285a; **15**:434a
    in water, **3**:541b–542a; **9**:148b
  sound vibrations in, **3**:258b; **13**:103b; **15**:486a
  specific gravity, **2**:588a; **5**:493a; **13**:373a
  specific heats, **1**:348a, 616b; **2**:169b–170a, 266b; **3**:82a, 317a*; **4**:241a; **5**:319a; **7**:161a, 526a; **8**:118a; **9**:218a, 220a, 223a; **10**:329b; **11**:150a, 292a, 530a; **14**:171a, 185a; **15**:95a, 363a, 449a
  spectra, **4**:497b; **7**:76a; **10**:345b–346a; **11**:47a; **12**:238a–b, 499a; **13**:473a; **14**:629b
    theories of, **8**:596a
  statistical mechanics of, **4**:578a; **10**:235b
  super-dense, **4**:280a
  surface effects, **9**:224b–225a
  theory of
    eighteenth-century, **8**:79b, 80a; **11**:144b
    nineteenth-century, **3**:83a; **5**:124a; **7**: 220b–221b, 526a; **8**:508b–510b; **13**: 104b; **15**:377a–379a
    twentieth-century, **4**:314a, 317b–318a; **8**:185b–186b; **12**:219a; **15**:41a
    vibratory theory, **13**:238b

see also Gases: kinetic theory of, quantum mechanics of, statistical mechanics of
thermal properties, **1**:348a–b; **3**:20a; **6**:470a; **11**:292a, 353a–b; **15**:486b
discontinuity, **12**:496a–b
heat conduction, **2**:263a; **7**:526a; **9**:222a–b; **13**:11a; **14**:170b
heat of combustion, **2**:69b
inversion temperatures, **10**:206a–b
see also Gases, combustibility
transport coefficients, **8**:185b, 186a; **9**:222b, 224b
transport processes in, **2**:262b–263b, 265b; **4**:375b–376b
mean-free-path method, **4**:376a
vibrational energy, **4**:414a
virial coefficients, **8**:185b, 186a
viscosity, **2**:263a; **3**:477b; **4**:376a; **8**:478b; **9**:217b–222b *passim;* **11**:290b–291a; **13**:156a
volcanic, **2**:588a
volume
pressure effects, **15**:355a–b
temperature effects, **1**:138a–b; **3**:208a; **5**:318a; **14**:76a; **15**:355b, 357b, 378b–379a
see also Chemistry, pneumatic; Coal gas; Eudiometry; Illuminating gas; Matter, states of; Mines, damps from; Natural gas; Pneumatics; Poisoning, by gas; Power gas; specific gases; Vapor(s)
Gaskell, Walter Holbrook (1847–1914), 5:279–284; **5**:84b*; **9**:506a
on heartbeat, **5**:82b
influence of, **8**:15b, 16a, 451a,b, 452b
influences on, **5**:80b, 83a; **7**:504b; **11**:518a
The Involuntary Nervous System, **8**:17a
students of, **12**:395b, 396a; **15**:104b
Gas-lighting
see Lighting, artificial
Gasoline
antiknock agents, **9**:375b–376a
cracking, **9**:608b; **14**:601b
ethyl, **7**:316a, 498a
fractions, **14**:602a
isolation of, **12**:435b
octane number, **7**:21b
shale
desulfurization of, **9**:608b
Gasometer, **8**:79a; **9**:152b
Gasometry, **13**:574a; **14**:123a
Gaspard, Marie Humbert Bernard, **9**:10b
Gassendi (Gassend), Pierre (1592–1655), 5:284–290; **1**:495a; **2**:83b, 201a; **5**:48a, 247b, 344a*; **6**:548b; **7**:308a; **10**:332b; **11**:486b
acquaintances of, **2**:348b, 353b; **4**:46a, 505b; **5**:408b; **6**:360b, 445b, 598b; **7**:375a, 537a; **9**:528a; **10**:595b; **13**:335a; **15**:69a
Animadversions on the Tenth Book of Diogenes Laertius
translations of, **3**:209a
astronomy, **7**:305a; **10**:153a,b; **12**:343a
and Beeckman, **1**:567a, 568a
and Bérigard, **2**:13a–14a
correspondence of, **3**:100b; **4**:505b; **6**:520a; **8**:267a–268b; **9**:598a; **13**:256b; **14**:254b
critics of, **7**:583b, 612a
and Descartes, **3**:321a; **4**:52a; **5**:285a–288a *passim*
Epicuri philosophiae Syntagma, **2**:378a
Exercitationes paradoxicae, **2**:13a
historiography, **11**:351b
hydrology, **10**:522a
influence of, **1**:484a; **2**:309a, 354a; **4**:96a; **5**:415b; **6**:515b; **10**:43b, 60b; **12**:581b, 585a, 600b; **13**:30b
kinematics, **8**:267b–268a
and Mersenne, **5**:285a,b, 287a, 288b; **9**:316b, 320a, 321a(n)
and Montmor, **9**:497b–498a
natural philosophy, **2**:13a; **4**:495b; **5**:354a; **6**:608b; **8**:152a, 173b, 436b, 538b
and Peiresc, **10**:489b–492a
philosophy, **7**:539b
Syntagma philosophicum, **2**:13a
La vie de Tycho Brahe, **9**:497b
Vita Peireskii, **14**:376b(n)
Gassendism, **1**:474a; **9**:498b
Gasser, Achilles Pirmin, **11**:395b, 396a; **13**:141a
Gasser, Herbert Spencer (1888–1963), 5:290–291; **4**:398b–399a; **8**:29b
Gassicourt
see Cadet de Gassicourt
Gassiot, John Peter (1797–1877), 5:292–293; **5**:341a; **12**:51a,b, 52b(n)
Gasteromycetes, **13**:489b
Gasterophilus, **7**:273a; **12**:224a
Gastraea theory, **1**:421a, 601a; **5**:447a; **6**:9a, 339a; **7**:475a; **11**:361a
critics of, **7**:402a
Gastraphetes, **6**:312b
Gastric juice, **15**:108a
analysis of, **11**:426a–b; **12**:558b
bacteria in, **11**:428b
carbohydrates, action on, **2**:28b
enzymatic factor in, **2**:28a
hydrochloric acid in, **1**:542b; **11**:173a, 426a–b
in vitro studies, **12**:558b
isolation of, **13**:46b
phosphoric acid in, **14**:558b
secretion of
see Secretion, gastric
Gastritis, **1**:544a
chronic, **3**:490a
hyperplastic
surgical treatment of, **1**:438b
Gastroenteritis, **2**:508a; **4**:404b
Gastropods, **6**:613a; **7**:460b
classification of, **5**:355a
embryogeny, **3**:389b; **11**:255a
fossil, **5**:486b; **9**:579a; **10**:221b
theories of, **11**:234b
Gastrotricha, **4**:290a
Gastrula, **4**:266b; **6**:9a; **14**:426a
epibolic, **1**:601a
Gastrula theory
see Gastraea theory
Gastrulation, **7**:474b, 475a; **11**:597a
in amphibians, **2**:384a
in mammalian embryo, **1**:601b
in vertebrates, **2**:384a
Gatenby, James Brontë, **4**:207a
Gates, Reginald Ruggles (1882–1962), 5:293–294; **14**:103b*, 104a*; **15**:513a*
Gattoni, Giulio Cesare, **14**:69b, 70a
Gatty Laboratory, **8**:609a
Gaub (or Gaubius), Jerome David, **3**:37b; **7**:11a; **8**:327b; **15**:6a
Institutiones pathologiae medicinalis, **14**:321a
Gaucher, Philippe Charles Ernest
Gaucher's disease, **1**:360a
Gaudichaud-Beaupré, Charles, **5**:224b; **11**:242b; **12**:550b; **13**:506b
Gaudin, Alexis, **5**:295a
Gaudin, Marc Antoine Augustin (1804–1880), 5:294–295; **1**:518b; **3**:48a
and Avogadro's hypothesis, **3**:47b
Gaudry, Albert Jean (1827–1908), 5:295–297; **1**:491a; **2**:346b; **5**:357b; **8**:43b; **15**:51b
Gauge
caliber, **6**:144a
Gauging, **4**:551b; **10**:572a; **11**:457b, 590b
Gaule, Justus Georg, **2**:141b; **4**:359b; **5**:463a; **7**:504b
Gaultheria procumbens
oil of, **3**:10b
Gaultier de Claubry, Henri-François (1792–1878), 5:297–298
Gaultier de La Valette, Joseph, **5**:284b; **10**:489a, 489b–490b, 491a
Gaupp, Ernst, **2**:383b
Gause, George Francis, **5**:545b
Gauss, Carl Friedrich (1777–1855), 5:298–315; **1**:14a; **3**:214a, 302a,b, 432b; **4**:1a, 127a; **5**:158b*, 261b; **6**:26a; **7**:602a; **9**:125b
acquaintances of, **2**:98a; **6**:326b; **7**:51a, 52a, 53a; **10**:573a; **12**:234a
Allgemeine Theorie, **12**:50b
astronomy, **2**:442a; **10**:592a; **11**:377a
and the Bolyais, **2**:268a,b, 269a, 270a–b; **5**:301a, 302b, 308a
celestial mechanics, **1**:429a; **4**:547b; **5**:299b–301a, 302a; **9**:449b; **10**:111a, 198b
correspondence of, **10**:199a; **11**:377a; **12**:234b; **13**:264b
and Dedekind, **4**:1b–2a, 3a
Disquisitiones arithmeticae, **4**:4b, 124a,b; **5**:299b–300b, 309b, 375a–b; **6**:307a; **7**:52a; **13**:219a
Disquisitiones generales circa seriem infinitam, **4**:341a; **5**:203b–204a
Disquisitiones generales circa superficies curvas, **8**:385b; **11**:611a
editions of works, **2**:442b; **4**:4b; **12**:599b; **14**:158b
and Eisenstein, **4**:340b–343a
and Encke, **4**:369a–b
Gauss curve, **14**:99a
Gauss-Bonnet theorem, **2**:288a
Gauss-Codazzi relations, **2**:288a
Gaussian coordinates, **4**:479b
Gauss-Krüger projection, **5**:304a
Gauss's theorem, **10**:248b; **14**:317a
geodesy, **5**:302b–304b, 305b, 306a, 309b; **7**:500b; **12**:234b
and Germain, **5**:375b
and Humboldt, **5**:301b, 304b–305a, 306a, 308a; **6**:552b
influence of, **1**:599b; **3**:559b; **4**:4a; **5**:108a, 572a; **7**:523a, 570a; **8**:429a; **10**:544b; **12**:469a
and Kaestner, **7**:206b
and Legendre, **8**:137b, 138a; **15**:365b
letters of, **2**:284a; **6**:522a
mathematics, **11**:52a
algebra, **2**:144b, 294b, 299a; **3**:136b, 137b, 138b, 164a, 166a; **5**:262b, 263a, 299a–300a, 307a, 309a; **7**:567a; **14**:279b
analysis, **1**:16b; **2**:470a; **3**:75b, 251a; **5**:302a, 304a,b, 306a, 309b, 572b, 573a; **7**:52b, 398b, 514b, 550a, 619a; **9**:127a; **10**:573b, 574a; **11**:409a, 449b, 452b
geometry, **1**:16a; **2**:271a, 288a; **4**:417b, 479b; **5**:299a, 301a, 302b, 304b, 309a; **7**:206b, 599b; **8**:137b, 432b, 433a, 434a; **9**:127a, 199b, 409b; **11**:455a; **13**:571b
logarithms, **3**:152b; **13**:431b
number theory, **1**:237b; **3**:137a–b; **4**:475a, 573a; **5**:298b–299b, 302a, 304b,

Gauss, Carl Friedrich (cont.)
307a; **6**:308b, 309a; **7**:494b, 565b, 616a; **8**:138b; **9**:173b, 411b–412a; **11**:55a; **12**:172b, 469b
reciprocity law, **1**:307a; **4**:474b; **5**: 299a,b, 304b; **6**:390b
and Olbers, **10**:198b–199a
personality, **5**:307b–309b
philosophy of mathematics, **7**:508a,b
physics, **5**:304b–306b; **8**:602b
electricity, **9**:209b
magnetism, **1**:521b, 595b; **5**:305a–306a, 309b; **6**:107a, 552b; **7**:608b; **9**:181b, 206b, 566b; **10**:192b; **11**:50a; **12**:49b, 50a–51a; **14**:204a–b; **15**:488a
optics, **5**:302a, 303a, 306b
publishing methods, **3**:134b; **5**:309a–b
statistics, **5**:302a, 307a, 309b, 477a; **9**: 453b; **15**:372a
law of errors, **5**:414a; **10**:458a; **13**:497b
least squares, method of, **2**:330a; **5**: 299a, 300a,b, 303b, 304a, 305a, 309a; **8**:187b; **10**:452b; **15**:363b, 365b, 367a
students of, **3**:59a, 495b; **5**:479a, 490b; **6**:230a; **9**:429a; **10**:110b, 543a; **11**: 448a; **12**:289b; **13**:4a–b
Theoria motus, **6**:49a; **8**:137b
"Theoria residuorum biquadraticorum," **1**:239b
translations of works, **12**:53a*
and Weber, **1**:7a; **5**:305a–306b, 308a, 309b; **7**:450b, 608b; **14**:204a–b
Gauss (ship), **4**:193b
Gauthey, Émiland, **10**:2a, 2b–3a, 3b
Gautier, Alfrède, **13**:127a; **14**:480b
Gautier, Armand E.-J. (1837–1920), 5:315; 3:10b; **4**:267a*; **13**:468a*; **14**:532a*
Gautier, Henri, **15**:56a
Gautier, Paul Ferdinand (1842–1909), 5: 315–316
Gavarret, Jules, **2**:366b
Gay, Frederick Parker (1874–1939), 5:316–317
Gay, John, **6**:139a–b
Gayant, Louis (*d.* 1673), 5:327–328; 4:37b, 267b; **10**:477a
anatomy, **10**:477b
Gay-Lussac, Joseph Louis (1778–1850), 5: 317–327; 1:200b; **2**:134a, 356b; **3**: 347a*; **4**:241b(n); **5**:157b
acquaintances of, **1**:201a; **3**:545b; **6**:325a; **9**:418b, 424b; **10**:499a, 511b; **11**:111a, 353a, 504b; **12**:523a; **13**:127a; **15**: 347b
analytical methods, **8**:331b; **9**:558b; **11**: 485b
on aqua regia, **1**:518b; **5**:323a
and Berthollet, **5**:317b–324a *passim*
capillarity, **15**:360a
on Charles's law, **3**:208a; **5**:78b, 318a
and Davy, **5**:320b–322b *passim*
elements, isolation of, **9**:451a
on gases, **7**:182a; **14**:74b, 75a, 76a–b
Gay-Lussac's law
see Charles, Jacques-Alexandre-César, Charles's law
Gay-Lussac's principle, **5**:323a; **6**:285a
on heat, **3**:316a; **5**:216a; **9**:237a; **15**:377b, 378a
high-altitude research, **2**:134b–135a; **5**: 318a–b
and Humboldt, **5**:318b–320a; **6**:551a–b
influences on, **15**:361a
on iodine, **3**:455b, 603a; **5**:322b
law of combining volumes
see Combining volumes, law of
and Liebig, **8**:330a–b, 331b
organic chemistry, **10**:357a
on oxymuriatic acid, **3**:602b; **5**:322a; **6**: 284b
photochemistry, **4**:182b; **5**:322a
physiological chemistry, **10**:367a, 370a; **12**:241a
on potassium, **3**:400a, 602b; **5**:320b–321b
students of, **2**:586b; **3**:79a; **5**:430a; **8**:277a; **12**:197a; **13**:127b, 499a
and Thenard, **4**:239b; **5**:320a–322a, 324b; **13**:311a–b, 312a
thermometry, **4**:240a; **15**:355b
on vapors, **3**:8b; **5**:320b; **15**:357a–b, 358a
on volcanism, **3**:586a
works translated, **4**:546b
Gay-Lussac, Jules, **8**:338b; **10**:499a
Gaze, Vera F., **12**:368b
Gazeta de literatura
founding of, **1**:127b
Gazette de santé
editors of, **10**:612a
Gazzetta chimica italiana
founding of, **12**:163b
Geanticlines, **3**:553b
Gears, **5**:387a; **7**:602a; **15**:519a
cycloidal, **8**:210a
differential, **12**:384b
epicycloidal profile, **7**:578a
hinge-lever, **3**:224a,b
segmental, **15**:253b
teeth, **8**:124a; **12**:131b; **14**:404a
classification of, **12**:211a
worm, **8**:210a
Geaster, **14**:193a
Geber
see Jābir
Ged, William, **13**:413b
Gedinnian (stratig. stage), **8**:463a
Geer, Charles de (1720–1778), 5:328–329
Geer, Gerhard Jakob de (1858–1943), 5: 329–330; 13:270b
Gegenbaur, Carl (1826–1903), 15:165–171; 6:7a, 10a; **7**:440a; **11**:382b; **13**:274a
students of, **4**:371a; **5**:51a, 453a, 502b; **8**:8a; **11**:211b; **12**:299a; **15**:123a
zoology, **3**:610b; **6**:535b
Gegenschein, **1**:465a
Gehlen, Adolf Ferdinand (1775–1815), 15: 171–173; 3:465b; **4**:134a; **11**:437a,b; **13**:310a
Gehler, Johann
Physikalisches Wörterbuch, **9**:578b
Gehrcke, Ernst, **8**:551a,b; **14**:171a
Gehring, Franz, **6**:534b
Geibel, Karl Georg Emil Georgevich, **4**:301b
Geic acid, **9**:558a–b
Geiger, Hans (Johannes) Wilhelm (1882–1945), 5:330–333; 9:261a; **11**:551a; **12**:29b
and Bothe, **2**:337b, 338a
Geiger counter, **2**:338a; **5**:330b–332a; **7**: 152b; **9**:261b; **13**:446b; **14**:172a
Geiger-Müller tube, **12**:34a
Geiger-Nuttall law, **5**:272a, 331a–b; **12**: 32a
Handbuch der Physik, **11**:265a
radiation, **11**:347b
radioactivity, **2**:398a; **3**:369b; **12**:30a–31b
Geikie, Archibald (1835–1924), 5:333–338; 3:158b*; **4**:549a*; **6**:587b*; **7**:186a; **8**:33b; **11**:34b; **12**:263b; **13**:273a
on contemporary scientists, **5**:68a*; **8**: 261a*; **9**:371a*; **11**:131b*, 277a, 372b*; **12**:545b*; **13**:149a*; **14**:627b*
Geikie, James (1839–1915), 5:338–339; 5: 39a, 334b, 337b
Geiser, Karl Friedrich (1843–1934), 5:339–340; 3:264a*; **11**:390b*, 480a(n); **13**:12b, 22a*, 155a
Geissler (pathol.), **1**:576a
Geissler, Friedrich Wilhelm Florenz, **5**:340a
Geissler, Johann Heinrich Wilhelm (1815–1879), 5:340–341; 11:46b
Geissler tubes, **5**:340b–341a; **6**:439b; **11**:446b
Geisslern, Ferdinand, **9**:278a
Geitel, F. K. Hans (1855–1923), 5:341–342; 12:506b(n)
and Elster, **4**:354b–357a; **13**:80b, 367b
Elster-Geitel activation number, **4**:356b
radioactivity, **3**:480b(n); **5**:394b; **12**:28a
Geitner, P., **1**:578b
Gelassenheit, **5**:120b
Gelatin, **3**:7b; **13**:161a
analysis of, **8**:275b
from bones, **3**:560b
in culture media, **2**:437b, 438a; **7**:423a
nutritive value, **9**:10b
in photography, **3**:333b; **4**:282b–283a
sulfuric acid, effects of, **2**:386a
technology of, **4**:165b
Gelfond, Alexandr Osipovich (1906–1968), 5:342–343; 4:475a; **6**:393b
Gélis, Amédée, **5**:72b–73a; **10**:499b
Gellibrand, Henry (1597–1636), 5:343–344; 2:462b; **5**:593b
Trigonometria Britannica, **5**:343b; **14**:51b
Gelmo, Paul Josef Jakob (1879–1961), 5: 344; 4:155b–156a
Gels, **5**:494b; **14**:633b
Gemeinnützige Abhandlungen, **13**:425b
Gemini (constellation)
novae in, **1**:598a; **6**:150a; **14**:521b
variable stars, **8**:539b
Geminus (*fl. ca.* 70 B.C.), 5:344–347; 1:247a; **3**:319a; **4**:391a, 433a(n); **10**:6a
calendar, **9**:339a(n); **10**:180a
commentaries by, **4**:415b–416a, 437b
Elementa astronomiae
editions of, **4**:213a
historiography, **4**:460a; **9**:272a,b; **15**:204a
influence of, **6**:406b; **9**:270a,b; **10**:529a; **13**:317b
Isagoge, **4**:172a
Liber introductorius Ptolemei ad artem spericam, **15**:178b
translations of works, **13**:401b
Geminus (also known as Lambrit or Lambert), Thomas (*ca.* 1510–1562), 5: 347–349; 4:487b
Compendiosa totius anatomie delineatio aere exarata, **14**:11b
Gemistus Pletho, Georgius, **6**:306a
Gemma Frisius, Reiner (1508–1555), 5:349; 1:179a; **2**:401b; **4**:5b, 97a; **9**:309b; **11**:591a; **13**:62a(n); **14**:594a
Arithmeticae practicae methodus facilis
commentaries on, **10**:494a
astronomy, **2**:403a
geodesy, **12**:500a
Gemmology, **4**:141b–142a, 597b; **11**:328b; **13**:407a–b, 511b
Gems, **12**:573a
artificial, **8**:213b; **15**:649b, 650b, 652a
coloring of, **4**:214b
assaying, **7**:340a, 342a, 343b, 346a, 347a–348a
collections of, **11**:304b
luminescence of, **4**:215a
magic virtue of, **7**:459a; **13**:348a
medical use of, **6**:396b–397a; **14**:585a
in Mesopotamian myths, **15**:641a
optical properties, **11**:266b–267a
values of, **7**:342a, 343a
Genderen Stort, A. G. H. van, **4**:372a
Genealogy
ancient Greek, **6**:213a

medieval, **14**:547a,b
seventeenth-century, **8**:150a–b, 268b
eighteenth-century, **8**:471a
General Electric Company, **3**:367b; **6**:546a; **8**:23a; **11**:306a; **13**:24a, 361b
The General Magazine of Arts and Sciences, **9**:141b
General Motors Research Corporation, **7**: 316a–b
Generalization, scientific, **1**:251b; **2**:304b; **3**:381b, 385a; **4**:228a–b, 232a; **8**:601b; **10**:486a; **11**:9a,b, 10b, 13b–14a; **13**: 585b; **15**:167a–b
Generation
see Embryogeny; Fertilization; Reproduction, biological; Spontaneous generation
Generation and corruption, **2**:394a; **15**:497a
Generators, electrical, **1**:284a; **10**:268b; **12**: 574b
alternating-current, **9**:77a
armature reaction, **8**:188a
efficiency of, **4**:284a
electrostatic, **7**:13a, 470a; **9**:607b
globe, **6**:170a–b
high-voltage, **13**:569b–570a
induction generator, **11**:460a
lightning type, **1**:284b
and motors, **8**:188a; **12**:131b
phase relationships, **8**:188a
polyphasic, **5**:597b
rotating field, **6**:223a
triboelectric, **6**:171a
turbine, **11**:588b
Generators, mathematical
G rule, **15**:688b–689b, 698b
Generelli, Giuseppe Cirillo, **9**:534a
Genes
alleles, **9**:522a
concept of, **7**:114a, 115a
contamination of, **3**:121b–122a
developmental role of, **13**:135a–b
exchange of, **3**:422b
genic balance, **2**:456a
interaction of, **9**:282a–b
lethal, **3**:493a; **9**:522a, 523a, 565a; **14**: 101a
complementary, **13**:138a
mechanism of, **7**:455b, 516b; **13**:135b
modifying factors, **3**:122a; **9**:564b; **10**: 570a–b; **13**:135b
mutation of
concept of, **9**:564b
discovery of, **5**:293b; **14**:96b
and evolution, **3**:572a; **5**:453b; **9**:520b; **14**:99b–101b
frequency, **3**:555b; **9**:565a; **13**:137b
induced, **7**:455b, 456a; **9**:565a; **10**:130a
mechanism of, **13**:136a–b
theories of, **3**:493a; **4**:271b; **6**:101a; **7**: 99a, 114b, 166b; **12**:76b
naming of, **7**:114a; **14**:99a
nature of, **5**:454a; **7**:455b; **9**:564b
polymerization, **4**:142b; **10**:130a
position effects, **9**:522a; **13**:136b–137a
recombination, **5**:11a; **10**:130a; **13**:135a, 137b
sex-linked, **9**:521a,b; **14**:432a
nondisjunction, **2**:456a
see also Characters, genetic; Chromosomes; Genetics; Linkage, genetic
Genesis, book of
see Bible
Genet, Edmond Charles, **9**:366a
Genethlialogy
medieval Arabic, **1**:35a, 38a, 39a; **11**: 226a; **13**:538b
Indian, **12**:576b, 599a; **13**:582a,b
Mesopotamian, **15**:644b–645a
Genetica, **8**:513a
Genetical Society (British)
founding of, **11**:211b
Genetics
early eighteenth-century, **9**:188a–b; **11**: 332b–334a
late eighteenth-century, **7**:409a
mid nineteenth-century, **7**:93b; **9**:277a–283a; **14**:34a
late nineteenth-century, **2**:362a–364a; **3**: 421b–422a, 572a; **5**:266b; **10**:449a–b, 451b, 461a–464a; **13**:477b–478a; **14**:96b–97a, 98b–100b
early twentieth-century, **1**:505a–506a; **2**:364b, 455b–456b, 584b; **3**:120b–122a, 259b, 422b–423a, 493a, 555b, 572b, 590b; **4**:133a*, 142b, 270b–272b; **5**:7b, 10b–11a, 225b, 293a–294a, 453a–454a, 598b–599a; **6**:121b; **7**:99a–b, 114a–115a, 286a; **8**:513a; **9**:520a–525b, 564b–565a; **10**:130a, 444b; **11**:211b–212b; **12**:416b–418a; **13**:133b–137a, 478a–479a; **14**:100b–101b, 251b–252a, 432a–433a; **15**: 506a–511b
mid twentieth-century, **3**:122b–123b, 555b, 564b; **6**:21a–23a; **7**:455b–456a, 516b–517a; **9**:565a–b; **10**:611a; **12**: 76b–77a; **13**:137b–138a; **15**:425b–426a
applied, **7**:455b; **15**:509b–510b
behavioral, **10**:435a
biochemical, **6**:498b; **7**:516b; **9**:525b; **13**:135b
botanical, **3**:259b; **12**:76b–77a; **15**:506a–510b
developmental, **13**:137a
experimental, **1**:505b; **5**:10b; **8**:3b
chemical mutagenesis, **12**:76b–77a
culture media, **2**:456a
X rays, **6**:338b
fate-mapping, **13**:137a
genetic ratios, **2**:58b
of geographic variation, **5**:453b
human, **3**:555b, 590b; **5**:11a, 598b; **6**:22a; **10**:464a; **11**:333b; **12**:76b; **13**:137b; **14**:230b
mathematics and, **2**:58a–b; **7**:99a, 115a; **14**:87a
medical, **12**:76b
molecular, **9**:525b
opposition to, **15**:425b–426a, 511a
physiological, **5**:454a
population, **2**:58a–b; **5**:453b–454a; **6**:22a; **7**:114a–115a; **9**:282a; **14**:230b
protozoan, **7**:99a–b
radiation, **9**:565a; **12**:77a
serological, **7**:623b
and society, **4**:271b–272b
statistical, **12**:461b
and taxonomy, **4**:308a
theoretical, **7**:455b; **13**:137b; **15**:506b
multiple-factor hypothesis, **3**:122a
pure line theory, **7**:114b
and quantum theory, **12**:222b–223a
as university discipline, **1**:505a
zoological, **3**:120b–123b
see also Biometrics; Dominance, genetic; Eugenics; Genes; Heredity; Hybridization; Linkage, genetic; Mosaicism; specific organisms
Genetics, **4**:271a
founding of, **3**:123a; **4**:272a
Genetics Society of America
founding of, **3**:123a
Geneva
Académie de Calvin, **3**:459a–b
Académie de Genève, **8**:35a–b, 37b–38a
botanical gardens, **3**:44a; **8**:38a
Conservatoire Botanique, **3**:44a
natural history museum, **3**:44a; **9**:538a
politics, **11**:134b, 135a
Société de Lecture, **8**:38a
Geneva, Lake, **4**:9b; **15**:158a–159a
Geneva, University of, **14**:57b
chemistry at, **9**:354b
Gengou, Octave, **2**:300b; **15**:522a–b
Genius, **15**:465b
Genocchi, Angelo, **3**:118a; **9**:268a; **10**:441a,b
Genotypes, **5**:266b
concept of, **7**:114a, 115a
Gentes herbarum
founding of, **1**:396b
Genth, Frederick Augustus (1820–1893), 5: 349–350
Gentiana, **5**:546a
Gentleman's Diary, **3**:538a; **5**:530b
Gentleman's Magazine, **3**:350b
Gentleman's Mathematical Companion, **5**: 466b*
Gentleman's Society of Spalding, **11**:579b
Gentzen, Gerhard (1909–1945), 5:350–351
Herbrand-Gentzen theorem, **5**:350b–351a
Genus, biological
concept of, **1**:262a–b, 278a, 306a, 524a; **8**:435b; **13**:443a
definition of, **11**:443b; **13**:443a
Genus, mathematical, **3**:315b; **11**:450a
definition of, **3**:315a
Geobotany
see Phytogeography
Geocentric theory
see Cosmology, geocentric theory
Geochemistry
eighteenth-century, **8**:147a, 469a
early nineteenth-century, **2**:158a–159a; **3**:292a–293b; **6**:54b–55a; **15**:457a
mid nineteenth-century, **3**:586b; **5**:72a–b; **6**:565a; **10**:265a; **11**:560b; **12**:318a, 435b
late nineteenth-century, **5**:89a, 161b–162a, 315b; **12**:140b
twentieth-century, **2**:370a; **3**:610a; **4**: 597b–600a; **5**:457a–458a; **8**:285b; **10**:126a, 136b; **11**:558b; **12**:273b; **13**:610b–611a, 617b–619b; **14**:183a–b
crenitic hypothesis, **6**:565a
living organisms, role of, **13**:619b
philosophy of, **13**:618a
regional, **4**:598b
of water, **9**:258a
Geochronology
see Earth, age of; Geology, time scale
Geöcze, Zoard de, **11**:260a
Geodesics, **1**:599b, 600a; **2**:288a; **3**:263b, 282a, 331a, 559b; **4**:103a, 323b, 327b, 328a, 336b–337a, 479b; **5**:202a; **6**: 240a; **7**:573b; **11**:408a,b, 454b; **14**: 311a
closed, **11**:58a; **12**:411a
on concave surfaces, **6**:4b
on convex surfaces, **2**:48b, 53b–54a
ellipsoid, **8**:385a–b
equation of, **8**:385b
on spheroids, **5**:472a
theory of, **4**:480a
Geodesy
medieval Arabic, **1**:38a; **2**:149a–b, 153a, 154b
Renaissance, **9**:191b
early seventeenth-century, **10**:490a

Geological Society of America
founding of, **3**:592b; **4**:361b, 366a; **7**:370b; **14**:392b, 439b, 443a, 517a
Geological Society of London, **3**:603b; **4**:10a, 541a; **5**:67a, 334b–335a; **8**:486a; **12**:277b, 488a; **14**:210b
founding of, **1**:359b; **2**:355b; **5**:518b; **10**:322b
Wollaston Medal, **14**:486b
Geological Survey, U.S.
in nineteenth century, **1**:552b; **2**:500a; **3**:483b, 536a, 593a; **4**:265b, 361b; **5**:49b; **6**:57b, 187a; **7**:45a, 62b–63a, 148b–149b, 370a,b; **9**:255b; **10**:273a; **11**:210a, 443a; **12**:83b, 229b–230a; **13**:572b; **14**:120b, 392a, 402a, 443a; **15**:92a
in twentieth century, **3**:483b, 520a, 536a; **4**:562b; **7**:112b; **8**:41b, 370a–b; **9**:104a; **13**:531b, 532b; **14**:297b
Alaska branch, **2**:500b
chemistry, **3**:292b–293a, 610a
critics of, **9**:100b
Division of Economic Geology, **4**:366b
establishment of, **4**:366a–b; **5**:395b; **6**:13b; **7**:89b; **11**:119a
glacial division, **3**:190a; **13**:269b–270a
hydrology, **9**:257a–b
and irrigation surveys, **11**:119b–120a
paleontology, **9**:134a
statistical division, **3**:609b
Geological Surveys
see under specific countries
Geologicheski vestnik
founding of, **1**:163b
Geology
ancient Greek, **11**:28b, 104a
medieval Arabic, **7**:342b; **13**:292a; **15**:265b
medieval Latin, **11**:469a; **14**:35b, 400a
Renaissance, **1**:78a, 97a; **10**:280b–281a; **11**:608a–b
seventeenth-century, **2**:612b–613b; **4**:406b; **7**:376a; **10**:259b–260a; **12**:256b–257a; **13**:30b; **14**:500b–501a
early eighteenth-century, **2**:347b, 490a; **5**:226b; **7**:495a; **9**:368b; **10**:509b; **11**:562b–563a; **12**:159a–b; **13**:86b–87a, 409b–410a, 530b, 564a; **15**:52b–56a
late eighteenth-century, **1**:233b–234a, 487b; **2**:347b–348a, 552b–554a; **3**:155b, 560b; **4**:27b–28a, 70a–72b, 149b–153a, 275a, 548b–549a; **5**:205a–b, 380a–b, 577a–579a; **6**:84a–b, 578a–584b; **7**:277b; **9**:418a; **11**:34b–36a, 272a–273a, 387a–b, 472a; **12**:517a–b, 560b, 561a–562a; **13**:494b, 565b; **14**:132a–133a
early nineteenth-century, **1**:327a–b, 359b, 413a–b; **2**:375a, 480b–481b, 482a–b, 493b–497a, 554b–557a, 566a–571a; **3**:210a, 395a–396a, 601a, 603b–604a; **4**:9b–11a, 234a, 273b–274a; **5**:11b, 13b–14b, 155b–156a, 430a, 537a–b; **6**:325a, 455b–457b, 509b–510a; **8**:245a–246a, 564a–569a, 593b–595a, 615a–616b; **10**:8b–9a, 164a, 287b; **12**:433a; **13**:448b, 466b–467a; **14**:88a–b, 240a, 462b; **15**:82b, 502a–503a
mid nineteenth-century, **1**:19b–20b, 73a–b, 95b, 209b–211a, 388a, 572b; **2**:108a–110a, 158a–159a, 172b, 341a–b, 356b, 497b–498a, 586b, 587a, 588a–b; **3**:70b, 297a–298a, 392a, 433a–434b, 470b–471a, 585b–586a, 607b–608b, 623b; **4**:73a–b, 289b, 308b–309b, 347b–350a; **5**:68b–69a, 71b–72b, 206b, 333b–338a, 538a–b, 599a–b, 610b–611b; **6**:137b, 288b, 436b–437a, 489b, 524a, 564a–565a; **7**:44b, 45a, 185b; **8**:578a; **9**:172b–173a, 420b–421a; **10**:170b, 258a–259a, 204a; **11**:500a–b, 504b–505b, 560b–561a; **12**:293a, 542b–544b; **13**:123b–124a, 467a, 581a–b; **14**:57b, 139a, 315b–316a, 397b–398b, 439a–b; **15**:151b–153a, 526a–527b
late nineteenth-century, **1**:129b–132a, 161b–163b, 300b, 397a, 552b, 585a–b; **2**:289a–b, 346b, 477a, 486a–b; **3**:160b–161a, 189b–191a, 191b–192a, 328a–b, 462b, 483b–484a, 536b, 592b; **4**:143b–145b, 361a–b, 365b–366b; **5**:214a–b, 338b–339a; **6**:13b–14a, 78b, 320a–b; **7**:30b–31a, 370b–371a; **8**:44b–45a, 122b–123a, 529a, 576b–577a; **9**:315a–b, 512b; **10**:32a–b, 203a, 272b; **11**:371b–372a; **12**:293a–b, 298a, 343b–344a, 519b–520a; **13**:272b–273b, 525a–526a, 572a–b; **14**:1a, 402a–b, 441a–b, 516b–517b
early twentieth-century, **1**:235b–237a, 393b–394b; **2**:373a–b, 489a, 548b–549b; **3**:532b–533a, 547b–548a, 591a–b; **4**:42b, 261b–263a; **5**:187a; **7**:160b–161b; **11**:575b–576a; **12**:83b–84a, 260b–261a, 279b; **13**:269b–271a, 345b
mid twentieth-century, **1**:393b–394b; **7**:266b–267a; **13**:275a–b
agricultural, **3**:161a
and archaeology, **2**:549b; **14**:183a
and biogeography, **1**:73a; **5**:67a–b
British school, **2**:567b–568a; **3**:395a
and chemistry
see Geochemistry
comparative, **2**:568b
dynamic, **1**:153b–154a; **2**:567b, 568b; **9**:532a–533a; **10**:166b
economic
Renaissance, **8**:242a
seventeenth-century, **14**:451a
eighteenth-century, **2**:292a; **4**:548b; **5**:205a, 428b, 568a; **6**:579a; **8**:376a; **13**:410a
early nineteenth-century, **2**:439b; **3**:411b–412a; **4**:10b–11a, 218a, 348a–349a, 541a; **5**:155b–156a; **12**:487b–491b
mid nineteenth-century, **1**:20b; **2**:158b, 570b; **3**:70a–b, 586b; **5**:441b, 611a; **7**:45a, 224a; **10**:258a–b, 265b; **11**:209a–b, 399a, 439a; **12**:434b–436b; **14**:315b, 439a, 625a
late nineteenth-century, **3**:606b, 609b; **4**:366a–b; **5**:458a, 498a, 547b–548b; **6**:187a; **7**:44a, 62b, 253a–b; **8**:577a; **9**:291b, 592a–b; **10**:167a; **11**:102b–103a, 210a, 439a–b; **13**:525b, 572b; **14**:300b–301a, 442b–443b, 599b; **15**:428a
twentieth-century, **2**:499b–501a; **3**:325a, 520a–b, 610a; **4**:7b–8a, 261b, 262a, 598a–600a; **5**:457a, 462a–b, 563a; **7**:267a, 530a; **8**:370a–b, 544b; **10**:169a; **11**:576a; **13**:525b–526a; **14**:60a, 298a–b, 441b–442a
and paleontology, **5**:467b
see also Mining, Ores
engineering, **2**:500a; **5**:533b; **10**:430a; **12**:140b
and evolution, **5**:357b; **6**:120b; **14**:215a
experimental
nineteenth-century, **1**:490b; **2**:158b–159a; **3**:586b–587a; **4**:141a; **6**:54a–56a; **7**:62b–63a; **8**:463b; **14**:402b
twentieth-century, **1**:51b–52a; **3**:484a; **13**:610a
wet clay technique, **3**:325b
and geometry, **1**:236a
handbooks, **4**:10a; **10**:169a
historical, **2**:567b; **4**:72a, 150b–152a; **5**:205a–b; **9**:593b; **12**:229b–230a, 230a–b, 272a–b, 272b–274b; **13**:64a, 572b; **14**:259b–263a
and geography, **1**:623a; **10**:29b
igneous, **1**:394a–b, 470a; **5**:336a–b; **6**:582b
kinematic, **1**:236b; **13**:148b
marine
see Oceanography, geological
mathematical, **4**:350a; **6**:502a–503b; **13**:572a
measurement in, **6**:79a
methodology, **8**:594b
military, **4**:598a, 599a
naming of, **4**:406b; **12**:122b
past and present causes in, **3**:162a; **12**:318a
philosophy of, **4**:599b–600a; **8**:565a, 568a–b; **10**:8b–9a
physical, **9**:593b–594b
and prehistory, **2**:346b
regional
eighteenth-century, **4**:493a; **5**:205a
nineteenth-century, **1**:163b, 397a, 491b; **4**:366a; **6**:238a–239a, 456a–b; **8**:260b–261a, 461b–462a; **9**:103a–b; **14**:627a; **15**:527a
twentieth-century, **6**:168b, 216a–b; **9**:197b–198a
and religion, **2**:568b–570b, 612b–613b; **3**:297a, 313b; **6**:586b–587a; **9**:389a–b, 533b; **10**:322a,b; **11**:372a; **12**:262a–b, 318b, 433b; **13**:547b; **14**:145a, 516b
statistical, **2**:158a
structural
early eighteenth-century, **9**:135a, 533a
late eighteenth-century, **5**:205a; **8**:68a–69a, 147b; **11**:35b, 302b–304b; **12**:119b–122b; **14**:132b
early nineteenth-century, **4**:402b–403a, 540a–541a; **5**:518b–519b; **14**:211a
mid nineteenth-century, **3**:529b–530a, 549b–554a, 566b–569a, 586b–587a; **6**:186b–187a; **7**:319b–320a; **8**:461a–462b; **9**:61a–b; **10**:258a–b; **11**:269b–270a; **12**:277a, 301a; **13**:143b–145b
late nineteenth-century, **1**:472a; **2**:89b; **3**:236a–237a, 593a–595a; **4**:265b–266a; **5**:329b, 395a–396a; **8**:30b–31a, 32b–33b; **9**:367b, 592b–594b; **10**:29b, 166b–168a; **11**:118b–120a; **12**:139b–140b; **13**:145b–148b; **14**:442b–443a; **15**:427a–428a
twentieth-century, **1**:153b, 235b–237a, 469b–470b; **2**:558a–b; **3**:325a–326a, 547b, 595a; **5**:221a–b; **8**:99a–b; **10**:128a–b, 168a–169a, 506b–507a; **11**:336a, 594a; **12**:274a; **13**:275b, 283a–285b; **14**:402b
experimental, **2**:558b
theories of
ancient Greek, **13**:85a–b
ancient Roman, **11**:39b
medieval, **10**:421a
Renaissance, **3**:66b; **8**:201b, 241a–243a; **13**:50a
seventeenth-century, **1**:550a; **2**:612b–613b; **6**:486b–487a; **8**:307b–308a, 416a; **13**:34a; **14**:501b–502a

Geology (cont.)
theories of (cont.)
early eighteenth-century, **2**:578a–579a; **8**:379a; **9**:26a–27b, 531b–534a, 539b; **13**:179a–b, 257b–258a, 564a; **15**:55a–56a
late eighteenth-century, **4**:70a–b, 151a–b; **5**:77a, 205a–b, 424b, 444a–445a; **6**:54a–b, 469a, 549a–b, 579b–584b; **7**:389a, 412a–b, 604a; **8**:68b–69a, 70b, 147a, 251b–252a, 373b; **10**:284a–b; **11**:302b–304b, 521a; **12**:121b–122b, 549a–b; **13**:441b–442a; **14**:60b, 119b, 259b–263a, 312a
early nineteenth-century, **2**:552b–557a, 568a–571a; **3**:395a–396a, 412a, 526a; **4**:10a; **5**:518b, 519a; **6**:55a–56a, 456b–457b, 509b; 7:69a–70b, 589a–b; **8**:565b, 594b, 615b–616b; **10**:322b–323a; **12**:514b; **13**:547b–548a; **15**:380a
mid nineteenth-century, **3**:550b–553b, 566b–569a; **5**:31b–32b, 533a; **6**:437b–438a, 454b, 502b–503a; **7**:45a, 185b–186a; **8**:530a, 575a; **9**:61a–b, 239a–b, 445b, 583a–585a; **10**:209b; **11**:133b–134a, 269b–270a, 276b–277a, 439a, 505b; **12**:261b–263b, 318a,525a; **13**:382b–383b, 620b–621a
late nineteenth-century, **9**:594a; **10**:171a–b; **11**:209b–210a, 439b; **14**:139a, 392a
twentieth-century, **2**:558b; **3**:162a, 191a; **5**:486b, 487a–b; **6**:188b, 475a–b; **8**:286a, 463b–464a; **12**:273a, 280b; **13**:270b–271a, 284b, 532a, 609a–611a; **14**:441b–442a
Chinese, **12**:380a–b
see also Catastrophism; Continental drift; Diluvialism; Evolution, theories of; Neptunism; Plutonism; Uniformitarianism; Vulcanism
time scale
Renaissance, **8**:241b
early eighteenth-century, **2**:204b, 579a, 581b; **8**:379a; **9**:26b; **15**:56a
late eighteenth-century, **1**:234a; **4**:72a, 151b; **5**:380a–b, 577b; **6**:580a, 581a–b; **8**:69a; **14**:260a
early nineteenth-century, **2**:494b, 496a, 496b–497a, 568b; **3**:526a–b; **4**:348b, 540b; **5**:356b, 357b; **6**:509b–510a; **7**:589b; **8**:569b; **10**:322b
mid nineteenth-century, **2**:108b–109a; **3**:297b; **4**:68a–b, 309a; **5**:533a; **7**:93b; **13**:383a–b
late nineteenth-century, **1**:397a, 491a; **3**:293b, 594a–b; **5**:336b–337a, 476a, 498a; **14**:392a, 443a, 503b
twentieth-century, **1**:285b; **4**:365a; **5**:329b–330a, 487a–b; **6**:475a; **7**:267a; **13**:107b, 532a
color code, **11**:375b
topographical, **3**:291b
see also Bibliography, geological; Congresses, scientific; Geophysics; Historiography, of geology; Hydrology; Maps, geological; Metamorphism, geological; Microgeology; Mineralogy; Mining; Oryctography; Paleobotany; Paleontology; Petrography; Petrology; Sedimentology; specific topics, countries, etc.; Stratigraphy; Tectonics; Terminology, geological
Geomagnetic index, **1**:478a
Geomagnetism
see Terrestrial magnetism
Geomancy, **1**:381a; **5**:49a*; **9**:364b, 435b; **12**:379b; **13**:509b; **15**:186a
Geometria Deutsch, **4**:260a
Geometrography, **8**:175b
Geometry
ancient Greek, **1**:171a–b, 180a–191a, 215a–230a, 231b–233a, 245a–b, 248b–249b, 338b–339a; **2**:549b–550a; **3**:391a; **4**:33b–34a, 103b–104b, 105a–b, 108b–110b, 171b–172a, 414a–429b, 437a–438a, 462b–463b, 466a–b; **6**:406b–408a, 410a–416b, 616a–b; **8**:189a–b, 192a–b; **9**:268b–276b; **10**:114b–115b, 180b–181a, 529a–530a, 587a–588a; **11**:26b, 28a–b, 221b–222b; **13**:296a–b, 302a–305a, 314b–317b, 319a–320a, 322a–b, 334a; **14**:603b–604a, 611a–b, 612b–613a; **15**:115b–118a, 503a–b, 671b–672a
ancient Roman, **1**:119b, 418a–b; **2**:230b–231a, 233b; **3**:176a; **9**:140b, 299b–300b; **10**:293b–300a; **11**:201b; **12**:314a–b, 579b–580a
Byzantine, **7**:29a–b
medieval Arabic, **1**:41b, 82a–83a, 437b, 444a–446a; **2**:154a, 619b; **4**:438a–443a; **6**:200a–203b; **7**:2b–3b, 79a–80b, 83a–b, 329a–330a, 335a, 352b–353a, 359b–360a; **9**:21b–22a, 85a, 556a–b, 615a; **10**:5b–6b; **11**:240a–b, 244b, 248b, 249a–b; **12**:91a, 431b, 432a, 447b; **13**:289a–291b, 510a–b
medieval Hebrew, **1**:22b–23a, 31a–b; **7**:360a; **8**:280a
medieval Latin, **1**:63a–b, 226a–228b, 381a; **2**:394b–395b, 435a; **4**:157b, 443a–448a, 605a, 608b–612b *passim;* **5**:360a–b, 365a,b, 401b; **6**:544b; **7**:129a, 175a–b; **10**:226a–228b; **14**:31a, 458b; **15**:163a
Renaissance, **1**:468a, 604b–605a, 607a; **2**:281a, 360a–361a, 534b, 462a, 618a–b; **3**:181a–b, 262b, 272b, 276b–277b, 311b, 515a–b; **4**:97b, 259a–261a, 596b, 619a–b; **5**:113a; **6**:144a; **7**:419a; **8**:193b, 223b–225b; **9**:264a, 488b; **10**:161b, 238a–b, 269b–271b, 416b, 494a–b; **11**:97b, 288a, 533a–b; **13**:48a, 259b, 260a–b; **14**:21a–24b, 29b; **15**:155a–b
early seventeenth-century, **2**:308a, 419a–b, 462b; **3**:128b, 149a–153a; **4**:46b–50a; **5**:382a–b, 408b–409a, 588b–589a; **6**:112b, 271b, 445a,b, 449b; **7**:194a, 269b, 301a, 557b–558a; **9**:598a–b; **10**:254b, 330b–331b, 342a–b; **11**:487a–488a, 489b; **12**:74b–76a, 206a–b, 332b, 500a–501a, 610b; **13**:235b, 433b–437a, 501a, 560b–561a
late seventeenth-century, **1**:164a–b, 292a, 474a–475a; **2**:47a, 308b–309a, 310a; **3**:182a–183a, 183b–184a; **6**:359a–b, 537a, 599b–602a; **7**:376b, 576b–577b, 583a–b, 611a; **9**:311a–b, 446b–447b, 599b; **10**:50a–53a, 264a–b, 314b, 335b–337a; **11**:404a–b, 512b–513a; **12**:459b–460b; **13**:480a; **14**:49a–b, 152a, 509b
early eighteenth-century, **2**:418a–b; **3**:284b; **4**:88b, 214b, 515b–516b; **5**:499a–500a; **8**:252a–253a; **9**:186a; **11**:328a; **12**:55a–56b, 445b–447a; **13**:54a–b, 266b–267a
late eighteenth-century, **3**:119b; **4**:515a; **5**:209b–210a; **7**:404b–405a, 598a; **8**:136b, 306a–b; **9**:55a–b, 156a–157b, 469b–476b; **11**:34a–b, 402a–b, 609a; **12**:283b, 444a–445a; **13**:54b–55a; **14**:589a
early nineteenth-century, **2**:214a–215b, 268a–269b, 270a–271a, 274a–b, 444a, 454b–455b, 473a; **3**:131b, 143a, 554b–555a; **4**:602a; **5**:299a, 302b, 367b–368b; **6**:2a–b; **8**:141a–142a; **11**:61b–62a; **12**:255a, 325a; **13**:127a–129a, 264a–b; **14**:140b–141a
mid nineteenth-century, **2**:470a; **3**:165a–169a, 212b–214a, 314a–315b, 322a–323a, 377a, 467b–469b; **4**:137a–b; **7**:163b–164a, 522a, 523b, 601a–b; **8**:429a–434b; **9**:199b; **12**:74a, 86b–87a; **13**:13a–21a, 219b, 221a; **14**:472a; **15**:66b
late nineteenth-century, **1**:425a; **2**:479a–480a; **3**:178a, 178b–179a, 428a–b, 559b–560a; **4**:102b–103a; **5**:198a, 211a–212a; **7**:396b–397b, 408a, 573b–575b; **8**:175b–176b, 542a–b; **9**:80a–b, 358a; **10**:24b–25a, 544a; **11**:390a, 509b–510a, 589a; **12**:195b, 213b, 225b–226a, 227b–228b, 246a–247a; **13**:250a–b, 414b, 623a–b; **14**:303a–b, 618b–619a
early twentieth-century, **2**:191a–192a; **3**:170b–171a, 399a; **4**:373a–b, 522b–523a; **7**:207a–b; **8**:504b–505a; **9**:358b; **10**:170a; **11**:394b; **12**:150a–b, 187b, 189b–190a, 532b–533b; **13**:23b, 124b–125a; **14**:283a–b, 336a–b
mid twentieth-century, **2**:191a–192a; **3**:171a, 222a
absolute, **2**:271a; **5**:555a; **12**:533a
abstract, **3**:166b
affine, **2**:192a; **3**:165a; **4**:281a, 331a; **5**:202a; **8**:161b; **9**:430b; **11**:260b, 362b
algebra and, **3**:451a; **5**:382a–b; **9**:475b; **14**:21a, 24a–b, 150a
algebraic
ancient Greek, **1**:183b–184b, 188a–189a
ancient Roman, **10**:297a
medieval, **4**:610b, 611a
Renaissance, **2**:280a–b; **3**:276b–277b
seventeenth-century, **4**:56b–58b; **8**:267b
eighteenth-century, **2**:114a; **14**:280a–b
early nineteenth-century, **5**:367b–368b
mid nineteenth-century, **1**:591a; **3**:167a–168a, 314a–b, 315a–b; **6**:357a–b; **11**:450a–b
late nineteenth-century, **2**:85b, 465a; **4**:373b, 618b; **5**:339b; **6**:547b; **8**:250a–b, 542b; **9**:358a; **10**:139b–141a, 343a, 594a–b; **11**:54b–55a, 509b–510a, 536a; **12**:246b–247a; **13**:133a
twentieth-century, **3**:117a–b; **4**:523a; **6**:402a*; **7**:530b; **8**:504b, 585a; **10**:138b–139a; **12**:228a, 330b–332a; **13**:23b
foundations of, **12**:331a
analytic
ancient Roman, **10**:296a
early seventeenth-century, **3**:64a, 615b–616a; **4**:46b, 51b, 56b–58b, 550b, 567b–569a; **5**:382b; **10**:331a; **11**:488a
late seventeenth-century, **5**:520b–521a, 526a–528b; **7**:577a–b; **8**:163b; **9**:303b–304a; **10**:52a–53a; **12**:460b; **14**:148a–150a, 465a–466b
early eighteenth-century, **3**:281a; **4**:516a–b; **5**:499b, 567a–b; **8**:609b–611b; **10**:319b; **13**:68b
late eighteenth-century, **3**:282a; **4**:107b, 479b; **7**:550b, 564b–565a; **8**:306a; **9**:343a–b, 470b, 475a–476b; **13**:418a
early nineteenth-century, **2**:214b, 444a; **3**:76a–78a; **5**:111a, 368a; **6**:2a; **9**:430a–b, 474a

mid nineteenth-century, **1**:590b–591a; **2**:424b; **3**:166b–167a, 314a,b, 331a; **4**:137a; **5**:523a; **6**:357a–358a; **7**:108b–109b; **8**:385a–b, 592a; **9**:431a, 553b; **11**:44b–46b
late nineteenth-century, **3**:559b; **5**:569a; **6**:75a–76a, 547b; **7**:397b; **8**:61b, 323b–326b, 367b, 542b; **9**:414a; **10**: 275a–b; **12**:246a–247a, 284a–285a; **13**:81a, 623a; **15**:195b
twentieth-century, **2**:320a, 593b; **3**: 117a; **11**:54b; **13**:125a
naming of, **7**:550b
applied
ancient Greek, **1**:222a; **6**:314a–315b
medieval Arabic, **1**:22b, 30b, 41b–42a
Renaissance, **1**:97a; **3**:515a; **4**:97a, 450a, 451a; **15**:155a–b
seventeenth-century, **2**:526b; **9**:13a
eighteenth-century, **3**:39a; **4**:166a; **8**: 611a
nineteenth-century, **9**:227a; **12**:325a–b; **14**:343a
and arithmetic, **1**:474b; **8**:267a–b
astral, **12**:255a
and astronomy, **1**:248b–249b, 256b; **2**: 348b; **13**:54b–55a; **14**:160b; **15**:208b
and atomism, **1**:381a; **2**:394b–395a
axiomatization of, **4**:9a; **6**:249a; **10**:343b–344b, 443a, 605b–606b; **12**:11b; **13**: 599b; **14**:304b, 612b
see also Geometry, foundations of
and biology, **1**:256b; **5**:470b
and chemistry, **3**:596b
Chinese, **3**:254b–255b; **8**:319a–320a, 419b–424b; **12**:375b, 378a–b; **13**: 484b–485a; **14**:543a, 545b
circular, **7**:397b
classification of, **11**:362b
complex, **10**:606a; **11**:77a, 79b; **12**:285a; **13**:125a
continuity of forms, **3**:513b–514a; **14**: 91a–b
and crystallography, **1**:591a; **5**:212a; **11**: 235b; **12**:195b–196a
of curves, **2**:418b; **12**:86b
definition in, **1**:22b; **2**:274b; **9**:311a
descriptive
ancient Greek, **15**:210b
Renaissance, **4**:259b, 260b
eighteenth-century, **6**:1a; **7**:550a, 599b–600a; **9**:470a–476a *passim*
nineteenth-century, **1**:590b, 591a; **2**: 350b, 473a; **3**:554b; **13**:132b; **14**: 342b–343a
twentieth-century, **5**:555a; **8**:504b
differential
eighteenth-century, **2**:330a; **4**:479b; **5**: 209b–210a; **11**:400b–401a, 401b–402b
early nineteenth-century, **3**:143a; **5**: 304a, 368b*; **7**:602a
mid nineteenth-century, **1**:599a, 599b–600a; **2**:88a, 287b–288a, 350a; **3**: 263b; **5**:309b; **7**:573b–575b; **8**:389b; **9**:199b, 409b–410a, 553b; **14**:231b; **15**:198b
late nineteenth-century, **2**:121b; **3**:179a; **7**:397b; **8**:287b, 324a–326b; **9**:80b; **10**:544b–545a; **11**:398a–b, 611a; **14**: 158b, 231b–232a
early twentieth-century, **2**:191b, 594a; **3**:95a, 170b, 171a; **4**:173b–174a, 287b–288a, 335b–337a, 373b; **5**:78a*, 201b; **11**:455b; **12**:150a–b, 214a–b, 285a; **13**:562a, 600a; **14**:356a–b, 437a
mid twentieth-century, **7**:207b; **14**: 122b, 311a; **15**:23a
absolute, **5**:555a
affine, **2**:192a; **5**:202a; **11**:260b
complex, **12**:285a; **13**:125a
conformal, **11**:260b
foundations of, **14**:311a
global, **3**:96a; **4**:337a; **6**:497a–b; **13**:600a
invariant methods, **1**:600a
local, **12**:214a
moving frames, method of, **3**:96a
projective, *see* Geometry, projective differential
of surfaces, **4**:257a–b, 480a; **9**:409b–410a
topological, **2**:192a
division of figures, **1**:22b; **4**:425b–426a; **6**:315a
Egyptian, **1**:31b; **13**:296b; **15**:670b–671a, 681a, 690b–693a, 696b–697b
elliptic, **3**:165b; **5**:201b; **7**:397a; **12**:533a; **13**:264b
enumerative, **3**:213b; **7**:164a; **12**:213b, 227b–228b, 331a; **13**:133a; **14**:618b–619a
Euclidean, **2**:352a; **3**:469a; **7**:397a,b; **8**: 432a
critics of, **8**:321b
see also Euclid
finite, **4**:523a
five-dimensional, **4**:331a
foundations of
Renaissance, **10**:416b
seventeenth-century, **13**:235b
eighteenth-century, **7**:206b, 226b; **9**: 157a
early nineteenth-century, **2**:269a; **5**: 154b, 301a; **13**:14a–b, 264b
mid nineteenth-century, **5**:307b; **6**: 243a; **8**:326b; **11**:448a, 454b–455b; **13**:16b–17b
late nineteenth-century, **6**:390b–391b; **8**:324a, 389a; **12**:10a, 364a–b; **13**: 414b
twentieth-century, **4**:101b, 373b, 560b; **6**:76b, 570a; **7**:207a–b, 313a; **8**:263a; **9**:358b, 502b; **11**:362b, 413a–b; **12**: 11a–b, 14a, 16a–b; **13**:125a; **14**:307b, 437a
see also Geometry, axiomatization of
four-dimensional, **3**:166b; **5**:78a*; **11**:46a; **12**:213b
in geology, **1**:236a
graphical methods in, **4**:48a–b; **9**:470b; **10**:226a–228a; **14**:343a
higher, **1**:187b; **5**:203b
hyperbolic, **3**:165b; **7**:397a; **9**:410a; **12**: 255a, 533a
imaginary, **8**:429b
Indian, **1**:309b; **2**:115b; **4**:608b; **9**:613b; **12**:598a
infinitesimal, **1**:229b, 449b; **2**:47a–48b, 351a; **4**:102b; **5**:55a*; **6**:599b–602a; **7**:446a, 615a; **8**:385a,b; **9**:470b–476a *passim;* **11**:206b–207a, 487a–b; **12**: 285a; **13**:130a, 417b
integral, **2**:191b, 192a
intrinsic, **3**:178a, 178b–179a; **5**:78a*, 210a; **11**:407b–409a
inversive, **8**:385b
Japanese, **1**:83b, 90b; **12**:291b–292b
kinematics and, **1**:445b; **5**:360a–b; **6**:379b; **7**:446b; **9**:80b; **11**:487b–488a, 488b; **12**:195b; **13**:290b
law of lever, **1**:219b, 220b–221a; **3**:182a,b
line, **4**:523a; **7**:396b–397a; **8**:542b; **11**: 46a–b; **13**:132b–133a
Lobachevskian, **1**:591a, 599b–600a; **4**: 371a; **7**:207a, 397a, 600a; **8**:429a–434b; **12**:533a; **13**:414b
measurement in, **13**:289b–290a
and mechanics, **3**:78a, 182a–183a, 195a; **7**:340a, 341a, 342a, 473a–b; **8**:124a; **10**:226a–228b; **11**:62a; **12**:534b; **14**: 149b, 620a
Mesopotamian, **1**:31b; **3**:276b; **4**:608b; **6**:315a; **11**:221b, 222a; **15**:670b–671b, 692a–693a
metric, **2**:121b; **3**:165a–b; **4**:336b–337a; **5**:202a; **12**:213b
models, **14**:343a
"modern," **3**:314a; **9**:475a
of *n* dimensions, **3**:166b, 323a; **6**:92a; **9**:502b; **12**:171b–172b, 532b, 533a–b; **15**:197a–b
non-Archimedean, **4**:523a; **13**:623b
non-Euclidean
ancient Greek, **14**:612b
eighteenth-century, **2**:331a; **4**:417a–b; **7**:405a, 600a
early nineteenth-century, **2**:269a, 270a–271a, 274a–b; **5**:301a, 302b, 304b, 309a; **7**:1b, 206b, 330a, 473a–b; **13**: 264a–b
mid nineteenth-century, **1**:591a, 599b–600a; **3**:128a–b, 165b, 322a, 323a; **5**:572b; **6**:249a, 522a
late nineteenth-century, **4**:137b, 618b; **5**:299a; **6**:391a; **12**:599b; **13**:414b; **14**:303a
early twentieth-century, **2**:121b; **3**: 111b, 179a, 399a; **4**:371a, 374b*, 523a; **12**:532b–533b
Cayley-Klein, **12**:533a
in function theory, **11**:52b, 53b
opposition to, **4**:101b; **7**:522b–523a; **8**:141a–142a; **12**:609a–b
precursors of, **12**:55a–56b, 255a; **13**: 510b
projective foundation of, **7**:397a–b
see also Euclid, parallel postulate; Geometry, Lobachevskian; Geometry, Riemannian; Lobachevsky; Riemann
non-Riemannian, **4**:336b
numerical, **2**:197b; **3**:314a; **9**:412a–414a
of paths, **8**:284b; **14**:311a
philosophy of, **6**:556b; **10**:51b–52a, 486a
and physics, **1**:226b; **3**:616a; **4**:51a, 53a; **5**:551b–552a; **7**:485b–486a, 487a; **8**: 39b, 196a–198a, 232a–233b, 434a; **11**:21b–22a
point, **11**:46a
of position, **10**:605b, 606a; **13**:4b
and probability, **3**:323a
problem-solving, **3**:78a; **4**:608b; **5**:209b, 382a–b; **8**:189a,b; **10**:271a; **13**:290a
projective
ancient Greek, **1**:191b; **4**:427a
seventeenth-century, **4**:47a–50a; **7**: 577a; **10**:330b–331b
eighteenth-century, **8**:252a–253a
early nineteenth-century, **2**:214b; **4**: 47a–50a; **5**:367b–368b; **6**:2a–b; **9**: 430b, 475a; **11**:76b–77b, 79b–80a; **12**:325b
mid nineteenth-century, **3**:165a–b, 214a, 314a, 467b; **5**:309a; **7**:163b; **11**:79a; **13**:4b–6a, 14b–21b, 221a
late nineteenth-century, **1**:194b; **3**:168a, 323a, 468a, 469a; **5**:569a; **7**:397a–b; **8**:287b, 433a–434a; **9**:80a–b, 358a; **10**:605b; **11**:390a, 460b; **12**:196a, 213b; **13**:133a, 623b
early twentieth-century, **2**:593b, 594a; **3**:117a, 170b, 399a; **4**:336b, 523a; **6**:473a; **7**:185a; **9**:173b; **11**:394b, 459b; **12**:330b–331a, 533a–b; **13**: 599b; **14**:212a, 437a, 559b
mid twentieth-century, **3**:96b*; **8**:457b
axiomatization of, **10**:605b–606b

Geometry (cont.)
projective (cont.)
complex, 10:606a; 11:77a, 79b
foundations of, 10:343b–344a; 11:413a; 13:5a–6a
hyperspatial, 8:504b
projective differential, 2:192a; 3:171a; 5:201b–202a; 14:356a–b
pseudo-Euclidean, 8:434a
of real world, 8:432a,b, 434b
Riemannian, 2:121b; 3:96a, 322a, 323a; 8:434a; 11:260b; 12:533a
generalization of, 4:336b–337a
of ruled spaces, 12:284a–b
solid, 2:395a; 4:422b–423b; 6:415a; 9:191b; 11:26b; 13:17a
analytic, 10:51b; 14:150b
space, 11:45b–46a
spherical
ancient, 9:297b–298a; 11:188a
Renaissance, 8:27b
seventeenth-century, 9:13a
eighteenth-century, 5:209b; 8:300a
nineteenth-century, 1:449b; 5:572b; 6:240a; 7:397b; 8:431b; 11:390a
Japanese, 1:313b
statics and, 1:214b, 219b–222a, 228b–229a; 3:150a, 182b; 5:369a*; 6:599b–600a, 601a; 10:336b–337a; 11:164a
of surfaces, 4:177b, 257a–b, 480a; 5:376a–b; 9:409b–410a; 12:86b
symbols in, 1:474a
symplectic, 8:324b
synthetic, 1:192a; 3:119b, 213a–b, 314a; 5:471b, 568b–569a; 6:2a; 7:3b, 163b; 9:80b; 11:390a, 394b; 12:86b, 225b–226a; 13:5b–6a, 13a–21a, 132b–133b; 14:222a; 15:195b
tensor, 7:207b
see also Circle, Square, etc.; Conics; Constructions, geometrical; Terminology; Trigonometry
Geomorphology
eighteenth-century, 2:347b; 4:492b–493b; 9:537b; 11:35b–36a; 13:257b–258a
early nineteenth-century, 6:456b–457a; 10:586a
mid nineteenth-century, 1:20b; 5:337b–338a; 6:437b–438a; 7:185b–186a; 8:260b; 11:276b–277a; 12:38b
late nineteenth-century, 1:174a–b; 3:593b–595a; 4:143b; 5:395b; 7:148b–149b, 496a–b, 511a; 8:31a; 9:103b–104a; 10:166b–168a, 429a–b, 502a, 504b–505a; 11:440a; 13:146b
twentieth-century, 2:191a, 548b–549b; 3:595a–b; 4:262a; 6:216a–b, 474b; 7:143b–145a; 9:149b–150b; 10:168a–169b, 503b, 505a–b, 507a–508b; 11:336a, 524b, 576a
Chinese, 12:380b
climatic, 10:505b
laws of, 11:35b–36a
mathematical, 10:504b
theories of, 6:55b
see also Erosion, Landforms, Physiography
Geophysical Research Corporation, 4:8a
Geophysics
medieval, 15:265b
seventeenth-century, 3:61a; 6:71b–72a
eighteenth-century, 2:329a–b; 12:159a–b; 15:295a–301b, 339b–341b
early nineteenth-century, 2:102a; 5:78b–79a; 6:106b–107a, 165b–166a
mid nineteenth-century, 1:364a; 4:590b–592a; 6:502a–503b; 7:608b–609b; 8:187a, 188b; 11:498a; 12:49a–52b, 209b–210a; 14:293a–b
late nineteenth-century, 1:300b, 521b, 522a, 552b; 3:583a–b; 5:573b; 7:595a; 12:187a–b; 14:503a–b
early twentieth-century, 1:154a, 521b; 2:168b, 372b–373b, 459b, 466b; 3:548a; 4:8a, 270a; 5:596a; 6:147a, 480a; 7:413b; 8:102b–103a; 9:180b–181b; 11:361b–362a, 412b; 13:83a–b, 166b–167a, 176a, 246b, 605a–608a; 14:215a–217a, 327a–b
mid twentieth-century, 1:478a; 2:558b; 5:161b; 6:474b–475b; 8:301a,b; 10:564a; 12:352b–353a; 13:608a–611a
and electricity, 1:547a
emergence of, 1:6a
experimental, 6:503a
philosophy of, 12:49b–50a
theoretical, 8:516a–b
see also Instrumentation, geophysical; Seismology; Volcanology
Geoponica, 2:256b
George I, King of England, 3:295b; 8:150b; 12:458a
George III, King of England, 1:435b; 2:420a; 4:565b; 6:130b, 234b, 494a, 567b, 569a; 7:11b; 8:321a; 9:595a; 11:148a, 499a; 12:226a, 365b; 14:419a
George, bishop of the Arabs, 1:275a
George of Ragusa, 10:489b
George of Trebizond, 1:274b; 11:351b; 15:474b
Georgia (Russia)
mathematics, 11:327a
Geospheres, 13:618b–619a
Geostatics, 4:46b, 47a
Geosynclines, 5:39a
concept of, 1:237a; 3:553b; 6:58a,b, 168b; 13:64a–b, 146b, 610a
Geothermal gradient, 3:412a
Geothermometry, 1:154a; 13:456a*
mathematical analysis of, 12:526a
Geotropism
see Plants, tropisms, geotropism
Gephyrea, 11:234b
Geraldy, Frank, 4:248b
Geranium, 4:401b; 8:303b
Gerard of Brussels (*fl.* first half of thirteenth century), 5:360; 1:226a; 7:178b*
De motu, 1:227b; 6:378b; 8:227b
Gerard of Cremona (*ca.* 1114–1187), 15:173–192; 2:283b
astronomy, 11:350a–b
Theorica planetarum (spurious?), 2:550b; 7:112a; 9:364b; 15:180a, 187b, 189a–b
translations by, 1:63b, 82a, 99b, 226a–227a, 228b–229a, 271a–b, 444a–b; 3:25a, 26a; 4:441a, 444b–445b, 542a, 609b, 611b–612a; 5:346b, 360a, 592a; 6:196b, 206a*, 617a; 7:23a, 38a,b, 82b, 141b–142a, 362b, 363a; 9:162a, 301b*, 362a; 10:5b; 11:202a, 245a, 349a; 13:293a*, 308a, 320a; 14:19a, 35b, 584b, 593a; 15:255a, 266b(n), 474b, 477a, 499a
Gerard of Feltre, 5:361a
Gerard of Sabbionetta, 1:512b; 2:283b, 435a; 7:112a; 15:186a, 189a
Gerard of Silteo (Sileto) (*fl.* thirteenth century), 5:361
Gérard, Frédéric, 5:357a
Gerard, John (1545–1612), 5:361–363; 9:313a
The Herball, 1:433a
editions of, 7:147a–b
Gerasimovich, Boris Petrovich (1889–1937), 5:363–364; 6:540a; 13:117b
Gerbec, Marko
see Gerbezius, Marcus
Gerbert (also Gerbert d'Aurillac, later Pope Sylvester II) (*ca.* 945–1003), 5:364–365; 1:62b; 3:214a; 4:157b, 605a; 6:301b, 302a, 544b
Gerbezius, Marcus (1658–1718), 5:366–367
Gerdy, Pierre Nicolas, 3:531a
Gergonne, Joseph Diaz (1771–1859), 5:367–369; 2:214b, 215a; 8:386b(n)
Annales de mathématiques pures et appliquées
see *Annales de mathématiques pures et appliquées*
correspondence of, 12:325b
geometry, 11:45a,b, 77b, 80a; 13:4b–5a
optics, 6:87b; 9:73a
Gerhard, Fridericus, 7:120a
Gerhardt, Charles Frédéric (1816–1856), 5:369–375; 3:193a
acid theory, 3:10b; 8:341b
acquaintances of, 2:598a; 4:394b; 14:394a
atomic weights, 6:463b
and Avogadro's hypothesis, 1:347a; 3:47b
and Chancel, 3:193b
chemical theory, 1:579a; 2:484b; 3:290a(n), 307a; 4:531b; 14:530a–b
critics of, 4:243a
influence of, 6:564b; 9:286b; 10:177a–b, 178a; 14:395a,b
and Laurent, 5:370a, 371b, 372b; 8:55a–b, 61a
organic chemistry, 3:193b; 14:396a
students of, 7:280a
supporters of, 2:620b; 5:315a; 9:287a
Traité de chimie organique, 7:280a
type theory, 2:621a–b; 3:193b–194a; 6:462b; 7:280a; 9:288a
Gerhardt, Karl, 4:297a, 403b
Gericke
see Guericke
Gerlach, Joseph von, 1:177a; 5:425a, 460b
Handbuch der allgemeinen und speciellen Gewebelehre des menschlichen Korpers . . . , 8:302a
Gerlach, Walther, 5:332b(n); 12:497a; 13:42b; 15:41a
Gerland, Ernst, 5:536b
Gerland, Georg Cornelius Karl, 1:174a
Gerling, Christian Ludwig, 5:301a, 307b; 12:255a
Germ cells, 14:234a,b, 235b; 15:54a
autonomy of, 1:417b
genesis of, 1:545b
vs. somatic cells, 3:121a–b
Germ layers, 4:266b; 5:283b; 7:95a*, 474b
formation of, 1:387a; 11:255a, 597a–b; 15:525a
in elasmobranchs, 1:420b
yolk-quantity theory, 1:421a
specificity of, 6:133a, 339a
Germ plasm, 11:518b; 14:233a, 234a–238a
Germ theory, 1:496b; 2:574b, 598b–599a; 3:588a; 8:405a, 406a, 408a; 10:352a, 368b–408a *passim;* 11:37a–38a; 14:249b
opposition to, 14:42a
Germain, Sophie (1776–1831), 5:375–376; 5:301a; 15:487b
physics, 3:259a; 15:363a–b, 376b, 485b, 487a
German Anthropological Society
founding of, 1:388a; 14:42a
German Botanical Society
founding of, 11:155a
German Geological Society, Berlin
see Deutsche Geologische Gesellschaft

German language, **11**:384b
in science, **15**:194b
German Metallographic Society
founding of, **5**:576b
German Research Association, **8**:52a,b, 53a
German Society for Electron Microscopy, **2**:318b
German Society for Technical Röntgenology
founding of, **5**:576b
German Society for the History of Medicine and Science
founding of, **13**:142a
Germanite, **12**:194b
Germanium
discovery of, **9**:290b; **14**:447a
prediction of, **9**:288b, 290a; **10**:37b
Germanus, Henricus Martellus (*fl. ca.* 1480–1496), 5:377
Germany
Americans studying in
astronomy, **5**:479a; **7**:270b; **12**:72b
biology, **2**:366b; **3**:29a, 187b; **4**:220a, 362a; **6**:77b, 121a, 131b, 264a; **8**: 555a; **9**:56a, 495b; **10**:252a, 276a; **11**:125b; **13**:96b; **15**:149a
chemistry, **1**:9b; **2**:258a; **3**:256b, 436b; **5**:393b; **6**:117a, 135b, 260a, 517a, 538b; **7**:208a; **8**:22b–23a, 289a; **9**: 360a; **10**:156a; **11**:576b; **12**:465a; **13**:429a; **15**:1a, 458b
geology, **2**:441a; **3**:483b; **4**:361a; **6**:13b; **7**:5a; **12**:260b; **14**:315b
mathematics, **2**:217a; **3**:345b, 399a; **4**: 618b; **6**:76a–b; **9**:502a; **10**:244b
medical sciences, **3**:447a; **11**:176a–b; **13**:300a; **14**:249a
physics, **1**:132b; **2**:382b; **4**:194b; **10**: 481b; **11**:531b; **12**:530a
psychology, **3**:247b; **6**:52b
astronomy, **2**:98b
biological sciences
anatomy, **15**:166a, 170a
anthropology, **14**:42a
botany, **6**:464b–465a; **7**:417a; **10**:13a–b
microscopy, **1**:476b
physiology, **4**:603b; **5**:171b; **7**:520b; **8**: 541b; **13**:460b
zoology, **15**:78b
chemistry, **3**:465a–b; **7**:395a; **11**:562b; **13**:70b, 466a; **14**:225a; **15**:434b–435a, 447b–448a, 456b–457a
education, **6**:294b; **7**:382b, 399b; **8**:367b; **10**:28b; **11**:7b, 123b, 344a–b, 613b; **12**:511b, 618b; **13**:40b; **15**:755a
geography, **9**:233a, 581a
geodesy, **5**:303a–b; **6**:239a–241a
geology, **1**:95b; **2**:341a–b, 557a; **3**:325a–b, 433b, 462b; **5**:155b–156a, 205a, 206b, 476a; **10**:29a, 502a; **11**:336a; **13**:64a
mineralogy, **5**:556a–b
oceanography, **6**:288a
paleontology, **2**:15a
stratigraphy, **2**:108b–109a
volcanoes, **5**:336a; **11**:336b; **12**:261b
historiography, **5**:119b
industry
chemical, **7**:21a; **9**:426a; **13**:580a
fishing, **6**:267b
mining, **3**:70a, 623a–624a; **5**:155a–156a; **6**:142b–143a; **7**:201b, 254a–b
optical, **5**:144a
pharmaceutical, **5**:99b; **13**:466a
mathematics, **4**:124b, 560b–561a; **6**:403b; **7**:50b–51a, 494b, 506a–b, 522a; **11**: 613a; **14**:325a
medical sciences, **4**:298b, 300b, 352b; **6**: 505b–506a; **9**:571b; **10**:558a–560a; **11**:586a; **12**:591b, 592a; **14**:39b–41a, 42b, 277a
pharmacy, **14**:639a–b; **15**:171b
nationalism, **4**:243a; **8**:181b–182b; **10**: 502b–503a
philosophy, **14**:483a; **15**:728a
physics, **6**:341a; **8**:52a–53a; **11**:8a, 612b; **12**:615a–b; **15**:490b
nuclear, **2**:338b; **15**:44a
politics, **7**:318a; **10**:11b–12a, 28b
scientific movement, **6**:17a, 252b; **7**:234b–235a; **10**:12b; **15**:447b, 449b–450a
sociology, **15**:133a
technology, **7**:379b; **8**:500b; **9**:578a; **11**: 344a, 354a–b; **12**:425a
theology, **14**:225b
university system
in nineteenth century, **15**:754b–755a
Germer, Lester Halbert, **3**:597b, 598a
Germination
of algae, **1**:70b
chemical studies, **11**:179a; **12**:124a, 309a
of legumes, **3**:44a
of mosses, **10**:575b
of orchids, **2**:34b, 35a
of seeds, **3**:43a; **7**:113b; **8**:425b; **11**:314b; **13**:332b
long-term, **1**:561b
of spores, **3**:30a; **4**:220a
Germs
see Bacteria; Diseases, infectious; Germ theory; Infection; Microbes
Gernardus, **5**:360a
Gernez, Désiré, **10**:360b, 373b, 374a,b
Gersh, Isidore, **1**:613b
Gersonides
see Levi ben Gerson
Gerstaecker, (Karl Eduard) Adolph, **15**:78b
Gerstner, Franz Joseph von, **2**:84b; **11**:63a
Gervais, Henri, **1**:130a
Gervais, Paul (François Louis Paul), **1**:130a; **2**:187b
Geschichte des Fixsternhimmels, **1**:340a
Gesell, Arnold Lucius (1880–1961), 5:377–378
Gesell Institute of Child Development, **5**:378a
Gesellschaft Deutscher Naturforscher und Aerzte
founding of, **10**:195b
Gesner, Konrad (1516–1565), 5:378–379; 11:184a; **13**:141a
acquaintances of, **1**:49b; **3**:12b; **11**:311b, 395b, 396b, 527b; **13**:501b
Cordus' work, publication of, **3**:414a
correspondence of, **3**:533b; **13**:463b*
editions of works, **12**:185a
entomology, **9**:440b
Historia animalium, **3**:13a
Historia plantarum, **1**:525a–b
illustrations, **13**:463a
influence of, **5**:380b; **7**:355a; **9**:122b, 179a; **14**:157b, 413b, 507b
influences on, **13**:535b
De rerum fossilium, **1**:526a
students of, **1**:525a
Gesneriaceae, **2**:521a
Gess, German Ivanovich
see Hess, Germain Henri
Gessard, Carle, **5**:73a
Gessen, Boris Mikhailovich
see Hessen, Boris Mikhailovich
Gessner, Johann Jakob, **5**:379b
Gessner, Johann Matthias, **5**:380b
Gessner (Gesner), Johannes (1709–1790), 5: 379–380; 7:596b
Gestaltism, **6**:299b
Gesù, P. Giovanni di, **2**:312a
Gesuati, Order of
see Jesuati
Geuther, Anton (1833–1889), 5:380–381
Geysers, **4**:65b, 564a
experimental, **7**:62b–63a
mechanism of, **2**:588b
al-Ghāfiqī, **1**:414a, 538b, 539a
Ghars al-Dīn Ibrāhīm al-Ḥalabī, **9**:604a
Ghaṭīgopa
commentaries of, **1**:308b
al-Ghazālī, Abū Ḥamīd Muḥammad al-Ṭūsī al-Šāfiʿī, **7**:138b, 336a; **8**:547b; **9**:29b; **11**:249b; **12**:1a, 7Ba, 8a; **13**: 514a; **14**:388a
Destruction of Philosophers, **1**:279b
translations of works, **13**:400b
Ghaznavid dynasty, **2**:148b
Ghent, University of, **9**:414b–415a
chemistry at, **7**:280b
Ghetaldi (Ghettaldi), Marino (*ca.* 1566–1626), 5:381–383; 12:500a; **14**:22b, 23a
Ghiberti, Lorenzo, **2**:534b, 535a; **6**:197a; **10**:475b; **14**:461a
Ghini, Luca (*ca.* 1490–1556), 5:383–384; 1:108b, 167a; **15**:80a
Ghiron, Guido Fubini
see Fubini, Guido
Ghisi, Martino (1715–1794), 5:384–385
Ghyben-Herzberg formula, **9**:258a
Giant's Causeway (Ireland), **4**:71b; **14**:312a
Giard, Alfred (1846–1908), 5:385–386; 3: 148a,b; **4**:147a; **7**:593b*; **8**:124b, 354a
Giauque, William Francis, **8**:47a
Gibbon, Edward, **3**:21a; **6**:383a; **14**:191a
Gibbs, George, **12**:433a, 434a
Gibbs, Josiah Willard (1839–1903), 5:386–393; 2:69a; **4**:228b, 231a; **6**:13b; **9**: 372b
on contemporary scientists, **3**:310b*; **10**:42b*
critics of, **2**:594a; **3**:166a; **13**:237a
Elementary Principles in Statistical Mechanics
translations of, **14**:614a
"Equilibrium of Heterogeneous Substances," **4**:229a
Gibbs free energy, **9**:394a
Gibbs function, **9**:394a
Gibbs-Donnan law, **13**:574b
Gibbs-Duhem equation, **4**:229a
Gibbs-Helmholtz equation, **6**:244b; **7**: 65b; **13**:578b; **15**:436b–441b
influence of, **11**:298a; **15**:462b–463a
influences on, **15**:198a
optics, **9**:213a; **13**:386b
phase rule, **11**:534a
proof of, **4**:229b
physical chemistry, **3**:310a; **4**:230a; **8**: 118b; **13**:244b, 245a
statistical mechanics, **2**:264b, 265a; **4**: 225b, 293a,b; **7**:493a; **9**:223b; **10**: 235b; **15**:195b
students of, **4**:6b; **14**:237a
thermochemistry, **15**:457a–b
thermodynamics, **2**:499a; **8**:289b; **9**:227a, 394a, 462a; **11**:10a; **15**:463b
translations of works, **15**:463a
vector analysis, **6**:92a, 212a; **9**:215b; **10**: 483b–484a
Gibbs, (Oliver) Wolcott (1822–1908), 5:393–394; 1:6a; **3**:292a; **5**:349b–350a; **6**: 565a; **12**:37a; **15**:84a
Giberti, Gian Matteo, **5**:105a
Gibson, Charles Stanley, **11**:86b, 87b
Giddy, Davies
see Gilbert, Davies
Gierster, Joseph, **6**:571b

Giese, Tiedemann, **10**:245b
Giesel, Friedrich Oskar (1852–1927), 5:394–395; 2:257b
  radioactivity, **4**:356a; **11**:280b; **12**:27a, 504b
Gieseler, Hilda, **1**:371a
Giessen, University of
  in seventeenth century, **7**:193b
  chemistry at, **9**:619b
  Hygienic Institute, **5**:220a
  physics at, **14**:348a
  zoology at, **8**:270a
Gigartinales, **7**:535b
Gilbert the Englishman, **1**:290a
Gilbert of Poitiers, **2**:19b; **13**:340a
Gilbert de la Porrée, **1**:91b; **2**:609a
Gilbert, Alfred, **11**:486a
Gilbert, Charles Henry, **5**:545a; **7**:170a
Gilbert, Davies, **3**:599a; **6**:385a, 510a
Gilbert, Grove Karl (1843–1918), 5:395–396; 3:593b; **10**:503b; **11**:120b*
  acquaintances of, **6**:57b; **7**:148b, 149a,b; **10**:502b, 506b; **13**:270a; **14**:503b
  geology, **4**:266a; **10**:509a*; **11**:118b
  glaciology, **13**:270b
  sixth-power law, **6**:503a,b
Gilbert, (Sir) Humphrey, **6**:20a
Gilbert, John, **5**:481a
Gilbert, Joseph Henry (1817–1901), 8:92–93; 8:347a
Gilbert, Ludwig Wilhelm, **2**:84b; **4**:394b; **5**:532a; **9**:429a; **11**:50b, 437a; **15**:172a
Gilbert, William (1544–1603), 5:396–401; 1:198b, 567b, 608a; **3**:545a; **6**:127b; **10**:153b; **12**:106a; **14**:365a, 367b, 461a
  biographies of, **13**:357a
  De magnete, **2**:543b; **10**:149b–150a, 537a–b; **14**:375a(n), 514b
    translations of, **13**:357a*
  magnetism, **1**:373a; **5**:48a, 246b, 575b; **6**:170b; **7**:295a; **10**:534b, 535a, 536b; **14**:364b
  De mundo nostro sublunari philosophia nova, **14**:377b(n)
Gilbert, William (of Melford), **5**:397a, 399b
Gilchrist, Percy Carlyle, **13**:346b; **15**:25a
Gilding, **3**:557a
Giles Colonna
  see Giles of Rome
Giles (Aegidius) of Lessines (*ca.* 1235–ca. 1304), 5:401–402; 1:99a; **5**:361a
Giles (Aegidius) of Rome (*b.* before 1247; *d.* 1316), 5:402–403; 1:276a
  astronomy, **15**:35b
  De erroribus philosophorum, **15**:263b
Giles de Thebaldis of Parma, **2**:435b
Giles, Robin, **6**:393b
Gilgameš Epic, **15**:639b–640a
Gill, David (1843–1914), 5:403–406; 4:351b*; **7**:239b; **8**:613b*; **12**:448b
  astronomy, **7**:238b; **9**:551b; **15**:493b
  Cape Photographic Durchmusterung, **7**:236a
  and Elkin, **4**:351a–b
Gill, Theodore Nicholas (1837–1914), 5:406–407; 7:170a, 238b; **15**:227b*, 228a*
Gill slits, **11**:307b
Gilliss, James Melville, **5**:479b; **9**:195b; **10**:33b
Gillot, Firmin, **8**:213a
Gillyflower
  genetics, **13**:478a,b
Gimbernat, Antonio de, **2**:443a; **13**:448b
Gimingham, Charles Henry, **3**:475a, 477a, 481b*
Gingrich, Newell Shiffer, **3**:370a
Gini, Corrado, **2**:319a
Ginkgo, **5**:442a
Ginzburg, Vitalii Lazarevich, **7**:618a; **8**:477b; **13**:241a
Giobert, Giovanni Antonio, **1**:586a
Giorgi, Giovanni (1871–1950), 5:407–408; 6:212a; **9**:532a
  Giorgi International System
    see M.K.S. system
Giorgini, Gaetano, **13**:5b
Giorgio, Francesco, **9**:316b, 319a
Giornale dei Letterati, **2**:311b, 312a
  editors of, **11**:404a
Giornale di geologia pratica
  founding of, **11**:576a
Giraffe, **5**:109a; **8**:274b
Girard, Albert (1595–1632), 5:408–410
  algebra, **4**:58a
  on catenary, **2**:47b
  commentaries of, **12**:446b
  Invention nouvelle en l'algèbre, **12**:205a
  notation, **11**:512a
Girard, Charles Frederic, **1**:405a; **7**:169b
Girard, Pierre-Simon (1765–1836), 5:410–411; 3:80a; **10**:4b*; **11**:63a, 164b; **12**:463a*
Girard College (Philadelphia)
  magnetic observatory, **1**:364a
  organizing of, **1**:363b
Giraud-Soulavie, J.L.
  see Soulavie, J.L. Giraud
Giridhara
  commentaries of, **2**:116a
Gironde River, **4**:614a
Girondins, **2**:321b; **3**:384a
Girtanner, Christoph (1760–1800), 5:411; 8:531a
  Anfangsgründe der antiphlogistischen Chemie, **2**:90b; **11**:438a
"Given," different geometrical senses, **4**:425a–b
Giving and taking, mathematical problems of, **4**:606b, 607b, 612a,b
Giza, Egypt, **12**:499a
Glacial deposits, **4**:262b; **7**:44b; **9**:241a; **10**:128a, 502a, 505b; **15**:427b
  classification of, **10**:429a
  drift sheets, **7**:267a
  mapping of, **5**:337b, 338b; **14**:516b–517a
  see also Erratic blocks, Moraines, Varves
Glacial theory, **1**:73a–b; **2**:568a, 570b; **3**:192a, 210a–211a; **4**:73a; **6**:582b; **10**:209b; **11**:277a; **14**:517a–b
  opposition to, **2**:556b; **3**:608b; **4**:27b; **7**:45a
Glaciation, **8**:33b
  Carboniferous, **4**:262a–b
  causal theories, **1**:623b; **3**:211a–b, 470b; **5**:487a
  cycles of, **11**:524b
  dating of, **1**:623b; **9**:241a
  and distribution of species, **5**:67b; **6**:492a; **15**:147b
  and ethnology, **11**:210b
  and evolution, **1**:73b; **5**:513a
  and landscape formation, **3**:593a; **6**:55b, 187a, 228a, 437b–438a; **7**:510b–511b; **9**:150b; **10**:169a, 502a, 505b; **11**:277a; **13**:270a; **14**:139a
  Pleistocene, **2**:549b; **3**:608b; **6**:116a; **7**:149a; **10**:504a; **12**:167b; **13**:270a–b; **14**:517a,b
  Quaternary, **6**:238b; **10**:429b
  and shoreline, **5**:329b
  see also Glaciers
Glacier Bay, Alaska, **14**:517a–b
Glaciers, **6**:187b
  and atmospheric moisture, **15**:159a
  energy balance, **13**:167a
  motion of, **6**:503a–b; **9**:61a, 180b, 220b; **12**:593b
    differential nature of, **5**:69a
    mechanism of, **3**:211b; **5**:69a; **13**:246b, 521b–522a
    rate of, **3**:211a; **5**:69a
  stratification in, **3**:211a
  valley, **2**:289a–b; **9**:149b, 150a–b
  variations of, **8**:544a; **9**:189b; **12**:122a
  veined structure, **5**:69a
  water-ice budget, **6**:78b
  see also Glaciation
Glaciology
  late eighteenth-century, **5**:445a; **11**:272a
  early nineteenth-century, **3**:210a–211b; **7**:70b; **13**:604b–605a
  mid nineteenth-century, **1**:20b, 73a–b; **4**:73a; **5**:69a, 337b, 599b; **9**:61a; **12**:167b, 300b, 301a
  late nineteenth-century, **2**:289a–b; **3**:189b–190b; **5**:329b, 338b–339a; **6**:14a, 78b–79a, 227a; **7**:370b, 510b–511b; **9**:592b, 593b; **10**:32b, 501b–502a; **11**:361b; **12**:83b; **13**:525b; **14**:443a, 516b–517b; **15**:159a
  twentieth-century, **1**:394b; **3**:204a, 533a; **4**:193b; **7**:149a–b; **10**:504a, 505b; **11**:524b; **13**:167a, 269b–270b, 517b; **14**:50b, 215a
  see also Glaciation, Glaciers
Gladstone, John Hall (1827–1902), 5:412–413; 7:620a; **9**:349a; **11**:354a*
Gladstone, William Ewart, **6**:25b; **7**:104b; **10**:260b; **11**:36b; **12**:180b; **13**:523b
Glagoleva-Arkadeva, Aleksandra Andreevna, **1**:284a; **8**:107a
Glaisher, James (1809–1903), 5:413
Glaisher, James Whitbread Lee (1848–1928), 5:413–414; 9:157b*, 613b*; **10**:61a; **12**:469a; **14**:51b*
  analysis, **12**:469b
  on Cayley, **3**:163a
Glanders, **10**:19a
  bacillus, **7**:425b; **8**:448b
  etiology of, **8**:448b
Glands, **4**:182a
  adrenal
    see Glands, suprarenal
  Bartholin's glands, **4**:268b
  castoreal, **10**:520a
  classification of, **5**:514b; **14**:286b
  conglobate, **8**:246b; **9**:64b
  cutaneous, **1**:545a
  digestive, **1**:544a; **8**:327b; **10**:567b–568a
    hormonal control of, **5**:43a
  electric currents in, **4**:203a
  embryogeny of, **5**:514b; **9**:570a–b; **11**:369b
  endocrine
    and morphogenesis, **3**:113a
    theories of, **10**:205b
  function of, **9**:64a
  gastric, **2**:165a; **6**:225b
    discovery of, **11**:216b
  hemolymph, **8**:294b
  intestinal, **2**:165a; **5**:226b–227a
  lacrimal, **1**:98b; **13**:31b–32a
  lymphatic
    function of, **13**:31b–32a
    structure of, **2**:165a
  mammary, **7**:612b
    hormonal influences on, **2**:345a
  mucous, **6**:225b
  nasal, **13**:32a
  neurophysiology of, **6**:225b–226a
  parathyroid, **4**:404a; **6**:78a
    hormones of, **3**:352a–b
    removal of, **8**:536a; **13**:589b

pineal, **1**:620a; **4**:63a–b; **12**:473a; **14**:407a
pituitary, **1**:620a; **3**:113b*
and carbohydrate metabolism, **8**:452b; **15**:228b–229a
disorders of, **3**:518b; **9**:109a
embryogeny, **12**:473a
extracts of, **9**:19b; **10**:205a,b; **12**:355b; **15**:105b, 228a, 229a
hormones of, **1**:11a; **2**:345b; **3**:352b–353a; **5**:225b
naming of, **12**:510a
tumors of, **3**:518b; **6**:518b
see also Hypophysectomy
prostate, **9**:165b–166a
salivary, **1**:46b, 369a; **10**:433b; **13**:31a,b
chromosomes of, **2**:456b; **10**:276b
histology, **6**:224a
innervation, **1**:452b; **5**:184b; **8**:15a,b; **10**:579b
metabolism, **1**:452b–453b, 454a
morphology, **6**:225b
scent, **13**:527b
serous, **6**:225b
sexual, **13**:527b; **14**:595b
accessory, **7**:32a; **8**:9a
structure of, **2**:226a, 301b; **9**:64a–b, 570a–b; **11**:216b; **12**:41a; **14**:199b
changes during secretion, **8**:15a
submaxillary, **5**:384b; **8**:14b; **14**:286b–287a
suborbital, **1**:545a
suprarenal, **4**:486b; **5**:470a, 514b; **8**:552b
extracts, **10**:205a–b; **12**:355b; **13**:53b; **15**:258a–b
functioning of, **15**:73a
and growth, **3**:113a
physiopathology, **1**:60a–b; **13**:589b
removal of, **2**:525b
role of, **1**:480b; **14**:287a
sweat, **2**:443a; **11**:216b
thymus, **5**:467b, 470a; **15**:473a
development of, **9**:56a
weight of, **3**:113a
thyroid, **5**:470a; **6**:78a
chemistry of, **9**:284b
embryogeny, **5**:514b; **6**:574a*
extract of, **2**:548a; **13**:589b; **15**:258a
function of, **10**:205b; **12**:165a; **14**:287a
and growth, **3**:113a; **7**:101a
histology, **3**:113a; **6**:224a
iodine content, **3**:113a
naming of, **14**:287a
pathology of, **8**:352b; **11**:99a; **14**:115a
perfusion of, **3**:91b
removal of, **8**:536a
transplantation, **6**:518b
urethral, **9**:323b
see also Endocrinology, Hormones, Secretion
Glans penis, **13**:527b
Glanvill, Joseph (1636–1680), 5:414–417; **10**:43b; **12**:586b*; **14**:210a
De proprietatibus rerum, **10**:515a
Glarus (Switzerland)
geology, **6**:227a–228a, 228b
Glaser, Christopher (*ca.* 1615–*ca.* 1672), 5:417–418; **3**:596b; **6**:380b; **8**:172b, 173a
Glaser's salt, **5**:418a
Traité de la chymie, **2**:334a
Glaser, Johann Heinrich (1629–1679), 5:418–419
Glaserian fissure, **5**:419a
Glaser, Karl, **5**:489a; **7**:281b
Glaserite, **5**:418a
Glasgow
Anderson's College, **3**:488b
botanical gardens, **6**:494a
Glasgow, University of
in nineteenth century, **13**:374a
astronomy at, **14**:415a
botany at, **1**:422a; **2**:371a; **6**:493a–b
chemistry at, **2**:173b; **3**:494a–b; **13**:372b
engineering at, **11**:294a
Hunterian Library, **4**:172b
Hunterian Museum, **1**:398b, 399a; **3**:486b, 488a; **6**:569b
mathematics at, **12**:445b, 446a
medical school, **8**:402a–b
observatory, **5**:562a
physics at, **13**:377b
Royal Infirmary, **8**:402a–405a
Glass
archaeology of, **13**:505b
coloring of, **5**:422b; **7**:525a; **10**:23b; **13**:327b; **14**:633a–b; **15**:652a
cooling of, **6**:54b; **7**:277b
inorganic inclusions, **14**:633a
manufacture of
medieval, **1**:5b; **13**:327b
Renaissance, **10**:23a–b
seventeenth-century, **7**:524b, 525a; **9**:313a
eighteenth-century, **1**:233b; **3**:200a; **4**:87a; **5**:602a; **8**:470a,b; **13**:67b
nineteenth-century, **1**:9a; **4**:243a; **5**:142b–143a; **11**:515b; **12**:211b–212a
twentieth-century, **11**:549a; **13**:454b, 505a–b
glassblowing, **5**:340a–341a
handbooks, **15**:652b
Mesopotamian, **15**:647b, 649b, 650a,b, 651a, 651b–652b
physical chemistry of, **13**:247a
properties of, **6**:546b; **13**:505b
compressibility, **2**:458b; **13**:130b
devitrification, **5**:576b*
electrical, **1**:547b; **3**:51b; **5**:131b–132a, 516a; **6**:169a, 170a–171b; **10**:146b; **11**:433a; **13**:471b; **14**:352b–353a
expansion, **4**:518a; **15**:312b
gas diffusion, **10**:289a
optical, **5**:143a, 144a; **6**:125b; **12**:281a,b; **13**:77b
solubility, **4**:214b
specific gravity, **7**:346a
specific heat, **2**:589b
specific volume, **13**:247a
structure of, **8**:117b; **15**:445a
thermodynamics of, **8**:118b; **15**:445a
types of
enamel (opaque), **10**:23b
flint, **8**:621a
heatproof, **11**:208a
laboratory, **12**:211b–212a
lead, **10**:23b
lithium, **12**:211b
milk, **14**:633a
optical, **2**:370a; **4**:532a, 563b; **5**:87a, 142b–144a, 271b; **9**:60b, 373b, 564a; **11**:377a, 580b; **12**:211b; **14**:182b, 515a–b
plate, **15**:24b
potassium, **13**:327b
pyrex, **3**:292b
ruby, **5**:422b; **12**:423a; **14**:633a
stressed, **5**:186a; **12**:281a,b
tempered, **9**:485b
Glass drops, **9**:485b, 498a; **11**:507b
see also Prince Rupert's drops
Glass harmonica, **3**:258a
Glauber, Johann Rudolph (1604–1670), 5:419–423; **4**:394a; **8**:130b; **13**:222b; **14**:508a
chemistry, **9**:424b; **11**:48b
Glauber's salt, **5**:351b, 420a, 422a–b; **6**:236b; **8**:70b; **9**:425a; **13**:245a; **14**:333a
mechanism of deposition, **7**:529b
Glaucoma, **2**:376b; **9**:75a
treatment of, **7**:273b
Glaucon, **11**:23b, 26b
Glauconite, **9**:589b; **11**:372a
Glazebrook, Richard Tetley (1854–1935), 5:423–424; **4**:501a*; **9**:227b, 229b*; **12**:366a, 367a*; **13**:102a, 365a
Light
translations of, **14**:614a
optics, **8**:593a*; **13**:386b
Glazes, **8**:118b
Gleason, Andrew Mattei, **2**:513a
Gleditsch, Ellen, **2**:259b; **3**:501a
Gleditsch, Johann Gottlieb, **1**:44b; **14**:386a; **15**:68a
Gleichen-Russworm, Wilhelm Friedrich von (1717–1783), 5:424–425; **2**:520b
Glen Roy, Scotland
Parallel Roads of, **3**:569a
Gley, (Marcel) Eugène (Émile), **10**:570a; **11**:427b; **13**:589b
Gley (soil), **14**:107a
Gliddon, George Robins, **9**:540b
Glider planes, **7**:247a, 465a; **8**:213b
Gliocadium, **9**:176a
Gliocephalis, **9**:176a
Gliomas, **3**:518b; **11**:465b
Glisson, Francis (*ca.* 1597–1677), 5:425–427; **5**:534b; **9**:98a; **14**:287a
anatomy, **6**:183b; **10**:478a
Glisson's capsule, **5**:426a
neurology, **6**:64b, 158a
De rachitide, sive morbo puerili . . . , **9**:243a
students of, **14**:146b
Gloag, James, **9**:199b
Globe, **2**:133a
celestial
ancient Greek, **1**:214a; **15**:219b
medieval Arabic, **11**:244b; **13**:149b
medieval Latin, **7**:119a
Renaissance, **2**:402b, 405b, 413b; **15**:478b
seventeenth-century, **2**:185b; **3**:421a; **6**:364a; **12**:163a
eighteenth-century, **4**:565b; **10**:255b
lunar, **6**:362a
relief, **11**:337b
terrestrial
Renaissance, **1**:574a; **3**:558a; **4**:260a; **5**:349a, 377a; **12**:199b
seventeenth-century, **2**:185b; **3**:421a
twentieth-century, **7**:265b
Globigerina
distribution of, **13**:360b
Globigerina ooze, **3**:161b; **9**:588b; **11**:372a
Globulins, **10**:242a–b
metabolic role, **12**:319b
molecular weight, **4**:522a
serum, **5**:149a; **13**:420b
Glomerulonephritis, **9**:95a; **14**:624a
Glossina, **2**:528b, 529a, 533b; **7**:429a,b
Glossocomus, **14**:366b–367a
Glossopetrae, **1**:482a; **12**:257a; **13**:33b
Glossopteris, **4**:262b; **14**:600a
Glottis, **4**:135b–136a; **15**:108a
Glover, John, **5**:326a
Glucal, **4**:603b; **15**:15b
Glucina, **13**:597a
Glückradt, Christopher (pseudonym)
see Hartmann, Johannes

Glucosamine, **13**:406b
Glucose
in blood, **13**:403b
cyclization of, **5**:6a
derivatives of, **5**:3b; **15**:13a
isolation of, **8**:520b
isomers of, **5**:3b
metabolism of, **3**:220a; **4**:360a–b; **8**:452b; **9**:359a–b
and phosphoric acid, **4**:485b–486a
phosphorylation of, **3**:416a
stereochemistry, **15**:13a
acid-base dependency, **6**:538b
tests for, **15**:109b
Glucosides, **12**:619b
constitution of, **12**:164a
structure of, **5**:4a; **9**:95b
synthesis of, **5**:2b, 4a; **9**:360a
Glutamic acid, **14**:174b
Glutamine, **13**:575a
Glutathione, **6**:501b–502a
Gluten, **7**:279b
reactions of, **7**:379a
Glycemia, **2**:26a, 29a; **9**:11a
traumatic, **2**:29b
Glycerides, **5**:6a; **6**:398a–b
Glycerin
derivatives of, **2**:65b–66a; **10**:575b
esterification of, **2**:66a; **5**:100a
optical properties, **11**:265b
Glycerol
derivatives of, **5**:6a
discovery of, **12**:148b
esterification of, **5**:5a
from soap, **3**:242a–243a
Glycerophosphoric acid, **5**:432b
Glycine, **2**:386a
observed in nature, **3**:256b
Glycogen, **3**:464a*
determination of, **10**:580a
discovery of, **2**:29a–b
isolation of, **2**:29b
in liver, **6**:287a
metabolism of, **3**:416a; **4**:360b; **8**:614b; **9**:359a; **10**:580a; **11**:51b, 565a; **14**:412b
in microorganisms, **4**:401b; **6**:111a
molecular structure, **3**:416a; **13**:3b
molecular weight, **13**:3b
phosphorolysis, **10**:326b
role of, **9**:506a
storage diseases, **3**:416a
synthesis of, **3**:416a
Glycogenesis, **2**:26a, 29a–b
as cellular process, **2**:29b
control by nervous system, **2**:29b; **8**:615a; **9**:506a
extrahepatic, **2**:29b
ferments in, **2**:29b
Glycolysis, **5**:6a
in air vs. nitrogen, **2**:184b
by cancer cells, **14**:176a–b
Glycoproteins, **8**:276a
Glycosides, **7**:518a; **14**:444a, 445b, 477a
in plants, **5**:582a; **15**:252b
synthesis of
in plants, **3**:279b
Glycosuria, **8**:452a, 555a
experimental, **8**:614b
traumatic, **2**:29b
Glycyrrhizin
discovery of, **11**:494b
Glyoxalase, **6**:502a
Glyphography, **3**:557a
Glyptodon, **10**:262a
Gmelin, Christian Gottlob, **5**:430a
Gmelin, Ferdinand, **5**:429b
Gmelin, Johann Friedrich, **5**:216a, 429b, 430b; **7**:366b
Gmelin, Johann Georg (1709–1755), 5:427–429; 6:380b; **7**:441a, 495a; **13**:28a; **15**:68a*
Flora sibirica, **13**:28b
supplements to, **5**:218a
Gmelin, Leopold (1788–1853), 5:429–432; 1:578b; **2**:587a; **5**:412a; **9**:578b; **14**:489a
chemistry, **5**:371b
Gmelin's salt, **5**:431b
Handbuch der Chemie, **5**:430b–431b
students of, **5**:349b; **7**:463a; **9**:445a; **14**:394a, 474a–b
and Tiedemann, **13**:403a–b
Gmelin, Philipp Friedrich, **7**:412a
Gmelin, Samuel Gottlieb, **5**:429a*
Gmelinite, **4**:597b
Gmunden
see John of Gmunden
Gneiss, **2**:289b; **3**:567b; **4**:565a*; **5**:556b; **9**:533a
origin of, **2**:553b, 555b; **9**:620b; **15**:427b
Gnetales, **8**:354a
Gnetum, **8**:512b
Gnomon, **1**:310b; **2**:115b, 417a; **3**:102b, 558b; **4**:390a; **12**:598b; **13**:440b; **15**:539b, 628a, 666b(n)
divisions of, **2**:153a–b; **15**:538a
meridian, **8**:22a
Gnomonics
ancient Roman, **15**:518b, 519b–520b
medieval Arabic, **1**:511b–512a; **2**:153a–b; **13**:291b; **14**:546b, 578a
medieval Latin, **15**:163b
Renaissance, **9**:191b; **15**:154a–b
seventeenth-century, **2**:333b; **3**:152b; **4**:47a–b, 48a,b; **6**:362b; **7**:577b; **12**:152a
eighteenth-century, **7**:405a
Egyptian, **15**:713b
Indian, **15**:538a, 539b–540a, 551b, 552b–553a, 559b–560a, 570b–572b, 576a, 592b, 610a, 628a
Mesopotamian, **15**:676b
see also Shadows, Sundials
Gnosticism, **2**:539b, 543a; **5**:107a; **6**:256b; **7**:1b, 42b; **10**:310a–b, 311a; **11**:250a, 324b–325a; **13**:513b; **14**:632b
psychology of, **7**:192b
Go (game), **12**:375b; **15**:749a
Goal-year-texts (astronomy), **15**:680b
Gobelins (tapestry workshop), **3**:240a,b, 419a, 560b; **4**:87a, 184b
Gobi Desert, **6**:217a; **11**:181b
Gobley, Nicolas-Théodore (1811–1876), 5:432–433
Gockel, Albert Wilhelm Friedrich Eduard, **6**:355a
Goclenius (or Goeckel), Rudolf, **6**:254a
Goddard, Henry, **5**:377b
Goddard, John Frederick, **5**:21a(n)
Goddard, Jonathan, **3**:482a; **5**:426b; **11**:532a; **14**:151a,b, 405a
Goddard, Robert Hutchings (1882–1945), 5:433–434; 11:420a
Gödel, Kurt, **2**:514a; **3**:57b; **6**:393a, 473b; **11**:107a; **14**:89b, 304a, 614b
Gödel's theorem, **5**:350b; **11**:61a
proof of, **3**:279a*
Godfrey, John, **5**:516b–517a
Godfrey, Thomas (1704–1749), 5:434; 6:6a; **8**:459b
Godin, Louis (1704–1760), 5:434–436; 4:23a; **7**:200b, 579a*; **15**:271b
geodesy, **2**:343a; **7**:183a, 200a; **15**:270a–271a
Godlewski, Tadeusz, **5**:394b
Godman, John Davidson, **10**:438b
American Natural History, **15**:472a
Godwin, Francis
Man in the Moone, **14**:364b
Nuncius inanimatus, **14**:368b
Godwin, William, **6**:140a; **9**:68a, 69a
Godwin-Austen, Robert Alfred Cloyne (1808–1884), 5:436–437; 1:472a; **5**:68a; **15**:51a
Goebel, Karl (Immanuel Eberhard Ritter von) (1855–1932), 5:437–439; 4:220a; **6**:467a, 468a*; **8**:513b; **9**:149b*; **12**:58b
Goebel, Severin, **9**:23a
Goedaert, Johannes (*ca.* 1617–1668), 5:439–440
entomology, **13**:172b
Goeppert, Heinrich Robert (1800–1884), 5:440–442; 3:336b, 337a; **10**:11b; **13**:462b; **15**:147a,b
Goethart, Jan Willem Christiaan, **8**:513a
Goethe, Johann Wolfgang von (1749–1832), 5:442–446; 2:72b; **4**:88a; **5**:595b, 624b*; **7**:430b; **9**:20a, 343a; **10**:287b; **11**:257b; **13**:244a
acquaintances of, **3**:258a; **4**:134a, 275b; **6**:218b, 456a, 550a; **8**:321a; **10**:11a; **11**:473b; **13**:455b, 506b; **14**:60b, 351b
botany, **2**:160a; **6**:10a; **10**:531a; **13**:506b; **15**:524b
comparative anatomy, **6**:10a; **10**:263a; **15**:168b
intermaxillary bone, **3**:38a
correspondence of, **11**:387b; **12**:588a; **13**:44b
criticism of, **4**:203b; **9**:463b; **10**:195b
editions of works, **13**:506b
embryology, **6**:592a
Faust, **1**:80b, 575b; **4**:203b; **7**:189b; **9**:561b; **13**:286b
geology, **5**:205b, 357a, 558b; **6**:84a; **11**:303b
influence of, **12**:154a, 281a, 607a; **13**:238a
metamorphosis doctrine, **2**:426b; **10**:13a
philosophy, **8**:160a
physiology, **9**:570a
religion, **2**:446a
Goette, Alexander Wilhelm (1840–1922), 5:446–447; 6:9b, 436a; **7**:402a
Goetz, Friedrich Wilhelm Paul, **11**:542b
Goeze, Johann August Ephraim, **5**:328b
Gohory, Jacques (Leo Suavius) (1520–1576), 5:447–448; 1:49b; **7**:536a; **12**:335a
Göhring, Oswald Helmuth, **6**:16b; **12**:505b
Góis, Benedetto de, **11**:403a
Goiter, **1**:485a; **8**:125b; **14**:248b
endemic, **3**:97a; **5**:502b; **8**:556b
theories of, **6**:433a–b; **10**:307a
and genetics, **12**:76b
removal of, **12**:165a
treatment of, **3**:113a; **4**:209a; **14**:115a; **15**:419a
Gold, **5**:595b; **15**:651b, 655a(n)–b(n)
Alaska, **2**:500a,b
alchemical theory, **2**:22b; **4**:550b; **10**:81a, 555a
alloys, **8**:312a; **11**:486a
analysis of, **5**:622b
assaying, **8**:621a
atomic weight, **13**:389b
Australia, **3**:297b
as catalyst, **13**:313a
colloidal, **9**:295b; **11**:437b; **13**:160b; **14**:633a–b
compounds of, **2**:171a; **10**:497b
organic, **11**:86b
deposits of, **6**:238b; **7**:267a; **9**:585a; **11**:608a; **12**:293a

origin of, **9**:421a; **10**:167a
placers, **2**:439b; **12**:293a
dissolving of, **2**:421b
economics of, **13**:145a
extraction of, **3**:475a; **5**:421a
from ashes, **12**:64a
from sea sand, **1**:549a
from sea water, **5**:622b
fulminating, **9**:148b
leaf, **13**:327b
medical use of, **5**:421a; **7**:208b
metallurgy, **4**:345b; **6**:79b–80a; **11**:328b; **13**:327a, 411b
amalgamation treatment, **4**:345b
refining, **1**:380a; **4**:394a; **11**:499a
separation from silver, **2**:143a; **8**:213b; **10**:557b; **13**:245b, 327a
mining of, **3**:462b; **8**:578a; **10**:167a, 555b; **13**:525b–526a; **15**:152b
physical properties, **11**:486a
potable, **4**:387b; **5**:448b*; **7**:525a; **8**:311b; **10**:281a
powder, **13**:327b
prospecting for, **5**:568a; **12**:293a, 436a–b, 514a; **13**:526a
specific gravity, **7**:346a
specific heat, **14**:354b
see also Transmutation of metals
Goldbach, Christian (1690–1764), 5:448–451; **2**:43b; **4**:469a; **8**:165b
calculus, **3**:228a
and Euler, **4**:475a, 476a, 479b; **5**:449a
Goldbach's hypothesis, **6**:393b; **12**:411a
theorem of, **3**:58a*; **5**:450b
Goldberger, Joseph (1874–1929), 5:451–453
Golden ratio, **4**:259a, 558a; **9**:310b; **10**:271a; **11**:221b; **15**:703b*
Golden section
see Golden ratio
Goldfish, **8**:395a
Goldfuss, Georg August, **2**:108a, 158a
Goldie, John, **11**:218b
Goldschmidt, Carl Wolfgang Benjamin, **4**:1b
Goldschmidt, Karl, **2**:4a
Goldschmidt, Richard Benedict (1878–1958), 5:453–455; **2**:625b; **9**:522a, 524a; **10**:610b
Goldschmidt, Victor (1853–1933), 5:455–456; **4**:597b, 599a; **7**:62b; **14**:515a
Goldschmidt symbols, **5**:455b
Goldschmidt, Victor Moritz (1888–1947), 5:456–458; **4**:409a; **5**:212b; **6**:561a; **14**:411b
Goldschmidt Company (Essen), **2**:4a
Goldsmith, Oliver, **1**:332b; **9**:481b
Goldsmiths
Renaissance, **2**:534b
Goldstein, Eugen (1850–1930), 5:458–459; **13**:365b; **14**:31b
Goldstine, Hermann, **14**:92a
Golfand, Iurii Abramovich, **13**:241b
Golgi, Camillo (1843–1926), 5:459–461; **2**:164b; **11**:273b; **12**:319a; **15**:430b
Golgi bodies, **14**:433a
Golgi-Mazzoni corpuscles, **11**:597a
Golgi's internal reticular apparatus, **1**:613b; **5**:461a, 585a; **15**:473a
Golgi's method, **9**:93b; **11**:274a
Golgi's organs, **9**:93b; **11**:597a
malaria, **5**:503a,b
neurohistology, **7**:439a; **11**:382b; **12**:355a; **15**:473a
students of, **4**:160b; **9**:93b; **10**:15a; **15**:473a
Golitsyn, Boris Borisovich (1862–1916), 5:461–462; **13**:243b
Golius, Jacobus, **1**:245b; **4**:57a, 59b, 542a, 543a; **5**:408b; **10**:296a
Golod, Evgenii Solomonovich, **1**:307a
Golovin, Mikhail Evsevevich, **4**:472b
Golovnin, Vasilii Mikhailovich, **8**:417b
Goltz, Bogumil, **5**:462b
Goltz, Friedrich Leopold (1834–1902), 5:462–464; **6**:60a, 440b, 441a; **8**:15a, 445a; **12**:355b, 395b; **14**:126a; **15**:20b
experiments of, **5**:463a
on semicircular canals, **2**:446b
Goltzius, Hendrik, **4**:183b
Golubev, Vladimir Vasilevich, **4**:287b; **11**:156b
Gomberg, Moses (1866–1947), 5:464–466; **13**:574a
Gomes, Bernardino Antonio, **10**:497b
Gomes, Diogo, **1**:573b
Gómez Ortega, Casimiro, **6**:310a
Gompertz, Benjamin (1779–1865), 5:466
Gompertz's law, **5**:466a
Gondolatsch, Friedrich Peter Max, **6**:540a
Gondwana Series, **4**:262b
Gondwanaland, **4**:263a
Goniatites, **2**:108a, 109a
Goniometer, **5**:557a; **8**:65b; **9**:234a, 379b; **12**:436b
contact, **6**:181a; **11**:521b; **14**:490a
reflecting, **11**:539a; **14**:490a–b
two-circle, **5**:212a–b, 455b; **14**:525b
Goniometry, **1**:358a; **3**:61b, 208a; **4**:597b; **5**:124b; **7**:559a, 599b, 601a; **8**:288b; **9**:393a; **13**:517b; **14**:241a
Goniophotometry, **2**:344a
Gonococcus
discovery of, **10**:17b–18a
Gonorrhea, **6**:73a
path of infection, **14**:278b
Gonsález, Domingo
see Gundissalinus, Dominicus
Gonseth, Ferdinand, **13**:14b
Gonville and Caius College, Cambridge, **3**:13b
astronomy, **13**:95b
Gonzáles, Ochoa, **1**:620b
Good Hope, Cape of
circumnavigation of, **9**:36b
Good Housekeeping, **14**:357b
Goode, George Brown, **1**:406a; **9**:315a
Goodrich, Edwin Stephen (1868–1946), 5:466–467; **8**:26b, 27b*
Goodricke, John (1764–1786), 5:467–469; **9**:486a; **10**:608a
Goodsir, Harry, **5**:470a
Goodsir, John (1814–1867), 5:469–471; **2**:598a; **5**:67a; **7**:416a*; **9**:588b; **13**:503b
nutrition, **1**:477b
Goodspeed, Arthur Willis, **11**:531a
Goodspeed, Thomas Harper, **12**:328a
Goodyear, John, **7**:147a,b
Gool, Jakob, **12**:205b, 610b
Göpel, Adolph (1812–1847), 5:471–472
analysis, **6**:308b; **11**:451a, 548a–b
Göpel relation, **5**:471b
Gopīrāja, **9**:23a
Goppelsroeder, Friedrich, **12**:198b; **13**:487b
Göppert-Mayer, Maria, **15**:41a, 43a
Goranson, Roy Waldemar, **5**:186a
Gordan, Paul Albert (1837–1912), 5:472–473; **2**:465a; **3**:164a; **7**:522b; **10**:137b; **13**:124b; **14**:202a
and Clebsch, **3**:315a; **6**:388b, 389a
invariant theory, **3**:164b, 165a, 314a,b; **5**:472a–b; **9**:358a
transformations, **12**:284b
Gordiacea, **1**:602b
Gordon, Alexander, **6**:521b; **14**:297a
Gordon, James Edward Henry, **9**:212b
Gordon, Lewis Dunbar Brodie, **11**:291b, 294a
Gordon, Walter (1893–1940), 5:473–474
Gore, George (1826–1908), 5:474; **9**:451a
Gorges
formation of
erosion theory, **13**:257b
Görgey, Arthur (1818–1916), 5:474–475
Gorgoniae, **13**:554b
Gorilla
anatomy, **14**:126b, 533b–534a
Gorky, Maxim, **1**:361a; **9**:125b
Gorny zhurnal, **12**:513b
Gorter, Cornelis Jacobus, **8**:475b
Gosling, Raymond, **5**:140a,b, 141a, 589b
Gossage, William, **12**:520b
Gosse, Louis, **11**:132b
Gosselet, Jules-Auguste (1832–1916), 5:475–476; **1**:471b, 472a; **3**:160b; **4**:147a; **8**:354a, 463b; **10**:210a*
Gossett, William Sealy (also "Student") (1876–1937), 5:476–477
correspondence of, **10**:457b, 465b(n)
statistics, **1**:154b; **5**:8a, 10a; **10**:454a–b; **14**:382a
Student's t-test, **5**:477a
Gossler, Gustav, **7**:425b, 426a,b
Gotch, Francis, **2**:599a; **7**:504b; **8**:533a; **11**:126a
Gotha, Duchy of, **6**:456a
Gotland
botany, **14**:116b
Göttingen
Institute for Aeronautical and Hydrodynamical Research, **7**:399b
Institute for Technical Physics, **8**:500b
Magnetische Verein, **14**:204b
Sozietät der Wissenschaften, **6**:457a
Göttingen, University of
in nineteenth century, **4**:127a, 380a; **7**:420b–421a; **12**:165a; **14**:126a
in twentieth century, **10**:137b–138a
chemistry at, **8**:502b–503a; **9**:355a; **13**:243b; **14**:444b, 445a, 477b
Dedekind at, **4**:1b–2a
geophysics at, **14**:327a–b
Göttingen Seven, **1**:7a; **5**:306a–b; **14**:204b
mathematics at, **7**:206a; **11**:447b–448a, 613a–b; **12**:195b
medicine at, **2**:225b
observatory, **5**:301a, 302a, 303a; **9**:234a; **11**:377a; **12**:249a–b
physical institutes, **5**:117b; **15**:435a
physics at, **11**:123a, 123b–124a, 446a; **12**:615a; **14**:61b; **15**:39b–40a, 41a–b, 43a
physiology at, **9**:259b; **14**:113b
zoology at, **7**:516b
Göttingische Zeitungen von gelehrten Sachen, **6**:66b
Göttingisches Magazin der Wissenschaften und Litteratur, **5**:75b
Göttling, Johann Friedrich August, **5**:217a; **12**:562b; **13**:465b; **14**:332b
Gottsche, Karl Moritz, **10**:13b
Gouan, Antoine, **2**:509a, 510a; **4**:156b, 178a
Goudin, Mathieu-Bernard, **4**:107b; **5**:567a
Goudsmit, Samuel Abraham, **1**:371a; **2**:248a; **9**:397b; **11**:420b
Gough, John, **3**:538a, 540a, 546b
Goujon, Jean, **1**:202b
Gould, Augustus Addison (1805–1866), 5:477–479; **6**:56b; **9**:536b; **13**:65b
Principles of Zoology, **5**:478a; **11**:113a
Report on the Invertebrata of Massachusetts, **3**:535b
zoology, **13**:66a
Gould, Benjamin Apthorp (1824–1896), 5:479–480; **1**:6a, 364b; **3**:194a; **4**:352a*, 369b, 590b; **10**:480a; **12**:37a,b*

Gould, Benjamin Apthorp (cont.)
astronomy, **14**:448b
Gould's belt, **5**:480a
Gould, Elizabeth Coxen, **5**:480b, 481a
Gould, John (1804–1881), 5:480–481; **2**:205b
Goupy, Joseph, **3**:129b
Goursat, Édouard Jean-Baptiste (1858–1936), 5:481–483; **7**:71b; **9**:182b
Cauchy-Goursat theorem, **5**:482a; **9**:502b
Cours d'analyse mathématique, **10**:20b
Gout, **1**:49b, 400a; **3**:205a, 244b, 578b; **4**:603a; **5**:137b; **10**:321b; **13**:215a, 282a; **14**:322a
purine metabolism, **4**:603a
treatment of, **2**:354a; **7**:15a*; **9**:242a
Gouy, Louis-Georges (1854–1926), 5:483–484
Government
and education, **6**:26a
and industry, regulation of, **4**:71a, 216a, 298a; **8**:621b; **14**:473b
and science
eighteenth-century, **6**:383b; **13**:494b–496a
early nineteenth-century, **1**:354b–355a, 364a,b; **3**:549b; **4**:11a
mid nineteenth-century, **10**:352b, 354a, 363a, 385a; **11**:118b, 120a; **12**:425b–426a
late nineteenth-century, **1**:406a; **4**:265b; **6**:243a–b; **10**:392b, 396a, 403a–b; **13**:523a
twentieth-century, **2**:318b; **3**:35b, 373a; **4**:284b; **6**:31a; **7**:248b; **11**:482b–483a; **13**:7b, 104b–105a, 159a, 419a; **15**:228b
aeronautics, **1**:133a; **8**:368b
atomic physics, **4**:580b, 581b, 582a
chemistry, **5**:202b; **8**:116b–117a; **11**:36b, 537a; **13**:429a; **15**:2a–b
Chinese, **12**:377b–378a, 384a–b, 386b–391a
education, **7**:26b
entomology, **5**:12a
geology, **15**:152b
horticulture, **15**:425b
in Japan, **15**:734b, 736a, 744a, 745b, 748a
mathematics, **14**:384a; **15**:745b, 748a
medicine, **1**:443b; **5**:180b; **12**:57b
physics, **13**:429b
see also Architecture, military; Exploration; Politics; Science(s), financial support of; specific countries; Technology, military; Warfare
Governors
for steam engines, **5**:87a, 387a
theory of, **9**:226b–227a
see also Regulators, mechanical
Govinda, son of Bhaskarabhaṭṭa, **2**:115a; **7**:331a
Govinda, son of Viśvanātha Tāmbe
commentaries of, **2**:416b
Govindasvāmin, **6**:115b
commentaries of, **2**:114b
Gowers, William Richard, **1**:496b; **4**:594a; **6**:518b
Gowers' tract, **1**:496b
Graaf, Regnier de (1641–1673), 5:484–485; **6**:509a; **8**:127a, 129a; **9**:497b; **13**:33b, 169a, 223b
anatomy, **9**:595a; **12**:40b; **13**:170b
experimentation, **1**:544a; **3**:487a
Graafian follicle, **5**:484b, 485a; **8**:127b–128a; **13**:563b
Grabau, Amadeus William (1870–1946), 5:486–488; **2**:171b; **13**:532a
Graben (geol.), **10**:168a
Gradient theory (embryology), **3**:247b
Gradients (math.), **2**:594a
Graebe, Karl James Peter (1841–1927), 5:488–489; **1**:390b; **2**:65a; **14**:247b
chemistry, **1**:168b; **3**:85a; **8**:328b–329a; **10**:516a; **12**:452b
Graefe, Albrecht von, **2**:129b, 376b, 422b; **4**:163a; **6**:434b
Graff, Kasimir Romuald (1878–1950), 5:489–490; **10**:213a*
Gräffe, Karl Heinrich (1799–1873), 5:490; **3**:555a; **8**:432b; **11**:611a; **14**:480b
Gräffe's theorem, **5**:490b
Graffenried, François Louis de, **1**:526a–b
Grafting, botanical, **2**:444b; **5**:582a; **12**:208a; **13**:390a
physiological effects, **15**:426a
Grafting, zoological, **13**:458a
chorioallantoic, **10**:105a
embryological, **10**:104b–105a
Grahacāranibandhanasaṅgraha, **15**:596a–b, 597a
Graham, Clarence Henry, **14**:201a
Graham, Duncan, **1**:443a
Graham, George (*ca.* 1674–1751), 5:490–492; **2**:140b, 388b; **6**:70a, 130b; **9**:156b
astronomical instruments, **2**:387b, 388a; **9**:156b, 464a
clocks, **2**:389a
Graham, Robert, **1**:423a; **5**:66a; **6**:489b
Graham, Thomas (1805–1869), 5:492–495; **3**:289a,b; **12**:478a; **14**:394b
acquaintances of, **5**:104a; **10**:510b; **11**:485b; **14**:394a
chemistry, **14**:489a,b
acids, **8**:339a, 340b, 341b
gases, **4**:79a; **9**:217b, 219b, 222a; **13**:11a
Graham's law, **5**:493a
students of, **5**:412a; **11**:36b, 536b; **14**:397b
Grahatanu (period)
definition of, **15**:593a
Grainger, Richard Dugard, **1**:477a; **9**:242a
Grains, **3**:259b
diseases of, **7**:323b
in nutrition, **4**:358a; **6**:135b
physiology of, **2**:290b; **12**:124a
preservation of, **3**:116b*, 436b*; **4**:224b*; **10**:326a; **11**:38a*
see also Cereal crops; specific grains
Gralén, Nils, **13**:163a
Gram, Hans Christian Joachim (1853–1938), 5:495–496
Gram's method, **5**:495a–b
Gram (unit of mass), **8**:85a
Gramicidin, **5**:42b
Gramineae, **1**:206a; **2**:322b, 519a; **5**:362a, 546a; **10**:13b; **11**:300a
Grammar
ancient Greek, **4**:392b
ancient Roman, **3**:175b; **5**:232a
medieval Arabic, **15**:232a
medieval Latin, **1**:564b; **6**:377a
seventeenth-century, **14**:151b–152a
eighteenth-century, **4**:172b; **11**:140a
algebra of, **10**:443b
Grammateus
see Schreiber, Heinrich
Gramme, Zénobe-Théophile (1826–1901), 5:496; **10**:268b
Gramme dynamo, **2**:439a; **5**:496a–b; **13**:286a
Gram-molecular weight, **3**:10b
Gramont, Antoine Alfred Arnaud Xavier Louis de (1861–1923), 5:497
Gran Paradiso massif, **11**:576a
Grand Canyon
geology, **4**:265b–266a
theories of, **10**:32b
Grande encyclopédie, **8**:61b; **9**:553b
Grande Soufrière (volcano), **10**:569a
Grand'Eury, Cyrille (1839–1917), 5:497–498; **2**:492b
Grandi, Guido (1671–1742), 5:498–500; **4**:515b; **12**:56b*
and Ceva, **3**:183b–184a
geometry, **1**:76b; **11**:402a; **14**:49b(n)
roses of Grandi, **5**:499b
Grandjean, François, **5**:186a,b
Grandjean de Fouchy
see Fouchy, Jean-Paul Grandjean de
Granger, Walter Willis (1872–1941), 5:500–501; **2**:172a
Granite, **3**:293b, 433b; **5**:518b, 519a; **7**:62b; **12**:520a; **13**:273a
age of, **1**:472b; **2**:496a–b
alterations of, **10**:125b
classification of, **12**:273b–274a
cleavage of, **3**:325b
formation of, **12**:543a; **13**:283b
fusion of, **12**:122b
graphic, **6**:583a
hybrid, **3**:434a
industrial use of, **1**:51b
joint planes, **5**:444b
and lava, **3**:567b
massifs, **3**:325a–b
origin of, **2**:553b, 555b; **4**:409b; **5**:444a–445a; **6**:55a, 475b, 582b–583b; **7**:69b, 70b; **8**:616a; **10**:9a; **11**:273a; **12**:121a, 273a; **14**:132b, 261a; **15**:55a
rapakivi, **12**:271b
veins, **15**:427a
Grant, Ronald Thomson, **8**:295a
Grant, Ulysses Sherman, **14**:443a,b
Granulation tissue, **2**:165a–b
Granulite, **3**:462b
Granulopexis, **9**:296a
Grape
coloring matter of, **5**:315b
diseases of, **3**:43a; **5**:504a; **12**:125a; **15**:130b
Grape sugar, **11**:171a
Graphing, **3**:460b; **10**:483b
theory of, **10**:544a; **14**:591b
Graphite, **11**:463b
analysis of, **12**:147b–148a
atomic structure, **9**:183b
crystal structure, **15**:17a
formation of, **5**:139b–140a, 141b
mining of, **14**:451a
in pencils, **15**:24b
Graphitic acid, **2**:484b
Graphium, **9**:176a
Graphometer, **2**:133a
Graptolites, **8**:32b–33a; **11**:594a; **13**:143b, 145b; **14**:312b
Grashof, Franz (1826–1893), 5:501–502; **11**:344b*
Grashof number of heat transfer, **5**:502a
Grasses, **3**:318a; **5**:218a
artificial, **12**:317a
pollution of, **6**:117a
taxonomy, **15**:224a–b
Grasshopper
chromosomes, **13**:157a
hearing in, **6**:287b
Grassi, Giovanni Battista (1854–1925), 5:502–504; **4**:198a; **9**:82b–83a; **12**:315a; **15**:166a
Grassi, Orazio, **5**:242b–243a
Grassmann, Hermann, Jr., **15**:195b
Grassmann, Hermann Günther (1809–1877), 15:192–199; **1**:207b
algebra, **3**:142b, 166a; **6**:95b; **8**:389a; **10**:138a; **14**:303b
critics of, **12**:284b; **13**:11b
editions of works, **4**:371a; **8**:542b
Grassmann coordinates, **15**:197b

Grassmann manifold, **15**:197b
influence of, **6**:95b; **12**:217a; **13**:125a
influences on, **1**:239b
number theory, **6**:522a
optics, **6**:246b; **9**:200a
vector methods, **10**:443a; **12**:74a; **15**:79b
Die Wissenschaft der extensiven Grösslehre oder die Ausdehnungslehre, **1**:591a; **2**:295b; **3**:166b; **5**:309a; **6**:92a; **9**:431a; **11**:46b
Grassmann, Robert, **15**:193b, 194b
Gratiolet, Louis Pierre (1815–1865), 5:504–506; **2**:60b, **8**:272a; **12**:551b*
Graunt, John (1620–1674), 5:506–508; **9**:454a; **11**:236b; **14**:179a
Natural and Political Observations Mentioned in a Following Index, and Made Upon the Bills of Mortality, **10**:565b–566a
Grauwacke, **9**:583b
Grave, Dmitry Aleksandrovich (1863–1939), 5:508–509; **3**:221b, 226a,b
school of mathematics, **5**:508b
Grave robbing
see Body snatching
Graves, John Thomas, **6**:90a, 91a, 572b
Graves, Thomas, **5**:66b
'sGravesande, Willem Jacob (1688–1742), 5:509–511; **3**:216a, 537b; **5**:130a; **6**:612b*
acquaintances of, **3**:459b; **4**:517a; **9**:595a; **10**:145b
electricity, **4**:215b
The Mathematical Elements of Natural Philosophy
translation of, **4**:45b*
mathematics, **1**:209a; **5**:567a; **7**:490b
mechanics, **1**:112b; **8**:159b
Physices elementa mathematica experimentis confirmata, **9**:595a
translations of, **5**:510a
students of, **3**:37b; **15**:6a
Gravier, Charles, **10**:523a, 570a
Gravimeter
bifilar, **12**:187a
torsion, **14**:515b
Gravimetry
eighteenth-century, **2**:329b; **3**:156b; **5**:78b; **8**:72a
nineteenth-century, **2**:433a; **4**:127b, 239b, 379b; **8**:504a–b; **10**:219b; **13**:45a, 76a
early twentieth-century, **1**:429b; **4**:379b; **6**:105b; **10**:231b–232b; **11**:291a; **13**:45b, 605a–611a; **14**:448a, 515b
mid twentieth-century, **2**:373a–b; **7**:497a; **10**:158b–159b
in economic geology, **13**:45b; **14**:298b
pendulum, **9**:180b–181a; **10**:159a–b
at sea, **13**:606a–608b
theoretical, **9**:449a
vacuum standard, **3**:476b
see also Balances, Weight
Gravitation
see Attraction, gravitational (Newtonian, universal)
Gravity, **8**:609b
absolute, **6**:241a
acceleration of, **2**:606a–b; **6**:240a
"accidental," **2**:590b; **8**:223a–b
apparent, **13**:608a
center of, **1**:222b, 229a, 256b; **2**:49a; **4**:46b; **5**:238b; **6**:205b*, 310b, 312a; **7**:345b; **10**:225a; **11**:488b; **13**:48b; **14**:149b–150a
and celestial mechanics, **5**:588a
definition of, **10**:298a
and equilibrium, **1**:419b; **6**:602a; **8**:223a–b
in geometry, **1**:164a,b, 219b, 221a; **3**:182a–b, 364a; **4**:516a, 570b–571a; **5**:588b; **6**:599b–600a; **7**:558a, 583b; **8**:223b–225b; **9**:191b, 430a–b; **10**:336b–337a; **13**:19a, 436a, 560b–561a; **14**:29b, 149b
motion of, **7**:341a
and oscillation, **6**:605b–606a, 608a
constant of, **11**:122b
definition of, **8**:227b–228a
diurnal variation, **9**:155a
geographical variation, **2**:372b; **9**:155a; **11**:412b; **13**:76a
and figure of the earth, **10**:483b; **13**:69b
and geological substructure, **2**:373a; **9**:180b–181a; **11**:127a
Vening Meinesz belts, **13**:609b–610b
and volcanic activity, **14**:515b
latitude, effects of, **2**:389a; **3**:282b; **6**:240b; **9**:164b; **15**:321a–b, 339a
measurement of, **2**:102a; **4**:379a–380a; **6**:240b–241a; **10**:483a
oceanographic variation, **2**:373a–b; **9**:181a
and plant growth, **3**:581b; **4**:135b
positional, **2**:193b–194a; **7**:172a–174b; **8**:215b, 216b
specific
ancient Greek study of, **1**:224a; **4**:431a
ancient Roman study of, **9**:300b; **15**:420b–421a
medieval study of, **2**:153a, 194a; **7**:325b, 340b–348a *passim*
seventeenth-century study of, **6**:125b, 127a, 255a, 608b
eighteenth-century study of, **6**:171b; **10**:109a*; **15**:335a
nineteenth-century study of, **3**:292b; **5**:321a
and acid strength, **6**:478a
and crystal structure, **7**:463b
determination of, **9**:445b; **11**:434b
of gases, **2**:588a; **5**:493a; **13**:373a
of liquids, **7**:341a; **8**:70b, 177a
of metals, **7**:346a
of minerals, **5**:455a
of solutions, **9**:291a–b
and specific volume, **7**:463b
and temperature, **6**:477b
and temperature, **6**:291b; **11**:122b
theories of, **3**:280b; **6**:485b, 606b–611a *passim;* **8**:497a; **14**:184a
vortex, **6**:610b; **12**:118a–b
and weight, **1**:541b
see also Acceleration, gravitational; Attraction, gravitational; Falling bodies; Gravimetry; Isostasy; Weight
Gravity gliding (geol.), **8**:543b, 544b
Grawitz, Paul Albert, **7**:424a
Gray, Andrew, **9**:173b
Gray, Asa (1810–1888), 5:511–514; **4**:274a; **8**:264b
acquaintances of, **1**:396a; **2**:103a; **3**:197a, 562b; **6**:492a; **13**:432b, 433a, 456a,b; **14**:192a, 533a; **15**:130b
botany, **3**:196b; **10**:285b
on contemporary scientists, **2**:521b; **6**:495a; **14**:534a*; **15**:131a
correspondence of, **3**:563a,b*; **10**:164b; **14**:135b
and Dana, **3**:549b, 550b, 553a
and Darwin, **5**:513a–b
Darwiniana, **14**:516b
and evolution, **3**:574a
Manual of the Botany of the Northern United States
revision of, **4**:583b; **14**:192b
Gray, Elisha, **1**:582b; **4**:248a
Gray, Henry (*ca.* 1825–1861), 5:514–515
Gray, Joseph Alexander, **3**:368a
Gray, Robert Whytlaw
see Whytlaw-Gray, Robert
Gray, Stephen (1666–1736), 5:515–517; **6**:169a
barometric level, **4**:44a,b
electricity, **4**:166b, 215a, 216a; **6**:38a; **10**:146a; **11**:433a; **14**:71b
Gray, Thomas, **7**:413b; **9**:407a
Graybill, Harry Webster, **12**:482b
Grayson, Eldred (pseudonym)
see Hare, Robert
Graz
Johanneum, **9**:448a
pediatric institute, **4**:404a
Graz, University of
chemistry at, **11**:129b
Grease (disease), **7**:96b
Great Barrier Reef, **3**:568b–569a; **7**:185b
Great Basin, **14**:192b
geology, **5**:395b
Great Britain
archaeology, **8**:442b–443a; **11**:6a
biology
anthropology, **13**:504a
botany, **1**:358b–359a; **5**:362b; **6**:163a, 265b, 491b; **7**:147a–b; **8**:372a; **11**:313b–316a; **12**:471b; **13**:341a–343a, 502a–b; **14**:189b, 190b, 463b
marine biology, **8**:609a
physiology, **12**:354a–b
zoology, **5**:31b; **6**:184a–b
Board of Longitude, **2**:141a; **6**:130b; **14**:563a
chemistry, **8**:186a; **11**:48b
colonies
botany, **6**:491b
Department of Scientific and Industrial Research, **3**:87b
education, **12**:618b–619a
English settlement of, **1**:564b
finance, **7**:318a
geography
biogeography, **5**:66b, 67b
ethnography, **1**:562a–b
exploration, **11**:225a–b
Geological Survey, **1**:393b–394a; **3**:470b; **5**:39a, 334a–335a, 336b, 337a, 436b; **6**:489b; **7**:185b; **8**:461a; **9**:585a; **10**:583b–584a; **12**:277b, 293a; **13**:273a
formation of, **4**:10b–11a
publications of, **5**:335a
geology, **4**:10b–11a; **5**:39a; **9**:583a–585a; **10**:259b–260a; **11**:34b–35b, 276b–277a; **13**:273a
hydrography, **5**:17a
meteorology, **5**:413a–b
mineralogy, **1**:359b
paleobotany, **6**:489b
paleontology, **8**:32b–33b, 528a; **9**:87a–b; **10**:40b, 322b
volcanism, **5**:336a–b
industry, **7**:78a
chemical, **6**:463a
mathematics, **8**:611b
medical sciences, **7**:147b; **13**:503b–504a
military organization, **6**:25b
National Institute for Medical Research, **15**:107a
natural history
Renaissance, **3**:13a
seventeenth-century, **2**:522a–523a; **3**:248a–b; **9**:313a; **11**:313a–318a; **14**:413a
eighteenth-century, **1**:433b–434b
nineteenth-century, **1**:358b–359a; **2**:18b, 566a–b
physics, **8**:592a,b; **14**:293a

Great Britain (cont.)
technology
instrumentation, **8**:21b
metallurgy, **10**:511b–512a
see also England, Scotland, Wales
Great Eastern (ship), **4**:492a
Great Inagua Island, **15**:89a
Great Lakes
geology, **4**:493a; **5**:395b; **13**:269b–270b
Great Plains
geology, **9**:255b
Great Salt Lake, **5**:395b
Great Soviet Encyclopedia, **7**:207b
Great Windmill Street School of Anatomy, **1**:398b, 583b; **3**:67b, 486b–487a; **6**: 569b
Greatest common measure, **13**:304b, 316a–317b; **15**:503b
Grebe, Ernst Wilhelm
Grebe point, **8**:175b
Greece
archaeology, **3**:162a; **6**:84a; **9**:489b; **12**: 180a–181b
calendar, **9**:337a–339a
chronology, **10**:550b
geology, **10**:29b
and Persia, **6**:316a
Greek Anthology, **4**:606b, 612a
Greek fire, **3**:20b, 21a
Green, Anne (criminal), **14**:405a
Green, George (1793–1841), 15:199–201; **3**:214a; **13**:75b, 78b; **15**:488a
Essay on the Application of Mathematical Analysis to the Theories of Electricity and Magnetism, **13**:376a–b
ether, **9**:211a; **13**:386a–b; **14**:67a
Green's formula, **3**:140b, 145a
Green's function, **2**:217b; **3**:145a; **4**:174a; **6**:392a; **8**:458b*; **9**:215b; **11**:454a; **15**:199b–200a
Green's theorem, **6**:294a; **10**:248b; **11**: 449b; **15**:199b
hydrodynamics, **13**:75a,b
optics, **8**:592a–b; **13**:77b
physics, **2**:105b; **3**:146a,b, 282b; **15**:316b, 487b, 488a
Green, Herbert Sydney, **15**:43b–44a
Green, Jacob (1790–1841), 5:517–518
Green, Thomas Hill, **6**:26a
Green, William, **3**:67b
Green Mountains, **11**:210a; **14**:516b
Green River
exploration of, **11**:118b
Greene, Edward Lee, **7**:102a–b, 285a; **9**:314a
Greenewalt, Crawford Hallock, **3**:371a–b
Greenhill, Alfred George, **2**:215b
Greenhouses, **13**:553b
Greenland
exploration of, **3**:204a; **4**:193b; **5**:225a; **14**:214b–215a; **15**:430b
geography, **10**:617b
geology, **4**:564b; **6**:78b
glaciation, **14**:517a
meteorology, **15**:430b
paleobotany, **12**:339b
Polar Year expedition, **3**:204a
Greenough, George Bellas (1778–1855), 5: 518–519; **2**:566b; **4**:65b; **6**:18b; **8**: 564b; **10**:322b; **14**:397b
geological map, **2**:482b; **4**:218a, 348b, 541a; **5**:537a; **9**:87a; **12**:491a
Greensand, **2**:341b, 495b; **5**:14a–b; **8**:564b
Greenstein, Jesse Philip, **3**:336a
Greenwich (England)
latitude of, **2**:389a; **4**:270a; **5**:24b
meridian of, **3**:106a
Seaman's Hospital, **2**:574b, 617a
Greenwich, Royal Observatory, **5**:404b; **12**:573b
in nineteenth century, **1**:85a–b, 86a; **4**: 354a–b; **9**:184a
founding of, **5**:22b–23a, 24b
Greenwich mean time, **3**:288b
instruments, **2**:141a, 388b; **3**:261a–b; **6**: 6a; **10**:207b
lunar observations, **2**:389a, 516b
meteorology, **5**:413a
reduction of observations, **4**:175b, 269b–270a
Greenwood, Isaac (1702–1745), 5:519–520; **14**:452a
Greenwood, Major, **14**:574a
Gregarinida, **4**:133a*, 142b, 207a; **7**:438a; **8**:145b–146a; **9**:328b
life cycle, **3**:148b, 492b; **12**:424a
Gregg, Alan, **5**:65a
Gregory XIII, Pope, **3**:312a, 558a,b; **5**:588a; **9**:308b; **10**:598a; **13**:587a; **14**:23a; **15**:45a
Gregory of Cittá di Castello (Tifernate), **1**:274a
Gregory Magistros, **4**:442b
Gregory of Nyssa (Saint), **4**:396b; **7**:25a; **10**:20b
Gregory of Rimini, **1**:199b; **2**:394b, 604b; **3**:420b; **10**:419b; **13**:349b
Gregory, David (1659–1708), 5:520–522; **5**:510b; **6**:68a; **14**:152b
acquaintances of, **3**:244a; **11**:1a–b
algebra, **8**:162a
analysis, **8**:162b, 163a
Exercitatio geometrica, **5**:528b
geometry, **5**:259b*
historiography, **4**:430b, 450b; **6**:71a; **10**: 303a*
influences on, **14**:637a
Memoranda, **3**:244b
and Newton, **5**:520b, 521b; **10**:44b, 45a, 51b, 52b, 53a, 56a, 61b, 83b–84a, 87b(n); **11**:1a–b
students of, **3**:458a,b; **7**:275a–b
Gregory, Duncan Farquharson (1813–1844), 5:522–523
Gregory, Frederick Gugenheim (1893–1961), 5:523–524; **11**:416a
Gregory, George, **6**:277a
Gregory (or Gregorie), James (1638–1675), 5:524–530; **3**:348b; **5**:25b, 520b, 522b*
acquaintances of, **8**:610a; **9**:484a
astronomy, **6**:69a, 331a
on contemporary scientists, **1**:475a
correspondence of, **2**:507b; **10**:48b; **13**: 265b
Exercitationes, **5**:521a
Geometriae pars universalis, **1**:474b; **5**: 521a
Gregory's series, **5**:527b; **15**:633a(n)
mathematics, **10**:45b; **12**:344b
and Newton, **5**:525b, 529a
optics, **3**:98a; **9**:159a
students of, **6**:555b
Tentamina de motu projectorum, **6**:127a
Vera . . . quadratura, **2**:49b; **6**:236a*
Gregory, John, **3**:494b; **12**:354a
Gregory, Olinthus Gilbert (1774–1841), 5: 530; **6**:577b*
Gregory, Sir Richard, **8**:443a
Gregory, William (1803–1858), 5:530–531; **8**:350a*; **10**:510b, 582a; **13**:500a*
Gregory, William King, **14**:410a
Greig, Edward David Wilson, **2**:529a
Greinacher, Heinrich, **3**:329b; **12**:34a
Greisen, **7**:249a
Gren, Friedrich Albrecht Carl (1760–1798), 5:531–533; **2**:565a; **7**:395a; **8**:520a; **13**:465b; **14**:78b
Grenoble, University of
geography at, **2**:190b
geological school, **1**:236a
Grenville Series, **1**:51a–52a
Gresham, Thomas, **14**:510b
Gresham College, London, **3**:482a; **4**:495b; **10**:200b–201a, 565a; **11**:532a; **14**: 151a, 371a, 501a, 510b, 513b
Gressly, Amanz (1814–1865), 5:533–534; **7**:43b; **14**:88a
Greville, Robert Kaye, **2**:19a
Grévin, Jacques, **5**:348b; **14**:11b
Grew, Nehemiah (1641–1712), 5:534–536; **14**:371b
cell formation, **12**:174b
influence of, **6**:130a; **8**:127a; **9**:65b; **13**: 526a
philosophy, **15**:54a
plant anatomy, **6**:41a,b; **9**:455b; **11**:315b
plant sexuality, **2**:188b; **9**:400b; **11**:83a; **15**:67b
Grey Friars
see Franciscans
Griesinger, Wilhelm, **2**:127a; **4**:198a; **5**:570a, 571a; **14**:277a
Griess, Johann Peter (1829–1888), 5:536–537; **3**:85a; **6**:461b; **7**:282a
Griffin, Donald Redfield, **12**:562b
Griffin, John Joseph, **3**:289a
Griffith, Alan Arnold, **12**:464b
Griffith, Frederick Reece, Jr., **1**:342b
Griffith, Richard John (1784–1878), 5:537–539
Griffith, William (1810–1845), 5:539–540; **6**:467b(n)
Micrographic Dictionary, **12**:259a
Griffiths, Ernst Howard, **3**:20a
Grignard, François Auguste Victor (1871–1935), 5:540–541; **12**:46b
Grignard reaction, **5**:540b; **7**:372b
Traité de chimie organique, **5**:100a
Grijns, Gerrit (1865–1944), 5:541–542; **4**: 311b; **10**:493a
Grimaldi, Claudio Filippo, **8**:150a
Grimaldi, Francesco Maria (1618–1663), 5: 542–545; **4**:506b; **6**:195a; **14**:461a
optics, **3**:100b, 528b; **10**:59b
selenography, **9**:485b; **11**:411b
Grimaldi, Honoré Charles
see Albert I of Monaco
Grimaux, Édouard, **2**:64b; **12**:147a
Grimm, Hans August Georg, **6**:304a; **12**: 530a
Grimm, Johann, **11**:102b, 103a
Grimm, Oskar Andreevich, **7**:410b
Grimmia, **6**:219a
Grinberg, Christopher, **5**:382b
Griner, Boris Matveevich, **12**:77a
Grinnell, Joseph (1877–1939), 5:545–546; **14**:402a
Grisebach, August Heinrich Rudolf (1814–1879), 5:546–547
Grisogono, Federico (also known as Federicus de Chrysogonis) (1472–1538), 5: 547; **4**:159a
Grobben, Karl, **5**:447b; **6**:9b; **7**:517a
Groddeck, Albrecht von (1837–1887), 5:547–548; **11**:103a
Grodzicki, Ladislaus, **10**:206b
Gróh, Gyula, **6**:365b
Groma, **6**:312b
Groningen, University of
astronomy at, **7**:236b
physics at, **3**:430a
Gronovius, Johann Friedrich, **1**:486b; **3**: 351a; **5**:408b, 607a; **8**:375a, 460a
Flora Virginica, **1**:431b, 485b; **3**:344a, 350b

Groombridge, Stephen
catalog of, **3**:261b*; **4**:269b; **7**:146a
observations of, **13**:110a–b
Groot, Jan Cornets de
see De Groot, Jan Cornets
Gros, Oskar, **15**:462a
Grosheintz, John Manfred, **5**:6a
Gross, Evgenii Feodorovich, **9**:76b
Grosse, Jean, **4**:223b
Grosseteste, Robert (*ca.* 1168–1253), 5:548–554; 1:100a, 272b, 337b; **4**:92b, 93a; **5**:402a; **7**:66b; **9**:362b; **10**:474a
astronomy, **1**:101a; **15**:35a
and (Roger) Bacon, **1**:377b–378a, 379a–b, 382b
on calendar, **1**:383a; **5**:553a; **10**:540a–b
critics of, **3**:26b
De impressione aëris, **1**:382b
influences on, **14**:388a; **15**:264a
De lineis angulis et figuris, **14**:458b
methodology, **1**:278b; **5**:551a–552a
multiplicatio specierum, **1**:379a
natural philosophy, **11**:42a; **14**:459a
optics, **1**:279a; **5**:549b, 550b–552b; **10**:475a; **14**:459b
Sphere, **3**:26a
translations by, **1**:271b–272a, 273b, 276a,b; **5**:549a–b; **12**:442b–443a
revisions of, **9**:437a
Grossi, Ernest von, **11**:177b
Grossmann, Ernst A. F. W. (1863–1933), 5:554
Grossmann, Marcel (1878–1936), 5:554–555; 4:327a; **11**:409a
Grote, Augustus Radcliffe (1841–1903), 5:555–556
Grote, George, **1**:404a; **3**:375a
Groth, Paul Heinrich von (1843–1927), 5:556–558; 5:457a; **9**:379b; **11**:540a; **12**:512a, 572b; **13**:617a; **14**:411b, 525a
Chemische Krystallographie, **11**:271a
crystallography, **4**:141b; **5**:455a
Fyodorov-Groth nomenclature, **5**:212a
Grothite, **5**:556b
Grotius, Janus
see De Groot, Jan Cornets
Grotrian Otto, **7**:449b
Grotthuss, Theodor (Christian Johann Dietrich) von (1785–1822), 5:558–559; 11:540b
electrolytic conduction, **2**:136b; **13**:364a
Grotthuss-Draper law, **4**:182b; **5**:558b
Grottoes, **13**:30b
prehistoric, **8**:44a
Grouchy, Sophie de, **3**:383b
Ground tilting, **7**:9b
Ground water, **2**:348a, 439b, 548b; **4**:262a; **5**:410b; **9**:257a–258a
theories of, **7**:149a; **9**:532a; **10**:128a, 522a
Grounding (elect.), **4**:129a
Group theory
early nineteenth-century, **3**:142b; **5**:473b; **11**:598b
mid nineteenth-century, **1**:370a; **3**:167b, 168a–169a; **7**:384a; **8**:385a
late nineteenth-century, **2**:615a–b; **4**:269a; **5**:198a–b; **7**:167b–168b, 397b–399a; **8**:324a, 325b–326b; **9**:502b; **10**:24a–b; **11**:589a; **13**:216a, 263b
early twentieth-century, **1**:307a,b; **2**:197a–b, 431b; **3**:135b, 345b; **4**:177a–b; **6**:472b, 572a; **8**:457a–b; **9**:173b, 358b, 387a–388a, 457b–458a; **12**:237a–b, 451b, 533b; **14**:283b–285a, 559b*
mid twentieth-century, **2**:352a–b; **4**:174a; **9**:71b, 183b; **11**:362b; **13**:498a; **14**:89b–90a
finiteness theorems, **3**:142b; **5**:198b; **6**:214b; **7**:168a,b
origins of, **5**:260b–264a *passim*
in physics, **1**:519b; **8**:434a
primitivity, **9**:387b
Groups (math.)
Abelian, **1**:307a; **3**:168b; **4**:125b; **5**:609a; **6**:309a; **7**:508a
abstract, **2**:352b, 615b; **3**:168b; **4**:2b; **9**:502b
concept of, **5**:198b
algebraic, **8**:326b
automorphic, **6**:572a; **8**:326a
characters, **2**:615b; **5**:198b
cohomology of, **6**:497a
collineation, **2**:197b
commutative, **5**:608b; **9**:387b
compact, **2**:513a; **3**:95b; **5**:608b–609a; **6**:572a; **14**:90a, 284b
complex, **14**:284b
composite, **6**:473a; **10**:24b
concept of, **8**:434a
congruence, **7**:398b
conjugateness, **3**:142b
continuous, **2**:513a; **4**:370b; **5**:201b; **7**:168b; **8**:326a, 458b*; **11**:55b; **14**:13b
analyticity of, **6**:393b
crystallographic, **12**:533b
decomposition of, **5**:263b; **14**:89b
definition of, **3**:168a–b; **10**:138b
dihedral, **7**:398a
discontinuous, **5**:201b; **6**:497a; **7**:168b
factor, **6**:472b
finite, **4**:83a; **12**:533b; **14**:284b
theory of, **2**:197b, 615a; **3**:168b; **5**:509a*; **9**:387b, 502b; **14**:285a
Fuchsian, **11**:53b, 54a
fundamental, **4**:9a; **11**:60a
Galois, **1**:307a; **7**:508a; **12**:237b
structure of, **7**:168a
hypoabelian, **7**:168a
icosahedral, **5**:472b
infinitesimal, **8**:326a
Lie, **2**:197b; **3**:95a–96b; **5**:201b; **8**:457b; **11**:53b; **12**:214a; **13**:125b; **14**:90a
representation of, **9**:71b; **14**:283b–285a
theory of, **3**:222a; **6**:76a; **7**:574a; **8**:326a–b; **9**:71b
linear, **2**:615b; **3**:96a; **5**:201b; **7**:168a; **8**:458b; **12**:237a–b
homogeneous, **2**:197b
Lorentz, **8**:434a; **15**:49b
Möbius, **8**:457b
of movements, **7**:168a
non-Euclidean, **8**:457b
odd order, **2**:615b
ordered, **2**:352b; **3**:142b
orthogonal, **3**:95b, 96a; **6**:572a; **14**:284a
parameter, **8**:326a
permutation, **2**:615b; **3**:142b; **5**:198a, 263b; **7**:168a–b, 567a; **10**:24a; **11**:598b
rationality group, **4**:177b
semigroups, **4**:420a; **8**:457a–b; **11**:107b; **13**:498a
semisimple, **14**:285a
simple, **6**:472b–473a
solvable, **2**:615b; **7**:168a; **9**:71b
subgroups, **2**:88a; **3**:142b; **7**:168a,b; **9**:387b; **13**:216a
symmetric, **2**:88a; **9**:183b
topological, **4**:9a; **6**:496b–497a; **7**:312b–313a; **9**:71b
transformations, **2**:197a; **4**:370b; **7**:397b–399a
theories of, **8**:324a, 325b–326b
transitive, **2**:615b
Weyl, **14**:284b
see also Group theory, Systems
Grove, William Robert (1811–1896), 5:559–561; 12:52a
Correlation of Physical Forces
translations of, **12**:289a
electricity, **5**:292a; **9**:215b; **11**:377b
Grove cell, **5**:559a–b
Groves, Leslie Richard, **10**:215a–b; **13**:429b
Growth, **4**:161a; **8**:282b; **9**:416a
of bones, **5**:45a; **6**:40b, 183b, 566a; **9**:546b
and diet, **9**:284b–285a; **10**:243a–b
endocrinology of, **3**:113a; **12**:474a
human, **2**:211b, 212b; **3**:113a, 590b; **9**:332b
theories of
ancient Greek, **1**:258b, 260b; **4**:384a
seventeenth-century, **5**:421b; **14**:405a
eighteenth-century, **15**:53b–54b, 525a
twentieth-century, **12**:410a
see also Morphogenesis; Plants, growth of; Plants, tropisms; Trees, growth of
Grubb, Howard (1844–1931), 5:561; 11:484b
Grubb, Thomas (1800–1878), 5:561–562
Grubbs, Frank Ephraim, **10**:460b–461a
Grubenmann, Johann Ulrich (1850–1924), 5:562–563; 1:236a; **4**:409a; **10**:124a,b, 126b; **12**:271b
Gruber, Ignaz von, **5**:563b
Gruber, Max von (1853–1927), 5:563–565; 4:299a, 304b*; **5**:184b; **7**:623a; **8**:412b*; **10**:563a*
cytology, **14**:430a
Gruber-Widal reaction, **5**:564b
Gruby, David (1810–1898), 5:565–566; 1:493b
Gruby's disease, **5**:566a
pathology, **12**:203a
Grünbaum, Adolf, **14**:608b–609b, 610b
Grünbaum, Albert Sidney, **5**:564b
Grundschollen (ground blocks), **3**:325b
Grünwald, Josef, **2**:192a; **4**:562a
Grüter, F., **2**:345b
Grynaeus, Simon, **4**:437b, 448a, 448b–449a; **13**:491b
Grzepski, Stanislaw, **2**:526b
Gua de Malves, Jean Paul de (*ca.* 1712–1786), 5:566–568; 3:281a; **4**:86a
algebra, **13**:129a
L'usage de l'analyse de Descartes . . . , **3**:460a
Guaco, **15**:430a
Guaiac wood
in syphilis treatment, **1**:620b; **5**:105b; **9**:75a, 179a; **10**:305a, 306b
Guaiacum, **10**:622a
Guam, **3**:551b
Guanidine, **4**:400a
derivatives of, **13**:337b, 338b, 406b
Guanine
synthesis of, **4**:604a; **5**:2b
Guano, **6**:553b; **8**:347a
deposits of, **10**:170b
Guarico, Luca, **1**:604b, 605b
Guasti, Cesare, **1**:612a
Guatemala
geology, **6**:13b
gravimetry, **14**:515b
Guayule, **4**:362a
Guccia, Giovanni Battista (1855–1914), 5:568–569
Gucker, Frank Thomson, **11**:418a
Gudden, Johann Bernhard Aloys von (1824–1886), 5:569–572; 4:160b; **5**:73b, 74a; **10**:131a
Gudden method, **5**:571a
Gudermann, Christoph (1798–1852), 5:572–573; 2:215b; **14**:219b, 220a

Guenther, Adam Wilhelm Siegmund (1848–1923), 5:573–574; 2:361a, 430a; **3**:59a; **9**:234b*, 248a*, 463b*; **12**:512b*; **14**:276b*, 336b
Guericke (Gericke), Otto von (1602–1686), 5:574–576; **2**:377b; **7**:375a; **8**:149b; **9**:16b; **10**:334a; **12**:210b
air pump, **6**:482a
experiments, **5**:516a; **6**:477b; **12**:210b
Guérin, Camille, **3**:22b
Guerrini, Guido, **8**:557a
Guertler, William Minot (1880–1959), 5:576–577
Guettard, Jean-Étienne (1715–1786), 5:577–579; **7**:199a; **12**:120a
geology, **1**:327a; **4**:71b, 348a, 548b; **9**:54b, 478b
and Lavoisier, **5**:579a; **8**:68a–b, 69a
marine biology, **10**:569a; **11**:334a
mineralogy, **11**:330b; **12**:67a,b(n)
paleontology, **15**:54b, 59a*
Guggenheim, Edward Armand, **2**:170b; **5**:103b; **8**:478b
Guggenheim Foundation, **5**:433b
Guglielmini, Domenico, **1**:616a; **3**:332a; **9**:77a, 370a, 486b, 511a; **11**:521b, 523a; **13**:566b
Guglielmini, Giovanni Battista, **1**:165a
Guiart, Jules, **11**:342a
Guibal, Théophile, **1**:327b; **3**:358b
Guibert, Nicolas (*ca.* 1547–*ca.* 1620), 5:579–580; **8**:310a; **15**:692b
chemistry, **7**:194b; **12**:79b
Guichard, Claude, **4**:336a
Guidi, Guido (also known as Vidus Vidius) (1508–1569), 5:580–581
Guidobaldo del Monte
see Monte, Guidobaldo, Marchese del
Guiducci, Mario, **5**:242b
Guiffart, Pierre
Discours sur le vide, **10**:332b
Guignard, Jean-Louis-Léon (1852–1928), 5:581–582; **2**:70b; **3**:422a
Guilandinus
see Wieland, Melchior
Guillaume, Charles Édouard (1861–1938), 5:582–583; **3**:239a
Guillemeau, Jacques, **10**:316a
Guillet, Léon Alexandre (1873–1946), 5:583–584; **11**:100b
Guilliermond, Marie Antoine Alexandre (1876–1945), 5:584–585
Guilt, **5**:178a,b
Guimet, Jean Baptiste, **3**:316a
Guinand, Pierre Louis, **5**:142b, 143a
Guinea pig
coat color
inheritance of, **3**:121a–b
embryology, **2**:161b
Guinea worm, **1**:26a, 496a
Guinier, André, **9**:312b
Guinness, Arthur and Sons (brewers), **5**:476b
Guinter, Joannes (*ca.* 1505–1574), 5:585–586; **4**:138b, 208b; **11**:527b; **12**:323b; **14**:4a
anatomy, **14**:4b
Guisnée, N., **3**:281a; **9**:186a; **11**:328a; **12**:118a
Guldberg, Cato Maximilian (1836–1902), 5:586–587; **9**:443a; **15**:458a
law of mass action, **2**:68b, 69a; **4**:412a; **9**:110a, 352b; **10**:440b; **14**:108b–109a
see also Mass action, law of
meteorology, **6**:98a; **12**:595b
solutions, **11**:298b
Guldin, Paul (1577–1643), 5:588–589; **9**:97a; **13**:561a
Centrobaryca, **3**:152a
Guldin's theorem, **3**:152a; **4**:110a; **5**:527a, 588b; **7**:583b; **10**:296b; **12**:292b; **13**:615b
mathematics, **1**:164a; **3**:151b; **7**:583a; **10**:336a
Gulf Stream, **1**:92a, 364a–b
charting of, **5**:136b–137a
Gulick, John Thomas, **11**:519a
Gull
speciation, **3**:572b
Gulland, John Masson (1898–1947), 5:589–590
Gulliver, Frederic Putnam, **7**:144a
Gullstrand, Allvar (1862–1930), 5:590–591; **7**:404a; **11**:215a
Gullstrand ophthalmoscope, **5**:590b
Gum senega, **4**:250a
Gümbel, Theodor, **12**:169a
Gumbotil, **7**:267a
Gums (bot.), **9**:79a
analysis of, **10**:497b
origin of, **14**:350b
sugar gum, **13**:405a–b
Guncotton, **3**:358b*
discovery of, **2**:340b; **7**:45a; **12**:198a–b
nitrogen in, **3**:488b
Gundissalinus, Dominicus (*or* Gundisalvo, *or* Gonsález) (*fl.* second half of twelfth century), 5:591–593; **9**:364b
translations by, **15**:175a,b, 181b–182b
De unitate et uno, **2**:231a
Gunn, Battiscombe George, **15**:697b
Gunpowder, **2**:77a–b; **14**:333a
chemical analysis of, **2**:455a; **3**:562a; **7**:379a; **8**:77a
energetic equivalent of, **8**:209b
force of, **13**:350b–351a
manufacture of, **2**:143a; **3**:388b; **5**:603a; **8**:133a, 507b; **9**:146b, 472b; **11**:616b; **13**:496a
Guns
Renaissance, **8**:213b, 214a
eighteenth-century, **7**:13a; **8**:328a; **11**:494a
nineteenth-century, **11**:5b, 6a; **12**:131b, 320a–b
inflammable air pistol, **14**:75a
Gunsight, **5**:561b; **11**:472a
Gunter, Edmund (1581–1626), 5:593–594; **5**:343b; **6**:271b; **10**:151b
Gunter's chain, **5**:593b
Günther (of Andernach), Johann
see Guinter, Joannes
Günther, Albert Charles Lewis, **7**:170a; **8**:609a
Gurney, Goldsworthy, **6**:292b
Gurney, Ronald Wilfrid, **5**:272a
Gurvich, Aleksandr Gavrilovich (1874–1954), 5:594–595; **9**:76a
Gutbier, Felix Alexander (1876–1926), 5:595–596
Gutenberg, Beno (1889–1960), 5:596–597; **9**:444a; **14**:327a,b
Guthrie, Frederick, **2**:547b; **15**:59a
Gutta-percha, **13**:450a
Guttmann, Paul, **4**:296b, 297b
Guy de Chauliac
see Chauliac, Guy de
Guy, William Augustus, **4**:593b
Guye, Charles-Eugène (1866–1942), 5:597–598; **10**:604b*
Guye, Philippe-Auguste, **5**:587b, 597a–b
Guyer, Michael Frederic (1874–1959), 5:598–599; **1**:498b
Guyonneau de Pambour, F. M.
see Pambour, F. M. Guyonneau de
Guyot, Arnold Henri (1807–1884), 5:599–600; **8**:264a; **10**:241a
Guyot (geol. term), **5**:600a
Guy's Hospital, **2**:464a–b; **10**:177a
Guy's Hospital Reports, **2**:464a
Guyton de Morveau, Louis Bernard (1737–1816), 5:600–604; **6**:2b; **11**:166b
acidity, **8**:76a
acquaintances of, **3**:315b; **4**:74a; **5**:202b; **6**:1a; **10**:497a
calcination, **3**:206b; **8**:74a
causticity, **9**:347a
chemical affinity, **7**:388a
chemical analysis, **2**:6a; **13**:309b
on chemical combination, **11**:167a, 435b
correspondence of, **8**:620b; **12**:67b(n)
critics of, **12**:68a(n)
crystallization theory, **2**:6b
disinfection, **3**:202a
and Lavoisier, **8**:80b, 81a
nomenclature, **2**:7b, 8a, 74a; **5**:90a; **6**:164b; **7**:389a
technology, **3**:317a*, 561a*
translations of works, **14**:225a
Guzman, Gabriel de, **1**:605b
Gwynne-Vaughan, David Thomas (1871–1915), 5:604–605; **2**:371a; **8**:5a,b
Gyldén, Hugo, **1**:429a; **2**:56b, 442a; **3**:18b; **11**:383a; **14**:481b
Gyllenhaal, Leonhard (1752–1840), 5:605; **12**:201a
Gymnosperms, **2**:491b
as distinct from angiosperms, **2**:517a, 520a; **10**:1b
fossil, **14**:462b
morphology, **3**:187b
Gymnotoca, **5**:385b
Gymnotus electricus, **4**:202b
Gynandria, **13**:392b
Gynandromorphism, **5**:453b; **10**:570a; **13**:135b, 137a
Gynecology
ancient Greek, **6**:318a, 422b, 425b, 427b
ancient Roman, **12**:539a–b
medieval Arabic, **12**:92a
medieval Latin, **13**:347b
Renaissance, **11**:33b
eighteenth-century, **1**:323a
nineteenth-century, **14**:278a–279a
twentieth-century, **10**:291b–292a
Gypsum, **3**:473b; **6**:19a; **8**:70b; **15**:662b(n)
analysis of, **8**:69a–b; **9**:106a
and anhydrite, **2**:159a
colors from polarized light, **2**:137b
crystallography, **14**:241a
deposits of, **10**:171a
fossils in, **2**:494b
and plaster of Paris, **8**:117a
Gypsy moth, **6**:524b; **13**:472b
Gyrations, **7**:446b
Gyrocompass, **12**:575a
Gyromagnetic effect, **11**:420b
Gyroscope, **4**:592a; **5**:64a
applications of, **12**:575a; **13**:483a; **14**:616a*
invention of, **5**:86b; **11**:62a
theories of, **5**:86b; **7**:399a; **10**:42b*; **15**:195b
Gyrostabilizer, **12**:575a

H

Haak, Theodore (1605–1690), 5:606–608; **3**:360a; **8**:267a; **10**:495b; **14**:365a, 368a–b

correspondence of, **2**:163a; **14**:368a–b, 373b
and Royal Society, **12**:583b–584a, 585a; **14**:364a
Haar, Alfréd (1885–1933), 5:608–609; 4:561b; **11**:459a
Haar's lemma, **5**:608b
Haar's measure, **5**:608b–609a; **6**:572b; **14**:90a
Haarlem
Teyler's Stichting, **8**:489a; **9**:152a,b; **13**:183b
Haarlem Academy
competitions of, **5**:441b
Haarmann, Wilhelm, **13**:406b
Haas, Arthur Erich (1884–1941), 5:609–610; 5:393b*
Haas, Wander Johannes de (1878–1960), 5:610; 7:272a, 516a*; **8**:52a, 499b*; **9**:211b
Einstein-de Haas effect, **5**:610a; **14**:172a
Haast, Johann Franz Julius von (1822–1887), 5:610–612
Ḥabash al-Ḥāsib, Aḥmad ibn ᶜAbdallāh al-Marwazī (*d. ca.* 864–874), 5:612–620; 1:36b; **14**:537b
astronomy, **7**:338a; **9**:84b; **13**:324a
trigonometry, **10**:6a; **14**:538a
zījes, **14**:537b, 575b
Haber, Fritz (1868–1934), 5:620–623; 5:117b; **6**:16a, 118a; **7**:553a; **8**:553a; **15**:159b, 447b, 451a
acquaintances of, **2**:3b; **6**:371a; **9**:354a; **14**:334a
ammonia synthesis, **2**:323b–324a; **3**:436b; **5**:621b–622a; **8**:118b; **9**:427a,b; **15**:435b
Haber Colloquium, **5**:622b
influences on, **15**:438b–439a
nitrogen fixation, **3**:479b
students of, **4**:545a; **6**:365a
thermodynamics, **15**:436b, 441a
Thermodynamik technischer Gasreaktionen, **15**:439a–b
Haberlandt, Gottlieb (1854–1945), 5:623–624; 3:581b; **6**:132a; **10**:577b*; **12**:256b*; **13**:542b*–543a*
Habert, Germain, **9**:497a
Habitus corporis, **1**:400a
Hachette, Jean Nicolas Pierre (1769–1834), 6:1–3; 3:327a*; **4**:74b*; **7**:615b*; **9**:476a; **11**:76b, 384b; **13**:128a
and Monge, **3**:450b; **6**:1a,b, 2b; **7**:550a; **9**:473b, 474a, 475b
thermodynamics, **3**:80b, 82b
Traité élémentaire des machines, **9**:477a
Hackforth, Reginald, **13**:315a
Hadamard, Jacques (1865–1963), 6:3–5; 2:359a*; **8**:563b*; **10**:275b; **11**:476a
acquaintances of, **2**:144a; **14**:13b
analysis, **3**:145a, 228a; **11**:54a, 454a–b
and Duhem, **4**:230a, 232a, 233b*
number theory, **4**:475a; **5**:299a; **7**:616a; **8**:139a; **9**:173b; **11**:453a,b; **13**:561b
Hadfield, Robert Abbott (1858–1940), 6:5; 4:532a
Ḥadīth, **1**:33a; **9**:88b
Hadley, George, **3**:540a; **12**:595b
Hadley, John (1682–1744), 6:5–6; 4:592a; **6**:176a; **8**:459b
quadrant of, **1**:326b; **5**:88a, 434a; **9**:163a, 371a*
Haeckel, Ernst Heinrich Philipp August (1834–1919), 6:6–11; 3:434b
acquaintances of, **5**:52b; **7**:440a; **8**:3a,b; **9**:569b; **11**:593b; **12**:422a; **14**:2b, 41a; **15**:466a
anthropology, **6**:8b, 9a–b
biographies of, **6**:340a*
on contemporary scientists, **4**:204a; **8**:4a*; **11**:571a; **15**:468b*
critics of, **6**:436a; **7**:402a; **9**:520b; **11**:308b, 361a; **15**:142b
cytology, **9**:333a; **13**:88b
ecology, **14**:181b
embryology, **1**:420b, 476b; **6**:338a
gastraea theory, **1**:421a, 601a; **6**:9a, 339a; **7**:475a
evolution, **3**:519b; **5**:444a; **6**:8b, 613b; **12**:316a, 337b; **14**:236b, 426b; **15**:168a, 169a,b
heredity, **1**:545b; **8**:589a; **14**:235a
influence of, **2**:564a; **5**:385b; **6**:53a, 167a,b; **11**:308b; **13**:88a; **14**:233a
influences on, **6**:573a; **13**:460b
Natürliche Schöpfungsgeschichte, **11**:254b
philosophy, **7**:498b; **13**:275b; **15**:167a
and Plate, **11**:19b
students of, **4**:186b; **5**:51a; **6**:336b, 337b, 338a; **7**:401b, 402a; **8**:8a; **11**:19a, 255a, 570a,b; **12**:297b; **13**:87b; **15**:123a
Ziele und Wege der heutigen Entwicklungsgeschichte, **5**:447a
zoology, **9**:573b
Haeften, Reinhard von, **6**:550a
Haemophilus, **11**:515b
Haemosporidia, **12**:141b
Haen, Anton de, **1**:333a; **11**:158b
Haenke, Thaddäus, **8**:507b; **11**:130a
Hafenreffer, Matthias, **1**:158b, 159a; **7**:300a
Haffkine, Waldemar Mordecai Wolfe (1860–1930), 6:11–13; 7:427b
Haffkine Foundation, **6**:12b
Hafnium
atomic weight, **6**:481a
discovery of, **2**:247a; **3**:429b; **6**:366a; **9**:544a; **13**:546b
spectrum, **3**:429b–430a
Hagecius, **2**:403b, 404a, 412b, 413a; **4**:213a; **14**:470a
Hagedorn, Hans Christian, **1**:443a; **7**:503a
Hagen, Ernst, **11**:582b
Hagen, Gotthilf Heinrich Ludwig, **11**:63b
Hagen, Johann Georg, **10**:321a; **13**:12a
Hagen, Karl Gottfried, **10**:27a; **15**:171b
Hagenbach, August, **11**:476b–477a, 479b
Hagenbach-Bischoff, Jacob Eduard, **1**:425b; **6**:434b, 435b; **11**:63b
Hagenow, Charles Frederick, **3**:369b
Hagfish, **3**:610b
Hägglund, Erik, **2**:4a
Hagins the Jew, **4**:502b; **6**:272b, 273a
Hague, Arnold (1840–1917), 6:13–14; 2:441a; **4**:365b, 367a*; **7**:5a, 62b
Hahn, Hans, **9**:248b; **11**:362b; **14**:614a
Hahn-Banach theorem, **1**:428a
Hahn, Martin, **2**:561b
Hahn, Otto (1879–1968), 6:14–17; 2:257b; **5**:341b; **8**:52b; **11**:348b*
acquaintances of, **6**:481a; **7**:517b; **15**:448a
on contemporary scientists, **5**:395a*; **10**:289b*; **12**:35b*, 505a
and Meitner, **9**:260a–b, 261b–262b
radiochemistry, **2**:259b; **12**:505b
actinium, **5**:394b
fission, **2**:251b; **4**:580a; **5**:161b; **7**:155a, 158b; **13**:227a
mesothorium, **2**:260a
radiothorium, **11**:281a
Hahn, Philipp Matthäus, **8**:166a
Hahnemann, Christian Friedrich Samuel (1755–1843), 6:17–18; 12:592a
Haidinger, Wilhelm Karl (1795–1871), 6:18–20; 6:293b; **13**:143b, 144a; **14**:625a
Haidinger's brushes, **9**:201b–202a; **13**:77a
mineralogy, **8**:551a; **9**:448a,b; **11**:359b; **14**:241b(n), 327a*
Hail
microscopy, **14**:143b
theories of, **1**:586a; **14**:74a
Haime, Jules, **1**:210b
Hainzel, Johann Baptist, **2**:402a–b
Hainzel, Paul, **2**:402a–b, 403b
Hair, **13**:551b
color of, **2**:255b
distribution of, **5**:197a
formation of, **2**:72b
histology of, **6**:589b
origin of, **11**:244b
as racial characteristic, **11**:178b
structure of, **1**:319a–b
see also Fibers
Haiti
botany, **13**:506a
discovery of, **15**:89a
geography, **10**:617a
natural history, **4**:67a
Ibn Ḥajala, **1**:437a
Ḥajj (pilgrimage)
medieval, **1**:516a
al-Ḥajjāj ibn Yūsuf ibn Maṭar, **11**:249b
translation of Euclid, **1**:63a; **4**:438a–447b *passim;* **10**:5b
Ḥājjī Khalīfa, **1**:30b, 35b, 37a; **5**:612a; **7**:336a, 338b, 352b, 362b; **9**:161a, 162a, 297a; **10**:300a; **11**:226b; **13**:539a
Hajos, Georg, **6**:394a
al-Ḥākim, bi-amr-Allāh-Manṣur, **6**:189a,b; **14**:574b, 575a–b
Hakluyt, Richard (*ca.* 1552–1616), 6:20–21; 10:239b
Principal Navigations, **14**:514a
Voyages, **4**:214a
Hakluyt Society, **9**:123b
al-Ḥalabī, **12**:362a
Halban, Hans von, **7**:155a,b
Halberstädter, Ludwig, **11**:175a
Halcyonarians, **11**:113b
Haldane, John Burdon Sanderson (1892–1964), 6:21–23; 10:472b*; **11**:212a; **14**:345a
genetics, **3**:572a,b; **9**:282b
Haldane's law, **6**:22a
probability, **15**:32a
Haldane, John Scott (1860–1936), 6:23–25; 6:21b, 26a, 264b; **10**:563a*; **14**:186b*
biochemistry, **1**:453a; **14**:175a
Haldane apparatus, **6**:24a
respiration, **11**:138b–139a
Respiration, **11**:139a
Haldane, Richard Burdon (1856–1928), 6:25–26
Hale, George Ellery (1868–1938), 6:26–34; 5:199a; **10**:157a
acquaintances of, **1**:55a,b, 597b; **4**:498a; **6**:529a; **7**:271a,b*; **9**:373b, 525a; **10**:156a,b; **11**:412a, 470a, 612b; **12**:72b, 265a, 346a
on contemporary scientists, **12**:239a*
correspondence of, **4**:279b; **10**:473b*; **12**:348a–b
International Astronomical Union, **7**:238b
National Academy of Sciences, **3**:130b–131a; **6**:30b–31b
spectroheliograph, **2**:424a; **4**:69b; **6**:28a,b, 32a
stellar spectra, **4**:251a; **10**:320b
students of, **6**:528b; **14**:521a
sun
magnetic field, **8**:582b, 583a*
sunspots, **1**:56a; **11**:379a

Hale, William (1797–1870), 6:34–35
Hale, William Ellery, **6**:27a, 29a
Halemaumau crater, **7**:63a
Hales, Stephen (1677–1761), 6:35–48; 3: 582a; **5**:134b; **6**:139a; **9**:245a,b
  botany, **7**:409a; **10**:520b
  chemical apparatus, **6**:42a–43a; **8**:74b
  evaporation, **8**:256a
  Haemastaticks, **6**:37a–38a; **11**:2b
  influence of, **2**:16b; **5**:130a; **6**:24b; **8**:73b, 74a; **11**:43a, 144a,b
  medicine, **14**:320a
  physiology, **13**:46b, 122b; **14**:321a
  pneumatic chemistry, **2**:176a, 329a, 524a; **8**:585a; **11**:144b, 562b; **13**:602b
  sounding gauge, **4**:44a
  Statical Essays, **6**:37a
  Vegetable Staticks, **4**:44b; **6**:36b–37a, 41a; **8**:72b
    translations of, **2**:576b
Hales, William, **9**:159a
Halévy, Élie, **2**:544b
Halft, Franz, **5**:605a
Halichoerus, **10**:16b
Halides
  crystallography of, **1**:460b–461a
  heats of formation, **1**:579b
Halite
  crystals of, **15**:251b
Hall, Asaph (1829–1907), 6:48–50; 4:265b; **5**:480b*; **11**:115a*
Hall, Basil (Captain), **5**:78b; **6**:263a; **14**:397b
Hall, Charles Martin (1863–1914), 6:50–51
  Hall-Héroult process, **6**:50b–51a, 319a–b
Hall, Chester More, **4**:148b–149a
Hall, Edwin Herbert (1855–1938), 6:51–52; 10:481b, 604b; **12**:54b*; **13**:473b*
  Hall effect, **6**:51a–b; **7**:221a, 553b; **8**:493a, 510b; **11**:446b
Hall, Granville Stanley (1846–1924), 6:52–53; 2:207b, 367a; **4**:160b; **5**:179b, 377b; **9**:56a–b
Hall, Harry Reginald Holland, **14**:504a
Hall, Henry Edgar, **8**:483a
Hall, Sir James (1761–1832), 6:53–56; 12: 262a; **14**:132b
  acquaintances of, **2**:174b; **6**:579a; **10**:8b; **12**:523a; **14**:120b
  chemistry, **2**:180a; **6**:585b
  critics of, **7**:389a; **9**:100b
  erratic blocks, **3**:210b
  Hall effect, **11**:460b
  heat of earth, **6**:584a
Hall, James, Jr. (1811–1898), 6:56–58; 5: 214a; **11**:593b
  acquaintances of, **1**:472a; **4**:364a, 365a*; **6**:517a; **8**:571a, 576b; **9**:255a; **10**:32a; **12**:229b; **14**:312b, 315b
  experiments of, **11**:539b
  geology, **3**:553b; **13**:146b, 581b
  influences on, **4**:274b
  paleontology, **13**:467a–b
  stratigraphy, **4**:364b, 365a(n); **11**:505a
  students of, **6**:186b
Hall, Sir John, **6**:578b
Hall, Marshall (1790–1857), 6:58–61; 1: 323a; **9**:241b, 242a; **11**:159b
  On the Reflex Function of the Medulla Oblongata and Medulla Spinalis, **9**:571a,b
Hall, Robert, **6**:58b
Halle, University of
  in eighteenth century, **7**:188a
  in nineteenth century, **9**:252b; **15**:172a
  founding of, **12**:600a
  medicine at, **6**:458b; **11**:363b, 364b
  observatory, **11**:545b
  physiology at, **15**:20b
Haller, (Victor) Albrecht von (1708–1777), 6:61–67; 6:283b*
  acquaintances of, **3**:365a; **5**:380a,b; **9**: 595a; **11**:272a, 503a; **12**:119b, 563a; **14**:450a
  anatomy, **9**:153a, 323a; **10**:582b
  botany
    taxonomy, **8**:377a
  and Caldani, **3**:15b
  on contemporary science, **1**:332b; **2**:225b; **5**:56a,b; **9**:306b*
  correspondence of, **2**:204a; **5**:429a; **7**:412b
  critics of, **8**:115a; **11**:159a
  Deux mémoires sur le mouvement du sang, **12**:556a
  Elementa physiologiae, **9**:245a; **13**:461a
  embryology, **2**:287a; **8**:160a; **15**:524b
  Historia stirpium indigenarum Helvetiae inchoata, **6**:533b
  historiography, **1**:611b; **3**:41a, 99b; **7**: 540a(n), 612b; **9**:368a; **10**:426b; **12**: 600b
  Icones anatomicae, **6**:533b
  influence of, **4**:88a; **5**:216b; **6**:59b; **7**:442b
  influences on, **11**:2b
  on irritability, **5**:55b, 427a; **7**:590b, 606a; **12**:138a; **14**:321a,b
    critics of, **3**:15a–b
  and La Mettrie, **7**:605b; **9**:187b–188a
  philosophy, **8**:506a
  physiology, **8**:352b; **9**:8a–b; **13**:123a
  Primae lineae physiologiae,
    translations of, **12**:510a
  on reproduction, **12**:559b
  students of, **6**:533b; **9**:252a
Halley, Edmond (*ca.* 1656–1743), 6:67–72; 5:23a, 520b; **7**:162b, 308a; **10**:91a(n), 255a*, 303a*
  acquaintances of, **2**:387a,b; **3**:104b, 244a, 459b; **4**:41a; **5**:435b; **6**:45b(n), 361a; **9**:452b, 499b; **11**:526b; **12**:458a; **13**: 178b
  astronomical tables, **4**:14b; **15**:289b, 290a, 324b–325a, 330b
    corrections to, **4**:23b; **15**:329b
    editions of, **3**:197b; **7**:580a; **8**:459b
  astronomy, **1**:241a; **6**:331b; **10**:199a; **14**: 519a
    cometary theory, **4**:166a; **10**:64b–65a
    eclipses, **3**:431a; **13**:295b–296a
    instruments, **5**:491a; **6**:6a, 363b
    lunar theory, **8**:179a; **15**:331b
    planets, **15**:324a, 331a
    transit studies, **4**:23b; **5**:404b; **7**:373b; **10**:72a; **11**:163a
  aurora borealis, **3**:540a
  biographies of, **4**:36b
  comet of
    see Comets, Halley's
  Cometography, **8**:179a
  correspondence of, **2**:388a; **5**:434a; **9**:465a
  critics of, **8**:610b
  earth, age of, **7**:161a
  and Flamsteed, **1**:403a; **5**:24b, 25a,b; **6**: 69b, 72a; **14**:511a
  horology, **10**:91b(n)
  hydrology, **10**:522a
  influence of, **6**:130a; **12**:50a, 446a
  influences on, **13**:96a
  magnetism, **1**:521b; **6**:71b, 106b–107a
  manuscripts of, **3**:348b
  mathematics, **1**:187b; **7**:163a; **12**:446a–b; **13**:266a
  mortality table, **2**:42b; **6**:71b; **9**:454a
  and Newton, **10**:44b, 61b–65a *passim,* 78b, 83b, 84a, 87b(n)
  optics, **9**:159a, 465a
  rainbow, **10**:89b(n)
  star catalog, **1**:403b*; **6**:69b; **7**:373b, 543b
  translations by, **9**:302a*
Hallier, Ernst Hans (1831–1904), 6:72–73; 3:338b; **7**:401b; **8**:296b; **10**:562b*; **12**:176a*
Hallmann, Eduard, **4**:200a, 204b
Hällström, Gustaf Gabriel, **6**:245a,b
Hallucination, **2**:448a
Hallwachs, Wilhelm Ludwig Franz (1859–1922), 6:73–74; 11:460b
  Hallwachs effect, **6**:74a
Halm, Jacob Karl Ernst (1866–1944), 6:74–75; 6:352b
Halma, Nicola (Abbé), **4**:17a
Halochromism, **10**:578b
Halogens, **1**:416b
  determination of, **14**:448a
  see also specific halogens
Haloragaceae, **2**:519b
Halos, **10**:59b; **14**:566b, 567a
Halosantho
  see Spermaceti
Halphen, Georges-Henri (1844–1889), 6: 75–76
  editions of works, **10**:593b
  geometry, **8**:324a; **9**:503a*; **14**:356b
  theorem of, **12**:228a
Halpin, James Garfield, **6**:136a
Halske, Johann Georg, **7**:56a; **12**:425a
Halsted, George Bruce (1853–1922), 6:76–77; 4:37a*, 458a*, 618b; **5**:314a*; **8**:434b*; **12**:57a*; **13**:218a
Halsted, William Stewart (1852–1922), 6: 77–78; 3:517a; **8**:447a; **14**:249b
  Halsted's suture, **9**:56a
Haly Abbas
  see al-Majūsī, Abu'l-Ḥasan ʿAlī ibn ʿAbbās
Ham, Johan, **8**:127b
al-Hamadānī, ʿAbd al-Jabbār, **15**:250a,b
Hamann, Carl August, **2**:171b
Hamberg, Axel (1863–1933), 6:78–79
Hamberger, Georg Erhard, **1**:66b; **6**:64a; **8**:327a,b; **9**:153b; **11**:432a,b; **12**:185a
Hamburg
  Museum of Natural History, **9**:431b
Hamburg, University of
  mathematics at, **2**:191b
Hamburger, Hartog Jakob, **3**:334a; **13**:579b
Hamburger, Johannes, **9**:126b; **11**:256a
Hamburger mathematische Einzelschriften
  founding of, **2**:191b
al-Hamdānī, Abū Muḥammad al-Ḥasan ibn Aḥmad ibn Yaʿqūb (also known as Ibn al-Ḥāʾik, Ibn Dhi 'l-Dumayna, *or* Ibn Abī 'l-Dumayna) (*ca.* 893–*ca.* 951), 6:79–80
Ḥamdānid culture, **4**:524b; **15**:250b
Hamer, Frances Mary, **9**:403b
Ḥāmid ibn Khiḍr al-Khujandi
  see al-Khujandi
Hamilton, Alexander, **6**:521b; **7**:89a, 319a; **13**:46a,b
Hamilton, Gilbert Van Tassel, **14**:550b
Hamilton, William (1730–1803), 6:83–85; 3:291a,b; **11**:303b; **12**:120a,b; **14**:60b, 259b
Hamilton, William (1745–1813), **11**:218a
Hamilton, William (1788–1856), 6:80–83; 3:88b, 448a; **4**:36a; **9**:199a, 206a, 210a
  epistemology, **9**:213b
  logic, **2**:295a; **6**:81b–83b
Hamilton, William Rowan (1805–1865), 6: 85–93; 1:207b
  acquaintances of, **4**:341a; **6**:325a; **7**:383b; **12**:86b; **13**:237a

algebra, **2**:467b; **6**:95b; **14**:303b; **15**:195b
analysis, **10**:250a; **14**:317b
celestial mechanics, **6**:88b–89b
on contemporary scientists, **1**:239b; **7**:103b*; **8**:592a
influence of, **2**:594a; **3**:166a; **8**:40b; **10**:480a
influences on, **15**:485b
geometry, **7**:523b; **11**:46a; **15**:197b
Hamiltonian *H*, **4**:579a,b; **6**:89a,b
Hamilton-Jacobi equation, **6**:89b; **12**:220a
Hamilton's principle, **6**:89b, 347b, 473a; **7**:453b; **12**:250a
generalization of, **9**:616b
least action, principle of, **6**:245a; **8**:602b
Lectures on Quaternions, **1**:240a
mathematical physics, **6**:87a, 88b–90b, 347b; **7**:54b; **8**:384b; **9**:223b; **12**:250b; **13**:130b
number theory, **1**:238b; **2**:271a; **6**:522a
optics, **6**:85b, 87a–88b; **8**:426a; **9**:73a, 226a; **12**:220a
quaternion theory, **4**:35b; **6**:90a–92b; **10**:138a; **15**:196a
Hamis ibn Zabara, **14**:592b
Hammer, Heinrich
see Germanus, Henricus Martellus
Hammers
hydraulic, **2**:444b
steam, **9**:615b
Hammett, Louis Plack, **6**:404b
Hammond, James, **8**:617a; **13**:218a
Hampson, William (*ca.* 1854–1926), 6:93–94; 8:365b; **10**:206b; **13**:454a
Hamster, **10**:16b
Hamy, Ernest-Théodore (Jules Théodore Ernest), **8**:45a; **11**:235a
Hamy, Maurice Théodore Adolphe (1861–1936), 6:94–95; 13:37b*
Ḥamza al-Iṣfahānī, **7**:362b
Hanbury, Daniel, **6**:492a
Hanckwitz, Ambrose Godfrey, **9**:105b
Hand
anatomy of, **3**:38a, 40b; **4**:520a
Handbooks
see specific sciences
Handbuch der biologischen Arbeitsmethoden, **3**:259b
Handbuch der Physik, **5**:332b*–333a*
Handbuch der physiologischen Botanik
editors of, **12**:58a
Handbuch für Astrophysik, **8**:539b
Handedness
see Left- and right-handedness
Hanford (Wash.), **3**:371b
Hankel, Hermann (1839–1873), 6:95–96; 4:205b
Hankel functions, **6**:95b
historiography, **4**:116a, 117b; **6**:416a; **7**:259a
mathematics, **12**:217a
Theorie der komplexen Zahlensysteme, **15**:195a
Hankel, Wilhelm Gottlieb (1814–1899), 6:96–97; 14:329b
Hanley, Sylvanus, **5**:67b
Han-lin Academy, **8**:314a; **12**:378b
Hann, Julius Ferdinand von (1839–1921), 6:97–99; 4:592b; **9**:107b, 149b
Handbuch der Klimatologie, **14**:52b
Hannequin, Arthur, **12**:609b
Hannover
mining, **13**:455b
politics, **8**:150b
triangulation of, **5**:303a–b
Hanoi
medical school, **14**:551b
Hansen, Emil Christian (1842–1909), 6:99–101; 7:100b
Hansen foundation, **6**:101a
Hansen, Gerhard Henrik Armauer (1841–1912), 6:101–103; 1:600b; **10**:18a
Hansen's bacillus, **6**:101b
Hansen, Peter Andreas (1795–1874), 6:103–104; 3:302a; **7**:51b, 480b
astronomy, **1**:257b, 339b; **2**:516b; **6**:399b; **9**:460b; **10**:481a*
Hansen's problem, **12**:501a
lunar tables, **6**:403a*; **10**:34a, 207b; **11**:115a
solar tables, **11**:114a,b
Hansen, William Webster (1909–1949), 6:104–105
Hansky, Aleksey Pavlovich (1870–1908), 6:105–106
Hansteen, Christopher (1784–1873), 6:106–107; 1:13a,b, 14b, 16b; **6**:277b, 474a; **12**:49b
geodesy, **13**:111a
Hantzsch, Arthur Rudolf (1857–1935), 6:107–109; 4:485a; **7**:623a; **9**:354a; **10**:578a; **14**:265a
diazo compounds, **1**:426b
physical chemistry, **15**:460b, 461a
stereochemistry, **1**:340b; **3**:300a; **9**:356b, 402b; **14**:265a–b
Hanzlík, Stanislav, **6**:98a
Hapelius, Nicolaus Niger (pseud. for Raphael Eglinus Iconius), **1**:550a
Happiness
measurement of, **9**:189a; **12**:178b; **15**:465b
Hapten, **7**:624b
Haptophore, **4**:298b–299a
Ibn ʿAbd al-Ḥaqq, ʿAbd al-Muʾmin, **14**:548a
al-Harawī, Aḥmad ibn Abī Saʿīd, **7**:335a; **9**:22a
Harborworks, **2**:326b
Harcourt, A. G. Vernon (1834–1919), 6:109–110; 4:130a, 411b–412a; **12**:418b
Harcourt, William Vernon, **6**:110a; **12**:25a; **13**:77b
Harden, Arthur (1865–1940), 6:110–112; 8:83a; **11**:538a
fermentation, **2**:562a; **4**:358a, 485a,b; **9**:359b
Hardening, metallurgical
precipitation, **9**:312b
theories of, **9**:312b
Harder, Johann Jacob, **10**:568a
Harding, Carl Ludwig (1765–1834), 6:112; 6:325a; **10**:199a; **12**:226a,b
Hardness
measurement of, **6**:342a, 534b; **11**:329a; **14**:132b
of metals, **2**:467a; **11**:329a
Hardy, Claude (*ca.* 1598–1678), 6:112–113; 4:46a; **6**:271b; **11**:486b; **14**:20a
Hardy, Godfrey Harold (1877–1947), 6:113–114; 2:238a, 490b; **5**:414a; **6**:451b; **7**:616b; **12**:10a, 426a; **13**:424a; **14**:344a
analysis, **4**:561b; **13**:263b–264a
Hardy-Weinberg law, **2**:58b; **14**:230b
historiography, **13**:318a(n)
and Littlewood, **7**:351b
number theory, **11**:453b; **14**:95a
and Ramanujan, **11**:256b, 268b, 269a
Hardy, William Bate (1864–1934), 15:201–202; 1:613b; **5**:81a
Hare, Robert (1781–1858), 6:114–115; 4:182a, 456b; **12**:432b, 433a
Hare, William, **7**:415a
Hare, William Kendrick, **1**:450a
Hare and hound problem, **4**:606a, 612b
Haridāsa
commentaries of, **2**:117a
Haridatta I (*fl.* A.D. 683), 6:115; 15:591b, 601a
Grahacāranibandhana, **15**:595b–596a
Mahāmārganibandhana, **15**:596b
Haridatta II (*fl.* 1638), 6:115; 4:100a
Jagadbhūṣaṇa, **15**:589a
Harihara
commentaries of, **2**:118a
Gaṇitacūḍāmaṇi, **15**:585a
Harikarṇa of Hisāranagara, **9**:42a
Harikṛṣṇa, **12**:598a
Harington, Charles Robert, **15**:11a, 107a, 258a
Hariot, Thomas
see Harriot, Thomas
Harker, Alfred (1859–1939), 6:116; 5:336b; **6**:475b
Harker, David, **4**:522a; **5**:186a
Harkhebi, **15**:725b–726b
Harkins, William Draper (1873–1951), 6:117–119; 1:321b; **9**:110a; **10**:107a*
Harkness, William (1837–1903), 6:119; 4:265b
Harlan, Richard (1796–1843), 6:119–121; 9:540a; **10**:438b
"Harmonia" theory, **1**:260b
Harmonic analysis (math.)
see Analysis, harmonic
Harmonics
ancient Greek, **1**:282a–283a; **4**:391b, 430b–431a, 462b
ancient Roman, **3**:14b; **10**:112b–113a
medieval Arabic, **4**:526a
Renaissance, **1**:606b; **2**:401a; **14**:29b
in art, **4**:259a
seventeenth-century, **1**:484a; **7**:301a; **9**:319b
eighteenth-century, **12**:128a–b, 477b; **13**:571a
nineteenth-century, **4**:265a*; **6**:245a–b
twentieth-century, **5**:456b
difference tones, **6**:245a–b
numerical, **1**:606b; **5**:249a–b; **12**:374b, 376a, 384b
spherical, **4**:126a; **6**:451b; **10**:26b
summation tones, **6**:245b
see also Music, theory of
Harmony, preestablished, **8**:157b–158a, 158b, 515a; **15**:53a
Harmony, universal, **2**:163b; **7**:290a, 300b–302a, 307b; **9**:1a; **13**:326a; **15**:85b
Harnack, Adolf von, **8**:181b; **11**:8b, 14a
Harnack, Carl Gustav Axel, **8**:111a
Harpagornis, **5**:611b
Harper, Robert Almer (1862–1946), 6:121–122
Harper, Roland McMillan (1878–1966), 6:122–123
Harper, Stanley Hugh, **4**:91a
Harpestraeng, Henrik (*d.* 1244), 6:123–124
Harpsichord
ocular, **3**:115a
al-Ḥarrānī, ʿAlī ibn ʿĪsā
Risāla fī ibṭāl ṣināʿat aḥkām al-nujūm, **1**:38a
Ḥarrānian science, **9**:160a; **11**:325a; **12**:447a,b, 448a; **13**:288b, 292b; **15**:249a
philosophy, **1**:33a–b; **7**:42b; **15**:182a
theology, **1**:33b–34a, 507b
Harries, Carl Dietrich, **13**:1b, 2a
Harriman Alaska Expedition, **9**:313b
Harrington, Bernard James, **1**:50b
Harriot (or Hariot), Thomas (*ca.* 1560–1621), 6:124–129; 1:541b; **4**:60b(n); **5**:397a, 527b; **14**:461a, 514b(n)
Artis analyticae praxis, **14**:147b, 150a

Harriot (or Hariot), Thomas (cont.)
astronomy, **2**:97b; **6**:361b; **12**:151b
mathematics, **3**:254a; **5**:409a,b
mechanics, **5**:529a; **6**:379a
navigation, **5**:593a; **9**:311a
optics, **6**:197a
Harris, Isaac Faust, **9**:284b; **10**:243b
Harris, John (*ca.* 1666–1719), 6:129–130
Lexicon chemicum, **14**:502a
Lexicon Technicum, **4**:86a; **6**:46b(n)
Harrison, James, **6**:130a,b
Harrison, John (1693–1776), 6:130–131; 6:70a; **7**:183b*; **12**:413b
chronometer, **4**:481b; **5**:491b; **9**:163a; **11**:495b
translations of works, **10**:572a*
Harrison, Ross Granville (1870–1959), 6: 131–135; 2:502a; **3**:389b, 390b; **8**: 447a; **9**:56b; **10**:276a
philosophy of biology, **14**:434b
zoology, **3**:91a; **10**:104b
Harrison, William, **6**:130b
Hart, Edwin Bret (1874–1953), 6:135–136; 1:357a; **4**:357b; **7**:209a*
Harteck, Paul, **12**:34a
Hartig, Georg Ludwig, **6**:136b
Hartig, Theodor (1805–1880), 6:136–137; 5:425a
critics of, **12**:100a
Harting, Pieter (1812–1885), 6:137–138; 1:136a; **3**:425a; **6**:535a,b, 613a*
Hartley, David (1705–1757), 6:138–140; 3: 598b; **12**:571a
editions of works, **11**:143a
Observations on Man, His Frame, His Duty, and His Expectations, **11**:140a
Hartlib, Samuel (*d.* 1662), 6:140–142; 1: 134a, 159b; **2**:163a; **7**:194a; **14**:368a, 376a(n), 377b(n)
acquaintances of, **2**:377a; **3**:360a; **5**: 606a,b, 607a; **10**:200b, 495a, 564b; **11**:314a; **12**:616b
Hartline, Haldan Keffer, **14**:201a
Hartman, Carl Gottfried, **13**:97a
Hartman, Philip, **14**:454a
Hartmann, Carl Friedrich Alexander (1796–1863), 6:142–143
Hartmann, Georg (1489–1564), 6:143–145
cosmology, **11**:351b
Hartmann, Johannes (1568–1631), 6:145–146; 3:471b; **8**:310a,b; **12**:307b
Hartmann, Johannes Franz (1865–1936), 6: 146–147; 1:352a; **12**:455a; **14**:56b
Hartree, Douglas Rayner (1897–1958), 6: 147–148; 15:63a
Hartsoeker (or Hartsoecker), Nicolaas (1656–1725), 6:148–149; 5:62a, 354a
microscope, **7**:110b
Hartt, Charles Frederick, **1**:397a; **7**:169a
Hartung, George, **8**:572a,b, 573b
Hartung, Nikolaj, **3**:529b, 532a
Hartwig, (Carl) Ernst (Albrecht) (1851–1923), 6:149–150
Hārūn al-Rashīd, **4**:438b, 445b; **7**:222b, 358b; **12**:432a
"Harvard Photometry," **1**:398a
Harvard University
in seventeenth century, **1**:134b
in eighteenth century, **9**:142a, 607b; **14**: 452a–b
in nineteenth century, **13**:350b
in twentieth century, **6**:260a; **12**:54a
astronomy at, **12**:349b
biology at, **14**:533a–b
botany at, **5**:512a, 513b; **10**:164a
Bussey Institution, **3**:121b, 123a,b; **4**: 272a; **14**:291b
chemistry at, **3**:397b–398a; **5**:393b–394a; **11**:418a
Cruft Laboratory, **10**:605a
geology at, **14**:315b–316a
Gray Herbarium, **14**:192a
history of science at, **12**:110a–b, 112b, 113b
Lawrence Scientific School, **6**:517a; **8**: 123a; **10**:599a
mathematics at, **4**:546b; **5**:520a; **10**:244b
Medical School
bacteriology at, **2**:367b
biochemistry at, **5**:53a
Laboratory of Surgical Research, **3**: 518a
physiology at, **2**:366b–367a; **15**:72a–b
mineral collection, **3**:397b–398a
Museum of Comparative Zoology, **1**:74a; **8**:264b; **9**:100a; **10**:272b; **11**:113a
observatories, **2**:284b, 285b; **3**:50b, 595b; **4**:180b; **9**:194b; **10**:479b, 599b–601b, 602a; **12**:348a–349b; **14**:448b–449a
southern hemisphere studies, **1**:397b, 398a; **12**:348b–349a
Peabody Museum of American Archaeology and Ethnology, **11**:219a; **14**:533b
physics at, **13**:473a
science at, **4**:546b; **5**:394a, 519b–520a
Harvey, Basil Coleman Hyatt, **1**:613b
Harvey, William (1578–1657), 6:150–162; 1:279a; **4**:388a, 585a; **5**:233b, 237b
acquaintances of, **11**:467a; **13**:474a
anatomy, **14**:287a
on chyle, **1**:316b; **10**:477b
circulation of blood, **1**:98a, 244b, 392a, 483a; **2**:337a; **5**:288b; **6**:155a–158a, 386b, 484a; **9**:323b; **10**:477a; **12**:324b; **13**:170a
De circulatione sanguinis, **2**:522b; **6**:157a
correspondence of, **12**:333a
critics of, **11**:467a
defenders of, **4**:377a; **11**:511a; **13**:222b
and Descartes, **4**:53b, 54b, 65a
editions of works, **4**:377b; **15**:4a
embryology, **1**:386b; **2**:595a; **3**:343a; **4**: 510a,b; **5**:427a; **6**:65a, 159a–161b; **7**:612b; **8**:580a; **13**:171a,b
entomology, **13**:172b
Exercitationes de generatione animalium, **1**:293b; **6**:157a–159b *passim,* 387a; **8**:127b
and Highmore, **6**:387a
influence of, **1**:482a, 484a; **2**:306a, 309a; **5**:415b; **9**:146a; **12**:600b; **14**:405a, 407a
library of, **4**:377b
Lumleian Lectures, **1**:523b; **6**:151a
manuscripts of, **6**:151b, 159b
methodology, **1**:110a; **6**:448b
De motu cordis (Exercitatio anatomica de motu cordis et sanguinis in animalibus), **3**:356b; **4**:62b, 509a; **5**:47b; **6**: 151b, 153a,b, 155a–157b; **8**:53b; **10**:490b; **12**:333b
neurophysiology, **5**:426b; **6**:158a–159a
physiology, **4**:64a; **13**:32a
precursors of, **11**:390b–391b
Second Disquisition to Jean Riolan, **4**:54b
Harvey, William Henry (1811–1866), 6:162–163; 2:19a; **6**:490a, 495a
Harvey Cushing Society, **3**:517b
Harz Mountains
geology, **5**:155b, 444b, 547b–548a; **8**: 511b; **13**:283a; **14**:60b
paleontology, **4**:401a; **11**:501b–502a
stratigraphy, **2**:109a
al-Ḥasan ibn Muḥammad al-Wazzān al-Zayyātī al-Gharnāṭī
see Leo the African
al-Ḥasan ibn Mūsā ibn Shākir
see Banū Mūsā
al-Ḥasan ibn al-Nakad, **7**:39b
al-Ḥasan al-Samarqandī, **7**:336a
Ibn Ḥasday
see Yūsuf ibn Ḥasday, Abū Jaᶜfar
Ḥasdāy ibn Shaprūṭ, **6**:186a; **7**:187a
Haselden, Thomas, **1**:475b*
Hasenöhrl, Friedrich (1874–1915), 6:163–164; 12:217b–218a, 530a
Haskins, Joshua, **4**:44a
Hasolle, James (pseudonym)
see Ashmole, Elias
Hassall, Albert, **13**:63a
Hassall, Arthur Hill, **3**:424b
al-Ḥaṣṣār, **1**:437b
Hasse, Helmut, **1**:306b; **2**:191b; **6**:286b, 394a; **10**:139a; **13**:23b; **14**:611a
Hasse, Karl, **7**:420b; **8**:514a; **11**:570a, 571b
Hasselbalch, Karl Albert, **7**:502a
Henderson-Hasselbalch equation, **6**: 260b–261a
Hasselberg, Bernhard, **12**:42b; **13**:115a
Hasselquist, Fredrik, **1**:59a, 124b; **8**:379b
Hassenfratz, Jean-Henri (1755–1827), 6: 164–165; **7**:14a(n); **8**:81a; **9**:472b; **10**:545b; **12**:66a, 67a
chemistry, **1**:64b; **7**:13a
mining, **4**:347b, 348a
Hassenstein, Bruno, **6**:215b
Hassler, Ferdinand Rudolph (1770–1843), 6:165–166; 1:364a,b; **7**:19b
Hastings, Charles Sheldon, **2**:424a; **7**:270b
Hastings, Edwin George, **6**:135b
Hastings Castle, **3**:606b
Hata, Sahachiro, **4**:298b, 300b; **7**:392b
Hatai, Shinkishi, **4**:161a
Hatch Act, **1**:325a–b
Hatchett, Charles (1765–1847), 6:166–167; 2:420a; **7**:69b; **10**:242a
Hatschek, Berthold (1854–1941), 6:167; 5: 109a,b; **7**:402a; **11**:174b
Hatschek, Emil, **5**:385b
Hattendorff, Karl, **14**:202b
Hauer, Franz von, **13**:143b
Haug, Gustave Emile (1861–1927), 6:168–169; 5:221b; **7**:252a; **8**:544a
Haug's rule, **6**:168a
Hauksbee, Francis (*ca.* 1666–1713), 6:169–175; 4:45a, 214b, 352b; **9**:141b
capillarity, **6**:40a; **15**:359b
electricity, **2**:324b; **4**:166b, 215b; **5**: 516a,b; **10**:92a(n), 146a; **11**:68a
Physico-Mechanical Experiments, **6**: 45b(n), 47a(n)
French edition of, **4**:70b, 518a(n)
Hauksbee, Francis (1688–1763), 6:175–176; 9:141b, 479b
Hausdorff, Felix (1868–1942), 6:176–177; 8:324a; **13**:548b
algebra, **11**:55b
analysis, **14**:573a
axiomatization, **12**:16b
Grundzüge der Mengenlehre, **12**:196a
Hausdorff-Banach-Tarski paradox, **14**: 89b
Hausen, Christian August, **2**:324b; **11**:433a
Hausen, Hans, **12**:272b
Häusermann, Emil, **2**:586a
Häusler, Hans F., **8**:452b
Hausser, Wilhelm, **7**:470a
Haussmann, K. G. F., **12**:187a
Hautefeuille, Paul Gabriel (1836–1902), 6: 177–178; 13:468a
Haute-Marne, **1**:572a
Haüy, René-Just (1743–1822), 6:178–183; 2:493b, 494a; **3**:412b, 445b, 523a; **5**:356b, 358a, 530b*; **8**:288b; **12**:67a
acquaintances of, **3**:291b; **5**:202b, 355b, 430a; **12**:523a; **13**:466b; **14**:239b; **15**:113b, 114b

on Aepinus, **1**:68a
capillarity, **15**:360a
classification system, **2**:94b; **3**:232b
critics of, **9**:448b; **11**:169a, 170a, 465a, 522a–b; **14**:490b
crystallography, **2**:6b, 355a–b, 431a, 482a; **3**:313a; **8**:56b; **9**:59a; **10**:357b; **11**:523a; **14**:240a, 490a
editions of works, **15**:114b
and education, **15**:342b
Iceland spar, **1**:557a; **9**:73a
influence of, **2**:356b; **4**:152b, 247a; **5**:295a; **12**:433a, 520a; **14**:258a
influences on, **15**:361a
metric system, **8**:85a
mineralogy, **11**:167b
notation, **4**:66a; **14**:241a
pyroelectricity, **2**:5a
students of, **10**:497b; **12**:317b; **13**:581a
Traité de minéralogie
translations of, **7**:254a; **14**:240a
Traité élémentaire de physique, **15**:361b
Haüynite, **5**:430a
Haverfield, John, **1**:88a
Havers, Clopton (*ca.* 1655–1702), 6:183–184
Hawaiian Islands
botany, **10**:164a
discovery of, **3**:397a
geology, **3**:484a; **7**:62b; **8**:578a
volcanoes, **3**:552a–b; **4**:266a; **7**:63a–b; **8**:286a
Hawksbee
see Hauksbee
Haworth, Adrian Hardy (1768–1833), 6:184
Haworth, Walter Norman (1883–1950), 6:184–186; **8**:276a; **15**:257b
Ibn Ḥawqal, Abu'l-Qāsim Muḥammad (*fl.* second half of tenth century), 6:186; **9**:36a
geography, **1**:29a; **9**:88b
Kitāb ṣūrat al-arḍ, **7**:8b
Hay, David Ramsay, **9**:199b, 200b, 201a
Hay
infusions of
bacilli in, **3**:339b; **7**:424a
sterilization of, **3**:339b
Hay fever, **2**:336a
Hayashi Dōshun, **8**:394b
Hayata (Hay'a), **15**:629a
Hayden, Ferdinand Vandiveer (1829–1887), 6:186–188; **4**:265b; **6**:57b; **9**:255a,b, 313b; **10**:273a
Hayden survey, **3**:438b; **10**:32a
Ibn Haydūr, **1**:437b
Hayek, Thadeus, **2**:293a
Hayem, Georges, **4**:277a
Hayes, Charles Willard, **2**:500a; **4**:7b
Hayford, John Fillmore (1868–1925), 6:188–189; **2**:373a; **6**:241a; **14**:503a
Ibn Ḥaylān, Yūḥannā, **4**:523b, 524a
Ibn al-Haytham, Abū ᶜAlī al-Ḥasan ibn al-Ḥasan, called al-Baṣrī, al-Miṣrī (965-*ca.* 1040), 6:189–210; **4**:117b; **7**:82b, 336a
algebra, **12**:92a
astronomy, **6**:197a–199b; **7**:334b; **8**:281a; **14**:577b
commentaries on, **7**:212b–213a
geometry, **1**:224b, 225b; **4**:441a,b; **6**:200a–203b; **7**:83a, 133b, 329a; **8**:429b; **9**:22a; **11**:240b, 402a
influence of, **4**:8b; **10**:474a; **11**:248b; **12**:91a, 358b, 501a; **14**:457b
influences on, **13**:291a
methodology, **6**:190b–191a, 195a, 198a, 199a, 203a–b
optics, **1**:81a, 170a, 191b, 379b, 380a, 475a; **2**:193b; **4**:93a; **5**:525a,b; **6**:190b–197a; **7**:212b–218b *passim;* **8**:280b; **10**:475a; **15**:264b
problem of Alhazen, **6**:194a, 200a–201b
statics, **7**:341a, 342b
translations of works, **13**:400b
On the Configuration of the World (Liber de mundo et coelo), **6**:197b–198a; **15**:475b
De crepusculis liber (spurious), **15**:180a
Discourse on Light, **6**:191b, 192a, 195b
The Halo and the Rainbow, **6**:192a, 195a
Kitāb al-Manāẓir, **11**:203a
On the Measurement of Parabolic Solids, **1**:445b
On the Measurement of the Sphere, **1**:445b
Optics (or *Perspectiva* or *De aspectibus*), **1**:227b; **6**:125b, 190b–197b *passim,* 200a–201b; **7**:175a
commentaries on, **4**:157b; **11**:250b
editions of, **11**:468a
translations of, **14**:458a–461a *passim;* **15**:187a, 188a
transmission of, **6**:196b–197a
On Paraboloidal Burning Mirrors, **14**:458a; **15**:118a
translations of, **15**:178a, 187a
On Spherical Burning Mirrors, **7**:213b
Ibn Ḥayyān, **1**:413b
Hazama, Shigetomi, **1**:310b–314b *passim;* **7**:18b
Hazard, Erskine, **5**:518a
Hazelius, Artur, **6**:477a
Ibn Ḥazm (Abū Muḥammad ᶜAlī ibn Aḥmad), **9**:39a; **14**:584a; **15**:263b
al-Hazmirī, **1**:437a
Head, Henry, **5**:80b, 283a; **8**:7b
Head
balancing of, **8**:125b
bones, **5**:419a; **14**:456b
embryogeny, **2**:383b
evolution of, **12**:337a
geometry of, **5**:113a
muscles, **1**:204a; **2**:376b, 531a; **3**:343a, 355b; **4**:520a; **12**:100b; **14**:450a
nerves, **4**:520b; **5**:419a; **9**:9a; **14**:143b, 406a
vascular system, **4**:520b
see also Brain, Craniology, Scalp, Skull
Headache, **9**:422a; **14**:408a
Health
and climate, **2**:619b; **4**:171a–b; **8**:554b; **9**:306b; **12**:289b; **13**:541a–b; **14**:464a
definition of
ancient Greek, **1**:104a; **4**:106a–b
medieval Arabic, **15**:242a
nineteenth-century, **2**:33a
and environment, **5**:452b; **6**:421a, 422a, 427a–b, 459b
and fresh air, **5**:137b; **6**:44b, 421b; **10**:558a,b; **12**:561a; **14**:638a
psychological influence of, **3**:1b–2a
and vision, **9**:75a
see also Hygiene, Public health
Hearing
Renaissance study, **3**:99b
seventeenth-century study, **1**:483b; **4**:268a
eighteenth-century study, **3**:437b; **13**:567a
nineteenth-century study, **2**:366b; **6**:287b; **8**:597a; **9**:571b; **12**:130a; **14**:492b
twentieth-century study, **5**:65b; **8**:101b–102a; **13**:104b
accommodation, **8**:597a
audibility limits, **14**:492b
binaural effect, **13**:104b
corpora quadrigemina, **2**:596a
in fishes, **3**:38a; **5**:355a; **6**:287b, 566b
frequency limits, **12**:130a
in insects, **6**:287b
in reptiles, **5**:355a; **7**:516b
and sight, reinforcement by, **8**:101b–102a
theories of, **8**:102b; **11**:267a; **14**:130b*
resonance, **3**:425a; **6**:245b–246a; **8**:101b
see also Acoustics, Deafness, Ear, Otology
Heart
anatomy
medieval Arabic discussion, **9**:603b
Renaissance discussion, **1**:619b; **4**:199a, 487a; **8**:202b, 204a; **9**:166a; **10**:598b; **14**:10a–b
seventeenth-century discussion, **4**:510b; **5**:288b; **7**:612b; **13**:32a–b
eighteenth-century discussion, **6**:63a–b; **15**:524b
ductus arteriosus, **2**:336b
fetal, **8**:171b
foramen ovale, **2**:336b; **6**:65b; **8**:53b
malformations, **7**:278b
pacemaker, **5**:281b; **7**:278b
septum, **1**:293a, 523b, 619b; **3**:356a–b; **5**:235a; **8**:54a; **10**:490b–491a; **13**:569a; **14**:10a, 11b
three-chamber theory, **1**:264b, 266b–267a; **4**:285a
beat
ancient Greek discussion, **1**:267a
medieval discussion, **1**:199b
Renaissance discussion, **3**:356b, 357a; **8**:204a; **14**:4b, 5a
seventeenth-century discussion, **6**:153b–154b, 157b–158a; **8**:526a; **9**:243b, 318b; **10**:568a
eighteenth-century discussion, **2**:2b; **4**:589b; **6**:63a–b; **9**:480a; **13**:122b, 231b
mid nineteenth-century discussion, **3**:220a; **8**:541b; **11**:368a
late nineteenth-century discussion, **5**:81b–83a, 279b–281b; **7**:504b–505a; **8**:14b; **11**:518a
early twentieth-century discussion, **5**:150a; **6**:526b; **12**:618a–b
blood pressure, influence of, **2**:366b; **9**:101b
chemicals, effects of, **11**:462b–463a
electrotherapy, **6**:94a*
in embryo, **3**:245b
and emotions, **9**:547a
heart block, **5**:281a,b
induced, **5**:82a
inhibition of, **5**:82a–b, 280a; **14**:200a–b; **15**:469a
irregularities of, **5**:150a; **8**:294b; **15**:101a–b
laws of, **12**:618a–b
myogenic theory, **5**:82a–83a, 280a–b, 281a–b; **8**:14b
neurogenic theory, **5**:82a–b, 280a–b; **7**:505a
pacemaker, **5**:281b; **7**:278b
refractory period, **7**:504b
sequence, **5**:280b–281a
shape changes, **4**:589b
in sleep, **14**:322a
and temperature, **9**:142b
and vagus nerve, **2**:111a; **5**:82a–b, 280a, 463a; **6**:225a; **8**:453a
cardiac cycle, **1**:536b; **6**:63a
see also Diastole, Systole
catheterization of, **2**:25a, 30b; **3**:219b, 220a
as center of life, **1**:262a, 266a–b, 324b; **11**:29b, 127b, 128a, 391a
opposition to, **13**:32a–b

Heart (cont.)
contraction of, **2**:237b; **12**:165a
all-or-none law, **7**:504b
extrasystoles, **14**:408a
salts, influence of, **6**:526b*
tonicity, **5**:280a
Treppe, **2**:366b
diseases of, **1**:360a–b; **3**:427a; **7**:556b, 614a; **8**:352b; **11**:559b; **14**:408a
abscesses, **14**:638a
auriculoventricular block, **5**:366b
decompensation, **1**:391b
diagnosis of, **1**:333a; **8**:295a; **12**:450b
infarction, **9**:95a
mitral stenosis, **3**:518a; **14**:26a
pericarditis, **1**:439a
rheumatic, **14**:253a
see also specific diseases by name
electrical phenomena, **1**:536b; **5**:282a
action current, **4**:333b–334b
resting heart, **4**:372a
velocity of conduction, **4**:372a
embryogeny, **9**:253a; **11**:255b; **12**:45a
enlargement of, **1**:438b
evolution of, **12**:45a
failure of, **9**:64b
treatment of, **1**:360b
fibrillation, **9**:506a
auricular, **4**:335a*; **5**:150a; **8**:294b; **15**:101a–b
ventricular, **5**:150a
force of, **7**:275a
histology, **8**:9a, 526a; **11**:216b–217a, 368a, 374a; **14**:233b
innervation of, **1**:537a; **5**:281b–282a; **10**:431b; **12**:138a, 165a
auriculoventricular bundle, **4**:398a
septum interatriale, **2**:111a
metabolism, **1**:453a
neurophysiology, **2**:111a; **5**:81b–83a; **8**: 294b–295a; **9**:142b; **12**:612a; **15**:101a
automaticity, **3**:68b, 245b; **6**:63a–b; **8**: 536a
conduction of excitation, **3**:68b–69a; **4**:371b–372a, 398a
isolated heart method, **7**:504b; **8**:451a, 453a–455a
suspension method, **5**:280a
types of activity, **4**:372a; **10**:432b
nutrition of, **11**:125b
pharmacology, **15**:100a–101b
position of, **1**:619b
protective reflex, **2**:111a
pumping rate, **2**:601a; **6**:154b–157a; **8**: 526a
role of
ancient Greek theories, **1**:262a, 266a–b; **4**:384b; **6**:318a
ancient Roman theories, **5**:234a–b
Renaissance theories, **3**:356b; **8**:201a, 204a; **12**:335b; **15**:80b
seventeenth-century theories, **6**:153a–157a; **7**:612b; **8**:525a–526b; **13**:223b
sounds, **2**:601a; **3**:219b, 220a; **12**:450b; **13**:300b
surgery of, **7**:502b–503a; **8**:552b
tumors of, **1**:612a
valves
lesions of, **3**:518a
role of, **5**:234a, 235a; **11**:238b
work done by, **2**:37a
see also Blood pressure; Cardiology; Cardiovascular system; Myocardium; Perfusion, of heart; Pericardium; Pulse; specific animals
Heart, **8**:296a
founding of, **8**:295a
Heart-lung preparations, **1**:534a
Heat, **6**:278b
absorption of
color effects on, **1**:364a; **3**:116b
atomic, **15**:439a–440b
and atomic weight, **4**:80b
and temperature, **15**:442b
of combustion, **2**:69a,b; **4**:241a, 554b; **6**:243b; **7**:462b; **8**:545b; **13**:351b; **15**:316a
conduction of
see Conduction, thermal
conservation of, **3**:156a; **5**:137a, 387b; **13**:381b; **15**:313a
constant summation law, **4**:241a; **6**:354a–b
cycle, **3**:75b, 80b–81b
of dilution, **4**:228b; **11**:418a, 577a
of dissociation, **1**:300a; **6**:519b–520a; **8**: 23b
distribution of, **13**:130b; **15**:360b, 363a
and electricity, **2**:123b; **3**:238b; **9**:205b; **10**:135a, 192a; **13**:374b–377a; **14**:75b, 354b; **15**:312b
and energy, **14**:360b
equivalence of transformations, **3**:305b–306a, 308b; **7**:181b–182a
"form" of, **1**:374b–375a
of formation, **3**:401a; **4**:78b; **7**:203b; **8**: 80a, 83b, 545b
free, **3**:304a, 316a; **15**:313a–b, 314b, 378a
of fusion, **2**:178a–b
of hydration, **1**:579b; **8**:48a*
latent, **3**:158a; **5**:216a; **14**:74a, 185b; **15**: 363a
in clouds, **4**:410b
concept of, **3**:304a; **4**:28b; **13**:378b, 379a,b, 381b, 383b; **14**:75b, 197b
discovery of, **2**:177b–179b; **3**:156a; **14**:353b–354b
and molecular weight, **13**:471b
and light, **5**:167a; **6**:586a; **9**:264b–265a; **13**:238b
mechanical equivalent of, **4**:40a; **5**:87a; **9**:238b; **11**:393a
and absolute temperature, **3**:308a–b
calculation of, **3**:20a, 505b; **6**:244a; **7**: 182a; **9**:236b–237a, 239b; **10**:518b; **11**:353b, 578b; **12**:288b; **14**:38a; **15**: 85b, 86a
discovery of, **6**:431b–432a
mechanical production of, **2**:379a; **7**:182a
molecular, **3**:304a; **4**:80b; **15**:440b
as motion, **1**:374b, 567b; **2**:180a, 378b, 379b; **9**:132b–133a; **11**:293a, 432b; **12**:287b, 288a; **14**:185a–b
motive power of, **3**:81a–82a; **5**:97b; **13**: 375a, 377b, 379a–382b
and temperature, **3**:82a; **5**:387b
of neutralization, **1**:299a; **8**:545b; **11**:418a, 577a; **13**:359a
of oxidation, **1**:579b
and physical state, **3**:11b–12a, 304a–b, 305b–306a
propagation of, **4**:222b
quantity of, **2**:178a; **4**:240a; **6**:292a; **13**: 382a; **15**:94b, 314a–315a, 486b
definition of, **4**:229b
radiation of
see Radiation, thermal
of reaction, **1**:299a; **4**:225b; **5**:353a; **6**: 354b; **8**:289b; **15**:314a,b, 437a–441b, 457a–b
reflection of, **7**:599a; **10**:603a
refraction of, **10**:603a
sensation of, **7**:599a
solar
Renaissance discussion, **1**:607b
mid nineteenth-century discussion, **8**: 1a, 2a; **9**:238b–239a; **14**:185b
late nineteenth-century discussion, **14**:53a
twentieth-century discussion, **8**:13a*
measurement of, **11**:111b
meteoric hypothesis, **9**:238b–239a
origin of, **7**:386b*
use of, **6**:326b
of solution, **4**:228b, 554b
specific, **5**:215b, 216a;
of air, **3**:156b; **9**:237a; **15**:86a, 358a
and atomic weight, **1**:348b; **2**:588b; **3**: 48b; **4**:240b–241a; **13**:411a
of compounds, **3**:156a–b; **7**:463b; **10**: 27a–b
concept of, **5**:313b
Debye equation, **3**:619a; **4**:414a
determination of, **2**:499a, 588b; **3**:20a; **4**:80a, 240b, 413b–414a; **8**:80a; **11**: 353a, 418a; **14**:354a–b; **15**:94b, 313b–314b
of electrolytes, **11**:577a
of fluids, **1**:128b–129a; **8**:545b
of gases, **1**:348a, 616b; **2**:169b–170a, 266b; **3**:82a, 317a*; **4**:241a; **5**:319a; **7**:161a, 526a; **8**:118a; **9**:218a, 220a, 223a; **10**:329b; **11**:150a, 292a, 530a; **14**:171a, 185a; **15**:95a, 363a, 449a
of glass, **2**:589b
at low temperatures, **15**:439a–448a
magnetic, **14**:245b
of minerals, **7**:161a; **10**:27a
Nernst-Lindemann theory, **8**:368a; **15**:442a
of platinum, **2**:589b
and pressure, **15**:378b
quantum nature of, **3**:367a; **4**:316a–b; **15**:40b
ratio of, **14**:185a
of solids, **10**:545b–546a
of solutions, **9**:110a
and spectrum, **2**:170a
of steam, **3**:20a
tables of, **9**:5b
and temperature, **2**:178b–179b; **4**:316a–b; **6**:271a; **7**:246b; **10**:27b; **11**:578b, 583a; **12**:437b, 438a; **13**:411a; **14**:39a; **15**:314a, 315a, 378b
theories of, **6**:164a; **7**:84b; **12**:218b; **13**:364a; **15**:487a
use of term, **9**:5b
of water, **2**:589b; **3**:20a; **6**:271a; **10**:27b; **11**:393a, 578b
vs. temperature, **2**:177b–178a; **3**:304a, 308b; **9**:132b–133a; **14**:155b
theories of
ancient Greek, **13**:92b, 330a–b
medieval, **1**:101a; **2**:149a; **5**:551a
Renaissance, **9**:132b–133a; **13**:279a
seventeenth-century, **9**:528a; **10**:59b
early eighteenth-century, **3**:345a; **4**: 352b; **8**:468b–469a; **9**:33b; **10**:79a; **11**:432b
late eighteenth-century, **2**:179b–180a; **5**:137b; **6**:585b; **7**:228b–229a; **8**:79b–80a; **9**:5b; **10**:602b–603a; **11**:134b–135a, 146b; **12**:286b; **14**:75b, 354a; **15**:94b–95b, 312b–316a
early nineteenth-century, **3**:81a–83b *passim,* 602a; **5**:97b; **6**:291b–292b; **8**:261b; **12**:287b, 288a
mid nineteenth-century, **3**:303b–310a; **4**:616a; **5**:387b; **8**:1b; **9**:19a, 236a, 264b–265a, 445a; **10**:27a–b, 28b; **11**:292a–293b, 454a; **12**:288a–b; **13**: 378a–384b

late nineteenth-century, **2**:356a; **9**:60a*; **11**:10a
twentieth-century, **2**:499a–b
caloric, **1**:348a, 349a; **2**:75b, 81a; **3**:81a–83b *passim,* 303b–306a *passim,* 541a, 599a,b; **5**:137b, 167a, 387b; **6**:178b, 284b, 354a; **8**:79b–80a, 81b, 82b; **9**: 471b, 477a; **10**:4a, 603a; **12**:223b, 287a; **15**:363a, 377a–379a, 486b
chemical, **10**:546a
corpuscular, **8**:261b
dynamical, **3**:304b–305a; **6**:244a
fluid, **2**:180a; **3**:376b; **5**:137b; **11**:134b–135a; **15**:313a
kinetic, **1**:563b; **3**:83b, 599a–b; **4**:314b; **5**:387b; **6**:291b–292b; **8**:468b–469a, 469b, 471b, 493a, 620b; **13**:11a–b; **15**:313a
mathematical, **3**:178b; **4**:126a, 222a,b; **8**:382a, 384b; **10**:247b–249a, 594b; **11**:409a; **12**:218b; **15**:486a–487a, 488b
mechanical, 3:304b, 358b*; **4**:258a, 313b; **5**:461b; **6**:164a; **7**:181b–182a; **9**:445b; **11**:8a, 353b; **13**:378b–384b, 615a; **15**:86b
particulate, **12**:289a
vibration, **3**:156a–b; **7**:181a; **9**:264b–265a; **13**:351a
wave, **14**:184b–185a
transfer of, **9**:462b; **10**:162a–163a; **15**: 360b
of vaporization, **2**:178b, 179a; **3**:287a; **6**:520a; **8**:545b
as vital quality, **1**:261b–262a
and work, **3**:303b, 304a–306a; **5**:86b–87a, 387b; **12**:288b; **13**:378a; **14**:123b; **15**:85b
conversion coefficient, **3**:83b
see also Calorimeter; Cooling; Earth, heat of; Insulation, thermal; Radiation, thermal; Temperature; Therapy; Thermodynamics; Thermometer
Heat, animal, **2**:226b; **3**:604b; **4**:241a; **6**:242a; **12**:287a; **15**:314a
brain and, **2**:483b–484a
production of, **1**:610b; **2**:30b–31a, 483b–484a; **4**:285b; **6**:119b
in heart, **4**:377a; **5**:235a,b; **6**:153a,b
in lungs, **2**:70b
and surface area, **11**:428a
theories of
Renaissance, **10**:598b
seventeenth-century, **4**:377a; **6**:157b–158a; **9**:244b; **14**:407b, 408b
eighteenth-century, **2**:3a; **8**:37b, 83b; **10**:80a; **15**:95a–b
nineteenth-century, **2**:483b–484a; **6**: 243b; **8**:133b, 348b; **14**:63b
chemical, **2**:483b; **4**:616a; **9**:237b–238b
see also Body temperature
Heat, innate, **5**:235a,b; **6**:153a,b
Heat balance, **6**:432a
Heat engine, **4**:501a*
analysis of, **7**:446a; **11**:293b; **13**:73b
Cagniard, **3**:8b, 9a
Carnot, **3**:81a–82b; **13**:378a
efficiency of, **11**:292b
reversibility, **13**:379b, 380b–381a, 381b–382b
theory of, **8**:212a; **13**:381b–382b
Heath, Josiah, **9**:591b
Heath, Thomas Little (1861–1940), 6:210–211; 12:108b; **13**:318a(n)
Aristarchus of Samos, **1**:249a
historiography, **1**:367b; **6**:95b; **13**:334a
of astronomy, **1**:247a, 248a; **6**:617a; **15**:204b(n)
of mathematics, **1**:245b; **4**:111a, 392b, 438b; **6**:315a, 408a, 411b, 412b; **9**: 275a; **10**:180b–181a, 294b, 299b, 344a; **13**:296b, 303b, 305b(n), 306b(n), 317b(n), 399b
A History of Greek Mathematics, **1**:249a; **4**:392a, 466a; **5**:347b(n); **8**:189b(n); **9**:192b(n), 276a(n)
mathematics, **1**:245a; **4**:104b, 113a, 419b; **9**:274b
Heating, artificial, **6**:383b; **10**:558a
physiology of, **7**:503a
steam, **13**:352a
theory of, **5**:97b
Heatley, Norman, **5**:42b
Heatstroke, **6**:24a
Heavens
mutability of, **2**:407a; **4**:168b; **5**:240b, 242a
Heaviside, Oliver (1850–1925), 6:211–212; 5:16a; **6**:350a*; **13**:364b
calculus of operators, **2**:491a; **15**:386b
critics of, **6**:92a; **13**:237a
electrical engineering, **7**:288b; **8**:502b
electrodynamics, **5**:15a,b; **6**:346b; **8**:494a; **9**:212a; **11**:9a
Kennelly-Heaviside layer, **1**:195b–196a; **6**:212a; **7**:288b
vector analysis, **9**:215b
Heberden, William, **5**:137b
Heberer, Gerhard
Die Evolution der Organismen, **4**:101b
Hébert, Edmond, **1**:472a, 491a; **4**:67b; **5**: 475b; **6**:168a; **9**:579a; **11**:375a; **14**: 626a
Hebra, Ferdinand von, **2**:129b; **12**:294b, 295a,b
Hebrew science, **4**:502a–b
mathematics, **1**:22b–23a, 31a–b; **7**:360a; **8**:279b–280a, 548b–550a
natural sciences, **2**:221b; **8**:280b
philosophy, **3**:469b–470a; **9**:27b–31b
physical sciences, **1**:383a; **8**:280a–281b
Hebrew University (Jerusalem), **4**:313a, 545a; **14**:248a
Hebrides (islands)
geology, **4**:549a; **6**:116a; **14**:132a
Hecataeus of Miletus (*fl. ca.* 500 B.C.), 6: 212–213; 4:389b, 467b
Hecht, Daniel Friedrich (1777–1833), 6: 213–214
Hecke, Erich (1887–1947), 6:214–215; 2: 191b; **6**:393b; **7**:398b, 508a; **11**:256b, 362a
Heckmann, Otto, **6**:532a
Hectocotyli, **1**:264a; **14**:57b
Hector, Sir James, **5**:611b
Hedgehog, **6**:535b
orbicular muscle, **3**:342b
Hedin, Sven Anders (1865–1952), 6:215–218; 9:104a; **10**:149a*
Hedin Foundation, **6**:217a
Hedonism, **4**:381b, 466a
Hedrick, Henry Benjamin, **2**:516b
Hedwig, Johann (1730–1799), 6:218–220; 1:46a; **5**:217a
Hée, Louis van, **3**:266b; **8**:314b
Heegaard, Poul, **4**:9a; **7**:312a; **8**:324b
Heer, Oswald (1809–1883), 6:220–222; 3: 530a; **5**:73b; **7**:43b, 437b; **8**:543a; **12**:255b
Heers, Henry van, **6**:254b
Hefner-Alteneck, Friedrich Franz von (1845–1904), 6:223
Hefner candle, **6**:223a
Hegel, Georg Wilhelm Friedrich
acquaintances of, **9**:345b; **12**:153b
critics of, **2**:545a; **4**:616b; **12**:198b; **15**: 409b, 411b–412a, 413a
epistemology, **9**:29a
influence of, **3**:312b; **4**:200a; **6**:564b; **7**: 68a, 189b, 505b; **8**:184a; **9**:456b, 572a, 578b; **12**:281a; **15**:135b
influences on, **5**:254b; **11**:493a
logic, **15**:142a–b, 424a
natural philosophy, **2**:552b–553a, 556a; **3**:335a; **5**:446a; **15**:415b
students of, **5**:256b; **7**:50b; **15**:526a
Hegeler, Edward Carl, **15**:79a
Hegelianism, **3**:110b; **5**:439a; **6**:26a; **7**:482b, 484b–485a; **9**:278a, 600a; **10**:184b; **12**:154a–b, 607a,b; **15**:132b, 133a, 139b, 143a, 404a–b, 405a
critics of, **12**:175a
Hegemonikon, **1**:324b
Heger, Paul, **15**:469a
Heiberg, Johan Ludvig, **3**:25a, 150a; **4**:416a, 424a, 431b, 437b–438a, 439b, 463b, 488b; **6**:71a, 310b, 311a, 412b; **9**: 275b(n); **10**:299b; **12**:579b; **13**:252b, 255a
Heidelberg, University of
in Middle Ages, **9**:136a
in nineteenth century, **7**:380a; **14**:170b
Anatomy Institute, **15**:166a
botany at, **6**:466b
Chemical Institute, **5**:430a–b, 431b
physics at, **8**:181b
physiology at, **7**:519b
Heidelberg man, **3**:606b
Heidelberger, Michael, **1**:343a*; **2**:300b; **13**:420a; **15**:252b
Heidenhain, Martin (1864–1949), 6:223–224; 8:15a
Heidenhain's stain, **6**:224a
Heidenhain, Rudolf Peter Heinrich (1834–1897), 6:224–227; 4:296a; **5**:279b; **6**:223b; **11**:217a,b*; **15**:22a*
acquaintances of, **10**:431b, 432b; **11**:574a; **12**:617b; **13**:452a
critics of, **15**:102a–b
physiology, **1**:452b, 544a; **10**:19a; **15**: 101b–102a
students of, **9**:284a; **10**:17b; **11**:571b; **14**:228a; **15**:20b
Heilbronner, Johann Christoph, **4**:612b; **9**: 501a
Heim, Albert (1849–1937), 6:227–228; 8: 543b; **9**:103b, 241a*; **11**:210a; **13**: 146a
Heim, Albert Arnold (1882–1965), 6:228–230; 6:228a
Heine, Heinrich Eduard (1821–1881), 6:230; 2:275b; **5**:97b; **7**:51a, 506b
Heine-Borel theorem, **2**:303b; **6**:230b; **13**:599b
Heinroth, Oskar, **14**:315a
Heintz, Wilhelm, **14**:454b
Heis, Eduard, **7**:400a
Heisenberg, Werner, **2**:245b; **5**:160b; **7**:617a; **9**:37b, 38a; **10**:213b; **12**:529b, 530b–531a, 615b; **15**:42a
matrix mechanics, **4**:318a; **5**:103a; **12**: 220b–221a
quantum theory, **2**:248a–249b; **7**:492a; **8**:52a,b, 474a,b, 475a; **9**:377a; **10**: 423a; **12**:221b; **15**:41b–42a, 42b, 43a
symbols, **4**:281b
uncertainty principle, **3**:369b; **8**:477a; **13**:275b; **15**:49a
Heister, Lorenz (1683–1758), 6:231–232; 3:464b; **8**:377a
Compendium anatomicum, **12**:303a*
De incrementis anatomie, **1**:318b

Heitler, Walter Heinrich, **8**:473a–474b; **12**: 419a, 530a; **15**:43a
Heitler, Werner, **15**:26b, 27a
Heitz, Emile, **2**:456a; **13**:136a
Hekat (unit of measure), **15**:688b, 694a, 694b–696a, 699a, 699b–700b
Hektoen, Ludvig (1863–1951), 6:232–233; **7**:171b*; **11**:177a*, 442b; **14**:252b
Helias, **1**:274b
Helicoid, **9**:343b
Helicon, **2**:549b; **9**:269a; **11**:28a
Heliconema, **12**:96b
Heliconidae, **1**:503a
Helicoprion, **7**:252a–b
Helicopter, **2**:439a; **7**:247a; **8**:213b
Hélie, Félix
  Traité de balistique expérimentale, **6**:545a
Heliocentric theory
  see Cosmology, heliocentric theory
Heliodorus of Larissa, **1**:481b*; **15**:550b, 563b
Heliograph, **8**:581a; **10**:119a
Heliometer, **4**:351a, 369b; **5**:144a, 404a,b, 405a; **6**:103b, 149b; **7**:146a; **10**:542b; **11**:378a; **12**:448b; **13**:482a
  invention of, **2**:343b
Heliometry, **1**:339b; **2**:98a, 101a
Helioscope, **6**:361a,b
Heliostat, **3**:430b; **5**:87a, 271b, 303a
Heliotrope (astron. instrum.), **11**:377a
  invention of, **5**:303a–b
Heliotropism, **8**:445b
Heliozoans
  cytology, **6**:337a
Helium, **2**:429a; **3**:478b
  determination of, **10**:288b, 289a
  discovery of, **3**:479b; **8**:441a; **9**:292b; **11**:279b
  fountain pressure, **8**:478a
  helium II, **7**:617b; **8**:478a–479a, 481a–482a
  ionized, **7**:268a
  ionization energy, **6**:614b–615a
  isotopes, **8**:478b
  liquid, **8**:482a–483a; **10**:206b; **15**:444b
    manufacture of, **4**:80a; **7**:221a, 272a–b; **12**:437b, 439a
    superfluidity, **8**:477a, 478a–479a, 481a–483a
  molecular ion, **9**:38a
  packing factor, **4**:281b
  phase transitions, **12**:438b–439a
  properties of, **13**:453b–454a
  quantum theory, **7**:491b
  in radioactive decay, **2**:260a; **3**:499b, 500a; **4**:80a; **11**:280b; **12**:28b, 30a,b, 504b
  solidification of, **7**:272a
  sources of, **3**:437a; **4**:80a; **11**:279b, 280b
  spectrum, **2**:244a; **4**:513b; **5**:102b, 126b; **7**:27a; **8**:503b, 579a*; **10**:346a, 347a–348a, 601a; **11**:279b, 446a, 612a
Helix, **8**:3b; **10**:16b
Helix (curve), **4**:259b
Hell (or Höll), Maximilian (1720–1792), 6: 233–235; **2**:599b; **7**:580b–581a; **9**: 231b, 326a; **10**:34b; **13**:39b
Helland-Hansen, Björn, **4**:344a,b; **15**:431b
Hellebore, **1**:75a, 212b; **4**:120a
Hellenism, **13**:308a
Heller, G., **8**:476a, 477b
Hellinger, Ernst (1883–1950), 6:235–236; **5**:530a*; **13**:428a
  Hellinger integral, **6**:235b
Hellner, Erwin Emil, **6**:303b
Hellot, Jean (1685–1766), 6:236–237; **8**: 621a,b; **11**:330b; **12**:65b
  phosphorus, **9**:105b
Hellriegel, Hermann (1831–1895), 6:237–238
Helly, Eduard, **1**:428a
Helm, Georg, **2**:266b; **5**:391b; **11**:9a; **15**: 463a, 464a
Helmersen, Grigory Petrovich (1803–1885), 6:238–239; **8**:187b; **10**:168b
Helmert, Friedrich Robert (1843–1917), 6: 239–241; **5**:8a
Helmholtz, August Ferdinand Julius, **6**: 241b, 242b, 243a
Helmholtz, Hermann von (1821–1894), 6: 241–253; **3**:437b; **4**:203b; **6**:350a*; **7**:449a, 519b; **12**:426a; **15**:22a*
  acoustics, **3**:9a; **6**:242b, 245a–246a; **7**: 470a; **8**:399a–b; **12**:128b; **13**:101b
  acquaintances of, **1**:7b; **4**:200a,b; **5**:196a; **7**:460a, 507b; **8**:540b; **9**:568a; **10**: 481b; **11**:8b, 578a,b; **12**:270a, 425a; **13**:243a; **14**:329b, 360a, 527a; **15**:20b
  conservation principle, **4**:557a; **6**:242a, 243b–244a, 250a–b
  on contemporary scientists, **14**:330b*; **15**:435b–436a
  critics of, **14**:338b; **15**:194a
  electricity, **11**:446a; **13**:11b
  electrodynamics, **4**:231a–232a, 536a; **6**: 243a, 250a–252b, 344a–345a, 348a–b; **8**:188a, 492a, 493b; **9**:155a, 205b–206a, 211a; **11**:9a; **13**:82b; **14**:205b, 207b
  energetics, **5**:462a; **6**:243b–245a; **9**:239b; **12**:288b; **15**:85b, 436b
    solar heat, **9**:239a
  epistemology, **6**:242b, 243a, 247b–249a; **11**:356a
  geometry, **6**:391a, 522a; **11**:455a; **14**:283b
    space problem, **8**:326b
  Gibbs-Helmholtz equation, **6**:244b; **7**: 65b; **13**:578b; **15**:436b–441b
  Handbuch der physiologischen Optik, **6**: 242b–243a, 247a–b, 248a
    editions of, **5**:590b; **7**:458a
  on hearing, **3**:425a; **6**:245b–246a; **8**:101b
  Helmholtz theorem, **6**:249b
  and Hertz, **6**:341a–342a, 342b–343a, 348a,b
  hydrodynamics, **2**:168a, 466a; **6**:249a–250a; **13**:385a
  influence of, **1**:536b; **2**:382b; **3**:32a; **4**: 225b; **7**:380a, 526a; **9**:101b; **10**:579b; **11**:264b, 298a, 612a; **13**:237a, 614b; **14**:185b
  influences on, **2**:18a; **12**:156b
  Die Lehre von den Tonempfindungen als physiologische Grundlage für die Theorie der Musik, **6**:242b, 246a
  mechanics, **6**:347b, 348a; **7**:54b
  meteorology, **6**:97b, 98a; **9**:108a; **11**:558a
  music, **11**:9a
  ophthalmology, **11**:215a
  ophthalmoscope, **2**:376b, 531a; **6**:246a
  optics, **8**:492a–b; **11**:460b, 582a
    critics of, **6**:299b–300a
    physiological, **4**:615b; **5**:445b; **6**:246a–249a; **8**:551b; **9**:200a,b; **14**:565a; **15**:194a
  philosophy of science, **7**:234b; **8**:515b; **12**:243b, 609b
  physiology, **4**:614b, 616b; **6**:241b–243b, 245a–249a; **11**:426b; **12**:241a; **14**: 492b
  psychophysics, **4**:558a
  students of, **1**:132b; **2**:265b, 272a; **4**: 378a,b; **5**:458b, 462b; **6**:28b, 51a, 52b, 286a, 470a, 476a; **7**:267b, 270b, 444a, 445a, 457b, 477b, 504a; **8**:180a, 551a; **9**:355a, 372b, 563b; **11**:8a, 149b, 213a, 551a, 611a; **12**:212b, 366a; **13**:79b, 416b; **14**:337b–338a; **15**:432b
  thermochemistry, **2**:69a; **5**:390b; **15**: 457a–b
  thermodynamics, **2**:261a; **3**:83a
  "Die Thermodynamik chemischer Vorgänge," **4**:228b
  Young-Helmholtz theory, **6**:246b–247b; **7**:458a
Helmholtz, Robert von, **11**:150a
Helmholtz-Gesellschaft, **12**:615a
Helminthologia
  founding of, **12**:453b
Helminthology
  seventeenth-century, **11**:342a
  eighteenth-century, **9**:575b; **12**:515b
  nineteenth-century, **1**:602b–603a; **2**:127b, 322a; **4**:198a, 236a; **5**:502b–503a; **8**:270b; **11**:592b; **12**:192b–193b
  twentieth-century, **5**:453a–b; **12**:453a–b; **13**:63a
  see also Worms
Helmont, Franciscus Mercurius, **6**:253b
Helmont, Johannes (Joan) Baptista van (1579–1644), 6:253–259; **1**:550a; **13**:559b
  acquaintances of, **9**:316b, 318a; **13**:30b
  botany, **6**:38b
  chemistry, **7**:194b; **8**:86a(n), 620b; **12**:79b
  critics of, **7**:459a; **13**:490b
  iatrochemistry, **2**:237b; **3**:596b; **4**:388a; **13**:234b
  influence of, **5**:420a, 427a; **7**:150b; **8**:130b; **9**:327a; **12**:616a–b; **14**:209b, 210a
  influences on, **10**:311a
  physiology, **4**:210b*; **10**:307b; **13**:235a; **14**:405b
  translations of works, **3**:208b
  vitalism, **1**:479a; **5**:426b
  willow tree experiment, **1**:549b; **6**:256b; **8**:71b; **14**:157b
Helmstedt, University of, **10**:573a
Heloderma
  venom, **8**:447b
Helsinki, University of
  botany at, **10**:165b, 166a
Helvetian (stratig. stage), **9**:240b
Helvetica physica acta, **5**:598a
Helvétius, Anne-Catherine de Ligneville, **3**:1a, 383b; **10**:611b; **14**:68b
Helvétius, Claude Adrien, **7**:581b; **9**:189a
Hemacytometer, **10**:205a
Hemagglutination
  see Agglutination, of blood
Hemarthrosis, **3**:204b*
Hematin, **5**:431b; **6**:505a
Hematology
  seventeenth-century, **9**:64b–65a
  eighteenth-century, **4**:352b; **6**:367a–b; **10**:326a
  nineteenth-century, **4**:296a–b; **9**:571b; **11**:133a; **12**:232a; **13**:77b
  twentieth-century, **3**:336a; **6**:233a; **9**: 417a, 545a–b; **15**:38a–b
  see also Blood
Hematoporphyrin, **6**:505b; **9**:96a
Hematosis, **9**:63b
Hematoxylin, **4**:395a; **8**:329b; **10**:517b
Hematozoa, **2**:528a; **5**:566a; **8**:296b
Hematuria, **2**:127b; **14**:253a
Hemel en dampkring
  editors of, **13**:12a
Hemeralopia, **10**:621b
Hemerology, **15**:644b, 729b, 733b
Hemianesthesia
  hysterical, **3**:205b
Hemianopia, **14**:492b–493a

Hemin, **9**:96a; **13**:274a; **14**:175a–b; **15**:157b
synthesis of, **15**:157b
Hemiplegia, **5**:207b; **7**:48b
Hemiptera, **2**:533b; **15**:123a
cytology, **9**:496a–497a
Hemisphere, surface of
area, **15**:697b
center of gravity, **3**:182b
Hemitonia
apoplectic, **1**:580a
Hemlock
poisoning by, **14**:256a
Hemmingsen, Axel Marius, **7**:503a
Hemocatheresis, **2**:339b
Hemochromogen, **6**:505a
Hemocuprein, **7**:273b
Hemocyanin, **5**:149a
molecular weight, **13**:161a–162a
radiation effects, **13**:163b
Hemodynamics, **4**:615b; **6**:63b; **9**:9a, 10b, 101b
Hemodynamometer, **11**:63a
Hemoglobin, **9**:96a; **11**:135b
aggregation of molecules, **1**:454a
as buffer, **13**:574b
and carbon monoxide, **6**:24a; **12**:546b
chemistry of, **6**:505a–b; **7**:273b
degradation products, **10**:22b
differentiation of, **7**:624b
discovery of, **9**:348b
genetic origin, **13**:136b
iron content, **2**:586a
molecular weight, **13**:2a–b, 160b–161a
and oxygen, **1**:453b; **6**:24b, 505a–b
physical chemistry of, **1**:454a
spectrum, **6**:505a; **12**:236b
structure, **4**:522a
synthesis of by organism, **2**:586a; **4**:357b
Hemoglobinemia, **9**:95a
Hemoglobinometer, **10**:205a
Hemoglobinuria, **9**:95a
paroxysmal, **4**:302b*; **9**:545b*
diagnosis of, **7**:623b–624a
Hemolysis, **2**:300b; **4**:299a; **5**:40a
eighteenth-century theory of, **2**:226a
by metals, **9**:295b
from radiation, **12**:88b
spleen and, **1**:439a–b
Hemophilia, **6**:526b; **9**:253a; **14**:584b
cause of, **9**:417a
genetics of, **2**:584b; **6**:22a; **9**:521a
Hemoptysis, **4**:386a
Hemopyrrole, **9**:96a; **10**:22b
Hemorrhage, **10**:612b
lacrimal, **6**:183b; **13**:262b
management of, **3**:518b; **8**:404b; **10**:316a
pathological, **9**:546a*
Hemostasis, **9**:12a; **10**:316a, 620b
Hemotherapy, **11**:429a
Hemp plant
see *Cannabis*
Hemprich, Wilhelm, **4**:288b–289a
Hemsalech, Gustave Adolphe, **5**:497a
Hemsterhuis, Siboldus, **11**:587a
Henbane, **1**:485a
Hench, Philip Showalter, **3**:353a; **15**:258b
Henckel, Elias Heinrich von, **3**:473b
Henckel, Johann Friedrich (1678–1744), 6:259–260; 8:468a; **9**:105a; **10**:26a
arsenic, **1**:529b; **2**:421b
Pyritologia, **9**:105b
works translated, **6**:468b
Hencky, Heinrich, **6**:534b
Henderson, Lawrence Joseph (1878–1942), 6:260–262; 3:335b; **11**:418a,b; **12**:110a–b; **13**:574b
Henderson-Hasselbalch equation, **6**:260b–261a
Henderson, Thomas (1798–1844), 6:263–264; 7:543b; **12**:498b
parallax, **2**:101b; **3**:404b; **8**:613b
Henderson, Velyian, **8**:452b
Henderson, Yandell (1873–1944), 6:264–265; 6:25b*
Hendriks, Eileen Mary Lind, **5**:39a
Henerus, Renatus, **4**:198b
Henfrey, Arthur (1819–1859), 6:265–267; 2:521b; **6**:597a*; **9**:602a*
and Hofmeister, **6**:466a–b
Micrographic Dictionary, **12**:259a
Henke, Karl, **7**:516b
Henking, Hermann (1858–1942), 6:267–268; 8:587b
Henle, Friedrich Gustav Jacob (1809–1885), 6:268–270; 9:572b; **11**:368a; **14**:127a*
acquaintances of, **13**:273b, 557a,b
contagion, **1**:493b; **7**:422b; **11**:38a; **12**:203a
Handbuch der systematischen Anatomie, **13**:274a
Henle's loop, **6**:270a
histology, **2**:165a; **7**:438a
students of, **2**:124a; **7**:420b, 421a, 437b, 519a; **9**:332a, 456b; **14**:126a, 233a,b
Henley, William, **3**:153b
Henna, **3**:347a*
Henne, Albert Leon, **13**:177a
Henneberg, Wilhelm, **4**:5a
Henneguy, Félix (Luis), **1**:418a
Hennell, Henry, **2**:66b; **8**:333b; **10**:510b
Henneman, Nicholaas, **13**:238b
Henrichsen, Sophus (1845–1928), 6:270–271
Henrion, Denis or Didier (*ca.* 1580–*ca.* 1632), 6:271–272; 11:3b
Henriques, Valdemar, **9**:284b
Henry II of France, **4**:584b–585a,b; **9**:11b; **10**:152b, 315b; **11**:286b; **14**:6a; **15**:45a
Henry III of France, **2**:337a, 540b, 541b; **10**:315b; **14**:18a
Henry IV of France, **1**:571a, 592a; **2**:542a; **6**:445a; **7**:536b; **8**:53b; **11**:288b, 533a; **12**:316b; **13**:507a; **14**:18a, 22a
Henry VIII of England, **5**:348a,b; **8**:360b; **13**:491b; **14**:47b, 507a
Henry (Aristippus), **1**:270b
Henry Bate of Malines (1246–*ca.* 1310), 6:272–275; 4:502b; **7**:133a*; **9**:435a
astronomy, **15**:35b
Magistralis compositio Astrolabii anno 1274 scripta, **1**:512b; **9**:435a
Henry of Denmark
see Harpestraeng, Henrik
Henry of Harclay, **2**:394a
Henry of Hesse (1325–1397), 6:275–276; 1:84a; **7**:119b
astronomy, **15**:35b
influences on, **1**:228b; **5**:361a; **14**:461a
optics, **10**:475b
Henry of Langenstein
see Henry of Hesse
Henry of Mondeville (*ca.* 1260-*ca.* 1320), 6:276–277; 1:107a; **2**:314b, 315a, 546b; **5**:348a; **10**:307a
Henry of Sinrenberg, **12**:62a
Henry, Aimé, **10**:12b
Henry, Charles, **4**:567b; **13**:252b, 253b, 254a
Henry, Joseph (1797–1878), 6:277–281; 6:546a
acquaintances of, **1**:364b; **4**:179a, 265b, 364b; **5**:1*1b; **8**:1a,b, 2a,b; **10**:33b, 480a; **13**:218a, 433a; **14**:52b
astronomy, **12**:266b
on contemporary scientists, **11**:22b*; **12**:132b*
correspondence of, **6**:57b
electricity, **11**:604a
electromagnetism, **4**:534b; **12**:26a; **13**:126a–b
influence of, **5**:479b; **7**:445a; **8**:122a; **9**:230b
meteorology, **4**:410a; **9**:197a
and National Academy of Sciences, **1**:365a
at Smithsonian Institution, **1**:405a, 406a; **9**:255b; **12**:495a
Henry, Paul Pierre (1848–1905), 6:281–282; 9:551b
Henry, Prosper Mathieu (1849–1903), 6:281–282; 9:551b
Henry, Thomas (1734–1816), 6:283–284; 3:539a
Henry, William (1774–1836), 6:284–286; 2:420b; **3**:545b, 546b; **12**:236a
Elements of Experimental Chemistry
translation of, **5**:297a
Epitome of Chemistry
editions of, **12**:433a
Henry's law (of partial pressures), **2**:588a; **3**:541b; **5**:493a; **6**:284b, 285a; **15**:434a
Henry, William Charles, **3**:547a; **6**:285b; **11**:173b
Henry Draper Catalogue, **3**:50a; **6**:350b; **10**:600b; **12**:348b
Henry Draper Fund, **10**:600a
Henry Mountains
geology, **5**:395b
Henschel, August Wilhelm Eduard Theodor, **5**:218a; **11**:215b
Henschel, Karl Anton, **2**:586b
Hensel, Kurt (1861–1941), 6:286–287; 7:383a*, 508a, 524a*; **11**:362b; **13**:23b
acquaintances of, **1**:370a; **4**:617a; **7**:507a; **14**:222a
algebra, **5**:108a; **7**:530b
Hensen, (Christian Andreas) Victor (1835–1924), 6:287–288; 2:29b; **6**:246a
Hensen's cells, **6**:287b
Hensen's duct, **6**:287b
Hensen's knot, **6**:287b
Henslow, John Stevens (1796–1861), 6:288–289; 1:358b, 359a; **2**:18b; **3**:565b
Henzi, Samuel, **9**:187b
Heparin
discovery of, **6**:526b
Hepatica, **12**:594b
fertilization in, **4**:546a
Hepatitis
immunization, **3**:336a
from overexertion, **11**:100a
Hepatoxylon, **2**:322a
Hepburn, John, **1**:442b
Hephaestio of Thebes, **8**:191b; **10**:547b, 548a; **15**:125a,b, 721b, 724a
Heptagon
construction of, **1**:224b–225b; **6**:207a*; **9**:598b; **11**:240a; **12**:432a
inscribed, **4**:551a; **6**:208a*; **14**:21a,b
stellated, **2**:360b
Heptane
compounds of, **11**:88b
Heracleides, **1**:213a
Heraclides Ponticus (*b. ca.* 390; *d.* after 339), 15:202–205; 1:315a; **11**:29a, 188a; **13**:305a; **14**:534b
cosmology, **1**:246b, 247a, 248b; **3**:15a; **6**:382a; **9**:141a; **11**:224b; **14**:162a
Heraclitus of Ephesus (*fl. ca.* 500 B.C.), 6:289–291; 6:315b; **10**:324b; **11**:25b; **14**:537a, 606a
Herapath, John (1790–1868), 6:291–293; 14:184a
Herapath, William, **6**:293a
Herapath, William Bird (1820–1868), 6:293–294; 6:291b

Herapathite, 6:291b, 293b
Herbals
ancient Greek, 4:107a
medieval Latin, 6:123b; 11:601b
Renaissance, 2:536a–537a; 4:138b–139a; 5:361b–362b; 8:483b–484a; 13:502a–b; 14:586a; 15:161a
seventeenth-century, 2:305b; 11:315b–316a
eighteenth-century, 6:400b
historiography of, 1:205b; 12:328b
regional
Flemish, 4:139a
Herbaria
Renaissance, 5:383b; 11:33a, 311b, 312a–b
seventeenth-century, 9:528b; 11:47b, 587b; 13:443b
eighteenth-century, 3:365b; 4:157a; 6:61b, 66a; 7:198b; 8:376a; 9:433a; 12:394a, 459a; 13:553b; 14:463b
early nineteenth-century, 2:517b–518b; 6:184b, 493a; 9:456a; 10:12a, 13b, 518a, 531a; 11:130a, 263b; 12:553a*, 591b; 13:390b; 14:386b; 15:471a
mid nineteenth-century, 1:614b; 2:19a; 3:42b; 5:539b, 546b; 6:6b, 494a; 7:165b, 196b, 533b; 10:166a, 575a; 12:594b; 13:154a, 432b, 464a; 14:95b
late nineteenth-century, 2:476a; 3:197a; 5:50a; 6:491a; 11:108b, 181b; 15:268b, 269a
early twentieth-century, 7:535b; 13:456b; 15:150b, 421b
Herbart, Johann Friedrich (1776–1841), 6:294–295; 2:1b; 5:152b; 8:507b, 508a, 514a; 9:572a; 11:454b; 12:607a
Herberstein, Siegmund von, 10:16b
Herbert, William (1778–1847), 6:295–297
Herbivores, 2:585b; 4:510b
Herbrand, Jacques (1908–1931), 6:297
Herbrand tautology, 5:351a
Herbrand theorem, 6:297a–b
Herbrand-Gentzen theorem, 5:350b–351a
Herbs
essences of, 5:421a
Herbst, Curt, 4:186b, 187a,b; 14:424b, 427b, 429b
Herbst, Ernst Friedrich Gustav, 6:575b
Herculaneum, 3:603b
Hercules (constellation)
alpha Herculis, 6:331b
Hercynian deformation, 1:394a; 2:89b; 3:325b; 8:463b, 511b
Herder, Johann Gottfried von, 5:251b; 7:225a, 367a–b; 8:514b
Heredity, 2:579a–b; 3:114a*
and antibodies, 5:599a
blending, 5:10b, 266b
vegetative, 4:271a; 15:426a
cryptomeric, 13:478b
and cytology, 5:447a; 14:431a–433a
and development, 8:589a
and embryogeny, 9:523b; 11:573b
and environment, 2:211a–b, 580b; 3:390a; 7:166a–b; 12:417a
and evolution, 1:505a–506a; 3:121a–122a, 590a; 5:10b–11a, 385b–386a; 7:114b; 9:519b–520b, 523b, 618b; 12:417a; 14:99a, 238a, 432b, 433b–434a
inbreeding, 2:58b; 4:270b–271b
intermediate forms, 9:282a
laws of, 5:266b; 6:8a; 10:449b, 461b–462a, 463b; 13:477b–478a; 14:96b
and memory, 12:298a–b
Mendelian, 1:505a–b; 3:421b, 422b–423a, 493a; 4:271a; 5:10b; 9:519b–525a; 11:19b; 14:574a
opposition to, 9:520a–b, 522a
outbreeding, 4:271a–b
adaptation to, 3:422a
quantitative, 4:271a
recessive inheritance, 2:58b
sex ratio, 3:120b; 8:588b, 589a; 9:518a–b, 520a
theories of
nineteenth-century, 5:51b; 11:573b; 13:89a; 14:34a, 233a, 234a–238a
twentieth-century, 3:149a*; 9:495a; 15:426a–b
chemical, 14:432a
chromosome theory
see Chromosome theory
cytoplasm theory, 2:363b; 7:95a
extrachromosomal, 14:433a
idioplasm, 6:338b; 9:601a–b; 14:234b–236b, 429b
orthogenesis, 5:386a; 11:535b; 13:276a; 14:314b
particulate, 2:190a; 3:574a; 5:10b, 266b; 7:115a; 8:589a; 9:187a, 188a, 280a
physiological, 5:453b–454a
plastid, 3:422b; 6:9a
vibratory, 1:505b–506a
see also Pangenesis
see also Acquired characteristics; Degeneration (biol.); Diseases, hereditary; Dominance, genetic; Eugenics; Genetics; Hybridization; Mendelism
Hérelle, Félix d' (1873–1949), 6:297–299; 5:270b; 13:520b
Heresy
analysis of, 5:121a
scientists accused of, 1:4b, 93b, 108b, 590a; 2:55a, 273b, 540b, 542a–b; 3:65a, 514b, 526b–527a; 5:244b; 6:254b, 449b; 9:16a, 309b; 10:172a, 420a; 11:42b, 74a; 12:105b, 322b–323a; 14:226a; 15:68b, 80b, 98b
see also specific heresies by name
Herglotz, Gustav, 1:306b; 2:191b, 192a; 11:613b; 14:327b
Héricourt, Jules, 11:429a
Hérigone, Pierre (*d. ca.* 1643), 6:299; 4:571a
Cursus mathematicus, 3:615b; 5:382b; 12:206a
Euclidean commentaries, 1:474a
Hering, Heinrich Ewald, 15:101a, 469a
Hering, Karl Ewald Konstantin (1834–1918), 6:299–301; 2:446a–b; 4:615b; 5:446a; 6:248a,b; 9:562b; 10:220b; 12:298a
Hering waves, 6:300a
Héritier de Brut
see L'Héritier de Brut
Herlinus, Christian, 3:585a; 4:451a
Herman, Robert (physics), 5:273a
Herman, Robert Alfred (math.), 4:278a
Hermann of Carinthia, 1:35b, 36a, 62a; 4:447b, 611b; 9:39b; 11:198a; 15:175a
De essentiis, 15:181a–b, 182a,b
Hermann the German, 1:272b
Hermann (Hermannus) the Lame (1013–1054), 6:301–303; 11:321b
Hermann of Reichenau
see Hermann the Lame
Hermann, Carl Heinrich (1898–1961), 6:303–304; 9:183b
Hermann, Jakob (1678–1733), 6:304–305; 2:49b; 4:24b*, 469a; 5:449a; 7:442b; 9:187b; 11:401b, 513b*; 15:52b
Phoronomia, 2:38a–b; 10:66a
Hermann, Ludimar, 1:303b; 4:201a, 202a, 204b; 5:37b; 15:20b
Hermannus Contractus
see Hermann the Lame
Hermaphrodites, 1:523b
self-sterile, 3:120b
Hermaphroditism, 3:148b; 4:89a, 590a; 9:185b; 12:137a
experimental, 10:570a
human, 2:72b; 6:63a
of oysters, 3:588b
of plants, 15:67b
of snails, 10:568a; 13:174a; 15:67b
Hermbstaedt, Sigismund Friedrich (1760–1833), 15:205–207; 3:465b; 4:41b; 7:395a; 8:87b*; 12:239b; 14:332b; 15:172a
critics of, 5:532b; 13:465b
Hermeias, 1:250a; 11:184a
Hermes Trismegistus (*n.d.*), 6:305–306
astronomical tables, 12:432a; 14:593a
influence of
in Byzantine era, 4:622a; 10:419a
in Middle Ages, 1:33a,b, 34b, 36b; 4:556a; 6:79b; 7:187a; 10:555b; 11:226a; 14:388a; 15:262b
in Renaissance, 2:23a, 539b, 540a, 543b; 5:121a; 14:225b
in seventeenth century, 1:571b; 9:509a
Hermeticism
ancient Greek, 2:256b
ancient Roman, 9:79b; 14:632a,b
Byzantine, 11:183b, 184b–185a
medieval, 1:33a, b, 91a; 15:182a, 249b
Renaissance, 1:79b; 2:539b–543b; 4:6a*, 170a, 208b–209b; 11:96b–97a, 606b
seventeenth-century, 1:317b; 2:305b, 317a,b; 4:75a, 377a; 5:47b–48a, 422b–423a; 6:145b, 256a; 8:311b; 9:23b–24a, 527b; 10:81a, 83b; 12:307b; 14:451a
opposition to, 8:310a–b
see also Naturalism
Hermite, Charles (1822–1901), 6:306–309; 3:164a; 6:547a; 7:506b; 8:141a, 382b; 13:79a*
acquaintances of, 1:194a; 2:299a; 8:367b; 9:426a; 10:594a; 11:414b; 13:217b; 14:630a
algebra, 5:264a; 7:103a–b, 398a; 13:129b, 250a
analysis, 2:468a; 7:602a; 11:53b, 207a
on Cayley, 3:163a
editions of works, 10:593b
Hermite interpolation, 4:562a; 6:308a
Hermite parameters, 14:94b
Hermitean polynomials, 3:229b; 6:308a; 9:127a; 15:385a, b
Hermite's method, 10:249b
Hermite's principle, 6:308a
influence of, 12:86b; 13:249b
number theory, 2:470a; 3:228a, 230a; 5:342b; 6:547b, 571b; 7:168b; 8:367b; 9:412a; 14:94b, 630b
students of, 3:177b, 178a; 4:102b; 5:481b; 8:124b; 11:55a
Hermits of St. Augustine, order of, 5:402a–b
Hermogenes, 1:170b; 15:516a
Hermotimus of Colophon, 4:414b; 6:414b; 13:305a, 334a
Hernández, Francisco (1517–1587), 6:309–310; 3:180a
Hernia, 3:219a*; 8:97a
atlas of, 12:137a
formation of, 13:171a
inguinal, 6:78a
strangulated, 2:478a; 11:254a; 13:448b
Hero of Alexandria (*fl.* A.D. 62), 6:310–315
commentaries by, 4:415b, 437a, 438a; 10:5b, 596b
editions of works, 3:585a; 13:337a*
geometry, 1:418b; 3:126a; 4:105a, 110a, 434a(n), 491a, 609a; 6:314a–315b;

8:300a; **10**:529a, 587b, 588a; **15**:691a,b
Hero's formula, **1**:226a, 227a, 445a; **3**:227a; **4**:612a; **6**:314b
proofs of, **7**:175a, 335a
historiography, **4**:109b
influence of, **1**:22b, 31b, 41b, 82b, 567a; **3**:276b; **15**:253b
influences on, **1**:148b, 223b; **5**:346b
mechanics, **8**:225a; **10**:298a; **14**:376b(n); **15**:519a
natural philosophy, **13**:92b–93a
optics, **6**:313a; **11**:201a
war machines, **1**:468a; **6**:312b
Automata, **1**:419a,b; **6**:312b
Belopoiika, **1**:419b; **4**:489b; **6**:312b
Catoptrica, **14**:458a, 460a
translations of, **9**:438a,b; **14**:457b
Dioptra, **6**:310a,b, 312b
Mechanics, **1**:224b; **4**:489b; **6**:310a,b, 312a, 313a
translations of, **11**:244b
Metrica, **1**:224a, 445a; **6**:314b–315a
Pneumatica, **1**:61b; **3**:491a; **6**:310a–312a; **9**:491b; **10**:588b; **13**:91a
translations of, **3**:364b
On Vaultings, **7**:29a–b
Herod the Great, King of Judaea, **10**:111a
Herodot, **1**:75a
Herodotus of Halicarnassus (*d. ca.* 425 B.C.), 6:315–316; **1**:259a, 263a,b; **4**:381b; **11**:75a
commentaries on, **11**:376b
geography, **4**:389b
historiography, **6**:212b, 213a; **13**:295b; **15**:646a, 676a, 678a, 703a*, 724b
zoology, **1**:264a, 265b
Heron of Alexandria
see Hero of Alexandria
Heron, David, **10**:454a; **14**:574a
Herophilus (*b. ca.* 330 B.C.), 6:316–319; **11**:128b; **13**:91a
anatomy, **1**:316a; **4**:383a–386a *passim;* **5**:234a; **11**:128a; **13**:490b
critics of, **12**:80b
influence of, **4**:105b; **5**:235b; **12**:539a,b, 540a
medicine, **6**:420b
psychology, **13**:94b
on pulse, **5**:231b, 235a
students of, **4**:383a–386a *passim;* **6**:318b; **10**:581b
Hérouard, Jean, **7**:536b
Héroult, Paul Louis Toussaint (1863–1914), 6:319
Hall-Héroult process, **6**:50b–51a, 319a–b
Herpes, **5**:317a; **8**:114b, 274a*
virus, **12**:492b; **14**:624a
Herpetology
seventeenth-century, **14**:13a
nineteenth-century, **6**:120a; **13**:551b–552a; **15**:91a, 126a–b, 227a–b
twentieth-century, **13**:25a–b
Herrera, Alfonso Luís (1868–1942), 6:319–320
Herrera, Gabriel Alonso de, **14**:112b
Herrick, Charles Judson (1868–1960), 6:320–322; **1**:613b; **3**:332a
Herrick, Clarence Luther (1858–1904), 6:320–322; **3**:120b, 331b, 332a; **5**:50a, 197a*; **7**:143b
Herrick, Edward Claudius, **10**:41a
Herring
year-class dominance, **6**:442a
Herrmann, Felix, **13**:576b
Herschel, Alexander, **6**:322a, 329b
Herschel, Alexander Stewart, **6**:323b
Herschel, August, **6**:329a
Herschel, Caroline Lucretia (1750–1848), 6:322–323; **5**:26a*; **6**:322a, 323b, 325a, 326b, 328a, 329a, 330a; **12**:524b
Herschel, John Charles William, **6**:322b
Herschel, John Frederick William (1792–1871), 6:323–328; **5**:492b; **6**:322a–b, 323a*, 330a; **10**:601a
acquaintances of, **2**:98a; **6**:90a, 92a; **8**:288b, 321a; **10**:591b; **12**:288a, 524a, 525a; **14**:204b, 562b
astronomy, **1**:53b; **8**:613b
double stars, **2**:100b; **12**:552a; **13**:110a
instruments, **4**:179a; **12**:226a
nebulae, **1**:295b; **13**:120b
Neptune, **8**:277b
observations, **10**:592a; **13**:110b–111a
photometry, **6**:70b
star catalog, **4**:185a–b
chemistry, **2**:515a; **10**:511b
on contemporary scientists, **14**:490b
correspondence of, **3**:93b, 481b*; **7**:106b; **8**:414b; **10**:608a–b; **12**:288a; **15**:388a*
critics of, **1**:178a
crystallography, **10**:357b
and education, **10**:437b; **14**:292b–293a
geosynclines, **3**:553b
Herschel effect, **6**:327a
Herschel's condition, **6**:325a
magnetism, **1**:203a; **7**:608b; **12**:49b–50a, 50b–51a
mathematics
probability, **9**:219a
teaching of, **1**:354b; **7**:551a
meteorology, **2**:147a
optics, **14**:492a
physical, **8**:592a; **13**:77a, 238a–b
physiological, **9**:200a,b
philosophy, **14**:293b
Treatise on Astronomy
translation of, **3**:450b
Herschel, William (1738–1822), 6:328–336; **5**:26a*; **6**:323b, 326b, 471a; **9**:330b, 371a
comet of (Uranus), **2**:216b, 221a; **6**:322a, 328b–329a; **8**:299b; **13**:425a; **15**:311b–312a, 354b
correspondence of, **10**:608a
critics of, **9**:2b
double stars, **6**:325a; **9**:231b; **12**:552a
editions of works, **4**:185b
influences on, **6**:70a; **14**:485a
nebulae, **1**:295b; **6**:322b–323a, 329b, 332a, 332b–333a, 541b, 543a; **7**:598b, 608a; **9**:330a–b; **12**:266b, 269b
optics, **5**:19a; **6**:586b; **11**:474b
star clusters, **6**:323a
stars, motion of, **4**:278b
sun's motion, **1**:241b
sunspots, **14**:415a
telescopes, **6**:112a, 322b, 324b, 328b–330a, 333b–334a, 335b; **8**:170b; **9**:330a
Herschel, William James, **6**:323b
Herschel family, **6**:322a–b, 323b; **12**:226a
Herter, Christian Archibald, **1**:10b; **4**:298b
Hertig, Arthur Tremain, **13**:97b
Hertwig, Günther, **1**:499a
Hertwig, Karl Wilhelm Theodor Richard von (1850–1937), 6:336–337; **6**:337b
acquaintances of, **6**:9b, 10a; **11**:174b; **15**:166a
on contemporary scientists, **12**:422b*; **14**:239a*
critics of, **7**:402a
embryology, **2**:626b; **4**:187a; **6**:338a, 339a
students of, **2**:361b, 363a; **4**:132b; **5**:453a; **6**:442a; **11**:19a
Hertwig, Wilhelm August Oscar (1849–1922), 6:337–340; **6**:336b–337a; **9**:381b
acquaintances of, **6**:10a; **7**:402a; **8**:282a; **9**:495b; **15**:166a
cytology, **1**:601b; **13**:88b–89a; **14**:235a
embryogeny, **4**:187a; **5**:51b, 52b
fertilization, **2**:626a; **8**:128a; **12**:192b, 193b
Handbuch . . . der Entwicklungsgeschichte der Wirbeltiere, **5**:36a
heredity, **9**:601b; **14**:235a, 236b
students of, **2**:363a; **7**:454b
Hertz, Gustav Ludwig, **2**:245a; **5**:117a–b; **6**:16a; **9**:260b; **14**:171b; **15**:448a
Hertz, Heinrich Rudolf (1857–1894), 6:340–350; **7**:268a; **8**:148b
acquaintances of, **6**:243a; **7**:267b; **11**:208a
on contemporary scientists, **14**:171b
correspondence of, **11**:9a, 581b, 582a
editions of works, **8**:180b
electrodynamics, **2**:167b; **4**:231b; **8**:444a, 493b–494a, 494b; **9**:98b, 212a; **11**:9a; **14**:339a
experiments of, **9**:376b; **13**:471b; **15**:42b
influence of, **2**:266a; **6**:74a; **11**:460b; **12**:26a; **14**:470a*
influences on, **6**:252b
mathematics, **6**:473a
on Maxwell's theory, **4**:190b–191a; **5**:15a; **6**:346b; **9**:216a, 217a
oscillators, **1**:304a; **2**:427b
Die Principien der Mechanik, **6**:245a, 347a–348a
radiation, **11**:530b; **13**:366a; **14**:348a
strength of materials, **6**:534b
students of, **14**:573b
thermodynamics, **8**:599a
Hertz, Paul, **5**:350b
Hertzsprung, Ejnar (1873–1967), 6:350–353; **9**:194b; **12**:251a, 253b*, 455a,b
astrophysics, **12**:248a, 250b
Hertzsprung-Russell diagram, **4**:280a; **5**:272b; **6**:351a; **8**:2a; **10**:317b; **12**:18a, 19a, 22b
Herzfeld, Karl Ferdinand, **6**:164a
Herzolith, **3**:211b*
Hesiod, **6**:290a, 291a; **13**:297a; **14**:536b
Hesperia (asteroid), **12**:160a
Hesperiidae, **4**:494a–b
Hess, Alfred Fabian, **14**:444b–445a
Hess, Christian, **15**:194b
Hess, Clemens, **9**:189b
Hess, Germain Henri (1802–1850), 6:353–354; **1**:161a; **9**:91b; **11**:437a
law of constant heat summation, **2**:69a; **4**:241a
Hess, Harry Hammond, **5**:600a; **13**:609b
Hess, Hermann Heinrich, **1**:20a
Hess, Moses, **15**:132b, 133a
Hess, Victor Franz (Francis) (1883–1964), 6:354–356; **9**:397b; **14**:422b
Hesse, Ludwig Otto (1811–1874), 6:356–358; **1**:294a,b; **3**:314a,b; **7**:51a; **8**:542b; **10**:25a; **12**:216b; **13**:22a*
algebra, **3**:164b; **11**:45b; **12**:171a
editions of works, **8**:542b
Hessel, Johann Friedrich Christian (1796–1872), 6:358–359
crystallography, **2**:431b, 441a; **12**:511b; **14**:240b
Hesselberg, Theodor, **2**:168b
Hessen, Boris Mikhailovich, **15**:138a
Hessian fly, **7**:89b
Heterakis, **12**:482b
Heterangium, **12**:259b
Heterochromatin, **10**:277a
Heterodiceras, **9**:579a
Heteroecism, **3**:613a–b

Heterogeneity, 12:570b–571a
Heterogenesis, 1:497a; 10:367a
Heterogenite, 5:162a
Heterolysin, 4:299a
Heteropods, 15:166b
Heterosis, 3:575a–b; 7:409a
  physiological interpretation of, 4:271b
Heterostylism, 3:422a, 575a; 4:401b
Hetper, Władysław, 3:278b
Hettner, Alfred, 1:174a
Hettner, Georg, 11:583a; 14:222a
Heubner, Otto, 1:569b, 576a, 578a; 4:403b
Heumann, Karl, 8:553b
Heuraet, Hendrik van (1633–*ca.*1660), 6:359; 10:43b, 45b
  geometry, 6:600a; 12:206b
Heurne, Jan van (1543–1601), 6:359–360
Heuser, Chester Henry, 13:97a
Hevea, 13:342a
Hevelius, Johannes (1611–1687), 6:360–364; 3:116a; 6:68a, 516a; 7:309b*, 373a, 375a
  astronomy, 1:169a; 6:515b; 11:532b
  correspondence of, 2:349b; 6:141b; 7:376a; 8:26a; 10:202a
  and Hooke, 6:361a, 363b, 487b
  manuscripts, 7:374a
  Prodromus Astronomiae, 8:459b
  star catalog, 1:403b*; 5:23a
Hevesy, György (1885–1966), 6:365–367; 2:245b; 4:599a; 6:371a; 8:503b*; 9:543a
  hafnium, 2:247a; 3:429b; 13:546b
  radiochemistry, 2:241a, 499a; 6:16b, 481a; 10:288b; 12:30b
Heweliusza
  see Hevelius, Johannes
Hewetson, John, 13:300b
Hewson, William (1739–1774), 6:367–368; 3:486b; 6:569a; 9:483a
Hexactinellida, 14:626b
Hexagon, 10:298a
  Brianchon's theorem, 2:455a–b; 3:555a
  inscribed, 6:415a; 8:240a; 10:330b
  Pascal's theorem, 2:455a; 3:555a; 10:330b
  skew, 3:555a
  stellated, 2:360b
  types of, 5:409a
Hexagram, 3:167b; 10:330b; 13:623a
Hexamera, 1:565a
Hexaphenylethane, 5:464b–465a
Hexosamines
  structures of, 8:276a
Hexosemonophosphate, 3:416a
Hexoses
  degradation of, 14:174a
  in fermentation, 6:111a
  in plants, 3:511b
Hey, James Stanley, 4:69b
Hey, Käte, 1:307b
Heymans, Corneille, 15:469b
Heyn, Emil (1867–1922), 6:368–369; 9:140a*
  metals, 9:139a; 10:247a*; 13:245b
Heynitz (Heinitz), Friedrich Anton von (1725–1802), 6:369–370
Heyrovský, Jaroslav (1890–1967), 6:370–376; 2:429a
Heytesbury, William (*fl. ca.* 1335), 6:376–380; 2:394a; 5:360b; 6:127a; 8:198a; 9:32b; 13:195a
  Regule solvendi sophismata, 13:185b, 197a, 206b, 208a
    commentaries on, 13:208b
  Tractatus de tribus praedicamentis, 4:237b; 13:349b
Hiärne, Urban (1641–1724), 6:380–381; 2:421b; 3:473a; 13:179a, 409b
Hibbard, Paul Llewellyn, 6:443b
Hibbert, George, 2:518b, 519a
Hibernation
  eighteenth-century study, 7:96a
  nineteenth-century study, 2:30b–31a; 13:557b
  twentieth-century study, 8:283b*
  and digestion, 6:566b
  of swallows, 3:129b, 350a; 8:379b
Ibn Hibintā (*fl. ca.* 950), 6:381; 9:160b
  Kitāb al-Mughnī, 7:222b
Hiccuping
  reflex nature of, 9:571a
Hicetas of Syracuse (*fl.* fifth century B.C.), 6:381–382
  cosmology, 1:246b; 10:590a; 11:224a
Hidalgo (asteroid), 1:352b
Hieracium, 5:191a, 546a; 9:281b–282a, 601b
Hieratics, 15:682a–b, 683a–b, 687a, 700b
Hieroglyphics, 7:374b
  translation of, 7:375a; 14:569b; 15:681a–687a, 700b
Hieron II of Syracuse, 1:213a,b
Higgins, Bryan (*ca.* 1737–1818), 6:382–384; 14:562b
  Experiments and Observations Relating to Acetous Acid . . . , 6:385a
Higgins, William (*ca.* 1762–1825), 6:384–386; 7:388b
High Voltage Engineering Corporation, 13:570a
Highlands, Scottish
  geology, 8:33a
Highmore, Nathaniel (1613–1685), 6:386–388; 3:343a; 9:64a, 243a
  antrum of Highmore, 6:386b
  corpus Highmori, 6:386b
High-pressure studies
  eighteenth-century, 5:78b
  nineteenth-century, 1:579a; 4:141a; 6:54b–55a; 12:593b; 13:522a
  twentieth-century, 2:3b–4a, 323b–324a, 458a–460a; 3:239a, 334a; 5:103b, 621b–622a; 7:21a–b; 8:42a; 9:427b; 11:530b; 12:437b, 438a,b; 13:245a
  biological, 8:608a
  leakproof pressure seal, 2:458a
High-temperature studies
  nineteenth-century, 4:77b; 5:295a; 6:54b–55a, 115a; 8:117b–119a; 9:451b; 12:227a; 14:39a, 338b
  twentieth-century, 2:3b–4a, 323b–324a; 3:239a; 5:103a,b; 7:59b; 8:23a–b, 42a; 9:427b; 15:438a, 449b
  see also Pyrometry
Hiketas
  see Hicetas
Hilāl al-Ḥimṣī, 1:446a
Hilbert, David (1862–1943), 6:388–395; 3:96b, 560a*; 10:606a; 11:363a
  acquaintances of, 7:616a; 8:367b; 9:411b; 10:137b–138a; 12:249a; 13:548b; 14:68a, 202a, 613b; 15:39b, 40a, 41a, 42a–b
  algebra, 2:144b; 3:221b; 7:444b; 10:141a; 11:510a
    invariant theory, 3:164b, 165a, 314b; 5:472b; 6:388b–389b; 9:358a
  analysis, 1:427b, 500a; 2:271b; 3:135b; 5:608b; 6:391b–392b; 7:398b; 11:452a, 477b; 12:188a–b; 14:89a
  axiomatization, 1:292a; 5:152b, 154b; 6:77a*, 390b–391b; 12:11b; 14:283a
  on contemporary scientists, 4:4b, 323b; 14:221a
  correspondents of, 2:593b
  critics of, 11:60b, 61a; 13:125a; 14:437a
  foundations of mathematics, 2:512b, 514a; 3:56b; 4:4a, 561a; 5:350b, 351a; 9:502b; 14:89b
    formalism, 3:57a,b; 6:393a; 12:15a
  Grundlagen der Geometrie, 7:530b; 10:344b; 11:60b; 12:364b; 13:599b; 14:67b
  Hilbert space, 1:427b; 5:150b, 152a; 6:392a; 8:457b; 12:187b, 189a–190a; 13:58a, 549a; 14:46b, 90a–91b, 282a, 346b, 454a
  Hilbert-Schmidt theorem, 12:189a
  and Hurwitz, 6:570b, 571a
  influence of, 11:327a, 413a, 459a,b; 13:428a
  lectures, 6:235b; 15:39b
  logic, 8:546a
  mathematical problems, 1:307b; 2:513a; 4:9a; 5:342a–b; 6:393b–394a; 10:554a; 11:452b; 12:228b, 331a; 13:497b; 14:90a; 15:23a
  Maxwell-Boltzmann equation, 4:375b
  Methoden der mathematischen Physik, 12:220a
  number theory, 4:327b, 474b, 475a; 6:214a, 390a–b
    class field theory, 1:307a; 4:5a; 6:297b
  physics, 6:392b; 9:377a; 11:150a
  students of, 2:58a, 191b; 3:62b; 4:9a, 100b; 5:206a, 350b, 473a, 608a; 6:235b; 7:71b, 246b, 483a, 485b; 11:355a, 475b; 12:188a, 526a; 13:35b; 14:281a–b, 344a
  topology, 7:312b; 10:553b; 14:283a
Hildebrand, Joel Henry, 8:47a, 290b
Hildebrandt, Georg Friedrich (1764–1816), 6:395
  Handbuch der Anatomie, 14:199b–200a
Hildegard of Bingen (1098–1179), 6:396–398
Hilditch, Thomas Percy (1886–1965), 6:398; 3:243a
Hill, Archibald Vivian, 1:453b, 454a; 5:37a, 103a, 291a; 8:534a; 9:359a, 404a
Hill, Arthur, 6:296a
Hill, Carl Johan Danielsson, 2:468a; 7:103b
Hill, George William (1838–1914), 6:398–400; 3:584a; 6:49b; 10:35b*; 11:204b*, 207a
  celestial mechanics, 4:176a; 10:34a; 11:57a
  Hill's equation, 6:399b
  lunar theory, 2:516a; 4:481b; 14:453b
Hill, John (*ca.* 1707–1775), 6:400–401; 5:54a; 13:175a*
Hill, Leonard, 8:294b, 614a
Hill, Lester Sanders (1890–1961), 6:401–402
Hillebrand, William Francis, 2:588b; 3:292b; 11:279b
Hills, Edmond Herbert, 8:40b
Himalayas, the, 3:533a
  birds, 5:480b
  exploration of, 6:489b
  geology, 6:229b; 10:203a
  paleontology, 2:110a
Hind, John Russell (1823–1895), 6:402–403; 8:279a*
Hindenburg, Carl Friedrich (1741–1808), 6:403–404; 2:56b; 7:494b, 566b; 8:140b; 10:573b; 12:325b
Hindley, Henry, 8:210a
Hinge mechanisms
  theory of, 3:228b–229a
Hinks, Arthur Robert, 7:472b; 12:17b, 18b
Hinshelwood, Cyril Norman (1897–1967), 6:404–405; 15:36b
Hipparchus (*d.* after 127 B.C.), 15:207–224; 1:205a, 283a,b; 2:405a; 5:345a

astronomy, **1**:510b; **3**:319b, 406a; **4**:467a; **5**:345b, 347a; **9**:338a; **11**:188a–198a *passim*, 201b; **13**:225b; **15**:562a
precession of equinoxes, **9**:297a; **15**: 543a
on calendar, **3**:22a; **4**:459b; **5**:553a; **10**: 180a; **15**:538b
cosmology, **1**:247b, 248a, 249b; **7**:143a; **13**:326a
on Eratosthenes, **4**:389b–391a *passim*
geography, **11**:200a, 225b
influence of, **3**:319a; **5**:347a; **11**:104a; **13**:84a; **14**:593b
influences on, **1**:192a
mathematics, **9**:297a–b, 299b; **15**:559a
mechanics, **5**:239a; **7**:136a
observations of, **14**:577a; **15**:349b
On the Rising of the Twelve Signs of the Zodiac, **6**:617a
Hipparion, **3**:264a
Hippasus of Metapontum, **11**:222a,b; **13**: 304a, 305b(n)
Hippias of Elis (*fl.* 400 B.C.), 6:405–410; **4**:103b, 104a, 464b(n)
Hippocampus minor, **6**:593a,b
Hippocrateans
pathology, **1**:104a
Hippocrates of Chios (*fl.* second half of fifth century B.C.), 6:410–418; **8**:189b; **11**:288b
duplication of cube, **1**:232a; **4**:105a, 489a; **8**:192b; **9**:270a
Elements, **4**:415b; **6**:414b–416b
quadrature of lunes, **4**:463a; **6**:202b; **11**: 589a
squaring of circle, **1**:119b, 171a; **4**:424a; **12**:441b
Hippocrates of Cos (460-*ca.* 370), 6:418–431
biographies of, **7**:187a; **8**:133a; **12**:540a
chemistry, **13**:234b–235a
commentaries on, **3**:218b; **4**:199a; **5**: 157a*, 232a; **11**:445a, 511b; **12**:540a; **13**:37b, 282a, 415b*
critics of, **1**:315a; **6**:317a
dietetics, **8**:506b*
editions of works, **1**:77b
Hippocratic oath, **6**:425a
influence of, **1**:235a, 267b; **2**:224b, 401b; **4**:105b; **6**:79b; **7**:539b; **9**:40b; **12**:312a
medicine, **1**:103b; **3**:577b
anatomy, **1**:316a
pathology, **12**:102a,b
pharmacology, **4**:119b
succussio, **1**:333a
on urine, **5**:231b
physiology, **7**:612b; **10**:598b
religion, **5**:232b
translations of works, **1**:107a; **3**:12b; **5**: 586a; **11**:245a; **15**:160b, 230b
Aphorisms, **1**:315b; **2**:23a; **11**:253b
commentaries on, **9**:604a; **10**:123a, 230b; **11**:238b; **12**:102a
translations of, **3**:393a
Epidemics
commentaries on, **9**:604b
Liber veritatis Ypocratis (spurious)
translations of, **15**:183a
De morbis popularibus
editions of, **5**:157a*
Nature of Man
commentaries on, **5**:230b, 231a; **9**:603a
Prognostic, **11**:253b
translations of, **3**:393a
De prognosticationibus aegritudinum secundum motum lunae (spurious), **9**:438a,b
In the Surgery, **1**:315b
see also Hippocrateans; Medicine, Hippocratic tradition
Hippopede, **4**:467a; **9**:300b; **10**:529b
Hippopotamus, **5**:504b
Hippuric acid, **10**:557a; **14**:233b
analysis of, **8**:331b
metabolism of, **7**:520a
synthesis of, **3**:511a; **4**:75b–76a
by organism, **2**:586a; **14**:233a
Hirayama, Kiyotsugu (1874–1943), 6:431; **7**:385b
Hirn, Charles Ferdinand, **6**:431b
Hirn, Gustave Adolfe (1815–1890), 6:431–432; **3**:306a; **6**:545b
Hirn's analysis, **6**:432a
Hiroshima, **10**:216a
Hirota, Shinobu, **9**:406b
Hirsch, August, **7**:530b
Handbuch der historisch-geographischen Pathologie, **3**:463b
Hirst, Edmund Langley, **6**:185a
Hirszfeld, Ludwig (1884–1954), 6:432–434; **7**:623b
Hirudinea
neuroanatomy, **1**:177a
Hirzebruch, Friedrich Ernst Peter, **6**:497a
Hirzgarter, Matthias
Detectio dioptrica corporum planetarum verorum, **6**:361b
His, Wilhelm (1831–1904), 6:434–436; **5**: 447a; **8**:542a*; **11**:255a
acquaintances of, **9**:380b; **15**:64b
critics of, **7**:402a, 439b
embryology, **11**:573b
histology, **11**:274a; **13**:590a
neuroanatomy, **5**:26b
students of, **7**:278b; **9**:56a, 57a; **13**:96b
Unsere Körperform und das physiologische Problem ihrer Entstehung, **11**: 571a
His, Wilhelm, Jr., **5**:281b
bundle of His, **4**:398a; **6**:434a
Hisinger, Wilhelm (1766–1852), 6:436–437; **2**:93a
analysis, **11**:437b
and Berzelius, **2**:91a, 92a–b
cerium, **3**:473b
electrochemistry, **3**:601b
Hispaniola
colonization of, **15**:89b
discovery of, **15**:89a
Hiss, Philip Hanson, Jr., **14**:624a–b
Hissarlik (Asia Minor), **12**:180a–181b; **14**: 42b
Histamines, **1**:11a; **8**:295a; **15**:10b
discovery of, **14**:445b
metabolism of, **15**:418a
pharmacology of, **15**:105b–106a
Histidine, **14**:445b
discovery of, **7**:467a
nutritional role, **6**:501a
Histiocytes, **9**:295b, 296a
Histochemistry, **3**:359a; **4**:162b, 603b–604a; **6**:434b; **9**:381a; **11**:300b
Histogenesis, **5**:36a; **13**:555b–556a
Histologische und Histopathologische Arbeiten über die Grosshirnrinde
editors of, **10**:131a
Histology
seventeenth-century, **1**:391b; **5**:426a; **9**: 63a–65a, 66a, 145b–146a; **13**:171a; **15**:29a
eighteenth-century, **1**:392a; **4**:26a, 589b, 590a; **9**:511a–b; **13**:465a; **14**:15a
mid nineteenth-century, **1**:476b, 477a, 539b–540a; **2**:124b–125a, 375b–376b, 532a; **3**:424b–425a; **4**:162b; **6**:269a–270a, 434a–435a, 589b; **7**: 437b–440a, 520a; **8**:254a, 302a–b, 400b–401b, 528a, 541b; **9**:570a–b; **10**:267a–b; **11**:216a–217a, 360b–361a, 367b–369b, 380b, 491a–492a, 565b–566b; **12**:224a, 231b–232b, 242a; **13**:461b, 555b–556b; **14**:40a, 113b, 233b; **15**:109a
late nineteenth-century, **1**:176b, 177b, 417b, 438b, 439a, 613b; **2**:164b–165b, 344b; **3**:184b, 187b; **4**:266b–267a, 296a–297a, 276a–277b; **5**:26b, 34b, 36a, 74a, 459b–460b; **6**:225b–226a, 287b; **7**:35b; **8**:3b, 8a–9a, 15b–16a, 145b, 556b; **9**:56a–57a, 79a; **11**:176a–b, 177a, 295a–296b, 374a, 382b, 517b; **12**:319a–b, 355a–356a; **13**:589a; **14**:126a–b, 228a–229b, 278a–b; **15**:201b
twentieth-century, **3**:112b; **4**:397b; **5**: 453b, 594a–b; **6**:224a; **7**:502b; **8**: 282b–283a, 447b, 552b; **12**:48a–b; **13**:590a; **14**:596b; **15**:473a
botanical, **6**:401a
comparative, **8**:254a, 302a–b; **9**:259a; **11**:234b, 592b
and evolution, **14**:596b
handbooks, **1**:177b
pathological, **2**:129b; **5**:565b–566a; **9**:93b, 95a, 572b; **10**:15a–16a; **11**:295a, 374a; **14**:252b; **15**:479b–480a
techniques
silver impregnation, **8**:92a
terminology, **1**:540a
see also Illustration, histological; Microbiology; Neurohistology; Staining, microbiological; Tissues
Histomonas, **12**:482b
Histone
composition of, **7**:467a
Historicism, **3**:383a
Comte's law of three stages, **3**:375b
Historiography
ancient Greek, **1**:250b; **4**:81b; **6**:315b–316a; **10**:111b; **11**:105a; **13**:83b–84a
ancient Roman, **11**:38b–40b
Byzantine, **10**:208a
medieval Arabic, **1**:28b; **2**:150a; **3**:549a; **6**:79a–b; **7**:321b–322b, 356b–357a, 362b; **9**:171a–b; **14**:547a
medieval Latin, **1**:565a; **14**:35a
Renaissance, **5**:448a; **7**:419a; **9**:32b, 190b, 254b
seventeenth-century, **2**:305b; **11**:587b; **12**:85a, 460b; **13**:583b
eighteenth-century, **8**:470b, 471a; **9**:159a; **11**:42b; **12**:549a; **13**:425b
nineteenth-century, **1**:28b; **6**:456b; **7**: 547b; **10**:448b; **12**:105a; **13**:238a, 239a, 372b; **14**:69a
twentieth-century, **7**:6a, 7a; **11**:431a
of agriculture, **15**:493a*
of alchemy, **2**:64b; **7**:464b; **10**:288b; **14**: 333a
of anatomy, **9**:479b–480a, 481b, 483a; **11**:100a
of architecture, **2**:201b; **4**:496a; **12**:545a; **14**:29a–b, 293a, 404a, 636a; **15**:516a
of art, **5**:447b; **7**:412a; **11**:39b, 304b; **14**:3a
astrological, **1**:34b–35a; **7**:222b, 362b; **9**:160a
of astrology, **12**:432a, 439b–440a; **15**:519b
of astronomy
ancient Greek, **4**:463b
medieval Arabic, **11**:246b–247b
seventeenth-century, **6**:362b
early eighteenth-century, **6**:70b; **9**:501a
late eighteenth-century, **1**:401a; **8**:

Historiography (cont.)
of astronomy (cont.)
143b–144b; **10**:572a*, 615b; **14**:485a; **15**:329a–b, 343a, 344a
early nineteenth-century, **1**:513a; **2**:134a; **3**:14a; **4**:17a; **7**:580b; **10**:199a; **15**:372a, 379b
mid nineteenth-century, **2**:533a; **4**:369b; **8**:369b; **9**:2b; **12**:170a; **14**:471b
late nineteenth-century, **4**:185b; **9**:81a, 184b; **11**:320a, 378a; **12**:162a–b; **13**:253a, 423b*; **14**:54b, 479b–480a
early twentieth-century, **1**:172b, 398a; **2**:126b; **3**:35b; **6**:147a, 431b, 548b; **7**:7a; **10**:290a; **13**:12a, 139a
mid twentieth-century, **13**:119a
autobiographical
ancient Roman, **5**:227b
medieval Arabic, **6**:190a; **7**:321b
of bacteriology, **2**:584b; **8**:448b
biographical
ancient Greek, **1**:282a
ancient Roman, **12**:540a
medieval, **2**:609b–610a; **7**:187a–b; **14**:547a–b
Renaissance, **1**:419a–b
seventeenth-century, **5**:287b; **14**:157a–b
eighteenth-century, **4**:166b; **5**:61a–b; **11**:100a
nineteenth-century, **2**:88b, 453a, 476a; **3**:265a, 525b; **4**:560a; **5**:45a, 335b, 359b; **6**:76b*; **7**:549a; **9**:557b; **11**:50b, 589a; **13**:373a
twentieth-century, **4**:133a; **5**:208a, 454a; **8**:505a; **10**:118a, 329b, 503a–b; **12**:465b; **13**:26b; **14**:324a–b
of biology
nineteenth-century, **10**:367a; **11**:110a, 382b
twentieth-century, **2**:221b; **3**:148b; **7**:273b, 517a; **10**:149a–b, 523a; **11**:257b; **12**:208a; **13**:138a
of botany
seventeenth-century, **7**:147b
eighteenth-century, **3**:177a*; **7**:586a; **13**:444a*
nineteenth-century, **2**:160a, 476a; **3**:562b–563a; **7**:417a; **10**:13b; **12**:60a; **13**:462b
twentieth-century, **1**:205b; **12**:328b; **13**:343a
of cartography, **10**:148b; **11**:394b; **15**:256a–b
of chemistry
Renaissance, **13**:141a
seventeenth-century, **2**:317b
eighteenth-century, **14**:333a
early nineteenth-century, **13**:373b
mid nineteenth-century, **3**:48a, 240b; **4**:242b–243a; **6**:461b; **7**:463a, 464a–b, 552a; **9**:347b; **12**:478b
late nineteenth-century, **2**:64b, 186b, 624b; **3**:85b; **9**:131a, 305a, 467b*, 557b; **11**:538a; **12**:209a; **13**:389b; **15**:466a
early twentieth-century, **1**:168a,b; **2**:169b; **3**:334a; **4**:227b; **5**:489a, 595b; **6**:398b*; **7**:59b; **9**:341a; **11**:319a; **12**:465b; **13**:411a*, 546b; **14**:125a–b
mid twentieth-century, **3**:111a; **4**:21a; **9**:428a; **10**:329b
and chronology, **1**:565b
of cosmology, **4**:227b; **7**:485b–486a, 487b–488a
of crystallography, **1**:463a*; **9**:341a
of culture, **3**:111b–112a; **4**:34a, 81b
of dentistry, **13**:142b
of embryology, **2**:161a; **6**:339b
of epistemology, **1**:365b–366a; **8**:517a–b
of exploration, **6**:20b
of geography, **1**:173a, 623b; **5**:77a; **6**:552a,b; **9**:104a; **10**:239b–240a, 284b; **11**:180b, 376b; **13**:124a
of geology
nineteenth-century, **1**:210b; **3**:435a*; **5**:13b–14a, 335b, 556b; **12**:514b; **14**:627a
twentieth-century, **1**:52a–b, 395a; **4**:599a; **5**:337a; **8**:286a; **9**:104a, 315b; **10**:169a, 430b
of heredity, **7**:115a–b
of hydraulics, **2**:39b
of instrumentation, **7**:2b*
of invention, **1**:554b, 555a
of Jesuits, **1**:483b
Jewish, **1**:323b
of logic, **3**:456b; **8**:546b; **10**:486a–b
of mathematics
ancient Greek, **1**:223b; **4**:462b–463b; **6**:408a; **13**:296a–b
Byzantine, **11**:160b–161a
medieval Arabic, **1**:508a
Renaissance, **1**:419a,b
seventeenth-century, **3**:152b; **5**:60b; **9**:500b–501b; **14**:150a
eighteenth-century, **1**:206b–207a; **2**:335a; **3**:284b, 459b; **6**:71a, 577a; **7**:206b, 570a; **15**:368a,b
early nineteenth-century, **4**:17a; **5**:490b; **7**:550b; **14**:500a
mid nineteenth-century, **1**:591b; **2**:283b–284a; **3**:212b–214a; **4**:36b; **7**:54b, 55a; **13**:427a–b; **14**:471b–472b
late nineteenth-century, **1**:600a; **2**:430a; **3**:59a, 135b, 512a–b; **5**:414a, 573b; **6**:95b; **7**:466a; **8**:250b, 368a; **9**:81a; **11**:589a; **12**:599b; **13**:81b, 155a–b, 250b, 251b–255a, 414b; **14**:343a, 619a–b
early twentieth-century, **2**:320a, 358a–b, 465b, 545b; **4**:84a, 373b; **6**:211a; **7**:207b, 408b; **8**:367a, 505a; **9**:387b; **10**:118a, 170b, 444a; **11**:394b; **13**:23b, 470a, 615b; **14**:336b–337a; **15**:256a–b
mid twentieth-century, **1**:584b; **3**:117b, 222a, 399a; **4**:9a, 337a–b; **5**:108a, 342b–343a; **6**:235b, 548b; **11**:363a; **13**:428b
of mechanics, **1**:419b; **7**:570a; **9**:501a
of medicine
ancient Roman, **12**:540a
medieval, **3**:218b; **7**:24b–25a, 187a–b
Renaissance, **1**:484b, 603b, 604a; **8**:483b
seventeenth-century, **1**:451b; **4**:135b
eighteenth-century, **3**:2a–b; **5**:157a; **8**:506a
nineteenth-century, **2**:336a; **3**:240b, 463b–464a, 517b; **4**:252a, 289a; **6**:618b; **8**:133a; **11**:301a; **12**:591b; **15**:110a
twentieth-century, **3**:113b*, 453b; **5**:208a–b, 495b; **8**:295b; **9**:296a; **12**:88b; **13**:142a–b; **14**:324a–b
of meteorology, **12**:366b
of microscopy, **6**:137b
military, **11**:67a
of mining, **11**:304b
of music, **1**:282a; **2**:354a; **4**:135b, 525a; **7**:376b; **9**:319b, 501a; **10**:521a
of nutrition, **8**:591b
of paleontology, **14**:119b
of philosophy
medieval, **1**:34b, 378b
Renaissance, **2**:542b–543a
eighteenth-century, **7**:443b*
nineteenth-century, **9**:382a; **12**:11a; **13**:253b; **15**:404a–b
twentieth-century, **2**:545b; **3**:110b–111b; **7**:488a–b; **11**:258a, 363a; **14**:636a
philosophes', **1**:114b
of photography, **4**:283a
of physics
medieval Arabic, **7**:340b, 341b, 343a
late eighteenth-century, **7**:494a; **9**:501a; **11**:141b–143a, 610a*
early nineteenth-century, **1**:202a; **2**:336a
mid nineteenth-century, **2**:283b–284a; **11**:50b, 116a; **13**:237a–b
late nineteenth-century, **8**:597b; **11**:543b–544a, 581b; **13**:81a; **14**:67b
early twentieth-century, **1**:519b; **2**:169b, 545b; **4**:227b–228a; **5**:609a; **7**:485b–488a, 516a; **8**:40b; **14**:158b, 317b
mid twentieth-century, **7**:86b; **9**:449b; **13**:599a; **14**:636a; **15**:251b
of physiology, **1**:617a; **4**:399a; **5**:83b, 207b; **15**:110a
of protozoology, **4**:290a
of religion, **1**:34b; **2**:591a; **4**:462b; **6**:469a, 556b; **15**:137b
comparative, **5**:60a
of science
ancient Greek, **4**:462b–463b; **13**:333a–b
medieval Arabic, **1**:34a–b; **4**:502b; **7**:322a
seventeenth-century, **4**:221a; **5**:415a; **12**:580b–586b
eighteenth-century, **1**:554a; **4**:25a*; **5**:195b; **8**:261b; **11**:472b
early nineteenth-century, **2**:134a, 187a; **3**:525a–b
mid nineteenth-century, **1**:178a; **3**:45b, 525a–b; **4**:183a–b; **5**:359b; **14**:293b–294a; **15**:133a, 134a–144b, 412b–413a
late nineteenth-century, **1**:175a; **2**:88b; **3**:42b; **4**:203b; **5**:211b; **8**:600a–603b; **9**:123b, 382a–b; **10**:148b, 486a–b; **11**:543b–544a; **13**:251b–255a, 357a, 550b; **14**:481a
early twentieth-century, **1**:175a; **2**:340a, 357b–358a, 358b; **3**:148b; **4**:359a; **5**:190a, 408a*; **7**:503b, 514b; **9**:341a–342a, 378a–b; **10**:594b–595a; **11**:388b–389a; **12**:95a, 108a–112a, 239a*; **13**:345b; **15**:423a, 424b, 466a–b
mid twentieth-century, **1**:365b–366a; **3**:148b, 359b; **4**:599a; **7**:482b–488b; **12**:112a–114a; **14**:636a–b; **15**:44a
institutionalization of, **9**:378a–b; **12**:108b–114a; **13**:254b
philosophy of, **1**:114b; **7**:486b; **8**:600a–603b
of surgery, **7**:279a
of technology, **1**:554b; **6**:2b; **9**:491b–492a; **11**:79a; **12**:383a; **13**:357a, 398b; **15**:406b–408a
of zoology, **3**:439a*, 525b; **5**:442b; **13**:601b; **15**:78b
see also Experiment, historical repetition
History
ancient Greek concept of, **6**:315b
ancient Roman concept of, **1**:335a
environmental interpretation of, **12**:344a
philosophical use of, **2**:545a–b; **5**:121a
philosophy of
medieval Arabic, **7**:321b–322a; **9**:171b
seventeenth-century, **5**:59b–60a
eighteenth-century, **9**:490b–491a

nineteenth-century, **4**:183a; **12**:155b, 591b; **14**:294a; **15**:136a–138b, 142a, 408b, 410b, 415a
twentieth-century, **2**:358b; **12**:108a–114a *passim;* **15**:424b
science and, **3**:452b–453b; **4**:203b; **7**:604b; **8**:183b; **11**:14a; **15**:730b, 732b
as social science, **14**:68b
theories of
cyclical, **4**:183a
psychological, **7**:191b, 192b
see also Historicism
History of Mathematical Astronomy in India, 15:533–633
History of nature, **2**:204b, 209a–211a, 502a, 579a–581a; **15**:135b, 139b, 144a, 408b, 412b–415b
History of Science Society (Cambridge, Mass.), **12**:109b
Hitchcock, Albert Spear (1865–1935), 15: 224–225
Hitchcock, Charles Henry, **4**:361b; **14**:516b
Hitchcock, Edward (1793–1864), 6:437–438; **4**:361a,b, 363b, 364b; **6**:57b; **12**:433b, 434a
geology, **4**:365a(n); **7**:252a
Hitler, Adolph, **4**:155a, 313a; **5**:183a; **8**:52b, 182b; **12**:530b
Hittites, **14**:504b
Hittorf, Johann Wilhelm (1824–1914), 6: 438–440; **9**:391b; **10**:345a–b; **11**:46b–47a; **12**:191b
electrochemistry, **7**:449b; **15**:459a
Hitzig, (Julius) Eduard (1838–1907), 6:440–441; **1**:569b; **4**:594a; **7**:49b*
neurophysiology, **4**:594b; **5**:196b, 255a, 463b; **6**:518b; **7**:48a
Hiuan-tsang, **2**:416b
Hjelm, Peter Jacob, **4**:346b; **12**:147b
Hjelmslev, Johannes Trolle, **4**:414a; **9**: 157a,b, 447a
Hjort, Johan (1869–1948), 6:441–442; **9**: 588a, 590a
Hjorth, Sóren, **12**:425b
Ho Ch'êng-tien, **13**:484b
Ho Yüan-shih, **3**:266a
Hoadly, Benjamin, **3**:296a; **14**:419a
Hoagland, Dennis Robert (1884–1949), 6: 442–444
Hoagland, Hudson, **10**:611a
Hoaxes, scientific
in archaeology, **9**:538b
in astronomy, **1**:467b*
in biology, **12**:561b–562a
in geology
Würzburg Lügensteine, **2**:15a–b
in paleontology, **3**:606b–607a
see also Forgeries, literary
Hobbes, Thomas (1588–1679), 6:444–451; **14**:178a*
acquaintances of, **1**:541a; **2**:13a; **4**:95b; **6**:151a; **9**:316b; **10**:200b; **12**:459b
and Bacon, **6**:445a–b
and Boyle, **2**:377b; **6**:450a
on contemporary scientists, **1**:383a; **9**: 322a(n)
critics of, **1**:474b; **3**:492a; **9**:509b; **12**:357b
and Descartes, **4**:52a; **6**:446a,b
Elementorum philosophiae, **8**:151a
geometry, **4**:419a; **10**:495b
influence of, **3**:208b; **8**:153b
influences on, **5**:553b
optics, **6**:445b–446a; **9**:319a
philosophy, **5**:416a; **6**:446b–449b; **8**:151b; **15**:139b
students of, **10**:564b
translations of works, **12**:585a
and Wallis, **6**:449b–450a; **14**:152a
Hobbie, R., **6**:366a
Hobbism, **6**:129b
Hobbs, William Herbert, **7**:149b
Höber, Rudolf, **15**:21b
Hobson, Ernest William (1856–1933), 6:451; **6**:114a; **9**:613b*
Hochart, Polydore, **13**:253a
Höchst Chemical Works, **4**:300b
Hochstetter, Ferdinand von, **4**:140b; **5**:611a
Hockin, Charles, **9**:209b
Hodge, Frederick Webb
Handbook of the American Indian, **11**: 119b
Hodgkin, Dorothy Crowfoot, **15**:18a
Hodgkin, Thomas, **8**:414b
Hodgkinson, Eaton (1789–1861), 6:452; **11**:294a
Hodgson, James, **6**:169b
Hodgson, Joseph, **2**:375a,b, 443a
Hodierna, Giovanni
see Odierna, Gioanbatista
Hodograph, **6**:90a
Hodometer, **8**:198a; **15**:519a
Hoefer, Ferdinand, **7**:464a,b; **9**:401b
Hoek, Martinus (1834–1873), 6:452–453
Hoëné-Wroński (Hoehne), Józef Maria (1776–1853), 15:225–226; **4**:84a
algebra, **7**:570b; **11**:600a*
method of, **12**:326a
Wronskians, **10**:249b, 442a; **15**:226a
Hoeven, Jan van der (1801–1868), 6:453–455; **6**:535a
Hoff, Karl Ernst Adolf von (1771–1837), 6: 455–458; **2**:204a; **12**:182b
Hoffa, Mauritius, **5**:150a
Höffding, Harald, **2**:240a; **15**:422b, 424b
Hoffman, Ernst Theodor Amadeus, **1**:389a
Hoffmann, Adolf Friedrich, **7**:597b
Hoffmann, Banesh
EIH (Einstein-Infeld-Hoffmann) method, **7**:10a
Hoffmann, Carl August Siegfried, **14**:259a
Handbuch der Mineralogie, **2**:440a
Hoffmann, Caspar, **1**:482b; **6**:152a, 157b; **9**:153b
Hoffmann, Erich, **10**:18b; **11**:174b; **12**:142a
Hoffmann, Friedrich (1660–1742), 6:458–461; **2**:2b, 226b; **7**:188a; **9**:483a; **10**: 25b; **13**:543a; **14**:212b
chemistry, **2**:175b; **9**:106a
Hoffmann's drops, **6**:460a
students of, **9**:104b; **11**:109a; **12**:185a
at University of Halle, **12**:600a,b
Hoffmann, Friedrich Albin, **8**:8b–9a
Hoffmann, Georg Franz, **10**:531a
Hoffmann, Karl, **6**:223a
Hoffmann, Paul-Erich, **8**:273b
Hoffmann
see also Hofmann
Hofmann, August Wilhelm von (1818–1892), 6:461–464; **3**:85a; **5**:125a, 341b*; **7**: 282b, 464b*; **9**:19a*; **10**:511a, 516a; **15**:11b
acquaintances of, **7**:507b; **8**:55a, 112b; **9**:178a; **14**:394a, 529b
chemistry, **14**:530b
on contemporary scientists, **14**:475a, 478b*, 622b*
dyestuffs, **5**:2a, 197b; **8**:349a–b
Hofmann's method, **5**:320b
Hofmann's violets, **6**:463a–b
influence of, **3**:84b; **5**:536b
students of, **1**:288a, 340b, 426a; **3**:474b, 476a; **5**:214b, 215a; **7**:65a; **9**:360a; **10**:37a, 177a, 515b; **11**:551a; **13**:406b; **14**:141b
theoretical chemistry, **5**:373a; **14**:395a
works edited by, **4**:19a; **5**:104a*; **8**:350a*
Hofmann, Moritz, **10**:567b
Hofmann
see also Hoffmann
Hofmeister, Friedrich, **6**:464a–b
Hofmeister, Wilhelm Friedrich Benedikt (1824–1877), 6:464–468; **12**:58a, 60a, 259a; **13**:88a–b
biological chain, **2**:492a
cryptogams, **2**:160a; **6**:219a; **9**:601a; **11**: 152a
The Higher Cryptogamia, **3**:29b
plant development, **2**:426a; **12**:256a
plant reproduction, **2**:520a; **5**:539b; **6**: 266a; **15**:67b–68a
students of, **2**:436b; **5**:437b; **13**:416b; **14**:96a, 97b
translations of works, **6**:267a*
Hog cholera, **2**:575b; **14**:249b
bacillus, **4**:620a; **12**:480b, 482a
diagnosis of, **2**:300b
vaccination, **3**:189a; **4**:211a
Hohe Tauern, **13**:285a–b
Hohenheim, Theophrastus Bombastus von
see Paracelsus, Theophrastus . . .
Hohmannite, **5**:162a
Hoho, Paul, **14**:218a
Hoki naiden, **15**:733b
Holbach, Paul Henri Thiry, Baron d' (1723–1789), 6:468–469; **7**:525b*; **8**:81a; **9**:148b; **10**:23b*; **11**:562b; **14**:68b
acquaintances of, **4**:85a; **11**:563a; **13**:565b
salon, **11**:563b
Holbein, Hans, **5**:447b; **6**:145a*; **12**:459a
Holborn, Ludwig Christian Friedrich (1860–1926), 6:469–470; **7**:528b; **11**:8b; **14**:338b
Holbrook, John Edwards (1794–1871), 15: 226–228
Holden, Edward Singleton (1846–1914), 6: 470–472; **1**:465b, 466a,b; **2**:285a; **4**: 180a,b; **6**:336a*; **7**:270b
Holden, Isaac, **12**:328a
Holden, Roy Jay, **6**:121b
Hölder, Otto Ludwig (1859–1937), 6:472–474; **9**:232b*; **11**:510a*; **14**:222a, 454a
Hölder continuity, **6**:472a
Hölder inequality, **6**:472b
Jordan-Hölder theorem, **6**:472b; **7**:168a
Holder, William, **14**:152a
Holism, **6**:408b(n); **12**:111b, 113b
Höll, Maximilian
see Hell, Maximilian
Holland, Robert, **2**:476a
Holland, Samuel, **3**:396b
Holländische Beiträge zu den anatomischen und physiologischen Wissenschaften, **4**:162b; **9**:456b
Hollandus, Isaac, **13**:559b
Hollandus, John Isaac, **5**:420a; **13**:559b
Holliday, Francis, **13**:68b
Holmboe, Bernt Michael (1795–1850), 6:474
and Abel, **1**:13a, 14a, 15a, 16a; **6**:474a
correspondence of, **11**:599a
Holmes, Arthur (1890–1965), 6:474–476; **1**:394b; **2**:260a; **13**:610b
Holmes, Kenneth Charles, **5**:141a
Holmes, Oliver Wendell, **14**:297a, 532b, 533a
Holmes, Robert, **12**:85b
Holmes, Timothy, **5**:45b, 515a
System of Surgery, **8**:402a
Holmgren, Albert, **6**:4a
Holmgren, Erik, **5**:152a
Holmgren, Frithiof (1831–1897), 6:476–477; **2**:531a; **7**:520b; **13**:393a; **15**:22a*
Holmium
discovery of, **3**:321b

Holmyard, Eric John, **7**:39b, 41a, 42b
Holocene epoch
  artifacts, **1**:131b
  paleontology, **3**:519b–520a
  stratigraphy, **1**:131a
Holocephali, **6**:535b
Holography, **14**:617a
Holothurians, **11**:113b; **12**:297b
  embryogeny, **7**:475a
Holt, Arthur, **13**:343a
Holtzmann, Karl Heinrich Alexander, **3**: 83a, 306a; **6**:243b, 244a–b; **9**:239b
Holwarda of Franeker, **7**:373b
Holweck, Fernand, **3**:502a
Holywood, John of
  see Sacrobosco, Johannes de
Holzmann, Wilhelm, **1**:367b; **4**:117b, 212b, 449b, 451a; **5**:409a; **7**:269b; **14**:20b
Homann Cartographic Bureau, **9**:233a, 234a
Homberg, Wilhelm or Guillaume (1652–1715), 6:477–478; **4**:352b; **5**:352a; **7**:111a; **8**:174b
  chemistry, **1**:451b; **2**:16b, 381b; **5**:353a; **6**:43a, 46a(n)
Hombres-Firmas, Louis Auguste d', **3**:264b
Home, Daniel Dunglas, **3**:475b
Home, Everard (1756–1832), 6:478–479; **1**:433a; **2**:420a, 483a,b; **3**:487a; **6**: 567a
  critics of, **14**:564b
  and Hunter's manuscripts, **3**:323b, 324a; **6**:479a
  Lectures on Comparative Anatomy, **1**: 521a
  optics, **14**:564a
Home, Henry, Lord Kames, **8**:256a; **14**:132a
Home economics
  nineteenth-century, **3**:7b
Homeomerism, **13**:546b
Homeo-osmosis, **2**:339b
Homeopathy
  Renaissance, **10**:307b–308a
  nineteenth-century, **14**:116b, 117b
  critics of, **12**:592a
  origin of, **6**:18a–b
Homeostasis, **1**:450a, 610b; **2**:33a, 448a; **9**: 187b; **11**:545b; **13**:530b; **15**:73b–74b
  of hormones, **3**:353a
Homer, **3**:296b; **4**:622a; **6**:316a, 327a, 450a; **7**:40b; **12**:179b–181b *passim,* 554a; **13**:297a, 331b; **15**:638a
  critics of, **14**:536b
Hominoidea, **2**:110a
Homme, **9**:539a
Homo erectus, **8**:105a
Homo ferus, **2**:204b
Homo habilis, **8**:105a
Homochrome phenomenon, **3**:492b
Homogeneity, mathematical principle of, **4**:506a; **5**:54b; **14**:19b
Homography, **3**:213a,b; **12**:213b
Homology (biol.)
  ancient Greek discussion, **1**:261b
  Renaissance discussion, **8**:205a; **11**:528a
  early nineteenth-century discussion, **5**: 356b; **11**:361a; **12**:131a
  mid nineteenth-century discussion, **6**: 466a, 592a; **7**:546a; **9**:572b; **12**:299a; **15**:123a, 166b
  late nineteenth-century discussion, **3**: 389b; **4**:266b; **6**:9a; **7**:475a; **8**:27a; **13**:405b; **14**:426b
  twentieth-century discussion, **5**:467b
  concept of, **7**:402b; **10**:263a
Homology (chem.)
  concept of, **5**:370b–371b; **6**:462a
Homology (math.), **3**:170b, 171a; **6**:496b–497a; **7**:577a; **11**:60a, 79b, 449b
Homomorphism (math.), **4**:420a; **10**:139a
  inverse, **6**:496b, 497a
Homoplasy, **8**:27a
Homotopy, **2**:513a; **6**:496b–497a; **14**:311a
Homotyposis, **10**:462a–463a
Homozygosity, **4**:271b
Homunculus, **7**:41b
Honda, Kotaro (1870–1954), 6:479–480; **9**: 606b
Honda, Rimei, **7**:19a
Hondius, Henry, **5**:408b
Hondius, Jodocus, **14**:514a
Honduras
  discovery of, **15**:89b
Honest Whigs, Club of, **3**:52a
Honey
  analysis of, **8**:174b
Honeybee
  see Bee
Honeycombs, **8**:612b*; **12**:533a–b
  mathematical analysis of, **8**:306a
  measurement of, **14**:534a
  structure of, **7**:442b; **11**:332b
Honeystone, **14**:476a
Hong Kong
  College of Medicine
    founding of, **9**:83a
Hönigschmid, Otto (1878–1945), 6:480–481; **11**:418a
  atomic weights, **11**:417b; **14**:319a
  radiochemistry, **12**:505b
Honing, Jan Antonie, **14**:97a, 101a
Hönl, Helmut, **8**:473a; **12**:529b
Honnecourt
  see Villard de Honnecourt
Hoogenberg, Frans, **10**:239a
Hoogewerff, Sebastian, **5**:215a
Hooke, Robert (1635–1702), 6:481–488; **2**: 614a*; **3**:528b; **6**:36a, 495b
  acquaintances of, **2**:46b; **8**:149b; **9**:310b; **10**:292b; **12**:581a; **13**:526b; **14**:374a, 405a
  Animadversions on the First Part of the Machina Coelestis of . . . Hevelius, **6**:363b
  astronomy, **11**:163a
  and Boyle, **6**:482a–b, 487b; **9**:245a
  celestial mechanics, **6**:485b–486a; **10**: 62a–64b, 72a
  chemistry
    air, **6**:40b, 41a, 47a(n)
    combustion, **6**:484b–485a
  collection of, **3**:362a
  on contemporary scientists, **14**:369b–370a
  correspondence of, **9**:465a; **10**:36b
  critics of, **14**:565a
  crystallography, **2**:431a; **11**:523a
  on Dee, **4**:6a*
  earth's revolution, **1**:616a
  editions of works, **4**:40b
  experiments, **6**:170a, 483a
  geology, **6**:486b–487a; **8**:416a; **11**:302b–303a, 304a
  and Haak, **5**:607a–b
  and Hevelius, **6**:361a, 363b, 487b
  Hooke's law, **3**:32a, 145b, 146a; **6**:482b, 485a, 487a; **10**:3a
  influence of, **8**:127a; **9**:534a
  influences on, **14**:374b, 377a(n)
  instrumentation, **5**:279a(n); **6**:487a; **13**: 336a
    air pump, **2**:377b
    clocks, **4**:40b; **6**:482b, 599a, 605a
    telescopic sights, **1**:341b
  and Leibniz, **5**:607a
  and Mayow, **6**:47a(n); **9**:245a
  Micrographia, **1**:411b; **2**:379b; **6**:483b–485a; **9**:66a, 498b; **10**:43b, 54b; **14**: 490b, 493b(n)
  microscopy, **5**:534b, 535b
  and Newton, **6**:68a, 484b, 486a, 487a,b; **10**:44b, 54a–55a, 56a, 60a, 61b, 62a–64b, 89b(n)
  and Oldenburg, **6**:484b, 487a; **10**:202a
  optics, **5**:544a; **6**:484a–b; **10**:59a
  parallax measurement, **2**:387b; **9**:463b–464a
  physiology, **4**:65a; **8**:525b, 526b; **9**:243a–245b *passim;* **14**:407b
  and Royal Society, **2**:507a; **6**:169b, 483a–b, 487a; **10**:201a,b; **14**:371b, 373b, 510b
  zoology, **13**:527a
Hooker, John Daggett, **6**:30a
Hooker, Joseph Dalton (1817–1911), 6:488–492; **3**:582a; **6**:289a, 494a, 592b; **12**:259a
  acquaintances of, **1**:412b; **2**:476a; **3**:576b; **5**:539a; **8**:528a, 573b; **13**:523a
  on contemporary scientists, **12**:594b; **14**:135b
  correspondence of, **5**:513a; **13**:341b
  critics of, **2**:476b
  and evolution, **3**:573b, 574a
  Flora of British India, **13**:342b
  Genera plantarum, **1**:615a; **2**:18b
  influences on, **5**:75b
  at Royal Botanic Gardens, **6**:494b; **13**: 342a
  taxonomy, **9**:340b
  and X Club, **5**:125a
Hooker, William Jackson (1785–1865), 6: 492–495; **1**:423a; **5**:540a*; **6**:488b, 489b; **8**:564a; **12**:472a*; **14**:189b
  acquaintances of, **2**:19a, 521b; **8**:372b
  British Flora, **1**:358b, 359a; **13**:341a
  at Kew Garden, **1**:89b; **6**:491a
  Synopsis Filicum, **1**:412b
Hookeria, **6**:493a
Hookworms, **7**:447a; **12**:142a; **13**:63a
Hoole, Charles, **3**:360b
Hope, Thomas Charles (1766–1844), 6:495–496; **2**:174a, 175a, 180a; **5**:492b; **6**: 84a; **13**:499a
  Hope's experiment, **6**:495b
  students of, **5**:13b, 537a
Hopf, Eberhard, **9**:405a
  Wiener-Hopf equation, **14**:346a,b
Hopf, Heinz (1894–1971), 6:496–497
  Hopf invariant, **6**:496b
Hopff, Heinrich, **9**:354a
Hopkins, Frederick Gowland (1861–1947), 6:498–502; **4**:310b; **5**:83a; **8**:452b; **15**:202a*, 418a
  biochemistry, **5**:37b–38a, 451b; **9**:381a
  students of, **4**:357b; **5**:41b, 42a; **6**:21a; **15**:417b
Hopkins, William (1793–1866), 6:502–504; **2**:570a; **9**:88a*, 199a; **13**:74b, 75a, 236b; **15**:200b
Hopkinson, John (1849–1898), 6:504; **7**:56a; **9**:212b
Hopkinson, Thomas, **5**:131a
Hoppe-Seyler, (Ernst) Felix (Immanuel) (1825–1895), 6:504–506; **7**:273b; **14**:41a
  biochemistry, **1**:362b; **2**:561b; **9**:348b; **12**:319a,b
  students of, **5**:148b, 149a, 269a; **7**:466b, 467a, 519b; **9**:380b
  Zeitschrift für physiologische Chemie
    see *Zeitschrift für physiologische Chemie*
Hops, **12**:317a

Horace (poet), **4**:30b, 172b; **5**:365a; **7**:27b; **14**:363b
Horbaczewski, Jan (1854–1942), 6:506
Horizon
naming of, **1**:339a; **4**:429b
refraction at, **14**:579a
Hörlein, (Philipp) Heinrich, **4**:154a
Hormones
adrenal, **10**:611a; **14**:445b
and carbohydrate metabolism, **3**:415b–416a
discovery of, **1**:537a
feedback mechanisms, **15**:229a
female, **3**:352b; **12**:474b
gonadotrophic, **8**:359a
placental, **3**:352b
and immunity, **5**:317a
inhibitory action, **10**:570a
lactogenic, **2**:345b; **5**:225b
male, **5**:209a
interstitial theory, **2**:344b
and sexual development, **8**:358b–359a
and morphogenesis, **8**:359a
naming of, **12**:617b
in plants, **3**:576a, 581b; **14**:255b
protein nature of, **1**:11b
purification of, **3**:352b
replacement therapy, **12**:472b, 474b–475a
sex, **14**:595b
and behavior, **8**:359a
steroid nature of, **10**:611a; **15**:258b
structure of, **14**:445b
therapeutic use of, **12**:384a
immune reactions, **3**:353a
trophic, **3**:352b–353a
and wound healing, **1**:303b
see also Endocrinology; Glands; specific hormones
Horn, George Henry (1840–1897), 6:506–507; 5:214a
Horn, Jakob, **2**:144b
Horn d'Arturo, Guido (1879–1967), 6:507–508; 9:78a*, 487a*
comets, **11**:412b
Horne (Hornius), Johannes van (1621–1670), 6:508–509; 13:31a
anatomy, **13**:169a, 170b
students of, **5**:484a,b; **11**:587a; **12**:40a; **13**:169a
Horne, John, **5**:337a; **8**:33b
Hornea, **8**:5b
Horner, Johann Friedrich, **6**:434b
Horner-Bernard ocular syndrome, **2**:30b
Horner, Johann Kaspar, **11**:377a
Horner, Leonard (1785–1864), 6:509–510; 5:334a, 436b; **8**:569a; **12**:522b; **13**:499a
Horner, William George (1786–1837), 6:510–511; 9:578b; **12**:522b
Horner's method, **3**:254b, 267b; **4**:610a; **6**:510a–511a; **7**:257a; **8**:315b, 423a; **12**:291a; **14**:540b, 541b, 544a,b
number theory, **13**:219a
Hornfels, **5**:39b*
Horns
jugular, **9**:66a
morphogenesis, **9**:66a
and teeth, compensation principle, **5**:443a–b
Hornsby, Thomas (1733–1810), 6:511–512
parallax determination, **6**:263b
Hornschuch, Friedrich, **10**:13b
Horocycle, **8**:430b
Horologium viatorum, **6**:302a
Horology
ancient Greek, **15**:209b–210b
medieval Arabic, **1**:5b, 152b; **12**:357b, 360b–362a; **15**:259a–b
medieval Latin, **1**:122a, 382a; **4**:164a–b; **6**:301b–302a; **9**:363b
Renaissance, **2**:403b, 406a, 534b; **4**:183b; **6**:144a; **8**:198a, 208a; **14**:273b
seventeenth-century, **1**:549b; **2**:349b; **3**:23a; **4**:40b, 184b; **5**:246b; **6**:363a, 482b, 604b–605a, 608a; **7**:376a, 578a; **8**:271b*; **9**:97b, 485b, 507b; **10**:314b; **11**:411b, 423b, 525b
eighteenth-century, **2**:43a, 389a; **3**:39a; **4**:565b; **5**:491a; **6**:130a–131a; **9**:163b; **11**:471b; **12**:118a; **14**:311b, 485a
nineteenth-century, **3**:388b; **4**:354a; **5**:86a, 224a*, 583a; **10**:133b; **12**:131b; **13**:111b
twentieth-century, **3**:239a
see also Chronometer, Clock, Gnomonics, Time
Horopter, **4**:615a; **6**:248b, 300a
Horoscopes, **4**:622a; **15**:125b, 540b, 644b, 674a, 678a, 679b, 680b, 719b, 723b, 725a, 726b
see also Astrology
Horosphere, **8**:431a
Horrebow, Christian (1718–1776), 6:512–513
Horrebow, Peder, **6**:512b
Horrebow, Peder Nielsen (1679–1764), 6:513–514; 6:512a–b; **11**:527a*
latitude determination, **13**:12a
translations of works, **13**:99b
Horrocks, Jeremiah (1618–1641), 6:514–516; 3:457b; **5**:526a; **7**:308a; **10**:60b; **12**:342b–343a
editions of works, **14**:147a
lunar theory, **4**:166a; **5**:23b–24a; **8**:178b
manuscripts of, **13**:445a
Opera posthuma, **5**:528b
publication of, **3**:348b
Horror vacui, **4**:383b–385a *passim,* **5**:234b, 575a; **6**:311a; **9**:136b; **13**:93a,b, 438a; **15**:33b
Horse
age of, **11**:159a, 604b
anatomy, **8**:204b; **15**:112b
artificial insemination, **7**:32a
breeding lines, **3**:122b
coat colors
inheritance of, **3**:122b; **13**:134a
diseases of, **3**:359b; **7**:96b, 100b, 286b–287a, 429a
domestication of, **1**:351b; **15**:649b, 650a,b
evolution of, **3**:573a; **5**:46b; **6**:595a–b; **7**:480a; **9**:134b
experimental use of, **1**:576a; **3**:219b
feeding of, **8**:29b
fossil, **5**:501a*; **6**:595a; **8**:169b
hybridization, **7**:32b–33a
maximum daily work, **4**:44a
motions of, **6**:445a
palomino breed, **3**:122b
physiology, **11**:390b–391b
Przhevalski's, **7**:33a
veterinary science
medieval Latin, **2**:314a; **11**:600b–601a
Renaissance, **11**:604b–605a
wild, **11**:181b
Horse buying, problem of (math.), **4**:606b, 607a, 612b
Horse typhoid, **10**:398b, 399b
Horsemeat, **5**:359b
Horseshoe crab, **8**:27a
Horsfield, Thomas, **1**:485a; **2**:520b–521a; **11**:262a
Horsford, Eben Norton (1818–1893), 6:517–518; 3:397b; **4**:274b, 364b; **6**:57a
Horsley, Samuel, **4**:450b; **5**:137a; **6**:577a; **7**:163a; **10**:87b(n)
Horsley, Victor Alexander Haden (1857–1916), 6:518–519; 2:584a, 599a; **7**:431a*; **13**:520a
neurology, **1**:569b; **3**:517b; **7**:49*; **12**:355b
Horst, Johann Daniel, **6**:387a
Horstmann, August Friedrich (1842–1929), 6:519–520; 13:578b
Horsts, **10**:168a; **11**:336b
Hortensius, Martinus (1605–1639), 6:520–521; 1:566b; **8**:28a; **12**:501a
Horticultural Society of London, **8**:371a–b
Horticulture
Renaissance, **4**:412b; **5**:361b; **11**:97a
seventeenth-century, **4**:495a–b; **12**:357a; **13**:451a
eighteenth-century, **1**:88a–89a, 435b, 486a–b; **2**:257a–b, 390a; **3**:172b, 349b, 350a, 351a; **8**:178a; **9**:365b–366a, 390a–b; **10**:283b
nineteenth-century, **1**:89a–b, 422b, 423a; **2**:444b; **3**:556b–557a; **5**:83b; **7**:408b–409b; **8**:371a–372b, 521a; **9**:279a, 282b, 618a–b; **14**:33b–34a
twentieth-century, **1**:422b; **3**:259b; **10**:130a; **15**:425a–426b
American, **1**:396a–b, 486a–b
British, **1**:88a–89a, 89a–b, 422b, 423a; **3**:350a, 351a, 556b–557a; **5**:361b
forcing, **3**:359b
French, **1**:243b; **2**:444b
humidity, role of, **3**:556b–557a
see also Agriculture; Arboriculture; Botany; Congresses, scientific; Education, horticultural; Forestry
Hortulanus, **2**:23a
Hosack, David (1769–1835), 6:521; 4:21b, 273b; **11**:218a; **13**:47a; **14**:457a*
Hosemann
see Osiander, Andreas
Hoshino, Sanenori, **3**:265b
Hosius
see Osius
Hospitals
medieval, **9**:602b
Renaissance, **7**:16b
eighteenth-century, **12**:458b
reorganization of, **3**:2a; **8**:258b
nineteenth-century, **8**:7b; **11**:132b
maternity, **3**:489b; **12**:294b–296a
military, **11**:148a, 363b–364a
outpatient, **11**:132b
see also Asylums; specific hospitals listed under cities, *e.g.,* Paris, La Charité
Hospitalstidende
editors of, **8**:7a
Hot springs
see Water, springs of
Hottentots, **2**:505a; **13**:282a–b
Houdijk, Alexander, **12**:497a
Hoüel, Guillaume-Jules (1823–1886), 6:522; 2:271a; **8**:433a; **13**:249b, 414b
Hough, George Washington (1836–1909), 6:522–523
Houghton, Douglass (1809–1845), 6:523–524; 4:274b
Houghton, William, **5**:491a
Hourglass, **2**:42b; **8**:198a
Hours, **15**:711a, 711b–712b, 723b
civil, **11**:189b; **15**:520a
equal, **15**:714a
seasonal, **15**:713a, 714a
Housefly, **5**:424b; **6**:525a
diseases of, **3**:340b*
Houses, celestial, **6**:273b; **7**:82b, 123b, 299b; **15**:154b
Houses of Life, **15**:723a

Houssay, Bernardo Alberto (1887–1971), 15:228–229; **3**:416a; **8**:452b
Houston, Edwin James, **7**:288a; **13**:361b
Houston, William, **12**:530a
Housz
see Ingen-Housz
Houtermans, Fritz Georg, **5**:272a
Houtgast, Jakob, **9**:415a
How, William
Phytologia Britannica, **7**:147b; **9**:313a
Howard, Leland Ossian (1857–1950), 6:524–525; **5**:71b*; **9**:537a*
Howe, James Lewis (1859–1955), 6:525; **3**:301b–302a*
Howe, William, **11**:314a
Höwelcke, Johann
see Hevelius, Johannes
Howell, William Henry (1860–1945), 6:525–527; **4**:398a; **9**:142b, 266a*; **12**:470a
Hrdlička, Aleš (1869–1943), 6:527–528; **2**:172a
Hsiao-ch'ang, **1**:312b
Hsieh Chuang, **12**:380a
Hsin hsiu pen-ts'ao, **8**:396a*
Hsi-yang hsin fa li shu, **14**:160a
Hsu Feng-k'ao, **3**:266a
Hsu Kuang-ch'i, **11**:403a
Hsueh Feng-tso, **14**:159b, 160b
Hua Heng-fang, **3**:266b
Huang, Kun, **15**:41a, 43b
Huang P'ei-lieh, **14**:540a
Huang, Su-Shu, **13**:118b
Huang Ting
T'ien-wen ta-ch'eng kuan-k'uei chi-yao, **15**:731b
Huang Yü-chi, **8**:419b
Ibn Hubal al-Baghdādī, **15**:241b
Ḥubaysh ibn al-Ḥasan al-Aʿṣam, **7**:25a; **15**:230b, 236a–242a *passim*
Hubble, Edwin Powell (1889–1953), 6:528–533; **12**:349a, 455a
galaxies, **1**:352b; **9**:405b
Hubble's constant, **6**:531b
Hubble's law, **6**:530b–531b
nebulae, **6**:30b; **8**:582b; **12**:455b, 456a
red shift, **4**:330a; **13**:429b
Huber, François, **2**:286b; **3**:45b; **11**:132b; **12**:309a
Huber, Johann Jacob (1707–1778), 6:533–534
Huber, Maksymilian Tytus (1872–1950), 6:534–535
Hübner, Hans, **1**:357a; **5**:12b; **14**:141b
Hubrecht, Ambrosius Arnold Willem (1853–1915), 6:535–536; **6**:138b*; **7**:94b; **14**:104a*
Hückel, Erich, **3**:618a, 620a; **5**:218a; **12**:419a–b; **15**:461a
Debye-Hückel theory, **1**:300a; **2**:170a, 499a; **7**:208b; **10**:156b; **13**:156a
Hückel, Walther, **5**:218a; **14**:445b
Hudde, Jan (1628–1704), 6:536–538; **1**:474b; **4**:571a; **10**:43b, 45b; **14**:467a–b
acquaintances of, **2**:46b; **5**:520b; **6**:359a; **8**:149b, 163a
algebra, **11**:512b; **12**:206b
instrumentation, **6**:148b; **13**:173b
Hudson, Claude Silbert (1881–1952), 6:538
Hudson, William (1733–1793), 6:538–539
Hudson, William Henry, **12**:258a
Hudson's Bay Company, **5**:76b
Hueppe, Ferdinand, **7**:171a, 423a, 426b
Huerto, Garcia del, **8**:120b
Hues, Robert, **5**:593a
Tractatus de globis coelesti et terrestri ac eorem usu conscriptus, **9**:123b
Hufeland, Christoph Wilhelm, **4**:288b; **5**:217b; **11**:363b, 364b
Hufnagel, Leon (1893–1933), 6:539–540; **5**:364b*
Hügel, Karl Alexander Anselm von, **11**:177b
Huggins, Margaret Lindsay Murray, **6**:540b–541a
Huggins, William (1824–1910), 6:540–543; **6**:70a; **11**:537a
acquaintances of, **3**:475b; **4**:498a; **14**:55b
on contemporary scientists, **8**:46b*
correspondence of, **9**:215a
instrumentation, **4**:168a, 180a; **10**:31b
and Lockyer, **6**:542b–543b
nebulae, **1**:295b; **11**:485a; **12**:266b
spectroscopy, **2**:146b; **3**:125b(n); **5**:20a, 561b*; **6**:350b; **9**:391b–392a; **12**:269a; **14**:55a
Hugh of Lucca, **2**:314a,b; **6**:276b
Hugh of St. Victor (*d.* 1141), 6:543–545; **12**:83a
Hugh of Santalla, **4**:543b; **9**:160b; **15**:186a
Hughes, Edward David, **14**:124b
Hughes, Robert, **6**:127b
Hugo, Victor Marie, **2**:544b; **5**:44a, 365b; **9**:327a
Hugoniot, Pierre Henri (1851–1887), 6:545–546; **4**:230a; **13**:104b
Rankine-Hugoniot relations, **11**:293a
Huguenots, **12**:134b; **13**:593a
Hūlāgū Khān, **9**:555a; **13**:509a, 510b
Hull, Albert Wallace (1880–1966), 6:546–547; **8**:25a*
Hull, Gordon Ferrie, **1**:301a; **11**:122b
Hull (Eng.)
Botanic Garden, **6**:184b
Human Biology
editors of, **10**:445a
Human nature
seventeenth-century ideas of, **6**:447a–b
Humanism
medieval Latin, **1**:4b, 91b
Renaissance, **1**:77b, 79b; **2**:536a, 590b; **3**:413a; **4**:212b; **5**:121a, 580b; **9**:264a, 581a; **10**:122b; **11**:457a; **12**:151a; **13**:209a, 210b; **14**:29b, 47a–b, 127b; **15**:160b–161a, 474a
in medicine, **1**:612a; **4**:285a; **8**:248a–249b, 361a, 554b; **9**:165b
seventeenth-century, **5**:284b, 286b
eighteenth-century, **4**:85a; **5**:380a; **7**:412b
nineteenth-century, **15**:132a–b
twentieth-century, **3**:112a; **12**:109b, 111b–112a
Humaria, **5**:584b
Humason, Milton Lasell, **6**:531a,b; **10**:473b
Humbert, Marie-Georges (1859–1921), 6:547–548; **4**:373b
Humbert, Pierre (1891–1953), 6:548–549; **10**:342a*; **11**:5b*
Humboldt, Friedrich Wilhelm Heinrich Alexander von (1769–1859), 6:549–555; **1**:381b; **2**:516b; **3**:302a; **5**:74b, 75a; **7**:368a; **9**:196b, 457a; **13**:583b; **14**:187a–b; **15**:267a
acquaintances of, **1**:72b, 201a, 203b, 388b; **2**:510a, 552b, 554a; **3**:16b, 210a; **4**:201b; **5**:75b, 77a, 155a–b; **6**:268b, 325a, 456a; **7**:319b, 506a; **8**:104a, 321a, 564b; **9**:177b, 252b, 345b, 570b; **10**:543a, 573a; **11**:367b, 376b, 464a,b, 473a, 539a,b, 541a; **12**:173b, 182a, 215a, 301a, 587b; **13**:13a, 14a, 109b, 127a, 466b, 506a, 557a; **14**:54a, 204a,b, 257a, 386b, 471b; **15**:82b, 123a, 430a
astronomy, **5**:256b; **6**:553b; **12**:239b
biographies of, **2**:533a
botany, **5**:546b; **6**:493b; **14**:117b, 386b
on contemporary scientists, **12**:524b
correspondence of, **7**:16a*, 533b; **11**:387b; **12**:50b
critics of, **1**:173b–174a; **12**:263a, 300b
and Ehrenberg, **4**:288b–291a *passim,* 292b*
and Eisenstein, **4**:341a–b, 342a–b
and Gauss, **5**:301b, 304b–305a, 306a, 308a; **6**:552b
and Gay-Lussac, **5**:318b–320a; **6**:551a–b
geology, **9**:620b; **12**:49b
historiography, **15**:90b*
Humboldt effect, **6**:553b
influence of, **2**:207a; **6**:6b; **9**:177a, 196a, 431b; **11**:180b, 181a, 337a; **12**:38a, 154b, 215a, 300a, 421a; **14**:133b, 190a
influences on, **14**:387b
Kosmos, **7**:55a; **8**:514b
commentaries on, **3**:434b
translations of, **15**:527b
magnetism, **2**:135a, 472a; **3**:260a; **5**:305a, 306a; **6**:107a; **7**:608b; **8**:187a; **12**:50a, 51a
meteorology, **1**:203a; **2**:557b; **4**:28b, 175a; **6**:551b, 552a
patronage of, **2**:98a, 108a, 356b, 422b; **3**:466b, 623b; **4**:124a,b, 200b, 289b, 369b; **6**:242a, 553a; **7**:51a, 521b; **8**:330a,b; **9**:344b; **11**:360b, 368a, 368b–369a; **13**:554a–b
physiology, **9**:570a
Plantae aequinoctiales . . . , **13**:506a
pneumatic chemistry, **14**:74b, 75a
Political Essay on the Kingdom of New Spain, **4**:345b
students of, **1**:19b; **15**:526a
translations of works, **12**:53a*; **13**:466b; **14**:472b(n)
Voyage aux régions équinoxiales du Nouveau Continent, **15**:267b
Humboldt, Wilhelm von, **6**:549a, 550b, 551b; **11**:363b; **14**:239b
Hume, David (1711–1776), 6:555–560
acquaintances of, **2**:174a; **9**:480b, 483b; **11**:499a
causation, **1**:141a; **3**:572b; **7**:233b–234a; **8**:261b
critics of, **12**:178a; **14**:287b–288a
influence of, **3**:598b; **4**:321b; **6**:596a
influences on, **1**:376b; **2**:18a; **5**:288b(n), 416a; **10**:173a
logic, **5**:9a; **11**:357b
on miracles, **3**:120a*
natural philosophy, **9**:189a
translations of works, **11**:492b
Hume-Rothery, William (1899–1968), 6:560–562
Humic acids, **9**:558a–b
Humidity, atmospheric
determination of, **12**:595b
by weight of wool, **3**:515b
in horticulture, **3**:556b–557a
see also Hygrometer
Hummingbird, **5**:481a; **8**:265b
Humors, medical theory of
ancient Greek, **1**:315a; **4**:106a, 385b–386a; **6**:317a, 422a, 427b–428a; **11**:127b
ancient Roman, **1**:212b; **5**:230b, 231a
Byzantine, **10**:417a–418a
medieval, **7**:28a; **9**:469a; **11**:244b, 601a; **12**:82a; **15**:240a, 498b
Renaissance, **1**:400a; **4**:387a,b, 585a; **5**:106b, 107a; **11**:604b–605a; **13**:535a
seventeenth-century, **5**:48a; **6**:458b–459b; **7**:165a; **10**:478a; **11**:467a; **12**:101b–102a, 102b, 311b
eighteenth-century, **8**:416b

nineteenth-century, **3**:463b; **5**:233b
fluxes of, **6**:423a–b, 428a
rejection of, **4**:209b; **10**:307b; **12**:335a, 458b, 538b
and seasons, **6**:426b
specificity of, **6**:426b
Humphrey, George Colvin, **1**:357a; **6**:135b
Humphry, George Murray, **5**:81b; **13**:504a
Humus, **4**:144b; **8**:178a
Ḥunayn ibn Isḥāq ibn Ḥunayn, 7:24b
Ḥunayn ibn Isḥāq al-ʿIbādī Abū Zayd (Johannitius) (808–873), 15:230–249; **1**:107a, 444a; **2**:619b; **5**:233b; **7**:24b, 187a; **13**:230b; **14**:458b
Ars medica, **9**:40b; **11**:238a; **15**:239a–242a
critics of, **11**:445a
Isagoge
commentaries on, **13**:415b*
medicine, **14**:638a; **15**:188b
optics, **1**:380a
Questions
commentaries on, **9**:604b
translations by, **1**:275a; **4**:121b; **6**:203b(n); **7**:25a, 40a; **9**:301b*; **10**:111b; **11**:32a, 238a; **15**:261b
translations of works, **13**:401b
Hund, Friedrich, **3**:4a; **8**:474b; **12**:22a, 419a; **15**:43a
Hundt (Hund, Canis), Magnus (1449–1519), 6:562–563
Hungarian Academy of Sciences
see Magyar Tudomanyos Akademia
Hungarian language
in science, **7**:22b
Hungary
botany, **7**:390b–391a
chemistry, **14**:447b–448a, 603a
geology, **2**:106a
National Museum, **7**:391a
politics, **5**:474b–475a; **9**:148b
travelers to, **2**:463b
Hunger mechanism, **3**:69a
Hunsrück Mountains
geology, **8**:511b
Hunt, Franklin Livingston, **4**:195b
Hunt, George, **2**:212a
Hunt, Henry, **5**:515b
Hunt, James (1833–1869), 6:563–564
Hunt, Reid, **4**:298b; **15**:106b
Hunt, Thomas, **6**:563a
Hunt, Thomas Sterry (1826–1892), 6:564–566; **2**:289b; **3**:478b; **4**:364b; **8**:461b, 462a; **12**:435a
A New Basis for Chemistry, **12**:607a
Hunter, Alexander Jardine, **8**:355a
Hunter, John (1728–1793), 6:566–568; **6**: 569a
acquaintances of, **1**:398b; **3**:67b; **4**:172b; **7**:11b; **12**:136b, 509b
anatomy, **6**:367a
critics of, **8**:97b; **12**:560a
evolution, **11**:137a
and Home, **6**:478b–479a
influence of, **3**:324a; **5**:470b
Lyceum Medicum Londinense, **3**:487b
medicine, **3**:487a
Observations on . . . the Animal Oeconomy, **7**:96a
optics, **14**:564a
physiology, **9**:8a
scientific collections, **3**:323b–324a; **6**:567b
students of, **2**:483a; **3**:486b; **7**:95b, 96a; **10**:321b; **12**:471a; **14**:296b, 297a, 456a
Hunter, John Stuart, **14**:554a
Hunter, William (1718–1783), 6:568–570; **1**:398b; **3**:486b, 487a, 494a, 577b; **4**:172b; **6**:566a, 567a; **7**:11b; **12**:136b
Anatomy of the Gravid Uterus, **1**:399a–b*
anatomy school, **1**:398b; **3**:487b; **6**:367a
and Douglas, **4**:173a
Medical Commentaries, **6**:566a–b; **9**: 481b, 482b–483a
students of, **9**:482b; **14**:296b
Hunterian Museum
see under Glasgow, University of; London
Huntington, Edward Vermilye (1874–1952), 6:570; **2**:298a; **10**:485a
Huntington, Ellsworth, **11**:210a
Huntington, Henry Edwards, **6**:32a
Huntington, Oliver Whipple, **3**:398a
Huntington, Robert, **9**:465a
Huntington Hospital (Boston), **4**:195a, 196a
Hunyady, Eugen von, **7**:530b
Huret, Grégoire, **4**:47b; **12**:85a
Hurewicz, Witold, **6**:496b–497a
Huronian orogeny, **2**:89b
Huronian System, **6**:565a
Hurricane
theories of, **4**:592b; **8**:174a–b
Hurst, Charles Chamberlain, **3**:121a
Hurter, Ferdinand, **1**:21b
Hurwitz, Adolf (1859–1919), 6:570–573; **4**: 2b, 5a; **6**:388a,b; **9**:411b; **12**:526a; **14**:222a
analysis, **4**:548a; **6**:389b; **7**:398b
correspondence of, **12**:228a
group theory, **14**:284a,b
Hurwitz' theorem, **6**:572b
stability theory, **13**:73a
al-Ḥusain, **4**:117b
Ibn al-Ḥusayn, **11**:240b
Huschke, Emil (1797–1858), 6:573–574; **6**:6b
Huschke's auditory teeth, **6**:573b
Husemann, Elfriede, **13**:3b
Huser, John, **10**:306b
Hūshank, **1**:34b
Husserl, Edmund, **2**:277b; **4**:100b; **7**:483a, 484b; **14**:222a, 613b
Hussey, William Joseph (1862–1926), 6: 574–575; **1**:87b; **3**:510a; **12**:139b*
Hutchins, Thomas, **9**:537b; **10**:509b
Hutchinson, Arthur, **15**:17b
Hutchinson, John (1811–1861), 6:575–576
Hutchinson, Jonathan, **7**:46b, 47b; **8**:407b
Hutten, Ulrich von, **1**:620b; **2**:535b
Hutton, Charles (1737–1823), 6:576–577; **5**:530a; **8**:271b; **9**:163b, 504b
historiography, **3**:621b; **7**:443a; **9**:159b*; **10**:264b; **12**:445a*; **14**:312a*
Hutton, Frederick Wollaston, **5**:611b
Hutton, James (1726–1797), 6:577–589; **1**: 422a; **2**:175a
acquaintances of, **2**:174b; **4**:250a; **11**: 304b; **14**:132b
agriculture, **6**:578b, 584b–585b
chemistry, **2**:180a; **6**:585b–586b
critics of, **2**:556b, 568a; **4**:28a, 72a, 151b; **5**:31b; **7**:69a, 389a; **10**:322b; **13**:448b
evolution theory, **6**:585a–b
geology, **5**:13b; **6**:578a–584b; **11**:376b
plutonism, **1**:413a; **2**:554b; **6**:583b
specimen collection, **6**:579b
theory of the earth, **3**:210b; **6**:580b–584b; **7**:69a, 185b; **8**:594b; **10**:209b; **15**:56a
uniformitarianism, **1**:413b; **6**:582b; **12**:273a
heat theories, **8**:261b
influence of, **12**:262a
influences on, **12**:122b
meteorology, **6**:585b
philosophy, **6**:586b–587a
Theory of the Earth, **1**:395a; **7**:389a; **9**: 534a; **11**:34b
defenders of, **6**:54a–55b
editions of, **5**:335b
forerunners of, **13**:441b–442a
Hutton, William, **8**:371b; **14**:397b, 462b
Huttonism, **10**:8b; **15**:55a
opposition to, **12**:276b; **13**:372b
Huxley, Julian Sorell, **5**:467a; **9**:542a, 564b
Huxley, Thomas Henry (1825–1895), 6:589–597; **1**:420a; **3**:475b; **5**:290b; **11**:331a, 537a; **13**:341b
acquaintances of, **5**:125a; **8**:528a; **9**:142a–b, 199a; **10**:40a; **12**:258a, 259a, 570a; **13**:521b, 522b; **14**:424a, 425b
anthropology, **5**:46b; **6**:593b
on contemporary scientists, **9**:346b*; **10**:263b*; **12**:468b; **13**:524a*
critics of, **1**:540a; **5**:513b; **12**:263a; **14**: 146a
and Darwinism, **3**:576b; **6**:592b
and education, **6**:595b–596a; **11**:514b
embryology, **1**:601a; **5**:444a; **6**:535b, 590a–592a; **7**:402a
on evolution, **2**:504a; **3**:571a, 572b, 574a; **6**:9b, 591a, 592a–594a; **13**:383b, 504a; **15**:166b
and Foster, **5**:80a,b, 81a; **6**:595b
Huxley's layer, **6**:339a, 589b
influence of, **6**:8b; **9**:428a–b; **14**:523b
influences on, **11**:380b
Lessons in Elementary Physiology, **5**:83b
Man's Place in Nature, **9**:538b
Manual of the Anatomy of Invertebrated Animals, **5**:386a*
marine zoology, **6**:589b–591a; **14**:146a
on mathematics, **13**:217b
and Owen, **6**:592a, 592b–593b; **10**:262b–263a; **11**:514a
paleontology, **6**:591b, 594b–595b; **12**:38a, 489b
philosophy, **6**:53a, 596a–b
students of, **2**:371a, 501b; **5**:128a; **9**:512b; **10**:241a; **12**:519b; **13**:341a
translations of works, **15**:78b
and Wilberforce, **4**:183b; **6**:592b–593a
Huygens, Christiaan (also Huyghens, Christian) (1629–1695), 6:597–613; **2**: 379a, 431a; **3**:103b, 105a, 111a; **5**: 54b, 240b, 575a; **8**:437b; **9**:485b
and Académie Royale, **3**:102a, 534b
acquaintances of, **2**:53a; **3**:615a; **4**:49b; **5**:520b; **6**:148b, 149a; **8**:149b, 161a,b; **9**:498a; **10**:201a, 292b, 293a, 315a, 339b; **11**:526b; **12**:584b; **13**:334b–335a, 479b, 480a,b
astronomy, **2**:592b; **4**:128a, 506a; **5**:287b; **6**:603b–604a
and Boulliau, **2**:349b
on contemporary scientists, **12**:460b; **14**:466b
correspondence of, **2**:281b; **3**:348b; **4**:75a, 505b; **5**:158b; **6**:520a, 538a; **8**:26a, 250b, 305a; **9**:114a, 465a, 498b, 595a, 599b; **10**:176b, 202a; **11**:423b; **12**: 205b–206a, 210b, 460a, 584a; **13**: 336a,b; **14**:49b(n), 373b; **15**:195b
Cosmotheoros, **6**:611a–b
critics of, **12**:118a–b; **14**:565a
crystallography, **11**:523a; **14**:490b
and Descartes, **4**:52b, 59b; **6**:598a, 599b, 602a–603a, 608b, 610a
Dioptrica, **12**:501a
earth, figure of, **3**:103a, 282b
editions of works, **5**:510b; **7**:466a
engineering, **8**:212a
and Fermat, **3**:64a; **4**:571a, 572b, 573a
and J. Gregory, **5**:524b, 526b, 527b, 529a
Horologium oscillatorium, **2**:49a; **9**:116b; **10**:72a; **11**:423b; **15**:271b

Huygens, Christiaan (cont.)
horology, **2**:507b; **3**:63b; **4**:40b; **5**:86a; **6**:130a, 482b, 604b–605a, 608a; **9**: 507b
Huygens ocular, **6**:603b
Huygens' principle, **1**:600a; **3**:144a, 145a; **4**:230a; **5**:167a, 168a; **6**:191b, 610a; **7**:262b; **9**:129b; **14**:567b; **15**:362b
hydrostatics, **5**:528b; **6**:602a
influence of, **2**:593a; **3**:62b; **4**:44a; **6**:611b–612a; **9**:596a; **12**:84b
instrumentation, **13**:336a
and Mariotte, **9**:114a–120b *passim*
mathematics, **1**:474b; **6**:599b–602a, 612a; **10**:45b
analysis, **2**:47a,b, 48a; **8**:164b, 268a; **12**:75b; **14**:148a
geometry, **5**:499b; **6**:359a,b, 537a; **10**: 115b, 336b; **11**:402a–b; **12**:206a–b; **15**:118a
probability, **3**:451b; **6**:537a; **9**:454a, 499b
mechanics, **4**:60a,b; **5**:528b; **6**:604b–611b; **8**:152b, 209a; **10**:61a, 62a, 70a; **12**: 118a; **14**:150a
central force, **2**:54b; **3**:431b; **7**:276a
dynamics, **8**:163b
gravitation, **6**:606b–611a *passim;* **9**: 115a
impact, **6**:602a–603a; **8**:151a, 152a, 153a; **9**:116a,b
pendulum, **6**:604b–606b
metrology, **9**:554b
and Newton, **6**:599a, 603b, 609a–b, 610b, 611a; **10**:44b, 45a, 54a
optics, **5**:483b; **6**:609b–610a; **9**:120b(n), 159a, 464a*; **10**:60a; **13**:78a
microscope, **7**:110b
rainbow, **1**:475a
refraction, **1**:481b; **2**:138b; **5**:167a, 170b; **6**:603a–b, 610a; **8**:460a; **9**: 73a,b; **10**:79b; **14**:491b; **15**:362a–b
telescope, **1**:341a–b; **4**:41a; **6**:6a, 137b, 363a; **9**:464a*
philosophy, **6**:608b–611a; **11**:507b
De ratiociniis in ludo aleae, **1**:208b; **4**: 574b; **9**:452a,b; **10**:335b; **12**:206b
commentaries on, **2**:50a
students of, **9**:446b
Systema Saturnium, **2**:349b, 592b
Traité de la lumière, **9**:117a; **15**:361b–362a
vortex theory, **2**:38b; **6**:604a, 609b, 610a–b; **9**:52a; **10**:91b(n)
Huygens, Constantijn *(père),* **4**:184b, 505b; **5**:408a,b; **6**:597b–598a, 599a; **8**:127a; **9**:322a(n); **12**:610b; **14**:254b, 368a, 376b(n), 377a(n)
Huygens, Constantijn *(fils),* **6**:148b, 597b–598a, 598b, 599a, 603a; **12**:205b
Hven, island of, **2**:404a–405a, 412a–b
Hyades (star cluster), **6**:351b
Hyatt, Alpheus (1838–1902), 6:613–614; **5**:50a, 486b; **9**:101a*, 536b; **10**:272b, 273b; **14**:1a
evolution, **15**:92b
Hybrid vigor
see Heterosis
Hybridization, **5**:385b; **7**:409a
bees, **9**:282b
bisexual *vs.* unisexual, **14**:100b
botanical, **2**:257a–b, 390a; **3**:351a, 422a; **5**:218a–b; **6**:121b; **7**:440b–442a; **8**: 377b, 460a, 513a; **9**:618a–b; **10**:130a; **13**:477b–479a; **14**:98b, 99b–101b; **15**:68a, 425a–426a
and evolution, **14**:190a
and sterility, **6**:296a; **7**:441b; **9**:561a
capacity for, **4**:512b; **5**:218b
chemical methods, **6**:337a
and evolution, **8**:513a–b; **9**:618b
experimental, **1**:505b; **2**:286b; **4**:286a; **5**: 218b, 359b; **8**:3b; **9**:277a–283a, 520b, 601b; **11**:212b; **12**:555b–556a; **13**: 281b–282a; **14**:33b–34a, 314b; **15**: 113a
and fertility, **6**:296a; **9**:282b; **14**:101a
racial, **11**:137b
theories of, **9**:188a
twin hybrids, **14**:100b, 101a
zoological
interspecies, **7**:32b–33a
and sexuality, **6**:22a
and sterility, **13**:137b
see also Selective breeding
Hydatid, **4**:621a
Hydra (constellation)
epsilon Hydrae, **3**:605b
Hydra (zool.), **10**:569a
animal nature of, **13**:457a–b
inversion of, **13**:457b–458a
physiology, **13**:457b–458a
regeneration of, **8**:580a; **11**:333b–334a; **13**:457a–458a
Hydrachnellae, **9**:575a
Hydrachnida, **7**:111a; **9**:575b
Hydracids
see Acids
Hydracrylic acid, **1**:578a; **14**:454b
Hydrargyrosis, **10**:307a
Hydrates, **1**:298b; **11**:534a,b, 541a; **14**:31b
Hydration
heat of, **1**:579b; **8**:48a*
ionic, **7**:203b; **8**:290b
surface films, **4**:76b
Hydraulic engine, **3**:8a
Hydraulic lift, **1**:607b
Hydraulic ram, **9**:493a; **10**:118b
Hydraulics
ancient Roman, **15**:514b
medieval Arabic, **4**:541b–542a; **6**:189a
Renaissance, **2**:279b; **8**:207a–b, 212b–213a, 214b; **9**:147a; **11**:405b; **13**:233b
early seventeenth-century, **3**:115b–116a; **4**:46a, 49a–b; **13**:437a
late seventeenth-century, **3**:101a, 182a; **9**:119b, 369b–370a, 485a, 486b, 529b–530a; **10**:293a; **12**:127b, 210b
early eighteenth-century, **2**:54b; **3**:39a; **4**:38b, 129a; **5**:499a, 511a; **11**:5a–b, 65b, 399b; **12**:283a
late eighteenth-century, **2**:326b; **3**: 387b(n); **4**:208a, 470b, 481a, 501b; **5**:195b; **9**:493a; **11**:164a, 401a–b; **12**:462b; **14**:494a–495a
early nineteenth-century, **1**:328a; **2**:473a; **3**:417b; **4**:402a–b; **5**:410b–411a; **7**: 499b; **10**:118b; **11**:164b–165a
mid nineteenth-century, **2**:469b; **3**:222b, 358a; **5**:101a; **8**:62b; **12**:131b–132a; **14**:232b, 617b–618a
late nineteenth-century, **2**:84b, 356a, 470a; **8**:287b, 464a, 465b–466a; **14**: 620b
early twentieth-century, **2**:466a; **7**:247a–b; **8**:489a
mid twentieth-century, **8**:301a,b
historiography of, **2**:39b
see also Backwater; Flow; Hydrodynamics; Outflow; Overflow; Resistance, in fluids; Underflow
Hydrazine, **13**:337b, 338b; **14**:334a
commercial production, **3**:511a
discovery of, **3**:511a
organic derivatives of, **5**:1a–b
Hydrazoic acid
discovery of, **3**:511a
structure of, **3**:511a–b
Hydrides
decomposition of, **10**:289a
synthesis of, **10**:288b
Hydriodic acid, **10**:530b
dissociation of, **6**:177b
Hydrobromic acid, **10**:530b
Hydrocarbons
nineteenth-century study, **2**:420b; **5**:125b; **12**:209a
acetylenic, **4**:553a–b
alicyclic, **1**:426a–b; **9**:131a–b
allene
optical activity, **11**:87a
aromatic, **1**:426a–b
boiling points, **5**:372b
classification of, **2**:68a; **8**:58a–60b
colored, **7**:518a
combustion of, **5**:621a
cracking of, **4**:553b
cyclic, **4**:553b
catalytic reactions of, **1**:414b–415a
decomposition of, **5**:621a
dehydrogenation of, **4**:553b
"Diels," **4**:91a
formulas for, **3**:542b; **5**:372b; **6**:284b–285a; **10**:178a
gaseous, **6**:284b
hydrogenation of, **1**:414b; **2**:4a; **8**:108a; **14**:156b
illuminating power of, **5**:128a; **6**:284b
liquid
manufacture of from coal, **2**:3b
nitrification of, **13**:468a
nomenclature, **6**:463b; **8**:60b
oily, **10**:498a
phenylated, **5**:464b–465a
polymerization of, **2**:624a; **4**:553b; **7**: 457a–b; **8**:108a; **9**:18b
quaternary, **7**:552a
radicals, **14**:530b–531a
reactions of, **5**:541a; **8**:56a–61a
reduction of, **2**:68a
series, **4**:244b–245a; **8**:56a, 58a–59b
structure of, **4**:91a
synthesis of, **2**:67a–68a; **5**:540b; **7**:21a; **14**:601b–602a
unsaturated, **8**:108a
see also specific hydrocarbons
Hydrocele, **6**:521a
Hydrocephalus, **5**:157a*; **8**:352b; **14**:584b
Hydrocharis, **2**:322b
Hydrochemistry, **4**:42b; **14**:50b
Hydrochloric acid
composition, **11**:146a
density, **9**:95b
gas
discovery of, **11**:144b
reactions of, **5**:320a, 321a; **12**:146a
naming of, **5**:323b
oxidation of, **6**:177b
preparation of, **2**:421b; **5**:420b, 422a; **6**: 255a
industrial, **3**:201a, 202a; **7**:277b
in stomach, **13**:403b
formation of, **3**:351b
vapor density, **13**:299a
see also Muriatic acid
Hydrocyanic acid, **3**:176b
analysis of, **2**:76b; **5**:324b
determination of, **5**:582a
discovery of, **12**:148b
in plants, **5**:582a
preparation of, **5**:323b; **8**:342b
reactions of, **3**:279b; **8**:333a; **10**:499a
toxicity, **9**:9a; **15**:172a

Hydrodemethylation, **7**:21b
Hydrodictyon, **11**:154a; **13**:596a
Hydrodynamics
  medieval Arabic, **7**:344b–345a
  Renaissance, **3**:66b; **8**:198b, 200b, 211a, 227a–b
  seventeenth-century, **1**:138a–b, 566b, 567b; **3**:182b; **9**:13b–14a, 116a, 119a–b; **13**:437a
  early eighteenth-century, **1**:113a, 113b–114a; **2**:39b–40b; **3**:73a–b; **8**:177a; **11**:66a*
  late eighteenth-century, **2**:335a; **4**:477a, 480b–481a; **9**:471b; **14**:353b, 589a
  early nineteenth-century, **1**:327b–328a; **3**:132a, 135b, 143b; **11**:78a; **14**:200a, 206a; **15**:225b, 226a
  mid nineteenth-century, **1**:128a–129a; **2**:88b, 470a; **3**:314a; **4**:5a, 615a; **5**: 101a, 200a–b; **6**:34b–35a, 249a–250a, 503a,b; **8**:384b; **9**:19a, 392b; **10**:247b; **11**:63a–b, 293b, 345a; **12**:73b; **13**: 75a–76b, 385a; **15**:85a,b, 200a
  late nineteenth-century, **1**:600a–b; **2**: 166a–167a, 167a–b, 356a; **3**:178b, 265a; **4**:230a–b; **8**:287b; **9**:107b; **10**: 551a–b; **11**:58a, 446a; **13**:73b, 362b–363a; **14**:338a, 356a
  early twentieth-century, **1**:500a; **2**:167b–168a, 466a, 565b, 594a; **4**:344a–b; **7**:85a–b, 246b–247a; **8**:13b*, 284b, 559b–562b; **9**:616b; **10**:563b; **11**: 123a–124b, 306a, 557b; **12**:496b, 527a, 612b, 613a
  mid twentieth-century, **6**:147b; **7**:436b–437a; **12**:414b–415b
  applied, **6**:2b
  continuity, principle of, **8**:227a–b
  continuity equation, **4**:480b; **9**:222a
  differential hydrodynamic equations, **1**: 114a; **4**:126b, 480b
  discontinuous motion, **6**:250a
  and earth's shape, **1**:113a
  and electricity, **9**:206a–b
  and electrodynamics, **2**:166b–167a
  equal circulation, principle of, **8**:227b
  graphical methods, **11**:293b
  hydrodynamic paradox, **1**:114a
  laws of, **6**:503a,b; **14**:200a
  and magnetism, **9**:207b
  mathematics of, **3**:137a; **6**:5a, 249a–250a; **7**:594b–595a; **8**:457b*; **9**:548a–549a; **13**:75a–76b; **14**:613b
  parallel cross sections, hypothesis of, **2**: 39b
  physical, **2**:168a
  physiological, **4**:398b; **6**:459a,b
  theories of, **2**:168a; **4**:126b
  velocity-potential equation, **4**:480b
  of viscous liquids, **9**:616b
  vortex motion, **6**:249b–250a
  see also Aerodynamics, Fluids, Gas dynamics, Gases, Hydraulics, Hydromechanics, Hydrostatics, Liquids, Outflow
Hydroelectricity, **5**:407a; **6**:79a; **8**:180b; **13**:73a
Hydrofluoric acid
  discovery of, **12**:146a
  preparation of, **5**:321b–322a, 474a; **13**: 311b
  vapor density, **13**:389b
Hydrogen
  absorption, by metals, **6**:177b; **13**:468a
  active, **3**:401a
  in air, **5**:315b
  atomic volume, **1**:129a
  atomic weight, **2**:135a; **8**:510a; **10**:158a
  as standard, **3**:48a; **9**:91b, 351b–352a; **11**:173a
  in ballooning, **3**:207b–208a; **5**:602a–b; **8**:78b–79a; **9**:492b
  in biological fluids
    measurement of, **4**:220a
  bonding of, **1**:519b; **5**:589b; **7**:622b; **8**:47a; **9**:38a; **12**:418b–419a
  combustion of, **2**:75a–b; **4**:241b; **8**:77b–78a; **15**:316a
  compressibility, **5**:610a
  determination of, **8**:331b
    microchemical, **11**:129b
  discovery of, **2**:380b
  dissociation of, **8**:83b
  electrochemical nature of, **4**:246b; **6**: 373a–b; **8**:113a
  electroconductivity, **5**:558b
  elementary nature of, **8**:82a,b; **11**:173a
  from illuminating gas, **3**:299a
  interstellar, **4**:280a; **13**:117b
  inversion temperature, **10**:206b
  ionic charge, **13**:82b, 369a–b
  liquefaction of, **4**:79b; **6**:94a; **7**:221a; **8**: 47a; **10**:206b, 604a; **13**:454b; **14**:522b; **15**:442a
  liquid
    energy from, **3**:299a
  in metabolism, **7**:406b; **13**:394a
  metallic nature of, **3**:599b, 602b; **8**:60a
  molecular forms
    equilibrium distribution, **4**:545a
  molecule
    diatomic nature of, **5**:294b; **8**:61a; **14**: 531a
    quantum mechanics of, **8**:473a–475a
    structure of, **2**:242b; **11**:422a
  negative ion, **6**:615a
  occlusion of, **5**:493a–b
  ortho form, **12**:438b; **15**:444b
  in periodic table, **2**:429a
  in phlogiston theory, **7**:388b
  in plants
    source of, **8**:345b
  production of, **8**:79a–b, 365b; **9**:343a, 471b; **13**:243b; **15**:316a
  as propellant, **3**:578b
  quantum theory, **7**:491b, 492b; **10**:214a; **11**:422a; **12**:31b, 219a–220a; **15**:42a, 49b
  reactions of, **5**:319b; **8**:118b
    with oxygen, **5**:621b; **6**:404a,b
    photochemical, **3**:197b; **5**:322a; **11**: 537b; **13**:311b; **15**:37a–b, 448a–b
  specific heat, **4**:414a
  spectrum, **2**:243a, 244a, 246b; **3**:420a; **5**:102b; **7**:27a, 268a, 553b–554a; **10**:217a; **11**:47a, 422a, 478b–479a; **12**:31b, 42b–43a; **13**:104a, 240a
    doublet structure, **2**:245b
    F line, **3**:36a
    fine structure, **11**:422b
    series, **1**:425b; **2**:145b; **8**:578b; **10**:347b; **11**:446a; **12**:235b
  in sun, **12**:22a
  valence, **8**:508a
  see also Air, inflammable; Deuterium; pH
Hydrogen bomb, **5**:273a; **7**:527b; **8**:95b; **10**:217b
Hydrogen bromide
  dissociation of, **15**:36b–37a
Hydrogen chloride
  as disinfectant, **3**:202a
Hydrogen cyanide
  formation of, **4**:79a
Hydrogen disulfide, **12**:46b
Hydrogen iodide
  decomposition of, **15**:36b, 37a–b
Hydrogen ion
  measurement of
    see pH
Hydrogen peroxide
  as antiseptic, **3**:189a
  discovery of, **13**:310b, 312a–313a
  preparation of, **13**:312b
Hydrogen persulfide, **13**:310b
Hydrogen phosphide
  discovery of, **7**:390a*
Hydrogen sulfide, **7**:390a*; **12**:145b
  as analytical reagent, **5**:164a; **11**:167a
  synthesis of, **12**:147b
Hydrogen telluride, **3**:602b
Hydrogenation
  of carbon, **2**:4a, 324a
  catalytic, **3**:86a; **4**:20b, 134a; **6**:398a; **7**: 21b; **12**:47a; 15:2b
  of hydrocarbons, **1**:414b; **2**:4a; **8**:108a; **14**:156b
Hydrogène bicarboné
  see Olefiant gas
Hydrogenium
  physical constants of, **4**:79a, 80a
Hydrogenolysis, **1**:415b; **14**:602a
Hydrogeology
  see Hydrology
Hydrography
  medieval Arabic, **9**:36a
  seventeenth-century, **6**:71b
  eighteenth-century, **1**:351a; **2**:343a, 551b; **3**:396b–397a; **5**:136b–137a; **9**:90b, 250a; **10**:571b–572a; **12**:114b
  early nineteenth-century, **1**:202a; **4**:256a; **8**:101a, 417b–418a; **11**:554a; **12**: 114b–115a
  mid nineteenth-century, **1**:210a; **2**:430b–431a; **4**:491b–492a; **5**:16a, 17a; **9**: 551a–b
  late nineteenth-century, **4**:143b, 266a; **7**:149a; **11**:181b
  twentieth-century, **3**:203b–204a; **4**:614a; **7**:184a, 416b; **9**:150b; **12**:279b, 411b–412a; **14**:634a
  handbooks, **12**:114b
Hydroids
  colonies, **4**:186b; **6**:590a
  embryogeny, **1**:601a; **7**:516b
  regeneration of, **4**:187a
Hydrology
  Renaissance, **10**:280b
  seventeenth-century, **3**:116b*; **10**:521b–522a; **13**:614a
  eighteenth-century, **12**:554a–b; **14**:132b, 144b
  mid nineteenth-century, **12**:168a; **13**: 144b; **15**:85a
  late nineteenth-century, **1**:162a, 174a–b; **6**:78b–79a; **7**:44a; **8**:466a–467a; **9**: 593b; **10**:430a, 505b; **12**:140b, 545a; **13**:413a; **14**:52b, 53a–b
  early twentieth-century, **4**:42b, 462a; **6**: 215b; **7**:149a, 410b; **8**:146a; **9**:257a–258a; **10**:128a; **13**:619b; **14**:50b, 107a
  mid twentieth-century, **12**:414b–415b
  see also Hydraulics, Limnology, Oceanography, Rivers, Water
Hydrolysis, **1**:297a–300b; **3**:279b; **5**:125b
  enzymatic, **5**:589b
  of nucleic acids, **4**:603b
  of proteins, **14**:602b
  of starch, **7**:378b–379a; **10**:436a, 532a
  of wood, **2**:4a
Hydromechanics
  Renaissance, **11**:97b
  seventeenth-century, **2**:311a; **10**:61b–62a, 75a–76b

Hydromechanics (cont.)
eighteenth-century, **2**:299b; **4**:216a, 480b–481a; **5**:410b; **7**:565a
mid nineteenth-century, **10**:248a
late nineteenth-century, **3**:195a; **6**:545a–b; **11**:393a–b; **13**:26b; **14**:620b–621a
twentieth-century, **5**:188a; **7**:437a; **10**:20a, 514a
see also Hydraulics, Hydrodynamics, Hydrostatics
Hydromedusae, **14**:234a
Hydrometer, **2**:194a; **3**:208a; **4**:518a; **6**:386a; **10**:109a; **12**:132a; **13**:225b
constant-immersion type, **8**:70a–b
Hydronium ion, **6**:109a
Hydrophobia, **1**:563b; **10**:401a; **14**:187b
see also Rabies
Hydroplaning, **7**:437a
Hydroponics, **6**:444a
Hydroquinone, **1**:21b; **5**:489a; **14**:477b
Hydroscope, **6**:616a
Hydrosphere, **13**:619b
thermodynamics of, **2**:168a
Hydrostatics
ancient Greek, **1**:223a–b, 228b, 229b
ancient Roman, **10**:300b–301a
medieval, **1**:94a; **7**:336b–337a, 339a–348a
Renaissance, **1**:607b; **5**:238b; **8**:225b–227a; **9**:264a; **13**:49a
seventeenth-century, **1**:165a; **2**:378a; **5**:241b, 242a, 382a, 525a, 528b; **6**:602a; **9**:13b–14a, 119a–b, 529b–530a; **10**:332b–334b; **11**:488b
eighteenth-century, **2**:51b*; **3**:282b, 431a; **4**:480b; **5**:511a; **15**:295b–296a
nineteenth-century, **2**:424b, 470a; **4**:230b; **8**:559b, 596a; **9**:392b; **12**:73b; **15**:66b
Archimedes' principle, **1**:223a, 228b
and geometry, **1**:214b, 222a, 222b–223b
hydrostatic paradox, **4**:60b; **9**:116a, 117a, 119b
laws of, **10**:334a
oriented body, **4**:230b
Stevin's paradox, **1**:223a, 607b; **13**:49a
Hydrotherapy
eighteenth-century, **2**:301b
nineteenth-century, **9**:242a–b
see also Balneology, Baths
Hydroxyl radical
naming of, **3**:47a
Hydroxylysine, **13**:575a
Hydroxyproline
discovery of, **5**:4b
Hydrozoa, **4**:290a; **5**:109a; **14**:234a
Hyena, **1**:263b; **2**:569b
fossil, **2**:569b; **3**:264a
Hygiene
ancient Greek, **4**:106b, 386a
ancient Roman, **12**:540a
Byzantine, **10**:417a–b
medieval Arabic, **9**:41a; **11**:238b, 445a; **14**:585a
Renaissance, **7**:16b; **13**:569a
seventeenth-century, **4**:495a
eighteenth-century, **1**:563b; **3**:7a, 97a; **5**:221a, 355a; **9**:62a; **10**:612a
mid nineteenth-century, **2**:386b; **6**:434b; **10**:557b; **12**:233a, 478a–b; **15**:85a
late nineteenth-century, **3**:189a; **5**:74a, 125a, 128b, 219b–220a, 564a–b; **7**:93b, 423a, 425b; **8**:7b, 449b–450b; **9**:86a; **10**:176b–177a; **11**:585b; **12**:481a–b
twentieth-century, **4**:212a*; **5**:270a,b; **9**:95a, 335a; **11**:595b; **12**:96a–b; **13**:394a; **15**:521b
experimental, **10**:558a–b
industrial
eighteenth-century, **3**:7a
nineteenth-century, **3**:6b, 237b
twentieth-century, **6**:25a; **11**:585b
military, **7**:428a,b; **8**:450b; **11**:148a
naval
eighteenth-century, **4**:224b*; **8**:363a
sanitary codes, **7**:16b
social, **14**:41b–42a, 45a
see also Congresses, scientific; Public health
Hygrometer, **1**:97a, 138a, 586a; **2**:586b; **3**:534b, 540a, 556b; **4**:565b; **6**:342a; **7**:599a; **8**:198a; **12**:103a
hair, **4**:28b; **12**:122b
ivory bulb, **4**:28b
wet and dry bulb, **5**:413a; **8**:262a
Hygrometry
eighteenth-century, **7**:599a; **10**:602b
nineteenth-century, **12**:366b
Hygroscope, **3**:581b; **9**:465a
Hyla, **2**:322a
Hylacomylus
see Waldseemüller, Martin
Hylaeosaurus, **9**:87b
Hylleraas, Egil Andersen (1898–1965), 6:614–615; **7**:492a
Hylobates, **2**:110a
Hylozoism, **1**:260b; **3**:492a; **5**:427a
Hymen, **4**:521a
existence proved, **15**:4b
Hymenium, **2**:19a
Hymenomycetes, **2**:583a
Hymenophyllaceae, **11**:130a
Hymenoptera, **6**:524b
Hyoscine, **14**:496a
Hyoscyamine, **15**:103a
Hyoscyamus, **10**:307a
Hypacusis, **13**:567b
Hypatia (*ca.* 370–415), 6:615–616; **4**:491a; **13**:321a–b
commentaries by, **1**:191b; **4**:111b
students of, **13**:225a–b
thermometry, **7**:376a
Hyperactivity, **8**:455a
Hyperalgesia, **8**:295b
Hyperbola, **3**:183a; **4**:104a; **5**:410a; **10**:295a; **14**:274b
area of, **5**:526b
asymptotic property, **7**:133b; **9**:270a, 273b–274a
conjugate, **1**:185b
construction of, **14**:465a–466a
meaning of word, **4**:421a
two branches considered as one, **1**:185b
see also Conics; Quadrature, of hyperbola
Hyperboloid, **4**:429a
center of gravity, **3**:182b
of revolution, **3**:555a
volume of, **3**:182b; **13**:435b
Hypercycle, **8**:430b
Hyperdeterminant, **3**:164a, 166b; **6**:357b
Hyperesthesia, **2**:525a
Hypergeometry, **3**:166b
Hyperion (satellite), **8**:46a; **10**:35a
Hypermastigina, **4**:142b
Hypermetamorphosis (zool.), **4**:504a
Hyperplane, **9**:414a
Hyperquadric, **3**:166b
Hypersensitivity
tuberculin, **14**:624a
Hyperspace, **2**:105b; **3**:166b, 169b; **4**:373b; **10**:35b*, 606a; **12**:284a–285a
recursion method, **13**:623b
Hypersurfaces, **5**:202a; **11**:57a
linear systems of, **12**:331b
Hypertension
see Blood pressure
Hypertrophy, **15**:642b
Hypha, **14**:93b
Hypnosis, **2**:132a; **5**:74a, 138a; **6**:226b; **9**:327b; **11**:425a
as anesthesia, **9**:242a
in psychoanalysis, **2**:447b
in psychotherapy, **3**:205b; **5**:174a; **11**:425b
repetition, effects of, **11**:425a
and spiritualism, **5**:531a
theories of, **8**:15a
in three stages, **2**:35b
see also Animal magnetism, Mesmerism
Hypoblast, **1**:601a
Hypochondria, **6**:387b; **8**:506b; **10**:613b; **14**:407b, 456a
Hypocycles (astronomical), **2**:412a
Hypogene processes, **4**:598b
Hypoglycemia, **15**:74a
Hypomagma, **7**:63a
Hyponitrous acid, **13**:338b
Hypophosphoric acid, **12**:64a
molecular weight, **11**:551b
Hypophosphorous acid, **14**:529b
discovery of, **4**:239b
Hypophysectomy, **3**:352b
effects of, **12**:472b–475a; **13**:589b
Hypophysis
see Glands, pituitary
Hyposulfites
as analytical agent, **3**:194a*
Hypothalamus
anatomical description of, **5**:74a
Hypothermia, **7**:502b
Hypothesis
Renaissance discussion, **3**:410b; **10**:123a
seventeenth-century discussion, **4**:52b–54b; **7**:294b, 295a, 488a; **9**:120a, 318a–b; **10**:77a, 85a–b
eighteenth-century discussion, **8**:321b–322a
nineteenth-century discussion, **2**:31b–32a; **3**:376b–377a, 452a–b; **6**:279b; **8**:596b; **10**:580b; **14**:434b; **15**:143b
twentieth-century discussion, **3**:32b–33a; **8**:294b; **9**:524a; **11**:14b
testing of, **5**:8a–9a; **14**:122a
see also Method, scientific; Theory
Hypothyroidism
replacement therapy, **6**:518b
Hypoxanthine, **7**:466b
synthesis of, **5**:2b
Hypoxylon, **13**:489b
Hypsicles of Alexandria (*fl. ca.* 175 B.C.), 6:616–617; **7**:81a
De ascensionibus signorum, **15**:179b
astronomy, **15**:210a
editions of works, **13**:509b
and Euclid's *Elements,* **1**:245b; **4**:415b, 438b
mathematics, **4**:116a; **9**:301a(n)
translations of works, **7**:25a; **11**:244a, 245a
Hypsometer, **9**:443a
Hypsometry, **13**:412b–413a
Hyrtl, Joseph (1810–1894), 6:618–619; **3**:424b; **7**:23b; **13**:274a
Hysteresis, **3**:239b*; **5**:597b; **6**:504b; **8**:480a; **11**:578a
magnetic, **3**:212a; **6**:5b; **12**:281b
discovery of, **4**:500a, 500b–501a; **11**:460b; **14**:171a
law of, **13**:24a
thermodynamics of, **4**:229a,b

Hysteria, **2**:483b; **3**:205b; **5**:27a, 172b, 176a; **6**:256a, 387b; **8**:506b; **9**:422b; **10**: 307a; **14**:407b
treatment of, **2**:35b; **5**:173b–174b, 178b; **9**:326a

I

I Chih-han, **3**:266a
I Ching
see *Book of Changes*
Iamblichus (*ca.* 250–*ca.* 330), 7:1
Collection of Pythagorean Dogmas, **11**: 185b
commentaries by, **4**:111a
commentaries on, **2**:229b, 230a; **11**:183b; **12**:442b
De communi mathematica scientia, **12**: 575b
historiography, **1**:245a; **4**:104a, 160a, 461a, 607a; **6**:407a, 410b; **10**:113b(n); **11**:221a,b; **13**:314b
influence of, **2**:539b; **6**:616a; **11**:161b; **12**:441a, 442a
De mysteriis, **11**:184b
Neoplatonism, **11**:160b
In Nicomachi . . . , **13**:399b, 400a
number theory, **9**:191b
Theologumena arithmetica
editions of, **11**:18a
Iatrochemistry
Renaissance, **1**:49a–b, 571a; **3**:471b–472a; **4**:208b–209b; **10**:307a–308a; **11**: 606a–b; **12**:334b–336a; **13**:535a–b
early seventeenth-century, **2**:308b; **4**: 170a; **5**:419b–423a; **6**:145b–146a; **8**: 311b; **13**:559a–b; **14**:209b
late seventeenth-century, **1**:451a, 548b, 549b; **2**:2b, 353a; **3**:596b; **7**:150b; **8**:130b, 174a; **9**:146a; **11**:40b; **12**: 601a, 616b; **13**:169b, 223a–b, 282a; **14**:26a, 212b
early eighteenth-century, **1**:322b; **5**:366b; **6**:459a–b
opposition to, **2**:226a–b, 237b; **4**:387b–388a; **7**:188a; **12**:604a; **13**:490b
Iatromathematics, **1**:392a; **2**:52a; **3**:244a–b; **11**:1b–2b; **12**:102a; **13**:231b–232a
opposition to, **7**:188a
Iatromechanics, **1**:592b–593b; **2**:225a, 226a–b, 237b, 306a, 309b, 312a; **3**:244a; **4**:136a, 589a; **6**:459a–460a; **7**:274b, 605b, 606a; **9**:64a–b, 243a; **10**:10a, 611b; **13**:122a–123a; **14**:587b
corpuscular, **1**:593a; **2**:2b
hydraulic, **1**:593b; **11**:2a–b
opposition to, **1**:479a–b
Iatrophysics, **1**:322b, 392a; **2**:37a; **3**:332a; **6**:37a; **8**:173b; **12**:101b, 103b, 601a
Iatrosophists, **10**:230a
Ibáñez de Prado, Martín, **7**:1b
Ibáñez e Ibáñez de Ibero, Carlos (1825–1891), 7:1–2
Ibex, **12**:169a
Ibis
founding of, **12**:257b
IBM, **15**:128b, 129a
Ibrāhīm ibn ʿAbdallāh, **1**:275b
Ibrahim Pasha of Parga, **10**:616b–617a
Ibrāhīm ibn al-Ṣalt, **15**:236b
Ibrāhīm ibn Sinān ibn Thābit ibn Qurra (908–946), 7:2–3; **6**:199b, 207b*; **7**: 335a
Ibrāhīm ibn Yaʿqūb al-Isrāʾīlī al-Turṭushi (*fl.* second half of tenth century), 7:4; **1**:414a
Icacinaceae, **15**:147a
Icarus (asteroid), **1**:352b
Ice, **15**:663b(n)
Antarctic
classification of, **1**:595b
Arctic, **9**:43b; **14**:50b
laws of, **14**:635a
classification of, **8**:471a
flexure of, **14**:635a
latent heat, **14**:353b–354a
as lens, **9**:115a
melting point
pressure effects, **13**:236b
molecular structure, **1**:519b; **2**:171a
oceanic, **2**:557b
drifting of, **4**:344a–b; **14**:635a
forecasting of, **14**:634b, 635a
physical properties, **9**:180b
polymorphism, **2**:459b; **13**:246b
in rivers, **8**:461a, 465b, 467a
thawing of, **4**:28b; **8**:73a
vaporization of, **10**:553a
volatilizing point
and pressure, **14**:560b
Ice Age, **13**:605a
causes of, **14**:139a
concept of, **1**:73a–b; **5**:337b; **12**:167b
dating of, **14**:443a, 517b
and loess, **1**:623a; **3**:462b; **7**:511b
man in, **5**:339a; **14**:443a, 517b
multiple glaciation, **3**:190b, 533a; **5**:338b–339a; **10**:429b, 502a, 504a
and plant distribution, **1**:623a
unitary theory, **14**:517b
see also Glaciers, Glaciology
Icebergs
origin of, **8**:471a
Iceland
botany, **6**:493a
exploration of, **2**:463b; **5**:225a
geology, **11**:135b; **14**:625a
volcanoes, **11**:336b
Iceland spar
see Calcite
Ichneumonidae, **3**:580a; **5**:439b; **10**:13a; **14**:413a
life cycles, **8**:415b
Ichnology, **6**:438a; **9**:617b; **12**:294a*
experimental, **9**:617b
Ichthyology
medieval Arabic, **1**:30a
Renaissance, **1**:596a; **9**:192a; **11**:527b–528b; **12**:89b–90a; **13**:502a
seventeenth-century, **11**:316b; **13**:527b; **14**:157b, 413b
eighteenth-century, **1**:305b–306a; **2**:509b; **3**:365a, 522b; **5**:76b; **12**:536b, 560b
nineteenth-century, **2**:282a, 423a; **3**:525a; **5**:196a, 406a–b; **7**:169b–170a, 547a–b; **8**:266b; **12**:421b–422a; **13**:551b–552a, 554b; **14**:57a–b; **15**:227a
twentieth-century, **1**:622a,b; **4**:309b–310a; **6**:267b; **7**:314a–b; **8**:146a; **9**: 331a–b, 550b; **13**:151a–b
see also Fishes, Paleoichthyology
Ichthyopsida, **6**:595a
Ichthyosaurus, **3**:395b; **4**:147b; **7**:60a; **9**:314a
coprolites of, **2**:569b–570a
Icilius, Quintus, **4**:2a
Icones plantarum
editors of, **6**:491b–492a
founding of, **6**:495a*
Iconius, Raphael Eglinus
see Hapelius, Nicolaus Niger
Iconographic Encyclopaedia of Science, Literature, and Arts, **1**:405a
Iconography
see Illustration
Icosahedron, **6**:92b
discovery of, **13**:303b–304a
inscribed, **6**:616b
transformations of, **7**:398a
Id, **5**:175b, 178a–b
Iddings, Joseph Paxson (1857–1920), 7:5–6; **3**:484a; **4**:564b; **6**:13b, 14a,b*; **7**: 266b; **9**:531a; **14**:183b
Idealism
seventeenth-century, **2**:613a; **6**:152b, 256b; **8**:158b, 437b–439a
eighteenth-century, **2**:223b; **3**:381b; **12**: 155a
nineteenth-century, **1**:70b–71a, 73a–b; **3**:90a; **6**:86b–87a, 90b, 247b; **7**:234a; **8**:514a; **12**:156a, 607a
twentieth-century, **3**:90a; **6**:26a,b; **7**:86a, 482b; **8**:257b; **11**:258b; **14**:124a
critical, **2**:545a–b; **3**:111b
empirical, **6**:596b
and evolution, **1**:73a–b
mathematical, **1**:337a; **7**:327a
rejection of, **2**:577b; **3**:278b; **8**:183b, 184a; **12**:59b; **15**:79b, 128a
subjective, **8**:603b
and taxonomy, **1**:70b–71a; **5**:192a, 443a
transcendental, **7**:227a, 229b, 230b, 235a; **9**:359a
Ideals (math.)
algebraic, **2**:90a; **6**:389a–b; **11**:55b
ascending chain condition, **10**:138b
concept of, **4**:4b–5a; **10**:138b
differential, **11**:471a
fundamental theorem, **6**:572a
invention of, **5**:306b
polynomial, **7**:444b; **8**:585a; **10**:138b, 140b–141a
primary, **10**:138b
prime, **1**:307a; **6**:390a,b
right
double-chain law, **1**:307b
theory of, **3**:228b; **4**:125b; **10**:138a–139a
Ideas
historiography of, **8**:517a–b
sociology of, **15**:137b, 138b–139a, 141a, 142a, 408b, 412b
Idelson, Naum Ilich (1885–1951), 7:6–7; **10**:31a*
Idempotent law, **2**:295b, 296a
Identity, mathematical, **4**:116a,b
theory of, **8**:263a
Identity, principle of (phil.), **8**:156b, 157a, 158b; **12**:155b–156a; **15**:423a–b, 424a
Idéologues, **1**:140b; **3**:1b, 326b, 383a; **5**:166a, 252b–253a; **7**:591b; **9**:7b; **10**:611b; **14**:68b; **15**:341b
critics of, **14**:44b
Ideology, **15**:135a, 138b–139a, 140a–b
Idiocy, **10**:613b
Idioplasm theory, **6**:338b; **9**:601a–b; **14**: 234b–236b, 429b
Ido, Yutaka, **10**:142a
Ido (language), **3**:456b; **15**:465b
al-Idrīsī, Abū ʿAbd Allāh Muḥammad ibn Muḥammad ibn ʿAbd Allāh ibn Idrīs, al-Sharīf al-Idrīsī (1100–1166), 7: 7–9; **1**:29a, 538b
Ieyasu, **8**:394b
I.G. Farbenindustrie, **1**:105b; **2**:324a; **4**:154a; **5**:344b; **6**:304a
Igneous activity, **3**:553b
Iguanodon, **1**:603a; **3**:606b; **4**:147b; **6**:595a; **9**:87b
Iḥṣāʾ al-ʿulūm (Enumeration of the Sciences), **4**:525b
I-hsing, **12**:375b, 378a

Ikenberry, Ernest, **9**:225b
Ikhshīdid culture, **4**:524b
Ikhwān al-Ṣafāʾ (*n.d.*), **15:249–251; 4**:525b; **7**:40a
Rasāʾil, **7**:358a; **9**:39a,b, 40a
I-ku chi, **8**:319a
Ilberg, Johannes, **3**:175a; **5**:227a–b; **11**:603a
Ilg, Albert, **13**:327a
Ilg, Frances Lillian, **5**:378a
Iliastrum, **10**:309b
Ilkovič, Dionyz, **6**:373a
Illiger, Johann Carl Wilhelm, **4**:504b
Illinois
geology, **4**:562b; **12**:229b; **13**:531b; **14**:315b
paleobotany, **8**:264a–b
paleontology, **9**:255b
State Natural History Survey, **5**:69b
Illinois, University of, **9**:387b
chemistry at, **15**:1a–2a
Illuminati, Order of, **5**:195b; **11**:497a
Illuminating gas, **4**:134a, 531b; **5**:126b; **12**:435b
comparative analysis of, **6**:284b
hydrogen in, **3**:299a
manufacture of, **3**:556b
nitrogen in, **5**:128a
Illumination, **14**:343a
effective use of, **7**:77b
photometric law
seventeenth-century, **3**:116a
eighteenth-century, **2**:328b
psychology and, **14**:550b
see also Lighting, artificial
Illumination, metaphor of, **1**:334a, 336a, 337b; **11**:250a; **13**:513a
Illustration
alchemical
eighteenth-century, **1**:451a
anatomical
medieval, **6**:276b
Renaissance, **1**:620b; **3**:40a,b, 99a,b, 343a; **4**:487a–488a, 508b; **5**:347b–348b, 580b; **6**:562b; **8**:54a, 202b; **13**:569a; **14**:4b, 8a–12a
seventeenth-century, **1**:316a, 480a, 482b, 523b; **4**:511a; **5**:484b–485b; **12**:577b; **13**:33b; **15**:29a–b
eighteenth-century, **3**:16a, 38a; **9**:153b; **12**:137a, 138a, 510a; **13**:566b; **15**:4b
nineteenth-century, **3**:490a; **5**:515a; **9**:241b; **11**:382a; **12**:510b
twentieth-century, **15**:64b–65a
Chinese, **15**:737b, 739a,b, 740a
historiography of, **13**:142a
Japanese, **15**:740b
Mesopotamian, **15**:642b
architectural
seventeenth-century, **2**:201b
astronomical
Renaissance, **1**:530b; **2**:407b, 412b
seventeenth-century, **6**:361b–362a, 364a
nineteenth-century, **1**:464b; **2**:284b; **4**:19a; **13**:472b
biological
seventeenth-century, **6**:483b
nineteenth-century, **8**:169b
twentieth-century, **13**:165b
botanical
ancient Roman, **4**:120b–121a
Renaissance, **1**:592a–b, 596a; **2**:219a–b, 536a–537a; **4**:138b–139a; **5**:362a; **8**:121a, 484a; **9**:179b, 180a; **15**:161a
seventeenth-century, **2**:334a; **10**:426b*; **11**:587b–588a; **13**:449a
eighteenth-century, **1**:520a–b, 520b–521a; **2**:322b; **3**:172b, 365b; **4**:99a, 295a; **5**:217a; **6**:66a; **7**:58b, 412a; **8**:303b; **11**:343a; **12**:587b
nineteenth-century, **1**:485b*, 521a; **2**:159b, 160a; **3**:44b; **6**:467a, 489a, 492b, 493a,b; **7**:534a; **9**:432b, 455b; **11**:263b*; **12**:60a, 552b; **13**:299b, 395b, 506a–b
twentieth-century, **3**:318a
historiography of, **13**:463a
Oriental, **8**:394b
chemical, **8**:311a
copperplates, **3**:40b; **4**:487b; **8**:71a; **13**:395b; **14**:11b
cosmological
medieval, **6**:396a
crystallographic
Renaissance, **5**:379b
cytological
nineteenth-century, **2**:626a–b; **11**:255b
embryological
seventeenth-century, **4**:510a, 511a–b
eighteenth-century, **12**:510a
nineteenth-century, **1**:476b; **6**:9b
entomological
seventeenth-century, **2**:591b; **13**:29a,b
eighteenth-century, **3**:320a–b; **5**:328b; **8**:580a
nineteenth-century, **8**:528b
ethnographical, **10**:323b
geological
nineteenth-century, **4**:10a; **6**:227b; **10**:258a; **11**:505a; **12**:543a
twentieth-century, **6**:116b
histological
nineteenth-century, **1**:540a; **2**:375b
twentieth-century, **2**:345a
ichthyological, **11**:316b
industrial
eighteenth-century, **4**:86b–87a
malacological
seventeenth-century, **2**:591a
nineteenth-century, **5**:478a,b
medical
medieval Latin, **10**:21b
eighteenth-century, **3**:97b; **6**:569b
nineteenth-century, **3**:323b, 324a
twentieth-century, **15**:64b–65a
historiography of, **13**:142a
metallurgical
Renaissance, **2**:143a
seventeenth-century, **8**:464b
microbiological
eighteenth-century, **6**:219b–220a
nineteenth-century, **4**:236b
twentieth-century, **6**:121b
mineralogical
eighteenth-century, **3**:61b
natural history
Renaissance, **1**:526a
eighteenth-century, **1**:488b; **2**:322a–b; **10**:323a–b; **12**:536a–537a
nineteenth-century, **8**:266a–b; **10**:439b; **12**:552b
ornithological
medieval Latin, **5**:148a*
eighteenth-century, **3**:129b; **8**:274b; **12**:537a
nineteenth-century, **1**:330a–331b; **5**:480b–481a; **14**:417a–b
paleontological
Renaissance, **5**:379b
eighteenth-century, **7**:412a–b; **14**:119b
nineteenth-century, **1**:491a; **3**:392a; **8**:570a; **9**:346a; **10**:322b; **12**:229b, 552b; **13**:531b; **14**:312b
twentieth-century, **2**:505a
pharmacological
Renaissance, **1**:47b
physics, **9**:566b; **10**:145b
scientific
eighteenth-century, **4**:26a
surgical
medieval Arabic, **14**:584b
Renaissance, **2**:546b
technological, **6**:312a; **9**:147a; **11**:111a; **13**:568a
topographical, **10**:323b
wood blocks, **4**:138b–139a; **5**:362a; **11**:587b; **13**:568a
historiography of, **13**:463a
zoological
medieval, **7**:64b
Renaissance, **12**:90a
eighteenth-century, **11**:503a
nineteenth-century, **1**:330b–331a; **2**:18b; **4**:12a; **9**:536b; **13**:168a; **15**:83a, 227a
see also Art, Paintings
Ilmenite
deposits of, **13**:519a
Images
diffraction, **1**:8a–b
perfect, **9**:225b
photographic
intermittency effect, **1**:21b
mutual repulsion of, **7**:472a
pinhole, **12**:381b; **14**:458b
resolution of, **1**:8a–b, 135a–b
see also Vision
Imagination
active, **7**:191a,b
theories of, **6**:559b; **9**:98a; **14**:406b
Imamura, Akitune (1870–1948), 7:9–10
Imidazole
derivatives of, **14**:445b–446a
Immersed bodies
in air vs. water, **4**:208a
Immersion
heat of, **6**:118b
Immigration, **3**:556a
Immortality of man, **1**:119a; **3**:470a; **7**:68a; **11**:71b, 74a
Immune reactions, **3**:353a
Immunity, **9**:328b; **11**:37b
antibacterial vs. antitoxic, **15**:522a
athreptic, **4**:299b
bacteriophages and, **6**:298a–b
duration of, **10**:396a, 398a
and evolution, **15**:507a
hormones and, **5**:317a
inherited, **4**:297a
local, **15**:522a
in lower vertebrates, **9**:328a
and nutrition, **8**:274a*
in plants, **3**:359a–b; **15**:505b
genetics of, **15**:507a
theories of, **3**:332b; **5**:270b; **10**:391a–b, 392b–394b, 406b–408a; **11**:428b–429a
humoral, **2**:300a–301a; **9**:334b
phagocytosis, **9**:333a–334b
side-chain theory, **4**:298b–299b; **5**:564b
in tissues, **5**:317a
transient, **4**:297a
vitamins and, **5**:317a
Immunization
active, **1**:576b–577a; **4**:297a,b; **7**:427b, 429a; **8**:556b; **12**:482a
passive, **1**:577a; **3**:336a; **4**:297a–b
see also specific diseases; Vaccination
Immunochemistry, **1**:301b, 342b; **8**:608a; **14**:523b, 624a
Immunology
eighteenth-century, **7**:96a–97b
late nineteenth-century, **4**:297a–299b; **5**:269a–270b, 564a–b; **7**:427a–b; **8**:

556b–557a; **9**:333b–334b; **10**:390a–408a; **11**:428b–429b; **12**:88a–b, 480b; **14**:250a, 511b; **15**:521b–522a
early twentieth-century, **1**:342a–b; **2**:300a–301a; **4**:499a; **5**:29b, 40a–b, 316b–317a, 599a; **6**:232b–233a; **7**:623a–624b; **9**:545b; **10**:141b–143a; **11**:271b–272a, 430a–431a, 443a; **12**:48b, 315a, 481b–483b; **14**:252b, 511b–512a, 551b, 623a–624b; **15**:522a–523a
mid twentieth-century, **1**:342b–343a; **12**:492b; **15**:454b
botanical, **15**:508b
"wet nurse" experiments, **4**:297a–b
see also Immunity, Serology
Impact
Renaissance discussion, **8**:199b, 230b–231b
seventeenth-century discussion, **4**:58b–59b, 60a–b; **13**:438a; **14**:150a
eighteenth-century discussion, **1**:111b–112a; **5**:509b; **7**:231a; **10**:80a
nineteenth-century discussion, **10**:250b
deformation, **9**:115b–116a
elastic
seventeenth-century analaysis, **6**:602a–603a; **9**:51b–52a
twentieth-century analysis, **1**:570b
of flowing water, **9**:116a, 119b
with friction, **10**:514b
inelastic, **10**:250b
laws of, **1**:424b; **6**:602b–603a; **9**:51b; **10**:62a, 314b; **14**:509b–510a
measurement of, **9**:115b
transmission of, **9**:116a
see also Collision, Percussion
Impatiens, **6**:490a
Impedance, **6**:278a
motional, **10**:605a
naming of, **6**:212a
Imperial Chemical Industries, Ltd., **1**:286b; **12**:437b; **15**:159b
Imperial College of Science and Technology (London)
plant physiology at, **5**:523b–524a
Impetigo, **12**:203a
Impetus
Byzantine discussion, **7**:136a
medieval Arabic discussion, **1**:28a; **15**:33b
medieval Latin discussion, **1**:94b, 100b, 198b; **2**:194a–b, 605b–607b; **4**:157b; **5**:113b–114a; **6**:275b; **9**:136b–137b; **10**:223b–224a, 226a, 421a
Renaissance discussion, **1**:606a–b, 607a,b; **3**:66b, 280b; **4**:237b; **7**:486b–487a; **8**:199b, 200a, 228a–230a, 231a,b; **10**:123a, 512b; **12**:136a, 548a
seventeenth-century discussion, **5**:247a; **8**:268a; **13**:438a
compound, **8**:229a–230a
conservation of, **8**:231b
definition of, **2**:606a
increments of, **1**:606a–b; **2**:606a, 607a
and local motion, **3**:420b
in optics, **7**:136a–b
rotational, **2**:606b
vs. rectilinear, **6**:275b
self-expending, **2**:590b
see also Inclination
Imponderables, **6**:278a–b
Impotence, **4**:67a*
Impregnation
see Fertilization, zoological
Impressionism, **11**:532a
Impulse (mech.), concept of, **9**:97a
al-ᶜImrānī, ᶜAlī ibn Aḥmad, **1**:23a; **4**:438a; **11**:32a, 226a
Inada, Ryukichi, **10**:142a
Inbreeding
see Heredity
Incandescence
spectra, **4**:183a
Incidence (geom.), **8**:324b
Inclination (mech.), violent, **1**:28a, 410a
see also Impetus
Inclination, magnetic
laws for, **2**:135a
measurement of, **6**:550b; **8**:426b
Inclinations (geom.)
see *Neusis*
Inclined plane
ancient Roman study, **10**:298a
medieval study, **1**:410a; **7**:173b–174a; **13**:260a
Renaissance study, **7**:175a; **8**:200b, 216b–218b, 233a–b; **13**:48b–49a, 260a
seventeenth-century study, **1**:424a; **5**:239a,b, 240b, 246a; **7**:487a; **9**:318a, 488a; **11**:488a; **13**:436b, 586a; **14**:366b
distribution of forces on, **3**:183a
friction on, **2**:39a
law of, **8**:215b; **13**:48b–49a
obliquity of, **8**:216b–218a
Inclusion, geological, **5**:205a
Inclusion, logical, **10**:485a
Incommensurability
ancient Greek discussion, **4**:419a,b, 424a, 447a, 466a; **11**:28a
medieval Latin discussion, **2**:393b, 394b, 395b
seventeenth-century discussion, **5**:409b; **8**:267a–b
nineteenth-century discussion, **3**:178a
definition of, **4**:422a; **7**:242a–b
discovery of, **11**:222a–b; **14**:611a–b
test of, **13**:316a–317b
Incompatibility (logic), **12**:14b
Incubation, **11**:333b
Incubator
seventeenth-century, **4**:184b
d'Arsonval's, **1**:304a
sunlight, **5**:147b
Indeterminacy, **2**:248b–249b, 356b; **3**:111b, 275a, 369b, 565a*; **8**:369a, 477a; **10**:442a; **13**:275b; **15**:49a
Index Kewensis, **6**:491b
Index Librorum Prohibitorum, **1**:484a; **5**:244b, 250b; **7**:303a; **8**:257a; **9**:49a, 168a; **12**:85b, 134b, 199b; **14**:160a,b
Indexing, **14**:507a
India
archaeology, **14**:505a
astrology, **2**:154b; **5**:275b; **6**:115b; **7**:222b–223b, 315a,b, 583a; **10**:314a*, 548a; **11**:264a; **12**:576b, 598a–599a; **13**:581b–583a; **14**:549a–b
astronomy, **15**:533a–633b
pre-tenth-century, **1**:34b, 308a; **2**:114b–115a, 416b–417b; **4**:555b–556b; **5**:615a; **7**:361a, 582b–583a; **8**:46b; **10**:425b–426a; **12**:576b; **13**:582a, 590b; **15**:209a, 211a, 215b–216a, 217a, 221a
tenth- through fifteenth-century, **1**:309b; **2**:117b–119b, 154a–b, 416a–b; **3**:584b; **9**:21a, 22b–23a, 42a, 580b; **10**:129a–b, 313a–b; **12**:115b–116a, 598b; **14**:28b
sixteenth-century, **1**:48b–49a; **4**:100a; **5**:274a–276a; **7**:314b–315b; **11**:264a
seventeenth-century, **6**:115b; **7**:219a–b, 512a–b; **9**:175b, 579b–580a, 607a; **11**:290a–b; **12**:343a
eighteenth-century, **7**:62a, 80b–81b; **8**:144a; **15**:329a–b
Drgganita system, **1**:48b; **10**:129a, 313b; **15**:602b, 613b–614a
Greco-Mesopotamian, **15**:538b–554a
Islamic influences, **15**:625b–629a
Mesopotamian, **15**:536a–538a
Parahita system, **15**:595b–596b, 601a, 602b, 613b
Vākya system, **15**:596a, 601a–b, 602b
Vedic, **15**:534b–536a
atomic energy, **15**:26a, 27b
atomism, **4**:31a
botany, **3**:45b; **5**:539a; **6**:489b–490a; **10**:236b–238a; **13**:342b
British development of, **9**:123a–b
calendar, **2**:154a,b; **3**:584b; **4**:100a; **5**:274a; **10**:426a; **12**:343a; **15**:534a–630a *passim*
chemistry, **11**:319a
cosmography, **2**:117b; **14**:548b
cosmology, **1**:308a; **7**:338a, 582b; **10**:129a; **13**:582a; **15**:554a–555a
earthquakes, **10**:203a, 204a
education, **15**:48a
entomology, **4**:493b–494a
epidemiology, **5**:219b, 220a; **6**:11b–12b
ethnology, **12**:537a
finance, **7**:317b
geography, **1**:309b; **2**:117b, 154a; **4**:556a; **7**:582b; **11**:376a–b; **13**:582b; **15**:626a–628b
Geological Survey, **10**:203a, 204a
geology, **5**:519a; **8**:577a; **13**:275b
glaciation, **3**:591a
industry, **11**:319a
mathematics, **1**:30b, 31a, 308a, 309b; **2**:115b–117b, 417a; **4**:117b, 502b, 605a, 608b, 612a; **5**:159a; **7**:325b–326a, 531b–533a; **9**:22a–b, 580a, 613b–614a, 614a–615a; **11**:18a; **12**:597b–598a; **13**:544b–546a; **14**:471a(n), 548b–549a; **15**:9b–10b
numeral system, **1**:40a, 41a, 62b–63a; **4**:604b, 605a,b, 612a; **6**:115b; **7**:360a–b, 362b–363a, 532b; **11**:18a, 458a; **13**:544b; **14**:472a
medieval visitors to, **1**:516a, 517a; **2**:150a–b, 154a–b
natural history, **2**:205a–b; **12**:536b
paleontology, **4**:518b–519a
pharmacy, **10**:616a
philosophy, **15**:47a
relief programs, **13**:167a
religion, **15**:535b–536a, 630a
science administration, **12**:71b
sociology, **10**:237b
statistics, **6**:22b–23a
surveying, **13**:76a
India Plague Commission, **14**:511b
Indian Academy of Sciences, **11**:266a,b
Indian Affairs, U. S. Office of, **12**:204a–b
Indian Association for the Cultivation of Science, **11**:264b
Indian Institute of Science, **15**:26a
Indian Journal of Physics, **11**:265b
Indian Medical Service, **11**:555b
Indian Ocean
exploration of, **3**:365b
geography, **9**:172a
navigation of, **9**:35b–37a
salinity, **8**:188b
Indiana
Geological Survey, **10**:258a,b
geology, **5**:49b
Indiana University, **7**:169a
Indians, American
see Amerindians

seventeenth-century concept of, **5**:60b; **6**:127b; **7**:483b–484a
eighteenth-century concept of, **4**:477a
nineteenth-century concept of, **10**:484a–b; **13**:250b
actuality of, **3**:57a; **4**:237b; **5**:114a; **8**:158a; **9**:32b–33a; **13**:291b–292a
denial of, **2**:577b; **5**:115b; **7**:138a–b, 508b; **15**:496a
geometric vs. metaphysical, **5**:61b, 510a
and God's power, **3**:420b; **10**:420b
inequality of infinites, **2**:395b
infinite sets, **2**:277a–b; **3**:56a; **8**:111a–b; **12**:364b
mathematical
opposition to, **7**:206b
theory of, **3**:55a, 456a; **5**:108a; **6**:114b*; **8**:305b; **9**:303b
relativity of, **7**:490b
successive, **5**:114a
see also Series, infinite; Space
Infinite descent, method of, **4**:474b, 573a, 574a–b; **5**:159a; **7**:566a; **8**:138a–b; **10**:484a; **11**:55b
Infinitesimals
ancient Greek discussion, **1**:215b, 229b; **14**:611a,b
medieval Arabic discussion, **1**:444b
medieval Latin discussion, **6**:378a
seventeenth-century discussion, **1**:164a,b; **2**:47a, 52a; **3**:149b–152a; **4**:58a, 506a; **5**:588b–589a; **6**:599b; **8**:164b–165a; **10**:47a–b, 48b–49a, 63a, 335b, 336a, 337a; **14**:148a
eighteenth-century discussion, **3**:76a–b; **5**:510a, 567a; **15**:368a
nineteenth-century discussion, **13**:81b
concept of, **3**:515b; **4**:222b, 477a; **6**:446a; **7**:490b; **9**:303b; **10**:120b
resistance to, **2**:274b–275a; **5**:259a
in statics, **7**:172b
see also Calculus, infinitesimal
Infinity
see Infinite
Inflammation
eighteenth-century study, **2**:226b; **6**:566b
nineteenth-century study, **2**:165a–b, 172b, 445a; **4**:276a; **5**:269b; **8**:401b; **10**:493a; **14**:229a
twentieth-century study, **5**:43a; **14**:596b
interstitial, **9**:95a
light-induced, **4**:620b–621a
theories of, **9**:333b–334a, 334b
treatment of
steroid, **3**:353a
vascular, **14**:40a
Inflection (math.), **3**:460b; **6**:357a,b, 359b*
theory of, **7**:185a
Influence lines, **9**:446b
Influenza, **3**:13a; **5**:317a; **7**:614a
epidemic, **7**:171a; **14**:405b
equine, **13**:229a
immunity, **15**:522a
virus, **7**:518a; **12**:492b; **15**:454b
Information, statistical, **5**:9a; **7**:352a
Information theory, **11**:546a; **14**:347a
Infusoria, **1**:417b; **2**:481a; **3**:337a; **4**:142b, 289b, 290a, 371b; **7**:110b; **9**:328b
chemistry of, **11**:541b
heat, effects of, **12**:557a; **14**:534a
marine, **3**:338a; **7**:455a; **12**:560b
origin of, **7**:111a; **12**:555a
polygastric theory, **4**:235a–b
reproduction, **11**:174b; **12**:557a
taxonomy, **4**:290b; **9**:575a,b
see also Microbes
Ingen-Housz, Jan (1730–1799), 7:11–16; **7**:97a; **9**:325b
physics, **5**:137b; **12**:427b; **14**:71b
plant physiology, **5**:277b; **8**:82b, 345b
smallpox, **13**:181b
Ingold, Christopher Kelk, **8**:32a, 485a; **9**:356b; **13**:388b; **14**:124b
Ingram, George Lewis Yeatman, **13**:520b
Ingrassia, Giovanni Filippo (*ca.*1510–1580), 7:16–17; **3**:41a, 355b; **4**:487a, 520a; **14**:6a
Inheritance
see Heredity
Inheritance, mathematical problems of, **3**:276b
Inhibition (neurol.), **13**:530a; **14**:105b
central, **12**:270a, 396b–397a, 399b
discovery of, **14**:200a–b
irradiated, **8**:15a
theories of, **2**:525b; **8**:533b; **9**:265b–266a
Injection (geol.), **4**:263a*
Injection (med.)
for anatomical study, **5**:484b; **6**:62b; **8**:8b, 327b, 328a; **9**:153b, 481b
intravenous, **4**:146a
experimental, **1**:391b, 392a; **10**:202a; **12**:210b
introduction of, **1**:360a–b; **4**:37b; **8**:524b
see also Inoculation
Ink
carbon, **12**:381a
sympathetic, **6**:237a*
Inliers (rocks), **8**:33b
Innervation
double, **2**:124b, 301b
reciprocal, **2**:30b; **12**:397a,b; **13**:530a–b
concept of, **2**:26b; **9**:266a
see also specific body parts and organs
Innes, Robert Thorburn Ayton (1861–1933), 7:17–18; **10**:34a
Thiele-Innes constants, **7**:18a; **13**:339a
Inō, Tadataka (1745–1818), 7:18–20
Inoculation
botanical, **1**:493b; **3**:613a
immunological
see Vaccination
intranasal, **12**:492b
see also Immunization, Injection
Inositol, **4**:358a; **5**:6b
Inquieti, Accademia degli
see Accademia degli Inquieti
Inquisition, **1**:468a, 589b–590a; **2**:310a, 540b, 542a–b; **3**:65a, 342a, 420b; **4**:158b, 505b; **9**:13a; **10**:237a,b; **11**:96a, 97a; **12**:105b, 134b, 333a, 428b; **15**:45a, 68b–69a, 98b
and Galileo, **3**:115b, 116b; **5**:242a, 244a–245a
and Helmont, **6**:254b
Insanity, **3**:577b; **12**:597a
colloidal phenomena, **1**:431b
heredity and, **10**:464a
see also Asylums, Mania, Psychiatry
Insecticides, **3**:185b
contact, **9**:576b–577b
ecological, **3**:580a
lead arsenate, **3**:69a
lime/sulfur mixture, **3**:580a
natural, **9**:577a
persistence of, **9**:577a–b
standardization of, **12**:99a
Insectivores
embryology, **6**:535b
Insects, **4**:236a
anatomy, **6**:484a; **14**:510b
eyes, **2**:481a; **6**:484a; **10**:176b; **12**:318a
mouth parts, **7**:286a
sense organs, **3**:247b
wings, **3**:184b
aquatic, **11**:102a
biological control of, **5**:70b; **6**:524b, 525a; **9**:333a
as cause of disease, **1**:526a; **5**:451b; **12**:315a, 468a, 481a
classification of, **1**:502a, 504a; **2**:4b; **4**:494a–b, 512a–b; **5**:73b–74a, 355a; **6**:507a, 524b; **8**:48b–49a; **11**:317a, 331b–332a, 503a; **13**:174a
collections of, **12**:201a
cytology, **8**:587b–588a; **14**:431b
deformities, **12**:612a
development, **13**:172a–174b
ecology, **5**:70b; **12**:98b
economic importance of, **9**:440b; **11**:332a
embryology, **6**:267a; **10**:273b; **14**:233b, 291a
ethology, **1**:329a; **4**:504a–b; **5**:73b; **8**:528b–529a; **11**:331b–332b
flight, **7**:503b
fossil, **6**:220b–222a; **12**:264b
germ strings, **2**:625b
identification of, **5**:440a
instinct in, **4**:504b
life cycles, **5**:12a, 439b
metabolism
temperature effects, **7**:503a
water, **1**:357a
metamorphosis, **5**:328b, 439b; **8**:580a; **14**:233b, 451b, 507b
disproof of, **13**:171a–174b
microanatomy, **4**:207a
mimicry, **1**:502b–504a; **9**:560b–561a
neurohistology, **14**:596b
physiology, **4**:505a*; **10**:39b
and plants
damage, **9**:278b
pollination, **2**:390a; **7**:417b, 441a; **12**:459a, 588a–b; **15**:130b
reproduction, **1**:265a–b; **2**:591b; **3**:180b; **5**:439b; **6**:159b; **11**:341b–342b; **12**:421a,b; **13**:563a,b
respiration, **7**:503a
social, **13**:134b, 137b; **14**:291b
sound generation, **6**:229b; **10**:605a
swarming of, **1**:526a
unity of type, **5**:443b
vision, **8**:529a
color preferences, **8**:518a
see also Entomology; specific insects
Insemination, artificial
see Artificial insemination
Insertion, mathematical problems of, **1**:219a; **5**:382a,b
Insolubility (math.), **3**:275a
Insomnia, **11**:244a
Instinct
nineteenth-century theories, **1**:404a; **8**:445b–446a, 446b; **14**:44b
in animals, **3**:521a; **5**:45a*; **11**:567b
in insects, **4**:504b
and intelligence, **9**:513a
and physical causation, **5**:193b–194b
predatory, **4**:504b
self-preservation, **3**:521a
sexual, **3**:521a; **5**:146a
social, **3**:521a–b, 574b
Institut Catholique
founding of, **3**:133a
Institut d'Astrophysique
founding of, **9**:411a
Institut d'Égypte, **5**:93b, 94a; **7**:615a
Institut de France
see Académie des sciences, Paris
Institut de la Morale Universelle, **10**:2b
Institut de Paléontologie Humaine (Paris)
founding of, **1**:92b; **2**:347a

- Institut für Serumforschung und Serumprüfung (Steglitz), **4**:297b–298a
- Institut International d'Embryologie
  - founding of, **6**:536a
- Institut national des sciences et des arts, Paris
  - *see* Académie des sciences, Paris
- Institut Océanographique (Paris)
  - founding of, **1**:92b
- Institut Pasteur
  - *see* Pasteur Institute, Paris
- Institute for Advanced Study (Princeton), **4**:313a; **7**:487b; **10**:216b–217b; **13**:600a; **14**:89a
- Institute for Infectious Diseases (Berlin), **4**:297b
- Institute for Research in Tropical America, **15**:224b
- Institute for the Study of Human Problems (Paris), **3**:91b
- Institute for the Study of Platinum and Precious Metals (U.S.S.R.), **3**:271a
- Institute of Applied Chemistry (U.S.S.R.), **3**:271a
- Institute of Chemistry, **5**:127a*
  - founding of, **5**:125a
- Institute of Industrial Chemistry, London, **12**:494a
- Institute of Mathematical Statistics, Berlin
  - founding of, **2**:58a
- Institute of Mathematical Statistics, U.S.
  - founding of, **14**:383b
- Institute of Metals (G. B.)
  - founding of, **3**:87b
- Institute of Radio Engineers, **1**:338a–b
  - founding of, **4**:7a
- Institute of Scientific Research (Birmingham), **5**:474b
- Institution of Electrical Engineers (G.B.), **3**:288b; **9**:155b
- Institution of Engineers, **11**:291b
- Instrumentalism
  - eighteenth-century, **2**:17b
  - twentieth-century, **2**:399a
- Instrumentation, scientific, **12**:283a; **14**:196a–b
  - anthropological, **9**:144a
  - astronomical
    - ancient Greek, **1**:248a; **4**:467a; **9**:338b; **13**:323a–b; **15**:219a–b
    - ancient Roman, **10**:299a; **11**:189b
    - medieval, **1**:122a–b, 382a, 507b, 509a; **2**:24a, 148b–153a *passim;* **3**:25b–26a, 27a, 217b–218a; **4**:556a; **5**:365b; **6**:273a–b, 301b–302a; **7**:38b, 118a–120a, 123b, 124b–126a, 142b–143a, 259b–260a, 338b, 353a–354a; **8**:280b; **10**:540a–541a; **11**:244b, 415a–b; **12**:357b, 361a–362a; **13**:149b, 400b, 510b–511a, 536a–b; **14**:389b, 400a–401a, 575a–b, 592b–593a, 594a–b; **15**:162b–163b
    - Renaissance, **1**:179a, 573b; **2**:401b–413b *passim*, 602a; **3**:402a; **4**:260a, 584b; **5**:348a, 349a; **6**:144a; **9**:191b–192a; **10**:160b–161a; **11**:396a; **13**:621b–622a; **14**:273b, 470b; **15**:154a–156a, 477b–478b
    - early seventeenth-century, **3**:262a, 615a; **5**:278b; **7**:294b; **9**:336a–b
    - late seventeenth-century, **1**:169a, 341a–b; **2**:310b, 593a; **3**:97b–98a, 100b–101a, 101b, 102b; **4**:168b; **5**:22b, 24b, 525b; **6**:68a, 360b–363b, 487a; **7**:577b; **9**:484b, 485b; **10**:546b, 595b; **11**:526b
    - early eighteenth-century, **2**:133a, 328a, 387b; **3**:284a, 430b; **5**:88a, 435a, 491a–b; **6**:5b–6a, 69b; **8**:21a–22b, 170b; **9**:77b, 463b–464a; **14**:589a
    - late eighteenth-century, **1**:310b–311a; **2**:140b–141a, 216b, 328b–329a, 388b; **4**:565b; **6**:112a, 328b–330a, 511b–512a; **7**:87a, 543a,b; **8**:178b, 350b, 470b; **9**:231b; **10**:198a, 591b; **11**:284b–285a; **12**:413a–b; **14**:452b, 485a–b
    - early nineteenth-century, **1**:202b, 241a, 568b; **2**:98a,b, 101a, 285b, 471b, 472a; **5**:144a, 271a–b, 302a–b; **6**:324b; **7**:263a; **11**:354b, 377a; **12**:552a; **13**:109b
    - mid nineteenth-century, **1**:136a, 242b, 243a; **2**:284b, 451b; **3**:92b, 93b, 187a, 238b, 288a, 605b; **4**:19a, 161b–162a, 369b; **5**:561b–562a; **7**:75a–b, 209b, 608a; **8**:46a; **9**:460a; **10**:328b–329a; **11**:377b–378a; **12**:36b–37a, 269a; **13**:109b; **14**:628a–629a
    - late nineteenth-century, **1**:54a, 87b, 172a, 398a, 465b–466a, 597a–b, 598a; **2**:332b, 423b–424a, 615a; **3**:35a, 194b, 261b, 288a, 298b, 366a–b; **4**:30a, 179a, 498a; **5**:102b*, 315b–316a, 404a–405b *passim;* **6**:27b–29a, 94b, 106a, 119a, 282a–b, 522b; **7**:270b–271a; **8**:387b–388a; **10**:31b, 35a, 279b, 528b, 600a; **11**:297a, 378a; **12**:139a, 247b–248a; **13**:344a, 481b–482a; **14**:54b–55a, 56b, 448b–449a
    - early twentieth-century, **1**:19a, 55a–56b; **2**:196b, 333a; **3**:298b; **4**:26b, 27a, 69a–b, 407b; **5**:199b, 489b–490a; **6**:29a–30b, 32a–b, 146b, 147a, 149b, 508a; **7**:472a; **8**:582a; **9**:45b–46a; **10**:234b, 473a, 585a; **11**:18b, 470a, 552b; **12**:186a–b, 249b; **13**:409a; **14**:482a
    - mid twentieth-century, **1**:352b–353b; **5**:183a–b; **8**:581a–b; **9**:46a–b; **10**:158b–159a; **11**:553a
    - Chinese, **12**:376b, 384b
    - constants, **4**:547b
    - corrections for, **2**:329a, 402a,b, 405a, 411b
    - division of, **2**:402a, 412b; **3**:120a*, 158a; **5**:491a; **6**:363a; **9**:156b; **11**:377b
    - India, **2**:118a; **5**:275b, 276a; **7**:81a–b; **15**:538a, 593b, 628a, 629b
    - theory of, **2**:98b; **6**:103b
  - aviation, **5**:188a
  - ballistic, **4**:39b
  - biological, **3**:581b
  - biophysical, **1**:303a–b
  - chemical, **5**:432b, 595b; **6**:372b; **9**:140a, 343b; **13**:262b; **14**:488b
  - crystallographic, **5**:455b, 557a; **9**:379b; **12**:436b; **14**:490a–b
  - drafting, **2**:141a, 419a
  - electrical
    - Renaissance, **5**:398a
    - eighteenth-century, **1**:67a, 547a; **3**:153b, 444b–445a; **14**:73a, 78a–b
    - early nineteenth-century, **1**:347b, 557a–b; **4**:535b; **10**:188a; **12**:254a–b
    - mid nineteenth-century, **1**:555b, 556a; **2**:438b–439a; **3**:260b; **4**:201b, 202b; **11**:50a–b, 111b; **14**:330b
    - late nineteenth-century, **1**:298a, 303b; **4**:40a; **6**:73b–74a; **10**:345b, 346a; **11**:122b
    - twentieth-century, **1**:284a,b, 287a–b, 292b, 304a, 320b–321b; **2**:1b, 2a; **5**:597b; **8**:294b, 533a; **10**:348a, 604b; **12**:33b, 95a; **14**:422a
  - electromagnetic, **1**:203a, 557b
  - electrophysiological, **1**:536b; **2**:1b, 367a; **4**:201b, 202a,b, 333b–334a; **8**:387b, 533a; **15**:21a–b
  - geodetic
    - medieval, **4**:609b
    - Renaissance, **1**:179a; **9**:309b; **13**:260a
    - seventeenth-century, **8**:65b; **9**:115b; **10**:596a; **13**:335b–336a
    - eighteenth-century, **2**:299b, 327b, 329b–330a; **4**:15a,b; **8**:22a, 350b–351a; **9**:234a–b, 250b, 330b; **15**:334b
    - nineteenth-century, **5**:303a–b; **6**:240a; **7**:262b–263a, 500a; **10**:9a; **11**:208a, 354b
    - twentieth-century, **2**:372b; **3**:298b; **10**:158b–159b; **12**:249b
    - rectification of, **2**:329b–330a
  - geological, **6**:79a, 229b; **9**:257b
  - geophysical, **1**:521b; **4**:379b; **7**:9b, 609a; **9**:180b; **10**:210b; **11**:291a; **13**:605b–608b; **14**:216a, 629a
    - temperature effects, **14**:503b
  - hydraulic, **14**:494b–495a
  - limnological, **2**:142a
  - mathematical
    - ancient Greek, **1**:228a
    - Byzantine, **4**:488b; **7**:29a
    - medieval, **1**:445b; **4**:157b; **6**:206b*
    - Renaissance, **2**:602a; **4**:260b; **9**:488b
    - seventeenth-century, **2**:83a, 132b–133a; **3**:60b, 183b; **4**:13a–b, 550a; **5**:240a, 593a–b; **6**:271b; **9**:336a–b
    - eighteenth-century, **2**:133a, 140b; **11**:471b; **13**:470b
    - nineteenth-century, **1**:147b; **10**:133b; **13**:470b
    - division of, **3**:120a*
  - medical
    - ancient Roman, **1**:212b
    - seventeenth-century, **12**:102a, 103a,b
    - twentieth-century, **3**:518b
  - metallurgical
    - twentieth-century, **3**:239a–b; **11**:549b–550a
  - meteorological
    - Renaissance, **12**:101b
    - seventeenth-century, **1**:138a; **3**:534b; **6**:487a; **7**:578a; **12**:102b–103a
    - eighteenth-century, **3**:158a, 435b, 540a; **4**:28b; **9**:5b; **12**:122b
    - nineteenth-century, **1**:586a; **3**:556b, 557a; **4**:29a, 410b, 546b; **6**:522b; **12**:595a,b; **14**:292b, 356b–357a
    - twentieth-century, **4**:171a; **9**:189b
  - mineralogical, **5**:212a–b
  - navigational
    - Renaissance, **3**:423b; **4**:5b; **5**:399a; **6**:126a; **9**:254b; **10**:149b
    - seventeenth-century, **4**:214a; **5**:593b; **8**:65a; **9**:527b
    - eighteenth-century, **2**:343b; **3**:397a; **5**:434a; **6**:6a; **8**:170b, 459b, 470b; **9**:5b, 163a, 234a, 607b; **10**:572a; **13**:184a
    - nineteenth-century, **7**:209b; **14**:491b
    - twentieth-century, **7**:265a–b; **11**:84a
  - oceanographic
    - Renaissance, **1**:97b
    - eighteenth-century, **6**:44a
    - nineteenth-century, **1**:202a; **2**:557b; **12**:132a
    - twentieth-century, **4**:43a, 344b; **15**:431b
  - optical
    - medieval Arabic, **6**:194a, 195a
    - seventeenth-century, **3**:23a–b; **4**:128a, 184a,b; **5**:542b; **6**:137b, 603a–b; **7**:73a–b; **9**:595a
    - eighteenth-century, **2**:328b, 344a; **4**:

148b–149a; **6**:148b–149a, 176a; **9**: 141b, 607b; **11**:471b
early nineteenth-century, **1**:202a–b; **2**: 139b; **5**:143a–144a; **14**:491b–492a, 570b (n)
mid nineteenth-century, **1**:556b; **4**: 236b; **6**:19a, 246a; **8**:414a–b; **9**:201a–b, 551b; **10**:133b–134a; **11**:95a–b; **12**:304b, 519a–b; **13**:22b, 77a
late nineteenth-century, **2**:382b; **4**:355b; **5**:557a, 561a–b; **7**:620a; **8**:551b; **9**: 372a, 372b–373a; **11**:208a
twentieth-century, **5**:590b; **7**:36b; **8**: 581a–b; **9**:373b–374a; **12**:352b–353a, 423a
binocular, **8**:144b
phase contrast, **14**:617a
resolving power, **13**:101b
see also specific instruments
philosophy of, **14**:482a, 490b
photochemical, **4**:182b–183a
in physics
medieval Latin, **10**:534a,b
Renaissance, **8**:198a
seventeenth-century, **2**:83b–84a; **7**:578a
early eighteenth-century, **4**:517a–518a; **5**:510b; **9**:595a–596b; **11**:5a–b
late eighteenth-century, **3**:208a; **5**: 78a–b; **7**:620b–621a; **8**:258b; **9**:607b; **11**:284b–285a; **12**:120a; **14**:354b, 452b
early nineteenth-century, **1**:586a; **3**: 497a; **5**:271a–b; **6**:55a
mid nineteenth-century, **2**:354b–355a; **4**:174b; **5**:306a, 340a–341a; **6**:96b; **7**:74a; **8**:2a; **12**:131b–132a; **13**:377b, 386b–387a; **14**:206b
late nineteenth-century, **2**:84b, 200b, 424a; **3**:498b, 504b, 506b; **6**:342a; **7**:160a; **8**:20a, 387b; **15**:59a–b
early twentieth-century, **1**:332a*; **2**: 458a, 600b; **3**:327a; **5**:330b–332a; **6**:470a; **7**:528b; **11**:290b–291a; **14**: 616b–617a; **15**:435a–b
mid twentieth-century, **1**:366a
physiological
nineteenth-century, **2**:184a, 366b–367a; **4**:371b, 615b, 616a; **5**:149b, 184b, 470b; **8**:540b, 541a; **9**:78b, 101b, 546b, 547a; **10**:175b–176a, 205a, 558a, 576b; **11**:425b, 427b; **15**:100a
twentieth-century, **1**:534a, 610b; **2**: 325b; **5**:37b, 65a; **7**:502a,b; **8**:532b–533a; **11**:126a
standardization of, **9**:102b
precision in, **2**:141a; **5**:491b–492a; **6**:240a
psychiatric, **2**:1b
topographical
ancient Greek, **6**:312b
Renaissance, **1**:97b
seventeenth-century, **13**:256b, 621b–622a
eighteenth-century, **11**:284b–285a, 471b
nineteenth-century, **9**:551b; **11**:95a–b
see also Apparatus; Calibration; Measurement; specific instruments
Instruments, musical
medieval, **4**:526a
eighteenth-century, **3**:115a; **4**:88b; **9**:186a
"denisdor," **4**:129a
drum, **11**:265b
enchanted lyre, **14**:289b
keyboard
barrel organ, **4**:88b
glass harmonica, **3**:258a
Neo-Bechsteinflügel, **15**:449b
water organ, **3**:491a
stringed, **12**:376a
acoustics of, **13**:76b
tuning of, **1**:606b; **5**:249a–b; **12**:127b–128a; **14**:152b; **15**:663b(n)
equal temperament, **1**:606b; **5**:249b; **13**:50b
standard frequency, **7**:445b; **8**:399a; **13**:11b
violin
acoustics of, **11**:265a; **12**:129a,b
bowing of, **4**:222b
strings, thickness of, **6**:541a
see also Temperament
Instruments, surgical
medieval, **6**:276b; **14**:584b
Renaissance, **10**:316a; **11**:254a; **14**:28a
seventeenth-century, **4**:511b
eighteenth-century, **3**:67b; **6**:44a; **8**:114b; **11**:112a
nineteenth-century, **2**:376b; **8**:34b, 402a; **11**:99a
Mesopotamian, **15**:645b
Insufflation
tracheal, **7**:502b; **9**:266a
Insulation, thermal, **13**:351a,b
Insulators, electrical, **1**:547b; **4**:215b; **5**:516b, 517a; **12**:427b
thermal conduction, **14**:208b
Insulin, Stephan, **2**:5a
Insulin
action of, **8**:452b
and carbohydrate metabolism, **3**:415b–416a; **15**:228b–229a
clinical trials of, **1**:442b–443a
commercial production of, **1**:442b, 443a; **7**:503a
discovery of, **1**:613b; **8**:614b
extraction of, **1**:440b–443a
isolation of, **1**:11a–b
naming of, **12**:356a
protamine addition, **1**:443a
purification of, **1**:442b; **3**:352a
standardization of, **7**:503a
structure of, **1**:11b
Insurance
eighteenth-century, **5**:209b; **9**:159b*; **14**: 179b; **15**:280a
nineteenth-century, **15**:31b, 66b, 373b
health, **10**:11b; **12**:458b
marine
theory of, **7**:549b
mathematics of, **2**:442a; **4**:470b, 473a; **13**:263b, 264a; **14**:618a
medical standards, **6**:576a
old age, **15**:172b
for risk, **2**:42a
see also Actuarial science, Annuities, Mortality
Integers
algebraic, **2**:90a; **14**:630a
congruence of, **5**:299b
Gaussian, **4**:124b, 125a
reciprocals, **15**:668a
representation of, **7**:54a,b; **11**:269a
theories of, **2**:352a; **15**:668a
Integrability, **6**:95b
Integrals, **6**:113b
Abelian, **2**:85a; **3**:315a; **5**:260a, 471b, 482a; **7**:398b, 477b; **10**:244b; **11**:54b, 548a–b
classification of, **5**:263b
periods of, **6**:214b; **10**:595a*
theory of, **5**:260a; **7**:478b, 479a; **10**: 25b*; **14**:223b
absolutely convergent, **8**:111a
algebraic, **11**:55b
asymptotics of, **3**:136b
bounds of, **9**:126a–b
Cauchy, **12**:513a
circle, **9**:311a
complex, **12**:526b
concept of, **3**:151a; **8**:110b, 111b, 384a, 562b; **10**:275a–b, 337a; **12**:461a
conditions of existence, **3**:385b
continuous, **8**:63b
definite, **1**:499b; **3**:132a, 136a, 137b, 138b, 230a; **4**:125b, 477b, 478a,b, 617b*; **8**:432a; **9**:303b; **11**:269a; **12**:535a; **15**:305a, 317b, 318b, 364b, 366a–b, 368a,b
mean-value theorem, **4**:205b–206a
theory of, **12**:512b
definition of, **3**:139b; **4**:126a; **8**:112a; **11**:402a, 454b
Denjoy, **7**:351b; **8**:558b; **12**:78a
determinant, **9**:449b
Dirichlet, **5**:201b; **11**:451b
divergent, **6**:5a
double, **3**:137b, 138a,b; **4**:478a; **8**:63b, 136a; **9**:409a
extrema, **4**:479b; **9**:471a; **10**:249a
theory of, **8**:111b; **10**:247b
variation of, **2**:349b; **5**:305a, 608b
elliptic, **1**:13a, 15b, 591b; **2**:271b, 299a; **3**:141a, 228b, 468b; **4**:478a; **5**:471b; **7**:52a–b; **8**:136a, 137b, 139b, 140a; **11**:268a, 401b; **13**:132a*, 221a, 238a, 291a; **14**:148b
applications of, **4**:480a; **10**:483b
classification of, **4**:516a; **7**:52b, 478b; **8**:299b
computation of, **7**:619a
multiplication of, **7**:53b
summation of, **7**:52b
transformation of, **1**:16a,b; **8**:139b, 140a–b
Eulerian, **1**:15b, 591b; **4**:2a, 478a; **8**:137b, 142a
extreme values of, **2**:53b; **10**:249a; **14**:614a
Fourier, **2**:238b; **3**:143b; **4**:578a; **5**:95a, 96b, 97a; **13**:424a; **14**:221b, 282a; **15**:385a,b, 386a
Fredholm, **14**:13b
Fresnel's, **5**:20b, 169a
gamma, **6**:214b
Hellinger, **6**:235b
hyperelliptic, **2**:271b, 272a, 299a; **3**:141a; **6**:307a; **7**:459b, 478b, 479a; **11**:449a
inversion of, **14**:220b, 221a
improper, **2**:320a; **3**:137a; **6**:472b; **9**:502b
definition of, **3**:136a
indefinite, **4**:478a
iterated, **8**:111b
Laplace, **6**:309a
Lebesgue, **2**:239a; **4**:547b–548a; **5**:201b; **8**:110a–112a, 558b; **11**:156b, 459b; **13**:424b, 562a,b; **14**:573a
limiting values of, **3**:224b, 230b
Mellin, **6**:214b
multiple, **2**:198a; **3**:137a; **4**:126a–b; **7**:55a, 530b; **8**:62b; **12**:246b
theory of, **3**:223a; **4**:478a; **10**:249a
naming of, **2**:47b
nonanalytic, **2**:356b
Poisson's, **15**:488b
Radon, **11**:261a
Riemann, **6**:95b; **8**:110a–b, 111a,b, 562b; **11**:454b
generalization of, **8**:110a–b
singular, **3**:136b, 138a; **4**:561b; **5**:54b; **7**: 568a; **9**:471a
Stieltjes, **6**:392a; **8**:112a, 562b; **9**:126b; **13**:55a, 57b–58a, 562a; **14**:573a
surface, **5**:201b
tables of, **10**:482a
theory of, **5**:262b; **7**:506b

Integrals (cont.)
- transcendental, **8**:163b
- Weierstrass', **5**:201b

Integraph, **8**:164a
Integration (math.), **5**:527b–528a; **7**:2b–3a; **13**:48b, 81b; **14**:513b–514a
- approximate, **3**:230a; **5**:302a; **7**:515a; **15**:294a–295a, 302b–303b, 307b–308a, 316b, 326a, 329b, 339b, 351a, 353a, 366a–b
- arbitrary closed paths, **3**:140b
- circular path, **3**:139a
- complex path, **3**:137b, 138b; **14**:188b
- continuous, **8**:165b
- definite, **1**:90b
- definition of, **11**:402a; **14**:573a
- of differential equations, **1**:110b; **2**:52b, 350a,b; **3**:385a, 457a, 615b; **4**:103a, 478b; **5**:209b, 449b; **7**:444a, 515a, 567b, 568a; **8**:136a, 139b; **9**:125a; **11**:56a, 399b–400b, 401b, 613a; **15**: 164b, 484b–485b, 488b
- double, **1**:90b
- of dynamical equations, **9**:324b
- by elliptic arcs, **8**:135b, 139b
- formulas of, **4**:515a
- of functional expressions, **1**:14a; **3**:432a–b; **5**:111a
- gradual, **8**:162b
- by halved separation, **11**:400a,b
- indefinite, **4**:478a
- by logarithms, **3**:223a
- "logical," **4**:177a–b
- multiple, **11**:454b
- of partial differential equations, **1**:142b; **3**:385a; **5**:110b, 111a, 482b*, 509a*; **7**:568a; **8**:136a; **9**:232b; **10**:573b–574b; **14**:87a
- by parts, **7**:561b, 574b; **10**:337a; **13**:266b
- of Pfaffian systems, **3**:314a
- rectangular path, **3**:138a–b
- of square-root expressions, **1**:15a
- symbol of, **3**:458b; **8**:163b
- term-by-term, **8**:110b–111a
- theory of, **1**:427b; **2**:351b; **6**:95b, 451b; **7**:54b; **8**:110a–112a, 324b–325b, 558a,b; **9**:232b; **11**:260a, 261a; **12**: 78a; **14**:345a

Integument (bot.), **5**:217a
Intellect, **3**:513b; **5**:106b; **12**:7a–7Ab
- active, **1**:409a
- active and passive, **1**:118a–b
- location of, **5**:44b; **11**:593a
- origin of, **14**:137a
- perfection of, **1**:409a
- and soul, **1**:27a, 260b

Intelligence
- early nineteenth-century study, **14**:44b
- late nineteenth-century study, **2**:132a
- animal, **3**:553a; **5**:45a*; **11**:518a–b
  - and age, **3**:521a; **5**:359a
- and brain convolutions, **4**:383a; **5**:505b; **11**:383a; **14**:407a
- and brain size, **2**:162a; **5**:359a
- evolution of, **11**:518a–b
- inheritability of, **3**:566b
- and instinct, **9**:513a
- localization of, **6**:441a
- measurement of, **2**:131b, 132a–b; **6**:53a
  - predictive, **5**:378b
- nature of, **5**:505a–b
- and race, **5**:505b

Intelligibles, **1**:118a–b
Intension and remission
- of forms, **2**:193a, 194a, 609a, 610b–611a; **3**:420b; **4**:237b; **5**:552a; **6**:378b; **10**: 62a; **13**:196b, 209b
- of knowledge, **7**:116a
- of light, **13**:200a–201a
- of motion, **13**:186a, 202a–204a
- of qualities, **13**:187a, 188b–193b; **14**:155b

Intensity
- degrees of, **13**:186b–187a, 187b, 188b–193b, 196b, 209b; **15**:265a
- vs. extension, **13**:190a,b, 196b–197a, 200a
- vs. magnitude, **9**:377b; **10**:62a

Intercalation, calendric
- *see* Calendar, intercalation

Interest
- compound, **2**:49a; **3**:272b, 276b; **11**:590b, 591a
- simple, **3**:272b, 348b*
- tables of, **3**:181b; **13**:48a

Interface (chem.)
- interfacial tension, **6**:118a–b

Interference phenomena, **2**:452b; **3**:430a; **4**:513b; **5**:86b; **6**:324b; **13**:101b, 103b; **14**:83b, 616b
- achromatism, **3**:420a
- concave mirror, **5**:483b
- coronas, **1**:358a; **9**:118b
- electron beams, **3**:598a; **7**:469b–470a; **8**: 52a
- fringes, **5**:19a, 20b; **8**:551a–b; **9**:155a; **14**:566a–567b
- intensities, determination of, **3**:420a; **8**: 426a–b
- Newton's rings, **1**:201b; **6**:333b; **10**:54b, 57a–59a, 60a; **11**:143a; **12**:477b
- photography, **8**:388a
- principle of, **14**:566a–568a
- theories of, **8**:50b, 51b–52a; **13**:115a
- and wave theory of light, **5**:143b, 167b–170a; **14**:566a–568a
- X rays, **2**:398b; **8**:51a–52a; **9**:543a
- *see also* Rainbow

Interferometer, **2**:424a; **4**:320a–b, 513b; **5**: 15a, 183b; **6**:30b, 94b; **8**:495b, 551b; **9**:215a, 372b, 374a, 531a; **10**:473a, 519a; **11**:583a; **12**:73a, 247b–248a
- rotating, **12**:69a–70a
- stellar, **9**:373b

Interferometry, **7**:555a; **9**:256a; **10**:518b, 519a; **13**:421a
Interglacial periods, **5**:339a
Interior Department, U. S., **11**:119a
Interlingua, **10**:443b–444a
Intermaxillary bone, **3**:38a; **5**:442b–443a, 444a
Intermédiare des mathématiciens
- founding of, **8**:176a

International Academy for the History of Science, **9**:378b
International Association of Academies
- establishment of, **5**:83b

International Association of Chemical Societies, **15**:466a
International Association of Seismology
- founding of, **14**:327b

International Association of Seismology and Physics of the Earth's Interior, **9**: 444b
International Astronomical Society, **11**:94b
International Astronomical Union, **2**:511b; **4**:270a; **6**:35a; **7**:238b; **9**:256a, 415a; **10**:317b; **12**:449b, 456a
- formation of, **5**:101b; **6**:31a
- *Scientific Delimitation of Constellations,* **4**:27a

International Brain Commission, **5**:26b
International Bureau of Latitudes, **6**:240b
International Bureau of Weights and Measures, **2**:102a; **3**:501a; **5**:582b, 583a; **7**:2a
International Catalogue of Scientific Literature, **5**:83b; **12**:573a
International Commission on Atomic Weights, **9**:351b–352a
International Commission on Intellectual Cooperation
- *see* League of Nations

International Commission on the Teaching of Mathematics
- American Committee on Graduate Work in Universities, **2**:217b

International Commission on Zoological Nomenclature, **13**:63a
International Conference on Weights and Measures, **2**:200b
International congresses
- *see* Congresses, scientific

International cooperation, scientific
- *see* Science(s), international cooperation

International Council for the Exploration of the Sea, **9**:590a
International Council of Scientific Unions, **6**:31a
International Critical Tables, **7**:323b
International Education Board, **2**:499a
International Electrical Engineering Commission, **5**:407b
International Electrotechnical Commission
- founding of, **7**:288b

International Encyclopedia of Unified Science, **5**:123a
International Epizootic Bureau, **12**:453b
International Federation of Electron Microscope Societies, **2**:318b
International Geodesic Association
- founding of, **7**:2a

International Geodetic and Geophysical Union, **6**:188a
International Geographical Union, **9**:150a
International Geophysical Year (1957), **1**: 196a; **13**:167a
International Hydrographic Organization, **12**:412a
International Illumination Commission, **2**: 200b
International Journal of American Linguistics
- founding of, **2**:212b

International Latitude Service, **1**:103a; **10**: 232a; **13**:45b
International Meteorological Committee, **2**:628b; **6**:97b
International Meteorological Institute, **11**: 558a
International Monetary Fund, **7**:318b, 319a
International Permanent Council for the Study of the Sea, **12**:191a
International Phytogeographic Excursion, **3**:318a
International Polar Commission, **14**:357a
International Polar Years, **2**:628b; **12**:412b; **14**:634b
International Psychoanalytic Association, **5**:178b–179a, 179b
International Radium Standard Commission (Paris), **5**:341b
International Research Council
- founding of, **6**:31a; **12**:238b

International Seismological Summary, **9**: 407a–b; **13**:500b
International Telecommunications Union, Geneva, **1**:288a
International Union for Cooperation in Solar Research, **1**:597b–598a; **6**:31a; **10**:600b
International Union of Geodesy and Geophysics, **9**:181a
International Union of Pure and Applied Chemistry, **11**:85a; **14**:182a

International Union of Pure and Applied Physics, **13**:601a
International Working Men's Association, **15**:133b
Internationale Erdmessung, **6**:239b, 240b, 241a
Internationale photographische Monatsschrift für Medizin und Naturwissenschaften, **5**:196b
Internationale Tabellen zur Bestimmung von Kristallstrukturen, **6**:303a,b
Internationale Zeitschrift für Metallographie
founding of, **5**:576b
Internationale Zeitschrift für Vitaminforschung, **12**:207b
Internationales Planeteninstitut, **2**:442b
Interoceptors
morphology of, **8**:92a
Interossei, **3**:40b
Interpolation (math.), **1**:42a–b; **2**:56b, 153b, 299a, 462b; **3**:126a, 273b–274a, 373b; **4**:478a; **5**:299a, 410a, 528a; **6**:309a; **7**:139b, 304a, 305a; **8**:142b*; **9**:127a, 555b, 556a, 612b; **10**:46b, 47a, 68b; **12**:377a, 378b; **13**:265b; **14**:148a–b, 149a
finite-difference, **6**:125a
Hermite, **4**:562a; **6**:308a
Lagrange's, **4**:562a; **7**:570a
mechanical device, **7**:260a
theory of, **3**:224b, 229b; **5**:342b; **7**:515a; **13**:55b; **15**:23b
Interrupter, electrical, **9**:166b; **14**:218a
mercury, **5**:87a
Intersection (math.), **4**:336a
Cayley's theorem, **3**:167a
Intersexuality, **5**:453b
Intestines
see Digestive system, intestines
Introspection, **9**:562a
Intrusion, geological, **1**:394b, 470a; **2**:554b; **3**:325a–b, 547b; **4**:262a, 409b; **5**:39a, 395b–396a; **6**:116a, 581a; **8**:99a; **11**:547b; **12**:273a–b; **14**:312a
Intuition, **5**:141b
role in science, **2**:32a, 356a; **6**:203a; **8**:257b; **11**:388b; **14**:343a
theories of, **2**:10a–b, 12a; **12**:155a, 156a
Intuitionism, **2**:304b, 512b, 513b, 514a; **6**:393a, 473a; **11**:60b; **12**:15a; **14**:285a
Intussusception theory, **3**:422a
Inulin, **5**:297b; **12**:470b
Invagination, **8**:27a
Invar (alloy), **5**:582b–583a
Invariance (math.)
algebraic, **3**:164a; **13**:219b
concept of, **3**:164a; **13**:219b–220a
foundations of, **11**:409a
and laws of physics, **4**:322b–323a, 331a–b
topological, **8**:542b; **12**:196a
Invariants, **3**:169b; **10**:137b
absolute, **11**:407b, 408b
binary, **3**:314b
birational, **3**:315b; **12**:330b
class, **12**:173a
differential, **3**:164a, 263b; **6**:75a,b, 76a; **7**:574a; **8**:324a, 326a
finite, **5**:472a–b
functional, **1**:294b
fundamental, **7**:398b
independent, **3**:164a
integral, **3**:96b*; **7**:446a
orthogonal, **6**:235b, 572a
projective, **3**:263b, 314a; **9**:358a; **12**:533a
reciprocity law, **6**:309a
ring basis, **6**:389a–b
semi-invariants, **8**:617a
simultaneous, **1**:294b
theory of
mid nineteenth-century, **2**:294b; **3**:163a, 164a–165a, 263a, 314a–b; **5**:472a–b; **6**:309a, 357a; **7**:397b; **12**:86b; **15**:198a
late nineteenth-century, **1**:294a–b; **2**:236b, 465a; **3**:169b; **6**:388b–389b; **8**:367b, 434a, 542b, 617a; **9**:358a; **10**:544a; **11**:536a; **12**:214b
early twentieth-century, **4**:83a; **6**:572a; **9**:358b; **13**:125a; **14**:283b
mid twentieth-century, **7**:492b
topological, **2**:513a; **12**:411a
Invention
biological foundation of, **15**:455a
philosophy of, **8**:257a–b; **11**:388b
trial and error method, **4**:284a
Inverse squares, law of
in astronomy, **2**:348b–349a
in chemistry, **5**:601a
in physics
electric force, **3**:157a, 444b; **5**:136a; **11**:142a, 496b
gravitation, **3**:431b; **5**:521a; **6**:486a; **8**:259a–b; **10**:61a–64b *passim,* 69a, 74a, 77b, 82b; **12**:288a; **15**:286b–287a, 359a
light, **2**:343b
magnetic force, **3**:444b; **9**:234b
see also Coulomb, Charles Augustin, Coulomb's law
Inversion, analytic, **4**:103a; **5**:521a, 528b; **6**:308b; **13**:15b
Cauchy, **3**:143b
Fourier, **3**:140a
Jacobian, **3**:315a; **5**:471b; **6**:307a, 308b; **11**:450b–451a, 452a, 548a–b; **14**:221a
Inversion, geometric, **3**:555a; **9**:156b
Inversion, polarimetric, **2**:139b
Inversion, stratigraphic, **1**:236a; **4**:402b
see also Faulting, geological; Nappes
Invertase, **6**:111b; **10**:378b
Invertebrates
absorption, **3**:492b
classification of, **3**:523a; **7**:475a, 546a, 588b; **10**:273a, 437a, 522b; **11**:67a, 317a, 331b–332a; **12**:133b, 453a; **13**:355b
digestion, **5**:149a
embryology, **5**:52a; **8**:357a; **9**:332a; **10**:273b; **14**:251a
evolution of, **10**:29b
excretion, **3**:492b
fossil, **4**:147b; **5**:50b, 214a, 488a*; **6**:57a; **9**:87a; **11**:386a
marine, **2**:502a; **9**:112b, 408a, 560b; **12**:479a–b; **13**:65b–66a; **15**:166a–b
biochemistry, **5**:149b
classification of, **1**:329a; **7**:546a; **14**:1b–2a
embryogeny, **7**:546a
life cycles, **12**:107a; **15**:166b
naming of, **7**:586b
neurology, **3**:88a; **14**:408a
neurophysiology, **3**:68b; **11**:517b–518a; **15**:20b
physiology, **10**:431b
psychology, **14**:550b
reproduction, **2**:617a; **12**:45b; **13**:461b
and vertebrates, transition hypothesis, **3**:526b; **5**:357a
see also individual families, genera, and species; Paleontology, invertebrate; Zoology, invertebrate
Investiganti, Accademia degli
see Accademia degli Investiganti
Investigator (ship), **2**:517b–518a
"Invisible College," **5**:415a, 425b; **6**:141a,b
Involution, geological, **8**:544a
Involution, geometrical, **1**:188b; **3**:213b; **4**:47a; **5**:339b, 569a; **10**:297b, 330b; **13**:128a
absolute, **3**:165a
classification of, **2**:85b
elliptical, **13**:5b–6a
theory of, **10**:296b–297a
Iodides
tests for, **10**:440b
Iodine
in air, **5**:315b
atomic weight of, **3**:398a; **7**:552b
compounds of, **2**:199b; **5**:322b; **10**:530b; **12**:321b
aromatic, **9**:357a
determination of, **2**:589b; **9**:95b; **10**:511a; **14**:448a
discovery of, **1**:143a; **3**:455a–b
as dyestuff, **5**:488b
and growth, **3**:113a
medical use of, **4**:209a; **9**:9b; **10**:620b; **14**:115a
metabolism of, **9**:284b
nature of, **3**:603a; **5**:322b
in nutrition, **6**:136a
optical properties, **3**:327b*
in organic substances, **5**:315b
poisoning
treatment of, **4**:297a
in sea water, **1**:416a–b; **3**:437a*; **5**:297a
and starch, **5**:297a; **12**:59b
test for, **1**:416a–b
Iodoform, **14**:478a
discovery of, **12**:321b
as disinfectant, **1**:574b–575a
formula of, **4**:245b
Iodol
synthesis of, **3**:279a
therapeutic use, **3**:279a
Iodonium compounds
discovery of, **9**:357a
Ioffe, Abram Fedorovich (1880–1960), 15:251–252, **4**:295b; **5**:160a; **7**:527a; **15**:8a
Ion exchangers, **12**:405a
Ionian Academy, **4**:257a
Ionians, **1**:247b; **2**:13b; **13**:295a–297b
Ionium, **2**:259b; **13**:504b
discovery of, **12**:29a
Ionization, **14**:98b
of air, **1**:559a; **4**:356a–b
amplification of, **12**:34a
atmospheric, **6**:355a–b; **8**:12b*
and altitude, **9**:397b–398a
by collision, **12**:30a
and condensation, **1**:490b; **14**:421a–422a
energy of, **6**:614b–615a
of fast particles, **7**:617b
of gases, **2**:397b; **3**:31b, 372b, 373a; **4**:355a,b; **12**:26b, 70b–71a; **13**:366a
ionization coefficients, **13**:446b
ionization potential, **5**:117b; **12**:21a, 44a, 71a; **13**:446b
measurement of, **4**:194b; **7**:152b, 153a
in nebulae, **5**:363b
positive, **11**:420b
secondary, **13**:446b
of solids, **5**:160b–161a
theories of, **1**:56a, 288b; **2**:398b, 487b; **5**:103b; **11**:298b–299a; **12**:20a–b, 26b; **13**:299a, 364a; **15**:460a–461a
collision, **13**:446a–b
thermal, **9**:405a; **12**:70b–71a
of vapors, **8**:47a
by X rays, **8**:10a; **12**:26b, 69a; **13**:366a–b
Ionones, **13**:406b

Ionosphere
  Appleton layers, **1**:196a
  electromagnetic properties, **15**:49b
  height of, **8**:94a
  Kennelly-Heaviside layer, **1**:195b–196a; **7**:288b
    prediction of, **6**:212a
  temperatures in, **5**:596b
  theory of, **8**:369a; **12**:238b
  see also Atmosphere
Ions
  activity coefficients, **2**:170a; **10**:156b
  association of, **2**:170a
  charge-to-mass ratio, **14**:598a
  complex, **14**:268a–269b
  degenerate states, **7**:493a
  density, **8**:24b
  diffusion of, **13**:446a
  distribution coefficients, **2**:170a
  electrostatic interactions of, **10**:156b
  entropy, **8**:47a
  excretion of, **9**:284b
  heavy
    acceleration of, **13**:570b
  hydration of, **14**:182b
  metal
    autodiffusion of, **6**:365b–366a
  migration of, **7**:449b; **14**:171b
  mobility of, **2**:487b; **8**:10a; **11**:117a
    theories of, **5**:117a–b; **9**:391b; **15**:433b
  positive, **13**:369a–b
    emission of, **3**:437a*
  quantum structure, **9**:256a
  radii, **14**:588a
  recombination of, **8**:10a
  specific interaction of, **2**:499a
  transport, **10**:525a
  transport numbers, **6**:439a; **7**:449b
  volume, **8**:503a
    and valency, **1**:462a
  see also Cations; Electrolytes; Ionization
Iowa
  Geological Survey, **6**:57b
  geology, **7**:267a
Ipatiev, Vladimir Nikolaevich (1867–1952), 7:21–22; 4:553a, 554a
  Ipatiev bomb, **7**:21a
Ipecacuanha, **3**:45a; **6**:491b; **15**:430a
  analysis of, **10**:497b
  medical use of, **9**:8a; **10**:622a
Ipomoea, **12**:259a
Iraq
  botany, **11**:311b
  geography, **7**:357a–b
Ibn ᶜIrāq
  see Manṣūr ibn ᶜIrāq
Ireland
  archaeology, **14**:518b
  astronomy, **11**:49b
  census of, **10**:565a
  chemistry, **6**:385a–386a
  economy, **10**:566a
  Geological Survey, **5**:538b; **10**:204a
  geology, **5**:537a–538b; **7**:185b–186a
  mapping of, **5**:537a–538b; **10**:565a
  paleontology, **5**:538a
  politics, **9**:465b
Ireton, John, **6**:183b
Iridium, **2**:589b; **3**:301b; **5**:197b
  analytical reagent for, **3**:272a
  as catalyst, **4**:241b; **5**:595b; **13**:313a
  complex compounds of, **8**:108b–109a
  discovery of, **3**:346b; **5**:92a; **13**:281a; **14**:487a
  melting point, **14**:39a
  stereochemistry, **4**:21a
Irinyi, János (1817–1895), 7:22
Iris (anat.)
  inflammation, **4**:285b
  movements, **5**:56a
  muscles, **8**:401a
  neurophysiology, **8**:16a
Iris (asteroid), **5**:405a; **6**:402b
Iris (bot.), **5**:83b
  genetics of, **13**:137b
Irish Corporation of Apothecaries, **6**:385a–b
Irish Linen Board, **6**:385b
Irkutsk province
  geology, **3**:529b
Iron, **15**:651b
  alchemical theory, **5**:353a
  biological occurrence
    in blood, **5**:226b; **9**:302b
    in enzymes, **14**:173b, 175a–b
    intestinal absorption, **10**:493a
    metabolism, **2**:585b–586a; **5**:226a–b; **8**:28b; **12**:95a
    nutritional role, **6**:136a
  chemistry of, **8**:171b, 312a; **11**:329b; **12**:64a
    carbon content, **3**:212a; **5**:601b; **6**:368b; **8**:119a; **9**:591b; **13**:246a–b
    corrosion, **9**:60b
    electrodeposition, **2**:340b
    isolation, **6**:395b
    passivity, **12**:197a–b
    reactions, **5**:324a
    solution in acid, **12**:146a
  compounds of, **3**:271b
    oxides, **11**:167b–168a, 329b; **13**:310b
    sulfides, **11**:168a
  detection of
    chemical, **8**:28b; **9**:105b
    with magnet, **5**:353a
  determination of, **2**:586a; **9**:445b; **10**:440b, 510b
  historiography of, **5**:190a
  magnetization of, **5**:398b; **11**:446a; **13**:80a–b; **14**:330a
    by electric current, **1**:203a; **11**:213a, 578a; **12**:26a; **14**:354b
  medical use of, **9**:417a
  metallurgy of, **2**:421b; **3**:87a; **4**:218a, 394a; **6**:479b, 480a; **7**:78b, 254b; **8**:119a, 260b; **9**:61a, 312b, 472a, 477b, 590b–591b; **10**:512a; **11**:463b, 499a; **13**:346a,b, 468a; **14**:263a, 451a; **15**:24b–25a, 651b
    alloys, **3**:239a; **5**:582b–583a; **9**:591a
    anthracite in, **9**:591b
    casting, **1**:233b; **2**:587b–588a; **3**:87a; **11**:329a–b
    manganese, separation from, **2**:565a
    polymorphic transformation, **3**:233a
    puddling process, **3**:87a
    smelting, **2**:588a; **3**:454b
    tempering, **3**:11b
  meteoric, **14**:477b; **15**:663a(n)
  ores of, **4**:243a
    analysis of, **3**:473b; **11**:169b
    arsenious, **3**:326b
    assaying, **5**:601b
    deposits of, **8**:102b–103a, 260b; **10**:512a; **11**:209b, 608a; **14**:441a, 442b; **15**:152b, 427b
    magnetic separation of, **4**:284a
    oolitic, **3**:161b
    origin of, **1**:623a; **3**:548a*; **14**:441b, 443a
    oxides, **2**:557b; **5**:157b; **9**:450b
    phosphoric, **13**:346a,b; **15**:25a
    pyrites, **5**:157b
    sedimentary, **6**:238b
  physical properties, **3**:454b; **12**:613a; **13**:78b
    combustibility, **9**:18b
    hardness, **8**:119a
    magnetism, **1**:283b–284b; **9**:566a; **14**:245b
    malleability, **11**:329a–330a
    permeability to gases, **3**:11b
    specific gravity, **7**:346a
    strength, **6**:452a–b; **10**:3a
    testing of, **14**:473b
  as spectroscopic source, **9**:256a
  spectrum, **12**:22a
  structure of, **9**:139a; **12**:543b–544a
    allotropy, **10**:246a–b
Iron Age, **3**:622b; **9**:489a; **14**:506a,b
  subdivisions of, **9**:539a
Iron carbonyl
  discovery of, **2**:70b
Irrationality
  definition of, **7**:242a–b
  proof of, **15**:503a–b
Irrationals
  ancient Greek discussion, **1**:189a; **4**:441b, 466a
  medieval Arabic discussion, **1**:445a; **7**:327a–b; **13**:510a; **15**:10a
  medieval Latin discussion, **4**:605b, 608a, 611b
  Renaissance discussion, **13**:59b
  nineteenth-century discussion, **8**:139a
  twentieth-century discussion, **1**:407a; **6**:393b
  in algebra, **3**:128b*; **7**:242a–b; **14**:325a
  approximate calculation of, **7**:558b
  classification of, **4**:421a–424a *passim;* **5**:263b; **10**:300a; **12**:93b; **13**:302b–303b
  compound, **4**:422a
  Dedekind cut, **4**:3b–4a
  discovery of, **13**:302a–b, 315a–317b
  natural, **6**:472b
  opposition to, **7**:507a,b
  quadratic, **7**:566b
  reduction of, **12**:206a
  representation of, **13**:81b
  theories of, **9**:307a–308a
Irreducibility, mathematical, **6**:389b
Irreversibility in physical systems, **4**:293b; **5**:392b; **7**:387a–b; **9**:616b; **11**:9a,b, 10b, 11a; **12**:514b; **14**:123b
  recurrence paradox, **2**:264a–b; **14**:614a
  reversibility paradox, **2**:263b–264a
  see also Reversibility
Irrigation
  ancient Greek, **1**:213b
  Renaissance, **14**:336a
  eighteenth-century, **2**:390a; **6**:219b; **11**:65b; **12**:283a
  nineteenth-century, **7**:148b; **9**:123b; **10**:430a; **12**:487b
  twentieth-century, **2**:538b, 548b; **11**:576a
  Mesopotamian, **15**:648a
Irrigation Survey, U.S., **7**:148b
Irritability
  seventeenth-century discussion, **4**:377a; **5**:426a–427a; **6**:158a, 159a, 256a; **10**:568a
  eighteenth-century discussion, **2**:226b, 302a; **3**:1b, 15a–b; **5**:55b–56b, 267b; **6**:64b; **8**:115a; **14**:321a,b
  nineteenth-century discussion, **2**:187a–b; **8**:515a; **9**:238b; **10**:434a
  of cardiac muscle, **2**:366b; **6**:63a–b; **12**:138a
  cytological studies, **7**:455b
  muscle, **7**:606a
  oxygen and, **5**:411a
  in plants, **3**:576a; **10**:575b–576a, 577a; **12**:471b; **15**:47a

specific, **10**:433a
theories of, **14**:105a–b
Irvine, Robert, **9**:589b
Irvine, William, **2**:174a, 179a–b; **3**:541a; **9**: 5b; **15**:94b–95a, 95b, 363a
Irving, John Duer, **7**:62b
Irving, Roland Duer, **11**:210a
Ibn ᶜĪsā, ᶜAbd al-Raḥman, **6**:189b
ᶜĪsā ibn Ibrāhīm al-Nafīsī, **1**:275b
ᶜĪsā ibn Yaḥyā, **7**:25a; **15**:230b, 236b, 238a–b
ᶜĪsā ibn Zurᶜa, **1**:275b
Isaac Israeli (*d. ca.* 955 A.D.), 7:22–23; 1: 23a*; **3**:394a
commentaries on, **15**:189a
Diete universales, **3**:393a; **13**:38a
De elementis
translations of, **15**:183a
Liber diffinitionum
translations of, **5**:592a; **15**:183a
Liber Jesod Olam, **15**:35b
works translated, **3**:393b
Isaac Judaeus
see Isaac Israeli
Isaac de Stella, **5**:592a
Isaacs, Charles Edward (1811–1860), 7: 23–24
Isaev, Vasilii Ivanovich, **2**:300a
Isatin, **4**:395a; **9**:95b
Ischemia, **2**:547b
Ischia
geology, **8**:566b–567a
Ischus, **1**:253a
Isentropic analysis, **11**:558a
Isethionic acid
discovery of, **9**:18b
Isfahan
observatory, **7**:324b, 325a, 336a–b
Isḥāq ibn Ḥunayn, Abū Yaᶜqūb (*d. ca.* 910), 7:24–26; **7**:187a; **15**:230b, 231a
translations by, **1**:224a, 275a–b; **4**:438a–447b *passim;* **6**:617a; **9**:301b*; **10**: 111b; **15**:236b–237a, 238b
Isḥāq al-Isrāᶜīlī
see Isaac Israeli
Ibn Isḥāq al-Tūnisī, **1**:438a
Ishikawa, Chiyomatsu, **14**:235b
Ishiwara, Jun (1881–1947), 7:26–27; 9:606b
Isidore of Seville (*ca.* 560–636), 7:27–28; 4:121a; **8**:242a; **9**:363b; **12**:81b; **14**: 35a
anatomy, **12**:81a
Etimologiae (Originum seu Etymologiarum), **1**:414a
Liber rotarum, **1**:565a
Isidorus of Miletus (*fl.* sixth century A.D.), 7:28–30; 1:169b, 223b; **4**:488b; **9**: 274b
Isidorus the Younger, **7**:29a
al-Isifīzarī, al-Muẓaffar, **7**:325b
Ising, Gustaf Adolf, **8**:95a
Ising model, **5**:103b
Isis, **12**:112b
editors of, **7**:447b; **9**:378a; **11**:257b
founding of, **12**:108b; **15**:466a
Isis oder enzyklopädische Zeitung von Oken, **10**:195b
Island arcs, **4**:309a
Island universe theory, **3**:509a–b; **6**:529b; **12**:347b; **15**:144b–145a
Islands
biogeography, **2**:320b–321a; **3**:569b; **14**: 135b, 139a
coral, **1**:595b; **3**:550a–552b, 595a; **8**:101a
discovery of, **2**:342b; **4**:256a; **8**:101a
elevation of, **3**:551b, 567b; **11**:303a
locations of, **8**:101a
origin of, **6**:457a; **11**:302b–303a, 304a; **13**:85a
subsidence of, **3**:551a–552a, 567b
volcanic, **2**:321a; **3**:551a–552a, 569a; **9**: 532b; **11**:104b, 133b, 336b; **13**:85a
see also specific islands
Islandus, Otto, **2**:406a
Ismāᶜīlī intellectual tradition, **1**:34b, 35a; **7**:40a,b, 42a; **13**:509a, 510a; **15**:249a–250b
Isoagglutinin, **6**:433a
Isobaric problems, **3**:178a
Isobutane, **2**:622b
Isobutylene, **2**:622b
Isobutyric acid, **4**:400a; **9**:131a
derivatives of, **5**:464a–b
Isochrone, **2**:47a–b, 48a, 507b; **4**:468b; **6**: 600b; **8**:163b
Isocyanines, **9**:403b
Isodimorphism, **5**:124a
Isoëtes, **8**:5b; **15**:130b
Isolation (med.)
concept of, **3**:489b
Isomerism, nuclear, **6**:16b; **7**:527a
Isomerism, optical
see Stereoisomerism
Isomerism, physical
see Enantiotropy
Isomerism, structural, **2**:65b, 68a, 485b; **4**: 91a, 400b; **5**:315a; **8**:508a; **9**:130b; **13**:468a; **14**:141b, 601b
acids and bases, **1**:168a; **6**:108b; **7**:281a; **14**:454b–455a
cis-trans, **1**:390a; **13**:577a; **14**:270a, 445b
concept of, **1**:340b–341a; **2**:95b; **8**:331a, 332a–b; **14**:474b–475a
opposition to, **7**:452b
coordination compounds, **3**:271b; **4**:21a; **13**:338b
diazo compounds, **1**:426b; **6**:108a
inorganic compounds, **14**:266a–272a
oximes, **1**:553b; **3**:300a; **9**:402b; **14**:265a
reversible, **2**:624b; **4**:553a,b
sugars, **5**:3b–4a; **6**:538b
theories of, **2**:620b, 622a–b; **4**:244b–245a, 247a, 553a,b; **5**:325a; **11**:173a
Isomerization, **2**:624a
catalytic, **7**:21b
mechanism of, **9**:131b
reciprocal, **14**:602a
Isometry, **6**:497a–b
Isomorphism (cryst.), **1**:20a; **3**:10b; **6**:179a; **8**:59a; **9**:286b, 289a–b, 423b–424b; **10**:357b; **12**:304b; **13**:476a, 517b
concept of, **5**:556b; **11**:539b; **13**:546b, 617b
discovery of, **2**:72a; **5**:203a; **6**:182a; **11**: 539a; **14**:588a,b
law of, **2**:93b, 106b; **5**:324a, 373a; **8**:56b–57a; **9**:424a–b
and optical activity, **10**:357a–358b
rejection of, **11**:522a
Isomorphism (math.), **4**:420a; **8**:434a
Isoniazid, **4**:155b
Isooctane
synthesis of, **7**:21b
Isoperimetry, **2**:48a, 53b, 58a, 395a; **3**:442b; **5**:195b; **6**:325b, 572a; **7**:178b*; **8**: 306b*, 611a; **9**:409b; **10**:250a, 295b; **11**:487a, 533a; **14**:603b–604a
of convex bodies, **2**:192a; **9**:414a
Isoprene
isolation of, **13**:410b–411a
synthesis of, **4**:553a–b; **5**:531a; **7**:21a, 457a; **13**:1a
Isoquinoline, **3**:347b; **5**:215a
Isorotation, **6**:538b
Isostasy, **1**:470b; **2**:329b; **3**:548a; **8**:241b, 516b; **10**:524b*; **14**:215b, 503b
compensation tables, **2**:373a
concept of, **4**:493a,b, 555b; **6**:58a, 188a–b
distribution of compensations, **5**:395b; **6**:188b; **8**:99a; **13**:532a, 609a–610a
and earthquakes, **4**:266a; **9**:181a
limiting depth, **6**:188b, 241a
oceans and, **2**:373a–b; **5**:329b
regional, **13**:609b
rejection of, **13**:271a
Isotherm, **6**:551b
Isotopes
concept of, **3**:479a; **6**:365b; **12**:32a, 504b–505a; **13**:159b–160a
experimental confirmation, **11**:417b
definition of, **2**:241a; **13**:546b
discovery of, **1**:321a–b; **13**:369b
distribution of, **1**:321b
isotope dilution method, invention of, **6**:366a
in minerals, **5**:457b
naming of, **12**:505a
radioactive, **1**:321a; **2**:259b; **4**:580a–581a; **6**:16a, 481a; **7**:158b
artificial, **6**:365b, 366b; **7**:154b
in biochemistry, **10**:326b
chemical identification of, **7**:158a
inseparability of, **6**:365a–366b; **10**:288b
production of, **13**:163b
ratios, **10**:236a
separation of, **2**:499a; **4**:545a, 581b; **6**: 481a; **8**:292b–293a, 369a; **12**:505a
by diffusion, **1**:321a; **6**:117b; **12**:438a
electromagnetic, **8**:95b; **15**:8b
by fractional distillation, **1**:321a; **8**:482a
by photochemical means, **7**:519a
in stellar atmospheres, **12**:368a
see also specific elements
Israel
scientific research, **4**:545b; **14**:248a
Issel, Arturo (1842–1922), 7:30–31; 11:575b
Issyk Kul, **12**:300b
al-Iṣṭakhrī (Abū Isḥāq Ibrāhīm ibn Muḥammad al-Fārisī), **1**:29a; **6**:186a; **9**:88b
Istanbul, University of
astronomy at, **5**:183a
philosophy at, **11**:355b–356a
Iṣṭifan ibn Bāsīl, **4**:121b; **15**:237a, 238b
Istituto Elettrotecnico Nazionale Galileo Ferraris (Turin), **4**:589a
Istvánffy, Gyula, **2**:437b
Italian Anatomical Society
founding of, **8**:552b
Italian Geological Society, **3**:328a
Italian Institute of Chemistry
founding of, **3**:46b
Italian Mathematical Union
founding of, **10**:610a
Italian Society of the History of Medicine
founding of, **1**:360b
Italy
archaeology, **9**:538a, 547a
astronomy, **13**:232b
botany, **1**:525b; **13**:449a
Consiglio Nazionale delle Ricerche, **4**: 580b
engineering, **13**:233b
geodesy, **1**:19a; **8**:504a
geography, **1**:19a; **3**:533a; **9**:13a
geology, **1**:234a, 491a–b; **2**:439b, 481a–b; **3**:328a–b, 533a; **5**:430a; **7**:30b–31a; **8**:564b
guide books, **7**:581a
mathematics, **3**:117a; **12**:285a–b
meteorology, **13**:232b

Italy (cont.)
  National Institute for the History of Science, **4**:373b
  natural history, **1**:587a; **12**:120a
  paleontology, **1**:587a; **2**:481a; **8**:566b–567a
  philosophy, **14**:581a
  physics, **4**:578a–580b
  Piedmont, Renaissance rehabilitation of, **1**:605a
  politics, **14**:86b
  scientific instrumentation, **8**:21b
  scientific societies, **2**:307b–312a *passim*
  stratigraphy, **2**:108b
  volcanology, **5**:336a; **6**:84a; **12**:562a
  zoology, **1**:130a; **3**:47a
Itching, **5**:184b
Iteration, **3**:126a, 128a, 143a, 222b, 385a; **4**:547b; **5**:151a, 410a; **6**:392a; **7**:446a; **9**:449b; **10**:68b, 72b, 553b, 610b; **11**:227b–228a, 228b
  Maxwellian, **9**:225b
Ithaca (Greece), **12**:180a,b
Ittmann, George Philip, **7**:492b
Ivanov, Ilya Ivanovich (1870–1932), 7:31–33
Ivanov, Piotr Pavlovich (1878–1942), 7:33–34; **7**:476b*
Ivanovsky, Dmitri Iosifovich (1864–1920), 7:34–36
  viruses, **8**:450a; **15**:14b
Ives, Frederic Eugene, **7**:36a
Ives, Herbert Eugene (1882–1953), 7:36–37; **11**:442a*
Ivory, James (1765–1842), 7:37; **8**:261b, 262a; **13**:219a; **15**:487b
  Ivory theorem, **1**:147a; **7**:37a
Ivory, **15**:648a
  from mammoths, **1**:30a; **7**:60a–b
  medical use of, **10**:237a
Ivy, Andrew Conway, **3**:69b
Izvestiya Akademii nauk SSSR
  editors of, **10**:169b
ᶜIzz al-Dīn Abū Isḥāq Ibrāhīm ibn Muḥammad ibn Ṭarkhān al-Suwaydī, **9**:604a

J

Jābir ibn Aflaḥ al-Ishbīlī, Abū Muḥammad (*fl.* first half of twelfth century), 7:37–39; **4**:441a; **8**:280a; **9**:264a; **15**:477a
  De astronomia
    translations of, **15**:178b
  Correction of the Almagest, **11**:249a
  Flores, **11**:415a
  Iṣlāḥ al-Majisṭī, **8**:280b; **15**:33b
  manuscripts of, **14**:273a
  translations of works, **13**:400b, 401b
Jābir ibn Ḥayyān (*fl.* late eighth and early ninth centuries?), 7:39–43; **2**:23a, 64b; **4**:441a; **7**:38a; **10**:555b; **13**:511a
  alchemy, **1**:33b
  Liber divinitatis de LXX
    translations of, **15**:185b
Jābir ibn Sinān al-Ḥarrānī, **1**:507b
Jaborandi
  physiological effects, **8**:14b
Jaccard, Auguste (1833–1895), 7:43–44; **12**:139b
Jackson, Charles Thomas (1805–1880), 7:44–46; **1**:542b; **4**:73b; **5**:478b; **7**:44b; **12**:434b; **14**:315b
Jackson, Elisabeth, **5**:589b
Jackson, John Hughlings (1835–1911), 7:46–50; **1**:496b; **4**:593a,b, 594a,b; **5**:173b, 196b; **6**:59a, 140a, 440b
  Jacksonian epilepsy, **3**:205b; **7**:48a
Jackson, Samuel, **6**:120a
Jacob of Cremona, **1**:229a,b; **3**:515b
Jacob of Edessa, **1**:275a
Jacob, Charles Edward, **9**:258a
Jacob, François, **13**:137a
Jacob staff
  see Cross staff
Jacobi, Carl Gustav Jacob (1804–1851), 7:50–55; **3**:224b, 302b, 466b; **4**:341a; **5**:261b, 262b, 301a; **7**:460a, 522b; **8**:63a
  and Abel, **1**:16b–17a, 17b; **7**:51a, 52a,b, 53a,b
  acquaintances of, **5**:307a, 308a; **7**:506a; **8**:382b; **10**:27a, 28b; **11**:447b; **13**:13b, 15a
  algebra, **3**:164a; **6**:357a, 403b; **7**:508a; **12**:171a; **13**:220a
  analysis, **2**:271b; **4**:475a; **5**:304b, 309b; **7**:408a, 444a; **8**:325b, 383b, 384a; **10**:249a,b, 250a, 573b, 574a–b; **11**:54b, 449a, 452a; **13**:238a; **15**:198b
  on contemporary scientists, **13**:18a
  correspondents of, **6**:552b; **7**:521b; **8**:136b, 137b, 138a, 139a, 140b–141a
  and Dirichlet, **4**:124b, 126a
  editions of works, **12**:599b; **14**:223a
  and Eisenstein, **4**:341b–342a, 342b
  Fundamenta nova theoriae functionum ellipticarum, **1**:16b; **3**:169a; **5**:572b; **14**:219b; **15**:488b
  Gesammelte Werke, **3**:314a
  Hamilton-Jacobi equation, **6**:89b; **12**:220a
  and Hermite, **6**:307b
  influence of, **12**:225b; **13**:4b, 422b; **14**:202a
  influences on, **15**:485b
  Jacobi condition, **2**:198a
  Jacobi polynomials, **12**:513a
  Jacobian inversion, **3**:315a; **5**:471b; **6**:307a, 308b; **11**:450b–451a, 452a, 548a–b; **14**:221a
  Jacobians, **3**:136b, 142b; **7**:55a; **10**:442a
  Jacobi-Liouville theorem, **7**:37b
  mathematical physics, **2**:470a; **3**:142b; **5**:508b; **6**:92b; **8**:384b, 561b, 562a; **9**:232b, 324b–325a, 566a; **11**:58a; **13**:130b
  number theory, **8**:138b
  school of mathematics, **3**:313b–314a; **4**:2a; **5**:472a; **7**:51a
  students of, **1**:294a; **2**:299a; **6**:230a, 356b; **7**:108b, 109a,b; **11**:548a; **12**:170b, 289b
  translations of works, **12**:170b
  Vorlesungen über Dynamik, **15**:485b
Jacobi, Maximilian, **5**:570a
Jacobi, Moritz Hermann von (1801–1874), 7:55–56; **14**:206b
  electric motor, **7**:181a; **8**:188a
  Mémoire sur l'application de l'Électro-Magnétisme au Mouvement des Machines, **10**:192b
Jacobins, **2**:321b; **3**:7a, 72a, 400a; **9**:343a; **12**:549a; **13**:571a
  opposition to, **11**:497a
Jacobites, **13**:67b, 68a
Jacobs, Walter Abraham (1883–1967), 15:252–253
Jacobs, Wilhelm, **6**:456a,b
Jacobsen, Carlyle Ferdinand, **4**:286b
Jacobsen, Jacob Christian, **6**:100b; **7**:393a; **10**:380a–b
Jacobson, Egbert, **15**:465b
Jacobson, Ludwig Levin, **2**:443a; **11**:379b
  Jacobson's organ, **2**:504a
Jacobson, Paul, **9**:357a
Jacquard, Joseph Marie
  Jacquard loom, **1**:356a
Jacquet, Pierre Armand (1906–1967), 7:56–57
Jacquier, le P. François, **2**:326b; **3**:216a, 282a; **4**:515b; **5**:193a
Jacquin, Josef Franz, Baron von, **7**:58a; **9**:152b
Jacquin, Nikolaus Josef (1727–1817), 7:57–59; **1**:520a,b; **5**:202b, 430a; **7**:12a; **8**:85b(n); **9**:152b
  botany, **13**:181b
  students of, **1**:17b; **9**:559a
Jadeite
  synthesis of, **2**:459b
Jadwal tables, **1**:508b
Jaeger, Frans Maurits (1877–1945), 7:59; **4**:185a*
Jaeger, Georg Friedrich (1785–1866), 7:59–60; **7**:368a, 369a*
Jaeger, Werner, **1**:258b; **4**:461a; **11**:28b; **12**:440b–441a
Jaekel, Otto (1863–1929), 7:61–62; **14**:627b*
Jaᶜfar al-Balkhī
  see Abū Maᶜshar
Jāᶜfar ibn Muḥammad al-Ṣādiq, **7**:39b, 40a; **15**:250a
Jaᶜfar ibn Yaḥyā, **7**:39b
Jagannātha (*fl. ca.* 1720–1740), 7:62
  translations by, **7**:81a,b; **15**:628b
Jäger, Gustav, **8**:511a; **14**:234b
Jäger, Karl Christoph Friedrich von, **10**:191b
Jaggar, Thomas Augustus, Jr. (1871–1953), 7:62–63; **6**:13b; **7**:474b*
al-Jāḥiẓ, Abū ᶜUthmān ᶜAmr ibn Baḥr (*ca.* 766–868/869), 7:63–65; **3**:548b; **11**:231a; **15**:261a
Jahn, Hans Max (1853–1906), 7:65
Jahnke, Eugen, **2**:594a; **11**:150a
Jahrbuch der Radioaktivität und Elektronik
  founding of, **12**:613b
Jahrbuch für Elektrochemie
  editors of, **15**:450a
Jahrbuch für psychoanalytische und psychopathologische Forschungen
  editors of, **7**:190b
Jahrbuch über die Fortschritte der Mathematik, **4**:560b
  editors of, **14**:453b
Jahrbücher für wissenschaftliche Botanik
  editors of, **10**:576b; **13**:90a
  founding of, **11**:155a
Jahresbericht über die Fortschritte der Chemie
  editors of, **1**:346b; **4**:19a; **7**:464a
Jahresberichte über die Fortschritte der gesammten Medicin in allen Ländern, **13**:557b
Jaillard, Pierre François, **3**:588a; **10**:386b–387a
Jains
  astronomy, **15**:538a, 554b
Jalāl al-Dīn Muḥammad ibn ᶜAbd al-Raḥmān al-Qazwīnī, **9**:604a
Jamaica
  discovery of, **15**:89a
  geology, **4**:9b–10a
  natural history, **3**:129a; **12**:457b
  sugar industry, **6**:383a–b
Jamāl al-Dīn Yūsuf al-Khiṭāᵓī, **12**:362b
Jambūdvīpa (continent), **15**:554a–b
James I, King of England, **2**:406b; **4**:183b; **5**:47a, 396b; **6**:150b; **7**:306a; **9**:609b, 611a; **13**:507b, 508a

James II, King of England, **8**:129b; **10**:44b, 565a, 566a; **12**:586a
James of Forlì, **4**:237b; **9**:132b; **13**:349b
James of Venice, also known as Iacobus Veneticus Grecus (*d.* after 1147), 7: 65–67
translations by, **1**:99b, 270a–b, 272a,b, 273b; **2**:230b, 231a; **15**:175a
revisions of, **9**:436a, 437a
James, Hubert Maxwell, **8**:474b
James, Reginald William, **15**:63a
James, Robert, **4**:89b*
James's powder, **10**:446a
James, Thomas (Captain), **5**:343b–344a
James, Thomas Potts, **8**:264a
James, William (1842–1910), 7:67–69; 3: 471a*; **4**:558b*; **8**:597b
acquaintances of, **2**:367a; **6**:52b; **10**:486b; **15**:151b, 467a(n)
correspondence of, **14**:528a
influence of, **8**:517a; **10**:485b
influences on, **1**:74b; **2**:12a; **14**:534a
James-Lange theory, **8**:7b
philosophy, **2**:461a; **14**:609a
students of, **3**:335a; **15**:71a
Jameson, Robert (1774–1854), 7:69–71; 1: 476b; **5**:67b; **6**:579b; **14**:132b
acquaintances of, **3**:291b, 565b; **4**:65b; **5**:31b, 32a, 71b; **9**:447b; **12**:523a; **13**:499a; **14**:257a; **15**:151b
students of, **2**:341a; **3**:585b; **5**:13b, 66a, 469a, 537a; **9**:540a; **10**:204a
Jamieson, Thomas Francis, **3**:569a; **5**:337b
Jamin, Jules Celestin, **1**:202b; **3**:420a; **5**:496a; **8**:387b; **9**:154b, 166b, 566b
Jamshīd ibn Maḥmūd al-Kāshī
see al-Kāshī
Janet, Pierre Marie Félix, **7**:67b, 190a; **9**: 382a
Janiszewski, Zygmunt (1888–1920), 7:71–73; 9:248a,b; **12**:426b
Jansen, Barend Coenraad Petrus, **14**:393a, 446a
Jansen, Zacharias (1588–*ca.* 1631), 7:73; 2:305b
Jansenism, **1**:292a; **4**:135a, 496a, 505b; **5**: 166a, 577b; **10**:332a, 333a, 335b, 337b, 614a; **11**:42b; **14**:69b, 301b
critics of, **9**:48a
Jansky, Jan, **9**:545a
Janson, Stanley Eric, **11**:88b
Janssen, Pierre Jules César (1824–1907), 7: 73–78; 12:248b; **13**:344b*, 408a
camera, **9**:102a
observatory, **4**:68b; **6**:105b, 106a
solar prominences, **6**:28a; **8**:440b; **11**: 320a; **14**:629a
spectroscopy, **11**:279b; **12**:268a, 269a
Janssens, Frans Alfons, **5**:35b; **9**:521b; **13**: 158a; **14**:432b
Japan
agriculture, **12**:116a–117a
astrology, **15**:729a,b, 730a, 731b
Astronomical Bureau, **15**:734b, 736a,b
astronomy, **1**:310a–314b; **7**:18b–20a; **12**:403a–404a; **15**:729a–737a, 739b, 743a, 745b, 747b, 749a, 753b
biological sciences, **9**:536b
anatomy, **15**:739b–742a
bacteriology, **7**:391b
botany, **7**:205a; **13**:391b–392a
zoology, **14**:314a
calendar, **1**:310a–313b; **7**:18b; **15**:729a, 731b, 732a, 734b, 735a–b, 737a, 746a, 750a, 752a
reform, **1**:310b–313b *passim;* **12**:403a–404a; **15**:730a, 731b, 734a–b, 735a,b
and China, **15**:729a, 732b–734a, 737a–739b, 756b
education, **15**:753b, 755b
han schools, **15**:748a
medical, **15**:742b–743a
engineering, **15**:744a
exploration of
eighteenth-century, **2**:14b
geography, **7**:204b; **9**:190a; **13**:583b
geodesy, **7**:18b–20b; **15**:744a, 745b, 746b, 747b
geology, **7**:474a; **8**:577a; **9**:181b
gravimetry, **9**:181a
mineral resources, **11**:209b
paleobotany, **9**:617b
seismology, **7**:9a–b; **9**:406b–407a; **10**: 210b–211a; **12**:341b
Institute for Infectious Diseases, **7**:392a–b
mathematics, **1**:83a–84a, 90a–91a; **3**: 254a, 265b–266a, 270b; **8**:314b, 419a; **12**:290b–292b; **15**:744b–755b
enri calculus, **15**:747b, 751b, 752a,b
idai, **15**:749a, 750a, 750b, 751a,b
number theory, **15**:751a
Saijo school, **15**:751b
Seki school, **1**:83b, 90a; **15**:747b, 751a–b, 753b
Takuma school, **15**:752a
tenzan algebra, **15**:751a,b
yojutsu, **15**:752a–b, 753a, 754a, 755b
medicine, **1**:310a,b, 312a; **15**:737a–744b, 749a
Koihō school, **1**:312a; **15**:739b–740a, 740b, 741a,b
surgery, **1**:312a; **6**:231b; **15**:741b
meteorology, **14**:52b
natural philosophy, **12**:406b–409a; **15**: 728b, 730a–b, 741a
philosophy, **15**:729a, 730a–b, 732b, 743a–b, 749a
physics, **9**:606b
politics, **7**:205a; **15**:733b, 734a
social structure, **15**:728b, 734b, 742b–743a, 743b–744a, 745b, 748a
Western science, availability of, **7**:19b–20a; **12**:117b; **14**:394b; **15**:730a, 734a–744a *passim,* 753b
Japan, Sea of, **4**:42b
Japan Institute
founding of, **5**:623a
Japan Trench, **9**:181a
Japanese Association for the Advancement of Science, **9**:181a
Japanese Scientific Thought, 15:728–758
Jappelli, Gaetano, **2**:339b
Jaquemet, Claude, **11**:392a
Jaquet, Alfred, **2**:586a
Jardin des Plantes (Paris)
see Jardin Royale des Plantes (Paris)
Jardin du Roi
see Jardin Royale des Plantes (Paris), Muséum National d'Histoire Naturelle (Paris)
Jardin Royale des Plantes (Paris)
in seventeenth century, **9**:92b; **13**:442b
in eighteenth century, **2**:576b; **3**:365a; **4**:216a–b; **13**:390a, 553b; **15**:111a,b, 113b, 342b
in nineteenth century, **9**:418b
anatomy at, **4**:267b; **9**:252b; **10**:477a; **11**:99b
botany at, **7**:197a, 199a,b
chemistry at, **3**:596b; **5**:90a, 352b; **8**:130a–b, 171a, 619b; **11**:562a, 563a, 564a
collections, **3**:365b; **4**:156b–157a; **7**:544a; **8**:274b
founding of, **7**:536a–537a
natural history at, **13**:565b
reorganization of, **5**:92a, 356a; **7**:198b, 547a–b, 586b, 588b; **12**:66b; **15**:111a–b, 113b
zoology at, **5**:355b
see also Muséum National d'Histoire Naturelle (Paris)
Jargonium, **12**:544a–b
Jarisch, Adolf, **2**:111a
Jarisch-Herxheimer reaction, **4**:301a
Jarrett, Alan, **5**:183b
Jars, Antoine Gabriel (1732–1769), 7:78; 8:70b; **12**:65b
Jasper
artificial, **8**:213b
Jastrow, Joseph, **5**:377b; **6**:53a; **10**:485a–b
Jaundice, **5**:226b
epidemic, **5**:270b
hemolitic, **9**:95a
hemorrhagic, **10**:142a
Java
botany, **2**:520b–521a; **8**:512b
natural history, **11**:261b
paleobotany, **5**:442a
Java General Agriculture Experiment Station, **13**:460a
Java man, **3**:606b
Javelle water
see Bleaching agents, chlorine
Jaw
movements of, **4**:589b
al-Jawharī, al-ᶜAbbās ibn Saᶜīd (*fl. ca.* 830), 7:79–80; 4:438a, 440b; **12**:91a
tables of, **14**:538a
Jayalakṣmaṇa
commentaries of, **2**:118a
Jayasiṃha (Savāī Jayasiṃha II) (1686–1743), 7:80–82; 7:62a; **15**:628b, 629a
al-Jayyānī, Abū ᶜAbd Allāh Muḥammad ibn Muᶜādh (*b. ca.* 989; *d.* after 1079), 7:82–83; 4:441a
Kitāb al-masālik wa'l-mamālik, **7**:8b
translations of works, **15**:180a
al-Jazarī, Badīᶜ al-Zamān Abu'l-ᶜIzz Ismāᶜīl ibn al-Razzāz (*fl.* 1206), 15:253–255
Ibn Jazla, Abū ᶜAlī Yaḥyā ibn ᶜĪsā
Taqwīm al-abdān, **1**:29a
Ibn al-Jazzār, Abū Jaᶜfar Aḥmad ibn Ibrāhīm ibn abī Ḫālid, **3**:393b, 394a; **11**:244a, 445a; **13**:401b
Jealousy, **5**:377b
Jeans, James Hopwood (1877–1946), 7:84–86; 2:244b; **3**:583a, 584a; **6**:34a*, 113a; **12**:10a; **14**:296b
astrophysics, **8**:562b; **12**:22b
and Eddington, **12**:23a
gas theory, **9**:222b
Rayleigh-Jeans law, **7**:84b–85a; **13**:102b, 104a
Jeaurat, Edme-Sébastien (1724–1803), 7:86–88; 1:400b
Jedlik, Ányos, **4**:378b; **12**:425b
Jefferson, Mark, **2**:373b
Jefferson, Thomas (1743–1826), 7:88–90; 11:218a; **12**:121a–b; **14**:456b
acquaintances of, **1**:485b; **3**:400a; **4**: 252a,b, **6**:551a
correspondence of, **3**:562b; **4**:503a; **9**:54a; **11**:263a; **13**:46b
entomology, **12**:133b
influence of, **9**:537b; **10**:438b; **11**:141b; **14**:417a
metrology, **11**:472a
vaccination, **7**:97a
Jefferson Medical College (Philadelphia), **4**:252a
founding of, **5**:518a

Jeffrey, Edward Charles (1866–1952), 7:90–91; 8:354b*
Jeffreys, Harold, **5**:596b; **7**:151a
Theory of Probability, **1**:532a
Jeffreys, John Gwyn (1809–1885), 7:91–92; **9**:542a; **12**:107b*
collection of, **3**:536a
oceanography, **13**:360b
Jeffries, Zay (1888–1965), 7:92–93; 3:367b; **9**:312b*
Jehuda ben Barsillae al-Barceloni, **1**:23a
Jelinek, Karl, **2**:533a; **6**:97a,b
Jellyfish
fossil, **14**:120b
Jena
glassworks, **12**:211b
Jena, University of
anatomy at, **6**:573a
botany at, **11**:152b
Faculty of Mathematics and Natural Sciences, **5**:595a
medicine at, **11**:511a; **14**:212b–213a
Phyletisches Museum, **11**:19b
reform of, **6**:456b
zoology at, **11**:570b; **15**:170a
Jenaische Zeitschrift für Medizin und Naturwissenschaft, **15**:166b
Jenkin, Henry Charles Fleeming (1833–1885), 7:93–94; 1:506a; **4**:500a, 501a*; **9**:209b, 227a
Jenkins, Louise, **12**:177a
Jenkinson, John Wilfred (1871–1915), 7: 94–95
Jenner, Edward (1749–1823), 7:95–97; 3: 220a; **6**:567b; **7**:11b; **8**:409a; **10**:396b
critics of, **3**:463b
and Pearson, **10**:445b–446a
smallpox, **10**:390a, 391b, 395b; **12**:57a
Jennings, Herbert Spencer (1868–1947), 7: 98–100; 8:45b, 446b; **9**:167a; **10**: 445b*
Behavior of the Lower Organisms, **8**:446a
Jennings, W. N., **11**:531a
Jensen, Carl Oluf (1864–1934), 7:100–101; **4**:299b; **12**:88a
Jensen, Ingeborg Hammer, **14**:632b
Jensen, Johan Ludvig William Valdemar (1859–1925), 7:101–102; 6:472b; **10**:117b
Jensen's theorem, **7**:102a
Jepson, Willis Linn (1867–1946), 7:102–103; **12**:329a*
Jequirity, **12**:88a
Jerrard, George Birch (1804–1863), 7:103; **2**:468a; **6**:92b
Mathematical Researches, **2**:467b
Jessop, Francis, **14**:413a,b
Jesuati, Order of, **1**:164a; **3**:115b, 149a
Jesuits
in Renaissance, **1**:587b; **3**:311a; **4**:506b(n); **10**:137a; **14**:29a
in early seventeenth century, **2**:131a; **3**:3a, 100a, 133b, 360a, 528a; **4**:505a, 506b(n); **5**:542a–b, 588a; **7**:557b, 583a; **11**:411a; **12**:74b, 152a; **13**:235b, 236a; **14**:636b
in late seventeenth century, **2**:591a; **3**: 183b; **4**:221a, 505a; **5**:57b–58a, 542a–b; **9**:50a; **10**:314a–b; **12**:55a
in early eighteenth century, **2**:326a,b; **3**: 114a; **4**:84b; **5**:60a; **6**:233b; **10**:571b; **11**:401a; **13**:39a–b
in late eighteenth century, **2**:315a, 327b, 334b; **5**:600b; **6**:234b; **9**:231b, 458a; **14**:69b
in nineteenth century, **10**:186b; **12**:266a–b
in twentieth century, **13**:274b, 276a–b
astronomy, **7**:19a,b, 20a; **10**:571b; **12**:85b, 151b; **15**:735a, 736a, 747b, 751b
calendric science, **7**:18b; **12**:403b–404a; **14**:160a
Ch'ung-chen li-shu, **1**:312b, 313b
critics of, **2**:526b; **6**:253b–254a; **9**:15b–16a; **10**:335b; **14**:301b
and Galileo, **5**:241b, 382b
historiography of, **1**:483b
medicine, **15**:737b
missionaries, **1**:310a, 312a; **3**:269a; **5**: 259b*; **8**:150a, 314b; **11**:402b–403a; **12**:210b; **14**:159b–165b *passim*
physics, **7**:375b; **10**:512b; **12**:210b; **14**:71b
and Prague University, **9**:96b–97a
Jesup, Morris Ketchum
North Pacific Expedition, **2**:212b
Jesus
birth date, **9**:168a
Jet
force of reaction of, **2**:40a–b
pressure of, **2**:40a
propulsion, **6**:34b–35a
Jet streams, theory of, **3**:195a–b; **10**:20a; **11**:558b
Jeu de paume, **2**:50b
Jevons, William Stanley (1835–1882), 7: 103–107; 3:451a
logic, **11**:94b
Pure Logic, **2**:297b
Jewett, Charles Coffin, **6**:280a
Jewett, Frank Baldwin (1879–1949), 7:107–108
Jewett, Frank Fanning, **6**:50b
Jewitt, Llewellyn, **8**:105a
Jibrāʾīl ibn Bukhtīshūʿ, **15**:230b, 235b, 237b
Jimeno, Pedro, **4**:487a, 520a; **7**:17a
Jīvanātha of Patna, **9**:42a
Jñānarāja
Siddhāntasundara, **9**:580a; **15**:614b
Joachim of Fiore, **1**:159a
Joachim, Georg
see Rheticus
Joachimsthal, Ferdinand (1818–1861), 7: 108–110; 7:521b, 522a; **8**:383b; **14**: 329b
Joachimsthal equation, **8**:385a
Joachimsthal surfaces, **7**:109b
theorems of, **7**:109b
Joachism, **1**:289b
Job of Edessa
see Ayyūb al-Ruhāwī
Joblot, Louis (1645–1723), 7:110–112; 2: 479b; **4**:133b*
Jodrell, Thomas Jodrell Phillips, **6**:491a
Joestellius, Melchior, **2**:413a–b
Johann Georg, Elector of Brandenburg, **13**:397a,b
Johannes
see also John
Johannes Lauratius de Fundis (*fl.* 1428–1473), 7:112
Johannes Leo
see Leo the African
Johannesburg (U. S. Afr.)
observatory, **7**:18a
Johannitius
see Ḥunayn ibn Isḥaq
Johannsen, Albert (1871–1962), 7:112–113; **3**:484a
Johannsen, Wilhelm Ludvig (1857–1927), 7: 113–115; 10:576b; **12**:298b
genetics, **3**:122a; **14**:99a; **15**:507b
Johansen, (Fredrik) Hjalmar, **15**:431a
Johansson, Carl Hugo, **13**:245b
John XXII, Pope, **5**:113b, 115a; **10**:172a
John II, king of Portugal, **14**:583b
John Actuarios
About the Urine, **1**:121b
John Afflacius
see Afflacius, Johannes
John Buridan
see Buridan, John
John of Damascus, **5**:549b; **10**:21a; **15**:188a
John Danko of Saxony
see John of Saxony
John de' Dondi
see Dondi, Giovanni
John of Dumbleton (*d. ca.* A.D. 1349), 7: 116–117; 13:206a
Summa logicae et philosophiae naturalis, **2**:393a–b; **13**:196a, 197a, 201a
John Duns Scotus
see Duns Scotus, John
John of Gmunden (*ca.* 1380–1442), 7:117–122; 11:415a; **15**:473b
Tractatus de sinibus, chordis et arcubus, **14**:19a
John of Halifax
see Sacrobosco, Johannes de
John of Holywood
see Sacrobosco, Johannes de
John of Italy, **1**:270a
John of Jandun, **4**:237a; **6**:378a; **15**:35b
John of Legnano, **6**:275b
John Lichtenberger
see Lichtenberger, Johann
John of Lignères, or Johannes de Lineriis (*fl.* first half of fourteenth century), 7:122–128; 7:120a, 139a, 140a, 141a
Canones tabularum primi mobilis, **14**:19a
equatorium, **3**:26a; **7**:119b; **10**:541a
tables of, **7**:129b–130a, 139b
John Marliani
see Marliani, Giovanni
John of Messina, **7**:142b*
John of Meun, **1**:91b
John of Mirecourt, **2**:604a
John of Murs (*fl.* first half of fourteenth century), 7:128–133; 7:123a, 141a; **8**: 281b; **12**:440a
De arte mensurandi, **1**:228a–b
Circuli quadratura, **1**:228a–b
Expositio tabularum Alphonsi regis Castelle, **7**:139b
Quadripartitum numerorum, **8**:226a
tables of, **7**:120a, 124b; **15**:155a
John of Palermo (*fl.* 1221–1240), 7:133–134; **4**:605a, 610a,b
John Peckham
see Peckham, John
John Philoponus (*fl.* first half of sixth century A.D.), 7:134–139
astronomy, **13**:323a
commentaries of, **1**:272a, 276a; **4**:433a(n); **9**:276a(n), 437a, 438b; **10**:112b; **11**: 184a, 245a
critics of, **4**:525a; **12**:441b, 442a
historiography, **6**:410a, 412a; **11**:27b
A History of Physicians, **7**:24b–25a
influence of, **2**:590b; **8**:238a; **14**:582a; **15**:262b
influences on, **1**:137a; **12**:440b
mechanics, **1**:254b, 409b, 410b
philosophy, **11**:161a; **12**:441a
John of Rupescissa, **4**:170a; **5**:448a; **10**:308a; **13**:535a
John of Sacrobosco
see Sacrobosco, Johannes de
John of St. Amand, **10**:535b
John of Salisbury, **2**:19b; **7**:65b, 66a; **13**:339b
John of Saxony (*fl.* first half of fourteenth century), 7:139–141; 7:123a, 124a,b, 130a
commentaries by, **7**:123b; **11**:226a,b; **15**:191a(n)
John Scotus Eriugena
see Eriugena, Johannes Scotus
John of Seville, **4**:607b; **7**:140b, 362b; **11**: 322b

translations by, **1**:35b, 36b, 37a, 62a; **3**: 25b; **4**:542a, 611b; **9**:161b, 162a; **11**:226a,b, 244b, 349b; **13**:538b; **14**: 593a; **15**:175a,b, 178a–182a *passim,* 187a, 189b, 190a(n)
John of Sicily (*fl.* second half of 13th century), 7:141–142; 7:123a, 128a*; **14**:401a
John Simonis of Selandia (*fl.* fifteenth century), 7:142–143
John of Spain
see John of Seville
John of Speyer, **7**:123a
John of Tinemue
De curvis superficiebus Archimenidis, **1**: 227b
John, son of David, **11**:31b
Johne, Albert
Johne's bacillus, **13**:520b
Johns Hopkins Hospital, **6**:77b, 78a; **9**:56a; **13**:300a–b
Johns Hopkins University
in nineteenth century, **3**:447a; **9**:142b–143a; **15**:64b
bacteriology at, **3**:94a
biology at, **3**:389b; **5**:80b; **14**:424a
chemistry at, **11**:370b–371a
endocrinology at, **1**:11b
Hunterian Laboratory, **3**:517b
mathematics at, **13**:218a
Medical School, **9**:57b; **12**:48a; **14**:249a–250a
anatomy at, **9**:56b
embryology at, **9**:57a
nutrition at, **8**:591a
pathology at, **3**:94a
pharmacology at, **1**:10a
physics at, **1**:133a; **11**:578a
psychology at, **6**:52b–53a
School of Hygiene and Public Health, **6**:526a; **14**:250a
Studies from the Biological Laboratory, **9**:143a
Johnson, Douglas Wilson (1878–1944), 7: 143–145; **3**:595b; **4**:562b
Johnson, George, **7**:147a
Johnson, Manuel John (1805–1859), 7:145–146
Johnson, Samuel, **3**:487b; **6**:383a
Johnson, Samuel William, **1**:325a; **3**:256b; **6**:13b; **10**:150b, 242a
How Plants Grow, **13**:341b
Johnson, Thomas (*ca.* 1600–1644), 7:146–148; **5**:362a,b
Johnson, Thomas Hope, **3**:370b
Johnson, Willard Drake (1859–1917), 7: 148–150
Johnson, William (*ca.* 1610–1665), 7:150
Johnson, William Ernest (1858–1931), 7:151
Johnston, James Finlay Weir, **10**:150a
Elements of Agricultural Chemistry and Geology, **12**:233b
Jointing (geol.), **3**:325b, 586b
Joints (anat.)
diseases of, **2**:483a–b; **8**:402b; **11**:602b
loose cartilages in, **3**:487a
mechanics of, **4**:615a; **5**:581a; **8**:202b
synovium
function of, **6**:183b
Joliot, Frédéric (1900–1958), 7:151–157; **3**:329b, 502b; **4**:580a; **7**:56b, 158a; **8**:12a
Joliot, Hélène, **7**:152a
Joliot, Pierre, **7**:152a
Joliot-Curie, Irène (1897–1956), 7:157–159; **3**:329b, 501b, 502a,b, 503a*; **4**:580a; **7**:152a–156a *passim;* **9**:262a; **12**:71b
Joliot-Curie family, **12**:34a
artificial radioactivity, **6**:366b
Jolly, Philipp Johann Gustav von (1809–1884), 7:160; **2**:161a, 430a; **4**:615a; **6**:341a; **11**:7b, 611a
Joly, John (1857–1933), 7:160–161; **4**:130b, 131a; **12**:520a; **13**:619a
Joly, Nicolas, **4**:21b, 210b; **10**:372a
Jolyet, Félix, **9**:506b(n)
Jombert, Charles Antoine, **9**:500b
Jones, Donald Forsha, **4**:271b, 272a
Jones, Ernest, **2**:446a; **5**:173a, 179a, 181a*
Jones, Harold Spencer
see Spencer Jones, Harold
Jones, Harry, **8**:478a
Jones, Harry Clary (1865–1916), 7:161–162; **6**:561a; **10**:577b*; **11**:299b*; **13**:67a*, 581a*; **15**:467a(n)
Jones, Inigo, **3**:209a; **14**:29b
Jones, John Edward
see Lennard-Jones, John Edward
Jones, Owen Thomas, **8**:33a
Jones, Samuel, **7**:22a
Jones, Thomas ap Catesby, **9**:195b
Jones, Thomas Wharton, **3**:424b; **6**:589b; **8**:400b
Jones, William (1675–1749), 7:162–163; **5**: 522a*; **8**:459b; **10**:87b(n)
Jones, Wood, **11**:595b, 596a
Jonquières, Ernest Jean Philippe Fauque de (1820–1901), 7:163–164
Jonston, John (1603–1675), 7:164–165
Jordan, (Claude Thomas) Alexis (1814–1897), 7:165–167
botany, **14**:99b
Jordanon species, **7**:166b; **8**:513a; **14**:99b–100a
Jordan, Camille (1838–1921), 7:167–169; **3**:164a; **4**:5a; **6**:76a*, 309b*, 547a, 548a*; **7**:397b; **8**:323b, 367b, 563b*
Cours d'analyse, **2**:238a; **6**:113b; **8**:110b
group theory, **2**:615b; **4**:83a; **5**:264a; **12**: 511b
Jordan curve, **2**:274b, 513a; **4**:174a, 562a; **7**:72a; **13**:562a, 599b
Jordan-Hölder theorem, **6**:472b; **7**:168a
measure theory, **8**:110b
number theory, **6**:309a; **11**:55a
students of, **3**:456a; **9**:387b
Traité des substitutions, **8**:325b
Jordan, David Starr (1851–1931), 7:169–170; **1**:74b, 405b; **4**:309b; **5**:406b, 545a; **7**:286a
Jordan's law, **7**:169b–170a
Jordan, Denis Oswald, **5**:589b
Jordan, Edwin Oakes (1866–1936), 7:170–171
Jordan, Hubert Fairlee, **6**:118a
Jordan, Louis, **1**:460b
Jordan, Marcus, **10**:239a
Jordan, Pascual, **4**:318a; **5**:118a; **10**:423a; **15**:41b, 42a, 43a
Jordanon species
see Jordan, (Claude Thomas) Alexis
Jordanus de Nemore (*fl. ca.* 1220), 7:171–179; **4**:228a; **8**:100b; **13**:62a(n), 349b; **15**:256a
Arithmetica, **1**:82b, 381b; **3**:25a; **4**:446b
critics of, **9**:487b, 488a
Elementa de ponderibus, **1**:63b; **8**:215b, 216a
Liber de ratione ponderis, **8**:215b, 216a–217a, 218b, 219a, 222a–b
Liber philotegni de triangulis, **5**:360a
De numeris datis, **11**:458a
Opusculum de ponderositate, **13**:260a,b
Planisphere, **3**:364a
statics, **1**:94a, 229a; **2**:193b; **5**:247a
De triangulis, **1**:227b, 381b; **2**:360b; **3**: 25a; **4**:609b
Jordanus of Saxony, **1**:99a; **7**:172a
Jørgensen, Sophus Mads (1837–1914), 7: 179–180; **2**:169b, 170b; **12**:546a
chemical theories, **2**:199b; **14**:267b, 269b, 270a–b
Josephus, Flavius, **1**:148b; **11**:312a; **15**:125a
Jossigny, Paul, **3**:365b
Jost, Adolph
Jost's law, **9**:562a
Jost, Friedrich, **15**:435b
Jöstel, Melchior, **11**:3b; **14**:471a
Jouan, C., **3**:189a
Joubert, Jules François, **9**:155a; **10**:382b, 385b
Jouguet, Émile, **4**:230a, 232a, 233b*
Joule, James Prescott (1818–1889), 7:180–182; **11**:36b; **13**:237b, 614b
biographies of, **11**:393b
on Carnot's function, **3**:306a
correspondence of, **9**:227b
electromagnetism, **9**:566a
gases, **5**:319b; **6**:292b–293a
geophysics, **6**:503a
heat, **3**:83a,b, 304a; **4**:554b; **5**:387b; **6**: 243b, 244a, 432a; **12**:288b; **13**:378b–382a *passim;* **15**:85b, 86a,b
Joule-Kelvin equation, **3**:20a
Joule's law (of electric currents), **1**:555b; **7**:181a–b
Joule's law (of gases), **7**:182a
Joule-Thomson effect, **4**:79b; **6**:94a; **7**: 182b; **8**:365b
and Mayer, **3**:306a; **7**:180b, 182a; **9**:239b
Joule (physical unit), **5**:407b
Jourdanet, Denis, **2**:61b
Journal de chimie physique
editors of, **7**:162a
founding of, **5**:597a
Journal de Chirurgie, **2**:122a
Journal de Genève
founding of, **10**:603a
Journal de géologie, **2**:341b
Journal de l'anatomie et de la physiologie normales et pathologiques de l'homme et des animaux, **11**:492a
Journal de l'École polytechnique, **15**:483a
Journal de Liouville
see *Journal de mathématiques pures et appliquées*
Journal de mathématiques pures et appliquées, **5**:367b, 482a
editors of, **8**:61b
founding of, **8**:382b–383a
Journal de médecine
editors of, **7**:556b
Journal de médecine, chirurgie et pharmacie, **3**:427a
Journal de médecine et de chirurgie de Toulouse
founding of, **11**:234a
Journal de Paris, **3**:7a
Journal de pharmacie et de chimie, **5**:432a
Journal de physiologie expérimentale et de pathologie
founding of, **9**:9b
Journal de physique, **1**:64b; **2**:321b
editors of, **7**:602b–603a
Journal de Trévoux, **3**:114a
Journal de Verdun, **4**:70b
Journal der Pharmacie, **13**:466a
editors of, **15**:172a
Journal des débats, **5**:84b
Journal des sçavans, **3**:283b; **5**:58b, 259a
editors of, **8**:621a; **12**:117b
founding of, **12**:84b–85b
Journal d'histoire naturelle
editors of, **10**:497a*
Journal encyclopédique, **3**:283b
Journal für Chemie, Physik und Mineralogie, **15**:171b

Journal für Chemie und Physik, **4**:395b*; **14**:489a
founding of, **12**:254a
Journal für die Baukunst, **3**:466b; **4**:501b
Journal für die reine und angewandte Mathematik
see *Crelle's Journal*
Journal für makromolekulare Chemie, **13**:3b
Journal für Physik und physikalische Chemie des Auslandes
editors of, **7**:509a
Journal für praktische Chemie, **4**:394b, 395b*
editors of, **7**:451a; **9**:92a*
Journal für technische und ökonomische Chemie, **4**:395b*
Journal helvétique, **15**:53a
Journal littéraire de la Haye
founding of, **5**:509a
Journal of Anatomy, **4**:354a*
editors of, **7**:279a
Journal of Anatomy and Physiology, **3**:463a
founding of, **13**:504a
Journal of Animal Behavior
founding of, **14**:550b
Journal of Applied Psychology
founding of, **6**:53b
Journal of Atmospheric Research
founding of, **1**:196a
Journal of Biological Chemistry, **5**:53b
editors of, **13**:574a
founding of, **1**:10b
Journal of Botany, **2**:476b
founding of, **6**:495a*
Journal of Botany and Kew Gardens Miscellany
founding of, **6**:495a*
Journal of Comparative Neurology, **4**:161a
founding of, **6**:320b
Journal of Economic Entomology, **12**:98b
Journal of Experimental Medicine, **5**:40b
editors of, **14**:250a
founding of, **1**:10b
Journal of Experimental Zoology, **11**:574a
editors of, **10**:104b
founding of, **3**:123a, 390b; **6**:134a
Journal of Genetics
founding of, **1**:505b; **11**:212a
Journal of Geology
editors of, **12**:84a
founding of, **3**:190a
Journal of Geophysical Research, **1**:522a
Journal of Heredity
founding of, **3**:123a
Journal of Infectious Diseases
editors of, **6**:233a; **7**:171a
Journal of Meteorology, **11**:558a
Journal of Morphology
editors of, **8**:589b; **9**:496a
founding of, **14**:314a
Journal of Natural Philosophy, Chemistry and the Arts
see *Nicholson's Journal*
Journal of Neurophysiology
founding of, **5**:208a
Journal of Organic Chemistry
founding of, **7**:323a
Journal of Pharmacology and Experimental Therapeutics
founding of, **1**:10b
Journal of Physical Chemistry
founding of, **1**:431a; **9**:394b
Journal of Physiology, **2**:367b; **9**:143a; **11**:139a
editors of, **2**:367b; **8**:17a–b
founding of, **5**:83a
Journal of Preventive Medicine
editors of, **7**:171a
Journal of Religious Psychology
founding of, **6**:53b
Journal of Sedimentary Petrology
editors of, **13**:519a
Journal of the American Chemical Society, **11**:371a
editors of, **10**:158a
Journal of the American Medical Association, **4**:154b
Journal of the Bombay Natural History Society, **4**:494a
Journal of the British Astronomical Association, **4**:176b*
editors of, **9**:184b
Journal of the Chemical Society, **12**:573a
Journal of the Franklin Institute
editors of, **7**:162a
Journal of the History of Ideas
founding of, **8**:517a
Journal of the Scientific Laboratories of Denison University, **5**:50a
Journals, scientific
see Periodicals, scientific; specific periodicals by title
Joy, Alfred Harrison, **13**:98b
Joyner, Mary Cross, **12**:265b
Juan y Santacilla, Jorge (1713–1773), 7:183–184; **5**:435b; **6**:512b; **13**:530b
Juan Yuan, **3**:266a; **8**:314a; **14**:540a; **15**:733a
Judaism
medieval, **9**:28a–31b; **12**:175a
twentieth-century, **9**:341b
Juday, Chancey (1871–1944), 7:184; **2**:142a–b
Judd, Orange, **1**:325a
Judgment
analytical theory of, **8**:155a
nature of, **2**:545a; **5**:106b
Juel, Sophus Christian (1855–1935), 7:184–185; **10**:544a*; **14**:279b
Juglans regia, **3**:588a
Juhn, Mary, **8**:357a, 359a
Jukes, Joseph Beete (1811–1869), 7:185–186; **3**:297b; **4**:492b*; **12**:263b
Student's Manual of Geology, **5**:335a
Julian the Apostate, **10**:230b; **13**:308a
Julin, Charles, **1**:601a
Julius, Willem Henri (1860–1925), 7:186–187; **9**:415a; **10**:235b, 236a; **11**:150a; **12**:73a
Julius suspension, **7**:186b
Ibn Juljul, Sulaymān ibn Ḥasan (944-*ca.* 994), 7:187–188; **4**:121b
historiography, **15**:230a, 263b
Jullander, Ingvar, **13**:163a
July Revolution of 1830, **3**:132a, 527a; **5**:261a, 367b
Ibn Jumaᶜ, **1**:539a
Juncaceae, **10**:13b
Juncker, Johann (1679–1759), 7:188; **9**:104b; **10**:26a; **13**:543a
Jung, Carl Gustav (1875–1961), 7:189–193; **5**:456b; **10**:312a*, 424a–b
alchemy, **4**:170a
critics of, **7**:190b
and Freud, **5**:178b–179a; **7**:190a–b
Junge, Joachim
Opuscula botanico-physica, **1**:293a
Jungfleisch, Émile Clément, **2**:64a, 65a, 67b, 69b–70a; **11**:426a; **15**:434a
Jungius, Joachim (1587–1657), 7:193–196; **1**:495a; **7**:355a, 375a; **11**:314b, 315b; **12**:79a,b; **15**:81a
Isagoge phytoscopica, **11**:314a
Juno (asteroid), **5**:404b; **6**:94b, 334b; **9**:429b
discovery of, **10**:199a
Juno Georgia (asteroid), **6**:112a
Jupille, Jean-Baptiste, **10**:404b
Jupiter (planet), **2**:146b; **10**:34a
atmosphere, **12**:267a
brightness, **2**:284b
and comets, **3**:283a–b; **10**:42a; **15**:355a
distance of, **3**:409a
flattening of, **1**:55b; **3**:101b; **15**:337b, 338a, 352b, 354a
motion of, **7**:598b
irregularities of, **2**:43a, 328b; **6**:515b; **8**:144b*; **15**:289b, 329a–331b
observation of, **3**:357b; **6**:105b; **7**:376a; **8**:465a; **13**:115a
oppositions of, **8**:114b*; **15**:356a
orbit, **5**:224a; **6**:399b; **7**:563a; **8**:277a, 279a
eccentricity, **7**:139b; **11**:525b
inequalities in, **7**:562a
perturbations of, **6**:103b; **9**:174b
photography of, **3**:366a; **12**:36b, 186b, 454a; **14**:521b
rotation
period of, **1**:597b; **3**:101b; **10**:180b, 585a; **12**:455a
satellites of, **1**:429a; **2**:511b*; **3**:23a; **4**:128a; **5**:24a; **6**:94b, 124b, 126b, 335a; **7**:18a, 299a–b, 543a; **10**:19b; **11**:472a; **14**:452a
capture hypothesis, **3**:18b
and cosmology, **5**:241b
discovery of, **1**:463b, 466a; **3**:457a; **5**:241a, 287b; **10**:107a–b, 489a, 526b
distances of, **15**:337b
dynamics of, **12**:448b–449a
eclipses of, **2**:388b; **3**:198a; **6**:263b; **9**:215a; **11**:525b–526a, 532b; **12**:95b; **13**:113b; **15**:337b
geographic use of, **4**:22b, 23a; **5**:242b; **6**:520b; **7**:374a; **9**:163a; **10**:490a; **11**:554a
irregularities in movements of, **3**:103b; **7**:562b–563a; **9**:90a
light of, **1**:401a; **6**:334b; **9**:267a; **12**:267a
mass of, **15**:337b, 338a, 354a
motions of, **1**:311a, 400b–401a; **2**:309b, 310b; **5**:515b; **6**:361b; **9**:90b; **14**:489b; **15**:325a–327a, 336b–338b, 354a–b
naming of, **9**:247a
orbits of, **3**:472a; **10**:64b, 69a, 77a–b, 159a; **11**:552b; **14**:178a–179a
periods of revolution, **10**:176b, 489a, 602a
tables of, **1**:400b–401a; **2**:307b; **3**:101b, 103a; **4**:14b; **9**:77b, 90b, 247a,b; **11**:114b; **12**:95b; **14**:178b–179a; **15**:325a,b, 326b, 327a, 337b
visibility of, **5**:88a
and Saturn, **2**:401b; **4**:481b; **7**:290b, 297b, 374a; **8**:179a; **10**:64b, 76b, 77b; **14**:389b; **15**:250a, 323a–325a, 329a–331b, 351a, 353b
sixty-year cycle, **3**:584b
size
diameter, **2**:388b
mass, **10**:34b–35a, 111a
spectrum, **4**:180a; **9**:392a; **12**:37a, 267a; **14**:55a
surface features, **6**:523a; **8**:522a
belts, **1**:464b; **3**:101b; **11**:411b; **14**:485a
changes in, **8**:50a; **10**:585a
equatorial white spot, **1**:464b
great red spot, **1**:464b; **2**:433a; **6**:522b; **10**:119b, 585a; **13**:45a; **14**:485a
south tropical disturbance, **10**:585a
spots, **3**:101b; **4**:128a; **9**:115b
synodic period, **4**:467a; **15**:219a
tables of, **2**:359b; **4**:14b; **7**:86b, 123a, 124a; **15**:331b, 356a
zodiacal period, **5**:345b
Jura, George, **6**:118b

Jura Mountains, **4**:27b
geology, **2**:496b; **3**:210b–211a, 325a; **4**: 73b; **5**:533a–b; **6**:168b, 229a; **7**:43b; **9**:104a; **12**:139b
paleontology, **7**:43b; **9**:99b
Jurassic period, **4**:9b; **9**:583a; **12**:489b
deformation, **7**:251b
orogeny, **9**:87a; **13**:64a
paleobotany, **3**:187b; **12**:104b, 260a; **13**: 345b
paleogeography, **2**:554a; **7**:250b; **12**:488b
paleontology, **1**:285a–286a, 393a; **2**:109b, 491b–492a; **3**:395b, 526b, 530a, 572b; **5**:500b; **6**:221a; **7**:43b, 320a, 370b; **8**:594a; **9**:87b, 99b, 100b–101a, 134b, 346a; **10**:29a,b, 40b, 128a, 211b–212a, 221b, 222a, 262a, 265b; **11**: 336b, 386a, 501b; **14**:626a; **15**:527a
stratigraphy, **1**:285a–286a; **2**:108b, 341b, 495a; **3**:297b, 530a; **4**:402b, 540b; **5**:14a–b, 335a, 519a, 533b; **6**:168b; **9**:240b, 579a; **10**:211b–212a, 428a,b, 584a; **11**:235b, 567a; **12**:140b, 276b, 487a; **13**:87a; **14**:210b–211a
Juries, **15**:31b
Jurin, James, **3**:216a; **7**:275a; **11**:494a; **12**: 446a
Jurine, Louis, **12**:562b
Jurisprudence
mathematical aspects of, **2**:618a
medical, **4**:252a
see also Medicine, forensic
al-Jurjānī, Seyyid al-Sharīf, **11**:227a; **15**:629a
Jussieu, Adrien Henri Laurent de (1797–1853), 7:196–197; **5**:20b; **7**:199a*; **10**:511b; **12**:72a*, 215a; **13**:489a; **15**:268a*
Jussieu, Antoine de (1686–1758), 7:197–198; **1**:58b; **4**:216b; **7**:58b, 199a; **8**:177b; **9**:17b; **10**:569a; **15**:58a*
Jussieu, Antoine-Laurent de (1748–1836), 7:198–199; **7**:196b, 199b, 200b
acquaintances of, **5**:218a; **10**:518a; **12**: 471b; **13**:495b
botany, **13**:444a*
on contemporary scientists, **13**:554a
correspondence of, **12**:536b
Genera plantarum, **1**:59a; **5**:217a; **7**:199b
influence of, **11**:234b; **12**:535b
students of, **2**:321b; **3**:154b; **4**:156b; **5**: 355b
taxonomic system, **1**:58b; **2**:519a; **5**:217a, 512b; **7**:585a, 586a; **8**:372a, 377a; **10**:163b; **13**:432b; **15**:267a
Traitement contre la ténia ou ver solitaire . . . , **8**:623a*
Jussieu, Bernard de (1699–1777), 7:199–200; **5**:354a*; **7**:198a, 200b
acquaintances of, **3**:365a; **4**:216b, 223b, 295a; **5**:577b; **7**:58b; **8**:177b; **9**:17b, 254a
botany, **1**:59a
correspondence of, **10**:569a
influences on, **13**:443a
marine biology, **10**:569a
students of, **1**:58b; **3**:172b; **5**:355a; **8**:67b; **9**:53b, 366a; **10**:325b; **13**:390a
taxonomy, **1**:143a; **7**:412b
Jussieu, Christophle, **7**:198a
Jussieu, Joseph de (1704–1779), 7:200–201; **2**:343a; **7**:183a; **15**:270b
Jussieu family
acquaintances of, **8**:375b
genealogy, **7**:196a
Justi, Johann Heinrich Gottlob von (1720–1771), 7:201–203; **3**:473b; **4**:471b
Justice, **6**:447a
natural, **1**:119a
Justinian I, Byzantine emperor, **1**:279b; **7**: 28b, 29a; **9**:532b; **13**:340a
Justus Liebig's Annalen der Chemie
editors of, **13**:337b; **14**:446a
Ibn Juzayy, **1**:516a
Jyeṣṭhadeva
Yuktibhāṣā, **1**:48b

K

Kabat, Elvin Abraham, **5**:6b
Kablukov, Ivan Alexsevich (1857–1942), 7: 203–204; **9**:130b
Kac, Mark, **9**:219a
Kaempfer, Engelbert (1651–1716), 7:204–206; **13**:392a
translations of works, **12**:407a
Kaempferia, **13**:354b
Kaestner, Abraham Gotthelf (1719–1800), 7:206–207; **1**:615b; **3**:62a*, 345a*; **5**:299a; **6**:403a; **7**:419b(n), 596b; **11**:305a*
Anfangsgründe der Mathematik, **2**:273b
and Gauss, **5**:308a; **7**:206b
geometry, **2**:268b
students of, **2**:268a, 420b; **6**:112a; **7**:404b; **8**:320b; **10**:197b, 572b; **12**:226a
Kāfūr, **4**:524b
Kagan, Benjamin Fedorovich (1869–1953), 7:207–208
Kahlenberg, Herman, **7**:208b
Kahlenberg, Louis Albrecht (1870–1941), 7: 208–209; **15**:467a(n)
Kähler, Erich Ernst, **12**:214a
Kahn, R. H., **1**:450a
Kaiser, Frederik (1808–1872), 7:209–210; **6**:453a; **8**:488a; **11**:163a
Kaiser Wilhelm Gesellschaft zur Förderung der Wissenschaften, **6**:17a; **9**:262a; **11**:347b; **15**:449b
biology institutes, **3**:421b; **7**:516b
chemistry institutes, **1**:553a; **5**:117b, 622a,b, 623a; **6**:16a; **7**:517a; **9**:260b; **15**:159b
founding of, **8**:181b
medical institutes, **7**:517b; **11**:585b; **15**: 521b
physics institutes, **3**:618a; **8**:52b
see also Max Planck Gesellschaft zur Förderung der Wissenschaften
Kaiserliches Gesundheitsamt (Berlin), **8**: 448a
Kala-azar, **15**:454b, 480a
Kalāṃśa (time degree), **15**:563a,b, 575b, 586a, 589b
Kalbe, Ulrich Rülein von
see Rülein, Ulrich
Kaleidophone, **14**:289b
Kaleidoscope
invention of, **2**:390a, 452b
Kalm, Pehr (1716–1779), 7:210–211; **4**:492b; **5**:76b; **8**:379b
and Bartram, **1**:487a,b
Travels in North America
translations of, **9**:54b
Kalmia, **8**:178a
Kalonymus ben Kalonymus, **10**:111b
Kalpa (period), **4**:556a; **15**:555a, 556a, 557a, 565a, 581a, 590a, 591a, 597a, 609b, 618a, 619b, 620a
Kaluza, Theodor Franz Eduard (1885–1954), 7:211–212; **4**:330b, 331a
Kalyāṇa
Khecaradīpikā, **15**:589a
Tithikalpadruma, **15**:589a
Kalyāṇavarman, **7**:223a
Kamāl al-Dīn Abu'l-Ḥasan Muḥammad ibn al-Ḥasan al-Fārisī (d.1320), 7:212–219; **6**:204a(n); **11**:251b
optics, **6**:195a,b, 200a
Tanqīḥ al-Manāẓir, **6**:196b; **11**:250b
Kamāl al-Dīn ibn Yūnus, **13**:509a, 514b
Kamalākara (b. ca. 1610), 7:219; **9**:580a; **11**:290a; **15**:555b
commentaries of, **5**:274b
Siddhāntatattvaviveka, **15**:613b, 615b–617b
sine tables, **15**:616b, 626a
Kamalākara Bhaṭṭa
Nirṇayasindhu, **9**:607a
Kamalanayana, **12**:115b
Kamchadals, **7**:495b
Kamchatka Peninsula
geography, **2**:14b; **13**:28b
natural history, **7**:495a–b; **13**:28b
Kamel, Georg Josef, **8**:395a
Kamerlingh Onnes, Heike (1853–1926), 7: 220–222
acquaintances of, **3**:312b; **5**:610a; **6**:163b; **7**:272a, 515b; **13**:243b; **14**:597b
on contemporary scientists, **7**:516a*; **10**:207a*; **14**:599a*
enthalpy, **9**:462a
low temperature studies, **5**:610b; **8**:475b; **14**:246b*
helium, **4**:80a; **8**:478a; **10**:206b
translations of works, **13**:12a
Kamiya Teirei, **1**:83a–b
Kammerer, Paul, **6**:339a
Kanai, Kiyoshi, **12**:341b
Kanaka (fl. ca. 775–820), 7:222–224; **1**:33a; **4**:555b; **6**:381b
Kanakācārya, **7**:223a
Kandel, David, **2**:219a
Kane, Robert John (1809–1890), 7:224
Kaneite, **7**:224a
K'ang Hsi, Emperor, **11**:403a; **14**:162a
Kankah al-Hindī
see Kanaka
Kansas
geological survey, **14**:410a
geology, **13**:518b
paleontology, **14**:409b
Kansas, University of
entomology at, **7**:285b–286a
medical school, **14**:409b
Kansas State Agricultural College, **14**:409a–b
Kant, Immanuel (1724–1804), 7:224–235; **2**:18a; **6**:82b; **7**:505b; **8**:159a
acquaintances of, **8**:321a
Allgemeine Naturgeschichte und Theorie des Himmels, **7**:598b
axiomatization, **6**:249a; **12**:11a–b
biology, **5**:75b; **7**:232b–234a
causation, **4**:101a; **7**:232b–234a; **8**:159b
cosmogony, **7**:85b, 231b–232b; **9**:552b; **10**:42a; **15**:344b
cosmology, **6**:331b, 332b; **14**:519a
critics of, **9**:2b; **10**:247a; **12**:155b, 178a; **15**:79b
dynamics, **6**:395a–b; **7**:227b–228b; **8**:159b
critics of, **9**:578b
epistemology, **4**:100b; **6**:26a,b; **7**:226a–227a, 233b–235a; **12**:156a
idealism, **2**:545a; **3**:110b, 111b; **7**:227a, 229b, 230b, 235a; **12**:154a, 155a
influence of, **4**:188a, 616b; **6**:342b, 564b; **7**:189b; **10**:482b; **11**:389a; **13**:466a; **14**:67b; **15**:86b
influences on, **11**:303a; **14**:483b, 484a
Kritik der reinen Vernunft, **6**:87a, 90b–91a; **10**:183b; **12**:155a
geometrical influence on, **4**:425a
and Lambert, **7**:597b

Kant, Immanuel (cont.)
matter, theory of, 7:227a–228b, 229b–230a; **8**:373b
Metaphysische Anfangsgründe der Naturwissenschaft, **4**:530b; **10**:183b–184a
Monadologica physica, **8**:160a
on motion, 7:230a–231a
ontology, **1**:140b–141a; **5**:286a
space, theory of, **2**:270b; 7:225b–226b, 229b–231a; **11**:455a; **13**:217b
Der Streit der Fakultät, **15**:754b
students of, **11**:434b
translations of works, **14**:480a
Kantianism
eighteenth-century, 7:367a, 368a; **11**:363a; **12**:153b
early nineteenth-century, **2**:221a; **5**:193a–194b, 532b; **6**:294b, 395a–b; **10**:182b–185b; **12**:156a, 592a; **14**:241b; **15**:225b
mid nineteenth-century, **6**:86b–87a, 90b, 241b, 243b, 247b; **9**:206a; **12**:175a, 607a,b; **14**:220b, 294a; **15**:167a
late nineteenth-century, **2**:31b; **9**:143b
twentieth-century, **9**:359a; **11**:14b
opposition to, **11**:356a–b
see also Neo-Kantianism
Kaolin, **5**:578b; **11**:330a–b; **13**:246b
analysis of, **2**:72a
Kapitsa, Petr Leonidovich, **3**:329a; **8**:12a, 478a; **12**:33b; **15**:251b
Kappler, Eugen, **12**:497a
Kapteyn, Jacobus Cornelius (1851–1922), 7:235–240; **4**:272b, 273a; **5**:406a*; **10**:290a; **11**:156a; **12**:346b
acquaintances of, **5**:405b; **8**:582a; **12**:448a–b; **14**:616b
astronomy, **4**:269b, 278b; 7:481b; **12**:250b, 265a, 347a; **13**:475a
Kara Sea, **14**:635a
al-Karābīsī, **4**:438a, 441b
al-Karajī (or al-Karkhī), Abū Bakr ibn Muḥammad ibn al-Ḥusayn (or al-Ḥasan) (*fl. ca.* 1000), 7:240–246; **1**:22b, 31b; **12**:94a; **14**:388a
algebra, **4**:609a, 610a, 611b; 7:362b; **12**:92b, 93a,b
Algebra
editions of, **14**:472a
arithmetic, **1**:41b; **15**:9b
al-Badīʿ, **12**:91b
critics of, **12**:92a
and Diophantus, **4**:117b
al-Fakhrī, **13**:540a; **15**:119a,b
De mensuratione terrarum
translations of, **15**:177b
Karakoram Range, **3**:533a
Karakum Desert, **4**:598b
Karam, Ahmed, **13**:18a
Karaṇa (time unit), **10**:426a
definition of, **15**:546a
Kardaga (arc), **14**:19a, 593a; **15**:478b
Karelian (stratig. stage), **4**:409b
Kari-Niazov, T. N., **13**:537a
al-Karkhī
see al-Karajī
Karlsruhe
gardens, 7:440b
Physical Institute, **6**:342b
Polytechnische Schule, **11**:344a–b
Karlsruhe Congress, 1860, **1**:347a; **2**:68a, 316a, 621b; **3**:46b, 48b–49b; **4**:394b; 7:280b–281a; **9**:110a, 287a, 288b, 347b, 348b; **10**:178a; **12**:163b
Kármán, Theodore von (1881–1963), 7:246–248; **5**:64b, 608b; **11**:124a, 478a; **15**:43a
aerodynamics, **11**:124b
Kármán vortices, 7:246b–247a
specific heat, **15**:40b, 442b
al-Karmānī, **9**:39b
Karolinska Institutet, **2**:91a; **11**:379b
Karpinski, Louis Charles (1878–1956), 15:255–257; **14**:119a*; **15**:177b
Karpinsky, Alexandr Petrovich (1847–1936), 7:249–254; **4**:598b, 599a; **9**:594b*
Karpov, Lev Yakovlevich, **1**:361b
Karrer, Paul (1889–1971), 15:257–258; **4**:301a, 486a; **6**:185a; **13**:487b; **14**:272b*
Karroo formation, **2**:489b, 504b, 505a; **4**:262a; **10**:40b
Karst topography, 7:512b; **12**:194b
water circulation in, 7:513a; **12**:140b
Karsten, Dietrich Ludwig Gustav, **2**:552b; **3**:623b; **11**:465a; **14**:239b, 240a
Karsten, Karl Johann Bernhard (1782–1853), 7:254–255; **14**:242a*
Karyogamy, **2**:626a
Karyokinesis, **3**:390b*
Karyology, **10**:1b
Karyotype, **10**:276b–277a
Kasha, Michael, **8**:293b
al-Kāshī (or al-Kāshānī), Ghiyāth al-Dīn Jamshīd Masʿūd (*d.* 1429), 7:255–262; 7:62a, 533a; **11**:227a; **13**:514b, 536a
astronomical tables, 7:377a; **12**:362a
Miftaḥ-al-ḥisab, **12**:93b
trigonometry, **9**:555b; **11**:227b, 228a,b; **13**:536b
Zīj-i Khāqānī, 7:81a
Kassel
observatory, **14**:359a
Kassel, University of
in seventeenth century, **6**:145b
Kasso, Lev Aristidovich, **1**:284a; **8**:102a, 107b; **13**:598a
Kassowitz, Max
Children's Clinic, **5**:172b, 173b
Kästner, Abraham Gotthelf
see Kaestner
Kastner, Wilhelm Gottlob, **8**:330a
Kaṭapayādi method (math.), **6**:115b
Katayama, Masao, **8**:503a
Kater, Henry (1777–1835), 7:262–263; **5**:79a; **12**:523b
Kater's pendulum, 7:262b–263a; **14**:571b*
A Treatise on Mechanics, **3**:450b
Kathlamet Indians, **2**:212a
Ibn Kaṭīr al-Farghānī
see al-Farghānī
Katmai, Mount, **4**:563b–564a
Kato, Genichi, **5**:65b
Katz, David, **9**:562b
Katzenbuckel (mt.), **11**:547a
Kauffman, Niklaus
see Mercator, Nicolaus
Kaufmann, Bruria, **4**:331a
Kaufmann, Herman Samuel, **4**:522b
Kaufmann, P. (physiol.), **15**:469b
Kaufmann, Walter (or Walther) (1871–1947), 7:263–265; **11**:475b
cathode rays, **13**:367a
electrodynamic theory, **1**:24a, 24b–25a; **2**:559b; **5**:597b; **8**:496b; **11**:13b; **14**:218a
Kausche, Gustav Adolf, **13**:3b
Kauṭilya
Arthaśāstra, **15**:538a, 541a
Kavraysky, Vladimir Vladimirovich (1884–1954), 7:265–266
Kavraysky grid, 7:265a
Kawamoto Komin
Kikai kanran kogi, **15**:742a
Kawamura, Rinya, **13**:263a
al-Kawm al-Rīshī, **12**:362b
Kay, George Frederick (1873–1943), 7:266–267
Kay, John
Edinburgh Portraits, **1**:452a; **2**:174b
Kayak Island, **2**:14b
Kāyasthas, 7:315b
Kayser, Heinrich Johannes Gustav (1853–1940), 7:267–268; 7:528a; **10**:345b, 347a; **11**:150b*, 476b–477a
Kayser formula, **12**:43a
and Runge, **11**:611b–612a
spectral series, **1**:425b; **11**:279b
Ibn al-Kayyāl, **12**:362a
Kazan University, **2**:623b; **4**:308a; **11**:610a
chemistry at, **3**:301a
neurology at, **1**:580a
Keckermann, Bartholomew (*ca.* 1571–1609), 7:268–270
Keeler, James Edward (1857–1900), 7:270–271; **3**:35b, 36a, 366b, 509a; **6**:28a,b; **9**:203b; **10**:526b; **11**:484b
Keen, William, Jr., **9**:422a
Keesom, A. P., **8**:478a
Keesom, Willem Hendrik (1876–1956), 7:271–272; **5**:610a; **8**:475a, 478a
Keewatin period, **8**:99a
Kehrer, Ferdinand, **15**:105b
Keibel, Franz, **9**:57a; **13**:97a
Keilin, David (1887–1963), 7:272–274
Keill, James (1673–1719), 7:274–275; **8**:416b; **10**:84a
Medicina statica britannica, **6**:37b, 39b
physiology, **6**:37a; **11**:2b; **13**:231b
Tentamina medico-physica, **6**:37b
Keill, John (1671–1721), 7:275–277; **4**:43b, 45a, 450b, 452a; **5**:156b, 510a, 522a*; **13**:265a
acquaintances of, **5**:510b; **8**:165a; **9**:400a; **10**:500b; **13**:67b
chemistry, **8**:621a; **12**:408a
correspondence of, **5**:509b; **6**:174a(n), b(n); **13**:265b
geodesy, **4**:44a
Introductio ad veram astronomiam, **1**:314a; **8**:179a; **12**:407a–b
natural philosophy, **2**:613b; **12**:407b
physiology, **13**:231b
Keir, James (1735–1820), 7:277–278; **11**:141a
Keith, Arthur (1866–1955), 7:278–279; **2**:171b, 375b; **3**:324b*, 606b; **5**:281b; **8**:294b; **11**:594a; **13**:504a
Keith, Thomas, **8**:406b
Kekule von Stradonitz (Kekulé), (Friedrich) August (1829–1896), 7:279–283; **3**:476a, 511a; **4**:395a; **8**:55b; **9**:110b*
acquaintances of, **2**:621a–623b *passim;* **3**:449b; **9**:347b; **10**:177a; **11**:92b; **14**:141b, 394b
benzene, **1**:390a; **3**:300a–b; **5**:488b; **8**:61a, 508a; **9**:352a
carbon, **1**:390b; **10**:359a; **14**:531a
and Couper, **3**:448a; **5**:374a
critics of, 7:452b, 552a; **14**:265b, 531b
and Dewar, **4**:78b
influence of, **14**:395b
Karlsruhe Congress, **3**:48b–49a; **9**:287a
Kekulé's rule, **11**:92b
Lehrbuch der organischen Chemie, **8**:508a
salicylic acid, **3**:448b
structural formulas, **2**:68a, 515a; **5**:1a–b, 371a; 7:451b, 452b, 453a; **8**:109b; **10**:178a; **12**:593a

students of, **1**:168a,b, 389b, 528b, 578a; **2**:59a; **3**:286a; **4**:399b, 400a; **7**:551b; **10**:578a; **12**:593a; **13**:389a, 575b
on valence, **2**:199b; **3**:448b, 449a; **5**:126a; **14**:266b–267a
Keller, Allen Dudley, **1**:450a
Kelley, Edward, **4**:6a; **8**:311a; **12**:307a
Kellner, David (*b.* mid-seventeenth century), 7:283–284
Kellogg, Albert (1813–1887), 7:285; 7:102a
Kellogg, Oliver Dimon, **2**:145b; **6**:472a
Kellogg, Vernon Lyman (1867–1937), 7:285–286; 8:356b; **14**:239a*
Kelp
metabolism, **6**:443a
Kelser, Raymond Alexander (1892–1952), 7:286–287
Kelvin, Lord
see Thomson, Sir William (Baron Kelvin of Largs)
Kemp, John, **6**:26a
Kempe, Alfred Bray, **10**:483b
Kempe, Harry Robert, **6**:211b
Kempf, Paul, **9**:563b, 564a; **14**:55b
Kendall, Edward Calvin (1886–1972), 15: 258–259; 3:353a
Kendall, Larcum, **6**:130b
Kendall, Maurice George, **14**:574a
Kendrew, John Cowdery, **7**:274a
Keng Shou-ch'ang, **8**:419a
Kennedy, Alexander Blackie William (1847–1928), 7:287–288; 10:448a
Kennedy's law of three centers, **7**:287b
Kennedy, Joseph William, **4**:581b; **8**:47b
Kennedy, William Quarrier, **1**:154a, 394b
Kennelly, Arthur Edwin (1861–1939), 7: 288–289; 1:195b; **10**:604b
Kennelly-Heaviside layer, **1**:195b–196a; **6**:212a; **7**:288b
Kennicott, Robert, **3**:535b
Kenny, Méave, **4**:154b
Kent, Albert Frank Stanley, **5**:281b
Kent (English county)
botany, **7**:147a
geology, **5**:437a
Kentish, Edward, **6**:575b
Kentucky
botany, **11**:263b
Geological Survey, **10**:273a; **12**:343b; **13**:531b
geology, **5**:49b; **10**:258b
paleontology, **1**:487b; **14**:456b
Kenya
prehistory, **8**:104b–105b
Kepler, Johannes (1571–1630), 7:289–312; 3:111a
acquaintances of, **1**:160a; **2**:602a,b; **3**:59b; **9**:247a; **12**:163a, 499b; **14**:359a, 636b
astrology, **3**:60a; **7**:289a, 289b–290a, 297b, 299b, 301a–b, 302b, 306b
biographies of, **2**:453a
and Brahe, **2**:413b, 414a; **7**:293a, 293b–295a, 298a, 302b
celestial mechanics, **4**:166a; **6**:514b–515a; **7**:291b–292b, 295a–b, 296b–297a, 303a–304a, 307b; **8**:152a,b; **9**:310b; **11**:28b, 288b
lunar theory, **5**:23b; **6**:516a; **7**:303b–304a
on comets, **2**:433b; **3**:528b; **7**:302a–b; **10**:64b
on contemporary science, **1**:607b; **2**:542b; **11**:366a
correspondence of, **1**:447b; **3**:262a, 528b; **4**:551b; **6**:124b, 125b, 127b; **8**:281b; **9**:611a, 613a; **12**:151b
cosmology
see Cosmology, Keplerian
editions of works, **4**:269a
ephemerides of, **4**:551b; **7**:306b
and Fludd, **5**:48a; **7**:302a
and Galileo, **5**:237b, 239b, 241b, 243b, 247b; **7**:293a, 299a, 302b, 307b–308a
and Hortensius, **6**:520b
influence of, **2**:306a; **3**:101a, 103a; **10**: 486b; **11**:489a; **12**:103b, 342b, 431a; **13**:30b, 96a–b, 433b
influences on, **1**:229b, 513a; **10**:153b; **11**:162a; **12**:136a; **13**:401a; **14**:461a
Kepler's equation, **3**:146b; **5**:521a, 528b, 615a; **7**:304a, 305a
Kepler's problem, **5**:528b; **7**:567a; **10**:68b, 72b; **13**:266a
kinematics, **5**:288a
magnetism, **5**:401a–b, 575b; **7**:295a–b, 296b–297a, 303b
manuscripts of, **6**:361a
and Mästlin, **7**:289b, 290a–b, 291b, 294b, 299b, 303a, 303b–304a; **9**:168b
mathematics
geometry, **7**:301a; **9**:303b; **11**:609a; **12**:75b; **13**:434b; **14**:471a
infinitesimals, **5**:588b–589a; **10**:335b–336a
logarithms, **7**:304b, 305a; **9**:159a
natural philosophy, **1**:178b; **2**:431a
optics, **1**:81a, 475a; **4**:63a; **6**:194b, 197a; **7**:298a–299a; **9**:97b; **10**:475b; **11**:526a
photometric law, **3**:116a
telescope, **5**:278b
parallax, **6**:515b
planetary laws, **1**:311a–b; **2**:413b; **5**:23a; **6**:514b; **7**:308a; **10**:60a–64b *passim,* 69b, 71a–78b *passim;* **12**:163a; **13**: 96a; **15**:337b
corroboration of, **10**:50a
distance law, **7**:295b, 303b
first law, **7**:297a; **14**:594b
second law, **5**:25b; **7**:295b; **12**:595a; **14**:177b–178a, 446b; **15**:350b
third law, **1**:314a–b; **3**:431b; **6**:69a, 486a; **7**:302a, 303b; **10**:73a, 76a–b; **15**:323b–324a
theology, **5**:588a; **7**:289b, 291b, 300a, 303a, 307b
Astronomia nova, **2**:348b, 465b; **5**:525b; **6**:126a; **7**:296a–297a, 301b, 302b; **10**:245b; **11**:202b
commentaries on, **4**:17a
Dioptrice, **4**:58b; **10**:43a
Dissertatio cum Nuncio Sidereo, **5**:241b
Epitome astronomiae Copernicanae, **7**: 295b, 300b, 301b, 302b–304a
Harmonice mundi, **7**:300b–302a; **11**:203a
Narratio de observatis a se quatuor Jovis satellitibus, **5**:241b
Prodromus dissertationum cosmographicarum . . . , **5**:239b; **7**:290a–293b
Stereometria doliorum vinariorum, **3**: 150a–b; **5**:588b
Tabulae Rudolphinae, **2**:411a, 413b; **3**: 457b; **7**:295a, 304a–306a, 373b; **9**: 527b; **12**:163a, 332b
corrections to, **3**:457b; **6**:514b, 515b
Ad Vitellionem paralipomena, **5**:525a; **12**:501a
Kerasin, **8**:276a
Keratin
structure of, **1**:319a–b
Keratosa, **6**:9a
Kerch Peninsula
geology, **1**:161b–163b
Kerékjártó, Béla (1898–1946), 7:312–313
Kerényi, Carl, **7**:192a
Keresan Indians, **2**:212a
Kermes mineral, **11**:48b, 494b
Kernel (math.), **12**:188a–189a, 214b; **14**: 591b–592a
Kerosene
isolation of, **12**:435b
Kerr, John (1824–1907), 7:313–314; 4: 191a
birefringence, **2**:203a
Kerr effect, **2**:203a; **7**:221a, 313b; **8**:510b; **11**:460b, 530a; **14**:597b
Kerr, John Graham (1869–1957), 7:314; 9: 590b*
Keśava (*fl.* 1496), 7:314–316; 5:274a
commentaries of, **2**:116a; **12**:116a
Grahakautuka, **15**:624a
Jātakapaddhati
commentaries on, **9**:579b
Muhûrtatattva, **5**:275b
Keśavārka
commentaries of, **2**:119a
Vivāhavṛndāvana, **5**:275b
Kessler, Friedrich, **6**:110a
Ketene group, **3**:347b
discovery of, **13**:1a
Ketones, **3**:193a–b
and aldehydes, **5**:1b; **7**:452b
aliphatic-aromatic, **3**:300a
hydrolysis of, **3**:279b
isomers of, **11**:92b
naming of, **5**:431b
nature of, **3**:193b
oxidation of, **11**:92b
photochemistry of, **14**:247b
synthesis of, **4**:553b; **12**:47a
Ketteler, Eduard, **11**:582a
Kettering, Charles Franklin (1876–1958), 7: 316; 9:375b, 376a,a*
Kettle holes
origin of, **6**:238b
Kettlewell, Henry Bernard Davis, **3**:572b
Keulen, Ludolph van
see Ceulen, Ludolf van
Keuper sandstone, **1**:95b
Kevalarāma Pañcānana
Gaṇitarāja, **15**:618a
Kew Bulletin
founding of, **13**:342b
Kew Gardens
see Royal Botanic Gardens (Kew)
Kew observatory, **5**:292a; **12**:51a; **13**:51b, 52a
Keweenawan (stratig. unit), **14**:440b
Key, Axel, **11**:382a,b
Keynes, John Maynard (1883–1946), 7:316–319; 11:285b–286a; **12**:10a
historiography, **7**:104b; **10**:82a
probability, **3**:453b(n); **7**:151a–b
Treatise on Probability, **9**:419b
Keynes, John Nevile, **7**:316b–317a
Keys, John
see Caius, John
Keyser, Cassius Jackson, **11**:107a
Keyserling, Alexandr Andreevich (1815–1891), 7:319–320; 13:620b
geology, **9**:585b*; **13**:621a
Keyzer, Pieter Dirckszoon, **1**:530b
Khaikin, Semen Emmanuilovich, **15**:7a
Khair al-Din Barbarossa, **10**:616b
Ibn Khalaf, ᶜAlī, **14**:400a, 594a
Ibn Khaldūn (ᶜAbd-al-Raḥmān ibn Muḥammad) (1332–1406), 7:320–323; 1:30b, 437b; **4**:439b; **7**:334b, 338b
on al-Battānī, **1**:512b
Muqaddima, **1**:350b; **15**:263a
Ibn Khaldūn, Abū Muslim, **9**:39b
al-Khalīl, ibn Aḥmad ibn ᶜOmar al-Farāhīdī, **1**:5b; **15**:230a

al-Khalīlī, Shams al-Dīn Abū ʿAbdallāh Muḥammad ibn Muḥammad (*fl. ca.* 1365), 15:259–261; 12:357b, 361a; 14:579a
Ibn Khallikān, 4:541b
Khaṇḍakhādyaka, 15:580b
Khanka, Lake, 10:167b
al-Kharaqī, 7:334b
Kharasch, Morris Selig (1895–1957), 7:323
al-Kharīzī, Jehuda ben Salomo, 15:233a
Kharkov University
in nineteenth century, 10:247a–b
Ibn al-Khaṭīb, 1:437b
al-Khaṭṭābī, 4:524a
Ibn Khayr, 1:414a
al-Khayyāmī (or Khayyām), Ghiyāth al-Dīn Abu'l-Fatḥ ʿUmar ibn Ibrāhīm al-Nīsābūrī (or al-Naysābūrī), also known as Omar Khayyam (*ca.* 1048–*ca.* 1131), 7:323–334; 6:471b
acquaintances of, 12:91b
algebra, 4:610a, 611b; 7:242a, 257a,b, 324a–b, 327b–329a, 362b; 9:21b; 13:291a, 515a
Algebra
editions of, 14:472a
arithmetic, 7:325b–327b
astronomy, 7:324b–325a, 336b, 337a
balance, 7:325b, 340a, 341b, 342b, 343a
calendar, 7:259b, 324b, 325a
on contemporary scientists, 11:240a,b
geometry, 4:439b, 441a,b; 6:202a; 7:329a–330a, 335a; 8:429b; 12:56b; 13:289b, 510b
influence of, 9:26a; 12:91a; 13:510a
philosophy, 7:330a–b
poetry, 7:330a–331a
ratios, 7:83a, 326a–327b
al-Khayyāṭ, (Abū ʿAlī) Yaḥyā ibn Ghālib, 1:23a; 9:161b; 11:32a
Ibn al-Khayyāṭ, 9:39b
al-Khāzimī, Abu'l-Fatḥ, 7:336a,b
al-Khāzin, Abū Jaʿfar Muḥammad ibn al-Ḥasan al-Khurāsānī (*d.* 961/971), 7: 334–335; 4:440b; 7:336a, 338b; 9:21b
astronomy, 15:35b
commentaries on *Almagest,* 1:510b; 4: 438a
al-Khāzinī, Abu'l-Fatḥ ʿAbd al-Raḥmān (sometimes Abū Manṣūr ʿAbd al-Raḥmān *or* ʿAbd al-Raḥmān Manṣūr) (*fl. ca.* 1115–1130), 7:335–351; 7: 334a
historiography, 15:420b
Mīzān al-ḥikma, 7:325b; 9:300b; 13: 293b*
Khibiny Mountains, 4:598b
Khinchin, Aleksandr Yakovlevich (1894–1959), 7:351–352; 2:145b; 3:228a; 5:342a; 8:558a; 9:129b; 10:553b; 12:410b
Khizhin, Pavel Pavlovich, 10:433a
Khramov, V. A., 15:8a
al-Khujandī, Abū Maḥmūd Ḥāmid ibn al-Khiḍr (*d.* 1000), 7:352–354; 2:148b–149a, 153b; 7:260a; 9:84b–85a
Khunrath (or Kunrath or Kuhnrat[h] or Cunradius or Conrathus), Conrad (*d. ca.* 1614), 7:354–355
Khunrath (or Kunrath), Heinrich (*ca.* 1560–1605), 7:355–356; 8:311b
al-Khūrī, Yūsuf, 15:236a–b
Ibn Khurradādhbih (or Ibn Khurdādhbih), Abu'l-Qāsim ʿUbayd Allāh ʿAbd Allāh (*ca.* 820–*ca.* 912), 7:356–357; 9: 88b
Kitāb al-masālik wa'l-mamālik, 7:8b
Khusrau Anūshirwān
see Anūshirwān
al-Khuwārizmī, Abū ʿAbd Allāh Muḥammad ibn Aḥmad ibn Yūsuf (*fl. ca.* 975), 7:357–358
Mafātīḥ al-ʿulūm, 9:84a
al-Khuwārizmī, Abū Bakr, 7:357b
Khwārazm, region of, 2:148a–152b *passim,* 155b
al-Khwārizmī, Abū Jaʿfar Muḥammad ibn Mūsā (*fl. ca.* 800–847), 7:358–365; 6:315b; 7:357b; 14:537b
algebra, 1:22b, 30b, 31a–b, 40a, 444a; 4:118a(n), 608b, 609a, 611b; 7:241a, 243b, 244a, 327b, 358b–360a, 362b; 12:93a; 13:291a; 15:669b
Algebra, 11:458a
commentaries on, 11:533b
translations of, 1:22b; 4:611b; 7:362b; 11:591a; 15:175a, 177b
arithmetic, 7:129a–b, 359a–b, 360b; 13: 545a
Arithmetic
translations of, 15:175b, 187b
astrology, 6:381b; 7:82b, 358b, 362b
astronomical tables *(Zīj al-Sindhind),* 1:61a, 62b; 3:26b; 4:542a, 543b; 5: 613b; 7:82b, 360b–361b, 363a; 11: 322a; 12:361a; 13:401a; 14:576b, 593a
revision of, 9:39a,b
translations of, 7:361a, 363a
geography, 7:361b–362a, 363a
geometry, 7:359b–360a
Kitāb Ṣūrat al-Arḍ, 11:202b
De numero Indorum, 1:63a; 7:358b, 360a–b; 15:10b
translations of works, 15:256a
trigonometry, 5:612a–b; 7:361a
Kidd, John (1775–1851), 7:365–366; 2: 566a,b, 568a,b; 3:291b; 8:56a
Kidinnu (Kidenas), 15:678b
Kidney(s)
anatomy
Renaissance discussion, 1:98b, 619a–b; 4:486b, 520b; 9:165b
seventeenth-century discussion, 1: 592b–593a; 2:311a; 9:64a; 10:520a
eighteenth-century discussion, 5:267a
nineteenth-century discussion, 6:270a–b; 7:23b
twentieth-century discussion, 15:65a
abnormalities, 4:520b
comparative, 6:224a
artificial, 1:11a
of birds, 5:267a
blood flow, 12:470b
calculi
see Calculi, renal
diseases of, 9:95a; 11:374a; 12:470b; 14: 638a*
Bright's, 2:464a
and lead poisoning, 5:219a
nephritis, 3:447b; 13:574b
nephrosis, 1:439a
treatment of, 7:375b
embryogeny, 9:570b; 15:525a
evolution of, 12:470a
function
clearance, 12:470b
measurement of, 1:11a; 12:470b; 13: 575a
theories of, 8:452b, 540a, 541b; 14:10a
fusion of, 1:619b; 2:337a
glomeruli
filtration rate, 12:470b
permeability of, 1:534a
histology, 2:376a; 4:590a; 6:226a; 8:541b
metabolism, 1:453a, 454a; 6:24b; 8:452a
perfusion techniques, 1:534a; 2:237b; 3: 91b
physiology, 6:226a–b; 7:23b–24a, 519a; 10:220b; 11:139a; 12:470a–b
polycystic, 9:66a
relative positions of, 4:285a, 486b; 9:165b; 14:9b–10a
role of, 8:204a, 584a; 12:470a
and hypertension, 12:470b
transplantation, 3:91a
see also Nephrectomy, Urine
Kidston, Robert, 3:30b, 608a; 5:498a, 605a; 8:5a,b
Kiel
zoology institute, 9:432a
Kiel, Treaty of (1814), 1:12b
Kiel, University of
observatory, 11:542b
physics at, 11:8a
Kiel Bay, 9:432a
Kielmeyer, Carl Friedrich (1765–1844), 7: 366–369; 3:522a,b, 523b, 527a; 5: 217b, 356b–357a, 429b; 12:155a
Kierkegaard, Sören Aabye, 12:156b
Kiev, University of
geology at, 1:162a–b
Kikuyu tribe, 8:104b
Kilauea (crater), 3:552a; 8:578a
Kilborne, Fred Lucius, 12:481a
Kilian, Charles-Constant-Wilfrid, 5:221b; 6:168b; 13:283b
Kiliani, Heinrich, 5:3a, 4a; 14:444a
Killing, Wilhelm, 3:95a–b; 4:337a; 8:326b; 14:222a
Kilwardby, Robert, 1:100a; 5:592b
Kimberlite, 6:475b
Kimura, Hisashi (1870–1943), 7:369–370
Kinckhuysen, Gerard, 9:310a; 10:44a
al-Kindī, Abū Yūsuf Yaʿqūb ibn Isḥāq al-Ṣabbāḥ (*ca.* 801–*ca.* 866), 15:261–267; 1:36b, 276a; 4:525b, 543b(n)
acquaintances of, 11:244a
De aspectibus, 14:458b
translations of, 15:178a
biographies of, 7:187a
on Euclid, 4:438b, 440b, 441a,b
De gradibus, 1:291b*
translations of, 15:185a–b
influence of, 1:34b, 290b; 4:526a; 10:474a; 11:226a; 13:511a
influences on, 9:160a
De intellectu
translations of, 5:592a; 15:181a
optics, 1:379a; 10:475a
pharmacology, 2:392b–393a; 14:156a
philosophy, 1:33a; 7:23a
De quinque essentiis
translations of, 15:181a–182b
De ratione
translations of, 15:187a
De somno et visione
translations of, 15:182b
students of, 1:33b
translations by, 11:244a
Kinematics
medieval Latin, 1:94b, 197b; 2:393b–394a, 551a; 5:116a, 360a–b; 6:376b, 378a–379b; 7:116a–b; 10:225b–228b
Renaissance, 1:605b–606b, 607a–b; 8: 210b, 231b–233b; 9:133a–b; 12: 548a–b
seventeenth-century, 5:238b–242b *passim,* 245a, 246a–247a, 287b–288a; 6:126b–127a; 8:267b–268b; 13:583a–b
eighteenth-century, 5:195b; 14:197b
nineteenth-century, 1:615b–616a; 2:215b; 3:560a; 5:502a; 7:446b; 8:287b; 11: 383b–385a; 12:74a, 195b, 534b; 13: 386a; 14:403b–404a, 620a,b

twentieth-century, 2:168b, 169a, 191b–192a; 4:321a–322b; 13:125b; 14:305a
articulated systems, 7:446b
"fastest moving point rule," 2:394a
geological, 1:236b; 13:148b
and geometry, 1:445b; 5:360a–b; 6:379b; 7:446b; 9:80b; 11:487b–488a, 488b; 12:195b; 13:290b
liquid, 14:620a,b
of mechanisms, 7:287a–b
Kinetic theory of gases
see Gases, kinetic theory of
Kinetics
chemical
nineteenth-century, 1:300a–b; 2:589a; 4:411b–412a; 5:587a–b; 6:110a; 7:462a–b, 620a; 9:304b–305a, 352b, 619b; 13:310a, 577b–578b; 14:109a; 15:458a, 461b–462a
early twentieth-century, 1:361a–363a, 414b–415b; 2:170a–171a; 3:176b, 197b; 4:414a; 7:493a–b; 8:24a, 369a; 12:218b, 404b, 418b; 13:429b; 15:36b–37b
mid twentieth-century, 1:519b; 2:68b–69a; 4:82b, 545b; 6:404a–405a; 7:617a; 13:7a
enzyme reactions, 1:537b; 6:22a
temperature effects, 1:300a
electrochemical, 6:373a–b
physical
twentieth-century, 8:11a
King, Alfred Freeman Africanus, 11:556a
King, Clarence Rivers (1842–1901), 7:370–371; 1:552b; 4:265b, 365b; 7:5a; 11:443a; 14:192a, 516b
influences on, 14:315b
and U.S. Geological Survey, 1:490b; 4:366a; 6:13b, 57b; 11:1119a,b
King, Edward, 9:534a
King, Harold, 14:335a
King, Philip Parker, 5:16a, 17a
Narrative of a Survey of the . . . Coasts of Australia, 2:520a
King, Victor, 14:271a
King's College, Aberdeen
observatory, 5:404a
Kings College, Cambridge
religion at, 10:447b
Kings College, London
chemistry at, 3:557b
King's College for Women, London, 15:417b
King's evil (disease), 4:495a
Kingsley, Charles, 6:563a, 593b, 596b, 616a
Kinnersley, Ebenezer (1711–1778), 7:371–372; 1:547a; 5:131a, 136a
Kino, Eusebio, 12:431a
Kinoplasm, 4:546a
Kinsey, Alfred Charles, 4:584a
Kipfer, Paul, 10:597b
Kipper, Aksel Ianovich, 12:368b
Kipping, Frederick Stanley (1863–1949), 7:372–373; 7:552b*; 8:31b; 10:517b; 11:85a–b
Kirby, William, 1:502b; 6:492b
Kirch, Christfried (1694–1740), 7:373–374
Kirch, Christine (*ca.* 1696–1782), 7:373–374; 2:221a
Kirch, Gottfried (1639–1710), 7:373–374; 4:169a
Kirch, Maria Margarethe Winkelmann (1670–1720), 7:373–374
Kircher, Athanasius (*ca.* 1601–1680), 7:374–378
acquaintances of, 2:83a; 6:360b; 9:25a, 97a; 10:103b
Ars magna lucis, 5:525a
commentaries on, 2:13a
correspondence of, 3:61a*; 8:149b; 10:314b; 11:342b*
critics of, 2:317b
influence of, 2:591b; 8:153b
Magnes, sive de arte magnetica, 2:84a; 5:575b; 14:376b(n)
Musurgia, 2:83b
optics, 5:525b
Primitiae gnomonicae catoptricae, 6:362b
students of, 2:591a; 12:210a–b
Kircherian Museum, 2:591a
Kirchhof, Konstantin Sigizmundovich (Gottlieb Sigismund Constantin) (1764–1833), 7:378–379
starch conversion, 4:134a
Kirchhoff, Gustav Robert (1824–1887), 7:379–383; 2:564a; 6:243a, 342b; 8:510b; 10:28b; 11:8b, 9a
acquaintances of, 2:587a; 6:342a, 356b, 357a; 7:460a; 8:387b; 9:178a; 12:511b; 13:80a; 14:55b
critics of, 4:616b
equations, 8:491a; 15:437a
Gesammelte Abhandlungen, 11:10b
hydromechanics, 3:195a; 14:620b
influence of, 12:268a
Kirchhoff function, 8:551b; 10:345b, 346a
magnetism, 13:80a
mechanics, 6:347b
elasticity, 3:146a
vibration, 3:258b
physical chemistry, 14:560b
radiation, 4:183a; 6:350b; 11:9a, 10b
law of, 4:180a; 5:305b; 11:149b–150a; 13:52a; 14:339b
spectroscopy, 1:166a–b; 2:589a; 3:476a; 5:20a, 86b, 143b; 6:541b; 7:74a–b; 8:440b; 11:47a; 12:37a; 14:55b
students of, 2:272a, 382b; 4:378a,b; 6:286a, 341b; 7:220a, 267b, 477b, 551b; 8:553a; 9:347b, 355a; 11:8a, 213a, 241b, 611a; 12:216b, 238a; 13:79b, 416b; 14:170b
translations of works, 11:538a
Kirkaldy, David (1820–1897), 7:383
Kirkby, Paul Jerome, 13:446b
Kirkham, William Barri, 3:493a
Kirkman, Thomas Penyngton (1806–1895), 7:383–384; 13:20a
Kirkwood, Daniel (1814–1895), 7:384–387; 13:270b
Kirkwood's gaps, 7:385a–b
Kirkwood, John Gamble (1907–1959), 7:387; 3:336a, 619b; 7:387a; 13:430a*; 15:44a
al-Kirmānī, Burhān al-Dīn Nafīs ibn ᶜAwaḍ, 9:604a
Kirschner, Aage, 2:171a
Kirtland, Jared Potter, 7:169b
Kirwan, Richard (*ca.* 1733–1812), 7:387–390; 1:65a*; 2:368a
acquaintances of, 5:215b; 7:69b; 10:108a; 12:494b; 15:94b
chemistry, 2:6a; 3:465b; 5:532b, 601a; 9:5b
correspondence of, 12:67b(n)
critics of, 9:472a
Elements of Mineralogy
translations of, 12:329a
Essay on Phlogiston
translations of, 8:80b; 10:108a
influence of, 6:385a–b; 7:69a; 14:75b
mineralogy, 2:552b; 5:537a; 6:579b, 583b
Kisch, Bruno, 15:469b
Kistemaker, Jacob, 7:493a
Kitaibel, Pál (or Paul) (1757–1817), 7:390–391
Kitasato, Shibasaburo (1852–1931), 7:391–393; 7:425b, 429b
antitoxins, 1:575a; 7:427a; 11:569b; 15:521b
plague, 7:427b; 14:551b
Kitasato Institute, 7:392b
Kitchen middens, 5:72b
naming of, 13:9b
Kites
Franklin, 3:153b; 5:134b
Kjeldahl, Johann Gustav Christoffer (1849–1900), 7:393–394; 7:113b; 12:546a
Kjeldahl apparatus, 2:258a
Kjeldahl flask, 7:393b
Kjeldahl method, 7:393a–b; 12:546b
Kjellman, Frans Reinhold, 7:535a; 13:165a
Kjerulf, Theodor, 2:486a; 14:59a
Klácel, Matthew, 9:277b–278a
Klamroth, Martin, 4:439a–b, 445a, 447b
Klaproth, Julius, 1:19b
Klaproth, Martin Heinrich (1743–1817), 7:394–395; 3:465b; 5:430a; 13:597b*; 14:332b
acquaintances of, 1:19b; 5:202b; 11:539a, 540b; 14:257b; 15:172a
chemical theories, 5:532b; 8:520a
critics of, 7:391a; 11:465a
elements discovered, 2:92b; 3:473b
influences on, 10:26a
mineral analysis, 6:495b; 7:378a; 9:106b, 559b; 11:437b
students of, 12:239b; 14:239b, 386a; 15:171b, 205b–206a
Klarer, Josef, 4:154a
Klatt, Virgil, 8:180a–b
Klau, Christoph
see Clavius, Christoph
Klebs, Arnold Carl, 3:517b; 5:208a
Klebs, Edwin, 4:276a,b; 12:58b; 14:41a
Klebs-Loeffler bacillus, 4:404b, 276b; 7:425b; 8:448b–449b; 11:569a; 14:551b
Klebs, Georg Albrecht (1857–1918), 7:395–396; 4:220a
Kleeman, Richard Daniel, 2:398a
Kleemann, Christian Friedrich Carl, 11:503a
Kleene, Stephen Cole, 2:514a; 13:497b–498a
Kleijn, Adrianus Paulus Huibertus Antonie de, 9:20a,b
Klein, Christian Felix (1849–1925), 7:396–400; 1:352a; 2:271a,b; 4:293a, 373b, 561a; 6:357a; 8:388b, 500b; 9:431a*, 461b; 11:568b
acquaintances of, 3:167a; 7:616a; 8:365b, 433b; 9:358a; 10:137b; 11:46a, 123b–124a, 407b, 613a,b; 12:249a; 13:243b; 14:222a
algebra, 3:221b
analysis, 7:602a; 11:53a–b, 451b, 452a,b; 12:247a
correspondence of, 2:594a; 10:443a; 11:53b; 12:172b, 228a, 284a
critics of, 13:133a
on Eisenstein, 4:342b
Elementarmathematik vom höheren Standpunkte aus, 6:235b
Encyklopädie der mathematischen Wissenschaften, 10:422a
Erlanger Programm, 2:513a; 4:522b; 7:397a–b; 8:325b, 434a; 9:358b; 11:362b; 12:214b; 13:599b
on Gauss, 5:306a, 309a, 311a*, 312a*
geometry, 1:160a; 2:192a; 3:165a,b, 223a; 7:446a, 574a; 8:324b, 434a; 9:127a; 10:343b–344a; 11:46a,b; 12:225b, 284a, 533a; 13:5b
group theory, 3:168b; 5:472b; 7:168a–b
and Hurwitz, 6:570b, 571a–b
influence of, 2:217a,b, 465a; 3:62b; 4:370b; 9:502a; 10:486a; 11:509b; 12:195b

Klein, Christian Felix (cont.)
  Kleinean functions, **11**:52b
  and Lie, **8**:323b, 324a, 325b
  mathematical physics, **13**:240a
  matter theory, **15**:463a
  number theory, **6**:214b
  and Sommerfeld, **12**:526b, 527a
  students of, **2**:58a, 121a, 191b, 272a, 430a; **3**:345b; **4**:100b, 269a, 618b; **5**:181b; **6**:214a, 472b; **7**:246b; **8**:367b; **9**:457b; **10**:244b, 275a; **11**:407a, 475b; **13**:81a; **15**:40a
  Theorie des Kreisels, **12**:527a
  Über Riemann's Theorie der algebraischen Funktionen und ihrer Integrale, **11**:449a–b
Klein, Edward Emanuel, **11**:555b; **12**:355b
Klein, Hermann Joseph (1844–1914), 7:400
Klein, Jacob Theodor (1685–1759), 7:401; **3**:350a
  Système du règne animal
    translation of, **2**:474a
  taxonomy, **13**:425b
Klein, Marthe, **3**:501b
Klein, Oskar Benjamin, **2**:245b; **3**:369b; **7**:211b
Kleine, Friedrich Karl, **7**:428b, 429a
Kleinenberg, Nicolaus (Nicolai) (1842–1897), 7:401–403; **9**:333b
Kleiner, Israel Simon, **1**:441a
Kleist, Ewald Georg von (*ca.* 1700–1748), 7:403; **2**:324b; **9**:596b
Klemensiewicz, Rudolf, **1**:544a; **14**:278a
Klevens, Howard Bliss, **6**:118b
Klingenstierna, Samuel (1698–1765), 7:403–404; **3**:282a, 284a; **14**:352a, 353a
  Inledning til naturkunnigheten, **14**:354a
Klinger, R., **6**:433a
Klippe (geol.), **5**:221b; **6**:229a
Klüber, Harald von, **5**:183b
Klug, Aaron, **5**:141a
Klug, Josef, **9**:247b
Klügel, Georg Simon (1739–1812), 7:404–405; **7**:206b; **10**:573a
Kluyver, Albert Jan (1888–1956), 7:405–407; **15**:15a*
Kluyver, Jan Cornelis, **7**:405b, 465b; **10**:461a
Klystron, **6**:104b, 105a; **11**:422a
Knab, Frederick, **6**:524b, 525a
Knapp, Georg Friedrich, **2**:318b, 319a
Knee jerk, **2**:367a; **12**:396a, 397a
Kneser, Adolf (1862–1930), 7:407–408; **2**:271b; **5**:350b; **7**:507a, 508a; **9**:410b*; **14**:222a
  mathematics, **2**:236b; **4**:9a
Knight, Gowin, **3**:51a; **4**:149a, 529b, 530a; **6**:384a; **7**:13b
Knight, Joseph, **2**:518b
  On the Cultivation of . . . Proteeae, **2**:519a
Knight, Thomas Andrew (1759–1838), 7:408–410; **2**:257a; **8**:445a; **13**:463a*; **14**:129b(n)
Knights of Malta, **4**:149b, 150a,b
Knipovich, Nikolai Mikhailovich (1862–1939), 7:410–411
Knipovich (ship), **14**:634b
Knipping, Paul, **4**:196a; **5**:117b; **6**:303a; **7**:468b; **8**:51a; **11**:531a; **15**:61b
Knobel, Edward Ball, **10**:543b
Knoblauch, Karl Herrmann, **13**:521b
Knöfler and Company, **6**:15b–16a
Knoll, Max, **2**:318a
Knoop, Franz, **4**:360a; **10**:22b; **14**:445b
Knopp, Konrad (1882–1957), 7:411; **7**:616b*; **13**:81b
Knorr, Georg Wolfgang (1705–1761), 7:411–413; **9**:534a; **14**:119b
Knorr, Ludwig, **5**:5b, 381b, 621a
Knorre, Georg, **7**:423a
Knots (math.), **3**:169a; **7**:384a; **11**:362b; **13**:237a
Knott, Cargill Gilston (1856–1922), 7:413; **2**:594a; **7**:314a*; **9**:606a; **13**:237b*
Knowledge
  see Epistemology,
Knowledge and Illustrated Scientific News, **11**:163a
Knox, George James, **9**:451a
Knox, Robert (1793–1862), 7:414–416; **1**:452a; **5**:469a
Knox, Thomas, **9**:451a
Knucklebones (game), **3**:66a
Knudsen, Martin Hans Christian (1871–1949), 7:416–417; **14**:635b*; **15**:83b
  Knudsen manometer, **7**:416b
  oceanography, **15**:431b
Knuth, Paul Erich Otto Wilhelm (1854–1900), 7:417–418
Kobeko, Pavel Pavlovich, **7**:527a
Köbel (or Kobel or Kobilin or Kiblin or Caballinus), Jacob (*ca.* 1460–1533), 7:418–420; **11**:457b
Koch, Eberhard, **15**:469b
Koch, Fred Conrad, **8**:359a
Koch, Heinrich Hermann Robert (1843–1910), 7:420–435; **9**:259b; **11**:585b
  acquaintances of, **10**:493a; **12**:87b, 481b; **14**:512a
  anthrax, **4**:276a; **7**:421b–424b *passim*; **10**:386a–b, 388a; **11**:70b
  bacteriology, **5**:564a; **9**:334b
  cholera, **5**:219b; **6**:11b; **7**:421a, 424b–425b, 427a–b; **10**:267b
  and Cohn, **3**:340a; **4**:296a; **7**:422a,b
  colleagues, **1**:575a; **4**:276b, 297b, 310b, 311a; **5**:219a,b, 220a; **14**:551b; **15**:521a–b, 522a
  correspondence of, **8**:407b–408a
  critics of, **1**:496b; **7**:100b; **8**:296b, 409a
  on disinfection, **8**:408a
  epidemiology, **7**:424b–425b, 427a–429b
  and Gaffky, **7**:423a–429a *passim;* **8**:448a
  influence of, **4**:403b; **5**:52a; **10**:383a; **11**:176b; **12**:480b; **14**:443b
  influences on, **14**:228b
  Koch's lymph, **7**:426a
  Koch's phenomenon, **7**:426a
  Koch's postulates, **7**:422b, 424a; **8**:448b
  Koch-Weeks bacillus, **4**:618a; **7**:425a
  malaria, **7**:428a–b
  methodology, **2**:438a, 575b; **3**:464a*; **6**:100b; **10**:388b; **12**:480b
  and Pasteur, **7**:423b, 424a–b, 426a; **10**:354b, 364a, 397a–400a *passim;* **12**:87b
  and Pettenkofer, **10**:559b–560a, 560b
  rabies, **10**:401b
  students of, **7**:100a, 391b, 392a,b; **8**:448a–b; **9**:94b; **10**:155a; **11**:176b; **12**:395b; **14**:249a
  tuberculosis, **2**:528a; **4**:296b; **7**:423b–430a *passim;* **12**:481b
    bovine, **3**:22b, 588a; **7**:423b–424a, 427a, 428b, 429a,b; **8**:409a
    tuberculin treatment, **4**:297a; **7**:425b–427a
  Ueber die Milzbrandimpfung . . . , **10**:398a
Koch, Helge von (1870–1924), 7:435–436; **9**:426b; **11**:59b
  curves of, **3**:178b
Koch, Wilhelm, **2**:159b
Kochin, Nikolai Yevgrafovich (1901–1944), 7:436–437; **5**:188b
Kodak Research Laboratories, **9**:202a
Kodaṇḍarāma
  commentaries of, **1**:308b
Koebe, Paul, **4**:560b; **6**:394a; **7**:399a; **11**:54a, 452a
Koechlin, Maurice, **3**:496a
Koechlin et Cie, **3**:9a
Koeck, Christian, **12**:510a,b
Koelliker, Rudolf Albert von (1817–1905), 7:437–440; **11**:256a*; **12**:422b*
  acquaintances of, **5**:460a; **6**:268b; **8**:302a; **11**:368a; **12**:421a; **13**:274a
  correspondence of, **3**:425a
  critics of, **1**:540a; **7**:402a
  cytology, **13**:89a
  handbooks, **2**:617b*; **3**:424b; **8**:302a
  heredity, **14**:235a
  histology, **8**:401a; **10**:267a; **11**:566a–b
  influence of, **3**:425b; **9**:332a
  students of, **4**:275b; **6**:7a, 224a, 287a, 434a; **12**:299a; **14**:331a; **15**:78a, 166a,b
Koelreuter, Joseph Gottlieb (1733–1806), 7:440–442; **2**:257a; **9**:280a,b
  on reproduction, **5**:218a; **12**:588a; **15**:67b, 68a
  translations of works, **15**:507a
Koenig
  see also König
Koenig, Dénes, **7**:444b
Koenig (König), Johann Samuel (1712–1757), 7:442–444; **3**:216a, 282a; **4**:471b
  Koenig's law, **7**:443a
  and Maupertuis, **4**:471b; **7**:442b–443a; **9**:186b–189a *passim*
  theorem of, **2**:38b
Koenig, Julius (1849–1914), 7:444; **7**:530b
Koenig, Karl Rudolph (1832–1901), 7:444–446
  tuning fork, **13**:11b
Koenig, Samuel, **7**:442a
Koenigs, Gabriel (1858–1931), 7:446; **3**:428b
Kofoid, Charles Atwood (1865–1947), 7:447; **13**:268a
Kögl, Fritz, **14**:255b
Köhler, Alexander Wilhelm, **14**:259a
Köhler, August Karl Johann Valentin (1866–1948), 7:447–449; **12**:423a
  Köhler principle of illumination, **7**:448a
Kohl-Larsen, Ludwig, **11**:336b
Kohlrausch, Friedrich Wilhelm Georg (1840–1910), 7:449–450
  acquaintances of, **1**:297b; **6**:73b, 74a, 343b; **7**:528a; **14**:172a, 338b; **15**:458b
  on contemporary scientists, **14**:330b*
  electricity, **4**:500b; **7**:380b; **14**:205a
  electrochemistry, **1**:298a; **6**:439b; **9**:391b; **14**:268b; **15**:459a
  magnetometer, **6**:470a
  students of, **1**:490a; **8**:106b; **11**:445b; **15**:433a
Kohlrausch, Karl Wilhelm Fritz, **12**:218a
Kohlrausch, Rudolph Herrmann Arndt (1809–1858), 7:450
  electrodynamics, **9**:205a, 209a; **14**:205a, 207a
Kohlrausch, Wilhelm Friedrich, **7**:449a
Kohlschütter, Arnold, **1**:56b; **8**:364a, 539b
Kohlstock, Paul, **7**:427b
Koide Shuki, **15**:734b–736a
Koihō school, **1**:312a; **15**:739b–740a, 740b, 741a,b
Kola Peninsula
  botany, **10**:165a
Kolb(e), Peter, **7**:373b
Kolbe, Adolf Wilhelm Hermann (1818–1884), 7:450–453; **14**:454b

acquaintances of, **2**:515a; **5**:536b; **9**:130b; **14**:478a–b
chemistry, **2**:587b, 621b; **5**:431b
acids, **1**:578a; **14**:169a
radicals, **2**:588b; **5**:372b; **14**:530b
on contemporary scientists, **4**:395b*; **13**:577b
critics of, **10**:177b; **11**:92b; **14**:531a
and Frankland, **5**:125a,b; **7**:451a, 452a
influence of, **2**:623a; **3**:511a
influences on, **4**:395a
students of, **1**:288a, 553a; **3**:299b; **5**:488b; **6**:107b; **9**:467a; **10**:502a
Kolhörster, Werner, **2**:338a; **6**:355b
Kolk, Schroeder van der, **2**:124b; **3**:425a
Kolle, Wilhelm, **7**:427b; **14**:512b; **15**:522b
Kolletschka, Jakob, **12**:295a
Kolloid Gesellschaft, **10**:251b
Kolloidchemische Beihefte
editors of, **10**:251b
Kolmogorov, Andrey Nikolaevich, **1**:366b; **3**:231a; **7**:351a,b; **8**:558a; **9**:71a, 129b, 420a; **10**:553b
Kolosov, Gury Vasilievich (1867–1936), 7:453–454
Kolthoff, Izaak Maurits, **6**:374a; **13**:226b
Koltzoff, Nikolai Konstantinovich (1872–1940), 7:454–457; 12:76b; **14**:595a
Koltzoff principle, **7**:455a
Kolyma River, **12**:279b
Komandorskiya Ostrova Islands, **13**:25a–b
Komensky, Jan Amos
see Comenius, John Amos
Kon, George Armand Robert, **4**:91a
Kondakov, Ivan Lavrentievich (1857–1931), 7:457; 2:623b
König
see also Koenig
König, Arthur (1856–1901), 7:457–458; 6:253a*, 350a*; **9**:201b
König, Emanuel (1658–1731), 7:458–459
Konig, Julius, **10**:140a–b
Königer, Hermann, **10**:291b
Königliche Materialprüfungsamt, **6**:368a–b; **9**:139a
Königsberg
in nineteenth century, **7**:379b
Königliche Anatomische Anstalt, **2**:595b
observatory, **2**:98a
seven bridges, problem of, **4**:479b
Königsberg, University of, **7**:50b–51a; **10**:27a; **11**:307b
chemistry at, **5**:488b
mathematics at, **3**:313b–314a; **6**:388a–b, 571a; **14**:202a
physics at, **10**:28b
Königsberger, Leo (1837–1921), 7:459–460; 7:444a, 477b; **11**:9a, 551a; **14**:222a, 223a, 224a*, 481b
Koninck, Laurent-Guillaume de (1809–1887), 7:460–461; 9:61b*; **12**:619b
Koningsberger, Jacob Christiaan, **13**:460a
Konkoly Thege, Miklós von (1842–1916), 7:461
Konovalov, Dmitry Petrovich (1856–1929), 7:461–462; 2:623b
Konovalov's laws, **7**:462a
Konrad of Megenberg, **10**:310a
Kontorova, Tat'iana Abramovna, **5**:161a
Koomans, Catharina Maria, **1**:394b
Koopman, Bernard Osgood, **2**:145b; **14**:90b
Kopfermann, Hans, **7**:554a
Kopp, Hermann (1817–1892), 7:463–464; 5:371a; **12**:209a; **15**:422b
Geschichte der Chemie, **11**:437a
historiography, **2**:64b; **7**:40a; **13**:560a*
laws of, **8**:509b; **11**:278b; **14**:560b; **15**:439a–b, 440b
mineralogy, **8**:245b
students of, **2**:59a; **5**:163b, 349b; **9**:355a; **12**:208b
Kopp, Johann Heinrich, **7**:463a; **8**:246a
Koppel, Ivan, **11**:552a
Koran, **3**:516a; **7**:339b; **9**:88b; **12**:7Ba; **15**:494b, 657b(n)
animals in, **3**:548b, 549a
commentaries on, **15**:262a
interpretation of, **12**:7Ab
Korana race, **2**:505a
Korea
geology, **7**:474a
mathematics, **3**:265b
Koref, Fritz, **15**:442a
Korkin, Aleksandr Nikolaevich, **3**:226a; **5**:508b; **7**:513b; **9**:124b, 126a, 324a, 410a; **14**:94b, 630a,b
Korn, Arthur, **6**:472a
Körner, Wilhelm, **3**:46b; **7**:281b, 282a, 552a; **14**:269b
Korolev, Sergey Pavlovich (1907–1966), 7:465
Korotkoff, sounds of, **4**:398b
Korschelt, Eugen
Vergleichende Entwicklungsgeschichte der Tiere, **7**:33b–34a
Korteweg, Diederik Johannes (1848–1941), 7:465–466; 2:512a; **6**:612b*; **12**:213b*, 501b*
Koschmieder, Hans Harald, **2**:344a
Kosmos
editors of, **7**:498b
Kossel, (Alexander Richard August) Hermann, **1**:576a; **4**:297b; **7**:427a, 428a
Kossel, Karl Martin Leonhard Albrecht (1853–1927), 7:466–468; 4:603b; **5**:53a; **6**:135b, 264a, 506a; **8**:275b
biochemistry, **6**:505b; **10**:242b
Kossel, Walther (Ludwig Julius Paschen Heinrich) (1888–1956), 7:468–470; 2:245a; **4**:201a; **7**:265a*; **9**:544b
Kossel effect, **7**:469b–470a
Kostanecki, Stanisław (1860–1910), 7:470–471; 12:423b
Liebermann-Kostanecki rule, **7**:471a
Köster, Werner, **13**:247a
Kostinsky, Sergey Konstantinovich (1867–1936), 7:471–473; 5:363a
Kostinsky effect, **7**:472a
Kotelnikov, Aleksandr Petrovich (1865–1944), 7:473
Kotelnikov, Piotr Ivanovich, **7**:473a
Kotelnikov, Semën Kirillovich, **4**:471a
Köthe, Gottfried Maria Hugo, **13**:428a,b
Kotō, Bunjiro (1856–1935), 7:473–474
Kotzebue, Otto von, **4**:407a; **8**:187a; **15**:82a
Kou Shou-ching, **1**:310b
K'ou Tsung-shih, **8**:391a, 396b*; **12**:383b
Kovalevsky, Aleksandr Onufrievich (1840–1901), 7:474–477; 7:32a; **9**:112a, 332a–b
comparative embryology, **1**:420b; **11**:380a
critics of, **15**:513b
marine zoology, **1**:601a; **9**:113b*
paleontology, **12**:38a
Kovalevsky, Maxim, **7**:479a–b
Kovalevsky, Sonya (or Kovalevskaya, Sofya Vasilyevna) (1850–1891), 7:477–480; 4:480b; **9**:426b; **11**:383a
Cauchy-Kovalevsky theorem, **3**:143b; **7**:478a–b
mechanics, **14**:620a
on Saturn's rings, **1**:597b
and Weierstrass, **9**:427a; **14**:223a–b
Kovalevsky, Vladimir Onufrievich (1842–1883), 7:480; 1:18a, 161b; **4**:147a, 148a; **7**:477b
Kovalsky, Marian Albertovich (Voytekhovich) (1821–1884), 7:480–482; 13:119a
Kowalewski, Gerhard
Bestand und Wandel, **3**:56a
Kowalewsky
see Kovalevsky
Kowarski, Lew, **7**:155a,b
Koyré, Alexandre (1892–1964), 7:482–490; 2:546a; **9**:342a; **12**:113b; **15**:422b, 424b
historiography, **3**:433a*; **5**:287a,b, 288b; **6**:174b(n), 488a*; **9**:510b; **14**:227b*; **15**:746b
Kozyrev, Nikolai Alekseevich, **7**:400a
Kraepelin, Emil, **1**:446b; **5**:571a; **7**:190a; **14**:277b
Krafft, Georg Wolfgang, **11**:432a–b; **13**:10b, 425a
Krafft, Johann Daniel, **7**:525a
Krafft, Wolfgang Ludwig, **4**:472b; **8**:299a
Krafft-Ebing, Richard von, **14**:115a
Lehrbuch der Psychiatrie, **7**:190a
Kraft, Jens (1720–1765), 7:490–491
Krakatoa, **11**:336b
botany, **13**:460a
Krakow, University of
in Renaissance, **3**:401b
in seventeenth century, **2**:526b
in World War II, **1**:428b
Krakow Observatory, **1**:429b, 430a
Kramers, Hendrik Anthony (1894–1952), 7:491–494; 2:245b–248a *passim;* **3**:429b; **4**:294b*; **5**:331b; **10**:236b*
absorption coefficient, **4**:279b
Kramers' theorem, **7**:493a
Kramers-Kronig relations, **7**:492a
quantum theory, **2**:338a; **3**:369a–b; **7**:518b; **15**:41b
W(entzel)-K(ramers)-B(rillouin) method, **7**:492a–b
Kramp, Chrétien or Christian (1760–1826), 7:494–495; 8:140b
Kramp's transcendental, **7**:494b
Krarup, Carl Emil, **8**:502b
Krasheninnikov, Stepan Petrovich (1711–1755), 7:495
Krasnov, Andrey Nikolaevich (1862–1915), 7:496
Krasovsky, Theodosy Nicolaevich (1878–1948), 7:496–497
Krassowski, Jan, **6**:539b
Kratzer, Adolf, **5**:609b
Kraus, Charles August (1875–1967), 7:497–498
Kraus, Paul, **7**:40a, 41a, 42a,b
Krause, Ernst Ludwig (also known as Carus Sterne) (1839–1903), 7:498–499; 10:196b*; **14**:58a*
Krause, Wilhelm, **4**:202b; **7**:421a; **11**:566a
Krauss, Heinrich Carl Wilhelm, **11**:215b
Kraut, Karl, **5**:430b, 431b
Krayenhoff, Cornelis Rudolphus Theodorus (1758–1840), 7:499–501; 13:468b, 469a
Krebs, Georg, **5**:388a
Krebs, Norbert
Deutsches Lebensraum in Mitteleuropa, **10**:503a, 505b
Kremer, Gerhard
see Mercator, Gerardus
Křepelka, Jindřich, **2**:429a
Kreps, Evgenii Mikhailovich, **10**:220b, 435a
Kretz, François Xavier, **11**:78a, 79a–b
Kreuzstein, **2**:555a
Krishnan, Kariamanikkam Srinivasa, **9**:76b; **11**:265a,b; **13**:240a

Kristineberg zoological station
founding of, 8:519b
Krogh, Schack August Steenberg (1874–1949), 7:501–504
Kroman, Kristian
Naturerkenntnis, 15:422b
Kron, Erich, 9:563b; 12:251a
Kronecker, Hugo (1839–1914), 7:504–505; 2:62b, 366b; 4:201a; 6:52b; 7:519b, 521a*; 8:542a*; 9:265b; 15:99b
Kronecker, Leopold (1823–1891), 7:505–509; 4:5a
acquaintances of, 4:2b, 3a, 127a; 6:230a
algebra, 2:470a; 7:398a; 13:220b, 571b
algebraic surfaces, 3:117a; 12:284b
and Cantor, 3:53a,b, 57b
on contemporary scientists, 4:4b, 206b*
editions of works, 6:286b
group theory, 3:168b; 7:167b
Grundzüge einer arithmetischen Theorie der algebraischen Grössen, 7:444b
historiography, 5:264a
influence of, 2:491a; 5:198b; 6:297b; 12:245b; 13:23b
Kronecker index, 11:57a
Kronecker-Castelnuovo theorem, 3:117a
Kronecker's theorem, 6:394a
proof of, 14:202b
and Kummer, 7:521b–523b *passim*
number theory, 4:125a; 6:388b, 390a, 571b; 7:523b
students of, 4:618b; 6:95a, 286b, 472a,b, 570b; 7:407b; 8:253a; 9:358a, 502a; 10:24a, 343a; 11:611a; 13:81a, 155a
theorems of, 14:282b
and Weierstrass, 7:506b, 507a–508a, 523a; 14:222b
Krönig, August Karl (1822–1879), 7:509–510
gas theory, 3:306b; 14:185b
Kronig, Ralph de Laer, 3:430a; 10:424a; 12:21b
Kropotkin, Petr Alekseevich (1842–1921), 7:510–512; 10:167a, 169a; 11:338a*
Krosigk, Baron von, 7:373b, 374a
Kṛpārāma Miśra of Ahmadabad
commentaries of, 2:117a; 7:315b; 9:42a
Kṛṣṇa (*fl.* 1653), 7:512
Karaṇakaustubha, 15:624b
Kṛṣṇa, son of Ballala, 9:579b–580a; 11:290a; 12:599a
Kruber, Aleksandr Aleksandrovich (1871–1941), 7:512–513
Krüger, Adalbert, 1:243a
Gauss-Krüger projection, 5:304a
Krüger, Johann Gottlob, 7:403a
Geschichte der Erde, 9:596b
Krüger, Peter, 6:360b, 362b
Krusch, Johann Paul, 14:59a
Krusenstern, Adam Johann von
see Kruzenstern, Ivan Fedorovich
Kruzenstern, Ivan Fedorovich, 1:594a, 595b; 7:20a; 8:187a; 11:377a
Krylov, Aleksei Nikolaevich (1863–1945), 7:513–514; 3:226a, 230a; 5:188b, 462b*; 7:7b*; 9:43b*; 15:165a*
Krylov, Nikolai Mitrofanovich (1879–1955), 7:515
Krypton
density, 11:280b
discovery of, 11:280a; 13:454a
penetrability, 15:491a
spectrum, 4:513b
Krystallin, 5:197b
Kṣepa (remainder), 15:567a–b
Ktesibios
see Ctesibius
Kü Yen, 6:217a
Ku Yen-wu, 14:159b, 163b, 164a
Kublai Khan, 8:313b, 314a
Kučera, Bohumil (Gottlieb), 6:370b, 371b, 374a
Küchenmeister, Gottlob Friedrich Heinrich, 1:602b; 8:270b
Kuei-shu (reduced number), 3:252b, 253a
Kuenen, Johannes Petrus (1866–1922), 7:515–516
Kuenen, Philip Henry, 1:394b; 13:609a, 610a, 611a
Kufarev, Pavel Parfentevich, 14:592a
Kuffler, Abraham, 4:184a,b
Kuffler, Jacob, 4:184a
Kuffler, Johannes, 4:184a
Kugler, Franz, 1:389a
al-Kūhī
see al-Qūhī
Kuhlmann, Frédéric, 10:499a
Kuhlmann, Wilhelm H. F., 11:129b
Kühn, Alfred (1885–1968), 7:516–517; 8:529a; 15:125a*
Kühn, Heinrich, 4:477b
Kuhn, Richard (1900–1967), 7:517–518; 13:487b
Kuhn, Werner (1899–1963), 7:518–519
Kühne, Wilhelm Friedrich (1837–1900), 7:519–521; 7:466b
acquaintances of, 2:366b, 547b; 6:506a; 8:387b
and Chittenden, 3:256b–257a
critics of, 1:540a
histology, 11:566b
influences on, 2:27a, 31a, 531a
physiology, 7:402b*; 14:113b
students of, 5:34b; 7:504a; 8:14b; 9:19b; 12:617b; 15:503b
Kühnen, Friedrich, 6:241a
Kuiper, Gerard Peter, 7:232a
Kukarkin, Boris Vasilievich, 1:353a; 10:317b
Kullenberg, Börje, 10:564a
Külpe, Oswald, 9:562a; 14:528b
Kulturkreislehre, 2:210b
Kumagai, Naoiti, 9:181a
Kummer, Ernst Eduard (1810–1893), 7:521–524; 4:341b, 342a, 573a; 6:472a; 7:109a; 12:245b; 14:221a
acquaintances of, 3:466b; 4:2b; 6:230a; 14:222a
algebra, 4:474b; 5:198b, 306b; 6:390a
analysis, 11:452b
astronomy, 2:299a
on contemporary scientists, 15:193b
correspondence of, 3:134b
Kummer points, 12:284b
Kummer surfaces, 5:471b; 6:547b; 7:397a, 523b; 11:509b, 589a
Kummerian development, 8:253b
Kummer's equation, 5:482a
logarithms, 7:619b
number theory, 4:4b, 125a
students of, 1:370a; 3:53a; 5:203b, 472a; 6:570b; 7:108b, 505b–508b *passim;* 9:358a, 563b; 10:24a; 12:195b; 13:81a, 155a; 14:630a; 15:65b
Kunckel (von Loewenstern), Johann (*ca.* 1630–*ca.* 1703), 7:524–526; 5:420a
Laboratorium chymicum, 12:144a
phosphorus, 6:477b; 9:105b
Kundt, August Adolph (1839–1894), 7:526; 7:449a; 14:171b
acquaintances of, 6:28b, 343b; 11:8b, 529a,b; 14:338a
gas theory, 9:224b, 225a; 12:496a–b
Kundt tube, 7:526a
Kundt's dust figures, 7:526a
optics, 4:191a; 11:530a, 581b
students of, 4:360b; 5:148b; 6:73b; 7:267b; 8:106b; 10:345a; 11:551a, 581a; 14:347b, 348a, 633a
and Warburg, 14:170b–171a
K'ung Chi-han, 8:419a,b
K'ung Kuang-shen, 8:314b
Kunitz, Moses, 1:11b
Kunkel von Löwenstern
see Kunckel, Johann
K'unlun
geography, 6:216b
Kunrath
see Khunrath
Kunsman, Charles Henry, 3:597b
Kunth, Carl Sigismund (1788–1850), 15:267–268; 3:336b; 11:152a
Kuntze, Carl Ernst Otto (1843–1907), 15:268–269
Kuo Shou-ching, 3:270b; 12:378b
Kupffer, Carl, 2:124b, 361b
Kuranishi, Masatake, 3:95b
Kuratowski, Kazimierz, 3:55a; 14:591a–b
Kurchatov, Igor Vasilievich (1903–1960), 7:526–527; 15:8a, 251b
Kurile Islands, 2:14b
Kurils, 7:495b
Kurlbaum, Ferdinand (1857–1927), 7:528; 11:8b
optics, 6:470a; 8:551b–552a; 11:582b
Kurnakov, Nikolai Semyonovich (1860–1941), 7:529–530; 3:234b; 4:599a; 14:123b
metals, 13:245b
Kürschák, József (1864–1933), 7:530–531
Kursk Magnetic Anomaly, 8:102b
Kurti, Nicholas, 12:437b, 438a; 15:444b
Kurz, Karl, 1:455b
Kusawa Makoto, 1:90a
Kushner, Samuel, 4:155b
Kushyār ibn Labbān ibn Bāshahrī, Abu'l-Ḥasan, al-Jīlī (*fl. ca.* 1000), 7:531–533; 5:612a; 7:325b
arithmetic, 9:614b
Uṣūl Ḥisāb al-Hind, 9:614b, 615a
Kussmaul, Adolf, 1:543a; 14:41a
Küster, Ernst, 11:574b
Küster, William, 15:157b
Küstner, Friedrich, 3:194b; 7:238b, 369b
Kutenai Indians, 2:212a
Kutorga, Stepan Semyonovich, 7:474b; 9:286a
Kutscher, Friedrich, 8:452a; 10:242b
Kützing, Friedrich Traugott (1807–1893), 7:533–534; 2:561a; 10:365b; 12:243b
Kuvera Miśra, 12:115b
Kuyper, Gerhard, 9:152a
Kuzmin, Rodion Osievich, 5:342b
Kwakiutl Indians, 2:212a
Kyanization, 9:458b, 459a
Kyanol, 5:197b
Kylin, Johann Harald (1879–1949), 7:534–536
Kymograph, 2:366b; 5:56a; 8:541a
Kynurenic acid
metabolism of, 9:284b
Kyschtymite, 7:249b

L

Laar, (Peter) Conrad, 2:624b
Labiatae, 1:615a
Labiograph, 2:531b
La Blancherie, Pahin de, 3:61b
Labor, 15:413b–415b
division of, 15:134b, 135a, 136b, 141b, 407a–408b
evolutionary significance of, 15:140b, 141b, 143b
Labor relations
nineteenth-century, 1:471b; 13:144a–b

Laborde, Albert, **3**:508a
Laborie, Louis Guillaume, **3**:7a; **10**:326a; **12**:63b–64a
Laboulbène, Jean-Joseph-Alexandre, **3**:588b
Laboulbeniaceae, **13**:299b
Labrador (province)
botany, **1**:433b
discovery of, **10**:617b
geology, **3**:548a
Labradorite, **11**:267a
La Brea tar pits
see Rancho La Brea tar pits
La Brosse, Guy de (*ca.* 1586–1641), 7:536–541; **6**:41a
Laburnum
chemical analysis of, **3**:159a
Labyrinthodonts, **6**:594b; **7**:60a; **10**:262a
Lac virginis, **8**:311b
Lacaille, Nicolas-Louis de (1713–1762), 7:542–545; **1**:400b; **2**:344a; **8**:67b, 179b; **10**:614b; **13**:39b
L'art de vérifier les dates
editions of, **10**:615a
Coelum australe stelliferum, **9**:90b
geodesy, **3**:108a; **6**:326a; **7**:579b; **8**:22a, 177a; **9**:187a; **10**:596a; **12**:498b; **15**:287b
star catalog, **1**:403b*; **10**:591b
Lacaze-Duthiers, Félix-Joseph Henri de (1821–1901), 7:545–546; **3**:184b, 492b; **4**:12a,b*; **5**:148b, 385b; **9**:113a; **10**:522b
Laccase, **2**:86b
Laccoliths, **5**:395b–396a; **7**:62b
Lacépède, Bernard-Germain-Étienne de la Ville-sur-Illon, Comte de (1756–1825), 7:546–548; **5**:355b–356a; **13**:554a; **15**:126b
acquaintances of, **2**:322a; **10**:518a; **15**:113b, 126a
on contemporary scientists, **4**:149b; **7**:569b; **15**:114a*
correspondence of, **3**:523a; **12**:536b
Histoire naturelle des poissons, **3**:365a
influence of, **7**:592b
students of, **10**:209a
La Cerda, Vicente de, **9**:432b
La Chambre, Marin Cureau de, **2**:334a
Lacoe, Ralph Dupuy, **8**:264b
La Condamine, Charles-Marie de (1701–1774), 15:269–273; **3**:281a; **8**:170b; **10**:145a
geodesy, **2**:343a; **5**:435b; **7**:183a, 200a; **9**:186b; **15**:287b, 334a
Lacordaire, Jean Théodore, **4**:504b
La Coste, Wilhelm, **7**:283a
Lacrimation, **14**:322a
Lacroix, Alfred (1863–1948), 7:548–549; **4**:153b*; **6**:168a
acquaintances of, **4**:597b; **13**:128a
on contemporary scientists, **5**:89b*; **6**:178a*; **7**:546b*; **8**:31b*; **9**:367b*; **15**:225a, 485b
historiography, **4**:218b*; **6**:182b*; **10**:569b*; **12**:537b*
petrology, **3**:161a; **9**:367a
students of, **2**:500a; **13**:127b; **14**:440a
Lacroix, Sylvestre François (1765–1843), 7:549–551; **7**:569b; **9**:476a
acquaintances of, **9**:471b, 473a; **15**:346a
on contemporary scientists, **5**:95a, 110b, 111a; **7**:567b, 572a*
historiography, **3**:284b
students of, **3**:450b; **11**:76b
Traité des différences et des séries, **10**:327b
Traité du calcul différentiel et du calcul intégral, **6**:399a
translations of, **1**:354b; **6**:324a
translations of works, **4**:546b; **10**:437b
Lactantius Firmianus, **2**:539b; **3**:343a; **6**:305b; **7**:27b
Lactation
physiology of, **5**:225b–226a; **15**:417b
Lacteals
see Lymphatic system
Lactic acid, **4**:554b; **8**:339a
in muscle metabolism, **2**:92b; **5**:37b–38a, 184b; **6**:225b, 501a–b
optical activity, **13**:576a–b
structure, **14**:454b–455a
Lactoflavin
see Vitamin(s), B_2
Lactones, **3**:347b; **4**:400a; **5**:13a; **14**:560b
Lactose, **5**:225b; **6**:185b
synthesis of, **6**:538b
Ladenburg, Albert (1842–1911), 7:551–552; **5**:493b
benzene, **1**:390a; **3**:300a; **7**:283a
Ladenburg, Erich, **7**:553a
Ladenburg, Rudolf Walther (1882–1952), 7:552–556; **15**:448a
Ladies' Diary, **3**:537b, 538a; **5**:530b
editors of, **6**:577a; **12**:444a
Laennec, Théophile-René-Hyacinthe (1781–1826), 7:556–557; **9**:511b
pathology, **2**:444b, 464a, 508a; **4**:197b; **9**:512a; **12**:450a
La Faille, (Jean) Charles de (1597–1652), 7:557–558; **7**:583a; **8**:26a; **13**:561a
geometry, **13**:436a
Theoremata de centro gravitatis, **12**:74b
La Faye, Georges de, **1**:322b
Laffont, Marc, **9**:506b(n)
Lafleur, Henri Amédée, **3**:447a–b
Lagadha, **12**:576b
Jyotiṣavedāṅga, **15**:535a, 536a–537a, 538a,b, 539b, 542b
La Galla, Giulio Cesare, **7**:375b
Lagasca y Segura, Mariano, **3**:154b; **9**:140a
Lagenidae, **11**:386a
Laghi, Tommaso, **3**:15b; **5**:267b
(Laghu)vasiṣṭhasiddhānta, **15**:612b
Lagny, Thomas Fantet de (1660–1734), 7:558–559
Lagrange, Eugène, **14**:218a
Lagrange, Joseph Louis (1736–1813), 7:559–573; **3**:386a; **5**:55a; **8**:85a, 137a, 178b; **11**:61b; **13**:140a*
acquaintances of, **1**:77a, 548a; **4**:515b; **5**:375a
and d'Alembert, **7**:560b–567a *passim,* 570b
and Berlin Academy, **3**:119b; **8**:305b
biology, **12**:563a
celestial mechanics, **4**:480b; **7**:562a–564b, 568b, 570a; **15**:286b, 289a,b, 290a, 293b–294a
perturbation theory, **1**:401a, 428b–429a; **7**:564a–b; **15**:484a–485a
planets, **2**:442a; **15**:289b, 323b–324a, 325a–b, 327a,b
and Condorcet, **3**:384b; **7**:563b–567b *passim*
on contemporary scientists, **1**:113b; **3**:131b, 562a; **4**:107b; **5**:111a; **11**:507a; **14**:180a; **15**:225a, 384a, 489b
correspondence of, **2**:284a; **3**:388a(n); **15**:276b, 297a, 303b, 315b, 391b*
critics of, **15**:225b
editions of works, **14**:158b
and education, **15**:342b, 347a
equations of, **3**:195a; **9**:211a–b, 213b–214a; **10**:3a; **11**:408b; **13**:363b
and Euler, **7**:560a–567b *passim,* 571a
influence of, **1**:13a, 200b; **3**:377a; **4**:340b; **5**:93b, 260a,b; **6**:452a; **7**:50b, 480b; **9**:308a; **10**:186b; **11**:134b; **13**:78b, 374a, 584b; **15**:310b, 368a, 383a
influences on, **15**:383a
Lagrange variation principle, **8**:592b
Lagrange's interpolation, **4**:562a; **7**:570a
Lagrange's remainder, **3**:136a; **7**:570b
Lagrange's series, **3**:140a; **7**:566b–567a
Lagrangians, **6**:89b; **11**:13b, 568b; **12**:599b; **13**:363b, 364a
and Laplace, **7**:561b–569b *passim;* **15**:293b–294a
mathematics, **7**:494b; **8**:142a
algebra, **3**:128a, 164a; **5**:262b, 263a; **7**:564b–567a, 570a; **8**:306b; **11**:471a, 598b; **13**:129a, 571b
analysis, **1**:207a, 240a; **2**:275a; **3**:385b, 418a; **4**:479a; **5**:54b, 95b, 159b, 299a; **7**:550a,b, 560a–571a *passim;* **8**:139b, 140b; **9**:343b, 471a, 476b; **10**:327b, 574b; **15**:276b, 277b, 303a–b, 306b, 383a
geometry, **3**:166b; **4**:174a; **5**:472a, 508b; **7**:564b–565a, 598a, 599b; **8**:136b, 306a; **9**:475a
number theory, **2**:294b, 299a; **3**:168b; **4**:474b, 478b; **5**:409b; **7**:565b–566a; **8**:138b; **9**:412b; **14**:438a
probability, **9**:55a; **15**:281a, 383a
problems, **2**:198a, 271b–272a; **3**:63a; **11**:260b, 407a
Mécanique analytique, **3**:72b; **6**:399a; **8**:135a; **11**:164a; **15**:192b, 485a
editions of, **2**:88a
Mémoires de Turin, **8**:141b
and metric system, **4**:14b, 16a; **15**:334a, 335b
on observational error, **2**:42b; **15**:282b
physics, **3**:379a; **5**:95a
acoustics, **6**:245a; **7**:561b, 565a; **15**:306b, 357a
hydromechanics, **7**:565a
mechanics, **2**:101b; **4**:126b; **5**:376a; **6**:89a, 347b; **7**:54b, 55a, 560b–561a, 561b–562a, 564b–565a, 568b, 569b–570a; **9**:323b; **11**:62a; **13**:386a, 585b
students of, **11**:6b; **15**:480b
Traité de la résolution des équations numériques de tous les degrés, **2**:573b; **6**:307a
commentaries on, **11**:61b
Traité des fonctions analytiques, **3**:131b
translations of works, **3**:466b
Laguerre, Edmond Nicolas (1834–1886), 7:573–576
geometry, **2**:192a; **3**:165a; **7**:397a,b
Laguerre functions, **7**:574b
Laguerre transformations, **7**:185b*; **8**:324a
Laguerre's equations, **7**:574a–575b
Laguerre's polynomials, **3**:229b; **7**:574b, 575a–b
Laguesse, Gustave Édouard, **1**:440b; **8**:8b
Laguna, Andrés de, **1**:98b
La Hire, Gabriel-Philippe (or Philippe II) de (1677–1719), 7:576; **8**:132a; **9**:90a, 532a; **13**:337a*
La Hire, Philippe de (1640–1718), 7:576–579; **1**:191b; **7**:576a
acquaintances of, **1**:581b; **4**:267b, 505b; **8**:151b; **9**:120b; **10**:596a,b
astronomy, **14**:415a
on contemporary scientists, **9**:119a,b
correspondence of, **13**:480b
critics of, **8**:132a
crystallography, **11**:523a
engineering, **4**:49b; **8**:213a; **11**:163b
geodesy, **3**:103b; **10**:596a
geometry, **7**:109b; **9**:598b; **10**:331b
influences on, **4**:47a; **14**:594a

La Hire, Philippe de (cont.)
mechanics, **2**:202a; **8**:211a; **9**:116b
Nouvelle méthode en géométrie, pour les sections des superficies coniques et cylindriques, **8**:253a
optics, **9**:118b
students of, **10**:319a
Tabulae Astronomicae, **5**:435a; **7**:81a; **8**:132a
Laidlaw, Patrick Playfair, **12**:492a,b; **15**:105b
Lairesse, Gerard de, **15**:29a,b
Laisant, Charles-Ange, **8**:62a, 176a
Ibn Laith, Abu'l-Jūd Muḥammad
see Ibn Layth
Lake Bonneville, **5**:395b
Lake dwellings, **12**:38b; **14**:506a
Lake Superior
geology, **7**:45a; **14**:443a
Lake Survey, U.S., **12**:479a
Lakes
bottoms, **15**:158b
chemistry of, **2**:142a
classification of, **5**:599a
color of, **8**:122a
density stratification, **14**:132b
drying up of, **1**:622b
ecology of, **5**:70a–b
free oscillations in, **3**:265a; **6**:480a; **12**:342a; **15**:158b
freezing of, **9**:189b
gases dissolved in, **2**:142a
"heat budget" of, **2**:142a
light penetration, **2**:142a–b
mineralization of, **1**:622b
origin of, **1**:622b; **8**:241b; **11**:277a; **14**:139a
oxygen in, **2**:142a
productivity of, **2**:142a; **7**:184a
profundal fauna, **15**:158b
as reservoirs, **1**:622b
salinity of, **1**:622b; **6**:71b; **9**:593b
temperature of, **1**:622b
seasonal variations in, **2**:558a
stratification of, **2**:558a
thermal exchange, **2**:142a
see also Limnology
Lakṣmīdāsa
commentaries of, **2**:115b, 118a
Lakṣmīdhara, **2**:115a
Śīghrasiddhi, **15**:588b
Lakṣmīnātha Miśra
commentaries of, **2**:118a
Lalande, Joseph-Jérôme Lefrançais de (1732–1807), 7:579–582; 1:508a; **4**:16b, 24a, 108a*; **5**:55a*; **10**:615b; **11**:411b*
acquaintances of, **2**:334b; **3**:6a, 198a, 283b, 365a; **4**:23b; **8**:179b; **9**:500b, 501a
astronomical tables, **7**:542b
astronomy, **1**:241b; **6**:234b; **7**:543b; **12**:239b; **14**:415a, 479b
on contemporary science, **3**:365b*, 446a(n); **4**:25b*; **6**:487a; **7**:545a, 572b*; **8**:70b, 88a*, 145a*, 611b; **9**:153a, 329b, 554a; **10**:572b*; **14**:180a
correspondence of, **1**:436b; **9**:250a; **13**:183b; **14**:179a; **15**:325a
critics of, **15**:310b
and Delambre, **4**:14a
Encyclopédie, **3**:383b
and French Revolution, **15**:335b
Histoire Céleste Française, **1**:403b*
influence of, **2**:97b; **7**:19b
observations of, **13**:110a
star catalog, **9**:551b
Tables astronomiques de M. Halley, **14**:179a
Traité d'astronomie, **1**:314a; **4**:14a,b; **7**:19a; **15**:337b, 340a, 735a
Lalande, Michel-Jean-Jérôme Lefrançais de, **7**:579b, 580b; **8**:382b
Lalla (*fl.* eighth century), 7:582–583; 2:416a; **12**:598a; **15**:609a
commentaries of, **2**:417b
Jyotiṣaratnakośa, **12**:598b
Śiṣyadhīvṛddhidatantra, **2**:119a; **15**:596b–598b
Lalouvère, Antoine de (1600–1664), 7:583–584; 4:573a; **10**:336b; **14**:149b
Lalovera, Antonius
see Lalouvère, Antoine de
Lamarck, Jean Baptiste Pierre Antoine de Monet de (1744–1829), 7:584–594; 2:123a; **4**:67b
acquaintances of, **1**:328b; **3**:43b; **5**:356a; **6**:493b; **8**:48b, 49a; **9**:366a; **12**:130b; **13**:554a; **15**:50b
biographies of, **10**:273b, 523a
botany, **7**:584b–586b; **14**:100a
chemistry, **7**:586b–587b, 590b
conchology, **5**:477b; **7**:588b; **8**:373b
correspondence of, **12**:536b
critics of, **12**:578b*
and Cuvier, **3**:522b, 523a, 526a–527a; **7**:592b, 593a
Dictionnaire de botanique, **1**:59a; **7**:585a
evolution theory, **2**:10b; **3**:523b, 526b, 573b, 579b; **5**:357b, 359b; **6**:592a; **7**:585b–593a *passim*
critics of, **8**:27a, 569a, 574a; **11**:567b
Flore française, **15**:113b
revision of, **3**:44b
geology, **7**:589a–b
Histoire naturelle des animaux sans vertèbres
editions of, **4**:68a
Hydrogéologie, **3**:526a; **5**:356b
influence of, **2**:492b; **3**:565b; **6**:120b; **10**:210a; **11**:491b, 567b; **12**:318a; **15**:50b
meteorology, **3**:436a; **7**:587b–588b
paleontology, **2**:481a, 494b; **7**:588b–589b; **10**:322b
Philosophie zoologique, **6**:10a; **8**:565b
translations of, **8**:3b
philosophy, **5**:358a
psychology, **7**:591a–b
students of, **2**:320b; **3**:45a, 154b; **12**:317b; **14**:351a
zoology, **15**:126b
Lamarckism, **3**:429a; **4**:12b; **5**:385b; **6**:9a, 10a, 339a; **8**:125a, 586a, 616a; **9**:460b–461a; **10**:287b, 523a; **11**:20a; **12**:298a, 570b, 571a; **14**:181b, 189b–190a, 291b
forerunners of, **2**:580b; **4**:512b
opposition to, **2**:286b, 505b; **3**:395b; **9**:389a–b
see also Neo-Lamarckism
Lamarle, Anatole, **13**:414b
Lamb, Horace (1849–1934), 7:594–595; 2:397b, 398b; **3**:135b, 329a; **4**:278a; **8**:185b; **11**:393b; **13**:364b; **14**:63a*
Lambert of Auxerre, **14**:391b
Lambert, Aylmer Bourke, **11**:218b; **15**:471a
Lambert, Johann Heinrich (1728–1777), 7:595–600; 7:404b–405a, 566b
acoustics, **15**:357a
acquaintances of, **2**:220b; **7**:563a
analysis, **11**:402a
celestial mechanics, **2**:328a; **15**:324b–325a
cosmology, **2**:221a; **6**:331b, 332b; **12**:250b
editions of works, **12**:599b; **14**:158b
Euler-Lambert theorem, **7**:481b; **13**:139a
geometry, **9**:157a; **14**:141a
influences on, **1**:138b; **7**:206b; **8**:160a; **13**:267a
and Kant, **7**:597b
Lambert's equation, **1**:429a
Lambert's fractions, **7**:599b
Lambert's laws, **2**:328b, 343b; **7**:599a
Lambert's series, **7**:599b
Lambert's theorem, **7**:598a; **8**:139a
number theory, **4**:475a
Photometria, **2**:343b
pneumatic chemistry, **14**:76a
trigonometry, **7**:258a; **8**:299b
Lambert, Thomas
see Geminus, Thomas
Lambrit, Thomas
see Geminus, Thomas
Lamé, Gabriel (1795–1870), 7:601–602; 2:88b; **6**:1b; **8**:382b; **11**:206b; **13**:128a
acquaintances of, **3**:287a; **15**:31a
analysis, **9**:175a
elasticity theory, **8**:510b
Examen des différentes méthodes employées pour résoudre les problèmes de la géométrie, **5**:368b
geometry, **3**:134a–b; **5**:368a
heat theory, **10**:248b
influence of, **1**:599b; **8**:287b
influences on, **9**:476a; **12**:130a
Lamé equations, **1**:600a; **6**:308a, 309a; **9**:127a
Lamé functions, **6**:230b; **7**:492b, 601b; **12**:513a; **14**:158b, 317a
Lamellibranchia, **1**:285a; **5**:109b; **7**:460b
embryology, **8**:355a
La Mer, Victor Kuhn, **2**:499a; **3**:83b(n)
Lamétherie, Jean-Claude de (1743–1817), 7:602–604; 3:62a*; **7**:15a*; **14**:79b
Journal de physique, **1**:345b
and Lavoisier, **8**:80a–b
La Mettrie, Julien Offray de (1709–1751), 7:605–607; 2:563b; **4**:88a, 203b, 204a; **6**:469b; **8**:506a; **9**:187b
and Haller, **9**:187b–188a
L'homme machine, **4**:64b; **9**:188a
La Michodière, Jean Baptiste François de, **15**:309b
Laminariaceae, **5**:582a; **8**:29b; **12**:125b
Lamont, Johann von (1805–1879), 7:607–609; 2:430a; **6**:98b; **10**:194a*; **12**:51b, 518b
Lamothe, Léon Jean Benjamin de, **4**:39a
Lamouroux, Jean Vincent Félix (1776–1825), 7:609–610; 1:69a
La Mouthe (cave), **2**:450b; **11**:484a
Lampadius, Wilhelm August, **3**:433b; **5**:202b; **14**:257b
Lampland, Carl Otto, **8**:522b
Lamprey
embryology, **3**:610b
evolution of, **15**:513b
light organs, **7**:438b
symbiosis, **1**:622b
Lamy, Bernard (1640–1715), 7:610–611
Les élémens de géométrie
editions of, **4**:70b
Traitez de méchanique, **13**:585b
Lamy, Claude Auguste, **3**:476b
Lamy, Guillaume (*fl. ca.* 1665–1685), 7:611–613; 4:38a*
Lancelet
see *Amphioxus*
Lanceolatus, **14**:218b
Lancet, **6**:59a; **8**:406a, 407a, 408a
Lancet fish, **11**:308a
Lanchester, Frederick William (1868–1946), 7:613; 9:414b*

Lancisi, Giovanni Maria (1654–1720), 7: 613–614; 1:391b; **4**:487b; **5**:503a; **9**: 66b*, 309a; **10**:266b; **13**:563a
De motu cordis et aneurysmatibus, **1**:318b
Lancret, Michel Ange (1774–1807), 7:614–615; 9:476a
Land, Edwin Herbert, **6**:291b
Land
nationalization of, **14**:134a
reclamation of, **12**:344a, 371b; **13**:437a
valuation of, **5**:537b, 538b
Landau, Edmund (1877–1938), 7:615–616; 4:5b*; **6**:114a, 214a; **11**:362a; **12**: 247a*, 426a; **13**:391a
Bohr-Landau theorem, **2**:238b
mathematics, **6**:417b(n); **13**:562a
students of, **2**:238a; **5**:350b; **11**:256a; **13**:35b; **14**:344a
Landau, Lev Davidovich (1908–1968), 7: 616–619; 8:477b, 478b, 481a; **13**:241a
Landé, Alfred, **1**:371a; **2**:245a; **9**:397b; **10**: 348a, 422b; **12**:21a
Landeck (region), **2**:553b–554a
Landen, John (1719–1790), 7:619; 3:78a; **5**:466b*, 572b; **7**:567b
Landen transformation, **7**:619a–b; **8**: 139b, 140a
Landforms
arid, **2**:549a
classification of, **10**:504b–505a; **11**:440a
genesis of, **4**:143b; **10**:504b–505b, 507b–508b; **11**:181b, 576a; **12**:38b; **13**:257b
Landolt, Hans Heinrich (1831–1910), 7: 619–620; 4:194b; **5**:412b; **7**:65a,b*; **9**:347b, 351b; **11**:551a; **15**:435b, 460b
Physikalisch-chemische Tabellen, **14**:448a
Landriani, Marsilio (*ca.* 1751–*ca.* 1816), 7: 620–621; 8:622a(n); **14**:74b
atmidomètre, **1**:586a
Landsberg, Georg (1865–1912), 7:621; 6: 286b; **7**:531a*
optics, **11**:265b
Landsberg, Grigory Samuilovich (1890–1957), 7:621–622; 9:76a,b; **13**:240a
Landscape analysis, **3**:593a–595a, 595b; **10**:504b–505a, 507b–508b
Landshut, University of, **5**:202b
Landslides, **6**:228a
causes of, **10**:430a
experimental evidence, **2**:159a
subaqueous, **6**:229a
Landsteiner, Karl (1868–1943), 7:622–625; 9:586b*; **13**:420a
blood groups, **6**:432b; **9**:545a–b
Donath-Landsteiner test, **7**:624a
Lane, Jonathan Homer (1819–1880), 8:1–3; 4:279a
Lane, Timothy, **2**:524a
Lane Fox, Augustus Henry
see Pitt-Rivers, Augustus Henry Lane Fox
Lanfranchi of Milan, **2**:537a, 546b; **3**:218b
Lang, Arnold (1855–1914), 8:3–4; 6:10b; **7**:498b; **14**:424a
Lang, Karl Nikolaus (1670–1741), 8:4; 3: 210b; **7**:412b
critics of, **15**:53a
paleontology, **15**:54b
Lang, William Henry (1874–1960), 8:4–7; 2:371b; **3**:30b, 608a; **5**:540a*
Lange, Carl Georg (1834–1900), 8:7–8; 7:68a
James-Lange theory, **8**:7b
Lange, Friedrich Albert, **14**:527b
Lange, Ludwig, **14**:528a
Langenbeck, Bernhard von, **2**:129b; **5**:196a; **11**:215a
Langenstein, Henry of
see Henry of Hesse
Langer, Carl (chem.), **9**:467a; **14**:175a
Langer, Rudolph Ernest, **2**:144b
Langer von Edenberg, Carl, **10**:267a; **11**:255a
Langerhans, Paul (1847–1888), 8:8–9
islets of, **1**:440b–442b, 613b; **8**:8a,b; **12**: 355b–356a; **13**:589a; **14**:219a
Langerhans' cells, **8**:8b
Langerhans' layer, **8**:9a
Langevin, Paul (1872–1946), 8:9–14; 10: 276a*; **12**:218b
acquaintances of, **1**:519a; **2**:120b, 466a; **4**:253a; **10**:525a; **11**:388b, 419b; **14**: 243b; **15**:422b
historiography, **2**:359a*
Langevin function, **3**:620a
Langevin sandwich, **8**:11b
magnetism, **3**:506b; **14**:244b–245a, 246a
piezoelectricity, **3**:505a
relativity theory, **12**:69b
students of, **2**:487b; **7**:151b, 152a, 157b; **15**:48a
Langguth, Johann, **7**:284a
Langley, Edward Mann, **1**:584b
Langley, John Newport (1852–1925), 8:14–19; 5:80a, 279b
acquaintances of, **5**:80b, 82b, 83a; **7**:519b; **8**:451b, 452b; **12**:355b
neurology, **1**:452b; **5**:282b; **8**:534b
students of, **9**:19b; **12**:395b; **15**:104b
Langley, Samuel Pierpont (1834–1906), 8: 19–21; 2:423b; **7**:270b; **10**:220b; **12**: 495a*; **14**:38b–39a, 121a
aviation, **1**:490a; **10**:35a
bolometry, **10**:345b
at Smithsonian, **1**:583a; **8**:2b
Langlois, Claude (*ca.* 1700–*ca.* 1756), 8:21–22; 8:170b
Langmuir, Irving (1881–1957), 8:22–25; 6: 546a
acquaintances of, **11**:422a; **15**:450b
chemistry, **15**:36b
Child-Langmuir space-charge equation, **8**:23b–24a
electronics, **4**:7a; **11**:421b
instrumentation, **4**:253b
Langmuir adsorption isotherm, **8**:24a
Langmuir probe, **8**:24b
Lewis-Langmuir theory, **8**:292a
molecular physics, **4**:76b; **6**:118a; **7**:469b; **9**:375b; **12**:419a
Langren, Michael Florent van (*ca.* 1600–1675), 8:25–26; 7:558b*; **11**:395a
Language
artificial
international, **1**:548b, 591b; **3**:384b, 456b; **5**:607b; **6**:112b, 141a; **7**:376b; **9**:320a, 415b; **10**:443b–444a; **15**:465b
philosophical, **2**:163b, 223b; **3**:362a; **8**:153b–154a; **9**:320a; **13**:180a; **14**: 365b, 372a–374a
symbolic, **6**:299a; **7**:598a
and culture, **2**:208b, 210a; **4**:34a; **15**:635a–636a
history of, **1**:274b, 564b
and race, **2**:208b; **11**:137b–138a
teaching of, **1**:377b; **3**:360a,b, 362a; **8**: 547a–b
theories of, **1**:2b, 292a; **2**:240a; **3**:381a–b; **6**:447a; **9**:319b–320a; **10**:172b, 174b; **14**:361b–362a, 365b, 367a, 368b, 468b–469a
correspondence, **13**:180a
in translating and interpreting, **2**:619b
see also Etymology; Grammar; Linguistics; Philology; Physiology, linguistic; Semantics; specific languages; Speech; Translation; Transmission of science
Lankester, Edwin Ray (1847–1929), 8:26–27; 2:529a; **8**:417b*; **9**:335b*; **12**: 175b*
acquaintances of, **2**:366b; **5**:467a; **6**:535a, 595b; **9**:542a
on contemporary scientists, **6**:597a*; **11**:520a*
embryology, **1**:420b, 601a
students of, **1**:535b; **13**:341b; **14**:251a
Treatise of Zoology, **1**:507a
La Noë, Gaston Ovide de, **9**:103b
Lansberge, Philip van (1561–1632), 8:27–28; 1:566b; **3**:262b*; **4**:619b; **6**:520a; **11**:4a
Commentationes, **6**:520b; **14**:365a
Cyclometria nova, **12**:500b
Tabulae motuum coelestium perpetuae, **6**:514b, 515b
Lanthanides
see Rare earth elements
Lanthanum, **3**:321b; **9**:262a
atomic weight, **2**:428b
discovery of, **9**:541b
isolation of, **2**:588b
isotopes of, **7**:158b
Laon, Ralph of
see Ralph of Laon
La Pérouse, Jean-François de, **5**:224b; **7**: 603a
Lapicque, Louis (1866–1952), 8:28–30; 4: 399a; **5**:291a; **11**:431b*
Lapidaries, **2**:256b, 293a
Lapis Bononiensis, **7**:375b
Laplace, Pierre-Simon, Marquis de (1749–1827), 15:273–403; 8:137a; **15**:481b
acquaintances of, **2**:74a; **3**:131b, 258a, 440b, 545b; **5**:94b, 318a; **6**:325a, 553b; **8**:135a; **9**:72b; **10**:247b; **12**: 523a; **13**:127a
biographies of, **7**:7a; **10**:470b*–471a*
and Biot, **2**:133b, 134b, 135b, 137b; **15**: 347a–b, 361a–b
celestial mechanics, **1**:54a; **2**:442a; **3**:169b; **4**:126b; **7**:479a; **8**:135b; **9**:202b, 203a–b, 239a; **13**:374b; **15**:286a–290b, 293b–301b, 316a–333a, 336b–339b, 343a–344a, 348a–356b, 379b–382b
comets, **2**:327b, 328a; **10**:198b; **15**: 290b–292b, 309b–312a, 316b, 323b, 351a, 352a, 354b–355a
Jupiter's satellites, **15**:325a–327a, 336b–338b, 354a–b
perturbation theory, **1**:313a; **2**:359b, 600a; **4**:16b; **8**:277a, 530a; **11**:207a; **13**:153a; **15**:323b–324a, 329a, 337a–338b, 343b, 348a, 349a, 350b–351a, 353b, 356b, 484a–b
chemistry, **8**:78a, 79b–80a, 80b; **12**:64b; **14**:75b; **15**:315a–316a
on contemporary scientists, **3**:527b; **4**: 107b, 474a; **5**:95a; **7**:569a; **15**:484a, 489b
cosmogony, **4**:555b; **7**:85b, 385a; **9**:107b, 552b; **10**:42a; **11**:21b, 498a; **12**:247b; **15**:344a–345b
critics of, **7**:385b
critics of, **7**:37b; **9**:2b; **11**:599a–b
and education, **15**:341b–342b, 346b–347a
Essai philosophique sur les probabilités, **15**:274a, 279a, 281b, 285a,b, 308b, 309b, 342b, 343a–b, 363b–364a, 367b, 373b, 374a–375b
Exposition du système du monde, **1**:240a; **3**:450a; **15**:274a, 335a, 336b, 342b–345b, 348a, 350a, 351a–b, 374b, 379a,b
editions of, **15**:343a–b

Laplace, Pierre-Simon, Marquis de (cont.)
formulas of, **1**:129a; **8**:384b
and Fourier, **15**:377a, 380a, 384a–385a
geodesy, **15**:352b, 376a–b
geophysics, **15**:295a–301b, 320a–321b, 339a–341b, 351b–352a, 354a, 372b, 376b, 380a
historiography, **15**:329a–b, 343a, 344a, 368a,b, 372a, 379b
influence of, **1**:13a, 66a, 200b; **2**:70b, 483b; **3**:377a, 379a, 583a; **4**:151b; **5**:96b; **6**:241b, 452a; **7**:480b; **9**:547b; **10**:186b, 546a; **11**:236b, 344a; **13**:78b, 374a; **15**:199b, 360b–363b, 376b–379a, 481a, 486a–489b *passim*
and Lagrange, **7**:561b–569b *passim;* **15**: 293b–294a, 383a
Laplace coefficient, **10**:34b; **13**:139b
Laplace equations, **2**:105b; **3**:559b; **4**: 126b; **5**:96a; **6**:4b; **7**:601a–b; **10**:553b; **11**:449b; **12**:285a; **13**:26b; **14**:317a
Laplace expansion, **7**:435b; **15**:295a
Laplace functions, **7**:561b, 601b; **11**:59a; **15**:319b
Laplace integrals, **4**:125b; **6**:309a
Laplace series, **4**:561b
Laplace theorem, **15**:317b
Laplace transformations, **1**:500a; **2**:58a, 144b; **6**:211b; **10**:610b; **15**:382b–386b
Laplace's angles, **15**:319a
and Lavoisier, **15**:312a–316a, 356b, 358a
and Legendre, **15**:316b–317a
mathematics
algebra, **11**:471a
analysis, **3**:137b, 229a; **4**:478b; **5**:97a; **7**:37a–b, 550a, 566b, 568a,b; **11**:599a; **15**:276b–283a, 292b–296a, 298a–300b, 302b–311b, 316b–333a, 339a, 352a, 366a–b, 382b–386b
error theory, **3**:141b; **9**:218b; **15**:282a–283a, 320b–321a, 325a, 329b–331a, 339a–b, 351b–352b, 363b–365b, 372a–b, 373a,b, 376a–b, 381b
geometry, **7**:598a
probability, **1**:532a; **3**:225a, 230b, 231a, 386a–b; **7**:106b; **8**:306a–b, 530a, 562b; **9**:127b, 219a, 419b, 453b; **15**: 31b, 279a–286a, 290b–292b, 303b–309b, 325b, 344b–345b, 363b–376b, 380a–382b, 385b–386a, 489a
statistics, **1**:209a; **3**:387b(n); **15**:302b, 304a–306a, 308b–309b, 371b, 373a,b, 394b*
Mécanique céleste, Traité de, **1**:201b; **2**: 134b, 360a; **3**:18b, 131b; **4**:590b, 591a; **6**:85b, 399a; **10**:33b; **13**:423a; **14**:219b; **15**:192b, 273b, 274a, 277b, 295a, 309b, 318a, 319a, 320a, 322b, 332b, 336b, 339a,b, 343a, 345b–356b, 358b, 359a–b, 361b, 377b, 378b, 379a–381a, 484b, 486b
translations of, **2**:368b; **10**:479a; **12**: 524a
Mémoire sur la chaleur, **15**:95b, 312b–315b, 358a, 363a
and metric system, **4**:14b, 15b, 16a; **15**: 312b, 320a,b, 333b–336a
physics
acoustics, **8**:122a; **10**:76a; **15**:306b, 356b–358a, 378a
calorimetry, **8**:620b; **15**:313b–316a
capillarity, **6**:169a; **15**:312b, 358a, 359b–360a, 379a
electricity, **2**:136b, 137a; **14**:74a; **15**: 312b, 360b–361a
heat theory, **3**:82a,b; **15**:312b–316a, 358a, 360b, 363a, 377a–379a
mechanics, **4**:480b; **7**:55a; **15**:343b, 348a–b, 350a–b, 368b
molecular forces, **15**:315b, 358b–363a *passim,* 379a
optics, **9**:73b; **15**:338b, 355a–356a, 359a–b, 361b–363a
physiology, **8**:83b, 133b; **9**:238a
students of, **15**:480b
Théorie analytique des probabilités, **15**: 31a, 273b, 274a,b, 306a, 346b, 363b–364a, 365b, 367a–376b, 380a–b, 381b–382a, 385a–b
Théorie du mouvement et de la figure elliptique des planètes, **2**:216b; **15**: 316a–318a, 358a
Lapland, **1**:388a; **2**:430b
botany, **8**:375a; **10**:165a; **14**:116b, 117a
ethnography, **10**:256a; **11**:383a
exploration of, **5**:225a
geodesy, **3**:281b, 283b–284a
geology, **6**:78b
La Plata
observatory, **6**:147a, 574b
Lapparent, Albert-Auguste Cochon de (1839–1908), 8:30–31; **1**:473a; **7**: 615b*; **9**:60a*, 103a, 579a
Leçons de géographie physique, **3**:595b
Traité de géologie, **3**:595b
Lapworth, Arthur (1872–1941), 8:31–32; **8**:292a; **10**:329b, 517b
Lapworth, Charles (1842–1920), 8:32–34; **5**:337a; **10**:504b
La Ramée, Pierre de
see Ramus, Petrus
Laramie Formation, **6**:14a
Lardner, Dionysius
Cabinet Cyclopaedia, **6**:325b–326a
Treatise on Mechanics, **3**:450b
Lardy, Henry, **5**:6a
Large numbers, law of
see Probability, law of large numbers
Larghi, Bernardino (1812–1877), 8:34–35
La Rive, Arthur-Auguste de (1801–1873), 8: 35–37; **8**:38b, 307a*; **10**:193b(n); **13**:127a, 128a
electricity, **11**:604a
ozone, **12**:198a
La Rive, Charles-Gaspard de (1770–1834), 8:37–39; **1**:202b; **4**:242a; **6**:350a*; **12**:523a
La Rivière, Estienne de, **4**:413a
Larix, **9**:390a
Larmor, Joseph (1857–1942), 8:39–41; **15**: 39b
acquaintances of, **13**:362b, 365a
on contemporary scientists, **3**:158b*; **8**: 499b*; **11**:569a*, 584b*
correspondence of, **10**:106a*
electrodynamics, **8**:10b, 494a, 497a; **9**: 217a, 227a
historiography, **9**:206a, 229a(n); **10**:107a*
Larmor precession, **8**:40b
Lorentz-Fitzgerald contraction, **2**:383a; **5**:15b
optics, **7**:313b; **8**:592b; **9**:213b, 214b, 372b
radioactivity, **3**:480b(n)
students of, **15**:40a
La Roche, Estienne de (*fl. ca.* 1520), 8:41; **2**:618b; **3**:272b, 275b, 277b
La Rochefoucauld, (Duke) Louis-Alexandre de, **4**:71b, 149b, 150a; **15**:333a
La Roque, (Abbé) Jean-Paul de, **5**:259a
Larsen, Esper Signius, Jr. (1879–1961), 8: 41–43; **3**:484a; **4**:564b
Lartet, Édouard Amant Isidore Hippolyte (1801–1871), 8:43–44; **15**:51b
Lartet, Louis (1840–1899), 8:44–45; **2**:346b; **11**:235a
Larvae
dipterous, **7**:437b
respiration, **7**:273a
of frogs
digestive organs, **3**:424b
of insects, **13**:171b
Larval body theory, **7**:33b
Laryngectomy, **2**:130a
Laryngology
nineteenth-century, **3**:530b–531a
Laryngoscopy
seventeenth-century, **2**:305b
nineteenth-century, **3**:530b–531a; **13**: 493a–b
Laryngotomy, **3**:99a; **8**:411a*
Larynx
anatomy, **3**:99a–b; **4**:488a, 520a; **6**:269b; **12**:384a
cartilages in, **1**:619b
innervation of, **2**:30a; **12**:80b, 81b
mechanics of, **14**:404a
La Sagra, Ramon de
Histoire naturelle de Cuba et des Antilles, **10**:221b
Lascaux (cave), **2**:450b
Laser, **3**:369b; **4**:317b
Lashley, Karl Spencer (1890–1958), 8:45; **7**:98a
Lasker, Emanuel, **7**:444b; **10**:140b
Lasos of Hermione, **11**:222b
Lassaigne, Jean-Louis, **13**:403b
Lassar-Cohn
see Cohn, Lassar
Lassell, William (1799–1880), 8:46; **9**:615b
Lassone, Joseph Marie François de, **3**:5a, 419a; **8**:352a,b, 353a
Lāṭadeva (*fl. ca.* A.D. 505), 8:46; **10**:425b, 426a; **12**:115b; **15**:543a, 545a
Sūryasiddhānta, **13**:582a; **15**:550b, 569b, 574b, 596a, 603a–605b *passim,* 608b
Latchford, James Kyran, **1**:442b
Laterite
origin of, **4**:141a
Latex, **2**:86b; **14**:349b
Lathe
elliptic, **8**:207b, 212b
Latimer, Wendell Mitchell (1893–1955), 8: 46–48
Latin (language), **13**:588b
teaching of
seventeenth-century, **3**:360a, 362a
Latin America
flora, **3**:154b
geography, **1**:48a
see also South America; specific countries
Latin square, **8**:617a
Latitude, determination of
ancient Greek, **11**:225a; **15**:220a
medieval, **2**:148a, 150a,b, 153a, 435a; **6**: 302a; **7**:353b–354a; **13**:536a; **14**:575a
Renaissance, **10**:150a, 161a, 513b; **14**: 275a, 359a; **15**:474b
seventeenth-century, **5**:593b; **7**:20a; **8**:65a; **11**:423a
eighteenth-century, **3**:198a, 284a; **5**:435a–b; **6**:234a, 513b; **7**:81a; **8**:350b–351a; **9**:250b–251b; **11**:609b
nineteenth-century, **1**:402b–403a; **2**:98b; **3**:14a; **6**:240b; **8**:487b, 504a; **11**:379a; **13**:109a, 120b
twentieth-century, **1**:103a; **3**:237a; **4**: 270a; **7**:266a; **10**:232a; **11**:552b; **13**: 12a, 45b
degree, length of
medieval Arabic measurements, **1**:38a; **2**:149a–b, 153a; **14**:538a
Renaissance measurements, **4**:584b; **9**:191b

early seventeenth-century measurements, **5**:542b–543a, 593b; **10**:151b; **12**:500a–b
late seventeenth-century measurements, **3**:103a–108a *passim;* **7**:577b; **9**:554a; **10**:151b, 595b–596a
early eighteenth-century measurements, **2**:343a; **3**:173b, 281b; **4**:15a; **7**:183a, 542b, 576a; **8**:22a, 178b; **9**: 34a, 186b–187a; **10**:256a; **15**:270a–271b
late eighteenth-century measurements, **2**:299b, 329b; **4**:132a; **5**:195a; **6**:550b; **9**:164b, 231b, 370b; **10**:614b; **13**: 184a; **15**:334a–335b
early nineteenth-century measurements, **5**:303a,b; **6**:103a; **7**:19a; **12**: 234a–b, 498b, 518b; **13**:111a
late nineteenth-century measurements, **5**:405b; **6**:239a
early twentieth-century measurements, **1**:103a; **7**:472a; **10**:523b–524a
at Greenwich, **2**:389a; **4**:270a; **5**:24b
laboratory method, **3**:367a
mean, **10**:232a
photographic, **12**:249b
at sea, **2**:140b; **5**:399a
variation of latitude, **5**:224a*; **6**:240b; **7**:472a; **8**:516b; **9**:174b; **10**:35a, 153b, 159a, 232a; **11**:552b; **13**:45a,b
annual term, **7**:369b
discovery of, **3**:194a–b
Latitude of forms, **2**:393b; **13**:186b–187a, 189b, 190a, 194b
Latitudinarianism, **14**:361a, 370b, 371b
Latreille, Pierre-André (1762–1833), 8:48–49; **1**:328b, 329a; **2**:322a; **6**:454a
zoology, **2**:493b; **3**:525a; **8**:265b; **15**:126b
Lattices, crystallographic, **2**:441a; **5**:557b; **6**:180b, 614b; **8**:485b; **9**:377a
classification of, **10**:125a
definition of, **5**:557a
dynamics of, **11**:266b
ionic theory of, **4**:141b; **15**:251b, 461a
law of Bravais, **2**:431a–b; **5**:124a, 186a–b, 213b*; **9**:59a; **12**:511b, 512a
law of mean indices, **5**:186a
motion within, **5**:161a; **15**:40b–41a, 41b
nomenclature, **6**:303b
vacancies in, **5**:161a
Lattices, mathematical theory of, **2**:298a–b; **9**:412b–414a; **10**:485a; **11**:269a; **12**: 451b
Latyshev, Georgii Dmitrievich, **15**:8a
Lau, Hans Emil (1879–1918), 8:50; **6**:351a
Laub, Johann Jakob, **8**:181b; **15**:41a
Lauchen, George Joachim von
see Rheticus, George Joachim
Laud, William, **5**:343b; **10**:151b
Laudanum
liquid, **13**:215a
Laue, Max von (1879–1960), 8:50–53; **5**: 455a; **12**:218b; **13**:226b
acquaintances of, **4**:312b, 357a; **6**:16a; **7**:468b; **12**:71a; **13**:41a; **15**:37a,b*, 41a, 448a
on contemporary scientists, **5**:332b*, 333a*; **11**:584b*; **12**:615b; **14**:342b*, 526b*; **15**:451a
crystallography, **2**:399b; **15**:61b–62b
X-ray, **2**:431b, 487b; **4**:196a; **5**:457a; **6**:303a; **9**:542b, 543a; **11**:531a; **12**:31b
electrodynamics, **11**:421b; **12**:69b–70a
historiography, **4**:292b*; **6**:349b*
Laue diagrams, **2**:399b; **8**:51a
Laue-Friedrich-Knipping phenomenon, **2**:399a
Laughing gas, **3**:334a
Laugier, Jean, **9**:92b
Laugier, Paul-August Ernest, **1**:202b; **8**: 64b
Laugier, Pierre, **9**:17b
Lauraceae, **10**:13b
Laurasia (supercontinent), **4**:263a
Laurel, oil of
rotation of light, **2**:139a
Laurencet, **3**:526b; **5**:357a; **8**:49a
Laurens, André du (Laurentius) (1558–1609), 8:53–54
Historia anatomica, **1**:523b
Laurent, Auguste (or Augustin) (1807–1853), 8:54–61; **1**:416a; **3**:10b
acquaintances of, **1**:518a; **3**:133b; **4**:242b; **10**:357a; **14**:394a
analysis (math.), **3**:139a; **14**:223b
and Avogadro's hypothesis, **1**:347a; **3**: 47b, 307a
and Berzelius, **8**:55a, 57b
and Chancel, **3**:193b
chemical theory, **1**:518b; **2**:96a; **4**:246a,b, 247a, 531b; **5**:372a, 374a, 431a; **6**: 462a; **14**:488b, 530a–b
critics of, **2**:96b; **4**:243a
and Gerhardt, **5**:370a, 371b, 372b; **8**: 55a–b, 61a
influence of, **2**:620b; **6**:463b, 564b; **10**: 177a,b; **14**:395a,b
Méthode de chimie
translations of, **10**:177a
organic chemistry, **1**:390a; **3**:193b–194a; **4**:395a; **5**:373a; **8**:55a; **10**:359a
Laurent, Matthieu Paul Hermann (1841–1908), 8:61–62
Laurent, Pierre Alphonse (1813–1854), 8: 62–64
Laurent series, **3**:138b; **8**:62a
theorem of, **3**:140b, 141b; **8**:63a
Laurentian disturbance, **8**:99a–b
Laurentian peneplain, **8**:99b
Laurentian System, **1**:51a; **6**:565a
Laurentianus
see Lorenzi, Lorenzo
Lauric acid
discovery of, **5**:475a
Laurillard, Charles Léopold, **3**:524a, 525a
Lauritsen, Charles Christian, **9**:397a
Lausanne
Musée Géologique Cantonale, **11**:375a
Lausanne, University of
geology at, **8**:543a, 544a
Laussedat, Aimé (1819–1907), 8:64–65
Lautemann, Eduard, **1**:578a; **2**:68a
Lautenschläger, Carl Ludwig, **13**:1a
Lautite, **5**:162a
Lava, **3**:530a; **7**:370b–371a; **13**:273a
analysis of, **12**:562a
floods, **8**:123a
flow rate, **12**:562a
formation of, **12**:263b
and granite, **3**:567b
Hawaiian Islands, **3**:484a
heat of, **4**:152a; **7**:63a
intrusive nature of, **8**:616b
mineralogical nature of, **3**:412a, 567b; **4**:152a
phonolite, **11**:269b
pillow, **1**:394b; **5**:39b*
subaerial, **11**:303b
trap, **8**:594b
see also Basalt
Lavanchy, Charles, **5**:598a
Lavanha, João Baptista (*ca.* 1550–1624), 8:65
Lavater, Johann Kaspar, **5**:252b; **8**:322a
Laveran, Charles Louis Alphonse (1845–1922), 8:65–66
malaria, **5**:461a; **9**:328b; **11**:555b, 556a
Laves, Fritz H., **5**:458a; **6**:561a
Lavoisier, Antoine-Laurent (1743–1794), 8: 66–91; **3**:7a; **4**:343b, 453a; **5**:91b, 138a; **7**:78a, 569a; **9**:327a
and Académie des sciences, **15**:333b–334a, 336a
acquaintances of, **2**:74a, 88b; **3**:5a,b, 327a*; **5**:78a; **6**:164b, 178b, 284b; **8**:612b; **9**:152a, 343a; **12**:64a, 147b, 286b; **13**:571a; **14**:14b
analytic chemistry, **11**:168a; **12**:64a; **13**: 311b
biographies of, **9**:379a*; **10**:372b
biology, **12**:555b
and Bucquet, **2**:572b; **8**:76a, 77b
chemical theory, **2**:7a, 74b–77a, 92a,b, 177a, 180a–b; **3**:201b; **5**:90a,b, 532b, 602b; **7**:388b; **8**:117a, 321a, 520a; **9**:341a
acids, **3**:200b, 602b; **4**:531b; **5**:323b, 372a, 411b; **8**:76a–77b, 82b; **10**:184b; **13**:312a
alum, **9**:106a
calcination, **8**:74a–75b, 76b, 77b; **11**: 389b
causticity, **3**:201a; **8**:76b; **9**:347a
elements, **1**:143a; **3**:544a, 602a; **8**:72a–73b, 80b, 81b–82b; **15**:12a
matter, states of, **8**:73a, 79b, 80a, 256a
water, **2**:186a; **8**:71b–73b, 78a–79b; **9**: 343a,b, 477a–b; **12**:64b; **14**:197a
combustion theory, **2**:75a–b; **5**:216a, 326b, 601a; **6**:395a, 485a; **8**:79b–80b, 373b, 530b, 531a, 620a,b; **9**:6a, 152a
critics of, **6**:585b; **9**:148a
critics of, **1**:142b; **3**:419b, 465b, 600a; **4**:29a; **6**:395a; **7**:603a; **11**:143b, 146a–b, 437b, 523a; **14**:75b
electricity, **14**:74a
geology, **5**:578b; **8**:68a–69a, 70b, 71a; **9**:478b; **12**:67b(n)
and Guettard, **5**:579a; **8**:68a–b, 69a
heat theory, **1**:616b; **2**:70b; **3**:599a; **5**: 137b; **8**:79b–80a, 81b, 82b, 620b; **14**:213b; **15**:95b, 312a–316a
influence of, **1**:563b; **2**:93b, 483b; **3**:158a, 198a, 540b; **5**:370b, 411a; **6**:35b, 54a, 284b; **9**:140a, 527a; **10**:552b, 603a; **11**:166b; **12**:155a, 329a; **13**:313b, 469a; **14**:600b
influences on, **11**:563a; **13**:494b
laboratory notebooks of, **2**:64b; **8**:74b
and Laplace, **15**:312a–316a, 356b, 358a
Mémoire sur la chaleur, **15**:95b, 312b–315b, 358a, 363a
method of, **8**:75b, 77a, 81a–82a
Méthode de nomenclature chimique
see *Méthode . . .*
and Monge, **8**:80b, 81a; **9**:471b, 472a, 477a
nomenclature, **1**:64b; **2**:292a; **5**:411b; **6**: 182a–b; **8**:77b, 80b, 81a, 82a
Opuscules
translations of, **6**:283b
organic chemistry, **5**:324b; **7**:393a; **8**:82b–83a
physiological chemistry, **2**:26a, 29a, 30b; **8**:75b, 82b–84a, 133b, 348b; **9**:238a, 245b, 246a; **12**:563a; **13**:393b
pneumatic chemistry, **2**:73b; **14**:75a,b
and Scheele, **12**:146b–147b
students of, **11**:464a
technology, **13**:496a
Traité élémentaire de chimie, **2**:572b; **3**:

Lavoisier, Antoine-Laurent (cont.)
7b, 12a, 465b, 382b, 598b; **8**:81a–83a; **14**:475a, 562a
translations of, **7**:395a; **15**:206a
translations of works, **14**:225a
Lavoisier, Mme. Marie Anne Pierette Paulze, **7**:388b; **8**:71a, 80b, 84a, 87b*; **12**:286b; **13**:350b
Lavrentiev, Boris Innokentievich (1892–1944), 8:92
Law
natural, **8**:437a, 439a; **12**:548b
natural vs. positive, **6**:406b, 447a,b
Lawes, John Bennet (1814–1900), 8:92–93; **8**:347a
Lawn tennis
mathematics of, **2**:491a
Lawrence, Ernest Orlando (1901–1958), 8: 93–96; **3**:329b, 371a; **6**:104a; **8**:293a; **10**:216a; **12**:33b
Lawrence, George Newbold, **1**:405a
Lawrence, Jason Valentine O'Brien, **6**:119b
Lawrence, John, **8**:96a
Lawrence, Thomas Edward, **14**:504a
Lawrence, William (1783–1867), 8:96–98
Lectures on Physiology, Zoology and the Natural History of Man, **11**:137b
Lawrence Scientific School, **6**:517a; **8**:123a; **10**:599a
Lawrencium, **8**:96b
Laws, scientific
nature of, **3**:32a–34b, 377a,b
see also Nature, laws of
Lawson, Alexander, **5**:478b; **14**:417b
Lawson, Andrew Cowper (1861–1952), 8:98–100; **8**:41b
Lax, Gaspar (1487–1560), 8:100; **3**:171b; **9**:32b, 264a; **14**:47a
Laxatives, **14**:637b
Laycock, Thomas, **3**:88a–b; **4**:593a; **7**:47b, 48b
Ibn Layth, Abu'l-Jūd Muḥammad, **7**:257b, 329a, 334b; **11**:240b; **12**:432a
Layth ibn Kahlān, **9**:36a
Lazarev, Mikhail Petrovich (1788–1851), 8: 101; **1**:594b–595a
Lazarev, Petr Petrovich (1878–1942), 8: 101–103; **8**:107b; **13**:548b, 598a,b; **15**:165a*
Lazarists, **2**:82b
Lazarus, Moritz, **6**:53a
Lazear, Jesse William, **4**:620a; **11**:346b
Lazzaroni, **1**:364b; **5**:394a*, 479b; **14**:448b
Lea, Arthur Sheridan, **5**:80b, 82b
Lea, Einar, **6**:442a
Lea, Isaac (1792–1886), 8:103–104
Lead, **15**:651b
alloys, **3**:560b; **8**:321a; **9**:106b; **11**:534b
atomic weight, **6**:481a; **13**:499b
calcination of, **8**:311b–312a; **11**:389a–b
compounds
oxides, **11**:168a, 169b, 170b
salts, **3**:202a; **4**:134a
solubility of, **10**:288b
detection of, **5**:73a; **6**:18a
in gasoline, **9**:375b–376a
isotopes, **6**:365a–366a; **10**:288b; **12**:505b
metallurgy of, **3**:473a; **15**:651b
mining of, **3**:358a; **5**:415a; **7**:78a; **12**: 204a,b; **13**:68a; **15**:152b
ores, **3**:190a, 232b, 346b; **12**:63b–64a
deposits of, **10**:258a; **11**:608a; **14**:315b
in plants, **6**:366a
poisoning by, **3**:203b; **4**:301b; **5**:73a, 137b, 219a; **7**:421b; **11**:373b; **15**:107b
treatment of, **9**:296a
in radioactive decay
and minerals, dating of, **2**:259b–260a; **8**:42a–b; **12**:20a, 29a
radioactivity of, **4**:356a
and atomic weight, **11**:417b
in rocks, **6**:366b
specific gravity, **7**:346a
specific heat, **14**:354b
valence of, **2**:622b
Leaders in Education, **3**:130b
Leaf
see Leaves
League of Nations, **12**:274b; **15**:431b
Health Organization, **15**:107a
International Commission on Intellectual Cooperation, **2**:8b, 120b; **3**:502a; **4**: 313a; **8**:489b; **10**:275a
Leakey, Jonathan, **8**:105a
Leakey, Louis Seymour Bazett (1903–1972), 8:104–105; **11**:336a,b
Leakey, Mary Douglas Nicol, **8**:105a
Leakey, Richard, **8**:105a
Learning
neurology of, **5**:175a–b
psychology of, **9**:562a–b
theories of, **8**:45b
Least action, principle of, **3**:226b, 562a; **4**: 471b, 480b, 482a; **6**:88b, 243a, 245a, 343a; **7**:26b, 442b, 561a,b, 562a, 565a; **8**:39b, 384b, 602b; **9**:174b, 187b, 189a, 232b; **12**:250a
in optics, **6**:87b, 88a; **9**:73b; **15**:362b
Least constraint, law of, **5**:305a; **8**:602b
Least squares, method of, **3**:141b, 142a, 229b, 432b; **4**:1b–2a; **5**:299a, 304a, 305a; **7**:7b*, 107a; **8**:187b; **9**:127b, 129b, 219a, 410b; **10**:452b; **12**:290a; **15**:31a, 31b–32a, 365b–367a, 372a, 376a, 380b
in astronomy, **1**:19a–b, 429b; **2**:330a; **3**: 374b; **5**:300a,b, 309a, 414a; **8**:137b; **15**:363b
in surveying, **2**:102a; **5**:303b; **6**:240a
Least time, principle of
see Fermat, Pierre de, Fermat's principle
Least work, principle of, **3**:118a–b; **9**:268a
Leather
technology of, **4**:165b
see also Tanning
Léau, Léopold, **3**:456b
Leaves
arrangement of
see Phyllotaxis
cuticle, **5**:277b
etiolated, **2**:286b; **10**:281b; **12**:59a
falling of, **13**:596a
guard cells, **5**:624a
hairs, **5**:624a
heterophylly, **12**:167b
hydathodes, **5**:624a
as model for plant, **2**:160a; **5**:443b; **13**: 506b
morphology, **5**:604b–605a
movement of, **2**:286b; **10**:577a
periodic, **10**:576a
nutrition of, **2**:286b
morphological effects, **9**:461a
oil glands, **5**:624a
osmotic properties, **5**:277b
permeability of cells, **4**:131b
photosynthetic efficiency, **5**:523b
senescent, **3**:359a
sensitivity, **3**:422a
stomata, **2**:184a; **3**:581b–582a; **4**:264a; **5**:277b, 535b; **6**:219b; **9**:78b
structure of, **12**:99b, 256b
vascular bundles, **8**:354a
water absorption, **5**:277b
see also Photosynthesis, Transpiration
Leavitt, Henrietta Swan (1868–1921), 8: 105–106; **6**:352a, 529b; **12**:346b
Lebailly, Charles, **15**:454b–455a
Lebanon
stratigraphy, **2**:341b
Lebeau, Paul [Marie Alfred], **9**:451a
Lebedev, Petr Nikolaevich (1866–1912), 8: 106–108; **1**:283b, 284a; **4**:305b; **8**: 102a; **13**:598a
physics, **1**:301a; **6**:163b; **9**:212b; **11**:122b
Lebedev, Sergei Vasilievich (1874–1934), 8: 108; **2**:624b; **4**:553a, 554a
Lebedev Physics Society, Moscow
see Moskovskoe fizicheskoe obshchestvo imeni I.I. Lebedeva
Lebedinsky, Vyacheslav Vasilievich (1888–1956), 8:108–109; **3**:271a
Le Bel, Joseph Achille (1847–1930), 8:109–110; **7**:452b; **11**:85b; **14**:530a
stereochemistry, **2**:485b, 622b; **10**:359a; **13**:575b, 576b; **14**:265a, 455a
Lebenskraft
see Vital force
Lebensphilosophie, **3**:112a
Lebensraum, **10**:502b–503b; **11**:309a–b
Lebert, Hermann, **11**:132b, 133a, 491a; **14**:228a
Lebesgue, Henri Léon (1875–1941), 8:110–112; **3**:63a; **6**:451b; **7**:71b, 72b*; **8**: 558b; **10**:595a; **14**:243b
analysis, **2**:304a; **11**:59a, 260a, 261a; **14**:572b–573a
on contemporary scientists, **1**:408a; **7**: 169a*
historiography, **11**:290a*; **13**:571a–b
influences on, **4**:206b, 561b; **5**:97b; **8**: 386b(n); **14**:86a
Lebesgue integral, **2**:239a; **4**:547b–548a; **5**:201b; **8**:110a–112a, 558b; **11**:156b, 459b; **13**:424b, 562a,b; **14**:573a
Lebesgue, Victor Amédée, **3**:251a
Le Bey, Raymond, **8**:354a
Le Blanc (pseudonym)
see Germain, Sophie
Le Blanc, Max Julius Louis (1865–1943), 8:112–113; **4**:242b; **13**:228a; **15**:433b, 448a, 456a, 468b*
Leblanc, Nicolas (1742–1806), 8:113–114
Leblanc process, **8**:113b–114a; **10**:497a
Le Blond, Jean Baptiste Alexandre, **1**:243b
Le Bouyer de Fontenelle, Bernard
see Fontenelle, Bernard Le Bouyer de
Le Breton, André François, **4**:86a; **5**:567b
Lebrun, Charles-François, **12**:66b
Lecanoric acid
synthesis of, **5**:6a
Lecanorin, **12**:236a
Le Cat, Claude-Nicolas (1700–1768), 8:114–116; **3**:16a*; **8**:506b; **10**:614b
Le Cazre, Pierre, **5**:287b
Le Châtelier, Henry Louis (1850–1936), 8: 116–120; **5**:389b
acquaintances of, **3**:498a; **5**:583b; **6**:319a; **13**:243b, 617a
chemistry, **5**:621b; **9**:59a–b; **13**:245b; **15**:463a, 464a
influence of, **3**:212a, 240a*; **7**:529a; **11**: 100b
Le Châtelier principle, **4**:228b; **8**:118a–b; **12**:543a; **13**:578b; **15**:438b, 439a, 441a
pyrometry, **1**:490b; **3**:234a; **10**:246a; **11**: 486a
Le Châtelier, Louis, **8**:116a
Lecher, Ernst, **4**:306a; **11**:9a
Lecithin
composition of, **6**:505b; **8**:276a
discovery of, **5**:432a–b
Leclanché, Georges
battery, **2**:439a
Leclerc, Daniel, **4**:135b; **5**:157a

Leclerc, Georges-Louis
  see Buffon, Georges-Louis Leclerc, Comte de
Leclerc, Sébastien, **10**:264b
L'Écluse (Clusius), Charles de (1526–1609), 8:120–121; **1**:596b; **10**:238a; **13**:451a
  acquaintances of, **4**:139a; **10**:489a; **11**: 311b, 312a, 527b
  correspondence of, **13**:463b*
  illustrations, **13**:463a
LeConte, John (1818–1891), 8:121–122; 8: 103b; **12**:257b
LeConte, John Eatton, **8**:121b
LeConte, John Lawrence, **6**:507a; **8**:121b; **12**:133b*
LeConte, Joseph (1823–1901), 8:122–123; 8:122b*, 463b; **9**:314a
Lecoq, Karl Ludwig Edler von, **5**:303a
Lecoq de Boisbaudran
  see Boisbaudran, Paul Émile Lecoq de
Lecornu, Léon François Alfred (1854–1940), 8:124; 13:286a*
Le Dantec, Félix (1869–1917), 8:124–125
Ledermüller, Martin, **5**:424b
Le Double, Anatole Félix (1848–1913), 8: 125–126; 11:254a
Leduc, Sylvestre-Anatole, **6**:51b; **11**:460b
  Leduc effect, **11**:446b
Lee, Alice, **10**:453a
Lee, John, **2**:147a; **5**:413b
Lee, Tsung Dao, **4**:582b; **9**:38b; **10**:425a
Leech, **9**:401b, 560a
  anatomy, **5**:504b; **14**:314b
  eggs, **2**:4b
  embryogeny, **14**:314b
  trypanosomes in, **2**:533b
Leech book, **6**:123b
Leeuwenhoek, Antoni van (1632–1723), 8: 126–130; 10:202a
  acquaintances of, **5**:484b; **8**:149b, 157b; **13**:169a
  biographies of, **4**:133a
  blood
    circulation of, **6**:63b
    corpuscles, **6**:64a
    hemolysis, **2**:226a
  correspondence of, **6**:68b; **12**:208a; **15**:29b
  critics of, **6**:148b–149a
  embryology, **4**:291a; **11**:333a
  histology, **6**:183b
  influence of, **5**:328b; **6**:603b
  microbiology, **3**:332b; **4**:235a; **7**:110b; **11**:37a,b; **12**:555a, 557b; **13**:457b
  microscopes of, **1**:412a; **6**:137b; **9**:595a
  physiology, **10**:365a; **13**:46b
    reproduction, **7**:111a; **12**:557a, 559b
  on seeds, **1**:561a
Leeward Islands
  discovery of, **15**:89a
Le Febure
  see Le Febvre
Le Febvre, Nicaise (*ca.* 1610–1669), 8:130–131; 3:596b; **4**:494b; **5**:417b, 418a, 420a; **8**:173a
Le Febvre, Robert, **6**:273a
Le Fèvre (or Le Fèbvre), Jean (*ca.* 1652–1706), 8:131–132; 7:576a
Le Fèvre, Nicolas
  see Le Febvre, Nicaise
Leffler, Fritz, **7**:479a
Lefschetz, Solomon, **6**:496a,b; **7**:312a; **10**: 140a; **11**:60a; **14**:310b
Left- and right-handedness, **1**:259a; **2**:255b; **6**:425b
Leg
  anatomy, **4**:520a
Legallois, Julien Jean César (1770–1814), 8:132–135; 5:149b; **6**:59b; **8**:506a; **10**:582b
Legendre, Adrien-Marie (1752–1833), 8: 135–143; 2:134b; **3**:134b, 137b; **4**: 137b; **5**:111a
  and Abel, **1**:15b, 16b–17a; **8**:138a, 141a
  acquaintances of, **7**:568b; **10**:247b; **13**: 184a
  algebra, **1**:239b; **2**:573b–574a; **8**:139a, 306b
  analysis, **2**:271b; **3**:432b; **4**:516a; **5**:572b; **7**:52a,b, 478b, 550a; **13**:238a
  celestial mechanics, **1**:429a; **8**:135a–b, 137b; **15**:316b–317a, 319a
  correspondence, **1**:237b; **7**:50b–53a
  Éléments de géometrie, **5**:260a; **14**:85b
  Essai sur la théorie des nombres, **15**:389a*
  and Gauss, **8**:137b, 138a; **15**:365b
  geodesy, **3**:106a; **4**:15a; **9**:250b; **12**:518b
  geometry, **7**:550b, 599b; **2**:274a; **8**:136b, 141a–142a, 429b
  influence of, **4**:124b; **10**:186b; **12**:325a; **13**:374a; **15**:485b
  and Laplace, **15**:316b–317a
  "Legendre conditions," **8**:135b
  Legendre functions, **6**:309a; **14**:317a
  Legendre polynomials, **6**:230b, 545b; **8**: 61b, 135b; **9**:127a; **15**:317a
  Legendre transformations, **7**:530b; **8**: 136a, 140b–141a
  Legendrian coefficients, **1**:54b
  number theory, **3**:227b; **4**:124a, 125a, 258a, 474b, 478a; **7**:494b, 565b
  statistics, **15**:363b, 365b, 367a, 372a
  theorems of, **4**:9a; **6**:240a; **8**:135b–136a, 137b
  Théorie des nombres, **5**:375a
  Traité des fonctions elliptiques et des intégrales eulériennes, **6**:399a; **7**:53a; **14**:219b
  trigonometry, **8**:135b–136a
  works translated, **3**:466b; **4**:546b
Legendre, René, **4**:399a; **8**:29b
Le Gentil de La Galaisière, Guillaume-Joseph-Hyacinthe-Jean-Baptiste (1725–1792), 8:143–145; 2:216b
Léger, Urbain-Louis-Eugène (1866–1948), 8:145–146; 4:207a; **12**:424a
Legislation
  antipollution, **3**:69a
  geological, **14**:439a
  medical, **8**:552b
  mining, **2**:292a; **3**:624a; **4**:346a; **5**:93a; **7**:254b; **14**:442a
  penal, **5**:74a
  psychiatric, **14**:115a–b
  public health, **8**:270b, 313a; **12**:48b
  scientific, **2**:303b
  social, **9**:67b
  vivisectionist, **3**:69a–b
  see also Patents, law movement
Legros, Georges Victor, **4**:504a
Le Gros Clark, Wilfred Edward, **3**:607a
Legumes, **3**:44b; **9**:253b
  germination, **3**:44a
  hybridization, **13**:478a–b
  nitrogen fixation, **2**:357a; **6**:237b; **8**:93a; **15**:14a
  storage of, **14**:45b–46a
  taxonomy, **2**:519b; **9**:434b
Legumin, **2**:386a
Lehfeldt, Robert Alfred, **2**:184a
Lehmann, Alfred Georg Ludvig, **2**:1b
Lehmann, Carl Gustav, **7**:519a
Lehmann, Johann Gottlob (1719–1767), 8: 146–148; 6:259b; **9**:106b*; **14**:259b
  stratigraphy, **9**:534a
  works translated, **6**:468b
Lehmann, Otto (1855–1922), 8:148–149; 5:186b; **9**:182b
Lehrbuch der organischen Chemie, **4**:400a
Lehrer, Erwin, **12**:497a
Leibenson, Leonid S.
  see Leybenzon, Leonid S.
Leibniz, Gottfried Wilhelm (1646–1716), 8: 149–168; 2:326a; **3**:110b; **6**:253a; **7**: 598b; **9**:303a; **11**:257b
  acquaintances of, **4**:506a,b; **5**:448b; **6**: 253b, 304b, 305a, 459b, 537a, 538a; **7**:559a; **9**:49b; **10**:315a; **13**:33a; **14**: 49b(n)
  and Bernoulli (Johann I), **2**:52a–54a *passim*
  on Bisterfeld, **2**:163b
  and Bruno, **2**:543b
  and Clarke, **3**:295b–296a, 450a; **7**:225b; **8**:151b, 152b, 158b, 159a; **14**:83b
  correspondence of, **1**:292a; **2**:57a; **3**:119b, 152a, 348b, 460a, 534b; **4**:84a, 505b, 517a; **5**:259b*, 449b; **6**:599a,b; **7**: 376b; **8**:160a, 164a, 165b, 305a; **9**: 120b, 187b, 447a; **10**:293a, 331a, 335b; **11**:400a; **13**:585a, 586a; **14**: 151a, 483a; **15**:52b
  cosmogony, **7**:232a
  critics of, **2**:358b, 577b; **3**:485a–b; **6**:148b; **10**:120b, 500b–501a; **11**:494a; **12**: 430a; **14**:84b
  and Descartes, **4**:52b, 64b; **8**:151a, 152b–153b
  Dissertatio de arte combinatoria, **8**:550a; **9**:554b
  dynamics, **2**:49a, 330a; **3**:454a; **4**:471b; **8**:151a–153a, 156b–159b *passim*
    critics of, **6**:608a
  editions of works, **7**:561a; **11**:303a
  Essay de dynamique, **9**:51b
  geology, **7**:412a; **9**:532b
  Historia inventionis phosphori, **7**:525a
  and Hooke, **5**:607a
  influence of, **2**:274a; **3**:56b; **4**:203b; **7**: 505b; **8**:514a; **10**:443b; **11**:493a–b; **12**:607a; **14**:483a,b, 484a; **15**:368a, 489b
  influences on, **1**:126a; **3**:284a, 362b; **6**: 459b; **7**:375a; **8**:547a; **10**:337a; **12**: 136a, 460b; **13**:210b
  Leibniz series, **8**:163a
  and Locke, **8**:151b, 153a–b
  logic, **2**:295a, 379a; **3**:452a, 456a; **7**:195b*; **8**:154a–155b, 159b–160a; **12**:55b, 283b; **13**:612a; **14**:372b, 373b
  and Malebranche, **9**:49b–52a *passim*
  mathematics, **5**:61a, 529a
    algebra, **2**:281b; **3**:461a; **7**:55a, 508a; **8**:161b–162a, 163a
    analysis, **2**:47a–48b *passim,* 274b; **3**: 76a; **5**:60b; **6**:600b; **7**:490b, 560a; **8**: 161b–166b, 304a, 384a; **9**:50a, 51a; **11**:402a, 512b; **13**:69a, 266b
    calculating machine, **8**:149b, 161a, 162a, 166a; **11**:66a*
    calculus, **1**:590b, 164b, 229b; **2**:43b, 47a, 53a, 54a; **3**:117; **4**:546b, 572b; **6**:600b; **8**:153b–155a, 161a–166b, 304b; **9**:50a–b; **10**:120a,b
    combinatorial analysis, **6**:403b; **8**:153b–154a, 160b–161a
    geometry, **3**:616a(n); **11**:402a; **12**:344b; **13**:571b; **14**:49b(n); **15**:193a, 195b–196a
    notation, **1**:354b; **3**:458b; **5**:54b; **8**:161b, 163b; **10**:437b; **14**:180b; **15**:99a
    number theory, **8**:161a, 162a, 165b–166a; **11**:256b; **14**:438a
    series, **2**:38a, 49a,b; **3**:615b; **4**:477a; **8**: 161a–163a *passim,* **9**:500a
  and Newton, **2**:54a, 57a,b; **3**:295b–296a, 431b; **5**:156b; **6**:72a; **7**:225b–226a, 276a; **8**:151a–b, 158b, 159a,b, 161b–

Leibniz, Gottfried Wilhelm (cont.)
165b *passim;* **9**:452b; **10**:44b, 46a,b, 48b, 52b, 68a, 84a–85a; **14**:150b
"Nova methodus pro maximis et minimis . . . ," **3**:458b
and Oldenburg, **8**:149b, 160a–164b *passim*
philosophy, **3**:111b, 451b; **9**:31b; **13**:278b
epistemology, **8**:151b, 153b–154a; **9**: 189a
metaphysics, **3**:456a; **7**:225b–226b, 229b, 597b, 598a; **8**:156b–160a
methodology, **3**:381b; **8**:153b–154a, 155b–156b, 159a
monadology, **3**:90a, 381a; **4**:471b; **7**: 227b; **8**:152a, 157a–159a, 160a, 468b
universal language, **13**:180a
psychology, **8**:157b
scala naturae, **2**:286b; **4**:290b
Theodicy, **3**:295b, 485b
and Tschirnhaus, **8**:160a, 162a,b; **13**: 479b–480a, 480b
Leibnizianism
eighteenth-century, **3**:73b, 215b–216a, 216b; **7**:225b–226b, 442b; **15**:54a,b
nineteenth-century, **12**:240a
Leiden, University of, **4**:139a, 293a
botanical garden, **2**:225a
botany at, **8**:121a; **14**:95b
engineering school, **3**:181a
Kamerlingh Onnes Laboratory of Experimental Physics, **5**:610a; **7**:272a
medicine at, **2**:225a–b; **6**:360a; **13**:223a; **15**:3b, 4a
observatory, **7**:209a–b; **12**:449b
physics at, **8**:488b, 489a
Leidy, Joseph (1823–1891), 8:169–170; 5: 478b; **8**:104b*; **12**:257b; **13**:63a
critics of, **14**:146a*
paleontology, **6**:187a; **14**:409b
Leigh Browne Trust, **3**:463a
Leighton, William Allport, **6**:491a
Leiomyogram, **12**:319b
Leipzig, University of
anatomy at, **6**:434b; **11**:255a
astronomy at, **9**:463a
botanical garden, **9**:340a
botany at, **6**:218b; **9**:340a
chemistry at, **4**:394b; **7**:451a; **15**:433a–b, 456a–b
Debye and, **3**:618a
geophysical institute, **2**:168b
history of medicine at, **13**:142a,b
observatory, **2**:532b; **9**:429a–b
physiology at, **8**:540b–541a
zoology at, **8**:270a
Leipziger Magazin für Naturkunde, Mathematik und Ökonomie, **6**:403b
Leipziger Magazin für reine & angewandte Mathematik, **2**:56b; **6**:403b, 404a
Leishmania, **15**:480a
Leishmaniasis, **15**:97b, 454b
Leitch, Isabella, **7**:503a
Lejeuneaceae, **15**:149b
Leland, John, **10**:259b; **11**:415b
Lemaire, Jacques (*fl.* 1720–1740), 8:170–171
Lemaire, Jules, **8**:404a
Lemaire, Pierre (*fl.* 1733–1760), 8:170–171; **7**:111a
Lemaître, Georges, **3**:370b; **4**:278b, 281b, 282a; **6**:531a; **12**:449a
Lemberg, Johann, **13**:243a; **15**:455b, 457a
Lembert, Max Ernest, **11**:417b
Lemery, Jacques, **8**:174b
Lemery, Louis (1677–1743), 8:171–172; 2: 16b; **4**:223b, 352b; **5**:226a, 352b; **8**: 174b
chemistry, **5**:353a,b; **6**:43a, 46a(n); **7**: 111b*
Lemery, Nicolas (1645–1715), 8:172–175; **5**:354a, 417b; **6**:41b; **7**:274a; **9**:341a; **12**:457a
chemistry, **5**:353b; **6**:237a*
Cours de chymie, **1**:571b; **5**:418a; **8**:130b, 619a; **11**:562a
Lemna, **7**:199a
Lemniscate, **8**:139b; **9**:55a
area of, **4**:516b
rectification of, **2**:48a; **4**:516a; **7**:52a
Lemoine, Émile Michel Hyacinthe (1840–1912), 8:175–176; 8:62a
Lemoine circle, **8**:176b
Lemoine point, **8**:175b
Le Monnier, Louis-Guillaume (1717–1799), 8:176–178; 3:172b; **4**:156b; **7**:198a, 199a; **8**:506a; **9**:366a
Le Monnier, Pierre-Charles (1715–1799), 8: 178–180; 3:174a*; **5**:567a,b; **7**:563b, 569a; **8**:22a; **10**:614b, 615b
astronomy, **2**:359b; **10**:256a
critics of, **3**:283b
geodesy, **3**:39a, 281b; **15**:270a
Institutions astronomiques, **5**:24a; **7**:581a
meteorology, **3**:51b
students of, **1**:58b; **7**:579b, 580a
Lemons
and scurvy, **8**:363a
Lemur, **6**:455a*
Lemuroidea, **9**:428b
Lenard, Philipp (1862–1947), 8:180–183; **7**:469b; **10**:347a; **14**:171b
acquaintances of, **4**:357a; **6**:340b, 343a; **11**:529b; **12**:615a
critics of, **15**:451a
law of absorption, **13**:367b
radiation, **7**:553a; **11**:530b, 531a–b; **13**: 366a,b, 367a; **14**:338b
spectroscopy, **11**:479b
students of, **7**:468b; **15**:490b, 491a
Length
measurement of
standardization of, **1**:612b; **4**:354a; **6**: 236b, 606b; **14**:510a, 569a; **15**:271b
telegraphic, **6**:240b
units of
medieval Arabic, **7**:257b–258a
medieval Latin, **4**:609a
Renaissance, **3**:276b–277a
seventeenth-century, **9**:554a–b; **12**:500b
eighteenth-century, **2**:299b
nineteenth-century, **2**:102a
Egyptian, **15**:689b, 696a
French, **3**:107b
see also Distance; Measurement
Lenin, Nikolai
see Lenin, Vladimir Ilyich
Lenin, Vladimir Ilyich (1870–1924), 8:183–185; 4:598a; **5**:122b; **8**:603b; **15**:425b, 506a
critics of, **15**:38a
Lennard-Jones, John Edward (1894–1954), 8:185–187; 8:475a
Lennard-Jones potential, **8**:186a
Lenoir, Étienne, **4**:15a; **8**:22a
Le Normand, Louis-Sébastien, **9**:492b
Lens(es)
achromatic, **1**:115b, 135a–b, 460a; **2**: 328b; **4**:128a, 148b–149a, 482a; **5**: 521a; **6**:511b; **7**:404a; **8**:414a; **11**:284b
Amici's, **1**:136a
anaclastic problem, **4**:58b, 59b
angle of sight, **4**:158b
annular, **12**:519a
apochromatic, **1**:6b, 9a
burning, **1**:379b; **5**:382a,b, 552a; **11**:145a; **13**:480a
color correction, **12**:37a
compound, **6**:325a
conical, **5**:525b; **6**:508a
Crookes, **3**:475a
curvature of, **3**:420a
cylindrical, **1**:86b
echelon, **5**:166b
fish-eye, **9**:199b, 225b
focal length of, **3**:153a; **6**:603b
foci of, **9**:465a
Fraunhofer, **3**:288a
grinding of, **3**:23a; **8**:126b; **9**:336a, 464a*
hemispheric, **1**:135b–136a
hyperbolic, **4**:59b
magnifying, **1**:379b
meniscus, **14**:491b–492a
numerical aperture, **1**:8b, 135a, 135b–136a
ophthalmological, **4**:60a; **5**:590b
perfection of, **3**:288a; **13**:437b
spherical, **6**:603a
systems of, **5**:144a, 306b; **9**:226a; **11**:97b, 552b
testing of, **5**:87a, 144a; **6**:361a; **9**:46a
theory of, **4**:58b
wide-angle, **11**:553a
Lenssen, Ernst, **9**:349a
Lenz, Emil Khristianovich (Heinrich Friedrich Emil) (1804–1865), 8:187–189; **1**:298a; **9**:566a; **10**:28a, 192b
Lenz's law, **6**:251b; **8**:187b–188a
Leo X, Pope, **1**:618a; **8**:190a, 194a, 242a; **10**:122b; **11**:74a; **13**:58b
Leo VI, Patriarch of Constantinople, **8**:190b
Leo (*fl.* first half of fourth century B.C.), 8: 189–190; 6:414b; **13**:334a
Leo the African (also known as al-Ḥasan ibn Muḥammad al-Wazzān al-Zayyātī al-Gharnāṭī) (*fl. ca.* 1485–1554), 8: 190; 13:583b
Leo de Bannolis or Balneolis
see Levi ben Gerson
Leo Choerosphactes, **8**:190b
Leo Hebraeus
see Levi ben Gerson
Leo Judaeus
see Levi ben Gerson
Leo the Mathematician (also known as Leo the Philosopher) (*fl. ca.* 790–869), 8: 190–192; 1:223b; **4**:415b, 491a
Leo Suavius (pseudonym)
see Gohory, Jacques
Leo of Thessalonica
see Leo the Mathematician
Leodamas of Thasos (*fl. ca.* 380 B.C.), 8:192; **6**:411a; **8**:189a; **11**:28a
Léon, Xavier, **2**:544b
Leonardo, **13**:550b
Leonardo da Pisa
see Fibonacci, Leonardo
Leonardo da Vinci (1452–1519), 8:192–245; **1**:97a; **2**:340a; **3**:40b; **4**:227b; **9**:147a
acquaintances of, **3**:64b; **9**:146b; **13**:430b
art, **8**:193b–194a, 201b, 206b, 241a–b, 242b
theory of, **8**:234b
biographies of, **13**:81a
biology
anatomy, **3**:114a*; **8**:194a, 202a–205b; **13**:430b
botany, **8**:205b–206b
physiology, **4**:590a; **8**:197a–b, 201b–202a, 203a–204a; **13**:122b
and Cardano, **3**:66a
chemistry, **8**:213b

Codice Atlantico, **2**:470b
cosmology, **8**:201a–b
editions of works, **14**:636a
geology, **8**:201b, 241a–243a, 416a
influence of, **1**:419b; **3**:116a; **10**:281a
influences on, **13**:568a; **14**:461a
manuscripts of, **3**:116a; **8**:194a–195a; **12**:111a; **13**:430b; **14**:29b
mathematics, **8**:234a–241a
algebra, **8**:198a, 234b
geometry, **3**:150a; **8**:193b, 196a–198a, 200a, 223b–225b, 234b–241a; **9**:157a; **10**:271a
perspective, **8**:234b
proportion, **4**:259a, 260b; **8**:235a, 236a, 237b
mechanics, **8**:197b–198a, 200a–b, 208a
dynamics, **1**:94b; **8**:227b–231b
hydrodynamics, **8**:198b, 200b, 211a, 227a–b
hydrostatics, **8**:225b–227a
kinematics, **8**:210b, 231b–233b
statics, **8**:212b, 215a–225b
natural philosophy, **8**:197b–198a, 199a–200a, 227b–228b, 240b–241a
optics, **8**:196a–197a; **9**:192a; **10**:475b
and Pacioli, **8**:193b, 198a, 235a–236a; **10**:269b
scientific method, **8**:194b–196a, 198a–200a, 202a
technical drawings, **13**:613b
technology, **8**:206b–214b; **15**:255a
Leonardo da Vinci publishing house, **9**:378a
Leonhard, Georg von, **2**:498a
Leonhard, Karl Cäsar von (1779–1862), 8: 245–246; 2:341b, 498a; **3**:433b; **5**: 349b, 430b; **6**:358b; **14**:626a; **15**:526a
Leonhardi, Johann Gottfried (1746–1823), 8:246–248; 5:532b
Leoniceno, Nicolò (1428–1524), 8:248–250; 3:413a; **8**:360b, 361a; **9**:74b; **15**:161b*
medicine, **3**:40b; **9**:75a; **14**:28a
Leopold I, Grand Duke of Tuscany, **5**:55b
Leopold de' Medici
see Medici, Leopoldo de'
Leopoldina
see Academia Caesarea Leopoldina-Carolina Naturae Curiosorum
Le Paige, Constantin (1852–1929), 8:250; 11:405a*; **12**:460b*
Lepaute, Nicole-Reine Étable de Labrière, **3**:283b; **7**:580a
Lepekhin, Ivan Ivanovich (1740–1802), 8: 251–252
Lepidocarpon, **12**:260a
Lepidodendra, **5**:498a
Lepidoptera, **6**:167a, 184a–b; **10**:439b
morphology, **12**:131a
scales, **7**:286a
taxonomy, **5**:555b
see also Butterflies, Moths
Lepidosiren, **7**:314a
Lepidotus, **3**:606a
Lepinay, Macé de, **4**:513b; **10**:519a
Leplat, Émile-Claude, **3**:588a; **10**:386b–387a
Le Pois, Nicolas, **1**:399b
Le Poivre, Jacques-Francois (*fl.* early eighteenth century), 8:252–253
Leppmann, Arthur, **4**:296b
Leprosy, **3**:97a; **7**:427b; **9**:81b; **11**:249a; **13**: 282b; **14**:638b*; **15**:645a
bacilli, **10**:18a
etiology of, **6**:101b–102a; **10**:18a
transmission of, **5**:107a; **6**:102a
treatment of, **4**:155a; **11**:121a
Lepsius, Johannes, **7**:597b
Leptalides, **1**:503a,b
Leptocephalus, **4**:12a
Leptosphaeria, **9**:78b
Leptospira, **10**:142a–b
Lerch, Mathias (1860–1922), 8:253; 14:222a
Lerch formula, **8**:253b
Lerch theory, **8**:253b
Lerche, Johann Jacob, **5**:428b
Lereboullet, Dominique-Auguste (1804–1865), 8:253–255
Le Rossignol, Robert, **5**:621b
Le Roy, Alphonse, **12**:428a
Le Roy, Charles (1726–1779), 8:255–256; 13:412a*; **15**:302b
and Académie des sciences, **15**:333b
Le Roy, Édouard (1870–1954), 8:256–258; 4:233b*; **13**:275a; **15**:423a
analysis, **13**:26b
Le Roy, Jean-Baptiste (1720–1800), 8:258–259; 3:440b, 562a; **8**:71b, 79a; **10**: 147a, 615b; **12**:147b
Le Roy, Julien, **8**:170b
Le Roy, Pierre, **3**:106a
Le Royer, Augustin, **9**:109b; **11**:132b, 133a
Lesage, George-Louis (1724–1803), 8:259–260; 8:305a–b; **11**:134b; **13**:52b
correspondence of, **3**:286a*
Lesbros, Joseph Aimé, **11**:78a
Le Seur, Thomas, **2**:326b; **3**:216a, 282a; **4**: 515b; **5**:193a
Lesgaft, Pëtr Frantsevich, **9**:535a; **10**:220b
Lesions, **3**:588b
classification of, **7**:556b
specificity of, **2**:444b–445a; **11**:560a
and symptoms, **9**:511b
venereal, **8**:362a
Leske, Nathanael Gottfried, **6**:403b
Lesley, J. Peter (1819–1903), 8:260–261; 4:74a*; **6**:564b; **8**:264b*, 576b; **10**: 259a; **14**:516b
Leslie, George Arthur James, **4**:493b
Leslie, John (1766–1832), 8:261–262; 3: 317a*
acquaintances of, **6**:263a; **9**:615b; **12**:523a
acoustics, **13**:76b
on erratic blocks, **3**:210b
Leslie cubes, **8**:20b
meteorology, **6**:585b
probability, **7**:106b
on radiation, **11**:115b; **13**:352a
students of, **5**:68b; **14**:184a
translation of works, **6**:3a*
Leśniewski, Stanisław (1886–1939), 8:262–263; 8:546a
Lesnoy zhurnal, **9**:535b
Lespinasse, Julie Jeanne Éléonore de, **1**: 110b, 117a; **3**:380b, 383b, 384a; **8**: 506a
Lesquereux, Leo (1806–1889), 8:263–265; 6:187a; **10**:32b; **12**:104b
Lesser, Friedrich Christian, **8**:579b
Lessines
see Giles (Aegidius) of Lessines
Lesson, René-Primevère (1794–1849), 8: 265–266; 4:256a; **5**:277a
Lesueur, Charles-Alexandre (1778–1846), 8:266–267; **7**:44b; **8**:615b; **10**:518a; **12**:133a; **13**:467a; **14**:328b
Le Tenneur, Jacques-Alexandre (*d.* after 1652), 8:267–269
Lethargy, **14**:408a
Letnikov, Aleksei Vasilevich, **8**:433a
Le Tonnelier de Breteuil
see Châtelet, Gabrielle-Émilie Le Tonnelier de Breteuil, Marquise du
Leucine, **7**:467a
discovery of, **2**:386a; **11**:171a
Leucippus (*fl.* 5th century B.C.), 8:269; 3: 492a; **4**:30b, 31a,b, 32b, 381b
Leucite, **3**:484a
Leuckart, Friedrich Sigismund, **5**:444a; **13**:403a
Leuckart, Karl Georg Friedrich Rudolf (1822–1898), 8:269–271; 6:455a*; **8**:449b; **9**:332a; **12**:193a
physiology, **14**:113b
students of, **2**:625b; **3**:247b; **6**:167a; **9**: 416a; **11**:255a; **13**:62b; **14**:233b, 314a, 424a
Leucocytes, **4**:296a; **12**:423b
amoebic movements of, **3**:588b
composition of, **4**:236a
damage to, **5**:29b
enzymes in, **4**:618a; **5**:29b
phagocytosis, **2**:165b; **9**:333a–334b
role of, **10**:493a
Leucocytosis, **6**:506b; **8**:274a*
Leucol, **11**:615b
Leuconostos, **13**:405b
Leucoscope, **7**:458a
Leucotomy, **4**:286b–287a
Leukemia, **5**:37a
causes of, **7**:159a
diagnosis of, **4**:296b
experimental, **9**:587a*
lymphocytic, **14**:623b
and metabolic rate, **14**:65b
pathology of, **1**:439a–b
Leuko-
see Leuco-
Leunis, Johannes
Synopsis der drei Naturreiche, **11**:502a
Leupold, Jacob, **8**:208b; **9**:147a
Leurechon, Jean (*ca.* 1591–1670), 8:271–272; 10:264a
Recréations mathématiques, **6**:271b
Leuret, François (1797–1851), 8:272–273; 13:403b
Levaditi, Constantin (1874–1953), 8:273–274
Levaillant (Le Vaillant), François (1753–1824), 8:274–275; 1:331a
Level (instrument), **5**:88a; **6**:312b; **7**:342b; **10**:596b
barometric, **4**:44a
bubble, **13**:336a
dioptric, **9**:485b
spirit, **4**:253b
water type, **2**:133a; **9**:115b
Le Vene, Clara Mae, **12**:230a
Levene, Phoebus Aaron Theodor (1869–1940), 8:275–276; 5:40a; **7**:466b; **13**:574a; **15**:252b
Lever, **13**:292a; **14**:366b; **15**:519a
bent, **8**:215b, 218b–219b
proof of, **7**:174a–b
law of, **5**:245a,b; **7**:343b–344a, 344b, 345b; **8**:215a, 223b, 224a,b; **11**:488b
in geometry, **1**:219b, 220b–221a; **3**: 182a,b
proof of, **1**:222a–b, 228b–229a; **2**:194a; **4**:431b; **7**:173a, 174a; **8**:215b
potential lever arm, **8**:217b–222a *passim*
theory of, **4**:431a–b; **8**:230b; **13**:48b
see also Equilibrium, physical
Leverett, Frank, **13**:270a
Le Verrier, Urbain Jean Joseph (1811–1877), 8:276–279; 2:288a; **5**:87a; **9**: 551b; **11**:206b; **13**:284b
acquaintances of, **1**:157b, 449a; **3**:18b; **8**:382b; **11**:319b, 320a, 414b, 498a; **13**:36b, 283b, 422b
astronomy, **1**:510b; **2**:359b; **3**:146b; **5**:86a; **10**:34b
Neptune discovery, **1**:53b, 202b, 295b; **3**:187a; **5**:257a–b, 258a; **8**:299b; **10**: 479b

Le Verrier, Urbain Jean Joseph (cont.)
biographies of, **2**:88b; **7**:7a
celestial mechanics, **13**:423a
critics of, **11**:114b
and Delaunay, **4**:20a
and Gaillot, **5**:223b–224a
at Paris observatory, **5**:21b; **7**:75b
Levi ben Gerson (1288–1344), 8:279–282; **1**:23a; **6**:197b; **7**:120a, 245a; **9**:28b
astrology, **7**:131a
astronomy, **1**:573b; **11**:203b(n); **13**:401a
Astronomy, **1**:512b
Sefer Tekunah, **15**:35b
Levi, Giuseppe (1872–1965), 8:282–283; **11**:598a*
Levi's law, **8**:282b
Levi, Hilde, **6**:366b
Levi-Civita, Tullio (1873–1941), 8:284–285; **2**:144a; **3**:263b; **4**:280b, 577b; **12**:214a; **13**:623b
analysis, **11**:408a–409a
geometry, **4**:336b; **5**:555a; **11**:455a
Levine, Philip, **7**:624b
Levinson-Lessing, Franz Yulevich (1861–1939), 8:285–287; **5**:214a*
Levitskaya, Maria Afanasevna, **8**:107a
Levity, positive, **2**:311a
Lévy, Maurice (1838–1910), 8:287–288; **2**:350a; **6**:546a*
Lévy, Paul Pierre, **7**:351b; **9**:129b
Levy, S. (Paul), **7**:554a
Lévy, Serve-Dieu Abailard (called Armand) (1795–1841), 8:288–289; **4**:65b, 66a
Lévy-Bruhl, Lucien, **9**:340b, 341b; **15**:422b, 424a
Lewes, George Henry, **3**:379a; **5**:255a; **15**:78b
Lewis, Exum Percival, **13**:107b
Lewis, Gilbert Newton (1875–1946), 8:289–294; **3**:329b; **7**:497b; **8**:47a; **11**:418a
chemistry, **2**:499a, 524a; **7**:469a–b
electrochemistry, **6**:372a; **15**:461a
Lewis acids, **8**:293a
Lewis-Langmuir theory, **8**:23b, 292a; **12**:419a
on relativity, **13**:429b
thermodynamics, **9**:394a; **15**:438b, 439a
Thermodynamics and the Free Energy of Chemical Substances, **5**:621b
Lewis, Henry Carvill, **14**:516b
Lewis, Margaret, **8**:283a
Lewis, Meriwether, **1**:485a; **7**:89b; **11**:218b
Lewis, Morris James, **9**:422b
Lewis, Thomas (1881–1945), 8:294–296; **4**:335b*; **6**:523b*; **7**:502b; **15**:106a
cardiology, **5**:281b; **15**:101a–b
Lewis, Timothy Richards (1841–1886), 8:296–297; **4**:133b*; **9**:81b–82a
Lewis, Warren Harmon, **8**:283a; **9**:56b
Lewis, Wilfrid Bennett, **3**:330a
Lewis, William (1708–1781), 8:297–299; **11**:143b
Lewis, William Cudmore McCullagh, **13**:429b
Lewis and Clark expedition, **1**:485b; **3**:439a; **7**:89b; **9**:540b; **10**:438b
botany, **11**:218a,b
Lewisian (stratig. division), **8**:33a
Lewisite, **10**:121b
Lewisohn, Richard, **7**:623b
Lexell, Anders Johan (1740–1784), 8:299–300; **4**:472b, 473a; **5**:209b; **15**:302a
analysis, **7**:566b
geometry, **8**:306a
Lexicography
see Dictionaries
Lexis, Wilhelm Hector Richard Albrecht, **2**:318b, 319a; **15**:31b
Lexis criterion, **2**:319a
Leybenzon, Leonid Samuilovich (1879–1951), 8:300–301; **10**:552a*; **14**:621b
Leybourn, Thomas, **14**:141a
Leybourn, William, **8**:459a
Leyden jar, **5**:268a, 292a; **7**:403b; **13**:224b
analysis of, **5**:131a–132a, 133a
discharge
electromagnetic vs. electrostatic value, **7**:450a–b
successive discharges, **2**:137a; **6**:279a
invention of, **2**:324b; **9**:596a–b
oscillations, **4**:559a; **8**:444a
theories of, **1**:67a; **8**:177a; **10**:146a; **14**:72b, 194a, 352b, 353a
Leydig, Franz von (1821–1908), 8:301–303
cytology, **12**:231b
gland of Leydig, **8**:302b
historiography, **11**:503b*; **12**:186a*
Leydig's cells, **8**:302a
students of, **4**:275b; **5**:446b; **6**:7a, 434a; **12**:299a; **14**:203a, 331a; **15**:166b
Leyst, Ernest Egorovich, **8**:102b
L'Héritier de Brutelle, Charles Louis (1746–1800), 8:303–304; **2**:321b, 510a; **3**:173a*
botany, **11**:343a; **14**:463b
Stirpes novae, **3**:172b
Lhermite, Michel, **10**:530b
L'Hospital (L'Hôpital), Guillaume-François-Antoine de (Marquis de Sainte-Mesme, Comte d'Entremont) (1661–1704), 8:304–305; **1**:111a; **3**:282b; **7**:558b; **10**:65a; **11**:392a
Analyse des infiniment petits pour l'intelligence des lignes courbes, **2**:49b; **3**:281a; **5**:60b; **9**:50b; **12**:444b; **13**:586a
and Bernoulli (Johann I), **2**:52a, 53b
commentaries on, **2**:467b
correspondence of, **1**:206b; **3**:461a; **6**:612b*; **8**:164a; **14**:49b(n)
engineering, **2**:49a
L'Hospital's rule, **8**:304a,b; **12**:118a
mathematics
analysis, **2**:47b; **3**:485b; **8**:163a; **9**:50a–b; **11**:512b
calculus, **2**:52a, 342b; **6**:600b
geometry, **3**:183b; **11**:402b, 513a
Traité analytique des sections coniques, **1**:75b; **3**:281a
L'Huillier (or Lhuilier), Simon-Antoine-Jean (1750–1840), 8:305–307; **13**:127a
analysis, **4**:516a; **13**:266a
geometry, **8**:300a; **13**:132a*
Lhwyd, Edward (1660–1709), 8:307–308; **8**:4a, 416a,b; **11**:41b*; **14**:500b–501a
Lithophylacii Britannici ichnographia, **14**:502b
Li Chih (also called Li Yeh) (1192–1279), 8:313–320; **3**:254a, 265b, 266b
algebraic equations, **3**:254b
geometry, **3**:255b
Ts'e-yüan hai-ching, **14**:538b
Li Jui, **3**:250a, 266a; **8**:314a–b; **14**:540a
Li Shan-lan, **3**:266a, 269a; **8**:314b; **11**:403a
Li Shih-Chen (1518–1593), 8:390–398
Li Shun-feng, **3**:269b; **8**:419a
Li Te-tsai, **3**:266b
Li Wen-i, **3**:266b
Li Yeh
see Li Chih
Li Yen, **3**:255a, 266a; **8**:314b, 317a, 419a, 423a; **13**:484b–485a; **14**:539a,b, 540a, 544b
Li Yen-wen, **8**:390b
Liapunov
see Lyapunov
Liassic (stratig. stage), **3**:395b; **9**:617b
Libavius (or Libau), Andreas (*ca.* 1560–1616), 8:309–312; **4**:209a; **5**:47a; **7**:355a; **13**:535b
Alchymia, **1**:571b
chemistry, **3**:201b; **7**:194b; **12**:79b
Defensio alchymiae transmutatoriae opposita Nicolai Guiberti, **5**:580a
Libbertz, Arnold, **7**:426a
Liber de ponderibus, **8**:215b
Liber trium fratrum de geometria, **3**:512a
Liberal arts
see Quadrivium, Trivium
Liberalism, **12**:570a
Libertins, **7**:537a
Libido, **5**:176b–178a *passim,* 180b
Libocedrus, **5**:442a
Libraries
medieval Arabic, **6**:189a; **13**:511a
astronomical, **10**:198a
biological, **15**:124a
botanical, **13**:456b
entomological, **12**:201a
geological, **12**:293b
historical, **4**:8a; **14**:481a
paleontological, **4**:67b
private, **14**:451b
public, **2**:173a
Library of Congress, **5**:406b
Librations, **2**:616a
Libri, Guglielmo, **1**:206b–207a; **2**:353b*; **4**:605b, 612b; **8**:242b; **9**:191a, 192a; **15**:482a–b, 483b
Libri-Carrucci della Sommaia (Count Guillaume Brutus Icilius Timoleon), **5**:111b; **7**:550b; **8**:382a
Liceaga, Eduardo (1839–1920), 8:312–313
Licent, Émile, **13**:274b, 275a
Liceto, Fortunio, **2**:13a
Lichen (disease), **10**:19a
Lichenology, **2**:290b; **5**:191a; **8**:512b; **10**:165b–166a; **12**:256a
Lichens, **3**:259b; **4**:560a; **5**:6a
antibiotic properties, **15**:150b
development of, **2**:160a; **3**:613b
dyestuffs from, **12**:236a
gonidia, **15**:14a
identification of, **10**:165b
nature of, **10**:166a
physiology, **3**:44a, 45a; **13**:489b
reproduction, **4**:288b; **13**:489b
symbiotic composition of, **2**:290b; **10**:166a; **12**:256a
taxonomy, **1**:46a; **5**:191a; **7**:197b; **13**:178a; **15**:150a–b
see also Cryptogams
Lichfield Botanical Society, **3**:578a
Lichtenberg, Georg Christoph (1742–1799), 8:320–323; **13**:466a
acquaintances of, **1**:615b; **3**:258a; **5**:75b, 411a; **14**:69b
on contemporary scientists, **7**:600b*; **14**:80b
correspondence of, **14**:73b, 75b
Lichtenberg figures, **8**:322a
Lichtenberg's alloy, **8**:321a
Opera inedita Tobiae Mayeri, **9**:234a
students of, **2**:420b; **7**:366b; **10**:197b, 572b
Lichtenstein, Leon, **4**:548a; **6**:472a; **14**:453b
Lichtenstein, Martin Hinrich Carl, **2**:422b; **4**:288b, 289a,b; **5**:77a; **9**:344b, 431b; **12**:420b, 421a
Lick, James, **6**:471a,b
Lick Observatory
in nineteenth century, **1**:465a–466a; **3**:35a, 366a–b; **6**:470b, 471a–b; **7**:270b; **12**:139a; **14**:521b
in twentieth century, **3**:35b–36b *passim,* 327a, 509a; **9**:503b; **10**:526a–b
establishment of, **10**:35a

Liddell, Edward George Tandy, **12**:399a–b
Lidén, Ragnar, **5**:330a
Lie, Marius Sophus (1842–1899), 8:323–327; **2**:191b; **4**:177b, 205b; **5**:265b*, 482b*; **6**:473a
acquaintances of, **13**:216a; **14**:58a, 222a
analysis, **9**:232b; **10**:573b, 574b
correspondence of, **14**:219b
editions of works, **4**:370b; **12**:150a
and Engel, **4**:370b
geometry, **2**:192a; **3**:143a; **7**:397a,b; **9**:476b; **11**:408b; **13**:125a; **14**:281a*
group theory, **3**:168b; **7**:168a–b; **8**:457a
Helmholtz-Lie space problem, **8**:326b
influence of, **9**:457b; **13**:125a
invariance, **3**:428a
and Klein, **8**:323b, 324a, 325b
Lie algebra, **3**:95a–b; **8**:326a,b; **9**:71b; **11**:55b–56a; **13**:219b
Lie groups, **2**:197b; **3**:95a–96b, 222a; **5**:198a, 201b; **6**:76a; **7**:574a; **8**:326a–b, 457b; **9**:71b; **11**:53b; **12**:214a; **13**:125b; **14**:90a
students of, **2**:197a; **9**:387b; **12**:150a; **13**:390b
theorems of, **11**:55b
Liébeault, Ambroise Auguste, **2**:35b; **5**:174a
Lieben, Adolf, **3**:46b; **12**:452a
Lieber, Thomas
see Erastus, Thomas
Lieberkühn, Johannes Nathanael (1711–1756), 8:327–328
Lieberkühnian glands, **5**:226b; **8**:327b
Liebermann, Carl Theodore (1842–1914), 8:328–329; **1**:390b; **5**:381b*; **9**:357b*
alkaloids, **5**:394a
anthracene, **1**:168b
dyestuffs, **3**:85a; **5**:488b, 623a*; **10**:516a; **14**:247a
Liebermann-Kostanecki rule, **7**:471a
students of, **1**:426a; **11**:551a; **15**:36b
Liebherr, Joseph, **5**:142b
Liebig, Justus von (1803–1873), 8:329–350; **2**:564a, 570b; **3**:49a; **4**:400b*; **5**:341a, 531b*, 622b; **7**:283a, 451a; **11**:34a*; **13**:500a*
acid theory, **5**:372a, 493b; **8**:338a–342b
acquaintances of, **2**:161a, 426a, 586b; **6**:553a; **8**:55a, 270a; **10**:510b, 558a; **11**:536b; **12**:227a
agricultural chemistry, **3**:601a; **6**:237b; **8**:344a–348a; **10**:242b; **12**:328b
on alcohol, **3**:193b
analytical method, **2**:95b; **4**:75b–76a; **8**:330a–333b; **14**:476a–477b
criticism of, **5**:371b
Animal Chemistry, **2**:124a
and Berzelius, **8**:336a, 338a–344b *passim;* **14**:478a
bromine, **1**:416b
Chemische Briefe, **7**:279b
critics of, **9**:457a
on contemporary scientists, **2**:66b; **3**:243a; **11**:359b; **14**:169a
correspondence of, **9**:566b
critics of, **11**:172b; **12**:175b, 466b; **13**:452b; **14**:66a
disputes, **2**:91b, 96b; **3**:9b; **5**:370b; **9**:238a–b, 425b, 558a
and Dumas, **8**:335a–342b *passim,* 349a; **14**:601a
dyestuffs, **8**:349a–b
ether theory, **3**:193b; **8**:333b–338a; **9**:91b; **14**:395a
and Gay-Lussac, **8**:330a–b, 331b
Giessen laboratory, **4**:394b; **8**:540b; **14**:622a
students and assistants, **1**:156a, 578a; **2**:484b; **3**:333a; **4**:400a; **5**:104a, 125a, 163b, 349b, 369b, 412a, 531a; **6**:517a; **7**:224a, 279b, 280a, 463a; **8**:93a; **9**:109b, 178a, 391a, 566a; **10**:557a; **11**:36b, 352b; **12**:236a, 478a; **13**:373b, 452a; **14**:57a, 315b, 394a, 529b
teaching methods, **3**:397b, 465b; **6**:461a; **8**:330b–331a
influence of, **1**:389b; **9**:445b, 619b; **12**:59a, 619b; **15**:134a, 136a
influences on, **13**:312a; **14**:487b
Jahresbericht
see *Jahresbericht über die Fortschritte der Chemie*
Liebig reflux, **9**:445b
nationalism, **4**:243a; **8**:336b
Die organische Chemie in ihrer Anwendung auf Agricultur und Physiologie, **2**:162a; **8**:93a
translations of, **11**:36b
and Pelouze, **8**:336a, 337a, 340a, 342a; **10**:499a
physiological chemistry, **4**:616a; **8**:342b–345b, 348a–349a, 402b; **9**:238a–b; **12**:563a; **14**:63b, 64a, 65b
fermentation, **2**:561a; **3**:9b; **8**:349b; **10**:361b, 365a,b, 376b, 378a–379b; **12**:243b
protein metabolism, **2**:162a; **5**:126b; **14**:454b
radical theory, **2**:96a, 587b; **5**:125b, 372b; **8**:332b–338a; **14**:530b
students of, **9**:259a; **11**:370b; **13**:243a; **14**:63a; **15**:457a
Die Thierchemie oder die organische Chemie in ihrer Anwendung auf Physiologie und Pathologie, **2**:162a; **6**:243b; **9**:91b
translations of works, **5**:369b, 531a; **12**:478a; **14**:529b
and Wöhler, **8**:331a–335a *passim,* 339a, 342b–343b; **11**:495a; **14**:474b, 476a–477b
Liebmann, Heinrich, **4**:336b
Liège, University of
in nineteenth century, **12**:244a
geology at, **8**:463a
Liénard, Alfred, **5**:186a; **8**:496a; **9**:217a
Liesganig, Joseph Xaver (1719–1799), 8:350–351
Lieutaud, Jacques, **4**:22b; **7**:86b
Lieutaud, Joseph (1703–1780), 8:352–354; **11**:99b
editions of works, **11**:100a
Lieutaud's body, **8**:353a
Life
altitude, effects of, **6**:221a
and asymmetry, **10**:360b–361b
autotrophic, **1**:623b
definition of, **2**:122b, 187a–b; **3**:453a; **5**:194a; **12**:571a; **14**:40b
extraterrestrial, **1**:301a; **3**:327b*; **5**:21b, 59a–b; **6**:333b, 334a, 611a–b; **7**:299a, 598b; **8**:520b, 521b–522a; **11**:116a; **14**:363a, 364b
signals from, **13**:287a
laws of, **11**:262b–263a
nature of, **2**:579b; **7**:587a,b, 588b, 590a, 591a, 592b; **8**:125a, 133b; **10**:576a, 580a; **11**:491b; **12**:222b–223a, 601b; **14**:426a; **15**:11b–12a, 47a, 127b–128a, 525a
and optical activity, **10**:359a–361b; **15**:103b
organization of
theories of, **2**:11a; **7**:589b–590b; **15**:12a
origin of, **1**:496b–497b, 623b; **2**:579b; **6**:23a, 320a; **7**:590a–b, 591a; **8**:169b; **9**:26b, 182b, 564b; **10**:195a, 355a; **15**:18a
archebiosis, **1**:497a
geological evidence for, **9**:583b, 584b
heterogenesis, **1**:497a; **10**:367a
panspermic theory, **1**:301a
see also Spontaneous generation, Vitalism
philosophy of, **8**:257a–b
prolongation of, **1**:380a; **8**:133b; **9**:335a
protoplasmic theory of, **3**:337a–b; **4**:235b–236a; **6**:596b
signs of, **13**:262b
spontaneous manifestations, **14**:3a
suspension of
by dehydration, **1**:561b
in nature, **1**:561b; **7**:447a
Life expectancy
medieval discussion, **2**:152b
seventeenth-century discussion, **10**:565b
eighteenth-century discussion, **1**:116b; **4**:38b–39a; **12**:444b
nineteenth-century discussion, **4**:36b; **15**:31a–b, 373b
twentieth-century discussion, **8**:512a
mean life expectancy, **4**:39a
see also Actuarial science, Longevity, Mortality
Lifshits, Evgenii Mikhailovich, **7**:617b, 618a
Lifting systems, **8**:209a–b, 211a
Ligaments, **3**:355a
Light
absorption of, **2**:120b; **11**:208a
and chemical structure, **4**:191b
and electrical conductivity, **4**:191a; **8**:502a; **9**:212b
interstellar, **5**:363b; **7**:237b; **9**:411a; **10**:199a, 317b; **12**:249a; **13**:110b–111a, 475a
laws of, **7**:599a
selective, **11**:582a
in solids, **5**:160b; **11**:10b
asymmetry of, **5**:169a–b
biological action of, **1**:536a; **3**:586a; **4**:620b–621a
chemical effects of
see Photochemistry
diffusion of, **13**:351b
at high altitude, **1**:519b
interstellar, **13**:408b
and electricity, **8**:36b; **10**:146a; **11**:530a
emission of
Lambert's law, **2**:328b
mechanism of, **10**:347a,b
energy of, **4**:322b; **14**:175b–176a
heat and, **5**:167a; **6**:586a; **9**:264b–265a; **13**:238b
intensity of, **1**:81b; **4**:182b
magnetic effects on, **3**:565a*; **4**:191a; **8**:40a; **14**:597a–598b
see also Faraday effect; Kerr effect; Optical activity, magnetic; Zeeman effect
and matter, **8**:494b–495a; **10**:59b, 80a, 525b; **13**:238a–b
momentum of, **11**:142b
and motion, **4**:378a; **5**:20a–b; **7**:26b; **8**:495a; **9**:154b; **12**:69a–b
penetration of into water, **2**:142a–b; **5**:52a; **10**:564a
periodicity of, **10**:59a, 60a
point source of, **5**:483b
polarized
analysis of, **13**:77a
colors produced, **2**:137a–139a
direction of vibration, **6**:19a; **13**:77a
interference phenomena, **5**:19a, 169b

Light (cont.)
polarized (cont.)
path-difference, **5**:186a
reflection of, **5**:169b; **7**:313b; **13**:77a
rotation of
see Optical activity
pressure of, **1**:301a; **3**:477a; **5**:363b, 461b; **8**:289b, 290b, 491b; **9**:212b; **10**:105b
measurement of, **8**:107a–b
propagation of
medieval Arabic theories, **6**:191a–192a; **15**:264a–b
medieval Latin theories, **10**:475a–b; **13**:199b–201a; **14**:459a–b
Renaissance theories, **8**:196a–197a
seventeenth-century theories, **1**:568a; **5**:543a, 544a; **9**:52b
eighteenth-century theories, **3**:486a; **8**:115a
nineteenth-century theories, **13**:384a–b
in crystalline medium, **6**:90a; **7**:601b
curvilinear, **4**:306a
mathematical analysis of, **7**:478a
rectilinear, **2**:328b; **14**:567b
temporal aspects, **14**:459b
red shift, **4**:279b, 280b, 325a, 329a, 330a, 498a; **5**:182a,b, 188b; **6**:531a,b, 532a; **10**:519a; **12**:280b; **13**:429b
non-relativistic, **5**:183b–184a
relativistic deflection of, **3**:472a; **4**:270a, 278a, 280b, 312b, 325b, 328b–329a, 359a; **5**:182a,b, 183b; **12**:449a; **13**:474b
resonance, **4**:578a
scattering of
see Scattering, of light
and sound, **14**:565a–566a
speed of, **4**:281a; **5**:245a; **8**:1a, 491a, 497a, 498a; **9**:205a, 209a,b; **10**:315a
in dispersive media, **5**:483a
and earth's rotation, **9**:373b
finite nature of, **3**:101b, 103b; **4**:322b, 369a, 575a; **9**:77b, 90a,b; **11**:526a
in flowing water, **8**:50b; **9**:372b
in glass vs. air, **10**:134a
measurement of, **2**:439a; **3**:420a; **4**:191b; **5**:19a–20b, 85a–86a; **7**:445b; **8**:502a; **9**:371b–372a, 372b, 373b; **10**:34b, 473b, 528b; **11**:526a–b; **12**:448b; **15**:325a
in moving media, **14**:598b–599a
and relativity theory, **7**:26b
uniformity of, **15**:338b
variations in, **4**:320a–b; **9**:214b, 215a
in water vs. air, **1**:202a; **5**:19a–b, 85b–86a
and wavelength, **3**:146b
spherical waves, **5**:483a–b
standing waves, **7**:36a,b
theories of
ancient Greek, **11**:42a; **13**:92b, 94a
ancient Roman, **4**:369a
Byzantine, **7**:136a–b
medieval Arabic, **2**:149a; **6**:191a–192a, 194b, 195b; **7**:213a
medieval Latin, **5**:403a, 549b, 550b–552b; **7**:116b; **14**:388a,b, 459a
Renaissance, **1**:81b
early seventeenth-century, **1**:373a; **4**:58b–60a, 96a, 575a; **6**:445b–446a; **7**:375b; **9**:318a–b, 319a
late seventeenth-century, **1**:481b; **5**:543a–544a; **6**:484a–b; **10**:53b–60a, 76a, 83a
early eighteenth-century, **2**:328b, 329a; **4**:482a; **8**:171b; **9**:33b; **10**:79b–80a; **12**:283b, 477a–b
late eighteenth-century, **5**:137a, 195a; **6**:585b; **8**:82b, 322a, 470a; **11**:142b, 438a; **13**:351b; **14**:565a–b; **15**:338b
early nineteenth-century, **5**:166a–171a; **7**:263a; **9**:72b–74a
mid nineteenth-century, **6**:87b, 88b; **8**:63a; **11**:292a; **14**:208b
late nineteenth-century, **5**:391a; **8**:491a–496b, 501a, 502a; **13**:386a–b
twentieth-century, **4**:196b–197a; **9**:396b; **11**:480a; **12**:528a–b
corpuscular, **1**:475a; **2**:137b; **3**:477a, 599a–600a; **4**:44b; **5**:287b; **6**:178b; **9**:214a, 267a; **10**:54a, 55b, 60a, 79b; **12**:106a; **15**:355a, 355b–356a, 359a, 361b, 362a–b, 377a
dynamic, **8**:592a–b
elastic, **3**:146a–b; **8**:490b, 501a; **11**:581b; **13**:101b
electrical, **8**:491a,b
electromagnetic, **2**:169b; **4**:189b, 190b–191b, 231b, 306a, 315b, 537b; **5**:15a, 390b–391a, 559a; **6**:344a; **8**:40a, 492a–496b, 502a, 592b; **9**:209b–215a; **11**:9a, 581b–583a
emission theory, **1**:201b; **2**:451a, 452a–453b; **5**:137a; **6**:445b–446a; **9**:73a–b; **14**:565b, 567a
fluid theory, **5**:543b–544a
mechanical, **4**:189b, 190a–b; **6**:484a–b; **14**:61b
mediumistic, **6**:446a
particulate, **2**:452a; **3**:619b; **5**:85b, 167a; **9**:73b; **12**:289a, 477a, 528a
quantum, **4**:315a–317b; **12**:464b; **13**:599a
ray theory, **6**:191b–192a
undulatory
see Light, theories of, wave theory
vibrational, **3**:83a–b; **4**:533a; **5**:167b; **9**:52a–b; **11**:241b; **14**:565a–566a
wave theory, **1**:7b, 8a–b, 201b–202a, 357b, 600a; **3**:619b; **4**:319a, 534b; **5**:19a, 85b, 86b, 137a, 143b, 166a–171a, 445b; **6**:88a–b, 484a, 609b–610a; **8**:40a, 50b–51a, 426a,b; **9**:74a, 213a, 214a, 264b, 372a–374a *passim;* **10**:27b, 60a, 546a; **11**:115b, 116a; **12**:519a; **13**:76b, 104a, 238b; **14**:491b, 565a–568b, 617a; **15**:355b–356a, 362a–b, 376b
thermal effects of, **1**:203a
and wavelengths, **12**:281a
transmission of, **1**:374a; **9**:244a; **13**:351b
Bouguer's law, **2**:343b
external effects on, **2**:382b
wavelength of, **1**:166b, 357b; **5**:143b; **8**:578b; **9**:154b; **10**:519a; **11**:149b, 579a, 582a; **14**:566a–567b
Newton's values, **2**:138a
shifts in, **10**:519a
standardization of, **4**:513b; **9**:256a–b
and standardization of meter, **1**:612b
white
nature of, **5**:483b
see also Aberration; Catoptrics; Color(s); Crystallography, optical; Dioptrics; Dispersion; Illumination; Optics; Photometry; Polarization; Radiation; Reflection; Refraction; Sunlight
Lighters, **2**:340a; **4**:134a; **7**:13a, 22a
Lighthouses
nineteenth-century, **2**:200b; **5**:166b, 170b; **6**:504b; **11**:377b; **12**:519a; **13**:105a
Lighting, artificial
coal-gas, **4**:250a
efficiency of, **13**:351b
electric, **2**:588b; **3**:475a, 478a; **4**:216a, 248b; **5**:496a; **10**:236a–b, 552b; **14**:217b
arc lamp, **2**:200b, 439a; **5**:87a; **6**:223a; **12**:574b; **13**:286b, 361b
filament lamp, **5**:560a; **8**:23a–b; **15**:449b
fluorescent lamp, **3**:367b
incandescent, **4**:284a; **11**:377b
mercury-arc, **13**:24b
sodium vapor lamp, **3**:367a
standards, **6**:504b
gas, **1**:44a; **2**:566b; **3**:316b*, 388b, 556b; **11**:355a; **14**:80b; **15**:472a
purity standards, **6**:110a
kerosene, **15**:526b
limelight, **6**:115a
oil, **6**:383b; **9**:343a; **13**:351b
street lighting, **8**:69b; **9**:343a
wick lamps, **8**:69b
miners', **3**:603a; **5**:558b; **7**:509b; **8**:118a
wood-gas, **10**:557b
Light-mill
see Radiometer
Lightning
eighteenth-century study, **2**:5a; **3**:51b; **10**:321b; **11**:433a–b, 472a
nineteenth-century study, **1**:300b; **6**:471a
artificial, **13**:287a
conductors of, **2**:5a; **4**:129a–b; **8**:470a
and electromagnetic induction, **6**:279a
photography of, **12**:201b; **15**:60a
protection against, **4**:248b
and ships, **3**:557b
spectra of, **7**:526a
theories of
ancient Greek, **11**:104b
eighteenth-century, **3**:535a; **4**:129a–b; **5**:134a–135b; **8**:174a–b, 459b; **10**:146b; **11**:66b–67a
nineteenth-century, **4**:248b
twentieth-century, **12**:201b
Lightning (ship), **7**:91b, 92a; **13**:360a–b
Lightning rod, **1**:435b; **2**:83a; **5**:130b, 134a–135b; **7**:499a–b; **8**:258b, 321a; **9**:153a; **11**:111b; **14**:69b, 194b
shape of, **7**:13a; **14**:419a
Lightships, **6**:504b
Light-years, **2**:328b
Lignères
see John of Lignères
Lignier, Élie Antoine Octave (1855–1916), 8:354
Lignification, **9**:418b
Lignin, **10**:436b; **12**:233b
analysis of, **3**:359a
Lignites
age of, **1**:210a; **10**:32b
formation of, **12**:593b
Lignoceric acid, **8**:276a
Lignum nephriticum, **2**:381a; **7**:375b
Liguria
geology, **7**:30b, 31a; **11**:575b, 576a
Li-hsiang k'ao-ch'eng, **1**:311a, 314a; **7**:18b, 19a; **14**:162a; **15**:734b, 735a,b
Likelihood, statistical, **10**:458a
method of, **5**:9a; **10**:456a, 457a, 466b(n)
Liliaceae, **3**:44b
Lilienthal observatory, **12**:226a–b
Lilium, **3**:422a; **4**:131a; **10**:1b
cytology, **4**:545b, 546a
Lille, Alain de
see Alain de Lille
Lille
Botanical Gardens, **8**:354a
Faculty of Sciences, **10**:361a–b
Lillie, Frank Rattray (1870–1947), 8:354–360; **3**:335b; **14**:315a*
Lilly, William, **1**:317a–b; **12**:342a–b, 343a; **14**:285b, 286a
Worlds Catastrophe, **1**:317b
Lily of the valley
Renaissance description, **2**:219a
Lima, Almeida, **4**:287a

Limbs, artificial, **13**:73b
Lime
  in building, **15**:662b(n)
  burning of, **7**:58a
  hydraulic, **9**:538a
  metabolic effects of, **2**:585b
  nature of, **5**:353b
  in plants, **8**:344b
  slaked
    heat of, **4**:214b
Limelight, **6**:115a
Limes
  and scurvy, **8**:363a
Limestone, **5**:155b, 475b, 538a; **14**:132b
  calcination of, **2**:175b, 176a,b
  in cement production, **7**:44a
  consolidation of, **6**:54b–55a
  fossils in, **2**:494a–b; **8**:567a; **11**:35b; **12**:543a
  freshwater, **8**:565a, 566a
  Magnesian, **12**:276a, 489a; **13**:620b
  occurrence of, **10**:259b–260a; **14**:440b
  oolitic
    misinterpretation of, **2**:347b
  organic, **1**:162b–163a
  origin of, **2**:496b; **3**:161b; **11**:439a
  undersea, **6**:229a
  see also Muschelkalk
Limewater, **2**:175a, 177a; **6**:44b
Limit(s) (math.), **3**:265a; **5**:201b
  central theorem, **3**:142a; **8**:563a
  concept of, **1**:115b–116a, 256a, 407a, 408a; **2**:275a; **6**:377b, 378a; **8**:306a; **9**:303b; **10**:65b–66a; **12**:461a; **13**:561a
  definition of, **3**:136a; **4**:35b
  method of, **3**:76a–b; **4**:115b; **5**:97a, 204a; **10**:63a, 65a–66a, 72a
  theorems of, **3**:230b–231a; **7**:352a; **10**:553b–554a
  theory of, **1**:444b; **5**:96b; **7**:411b; **9**:303b, 304a, 503a*; **10**:51a, 443a; **12**:446b
  variable, **5**:305a
Limnobium, **2**:322b
Limnology
  eighteenth-century, **9**:135b
  nineteenth-century, **1**:174b; **2**:142a–b, 558a; **4**:9b; **5**:70a–b; **15**:158a–519a
  twentieth-century, **1**:622b; **2**:142a–b; **7**:184a–b; **13**:519a
  founding of, **15**:158b, 159a
  see also Lakes
Limonene
  structure of, **13**:550a
Limpricht, Heinrich, **5**:12b; **8**:329a
Limulus, **10**:273b
  heart, **3**:68b
Lin, Chia-Chiao, **8**:364b
Linacre, Thomas (1460–1524), 8:360–361; **8**:248b; **14**:47a
  Linacre lectures, **8**:361a
Linaria, **3**:575a; **8**:377b
Lincei, Accademia dei, **2**:306b; **3**:117a, 118b; **4**:30a; **5**:241b, 242a, 243a, 244a, 382b; **7**:193b; **12**:106a, 257a; **13**:560b
  founding of, **3**:179a–b; **11**:96a–b; **13**:29a–30a
  Pontifical Academy of the New Lincei, **2**:284a
  reconstitution of, **3**:180a
Lincoln (England)
  Mechanics Institution, **2**:294a
Lind, James (1716–1794), 8:361–363; **1**:98b; **14**:81a(n)
Lindau, Gustav, **2**:437b
Lindbergh, Charles Augustus, **3**:91b; **4**:253a; **5**:433b; **8**:24b
Lindblad, Bertil (1895–1965), 8:364–365; **6**:532a; **7**:239a; **12**:348a,b; **13**:98b, 99b*; **14**:179b*
Lindblad, Per Olof, **8**:364b
Linde, Carl von (1842–1934), 8:365–366; **4**:269a; **6**:94a; **9**:461b; **10**:206b; **14**:618a
Linde, Jonas Otto, **13**:245b
Lindelöf, Ernst Leonhard (1870–1946), 8:366–367
Lindemann, Adolphus Frederick, **14**:482a
Lindemann, Carl Louis Ferdinand (1852–1939), 8:367–368; **3**:62b, 314b; **6**:388a–b, 570b; **12**:247a*, 527b; **14**:67b
  number theory, **4**:475a; **5**:342b; **6**:308a
  students of, **12**:526a; **14**:336a
Lindemann, Frederick Alexander (1886–1957), 8:368–369; **1**:320b, 321a; **9**:544b; **12**:222a, 437b; **15**:450b, 451a*
  Nernst-Lindemann theory, **8**:368a; **15**:442a
  physics, **2**:170a; **6**:404b; **11**:583a
Lindenau, Bernhard August von (1779–1854), 8:369–370; **5**:301b; **10**:110b; **13**:109b
Lindenberg, Johann Bernhard Wilhelm, **10**:13b
Linderstrøm-Lang, Kaj, **13**:3a
Lindgren, Waldemar (1860–1939), 8:370–371; **3**:484a; **9**:316a*
Lindhard, Johannes, **7**:502a,b
Lindley, John (1799–1865), 8:371–373; **1**:89b, 359a, 615a; **2**:521a,b; **5**:539a; **6**:493a, 494a; **14**:397b
  Fossil Flora of Great Britain, **14**:397b, 462b
  Introduction to the Natural System of Botany
    editions of, **10**:13b; **13**:342b
  translations of works, **13**:463b*
Lindos (city), **4**:463b–464a
Lindsay, James Ludwig, 26th earl of Crawford, **5**:404a,b
Lindstedt, Anders, **4**:19b; **10**:249b
Linea gypsea, **9**:135a
Linear programming, **3**:141b
Linearity, **2**:215a
Lines
  asymptotic, **4**:257a
  complexes of, **11**:46a–b, 509b; **13**:133a
    quadratic, **11**:46b
  congruences of, **4**:257b; **9**:471a; **11**:46a–b
    concept of, **9**:476a
  conjugate, **4**:257a
  division of, **4**:569a–b
  intersection of, **4**:426b–427a; **7**:601a; **9**:156b, 447a; **15**:198a
  Segre, **12**:285a
  signed, **3**:469a
  in space, **2**:350a
  systems of, **4**:336a; **6**:85b; **12**:285a
  theory of, **2**:274a
  third-order, **9**:431a
  transverse, **3**:182a
Linfoot, Edward Hubert, **5**:183a
Ling lung k'ai fang, method of, **3**:254b
Linguistics
  medieval Arabic, **9**:89b
  seventeenth-century, **1**:292a; **2**:305b; **7**:376b; **9**:3b, 319b–320a
  eighteenth-century, **7**:495b; **8**:471a; **11**:140a
  nineteenth-century, **2**:209b; **4**:147b, 285b–286a; **6**:454a–b; **7**:389b; **11**:137b–138a, 149b, 263b; **14**:68b–69a, 569b; **15**:83a, 193b–194a, 194b–195a
  twentieth-century, **2**:210a, 212b; **3**:456b
  Amerindian, **2**:212b; **15**:761b–762a
  Celtic, **8**:308b
  Mesopotamian, **15**:635a
  see also Acoustics, linguistic; Etymology; Language; Philology; Phonetics; Physiology, linguistic; Semantics; specific languages
Link, Heinrich Friedrich (1767–1851), 8:373–374; **4**:288b, 289b; **7**:254a; **8**:87a*; **11**:592a, 593a*; **12**:592a
  students of, **9**:344b, 423a; **12**:420b
Linkage, genetic, **2**:58b; **6**:22b; **9**:282a, 520a, 564b; **11**:212b; **13**:134b–135a
  and age, **6**:22a
  autosomal, **11**:212a
  prediction of, **1**:505b; **13**:158a
  sex, **9**:521a,b; **11**:212a; **14**:432a
  in vertebrates, **6**:21b
Linnaea, **15**:82b
Linnaeus (or von Linné), Carl (1707–1778), 8:374–381; **2**:4b, 516b; **8**:460a; **9**:529a
  acquaintances of, **1**:434a, 486b; **2**:225a; **3**:365a; **4**:99a; **7**:199a; **8**:177b; **12**:143b
  and Artedi, **1**:305b–306a
  on Banks, **1**:433b
  biographies of, **11**:382b
  botany, **1**:124b; **5**:379b; **6**:219a; **8**:177a, 395a, 436a, 483b; **9**:618b; **11**:83b, 218a; **12**:515a; **13**:450b
  and Colden, **3**:344a, 351a
  correspondence of, **3**:350a,b; **5**:76b, 380a,b, 429a; **7**:199b; **12**:471a; **13**:391b, 565b; **14**:132a; **15**:429b
  critics of, **1**:58b, 59a; **2**:577b, 580a; **3**:579b; **7**:585a; **9**:140a, 253b, 254a, 434a; **10**:284a; **12**:588b; **14**:387a; **15**:112b
  crystallography, **11**:520b, 521a,b, 522a
  defenders of, **1**:70b; **2**:509a; **3**:320a, 579a
  Genera plantarum, **7**:199b
    translation of, **3**:578a
  geology, **3**:210b; **7**:412a; **8**:147a, 376a, 379a
  herbarium, **6**:521a; **7**:198b; **12**:471a; **13**:177b
  Hortus Cliffortianus, **4**:295a
  influence of, **2**:257a; **10**:509b; **12**:108b, 202a, 471a–b; **13**:391b, 409b, 425a
  influences on, **11**:588a; **13**:215a, 443a; **15**:81a
  Mantissae Plantarum
    translation of, **3**:578a
  medicine, **8**:374b–375a, 379a–b
  nomenclature, **1**:524a; **3**:535a; **6**:182a; **7**:58b, 547b; **8**:375a, 377b, 379a; **9**:390a, 400b; **10**:116b, 284b; **15**:268b
  Philosophia botanica, **6**:454a; **7**:199b; **14**:386b; **15**:68a
  Species plantarum, **1**:431b; **7**:205a, 210b; **8**:435b; **9**:390a
    editions of, **14**:386b–387a
  students of, **1**:45b, 554a; **4**:512a,b; **5**:74a, 77a; **7**:210a; **8**:379b; **12**:515a
  Systema naturae, **1**:306a; **3**:119b*, 320a; **8**:460a; **12**:185a
  Systema vegetabilium
    editions of, **10**:531b
  taxonomy, **1**:69a, 359b; **2**:422b, 440a, 519a; **3**:351a, 495a*; **6**:66a, 400b; **7**:197a; **8**:48b; **9**:368a; **12**:515b; **13**:174a, 425b
    and chemical nomenclature, **2**:7b
    nonscientific use of, **2**:315b
    overthrow of, **1**:520b
  theology, **8**:379b
  translations of works, **12**:471a; **14**:463b
  on Vaillant, **13**:554a
  zoology, **3**:129b; **5**:355a; **8**:378b–379a; **14**:299b
  see also Taxonomy, botanical, Linnaean
Linné, Carl von (the younger), **8**:376a; **12**:536b; **13**:177b, 392a

Linnean Society of London, **14**:135b
founding of, **1**:435a; **12**:471b
Linnemann, Eduard, **7**:281a
Linneon species, **8**:513a,b; **14**:100a; **15**:508a
Linseed oil, **13**:327b
Linth River, **4**:402a–b
Linus, Francis, **2**:377b
Lion
anatomy, **1**:263a; **4**:519b; **8**:204b
Liouville, Joseph (1809–1882), 8:381–387; **2**:288a; **4**:222b
acquaintances of, **2**:299a; **3**:559b; **13**:128a,b; **15**:200b
on contemporary scientists, **14**:221a
correspondence of, **2**:105a; **13**:250b
influence of, **5**:261b, 262a; **6**:307a, 308b; **9**:474a; **11**:471a
Jacobi-Liouville theorem, **7**:37b
Liouville's equation, **7**:387a
Liouville's Journal
see *Journal de mathématiques pures et appliquées*
mathematics
algebra, **3**:134a–b; **7**:167b; **13**:129b
analysis, **3**:228a; **5**:96b, 204a; **6**:391b–392a; **10**:249b; **13**:130a
geometry, **4**:102b; **11**:57b
number theory, **3**:128a; **8**:367b; **15**:65b
mechanics, **12**:118a
students of, **13**:375a; **14**:472a; **15**:65b
Sturm-Liouville theory, **2**:144a–b; **5**:608a; **7**:408a; **8**:384b–385a; **11**:59a; **14**:281b–282a
theorems of, **3**:54a, 140b; **8**:386a
Liouville's Journal
see *Journal de mathématiques pures et appliquées*
Lipase, **4**:618a
Lipids
in blood plasma, **8**:608a
composition of, **8**:276a
metabolism, **4**:604a
Lipkin, Libman Izrailevich, **3**:229a
Lipmann, Fritz Albert, **9**:359b
Lipogenesis, **1**:610b
Lipoproteins, **12**:546b
Lipotropism, **4**:297a
Lipperhey (Lippershey), Hans, **2**:305b; **5**:241a; **9**:336b; **12**:105b
Lippmann, Gabriel Jonas (1845–1921), 8:387–388; **3**:498a, 501a,b; **9**:372a; **11**:388a; **12**:496a
electricity, **1**:303b; **6**:371b
electrometer, **1**:536b; **4**:333b
photography, **7**:36a; **14**:348b
piezoelectricity, **3**:504b
Lips, Henricus Conrad, **7**:355b
Lipschitz, Rudolf Otto Sigismund (1832–1903), 8:388–390; **4**:4a; **6**:286a, 356b; **11**:407a; **13**:124b
analysis, **3**:263b; **11**:407b; **13**:80a
Cauchy-Lipschitz method of approximation, **3**:143a; **5**:204a
Lipschitz algebra, **8**:389a
Lipschitz condition, **8**:389a; **14**:345a
Lipschütz, Alexander, **2**:344b
Liquefaction
of air, **3**:299a; **4**:79b; **6**:94a; **10**:206a; **13**:454b; **14**:522b
of gases, **1**:129a, 161a, 304a; **2**:65a; **3**:603b; **4**:79a–80b; **6**:94a; **7**:220b–221a, 272a–b; **8**:47a; **9**:291a, 477b; **10**:206a–b; **14**:109b, 522a–b
critical temperature, **3**:11b
temperature of, **1**:586a
theories of, **7**:587a
Liquids
associated, **11**:278b–279a
compressibility, **10**:185b; **11**:530a; **13**:127a–b, 130b, 244b, 246b
pressure effects, **1**:128a
temperature effects, **1**:128a; **15**:85b
critical states, **11**:278b–279a
critical temperatures, **13**:389b
diffusion of, **4**:615a–b; **5**:494a–b
equation of state, **5**:587b; **8**:186b
expansion of, **2**:136a; **4**:240a
Fermi-liquid, **7**:617b
fluidity of, **8**:177a
friction in, **4**:591b; **6**:249a, 250a; **10**:551a–b; **12**:595a, 613a
intermolecular distance, **7**:272a
internal field, **3**:619b
magnetic properties, **6**:271a; **11**:241b
mixtures of, **4**:228b; **7**:462a, 529b; **11**:534b
and vapor pressure, **9**:107b
molecular volumes, **13**:389b
molecular weights, **11**:278b–279a
optical properties
scattering, **7**:622a; **11**:265b–266a
refraction in, **1**:201b; **6**:453a
refractive index, **5**:412b; **6**:355b
rotation of, **1**:598b; **3**:225a, 583b; **7**:54b–55a; **8**:384b, 559b–562b; **11**:498a
forms of, **7**:85a–b
solidification of, **1**:129a
spectroscopy of, **5**:497a
structure of, **15**:19a
surface films, **8**:24a
surface tension
and temperature, **11**:279a
surfaces, **8**:596a, 604b; **15**:359b, 360a
tensile strength, **4**:131a
theories of, **11**:507b
kinetic, **2**:466a; **5**:161a; **14**:620b; **15**:44a
quantum mechanics, **8**:478b
thermal motion in, **5**:161a
vaporization of, **3**:8b–9a
viscosities, **13**:389b
volume of
temperature effects, **1**:138b
weightless, **11**:21b
see also Fluids; Hydrodynamics; Matter, states of; Motion, in a liquid; Solutions; Superfluidity
Liris nigra, **4**:504b
Lisboa, João de (*d.* before 1526), 8:390
L'Isle, Armet de, **3**:500b
Lislet-Geoffroy, Jean-Baptiste, **3**:365b
Lissajous, Jules Antoine (1822–1880), 8:398–399; **11**:460b
acoustics, **14**:290a
Lissajous figures, **8**:398a, 399a
Lissauer, Heinrich, **14**:229a
List, Karl, **5**:430b
Lister, Joseph (1827–1912), 8:399–413; **2**:375a, 528a; **5**:564a; **7**:422b; **10**:408a; **12**:354b
antiseptic methods, **3**:185a; **7**:423a,b; **10**:384b
on contemporary scientists, **6**:13*; **7**:426a, 428b, 434b*; **12**:478b
Lister, Joseph Jackson (1786–1869), 8:413–415; **8**:400a
Lister, Martin (*ca.* 1639–1712), 8:415–417; **3**:181a*; **5**:439b, 535b; **12**:104a*
acquaintances of, **1**:318a; **8**:307b; **9**:323a, 539b; **11**:313a
and Banister, **1**:431b–432a
Historia sive synopsis methodice Conchyliorum, **1**:431b; **8**:307b
taxonomy, **13**:174a
zoology, **2**:591b; **11**:317a
Lister Institute
founding of, **11**:538b
Listing, Johann Benedict, **4**:1b; **6**:247a; **9**:431a; **10**:484b; **14**:205a
Listing's law, **6**:248a
Liston, Robert, **3**:531a; **8**:401a; **12**:502b; **13**:493a; **14**:397b
Listwanite, **7**:249a–b
Literary Gazette
founders of, **14**:128b
Literature
influence of science upon, **1**:489b
primitive, **2**:212a; **15**:642a
statistical studies, **14**:574a
see also Philology, Poetry
Litharge
medical use of, **14**:585a
Lithistida, **14**:626b
Lithium
bombardment of, **3**:329b; **7**:159b*; **12**:33b
discovery of, **2**:94b; **15**:7b
disintegration of, **8**:95a
isolation of, **2**:588b
pharmacology of, **9**:422b
preparation of, **9**:178a
salts of, **13**:468a
Lithium hydride, **6**:615a
Lithocysts, **11**:517b, 518a
Lithography
eighteenth-century, **2**:355b
nineteenth-century, **10**:118b, 583b; **13**:168a; **15**:472a
Lithology
eighteenth-century, **1**:244a; **4**:152a; **5**:205a; **8**:68b–69a; **11**:35a; **12**:517a–b; **13**:87a
nineteenth-century, **2**:495a,b; **4**:274a, 309a, 363b; **5**:538a, 548a; **6**:178a; **8**:594a; **10**:265b; **12**:140b
twentieth-century, **6**:229a
see also Petrography, Petrology, Rocks
Lithopedion, **1**:323a
Lithosphere, **1**:470b
Lithotomy, **4**:173a; **8**:114b; **9**:322b, 508a*; **12**:577b*; **14**:584b
see also Cystotomy
Lithotripsy, **11**:99a
Litke, Fyodor Petrovich (1797–1882), 8:417–418; **12**:114b
Litmus, **5**:325b; **7**:224a
Litorella, **7**:199a
Little, Ralph Bulkley, **12**:482b
Little, William Arthur, **8**:479a
Little Astronomy
see *Treasury of Astronomy*
Little Tower Street Academy, **13**:68a
Littlewood, John Edensor, **2**:238a; **4**:561b; **6**:113b, 451b; **7**:351b; **8**:584b; **11**:268b; **13**:263b–264a; **14**:95a, 344a
Littré, Émile, **3**:379a; **10**:354b; **11**:491b, 492a
Littrow, Joseph Johann von, **2**:600a; **9**:578b; **11**:610a; **14**:480b
Littrow, Karl Ludwig von, **6**:234b; **9**:201b; **10**:592b; **14**:242b
Liu Ch'uo, **3**:269b
Liu Hsin, **8**:419b; **13**:484b
Liu Hui (*fl. ca.* A.D. 250), 8:418–425; **3**:268b
Chiu-chang suan-shu, **3**:250a, 255a
commentaries on, **14**:539a–541b
geometry, **8**:320a; **13**:484b
Hai-tao suan-ching, **14**:545b–546a
Liu I, **3**:265b; **14**:539a, 544a
Liu Pi-chien, **14**:539a,b, 544b
Liu Ta-chien, **3**:266b
Liu Yu-chieh, **3**:266b
Liusternik, Lazar Aronovich, **12**:411a
Live force
see *Vis viva*
Liveing, George Downing, **4**:79a; **11**:611b

Liver
anatomy, **6**:317b; **9**:165b, 468b; **12**:81b, 82b; **14**:9b
and anemia, **3**:335b
blood-capillaries, **3**:112b
cancer, **15**:418a
as center of venous system, **1**:316a–b
connective tissue, **3**:112b
degeneration of, **4**:297a
fatty, **14**:252b
embryogeny, **2**:383b
excision of, **14**:587a
as food, **9**:417a
function of, **3**:356b; **5**:234a, 426a; **6**:300a; **9**:9a; **11**:467a–b
functional tests, **4**:618a
glycogen in, **6**:287a
hematopoiesis, **7**:438a
histology, **4**:276a, 589b; **5**:426a; **8**:254a; **9**:56b–57a, 416a; **12**:41a
structural units, **9**:57a
inflammation of, **4**:589b
metabolism, **1**:454a; **4**:360a; **8**:614b; **10**:22a
glycogenesis, **2**:26a, 29a–b; **8**:123a, 615a; **9**:506a; **12**:165a; **15**:109b
parasites, **10**:477a
pathology, **2**:464a; **8**:254a; **11**:100a
perfusion of
nineteenth-century, **2**:29a
twentieth-century, **4**:360a
sugar in, **1**:471a; **2**:29a; **15**:109b
Liver fluke, **6**:219b; **8**:270b; **11**:342a
Liverpool
Botanic Garden, **2**:335b
Fever Hospital, **2**:335b
Liverworts, **6**:493a; **10**:13b; **13**:416a
culturing of, **10**:13b
evolution of, **15**:149b
and ferns, **3**:30a–31a
and mosses, **6**:219b
reproduction, **2**:160a; **3**:422a
taxonomy, **15**:149a–150a
see also Cryptogams
Livingston, Burton Edward (1875–1948), 8: 425
Livingston, Milton Stanley, **3**:329b; **8**:94b
Livingstone, David, **2**:528b; **8**:103b, 613b; **9**:81a, 585a
Livy, **1**:213a, 214a
Lizards, **2**:322b
paleontology, **2**:504b; **4**:549a
Llobet of Barcelona, **11**:321b
Lloyd, Bartholomew, **6**:86a
Lloyd, Humphrey (1800–1881), 8:426–427; **5**:137a; **6**:88b; **8**:591b; **12**:86b
geomagnetism, **1**:364a; **11**:555a*; **12**:49a, 50a, 50b–51a
Lloyd, John Uri (1849–1936), 8:427–428
Lloyd's reagent, **8**:427b
Lloyd Brothers, **8**:427a,b
Llwyd, Edward
see Lhwyd, Edward
Lo Shih-lin, **3**:266a, 269a; **14**:540a
Lob Nor (lake), **11**:181a
Lobachevsky, Nikolai Ivanovich (1792–1856), 8:428–435; **3**:224b; **5**:298b, 307a; **7**:206b, 473a
algebra, **3**:555a
biographies of, **7**:207b
commentaries on, **7**:7a
editions of works, **7**:207b, 473b
Geometrische Untersuchungen zur Theorie der Parallellinien, **2**:271a
geometry, **1**:591a, 599b–600a; **2**:274b; **3**:322b; **4**:371a, 417b; **5**:302b; **7**:207a, 397a, 600a; **8**:429a–434b; **12**:533a; **13**:414b
Lobachevskian function, **8**:430a
Lobachevskian spaces, **8**:434a; **10**:486a; **12**:609a,b; **13**:414b
translations of works, **4**:371a; **6**:76b, 522a
trigonometry, **9**:410a
L'Obel (or Lobel), Mathias de (1538–1616), 8:435–436; **4**:139a; **5**:362a; **8**:121a; **10**:426b*, 489a; **11**:312a, 527b; **13**:463a; **14**:586b
Lobelia, **8**:436a
Lobotomy, **4**:287a(n)
Lobster, **12**:479b
anatomy, **14**:408a
Location
see Place
Lock, Charles George Warnford, **10**:37a
Lock, Robert Heath
Recent Progress in the Study of Variation, Heredity and Evolution, **9**:564b; **14**:432b
Locke, John (1632–1704), 8:436–440; **5**:62a, 120a, 286b; **14**:380a
acquaintances of, **10**:45a; **12**:457b; **13**:214a–b; **14**:405b
Condillac on, **3**:381a
correspondence of, **9**:465a–b
critics of, **1**:322b; **9**:383b; **15**:57a*
Elements of Natural Philosophy
translations of, **3**:119b
Essay Concerning Human Understanding, **2**:614a*; **3**:380b; **6**:139a; **9**:465a; **12**:106a; **13**:214b
influence of, **2**:16a, 577b; **3**:598b; **4**:86a; **6**:37a, 555b; **7**:596a, 597b; **9**:463b; **10**:611b, 613b; **11**:140a; **14**:83a, 288a–b
influences on, **10**:173a
on language, **14**:380a(n)
and Leibniz, **8**:151b, 153a–b
medicine, **13**:214b
physiology, **6**:46a(n)
sensationalism, **1**:111a, 114b; **8**:437b–438a; **14**:367a–b
Lockert, George, **2**:603b
Locks, hydraulic, **2**:104a; **3**:388b, 578b; **5**:101a, 410b; **13**:50a; **15**:662a(n)
Lockyer, Joseph Norman (1836–1920), 8: 440–443; **5**:127a*; **6**:565a
acquaintances of, **10**:483a; **11**:297a, 485b; **12**:238a
astronomy, **12**:19a; **14**:629a
chemistry, **3**:478b
and Fowler, **5**:101b–102a
and Huggins, **6**:542b–543b
spectroscopy, **4**:270a
dissociation hypothesis, **4**:79a, 180a; **5**:102a
enhanced lines, **1**:56a
helium, **5**:126b; **11**:279b
solar prominences, **6**:28a, 542a; **7**:75a; **12**:268a
Locomotion
ancient theories, **1**:118b, 264b
seventeenth-century theories, **6**:158a–159a
eighteenth-century theories, **10**:79b
and brain, **9**:10a
mechanics of, **9**:102a; **11**:571b
Locomotives, **8**:365a; **12**:527a
steam, **3**:287a; **10**:286a; **12**:287b, 288a
Locus, **5**:569a; **8**:306b; **13**:305a
classification of loci, **3**:322a–b; **4**:429a; **10**:295b–296b
definition of, **4**:429a
for four lines, **1**:185b, 245a,b; **4**:427a, 427b–428a; **10**:295b, 296a–b
linear, **4**:429a
plane, **1**:188b–189a; **4**:567b–568a; **5**:408b
solid, **1**:245a; **4**:427b–428b, 568b–569a; **5**:408b; **9**:276a(n); **10**:331a
surface, **1**:245a; **4**:429a; **9**:300a; **10**:295a, 298a
for three lines, **1**:185b, 245a,b; **4**:427a, 427b–428b; **10**:295b
Locust (bot.), **8**:178a
Locust (insect), **6**:298a
as food, **9**:441a
Loder, Just Christian von, **5**:442b
Lodestone, **5**:397b–400b *passim;* **10**:534b
Lodge, Oliver Joseph (1851–1940), 8:443–444; **1**:297a, 456a; **3**:481b*; **9**:373a; **12**:238a
acquaintances of, **6**:343b; **13**:356b
atomic theory, **13**:364a
on contemporary scientists, **6**:212b*; **11**:615b*; **13**:523a,b
electromagnetism, **5**:15a; **6**:345a; **11**:93b, 122b
radiotelegraphy, **13**:357a
Lodoicea, **12**:536a
Loeb, Jacques (1859–1924), 8:445–447; **3**:390a; **5**:40a; **9**:525a; **10**:155a, 251b; **15**:201b
acquaintances of, **4**:616b; **5**:463a; **9**:516b; **12**:108b; **15**:466b(n)
biology, **4**:12b; **6**:337a; **7**:98b; **8**:102a–b; **9**:10a; **14**:434b
psychology, **11**:518b
Loeb, Leo (1869–1959), 8:447–448; **6**:132a; **15**:75b
Loeffler (Löffler), Friedrich August Johannes (1852–1915), 8:448–451; **4**:276b; **5**:219a; **7**:423a, 424a, 425b, 434a*
Klebs-Loeffler bacillus, **4**:404b, 276b; **7**:425b; **8**:448b–449b; **11**:569a; **14**:551b
Loeffler, Gottfried Friedrich Franz, **8**:448a
Loefling, Pehr, **8**:379b
Loeser, Hanns, **13**:425a
Loess, **9**:593b; **10**:169a; **12**:168a
distribution, **10**:167b
origin of, **1**:623a; **3**:462b; **4**:145b; **7**:511b; **10**:128a, 166b, 167b, 429a; **11**:209b–210a, 439b
Loew, Ernst, **7**:417b
Loewi, Otto (1873–1961), 8:451–457; **12**:319b; **15**:73b, 105a, 106b, 107a
Loewinson-Lessing
see Levinson-Lessing, Frants Yulevich
Loewner, Charles (Karl) (1893–1968), 8: 457–458
Loewy, Alfred (1873–1935), 8:458–459
Loewy, Maurice, **13**:37a
Atlas de la lune, **5**:316a
Log (naut. instrument), **2**:343b; **5**:593b; **10**:151b
Logan, James (1674–1751), 8:459–461; **3**:344a; **5**:434a
and Bartram, **1**:486a–b, 487a
hybridization, **3**:351a
Logan, William Edmond (1798–1875), 8: 461–462; **2**:128b; **3**:607b; **4**:364b; **6**:57b, 564a, 565b*; **11**:505a; **12**:293a,b
Logan's line, **2**:129a; **8**:462a
petrography, **1**:51a; **6**:565a
Logarithmica, **6**:600b, 607b
Logarithms
Renaissance use of, **11**:591a
seventeenth-century use of, **2**:527a, 602b; **3**:152b; **5**:588b; **6**:271b; **10**:151b; **12**:75b
eighteenth-century use of, **9**:159a; **10**:109b*
nineteenth-century use of, **2**:441b
twentieth-century use of, **13**:431b
of algebraic numbers, **5**:342a

Logarithms (cont.)
antilogarithms, **2**:602b
base 2, **1**:84a
base 10, **3**:61a, 432a
base 10^9, **3**:61a
Briggs, **2**:462a–463a; **3**:432a; **4**:551a
calculation of, **2**:462b; **3**:459a; **6**:600a; **10**:48a; **11**:472a; **12**:229a, 345a; **13**:60a, 266b
of complex numbers, **4**:477a
construction of, **9**:311a
definition of, **9**:612a
hyperbolic form, **2**:462a–b; **8**:306b*
of imaginary numbers, **4**:515b
in integration, **3**:223a
invention of, **9**:610a–613a
iterated, **7**:351b
Kepler's, **7**:304b, 305a
multivalued, **3**:137a
naming of, **9**:610a
natural, **2**:507a, 602b; **3**:432a; **7**:559a; **8**:164a
of negative numbers, **4**:477a; **8**:166a
Oriental, **1**:84a, 90b
parabolic, **5**:573b*
popularization of, **4**:551a
of a sum or difference, **3**:152b
theory of, **12**:446b
arithmetic, **9**:304a
see also Tables, logarithmic
Logic
ancient Greek, **4**:423b, 461b–462b; **10**:324b–325a; **11**:24a–b; **14**:607a–b, 613a
ancient Roman, **2**:229a–233a; **5**:232a; **12**:340b
medieval Arabic, **8**:547b; **9**:28a, 604b; **12**:91a; **13**:292b, 509a,b, 510a; **15**:494a, 497a
medieval Latin, **1**:94a, 94b; **2**:391a–b, 603b–604b, 609a–b; **4**:447a; **6**:376b–378a; **7**:66a; **8**:547a–550a; **9**:136a–b; **10**:172b–174b, 419b–420b; **13**:184b, 185a, 193b; **14**:391a–b
parva logicalia, **2**:609b
systematization of, **1**:2a
Renaissance, **3**:171b, 172a, 262a, 280b, 420b; **4**:237b, 450b–451a; **8**:100b, 309b; **9**:32b–33a; **11**:286b; **14**:47a, 580b–581b
seventeenth-century, **2**:47a, 601b; **3**:60a; **7**:193a–b, 194b; **8**:154a–155b; **12**:55a–b; **13**:613b
eighteenth-century, **3**:387a, 462a*, 485a; **6**:556a–b; **7**:597b–598a; **11**:392b*; **14**:482b–483a
early nineteenth-century, **1**:614b; **2**:277b–278a; **7**:389b; **14**:288a–289a
mid nineteenth-century, **2**:295a–298b; **3**:452a–b; **4**:35b–36b; **6**:81b–83b; **7**:105a–b; **9**:383b–384b
late nineteenth-century, **1**:404a; **4**:136b, 137b–138a; **5**:152b–154b; **8**:542b; **10**:484a–486b; **13**:612a–613a
early twentieth-century, **2**:512b, 514a; **3**:455b–456b; **4**:102a, 188a, 226b; **5**:350b–351a; **6**:5a; **7**:151a–b; **11**:107a–b, 357b
mid twentieth-century, **3**:278a–b, 335b*; **11**:260a; **14**:89b
Aristotelian, **2**:295a, 296a–b; **3**:513b; **6**:81b
associative principle, **14**:304a
axioms of, **3**:278b
Boolean, **7**:151a
of classes, **4**:137b; **7**:105b; **8**:154b, 159b; **11**:94b
classical, **5**:350b
deductive, **10**:442b
diagrammatization, **4**:137b; **8**:548b–549b; **10**:483b; **13**:612a–613a
Eleatic, **1**:149a
equational, **7**:105b
formal analysis, **14**:288a–289a
foundations of, **3**:278b; **5**:350b–351a; **7**:151a; **14**:288a–289a, 468a,b
infinitary, **5**:351a
intuitionist, **5**:350b; **8**:546b
many-valued, **8**:546a
mathematical
eighteenth-century, **7**:598a; **12**:283b; **13**:39a
mid nineteenth-century, **2**:294b; **6**:81b–83b; **8**:590a–b
late nineteenth-century, **2**:593b; **5**:152b; **10**:442a–443a; **11**:94b–95a; **12**:216b–217a; **13**:550b, 551a
early twentieth-century, **5**:10a, 350b–351a; **6**:297a–b; **8**:262b–263a, 546a–b; **10**:274a; **11**:285b; **12**:12a–16b; **14**:304a–b
mid twentieth-century, **9**:71a–72a; **13**:497b–498a
ordinal, **13**:498a
modal, **8**:546a,b
predicate, **5**:350b; **14**:304a
propositional, **5**:153a; **8**:262b–263a; **11**:94b; **14**:607a
multivalued systems, **11**:107a
Ramist, **1**:126a–b, 134a
and science, **8**:81a–b
sentential, **14**:304a
symbolic, **2**:295a–298b; **4**:138a; **7**:105a–b; **8**:154a–155a, 549a; **10**:606a–b; **11**:107a–b; **12**:12a, 14a–15a, 217a, 283b
terminist, **8**:100b; **14**:391a
triadic, **10**:485b
see also Calculus, propositional; Dialectic; Historiography, of logic; Mathematics, and logic; Notation, logical; Propositions; Syllogism; Synthesis, logical
Logica moderna, **6**:376b–377b; **10**:174b
Logical piano, **7**:105b
Logical positivism, **3**:379a; **5**:122b; **11**:355b, 356b–357a; **12**:177b
Logicism, **2**:304b; **14**:304a
Logistic (arithmetic), **4**:111a
Logos, **5**:439a; **6**:289b, 290a; **12**:311b; **14**:607a
Lohest, Marie Joseph Maximin (called Max) (1857–1926), 8:462–464; 5:109b, 110a*
Lohmann, Karl Heinrich Adolf, **6**:501b
Löhneyss (or Löhneis or Löhneyssen), Georg Englehardt von (1552-*ca.* 1625), 8:464–465; 4:394a
Lohrmann, Otto, **6**:303b
Lohrmann, Wilhelm Gotthelf, **1**:568b; **9**:2a
Lohse, Wilhelm Oswald (1845–1915), 8:465; 6:28a; **12**:579b*; **14**:55a, 57a*
Loire (river), **2**:347b
Loire basin
geology, **5**:497b
Lokhtin, Vladimir Mikhaylovich (1849–1919), 8:465–467
Lombard, Jean, **10**:489a, 491a
Lombard, Peter
see Peter Lombard
Lomonosov, Mikhail Vasilievich (1711–1765), 8:467–472; 1:67b; **4**:471a, 472b; **6**:259b; **8**:251b; **9**:17a; **10**:102a*; **11**:609a,b; **12**:514a
astronomy, **2**:5a
electricity, **11**:433a–b
Lomonosov's phenomenon, **12**:353b*
philosophy, **12**:514b
works translated, **5**:75a
London, Fritz (1900–1954), 8:473–479; 4:579a; **8**:52a, 186a, 479b–483a *passim;* **12**:419a; **15**:448a
London forces, **14**:111b
London, Heinz (1907–1970), 8:479–483; 8:52a, 292a, 475b–479a *passim*
London order parameter, **8**:476a
London
Adelaide Gallery, **12**:131b
Apothecaries' Garden, **6**:539a
cholera, **12**:503a
College of Physicians, **3**:12b–13a, 13b, 208b, 209a; **4**:377a,b; **5**:157a, 348b, 425b; **9**:400b; **14**:194b, 253a, 507a
founding of, **8**:361a–b
library, **9**:313a
Pharmacopoeia, **4**:209b; **5**:396b–397a; **12**:458a
physic garden, **5**:361b
College of Surgeons, **3**:323b
Hunterian museum
see London, Hunterian museum
research institute, **7**:279a
Geological Museum, **5**:39a
Great Fire, **14**:510a
Guy's Hospital, **2**:464a–b; **10**:177a
Houses of Parliament, **12**:489a
Hunterian Museum, **3**:323b–324a; **6**:478b–479a, 568a; **7**:278b; **10**:260b
London Exhibitions, **1**:9a; **2**:355a; **3**:476b, 586b; **6**:237b; **8**:462a, 572a; **11**:36b, 79a; **12**:26a, 132a, 519b
Middlesex Hospital, **6**:569a
Medical School, **9**:241b
Museum of Practical Geology, **2**:567a; **4**:10b–11a; **6**:591b
Ophthalmic Infirmary, **8**:97b
population of, **5**:507b; **15**:304b, 305b
rebuilding of, **4**:496a; **6**:483b
St. Mary's Hospital, Paddington, **5**:29a; **14**:511b–512a
St. Paul's Cathedral, **4**:496a; **14**:419a, 510a
surveys of, **13**:96a
water supplies, **5**:125a, 128b; **14**:169b, 513b
see also Royal Society of London
London, University of
in nineteenth century, **3**:88b–89a; **8**:529a
founding of, **5**:530b
Imperial College of Science and Technology
founding of, **6**:26a
library, **4**:36b
reorganization of, **5**:83b
London Basin
paleontology, **10**:322b
London Geological Society
founding of, **10**:585b
London Journal of Botany
founding of, **6**:495a*
London Mathematical Society, **13**:217b
founding of, **4**:35b; **11**:297a; **13**:427a
London Pharmacopoeia, **12**:458a
London School of Tropical Medicine
founding of, **9**:83a
Long, Perrin Hamilton, **4**:154b
Long, Stephen Harriman, **6**:120a; **8**:103a; **10**:439b; **12**:133a
Longet, François Achille, **2**:30a; **12**:164b; **15**:108a,b, 109a
Longevity, **1**:582b
human, **5**:45a*; **10**:445a
Longfellow, Henry Wadsworth, **12**:204b

Longitude, determination of
ancient Greek, **13**:85a
ancient Roman, **11**:200a
medieval, **1**:29a; **2**:149a, 153a, 435b; **7**: 119b
Renaissance, **1**:179a; **5**:349a; **14**:275a; **15**:88a
seventeenth-century, **1**:541a; **3**:101b, 103a; **4**:214a; **5**:23a; **6**:299a, 520b; **7**:375b; **8**:25b; **9**:527b–528a; **10**:489a, 490a; **11**:424a, 525b
eighteenth-century, **1**:547b; **2**:43a, 388b, 389a; **3**:104b, 198a; **4**:23a,b, 603a; **5**:88a, 435a; **7**:374a; **10**:572a, 591b, 607b, 615a; **11**:609b; **12**:414a*; **15**: 349b
nineteenth-century, **1**:364a; **2**:97b, 285b, 533a; **5**:258a, 479b; **6**:263a, 325b; **7**:19a; **8**:487b; **9**:421b, 551b; **10**:219b, 479b–480a, 528b*; **12**:234b; **13**:37a, 109a, 111b, 120b
twentieth-century, **4**:278a, 408a*
chronometric, **6**:402b
eclipse observations, **2**:149a, 153a; **3**:198a; **5**:343b–344a, 435a; **9**:233b; **10**:490a; **14**:275a; **15**:220a
by lapsed time, **5**:349a
lunar observations, **10**:110b–111a, 490a; **12**:518b
magnetic, **1**:364a
photographic, **8**:388a
at sea
Renaissance, **5**:399a; **8**:390a; **13**:49b
seventeenth-century, **1**:138a; **3**:63b; **5**: 23a, 242b, 343b–344a; **10**:176b; **11**: 532b; **12**:611a; **13**:96a
eighteenth-century, **2**:43a, 389a; **4**: 481b, 517a; **5**:491b; **9**:234b; **10**:614b; **11**:495b–496a; **13**:179a, 184a; **15**: 325a
nineteenth-century, **5**:17a; **11**:554a; **15**:225b
chronometric, **6**:130b, 482b, 604b, 605a; **9**:163a
Jupiter's satellites, **6**:520b; **9**:163a
lunar observation, **2**:389a; **6**:70a, 299a; **7**:544b; **8**:25b; **9**:162b–163a, 527b–528a
systematic observations, **2**:343a
telegraphic, **1**:364a; **3**:261b*; **4**:354a, 370a*; **5**:479b; **8**:388a, 504a; **13**:37a
Longolius, Gisbert, **13**:502a
Longomontanus
see Severin, Christian
Lonicera, **8**:483b
Lonicerus (Lonitzer), Adam (1528–1586), 8: 483–484
Naturalis historiae opus novum, **14**:507b
Lonicerus, Johann Adam, **8**:484a
Lonitzer, Adam
see Lonicerus, Adam
Lonsdale, Henry, **7**:414b, 415b
Lonsdale, Dame Kathleen Yardley (1903–1971), 8:484–486; **1**:319a; **6**:304b*
Lonsdale, William (1794–1871), 8:486; 5: 436b; **9**:584a
Loomis, Alfred Lee, **14**:498a; **15**:60a
Loomis, Elias (1811–1889), 8:487; 4:410a
cold waves, **9**:530b
Loomis, Francis Wheeler, **5**:609b
Loosestrife
heterostylism, **3**:575a
Lopadorhynchus, **7**:402a
Lophius, **4**:377b
Loranthaceae, **5**:539b
Lorentz, Hendrik Antoon (1853–1928), 8: 487–500; 4:168a*; **6**:612b*; **8**:11b; **12**:527b
acquaintances of, **2**:466a; **3**:430a; **4**:293a; **5**:610a; **11**:422a; **14**:597b
on contemporary scientists, **7**:221a, 466a*, 516a*; **14**:599a*
correspondence of, **11**:9b
and Einstein, **4**:312b; **8**:489a, 490a, 498a–499a
electromagnetic theory, **1**:24b–25a; **2**: 266a, 559b; **3**:310a; **4**:306a, 315a, 319b, 321a; **6**:347a; **9**:213a, 377a; **11**:479b, 530a–b
electron theory, **4**:191a, 315a, 317a, 319a–b, 320b, 323b; **5**:391a, 597b–598a; **7**:264a,b; **8**:39b, 488b, 493a–499a; **11**:58b, 421a, 447a; **13**:367a, 368a; **14**:208b
equations of, **1**:500a; **8**:495a
ether, **8**:492b–496a; **9**:215a–b, 372a
heat theory, **8**:493a
influence of, **2**:265b; **9**:616b; **14**:340b; **15**:39b
invariance, **4**:322b–323a
Lorentz force, **8**:495b; **9**:212a; **13**:364b
Lorentz group, **9**:38b
Lorentz transformations, **2**:594a; **4**: 322a,b, 323b, 331b; **8**:40b, 495b, 496a, 498a; **11**:58b; **14**:62a, 305a, 317b
derivations of, **5**:122b
Lorentz-FitzGerald contraction, **2**:382b–383a; **4**:320a–b; **5**:15a–b, 597b; **8**:10b, 40b, 495b–496a, 498a; **9**:215b; **13**: 471b
Lorentz-Lorenz formula, **7**:620a; **8**:493a, 501b
optics, **9**:372b
relativity, **8**:498b; **11**:13b; **14**:317b, 598b–599a
students of, **10**:235b; **11**:476b; **13**:12a
on time, **4**:321b
Zeeman effect, **3**:420a; **8**:496a–b; **11**:612b; **14**:597a–b, 598a
Lorenz, Hans (1865–1940), 8:500–501; 5: 502a*; **11**:478a; **14**:618a
Lorenz, Ludwig Valentin (1829–1891), 8: 501–502; 15:83b
electrodynamics, **6**:344a; **7**:381a,b; **8**: 491a; **9**:217b; **10**:27b; **14**:33b*
Lorentz-Lorenz formula, **7**:620a; **8**:493a, 501b
Lorenz law, **8**:502a
Lorenz, Richard (1863–1929), 8:502–503; 6:365a
Lorenzen, Paul, **6**:473a
Lorenzi, Lorenzo, **1**:274a, 611b
Lorenzini, Antonio, **3**:60a
Lorenzoni, Giuseppe (1843–1914), 8:503–504
Loria, Gino (1862–1954), 8:504–505
correspondence of, **12**:228a
geometry, **12**:284a
historiography, **3**:215a*, 467a; **4**:103b*, 113a, 516b*, 605b; **5**:112b*; **6**:112b, 411a; **7**:164a*; **8**:381b, 386a; **9**:80b*, 157b*, 191a, 378a; **10**:529b; **12**: 173a*; **13**:132a*, 255a; **15**:702a*
translations of works, **14**:336b
Lorraine
fauna of, **3**:492b
Lorraine Shales, **4**:363b
Lorry, Anne Charles (1726–1783), 8:505–507; 12:104a*
Lortet, Marius, **8**:354a
Lory, Charles, **13**:283b
Los Alamos Scientific Laboratory, **4**:582a; **5**:273a; **8**:95b; **10**:215b
Loschmidt, Johann Joseph (1821–1895), 8: 507–511; 1:168a; **9**:222b
gas theory, **13**:11a
irreversibility, **2**:263b; **4**:293b; **11**:11a
Loschmidt's constant, **8**:510b
molecular size, **2**:261a; **14**:185b
Losh, William, **4**:250a
Lossen, Karl August (1841–1893), 8:511–512; 5:548b*
Lotka, Alfred James (1880–1949), 8:512
Lotsy, Jan Paulus (1867–1931), 8:512–513; 7:166b; **10**:254a*; **15**:508a
Lotze, Hermann Rudolph (1817–1881), 8: 513–516; 2:1b; **6**:25b, 342b, 434a; **9**:561b; **14**:233a
logic, **12**:217a
philosophy, **4**:203b; **12**:243a
physiology, **6**:248a; **14**:113b
Loudon, John
Encyclopaedia of Plants, **8**:371b
Louis IV of Bavaria (king of Germany and Holy Roman emperor), **5**:113b; **10**: 172a
Louis IX, King of France, **14**:34b
Louis XIII, King of France, **7**:536a,b; **10**: 489b; **13**:613b; **15**:69a
Louis XIV, King of France, **2**:201a; **3**:71a, 97b, 267b, 421a; **4**:135b, 603a; **5**: 158b, 417b; **6**:598b, 599a; **8**:149b; **9**:118a; **10**:521b; **11**:47a, 423b, 525b, 526b; **12**:127a, 584b, 585b; **13**:553b, 591a; **14**:49a
Louis XV, King of France, **2**:576b; **3**:6a, 426b; **4**:22b, 515b; **5**:491b; **7**:183a, 199b; **8**:115a, 177b, 178b, 352a, 506a; **11**:68a; **12**:63b, 302b, 429b
Louis XVI, King of France, **3**:72a, 106a; **4**:14b; **5**:603a,b; **9**:53b; **11**:99b, 522b; **12**:64b; **13**:494a, 495b
Louis XVIII, King of France, **2**:106a, 355b; **3**:106b; **5**:94a, 603b; **6**:2a; **11**:100a, 166b; **15**:346b
Louis, Pierre Charles Alexandre, **2**:366a, 574b; **6**:59a; **7**:44b
Louisiade Archipelago, **2**:342b
Louse, **7**:286a
anatomy, **6**:484a
and disease, **5**:270b, 451b; **11**:442b; **14**: 623b; **15**:454a–b
microanatomy, **8**:128a
plant, **15**:52b
reproduction, **8**:128a
Louvain University, **14**:4a
Louville, Jacques Eugène, **6**:512b
Louyet, Paulin, **9**:451a
Love, Augustus Edward Hough (1863–1940), 8:516–517; 3:135b; **6**:113a; **8**:599a
Love waves, **8**:516b
Love's numbers, **8**:516b
Lovejoy, Arthur Oncken (1873–1962), 8: 517–518; 9:189a
Lovelace, Lady, **1**:356b*
Lovell, John Harvey (1860–1939), 8:518
Lovell, Robert
Panzoologia, **9**:441a
Lovén, Sven (1809–1895), 8:518–519; 11: 380a, 381a*, 593a; **12**:107a
Lovits, Georg Moritz, **8**:519b
Lovits (Lowitz), Johann Tobias (Tovy Yegorovich) (1757–1804), 8:519–520; 7:378a
Low, George Carmichael, **9**:83a
Lowell, John Amory, **8**:521a; **14**:533a
Lowell, Percival (1855–1916), 8:520–523; 10:602a; **12**:454a,b, 456a; **13**:474b
critics of, **14**:139b
Lowell Observatory (Flagstaff, Arizona), **8**:521b; **12**:454b, 456a

Löwenheim, Leopold
  Löwenheim-Skolem paradox, **6**:393a
Löwenstein, Frederick, **1**:287a; **4**:7a
Lower, Richard (1631–1691), 8:523–527; 9:245b, 400a; **14**:405b, 407a
  anatomy, **14**:406a, 407a,b
  circulation of blood, **9**:323b
  on combustion, **6**:484b
  physiology, **4**:37b; **9**:243a; **14**:407b
  Tractatus de corde, **9**:243b; **14**:407b
Lower, William, **6**:126a, 127b
Löwig, Carl Jacob, **1**:416b, 578a; **2**:620b; **7**:619b; **8**:553a
Lowitz, J. T.
  see Lovits, Johann Tobias
Löwl, Ferdinand, **3**:593b
Low-pressure studies, **7**:416b; **8**:23a–b
Lowry, Thomas Martin, **2**:499a; **8**:292a
Lowson, James Alfred, **7**:392a
Low-temperature studies
  mid nineteenth-century, **8**:1a–b; **12**:227a
  late nineteenth-century, **7**:220b–222a; **10**:206a–b, 604a; **14**:338b, 522b
  early twentieth-century, **5**:33a, 610a–b; **7**:272a–b; **11**:418a; **12**:437b–439a; **15**:201b, 438b–448a
  mid twentieth-century, **7**:618a; **8**:47a, 475b–479a, 479b–483a
  see also Cryogenics
Loxodrome, **10**:161a; **13**:49b
Loxomma, **6**:594b
Loys de Chéseaux, Jean Philippe, **3**:485b; **6**:331b; **10**:199a; **13**:110b
Lubbock, Sir John (Lord Avebury) (1834–1913), 8:527–529; 11:6a
  Lubbock nest, **8**:528b
  Prehistoric Times, **9**:538b
  prehistory, **9**:539a
Lubbock, John William (1803–1865), 8:530; 14:185a
  tides, **4**:591b; **14**:293a
Lubbock, Richard (*ca.* 1759–1808), 8:530–531
Lubieniecky, Stanislaus, **5**:575a
Lublin, University of
  founding of, **6**:433a
Lubricants, **13**:398b
  hydrodynamic, **8**:301a
  liquid nitrogen, **3**:299a
Lubrication, **12**:613a
  film, **11**:393a
  theory of, **10**:551a–b; **11**:393a; **12**:527a; **14**:620b; **15**:201b
Lubs, Herbert, **3**:290b
Luca, Sebastiano de, **2**:66a; **3**:47a, 48a
Lucas, Francis Robert, **8**:532a
Lucas, François-Édouard-Anatole (1842–1891), 8:531–532; 4:613b*
Lucas, Keith (1879–1916), 8:532–535; 5:64b, 65a, 81a; **11**:126a
Lucas-Championnière, Just Marie Marcellin, **8**:406b
Lucasian professorship, **1**:354b, 473b, 474a,b, 475b
Lucca
  observatory, **11**:82b
Luciani, Luigi (1840–1919), 8:535–536; 2:339a; **9**:93b, 94b*; **10**:175b
  Luciani triad, **8**:536b
  Luciani's phenomenon, **8**:536a
Lucifer, **2**:502a
Lucius Apuleius of Madauros, **6**:306a
Luck, **1**:251b
Luckhardt, Arno Benedict, **3**:69a
Lucretius (*ca.* 95–*ca.* 55 B.C.), 8:536–539; 1:380a; **4**:368b; **10**:44b
  atomism, **5**:107a, 286b
  influence of, **1**:567a; **2**:542a; **7**:27b, 28a, 607a, 612a; **10**:494a
  Renaissance interpretation of, **2**:543a
  De rerum natura, **4**:26a, 382b; **5**:106a; **8**:259b; **10**:82b; **14**:69b; **15**:33b
    translations of, **4**:495b
Lud (Ludd), Gauthier, **14**:127b
Ludendorff, F. W. Hans (1873–1941), 8:539–540; 3:262b*; **9**:564a*
Ludolph's number, **3**:181a
Ludwig II, king of Bavaria, **5**:570b; **10**:558b; **12**:422a
Ludwig, Carl Friedrich Wilhelm (1816–1895), 8:540–542; 6:300a,b; **15**:22a*
  acquaintances of, **2**:531a; **4**:200a,b, 204b, 614b, 615a; **6**:434a, 435b; **10**:431b; **12**:270a; **14**:199b
  on contemporary scientists, **14**:201b*
  Leipzig laboratory (Physiologische Anstalt), **4**:201a; **8**:8a
    students and assistants, **1**:9b; **2**:141b, 366b, 547b; **5**:26b, 184b, 279a,b, 541b, 563b; **6**:52b, 434b, 476a; **7**:504a–b; **8**:535b; **9**:56a, 57b, 380b, 416a; **11**:585a; **14**:496a
  philosophy, **4**:203b, 616b; **6**:226a, 242a; **8**:270b, 515b; **12**:243b
  Physiologie des Menschen, **13**:557b
  physiology, **1**:537a; **2**:62a; **13**:493b; **14**:113b
    cardiac function, **5**:150a; **11**:63a
    kidney function, **4**:264b; **12**:470b; **15**:102a
    techniques, **9**:101b; **11**:425b, 462b
  students of, **4**:614b; **6**:434a; **7**:519b; **9**:347b, 546b; **13**:10b; **14**:249a, 424a
Ludwig, Ernst, **7**:65a, 623a; **13**:476a
Ludwig, Karl (pseudonym), **2**:564a*
Ludwig-Maximilian University (Munich)
  chemistry at, **5**:203a
Lueroth (or Lüroth), Jakob (1844–1910), 8:542; 4:103a*; **12**:217b*; **14**:222a
  Clebsch-Lueroth method, **8**:542b
  Lueroth quartic, **8**:542b
  Lueroth theorem, **8**:542b
Lues
  see Syphilis
Lufthansa Corporation, **6**:216b
Lugeon, Maurice (1870–1953), 8:543–545; 1:235b; **5**:222b*; **6**:228b*; **11**:375a, 376a*; **13**:146a
  geology, **5**:221a,b, 222a*; **6**:228a; **12**:140b; **13**:285a
Luggin, Hans, **5**:621a
Luginin, Vladimir Fedorovich (1834–1911), 8:545; 7:204a*
Luillier, François, **5**:285a
Łukasiewicz, Ignacy, **15**:526b
Łukasiewicz, Jan (1878–1956), 8:546–547; 8:262b; **11**:107a
  Łukasiewicz system, **8**:546a
Lukens, Isaiah, **12**:131b
Lull, Ramón (*ca.* 1232–1316), 8:547–551; 1:80a
  alchemy, **10**:555a
  Ars magna, **1**:126a; **2**:163b
  commentaries on, **3**:513a; **5**:59a; **13**:210b
  influence of, **1**:158b, 571b; **2**:23a, 317b, 540a; **3**:359b; **5**:448a; **7**:376b; **8**:153b, 242a; **13**:535a
  on language, **14**:372a
  mathematics, **3**:277b
  De quadratura et triangulatura circuli, **2**:360b
Lullism, **2**:541b, 543b; **10**:310a
Lumbar puncture, **2**:529a
Lumbricales
  see Earthworm
Lumen (phys. unit), **2**:200b
Luminal, **5**:2b
Luminescence, **1**:519a, 556a–b; **12**:191a,b, 232a
  bacterial, **10**:236b
  and electricity, **5**:516a–b; **11**:68a
  in eyes, **2**:531a
  fluorescent, **1**:556b; **4**:356a; **6**:327a; **7**:375b; **10**:525b; **14**:497a
    alkali metals, **4**:253b
    energy output, **13**:598b
    explanations of, **4**:315b
    polarization of, **13**:599a
    spectra, **13**:599a
    Stokes's law of, **1**:458b; **13**:77a–b
  oxyluminescence, **4**:20b
  phosphorescent, **1**:67b; **3**:52a, 271b; **4**:214b, 215a; **5**:493a; **6**:278a, 477b; **8**:180a–b; **14**:419b*
    artificial, **7**:375b
    fungi, **4**:505a*
    of sea water, **3**:52a; **4**:157a*, 290a; **7**:375b; **12**:560b
    temperature effects, **1**:556b; **10**:552b
    theories of, **3**:600a; **5**:558b; **8**:256b, 293b; **10**:552b
    in vacuum, **1**:67b; **3**:477b; **4**:214b; **6**:170a–171b, 173b
    and X rays, **11**:58b
  radioactive, **5**:394a
  theories of, **5**:559a; **10**:552b
  and X rays, **1**:558b–559a
Luminosity, stellar
  see Stars, luminosity
Lummer, Otto Richard (1860–1925), 8:551–552; 8:50b; **10**:346a; **11**:8b, 149b; **13**:40b
  Lummer fringes, **8**:551b
  radiation, **7**:528a; **11**:12a, 150a–b
Lumpfish, **13**:527b
Luna, Emerico (1882–1963), 8:552–553
Lunar Society (Birmingham), **3**:578a, 579a; **7**:277a,b; **11**:141a; **14**:213b, 463a
Lunar theory
  see Moon, motion of
Lund
  observatory, **9**:460a
Lundmark, Knut, **6**:531a; **8**:582b; **12**:455b
Lundsgaard, Einar, **9**:359b
Luneberg, Rudolf Karl, **9**:225b
Lunes, **2**:37b; **3**:222a; **8**:240b
  see also Quadrature, of lunes
Lunge, Georg (1839–1923), 8:553; 5:621a; **9**:95b, 357b*; **14**:265a
Lungfish, **4**:147b; **7**:314a; **12**:298a
  physiology, **12**:470a
Lungs
  anatomy, **1**:267a, 619b; **9**:63a–b; **11**:510a; **12**:510b; **14**:408a
  blackening of, **10**:446a
  circulation in, **1**:303a; **2**:367a; **7**:502a; **9**:20a, 603b–604a
  diseases of
    diagnosis, **6**:576a
  drug absorption, **9**:296a
  elasticity of, **1**:303a
  embryogeny, **9**:253a
  epithelium
    oxygen secretion, **1**:454b
  gas exchange, **5**:149a, 150a; **7**:501b–502a; **8**:541b
  glucose in, **3**:220a
  histology, **8**:552b
  role of, **3**:356a–b; **8**:525b–526b; **9**:63a–b
  thrombocyte formation, **6**:526b
  volume of, **6**:575b–576a
Lupine, **6**:237b
  genetics, **14**:34a

Lupulin, **10**:530b
Lupus
treatment of, **4**:621a
and tuberculosis, **10**:18a
Lüroth
see Lueroth
Lusin, N. N.
see Luzin, N. N.
Lusitanus, Amatus (Rodrigues, João) (1511–1568), 8:554–555; 3:41a; **4**:508b; **9**:179a
Lusk, Graham (1866–1932), 8:555–556; 1:325b; **8**:90b*; **14**:66a, 67a*
Lustig, Alessandro (1857–1937), 8:556–557; 2:164b
Lusus naturae, **7**:412a
Lute, **4**:526a
Lutetia (mineral), **9**:110b
Lutetium
discovery of, **9**:544a; **13**:546a
Luther, C. Robert (1822–1900), 8:557; 1:616a; **2**:169b
Luther, Martin, **1**:80a; **5**:119a, 120b, 121a; **10**:448a; **13**:58b–59a; **14**:225b, 226a
Luther, Wilhelm, **8**:557b
Lutheranism, **1**:158b–159a, 480b; **2**:222a,b, 223a, 541b; **4**:212b; **10**:245a; **11**:457a; **13**:58b; **14**:529b; **15**:160b
critics of, **6**:254b; **10**:305b; **14**:225a–226b
Lutidones, **3**:347a–b
Lutugin, Leonid Ivanovich, **3**:236b; **9**:594a
Lutz, Anne Mae, **14**:101a
Lutz, Frank Eugene, **9**:520b
Luxation
see Dislocations
Luynes, Honoré-Théodoric-Paul-Joseph d'Albert (duc) de, **8**:44b; **10**:255b
Luyten, Willem Jacob, **5**:363a,b
Luzin, Nikolai Nikolaievich (1883–1950), 8:557–559; 4:287b; **8**:112a; **10**:102b*; **13**:35b
analysis, **4**:548a; **11**:156b–157a
students of, **7**:351b; **12**:410b; **13**:548b
Luzzi
see Mondino de' Luzzi
Lyapunov, Aleksandr Mikhailovich (1857–1918), 8:559–563; 3:225a, 226a; **4**:231a; **9**:125b; **10**:251a*; **13**:26a
analysis, **10**:249b; **13**:26b
on Chebyshev, **3**:223b
mechanics, **3**:584a; **7**:85a; **9**:450a*; **13**:26b; **15**:6b–7a
probability, **3**:231a; **9**:127b, 128a–b; **15**:32b
works edited, **7**:437a
Lycée des Arts (Paris), **5**:91b
Lyceum, Athens
see Athens
Lyceum Medicum Londinense, **3**:487b; **6**:567b
Lyceum of Natural History
founding of, **1**:330a
Lycopene
structure of, **15**:257b
Lycopodiales, **2**:492a
reproduction, **8**:5a–b; **15**:67b
Lydall, John, **14**:405a
Lydekker, Richard, **5**:46b
Lydus, John, **8**:191b; **10**:547b; **11**:18a
Lye
manufacture of, **1**:44b
Lyell, Charles (1797–1875), 8:563–576; 2:288b, 564a; **6**:437b; **9**:534a; **12**:517b
acquaintances of, **1**:72b; **2**:567a; **3**:607b, 608b; **4**:364b; **5**:71b, 334a, 436b; **6**:437b; **8**:104a, 169b, 260b, 528a; **10**:261b; **11**:133b; **12**:524a; **14**:397b
anthropology, **5**:207a; **11**:137b; **15**:51b
Antiquity of Man, **6**:510a; **9**:538b; **14**:516b
archaeology, **2**:341a
biogeography, **1**:73b
biographies of, **1**:395a
on contemporary scientists, **4**:519a*; **8**:595a; **12**:433b
correspondents of, **2**:375a; **6**:57b
critics of, **9**:583b, 584b; **12**:262b, 263b; **13**:383b
and Darwin, **6**:490b, 592b; **14**:135a–b
editions of works, **7**:480a
Elements of Geology, **5**:14a; **14**:136b–137a
erratic blocks, **3**:210b
and evolution, **3**:573b; **5**:513b; **8**:98a; **14**:190a
Geological Evidences of the Antiquity of Man, **14**:137a
geomorphology, **7**:185b, 186a; **11**:277a, 376b
and Hutton, **6**:584b
influence of, **1**:129b; **6**:8b; **9**:61a; **12**:525a, 570b; **14**:133b
influences on, **2**:617b; **3**:324a; **12**:262a
and Murchison, **9**:583a
paleoclimatology, **6**:503b
paleogeography, **5**:337b
paleontology, **1**:469a; **2**:481a; **4**:234a
Principles of Geology, **2**:497b, 570a,b; **3**:395b, 567a–b; **4**:68a; **5**:66b, 335b; **6**:457a–b, 509b, 584b; **8**:242b; **12**:184b, 262a, 276b, 514b; **14**:136b–137a, 138a
translations of, **11**:133b
stratigraphy, **4**:68a; **13**:145b
students of, **14**:315b
tectonics, **2**:496a; **3**:550b, 568a,b; **9**:584a
uniformitarianism, **1**:130a; **3**:434b, 552b; **5**:32a; **12**:276b
volcanology, **3**:552a; **11**:134a; **12**:263a
Lyencephala, **10**:262b
Lyginodendron, **12**:259b
Lyman, Benjamin Smith (1835–1920), 8:576–577; 8:261a*
Lyman, Chester Smith (1814–1890), 8:577–578
Lyman, Theodore (1874–1954), 8:578–579; 11:114a*; **12**:236a*; **13**:473b*; **14**:499b*
Lyman region, **8**:578b
Lyman series, **1**:425b
spectroscopy, **12**:235b
Lymantria
genetics, **5**:453b–454a
Lymph
formation of, **3**:69a
nature of, **9**:571b
toxicity of, **15**:469a
Lymph heart, **11**:126a, 296b; **12**:612b*
Lymphatic system
ancient Greek study, **6**:317b
seventeenth-century study, **5**:426a; **10**:477a–478a, 490b; **11**:586b–587a; **12**:40a; **13**:31a–32a; **14**:13a
eighteenth-century study, **3**:487a–b; **6**:63b, 367a,b, 566a; **9**:153a–b, 482b–483a
nineteenth-century study, **8**:541b; **11**:295b, 296b, 383a; **12**:617b; **13**:274a
twentieth-century study, **8**:552b
capillary system, **11**:374a
cisterna chyli, **1**:482b;**10**:477a
discovery as separate system, **1**:482b; **6**:367a; **9**:482b–483a; **11**:467a–b, 588a
ducts, **2**:237b; **4**:589b; **13**:31a–b
embryogeny of, **12**:48a–b
histology, **6**:434b
lacteals, **1**:316a–b; **6**:317b; **14**:13a
lymph nodes, **9**:64b
reticular cells, **2**:165a,b
sinuses of, **2**:165a
origin of, **9**:153b
thoracic duct, **1**:316a,b, 482b; **10**:477b; **11**:586b; **14**:13a
valves, **12**:40a; **13**:170b
Lymphocytes, **2**:531b; **6**:367b; **9**:586b*
aggregation of, **11**:374a
role of, **9**:550a
Lymphosarcoma
transmission of, **4**:499a
Lyonet, Pierre (1706–1789), 8:579–580; 12:556a,b; **13**:458a
Lyons, Israel, **1**:433b
Lyons, Councils of, **1**:99a; **5**:550a; **9**:435a
Lyons, University of
medicine at, **11**:373b
paleontological collections, **4**:39b
Lyophilization, **9**:586a
Lyot, Bernard (1897–1952), 8:581–582; 6:28b, 106a
Lyra (constellation)
alpha Lyrae
see Vega
beta Lyrae, **10**:608a
spectrum, **5**:199a; **9**:195a
variability of, **2**:147a; **5**:468b
Lysenko, Trofim Denisovich, **6**:22b; **9**:565a,b; **15**:19a, 425b, 426a, 510b–511a
Lyser, Michael, **1**:482b; **11**:587a
Lysergic acid
structure of, **15**:252b
Lysine, **7**:467a
nutritional role, **9**:285a; **10**:243a
Lysocythin, **5**:100a
Lysozyme, **5**:317a
discovery of, **5**:29b–30a
purification of, **5**:42a

M

Maanedsskrift for Dyrlaeger
founding of, **7**:101a
Maanen, Adriaan van (1884–1946), 8:582–583
nebulae, **3**:509b; **6**:529b; **12**:347b, 456a
van Maanen's star, **8**:582b
Mably, Abbé de, **3**:380a
MacAlister, Donald, **9**:226b, 227b
Macallum, Archibald Byron (1858–1934), 8:583–584; 3:351b; **8**:355a
McAtee, Waldo Lee, **12**:417a
Macaulay, Francis Sowerby (1862–1937), 8:584–585; 7:444b
Macbride, David (1726–1778), 8:585–586; 2:572a
critics of, **13**:46b
Experimental Essays on Medical and Philosophical Subjects, **8**:73b–74a
medicine, **10**:612b
pneumatic chemistry, **2**:177a; **6**:283a; **11**:144a
MacBride, Ernest William (1866–1940), 8:586
McCallien, William John, **1**:394b
MacCallum, William George, **11**:556a
McCarthy, Joseph Raymond (senator), **7**:247b; **10**:217b; **12**:350b
McCarty, Maclyn,**4**:604a
McClean, Frank, **5**:405b; **10**:31b
MacClelland, John, **5**:539a
McClelland, John Alexander, **13**:366a, 367b; **14**:420b

McClintock, Barbara, **14**:432b
McClung, Clarence Erwin (1870–1946), 8: 586–590; 7:286a*; **9**:518b; **13**:156b, 157a; **14**:410b, 424b, 431a, 432b
McColl, Hugh (1837–1909), 8:590; 10:442b
McCollum, Elmer Verner (1879–1967), 8: 590–591; 1:357a; **6**:135b, 136a, 443a; **9**:284b, 285a; **10**:243b
MacCormac, William, **8**:407b
McCormick Institute of Infectious Diseases, **6**:232b
McCoy, Frederick, **3**:297b; **5**:538a; **7**:460b
McCoy, Herbert Newby, **2**:257b, 259a,b; **12**:504b; **13**:262b
McCrea, William Hunter, **9**:406a
MacCullagh, James (1809–1847), 8:591–593; 2:452b; **6**:88b; **8**:40b, 501a, 591b; **12**:86b
  optics, **8**:40a; **9**:74a, 213a–b; **13**:77b; **14**:67a–b
Macculloch, John (1773–1835), 8:593–595
MacCurdy, Hansford M, **3**:121b
MacDonald, Donald Laurie, **5**:6b
McDonald Observatory (Texas)
  founding of, **13**:116a–b
McDougall, Alexander, **12**:478b
McDowell, James, **2**:424a
Macellama
  see Māshāʾallāh
Macer Floridus
  see Odo of Meung
Macewen, William, **3**:517b; **8**:406a
Macfarlane, Alexander, **2**:594a; **10**:604b
McGee, William John, **7**:149a; **14**:517b
Macgill, John, **9**:479b
McGill University (Montreal)
  Dawson and, **3**:607b
  Institute of Endocrinology, **3**:353a
  library, **3**:607b
  physics at, **12**:27a
  zoology at, **8**:586a
MacGillivray, William, **1**:330b; **5**:469b; **14**:462b
Macgregor, Malcolm Evan
  Mosquito Surveys, **11**:556b
Mach, Ernst (1838–1916), 8:595–607; 10: 57b; **13**:437a
  acquaintances of, **2**:446a, 449a; **10**:422a; **14**:123a
  correspondence of, **6**:349b*; **15**:456b
  critics of, **4**:374a; **5**:213a; **10**:485a; **11**:9a, 14a
  influence of, **4**:321b; **5**:609b; **6**:342b; **11**: 124a; **12**:218a; **14**:67b; **15**:463b
  Mach angle, **8**:597a
  Mach criterion, **8**:602b
  Mach number, **8**:597b
  Mach's bands, **8**:596b, 597a
  Mach's principle, **2**:17b; **4**:326b–327a, 329b–330a; **8**:599b–600a; **14**:306b
  mathematics, **15**:681a–b
  Die Mechanik in ihrer Entwicklung, historisch-kritisch dargestellt, **1**:222b; **4**:227b, 313b
  optics, **5**:445b
  philosophy, **2**:18a; **3**:32a, 33a, 34a; **4**: 233a*; **5**:391a; **11**:357a; **12**:609a; **15**:38b
    positivism, **2**:461a; **3**:379a; **5**:123b*; **8**:183b
  physics, **1**:112a; **2**:266b, 267a; **4**:225b; **6**:347b; **8**:159b; **15**:60a, 464a
  physiology, **1**:446b; **2**:446b, 515b; **5**:45a; **11**:215b
  psychology, **4**:588a
  and Stallo, **12**:609b
  students of, **9**:443b; **11**:174b
  translations of works, **15**:79b
Mach, Ludwig, **8**:597a
Macheboeuf, Michel (1900–1953), 8:607–608
Machin, John, **7**:163a; **10**:78a, 501a; **12**: 344b–345a; **13**:68a, 265a, 266a
Machine tools, **8**:212b, 214a; **9**:615b
Machines
  ancient Greek discussion, **6**:312a
  ancient Roman discussion, **10**:298a; **15**: 518b–519a
  medieval discussion, **1**:380b
  seventeenth-century discussion, **14**:150a, 366b–367a
  eighteenth-century discussion, **2**:40a; **5**: 511a
  nineteenth-century discussion, **15**:407a–b
  analysis of, **11**:384b; **14**:403b–404a
  classification of, **6**:2b; **14**:403b–404a
  as corpuscular systems, **3**:73b
  definition of, **3**:72b
  design of, **8**:208b; **11**:384b
  drive mechanisms, **8**:210b
  efficiency of, **3**:73a, 417a; **8**:210b–211a, 214a
  electrical, **3**:52a; **8**:258b; **9**:152a, 155a, 607b; **11**:50a; **12**:427b
  friction in, **1**:138b; **3**:71a, 358b*; **6**:431b; **8**:209a, 209b–210b; **10**:319b, 521a, 551a–b
  hydraulic, **3**:417b, 418a; **4**:481a; **5**:101a; **8**:211a–b; **11**:164a; **13**:93a; **14**:568b
  logic, **7**:105b
  lubrication theory, **12**:527a
  multiple, on single power source, **8**:214a–b
  optimal conditions, **3**:72b
  perpetual motion, **3**:388b; **5**:510a; **6**:141b; **8**:208b–209a; **10**:536b
  see also *Perpetuum mobile*
  pneumatic, **14**:568b
  relative motion in, **3**:417b–418a
  speed regulation, **8**:124a; **13**:73a; **14**:620a
  theory of
    eighteenth-century, **3**:71a; **4**:44a; **9**: 477a
    nineteenth-century, **3**:72b–75a, 82b, 224a, 226b, 228b–229a; **5**:501b; **6**:2b; **9**:461b; **10**:3b–4a, 546a; **11**:78a, 80a; **13**:73a–b
  work done by, **3**:82b; **8**:209a–b, 230b
  see also Engineering, Technology
Macintosh, Charles, **3**:289a; **8**:612b
McIntosh, William Carmichael (1838–1931), 8:608–609
McIntyre, Donald B., **13**:441b–442a
McIver, Monroe Anderson, **15**:74a
Mack, Karl Friedrich, **12**:187a
McKay, Alexander, **5**:611b
Mackay, James Townsend
  Flora Hibernica, **6**:162b
Mackay, John Sturgeon, **14**:141a
McKendrick, John Gray, **6**:476b
McKenzie, Doris, **4**:155b
Mackenzie, James, **5**:281b, 283a; **7**:278b; **8**:294b, 295a
Mackie, William, **8**:5b
MacKinney, Loren, **6**:276b
MacLane, Saunders, **6**:497a
Maclaren, Charles, **5**:336a
Maclaurin, Colin (1698–1746), 8:609–612; 2:18a, 418a; **5**:511b, 522a*; **13**:266a
  Account of Sir Isaac Newton's Philosophical Discoveries, **9**:481a
  acquaintances of, **4**:565b; **9**:480b; **12**:446a
  celestial mechanics, **15**:317b
  correspondence of, **3**:458a, 461b, 486a; **13**:68a
  Euler-Maclaurin formula, **4**:475b–476a; **10**:249b; **12**:535a
  Geometria organica, **2**:418b
  hydrodynamics, **8**:561b
  influence of, **3**:78a, 460a; **4**:45a; **5**:466a; **11**:79b
  Maclaurin's series, **8**:611a; **13**:266a
  mathematical physics, **7**:54b–55a; **8**:562a; **11**:58a
  mathematics
    analysis, **7**:490b; **8**:139b; **14**:180a
    geometry, **4**:480a; **7**:565a; **10**:66a; **15**: 198a
  students of, **6**:555b, 578a; **12**:413a; **13**:54a
  theorem of, **2**:297a
  tides, **3**:282b; **4**:591a; **15**:296b
  translations of works, **10**:572a*
  Treatise of Algebra, **3**:461a
Maclean, John (1771–1814), 8:612–613; 2: 174a; **11**:147b*; **12**:432b
McLean Hospital for the Insane (Waverley, Mass.), **5**:53a
Maclear, Thomas (1794–1879), 8:613–614; 6:326a,b; **12**:498b
Macleay, William Sharp, **3**:192b; **5**:192a; **6**:589b; **13**:168a
Macleod, John James Rickard (1876–1935), 8:614–615; 2:584b*; **3**:354a*
  insulin, **1**:441b, 442b, 443a; **3**:352a
MacLeod, Jules, **4**:401b; **14**:99a
Maclure, William (1763–1840), 8:615–617; 7:44b; **8**:266a–b; **10**:258a, 439b; **12**: 133a; **13**:467a
  stratigraphy, **3**:313b; **4**:274a, 363b
M'Mahon, Bernard, **11**:218a
MacMahon, Percy Alexander (1854–1929), 8:617; 13:222a*
  number theory, **11**:269a
McMeekin, Thomas Leroy, **3**:336a
McMillan, Edwin Mattison, **8**:95b, 96a; **13**:600b
Macmillan, William Duncan (1871–1948), 8:617–618; 9:263a, 553a*
MacMunn, Charles Alexander, **7**:273b
McNab, James, **1**:423a
McNab, William, **1**:423a
Macquer, Philippe, **8**:619b
Macquer, Pierre Joseph (1718–1784), 8: 618–624; 3:560b; **5**:90a, 600b; **11**: 144a, 330b
  acquaintances of, **2**:73b; **3**:5b, 419a; **9**: 54b; **11**:563a, 616b
  chemistry, **1**:527a; **2**:7a; **5**:601a,b; **8**:76a; **12**:64a, 427b
  on contemporary scientists, **9**:478b; **11**: 564b*; **12**:64b
  correspondence of, **12**:146a, 303a*
  critics of, **8**:80a
  Dictionnaire de chymie, **1**:345a; **3**:6a; **8**: 86a(n)
    translations of, **7**:277a; **8**:247a; **14**:75b
  dyeing, **3**:489a
  fermentation, **8**:82b
Macrauchenia, **10**:262a
Macrobius, Ambrosius Theodosius (*fl.* early fifth century A.D.), 9:1–2; 2:21a; **6**:544b; **7**:27b; **10**:179b
  Commentum in somnium Scipionis, **12**: 62a
    translations of, **11**:18a
  cosmology, **15**:203a
  historiography, **4**:391b; **6**:301b
  Saturnalia, **1**:269b
Macrocosm
  see Microcosm and Macrocosm
Macromolecules, **3**:620b; **7**:493a, 518b–519a; **13**:1a–3b, 421b
  shape of, **13**:2b

Macrophages, **2**:165b; **8**:8b–9a
McVittie, George Cunliffe, **6**:532a
Madagascar, **3**:365b
  botany, **9**:366b
  geology, **2**:569a
  mineralogy, **7**:548b
Madder, **4**:224b*, 504a; **5**:226a, 297b
Madeira (islands), **8**:9a
  biology, **8**:572a–b, 573a
  geology, **8**:572a–b
Mādhava Miśra, **12**:115b, 599a
Mādhava of Saṅgamagrāma, **9**:613b; **10**: 313a; **15**:602a, 632b(n)
  Veṇvāroha, **1**:49a
Mädler, Johann Heinrich (1794–1874), 9: 2–3; **7**:481b
  and Beer, **1**:568b–569a
  lunar map, **2**:147b; **8**:321b
Madrepores, **10**:569a
Madrid
  Botanical Gardens, **3**:154b
  Madrid Exposition, **1**:573a
Madrisio, Niccolò, **9**:532b
Madroño
  editors of, **7**:103a*
Madsen, John, **12**:31b
Madsen, Thorvald, **1**:301b; **4**:299a; **10**:141b; **12**:88a,b; **15**:107a
Maestlin, Michael
  see Mästlin, Michael
Magalhães
  see Magellan
Magalotti, Lorenzo (1637–1712), 9:3–4; **2**: 309b; **11**:342b*; **13**:172a
Magati, Cesare (1579–1647), 9:4–5; **9**:12a; **10**:307a
Magati, Giovanni Battista, **9**:4b
Magazin for Naturvidenskaben
  founding of, **1**:13b–14a
Magazin für die gesamte Mineralogie, Geognosie und mineralogische Erdbeschreibung, **6**:456a
Magazin für die Oryktographie von Sachsen, **5**:156a
Magazin für die Technologie, **15**:206a
Magazin von merkwürdigen neuen Reisebeschreibungen, **5**:77a
Magazin vor Aerzte, **3**:465a
Magdalenian remains, **8**:44a
Magdeburg hemispheres, **5**:575a
Magellan, Jean-Hyacinthe (Magalhães, João Jacinto de) (1722–1790), 9:5–6; **3**:474a*; **8**:74b; **10**:108a
  Nouvelle théorie du feu élémentaire, **14**: 354a; **15**:95b
Magellan, Strait of, **1**:574a
Magellanic Clouds, **6**:326b
  distance of, **6**:352a–b; **12**:348b
  nebulae, **12**:349a
  variable stars, **8**:106a; **12**:346b
Magendie, François (1783–1855), 9:6–11; **1**:479b; **2**:122b; **12**:164b
  acquaintances of, **5**:566a; **10**:497b; **13**: 557a; **15**:382b
  Bell-Magendie law, **1**:584a; **8**:17a; **9**:9b; **14**:128b
    confirmation of, **9**:570b
  and Bernard, **2**:25a–b; **9**:7b–8a, 8b, 9b, 10a
  critics of, **9**:572a, 573b; **15**:109a
  empiricism, **2**:31b
  Formulary, **4**:252a
  methodology, **12**:240b
  neurology, **2**:29b, 30a; **6**:59b; **9**:20a, 241b, 242a; **10**:500a; **14**:128b
  Précis élémentaire de physiologie, **4**:285b
Maggi, Bartolomeo (1477–1552), 9:11–12; **1**:204a; **14**:6a
Maggi, Vincenzo, **13**:278a
Maggio, Giulio Cesare Aranzio
  see Aranzio, Giulio Cesare
al-Maghribī
  see Muḥyi 'l-Dīn al-Maghribī
Magic
  ancient Greek, **2**:256b
  ancient Roman, **1**:150a–b; **5**:231b; **7**:1b
  medieval Arabic, **1**:122b; **14**:118a
  medieval Latin, **9**:364b
  Renaissance, **1**:79b, 80a; **3**:66a; **5**:448a; **6**:306a
  seventeenth-century, **12**:211a; **15**:69a,b
  analysis of, **5**:145b
  astral, **2**:539b, 541a
  critics of, **14**:389a
  Egyptian, **6**:305b, 306a
  Mesopotamian, **15**:645b
  natural
    medieval, **1**:380a–b; **14**:388b–389a
    Renaissance, **1**:80a, 373a; **2**:539a–543b; **5**:580a; **10**:308a–b, 309a, 310a; **11**: 96b–98a
    seventeenth-century, **6**:254a–b; **7**:355b–356a; **14**:209b–210a
    eighteenth-century, **14**:333b*
    critics of, **4**:387a,b; **10**:512b
  religious
    see Theurgy
  sympathetic, **5**:190a
  see also Necromancy; Sorcery
Magic circles, **14**:544b
Magic lantern, **7**:376a; **9**:595a
Magic squares, **1**:368a; **4**:550a; **5**:129b, 159b; **7**:577b; **9**:614a*; **13**:60a,b; **14**:544b–545b
Magini, Giovanni Antonio (1555–1617), 9: 12–13; **1**:244b, 447b; **2**:414b*; **3**: 149b; **4**:619b; **5**:238b, 239b; **11**:533a
  planetary tables, **10**:153b
Magiotti, Raffaello (1597–1656), 9:13–14; **2**:83a,b, 84a; **9**:25a; **10**:103b
  hydrostatics, **7**:376a; **10**:332b
Magirus, Johann, **1**:126b; **10**:43b, 60a
Magisterium, **8**:311b
Magma
  assimilation, **4**:564a
  crystallization, **10**:126a
  Devonian, **1**:394b
  differentiation of, **2**:370a, 486a; **3**:547b; **4**:141a, 564a–b; **8**:285b; **10**:126a; **14**:58b–59a
  flow textures, **3**:325b
  fluidity, **5**:395b
  immiscibility, **4**:564a
  inclusions in, **2**:370b
  physical chemistry of, **7**:5b
  stoping, **1**:470a; **3**:547b
  succession through time, **13**:64b
  Tertiary, **1**:394b
  theories of, **1**:470a; **2**:369b–370a; **4**:409b; **8**:285b; **13**:64b
  volatile constituents, **7**:5a; **10**:125b–126a, 127a
  see also Basalt; Lava; Plutonism; Rocks, igneous
Magmatites, **3**:434a
Magnenus, Johann Chrysostom (*ca.* 1590–*ca.* 1679), 9:14–15
Magnesia, **5**:421a; **9**:106a
  in plants, **8**:344b
Magnesia alba, **2**:175b–176b
  manufacture of, **6**:283a
Magnesia usta, **2**:175b–176a, 176b
Magnesium
  alloys, **11**:550a
  commercial manufacture of, **2**:588b
  compounds of, **2**:175b–176b; **3**:246b; **4**: 554a–b; **5**:90b
    organic, **5**:540a–541a
  as illuminating agent, **2**:588b
  isolation of, **2**:588b; **3**:602a
  pharmacological effects, **9**:266a
  preparation of, **9**:178a
  in sea water, **5**:72a
  separation from calcium, **2**:565a; **4**:134b
  spectroscopy of, **9**:154b
  tests for, **14**:486b
Magnetic analysis, **6**:480a
Magnetic coefficients
  and temperature, **6**:480a
Magnetic density, **3**:445a
Magnetic dip
  discovery of, **6**:144a; **10**:149b–150a
  see also Compass, magnetic
Magnetic moment, **1**:371a; **3**:445b; **9**:211b
  atomic, **9**:38a; **13**:42a–b; **14**:243b, 244b–246a
    equation for, **3**:4a
    vs. atomic number, **3**:4a
  of deuterons, **13**:42b
  of electrons, **13**:43a
  and mechanical moment, **5**:610a–b
  molecular, **8**:10b–11a
  of neutrons, **13**:240b
  of protons, **13**:42b, 43a
Magnetic poles
  existence postulated, **3**:424a; **9**:578b
  movement of, **6**:240b; **9**:254b
  position of, **4**:256b*; **14**:354b
    north pole, **11**:554b
    south pole, **1**:595b; **3**:591a; **9**:198a
Magnetic susceptibility, **3**:4a; **6**:480b
  and temperature, **8**:11a; **14**:245b
Magnetische Verein, **5**:306a, 309b; **7**:608b
Magnetism
  medieval Arabic study, **1**:410a; **2**:619b
  medieval Latin study, **1**:380a,b; **10**:532b–537b
  Renaissance study, **5**:397a–401b; **9**:192a; **10**:149b–150a; **12**:106a
  early seventeenth-century study, **3**:3a; **5**:48a; **7**:375b
  late seventeenth-century study, **2**:311a; **3**:534b; **4**:222a*; **5**:543a, 575b; **7**: 578a; **11**:121b, 506b; **13**:335a
  early eighteenth-century study, **4**:214b–215a; **6**:172a; **7**:111a; **8**:177b, 179b; **13**:265b
  late eighteenth-century study, **2**:474b; **3**:153b, 155b, 440a, 444b–445a, 445b–446a; **6**:233b; **7**:13a–b; **11**:472a
  early nineteenth-century study, **2**:135a; **6**:2b; **15**:485b
  mid nineteenth-century study, **1**:556a; **9**:19a, 566a; **13**:521b; **14**:330a
  late nineteenth-century study, **3**:265a, 505b–506b; **4**:500a–501a; **9**:155a; **11**:393a
  twentieth-century study, **3**:3b–4a; **4**:253a; **5**:610a–b; **6**:480a–b; **7**:316b; **14**: 243b–246a
  anisotropic, **6**:480b
  "Archimedean law," **1**:556a
  of chemical compounds, **14**:330a
  Chinese study, **12**:379b–380a
  distribution of, **1**:147a; **3**:445b
  and electricity
    see Electricity, and magnetism
  in electromagnetic field
    frequency effects, **1**:283b–284b
  in electrostatic field, **4**:305b
  and hydrodynamics, **9**:207b
  in iron, **1**:283b–284b; **9**:566a; **14**:245b
    limits of intensity, **9**:566a

Magnetism (cont.)
in liquids, **6**:271a; **11**:241b
measurement of, **5**:305b, 306a, 309b; **6**: 342b; **7**:375b, 449a; **8**:188a; **14**:204a–b
mechanical production of, **2**:379a
medical use of
see Animal magnetism
Mesopotamian study, **15**:651b
as nuclear force, **4**:196b
optical effects of, **4**:191a; **8**:40a; **14**:597a–598b
rotation
see Optical activity, magnetic
in oxygen, **1**:556a; **8**:11a
philosophy of, **1**:373a; **5**:194b
rotation, influence of, **3**:260a
solar, **1**:56a, 478a; **6**:29b–30a, 32a; **8**: 582b–583a
and temperature, **3**:505b–506b; **12**:438a; **14**:245b
theories of
ancient Greek, **4**:32b
ancient Roman, **1**:119a
medieval Latin, **1**:198b; **10**:535a–536b
Renaissance, **2**:543b; **5**:397b–401b
seventeenth-century, **1**:568a; **6**:609a
eighteenth-century, **1**:67b; **2**:38b, 43a; **3**:440a, 445b–446a; **4**:482a; **6**:178b; **7**:111a, 587a; **9**:234b; **11**:66b–67a, 134b; **13**:183b
early nineteenth-century, **1**:146a; **4**: 532b, 534a; **8**:38a–b; **9**:204b
mid nineteenth-century, **4**:538b–539a; **6**:278b; **9**:205a–b, 206b, 207b; **13**: 377a–b, 384a–b; **14**:208a, 330a
late nineteenth-century, **3**:506b; **4**:225b; **5**:496b; **13**:363a
twentieth-century, **1**:455b; **3**:506b; **6**: 480a; **7**:622b; **8**:11a
domain theory, **1**:455b
lines of force, **4**:538b–539a
mathematical, **7**:84a; **13**:183b; **15**:199b
mechanistic, **1**:568a
molecular polarization, **1**:145a; **3**:445b–446a
one-fluid theory, **1**:67b; **3**:445b; **5**:136a–b; **6**:178b
particulate, **12**:289a
two-fluid theory, **3**:445b
see also Animal magnetism; Diamagnetism; Electromagnetism; Ferromagnetism; Fields (phys.), magnetic; Force, magnetic; Magnetization; Magnets; Permeability, magnetic; Viscosity, magnetic
Magnetism, terrestrial
see Terrestrial magnetism
Magnetite, **14**:244b
deposits of, **13**:519a
Magnetization, **10**:481b
and electric discharge, **14**:354b
field effects, **13**:80a–b
hysteresis, **4**:500b–501a; **14**:171a
intensity distribution, **8**:188a
of iron, **5**:398b; **11**:446a; **13**:80a–b; **14**: 330a, 354b
by electric current, **1**:203a; **11**:213a, 578b; **12**:26a; **14**:354b
laws of, **14**:244b–245b
reversible polarity, **10**:535b–536a
shape effects, **1**:284b
of single crystals, **6**:480b
spontaneous, **14**:245a–b
by sunlight, **12**:523b
temperature effects, **3**:620a; **6**:480a; **14**: 206b, 244b–245b, 330a
thermoremanence, **12**:379b
Magnetochemistry, **3**:3b–4a
Magnetodynamics, **1**:284b
Magnetoelectricity, **3**:260b; **4**:192a
Magnetograph
solar, **6**:30a
Magnetometer, **5**:305b; **11**:291a; **14**:206b
bifilar, **5**:306a
torsion, **6**:470a
Magneton theory, **3**:4a; **14**:246a
Magnetostriction, **6**:480a; **9**:606a; **10**:605a
Magnetron, **6**:546b; **11**:422a
Magnets
artificial, **1**:67b; **3**:51a; **7**:13b, 111a; **9**: 371a*, 607b; **11**:432b
cooperative phenomena, **5**:103b
damping of, **12**:281b
experimental use of, **2**:84a
floating, **9**:230b–231a
and magic, **2**:543b
permanent, **3**:329a; **4**:231b
polarity of, **5**:400b; **9**:155a
determination of, **10**:534b–535a
law of, **10**:535a
spherical, **5**:398a; **10**:534b–535a, 536a–b
surgical use of, **6**:276b
temperature effects on, **7**:609a
see also Compass; Electromagnets
Magni, Domenico, **2**:310a
Magni, Valeriano (1586–1661), 9:15–17; 10:332b
Magnification
medieval use of, **1**:379b; **5**:552b
theory of, **6**:603b
see also Megascope, Microscope, Telescope
Magnitsky, Leonty Filippovich (1669–1739), 9:17; 8:467b
Magnitude (math.), **4**:466b
absolute, **1**:238a–b; **8**:141b
divisibility of, **4**:447a; **7**:83a
extended, **3**:166b
geometrical, **7**:83a
homogeneous, **8**:141b
and number, **4**:447a
sensed, **3**:213a,b
theory of, **4**:420a, 421b–422b, 424a; **10**: 344a
Magnol, Pierre (1638–1715), 9:17–18; 7: 197a; **12**:457b; **13**:442b, 443a
Magnolia, **9**:112a
Magnoliaceae, **5**:441b; **9**:18a
Magnus, Heinrich Gustav (1802–1870), 9: 18–19; 3:468a; **6**:243a; **11**:360b*
acquaintances of, **5**:556a; **7**:509a; **12**:425a; **14**:170b
chemistry, **8**:333b–334a, 335a–b; **9**:348b
Magnus effect, **9**:19a
Magnus' green salt, **9**:18b
Magnus' rule, **9**:19a
optics, **11**:582a
students of, **1**:294a, 389b; **3**:303b; **7**:526a; **12**:236a; **13**:79b; **14**:329a,b, 628a
Magnus, Olaus
see Olaus Magnus
Magnus, Rudolf (1873–1927), 9:19–21
Magnus, Valerianus
see Magni, Valeriano
Magnus Ephesius, **12**:538a
Magrou, Joseph, **3**:429a
Magyar Tudomanyos Akademia, Budapest, **4**:380b
Mahādeva (*fl.* 1316), 9:21
Mahādevī, **4**:100a; **15**:588b, 589a
Mahādeva, son of Luniga
commentaries by, **3**:584b; **12**:598b–599a
al-Māhānī, Abū ʿAbd Allāh Muḥammad ibn ʿĪsā (*fl. ca.* A.D. 860), 9:21–22
astronomy, **14**:575b
commentaries by, **4**:438a, 441a
geometry, **7**:3a, 326b, 327b–328a, 335a; **10**:6a; **13**:307a(n)
Mahāpātas, **3**:584b
Maharam, Dorothy, **14**:89b
Mahāsiddhānta, **15**:580b
Mahāvīra (*fl.* ninth century), 9:22; 12:597b
Gaṇitasārasangraha, **8**:419b
Mahāyuga (period), **15**:590a,b, 599b–600a, 603a–611a *passim*, 622a
Mahendra Sūri (*fl.* 1370), 9:22–23
sine table, **15**:626a–b
Yantrarāja, **15**:626a–628a
Maheśvara, **2**:115a,b
commentaries of, **2**:118a
Mahīdhara
commentaries of, **2**:116a
Maḥmūd al-Ghaznawī, **2**:148b–151a *passim*; **9**:83b
Maḥmūd ibn Masʿūd al-Shīrāzī
see Quṭb al-Dīn al-Shīrāzī
Maier, Michael (*ca.* 1568–1662), 9:23–24; 10:83a; **12**:307a
Maievsky, Nikolai Vladimirovich, **3**:195b, 224b
Maige, Albert (1872–1943), 9:24–25
Maignan, Emanuel (1601–1676), 9:25; 2:83b; **10**:103b, 314b
Maillet, Benoît de (1656–1738), 9:26–27; 6:469a; **7**:592b
geology, **15**:55b
Telliamed, **8**:242b
Maimonides, Rabbi Moses ben Maimon (*ca.* 1135–1204), 9:27–32; 6:189b; **11**: 326b
astronomy, **5**:619a–b
Cresca on, **3**:470a
The Guide of the Perplexed, **1**:409a; **7**:38a
influence of, **10**:474a; **13**:400b, 401b
influences on, **1**:23a*; **4**:121b, 524a; **11**: 325a
Mishnah Torah, **1**:512a–b
Sharḥ asmāʾ al-ʿuqqār, **1**:539a
translations of works, **13**:401b
Main, Robert, **11**:163a
Mainardi, Gaspare, **3**:331a; **10**:544b
Maine de Biran, Marie-François-Pierre, **1**: 141a; **5**:167a
Maine (state)
Geological Survey, **7**:45a
Maior (or Maioris), John (or Jean Mair) (1469–1550), 9:32–33; 2:395b; **3**: 171b, 420b; **4**:237a,b; **8**:100a; **12**: 547b; **13**:349b
Mair, Alexander, **1**:530b
Mair, Jean
see Maior, John
Mair, Simon
see Mayr, Simon
Mairan, Jean Jacques d'Ortous de (1678–1771), 9:33–34; 3:216a, 486a; **5**:61b
acquaintances of, **1**:323a; **2**:327a; **3**:459b; **5**:58a
Dissertation sur la glace, **2**:578b
geodesy, **3**:105a
Nouvelles recherches sur la cause générale du chaud en été et du froid en hiver, **2**:578b
physics, **2**:343b; **3**:282b
Maire, Christopher, **2**:329b; **13**:39b
Maire, René-Charles-Joseph-Ernest (1878–1949), 9:34–35
Maitland, Peter, **9**:403a
Maize, **1**:325a; **3**:7a; **12**:317a; **14**:452a*
breeding experiments, **4**:270b–271a, 271b; **8**:459b–460a
genetics, **3**:422b; **6**:121b
xenia effect, **3**:422a; **13**:478b

in nutrition, **4**:358a
see also Corn
Ibn al-Majdī, **12**:362b
Ibn Mājid, Shihāb al-Dīn Aḥmad ibn Mājid (*fl.* fifteenth century A.D.), 9:35–37
Majorana, Ettore (1906–1938), 9:37–39; 4:578b
al-Majrīṭī, Abu'l-Qāsim Maslama ibn Aḥmad al-Faraḍī (*d. ca.* A.D. 1007), 9:39–40; **7**:42a; **9**:160b
spurious works, **7**:223b
translations by, **1**:61a; **7**:361a; **11**:198a
al-Majūsī, Abu'l-Ḥasan ᶜAlī ibn ᶜAbbās (Haly Abbas) (*d.* A.D. 994), 9:40–42; **1**:382a, 620a; **3**:394a; **13**:229a
Kāmil al-ṣināᶜa al-ṭibbiyya, **15**:498a–b
Kitāb al-Mālikī
translations of, **3**:393b; **12**:81a; **13**:38a–b
medicine, **15**:246b(n)
Makaranda (*fl.* 1478), 9:42
Makaranda, **15**:614a
Makarov, Stepan Osipovich (1849–1904), 9:42–43; 8:188b
Makerstoun, observatory at, **2**:472a
Makkibhaṭṭa, **12**:598b
Makower, Walter, **6**:16a
Makromolekulare Chemie, **13**:3b
Maksimov, Alexander A., **1**:613b; **2**:165a
Maksimov, Nikolay Aleksandrovich (1880–1952), 9:43–45; 10:281b
Maksutov, Dmitry Dmitrievich (1896–1964), 9:45–47
Mal de la rosa, **3**:97a,b
Mal del segno, **1**:492b–493b
Malachite green, **3**:272a
Malacology
seventeenth-century, **1**:431b–432a; **2**:591a–b; **8**:307b
eighteenth-century, **4**:178a; **7**:401a; **9**:575b; **11**:67a, 331a–b; **12**:215b
early nineteenth-century, **2**:187a, 481a; **8**:373b; **11**:243a
mid nineteenth-century, **3**:392a, 535b–536b; **4**:67b–68b; **7**:91b–92a, 545b–546a; **8**:103a–104a; **12**:299a–b, 550b–551a; **13**:65b
late nineteenth-century, **2**:486a; **7**:30b, 448a; **9**:112b, 536b
see also Conchology, Shells
Malaguti, Faustino Jovita Marianus, **1**:518b; **3**:10b; **4**:246a
Malapterurus electricus, **4**:202b
Malaria
ancient Roman discussion, **5**:231b
seventeenth-century discussion, **1**:25b; **12**:577b
eighteenth-century discussion, **5**:366b; **11**:148a
nineteenth-century discussion, **1**:360a; **2**:481a; **4**:252a; **8**:313a; **11**:346a, 555b
twentieth-century discussion, **4**:499b*
avian, **7**:428a
control of, **3**:185a–b; **5**:503b; **7**:16b, 428a–b, 614a; **8**:557a; **9**:83a, 95a, 577a; **11**:556b; **14**:551b
diagnosis of, **3**:185a; **5**:461a–b
etiology of, **7**:428a; **11**:555b–556b; **12**:141b–142a
induced, **14**:115b
pathology of, **3**:447a; **5**:461a–b, 503a–504a; **8**:65b–66a; **9**:94b–95a; **12**:482a; **13**:300a–b
transmission of, **5**:503a–b; **6**:524b; **7**:428a, 614a; **8**:66a; **9**:82a–83a; **11**:555b–556b; **12**:315a, 482a
treatment of, **1**:360b; **2**:534a; **5**:100a; **6**:494b; **7**:614a
with methylene blue, **4**:296b
with quinine, **5**:461a–b, 503b
Malay Archipelago
anthropology, **9**:144a
exploration of, **14**:134a, 134b–135a
genesis of, **13**:610a, 611a
zoogeography, **14**:138b
zoology, **14**:203a–b
Malayendu Sūri
commentaries by, **9**:22b; **15**:626a–628a
sine table, **15**:616b
Maldive Islands, **1**:516a, 517a
Malebranche, Nicolas (1638–1715), 9:47–53; **1**:322b; **3**:381b; **4**:64b; **5**:61a; **8**:151a
acquaintances of, **2**:52a; **5**:226a; **7**:611a; **11**:392a, 512b
critics of, **1**:292a; **2**:577b
influence of, **2**:16a, 46b; **4**:228a; **7**:596a; **9**:491a; **15**:54a
influences on, **13**:174b
natural philosophy, **11**:158a
Recherche de la vérité, **4**:515b; **5**:59b
students of, **9**:33b, 499b; **11**:157b
Maleic acid, **2**:386a; **10**:516b
discovery of, **7**:281a
structure of, **14**:455a
Malesherbes, Chrétien-Guillaume de Lamoignon de (1721–1794), 9:53–55; **5**:578a; **7**:199a; **8**:303b; **9**:478a; **11**:563a; **13**:390a, 497a; **15**:113b
and *Encyclopédie,* **4**:86a
Malfatti, Gian Francesco (1731–1807), 9:55; **1**:91a; **11**:598b
Malfatti resolvent, **9**:55a
Malfatti's problem, **9**:55a; **13**:15a–b, 259b
Malic acid, **4**:76a; **8**:339a, 340b; **10**:576a
discovery of, **12**:148b
optical properties, **10**:358b
preparation of, **10**:359a–b
Malinowski, Bronislaw, **5**:145b
Ibn Malkā
see Abu'l-Barakāt
Mall, Franklin Paine (1862–1917), 9:55–58; **8**:447a; **12**:48a; **13**:96b, 97a; **14**:249b; **15**:64b
Manual of Human Embryology, **13**:97a
Mallard, (François) Ernest (1833–1894), 9:58–60; **5**:185b; **8**:31a, 118b, 119a; **13**:283a, 284a
crystallography, **1**:551b; **5**:186b
mining, **8**:118a
Mallâri
commentaries of, **5**:274b; **7**:219a, 315b; **15**:624b
Mallery, Garrick, **11**:119b
Mallet, Frederick, **2**:5a; **7**:404a
Mallet, John William, **9**:61a
Mallet, Robert (1810–1881), 9:60–61
Mallikārjuna
commentaries by, **15**:608b
Mallophaga, **7**:286a
Mallory, Frank Burr, **3**:447b
Malmquist, Karl Gunnar, **6**:539b
Malone, Luke, **4**:155b
Malouin, Paul-Jacques (1701–1778), 9:62; **8**:619b; **13**:602b
Malpighi, Marcello (1628–1694), 9:62–66; **1**:110a; **3**:343a; **4**:487a; **8**:157b
acquaintances of, **2**:309a; **10**:266b; **13**:33b
Anatome plantarum, **2**:591a; **5**:535a; **11**:314b; **13**:563b
anatomy, **1**:593a
bone, **6**:183b
circulatory system, **6**:63b
glands, **2**:226a; **14**:199b
kidney, **4**:590a
and Baglivi, **1**:391b
botany, **5**:535b, 536a; **6**:41a,b, 219b; **9**:455b; **11**:315b, 342a; **15**:67b
correspondence of, **1**:593b; **13**:169b
critics of, **12**:41a; **14**:15a
embryology, **1**:293b; **3**:482b; **6**:65b
entomology, **13**:172a
hematology, **8**:128b
influence of, **2**:237b; **5**:366a; **8**:150a; **9**:511b; **13**:179b
Malpighian corpuscle (body, gland, nodule), **2**:376a; **7**:23b; **9**:64a, 66a
Malpighian layer, **2**:165a; **8**:8b, 9a; **9**:63b
Malpighian tuft (glomerulus), **7**:23b
Malpighian vesicles, **9**:456b
De polypo cordis, **9**:302b
De pulmonibus, **1**:483a
reproduction, **12**:559a,b
respiration, **9**:243a
students of, **9**:3b, 511a; **13**:563a, 566a–b
Malpighiaceae, **5**:546a
Malt
reactions of, **7**:379a
Malta
geology, **7**:30b
Malta fever, **2**:528a; **9**:546a*
Maltase, **5**:4a
Malthus, Thomas Robert (1766–1834), 9:67–71; **2**:570b; **3**:291a; **11**:236b; **14**:133b, 136b; **15**:139b
Essay on the Principle of Population, **3**:570b–571b; **4**:271b
translations of, **11**:135a
influences on, **5**:129a; **13**:448a
population, **9**:385a; **15**:139b
Malthusianism, **3**:564b; **7**:104b; **12**:570b; **13**:616a
predecessors of, **1**:424b
Maltose
derivatives of, **5**:4b
Maltsev (or Malcev), Anatoly Ivanovich (1909–1967), 9:71–72
Maltsev algebra, **9**:71b
Maltsev-Kolchin theorem, **9**:71b
Maltsov, Nikolai Sergeevich, **6**:105b
Maltus, **2**:202a
Malus, Étienne Louis (1775–1812), 9:72–74; **1**:202a; **5**:110b; **6**:1b; **15**:481a
acquaintances of, **1**:201a; **5**:111a; **7**:615a; **15**:347b, 361a
Malus theorem, **4**:257b; **9**:73a
optics, **1**:616b; **2**:451b, 452a; **6**:87b; **9**:476a
polarization, **1**:201b; **2**:137a,b; **5**:169a–b; **12**:281a; **14**:567a; **15**:362a
refraction, **1**:358a; **15**:361b, 362a,b
Malvaceae, **9**:253b
Malvasia, Cornelio, **1**:341b; **3**:100a; **9**:485a,b
Mamlakat al-Islām, **9**:88b
Mammals, **1**:263a; **3**:439a; **5**:358b
anatomy, **1**:263a; **2**:422b; **5**:442b; **14**:203b
body temperature, **5**:149b
in captivity, **3**:521a
classification of, **5**:46a–b, 359a; **6**:592b–594a, 595a; **9**:313b; **13**:402a; **14**:203b
by cerebral anatomy, **10**:261b, 262b
by dentition, **3**:521a
coat colors
inheritance of, **3**:122b
dentition, **5**:469b; **12**:38a
eggs of
see Eggs, mammalian
embryology, **1**:601b; **6**:535b–536a; **10**:105a; **13**:97b
endocrinology, **12**:474a–475a
evolution of, **2**:504a–505b; **5**:46b; **6**:536a; **7**:480a; **12**:38a–b
fossil, **2**:423a, 494a; **3**:532a; **4**:401a, 518b; **5**:109b, 156a, 295b–296a, 357b, 500b;

Mammals (cont.)
7:60a, 588b–589a; **8**:43b, 169b, 573a; **9**:134b, 314a; **10**:16a–17a; **11**:336a–b, 500b; **12**:260b–261a, 278a; **13**:72a, 145b, 257b–258a; **15**:50b–51a
fur-bearing, **5**:545b*
genetics, **3**:120b–123b
heart, **5**:281b
hybridization, **5**:359b
lifespans, **11**:586a
marine, **1**:491a; **13**:28b
migrations
ancient, **8**:43b
morphology, **5**:481a
osteology, **5**:504b
phylogenesis, **10**:16b–17a
physiology
epithelial mechanism, **7**:502b
reproduction, **5**:484b–485b; **8**:127b–128a; **10**:611a
artificial insemination, **7**:31b–33a
size
inheritance of, **3**:122b
social life, **3**:521a–b
see also specific mammals
Mammoth, **2**:423a; **5**:222a*; **12**:184a
Big Bone Lick fossils, **1**:487b
Mammoth Cave, **10**:273a
al-Maʾmūn, ʿAbdallāh, **1**:33a, 275a, 444a, 510b; **2**:149a; **4**:439a, 541a, 542b; **5**:612a; **6**:189a; **7**:24b, 79a, 358b, 361b, 362a; **8**:191a; **11**:202a,b; **13**:149b, 324a, 538a; **14**:537b; **15**:88a, 230b, 261a
al-Maʾmūn, Abu'l-ʿAbbās, **1**:350b; **2**:149b–150a; **9**:83b; **14**:592b, 594a
Man
antiquity of, **6**:510a; **8**:573b–574a; **11**:131a, 399a; **14**:516b, 517a,b; **15**:51a–52a
apes and, **3**:574a–b; **5**:356b, 359a–b; **6**:592b–594a; **7**:278b, 606b; **8**:378b; **11**:514a, 592b; **13**:527b–528a
neoteny theory, **2**:255b; **5**:359a
average, **11**:237a
brain
growth of, **4**:160b–161a
size of, **2**:162a; **3**:524b; **5**:359a–b; **11**:514a
classification of, **11**:235a
by physical characteristics, **2**:320b, 321a; **11**:381a
by primitiveness, **1**:388a
creation of, **2**:21a, 321a, 568b; **3**:392b; **6**:454a–b; **9**:48b; **10**:195b; **14**:114a, 296a
embryogeny, **3**:574b; **9**:57a, 416a; **11**:383a; **13**:96b–97b
equality of, **6**:406a,b, 447b, 553a; **11**:323a; **12**:109a
evolution of, **2**:347a; **3**:390a, 553b, 556a, 573b–574b; **5**:221b–222a, 354a*; **6**:9a–b, 469a, 527b–528a; **7**:279a, 590b; **8**:125b, 574a–b; **10**:290b; **13**:275a–276b, 504a; **14**:439b; **15**:93a, 140a, 140b–142a, 144a–b
dating of, **6**:8b
"nature" vs. "nurture" controversy, **3**:390a
experimental use of, **9**:547a
fossil, **1**:130a; **2**:171b–172a, 189a–b, 341b, 347a, 505a–b; **3**:264a–b, 328b, 606b–607a, 608b; **5**:109b–110a, 206b–207a, 500b; **8**:43b, 105a; **10**:17a, 40b; **11**:235a, 336b; **12**:184a, 318a; **15**:51b–52a
and diluvialism, **2**:568a
future of, **3**:564b; **15**:455a
genetics, **3**:555b, 590b; **5**:11a, 598b; **6**:22a; **10**:464a; **11**:333b; **12**:76b; **13**:137b; **14**:230b
chromosomes, **10**:276b–277a
spontaneous mutations, **6**:22a; **9**:565b
life-span, division of, **14**:584b–585a
longevity, **5**:45a*; **10**:445a
as a machine, **7**:606a–607a; **8**:202a, 210b–211a; **14**:204b, 206a
morphology
constitutional types, **3**:113a–b
cultural influences on, **9**:144b–145a
erect posture, **5**:110a; **15**:112b
and nature
eighteenth-century discussion, **11**:43a–44a; **13**:258a
nineteenth-century discussion, **3**:453a; **6**:139b, 596a; **15**:135a, 140a, 140b–142a, 144a–b, 412b–415b
twentieth-century discussion, **6**:228b; **8**:512a; **14**:107a
Chinese discussion, **12**:371a, 374b, 390a–391a
ecological effects, **8**:512a
exploitation, **12**:371a, 390a–391a; **15**:413b–415a
geological effects, **4**:600a; **5**:395b; **12**:344a; **13**:619b
man as master, **2**:581b; **3**:362b; **5**:135a, 213b; **7**:366a; **15**:406b, 634a–b, 635a, 646b
separation of, **3**:453a
perfectibility of, **3**:1b, 470a
pigmentation, **15**:5a*
prehistoric, **1**:173a; **4**:519a; **5**:296a; **7**:31a; **8**:44a
Australopithecus, **2**:505a; **6**:9b; **8**:105a; **13**:276a
Cro-Magnon, **8**:44b; **12**:184a, 520b
Heidelberg man, **3**:606b
Homo erectus, **8**:105a
Homo ferus, **2**:204b
Homo habilis, **8**:105a
Ice Age, **5**:339a; **14**:443a, 517b
Java man, **3**:606b
Neanderthal man, **2**:189a–b, 347a, 617b; **3**:606b; **5**:109b, 206b–207a; **6**:9b, 527b, 593b; **8**:464a, 574a; **12**:184a
Pampean man, **1**:130a, 131a–b
Peking man, **2**:171b–172a; **5**:486b, 500b; **13**:275a
Piltdown man, **3**:606b–607a; **4**:353b
Pithecanthropus, **6**:8b, 9b, 527b; **8**:29a; **10**:17a; **13**:275a, 504a
Plesianthropus, **2**:505a
primitive, **2**:189b; **8**:528b; **9**:144a
osteology of, **5**:46b
as social animal, **14**:291b–292a
uniqueness of, **5**:505a–b; **6**:593a; **8**:98a; **10**:262b, 263a; **11**:235a; **14**:136b–137a, 367a; **15**:140b, 143a–b
work done by, **8**:211a
maximum daily, **2**:37b, 43a; **3**:442b; **4**:44a
see also Anatomy, human; Anthropology; Culture; Demography; Ethnography; Ethnology; Evolution; Immortality of man; Paleolithic man; Paleontology, human; Population; Race; Reproduction, human; Sociology
Man and Nature in Mesopotamian Civilization, 15:634–666
Management (econ.), **12**:274b
Manardo, Giovanni (1462–1536), 9:74–75; **8**:248b; **11**:253b; **15**:161b*
Manardo, Timoteo, **9**:74b, 75a
Manasse(h)
see Māshāʾallāh
Manatee, **11**:528b
Manceaux, Louis, **15**:454b
Manchester (England)
School of Medicine and Surgery, **3**:546a
Manchester, University of
chemistry at, **12**:236b; **14**:247b
crystallography at, **9**:380a
on Dalton, **3**:547a
physics at, **12**:29b, 30b, 238b
School of Combustion Research, **4**:130a
Manchester Academy, **3**:538b; **6**:283a
Manchester Literary and Philosophical Society (England), **3**:399b–400a, 539a, 541a,b, 544b–545a; **13**:126a
founding of, **6**:283a; **14**:297a
Manchester Memoirs, **3**:542b, 545a
Mandelshtam, Leonid Isaakovich (1879–1944), 9:76–77; **13**:239b, 240a, 241b; **15**:6b
Mandelshtam-Brillouin effect, **9**:76b
optics, **7**:622a; **10**:102a*; **11**:265b
Mandeville, Bernard de
Fable of the Bees, **3**:215b
Mandible
resection of, **8**:34b
Mandocca (astron.), **15**:568b, 580a, 584a, 587a–611b *passim,* 614a, 618a–624b *passim*
Manetti, Gianozzo, **1**:274a
Manfredi, Eustachio (1674–1739), 9:77–78; **3**:183b*; **9**:90a, 485a; **13**:566b; **14**:589a
astronomy, **6**:512b
Elementi della geometria, **14**:589a
ephemerides, **14**:589b
Manfredi, Gabriele, **5**:526a; **9**:55a; **11**:400a
Manfredi, Michele (pseudonym), **11**:411b*
Manganese, **5**:157b
atomic weight, **13**:499b
compounds of, **9**:424b; **10**:530b
arsenide, **7**:224a
oxides, **15**:7b
deposits of, **1**:20b; **3**:609b
oceanic, **2**:557b; **13**:360b
discovery of, **5**:223a
and iron, **13**:468a
separation of, **2**:565a
isolation of, **9**:451b
lead dioxide test, **3**:488b
in metallurgy, **15**:25a
nodules, **9**:589b; **11**:372a
spectrum, **3**:124b–125a
Mange, **5**:270b
etiology of, **3**:180b, 332b
Mangin, Louis Alexandre (1852–1937), 9:78–79; **2**:290b
Mangold, Hilde, **12**:568b
Mangold, Otto, **12**:568b
Mangoldt, Hans von, **7**:411b; **11**:453a; **14**:222a
Mangonel, **8**:213b
Mangrove, **12**:166a, 190b
Manhattan Engineer District (MED), **2**:460a; **3**:371a–b; **4**:582a; **7**:498a; **8**:47b, 95b; **10**:215a–216a; **13**:227a–b, 429b; **14**:92a
Manhattan project
see Manhattan Engineer District (MED)
Mania, **10**:613b
dancing, **10**:307a
treatment of, **1**:332b
see also Hysteria, Insanity, Psychiatry
Manichaeism, **1**:333b, 334b, 335a–b; **7**:41b; **11**:325a; **12**:442a
opposition to, **14**:388b

Manifolds (math.), **3**:56a; **11**:390a
algebraic, **12**:171a
almost-complex, **6**:497a
analytic, **8**:324b; **14**:310b
compact, **6**:497a
completeness, **14**:311a
complex
definition of, **14**:283a
curvature of, **12**:172b
differentiable, **3**:95a; **4**:336b; **14**:283a
definition of, **13**:600a; **14**:311a
fibering of, **8**:325a
Grassmann, **15**:197b
homology, **11**:60a
intersection of, **2**:114a
Kähler, **12**:214a
mappings of, **6**:496b
n-dimensional, **6**:249a; **8**:389b; **11**:454b
nilpotent, **9**:71b
Riemannian, **8**:389b; **12**:214a; **14**:283a–b
topological
Dehn's lemma, **4**:9a
triangulation, **11**:60a
unirational, **4**:523a
Manilius, Marcus (*fl.* beginning of first century A.D.), 9:79–80
Astronomica, **2**:349a; **4**:622b*
editions of, **10**:615b
Manis, **6**:535b
Manitoba, University of
botany at, **2**:582a
Mankah al-Hindī, **7**:223a,b
Mankind Quarterly, **5**:294a
Mann, Frederick George, **11**:88a
Mannheim, Victor Mayer Amédée (1831–1906), 9:80; 11:398a
Mannheim (Germany)
Academia Theodoro-Palatina, **9**:253b, 254a
botanical gardens, **9**:253b, 254a
observatory, **9**:231b
Mannitol, **4**:554b
Manometers, **1**:128a–b; **3**:12a; **6**:39a,b, 44a; **13**:574a
absolute, **7**:416b
Bourdon's, **2**:355a
flame, **7**:445a
Knudsen's, **7**:416b
metal, **2**:439a
spring, **4**:615b
Manometry, **14**:174a
Mansfield, Charles Blackford, **6**:461b, 463a
Mansion, Paul (1844–1919), 9:80–81; 10:25a; **11**:398b*, 414a*; **13**:415a*
historiography, **5**:314a*; **6**:308a; **12**:57a*, 469b*; **15**:687a, 702a*
Manson, Patrick (1844–1922), 9:81–83; 2:127b, 528b; **8**:66a, 296b; **11**:557a*
malaria, **5**:503a; **11**:555b
and Ross, **5**:503b; **11**:555b
al-Manṣūr (Abū Jaᶜfar) (caliph), **4**:438b, 555b; **7**:223a, 360b; **9**:160a; **11**:32a; **13**:538a
Manṣūr ibn ᶜAlī ibn ᶜIrāq, Abū Naṣr (*d. ca.* A.D. 1036), 9:83–85; 7:334b; **9**:22a; **11**:240b; **14**:577b
geometry, **13**:291b
students of, **2**:148a; **12**:431b
trigonometry, **7**:352b; **9**:298a; **13**:510b
Manṣūr ibn Athānās, **15**:236b
al-Manṣūr Sayf al-Dīn Qalāwūn al-Alfī, **9**:602b
al-Manṣūr Yaᶜqūb ibn Yūsuf, **12**:1b, 2b, 8a
Mansurova, V. V., **12**:76b
Mansvelt, C. G. van, **7**:391a–b
Mantegazza, Paolo (1831–1910), 9:86–86; 2:164b
Mantell, Gideon Algernon (1790–1852), 9:86–88; 3:324a; **8**:564b, 566a; **15**:151b, 152a
Medals of Creation, **14**:398a
Pictorial Atlas of Fossil Remains, **10**:322b
Manten, Adriaan, **7**:406b
Manticore, **1**:263a; **13**:528a
Mantinea (Greece), **1**:282a
Manure, **4**:224a; **7**:15b*, 389b, 538a; **8**:93a; **9**:609b; **10**:436a
Mao Chin, **14**:540a
Mao I, **8**:419b
Mappae clavicula, **1**:62a
Mappings (math.), **3**:136b, 139a, 141a, 143a; **4**:4a, 420a; **5**:477a; **8**:325a; **10**:244b; **11**:256a
bounded families of, **1**:428a
classification of, **6**:496b–497a
conformal, **1**:500a; **2**:144b; **3**:263a; **4**:477b, 560b; **7**:398a; **8**:367a, 433b, 458a*; **11**:54a,b, 259b, 394b, 449b, 454b; **12**:212b; **14**:592a
fundamental theorem of, **4**:562a; **11**:449b–450a, 451b–452a; **12**:246a–b
continuous, **2**:513a
fixed points, **2**:236b
degree of, **2**:513a; **6**:496b
differentiable, **2**:192a
discontinuous, **3**:54b–55a
equiareal, **3**:331a
equicontinuous, **7**:312b–313a
homographic, **7**:574a
inverted, **6**:496b
kinematic, **2**:192a
monotonic, **8**:457b
nonorthogonal, **2**:153a
nonstereographic, **2**:153a
of surfaces, **7**:312b
theory of, **1**:427b; **5**:303b–304a
see also Geometry, projective; Projection (geom.)
Maps
coloring of, **2**:145b; **3**:169a; **9**:430b–431a; **10**:483b
fishing, **3**:204a
geological
Renaissance, **8**:242a
seventeenth-century, **4**:348a; **8**:416a; **10**:260a
early eighteenth-century, **9**:135a; **13**:87a, 409b
late eighteenth-century, **4**:71b, 348a, 493a; **5**:205b, 578a–b; **8**:68b; **12**:549a; **14**:68b
early nineteenth-century, **2**:482a,b, 494a, 495a; **3**:396a*; **4**:218a, 348a–349b, 540b, 541a; **5**:14a, 518b–519a, 537a–b; **8**:565a, 593b–594a, 594b, 615b; **9**:87a; **10**:9a, 209a, 583b; **11**:35b; **12**:487a–491b; **13**:467a; **14**:88a, 211a; **15**:502b
mid nineteenth-century, **1**:20b, 210a, 572a; **2**:109a, 557a; **3**:328a, 392a, 433b, 586b, 623b; **4**:10a, 73b, 364b, 401a; **5**:537b–538b; **6**:238b, 436b; **7**:43b–44a; **8**:571b; **9**:99b–100a, 100b, 421a; **10**:258a, 584a; **11**:276b, 500b, 539b; **12**:293a, 304a, 514a–b; **13**:124a, 467a; **15**:527a
late nineteenth-century, **1**:472a–b; **2**:89b, 486a; **3**:236b, 462b, 529b, 587a; **4**:39a, 143b–145b *passim;* **5**:547b; **6**:14a, 187a, 227b; **7**:250a, 253a; **8**:30b, 32b, 511b, 543b, 577a; **9**:592b, 593a; **10**:29a, 502a; **11**:439b; **12**:272a; **13**:283b–284a; **15**:427b, 428a
early twentieth-century, **1**:52a, 236a–237a; **3**:483b–484a, 591b; **4**:261b, 262a, 598a; **6**:168b, 228a, 229a; **8**:544a; **10**:124a; **12**:230a, 273a; **13**:532a
mid twentieth-century, **1**:394a,b; **5**:221b; **8**:544b; **9**:104a
marine, **3**:204a
relief, **8**:544b
geomagnetic, **1**:521b; **4**:492a; **6**:107a; **7**:413b, 480b, 609a; **8**:426b; **9**:155a; **11**:412b, 554b; **12**:187a; **13**:517b; **14**:206b, 354b
geotechnical, **10**:127a
historical, **10**:239b–240a
hydrographic, **4**:614a; **8**:101a
hydrological, **7**:44a
hypsometrical, **13**:412b
meteorological, **4**:175a; **11**:525a
synoptic, **8**:487a
mineralogical, **14**:60b
oceanographic, **4**:70b; **9**:104a
paleogeographical, **7**:250a–b
phytogeographical, **9**:35b
relief, **6**:550b; **7**:8a
seismic, **9**:61a
tectonic, **1**:236b, 237a
topographical, **4**:217b; **7**:148b, 370b; **8**:64b–65a, 65a–b; **12**:610b
three-dimensional, **12**:380a
see also Atlases, Cartography, Geography, Projection (geom.), specific countries and areas
al-Maqdisī (or Muqaddasī), Shams al-Dīn Abū ᶜAbdallāh Muḥammad ibn Aḥmad ibn Abī Bakr al-Bannāᵓ al-Shāmī al-Maqdisī al-Bashshārī (*b. ca.* A.D. 946), 9:88–89
al-Maqsī, **14**:578b
Marāgha
observatory, **7**:353b; **13**:509a–b, 513b
Maraldi, Giacomo Filippo (Maraldi I) (1665–1729), 9:89–91; 3:103b, 107a; **4**:23a
Maraldi, Giovanni Domenico (Maraldi II) (1709–1788), 9:89–91; 3:198a; **7**:542b, 580a; **8**:22a
astronomy, **11**:525b–526a
students of, **3**:106a; **8**:143b
Marat, Jean-Paul, **5**:91b; **8**:85a; **15**:302a, 336a
Marattiales, **5**:605a; **11**:130a
Marble, **11**:539b; **14**:132b
ammonitico rosso, **8**:242b
formation of, **9**:533a
industrial use of, **1**:51b
source of, **14**:183a
Marburg school, **3**:110b
Marcasite
deposits of, **13**:566a
Marcellus, Marcus Claudius, **1**:214a
Marcet, Alexandre, **2**:335b, 336a; **8**:37b; **11**:172a; **12**:523b; **14**:492a
Marcet, Jane, **4**:528b–529a
Marcgraf, Georg
see Markgraf, Georg
Marchand, Richard Felix (1813–1850), 9:91–92; 4:394b, 395a
Marchant, Jean (1650–1738), 9:92–93
Marchant, Nicolas (*d.* A.D. 1678), 9:92–93
Marchantia, **9**:418b; **10**:13b
Marchi, Vittorio (1851–1908), 9:93–94
Marchi's globules, **9**:93b
Marchi's method, **8**:552b; **9**:93b
Marchi's tract, **9**:93b
Marchia
see Francis of Marchia
Marchiafava, Ettore (1847–1935), 9:94–95; 5:461a, 503a
Marchiafava-Bignami syndrome, **9**:95a

Marchiafava, Ettore (cont.)
Marchiafava-Micheli syndrome, **9**:95a
Marchiafava's postpneumonic triad, **9**: 95a
Marchlewski, Leon Paweł Teodor (1869–1946), 9:95–96; 10:22b; **12**:236b
Marci of Kronland, Johannes Marcus (1595–1667), 9:96–98; 7:377b*; 9:96b
Marckwald, Willy, **11**:85b, 87a; **12**:504b
Marco Polo
see Polo, Marco
Marcolongo, Roberto, **2**:594a; **8**:216b–217a, 225b, 237b, 238a; **10**:444a, 606a
Marconi, Guglielmo (1874–1937), 9:98–99; 4:6b, 284a; **6**:348b–349a
radiotelegraphy, **1**:195b; **2**:427b; **5**:33a; **11**:94a; **13**:357a
Marcou, Jules (1824–1898), 9:99–101; 1: 469b; **3**:392b*; **4**:364b, 365a*,b*; **6**: 57b; **10**:32b
Marcus Graecus
Liber ignium, **2**:64b
Marcus, Johannes
see Marci of Kronland
Maréchal, Sylvain, **7**:581b
Maret, Hugues, **5**:601a, 602b; **14**:225a
Marey, Étienne-Jules (1830–1904), 9:101–103; 11:431a
acquaintances of, **1**:304a; **5**:149a; **9**:546b; **11**:425b; **13**:288a
cardiac physiology, **3**:220a; **7**:504b
Marey's tambour, **9**:102a
physiological instrumentation, **1**:303b; **8**:541a; **9**:505b; **11**:425b, 426b
students of, **2**:366a; **3**:184a; **15**:504a
Margaric acid, **8**:58a
Margarine, **6**:499b
Margerie, Emmanuel Marie Pierre Martin Jacquin de (1862–1953), 9:103–104; 6:169a*, 227b; **8**:31b*
Marggraf, Andreas Sigismund (1709–1782), 9:104–107; 2:221a; **6**:259b; **7**:394a; **10**:26a; **11**:109b
chemistry, **1**:44b, 529b; **8**:69b; **14**:332b
Margó, Theodor, **1**:176b
Marguerite, Frédéric, **10**:440b
Margules, Max (1856–1920), 9:107–108; 6:97b; **7**:462a
Duhem-Margules equation, **4**:228b; **9**: 107b
Margulies, Benedikt, **8**:507b
Maria Theresa (Empress), **1**:332b; **2**:315b; **7**:12a; **8**:351a; **9**:325b; **11**:37a; **13**:39b, 181b
Mariana Islands, **11**:175b
botany, **11**:130a
gravimetry, **9**:181a
Marianas Trench, **10**:598a
Mariani, Andrea, **9**:62b
Marianini, Stefano, **8**:36a
Mariano, Jacopo
see Taccola, Jacopo Mariano
Maricourt, Pierre de
see Peter Peregrinus
Marie, (Abbé) Joseph-François, **8**:135a
Marie, Pierre (1853–1940), 9:108–109; 3: 205b; **4**:286a
on aphasia, **1**:439a
Marienbad, **11**:387b
Marignac, Jean Charles Galissard de (1817–1894), 9:109–111; 3:307a
atomic weight, **1**:321b; **2**:429a; **11**:173b; **12**:619b–620a
chemistry, **5**:216a; **12**:198a; **13**:546a
Marijuana
see *Cannabis*
Marine acids, **2**:76a–b
Marine Biological Association (U.K.), **3**: 89a–b
founding of, **7**:92a; **8**:26b
Marine ether, **14**:508b
Marinelli, Olinto, **3**:533a
Marinescu, Gheorgi, **13**:589b
Marinoni, Johann Jacob von, **6**:233b
Marinus (*fl.* second half of fifth century A.D.), 9:111; 6:271b; **10**:6a
commentaries of, **4**:425b; **10**:300a
historiography, **4**:159b, 430b; **11**:160a, 161a,b
Marinus of Tyre, **11**:198b; **15**:88a
Marion, Antoine Fortuné (1846–1900), 9: 111–114; 7:475b; **8**:145b; **12**:104b
Mariotte, Edme (*d.* 1684), 9:114–122; 12: 118b
astronomy, **9**:115b
botany, **9**:114a–b
correspondence of, **1**:169a
editions of works, **7**:578b
Essai de logique, **8**:86b(n)
geology, **9**:532a
hydrodynamics, **4**:208a; **9**:14a, 116a, 119a–b
hydrology, **10**:522a
Mariotte's law
see Boyle, Robert, Boyle's law
Mariotte's paradox, **9**:119b
mechanics, **7**:578a; **9**:114b, 115b–117a
meteorology, **9**:117b–118a, 119a, 120b
methodology, **9**:118a–b, 120a–b
observations of, **12**:123b
optics, **9**:114b–115a, 118a–119a; **10**:477a; **14**:570b(n)
philosophy, **11**:507b
pneumatics, **9**:117a–118a
Traité de la percussion ou chocq des corps, **9**:51b–52a
Traité du mouvement des eaux et des autres corps fluides
translation of, **4**:45b*
Marius, Simon
see Mayr, Simon
Marivaux, Pierre, **1**:323a; **5**:58a
Mark, Herman Francis, **4**:522a; **6**:303b; **9**: 354a; **13**:227a
Markgraf (or Marcgraf), Georg (1610–1644), 9:122–123; 10:621b; **13**:451a
Markham, Clements Robert (1830–1916), 9:123–124; 6:494b
Markov, Andrei Andreevich (1856–1922), 9:124–130; 3:225a, 226a, 230b; **5**: 508b; **8**:560b; **11**:107b; **12**:535a
Markov chains, **7**:352a; **9**:128b–129b
Markov processes, **9**:129b
probability, **1**:366b; **3**:231a; **8**:562b, 563a
Markov, Vladimir Andreevich, **9**:127a
Markovnikov, Vladimir Vasilevich (*ca.* 1837–1904), 9:130–132; 2:621a,b, 622b, 623b; **9**:291b
Markovnikov rule, **9**:131b
students of, **3**:246a; **7**:203b; **10**:1a; **12**: 336b
Marl, **5**:475b; **8**:565a, 567a; **14**:463b
Marliani, Giovanni (*d.* A.D. 1483), 9:132–134; 2:393b; **13**:209a
Probatio cuiusdam sententie calculatoris de motu locali, **13**:210a
Marliani, Girolamo, **9**:132b
Marliani, Paolo, **9**:132b
Marliani, Pietro, **9**:132b
Marmara, Sea of
density of, **9**:43a
Marmot
hibernation, **13**:557b
Marne (river), **2**:347b
Marneck, F. H. (pseudonym)
see Bolza, Oskar
Marolois, Samuel, **5**:408b; **12**:205a
Marquart, Ludwig, **5**:163b
Marquesas Islands, **3**:397a
Márquez-Miranda, Fernando, **1**:131b
Marr, John, **5**:343b
Marr, John Edward, **2**:288b; **6**:116a; **8**:33a; **12**:339a
al-Marrākushī, Abū ᶜAlī, **9**:35b; **12**:2a, 358a, 361a; **15**:260a
Marre, Aristide, **3**:272a,b, 273a, 275b; **8**:41a
Marrot, Louis, **1**:327b
Marrow (anat.)
erythrocytopoiesis, **2**:165a
disruption of, **9**:295b–296a
lymphoid tissue, **15**:479b
phagocytosis, **2**:165b
role of, **8**:172a; **9**:146a
Mars (planet), **12**:187b
atmosphere of, **1**:57a; **3**:36a; **7**:74b; **12**: 267a
distance of, **3**:409a
drawings of, **7**:209b
inclination of axis, **6**:334b
life on, **8**:520b, 521b–522a; **14**:139b
motion of, **2**:411a; **13**:481a; **15**:353b
observation of, **1**:172a–b, 568b; **2**:387a, 413b; **5**:21b, 404b; **7**:112a, 271b*; **8**:465a, 521b, 613b; **11**:424a; **12**: 161a–162a; **13**:115a
occultations of, **11**:28b
orbit, **7**:294a–b, 295b–296b; **11**:552b–553a
photography of, **10**:602a; **11**:553a; **12**: 454a–b; **13**:408b; **14**:521b
relief map of, **3**:357b
rotation, **14**:637a
axis of, **12**:161a
period of, **1**:172b; **3**:101b; **6**:334a–b, 604a; **11**:163a; **12**:455a
satellites of, **7**:385b, 472a
discovery of, **6**:49a–b
size
diameter, **2**:388b
mass, **6**:50a*
and solar parallax, **1**:157b; **3**:103a; **5**:24b, 258a, 479b; **6**:263b; **10**:34b; **11**:424a
space probes, **7**:465b
spectrum, **3**:36a; **4**:180a; **9**:392a; **12**:37a, 73a; **14**:55a
surface features, **2**:433a; **5**:490a; **8**:581a; **9**:2a; **10**:585a; **11**:163a; **12**:161a–162a, 226b
canals, **1**:172b, 467a; **3**:357b; **8**:521b–522a; **10**:119b; **12**:161b, 267a; **13**: 474b
changes in, **8**:50a; **12**:454a
nature of, **12**:353a
oases, **10**:602a
spots, **3**:101b; **5**:21b; **6**:334b; **14**:637a
synodic period, **4**:467a; **10**:180b; **15**: 219a
tables of, **7**:305a; **8**:369b; **13**:4a
vegetation on, **13**:408b–409a
zodiacal period, **5**:345b
see also Parallax, of Mars
Marsden, Ernest, **5**:331a; **12**:31a,b, 32b
Marseilles
Academy of Marseilles, **10**:568b, 569b
latitude of, **10**:572a
marine biology, **9**:112b
Museum of Natural History, **9**:112b
observatory, **8**:279a; **10**:571b; **11**:82b; **13**:36b
Marseilles, University of
zoology at, **9**:112a
Marsh, Adam, **1**:377b; **5**:549a

Marsh, Othniel Charles (1831–1899), 9:134; 6:13b
  acquaintances of, **14**:409b, 410a
  biographies of, **12**:230a
  and Cope, **15**:92a
  Marsh's test, **10**:557a
  paleontology, **6**:595a–b; **11**:176a; **12**:38a
Marshall group (stratig. unit), **14**:439a
Marshes
  draining of, **2**:279b, 326b; **5**:499a
  gases from, **14**:74b, 638a
  utilization of, **9**:54a
Marsili, Antonio Felice, **2**:591b
Marsili (or Marsigli), Luigi Ferdinando (1658–1730), 9:134–136; **9**:511a, 532a, 533a; **10**:568b
Marsilius of Inghen (or Inguem or de Novimagio) (*d.* A.D. 1396), 9:136–138; **1**:199b; **5**:403a; **8**:223b; **13**:349b
  Abbreviationes libri physicorum, **6**:275b
Marsupials, **2**:342b; **5**:356a
  anatomy, **2**:504a; **13**:527b
  evolution of, **4**:148a
  fossil, **10**:262a
  reproduction, **10**:261a–b
  taxonomy, **5**:46a; **10**:262b
Marta, Giacomo Antonio, **15**:68b
Martens, Adolf (1850–1914), 9:138–140; **10**:246b
  metallurgy, **6**:368a–b; **12**:544a
Martensite, **10**:246b
Marth, Albert, **8**:46a
Martí Franqués (or Martí d'Ardenya), Antonio de (1750–1832), 9:140
Martialis, Gargilius, **4**:121a; **5**:229b
Martianus Capella (*fl. ca.* 365–440), 9:140–141; **5**:365a; **9**:1b
  commentaries on, **2**:21a,b
  cosmology, **1**:244b; **15**:204b(n)
  De nuptiis philologiae et Mercurii, **1**:91b; **4**:396a, 443a
Martin of Bohemia
  see Behaim, Martin
Martin, André (philos.), **7**:611a
Martin, André (vet. med.), **1**:576a
Martin, Benjamin (*ca.* 1704–1782), 9:141–142
  instrumentation, **14**:452b
  translations of works, **12**:407a
Martin, Hans, **7**:519a
Martin, Henry Newell (1848–1896), 9:142–143; **8**:14b; **9**:516b
  acquaintances of, **1**:9b; **2**:501b; **3**:447a; **5**:80b; **6**:526a; **14**:249b
  students of, **3**:389b; **9**:515b; **14**:424a, 426a
Martin, James, **8**:309b
Martin, Lillien Jane, **9**:562a
Martin, Louis, **11**:569b
Martin, Rudolf (1864–1925), 9:143–145
Martin, Thomas Henri, **9**:269b; **15**:203b
Martin, William Ted, **14**:347a
Martine, George, **2**:179a; **13**:10b
Martineau, Harriet, **3**:379a; **6**:521b; **9**:67b; **12**:523b
Martínez, Crisóstomo (1638–1694), 9:145–146
Martini, Francesco di Giorgio (1439–1501), 9:146–148; **8**:208a; **9**:488b*; **13**:233b, 568a
Martini, Martin, **7**:375b
Martinique, **7**:63a
Martinovics, Ignác (1755–1795), 9:148
Martins, Charles, **2**:430b, 431a, 432a–b*
Martius, Carl Alexander, **3**:85a; **6**:461b, 463a
Martius, Karl Friedrich Philipp von (1794–1868), 9:148–149; **8**:371b; **12**:215a; **14**:181a
  botany, **10**:13a,b
  on contemporary scientists, **2**:517a, 519b, 521b; **8**:374a*; **10**:212a*, 499b*; **12**:254b*
  Flora Brasiliensis, **1**:412b; **4**:307a; **9**:417b; **15**:147a
  historiography, **9**:123a*
  Nova genera et species plantarum, **2**:159b
  and Spix, **1**:501b; **12**:578a–b
Martius yellow, **3**:85a; **5**:489a
Martonne, Emmanuel-Louis-Eugène de (1873–1955), 9:149–151; **1**:208a
Martyn, John, **9**:311a; **14**:193a
Martynia, **7**:441a
Marum, Martin (Martinus) van (1750–1837), 9:151–153; **7**:16a*, 499b; **12**:197b; **13**:183b, 468b, 469a
al-Marwazī
  see Ḥabash al-Ḥāsib
Marx, Ernst, **4**:298b
Marx, Karl (1818–1883), 15:403–417; **3**:497b; **5**:254b; **10**:448a; **11**:258b; **12**:209a; **13**:551a
  economics, **7**:104b
  editions of works, **15**:133b
  and Engels, **9**:70a; **15**:132b–144a *passim*
  historiography, **5**:285b
  Das Kapital, **15**:133b
Marxism, **1**:361a; **8**:183b–184b, 603b; **10**:289b, 290a; **12**:108a; **15**:17a, 18b, 38a
  and evolution, **13**:276b
  and science, **5**:213a; **15**:38b
Marysville mining district, **1**:470a
al-Marzūqī, **11**:246b
Māsa (month)
  definition of, **15**:535a
Masaryk, Tomáš Garrigué, **11**:257a–b, 258a
Masatada Irie, **1**:90a
Ibn Māsawayh (Mesuë), Yūḥannā, **7**:187a; **11**:246b; **13**:230b; **15**:230a–b, 236a, 237a,b, 242a
  Aphorisms
    translations of, **15**:188a
Mascagni, Paolo (1755–1815), 9:153–154; **11**:510a
Mascarene Islands, **2**:321a
Mascart, Éleuthère Élie Nicolas (1837–1908), 9:154–156; **1**:613a*; **2**:465b; **3**:420a*; **4**:254a(n); **8**:10a, 551a; **9**:372a
Mascheroni, Lorenzo (1750–1800), 9:156–158; **1**:42a; **9**:447a; **13**:18a
  Geometria del compasso, **12**:325a
Maschke, Heinrich, **2**:198b, 272a; **9**:502a; **11**:55b
Mas-d'Azil, **10**:607a
MASER, **11**:12a
Maseres, Francis (1731–1824), 9:158–159; **6**:577a
Māshā'allāh (*fl. ca.* 762–815), 9:159–162; **4**:555b; **7**:362b; **13**:538a; **15**:125a
  astrology, **7**:222b
  De elementis et orbibus celestibus
    translations of, **15**:178b–179a
  Epistola de rebus eclipsium, **15**:179a
  Fī 'l-qirānāt wa 'l-adyān wa 'l-milal, **6**:381b
  influence of, **1**:33a, 34b, 382a; **2**:435a; **3**:217b; **11**:226a; **13**:538b
  De scientia motus orbis
    translations of, **15**:179a
Masing, Georg, **13**:245b
Maskelyne, Nevil (1732–1811), 9:162–164; **6**:131a*, 577a
  acquaintances of, **2**:468a; **6**:329a; **10**:591b
  astronomy, **6**:328b, 331a; **8**:593b; **9**:164a, 234b
  geodesy, **9**:164b
  observations of, **2**:99a–b; **10**:207b; **15**:354a
  star catalogs, **4**:16b
Maslama ibn Aḥmad al-Majrīṭī
  see al-Majrīṭī
Mason, Charles (1728–1786), 9:164–165; **15**:339b
  Mason and Dixon line, **4**:131b–132a
Mason, Max, **2**:198a
Masonry, **3**:441b; **9**:507b*; **14**:31a
  rupture of piers, **3**:441a, 442a, 444a
Mass
  atomic, **4**:314b; **13**:369b
  concept of, **1**:424a
  definitions of
    seventeenth-century, **10**:70a
    eighteenth-century, **1**:111b–112a; **2**:330b
    nineteenth-century, **8**:599a; **10**:449a
  electric, **3**:445a
  electromagnetic, **4**:323b; **8**:10b, 497a, 498a; **13**:364b
  of electron, **4**:281a,b; **5**:102b
  and energy, **4**:323a–b, 324a–325a; **6**:164a; **7**:37a*; **8**:10b, 95a, 290b
  gravitational, **4**:325a, 326b–327a, 329b–330a, 331b, 380a; **14**:599a
  inertial, **4**:324b–325a, 331b, 380a; **8**:599a; **10**:75b; **14**:599a
  measurement of, **1**:321a–b; **8**:85a, 95a, 470a; **10**:519a
  and velocity, **5**:15a; **7**:264a; **8**:496a–b, 498a
  vs. weight, **6**:608b; **10**:62a, 75b, 77b
  see also Chemical mass
Mass action, law of, **1**:299a, 300a, 301b, 579a; **2**:68b, 69a; **4**:411b–412a; **5**:412a–b, 587a–b; **6**:538b; **8**:503a; **9**:110a, 352b, 619b; **10**:440b; **11**:541a–b; **13**:578a; **14**:108b–109a, 395b; **15**:457b, 459a,b, 461a, 462a
  confirmation of, **9**:616a; **13**:359a
  discovery of, **6**:110a
  and gaseous explosions, **4**:130a
Mass production
  nineteenth-century, **1**:355a
Mass spectrograph
  see Spectroscopy
Massa, Niccolo (1485–1569), 9:165–166; **1**:47a, 619b
  Anatomiae liber introductorius, **4**:199a
Massachusetts
  Geological Survey, **5**:477b; **6**:437b; **7**:148b
  geology, **4**:361b
  mollusks, **5**:477b–478a
  State Board of Health, **12**:481b
Massachusetts Institute of Technology
  chemistry at, **10**:156b
  Compton and, **3**:373a
  founding of, **11**:504b–505a
  mathematics at, **14**:344a
  physics at, **10**:599b
  Radiation Laboratory, **8**:95b
Massage
  ancient Greek use of, **4**:106b
Massari, Julius, **14**:278b
Massaria, Alessandro, **1**:360a–b
Massat (prehistoric site), **8**:44a
Massau, Junius, **10**:170a
Massey, Harrie Stewart Wilson
  Theory of Atomic Collisions, **15**:42b
Massieu, François Jacques Dominique, **4**:228b

- Massif Central
  - geography, 1:208a
  - geology, 9:367b, 583a; 13:283b
  - paleobotany, 5:498a
- **Masson, Antoine-Philibert (1806–1860), 9: 166–167;** 2:438b–439a; 9:566b
- Masson, Francis, 1:88b; 13:391b
- **Mast, Samuel Ottmar (1871–1947), 9:167;** 7:99a; 8:45b
- Mast cells, 4:296a
- Mastectomy, 6:509a
  - radical, 8:402a
- *Mastigocladium,* 9:176a
- **Mästlin, Michael (1550–1631), 9:167–170;** 4:549b; 11:367a*; 12:499b
  - *Ephemerides,* 7:289b
  - and Kepler, 7:289b, 290a–b, 291b, 294b, 299b, 303a, 303b–304a; 9:168b
  - students of, 1:158b, 159a; 12:163a
- Mastodon, 3:324b*; 8:43b; 10:438b; 12: 437a*; 15:472a–b
- Masᶜūd ibn Maḥmūd, 2:151a
- **al-Masᶜūdī, Abu ʾl-Ḥasan ᶜAlī ibn al-Ḥusayn ibn ᶜAlī (*d. ca.* A.D. 957), 9:171–172;** 4:524a
  - algebra, 13:514b–515a
  - historiography, 7:357a, 362a; 12:447a,b; 15:232b
- Masurium, 10:136b
- Matches, 2:340b, 620b; 4:42a*; 5:474a; 7: 13a, 89a; 12:227a
  - safety, 7:22a–b
- *Matematicheskii sbornik,* 2:425a; 3:222b
  - editors of, 10:553b
- *Matematisk Tidsskrift*
  - editors of, 7:185a
- *Matematyka,* 14:590b
- Materia medica
  - ancient Roman, 1:167a–b, 212b; 11:39a
  - Byzantine, 10:418b
  - medieval Arabic, 9:40b–41a; 13:38b; 15:499a
  - medieval Latin, 6:123b
  - Renaissance, 3:414b; 8:554a; 9:179b–180a; 10:237a–b
  - seventeenth-century, 1:369a; 6:145b; 7: 164b; 14:27a
  - eighteenth-century, 2:173b; 3:465a; 5: 352b; 8:297a–b, 352b–353a, 379a; 13:602b, 603a
  - nineteenth-century, 2:547b, 548a; 4:252a; 6:18a; 14:44b
  - astrological, 10:548b
  - Chinese, 12:374a, 383b–384a, 384b
  - development into pharmacology, 1: 10a–b; 2:548b
  - Indian, 12:576b
  - Japanese, 15:742a
  - Mesopotamian, 15:645a–b, 646b
  - separation from botany, 1:524a; 13:443b; 15:81a
  - *see also* Botany, medical; Pharmacopeia; Pharmacy
- Materialism
  - ancient Greek, 1:259a, 260a, 315a, 324b; 4:369a
  - medieval, 2:193a
  - seventeenth-century, 5:285a–288b; 6: 446b–447a; 7:612b
    - opposition to, 3:492a; 9:509b–510a
  - eighteenth-century, 3:114b–115a, 295a–b, 540a; 4:85a, 87b; 6:468b–469b; 7: 605a–607a; 9:26a, 148b, 490b; 11: 143a, 334a; 13:179b; 15:54a
    - opposition to, 5:577b; 11:497a
  - early nineteenth-century, 7:587b, 590a–b, 591a,b; 8:97b–98a; 11:560a; 12:154a–b; 15:127b
  - mid nineteenth-century, 2:563a–564a; 3:378a, 434b; 6:7a, 60a, 596b; 8:445a; 9:456b, 457a; 10:266a, 371b; 14:57b, 114a; 15:408b–409a, 412a–b, 413a
  - late nineteenth-century, 1:173b; 3:90a; 5:211a, 439a; 12:209a; 13:417a; 15: 143a, 145a
  - nineteenth-century
    - opposition to, 3:110b, 526b; 4:556b; 7:384a; 8:515b, 601a; 12:224b, 278a, 607a–b; 15:79b
  - twentieth-century, 5:213a; 8:603b; 12: 337a, 356a; 13:530a; 14:3a
    - and heredity theories, 1:505b, 506a
    - opposition to, 11:258b; 14:302b; 15:38a
  - and Christianity, 6:7a; 13:523a
  - dialectical, 8:183b–184b
  - historical, 15:133a, 134a, 136a–138b, 140a, 143b–144a
- *Matériaux pour l'histoire positive et philosophique de l'homme*
  - founding of, 9:538b
- Mathematical and Physical Society (Hungary)
  - founding of, 4:380b
- Mathematical Association of America
  - founding of, 9:502a; 14:559b
- Mathematical Society (Berlin), 3:53a
- Mathematics
  - ancient Greek, 4:391b–392b; 5:159a, 346a–347a; 13:301b–305a, 615b; 14:608a–611b
  - ancient Roman, 2:229a, 233a–234a; 4: 111a–117a; 9:140b–141a; 13:325b–326a
  - Byzantine, 4:159b–160a, 488a–491b; 6: 615b–616a; 8:191a–b; 11:185b; 13: 37b
  - medieval Arabic, 1:444a; 4:611b–612a; 5:159b; 7:25a, 256b–259a; 12:91b–94a; 13:288b–291b, 509b–513b; 14: 471b–472b; 15:497b–498a
  - medieval Hebrew, 8:279b–280a
  - medieval Latin, 2:393a–b; 3:23b–27b; 4: 604b–612b; 5:364b–366a; 7:117b–118a, 128b–129a; 10:224a–b; 13: 340a; 14:35b
  - Renaissance, 2:602b–603a, 618a–b; 3: 125b–126a, 262a–b, 280a–b, 311a–312a, 363b–364b, 401b, 585a; 4:8b, 212b; 5:238a–b, 250a; 9:12b, 32b, 190b–192a, 264a; 10:7b–8a, 160b–161b; 11:402b–403a, 405b–406a, 468a, 532b–534a; 12:199b; 13:51a, 349a–350a; 14:513b–514b
  - early seventeenth-century, 1:541a–b; 2: 83a, 349a, 358b, 526a–527a; 3:3a, 63b–64b, 115b, 126a–128b, 614b–615a; 4:549b–551b; 5:54b–55a, 343a–344a, 588a–589a, 593b; 6:124b–125b, 145b, 271a–b; 7:193b–194a; 8:271b; 9:12b–13a, 336a; 10:495a–b; 12:163a, 205a–206b, 499b–501a; 13:235b–236a; 14:51a–b
  - late seventeenth-century, 2:349a, 358b, 506b–507b; 3:61a, 348a–b; 4:95b, 505b–506a; 5:59a, 60b, 158b–159b, 259a–b, 520b–521b, 524b–529a; 6: 71a, 536b–538a; 7:583a–584a; 9: 310a–311b, 529b, 599a–b; 13:585a–586a
  - early eighteenth-century, 2:56a, 57b, 326a–b, 418a–b, 576a,b, 577a–b; 3: 39a, 244b, 281a–282b, 431b–432b, 458a–459a, 459a–461b; 4:88a–89a, 166a–b, 468a–479b; 5:61b, 434a–b, 448b–450b, 509b–510a; 6:129b, 304b–305a; 7:162b–163a, 442a–443a, 542b, 558b–559a; 9:17a; 11:67b, 494a; 12:117b–118a; 13:39a–b, 265a–267b; 14:295b
  - late eighteenth-century, 2:51b, 56a, 326a–b, 342b; 3:41b, 119b, 155b, 208a, 383b–387b, 538a; 4:88a–89a, 468a–479b; 5:195b; 6:403a–404a, 576b–577a; 7:206a–b, 404b–405a, 559b–571a, 599a–600a; 8:305a–306b; 11: 579b; 13:184a; 14:179b–180b; 15: 98b–99a
  - early nineteenth-century, 2:133b, 443b–444a; 3:131b–146b, 466a–b; 5:110a–111b, 298a–310a, 375a–376b, 466a–b; 6:85b–86a, 474a–b; 7:494a–b, 549b–551a; 13:237b; 15:270a
  - mid nineteenth-century, 2:349b–350a, 573a–574a; 3:163a–170a, 222b–231a, 302a–b, 591b–592a; 4:35a–36b, 123b–127a, 167a–b, 340b–343a; 5: 158a–b, 259b–264a, 471a–b, 572a–573a, 587a; 6:95a–b, 230a–b, 356b–357a, 502a–b; 7:383b–384a, 505b–508b, 521b–523b; 10:479a–480b; 11:568a–b; 12:86a–87a, 170b–173a; 13:216b–221a, 426b–427b; 15:192a–198b
  - late nineteenth-century, 2:302b; 3:52b–57b, 177b–179a, 263a–b, 345b–346a, 456a, 583a; 4:1a–5a, 84a, 137a–138a, 269a; 5:77b–78a, 198a–b, 413b–414b, 481b–482a, 568b–569b, 573a–b; 6: 3a–5a, 76a–b, 341a, 388a–391b, 547a–b; 7:167a–168b, 396a–399b; 8:388b–389b; 12:245b–247a, 511b; 13:249b–250b
  - early twentieth-century, 2:217a–b, 357b, 358a, 490b–491a, 512a–514a; 3:196a, 328b–329a; 4:83a–b, 278a, 370a–371a, 560b–561a, 617a–b, 618b; 5: 187b, 200b–202a, 206a, 407a; 7: 407b–408b, 411a–b, 530a–531a; 13: 390b–391a; 14:344a–345a, 559a–b; 15:256a
  - mid twentieth-century, 6:402a, 496a–497b; 12:410b–411a; 14:89a–92b
  - analogy in, 11:52b; 14:347a
  - applied
    - ancient Greek, 6:314a–315b
    - Renaissance, 4:5b, 97a; 13:48b
    - seventeenth-century, 3:621b; 4:550a; 6:612a; 10:264a; 12:127a–b
    - early eighteenth-century, 1:116a–b; 2: 37b–38a, 54a; 4:473a–b; 9:233a; 10:320a; 15:754a
    - late eighteenth-century, 1:326b; 2:327a; 4:473a–b, 501b; 5:209b; 7:599b–600a; 11:434b
    - early nineteenth-century, 1:354b; 4: 257a; 5:530b; 6:213b; 7:50b; 15: 480b–490a
    - mid nineteenth-century, 2:166a, 424b–425a, 469b; 3:224a, 226a–227a, 450a–452a; 4:5a, 125b–126b, 369b; 8:63a–b; 11:568b; 14:105b
    - late nineteenth-century, 1:599b, 600a–b; 2:236b, 515a; 3:322b, 583b–584a; 5:150b–152a; 6:92b; 7:399a–b, 465b–466a, 594b–595b; 8:542b; 9:324a–325a; 10:447a–448b; 12:238a–b; 13: 339a; 14:614a, 619b–621a; 15:65b
    - early twentieth-century, 2:303a–304b, 491a, 594a, 616a; 3:63a, 564a; 5: 103a, 508b; 6:570a–b; 7:413b, 514a–b, 574b; 8:301a–b, 617b; 9:419a–b; 10:20a–b; 11:613a–b; 13:24b, 391a; 14:614a

mid twentieth-century, **3**:63a; **6**:147a–148a; **7**:211b; **13**:36a; **14**:345b–347a
Egyptian, **15**:688b, 694a–697a, 699a, 699b–700b
Japanese, **15**:746b, 750b
philosophy of, **5**:193b; **6**:87a; **11**:288a–b
arithmetization of, **7**:523b
and art, **4**:258b–261a
and astronomy
see Astronomy, mathematical
Chinese, **15**:745b–746a
consistency of, **2**:512b; **6**:392b–393a
critics of, **4**:88b–89a; **5**:347a
definition of, **3**:57b; **15**:681a–b, 745a
Egyptian, **2**:320a; **15**:681a–705b
engineering, **5**:202a
epistemology of, **4**:52b–55a; **7**:226b
evolution of, **2**:545b
foundations of, **2**:512b, 513b–514a, 593b; **3**:278a–b; **4**:373b; **5**:108a–b, 152b–154b, 350b–351a; **6**:392b–393b, 473a–b, 570a; **7**:72a, 444b; **8**:262b–263a, 389a, 432b, 546a; **9**:502b; **10**:343a–344b, 442a–443a; **11**:60b–61a, 285b; **12**:11a–16b, 216b, 364a–b, 451b–452a; **14**:285a, 303b–304b; **15**:225a, 573a
and genetics, **2**:58a–b; **7**:99a, 115a; **14**:87a
in geology
see Geology, mathematical
geometrization of, **4**:424b–425a
and idealism, **1**:337a; **7**:327a
internationalization of, **3**:53a
intuitionistic, **2**:512b, 514a
Japanese, **3**:254a, 265b–266a, 270b; **8**:419a; **15**:744b–755b
Seki school, **1**:83b, 90a; **15**:747b, 751a–b, 753b
as language of science, **4**:615b
and logic, **1**:381a; **2**:163b–164a, 295a–298b; **4**:137b–138a; **8**:159b–160a; **10**:442a–b; **12**:364b; **13**:510a; **14**:304a; **15**:369a, 755a
meaning of word, **6**:544a
and mechanics, **1**:582a; **7**:116a–b; **13**:584b, 586a; **14**:149b–150a
and medicine, **1**:290b; **5**:228b, 231a–b; **10**:611b, 613a; **12**:335a
Mesopotamian, **2**:320a; **4**:111b, 117a, 612a; **6**:315a; **13**:296b; **15**:10a, 208b, 667a–672b, 692a–693a
and metaphysics, **3**:57a; **7**:597b; **12**:575b–576a
methodology of, **2**:465a; **7**:107a, 194b; **10**:181a
and music, **1**:282a–283a, 599a; **2**:233b; **5**:249a–b; **7**:128b, 129a; **11**:222b
and nature, **8**:148a
"new mathematics," **2**:352a
and optics, **1**:81b, 279a; **5**:483a; **6**:88a; **14**:568a
philosophy and, **2**:395b, 545b; **7**:597b–598a; **11**:26b; **13**:428b
philosophy of
ancient Greek, **1**:256a–258a; **5**:346b; **9**:269a–b
ancient Roman, **7**:1a
Byzantine, **11**:160b
medieval Arabic, **7**:3b; **11**:250a
medieval Latin, **1**:100a,b, 379a, 380b–381b; **5**:549b
Renaissance, **1**:97a; **3**:513b–514a; **10**:512b; **12**:97b
seventeenth-century, **1**:373a, 474b; **10**:51a–52a
eighteenth-century, **1**:112b; **2**:17a, 577b; **6**:557b–558a
early nineteenth-century, **5**:97b, 193a–b; **8**:142a; **14**:500a; **15**:225b–226a
mid nineteenth-century, **3**:226a–b; **5**:263b–264a; **7**:522b; **9**:384a; **13**:217b
late nineteenth-century, **4**:206a; **6**:4a–5a; **7**:508a–b; **9**:382a; **10**:484b
twentieth-century, **2**:358a–b; **3**:111a; **4**:374a; **8**:262b–263a; **11**:107b, 261a, 362b, 413a–b; **12**:14a; **13**:125a; **14**:304b–305a, 469b; **15**:424a
and physics
see Physics, mathematical
preeminence of, **1**:251a–b, 279a, 337a, 381a; **3**:382a–b, 452a; **5**:213b, 243a, 247a; **6**:280a, 544a; **7**:55a, 193b; **8**:194a, 199b, 234a, 240b–241a; **9**:111b; **10**:482b, 486a, 495a–b; **11**:42a, 289a; **14**:164b, 221b
and reality, **4**:330a
"social," **3**:386b–387b; **15**:374b, 375b
surveys of, **2**:351b–352b, 358a–b
and theology, **1**:91a
see also Algebra; Analysis (math.); Arithmetic; Calculation; Congresses, scientific; Education, mathematical; Geometry; Historiography, of mathematics; Instrumentation, mathematical; Notation, mathematical; Number(s); Recreations, mathematical; specific countries; specific topics; Terminology, mathematical
Mathematics and Astronomy in Mesopotamia, 15:667–680
Mathematics of Ancient Egypt, 15:681–705
Mathematics Teacher, **14**:559a
Mathematische Annalen
editors of, **14**:202a
founding of, **3**:314a; **10**:25a
Mathematische Nachrichten
founding of, **12**:188a
Mathematische Zeitschrift
editors of, **14**:453b
founding of, **7**:411b
Mathematischer Verein (Berlin), **11**:611a
Mather, Cotton, **1**:126b
Mather, William Williams (1804–1859), 9:172–173; **4**:364a,b; **13**:581b; **15**:152b
Matheronia, **9**:579a
Mathers, Patrick (pseudonym)
see Sanders, William
Mathesis
editors of, **10**:25a
founding of, **9**:81a
Mathews, Albert Prescott, **7**:466b; **14**:424b, 429a
Mathews, George Ballard (1861–1922), 9:173–174
Mathieu, Charles, **1**:202b
Mathieu, Claude Louis, **1**:201a; **2**:135b; **4**:17a
Mathieu, Émile Léonard (1835–1890), 9:174–175
Mathieu functions, **9**:175a; **14**:317a
Mathieu's equation, **9**:175a
Mathieu, Félix, **10**:333b
Mathurānātha Śarman (*fl.* 1609), 9:175; **11**:264a
Ravisiddhāntamañjarī, **15**:614a
Matijasevič, Iurii Vladimirovich, **6**:394a
Maton, William George, **11**:35a
Matrices (math.), **1**:429a; **3**:165b–166b; **5**:151a–152a; **6**:92b, 402a; **7**:530b; **8**:457a; **10**:139a; **12**:237b; **14**:211b, 454a; **15**:195b
infinite, **5**:96a; **6**:235b*; **7**:435b–436a; **9**:426b; **11**:459b; **12**:188a,b; **13**:428a
inverse of, **3**:142b; **7**:565a; **14**:92a
normal form, **7**:435b, 436a
notation, **4**:137b
nullity, **13**:220b
theory of, **3**:169a; **4**:281b; **10**:483b–484a; **14**:90b
Matrix, anatomical
naming of, **9**:468b
Matrix mechanics, **7**:518b; **10**:214a; **15**:41b–42a, 48a
opposition to, **4**:318a; **12**:220b
and wave mechanics, **11**:422a; **12**:220b–221a; **15**:49a
Matruchot, Louis (1863–1921), 9:175–176; **5**:584a
Matsunaga, Ryohitsu, **15**:751a
Matsunaga Yoshisuke, **1**:90a,b
Matsuyama
see Matuyama, Motonori
Matter
as cause of spatial structure, **3**:322b–323a
cohesion of, **2**:330b; **3**:441b, 443b–444a; **4**:506b; **5**:124a, 245b; **6**:609a
conservation of, **4**:369a; **7**:194b; **8**:78a, 83a, 469a, 470b, 471b
divisibility of, **2**:484b; **7**:229b–230a, 276a; **8**:374a; **14**:489a–b
electricity and, **3**:499b, 501a; **8**:497a; **9**:207a; **10**:146a; **11**:421a
and energy, **8**:618a
and ether, **4**:319a; **6**:347a; **8**:40a–b, 493a,b, 494a–b, 497a, 508b; **9**:213b, 548a–b
extension of, **2**:330b; **4**:52a
and fields, **4**:315a, 325b–326a; **8**:492b, 494b, 498b; **9**:377a
and force, **2**:563b; **4**:203a–b, 204a, 367b–368a, 529b, 537a; **5**:193b; **6**:243b–244a, 247b, 335b, 395b; **7**:227b–228b
and form, **1**:254b–255b, 277b, 278a; **9**:252b; **11**:244b; **12**:7a–b, 59b; **15**:495a
heterogeneity of, **3**:544a–b
impenetrability of, **2**:330b; **4**:529b; **7**:228b; **8**:152b
indestructibility of, **6**:255a
interstellar, **4**:280b; **6**:146b, 540a; **9**:376b; **10**:199a, 290a; **11**:320a; **12**:265b, 455a; **13**:111a, 475a; **14**:482a
origin of, **9**:449a
and radiation pressure, **11**:123a
and light, **8**:494b–495a; **10**:59b, 80a, 525b; **13**:238a–b
mind and, **4**:557a, 558a; **6**:596a; **12**:601a
and motion, **4**:52a, 54a; **9**:214a; **12**:601b; **15**:127b, 143a
penetrability of, **2**:330b
and radiation, **4**:196a; **7**:85b; **9**:404b; **15**:49a,b
radiation, conversion to, **4**:280a
and space, **2**:326a; **3**:322b–323a
and spirit, **6**:256a–b
states of, **3**:486a; **5**:494b; **7**:195a; **9**:182b
changes in, **3**:599b–600a; **9**:33b; **13**:244b–245a, 578b; **15**:315a
continuity between, **1**:161a; **4**:228b; **9**:222b
critical phases, **5**:389a, 390b
critical states, **8**:186b; **13**:80b–81a
equation of, **2**:40a; **3**:308b, 309a–b; **7**:220b, 221a, 272a, 515b; **8**:503a; **11**:417b; **14**:111a, 561a; **15**:441a, 486b
fourth state, **3**:477b–478a, 599b
glass state, **13**:246b–247a
mesomorphic, **5**:186b; **8**:148b
polytropic change of, **4**:361a
temperature discontinuity, **8**:102a
temperature effects, **8**:118a–b
theories of, **5**:103b, 161a; **8**:73a, 79b,

Matter (cont.)
states of (cont.)
256a, 374a; **11**:9b–10a; **13**:245a, 246b–247a
see also Phase (chem.), etc.
subtle
vortex theory, **2**:38b; **9**:52a
theories of
ancient Greek, **1**:277a; **10**:208a; **11**: 104a
ancient Roman, **3**:15a
medieval, **2**:20b; **5**:403a, 549b; **7**:346a; **11**:324b–325a; **15**:495a
Renaissance, **13**:278b–280a
seventeenth-century, **5**:422a; **6**:256a–b; **7**:165a, 611b; **9**:51b–52b, 117b; **10**: 59a; **11**:507a–b; **13**:235a
early eighteenth-century, **1**:111b; **5**: 353a, 511a; **6**:173b; **7**:188b; **9**:33b; **10**:80b–81a; **12**:601a, 604b; **15**:54a–b
late eighteenth-century, **2**:330b; **3**: 599b–600a; **5**:601a; **6**:335b, 586a; **7**:227a–228b, 229b–230a; **8**:620a; **11**:142b–143a, 146b–147a; **13**:602b–603a; **15**:313b
early nineteenth-century, **6**:395a–b; **11**:172b, 173a–b; **15**:127b
mid nineteenth-century, **2**:563b–564a; **4**:529b–531a; **6**:8b–9a; **9**:207b–209a, 210a, 548a–b; **11**:291b–293a; **13**: 380a–b, 384a–386a
late nineteenth-century, **15**:462b–463a, 464b
twentieth-century, **4**:316a, 317a, 318a; **9**:377a–b
dynamic, **6**:395a–b
ionic, **9**:213b
optical, **8**:493a
vortex theory, **9**:207b–209a, 210a, 211b; **11**:291b–293a; **13**:362b–363b, 365b, 385a–386a
Matter, structure of
ancient Greek theories, **1**:149a–b; **4**:367b–368b; **13**:92b–93a
medieval theories, **1**:199a–b; **4**:94b
minima naturalia, **1**:101b, 199b; **5**: 402b; **8**:268a; **10**:123a
seventeenth-century theories, **5**:245b, 422a; **6**:608b–610a
eighteenth-century theories, **3**:74a, 539b–540a; **5**:353a–b; **8**:468b–469a
nineteenth-century theories, **3**:602a–b; **5**:494b; **8**:63a, 374a; **9**:230b–231a, 531a; **11**:10a; **12**:288a, 525a; **14**:209a, 488a–490a
twentieth-century theories, **9**:606b; **15**: 464a–b
corpuscular theory
ancient Greek, **2**:13b; **4**:383a–b
seventeenth-century, **1**:567a; **2**:13a–14a, 226b, 378a–381b; **7**:194b–195a, 459a; **14**:405a–b, 408b
eighteenth-century, **1**:113a, 115b, 451a; **3**:344b; **6**:459a–460a; **14**:353a, 484a
nineteenth-century, **7**:220a–b; **10**:4a; **11**:173a–b; **15**:487a–b
electron theory
see Electron theory
structural types, **5**:203a
nematic, **5**:186b
universal law of forces, **2**:330a–b
see also Atomism; Atomic theory; Molecular theory; Physics, atomic; Physics, nuclear
Matterhorn, **10**:575a; **12**:121a
Matteucci, Carlo (1811–1868), 9:176–177; **4**:200b, 201b–202a; **9**:570a; **12**:163b, 164b; **15**:108b
Matteucci, Petronio, **14**:589b
Matthaei, Gabrielle Louise Caroline, **2**:184a
Matthes, François Émile, **4**:562b
Matthew, George Frederick, **1**:397a
Matthew, William Diller, **2**:171b; **5**:501a*
Matthews, Bryan Harold Cabot, **14**:201a
Matthiessen, Augustus (1831–1870), 9:178; **2**:588b; **3**:251a
Matthiola, **3**:422a; **9**:281b
Mattioli, Pietro Andrea Gregorio (1501–1577), 9:178–180; **1**:167a; **5**:383b, 384a; **13**:463a; **14**:586b
critics of, **8**:436a, 554a; **14**:336a
Opera quae extant omnia, **1**:524b*
translations by, **4**:122a; **13**:502b
Matunaga, Yosisuke, **12**:291a
Matuyama (Matsuyama), Motonori (1884–1958), 9:180–182
Matuyama Epoch, **9**:181b
Maud (ship), **13**:166b–167a
Maudith, John, **2**:435a; **11**:415a
Maufe, Herbert Brantwood, **1**:394a
Mauguin, Charles Victor (1878–1958), 9: 182–183; **6**:303a, 304a*
Maui (island), **3**:551b
Maule, George, **6**:463a
Mauna Loa (volcano), **3**:552a,b
Maunder, Annie S. D. Russell, **9**:184a
Maunder, Edward Walter (1851–1928), 9: 183–185; **3**:261a; **13**:344b*
Maunder, Thomas Frid, **9**:184b
Maupas, François Émile (1842–1916), 9: 185–186; **2**:626b
Maupas's law, **9**:185b
Maupertuis, Pierre Louis Moreau de (1698–1759), 9:186–189; **2**:577b; **3**:281a–b, 454b; **4**:44a, 203b; **7**:276b*, 591a
acquaintances of, **2**:56a; **3**:39a, 173b, 281a–b, 459b; **5**:54b; **15**:272a
and Berlin Academy, **4**:470b; **7**:561a,b, 579b
correspondence of, **3**:461b, 486a; **11**:65b
critics of, **2**:580a; **3**:561b, 562a
geodesy, **3**:105a; **5**:491b; **6**:469a; **8**:178b; **11**:545a; **13**:69b; **15**:270a, 287b
and Koenig, **4**:471b; **7**:442b–443a; **9**: 186b–189a *passim*
least action principle, **3**:74b; **4**:480b; **6**: 473a; **8**:602b; **9**:187b
and Mme. du Châtelet, **3**:215a,b; **9**:186b, 187a, 188b
philosophy, **3**:114b; **4**:471b; **7**:591a
and Voltaire, **9**:186b–189a *passim;* **14**: 83a–b
Maurepas, Jean Frédéric Phélippeaux, **2**: 576b; **7**:200a,b; **15**:111b
Maurer, Joseph (Josias Murerus), **12**:577b
Maurer, Julius Maximilian (1857–1938), 9: 189
Maurer, Ludwig, **8**:326a
Maurice of Nassau, **3**:181a; **5**:241a; **13**:47b, 50a
see also Nassau-Siegen, Count Joan Mauritz van
Mauritius (island), **3**:365b; **7**:544a–b
botany, **9**:366b; **11**:64b; **13**:354b
natural history, **12**:536b
Mauritz van Nassau, Johan
see Nassau-Siegen, Count Joan Mauritz van
Mauro, Fra (*d. ca.* A.D. 1459), 9:189–190
Maurolico, Francesco (1494–1575), 9:190–194; **1**:383a; **2**:311b; **6**:361a; **13**:561a; **14**:461a
Emendatio et restitutio conicorum Apollonii Pergaei, **2**:308b
geometry, **8**:224b, 225a
optics, **10**:475b
translations by, **1**:229b; **3**:149b; **9**:302a*; **13**:320b*; **15**:177a–b, 179b, 180a
Maurus of Salerno, **12**:82b
Maury, Antonia Caetana de Paiva Pereira (1866–1952), 9:194–195; **6**:350b, 351a; **10**:600a
Maury, Matthew Fontaine (1806–1873), 9: 195–197; **4**:291b; **12**:266b
Physical Geography of the Sea, **4**:591b
Maw, George, **6**:492a
Mawdūd, ibn Masᶜūd, **2**:151a
Mawson, Sir Douglas (1882–1958), 9:197–198
Max Planck Gesellschaft zur Förderung der Wissenschaften, **6**:17a; **7**:517b; **11**:9b, 347b
biology institutes, **7**:516b
chemistry institutes, **5**:117b
medical institutes, **2**:338b
physics institutes, **3**:618a
see also Kaiser Wilhelm Gesellschaft zur Förderung der Wissenschaften
Maxima and minima
ancient Greek discussion, **1**:188b
medieval discussion, **13**:197a–198a, 515a–b
seventeenth-century discussion, **4**:57b, 58a, 569a–571a, 575b; **6**:537b; **8**: 304b; **9**:447b; **10**:49a; **11**:404a; **12**: 460a; **13**:436a; **14**:49a
eighteenth-century discussion, **3**:460a,b; **5**:195b*; **7**:560b–561a, 561b; **8**:135b, 306a, 611b; **9**:186a; **15**:276b–277a
nineteenth-century discussion, **2**:271b, 299a; **3**:226b, 428b*; **4**:591b; **13**:19a–20a; **14**:630b
method of, **3**:441b; **4**:57b, 58a, 569a–571a, 575b; **6**:537b–538a; **8**:156a, 163b; **9**:232b; **10**:49b
Maximilian I Joseph, Elector of Bavaria, **5**:142a; **9**:149a; **15**:172a
Maximilian II, King of Bavaria, **10**:558a
Maximilian zu Wied
see Wied, Maximilian zu
Maximow
see Maksimov
Maximum
concept of, **3**:514a
Maximum solubility principle, **6**:538b
Maximum work, principle of, **2**:69a; **6**:354b
critics of, **4**:225b
Maxwell, James Clerk (1831–1879), 9:198–230; **1**:347b; **4**:482a; **5**:424a*; **6**:249b, 502a; **13**:101b
acquaintances of, **6**:542b; **7**:93b; **11**:122a, 568a–b, 578a; **12**:238a; **13**:236b
and Clausius, **3**:307b–308a
on contemporary scientists, **3**:163b; **11**: 293b; **12**:524b; **14**:109b
correspondence of, **14**:186a
critics of, **4**:227a; **11**:267a, 479b; **12**:570b
edition of Cavendish's work, **3**:155b
elasticity, **9**:199b–200a; **13**:385b
electricity, **8**:491b; **9**:204a–206b, 208b–209b, 215b–217b; **14**:32b
measurement of, **9**:209b; **13**:102a
electromagnetic theory, **1**:347b, 458a, 600b; **2**:166b; **3**:477a; **5**:63b; **6**:243a, 341b–348a *passim;* **8**:491a–495b *passim;* **9**:155a, 206b–217b, 377a; **10**:28a; **11**:10b, 122b, 407a; **13**:356b, 364b; **14**:205a; **15**:43b
confirmations of, **8**:107b; **11**:13a, 581b–582a
critics of, **4**:231a–232a; **14**:339a
equations, **6**:345a, 346a–b; **8**:476a, 477a; **9**:207a,b; **11**:461a, 479b
field equations, **1**:23b, 24b, 284a; **2**: 202b, 265b–266a; **4**:313b, 315a, 319a–b, 322b, 324a, 326a, 327b, 328b, 539b, 579a; **6**:251a–b, 252a;

8:493b, 494a, 495b; **9**:211b–212a, 215b; **14**:207b–208a
and light, **1**:301a; **4**:189b, 190b–191b, 306a; **5**:15a; **8**:492a–b, 502a; **9**:209b–215a; **11**:9a; **14**:598a
practical applications of, **6**:211b, 504b
predictions of, **8**:107a, 444a; **10**:105b
tests of, **11**:530a
engineering, **9**:226b–227a; **15**:7a
on ether, **9**:209a–b, 213b, 214b–215a
gas theory, **2**:261a–263b; **3**:307b–308a, 477b; **6**:293a; **7**:493a; **8**:509a,b, 510b; **9**:217b–225b; **13**:11a; **14**:170b
distribution equations, **7**:416b; **8**:284b; **9**:218b, 616b; **14**:340a
ensemble concept, **2**:265a; **9**:223b–224a
ergodic hypothesis, **9**:223b
radiometer, **9**:224a–225a
transfer equation, **9**:222a, 225a
and Gibbs, **5**:389a, 390b
influence of, **3**:31b, 179a; **5**:391b; **8**:40b; **13**:363b, 365b
influences on, **13**:78b
magnetism, **9**:206b, 207b, 213b
mathematics, **9**:204a, 211a–212a, 215b, 221b–222a, 225a
geometry, **9**:199b; **13**:571b
Matter and Motion, **4**:320b
Maxwell bridge, **9**:215b
Maxwell relations, **9**:227a
Maxwell spot, **9**:201b–202a
Maxwell-Boltzmann equations, **4**:375b–376b; **9**:222b
Maxwellian iteration, **9**:225b
Maxwellian molecules, **9**:221b, 222a,b
Maxwell's demon, **3**:306b, 479b; **5**:391b; **9**:227a
methodology, **13**:362b, 363a
optics
geometrical, **9**:199b–200a, 225b–226a
physical, **5**:390b–391a; **8**:491b; **9**:212b–213a
physiological, **6**:246b; **9**:200a–202a; **14**:565a
philosophy, **9**:206a, 213b–214a; **12**:609a
on Saturn's rings, **1**:597b; **7**:270b, 479a; **9**:202a–204a, 606b
statics, **3**:468b–469a; **9**:226a–b, 446b; **11**:294a
students of, **3**:264b; **5**:32b, 403b, 423b; **7**:594b; **10**:447a; **12**:366b
Theory of Heat, **3**:306b; **5**:388a, 389a
thermodynamics, **9**:210a, 227a–b
and Thomson, **9**:202b, 205b–211b *passim,* 220b, 221a, 222b, 224b; **13**:375a, 377b, 384b, 385b–386a
Treatise on Electricity and Magnetism, **2**:397b; **5**:391a; **6**:51b, 345b; **8**:494a; **9**:204a–b, 211a–213b; **13**:438a
translations of, **11**:8b
Maxwell, William, **9**:327a
Maya Numeration, Computation, and Calendrical Astronomy, 15:759–818
Mayall, Nicholas Ulrich, **6**:531b, 533a
Mayas
almanac, **15**:760a, 777a–b, 782b, 790b, 807a, 810a, 813b
almanac interval, **15**:771a–773b, 783a–784a
one-day corrections, **15**:802b–804a
astrology, **15**:777b, 804b
astronomy, **8**:539b; **15**:579a–818b
calendar, **15**:759a–814b
calendar round, **15**:760a, 765b, 766a, 769a–771a, 780a, 782a–783a, 813a
and eclipse reckoning, **15**:774b, 775b, 776b
trecena, **15**:760a, 764b, 765b, 766a, 769a,b, 770a, 772b, 777a
veintena, **15**:760a, 764b, 765b, 769a,b, 770a, 772b, 777a, 781a
Venus calendar, **15**:776b–789a, 812a–b
chronology
dates, **15**:767a–768b, 786b, 787a, 805a–b, 806a, 807b, 813a
day, **15**:764b–765a, 765b, 796a
day count, **15**:760a, 766a–767a, 769a, 773b, 812b–813a, 814a
Goodman-Martinez-Thomson correlation, **15**:808b–809a
lunar series, **15**:774a–776b, 810b–812b
month, **15**:765b, 781b
Venus epoch, **15**:782b, 786a
year, **15**:760a,b, 765a–b, 766a–b, 773a, 774a–775a, 777a, 780a–789a *passim,* 810b
Codex Dresdensis, **15**:759b, 773a, 775a, 776b, 806a–b, 809b, 813a–b
eclipse tables, **15**:774b, 789b–799a, 800a–802b, 811a,b, 812a
Venus tables, **15**:778a–b, 779a, 784a–b, 785a–b, 788a–b
eclipse reckoning, **15**:789b–799a
languages, **15**:759b, 760b–763b
Leiden Plaque, **15**:809a–b
Madrid Codex, **15**:813a–b
mathematics, **15**:769a
numeration, **15**:760b–764a, 779b, 799a–800a, 809b, 810a,b
numerology, **15**:804a–808b
Paris Codex, **15**:769b, 801a, 813a
writing, **11**:263b
Mayer, Alfred Marshall (1836–1897), 9: 230–231
magnetism, **13**:363a, 367a–b
Mayer's law, **9**:230b
Mayer, Christian (1719–1783), 9:231–232
Mayer, Christian Gustav Adolph (1839–1908), 9:232; **4**:370a; **8**:324a, 325b
Mayer problem, **2**:198a, 271b–272a; **7**: 408a
Mayer, Johann Tobias (1723–1762), 9:232–235; **2**:141a; **5**:532b; **6**:395a; **7**:596b; **8**:299b
editions of works, **8**:321b; **9**:163a
longitude, **9**:162b–163a
lunar tables, **2**:389a; **4**:481b; **6**:516a; **8**: 179a; **15**:289a
magnetic theory, **1**:67b
Mayer circle, **9**:234b
observations of, **2**:98b; **15**:312a
star catalog, **1**:241a, 403b*; **7**:87a
Mayer, Julius Robert (1814–1878), 9:235–240; **7**:235a; **13**:237a; **14**:360b
and Joule, **3**:83a, 306a; **7**:160a, 180b, 182a; **9**:239b
thermodynamics, **6**:432a; **14**:185b; **15**: 85b, 86b
Mayer, Maria Goeppert, **4**:582b
Mayer, Sigmund, **5**:150a
Mayer, Walter, **4**:331a
Mayer-Eymar, Karl (1826–1907), 9:240–241
Mayerne, Theodore Turquet de
see Turquet de Mayerne
Mayers, William Frederick, **8**:395a
Mayfly, **13**:171a,b, 173a
Mayl qasrī (impetus), **1**:28a
Ibn Maymūn al-Qaddāḥ, ᶜAbd Allah, **15**: 250a
Mayo, Frank Rea, **7**:323b
Mayo, Herbert (1796–1852), 9:241–242
Mayo Clinic (Rochester, Minn.), **15**:258a
Mayow, John (1641–1679), 9:242–247
alchemy, **12**:308a
chemical affinity, **1**:451b
combustion, **6**:42b, 43a, 484b
critics of, **1**:451a; **8**:171b
and Hooke, **6**:47a(n); **9**:245a
influences on, **14**:408b
nitro-aerial spirits, **6**:40b, 485a; **7**:459a; **14**:407b
respiration, **4**:377a,b; **8**:526b; **13**:169b
Mayr (Marius), Simon (1573–1624), 9:247–248; **3**:59b, 60a,b; **5**:240b
Mazenta, Giovanni Ambrogio, **13**:613b
Mazer wood, **13**:450a
al-Māzinī
see Abu Ḥāmid
Mazur, Stanisław, **1**:427a
Mazurkiewicz, Stefan (1888–1945), 9:248–250; **7**:71b; **12**:78a, 426b
Mazurkiewicz-Moore theorem, **9**:248b
Ibn al-Māzyār, **1**:37a
Mazzoni, Jacopo, **1**:608a; **5**:238b, 239b
Mazzuchelli, Giovanni Maria, **1**:484b
Mazzuoli, Lucio, **7**:31a
Mea Kaumen (island), **9**:532b
Mead, Margaret, **2**:210a
Mead, Richard, **1**:594a; **2**:188b; **3**:244a; **4**: 172b, 295a; **5**:157a; **9**:326a; **10**:500b, 501a; **11**:2b; **14**:501a
Meadows, William, **6**:263b
Meakins, Jonathan, **8**:295a
Mean speed theorem, **2**:194a
Mean value theorem, **3**:136a, 152b
Meaning, theory of, **2**:604b; **11**:356b–358a
Means
arithmetic, **1**:231b, 232b–233a; **3**:126a; **4**:160a; **5**:9a; **6**:472a–b; **10**:300a; **11**:221b; **13**:302b
method of, **10**:25a
arithmetic-geometric, **2**:299a; **5**:298b, 299a; **7**:568a
between vectors, **1**:239a
geometric, **1**:231b, 233a, 238a–b, 239a; **2**:393a; **3**:126a; **4**:160a, 421b; **9**:612b; **10**:300a; **11**:221b; **13**:302b
geometric-harmonic, **3**:183b
harmonic, **1**:231b, 232b; **4**:160a; **10**:300a; **11**:221b; **13**:302b
centers of, **11**:77b
subcontraries, **4**:466b
theory of, **1**:231b–233a; **4**:160a, 392b; **8**:458b*; **10**:294b; **11**:221b; **13**:326a; **15**:32b, 282a–283a
Measles, **1**:399b; **6**:73a; **11**:325b; **13**:292b; **14**:322a, 405b
transmission of, **5**:451a–b; **6**:233a
vaccination, **3**:336a, 577b; **15**:454b
Measure theory, **1**:427a,b; **2**:303b–304a, 352b; **3**:63a; **5**:608b–609a; **6**:95b, 176b, 230b, 451b; **8**:110b, 112a, 457b; **9**:420a; **10**:443a; **11**:156b, 256a, 260a; **13**:562a
constructive, **2**:513b
disintegration, **14**:89b–90a
see also Dimension, theory of
Measurement
Egyptian, **15**:694a–697b, 699a, 699b–700b
instruments for, **4**:551b; **11**:377b
microscopic, **8**:127b
philosophy of, **3**:34b; **7**:287b
rectilinear
see Length
standardization of
ancient Roman, **1**:418a
medieval Arabic, **11**:238b
medieval Latin, **9**:469a
Renaissance, **10**:160b
seventeenth-century, **9**:554a–b
early eighteenth-century, **3**:39a; **5**: 54a(n), 491b–492a; **6**:236b; **8**:22a
late eighteenth-century, **2**:141a, 299b; **3**:441a; **4**:14b–16a, 501b, 503a; **5**:78b; **7**:89a; **8**:351a–b; **9**:250b–251b; **13**: 411b, 495b–496a; **14**:312a*; **15**:312b

Measurement (cont.)
standardization of (cont.)
early nineteenth-century, **2**:135b; **6**:107a; **7**:262b–263a; **9**:459b; **14**:486b
mid nineteenth-century, **2**:102a; **5**:306a; **6**:165b, 239b; **7**:160a; **8**:613b; **9**:392a, 393a; **11**:377b; **12**:132a, 425b; **13**:23a
late nineteenth-century, **1**:403a, 612b; **3**:19b, 288b; **6**:223a; **7**:2a; **8**:30b; **9**:102b, 292b; **13**:102a
twentieth-century, **1**:612b; **2**:200b; **3**:327b, 565a*; **5**:407b, 423b–424a; **13**:517b
astronomy, **12**:265a–b
electricity, **3**:288b; **4**:588b, 589a; **5**:407b, 423b–424a; **6**:212a; **7**:93b, 182b, 288b; **9**:155b, 178a, 209b; **11**:579a; **13**:102a; **14**:205b
historiography of, **14**:480a
M.K.S. system, **5**:407b; **6**:212a; **7**:288b
optics, **8**:551b
spectroscopy, **1**:166b; **4**:513b; **9**:256a–b; **12**:73a
theory of, **8**:473a; **14**:67a, 305a–b
see also Anthropometry; Area; Density; Distance; Errors; Geodesy; Latitude; Length; Longitude; Mensuration; Metric system; Metrology; Micrometer; Photometry; Polarimeter; specific sciences, e.g., Astronomy, measurement in; Volume
Meat
chemistry of, **10**:557a
extract, **8**:330a; **10**:558a
infection of, **10**:527a
inspection of, **8**:270b
preservation of, **10**:326a
Méchain, Pierre-François-André (1744–1804), 9:250–252; 2:216b; **4**:16a; **7**:580b
celestial mechanics, **9**:330a; **15**:311b, 312a
geodesy, **3**:106a; **4**:15a–b; **7**:500a; **8**:135b; **9**:234b; **15**:335a, 376a
standard meter, **2**:135b; **8**:136b
Mechanical advantage, **8**:209a; **9**:114b
Mechanics
ancient Greek, **1**:28a, 100b–101a, 213b–214b, 229b; **4**:431a–b; **10**:586b–589a
ancient Roman, **9**:300b; **10**:298a; **11**:201b
medieval Arabic, **13**:292a; **15**:497b–498a
medieval Latin, **1**:100b–101a; **5**:113b–114a; **7**:116a–b, 129a; **13**:260a
Renaissance, **1**:419b, 607a; **2**:590b; **3**:66a–b; **4**:8b; **8**:200a–b, 208a; **9**:192a, 487b–488b; **12**:106a–b; **13**:48b–49a, 260a–b; **14**:29b
early seventeenth-century, **1**:424a; **2**:348b, 607b; **4**:47a, 60a–b; **7**:486b–487b; **9**:97a–b; **11**:488a–489b; **13**:436a–437a, 614a; **14**:254a–b, 362b–363a, 365b–367a
late seventeenth-century, **2**:47a, 49a, 306a, 311a, 592a–b, 613b; **3**:182b–183a; **5**:527b; **6**:481b–483b, 487a, 602a–603a, 604b–611b; **7**:611a; **9**:51a–52b, 114b, 115b–117a; **11**:508b; **13**:584b–586a; **14**:149b–150a, 371b, 509a, 509b–510a
early eighteenth-century, **1**:111a–114a; **2**:54a–55a, 57a, 330a; **3**:39a, 282b–283a, 454a; **4**:44a, 45b; **5**:499a, 511a; **9**:33b–34a; **10**:319a–320a, 500b–501a; **11**:65b, 400b; **12**:283b
late eighteenth-century, **1**:326b; **2**:335a; **3**:41b; **5**:209b; **7**:490a–b; **8**:135a; **15**:343b
early nineteenth-century, **2**:214a, 215b; **3**:8a–b; **5**:530a–b; **6**:213b; **11**:80a; **13**:13a–b; **14**:568b–569a; **15**:487b–489b
mid nineteenth-century, **2**:424b–425a, 470a; **5**:587a; **9**:19a; **11**:344a; **12**:73b–74a; **13**:128a–b, 130b
late nineteenth-century, **1**:194b; **2**:356a; **3**:194b–195a; **6**:343a; **13**:73b, 102b; **14**:86a–b
early twentieth-century, **1**:570a–b; **2**:594a; **5**:508b; **8**:124a, 500b; **11**:413b; **14**:437a
mid twentieth-century, **3**:220b
analytical
see Mechanics, rational
applied
ancient Greek, **6**:312a–313a
Renaissance, **13**:49a
eighteenth-century, **3**:441a–442b; **11**:496a–b
early nineteenth-century, **4**:257a; **5**:530b; **6**:2b; **10**:551a–b; **11**:77a, 77b–78b, 80a
mid nineteenth-century, **3**:224a, 417a; **6**:452a–b; **15**:66b
late nineteenth-century, **2**:167b; **5**:501b–502a; **7**:446a–b; **8**:287b; **14**:620a–621a
Cartesian, **2**:54b, 613b; **4**:60a–b
of changing mass, **13**:483b
classical, **2**:356a; **3**:308b, 417a, 418b
and relativity theory, **4**:322b–323a
collision theory of, **2**:49a
conservative systems, **2**:54b
continuum, **3**:73a, 146a; **4**:228a,b; **9**:200a
of elastic bodies, **2**:40b
and electromagnetism, **4**:315a, 321a, 323a, 330b
engineering
Renaissance, **13**:233b
seventeenth-century, **5**:245b; **7**:578a
eighteenth-century, **1**:581b–582a; **3**:70b–75a, 441b, 442a–b; **4**:44a, 45b; **9**:446b, 477a; **11**:163b–164b
nineteenth-century, **3**:78a, 358a–b, 417a–418a; **5**:530b; **7**:287b; **9**:138b–139a; **10**:2b–4b; **11**:165a, 345a; **12**:73b–74a
twentieth-century, **5**:64a; **6**:534a–b; **12**:612b–613a
epistemology of, **7**:226b
ergodic hypothesis, **2**:265a–b; **4**:293b, 577b; **9**:129b, 223b; **11**:56b
of flexible bodies, **2**:40b
foundations of, **4**:101b; **6**:245a, 347a–348a, 348b; **8**:151b–152a; **11**:384b; **13**:585b–586a
geometry and, **3**:78a, 182a–183a, 195a; **7**:340a, 341a, 342a, 473a–b; **8**:124a; **10**:226a–228b; **11**:62a; **12**:534b; **14**:149b, 620a
indeterminate problems, **1**:570b
industrial, **11**:77b, 78a, 80a
Lagrangian, **1**:195a
least constraint, law of, **5**:305a; **8**:602b
and mathematics, **1**:582a; **7**:116a–b; **13**:584b, 586a; **14**:149b–150a
minimum action, principle of
see Least action, principle of
n-body problem, **4**:371a; **10**:275b
non-Euclidean, **13**:414b
nonholonomic systems, **1**:570b
nonlinear, **7**:515a
and optics, **6**:89b; **12**:220a–b
philosophy of, **1**:111b; **6**:347b–348a; **8**:598a–603a; **10**:449a
physiological, **3**:184a–b; **11**:571a; **14**:204b, 206a
punctual, **4**:480a; **15**:196a
rational
early eighteenth-century, **2**:38a–39b; **3**:561b–562a; **4**:479b–480b; **6**:305a; **8**:611a–b; **12**:118a–b, 444b; **13**:265b, 266b–267a
late eighteenth-century, **7**:560b–561a, 561b–562a, 564b–565a, 568b, 569b–570a; **11**:164a; **15**:350a–b
early nineteenth-century, **3**:145b, 417a–418b; **5**:95b–96a, 97a; **10**:2b–4a; **11**:61b, 62a
mid nineteenth-century, **1**:591b; **2**:350b–351a, 469b; **3**:496a, 591b; **4**:19b–20a, 126a–b; **5**:86b, 305a; **7**:54b–55a, 109a; **8**:384b; **9**:268a–b, 410a; **10**:247b–250b, 480b; **11**:207a, 453b–454a; **12**:534b; **13**:130b, 236b
late nineteenth-century, **1**:599b; **2**:288b; **3**:225a, 309a, 428a–b, 559b; **7**:446a–b, 459b–460a, 466a, 478a, 479a; **8**:62a, 287b, 367b, 389a, 542b; **9**:174b–175a, 232b, 324a–325a; **10**:275b, 443b, 447b, 448b; **11**:408b, 568b; **12**:599b; **13**:26b
early twentieth-century, **1**:570b; **2**:320a, 465a; **5**:188a–b; **6**:5a; **7**:84a, 453a–b, 473a–b, 514a–b; **8**:284a–b, 559b–562b, 617b; **10**:514b; **11**:477b–478a; **13**:125b; **14**:13b–14a; **15**:164a–165a
mid twentieth-century, **3**:220b; **7**:515a; **10**:20a–b, 554a; **13**:36a; **15**:6b–7a
conservation of areas, principle of, **2**:39a; **3**:561b–562a
system of reference, **3**:417b–418b
and thermodynamics, **2**:471a
of rigid bodies, **2**:38b; **3**:145a; **7**:621b
of solid bodies, **7**:453b
statistical, **2**:261a–263b, 264b–265b; **4**:314a–318b, 325b–326a, 361a, 577b–578a; **5**:103a–b, 123a*; **6**:164a; **7**:352a; **9**:219a, 220a; **10**:235b–236a; **13**:176a, 226b–227a, 429a–b; **15**:44a
Darwin-Fowler method, **3**:564a; **5**:103a
foundations of, **4**:293a–b
and thermodynamics, **5**:391b–392b
structural
eighteenth-century, **3**:441a–442b
mid nineteenth-century, **6**:452a–b
in botany, **12**:256a
theoretical
see Mechanics, rational
transformability principle, **2**:245b–246a
see also Ballistics; Dynamics; Engineering; Gravity; Historiography, of mechanics; Hydromechanics; Kinematics; Machines; Matrix mechanics; Motion; Quantum mechanics; Statics; Wave mechanics
Mechanics, celestial
early seventeenth-century, **1**:424a; **2**:348b–349a; **5**:588a; **6**:514b–515a; **7**:291b–292b, 295a–b, 296b–297a, 303a–304a, 307b; **12**:342b–343a
late seventeenth-century, **2**:306a; **5**:23b–24b; **6**:68b–69a, 485b–486a; **8**:152a–b; **10**:61a–65b, 69a–78b; **11**:507a; **14**:446b
early eighteenth-century, **1**:113b; **2**:54b, 330a; **3**:281b, 282b–283b, 562a; **4**:471a, 480a–b, 481a–b; **5**:521a–b; **6**:36a–b; **8**:178b–179a
late eighteenth-century, **1**:400b–401a; **2**:578b; **4**:14b; **7**:544b, 562a–564b, 568b, 570a, 580a; **8**:135a–b, 299b; **15**:286a–290b, 293b–301b, 309b–312a, 316a–333a, 336b–339b, 343a–b, 348a–356b

early nineteenth-century, **2**:97b, 368b, 600a; **4**:16b; **5**:299b–301a, 302a; **6**: 291b; **7**:37b; **8**:137b; **10**:111a; **11**: 6b–7a; **15**:226a, 356b, 379b–382b, 484a–485b
mid nineteenth-century, **1**:53a–54a, 241a–242a, 339b; **2**:350b, 351a; **3**: 146b, 169b–170a; **4**:19b–20a, 369b, 591a; **6**:88b–89b, 103b, 522a; **7**:480b–481b; **8**:277a–279a, 384b; **9**:202a–204a, 239a–b, 430a; **10**:479b; **11**: 206b, 207a, 498a–b; **12**:524a
late nineteenth-century, **1**:156b; **2**:123b, 442a, 616a; **3**:18b–19a, 583a–584a; **5**:224a; **6**:94b, 399a–400a; **7**:17b; **8**: 560b; **9**:174b, 324b; **10**:33a–35b, 212b, 219a–b, 528b; **11**:57a–58a; **12**:247b–248a; **13**:114a, 422b–423a; **14**:481b, 620b
early twentieth-century, **1**:156b–157b, 428b–429b; **2**:144a, 145a, 516a–b; **3**:191a, 457a, 472a–b; **4**:328b, 547b; **5**:363b; **6**:431a–b, 539b; **7**:7a, 238b–239a; **8**:284b; **9**:553a*; **10**:19b, 159a; **11**:49a, 84a, 552a–553a; **12**:17b–18b, 95b, 448b–449b; **13**:95a, 99a, 153a–b; **14**:13b, 87a
mid twentieth-century, **2**:511a–b; **3**:220b–221a; **7**:10b; **9**:410b–411a, 449a–b, 450a*; **10**:317b; **13**:139a–b; **14**:453b–454a; **15**:128b–129a
and geodesy, **15**:351b–352a, 354a
plane of reference, **15**:350a–b
recurrence theorem, **11**:58a
see also Perturbation theory, Three-body problem, Two-body problem
Mechanistic philosophy
ancient Greek, **1**:315a; **4**:31a–33b
medieval, **2**:607b–608a
Renaissance, **5**:401b
early seventeenth-century, **1**:373b–374a, 568a; **3**:492a; **4**:52a–54b, 61b–65a, 96a; **6**:158a; **9**:15a–b, 316b–320a; **12**:101b–104a
late seventeenth-century, **2**:378a–382a; **5**:416a; **6**:482a, 484a, 487a, 608b–611a; **7**:459a, 612b; **8**:151a, 173a–174b; **11**:506b, 507a; **13**:169b–170b, 174b; **14**:26a
seventeenth-century
opposition to, **9**:509b–510a
early eighteenth-century, **1**:451a; **3**:295a–296a; **6**:43b, 139b–140a; **7**:606a–607a; **8**:114b–115a, 255b, 468a; **9**: 187a, 490b; **10**:80b; **11**:157b–158a; **12**:601a; **13**:543a
late eighteenth-century, **2**:579b; **7**:604a; **8**:259a–260a, 297a,b; **11**:145a–b
eighteenth-century
opposition to, **1**:479a–b; **7**:232b–233a; **8**:298b, 580a; **10**:26a; **12**:602b–603a, 604a–b
early nineteenth-century, **3**:378a, 541b, 542a,b
mid nineteenth-century, **2**:356a, 561a; **6**:7a,b, 226a, 242a, 243b, 245a, 252b; **7**:382b; **8**:514b–515b, 596a; **9**:571b; **12**:240b–243b
late nineteenth-century, **2**:626b–627a; **3**: 531a; **4**:188a, 313b; **5**:447a; **8**:445a–446a, 494a, 495a; **9**:516b; **10**:580b; **11**:571a, 573b–574a; **13**:362b–363a; **14**:434b; **15**:21b
nineteenth-century
opposition to, **1**:540a; **2**:9a–12a, 266b, 426b, 532a; **5**:391a–b; **8**:601a; **12**: 155a–b, 609a–b
twentieth-century, **1**:392a; **3**:31b–32a; **4**:225b; **5**:22a*; **8**:125a, 446b; **9**:525a; **12**:23b; **15**:11a
opposition to, **3**:90a; **4**:227a; **14**:124a
precursors of, **1**:495b; **7**:137b
Mechnikov, Ilya
see Metchnikoff, Elie
Meckel, Johann Friedrich, the Elder, **9**:252a; **12**:137b
Meckel, Johann Friedrich (1781–1833), 9: 252–253; 2:443a; **11**:364a; **15**:525b*
anatomical theory, **1**:387a,b; **12**:241b
embryology, **12**:316a
Meckel's diverticulum, **9**:253a
Meckel, Phillip Friedrich Theodor, **9**:252a, 482b; **11**:363b
Meconic acid, **8**:339a–341b *passim*
Meconium ileus
etiology of, **7**:624a
Médecine éclairée par les sciences physiques, **5**:90b
Mededeelingen, **13**:459b
Medial (geom.), **4**:422a
Medicago, **5**:383b, 384a
Medical Bureau to Aid Spanish Democracy, **15**:75b–76a
Medical Essays and Observations, **9**:480b, 481a; **11**:48a
Medical Research Committee (Great Britain), **5**:37a; **15**:418a,b, 419b–420a
Medical Research Society (London)
founding of, **8**:296a
Medici, Cosimo I de', **3**:354b, 558a,b; **5**:499a, 580b; **14**:5b
Medici, Cosimo II de', **13**:613b
Medici, Cosimo III de', **9**:368a; **13**:170a, 175a(n)
Medici, Leopoldo de', **2**:308a–311a, 349b, 592a,b; **5**:246b; **9**:3a, 14a, 369a,b, 484b
Medici, Lorenzo de', **1**:618b; **9**:14a
Medici family, **1**:621a; **2**:308b; **11**:341b
instrument collection, **5**:55b
patronage, **1**:618a
Medici planets
see Jupiter, satellites of
Medicine
ancient Greek, **1**:75a, 103b–104a, 212a–b, 261b–262a, 314b–315b, 324a–b; **3**: 174b–175a; **4**:105b–107a, 367a, 382b–386a; **6**:305b, 316b–318b, 418b–429a; **10**:581b, 589b; **11**:127a–128b
ancient Roman, **1**:234b–235a; **5**:227b–232a; **10**:230a–231a; **11**:602a–603a; **12**:538a–540b
Byzantine, **1**:68b–69a, 121a; **10**:417a–418b; **11**:186a; **12**:540b
medieval Arabic, **1**:399b; **2**:619a–b; **3**: 393a–394a; **6**:79b; **7**:22b–23a, 24b–25a, 42b, 187a–b; **8**:248a, 547b; **9**: 28a, 31a–b, 40a–41b, 602b–604b; **11**:238a–239a, 244a–b, 247b–248a, 249a, 251b–253a, 325b–326b, 444a–445b; **12**:1b, 7Bb–8a, 90a–b, 91b–92a; **13**:229a–b, 230b, 292b, 415a–b, 488b, 509a, 511b; **14**:112b, 584a–585a, 637a–638a; **15**:230a–231b, 238b–242a, 498a–499b
medieval Latin, **1**:4b, 107a, 289b–291a; **2**:314a–315a, 435b; **3**:218a–219a; **4**: 164a; **6**:123a–b, 276b, 396a–397a; **9**:363a, 364b, 467b–469a; **10**:21b; **11**:18a; **13**:347b; **14**:399b; **15**:188b
Renaissance, **1**:46b–47a, 98a–b, 125a, 399b–400a, 484b, 522b–524a, 525a, 526a; **2**:283a, 293a, 336b–337a, 535b, 537a, 546b–547a; **3**:12b–13b, 40a–41a, 64b, 342a–b, 363b, 413a, 471b–472a, 533b–534a; **4**:138b–139b, 170a, 208b–209b, 212b, 284b–285a, 387a–b, 486b–488a, 507a–508a, 584b–586a; **5**:104b–107b, 349a, 396b–397a, 447b–448a, 579b, 580b–581a, 585b–586b; **6**:309b; **7**:16b–17a; **8**:248a–249b, 360b–361b, 435b, 483b–484a, 554a–b; **9**:74b–75b, 132a–133b, 165a–166a, 178b–180a, 308b–309a, 440a–441a, 466a–b; **10**:123a, 152a–b, 281a, 304b–311a, 316a, 494a, 515a, 598a; **11**:253b–254a, 338a, 339b, 396b, 527b, 532b, 594b, 607b, 608a; **12**:89b, 97a–b, 134a–135a, 151a, 323b–324a, 334b–336a; **13**:100a–b, 141a, 397a–b, 430a, 501b–502b, 534b–535b, 569a, 587a; **14**:4a–12a *passim,* 27b–28a, 507a, 586a; **15**: 80a–b, 160b, 491b–492a
early seventeenth-century, **1**:480a–b, 482a, 571a; **2**:308a–b, 353b, 522a–523a; **3**:596a–b; **4**:377a–b, 388a, 511b; **5**:47a–48a, 425b; **6**:145b, 150b–151a, 254a, 255b–256a; **7**:164b–165a, 354b–355a, 355b, 536a–539b; **8**:53b–54a, 309a–310a, 311b; **9**:14b, 23a–b, 96b–97a, 97b–98a; **10**:426a–b; **11**: 466a–467b, 511a–b; **12**:78b–79a, 311a, 577a–b; **13**:222b, 473b–474a, 490a–b, 507a–508a
late seventeenth-century, **1**:481a, 482b–483a; **2**:224b, 237a–b, 291a–b, 305b, 317a–b, 354a; **3**:208b; **4**:37a–38a, 135a–136a, 495a; **5**:418b–419a, 484a; **6**:183a, 380a–b, 387a–b; **7**:283b–284a, 376b, 459a, 612a–613a; **8**:437a, 524a; **9**:62b–63a, 243a, 369b, 400a–b, 581b–582a; **11**:1a–2b, 341a–b; **12**: 40a–b, 457a–b, 616a–b; **13**:168b–171a, 213b–215a, 222b–223b, 282a–b, 526b, 563a; **14**:25a–26a, 212b–213a, 255b–256b, 404b–408b, 451b, 586b–587a; **15**:3b
early eighteenth-century, **1**:318b, 322b–323a, 393a; **2**:15a, 37a–b, 188b, 225a–227a, 390a, 490a; **3**:97a–b, 244a–245a, 332a–b, 344a,b, 494a–b; **4**:172b–173a, 352a–b, 589a–590a; **5**:156b–157a, 220b–221a, 226a–b, 352a–353a, 355a, 366b–367a, 379b–380a, 384b–385a; **6**:139a, 231a–b; **7**:200a–b, 605a–b, 614a–b; **8**:171a, 374b–375a, 618b; **9**:62a, 302b; **11**: 48a–b; **12**:159a–b, 302b–303a, 365b, 458a–b, 599b–604a; **13**:121b–122a, 181a–b, 567a–b; **14**:319b–320b, 449b–450b, 501a
late eighteenth-century, **1**:310a,b, 312a, 332b–333a, 398b–399a, 563b; **2**: 173b–174a, 301a–302a, 509a, 595a; **3**:1a–2b, 37a–38a, 67b, 153b, 199a, 206a, 419a, 437a–b, 464b–465a, 486b–488a, 577b–579b; **5**:89b–90a, 137b–138a, 217b–218a, 250b, 267a, 411a–b; **6**:17b–18a, 218a–b, 401a; **7**:11a–12a, 13a, 95b–97b, 494a; **8**: 177b, 246b–247a, 255a–b, 352a–353a, 361b–363a, 379a–b, 506a–b; **9**:305b–306b, 325b–327b; **10**:198a, 445b–446a, 611b–613b; **11**:66b, 99b–100a, 363a–365a, 387a–b, 616a–617b; **12**:24b, 185a–b, 215b, 591b; **13**:46b–47a, 441a, 448a–b, 465a, 543a; **14**:14b, 15b–16a, 194b, 224b, 253a–b, 322a–b, 456a–b, 463a–464b; **15**:113b
early nineteenth-century, **1**:485a–b, 617a; **2**:172b, 335b–336a, 366a–367b,

Medicine (cont.)
444b–445a, 463b–464b, 483a–b, 507b–508b; **3**:489a–490a, 562b; **4**:21b, 67a, 238a, 251b–252b, 263b, 288a–b, 307b–308a; **5**:7a, 13b, 276b–277a, 429b–430a, 440b–441a, 511b–512a; **6**:119b, 284a, 521a–b; **7**:59b–60a, 556a–557b; **9**:7a, 9a–b, 344b, 527a; **10**:321b–322a, 582a–b; **11**:136b–137a, 172a, 510a, 599b; **12**:57a–b, 223b–224a, 315b, 591b–592a; **13**:354a–b, 499a, 541a–b; **14**:562b–563a; **15**:226b–227a
mid nineteenth-century, **1**:417b; **2**:375a–376b, 386b, 477b–478a, 524b–525b, 563a–b, 574b–575b, 598a–b, 616b–617a; **3**:88a, 196b, 205a–b, 587b–588b; **4**:178b, 593a–b; **5**:11b–12a, 45b, 79b–80a, 203a, 477b, 478b, 504b, 565a–566b; **6**:241b–242a, 270a, 293b; **7**:23b–24a, 44b; **8**:121b–122a; **9**:85b–86a, 540a, 571b; **10**:39a–b, 68b–70b; **11**:177b, 301a, 419a, 565a–b; **12**:294a–296b, 420b–421b, 450a–451a, 465b–466a, 502a–503a; **13**:493a–b; **14**:39b–42b, 397a–398a; **15**:107b
late nineteenth-century, **2**:35b, 445b–449a; **3**:185a, 203b, 205a–b, 516b–517a; **4**:295b–302a, 403a–405a, 619b–620b, 620b–621a; **5**:219a–220a, 495a–b; **7**:46a–48b; **8**:7a–b, 312b–313a; **9**:422a–b; **10**:165a, 175b–176a, 204b–205b, 385b–408a, 492b–493a; **11**:345b–347a, 373b–374a, 462b; **12**:87b–88b; **13**:141b–142a; **14**:230a–b, 248b–250a; **15**:453b–454a
twentieth-century, **3**:336a, 415b; **4**:153b–156a; **5**:29a, 36b–37a, 290b–291a; **7**:622b–624b; **8**:273a–274a; **9**:417a; **15**:228a–b
astrological, **1**:49b, 244b, 382a–b, 399b; **2**:337a; **4**:585a; **5**:48a, 547a; **6**:123b; **7**:142b, 459a; **8**:192a*; **9**:74b; **10**:153b; **12**:324a; **14**:399b; **15**:154b
and aviation, **5**:208a
and biology, **5**:315a; **10**:195b
Brunonian system, **2**:595a; **12**:155b, 592a; **14**:351a
case histories, **2**:483a–b; **3**:343a; **4**:172b, 352b; **6**:423a; **8**:352b, 554a–b; **9**:511b–512a; **13**:508a*; **14**:407b, 408a, 464a
chemistry and, **1**:10a, 563b, 571a; **2**:317b; **4**:170a, 208b–209b; **5**:90a,b; **6**:293b; **9**:62a; **11**:51b; **12**:311a–312b, 604a; **13**:507a–508a, 535a–b
Chinese, **1**:323a; **8**:390b; **12**:374a, 383a–384a, 384b; **15**:737a–739b, 743b
Christianity and, **1**:235a; **10**:20b
clinical
ancient Roman, **11**:602a–b
medieval Arabic, **11**:444b
Renaissance, **3**:12b; **13**:100b
seventeenth-century, **1**:391b; **7**:164b; **13**:214b–215a
eighteenth-century, **1**:479a; **2**:225a–b, 301b–302a; **3**:426b–427b; **7**:605b; **9**:306b
nineteenth-century, **1**:60a–b, 360a–b; **2**:464b; **3**:531a; **4**:197b–198a; **6**:58b–59a, 60a–b; **8**:296a, 555b; **11**:482a, 559b–560a; **12**:202a–203a; **13**:300a–b
twentieth-century, **4**:618a; **8**:294a–296a; **13**:300b–301a
Japanese, **15**:742a,b, 743b, 749a
empirical, **3**:2a–b; **10**:581b; **12**:340a, 383b
ethics of
see Deontology
experimental
medieval Arabic, **14**:637b
seventeenth-century, **9**:369b; **12**:102b–104a
nineteenth-century, **2**:26b, 27a; **9**:9b; **12**:240b
twentieth-century, **3**:17a; **5**:150a; **9**:549b–550a, 586a–b
forensic
Renaissance, **7**:16b
seventeenth-century, **2**:237b; **12**:40b
eighteenth-century, **3**:37b
nineteenth-century, **2**:563b; **3**:88a, 237b–238a; **4**:163a, 593b; **5**:297b; **6**:498b; **13**:274a
twentieth-century, **4**:311b, 499a
French Revolution and, **3**:1a–2b; **10**:612a; **14**:16a
Galenic tradition
Byzantine, **10**:417a–b, 418a
medieval, **1**:290a; **5**:233b; **6**:396b; **9**:31a–b; **11**:601a; **12**:7Bb; **13**:229b, 292b; **14**:637b; **15**:239b–242a, 265a
Renaissance, **3**:12b, 342a,b, 354b, 355b; **4**:139a, 198b–199b, 486b; **5**:233b, 581a; **6**:360a, 562b; **8**:248a–249b, 360b–361a, 554b; **11**:253b; **12**:323b
seventeenth-century, **4**:511b; **7**:150b, 164b, 537b; **12**:603a
opposition to, **5**:47a, 48a; **10**:305a, 310b; **12**:79a
handbooks, **13**:448a; **15**:643b, 646a–b
Hippocratic tradition
ancient Greek, **4**:106a, 368b; **6**:418b–429a
ancient Roman, **3**:175a; **5**:228b, 230b–232b; **11**:602a
medieval, **1**:290a,b; **9**:41a; **11**:238a, 444a–b; **12**:7Bb; **13**:229b; **14**:584b–585a, 637b; **15**:500a*
Renaissance, **1**:399b; **4**:139a; **6**:360a; **8**:554b; **11**:253b; **12**:97b
seventeenth-century, **1**:391b; **4**:511b; **13**:490b
eighteenth-century, **1**:479a; **2**:227a; **3**:2a, 97a; **5**:366b, 384b; **9**:306b
nineteenth-century, **2**:444b; **7**:556b
industrial, **11**:585b
internal, **2**:445b, 464a; **15**:741b
Japanese, **1**:310a,b, 312a; **15**:737a–744b, 749a
magical, **5**:231b; **14**:118a; **15**:645b
critics of, **6**:422b, 424b
magnetism in
see Animal magnetism
malpractices in
eighteenth-century, **2**:15a
and mathematics, **1**:290b; **5**:228b, 231a–b; **10**:611b, 613a; **12**:335a
Mesopotamian, **15**:643a–b, 645a–646b
military
seventeenth-century, **15**:28b
eighteenth-century, **4**:589b; **6**:568a; **11**:112a, 148a; **13**:181b
nineteenth-century, **2**:527b–528b; **3**:604b; **4**:162b, 310b–311b; **7**:414a, 421b; **8**:406b, 448a; **9**:422a; **11**:177b, 242b, 345b–347a
twentieth-century, **2**:529a; **4**:618a; **8**:295a; **11**:577a–b; **13**:300b–301a; **14**:512a; **15**:73a–b, 420a
music and, **1**:332b
and natural sciences, **1**:393a; **4**:178a*
naval, **8**:362a–363a
Parisian school, **2**:444b; **7**:536a–b
and pharmacy, **9**:313a
and philosophy, **1**:4b, 480a; **3**:1b–2a; **5**:232a; **6**:422b
philosophy of
ancient Greek, **1**:290a, 315a; **6**:420a–425a *passim*
medieval, **1**:290a–291a; **3**:394a; **9**:40b–41b; **11**:444b; **14**:584a
Renaissance, **3**:471b; **14**:6b–7a
seventeenth-century, **1**:391b–392a; **9**:66b; **12**:333a
eighteenth-century, **1**:451b; **3**:2a–b; **5**:221a, 251a; **7**:605b–606a; **8**:352a; **9**:306a–b; **12**:458b, 602a, 603b–604a
nineteenth-century, **2**:31b–33a; **9**:10b; **12**:592a, 617b; **14**:40a–b, 41a, 42a
twentieth-century, **8**:295b–296a
Chinese, **15**:738b–739b
humoralism vs. solidism, **1**:315a; **15**:738b
Japanese, **15**:739b
physics and, **2**:601a
pneumatic, **13**:448b*
preventive
eighteenth-century, **1**:563b
twentieth-century, **3**:186a; **8**:295b–296a
professionalization of, **3**:12b–13a
and psychology, **9**:41a
rationalization of, **6**:422b, 424b–429a
regulation of
medieval, **12**:447b; **13**:415a
Renaissance, **1**:525a; **4**:585a–586a; **7**:16b; **8**:361a–b
eighteenth-century, **3**:2a–b
nineteenth-century, **7**:366a; **14**:409b
and science, **3**:244a–245a; **7**:520b; **9**:325b
socialization of, **6**:519a
statics in, **12**:102b–103a, 103b
and statistics, **1**:563b; **2**:42a–b; **3**:2b, 6b; **5**:137b; **6**:59a; **7**:44b; **8**:122a; **10**:404b, 405b; **11**:33b, 237a; **12**:294b–295a, 296b, 450b, 503a; **14**:139a, 230b–231a
and surgery, **8**:114b; **10**:304b
sympathetic, **5**:48a
theoretical
ancient Greek, **6**:317a
ancient Roman, **5**:230b, 231a–b; **12**:538b–540b
medieval Arabic, **15**:240a–242a, 265a–b, 498b
medieval Latin, **13**:186a, 189a, 195b
Renaissance, **10**:307b–308a, 309b–310a; **15**:160b–161a
seventeenth-century, **1**:592b–594a; **7**:150b, 537b; **11**:1b–2b, 466b–467b; **12**:79a, 101a–103b, 311a–312b, 601a
early eighteenth-century, **1**:392a, 594a; **2**:226a–b, 302a; **3**:244b–245a; **6**:458a–460a; **7**:188a; **8**:114b; **9**:511a–b; **12**:601a–604a; **13**:231b–232a, 563a
late eighteenth-century, **3**:15b, 37b; **8**:585a–b; **9**:325b–327b; **11**:37a–38a, 364a–b, 617a
nineteenth-century, **6**:18a–b; **10**:22a; **11**:560a; **15**:128a
tropical
Renaissance, **10**:236b–238a
seventeenth-century, **1**:25b–26a; **10**:621a–622a
eighteenth-century, **8**:362b
nineteenth-century, **4**:310b–311b; **7**:424b–425a, 427b–429b; **8**:65b–66a, 296b; **9**:81b–83a; **11**:555a–556a
twentieth-century, **3**:185a–186a; **9**:328b; **10**:142a–b; **15**:96a–97b, 454a–455a
see also Anatomy; Botany, medical; Congresses, scientific; Diagnostics;

Disease(s); Education, medical; Epidemiology; Gynecology; Historiography, of medicine; Iatrochemistry; Instrumentation, medical; Pathology; Physician, concept of; Pneumatic school; Prognostics; Psychosomatic medicine; specific countries and topics; Surgery; Terminology, medical; Therapeutics; Veterinary science
Medicines
  see Drugs
Medicus, Friedrich Casimir (Medikus, Friedrich Kasimir) (1736–1808), 9: 253–254; 9:434a,b
Medicus, Ludwig, 5:2a
Medina, Pedro de (1493–1576), 9:254–255; 1:530b
Mediterranean basin
  botany, 9:34b–35a
  geology, 7:30b–31a; 8:241b
Mediterranean fever, 2:528a
Mediterranean Sea
  biology, 9:112b
  currents, 10:568b, 569a
  geography, 10:619a; 11:251b; 13:85b
  geology, 10:29b
  length of, 10:490a; 11:200b
  mean sea level, 3:162a
  naming of, 11:75b(n)
  natural history, 5:67a
  navigation, 5:287b
  paleogeography, 13:147a
  paleontology, 1:491b; 12:517a–b
  zoology, 15:78a
Die medizinische Reform, 14:40b
Medulla oblongata, 5:251a
  histology, 5:172a; 12:224a
  respiratory center, 5:45a, 149b; 8:134a
Medusa (ship), 9:589b–590a
Medusae, 12:561a
  anatomy, 4:289a; 6:589b, 590a
  classification of, 4:290a; 6:8b
  embryology, 6:7b, 590a
  germ layers, 6:339a
  histology, 11:517b
  neurophysiology, 4:290b; 11:517b–518a; 12:355b
Meek, Fielding Bradford (1817–1876), 9: 255–256; 6:57a, 186b, 187a; 10:258b; 14:312b
Meerwein, Hans, 13:550a
Megakaryocytes, 2:165a
Megaliths, 2:450b; 9:490a, 539a
Megalonyx, 7:89b
Megalosaurus, 2:569b; 9:87b
Megaphone, 3:97b
Megascope, 3:208a
Megatherium, 2:570a; 3:324b*
Meggers, William Frederick (1888–1966), 9:256–257
Mégnié, Pierre, 8:79a
Mei Ku-ch'eng, 3:266a; 8:314b
Mei Wen-ting, 3:270b; 8:419b; 14:159b, 160b, 161a–b; 15:745b
Meickleham, William, 13:374a,b
Meigs, Charles Delucena, 6:120a
Meijering, Jan Laurens, 13:611a
Meiji era, 15:728a,b, 738b, 744a, 753b
Meillet, Alphonse, 9:538b
Meinesz, F. A. Vening
  see Vening Meinesz, Felix A.
Meinzer, Oscar Edward (1876–1948), 9: 257–258
Meinzo of Constance, 6:301b
Meiosis, 2:626b; 4:546a; 5:35b, 453b; 13: 157a–158a; 14:235b, 430b–431a
  in plants, 5:582a
Meisenheimer, Jacob, 1:553b
Meissen porcelain works, 8:621b; 11:109a
Meissner, Georg (1829–1905), 9:258–260; 2:129b; 6:434b; 7:420b, 421a; 13: 452b; 14:2b, 113b
  embryology, 1:477a
  Meissner's corpuscles, 8:9a; 9:259a
  Meissner's plexus, 9:259b
Meissner, Karl, 3:159b; 10:497b
Meissner, Walther, 8:52a, 476a, 478a
Meister, Joseph, 10:403b–404a, 405a
Meitner, Lise (1878–1968), 9:260–263; 4: 579b; 10:424b; 11:17b*; 12:35b*
  radiochemistry, 2:281b; 4:580a; 6:16a–17a; 7:158b; 12:505b
Mela, Pomponius
  see Pomponius Mela
Melancholy, 10:613b; 11:602a,b, 603a; 12: 92a
  see also Depression
Melanchthon (or Melanthon), Philipp Schwarzert, 2:401a; 3:414a; 4:431b; 5:28b, 121a; 10:8a; 11:365b; 12:322b; 13:59a
Melanism
  industrial, 5:453b
Melanophores, 7:455a
  and epithelial cells, 1:545a
Melanosporum, 9:176a
Melbourne
  Melbourne Exhibition, 11:384b
  observatory, 5:562a
Meletius (monk), 10:21a
Meliorism, 15:79a
Melissic acid, 2:484b
Meliteniotes, Theodore, 7:337a; 11:186b
Mellanby, Edward (1884–1955), 15:417–420; 5:41b; 6:499b
Mellanby, May Tweedy, 15:417b
Mellitic acid, 1:390a; 14:476a
  discovery of, 7:394b
Mello, Francisco de (1490–1536), 9:264
Mello, Francisco Manuel de, 9:264a
Melloni, Macedonio (1798–1854), 9:264–265; 3:46a; 5:68b; 13:52a
  physics, 9:166b; 13:238b; 14:185a
Melnikowite, 2:159a
Meloidae, 4:504a
Melotte, Philibert Jacques, 10:107b
Melsens, Henri, 8:597a
Melting point, 3:292b
  as analytic tool, 3:242b
  of crystals, 8:368a
  curves of, 13:245a
  determination of, 4:141a
  pressure effects, 13:236b
Meltzer, Samuel James (1851–1920), 9:265–266; 5:40a; 7:504b, 505a*
Melun
  Agricultural Society, 10:395a, 396b
Melvill, Thomas (1726–1753), 9:266–267
Melzi, Francesco, 8:194b, 207b, 234b; 13: 430b
Membrane(s)
  botanical
    composition of, 9:79a
  Bowman's, 2:376b
  extraembryonic, 1:386b
  fetal, 2:161a; 4 :264a
  fibrohyaline, 11:374a
  glandular, 9:64b
  mucous
    diseases of, 2:444b–445a
    regeneration of, 13:589a
  nictitating, 10:520a
  peritoneal, 2:165a; 4:172b
  pleural, 2:165a
  precipitation, 13:453a
  semipermeable, 5:7a; 13:453a
  serous, 2:165a
  synthetic, 10:577a
    permeability of, 9:354b
Mémoires de la Société ethnologique de Paris, 4:286a
Mémoires de la Société géologique du Nord
  founding of, 1:472a
Mémoires pour servir à l'histoire des plantes, 13:526b–527a
Mémoires pour servir à l'histoire naturelle des animaux, 4:267b, 268a; 13:526b, 528a
Mémorial de l'officier du génie, 11:78b
Memorial Sloan-Kettering Cancer Center, 4:499a
Memorie del Regio osservatorio di Firenze ad Arcetri, 4:162a
Memorie della Società degli spettroscopisti italiani
  founding of, 12:269b
Memory
  Renaissance discussion, 14:47b
  seventeenth-century discussion, 14:406b
  eighteenth-century discussion, 6:139b
  nineteenth-century discussion, 2:9b–10a, 448a
  twentieth-century discussion, 3:90a; 13: 227b
  and heredity, 12:298a–b
  latent, 1:27a
  location of, 10:21a
  in psychoanalysis, 2:449a
  theories of, 9:98a
  see also Mnemonics
Memory drum, 9:562a
Menabrea, Luigi Federico (1809–1896), 9: 267–268; 1:356b*; 9:578a
  and Castigliano, 3:118a–b; 9:268a,b
  principle of least work, 3:118a–b; 9:268a
Menaechmus (*fl.* middle of fourth century B.C.), 9:268–277; 4:415a
  acquaintances of, 4:103b; 11:22b; 13:334a
  critics of, 11:28a
  geometry, 1:245a; 4:104a, 105a, 489b; 6:407b; 14:49a; 15:116b
Mencarelli, Edmond, 7:57a
Mencius, 12:386a, 387b
Mendel, Johann Gregor (1822–1884), 9:277–283; 3:121a, 422a; 6:9a; 7:114a, 442a
  biographies of, 7:517a
  heredity, laws of, 2:363a, 364a; 7:623b
    confirmation of, 8:3b, 589a
    exceptions to, 13:134b
    independent assortment, 3:422b; 9:280b
    rediscovery of, 1:505b; 3:421b, 422a–423a; 5:10b, 266b; 6:21b; 7:115a; 9: 496b; 13:477b–478a; 14:96b, 99b, 430b–434a
    segregation, 3:572a; 9:280a, 618b
  influences on, 5:218b; 13:542b
  meteorology, 9:279a–b
  and Naegeli, 9:601b
  translations of works, 15:507a
Mendel, Lafayette Benedict (1872–1935), 9: 284–286; 10:243a–b; 14:252b
Mendeleev, Dmitry Ivanovich (1834–1907), 9:286–295; 1:528b; 2:624b, 625a; 3: 49b; 4:599a
  acquaintances of, 2:316a, 587a; 3:498a; 12:270a; 13:243b
  atomic weights, 11:173b
  mathematics, 9:127a
  periodic law, 2:428b–429a; 3:479a
    confirmation of, 9:290b–291a, 292b
  periodic table, 1:288b; 4:134b; 9:288b–289a, 348a,b, 349b, 351a; 10:38a; 11:136a; 12:513b

Mendeleev, Dmitry Ivanovich (cont.)
periodic table (cont.)
confirmation of, 2:247a, 254b, 428b–429a
predicted elements, 2:254b; 3:321b; 14:447a
Principles of Chemistry, 2:429a
and St. Petersburg Academy of Sciences, 2:624a; 9:292a
solutions, 7:162a, 203b, 462a; 13:244a
students of, 4:597b; 7:34b, 461b; 13:416a, 616b
Mendelism, 1:505a,b, 506a; 3:121a–123b, 590b; 8:446b, 589a; 9:518a–b; 10:130a, 463a–464a; 11:212b; 12:417a; 13:134a, 157a; 15:148a
and animal kingdom, 3:493a
and Darwinism, 5:10b; 9:282b–283a
see also Heredity, Mendelian
Mendelssohn, Kurt, 12:437b; 15:449b
Mendelssohn-Bartholdy, Felix, 4:2a; 6:230a, 286a; 7:506a; 14:61a; 15:122b
Mendoza, Diego Hurtado de, 12:548b
Mendoza, Eric, 8:482b, 483a
Meneghetti, Egidio (1892–1961), 9:295–296
Meneghini, Giuseppe, 1:491a; 3:328a; 7:30b
Menelaus of Alexandria (*fl.* A.D. 100), 9:296–302; 15:420–421; 1:22b, 227a, 445b; 7:81a; 13:319b
editions of works, 9:190b; 13:509b
geometry, 1:227a, 445b
hydrostatics, 7:340b, 341a,b, 342b, 343a
manuscripts of, 14:273a
Menelaus configuration, 11:189a
Sphaerica, 6:203b(n); 7:38b
editions of, 9:22a, 555b
translations of, 6:71a; 7:25a; 9:85a; 13:400b; 15:177a–b
theorem of, 1:42b; 2:153b, 154a; 3:182a; 4:611b; 7:38a,b, 352b; 9:39b, 298a–299b, 555a, 615a; 11:415a; 15:209b
proof of, 13:289b
Meng Shen, 8:396a*
Mengenlehre
see Sets, theory of
Menger, Karl, 7:105a; 13:549a–b; 14:121b
problems of, 9:249a
Menghini, Vincenzo Antonio (1704–1759), 9:302–303; 5:226b
Mengoli, Pietro (1625–1686), 9:303–304; 13:566a
mathematics, 2:49b, 53a; 3:151a
Ménière, Prosper, 1:446b; 14:398a
Meninges, 4:385b
cells of, 2:165a
meningiomas, 2:165a; 3:517b, 518b
Meningitis, 2:596a; 4:252a; 5:317a
cerebrospinal, 3:447b; 4:154b
treatment of, 5:40b
epidemic, 11:148a
etiology of, 9:95a
spinal, 8:7a
treatment of, 5:31a*
tuberculous, 14:322a
Meningococcus, 9:95a
antiserum, 5:40b; 12:482b; 15:522b
discovery of, 14:218b
Meniscus (liquid)
measurement of shape, 6:173a
Meniscus system (optics), 9:46a,b
Menologies, 15:644b
Menon, 4:462b; 6:421a–b, 424a; 10:589b
Menshutkin, Nikolay Aleksandrovich (1842–1907), 9:304–305; 2:624b; 9:288b
on esterification, 2:69a
Menstruation
and health, 6:318a
theories of, 11:565b
Mensuration, 6:76b*; 7:532a–b; 9:610b; 13:540b; 15:9b, 751b
Mentelius, Jacobus, 6:617a; 10:477a
Menten, Maud Lenora, 1:537b
Menthol
formula, 14:156b
Menuret de Chambaud, Jean Jacques (1733–1815), 9:305–307
Menzel, Donald Howard, 5:363a; 9:504a; 12:250a, 349b
Mephitic air (carbon dioxide), 12:24b
Méray, Hugues Charles Robert (1835–1911), 9:307–308; 1:407a; 6:3b; 11:413a
Mercantilism
seventeenth-century, 1:549a
eighteenth-century, 7:202a
Mercaptan, 14:600b–601a
Mercati, Michele (1541–1593), 9:308–309
Metallotheca, 1:318b
Mercator, Gerardus (or Gerhard Kremer) (1512–1594), 9:309–310; 4:5b, 166a; 10:239a; 13:49b
maps, 5:349a; 11:202b; 14:275b
Mercator projection, 4:214a; 5:304a; 7:20a; 9:309b; 14:513b–514a
theories of, 9:310a, 311a
Mercator, Nicolaus (Kauffman, Niklaus) (*ca.* 1619–1687), 9:310–312; 5:525a
Institutionum astronomicarum, 7:308a
Logarithmotechnia, 5:521a, 524b; 8:161a; 10:48a; 11:404a
Mercator series, 5:528a; 8:161b; 10:87a(n)
Mercure galant, 5:57b, 58b, 60b
Mercure suisse, 15:53a
Mercuriale, Girolamo, 3:98b; 4:507b; 5:238b; 9:309a; 13:587b
Mercurialis, 15:67b
Mercuric chloride
as disinfectant, 7:423a; 8:408a
Mercuric fulminate
discovery of, 4:185a
Mercurio Peruano, 13:541a
Mercurous chloride
vapor density, 3:617b
Mercurous nitrate, 11:318b
Mercury (Hg), 3:419b; 5:595b
in alchemical theory, 1:549b; 2:22b, 227a; 3:596b; 4:209b, 387b; 7:356a, 524b; 8:131a, 174a, 620a–b; 10:81a, 309b, 555a; 12:311b; 13:511b
atomic weight, 3:544a; 13:499b
calcination of, 1:529b; 8:75b
compounds of, 6:395b
alkyl, 7:323
oxides, 1:129a; 2:75a; 3:5a; 5:92a; 6:18a; 7:395a; 8:75a,b; 9:39b; 12:64a; 13:310b, 465b
reactions of, 1:529b; 9:401b
salts, 5:92a; 8:312a; 11:318b; 13:310b
deposits of, 4:346b; 11:608a
detection of, 13:71a
electrical properties, 8:387b
resistance, 1:303b; 2:458a; 7:221b
germs in, 10:371a
hydrodynamics of, 11:63a–b
isotopes of, 2:499a
medical use of, 3:206a, 487a, 489b; 5:385a*; 6:18a, 255a; 11:48b; 13:508a
dropsy treatment, 10:307a, 309a
ointments, 2:314b
syphilis treatment, 1:620b, 621a; 5:105b; 9:334b; 10:305a–b, 307a; 13:181b; 14:28a
tetanus treatment, 14:559a*
mining of, 7:200b
physical properties
capillary constant, 11:241b
compressibility, 1:128b; 2:458b
expansion, 2:177b; 3:20a; 4:240a; 15:312b
freezing point, 2:186b; 3:158a
refractive index, 7:554b
specific gravity, 7:346a
specific heat, 2:178b
vapor density, 3:47b; 5:320b
vapor pressure, 1:349a; 7:416b
physiological effects, 10:582b
poisoning by, 3:206a; 5:366b; 13:70b–71a; 14:256b
treatment of, 9:296a
purification of, 2:227a
spectrum, 7:554b
in thermometry, 4:518a
vapor, 5:294b; 7:526a
Mercury (planet)
conjunctions of, 9:508a*
distance from earth, 6:69a; 11:197a–b
distance from sun, 3:409a
halo, 3:261a
motion of, 7:338a; 10:5b; 11:224a; 15:353b
observations of, 9:122b; 11:472a; 13:325b; 15:218b
orbit, 1:509b; 3:407b–409a; 8:277a; 10:34b; 13:537a; 14:594b; 15:202b–204a
eccentricity, 11:251a; 12:360b; 13:472b
elongation, 13:326a
maximum equation, 14:576b
and relativity theory, 3:221a; 4:328b; 5:181b, 182b, 554b; 8:278a–b; 11:84a*
position of, 12:547a
latitude, 5:618a–b
rotation, 12:162a
period of, 8:522a
size of, 4:543a
apparent diameter, 4:25a*
spectrum, 14:55a
synodic period, 4:467a; 15:219a
tables of, 8:369b
transit of 1631, 3:528b; 5:285a, 287b; 7:305a; 10:490a; 12:163a
transits of
seventeenth-century, 6:69a; 7:373a
eighteenth-century, 2:328a; 3:198a; 4:14b; 5:88a; 8:178b; 9:78a, 330b; 10:607b, 614b; 14:452b; 15:589b
nineteenth-century, 6:263b, 334a; 9:459b; 13:472b; 14:479b
and longitude, 12:414a*
predictions of, 12:343a
and solar parallax, 4:23b
zodiacal period, 5:345b
Mereology, 8:263a
Merian, Johann Rudolf, 15:158b
Merian, Maria Sybilla
Metamorphosis insectorum Surinamensium, 11:503a
Merica, Paul Dyer (1889–1957), 9:312
Meridian, measurement of
see Latitude, determination of, degree, length of
Meridian, prime, 13:120b–121a
localities on
Indian determinations, 15:582a–589a *passim,* 593b, 597a, 607b, 610a, 612a, 614b, 616a
Meridian circle (astron. instrum.), 2:148a, 149b, 150a, 196b, 332b, 333a, 468b, 532b; 3:14a, 100b–101a, 558b; 4:30a; 5:405a, 554b; 7:81b, 209a–b, 481a, 608a; 8:178b, 504a; 9:77b, 460a; 12:518b; 14:448b, 485a–b, 575a
constants of, 6:94b
errors in divisions of, 4:26b
vertical, 11:377b
Mérimée, Léonor, 13:310a–b, 313a

Mering, Joseph von, **1**:441a; **5**:463a; **12**:355b
Merini, Michele, **5**:383b
Meriphyte theory, **8**:354b
Merkle, Karl, **15**:232b–233a
Merlan, **12**:575b–576a
Merogenesis, **5**:385b
Merogony, **4**:12a
Merohedrism, **9**:379b
Merons, **1**:518b
Merotomy, **1**:417b; **11**:175a
Merrett, Christopher (1614–1695), 9:312–313; **2**:523a; **6**:46a(n); **7**:525a; **11**:316a
Merriam, Clinton Hart (1855–1942), 9:313–314; **1**:406a; **3**:537a*
Merriam, John Campbell (1869–1945), 9:314; **8**:356b; **13**:72a
Merrill, Elmer Drew (1876–1956), 15:421–422
Merrill, George Knox, **15**:150a
Merrill, George Perkins (1854–1929), 9:314–316; **3**:392b*; **4**:365b*; **6**:438b*; **8**:261a*; **12**:433a
Mersenne, Marin (1588–1648), 9:316–322; **1**:245b; **2**:349a, 542b; **5**:48a, 247b; **13**:320b*
  Academia Parisiensis, **2**:353b; **4**:46a; **6**:112b; **9**:316b, 599a; **10**:330b, 342a; **12**:585a; **14**:368a,b
  acquaintances of, **1**:541a, 567a, 568a; **2**:163a; **3**:615a; **4**:95b, 571a; **6**:445b, 597b; **7**:537a; **9**:599a; **10**:546b, 564b; **11**:486b; **12**:205b; **13**:439a–b
  astronomy, **10**:490a
  Cogitata physico-mathematica, **14**:366b–367a, 368b, 379a(n)
  on contemporary scientists, **11**:487a
  correspondence of, **1**:425a*, 541b; **2**:83b; **3**:64a, 615b, 616a; **4**:505b, 567b, 574b; **5**:158b, 606b–607a; **6**:520a, 597b, 598a; **7**:540a(n); **8**:267a–268b; **9**:13b, 16a, 528a; **10**:103b, 489b, 490b, 495b; **12**:583b–584a; **13**:253a, 255a, 434a; **14**:254b, 368a–b, 379b(n)
  and Desargues, **4**:46a,b, 47b
  and Descartes, **4**:51b, 53a, 58a, 62a; **9**:316b, 318a–b, 320a, 598a; **13**:436a
  and Gassendi, **5**:285a,b, 287a, 288b; **9**:316b, 320a, 321a(n)
  geometry, **1**:191b, 474b; **10**:331a
  Harmonie universelle, **4**:46b; **9**:498b; **10**:342a
  hydrostatics, **10**:332b, 333b, 334a,b; **13**:438b
  influence of, **1**:484a; **2**:13a; **3**:362b; **4**:228a; **11**:488a; **14**:365b, 366b–368b, 375b(n)
  on language, **14**:372a, 373b
  manuscripts of, **1**:206b–207a
  Les méchaniques de Galilée, **14**:366b
  Mersenne's numbers, **8**:531b
  music theory, **4**:88b
  optics, **1**:81b; **6**:446a
  philosophy of science, **14**:367a–b, 374b(n)
  Reflexiones physico-mathematicae, **10**:333a
  students of, **10**:103a
  Synopsis mathematica, **14**:29b
  translations by, **5**:239b; **9**:302a*
  Universae geometriae . . . synopsis, **6**:128a
Mertens, Franz (math.), **6**:572a; **14**:222a
Mertens, Franz Carl (bot.), **7**:610a
Merton College, Oxford, **1**:94a; **2**:393a,b, 394a, 435a, 550b–551a, 609a; **3**:172a; **5**:247a; **8**:198a; **9**:32b; **13**:185a
  astronomy at, **3**:217b
  kinematics
    Merton rule, **1**:290b; **5**:240b; **6**:376b, 378b–379a; **10**:225b–226a
    mean speed theorem, **5**:360b; **13**:206b, 350a
      proof of, **7**:116b; **13**:203b, 210a
    natural philosophy, **6**:376a–379b; **7**:116a–b; **9**:611b
Mertz, Xavier Guillaume, **9**:198a
Meru (mountain), **15**:554a–b, 555a
Merules, **1**:518b
Merv (city), **7**:336a
Merwin, Herbert Eugene, **8**:41b
Méry, Jean (1645–1722), 9:322–323; **6**:65b; **7**:111a
Merz, Alfred, **2**:557b; **6**:97b
Merz, Georg, **3**:605b; **4**:30a; **12**:268a
Merz, Karl Friedrich, **8**:245b
Merz, Sigmund, **5**:144a
Mesabi Range, **14**:441a,b
Mesembryanthemum, **4**:99a
Mesencephalon, **3**:112b–113a
Mesenchyme, **4**:187b
Mesentery
  naming of, **9**:468b
Meshchersky, Ivan Vsevolodovich (1859–1935), 9:323–325; **5**:189a*; **13**:28a*
  Meshchersky's equation, **9**:324b
Mesiatsev
  see Mesyatsev
Mesitylene, **5**:13a; **7**:224a
Meslans, Maurice, **9**:451a
Mesmer, Franz Anton (1734–1815), 9:325–328; **5**:138a
  "animal magnetism," **1**:401b; **8**:84a
Mesmerism, **5**:138a; **9**:325b–327b; **10**:11a, 198a; **14**:137a
  see also Animal magnetism, Hypnosis
Mesnil, Félix (1868–1938), 9:328–329; **3**:148b; **15**:455a*
Mesoblast, **1**:601a
Mesoderm, **6**:435a; **11**:369a,b
  denial of, **7**:402a
  formation of, **6**:167b; **11**:255a
  origin of, **3**:120b; **6**:336b, 339a; **14**:426a–b
Mesographicum
  see Proportionals, between two lines
Mesolithic
  industries of, **10**:607a
Mesomeres, **15**:513b
Meson(s), **8**:96a; **10**:214b; **13**:241a,b; **15**:27a
  radioactivity of, **3**:370b
Mesonephros, **8**:302b
Mesopithecus, **2**:110a
Mesopotamia
  agriculture, **15**:648a, 649b
  alchemy, **15**:649b–650a
  archaeology, **14**:504b; **15**:646b–647b, 649a, 651b
    iconographic remains, **15**:647a, 649a
    "tells," **15**:647a
  architecture, **15**:648b, 650b, 652b, 658b(n)
  art, **15**:649a, 651b
  astrology, **15**:635b, 644a, 672a,b, 673b–674a, 677b, 678a–b
  astronomy, **6**:616b; **10**:290a, 426a; **11**:188a,b, 191b, 195b, 223a–224a; **12**:547a, 576b; **13**:295b–296a, 582a; **14**:593b; **15**:209b–212b *passim,* 216a, 218b, 219a, 220b, 331a, 640b, 643b–644a, 672a–680b
  calendar, **9**:337b, 338b–339a; **15**:535a, 536a–554a, 589a
  cartography, **15**:637b–638b
  commerce, **15**:650a–b
  cosmogony, **15**:638b–639a
  cosmography, **15**:638b, 639a–b
  cosmology, **14**:549a; **15**:638b–641b, 643b, 645a
  geography, **15**:637a–638a
  mathematics, **2**:320a; **4**:111b, 117a, 612a; **13**:296b; **15**:10a, 208b, 667a–672b, 692a–693a
    algebra, **1**:30b, 31b; **4**:115a; **15**:668b–670b
    arithmetic, **3**:126a; **13**:540a; **15**:668a, 670b
    geometry, **1**:31b; **3**:276b; **4**:608b; **6**:315a; **11**:221a,b, 222a; **15**:670b–671b, 692a–693a
  medicine, **15**:643a–b, 645a–646b
  metallurgy, **15**:651a–b
  meteorology, **15**:677a
  poetry, **15**:636a–637a
  raw materials, **15**:648a, 651a
  sociology, **15**:637a
  technology, **15**:646b–653b
Mesostomum, **12**:192b
  life cycle, **12**:193a–b
Mesothelium, **2**:165a
Mesothorium, **12**:504b
  discovery of, **2**:260a; **6**:15b–16a
Mesoxalic acid, **8**:343b
Mesoxylon, **12**:260a
Mesozoa
  proposed as phylum, **1**:601a
Mesozoic era, **2**:556b; **3**:161a,b; **5**:476a; **10**:168a
  naming of, **10**:584a
  paleobotany, **12**:104b, 339b
  paleontology, **1**:95b, 491b; **2**:489b; **5**:214a; **6**:14a, 57b, 222a, 594b–595a; **7**:61b; **9**:87b, 255b, 617b; **10**:128a, 222a, 429b
  stratigraphy, **1**:95b, 210a, 234a, 286a; **2**:108b, 496b; **3**:297b, 530a; **6**:168b; **7**:253a; **10**:428b; **12**:140a
  see also Cretaceous period, Jurassic period, Triassic period
Messahala
  see Māshā᾽allāh
Messenger of Mathematics, **5**:414a,b
Messier, Charles (1730–1817), 9:329–331; **1**:400b; **4**:14a; **13**:496a
  catalogue, **9**:250b
  comets, **2**:216b; **4**:23b; **8**:299b; **11**:82b
  nebulae, **6**:330a, 332a; **7**:87a
Messina (Sicily), **2**:311b
Messina, University of, **2**:307b–308a
Mesuë, Johannes, the elder
  see Ibn Māsawayl (Mesuë), Yūḥannā
Mesue, Johannes, the younger, **9**:75b
  Canones, **9**:469a
Mesyatsev, Ivan Illarionovich (1885–1940), 9:331
Metabolism
  aerobic, **6**:443b
  of amino acids, **4**:360a; **5**:53b; **8**:555a
  animal vs. vegetable, **4**:243a
  calorimetric study, **1**:325b, 610a–b
  of carbohydrates, **2**:28b–29b; **4**:360a–b; **7**:517b, 519a; **8**:452a,b, 614b–615a; **10**:580a; **14**:454b
    hormones and, **3**:415b–416a
    insulin and, **3**:352a; **15**:228b–229a
    in muscle, **2**:29b; **6**:501b
  and climate, **4**:311b; **10**:580a
  concept of
    early twentieth-century, **1**:610a–b
  endogenous vs. exogenous, **5**:53b
  and energy conservation, **1**:610a
  enzymatic systems, **2**:87a; **13**:393b–394a
  of fats, **2**:28a, 124a–b; **4**:243a; **8**:555a; **9**:285b; **13**:403b; **14**:454b
  of glucose, **3**:220a; **4**:360a–b; **8**:452b; **9**:359a–b
  hydrogen transfer, **7**:406b; **13**:394a
  inorganic, **2**:585a–586a
  intermediate, **2**:28b–29b, 124a–b; **4**:360a

Metabolism (cont.)
of iodine, **9**:284b
light, effects of, **9**:456b–457a
measurement of
nineteenth-century, **1**:610a; **5**:149b; **12**:466b–467a; **14**:63a–66b
twentieth-century, **1**:610b
of minerals, **2**:585a–586a; **4**:357b; **6**:136a
and muscle activity, **14**:64a–b, 65b–66a
of nitrogen, **9**:284b; **10**:326b; **11**:585b–586a; **14**:63b–66b
in plants, **3**:359a; **11**:179b; **12**:208a
of proteins, **2**:162a; **3**:220a; **8**:451b, 452a–b; **10**:580a; **14**:65b–66b, 454b
carbohydrates, influence of, **8**:555a
temperature effects, **7**:503a; **11**:585b
see also Biochemistry; Nutrition; Oxidation; Respiration; specific organs; Sugar, metabolism of
Metalamines, **2**:199b
Metallogenetic provinces, **8**:370a
Metallographist, **12**:126b
Metallography
nineteenth-century, **9**:139a; **10**:246a–b; **12**:126b, 543b–544a; **13**:8a
twentieth-century, **1**:609a–b; **3**:87b; **5**:576b, 584a*; **6**:368a–369a, 560b–561b; **7**:57a, 92b; **11**:549a–b; **13**:245b
methodology, **6**:561b; **10**:246b
nondestructive, **7**:57a
quantitative, **12**:126b
Metalloids, **7**:21b
spectra, **1**:166b
Metalloporphyrins, **3**:290b
Metallurgy
medieval Arabic, **6**:79b–80a
medieval Latin, **11**:504a; **13**:327a–b, 348a–b
Renaissance, **1**:77b–78a; **2**:142b–143b; **4**:393a–394a; **10**:515a; **13**:396b–397a; **15**:491b
early seventeenth-century, **1**:448a–b; **7**:355a; **8**:464b; **12**:307b
late seventeenth-century, **1**:451a, 548b; **5**:420a, 421a, 422b; **7**:284a–b; **9**:507b; **14**:210a
early eighteenth-century, **2**:421b; **3**:41b, 350b, 454b, 473a; **6**:236b; **8**:468a; **11**:328b–330a; **13**:179a; **14**:144b; **15**:93b–94a
late eighteenth-century, **1**:233b; **2**:292a, 315b; **3**:326b, 346b, 560b; **4**:345a–b, 346b; **5**:222b, 601b–602a; **6**:370a, 468b; **7**:78a–b; **8**:146b–147b, 321a, 470b; **9**:471b, 472a,b, 473b, 477a,b, 590b; **10**:446a, 496b; **11**:463b, 499a; **13**:411b; **14**:263a
early nineteenth-century, **4**:218a; **5**:156a; **6**:164b; **7**:254a–b; **9**:18b, 591a–b; **11**:359a; **12**:117a; **14**:486b–488a; **15**:502b
mid nineteenth-century, **2**:72a; **3**:586b, 617b; **4**:77b, 243a; **6**:142b–143a; **7**:383a; **8**:260b; **9**:60b–61a, 178a; **10**:240a, 511b–512a, 557a–b; **11**:33b–34a; **12**:543b–544a; **15**:24b–25a
late nineteenth-century, **2**:467a; **3**:11b, 212a, 233a–234b; **5**:582b–583a; **6**:5a–b, 319a–b; **8**:118b–119a; **9**:451b; **10**:246a–b; **11**:485a–486b, 534b; **12**:126a–b, 544a; **13**:8a, 346a–347a; **14**:123a
early twentieth-century, **1**:581a; **3**:87a–b, 212a, 235a, 239a–b, 436b; **5**:576a–b, 583b; **6**:368b, 479b–480b; **7**:529a–530a; **8**:119b; **9**:312a–b; **11**:100b–101a, 549a–550a; **13**:243b–244a, 245a–247b
mid twentieth-century, **6**:546b, 560b–561b; **7**:57a; **8**:109a
chemical, **7**:57a
Chinese, **12**:382b
historical, **6**:5b; **10**:446a, 512a; **11**:486a
Mesopotamian, **15**:647b, 651a–b
powder, **14**:487b
precision, **3**:239a–b
see also Alloys; Illustration, metallurgical; specific metals
Metalogic, **8**:546a
Metals
activity series, **1**:579a; **10**:81a
alchemical theory, **5**:353a; **13**:511b, 559a–b
analysis of
flame tests, **2**:589a
in solution, **3**:193a
calcination of, **3**:206a–b; **6**:165a*, 381a; **8**:74a, 470a; **12**:605a; **13**:47a*
catalysis, **3**:197b; **4**:20b; **7**:457b; **12**:47a–b; **13**:313a
chemical properties
causticity, **2**:75a
electromotive force, **15**:434a
passivity, **1**:160b; **6**:439a; **12**:191b
permeability to gases, **3**:11b
replaceability, **6**:478a
valence, **5**:125b–126a
colloids, **5**:595b; **13**:160a–b
electric synthesis of, **13**:159a–b
colors, **5**:422b; **11**:486b
compounds of, **1**:49a–b; **10**:496b; **14**:267a–271b
colored derivatives, **14**:478a
complex, **3**:321b; **10**:578b; **11**:87b–88b
intermetallic, **8**:119a; **13**:245a–b
organic, **5**:125b–126a, 321a, 465a; **7**:552a; **10**:578b; **11**:86b; **14**:169a
oxides, **9**:152b, 450b; **11**:167b–170b; **13**:310b
sulfides, **11**:168a–169b
condensation of, **5**:160b
corrosion of, **3**:176b; **5**:189b, 190a; **7**:57a; **9**:60b
crystal growth
recrystallization, **13**:246a
crystallography of, **1**:581a; **3**:87a, 619b; **4**:522b; **11**:486a, 539b, 549a; **13**:8a, 246a; **15**:63b
deformation of, **11**:549a–b
diffusion of, **12**:593b
dissociation of, **11**:281a
electric conduction, **6**:51b; **8**:475b–479a, 480a–481a, 491b, 502a; **9**:178a; **10**:188a–b, 500a; **11**:142a, 420b, 446b–447a, 474a; **13**:246a, 365a, 429a–b; **14**:330a
theories of, **14**:33a
electrical resistance
and temperature, **3**:19b; **4**:80b; **7**:221b; **8**:20a, 188a, 480b
electrodeposition of, **1**:431a; **2**:340b; **5**:474a; **7**:56b; **13**:23a
electron bombardment of, **3**:597b
electron theory of, **2**:239b; **4**:191b–192b, 355b; **5**:160b; **6**:561a; **7**:26b; **8**:497b, 498b; **11**:420a–421b; **12**:529b, 530a; **13**:364a; **14**:205a–b
electrons in, **4**:376a; **5**:160b; **7**:617b; **12**:529b, 530a; **15**:444a
in enzymes, **14**:174b
expansion of, **15**:312b
galvanic pairs, **1**:579a; **5**:268a; **14**:77b–79b
geochemistry of, **5**:457b
hardening of
theories of, **7**:92b
heat conduction, **2**:136a; **5**:137a–b; **7**:13b; **8**:502a; **14**:208b, 218b, 330a
interconvertibility, **12**:495b
isolation of, **2**:588b, 589b; **9**:451b
magnetization of
see Magnetization
medicinal uses, **10**:308a
microscopy of, **1**:609a; **7**:57a; **11**:121b, 549a–b
occlusion by, **4**:80b; **5**:493a–b
optical properties, **4**:189b; **8**:491b; **11**:581b, 582b
reflection, **2**:452b; **3**:420a; **9**:74a; **14**:348a–b
thin plates, **14**:348a–b
organic effects of, **2**:87a; **13**:394a
origin of, **4**:209a; **5**:422b; **8**:123a; **10**:280b; **11**:521a; **13**:559b
physical properties
coercive force, **5**:160b
contact potential, **5**:160b
density, **7**:340b, 341b
ductibility, **1**:518b
hardness, **2**:467a; **11**:329a
malleability, **1**:518b; **13**:246a–b
melting points, **2**:589a
specific gravity, **7**:346a
superconductivity, **2**:252a; **7**:221b
surface tensions, **5**:160b
tenacity, **1**:518b
in plants, **2**:86b–87a
polishing of
electrolytic, **7**:56b–57a
precipitates of, **2**:6a; **4**:394a; **5**:323a–b; **6**:255a, 257a; **11**:435b
radiation studies, **5**:459a; **11**:582b
reduction of, **1**:579a–b; **3**:317a*; **5**:7a; **12**:46b–47a
in rocks, **5**:72b; **8**:616a–b
separation methods, **5**:595b; **7**:21b; **8**:109a
solution of in acids, **2**:6a; **3**:176b; **7**:278a*; **8**:77b
spectroscopy of, **4**:79a; **7**:268a, 381b; **9**:154b, 391b
structure of, **15**:18b
amorphous state, **11**:594a–b
gaseous state, **1**:345b, 347a
grain size, **7**:92b
inner stresses, **6**:368b
polygonization, **7**:57a
sulfuration of, **6**:586b
tarnishing of, **13**:247a
temperature effects, **3**:87a; **6**:5b
theory of, **7**:617a; **8**:620a–b; **11**:437b–438a
granular theory, **11**:446b–447a
quantum, **13**:240a–b
thermoelectric order, **3**:497a
vaporization of, **9**:451b
work functions, **3**:597b; **7**:36b; **9**:396b
work-hardening, **11**:549a
see also Alloys; Amalgams; Ferromagnetism; Metallurgy; Mineralogy; Ore deposits; Salts, metal; specific metals; Transmutation of metals
Metamathematics, **3**:279a*; **4**:446a; **5**:350b; **6**:393a; **11**:107a; **14**:89b
Metamerism, **7**:454b; **8**:3b; **13**:243a
neural, **1**:545b
Metamorphic aureoles, **4**:262a
Metamorphism, biological, **2**:426b; **3**:180b; **5**:385b, 443b; **11**:307b; **13**:171a–174b; **14**:351b
of Anguillidae, **4**:12a
of crustacea, **13**:355a–b
epigamic, **3**:148b
of insects, **5**:328b, 439b; **8**:580a; **13**:171a–174b; **14**:233b, 451b, 507b

paleontological, **1**:469a
and phagocytosis, **9**:333b
of plants, **5**:438b; **14**:350a; **15**:524b
of sponges, **4**:12a
Metamorphism, geological, **1**:472a–b; **3**:608b; **11**:35b
eighteenth-century discussion, **2**:579a
nineteenth-century discussion, **2**:555b; **3**:567b, 568a; **4**:361b; **8**:33a–b, 463b; **9**:584a; **11**:561a; **12**:277a
twentieth-century discussion, **1**:236a, 394a; **3**:548a*; **6**:116b; **13**:572b; **14**:59b
alkali and lime, role of, **3**:434a
contact, **4**:366b, 565a*; **5**:457a; **6**:456b; **7**:249b; **8**:370a; **11**:372a, 547b, 593b; **14**:60b, 312a
dynamometamorphism, **8**:511b; **13**:284b
experimental, **1**:51b–52a; **2**:459b; **3**:587a*
hydrothermal, **4**:565a*
mass movement, **1**:552a
regional, **5**:457a; **11**:372a; **13**:283b–284a, 284b
regressive, **1**:552a
terminology, **1**:551b–552a
theories of, **6**:438a
magmatic, **1**:470a; **10**:126b
neptunist, **2**:158b–159a
transformationism, **2**:159a
thermal, **6**:116a
zonal principle, **5**:563a; **12**:271b; **13**:148b
see also Rocks, metamorphic
Metamorphites, **3**:434a
Metaphosphoric acid, **12**:46b
Metaphysical Club (Cambridge), **7**:68b
Metaphysical Society (London), **6**:597a; **13**:522b
Metaphysics
ancient Greek, **1**:255b; **4**:367b; **13**:329a–330b
ancient Roman, **11**:41b–42a
medieval, **1**:1a, 334a–337b; **4**:92b; **11**:445a; **12**:5a–7Bb; **15**:181b–182b, 494a–496b
Renaissance, **3**:280b, 514b
early seventeenth-century, **2**:163b, 601b; **15**:69a–b
late seventeenth-century, **5**:426b–427a; **8**:156b–160a; **11**:493a–b
early eighteenth-century, **3**:245a, 294b–296a, 381a–382a; **5**:510a, 510b–511a; **13**:180a; **14**:84b, 483a–484a
late eighteenth-century, **2**:328a, 330b–331a; **7**:598a; **11**:143a
early nineteenth-century, **1**:139a,b, 140b–142a; **7**:389b; **10**:182b–184a
mid nineteenth-century, **5**:470b; **15**:85a–b, 86b
late nineteenth-century, **3**:57a, 322b, 456a; **7**:68a–b
early twentieth-century, **2**:10a–11b; **4**:188a
mid twentieth-century, **7**:488a–b
and astronomy, **12**:2a
Cartesian, **4**:51a–55a
critics of, **12**:177b, 178a–b; **14**:468b–469a
mathematics and, **3**:57a; **7**:597b–598a; **12**:575b–576a
and psychology, **5**:193a
and science, **4**:51a–55a, 204a
see also Epistemology, Ontology
Metapsychics, **11**:431a
Metasomatism (geol.), **4**:361b, 409b; **6**:475b
classification of, **8**:370a
Metazoa, **11**:501b
classification of, **6**:167b; **8**:270b
Metchnikoff, Elie (1845–1916), 9:331–335; **2**:165b; **3**:588b; **7**:430a; **10**:415b*
acquaintances of, **1**:576b; **6**:11b; **8**:273a; **12**:270a; **14**:512a
bacteriology, **5**:269b
immunology, **4**:301b
pathology, **5**:270a
phagocytosis theory, **1**:439a, 575b; **5**:270b; **8**:408a; **10**:407b
students of, **1**:161b, 162a; **2**:300a, 584a; **3**:203b; **5**:269a; **6**:11a; **9**:328a; **11**:595b; **12**:423b; **15**:454a
on syphilis, **4**:300a; **10**:18b
Metempsychosis, **2**:539b; **4**:367a, 369a; **11**:220b
Meteor Crater (Arizona), **9**:315b
Meteorites
eighteenth-century study, **8**:321b; **12**:517b; **13**:40a*, 456a*
nineteenth-century study, **1**:20b, 528a; **2**:134b, 431a; **3**:587a; **6**:19a–b, 326b; **8**:441b–442b; **10**:42a; **11**:234a, 539b; **12**:216a
twentieth-century study, **3**:548b*; **9**:315b; **10**:289a–b; **12**:572b; **13**:246b
chemical analysis of, **2**:134b; **5**:457b, 558b; **11**:270b, 359b; **12**:433a; **13**:310b–311a; **14**:477b
Chinese study, **12**:381a
collections of, **3**:587a; **10**:42a, 288b
dating of, **10**:288b, 289a
light and sound emitted by, **6**:19b
mineralogical composition, **6**:19b; **9**:315b; **13**:476b
origin of, **1**:615b; **3**:210b, 258a; **5**:558b–559a; **6**:19b; **10**:199a, 289a; **12**:517b; **13**:476b
prediction of, **13**:296a
reality of, **2**:134b
spectra, **7**:268b, 461b*; **8**:441b–442b; **12**:269a
structure of, **12**:543b
Widmannstätten figures, **1**:581a; **14**:326b
Meteorological Office (Great Britain), **12**:366b
creation of, **5**:17b
Meteorologicheskii vestnik, **14**:52b
Meteorologische Zeitschrift, **6**:97a,b
Meteorology
ancient Roman, **12**:309b–310a; **15**:518a–b
ancient Greek, **1**:205a, 256b; **2**:283a; **3**:391a; **4**:32b, 172a, 460a; **5**:345b–346a, 346b; **11**:104b; **12**:310a; **13**:93b
medieval Arabic, **9**:41a; **10**:5b; **11**:246b, 247a, 251b; **12**:447b; **15**:232a, 265b, 497a
medieval Latin, **2**:603b; **4**:92b; **5**:402a; **9**:363a,b; **11**:469a; **14**:35b, 400a
Renaissance, **1**:468a, 607b; **2**:283a, 410a–b; **4**:158b–159a; **5**:400b–401a; **9**:192a; **11**:97b; **12**:101b; **14**:273a, 275b
early seventeenth-century, **3**:3a; **5**:288b; **6**:126b; **10**:176b; **12**:163a; **13**:438a–439b; **14**:254b
late seventeenth-century, **3**:61a, 101a–b, 248a,b; **4**:506a–b; **5**:25b, 575a; **6**:71b, 487a; **7**:577b; **9**:117b–118a, 119a, 120b, 244a, 486a; **11**:41a, 507a; **13**:445a; **14**:510b
early eighteenth-century, **2**:327a, 329a; **3**:173b; **4**:23a–b, 40b; **5**:133b–135b, 366b, 428b, 434b, 520a; **7**:576a; **8**:469a; **9**:77b, 90b, 266b; **11**:65b
late eighteenth-century, **1**:544a; **2**:4b–5a, 530b; **3**:51b, 52a, 153b, 435b–436a, 538a, 540a–541a; **4**:28a–29a, 71a, 224b*; **5**:88a, 137a; **6**:234a, 512b, 585b; **7**:389a, 587b–588b, 596b, 599a; **8**:67b, 144a, 256a, 262a, 321a, 470a; **9**:140a, 472a; **10**:256a, 607b; **11**:66b–67a, 135a, 472a; **12**:119b–120a, 120b–121a, 123b, 560b; **13**:183b–184a; **14**:73a–74a, 353b, 452b–453a, 456a, 464a, 485a
early nineteenth-century, **1**:203a, 586a; **2**:356b; **3**:313a, 556b–557a, 557b; **4**:256a, 546b; **5**:78b–79a, 318a–b; **6**:325a, 551b, 552a; **8**:101a; **9**:176b; **10**:553a; **13**:311a, 541a; **14**:253b, 292b; **15**:83a
mid nineteenth-century, **1**:20b, 210a, 364a, 528a; **2**:147a, 430b, 431a, 628a–b; **3**:238b, 297a,b; **4**:174b–175a, 248a, 400b, 410a–b, 591b–592b; **5**:16b, 17a–b, 158a, 266a, 413a–b, 599b–600a; **6**:277b, 280a–b, 326a, 326b–327a, 327b, 402b, 489b; **7**:104a, 146a, 608b; **8**:278b, 487a–b, 613b; **9**:196a–197a, 279a–b, 421a; **10**:500a–b; **11**:111b, 174a, 340a–b, 390a; **12**:50b, 168a, 498b–499a; **13**:522a; **14**:329b; **15**:85a, 86b–87a, 130b
late nineteenth-century, **1**:6a, 398a; **2**:466a, 479a, 533a; **3**:12a, 592b–593a; **4**:144a, 354a–b; **5**:21b, 257b, 258b; **6**:97a–99a, 243a, 342b, 432a; **7**:160a–b, 400a–b, 461b; **9**:107b, 155a, 442b–443a, 443b, 444b, 530b–531a; **11**:181b, 320a, 498b; **12**:187a–b, 411b, 512a, 595a–b; **13**:52a–b, 232b, 413a, 482a; **14**:52a–53b, 356b–357a; **15**:158b–159a, 430b, 493b
early twentieth-century, **1**:398a, 478a; **2**:168a–b; **3**:204a, 327b*; **4**:171a–b; **5**:187b–188a, 573b; **9**:107a–108a, 189b, 404a; **11**:557b–558a; **12**:279b, 366b; **13**:166a–167a; **14**:50b, 107a, 214b–215a
mid twentieth-century, **5**:161b, 596b; **7**:436b; **8**:24b; **11**:125a, 557b–558b; **12**:201b
and agriculture, **6**:281a*
astrological, **2**:410a–b; **5**:345b–346a; **11**:415a; **12**:447b
and chemistry, **1**:300b
China, **3**:256a; **12**:376a, 380b–381a
cosmic, **4**:161b
dynamic, **2**:168b; **5**:187b, 188a; **7**:437a; **9**:443a; **12**:595a–b
and electricity, **1**:547a
experimental, **4**:410b
law of areas, **12**:595a
law of rotation, **4**:175a
Mesopotamian, **15**:677a
of ocean regions, **1**:92a; **9**:196a–197a; **12**:411b
practical applications of, **3**:435b
solar, **13**:232b
statistical, **6**:98a,b; **8**:487a; **9**:279b
synoptic, **5**:187b; **9**:196a–b
systematic observations, **3**:435b; **4**:144a, 410a; **6**:326b–327a, 552a; **7**:588a–b, 596b, 608b; **9**:392a; **13**:183b–184a; **14**:52b; **15**:381a–b
theoretical, **5**:188a; **7**:588a; **8**:1b; **9**:443a; **10**:564a; **14**:74a
tropical, **11**:557b–558a
see also Atmosphere; Climatology; Clouds; Congresses, scientific; Historiography, of meteorology; Instrumentation, meteorological; Optics, atmospheric; Physics, atmospheric; Weather
Meteoroscope, **14**:273b
Meteors, **3**:432b*; **4**:222a*; **5**:434b; **6**:553b; **8**:277a; **10**:233b; **11**:104b; **15**:640b
Andromedid, **2**:126a

Meteors (cont.)
  Bielid (Belid), **2**:126a; **10**:42a
    perturbations of, **4**:176a*
  comets and, **4**:555a; **5**:258a; **7**:386a; **10**:41b–42a; **12**:160a–161a
  height of, **1**:615b; **12**:268b
  Leonid, **1**:54a; **4**:351b; **6**:105b; **7**:76b–77a; **12**:160b; **14**:242b
    period of, **10**:41b
    perturbations of, **4**:175b
  observations of, **6**:105b; **8**:487b; **11**:472a; **12**:163a
    daylight, **6**:523a
    photographic, **4**:351b
    simultaneous, **1**:615b; **2**:421a
    systematic, **3**:19a; **10**:41b
  orbits of, **2**:236a; **5**:258a; **10**:41a–b
  paths of, **4**:169a; **11**:542a
  Perseid, **6**:523a; **12**:160b, 367a
  showers of, **1**:364a; **2**:368b, 433a, 434a; **7**:386a; **10**:41a–b; **12**:160b–161a; **13**:408a
    orbits of, **14**:242b
    periodic, **2**:126a, 421a; **10**:41b; **13**:476b
  spectra, **2**:196a–b; **6**:508a; **12**:269a
  statistical studies, **10**:41b–42a
  Taurids, **12**:161a
  theory of, **3**:18b; **9**:449a; **12**:160b–161a, 266b; **14**:242b, 400a
  velocity of, **1**:615b; **4**:351b; **13**:482a
Meter (unit of length)
  see Metric system, standard meter
Meth, Richard, **11**:87a
Methane, **2**:420b; **3**:603a
  composition of, **6**:284b
  derivatives of
    optical activity, **11**:87a
  discovery of, **14**:72a, 74b
  formula of, **10**:178a
  synthesis of, **2**:66b; **7**:21a; **10**:532a; **12**:47a
Methanol
  see Alcohol(s), methyl
Method, scientific
  ancient Greek, **1**:188a, 220a–222a, 251a–258a, 263b–265a, 277a–279b, 444b, 474a, 589a–b; **3**:515b; **4**:52b, 416a–b; **6**:411a, 424b–425a; **9**:270b–271a, 317a; **10**:344a; **13**:94b, 332a–b
  ancient Roman, **2**:232a
  Byzantine, **1**:121b
  medieval Arabic, **6**:190b–191a, 195a, 198a, 199a, 203a–b; **7**:213a,b, 217b–218a; **9**:604b
  medieval Latin, **1**:4b, 197b, 278b; **2**:604a–608a; **4**:92b–94b; **5**:551a–552a
  Renaissance, **1**:98a, 109b–110a; **3**:40b, 410b; **5**:106a–b; **8**:194b–196a, 198a–200a, 202a, 249a; **10**:8a; **11**:96b, 287b–289a; **12**:151a; **14**:5a, 7a–b, 11a, 581a–b; **15**:45a,b
  early seventeenth-century, **1**:373a, 374a–376b, 447b, 567b–568a; **4**:51b–55a; **5**:238b, 247a; **6**:448b; **7**:194a–b, 269a; **9**:98a, 318a–b; **10**:333a, 334a; **12**:102a; **14**:209b
  late seventeenth-century, **3**:248a–b; **4**:506b; **5**:416a–b; **6**:487b, 608b; **8**:127a, 153b–154a, 155b–156b, 159a; **9**:48a, 49a–b, 51b, 65a–66b, 118a–b, 120a–b; **10**:51a–52a, 55b–56a, 65a, 85a–86a, 335b; **12**:582a,b; **13**:31a, 34a, 566b
  early eighteenth-century, **2**:287a; **3**:381b, 431a–b; **5**:62a; **9**:135a, 596a; **10**:120b–121a; **13**:564b; **14**:296b
  late eighteenth-century, **6**:549b; **7**:234a–b, 367a, 598a; **8**:75b, 77a, 81a–82a, 256a; **11**:521a; **12**:308b, 309a, 549b
  early nineteenth-century, **2**:274a,b, 275b; **4**:528a; **5**:167a; **8**:134a–b; **10**:187b–188a, 189a; **14**:351b, 493a
  mid nineteenth-century, **3**:240b, 376a–378b; **6**:242a, 279b; **7**:381a, 382a–b; **8**:592a; **9**:206a, 383b–384b, 573b–574a; **11**:209a; **12**:175a; **14**:41a; **15**:167a–b, 409b–412b
  late nineteenth-century, **2**:208a, 304b; **5**:396a; **6**:224b; **9**:516a–517a; **10**:580b; **11**:210b, 573b–574a; **13**:362b–365a; **14**:434b–435a; **15**:60b
  early twentieth-century, **2**:239b, 245b, 249a–250b; **4**:100b–102a; **5**:102a,b, 145b; **6**:389b; **7**:143b, 191b, 501b; **9**:524a–525b; **10**:125b; **14**:173a–174a, 448a
  mid twentieth-century, **3**:335a; **5**:141b
  agreement and difference, **1**:101a, 375a
  multiple working hypotheses, **1**:469b–470a; **3**:191a; **5**:396a; **7**:143b
  reproducibility, **14**:448a
  see also Analysis, philosophical; Approximation; Axiomatization; Deduction; *Demonstratio per impossibile;* Empiricism; Experiment; Explanation, scientific; Generalization; Hypothesis; Induction, logical; Observation; Positivism; *Reductio ad absurdum;* Synthesis, logical
Méthode de nomenclature chimique, **2**:74a; **5**:602b; **8**:77b, 80b, 81a, 82a; **10**:446b; **12**:65a
Methodist school of medicine
  ancient Greek, **1**:315a
  ancient Roman, **3**:175a; **12**:340a, 538a–540b
  eighteenth-century, **1**:392a
Methyl radical, **5**:465a; **10**:289a
Methyl salicylate, **3**:10b
Methyl series
  discovery of, **4**:245a
Methylamines
  separation of, **4**:20b
Methylene
  derivatives of, **2**:621a; **11**:419a; **13**:1a–b
Methylene blue
  see Dyestuffs, methylene blue
Methylene radical, **8**:335a; **14**:602a
Metius, Adriaen (1571–1635), 9:335–337
Metius, Jacob (*d.* A.D. 1628), 9:335–337
Meton (*fl.* second half of fifth century B.C.), 9:337–340; 11:197b
  calendar, **3**:21b; **4**:459b–460a; **10**:179b
  observations of, **11**:189b; **15**:218b
  see also Calendar, Metonic cycle
Metric system, **7**:313b*
  and cartography, **1**:573a
  in France, **1**:206b; **2**:475a; **3**:106a; **4**:14b–16a; **5**:91b; **6**:178b; **8**:85a; **9**:156a; **15**:312b, 320a,b, 333b–336a
  in Italy, **1**:344a
  micron, **6**:137b
  standard meter, **1**:612b; **2**:135b, 299b, 424a; **3**:617b; **4**:16a; **5**:435b, 582b; **7**:160a; **8**:136b; **9**:250b–251b, 373a; **10**:483b; **13**:121a, 184a; **14**:357a, 553b–554a
  in United States, **4**:156b; **10**:42a
Metrics (prosody)
  Indian, **5**:275b
  Roman, **3**:176a
Metrodorus, **3**:275b; **4**:382b–383a
Metrology
  medieval Arabic, **6**:206a*
  Renaissance, **10**:8a
  seventeenth-century, **9**:554a–b
  eighteenth-century, **9**:472a,b; **10**:501b; **11**:472a, 522b
  nineteenth-century, **5**:582b; **6**:327a; **9**:372b–373a; **11**:377b; **13**:359b
  twentieth-century, **5**:598a
  Biblical, **13**:401b
  historical, **11**:522b
  see also Measurement
Metropolitan Museum of Art (N. Y.), **3**:611a
Metropolitan Vickers Electrical Company, **3**:329a
Mettayer, Jamet, **14**:21a, 23b
Mettayer, Pierre, **14**:23b
Mettenius, Georg Heinrich (1823–1866), 9:340; 2:425b; **6**:466a
Metzger, Hélène (1889–1944), 9:340–342; 3:243b*; **11**:388b; **15**:422b
Meudon
  observatory, **4**:68b; **7**:75b
Meusnier de la Place, Jean-Baptiste-Marie-Charles (1754–1793), 9:342–344; 1:582a; **4**:102b; **8**:78b, 79a; **9**:476a
  Meusnier's theorem, **9**:343b
Mexico
  archaeology, **2**:212b
  biology, **6**:319b
  botany, **1**:127b; **3**:177a; **9**:432b–433a; **12**:326b–327a
  exploration of, **9**:432b; **10**:285b
  geography, **6**:551a,b; **12**:431a
  geology, **8**:616a–b
  mining, **4**:345b–346a
  mollusks, **5**:478b
  National Polytechnical Institute, **11**:546a
  natural history, **1**:127a–b; **3**:180a; **6**:309b–310a
  petroleum industry, **4**:7b
  politics, **11**:464b
  public health, **8**:313a
Mexico City
  botanical garden, **3**:177a
  Colegio de Minería, **11**:464a–b
  National Museum, **6**:319b
Mey, Johannes de, **5**:439b
Meyen, Franz Julius Ferdinand (1804–1840), 9:344–345
Meyendorff, Aleksandr Kazimirovich, **7**:319b
Meyer, Adolf, **4**:161a; **5**:180a; **6**:321b; **15**:75b
Meyer, Christian Erich Hermann von (1801–1869), 9:345–346
Meyer, Ernst (Sigismund Christian) von, **1**:553a; **2**:623a
Meyer, Georg Friedrich Wilhelm, **1**:46a; **5**:546a
Meyer, Georg Hermann von, **3**:496b
Meyer, Hans Horst, **8**:451b; **10**:257b
Meyer, Johann Friedrich (1705–1765), 9:346–347; 2:177a; **8**:74a, 76b; **11**:523a; **12**:65a; **14**:224b
Meyer, Julius Lothar (1830–1895), 9:347–353; 1:297a, 528b; **2**:623a,b; **9**:290a
  atomic weights, **11**:173b
  Die modernen Theorien der Chimie
    Cannizzaro's influence on, **3**:49a–b
  periodic laws, **2**:587a; **11**:136a
Meyer, Karl Friedrich, **2**:528a
Meyer, Kirstine Bjerrum, **2**:169b; **10**:185b
Meyer, Kurt Heinrich (1883–1952), 9:354
Meyer, Ludwig, **5**:570a
Meyer, Oskar Emil, **2**:266b; **8**:509b
Meyer, Richard Emil, **1**:390b; **9**:19b; **11**:551a
Meyer, Richard Joseph, **6**:481a,b
Meyer, Stefan, **5**:395a*; **6**:355a; **9**:260a; **10**:288a
Meyer, Victor (1848–1897), 9:354–358; 3:511a; **8**:553b; **11**:298b, 299b; **14**:141b, 265a

acquaintances of, **1**:340b, 390b; **2**:589b; **14**:601b
iodine compounds, **5**:315a
physical chemistry, **15**:463a
students of, **1**:610a; **2**:59a; **5**:464a; **15**:36b
synthetic chemistry, **5**:464b; **6**:107b
Meyer, Wilhelm Franz (1856–1934), 9:358; **3**:164a
Meyerhof, Otto (1884–1951), 9:359; **5**:6a; **6**:501b
Meyerhoffer, Wilhelm, **13**:579b, 580a; **15**: 433b
Meyerson, Émile (1859–1933), 15:422–425; **7**:488b; **9**:341b; **11**:389a; **12**:609b, 610a
Meynert, Theodor Hermann, **1**:580a; **4**:160b; **5**:27a, 73b, 174b–175b, 178a, 571a; **14**:277a
Meyranx, Pierre-Stanislas, **3**:526b; **5**:357a; **8**:49a
Meyronnes, Francis of
see Francis of Meyronnes
Mézeray, François de, **12**:85a
Mézières
École Royale du Génie, **6**:1a; **9**:470a–b, 471b; **15**:342a
Miaskite, **7**:249b
Miasmata, **3**:463b; **6**:269b
Mica, **3**:293b; **4**:141a; **11**:540b
chemical composition, **9**:183b
crystallography, **5**:457b; **9**:183b; **12**:436b
deposits, **10**:167a
optical crystallography, **2**:139a; **9**:183a
polarization
of light, **2**:451b–452a
of thermal radiation, **5**:68b
Mice
see Mouse
Micelles, **3**:620b; **6**:118b; **9**:601a–b
Michael of Ephesus, **1**:270a; **4**:429a–b
Michael Psellus
see Psellus, Michael
Michael Sars (ship), **9**:590a
Michael Scot (*d. ca.* A.D. 1235), 9:361–365; **4**:605a,b; **5**:147b, 592b; **12**:62a
alchemy, **10**:310a
translations by, **1**:266a, 272b, 273b; **5**: 148a; **15**:33a
Michael, Arthur (1853–1942), 9:360; **5**:215a
Michaelis, Leonor, **1**:537b, 613b; **13**:420a
Michaux, André (1746–1802), 9:365–366; **1**:489b; **3**:172b; **14**:417b
botany, **11**:218a, 343b*
Flora boreali-americana, **4**:21b; **11**:263a
Michaux, François André, **9**:366a
North American Sylva
editions of, **10**:164a–b
Michel, Gerhardt, **11**:583a
Michelangelo, **3**:354b, 355a
Michele, Agostino
Trattato della grandezza dell'acqua e della terra, **1**:607a
Micheli, Pier Antonio (1679–1737), 9:368–369; **12**:394b; **13**:257a, 449a
Michelini, Famiano (1604–1665), 9:369–370; **2**:592a; **9**:484b, 486b
Michell, John (*ca.* 1724–1793), 9:370–371; **3**:156a; **5**:335b; **6**:335b; **11**:142b
magnetism, **1**:67b; **3**:444a
optics, **11**:143a
probability, **1**:209a; **6**:330b; **9**:219a
stratigraphy, **11**:35a
torsion balance, **3**:158a, 446a(n); **4**:379a
Treatise of Artificial Magnets, **3**:51a–b
Michel-Lévy, Auguste (1844–1911), 9:366–367; **1**:472a; **2**:89b; **3**:161a; **5**:89a–b*; **13**:283b, 284a
Michels, Antonius, **8**:481b
Michelson, Albert Abraham (1852–1931), 9:371–374; **5**:86a; **7**:107b; **8**:11b; **9**: 395a; **13**:517b
acquaintances of, **3**:369b; **8**:94b; **11**:612b; **15**:40a
astronomy, **13**:37a
early experiments, **13**:104a
ether-drag experiment, **8**:495a; **12**:69b–70a
Michelson interferometer, **2**:424a; **4**:513b; **10**:473a–b, 247b–248a
Michelson-Morley experiment, **1**:365a; **4**:320a–b; **5**:15a–b, 20b; **8**:40b, 444a, 495b–496a; **9**:214a–215b, 372b–373a, 531a; **11**:13b; **13**:471b
critics of, **11**:461a
repetitions of, **9**:373b–374a, 386b; **10**: 473b; **14**:598b–599a
optics, **5**:483a; **7**:445b
on standard meter, **1**:612b; **10**:483b
Michelson, Vladimir Aleksandrovich, **10**: 346a; **11**:11a; **14**:340a
Michigan
geography, **15**:256a–b
Geological Survey, **6**:524a
geology, **5**:488a*; **14**:439a
Michigan, University of
medical school, **10**:155a
observatory, **3**:509b; **6**:575a
Michigan Territory
settling of, **12**:204b
Michurin, Ivan Vladimirovich (1855–1935), 15:425–427; **2**:257a
Michurinism, **15**:425b, 426a
Microanalysis, chemical
see Chemistry, analytical, microchemical
Microbacillus, **9**:176a
Microbacteria, **3**:338b
Microbalance, **1**:321a; **14**:318b, 319a; **15**: 435a
Microbes, **4**:210b
in air, **1**:497a; **3**:189a, 359b; **10**:362b, 367b, 368b–372a, 383a–b; **11**:37b, 110a–b; **12**:555a
attenuation of, **11**:429a
biochemistry
enzymes, **4**:211b
vitamins, **12**:207b–208a
heat resistance, **10**:383a
alkalinity, effect of, **10**:371a, 382b
naming of, **10**:389b
prediction of, **5**:107a–b
respiration in, **10**:155b
self-toxicity, **10**:407a–b
and soil fertility, **4**:211b
specificity of, **10**:390b, 399a; **11**:37b
variation, **14**:37a; **15**:14a
in water, **10**:382b
see also Bacteria, Bacilli, Infusoria, etc.
Microbiology
seventeenth-century, **5**:426a; **8**:127a–b
early eighteenth-century, **1**:411a; **3**:332b; **15**:52b
late eighteenth-century, **3**:426a–b; **5**:56b, 424b–425a; **7**:13b; **9**:575a–b; **12**: 554b–555a, 556a–b, 557a–558a, 563b
early nineteenth-century, **1**:135b–136b, 493a–b; **4**:146b–147a
mid nineteenth-century, **3**:336b–340a, 424a–425a, 587b–588b; **4**:289a–291b; **5**:531a; **10**:361b–376b; **11**: 110a–b; **12**:233a–b; **14**:350a–b, 534a
late nineteenth-century, **1**:397b; **2**:437a–438a, 527b–528b; **3**:220a, 321b–322a; **4**:210b–211a, 401b–402a; **7**:34b–35b; **9**:328a, 494b; **10**:155a–b, 376b–408a; **11**:428b–429a; **12**:480a–481b; **14**: 36b–37b; **15**:13b–15a
early twentieth-century, **1**:497a; **2**:301a, 528b–529b; **3**:359b; **4**:132b, 207a; **6**:224a; **7**:405b–407a; **9**:176a, 328b; **11**:174b–175b, 442a–b; **12**:481b–483b; **13**:519b–520b; **14**:36b–37b; **15**:97b, 492b
mid twentieth-century, **12**:207a–208a, 492a–493a
applied, **7**:406a–b; **8**:450a; **14**:247b
chemical, **7**:406a–b; **10**:22a
cultural methods, **5**:219a–b, 270b; **6**:73a; **7**:421b, 423a–b; **8**:407b; **9**:176a; **10**: 362b
anaerobic, **2**:584a; **10**:155a–b
capillary culture chambers, **2**:437a–b; **12**:87b
chemostat, **13**:227b
enrichment, **15**:14a
hanging drop, **2**:437b; **6**:131b–132a; **14**:37a
nutrient media, **2**:437b, 438a; **7**:423a; **8**:408a; **10**:363a–b, 365b; **11**:310b–311a, 428b; **14**:37a; **15**:14a
single-spore transfers, **2**:437a
solid media, **7**:391b; **8**:449a
sterilization of media, **3**:188b–189a
submerged, **7**:406b
successive dilution, **10**:386b, 390b–391a, 392a
in vitro, **8**:448b; **12**:208a
in vivo, **5**:40b
ecological, **14**:37b
and ontology, **8**:157b
techniques, **1**:496b; **7**:13b
dissection, **6**:219b
fixation, **6**:224a; **11**:216a; **12**:232b
injection, **2**:375b–376a; **5**:426a
mounting, **3**:425a; **7**:422a; **12**:545a
paraffin sectioning, **2**:141b; **3**:30a
sterilization, **10**:383b
see also Staining, microbiological
see also Apparatus, microbiological; Bacteriology; Cytology; Histology; Microscopy
Microcephaly, **5**:506b*
Microcinematography, **12**:423a
Microclimatograph, **7**:503a
Micrococcin, **5**:43a
Micrococcus, **4**:311a; **10**:375b, 390a
M. lysodeikticus, **5**:29b
M. melitensis, **2**:528a
Microcosm and macrocosm
ancient Greek discussion, **2**:223b; **6**: 421a
medieval Arabic discussion, **7**:64b; **15**: 497b
medieval Latin discussion, **5**:549b; **6**: 396b; **7**:27b
Renaissance discussion, **1**:80a; **3**:471b; **4**:209b, 387b; **5**:120a; **8**:201a, 241a; **10**:308b; **12**:335a; **14**:225b, 226b
early seventeenth-century discussion, **1**: 159b; **3**:361b, 596b; **5**:47a–48a; **6**: 257a, 397b; **12**:333a; **13**:559b; **14**: 286a; **15**:69b
late seventeenth-century discussion, **5**: 421b, 422a; **14**:373a
nineteenth-century discussion, **8**:514b
Microdissection, **13**:173b
Microfilariae, **8**:296b
periodicity, **9**:82a
Microgeology, **4**:291a
Microglia, **11**:465b
Microincineration, **13**:556b
Micromachines, **3**:239a
Micromanipulation, **3**:184b; **5**:56b; **13**:268b, 269a

seventeenth-century observation of, **5**: 241a
eighteenth-century observation of, **6**: 332a–b
nineteenth-century observation of, **6**: 282a; **12**:269b
twentieth-century observation of, **1**:398a
brightness of, **5**:363b; **12**:265b
center of, **4**:273a; **9**:2b
Mesopotamian observations of, **15**:640b
photography of, **1**:465b, 466b; **4**:273a; **10**:290a
size of, **6**:529b; **12**:346b, 347a–348a
structure of, **4**:272b–273a, 279a; **6**:332b, 531b; **11**:470a; **12**:346b–347b, 448b; **13**:110b
theories of, **5**:400b; **6**:331b–332b, 416b; **7**:231b, 598b; **12**:406a; **13**:120b; **14**: 519a
see also Galaxy, the
Mill, James, **6**:140a; **9**:383a
Mill, John Stuart (1806–1873), 9:383–386; **2**:132a; **3**:88b; **6**:140b*
acquaintances of, **1**:404a; **2**:18a; **4**:504a; **12**:525a; **14**:394a
and Comte, **3**:375a, 376b, 379a
critics of, **1**:178a; **3**:32a; **4**:616b; **6**:563b; **7**:105a, 106a; **12**:571a
method of, **1**:375a, 376b
philosophy, **7**:234a; **14**:293b
System of Logic, **3**:377a; **4**:100b; **14**:527a
utilitarianism, **6**:140a
Miller, Dayton Clarence (1866–1941), 9: 386–387; **9**:373b–374a, 531a
Miller, Frederick Robert, **1**:440b, 441b
Miller, George
Logarithmicall Arithmeticke, **14**:51b
Miller, George Abram (1863–1951), 9:387–388; **2**:197b; **3**:168b; **5**:589a*; **14**: 559b*
Miller, Hugh (1802–1856), 9:388–390; **5**: 334a
Old Red Sandstone, **5**:333b
Miller, Philip (1691–1771), 9:390–391; **1**: 486b; **6**:46a(n), 539a; **12**:459a
Gardener's Dictionary, **1**:88a; **9**:254a; **13**:497b*
Miller, William Allen (1817–1870), 9:391–392; **1**:161a; **6**:541a,b
Miller, William Hallowes (1801–1880), 9: 392–393; **2**:441a; **4**:66a; **5**:212a; **6**: 289a; **8**:288b
Millerian indices, **9**:393a; **14**:241a
Miller, William Lash (1866–1940), 9:393–395; **15**:467a(n)
Millet (grain), **15**:648b
Milliet de Chales
see Dechales, Claude François Milliet
Millikan, Robert Andrews (1868–1953), 9: 395–400; **3**:369b; **4**:193a*; **7**:107b, 247b; **10**:156b, 157a; **11**:265b
on contemporary scientists, **9**:374a*; **12**:35b*
on cosmic rays, **2**:338a; **6**:355b; **8**:618a
droplet technique, **9**:261b; **11**:348a
electronic charge, **3**:370a; **13**:447a(n)
students of, **1**:292b; **6**:528b
Millington, Thomas (1628–1704), 9:400–401; **7**:275b, 276a; **14**:405a
anatomy, **14**:406a
Million, **3**:273a; **5**:409a
Millon, Auguste-Nicolas-Eugène (1812–1867), 9:401–402
Millon's reaction, **8**:312a
Mills, Darius Ogden, **3**:35a,b, 36a
Mills, Edmund James, **5**:412a
Mills, William Hobson (1873–1959), 9:402–404; **11**:87b
Mills (tech.), **8**:207b, 211a–b, 213a, 214b; **13**:614a
Milne, Edward Arthur (1896–1950), 9:404–406; **3**:510b*; **4**:278b; **5**:103b*; **7**: 84a,b; **10**:32a*; **12**:71a; **13**:119b
astronomy, **6**:531b
Milne's equation, **9**:404b
Milne, John (1850–1913), 9:406–407; **7**: 413b; **10**:203a,b, 210b; **13**:500b
seismograph, **14**:327a
Milne-Edwards, Henri (1800–1885), 9:407–409; **4**:68a, 285a, 286a*; **5**:360a*, 566a; **7**:540b(n)
acquaintances of, **1**:602b; **11**:234a
anatomy, **2**:443a
critics of, **12**:316a
embryology, **1**:418a; **8**:254b
influence of, **12**:551a, 570b
Introduction to General Zoology, **2**:60b
marine zoology, **1**:328b–329a
students of, **7**:545b, 546a; **12**:469a
translations of works, **7**:416a*
Milner, Samuel Roslington, **10**:156b; **13**: 156a
Milnor, John Willard, **6**:572b; **14**:311a
Milton, John, **2**:377a; **5**:49a*, 247b, 607b; **6**:141a; **10**:200a,b; **14**:51b
Mimicry, **5**:385b
Batesian, **1**:501b, 503a–504a; **9**:560b; **14**:136a
by insects, **1**:502b–504a; **3**:572b; **9**:560b–561a; **11**:212a; **14**:136a
Müllerian, **1**:502b–504a; **9**:559b, 560b–561a; **14**:136a–b
and natural selection, **1**:503b; **9**:560b–561a; **12**:417a–b; **14**:136a–b
Mimosa pudica, **4**:216a; **7**:538a
movements of, **2**:61b; **10**:575b; **11**:315b
stimulus transmission, **5**:624a
Mīnarāja, **12**:576b
Mind
and body
see Dualism, philosophical
concept of, **1**:404a; **8**:515a
and cultural development, **2**:207b, 209a–210a
evolution of, **9**:513a
location of, **10**:21a–b; **15**:737a, 738a
and matter, **4**:557a, 558a; **6**:596a; **12**:601a
origins of, **3**:332a; **6**:321a
see also Intellect, Intelligence, Psychology, Reasoning process
Mind
founding of, **1**:404a
Minding, Ernst Ferdinand Adolf (or Ferdinand Gotlibovich) (1806–1885), 9: 409–410; **2**:215b; **10**:544a,b
Minding's theorem, **9**:409b
Mine
see Mines
Mineral waters
Renaissance discussion, **1**:526a
seventeenth-century discussion, **6**:257a
eighteenth-century discussion, **2**:3a, 301a, 523b; **5**:428a; **11**:565a*
nineteenth-century discussion, **9**:593a; **10**:582b
analysis of
Renaissance, **4**:387a; **13**:397a
seventeenth-century, **2**:353a, 381b; **6**: 380b; **8**:311a, 312a; **11**:40b
eighteenth-century, **1**:529a–b; **2**:5b, 291b–292a; **3**:5a, 561a*; **4**:156b; **5**: 90b; **6**:395b, 460a; **7**:390a*, 390b; **8**:68b, 70a–b, 177a, 618b; **9**:478a; **10**:446b*; **11**:48b, 387b, 616b; **13**: 602b, 603b; **14**:144a, 464a
nineteenth-century, **2**:91a, 589a–b; **5**: 165a, 203a, 315b; **6**:293b; **9**:131a, 355a, 401b; **11**:51a; **14**:448a
artificial
eighteenth-century, **2**:5b
nineteenth-century, **2**:91a
effervescent
artificial, **13**:602b
gases in, **2**:523b–524a; **6**:42a; **13**:602b; **14**:448a
medical use of, **5**:366b; **6**:383b; **8**:416b*; **10**:307a
assessment of, **10**:204b
springs, **5**:380b; **6**:387b; **10**:167a; **11**: 387a–b
Mineralization, **3**:586b
Mineralogical Abstracts, **12**:573a
Mineralogical Magazine, **12**:573a
Mineralogical Society (St. Petersburg), **12**: 513b
Mineralogische Mitteilungen
editors of, **1**:552a
founding of, **13**:476b
Mineralogy
ancient Greek, **2**:256b; **13**:330b–331b
medieval Arabic, **1**:533a; **2**:153a, 155a; **7**:343b; **11**:232a; **13**:407a–b, 511a–b; **15**:232a
medieval Latin, **1**:101b; **6**:123b; **14**:35b
Renaissance, **1**:78a, 611b; **2**:293a–b; **9**: 309a
seventeenth-century, **6**:380b; **9**:508a*; **14**:210a
early eighteenth-century, **1**:305b, 393a; **2**:490a, 579a; **6**:259b; **8**:468a, 469a; **14**:144a–b, 501a, 502a–b
late eighteenth-century, **1**:359b; **2**:5a–6a, 8a, 291b, 292a, 315a–b, 552b–553a; **3**:41b, 61b, 473a–474a, 560b; **4**:149b–153a, 343b–344a, 348a, 549a*; **5**:76b, 215b, 216a, 223a, 568a, 578a–b, 602a; **6**:385a,b, 401a, 468b; **7**:202a, 389a, 390b, 604a; **8**:68a, 71a, 147a–b; **9**:5b–6a, 478b, 559a–b; **11**:166b, 272a, 304b, 387a–b, 464a–b, 520b–523a; **12**:329a–b; **13**:565b–566a; **14**: 60b, 132a, 224b, 256b–263a, 463b–464a; **15**:113a–b
early nineteenth-century, **1**:359b, 557a; **2**:72a, 91a, 94a, 106a–b, 355a–b, 420a, 440a–441a, 481a, 482a, 493b, 494a, 566b; **3**:313a–b, 232b, 291a–b, 400a; **4**:217b; **5**:155b–156a, 202b–203a, 430a,b; **6**:178a–182b, 354a, 358b, 436b–437a, 456a, 552a; **7**:69b, 254a, 365b; **8**:103a, 245b–246a, 288a–289a, 594a; **9**:423a–424b; **10**: 8b–9a, 27a, 585b–586a; **11**:464b–465a; **12**:204b, 329b, 433a, 513b–514a, 552b; **13**:466b; **14**:88a, 239b–240a, 293a, 326b
mid nineteenth-century, **1**:20a, 572a–b; **2**:199a–b, 431b, 440a–441a; **3**:397b–398a, 411b–412a, 549b; **4**:65b–66b, 182a, 309a–b; **5**:89a, 610b; **6**:18b–19b, 142b–143a; **7**:45a; **8**:103b; **9**: 110a, 392b–393a, 425b, 541b, 620a–b; **10**:148a, 258b, 265a–b; **11**:385b–386a, 539a–540a; **12**:304a–306b, 544a–b; **13**:373b, 467a, 475b–476b; **14**:108b, 625a–b; **15**:526b
late nineteenth-century, **1**:461a, 528b–529a, 551b; **2**:289a, 486a–b; **3**:237a, 292a–293b, 321a; **4**:65b–66b, 140a–141a, 361b; **5**:161b–162a, 185b–186a, 211a–213b, 455a, 547b–548a, 556a–557b, 562b–563a; **6**:177b–178a; **8**: 31a, 463a; **9**:58b–59a, 367a–b, 379b,

- Mineralogy (cont.)
  - 592a–593b; 11:372a, 547a–b; 12:520a; 13:476b; 14:58a, 525a–b; 15:427a
  - early twentieth-century, 2:369b–370b; 3:204a; 4:408b–409b, 563b, 597a–600a; 5:457a–b, 497a–b; 7:161a–b, 548b–549a; 8:42a–b; 9:315b; 10:124a–127a; 12:194b, 572b–573a; 13:246b, 616b–618a; 14:59b, 183a–b, 440a–b, 516a*, 588a–b
  - mid twentieth-century, 12:194b–195a
  - chemical
    - eighteenth-century, 2:5a; 3:5b, 473a–474a; 8:69a–b; 12:63b–64b, 494b
    - nineteenth-century, 4:217b; 5:349b–350a; 6:178b, 181b, 436b; 11:270b
    - twentieth-century, 3:400b–401a; 8:42a
  - and chemistry, 2:552b; 5:203a; 14:257b–258a
  - crystallographic, 4:217b; 6:178b
  - experimental, 4:141a–142a
  - geometric, 5:211a, 212a
  - handbooks, 5:162a; 14:144b; 15:526b
  - and industry, 6:164b
  - methodology of, 5:211b–212a; 8:42a
  - optical, 2:452b; 5:212b; 7:112b; 8:42a, 103b; 9:393a; 10:110a; 11:266b–267a; 14:440a–b, 443a
  - philosophy of, 7:549a; 10:8b–9a; 13:617b
  - systematic, 2:440a–b; 9:447b–448b
  - theories of
    - seventeenth-century, 1:549b
    - eighteenth-century, 4:152b; 11:167b
    - nineteenth-century, 6:564b–565a; 11:169a, 170a
    - early twentieth-century, 13:617b–619a
  - *see also* Apparatus, mineralogical; Chemistry, mineral; Crystallography; Petrography; Petrology
- Minerals, 5:205a
  - alteration products, 3:293b
  - ammonia analogues, 3:293b
  - analysis of
    - early eighteenth-century, 3:473a–474a; 6:259b; 11:109a
    - late eighteenth-century, 1:529b; 2:5a–b, 572a; 5:602a; 7:394b; 8:147a,b; 14:214a*, 257b, 333a
    - early nineteenth-century, 2:92b, 94a, 565a; 3:291b, 412a; 5:430a; 11:169b, 437b, 464b; 13:311a, 597a; 14:476a; 15:7b
    - mid nineteenth-century, 2:199b; 5:203a; 6:564a; 9:18b; 10:265a; 11:270b, 541a; 12:544b; 13:476a; 14:477b
    - late nineteenth-century,2:199b, 585b; 3:292a,b, 293a, 321a; 5:162a, 455b; 11:318b; 12:520a; 14:447a
    - early twentieth century, 10:136b; 14:588a
    - in biological materials, 2:627b
    - blowpipe method, 2:5a–b; 3:474a; 5:223a
    - immersion method, 8:41b
    - wet methods, 2:5b; 8:69a–b
  - association, 7:249b; 12:329b, 514a
    - age relations, 2:440b
  - birefringence, 9:367a
  - classification of
    - Renaissance, 2:293a
    - eighteenth-century, 2:8a; 3:474a, 560b; 4:152a–b; 6:178a–b, 401a; 8:379a; 11:521b–522a; 13:425b; 14:132a, 502a–b
    - nineteenth-century, 2:94a–b, 494a; 3:292a*; 5:216a*, 431b; 6:18b; 7:70b*; 11:539b; 13:476a; 14:293a; 15:115a
    - chemical, 2:490a; 3:232b; 6:181b–182a; 7:389a, 604a; 12:329a, 513b; 13:373b
    - Linnaean system, 1:359b; 11:521a
    - "natural" system, 2:440a–b; 6:565a; 10:8b–9a
    - physical, 3:232b; 6:181b–182b; 9:448a–b; 12:329a–b; 13:565b
    - Wernerian system, 1:359b; 2:552b; 14:257b–258b
  - collecting of
    - late seventeenth-century, 11:40b; 13:443b
    - early eighteenth-century, 12:459b; 14:144a, 502a
    - late eighteenth-century, 3:291a; 4:150a; 7:389a, 391a; 9:54b; 11:520b; 12:63b, 65b, 66a, 329b; 14:132a, 258b–259a
    - early nineteenth-century, 2:552b; 4:10b, 152b; 6:18b; 9:152b, 447b, 448a,b; 12:433a; 15:152a
    - mid nineteenth-century, 11:501b, 502a
    - late nineteenth-century, 5:162a, 556a; 7:548b
    - early twentieth-century, 4:597a
  - crystal forms, 2:293a; 3:596b; 5:211a; 10:281a
    - classification of, 2:293a; 9:448a–b
    - dihedral angles, specificity of, 3:61b
  - crystallization of, 6:177b–178a; 14:58b
  - dating of, 12:29a
  - dehydration in, 14:588a
  - deposits of, 3:609b; 4:598a; 5:223a, 578a–b; 7:530a; 8:147a; 9:593b; 10:167a; 13:34a, 409b, 618a; 14:60b; 15:152b
  - descriptions of, 5:162a; 6:401a; 11:520b
  - discovery of, 5:162a, 223a, 350a; 8:288b–289a; 14:259a
  - equivalent weights, 6:565a
  - genesis of, 2:579a; 3:434a; 4:309a; 6:565a; 7:586a, 589b; 8:42a; 10:9a, 265b; 12:514a
    - Becke volume rule, 1:551b
  - geological distribution of, 1:572a
  - heat, effects of, 3:560b; 4:141b–142a
  - industrial processing of, 2:292a
  - medical use of, 1:77b; 5:203a
  - in metabolism, 2:585a–586a; 4:357b; 6:136a
  - models of, 3:61b
  - molecular weights, 6:564b
  - naming of, 2:440b; 5:602a; 12:572b, 573a; 14:259a
  - orthoclase, 4:66b
  - paragenesis of, 2:440b–441a; 5:556b; 11:465a, 521a; 12:514a; 13:476a; 14:263a
    - laws of, 13:617b–618a
  - phase rule, 5:457a
  - physical properties, 2:293a, 552b
    - and chemical composition, 12:514a
    - coloration, 4:141b–142a
    - hardness, 7:62b; 9:448a; 14:132b
    - melting points, 4:141a
    - specific gravity, 5:455a
    - specific heats, 7:161a; 10:27a
    - thermoelectricity, 1:66b–67a, 557a
    - viscosity series, 4:141a
    - volatility, 9:106b
  - in plant nutrition, 2:257a; 5:523b; 8:346b–347b; 9:461a; 11:310b–311a, 416a; 12:59a, 123b
  - and plants, similarities between, 6:259b
  - pleochroic halos, 7:161a
  - prospecting for, 9:180b; 10:158b
    - seismic, 5:462a–b
  - radiation, effects of, 4:141b–142a
  - radioactive, 12:505b
  - recrystallization, 4:141a
  - separation methods, 7:21b
  - structure of, 1:529a; 4:66a; 5:203a, 457b; 10:110a; 13:618a
    - allotropy, 8:119a
    - amorphous, 2:440b; 5:124b
    - atomic, 5:212a
    - and chemical composition, 14:257b
    - fluid inclusions, 14:625a
    - origin of, 11:521a, 522a
    - polymorph transformations, 4:563b
    - pseudomorphs, 3:293b; 6:19a
  - substitution in, 14:588a
  - synantectic, 12:274a
  - synthesis of, 3:293a,b; 4:77b, 141a; 5:185b–186a, 497a; 9:425b; 12:305b–306a; 14:440b
  - *see also* Crystallography; Geology; specific minerals
- Mines
  - air, motion of, 8:469a
  - atmosphere in
    - analysis of, 12:478b
  - damps from, 2:523b; 6:236b; 9:508a*
  - development of, 3:358a
  - draining of, 9:609b; 12:244b
  - explosions in, 3:603a; 8:117b–118a; 9:59a–b
  - flooding of, 10:128b
  - mapping of, 6:370a; 13:409b
  - "miners' mine" of Rancié, 1:327b
  - prehistoric, 2:346b
  - temperature constant of, 1:327b
  - ventilation of, 3:358a; 7:78b*; 11:305b
- Mines, magnetic
  - defense against, 7:527a–b
- **Mineur, Henri (1899–1954), 9:410–411**
- Ming dynasty, 14:159b, 163a–165a
- Minima
  - method of
    - *see* Maxima and minima, method of
- *Minima naturalia,* 1:101b, 199b; 5:402b; 8:268a; 10:123a
- Minimax principle, 2:145b, 304a–b; 14:92b
- Minims (religious order), 1:547a; 4:602b; 9:25a, 316a–b; 10:103a; 11:47a
- Mining
  - medieval, 2:155a; 10:555b
  - Renaissance, 1:77b–78a; 4:393a–b; 11:338a, 457a, 607b–608a; 13:397a
  - seventeenth-century, 1:448a–b, 549a, 571b; 2:592a; 5:415a; 8:150a, 464b
  - early eighteenth-century, 2:421b, 523b; 3:346b, 473a; 5:428b; 6:236b; 7:200b; 12:116a; 13:68a, 87a, 409b
  - late eighteenth-century, 1:233b–234a, 554a,b; 2:292a, 315a–b, 439b; 3:41b; 4:346b; 6:164b, 369b–370a, 549b; 7:78a, 183a, 201b–202a, 389a; 8:146b–147b, 470b; 9:478b; 11:304b; 12:116b, 487a; 13:455a–b; 14:60b, 263a
  - early nineteenth-century, 1:327a,b; 2:72a, 480b; 3:210a, 358a, 411b–412b; 4:218a, 345b–346a, 348a–349a; 5:101a, 155a–156a, 537a; 6:213b; 7:254b; 12:117a, 204a,b; 14:88a; 15:502b
  - mid nineteenth-century, 1:95b; 2:166a, 350b; 3:70a–b, 358a, 434a, 586b, 623a–624a; 4:10b, 365b; 5:441b; 6:142b–143a, 238a, 576a; 8:54b; 11:399a; 12:433b; 14:315b; 15:152b
  - late nineteenth-century, 1:552b, 573a; 3:462b; 5:161b, 211a, 213a; 7:253a, 370a; 8:578a; 9:291b; 10:167a; 14:59a, 441b
  - twentieth-century, 1:472b; 4:598b; 6:320b; 8:99b; 10:128b; 13:525b–526a
  - economics of, 8:464b
  - hazards of, 6:24a, 265a, 576a
  - hydraulic, 5:395b

legislation, **2**:292a; **3**:624a; **4**:346a; **5**:93a; **7**:254b; **14**:442a
surveying, **7**:265a; **14**:232b
see also Engineering, mining; specific localities, materials, metals, minerals
Minium
see Red lead
Minkh, Grigorii Nikolaevich, **5**:270b
Minkowski, Hermann (1864–1909), 9:411–414; 3:228a; **4**:2b; **6**:26a, 309a, 388b, 473a, 571a; **7**:26b, 616a
acquaintances of, **6**:343a, 394b; **12**:249a, 526a; **14**:94b, 202a, 222a; **15**:40a
number theory, **6**:390a; **11**:55a; **12**:469b
on relativity, **4**:323b, 327a, 331b; **8**:431b
students of, **3**:62b; **4**:100b; **7**:71b; **11**: 480a(n); **15**:39b
theorem of
proof of, **6**:572a
topology, **6**:394a; **14**:283a
Minkowski, Oscar, **1**:441a; **8**:452a; **9**:411b; **12**:355b
Minkowski, Rudolf, **1**:352b
Minnaert, Marcel Gilles Jozef (1893–1970), 9:414–416; 10:290a
Minnesota
Geological Survey, **14**:441a
geology, **12**:229b; **13**:531b; **14**:439a, 442b–443a
Minnesota Academy of Sciences, **14**:443a
Minorites, **3**:421a
Minot, Charles Sedgwick (1852–1914), 9: 416; 2:367a; **9**:57a, 58a
cytology, **14**:235b
Minot, George Richards (1885–1950), 9: 416–417; 3:335b
Mints (coinage)
medieval Arabic, **6**:79b
Renaissance, **2**:142b, 143a; **4**:393a–b; **8**: 214b; **11**:338a–b
seventeenth-century, **6**:68b; **9**:485b
eighteenth-century, **2**:315a,b; **3**:5a, 41b, 383b, 560b; **6**:370a; **10**:83b; **11**:472a; **13**:411b
nineteenth-century, **2**:420a; **3**:10b, 617b; **5**:492b, 603b; **6**:327a–b; **7**:103b; **10**: 557a–b; **11**:485b
Miocene epoch
naming of, **8**:569b
paleobotany, **12**:104b
paleontology, **1**:491b; **6**:221b, 492a, 595a; **8**:105a; **9**:346a; **14**:313a
stratigraphy, **2**:497b; **3**:536b; **4**:68a; **11**: 269b, 270a
Miolati, Arturo, **11**:551b; **14**:265a, 268b–269b
Miquel, Auguste, **14**:141a
Miquel, Friedrich Anton Wilhelm (1811–1871), 9:417; 9:558b*; **10**:253b
Mira Ceti (star), **7**:194a; **8**:539b
parallax, **7**:472b
periodicity of, **2**:349a; **7**:373b; **9**:486a
spectrum, **12**:368a
Mirabeau, Honoré Gabriel Riqueti, **1**:323a; **3**:1a, 7a; **7**:568b
Mirabilis, **7**:441a; **8**:178a; **9**:281b
Miracles
Renaissance discussion, **11**:71b–72b
critical analysis of, **6**:559a–b; **11**:116a; **14**:287b–288a
Mirages, **2**:135a; **9**:473b, 477a; **14**:216a, 491a
Mirbel, Charles François Brisseau de (1776–1854), 9:418–419; 5:539a; **6**:493b; **10**:13a
anatomical theory, **2**:491b; **12**:241b
botany, **6**:465a; **9**:455b
Histoire naturelle des végétaux, **7**:586a
works translated, **10**:13b
Mīrīm Chelebī, Maḥmūd ibn Muḥammad, **7**:259a; **11**:227b
Miriquidite, **5**:162a
Mirrors
aluminized, **4**:253b
burning, **1**:170a, 186b, 187a, 189b, 191b, 214a; **2**:474b; **3**:153a; **4**:105a–b; **5**: 526a; **6**:195a, 477b; **7**:376a; **9**:610a; **10**:533b, 534a; **12**:381b; **13**:480a; **15**:116a–b, 118a, 264b–265a
casting of, **10**:328b
cellular, **11**:470a
circular, **13**:480a
concave, **8**:214b; **12**:381b
focal point, **10**:475b
conical, **5**:525b
convex, **12**:381b
cylindrical, **6**:201a,b
flat, **3**:366a
grinding of, **10**:329a
illusions, creation of, **6**:313a
magic, **12**:381b–382a, 383a
making of, **6**:313a
metal, **2**:424a; **4**:179a; **5**:419b, 422b; **9**: 46b; **12**:413a,b
parabolic, **4**:105b; **11**:98a; **14**:274b, 458a; **15**:118a
construction of, **3**:366a; **6**:5b; **13**:480a; **15**:116a–b
polishing of, **8**:46a, 212b
silvering of, **2**:424a; **3**:366b*; **5**:87a; **13**: 22b
spherical, **2**:328b; **6**:125b, 201a; **15**:116a
concave, **4**:105b; **6**:200a, 201a; **14**:460b
tasselate, **6**:508a
testing of, **5**:87a; **9**:46a–b
see also Reflectors
Miscarriage
allergic theory of, **6**:433a
Miscegenation, **6**:563b
Miscellanea curiosa mathematica, **13**:68b
Misener, Austin Donald, **8**:478a, 480b
Mises, Richard von (1883–1953), 9:419–420; 5:122b; **11**:384a, 615a*; **12**:196b*, 247a*; **14**:590b; **15**:451a
elasticity, **6**:534b
probability, **13**:612a; **15**:32b
Mīshā
see Māshāʾallāh
Mishnah, **9**:28a
Mishnat ha-Middot, **1**:22b, 31a–b; **7**:360a
Missiles
chemical, **5**:421a
spherical
range of, **3**:224b
see also Ballistics
"Missing links"
see Paleontology, "missing links"
Mississippi (river), **8**:572a
exploration of, **12**:204b
geology, **14**:443a
source of, **12**:204a
Mississippi (state)
surveying of, **7**:89b
Mississippi Valley, **8**:266b
geology, **10**:164a; **13**:519a, 532a
Mississippian period
paleontology, **6**:594b; **9**:255b
Missouri
Geological Survey, **6**:57b
geology, **4**:563a
lead mines, **12**:204a,b
Missouri Botanical Garden (St. Louis), **13**:456b; **15**:130a, 224a
Mistletoe, **13**:456b
Mitchel, Ormsby MacKnight, **6**:522b
Mitchell, Andrew, **13**:224a,b
Mitchell, Elisha (1793–1857), 9:420–421
Mitchell, George, **9**:447b
Mitchell, John, **1**:487a; **3**:350b
Mitchell, John Kearsley, **4**:182a, 419b, 620a; **9**:422a
Mitchell, Maria (1818–1889), 9:421–422
Mitchell, Samuel Alfred, **12**:72b
Mitchell, Silas Weir (1829–1914), 9:422–423; 2:27a, 367b; **4**:619b; **7**:48b*; **10**:141b; **14**:534a
on contemporary scientists, **15**:107b, 109a, 110a
Mitchell, Walter Mann, **6**:29b
Mitchell, William, **9**:421b
Mitchell, Mount, **9**:421a
Mitchill, Samuel Latham, **1**:330a; **3**:600a; **8**:613a, 616a; **11**:263a; **14**:132b
Mites
and disease, **5**:451b
microanatomy, **8**:128a
reproduction, **8**:128a
in scabies, **2**:291a–b; **5**:570b; **8**:125b
Mithridate, **10**:281a
Mitochondria, **5**:584b–585a; **14**:175b, 433a
microchemical analysis, **1**:613b
in vitro studies, **8**:282b–283a
see also Chondriosomes
Mitosis
nineteenth-century study, **1**:601b; **2**:626a; **4**:131a; **5**:35a–36a, 51a–b, 581b–582a; **12**:193b; **14**:235a, 429a
twentieth-century study, **3**:390a; **5**:584b, 594a–b, 624a
anaphase, **5**:35a
bipolar, **1**:498b, 499a
karyogamic reduction, **1**:601b
monopolar, **1**:498b, 499a
radiation, effects of, **5**:594b
reduction division, **4**:131a
see also Cell(s), division of; Cytology
Mitouard, Pierre, **3**:199a; **8**:74b
Mitscherlich, Alexander
Mitscherlich process, **9**:426a
Mitscherlich, Eilhard (1794–1863), 9:423–426; 2:587a; **5**:373a; **6**:461b; **11**:360b*
acquaintances of, **2**:556b, 586b; **4**:289b; **5**:197b; **6**:553a; **11**:540b; **14**:629a
and Berzelius, **2**:91b, 93b, 95b; **9**:423b–425b
chemical theories, **14**:395a
crystallography, **6**:18b; **11**:539a; **14**:241a, 490b
isomorphism, **2**:106b; **6**:18b, 182a; **8**: 56b–57a; **11**:539a,b
editions of works, **11**:560b
geology, **7**:70b
organic chemistry, **10**:357a–b
students of, **3**:336b; **6**:241b; **7**:505b, 619b; **9**:18b, 431b; **11**:270b; **14**:329b
Mittag-Leffler, Magnus Gustaf (Gösta) (1846–1927), 9:426–427; 3:57a; **5**: 150b; **7**:102a, 480a*; **14**:224a*
acquaintances of, **3**:53b; **7**:478a; **11**:611b; **14**:222a
analysis, **4**:103a; **8**:367a
theorem of, **5**:201b
Mittasch, Alwin (1869–1953), 9:427–428
Mitteilungen zur Geschichte der Medizin und der Naturwissenschaften, **13**:142a
Mitteleuropäische Gradmessung, **6**:239a, 240b
Mittelgebirge, **11**:387b
Mittheilungen aus dem Kaiserlichen Gesundheitsamt, **10**:397b
founding of, **7**:423a
Mivart, St. George Jackson (1827–1900), 9: 428–429
evolution, **1**:506a; **15**:507b
Mixture
concept of, **6**:427b
Mīzān (balance), **7**:41b

Mizar (star)
period of, **9**:194b
al-Mizzī, **12**:357b, 358a, 361a; **15**:259b
M.K.S. system, **5**:407b; **6**:212a; **7**:288b
Mneme, **12**:298a
Mnemonics
ancient Greek, **6**:405b
medieval, **1**:565a
Renaissance, **2**:539a; **3**:65a; **11**:97a
seventeenth-century, **1**:126b; **5**:49a*; **14**:380a(n)
nineteenth-century, **7**:384a
magical, **2**:540b–542a, 543b; **15**:642a
see also Memory
Mo Jo, **3**:266b
Moa, **5**:611a; **10**:262a
Moberg, Karl Adolf, **12**:272a
Möbius, August Ferdinand (1790–1868), 9:429–431; **2**:470a; **15**:193a
acquaintances of, **1**:295b; **15**:194a
barycentric calculus, **1**:590b, 591a; **5**:309a; **6**:95b; **15**:196a, 197a
Der barycentrische Calcul: Ein neues Hülfsmittel zur analytischen Behandlung der Geometrie, **3**:213b; **5**:368a
critics of, **12**:284b
geometry, **2**:192a; **3**:166b; **4**:551a, 602a; **7**:397b; **11**:45a, 46a, 80a; **13**:5b, 17b; **15**:193a
influence of, **10**:24b, 486a
influences on, **1**:239b; **3**:466b
Möbius function, **9**:430b
Möbius groups, **8**:457b
Möbius net, **9**:430b
Möbius strip, **9**:431a
students of, **1**:178a; **6**:95a
Möbius, Karl August (1825–1908), 9:431–432; **2**:626a
on oysters, **5**:70b
Mōchus, **4**:31a
Mociño, José Mariano (1757–1820), 9:432–434; **12**:327a
collections of, **15**:471a
Models, mathematical
theory of, **9**:71b
Models, molecular structure
see Stereochemistry
Models, technological
Renaissance, **8**:198a–b
Modena
Collegio dei Nobili, **3**:426a
Modena University
in nineteenth century, **11**:599b
anatomy at, **12**:136b
Modulation, musical
ancient Greek use of, **1**:282b
Modulo arithmetic, **15**:10a
Modulus (math.), **1**:239b; **5**:299b
law of, **6**:91a
naming of, **1**:238b
singular, **14**:189a
theory of, **4**:5a; **12**:173a
Moench, Conrad (1744–1805), 9:434
Methodus plantas horti botanici et agri Marburgensis, **9**:254a
Moenkhaus, William J., **9**:520b
Moerbeke, William of (*ca.* 1220–1286), 9:434–440; **3**:24b; **6**:272b, 273b; **12**:440a; **14**:457b; **15**:176a
translations by, **1**:199b, 223b, 227b–229b, 268b, 271b, 272a,b, 273b; **4**:105b, 491b; **5**:402a; **6**:275b; **7**:66a, 128b, 129a; **8**:224b, 225b; **9**:362b; **11**:197b; **12**:442b, 443a; **13**:259b, 260a, 308a; **14**:21b, 457b
Moffett (Moufet, Muffet), Thomas (1553–1604), 9:440–441; **12**:334b; **13**:451a; **14**:507b
entomology, **11**:317a; **13**:171a, 172b
Theatrum insectorum, **13**:508a; **14**:507b–508a
Mohanadāsa
commentaries of, **2**:118a
Mohism, **12**:381a–b
Mohl, Hugo von (1805–1872), 9:441–442; **3**:611b; **5**:536a; **6**:465a, 467a; **9**:149a,b*; **12**:215a, 259a
critics of, **11**:153a; **12**:175b
cytology, **3**:337a,b; **6**:266a; **9**:600b; **11**:153a, 300b; **12**:174b
works translated, **6**:267a*
Mohn, Henrik (1835–1916), 9:442–443
meteorology, **6**:98a; **12**:595b
Outline of Crystallography, **14**:108b
Mohole project, **9**:444b
Mohorovičič, Andrija (1857–1936), 9:443–445
Mohorovičič discontinuity, **9**:444a,b
Mohr, Carl Friedrich (1806–1879), 9:445–446; **5**:326a; **9**:567a*
Mohr balance, **9**:445b
Mohr salt, **9**:445b
Mohr test, **9**:445b
Mohr, Christian Albert, **7**:60b(n)
Mohr, Christian Otto (1835–1918), 9:446; **3**:496b
Mohr, Georg (1640–1697), 9:446–447
constructions, **9**:157a–b
Euclides danicus, **9**:157a
Mohs, Friedrich (1773–1839), 9:447–449; **2**:440a; **6**:18b; **8**:245b; **12**:227a; **14**:257a, 293a
crystallography, **9**:620a; **14**:241b(n), 490b
mineralogy, **7**:70b*; **9**:620b
Mohs scale, **9**:448a
students of, **14**:232a
Moigno, François, **12**:289a
Moine schists, **9**:585a
Moine thrust belt, **5**:337a–b
Moinian (stratig. division), **8**:33a–b
Moiseev, Nikolay Dmitrievich (1902–1955), 9:449–450
Moissan, Ferdinand-Frédéric-Henri (1852–1907), 9:450–452; **3**:333a, 400b; **5**:595b; **6**:480b; **8**:116b; **12**:46b; **13**:70a
chemistry, **4**:80b; **5**:474a; **12**:46b, 47a
Traité de chimie minérale, **3**:401a
Moivre, Abraham de (1667–1754), 9:452–455; **3**:244b; **10**:86b(n), 87a,b(n), 95b*
acquaintances of, **3**:459b; **5**:448b
actuarial science, **6**:71b
calculus, **3**:432a
central limit theorem, **3**:225a, 230b, 231a; **8**:165a; **9**:127b
combinatorial analysis, **6**:403b
correspondence of, **2**:54a; **3**:461b; **13**:265b
critics of, **9**:500a
Doctrine of Chances, **1**:531a; **2**:50b; **7**:561a; **9**:500a; **12**:444b; **15**:279b
geometry, **11**:402a
de Moivre's theorem, **1**:239a; **4**:477a; **8**:162a; **9**:454b
probability, **2**:42a; **8**:306a; **15**:281a, 282a
series, **2**:37b; **13**:68b, 69a
statistics, **10**:470b*
students of, **4**:35a; **13**:224a
trigonometry, **3**:120a
Mojsisovics von Mojsvar, (Johann August Georg) Edmund, **3**:530a
Molasse, **6**:228b, 229a; **13**:123b
oil in, **7**:44a
Moldenhauer, Wilhelm, **1**:578b
Moldenhawer, Johann Jacob Paul (1766–1827), 9:455–456
Molds (biol.)
antibacterial, **5**:30a, 42a–b
reproduction, **12**:557b–558a
Mole (animal), **3**:7b
eyes of, **1**:262b
Molecular beams, **3**:329a; **4**:253a–b; **13**:42a–b
Molecular theory
late eighteenth-century, **2**:79a, 80a–82b, 181a
early nineteenth-century, **1**:141b, 145a
mid nineteenth-century, **1**:518a–b; **2**:620b, 621b, 628a; **7**:281a; **14**:184b–186a
late nineteenth-century, **8**:492b–493a; **14**:61b
early twentieth-century, **13**:41a–42a
in biology, **14**:184b
Molecular volume
of gases, **13**:299a
Molecular weight, **3**:289a
and density, **1**:347a
determination of, **1**:345b, 347b; **4**:79a; **11**:370b
Beckmann thermometer, **1**:553b
by electric conductivity, **1**:297b
by freezing point, **11**:298b; **15**:435a–b
by light scattering, **3**:620b
by surface film, **4**:76b
by ultracentrifuge, **13**:160b–163a
and vapor density, **1**:344b; **3**:48b; **4**:243b; **13**:243a; **14**:531a; **15**:435b
by viscosity, **13**:2b
hydrogen standard
molecular vs. atomic, **1**:344b
and latent heat, **13**:471b
and osmotic pressure, **13**:579b
and viscosity, **13**:2b
Molecules
atoms, distinct from, **1**:344a, 347a; **2**:242a; **3**:47a–49a; **4**:243b–244a; **14**:531a
collisions of, **2**:261a–265b; **3**:307b–308a; **9**:221a–b
distribution of velocities, **3**:308a
persistence of velocity, **7**:84b
concept of, **1**:345a, 346b, 348b; **3**:49a; **8**:56b; **13**:160a
diatomic, **1**:345a, 347a; **3**:307a, 430a; **5**:117b, 372b; **6**:615a; **8**:61a
dipole moment, **3**:619a; **8**:475a
distance between, **8**:510a
distribution of, **14**:617a
energy of, **2**:262a; **8**:473b–474b
energy of dissociation, **5**:118a
interaction of, **3**:443a–444a; **7**:387a; **8**:57a, 475a–b; **10**:532a; **13**:241a
Maxwellian, **9**:221b, 222a,b
mean free path, **3**:307a–b, 477a–b; **9**:218a, 219b, 220a, 221a
microstates, **2**:264b–265a
molecular ray studies, **4**:253b
motion of, **3**:306b–307a
mutual attraction, **13**:156a
numbers per c.c., estimates of, **8**:509b–510b
polar, **3**:336a; **8**:475b; **9**:376b
quantum theory of, **4**:317a
quasi-molecules, **5**:118a
reactive centers, **8**:32a
rearrangement of, **1**:341a; **8**:32a
relaxation time, **3**:619b
resonance, **12**:419a–b
rotation of, **7**:491b; **9**:207b–209a
sizes of, **2**:261a; **5**:597b; **8**:508b–510a; **9**:222b
spin of, **13**:241a

structure of, **3**:619a; **4**:69a, 312b; **8**:57a–b, 59b–60b, 61a; **9**:230b; **15**:18a
and crystallography, **1**:288b; **5**:455b–456a, 557b
electron diffraction method, **3**:619b
at high temperatures, **9**:356a
molecular orbital method, **8**:186a–b
octet rule, **8**:291b
and optical activity, **1**:340b–341a; **2**:139a–b; **5**:128b; **9**:223a, 356b, 403a; **10**:357b–361a; **11**:85b–88b; **13**:576b–577b; **14**:124b
and physical properties, **2**:622a; **3**:336a
restricted rotation, **1**:340b–341a; **9**:356b, 403a
space groups, **8**:485a
submolecular units, **1**:518b
and surface properties, **6**:118a
symmetry, **5**:295a; **8**:485a
swarms of, **10**:235b, 236a
velocity, **6**:292b–293a
distribution of, **9**:218a–220a, 221b, 224b
relative, **3**:307b, 308a
vibration, **5**:117b–118a; **8**:596a; **11**:266a
frequency of, **8**:474a
see also Macromolecules, Polymers, Stereochemistry
Molécules intégrantes, **6**:179a–181b
Moleschott, Jacob (1822–1893), 9:456–457; **4**:162b, 164a*; **9**:546b
Kreislauf des Lebens, **2**:563b
Molières, Joseph Privat de
see Privat de Molières, Joseph
Molin, Fedor Eduardovich (1861–1941), 9:457–458
Molina, Juan Ignacio (1740–1829), 9:458
Molk, Jules, **13**:250a
Moll, Carl L., **11**:384a
Moll, Friedrich Rudolf Heinrich Carl (1882–1951), 9:458–459; **2**:600b; **9**:415a; **11**:479b
Moll, Gerard (1785–1838), 9:459–460; **5**:79a(n); **6**:137b, 278a; **13**:184a*
Moll, Jan Willem, **10**:253b
Moll, Willem Jan Henri, **10**:235b
Moll's thermopile, **9**:460a*
Möller, Didrik Magnus Axel (1830–1896), 9:460
Møller, Jens Peter, **13**:99a
Molliard, Marin (1866–1944), 9:460–461; **3**:359a; **9**:176b*
Mollier, Richard (1863–1935), 9:461–463; **10**:162b; **14**:618a*
Mollier diagrams, **9**:462b; **13**:73b
Mollusks, **2**:322a; **4**:289a, 290a
anatomy, **1**:264b; **3**:525a; **5**:34b, 504b; **6**:535b; **7**:546a; **8**:415b, 519a; **12**:551a; **13**:465a, 554b
classification of, **1**:58b; **3**:536a–b; **4**:42a; **5**:355a, 406b; **6**:590b–591a; **7**:546a; **8**:415a–b, 519a; **9**:575b; **11**:67a; **12**:550b–551a; **14**:1b; **15**:111a
collections of, **4**:67b; **7**:92a
crag, **7**:91b
distribution of, **5**:66b, 67a
embryology, **5**:51a; **6**:167a; **8**:103b, 519a
evolution of, **1**:587a; **8**:519a
freshwater, **2**:106a; **8**:103a–b
life cycles, **12**:107a
paleontology, **1**:163a, 210a, 587a; **2**:494b–495a, 570b; **3**:391b–392b, 536a; **4**:67b–68b, 147b; **9**:240b–241a, 314a; **10**:29a–b; **11**:386a; **13**:531b; **14**:88a, 312b
physiology, **8**:103b
reproduction, **11**:133a
spontaneous generation controversy, **2**:591a–b
terrestrial, **5**:478a
see also Malacology, Shells
Mollweide, Karl Brandan (1774–1825), 9:463; **7**:405a; **9**:429a, 430a
Molodenski, Mikhail Sergeevich, **7**:497a
Molther, Johann
Problema deliacum, **10**:115b
Moluccas (islands), **3**:365b
natural history, **2**:342b; **12**:536a
Molybdates, **14**:588b
Molybdenite, **12**:147b
Molybdenum
compounds of, **2**:565a; **3**:400b–401a; **5**:394a
isolation of, **9**:451b
naming of, **12**:147b
spectrum, **3**:125a
Molyneux, Emery, **14**:514a
Molyneux, Samuel (1689–1728), 9:463–464; **5**:26a*, 491b; **6**:6a; **9**:309b
parallax measurement, **2**:387b–388a
Molyneux, William (1656–1698), 9:464–466; **5**:279a(n); **9**:120b(n)
Molyneux problem, **9**:465b
Moment, physical, **4**:506a
composition of, **8**:222a; **11**:61b
equation of, **11**:488b
of a force, **3**:66b; **10**:4a, 250a
inertia, **3**:230a; **5**:241b; **6**:606a
momental ellipsoid, **3**:145a
theory of, **3**:224b, 229b–230a, 230b
three-moments equation, **9**:446b; **14**:473b
turning, **8**:301a
see also Magnetic moment
Moments, method of, **5**:9a; **6**:176b; **8**:562b, 563a; **9**:126a–129a *passim;* **10**:47b, 48b–49a, 66a–67b, 450b, 452b, 456a, 457a; **13**:57a–b, 58a; **15**:32a
Momentum, **2**:330a; **3**:296a; **4**:44a; **11**:488b
angular
conservation of, **5**:242a; **15**:343b, 350a–b
nuclear, **1**:371a
conservation of, **1**:111b–112a; **3**:74b; **4**:58b–59b, 60a–b, 323a, 327a,b; **6**:292a, 603a; **8**:153a; **10**:60b; **11**:493b; **15**:8a
general covariance, **4**:329a
and quantum theory, **2**:338a
definition of, **2**:606a; **10**:70a; **15**:343b
electrokinetic, **9**:211a
electromagnetic, **13**:365a
reduced, **9**:209a, 211a
transmission of, **2**:54b
Monaco
Musée Anthropologique, **1**:92b
Musée Océanographique, **1**:92b
Monadelphiae, **3**:154b
Monades, **3**:337a
Monadology, **3**:90a, 381a; **4**:291a; **7**:225b–226b; **8**:152a, 157a–159a, 160a, 514a; **9**:576a; **15**:54b
critics of, **4**:471b; **7**:227b, 229a–230a
materialistic, **8**:468b–469a
Monardes, Nicolás Bautista (*ca.* 1493–1588), 9:466
works translated, **8**:120b
Monatliche Correspondenz zur Beförderung der Erd- und Himmelskunde
editors of, **8**:369a–b
founding of, **14**:583a
Monatshefte, **12**:175a
Monazite, **4**:80a
Monazite sand, **12**:505b
Monceau, Henri-Louis
see Duhamel du Monceau, Henri-Louis
Mond, Ludwig (1839–1909), 9:466–467; **11**:282a*; **12**:521a
chemistry, **2**:70b; **11**:282b*; **14**:175a
Mondeville, Henry of
see Henry of Mondeville
Mondino de' Luzzi (Liucci, Liuzzi) (*ca.* 1275–1326), 9:467–469; **1**:47a, 98b, 107a, 620a; **4**:285a; **5**:585b; **12**:82a; **14**:8b
anatomy, **9**:165b
Anothomia
editions of, **1**:618a, 618b–619a
critics of, **8**:248b; **13**:430b
Money
see Coinage, Currency, Mints
Monge, Gaspard (1746–1818), 9:469–478; **3**:441a; **4**:14b, 49b; **5**:601b; **7**:571a; **8**:89b*
and Académie des sciences, **3**:132a; **5**:95a; **9**:471b–472a; **15**:334a
acquaintances of, **3**:326b; **9**:72b, 343a; **11**:163b; **13**:571a; **15**:302a
analysis, **7**:550a,b; **8**:323b; **9**:471a, 476a–b
Application de l'analyse à la géométrie, **8**:385b
and Berthollet, **3**:346a; **9**:472a,b, 473a, 477b
and Carnot, **3**:70b–71a
chemistry, **8**:89b*; **9**:343b, 471b–472a, 477a–b; **14**:75b
editions of works, **8**:383a
and education, **9**:472a–473a; **15**:342a–b, 347a
École Polytechnique, **2**:74a, 133b; **5**:93b; **7**:569a,b; **9**:473a, 474a; **15**:481a
École Royale du Génie, **9**:470a–b, 471b
Géométrie descriptive, **2**:473a
geometry, **2**:288a, 418b; **3**:213a; **4**:350a, 479b; **6**:2a; **7**:599b; **9**:469b–476b; **11**:46a, 77a; **12**:172a,b
"cutting and filling," **1**:194b; **4**:257b; **9**:471a, 475b–476a
theory of characteristics, **4**:205b; **6**:545b
and Hachette, **3**:450b; **6**:1a,b, 2b; **9**:473b, 474a, 475b
influence of, **2**:214b; **3**:559b; **5**:96a, 367b; **7**:615a; **8**:136a; **11**:61b, 79b
and Lavoisier, **8**:80b, 81a; **9**:471b, 472a, 477a
Monge curves, **8**:325a
Monge equations, **9**:476b
physics, **9**:471a–472a, 477a
politics
French Revolution, **9**:472a–473a
Napoleon, **9**:473a–474b
students of, **2**:72a, 454b; **4**:257a; **7**:549b, 550a; **9**:342b; **11**:76b; **13**:417b, 418a
technology
machines, **6**:2b; **11**:80a
metallurgy, **9**:471b, 473b, 477a,b
military, **9**:470a, 472b
and Vandermonde, **9**:471a,b, 472a,b, 477b
Mongez, Jean-André, **7**:603a; **9**:5b
Mongolia
exploration of, **11**:180b
geography, **6**:216b; **11**:108b
geology, **10**:167b
paleontology, **5**:488a*, 500b
Mongols, **9**:332b
Mongré, Dr. Paul (pseudonym)
see Hausdorff, Felix
Monilia, **2**:562b
Monism, **3**:493a; **6**:7a–10a *passim,* 152a, 256b, 257b; **7**:41b, 498b; **9**:456b, 495a; **11**:570b; **12**:297b, 407b, 408b; **15**:79a–80a, 462b, 466a
psychomonism, **14**:3a

Monist, 15:79a
Moniteur scientifique, 1:361a
Monium, 3:478b
Moniz, Egas
 see Egas Moniz, A. A. F.
Monkey
 cattarrhine, 2:110a
 cynomorph, 2:110a
 embryogeny, 13:97a–b
 experimental use of, 7:624a
 fossil, 3:328a
 hominoid, 1:173a; 2:110a
 hybridization, 7:33a
 parasites of, 4:132b–133a
 rhesus, 12:475a
Monnet, Antoine-Grimoald (1734–1817), 9:478–479; 3:473b; 8:69a; 9:54b; 12:65b
 Atlas et Description minéralogiques de la France, 4:348a; 5:578b
 on contemporary scientists, 11:562b
Monoblepharis, 14:93b
Monoceros (constellation)
 nebulae, 11:553b*
Monocotyledons, 5:362a; 9:418b; 15:148a
 aquatic, 12:125a
 derivation of, 2:103b
 vs. dicotyledons, 11:314b
 morphology, 1:205b–206a
 vascular system, 9:78b
Monocytes, 12:48b
Monod, Jacques, 13:137a
Monoecia, 13:392b
Monoglycerides
 preparation of, 15:15b
Monografie matematyczne
 editors of, 9:248b
Monophysites, 1:275a
Monotonicity (math.), 2:276a
Monotremes
 anatomy, 2:504a
 classification of, 5:46a; 10:262b
 reproduction, 10:261a–b; 12:298a
Monotropy, 3:333b–334a
Monro, Alexander (Primus) (1697–1767), 9:479–482; 2:173b; 6:568b; 7:11b; 8:612a; 9:483a; 11:48a; 12:354a; 14:320a,b
Monro, Alexander (Secundus) (1733–1817), 9:482–484; 6:569b; 7:11b; 9:481a,b; 10:582a; 12:354a, 509b; 14:456b
 anatomy, 6:367a,b
 foramen of, 9:483a; 14:322a
 medicine, 14:322a
Monro, Alexander (Tertius), 5:470a; 7:414a, 415a; 9:484a; 10:260b
Monro, Cecil James, 9:201a, 202a
Monro, Donald, 9:480a
Monro, John, 9:479b, 480b
Monsoons, 14:53a
Monstrillidae, 9:328b
Mont Blanc, 2:431a; 12:119b–120a, 120b, 123b; 14:38b
Montagne of Limoges, 9:329b
Montagu, Lady Mary Wortley, 12:458b
Montague, Charles, 10:44b, 45a, 83b; 13:528a
Montaigne, Michel Eyquem de, 9:316b
Montalbani (Ovidio?), 5:543a
Montalte, Louis de (pseudonym), 10:335b
Montaña de Monserrate, Bernardino
 Libro de anathomia del hombre, 14:11b
Montanari, Geminiano (1633–1687), 9:484–487; 2:592a; 5:468b; 7:373a
Montbéliard, Guéneau de, 2:107a,b
 Histoire des oiseaux, 2:474b
Montbéliard (principality)
 College of Medical Practitioners, 1:525a
Mont-Beuvray (city), 3:622b
Montcarville, Robert Benet de, 1:400b
Montdidier, Obert de, 6:272b
Monte, Giambattista da, 3:12b; 4:519b; 8:248b
Monte, Guidobaldo, Marchese del (1545–1607), 9:487–489; 3:364b
 geometry, 5:240a; 9:598b
 influence of, 1:607b, 608a; 5:238b, 239a, 246b; 14:29b
 influences on, 1:229b
 Liber mechanicorum, 7:174b; 14:366b
 mechanics, 1:419b; 5:239b; 7:174b–175a; 14:366b
 students of, 1:419a
Monte Ferru
 geology, 4:140b
Monte Rosa, 12:121a, 123b
Montefeltro, Federico di, 1:419b; 9:146b
Montelius, Gustav Oscar (1843–1921), 9:489–490
Montesquieu, Charles-Louis de Secondat, Baron de la Brède et de (1689–1755), 9:490–491; 3:114a, 378a, 560b; 5:58a; 7:202a; 9:188b
Montet, Jacques, 13:603a
Montgéry, Jacques-Philippe Mérigon de (1781–1839), 9:491–492; 3:389a*; 9:505a*
Montgolfier, Étienne Jacques de (1745–1799), 9:492–494; 3:7a, 78b*, 207b; 8:78b; 10:609a
Montgolfier, Michel Joseph de (1740–1810), 9:492–494; 3:78b*; 4:74a; 12:287b, 288a
 balloons, 3:7a, 207b; 8:78b; 10:609a
 mechanics, 10:3b
 Montgolfier process, 3:317a*
Montgomery, Deane, 2:513a
Montgomery, Edmund Duncan (1835–1911), 9:494–495
Montgomery, Thomas Harrison, Jr. (1873–1912), 9:495–497
 cytology, 8:588b, 589a; 13:157b; 14:431a
Monthly American Journal of Geology and Natural Science, 15:152a
Monthly Notices of the Royal Astronomical Society, 3:472b
Months, 15:723b
 anomalistic, 15:211b–212a, 540b, 541a, 601b
 draconitic, 15:211b
 length of, 6:302a; 15:708a
 lunar, 4:459b–460a; 5:553a; 12:378b; 15:707b, 708a, 709a,b, 711b, 723b, 789b
  length of, 5:345b, 347a; 9:337a; 10:180a–b; 15:775a,b, 790a, 800a, 803a
 names of, 2:152b; 9:337b; 15:707a, 709a, 716b
  Indian, 15:535a
  Mayan, 15:765b, 781b
  Syrian, 1:512a
 sidereal, 15:640b
  length of, 15:538b
 solar
  length of, 15:538b
 synodic, 3:251a; 15:568a
  length of, 9:337a; 10:180a; 11:191b; 12:378b; 15:211b–212a, 535a, 536a,b, 537a, 538b, 590b, 679a
 zodiacal, 4:460a; 9:338b
Montmor, Henri Louis Habert de (*ca.* 1600–1679), 9:497–499; 5:285a; 9:501a
 Académie Montmor, 4:49b; 9:498a–b; 10:492a, 546b; 11:506b; 12:585a–b; 13:334b–335b, 336a
 blood circulation experiments, 4:37b–38a
Montmort, Pierre Rémond de (1678–1719), 9:499–500; 2:42a; 3:281a; 8:151a; 11:392a; 13:265a–b
 Essai d' analyse sur les jeux de hazard, 2:50b, 57a; 9:452b
 probability, 9:454a
Montpellier, University of
 in Renaissance, 10:152a
 in eighteenth century, 1:617a
 botanical garden, 1:592a; 2:510b; 3:173b*; 7:536b; 9:17b–18a
 medicine at, 1:289b, 290a,b, 592a; 2:122b; 6:276a; 9:306a,b; 11:311a–b, 527b
Montreal Normal School
 founding of, 3:607b–608a
Montregian Hills, 1:51b
Montucla, Jean Étienne (1725–1799), 9:500–501; 2:334b, 335a; 6:577a; 8:271b; 10:264b
 Histoire des mathématiques, 5:383a; 7:550b, 580b; 8:306a
 historiography, 1:508a; 4:612b; 6:128a, 407a; 7:584a; 9:275b(n); 10:264a; 11:227a; 12:61a; 13:326b, 585a–b
Monzonite, 8:30b
Mookerjee, Asutosh, 15:47b–48a
Moon, 9:616a
 and astrological prediction, 15:672b
 atmosphere of, 2:329a; 4:24a*; 5:25b; 9:233b; 13:82a
 brightness, 2:284b
  scale of, 2:147b
  secular variations of, 2:147b
 color, 1:607b; 2:147b; 12:352b; 15:724b
 conjunctions of, 15:677b
 crescent, 1:309b; 2:114b, 115a, 117b, 119a, 416a, 417a,b; 3:584b; 5:274b; 7:219b, 512b, 582b; 10:129b; 12:598b; 14:28b
  visibility of, 1:512b; 2:115a, 118a, 155a; 5:619a–620a; 13:292a; 15:260b, 774a
 culminations, 6:49a; 10:110b–111a
 duration of, 15:760a, 774a, 775a, 776a, 810b
 and earth
  distance between, 1:189b, 248b–249b, 509a; 3:25a–b; 10:129a, 299a; 11:104a, 188a, 194a, 197a; 12:406a; 15:213b–215b, 561a, 594a, 604b, 605b
  electric influence, 1:300b
  meteorological influence, 3:436a; 10:199a; 15:382a
  mutual attraction, 5:400a, 401a; 6:291b; 10:61a; 15:325b–326a, 332a–b, 348a
  mutual illumination, 3:116a
  nutation effect, 2:388a–b
  tides, 3:248b, 423b; 4:506a–b, 591a; 5:400a, 401a, 551a; 11:104a
 far side, 9:415a
 heat reflected by, 7:578a
 latitude of, 5:616b; 10:426a; 14:577a; 15:544b, 548b, 560a–b, 573b, 587b–588a, 594b, 679a
 libration of, 6:149b, 361b, 362a; 7:562a–b; 9:233b; 15:354a–b
  heliometric observation, 1:429a,b
  photographic observation, 11:156a
 life on, 5:59a–b; 6:334a; 7:299a; 14:363a, 364b
 light of, 4:368a; 6:195a, 197b; 12:406a; 13:326b, 409a, 424b; 14:629a
 longitude of, 1:49a; 5:274a, 616a; 7:362a, 512b; 9:607a; 11:553b*; 15:537a,

539a–b, 540b–541a, 543a, 545b, 624a, 679a
inequality in, **2**:412a, 516b; **6**:516a
secular acceleration of, **3**:457a
mapping of
see Selenography
and meteorite theories, **1**:615b
moon-age, **15**:760a, 774a–b, 776a,b, 803a, 809a, 810b
moon-number, **15**:760a, 775a, 776a, 812a, 814a
motion of
ancient Greek study, **15**:211b–213b
ancient Roman study, **11**:191b–194a
medieval Arabic study, **5**:616a–b; **6**: 198b; **7**:259a; **11**:251a; **12**:359b–360a; **13**:292a; **15**:34b
medieval Hebrew study, **8**:281a
medieval Latin study, **2**:24a
Renaissance study, **2**:411b–412a; **3**: 406a–b; **5**:400a
seventeenth-century study, **5**:23b–24a, 25b; **6**:610b; **7**:303b–304a; **9**:310a; **12**:342b–343a
eighteenth-century study, **1**:114b; **3**: 431a,b; **4**:478b, 481a–b; **5**:195a; **6**: 69b–70a; **7**:562a–b, 563b–564a; **9**: 233b–234a; **15**:317b–318a, 348a
nineteenth-century study, **1**:53b–54a, 86b; **4**:19b; **6**:103b; **8**:530a; **10**:219b; **11**:7a; **13**:422b
twentieth-century study, **1**:156b, 429b; **2**:511a, 516a–b; **3**:457a; **12**:574b*; **15**:129a
annual equation, **2**:412a; **5**:23b; **7**:293b, 304a
Chinese study, **12**:378a
Indian study, **9**:580b; **15**:536b, 557a
irregularities, **2**:411b–412a, 600a; **6**: 399a,b, 516a; **7**:304b; **8**:178b–179a; **9**:239b; **10**:34a, 73b, 77a, 78a, 298b; **11**:207a; **12**:377b; **15**:348a, 349a, 353b–354a
in latitude, **15**:213a, 216a–b
in longitude, **1**:511a
maximum equation, **12**:360a
Mesopotamian study, **15**:678b, 679a
nineteen-year cycle, **3**:436a
secular acceleration, **1**:54a; **3**:169b–170a; **4**:20a, 591a; **7**:564a; **14**:576a; **15**:286b, 288b–289a, 290a–b, 331b–332b, 344a, 349a–b, 356a
speed of, **15**:539b, 545a–b
mean motion, **11**:192a; **15**:540b–541a, 570a, 586a, 611a–b, 613a–b, 623a–b, 624a–b
nature of, **15**:202b
occultations, **6**:325a; **7**:18a
orbit of, **6**:69b–70a; **7**:304a; **9**:141a; **15**: 676a
apogee, **3**:216a, 283a; **15**:543a
dimensions of, **15**:586a, 608a–b
eccentricity of, **11**:192b; **14**:576b
elongation from sun, **15**:219a, 548b–549a
equation of the center, **2**:119b
evection, **15**:582b–583a, 619a
inclination of, **2**:411b–412a; **10**:78a; **11**:192b; **12**:359b; **15**:354a, 544a, 550a
nodes, **2**:412a; **3**:457a; **7**:562a; **15**:543a, 605b, 611a–b, 613a, 614a, 623a–b, 624a–b, 625a–b
perigee, **2**:468b; **4**:481a; **6**:399b
origin of
cometary theory, **13**:270b
resonance theory, **3**:583a
penumbra, **11**:319b
phases of, **3**:319b; **4**:543a; **5**:345b; **10**: 419a; **15**:535b, 575b, 595a, 643b, 679a, 706b, 708a
theories of, **15**:519b
photography of, **4**:19a, 179a, 180b, 182b; **5**:21a(n); **10**:329a, 602a; **12**:20a, 36b, 37a, 186b
photometry of, **9**:415a; **11**:542b
position of, **15**:536b, 729b
altitude, **15**:576a
radiant heat of, **12**:499a; **15**:59b
rising and setting of, **15**:676b, 679a
rockets to, **5**:433b
shape of, **10**:78a
size, **3**:423b; **10**:299a
apparent diameter, **1**:248b, 249a, 511a; **8**:459b; **9**:233b; **11**:349a; **15**:213b, 219a
circumference, **15**:608a
diameter, **1**:249a–b; **2**:403b; **10**:595b; **11**:156a; **15**:215a–b, 544b, 550a, 560b, 579a, 594a–b, 598b, 605a, 608a, 798a
mass, **4**:591b; **12**:574b*; **15**:353b
radius, **13**:114a; **15**:608a
spectra, **4**:180a; **9**:392a; **12**:37a
in eclipse, **9**:504a
and sun
conjunctions of, **7**:130a–b, 139b; **10**: 299a; **15**:590b
mutual attraction of, **15**:332a, 380b–381b
surface, **2**:146b; **3**:103a; **6**:207a*; **8**:581a
changes in, **7**:400a; **10**:602a
craters, **1**:569a; **5**:396a; **6**:35a; **12**:267a
erosion of, **12**:280b
genesis of, **10**:430a
light absorption, **14**:499b*
and longitude determination, **5**:88a
mountains, **1**:569a; **5**:542b; **6**:333b–334a
nature of, **5**:241a; **12**:353a
petrology, **14**:515b
photometry of, **12**:353a
volcanoes, **6**:334a; **7**:263a, 400a
tables of
ancient Greek, **1**:191a
ancient Roman, **11**:191b–192a, 193b
Byzantine, **12**:406a
medieval Arabic, **5**:616a,b; **7**:259b; **14**:578b
medieval Latin, **5**:553a; **7**:118b, 119b, 123a, 130b; **10**:540b; **14**:390b
seventeenth-century, **5**:23b–24a, 25b; **6**:115b; **7**:304a, 577b; **9**:527b; **13**:96a
eighteenth-century, **2**:389a; **3**:283a; **4**: 22b, 481b; **8**:178b–179a; **9**:162b–163a, 164b, 233b, 234b; **15**:349a,b
nineteenth-century, **2**:600a; **6**:103b, 403a*; **10**:219b, 480b*–481a*; **11**:7a, 114b, 115a; **12**:170a
twentieth-century, **2**:516a–b; **15**:129a
Mayan, **15**:774b, 812b
temperature of
in eclipse, **10**:107b
visibility of, **7**:259b, 338a; **10**:129b; **15**: 553a–554a, 561b–564b
zodiacal period, **5**:345b
see also Eclipses, lunar; Parallax, lunar; Selenography
Moonstone
optical properties, **11**:267a
Moore, Charlotte, **12**:21b, 23a, 73a
Moore, Eliakim Hastings (1862–1932), 9: 501–503; **2**:198a; **6**:235b, 398b; **13**: 599b; **14**:211b
general analysis, **1**:427b; **2**:144a; **12**:364b
number theory, **14**:212a
Moore, George Edward, **7**:317b; **12**:10a; **14**:468a
Moore, John Edward Shorec, **4**:546a
Moore, John Hamilton, **2**:368b
Moore, Sir Jonas, **5**:22b–23a, 24b
Moore, Joseph Haines (1878–1949), 9:503–504; **3**:35b, 510b*; **14**:499a*
Moore, Richard, **11**:280b
Moore, Robert Lee, **9**:249a; **13**:600a; **14**: 382a, 591a
Moore, Stanford, **15**:15b
Moore, William (*fl. ca.* 1806–1823), 9:504–505
Mooser, Hermann, **14**:623b
Moraceae, **9**:417b
Moraines, **10**:504a; **12**:83b; **13**:270a; **14**:627a
Moral sense, **11**:617b
origin of, **3**:574b; **6**:139a–b
science and, **2**:63b–64a, 65a, 596b; **5**: 474b*; **14**:343b
see also Ethics
Morat, Jean-Pierre (1846–1920), 9:505–507
Moravia
stratigraphy, **2**:108b
Moravian Brethren, **3**:360b
Moray (or Murrey or Murray), Sir Robert (*ca.* 1608–1673), 9:507–509
acquaintances of, **5**:524b, 525a; **6**:482b, 483a; **12**:584b
correspondence of, **9**:498b; **12**:584a; **14**: 373b
and Royal Society, **6**:598b; **8**:130b; **14**: 371a
Morbidity
tables, **14**:231a
Morbus articularis, **1**:400a
Morbus hungaricus, **11**:606a
Mordants
see Dyeing
Mordell, Louis Joel, **11**:55b
More, Henry (1614–1687), 9:509–510; **4**: 64b; **5**:417a
acquaintances of, **2**:612b; **10**:43b, 44b; **14**:379a(n)
influence of, **5**:415b; **10**:60b; **11**:318a; **14**:378b(n)
philosophy, **3**:492a; **7**:488a
Morehouse, Daniel Walter, **10**:107a
Morehouse, George, **9**:422a
Moreno, Francisco, **1**:130a,b
Morey, George Washington, **4**:564a
Morgagni, Giovanni Battista (1682–1771), 9:510–512; **1**:318b; **3**:437a; **9**:94b
Adversaria anatomica, **13**:566b
influences on, **1**:604a, 611b; **3**:41a; **5**:366b
pathology, **2**:122b, 464a; **4**:197b; **13**: 567b*
De sedibus et causis morborum per anatomen indagatis, **9**:66b; **13**:567b
students of, **3**:15b; **11**:37a, 66b; **12**:136b, 137a
Morgan, Conwy Lloyd (1852–1936), 9:512–513; **6**:260b; **11**:520a*
psychology, **8**:446a; **11**:518b
Morgan, Edward James, **6**:501b
Morgan, Herbert Rollo (1875–1957), 9:513–514
Morgan, Lewis Henry, **11**:120a; **15**:144b
Morgan, Thomas Hunt (1866–1945), 9:515–526; **2**:361a; **3**:390a,b; **4**:189a*; **6**: 131b; **11**:575a*; **12**:569a; **14**:436b*
acquaintances of, **3**:389b; **8**:445b; **14**:424a
critics of, **7**:115b
embryology, **8**:446a; **14**:428b
genetics, **3**:122a, 123b, 422b; **4**:271b; **6**: 338a; **10**:276b; **13**:134b–135a, 136a, 158a; **14**:99a, 101a, 432a–433a
influence of, **2**:455b; **12**:416b

Morgan, Thomas Hunt (cont.)
influences on, **2**:502a; **3**:121a
philosophy of biology, **14**:434b
students of, **9**:564b; **13**:134a
translations of works, **15**:507a
Morgenroth, Julius, **4**:297b, 298b, 299a
Morgenstern, Oskar, **14**:92b, 121b
Morgulis, Sergius, **14**:550b
Mōri, Sigeyosi, **12**:290b; **15**:745b
Morichini, Domenico Lino (1773–1836), 9: 526–527
Morienus (or Marianos), **2**:23a; **10**:555b
Morimura, Ichizaemon, **7**:392a
Morin, Antoine, **11**:132b, 133a
Morin, Arthur Jules, **10**:286a,b; **11**:78a
Morin, Francis Joseph
Morin temperature, **6**:480b
Morin, Jean-Baptiste (1583–1656), 9:527–528; **3**:596a; **4**:55a(n); **5**:285b, 287b; **7**:308a
Famosi et antiqui problematis de telluris motu vel . . . , **8**:28a
longitude, **1**:541a; **6**:299a; **10**:342a
Morison, Robert (1620–1683), 9:528–529; **9**:92b; **11**:312a, 314b, 315b
Historia plantarum, **1**:431b
Morisot-Deslandes, Pierre Joseph, **9**:500b
Morland, Samuel (1625–1695), 9:529–530; **8**:211b
Morley, Edward Williams (1838–1923), 9: 530–531
ether-drag experiment, **5**:20b; **8**:495a; **9**: 372b–373a
Michelson-Morley experiment
see Michelson, Albert Abraham, Michelson-Morley experiment
Morlot, Adolph von, **6**:19a
Morningfirst (astron.), **15**:673a, 679a
Moro, Antonio-Lazzaro (1687–1764), 9: 531–534; **7**:412b; **14**:259b
Morocco
botany, **2**:510a; **6**:492a; **9**:34b–35a
Morosini, Andrea, **12**:101a,b, 106a
Morozov, Georgy Fedorovich (1867–1920), 9:534–536; **14**:107a
Morozov, Nikolai Aleksandrovich, **7**:6a, 7a
Morozzo, Carlo Ludovico, **8**:620b
Morphine, **3**:246a
addiction to, **5**:172a
analysis of, **3**:160a
discovery of, **3**:159b; **12**:320b
medical use of, **9**:9b
structure of, **5**:589a
synthesis of, **5**:421a, 531a
toxicity, **11**:430a
Morphogenesis, **8**:282b
botanical, **6**:467a
experimental, **5**:624a–b
comparative, **13**:173b
endocrine gland, influence of, **3**:113a
environmental effects
temperature, **8**:357a; **11**:333a
feathers, **8**:359a
functional adaptation theory, **11**:570b–571b
insects, **13**:171a–174b
levels of organization, **6**:224a
mathematical study, **13**:353a, 498a–b
theories of, **3**:247b; **5**:594b; **7**:33b–34a
vertebrate, **1**:386b
see also Embryogeny, Growth
Morphologisches Jahrbuch
founding of, **15**:166b
Morphology
botanical
see Plants, morphology
crystallographic, **1**:461a, 462a, 529a; **4**: 66a; **9**:58b; **10**:125a–b; **11**:521b–523a
ideal shape, **3**:505a
and physical properties, **15**:115a
statistical, **10**:125b
see also Lattices
cytological, **1**:545b; **2**:362b; **7**:455a; **8**: 282b, 302b; **14**:41b
embryological, **3**:389b; **4**:147a, 510a; **11**:255a–b
entomological, **3**:247b, 596b; **5**:328b
human, **3**:113a–b, 555b; **7**:278b
microbiological, **3**:519b; **7**:406b, 422a; **10**:399a; **14**:37a
mineralogical, **1**:529a; **5**:211b, 548a
naming of, **5**:442b
neurological, **5**:282a, 571a
paleoanthropological, **2**:172a
paleontological, **2**:204b; **6**:594b; **7**:252b
philosophy of, **5**:193b
zoological
eighteenth-century, **2**:204a, 509b; **13**: 458b
mid nineteenth-century, **2**:282a; **3**:89a; **4**:200a; **5**:462b–463a; **8**:270b, 302a–b; **12**:192b, 193b–194a; **15**:123a
late nineteenth-century, **1**:420a–420b, 507a; **5**:109a, 385a–b, 447a; **6**:131b; **9**:496a
twentieth-century, **1**:506a; **3**:112b–113a; **5**:453b; **7**:314a; **11**:211b, 212b
comparative, **1**:601a–b; **2**:383b, 504a–b; **5**:442b–444a; **6**:336b, 591b–592a; **8**: 415b; **9**:252b–253a; **12**:130b–131a; **15**:166a–170a, 473a
developmental, **5**:467a–b; **12**:410a; **14**:425a–430b
evolutionary, **2**:501b–502a; **4**:142b; **5**: 283a–b, 467a–b; **6**:7b–9b; **9**:516a–b; **12**:336b–338b, 410a–b
experimental, **3**:590a
invertebrate, **14**:251a
microscopical, **2**:164b–165b; **8**:128a–b
philosophy of, **5**:470b; **9**:65a, 516a; **10**:278b; **13**:353a; **15**:78b, 168a–169a
see also Anatomy, Geomorphology
Morrey, Charles Bradfield, Jr., **6**:472a
Morse, Edward Sylvester (1838–1925), 9: 536–537; **9**:416b*; **10**:272b; **14**:315a*
Morse, Jedidiah (1761–1826), 9:537–538
Morse, Samuel Finley Breese, **5**:305b; **6**: 280b; **7**:45a–b, 56a; **9**:537a
Mortality
Gompertz's law, **5**:466a
infant, **5**:137b; **10**:464a
statistics, **2**:42a–b; **3**:13a; **4**:39a; **5**:507a–b; **6**:71b, 453a, 538a; **9**:454a–b; **10**:565b; **12**:467a; **13**:263b; **14**:179a–b, 230b, 467a; **15**:31b, 302b, 373b
see also Actuarial science, Life expectancy
Mortar (masonry), **3**:8b
Mortillet, Louis-Laurent Gabriel de (1821–1898), 9:538–539; **1**:130a; **9**:489b; **10**:607a; **11**:270a, 484a
Mortimer, Cromwell, **5**:517a; **12**:458b
Morton, Henry, **4**:180a; **13**:398b
Morton, John (*ca.* 1670–1726), 9:539–540; **7**:412a
Morton, Richard, **6**:183a; **13**:526b
Morton, Samuel George (1799–1851), 9: 540–541; **9**:495b; **15**:153a
Morton, William Thomas Green, **5**:478b; **7**:45b
Morula, **1**:476b; **2**:161b
Morus, **15**:67b
Morveau
see Guyton de Morveau, Louis Bernard
Moṣadeva
commentaries of, **2**:116a
Mosaic gold (stannic sulfide), **14**:508b
Mosaicism (genetics), **13**:135b, 137a
Mosander, Carl Gustaf (1797–1858), 9:541; **2**:94b; **3**:321a, 473b
Mosandrite, **9**:541b
Mosasaurus, **4**:147b, 549a; **9**:152b; **14**:410a
Moscati, Pietro, **12**:57a
Moschopulous, Manuel, **1**:368a
Moscow (Russia)
Institute of Railway Engineers, **4**:305b
Navigation School, **9**:17a
observatory, **13**:481b
Timiryazev Institute, **10**:1a
zoo, **14**:595b
Moscow, University of, **8**:102a, 107b; **13**: 598a, 617a
astronomy at, **10**:317a
biochemistry at, **11**:179a
biology at, **7**:454b
chemistry at, **8**:545b; **9**:130b, 131a
founding of, **8**:470b
mathematics at, **3**:222b; **7**:207b
physics at, **1**:284a; **13**:80a, 241b
Moscow Basin, **6**:238b; **11**:567a
Moscow Mathematical Society
see Moskovskoe matematicheskoe obshchestvo
Moscow Physics Society
see Moskovskoe fizicheskoe obshchestvo, imeni I. I. Lebedeva
Moscow school of mathematics, **4**:287b
Moseley, Henry Gwyn Jeffreys (1887–1915), 9:542–545; **3**:564a; **12**:504a
atomic theory, **6**:365b; **7**:469a; **12**:31b–32a; **13**:546b
radioactivity, **12**:505b
spectrometer, **11**:531a
spectroscopy, **2**:245a, 399b, 488a; **12**:44a
Moseley, Henry Nottidge, **9**:542a, 588b; **13**:341b
Moser, Albert, **15**:71b
Moser, James, **6**:244b
Moser, Paul, **4**:404b
Moses (prophet), **9**:30b, 509a; **11**:44a; **14**: 296a
Moses of Khoren, **10**:300a–b
Moses of Vercelli, **7**:65b
Moskovskoe fizicheskoe obshchestvo imeni I.I. Lebedeva, **4**:305b
Moskovskoe matematicheskoe obshchestvo, **3**:222b; **15**:65b
founding of, **2**:425a; **3**:592a; **10**:544b
Mosquito, **2**:591b
and disease
equine encephalomyelitis, **7**:286b–287a
malaria, **5**:503a–b; **6**:524b; **7**:428a, 614a; **8**:66a; **9**:82a–83a; **11**:555b–556b; **12**:315a, 482a
yellow fever, **3**:94b; **4**:620a–b; **6**:524b; **11**:346a–347a
life cycle, **9**:82a
taxonomy of, **6**:524b
Moss, Herbert, **3**:20a
Moss, William Lorenzo (1876–1957), 9:545–546
Mosses, **6**:489a, 492b–493a; **15**:149a
anatomy, **9**:368b
classification of, **6**:219b; **13**:178a
development, **10**:13a
and liverworts, **6**:219b
reproduction, **3**:422a; **6**:219a–b; **11**:154a–b; **13**:542b
spermatozoids, **10**:576a
water transport, **12**:167b
see also Bryology, Cryptogams
Mosso, Angelo (1846–1910), 9:546–547; **2**: 2a, 27a, 62b, 339a, 366b; **7**:504b

Mossotti, Ottaviano Fabrizio (1791–1863), 9:547–549; **1**:135b; **2**:105a; **4**:161b; **9**:216b
electricity, **8**:106b; **14**:33a, 209a
gravitation, **11**:173b
Mossotti-Clausius equation, **6**:163b; **9**: 548b
Moths, **3**:320a; **9**:575b; **10**:273a
clothes moth, **11**:332a
genetics, **6**:22a
melanism, **3**:572b; **5**:453b
winter, **2**:4b
see also Lepidoptera
Motility, **4**:264a
extrapyramidal, **1**:546a
Motion
absolute, **8**:158b–159a; **10**:70a
arguments against, **2**:17b
vs. relative, **2**:330b–331a; **5**:64a
categories of, **13**:185b–186a, 188a
causes of, **1**:260a; **2**:590b; **13**:206b
central, **2**:145a
centrally accelerated, **2**:49a, 54b
circular
ancient discussion, **1**:256a
medieval discussion, **15**:496a
Renaissance discussion, **1**:607a; **3**: 403b–404a
seventeenth-century discussion, **5**:243b; **6**:607a, 485b, 486a, 487b; **10**:62b, 72a
eighteenth-century discussion, **7**:230b–231a
nineteenth-century discussion, **2**:168a
velocities of, **2**:168a
communication of, **3**:74a–b
composite, **1**:474b, 607a; **3**:183a; **5**:243b, 288a; **8**:229b–230a; **9**:174b; **11**:487b, 488b
acceleration in, **3**:418a
of compressible fluids, **3**:195b
concept of, **6**:602b–603a; **7**:485b
conservation of, **5**:243b; **6**:602b–603a; **8**:152b–153a, 470b, 471b; **9**:115b; **11**:507a
constraints upon, **1**:112a
continuance of, **1**:419b
continuous, **2**:512a
curvilinear
medieval analysis of, **1**:94a
cycloidal, **3**:432b
elliptical, **2**:101b; **10**:72a–b; **15**:350b
equations of, **1**:194b–195a; **4**:324a, 326b, 328a, 329a–b, 480a; **6**:89a–b; **7**:10a; **9**:324b; **11**:13b; **13**:130b, 139b; **14**: 306a; **15**:286b, 288b, 326a, 337a, 339a, 351a, 353b, 484b–485b
canonical, **6**:89b; **7**:54b
and ether, **5**:20a–b; **6**:346b–347a, 453a; **8**:40a–b, 444a, 493b; **9**:215a–b
of fluids
see Hydromechanics
and force, **9**:51a
forced
see Motion, violent
gravitational, **3**:158a; **14**:149b
infinitesimal, **3**:145a
at an instant, **1**:256a; **4**:480a
Keplerian, **10**:74b
quantization of, **2**:245b
laws of
medieval study, **7**:116a–b
Renaissance study, **8**:230a–b; **9**:133a–b; **10**:161a–b
seventeenth-century study, **4**:52b–60a; **7**:194a; **10**:62a, 70a–b, 72a
eighteenth-century study, **1**:111b–112b; **4**:480a; **5**:511a; **9**:187b; **15**:350a,b
nineteenth-century study, **2**:214a; **3**: 74b–75a, 78a; **6**:347b; **10**:449a
and light, **4**:378a; **5**:20a–b; **7**:26b; **8**:495a; **9**:154b; **12**:69a–b
in a liquid, **7**:344b–345a; **14**:621a
geometric interpretation, **3**:195a
resistance to, **8**:284b
local, **3**:420b; **10**:314b; **13**:197b, 201a–207b
in mathematics, **10**:48b
reality of, **1**:100b; **4**:237b
velocity of, **4**:237b; **6**:378b–379a; **13**: 186a, 187a, 188a
and mass, variation of, **9**:324a–325a
and matter, **4**:52a, 54a; **9**:214a; **12**:601b; **15**:127b, 143a
measurement of, **2**:590b; **9**:236a; **13**:185b–187a, 195a, 206b, 207a–b
medium, effects of, **1**:254a; **7**:135b, 487a
multiply periodic, **2**:246a
natural
ancient Greek discussion, **1**:260b
Byzantine discussion, **7**:135b–136a
medieval discussion, **7**:116b; **9**:136b; **13**:186a
Renaissance discussion, **1**:607a; **8**:228b, 229a
and fatigue, **1**:409b
laws of, **1**:254a–b; **4**:382a
neutral, **5**:239a, 247a
nonholonomic systems, **3**:195a
optics of, **6**:453a
original time of, **1**:410a
parallelogram of, **1**:111b; **9**:116a
perception of, **3**:116a
perpetual, **3**:39a, 404a; **14**:366a
impossibility of, **3**:66b, 75a, 81a; **4**:44a; **8**:209a, 230b; **13**:379a, 382a–b
machines, **3**:388b; **5**:510a; **6**:141b; **8**: 208b–209a; **10**:536b
magnetic, **10**:536b
see also *Perpetuum mobile*
photographic analysis of, **7**:77a
of a point, **1**:256a,b; **2**:330a, 424b; **4**: 480a,b; **11**:488b; **12**:599b; **13**:436b
projectile
ancient Greek discussion, **1**:198b, 229b, 258a; **7**:486b
Byzantine discussion, **7**:136a
medieval discussion, **1**:28a, 94b, 100b, 198b, 410a; **2**:605b–607a; **4**:157b; **5**:113b–114a, 116a; **7**:116b; **9**:137a–b; **10**:223b–224a, 420b–421a
Renaissance discussion, **1**:607a, 495b; **2**:590b; **3**:66b; **4**:237b; **8**:228a–230a; **9**:488a–b; **12**:106b; **13**:260a
seventeenth-century discussion, **5**:246a, 247a, 529a; **6**:379a; **10**:72a,b; **13**: 433b, 436a–437a
nineteenth-century discussion, **13**:237a; **15**:489b
and earth's rotation, **15**:355b, 489b
initial acceleration, **2**:590b
photographic study, **8**:597a
in resisting media, **6**:607b; **9**:119a,b; **14**:621b
see also Ballistics
quantity of, **10**:70a
quantum theory, **7**:27a
random, **5**:483b
rectilinear, **1**:279b, 326b; **3**:403b–404a; **7**:230a–b
relative, **3**:417b–418a; **4**:592a; **5**:86b, 243b; **6**:602b, 607a; **7**:230a–231a, 303a; **8**:152b, 158b, 599a–b; **10**:225b, 449a; **14**:611a
resistance to
see Resistance, mechanical
rigid, **3**:323a
of rolling bodies, **9**:325a
rotational, **2**:393b–394a; **3**:195a; **4**:480a; **5**:195b, 239a, 360a–b; **8**:227b
transformation into rectilinear motion, **3**:228b–229a; **13**:234a
sets of, **2**:145a
of a solid body, **3**:195a
about a fixed point, **4**:480a; **13**:130b; **14**:620a
and sound, **1**:233a; **4**:167b; **5**:20a; **8**:596a
spiral, **3**:145a, 516a; **10**:72a
stability of, **8**:284b, 560a–562b; **14**:620a
stationary, **8**:284b
terminus of, **2**:610b–611a
theory of
ancient Greek, **1**:100b–101a, 149a, 253a–256b *passim,* 258a, 260a,b, 279b; **3**:403b–404a; **4**:382a; **6**:312a; **7**:1a; **13**:93a–b; **14**:606b, 608b–609b, 610b–611a
Byzantine, **7**:135b–137a
medieval, **1**:94a, 409a–410b; **5**:115b–116a, 402b; **9**:136b–137b; **12**:429a; **13**:185b–187a, 197a–199b, 201a–207b; **15**:495b–496a
Renaissance, **3**:172a, 403b–404a, 420b; **8**:199b, 228a–230a; **12**:106a, 135b–136a; **13**:349b–350a
early seventeenth-century, **4**:52b–60a; **5**:238b–239a, 240b–241a, 288a; **6**: 445b
late seventeenth-century, **8**:151b–153a, 158b–159a; **10**:61b–76b; **11**:507a
eighteenth-century, **3**:486a; **7**:230a–231a; **10**:80a; **12**:601b–602a
nineteenth-century, **5**:193a
fluxus formae, **1**:100b; **2**:610b–611a; **5**:116a
forma fluens, **1**:100b; **2**:610b–611a; **5**: 116a
transformation of, **2**:245b–246a; **6**:2b; **8**:210b; **9**:477a; **12**:211a
transitive, **2**:145a
types of, **2**:590b
uniform, **1**:253a–b; **2**:393b–394a; **5**:246a; **10**:228b; **15**:731b
circular, **2**:348b, 349a; **6**:199a; **7**:137a
uniformly accelerated (uniformly difform), **1**:326b; **6**:127a, 376b, 378b–379a; **10**:228a–b; **12**:548a; **13**:206b
in vacuum, **1**:198a, 254a–b, 255a, 410a, 424b; **2**:606b–607a; **5**:402b; **7**:135b; **9**:16a
violent, **1**:253a–b, 258a, 607a; **3**:404a; **7**:136a; **9**:137a
vortical, **5**:188a; **8**:227b; **10**:76a–b; **13**: 237a
wandering, **2**:145a
see also Acceleration, Falling bodies, Gravitation, Intension and remission, Kinematics, Rotation, Velocity
Motion pictures
sound, **4**:7a, 253b; **7**:108a
stereoptics, **11**:306a
see also Cinematography
Motion sickness, **3**:353a
Motoneuron
concept of, **12**:399b–400a
Motors
efficiency of, **9**:155a
electric, **4**:248a,b; **7**:56a; **9**:205a; **10**:268b; **12**:131b; **14**:628a
alternating-current, **4**:588b–589a; **13**: 286b, 361b
compound winding, **4**:40a
and generator, **8**:188a

Motors (cont.)
  induction, **13**:286a–b
    power of, **8**:188a–b
  electromagnetic, **11**:46b
  thermic, **1**:138a
  traction
    coupling of, **6**:504b
  usefulness of, **10**:3b
  work delivered by, **11**:165a
  see also Engineering, mechanical; Engines
Mott, Nevill Francis
  Theory of Atomic Collisions, **15**:42b
Motte, Andrew, **10**:53a, 501a
Mottram, James Cecil (1879–1945), 9:549–551
Mouchez, Ernest Barthélémy (1821–1892), 9:551–552; **4**:68b; **8**:279b*; **13**:423a
Moufet
  see Moffett, Thomas
Moulton, Forest Ray (1872–1952), 9:552–553; **2**:198a; **3**:191a; **6**:34a*, 399b; **8**:350b*, 617b; **13**:269a
  cosmogony, **7**:85b; **8**:522a
  cosmology, **12**:280b*
Mount Egmont (New Zealand)
  height of, **8**:101a
Mount Hood, **6**:14a
Mount Lassen
  zoology, **5**:545b
Mount Rainier National Park, **14**:402a
Mount Wilson Observatory, **6**:529a, 574b; **10**:473a; **11**:470a
  early development, **1**:55a–56a
  founding of, **6**:29a–b, 32b
Mountain sickness
  theories of, **7**:505a
Mountaineering, **6**:551a
Mountains, **2**:551b
  ages of, **2**:341b; **3**:594a; **4**:349b; **5**:380a
  in Mesopotamian myths, **15**:637a, 640a
  Paleozoic, **1**:394a
  snow line, **12**:300b–301a
  structure of, **2**:191a; **5**:444a–b; **8**:147a; **9**:135a, 533a; **11**:272b, 505a–b; **12**:121b–122b, 262b; **13**:146a; **15**:55b
  vegetation, **14**:117a
  see also Altitude; Orogenesis; specific mountains and mountain ranges; Tectonics
Moureu, Charles, **4**:20b; **5**:99b; **12**:46b
Mourey, C. V.
  La vraie théorie des quantités négatives, et des quantités prétendues imaginaires, **1**:240a
Mouse, **10**:16a; **13**:354b
  dancing, **14**:550b
  diseases of, **8**:450a
  embryology, **7**:94b
  experimental use of, **14**:236a
    psychology, **14**:550b
  genetics, **3**:493a, 555b; **5**:10b
  yellow, **3**:493a
Mousnier, Pierre, **4**:505b
Mousterian remains, **2**:346b; **5**:109b; **8**:44a, 464a
Moutard, Théodore Florentin (1827–1901), 9:553–554
  transformation of, **4**:338a*
Mouton, Gabriel (1618–1694), 9:554–555
Moxa, **1**:125a; **7**:205a
Mozart, Wolfgang Amadeus, **7**:58a; **9**:326a
Moziño
  see Mociño, José Mariano
Ibn Muʿādh al-Jayyānī
  see al-Jayyānī
Muʾayyad al-Dīn al-ʿUrḍī, **13**:510b
Mucha, Viktor, **7**:624a
Mucic acid, **8**:340b
  discovery of, **12**:148b
Mucilages, **9**:79a
Mucor, **6**:100b; **10**:381b; **12**:208a
Mucosa
  gastric, **1**:613b
Mucus
  formation of, **5**:43a; **6**:255b, 269a; **14**:143b
  nasal
    lysozyme in, **5**:29b
  protective function, **5**:43a
  secretion of, **5**:43a
Mud
  analysis of, **7**:184a
  saline, **4**:598b
Mud volcanoes, **1**:20b, 163b; **6**:238b
Mudge, John, **12**:413b
Mudge, Richard Zachariah, **15**:152b
Mudge, Thomas, **9**:163b
Mueller
  see also Müller
Mueller, Rudolf, **7**:624a
Muffet
  see Moffett, Thomas
Mügge, Otto, **13**:246a
Muḥammad ibn ʿAbd al-Bāqī, **4**:445b
Muḥammad ibn ʿAlī Sipāhīzāde, **1**:29a
Muḥammad ibn Ibrāhīm al-Abūlī, **1**:437b
Muḥammad ibn Mūsā al-Khwārizmī
  see al-Khwārizmī
Muḥammad ibn Mūsā ibn Shākir
  see Banū Mūsā
Muḥammad ibn Shādān, **9**:36a
Muḥammad ibn Yaḥyā al-Sharīf, **1**:437a
Muḥammad ibn Yūsuf al-Warrāq, **1**:414a
Mühl, Karl von der, **15**:158b
Mühlenfells, Heinrich, **12**:307a
Muhūrta (time unit), **15**:539b
  definition of, **15**:536b–537a, 541a
Muḥyi ʾl-Dīn al-Maghribī (Muḥyi ʾl-Milla wa ʾl-Dīn Yaḥyā ibn Muḥammad ibn Abi ʾl-Shukr al-Maghribī al-Andalusī) (*fl. ca.* 1260–1265), 9:555–557; **7**:80a; **13**:510b, 511a
  editions by, **4**:440b, 441b, 444b, 453b*
  Zīj, **14**:578a
Muir, Matthew Moncrieff Pattison (1848–1931), 9:557; **3**:341b*; **13**:389b*; **15**:458a
Muir, Thomas, **13**:571b; **15**:226a, 295a
Muirhead, Alexander, **8**:444b
Mukunda
  commentaries of, **2**:116a
MUL•APIN (astronomical compendium), **15**:676a–b, 677b–678a
Mulberry tree, **12**:317a
Mulder, Gerardus Johannes (1802–1880), 9:557–559; **2**:95b; **3**:334a; **4**:162b; **6**:137b; **9**:456b; **10**:150b
  botany, **6**:138a; **8**:348b
Mulders, Gerardus Franciscus Wilhelmus, **9**:415a
Mule, **2**:580a; **4**:106a
Müller
  see also Mueller
Müller, Christian Frederik, **9**:575b
Müller, Franz (Ferenc), Baron de Reichenstein (1740–1825), 9:559; **5**:340b; **7**:391a
  Müller glass, **9**:559b
Müller, Friedrich, **8**:519a
Müller, Fritz (Johann Friedrich Theodor) (1822–1897), 9:559–561; **11**:217b*; **14**:233a
  botany, **7**:417b
  Für Darwin, **7**:475a; **9**:332a
  embryology, **6**:436a
  evolution, **14**:236b
Müllerian mimicry, **1**:502b–504a; **9**:559b, 560b–561a; **14**:136a–b
Müller, Georg Elias (1850–1934), 9:561–563; **4**:557b; **11**:475b
Müller, Gerhard Friedrich, **4**:472a; **5**:428a,b; **7**:495a; **8**:470b; **13**:28a
Müller, Gustav (1851–1925), 9:563–564; **11**:542a; **14**:55b, 57a*
  Geschichte und Literatur der veränderlichen Sterne, **6**:150a
Müller, Heinrich, **4**:275b, 276a; **11**:214b; **15**:166a
Müller, Helmut Otto Theodor, **6**:540a
Müller, (Heinrich Ludwig) Hermann, **9**:559b
  Fertilization of Flowers, **7**:417b; **8**:518a; **13**:352b
Muller, Hermann Joseph (1890–1967), 9:564–565; **3**:122a; **13**:134a; **14**:424b, 432a, 433b, 435a; **15**:506b
  genetics, **9**:521b–523b *passim;* **10**:276b; **13**:136a–b, 137a; **14**:101a
  translations of works, **15**:507a
Müller, Iwan von, **5**:228a
Müller, Johann
  see Regiomontanus
Müller, Johann (Heinrich Jacob) (1809–1875), 9:565–567; **11**:111b*; **13**:274a
  Lehrbuch der Physik, **12**:218b
Müller, Johannes, **2**:412b, 413a
Müller, Johannes Peter (1801–1858), 9:567–574; **2**:162a; **3**:88a; **4**:289b–290a; **7**:368a; **15**:170a
  acquaintances of, **1**:476b; **6**:269a, 553a; **9**:259a; **11**:368b; **12**:270a; **13**:556a; **14**:40b; **15**:165b
  anatomy, **7**:23b; **8**:302a; **11**:380a
  on contemporary scientists, **11**:592b
  correspondence of, **11**:382b; **12**:202b
  critics of, **13**:556b
  and (Emil) du Bois-Reymond, **4**:200a,b, 201a, 203a,b; **9**:570a
  embryology, **5**:514b; **13**:33b
  Handbuch der Physiologie, **9**:177b; **12**:240b
    translations of, **12**:354a
  influence of, **2**:109b, 596a; **6**:7b; **14**:2b; **15**:503b
  influences on, **1**:479b, 543a; **6**:140a; **7**:235a; **11**:593a; **12**:156b
  Müllerian duct, **1**:421a–b; **9**:570b
  neurophysiology, **1**:583b, 584a; **5**:445b; **6**:245b–246a
    reflexes, **6**:60a
  physiology, **1**:404a
    vision, **6**:248a–b; **11**:565b
  Pouillet's Lehrbuch der Physik und Meteorologie, **11**:570a
  students of, **2**:124a, 160b, 531a; **3**:336b, 611b; **4**:614b; **6**:7a, 241b–242a, 268a–b, 269b, 434a; **7**:437b, 498b; **8**:515b; **9**:431b, 560a; **10**:579a, 580b; **11**:360b, 367b, 368a; **12**:164b, 173b, 192b–193a, 231a, 240a–b; **14**:40a, 527a
  vitalism, **12**:240b, 242b
Müller, Otto Frederik (1730–1784), 9:574–576; **12**:107a, 555b
Müller, Paul (1899–1965), 9:576–577
Müller, Walther
  Geiger-Müller counter, **5**:332a
Müller, Worm, **8**:324a
Müller-Breslau, Heinrich (Franz Bernhard) (1851–1925), 9:577–578; **3**:496b
Mullet, **1**:263b
Mulliken, Robert Sanderson, **8**:474b; **12**:419a–b
Multinomial theorem, **8**:165b

Multiple proportions, law of, **8**:468b; **11**: 170b, 171a
experimental proof, **13**:373a; **14**:488a
and organic chemistry, **4**:243a; **8**:57b
Multiples (math.)
least common, **4**:421b; **8**:420b
Multiplicatio specierum, **1**:379a; **6**:191b
Multiplication, **4**:489a; **7**:176a–b; **13**:544b–545a; **14**:539a,b, 542a–543a
algebraic, **7**:241b, 243a; **15**:197a,b
complex, **1**:16a; **5**:206a; **6**:214a–b, 390a, 547b, 572a; **13**:216a; **14**:202b
definition of, **12**:15b–16a
distributive, **15**:196b
Egyptian techniques, **15**:684a–685a, 690a–b
geometric, **4**:56b; **9**:431a
logical, **2**:295b–296a; **8**:154b
Mayan techniques, **15**:769a
table of, **2**:233b; **7**:533a; **10**:540a; **15**:155a, 668a, 693a
Multiplier, electrical, **1**:347b; **3**:329b; **4**:201b, 202a,b; **10**:188a, 190a; **11**:50a; **12**: 254a–b; **14**:73a, 78a–b
mathematical theory of, **10**:189b
Mu-meson
see Muon
Mumia, **10**:307a
Mummies
animal, **3**:526b; **5**:356b; **10**:16b
human, **4**:353b
histology, **11**:596a
Mummification, **1**:596a–b
al-Munajjim, Abu'l-Ḥasan ᶜAlī ibn Yaḥyā, **15**:231b, 234b, 236a,b, 237a
Muncke, Georg Wilhelm (1772–1847), 9: 578–579
Munich
Bayerische Akademie, **7**:597a
Bogenhausen Observatory, **7**:608a
botanical gardens, **5**:438a; **9**:149a
herbarium, **9**:149a
mental hospital, **5**:570b
nutrition, **9**:145a
Polytechnikum, **4**:269a
public health, **10**:559a–b
Royal Institute for Experimental Forestry, **10**:427a
State Paleontological Collection, **2**:489a–b
Technische Hochschule, **5**:64a; **8**:365a–b
Munich, University of
in nineteenth century, **12**:154a
in twentieth century, **12**:530a–531a
anatomical amphitheater, **4**:146b
botany at, **9**:149a
chemistry at, **1**:390b; **6**:481a; **9**:354a; **14**:444b
paleontology at, **14**:626a, 627a
physics at, **11**:7b; **12**:527b, 529b–530a; **14**:339a
physiology at, **12**:421b
Munier-Chalmas, Ernest Charles Philippe Auguste (1843–1903), 9:579; **6**:168a; **12**:183a
Munīśvara Viśvarūpa (*b.* 1603), 9:579–580; **11**:290a; **15**:626a
commentaries of, **2**:116a, 118a
critics of, **7**:219b; **15**:616b
Siddhāntasārvabhauma, **15**:614b–615a
sine table, **15**:614b
Muñjāla (*fl.* 932), 9:580; **10**:314a*; **15**:625b
astronomy, **15**:582b, 586b
Bṛhanmānasa, **15**:618a
Laghumānasa, **15**:618a–619a
sine table, **15**:614a, 618b
Munk, Hermann, **4**:201a; **6**:440b, 441a; **10**:580a
Münster, Sebastian (1489–1552), 9:580–581
Compositio horologiorum, **6**:144a
Cosmographia, **1**:158b
Münster, University of
in nineteenth century, **6**:438b
Muon, **3**:371a; **6**:356a; **11**:117a–b; **15**:27a
Muqaddasī
see al-Maqdisī
Mural circle (astron. instrum.), **2**:471b, 472a; **5**:271b; **6**:263b; **10**:591b
Muralt, Johannes von (1645–1733), 9:581–582
Murchison, Charles, **2**:575b
Murchison, Roderick Impey (1792–1871), 9:582–585; **2**:569a; **4**:27b
acquaintances of, **2**:288b; **3**:604a; **5**:436b, 437a; **7**:319b; **8**:103b, 104a; **12**:524a; **13**:620a,b; **14**:397b; **15**:151b, 527b
correspondence of, **3**:297b; **9**:388b; **12**: 38a; **15**:153a
critics of, **2**:570b; **4**:143b; **7**:45a
erratic blocks, **3**:210b
and Geological Survey, **4**:11a; **5**:334a–338a *passim*
The Geology of Russia in Europe and the Ural Mountains, **7**:320a
translations of, **10**:266a
influence of, **4**:365a; **8**:461a, 462b; **11**: 501b; **12**:525a
and Lyell, **8**:566a–b
paleontology, **11**:500b
and Sedgwick, **9**:583a, 583b–584b; **12**: 276b, 277a–b
Silurian System, **1**:468b–469a; **5**:14a
stratigraphy, **3**:236b; **6**:238b; **7**:250b; **8**: 33a, 486a,b; **9**:389b; **12**:514b; **13**: 620b–621a
tectonics, **2**:570a
Murdoch, Patrick, **13**:224a
Murexid, **8**:343b
Muriatic acid, **1**:143a; **2**:552b
composition of, **3**:602b; **5**:411b; **6**:284b; **8**:77b
see also Hydrochloric acid
Murlin, John Raymond, **1**:441a–b
Murmansk
biological station, **4**:42b
expeditions to, **7**:410b
Murmur
cardiac, **3**:220a; **12**:450b
respiratory, **3**:220a
Murphy, James Bumgardner (1884–1950), 9:586–587
Murphy, Robert, **13**:80a; **15**:200b, 386a
Murphy, William Parry, **9**:417a
Murrain, **13**:495b
Murray, Alexander, **8**:461b
Murray, Francis Joseph, **14**:90b, 91a,b
Murray, George Redmayne, **13**:589b
Murray, George Robert Milne (1858–1911), 9:587–588
Murray, Johan Anders, **10**:531a,b; **14**:224b
Murray, John (1841–1914), 9:588–590; **1**: 71b; **3**:161a; **6**:442b; **11**:372a; **13**:360b
Murray, John (chemist), **3**:604b; **6**:588a*; **11**:437b
Murs
see John of Murs
Mus
see Mouse
Mūsā ibn Muḥammad ibn Maḥmūd al-Rūmī Qāḍīzāde
see Qāḍī Zāda al-Rūmī
Mūsā ibn Shākir, **1**:444a
Mūsā ibn Shākir, sons of
see Banū Mūsā
al-Musabbiḥī, **14**:579a–b
Muscardine, **1**:329a, 492b–493b; **10**:373a; **12**:202b–203a
Muscarine, **15**:100a
Muschelkalk (shell limestone), **1**:95b; **2**: 109a,b
Muscles
anatomical description
Renaissance, **1**:204a, 619a, 620b; **3**: 40a,b, 99a,b, 343a, 355b; **4**:520a; **8**: 202b; **13**:569a; **14**:9a
seventeenth-century, **13**:32b
eighteenth-century, **5**:226b; **13**:122b; **14**:450a
nineteenth-century, **2**:531a; **7**:278b, 438b; **8**:125b
atrophy
see Atrophy, muscular
chemical composition
lactic acid, **2**:92b; **6**:501a–b
nitrogen, **5**:90b–91a
proteins, **9**:359b; **13**:53b
respiratory pigments, **7**:273a–b
contraction of
ancient theories, **4**:385b
seventeenth-century theories, **3**:482b–483b; **4**:63a; **5**:426a–b; **6**:256a; **9**: 243b, 244b; **10**:54b, 520b; **13**:32b–33a, 170a–b; **14**:407b
eighteenth-century theories, **2**:3a, 37a; **5**:56a; **6**:38a; **7**:274b; **10**:278b; **13**: 122a–123a; **14**:450a
nineteenth-century theories, **4**:248a*, 372a; **6**:242a; **10**:257a; **11**:566a; **13**: 452b; **14**:492a–b
early twentieth-century theories, **8**: 533a–534a; **9**:354b; **11**:126a
all-or-none law, **2**:366b; **7**:504b; **8**: 533a–b; **11**:125b–126a
chemicals, effects of, **11**:463a
energy relations, **9**:359b
heat production, **3**:220a; **4**:264b, 616a; **5**:184b; **6**:225a–b, 243b; **9**:237b, 359a, 547a
induced, **4**:201b, 202a; **9**:177b; **11**:133a, 296b, 474a; **13**:33a; **14**:253a
latent period, **11**:427a
measurement of, **4**:616a; **5**:184b; **11**: 426b–427a; **12**:240b–241a
mechanics of, **9**:101b
refractory period, **5**:56a
reversibility of chemical processes, **4**: 360b
secondary, **11**:427a
thermodynamic analysis, **11**:446a
torsional movement, **4**:614b, 615a
total energy output, **6**:225a–b
treppe, **5**:56a
types of, **2**:339b
and vasodilatation, **5**:280a
and volume, **13**:170a–b
efficiency of, **9**:359b
electric currents in, **1**:303b; **2**:599a; **4**: 203a; **7**:438b
autogenous, **4**:202a
injury current, **4**:201b, 202a; **9**:177a–b
molecular interpretation, **4**:202a
resting potential, **9**:177a–b
electric stimulation of, **3**:15b; **4**:202a–b, 616a; **5**:267b; **7**:505a; **8**:533a; **9**:505b, 570a; **13**:547b
laws of, **2**:532a
fibers of, **8**:9a; **13**:231b
cellular nature of, **7**:438a
consistency of, **7**:520a
models of, **4**:372a
stimulation of, **11**:125b–126a
sulfuric acid, effects of, **2**:386a

Muscles (cont.)
histology
ancient theories, **4**:385b
seventeenth-century, **3**:482b; **13**:32b
eighteenth-century, **1**:392a
nineteenth-century, **1**:477a, 540a; **2**:376a,b; **6**:287b; **7**:438a, 520a; **8**:401a,b, 556b; **12**:231b; **13**:555b–556b
twentieth-century, **2**:339b; **6**:224a; **7**:502b
measurements of, **2**:376a
metabolism, **4**:360a–b; **5**:37b–38a, 126b; **7**:503a; **8**:614b; **9**:359a–b; **10**:326b; **11**:585a; **13**:452b; **14**:454b
glucose, **3**:220a
inogen theory, **5**:37b, 38a
inorganic ions, **5**:184b
lactic acid, **2**:92b; **5**:37b–38a, 184b; **6**:225b, 501a–b
oxygen consumption, **5**:184b; **7**:502b
theories of, **9**:238a–b
neurophysiology
eighteenth-century study, **6**:38a, 64b–65a
nineteenth-century study, **2**:111a, 532a; **4**:202b, 616a; **5**:38a*; **6**:225a–b; **7**:505a, 520a; **9**:505b, 547a, 562a; **12**:240b–241a, 319b
twentieth-century study, **1**:569b; **4**:397b; **5**:208a; **8**:532b–535a; **9**:19b–20b
antagonistic action, **12**:396b, 397b
comparative, **8**:534b
end plates, **11**:566b
excitability, **7**:505a, 520a; **14**:200b–201a
fatigue, **5**:37b–38a, 56a; **7**:504a; **9**:547a; **14**:105b
heat rigor, **7**:520a
irritability, **10**:257a
pseudomotor phenomenon, **6**:225b
sensory receptors, **5**:185a; **12**:396b
summation, **8**:533b–534a
tonus, **6**:60a, 225a, 265a; **9**:20a
vasomotor reactions, **5**:279b–280a; **6**:225b
nomenclature, **1**:523a; **4**:199a
perfusion of, **5**:184b
pharmacology, **9**:19b
preservation of, **5**:38a
relaxation of
measurement of, **11**:427a
sarcoplasm, **2**:339b
smooth, **1**:392a; **4**:616a; **7**:438a; **9**:456b; **12**:319b
see also Myography; Myology; Neurophysiology; specific muscles and parts of body
"Muscular sense," **1**:496b
Musée (de Monsieur), **10**:609a,b; **11**:166a, 167a
Musée Houiller
founding of, **1**:472b
Muséum National d'Histoire Naturelle (Paris)
in eighteenth century, **5**:92a, 356a; **7**:198b, 547a–b, 586b, 588b; **15**:111a–b, 113b
in nineteenth-century, **2**:282a,b, 346b; **4**:256a; **6**:553b; **13**:554a–b
anatomy at, **3**:520b–521a, 524b–525a; **5**:504b
botany at, **7**:196b, 198b; **13**:390b, 443b
entomology at, **8**:48b; **9**:408a
menagerie, **5**:359b
mineralogy at, **6**:178b
paleontology at, **1**:210b–211a; **5**:295b, 296b; **10**:221b
physics at, **1**:557a
zoology at, **5**:356a; **7**:547a–b, 586b, 588b; **11**:243a,b; **13**:551b; **15**:126b
see also Jardin Royale des Plantes (Paris)
Museums
anatomical
eighteenth-century, **6**:367b; **12**:41a
nineteenth-century, **7**:414b; **14**:533b
archaeological, **9**:538b; **13**:357b; **14**:533b
biological, **6**:567b
botanical, **6**:289a, 493b, 494b
ethnographical, **13**:358a; **14**:42b
geological, **7**:224a; **12**:276a; **14**:501a
maritime, **4**:257a
medical
eighteenth-century, **1**:398b, 399a; **3**:323b, 486b, 488a; **6**:569a
nineteenth-century, **3**:323b–324a, 488a; **6**:478b–479a; **11**:99a
twentieth-century, **7**:278b
mineralogical, **6**:385b; **12**:215b–216a; **14**:240a
natural history
Renaissance, **13**:397b
seventeenth-century, **7**:375a–b; **13**:450a, 451a–b; **14**:505b
eighteenth-century, **5**:55b; **10**:436b–437a, 438b–439a; **12**:556b, 560b; **14**:69b; **15**:111b
nineteenth-century, **4**:144a; **6**:184b; **7**:70a, 320a, 391a; **9**:432a, 588b; **13**:467a; **14**:88a; **15**:472a,b
oceanographic, **10**:503a, 522b
paleontological, **5**:295b, 296b
physics, **5**:55b
public, **1**:318a; **4**:269a; **13**:357b–358a; **14**:505b
purpose of, **5**:46a; **10**:438b–439a
technological, **13**:351b
zoological, **12**:215b–216a
see also individual museums under their titles or under specific city or university
Mushet, David (1772–1847), 9:590–592
Mushet, Robert Forester, **9**:591a,b; **15**:25a
Mushketov, Ivan Vasilievich (1850–1902), 9:592–594; 10:167a, 168a
Mushrooms, **8**:121b*; **11**:592b
cultivation of, **3**:429a; **9**:176a
fossil, **3**:45a
as fruiting bodies, recognition of, **4**:264b
membranes in, **9**:79a
parasitic, **3**:45a
poisonous, **8**:177b
poisons of, **14**:334b
sexual generation of, **4**:288b
see also Mycology
Music
ancient Greek, **1**:282a–283a
ancient Roman, **2**:230b; **3**:109b, 175b, 176a; **9**:141a
Byzantine, **11**:185b
medieval Arabic, **4**:525b–526a; **7**:357a; **15**:498a
medieval Latin, **6**:544a; **9**:363a; **14**:155b
Renaissance, **2**:401a; **3**:280b; **5**:238a; **9**:192a
seventeenth-century, **2**:507b
eighteenth-century, **1**:332b; **3**:258a
nineteenth-century, **2**:316b
twentieth-century, **5**:222a
aesthetics of, **4**:88b
and cosmology, **5**:47b, 48a
international pitch, **7**:445b
and mathematics, **1**:282a–283a, 599b; **2**:233b; **5**:249a–b; **7**:128b, 129a; **11**:222b
and medicine, **1**:332b
philosophy of, **5**:550b; **9**:319b
theory of
ancient Greek, **1**:231b, 232b–233a; **11**:26b–27a, 220b, 222b–223a; **13**:315b
ancient Roman, **2**:233b; **10**:112b–113a, 300b; **11**:201a–b; **13**:325b–326a
medieval Arabic, **1**:446a; **4**:525b–526a; **7**:326a–b; **11**:249b; **13**:292a; **15**:265b
medieval Latin, **6**:302a; **7**:128b; **11**:203a
Renaissance, **1**:606b; **5**:238b, 249a–250a; **13**:50b
seventeenth-century, **4**:46b; **5**:47b; **9**:319a–b; **14**:152b
eighteenth-century, **1**:115a; **3**:153b–154a; **6**:328a; **12**:127b–128b; **13**:265a
nineteenth-century, **6**:246a, 327b; **11**:9a
Chinese, **12**:376a
see also Acoustics, musical; Harmonics; Historiography, of music; Instruments, musical; Pitch; Terminology, musical
Musk
synthesis of, **3**:246b
Musket, **13**:592b
Muskhelishvili, Nikolai Ivanovich, **7**:453b
Muskūya (Miskawayh), **13**:513a
Muṣliḥ al-Dīn Muṣṭafā ibn Shaᶜbān al-Surūrī, **9**:604a
Muspratt, James, **12**:520b
Musschenbroek, Jan van, **5**:510b; **9**:595a,b; **12**:500b
Musschenbroek, Johan, **9**:595a
Musschenbroek, Joost Adriaensz, **9**:594b–595a
Musschenbroek, Petrus van (1692–1761), 9:594–597; 3:37b; **5**:510b; **6**:149a; **8**:460a; **9**:3b*; **13**:235b; **14**:70a, 333b*
acquaintances of, **4**:517a; **7**:596b
correspondence of, **3**:216a; **10**:147a
electricity, **2**:324b; **5**:131b; **7**:403a; **10**:146a; **11**:334b
Elementa physicae
translations of, **14**:354a
Natural Philosophy, **3**:537b
pyrometer, **4**:44a
students of, **7**:11a
Musschenbroek, Samuel, **9**:595a; **13**:173b
Mussel
anatomy, **12**:45b
culture of, **9**:432a
toxicity, **11**:430a
Musset, Charles, **4**:210b; **10**:372a
Mustard gas, **5**:128a
synthesis of, **5**:465b; **11**:87b; **14**:601b
Mustel, Evald Rudolfovich, **13**:408b
Mustelidae, **3**:439a*
al-Muᶜtaḍid (caliph), **7**:24b; **10**:5b, 6b; **12**:447a,b; **13**:289a
Mutation, genetic
see Genes, mutation of
al-Mutawakkil (caliph), **1**:36b; **4**:541b; **7**:64a; **13**:230b; **15**:231a,b, 261a
Mutawassiṭāt, **6**:189b
Muᶜtazilism, **9**:171a
Ibn al-Muthannā ibn ᶜAbd al-Karīm, Aḥmad, **4**:543b; **7**:223b, 361a
Muthmann, Friedrich Wilhelm, **5**:557a; **12**:572b
Mutis y Bossio, José Celestino Bruno (1732–1808), 15:429–430; 3:16b; **4**:346b; **6**:551a; **11**:605b; **13**:463b
Historia de los árboles de quina, **13**:464a–b
Quinología, **13**:464b
Mutoli, Pier Maria (pseudonym), **2**:310b
al-Muwaffaq al-Dīn Yaᶜqūb al-Sāmarrī, **1**:33a; **11**:251b; **15**:239a
Muwaḥḥids, **14**:637b

Muẓaffar al-Asfuzārī, **4**:439b
Muẓaffar al-Dīn Abu'l-Thanāʾ Maḥmūd ibn Aḥmad al-ᶜAyntābī ibn al-Amshāṭī, **9**:604a
Myasthenia gravis
  clinical description, **14**:407b
Mycelium
  anatomy of, **9**:368b
Mycenaean civilization, **10**:550b; **12**:180a,b
Mycobacteria, **6**:101b–102a
  chemotherapy, **4**:155b
Mycoderma, **6**:100a; **10**:364b, 365b–366a, 376b, 377a, 381b–382a
Mycology
  eighteenth-century, **6**:219b; **9**:368b
  early nineteenth-century, **4**:67a*, 264b, 288b; **5**:190b–191b, 192b; **10**:531a–b
  mid nineteenth-century, **2**:19a, 436a–438a; **3**:611a–614a; **4**:21b; **6**:73a; **10**:13a–b; **13**:489b–490a
  late nineteenth-century, **2**:436a–438a; **3**: 317b, 611a–614a; **4**:21b; **6**:100a–101a; **9**:79a, 587a–b; **10**:1a–b, 253b; **13**:299b; **14**:93a–b
  early twentieth-century, **2**:34b–35a, 534a, 582a–583b; **3**:318a, 429a; **4**:220a–b; **5**:584a–b; **9**:34b–35a, 175b–176a, 182b; **12**:125a
  mid twentieth-century, **2**:582a–583b
  cultural methods, **2**:437a–b
  medical, **5**:566a
  see also Fungi
Mycomelinic acid, **8**:343b
Mycorrhiza, **9**:78b
  in orchids, **2**:34b–35a
Mycosis, **10**:19a
Mydorge, Claude (1585–1647), 9:598–599; **1**:245b; **3**:622a; **6**:271b, 548b*; **8**: 271b; **10**:264a
  acquaintances of, **4**:46a; **5**:285a; **6**:112b; **10**:342a; **11**:486b
  astronomy, **10**:490a
  critics of, **6**:299a; **9**:528a
  geometry, **14**:148a
  optics, **4**:59b
Myelitis
  chronic, **8**:7a–b
Myelogenesis, **5**:27a–b
Myhill, John Renfred, **3**:278b
Mylon, Claude (*ca.* 1618–*ca.* 1660), 9:599
Mylonite, **5**:187a
Myocardiograph, **15**:100a
Myocarditis, **11**:374a
Myocardium
  histology, **8**:556b
  hypertrophy, **5**:366b
Myodynamics, **7**:519b, 520a
Myoglobin, **7**:520a
Myograph, **11**:426b; **12**:400b
  invention of, **6**:245a
  string, **4**:334a
Myography, **1**:303b
  isometric, **12**:399a,b
Myology, **3**:40b; **4**:172b; **5**:471a*; **10**:501a
  see also Muscles
Myosin, **7**:520a
Myrbäck, Karl David Reinhold, **4**:485b
Myriapods, **2**:422b; **4**:505a*; **5**:502b; **14**:496a
  spermatogenesis, **2**:345a
Myrmica, **8**:528b
Myrsinaceae, **3**:43a
Myrtaceae, **2**:519b; **8**:354a
Mysteries (Byzantine), **1**:38b*, 39b*
Mysticism
  ancient Roman, **14**:632a
  medieval Arabic, **11**:249b; **12**:3a; **13**:510a
  medieval Latin, **4**:92b
  Renaissance, **2**:222a–223b; **4**:6a; **5**:121a; **11**:97a; **14**:225a–226b
  seventeenth-century, **4**:550a–551b; **6**: 253b, 257b; **7**:485a, 488a; **9**:98a; **13**:174b
  nineteenth-century, **11**:474b
  twentieth-century, **2**:512b; **4**:282a; **5**:456b
  mathematical, **7**:307b
Mythology
  nineteenth-century study, **11**:137b; **12**: 154a
  twentieth-century study, **2**:210b
Myths
  ancient Greek, **6**:213a
  Amerindian, **2**:209a, 210b
  Mesopotamian, **15**:637a–641b
  psychology of, **7**:190b–191a
  and religion, **5**:145b
  and science, **3**:580a; **4**:159b; **5**:60a
Myxedema, **6**:518b
  congenital, **3**:415b
  treatment of, **2**:548a
Myxine, **11**:379b–380a
Myxobacteriaceae, **13**:299b
Myxomatosis, **12**:96b
Myxomycetes, **2**:437a,b
  developmental history, **3**:612a–b
  see also Cryptogams
Myxosporidia, **9**:328b
Myzostomum, **13**:355a; **14**:291b; **15**:430a

N

N rays, **2**:203a; **11**:481b*; **12**:29a; **14**:499a*
Nabarro, David Nunes, **2**:528b–529a
Nabataeans, **1**:350b; **14**:118a
al-Nabātī
  see Abu'l-ᶜAbbās al-Nabātī
al-Nabulusī, **12**:362b
Nabu-Rimannu, **14**:593b; **15**:678b
Nachtansicht, **4**:556b
al-Nadīm, Muḥammad ibn Isḥāq ibn Abī Yaᶜqūb, **4**:117a; **5**:612a; **7**:79a, 357a; **9**:501a; **10**:5b; **11**:244a; **13**:538a; **15**:125b, 237b, 263b
  on Euclid, **4**:438a–442b *passim*
  al-Fihrist, **1**:30b, 35a*–39b* *passim,* 226a, 507b, 508b, 509a, 513a; **4**:542a; **7**:39b, 223a, 334a, 531b; **9**:160a, 297a; **12**:447b; **13**:37b, 326b; **14**:118a; **15**:119a, 179a, 182a, 231a, 420b
Naegeli, Carl Wilhelm von (1817–1891), 9: 600–602; **3**:421b, 611b; **12**:58b, 60a, 259a
  acquaintances of, **2**:560a; **6**:221a; **7**:437b; **12**:255b, 421a; **15**:147a
  bacteriology, **5**:219a, 564a
  botany, **6**:219b; **11**:154a
  critics of, **7**:422a, 424a; **14**:96b
  cytology, **3**:337b, 423a
  embryology, **6**:338b, 339a
  heredity, **14**:234b–235a, 429b
  and Mendel, **9**:281b, 282a,b, 601b
  Das Mikroskop, **12**:255b
  students of, **3**:422a; **14**:181a
Ibn al-Nafīs, ᶜAlā al-Dīn Abu'l-Ḥasan ᶜAlī ibn Abi'l-Ḥazm al-Qurashī (or al-Qarashī) (*d.* 1288), 9:602–606; **12**: 324b
  commentaries by, **11**:251b
  Mūjiz al-qānūn, **15**:246b(n)
  Sharḥ al-Qānūn, **15**:498b, 499a
Nafīs ibn ᶜIwaḍ al-Kirmānī, **12**:90b
Nagakubo, Sekisui, **7**:20a
Nagana, **2**:528a–b
Nagaoka, Hantaro (1865–1950), 9:606–607; **6**:480a
Nagata, Masayoshi, **6**:394a
Nagel, August, **6**:239a,b
Nagendra Nath, S., **11**:266a–b
Nageotte, Jean, **8**:7a
Nāgeśa (*fl. ca.* 1630), 9:607; **15**:625a
Naiads
  propagation of, **9**:575b
Ibn Nāᶜima, ᶜAbd al-Masīḥ, **1**:275a,b; **15**: 263a
al-Nairīzī
  see al-Nayrīzī
Nairne, Edward (1726–1806), 9:607–608; **11**:284b
  critics of, **14**:419a
  instrumentation, **14**:452b
Najīb al-Dīn
  see al-Samarqandī, Najīb al-Dīn
Najm al-Dīn Dabīrān al-Kātibī al-Qazwīnī, **11**:248a, 249b; **13**:510b
Nakanishi school, **1**:90a
Nakṣatras (constellations), **2**:417b; **3**:584b; **4**:100a; **5**:275a; **15**:536b, 554a–b, 556b, 570a, 588b, 590b, 607b, 608a–b, 625a
  tables of, **15**:535a, 535b–536a, 537b, 541a, 565a–b, 578a–b, 594a–b, 599a–b, 610b, 613a, 617b, 621b, 625b
Nallino, Carlo Alfonso, **1**:507b–514a *passim;* **7**:337b, 358b, 361b, 362a; **9**:84a; **12**:162a; **15**:173a, 421a
Nametkin, Sergey Semenovich (1876–1950), 9:608–609
  Nametkin rearrangement, **9**:608b
Namur, basin of, **5**:476a
Nan Shan (Kunlun), **11**:181b
Nansen, Fridtjof (1861–1930), 15:430–432; **2**:168b; **5**:460a; **13**:166a; **14**:50b; **15**:124b
  Fram expedition, **4**:344a–b; **9**:443a; **15**: 431a
  Nansen ridge, **15**:431b
Nantes, Edict of, **5**:60a
Nantucket
  Maria Mitchell Association, **9**:421b
Naphtha
  distillation of, **14**:156b
Naphthalene, **2**:420b; **3**:347b; **5**:13a; **8**:31b; **9**:377a
  derivatives of, **1**:426a–b; **3**:321b; **8**:55b–56b, 59a, 329a; **9**:109b; **10**:515b
  discovery of, **13**:243b
  formula, **1**:288b, 426a–b; **4**:400a
  natural compounds of, **3**:46b
  synthesis of, **2**:65b; **8**:56a
Naphthenic acids, **3**:246b; **8**:56b, 57b
Naphthol yellow, **3**:85a
Napier, Archibald, **9**:609b
Napier, John (1550–1617), 9:609–613; **4**: 552b*, 619b
  and Briggs, **2**:462a–b
  logarithms, **4**:551a; **9**:159a; **10**:151b
  Mirifici logarithmorum canonis descriptio, . . . , **2**:462a; **7**:304b; **14**:514b
  Napier analogies, **9**:610b
  Napier's bones, **9**:610b
  Rhabdologia, **2**:462a
  translations of works, **14**:51a
Napier, Mark, **9**:610a
Napier, Robert, **9**:609b, 610a
Naples
  academies of, **11**:96a
  botanical garden, **11**:66b
  Geological Institute, **1**:491a
  observatory, **10**:592b–593a
  Stazione Zoologica, **9**:517a; **13**:63a; **14**: 424a, 434b
    founding of, **7**:401b; **15**:123a–125a
Naples, University of
  in late eighteenth century, **11**:66b
  founding of, **5**:147a

Napoleon I, Emperor
and archival materials, 2:542a; 4:437b
career, 3:526b; 5:166a, 301b, 356a
and education, 2:133b; 3:465b, 523b; 5:93a; 6:178b; 15:341b, 347a
military tactics, 3:131b; 7:499b
and science, 4:348b; 5:320b; 7:97b; 9:477a; 11:236b; 15:335b
scientists, relations with, 1:107b; 2:74a; 3:1b, 6a, 79a, 106b, 199a, 200a, 258b, 427a–b, 603a; 4:16a; 5:93b, 94a, 250b; 6:324a; 7:569b, 581b, 588b; 9:252b, 473b–474a; 10:327a; 11:165a; 12:137a, 286b; 13:418a; 15:342b, 345b–347a
political appointments, 3:72a; 5:92a, 603b; 7:547b; 9:473b–474a; 10:602b; 11:272b; 14:68b, 79b–80b; 15:346a–347a
see also Egypt, Napoleonic expedition to
Napoleon III (Louis Napoleon), 3:132b, 451a; 5:85a, 496b; 10:352b, 353b–354a, 384b–385a; 14:476a
Napoleon, Prince (Bonaparte, Joseph Charles Paul), 1:573a
Napoleonic Wars, 1:12b, 45a, 358b, 436a–b
Napp, Cyrill Franz, 9:277b, 278a
Nappes, 1:236a; 4:402b; 5:221b; 6:227a, 228b; 8:543b, 544a; 11:575b, 594a; 13:285a
facies relationships, 6:229a
Helvetic, 1:235b; 6:229a
involuted, 6:228a
origin of concept, 2:87b; 12:140a–b; 13:146a
Naquet, Alfred Joseph, 2:485b; 14:529b
Nārāyaṇa (*fl.* 1356), 9:613–614; 10:313a
Nārāyaṇa, son of Govinda, 9:579b
commentaries of, 2:115b; 5:274b; 7:315a
Nārāyaṇa Bhaṭṭatiri
Prakriyāsarvasva, 1:48b
Narceine
discovery of, 10:498a
Narcissus, 6:296a
Narcolepsy, 14:408a
Narcosis
effects on brain, 5:65b
theories of, 10:257b
Narcotine, 9:178a
discovery of, 11:494b
Nares, George Strong, 7:92a
Narwhal, 13:450a; 14:505b; 15:662a(n)
Naryn River, 12:300b
al-Nasawī, Abu'l-Ḥasan, ʿAlī ibn Aḥmad (*fl.* 1029–1044), 9:614–615; 4:607b, 608a; 7:325b, 335a, 531b, 532b
Näsijärvi (lake), 12:272b
Naṣīr al-Dīn al-Ṭūsī
see al-Ṭūsī, Muḥammad ibn Muḥammad ibn al-Ḥasan
Nāṣir-i-Khusraw, 11:325a; 13:513a
Nasmyth, James (1808–1890), 9:615–616; 4:19a
Nassau-Siegen, Count Joan Mauritz van, 9:122a–b; 10:621a; 13:169a
Natalis, Stephanus
see Nöel, Étienne
Natanson, Edward, 9:616a
Natanson, Władysław (1864–1937), 9:616–617; 2:120b; 4:562a; 12:498a*
Natanson's rule, 9:616b
Natanson's thermokinetic principle, 9:616b
Nathorst, Alfred Gabriel (1850–1921), 9:617–618; 6:78b; 13:345a,b
National Academy of Sciences (U.S.)
in nineteenth century, 4:179b, 180b; 6:86a; 12:434b
in twentieth century, 3:86a, 130b–131a, 484a, 509b; 6:30b–31b; 8:356b; 12:347b–348a
as advisory agent to government, 3:35b
founding of, 1:74a, 364b; 4:182a; 6:57b, 437b; 10:480a; 11:505a
Uranium, Committee on, 3:371a
National Advisory Cancer Council (U.S.), 4:499b
National Advisory Committee for Aeronautics (U.S.), 1:133a
National Aeronautics and Space Administration (U.S.), 5:433b
National Bureau of Standards (U.S.), 1:364b; 3:327a; 10:158a; 12:132a
and physics study, 1:133a
statistical methods, 14:553b–554a
National Commissariat of Health (USSR) Biochemistry Institute
founding of, 1:361b
National Conference of Standard Laboratories (U.S.), 14:554a
National Defense Research Committee (U.S.), 3:373a; 7:108a; 15:2b
National Electric Light Association (U.S.), 10:105b
National Electric Signalling Company (U.S.), 4:601b
National Geographic Magazine, 3:318a
National Geographic Society (Washington)
expeditions, 2:500a; 4:563b; 8:105a
founding of, 1:583a; 9:313b
National Herbarium (U.S.), 13:432b
National Institute of Physics and Chemistry (U.S.), 3:4a
National Medical College (Washington, D.C.), 12:481b
National Museum
see Smithsonian Institution
National Physical Laboratory (G.B.), 3:564b; 11:486a; 12:612b–613a; 13:102a, 105a
establishment of, 5:423b–424a
metallurgy, 11:549b–550a
National Radio Astronomical Observatory (Green Bank, W.Va.), 13:117a
National Research Council (Canada), 13:7a–b
founding of, 8:583b
National Research Council (U.S.), 3:372b; 5:40b; 8:117a, 356b; 9:396b, 515b; 10:104b, 105b, 157a; 13:519a; 14:298b, 550a; 15:73a
biology, 8:589b
founding of, 3:484a; 6:31a; 14:121a
National Society for Medical Research (U.S.), 3:69a–b
Nativism, 6:248a–b, 299b
Natural gas, 3:606b; 11:469a; 12:562a
analysis of, 9:608b; 10:289a
helium from, 3:437a
search for
anticlinal theory, 14:301a
technology of, 8:301b
Natural history
ancient Greek, 1:258b–266a; 2:256a–b
ancient Roman, 11:38b–40a
medieval Arabic, 1:29b–30a; 7:64a–b; 14:585a
medieval Latin, 1:565a; 11:469a–b; 13:347b–348b
Renaissance, 1:47b, 48a, 108b–109b, 526a; 3:13a; 4:412b–413a; 6:309b–310a; 8:190a, 391a; 10:280a–281a; 13:501b–503a
early seventeenth-century, 7:164b; 10:621b; 13:449b–450b, 451a; 14:505b
late seventeenth-century, 1:431b–432b; 2:305b, 522a–523a; 3:248a–b; 4:135b; 5:415a; 7:204b–205a, 458b–459a; 9:313a; 10:202a; 11:40a–41b, 47a–b, 313a–318a; 12:457a–b; 13:282a–b, 443b; 14:157a,b, 413a–b
early eighteenth-century, 1:243b–244a, 305a–306a, 434b; 2:15a; 3:97a, 129a–b, 344a, 486a; 4:40b–41a, 172b, 173a; 5:428a–429a, 577b; 7:200a–b, 210a–b, 495a–b; 8:374b–376b; 9:135a–b, 490b, 539b, 582a; 10:568b–569b; 11:42b–44a; 12:159a–b, 459a; 13:28a–b, 257a–258a, 563a; 14:84a, 193a–b; 15:111a–b, 271a–b
late eighteenth-century, 1:58b–59a, 351a–b, 433b–434b, 487a–488a, 488b–489b, 554a; 2:4b, 107a, 203b–204b, 315b, 321b, 322a–b, 474a–b, 576a–581b; 3:154b*, 198a, 365a–b, 397a, 522a–b; 4:156b; 5:75a–76a, 76b–77a; 6:539a, 566b; 7:57b–58a, 88b, 95b–96a, 544a, 547a–b; 8:251a–252a, 274b–275a; 9:458a; 10:283a–284b, 509b–510a; 11:472a, 564b–565a; 12:120a, 185a–b, 515a–516b, 535b–537a, 554a–b, 556a–b, 560a–562b; 13:530b–531a, 565a–566a; 14:299b–300a
early nineteenth-century, 1:328b–329a, 330a–331b, 484b–485b; 2:320b–321a, 566a–b; 4:67a, 288b–289b, 406b–407a; 5:224b, 276b–277a, 355b–358b, 518a; 7:69a–70a; 8:265a–b, 266a–b; 10:195a–b, 439b, 518a; 11:242b–243b, 261b–262a, 262b–263b; 12:72a, 202a; 13:353b–356a; 14:44a–b, 186b–187b, 328a–329a, 351b; 15:82a–83a, 472a–b
mid nineteenth-century, 1:72b–74b, 358b–359a, 387b–388a, 501a–b, 587a; 2:18b, 205a–206a, 282a; 3:549b–550b, 569b–570a; 4:308a–b; 5:51a, 66a–68a, 206b, 406b, 469b, 480b–481a; 6:280b, 437a; 7:420b; 9:344b, 428a–b; 10:221a–222a, 240a–b, 439b–440a; 11:118a; 13:360a; 14:133b–135a, 397a–398b
late nineteenth-century, 1:585a–b; 3:439a, 535b–536b; 4:144a; 5:69b–71a, 162a, 214b; 6:320a–b; 7:314a, 475b; 10:285a–b; 11:180b–182a; 13:352b, 472b, 525b
early twentieth-century, 5:150a; 14:552a
and chemistry, 2:572a; 5:90a
and medicine, 4:178a*
methodology, 2:581a
philosophy of, 2:187a–b; 4:512a; 5:213a; 9:100a, 101a; 10:9a, 284a; 13:526b–528a
as university discipline, 1:74a, 109a
see also Botany, etc.; Illustration, natural history; individual countries and geographic areas
Natural History Society of Edinburgh, 14:132a
Natural law, 6:406b, 447a,b; 8:437a, 439a; 12:548b
Natural philosophy
ancient Greek, 1:103b–104a, 149a–b, 151a–152a, 247b, 251b–262b; 4:30b–32b, 367a–369a, 381b–382b, 466a; 6:290a–291a; 8:269a–b; 10:181a; 11:29b, 103b–104a; 13:91a–95a, 295a–297b, 330a–b; 14:606a–607b, 608a–611b; 15:202b
ancient Roman, 1:118b–119a; 8:537a–538b; 11:42a; 12:310a–b; 15:518a

Byzantine, **7**:134a–138b
medieval Arabic, **1**:408b–410b; **2**:619b; **6**:191a–192a; **7**:343b–345b; **11**:231b–232b, 251b, 324a–325b; **12**:6b; **13**:511b; **15**:494b–496b
medieval Latin, **1**:61b–62a, 93b, 378b; **2**:192b–194b, 391a–395b, 603a–608a, 609b–611a; **3**:470a; **4**:157b, 447a; **5**:113b–114a, 115b–116a, 402b–403a, 549a–553a; **6**:275b, 376a–b, 378a–379b; **7**:27b–28a, 116a–b; **9**:136a–137b; **10**:223a–228b, 419b–421b; **12**:428b–429b; **13**:184b–211a, 340b, 534a–b; **14**:457b–461a
Renaissance, **1**:80a–b, 495a–b; **2**:590a–b; **3**:171b–172a, 280b, 420b; **4**:237a–238a; **5**:105a–107b, 397a–401b; **8**:100b, 197b–198a, 199a–200a, 227b–228b, 240b–241a; **9**:32b–33a, 132a–133b; **10**:122b–123a, 308b–311a, 416b, 512a–b; **11**:72b, 96a–98a, 606a–b; **12**:106a–b, 135b–136a, 151a, 548a–b; **13**:277b–280a; **14**:580b–582a; **15**:45a–b
early seventeenth-century, **1**:126a–b, 134a–b, 373a–376b; **2**:601b; **4**:51a–54b, 285b–288b; **5**:606b; **6**:156a, 254a–255a, 256a–257b, 445b–446b; **7**:164b–165a, 269a–b; **9**:14b–15b, 16a–b, 25a–b, 98a, 316b–320a; **10**:137a; **13**:559b; **14**:302a, 367b–369a; **15**:68b–70a
late seventeenth-century, **1**:317b, 550a; **2**:306b–312a, 317b, 378a–382a; **3**:183b, 208b–209b, 360b; **4**:95b–96a, 221a, 505b–506b; **5**:415a–417a; **6**:129b, 380a, 608b–611a; **7**:611b–613b; **8**:151a–153a, 173a–b, 416a; **9**:51a–53a, 464b–465b, 498a–499a, 509a–510b; **10**:43b–86a *passim*, 520b, 565b; **11**:493a–b, 506a–508b; **13**:335a–336b; **14**:157a–b, 209b–210a, 369a–374b, 405a–b, 451b
early eighteenth-century, **1**:75b, 411a–412a; **2**:324b, 326a,b; **3**:51a–52a, 215a–216b, 294b–296a, 431a–b, 486a; **5**:62a, 509a–511b, 519b–520a; **7**:225b–234a, 275b–276a; **8**:467b; **9**:187a–b, 595a–596b; **10**:319b–320a; **11**:43b, 157b–158a; **13**:39a, 179b; **14**:83b–84b, 484a; **15**:54a–b
late eighteenth-century, **2**:327b, 330a–331a; **3**:155a–158b, 344b–345a, 537a–540a, 579a; **4**:471b, 481b–482a; **5**:276a–b; **6**:335b, 383b–384a, 550a, 580b, 582a, 587a; **7**:367a–368a, 587a–b, 602b–604b; **8**:471a–b; **9**:607b; **11**:139b–147a, 496b–497a; **12**:407b–409a, 429b–430b; **13**:54b, 184a
early nineteenth-century, **3**:17b, 556a–557b; **4**:527b–539b, 546b; **5**:191a–b, 193a–195a; **6**:395a–b; **8**:261b–262a, 373b–374a; **10**:182b–185b; **11**:473b–474b; **12**:155a–156b; **14**:44b, 562a–569b; **15**:199b–200b
mid nineteenth-century, **5**:68b; **6**:86b–87a; **10**:266a, 360b–361a; **11**:344a; **12**:224b, 514b, 523b, 569b–570a; **13**:238a–b, 521b–523b
critics of, **3**:376b, 377b
late nineteenth-century, **11**:135b–136a; **12**:297b, 607b–610a
twentieth-century, **1**:206a; **2**:120b–121a, 145b; **5**:213a–b; **8**:11b; **14**:304b–308b, 348a, 349b
experimental
medieval Latin, **10**:532b–537b
Renaissance, **5**:399b
seventeenth-century, **1**:317b; **5**:415a–417a; **10**:103b, 546b; **11**:489a–b, 506a–507b; **13**:445a; **14**:151a–b
eighteenth-century, **4**:43b–45b; **9**:141b, 595a–596b, 607b; **11**:67b–68a, 472a–b, 579b–580a; **12**:462b; **13**:265b; **14**:418b, 419a
Japanese, **12**:406b–409a; **15**:728b, 730a–b, 741a
see also Idealism, Materialism, *Naturphilosophie,* Physics, Romanticism
Natural place, **1**:248a, 254a, 336b; **3**:404a; **12**:4b, 429a; **15**:495a, 496b
as cause of movement, **1**:198b; **13**:198a; **15**:496a
as cause of weight, **8**:199b
critics of, **13**:92a
Natural Science Society (Switzerland), **3**:211a
Natural selection
ancient Greek discussion, **4**:368a
eighteenth-century discussion, **3**:579b; **4**:512b; **12**:588b
early nineteenth-century discussion, **14**:253b
mid nineteenth-century discussion, **1**:501b, 502a–b; **2**:205b–206a; **3**:88b, 570b–573b; **6**:280b, 594a; **8**:572b–575b *passim;* **9**:283a, 601b; **10**:210a; **14**:58a, 135b–138a, 534a
late nineteenth-century discussion, **3**:42b; **4**:504b; **5**:385b, 624a; **6**:613a–b; **7**:371a, 480b; **8**:27a, 125a; **13**:343a, 417a; **14**:233a, 234a,b, 236b–237a
early twentieth-century discussion, **3**:390a; **5**:467b; **7**:114a, 253a; **9**:523b; **11**:19a–20a; **14**:432b, 434a; **15**:507b–509b
at cellular level, **11**:571a–b; **14**:234b, 236b, 237a–238a
and coloration, **9**:550a–b
and ecology, **5**:70a–b
experimental study, **3**:120b–122a; **5**:10b–11a
inorganic, **3**:594a
and man, **14**:136b–138a
mathematics of, **6**:22a; **10**:449b
and mimicry, **1**:503b; **9**:560b–561a; **12**:417a–b; **14**:136a–b
opposition to, **1**:506a; **2**:10b–11a, 426b; **3**:493a; **5**:192b, 447a; **6**:222b; **7**:439a; **9**:519a–b; **10**:149a, 262b; **11**:308b; **12**:38a, 167b, 567b; **15**:92b
proof of, **3**:572b
types of, **12**:410a
see also Darwinism
Naturalism (phil.)
medieval Latin, **10**:225b
Renaissance, **5**:120a; **8**:241b; **10**:310b, 311a; **11**:73b, 74a; **13**:279b–280a
eighteenth-century, **4**:85a–b, 89a; **6**:469a–b, 558b
twentieth-century, **3**:335a
Neoplatonic, **1**:91b
scientific, **5**:252a–255b *passim;* **13**:522b
see also Animism; Hermeticism; Magic, natural; Neoplatonism
Nature
balance of, **1**:260a; **3**:525b–526a, 527b; **5**:70a; **8**:569a
concept of, **8**:517b
constancy of, **7**:164b–165a
convention, opposed to, **1**:171b–172a
disorder of, **2**:577b; **15**:731a
economy of, **1**:278b; **14**:490a
form in, **1**:374b–375b; **10**:11b
historical approach to, **2**:204b, 209a–211a, 502a, 579a–581a; **15**:135b, 139b, 144a, 408b, 412b–415b
irreversibility of, **4**:293b; **5**:392b; **7**:387a–b; **9**:616b; **11**:9a,b, 10b, 11a; **12**:514b; **14**:123b
language of, **2**:223b
laws of
Byzantine discussion, **7**:135b
medieval Latin discussion, **1**:380a
Renaissance discussion, **5**:105b; **8**:214a
seventeenth-century discussion, **4**:52a,b; **8**:155b, 156a, 158a; **9**:48b; **11**:489a
eighteenth-century discussion, **2**:578a; **5**:511a; **8**:469a; **9**:187a; **10**:120b; **11**:523a; **13**:564b
nineteenth-century discussion, **2**:564a; **3**:192a,b; **5**:193a; **6**:341b; **7**:588a; **8**:134b; **10**:183b, 485a; **11**:9a; **12**:261b; **13**:460b; **15**:79b, 142b
twentieth-century discussion, **3**:335a; **11**:15a; **13**:247a; **14**:348a; **15**:755b
constancy of, **5**:61a; **14**:361b
simplicity of, **3**:283a, 451b; **7**:181a; **13**:171b; **14**:341a; **15**:344a
and man
see Man, and nature
order of, **3**:361b–362a, 362b; **6**:261a–b; **8**:57b, 378a; **9**:316b; **11**:168b–169a; **15**:371b, 731a
reverence for, **2**:223a–b
state of, **6**:447b, 448b
unity of
medieval view of, **3**:470a
Renaissance view of, **4**:170a
seventeenth-century view of, **8**:127a; **11**:493a–b
eighteenth-century view of, **4**:89a–b; **5**:442b–444a; **6**:549b; **8**:322a, 468b, 471a; **9**:532b
nineteenth-century view of, **1**:476b–477a, 477b; **4**:145a–b, 265a; **5**:167a, 191b; **6**:8b, 564b; **7**:585b, 586a, 587b; **9**:252b, 281a, 287b; **11**:116a, 473b–474b; **15**:728a
twentieth-century view of, **6**:530b; **8**:31a; **11**:14a
Chinese view of, **15**:732b, 738a
Japanese view of, **15**:732b
universal, **1**:380a
variability of
see Variation, biological
views of, **15**:728a–b, 729a, 732b, 738b, 746a, 754a, 757a(n)
cyclical, **15**:732a
war of, **3**:571a–b; **8**:378a; **11**:567b; **15**:139b
zones of, **4**:145a
see also Book of Nature; *Scala naturae;* Theology, natural
Nature, **4**:176b*; **13**:522b
founding of, **8**:443a
Nature printing, **12**:543b
Naturforscher,
founding of, **14**:120a
Naturphilosophie, **1**:69b, 387a, 388b; **2**:159b–160a, 160b, 425b, 426b, 545a, 552b–553b, 595a; **3**:522a; **4**:102a*, 146a,b, 264a, 530b–531a; **5**:191a–b, 192a, 193a, 194b, 218a, 251b, 444a, 446a; **6**:270a, 564b, 573a; **7**:367a, 368a; **9**:252a,b, 568a, 569b, 572a; **10**:11b, 13a, 182b–185b, 193a, 194b–196a, 263a; **11**:159b, 257b, 363a, 364b, 473b; **12**:38b, 154a–156b, 197a, 202a, 510a, 570a, 578b*, 607a; **13**:460b,

Naturphilosophie, (cont.)
462b, 466a, 542a; 14:351a–b; 15:50b, 85b, 167a, 168b, 462b, 463b
critics of, 3:232b; 7:235a; 8:373b–374a; 9:442a, 573b, 578b; 11:379b; 12:174b–175a, 592a; 15:127b
see also Romanticism
Naucrates, 1:180a; 4:438a
Naudé, Philipp, 4:474b
Naudin, Charles (1815–1899), 9:618–619
evolution, 15:507b
Naumann, Alexander (1837–1922), 9:619–620
Naumann, Karl Friedrich (1797–1873), 9:620; 3:462b; 14:625a
crystallography, 9:392b–393a, 448b; 14:241a
editions of works, 14:625b*
geology, 3:433b
Nausiphanes, 4:30b, 381b
Nautical Almanac (G.B.), 2:600a; 3:373b; 4:175b–176a; 6:402b; 9:164b–165a, 234b; 14:563a
criticism of, 12:552a
founding of, 9:163a
Nautical Almanac (U.S.), 1:6a; 6:119b; 9:421b; 10:34a; 15:128b–129a
Nautilidae, 2:556a; 7:314b, 460b; 10:221b, 261a
Navagero, Andrea, 5:104b, 106a
Naval Academy, U.S., 13:398b
observatory, 8:19b
Naval Consulting Board (U.S.), 12:575a
Naval Observatory (U.S.)
in nineteenth century, 5:480a; 6:49a, 119a–b, 471a; 10:33b
in twentieth century, 5:554b; 11:470a, 552b
founding of, 9:195b
Naval Research Laboratory (U.S.), 1:338a
establishment of, 4:284b
Naval strategy, 4:214a; 9:43b
Navāṃśas (ecliptic arcs), 15:125b
Navarro Martin, Augusto, 8:273b
Navashin, Sergey Gavrilovich (1857–1930), 10:1–2; 3:422a; 5:582a; 14:94a*
Nave, Annibale della, 1:108a; 4:587a, 595a,b
Navier, Claude-Louis-Marie-Henri (1785–1836), 10:2–5; 3:72b, 417a; 5:98a–b; 8:383a, 384b; 11:344a; 12:74a*
on Bélidor, 1:582a
elasticity, 3:259a; 15:487a,b
equations of, 3:146a; 8:502a
equilibrium, 3:145a,b; 146a
hydrodynamics, 3:80b; 4:591b; 13:75a
Navier-Stokes fluids, 4:230a,b
Navigation
ancient Greek, 13:295a
medieval Arabic, 9:35b–37a
Renaissance, 2:462a; 3:423b–424a; 4:5b, 97a; 5:397a–b, 398b–399b; 6:20b; 8:390a; 9:254b, 264a, 309b; 10:150a, 160b, 161a, 513a–b, 619a; 13:47b, 49b; 14:275a, 513a–514b, 583b; 15:88b–90a
early seventeenth-century, 2:185b; 4:214a; 5:287b, 343b–344a, 593a–b; 6:520b; 8:65a; 10:151a–b, 255a*; 12:501a
late seventeenth-century, 3:348a, 621b; 5:23a; 6:604b; 11:423b, 532b; 13:336a
early eighteenth-century, 2:43a, 54a–b, 343a–b; 4:469b, 470b; 5:434b, 515b; 6:130b, 513b; 7:162b; 8:179b; 9:17a; 10:571b–572a; 11:65b, 392b*; 14:518b
late eighteenth-century, 2:299b, 342b–343a, 389a; 3:383b, 396b–397a; 4:472b, 481b; 5:136b–137a; 7:183b, 580b; 8:144b, 470b–471a; 9:234a–b; 11:495b–496a; 13:184a; 14:354b
nineteenth-century, 1:460a, 594a–595b; 2:368b; 4:492a; 5:17a; 8:101a; 9:195b–196a
twentieth-century, 3:374b; 4:614a; 5:65a; 7:265b–266a; 11:84a; 14:50b, 614a
aerial, 4:253b; 8:532b; 12:18a
automatic piloting, 1:570b
celestial
Renaissance, 1:573b
great circle, 10:151b, 161a
Mesopotamian, 15:637a
position lines, 7:266a
rhumb line, 10:161a
see also Hydrography, Latitude, Longitude
Navratil, Ernst, 8:454a,b
Navy, U.S.
in nineteenth century, 9:195b
exploration, 9:195b
radiotelegraphy, 1:338a
Nawaschin
see Navashin
Nayanasukhopādhyāya, 7:81a,b; 15:628b
al-Nayrīzī, Abu'l-ᶜAbbās al-Faḍl ibn Ḥātim (*fl. ca.* 897–922), 10:5–7; 5:592a; 6:315b; 7:79b; 9:21b
commentaries of, 3:25a, 512a; 4:437a–441b *passim,* 445a–b, 448a, 611b; 5:346b; 6:199b; 13:307a(n)
Super Euclidem
translations of, 15:178a
Zīj, 14:575b
al-Nayrīzī, Abū Manṣūr, 7:343b
Naẓīf ibn Yumn, 4:438a, 439b
Nazism
opposition to, 1:519b; 2:238b; 5:206a, 623a; 6:304a; 7:555a; 8:52a–b; 12:34b; 14:613b; 15:75b, 451a
psychology of, 7:192a; 11:309b
and scientists, 3:618a; 4:154b–155a, 188a; 6:355b; 8:182b; 10:348b; 11:362b; 12:78a, 424a, 615a–b; 13:229a; 14:14a, 444a, 590b; 15:159b, 490b
Jews, 3:333a; 4:9a, 313a; 5:118a, 173a, 179b, 453a, 457b, 458a, 473a; 6:176b, 235b, 433a; 8:451b, 473a; 9:262a, 341b, 359a; 10:138a; 11:148b–149a, 355b, 551a; 12:237a, 530b–531a, 615b; 13:43a, 227b, 428a; 14:411b; 15:15b, 43b
Neander, Michael (1529–1581), 10:7–8
Neanderthal man, 2:189a–b, 347a, 617b; 3:606b; 5:109b; 6:9b, 527b, 593b; 8:464a, 574a; 12:184a
discovery of, 5:206b–207a
Nebraska
geology, 6:187a
Nebulae, 2:221a, 387a, 8:364b; 10:526b; 12:192a, 266b
angular velocities, 6:529b
catalogs of, 2:126b; 4:185a–b; 6:325b, 326b, 327b, 330a; 7:87a, 544a; 9:330a–b; 12:200b, 368b
classification of, 6:528b–529a; 12:269b
Crab, 1:352b; 9:330a
expansion of, 4:249a
dark, 1:466b; 10:290a; 14:482a
diffuse, 12:368b, 455a
discovery of, 1:465a,b, 466a; 4:506a; 6:322b, 402b; 7:598b; 8:46a, 143b; 9:247a, 250a–b, 330a; 10:489a, 607b; 13:36b
distance of
and velocity, 6:353a; 12:455b–456a
distribution of, 6:507b; 11:163a; 12:368b
drawings of, 6:326b
emission, 12:349a, 368b
filamentary, 12:368b
galactic, 6:30b; 7:231b; 12:349a
origin of radiation, 6:529a
gaseous, 1:295b; 2:433a; 4:279a; 5:363b; 6:507b; 9:415a; 12:269b; 14:482a
discovery of, 6:542a
radial velocities, 12:153a
heat radiated from, 3:327a
nature of, 6:70a, 329b, 332a, 332b–333a, 541b–542a, 543a; 13:120b; 15:345b
observations of, 6:471a; 14:54b, 56b
origin of, 2:123b; 5:272b
photography of, 4:249a,b; 6:528b–529a; 7:271a, 472a; 11:470a–b, 484b–485a; 12:139a, 368b; 14:481b; 15:493b
planetary, 4:279a; 6:507b, 529a; 8:539b; 14:482a, 521b
absolute magnitudes of, 8:582b
central stars of, 5:363b
distances of, 8:582b
equilibrium studies, 5:363b
ionization in, 5:363b
proper motions of, 8:582b
spectra, 12:368a
radial velocities, 7:270b; 9:503b
rotation of
rate of, 8:582b
spectra, 1:295b; 2:485b; 4:280b; 6:542a, 542b–543a; 7:270b; 9:184a; 12:153a, 269b, 455a; 13:117b; 14:55a, 482a
theories of, 10:106a–b; 12:368b
spiral, 4:279a; 6:529a; 10:329a; 15:493b
abundance of, 7:271a
distances of, 8:582b; 11:470b; 12:347b, 348a, 349a
distribution of, 14:482a
formation of, 7:85b
island universe theory, 3:509b; 12:347b
knots in, 9:552b
motion of, 8:582b
nature of, 6:529b; 12:455b
novae in, 12:347b
proper motions of, 8:582b
radial velocity, 12:455a–456a
rotation of, 10:473b; 12:456a
structure of, 11:485a
variable, 6:402b
"white"
island universe theory, 3:509a–b
see also Orion, etc., for specific nebulae
Nebular hypothesis, 3:192a; 7:385a; 9:552b, 594a; 11:21b; 15:343a, 344a–345a
critics of, 3:190b, 191a; 7:385b
Nebulium, 4:280b; 6:542a, 543a; 7:270b; 10:106b*; 11:280b; 12:44a; 14:521b
Necessity, 2:231b; 3:296a, 452a; 5:510a; 14:124a
and chance, 15:139b, 143a, 144b
"hypothetical," 1:260a, 261a
of natural causes, 1:259b–260a
random, 1:259b
Nechepso, 4:622a; 10:547b–548b; 15:724a
Neckam, Alexander, 13:348a
Necker (Necker de Saussure), Louis-Albert (1786–1861), 10:8–9; 13:131b*
Necromancy, 2:540b; 9:364b; 11:161b
Necrosis, 14:228b, 252b
Nectaries, 2:290a–b
Neddermayer, Seth Henry, 6:356a
Nederlandsch kruidkundig archief, 13:154a
Nederlandsch Lancet, 4:162b
Nederlandsche Vereeniging voor Koeltechniek
founding of, 7:222a
Needham, Dorothy Mary Moyle, 6:501b
Needham, John Turberville (1713–1781), 10:9–11
critics of, 12:556a–b, 557a, 559b
microbiology, 2:520b; 12:557b

Nouvelles observations microscopiques . . . , **12**:554b
on reproduction, **2**:579b; **10**:367a; **12**: 241a, 554b–555a
Needles, magnetic
damping of, **7**:13b
Nees von Esenbeck, Christian Gottfried (Daniel) (1776–1858), 10:11–14; 1: 69b; **2**:158a; **4**:288b, 291b; **12**:175b; **13**:462b
on fungi, **5**:191b
students of, **2**:160b; **3**:336b; **13**:555b
Nees von Esenbeck, Theodor Friedrich Ludwig, **10**:11b, 13a
Neesia, **10**:13a
Nef, John Ulric (1862–1915), 10:14–15
Nef reactions, **10**:14b
Negative numbers
see Numbers, negative
Nègre, Léopold, **3**:22b
Negri, Adelchi (1876–1912), 10:15–16
Negri bodies, **10**:15a–16a
Negrín, Juan, **15**:75b, 76a
Negro, **6**:563b
albinism, **9**:187a
anatomy, **12**:509b; **14**:534a
brain size, **13**:404a
distribution of, **8**:28b–29a
heredity, **5**:294a
morphology, **6**:454a
presumed inferiority of, **4**:272a; **9**:541a
serology, **7**:624b
Neher, Henry Victor, **9**:398b
Nehring, Alfred (1845–1904), 10:16–17
Nehru, Jawaharlal, **15**:27b
Neighborhoods (math.), **6**:4b
Neil, William, **1**:474b; **5**:527a; **6**:359b
Neilson, James Beaumont, **9**:591a
Neisser, Albert Ludwig Sigesmund (1855–1916), 10:17–19; 4:300b, 304a*,b*; **11**:175a; **12**:87b; **15**:522b
Neisser, Max, **4**:297b, 298b
Nekrasov, Aleksandr Ivanovich (1883–1957), 10:19–20; 14:621b
Neleus, **13**:328b, 329a
Nelson, Alfred, **1**:477a
Nelson, Horatio, **3**:487b; **6**:84a
Nematodes, **1**:496a; **2**:322a, 626a; **5**:453b; **7**:110b; **9**:112b; **12**:193b
anabiosis, **3**:588b
anatomy, **9**:112a
digestion
intracellular, **9**:333a
parasitic, **1**:602b; **3**:588b; **8**:296b; **9**:185b; **12**:482b
reproduction, **9**:185a,b, 332a
see also Helminthology, Parasitology
Němec, Bohumil, **3**:581b; **5**:624a
Nemerteans, **6**:535a–b; **9**:112b, 495b
morphology, **11**:211b, 212b
Nemesius (*fl.* 390–400), 10:20–21
On the Nature of Man
translations of, **7**:25a
Nemminski, W.
see Pravdich-Neminskii, Vladimir Vladimirovich
Nemophila
editors of, **7**:103a*
Nemore, Jordanus de
see Jordanus de Nemore
Nencki, Marceli (1847–1901), 10:22–23; 1: 9b; **7**:471a; **9**:96a
Neocene epoch
paleontology, **1**:162a, 163a–b
see also Miocene epoch, Pliocene epoch
Neoclassicism, **12**:135a
Neo-Confucianism, **14**:163a–165b
Neo-Darwinism, **2**:190a; **5**:453b; **6**:9a; **7**: 166b; **9**:143b; **11**:20a; **12**:571a; **13**: 276a; **14**:233a, 237a–238b; **15**:93a
opposition to, **9**:519a–b; **12**:417b
Neodymium
atomic weight, **2**:428b
discovery of, **3**:321b, 478a; **9**:541b
Neo-Kantianism, **1**:178a; **3**:110b–112a; **5**: 122b; **12**:112a; **15**:422b
Neo-Lamarckism, **3**:318a; **6**:613b; **7**:371a, 593a; **10**:273b, 427b; **13**:276a; **14**: 237a, 313a; **15**:92b–93a
Neolithic, **5**:109b; **9**:538a
fossils, **5**:110a
naming of, **8**:528a
subdivision of, **9**:489b, 539b
transition from Paleolithic, **2**:347a
see also Stone Age
Neon
discovery of, **6**:94a; **11**:280a; **13**:454a
ions, **13**:369b
isotopes, **1**:321a; **12**:505a
liquefaction of, **13**:454b
spectrum, **4**:513b; **7**:554a; **10**:348a
Neoplatonism
ancient Greek, **1**:268a, 269b; **6**:305b; **11**:105b; **12**:575b; **14**:535b
ancient Roman, **1**:255b, 333b, 337a; **2**: 229a,b, 234a; **5**:232b; **9**:1a; **11**:41b–42a; **14**:607b
Byzantine, **4**:159a–b; **6**:616a; **7**:137b; **9**: 111a–b; **11**:160a–162a, 183a–184a; **12**:440a–443a
medieval Arabic, **1**:33b; **5**:592a; **7**:23a, 41b; **9**:28b; **13**:229b, 488b; **15**:181b–182b, 249a–250b, 262a, 494b
opposition to, **12**:5b, 7Ba
medieval Latin, **1**:91a–b, 99b–100a, 272a; **3**:470a; **4**:92b, 396b; **5**:549b, 592a; **8**:547b, 549a–b; **9**:436a, 438b; **13**: 534b; **14**:459a; **15**:181b–182b
Renaissance, **1**:80a; **2**:539b–540a; **3**:514b, 516a; **4**:170a; **5**:107a, 121a, 448a, 580a; **10**:309b, 310a, 311a, 598b; **11**:96b–97a, 606b; **14**:225b
seventeenth-century, **2**:222b, 590b; **3**: 492a, 596b; **5**:47b; **6**:256b; **7**:355b–356a; **8**:130b; **9**:509a; **12**:311b, 333a; **15**:69a,b
eighteenth-century, **6**:459a; **11**:140a
and Christianity, **1**:334a, 336a; **4**:396b; **11**:161a
mathematics, **3**:127b–128a
Syriac, **1**:33a; **7**:1a–b
see also Naturalism
Neoprene, **3**:86b; **10**:122a; **13**:66b
Neo-Pythagoreanism, **2**:233a; **4**:425a; **7**:1a
Neosalvarsan, **4**:301a; **9**:546a*
Neotectonics, **10**:169a
Neoteny, **2**:255b; **15**:126b
theory of, **5**:359a
Nepenthes, **6**:492a
Nephelescope, **4**:410b
Nepheline, **4**:141a; **5**:186a
Nephelinite, **11**:547a
Nephelometer, **11**:417a
Nephrectomy, **13**:567b
unilateral, **14**:587a
Nephridia, **5**:467a–b
Nephrite
archaeology, **9**:592b
Nephritis
see Kidney(s), diseases of
Neptune (planet), **2**:516b
discovery of, **1**:53a–b, 86b, 202b, 295a–b; **2**:359b; **3**:186b–187b; **4**:369b; **5**: 257a–b, 306b–307a; **6**:325b; **8**:277a–b
distance of, **13**:425a
existence predicted, **8**:299b
motion of, **7**:481a
observation of, **10**:542b; **13**:115a
orbit, **5**:224a, 257b; **7**:481a; **8**:277b, 522b; **10**:479b
pre-discovery observations of, **7**:608a
rotation
period of, **9**:504a
satellites of, **6**:49b, 471a; **8**:46a; **10**:35a
measurement of, **10**:30b
spectrum, **12**:267a
tables of, **7**:481a; **10**:34a
Neptunism
Renaissance, **8**:201b, 241a–b
eighteenth-century, **1**:413b; **2**:578a,b; **4**: 70a, 151a; **5**:444a–445a; **6**:549b; **7**: 589b; **8**:70b; **9**:26a–27b; **11**:521a; **12**:121a, 121b–122a, 549a,b; **14**:60b, 259b–263a, 387a; **15**:55a–56a
nineteenth-century, **1**:327a; **2**:158b–159a, 481a, 552b–557a *passim;* **3**:313b, 412a, 433b; **4**:549a; **5**:518b; **6**:565a; **7**:69b; **8**:245b; **10**:322b; **11**:387b; **12**:514a; **15**:526b
critics of, **6**:456a,b; **9**:532a–534a; **14**:240a
Neptunium, **9**:263a
Nereis, **3**:389b; **8**:357b; **14**:426a–b, 428a
Neri, Antonio (1576–*ca.* 1614), 10:23–24
L'arte vetraria, **7**:525a
translations of, **9**:313a
Neri, Filippo de, **9**:308b, 309a
Nerine, **4**:173a
Nernst, Hermann Walther (1864–1941), 15: 432–453; 10:348b; **13**:244a
acquaintances of, **1**:297b, 519a; **2**:3b; **4**: 194b, 312b, 485a; **6**:16a; **8**:503a; **9**: 260b; **11**:148b; **12**:71a; **13**:243a; **15**: 37a, 456a, 458b
ammonia synthesis, **5**:621b
chemical warfare, **5**:622a
electrochemistry, **1**:537b; **8**:290b
heat theorem, **4**:316b, 414a; **8**:118b, 289b, 291a; **11**:10a, 418a; **12**:437b, 438b; **13**:41b; **15**:432b, 436a–448a
influence of, **4**:413b; **11**:298a; **13**:40b
laboratory, **13**:41b
mathematics, **12**:196a
Nernst effect, **6**:51b; **11**:446b
Nernst lamp, **15**:449b
Nernst-Lindemann theory, **8**:368a; **15**: 442a
neurophysiology, **8**:534a
physical chemistry, **2**:70a, 498b; **13**:243b; **15**:460a,b
specific heats, **2**:170a; **8**:368a; **11**:583a; **15**:40b, 442a
students of, **2**:169b; **6**:538b; **8**:23a, 289a; **9**:395a; **10**:136a, 329b; **11**:417b, 576b; **12**:437b; **14**:173a; **15**:36b
Theoretische Chemie, **13**:159a
Nero, **1**:381b; **11**:38b; **12**:309b
Nerves
cytology, **4**:385b; **5**:461a; **8**:30a, 282b; **11**:465b
axis cylinder, **4**:372a; **11**:296a
corneal spires, **5**:460b
fibrils, **5**:460b; **8**:283a; **11**:296b, 368a
sheaths, **11**:382b, 597a; **14**:229a
degeneration of, **15**:419a–b
discovery of, **6**:317a; **11**:128a
electric current in, **2**:161a; **4**:202a, 334a; **5**:291a
measurement of, **4**:202b
electric stimulation of, **4**:616a; **5**:267b; **9**:483b; **13**:529a
laws of, **10**:579a
polarization, **4**:202a, 310b
endings of, **5**:460b–461a; **7**:520a; **8**:92a; **10**:579b, 598b; **11**:274a–b, 296a,b, 382b, 566a–b; **12**:355a–b; **14**:113b
proprioceptors, **11**:596b–597a

Nerves (cont.)
endings of (cont.)
in skin, **8**:8b
in voluntary muscles, **1**:540a
excitability, **4**:399a; **5**:291a,b; **9**:480a–b
theories of, **8**:102a–b
time factor, **8**:29a–b
fibers of, **8**:9a; **15**:430b
bundles, **11**:296a
characterization of, **5**:291b
conductibility of, **7**:520a; **14**:200b–201a
degeneration of, **9**:93b
development of, **6**:131b–132a, 435a
distribution of, **11**:216b
efferent vs. afferent, **8**:16b, 17a
experimental cross unions, **8**:16a–b
morphology, **5**:56b, 282a, 291b
postganglionic, **2**:124b
spinal, **2**:124b
structure, **1**:177a; **2**:361b; **5**:172a; **11**:216b, 368a,b; **13**:555b; **14**:143a
unmyelinated, **5**:291b
function of
ancient Greek theories, **6**:317b
eighteenth-century theories, **2**:3a; **11**:159a–b; **13**:122a–123a
nineteenth-century theories, **1**:584a; **5**:460b; **9**:241b–242a; **11**:274a–b; **12**:355b
twentieth-century theories, **8**:16b
Japanese theories, **15**:738a
impulse, **5**:65a–b; **9**:564b; **14**:113b
action potential, **4**:398b–399a; **5**:291a–b
after-potential, **5**:291a–b
all-or-none law, **5**:65b; **8**:533b
chemical transmission, **8**:453a–455a; **15**:73b, 74a–b, 106b–107a
excitation wave, **15**:21a,b
feedback mechanisms, **2**:446a; **6**:300a
and fiber diameter, **4**:399a; **5**:291a–b
inhibition, **8**:533b
interruption of, **8**:15a–b
recurrent sensitivity, **2**:25a, 26a, 30a
summation, **8**:533b–534a
synapse time, **15**:21a
temperature coefficient, **8**:533b
theories of, **3**:483a–b; **5**:65b; **6**:245b–246a, 247b; **8**:16b–17a, 29a–b, 453a–455a, 533b–534b; **9**:64a, 571a; **10**:257a; **11**:274b; **12**:95a–b; **13**:263a; **15**:21b
velocity of, **3**:68b; **5**:149a; **6**:242a, 245a; **8**:29b
ligation of, **2**:237b
metabolism, **13**:262b
origin of, **4**:385b; **5**:251a; **6**:317a; **9**:242a; **12**:509b, 510a
respiration in, **13**:394a
sectioned, **8**:92a; **9**:86b*
degeneration of, **11**:296a; **14**:142b–143a
regeneration of, **2**:124b; **3**:487a; **5**:45a, 56b; **6**:526b*; **8**:65b; **11**:274b–275a, 296a; **12**:356a
types of
afferent, **13**:543b
centrifugal vs. centripetal, **3**:46a
classification of, **5**:571a
inhibitory, **2**:124b
involuntary, **1**:584a; **5**:282b
medullated, **11**:296a
mixed, **9**:571a
pain, **8**:295a–b
peripheral, **1**:583b–584a; **2**:124b; **8**:16b; **14**:26a
pilomotor, **8**:15b
secretory, **2**:301b; **6**:225b; **10**:579b
sensory vs. motor, **4**:385b; **6**:317b; **9**:241b, 242a, 570b–571a; **11**:566b
spinal, **1**:584a; **8**:15b; **9**:9b–10a; **14**:128a–b
trophic, **2**:73a; **6**:225b; **9**:10a; **10**:432b–433a
vasoconstrictive, **2**:525a–b; **14**:143b
vasodilator, **9**:505a, 505b–506a
vasomotor, **9**:56b
see also Ganglia, Innervation, Nervous system, Neuroanatomy, Neurohistology, Neurology, Neurons, Spinal cord, Vagus nerve, Vasomotor system
Nerville, Ferdinand Guillebot de, **1**:613a; **9**:155a
Nervous system
autonomic, **5**:281b–283a
function of, **9**:571a; **14**:408b; **15**:73b–74b
histology, **2**:124b; **8**:15a–17a, 92a; **14**:143a–b
naming of, **8**:16a
physiology, **8**:452b–455a; **9**:19b, 505a–506a; **10**:220b; **12**:165a
crossings, **14**:450a
degeneration of
from decortication, **8**:15a
embryogeny, **3**:245b, 331b–332a; **6**:435a; **8**:586a; **9**:253a; **10**:105a; **11**:274b, 369a, 510b; **12**:473a, 568b–569a; **13**:96b–97a
evolution of
theories of, **7**:590b, 591a–b
evolution and dissolution of, **7**:48b
extrapyramidal pathways, **3**:113a
fasciculi, **2**:595b–596a
fluid in, **13**:122a–123a
circulation of, **1**:392a
functional analysis, **11**:546a
and glycogenesis, **2**:29b; **8**:615a; **9**:506a
gray matter, **5**:460b; **10**:130b, 131a–b
interstitial cells, **11**:465b
lesions of
diagnosis of, **2**:525a
and epilepsy, **2**:525b
mapping of, **8**:15b
morphogenesis, **1**:386b; **5**:283a–b; **7**:48b; **11**:274b; **14**:596b
nerve trunk
indefatigability of, **2**:367a
nocifensor system, **8**:295b
oncology, **2**:597b; **3**:518b; **11**:465b
parasympathetic, **6**:225b; **14**:406b–407a; **15**:106b
naming of, **8**:16a
peripheral, **11**:296a
phase states, **10**:434b
supporting tissue, **14**:229a–b
sympathetic, **2**:111a, 124b; **4**:487b; **5**:282b; **6**:225b; **7**:437b; **8**:15a; **9**:505b–506a, 571b; **13**:556b; **14**:406b–407a, 450b; **15**:73b
cytology, **11**:368a
emergency function, **15**:73a
extirpation studies, **11**:545b–546a
galvanization, **2**:30a
histology, **14**:143a–b
origin of, **11**:112a–b
paralysis, **2**:30b
role of, **10**:220b
sectioning, effects of, **2**:30a, 525a–b; **11**:112a–b
trophic effect, **8**:92a
see also Brain; Disease, neurological; Nerves; Neuroanatomy; Neurohistology; Neurology; Spinal cord
Nesbitt, Robert, **6**:183b
Nesfield, William Andrews, **6**:494a
Nesting, mathematical, **4**:606b, 607a
Nestorians, **1**:275a
Netherlands
botany, **10**:253b
economics, **14**:466b
entomology, **5**:439b
geodesy, **7**:499b–500b
geology, **6**:137b
mathematics, **7**:466a
medicine, **13**:490b
Meteorological Institute (K.N.M.I.), **2**:628a
military science, **13**:50a
politics, **14**:465a
Netherlands Botanical Society, **13**:154a
Netherlands East Indies
botany, **9**:417b
Netherlands Society of Sciences (Haarlem)
competitions of, **5**:218a
Nets, theory of, **3**:225a
Nettesheim
see Agrippa, Heinrich Cornelius
Netto, Eugen (1848–1919), 10:24; **5**:311a*; **12**:451b; **14**:222a
Theory of Substitutions. . .
translation of, **3**:346a*
Neuberg, Carl, **6**:111b
Neuberg, Joseph (Jean Baptiste) (1840–1926), 10:24–25; **9**:81a
Neuchâtel, University of
geology at, **12**:140a
Neue Gesellschaftliche Erzählungen, **13**:425b
Neues allgemeines Journal der Chemie, **3**:465b; **15**:171b
Neues Berliner Jahrbuch für die Pharmazie, **4**:134a
editors of, **15**:171b
Die neuesten Entdeckungen in der Chemie, **3**:465a
Die neuesten Forschungen im Gebiete der technischen und ökonomischen Chemie
see *Journal für technische und ökonomischen Chemie*
Neufeld, Fred, **7**:426b, 428b; **12**:142a
Neuilly
bridge at, **11**:163b–164a
Neumann, Carl Gottfried (1832–1925), 10:25; **6**:97a*; **10**:27b, 29a*; **12**:172b
acquaintances of, **3**:314a; **6**:230b, 356b
analysis, **4**:561b; **5**:151a, 388a; **11**:58b; **13**:26b
correspondence of, **11**:15b(n)
method of, **5**:201b; **11**:59b
natural philosophy, **12**:608b–609a
physics, **14**:208b
potential theory, **6**:392a; **7**:398a
students of, **9**:457b; **11**:445b
Vorlesungen über Riemanns Theorie . . ., **11**:451a,b, 452a, 455b
Neumann, Caspar (1683–1737), 10:25–26; **9**:104b; **11**:109a
chemistry, **9**:105b, 106a
lectures of, **8**:297b, 298a; **11**:143b
Praelectiones chemicae, **12**:144a
Neumann, Franz Ernst (1798–1895), 10:26–29
acquaintances of, **7**:50b; **14**:240a
biographies of, **14**:67b
critics of, **6**:242a; **11**:446a
crystallography, **9**:393a; **14**:61b, 241a
and education, **7**:51a
electrodynamics, **6**:251a,b, 252a, 343b; **7**:380b; **8**:490b–491a, 492a; **9**:205a,b, 207a, 217a
on ether, **8**:494a
influence of, **14**:61a–b, 241b
mathematical physics, **9**:200a; **11**:63b
Neumann's laws, **1**:348b; **7**:463b; **8**:188a; **10**:27a–b; **11**:353a
optics, **4**:190a; **8**:490b, 592a; **13**:77b

students of, **1**:147a, 294a; **2**:299a; **3**:314a; **4**:205a, 378a,b; **6**:230a; **7**:380a; **8**: 388b; **9**:232a, 347b; **11**:241b; **12**: 216b, 289b, 511b; **14**:158a, 202a
Neumayer, Georg von, **4**:193b; **12**:595a
Anleitung zu Wissenschaftlichen Beobachtungen auf Reisen, **11**:440a
Neumayr, Melchior (1845–1890), 10:29–30; **1**:162a, 491a
Neural crest, **6**:133a
Neuralgia, **4**:296b
crural, **1**:400a
traumatic, **9**:422a
trigeminal
surgery for, **6**:519a
Neurasthenia
Beard's, **3**:205b
Neurath, Hans, **1**:319b
Neurath, Otto, **15**:140a
Neurine
synthesis of, **14**:531b
Neuritis
optic, **7**:47b
Neuroanatomy
ancient Greek, **4**:385b; **6**:317b
medieval, **1**:4b
Renaissance, **4**:413a, 487b–488a, 520b; **8**:203a; **14**:9a–b
seventeenth-century, **1**:480b; **3**:99a; **9**: 63b; **14**:25b–26a, 256a, 406a–407a
eighteenth-century, **2**:37a; **6**:533b; **9**:483a; **11**:112a–b; **12**:137b–138a, 509b–510b; **14**:14b–15a
early nineteenth-century, **1**:583a–584a; **2**:595a; **3**:437b; **5**:251a–b; **9**:241b–242a; **11**:380a; **12**:596a–597a; **14**: 129a–b
mid nineteenth-century, **2**:26a, 29b, 30a, 124b, 160b; **3**:424b, 425a, 489b, 490a; **5**:571a–b; **12**:224a–b
late nineteenth-century, **1**:580a; **3**:245b; **4**:266b; **5**:26b–27b, 74a, 282b–283a, 459b–461a, 571a; **9**:422a; **11**:274a–b, 382b–383a; **12**:396a–b; **14**:249a
twentieth-century, **3**:113a; **5**:291b; **6**: 321a; **8**:552b; **13**:590a; **15**:473a
comparative, **2**:160b; **5**:505a; **8**:272b; **11**:510b; **12**:612a; **14**:407a
see also Brain, anatomy; Nerves; Nervous system; Neurohistology; Neurology
Neuroblasts, **6**:435a
Neuroglia, **11**:275a, 374a, 465b, 492a
existence demonstrated, **14**:229b
Neurohistology
seventeenth-century, **9**:63b
mid nineteenth-century, **2**:124b; **7**:437b–438a; **9**:259a–b; **11**:216b, 367b–369b, 491b, 492a, 566a–b; **12**:224a–b; **13**: 555b, 556b; **14**:113b, 142b–143b
late nineteenth-century, **1**:177a–b; **2**:584a; **5**:74a; **7**:439a,b; **8**:15a–16a, 556b; **9**:93b; **10**:130b–131b; **11**:273b–274b, 296a–b, 374a, 382b, 596b–597a; **12**:355a–b; **14**:126b, 229a–b; **15**:430b
twentieth-century, **8**:92a, 282b; **11**:274b–275a, 465b, 597a; **14**:596b
Neurology
ancient Greek, **6**:317a–b
medieval Arabic, **11**:444b
Renaissance, **1**:618b; **2**:336b; **14**:9a–b
seventeenth-century, **2**:312a; **3**:483a–b; **9**:63b–64a, 97b–98a; **14**:406a–408a; **15**:29b
eighteenth-century, **2**:3a; **3**:437b, 561a*; **5**:55b–56b; **9**:326a–327b, 480a–b; **14**:321b–322a, 450a
early nineteenth-century, **5**:251a–252a; **15**:128a
mid nineteenth-century, **1**:404a; **2**:25a, 26a, 29b–30b, 525a–b; **3**:46a, 88a, 205a–b; **7**:421a; **11**:369b; **13**:493a; **14**:57b
late nineteenth-century, **1**:580a; **3**:203b; **4**:12a, 160b–161a; **5**:459b–461a; **9**: 108b–109a, 422a–b; **13**:589a
twentieth-century, **1**:545b; **2**:597b–598a; **3**:112b–113a; **4**:286a–287a, 353a–b; **6**:320b–321a; **9**:108b–109a; **11**:257b; **12**:356a; **15**:473a
clinical
nineteenth-century, **1**:496a–b, 569a–b; **4**:593b–594a; **5**:172a; **7**:47a–48b
mid twentieth-century, **8**:455a
clinical anatomical method, **3**:205b
comparative, **5**:207b–208a; **6**:320b–321a
continuity theory, **10**:580b
dominance theory, **13**:529b–530b
electric fluid theory, **5**:56a
nervous fluid theory, **9**:10a
and psychology, **5**:173b
reflex theory, **5**:174b–177a
as a science, **1**:496b
sexual, **9**:571b
see also Brain; Disease, neurological; Nervous system; Neuroanatomy; Neuron theory; Neurophysiology; Reflexes
Neuron theory, **1**:177b, 545b; **2**:124b; **5**: 282b, 460b; **7**:439a; **8**:7a, 92a; **9**:63b–64a; **11**:274a–b, 296a–b, 382b, 597a; **12**:355b, 397a–b; **14**:126b, 143a
opposition to, **10**:131b
Neurons
classification of, **11**:274b
differentiation of, **8**:283a
size of, **8**:282b
structure of, **8**:283a
Neuropathology
seventeenth-century, **14**:407b, 408a
early nineteenth-century, **10**:321b–322a
late nineteenth-century, **4**:593b; **5**:172a–174a; **8**:7a–b; **9**:93b, 109a, 422a–b; **10**:130b–131b; **14**:277a–b
as tool of neurology, **7**:47a–48b
see also Disease(s), neurological
Neurophysiology
ancient Roman, **5**:230b
seventeenth-century, **4**:62b–63b; **5**:419a, 426a–b; **6**:158a–159a; **14**:406a–408b, 587b
eighteenth-century, **6**:64a–65a, 139b–140a, 459a–b; **8**:506a; **9**:483b; **11**: 112a–b, 159a–b; **13**:122a–123a, 543b–544a; **14**:321a–322a
early nineteenth-century, **2**:595b–596a; **5**:44a–45a, 254b; **8**:134a; **9**:9b–10a, 241b–242b, 570b–571b; **10**:582b; **14**:128a–b
mid nineteenth-century, **2**:111a–b, 124b, 532a; **3**:88a–b; **4**:616a; **6**:59a–60a, 140a, 245a–249a, 440a–441b; **8**:122a; **9**:456b–457a; **10**:579a–b; **11**:133a, 368a; **12**:164b–165a, 244b, 270a–b; **13**:556b–557a; **14**:142b–143b, 184b, 200a–201b; **15**:109a
late nineteenth-century, **1**:569a–570a; **2**: 366b, 367a; **4**:371b–372a, 594a–b; **5**:38a*, 171b–172a, 255a, 279b–283b, 463a–b; **6**:287b, 299b–300b, 518b; **7**:47a–48b; **8**:7a–b, 14b–16a, 27a, 445a, 536a–b; **9**:265b–266a, 505a–506a; **10**:252a–b, 431a–433a; **11**: 382b, 425b–426a; **12**:270b, 355b, 395b–397b; **14**:2b, 105a–b; **15**:21a–b
early twentieth-century, **1**:440b, 580a; **3**:68b–69a; **4**:397b–399a; **5**:65a–b, 291a–b, 541b; **8**:16a–17b, 29a–30a, 452b–455a, 532a–535a; **9**:19b–20b, 564b; **10**:220a–b, 433a–435a; **12**: 397b–402a; **13**:262b, 263a, 393a, 394a, 529a–530b; **15**:73a–74b, 105b–107a, 435a
mid twentieth-century, **5**:207b–208b; **8**: 45b, 295a–b; **11**:545b–546a
chemical mediation theory, **11**:546a
comparative, **4**:594a–b; **11**:510b; **12**:396b
inhibition theory, **9**:265b–266a
integrative action concept, **12**:397b–398b, 400b–401b
methodology, **12**:400b
quantitative, **11**:216b
sensory-motor interpretation, **6**:139b
see also Electrophysiology, Reflexes
Neuropsychiatry, **1**:580a; **14**:277a–b
Neurosis
classification of, **10**:612b–613a, 613b
etiology of, **5**:176a
experimental, **10**:434b
and normality, similarity of, **5**:178a
treatment of, **5**:178b
Neurospora, **14**:255a
Neurosurgery
Renaissance, **1**:618b
late nineteenth-century, **4**:594b; **5**:255a; **6**:518b–519a
twentieth-century, **2**:597b–598a; **3**:336a, 517a–518b; **5**:207b–208a
see also Psychosurgery
Neurosyphilis, **5**:27a
Neurotropism, **4**:297a
Neusis, **1**:188b, 215a, 218b–219a, 224b–225b, 227b, 228a, 245a; **5**:408b, 525b; **6**:413a, 414b–415a; **9**:447b; **10**:295a; **14**:21b
Neutralization, chemical, **6**:255a, 478a
heat of, **1**:299a; **8**:545b; **11**:418a, 577a; **13**:359a
law of, **11**:435a–437a, 541a
volume changes, **15**:457b
Neutrino
emission of, **5**:272b; **7**:154a,b
existence predicted, **9**:261a; **10**:424b–425a
theory of, **4**:579b; **9**:38b
Neutron(s), **4**:82b; **8**:47b*; **11**:14b
bombardment by, **4**:580a–582a; **6**:16b–17a, 118a; **7**:154b, 158b; **9**:261b–263a; **10**:289a; **12**:34a
capture of, **2**:251a; **15**:8a
diffusion of, **4**:376a
discovery of, **2**:338b; **3**:329b; **7**:153b; **8**: 95a; **15**:27a
emission of, **2**:252a; **7**:154a–b, 155a; **15**:8b
gamma-ray-induced, **13**:227a
existence predicted, **6**:118a
existence proved, **12**:34a
impact of, **2**:250b–252a; **8**:13a*
magnetic moment, **13**:240b
production of, **7**:554b; **8**:482a; **13**:163b
scattering of, **4**:582a; **8**:482a
secondary, **4**:580b–582a
slow, **4**:580b–582a
spin, **9**:38a
transport theory, **2**:338b
Neutron activation analysis
invention of, **6**:366b
Neutronology, **4**:582a
Neuymin, Grigory Nikolaevich (1886–1946), 10:30–31; **6**:105b
Neva (river), **8**:465b
Nevada
paleontology, **14**:312b
Nevrologicheskii vestnik
founding of, **1**:580b
New American Cyclopedia, **15**:133b
New Brunswick (Can.)
geology, **14**:120b

New Brunswick (Can.), University of
natural science at, 1:397a
New Caledonia (island), 3:397a
New England
geology, 1:397a; 3:593a; 4:361b; 6:438a; 11:210a; 14:516b
medicine, 14:451b
New Guinea
ethnology, 5:277a
New Hampshire
Geological Survey, 7:45a
New Harmony, Indiana, 8:266b, 615b; 12:133a; 13:467a; 14:328b
New Hebrides (islands), 3:397a
geology, 9:198a
New Holland
see Australia
New Jersey
geography, 4:492b
Geological Survey, 6:57b
geology, 3:593a,b; 4:564a; 11:505a
paleontology, 6:120a
New Mexico
geology, 4:265b
paleontology, 14:410a
New Red Sandstone, 12:276a
New School movement, 14:516a
New South Wales, 4:492a
New Spain
see Mexico
New World
ethnography, 1:48a
metallurgy, 10:515a
New York (state)
botany, 5:518a; 13:432b
Elgin Botanic Garden, 6:521a
entomology, 5:12a
geography, 4:492b
Geological Survey, 3:392a; 4:274a, 363b–365a; 6:57a–b, 437b, 517a; 9:172b, 173a; 13:581b
geology, 4:273b–274a; 5:487a; 9:100b; 11:505a, 594a; 13:621a; 14:120b, 392a
limnology, 7:184a
paleontology, 6:57a–b; 11:593b–594a; 14:312b
New York Academy of Medicine, 7:24a
New York Academy of Sciences, 10:32b
New York Botanical Garden, 6:121b; 13:432b
founding of, 2:477a
publications of, 2:477a
New York City
Bellevue Hospital, 6:521b; 14:249a
Memorial Hospital, 4:499a
Metropolitan Museum of Art, 3:611a
New York Industrial Exhibition, 8:572a
New York Mathematical Society
founding of, 5:414b
see also American Mathematical Society
New York Medical Journal
editors of, 7:23b
New York Pathological Society
founding of, 7:24a
New York State College of Agriculture, 1:396b
New York State Museum
founding of, 6:57b
New-York System (geol.), 4:363b; 6:57b
New York University
Ph.D. degree, 4:182a
School of Medicine, 4:182a
New Zealand
botany, 6:489a; 13:168a
charting of, 3:396b–397a
exploration of, 5:611a
Geological Survey, 5:611b
geology, 5:611a–b
hydrography, 4:492a
natural history
eighteenth-century, 1:434b
paleontology, 5:611a–b
Newall, Hugh Frank (1857–1944), 10:31–32; 6:34a*; 9:404a,b; 12:95b; 13:95b, 371a*
Newberry, John Strong (1822–1892), 10:32–33; 3:610b; 5:395a; 6:187a; 7:252a; 8:463a; 14:442b
Newcastle Chemical Society
founding of, 8:553a
Newcastle University, 3:238b
Newcomb, Simon (1835–1909), 10:33–36; 3:35b; 4:593a*; 6:119b; 9:372a; 11:554a*
acquaintances of, 4:265b; 5:405b; 8:2b; 11:552a
celestial mechanics, 6:399b; 13:422b; 14:576a
latitude variation, 3:194b
planets, 7:481a; 11:552b–553a
precession formula, 1:313a–b; 13:120a
on contemporary scientists, 2:98b; 3:169b; 6:234b, 398b
critics of, 3:221a
influence of, 6:29a, 471a
light velocity, 5:86a; 9:372b
at Naval Observatory, 6:49b
Newcomb operators, 10:34b
observations of, 5:258a
Popular Astronomy, 3:35a; 10:479b
Reminiscences of an Astronomer, 6:103b–104a
students of, 1:464b; 6:471b; 11:83b
tables of, 4:176a
Newcomen, Thomas (1663–1729), 10:36–37
Newcomen engine, 2:179a; 4:44a; 10:293a; 11:496a; 12:463a
Newcomer, Harry Sidney, 5:291a
Newfoundland
botany, 1:433b
geology, 7:185b
survey of, 3:396b
Newlands, Benjamin E. R., 10:37a
Newlands, John Alexander Reina (1837–1898), 10:37–39; 3:476b; 5:412a; 6:461b; 9:349a; 11:136a
Newman, Maxwell Herman Alexander, 14:311a–b
Newport, George (1803–1854), 10:39–40
embryology, 1:477a
Newt
embryogeny, 12:567b–569b
reproduction, 12:559a
Newton, Alfred, 1:505a; 5:80a; 12:258a; 13:168a
Newton, Edwin Tulley (1840–1930), 10:40–41
Newton, Hubert Anson (1830–1896), 10:41–42; 8:478b*; 9:501b; 10:481a*
Newton, Isaac (1642–1727), 10:42–103; 2:294a, 418a; 3:111a; 5:308a, 375a, 466a; 6:35b; 8:437b; 12:458b
acoustics, 7:561b; 10:76a; 12:128a; 15:356b–357a, 358a, 377b
acquaintances of, 4:41a; 5:54a; 6:36a,b, 37a; 8:307b, 609b, 610a; 9:135b, 452b, 499b; 10:500b–501a; 11:1a–b, 526b; 12:458a; 14:27a, 295b
alchemy, 7:319a; 10:81a–83b
algebra, 3:222b; 4:475b; 5:409b; 6:511a; 8:162a; 10:51b; 11:611a; 12:291b, 611a(n); 13:217b
notation, 10:254b
analysis, 1:591a; 2:53b; 3:152b, 167a, 282a; 5:528a; 7:490b, 514b; 8:304a; 10:45a–53a, 65a–68b; 13:266a; 14:150b; 15:99a, 294a
binomial theorem, 10:45b–47a, 48a; 14:149a
calculus, 1:229b, 531a; 2:43b, 54a, 274b; 3:76a, 117b, 228a, 432b, 458a–b; 4:506a,b, 618b; 5:54b; 10:47a–51a; 11:402a; 13:436b; 15:368a
fluxions, 8:610b–611b; 10:47b–51a, 53a, 66a, 67b, 68a
notation, 1:354b; 5:54b; 8:611a; 10:47b, 52a; 13:265b; 14:180a
astronomy, 6:331a,b; 7:373b
and Barrow, 1:475a–b; 10:43b, 44a, 45b, 48a, 49b, 55a
and Bentley, 2:382a; 10:45a
biographies of, 2:453a; 4:36b; 5:61a; 7:7a; 8:505a, 559a; 11:544a; 13:26b, 81a, 599a; 14:67b
biology, 6:45b; 10:54b, 83a
Biot on, 2:134a
celestial mechanics, 4:52b; 5:301a; 6:70a; 9:311a; 10:61a–65b, 69a–78b
comets, 6:68b; 10:64a, 64b–65a, 78b; 15:310a, 311b
earth's revolution, 1:616a
lunar theory, 4:166a; 5:24a, 521a; 6:516a; 8:178b; 10:44b, 78a–b, 84a
chemistry, 5:354a; 8:621a; 10:59b
acidity, 8:76a; 10:81a
crystallization theory, 2:6b
and Cheyne, 3:244b
on contemporary scientists, 10:254b; 14:509b
correspondence of, 3:214a; 9:301a, 465a, 508b*; 10:44b, 315a; 13:67b, 480a
cosmogony, 2:578a; 7:232a
cosmology, 6:611a; 7:228a, 231b, 232b; 10:76b–78b
see also Cosmology, Newtonian
and Cotes, 3:430b, 431a
critics of, 3:114a–115a; 4:88b; 6:148b; 13:270b
earth, figure of, 3:103a, 282b; 7:183a; 9:164b; 10:73b, 76b, 78b; 11:127a; 13:69b; 15:270a, 271b
economics, 7:319a
editions of works, 11:49b
experiments, 8:459b; 10:75b, 77b, 80b
duplicators of, 4:43b, 44a–b; 10:55b
and Flamsteed, 5:25a,b, 516a; 10:44b, 64a,b, 78b, 84a
on Galileo, 5:247b; 10:69b, 72a
gas pressure, 6:292a
geography, 13:583b
geometry, 1:188b, 191b; 4:572b; 5:60b; 6:112b; 8:609b–610a; 9:598b; 10:50a–53a, 72b, 115b, 296a–b, 297a; 13:17a, 68b, 266b; 15:118a, 632b(n)
gravitation, 2:389a; 3:103a, 295b; 7:580a; 10:44a, 54b, 61a–65a, 71a–78b, 596a
and the Gregorys, 5:520b, 521b, 525b, 529a; 10:44b, 45a, 51b, 52b, 53a, 56a, 61b, 83b–84a, 87b(n); 11:1a–b
and Hauksbee, 6:169a–173b *passim;* 10:92a(n)
heat theory, 10:59b, 79a
historiography, 10:82a–b
and Hooke, 6:68a, 484b, 486a, 487a,b; 10:44b, 54a–55a, 56a, 60a, 61b, 62a–64b, 89b(n)
and Huygens, 6:599a, 603b, 609a–b, 610b, 611a; 10:44b, 45a, 54a
hydromechanics, 2:299b; 7:565a; 8:561b; 10:61b–62a, 75a–76b, 551b; 15:295b, 360a
influences on, 1:187b; 2:179b, 379a; 3:209a; 4:47a; 5:288a, 594a*; 9:510b;

10:45b; **13**:96a; **14**:29b, 149a, 377a(n), 637a
and Kepler's laws, 7:308a; **10**:60a–64b *passim,* 69b, 71a–78b *passim*
law of cooling, **2**:136a–b, 178a; **4**:240a; **13**:10b
confirmation of, **11**:432b
laws of motion, **1**:111b; **3**:344b; **6**:347b; **9**:189a; **10**:62a, 70a–b
second law, **1**:112a; **10**:70a–b
third law, **5**:239a; **6**:35a, 250b; **8**:200a–b; **9**:291a, 504b; **10**:62a; **12**:609a
lectures, **10**:44a–b, 51b, 55a–b
and Leibniz, **2**:54a, 57a,b; **3**:295b–296a, 431b; **5**:156b; **6**:72a; **7**:225b–226a, 276a; **8**:151a–b, 158b, 159a,b, 161b–165b *passim;* **9**:452b; **10**:44b, 46a,b, 48b, 52b, 68a, 84a–85a; **14**:150b
manuscripts of, **1**:54b; **3**:348b; **7**:162b–163a, 319a
matter, theory of, **6**:173b; **7**:227b, 228b; **10**:59a, 80b–81a
mechanics, **2**:306a; **5**:528b; **6**:347b, 611b; **8**:599a–b
circular motion, **6**:486a
critics of, **6**:609a–b
dynamics, **8**:163b; **10**:60a–78b
parallelogram of forces, **2**:38b; **3**:460a
pendulums, **5**:528b; **10**:61a, 70a, 75b
vortices, **10**:76a
music, **13**:265a
natural philosophy, **2**:54b; **7**:225b–232b *passim;* **11**:158a; **12**:608b; **14**:418b, 419a
Newton-Cotes formula, **3**:432b
Newton-Stirling formula, **13**:68b
Newton's rings, **1**:201b; **6**:333b; **10**:54b, 57a–59a, 60a; **11**:143a; **12**:477b
central spot, **13**:76b
secondary, **13**:11b
Newton's rule, **13**:217b, 218b–219a
optics, **1**:481b; **2**:328b, 379b, 451a; **3**: 215b, 284a, 454b; **6**:192a; **9**:33b; **10**:44a,b, 53a–60a, 74a, 76a, 79a–81a, 83a; **12**:283b; **14**:565a, 566a, 566b–567a, 567b
colors, **2**:452b–453a; **6**:603b; **9**:118b, 200a, 201a, 463b; **10**:53a–60a *passim*
diffraction, **5**:544b
fits, theory of, **2**:137b, 138a; **10**:59a, 79a,b
opposition to, **5**:445a–b, 446a; **11**:142b
refraction, **7**:404a; **9**:73a, 267a; **15**:359a, 361b
telescope, **3**:97b–98a; **4**:46a*, 148b; **5**: 525b, 529a; **6**:5b; **10**:53a, 54a, 56b
vision, **10**:59b–60a, 79a–b
and Royal Society, **10**:84a,b, 202a
scientific method, **3**:431a–b; **6**:195a; **7**: 488a; **10**:51a–52a, 55b–56a, 65a, 68a, 85a–86a
rules of reasoning, **5**:511a; **6**:557a; **10**:77a, 85b
theology, **2**:134a; **7**:275b, 488a; **8**:612a; **10**:81b–82a
on tides, **2**:43a; **10**:71a, 73b, 78a; **15**:296b
on time, **10**:49b–50a, 70a
vacuum, **3**:295a, 296a
De analysi per aequationes numero infinitas, **3**:458a–b; **7**:163a; **8**:163a; **9**: 311a; **10**:48a,b, 49a
Arithmetica universalis, **5**:510a; **10**:115b; **12**:206a
commentaries on, **3**:119b
editions of, **5**:510b
Commercium epistolicum, **2**:54a; **10**:49b, 51a–b, 68a, 84b
Enumeratio linearum tertii ordinis, **3**: 460a; **5**:567a; **8**:165a; **10**:52a–b
Lectiones opticae, **1**:475a; **10**:53a, 55a–b, 56a
Methodus fluxionum et serierum infinitarum, **7**:163a; **10**:48b–50a, 52a
translations of, **2**:576b
De motu, **10**:62a, 64a, 71a, 74b
De natura acidorum, **10**:80b, 81a; **11**:1b
Opticks, **2**:180b, 326a; **3**:344b, 544b, 599a; **4**:44b; **5**:130a, 509b, 510b; **6**:36a, 40a, 41a, 68b, 139b, 383b, 478a, 484a–b; **9**:266b; **10**:53a–60a *passim,* 85a,b; **11**:68a; **14**:492b, 562a; **15**:358b
commentaries on, **14**:83b, 84a
editions of, **2**:330b; **3**:244b; **6**:41b, 172a, 173a–b; **10**:79a–81a, 83a; **11**:146b; **14**:415a
translations of, **3**:295a; **11**:508a; **13**: 599a
Philosophiae naturalis principia mathematica
see *Principia mathematica*
Principia mathematica, **10**:44a–85b *passim;* **14**:254b
analysis, **1**:164b; **13**:265b, 561a
celestial mechanics, **4**:591a
commentaries on, **3**:282a, 296b(n); **5**: 193a; **8**:611a; **14**:83b; **15**:138a
criticism in, **2**:54b; **3**:152b
editions of, **3**:431a; **6**:173a–b; **7**:488a; **9**:465b*; **10**:67b, 68a, 85a, 501a
gravitation studies, **3**:156a; **5**:135b, 521a; **11**:424b
inaccuracies in, **5**:247b
influence of, **2**:326a; **3**:294b, 458b, 537b, 541a; **4**:169a, 283b, 590b; **5**: 499a, 509b, 510b; **6**:85b; **7**:442a; **8**: 459b, 610a; **9**:452b; **13**:179b, 183b; **14**:562a; **15**:317a, 350b
influences on, **2**:349a; **4**:425a; **5**:24a, 528b; **9**:120b
interpretations of, **3**:90a
motion studies, **1**:111b, 113a; **2**:39a, 49a; **3**:344b; **4**:480a
publication of, **6**:68a, 487a; **13**:584b, 585b
reviews of, **7**:308a
translations of, **3**:216a–b, 281b; **7**: 514b*; **10**:501a; **14**:83a
Tractatus de quadratura curvarum, **3**: 152b; **8**:165a; **10**:50a–51a, 66a, 68a, 72b; **11**:494a
Newton meter
see Joule (physical unit)
Newtonianism
late seventeenth-century, **2**:613b; **5**:520b–521b, 524b; **6**:129b
early eighteenth-century, **1**:111a; **2**:16a, 226b; **3**:104b, 105a,b, 107a–b, 173b, 215a–216b, 244a–b, 281a–284a, 294b–296b, 380b, 381b, 382b, 485b; **4**:44a–b, 45a, 88a–b, 203b; **5**:62a, 156b, 509a–511b, 519b; **6**:36a, 555b; **7**:274b, 275a–276a; **8**:159b, 611b; **9**:33b–34a, 186a–b, 532b, 595b, 596a; **10**:145b, 500b–501a; **11**:2b, 158a, 399b, 400b, 494a, 508a; **13**:224a, 231b; **14**:83a–84b, 296b, 589a
late eighteenth-century, **1**:67a, 326b; **2**: 181a, 326a, 327b, 330a–b, 577b, 579b; **3**:73b, 155b–157a, 206a, 444a–b, 446a, 539b; **4**:26a, 529b, 565b; **5**:195b; **6**:383b–384a, 386a; **7**:225b–232b *passim,* 490a–b; **8**:305a; **9**:234b, 325b–326a; **10**:108a; **11**:140a, 142a, 144b, 497a; **12**:429b–430b; **13**:183b, 224b, 494b; **14**:452b
nineteenth-century, **3**:542b, 600a; **5**:193a–b, 299a; **6**:178b, 279a; **8**:159b; **9**:204b; **12**:288b, 608b–609a; **13**:541a; **15**: 358a,b, 360a, 368a, 377a, 384a
and Cartesianism, **11**:544a
in chemistry, **6**:41b
critics of, **2**:16b–18a; **3**:295b–296a; **5**: 156b, 166b, 167a; **7**:276a; **11**:157b, 580a; **12**:118b, 609a; **13**:603a
earth, figure of, **3**:104b, 105a, 107a, 173b; **7**:542b
in Japan, **1**:311a; **12**:407a–408b; **15**: 736a,b
and medicine, **11**:2b
and theology, **3**:114b–115a, 244b, 295a–296a; **7**:275b
Neyman, Jerzy, **5**:10a, 477a; **14**:383a,b, 437b
Neystett, André Reusner de, **3**:63b
Neyt, Adolphe, **1**:601b
Niagara Falls
geology, **8**:571a; **13**:270a–b
harnessing of, **13**:286b–287a
Nicaea, Council of, **3**:513b
Nicaise, Jules Édouard, **3**:218b; **6**:276b
Nicaragua
discovery of, **15**:89b
proposed canal, **4**:266a
Nicarchus, **11**:127a,b
Niccolite, **3**:473a–b
Nice (France)
observatory, **10**:528b; **13**:344a
Niceron, Jean-François (1613–1646), 10: 103–104; 13:434a–b
Nicéville, Lionel de, **4**:493b
Nicholas V, Pope, **1**:229a; **3**:513b, 515a
Nicholas (monk), **7**:187a
Nicholas of Autrecourt, **2**:603b, 605a
Nicholas Chuquet
see Chuquet, Nicolas
Nicholas of Cusa
see Cusa, Nicholas
Nicholas of Damascus
see Nicolaus of Damascus
Nicholas of Dinkelsbühl, **7**:118a, 119b–120a
Nicholas of Lyra, **2**:23b
Nicholas of Reggio, **5**:233b
Nicholas of Sicily, **1**:272b
Nicholas, John Spangler (1895–1963), 10: 104–105; 6:135a*
Nichols, Ernest Fox (1869–1924), 10:105–106
Nichols radiometer, **10**:105b
optics, **1**:301a; **11**:122b, 582a
Nichols, George, **15**:149b
Nichols, James Burton, **13**:160a
Nicholson, Edward Chambers, **6**:461b, 463a
Nicholson, Henry Alleyne, **8**:33a; **13**: 532b(n)
Nicholson, John William (1881–1955), 10: 106–107; 14:310b
quantum theory, **12**:528b
Nicholson, Seth Barnes (1891–1963), 10: 107; 6:33b*; **8**:582b; **12**:73a
Nicholson, William (1753–1815), 10:107–109; 1:43b; **5**:537a
Dictionary of Chemistry, **2**:7a; **3**:598b; **13**:547b
electricity, **14**:73a
electrochemistry, **3**:68a, 600b
Journal of Natural Philosophy, Chemistry and the Arts
see *Nicholson's Journal*
translations by, **3**:203a*; **7**:388b
Nicholson's Journal, **3**:541a, 544a, 600b; **10**:108b; **13**:413b–414a
Nickel, **4**:394b
alloys, **3**:239a; **5**:582b–583a; **9**:312b; **11**: 550a

Nickel (cont.)
atomic weight, **1**:458a–b; **11**:417a
as catalyst, **12**:46b–47a
compounds of, **13**:310a
complex, **11**:88a
salts, **3**:271b
discovery of, **3**:473a–b
divalent
analytical reagent for, **3**:272a
electrodeposition of, **2**:340b; **7**:56b
extraction of, **9**:467a
magnetic properties, **1**:283b–284b; **11**:578a; **14**:245b
in marine sediments, **10**:564a
ores
analysis of, **3**:473a–b; **11**:437b
deposits of, **4**:262a; **7**:267a
formation of, **14**:58b
physical properties, **3**:401a*
purification of, **11**:495a
separation from cobalt, **2**:565a; **5**:7a
wire
embrittlement of, **3**:87a
Nickel carbonyl
discovery of, **9**:467a
Nicklès, François Joseph Jérôme, **2**:340b; **8**:55b; **9**:401b, 451a
Nicod, Jean, **12**:14b, 15a
Nicol, James, **5**:337a; **8**:33a
Nicol, William (1768–1851), 10:109–110; **2**:492a; **5**:455b; **9**:199b; **14**:462b
Nicol prism, **9**:201b; **10**:109b–110a; **13**:273a
Nicolai, Christoph Friedrich, **1**:555a; **2**:279b
Nicolai, Friedrich Bernhard Gottfried (1793–1846), 10:110–111; **10**:573a; **12**:518b
Nicolaus of Damascus (*b.* 64 B.C.), 10:111–112
De plantis, **1**:268a, 271b
editions of, **12**:135a
translations of, **7**:25a; **11**:504a
Nicole, François, **9**:499b; **13**:68b
Nicole, Pierre, **1**:292a; **3**:460a; **10**:335b
Nicolle, Charles Jules Henri (1866–1936), 15:453–455; **14**:623b
Nicollet, Joseph Nicholas, **11**:82b; **13**:4a
Nicolson, John Thomas, **1**:52a; **3**:19a,b, 20a
Nicomachus of Gerasa (*fl. ca.* A.D. 100), 10:112–114; **4**:159b
commentaries on, **4**:160a; **11**:221b
influence of, **1**:91b, 267b; **2**:233a, 590b; **3**:272b; **5**:365a; **13**:349b; **14**:35b
Introduction to Arithmetic, **4**:111a, 392a; **9**:140b–141a; **13**:540a
number theory, **9**:191b; **15**:10a, 670a, 671b
Nicomedes (*fl. ca.* 250 B.C.), 10:114–116
conchoid, **4**:489b; **6**:406b; **10**:295a, 300a
mean proportionals, **1**:288a; **4**:105a
quadratrix, **4**:103b, 104a; **6**:407b
Nicot, Jean (*ca.* 1530–1604), 10:116–117
Nicoteles of Cyrene, **1**:185b; **3**:391a
Nicotiana, **7**:441b; **12**:328a
genetics, **4**:271a
Nicotinamide-adenine dinucleotide (NAD), **15**:257b
Nicotinamides, **4**:155b, 486a; **14**:174a
Nicotine, **3**:246b
detection of, **12**:620a
neurological effects, **8**:15a–b, 16b
Nicotinic acid, **5**:209a
metabolism of, **4**:358a
Niebuhr, Carsten (1733–1815), 10:117; **5**:74b; **6**:455a*
Niello, **13**:327b
Nielsen, Niels (1865–1931), 10:117–118
historiography, **14**:279b
Niemann, Carl, **15**:15b
Niementowski, Stefan, **7**:471a
Niepce, Claude, **10**:118a–119a
Niepce, Joseph (later Nicéphore) (1765–1833), 10:118–119; **1**:203a; **14**:492a
engine, **3**:78b*
Niepcotypy, **4**:283a*
Niesten, Jean Louis Nicolas (1844–1920), 10:119–120
Nietzsche, Friedrich Wilhelm, **7**:189b; **15**:80a
Thus Spake Zarathustra, **7**:191b
Nieuw archief voor wiskunde
editors of, **7**:466a; **12**:213b
Nieuwentijt, Bernard (1654–1718), 10:120–121; **2**:53a; **8**:164b
critics of, **6**:304b
Nieuwland, Julius Arthur (1878–1936), 10:121–122; **3**:86b
Nieuwland, Pieter, **13**:184a, 469a
Nife, **13**:148a
Nifo, Agostino (*ca.* 1469–1538), 10:122–124; **8**:228b; **11**:74a; **12**:134b
Niger Hapelius, (Nicolaus)
see Hapelius, Nicolaus Niger
Nigg, Clara, **7**:624b
Niggli, Paul (1888–1953), 10:124–127; **5**:563a,b*; **8**:485a; **12**:195a
chemistry, **5**:455a; **6**:303b; **13**:2a
geology, **4**:409a
glaciers, **13**:246b
Niggli values, **10**:126a
Versuch einer natürlichen Klassifikation der im weiteren Sinne magmatischen Erzlagerstätten, **12**:195a
Night
divisions of, **15**:711a
Nightingale, Florence, **2**:376b; **8**:406b; **11**:237b; **13**:217a
Nightmare, **14**:408a
Nihil, Johann, **15**:474a,b
Nijānanda
commentaries of, **2**:117a
Nikander of Colophon, **3**:414a
Nikiforovskii, Petr Mikhailovich, **10**:435a
Nikitin, Sergey Nikolaevich (1851–1909), 10:128; **5**:212b; **7**:250a, 320b*; **15**:428a
Nikodým, Otton
Radon-Nikodým theorem, **1**:427b; **11**:261a
Nīlakaṇṭha (*b.* 1444; *d.* after 1501), 10:129; **10**:313a
commentaries of, **1**:308a,b
Golasāra, **15**:622a–b, 623a
Siddhāntadarpaṇa, **15**:622b, 623a–b
Tantrasaṅgraha, **1**:48b; **15**:622a–623a
Nīlāmbara Jhā, **15**:629a
commentaries of, **2**:116a; **9**:42a
Nile (river), **6**:189a,b
engineering study, **5**:410b
floods
theories of, **10**:181a
source of, **2**:530b
Nilson, Lars Fredrik, **2**:589b; **3**:321b; **9**:110b, 290a,b, 291a
Nilsson, Nils Ragnar, **4**:485b
Nilsson, Sven, **8**:518b–519a; **9**:489a
Nilsson-Ehle, Herman (1873–1949), 10:129–130; **4**:271a; **13**:478a
Ninnis, Belgrave Edward Sutton, **9**:198a
Niobium, **13**:468a
determination of, **10**:242a
discovery of, **6**:166b; **11**:541a
Nippoldt, Wilhelm August, **7**:449b
al-Nīsābūrī, Ibn Abī Ṣādiq, **4**:440b; **15**:239a–b
Nishimura Tosato, **15**:746a–b, 748a
Nishina, Yoshio, **3**:369b; **9**:606b; **13**:240a
Nissl, Franz (1860–1919), 10:130–132; **5**:571a
Nissl bodies, **8**:282b; **10**:130b, 131a
Nisus formativus, **2**:204a
Nitella, **6**:443a–b
Niter, **2**:380a; **3**:21a; **4**:96a, 156b; **8**:77a; **11**:564a
aerial
see Nitro-aerial spirits
analysis of, **10**:510b
artificial beds of, **3**:201b
deflagration of, **6**:395b
detonation of, **6**:42b, 115a
extraction of, **3**:201b
from wood, **5**:421a
formation of, **8**:171b, 173b
manufacture of, **2**:143a; **3**:419b*; **4**:393b; **5**:602a, 603a; **8**:122a; **9**:472b; **11**:472a, 616b; **12**:64a, 329b; **13**:496a, 596b
nature of, **8**:469b; **9**:244a
origin of, **4**:149b; **10**:265b
refining of, **11**:563a
three forms of, **5**:421b–422a
as universal salt, **5**:421b–422a
Nitophyllum, **13**:165a
Nitramide, **13**:338b
Nitrates, **8**:171b
analysis of, **3**:488b; **9**:105b
production of, **8**:24b
Nitration, **3**:246a; **9**:608b
of aromatic compounds, **14**:334a
Nitric acid
composition of, **8**:77a
density, **9**:95b
explosive reactions of, **6**:41a
preparation of, **2**:421b; **6**:255a
industrial, **3**:201a, 202a; **4**:165b; **8**:553b; **15**:462a
reactions, **2**:386a; **9**:401b
in voltaic pile, **3**:601a
Nitric oxide, **3**:600b
discovery of, **11**:144b
formation of, **8**:23a
preparation of
industrial, **3**:437a; **15**:438a
Nitrides, **9**:401b
Nitrification, **2**:356b, 357a; **9**:401b
of hydrocarbons, **13**:468a
Nitriles, **5**:125b
condensation, **13**:388b
conversion of, **7**:451a, 452a
isonitriles, **5**:315a
Nitrites
tests for, **2**:340b; **10**:440b
Nitro-aerial spirits, **1**:451a; **5**:47b; **6**:40b, 485a; **7**:459a; **8**:526b; **9**:244a–b, 246a; **12**:307b–308a; **14**:407b
Nitrobenzene, **5**:197b
reduction of, **14**:622a
Nitrocellulose
discovery of, **10**:499a
Nitrogen, **5**:411b
active, **13**:107b
in atmosphere, **11**:279a
atomic weight, **3**:476b; **10**:510b; **11**:279a, 417b; **12**:620a; **13**:499b; **14**:318b
in blood, **2**:62a
bonding of, **1**:553b; **6**:108a
compounds of, **2**:562b; **5**:622a; **7**:372b; **13**:429a
azo, **5**:215a; **7**:471a
colorimetry of, **5**:53a
diazine, **5**:215a
diazo, **3**:511a–b; **5**:1a–b, 536b; **6**:108a–b; **7**:282a; **9**:354a

heterocyclic, **5**:215a
imino, **13**:388b
isodiazo, **1**:426b; **6**:108a
isonitroso, **1**:553b
nitrite, **9**:355b
nitro, **9**:355b
organic, **6**:462a–b; **12**:418b, 419b; **13**:338b; **14**:334a
oxidation and reduction of, **1**:426b
oxides, **3**:542b, 600b; **4**:239b; **5**:197b, 321a, 323b; **6**:285a, 550b
stereochemistry of, **6**:108a; **8**:109b; **9**:356b; **14**:265a–b, 267a–271b
thermochemistry of, **2**:70a
see also Ammonium compounds
compressibility, **1**:128a
density, **13**:102b–103a
determination of, **8**:331b, 332a; **9**:91b; **12**:164a; **14**:169a–b
Dumas method, **3**:341b; **7**:393a; **9**:91b
Kjeldahl method, **7**:393a–b; **12**:546b
discovery of, **12**:24b–25a
disintegration of, **11**:281a
elementary nature of, **2**:96a; **8**:82a,b
fixation of
anaerobic, **14**:37b
by bacteria, **15**:14a
in laboratory, **2**:70b; **3**:436b, 479b; **5**:621b, 622a; **9**:427b; **15**:438a
by plants, **4**:220a; **8**:344b–347b *passim,* 512b; **14**:45b, 46a
by soil, **2**:356b–357a
symbiotic, **14**:37b
inversion temperature, **10**:206a
isotopes
radioactive, **7**:154b
liquefaction of, **10**:206a; **14**:522b
as lubricant, **3**:299a
metabolism of, **9**:284b; **10**:326b; **11**:585b–586a; **14**:63b–66b
in plants, **3**:359a; **11**:179b; **12**:208a
metallic nature of, **3**:599b, 602b
in muscles, **5**:90b–91a
naming of, **10**:446b
nuclear bombardment, **6**:117b–118a
oxidation of, **10**:329b
in plants, **5**:91a
source of, **8**:344b, 345b, 347b
production of, **8**:365b
pulmonary exchange, **7**:501b
quantum mechanics, **10**:214b
reactions of, **8**:118b
spectrum, **12**:238a
valence, **1**:553b; **8**:508a; **14**:267a
pentavalence, **1**:426b
stereochemistry of, **6**:108a
Nitrogen dioxide
discovery of, **11**:145b
Nitrogen pentoxide
synthesis of, **4**:77b
Nitrogen trichloride
discovery of, **4**:239a
Nitroglycerine, **14**:622b
production of, **10**:132b
Nitrolic acids, **9**:355b
Nitrometer, **3**:341b; **12**:164a
Nitroparaffins
preparation of, **9**:355b
Nitrophenols, **5**:197b
Nitroprussides, **11**:36b
Nitrous acid, **12**:145a
action of, **9**:178a
composition of, **9**:471b
freezing point, **3**:158a
reactions of, **9**:355b
Nitrous air, **14**:74b
Nitrous ether, **14**:508b
Nitrous oxide
anesthetic use of, **2**:62b; **3**:600a–b
discovery of, **11**:145b
Nitti, Frederico, **4**:154b; **5**:100a
Nivation, **7**:149a; **9**:150b
Niven, David (bot.), **2**:518b
Niẓām al-Dīn al-Naishāpūrī, **11**:251b
Niẓāmī ᶜArūḍī Samarqandī, **7**:324a, 325a
Nobel, Alfred Bernhard (1833–1896), 10:132–133; 9:291b
Nobel Foundation, **13**:419a–b
Nobel Institute, **9**:262b
Nobel Prize, **10**:133a; **14**:632b–633a
committees, **11**:382b
disputes over, **5**:622a
Mainau Declaration, **6**:17a
Nobert, Friedrich Adolph (1806–1881), 10:133–134
Nobili, Leopoldo (1784–1835), 10:134–136; 4:201b; **10**:500a
Noble, Edward Clark, **1**:441b–442a, 442b
Noble, Henry Robertson, **8**:496b; **13**:471b
Nocard, Edmond Isidore Étienne, **7**:426b
Noctilabium (astron. instrum.), **2**:24a
Noctuidae, **5**:555b
Nocturnal (astron. instrum.), **5**:344a; **7**:119b
Noddack, Ida, **9**:262a
Noddack, Walter (1893–1960), 10:136; 15:448b
Nodder, Charles Reynolds, **9**:402b
Nodes (math.), **1**:307b; **3**:167a
Nodzu, Ryuzaburo, **13**:2b
Noël, Étienne (1581–1659), 10:137; 7:485b; **10**:333a–b, 342b
Noelting, Emil, **7**:471a; **14**:265a
Noether, Amalie Emmy (1882–1935), 10:137–139; **5**:472b–473a; **6**:297a, 496a,b; **13**:23b; **14**:90b
algebra, **4**:5a; **7**:444b; **10**:140b
Noether, Max (1844–1921), 10:139–141; 3:315a; **10**:244b
algebraic geometry, **2**:465a,b; **3**:167a, 315b; **6**:75b, 358a*; **11**:536a; **12**:330b
transformations, **3**:323a; **5**:569a; **12**:284b
analysis, **11**:451b
on contemporary scientists, **5**:473a*; **6**:309b*; **8**:327a*, 542b*; **13**:222a*; **14**:619b*
Die Entwicklung der Theorie der algebraischen Functionen . . ., **11**:448b
historiography, **3**:135b, 167b; **4**:269a*; **11**:448b; **13**:6b*
Noether theorem, **8**:585a; **10**:140a–b
Noetherian conditions, **10**:140a–b
Skolem-Noether theorem, **12**:451b
Noetics, **1**:118a–b; **15**:262a
Noeud vital, **8**:506a
Noguchi, (Seisaku) Hideyo (1876–1928), 10:141–145; 5:40a
Noll, Friedrich Carl, **5**:453a
Noll, Fritz, **12**:58b; **13**:90a
Nollet, Jean-Antoine (1700–1770), 10:145–148; 4:216a; **5**:138b*; **9**:470a
acquaintances of, **3**:281a; **8**:115a; **9**:595b; **13**:565b
botany, **10**:577a
correspondence of, **3**:486a; **9**:596a–b; **14**:70a
critics of, **8**:177b
electricity, **1**:547a,b; **2**:324b; **5**:516a; **8**:258b; **13**:224b–225a; **14**:194a, 352b
influence of, **2**:474b; **4**:216b; **12**:427a; **14**:70a
physics, **11**:432b
physiology, **12**:559a
students of, **3**:106a; **10**:325b; **12**:63b
Nomenclature
anatomical
Renaissance, **1**:523a; **3**:343a; **4**:199a, 285a; **5**:586a
seventeenth-century, **10**:477a
eighteenth-century, **14**:14b
nineteenth-century, **2**:596a; **6**:436a; **10**:261b, 263a
historiography of, **6**:618b
astronomical, **1**:242a, 530a; **9**:23a; **13**:149b; **15**:673a
biological
binomial, **3**:129b; **7**:547b; **14**:133a
botanical
ancient Roman, **4**:120b
Renaissance, **1**:525b–526a; **2**:218a–b, 536b; **13**:502a,b
seventeenth-century, **1**:524a; **11**:313b–314a; **13**:443a
eighteenth-century, **4**:99a; **6**:65b–66a; **7**:585a; **8**:375a; **9**:390a, 434a–b
nineteenth-century, **2**:477a; **3**:43a; **6**:296a; **8**:372b; **10**:531b; **15**:194b, 268b–269a
twentieth-century, **5**:192b, 498a; **15**:421b
binomial, **1**:369b, 524a, 592b; **7**:58b; **8**:377b
historical, **13**:343a
saints' names, **1**:525b–526a
synonymy, **9**:179a,b; **11**:595a, 601b; **13**:502a,b
calendric, **1**:512a; **2**:152b
chemical
Renaissance, **10**:310b
late eighteenth-century, **1**:64b; **2**:7b–8a, 74a, 180b, 292a; **3**:201b, 382b; **5**:90a, 411b, 601b, 602b; **7**:389a; **8**:77b, 80b, 81a, 82a, 620b; **10**:446b*; **12**:65a
early nineteenth-century, **3**:6a, 159a, 232b; **14**:156b*
mid nineteenth-century, **1**:348a; **3**:49a; **6**:463b; **7**:281a; **8**:60b; **10**:37a; **14**:396a
early twentieth-century, **11**:551b; **13**:71a
binary, **2**:7b–8a
Russian, **6**:354b
crystallographic, **5**:212a; **6**:303a,b
geographical, **15**:637a
geological, **7**:63a; **10**:504a; **11**:375a–b; **14**:392b*
geographical reference, **4**:363b
mineral facies, **4**:409a
petrographic, **5**:212b, 486b–487a, 563a; **7**:5a–b; **8**:286a
strata, **4**:274a; **7**:267a; **12**:277b, 490a
time periods, **2**:496b–497a
mineralogical
eighteenth-century, **5**:602a; **6**:178b
nineteenth-century, **2**:440b; **6**:182a–b
twentieth-century, **5**:576b; **14**:183b
paleontological, **12**:182a–b
zoological, **2**:110a; **6**:401a; **9**:432a; **11**:443b; **12**:264a; **15**:78b
entomological, **3**:320a; **5**:328b; **8**:379a; **12**:201a
trinomial, **11**:443b
see also *Méthode de nomenclature chimique;* Terminology
Nominalism
medieval, **1**:1b, 84a, 100b, 199b; **2**:193b, 395b, 604a–b; **9**:136a–b; **10**:171b, 173b–174b
Renaissance, **3**:172a; **4**:237b; **9**:32b–33a; **12**:97b, 547b, 548a
seventeenth-century, **5**:286a; **6**:446b–447a

Nominalism (cont.)
  nineteenth-century, **15**:168b
  twentieth-century, **3**:278a
  opposition to, **10**:485a
Nomogram, **6**:261b; **7**:260a, 265b; **10**:170a–b; **11**:197b; **14**:552b
Nonconformists (religious), **1**:531a; **14**:209b, 371b
Noncontradiction, principle of, **8**:546a
Nonius (astron. instrum.), **10**:161a
Nonnaturals, medical theory of, **15**:241a–242a
Nontronite, **13**:246b
Noorden, Carl von, **1**:446b; **4**:359b; **8**:451b, 452a
Noosphere, **13**:275a, 276b
Norddeutsche Zeitung, **15**:193b
Nordenflicht, Thaddeus von, **4**:345b
Nordenskiöld, (Nils) Adolf Erik (1832–1901), 10:148–149; **1**:528a; **6**:78b; **11**:383b*; **12**:144b–145a, 232b; **15**:430b
Nordic Institute for Theoretical Atomic Physics, **2**:252b
Nordisk astronomisk tidsskrift
  editors of, **13**:99b
Nordisk Insulinlaboratorium, **7**:503a
Nordmann, Charles, **4**:69b
  Tikhov-Nordmann effect, **13**:408b
Nordström, Gunnar, **4**:327a; **7**:26b
Nörlund, Niels Erik, **2**:144a; **10**:117b
Norm (geol.), **3**:484a
Normal (math.)
  affine, **4**:257b
  binormal, **12**:74a
  concept of, **9**:476a
  construction of, **4**:57b; **10**:180b
  systems of, **11**:46a
Normality, chemical
  concept of, **13**:547b
Norman, Robert (*fl.* late sixteenth century), 10:149–150; **5**:593a
  The Newe Attractive, **5**:399a; **6**:144a
North, Marianne, **6**:491a
North America
  anthropology, **11**:219a
  astronomy, **14**:451b
  botany, **6**:163a, 492a; **7**:210b; **9**:54a; **10**:163b–164b; **11**:218a–b; **15**:149b–150a
  colonization of, **6**:20a–b
  education, **11**:464a
  ethnography
    sixteenth-century, **1**:48a; **6**:124a
    eighteenth-century, **1**:489a,b
    nineteenth-century, **11**:308b
    twentieth-century, **6**:527b
  exploration of, **3**:439a; **4**:361b; **5**:512b; **10**:285a–b
  geodesy, **9**:164a–b
  geography, **2**:191a; **4**:492b–493a; **10**:617b
  geology, **3**:462b, 483b–484a; **4**:492b–493a; **8**:99a–b, 571a–572a; **9**:104a; **12**:229b–230b, 274b; **13**:518a, 572b
    glaciation, **11**:113b; **14**:517a
    stratigraphy, **10**:164a; **13**:518b–519a, 532a–b, 581a–b
  natural history
    Renaissance, **1**:48a
    seventeenth-century, **1**:431b–432b; **4**:135b
    eighteenth-century, **1**:487a–488a, 488b–489b; **2**:322a–b; **3**:129a–b, 198a, 344a; **5**:76b
    nineteenth-century, **1**:330a–331b, 484b–485b; **3**:439a, 535b–536b
  paleogeography, **12**:230a; **13**:531b
  paleontology, **8**:169b; **9**:134a–b; **13**:531b; **14**:120b
  philosophy, **8**:460b
  zoology, **5**:76b; **6**:120a; **15**:227a
    entomology, **6**:507a
    ichthyology, **8**:266b
    ornithology, **3**:438b; **10**:164a; **11**:443a–b
  see also America; Canada; Mexico; United States; individual states and regions
North American Entomologist, **5**:555b
North American Flora, **2**:477a
North Atlantic Treaty Organization (NATO)
  Advisory Group for Aeronautical Research and Development (AGARD), **7**:248a
North Carolina
  geology, **9**:420b–421a
North Pacific Exploring and Surveying Expedition, **5**:478b; **13**:65b
North Pole
  exploration of, **7**:266a; **12**:279b; **13**:166b
  geography, **12**:412b
North Sea
  gravimetry, **13**:611a
North Star
  see Polestar
Northeast Passage, **13**:165a
Northern Boundary Commission, **3**:438b
Northern lights
  see Aurora borealis
Northern Pacific survey, **3**:592b
Northern Transcontinental Survey, **11**:210a
Northrop, John Howard, **1**:11b; **13**:152b, 420a
Northumberland
  geology, **14**:462b
Northwest Coast
  ethnology, **2**:207b, 209a
Northwest Passage, **2**:463a; **3**:397a; **6**:20a; **11**:554a–b; **12**:49a
Northwestern University
  Dearborn Observatory, **6**:522b
Norton, John Pitkin (1822–1852), 10:150–151; **10**:242a; **12**:435a
Norton, Thomas Herbert, **2**:588b
Norway
  astronomy, **6**:107a
  fisheries, **6**:442a
  Geological Survey, **2**:486a
  geology, **2**:486a; **4**:406b; **5**:457a; **9**:620a; **13**:147b; **14**:59a,b
  glaciers, **5**:69a
  leprosy in, **6**:101b–102a
  meteorology, **9**:442b–443a
  physics, **6**:615b
  surveying of, **6**:107a
  zoology, **9**:575b–576a
Norwegian Polar Institute, **13**:167a
Norwegian Sea
  oceanography, **15**:431b
Norwood, Richard (1590–1665), 10:151–152
  Trigonometrica, **5**:344a
Nose
  morphology, **2**:504a
  see also Rhinoscopy
Nosology
  ancient Greek, **6**:422a, 423a,b
  Renaissance, **1**:399b
  eighteenth-century, **3**:2a, 495a; **8**:379a–b; **10**:612b–613a
  nineteenth-century, **2**:508b*; **5**:571a
  twentieth-century, **1**:439a
  comparative, **11**:178a
  see also Disease, Pathology
Nostradamus, Michael (1503–1566), 10:152–153; **12**:135a
Notation
  chemical
    seventeenth-century, **7**:459a
    eighteenth-century, **1**:64b; **3**:495a; **6**:164b; **10**:446b*
    early nineteenth-century, **2**:94a; **3**:47b, 289a–b
    mid nineteenth-century, **2**:484b–485a, 515a; **3**:448b–449a; **4**:400a; **5**:126a, 370b–373b; **6**:463b; **7**:280b, 451b; **8**:508a; **10**:177b–178a; **14**:396a
    late nineteenth-century, **2**:67b–68a; **5**:2b, 315a; **13**:578a
    twentieth-century, **1**:321b
    philosophy of, **5**:370b
  crystallographic, **1**:463a*; **4**:66a; **6**:303a; **8**:288b; **9**:183b, 392b–393a, 620a; **14**:241a
  engineering, **1**:355a
  logical
    seventeenth-century, **6**:299a
    nineteenth-century, **4**:36a–b; **5**:152b–153a; **10**:442b
    twentieth-century, **8**:546a; **12**:14a–b
  mathematical
    ancient Greek, **4**:112a–b
    Byzantine, **8**:191a
    medieval Arabic, **1**:31a, 437b; **11**:230a; **13**:539b, 545b
    medieval Latin, **4**:607a–b, 609a, 611b, 612a; **13**:340a
    Renaissance, **2**:281a–b; **3**:274a, 277b, 311b–312a; **8**:41a; **11**:590a, 591a; **13**:48b, 60a; **14**:19b, 325a
    early seventeenth-century, **4**:56a–b, 61a(n); **5**:54b, 382b, 408b–409a, 410a*; **6**:124b, 125a–b, 299a; **7**:194a; **10**:151b, 254b–255a; **11**:533b; **14**:20b–21a, 24b
    late seventeenth-century, **4**:451b–452a, 576a(n); **5**:410a*; **8**:161b, 163b; **10**:47b, 52a, 495b; **11**:512a; **14**:148a
    early eighteenth-century, **3**:458b, 460b–461a; **4**:474a; **7**:163a
    late eighteenth-century, **2**:113a; **6**:403b; **7**:570b; **13**:571a–b; **14**:180a–b; **15**:278a, 293a, 328a
    early nineteenth-century, **1**:240a, 354b; **2**:214b; **3**:136a; **4**:546b; **5**:96a, 97a; **6**:324a; **7**:494b; **12**:325b; **14**:500a
    mid nineteenth-century, **3**:314b; **4**:137a,b; **5**:573a
    late nineteenth-century, **3**:165b–166a; **5**:391a, 414a; **6**:346b; **7**:444b; **8**:250b; **10**:484a; **11**:409b(n)
    twentieth-century, **7**:492b; **9**:502b; **12**:214b
    abridged, **2**:214b; **5**:368b
    Chinese, **8**:315a–319b; **12**:376a
    Egyptian, **15**:682a–684b, 687a, 700b
    infinity, **14**:148a
    Japanese, **1**:83b, 90b; **12**:291a
    Mesopotamian, **15**:667b
    philosophy of, **5**:391a; **14**:500a
    see also Algebra, Analysis, Arithmetic, Calculus, Combinations, Decimal system, Determinants, Fractions, Numeration, Powers, Quaternions, Radicals, Trigonometry, Vector notation
  musical
    medieval Latin, **6**:302a
    twentieth-century, **6**:229b
  in physics, **9**:209b; **11**:476b
  statistical, **5**:9b

Notes From the Royal Botanic Garden, Edinburgh
founding of, **1**:422b
Notgemeinschaft der deutschen Wissenschaften, **5**:622b; **12**:615a,b
founding of, **4**:269a
Nothnagel, Carl Wilhelm Hermann, **8**:7a
Notiometer
see Hygrometer
Notochord, **1**:421b
discovery of, **1**:386b
Nott, Josiah Clark, **4**:620b; **9**:540b–541a; **11**:346b
Nottingham
Bromley House Subscription Library, **15**:199b
Nous, **1**:118a, 149b; **9**:111b
Nouveau cours complet d'agriculture théorique et pratique, **2**:322a
Nouveau dictionnaire d'histoire naturelle, **2**:322a; **10**:326a
Nouvelle correspondance mathématique, **3**:177b
editors of, **10**:25a
founding of, **9**:81a
Nouvelles annales de mathématiques, **3**:467b
editors of, **8**:61b; **10**:514b*
Nova acta Academiae Caesareae Leopoldino Carolinae germanicae naturae curiosorum
editors of, **10**:12b
Nova Scotia (province)
geology, **3**:607b, 608a; **8**:571a–b
Novae, **1**:466b; **4**:249b
of 1572 (Brahe's), **2**:402b–404a, 408a, 410b; **4**:6a*, 97b; **7**:297b; **9**:167b
of 1604, **3**:60a; **5**:240b; **7**:297b–298a; **9**:247a
and astrology, **7**:297b
discovery of, **2**:402b–404a; **3**:509a; **5**:34a; **6**:542a
evolution of, **11**:319b
fluctuations in, **4**:167b; **6**:70a
photography of, **11**:470a
reactions in, **5**:272b
spectra, **1**:57a, 598a; **2**:146b; **3**:36a, 510b*; **5**:199a,b; **8**:50a, 465a; **9**:184a, 504a; **11**:319b, 542b; **14**:479b, 521b
changes in, **1**:467a; **14**:55b
theories of, **2**:123b; **7**:268b, 297b; **8**:50a; **9**:405b; **12**:282b
see also specific novae under individual constellations; Supernovae
Novara, Domenico Maria (1454–1504), 10:153–154; 11:352a
Novaya Zemlya
exploration of, **12**:279b
natural history, **1**:387b–388a
surveying of, **8**:417b; **14**:634a
Novick, Aaron, **13**:227b
Novitski, Edward, **13**:137b
Novy, Frederick George (1864–1957), 10:154–155
Novy jar, **10**:155a–b
Nowell, John, **1**:412b
Noyes, Arthur Amos (1866–1936), 10:156–157; 1:11a; **8**:289a; **13**:429a; **14**:182a; **15**:467a(n)
Noyes, William Albert (1857–1941), 10:157–158; 11:371b*; **14**:183a*; **15**:1a,b
Nozu
see Nodzu
Nṛsiṃha, **7**:315a; **9**:580a; **15**:625a,b
commentaries of, **5**:274a,b; **15**:612a
Nṛsiṃha of Golagrāma, **11**:290a
commentaries of, **2**:119a; **5**:275a; **7**:219a
Nṛsiṃha of Nandipura, **9**:21a
Nuck, Anton, **2**:224b; **15**:28b
Nucleal reaction, **4**:603b, 604a
Nuclear moment, **14**:599a
Nucleic acids, **10**:277a
in cancer, **4**:486a
composition of, **4**:603b; **7**:466b; **8**:275b–276b; **9**:381a
discovery of, **9**:381a
distribution of, **4**:603b–604a
hydrolysis of, **4**:603b
structure of, **4**:603b–604a; **5**:589a–b; **6**:505b
sugars of, **4**:603b; **7**:466b; **15**:252b
see also DNA, RNA
Nucleins
chemical characterization of, **7**:466b–467a
discovery of, **6**:505b
function of, **7**:467a
Nucleogelase, **4**:604a
Nucleolus, **5**:34b, 56b; **9**:496b–497a; **12**:193b; **13**:556b
Nucleon(s), **4**:583a
interaction of, **13**:241a
Nucleonics, **7**:93a
Nucleoplasm, **13**:88b
Nucleoproteins
discovery of, **9**:381a
Nucleosides, **8**:275b; **15**:252b
Nucleotides, **5**:273a
composition of, **4**:486a, 604a; **8**:276a
synthesis of, **5**:2a,b
Nucleus, atomic
alpha decay, **5**:272a
angular momentum, **1**:371a
beta decay, **4**:579b–580a; **5**:272b; **9**:261a
binding energy, **8**:95a
bombardment of, **13**:600b
compound state, **2**:251a–b
disintegration of, **8**:95a; **11**:281a
ejection of, **7**:153b
fusion of, **6**:117b; **12**:34a
heavy
statistical theory of, **5**:161a
light
theory of, **9**:38a
magnetic moment, **1**:371a; **4**:579a
models of, **2**:240b–241a, 250b–251b; **7**:211b
spin, **12**:438a
theory of, **1**:371a; **5**:160b, 161a, 272a; **7**:618a
statistical, **7**:617a
transmutation of, **2**:338a–b; **3**:329b, 501a; **7**:154a–b
see also Fission, nuclear; Neutrons; Physics, nuclear; Proton(s)
Nucleus, biological, **9**:600b; **13**:556b
constitution of, **5**:35a; **6**:505b
continuity of, **5**:51b
cytoplasm, influence of, **2**:364a–b; **4**:187b; **13**:89a
discovery of, **2**:517a; **12**:174a
division of, **2**:626a–b; **4**:545b–546a; **5**:34b–36a; **6**:121b; **11**:573a–b; **12**:193a–b; **13**:88a–89a
equivalence of nuclei, **4**:187b
fusion of, **6**:121b
genetic primacy of, **6**:338b
histological change, **14**:252b
naming of, **2**:520a
of nerve cells, **8**:282b
nucleic acids in, **4**:603b; **9**:381a
role of, **1**:417b; **3**:29b; **4**:187b; **5**:51a–52b; **7**:95a, 439a, 440a; **9**:495a; **13**:89a; **14**:235a–b, 430a–b
see also Chromosomes; Cytology; Egg, nucleus; Mitosis
Nuée ardente, **7**:548b–549a
Nueva Granada, **15**:429b–430a
Null, Jacques Friedrich von der, **9**:447b, 448b
Null hypothesis, **5**:8a–b, 9a; **14**:382b, 383a
Null space, **5**:151b
Null system, **9**:430b
Number(s)
absolute, **1**:238a–b
naming of, **1**:238a,b
abundant, **4**:83a; **15**:10a
algebraic, **3**:54a,b; **6**:308b
approximation of, **13**:391a
criterion for, **9**:412b
divisibility of, **12**:513a
expansion of, **8**:383b
logarithms of, **5**:342a
theory of, **3**:221b; **6**:286b; **7**:508a; **11**:256b
algebraic couples, **6**:90b
altar, **11**:221a
amicable, **2**:527a; **11**:221a; **13**:289a
Bernoulli, **2**:50a; **4**:475b; **6**:572a; **11**:268a; **12**:291b
computation of, **1**:54b, 356b*
theory of, **3**:302b; **13**:6a–b
Betti, **2**:105b; **3**:95b; **11**:57a, 60a
bicomplex, **12**:285a
biquadratic, **8**:386b(n)
cardinal
definition of, **3**:55b–56a; **4**:4a–b; **12**:12b
infinite, **2**:277a–b; **7**:138b
classification of, **3**:56a; **4**:160a; **10**:113a; **13**:302a
complex, **2**:269a; **4**:475a; **12**:150a
algebraic interpretation of, **2**:280b; **3**:135b, 142b; **4**:474b; **6**:90a, 95a–b
application of, **7**:473b, 601b
generalization of, **15**:196a
geometric representation of, **1**:237b–240a; **3**:137a–b; **4**:35b; **5**:111a–b; **6**:90a; **8**:139a; **13**:125a; **14**:279b–280b; **15**:196a
logarithms of, **4**:477a
as ordered pairs of real numbers, **1**:238b
theory of, **1**:591b; **3**:137b, 142b; **4**:4b, 477a–b; **6**:522a; **9**:454b; **10**:437b; **13**:6a
composite, **2**:233a; **4**:573b; **10**:113a; **11**:269a
compound, **3**:274a
computability, **13**:497b
concept of, **3**:111a; **4**:611a–b; **7**:226b, 327a–b; **13**:289b, 291b
congruum, **4**:610b–611a
conjugate, **2**:280b
conservation principle, **12**:228b; **13**:125a
cubic, **5**:588b; **10**:113a; **11**:221a; **15**:10a
aliquot parts, **5**:159a; **14**:149a–b
rules of, **13**:540b
sum of, **2**:233a–b; **7**:353a
decomposition of, **4**:574a; **5**:159b, 409a–b, 414b; **6**:214b; **7**:523b, 565b; **8**:135b; **9**:126a, 411b; **12**:469a–b; **14**:592a
definition of, **5**:152b, 153b; **7**:242a; **10**:113a; **12**:12b
diameter, **4**:419a
equal, **4**:419b
equipollent, **3**:274b
factoring of, **4**:608b, 611a; **11**:269a
figurate, **2**:233a; **4**:550a; **6**:125a; **11**:221a–b
friendly, **4**:573b
Galois imaginaries, **5**:261a, 263b
geometry of, **2**:197b; **14**:94b–95a
heteromeke, **11**:221a
hexagonal
central, **5**:159a

Number(s) (cont.)
hypercomplex, **1**:307b; **6**:90a–b, 91a–b, 92a, 95b; **7**:384a; **8**:389a; **10**:138a; **14**:211b–212a, 280a
ideal, **4**:4b; **7**:506a, 523a–b; **14**:630b
imaginary, **3**:65b, 560a; **4**:477b; **5**:302a, 466b*; **8**:166a; **14**:500a; **15**:366a–b, 368b
geometric representation of, **1**:238a–240a; **2**:271a; **3**:76b–77b; **12**:325b; **14**:150a
interpretation of, **2**:280b; **4**:515b; **8**:165b
naming of, **1**:238a
in physics, **1**:573a
infinite, **12**:13a
irrational
see Irrationals
and magnitude, **4**:447a
Mersenne, **8**:531b
natural, **1**:148b, 257a; **3**:221b; **5**:153b; **7**:328b; **11**:60b
logic of, **10**:442b, 443a
representation of, **5**:471b; **6**:214b; **11**:256b
sum of powers, **4**:550a
negative
medieval Arabic use of, **1**:41a
medieval Latin use of, **4**:607b, 610a, 611b
Renaissance use of, **3**:273b, 274a; **13**:60a
seventeenth-century use of, **3**:116b
eighteenth-century use of, **1**:66b; **8**:166a
nineteenth-century use of, **2**:116b; **3**:76b–77b
Chinese use of, **3**:254a; **8**:422b
generation of, **1**:238a
geometric significance of, **5**:410a
Indian use of, **1**:41a
logarithms of, **4**:477a
opposition to, **9**:158b–159a
termed "imaginary," **1**:238a
theory of, **10**:437b
odd
succession of, **9**:191b
ordinal
definition of, **14**:89b
theory of, **2**:593b; **3**:55a,b, 56a–b, 111a; **4**:4a–b
p-adic, **5**:108a; **6**:286b, 390b; **7**:530b; **13**:23b
partition of
see Partitions (math.)
pentagonal, **11**:221a
perfect, **2**:527a; **3**:128b*; **4**:573b; **9**:191b; **10**:479a; **15**:10a
computation of, **4**:606b, 611a; **11**:221b; **13**:289a
multiply, **5**:159a
theory of, **2**:361a; **4**:83a; **11**:220b–221a, 222a
polygonal, **2**:233b; **3**:131b–132a, 142b–143a; **4**:117a; **8**:138b; **9**:191b; **10**:113a; **15**:10a
definition of, **4**:116a–b; **6**:617b
reciprocal, **8**:161a
sequences of, **13**:60a,b, 539b–540a
prime, **2**:233a; **3**:178b, 221b, 302b, 345b; **4**:5a, 125a, 421b, 474b, 573a–574a, 611b; **5**:159a–b; **6**:214b, 393b; **7**:565b; **8**:135b, 385b, 386a; **9**:173b; **10**:113a; **11**:453b; **12**:599b; **13**:399b; **14**:438a
distribution of, **3**:223b, 227a–228a; **4**:475a; **5**:299a; **6**:4a; **7**:616a; **8**:138b–139a, 142a, 502b; **11**:268b; **13**:561b
partition of, **4**:342a
sieve method, **4**:392a–b, 573b; **8**:162a; **10**:113b(n)
Staudt's, **13**:6b
tables of, **7**:54b; **12**:345a
testing of, **5**:450a–b; **8**:531b; **15**:10a
see also Prime number theorem
pure, **4**:281b
pyramidal, **4**:550a
sequences of, **13**:539b–540a
Pythagorean, **1**:22b; **4**:610b, 612a; **11**:221a
Pythagorean triples, **4**:610b, 612a; **15**:503a–b, 671b
rational, **1**:306b, 307b; **2**:276b, 358a; **3**:273a
definition of, **4**:3b, 419b; **9**:310b
sequences of, **3**:53b–54a
real, **3**:53a; **13**:289b
algebraic structure of, **7**:242a
construction of, **4**:4b
definition of, **3**:54a; **4**:3b, 466a; **12**:12b–13a
distribution of, **14**:282a–b
expansion of, **7**:351b
order relation, **11**:512a
representation of, **3**:54a
theory of, **2**:274b, 275a–b, 276b–277a; **3**:53b–54a; **4**:206a; **9**:307b; **11**:409a–b
relatively prime, **4**:474a
side, **4**:419a
solid, **15**:119a
square, **2**:435a; **10**:113a; **11**:221a
aliquot parts, **5**:159b; **14**:149a–b
reciprocals of, **4**:475b
tables of, **5**:588b; **9**:12b
theory of, **4**:610b–611a; **13**:540b
stereometric, **11**:221a
transcendental, **3**:54a, 128a; **5**:472b; **6**:393b; **7**:559a; **8**:165b; **15**:368b
theory of, **4**:475a; **5**:342a–b; **8**:367b–368a, 383b
transfinite, **1**:407b; **2**:593b; **3**:53b, 55b; **6**:570a; **7**:507b
triangular, **2**:233b; **6**:125a; **8**:161b; **11**:221a; **13**:60b; **14**:592a
reciprocals of, **8**:161a; **9**:303b
see also Magic squares, Numerals, Numeration
Number theory
ancient Greek, **1**:231b; **4**:421b, 466a; **6**:411b, 617b; **11**:220b–223b; **13**:399b–400a
ancient Roman, **4**:115b–117a; **7**:1a; **10**:113a; **13**:325b; **15**:119a, 670a
Byzantine, **4**:160a
medieval Arabic, **1**:22b; **4**:502a–b; **7**:242a–b, 243a–b; **12**:93a–b; **13**:289a, 539b–540b; **15**:10a
medieval Hebrew, **8**:280a
medieval Latin, **4**:605a, 610b–611a; **10**:474b
Renaissance, **3**:275a, 514a; **9**:191b; **13**:48a, 59b–60a, 60b
early seventeenth-century, **1**:367b; **3**:116b; **4**:117b–118a, 550a, 612b; **5**:409a–b, 410a
late seventeenth-century, **2**:131a, 527a; **3**:64a; **4**:567b, 573a–574b; **5**:158b–159b; **7**:577b; **8**:161a, 162a, 165b–166a; **9**:599b; **11**:512a; **14**:149a–b
eighteenth-century, **2**:577a; **3**:285a; **4**:474a–475a, 515b; **5**:450a–b; **7**:565b–566a, 599b; **8**:135b, 137a–b; **14**:180a, 438a–b; **15**:391b*
early nineteenth-century, **1**:459b; **2**:275a–b, 276b–277a; **3**:142b–143a, 466b*; **5**:298b–299b, 302a, 304b, 375a–b; **8**:137b, 138a–139a, 142a; **11**:61b
mid nineteenth-century, **1**:449b, 591b; **3**:223a–b, 227a–228a, 302b; **4**:3b–5a, 124a–126a, 258a; **5**:203b, 260b, 261a, 263b, 307a, 309b; **6**:307b, 308a, 308b–309a; **7**:52a, 54a,b, 164a, 384a, 505b–506b, 522a, 523a–b, 601b–602a; **8**:383b, 385b–386a; **9**:307a–308a, 430b; **10**:479a; **11**:22a, 453a–b; **12**:170b, 172b, 468b–469b; **13**:6a–b, 219a–b; **15**:66b
late nineteenth-century, **1**:370a; **3**:53a–56b, 178a, 456a; **4**:1b, 177b; **5**:153b, 198a, 414a,b; **6**:4a; **7**:398b, 508a; **8**:62a*, 367b–368a, 389a–b, 502b, 531b; **9**:125b–126a, 173b, 411b–414a; **10**:24a, 344b, 442b, 443a, 484a–b; **11**:53b–54a, 55a–b; **12**:513a; **13**:83a, 218a, 561b–562a; **14**:202b, 630a–b; **15**:65b
early twentieth-century, **1**:584b; **2**:58a, 90a, 145b, 197b, 236b; **3**:96b, 345b, 346a; **4**:83a, 288a; **5**:206a, 350b, 508b; **6**:113b, 114a, 214a–b, 286b, 473b, 571b–572b; **7**:351b, 616a–b; **8**:253b; **10**:118a; **11**:256a–b, 260b, 268a–269a, 394b; **12**:237b, 426a–b, 451b; **13**:391a; **14**:94a–95a, 188a, 212a, 282a–b, 317a
mid twentieth-century, **1**:584b; **4**:83a; **5**:342a–b; **6**:497b; **12**:411a, 426b; **14**:592a
additive, **4**:83a; **14**:95a, 282b
algebraic, **1**:16a; **2**:90a; **4**:124a–125a; **6**:390a–b, 572a; **8**:389a; **11**:362b; **14**:94a–b
analytic, **1**:584b; **4**:124b–125a, 474b–475a; **5**:342a–b; **6**:214a–b, 547b; **7**:616a–b; **8**:138b, 386a; **11**:256b, 268b; **14**:94b
applied, **15**:66b
Egyptian, **15**:692b
foundations of, **5**:350b
geometric, **9**:412a–414a
and invariant theory, **4**:83a
Japanese, **1**:84a; **15**:751a
Mesopotamian, **15**:671b–672a
neo-Pythagorean, **2**:233a
Pythagorean, **5**:365a; **6**:302a; **9**:1a; **15**:10a
Numerals
Arabic, **7**:176b, 360b, 418b
origin of, **3**:214a
Greek, **1**:189a; **4**:112b
Indian, **1**:40a, 41a, 62b–63a; **4**:604b, 605a,b, 612a; **6**:115b; **7**:360a–b, 362b–363a, 532b; **11**:18a, 458a; **13**:544b; **14**:472a
Roman, **4**:605b; **7**:418b
Numeration
Amerindian, **15**:761b–762a
Aztec system, **15**:763b
base *n*, **3**:61a
binary, **6**:125b; **7**:559a; **8**:165b; **9**:610b; **10**:484a; **15**:52b
decimal
see Decimal system
Egyptian, **15**:682a–b
large numbers, **1**:222a
Mayan, **15**:760b–764a, 779b, 799a–800a, 809b, 810a,b
place-value, **1**:189a; **7**:360b; **13**:544b
positional-principle, **1**:40a, 41a; **8**:280a; **15**:667b
primitive, **15**:761a–762a
sexagesimal, **4**:610a,b; **7**:118a, 123b, 124a, 126a, 128b, 139b, 142a, 258a–b, 259a, 360b, 532a–533a; **9**:614b–615a; **10**:540a; **11**:188b; **12**:94a; **13**:323b, 544b, 545a; **15**:9b, 155a, 667b, 714a
vigesimal, **15**:760a, 761b, 762a, 770a

Numerology
ancient Greek, **10**:548b, 590a; **11**:220b
ancient Roman, **13**:326a
medieval, **1**:38b; **7**:40b, 41b–42a; **9**:363a; **10**:474b
Renaissance, **13**:58b–59a, 59b, 60b
seventeenth-century, **2**:526b
eighteenth-century, **8**:379a
Biblical, **4**:550a–551b; **13**:58b–59a, 60b
Chinese, **12**:374b, 385b
Japanese, **15**:745a
Mayan, **15**:804a–808b
Numerov, Boris Vasilievich (1891–1943), 10:158–160; **7**:6b
Numismatics, **2**:591a; **4**:496a; **5**:53b–54a; **10**:239b; **11**:67a, 593a; **13**:357b; **15**: 52a
Nummulites, **1**:210b; **6**:229a; **11**:375a; **12**: 38a; **15**:52b
dimorphism, **12**:183a
structure of, **15**:53b
Nunataks
persistence of plants, **4**:583b, 584a
Nuñez Salaciense, Pedro (1502–1578), 10: 160–162; **7**:39a; **12**:501a; **13**:49b, 621b–622a; **15**:180a, 476a
De erratis Orontii Finaei, **15**:155b
Nuovo cimento, **3**:47a
Nuremberg
Cosmographical Society, **9**:233b
fortification of, **4**:260b
Nursing
handbooks
nineteenth-century, **2**:130a
Nussbaum, Moritz, **6**:131b; **9**:601b; **13**:458a; **14**:234b, 430a
Nussbaum, N., **2**:73a
Nusselt, Ernst Kraft Wilhelm (1882–1957), 10:162–163; **9**:462a,b
Nutation, **8**:504a
constant of, **2**:99a, 469a*, 616b; **8**:369b; **9**:514b; **10**:542b; **12**:449a, 574b*; **13**:112a*
lunar, **15**:354a
solar, **2**:469a*
tables of, **2**:388b
terrestrial, **2**:388a–b; **4**:480b; **8**:179a; **9**: 226a; **10**:119b, 232a; **15**:287b, 348b
and tides, **15**:297a–b, 301a–b
Nutka, **9**:433a
Nutmeg
cultivation of, **11**:64b
oil of, **11**:121a
Nutrition
ancient Greek, **4**:106b
medieval, **3**:393b
seventeenth-century, **4**:136a
eighteenth-century, **2**:301b; **8**:171b–172a; **10**:325b–326a; **13**:351a–b
early nineteenth-century, **3**:7b; **9**:9a
mid nineteenth-century, **1**:477b; **2**:28b, 596b; **4**:163b*; **6**:517b; **9**:10b, 457a; **10**:558a–b, 621a; **12**:466b–467a; **14**: 63b–66b
late nineteenth-century, **1**:325b; **2**:162a, 366b, 585a–586a; **4**:403b, 405a; **8**: 536a; **10**:493a, 580a; **11**:585a–586a; **14**:109a
early twentieth-century, **1**:610a; **2**:339b; **3**:69a; **5**:37a, 451b–452b, 541b–542a; **6**:135b–136a, 499a,b, 501a; **7**:502a; **8**:29b, 452b, 555a–556a, 590b–591b; **9**:284b–285b, 549b; **10**:243a–b, 493a–b; **11**:586a; **14**:524a; **15**:201b, 418a–b
mid twentieth-century, **3**:257a–b; **4**:357b–358a; **14**:46a, 393b, 524a–b; **15**:418b–420a, 493a
agricultural, **6**:135b; **8**:591a; **13**:66b
of aquatic organisms, **7**:503a–b
and development, **9**:144b, 145a
and disease, **1**:443a; **10**:558b, 621b
experimental studies, **8**:349a; **9**:284b–285a; **14**:524a
and genetics, **11**:211b
and immunity, **8**:274a*
infantile, **4**:403b, 405a
medical applications of, **9**:285b
nutrient transport, **8**:128a–b
theories of, **8**:348b–349a; **10**:580a
thermogenesis, margin of, **8**:29b
trace elements, **9**:285a
see also Deficiency diseases; Dietetics; Food; specific organisms; Vitamins
Nuttall, George Henry Falkiner, **7**:273a; **14**:249b
Nuttall, John Michael
Geiger-Nuttall law, **5**:272a, 331a–b; **12**: 32a
Nuttall, Thomas (1786–1859), 10:163–165; **1**:485a, 489b; **3**:391b; **6**:120a
botany, **14**:456b
The Genera of North American Plants . . . , **11**:218b
Nux vomica, **9**:8a
Nychthemeron, **15**:535b, 554b, 581a
Nyctalopia, **5**:137b
Nycticebus, **6**:535b
Nylander, Fredrik (1820–1880), 10:165
Nylander, William (1822–1899), 10:165–166
Nylon, **3**:86b; **13**:66b
Nymphaeaceae, **9**:24b
anatomy, **5**:604b
Nystagmus, **11**:215b
caloric, **1**:446b
and labyrinthine reflexes, **2**:447a
Nysten, Pierre-Hubert
Dictionnaire de médecine, **11**:492a
Nyt Tidsskrift for Fysik og Kemi
editors of, **6**:271a

O

O., Anna, **2**:447b, 448b, 449a; **5**:174a
Oak Ridge, Tenn., **3**:371b; **5**:493a; **8**:95b
Oak tree, **9**:366b; **13**:456b
galls, **2**:591a
Oakley, Kenneth Page, **3**:607a
Oases
evolution of, **11**:210b
Oats
chemical constitution of, **10**:150a, 242a
Obelisks, **2**:326b
Oberlin College, **14**:516a
Obermeier, Otto, **7**:422a
Oberndorfer, Johann, **11**:606b
Oberon (satellite), **13**:113b
Oberth, Hermann, **5**:433a; **13**:484a
Oblique ascension, **14**:577b
Obozrenie psikhiatrii, nevrologii i eksperimentalnoy psikhologii
founding of, **1**:580b
Obruchev, Vladimir Afanasievich (1863–1956), 10:166–170; **1**:162b; **7**:254a*; **9**:593a, 594a,b*; **11**:108b*, 182b*; **12**:301a
Observaciones sobre la fisica, historia natural y artes utiles
founding of, **1**:127b
Observation
ancient use of, **1**:263b–265a; **6**:424a,b, 428b, 429a
medieval use of, **1**:278a–b; **5**:551b–552b
seventeenth-century use of, **6**:151b–152a
eighteenth-century use of, **2**:287a; **9**:135a, 306a; **12**:308b
nineteenth-century use of, **2**:447a, 464a
Mesopotamian use of, **15**:641b, 642b, 643a–b
and theory, **5**:102b
see also Error, theory of
Observations sur la physique, sur l'histoire naturelle, et sur les arts, **5**:411a; **8**:72a
editors of, **8**:80a–b
Observatories
medieval, **7**:255b–256b, 324b, 325a, 336a–b, 353b; **11**:240a; **13**:509a–b, 510b–511a, 513b, 535b–536a; **14**: 537b–538a, 575a–b
seventeenth-century, **9**:485a
eighteenth-century, **3**:13b; **7**:81a, 183a, 373b, 543a; **9**:231b; **10**:614b; **11**:472a; **12**:226a; **13**:513b; **14**:178b, 583a
nineteenth-century, **3**:14a, 93a–b; **6**:522b; **7**:75b, 76a, 400a–b; **11**:320a; **12**:234b, 518b; **14**:242b
twentieth-century, **2**:424a; **6**:574b; **8**:440b
astrophysical, **9**:563b
cosmic ray, **6**:355b
geophysical, **12**:51a, 187a; **14**:327b
magnetic, **2**:472a; **3**:260a, 261b, 557b; **5**:305b–306a; **6**:107a, 552b; **7**:461b, 609a; **8**:426b; **14**:204b–205a
meteorological, **3**:557b; **4**:171a,b, 174b–175a; **6**:98b, 432a; **7**:461b; **9**:108a, 264b, 590a; **13**:52a; **14**:356b, 357a; **15**:493b
optimum location of, **8**:521a,b
private
Renaissance, **14**:359a–b
seventeenth-century, **3**:100a; **6**:360b–361a, 363b
eighteenth-century, **4**:14a; **6**:112a; **7**: 373b; **10**:198a, 607b; **14**:485a
nineteenth-century, **1**:616a; **2**:97b, 147a–b, 285a, 471b, 503a; **3**:92b, 93b, 605b; **4**:19a, 29b–30a; **5**:480a, 561b; **6**:27b, 28a, 103b, 402b, 540b; **7**:270a, 400a, 461a–b; **8**:46a, 465a, 521b; **9**: 2a; **10**:218b–219a; **11**:114a; **12**:36b, 498b, 551b; **14**:54b–55a, 481b, 628b
twentieth-century, **8**:50a; **11**:542b
seismological, **9**:444b
southern hemisphere, **3**:16b; **5**:404b–405b; **6**:326a, 574b; **7**:18a, 543a; **8**: 539b, 613b; **9**:122b; **10**:526b; **12**: 251a; **14**:521b
see also individual observatories by location or name
Observatory
editors of, **9**:184b
Obsessive states (psych.), **1**:580a
Obstetrics
ancient Greek, **6**:318a–b
ancient Roman, **12**:539a
Byzantine, **10**:417a, 418a
medieval Arabic, **14**:584b
medieval Latin, **13**:347b
Renaissance, **5**:586b; **10**:316a
seventeenth-century, **9**:581b; **12**:40b
eighteenth-century, **4**:173a; **12**:100b, 427b–428a; **14**:296b–297a
nineteenth-century, **1**:476a–b; **5**:277a; **12**:294b–296b, 502b
twentieth-century, **6**:433a; **11**:374b*
anesthesia, **12**:502b
Walcher position, **14**:584b
see also Midwifery, Pregnancy
Obturator, **8**:554b
Ocagne, Philbert Maurice d' (1862–1938), 10:170; **6**:261b; **11**:398b*
Occasionalism, **8**:514a, 515a; **9**:49a–b

Occhialini, Giuseppe P. S., **4**:578b; **12**:33a; **15**:26b, 27a
Occlusion (chem.), **4**:80b; **5**:493a–b
Occult phenomena, **1**:198b; **5**:62a; **10**:80a; **13**:348a
see also Spiritualism
Occultations, **9**:330b; **11**:188a
computation of, **7**:494a
of the moon, **6**:325a; **7**:18a; **10**:110b
of planets, **11**:28b
of stars, **1**:429b; **6**:263b; **9**:233b, 429a; **11**:114b; **13**:114a; **14**:589b
computation of, **7**:481b
and longitude determination, **3**:198a
prediction of, **1**:429b; **6**:263a
Occupational diseases, **1**:563b; **7**:421b; **14**:256b
miners', **1**:77b; **2**:523b; **4**:198a; **5**:366b; **10**:305b, 307a, 527a
radiation effects, **6**:355b
Oceania, **5**:224b
Oceanography
late eighteenth-century, **3**:198a; **5**:76b–77a, 136b–137a
early nineteenth-century, **1**:594b–595b; **5**:79a, 224b; **15**:83a
mid nineteenth-century, **4**:73b, 291a–b; **5**:66b–67a, 68a; **6**:326a; **7**:91b–92a; **8**:188b; **9**:196a–197a; **11**:113a–b; **15**:85a
late nineteenth-century, **1**:71b–72a, 162a,b; **2**:557b–558a; **3**:89a–b, 322a; **4**:181a*; **9**:443a, 587b, 588a–590a; **12**:544b–545a; **13**:360a–361a, 552a; **15**:430a–431b
early twentieth-century, **1**:92a–b; **2**:373a–b; **3**:203b–204a; **4**:42a–43a; **9**:331a–b; **10**:503a; **12**:411b–412b; **13**:166b–167a, 352b
mid twentieth-century, **6**:134a; **8**:356b; **11**:558b; **13**:167a; **14**:50b
chemical, **1**:416a–b; **4**:127b; **5**:72a
geological, **1**:394b; **3**:204a; **5**:72a, 437a; **6**:229a; **7**:144a–b; **9**:26b–27b, 181a, 588a–589b; **11**:113a–b, 372a; **15**:380a
methodology, **4**:43a
physical, **3**:89b; **4**:344a–345a; **9**:42b–43b, 135a–b; **10**:563b–564a, 597b–598a; **11**:131a; **12**:414b–415b; **14**:634a–635b; **15**:431a–b
regional, **12**:412b
theoretical, **10**:564a
see also Bathyscaphe; Biology, marine; Instrumentation, oceanographic; Paleo-oceanography; Zoology, marine
Oceanology, **14**:635a–b
Oceans
aeration of, **14**:634b
circulation in, **2**:168a; **12**:412a; **13**:167a, 361a
general, **3**:89b; **4**:591b–592a; **9**:196b–197a
horizontal, **12**:415a
and islands, **12**:415b
mathematical theory, **4**:344b
thermohaline, **2**:557b; **9**:196b–197a
vertical, **14**:634b, 635a
and winds, **4**:344a–b
and climate, **10**:564a
currents in, **2**:43a; **4**:256a; **11**:376b
biological effects, **4**:42b
charts of, **9**:196a
and climate, **14**:634b
and Coriolis force, **9**:43a
countercurrents, **12**:415a–b
and density, **3**:471a; **14**:634b
and earth's rotation, **4**:591b–592a
geostrophic, **12**:415a; **15**:431b
measurement of, **4**:344b; **12**:414b, 415b
and plankton, **3**:322a
speed of, **4**:12b
turbidity, **1**:394b
vertical, **2**:557b
and wind, **3**:471a; **4**:592b; **12**:415a–b
dead water phenomenon, **15**:431b
depth, **15**:296b–301b *passim*
distribution of, **7**:251a
dredging of, **12**:107a; **13**:552a
dynamic equilibrium, **15**:339b, 352b
energy from, **3**:299b; **4**:600a
floor of
contraction of, **3**:552b
deformations of, **13**:147b
heat of, **3**:552b
nature of, **13**:360b
ridges, **9**:589a–b; **14**:216b
rise of, **3**:66b
sediments, **9**:588b–589b; **11**:372a
soundings, **9**:589b
tilting of, **9**:181a
types of, **9**:533a
gelatinous deposits, **2**:558a; **14**:146a
gravimetry, **2**:329b, 373a–b
and isostasy, **2**:373a–b
life in
and depth, **5**:66b, 57a; **7**:91b–92a; **13**:360a–b; **14**:145b–146a
mutual dependence of, **12**:412a
origin of, **13**:148a
paleogeography, **9**:589b
permanence of, **3**:552b
salinity of, **14**:191b
changes in, **6**:71b–72a; **7**:161a
mapping, **2**:557b; **4**:127b; **8**:188b; **15**:431a–b
theories of, **13**:292a; **15**:232a
temperature-salt curves, **12**:415a
thermometry, **1**:595a; **3**:204b*; **5**:137a; **7**:91b, 92a, 414a; **8**:188b; **9**:43a; **10**:518a*; **11**:131b*; **13**:237a, 360b–361a; **14**:191b; **15**:431a–b
turbulence in, **11**:558a; **12**:414b
underthrusting, **13**:64b
waves
height of, **2**:84b; **13**:167a
oil on, **5**:136b
shape of, **11**:293b
theories of, **14**:338a
zoogeography of, **1**:623b; **2**:320b, 321a; **12**:479b; **13**:552a
see also Bathymetry; Biology, marine; Ice, oceanic; Sea water; Shoreline; specific oceans; Tides, marine
Ochiai, Eiji, **13**:2b
Ochoa, Severo, **9**:359b
Ochsenfeld, R., **8**:476a, 478a
Ochsenius, Carl (1830–1906), 10:170–171
Ockenfuss, Lorenz
see Oken, Lorenz
Ockham, William of (*ca.* 1285–1349), 10:171–175
critics of, **2**:609b, 611a; **9**:137b; **10**:421a
editions of works, **1**:94a
influence of, **2**:604b; **12**:105b
influences on, **1**:199b; **4**:255a,b; **6**:194a; **7**:66b
on local motion, **1**:100b; **2**:610b
logic, **1**:94b
natural philosophy, **12**:548a
Ockham's razor, **1**:278b; **5**:511a; **10**:173b; **11**:357a
Summa logicae, **2**:604b, 609b
theology, **5**:113b
Ockhamism, **3**:172a; **7**:116a; **10**:419b
Ocreatus, Johannes, **1**:61a, 63a; **4**:448a
Octahedron
discovery of, **13**:303b–304a
Octant (astron. instrum.), **2**:141a; **3**:102b; **6**:6a, 363a; **8**:170b; **9**:5b, 607b*; **10**:572a
Octave, mathematical, **6**:572b
Octave, musical
division of, **1**:282b; **12**:127b
Octopus
anatomy, **13**:9b–10a
physiology, **5**:149a
Oddi, Marco de, **1**:522b
Oddi, Ruggero (1864–1913), 10:175–176
Oder (river), **10**:505b
Odessa
bacteriological institute, **9**:334a
Odic force, **11**:359b
Odierna (or Hodierna), Gioanbatista (1597–1660), 10:176
Odington, Walter of
see Walter of Odington
Odling, William (1829–1921), 10:177–179; **1**:161a; **3**:481a*; **8**:55b; **9**:349a; **10**:517a; **14**:394b, 395b
Odo of Meung, **6**:123b; **11**:601b
Odometer, **1**:97b; **6**:311a, 312b; **12**:384b
Odontograph, **14**:404a
Odontography
comparative, **12**:38a
Odors, **9**:3b
Odstrčil, Johann, **8**:597b
Oecolampadius, Johannes, **10**:305a; **12**:322b
Oeder, Georg Christian, **9**:576a
Oedipus complex, **5**:176b, 178a,b, 179a
Oedogonium, **11**:153b, 154a
Oehl, Eusebio, **2**:164b; **5**:459b; **12**:319a,b
Oenology
Renaissance, **10**:306a
eighteenth-century, **10**:283b
nineteenth-century, **3**:7b; **4**:135a; **6**:359a*
see also Viniculture, Wine
Oenopides of Chios (*fl.* fifth century B.C.), 10:179–182; **4**:463a–b; **6**:410b, 416b; **9**:156b; **11**:223a; **14**:603b
Oenothera, **8**:354a
genetics, **5**:293a–b; **9**:520b; **13**:137a; **14**:100a–101b
Oersted, Hans Christian (1777–1851), 10:182–186; **1**:347b; **11**:16a(n)
acquaintances of, **2**:91a; **5**:94b; **6**:106b; **11**:474b; **12**:169b; **14**:600b; **15**:84a
chemistry, **14**:475b
correspondence of, **6**:107a
editions of works, **2**:169b
electromagnetism, **1**:202b, 143b; **4**:530b; **5**:407b; **9**:204b, 459b
geology, **5**:71b
hydrodynamics, **15**:85b, 86a
influence of, **3**:497a; **6**:2b; **10**:186b; **15**:84b–86b
Materialien zu einer Chemie des neunzehnten Jahrhunderts, **3**:232b
students of, **11**:379b
thermoelectricity, **5**:98b*
translations of works, **12**:317b–318a
Oertel, Max, **8**:449a
Oettingen, Arthur Joachim von, **4**:559b; **9**:616b; **13**:243a; **15**:455b, 456a, 463a
Oeynhausen, Karl von, **3**:623a–b
Offret, Albert, **1**:472a
Ogston, Alexander, **8**:406a
Ohain, Karl Eugen Pabst von, **14**:257a, 258b
Ohio (state)
Geological Survey, **6**:57b; **9**:172b; **10**:32a; **13**:531b
geology, **4**:562b, 563a; **5**:49b, 395a; **6**:565b*; **13**:621a

Ohm, Georg Simon (1789–1854), 10:186–194; **4**:123b; **8**:262a
law of acoustics, **4**:222b; **6**:245b, 246a
Ohm's law, **3**:32a, 557a; **4**:556b; **6**:278a; **7**:380b, 449a, 450a; **8**:187b, 502a, 510b; **10**:189b, 191b; **14**:67b, 206b
verification of, **11**:111b; **14**:290a
Ohm, Martin, **2**:275a; **10**:186a, 193a
Ohm (elect. unit)
absolute value of, **11**:578b
standardization of, **1**:612b; **7**:93b; **8**:502a; **9**:155a,b; **12**:238a; **14**:330a
Oidium, **5**:503a, 566a; **11**:491b
Oikoumene, **4**:389a, 391a–b; **9**:172a; **13**:84b–85b; **15**:637a–b, 638b
Oil(s)
cod liver, **5**:432b; **10**:243b; **15**:418b, 419a
drying, **9**:558b
essential, **5**:374b*, 412b, 420b; **14**:141b–142a
constitution of, **3**:279b
inflammation of, **3**:419b; **4**:531b
mineral
origin of, **3**:206a
nature of
eighteenth-century discussion, **3**:206a
purity of, **5**:432b
ray liver, **5**:432b
solubility of, **8**:618b
vegetable
processing of, **7**:378b; **15**:651a
see also Petroleum
Oil wells
wastes from, **2**:374a
Okapia, **5**:109a
Okazaki Yasuyuki, **1**:83a
Oken (Okenfuss), Lorenz (1779–1851), 10:194–196; **10**:12b
acquaintances of, **12**:197a
anatomical theory, **5**:443b, 444a; **12**:241b
critics of, **2**:595a; **11**:592b; **12**:198b
embryology, **6**:592a
influence of, **1**:69b, 73a; **2**:425b; **6**:564b, 573a; **9**:578b; **12**:281a, 607a
influences on, **12**:155b, 156b
Lehrbuch der Naturphilosophie, **5**:191b
translations of, **10**:263a
students of, **7**:437b; **9**:600a
Okhotsk, Sea of, **4**:42b
Oklahoma, University of
library, **4**:8a
Olacaceae, **15**:147a
Olaus Magnus (1490–1557), 10:197
Olbers, Heinrich Wilhelm Matthias (1758–1840), 10:197–199; **3**:302a
Abhandlungen . . . die Bahn eines Kometen zu berechnen
editions of, **5**:257b, 258a
acquaintances of, **3**:258a; **5**:301b; **6**:112a, 326b; **11**:545b; **13**:109b
astronomy, **2**:433b; **13**:110b
and Bessel, **2**:97b
celestial mechanics, **1**:429a; **2**:328b, 442a; **6**:70b
correspondence of, **2**:98a; **5**:303b, 304b, 305a; **12**:234b
geometry, **7**:598a
on meteorites, **1**:615b
paradox of, **7**:263a; **8**:617b–618a; **10**:199a; **14**:629a
Old Red Sandstone, **5**:39b*, 334b; **12**:277a
deposition of, **5**:336b–337a
fossils in, **5**:32b; **8**:104a; **9**:388b, 389a
fractures in, **11**:35b
and stratigraphy, **8**:486b; **9**:583b, 584a; **11**:35a
Oldenburg, Henry (*ca.* 1618–1677), 10:200–203; **5**:529b*
acquaintances of, **2**:379a; **3**:534b; **4**:221b; **6**:598b; **9**:498a, 554b; **12**:585a; **13**:334b, 479b; **14**:369b
on contemporary scientists, **9**:245a
correspondence of, **2**:523a; **3**:348b; **5**:22b, 415a; **6**:141b; **7**:111a–b(n); **8**:250b; **9**:447a; **10**:44b, 46a,b, 54b, 81a, 254b, 546b; **11**:313a; **12**:85b, 460a, 583a, 584a–b; **13**:336b, 480a; **14**:371a
and Hooke, **6**:484b, 487a; **10**:202a
and Leibniz, **8**:149b, 160a–164b *passim*
and Royal Society, **5**:416a; **12**:84b, 584a, 585b; **14**:151b, 509b
translations by, **1**:481b; **13**:34b*
Oldham, Richard Dixon (1858–1936), 10:203; **5**:596a–b; **9**:61a, 444a
Oldham, Thomas (1816–1878), 10:203–204; **10**:203a
Oldhamia, **10**:204a
Olduvai Gorge, **8**:105a; **11**:336a–b
Olefiant gas (ethylene), **2**:420b; **5**:322a, 325a; **8**:332b–337b *passim;* **13**:469a
Olefins, **3**:246b; **14**:334a, 602a
oxidation of, **11**:92b
synthesis of, **7**:457a
Olekma (river), **7**:510b
Olenellus, **8**:33b
Oleson, Jerome Jordan, **4**:155b
Oleum, **14**:447a
Olfaction, **2**:124b
Olfactory apparatus, **12**:137b
Oligocene epoch
conception of, **2**:108b
paleontology, **2**:110a; **5**:357b; **6**:222a; **12**:261a; **13**:72a
stratigraphy, **3**:536b; **14**:210b–211a
subdivision of, **2**:108b–109a
Oligochaetes, **10**:522b; **11**:535b
freshwater, **9**:185b
reproduction, **9**:185a; **13**:458a–b
Oligosaccharides, **7**:518a
production of, **14**:603a–b
Oliphant, Marcus Laurence Elwin, **3**:329b; **12**:33b, 34a
Oliva, Antonio, **2**:309b, 310a
Olive tree, **7**:89a
Oliver, Daniel, **2**:371a, 476a; **12**:259a
Oliver, George (1841–1915), 10:204–206; **2**:322a; **12**:355b; **15**:106b
Olivi, Peter John, **1**:289b
Olivier, Louis, **11**:413b, 428b
Olivine
role of, **13**:476a
Olmsted, Denison, **1**:364a; **8**:1a, 487b; **9**:8a; **10**:41a, 150a; **11**:340a; **12**:433b; **14**:315b
Olsen, Elias, **2**:406a, 409a
Olsen, Olav Johan, **2**:437b
Olszewski, Karol Stanisław (1846–1915), 10:206–207; **4**:79b; **14**:522a,b
inert gases, **11**:279a–b
Olufsen, Christian Friis Rottbøll (1802–1855), 10:207; **11**:114b; **12**:235a*
Olympiads, **4**:389a, 392b
Olympic festival, **6**:405a,b
Olympiodorus (*fl. ca.* 360–425), 10:208; **1**:137a; **6**:305b, 416b; **10**:419a; **11**:160a, 186b; **12**:440b; **14**:582a
commentaries of, **7**:25a; **11**:184a; **14**:631b
Omalius d'Halloy, Jean Baptiste Julien d' (1783–1875), 10:208–210; **4**:348b
Omar Khayyam
see al-Khayyāmi
Omboni, Giovanni, **1**:491a
Omens, **15**:536a, 549b, 550a, 640b, 641b, 642a–643b, 645a, 646a, 673a–674a, 724a,b, 725a, 726b, 729b, 730a,b, 731b
apodosis, **15**:642a–b
protasis, **15**:642a, 646a
Omentum
embryogeny, **9**:570b
Omori, Fusakichi (1868–1923), 10:210–211
Onagraceae, **6**:465a
Onchocerca, **8**:270b
Oncley, John Lawrence, **3**:336a
Oncology
nineteenth-century, **5**:515a; **9**:572b; **14**:41b
early twentieth-century, **2**:597b; **3**:518b; **4**:498b–499b, 546a; **8**:557a; **9**:586a–b; **11**:465b
mid twentieth-century, **2**:597b; **4**:498b–499b; **6**:433a; **9**:550a
experimental, **4**:499a
see also Tumors
100 birds, problem of, **4**:606a, 610a–b, 612a,b
Oneiromancy, **1**:38a, 120b, 290b; **4**:81b; **9**:364a; **15**:497b
critics of, **10**:512b
see also Dreams
Oniscidae, **2**:422b
Onnes, Heike Kamerlingh
see Kamerlingh Onnes, Heike
Ono Ranzan, **8**:394b
Ono, Takashi, **6**:394a
Onomastica, **15**:634b–636a, 649a
Onsager, Lars, **3**:619b; **7**:493a; **8**:482a; **15**:461a
Ontario (lake), **8**:571a
Ontogeny
eighteenth-century study, **7**:367b–368a
nineteenth-century study, **3**:245b; **5**:444a, 447a, 604b–605a; **6**:8a–9a, 9b, 10a
twentieth-century study, **2**:255b; **14**:596b; **15**:14a
and heredity, **8**:589a
and phylogeny, **6**:613b; **9**:65a; **12**:45a, 337b–338b, 410a; **15**:92b, 169b–170a
variability of, **12**:337b
see also Embryogeny
Ontology
ancient Greek, **10**:324b; **12**:575b–576a; **14**:536b
Byzantine, **12**:441b–442a
medieval, **1**:33b; **4**:255a–b; **7**:484a–b; **10**:173b–174a; **11**:249b; **12**:6b–7a; **15**:263a
seventeenth-century, **4**:52a, 53a, 53b–54a; **5**:286a–287b; **7**:483b–484a; **8**:157a–159a; **9**:509b–510a
eighteenth-century, **13**:180a; **14**:483a–484a
nineteenth-century, **1**:140b–141a; **5**:193b–195a, 252b–253a; **6**:294b; **15**:143a
twentieth-century, **8**:263a; **9**:495a; **14**:307a–308b
rationality principle, **3**:335a
see also Creation, Metaphysics, Reality
Onychogryphosis, **9**:66a
Oogenesis
see Ovogenesis
Oort, Jan Hendrik, **7**:481b; **8**:364b; **13**:98b, 99a
Oozooid, **3**:148b
Opal, **9**:559b
optical properties, **11**:267a
structure of, **5**:203a
Opalescence
critical, **14**:616b
of fluids, **4**:315a; **10**:235b
of gases, **12**:497a–b
Oparin, Aleksandr Ivanovich, **1**:361b; **6**:23a

The Open Court, 15:79a
Operational research (indust.)
  nineteenth-century, 1:355a
Operationalism, 2:459b, 460b–461a; 4:102a
Operations, mathematical, 1:207b; 2:105a, 115b; 12:217a
  distributive, 2:320a
  infinite, 3:151b
  symbols of, 8:250b
  see also Calculus, operational
Operators (math.), 2:294a; 6:92a, 214b
  compact, 14:285a
  continuous, 2:513a
  difference, 15:41b
  differential, 2:294b; 3:145a
    linear, 3:169b; 11:470b; 13:428a–b; 14:283a
  Hermitian, 14:90a–b, 91b, 282a
  integral, 14:282a
  nonself-adjoint, 2:144a–b
  rings of, 14:90a–91b
  unbounded, 14:90a,b
Ophiobolus, 9:78b
Ophioglossales
  reproduction, 8:5a–b
  vascular organization, 3:30a
Ophiolites, 2:496a; 11:576a
Ophiuchus (constellation)
  novae, 1:352b
  P Ophiuchi, 13:110a
  70 Ophiuchi, 8:50a
Ophris, 4:295b*
Ophryoscolecidae, 4:142b
Ophthalmia, 12:88a
  contagious, 11:462a
  Egyptian, 7:425a
  nutritional, 10:243b
Ophthalmology
  ancient Greek, 6:316b
  ancient Roman, 12:540a
  Byzantine, 10:418b
  medieval Arabic, 9:604b; 13:292b; 14:112b, 584b; 15:500a*
  Renaissance, 9:75a; 15:161b*
  seventeenth-century, 11:511a
  eighteenth-century, 6:231a; 8:114b; 10:198a; 11:112a–b, 158b; 13:465a, 567b
  nineteenth-century, 2:376a–b; 4:162b–163b, 285b; 7:47b; 8:97b; 11:177b, 215a
  twentieth-century, 5:541b, 590b
  see also Eye; Optics, physiological; Vision
Ophthalmometer
  invention of, 6:247b
Ophthalmoscope, 2:376b; 7:74b; 11:215a
  invention of, 2:531a; 4:163a; 5:590b; 6:242a, 246a; 7:47b
Ophthalmotonometer, 4:615b
Opium, 3:246a; 14:585a, 587b
  addiction to, 6:62a
  chemical analysis of, 2:572b*; 3:159b; 4:41a; 11:494b–495a
  effects of, 2:31a; 6:146a*; 9:41a, 557b; 10:582b; 14:322b*
Oporinus, Johannes, 10:306a; 14:5b, 6a
Opossum
  anatomy, 13:527b
  fossils, 3:525b
  reproduction, 15:472b
Opotherapy
  see Organotherapy
Oppel, Albert (1831–1865), 10:211–212; 10:29a; 14:626a
  Die Juraformation Englands . . . , 1:285a,b
Oppenheim, Samuel (1857–1928), 10:212–213; 12:253b*
Oppenheim, Stephanie, 9:144b
Oppenheimer, J. Robert (1904–1967), 10:213–218; 4:294b, 582a; 7:487b; 8:94b, 95b–96a; 9:397a; 15:43a
  Born-Oppenheimer method, 10:214a
Oppolzer, Johann von, 2:129b, 445b; 10:218b, 219b
Oppolzer, Theodor Ritter von (1841–1886), 10:218–220; 7:6a; 8:504a; 10:279b
Opposites
  coincidence of, 3:513b–514b, 515b–516a
  interaction of, 13:330b
  and physiology, 1:171b
  as primary substances, 1:171b; 6:395b
  tension of, 6:289b–290a, 291a, 294b; 7:367b; 12:155a–b
Opsonic index, 5:29b
Opsonins, 2:584a; 9:334b, 545b*; 14:512b
Optic box, 2:535a–b
Optical activity, 15:11b
  in crystals, 2:137b–139b; 5:170a; 6:303b; 10:357a–358b; 12:304a–b; 14:525a
  in gases, 7:526a; 11:530a
  and life, 10:359a–361b; 15:103b
  magnetic, 1:558b; 2:382b; 4:538a–b; 7:492b, 526a; 8:493a; 10:516b; 13:615a; 14:329b
    theories of, 4:191a; 7:553b; 8:491b, 496b; 9:207b, 209b, 211b; 13:377a, 384a; 14:208b, 597b–598a
  and molecular structure, 1:340b–341a; 2:139a–b; 5:128b; 9:356b, 403a; 10:357b–361a; 11:85b–88b; 13:576b–577b; 14:124b
  theories of, 3:271b; 7:518b
  see also Compounds, organic, optical properties; Stereoisomerism
Optical illusions, 10:103b–104a; 11:21a, 200b, 472a–b
  in psychology, 2:132a
Optical systems, 12:250b
  achromatic
    theory of, 7:403b–404b
  aplanatic, 1:7b
    theory of, 7:403b–404b; 9:46a
  system laws, 5:591a
  testing of, 9:46a
  see also Instrumentation, optical; Lenses
Optics
  ancient Greek, 1:189a–b, 481b*; 4:430a–b; 6:416b; 13:322b–323a; 15:220a
  Byzantine, 7:136a–b
  medieval Arabic, 2:153a; 6:190b–197a; 7:212b–218b; 11:203a, 248b, 250a–b, 251b; 15:232a, 263b–265a
  medieval Latin, 1:101a, 279a, 337b, 379a–380a; 2:193a–b; 4:157b; 5:549b, 550b–552b; 6:275b; 7:116b; 10:474a–476a; 11:203a; 13:199b–201a; 14:35a, 457b–461a
  Renaissance, 1:81a–b, 97a; 5:28b, 401a; 9:192a; 11:97b–98a, 468a; 12:106a
  early seventeenth-century, 3:116a; 4:46b, 58b–60a, 574b–575b; 7:269b, 375b–376a; 10:176b; 11:489b; 13:30b
  late seventeenth-century, 1:424b, 475a, 481b; 2:379b–380a; 3:104b; 4:506b; 5:543a–544b; 6:68b; 9:465a–b; 11:526a–b; 14:83b
  early eighteenth-century, 2:343b–344a; 3:115a, 344b; 4:214b, 482a; 5:511a; 9:464a
  late eighteenth-century, 3:155b; 5:137a, 276a; 6:384a; 7:405b*; 11:472a–b
  early nineteenth-century, 1:201b–202b, 460a; 4:234a; 5:302a; 9:159a; 10:109b–110a; 11:111a; 15:361b–363a
  mid nineteenth-century, 1:135a–136a, 357b–358a; 3:144a–b; 5:305a; 6:278b; 13:22b; 14:290a
  late nineteenth-century, 1:6b–9a, 158a; 3:135b
  twentieth-century, 2:594a
  applied
    nineteenth-century, 13:77a–b
  atmospheric
    medieval Latin, 14:459a
    Renaissance, 5:28b
    seventeenth-century, 9:118b
    eighteenth-century, 2:344a
    nineteenth-century, 1:358a; 2:421a; 3:303b; 5:144a; 6:165a*, 240b; 10:9a; 12:289b; 13:522a; 14:343a, 491a–b
    twentieth-century, 1:519b; 12:352b–353a; 13:107b, 409a; 14:216a
    Chinese, 12:380b–381a
  and electricity, 4:189b–191b, 588b; 6:344a, 345b–346a; 9:212b–213a
  electronic, 15:8b
  as fundamental science, 5:549b; 14:459a
  geometrical
    ancient Greek, 1:256b
    ancient Roman, 11:200b–201a
    medieval, 1:379a; 4:92b–94b; 5:551b–552b; 6:200a–201b; 7:2b; 14:458b–461a
    Renaissance, 4:158b; 14:274b
    early seventeenth-century, 3:149b, 152b–153a; 4:184a,b; 5:278b; 6:125b, 128a; 7:298a–299a; 9:25b, 528a, 598a; 10:103b–104a; 11:395a; 12:501a–b; 14:636b–637a
    late seventeenth-century, 3:97b–98a; 5:515b, 521a, 524b, 525a–526a, 529a; 6:538a, 603a–b; 9:114b; 10:53a–60a, 74a; 13:480a
    early eighteenth-century, 6:5b–6a; 7:110b; 8:460a; 12:477b
    late eighteenth-century, 2:327a–b, 328b–329a; 3:284a; 4:148b–149a; 7:403b–404b, 596b; 9:477a; 12:413a–b
    early nineteenth-century, 1:202a–b, 460a; 4:257b; 5:142a–144a, 303a; 6:2b, 325a; 7:263a; 9:72b–74a; 14:491a–492a
    mid nineteenth-century, 1:135a–136a; 5:87a–b; 6:85b, 87a–88b; 7:509b; 8:413b–414b; 9:199b–200a, 225b–226a; 12:289b; 13:130b, 615a
    late nineteenth-century, 1:6b–9a, 158a; 3:420a; 5:102b*; 6:92b, 119a, 282a–b, 504b; 9:564a; 11:208a, 460b; 13:101b
    twentieth-century, 3:63a; 5:122b, 590b–591a; 9:45b–46a; 11:553a; 12:186a–b, 250b
    Chinese, 12:381a–382a
  historiography of, 14:158b
  ionic, 15:8b
  magneto, 3:564a; 4:190a
  mathematics and, 1:81b, 279a; 5:483a; 6:88a; 14:568a
  and mechanics, 5:550a; 6:89b; 12:220a–b
  physical
    seventeenth-century, 5:422b; 6:445b–446a, 484a–b, 609b–610a; 9:52a–53a, 97b, 118a–119a, 318a–319a; 10:315a
    early eighteenth-century, 2:328b–329a; 3:283b–284a, 486a; 4:23b, 44a–b; 10:79a–80b; 11:68a; 12:477a–b
    late eighteenth-century, 3:281b, 454a–b; 6:586a–b; 7:599a; 9:266b–267a; 11:142b–143a; 13:351b
    early nineteenth-century, 2:137a–139b, 451a–453b; 3:145b–146b, 489a, 599a–600a; 5:166a–171a, 445a–b; 6:324b, 333b; 7:263a; 9:72b–74a; 10:546a; 11:115b–116a, 474b; 12:281a–b; 14:491b, 565a–568b

Orion (cont.)
nebulae (cont.)
photography of, **3**:366a; **4**:180b; **11**: 484b, 553b*
spectroscopy of, **4**:180a–b
Orkney Islands, **5**:39a; **11**:75b
Orleans, Louis-Philippe-Joseph d', **8**:113b
Orlicz, Władysław, **1**:427a
Orlov, Aleksandr Petrovich, **9**:593a
Orlov, Aleksandr Yakovlevich (1880–1954), 10:231–233; 7:472a
Orlov, Sergey Vladimirovich (1880–1958), 10:233–235; 2:433b
Ornithodoros, **7**:429a
Ornithology
ancient Greek, **1**:120b, 263a, 264a
medieval, **5**:147b–148a
Renaissance, **1**:596a; **3**:342b–343a, 534a; **9**:441a; **13**:502a
seventeenth-century, **11**:316a–b, 588a; **14**:413b
eighteenth-century, **2**:474a–b; **3**:129b; **5**: 76b; **7**:96a, 200b; **8**:274b–275a; **13**: 601b; **14**:299b
early nineteenth-century, **1**:329b–331b; **2**:282a; **7**:97b; **8**:265b; **10**:9a; **12**:216a; **14**:187b, 416b–418a
mid nineteenth-century, **5**:480b–481a; **6**:295b; **10**:164a, 221b–222a; **12**:51b, 257b–258b
late nineteenth-century, **1**:405a–b; **3**: 438b–439a; **5**:162a–b, 545a; **9**:313b, 432a; **10**:427a; **11**:443a–b; **13**:25a–b
twentieth-century, **10**:244a
collecting manuals, **3**:439a
tropical, **8**:274b
see also Bibliography, ornithological; Birds; Illustration, ornithological; specific birds
Ornithopters, **6**:167a
Ornithorhynchus, **6**:454b*
Ornithoscelida, **6**:595a
Ornstein, Leonard Salomon (1880–1941), 10:235–236; 2:600b; 7:407a; 9:415a; **14**:616b
Orogenesis
medieval Arabic theories, **13**:292a
seventeenth-century theories, **2**:613b; **13**:34a
eighteenth-century theories, **2**:329a; **3**: 395b; **4**:493a; **5**:380a; **8**:252a; **10**: 283b, 284a–b; **11**:272b–273a
nineteenth-century theories, **1**:20a, 174b; **2**:555b, 569a; **4**:265b; **5**:395b; **8**:123a; **11**:118b, 210a; **12**:262b–263a, 276b, 300b; **13**:145b–146a, 284b–285b
twentieth-century theories, **1**:52a, 236a–237a; **2**:373a; **4**:598b; **8**:286a; **10**: 168a–169a; **12**:496b; **13**:64a,b, 271a, 610a; **14**:215b
contraction theory
see Earth, contraction hypothesis
erosion theory, **3**:66b
geosynclinal-contraction theory, **3**:553b; **6**:58a
gravity, role of, **2**:558b
horizontal stresses, **6**:187a
isostatic adjustment, **8**:99a
Neptunist theory, **2**:553b; **5**:444a–b; **8**: 241a–b; **15**:55b–56a
phases of, **13**:64a,b
as slow process, **4**:309a
tangential forces, **11**:505b
volcanic theory, **9**:532b–533b
wave theory, **2**:89b
see also Mountains
Orography, **1**:173a, 174b; **4**:143b; **6**:551b; **7**:511a; **11**:181b; **12**:301a; **13**:412b
Orohippus, **6**:595a
Orosius, Paulus, **7**:27b, 187a
History (Adversus Paganos Historiarum), **1**:414a
Oroya fever, **10**:142b
Orpheus, **15**:674a
Orphism, **4**:367b; **11**:161b
Orrery, **5**:491a, 520a*; **6**:36b; **11**:104a, 471b
naming of, **6**:45a(n)
see also Planetarium
Orta, Garcia d' (*or* da Orta) (*ca.* 1500-*ca.* 1568), 10:236–238; 1:47b; 11:312a
Coloquios dos simples e drogas he cousas medicinais da India
translations of, **9**:123b
Ortega, Casimiro Gómez, **2**:510a; **3**:154b, 177a; **11**:605b; **15**:470a
Ortega, Juan de (*ca.* 1480-*ca.* 1568), 10:238; 1:437b
Ortelius (or Oertel), Abraham (1527–1598), 10:238–240; 6:20a; 8:190b
Theatrum orbis terrarum, **9**:192a
Ortensius
see Hortensius, Martinus
Ortheziinae, **2**:321b
Orthmann, Wilhelm, **9**:261a
Orthobiosis, **9**:335a
Orthodrome, **10**:161a
Orthoepy, **4**:172b
Orthogenesis, **5**:386a; **11**:535b; **13**:276a; **14**:314b
Orthogonality, **5**:482a
Orthography, **3**:109b
Orthonectida, **5**:385b; **9**:328b
evolutionary cycle, **3**:148b
Orthopedics
seventeenth-century, **4**:511b; **6**:253b; **7**: 284a
nineteenth-century, **11**:571b
twentieth-century, **7**:279a
Orthoptera
anatomy, **12**:318a
taxonomy, **12**:264b
Orthostathmescope, **3**:299a*
Orton, James (1830–1877), 10:240–241
Oryctognosy, **14**:257b
Oryctography, **1**:393a
Osann, Gottfried Wilhelm, **3**:301a–b
Osazones, **5**:1b, 3a–b
Osbeck, Petrus, **5**:76b; **8**:379b
Osborn, Henry Fairfield (1857–1935), 10: 241; 1:18a; 3:610b; 5:500b; 6:321a; 7:480a; 12:260b; 13:151a
on contemporary scientists, **8**:169b; **13**: 277a*; **15**:93b*
Osborne, Thomas Burr (1859–1929), 10: 241–244; 3:335b; 8:591a; 9:284b–285a; 14:252b
Osborne beaker method, **10**:242a
Oscana, **2**:322a
Oschatz, Adolph, **11**:216a
Oscillaria, **3**:337b, 338a
Oscillation, **12**:534b; **14**:170b
autooscillations, **15**:7a
average energy, **4**:317a
center of, **2**:49a; **6**:605a, 605b–606b; **9**: 116b, 117a; **11**:488b; **13**:265b, 266b
definition of, **4**:60b
of elastic sphere, **7**:595a
harmonic, **6**:164a, 607b–608a
mathematics of, **15**:6b–7a
non-harmonic, **2**:41a
quantization of, **2**:241b; **4**:316a
relaxation, **11**:461a
theories of, **2**:217b; **8**:384a; **12**:218b
torsional, **3**:443a
see also Vibration
Oscillation, electronic, **4**:559a
spontaneous, **1**:455b
virtual, **7**:492a
Oscillators
Barkhausen-Kurz, **1**:455b
crossed-field, **6**:546b
Hertz, **1**:304a; **2**:427b
Oscillograph, **2**:84b, 318a; **14**:217b
bifilar, **2**:200b
cathode-ray, **4**:398b; **5**:291a,b; **6**:373b
invention of, **2**:200b, 427b–428a
"soft iron," **2**:200b
Oscilloscope
see Oscillograph
Osgood, William Fogg (1864–1943), 10:244–245; 2:359a*; 3:345b
Osiander, Andreas (1498–1552), 10:245–246; 3:410b; 13:59b
Osimo, Marco, **10**:373a
Osipovski, Timofei Fiodorovich, **10**:247a,b
Osiris, **12**:112b
Osius, **3**:14b
Osler, William, **1**:613b; **2**:599a; **3**:447b, 517a,b; **5**:208a; **6**:77b, 232b; **8**:447a; **9**:422b; **12**:483b; **13**:300a,b; **14**:249b; **15**:65a
Bibliotheca prima, **1**:323b
Oslo, University of
in twentieth century, **6**:615a; **13**:167a
founding of, **1**:12b
observatory, **11**:377b
Osmic acid, **2**:620b
Osmium, **3**:301b; **5**:197b
analytical reagent for, **3**:272a
as catalyst, **13**:313a
discovery of, **3**:346b; **5**:92a; **13**:281a; **14**:487a
Osmometer, **4**:264b
Osmond, Floris (1849–1912), 10:246–247; 3:234b; 7:529a; 13:245a
metallurgy, **2**:467a; **3**:212a, 233b–234a; **9**:139b*; **12**:544a; **13**:245b
Osmosis, **4**:182a, 203a, 211b; **5**:7a, 494a
in algae, **8**:29b
in blood, **4**:311b; **5**:149a–b; **7**:503a
collodion membrane, **4**:615b
electrical equilibria, **1**:537b
isotonic coefficient, **13**:579a–b; **14**:98a–b
in plants, **10**:576a, 577a–b; **14**:96a, 97b–98b
regulation of, **7**:503b
theories of, **2**:531a; **4**:264a–b; **7**:160a; **13**:579a–b
see also Pressure, osmotic
Osmundaceae
fossil, **8**:5b
phylogeny, **5**:605a
Osservazioni e memorie dell'osservatorio astrofisico de Arcetri, **4**:162a
Ossification, **3**:342b; **4**:519b–520a
Osteoarthritis, **11**:596a
Osteoclasts, **7**:439b; **11**:492a
Osteology
ancient Roman, **12**:540a
Renaissance, **7**:17a; **14**:8b–9a
seventeenth-century, **5**:220b; **6**:183a–b
eighteenth-century, **5**:267a, 442b–443a, 443b–444a; **8**:115b*; **9**:480a; **12**:137a; **13**:465a
nineteenth-century, **3**:264a; **5**:504b; **6**: 594b; **10**:194b
twentieth-century, **7**:279a
comparative
nineteenth-century, **2**:187a; **5**:46a–b; **10**:287b; **12**:38b, 45a
vertebrate, **2**:423b; **5**:357b
see also Bones, Skeletal system
Osteoma, **4**:499b

Osteomyelitis
etiology of, **10**:389b
Ostertag, Robert von, **15**:522a
Ostraca
mathematical, **15**:681a–b, 690a
tables of, **15**:704a*–b*
Ostracoda, **11**:386a, 501b, 502a
Paleozoic, **1**:494b; **13**:531b, 532a
Ostracodermi, **12**:337a
Ostrich, **3**:569b; **13**:528a
Ostrogradsky, Mikhail Vasilievich (1801–1862), 10:247–251; **3**:224b, 225b; **12**:534a
algebra, **3**:223a
on contemporary scientists, **2**:424b; **8**:432b–433a
influence of, **3**:228a; **7**:477b
students of, **10**:551a; **14**:105b
Ostromyslensky, Ivan Ivanovitch, **2**:316b
Ostrovski, Alexander Markovich, **3**:221b; **5**:508b; **7**:530b
Ostwald, Carl Wilhelm Wolfgang (1883–1943), 10:251–252; **8**:427b; **15**:447b
Ostwald, Friedrich Wilhelm (1853–1932), 15:455–469; **5**:446b*; **8**:598a; **10**:424a; **11**:283a*; **14**:329b
acquaintances of, **4**:165a,b; **8**:112b; **11**:279a; **12**:108b; **13**:243a, 576a; **14**:123a; **15**:36b, 433a
and Arrhenius, **1**:297a,b, 299a
chemistry, **5**:389b, 390b; **13**:228a, 244b; **14**:360a
ammonia, **2**:323b
atomic weight, **6**:481a–b; **9**:351b
catalysis, **7**:462b; **9**:352b; **12**:47a
dilution law, **1**:300a; **15**:459b
electrochemistry, **5**:559a; **11**:475a*
law of mass action, **5**:587a; **14**:109a
law of successive reactions (or stages), **5**:324a; **15**:461b
on contemporary scientists, **10**:578a*; **11**:283b*; **13**:581a*; **14**:330b*
critics of, **1**:288b; **2**:267a; **11**:9a; **12**:526b
crystallography, **9**:379b
editions by, **7**:381b
on genius, **7**:179b
glaciers, **13**:246b
influence of, **2**:498b; **5**:609b; **15**:435a, 436a
Ostwald process, **9**:427a
Ostwald-Walden rule, **14**:124b
physics, **11**:63b
atomic theory, **2**:266b; **4**:315a
energeticism, **4**:225b, 313b; **5**:391a
thermodynamics, **8**:289b; **9**:227a, 394a
students of, **1**:431a, 553a; **2**:258a, 559a; **3**:176a, 436b; **6**:350a; **7**:162a, 208a; **8**:289a; **9**:354a, 378a, 393b, 427a; **10**:156a, 578a; **11**:129a, 417b; **12**:404b; **14**:124b; **15**:159b
translations of works, **9**:557b
Zeitschrift für physikalische Chemie, **1**:536b; **13**:576a; **15**:456a, 459b–460a
Otho, Lucius Valentinus, **3**:262a; **5**:409a; **11**:4a, 397a–b
Otiorhynchidae, **6**:507a
Otitis, **14**:638a
Otocysts, **4**:12a
Otology, **1**:446b–447a; **4**:268a; **6**:453b; **13**:567a; **14**:398a
see also Ear, Hearing
Otoneurology, **1**:447a
Otsuki Gentaku
Chotei kaitai shinsho, **15**:738a
Ott, Isaac (1847–1916), 10:252–253; **2**:367a
Otter, William, **3**:291a; **9**:67a–b
Otto III, Holy Roman emperor, **5**:365b
Otto, Adolph Wilhelm, **5**:441b
Otto, Friedrich Julius, **5**:493a; **12**:198b
Otto, Robert, **1**:553a; **12**:482a
Oudemans, Corneille Antoine Jean Abram (1825–1906), 10:253–254; **6**:453a; **13**:154a; **14**:96b
Oudin, Paul, **1**:558b
Oughtred, William (1575–1660), 10:254–255; **2**:463a; **5**:278b(n); **9**:369a
acquaintances of, **9**:310a; **14**:287a
Apologeticall Epistle, **4**:13b
Circles of Proportion, **4**:13b
Clavis mathematicae, **14**:147b, 150a
and Delamain, **4**:13a–b
editions of works, **14**:147a
Euclidean commentaries, **1**:474a
geometric symbolism, **1**:474a; **4**:451b–452a; **10**:338b(n)
influence of, **10**:43b, 45b; **14**:148a
translations by, **8**:272a*
Ourcq (river)
canal, **5**:410b
Ousia, **1**:259b
Outbreeding
see Heredity
Outflow, **11**:164b, 306a; **13**:247b
constant level, **2**:39b
falling level, **2**:39b
force of, **9**:116a
of gases, **6**:545b
laws of, **11**:63a, 78a
non-stationary flow, **2**:39b
radial contraction of stream, **2**:39b, 299b
relation of velocity and head, **3**:116a
speed of, **9**:13b–14a, 116a, 119b; **13**:437a
stationary currents, **2**:40a
velocity law, **1**:567b
Outhier, Réginald (1694–1774), 10:255–256; **3**:281b
Ova
see Egg(s)
Oval, **11**:488a, 489b
generation of, **9**:199b
Ovals of Cassini, **3**:103b; **5**:521b
Ovariotomy, **2**:386b; **8**:406b
Ovary
anatomy, **4**:521a; **6**:509a; **11**:565b
embryogeny, **10**:579b
histology, **11**:465b
hormones of, **1**:123a–b
morphology
changes in, **5**:484b–485a
naming of, **5**:484b
nature of, **13**:33b, 170b–171a
transplantation of, **3**:121b
tumors of, **2**:386b
weight, **3**:113a
see also *Corpus luteum*
Overflow, **4**:208a
Overthrusts, **3**:162a, 433b; **13**:146a, 147a
Overton, Charles Ernest (1865–1933), 10:256–257; **12**:232a
Overvoltage, **15**:435a
Ovid, **2**:21a; **4**:446a; **7**:27b
Oviduct
discovery of, **13**:33b
Ovipara, **1**:264b, 265a; **2**:591b; **4**:509a; **6**:159b–161a; **7**:547a
fetal envelopes, **4**:264a
Ovism, **2**:287a; **7**:441a; **9**:52b; **11**:317a, 333b; **12**:559b; **13**:564a
opposition to, **2**:579b; **8**:127b–128a, 580a
Ovogenesis, **1**:123a–b, 545b; **6**:338b; **11**:565b; **12**:45b
Ovovipara, **1**:264b, 265a
Ovulation
hormonal control of, **2**:345a; **10**:611a
human, **11**:110a
periodic nature of, **2**:161b
rabbit, **2**:345a; **3**:487a
Ovule, **5**:539b; **13**:405b; **15**:513b
degeneration of, **13**:406a
Ovum
see Egg(s)
Owen, David Dale (1807–1860), 10:257–259; **4**:363b; **9**:255a
Owen, George (1552–1613), 10:259–260
Owen, Richard (1804–1892), 10:260–263; **5**:45b
acquaintances of, **1**:501b; **3**:324a, 424b; **6**:120b; **9**:428a; **14**:533a; **15**:78a
anatomy, **2**:503b, 504a,b; **5**:444a; **6**:591a
correspondents of, **3**:298a; **14**:128b
embryology, **1**:477a; **6**:592a
evolution, **2**:617b; **3**:574a; **5**:46b
and Huxley, **6**:592a, 592b–593b; **10**:262b–263a; **11**:514a
influences on, **11**:380b
paleontology, **7**:252a; **8**:572a; **9**:87a–b
taxonomy, **6**:593a–b
Owen, Robert Dale, **3**:539a; **8**:266b, 615b; **10**:257b; **11**:505a; **12**:133a; **13**:467a; **14**:133b, 328b
Owens College, Manchester, **14**:398a, 399a
chemistry at, **11**:536b–537b
engineering at, **11**:393a
physics at, **13**:52b
Oxalates, **5**:493a
composition of, **14**:488a
heat, effects of, **4**:239a
Oxalic acid, **1**:168b; **5**:125b; **13**:373a
composition of, **4**:239a; **8**:338b
determination of, **10**:440b
isolation of, **12**:148a–b; **14**:333a
manufacture of, **3**:201a, 202a
Oxalic ester
composition of, **4**:246b
Oxamic acid
discovery of, **1**:416b
Oxamide, **8**:334b
composition of, **4**:244b
Oxazole, **5**:215a
Oxford Philosophical Society, **11**:40b
Oxford University
in Middle Ages, **1**:377b; **12**:62b
in seventeenth century, **6**:387a, 482a, 483a; **7**:147a; **13**:501a; **14**:147a–148a, 368b–370a, 405a
in eighteenth century, **6**:385a
in nineteenth century, **3**:586a; **7**:365b; **8**:565b
archives, **14**:152b
Ashmolean Museum, **1**:318a; **8**:307b; **11**:40b
astronomy at, **11**:156a
botanical garden, **3**:586a
botany at, **1**:422b; **4**:98b; **9**:528b; **12**:394a–b
"calculators," **3**:420b; **4**:237b; **6**:376a–379b; **9**:133b; **12**:548a; **13**:185a–211a, 349a–350a
"Caroline creations," **6**:387a; **7**:147a
chemistry at
seventeenth-century, **1**:318a; **11**:40b
nineteenth-century, **6**:109b
twentieth-century, **10**:517a; **12**:505b
Clarendon Laboratory, **8**:368b, 369a; **12**:437b–438a
criticism of, **2**:540b–541a
Dunn School (Sir William), **5**:41b, 42a
geography at, **9**:123b
geology at, **2**:566a–567a
mathematics at, **14**:147b–148a
medicine at, **5**:40b
metallurgy at, **6**:560b
pathology at, **5**:41b

Oxford University (cont.)
physics at, 9:544a
physiology at, 6:25a; 8:524b–525a; 9:243b; 12:399a, 400a
Honours Physiology School, 5:41a–b
Radcliffe Observatory, 2:141a; 7:146a
founding of, 6:511b–512a
removal of, 4:270a
science at, 9:542b; 11:116b
Sheldonian Theater, 14:510a
University Museum, 10:584a
zoology at, 9:542a; 11:514b
see also individual colleges, e.g., Merton College
Oxfordshire
natural history, 11:40b
Oxidases, 2:86b; 7:273a–b; 10:282a; 13:393b–394a
naming of, 2:86b
Oxidation, 4:134a
agents for, 2:70a; 9:352b
autoxidation, 5:621b; 12:198a; 13:453a
biological, 2:25a; 5:126b; 7:406a–b; 10:22a
theories of, 1:361a–363a; 4:296b; 10:22b; 13:393a–394a, 452b; 14:334b
catalytic, 4:357b
chain theory of, 1:362b
conjugate, 12:404b
destructive, 9:131a; 11:92b
energy of, 5:621b
enzymatic, 1:362a, 362b–363a; 2:86b; 6:501b; 13:393b–394a, 452b
heat of, 1:579b
inorganic, 2:75a
intracellular, 6:505b
of metals
theories of, 13:247a, 310b
of nitrogen compounds, 1:426b
of organic compounds, 2:86b, 620b; 4:76a; 9:576b; 11:370b, 371a; 13:549b–550a; 14:169b
peroxide theory of, 1:361a, 362a–363a
quantitative study, 11:436b–437a
slow, 1:361a, 362a–b
spontaneous, 4:20b
Oxidation potential, 8:47a,b
Oxidation-reduction reactions, 1:362a, 363a; 3:47a, 235b, 279b, 290b; 14:173a
Oxides, 9:401b
classification of, 9:289a
composition of, 14:488b
inorganic vs. organic, 14:531b
metallic, 9:152b, 450b; 11:167b–170b; 12:47a; 13:310b
Oximes
isomers of, 1:553b; 3:300a; 9:402b
preparation of, 9:355b
stereochemistry of, 1:340b; 9:355b
Oxmantown
see Parsons, William
Oxonium compounds, 1:390b; 14:412a
Oxygen
and acid theory, 1:142b–143a; 8:58a
activation of, 1:362a–363a
in air
variability of, 7:160a–b
allotropy, 6:177b; 12:198a; 13:468a
in animal tissues
distribution of, 4:296b
in atmosphere, 5:318b; 9:140a, 530b–531a
atomic volume, 1:129a
atomic weight, 3:398a, 449a, 476b; 9:531a
as standard, 2:429a; 6:463b; 9:110a, 351b–352a
basic properties, 1:390b
in blood, 2:161a; 4:615b; 6:165a*, 261b; 9:18b
and atmospheric pressure, 2:61b–62b
dissociation curves, 1:453b–454a, 454b
saturation of, 1:453a
tension of, 1:453a; 2:62a; 7:502a
combustion of, 8:78a
critical temperature, 1:128b; 6:470a
determination of
in air, 6:395b; 7:620b
in organic substances, 1:528a
in water, 14:448a
discovery of, 6:484b; 8:75a–b; 11:145a–b, 146a; 12:144b, 147a–b
electric discharge through, 13:468b
inversion temperature, 10:206a
and irritability, 5:411a
isotopes of, 6:118a
in lakes, 2:142a
linkages of, 1:390b
liquefaction of, 3:11b; 4:79b; 7:221a; 10:206a, 604a; 14:522b
liquid
magnetic properties of, 4:79b
magnetic properties, 1:556a; 8:11a
in metabolism, 5:37b–38a
naming of, 8:77b
nature of, 3:599a,b; 11:438a
diatomic, 3:307a; 4:554b; 5:294b
elementary, 8:82a,b
optical properties, 13:107b
refractive index, 7:554b
poisoning by, 2:62a
production of, 4:134b; 7:13a; 8:47a, 365b
as propellant, 3:578b
prophylactic use of
at high altitudes, 2:62a
reactions of
with hydrogen, 6:404a,b
with mercury, 1:129a
in respiration, 8:75b, 83b–84a; 14:65a
role of, 8:73b, 75a–78b, 79b–80a, 82b
in compounds, 2:92a,b
in saliva, 1:453b
in solar spectrum, 4:179b–180a
spectrum, 11:612a
therapeutic use of, 1:360a; 5:57a; 13:448b*
invention of, 7:13a
valence, 1:390b; 8:508a
Oxyhemoglobin, 6:505a,b
Oxyluminescence, 4:20b
Oxymuriatic acid
elementary nature of, 3:602b–603a; 5:321b–322b; 6:284b
see also Chlorine
Oxytocic, 15:105b
Oyster, 2:502b, 591b
anatomy, 14:408a
coloration, 12:125a
culture of, 3:610b; 6:267b; 9:432a
ecology of, 5:70b
food supply of, 8:512b
fossil, 2:568b
hermaphroditism, 3:588b
physiology, 10:493a
Ozanam, Jacques (1640–*ca.* 1717), 10:263–265; 2:131a; 3:622a*; 8:271b; 9:452a
mathematical problems, 5:528a; 8:161b; 11:512a
Recreations in Mathematics and Natural Philosophy
editions of, 6:577a; 9:500b
A Treatise on Fortification, 4:43b
Ozark Mountains
geology, 15:152b
Ozarkian (Sierran) epoch, 8:123a
Ozarkian System, 13:532a,b
Ozersky, Aleksandr Dmitrievich (1813–1880), 10:265–266
Ozonam (pseudonym), 10:264b
Ozone, 9:152b; 13:468b; 14:394a
chemical properties, 12:197b
constitution of, 1:161a; 8:508a; 12:197b–198a
density, 13:236b
equilibrium with oxygen, 11:347b
formula, 3:307a; 7:552b; 10:178b*
liquefaction of, 6:177b
magnetic properties, 4:79b
thermal decomposition of, 3:197b
in upper atmosphere, 3:328a*; 4:514a, 600a; 5:102a; 13:107b
Ozonization, 5:541a
Ozonizer, 2:70a

P

Paalzow, Carl Adolph, 7:528b; 11:8b, 581b
Paauw, Jan, 9:595b
Pacal (Maya ruler), 15:804b, 805a, 807a,b
Pacchioni, Antonio (1665–1726), 10:266
Pacchioni granulations (Pacchionian bodies), 10:266b; 11:383a
Pachymeres, Georgius, 4:117b
Pacific Coast mountain ranges, 1:552b
Pacific Ocean
biology, 7:447a
ethnography, 5:225a*
exploration of, 3:396b–397a; 4:491b–492a; 5:75a–b, 76b–77a; 13:65b; 15:82a
geology, 6:229a
oceanography, 4:42b
salinity, 8:188b
subsidence in, 3:551b–552a
surveying, 12:114b
temperatures of, 1:595a–b; 9:43a
see also South Pacific
Pacific Railroad Surveys, 5:512b; 10:32a
Pacifism, 4:312b–313a; 7:151b, 156a, 159a; 8:11b–12a; 11:355b, 431a; 12:10a–b; 15:19b–20a, 465b–466a
Pacini, Filippo (1812–1883), 10:266–268; 3:113b*; 11:566a
Pacini corpuscles, 10:267a–b
Pacinotti, Antonio (1841–1912), 10:268–269
Pacinotti dynamo, 4:248b; 5:496a
Pacinotti, Luigi, 10:268b
Pacioli, Luca (*ca.* 1445–1517), 10:269–272; 1:82b; 4:259a, 448b; 5:113a
algebra, 12:92b
arithmetic, 8:41a–b
Divina proportione, 4:259a; 8:193b, 235a, 240a
and Leonardo da Vinci, 8:193b, 198a, 235a–236a; 10:269b
Summa de arithmetica, geometria proportioni et proportionalita, 2:279b; 3:273a, 275b, 276a, 277a,b; 4:586b, 587b, 595a, 612a,b; 8:235b, 236a, 238a
De viribus quantitatis, 3:276b
Packard, Alpheus Spring, Jr. (1839–1905), 10:272–274; 9:536b; 14:534b*
Packe, Christopher, 5:205b
Padmanābha, 2:416b
commentaries of, 2:119a
Yantraratnāvalī, 5:276a
Padoa, Alessandro (1868–1937), 10:274; 10:442b
Padtbrugge, Robertus, 13:169a
Padua
botanical garden, 1:167a; 14:13a, 335b, 336a

Padua, University of
in Renaissance, **3**:98a–99a; **12**:134b
in eighteenth century, **13**:563a
anatomy at, **14**:4a–b, 5a, 7a, 13a
and Fabrici, **4**:507a–b
medicine at, **6**:360a
methodology, **6**:448b
Padua Observatory
gravimetry, **8**:504a–b
Páez, Pedro, **2**:530b
Pagan, Blaise, **13**:592a
Pagano, Giuseppe (1872–1959), 15:469–470
Page, Charles Grafton, **8**:1a; **11**:604a
Paget, James, **3**:424b; **8**:7b, 407b; **13**:503b
Pagliani, Luigi, **4**:198a
Pahlavi intellectual tradition, **1**:32b, 33a; **13**:538a
Paige, Sidney, **2**:549a; **3**:484a
Paillette, Adrien, **1**:210b
Pain, **5**:291b; **8**:295a–b; **10**:220b
diffusion of, **1**:400a
double response, **8**:295a; **13**:393a
projected, **8**:7b
referred, **5**:283a; **8**:15b, 295b; **14**:322a
reflex, **8**:7b
relief of, **3**:517b
dye therapy, **4**:296b
lobotomy, **4**:287a(n)
traumatic, **11**:425b
visceral, **8**:295b
and weather, **9**:422a
see also Anesthesia
Painlevé, Paul (1863–1933), 10:274–276; **2**:303a; **6**:547b*; **8**:14a*; **11**:478a; **13**:250a
analysis, **2**:358a; **3**:220b; **11**:56a
on contemporary scientists, **11**:431b*; **13**:251a*; **14**:480b*
paradox of, **1**:570b
Paint
chemistry of, **5**:189b
gold, **15**:24b
manufacture of, **4**:250a
protective efficiency of, **5**:190a
quick-drying, **7**:316a
Paintal, Autar Singh, **2**:446b
Painter, Theophilus Shickel (1889–1969), 10:276–277; **2**:456b; **9**:522a; **13**:136a
Painting
encaustic, **15**:472a
three-color palette, **7**:36a
Paintings
optic box, **2**:535a–b
restoration of, **2**:74a; **10**:557b–558a; **13**:310b, 313a
see also Art, Illustration
Paitāmahasiddhānta, **13**:582a; **15**:537a,b, 538a, 555b–579a *passim,* 583b, 585a,b, 586a, 590a–592b *passim,* 597a, 604a, 609b, 610a,b, 620a, 625a
Pakṣas (half-months)
definition of, **15**:535b
Pakṣas (schools of astronomy), **15**:534a–b
Palache, Charles, **5**:455b; **7**:62b
Palaeobiologica
founding of, **1**:18b*
Palaeontographica
founding of, **9**:346a
Palaeontologia Indica, **10**:204a
Paläontologische Gesellschaft (Berlin), **7**:61b
Paläontologische Zeitschrift
founding of, **7**:61b
Palassou, Abbé Pierre-Bernard, **1**:327b; **3**:526a
Palate
anatomy, **13**:31b
cleft, **8**:554b
morphology, **2**:504a
Palau Islands, **12**:299a
Paleethnology, **2**:189b; **3**:328a–b; **7**:31a
Paleoanthropology
see Paleontology, human
Paleobiogeography
nineteenth-century, **10**:29b; **12**:38b; **14**:117a–b, 139a
twentieth-century, **12**:339b
Paleobotany
Renaissance, **5**:379b
eighteenth-century, **2**:204b; **4**:380a; **7**:197b; **10**:283b; **12**:159b
early nineteenth-century, **2**:491b–492b; **3**:45a; **10**:322a–b; **12**:182a, 202a; **13**:44a–b
mid nineteenth-century, **3**:608a; **5**:441a–442a; **6**:221a–b, 489b; **7**:43b; **8**:264a–b, 371b; **10**:110b; **11**:270a; **12**:104b–105a, 169a, 277a; **13**:9b, 542a–b; **14**:397b, 462b, 495b
late nineteenth-century, **2**:109b; **3**:187b, 322a; **5**:497b–498b; **6**:100a; **8**:354a–b; **9**:112a, 617b; **10**:32a–b; **11**:373a, 439b; **12**:259a–b; **14**:398b–399a, 599b–600a; **15**:147b
twentieth-century, **2**:371b; **3**:30b–31a, 188a; **5**:605a; **7**:90b–91a, 252b–253a; **8**:5a–6a; **12**:260a, 339a–b; **13**:345a–b; **14**:297b–298b
of coal, **2**:86a; **3**:608a; **4**:262a; **5**:441b, 497b–498b; **10**:32a,b; **13**:345a; **14**:398b, 600a
and stratigraphy, **14**:298a
thin sections, **2**:492a; **7**:90b
see also Paleontology; Plants, fossil; specific countries; specific geological time divisions
Paleocene epoch
deformation, **7**:251b
Paleochemistry, **8**:584a
Paleoclimatology
early nineteenth-century, **2**:569b; **5**:32a; **8**:568a; **12**:182a; **13**:604b
mid nineteenth-century, **2**:492a; **3**:568b; **6**:221b, 503b; **8**:519a; **9**:583b; **10**:29b; **11**:567a; **12**:104b, 168a; **13**:9b
late nineteenth-century, **3**:190b; **7**:511a; **10**:502a; **14**:53b, 316a
early twentieth-century, **1**:470a; **2**:549a–b; **10**:504a, 505b; **12**:339b; **14**:216b, 298b
mid twentieth-century, **1**:623a–b
Chinese, **12**:380b
hydrological, **6**:229a
theories of, **14**:298b
polar control, **5**:486b, 487a
Paleoecology, **1**:163b, 507a; **3**:520a; **4**:309a; **5**:486b, 533b–534a; **6**:222a; **10**:17a; **11**:386a; **13**:519a
Paleogeography
ancient Greek, **13**:85a–b
eighteenth-century, **10**:283b; **13**:179a, 257b
early nineteenth-century, **8**:566a
mid nineteenth-century, **4**:308a–309b; **5**:66b–68a, 437a, 534a; **7**:320a; **8**:519a–b; **12**:300b
late nineteenth-century, **1**:162b, 163a; **5**:476a; **7**:30b, 250a–253a, 510b–511b; **8**:463b; **10**:29b, 429a; **13**:147a–148a
early twentieth-century, **1**:623a; **3**:533a; **6**:168b, 216b; **12**:230a; **13**:270a–b, 531b
pulsation theory, **5**:486b, 487a–b
and tectonics, **6**:168b; **7**:251a–b
see also specific geological time divisions
Paleography, **3**:512a–b; **4**:212b–213a; **11**:149b
Paleohistology, **11**:595b–596a
Paleoichthyology
nineteenth-century, **1**:73a, 491a–492a; **3**:328a–b, 529a–530a, 606a–b; **4**:147b; **5**:32b, 469b; **6**:594b–595b; **7**:43b–44a; **8**:266b, 463a; **9**:388b–389a, 568b; **11**:340b
twentieth-century, **1**:622b; **3**:606a, 610b; **5**:467b; **7**:61a–b
Paleolithic, **4**:263a*
archaeology, **5**:109b; **10**:606b–607a
art, **2**:450b; **9**:539a; **10**:607a; **11**:484a
artifacts, **5**:163a; **10**:607a
industries, **2**:450b
naming of, **8**:528a
paleontology, **13**:275a
subdivisions of, **9**:539a; **10**:607a
transition to Neolithic, **2**:347a
see also Stone Age
Paleolithic man, **2**:189b; **3**:532b; **7**:278b; **14**:443a
Paleomagnetism, **9**:181b; **14**:216b
Paleontological Society (U.S.)
founding of, **5**:50a
Paleontology
Renaissance, **8**:242a; **9**:309a
seventeenth-century, **6**:486b; **8**:307b–308a; **11**:317b; **12**:257a; **13**:34a
early eighteenth-century, **1**:393a; **2**:490a; **7**:401a; **8**:4a–b; **9**:26b–27a, 368b; **12**:159b; **13**:179a, 257b–258a; **14**:502a–b; **15**:52b–55b
late eighteenth-century, **1**:234a; **2**:580b; **3**:43b; **6**:401a; **8**:147b; **9**:152b; **12**:487a–b, 560b; **14**:119b–120a, 133a
early nineteenth-century, **2**:187a, 481a, 556a; **3**:324a, 325a, 525b–526b; **4**:234a; **6**:358b*; **8**:564b–568b; **9**:86b–87b; **10**:287b, 322a–323a; **12**:182a–b, 488a–491b; **13**:44a, 354b–355a, 448b; **14**:88a, 456b; **15**:472a–b
mid nineteenth-century, **1**:73a–b; **2**:497b–498b; **3**:87b; **7**:320a; **9**:568b; **10**:211b–212a, 258b, 265b, 584a; **11**:269b, 386a, 500a–b; **12**:167b, 276b–278a, 318a–b; **13**:143b–144a, 145b, 467a–b, 554b, 620b–621a; **14**:439a; **15**:50b–52a, 151b–153a, 527a
late nineteenth-century, **1**:129b–132a, 162a–163b, 468b–469a, 471b–473a, 507a; **2**:172b, 346b; **3**:492b; **4**:147a–148a; **5**:475b–476a; **7**:251b–253a; **8**:543a; **9**:579a–b; **10**:606b–607b; **12**:229b–230a, 298a, 520a; **13**:145b; **14**:120a–b, 392a–b, 626a–b
early twentieth-century, **1**:18a, 397a; **3**:533a; **4**:39a–b; **8**:587a; **11**:336a–b; **13**:518b; **14**:120b, 215a
mid twentieth-century, **1**:285a–286a
comparative, **5**:577b; **10**:16b; **11**:500a–b, 501b; **14**:199b; **15**:112b
ethological, **4**:147b–148a
and evolution, **1**:473a, 491b–492a, 587a; **3**:569b–570a, 572b–573a; **4**:147b–148a; **5**:295b–296b, 357b; **7**:411b–412b, 480a–b, 588b–589b; **8**:486b; **9**:134b; **12**:38a–b; **14**:409b, 626b; **15**:169b
and geology, **1**:95b, 201a–211a; **2**:108a; **3**:192a, 236b, 297b, 433b; **4**:363b–365a, 493a; **5**:437a, 533a–b, 583a; **6**:186b–187a, 238a–239a, 437a; **7**:

Panum, Peter Ludwig, **6**:100a, 287b; **12**:87b
Pao Ch'i-shou, **14**:544b
Pao Huan-chih, **8**:419a
Pao T'ing-po, **14**:540a
Pap test, **10**:291a
Papakyriakopoulos, Christos Dmitriou, **4**: 9a; **14**:311b
Papaleski, Nikolay Dmitrievich, **9**:76a, 76b–77a
Papanicolaou, George Nicholas (1883–1962), 10:291–292
Pap test, **10**:291a
Papaver, **14**:99b
Paper industry, **2**:104b; **4**:71a; **5**:344b; **7**: 204b; **9**:60b
chlorine, use of, **3**:202a
repulping, **10**:497a
Papillae, **9**:63b
Papilledema, **7**:47b
Papin, Denis (1647–*ca.* 1712), 10:292–293; **6**:169b, 599a; **8**:151a; **10**:521b; **13**: 351b
experiments, **6**:46b(n), 170a; **8**:212a
Papin's digester, **9**:153a; **14**:354b
Pappus of Alexandria (*fl.* 300–350), 10:293–304; **9**:296b, 299b, 300a
acquaintances of, **12**:579b
astronomy, **15**:219a
Collectio, **1**:245a, 468a; **4**:55b, 212b, 489b, 567a,b, 569a; **5**:588b; **14**:19b, 603b, 604b; **15**:210a, 213b
commentaries on, **3**:364b; **12**:446a–b
commentaries of, **1**:223b, 249a; **4**:103b, 392a,b; **6**:310a, 313a, 407a, 617a; **13**:320a
on Apollonius, **1**:179b–191b *passim;* **5**:382a; **12**:206b, 500a
on Archimedes, **1**:213b, 214a, 223b; **3**:391a
on Aristaeus, **1**:245a–b
on Euclid, **1**:388b; **4**:414b, 415b, 416a, 421b, 424a–430a *passim,* 437a–b, 439b; **13**:302b; **15**:177b
on Ptolemy, **4**:491a; **11**:202a; **13**:321b
editions of works, **10**:573a; **12**:447a
geometry, **1**:219a; **4**:104a,b, 110b, 426b–427a, 429a; **9**:270b, 297b; **10**:51b, 115b; **12**:580a; **15**:118a(n)
influence of, **1**:41b; **3**:149a, 515a; **4**:489a; **5**:408b, 410a; **6**:598a; **11**:288a; **13**: 15a; **14**:458b
influences on, **15**:220b
lemmas, **3**:214b*; **6**:71a; **12**:314a
mechanics, **1**:419b; **7**:174b–175a, 341a; **8**:218b; **9**:488a; **14**:366b
Pappus' problem, **3**:461b; **4**:47a–b; **5**: 209b; **8**:306a; **10**:296a, 331a
theorems of, **3**:152a; **5**:527a; **10**:298a; **12**:292b; **13**:15b–16a
translations of works, **9**:487b; **15**:178b
on *Treasury of Analysis,* **4**:416a, 425b, 426b, 429a
Paprika, **6**:185b
Papyri, **3**:603b
mathematical, **15**:681a–b, 691a–692a, 697a–b
demotic, **15**:690b–694a, 698b, 699b
Greek, **15**:687a
Hieratic, **15**:682a–691a, 693a, 694a–700b
tables of, **15**:704a*–b*
translations of, **15**:687a, 690b, 698b–699a
Para-aminobenzoic acid, **7**:518a
Parabiosis, **14**:105b
Parabola, **4**:104a; **14**:274b
area of, **1**:218a–b; **6**:315a
construction of, **5**:382a; **14**:465a–466a; **15**:116a–b
cubic, **6**:359a
directrix, **14**:465b
evolute of, **2**:47b
focus of, **1**:186b–187a; **3**:554b–555a; **14**: 466a
meaning of word, **4**:421a; **9**:274a
rectification of, **2**:49b; **4**:516b; **6**:359a–b, 600a, 602a; **11**:240b
segments of, **13**:289b–290a
center of gravity, **12**:206a
semicubic, **6**:359b
rectification of, **2**:507b
spiral, comparison with, **3**:183a; **11**:487b; **12**:75b
tangents to, **11**:404a
Torricelli's, **13**:437a
see also Conics; Mirrors, parabolic; Quadrature, of parabola
Paraboloid, **4**:429a, 570b; **13**:290a
of revolution, **9**:191b
surface area, **6**:600a
Paraboloidal solid
center of gravity, **3**:182b, 364a
volume of, **3**:182b
Paracelsianiam
medieval, **6**:397a
Renaissance, **4**:169b–170a, 208b–209b; **5**:448a; **9**:440a; **11**:606a–b; **12**:334b–336a; **13**:141a, 397a; **14**:225b–226b
early seventeenth-century, **3**:596b; **4**: 388a; **6**:254a–257b; **7**:165a, 355b, 537b, 538b–539b; **8**:309a–312a; **9**: 23b; **12**:79a, 311a–312b, 333a; **13**: 507b, 559b
late seventeenth-century, **8**:130b
opposition to, **11**:466b
Paracelsus, Theophrastus Philippus Aureolus Bombastus von Hohenheim (*ca.* 1493–1541), 10:304–313; **5**:421b; **7**: 485a; **9**:327a
acquaintances of, **11**:395b; **13**:30b
alchemy, **1**:549b; **5**:580a
Archidoxes, **1**:571b
biographies of, **11**:257b
critics of, **2**:293a; **4**:209a, 387b; **12**:311b; **13**:559b; **14**:321a
editions of works, **1**:49b, 50a*; **12**:335a; **13**:142a
geology, **8**:416a
influence of, **1**:49a, 126b, 158b, 159a; **2**: 222b, 223a, 540a; **5**:420a; **7**:355a; **11**:493b; **12**:307b, 600b; **14**:209b, 210a
influences on, **13**:535a
medicine, **6**:145b
pathology, **3**:471b
physiology, **6**:257a
translations of works, **4**:170a; **5**:579b
De vita longa
commentaries on, **5**:448a
Paracentesis
of thorax, **6**:367a; **9**:482b, 483a; **12**:103a
Parachute, **8**:213b; **9**:492b; **13**:614a
size of, **4**:208a
Paradies, Maria-Theresa von, **9**:326b
Paradoxes
Achilles and the tortoise, **3**:456a; **14**: 608a–b
collections of, **4**:36b, 138a
Cramer's paradox, **3**:460a–461b *passim,* **11**:45b
falling millet, **14**:610b
flying arrow, **14**:608b–609b
liar paradox, **2**:604b
moving blocks, **14**:610b–611a
Petersburg, **2**:42a, 57a, 577b; **8**:321a–b; **13**:612a
Russell's, **3**:278a; **5**:154b
self-reference, **14**:391a–b
see also Zeno of Elea, paradoxes
Paraffins, **9**:608b; **11**:359b
dehydrocyclization, **7**:21b
isolation of, **12**:435b
isomerization, **7**:21b
Paragenesis, mineral, **2**:440b–441a; **5**:556b; **11**:465a, 521a; **12**:514a; **13**:476a; **14**:263a
Paraguay
natural history, **1**:351b; **7**:314a
Parahita system (astron.), **6**:115b
Paralbumin, **5**:149a
Parallax
ancient Roman discussion, **10**:298b
medieval discussion, **4**:543a; **5**:618b–619a; **6**:197b; **7**:259b; **10**:5b
seventeenth-century discussion, **7**:298b
of comets, **4**:213a; **6**:362a; **12**:500b
computation of
Indian, **15**:574b–575a, 592a–b, 605a, 610b
dynamical, **6**:352b
horizontal, **15**:631b(n)
lunar
ancient Greek study, **15**:215a–b
ancient Roman study, **10**:299a; **11**: 188a, 194a,b
medieval study, **1**:509a; **2**:24a; **5**:618b–619a; **10**:540b
Renaissance study, **3**:406a
seventeenth-century study, **5**:526a
eighteenth-century study, **5**:435a; **7**: 543b, 579b, 580a; **8**:179b; **14**:589b; **15**:348a, 353b
nineteenth-century study, **6**:263b; **10**: 207a
twentieth-century study, **3**:472a
Indian study, **15**:543b–544b, 561b, 592b
of Mars, **3**:103a; **7**:543b; **10**:596b
planetary, **6**:515a; **8**:504a; **15**:577a–b, 588a
solar
ancient Greek study, **15**:213b, 214b, 215a–b
ancient Roman study, **11**:194a–b
medieval Arabic study, **14**:576b
Renaissance study, **2**:411b; **3**:528b
seventeenth-century study, **3**:103a; **5**: 24b, 526a; **6**:515a–b; **11**:424a
eighteenth-century study, **2**:388b; **3**: 198a; **4**:23b, 481a; **5**:435b; **6**:234b, 511b; **7**:543b, 580b–581a; **8**:299b; **9**:164a; **10**:615a,b; **11**:472a, 609b; **12**:413b–414a; **15**:353b
nineteenth-century study, **1**:157b–158a, 340a; **4**:176b*, 351a–b; **5**:86a, 258a, 404b, 405a,b, 479b; **6**:119b, 263b; **10**:34b; **11**:114b, 115a
twentieth-century study, **4**:251a; **6**: 147a; **7**:472b; **10**:526a, 529a*; **12**: 449a, 574a
Indian study, **15**:543b–544b
spectroscopic, **1**:56b; **6**:352b; **13**:98a
stellar
ancient Greek study, **1**:247b
Renaissance study, **2**:403a, 410a; **3**: 403b; **4**:6a*
seventeenth-century study, **5**:243b, 526a
eighteenth-century study, **2**:328b, 387b–388a; **3**:285b*; **6**:512b, 513b
early nineteenth-century study, **2**:100a–101a, 468b; **3**:14a; **5**:144a; **6**:263b; **13**:110b
mid nineteenth-century study, **1**:339b; **3**:404b; **10**:542b; **13**:120b

Parallax (cont.)
stellar (cont.)
late nineteenth-century study, **1**:466b, 597b; **4**:351a; **5**:404a; **6**:49b
early twentieth-century study, **3**:509b; **5**:554b; **6**:149b; **7**:237a, 239a,b; **8**: 582a–b; **12**:448b
"absolute" vs. "relative," **2**:100b
average parallax, **11**:156a
definition of, **12**:176b
discovery of, **5**:25b
photographic determination, **7**:472a; **12**:17b, 18b, 176a–b
of Venus, **7**:543b
Parallax instrument, **2**:403b, 404a; **3**:402a; **7**:123b; **10**:299a
Parallel lines, **5**:302a
construction of, **1**:468a
definition of, **5**:346b; **6**:201b–202a; **10**:6a; **12**:314a
coplanarity, **3**:323a
Lobachevskian, **8**:430a
theories of, **2**:268b–269a, 270a, 274a–b; **6**:201b–202b; **7**:329a–330a; **8**:141a–b; **11**:34b; **12**:255a; **13**:264b
distant parallelism, **4**:331a
foundations of, **7**:206b
see also Euclid, parallel postulate
Parallel mechanism, **3**:228b
theory of, **3**:229a
Parallelepiped, **14**:94b
transformation of, **14**:274b
Parallelism, **4**:336b; **10**:514b
in curved spaces, **8**:284a
Parallelogram
analytic, **5**:567a
Paralysis
eighteenth-century study, **11**:112a
nineteenth-century study, **1**:496b; **3**:588b
twentieth-century study, **5**:207b
bulbar, **8**:7a
cerebral, **5**:173b
creeping, **15**:522b
diphtheric, **5**:384b
Duchenne's, **3**:205b
electric shock treatment, **5**:137b–138a
generation of, **1**:392a
infantile, **3**:205b; **5**:173b; **6**:233a*
paralysis agitans, **6**:60a
progressive, **3**:490a
pseudohypertrophic, **3**:205b
theories of, **11**:127b; **14**:408a
vascular, **2**:30a
see also Hemiplegia; Paraplegia
Paramagnetism, **7**:13b, 492b–493a
measurement of, **11**:241b
and temperature, **3**:506a; **12**:438a
Curie's law, **3**:506b
theories of, **8**:10b–11a; **14**:244b
quantum theory, **13**:240a
Paramakāraṇa, **12**:599a
Paramaśukla
commentaries of, **2**:117a
Paramatta, observatory at, **2**:472a
Paramecium, **6**:401a; **9**:550a
behavior of, **7**:98b
sexuality in, **7**:99b
Parameśvara (*ca.* 1380–*ca.* 1460), 10:313–314; **1**:48b; **10**:129a; **15**:622a
astronomy, **15**:602b, 630a
commentaries by, **1**:308a,b; **2**:114b, 115a, 115b; **9**:580b; **12**:599a
Dṛgganita, **15**:613b–614a
Parameters, **7**:109a
auxiliary, **8**:253b
variation of, **4**:478b; **9**:476b
Paranucleins, **7**:466b–467a
Parapegma, **3**:391a; **4**:172a, 460a; **5**:345b; **9**:339a(n)
Paraplegia
surgery for, **6**:519a
Parapsychology, **4**:188a; **5**:22a; **7**:189b–190a, 192b
see also Psychic research, Spiritualism
Parāśara, **15**:619a
Parāśarasiddhānta, **1**:309b
Parasilurus, **1**:265b
Parasitism, **3**:332b; **5**:385b; **7**:60a, 171a; **11**:491a
chemotherapy, **4**:299b–301b
facultative, **9**:176a
fungus, **1**:493a–b; **2**:437b; **3**:45a, 612a–613b; **5**:56b, 566a; **6**:73a; **10**:1a–b, 253b; **11**:132a, 491b; **12**:202b–203a; **13**:299b, 489b
hematozoic, **2**:528a; **5**:566a; **8**:296b
heteroecism, **3**:613a–b
intestinal, **3**:97a, 588a; **4**:132b–133a, 198a; **13**:563b
life cycles, **4**:207a; **8**:415b; **9**:82a–83a; **12**:424a
malaric, **5**:461a–b, 503a–b; **9**:82b–83a
nature of, **1**:493b; **12**:482b
pathogenic, **1**:493a–b; **5**:570b; **9**:83a; **10**:372b–376a, 384a, 388b
of plants, **3**:588b; **4**:223b; **9**:461a–b; **10**: 1a–b; **11**:131b, 132a; **13**:406a, 563b
transformations caused by, **3**:148b
Parasitology
Byzantine, **1**:121a
seventeenth-century, **1**:26a; **8**:415b; **11**: 342a–b; **15**:29b
eighteenth-century, **2**:204b
mid nineteenth-century, **1**:602b; **2**:127b, 617a; **3**:588a–b; **4**:236a, 275b, 289a; **5**:566a; **6**:73a; **8**:169a, 270a–271a; **11**:301a
late nineteenth-century, **2**:528a–529a; **5**: 502b–504a; **8**:65b–66a, 296b; **9**:81b–83a, 94b–95a; **10**:527a–b; **12**:480a–481b; **13**:62b–63a; **14**:2a
twentieth-century, **2**:533b–534a; **3**:148b; **4**:132b–133a; **6**:319b; **7**:272b–273a, 447a–b; **8**:145b–146a; **9**:328b; **11**: 175a; **12**:142a, 207a, 453a–b, 481b–483b; **13**:63a; **15**:454b, 480a
ecological, **4**:142b
experimental
twentieth-century, **14**:595b
medical, **2**:533b
see also Helminthology
Parasitology
editors of, **7**:273a
Paraskevopoulos, John Stefanos, **12**:349a
Paraśurāma Miśra
commentaries of, **2**:115b
Parathyroid
see Glands, parathyroid
Paratyphoid, **5**:270b
diagnosis of, **6**:298b
etiology of, **6**:433a; **12**:96b
Pardies, Ignace Gaston (1636–1673), 10: 314–315; **7**:583a; **10**:54a
works translated, **6**:130a*
Paré, Ambroise (*ca.* 1510–1590), 10:315–317; **2**:315a; **3**:534a; **5**:448a; **11**:254a; **14**:6a
Anatomie universelle, **14**:11b
medicine, **2**:336b; **9**:12a
neurosurgery, **1**:618b
translations of works, **7**:147b
Pareiasaurs, **10**:40b
Parenago, Pavel Petrovich (1906–1960), 10: 317–319
Parenchyma
atrophy of, **14**:229b
naming of, **5**:535b
pulmonary
structure of, **9**:63a, 64a
wood
structure of, **12**:99b
Parent, Antoine (1666–1716), 10:319–320
geometry, **13**:418a
mechanics, **3**:145b; **6**:452a; **8**:209a; **11**: 496b; **12**:118a
Parerga, **3**:491b
Paresis, **5**:208a; **10**:131a
etiology of, **10**:142a
treatment of, **14**:115a–b
Pareto, Vilfredo, **6**:262a
Parhelia, **4**:216a; **5**:287b; **9**:118b; **10**:9a, 490a
Parigiano, Emilio, **6**:155b
Parija, Prankrishna, **2**:184b
Paris
Bicêtre, **10**:612a
botany, **13**:506a, 553b
Catholic University, **8**:31a
cemeteries, **3**:7a
Centre National de la Recherche Scientifique
founding of, **2**:303b
La Charité hospital, **2**:301a,b; **3**:427a
Cité Universitaire, **1**:194a
Conservatoire des Arts et Métiers, **3**:316b
École de Médecine
see Paris, University of, Faculty of Medicine
hospitals, **15**:302b
Hôtel-Dieu, **1**:401b
surgical clinic, **2**:122a
Institut de Chimie Physique, **4**:253b–254a; **13**:546a
Institut Henri Poincaré
founding of, **2**:303b
Jardin d'Acclimatation
founding of, **5**:359b
Lodge of Nine Sisters, **7**:581b
mapping of, **8**:64b
meridian of, **3**:103b, 104b, 105a, 106a, 107a–b, 108a; **7**:576a, 577b; **9**:90a,b, 186b; **15**:334a
modernization of, **4**:242b
observatory
in seventeenth century, **3**:102a–b, 103b–104a; **9**:90a; **10**:520b–521a, 596a; **11**:525b
in eighteenth century, **2**:141a; **3**:104b, 106a, 108a, 443a; **8**:21b, 22a; **15**: 339b–340a
in nineteenth century, **4**:20a, 68b, 69a; **5**:86a, 87a, 257a, 271b; **6**:282a; **9**: 551b; **10**:34a; **11**:319b, 320a; **15**:380b, 381b
in twentieth century, **11**:470a
historiography of, **14**:480a
meteorology, **8**:278b
Palais de la Découverte, **10**:525b
Paris Exhibitions
early nineteenth-century, **3**:200a; **5**: 271b
mid nineteenth-century, **1**:573a; **2**:26b; **3**:328a, 586b; **4**:257b; **5**:87a; **6**:494b; **8**:30b, 399b, 462b; **9**:538b; **11**:79a, 235a, 384b; **13**:464b; **14**:473b
late nineteenth-century, **1**:130a, 147b, 173a; **2**:166b, 439a; **3**:207a; **8**:407b; **9**:155b; **11**:579a
twentieth-century, **2**:467a; **5**:316a
population of, **15**:304b, 305b
public health, **9**:306b
Salpêtrière, **2**:35b; **3**:205a; **8**:272b; **10**: 612a, 613a
water supplies, **4**:38b, 46a; **5**:410b; **8**: 71a–b

Paris, University of
in Middle Ages, 1:228a–b; 2:603a–b; 9:361b; 12:62b; 15:189a
in Renaissance, 11:286b–287a
Faculty of Arts
in Middle Ages, 6:274a; 9:136a
Faculty of Medicine
in Renaissance, 4:209a, 585b; 5:585b; 10:316b
in seventeenth century, 1:571b; 7:536a–b; 8:310a; 11:466a–b; 13:507b
in eighteenth century, 5:352b; 8:353a
in nineteenth century, 9:8b; 11:560a
anatomy at, 2:442b–443a; 3:490a; 11:466b
chemistry at, 14:529b
Faculty of Sciences, 9:418b
geological collection, 6:168a
Institut du Radium, 3:501b, 502a; 7:152a, 156a, 158a, 159a
Laboratory of Physiological Psychology, 2:131b
logic at, 6:376b
Napoleonic reorganization, 2:133b; 3:523b
physics at, 10:145b
women teaching at, 3:500b
zoology at, 7:546a
Paris Basin
geology, 3:161a,b, 162a; 8:30b, 564b–565a, 567b; 9:579a; 10:209a
paleontology, 2:481a, 494a–495b; 4:67b–68a; 10:221b, 222a
stratigraphy, 4:68a; 5:476a; 11:133b; 14:211a
Paris Chemical Society
see Société chimique de France, Paris
Paris green, 1:553a
Paris Peace Conference, 11:524b
Parisian anatomists, 4:267a–b
Parity (phys.)
nonconservation of, 10:425a–b; 11:117b
in weak interactions, 7:617b
Parkhurst, John Adelbert (1861–1925), 10:320–321; 6:28b
Parkin, John, 8:354b
Parkinson, James (1755–1824), 10:321–323; 7:401a
Parkinson's disease, 10:321b–322a
treatment of, 6:518b
Parkinson, John, 7:147a,b; 9:313a
Parkinson, Samuel, 9:202b
Parkinson, Sydney (*ca.* 1745–1771), 10:323–324; 1:434a
Parkinsonism
postencephalitic, 3:112b–113a, 205b
Parks, national, 6:14a
Parmenides of Elea (*ca.* 515–*ca.* 450), 10:324–325; 8:269a; 11:24a; 12:340b; 14:536b, 607b
logic, 1:149a
"On Nature," 1:170b–171a
philosophy, 4:31a, 32a, 367b–368a; 10:208a; 14:608a, 610b, 611a,b
Parmentier, Antoine-Augustin (1737–1813), 10:325–326; 3:7a,b; 11:165a*
Parnas, Jakub Karol (1884–1949), 10:326–327; 6:501b
Parona, Corrado, 4:198a
Parona, Ernesto, 4:198a
Parrington, Francis Rex, 2:504b
Parrish, Joseph, 6:119b; 9:540a
Parrot, Georg Friedrich, 8:187a, 188b; 13:109b
Parrot (bird)
anatomy of, 3:342b
Parry, William Edward, 5:78b; 7:70b; 11:554a–b; 12:49a; 14:491b
Parseval des Chênes, Marc-Antoine (1755–1836), 10:327–328; 13:562a
analysis, 7:566b; 13:27a
theorem of, 10:327a–b; 11:459a
Parsley
oil of, 3:279b
Parson, Alfred, 8:291a–b
Parsons, Charles, 11:306a
Parsons, William, third Earl of Rosse (1800–1867), 10:328–329; 4:178b, 185a; 5:562a; 9:88a*; 10:584a; 13:82a
Parthenogenesis, 1:265b; 5:624a; 8:270b; 9:185a,b
of aphid, 2:286a; 3:180b; 8:580a; 11:333a–b
artificial, 4:12a–b; 5:385b; 6:337a; 7:455b; 8:446a
by irritation, 13:478b
traumatic, 1:498a–499a; 2:384b
discovery of, 2:286a, 287a; 11:333a–b; 12:421b
of rabbits, 10:611a
Partial pressures, law of
see Gases, partial pressures, law of
Particle, integral, 5:473a–b
and molecular theory, 1:141b
Particles, subatomic
and antiparticles, 9:38b
collisions of, 4:577b–578a, 583a; 7:617b
counting of, 12:30b, 34a
detection of, 6:356a; 11:117a–b
interaction amplitude, 7:617b; 8:96a
mathematics of, 15:27b
orbits of, 7:466a
relativistic theory, 7:493b; 9:38a–b
resonances, 4:583a
single-domain, 5:160b
spinor analysis, 7:10a
wave nature of, 13:42b
wavelength *vs.* momentum, 15:49b
see also Accelerators, particle; specific particles
Partington, James Riddick (1886–1965), 10:329–330
on contemporary scientists, 12:47b*; 13:518a*
historiography, 3:48b, 206b*; 4:394a; 6:386a, 496a*, 586a; 8:86b(n), 87b*, 90b*; 9:246a; 11:284a*; 12:63b, 64a, 148a,b, 311a–b; 14:197a, 508b
Partition law (chem.), 2:69b–70a
Partitions (math.), 4:474b; 6:114a; 8:617a; 11:269a; 13:218a
theory of, 3:169a–b; 13:219a–b
Partridge (bird), 1:265b
Parturition, 6:60a; 9:41a
Parvan (period), 15:582a, 598b
definition of, 15:575b
Pasadena (Cal.)
as research center, 6:32a
Pascal, Blaise (1623–1662), 10:330–342; 3:284a, 384b, 622a; 9:599a
acquaintances of, 2:353b; 3:63b; 4:46a; 6:598b; 9:498a; 10:476b, 546b; 11:486b
biographies of, 2:88b
and Carcavi, 3:64a–b
correspondence of, 3:66a, 214a; 8:250b; 9:316b; 12:460a
critics of, 7:448b; 10:137a; 11:507a
editions of works, 2:335a, 358b, 544b, 545a,b
Équilibre des liqueurs, 5:288a; 9:117a
Essay pour les coniques, 4:47a
experiments of, 11:489a; 13:445a
Histoire de la roulette, 7:583b
influence of, 3:226a; 6:36a; 7:550b; 11:79b
influences on, 1:191b; 4:49b; 13:236a
and Lalouvère, 7:583b
Lettres de "A. Dettonville" (=Pascal) contenant quelques unes de ses inventions de géométrie . . . , 2:349b; 3:152a; 8:161b
manuscripts of, 13:479b
mathematics, 1:474b; 6:112b
algebra, 8:162a
calculating machine, 3:64a; 11:66a*
calculus, 3:152a
game theory, 3:66a; 9:599b
geometry, 4:506a; 6:598a, 605b; 8:304a; 10:115b; 11:487b; 12:460a; 13:623a; 14:149b, 509b
number theory, 13:540a
probability, 3:451b; 4:574b; 6:600a; 15:280a
Méthode, 1:292a
Pascal configuration, 8:542a
Pascal's law, 8:227a; 9:64a
Pascal's triangle, 3:268a–b; 7:257a; 10:335a; 12:93a; 13:259b, 510a; 14:540b
philosophy, 11:507b; 15:375a
physics, 3:114b
atmospheric pressure, 3:64a; 8:267b; 13:439a
hydrostatics, 2:378a; 4:60b; 9:369b
vacuum, 9:16a
Repertorium, 8:458b
editions of, 14:336b
theology, 5:62a
theorem of, 2:215b, 455a; 3:167b, 555a; 5:368a; 10:330b–331a; 14:212a
proof of, 5:368b
Traité du triangle arithmétique, 11:404a
Pascal, Étienne (1588–1651), 10:342–343; 6:299a; 10:333a
acquaintances of, 1:541b; 2:353b; 4:46a,b; 9:598a; 11:486b
critics of, 9:528a
geometry, 4:47b
limaçon of, 10:342b
physics, 10:332b
Pasch, Moritz (1843–1930), 10:343–345; 2:269a; 9:502b; 10:443a; 13:125a, 599b
geometry, 6:391a; 7:206b; 11:536a
Pasch's axiom, 10:343b
Paschen, Louis Carl Heinrich Friedrich (1865–1947), 10:345–350; 1:370b; 12:529a
acquaintances of, 1:371a; 11:478b; 15:448a
on contemporary scientists, 7:268b*; 12:615b
Paschen series, 1:425b; 10:347b
Paschen-Back effect, 1:370b; 10:347b, 423a
Paschen's law, 10:345a
radiation, 8:551b; 11:150a, 582a; 14:339b
and Runge, 11:612a–b
spectroscopy, 2:245a; 11:479b
Passignani, Domenico, 12:151b
Passions
seat of, 5:253a
Pasteur, Louis (1822–1895), 10:350–416; 2:438a, 471a; 4:66a; 7:281b, 518a, 545b; 9:333a; 10:594a; 12:124a
acquaintances of, 1:417a, 498a; 2:27a; 3:22a, 189a; 6:307b; 8:7b, 124b, 407b, 408a; 9:328a, 334a,b; 10:522b, 593b; 11:428b; 12:87b; 14:36b; 15:11b
and Bastian, 10:372b, 382a–383a
and Biot, 10:352b, 357a, 361a,b, 366b
chemistry, 10:357a–361a
isomerism, 7:517b; 8:109b; 13:576b
optical activity, 2:139b
racemic compounds, 7:552a
on contemporary scientists, 3:588a; 4:78a*; 6:11b

Pasteur, Louis (cont.)
correspondence of, **2**:436b; **8**:406a
critics of, **1**:496b; **2**:33a, 70b; **8**:296b; **11**:492a; **15**:11b–12a
and Duclaux, **4**:210b, 211a; **10**:355b, 373b, 380a, 385b, 389b
and Dumas, **10**:352b, 354a, 356b, 366b, 372b, 373b, 377a, 400a
and Gamaleya, **5**:269a–b, 270a
influence of, **3**:207a, 219b, 220a; **5**:52a, 128b; **6**:100b; **7**:100b; **8**:273b, 409a; **11**:70a, 176b; **13**:522a; **14**:36b, 443b, 534a; **15**:103b
influences on, **12**:563b; **15**:115a
jubilee, **8**:408b
and Koch, **7**:423b, 424a–b, 426a; **10**:354b, 364a, 397a–400a *passim;* **12**:87b
microbiology, **2**:575b; **7**:100a
antibiosis, **5**:42a
fermentation, **2**:31a–b, 560b, 561a–b; **3**:9b, 587b; **8**:349b; **10**:361b–370a, 376b–382b
methodology, **2**:528a; **9**:176a; **11**:428b
nomenclature, **3**:338a,b
putrefaction, **8**:403a; **10**:364b–365a, 370a
spontaneous generation, **3**:188b, 339b, 614a; **7**:111a, 421a; **10**:10b, 355a, 366b–372b, 380b–383b, 389b; **11**:110b; **12**:563b
Pasteur's fluid, **10**:363b
patents, **10**:352b–353a, 366a,b, 380a
pathology, **10**:383b–408a
anthrax, **3**:588a; **10**:384b–385a, 386a–388b, 392b–398b
chicken cholera, **8**:408a; **10**:390b–392b
pest control, **8**:450a; **10**:406a–b
rabies, **6**:518b; **10**:398b, 400b–408a, 527a
silkworm diseases, **1**:418a; **10**:372b–376b
swine erysipelas, **10**:399b–400b
and Pouchet, **10**:371b–372a
students of, **11**:310b, 311a, 595b; **12**:96a; **13**:405a
Pasteur Institute, Algiers, **12**:315a
Pasteur Institute, Paris
in nineteenth century, **4**:211a–b; **6**:11b; **8**:66a; **9**:334a–b; **10**:414b*; **11**:569a,b
in twentieth century, **4**:154b; **8**:273a; **14**:36b, 551b
offshoots, **3**:22a; **9**:334a; **10**:355b
organization, **3**:22b, 189a; **4**:211a–b; **5**:99b–100a; **8**:66a; **10**:352b, 384b–385a, 403b, 404b, 408a
Pasteur Institute, Tunis, **15**:454a
Pasteurella, **3**:189a; **7**:392a
Pasteurization, **10**:366b, 380a
Pastor, Julio Rey
see Rey Pastor, Julio
Pasumot, François, **4**:71b
Patagonia, **12**:261a
Pātas (configurations), **2**:417b; **5**:274b, 275b; **7**:219b, 512b, 582b; **10**:129b, 426a; **14**:28b
definition of, **15**:546a, 564b
Patella
fracture of, **8**:402a, 407b
Patent Office, U.S., **8**:1a, 2b
Patents
chemistry, **2**:59b; **4**:250a; **5**:489a; **10**:516a; **12**:47b, 521a; **14**:247b, 393a
conflict over, **1**:287a,b, 288a; **4**:6b, 7a, 148b, 601b; **6**:51a; **8**:93a; **9**:507b; **14**:196b
electricity, **4**:284a; **13**:286b, 361b; **15**:449b
engineering, **3**:436b; **10**:36b; **11**:499b, 588b; **13**:431b; **14**:196b, 520a,b
income from, **14**:121a
law movement, **2**:59b; **3**:85b; **5**:559b; **7**:89a–b; **11**:384b
metallurgy, **9**:591a–b; **13**:346b
microbiology, **10**:352b–353a, 366a,b
physics, **10**:604a, 605a; **15**:60a
printing, **13**:413b
Paternity disputes, **7**:623b
Paternò, Emanuele, **9**:378a; **11**:298b
Pathognomy, **8**:322a
Pathology
ancient Greek, **1**:104a, 324b; **4**:385b–386a
ancient Roman, **1**:212a–b; **3**:174b
medieval Arabic, **15**:499a
medieval Latin, **11**:601a
Renaissance, **3**:342a, 355a–b, 471b; **5**:105b–106a, 379b; **9**:75a; **10**:307a; **11**:33a–b, 604b–605a
early seventeenth-century, **1**:523b; **5**:425b
late seventeenth-century, **1**:391b; **5**:220b; **9**:323a; **13**:214b–215a; **14**:256a, 408a
early eighteenth-century, **3**:437b; **5**:366b–367a; **7**:614a
late eighteenth-century, **1**:399a; **3**:579a; **8**:352a–b; **12**:137a
early nineteenth-century, **1**:492b–493b, 544a; **2**:444b–445a, 463b–464a; **7**:556b–557a; **11**:510b; **12**:202b
mid nineteenth-century, **2**:164b, 478a, 575a–b, 598b–599a, 617a; **3**:489b, 587b–588b; **4**:198a; **5**:515a, 565b–566a; **6**:269b, 270a; **9**:10b; **10**:372b–376b, 383b–385a; **11**:300b–301a, 515b; **12**:202b–203a, 295a–296b, 612a; **14**:40b–42a
late nineteenth-century, **1**:438b–439b, 574b–577a; **2**:528a; **3**:17a, 94a–b, 447a–b, 463a–b; **4**:275b–277b; **5**:461a–b; **6**:101b–102a; **7**:100a–b; **8**:9a, 124b, 296b–298a, 556b–557a; **9**:93b–94a, 94b–95a, 333a–334b; **10**:386a–408a; **11**:176a–177a, 345b–347a, 428b–429b, 482a; **12**:87b–88b, 480a–481b; **13**:300a–b; **14**:115a, 126a–b, 218b–219a, 228a–229b, 248b–250a, 511b; **15**:521a–522a
twentieth-century, **2**:597b; **4**:498b–499b; **5**:39b–41a, 41a–43b, 316b–317a; **6**:232b–233a; **7**:100b–101a, 286b–287a; **8**:557a; **9**:176a; **10**:15a–16a, 141b–143a, 291a–b; **11**:175a, 442a–443a; **12**:96b, 141a–142b, 481b–483b; **14**:115b–116a, 252b; **15**:96b–97b, 454a–455a, 480a
botanical
see Plants, pathology
cellular, **11**:300b–301a, 492a
chemical, **14**:252b
classification, **12**:202b
comparative, **12**:480a–483a
constitutional, **14**:219a
embryological, **3**:245b; **13**:97a
experimental
eighteenth-century, **11**:99b
mid nineteenth-century, **2**:129b; **5**:566a; **12**:295b; **13**:451b
late nineteenth-century, **2**:599a; **4**:296b; **5**:269b; **9**:82a–83a, 86a; **11**:346b–347a
early twentieth-century, **4**:154a–155b; **5**:41b; **8**:447a–b; **15**:454a–455a
mid twentieth-century, **5**:43a
geographical, **11**:178a
handbooks
nineteenth-century, **3**:463b; **4**:277a–b
theories of
seventeenth-century, **12**:101b–102a
eighteenth-century, **2**:122b; **12**:602a–b; **13**:564a
nineteenth-century, **2**:508a–b; **6**:73a
cellular, **14**:41a–b
Chinese, **15**:739a
monistic, **2**:508a
see also Anatomy, pathological; Disease; Epidemiology; Histology, pathological; Neuropathology; Nosology; Paleopathology; specific diseases
Patin, Gui, **2**:317a; **4**:135a; **7**:536a; **9**:497b, 498a; **11**:466b; **12**:84b
Patricius, Franciscus
see Patrizi, Francesco
Patrizi, Francesco (also Patrizzi or Patricio) (1529–1597), 10:416–417; **1**:447b; **4**:158b; **9**:316b; **13**:278b
Patterson, Sydney Wentworth, **12**:618a
Patterson, Thomas Stewart, **9**:245b, 246a
Pattinson, John, **13**:7b, 8a
Paul V, Pope, **5**:242b; **12**:105b
Paul of Aegina (*fl.* A.D. 640), 10:417–419; **1**:604a; **4**:106b; **11**:312a, 444b; **12**:542a(n)
editions of works, **3**:534a
Epitome, **12**:81a
De re medica, **1**:68b; **3**:534a
works translated, **2**:537a; **5**:586a; **8**:361a
Paul of Alexandria (*fl. ca.* A.D. 378), 10:419; **10**:425b; **15**:631b(n)
Introduction, **8**:191b
Paul of Middelburg, **1**:382b; **12**:134a,b
Paul of Venice (*ca.* 1370–1429), 10:419–421; **13**:349b
De compositione mundi
editions of, **4**:237b
on local motion, **1**:100b
Summa philosophiae naturalis
editions of, **4**:237b
Pauli, Simon
see Paulli, Simon
Pauli, Wolfgang (1900–1958), 10:422–425; **4**:294b*; **12**:44b*, 69b
acquaintances of, **1**:352b; **2**:245b, 249a; **7**:617a; **10**:214a; **12**:529b; **15**:25b, 43a
atomic structure, **2**:247b; **5**:160b
exclusion principle, **2**:247b, 248a; **4**:578a; **15**:49b
beta decay, **4**:579b; **9**:261a
electrodynamics, **4**:321a; **9**:377a
philosophy of science, **7**:192a
quantum theory, **2**:248a; **4**:294a; **15**:42a,b
Pauling, Linus, **1**:319b; **5**:589b; **7**:469b; **8**:47a, 186a, 292a; **12**:419a,b; **14**:440a; **15**:63b
Pauliśa (*fl.* fourth or fifth century A.D.), 10:425–426; **10**:419b
Pauliśasiddhānta, **7**:363b(n); **8**:46b; **13**:582a; **15**:540b, 543b, 545a–550b *passim,* 580b, 597a, 598b, 605b, 607b–608a, 609a,b
Paulli, Simon (1603–1680), 10:426–427; **2**:317a; **13**:30a
Paulsen, Friedrich, **11**:611a
Pauly, August (1850–1914), 10:427–428
Pavesi, Angelo, **3**:49a
Pavia, University of
in late eighteenth century, **14**:72a, 79a
anatomy at, **12**:136b
medicine at, **11**:99a
museum, **15**:473a
natural history at, **12**:556a–b, 560a–b
pathology at, **9**:86a
Pavlov, Aleksei Petrovich (1854–1929), 10:428–431; **8**:472b*
Pavlov, Ivan Petrovich (1849–1936), 10:431–436; **8**:472b*; **10**:621a
acquaintances of, **7**:32a; **10**:22a; **12**:95a; **15**:74b
digestion, **1**:537a–b, 544a; **6**:225b–226a
influence of, **2**:597b; **6**:226b; **14**:550b
influences on, **12**:270b, 271a

neurophysiology, **3**:89a
students of, **10**:220a
Pavlovsk
observatory at, **14**:356b
Pavón y Jiménez, José Antonio (1754–1840), 15:470–471; **4**:156b; **11**:605b
Pavy, Frederick, **2**:27a, 29a; **15**:109b, 110a
Payen (surgeon), **5**:288b; **10**:490b
Payen, Anselme (1795–1871), 10:436; **3**: 238a*, 317a*; **8**:117a, 342b, 349a; **10**:532a
Payer, Julius von, **7**:511a
Paykull, Gustaf (1757–1826), 10:436–437
Payne, Fernandus, **9**:520b
Payne, William Wallace, **6**:28b
Payne-Gaposchkin, Cecilia, **3**:50b; **12**:22a, 349b
Pea
genetics, **3**:422a; **7**:409a; **9**:278b–282a
Peach, Benjamin, **5**:337a; **8**:33b
Peach tree
diseases of, **12**:468a
Peachey, Stanley John, **11**:85b, 86a–b
Peacock, George (1791–1858), 10:437–438; **4**:35a; **6**:324a; **7**:551a
algebra, **6**:95b; **14**:303b
on contemporary scientists, **14**:493a, 565a
and education, **1**:354b; **14**:293a, 500a
Peale, Charles Willson (1741–1827), 10: 438–439
paintings, **15**:472a
Peale, Rembrandt (1778–1860), 15:471–472
Peale, Titian Ramsay (1799–1885), 10:439–440; **6**:120a; **8**:615b; **12**:133a
Péan de Saint-Gilles, Léon (1832–1862), 10: 440–441; **2**:68b–69a; **15**:461b
Peano, Giuseppe (1858–1932), 10:441–444; **15**:199a*
acquaintances of, **2**:593b; **10**:605b; **13**: 550b
analysis, **1**:590b
Calcolo geometrico, **15**:195b
on contemporary science, **10**:606a
correspondence of, **3**:58a*; **11**:413b
critics of, **11**:60b, 61a
geometry, **1**:76b; **10**:344b
axiomatization, **7**:616a
vectors, **2**:594a
historiography, **13**:266b
influence of, **2**:513a; **8**:473a; **9**:502b; **12**: 217a; **13**:599b; **14**:304a
logic, **5**:152b; **8**:159b, 590b; **10**:274a; **12**:12a,b, 14a
measure theory, **7**:167b
number theory, **2**:352a; **12**:15b
Pearce, Richard Mills, **3**:447b
Pearl, Raymond (1879–1940), 10:444–445; **10**:472a*–b*
Pearls, **13**:331a; **15**:663a(n)
artificial, **11**:331a
medical use of, **10**:237a
optical properties, **11**:267a
Pearson, Egon Sharpe, **5**:477a; **10**:465a; **14**:382b, 383a
Pearson, Fred, **9**:374a; **10**:473a
Pearson, George (1751–1828), 10:445–447; **5**:92b; **6**:521a
Pearson, Karl (1857–1936), 10:447–473; **2**: 211b; **3**:323a*; **8**:599a, 603b; **15**: 402a*
acquaintances of, **10**:481b; **14**:573b
and Bateson, **1**:506a
biometrics, **14**:251b, 252a
correspondence of, **15**:387b*
critics of, **10**:485a; **14**:574a
and Galton, **5**:266a
heredity theory, **5**:10b, 266b
influence of, **12**:461a, 532b
philosophy of science, **4**:374a
statistics, **3**:590b; **5**:8a
students of, **5**:476b; **10**:444b; **14**:382b
Pearson, Richard, **6**:577a
Pease, Francis Gladhelm (1881–1938), 10: 473; **4**:279b; **6**:30b; **9**:374a
Pease, Michael, **11**:212a
Peat
analysis of, **8**:147b; **14**:132b
effects on man, **6**:563b
formation of, **8**:264a; **13**:9b
fossils in, **15**:472a
origin of, **10**:322a–b; **14**:298a
Peaucellier, Charles, **3**:229a; **9**:157a
Pebrine, **1**:417b–418a; **10**:372b–376a
Pecham, John (*ca.* 1230–1292), 10:473–476; **1**:197a; **4**:157b; **12**:62a, 428b; **15**: 264a
optics, **2**:193b; **14**:459b, 460a
Perspectiva communis, **6**:197a, 275b; **14**:458b, 461a
editions of, **3**:280b
Pechlin, Johannes Nicolaus, **10**:567b
Pechmann, Hans von, **9**:402a; **12**:418b
Peck, William Dandridge, **1**:43b; **11**:218a
Peckham, John
see Pecham, John
Pecopteris, **5**:498a
Pecquet, Jean (1622–1674), 10:476–478; **2**: 379b; **5**:288b; **9**:498a; **10**:334b; **11**: 507b
anatomy, **1**:316b, 482b; **4**:267b; **5**:327b; **9**:63b; **11**:467a–b; **13**:31a
experiments of, **13**:445a
optics, **9**:115a
physiology, **13**:46b
Pectic acid, **2**:386a
Pectin, **9**:79a
Pedagogy
see Education
Pediatrics
medieval Arabic, **11**:239a
Renaissance, **11**:33b
eighteenth-century, **5**:411a,b; **14**:322a
nineteenth-century, **4**:403a–405a
twentieth-century, **5**:378a
Pediments (geol.)
origin of, **2**:549a; **7**:144b
Peebles, Phillip James Edwin, **5**:273a
Peel, Robert, **2**:567a; **7**:146a; **11**:36b; **12**:524b
Pegmatite, **2**:486b; **4**:597b, 598b, 599b; **7**: 249b
Pegram, George Braxon, **4**:581b
Pei Wen-chung, **2**:172a; **13**:275a
Peibla, **7**:13a
Peierls, Ronald Frank, **5**:103b; **7**:617a
Peirce, Benjamin (1809–1880), 10:478–481; **2**:285b, 366a; **14**:295a*
acquaintances of, **2**:368b; **8**:2b; **14**:448b
algebra, **3**:166a; **10**:483b
critics of, **2**:284b
and *Lazzaroni,* **1**:364b; **5**:479b
Linear Associative Algebra, **10**:483b
Peirce's criterion, **10**:480a
students of, **5**:479a; **10**:33b
and U.S. Coast Survey, **4**:590b–591a; **11**:113b
Peirce, Benjamin Mills, **10**:479a
Peirce, Benjamin Osgood, II (1854–1914), 10:481–482; **2**:217a, 460a; **10**:244b
Peirce, Charles Sanders (1839–1914), 10: 482–488; **6**:400b*; **8**:604b; **10**:538b*
acquaintances of, **13**:218a
influence of, **12**:217a; **13**:550b
influences on, **3**:376b
logic, **2**:297b; **4**:36b; **6**:83b; **8**:159b; **10**: 442b; **11**:107a; **12**:12a, 14a,b
philosophy, **2**:461a; **7**:68a–b
Peirce, James Mills, **2**:217a; **10**:479a
Peiresc, Nicolas Claude Fabri de (1580–1637), 10:488–492; **5**:247b, 284b, 285a; **6**:548b
acquaintances of, **2**:83a; **5**:285b; **7**:375a; **9**:528a; **15**:69a
astronomy, **3**:528b
correspondence of, **2**:13a; **5**:408b; **9**:316b, 598a
Peithon, **12**:314a
Pekelharing, Cornelis Adrianus (1848–1922), 10:492–493; **4**:311a; **9**:457a*; **12**:225a*
Peking man
see *Sinanthropus pekinensis*
Peking Union Medical College, **2**:171b; **14**:523b
Pelagianism, **2**:395b
Pelargonin, **14**:412a
Pelecus, **10**:16b
Pelecypods
classification of, **3**:536a–b
Pelée, Mount, **7**:548b–549a
Peletier, Jacques (1517–1582), 10:493–495; **2**:618b; **4**:449a,b; **14**:20b
Commentarii tres, **7**:133b
In Euclidis elementa demonstrationum, **3**:311b
geometry, **10**:338b(n); **14**:21b
Pelican, **10**:520a
Péligot, Eugène, **3**:10b; **4**:554a; **8**:335a
Pell, John (1611–1685), 10:495–496; **6**:598a; **14**:376a(n)
acquaintances of, **3**:360a; **5**:606b; **6**:141a, 445b; **8**:149b, 161a; **9**:310a, 554b; **14**:365a, 377b(n)
correspondence of, **6**:128a, 141b; **7**:195b*; **14**:51b, 377b(n)
critics of, **3**:63b
Idea of Mathematics, **5**:607a
mathematics, **8**:162b; **14**:150a
Pellian equations, **4**:117a, 125a; **10**:238b
Pellacani, Paolo, **10**:205b
Pellagra, **1**:360a; **3**:97a,b; **13**:448b
elucidation of, **5**:451b–452b
treatment of, **4**:358a; **5**:209a
Pelletier, Bertrand (1761–1797), 10:496–497; **10**:326a; **12**:287b*
chemistry, **8**:331b, 332a
metallurgy, **3**:560b
pharmacy, **11**:562a
Pelletier, Félix, **3**:606b
Pelletier, Pierre-Joseph (1788–1842), 10: 497–499
and Caventou, **3**:159b
chemistry, **4**:243a; **5**:91a; **9**:8a; **11**:494b; **14**:156a–b
students of, **13**:499a
Pelmatozoa, **1**:507a; **7**:61a
Pelops, **5**:228a,b
Pelouze, Théophile-Jules (1807–1867), 10: 499; **1**:416a; **8**:55b
acid theory, **8**:334a, 339a, 340a,b, 341b
acquaintances of, **2**:25a, 586b; **3**:193a; **5**:157b, 158a; **9**:401b
and Dumas, **8**:340a, 342a, 344a
and Liebig, **8**:336a, 337a, 340a, 342a; **10**:499a
metallurgy, **14**:493a(n)
students of, **1**:471a; **2**:63b; **10**:440b
Peltier, Jean Charles Athanase (1785–1845), 10:499–500
Peltier effect, **6**:51b; **7**:65b; **8**:188a; **11**: 50b, 446b
Peltogaster
neurology, **4**:12a
Pelvis
anatomy, **4**:173a; **14**:126b
female vs. male, **1**:619b; **10**:598b
Pelycosaurs, **2**:504b

Pemberton, Henry (1694–1771), 10:500–501; **3**:537b; **4**:45a; **5**:511b; **10**:61a, 85a, 85a(n)
View of Sir Isaac Newton's Philosophy, **5**:130b; **10**:88a(n)
Pemphigus, **10**:19a
Pena, Jean, **1**:81a
Pena, Pierre, **8**:435b
Pénaud, Antoine, **2**:439a
Pencil (math.), **3**:165a, 213b; **5**:368b; **8**:430a–b; **10**:297b; **13**:16b, 18b
conic, **3**:153a; **8**:250b; **10**:573b; **11**:44b; **12**:75b, 225b; **13**:17a; **14**:465b
Penck, Albrecht (1858–1945), 10:501–506; **5**:339a; **7**:144b; **9**:149b; **11**:439b, 440a, 441a*, 524b; **13**:477b
Morphologie der Erdoberfläche, **3**:595b
Penck, Walther (1888–1923), 10:506–509; **7**:144b; **10**:503b, 505a
Pendulum
early seventeenth-century study, **1**:424a; **5**:240a,b, 246a; **6**:255a; **11**:411b
late seventeenth-century study, **2**:311a; **3**:183b; **5**:528b–529a; **9**:114b, 115b–116b; **10**:70a, 75b; **14**:510a
eighteenth-century study, **2**:40b–42a; **3**: 282b; **4**:88b
nineteenth-century study, **5**:112a*; **10**: 483a; **13**:75b–76a, 78a; **14**:569a
twentieth-century study, **8**:124a
as analogy for planetary motion, **6**:515a
astronomical, **9**:554b
ballistic, **9**:504b; **11**:493b, 494a; **13**:350b
compound, **6**:605a, 605b–606a; **9**:116a–b; **11**:488b
conical, **10**:61a; **11**:207a
continuous oscillation, **8**:388a
convertible, **10**:483a
cycloidal, **2**:507b; **6**:605a; **10**:315a; **11**: 400b
damping of, **9**:217b
double, **1**:194b; **2**:42a
and electricity, **5**:517a
finite amplitude, **6**:164a
in fluid media, **15**:200a
Foucault, **3**:404b; **4**:592a; **5**:86a–b, 307b
theory of, **14**:617b
friction effects, **13**:75b–76a
in gravimetry, **9**:180b–181a; **13**:605b–608b
horizontal, **10**:231b; **14**:629a
in hydrodynamic measurement, **3**:182b; **9**:119b
isochronism, **3**:615b; **5**:246a; **6**:255a
laws of, **3**:182b, 183a; **9**:319b
in machines, **8**:210b
mercurial, **13**:470b
momentum, **3**:562a
periods of oscillation, **1**:194b; **11**:472a
and amplitude, **6**:604b–605a; **14**:254b
and latitude, **9**:119b
and length, **3**:182b; **6**:605a, 607b; **9**:97b, 554b
measurement of, **8**:388a
and temperature, **14**:254a–b
and time of free fall, **6**:605a,b
Repsold, **8**:504a
resistance to, **10**:75b; **14**:569a
reversible, **6**:241a; **7**:262b–263a; **10**:219b
seconds', **2**:299b, 329b; **6**:44b; **8**:351a; **9**:318a, 319b
latitude corrections, **2**:389a
length of, **2**:102a, 135b; **3**:103a, 116a; **5**:79a, 435b; **6**:606b, **7**:263a; **8**:504a; **9**:554a; **10**:542a; **11**:424a–b; **12**:234b; **13**:495b–496a; **14**:569a; **15**:271b, 287b, 320b–321b, 334a, 339a, 348b, 352b
spring, **11**:378a
standardization of, **1**:403a; **2**:299b
support flection, **8**:504b; **10**:483a
tautochronism, **6**:604b–605b; **9**:115b; **10**:315a
temperature effects, **5**:491a
theory of, **5**:86b; **12**:118a; **15**:484a
from two points, **2**:368b
see also Clock, pendulum
Peneplain, **3**:593b, 594b, 595b; **8**:99b; **9**:149b, 150b
Peneroplidae, **11**:386a
P'eng Che, **8**:314b
Penicillin, **4**:154b, 156a
clinical trials, **5**:42b
discovery of, **5**:30a
manufacture of, **5**:30a, 42b–43a
purification of, **5**:42a–b
Penicillium, **2**:437a; **9**:78b
cultivation of, **5**:30a, 42b; **10**:363b
and fermentation, **10**:360b, 377a, 381b
Penington, John, **11**:616b–617a
Peninsular War, **9**:140a
Penn, William, **8**:459a
Pennant, Thomas (1726–1798), 10:509–510; **3**:291b; **5**:76b; **12**:516b; **14**:299b
Indian Zoology
editions of, **5**:77a
Second Tour of Scotland . . . , **4**:549a
Penney, William, **3**:430a
Pennsylvania
agriculture, **5**:350a
geography, **4**:492b
Geological Survey, **5**:350a; **6**:564b; **8**: 260b–261a
geology, **3**:593b; **11**:504b, 505a
paleobotany, **8**:264a–b
Pennsylvania, University of
in eighteenth century, **5**:130a; **11**:616b; **14**:456a
medical school
chemistry at, **6**:115a
museum, **14**:504a
pathology at, **5**:40a
veterinary medicine, **7**:287a
zoology at, **9**:496a
Pennsylvanian period
paleobotany, **14**:298a
paleontology, **6**:594b; **9**:255b
Penny, Frederick (1816–1869), 10:510–511; **13**:500a
Penny, Thomas, **9**:440a,b; **14**:507b
Penny Cyclopaedia, **4**:35b
Pensa, Antonio (1874–1970), 15:473
Pentacrinus, **13**:355a
Pentagon, **3**:128b*
center of gravity, **8**:224a–b
diagonal of, **11**:222a
inscribed, **4**:610b
perimeter of, **13**:19b
sides of, **4**:609b–610a
stellated, **2**:360b; **9**:610b
types of, **5**:409a
Pentagram, **11**:221b
Pentamethylethanol, **2**:624a
Pentanes, **2**:622b
Pentastemon, **4**:401b
Pentastomum, **8**:270b
Pen-ts'ao p'in hui ching yao, **8**:392b, 394a, 397a*
Penyafort, Saint Ramon de, **8**:547b, 548a
Penzias, Arno Allan, **5**:273a
Pepsin, **7**:466b; **8**:342b; **10**:220a; **15**:15b
discovery of, **12**:241a
isolation of, **13**:152b
protein nature of, **10**:493a
purification of, **2**:531b–532a
Pepsis, **1**:261b
Peptides
analysis of, **13**:574a
physical chemistry, **3**:336a; **13**:421a
structure of, **15**:257a
synthesis of, **14**:247b; **15**:15b
Peptones, **14**:66b
Pepys, Samuel, **2**:507b; **4**:496a; **5**:520b; **6**: 141a, 487b; **10**:45a; **12**:445b–446a, 457b, 583a; **14**:147b, 362a, 371b–372a
Pérard, Albert, **5**:583a
Perception
eccentric, **8**:7b
natural, **5**:426b
physiology of, **2**:367a
psychology of, **2**:193b; **9**:562b
theories of
ancient Roman, **8**:538a
seventeenth-century, **5**:426b; **8**:157b, 437b–438b; **9**:98a; **12**:312a
eighteenth-century, **6**:557a–558a
nineteenth-century, **6**:81a, 248a; **2**: 448a; **15**:127b
twentieth-century, **7**:68b; **2**:10a–b
see also Sensation
Perchloric acid, **6**:109a
Percival, John, **15**:505b
Percival, Thomas, **3**:539a; **6**:283a–b, 284a; **10**:108a
Percussion, mechanical, **2**:311a; **8**:610a; **10**:250b; **14**:150a
center of, **9**:116a–b, 117a; **13**:266b
heat of, **10**:59b
Lagrange's equations, **1**:570b; **7**:565a
see also Collision, Impact
Percussion, medical, **3**:427a; **10**:613a; **12**: 202b, 451a
development of, **1**:332b–333a; **12**:450a–b
Percy, John (1817–1889), 10:511–512; 11: 485b
Pereira (or Pererius), Benedictus (1535–1610), 10:512–513; **2**:601b
Pereira, Duarte Pacheco (*ca.* 1460–1533), 10:513
Pérès, Joseph Jean Camille (1890–1962), 10:513–514; **2**:466a
Pérez de Vargas, Bernardo (*ca.* 1500–1533), 10:514–515
Perfection, principle of, **8**:156a,b
Perfumes, **9**:401b; **13**:511b; **15**:649b, 650a,b
Perfusion, **8**:541a
chemical effects, **6**:38a, 43b
concept of, **8**:133b
of heart, **9**:142b, 295b; **11**:462b
of kidneys, **1**:534a; **2**:237b; **3**:91b
of liver, **2**:29a; **4**:360a
of muscles, **5**:184b
of severed head, **2**:525b
of thyroid gland, **3**:91b
Pergamum, **5**:227b–228a
Pericarditis, **1**:399b
treatment of, **4**:621a
Pericardium, **1**:619b; **15**:479b
Pericarp, **5**:217a; **13**:332b
Pericles, **1**:149a; **11**:23b
Pericranium, **5**:586a
Peridinia, **4**:207a
Périer, Florin, **8**:267b; **10**:330b, 333b, 334a
Périer, Jacques-Constantin, **2**:104a
Périgord
prehistory, **8**:44a
Periodic acid
discovery of, **9**:18b
Periodic table of the elements
see Elements, chemical, periodic system
Periodicals, scientific
abstracts, **12**:573a
agricultural, **2**:322a

anatomical
nineteenth-century, **5**:470a; **9**:456b; **11**:295b, 492a; **13**:556b; **15**:166b
twentieth-century, **2**:255b; **9**:57a
anthropological
nineteenth-century, **2**:479a; **4**:286a; **9**:86a, 538b, 539a
twentieth-century, **2**:347a; **5**:294a; **6**:527b; **9**:145a
archaeological, **14**:517a
astronomical
eighteenth-century, **7**:598b
nineteenth-century, **4**:162a, 176b*, 370a; **5**:199b, 479b; **6**:28b, 103a; **8**:369b; **9**:551b; **10**:119b; **12**:234b; **14**:583a
twentieth-century, **1**:430a; **3**:472b; **4**:27a; **6**:508a
astrophysical, **13**:232b
biochemical
nineteenth-century, **6**:505b
twentieth-century, **1**:10b
biological
nineteenth-century, **1**:418a; **4**:12b; **9**:143a, 546b; **10**:558b; **12**:193a; **15**:93a
twentieth-century, **3**:148b, 245b; **4**:220b; **7**:456a; **8**:589b; **10**:445a; **11**:19b, 374a
botanical
nineteenth-century, **2**:290b; **3**:611b–612a; **6**:491b–492a, 494b, 495a*; **9**:345b, 442a, 600a; **11**:155a; **13**:459b, 460a; **15**:82b, 147b
twentieth-century, **1**:396b; **2**:477a; **3**:188a; **4**:584a; **7**:103a*; **8**:512b–513a
chemical
eighteenth-century, **1**:44b; **3**:465a–b; **5**:90b, 602b; **8**:81a
early nineteenth-century, **1**:64b, 346b; **2**:95b, 96b, 564b–565a; **5**:558b; **12**:254a; **15**:171b–172a
mid nineteenth-century, **2**:95b, 96b, 340b*; **3**:47a, 475a; **4**:394b, 395b*; **5**:12b, 164b–165a, 370a, 432a; **7**:464a; **8**:55b; **9**:401b, 558a; **14**:394b
late nineteenth-century, **1**:536b; **3**:476b; **10**:157a; **11**:371a; **12**:163b; **13**:576a; **15**:456a
twentieth-century, **1**:361b, 431a; **2**:429a; **6**:375a; **7**:323a; **13**:3b
engineering
eighteenth-century, **4**:501b
nineteenth-century, **14**:617b
general
seventeenth-century, **2**:311b; **3**:183b; **5**:259a; **6**:487a; **10**:201b–202a; **12**:84b–85b
eighteenth-century, **3**:114a, 283b, 350b, 542b, 545a; **5**:58b, 75b, 509a–b; **8**:72a; **9**:141b; **10**:108b, 319b, 603a; **13**:179a, 425b; **14**:120a; **15**:53a, 349b
nineteenth-century, **1**:405a, 502a, 583a; **2**:60b, 103a, 420a, 439a; **3**:16b, 80a,b, 130a,b, 148b, 475b, 556b; **4**:242b, 401a; **5**:50a, 386a, 389a; **6**:1b; **8**:35b–36a, 443a; **9**:538b; **10**:108b, 195b, 272b; **11**:163a; **12**:175a, 289a; **13**:413b–414a; **15**:483a, 484a
twentieth-century, **3**:130b; **4**:597b; **5**:201a
genetics, **1**:505b; **4**:271a; **11**:212a
geological
nineteenth-century, **1**:173b; **2**:93a, 128a, 341b; **3**:70a, 623b; **5**:14b; **6**:143a, 456a; **8**:245b; **11**:103a; **14**:439b; **15**:152a
twentieth-century, **2**:191a; **3**:190a; **4**:144b; **8**:99b, 370b; **14**:442a
history of science
nineteenth-century, **2**:284a
twentieth-century, **12**:108b; **13**:142a
linguistics, **2**:212b
mathematical
eighteenth-century, **2**:56b; **3**:537b, 538a; **6**:403b–404a
nineteenth-century, **1**:13b, 14b, 65b; **2**:299a, 425a, 469b, 479a–b; **3**:53b, 134a, 177b, 222b, 314a, 466a–b; **4**:84a; **5**:367b, 414a,b, 522b, 569b; **8**:176a, 324a, 382b; **9**:81a, 426b–427a; **10**:25a, 442b; **11**:236a; **13**:218a
twentieth-century, **1**:427b; **2**:191b; **4**:373b, 560b; **5**:201a, 608a; **7**:72a, 411b; **9**:72a, 248a, 419a; **11**:394b; **12**:188a
medical
seventeenth-century, **1**:483a
eighteenth-century, **2**:122a; **3**:427a, 465a; **5**:90b; **11**:48a
nineteenth-century, **1**:10b, 360a, 485b*; **2**:464a, 508a; **3**:463a; **4**:162b; **6**:268b; **11**:301a; **14**:40b
twentieth-century, **5**:40b; **6**:233a; **8**:295a, 296a; **15**:97b
metallurgical, **5**:576b; **8**:119b; **12**:126b
meteorological, **6**:97a; **11**:558a
microbiological
nineteenth-century, **2**:164b; **3**:338a; **4**:211a; **6**:73a; **12**:232a
twentieth-century, **5**:453b; **7**:171a; **9**:328b
mineralogical
nineteenth-century, **5**:557b; **13**:476b
twentieth-century, **5**:456b*, 562b; **12**:573a
mining, **12**:513b
natural history, **1**:127b; **3**:154b; **9**:536b; **10**:122a
navigational, **9**:163a
neurological
nineteenth-century, **4**:593b; **6**:320b
twentieth-century, **5**:208a
oceanographic
eighteenth-century, **5**:77a
twentieth-century, **1**:92b
paleontological, **1**:18b*; **2**:189b; **3**:520a; **7**:61b; **9**:346a
parasitological, **2**:534a
pharmaceutical, **3**:6a; **4**:134a; **13**:466a; **15**:172b
philosophical
eighteenth-century, **10**:182b
twentieth-century, **8**:517a
photographic, **4**:283a; **5**:196b
physics
eighteenth-century, **1**:44b; **5**:532a
nineteenth-century, **1**:346b; **4**:189b; **13**:126b
twentieth-century, **1**:196a; **4**:522a; **5**:332a; **8**:103a; **11**:265b; **12**:613b; **15**:490b
physiological
eighteenth-century, **11**:364a
nineteenth-century, **4**:201a; **5**:83a; **9**:9b, 456b; **13**:556b
twentieth-century, **4**:372b*; **12**:356a; **14**:2b
private
nineteenth-century, **3**:134a; **13**:372b
twentieth-century, **3**:520a
psychological
nineteenth-century, **2**:131b; **6**:53b
twentieth-century, **1**:580b; **6**:53b; **14**:550b
public health
nineteenth-century, **7**:423a, 425b; **8**:272a–b; **10**:559b
twentieth-century, **5**:220a
veterinary, **7**:101a
zoological
nineteenth-century, **5**:52a; **7**:439b, 546a; **8**:271a; **12**:257b, 421a; **14**:314a; **15**:78b
twentieth-century, **3**:123a, 247b, 390b; **6**:134a
see also individual periodicals by title in main body of Index, and combined listing of all titles at end
Periodico di matematiche, **4**:373b
Periosteum, **6**:65b; **8**:34a–b; **9**:146a
role of, **5**:45a; **6**:183b
Peripateticism
ancient Greek, **1**:250a–b, 268a, 282a; **2**:590a; **4**:81b, 383a,b, 388a, 460b, 462b, 463b; **13**:94b, 328b–333b
ancient Roman, **1**:117b; **14**:607b
Byzantine, **13**:307b
medieval Arabic, **6**:191a, 192a; **7**:339b; **11**:249b, 250a; **13**:511b–512a
opposition to, **1**:27a, 410a
Renaissance, **8**:215b, 230a, 283a; **12**:151a; **15**:45b
opposition to, **1**:495a; **3**:60a; **12**:106a
seventeenth-century, **1**:126b, 391a; **7**:611b; **9**:47b, 62b
opposition to, **7**:195a
botany, **9**:455a; **10**:111b
dynamics, **1**:410a; **8**:230a
historiography, **4**:462b, 463b
meteorology, **2**:283a
optics, **6**:192a
philosophy, **14**:607a
physics, **9**:160a
Peripneumonia, **4**:386a
bovine, **2**:300b
Periscope, **5**:561b
Perisodactyls, **12**:38b
Perisporaceae, **9**:176a
Peristalsis, **4**:371b, 586a
mechanism of, **1**:537a; **15**:72b
Peristome, **6**:219b
Peritectonics, **5**:576b
Peritoneum
anatomy, **1**:619a; **3**:99b; **4**:172b
"gelatinous disease," **3**:490a
peritonitis, **1**:612a; **7**:556b
Perkin, William Henry (1838–1907), 10:515–517; **3**:84b; **6**:461b, 463a,b*; **14**:247b, 455b*
Perkin's mauve, **3**:84b; **10**:515b–516a
Perkin's synthesis, **10**:516b
Perkin, William Henry, Jr. (1860–1929), 10:517; **9**:131b; **15**:11b
acquaintances of, **5**:589a; **8**:31b; **13**:388a
chemistry, **11**:87a; **13**:388b
on contemporary scientists, **4**:400b*; **12**:237a*
students of, **3**:87a; **6**:185a,b; **7**:372a
Perkins, Jacob
steam engine, **3**:82b
Perkins, William (Puritan), **1**:133b, 134a
Permafrost, **10**:169b
depth of, **9**:375a
Permanganates
reduction of, **6**:110a
Permeability, magnetic, **11**:578a
complex, **1**:284a
Permian period
deformation, **7**:251b
glaciation, **11**:277a
naming of, **9**:584a

Permian period (cont.)
paleobotany, **3**:433b; **11**:373a; **12**:182a; **14**:600a
paleogeography, **7**:250a,b
paleontology, **2**:110a, 489a, 505a; **5**:296a, 486b, 441b, 486b; **9**:255a, 346a; **10**: 40b, 128a; **11**:386a; **14**:410a
rocks of, **2**:486a; **6**:457a
stratigraphy, **2**:109a, 505a; **4**:262a,b, 402b; **5**:155b, 156a; **6**:187a, 238b; **7**:253a; **8**:33b; **12**:514b; **13**:620b–621a
Permutations (math.), **3**:168a,b; **4**:502b; **5**: 262a, 263a; **9**:159a; **12**:375b–376a; **13**:216b
Perny de Villeneuve, Jean, **7**:500a
Peromyscus, **13**:151a
Péron, François (1775–1810), 10:517–518; **8**:266a
Peronospora, **3**:612b–613a
Pérot, Jean-Baptiste Gaspard Gustav Alfred (1863–1925), 10:518–519; **1**:612b; **4**:513b; **8**:551b, 581a
Peroxidase, **3**:259b; **6**:111b; **7**:273b; **10**:282a; **13**:152b
Peroxides
in green plants
assay method, **1**:363a
in oxidation mechanism, **1**:362a–363a
peroxide effect, **7**:323b
preparation of, **13**:313a
Perpendicular
see Normal (math.)
Perperot, Henri, **3**:401a
Perpetuum mobile, **4**:184a,b; **6**:608a; **12**: 497b; **14**:31a; **15**:443b, 464b
Perquin, Louis Henri Cyril, **7**:406b
Perraudin, Jean-Pierre, **3**:210a; **13**:604b
Perrault, Charles, **4**:49b, 135b; **5**:60a; **13**: 335b
Perrault, Claude (1613–1688), 10:519–521; **5**:61a; **10**:477a
acoustics, **8**:162b
anatomy, **3**:524b; **4**:267a–b; **5**:327b; **9**: 323a; **10**:478a
architecture, **3**:102a; **4**:37b
botany, **9**:92b, 114b
freezing, **9**:115a
influence of, **4**:26a, 135b
manuscripts of, **4**:267b
medicine, **4**:136a
optics, **9**:115a
zoology, **13**:526b
Perrault, Pierre (1611–1680), 10:521–522; **9**:119a, 532a
Perret, Auguste, **11**:430a
Perret, Frank Alvora, **7**:63a
Perrier, Carlo, **10**:136b
Perrier, Edmond (1844–1921), 10:522–523; **11**:235a*, 283b*
Perrier, François, **13**:37a
Perrier, Georges (1872–1946), 10:523–524
Perrimond, I. I., **15**:8b
Perrin, Jean Baptiste (1870–1942), 10:524–526; **2**:169b, 466a; **7**:157b; **8**:10a; **10**:276a*; **14**:633a
acquaintances of, **1**:519a; **7**:152a–b; **8**:9b; **14**:243b, 244a
on Brownian motion, **2**:170b; **4**:315a; **15**:464a,b
cathode rays, **8**:180b; **13**:366a
chemical kinetics, **13**:429b
on electrons, **3**:499b
Perrine, Charles Dillon (1867–1951), 10: 526–527; **7**:271b*
Perron, Oskar, **9**:126b
Perroncito, Edoardo (1847–1936), 10:527; **4**:198a
Perronet, Jean-Rodolphe (1708–1794), 10: 527–528; **3**:442b; **11**:163b
Perrotin, Henri Joseph Anastase (1845–1904), 10:528–529
Perseus (*fl. ca.* third century B.C.), 10:529–530
Perseus (constellation)
alpha Persei, **10**:232a
beta Persei
see Algol (star)
nebulae, **11**:553b*
novae, **1**:598a; **10**:526a
star clusters, **8**:582b
Perseus (ship), **9**:331a
Persia
botany, **7**:204b
geography, **6**:215a, 216a; **7**:204a–b
and Greece, **6**:316a
theology, **11**:250a
Persico, Enrico, **4**:577a,b, 578a
Person, Charles Cléophas, **11**:418a
Persona
definition of, **2**:233a
Personality
and culture, **2**:210a
faculty theory, **5**:253a–b
Personne, Jacques (1816–1880), 10:530
Persoon, Christiaan Hendrik (1761–1836), 10:530–532; **10**:365b
Persoz, Jean-François (1805–1868), 10:532; **2**:139b; **5**:297b; **8**:342b; **10**:436a
Perspective
ancient Greek, **4**:430a
medieval, **2**:193b, 435a; **4**:157b; **10**:475b
Renaissance, **1**:97a, 607a; **2**:534b–535b; **3**:559a*; **4**:259a–261a; **5**:113a; **8**: 234b; **9**:488b; **13**:48b
seventeenth-century, **2**:333b, 419a; **4**: 46b–50a; **9**:447b; **10**:103b–104a, 331a; **11**:508b
eighteenth-century, **7**:86b, 599b; **11**:142a–b; **13**:267a; **14**:589a
nineteenth-century, **2**:473a; **4**:248a, 502a*; **8**:62a*; **9**:474a
anamorphosis, **10**:103b–104a
curvilinear surfaces, **2**:333b
drawing machine, **14**:196b
inverse problem, **13**:48b, 267a
linear, **4**:259a
monocular, **2**:535a
theory of, **1**:373a; **13**:267a
triangle theorem, **2**:333b
Perspective pratique, **4**:48a
Perspicilla (astron. instrum.), **9**:336b
Perspiration, **6**:37b; **14**:322a
and age, **4**:136a
composition of, **4**:554b
insensible, **1**:610b; **3**:487a; **4**:136a; **6**:37b; **7**:503a; **12**:102b, 103b; **13**:557a
physiology of, **6**:24b
Persulfuric acid, **3**:85a–b; **5**:189b
Persulfuric anhydride
discovery of, **2**:70a
Perturbation theory
seventeenth-century, **10**:73b, 77b
eighteenth-century, **1**:313a, 401a; **4**:481b; **15**:350b–351a
early nineteenth-century, **4**:16b; **5**:302a; **15**:226a
mid nineteenth-century, **1**:53a; **4**:19b; **6**: 89a; **8**:384b, 530a; **10**:41b; **11**:207a
late nineteenth-century, **1**:156b; **3**:18b, 146b; **6**:522a; **10**:42a
twentieth-century, **1**:429a; **2**:236b; **9**: 449b; **10**:159a; **13**:139a, 153a; **14**:13b
asteroids, **1**:157a; **2**:442a; **4**:369b; **6**:539b; **9**:460b; **10**:528b; **11**:552a; **12**:17b; **13**:422b–423a
in atomic theory, **2**:246a; **4**:579a; **15**: 6b–7a
comets, **2**:433a; **3**:283a–b; **4**:369b, 514b; **5**:210a; **7**:480b, 564a–b, 581a; **10**:30b, 42a; **13**:422b–423a; **15**:354b–355a
direct vs. indirect effects, **2**:101b
in geodesy, **6**:240b
moon, **8**:530a
earth, oblateness of, **2**:600a
sun, influence of, **2**:600a; **10**:76b, 78a
planets, **4**:547b; **6**:94b, 103b; **7**:481a, 564a; **8**:277a–278a, 299b, 530a; **9**: 174b; **10**:34b–35a, 76b, 479b; **11**: 84a*; **13**:45a; **15**:323b–324a, 329a, 337a–338b, 343b, 348a, 349a, 353b, 356b, 484a–b
satellites, **12**:448b
sun
moon, influence of, **7**:544b
see also Mechanics, celestial
Peru
botany, **4**:156b, 603a; **11**:605b; **15**:470a–471a
conquest of, **5**:448a
exploration of, **4**:603a
geodesy, **7**:183a; **15**:270a–271b
medicine, **13**:541a–b
metallurgy, **1**:448a–b
meteorology, **1**:398a
Perutz, Max Ferdinand, **4**:522a; **7**:274a; **15**:63b
Peruvian bark
see Cinchona
Pessimism, **11**:325a
Pessina (Pešina), Ignaz, **11**:159a
Pest, University of
medicine, **12**:295b–296a
Pest control
biological methods, **8**:450a; **10**:406a–b
Pestalozzi, Johann Heinrich, **6**:294b; **11**: 214a; **13**:12b–13a, 13b, 16a
Pesticides, **9**:576b
Pesu (unit of measure), **15**:688b, 694b–696a
Petalite, **15**:7b
Peter I (the Great) of Russia, **2**:14b; **4**:23a; **6**:68a, 148b; **7**:374a; **8**:129b, 150b, 471a; **9**:17a; **11**:399b; **12**:41a
Peter Abano
see Abano, Pietro d'
Peter Abelard
see Abailard, Pierre
Peter of Ailly
see Ailly, Pierre d'
Peter of Alexandria, **7**:131a; **8**:280b, 281a
Peter Bonus
see Petrus Bonus
Peter of Dacia
see Peter Philomena of Dacia
Peter Deacon, **3**:393a
Peter Lombard, **7**:117b
Sentences, **10**:171b, 172a,b
commentaries on, **2**:609a; **3**:171b; **4**: 254b; **5**:113b, 115a, 403a; **9**:33a, 136b; **13**:534a
Peter of Mantua, **4**:237b; **13**:349b
Peter Nightingale
see Peter Philomena of Dacia
Peter Peregrinus (Pierre de Maricourt) (*fl. ca.* 1269), 10:532–540; **1**:380b, 381b, 606a; **3**:24b
Letter on the Magnet, **5**:397b, 399b; **6**: 145a*
Peter Philomena of Dacia (Petrus Dacus, Petrus Danus, Peter Nightingale) (*fl.* 1290–1300), 10:540–542; **6**:123b
commentaries of, **3**:512a; **12**:63a
Peter of St.-Omer, **10**:541a

Peter of Spain, **2**:603b, 604b; **4**:462a; **7**:117a; **10**:420b; **14**:391b
Peters, C., **5**:458a; **6**:304a
Peters, Carl F. W. (1844–1894), 10:542
Peters, Christian August Friedrich (1806–1880), 10:542–543; **6**:513b; **11**:378a
Peters, Christian Heinrich Friedrich (1813–1890), 10:543; **5**:199b; **11**:204a*
comets, **12**:160b–161a
Peters, John, **13**:574a
Petersburg mathematical school, **3**:225b–227a; **5**:508b; **7**:514b; **9**:126a
Petersburg paradox, **2**:42a, 57a, 577b; **8**:321a–b; **13**:612a
Petersen, Carl, **9**:290a,b
Petersen, Julius (1839–1910), 10:544; **10**:117b
Peterson, Elbert Axel, **13**:421b
Peterson, Karl Mikhailovich (1828–1881), 10:544–545; **8**:559a; **9**:409b
Peterson surfaces, **10**:545a
Peterson's problem, **4**:287b
Petersson, Wilfried Hans Henning, **6**:214b; **7**:399a
Petit, Alexis Thérèse (1791–1820), 10:545–546; **1**:201a; **3**:80b; **4**:238b; **5**:271b
and Dulong, law of, **1**:348b; **2**:93b; **3**:48b; **4**:239b–241a, 316a; **9**:347b; **10**:545b–546a; **13**:10b–11a; **14**:38b, 39a; **15**:377a
approximate nature of, **11**:353a,b
physics, **1**:201b; **4**:242a*; **10**:3b
Petit, Pierre (*ca.* 1594–1677), 10:546–547; **1**:341b*; **2**:317a; **9**:498b; **13**:335a
vacuum, **9**:16a,b; **10**:332b, 342b
Petit mal, **4**:387a
Petiver, James, **1**:432a,b; **9**:17b; **13**:449a
Petosiris, Pseudo- (*fl.* second and first centuries B.C.), 10:547–549; **1**:38b; **4**:622a
Petre, Robert James, **3**:350a
Petreius, Johannes, **10**:245b
Petrie, (William Matthew) Flinders (1853–1942), 10:549–550; **6**:110b; **9**:489b; **11**:6a; **12**:181a; **13**:218b; **14**:504b; **15**:727b*
Petrifaction, **8**:242a
theories of, **10**:280b
Petrochemical industry, **7**:21b
Petrogenesis, **3**:434a; **4**:141b, 361b, 409b; **12**:543a; **13**:331a
assimilation, **4**:564a
enrichment, **4**:564a–b
reaction principle, **2**:370a
Petrographic provinces, **7**:5b; **8**:42a; **10**:126a
Petrography
late eighteenth-century, **14**:60b, 261a–262a
early nineteenth-century, **1**:95b, 210a; **2**:568b; **3**:411b, 412a; **4**:10a; **12**:514a
mid nineteenth-century, **1**:20a; **3**:434a; **5**:533a–b; **9**:620a–b; **11**:269b; **13**:476a–b
late nineteenth-century, **1**:51a–52a, 472a, 551a–552a; **2**:289b, 346b, 486a–b, 500a; **3**:237a, 293a–b; **5**:49b, 89a, 212a–b, 562b–563a; **6**:14a; **7**:5a–b, 249a–250a; **8**:463a; **9**:367a–b, 593b; **11**:372a, 534b, 547a–b, 593b; **13**:148b, 273a; **15**:427b
twentieth-century, **1**:394b; **3**:161a–162a; **4**:42b; **6**:475b; **7**:112b–113a; **8**:285a–286a; **10**:125b–127a; **14**:59b, 60a
architectural, **3**:162a
experimental, **8**:285b
graphic representation of analyses, **1**:552a
laws of, **7**:249b
methods, **7**:249b
optical, **1**:472a; **5**:89a; **7**:5a, 112b–113a, 249a, 370b; **8**:41b–42b, 370a, 511b; **9**:315a, 367a; **10**:110a; **11**:209b, 539b, 547a–b; **12**:542b–543b **13**:572a; **14**:625a–b
regional, **8**:285a
sedimentary, **3**:161a–162a
theoretical, **8**:285a
see also Petrology; Rocks; specific geographic areas
Petroleum, **3**:89a; **4**:181a*; **9**:287b; **10**:499b; **13**:410a
artificial, **2**:3b–4a; **14**:602a
chemistry of, **8**:108a; **9**:131a–b, 148b, 608a–b; **12**:209a, 435b; **13**:243b
drilling for, **1**:528a; **8**:301a,b
false equilibrium, **4**:230a
medical applications, **9**:291b
occurrence of, **6**:238b; **8**:252a; **10**:429b; **12**:381a; **14**:132b
oil shale, **3**:610a; **6**:238b; **14**:298a
origin of, **2**:3b; **10**:171a
distillation theory, **1**:20b
inorganic theory, **9**:291b
organic theory, **2**:579a; **7**:44a; **10**:265b; **14**:124b, 298a, 602a
processing of, **2**:4a; **3**:437a*; **4**:554b; **9**:291b; **12**:47b, 435b; **14**:248a, 602a
byproducts, **14**:393a
cracking process, **3**:610a; **14**:601b, 602a
distillation, **15**:526b
fractionation, **14**:182b, 561a
reserves of, **3**:610a; **13**:525b
search for, **2**:466a; **4**:7b–8a; **6**:229b; **7**:44a; **8**:577a; **12**:436a–b
anticlinal theory, **1**:20b; **14**:301a
carbon-ratio theory, **14**:298a
Foraminifera, **3**:520a–b
seismography, **4**:8a
sulfur compounds of, **9**:608b
Petrologists' Club of Washington
founding of, **3**:484a; **4**:564b
Petrology
ancient Greek, **13**:330b–331b
late eighteenth-century, **1**:234a–b; **4**:549a; **12**:329a–b; **14**:132b
early nineteenth-century, **8**:246a, 594a–b, 615b–616a; **12**:514a
mid nineteenth-century, **2**:567b; **4**:66a, 140b–141a; **8**:30b; **11**:560b–561a; **12**:276b–277a; **13**:475b–476b
late nineteenth-century, **2**:289b; **3**:292b–293b; **4**:361b; **5**:89a–b; **6**:116a–b; **7**:370b, 473b–474a; **8**:511b; **10**:32b; **11**:181b; **12**:271b–272b, 520a; **13**:283a–284a, 476b; **14**:58a
early twentieth-century, **2**:369b–370b; **3**:484a; **4**:141a–142a, 262a, 408b–409b, 563a–564b; **5**:39a, 457a, 486a–487b; **7**:112b–113a, 548b–549a; **8**:41b–42b; **10**:124a–b, 125b–127a; **12**:194b, 272b–274b; **13**:285a; **14**:58b–59a, 440b, 515a–b
mid twentieth-century, **1**:394b; **6**:475b; **12**:195a
applied, **10**:126b–127a
experimental, **2**:369b–370b, 588a–b; **4**:141a–b, 563a–564b; **5**:89a–b; **9**:367a–b; **11**:539b; **12**:122b
handbooks, **11**:561a
phase-equilibrium studies, **2**:369b; **10**:125b
sedimentary, **5**:486a–487b
statistical, **9**:367a
structural, **1**:153b–154a; **3**:325b
theoretical, **8**:285b
see also Petrography, Rocks, Sedimentology
Petrov, Nikolay Pavlovich (1836–1920), 10:551–552; **10**:248a; **14**:106a
Petrov, Vasily Vladimirovich (1761–1834), 10:552–553
Petrova, Maria Konstantinovna, **10**:434b
Petrovsky, Ivan Georgievich (1901–1973), 10:553–554
Petroz, Claude Henri, **10**:498a
Petrus
see also Peter
Petrus Alphonsus, **1**:61a, 62b
Petrus Bonus (Bonus Lombardus or Buono Lombardo of Ferrara) (*fl. ca.* 1323–1330), 10:554–556
Petrus Camper
founding of, **2**:255b
Petruschky, Johannes, **7**:426b
Petrzhak, K. A., **7**:527a
Pettenkofer, Max Josef von (1818–1901), 10:556–563; **4**:403b; **9**:351b
biochemistry, **13**:452b
chemical elements, **9**:348b–349a
editions of works, **9**:348b
pathology, **2**:575b; **7**:425a, 427a
physiology, **14**:64b–66a; **15**:108b
students of, **5**:563b; **14**:249a
Pettersson, Hans (1888–1966), 10:563–564
Pettersson, Sven Otto, **1**:297a
Pettit, Edison, **10**:107b
Petty, William (1623–1687), 10:564–567; **9**:454a; **12**:461b*; **14**:179a
acquaintances of, **3**:482a; **5**:507a; **6**:141b, 482a; **14**:369a, 405a
and Royal Society, **14**:510b
technology, **12**:85a–b; **14**:371b
Petuntse, **5**:578b; **11**:330a–b
Petzetakis, Michel, **9**:506a
Petzval, Joseph Max, **4**:168a; **5**:144a
Peucer, Caspar, **2**:402a, 410b; **11**:366b; **13**:59a
Peurbach (Peuerbach), Georg (1423–1461), 15:473–479; **11**:349b
astronomy, **13**:324a
influence of, **3**:27b; **7**:118b, 120b, 269b; **10**:161b
influences on, **7**:119b, 120a; **14**:19a
and Regiomontanus, **11**:348b–349a, 350a
tables of, **1**:165b; **14**:275a
Theoricae novae planetarum, **1**:512b; **6**:197b; **7**:112b; **11**:351a
commentaries on, **11**:365b
editions of, **15**:154a
trigonometry, **7**:118a; **14**:471a(n)
Pevtsov, Mikhail Vasilevich, **11**:181a; **12**:301a
Peyer, Johann Conrad (1653–1712), 10:567–568
Peyer's patches, **2**:445a, 531b; **10**:567b; **11**:148a; **13**:32a
Peyrard, François, **4**:416a, 437b; **13**:322a–b
Peyssonnel, Jean André (1694–1759), 10:568–569; **14**:193b
Pezard, Albert (1875–1927), 10:569–571
Pezenas, Esprit (1692–1776), 10:571–572; **9**:555a*
Histoire critique de la découverte des longitudes, **10**:595b
Peziza, **6**:100a
Pfaff, Christoph (Christian) Heinrich, **3**:522b–523a, 523b; **7**:368a; **9**:578b
electric currents, **1**:347b; **8**:36b; **9**:152a; **10**:572b
Pfaff, Johann Friedrich (1765–1825), 10:572–574; **5**:301a, 304b; **9**:476b; **11**:545a; **12**:234a
Pfaffian equation, **3**:143a; **4**:370b; **8**:324b; **10**:574a; **12**:214a
integration of, **15**:198b

Pfaff, Johann Friedrich (cont.)
Pfaffian forms, **10**:573b–574b
problem of, **15**:198b
students of, **5**:299b; **9**:429a; **14**:204a
see also Systems, mathematical, Pfaffian
Pfalzpaint (or Pfolspeunt), Heinrich von, **2**:546b
Pfaundler, Leopold von, **6**:355a; **13**:578a
Pfeffer, Wilhelm Friedrich Philipp (1845–1920), 10:574–578
acquaintances of, **3**:421b; **7**:535a; **12**:58b; **13**:90a
cell physiology, **10**:257a; **13**:453a, 579a
students of, **2**:184b; **3**:30a; **4**:220a; **7**:395b
Pfeifer, Richard Arwed, **1**:570a; **5**:26b, 27a
Pfeiffer, Ludwig
Die Protozoen als Krankheitserreger, **1**: 439a
Pfeiffer, Paul (1875–1951), 10:578; 14:272b*
Pfeiffer, Richard Friedrich Johann, **2**:300a; **5**:564b; **7**:425b, 427a,b, 428a,b; **14**: 512b; **15**:521b
Pfitzner, Wilhelm, **13**:89a
Pfleiderer, Christoph Friedrich, **8**:305b
Pflüger, Eduard Friedrich Wilhelm (1829–1910), 10:578–581; **4**:201a; **5**:340a; **6**:505b; **11**:159b, 571b; **14**:2b
critics of, **6**:338b
electrophysiology, **4**:202a
embryology, **11**:572a, 573b, 575a*
metabolism, **5**:37b; **11**:427a; **14**:66a
Pflüger's law, **2**:111a; **10**:579a
philosophy, **5**:463a; **6**:59b
respiration, **4**:296b; **5**:150a; **13**:393b
Pflüger's Archiv
see *Archiv für die gesamte Physiologie des Menschen und der Tiere*
Pfuhl, Eduard, **7**:426a
Pfund, August Herman
Pfund series, **1**:425b
pH
of blood, **1**:534a
concept of, **12**:546b; **13**:228b
of lakes, **7**:184a
measurement of
apparatus for, **1**:534a; **5**:621a
errors in, **2**:170b
of soil, **2**:170b
Phacochoerus, **3**:521a
Phaeinos, **9**:339a
Phaeophyceae, **12**:125a–126a
Phagocytes, **1**:439a; **12**:423b
embryogeny, **9**:333b
naming of, **9**:333b
Phagocytosis, **1**:575b; **3**:588b; **5**:29b; **7**:455a; **8**:273a, 408a; **10**:407b, 493a
role of, **2**:165b; **9**:333a–334b
Phagotherapy, **6**:298b
Phalangers, **2**:504a
Phanerogams, **8**:354b; **10**:531b
analogies with cryptogams, **3**:29b; **6**:219a
embryogeny, **9**:600a
fertilization, **13**:89a
seeds of, **5**:217a
taxonomy, **13**:406a; **15**:148a
Pharmacognosy
Renaissance, **3**:413b–414b
Oriental
sixteenth-century, **8**:390b–395a
Pharmacology
ancient Roman, **3**:174b; **5**:231b
medieval Arabic, **1**:414a, 538b–539a; **2**: 155a–b; **11**:445a; **14**:112b; **15**:265a–b
medieval Latin, **2**:392b–393a
Renaissance, **1**:109a, 596a–b; **2**:536b, 537a; **10**:236b–238a, 307a; **12**:323b; **15**:162a*
seventeenth-century, **1**:524a; **4**:184a; **6**: 255a, 380b; **11**:314a; **14**:256a
eighteenth-century, **5**:352b; **6**:395b; **14**: 464a–b
early nineteenth-century, **2**:422b; **3**:45a; **9**:8a–b, 9b; **11**:364b, 615b; **12**:320b; **15**:127b
mid nineteenth-century, **2**:111a; **9**:422a; **11**:419a; **12**:612a; **14**:350b
late nineteenth-century, **2**:547b–548a; **5**: 495b; **8**:329a; **9**:422b; **10**:252b, 257b; **11**:462b; **14**:448a
early twentieth-century, **1**:9b–11b; **8**: 451a–455a; **9**:19b, 265b–266a; **15**: 11a, 99b–103b, 105a–107a, 418b
mid twentieth-century, **5**:42b; **8**:485b
experimental, **9**:295b–296a, 466b; **10**: 613a; **11**:215b; **13**:461a; **14**:495b–496b
Galenic doctrine, **2**:536b
Mesopotamian, **15**:645b, 654b(n)
molecular, **2**:548a
Oriental
sixteenth-century, **1**:47b; **8**:390b–395a
see also Drugs; Herbals; Illustration, pharmacological; Pharmacy; specific drugs; Terminology
Pharmacopoeia
ancient, **4**:119b–122a; **6**:428b
medieval Arabic, **1**:538b–539a; **2**:155a–b; **11**:445a; **14**:112b
Renaissance, **1**:109a; **2**:547a; **3**:413b, 414a–b; **4**:209a
seventeenth-century, **1**:483a; **4**:209b; **5**: 420b–421a; **8**:174a; **10**:310b; **13**: 490b, 507b–508a; **14**:639a–b
eighteenth-century, **5**:352b; **8**:297b–298a; **9**:582a; **10**:501a, 613a; **12**:458a; **15**: 206b
nineteenth-century, **2**:336a; **8**:247b*; **11**: 301a; **14**:448a
twentieth-century, **5**:495b; **14**:496a
commentaries on, **14**:639a–b
Mesopotamian, **15**:636a, 645a
Oriental, **8**:391a–397b; **15**:741b
Pharmacy
ancient Greek, **4**:106b–107a; **6**:427b
ancient Roman, **4**:119a–121b; **12**:539b–540a
medieval Arabic, **11**:238b; **12**:90b; **13**: 415b; **14**:585a, 638b*; **15**:499a
medieval Latin, **1**:290b; **6**:123a–b; **9**:469a
Renaissance, **2**:546b; **3**:413b–414b; **4**: 199a, 209a, 521a; **5**:448a; **9**:466a–b; **10**:616a; **13**:535a
early seventeenth-century, **5**:419b, 439a–b; **6**:145b–146a; **7**:146b–147a, 354b–355a; **8**:311b; **12**:78b–79a; **13**:507b–508a
late seventeenth-century, **1**:451a; **3**:180b; **5**:352a, 417b–418a; **7**:150a–b; **8**: 172a–174a; **9**:507b; **12**:39b–40a; **13**: 234b; **14**:639a–b
early eighteenth-century, **1**:529a–b; **2**: 564b; **5**:351b; **10**:25b–26a; **11**:48b, 562a; **14**:193a
late eighteenth-century, **1**:527a–b; **3**:5a, 7a, 177a; **4**:41a; **5**:531b–532a; **7**:11a, 378a; **8**:297a–298a, 352b–353a; **11**: 563a; **12**:143a–144a, 146b; **13**:465b; **14**:332a–b; **15**:205b–206b
early nineteenth-century, **2**:420a; **3**:6a, 159a, 237b, 301a; **4**:560a; **8**:265a,b; **9**:9b; **10**:497a–498a; **11**:494b–495a; **12**:286b, 320a–b; **13**:466a; **14**:44a–b, 600b; **15**:171b–172b
mid nineteenth-century, **5**:72b–73a, 432a, 475a–b; **7**:224a; **8**:296a; **9**:445a–b; **10**:530a–b, 556b–557a, 558a, 575a–b; **14**:111b–112a
late nineteenth-century, **4**:133b–134a; **5**: 99b; **8**:427a–b; **11**:120b; **12**:166a
early twentieth-century, **5**:209a; **6**:319b; **7**:208a–b
Chinese, **8**:390b–397b
handbooks
Renaissance, **2**:537a, 546b
mathematical, **1**:290b
military, **3**:159a; **9**:401b; **10**:325b; **11**:51a; **12**:321a–b; **15**:206b
see also Chemistry, pharmaceutical; Drugs; Pharmacology; Teminology, pharmaceutical
Pharmazeutischer Verein (Baiern), **15**:172b
Pharyngodopilidae, **3**:328a
Pharynx
anatomy, **14**:126b
histology, **13**:590a
paralysis of, **14**:638a
Phase (chem.)
conditions of separation, **7**:529b
fugacity, **8**:289b
Phase contrast (optics)
discovery of, **14**:617a
Phase diagrams, **7**:529a
Dalton points, **7**:529b
see also Matter, states of
Phase equilibria, **1**:431a; **2**:369b; **7**:515b; **8**:42a; **10**:125b; **11**:534a–b; **12**:438b; **13**:580a; **15**:434a
pressure effects, **12**:593b
Phase rule, **1**:431a; **5**:390a; **6**:561a; **7**:204a*, 529a; **11**:534a–535b; **15**:463b
metallic systems, **8**:119a
proof of, **4**:229b
Phase transitions, **8**:482b–483a; **11**:123a; **12**:438b–439a; **14**:31b; **15**:464a
second-order, **7**:617a
theory of, **12**:495b; **13**:245a; **14**:123b
Phaseolus, **3**:422a; **9**:281b
toxicity, **5**:582a
Phenacetin, **3**:246a
Phenanthrene
discovery of, **5**:13a
Phenazine
synthesis of, **3**:300a
Phenobarbital
synthesis of, **5**:2b
Phenol, **2**:66a; **11**:615b
and aniline, **6**:462a
as antiseptic, **8**:403a–408b
spray, **8**:405a–b, 408a,b
derivatives of, **7**:282a
as disinfectant, **3**:479b; **4**:299b; **7**:423a–b; **8**:403a–b, 408a; **12**:478b
etherification of, **3**:10b
oxidation of, **2**:86b
synthesis of, **5**:420b; **7**:282a; **9**:354a; **14**: 531b
Phenology, **2**:219b; **14**:300a
Phenolphthalein
discovery of, **1**:389b
Phenomena, **1**:277b, 278b
saving of, **1**:252b; **4**:226b
in astronomy, **1**:313b; **3**:410a; **15**: 645a
in medicine, **1**:290b
in physics, **2**:605a
Phenomenalism, **8**:158a–b; **11**:357a; **12**:607a
opposition to, **15**:423a
Phenomenology
eighteenth-century, **7**:598a; **11**:432a,b
nineteenth-century, **4**:190b–191a; **6**:299b; **7**:381a, 382a–b; **8**:592a; **11**:9a, 10a, 14b; **13**:363b, 364a
twentieth-century, **3**:112a; **7**:483a, 484b
experimental, **9**:562b
Phenomenotechnology, **1**:366a

Phenotype
concept of, **5**:266b; **7**:114a, 115a
phenocopies, **5**:454a
Phenyl salicylate, **3**:246a
Phenylation, **6**:463a
Phenylhydrazine
formula of, **5**:1b
reactions of, **4**:603b; **5**:1b, 3a–b
Phenylurea
discovery of, **3**:193b
Pheochromocytoma, **15**:105b
Phidias (astronomer), **1**:213a
Philadelphia
Academy of Natural Sciences
founding of, **8**:616b; **12**:133a; **13**:467a
Association for Medical Instruction, **9**: 540a
Central High School, **1**:363b–364a
Independence Hall, **12**:131b
Library Company, **3**:350a; **5**:130a,b; **8**: 460b
Medico-Chirurgical College, **10**:252b
Mint, **12**:132a
observatory, **11**:472a
Philadelphia Centennial Exposition, **4**: 179b; **7**:445a; **10**:551a; **11**:384b; **12**: 574b
Philadelphia Museum, **10**:438b–439a, 439b
Philadelphia, College of
see Pennsylvania, University of
Philadelphia Medical and Physical Journal
founding of, **1**:485b*
Philalethes, Eirenaeus (pseudonym), **12**: 616b; **14**:451b
Philanthropy, **3**:44a
Philinus of Cos (*fl. ca* 250 B.C.), 10:581; **6**:318b(n)
Philip II, King of Spain, **6**:309b; **7**:16b; **10**:239b; **14**:6b, 18a, 29b
Philip, Alexander Philips Wilson (1770–*ca.* 1851), 10:581–583
Philippine Islands, **5**:40a
botany, **11**:130a; **15**:421a–b
exploration of, **12**:299a
natural history, **12**:536a
Philipps-Universität (Marburg), **9**:434a
Philippus (medicine), **5**:228a
Philippus of Medma, **4**:414b; **13**:334a
Philistion of Locri, **4**:106a, 384a, 385a, 466a; **11**:25a, 29b, 127b
Phillip, Kurt Hugo Max, **9**:261b
Phillips, John (1800–1874), 10:583–584; **7**: 460b; **12**:52b(n), 488a,b; **14**:396b
Index Geological Map of the British Isles, **5**:537b–538a
Phillips, John Charles, **3**:121b
Phillips, Paul, **6**:136a
Phillips, Peregrine, **11**:34a
Phillips, Richard, **4**:533a
Phillips, Theodore Evelyn Reece (1868–1942), 10:584–585; **6**:353b*
Phillips, William (1775–1828), 10:585–586; **3**:395a; **9**:583a; **12**:490b
crystallography, **14**:490b
An Elementary Introduction to Mineralogy, **4**:66a
editions of, **9**:393a
Outlines of the Geology of England and Wales, **5**:519a
Phillipsite, **10**:585b
Philo (prefect of Ptolemy), **4**:390a
Philo of Byzantium (*fl. ca.* 250 B.C.), 10: 586–589; **3**:491b; **6**:312b; **15**:253b
geometry, **1**:227b; **4**:105a, 489b, 609a–b
Pneumatica, **1**:61b; **3**:491a; **6**:310b, 311a,b
Philo of Gadara, **4**:491a; **12**:579b
Philo Judaeus, **1**:148b
Philo of Tyana, **9**:300a
Philodemus of Gadara, **8**:537a; **14**:612b
Philolaus of Crotona (*fl.* second half of fifth century B.C.), 10:589–591; **11**:25a
cosmology, **1**:246b, 248b; **3**:402b; **6**:381b; **11**:224a; **15**:204b(n)
Philology
ancient Roman, **5**:232a; **13**:588b
medieval Arabic, **2**:155a–b; **6**:79a,b, 80a; **11**:246b–247b
medieval Latin, **9**:436a
Renaissance, **4**:212b, 412b; **5**:379a; **8**: 248b–249b, 554a; **9**:581a; **10**:116b; **13**:613b
seventeenth-century, **1**:549a; **2**:317a–b; **4**:566b; **13**:528a
eighteenth-century, **1**:554a; **6**:403a; **8**: 308b; **9**:455a–b; **10**:324a; **14**:119a–b; **15**:53b, 56a–b
nineteenth-century, **4**:452a; **12**:513b, 588b; **13**:238a, 239a, 252b; **14**:95b; **15**:133a
twentieth-century, **10**:443b–444a; **13**: 343a
Biblical, **1**:323a
Chinese, **12**:374a, 388a
Mesopotamian, **15**:635a–636a, 642a
see also Linguistics, Literature
Philonides, **1**:179b; **4**:108b; **14**:604a
Philoponus
see John Philoponus
Philosopher's egg, **2**:22b–23a
Philosopher's stone, **2**:317b; **4**:6a; **5**:420a, 421b, 423a; **6**:257a; **8**:310a,b, 311a; **11**:98a, 606b; **12**:307a,b
mercury theory, **2**:22b; **10**:555a
recipes for, **1**:549b
sulfur theory, **5**:422b
Philosophes, **1**:110b–117a *passim;* **3**:379a, 380a,b; **4**:64b, 84b; **6**:468b; **7**:586a, 605a; **8**:84b, 445a; **11**:563b
Philosophical and Literary Society (Liverpool)
founding of, **2**:335b
Philosophical Collections, **6**:487a
Philosophical Magazine, **4**:182b; **10**:108b
editors of, **7**:224a
founding of, **13**:414a
Philosophical Society (London)
see Chapter Coffee House Society
Philosophical Society of Edinburgh, **6**:578b; **9**:480b, 483b
founding of, **12**:413a
Philosophical Society of Glasgow, **13**:373b
Philosophical Society of Washington (D.C.)
founding of, **3**:536a
Philosophical Transactions
see Royal Society of London, *Philosophical Transactions*
Philosophische Studien, **14**:527b
Philosophisk repertorium for faedrelandets nyeste litteratur, **10**:182b
Philosophy
ancient Greek, **2**:229a; **4**:381a–382b; **6**: 289b–291a, 305b–306a, 406a–b; **10**: 324a–325a; **11**:22b–30a, 103b–105b, 128b; **12**:575b–576a; **14**:534b–535b, 606a–607b, 608a–611b; **15**:202b
ancient Roman, **1**:117b–119a; **5**:227b–229b, 232a; **7**:1a–b; **11**:41b–42a; **12**:309b, 340a–341a
Byzantine, **4**:159b; **6**:616a; **9**:111a–b; **11**:160a–162a, 183a–185a; **12**:405b, 440a–443a; **13**:307b–308b
medieval Arabic, **3**:469b–470a; **4**:523b–525b; **5**:592a–b; **6**:190a, 194a; **7**:23a, 40b–41a, 330a–b, 339b; **9**:27b–31b; **11**:230b–231b, 238b, 248a, 249a–250a, 253a, 323a–326b, 444b; **12**: 1b–3a, 5b–7Bb; **13**:229b, 230b, 292b, 488b, 509a–513b; **15**:231b–232a, 249a–250b, 261a–264a, 494a–496a
medieval Hebrew, **3**:469b–470a; **9**:27b–31b
medieval Latin, **2**:435a; **4**:396a–b; **7**:65b–66b; **9**:364a, 435b–438b; **10**:171b–174b; **12**:428b–429b; **14**:388a–389a
Renaissance, **2**:539a–543b; **3**:512b–516a; **4**:170a; **5**:104b–107b, 119b–121b; **9**:32b–33a; **10**:122b, 416a–b; **11**:71a–74b, 286a–287b; **12**:97a–b, 150b–151a; **14**:47a–48a; **15**:80b
early seventeenth-century, **1**:447b; **2**: 163b–164a, 601b; **3**:492a; **5**:284b–287b; **6**:145b, 446b–449b; **7**:268b–269a; **9**:15b–16a; **12**:333a; **13**:573b; **15**:68b–70a
late seventeenth-century, **5**:59b–60b, 535a; **7**:611a, 611b–612a; **8**:149a–160a, 436a–439b; **9**:47a–53a, 509a–510b; **10**:314b–315a; **13**:214b, 479b, 480b
early eighteenth-century, **3**:484b–485b, 598b; **4**:84b–86a, 468a, 471b; **5**:510a, 567b; **6**:555b–559b; **7**:442b, 605a–607a; **8**:460a–b; **9**:189a, 490b–491a; **10**:80a; **12**:599b–604a; **13**:265b; **14**: 84a–b, 482b–484a; **15**:54a–b
late eighteenth-century, **3**:1b–2a, 119b; **4**:86a–89b; **6**:468b–469b; **7**:490b, 596a–600a; **9**:148b; **13**:494b–496a
early nineteenth-century, **5**:193a–195a; **6**:80b–81a, 294b; **11**:599a–b; **12**: 153b–157a, 596b–597a; **15**:225a–226a
mid nineteenth-century, **2**:563a–564a; **3**:240b, 322b, 434b, 588b; **6**:241b, 564b, 596a–b; **7**:505b; **8**:507b, 514a–515b; **9**:383a–385b, 494b; **10**:354b–355a; **12**:607a–b; **13**:251b–252a; **14**: 57b, 113b–114a; **15**:50b, 404a–415b
late nineteenth-century, **2**:8b–10a, 544b–545b; **5**:153b, 439a; **6**:176b, 342b; **7**:67a–68b, 498b; **8**:445a; **9**:143b, 292a, 561b; **10**:448a, 484b–486b; **11**:120a, 309a–b; **14**:343a,b; **15**:422b
early twentieth-century, **2**:545b–546a; **3**:90a–b, 110b–112a, 278a–b, 312b, 335a–b; **4**:186b, 188a; **5**:456b; **6**: 26a–b; **8**:183b–184b, 257a–b; **9**:335a, 495a; **11**:14a–15b, 257b–258b, 388a–389a; **12**:11a, 177a–178b, 218a; **13**: 73b–74a, 274b–275a, 530a, 550b–551a; **14**:302b–308b, 468b–469a; **15**:465b–466b
mid twentieth-century, **7**:482b–488b; **9**: 427b–428a; **11**:363a; **12**:222b–223a, 401a–b; **13**:275b–276b; **14**:467b–469b
and astronomy, **14**:165a–b
Chinese, **14**:163a–165b; **15**:733b
definition of, **13**:307b
divisions of
ancient Roman, **2**:231a–b; **12**:340b
Byzantine, **1**:137b
medieval, **5**:592a–b
Ḥarrānian, **1**:33a–b; **7**:42b; **15**:182a
Japanese, **15**:729a, 730a–b, 732b, 743a–b, 749a
and mathematics, **2**:395b, 545b; **7**:597b–598a; **11**:26b; **13**:428b
and medicine, **1**:4b, 480a; **3**:1b–2a; **5**: 232a; **6**:422b
and politics, **1**:409a; **3**:2a; **4**:81b; **6**:446b–449b
and science, **3**:452b; **5**:122b–123a; **6**:7b, 24b–25a, 26a; **7**:581b, 612a; **8**:514a;

Philosophy (cont.)
9:382a; 11:543b–544a; 13:425b; 15:138b, 728a, 730b
and theology, 2:163b, 232b; 6:448a–449b; 8:436b; 9:30a, 48a–49b; 11:71a–74a, 257b, 258b; 12:2b, 7Ab, 8a
transcendental, 3:110b; 12:155b
see also Congresses, scientific; Education, philosophical; Epistemology; Historiography, of philosophy; Metaphysics; Natural philosophy; *Naturphilosophie;* Ontology; Science, philosophy of; specific sciences, philosophy of; Terminology
Philosophy of Science Association, 5:123a
Phlébentérisme, 11:234b; 12:551a
Phlebitis, 2:443a; 3:489b
naming of, 2:443a
Phlebotomy
ancient Greek, 1:315a; 4:386a
Byzantine, 10:417b–418a, 418b
medieval, 1:565a; 2:619b; 6:123b; 7:118b; 9:41a; 11:244a
Renaissance, 2:337a; 5:233b; 9:466b; 14:4b–5a
seventeenth-century, 1:25b
nineteenth-century, 2:508a; 3:562b
opposition to, 6:58b–59a; 8:134b
Phlegm
as cause of disease, 6:396b
Phlegmasia, 2:508a; 10:612b
Phlegrean fields, 2:439b
Phloem, 9:418b
transport mechanism, 4:131b
Phlogiston, 1:451a; 3:153b, 156b, 157b, 495a; 4:151a; 6:384a; 7:587a; 8:298a
early reference to, 1:550a
identity of, 3:465b; 4:352b; 5:353a; 7:388b
negative weight, 5:532a–b; 13:603a; 14:333a
theory of
early eighteenth-century, 1:550b; 2:226b; 3:454b; 7:188b; 9:105a; 10:26a; 11:562b; 12:303a, 604b–605b
late eighteenth-century, 1:527b, 529b; 2:6a, 74b–76b, 180a, 292a; 5:215b–216a, 411b, 532a–b, 600b–601a; 6:385a, 585b–586b; 7:277b, 388b–389a, 603b; 8:73b, 75b, 247a, 619a, 620a–b; 9:6a, 148a–b, 477a, 478b; 11:109b, 146a–b, 437b–438a, 523a; 12:64b, 144a–147b; 13:469a, 602b–603a; 14:75a–b, 191a, 333a; 15:94b, 95a–b
nineteenth-century, 4:345b; 6:395a
opposition to, 6:283b, 460a; 8:79b, 80a, 373b, 468b, 530b–531a; 9:343a; 12:144a–b, 145a, 365b; 13:465b–466a; 14:464a; 15:206a, 312a
quantification of, 2:6a
weight of, 3:206a–b; 11:434b
see also Chemistry, antiphlogistic
Phlorizin, 9:95b; 12:619b
Phobias, 1:580a
Phoebe (satellite), 11:552a–b
discovery of, 10:602a
Phonation
Renaissance study, 3:99a
seventeenth-century study, 4:135b–136a
eighteenth-century study, 4:590a
experimental study, 4:590a
see also Physiology, linguistic; Speech; Voice
Phonautograph, 7:445b
Phonetic transcription, 2:531b
Phonetics, 8:536a; 15:194b
physiological, 3:531a; 4:264a; 11:215b; 14:152a
Phonodeik, 9:386b
Phonograph, 4:248b, 500a; 7:108a
Phonon(s), 13:240a
Phonoptometer, 8:399a–b
Phora, 8:528b
Phoronis
embryogeny, 7:475a
Phosgene gas
preparation of, 3:604b
Phosoxygen, 3:599a,b
Phosphatase, 6:111a
Phosphates, 5:90b
in animal matter, 5:92b
concretions, 11:372a
deposits of, 2:72a; 5:437a; 6:229a; 8:463a; 13:272b
in fermentation, 6:111a
as fertilizers, 8:92b–93a; 11:179b; 13:347a
in igneous rocks, 5:104a
manufacture of, 8:553b
mining of, 9:590a
organic, 5:6a
polybasicity of, 5:493b–494a
in soil
source of, 5:104a
Phosphatides, 4:604a; 5:432a
Phosphides
preparation of, 10:496b
Phosphine, 5:493a; 10:446a
Phosphites, 5:92a
Phosphocreatine, 9:359b
Phosphoproteins, 8:608a
Phosphorescence
see Luminescence
Phosphoric acid, 2:381b; 3:289b
accumulation in organisms, 2:159a
composition of, 4:239b; 8:76b; 10:177b
in fermentation, 4:485b
in plants, 8:346a
polybasicity of, 5:493b–494a; 8:339a, 340b, 545b
preparation of, 3:401a
purification of, 2:292a
Phosphorites
see Phosphates
Phosphorolysis, 3:416a
Phosphoroscope, 1:556b
Phosphorous acid, 13:310b
composition of, 4:239b; 9:304b
Phosphors
hermetic, 4:215a
production of, 4:215a
Phosphorus, 3:8b; 6:236b; 8:174b
allotropy, 6:177b, 438b; 12:227a; 13:468a
in bones, 5:223a
bottles, 4:42a*
calcination of, 2:75a; 9:105b
combustion of, 6:42b; 8:74b, 76b, 77a; 12:562b
compounds of, 3:10b; 4:243a; 8:277a; 13:70a, 71a, 389b
acids, 4:239b; 9:105b, 106a; 10:496b; 11:541a, 551a
organic bases, 6:462b
oxides, 13:517b
phosphonium salts, 3:347a
discovery of, 7:525a
in eudiometry, 1:586a
in match heads, 2:340b; 5:474a; 7:22b; 12:227a
mercurial, 1:67b; 3:477b; 4:214b; 6:170a–171b, 173b
metabolism of, 4:357b; 6:136a, 366b
and carbohydrates, 10:326b
hormonal control of, 3:352a–b
molecular weight, 4:243b
nature of, 2:381b; 9:105b–106a
oxidation of, 2:75a; 4:241b*
in plant nutrition, 4:250b; 11:416a
preparation of, 6:477b; 7:525a; 9:105b–106a; 12:146a; 14:478a
radioisotopes, 7:154b; 13:163b
in biochemistry, 6:366b; 10:326b
reactions of, 12:64a
red, 10:530b
in steel, 13:8a
valence, 13:389b; 14:267a
vapor density, 3:47b
white
conversion to red form, 5:341a; 12:227a
detection of, 9:425b
Phosphorylase
discovery of, 3:416a
hormonal control of, 3:416a
Phosphorylation, 3:416a; 4:360b; 7:406a; 14:174b
Phosphovinic acid, 8:334a,b
Photius, 1:69a, 120a; 10:208a; 14:631a
Photobiology, 9:414b
Photocathode, 4:355a
Photochemistry
seventeenth-century, 12:79b
eighteenth-century, 12:145b
early nineteenth-century, 1:203a; 5:7a, 322a; 11:474b; 12:281a; 13:311b, 312b
mid nineteenth-century, 1:556a; 2:588b, 589a; 4:181b–183a; 11:536b; 13:238b–239a
late nineteenth-century, 3:279b; 4:282b–283a; 7:520b; 11:149b
early twentieth-century, 4:545a; 6:350a; 8:102a,b; 10:136b; 13:160a, 598a; 14:172a, 173a, 175a–b, 247b, 444b–445a; 15:37a–b, 448a–b
mid twentieth-century, 3:197b; 4:82b, 362b–363a; 7:518b, 519a; 8:293a–b; 13:7a
applications of, 3:279b
laws of, 1:21b; 10:136a; 11:537b; 12:248b; 13:416b
Grotthuss-Draper law, 4:182b; 5:558b
quantitative, 14:172a
Photoelasticity, 2:452b; 9:200a
Photoelectric cells, 3:31b; 4:253b, 356b; 11:542b
Elster-Geitel, 4:355b; 13:80b
Photoelectric effect, 6:96b; 11:420a,b, 441a, 460b; 12:191a, 528b; 13:364b; 14:218a
equation for, 9:396a–b
experimental study, 4:355a–b; 5:341b; 6:74a, 345b; 13:80b, 240b; 15:251b
laws of, 13:80b
in potassium vapor, 8:94b
quantitative studies, 7:36b; 8:181b
and temperature, 9:395b
theories of, 3:372b; 4:315b–316a; 7:553a
Photoelectrons
distribution of, 12:529b
emission of, 3:372b
velocities of, 2:488a
Photoengraving, 5:18b; 13:239a
Photoglyphy, 13:239a
Photogrammetry, 5:65a, 329b; 8:504b; 13:267a
Photographic zenith tube (PZT), 11:552b
Photographische Himmelskarte, 8:539a
Photography
nineteenth-century, 2:453a; 4:182a–183a; 5:52a, 196a,b, 295a; 6:327a; 8:64b–65a; 10:118b–119a; 11:359b; 13:238b–239a; 14:492a
twentieth-century, 5:118a; 7:472a; 9:403a–b

aerial, **5**:65a; **7**:36a; **8**:64b–65a; **9**:403b; **13**:345a–b, 408a
in anatomy, **6**:436a
in astronomy
mid nineteenth-century, **2**:284b–285a, 285b, 616b; **4**:19a, 178b–180b, 182b, 555a; **5**:18b, 86a; **10**:329a, 584b; **11**:95b, 156a; **12**:36b–37a, 289b
late nineteenth-century, **1**:21a–b, 398a, 465b–466a, 466b, 597a,b; **2**:616a; **3**:261a, 366a–b; **4**:269b, 351b; **5**:33b–34a, 405a, 480a; **6**:105b, 106a, 119a, 282a–b, 350b–353a; **7**:76a, 77a, 236a–b, 461a, 471b–472b; **8**:465a; **9**:184a, 230b, 551b; **10**:279b, 600b, 602a; **11**:320a,b, 484b–485a; **12**:139a, 153a, 248a–b; **13**:339a, 481b; **14**:56a, 449a, 479b, 481b, 557b; **15**:493b
early twentieth-century, **1**:398a; **2**:196a; **3**:50a–b, 428b, 472a, 509a–b; **4**:249a–b, 251a; **6**:150a, 350b–353a, 507b–508a; **7**:239a; **8**:50a; **10**:30b, 234b; **11**:470a, 552b, 553a; **12**:17b, 18b, 20a, 176b, 186b, 249b–250a, 251a, 448b, 574a; **13**:45a,b, 408a,b; **14**: 521a–b
mid twentieth-century, **8**:583a; **12**: 368b, 453b–454b
international photographic surveys, **3**:261b; **8**:539a; **13**:232b
see also *Carte du Ciel*
in biology, **6**:122b; **8**:532a, 533a
chemistry of, **1**:431a, 471b; **3**:333b; **4**: 282b–283a; **6**:325a, 327a; **12**:235a–b, 404b; **13**:160a
sensitizers, **11**:87b
color, **1**:614a; **7**:36a, 161a; **8**:388a; **9**:202a; **11**:18b; **14**:348b, 497a
in geodesy, **8**:64b–65a
handbooks, **4**:283a
infrared, **11**:477a
invention of, **10**:118b–119a
in measurement, **8**:388a
optical sensitivity, **3**:476a; **9**:403a–b; **10**: 136b
papers for, **13**:238b–239a
"gaslight," **1**:385a
in physics, **1**:320b, 321b; **4**:195b, 559a; **8**:597a; **11**:486a, 553a
portraiture, **4**:182b
processes in, **1**:284a, 597a,b; **5**:18b; **6**: 327a; **15**:462a
albumen beer, **1**:21b
dry collodion, **3**:476a
wet collodion, **3**:476a; **9**:202a
quenching effect, **6**:327a
in science, **3**:476a; **5**:18b; **8**:518a
spark, **15**:60a
in spectroscopy, **9**:391b–392a
stereoscopic, **4**:19a
in surveying, **11**:95b
techniques
long exposures, **11**:484b
short exposures, **7**:77a
see also Cinematography, Daguerrotypy, Emulsions, Motion pictures, Photomicrography, Telephotography
Photogravure, **4**:283a*
Photoheliograph, **2**:196a; **3**:261b; **4**:19a; **7**: 76a
Photoionization
explanations of, **4**:315b
Photolysis, **13**:7a
Photometer, **5**:87a; **8**:551b; **10**:28b, 600a; **13**:22b
invention of, **2**:344a
polarizing, **4**:219a; **11**:542b; **12**:346a
recording, **11**:542b
shadow, **13**:351b
split-field, **15**:465a
stellar, **5**:489b–490a; **13**:481b
step, **11**:208a
universal, **6**:146b
wedge, **11**:542a; **12**:352b
Photometry
Renaissance, **1**:81b; **9**:192a
eighteenth-century, **2**:343b–344a; **7**:599a
nineteenth-century, **1**:202a–b; **2**:200a–b; **5**:33a, 573b; **6**:110a; **10**:483a; **11**:208a
twentieth-century, **7**:36b; **12**:352b–353a; **13**:598a
in astronomy, **1**:398a, 429b–430a; **4**:514a; **5**:489b–490a; **6**:74b, 146b, 326b; **9**: 415a, 563b–564a; **10**:317a, 320b–321a, 600a; **11**:156a; **12**:96a, 289b, 352b–353a, 448b; **13**:22b, 481b; **14**: 414b, 628b–629a
photographic, **6**:352a; **8**:50a, 105b–106a; **10**:600b; **11**:542a–543a, 553b*; **12**:153a, 248a–b, 265a–b, 454a
in botany, **14**:349b
electrochemical, **1**:556a
flicker, **11**:531b–532a
geographic, **14**:349b
photoelectric, **3**:31b; **4**:355b; **5**:341b; **11**: 441a, 542b
photographic, **14**:349b
units of measurement, **14**:338a
Blondel, **2**:200a–b
Hefner candle, **6**:223a
standardization of, **14**:39a
see also Spectrophotometry
Photomicrography, **1**:614a; **2**:627b; **4**:178b, 182b, 522b; **5**:196b; **7**:422a, 448a; **8**:589b; **11**:486a; **12**:423a
Photomicroscope, **4**:522b; **7**:448a; **12**:423a
Photons, **7**:154a; **9**:398a,b; **15**:49a
high-energy beams, **3**:369b
linear momentum, **3**:368b, 369a
scattering of, **2**:338a; **14**:420b
statistics, **4**:361a
Photosensitization, **13**:7a
Photosynthesis
eighteenth-century study, **2**:286b; **7**:12a–13a; **11**:146a; **12**:308b–309a
nineteenth-century study, **2**:290b; **3**:43b; **12**:59a,b; **13**:416a–417a; **14**:255a
twentieth-century study, **3**:30b; **4**:362a–363a; **5**:118a,b, 523b; **7**:35a, 316b; **9**:44b; **10**:236b; **14**:174b–175a
discovery of, **7**:12a–13a; **11**:144b
and light intensity, **13**:417a
light vs. dark stage, **4**:363a; **14**:176a
in marine vegetation, **3**:600a
oxygen formation in, **1**:362a
photosynthetic unit, **4**:363a
quantum requirement, **4**:362b; **14**:173a,b, 175b–176a
red drop
Emerson effect, **4**:362b–363a
two-step model, **5**:118b
see also Plants
Phototaxis, **13**:88b
Phototheodolite, **8**:64b
Phototherapy, **4**:621a
Phototropism, **1**:124b; **3**:45a, 575b–576a, 581a–b; **4**:188b*, 560a; **6**:401a; **7**: 538a; **12**:357b; **14**:255a–b; **15**:71b
in animals, **11**:257b
in fungi, **6**:100a
in microscopic organisms, **3**:338a
photochemical nature of, **14**:255b
Phragmén, Edvard, **8**:366b; **9**:426b
Phreatophyte, **9**:257b
Phrenitis, **4**:386a
Phrenological Journal
editors of, **14**:189b
Phrenology, **2**:478a, 508a; **3**:192b; **5**:196b, 531b*; **6**:440b; **10**:500a; **12**:571a; **14**:137a, 189a–b
critics of, **5**:44b, 45a; **8**:272b; **11**:593a
origins of, **5**:252a–255b; **12**:596b–597a
Phtanites, **1**:472b
Phthalazine, **5**:215a
Phthalein, **1**:389b
Phthalic acid, **1**:389b
Phthalylacetic acid, **5**:215a
Phthisis, **1**:562a; **6**:183a; **9**:75a
contagion, theory of, **5**:107a; **11**:100a
diagnosis of, **11**:462b
treatment of, **7**:426a
and tuberculosis, **2**:444b; **7**:423b; **8**:9a; **13**:223a
Phtyriose, **9**:78b
Phycocyanin, **7**:534a
Phycoerythrin, **7**:534a
Phycology
see Algology
Phycomyces
biochemistry, **12**:207b–208a
Phycomycetes, **13**:299b
sporangiferous filaments, **4**:402a
Phyllocyanin, **9**:96a
Phylloerythrin
discovery of, **9**:96a
Phylloglossum, **2**:371a
Phyllophylin, **9**:96a
Phylloporphyrin, **9**:96a
Phylloscopi, **14**:299b
Phyllotaxis, **11**:416a; **12**:357b; **15**:14a
spiral, **2**:426a
theory of, **8**:206a–b; **12**:167b, 256a
Phylloxera, **1**:587a; **5**:504a; **9**:78b, 112b; **10**:406a, 436a
Phyloembryogeny, **12**:337b–338b
Phylogeny
botanical
eighteenth-century, **9**:93a
nineteenth-century, **2**:86a, 103a–b, 583a; **3**:29b, 318a; **4**:401b; **13**:542a; **14**:181b; **15**:147b–148a
twentieth-century, **3**:188a; **5**:438a, 584b, 605a; **15**:149b, 224b
adaptations, **2**:492a; **3**:575a–b, 576a; **5**:438b; **7**:409b; **9**:24b, 44b
variation, **7**:395b–396a; **9**:75b
see also Evolution, of plants
indicators of
salt content, **2**:585b
invertebrate, **8**:3b; **14**:1b
and ontogeny, **6**:613b; **9**:65a; **12**:45a, 337b–338b, 410a; **15**:92b, 169b–170a
paleontological, **2**:505b; **5**:486b; **7**:252a; **10**:429b; **11**:235b; **12**:38a
phylogenetic acceleration, **10**:429b
vertebrate, **7**:61b; **9**:134b; **12**:337a, 409b–410b
zoological
eighteenth-century, **5**:442b–444a
early nineteenth-century, **5**:356b–357a
mid nineteenth-century, **6**:7b–9b, 10a; **12**:299a; **15**:123a
late nineteenth-century, **1**:420b–421b; **2**:339b, 502a; **4**:207a; **5**:385a–b; **6**: 167a–b, 336b, 535a–b, 536a; **7**:314b; **9**:496a; **10**:128a; **14**:425a,b; **15**:169a–170a
twentieth-century, **8**:586a; **13**:137b
see also Evolution, Taxonomy
Phylotimus, **11**:128a,b
Physalia, **1**:92b; **11**:101b
Physical Review, **3**:370b, 372b
editors of, **9**:396b; **10**:105b, 482a

Physical Society of London, **4**:280b
founding of, **5**:412a
Physician, concept of
medieval, **1**:4b; **14**:585a
Renaissance, **2**:337a
nineteenth-century, **7**:520b
Japanese, **15**:743a–b
Mesopotamian, **15**:643a, 646b
Pneumatic school, **1**:235a
Physicians, Royal College of
see London, College of Physicians
Physicians and Surgeons, College of
see Columbia University, College of Physicians and Surgeons
Physicotheology, **5**:443b
Physics
Byzantine, **11**:185a–b; **13**:225b
medieval Arabic, **11**:445a
medieval Latin, **5**:402b–403a
Renaissance, **1**:97a
early seventeenth-century, **2**:592b, 593a; **4**:52a–54a; **5**:158b, 238a–247b, 382a; **13**:436a–439b
late seventeenth-century, **1**:483b–484a; **6**:481b–487b; **7**:376a; **10**:314a–315a; **12**:210b–211a
early eighteenth-century, **3**:114a–115a, 296a, 562a; **4**:26a; **5**:435b; **6**:148b; **7**:111a
late eighteenth-century, **2**:51b, 299b–300a; **3**:258a, 426a, 443a–446a; **4**: 503a–b; **5**:195a; **6**:178b, 403a; **7**: 598b–599a; **11**:66b–67a; **13**:425a; **14**:352a–355a; **15**:94b–95b, 312a–316a
early nineteenth-century, **2**:133b–139b; **3**:8b–9a, 258b–259a, 260a–b, 497a; **4**:29a, 239b–241b, 546b; **6**:178b, 324b–325b; **9**:578b; **10**:545b–546a, 552a–553a; **12**:253b–254b; **15**:376b–379a
mid nineteenth-century, **3**:18a, 238b; **4**: 167a–168a, 258a; **5**:68a–b, 124a, 304b–306b, 559a–560b; **6**:96a–b, 277b–279b; **7**:379b–382b, 509a–b; **8**:122a; **9**:18b–19a, 176b–177a, 566a–567a; **13**:614b–615a
late nineteenth-century, **2**:200a–b, 202b–203a, 397b, 427a–b, 465b–466a, 559a; **3**:11b–12a, 179a, 265a, 498a, 503b–506b; **4**:68b–69a, 500a–501a; **5**:63b, 423b; **6**:73b–74a, 270b–271a, 341a–349a; **7**:313a–b; **8**:148a–b; **11**: 122a–123a, 241b–242a, 445b–447a; **12**:238a–b; **13**:51b–52b, 79b–81a, 471a–472a; **14**:170b–171b
early twentieth-century, **2**:487b–488b, 559b; **3**:3b–4a, 367a–b, 429b–430a; **4**:253a–254a, 305b–306b; **5**:64a–b, 423b–424a, 610a–b; **6**:51a–b; **7**:84a–85a, 160b–161a, 271b–272b; **8**:368a–369a; **11**:290b–291a; **13**:471b
mid twentieth-century, **2**:457b–461a, 600b–601a; **3**:312b, 328b–330b
analytic, **3**:379a
applied
nineteenth-century, **13**:356b–357a, 471b; **15**:449b
early twentieth-century, **4**:408a–b*; **7**: 446b; **8**:23a–24b, 365b, 489a; **10**: 604b–605a; **11**:213a; **15**:60a–b, 251a–b
Aristotelian, **2**:378a; **7**:486b
atmospheric
nineteenth-century, **1**:6a, 300b, 521b; **2**:136b; **12**:595a–b
twentieth-century, **1**:195b–196a; **6**: 147b, 355a–356a; **12**:366b, 496b; **14**:422a–b
atomic, **1**:370b–371a; **2**:239b–248a, 466a, 488a, 559b; **3**:4a, 124a–125a, 367b, 373a, 597a–598a; **4**:195b, 577b–579a; **5**:117a–118b, 160b–161b, 609a–b, 610b; **7**:468b–470a; **8**:284b; **9**:38a, 256a–b; **10**:106b, 213b–218b, 236a; **12**:27b–32b, 528b–529b; **13**:82a–b, 367a–369b; **14**:327a, 420a–422a; **15**:251b
axiomatization of
see Axiomatization, in physics
chemical, **2**:169b–170a; **3**:474b, 476a–480a, 617b–620b; **4**:211b; **5**:597a–b; **8**:102a
cosmic, **1**:300b–301a; **9**:260b, 566b
crystal, **1**:461a; **3**:598a; **5**:160b–161a; **14**:61b–62a; **15**:40b–41a, 41b
electron, **3**:372b–373a; **5**:160b–161b; **6**: 546a–b; **7**:263b–264b; **8**:498b; **9**: 395b–397a; **11**:419b–422b; **12**:527b–528a
and engineering, **6**:105a
experimental
Renaissance, **3**:66b; **5**:249b
early seventeenth-century, **1**:566b–568a; **2**:83b–84a, 523a; **5**:288a–b; **7**:194a; **9**:13b–14a, 97a–b; **11**:411a–b
late seventeenth-century, **2**:377b–379b, 523a; **3**:482b; **5**:574b–575b; **6**:477b; **7**:578a; **9**:3b, 114a–120b, 485b; **11**: 121b
early eighteenth-century, **2**:43b, 55a; **3**:51a–52a; **4**:166a–b, 214b–216a, 517a–518a; **5**:130a–138a, 226a, 509b–511b; **6**:36a, 169a–173b, 175b–176a; **8**:177a–b; **10**:145a–147a; **11**: 65b, 432a–433b; **15**:271a
late eighteenth-century, **2**:474b–475a; **3**:153b–154a, 207b–208a, 535a–b; **6**:233b; **7**:13a–b, 547a; **8**:70a, 322a; **9**:5b, 152a, 471a–472a, 477a; **10**: 602b–603a; **12**:427a–b; **13**:350b–352a; **14**:69b–79b
early nineteenth-century, **8**:38a–b; **9**: 459b; **10**:184b–185b, 186b–189b; **12**:129a–130a, 281a–b; **13**:126a–b, 127a–b, 130a–b; **14**:204a
mid nineteenth-century, **4**:174b, 200b; **5**:84b–87b, 340b–341a; **7**:160a, 180b–182b, 450a–b; **8**:1a–b, 35a–36b, 187a–188b, 398a–399b, 596a; **9**: 166a–b; **11**:21b–22a, 46b–47a, 50a–b, 111a–b, 353a–b, 559a–b; **13**:236b–237a, 522a; **14**:204a–207a, 289a–290a, 329a–330b, 360a, 479a–b
late nineteenth-century, **3**:19a–20a; **4**: 194b, 354b–357a; **5**:483a–484a; **6**: 438b–439b; **7**:220a–222a, 267b–268a, 449a–b, 458a, 526a–b; **8**:387b, 502a, 597a–b; **9**:154a–155b, 230a–231a, 291a, 371b–373a, 531a; **10**: 345a–346b, 518b–519a; **11**:149b–150b, 393a–b, 460a–461a, 529a–531b, 531b–532a, 578a–579a; **12**: 26a–27a, 366a–b, 496a–b; **13**:10b–11b, 100b–104b, 237a, 365a–368a, 473a–b; **14**:32b–33a, 38a–39a, 67a–b, 217b–218a, 338a–340b, 340b–341a, 347b–348b, 522a–b, 597a–598b; **15**: 59a–60b, 83b, 432b–433a
early twentieth-century, **3**:31b; **4**:577a–579b; **5**:341b, 433a–b, 610b; **7**:416a–b, 515b–516a, 527a, 621b–622b; **8**: 102a, 106b–107b; **9**:76a–77a, 373a–374a, 376b–377b, 386a–b, 395b–399b, 542a–544a, 606a–b; **10**:236a, 347a–348b; **11**:264b–266a, 441a–b, 580b–581a; **12**:218a–b, 613b–615b; **13**:40b–43a, 104a–105b, 175b–176a, 368b–369b, 445b–447a; **14**:171b–172b, 218a–b, 243b–246a, 340b–341a, 497a–498a, 598b–599a; **15**: 61a–64a, 251a–b, 491a
mid twentieth-century, **4**:579a–583a; **8**:94a–96b; **11**:117a–b, 266a–267a; **13**:569b–570b
and physiology, **4**:200b
foundations of, **2**:460a–461a; **8**:497a–499a; **12**:16a–b
fundamental units, **13**:82b
of gases, **4**:280a
geometrization of, **4**:323b, 330a–331b; **7**:213b–218b
and geometry, **1**:226b; **3**:616a; **4**:51a, 53a; **5**:551b–552a; **7**:485b–486a, 487a; **8**: 39b, 196a–198a, 232a–233b, 434a; **11**:21b–22a
handbooks, **5**:332b*; **7**:449b
high-energy, **8**:95a; **11**:117b
low-temperature studies
see Low-temperature studies
marine
see Oceanography, physical
mathematical
ancient Greek, **1**:251b–252a, 256a–258a
medieval, **1**:337a; **13**:184b–211a
Renaissance, **13**:349a–350a
seventeenth-century, **1**:566b–568a; **4**: 52b–60a; **8**:267b–268b; **9**:116a–b
early eighteenth-century, **4**:471a; **8**: 609b, 610a
late eighteenth-century, **1**:110b–116a; **3**:155b–158a; **4**:476b–477a, 478b–479a; **14**:355a; **15**:293a
early nineteenth-century, **1**:343b; **2**: 134b, 368a–b; **3**:135b, 143b–146b; **5**:94b–97b; **7**:37a–b; **9**:547b–548a; **10**:4a, 187b–193a, 546a; **11**:6b; **13**: 130a–b; **15**:356b–363b, 480b–490a
mid nineteenth-century, **1**:147a–b; **2**: 88a–b, 166b, 288b; **3**:135b, 143b–146b, 186b, 303a–310a, 314a; **4**:124a, 126a–b, 205a–b, 222a–b; **6**:243b–245a, 249a–250a, 291b–293a; **7**:54b–55a, 379b–382a, 601a–602a; **8**:1a–2b, 382a–b, 384b, 385b, 591b–592b; **9**: 206b–207b, 215b, 548a–549a, 566a; **10**:27a–28b, 247b–250b; **11**:21b–22a, 46b, 126b–127a, 389b–390a, 453b–454b; **13**:74b–78b, 217a, 374a–387b; **14**:221b; **15**:192b–193a, 755a
late nineteenth-century, **2**:105b, 167b, 356a, 471a, 615a; **3**:178b; **4**:407b; **5**:150b–152a, 461b–462a, 483b; **6**: 211b–212a, 342a–b, 451b, 473a, 545b; **7**:235b, 288b, 399a, 594b–595a; **8**:39b–40b, 502a, 516a, 562b; **9**:174a–175a; **10**:25a, 481b; **11**:7b–12b, 58a–59b, 407a–409b; **12**:526a–527a; **13**: 82a–b, 237a, 362a–368a; **14**:61b–62a, 86a–b, 158a–b, 202a–b, 503a–b, 613b–614a
early twentieth-century, **1**:320b, 365b, 499b–500a; **2**:144b, 303a, 466a, 565b–566a; **4**:375b–376b, 577a; **5**: 202a, 407a, 473a–b; **6**:163b–164a, 214b, 392b; **7**:211a–212a, 515a, 574b; **9**:173b, 414a–b, 461b–462b; **10**: 106a–b, 594b; **11**:12b–13b, 475b–480a; **12**:496b–497b, 527b; **13**:26a–27b, 83a, 368a–369b, 429a–b; **14**:87a, 283b, 317a–b, 345a–b
mid twentieth-century, **5**:122b; **6**:147b–148a, 614a–615b; **7**:387a–b, 491b–493a; **8**:457b; **14**:91b–92a; **15**:7a, 27b
critics of, **8**:182a–b
unitary problem, **7**:211a–b

measurement in
ancient Greek, **1**:253b
nineteenth-century, **9**:227b
twentieth-century, **6**:469b–470a
apparatus, **2**:600b
medical, **2**:600b, 601a; **9**:566b
molecular
mid nineteenth-century, **4**:203a, 615a; **8**:508b–510b; **9**:217b–225b; **14**:184a–186a
late nineteenth-century, **3**:477a–478a; **5**:462a; **8**:148b; **13**:156a; **14**:109b–111b, 343b
early twentieth-century, **2**:337b; **3**:329a; **4**:76b, 253a–254a; **5**:117b–118a, 597a–598a; **7**:84b–85a; **8**:102a, 473a–475b; **9**:38a; **10**:235b; **14**:497a; **15**:19a
mathematical analysis and, **1**:143a; **14**:617a; **15**:358a–360b
neutron, **3**:430a; **7**:527a
nuclear
nineteenth-century, **1**:559b–560a
early twentieth-century, **2**:338a–b, 488a; **3**:329b–330b, 430a; **4**:579a–583a; **6**:117a–118a; **9**:260a–261b; **10**:525b; **11**:347b; **12**:32b–34b, 71a–b; **13**:569b–570b
mid twentieth-century, **2**:250b–252b; **3**:371a–b; **5**:272a–b; **6**:104a–105a, 615a; **7**:153a–156a, 158b–159a, 527a–b, 554b–555a, 617a; **8**:95a–96a, 293a; **9**:37b–38b, 261b–263a; **10**:214b–218a; **12**:438a; **13**:227a–b, 240b–241b, 600b–601a; **15**:8a–9a, 26a–27b, 251b
and astrophysics, **5**:272b
biomedical applications, **8**:96a
critics of, **8**:182b
particle, **4**:582b–583a; **7**:617b; **9**:38a–b; **11**:117a–b, 446a; **12**:613b–614a; **15**:26b–27b
philosophy of
ancient Greek, **1**:251b, 252b, 254b; **5**:346b
eighteenth-century, **3**:114b; **8**:321b–322a; **10**:146a; **11**:432b
early nineteenth-century, **5**:446a
mid nineteenth-century, **2**:563b; **7**:381a–b, 382b; **8**:122a, 596a–b; **9**:206a, 213b–214a, 566a
late nineteenth-century, **5**:391a–b; **6**:243a, 347b–348a; **8**:116b, 490a, 599a–605b
early twentieth-century, **3**:31b–34b, 111b; **4**:281a–282a; **5**:122b–123a; **8**:184a; **9**:406a; **11**:13b–15b, 388b; **13**:176a; **14**:341a; **15**:424a
mid twentieth-century, **10**:425b; **11**:356b; **12**:222b–223a; **14**:318a; **15**:251b
plasma, **15**:8b–9a
radiation
early nineteenth-century, **11**:115b–116a
mid nineteenth-century, **4**:183a; **13**:522a
late nineteenth-century, **3**:477a–478a, 498b–502b, 506b–508a; **5**:458b–459a, 461b–462a; **7**:528a–b; **8**:180a–181b, 551b–552a; **10**:345b–348b, 525a; **11**:149b–150b, 460b–461a, 581a–582b; **12**:69a; **14**:31b–32a, 338a–341a; **15**:46b–47a
early twentieth-century, **1**:320b–321a, 490b, 519a–b; **2**:337b–338b, 398a–399b, 487b–488b; **3**:31b, 312b, 367a–371a, 564a; **4**:194b–197a, 253b, 330b–332a, 355a–356b; **6**:355a–356a; **7**:84b–85a, 152b–154b, 158a, 186b, 468b, 553a–554a; **8**:9b–10a, 181b–182b; **9**:542b–543a; **10**:105b–106a, 236a; **11**:347b–348a, 582b–583a; **12**:191b, 613b–615a; **13**:548b; **14**:497b
mid twentieth-century, **6**:104a; **7**:469b–470a; **13**:241a
Scholastic, **2**:378a
solid-state, **2**:325a, 431b, 459a; **6**:147b, 303a–304a; **7**:617a; **9**:213a; **12**:464b; **15**:40b
statistical
see Statistics, in physics
stratospheric, **11**:347b
textile, **1**:319a
theoretical
early eighteenth-century, **8**:468a–469a; **9**:33a–34a
late eighteenth-century, **3**:72b–75a; **7**:490a–b, 587a; **8**:255b–256b, 259a–260a; **9**:234b; **11**:134b–135a
early nineteenth-century, **5**:97a–b, 167a, 558b–559a; **6**:291b–292b; **9**:578b; **10**:546a; **12**:287b–288a; **13**:127a–b
mid nineteenth-century, **2**:261a–267b; **6**:278b; **7**:181a–182b; **8**:508a–511a, 596a; **9**:204a–227b, 235a–240a, 264b–265a, 445a–b, 548a–549a; **11**:291b–294a; **12**:288a–289a; **13**:374b–386b; **14**:184a–186a, 206b–209a, 360b; **15**:85a–86b
late nineteenth-century, **3**:179a, 428a–b; **4**:189b–192b, 194b, 228a–232a, 378a–380b; **5**:15a–16a, 386b–392b; **7**:220a–b; **8**:444a, 488a–497a, 501a–502b, 598a–603a; **12**:42a–44a, 511b; **13**:100b–104b; **14**:61a–b, 67a, 231b, 337b–340b, 597a–598b
early twentieth-century, **3**:564a; **4**:292b–294b, 312a–332a, 408a, 577a–579b; **5**:103a–b, 122a–123a, 160a–161b, 188a–b; **7**:26b–27a; **8**:10b–11b, 50b–53a, 496b–499a; **9**:182b, 377a–b, 606b, 616b; **10**:235b–236a, 422a–423a; **11**:475b–480a; **12**:34a, 69b, 217b–221b, 463b–464b, 527b–530a; **13**:40b–42a, 104a–105b, 175b–176a; **14**:304b–308b, 340b–341a, 437a, 616b–617a; **15**:39b–42b, 48a–49b
mid twentieth-century, **2**:120b–121a, 168b–169a, 239b–252b; **3**:367b–370a; **7**:10a–b, 36b, 491a–493b, 617a–618a; **8**:185a–186b, 473a–479a, 479b–483a; **9**:405b–406a; **10**:213b–218a, 423b–425b; **12**:222b–223a; **13**:239b–241b; **15**:6b–7a, 25b–27b, 42b–44a
and reality, **4**:318b–319a
vacuum, **3**:477b–478a
see also Acoustics; Apparatus, physics; Biophysics; Congresses, scientific; Crystallography; Dynamics; Electricity; Engineering; Ether, physical; Historiography, of physics; Instrumentation, physical; Magnetism; Mechanics; Natural philosophy; Optics; Spectroscopy; Statics; Terminology
Physikalisch-Medizinische Gesellschaft, Würzburg
founding of, **7**:439b
Physikalische Blätter
founding of, **15**:490b
Physikalische Gesellschaft (Berlin)
see Deutsche Physikalische Gesellschaft, Berlin
Physikalische-technische Reichsanstalt, **12**:426a, 615a
Physikalische Zeitschrift
editors of, **11**:446a
Physikalischer Verein (Frankfurt am Main), **1**:7a
Physiocrats, **3**:386b; **7**:202a; **13**:494b, 495b; **14**:15a–b
Physiognomy, **4**:266b; **5**:252b, 506b*; **6**:562b, 573b; **8**:322a; **9**:364a; **11**:97a–b, 238b; **12**:596b; **14**:128b; **15**:497b
Physiography, **2**:374a, 500b; **3**:595a; **4**:562b–563a; **8**:543b
see also Geography, physical; Geomorphology
Physiological Abstracts
editors of, **11**:139a
Physiological selection, **11**:519a
Physiological Society (London)
founding of, **4**:593b; **12**:356a
Physiological Zoology
founding of, **3**:247b
Physiology
ancient Greek, **1**:266a–267a; **4**:106a–b, 383a–386a; **6**:317b–318b, 428a; **11**:29b, 127b–128b; **13**:92b, 297a
ancient Roman, **5**:230b, 233b–235b
medieval Arabic, **11**:238b; **13**:292b, 407b; **14**:638b*; **15**:498b
medieval Latin, **6**:396b; **9**:364a
Renaissance, **1**:611b; **3**:355b, 356a–357a; **4**:138b, 585a–586a; **5**:379b; **8**:197a–b, 201b–202a, 203a–204a; **10**:598b–599a; **11**:390b–391b; **12**:335b; **14**:9a–11b
synoptic tables, **4**:139a
early seventeenth-century, **1**:315b–316b; **2**:306a, 308a–b; **4**:61b–65a, 377a–b, 510a–511a; **6**:151b–159a, 255a–256a, 257a; **8**:53b–54a; **11**:467a–b
late seventeenth-century, **1**:391b–392a, 482b, 592b; **2**:237a–b; **3**:209a, 482b–483b; **4**:37b, 135b–136a; **5**:418b, 426a–427a; **7**:459a; **9**:243a–246a; **10**:520a–b; **11**:507b; **13**:32a–33b, 169b–174b; **14**:405a–b, 407b–408b
early eighteenth-century, **2**:2b–3a, 37a–b, 226b–227a; **4**:589b–590a; **6**:63a–65a; **7**:274b–275a, 606a–b; **8**:416b; **10**:501a; **12**:302b–303a; **13**:121b–123a, 179b–180a, 231b–232a, 563a–564b, 567a
late eighteenth-century, **1**:479a–b; **2**:122a, 581a–b; **3**:1b–2a, 15a–b, 579a; **5**:55b–57a; **6**:550a; **8**:37b, 506a; **9**:306a; **11**:158b–159b; **12**:287a; **15**:4b, 6a, 95a–b
early nineteenth-century, **1**:386a–387b, 518b, 542b–544a; **2**:187b, 335b–336a, 366a–367b, 483b–484a, 594b–596b; **4**:146a–147a, 243a, 252a, 264a–265a, 285b; **5**:44a–45a, 430b; **6**:119b–120a; **7**:60a; **8**:97b–98b; **11**:133a, 173b, 510a–b, 592b–593a; **12**:223b–224a; **13**:461a–b; **14**:44b, 351a–b, 492a–493a
mid nineteenth-century, **1**:471a; **2**:25a–33a, 61a–62b, 72b–73a, 110b–111b, 124a–125a, 161a, 162a, 375b–376a, 525a–b, 531a–532a; **3**:219b–220a, 530b–531a, 550b, 604b; **4**:75b, 162b–163b, 182b, 554a; **6**:137b–138a, 241b–243b, 252b–253a, 270a, 540b, 575a–576a; **7**:438a–b, 519a–520b; **8**:103b, 596a–597a; **9**:235b, 237b–238b, 422a; **11**:63a; **12**:224b, 354a–b, 421b; **13**:555b; **14**:57b, 113a–114a, 128b, 199b–200a, 204b, 206a
late nineteenth-century, **1**:325b, 535b–537a; **2**:25a–33a, 70b, 446a–447a, 515b, 547b–548a, 585a–586a, 598b–

Physiology (cont.)
599a; **3**:219b–220a, 492b–493a; **4**: 12a, 79a, 266b–267a, 371a–372a; **5**: 80a–83b, 148b–150a, 184b–185a, 196a–197a, 279a–283b, 462b–463b; **6**:78a, 225a–226b, 287a–b, 476a–477a, 596a; **7**:504a–505a; **8**:14a–16a, 535b–536b, 556a–b, 597a–598a, 603a–605a; **9**:142a–143a, 380b–381b; **11**:135b–136a, 517a–518a, 585a–586a; **12**:319a–b; **13**:53a–b; **14**:2b–3a; **15**:20b–21b
early twentieth-century, **1**:449b–450b, 452b–455a, 610a–b; **2**:339a–b, 429a, 597b; **3**:113a, 247b, 518a–b; **4**:311b, 333b–334b, 398a–b, 618a; **5**:36b–38a, 541b–542a, 598b; **6**:23b–25a, 260b–262a, 264a–265a, 526a–b; **7**:501a–504a; **8**:29b–30a, 101b–102b, 294b–295b, 451b–455a, 555a–556a, 583b–584a; **11**:138b–139a; **15**:99b–103b, 104b–107a, 417a–b
mid twentieth-century, **1**:533b–534b; **5**: 41a–43a; **6**:21a–22b; **7**:455b–456a, 516a–517a, 519a; **8**:359a
anatomical, **8**:526b
and anatomy, **1**:535b; **6**:62a; **9**:345a–b
applied, **6**:476b; **7**:503a
botanical
ancient Greek, **13**:331b–332b
Renaissance, **8**:205b–206b
seventeenth-century, **3**:180a; **5**:535a–b; **9**:114a–b; **10**:520a–b; **11**:314b–315b; **14**:501b
eighteenth-century, **2**:286b; **3**:45a; **6**: 36b, 38a–41b, 401a, 549b, 550a; **7**: 12a–13a; **9**:151b; **12**:308b–309a, 408a; **15**:113b
early nineteenth-century, **1**:69b; **2**:160a; **3**:43b; **4**:264a–b; **5**:218a–b, 441a; **9**: 418a–b; **11**:216a; **12**:123b–124a; **13**: 462a–463a, 595b–596a
mid nineteenth-century, **3**:338a; **4**:560a; **5**:277b–278a; **6**:136b, 138a; **9**:345a–b, 442a, 600a; **11**:310b–311a; **12**:59a; **13**:542a; **14**:350a
late nineteenth-century, **2**:27b, 61a–b, 183b–185a, 290a–b; **3**:340a; **4**:76b, 401b–402a, 545b; **5**:623b–624a; **6**: 237b; **7**:34b–35b, 113b–114a, 395b–396a; **9**:78b, 450b; **10**:281b–282b, 575b–577b; **11**:154b–155a; **12**:58a–60b, 166a–b; **13**:88b, 243a, 405b–406a, 416a–417a; **14**:95b–98b, 255a, 349a–b
early twentieth-century, **3**:359a; **4**: 220a–b; **5**:438b, 523b–524a; **8**:425a–b; **9**:24b–25a, 44a–45a, 460b–461b; **13**:486a–488a; **14**:255a–b; **15**:47a
mid twentieth-century, **4**:362a–363a; **6**:443a–444a; **11**:416a
techniques, **12**:59b–60a
cellular, **4**:371b; **5**:585a; **9**:442a; **10**:256b–257b; **13**:486a,b; **14**:2b–3a
cerebral, **5**:196b
chemical, **6**:498b; **8**:446a
comparative
Renaissance, **11**:528a
seventeenth-century, **9**:323a
eighteenth-century, **4**:178a; **7**:367b–368a; **13**:543b
nineteenth-century, **3**:88a; **8**:123a; **9**: 569b; **11**:517b–518a; **15**:123b, 124b, 126b
twentieth-century, **1**:610b; **2**:325a–b, 339b; **3**:68b–69a; **7**:503a–b; **8**:534a–b; **10**:433b, 434b–435a; **11**:101b–102a; **12**:470a–b
cosmic influences upon, **1**:300b
developmental, **5**:447a; **7**:395b–396a, 516a–517a
evolutionary approach, **5**:82b–83a, 280b, 281a; **8**:534b; **10**:220b, 435a; **11**:517a, 518a
experimental
ancient Roman, **5**:230b, 235a
medieval Arabic, **9**:604b
early seventeenth-century, **12**:103a; **13**:222b
late seventeenth-century, **5**:484b–485a; **6**:485a; **8**:524b–527a; **10**:567b–568a; **14**:509b, 587a–b
early eighteenth-century, **6**:36a–38a; **11**:334a; **14**:26a
late eighteenth-century, **8**:83a–84a; **11**:99b; **12**:553a–560a, 563a; **13**:46b–47a, 183a–b; **14**:320b–321b
early nineteenth-century, **8**:132b–134b; **9**:7a–10a; **10**:582b; **11**:68b, 214a–216a; **13**:403a–b; **14**:558b
mid nineteenth-century, **3**:531a; **5**: 514b; **7**:23b–24a; **8**:401a, 540a–541b; **9**:101b, 408a, 456b–457a, 567b–574a; **10**:558a–b; **12**:164b–165a, 240a–244b, 466b–467a; **13**:556b–558a; **14**:63a–66b, 527a; **15**:107b–110b
late nineteenth-century, **2**:599a; **3**: 219b–220a; **5**:80a–83b; **6**:518b; **8**: 445a–446a, 536a–b; **9**:102a–b, 265b–266a, 505b, 546b–547a; **10**:175b–176a, 204b–205b, 252a–b, 431a–433a, 579b–580b; **11**:295b–296b, 425a–429b, 462b–463a, 570b; **12**: 271a, 355a–356a; **14**:495b
early twentieth-century, **6**:21b–22a; **8**: 295a–b, 614a–615a; **9**:19b, 586a–b; **10**:220a–b; **11**:125a–126a; **12**:617b–618b; **15**:71a–77a, 228a–b, 469a–b
mid twentieth-century, **14**:595b–596a
critics of, **6**:618b
cross-circulation technique, **5**:149b–150a
long-term studies, **10**:432a
philosophy of, **9**:571b–572a, 573b
primate studies, **5**:207b–208a
fetal, **1**:455a; **4**:510a–511a
Galenic tradition, **4**:508b, 510b, 585a; **9**:468b; **11**:391a, 604b; **12**:101b–102a; **13**:100a; **14**:9a–11b, 587b
opposition to, **13**:32a
gastric, **3**:69a
handbooks
eighteenth-century, **6**:66a–b
nineteenth-century, **2**:547b, 599b*
high-altitude
see Altitude, physiological effects
histology and, **2**:375b; **14**:113b
human, **9**:547a
Japanese study, **15**:740b
linguistic, **2**:531b; **3**:530b–531a; **4**:508a; **6**:563a; **7**:48a–b
see also Phonation, Voice
and mathematics, **10**:198a
methodology, **10**:432a–b
microbiological, **14**:37a–b
muscle
see Muscles, neurophysiology
naming of, **4**:585a
nerve
see Neurophysiology
ovarian, **2**:345a
philosophy of
seventeenth-century, **4**:61b–65a; **5**: 426a–427a; **6**:151b–152b, 158a
eighteenth-century, **6**:549b; **8**:114b–115a; **9**:306a; **11**:364a–b; **13**:543a–b
early nineteenth-century, **2**:596a–b; **8**: 97b–98a; **9**:7a–9a; **11**:216a, 592b; **14**:351a–b; **15**:127b–128a
mid nineteenth-century, **2**:27a–b, 31b–33a; **6**:242a, 243b, 245a, 266a; **8**: 514b–515b; **9**:10b, 571b–572a, 573b–574a; **12**:240b–243b
late nineteenth-century, **2**:62b–63a; **4**: 616a–b; **6**:300a; **9**:102b, 516b; **10**: 431b–432b, 576a, 580b; **14**:2b–3a
twentieth-century, **1**:455a; **2**:239b–240a; **9**:359a; **11**:430b–431a; **12**: 298a–b, 356a; **14**:3a
physical, **2**:531a; **4**:200b–203a, 614b–616b
professionalization of, **5**:83a
and psychology
see Psychology, physiological
quantitative, **2**:239b; **13**:557a
recording techniques, **9**:101b–102b, 505b; **11**:425b, 426b
sexual
ancient Roman, **11**:602b, 603a
medieval Arabic, **12**:92a; **13**:407b
nineteenth-century, **9**:86a; **11**:565b
early twentieth-century, **2**:344b; **4**: 287a*; **7**:31b–33a; **8**:356b, 359a; **10**:291b; **14**:595b–596a
theories of, **7**:274b; **8**:53b–54a, 348b; **9**: 237b–238b; **10**:598b–599a; **12**:224b, 602a–603b; **13**:231b–232a; **14**:26a; **15**:504a–b
mechanistic, **5**:157a; **8**:203a–204a, 540b–541b; **9**:511b; **12**:302b
see also Biochemistry; Congresses, scientific; Digestion; Education, physiological; Electrophysiology; Endocrinology; Historiography, of physiology; Homeo-stasis; individual species and groups of animals; Instrumentation, physiological; Mechanics, physiological; Metabolism; Neurophysiology; Nutrition; Respiration; Sensation; Terminology
Physiotherapy, **1**:304a–b
Physostigmine, **8**:454b
Phytic acid, **15**:419a
Phytochemistry
seventeenth-century, **4**:135b
nineteenth-century, **3**:45a, 159a–160a; **5**:394b, 582a; **8**:427b; **9**:79a, 95b–96a, 418b; **10**:497b–498a; **12**:124a, 619b; **14**:112a
twentieth-century, **3**:279b, 511b; **11**: 120b–121a; **12**:452a
Phytoecology, **14**:181a–182a
Phytogenesis, **12**:174a
Phytogeography
eighteenth-century, **5**:74b; **8**:378a; **14**: 387a–b
early nineteenth-century, **2**:341a, 491b, 518b, 520b–521a; **3**:30a, 44b, 45a–b; **6**:551a,b; **8**:567b; **12**:215a; **14**:116b–117b
mid nineteenth-century, **1**:20b; **2**:431a; **3**:42b, 192b; **5**:513a, 546a–b; **6**:221a–b, 288b, 489a, 490b–491a, 492a; **7**: 196b; **8**:572a–b; **9**:345b; **12**:300a; **14**:189b–190b
late nineteenth-century, **3**:317b; **4**:145b; **7**:496a–b; **12**:166a–b; **13**:248b–249a, 343a; **14**:181b, 349b; **15**:147a–148a
early twentieth-century, **1**:623a; **4**:583b–584a; **6**:122a–b; **9**:24b, 34b–35a, 535b–536a; **12**:328a–b; **15**:508b–509b
mid twentieth-century, **15**:421b
centers of origin, **15**:508b–509b

and evolution, **14**:190a
temperature, role of, **12**:328b
see also Biogeography
Phytol, **9**:558b
Phytologia Britannica, **9**:528b
Phytopaleontology
see Paleobotany
Phytopathology
see Plants, pathology
Phytophthora, **3**:613a
Phytosaurs, **7**:60a
Phytospora, **9**:176a
Phytotaxonomy
see Taxonomy, botanical
Pi
approximations to, **11**:268b–269a
construction of, **11**:268b
expression of
as continued fractions, **2**:507a
as infinite product, **2**:507a; **14**:21b
introduction of symbol, **7**:163a
irrationality of, **4**:475a; **7**:599b; **8**:139a
"pi-theorem," **2**:565b
transcendence of, **4**:475a; **5**:472b; **6**:308a, 388a; **8**:165b, 367b–368a; **15**:368b
value of
ancient calculations, **1**:189a, 222a, 223b, 224a, 227a, 445a; **2**:618a; **6**: 301b, 314b; **7**:258b; **9**:556a
Byzantine calculations, **4**:491a
medieval calculations, **1**:445a; **2**:154a; **4**:608b, 609b; **5**:366a; **7**:256b, 257b–258b, 360a; **9**:556a; **11**:228a
Renaissance calculations, **3**:181a, 277a,b; **4**:8b, 260a; **5**:409a; **9**:335b; **11**:533a,b; **15**:155b
seventeenth-century calculations, **2**: 507a; **6**:600a; **8**:28a; **12**:332b, 500b–501a
eighteenth-century calculations, **7**:559a
nineteenth-century calculations, **12**: 344b–345a; **15**:373a
twentieth-century calculations, **12**:345a
Chinese calculations, **3**:254b; **8**:319a, 320a, 419b–420a; **13**:484b–485a; **14**:543a
Egyptian calculations, **15**:691a, 696b
Indian calculations, **2**:361a, 618a; **7**: 360a; **15**:557a, 579a, 586a,b, 591b, 593b, 597a, 607b, 609b, 611a, 614b, 616a
Japanese calculations, **12**:292a
Pi Sheng, **12**:382a–b
Pianese, Giuseppe (1864–1933), 15:479–480
Piarists, **1**:546b–547a; **9**:369a; **11**:213b
Piatti, Tommaso, **3**:64b
Piazzi, Giuseppe (1746–1826), 10:591–593; **6**:325a; **7**:580b; **11**:285a; **12**:498b
observations of, **2**:99a; **5**:299b; **10**:198b; **13**:110a, 425a
Praecipuarum stellarum inerrantium positiones . . . , **13**:149b
Picard, Casimir, **15**:50b
Picard, Charles Émile (1856–1941), 10:593–595; **4**:373b; **8**:324a, 599a; **9**:410a; **11**:478a
acquaintances of, **1**:194a, 407a; **2**:88b; **4**:253a; **5**:481b; **6**:307b; **13**:249b
analysis, **4**:177b; **8**:367a; **11**:56a; **14**:13b; **15**:23a
differential equations, **2**:358a; **4**:177a; **8**:326a; **11**:59a; **12**:246b
on contemporary scientists, **5**:484a*; **6**: 95a*; **8**:288a*; **13**:250b, 251a*, 388a*; **14**:87b*
geometry, **11**:54b, 55a; **12**:331a,b
historiography, **4**:233b*; **5**:21a*, 265b*; **6**:76a*
students of, **3**:456a; **7**:71b; **9**:182b; **11**: 388a
theorems of, **3**:63a; **8**:162b, 366b; **11**:54a; **12**:212b, 331b
proofs of, **2**:304a
Picard, Jean (1620–1682), 10:595–597; **9**: 554b
acquaintances of, **8**:132a; **9**:554a
astronomy, **2**:592b; **3**:103a
editions of works, **7**:578b
geodesy, **3**:39a, 63b, 103b, 105a; **4**:584b; **7**:542b, 577b; **9**:90a, 191b; **10**:614b; **11**:525b
meteorology, **9**:117b
micrometer, **1**:341b; **5**:278b
triboluminescence, **2**:55a; **6**:170a
Picardet, Claudine, **5**:601b, 603b
Picatrix, **7**:42a
Piccard, Auguste (1884–1962), 10:597–598; 14:246b*
Piccard, Jacques, **10**:598a
Piccard, Jean Félix, **5**:465a; **10**:597b
Picchi, Camillo, **2**:591b
Piccolomini, Alessandro, **1**:419b
Piccolomini, Arcangelo (1525–1586), 10: 598–599; **1**:495a; **3**:40a
Piccolomini, Ascanio, **5**:244b
Pickering, Edward Charles (1846–1919), 10:599–601; **2**:366a; **5**:34a*; **12**:348a
at Harvard Observatory, **1**:398a; **3**:50a; **5**:33b; **6**:28a
observational program, **7**:238b
photometry, **4**:181b*, 351b; **8**:105b–106a; **12**:176b; **14**:449a*
spectroscopy, **7**:268a; **9**:194b; **12**:251a; **14**:56b
Pickering, William Henry (1858–1938), 10: 601–602; **2**:366a; **8**:521b
Pickles, Samuel, **13**:1b
Pico della Mirandola, Giovanni, **1**:158b; **3**:280b, 470a; **5**:361a, 448a; **8**:248a; **9**:74b, 132b; **10**:512b
Disputationes . . . adversus astrologiam divinatricem, **3**:100a
Picot de Lapeyrouse, Philippe-Isidore, **4**: 150a; **11**:272b
Picric acid, **4**:243a; **5**:197b; **14**:508b
Picromel, **13**:309b
Picrotoxin, **10**:498a
Pictet, Adolphe, **11**:137b–138a
Pictet, Jean François, **3**:44b
Pictet, Marc-Auguste (1752–1825), 10:602–603; **2**:134b; **4**:242a; **6**:56a*, 550a; **8**:35a, 37b; **13**:127a
Essai sur le feu, **11**:134b
Pictet, Raoul-Pierre (1846–1929), 10:603–604; **3**:11b; **4**:79a; **7**:220b
Pien-ku t'ung-yuan, **14**:545b
Piérart, Désiré, **3**:160b
Pierce, George Washington (1872–1956), 10:604–605; **4**:197b*
Pieri, Mario (1860–1913), 10:605–606; **10**: 442b, 443a, 444a
Piero della Francesca
see Francesca, Piero della
Pierre
see Peter
Pietism, **1**:160a; **2**:223b; **3**:361a; **5**:120b, 121b; **12**:600b; **14**:226a
opposition to, **15**:132a
Piètre, Simon, **8**:53b; **11**:466a
Piette, Louis-Édouard-Stanislas (1827–1906), 10:606–607; **10**:222a
Piezochemistry, **3**:334a
Piezoelectricity, **1**:557a; **3**:504b–505b; **4**: 231a; **6**:96b, 178b, 546a; **7**:526b; **11**:530a,b
amplification by, **8**:11b
discovery of, **3**:504b–505a
industrial uses of, **3**:505a
inverse phenomenon, **3**:504b, 505a
of quartz, **3**:504b, 505b; **14**:525a
instrumental use of, **3**:498b, 504b, 506b, 507a; **10**:605a
theory of, **11**:446b
of tourmaline, **1**:557a; **3**:504b
Piezometer, **3**:505a
Pig
anatomy, **9**:468b; **12**:80b, 81a–83a
embryogeny, **13**:97a–b
evolution of, **7**:480a
Pigafetta, Antonio, **7**:174b
Pigeon, **14**:299b
behavior, **14**:315a
hybridization, **14**:314b
Pigmentation
chemistry of, **5**:430a; **13**:151b
embryogeny of, **8**:115a
heredity of, **3**:493a; **7**:455b, 516b; **11**: 211b–212a
theories of, **15**:5a*
see also Animals, coloration of; Variegation
Pigments (chem.), **5**:602a; **13**:508a; **15**:472a
chemical identification of, **6**:500b
dispersion of, **6**:118b
fading of, **8**:102a
manufacture of, **13**:327a–b, 331a
ommochrome, **7**:516b
in plants, **4**:362b; **9**:95b–96a, 576b; **10**: 497b; **15**:257a–b
flowers, **14**:350a, 412a
leaves, **7**:35a; **10**:497b; **13**:486b–487b
roots, **14**:46a
respiratory, **6**:505a–b; **7**:273a–b; **10**:282a
degradation products, **10**:22b
spectra, **12**:544a
structure of, **15**:257a–b
synthesis of, **13**:310a
visual, **8**:102b; **10**:136b
Pignano, Francis of
see Francis of Marchia
Pigorini, Luigi, **7**:31a
Pigott, Edward (1753–1825), 10:607–608
Pigott, Nathaniel (*d.* 1804), 10:607–608
Pike, Zebulon Montgomery, **3**:439a; **7**:89b
Pilatre de Rozier, Jean-François (1754–1785), 10:608–610; **4**:149b; **5**:90a; **11**:166a
aerostatics, **5**:602a; **8**:78b; **9**:492b; **11**: 166b
Pile drivers, **8**:211a, 213a
Pileworms, **6**:137b–138a
Pilgram, Anton, **6**:234b
Pilling, James, **11**:119b
Pilocarpine, **3**:246b
Pilomotor mechanisms, **8**:15b
Piltdown man, **3**:606b–607a; **4**:353b
Pilularia, **7**:199a
Pilzecker, Alfons, **9**:562a
Pi-meson
see Pion
Pinacol, **2**:624a
discovery of, **5**:13a
Pinacolone alcohol, **2**:624a
Pinacyanol, **9**:403b
Pincherle, Salvatore (1853–1936), 10:610; **2**:320a; **10**:605b; **12**:326b*; **14**:47a*
Pinckneya, **9**:366a
Pincus, Gregory Goodwin (1903–1967), 10: 610–611; **3**:123a
Pine tree, **3**:351a; **9**:54a; **12**:100a
Araucanian, **4**:156b
Pineau, Sévérin, **1**:522b

Pinel, Philippe (1745–1826), 10:611–614; **2**:122a, 508a; **3**:2b; **4**:67a*; **5**:251b; **8**:272b; **11**:559b, 617b
Nosographie philosophique, **2**:122b
Pinelli, Giovanni-Vincenzo, **4**:212b; **5**:239b; **10**:488b; **12**:106a
Pinene
structure of, **13**:550a
Pingré, Alexandre-Gui (1711–1796), 10: 614–616; **2**:126b, 414a; **3**:39a; **5**:402a; **7**:581a
Annales de l'astronomie, **7**:544b
Cométographie, **15**:311b
Pini, Pier Matteo, **4**:487a
Pinkus, Albin George, **3**:448b
Pinzón, Vicente Yáñez, **10**:617a; **15**:88b
Piobert, Guillaume, **11**:63a
Pion, **4**:580a, 583a; **13**:240b, 241a
discovery of, **11**:117a–b
Pioneering
twentieth-century, **2**:374a–b
Piotrowski, Gustav von, **6**:250a
Piperaceae, **9**:417b; **13**:456b
Piperidine, **5**:215a
Piperine, **10**:497b
Pipes, sound in, **12**:128a,b; **14**:204a, 206a, 565a
coaxial, **2**:41b
conical, **2**:41b; **4**:222b
cylindrical, **4**:222b
nodes, **12**:129b
open ends, **6**:242b; **14**:289b
Pipette
medieval discussion of, **1**:61b–62a
naming of, **5**:325b
Pipinus, Franciscus, **15**:173b, 174a
Pippard, Alfred Brian, **8**:480b; **13**:454b
Piqûre, **3**:184b
Piqûre sucrée
see Diabetes, puncture
Pires, Andre
Livro de marinharia, **14**:583b
Pires, Tomé (*ca.* 1470-*ca.* 1540), 10:616
Pirī Rais (or Re'is), Muḥyī al-Dīn (1470–1554), 10:616–619
Piria, Raffaele, **3**:46a, 47a; **4**:246a; **5**:536b
Pirie, Norman Wingate, **13**:152b
Pirogov, Nikolay Ivanovich (1810–1881), 10:619–621; **2**:597b
Piroplasmas, **2**:534a; **11**:442b
life cycles, **7**:429a
Piroplasmosis, **7**:428a
control of, **7**:428b–429a
Pirquet, Clemens von, **4**:404b, 405a; **5**:564b; **7**:427a
Pirsson, Louis Valentine, **3**:484a; **7**:5a–b; **12**:230a; **14**:183b
Pisa
botanical garden, **5**:383b
Ducal College, **14**:586b
tower of, **3**:3b; **4**:8b; **5**:238b
Pisa, University of
in Renaissance, **5**:238a,b; **10**:270a; **15**:45a
engineering at, **4**:102b
medicine at, **9**:62b
Pisano, Leonardo
see Fibonacci, Leonardo
Pisarev, Dmitrii Ivanovich, **1**:361a; **5**:211a; **10**:431a
Pisciculture, **1**:388a, 405a; **8**:146a; **9**:550b
Pisenti, Gustavo, **13**:589b
Piso, Willem (*ca.* 1611–1678), 10:621–622; **9**:122a
Pissarro, Camille, **3**:241b
Pistil, **2**:219a; **13**:405b; **14**:586a
Pisum
see Pea
Pitcairn, Archibald (1652–1713), 11:1–3; **3**:458b; **5**:520b; **6**:37a
on contemporary scientists, **1**:593b; **9**: 245a, 529a*
critics of, **8**:416b
physiology, **13**:46b
students of, **1**:594a; **3**:244a; **9**:479b
translation of works, **4**:45b*
Pitch (musical)
standard, **7**:445b; **8**:399a; **13**:11b
Pitchblende, **2**:565a
radioactivity, **3**:499a, 507a,b; **4**:356a
Pitchstone, **13**:273a
Pitfeild, Alexander, **13**:528a
Pithecanthropus, **6**:8b, 9b, 527b; **8**:29a; **10**: 17a; **13**:275a, 504a
Pithiatism, **2**:35b
Pitiscus, Bartholomeo (1561–1613), 11:3–4; **4**:619b
Thesaurus mathematicus, **3**:262a; **11**: 397b; **14**:51b
Trigonometria, **5**:594a*
trigonometry, **14**:471a
Pitocephalus, **9**:176a
Pitot, Henri (1695–1771), 11:4–5; **3**:282a; **4**:208a
Pitt, Moses, **9**:465a
Pitt, William, **1**:326a; **9**:67b, 68a
Pittacal, **11**:359b
Pittosporaceae, **2**:519b
Pitt-Rivers, Augustus Henry Lane Fox (1827–1900), 11:5–6; **1**:562b; **8**:529b; **14**:504a,b
Pittsburgh, University of
observatory, **8**:19b
Pituitary
see Glands, pituitary
Pituitrin, **1**:123b
Pitus, **14**:462b
Pixell, Helen Lucia Mary, **5**:467a
Pixii, Hippolyte, **8**:188a
Place
ancient discussion, **1**:252b–253a, 256b, 257a; **13**:93a
medieval discussion, **1**:27b–28a; **2**:619b; **3**:470a; **6**:208a*; **10**:420b; **15**:496a–b
see also Natural place
Placenta, **11**:514b; **14**:287a
as bacterial filter, **3**:587b
blood supply to, **6**:569b
development of, **6**:535b; **7**:95a
function of, **4**:511a
glycogenesis in, **2**:29b
hormones of, **3**:352b
naming of, **4**:511a
structure of, **9**:416a
types of, **4**:267a*, 511a
Placers, **2**:439b; **12**:293a, 514a
Placoderms, **7**:61a
Plagiarism
scientists accused of
ancient Greek, **2**:550a
medieval, **2**:314b; **3**:393b–394a; **13**:38b
Renaissance, **3**:355b; **4**:587a; **5**:362a–b, 580b; **10**:271a–b, 281a
seventeenth-century, **1**:541b; **4**:48a,b, 59b; **5**:522b*, 588b; **7**:294b, 583b; **8**:132a, 164b–165a; **9**:16a, 116b, 247a–b; **10**:84a–b; **11**:587a; **12**:151b, 501b; **13**:559b
eighteenth-century, **2**:54b; **3**:51a–b; **4**: 87a; **5**:411a; **7**:276a, 576a; **9**:188b, 454b, 500a; **12**:444a, 561b
nineteenth-century, **2**:206a; **3**:550a; **4**: 36a, 341b–342a; **6**:59b, 82b–83a, 247b; **7**:424a; **8**:260b, 349a, 404a; **9**:238a, 448b; **11**:384a, 565b; **15**:362b
twentieth-century, **2**:319a–b; **3**:606b
scientists guilty of
medieval, **10**:421a; **12**:82a; **15**:35a
Renaissance, **1**:94a, 606a, 620b; **3**:272b; **5**:348a–b; **7**:39a; **9**:75b; **10**:515a; **13**:260a; **14**:11b; **15**:161a
seventeenth-century, **3**:59b, 60b; **5**: 240b; **8**:464b, 525a; **13**:480a–b; **15**: 29a–b
eighteenth-century, **9**:481b
nineteenth-century, **3**:323b, 324a; **6**: 479a; **12**:536b; **13**:547b–548a
see also Patents, Priority disputes
Plagioclase, **2**:369b; **13**:476a
substitution in, **14**:588a
Plagiograph, **13**:220b–221a
Plagiolophus, **6**:595a
Plagiostomi, **9**:572b
Plague
medieval study, **3**:218b–219a; **4**:164a
Renaissance study, **1**:78a–b, 399b, 604a, 611b; **5**:107a, 586a; **7**:16b; **8**:483b; **10**:152a; **11**:33b, 608a; **13**:534b
seventeenth-century study, **7**:537b; **14**: 405b
eighteenth-century study, **9**:582a; **10**: 568b; **12**:303a, 458b
nineteenth-century study, **8**:313a
twentieth-century study, **3**:463a
bacteria, **14**:551b
bubonic, **3**:218b; **5**:220a, 270b, 316b
bacillus, **5**:40a; **7**:392a, 427b
prevention of, **5**:270a; **14**:551b; **15**:96b–97a
control measures, **1**:493b; **2**:390a, 524a*; **7**:60b(n), 427b
endemic, **7**:427b, 428a
epidemics, **5**:230a, 379a; **6**:254a; **7**:60b(n), 150b; **8**:281b; **14**:505b; **15**:96b
etiology of, **1**:424a–b; **3**:463b; **7**:427b; **11**:445a
historiography of, **13**:142b
phagotherapy, **6**:298b
pneumonic, **3**:218b
serum therapy, **8**:556b
Siberian, **5**:269b
transmission of, **1**:493b; **3**:219a; **4**:495a; **5**:107a; **7**:16b, 150b
vaccination, **6**:11b–12b; **8**:556b; **14**:551b; **15**:96b
Plains
structure of, **7**:149a
Plaisancian (stratig. stage), **9**:240b
Plana, Giovanni (1781–1864), 11:6–7; **10**: 481a*
hydrodynamics, **13**:75b
Planaria, **10**:444b
regeneration in, **3**:247b; **8**:169b, 357a
Planation, **5**:395b
marine, **11**:277a
Plancherel, Michel, **2**:265a
Planchon, Jules, **13**:464a
Planck, Max Karl Ernst Ludwig (1858–1947), 11:7–17; **4**:357a; **7**:383a*; **12**:530a; **15**:451a
acquaintances of, **2**:466a; **4**:294b, 312b; **6**:28b, 342b, 349b*; **8**:499b*; **9**:260b; **11**:422a, 582b, 611a, 613a; **12**:71a, 221b; **14**:171b; **15**:37a, 41a, 251b, 448a, 450a
on contemporary scientists, **3**:423a*; **11**:584b*; **12**:615b; **14**:330b*
correspondence of, **4**:197b*
critics of, **14**:338b
electrodynamics, **11**:420a
electron theory, **7**:264a
influence of, **4**:617a; **5**:117a; **11**:612a
influences on, **11**:150a
optics, **12**:528a

philosophy of science, **14**:173b
physical chemistry, **7**:462a; **15**:435b, 459a, 460a, 463a
Planck's constant, **2**:241b, 242b, 246a, 248b, 399a; **3**:597b; **4**:195a, 281a, 317a; **7**:211b; **10**:106b; **11**:12b–13a; **12**:43b, 528b; **15**:49a
determination of, **4**:195b; **9**:396b
Planck's law, **3**:327a; **4**:196b; **11**:150a–b
quantum theory, **2**:241b, 243b, 262b; **4**: 293b–294a, 316a; **6**:351a; **7**:382a, 528a; **8**:52a, 497b; **12**:43b, 251b; **15**:48b–49a
confirmation of, **11**:583a
radiation distribution law, **4**:316a, 317a–b; **6**:164a; **7**:84b–85a; **8**:20b, 478a, 552a; **10**:346b; **11**:12b, 582b; **13**:104a; **14**:338b; **15**:41b
derivation of, **15**:48a,b
testing of, **14**:172a
relativity, **4**:322b; **8**:182a
students of, **1**:23b; **2**:337b; **4**:194b, 485a; **6**:16a; **8**:50b; **9**:260a, 395a; **11**:355a; **12**:177a; **14**:613b
thermodynamics, **5**:390b; **8**:11a, 598b; **11**:149b; **15**:442b–443a, 447b
Vorlesungen über Thermodynamik, **4**: 576b
Plancy, Guillaume, **4**:585a,b, 586a
Plane (geom.)
Cartesian, **2**:513a
correlation, **13**:5a–b
cuttings of, **14**:590b, 591a
decomposition of, **7**:167b
division of, **7**:206b; **13**:16a–b
Gaussian, **15**:196a
principal lines of, **14**:343a
projective, **12**:172b
complex, **13**:6a
quadratic, **6**:357a
tangent, **13**:418a
topological characterization of, **7**:72a; **9**:249a
Plane translation theorem, **2**:513a
Planetarium
ancient Greek, **1**:214a
medieval, **1**:5b
seventeenth-century, **7**:375a
eighteenth-century, **4**:565b
see also Orrery
Planetesimal hypothesis, **3**:191a; **9**:552b
Planetoids
see Asteroids
Planetology, **5**:212b
Planets, **15**:640b
apparent angular diameters, **3**:528b; **5**: 24a; **6**:515a, 604a; **15**:563a–b, 576a–b, 586a–b, 591a, 621a–b
atmospheres of, **7**:74b; **13**:82a
axial rotation, **1**:597b; **7**:385a
laws of, **14**:620b
colors, **12**:352b
conjunctions of, **1**:309b; **2**:114b, 115a, 117b, 119a, 416a, 417a,b; **3**:584b; **5**:274b, 526a; **6**:273a; **7**:112b, 131a, 139b, 512b, 582b; **8**:281a–b; **14**:28b, 576a; **15**:555b, 576a, 576b–578b, 590a, 595a, 597a, 610b, 677a,b, 680b
densities, **10**:78a
distances of
Bode's law, **2**:221a; **5**:298b; **10**:198b; **13**:425a
from earth, **1**:511a; **2**:100a; **3**:25a–b; **4**:542b, 543a; **7**:219b, 259b; **11**:187a, 197a–b; **12**:406a; **15**:604b
from sun, **3**:409a; **6**:359a*
ratios of, **7**:292a–b, 301b–302a
evolution of, **8**:522a–b
heat radiated from, **3**:327a
hypothetical, **5**:258a
life on, **3**:327b*; **5**:21b, 59a–b; **6**:611a–b
signals from, **13**:287a
light of, **6**:195b; **9**:563b
polarization of, **8**:581a
Medicean
see Jupiter, satellites of
minor
see Asteroids
motion of
ancient Greek discussion, **3**:21b; **5**: 345b, 346a
ancient Roman discussion, **11**:195a–196b
medieval Arabic discussion, **7**:338a; **11**:250b–251a; **12**:360a–b; **15**:34b–35a
medieval Latin discussion, **2**:24a, 551a; **7**:124a, 130a, 139b, 142a
Renaissance discussion, **2**:405b; **3**:423b; **5**:400a; **11**:350a–b; **13**:49b; **15**:154b
seventeenth-century discussion, **7**:219b; **8**:152a–b; **10**:426a; **12**:332a–b, 342b; **13**:96a–b
eighteenth-century discussion, **2**:335a; **15**:353a–b
nineteenth-century discussion, **2**:349a; **7**:481a
in aphelion, **2**:54b; **3**:460a
Chinese discussion, **12**:378a; **14**:162b
Indian discussion, **2**:118a; **10**:313b; **15**:556a, 568b–569a
irregularities of, **5**:243b; **10**:207a; **11**: 190a; **12**:376b; **15**:34b, 289a–290b, 557b
Japanese discussion, **15**:750a
maximum equations, **4**:556a
mean motion, **4**:556a; **7**:512b; **10**:129a; **15**:555b, 556a, 567a, 569b, 579b–590b *passim*, 595a–613b *passim*, 617b–625b *passim*
Mesopotamian discussion, **11**:223a; **15**:643b, 644a, 679a–b
number of, **12**:3b
retrogradations, **1**:247b–248a; **3**:407b, 408b; **4**:467a, 542b; **11**:195a, 196b; **15**:218b, 519b, 542a, 584a, 601a, 679a
secular inequalities, **15**:286a, 293b–295a, 323a–333a, 355b
speed of, **15**:519b, 556a, 577a, 604a–b
spiral, **12**:3b–4a
zodiacal anomaly, **12**:4b
names of, **2**:152b; **15**:719a–b
orbits, **2**:236a; **4**:542b; **5**:526a, 529a, 576a; **6**:263b, 515a; **8**:278a; **9**:141a, 310b–311a, 514b; **10**:77a, 479b; **11**:472a; **12**:325b; **15**:676a
apogees, **4**:556a; **14**:576b, 577a; **15**: 599a–b
axes, **11**:206b; **13**:153b
dimensions, **3**:103a; **15**:556b–557a, 560b–561a, 579a, 591a, 608a–b, 609a, 616a
eccentric-epicyclic model, **1**:189b–190a, 247b–248a; **2**:118a; **7**:124b, 143a; **11**:195b–196a
eccentricities, **2**:123b; **5**:88a; **7**:123a; **8**:277a; **11**:196a; **15**:327a–329a
ellipticity, **1**:311a; **2**:216b, 310b, 348b, 349a; **3**:457b; **7**:481b; **14**:446b, 594b
elongations from sun, **15**:540b, 545b, 546a, 679b
inclination, **2**:43a, 54b; **8**:143b, 277a; **11**:196b; **13**:95a; **15**:327a–329a, 560a–b, 583a, 591b, 607b, 610a, 620b
nodes, **4**:556a; **5**:435a; **7**:564a; **15**:556a–b, 568b, 584a, 588a–b, 589a, 591b, 599a–b, 604a, 607b–611b *passim*, 618a–623a *passim*
oppositions, **15**:679a,b
secular variations, **2**:299a; **7**:564a,b; **9**:174b; **15**:484b
theories of, **2**:99b; **3**:407b–409a; **5**: 299b–301a, 435a; **6**:514b–515a; **7**:7a, 38b, 295a–297a, 303b; **8**:522b; **10**: 34b, 61b, 62b–64b; **12**:280b; **15**: 557b–558a
vectorial elements, **1**:429a
order of, **9**:1b; **11**:197a, 223b; **12**:5a; **13**:326a; **14**:637a; **15**:33b, 35a,b, 203a, 591b, 719a
origin of, **7**:385b; **8**:618a; **13**:179b
condensation theory, **7**:232a
periods, **11**:195a–b
sidereal, **15**:519b
synodic, **3**:319b; **4**:467a; **15**:540a, 541b–542a, 584b, 598a–b, 601a, 602a, 606a
zodiacal, **3**:319b; **5**:345b
photography of, **12**:139a
photometry of, **12**:353a
positions of, **2**:511a; **4**:547b; **6**:149b; **7**: 123b, 125a–b, 130a–b, 140a–b, 472b; **8**:504a; **11**:322a; **14**:390a,b; **15**:678b, 725b, 729b
altitudes, **5**:274b
culminations, **14**:593a
latitudes, **2**:117b; **5**:617b–618b; **7**:118b, 124a, 139b, 512b; **11**:187a, 196b, 197a, 504a; **14**:273b; **15**:560a–b, 589b
longitudes, **1**:48b, 309b; **2**:114b, 117b, 119a, 416a,b, 417b; **4**:100a; **5**:274a, 617a–b; **6**:273b; **7**:219b, 361a, 512b, 582b; **9**:607a; **10**:129a, 540b, 541a; **12**:598b; **14**:28b, 576b; **15**:125b, 541b, 555b, 557b, 566b–568b, 576b–589a *passim*, 593a,b, 601a–602a, 604a–b, 610a, 614a, 619a–623a *passim*
risings and settings of, **15**:578b
heliacal, **1**:309b; **2**:114b, 115a, 117b, 119a; **7**:582b; **10**:419a; **11**:196b; **15**: 588a, 595a, 643b
satellites of, **4**:176a; **6**:604a; **10**:602a
orbits of, **3**:105a, 583a–b; **6**:49b
shape, **3**:460a; **6**:94b; **8**:135b; **15**:295b–296a, 319b–321b
size, **3**:25b; **4**:543a; **7**:259b; **11**:187a
diameters, **1**:466b; **2**:403b; **6**:94b, 149b, 363b; **7**:219b; **10**:595b; **11**:197b; **15**: 560b, 591a, 608a, 616b, 621a
mass, **5**:224a; **6**:335b; **8**:278a; **10**:77b; **15**:353b
volumes, **11**:197b
spectra, **4**:69a; **5**:143a; **9**:184a; **12**:454b–455a
stations, **1**:189b, 247b–248a; **2**:551a; **3**: 407b; **7**:582b; **10**:419a; **11**:188a; **15**: 563b
surfaces, **3**:101b; **8**:581a; **9**:2a; **10**:119b
synodic arc, **15**:545b
tables of, **1**:36a, 510a, 512b; **2**:155a, 359b; **3**:430b; **4**:14b, 16b; **5**:224a, 617a–618a; **6**:115b, 273b–274a; **7**:86b, 118b, 123a,b, 130a–b, 139b–140b, 259b, 304b–305a, 577b; **8**:144b*, 278a, 369b; **9**:21a, 42a, 77b, 563b; **10**:34a–b, 540b; **11**:114b, 195b; **12**: 358b, 362a–b; **13**:536b–537a; **14**: 273b, 359a, 390b, 575b–579a, 593a; **15**:218b–219a, 478a, 540a–625a *passim*, 679a–b
temperatures, **10**:107b; **11**:123a
terrestrial influence of, **10**:548a; **13**:348a

Planets (cont.)
transits of, **7**:219b; **12**:598b
black drop, **14**:479b
trans-Neptunian, **8**:50a
visibilities, **2**:118a; **7**:338a; **10**:129b; **14**:28b; **15**:540a, 561b–564b, 575b, 576a–b, 584a,b, 586a, 589b, 592a, 595a, 598a,b, 601a, 607a–b, 620b, 621a, 625a, 677a,b, 680b
see also Astrology; Cosmology; Mechanics, celestial; Parallax, planetary; Perturbation theory; specific planets
Planimeter, **4**:40a
polar, **1**:147b
Planimetry, **6**:544b
Planisphere, celestial, **2**:133a; **3**:559a*; **4**:565b; **6**:68a, 69b; **13**:225b
Plankton, **1**:92b; **2**:142a; **3**:321b–322a; **4**:289b; **6**:9a; **7**:184a, 447a; **12**:166b; **15**:158b
collection net, **7**:447b; **13**:355a
composition of, **9**:78b
distribution of, **3**:322a
fossil, **11**:594a
life-cycles, **8**:519a
migration of, **13**:360b
naming of, **6**:287b
quantitative study, **6**:287b–288a
seasonal variations, **9**:587b
as stratigraphic tool, **3**:520a
suspension of, **10**:251b
taxonomy, **6**:589b
Plantamour, Émile, **1**:147a; **10**:483a
Plantin, Christophe, **4**:139a; **8**:121a, 435b; **10**:239a,b
Plants
adaptations
see Phylogeny, botanical, adaptations
anatomy
ancient Greek study, **13**:332b
Renaissance study, **2**:218b–219a; **15**:80b
seventeenth-century study, **8**:128b–129a; **9**:65b–66a
eighteenth-century study, **2**:286b; **7**:58b
nineteenth-century study, **1**:69b; **2**:161a, 290a, 371a, 492a; **3**:259b; **5**:441a–b, 539a,b; **8**:354a; **9**:78b, 455a–456a, 600b–601b; **10**:164a; **12**:125a, 255b–256a, 256a–b, 259a–b; **13**:405b, 462b, 462b–463a, 486a, 542a; **14**:349a–b
comparative, **1**:205b–206a; **2**:86a; **5**:217b, 534b–536a; **9**:418a
contractile tissues, **3**:338a
microscopic, **3**:180a; **5**:604b–605a; **6**:219b; **9**:79a, 345a–b, 418a–b, 441b–442a, 455b; **11**:215b–216a; **12**:99b–100a; **13**:555b
milk vessels, **12**:259a
organization, **4**:183b*
organography, **2**:430b; **5**:438b; **9**:345a, 418a–b; **13**:405b, 506b
physiological approach, **5**:623b–624a
preparation methods, **9**:455b
symmetry, **13**:405b
theories of, **9**:455b
vascular tissue, **9**:455b
see also Plants, morphology
animals, comparisons with
see Animals, plants, comparisons with
archetype, **13**:506b
chemical composition
acids, **12**:148a
alkalis, **14**:332b
analysis of, **2**:353a; **5**:352b–353a; **8**:171b, 344b, 346a, 584a; **9**:79a; **13**:619b; **14**:97b
enzymes, **5**:582a; **14**:45b, 255a
and flower structure, **15**:68a
glycosides, **5**:582a; **15**:252b
hormones, **3**:576a, 581b; **14**:255b
peroxides, **1**:363a
pigments, **4**:362b; **7**:35a; **9**:95b–96a, 576b; **10**:497b; **13**:486b–487b; **14**:46a, 350a, 412a; **15**:257a–b
proteins, **10**:150b
starch, **12**:59b
sugars, **1**:389b; **14**:255a
theories of, **5**:351b
definition of, **11**:315b
and environment, **1**:412b; **2**:257b; **3**:317b; **8**:378a, 425b; **12**:124a; **13**:542a
altitude, **3**:16b, 45a, 359a; **6**:221a, 550a; **11**:272a
chemistry, **9**:460b–461b
climate, **5**:546b; **6**:551b; **7**:585b; **13**:405b; **14**:117a, 387a
communities, **5**:441a; **14**:181b
drought, **9**:44b
habitats, **2**:290b; **3**:318a, 429a; **5**:380a
irritability, **3**:576a; **10**:575b–576a, 577a; **12**:471b; **15**:47a
light, **3**:359a; **10**:575b; **12**:59a, 168a; **13**:390a; **14**:349b
salt, **12**:166a–b
temperature, **5**:441a; **9**:44a–b; **12**:59a; **13**:405b, 596a; **14**:97a, 117a
water, **3**:359a; **14**:181b
see also Soil(s), and plants
evolution
see Evolution, of plants
fluids, **9**:418a
flow of, **5**:535b; **9**:151b, 345a; **12**:59b
osmosis, **10**:576a, 577a–b; **14**:96a, 97b–98b
water relations, **3**:581b–582a; **4**:546a; **8**:425b; **9**:44b
see also Sap
fossil, **2**:371b; **5**:379b, 380a, 441b–442a; **6**:489b; **7**:60a; **8**:572a–b
reconstruction of, **5**:498a
see also Paleobotany
as ground-water indicators, **9**:257b
growth of
apical, **9**:600b
circumnutation, **3**:575b, 581a; **14**:97b
electricity and, **2**:83a; **7**:13a
external stimuli, **5**:523b; **10**:575b; **14**:349b
measurement of, **2**:325b; **5**:523b; **14**:97b
pattern of, **6**:40a–b; **14**:97b
photoperiodism, **9**:44b
rate of, **11**:416a
stimulators, **9**:44b
theories of, **8**:205b–206a; **9**:490b; **12**:59a–b; **14**:97a–98a, 255b; **15**:524b
see also Plants, and environment; Plants, tropisms
identification of, **2**:219a; **5**:384a; **7**:585a; **9**:18a, 179a
life histories, **3**:29b; **6**:465b–466a
metabolism, **3**:279b, 511b; **11**:179a
artificial stimulation, **3**:279b
calcium, **14**:97b
carbon assimilation, **1**:361a, 362a, 389b; **7**:13a; **8**:345a–b
ion exchange, **4**:76b
nitrogen, **3**:359a; **4**:220a; **8**:344b–347b *passim,* 512b; **12**:208a; **14**:45b, 46a
organic acids, **9**:461a
proteins, **10**:281b, 575b
salt absorption, **6**:443a–444a
starch conversion, **12**:124a
vitamin synthesis, **14**:46a
morphology
Renaissance, **1**:526a; **2**:218b–219a; **9**:180a
seventeenth-century, **3**:180a, 596b; **7**:194b; **11**:314a–316a
nineteenth-century, **2**:19a, 103a, 160a, 426a–b, 518b; **3**:187b; **4**:545b; **5**:218a, 437b–438b; **6**:492a; **7**:196b; **11**:592a–b; **12**:255b; **14**:181b, 350a; **15**:267a
twentieth-century, **1**:205b–206a; **3**:188a; **7**:535a–b; **9**:461a; **15**:149b, 505b, 508b
comparative, **2**:371a–372a, 436a, 491b; **5**:535b–536a; **6**:265b; **7**:90b–91a; **13**:332b, 345b; **15**:147a
developmental, **2**:426a; **10**:575b; **13**:88a–89b
floral, **3**:422a; **11**:83a–b; **13**:88b; **14**:181b
and nutrition, **9**:461a
phyllode theory, **1**:206a
surface/bulk ratio, **2**:371b–372a
theoretical, **2**:371b; **5**:443a; **12**:59b
see also Plants, anatomy
movement of
see Plants, growth of; Plants, tropisms
nutrition of, **5**:277b; **7**:538a; **8**:71b, 205b–206a, 344b–348a; **9**:25a; **10**:83a; **11**:180a; **14**:501b
alkalies, **6**:381a; **8**:344b, 346a
carbon, **8**:345a
and environment, **11**:416a
gases, **1**:325a; **6**:237b, 283a–b
humus theory, **7**:13a
liquid culture, **12**:59a
minerals, **2**:257a; **5**:523b; **8**:346b–347b; **9**:461a; **11**:310b–311a, 416a; **12**:59a, 123b; **13**:417a
nitrogen, **7**:459a; **11**:179b
phosphorus, **4**:250b; **11**:416a
potassium, **11**:416a
rubidium, **11**:416a
salts, **2**:86b–87a; **4**:220a; **7**:459a
trace elements, **6**:444a; **11**:311a
see also Soil(s), and plants
pathology
seventeenth-century, **9**:66b
eighteenth-century, **4**:223b; **13**:411b
early nineteenth-century, **1**:521a; **2**:160a; **3**:45a; **11**:131b–132a
mid nineteenth-century, **2**:19a; **3**:43a, 588b, 612a–613b; **9**:345b; **12**:203a; **15**:130b
late nineteenth-century, **2**:103a; **5**:441a; **7**:34b–35b; **9**:112b; **13**:459b; **14**:93b, 96b; **15**:14b
twentieth-century, **4**:220a–b; **5**:504a; **6**:121b, 444a; **9**:24b, 35a, 78b–79a; **12**:125a
bacterial, **12**:468a
galls, **9**:461a; **11**:317a, 342a; **13**:173b, 563b; **15**:13b, 14a
tumors, **12**:468a
phylogeny
see Phylogeny, botanical
physical characteristics
color, **5**:351b
strength, **6**:467a
taste, **5**:535a
variegation, **1**:505b; **3**:422b
physiology
chemical influences on, **13**:462b
dormancy, **3**:43a; **8**:378a
electrophysiology, **11**:474a
freezing, **5**:441a; **9**:44a–b; **12**:59a
heat production, **4**:264b; **5**:278a, 441a
limiting factors theory, **14**:255a
poisons and, **7**:60a

secretion, **5**:582a; **9**:345b
sensation, **1**:260b; **5**:624a; **7**:538a
psychology of, **4**:557a
reproduction
see Reproduction, botanical
respiration, **2**:184a–185a, 290b; **4**:264a; **5**:277b–278a; **6**:40b–41b; **9**:44a, 450b; **12**:563a
biochemistry of, **10**:281b–282b
denial of, **8**:345a
discovery of, **7**:12b
and environment, **2**:290b; **12**:562b
gas ratio, **2**:290b; **7**:12b; **9**:78b; **12**:309a
and heat production, **5**:278a
and injury, **9**:44a
rate of, **9**:24b
and turgor, **9**:24b
sexuality, **1**:293b; **2**:188b, 257a, 390a, 491b; **3**:180a, 575b; **4**:98b, 288b; **7**: 440b–442a, 538a; **8**:374b, 376b–377a; **9**:496b; **14**:193a, 586a
denial of, **5**:218a–b; **11**:83a–b
discovery of, **9**:400b
and evolution, **4**:271a; **11**:154a
proof of, **3**:422b; **13**:553b; **15**:67a–68a
see also Reproduction, botanical
tropisms, **3**:43a; **8**:445a; **11**:315b; **12**:59a; **14**:255a–b
geotropism, **3**:581a–b; **4**:135b; **6**:467a; **7**:409a–b
sleep movements, **10**:577a
traumotropism, **3**:581a
twining, **3**:575b; **5**:513b
see also Phototropism
types
aquatic, **1**:205b–206a; **12**:167b
climbing, **3**:575b; **9**:24b, 441b; **14**:97b
creeping, **9**:24b
cultivated, **3**:43a; **7**:585b; **13**:249a; **15**: 508b–509b
dioecious, **11**:83b; **15**:67b
edible, **4**:584a; **10**:237b
exotic, **4**:99a
hermaphroditic, **12**:559b
insectivorous, **2**:599a; **3**:576a, 613b; **5**:513b; **12**:166a; **14**:98b
ligneous, **3**:359a; **9**:44a
littoral, **12**:166b
marine, **3**:338a
monoecious, **14**:586a
parasitic, **9**:461a–b, 490b; **11**:261b
subterranean, **6**:549b
toxic, **5**:582a
see also Botany; Chemistry, vegetable; Cytology; Embryology; Fertilization, botanical; Flowers; Genetics; Horticulture; Hybridization; Leaves; Paleobotany; Photosynthesis; Physiology, botanical; Phytochemistry; Phytogeography; Roots; Sap; specific plants and groups of plants; Stems; Taxonomy, botanical; Transpiration; Trees; Vegetation
Planudes, Maximus (*ca.* 1255–1305), 11:18; 1:367b; **4**:117b
translations by, **10**:111b–112a
Plaskett, Harry Hemley, **12**:349b; **13**:117b
Plaskett, John Stanley (1865–1941), 11:18–19; 4:280b; **12**:455a
Plaskett's twins, **11**:18b
Plasma, blood
see Blood, plasma
Plasma, physical, **15**:8b–9a
and magnetic fields, **2**:434a
naming of, **8**:24a–b
oscillations, **7**:617b
Plasma cells, **4**:296a
Plasma membrane, **12**:232a
Plasmal reaction, **4**:604a
Plasmaphaeresis, **1**:11a
Plasmodesmata, **12**:232a
Plasmodiophora, **14**:93b
Plasmodium, **2**:534a; **5**:503a
latency period, **14**:115b
Plasmolysis, **3**:259b; **14**:98a
Plaster of Paris, **8**:117a
Plastic nature, **3**:492a
Plastic surgery
see Surgery, plastic
Plastic technology
Renaissance, **8**:213b
Plasticity, **2**:466a; **9**:419b
of rocks, **1**:52a
Plastids
and heredity, **3**:422b; **6**:9a
Plastom theory, **3**:422b
Plate, Ludwig Hermann (1862–1937), 11: 19–20
Plate of heavens (astron. instrum.), **7**:260a
Plate tectonics, **13**:271a
Platearius, Matthaeus, **12**:83a
Plateau, Félix, **8**:518a; **11**:21a
Plateau, Joseph Antoine Ferdinand (1801–1883), 11:20–22; 6:247b
problem of, **4**:173b, 174a; **5**:608b; **11**:22a, 260a
Plateaus
basalt, **5**:336b
concept of, **6**:551b
Platelets
see Thrombocytes
Plater, Félix
see Platter, Félix
Plates, physical
orthotropic, **6**:534b
Platina, **3**:473b
Platinum, **2**:524a; **4**:156b; **5**:595b; **8**:619b; **14**:193a
alloys, **1**:44b; **3**:617b
analytical reagent for, **3**:272a
atomic weight, **4**:127b
as catalyst, **1**:586a; **3**:86a; **4**:134a, 241b; **5**:595b; **6**:285a; **13**:313a; **14**:602a
characterization of, **8**:298a
compounds of, **3**:271b; **4**:134a; **9**:18b
complex, **3**:235a–b, 321b; **11**:88a
organic, **11**:86b; **14**:600b–601a
stereochemistry, **9**:403a
coordination compounds of, **7**:179a, 529a
discovery of, **13**:530b
divalent, **3**:271b; **7**:529a
fusion of, **8**:78a
malleable, **10**:497a
melting point, **14**:39a
metallurgy of, **3**:235a, 346b; **4**:77b, 345b, 346b; **13**:280b–281a; **14**:486b–488a
occurrence of, **7**:183b*, 253b
ores, **3**:232b; **13**:280b–281a
paramagnetism, **7**:13b
platinum black, **4**:134a–b
radiation emitted by, **14**:39a
specific heat, **2**:589b
spongy, **1**:586a; **4**:134a, 241b
tetravalent, **3**:271b
in thermometry, **3**:19b; **5**:603a
use of, **13**:281a
Platinum metals, **5**:394a
analytical chemistry of, **3**:235b, 301a–b; **5**:595b
compounds of, **3**:271b, 617b; **6**:525b
separation of, **3**:271b
volatility of, **3**:482a*
Plato (427–348/347 B.C.), 11:22–31; 1:33a; **2**:577b; **4**:414a; **6**:410b
Academy
see Athens, Academy
acquaintances of, **1**:231b; **9**:269a; **12**: 575b; **13**:314b
biographies of, **7**:187a
biology, **1**:261b; **4**:384a, 385b; **13**:331b; **15**:737b
commentaries on, **1**:137a–b; **3**:456a; **5**: 232a; **10**:208b; **11**:106a*, 184a; **13**: 308a; **14**:535a–b
cosmology, **1**:259b; **2**:543a–b; **3**:21b; **9**:1b; **10**:590b; **11**:29b; **15**:34b
critics of, **2**:550a; **7**:41a; **11**:288b; **13**:91b, 292b
editions of works, **8**:191a
epistemology, **1**:589a–b; **11**:29b, 41b
education, philosophy of, **1**:251a; **11**: 24b, 26a–27a
Ideas (Forms), **1**:251a, 255b, 259b, 262a; **5**:144b; **11**:25b–26a; **12**:441a
ethics, **11**:24b, 29b, 325a–b
family of, **11**:23b
geology, **11**:28b; **13**:330b
geometry, **1**:227a; **4**:105a, 433a(n), 466b, 489b; **8**:192b, 238a; **9**:274a, 274b–275a; **11**:28a–b
on Hippias, **6**:405a,b, 406b, 407a
on Hippocrates of Cos, **6**:419a,b, 421a
influence of, **1**:2a, 82b, 258a, 259a, 268b, 495a; **2**:545a; **3**:516a, 579a; **4**:226b, 415a, 466a; **5**:121a; **9**:364a; **10**:590a; **11**:288b; **13**:307b, 325b–326a; **14**: 457b, 606a
music theory, **1**:282a,b
natural philosophy, **1**:254b; **4**:368b; **11**: 29b
politics, **9**:28b
on Pre-Socratics, **13**:295b
prisca theologia, **2**:539b; **3**:492a
psychology, **7**:137a
students of, **1**:246b; **4**:103b, 414b; **13**: 301b, 328b; **14**:534b; **15**:202a
theology, **11**:161a
translations of works, **5**:586a; **9**:437b; **15**:230b, 231b
Cratylus, **6**:290b
Dialogues, **3**:456a
commentaries on, **11**:161a–b
translations of, **14**:292b
Epistles, **2**:549b; **11**:23a
Erastae, **10**:179a,b
Gorgias, **11**:24b–25a
Laws, **4**:525b; **13**:334a
Meno, **1**:270b; **5**:409b; **8**:189b; **13**:290b
Parmenides, **1**:257b; **2**:550a; **3**:513a; **9**: 111b; **10**:324a; **14**:607b, 608a, 610b, 611b
Phaedo, **1**:270b; **6**:315b; **8**:538a; **10**:589b
critics of, **13**:94a
Phaedrus, **6**:419a
Philebos, **4**:466a
Protagoras, **6**:406b
Republic, **4**:525b; **6**:411a; **9**:30b; **12**:447b; **15**:249b
commentaries on, **9**:269a,b; **13**:292b, 326b
Sophist, **10**:325a; **13**:302a
Theaetetus, **4**:424a; **6**:290b, 291a; **11**: 222a; **13**:301b–302a, 303a, 307b, 314b–316b
commentaries on, **13**:315b
Timaeus, **1**:61b, 260a, 261a; **2**:19b, 21a; **3**:14b–15a; **4**:423a; **6**:411b, 421b; **8**: 537b; **11**:29b, 324b; **13**:303b, 333b; **14**:606a; **15**:746b
commentaries on, **9**:1a, 270a; **15**:203b
Plato Tiburtinus
see Plato of Tivoli
Plato of Tivoli (*fl.* first half of twelfth century), 11:31–33; 2:283b

Plato of Tivoli (cont.)
Liber embadorum, **4**:609a,b, 611b
translations by, **1**:22b, 23a, 226a, 227a, 508b, 509b, 512a,b, 513a; **4**:611b; **11**:349b; **13**:320a; **15**:175a, 177a
Platonism
ancient Greek, **4**:414b, 415a, 423b, 430b, 460b; **6**:305b; **14**:607b
opposition to, **4**:381b
ancient Roman, **3**:14b–15a; **5**:228b, 232a, 235b; **13**:326b
Byzantine, **1**:137a–b; **4**:159b
medieval, **1**:100a; **2**:19b, 21a; **4**:524a–525b; **7**:330b; **11**:323b, 324b, 444b; **13**:513a; **14**:388b; **15**:262a,b
Renaissance, **8**:240b; **10**:416b
opposition to, **15**:45b
seventeenth-century, **2**:612b, 613a; **5**:427a; **8**:304b; **9**:98a, 509a–b; **11**:313b, 317a, 317b–318a; **14**:370b
nineteenth-century, **3**:56b–57a; **14**:294a
twentieth-century, **7**:483a
Cambridge, **2**:612b, 613a; **3**:492a; **9**:509a–b; **14**:370b
and Christianity, **1**:137a; **10**:20b
in Japan, **15**:730b, 731a, 746b, 753b, 754b
see also Neoplatonism
Platter, Félix (1536–1614), 11:33; 8:310b
acquaintances of, **3**:342a; **5**:418b; **8**:309a; **10**:152b
anatomy, **6**:153a
natural history, **1**:526a
physiology, **1**:399b
students of, **1**:480a, 522a
Plattner, Karl Friedrich (1800–1858), 11: 33–34
Platyhelminthes, **2**:322a
life history, **12**:192b, 193a–b
nervous system, **8**:3b
Platyphora, **8**:528b
Platypus, **2**:504a; **3**:192b
Plaugiaulax, **3**:606b
Plaut, Felix, **15**:522b
Plavmornin (Plavchy morskoy nauchny institut), **9**:331a
Playfair, John (1748–1819), 11:34–36; 8:261a; **11**:497a; **12**:262a, 445a(n); **13**:54b; **14**:132b
acquaintances of, **2**:174b; **10**:8b; **12**:522b, 523a; **14**:169a
critics of, **4**:28a
geology, **3**:210b; **5**:445a; **9**:163b; **11**:376b
on Hutton, **6**:578a,b, 579a, 586a,b; **7**:389a
hydrodynamics, **6**:503a
Illustrations of the Huttonian Theory of the Earth, **6**:584b; **7**:69b; **9**:534a
Playfair's axiom, **6**:202b; **11**:34b
Playfair's law, **11**:35b–36a
Playfair, Lyon (1818–1898), 11:36–37; 3:448b; **4**:78a,b; **5**:125a; **6**:463b*; **8**:93a, 346b; **12**:478a
on gases, **2**:588a; **7**:451a
Pleasure principle, **5**:175b
Plebański, Jerzy, **7**:10b
Pléiade (poets), **10**:494a
Pleiades (star cluster), **4**:219a; **8**:582b
mapping of, **6**:49b; **7**:87b*; **10**:256a; **14**:479b
nebulae, **12**:455a
occultation of, **7**:209a; **11**:545a
photography of, **4**:351b; **5**:480a; **6**:282a; **11**:484b; **12**:37a
proper motion, **13**:474b
spectra, **6**:351a,b; **11**:542b
Pleione (star), **13**:118a
Pleistocene epoch
alluvial chronology, **2**:549a
glaciers, **2**:549b; **3**:608b; **5**:338b; **6**:116a; **7**:149a; **10**:504a; **12**:167b; **13**:270a–b; **14**:517a,b
paleobiogeography, **10**:16b
paleoclimatology, **2**:549a–b; **4**:565a*; **12**:168a
paleogeography, **3**:595a; **7**:144a; **13**:257b
paleontology, **1**:491b; **2**:110a, 505a; **3**:607a; **4**:519a; **5**:500b, 537b; **6**:221b; **7**:30b, 278b; **9**:134b; **10**:16a–17a, 40b; **11**:336a–b, 483b–484a, 500b; **13**:44a, 72a, 275a
stratigraphy, **1**:131a, 286a; **4**:361b, 402b–403a; **5**:329b, 395b; **7**:267a
Plektoids, **9**:300a
Plemelj, Josef, **2**:144b; **11**:452b; **12**:513a
Plempius, Vopiscus Fortunatus, **4**:64b–65a; **14**:254b
Plenčič (or Plenciz), Marcus Antonius (1705–1786), 11:37–38
Pleochroism, **13**:476a
artificial, **12**:305b
theories of, **6**:19a; **12**:520a
Pleomorphism, **8**:406a
in fungi, **3**:612a
Plesianthropus, **2**:505a
Plesiosaurus, **3**:395b; **14**:410a
Plethora (medic.), **4**:386a
Plethysmography, **2**:1a, 367a; **4**:615b; **9**:546b, 547a
Pleura
diseases of, **1**:360a; **4**:106a
extent of, **7**:24a*
Pleurisy, **4**:386a; **5**:228a–b
description of, **9**:40b
serous, **12**:450b
treatment of, **9**:291b; **14**:5a
Pleuritis
see Pleurisy
Pleuropneumonia, **6**:298a
Plica polonica, **3**:596b
Plimmer, Henry George, **2**:528b
Pliny (Gaius Plinius Secundus) (*ca.* A.D. 23–79), 11:38–40; 1:266a; **7**:28a
botany, **4**:560a; **5**:384a; **13**:343a; **14**:586a
critics of, **2**:293a,b; **8**:248b, 249a–b; **15**:161b*
editions of works, **3**:534a; **15**:113b
geography, **1**:381b; **8**:190a; **9**:140b, 532b; **11**:75b; **15**:88a
historiography, **4**:105b; **12**:547a; **15**:208b, 216a, 217a–b, 220b
influence of, **1**:109b; **2**:536b; **3**:401b; **11**:288b, 312a; **13**:565a; **14**:35a, 586b
influences on, **2**:256b; **3**:174b; **4**:119b; **11**:105b(n)
Natural History, **1**:150a, 565a; **3**:513a; **4**:108b, 390a,b
editions of, **1**:603b
translations of, **2**:336a; **5**:99a*, 579a*; **12**:329b
zoology, **2**:591b; **13**:502a
Pliny the Younger, **7**:27b; **11**:38a,b, 39a
Pliocene epoch, **3**:533a; **7**:144a
naming of, **8**:569b
paleontology, **1**:491b; **2**:110a, 481a; **4**:308b; **10**:40b
stratigraphy, **1**:234a, 286a; **2**:497b; **4**:68a; **5**:335a; **9**:241a; **11**:269b; **12**:517a–b
Pliopithecus, **8**:43b
Plössl, Simon, **1**:135b; **11**:69a, 216a; **13**:109b
Plot, Robert (1640–1696), 11:40–41; 1:318a; **7**:412b; **8**:307b, 416a; **13**:526a; **14**:500b, 501a
Natural History of Oxfordshire, **1**:432a; **3**:248b; **14**:379b(n)
Plotinus (*ca.* 204–270), 11:41–42; 1:275b
Enneads, **1**:27b, 276b
commentaries on, **11**:161a, 184a
influence of, **1**:269b, 333b; **2**:10b, 229a; **5**:121a; **6**:616a; **9**:111a; **11**:160b, 184a; **12**:442a; **14**:225b; **15**:494b
natural philosophy, **7**:1a; **14**:459a
philosophy, **1**:27b; **12**:576a
Plow, **7**:89a; **15**:653a
theory of, **5**:64a
Pluche, Noël-Antoine (1688–1761), 11:42–44; 7:602b
Le spectacle de la nature, **2**:286a; **8**:579b
Plücker, Julius (1801–1868), 11:44–47; 2:214b; **3**:466b; **5**:301a, 367b; **15**:195a
acquaintances of, **5**:340a
Analytisch-geometrische Entwicklungen, **5**:368b
critics of, **15**:198a
editions of works, **12**:196a
formulas of, **5**:368b
geometry, **1**:294b; **2**:215a; **3**:167a, 168a, 314a, 428a, 461b, 468a; **5**:368b; **6**:357b; **10**:139b; **11**:77b, 80a; **13**:133a
influence of, **7**:446a; **8**:323b; **12**:284a
Neue Geometrie des Raumes . . . , **7**:396b
physics, **5**:340b, 458b; **6**:439a,b; **13**:521b
Plücker-Clebsch principle, **3**:314a; **11**:45b
Plücker's coordinates, **3**:167a; **9**:430b; **11**:44b, 46a
Plücker's equations, **11**:45b
students of, **2**:299a; **6**:438b; **9**:566a
translations of works, **9**:81a
Plukenet, Leonard, **1**:431b, 432a,b; **11**:312a
Plumbago
see Graphite
Plumbing
medieval Latin, **13**:348a–b
eighteenth-century, **3**:578b
Plumeria, **11**:47b
Plumier, Charles (1646–1704), 11:47–48; 4:67a; **13**:442b
taxonomy, **1**:524a, 596b; **2**:218b, 536a; **8**:436a; **9**:18a; **11**:312b, 595a
Plummer, Andrew (*ca.* 1698–1756), 11:48–49; 3:494a,b; **6**:578a; **9**:482b; **14**:320a,b
Plummer's pills, **11**:48b
Plummer, Henry Crozier (1875–1946), 11:49; 6:516b*; **14**:155a*
Pluralism, **7**:68a,b
Plurality of worlds, **5**:59a–b, 403a; **8**:537b, 538b; **9**:26a–b; **10**:224b–225a; **12**:250b; **14**:364b, 367a
critics of, **9**:317a
Plutarch, **1**:152a, 247a, 248a,b; **4**:367a; **9**:274b, 532b
De communibus notitiis contra Stoicos, **9**:275b(n)
influence of, **11**:312a; **13**:92a
influences on, **11**:105b(n)
Moralia, **9**:274a, 296b
De Pythiae oraculis, **1**:283b
Quaestiones conviviales, **4**:433b(n); **11**:28a
translations of works, **12**:134b
Vitae parallelae, **1**:213a,b, 214a–b; **4**:392b; **6**:406a, 410b; **12**:547a; **14**:21b
Plutarch of Athens, **11**:160a,b
Pluteus paradoxus, **9**:573b
Pluto (planet)
discovery of, **8**:521a, 522b–523a; **10**:602a; **12**:456a
distance of, **13**:425a
existence predicted, **8**:522b
gravitational attraction of, **2**:516b
Pluton (geol.), **3**:325a
Plutonism, **1**:413a; **2**:158b, 439b, 554b, 578b–579a; **3**:412a; **4**:70b; **5**:518b; **6**:583b; **9**:532a–534a; **10**:265b; **12**:122a, 329a, 514a
critics of, **5**:31b

Plutonium, 10:215a
  in atomic bomb, 9:263a; 10:215b
  compressibility, 2:460a
  discovery of, 8:47b
  fission of, 4:581b–582a
  production of, 3:371a,b; 13:227a
Pluviometry, 7:577b; 12:595b; 13:445a; 14:52b
  see also Rain gauge
Pneuma, 1:261b, 262a, 264b, 315a, 324a–b; 3:319b; 4:106a–b; 7:28a, 137b; 11:29b; 14:606b, 607a; 15:498b, 738b
  in arteries, 4:383a–386a; 5:234b–235b; 6:317a–b; 11:127b–128a
  in muscle, 3:482b
  psychic, 4:383b–386a *passim;* 5:235b; 13:94a,b
Pneumatic school of medicine, 1:75a, 212a, 235a, 261b; 3:175a; 12:541a(n)
  founding of, 1:324a
Pneumatic trough, 6:42b–43a, 46b(n)
Pneumatics
  ancient Greek, 3:491a; 6:310b–312a; 10:588b
  medieval Arabic, 1:61b–62a, 446a
  seventeenth-century, 2:377b–378a, 379b; 3:482b; 5:574b–575a; 9:117a–118a; 10:292b–293a; 12:210b; 13:445a
  eighteenth-century, 3:153b, 431a, 600a–b; 5:511a; 6:36a; 11:67b; 14:72a, 74a–76b
  see also Chemistry, pneumatic; Eudiometry; Gases
Pneumococcus, 10:398b; 14:624a
  chemotherapy, 4:154b; 5:100b*
  serology, 1:342a–b
  staining of, 5:495a–b
Pneumograph, 4:615b
Pneumonia, 4:106a; 5:220b, 317a, 384b; 8:297a
  diplococcus
   discovery of, 4:276b
  hematogenic pathogenesis of, 1:438b
  hemolytic, 1:438b
  lobar, 1:342a
  mechanism of, 10:155a
  pneumococcal, 4:154b
  prevention of, 14:512a
  vaccination, 14:512b
  and vagus nerve, 2:30a
  viridans, 1:438b
Pneumothorax
  artificial, 11:482a
  traumatic, 9:483a, 545b*
Pochvovedenie, 4:144b
Pockels, Agnes, 4:76b
Podostemeae, 13:489b
Podostemonaceae, 14:255a
Poecilogony, 5:385b
Poetry
  philosophy of, 5:107a
  physiological analysis of, 2:531b
  Mesopotamian, 15:636a–637a
  scientific
   ancient Greek, 4:367a, 467a
   ancient Roman, 8:536b–538b; 9:79b; 11:602b
   medieval Arabic, 4:556a; 11:246b–247b; 13:150a, 488b; 15:500a*
   Renaissance, 5:105b–106a; 10:494a
   seventeenth-century, 11:121b; 12:616b
   eighteenth-century, 3:578b–579a, 580a; 5:355a; 8:469a; 11:67a; 13:224a; 14:69b–70a, 464a; 15:472a
   nineteenth-century, 10:557a; 12:168a
   twentieth-century, 14:498a
   Indian, 15:533b–534a
   Mesopotamian, 15:643b
  see also Prosody
Poetzl, Otto, 7:624a
Poggendorff, Johann Christian (1796–1877), 11:49–51; 10:148a*; 11:360b*; 14:330a
  acquaintances of, 6:553a
  Biographisch-literarisches Handwörterbuch, 4:559b
  editorships, 7:451a, 533b; 9:236a
  electricity, 3:497a; 5:341a; 10:188a,b; 11:604a; 12:254a
  translations by, 3:287b(n)
Poggiale, Antoine-Baudoin (or Baudouin) (1808–1879), 11:51
Pogonofora, 3:148b
Pogson, Norman, 7:146b*; 9:563b; 10:600a
Pohl, Robert Wichard, 4:357a; 5:117b; 14:445a; 15:41a
Poincaré, Jules Henri (1854–1912), 11:51–61; 1:558b; 2:357b; 6:31a; 8:599a; 10:593b; 11:481a(n); 13:140b*
  acquaintances of, 1:194a; 4:231b
  celestial mechanics, 12:247b; 13:423a
   recurrence theorem, 2:264a; 4:293b; 11:58a; 14:614a
   three-body problem, 1:429a; 2:145a
  on contemporary scientists, 3:96b, 420a*, 454a*; 5:316a*; 6:76a*, 399a; 7:576a*; 9:103a*; 13:250b, 422b, 424a*; 14:224a
  correspondence of, 6:350a*; 7:399a; 8:563b*
  critics of, 10:485a; 11:356b, 413b; 13:551a
  Euler-Poincaré formula, 11:60a
  influence of, 1:366b; 2:144b, 236b; 3:32a, 584a; 4:100b, 576b; 5:108b; 11:478a; 12:108a,b; 15:39b, 423a
  influences on, 2:105b
  mathematics
   algebra, 7:435b
   analysis, 6:392a; 7:399a, 446a; 10:248b, 249b; 11:452a–b; 12:247a; 13:26b, 27a; 15:386b
   differential equations, 2:358a; 6:75b
   foundations of, 2:512b; 4:226b; 12:13b, 609a
   function theory, 8:433b–434a
   geometry, 11:455a
   partial differential equations, 5:152a
   probability, 9:129b
   topology, 4:9a, 479b; 6:4b; 11:449b, 452a; 12:411a; 13:599b; 14:283a, 310b
  philosophy of science, 3:278a; 4:225b; 5:122b
  physics
   electrodynamics, 2:167b; 8:496b
   hydrodynamics, 7:85a
   mechanics, 4:19b; 8:561a,b, 562b; 15:6b
   optics, 9:372b; 12:526b
   potential theory, 10:25a
   quantum theory, 7:85a
   radiation, 3:498b; 11:476a
   relativity, 4:320b, 322a; 14:317b
   tides, 4:614a
  Poincaré series, 7:399a
  publication of works, 4:177b
  students of, 2:144a; 3:456a; 7:71b; 8:257a; 9:182b, 395a; 11:388a
  translations of works, 6:76b; 8:368a; 14:202a
  La valeur de la science, 8:257a
Poinsot, Louis (1777–1859), 11:61–62; 2:88b; 7:522b; 11:76b; 12:319a; 15:482b
  geometry, 7:301a; 11:46a
  mechanics, 3:145a; 8:592a; 9:430b, 477a; 13:130b
  Poinsot motion, 11:62a
Point
  bicomplex, 12:285a
  characteristic points, 14:311a
  circular, 3:165a; 11:79b
  complex, 12:284b
  concept of, 5:154b; 11:28b; 12:284b–285a
  conjugate, 7:408a
  definition of, 2:619b; 8:195b–196a; 12:16b
  existence of, 4:49b
  imaginary, 8:431a
  motion of, 1:256a,b; 2:330a, 424b; 4:480a,b; 11:488b; 12:599b; 13:436b
   accelerations of, 12:534b
  range of, 13:16b
  systems of, 11:536a
Point Lobos Reserve, 7:102b
Point sets, 3:55a
  measure of, 3:63a; 14:572b
Poiseuille, Jean Léonard Marie (1797–1869), 11:62–64; 5:432b; 6:37b; 9:10b
  Poiseuille's law, 8:510b
Poisoning, 7:42b
  arrow poisons, 13:558a
  by arsenic, 1:529b; 3:160a; 5:165a; 6:17b; 15:172b
  by carbon dioxide, 2:61b
  by carbon monoxide, 2:26a, 31a; 6:460a
  criminal cases, 3:257a; 5:417b
  by gas, 5:465b; 8:557a; 10:121b
   aftereffects, 1:454a–b; 11:139a
   protective measures, 5:190a; 8:29b
  by iodine, 4:297a
  by lead, 3:203b; 4:301b; 5:73a, 137b, 219a; 7:421b; 11:373b; 15:107b
  by mercury, 3:206a; 5:366b; 13:70b–71a; 14:256b
  in microorganisms, 3:271b
  by oxygen, 2:62a
  use in research, 2:31a
  vegetable, 1:575b; 5:57a*
  see also Antidotes; Toxicology; Venoms; Warfare, chemical
Poisson, Siméon-Denis (1781–1840), 15:480–490; 3:143b; 6:1b, 3a*; 8:382b; 13:128b
  acquaintances of, 3:79a; 6:325a; 7:51a; 8:288b; 9:418b; 10:247b; 12:523a; 13:127a, 129a; 15:347b, 380a
  celestial mechanics, 9:174b
  on contemporary scientists, 3:451a; 5:94b, 96b–97a, 261a–b, 262b; 9:55a; 15:384a, 401b*
  and Cournot, 3:450b
  critics of, 10:5a*; 15:31a
  equations of, 2:87b; 3:304a, 305a; 4:324a, 327b, 328a; 6:472a; 10:3a, 247b
  geophysics, 5:86b; 6:503a; 8:516b
  influence of, 2:214a; 3:146a, 418a, 451b; 4:222b; 7:480b; 9:547b; 10:186b; 11:344a; 13:78b; 15:199b
  influences on, 12:130a; 15:360b, 361a, 385a
  mathematical physics, 1:365b; 5:169a; 9:548a,b, 566a
   acoustics, 3:258b; 13:76a–b, 127a–b; 15:200b
   attraction theory, 1:147a; 3:446a; 15:316b
   capillarity, 9:174b; 15:360b
   elasticity, 3:259a; 9:220b; 10:249a; 14:61b, 329b; 15:363a–b
   electrostatics, 3:445b; 6:278b; 13:375a–376b; 15:360b–361a
   heat theory, 3:82a,b, 304a, 305a; 15:378b–379a, 384b
   hydromechanics, 3:591b; 4:591b; 13:75a,b; 14:206a
   matter theory, 10:4b*
   mechanics, 4:126b; 7:54b

Poisson, Siméon-Denis (cont.)
mathematics
analysis, **3**:138b, 231a; **4**:125b; **5**:96a,b; **7**:550a; **8**:558b; **10**:248b, 327b; **11**:471a; **15**:385b
probability, **2**:319a; **3**:223a; **7**:106b; **9**:419b
Nouvelle théorie de l'action capillaire, **15**:379a
Poisson approximation, **9**:454a
Poisson distribution, **2**:319a; **15**:489b
Poisson formula, **11**:58b
Poission-Bracket relations, **5**:103a
Poisson's integral, **15**:488b
Poisson's law, **6**:545b
Poisson's ratio, **3**:619a
Recherches sur la probabilité des jugements . . . , **15**:31b
theorem of, **3**:223a, 230b; **9**:324b; **13**:153b*; **15**:485b
proof of, **13**:19a
Théorie mathématique de la chaleur, **15**:379a
Traité de mécanique, **4**:222b, 576b, 577a; **6**:399a
translations of works, **4**:546b
Poiteau, Alexandre, **4**:67a; **13**:506a
Poivre, Pierre (1719–1786), 11:64–65, **3**:365b; **7**:199a; **12**:535b, 536a
Pokeweed
medical use of, **3**:344b
Poland
climatology, **11**:524b–525a
early science, **2**:526b
education
nineteenth-century, **4**:84a; **15**:502b
geodesy, **8**:351a
geography, **11**:524b
geology, **2**:108b; **15**:502b, 526b–527a
insurrection of 1863, **3**:529b, 531b
mathematics, **7**:71b–72b; **9**:248a; **12**:426b
mining, **15**:502b
physics, **7**:10b
religion, **9**:15b
Polarimeter, **1**:202a; **7**:620a; **9**:425b; **10**:28b; **14**:360a
photoelectric, **8**:581b
Polarimetry, **9**:601a
Polaris (star), **5**:365b
magnitude, **10**:600a
variability, **6**:352a
Polariscope, **1**:202a; **2**:139b; **5**:557a; **8**:581a; **12**:304b
Polarity (geom.)
plane, **13**:5b
Polarity, principle of, **3**:335a; **5**:194b, 446a; **11**:394b, 473b–474b; **12**:155a–b
apolarity, **11**:390a
Polarization
of X rays, **1**:457b–458a; **2**:203a; **3**:369b
Polarization, electrical
of dielectrics, **1**:347b; **6**:344b
electrochemical, **2**:94b–95a; **7**:65b; **8**:112b–113a
galvanic, **1**:297a; **7**:458a
and quantum theory, **7**:492a
and surface tension, **8**:387b
theories of, **15**:435a
waves of, **8**:492a,b
Polarization, magnetic
rotatory, **2**:203a; **3**:505b
Polarization, optical, **1**:558b; **4**:174b; **5**:87a, 167a; **6**:610a; **9**:174a
agents for, **6**:293b; **10**:110a
angle of
and index of refraction, **2**:452a
chromatic, **1**:201b
periodic effects, **5**:168b
theories of, **5**:169b
circular, **4**:66a; **5**:68b, 169b–170a; **9**:74a; **12**:281a–b
theory of, **5**:166b–170a; **9**:59a
compensated, **2**:139a
complementary, **2**:137b
discovery of, **15**:362a
elliptical, **2**:452b; **5**:186a; **7**:313b; **9**:616b; **13**:77a
lamellar, **2**:137b
laws of, **2**:137b; **3**:146b
naming of, **9**:73b
in outer space, **1**:353b
of radiant heat, **5**:68b; **9**:265a; **11**:115b
by reflection, **2**:137a–139b, 451b, 452a–b; **5**:169b; **9**:73b–74a
reversal of, **11**:582a, 583a
rotary, **1**:201b; **5**:597b
rotation of plane of
see Optical activity
of skylight, **1**:358a; **7**:265a
by successive refractions, **2**:452a
theories of, **1**:201b; **4**:190a; **5**:169b–170a; **8**:63a; **9**:74a; **10**:60a; **14**:567a–b
mobile polarization, **2**:137b
repulsive force, **2**:137b
see also Light, polarized
Polarography, **6**:371a–372b, 373a–374b
Polaroid, **6**:291b
Polars, **4**:49b; **11**:45b
projective definition of, **5**:569a
reciprocal, **2**:214b; **3**:213a; **5**:368a,b; **11**:79b
theory of, **11**:77b, 79b–80a
successive, **2**:214b–215a
Polder, D., **8**:475b
Polemarchus, **1**:338b; **3**:21b; **9**:269a
Polemoscope, **2**:305b; **6**:361a
Poleni, Giovanni (1683–1761), 11:65–66; **2**:43a; **3**:105a, 107a
Poles, celestial
motion of, **7**:472a
Poles, geometrical, **4**:49b
Poles, magnetic
see Magnetic poles
Poles, terrestrial
displacement of, **12**:187a
motion of, **6**:240b; **10**:153b–154a, 232a; **13**:45a,b; **14**:87a, 216b
position of, **1**:623b
see also North Pole, South Pole
Polestar, **1**:283b; **10**:542b
displacement of, **10**:596a
elevation of, **9**:36a
latitude of, **1**:573b; **12**:377a–b, 379b, 518b
position of, **6**:126a; **10**:536a; **15**:88b
Polhelm, Christopher, **13**:179a
Poli, Giuseppe Saverio (1746–1825), 11:66–67; **13**:465a
Policlinico
founding of, **1**:360a
Polinière, Pierre (1671–1734), 11:67–68; **10**:145a,b
Poliomyelitis, **5**:317a
diagnosis of, **7**:624a
etiology of, **7**:624a; **8**:273b; **10**:142a
vaccination, **11**:271b
virus, **5**:40b; **7**:624a; **8**:273b
Polish Astronautical Society, **14**:590b–591a
Polish Physical Society
founding of, **9**:616b
Politecnico
founding of, **2**:469b
Politi, Giambattista, **4**:451a
Politian, **8**:248a, 249a
Politics
ancient Greek, **4**:34b, 81b; **6**:406b
Byzantine, **13**:308b
medieval, **4**:525b; **9**:30b–31a; **10**:172b; **12**:447b
Renaissance, **12**:548b; **13**:50b–51a
seventeenth-century, **7**:204a, 205a; **8**:149b; **15**:69a
eighteenth-century, **4**:86a–b; **6**:66b, 468b, 469b; **7**:202a; **8**:133a; **9**:490b–491a; **10**:501b
nineteenth-century, **6**:551b; **7**:601a; **9**:385a–b; **12**:117a–b
twentieth-century, **11**:258a; **13**:551a
and philosophy, **13**:308b
psychology of, **7**:192a; **9**:491a; **11**:309a–b
scientists and
ancient Greek, **1**:231b; **6**:212b; **11**:23b–24a, 28b–29a, 219b–220a; **14**:534b–535a, 536a
ancient Roman, **10**:230b; **12**:309b; **13**:588a
Byzantine, **11**:183a; **13**:308a–b
medieval, **2**:149b; **5**:113b, 365b; **6**:79a, 275a–b; **7**:321a–b; **11**:600b; **15**:494a
Renaissance, **1**:77b; **3**:471b, 513a–b, 514b, 515a; **4**:97b, 212b; **5**:447b; **9**:264a, 609b–610a; **11**:287a, 338a–b, 608a; **14**:18a
early seventeenth-century, **4**:494b–495a, 566a–b; **6**:124b; **9**:23b, 507a; **10**:546b; **13**:613b; **14**:286a
late seventeenth-century, **2**:311b, 377b; **5**:520b; **6**:537a; **9**:465b, 497b, 529b; **11**:1a; **12**:256b, 580b, 581b, 586a; **13**:213b–214a; **14**:51b, 465a; **15**:29a
early eighteenth-century, **3**:344a; **4**:492b–493a; **5**:157a, 449a, 521a; **7**:442b; **8**:459a; **9**:464a; **11**:527a; **12**:429b; **13**:67b, 68a
late eighteenth-century, **1**:64b, 563b; **2**:321b, 510a; **3**:1b, 38a, 399b–400a, 579a; **4**:27b, 107b, 275a, 402a; **5**:75b, 110b, 130a, 138a–b, 267a, 411a, 600a–b; **6**:61b; **7**:88b, 89a; **8**:305b–306a; **9**:53b, 148b, 158b, 418a; **10**:327a, 438a, 509b; **11**:134b, 135a, 140b, 141b, 272b, 471b, 497a, 617b; **12**:120a–b, 549a; **13**:46b–47a, 350b, 410a, 417b–418a, 494a–496a, 571a; **14**:79a–b, 332b
see also French Revolution
early nineteenth-century, **1**:70a, 615b; **2**:320b; **3**:16b, 44b, 80a, 199a–200a, 375a, 400a, 450a, 523b–524a, 562b; **4**:402b; **5**:93b–94a, 166a, 193a, 216a, 301b; **6**:1b–2a; **7**:547b; **8**:37b, 245b, 369b; **9**:418b, 473b–474b; **10**:194b, 209a, 247a–b, 602b; **11**:464b; **13**:309b, 541b, 597a; **14**:79b–80a, 88b; **15**:375b, 376b
mid nineteenth-century, **1**:201a; **2**:281b–282a; **3**:42b, 46a,b, 132a–b, 336b–337a, 451a, 529b; **4**:73b, 74a, 242b, 257b; **5**:260a, 261a–b, 306a–b, 307a, 318a, 474b–475a, 506a; **6**:553a–b; **7**:22b, 51b, 505b, 522b; **8**:35b, 55a, 264a, 279a, 381b–382a, 545b; **9**:177a, 235b, 267b–268a, 383b, 538a; **10**:11b–12a, 28b, 328a–b, 353b–354b, 543a, 579a; **11**:61b, 78a–79a, 111a, 300a–302a, 337a, 398b–399a, 514b, 548a; **12**:38a, 200a, 261a–b, 276a, 296b, 570a; **13**:143a–b, 238a; **14**:40b–41a, 57b, 204b, 205a, 220b, 522a, 527a; **15**:133a–b, 193b, 405a–b, 526b
late nineteenth-century, **1**:360a, 361a–b; **2**:60a, 64a–b, 105a; **3**:184a, 497b; **4**:102b, 203b, 212a; **5**:211b, 385a, 573b; **7**:166b, 475b, 510a; **8**:529a; **9**:279a, 292a, 538b; **10**:165a, 321b; **11**:36b, 409b, 492a, 538b; **12**:271a;

13:24a, 394a, 523a–b, 623a; **14**:59a; **15**:38a
early twentieth-century, **2**:303b; **3**: 299b; **4**:286b, 312b–313b; **5**:83b, 329b, 363a, 460a, 573b; **6**:25b–26a, 217a, 264b; **7**:71b, 208b, 454b; **8**: 181b–182b, 183b–184b, 489b, 552b; **9**:125b; **10**:275a; **11**:19b, 613b; **12**: 10a–11a, 274a–b, 530a–b; **13**:45a–b, 416b, 529b; **14**:46b, 86b–87a, 302b
mid twentieth-century, **3**:112b; **5**:206a; **7**:155b, 158a, 248a, 486a; **8**:11b–12a, 52a–b; **9**:399b; **10**:215a, 525b, 526b; **12**:71b, 222a, 223a, 350a–b; **13**:71b, 227b; **15**:18b–20a, 74b–76b, 228b, 510b–511b
Politus, Bassanus, **13**:209b–210a
Tractatus proportionum introductorius ad calculationes Suisset, **13**:349b
Politzer, Adam, **1**:446b
Pollen, **5**:539b; **10**:9b
Brownian movement, **2**:520a–b
as climatic indicator, **2**:549a
composition of, **11**:68b
formation of, **5**:582a; **6**:465b; **9**:600a–b
function of, **6**:266a–b; **7**:441a; **11**:83b; **12**:559b–560a; **15**:68a
male vs. female, **3**:422b
membranes of
constitution of, **9**:79a
mitosis, **9**:24b
size of, **3**:422a
structure of grains, **1**:521a; **7**:441a
taxonomic use of, **2**:519a,b; **5**:424b
xenia phenomenon, **3**:421b, 422a; **13**: 478a–b
Pollen tube, **2**:492a, 519b; **5**:424b
function of, **5**:539b; **6**:266b, 465a; **9**:600a; **12**:174b
generative nuclei, **3**:422a
Pollender, Aloys (1800–1879), 11:68–71
Pollenia, **7**:272b
Pollination, **2**:390a; **13**:456a
cross-pollination, **2**:257a, 390a; **3**:575a–b; **7**:409a, 441a; **12**:588b
ecology of, **8**:378a, 518a
experimental, **3**:318b*; **8**:460a; **9**:278a,b, 280b–281a
insects, role of, **2**:390a; **7**:417b, 441a; **12**:459a, 588a–b; **15**:130b
mechanism of, **12**:587b–588b
see also Fertilization, botanical; Pollen
Pollinia, **13**:395b
Pollock, Frederick, **14**:438a
Pollution
air
seventeenth-century, **4**:495a
eighteenth-century, **9**:152b–153a
nineteenth-century, **12**:424b, 478b
twentieth-century, **12**:482a; **13**:24b
airborne, **6**:117a
industrial
early nineteenth-century, **3**:203a*
early twentieth-century, **3**:341b, 436b; **6**:117a
water, **12**:478b
ponds, **9**:176a
rivers, **5**:125a, 395b; **12**:478b; **14**:409b
Polo, Marco, **9**:190a; **11**:403a; **15**:88a
Polonium, **4**:356a
alpha particles of, **3**:502a; **7**:152b, 153a, 154a, 159a*
discovery of, **1**:559b; **3**:499a, 507a–b
electrochemical properties, **7**:152a
period of decay, **3**:500a
radioactivity, **3**:499b, 507b
Polovtsev, Valerian Viktorovich, **7**:34b
Polski przegląd kartograficzny, **11**:524b
Polverini, Giovanni, **8**:556b
Pólya, George, **4**:548a, 561b, 562a; **6**:114a, 571a; **9**:128b
Polybius, **1**:214a; **4**:81b; **6**:420a, 421b; **11**: 225b; **13**:83b–84a
Polychaetes, **13**:355a
embryogeny, **7**:402a,b
morphology, **9**:328b
phylogenetics, **9**:328b
reproduction, **3**:148b; **9**:328b
Polyclads, **8**:3b
Polyclitus, **4**:259a
Polycrystal Book Service, **4**:522a
Polydactyly, **9**:188a,b; **11**:333b
Polydora, **2**:322a
Polyembryony, **3**:493a; **13**:88b
Polyenes, **7**:518a
Polyethylene
synthesis of, **7**:21b
Polygalaceae, **3**:259a
Polygamia, **13**:392b
Polygastrica, **4**:235a
Polygonometry, **8**:299b–300a, 306b*
parallactic, **13**:111b
spherical, **1**:429a
Polygons, **5**:111a
area of, **1**:591a; **4**:609b
circumscribed, **7**:258a–b, 530b
Farey, **6**:571b
geodesic, **8**:385a–b
inscribed, **3**:277b; **4**:419a; **7**:258a–b, 530b; **10**:299b; **13**:289b–290a; **14**:19a, 21b–22a
of maximal area, **7**:561b
regular
area of, **6**:314b–315a
construction of, **4**:260a; **5**:299a
rule of sines, **7**:352b–353a, 500a,b; **8**:27b
stellated, **2**:360b, 395a, 527a; **11**:62a
Polygordius, **5**:109a
Polyhedra, **3**:143a; **5**:111a, 567b; **7**:384a
angles of, **3**:131b
congruence of, **4**:9a
Euler's theorem, **4**:55b, 479b
inscribed, **1**:474a; **13**:290b
molecular, **2**:431b
regular
see Solids, geometrical, regular
semi-regular, **10**:271a; **13**:48a
space-filling, **2**:395a
stellated, **7**:301a; **10**:271a; **11**:62a
theory of, **1**:449b; **5**:211a; **6**:92b; **7**:185a; **8**:306b; **9**:431a; **13**:23b
transformation of, **7**:207b
volume of, **1**:591a
see also Solids, geometrical
Polyhymnia (satellite), **10**:35a
Polyketides, **3**:347a
Polymastigina, **4**:142b
Polymerization, **2**:67b, 624a–b; **3**:86a–b, 246b, 279b; **5**:315a; **7**:21b; **9**:18b; **13**:3a
of aldehydes, **2**:316b, 621a; **7**:282b
of amino acids, **12**:164a
emulsion, **6**:118b
of rubber, **1**:106b; **6**:118b; **7**:323b, 457a–b; **8**:108a–b
of silicones, **7**:372b
theories of, **4**:553b; **9**:354a; **10**:15a
Polymers
diffusion of, **3**:620b
molecular weights, **3**:620b
natural, **9**:354a
sedimentation rate, **3**:620b
structure of, **13**:1b–3b
synthesis of, **4**:553a; **13**:66b, 163a–b
viscosity, **3**:620b; **13**:1b, 2a
Polymorphism, biological, **3**:148b; **8**:270b
of algae, **3**:259b
Polymorphism, crystal, **11**:465a; **13**:246b
at high pressures, **2**:459a
Polynesia
exploration of, **3**:549b
Polyneuritis, **4**:311a; **5**:209a*
Polynomials, **3**:120a*, 221b; **4**:124a; **11**:590b
algebraic
theory of, **2**:444a; **3**:142a,b; **8**:585a
Bernoullian, **12**:535a
Chebyshev, **3**:229b; **9**:127a; **13**:26b
complete
definition of, **2**:113a
derivatives of, **4**:569a–570a
deviating least from zero, **3**:229a–b; **9**: 127a
distribution of, **14**:282b
exponential, **8**:458a*
Fourier, **3**:142a
Hermitean, **3**:229b; **6**:308a; **9**:127a; **15**: 385a,b
irreducible, **6**:389b
Jacobi, **12**:513a
Laguerre's, **3**:229b; **7**:574b, 575a–b
Legendre, **6**:230b, 545b; **8**:61b, 135b; **9**: 127a; **15**:317a
*n*th-degree, **14**:630b
orthogonal, **3**:224b, 229b; **9**:127a; **12**:535a
roots of, **4**:475a–b; **7**:241b–242a; **10**:51b; **13**:56a
Zernike, **14**:617a
see also Algebra, of polynomials
Polyp (pathol.), **14**:584b
Polypeptides, **3**:511b; **5**:42b
synthesis of, **5**:4b–5a
Polypetalae, **14**:192b
Polypharmacy, **1**:290a
Polyploidization, **12**:76b–77a; **13**:478b
Polypnea
thermal, **11**:427b, 428a
Polyps (aquatic)
see Coelenterates
Polypterus, **13**:151a
Polysaccharides, **5**:4a; **7**:393b
in bacteria, **1**:342b
metabolism of, **3**:416a
physical chemistry of, **13**:163a
structures of, **6**:185b
Polystereoscope, **11**:460b
Polystigma, **3**:45a
Polystomum, **5**:453a
Polytechnisches Notizblatt
founding of, **2**:340b*
Polyteknisk tidsskrift, **5**:587b
editors of, **14**:109a
Polythionic acids, **12**:593a
Polytopes, **12**:171b–172a, 213b
Polyxenus, **2**:549b, 550a
Polyzoa
see Bryozoa
Pomeranchuk, Isaak Yakovlevich, **7**:617b, 618a; **15**:8b
Pomet, Pierre, **8**:174a
Pommard, Leblanc de, **4**:17b
Pomoideae, **8**:371b
Pomology, **1**:396a; **3**:43a, 173a*; **4**:225a*, 495b; **9**:278a, 282b, 455a–b
Pompeii, **6**:85a*
Pomponazzi, Pietro (1462–1525), 11:71–74; **4**:212b; **5**:104b; **10**:122b; **12**:134b
astrology, **1**:495b
historiography, **9**:133b; **13**:209a,b
De intensione et remissione formarum, **13**:209b
Super libello de substantia orbis, **13**:209b
Pomponius Mela (*fl.* A.D. 44), 11:74–76
Poncelet, Jean Victor (1788–1867), 11:76–82; **3**:169a; **5**:111b, 367b; **11**:166a*; **12**:73b

Poncelet, Jean Victor (cont.)
acquaintances of, **2**:214b; **14**:617b
analysis, **3**:229a
Applications d'analyse et de géométrie, **9**:553b
biographies of, **2**:88b
on contemporary scientists, **12**:325a–b
critics of, **10**:286a
on Desargues, **1**:541b
geometry, **1**:191b; **2**:214b, 418b; **3**:165a, 213a, 314a; **5**:368b; **9**:475a; **10**:331b; **11**:45b; **12**:228a; **13**:4b–5a, 18a
constructions, **1**:42a; **2**:455a; **4**:602a,b*; **7**:599b; **9**:157a
duality, **11**:45a
transformations, **3**:468a; **9**:598b
and Gergonne, **5**:368a
influence of, **3**:495b, 496a, 560a; **6**:357b; **7**:163b; **8**:323b; **9**:80a; **11**:344a, 414a; **13**:15a, 132b
influences on, **9**:476a
Introduction à la mécanique industrielle, **3**:82b
mechanics, **1**:582a; **3**:72b; **10**:3b, 286b
theorems of, **11**:45a
Traité des propriétés projectives des figures, **4**:47a; **5**:368a
Pond, John, **2**:468b; **6**:263a
Pongidae, **5**:110a
Ponograph, **9**:547a
Pons, Jean-Louis (1761–1831), 11:82–83
comet of, **3**:472b; **4**:369b; **9**:504a; **11**:544b; **13**:4a
Pons asinorum, **4**:417b, 418a
Pons Varolii, **13**:587b
Pont du Gard, **11**:5a
Pontecorvo, Bruno, **4**:578b, 580a
Pontécoulant, Philippe de, **3**:169b; **6**:399a
Pontedera, Giulio (1688–1757), 11:83
Pontine Islands
geology, **4**:140b
Pontine Marshes, **11**:164b
Pontriagin, Lev Semenovich, **5**:608b, 609a
Poor, Charles Lane (1866–1951), 11:83–84
Poor Laws, **9**:67b, 69b–70a
Poor Richard's Almanack, **1**:487a
Pope, Alexander, **2**:16b; **4**:172b; **6**:44b–45a
Essay on Man, **6**:558b; **9**:452b
Pope, William Jackson (1870–1939), 11:84–92; 5:623a; **8**:110a*; **9**:379b
crystallography, **1**:461a, 462a
stereochemistry, **8**:109b; **9**:403b
Popov, Aleksandr Nikiforivich (*ca.* 1840–1881), 11:92–93; 2:622b, 623b
Popov, Aleksandr Stepanovich (1859–1906), 11:93–94; 5:26b
Popov, N. A., **10**:435a
Poppe, Johann Heinrich Moritz von, **15**:406b
Poppius, Hamerus, **1**:550a; **8**:131a
Popular Science Monthly, **3**:130b
Popularization of science
ancient Roman, **9**:1a; **12**:309b
medieval Latin, **10**:223b, 475b
Renaissance, **15**:156a
seventeenth-century, **5**:58b–62b
eighteenth-century, **2**:42b; **4**:565b; **5**:221a; **9**:142a; **10**:572a; **11**:42b–44a; **12**:427b; **13**:425b, 601b; **14**:83a–84b; **15**:342b, 350a
early nineteenth-century, **3**:557b, 601a; **5**:276a–b; **10**:183a, 195b; **12**:521b; **13**:414a; **14**:128b, 562b, 568b
mid nineteenth-century, **3**:89a; **4**:532a; **5**:84b; **6**:138a, 242b, 252b, 280a, 552b, 595b–596a; **10**:11b; **11**:217a, 537a; **12**:173b, 215a, 289a; **13**:522b; **14**:439b; **15**:134a
late nineteenth-century, **2**:60b; **4**:144b, 147b, 504b; **6**:102b; **7**:498b; **9**:86a; **11**:382a; **13**:80a
twentieth-century, **1**:301b; **5**:454a; **6**:21b, 23a, 94a; **10**:217a–b; **11**:9b, 267a; **15**:50a, 61b
anthropology, **5**:145b
archaeology, **9**:490a; **12**:181b; **13**:357b–358a; **14**:504b
astronomy
ancient Greek, **3**:319a; **5**:345b
Renaissance, **1**:178b
eighteenth-century, **15**:342b – 345b, 736b
early nineteenth-century, **1**:200b; **2**:421a; **5**:518a; **11**:236a; **13**:109a
mid nineteenth-century, **1**:202b, 358a; **3**:377a; **7**:209b, 609a–b; **9**:2b
late nineteenth-century, **4**:176a; **5**:21b–22a, 102b*; **7**:400b; **11**:162b; **12**:248a; **13**:423b*
twentieth-century, **1**:87b; **3**:358a*; **6**:74b; **7**:86a, 400b, 472b; **8**:539b; **9**:184b; **10**:35a; **12**:23b, 153b, 250b, 349b; **13**:119a
bacteriology, **5**:128b; **11**:177a; **14**:623a
biology
nineteenth-century, **8**:170a; **9**:334b; **11**:110a; **12**:58a, 173b, 175a
twentieth-century, **7**:99b, 278b–279a, 286a*, 503b; **8**:27b*, 591b; **10**:523a; **15**:74a
botany, **2**:290b; **3**:340a; **4**:273b; **6**:221a; **9**:18a; **11**:595a; **12**:471b; **13**:88a, 542a; **15**:224a
chemistry, **3**:341b, 398b; **8**:330a; **11**:616b; **12**:433b; **13**:373b; **14**:492a
evolution, **13**:276a–b; **14**:233a, 236b
genetics, **7**:99a; **11**:212b
geography, **1**:178b, 203a; **9**:581a; **12**:301b
geology, **1**:358a, 413a; **2**:158b; **3**:434b; **4**:599a; **8**:246a; **9**:585a; **10**:29b, 430b; **11**:236a*; **12**:230a, 274a–b, 344a, 433b, 514b–515a; **14**:439a; **15**:92a, 152a
mathematics
Renaissance, **7**:419a
seventeenth-century, **10**:264a; **12**:206b
eighteenth-century, **1**:84a
nineteenth-century, **1**:194b, 459b; **2**:469b; **3**:322a; **6**:76b; **10**:479b; **11**:60b, 116a, 236a; **15**:366b, 374a–375a
twentieth-century, **1**:584b; **2**:303a; **11**:256b; **12**:16a; **13**:26b, 357a, 424b, 428b; **14**:347a, 559b
medicine, **9**:423a*, 582a; **10**:321b; **15**:271b–272a
meteorology, **1**:203a, 358a; **4**:410b; **7**:400b; **12**:201b, 366b, 512a
microscopy, **1**:411b
mineralogy, **1**:358a
natural history, **1**:587a; **2**:128a; **7**:458b–459a; **10**:509b; **13**:565a–566a; **14**:397b
nature study, **1**:72b
oceanography, **9**:590a; **10**:564a*
paleontology, **5**:221b, 296b; **10**:262a
pathology, **12**:482a
philosophy, **14**:483b
photography, **1**:203a
physics
seventeenth-century, **1**:483b, 484a; **11**:506a–b; **12**:211a; **14**:363a
eighteenth-century, **3**:215b–216b, 537b–538a, 546a; **4**:45a; **7**:371b; **10**:145b, 501a; **11**:67b–68a; **12**:429b–430b
nineteenth-century, **8**:596a, 597b; **11**:111a, 116a; **13**:52b
twentieth-century, **3**:125a–b; **4**:514a; **5**:272b, 364a, 407a; **6**:615b; **7**:10b, 26b, 86b, 516a; **8**:489a; **9**:399a; **10**:525b; **12**:16a; **14**:171a, 498a
statistics, **14**:554b
technology, **6**:468b; **13**:351b–352a, 398b–399a; **14**:106b*, 363a
zoology, **6**:454b; **11**:443b; **15**:126a, 472a–b
Population
dynamics of, **8**:512a
human, **9**:383a, 385a
as economic factor, **10**:566a; **13**:448a
"explosion" of, **3**:564b
and food supply, **4**:272a; **9**:67b–70a; **10**:444b
increase of, **1**:424b; **12**:570b; **13**:616b
limits on, **3**:570b–571a; **9**:68a,b; **13**:616b
statistics of, **2**:319a; **14**:179a–b, 230b; **15**:304b–306a, 371b, 373a, 394b*
statistical studies of, **3**:590a; **8**:512a; **12**:461b; **14**:437b
theories of, **8**:512a
see also Demography
Populin, **2**:386a
synthesis of, **12**:164a
Porath, Jerker, **13**:421b
Porcelain, **3**:560b; **6**:370a; **8**:621b; **11**:109a, 330a–b; **12**:615a; **13**:480a
see also Ceramics, Meissen porcelain works, Sèvres porcelain works
Porcupine (ship), **7**:91b, 92a; **8**:609a; **13**:360b
Poretsky, Platon Sergeevich (1846–1907), 11:94–95
Porisms, **1**:367b; **3**:78a; **5**:466b*; **10**:52b; **13**:54a,b
definition of, **4**:426b–427a
Porometer, **3**:581b; **5**:523b
Porosity, geological, **9**:257b
Porospora, **8**:146a
cycle of, **4**:207a
Poroxylon, **8**:354b
Porphyria, **15**:157b
Porphyrin, **6**:505b
derivatives of, **14**:412a; **15**:157b
Porphyrite, **3**:434a
Porphyry (philosopher), **5**:232b; **11**:160b, 161b
commentaries of, **4**:415b, 430b, 462b; **8**:191b; **9**:1a; **10**:300b; **11**:201a; **13**:91a; **15**:231b
commentaries on, **2**:229a–233b *passim;* **4**:437b
editions of works, **4**:391b; **14**:152b
historiography, **10**:112b; **11**:41b
influence of, **1**:333b; **2**:539b, 619b; **3**:14b; **5**:592a; **12**:441a; **15**:262b
influences on, **10**:112a
Isagoge, **1**:2a,b, 269b, 270a, 272b, 274b, 275a, 278a, 409a; **2**:229b, 230a,b, 231b, 232a, 609a; **11**:445a; **15**:231b
Predicables, **1**:94a; **10**:172b
Porphyry (rock), **5**:445a; **10**:557b
fusion of, **12**:122b
Porpoise, **11**:316b
Porro, Ignazio (1801–1875), 11:95
Port-Royal school, **1**:292a; **4**:135a–b; **7**:550b
Port-Royal Géométrie, **3**:284a,b
Port-Royal Grammar, **1**:292a
Port-Royal Logic, **1**:292a; **2**:50b, 545a; **3**:284b
Porta, Giambattista della (1535–1615), 11:95–98; 8:311a; **14**:461a
acquaintances of, **3**:179b; **5**:579b; **12**:101b, 105b, 106a; **13**:29a; **15**:68b

De humana physiognomia, **13**:29b
influence of, **2**:539a; **12**:333a; **15**:69a
Magia naturalis, **7**:299a
optics, **7**:376a; **10**:475b; **14**:461b(n)
Porta, Luigi (1800–1875), 11:98–99
Portal, Antoine (1742–1832), 11:99–100; **5**:251b, 484a; **8**:352b; **14**:14b
Histoire de l'anatomie et de la chirurgie, **5**:484b
Porter, Thomas Cunningham, **9**:542a
Porter, William Townsend, **15**:72b
Portevin, Albert Marcel Germain René (1880–1962), 11:100–101; **3**:234b, 239b; **5**:583a, 584a*; **13**:248b*
Portier, Paul (1866–1962), 11:101–102; 9: 176b*
toxicology, **11**:429b–430a
Portlandian (stratig. stage)
see Tithonian
Portlock, Joseph Ellison, **10**:204a
Portolano chart, **10**:617a
Portugal
botany, **8**:373b
Council of Mathematicians, **1**:573b
geological survey, **11**:399a
mathematics, **10**:160b
navigation, **1**:573b; **9**:36b–37a
Portuguese Guinea
geology, **4**:140b–141a
Portuguese man-of-war
toxicity, **11**:429b–430a
Portulacaceae, **3**:44b
Pose, Heinz, **5**:272a
Pošepný, Franz (1836–1895), 11:102–103; **5**:548a–b
Posidonius (*ca.* 135–*ca.* 51), 11:103–106; **3**:318b; **5**:345a
acquaintances of, **13**:83b
commentaries by, **9**:1a
geography, **3**:320a
geology, **11**:39b
geometry, **4**:433a(n); **8**:429b
influence of, **1**:75a; **3**:319b; **11**:444b; **12**: 441a; **13**:84a, 85a
Meteorologica
commentaries on, **5**:346b–347a
meteorology, **12**:310a
natural philosophy, **4**:30b–31a
psychology, **10**:21a
statics, **6**:310b
students of, **1**:235a
Positive rays, **5**:458b
analytic use of, **1**:320b–321b
decay time, **14**:340b
nature of, **13**:369a–b; **14**:340a
velocity of, **14**:340a
see also Canal rays
Positivism
ancient Greek, **1**:257a
seventeenth-century, **6**:477a; **11**:489a,b
eighteenth-century, **1**:112b; **3**:383a
nineteenth-century, **2**:65a, 485b, 545a, 624b; **3**:375a–379a; **4**:183a; **5**:97a, 255a, 370b; **8**:116b, 603b; **10**:2b, 4b, 177b; **11**:491b, 492a, 560a; **13**:252a, 499b; **14**:394a, 488b; **15**:741b
opposition to, **2**:26b, 31b; **10**:354b–355a, 485a
twentieth-century, **2**:461a; **4**:225a, 374a; **9**:419a; **11**:389a, 480a; **12**:109a–b, 113b; **14**:3a, 468b
opposition to, **5**:213a; **8**:183b; **11**:14a; **15**:423a
Chinese, **15**:730b
see also Logical positivism
Positron, **9**:261b
annihilation of, **4**:280a; **7**:154a
discovery of, **6**:356a; **8**:95a; **9**:399a; **12**: 33a; **15**:26b
emission of, **7**:153b–154b
prediction of, **10**:214b
scattering of, **15**:26b
secondary, **15**:26b
symmetrical theory of, **9**:38b
Possibility, **14**:483b
Post, Benjamin, **4**:522b
Post, Emil Leon (1897–1954), 11:106–108; **13**:498a
Postage stamps, **3**:42b
Post-mortems
ancient Greek, **4**:383a
medieval, **3**:218b
Renaissance, **1**:611b; **3**:354b
seventeenth-century, **1**:391b; **5**:419a; **7**: 150b; **10**:316a; **13**:222b–223a, 474a; **14**:256a, 407a,b, 408a
forensic, **2**:237b
eighteenth-century, **2**:225b–226a; **3**:2b, 332a; **5**:221a, 384b; **6**:567b; **7**:605b; **8**:352a
nineteenth-century, **2**:463b–464a; **6**: 233a*; **7**:44b; **12**:202b, 295a, 296a–b
Postulates, **4**:416b–417b
Posture
reflex control, **9**:19b–20b
Potamogeton, **4**:584a
Potanin, Grigory Nikolaevich (1835–1920), 11:108; **3**:531b; **12**:300b, 301a
Potash, **2**:176b; **3**:484a
bleaching of, **9**:60b
electrolysis of, **3**:602a
manufacture of, **6**:383b; **13**:580a
mining of, **10**:171a
in plants, **8**:344b
Potassium
alloys, **12**:321b
atomic weight, **9**:110a; **10**:510b; **11**:417b
compound nature of, **3**:602b; **5**:321a; **13**:311a
deposits of, **7**:530a
discovery of, **3**:601b–602a
isolation of, **5**:320b–321a
oxides, **13**:312a
photoelectric effect, **5**:341b
physiological effects, **11**:462b–463a
in plant nutrition, **11**:416a
preparation of, **3**:400a; **13**:311a
radioactivity of, **3**:31b; **6**:16a
reactions of, **5**:321a–b; **13**:311a–b
salts of, **3**:202a, 455a; **5**:197b
crystallography of, **5**:124a; **9**:424b; **15**:62b–63a
organic, **3**:341b; **8**:339a
physiological characteristics, **11**:179b
in sea water, **5**:72a
in solar spectrum, **12**:20b
specific gravity, **5**:321a
spectrum, **11**:476b, 580b
in tissues, **9**:549b
vapor
monatomicity, **4**:79a
Potassium bicarbonate, **13**:298b–299a
Potassium carbonate, **5**:420b
Potassium cyanide
analytic use of, **5**:165a
Potassium ferricyanide, **5**:431b
as indicator, **10**:510b
Potassium ferrocyanide
preparation of, **6**:395b; **12**:64a
Potassium permanganate, **10**:440b
Potassium silicate, **5**:420b
Potassium sulfate, **5**:418a
Potato, **8**:121b*; **11**:314a; **12**:317a; **13**:351a, 495a, 596a; **14**:96a
alcohol from, **3**:316b; **4**:250a
degeneration of, **3**:429a–b
nutritive value, **10**:325b
selective breeding, **4**:271a,b
tuberization, **2**:35a; **3**:429a; **9**:461a
Potato blight, **2**:19a, 437b; **3**:612a, 612b–613a; **6**:138a*; **9**:24b, 149b*, 176a; **10**:436a; **11**:155a
Potato oil, **3**:10a–b
Potential, mathematical
binary, **11**:408b
logarithmic, **10**:25a; **11**:261a; **13**:562b
Potential, physical
chemical, **5**:390a; **9**:227a, 394a
elastic, **2**:356b; **3**:146a
electrochemical, **4**:229a; **6**:371a–372b; **7**:493a; **8**:289a, 290a–b; **10**:345b; **11**:474a
electrodynamic, **6**:251a–252a, 343b; **8**: 492a; **9**:205a, 207a, 217a; **10**:28a; **13**:376a–b
retarded, **8**:490b–491a, 502a; **9**:217a
electromagnetic, **7**:211b
gravitational, **6**:240a; **7**:211b; **9**:180b; **15**:485b, 487b
concept of, **15**:316b, 318a–319a, 322a, 348a
interatomic, **8**:186a
magnetic, **6**:249b
thermodynamic, **4**:225b, 228b–229b, 230b, 231a; **8**:289b, 503a; **11**:446a
definition of, **4**:229b
Potential equation, **1**:499b; **4**:324a–b, 326b, 327b
Potential theory
eighteenth-century, **3**:282b
early nineteenth-century, **5**:303b; **15**: 199b–200a
mid nineteenth-century, **3**:315a; **4**:126b; **5**:306b; **6**:392a; **8**:592a,b; **10**:28a, 247b, 480b; **11**:453b; **15**:200b
late nineteenth-century, **1**:600a; **2**:105b; **3**:169b; **4**:269a, 378b; **6**:472a; **7**:398a; **8**:389a, 562b; **9**:225a; **10**:25a; **12**: 246a; **13**:263b, 422b, 423a; **14**:67a, 158a,b, 231b, 503b
early twentieth-century, **4**:561b; **6**:548b*; **8**:617b; **9**:358b; **11**:59a–b, 459b; **14**: 317a, 345b
mid twentieth-century, **7**:7a; **13**:562b
first vs. second potential, **9**:174b
Potentiality, **1**:258b, 277a–b; **7**:137a
Pothenot, Laurent, **9**:447b; **12**:500b
Potier, Alfred, **1**:558b; **9**:372a
Potsdam
observatory, **1**:340a; **6**:146b; **9**:563b, 564a; **12**:251a, 579a, 595a,b; **14**:55a–b, 56b
photometry, **14**:629a
Potsdam Sandstone, **4**:363b, 364a; **6**:187a
Pott, Johann Heinrich (1692–1777), 11:109; **3**:560b; **4**:223b; **6**:259b; **8**:69a; **9**:105a, 106a; **10**:26a,a*
Pottery
see Ceramics
Pottsville formation, **14**:298a,b
Pouchet, Félix Archimède (1800–1872), 11: 109–110; **2**:187b; **4**:210b; **10**:371a; **13**:288a; **14**:534a
Hétérogénie ou traité de la génération spontanée, **10**:367b
and Pasteur, **10**:371b–372a
Pouchet, (Charles Henri) Georges, **3**:184b; **5**:148b
Pouillet, Claude-Servais-Mathias (1790–1868), 11:110–111; **4**:410b; **13**:11a; **14**:38b
Éléments de physique expérimentale et de météorologie, **9**:566b

Pouillet, Claude-Servais-Mathias (cont.)
  Lehrbuch der Physik, 12:218b
  light theory, 15:377a
Poulenc, Camille, 5:99b
Poulton, Edward Bagnall, 1:504a; 9:561a; 14:136b
Poultry, 15:648a
  artificial insemination, 7:32a
  endocrinology, 10:570a–b
  genetics, 5:10b; 11:212a,b
Pound, James, 2:387a; 5:23a; 6:6a
  astronomical tables, 14:178b
Pound, Roscoe, 3:317b
Pourfour du Petit, François (1664–1741), 11:111–113
Pourquoi Pas? (ship), 3:203b–204a; 9:78b
Pourtalès, Louis François de (*ca.* 1823–1880), 11:113–114; 13:66a
Pourtalesia, 11:113b
Powalky, Karl Rudolph (1817–1881), 11: 114–115
Powders
  adsorption on, 6:118b
Powell, Baden (1796–1860), 11:115–116; 14:185a
Powell, Cecil Frank (1903–1969), 11:117–118; 15:27a
Powell, John Wesley (1834–1902), 11:118–120; 14:120b
  acquaintances of, 3:609b; 7:148b, 370a
  critics of, 9:100b; 14:517b
  geology, 3:593b, 594a; 4:266a
  Powell survey, 4:265b; 5:395b; 10:32a; 11:118b
  Report on the Lands of the Arid Region of the United States, 12:344a
Power, Frederick Belding (1853–1927), 11: 120–121; 14:496a
Power, Henry (1623–1668), 11:121–122; 13:445a; 14:379a(n)
Power
  boundaries of, 6:378a
  concept of, 1:253a–b; 8:209a; 13:195b, 196b, 197b, 200a
  conservation of, 8:231b
  pneumatic, 8:500a,b
  transmission of, 6:431b; 8:214a–b, 500b
    continuity principle, 3:73a
  see also Electricity, Energy, Force
Power gas, 9:467a
Powers (math.)
  ancient Roman use of, 15:119a–b
  Arabic use of, 1:31a, 437b
  and conics, 1:245b
  definition of, 3:55b–56b
  division of, 3:274b; 12:93a
  of imaginary numbers, 1:238b
  Indian use of, 1:31a
  integration of, 11:487a
  linear representation of, 2:281a
  multiplication of, 3:274a–b; 12:93a
  notation of, 2:281a–b; 8:41a; 11:590a; 14:19b
  summation of, 1:437b
  see also Exponents
Poynting, John Henry (1852–1914), 11:122–123; 9:227b; 12:238a; 13:364b, 369b
  glaciers, 13:246b
  Poynting flux, 11:122a,b; 14:339a
  Poynting vector, 3:368b
  students of, 1:320b; 5:189b
Pozzo, Francesco dal, 14:12a
Prabhākara, 1:308a; 2:115a
Prace matematyczno-fizyczne
  founding of, 4:84a
Prades, Jean-Martin (Abbé de), 4:86a
Praesepe (star cluster), 6:351b
  photography of, 5:480a; 12:37a
Praetorius, Johannes, 2:404a; 4:212b; 11: 561b
Prager, Richard, 9:564a; 12:349b
Pragmatism
  seventeenth-century, 5:286b
  eighteenth-century, 11:43a–44a
  nineteenth-century, 10:485b, 486a; 12: 591b
  twentieth-century, 7:68a–b; 13:550b
Prague
  Charles University, 3:171a
  Mathematical Institute, 3:171a
  National Museum, 11:130a; 13:44b
Prague, University of
  in late eighteenth century, 13:39b
  in mid nineteenth-century, 11:217a
  and Jesuits, 9:96b–97a
  philosophy at, 13:39b
Prain, David, 13:342b
Prandtl, Ludwig (1875–1953), 11:123–125
  acquaintances of, 5:64b; 7:247a; 11: 588b
  on contemporary scientists, 7:400a*; 11:615a*
  and Institute for Technical Physics, 8: 500b
  Prandtl number, 9:222a–b
  students of, 7:246b; 15:40a
  wave theory, 4:230b; 8:597a
Prantl, Karl Anton Eugen, 15:147b
Praśastidhara, 9:580b
Praseodymium, 3:321b, 478a
  atomic weight, 2:428b
  discovery of, 9:541b
Pratensis, Johannes, 2:401b, 403a; 12:334b
Pratt, Anne
  Flowering Plants, Grasses, Sedges and Ferns of Great Britain, 2:475b
Pratt, Frederick Haven (1873–1958), 11: 125–126
Pratt, John Henry (1809–1871), 11:126–127; 2:373a; 6:188a, 241a
Pratt, Joseph Hyde, 2:258b
Pravdich-Neminskii, Vladimir Vladimirovich, 2:2a
Praxagoras of Cos (*b. ca.* 340 B.C.), 11: 127–128; 4:105b, 383b, 384b, 511a; 5:234a; 6:316b, 317a–b
  school of, 4:383a, 385b
Precambrian era, 1:237a, 472b, 623a; 2:289b, 486b; 9:198a; 11:210a
  basements, 12:272a–274b
  paleontology, 1:472b; 3:591a; 14:298b
  rocks, 1:51a; 7:62b; 8:33a,b, 99a; 13: 572a–b
  stratigraphy, 2:89b; 4:262a–b, 409b; 6: 187b; 14:59b
  see also Archeozoic era, Primary era, Proterozoic era
Precession, 10:78a–b
  constant of, 10:35a, 76b; 12:170a; 13: 114a, 120a, 149b; 14:162a
  of electrons, 8:40b
  of a ring, 10:78b
  of viscous spheroid, 3:583a
Precession of equinoxes, 4:480a; 7:562a; 8: 504a; 9:514b; 15:542b–543a
  discovery of, 1:283b; 9:297a; 11:188a, 195a; 15:217b–218b
  rate of, 1:511a; 2:99a,b, 468b; 3:405b, 431a; 4:176a, 185b, 270a*, 542b; 10:592b; 13:485b, 534b, 537a; 14: 390b; 15:339a, 348b, 583b, 586b, 587b, 600b, 602a, 610a, 628a
    variation of, 1:312b, 313a–b; 2:99b; 3:318b–319a; 7:304a; 15:34a–b, 732a
  tables of, 2:388b
  theories of, 1:101a, 113b, 510b–511a; 2: 20b; 3:405a–b, 562a; 5:399b; 7:112b; 10:73b; 11:62a; 13:292a, 323b–324a; 15:33b, 287b, 348a
  and tides, 15:297a–b, 301a–b
  see also Trepidation
Precht, Julius, 11:612b; 12:614a
Precipitation, chemical
  as analytic tool, 2:5b–6a; 5:164a, 165a
  Chancel's method, 3:193a
  emanation studies, 6:16b
  explanations of, 6:257a
  fractional, 1:442b
  isoelectric, 1:443a
  of proteins, 12:546b
Precipitation, meteorological, 3:436a, 593a
  naming of, 8:256a; 9:486a
  remote recording of, 12:595b
  see also Hail, Rain, Snow
Precipitator
  electrostatic, 3:436b
Precipitins, 15:522a
Predators
  natural protection from, 9:560b–561a
  predator-prey relation, 5:70a–b
Predestination, 2:395b; 4:396a
Predicates
  classes of, 2:232b
  quantification of, 4:36a; 6:81b–83b
Predicativism, 5:108b; 14:614a–b
Prediction
  as aim of science, 3:377a
Prediction, mathematical
  linear, 14:346b
Preece, William Henry, 4:248b; 6:211b
Preformation, theory of, 1:260b, 293a–b, 387a; 2:204a, 287a, 384b; 3:247b, 482b, 611a*; 4:135b, 291a; 6:65a, 149a; 7:441a; 8:157b, 160a; 9:65b, 187a, 576a; 10:520a,b; 11:332b–334a, 597b; 12:559b; 13:174a–b, 564a; 15:54a
  opposition to, 2:579a; 7:441b; 8:580a; 9:187a, 520a; 11:159a; 15:524b
Pregl, Fritz (1869–1930), 11:128–129; 12: 452a
Pregnancy
  biochemistry of, 11:577a
  extrauterine, 1:323a; 3:487a
  length of, 2:72b; 15:660a
  multiple, 14:595b–596a
Prehistory
  nineteenth-century study, 3:264a–b; 8: 44a, 528a–b, 573b–574a; 9:538b, 539a; 15:50b–52a
  twentieth-century study, 2:450a–b; 3: 622b
  classification in, 9:539a
  three-age system, 14:506a–b
  see also Archaeology; Congresses, scientific; Paleontology, etc.
Premamiśra, 5:274b
Premaxillary bone
  see Intermaxillary bone
Premunition, 12:315a
Prenant, Auguste, 2:344b, 345a
Prentiss, Albert Nelson, 2:102b; 7:169a; 13:456a
Presbyopia
  see Vision
Presbyterianism, 2:453b
Presl, Karel Boriwoj (1794–1852), 11:130; 9:340b
Pre-Socratics, 1:149a–b, 150b–152a, 170b–172a, 495a–b; 2:13a–b; 4:32b; 7:539a; 10:324a–325a; 13:295a–297b; 14: 536a–537a, 607b–611b
  atomism, 4:30b–31a

Press (mech.), **13**:614a
hydraulic, **11**:603b
wedge, **8**:212b, 213a
Pressure
atmospheric
ancient Greek study, **6**:311a–b
seventeenth-century study, **1**:424b, 483b, 567b; **2**:83b–84a; **5**:575a; **9**:319b; **11**:121b
eighteenth-century study, **2**:40a
nineteenth-century study, **12**:595b
and altitude, **2**:40a; **3**:64a; **5**:428b, 528b; **6**:512b; **8**:267b; **9**:117b; **10**:333a,b; **13**:267a; **15**:355a–b
and boiling point, **4**:28b, 518a
and combustion, **5**:126a–b
distribution of, **4**:591b; **8**:188b
physiological effects, **1**:454a,b; **2**:61b–62b; **6**:24b; **9**:546b, 547a; **12**:159b; **13**:170a
theory of, **13**:439a
variation of, **3**:436a, 540a; **4**:28b–29a; **5**:288a–b; **11**:558a; **13**:438b; **15**:372b
and wind, **2**:628b; **4**:175a; **9**:117b
see also Barometry
concept of, **10**:334a
distribution of, **7**:437a
fluid, **2**:39b, 378a
distribution of, **4**:208a
hydrostatic, **1**:223a, 607b; **4**:230b; **8**:227a, 559b
of light rays, **1**:301a
measurement of, **1**:203a; **2**:458a,b
molecular, **3**:477b
osmotic, **1**:299a–b, 300a; **2**:170a; **3**:20a, 351b; **4**:131a, 228b, 264b; **8**:13a*, 425b; **11**:123a, 213a, 279a; **12**:617b–618a; **13**:453a, 579b; **14**:330a
body fluids, **2**:339b
colloids, **1**:537b
determination of, **13**:163a, 472a*
discovery of, **10**:147a
membrane equilibrium, **4**:165b
and molecular weight, **13**:579b
in plants, **10**:576a, 577a
thermodynamic definition of, **7**:462a
and vapor pressure, **13**:579a
of radiation, **2**:120b, 266a; **4**:279a, 322b, 361a; **9**:212b; **11**:122b–123a; **13**:106b*
of saturated vapor, **2**:136b
of steam, **1**:203a; **3**:541a; **15**:487a
and temperature, **2**:104a; **3**:80b; **4**:241b
of water, **6**:311a
see also Barometry, High pressure studies, Manometer, Vapor pressure
Pressure cooker, **10**:292b; **13**:351b; **14**:354b
Prestet, Jean, **7**:611a; **11**:392a
Élémens de mathématiques, **2**:50a; **9**:48a, 49b–50a
Preston, George Dawson, **9**:312b
Preston, Thomas, **10**:347b; **11**:612b
Preston's rule, **1**:370b
Prestwich, Joseph (1812–1896), 11:130–131; **5**:163a*; **8**:573b; **15**:51b
Prévost, Isaac-Bénédict (1755–1819), 11:131–132
Prevost, Jean-Louis (1790–1850), 11:132–133
Prévost, Louis-Constant (1787–1856), 11:133–134; **1**:210b; **2**:341b; **4**:67b; **5**:475b; **12**:317b, 318a
geology, **3**:526a; **8**:564b–565a
Prevost, Pierre (1751–1839), 11:134–135; **8**:35a, 260a*, 306a; **11**:132a,b*; **12**:523a; **13**:127a
heat, **9**:265a*; **13**:51b–52a
Prey, Adelbert, **13**:610b
Preyer, Thierry William (1841–1897), 11:135–136; **7**:519b; **11**:570a,b; **14**:2b
critics of, **15**:22a*
Spezielle Physiologie des Embryo, **11**:570b
Price, Richard, **1**:532a; **11**:140b, 143a; **14**:179b; **15**:281b, 304a–b
Prices
fluctuation of, **11**:123a
index-numbers, **7**:105a
tables of, **11**:457b
Prichard, James Cowles (1786–1848), 11:136–138; **8**:98a
Priest, Robert, **5**:362a–b
Priestley, John Gillies (1879–1941), 11:138–139; **6**:24a,b
Priestley, Joseph (1733–1804), 11:139–147; **2**:177a, 524a; **3**:538b; **5**:76b; **6**:283b; **7**:405b*; **8**:70a; **11**:579b; **12**:435b
acquaintances of, **6**:383a; **7**:11b, 277a; **9**:5b; **10**:108a; **11**:137a; **14**:194b, 197a, 463a, 464a, 508b; **15**:94b
biographies of, **13**:389b*
combustion, **8**:80a, 613a; **14**:213b
Considerations on the Doctrine of Phlogiston . . . , **1**:64b
on contemporary scientists, **1**:547b; **3**:599b
and Cooper, **3**:400a
correspondence of, **14**:71a–b
Directions for Impregnating Water with Fixed Air, **8**:74a
Disquisitions Relating to Matter and Spirit, **6**:335b
electricity, **5**:135b–136a; **6**:169a; **11**:496b
Experiments and Observations on Different Kinds of Air, **8**:74a, 75b
Hartley's Theory of the Human Mind, **6**:140a
and Higgins, **6**:384a
History and Present State of Electricity, **1**:68a; **3**:52a; **5**:135a
translations of, **2**:474b
influence of, **2**:483b; **3**:153b, 157a,b; **9**:140a; **12**:570a; **14**:70a
natural philosophy, **3**:539b, 540a
plant physiology, **7**:12a–b, 13b; **8**:345b
pneumatic chemistry, **14**:74b, 76a
ammonia, **2**:77a
carbon dioxide, **6**:283a
nitrogen, **12**:25a
oxygen, **7**:620b; **8**:75a,b, 76a; **12**:147b
Priestley's rings, **11**:142a
students of, **2**:335b
translations of works, **8**:85b(n)
water, **8**:78a
Primary era, **2**:492a
stratigraphy, **1**:472a; **4**:363b
Primates
anatomy, **7**:278b; **10**:261b; **14**:533b–534a
classification of, **10**:261b
comparative anatomy, **5**:505a; **9**:428b; **13**:527b–528a
comparative neurology, **5**:207b–208a
fossil, **9**:134b
persistence of infantile characteristics, **5**:359a–b
see also Apes, Chimpanzee, etc.; Evolution, of man; Monkeys
Prime meridian
see Meridian, prime
Prime mover
see *Primum mobile*
Prime number theorem, **3**:228a; **7**:599b, 616a; **13**:561b–562a
proof of, **8**:369a; **9**:173b
Primordia, **3**:422a
Primrose
see *Primula*
Primula, **1**:422b
genetics, **6**:22b
heterostylism, **3**:422a, 575a; **4**:401b; **5**:605a
Primulaceae, **6**:296a
Primum mobile, **1**:137b, 250b, 252b, 253b; **3**:424a; **7**:123b, 140a, 142a, 269b; **9**:136b, 137b; **11**:72b; **12**:5a–b; **14**:162b; **15**:33b, 34a, 154b
Prince Rupert's drops, **6**:477b; **9**:508a*
see also Glass drops
Princeton University
in late eighteenth century, **8**:612b
biology at, **3**:390a
library, **15**:343a
mathematics at, **4**:335b, 618b; **14**:211b
physics at, **7**:554b
statistics at, **14**:383b–384a
Principium sorbile, **8**:531a
Principles, Paracelsian
see *Tria prima*
Pringle, John (1707–1782), 11:147–148; **6**:54a; **7**:11a,b; **8**:585b; **14**:194b, 322a, 323a*
critics of, **13**:46b
and Franklin, **1**:435b–436a
Pringsheim, Alfred (1850–1941), 11:148–149; **5**:313b*; **10**:441b; **13**:268a*
analysis, **13**:266b
theorem of, **5**:201a
Pringsheim, Ernst (1859–1917), 11:149–151; **11**:8b; **13**:40b
radiation, **7**:528a; **8**:551b, 552a; **10**:105b, 346a; **11**:12a
Pringsheim, Nathanael (1823–1894), 11:151–155; **4**:206b; **13**:90a
acquaintances of, **10**:575b, 580b; **12**:193a
algology, **3**:337b; **6**:338a; **13**:395a
students of, **12**:99a; **13**:87b, 88a
Printing
Renaissance, **5**:347b–348a; **8**:213a; **11**:351a; **12**:199b; **13**:397a
seventeenth-century, **2**:185b; **8**:464b
eighteenth-century, **5**:129b–130a; **13**:413b
nineteenth-century, **2**:284a; **4**:18b–19a
early twentieth-century, **4**:283a
Chinese, **15**:730b
color, **11**:343a
movable-type
invention of, **12**:382a–b
see also Lithography, Typography
Printing press, **8**:213a; **9**:198b
Priority devices, **6**:604a; **8**:74b
Priority disputes
anatomy
Renaissance, **3**:355b–356a
seventeenth-century, **1**:482b; **5**:484b; **6**:509a; **11**:587a; **13**:31a, 33b, 171a
eighteenth-century, **6**:367a, 569b; **9**:482b–483a
nineteenth-century, **1**:584a; **10**:195b, 267a; **11**:510b
archaeology, **13**:357b
astronomy, **1**:53b; **2**:327b; **5**:257b; **6**:325b; **8**:277b; **10**:479b; **12**:151b
botany, **2**:426a; **11**:605b
chemistry
seventeenth-century, **7**:525a
eighteenth-century, **2**:186a; **6**:384a, 386a–b; **7**:391a
early nineteenth-century, **5**:322a–b; **8**:113b; **10**:498a; **11**:464b
mid nineteenth-century, **1**:518b; **2**:623a–b; **3**:448a, 617b; **4**:246a; **7**:464a; **8**:508a; **9**:290a; **10**:38a–b, 436a; **12**:544a

Priority disputes (cont.)
chemistry (cont.)
late nineteenth-century, 6:94a
twentieth-century, 14:476a; 15:439a
crystallography, 14:241b(n)
cytology, 11:255b
embryology, 2:161b
geography, 9:421a
geology, 5:69a, 578a; 9:255a, 584b; 12:167b, 277b; 15:92a
historiography, 7:464b
linguistics, 14:569b
mathematics
Renaissance, 9:610b–611a; 11:561b
seventeenth-century, 4:13a–b, 95b, 571a; 6:359b; 8:164b–166b; 9:613a; 10:52b, 84a–85a, 255a, 339a(n); 13:434b; 14:466a–b
eighteenth-century, 2:54a, 57a,b, 418b; 4:516a–b; 5:54b, 529a; 8:304b; 9:452b, 454b; 10:67b–68a
nineteenth-century, 2:214b; 4:341b–342a; 5:98a, 302b, 309a–b, 368a; 8:137b; 9:126b; 10:443b, 484a; 11:77b; 12:74a; 13:218a; 15:194a, 365b–366a
medicine, 1:576a; 3:578a; 5:478b, 564b; 7:45b; 11:70a, 556a; 13:493b
meteorology, 12:51b
microbiology, 5:564b; 10:18a, 397a–b; 13:520b
neurology, 9:241b, 259a
optics, 3:98a; 5:278b; 6:88b
parasitology, 2:529a; 5:503b–504a; 10:527b
physics
seventeenth-century, 6:68a, 485b–486a; 9:16a; 10:61b, 332b–334a
eighteenth-century, 1:114a; 8:258b; 9:596b; 14:353b
nineteenth-century, 3:306a, 309a; 5:15b, 388a; 6:242a, 280b; 8:2b, 180b–181a; 9:73b, 239b, 445a; 12:288b; 13:52a, 103a; 14:568a; 15:86b
physiology, 9:9b; 14:26a
technology, 3:118b; 4:589a; 5:434a; 6:604b, 605a; 7:45a; 8:459b; 10:366b; 11:94a; 12:103a; 13:357a
zoology, 1:58b; 8:104a, 270b
see also Patents, Plagiarism
Priroda, 4:597b
founding of, 7:456a
Prisca sapientia, 10:82b
Prisca theologia, 2:539b; 3:492a
Prismatic circle (astron.), 7:209b; 11:95b
Prismoid, 6:76b*
Prisms, geometrical
volume of, 8:421b; 15:670b
Prisms, optical, 1:424b; 5:302a, 445b
birefringent, 5:87a
color theory, 3:489a; 6:125b; 10:53a–b, 55b, 56a–b
direct-vision, 1:136a
Nicol prism, 9:201b; 10:109b–110a; 13:273a
polarizing, 10:109b–110a
Prisons, 3:7a
Pritchard, Charles (1808–1893), 11:155–156; 1:135b; 3:582b–583a; 8:279a*; 13:221b*
Privalov, Ivan Ivanovich (1891–1941), 11:156–157
Privat de Molières, Joseph (1677–1742), 11:157–158; 5:567b; 12:430a
electricity, 10:146a
vortex theory, 5:62a; 11:43b
Private College of Amsterdam, 13:169a, 171a
Probabilism
medieval Latin, 10:225b
seventeenth-century, 11:506b, 507b
eighteenth-century, 3:384b
nineteenth-century, 3:452b; 7:106a–b
twentieth-century, 1:366a; 9:129b
Probability, 4:51b; 5:98b; 6:402a*, 403b; 8:135a; 13:221a
a priori, 1:531b–532a; 15:303b, 304b, 369a, 371b, 373a
in astronomy, 15:280b, 282a–283a, 285b, 286a, 290b–292b, 325b, 344b–345b, 364a–365a, 370b, 372a, 380a–382b
biological, 11:599b
of causes, 8:306a–b
central limit theorem, 3:225a,b, 230b; 9:125b; 15:363b, 364a, 365a, 371b, 381a
proofs of, 9:127b–129a
chains of, 1:366b
of complex events, 15:303b–304a, 308b, 369a
concept of, 3:66a; 7:317b; 9:453a
continuous, 1:366b; 2:304a
Bertrand's paradox, 2:88b; 7:481b
definition of, 1:532a; 5:10a; 13:612a; 15:279b, 369a
denumerable, 2:303b
and entropy, 2:264b
and epistemology, 15:304a, 308b, 369a
finite, 2:304a
frequential theory of, 7:106b; 9:419b–420a; 11:357a–b
and geodesy, 15:376a–b
geometrical, 2:192a, 304a; 3:323a; 8:176a; 15:372b–373a
interpretation of, 3:386a; 4:314a
inverse, 7:106a; 10:456a; 15:280b–284b, 303b–306a, 365a–b, 373a, 375a,b
judicial, 15:370b–371a, 375a–b
law of large numbers, 2:47a, 304a, 319a; 3:66a, 223a, 225a, 230b, 231a; 7:351b; 9:125b, 127a–129a, 249b, 453a; 15:31a, 280a, 282a, 283b, 304b, 364a, 365a, 366b, 372a, 489a
lottery, 15:369b
movement of, 1:366b
in nature, 15:371b
objective, 3:451b
philosophical, 3:451b; 11:599a–b
posterior, 1:531b–532a
power law, 3:66a
and reality, 11:355b, 357a
rule of succession, 7:151a
sequential analysis, 14:121b–122a
stationary processes, 7:351b
statistical, 5:8b–10a
stochastic methods in, 1:366b; 6:600a
subjective, 3:451b
theory of
Renaissance, 3:66a
seventeenth-century, 2:47a, 50a–b; 4:574b; 6:600a; 8:162a; 9:599b; 10:334b–335a; 12:206b
early eighteenth-century, 1:208b; 2:42a–b, 56b, 57a, 576b, 577a–b; 3:461b; 9:452a–454b, 499b–500a; 12:444a,b; 13:265b
late eighteenth-century, 1:116a–b, 531a–532a; 3:386a–387b; 7:549b; 8:306a–b, 321a–b; 9:55a; 11:135a; 15:279a–286a, 303b–309b
early nineteenth-century, 1:140a,b; 3:451a–452a, 555a; 4:36b; 5:193b; 9:463a; 15:363b–376b, 385b–386a
mid nineteenth-century, 2:294b, 298a; 3:223a–225a *passim,* 230a, 230b–231a; 8:530a; 9:218b–219a; 10:249b; 12:289b; 13:611b–612a; 15:31a–32b, 489a–b
late nineteenth-century, 3:178a, 310a; 5:391b–392b; 7:106a–b; 8:562b–563a; 9:125a–129b; 10:450a; 11:94b
twentieth-century, 1:366b–367a; 2:304a, 319a, 352b, 615a; 3:399a; 5:477a; 6:5a; 7:151a–b, 317a–b, 351a–352a; 9:128a–129b, 248b, 249b, 419b–420a; 10:553b–554a; 11:49b, 357a–b; 13:139a; 14:454a; 15:23b
applications of, 3:451b–452a; 12:289b; 14:345a, 466b; 15:23b, 304a, 372a–376b
foundations of, 7:151a–b, 317a–b; 9:249b; 11:286a; 15:23b, 66b*–67a*
and statistics, 3:592a
urn models, 2:42b; 4:293b; 9:129b; 11:599b; 15:281a, 281b–282a, 366b, 369b–370a, 371b, 375a
see also Calculus, of probabilities; Chance; Distribution, statistical; Errors; Game theory; Statistics
Problems, mathematical
see Paradoxes; Recreations, mathematical; specific problems by name
Proboscideans, 10:241b
Probus (Probha), 1:275a; 3:391a
Procardia, 1:601a
Proceedings of the Edinburgh Mathematical Society
editors of, 14:211b
Proceedings of the Geological Society of London, 5:14b
Proceedings of the National Academy of Sciences, 6:31b
editors of, 9:396b, 502a
Proceedings of the Royal Society of London
see Royal Society of London, *Proceedings*
Proceedings of the Zoological Society of London, 10:261a
Processus vaginalis, 3:38a
Procháska, Georgius (Jiří) (1749–1820), 11:158–160; 4:146a; 6:59b; 9:571a; 12:241b; 13:543b
Proclus (*ca.* 410–485), 11:160–162; 4:159a–b; 11:113a(n), 183b
astronomy, 15:219a
biographies of, 9:111a
commentaries of, 4:160a; 10:547b; 11:184b
on Euclid, 4:414b, 419a, 423b, 424b, 425b–426a, 449a, 463a; 9:299b; 10:299a,b; 13:334a; 14:612b
on Plato, 9:438b; 11:184a; 13:315b; 14:608a
critics of, 7:134a
Elements of Theology (Liber de causis), 1:271a–b, 275b; 9:438b
geometry, 4:417a, 466b; 8:429b; 10:529b; 13:317b
historiography, 4:103b, 104a, 415a,b, 424a, 426b, 429a–b, 430a–b, 437a, 462b; 5:346b, 347a; 6:406b–408a, 410b, 411a, 414a–b; 8:189a, 192a,b; 9:156b, 268b–269b, 270a, 296b, 501a, 598b; 10:179a,b, 180b, 181a, 529a; 11:28a, 30a(n), 221a, 288a; 13:296a, 302a, 303b, 305a, 314b; 14:603b
Hypotyposis, 1:511b
influence of, 3:513a, 516a; 5:121a; 7:194b, 300b; 8:191b; 11:184a,b, 339a; 12:442a; 14:225b; 15:262b, 494b
influences on, 2:230a; 4:437b; 11:105b(n)
In Platonis Timaeum commentaria, 12:442b
Sphere
translations of, 3:558b; 8:360b

students of, **4**:425b; **10**:6a
translations of works, **1**:468a; **9**:436a, 437b, 438b
Proconsul africanus, **8**:105a
Proctor, Richard Anthony (1837–1888), 11: 162–163; **6**:352a
Procyon (star), **13**:114a
companion of, **12**:139a
orbit, **1**:339b
proper motion, **6**:71a
spectrum, **14**:56a
variation in motion, **2**:99b
Product (math.)
indeterminate, **15**:195b
infinite, **2**:105a, 507a; **3**:54a; **4**:476a, 477a; **5**:572b–573a; **14**:148b
convergence, **13**:69a
doubly, **3**:169a; **4**:342b
transformation of, **4**:476a
Product theorem, **3**:142b
Profatius Tibbon
see Ibn Tibbon, Jacob ben Machir
Progesterone, **1**:123b; **2**:345a
Prognostics
ancient Greek, **6**:422b
ancient Roman, **5**:231a–b
Byzantine, **10**:417b
medieval, **2**:435b
Renaissance, **1**:125a; **4**:285a
Programming
linear, **5**:98b
Progress
medieval discussion, **2**:19b
eighteenth-century discussion, **7**:586a, 593a
nineteenth-century discussion, **3**:378b, 453b
as aim of mathematics, **2**:358b
science and, **9**:477a
theories of, **12**:570b
Progressionism, **3**:395a,b
Progressions, mathematical, **2**:115b; **4**:391b
arithmetical, **3**:221b; **6**:617a; **8**:420a; **9**: 611b–612b; **14**:543a
prime numbers in, **3**:228a; **4**:125a; **8**: 135b, 139a; **13**:561b
summation, **12**:62b–63a; **13**:539b; **15**: 10a, 670b, 693a–694a, 698b–699b
geometric, **3**:276b; **4**:421b; **8**:420a; **9**: 611b–612b; **14**:545b
summation, **2**:276b; **13**:539b; **15**:10a, 670b, 684a–b, 694a–b, 699a
linear, **15**:751b
theory of, **4**:550a
see also Series
Progressivism, **12**:109a
Progressus rei botanicae, **8**:512b
Projectile motion
see Motion, projectile
Projection (geom.), **11**:198a, 198b–200a
azimuthal, **7**:265a
Bonne, **11**:200a, 203a
central, **9**:475a; **11**:77a, 79b
complex, **7**:573b, 574a
conformal, **3**:226b; **6**:124b; **9**:463a; **10**: 483b
conic, **2**:535b; **11**:200a; **14**:275a
cordiform, **14**:275a–b
cylindrical, **3**:424a; **9**:309b, 475a
equal-surface, **12**:499b*; **14**:275b
Gauss-Krüger, **5**:304a; **7**:265b
gnomonic, **1**:573a; **5**:455b; **9**:393a
homolographic, **5**:508b
imaginary, **14**:343a
mathematical conditions for, **7**:599b–600a
Mercator, **4**:214a; **5**:304a; **7**:20a; **9**:309b; **14**:513b–514a
orthographic, **1**:512a; **3**:364a; **8**:64b; **14**: 400b, 577a–b, 594a
parallel, **2**:394b
quincuncial, **10**:483b
radial, **2**:394b
skew, **13**:18a
spherical, **9**:393a; **10**:27a; **14**:275a
stereographic, **1**:81a–b, 573a; **2**:152b, 153a; **3**:27a, 312a, 364a, 555a; **6**: 124b; **7**:124b–125a, 126a, 178b*, 405a; **8**:433b; **9**:393a; **10**:302b(n); **13**:14b, 323b; **14**:273b, 275a, 400a,b, 525b, 594a,b; **15**:211a, 219b, 518b, 520a(n)
double, **3**:468a
theory of, **10**:52b
see also Geometry, projective; Mappings (math.)
Prokofeva, Aleksandra Alekseevna, **9**:522a; **13**:136a–b
Prolacerta, **2**:504b
Proline
discovery of, **5**:4b
Proluvium, **10**:429a
Promethium
isotopes of, **7**:158b
Promptov, Aleksandr Nikolaevich, **10**:435a
Promptuary, **9**:610b
Proneomenia, **6**:535b
Prontosil rubrum, **4**:154a–b; **5**:30a, 344b
Pronucleus, **2**:626a
Prony, Gaspard-François-Clair-Marie Riche de (1755–1839), 11:163–166; **3**:441b; **4**:501b
acquaintances of, **3**:440b; **9**:471b; **10**:528a
analysis, **10**:327b
on contemporary scientists, **3**:317a*; **10**:4b*, 616a*
engineering, **3**:287b(n); **15**:335b
historiography, **12**:128b(n)
Prony brake, **5**:101a; **11**:165a
tables of, **4**:16a; **8**:137a
Proof, in mathematics, **2**:274b; **4**:342b; **5**: 299a–b; **7**:167b
per impossibile, **4**:104b
theory of
ancient Greek, **1**:251a–b
seventeenth-century, **7**:194b
twentieth-century, **11**:107a; **14**:615a
Proof, in science, **2**:379a; **5**:232a; **8**:602a–b; **9**:383b
Proof plane, **3**:445a,b
Propellants, **3**:578b; **7**:13a
Propellers
theory of, **9**:343b; **14**:520b
thrust, **14**:621a
Prophatius Judaeus
see Ibn Tibbon, Jacob ben Machir
Prophecy
philosophy of, **9**:30b; **10**:81b, 82a; **12**: 323a; **15**:262b–263a
scientific, **13**:287b
Prophylaxis
ancient Greek, **6**:422b
seventeenth-century, **7**:164b
nineteenth-century, **2**:548a; **12**:295a–296a; **14**:42b
twentieth-century, **15**:96b–97a
Propionic acid, **5**:125b, 371a
synthesis of, **14**:169a
Proportion
ancient Greek discussion, **1**:253a–254b, 257a; **4**:419a–421b, 424b, 441b, 466b; **11**:222a–b; **13**:303a–b, 304a–b; **14**: 611b
ancient Roman discussion, **2**:233a; **10**: 113a; **13**:326a
medieval Arabic discussion, **1**:82a–b, 445a–b; **7**:83a–b, 326b–327a, 343a; **9**:21b–22a; **10**:6a
medieval Latin discussion, **7**:176b; **13**: 210a
Renaissance discussion, **8**:235a, 236a, 237b, 240a; **9**:147a, 488b; **13**:349b; **14**:29b; **15**:155a
seventeenth-century discussion, **2**:310a; **4**:451b
eighteenth-century discussion, **1**:238a–b; **9**:156b
nineteenth-century discussion, **4**:4a, 391b
anatomical, **4**:259a; **15**:4b
arithmetical, **8**:200a, 232a
in art, **4**:259a–b, 260b
between two lines, **1**:223b, 227a,b, 228a–b, 232a–b; **4**:105a, 392a, 451b, 466b, 489a–b, 612a; **6**:112b, 411a–b; **7**:29a, 175a–b; **8**:238a, 239a; **9**:270a–275a, 556b; **10**:115a,b, 294b, 587a–588a; **13**:291a; **14**:21a,b; **15**:116b
canons of, **4**:260b, 261a
Chinese discussion, **8**:420a
compounding of, **13**:194a, 201b, 202a
continuous, **2**:393a
definition of, **4**:419b, 441a,b, 446b, 451a, 466a; **7**:80a–b
expansion of, **3**:182a
geometric, **4**:516a; **10**:271a
in dynamics, **2**:392a; **13**:201a
in pharmacology, **2**:392b–393a
harmonic, **1**:481b*
metric, **2**:214b
nomenclature of, **3**:272b
pyramidal, **8**:197b–198a, 200a, 232a
rule of three, **7**:359b, 360a
tables of, **7**:118a, 120a, 128b
transformations of, **4**:608a–b; **12**:314a
see also Ratio
Proportional parts, method of, **3**:65b; **11**: 457b; **13**:192a–193b, 198b–199b
Propositions
abstract, **2**:277b–278a
categorical, **2**:604b; **6**:81b–83b
class-identity analysis, **6**:82a–b
conditional, **1**:3a
contingent, **4**:462a–b; **8**:155b
conversions of, **4**:423b
formal divisions of, **4**:416a
necessary, **8**:155b
particular, **2**:296b
primary vs. secondary, **7**:151a
and propositional functions, **2**:297b; **11**: 285b; **12**:14a
relations between
convertibility, **6**:82a
derivability, **2**:278a
entailment, **2**:278a
self-referential, **2**:604b
semantic status of, **3**:34b
syncategorematic, **6**:377a–b
unanalyzed, **2**:609b
universal, **2**:296b
see also Calculus, propositional; Logic, propositional
Proskauer, Bernhard, **7**:423a, 427a; **15**:521b
Prosneusis, **10**:298b–299a
Prosody, **8**:468a; **10**:178b*–179a*; **13**:217b
Prosthaphaeresis, method of, **2**:406a–b, 413a, 602b; **3**:262a–b, 312a; **9**:611a; **11**:561b; **14**:274a–b, 470b–471a
Prostitution
regulation of, **10**:18b
Protactinium, **6**:16b; **9**:260b; **12**:505b
Protagoras, **1**:171a,b; **4**:30b–34a *passim;* **6**: 405a, 408b(n); **13**:314b

Protamine
  composition of, 7:467a
Proteaceae, 2:517b, 518b–519a
Protease, 4:618a
Proteinase, 6:111a
Proteins
  amino acids in
    quantitative analysis of, 15:15b
  analysis of, 8:276a; 10:242a–243b; 12:546b; 13:574a; 14:252b
  in blood, 1:11a; 5:149a; 12:319b; 13:419b, 420b
  colloidal properties, 15:201b
  conjugated, 6:505b; 8:275a–276b
  denaturation of, 1:319b; 14:524a
  detection of, 9:401b
  digestion of, 7:520a; 9:259b; 10:493a; 13:403b
    toxicity from, 3:257a
  electrochemistry, 6:373b
  and enzymes, 1:11b; 13:152a–b; 14:412b
  and hormones, 1:11b
  hydrolysis of, 14:602b
  isolation of, 6:500b; 8:276a
  metabolism of, 2:162a; 3:220a; 8:451b, 452a–b, 555a; 10:580a; 14:65b–66b, 454b
    in plants, 10:281b–282b
  mobility of, 13:419b
  molecular weight, 13:161a–163a
  molecules
    uniformity of, 13:161a–162b
  naming of, 2:95b
  nature of, 3:257a; 6:339a–b; 8:446a; 9:558a
  nutritional role, 9:284b–285a; 10:243a–b, 558b; 11:585b–586a; 14:66b
    quantitative studies, 3:257a
  phosphorus-containing, 6:505b
  photochemistry of, 14:247b
  physical chemistry of, 3:335b; 8:446a; 13:160b–163a
  plant, 10:150b, 575b
    origin of, 13:460a
    specificity of, 10:243a
    toxic, 4:297a–b
  precipitation of, 12:546b
  putrefaction of, 6:505b
  reabsorption of, 2:531b
  reactions of, 6:500b
    dissociation-association, 13:161b–162a
    theories of, 7:467a–b
  separation of, 7:387a
  structure of, 1:11b, 319a–b; 4:522a; 5:141a; 8:276a; 13:3a; 14:602b; 15:63b, 257a
    and biological specificity, 7:467b
    folding, 1:319b
  surface films, 4:76b
  synthesis of, 7:467a; 13:24b
  therapeutic use of, 13:229a
  X-ray crystallography of, 7:274a
Proteinuria, 14:253a
Proteolysis, 9:284a
Proteosoma, 5:503a
  life cycle, 5:503b; 9:82b
Proterozoic era
  stratigraphy, 1:469a; 4:262b
Protestantism
  seventeenth-century, 6:141a
  nineteenth-century, 3:523b–524a, 526b, 527a
  see also Religion; specific religions by name; Theology
Proteus, 5:31a*; 6:455b*; 12:216a
Prothallia, 8:5a–b; 12:125b
Prothomo, 1:131b
Protista, 3:17a, 492b; 4:207a; 7:273a
  physiology, 14:2b–3a
Protistology
  see Microbiology
Protobiology, 6:298b
Protoceratops, 5:500b
Protochordates, 1:421b
Protococcus, 3:337a
Proton(s), 9:399a
  acceleration of, 3:329b–330a; 4:583a; 5:272a; 12:33b
  annihilation of, 4:280a
  ejection of, 7:153a–b; 12:32b
  impact, 2:250b–251a
  magnetic moment, 13:42b, 43a
  mass, 4:281b; 5:102b
  scattering of, 7:617b
  sources of, 12:33b
  total number, 4:281b
  transfer of, 4:553b
Protonema, 9:601a
Protophytes
  development of, 4:207a
Protoplasm, 9:418b; 13:457b
  biochemistry of, 9:44b
  chemicals, effects of, 4:235b
  circulation of, 1:135b
  colloidal chemistry of, 4:236a; 6:467a; 13:268b
  concept of, 9:442a
  differentiation in, 13:268b–269a
  identity of, in plants and animals, 3:337a–b; 4:236a
  naming of, 12:231b–232a
  nature of
    nineteenth-century theories, 1:540a; 3:337b, 612b; 5:36a; 6:266a, 596b; 7:519b; 9:600b; 11:136a; 12:232a; 13:89a–b
    twentieth-century theories, 6:498a; 9:494b
  streaming of, 2:517a, 532a; 3:426a–b
  structure of, 6:121b, 133b; 14:98b; 15:63b, 202a*, 257a
    alveolar theory, 2:627a
  vacuoles in, 4:235a–b
  vegetable
    freezing of, 1:561b; 9:44a–b
  see also Sarcode
Protothetic, 8:262b–263a
Protozoa, 2:627a; 5:109a, 317a, 329a; 8:127b
  anabiosis, 3:588b
  colony formation, 13:458a
  contractile vacuole, 7:110b
  digestion in, 8:124b
  division processes, 7:516b
  electric excitability, 7:520a
  evolution of, 4:142b
  fibrillar systems, 13:268b
  flagellating movements, 4:371b
  fossils, 4:291a,b
  freshwater, 8:169b
  genetics, 7:99a–b
  life cycles, 6:337a
  parasitic, 3:148b; 4:132b–133a; 5:503a; 7:447b; 8:145b–146a; 9:328b
  pathogenic, 3:17a; 5:503a; 8:66a; 12:141b–142b
  physiology, 7:519b, 520a
  protoplasm of, 13:268b
  regeneration of, 3:17b
  reproduction, 1:417b; 3:17b; 9:185a; 12:141b; 14:234b
  sensitivity to light, 4:372a; 9:167a
  social phenomena, 7:99a
  taxonomy of, 3:17a; 4:142b, 235b, 290a–291a; 8:270b; 12:421b; 13:268b
  see also subordinate groups
Protozoology
  eighteenth-century, 7:110b–111a
  nineteenth-century, 1:417b; 4:234b–236b, 291a; 6:337a; 7:438a; 8:169b; 9:185a–b; 12:231a–b; 14:146a
  twentieth-century, 3:17a–b; 4:132b–133a, 142b–143a; 7:98b–99b, 447a–b; 8:145b–146a; 9:167a, 328b; 12:141a–142b, 207a, 423b–424a; 13:268a–269a
Protyle, 3:478b, 479a,b
Proust, Joseph Louis (1754–1826), 11:166–172; 10:609a
  and Berthollet, 2:81b–82b, 93a; 11:168a–171a *passim*
  chemistry, 1:65a; 5:323a; 13:281a, 310b
  definite proportions theory, 2:81b–82b, 93a
  on Elhuyar, 4:345b
Proust, Marcel, 2:12a, 544b
Prout, William (1785–1850), 11:172–174; 5:71b; 13:499b
  chemical analysis, 1:572b; 8:343b; 13:403b
  chemical theory, 1:346a; 2:485b; 5:324b; 9:351b
  Prout's hypothesis, 1:321b; 2:135a, 624b; 3:476b, 478b; 4:247a, 395a; 6:481a; 9:91b, 110a; 10:510b; 11:173a; 13:367b, 373a, 499b
    disproof of, 11:416b; 12:619b–620a
    testing of, 9:531a
  specific gravity, 13:373a
Provana di Collegno, Giacinto, 3:328a
Provence
  botany, 8:121a
  geology, 2:89b
Proverbs, 3:548b; 5:119b, 121a
Providence, 1:119a; 5:167a
Provinces, floristic, 7:102b
Prowazek (Provázek), Stanislaus von Lanov (1875–1915), 11:174–175; 12:142a,b*, 143a*
  Prowazek's bodies, 11:175a
Proxima Centauri (star)
  discovery of, 7:18a
Pṛthūdakasvāmin, 15:615b
  commentaries of, 2:416b, 417a,b; 8:46b
Prudden, Theophil Mitchell (1849–1924), 11:175–177; 4:498b; 7:425b; 12:483b
Pruner, Georg, 7:120a
Pruner Bey, Franz Ignace (1808–1882), 11:177–179; 4:198a
Prunier, Fernand, 12:69b
Prussia
  geography, 11:396a
  Geological Survey, 2:108a, 109a; 8:511b
  industry, 15:206b–207a
  mining, 2:108a; 6:370a; 7:254b
Prussian Academy
  see Akademie der Wissenschaften (Berlin)
Prussian blue, 3:326b*; 5:216a, 324a; 6:165a*; 8:618b–619a; 9:105b; 12:145b, 148b; 14:394a
Prussian Geodetic Institute, 6:239b
Prussic acid
  see Hydrocyanic acid
Prutenic (Prussian) Tables
  see Reinhold, Erasmus, *Prussian Tables*
Pryanishnikov, Dmitry Nikolaevich (1865–1948), 11:179–180; 9:130b; 15:505b, 511b
Prym, Friedrich Emil, 4:2b; 11:449a
Przhevalsky, Nikolay Mikhaylovich (1839–1888), 11:180–182; 6:215b, 216a; 12:301a; 14:54a*
Psaronius, 14:600a

Psellus, Michael (Constantine) (1018–1078), 11:182–186; **1**:270a; **4**:111a, 117b; **6**:306a
Pseudo-Alexander, **1**:117b
Pseudo-Archimedes, **15**:253b
Psuedo-Aristotle, **1**:99b, 350b
see also Aristotle, spurious works
Pseudobranchia, **2**:509b
Pseudo-Democritus, **2**:256b
commentaries on, **13**:225b
Pseudo-Dionysius (also Pseudo-Areopagite and Dionysius the Areopagite), **4**: 396b; **5**:549b
Pseudo-Isidore
De ordine creaturarum, **1**:565b
Pseudoklossia, **4**:207a
Pseudomonas, **5**:73a
Pseudomorphs (mineral), **2**:440b
Pseudonitroles, **9**:355b
Pseudo-Plutarch, **7**:42b
Pseudoscience, **5**:252a
Pseudosphere, **1**:591a, 600a; **8**:433a
Psilophytales, **8**:5b
Psilophyton, **3**:30b–31a, 608a
Psora theory, **6**:18a
Psoriasis, **9**:75a; **10**:19a
Psyche
founding of, **12**:264a
Psychiatry
medieval Arabic, **11**:444b; **14**:585a
Renaissance, **8**:554b; **10**:307a, 309a; **11**: 33b
eighteenth-century, **7**:606b; **8**:506b; **10**: 612a–613b; **11**:617a–b; **13**:567b; **14**: 321b–322a
early nineteenth-century, **8**:37b; **11**:365a; **12**:597a; **15**:128a
mid nineteenth-century, **5**:570a–571a; **8**:272a–b; **10**:582b; **12**:224b; **13**: 396a–b
late nineteenth-century, **5**:26b, 73b, 74a; **6**:440a–441b; **8**:609a; **10**:131a; **14**: 114b–115a
twentieth-century, **2**:1a–2a; **5**:180b; **7**: 190a–193a; **10**:434b; **14**:115b–116a
dynamic, **5**:180a
and psychoanalysis, **5**:179a, 180b
somatist, **5**:255a
statistics in, **13**:396a
see also Hypnosis, Neuropsychiatry, Psychoanalysis, Psychotherapy
Psychic energy (*P-Energie*), **2**:1a–b; **7**:191b
Psychic research
Renaissance, **4**:6a
nineteenth-century, **7**:68a; **9**:416a; **10**:35a; **11**:359b; **13**:51b
twentieth-century, **8**:444b
see also Parapsychology, Spiritualism
Psychoanalysis
nineteenth-century, **2**:447b–449a; **5**: 172b–177a; **9**:327b
twentieth-century, **5**:177a–181a; **7**:190b, 193a
constancy principle, **2**:448a
historiographic use of, **1**:365b–366a
institutionalization of, **5**:178b–181b
and medical training, **5**:179b–181a
and psychiatry, **5**:179a, 180b
sexuality in, **2**:448b; **5**:172b, 176a, 179a; **7**:190b
see also Congresses, scientific
Psychobiology, **3**:332a; **6**:320b–321b; **10**: 427b
Psychogenesis, **6**:53b
Psychological Corporation
founding of, **3**:131a
Psychological Review
founding of, **3**:130a–b
Psychology
ancient Greek, **1**:26b–27a, 260a–b; **6**: 422b; **11**:104b–105b; **13**:94a–b
ancient Roman, **7**:1a–b; **8**:537a–b; **12**: 540a
Byzantine, **10**:20b–21b
medieval Arabic, **1**:26b–27a; **5**:592a; **11**:244b; **12**:92a; **13**:292b, 513a; **15**: 262a, 497a
medieval Latin, **7**:28a; **10**:474b, 475a; **12**:428b; **14**:458b
Renaissance, **8**:197a–b; **10**:21b, 122b–123a, 309a, 598b–599a; **14**:47b
seventeenth-century, **4**:63b; **8**:157b; **9**: 97b–98a; **14**:406a–b
early eighteenth-century, **2**:16b; **3**:380b–383a; **4**:85b; **6**:139a–140b, 555b–559b; **7**:606a–607a; **12**:603a–b; **13**: 179b–180a
late eighteenth-century, **2**:286b; **3**:1b–2a; **6**:469b; **8**:81b, 322a; **11**:135a
early nineteenth-century, **5**:250b–255b; **6**:294b–295a; **11**:617b; **12**:596b–597a; **14**:44b
mid nineteenth-century, **3**:240b, 575a; **4**:556b–558b; **8**:515a; **9**:572a; **12**: 571a–b; **14**:527a
late nineteenth-century, **1**:404a; **2**:35b–36a; **3**:130a; **5**:73b–74a; **6**:52b–53b; **7**:67b–68a; **8**:7b, 603a–605a; **9**:561b–562b; **11**:135b, 425a; **12**:270b–271a
early twentieth-century, **1**:580a; **2**:272b; **4**:12b; **5**:317a; **7**:189a–193a; **8**:45b, 446b; **9**:562b; **10**:433b–435a; **15**:465b
alchemical symbolism and, **4**:170a
analytical, **7**:190b–193a
animal, **1**:265b; **2**:581a; **3**:521a–b; **6**:229b; **8**:45b, 529a; **9**:432a, 513a–b; **14**: 549b–551a
and anthropology, **2**:209a–210a
associationist, **2**:16b; **5**:175b–176a, 178a; **6**:139b, 248a; **7**:190a; **11**:140a, 143a, 617b; **14**:47b, 83b
critics of, **12**:571a
behaviorist, **7**:607a
as biological science, **5**:252a–255b
of children, **2**:131b–132a; **5**:377b–378b; **6**:53a
observation methods, **5**:378a
comparative, **5**:253b; **7**:98b, 606a,b; **9**: 512b–513b; **11**:137b, 257b, 518a–b; **14**:408b, 549b–551a
complementarity in, **2**:253a–b
developmental, **11**:135b
dynamic, **7**:193a
elemental, **1**:366a
experimental
late nineteenth-century, **9**:513a; **10**: 485a–b; **14**:527b–528b
early twentieth-century, **2**:132a; **3**:130a; **9**:562b; **11**:388b; **14**:550a–b
mid twentieth-century, **4**:286b–287a
foreshadowing of, **1**:102a; **4**:558b
faculty theory, **5**:253a–255b; **11**:617b
Gestalt, **6**:299b; **9**:562b
mathematical, **8**:507b
and medicine, **9**:41a
metaphysics and, **5**:193a
methodology, **9**:562b
naming of, **6**:139b
neurology and, **5**:173b
objective, **1**:580a; **12**:270b–271a
pathological, **7**:67b
philosophical, **1**:404a
physiological, **1**:404a; **2**:131b; **3**:205b; **4**: 52a, 374a; **5**:254a; **6**:52b–53a, 573b; **7**:606a–b, 607a; **8**:45b; **11**:21b; **12**: 571a–b; **14**:44b, 527b–528b
of plants, **4**:557a
religious, **2**:208b–209a, 272b; **5**:145b–146a, 178b; **6**:53b; **7**:190b, 191b, 192b
sensationalist, **3**:1b–2a; **6**:140a; **7**:591b, 607a
statistical studies, **3**:130a; **5**:266a; **10**: 454a; **14**:382b
subjective, **10**:433b
theories of, **5**:173b, 174b–176a; **7**:591a–b; **14**:527a; **15**:128a, 133a
see also Intelligence; Mind; Soul, theories of; Testing, psychological
Psychometrics, **2**:132a–b; **6**:53a
Psychoneurological Institute (Russia)
founding of, **1**:580a
Psychopathology, **2**:448a–b; **10**:613b
Psychophysics
eighteenth-century, **7**:599a
nineteenth-century, **4**:557a–558b; **7**: 457b–458a; **8**:596a–597a, 605a; **9**: 561b–562b; **14**:200b–201b
Psychophysiology, **2**:1a–2a, 129b; **5**:252b; **6**:139b; **8**:7b; **9**:490b; **12**:175b; **14**:3a
Psychoses
metabolic study of, **5**:53a
toxic, **4**:286b
Psychosomatic medicine, **2**:36a; **5**:565a; **13**:511b
Psychosurgery, **4**:286b–287a
Psychotherapy
medieval Arabic, **9**:41a; **13**:229b, 230b
eighteenth-century, **10**:613b
nineteenth-century, **2**:36a, 447b–448b; **5**:174a–b; **8**:272b; **11**:137a, 365a
twentieth-century, **5**:178a–b, 180a–181a; **7**:192a
catharsis theory, **2**:447b–448b; **5**:174a
free association method, **5**:174a–b, 180a
hypnosis, **3**:205b; **5**:174a
punitive, **8**:272b
Psychrometer, **12**:595b
Pteraspis, **8**:27a
Pteridology, **9**:340a–b
Pteridophytes, **2**:371b; **12**:260a; **15**:130b
anatomy, **5**:604b
eusporangiate, **3**:30b–31a
evolution of, **14**:398b
reproduction, **8**:5a–b
see also Cryptogams
Pteridosperms, **13**:345b
Pterin, **14**:334b
Pterodactyl, **6**:594b; **10**:40b; **11**:336a; **14**: 410a
Pteropod ooze, **9**:588b; **11**:372a
Pteropoda, **12**:550b–551a; **15**:166b
Ptolemy I Soter, **3**:491a; **4**:414b, 415a; **6**: 316b; **9**:269a
Ptolemy II Philadelphus, **3**:491a; **4**:383a,b, 388b; **6**:316b; **13**:91a
Ptolemy III Euergetes, **3**:391a; **4**:388b, 389a, 392a; **6**:411a; **9**:270a, 274a; **15**:709a
Ptolemy (or Claudius Ptolemaeus) (*ca.* 100–*ca.* 170), 11:186–206; **1**:380b; **2**:405a; **3**:585b*, **5**:239b, 243b; **6**:313a; **9**: 162a, 299b; **11**:27b
astrology, **1**:382a; **7**:82b; **15**:125b
tables, **12**:432a
astronomy, **3**:60a; **4**:53a; **5**:612a; **9**:296b–297a, 338b; **11**:322a; **13**:323a; **15**: 349a,b, 519b, 533b, 543a,b, 594a–b, 619a
star catalog, **1**:403b*, 511a, 530a; **2**: 155a; **3**:318b; **6**:70b, 364a; **10**:486b, 543b; **12**:170a; **13**:149a–b, 537a; **15**:217a–b, 554a, 564b, 628a–b
tables, **7**:361a, 363a; **15**:331a
zodiac, **6**:617a

Ptolemy (cont.)
biographies of, **7**:187a
on calendar, **3**:22a; **4**:459b; **5**:553a; **15**:538b
criticism of, **3**:26b
commentaries on, **10**:123b*
on contemporary scientists, **13**:325b
cosmology, **1**:192a, 199a, 247a,b, 248a, 509a–510a; **2**:403b; **3**:405a–410a *passim*, **5**:403a; **7**:142a; **11**:224b; **13**:291b
earth's motion, **3**:404a–b
lunar theory, **2**:411b; **3**:406a–b; **10**:298b–299a; **15**:583a
see also Cosmology, Ptolemaic
critics of, **6**:80a, **7**:38b; **11**:288b; **14**:22b, 388b
editions of works, **9**:309b; **11**:31b; **14**:152b
errors of, **7**:362a; **11**:349a
geography, **1**:381b; **2**:153a; **4**:391a; **7**:8a–b; **9**:36b, 172a; **10**:300b; **15**:88a
commentaries on, **9**:13a
place names, **10**:239b
geometry, **1**:226a, 474a; **2**:618a; **3**:272b, 277b; **4**:157b, 417a, 609b; **8**:429b; **11**:201b; **12**:500b; **15**:155a
pi, value of, **7**:258b
Ptolemy's theorem, **1**:239a; **11**:227b; **14**:280b
and Hipparchus, **15**:208a–221a *passim*
influence of, **1**:33a, 36b, 39a, 91b, 267b; **9**:35b, 363b; **10**:419a, 474a; **11**:231a, 312a, 339a, 444b; **12**:61a; **13**:62a(n), 433b, 534b, 538b; **14**:286a, 593b; **15**:155b, 563b
mechanics, **11**:201b
method, **6**:199a
music, **1**:282b–283a; **5**:249a
optics, **1**:81b, 379b; **4**:430a; **5**:552a,b; **6**:125b, 144a
philosophy, **11**:201b
spurious works, **9**:438a; **11**:186b, 201b
translations of works, **2**:283b; **9**:438a; **15**:261b
trigonometry, **1**:42b; **8**:300a; **9**:298a–b, 555b; **14**:18b; **15**:163b, 209a, 478b, 550b
Almagest, **1**:189b–190b, 246a, 283b, 530a; **2**:401b; **3**:21b, 402b, 409a; **4**:443a; **6**:197b–199b *passim,* **9**:338a; **10**:180a; **15**:203a–b, 421a, 555b, 677b
commentaries on, **2**:435b; **4**:489a, 491a, 525a; **6**:199b, 615b; **7**:335a; **10**:5b, 6b, 293b, 294a, 298b–299a; **11**:202a, 249a, 351b, 365b; **13**:253a, 289a, 321a–322a, 323b; **14**:603b, 604b; **15**:178b
critics of, **6**:198b
editions of, **6**:189b; **8**:191a; **13**:509b
emendations of, **1**:508a–513b *passim;* **7**:38a–b; **11**:349a; **15**:474b–475a, 476b–477b
influence of, **1**:42a, 82b; **2**:24a–b; **3**:25a–b, 27b; **4**:93a, 442a; **5**:345a, 614b–618b *passim;* **7**:7a, 259a, 363a; **11**:226a, 415a, 504a; **12**:162a, 360b; **13**:400b; **14**:577a; **15**:33b, 497b
influences on, **1**:258a; **5**:345a; **15**:208a–221a *passim*
summaries of, **4**:542a–543a; **12**:62a
translations of, **1**:61b, 63b, 312b; **3**:402b, 409a; **4**:17a, 444a; **7**:25a, 62a, 81a; **9**:362a; **11**:97b, 349a; **15**:174a,b, 178b, 189a, 474b–475a, 477a, 626a, 628b
transmission of, **11**:202a–b
Analemma, **3**:364a; **11**:197b
Apotelesmatics
commentaries on, **8**:191b
Centiloquium (spurious), **1**:82a
Geography, **1**:29a, 84a, 511a; **5**:377a; **6**:79b; **7**:8b, 361b–362a, 363a; **10**:153b; **11**:187a, 198b–200b
commentaries on, **14**:274b, 275a
editions of, **11**:18a; **12**:323a–b; **14**:127b
translations of, **7**:357a; **9**:179a, 581a; **11**:349a–b
transmission of, **11**:202b–203a
Handy Tables, **10**:301a, 419a; **11**:187a, 196b–197a, 202a; **14**:576b, 577a
commentaries on, **13**:321a, 322a, 323b–324a
editions of, **13**:323a,b
Harmonica, **4**:430b; **7**:300b; **11**:201a–b, 203a
commentaries on, **10**:300b; **13**:91a
De judicandi facultate (spurious), **2**:349a
Optics, **6**:190b, 194a, 200a; **11**:187a, 203a; **13**:323a; **14**:458a, 460b
critics of, **6**:198b
Organum, **6**:144a
Phaseis, **3**:391a; **4**:172a, 391b, 460a; **11**:197b; **15**:208a
Planetary Hypotheses, **1**:511a–b; **3**:25b; **6**:198a, 381b; **7**:112b; **11**:187a, 196b, 197a–b; **15**:219a
critics of, **6**:198b
Planisphaerium, **7**:178b*; **11**:197b–198a; **13**:323b
commentaries on, **3**:364a; **10**:300a
translations of, **9**:39b; **13**:340a
Quadripartitum
see Ptolemy, *Tetrabiblos*
Syntaxis
see Ptolemy, *Almagest*
Tetrabiblos, **1**:23a, 511a; **4**:213a, 622a; **11**:161a, 198a–b
commentaries on, **1**:508a, 513b; **10**:5b; **11**:445a
translations of, **2**:435b; **11**:32a; **13**:538a
transmission of, **11**:202b
Turnings of the Scale, **4**:431b
Ptychopoda, **7**:516b
Public health
Renaissance, **1**:110a; **7**:16b; **8**:483b
seventeenth-century, **9**:582a; **13**:490b
early eighteenth-century, **3**:6b; **6**:44b
late eighteenth-century, **2**:83a; **3**:2a, 7a–b, 577b; **9**:306b; **10**:326a; **11**:100a, 387b; **14**:16a
early nineteenth-century, **2**:444b; **11**:559b; **12**:57b
mid nineteenth-century, **2**:575a; **3**:237b–238a; **4**:252a; **5**:297b; **6**:60b, 120a, 434b; **8**:272a–b; **10**:253a–b, 558b–559b; **11**:51a–b, 419a, 514a; **12**:451a, 465b–467a, 503a; **14**:41b–42a
late nineteenth-century, **1**:360a; **2**:164b, 548a; **3**:188b, 189a; **5**:73a, 219a–220a, 564a; **7**:420a–430b *passim;* **8**:313a, 409a, 449b; **10**:559b–560b; **11**:176b–177a, 514b, 516a, 555b; **14**:109a, 169b, 249b–250a, 409b
early twentieth-century, **3**:22b; **4**:311b; **5**:83b, 451a–452b; **7**:170b–171a; **8**:552b, 557a; **10**:18b–19a; **11**:556b; **12**:453a–b; **13**:63a; **14**:115a; **15**:96a–97b
mid twentieth-century, **12**:48b
legislation, **8**:270b, 313a; **12**:48b
Public Health Service, U.S., **5**:451a–b, 452a
Hygiene Laboratory, **3**:290b
Public service, **3**:440a; **5**:441a
Publication
Renaissance, **4**:412b–413a; **11**:351a
seventeenth-century, **3**:179b, 348b; **13**:29a–b; **14**:51a–b
Japanese, **15**:749a
Puccinia, **3**:613a–b; **14**:93b
Puccinotti, Francesco, **1**:611b, 612a
Pucherite, **5**:162a
Puerperal fever, **1**:323a; **10**:613a; **14**:42b
etiology of, **10**:389b–390a; **12**:294a–296b; **14**:297a
treatment of, **4**:154b; **10**:390a
Puerto Rico, **10**:617a
botany, **12**:326b
discovery of, **15**:89a
Puget, Louis de, **7**:111a
Pugwash movement, **3**:330b; **7**:10b; **11**:117b; **13**:227b, 241b; **15**:9a
Puiseux, Victor (1820–1883), 11:206–207; **3**:18b
Atlas de la lune, **5**:316a
mathematics, **2**:288a; **3**:141a; **11**:449a
Pulfrich, Carl (1858–1927), 11:207–209; **14**:482a
Pulkovo, Russia
observatory, **1**:6a; **2**:434a–b; **7**:6b, 209b, 471b, 472a,b; **8**:418a; **10**:30a–b, 198a; **12**:367b
astrophysics at, **1**:597b; **6**:105b; **13**:121a
founding of, **13**:109b–110a
Pullar, Frederick, **9**:590a
Pullar, Laurence, **9**:588b
Pulley, **8**:225b; **9**:488a; **14**:366b; **15**:518b
block and tackle, **13**:584b
compound, **1**:213b; **8**:198a, 230b
Pulmonary artery, **1**:267a, 619b; **3**:356a
Pulmonary vein, **1**:619b; **3**:356a–b
Pulsars, **4**:249a
Pulse
ancient Greek study, **1**:315a; **6**:317b–318a; **11**:128a
ancient Roman study, **1**:75a, 212b; **5**:231b
Byzantine study, **10**:417b
Renaissance study, **13**:100a–b
seventeenth-century study, **1**:523b; **5**:366b; **9**:97b
eighteenth-century study, **2**:302a; **6**:37b–38a
nineteenth-century study, **5**:184b; **9**:547a; **14**:200a
arterial
theories of, **3**:356b; **5**:234b, 235a; **6**:154b, 157a, 318a; **12**:556b
Japanese study, **15**:741a
measurement of, **5**:231b; **12**:103a
Mesopotamian study, **15**:643a
pulse waves, **2**:601a
crest of, **4**:398b
volume of, **4**:398a
respiration, influence of, **8**:294b
rhythm, **6**:318a
see also Heart, beat; Sphygmology
Pummerer, Rudolf, **13**:1b
Pumpelly, Raphael (1837–1923), 11:209–211; **2**:441a; **3**:434a, 592b, 594a; **5**:49b; **8**:370a
Pumpelly, Raphael Welles, **11**:210a
Pumping
seventeenth-century study, **1**:567b; **2**:311a; **5**:574b; **10**:36b
epicycloidal wheels, **4**:49a–b
eighteenth-century study, **2**:40a; **3**:39a; **4**:517a; **11**:499b
Pumpkin, **13**:478a
Pumps
air, **2**:378a, 379b; **3**:8a,b, 153b; **4**:44a, 166a; **6**:169b, 482a, 598b, 599a; **9**:152b, 498b, 595a; **10**:292b

double, **12**:502a
invention of, **2**:377b; **3**:491a; **5**:574b–575b
mercury, **5**:340b–341a; **7**:160a; **11**:50b
ballistic, **10**:293a
blood, **8**:540b
centrifugal, **8**:212b; **11**:306a
condensation, **8**:24a
curved-cylinder, **3**:8b
diffusion, **4**:253b; **14**:340b
electromagnetic, **13**:227a
fire, **4**:214b; **6**:311b; **11**:378a
force, **4**:44a
invention of, **3**:491a
lift pump, **1**:424a
pneumatic, **5**:295a; **8**:328a
steam, **10**:293a
suction, **2**:377b; **13**:233b, 438a,b
vacuum, **9**:607b
water, **6**:403a; **9**:609b; **11**:393b; **12**:244b; **15**:83b, 253b
Punnett, Reginald Crundall (1875–1967), 11:211–212; **1**:505b; **3**:121a; **5**:485b*; **9**:520a; **13**:134a; **15**:505b, 507a
Pupa
concept of, **13**:172b–173b
Pupin, Michael Idvorsky (1858–1935), 11: 213; **1**:287a; **6**:211b; **9**:395a
Puppis (constellation)
zeta Puppis
spectrum, **7**:268a; **10**:601a
Purāṇas, **15**:554a–b
Purbeckian (stratig. stage), **9**:87a
Purgatives, **2**:619b; **3**:562b; **14**:637b–638a
Purines, **15**:252b
formulas for, **15**:13a
metabolism of, **4**:603a; **6**:506b*; **8**:614b; **9**:284b
reactions of, **5**:2a
structure of, **5**:2a–b
synthesis of, **5**:2a–b
Puritanism, **1**:133b, 134b; **5**:343b; **14**:177b
Purkyně (Purkinje), Jan Evangelista (1787–1869), 11:213–217; **1**:477a, 543a; **6**: 300a; **12**:421b
acquaintances of, **11**:380b; **12**:58a; **13**: 556a
anatomical theory, **12**:241b
histology, **8**:302a; **9**:442a; **11**:257b, 368a; **14**:113a
influence of, **6**:375a
influences on, **11**:593a
physiology, **14**:113b
Purkyně shift, **11**:214b; **12**:353a*
Purkyně's cells, **11**:216b
Purkyně's fibers, **11**:216b–217a
Purkyně's tree, **11**:214b
Purkyně's vesicle, **11**:215b
students of, **3**:530b; **11**:152a; **13**:555a,b
Purple of Cassius, **14**:633b
Purpura, **6**:527a*
Purpuric acid, **5**:197b; **8**:343a
Purpurin, **8**:329a
Pursh, Frederick (Friedrich Traugott Pursch) (1774–1820), 11:217–219; **1**:485a; **10**:164b
Puruṣottama, **9**:42a
Purvis, Olive Norah, **5**:523b
Pus, **6**:479a*
bacteria in, **12**:88a
bactericidal power, **5**:29b
composition of, **9**:380b, 381a
formation of, **6**:255a, 269a,b
healing theory of, **2**:314b–315a; **3**: 219a
microscopic differentiation of, **5**:565b
origin of, **14**:143b
Putnam, Frederic Ward (1839–1915), 11: 219; **1**:74b; **2**:207b, 208a; **9**:536b; **10**:272b; **14**:517b
Putnam, James Jackson, **2**:367a; **5**:179b
Putrefaction, **6**:505b; **10**:80a
alchemical view of, **1**:550a
bacteria and, **3**:338b, 339b, 589b*; **12**:87b
gaseous products, **3**:156b
heat of, **10**:59b
theories of, **3**:345a; **5**:107a; **6**:283a; **8**: 402b–407b *passim;* **10**:364b–365a, 370a; **12**:241a
Putrid fever, **8**:74a
Pütter, August, **7**:503b; **14**:2b
Puydt, Marcel de, **8**:464a
Puységur, Jacques-François de Chastenet, Marquis de, **9**:327b
Puzzles, mathematical
see Recreations, mathematical
Pycnia, **2**:583b
Pycnidia, **13**:489b
Pycnogonida, **9**:516a–b; **14**:425b
Pycnometer, **7**:341b, 346a
Pyemia, **2**:598b; **8**:403a; **12**:87b
Pyman, Frank Lee, **11**:85b
Pyocyanine
isolation of, **5**:73a
Pyramid
center of gravity, **8**:200a, 224b–225b
triangular, **4**:602b*; **7**:564b–565a
volume of, **1**:220a; **3**:277a; **4**:424a, 466b; **8**:421b; **12**:375b; **15**:670b–671a, 697a–b
see also Tetrahedron
Pyramids, Egyptian, **6**:327a; **15**:697a–b
Pyrazole, **2**:560b, 562b
Pyrene, **5**:489a
discovery of, **8**:58b
Pyrenees
geology, **2**:89b; **3**:210a, 560b; **4**:218a; **8**: 45a; **9**:104a, 418a
glaciology, **11**:272a
natural history, **11**:272a–b
Pyréolophore, **10**:118b
Pyrethrum
as disinfectant, **3**:185b
Pyrexia, **2**:445a
Pyrheliograph, **4**:171a
Pyridine, **3**:246a–b, 347b
compounds, **6**:107b; **11**:278b
structure, **4**:78b; **7**:283b
synthesis of, **3**:246b; **11**:278b
and thiazole, **6**:107b–108a
Pyridine bases, **1**:156a; **11**:278b
physiological action of, **4**:79a
Pyrimidine, **5**:209a
metabolism of, **9**:284b
synthesis of, **5**:215a
Pyrites, **6**:259b; **11**:169b
decomposition of, **6**:165a*
deposits of, **13**:566a
distillation of, **6**:42a
in iron production, **5**:157b
medical use of, **14**:585a
thermoelectricity, **11**:539b
Pyrlimnometer, **2**:142a
Pyrochemistry, **2**:65b, 67b; **9**:356a
Pyrodynamics, **4**:208a
Pyroelectricity, **1**:461a; **3**:504a–b, 505a; **4**: 228b, 231a; **5**:497a; **6**:178b; **11**:530a,b
and morphology, **11**:539b
and piezoelectricity, **3**:505b
prediction of, **3**:505b
of quartz, **14**:525b
reversal of polarity, **2**:5a
theory of, **11**:446b
of tourmaline, **2**:5a; **3**:505b; **6**:178b
Pyrogallic acid, **2**:386a
Pyrolusite, **5**:223a
as catalyst, **4**:134b
chemical properties, **12**:146a–b
Pyrolysis, **8**:108a
inhibition of, **6**:404b
mechanism of, **13**:7a
Pyromagma, **7**:63a
Pyromeconic acid, **8**:341a,b
Pyrometer, **4**:44a; **5**:603a; **8**:328a; **10**:246a; **12**:131b, 462a; **15**:392b*
clay, **6**:55a; **12**:562a; **14**:213b
invention of, **9**:595b–596a
optical, **6**:470a; **7**:528b; **8**:117b
recording, **11**:486a
self-registering, **7**:529a
Pyrometry, **1**:490b; **3**:212a, 240a*, 234a, 557a; **6**:561a; **7**:599a; **11**:150b
see also High temperature studies
Pyrones, **3**:347a,b
Pyros (alloy), **3**:239b
Pyrosoma, **12**:481a
Pyrotartaric acid, **8**:339a
Pyrotechnics
Renaissance, **2**:143a
seventeenth-century, **3**:596b
eighteenth-century, **2**:577a
nineteenth-century, **9**:491b
Pyroxenes, **3**:211b*; **4**:141a
Pyrrhocoris, **6**:267a; **8**:587b
Pyrrhonism, **3**:485a; **12**:340a–b
Pyrrhotite, **14**:244b, 245a
Pyrrole, **11**:615b
derivatives of, **3**:279a; **15**:157b
ring structure, **3**:279a
synthesis of, **3**:246b, 279a
Pyrrolidine, **5**:215a
Pythagoras of Samos (*ca.* 560–*ca.* 480), 11: 219–225; **1**:103b, 245a; **6**:315b; **15**: 202b, 672a
astronomy, **1**:246b; **10**:179b; **15**:204b(n)
critics of, **6**:291a
geometry, **4**:104b, 433b(n), 605a; **8**:189b
influence of, **13**:307b; **14**:537a
music theory, **1**:282a; **5**:249a,b; **10**:112b–113a
number theory, **5**:365a; **6**:302a; **9**:1a; **15**:10a
numerology, **1**:38b
Pythagorean numbers, **1**:22b; **4**:610b, 612a; **11**:221a; **15**:503a–b
Pythagorean theorem, **4**:608b; **8**:238a, 422b–423a; **11**:221b; **14**:466a; **15**: 209a, 518b, 671a–b, 690b
Euclid's proof, **4**:418b–419a, 424b
generalization of, **4**:419a, 421a–b, 424b; **10**:294b; **13**:290b, 418a
proofs of, **13**:290b
stereometric analogue, **4**:551a
Pythagorean triples, **4**:610b, 612a; **15**: 503a–b, 671b
Pythagoreanism
ancient Greek, **4**:367b, 421b; **6**:406a; **11**:25a
ancient Roman, **15**:670a, 671b
medieval, **11**:250a
Renaissance, **8**:240b
seventeenth-century, **7**:488a
Oriental, **15**:745b, 746a
critics of, **12**:340b
Pythagoreans, **1**:231b, 333b; **4**:117a, 368a, 419a, 424a–b, 467b; **5**:159a; **6**:381b; **10**:589b–591a; **11**:25b; **13**:314b, 399b
algebra, **10**:300a
astronomy, **6**:416b; **11**:26b–27a; **13**:319b, 326a; **15**:33b
cosmology, **1**:246b; **3**:402b; **10**:590b
factions of, **11**:220a

Pythagoreans (cont.)
geometry, **4**:463a; **6**:410b, 411b, 414b–415a; **13**:303b, 304a,b; **14**:611a
music theory, **1**:282a–283a; **4**:430b, 462b; **11**:26b, 201a
number-atomism, **4**:31a
Pytheas of Massalia (*fl.* 330 B.C.), 11:225–226; **4**:390a; **13**:84b
Pythion, **14**:604a; **15**:116a
Pythius, **15**:516a
Pythogenic theory, **2**:575b

Q

Q quotient, **2**:319a
al-Qabīṣī, Abū al-Ṣaqr ʿAbd al-ʿAzīz ibn ʿUthmān ibn ʿAlī (*fl. ca.* 950), 11:226; **4**:438a, 542a
influence of, **12**:439a
al-Madkhal ilā ṣināʿat aḥkām al-nujūm, **1**:38a; **7**:140b
translations of works, **15**:175b, 187a
Ibn al-Qāḍi, **1**:437b
Qāḍī Ṣāʿid
see Ṣāʿid (Abū al-Qāsim) ibn Aḥmad ibn ʿAbd al-Raḥmān ibn Muḥammad ibn Ṣāʿid al-Qurṭubī al-Andalusī
Qāḍī Zāda al-Rūmī (Ṣalah al-Dīn Mūsā Pasha) (*ca.* 1364–*ca.* 1436), 11:227–229; **7**:255b, 256a,b; **13**:536a
algebra, **7**:259a
commentaries by, **12**:91a
al-Qalaṣādī (or al-Qalṣādī), Abu'l-Ḥasan ʿAlī ibn Muḥammad ibn ʿAlī (1412–1486), 11:229–230; **1**:437b
Qalonymos ben David, **15**:33a
al-Qarashī
see Ibn al-Nafīs
Qayṣar ibn Abi'l-Qāsim ibn Musāfir, ʿAlam al-Dīn, **6**:189b; **7**:79b
al-Qazwīnī, Ḥamdallāh, **7**:337a
al-Qazwīnī, Zakariyā ibn Muḥammad ibn Maḥmūd, Abū Yaḥyā (*ca.* 1203–1283), 11:230–233; **1**:29b; **3**:458b; **7**:4a, 343b
Qibla, **9**:85a; **10**:6a; **12**:361b, 431b; **14**:577b; **15**:259a, 260a
al-Qifṭī, Abu'l-Ḥasan ʿAlī ibn Yūsuf, **1**:33a; **2**:619a; **4**:541b, 542a; **5**:612a; **6**:189a–b, 190a, 199b; **7**:24b, 38b, 79a, 223a–b, 325a, 330b, 334a, 335a; **9**:159b; **10**:5b; **11**:226a, 444a; **12**:447b, 448a; **13**:292b, 538a; **14**:547a, 592b; **15**:231a, 261b, 263b
Taʾrīkh al-ḥukamāʾ, **1**:35b–39b *passim,* 507b, 508b, 513a; **9**:160a; **14**:537b
Quackenbush, Leonard Schuyler, **9**:520b
Quackery, **5**:252a; **7**:187b
Quadrangles
area of, **1**:40a; **3**:254b–255a
complete, **13**:16b
inscribed, **13**:18b
trigonometrical solution of, **8**:299b
Quadrant (astron. instrum.)
medieval Arabic, **1**:122a–b, 509a; **2**:150a; **7**:338b
medieval Latin, **3**:27a; **4**:609b; **7**:119a, 123b, 126a; **10**:540b; **12**:61a
Renaissance, **1**:179a, 573b; **2**:404a; **3**:276b–277a, 402a, 558b; **6**:144a; **9**:192a; **11**:339a; **15**:154a
early seventeenth-century, **5**:344a; **10**:495a; **12**:500b; **13**:256b
late seventeenth-century, **2**:133a; **3**:102b, 348b*; **4**:168b; **6**:362b–363a; **7**:577b; **11**:423a
early eighteenth-century, **2**:140b, 141a, 388b; **5**:88a, 491a–b; **8**:178b, 459b
late eighteenth-century, **1**:326b; **7**:579b; **8**:351a; **9**:231b
almucantar, **12**:361b
azimuthal, **14**:359b
Brahe's, **2**:402b, 403b, 405b–406a, 411a
with cursor, **6**:301b, 302a
double-reflecting, **5**:434a
Hadley's, **13**:470b
horizontal, **4**:13b
leveling of, **2**:329a
mural, **5**:435a; **6**:69b, 511b, 512a; **7**:543b; **8**:21b, 22a; **9**:234a; **13**:510b; **14**:589a
perfect, **12**:358a, 361b–362a
pocket, **6**:129b
portable, **8**:22a
quadrans novus, **13**:400b
reflecting, **9**:163a
shakāzi, **14**:594a–b
sine, **12**:361b; **14**:594a
with trigonometric grid, **15**:260b
trigonometric uses of, **15**:156a
universal, **15**:155b–156a
Quadratics, **3**:164a; **11**:407b
theory of, **6**:357a; **9**:475b
Quadratrix, **4**:56a, 429a, 451b; **8**:28a; **10**:295a; **12**:580a; **14**:21b
discovery of, **4**:103b–104b; **6**:406b–408a
Quadrature
ancient Greek methods, **1**:229b, 474a
Renaissance methods, **8**:236a–237b, 239a–b
seventeenth-century methods, **1**:474b; **3**:458b–459a; **4**:571a–573a; **5**:521a, 526a–b, 527a,b; **6**:599b–600a; **8**:161b–165b *passim;* **9**:599b; **10**:46a–47a, 48b, 50a–51a, 72b, 336a; **11**:487a; **13**:480a, 561a
eighteenth-century methods, **2**:56b; **3**:244b, 282a, 432a; **5**:499b–500a; **7**:577b; **9**:186a; **10**:256b*; **13**:418a
nineteenth-century methods, **9**:127a; **10**:249b, 442a, 443a
of algebraic curves, **6**:359a, 537a; **10**:103b; **13**:480b
approximation of, **3**:224b; **6**:599b–600a; **9**:127a
arithmetical, **8**:161b–164a *passim*
Chebyshev's formulas, **3**:230a
of circle
ancient Greek, **1**:171a–b, 226a–227a, 228a; **2**:549b, 550a; **4**:103b–104b; **6**:407a,b, 408a, 410b, 412b, 413b–414a
ancient Roman, **1**:119b; **10**:295a; **12**:579b–580a
Byzantine, **12**:441b
medieval, **1**:94a; **2**:435a; **3**:27b; **6**:202b; **7**:128b, 129a; **14**:400b
Renaissance, **1**:97a; **2**:360b, 618a; **3**:181a, 262b, 277b, 280b, 515b; **4**:260a; **8**:236b–237b; **9**:335b; **10**:161b; **11**:97b, 533b; **14**:21b; **15**:155b
seventeenth-century, **1**:341a; **3**:63b, 128b*; **6**:449b, 599b–600a; **7**:557b, 558a; **8**:28a, 161b–164a *passim;* **9**:312a*; **10**:46a–47a; **12**:75a, 75b–76a, 332b; **14**:148a, 152a
eighteenth-century, **4**:469b; **5**:449b; **7**:559b*; **9**:55b
nineteenth-century, **4**:36b; **8**:367b–368a; **11**:589a
twentieth-century, **6**:451b*
historiography of, **9**:500b
proof of, **7**:175a–b
of cone, **11**:487a
of cycloid, **2**:48b, 54a; **4**:58a, 506a; **8**:161b; **11**:487a; **13**:434b
of cylinder, **11**:487a
of hyperbola, **2**:49b, 507a; **4**:516b; **6**:599b, 600a,b, 602a; **8**:162a–165b *passim;* **9**:311b*; **10**:46a, 47a; **12**:75b; **13**:69a
of lunes, **1**:97a, 171a; **3**:222a; **4**:463a; **6**:202b, 411b–414a, 415b, 416a; **14**:21b
mechanical, **4**:562a; **15**:23b
of parabola, **1**:219b, 221b; **7**:2b–3a; **8**:163a; **13**:435a, 561a
of rectangle, **6**:414b
of spiral, **1**:219b; **4**:571b
theory of, **13**:26b
Quadratures (astron.), **15**:380b–381a
Quadrics, **2**:215a; **3**:560a; **12**:284a
classification of, **13**:6a
confocal, **4**:257b
definition of, **13**:5b
generatrices, **11**:79b
intersections of, **12**:225b, 284b
metric properties, **12**:225b
projective, **9**:173b
rotational, **3**:282a; **6**:359b
theory of, **2**:444a; **9**:475b; **11**:390a; **14**:282a
Quadrilaterals
area of, **5**:409a; **6**:314b; **12**:500a
complete, **13**:16b, 18b
concave, **4**:609b
cyclic, **12**:206a
inscribed, **3**:277b; **5**:409a
Saccheri, **6**:202a; **7**:329b, 600a; **12**:56a; **13**:510b
spherical, **13**:289b
types of, **5**:409a
Quadrillion, **3**:273a
Quadrivium, **1**:104b, 474a, 565a; **2**:229a, 230b, 233a–b, 395a, 401a; **3**:109b; **4**:443b; **5**:365a, 592b; **6**:544a; **8**:548a; **9**:140b; **11**:287b; **13**:340a, 588b; **15**:188a–b
Quaestiones, **1**:275b–276b
Quain, Jones, **5**:514b
Elements of Anatomy, **12**:354a
Quakers
see Friends, Society of
Qualities, primary
ancient Greek theories, **1**:261a–b, 277b, 278a, 324a,b; **4**:32a–b, 106a; **13**:92a–b, 93a
ancient Roman theories, **8**:537b–538a
Byzantine theories, **7**:136b
medieval theories, **1**:199a–b, 290b; **2**:610b; **6**:27b–28a; **7**:41b, 42a; **9**:137a; **13**:186a–193b, 195b–197a; **14**:155b–156a; **15**:265a–b, 496b
Renaissance theories, **1**:373b, 374b–375a, 375b; **3**:66a; **9**:133a; **11**:605a; **13**:279a
seventeenth-century theories, **2**:13b–14a, 237b; **7**:537b, 611b
and atomism, **1**:373b
maximum degree, **13**:205b–206a, 207b
mutations of, **9**:137a
quantification of, **1**:277b; **13**:186b–187a, 188b–193b, 196b–197a; **14**:155b–156a; **15**:265a–b
vs. secondary, **5**:252b; **8**:438b
Quantics, **3**:164b–165a, 169a, 314b
binary, **3**:169b
Quantification
logical, **3**:278b; **5**:153a; **6**:297a–b, 376b, 377b; **7**:151a
of motion, **13**:185b–187a, 195a
of qualities, **1**:277b; **13**:186b–187a, 188b–193b, 196b–197a; **14**:155b–156a; **15**:265a–b
Quantity
medieval theory of, **5**:402b; **10**:172b
twentieth-century theory of, **12**:364b
classification of, **3**:75b–76a

concept of, **1**:277a, 278a
definition of, **5**:152b
negative, **3**:76b–77b
sequence of, **10**:485a
Quantum chemistry, **8**:292a
Quantum electrodynamics, **3**:367b; **7**:617a,b; **11**:420a; **15**:43a
nonrelativistic, **7**:493b
and quantum mechanics, **3**:369b
reformulation of, **4**:579a
relativistic, **10**:423b
Quantum mechanics, **1**:519b; **2**:144b, 248a–250b; **3**:125a, 564a; **4**:425a, 578a; **5**:160b, 473a–b; **6**:92b, 147b, 614b–615a; **7**:518b, 554a, 617a; **8**:292a, 473a–479a, 480b–482a; **9**:77a, 217b, 397a; **10**:213b–214b, 236a; **11**:421b, 422a, 581a; **12**:21a–b, 33b, 419a, 438a–b, 528a–529b, 614b; **13**:176a, 240a–241b; **15**:8b, 444a, 446b–447b
approximation method, **15**:42b–43a
and classical physics, **4**:294a; **6**:89b; **7**:492b
confirmations of, **9**:544b
group-theory questions, **5**:198b
mathematization of, **7**:491b–493b; **14**:91b–92a, 454a; **15**:41b–43a
misinterpretations of, **5**:122b–123a
nonrelativistic, **5**:473a; **13**:240a; **14**:91b
objections to, **4**:318b; **12**:219a
and quantum electrodynamics, **3**:369b
relativistic, **3**:368b, 369b; **5**:161a; **12**:219a–b; **13**:240a
statistical, **3**:564a; **5**:103a,b; **8**:185b–186b, 473b; **14**:245a; **15**:6b, 48b–49a
see also Wave mechanics
Quantum numbers, **2**:245a–248a *passim*, 466a; **3**:125a, 430a; **4**:69a, 578a; **10**:423a; **12**:21a
Quantum theory, **1**:500a; **2**:169b, 240b–250b, 338a; **3**:368a–370a, 619a,b; **4**:293b–294a, 315a–318b, 376a, 414a, 577b; **5**:103a–b, 271b, 272a, 609a–b; **7**:26b–27a, 84a, 211b, 387a, 553a,b, 554b; **8**:52a, 292b, 293b, 552a, 578b; **9**:223a, 261a; **10**:106b, 345b, 422b–423b; **11**:12b–13a, 583a; **12**:218b, 251b–252a, 528a–529b; **13**:41a–b; **14**:317b; **15**:41a, 48b, 432b
and biology, **12**:222b–223a
and chemistry, **5**:621b; **12**:31b; **15**:441b
and classical theory, **2**:242a–250b; **7**:85a
correspondence argument, **2**:246b–247b, 248a–b, 250b; **7**:491b, 493b, 553b; **11**:13a; **13**:240a; **15**:41b
diffraction, **4**:196b
directionality of quanta, **4**:317b
and electron theory, **8**:497b–498a
experimental proof of, **5**:117a–b; **11**:266a; **12**:614a; **13**:599a; **15**:442b
and field theory, **4**:326a–b; **7**:617b; **8**:498b; **9**:377a
fine structure constant, **10**:423b
foreshadowings of, **2**:262a–b, 266a
foundations of, **6**:164a; **7**:235a; **12**:464a
generalization of, **4**:294a
indeterminism of, **3**:111b
interpretation of quantum, **7**:26b–27a
and magnetism, **14**:243b, 246a
mathematization of, **2**:248a–b; **7**:491b–492a; **8**:434a
opposition to, **4**:101b; **11**:481a(n); **13**:103b–104a, 364a–b
philosophical interpretation of, **2**:121a; **11**:356b, 358a
and relativity, **4**:281a–282a; **5**:182b; **10**:423b; **11**:13b
spatial quantization, **13**:42a–b
and specific heat, **2**:170a; **13**:41a; **15**:442a–448a
statistical, **8**:478a–479a; **12**:219a, 221b, 464a, 528a; **14**:91b; **15**:43a, 432b, 442a, 444a
stimulated emission, **4**:317b
and X-ray emission, **2**:398b; **3**:368a–370a; **7**:468b–469a; **10**:214a; **12**:614a; **14**:341a
Quantum thermodynamics, **14**:91b
Quarantine, **1**:493b; **3**:605a*
Quarries
nineteenth-century, **2**:439b; **5**:475b
Quartan fever, **9**:95a
Quarterly Journal of Experimental Physiology
founding of, **12**:356a
Quarterly Journal of Mathematics, **5**:414a
Quarterly Journal of Microscopical Science
editors of, **8**:27a*
Quarterly Journal of Pure and Applied Mathematics
editors of, **13**:79a*, 217b
Quarterly Journal of Science, **3**:475b
Quarterly Journal of Science, Literature and the Arts, **2**:420a; **6**:263a
Quarterly Review of Biology
founding of, **10**:445a
Quartics, **12**:284b
Quartz
fused, **5**:295a
as insulator, **15**:59b, 60b
optical properties, **11**:265b
circular polarization, **1**:461a; **12**:305b
colors produced, **2**:139a
optical activity, **2**:137b–139b; **5**:170a; **10**:357b
refractive index, **11**:582a, 583a
pebbles, **2**:569a
physical properties
actinoelectricity, **6**:96b
elasticity, **15**:251a
electrical, **15**:251a
piezoelectricity, **3**:498b, 504b, 505b, 506b, 507a; **10**:605a; **14**:525a
pyroelectricity, **14**:525b
rutilated, **5**:563a
structure of, **1**:461a; **5**:203a; **11**:539b; **14**:525b
allotropy, **8**:119a
and temperature, **4**:563b
twinning, **5**:186b
Quasars, **1**:353b
Quaternaria, **2**:189b
Quaternary period, **2**:486b; **4**:143b; **9**:173a; **10**:429a, 502a, 504a
formations, **8**:519a
glaciation, **6**:238b; **10**:429b
naming of, **12**:318a
paleobiogeography, **10**:16b
paleogeography, **7**:510a
paleontology, **2**:189a–b; **3**:264a, 532a; **4**:308b; **5**:109b, 221b, 296a; **8**:43b–44a, 464a; **10**:429b; **12**:184a; **13**:275a
stratigraphy, **4**:39a; **5**:329b; **6**:187b; **10**:128a; **11**:269b, 567a
see also Holocene epoch, Pleistocene epoch
Quaternions, **1**:240a; **3**:163b, 166a,b, 168a, 169a, 323a; **5**:309a, 391a; **6**:95b, 571a; **9**:215b; **13**:125b, 237a; **15**:17a
biquaternions, **3**:322a, 323a
concept of, **2**:295b; **4**:35b; **6**:90a–92b
definition of integrity, **6**:572a
pluquaternions, **7**:384a
theory of, **7**:473b; **15**:196a, 198a
Quatre Nations, Collège de, **1**:110b
Quatrefages de Bréau, Jean-Louis-Armand de (1810–1892), 11:233–235; **9**:409a*; **13**:288a
zoology, **1**:602b; **10**:372b, 523a; **11**:566b; **12**:551a
Quebec Group (geol.), **2**:129a; **8**:462a
Quenstedt, Friedrich (1809–1889), 11:235–236; **10**:211b; **11**:560b; **14**:240a
Quercetanus, Josephus
see Duchesne, Joseph
Quervain, Francis de, **10**:126b
Quesnay, François, **8**:115a
Quesneville, Gustave Augustin, **5**:370a; **9**:290b
Quet, Jean Antoine, **2**:67b; **5**:341a
Quetelet, Lambert-Adolphe-Jacques (1796–1874), 11:236–238; **2**:214b; **3**:113a, 555a*; **6**:3a*; **9**:460a*
acquaintances of, **3**:554b; **5**:305a; **11**:22a
Correspondance mathématique et physique, **2**:214a; **5**:367b
founding of, **11**:236a
correspondence of, **15**:31a
optics, **6**:87b; **9**:73a
statistics, **5**:266a; **9**:69a; **10**:465a(n); **14**:99a; **15**:352b
students of, **13**:616a
Theory of Probability as Applied to the Moral and Social Sciences
reviews of, **9**:219a
translations of works, **7**:416a*
Queuing theory, **7**:352a
Ibn al-Quff, Amīn al-Dawlah Abū al-Faraj ibn Muwaffaq al-Dīn Yaᶜqūb ibn Isḥāq al-Masīḥī al-Karakī (1233–1286), 11:238–239; **14**:584b
commentaries by, **11**:251b
al-Qūhī (or al-Kūhī), Abū Sahl Wayjan ibn Rustam (*fl. ca.* 970–1000), 11:239–241; **6**:206a*, 208a*; **7**:353b; **9**:21b; **12**:447b
geometry, **12**:432a
statics, **7**:341a, 342b
Quibell, Thomas Henry Hall, **9**:403a
Quicklime, **2**:175a–b, 176a; **5**:411a; **6**:395b
solubility of, **2**:175b, 176b–177a
Quicksand
classification of, **9**:593b
mobility of, **10**:166b
Quinazoline, **5**:215a
Quincke, Georg Hermann (1834–1924), 11:241–242; **2**:427a; **8**:180a, 181b; **14**:360b*
chemistry, **2**:70b
students of, **4**:354b; **7**:270b; **9**:372a; **10**:447a; **14**:338a
Quincy, John, **6**:37a,b
Complete English Dispensatory, **8**:297b
Quine, Willard Van Orman, **11**:107b; **12**:14b
Quinhydrone
discovery of, **14**:477b
Quinic acid
structure of, **5**:6a
Quinidine, **10**:288b
Quinine
botanical study, **4**:518b; **8**:512b; **13**:464b
determination of
in urine, **10**:530b
discovery of, **3**:159b; **10**:497b
extraction of, **5**:91a; **9**:417b
salts of, **6**:293b; **10**:498a
synthesis of, **10**:515b–516a; **12**:452b
as therapeutic agent, **3**:159b; **5**:461a–b, 503b; **7**:428a, 614a; **9**:9b; **12**:458b
Quinoline, **3**:246b, 300a; **5**:489a
physiological action of, **4**:79a
structure of, **4**:78b
Quinones, **5**:488b–489a; **14**:247a–b, 411b, 477b
bonding in, **5**:489a
in living cells, **3**:359a

Quinsy, **10**:613a
Quintana, Juan de, **12**:322b
Quintessence, **1**:80a, 250b, 571b; **4**:170a, 387b; **6**:417b(n); **7**:356a; **10**:308a, 309a, 310a; **11**:98a; **13**:535a
  see also Ether, Aristotelian
al-Qumrī, al-Ḥusayn ibn Nūḥ, **9**:40b
Ibn Qunfudh, **1**:437b; **11**:230a
Quotient
  differential, **4**:370b
    limit origin of, **3**:136a
Quoy, Jean-René-Constant (1790–1869), 11:242–244; **5**:224b; **13**:554b
Ibn Qurra
  see Thābit ibn Qurra
Qūshjī, ᶜAlī, **7**:257a
Qusṭā ibn Lūqā al-Baᶜlabakkī (*fl.* 860–900), 11:244–246; **7**:187a; **10**:21a
  algebra, **12**:92a
  commentaries of, **4**:117a
  De differentia spiritus et animae, **5**:592a
    translations of, **15**:175b
  De physicis ligaturis, **1**:291a–b
  translations by, **4**:439b; **6**:617a; **11**:32a; **13**:320a; **15**:119a, 261b
  translations of works, **13**:400b
Ibn Qutayba, Abū Muḥammad ᶜAbdallāh ibn Muslim al-Dīnawarī al-Jabalī (828-*ca.* 889), 11:246–247
  Kitāb al-anwāᵓ, **1**:37b
Quṭb al-Dīn al-Maṣrī, **13**:509a
Quṭb al-Dīn al-Rāzī, **11**:251b
Quṭb al-Dīn al-Shīrāzī (1236–1311), 11:247–253; **6**:196b; **12**:358b
  astronomy, **6**:199a; **7**:257b; **13**:510b; **15**:34b
  cosmology, **12**:358b–359a
  influences on, **7**:38b, 336a, 337a,b; **13**:511a, 513b
  optics, **7**:212b, 213a,b, 216a–b, 218a
  philosophy, **13**:513a
  translations by, **4**:442a
Quṭrub, **11**:246b

R

Rabaud, Étienne, **4**:504b
Rabbit
  embryogeny, **2**:161b; **5**:485a
  ovulation, **2**:345a; **3**:487a
  population, limitation of, **10**:352b, 406a–b
  size differences, inheritance of, **3**:122b
Rabbit problem (math.), **4**:606b–607a, 611a
Rabe, Eugene Karl, **12**:574a
Rabelais, François (*ca.* 1494-*ca.* 1553), 11:253–254; **8**:125b; **9**:75b; **11**:527b; **12**:135a
  Gargantua and Pantagruel, **12**:135a
Rabies
  contagious nature of, **9**:10b
  diagnosis of, **10**:15b–16a
  etiology of, **10**:15a–b, 142a, 398b
  paralytic, **5**:270a
  prevention of, **11**:569a
  serum therapy, **14**:551a–b
  transmission of, **10**:401b–402a
  treatment of, **10**:400b, 403b–406a; **11**:100a
  vaccination, **3**:189a; **4**:211a; **5**:269a–b, 270a; **6**:518b; **7**:427b; **8**:313a; **10**:400b–406a, 407a
  see also Hydrophobia
Rabl, Carl (1853–1917), 11:254–256; **1**:602a*; **6**:338b; **14**:236a
Racah, Giulio, **4**:578b
Race, **2**:204a, 538b; **3**:555b, 556a; **6**:563b; **7**:415b
  and brain size, **13**:404a
  Caucasian, **3**:192a
  and climate, **11**:243a–b
  and culture, **2**:208b, 211a–b; **11**:137b–138a
  fixed features, **4**:286a
  and inherent capacities, **4**:272a
  and intelligence, **5**:505b
  and language, **2**:208b; **11**:137b–138a
  and serology, **6**:433a
  theories of, **1**:175a; **5**:196b, 505b–506a; **6**:454a, 527b; **7**:278a,b; **8**:98a; **9**:540b; **10**:209b; **11**:136b–138a; **14**:253b
  three-fold division, **5**:46b
  variation within, **11**:137b
  see also Anthropology; Ethnology; specific races
Racemic acid
  naming of, **5**:325a, 431b
  optical properties, **10**:357a–360b
  production of, **10**:359b–360b
  separation of, **10**:360a–b
  and tartaric acid, **8**:332a; **10**:357a–360b
Racemic compounds
  preparation of, **10**:359a–360b
  pseudoracemism, **11**:85b
  separation of, **7**:552a–b; **10**:360a–b; **11**:85b–88b *passim;* **14**:173a, 271a
Racemization, **8**:276a; **11**:85a, 86a
  autoracemization, **14**:124b
Radar, **1**:196a, 287b; **2**:427b; **3**:330a, 373a; **6**:104b, 105a; **11**:422a; **12**:201b; **15**:386b
  in astronomy, **5**:405a
  centimetric, **8**:95b
Raddi, Giuseppe, **10**:13b
Rademacher, Hans (1892–1969), 11:256–257; **11**:269a; **13**:23b, 428b
Radiata
  taxonomy of, **8**:270a; **12**:421b
Radiation
  acoustic
    pressure of, **5**:202a
  amplitudes, **2**:246b, 248a
  black-body, **1**:519a; **2**:565b; **4**:293b–294a; **5**:273a, 609a; **6**:350b–351a; **7**:382a; **8**:289b; **11**:10b, 481a(n)
    constants of, **3**:327a
    distribution law, **4**:315b, 316a,b, 317a–b; **7**:84b–85a; **8**:497b; **13**:102b
    dynamics of, **11**:13b
    energy fluctuations, **4**:314b
    laws of, **8**:551b–552a
    long waves, **7**:528a
    measurement of, **7**:528a–b; **11**:123a
    and temperature, **7**:528a; **10**:346a–b; **11**:150a; **12**:153a
    theory of, **9**:212b
  cavity, **7**:26b; **11**:11a; **14**:340a
  continuous spectrum, **5**:68a–b
  cosmic, **2**:338b, 488a; **7**:153b; **12**:71b; **13**:176a
    concept of, **1**:301a; **14**:422b
    energy of, **9**:399a
    geographical variation in, **3**:370b; **13**:83a
    intensity of, **11**:348a
    latitude effect, **3**:312b, 370b; **9**:398a, 398b–399a
    meteorological variation in, **3**:370b
    mutagenic effects, **4**:131a
    nature of, **2**:338a; **3**:370b–371a; **6**:355a–356a; **9**:397b–399a; **10**:214b; **13**:241a; **15**:27a
    origin of, **4**:583a; **8**:618a; **9**:398a–b; **11**:348a
    particles in, **7**:617b; **11**:117a–b; **12**:33a; **13**:600b; **14**:422a
    showers of, **5**:332a; **10**:214b; **15**:26b–27a
    sun, influence of, **3**:370b
  damping, **11**:10b
  detection of, **2**:600b
  directional dependence, **12**:614a
  electromagnetic, **7**:518b; **14**:208a
    absorption of, **4**:191a–b
    and atoms, interaction of, **12**:528a–b
    diffraction of, **2**:399a
    displacement law, **14**:339b
    energy of, **14**:339a–b
    law of, **7**:379b
    in moving cavity, **6**:163b–164a
    polarization of, **15**:46b
    propagation of, **8**:444a
    reflection of, **2**:399a
  energy of
    distribution of, **11**:150a, 582b; **14**:338b, 339b–340a
    measurement of, **8**:20a–b; **12**:29a
  entropy of, **14**:339a
  equilibrium, law of, **7**:382a; **14**:346a
    theory of, **9**:404b
  flame, **1**:519b; **7**:186b
  genetic effects of, **13**:137b
  gravitational, **7**:10b
  infrared, **1**:519b; **13**:51b
    detection of, **6**:333b; **15**:59a–b
    interference phenomena, **5**:19a
    measurement of, **4**:81a; **8**:20b; **9**:566b; **10**:105b; **11**:149b, 582a
    penetration of, **11**:582b
    polarization of, **1**:616b
    production of, **11**:582b–583a
  intensities
    measurement of, **7**:528a; **9**:566b
  interstellar, **15**:449a
  ionizing
    absorption of, **2**:398a
    scattering of, **2**:398a
  laws of, **7**:528a; **10**:345b, 346b; **11**:149b–150b; **13**:11a, 52a, 102b, 104a
  magnetic deflection of, **4**:356a
  mathematics of, **6**:392b; **11**:10a–13a
  and matter, **4**:196a; **7**:85b; **9**:404b; **15**:49a,b
  metallic vapor, **1**:519b
  multipole, **7**:492b
  N rays, **2**:203a; **11**:481b*; **12**:29a; **14**:499a*
  naming of, **6**:446a
  natural, **11**:11a
  penetrating, **7**:153a–154a
    discovery of, **14**:32a
    ionization effects, **1**:559a–b
    reflection of, **1**:559a
    refraction of, **1**:559a
    see also X rays
  pressure of, **2**:120b, 266a; **4**:279a, 322b, 361a; **9**:212b; **11**:122b–123a; **13**:106b*
  residual, **11**:582a–b
  resonance, **5**:117a–b; **11**:478b
  secondary, **1**:559a; **3**:31b, 372a*; **6**:366a; **7**:153a–154a; **11**:265b; **12**:69a; **14**:32a
  solar
    and atmosphere, **1**:21b; **5**:68b; **13**:522a; **14**:38b–39a, 53a
    effects of, **12**:523b
    and geochemistry, **13**:619b
    infrared, **7**:270b; **11**:583a
    measurement of, **12**:266b; **14**:185b
    particle, **1**:478a
    pressure of, **1**:301a; **4**:555a; **6**:98b; **8**:107a

solar constant, **6**:106a; **7**:76b; **8**:20a; **11**:559b; **12**:353a; **14**:38b
and thickness of atmosphere, **2**:136b
total, **9**:89b
variation in, **4**:171a–b; **10**:35a; **11**:498b; **12**:250a; **14**:343a
wave, **1**:478a; **4**:69b
stellar, **8**:618a
stimulated emission, **11**:12a
temperature of, **11**:12a, 150b
theories of
eighteenth-century, **2**:330b
nineteenth-century, **7**:382a; **11**:9a, 115b; **13**:103b–104a
twentieth-century, **2**:243a–248b, 338a, 398a; **7**:493b, 553a; **8**:290b, 292b; **11**:480a
corpuscular, **3**:357b
field, **4**:316b–317a
quantum, **3**:369a; **4**:315b–318b; **10**:214b; **11**:12b–13a
wave, **3**:357b
thermal, **1**:519b; **9**:60b, 192a; **15**:83b
and atmosphere, **13**:522a
color effects on, **1**:364a
earth's heat loss, **9**:189b
energy of, **14**:339a
exchange theory, **13**:51b–52a
interference phenomena, **5**:86b
laws of, **4**:240a; **8**:261b; **12**:496a
mathematical treatment of, **5**:95a, 96b, 97b; **15**:488b
measurement of, **7**:74a; **9**:264b; **13**:352a
polarization of, **5**:68b; **9**:265a; **11**:115b
refraction of, **5**:68b; **6**:333b; **10**:603a
spectrum of, **9**:566b
and temperature, **5**:461b–462a; **11**:150a; **13**:10b–11a; **14**:39a
theories of, **3**:83a–b; **11**:10a–13a, 115b
ultraviolet, **9**:566a–b
absorption of, **3**:207a
biological effects, **3**:327b*; **5**:594b; **14**:444b–445a
and condensation, **14**:422a
discovery of, **11**:474b
Dorno's radiation, **4**:171b
ionization, **12**:26b
and metal dissociation, **11**:281a
polarization of, **1**:616b
therapeutic, **3**:327b
see also Alpha particles, Beta particles, Canal rays, Cathode rays, Gamma rays, Light, Luminescence, Positive rays, X rays
Radical theory (chem.), **1**:518b, 578a; **2**:96a, 587b, 620b, 621b, 623a; **5**:125b, 295a; **7**:451b–452b; **8**:56a–61a, 332b–338a, 341b; **14**:476b–477a
precursors of, **4**:239a
Radicals (chem.), **5**:373b; **13**:7a
concept of, **4**:244b–245a; **12**:546a
metallic, **7**:498a
notation, **5**:408b–409a
organic, **1**:362b; **3**:48b; **5**:125b, 324b, 372b; **7**:452a; **8**:82a, 332b–338a, 349b; **14**:476b–477a, 530b–531a
acidic properties, **9**:356b
free, **5**:464b–465a; **10**:15a, 288b, 289a; **14**:334a
and metals, **7**:498a
physiological effects, **11**:419a
polyvalent, **7**:280b
regrouping of, **2**:622a
stability of, **4**:553b
Radicals (math.)
algebra of, **7**:242b–243a; **8**:161b–162a
compound, **8**:161b
concept of, **9**:457b
Radiculites, **8**:354b
Radio, **1**:24a, 195b–196a; **2**:427b; **3**:298b; **4**:334a–b, 559b; **7**:108a, 468b; **8**:40b; **9**:76a; **12**:29a; **13**:357a, 598b
broadcasting, **4**:7a, 601b
educational, **9**:553a
circuits
heterodyne, **1**:287b; **4**:610b
regenerative, **1**:287a
resonant, **10**:604b
scatter propagation, **9**:99a
superheterodyne, **1**:287b
triode feedback, **1**:287a,b
detectors, **12**:26a
coherer, **2**:325a; **8**:444a; **9**:98b; **11**:93b
crystal, **2**:427b; **10**:604b
diode, **5**:33a; **6**:546b
electrolytic, **4**:601a
triode, **4**:6b–7a
distance measurement, **1**:196a
fading of signals, **1**:195b
feedback
triode circuit, **1**:287a,b
frequency modulation (FM), **1**:287b–288a
frequency stabilization, **3**:505a
generating apparatus, **2**:325a; **4**:601a–b
interference with
static, **1**:338a
Kennelly-Heaviside layer, **1**:195b; **6**:212a; **7**:288b
long-range transmission, **1**:338a; **9**:98b–99a
maritime, **9**:99a
microwave, **9**:99a
oscillator, **10**:605a
shortwave, **9**:99a
strength of signals
prediction of, **1**:338a
superregeneration, **1**:287b
time service, **1**:429b–430a
tuning, **8**:444a–b; **13**:357a
Radio Corporation of America, **1**:287b, 288a; **4**:601b; **9**:99a
Radio waves
see Wave(s), Hertzian
Radioactinium, **6**:15b
Radioactivity, **3**:176b*; **8**:47a; **9**:542b, 606b; **10**:347a
atmospheric, **4**:356b; **5**:341b; **12**:29a, 218a
and condensation, **1**:490b
dating methods, **2**:259b–260a; **3**:508a; **6**:16a, 474b, 475a; **7**:161a–b
helium method, **2**:260a; **10**:289a; **12**:29a; **13**:107b
lead method, **2**:259b–260a; **8**:42a–b; **12**:20a, 29a
see also Carbon, carbon-14, dating techniques
discovery of, **1**:559a; **3**:498b, 507a; **11**:58b
end products, **2**:259b; **12**:505b
heat production, **4**:195a; **13**:619a
identification of source
alpha particle ranges, **2**:398a
induced, **3**:329b–330a, 500a, 502b, 507b–508a; **4**:580a–b; **5**:332a
diminution of, **3**:508a
discovery of, **7**:154a–b
magnetic deflectability, **5**:394b
measurement of, **3**:312b, 498b–499a, 501a, 507a,b; **5**:330b–332a; **9**:260b
coincidence method, **5**:331b
scintillation-counting method, **3**:480a; **5**:331a
naming of, **4**:355b
physiological effects of, **7**:159a
cancer, **7**:157a*
genetic, **13**:138a
lesions, **3**:502b, 508a
toxicity, **3**:508a
Schweidler fluctuations, **5**:330b
statistics and, **2**:319a
theories of, **3**:479b–480a, 507a; **4**:355b–356a; **5**:394b; **7**:264a; **9**:292b; **12**:27b–28b; **13**:368b; **15**:464a
disintegration, **12**:504b–505b
transmutation, **1**:500a, 560a; **2**:241a, 258b–260a; **3**:480a, 499b–500a, 501a; **11**:280b; **12**:28a–29a
tracer techniques, **6**:16b; **7**:157a*; **10**:288b–289a
discovery of, **6**:365b–366b
types of, **7**:154b; **12**:26b–27a
alpha decay, **12**:505a
beta decay, **7**:158a,b; **12**:505a
weak, **6**:16a
see also Geiger counter; specific elements
Radiochemistry
late nineteenth-century, **12**:191b
early twentieth-century, **2**:257b, 258b–260b; **5**:394a–395a; **6**:14b–17a, 365a–366b; **7**:158a–b; **9**:260b, 262a–263a; **10**:288a–289a; **11**:280a–281a; **12**:27a–34a, 504b–506b; **13**:159b–160a
mid twentieth-century, **6**:365a–366b; **7**:154b–155a; **10**:288a–289a; **13**:163b–164a
applications of, **6**:16b
Radioelements
atomic weights, **6**:481a
beta radiations, **6**:16a
creation of, **7**:154a–b, 158b
decay series, **2**:257b, 258b–260a; **3**:479b; **6**:15b, 16b; **12**:28a–29a
displacement laws, **2**:241a; **6**:16a, 481a; **12**:32a, 505a–506b
gaseous products, **12**:504b
half-life, **6**:15b; **12**:27b, 28a
influence on environment, **3**:502b; **13**:619a
inseparability of, **2**:259b; **10**:288b; **12**:504b; **13**:159b
isolation of, **9**:262a
transmutation of, **3**:508a; **6**:365a–b
see also specific elements
Radiogeodesy, **9**:77a
Radiogeology, **13**:619a
Radiography, **3**:234a; **6**:366a; **13**:570a; **14**:218a; **15**:18a
in physiology, **15**:71b–72a, 72b
Radiolaria, **1**:394b; **6**:7a–b, 8b, 9a; **9**:573b
Radiolarian ooze, **9**:588b; **11**:372a; **13**:360b
Radiology, **4**:405a, 499a; **6**:94a; **11**:531a; **13**:361b
dosage measurement, **7**:470a
in veterinary medicine, **7**:100b
in World War I, **3**:501b; **7**:157b–158a, 159a; **8**:102b
Radiometer, **10**:105b; **11**:149b, 393b
Crookes radiometer, **3**:476b–477b; **13**:237a
electric, **3**:477b
invention of, **3**:476b–477a
Julius suspension, **7**:186b
theories of, **3**:477a–b; **7**:416b; **9**:212b, 224a–225a; **13**:237a
Radiometry, **3**:327a–b
standards of, **3**:327b
Radiomicrometer, **11**:583a; **15**:59a–b
Radiotelegraphy
see Radio
Radiotelephone, **9**:99a
Radiotelescope, **1**:353a, 430a; **13**:117a
Radiotherapy, **3**:331a*; **4**:195a, 195b–196a, 499a; **13**:164a, 570a
Dublin method, **7**:161b

Radiotherapy (cont.)
hematological effects, **9**:417a
measurement of dosage, **4**:196a; **9**:550a
safety programs, **9**:550a
see also Radium, therapeutic use of
Radiothorium, **11**:281a; **12**:504b
discovery of, **6**:15a–b
Radish
metabolism, **9**:461a
Radium
atomic weight, **3**:499b, 500a; **6**:15a; **13**: 389b; **14**:318b–319a
decay series, **3**:499b–500a; **12**:29a
discovery of, **1**:559b; **3**:499a–b, 507a–b; **4**:356a
disintegration of, **9**:542b; **14**:319a
gases occluded by, **4**:80b
half-life, **2**:259b; **12**:28b
heat release, **3**:508a
helium production, **5**:394b; **12**:504b
official standards for, **3**:501a
physiological effects of, **6**:355b; **12**:88b
lesions, **3**:502b, 508a
toxicity, **3**:508a
production of, **2**:260b; **3**:500b
radioactive constant, **12**:28b
radioactivity, **1**:559b–560a; **3**:479b; **5**: 331a,b; **7**:159a*; **12**:28a; **14**:32a
salts of
production of, **5**:394b
separation of, **2**:258b; **3**:507b; **6**:15a
therapeutic use of, **3**:501a, 502a; **6**:15b–16a; **9**:550a
in uranium decay, **2**:258b–259b; **9**:262a,b
Radium bromide, **5**:394b
Radium emanation (radon), **2**:258b–259a; **3**:499b, 500a, 501a; **4**:80b; **12**:29b–30a
atomic weight, **14**:318b–319a
distribution of, **4**:356b
nature of, **12**:28b
physical properties, **3**:507b–508a; **11**: 280a–b; **12**:30a
density, **11**:280b; **14**:318b
radioactive constant, **7**:159a*
radioactive lamps, **4**:195a
spectrum, **11**:280b; **12**:30a
Radius (astronomical instrument)
see Cross staff
Rádl, Emanuel (1873–1942), 11:257–259; 5:52a–b
Radó, Tibor (1895–1965), 11:259–260; 4: 174a, 562a*; **8**:457b*
Radon, Johann (1887–1956), 11:260–261; 8:112a
Radon curves, **11**:260b
Radon-Nikodym theorem, **1**:427b
Radon (element)
see Radium emanation
Raffeneau-Delile
see Delile, Alire
Raffinose
molecular weight, **14**:98b
Raffles, Thomas Stamford Bingley (1781–1826), 11:261–262
Rafflesia, **11**:261b–262a
Rafinesque, Constantine Samuel (1783–1840), 11:262–264; 3:391b, 563a; **6**: 120a; **13**:167b
critics of, **8**:104a
ichthyology, **7**:169b
Rāghava
commentaries of, **2**:116a
Rāghavānanda Śarman (*fl.* 1591–1599), 11: 264; 9:175b
Raghunātha
commentaries of, **1**:308b; **12**:599a
Ibn Rāhiwayh al-Arrajānī, **4**:438a
Rahn, Johann Heinrich
Teutsche Algebra oder algebraische Rechenkunst, **10**:495b; **14**:150a
Railroads
eighteenth-century, **4**:44a, 45b
nineteenth-century, **11**:340b, 399a; **14**: 473a; **15**:151b
France, **2**:439a; **3**:287a, 411b, 412b; **5**:101a; **7**:601a; **8**:31a, 116a; **12**:287b–288a
Germany, **3**:466a; **9**:138b
Great Britain, **1**:460a; **6**:292b, 452b; **11**:291b; **12**:570a; **15**:24b
Russia, **10**:551a
twentieth-century
Alaska, **2**:500b
Great Britain, **6**:148a
dynamometer car, **4**:40a
electric, **6**:504b; **13**:24b
steam-generated, **5**:407a
funicular, **13**:432a, 614a
and geology, **10**:128a
signaling, **6**:223a
wheels, **11**:291b
Railway Magazine and Annals of Science, **6**:292b
Rain
ammonia in, **2**:356b, 357a; **8**:344b, 345b
artificial production of, **4**:410b
electricity in, **3**:153b; **4**:355a
records, **10**:585a
theories of, **3**:540a; **6**:585b; **8**:321a; **15**: 640a–b
latent heat, role of, **4**:410b
see also Pluviometry
Rain forest, **13**:460a
Rain gauge, **3**:116a, 255b, 256a, 540a; **6**: 126b; **14**:509a
Rainaldi, Giovanni Battista
see Danti, Giovanni Battista
Rainaldi, Pier Vincenzio
see Danti, Pier Vincenzio
Rainbow
medieval Arabic study, **6**:195a; **7**:213a–218b; **11**:233b(n), 250a–b
medieval Latin study, **1**:101a, 380a; **4**: 92b–94b; **5**:552a; **6**:275b; **14**:459a
Renaissance study, **9**:192a
seventeenth-century study, **5**:543a–b; **6**: 125b; **9**:97b, 118b
eighteenth-century study, **2**:4b
nineteenth-century study, **1**:358a
altitude of, **6**:275b
artificial, **11**:506b
secondary, **4**:94a, 158b; **14**:567a
theories of
ancient Greek, **11**:104b
medieval Latin, **10**:475b
Renaissance, **4**:158b; **5**:28b, 401a
seventeenth-century, **1**:475a; **4**:52b, 60a; **10**:56b–57a, 176b
nineteenth-century, **9**:155a
Chinese, **12**:380b–381a
Rainier, Mount, **14**:402a
Rainpools
ecology of, **7**:102b
Rainwater
general infiltration, **10**:522a
Rainy, Harry, **13**:373a
Raistrick, Harold, **5**:30a
Rājāditya, **2**:116b
RaLBaG
see Levi ben Gerson
Raleigh, Walter, **6**:124a; **8**:131a
Rales, **7**:557a
Rāma, son of Ballāla, **7**:315a; **9**:579b; **10**: 129a; **15**:583b
Rāma of the Bhāradvājagotra, **7**:219a
Rāmacandra, son of Ananta
Rāmavinoda, **15**:614a
Rāmacandra, son of Ṣoṣaṇabhaṭṭa
commentaries of, **2**:116a
Rāmacandra, son of Vidyādhara
commentaries of, **2**:116a
Rāmadatta
commentaries of, **2**:116a
Rāmakṛṣṇa, **12**:115b
commentaries of, **2**:116a, 117a, 118a
Rāmakṛṣṇadeva
commentaries of, **2**:116a
Raman, Chandrasekhara Venkata (1888–1970), 11:264–267; 15:26a, 48a
Raman effect, **7**:492a, 622a; **9**:76b; **10**: 236a; **11**:265b–266b; **13**:240a; **14**: 499b*
prediction of, **12**:464a–b
Ramanathan, Kalpathi Ramakrishna, **11**: 265b
Ramanujan, Srinivasa Aaiyangar (1887–1920), 11:267–269; 6:113b–114a; **11**:256b
notebooks of, **14**:188b–189a
RaMBaM
see Maimonides, Rabbi Moses ben Maimon
Rameau, Jean Philippe, **1**:115a; **12**:128b
Ramée, Pierre de La
see Ramus, Peter
Ramelli, Agostino, **8**:211b; **9**:147a
Rames, Jean Baptiste (1832–1894), 11:269–270
Rāmeśvara, **4**:100a
commentaries of, **2**:116a; **12**:116a
Ramism, **1**:126a–b, 134a; **2**:163a, 601b; **7**: 269a
Rammelsberg, Karl Friedrich (1813–1899), 11:270–271; 1:426a; **4**:289b; **7**:620a; **11**:540a*, 542a*; **14**:240a
atomic weights, **9**:290a
students of, **8**:511b; **11**:551a; **14**:315b
Ramon, Gaston (1886–1963), 11:271–272; 9:329a*; **11**:569b
Ramón y Cajal, Santiago (1852–1934), 11: 273–276; 5:460a; **11**:465b; **14**:126b
critics of, **1**:177a
influences on, **11**:295b, 382b
neurohistology, **5**:460b; **10**:131b; **12**:355a, 397a; **15**:473a
Ramond de Carbonnières, Louis François Élisabeth (1755–1827), 11:272–273; 3:526a; **9**:418a; **15**:381b, 382a
Ramorino, Juan, **1**:130a
Ramsauer, Carl Wilhelm (1879–1955), 15: 490–491
Ramsauer circular method, **15**:491a
Ramsauer-Townsend effect, **13**:446b–447a; **15**:491a
Ramsay, Andrew Crombie (1814–1891), 11: 276–277; 5:334a–337b *passim,* 338b; **7**:186a; **12**:293a, 519b
Ramsay, William (1852–1916), 11:277–284; 2:175a; **4**:165b; **9**:452a*; **11**:299b*; **14**:109b*
acquaintances of, **3**:347a; **5**:13a; **6**:94a; **7**:515b; **9**:354a; **12**:108b, 504a; **14**: 560b
critics of, **12**:30a
inert gases, **2**:267a; **3**:479b; **9**:292b; **10**: 346a; **13**:103a, 453b–454a
spectroscopy, **8**:441a
influence of, **6**:15b, 370b; **15**:464a
manuscripts of, **13**:454b
organic chemistry, **13**:70b
physical chemistry, **14**:318b–319a, 560b
radioactivity, **5**:394b; **6**:15a; **12**:28b, 29b, 30a, 504b

students of, **4**:165a; **6**:398a; **10**:563b; **14**:318b
Ramsden, Jesse (1735–1800), 11:284–285; **5**:143a, 271b; **9**:231b; **10**:591b; **11**: 354a; **14**:485a
dividing engine, **2**:141a, 216b; **11**:284b, 354b
electricity, **12**:427b
optics, **3**:98a; **14**:564a
Ramses IV, **15**:706a
Ramsey, Frank Plumpton (1903–1930), 11: 285–286; **3**:453b(n); **12**:13b, 15a
Ramspoeck, Joseph, **8**:350b, 351a
Ramtilla, oil of, **3**:45a
Ramus, Peter (Pierre de La Ramée) (1515–1572), 11:286–290; **8**:162a
acquaintances of, **11**:468a
Arithmetica
editions of, **12**:500a
astronomy, **2**:402b
commentaries of, **4**:450b; **7**:119b
correspondence of, **11**:396b, 397a
critics of, **2**:527a; **8**:309b; **10**:8a; **12**:151a
Dialecticae libri duo, **1**:134a; **7**:193a
Geometria, **4**:619a; **11**:533a
translations of, **12**:500a
influence of, **3**:359b; **14**:20b, 363a
mathematics, **3**:585a
algebra, **4**:55b
geometry, **3**:284b; **10**:494a
optics, **1**:81a
Scholarum mathematicarum, **7**:269b
Ramusio, Giovanni Battista, **5**:104b; **8**:190b
Delle navigationi e viaggi, **6**:20b
Rana, **2**:322b; **4**:377b
Rancho La Brea tar pits, **9**:314a; **13**:72a
Rancié
mine of, **1**:327b
Rand Corporation, **7**:248a
Randall, Merle, **8**:291a, 292a
Randomization, **5**:8b; **14**:92a; **15**:371b, 372a
random walk, **10**:461a
statistical, **14**:554a
test for randomness, **15**:32a
Rang des Adrets, Paul Charles Alexandre Léonard, **12**:551a
Rangakusha, **1**:311b
Raṅganātha, son of Ballāla (*fl.* 1603), 11: 290
commentaries of, **2**:119b; **9**:579b
Raṅganātha, son of Nṛsiṃha, **7**:219a; **9**:580a
commentaries of, **2**:116a, 118a
Sūryasiddhānta, **15**:608b–618a *passim*
Rangefinder
stereoscopic, **11**:208a
Rank, Otto, **5**:179a
Rankine, Alexander Oliver (1881–1956), 11:290–291; **13**:471b
Rankine, William John Macquorn (1820–1872), 11:291–295; **11**:496b; **14**:569a
acoustics, **13**:104b
engineering, **9**:209a, 226a
matter theory, **9**:207b–208a, 210a; **14**: 185b
optics, **13**:78a
Rankine cycle, **10**:286b; **11**:293b
Rankine's method, **11**:291b
steam, **13**:381a–b
thermodynamics, **3**:306a, 308b
Ranunculaceae, **11**:315a
taxonomy, **7**:198a
Ranvier, Louis-Antoine (1835–1922), 11: 295–297; **11**:275a, 382b
acquaintances of, **1**:418a; **2**:27a; **13**:288a
neuroanatomy, **14**:249a
nodes of Ranvier, **4**:372a; **11**:296a
Ranvier's disks, **1**:303b
students of, **2**:366a; **5**:148b; **9**:416a; **11**: 373b
Ranyard, Arthur Cowper (1845–1894), 11: 297; **4**:498a; **11**:163b*
Rao, Ramachandra, **11**:268a
Raoult, François Marie (1830–1901), 11: 297–300
freezing point depression, **1**:553b; **13**: 579b
law of, **11**:299a
physical chemistry, **1**:299b; **2**:186b; **15**: 459a
Raphides, **3**:43a
Rare earth elements, **2**:258a,b, 429a; **3**:429b, 478a–479a; **9**:290a
atomic weights, **2**:428b; **9**:290b; **13**:546a
compounds of, **3**:321b; **14**:588a,b
Curie-Weiss law, **3**:4a
discovery of, **2**:94b; **5**:216a; **9**:541b
origin of, **3**:478b
salts of
spectroscopy of, **4**:79a
separation of, **2**:588b; **9**:110b; **10**:136b; **13**:546a
spectroscopy of, **2**:254b, 589b; **3**:478a–b; **9**:544a
valence, **3**:321b
Rarefaction
medieval discussion, **2**:394a
Renaissance discussion, **3**:66b
eighteenth-century discussion, **12**:408b
of air, **1**:152a
theories of, **13**:92b–93a, 194a–b
Rarity and density, **13**:193b–195b
Raschig, Friedrich, **3**:511a
Rasetti, Franco, **4**:577b–580a *passim*
Rask, Rasmus Kristian, **10**:117a
Rasori, Giovanni, **1**:492b, 493a
Raspail, François-Vincent (1794–1878), 11: 300–302; **2**:443a; **5**:261a; **9**:442a
cytology, **12**:174b, 241b
Raspe, Rudolf Erich (1737–1794), 11:302–305
basalt, **6**:84a–b
Beytrag zur allerältesten und natürlichen Historie von Hessen, **6**:84b
Rat
brain, **8**:45b
coat color
inheritance of, **3**:121b–122a
and disease, **5**:270b
experimental use of, **4**:161a; **8**:591a
nutrition, **8**:590b; **9**:284b–285a; **10**: 243a–b; **14**:66b
psychology, **14**:550b
hypophysectomy, **3**:352b; **12**:474a–475a
and man, age equivalents, **4**:161a
Ratdolt, Erhard, **1**:35b, 36b, 37a; **4**:448a; **6**:273a; **7**:123b, 139b; **10**:271a
Rateau, Auguste Camille Edmond (1863–1930), 11:305–307
Rathenau, Walther, **8**:182a
Rathke, Martin Heinrich (1793–1860), 11: 307–308; **2**:443a, 596a; **8**:270b
embryology, **2**:596b; **6**:592a; **9**:570b
Rathke's pocket, **11**:307b
zoology, **9**:575b; **13**:355a
Ratio
ancient discussion, **10**:113a
Byzantine discussion, **4**:160a
medieval Arabic discussion, **1**:82a–b; **7**: 83a–b, 326b–327b; **10**:6a; **12**:94a; **13**:510a
medieval Latin discussion, **1**:381a; **2**: 391b–393b; **5**:365b; **10**:224a–b
Renaissance discussion, **8**:100b; **11**:590a; **13**:349b; **15**:155a
eighteenth-century discussion, **8**:610b
anharmonic, **3**:213b; **4**:426b; **9**:299a–b, 430b; **10**:297a–b
Chinese discussion, **8**:421b
composite, **2**:393a,b; **4**:451b; **7**:327a–b; **10**:296a; **13**:289a–b
definition of, **4**:419b, 466a; **7**:83a; **9**:21b
geometrical, **1**:215a–b, 217a–b
limits of, **10**:51a, 65a–66a
indeterminate, **13**:81b
irrational, **10**:224a
logarithmic
definition of, **9**:304a
musical, **1**:282a–283a
superparticular, **1**:231b, 232b–233a, 282a; **5**:249a
transformations of, **4**:419b–420a
trigonometric, **4**:137a
see also Proportion
Rational intercepts, law of, **2**:431a
Rationalism
ancient Greek, **4**:367b; **11**:160b
medieval, **10**:225b, 420a
Renaissance, **5**:120a–b
seventeenth-century, **2**:379a; **8**:153a–b; **9**:47a, 53a, 316b; **10**:520a; **13**:33a; **14**:361a–b, 367a, 370b
eighteenth-century, **3**:114a–115a, 381a, 382a–b; **14**:483b
nineteenth-century, **8**:57b; **9**:383a–b; **12**:240a, 244a; **15**:132a, 502a
twentieth-century, **2**:358b; **3**:335a; **11**: 363a
Chinese, **15**:730b
experimental, **2**:31b–32a
rejection of, **3**:516a; **10**:120b; **11**:43b; **12**:156a; **15**:127b
utilitarian, **1**:233b
Ratke, Wolfgang, **3**:360a; **7**:193b
Ratnowsky, Simon, **5**:598a
Rattlesnake
anatomy of, **13**:527b
Rattlesnake (ship), **6**:589b–590a
Ratzeburg, Julius Theodor Christian, **2**:422b
Ratzel, Friedrich (1844–1904), 11:308–310; **2**:209a, 538b; **10**:503a
Rau, Johann Jacob, **4**:352b; **12**:41a; **14**:449b; **15**:4a
Rauber, August Antinous, **5**:447a; **11**:573b; **14**:234b
Rauchfuss
see Dasypodius, Cunradus
Raulin, Jules (1836–1896), 11:310–311; **10**:373b; **12**:208a
Raulin's medium, **11**:310b
Rauwolf, Leonhard (1535–1596), 11:311–312; **1**:525b
travel diary, **11**:317b
Rauwolfia, **11**:312b
Ravenelia, **4**:220a
Ravenna, Ciro, **3**:279b
Ibn al-Rāwandī, **4**:525a
Ray, John (1627–1705), 11:313–318; **7**:539a; **9**:529a; **13**:450a, 527b; **14**:379a(n)
acquaintances of, **3**:129a; **8**:307b–308a, 416b; **9**:539b; **12**:457a
botany, **2**:188b; **6**:35b, 38b, 41b, 130a
correspondence, **1**:432a; **2**:523a; **9**:17b
editions of works, **1**:358b; **4**:40b
geology, **7**:412a; **8**:416a
influence of, **10**:509b
influences on, **7**:147b; **11**:312a; **13**:502b, 503a; **14**:373b–374a, 374b, 413b
taxonomy, **1**:369b; **4**:99a; **8**:376b, 377a, 378b; **13**:174a, 442b–443a; **14**:373b–374a
and Willughby, **11**:313a–b; **14**:412b–413b
Catalogus plantarum circa Cantabrigiam nascentium, **7**:194b

Ray, John (cont.)
Historia plantarum, **1**:431b; **12**:394b; **14**:413b
Miscellaneous Discourses Concerning the Dissolution and Changes of the World, **8**:308a
Philosophical Letters, **4**:40b
Physico-Theological Discourses, **4**:40b
Stirpium Europaearum . . . sylloge, **12**: 394a
Synopsis methodica avium et piscium, **4**:40b
Synopsis methodica stirpium Britannicarum, **6**:539a; **8**:307b; **9**:539b; **12**:394a
Synopsis plantarum, **4**:98b
editions of, **4**:99a
Wisdom of God, **4**:40b; **9**:509b
Ray, Louis Lamay, **2**:549b
Ray, Prafulla Chandra (1861–1944), 11:318–319; 15:47b
Ray (fish)
anatomy, **9**:572b; **13**:32a, 33b
electric, **4**:289a; **6**:269a
Ray (geom.)
systems of, **7**:522a, 523b
Ray Society
founding of, **1**:358b
Rayer, Pierre-François-Olive, **1**:471a; **2**:25a, 443a, 524b; **3**:587b, 588b; **11**:69b–70a, 491b
Rayet, Georges Antoine Pons (1839–1906), 11:319–321; 14:479b
Wolf-Rayet stars, **3**:36a; **14**:479b
Rayleigh, Lord
see Strutt, John William
Rayleigh (phys. unit), **13**:107b
Raymond, archbishop of Toledo, **5**:591b
Raymond of Marseilles (*fl.* first half of twelfth century), 11:321–323; 15: 178b, 186a
Raymond of Sabunde, **3**:362b
Raymond, Albert L., **4**:82b
Raymond, Percy Edward (1879–1952), 11: 321; 13:534a*
Raymond, Rossiter Worthington, **11**:102b
Raynalde, Thomas, **5**:348b
Raynaud, François de, **4**:505b
Raynaud, Maurice
Raynaud's disease, **8**:295a
Raynaudet, Jacques, **1**:525b
Raynolds, William Franklin, **6**:186b
Rayon, **3**:207a
Rays
see Radiation
Razenkov, Ivan Petrovich, **10**:434b
al-Rāzī, Abū Bakr Muḥammad ibn Zakariyā (Rhazes) (*ca.* 854–*ca.* 935), 11: 323–326; 2:151a; **7**:334b; **9**:31b; **11**: 511b
alchemy, **14**:118a
balance, **7**:340a, 341b, 343a
biographies of, **7**:187a
commentaries on, **13**:415b*
critics of, **4**:525a
on Galen, **15**:235a, 238b
influence of, **1**:538b; **2**:23a, 546b, 619b; **3**:218b; **4**:157b; **6**:123a; **9**:40b; **10**: 555b; **11**:231a, 238a, 311b, 312a, 444b; **13**:511a; **14**:4a, 638b*
influences on, **10**:418b; **11**:244a; **13**:230b
medicine, **1**:620a; **9**:41a; **10**:21a; **15**: 246b(n), 498a,b
philosophy, **1**:27b
translations of works, **11**:445a; **13**:401b
De aluminibus et salibus (spurious), **14**: 35b
translations of, **15**:185b
Antidotarium
translations of, **15**:184a–b
Aphorismi Rasis
translations of, **15**:184b–185a
Continens
translations of, **15**:183b
De egritudinibus puerorum
translations of, **15**:184a
Experimenta Rasis, **15**:184a
al-Ḥāwī, **15**:498a
De iuncturarum egritudinibus
translations of, **15**:183b–184a
Kitāb Nuzhat al-mulūk (spurious), **15**: 184a
Liber Almansorius
translations of, **15**:183a
Liber divisionum
translations of, **15**:183a–b
Liber introductorius in medicinam parvus
translations of, **15**:183b
Liber luminis luminum (spurious)
translations of, **15**:186a
De preservatione ab egritudine lapidis
translations of, **15**:184b
al-Rāzī, Muḥammad ibn ᶜUmar, **11**:231b, 238b
Razmadze, Andrei Mikhailovich (1889–1929), 11:326–327
Reaction time, **14**:528a–b
Reactions, chemical
active transitional state, **1**:414b–415a
color changes, **2**:380b–381a
contact theory, **9**:425a
diagramming of
eighteenth-century, **2**:7b
electromotive force, **3**:333a,b
explanations of, **2**:79a–82b; **8**:32a, 173b, 475a; **9**:360a–b; **12**:312a–b; **13**:338a, 429b; **14**:125a
factors affecting, **2**:80a; **11**:541a
acidity, **6**:110a
concentration, **9**:352b; **11**:541a
light, **8**:102a
pressure, **6**:404b; **7**:21a–b
quantity, **2**:79b–80a; **5**:324a–b
solvents, influence of, **9**:305a
steric hindrance, **9**:357a, 360b
temperature, **1**:300a; **4**:80b, 412a; **6**: 110a; **9**:352b; **11**:541a; **13**:578a–b; **14**:360b; **15**:458a
time, **9**:352b
flow systems, **1**:415b
in nature, **3**:293a
prediction of, **1**:415a
reversibility of, **1**:579a; **4**:78a, 238b; **5**: 324a
definition of, **4**:229a
spontaneity of, **4**:225b; **15**:436a–b, 437b–438a
stepwise processes, **1**:415a
successive reactions, law of, **5**:324a
transition points, **3**:333a,b, 334a; **6**:519b–520a; **10**:15a
types of
addition, **2**:621b; **7**:323a–b; **9**:360a–b
Cannizzaro, **3**:47a
chain reactions, **15**:37a–b, 448b
classification of, **1**:415a; **2**:7a–b
endothermic, **2**:69a; **6**:244b; **13**:359a; **15**:437b
exothermic, **3**:603a; **13**:358b–359a; **15**:437b
heterogeneous, **13**:454b
induced, **12**:198a
inversion, **5**:5a
isenthalpic, **15**:437b
mediated, **6**:259b
monomolecular, **1**:415b; **13**:429b
nuclear, **7**:527a
photochemical, **1**:556a; **3**:197b; **13**:429b
quasi-unimolecular, **6**:404b
replacement, **5**:580a
secondary, **15**:448a–b
substitution, **4**:245a–b, 246b, 531b; **5**: 5a, 325a, 373a; **9**:131a; **14**:124b
thermonuclear, **5**:272a,b; **7**:527b; **13**: 241b
velocity of
absolute, **1**:415a
classification of, **13**:577b–578a
law of, **4**:412a; **14**:360a–b
see also Affinity, chemical; Catalysis; Compounds; Condensation, chemical; Equilibrium, chemical; Esterification reactions; Heat, of reaction; Kinetics, chemical; Oxidation; Reduction, chemical; Substitution theory, chemical
Reactivity
solid-state, **5**:223a
Reactors, nuclear, **3**:371a–b; **4**:376a, 582a; **7**:527b
cooling of, **13**:227a
design of, **3**:371a–b; **4**:581a–b; **15**:8b–9a
NRX, **3**:330a
Read, John, **11**:85b–86a, 87a
The Reader, **13**:522b
Reagh, Arthur Lincoln, **12**:482a
Real, Carlos, **10**:617b
Real, Miguel, **10**:617b
Real Escuela Metalúrgica
founding of, **4**:345a
Real Sociedad Económica Vascongada de Amigos del País, **11**:166a
Realism
medieval, **1**:1b, 100b
Renaissance, **3**:172a; **4**:237b; **9**:32b–33a
eighteenth-century, **7**:227a
nineteenth-century, **6**:247b; **10**:485a; **12**: 156b; **15**:79b
twentieth-century, **8**:517b*
rejection of, **10**:173b
Reality
plurality of, **3**:278a
structure of, **4**:318b–319a; **6**:294b
see also Ontology
Rearrangement, chemical, **3**:286b, 511b; **6**: 108a; **10**:288b; **13**:550a
Reason, **2**:581a–b; **3**:452a; **8**:257b
and authority, **1**:62a; **10**:334a
levels of, **15**:249b
sufficient reason, principle of, **8**:155b–156a, 158b, 159b; **14**:483b
see also Faith and reason, Rationalism
Reasoning process, **2**:132a
Réaumur, René-Antoine Ferchault de (1683–1757), 11:327–335; 1:58b; **2**: 473b; **3**:281a, 454b; **8**:580a
and Académie des Sciences, **4**:87a, 214b
acquaintances of, **3**:459b; **5**:226a; **10**:145a; **11**:5a
ceramics, **6**:54b
on contemporary scientists, **11**:503a
correspondence of, **3**:461b, 486a; **9**:596a; **10**:569a; **13**:457b, 458a
electricity, **10**:146a
influence of, **2**:286a; **5**:328b; **7**:442b; **8**: 579b; **9**:596a
marine biology, **10**:569a
Mémoires pour servir à l'histoire des insectes, **2**:474a; **12**:133b
natural history collection, **2**:474a; **5**:577b
physiology, **1**:543a–b; **12**:555b, 558a, 559a; **13**:46b
Reaumuria, **3**:45a

Rebound
Renaissance discussion of, **8**:231a–b
Recapitulation theory, **1**:387b; **3**:519b; **5**: 356b–357a; **6**:8b, 338a, 613b–614a; **7**:367b–368a; **9**:252b; **12**:316a, 337b, 338a; **15**:92b
opposition to, **8**:254b; **13**:97a; **15**:169b–170a
Recchi (Recho), Leonardo, **6**:309b
Recent Progress in Hormone Research, **10**: 611a
Recession problem (math.), **12**:500b
Recherche (ship), **5**:225a
Recherches physicochimiques, **13**:469a
Reciprocants
theory of, **13**:218a
Reciprocity (math.), **4**:341b; **11**:45a; **15**: 364b–365a
cubic, **4**:342b
laws of, **6**:390a,b; **15**:690a–b
biquadratic law, **4**:124b, 342a,b; **5**:304b
general law, **1**:307a; **3**:222a; **4**:474b; **6**:393b; **7**:523b
quadratic law, **4**:124b, 474b; **5**:299a,b; **8**:385b; **14**:630b; **15**:66b
polar, **2**:214b; **3**:213a; **5**:368a,b; **11**:79b
Reck, Hans (1886–1937), 11:335–337
Reclamation, geological, **12**:344a, 371b; **13**:437a
Reclus, Élisée (1830–1905), 11:337–338
General Geography, **7**:511b
Recoil, radioactive, **6**:16a
Recorde, Robert (*ca.* 1510–1558), 11:338–340; 10:151a; **11**:512a
The Pathway to Knowledge, **4**:449b–450a
Records of the Past, **14**:517a
Recreations, mathematical
Renaissance, **7**:418b; **10**:270a, 271a; **11**: 457b, 590b
seventeenth-century, **8**:271b; **10**:264a–b
nineteenth-century, **8**:531b; **11**:22b*; **12**:229a
twentieth-century, **8**:617a
Egyptian, **15**:694b–696a
games, **6**:92b
puzzles
medieval, **1**:105a; **4**:606a–607b, 610a–b, 612a–b
Renaissance, **3**:276b
seventeenth-century, **1**:367b, 368a
nineteenth-century, **4**:137b–138a
twentieth-century, **5**:190a
Japanese, **15**:749b
rythmomachia, **1**:468a; **6**:302a
yojutsu (packing problems), **15**:752a–b, 753a, 754a, 755b
Recrystallization, **3**:161b, 162a
secondary, **7**:92b
Rectangle
area of, **4**:609b
Rectangulus (astron. instrum.), **11**:415b
Rectification (math.)
seventeenth-century methods, **1**:474b; **2**:507b; **4**:572b–573a; **5**:521a, 527a–b; **6**:359a–b, 600a; **10**:337a; **12**:206b
rejection of, **4**:58a
eighteenth-century methods, **2**:56b; **3**: 282a; **9**:186a
of an arc, **8**:165b; **12**:292a; **13**:436a
of circle, **1**:219b, 228a; **4**:104b; **12**:291b–292a
of conic sections, **11**:401b
of cycloid, **10**:337a; **11**:487b; **14**:509b
of *elastica,* **2**:48a
of ellipse, **4**:516a; **8**:139b; **11**:487b, 609a
of lemniscate, **2**:48a; **4**:516a; **7**:52a
of logarithmic curve, **2**:49b; **8**:505a
of parabola, **2**:49b; **4**:516b; **6**:359a–b, 600a, 602a; **11**:240b
theory of, **8**:111b
Rectifier, electrical, **4**:284a; **5**:202a
Recueil zoologique suisse
founding of, **5**:52a
Red auxiliaries (math.), **15**:688a, 690a, 698b
Red clay (oceanic deposit), **9**:589a; **11**:372a
Red Cross, **4**:217b; **14**:252b
Radiological Service, **3**:501b
Sanitary Commission, **14**:623a
Red lead, **4**:243a; **7**:78a; **8**:74b
distillation of, **6**:42a
Red Sea
coloring of, **4**:289a
coral, **4**:289a, 290a
malacology, **7**:30b
navigation of, **9**:35b–36a
origin of, **7**:30b
Rede, William, **2**:435a
Redfield, William C. (1789–1857), 11:340–341; 4:410a,b; **8**:487a
Redi, Francesco (1626-*ca.* 1698), 11:341–343; 1:483b; **8**:127a, 128a; **14**:586b–587a
acquaintances of, **1**:391b, 593a; **2**:309b, 317a; **3**:180b
on digestion, **12**:558a
Esperienze intorno alla generazione degli insetti, **2**:591a; **5**:439b
experiments, **9**:3a, 369b; **10**:367a
Osservazioni, **2**:591b
on spontaneous generation, **2**:291a; **7**: 111a; **11**:317a; **12**:555a; **13**:173b, 563b
Redouté, Pierre-Joseph (1759–1840), 11: 343
illustrations by, **3**:172b; **8**:303b; **9**:366b
Liliacées, **3**:44b, 172b
Redtenbacher, Ferdinand Jakob (1809–1863), 11:343–344; 5:501b; **11**:384a; **13**:298b; **14**:342b
Reducibility, axiom of, **11**:471a; **12**:13b, 15a
critics of, **3**:278b; **11**:285b
Reductase, **10**:282a
Reductio ad absurdum
ancient Greek use of, **1**:215b, 218a, 219b, 226b, 227b, 257a; **4**:416b; **6**:415b
Byzantine use of, **7**:138b
medieval use of, **5**:360a–b, 551a; **6**:198b; **7**:3a
seventeenth-century use of, **6**:599b; **12**: 55b
Reduction, of astronomical observations, **2**:97b–99a *passim,* 616b; **4**:175b, 269b; **5**:405a, 480a; **7**:544b; **8**:557b; **11**:114b; **12**:176b
Reduction, chemical, **3**:246a; **4**:76a
agents for, **2**:68a; **14**:622a–b
of metals, **1**:579a–b; **3**:317a*; **5**:7a; **12**: 46b–47a
aluminothermal, **1**:579b
of nitrogen compounds, **1**:426b; **14**:622a–b
of organic compounds, **2**:68a; **14**:622a
Reduction, mathematical, **2**:271b, 467b; **13**:480a; **14**:23b
Christoffel's theorem, **3**:263b
Reductionism, **1**:10b; **2**:17b; **5**:374a, 598a; **9**:9a, 381a,b; **11**:140a, 143a, 147a; **15**:740a, 741b, 742a
critics of, **3**:376a; **8**:601a
Reech, Ferdinand (1805–1884), 11:344–345; 5:200a; **11**:353b
Reed, Walter (1851–1902), 11:345–347; 3: 94a
yellow fever, **3**:94b; **4**:620a; **8**:450b; **15**: 96b
Reed, William, **7**:124a
Reefs, **5**:77a
barrier, **3**:568b–569a
theories of, **3**:550b–552a
formation of
algae and, **12**:328a
fringing
theories of, **3**:551a, 569a
geographical distribution of, **3**:551a–b, 569a; **9**:181a
onkoid, **1**:162b–163a
origin of, **3**:595a; **9**:589b
see also Atolls
Rees, Abraham
Cyclopaedia, **4**:540a,b
Rees, Max Ferdinand Friedrich, **10**:166a
Refining
wastes, processing of, **8**:109a
by zone melting, **2**:459a
see also Metallurgy; Metals; specific metals
Reflecting circles, **2**:141a; **13**:470b
Reflection
atmospheric, **5**:28b
crystalline, **3**:419b
and electromagnetism, **4**:306a
in general surface, **5**:525b
at interface, **8**:490b, 492b
laws of, **1**:475a; **4**:58b–59a, 575a; **5**:552a; **6**:194a; **8**:426a–b, 592a; **10**:475b; **11**:200b; **14**:460a–b; **15**:265a
from magnetic substances, **4**:191a
measurement of, **2**:328b; **6**:194a
metallic, **2**:452b; **3**:420a; **9**:74a; **14**:348a–b
molecular theory of, **2**:337b
partial, **5**:170b; **14**:567b
of penetrating radiation, **1**:559a
phase inversion, **14**:348b
points of, **6**:200a–201b
from rough surfaces, **2**:344a
specular, **6**:194a
theories of
ancient Greek, **6**:313a
Byzantine, **7**:136b
medieval Latin, **14**:458a
seventeenth-century, **6**:610a; **10**:54b, 59b
nineteenth-century, **3**:146a–b
Chinese, **12**:381b
thermal, **7**:599a; **10**:603a
of X rays, **3**:370a, 564b*; **4**:196a–b; **9**: 543a; **15**:8a
see also Catoptrics
Reflection, mathematical, **12**:246b
Reflection gratings, **10**:345b
Reflectometer, **11**:208a
Reflectors
elliptical, **1**:170a
parabolic, **1**:170a
spherical, **1**:189b
see also Mirrors
Reflexes
axon, **8**:16a
Bezold-Jarisch, **2**:111a–b
carotid, **15**:469a–b
compounding of, **12**:396b
conditioned, **8**:7b; **10**:220a–b, 433b–435a; **13**:530a
coordination of, **12**:95b
cremasteric, **9**:422b
deep
sensory reinforcement of, **9**:422b
deglutition, **7**:504b
drugs, effect of, **6**:60a
flexion, **5**:65a
grasp, **6**:60a
Hering-Breuer, **2**:446b; **6**:300a
hind limb, **12**:396b, 397b–398a

- Reflexes (cont.)
  - metabolic, **8**:452b
  - physiology of, **6**:59a–60a; **9**:20a–b; **12**:395b–401b; **14**:321a–b
  - postural, **9**:19b–20b
  - proprioceptive, **12**:398a
  - scratch, **12**:397b–398a
  - spinal, **5**:65a–b, 175a
    - inhibition of, **12**:270a
  - stretch, **12**:399a–b; **14**:321a
  - theories of, **1**:323a; **9**:241b, 571a–b; **11**:159a–b; **12**:95b, 270b; **13**:530a, 543b; **14**:406b
  - time relations of, **5**:65a
  - and vagus nerve, **5**:463a
  - vasomotor system, **2**:367a
  - Whytt's, **14**:321b
- Reflexology, **1**:580a; **5**:463a–b
- Reformation, the, **3**:413a; **10**:173a
  - criticism of, **2**:540b–541a
- Refraction
  - astronomical, **2**:468b; **3**:14a; **4**:361a, 408a*, 555a; **7**:298b, 338a, 481b, 494b; **9**:234a; **11**:187a, 201a
  - atmospheric, **1**:358a; **2**:135a, 343b, 404a, 405a, 421a; **3**:319b; **5**:28b, 79a; **6**:364a; **7**:37b, 303a; **8**:144b; **11**:555a*; **12**:152a; **13**:111a; **14**:216a, 491a–b, 579a; **15**:355a–b, 359a–b
    - constants of, **3**:374b; **5**:554b
    - formula for, **1**:201b; **2**:40a; **3**:374b; **15**:359b
    - and sine law, **3**:101a
    - tables of, **2**:98a, 411b; **3**:103a, 374b; **5**:25b; **7**:544b; **8**:144a, 178b; **9**:162b; **15**:355b
    - and temperature, **10**:596b
  - conical, **1**:166a; **5**:525b; **8**:426a
    - prediction of, **6**:88a–b
  - constant of, **5**:224a*; **8**:501b
  - double
    - medieval Arabic study, **7**:214b
    - seventeenth-century study, **1**:481b; **6**:609b, 610a
    - eighteenth-century study, **7**:494a; **9**:472a
    - early nineteenth-century study, **1**:202a; **2**:138b, 451b, 452a,b; **3**:146a; **6**:178b; **9**:73a–74a; **15**:361b–362b
    - mid nineteenth-century study, **5**:68b; **7**:601b; **12**:304a–305b
    - late nineteenth-century study, **2**:382b; **5**:391a; **8**:107a; **14**:491b
    - twentieth-century study, **6**:614a; **8**:148b
    - in electrical field, **2**:203a; **7**:313b
    - induced, **9**:199b–200a, 221a
    - laws of, **5**:170a–b; **9**:73a–b; **10**:27b
    - in liquids, **9**:616b
    - and relativity theory, **13**:104a
    - and temperature, **9**:424b
    - theories of, **8**:501b; **9**:73a–74a, 213a; **10**:79b; **11**:292a; **13**:77b–78a; **15**:361b–362b
  - and electromagnetism, **2**:265b–266a
  - extraordinary vs. ordinary rays, **2**:138b; **15**:362b
  - in gases, **1**:348b; **2**:135a; **4**:241a; **7**:620a; **8**:492b; **10**:546a; **11**:142b
  - in glass, **2**:135a
    - temperature effects, **9**:564a
  - index of, **5**:143a; **6**:125b, 171b, 603b; **8**:42a; **9**:154b–155a, 256b*, 372b; **11**:142b; **12**:305b; **14**:491b
    - and chemical affinity, **1**:348b; **15**:361a–b, 457b
    - and color, **10**:53b, 56a–57a, 58b–59a
    - and density, **7**:620a; **8**:493a; **11**:201a; **15**:355a,b, 359b
    - and dielectric constant, **2**:265b; **9**:212b
    - and molecular weight, **4**:241a; **7**:620a
    - and negative dispersion, **7**:554a–b
    - and polarizing angle, **2**:452a
    - specific, **5**:412b
    - and velocity of light, **6**:610a
  - irregular, **9**:415a
  - laws of, **1**:379b; **2**:53b; **3**:599b; **4**:58b, 59a–60a, 158b, 575a–b; **5**:170b, 525b, 552a,b; **6**:128a, 194b, 603b; **7**:298a; **8**:156a, 163b, 592a; **10**:104a; **12**:187a, 501a–b
  - in liquids, **1**:201b; **6**:453a
    - refractive index, **5**:412b; **6**:355b
  - measurement of, **1**:358a, 551b; **2**:388b; **6**:125b, 194a; **7**:212b; **8**:460a; **11**:201a, 208a
  - of penetrating radiation, **1**:559a
  - sine law of, **1**:475a; **4**:59b; **5**:526a; **6**:128a; **9**:52b, 121a(n); **10**:56b; **12**:501b
  - in solids, **1**:201b
  - in sun, **7**:186b
  - tables of, **14**:458a, 460b
  - and telescope lenses, **3**:284a
  - theories of
    - medieval, **6**:194a–b; **14**:461a
    - seventeenth-century, **6**:610a; **10**:54b, 59b, 60a
    - eighteenth-century, **2**:474b; **3**:281b; **4**:481a; **9**:267a; **10**:79a
    - nineteenth-century, **3**:146a–b; **7**:494b; **9**:214b
    - twentieth-century, **2**:337b
  - of thermal radiation, **5**:68b; **6**:333b; **10**:603a
  - *see also* Dioptrics
- Refractometer
  - Abbe refractometer, **1**:7a; **11**:208a
  - double, **6**:74a
  - interferential, **9**:372a, 373a
- Refractometry, **13**:420b, 421a
- Refractory materials, **8**:118b; **9**:451b; **11**:549a, 550a
  - fusion of, **12**:433a
- Refrigerants, **9**:376a
  - *liquide Pictet,* **10**:604a
  - *see also* Coolants
- Refrigeration
  - eighteenth-century, **3**:202b*
  - early nineteenth-century, **8**:262a
  - late nineteenth-century, **8**:365b–366a, 500a; **10**:604a
  - early twentieth-century, **4**:501a*; **7**:221b–222a; **9**:462b; **13**:227a
  - mid twentieth-century, **8**:482a–483a
  - He^3-He^4 dilution refrigerator, **8**:482a–483a
  - *see also* Cooling
- Regel, Eduard-August, **6**:221a
- Regemorter, Ahasuerus, **5**:425b
- **Regener, Erich Rudolph Alexander (1881–1955), 11:347–348;** **4**:357a; **5**:330b; **6**:16a, 17b*; **11**:584b*
- Regeneration, botanical, **15**:14a
- Regeneration, zoological, **1**:59a; **3**:247b, 492b, 574a; **8**:169b; **9**:519a; **12**:567b; **14**:3a
  - and age, **12**:555b
  - in ascidians, **4**:187a
  - of bone, **7**:414b; **8**:34a–b; **13**:465a
  - in coelenterates, **5**:577b; **9**:408b
  - in crustaceans, **11**:333b
  - and embryogeny, similarity of, **9**:517b–518a
  - embryonic, **8**:357b; **11**:573a
  - of feathers, **8**:359a
  - in fish, **2**:510b*
  - of head, **1**:59a; **12**:555b
  - in hydra, **4**:187a; **8**:580a; **11**:333b–334a; **13**:457a–458a
  - hypotypic, **5**:385b
  - invertebrate, **8**:357a
  - of membranes, **13**:589a
  - ontogenetic vs. phylogenetic, **3**:17b
  - in salamander, **5**:447b*; **12**:555b
  - in snail, **1**:59a; **12**:555b
  - in tadpole, **12**:555b
  - theories of
    - larval body theory, **7**:33b–34a
  - in worms, **2**:286a; **7**:33b; **12**:555b
- Reggio Emilia, University of, **12**:554a
- Regimen, medical
  - ancient Greek, **1**:315a; **6**:318b, 422a–b
  - Byzantine, **10**:417a–b
  - medieval, **1**:289b–290a, 565a; **3**:219a
  - eighteenth-century, **3**:244b
- *Regimen sanitatis,* **6**:123b
- **Regiomontanus, Johannes (1436–1476), 11:348–352; 11**:366b*, 415b
  - astrarium, **4**:164b
  - astrology, **3**:27a
  - astronomy, **3**:27b, 277a; **14**:273b; **15**:35b, 189a
  - on al-Battānī, **1**:512b
  - *De cometae magnitudine . . . problemata XVI,* **12**:199b
  - on contemporary scientists, **2**:360b; **7**:119b, 120a–b, 175b; **13**:440b
  - cosmology, **11**:250b
  - *Disputationes contra Cremonensia in planetarum theoricas deliramenta,* **15**:475b
  - editions of works, **12**:199b; **15**:475a
  - *Ephemerides,* **1**:573b; **9**:169a*
  - geometry, **5**:409a; **7**:129a; **13**:291b; **14**:22b, 274a
  - historiography, **4**:111b, 117b; **9**:501a
  - influence of, **4**:619b; **7**:118b, 269b; **14**:273b
  - influences on, **13**:509b; **14**:19a, 461a, 593b
  - manuscripts of, **1**:229a; **6**:144a; **7**:134a*, 142b, 178b*; **11**:32b*
  - optics, **6**:144a
  - and Peurbach, **15**:473b–478a *passim*
  - students of, **1**:165b; **4**:260a
  - *Tabula directionum,* **1**:573b; **2**:401b
  - *De triangulis omnimodis,* **3**:273a, 277b; **7**:38b; **8**:281b
  - trigonometry, **1**:512a; **14**:19a, 470b; **15**:155b, 478b
- Regis, Emmanuel, **4**:286b
- Régis, Pierre-Sylvain, **1**:474a; **8**:173a–b; **11**:506b; **14**:25b
- Regius (de Roy), Henricus, **4**:65a; **13**:30b
- **Regnault, Henri Victor (1810–1878), 11:352–354;** **1**:347a
  - at Académie des Sciences, **11**:63a
  - acquaintances of, **2**:586b; **9**:154b, 401b; **11**:559a; **14**:617b
  - *Cours élémentaire de chimie*
    - translation of, **5**:214b
  - heat, **14**:360b
  - influence of, **1**:572b; **2**:65a; **3**:10b; **4**:246a, 247a; **14**:618a
  - organic chemistry, **1**:578a; **8**:335b–336a, 337b
  - students of, **3**:397b; **5**:18b; **9**:230b; **13**:375a; **14**:360a
  - thermometer, **3**:19b; **8**:117b; **13**:378a
  - vapor pressure, **1**:128a, 349a; **4**:240a; **7**:462a
- Regnault, Jules, **5**:86b
- Regress (logic), **14**:581b
- Regression, biological
  - *see* Degeneration (biol.)
- Regression, mathematical, **5**:8a, 9a, 266a; **7**:114b; **10**:449a–464a; **15**:330b, 367a
  - coefficient of, **10**:453a, 457b
  - multiple, **10**:461b, 463b

nonlinear, **10**:450a, 452a–b
partial, **10**:451b, 461b
Regression, psychological, **5**:177b
Regula falsi, **2**:602b; **11**:457b; **13**:61a(n)
Regulators, mechanical, **5**:87a
theory of, **14**:106a–b; **15**:7a
Regulus (metall.), **2**:421b
Regulus (star), **14**:576a, 577a, 593a; **15**:218a
Reiche, Fritz, **5**:117b
Reichenbach, Eugen, **1**:578a
Reichenbach, Georg Friedrich von (1771–1826), 11:354–355; **3**:14a; **5**:142b, 271b, 302a
instrumentation, **13**:109b
Reichenbach, Hans (1891–1953), 11:355–359; **4**:368b; **12**:179b*
Reichenbach, Heinrich Gottlieb Ludwig, **6**:464a, 465a
Reichenbach, Karl Ludwig (1788–1869), 11:359–360
animal magnetism, **5**:531a
Researches on Magnetism . . .
translation of, **5**:531a
Reichert, Karl Bogislaus (1811–1883), 11:360–361; **4**:200a; **5**:196a; **14**:126a; **15**:78a, 170a
comparative anatomy, **2**:504a–b
cytology, **12**:231b
histology, **6**:269b
Reichstein, Tadeus, **15**:258b
Reid, Harry Fielding (1859–1944), 11:361–362; **8**:99b
Reid, Thomas, **1**:404a; **3**:598b; **6**:81a; **14**:288b; **15**:95a
Reidemeister, Kurt Werner Friedrich (1893–1971), 11:362–363; **2**:192a
Reil, Johann Christian (1759–1813), 11:363–365; **5**:255b*; **9**:252a; **15**:172a
Reil's journal
see *Archiv für die Physiologie*
Reilhac, Grotte de, **2**:347a
Reimarus, Johann Albert Heinrich, **2**:220b; **8**:321a; **15**:205b
Reimer, Karl Ludwig
Reimer-Tiemann reaction, **13**:406b
Reincarnation, **10**:113a(n)
Reinhardt, Benno, **14**:40b
Reinhardt, Curt, **9**:431a
Reinhold, Erasmus (1511–1553), 11:365–367; **11**:351b; **15**:476a
influences on, **13**:401a
Prussian Tables, **1**:244b; **2**:401b–402a, 404a; **3**:312a; **6**:515b; **7**:269b; **9**:611a
editions of, **7**:289b; **9**:167b
Reinitzer, Friedrich, **8**:148b
Reinke, Johannes, **12**:125a
Reis, Johann Philipp, **1**:582b; **13**:357a
Re'is, Pirī
see Pirī Rais
Reisch, Gregor, **9**:160b, 581a; **13**:60b; **15**:154a
Reiset, Jules, **2**:586b; **9**:401b
Reissner, Ernst
Reissner's cell, **5**:172a
Reiter, Hanns, **14**:181b
Reive, Richard, **5**:525b
Reiyaku zyutu, **12**:292a
Rejuvenation, attempts at
testicular extracts, **2**:525b
Rejuvenescence, concept of, **2**:426a–b
Relapsing fever, **7**:422a,b
etiology of, **7**:429a; **8**:296b; **10**:155b
transmission of, **5**:270b; **12**:315a
treatment of, **4**:300b, 301a
Relation, mathematical, **4**:56b
dyadic, **10**:485a
Göpel relation, **5**:471b
logic of, **4**:36b
Relativism, **6**:447a; **15**:413a
opposition to, **2**:356a
Relativity, theory of, **1**:194b; **2**:145b, 169b, 266a, 460b; **3**:111b, 117b, 221a, 428b; **4**:306a; **5**:122b; **6**:392b; **7**:207b, 491b; **8**:11b, 184b, 290b, 292b, 444a, 617b; **9**:214a, 373a,b, 414b; **10**:422a–b; **11**:461a; **12**:218b; **13**:104a; **14**:437a; **15**:39b–40b
applications of, **12**:528a
and astronomy, **12**:449a
confirmations of, **1**:56b–57a; **3**:221a*, 472a; **4**:270a, 278a, 279b, 280b, 312b, 328b–329a, 498a; **5**:181b–184a *passim,* 554b; **8**:278a–b; **11**:13b; **12**:73a; **14**:598b–599a
consequences of, **3**:221a; **12**:528a
and Doppler principle, **4**:168a
and epistemology, **6**:26a–b; **7**:211b; **11**:358a
foundations of, **4**:374a; **7**:235a; **12**:16b; **14**:305a–306b, 308a–b
general relativity, **3**:96a; **4**:280b–281b, 336b, 577b; **5**:555a; **7**:26b, 211a–b; **8**:51a, 182a, 284a–b, 434a–b, 496b, 498a–499a, 599b–600a; **11**:455b; **12**:218a, 222b, 251b; **14**:13b
cosmological problem, **5**:188a–b; **12**:449a
critics of, **4**:408a; **7**:264a; **12**:69b
evolution of, **4**:323b–328b
experimental predictions, **4**:328b–329a; **5**:181b–182b
heavy bodies, motion of, **7**:10a–b
light waves, **7**:10a; **10**:519a
mathematical structure of, **4**:329a–b; **6**:453a; **10**:137b–138a; **11**:13a–b, 260b, 407b, 409a; **13**:548b; **14**:283b, 317a
precursors of, **4**:380a
testing of, **8**:478a; **13**:474b; **14**:306a
geometry of, **7**:397b, 473a–b; **12**:214a
kinematic, **9**:405b
misinterpretations of, **5**:122b
opposition to, **1**:25a; **2**:594a–b; **4**:101b, 232a; **7**:36b; **8**:52a–b, 182b, 600a; **9**:386b; **11**:84a; **12**:69b; **15**:80a
philosophical implications of, **3**:90a; **12**:177b
popularization of, **4**:281a; **5**:272b, 364a; **12**:16a
precursors of, **4**:378a–b; **6**:602b; **10**:449a
and quantum theory, **4**:281a–282a; **5**:182b; **10**:423b; **11**:13b
special relativity, **2**:145b, 383a, 559b; **3**:63a, 369a; **4**:315a, 317a, 319b–323b, 324a–331b *passim;* **5**:597b–598a; **7**:26b; **8**:10b, 431b, 434a, 498a; **9**:212b, 215b, 377a, 405b–406a; **11**:58b; **13**:429b; **14**:317b
confirmation of, **7**:36b; **8**:50b–51a; **12**:614a; **15**:27a
criticism of, **8**:517a
electrodynamic consequences, **1**:24b–25a; **4**:322b; **11**:479b–480a; **12**:528a, 529b; **13**:176a, 239b
spin extension of, **4**:281b
statistical extension of, **4**:281b–282a
see also Einstein, Albert, relativity
Relaxation time
gases, **9**:220a–221a, 224b
Religion
church and state, **5**:550a; **6**:448a; **8**:98a; **9**:388b; **10**:173a, 174b; **11**:141b; **12**:586a
comparative study
medieval, **2**:152b
Renaissance, **3**:515b–516a
critics of, **11**:323a–b
Egyptian, **2**:539b, 540a; **4**:467b; **6**:305b, 306a
philosophy of, **8**:257a; **11**:72b; **12**:156b–157a
psychology of
see Psychology, religious
reform of
medieval Latin, **3**:512b; **10**:474a
Renaissance, **3**:515a; **10**:310b
and science
ancient Roman, **1**:333b–338a; **12**:310b
medieval, **4**:525b
Renaissance, **3**:407a; **9**:33a; **10**:316b
early seventeenth-century, **1**:374a; **3**:115b, 116b; **7**:269b, 293a; **9**:316b–318b; **10**:334a; **14**:49a, 302a, 361b, 362a–b; **15**:69b
late seventeenth-century, **3**:492a; **5**:416b–417a; **6**:129b; **9**:53a; **11**:317b; **12**:582b–583a, 600b–601a; **14**:157a
eighteenth-century, **2**:107b; **6**:43b; **8**:579b; **12**:430a; **13**:425b
early nineteenth-century, **7**:366a; **15**:375a
mid nineteenth-century, **2**:63b–64a, 596b; **3**:88a, 553a, 608a; **4**:183b; **5**:16b–17a; **6**:438a; **8**:260b; **9**:388b–389b; **11**:116a, 127a; **12**:278a, 570a–b
late nineteenth-century, **2**:65a; **3**:190a, 608a; **5**:32a–b, 474b*, 555b–556a; **11**:520b*; **13**:9a, 52b, 74b, 78b, 523a; **14**:343b; **15**:79b
twentieth-century, **8**:355a; **9**:399b; **11**:15a–b; **12**:23b, 53b, 349b; **13**:284a; **14**:318a
alchemy, **4**:75a, 170a
anatomy, **14**:10b; **15**:737b
astrology, **4**:213a
biology, **1**:423b; **7**:99b; **10**:9b–10a, 11b, 210a, 371b; **14**:350b
Cartesianism, **11**:508a–b
chemistry, **3**:398b; **5**:104a
cosmology, **1**:248a, 336b, 588a–590a; **2**:540a; **3**:179b; **4**:506a; **5**:242a–245a *passim,* 248a, 286b–287a; **7**:291b; **9**:317a, 406a*; **13**:236a; **14**:254a, 365a, 518b–519b; **15**:638b–641b, 643b
Egypt, **15**:708b, 710b–711a, 716a–b, 724b
evolution, **3**:88b, 553b, 573b–574a, 576a–b; **5**:16b–17a, 513b; **7**:498b; **8**:98a, 123a–b; **9**:143a, 389a–b, 428b, 533b–534a; **12**:45b, 570a–b; **13**:274b–276b, 542a; **14**:137a–138a, 439b; **15**:158a
geography, **9**:172a
geology, **2**:568b–570b, 612b–613b; **3**:297a, 313b; **6**:586b–587a; **9**:389a–b, 533b; **10**:322a,b; **11**:372a; **12**:262a–b, 318b, 433b; **13**:547b; **14**:145a, 516b
India, **15**:535b–536a, 630a
Japan, **15**:755b, 756b
Newtonianism, **3**:114b–115a, 244b, 295a–296a; **7**:275b
separation of, **2**:577b, 581b, 604a; **3**:297a; **4**:162b, 524b–525a; **6**:586b–587a; **7**:225a; **10**:354b
spiritualism, **3**:475a–b; **7**:68a; **13**:52b; **14**:137a–138a, 629b
scientists and
ancient Greek, **4**:33b
ancient Roman, **5**:232b; **10**:301a
Byzantine, **11**:161b, 183a
medieval, **7**:325a, 330b; **12**:92a; **13**:230b
Renaissance, **10**:197a, 280a; **11**:96a, 286b, 312a, 528b; **12**:316b; **13**:501b; **15**:160b

Religion (cont.)
scientists and (cont.)
seventeenth-century, **2**:522b; **3**:492a; **5**:419b; **7**:293a; **8**:172b–173a, 437a; **10**:36a, 151b, 332a–b, 335b, 337a; **12**:117b; **13**:30b, 33a, 169b, 573a–b; **14**:209b, 449b
eighteenth-century, **11**:616a; **13**:180a, 448a; **14**:69b
early nineteenth-century, **2**:133b; **5**: 250b, 530b; **11**:172b; **13**:413b; **14**: 186b; **15**:127b
mid nineteenth-century, **2**:446a; **4**: 527a–b, 531a; **5**:412b; **8**:260b; **9**:560a; **10**:11b; **12**:240a–b, 244a, 465b–466a; **14**:137a–b, 350a; **15**:85a–b, 132a–b, 167b–168a
late nineteenth-century, **7**:67a, 68a, 189a–b; **9**:428b–429a; **10**:447b; **11**: 372a; **12**:544b; **13**:249b; **14**:109a
twentieth-century, **2**:240a; **8**:257a, 355a; **13**:529a,b; **14**:318a; **15**:16b–17a
sociology of, **9**:491a
see also Atheism; Bible; Heresy; Historiography, of religion; specific religions by name; Theology
Remainder problems, **4**:606b, 612a
Remainder theorem, **3**:136a
Remak, Robert (1815–1865), 11:367–370; **6**:434a; **7**:437b; **13**:556b
cytology, **6**:270a; **12**:231b, 242a
embryology, **6**:435a
pathology, **12**:203a
Remeis, Carl, **6**:149b
Remelé, Adolf, **9**:349b, 351a; **11**:541b
Remlinger, Paul, **10**:15b
Remmelin, Johann, **6**:183b
Remora (mythical creature), **2**:591b
Remote control, **13**:287b, 431b
Remsen, Ira (1846–1927), 11:370–371; **2**: 559a; **3**:609b; **5**:13a; **10**:157b; **14**:357b
Remsen's law, **11**:370b
Rémy, Joseph, **7**:31b
Renan, Joseph Ernest, **2**:25a, 63b, 545a; **12**:2a
Renard, Alphonse François (1842–1903), 11:371–372; **9**:589a,b
Renault, Bernard (1836–1904), 11:372–373; **2**:492b; **13**:405b
Renaut, Joseph-Louis (1844–1917), 11:373–374; **11**:295b
Renaut's layer, **11**:374a
Rendiconti del Circolo matematico di Palermo, **5**:569b
editors of, **9**:502a
Rendle, Alfred Barton, **2**:476a; **15**:224b
Renevier, Eugène (1831–1906), 11:374–376; **7**:43b; **8**:543a
Renieri, Vincenzo, **2**:306b, 307b; **3**:3b, 101b; **7**:308a; **9**:369a
Renilla, **14**:425b
Rennell, James (1742–1830), 11:376; **9**: 124a*
Renner, Otto, **14**:101a
Rennes, University of
geography at, **9**:149b
Rennie, John, **11**:496b
Renouvier, Charles Bernard, **2**:545a; **11**: 389a; **15**:422b, 423a
Essais de critique générale, **7**:67b
Rensselaer Polytechnic Institute (Troy, N.Y.), **5**:11b–12a; **6**:56b
founding of, **4**:274a–b
Renwick, James, Sr., **6**:277b
Renz, Franz Robert, **7**:472a; **10**:30a
Repeating circle (geodet. instrum.), **9**:234a–b, 250b
Repertorium der Physik, **4**:174b
Repertorium für Anatomie und Physiologie, **13**:556b
Repertorium für die Pharmacie, **15**:172b
Reppe, Walter, **4**:553b
Representation, mathematical, **3**:54a; **6**: 547b; **7**:53b–54a
of abstract groups, **5**:198b
conformal, **3**:63a; **4**:103a; **13**:562b
of groups, **2**:615b; **6**:214b; **9**:457b–458a
linear, **3**:95b; **12**:237a–b
theory of, **10**:139a; **11**:459a
of a variety, **10**:140b–141a
weight of, **3**:95b
Repression (psychol.), **5**:174a–b, 177b, 178b; **7**:190a
Reproduction, botanical
ancient Greek study, **1**:260b
Renaissance study, **15**:81a
seventeenth-century study, **1**:293a–b; **4**: 135b; **8**:128a; **15**:67a–68a
early eighteenth-century study, **4**:98b; **8**:459b–460a; **11**:83b
late eighteenth-century study, **5**:217a; **6**:218b–219b; **12**:559b–560a
early nineteenth-century study, **11**:216a; **13**:463a
mid nineteenth-century study, **2**:160a, 492a, 520a; **5**:539b; **6**:266a–b, 465b, 466a; **9**:561a, 600a–601b; **11**:152a, 153a–154b; **13**:462b; **14**:350a
late nineteenth-century study, **3**:422a; **5**:581b–582a; **8**:5a–b, 354a; **13**:88a–89b
early twentieth-century study, **2**:371b; **3**:30a,b; **5**:582a; **8**:5a–b; **12**:125a–126a; **13**:165a–b
asexual, **14**:398b
bisexual, **2**:257a
heterosis, **3**:375a–b
sexual, **9**:140a
biochemistry of, **12**:207a
see also Alternation of generations; Fertilization; Parthenogenesis; Plants, sexuality; Pollination; Seeds
Reproduction, graphic
eighteenth-century, **3**:578b; **14**:196b
nineteenth-century, **2**:340a
Reproduction, human
ancient Greek study, **4**:106a
ancient Roman study, **3**:175b
medieval Arabic study, **11**:232b
medieval Latin study, **9**:362a, 364a
seventeenth-century study, **13**:170b–171a
nineteenth-century study, **4**:147a*; **11**: 110a; **15**:107b
twentieth-century study, **5**:43a; **10**:611a
see also Ovulation
Reproduction, zoological
ancient Greek study, **1**:260b–261a, 265a–b; **13**:92b
medieval study, **1**:102a; **6**:275b; **7**:64b
seventeenth-century study, **2**:591a–b; **3**: 180b; **4**:64a–b, 509a–511a; **5**:484b–485b; **6**:159a–161b; **8**:127b–128a; **11**:341b–342b; **13**:174a–b
early eighteenth-century study, **2**:576b, 579a–580a; **10**:10a–b; **13**:457b, 458a–b, 563a–564b; **15**:52b, 54a
late eighteenth-century study, **3**:487a; **12**:554b–560a, 562a; **15**:524b
early nineteenth-century study, **11**:132b–133a; **15**:82b
mid nineteenth-century study, **1**:417b; **2**:73a, 617a; **6**:590b; **8**:270b; **9**:332a; **11**:565b; **12**:107a, 421a,b; **13**:9b; **14**:113a,b; **15**:78a
late nineteenth-century study, **6**:287b; **9**:86a, 185a–b; **12**:141b
twentieth-century study, **1**:622b; **2**:344b–345b; **3**:17b, 247b; **4**:101b–102a, 207a; **5**:502b; **10**:611a
female cycle, **10**:291b
hormonal control of, **1**:123a–b
see also Ovogenesis
hormonal control of, **1**:152b–153a; **10**: 611a
rates of
and ecology, **5**:70a
see also Artificial insemination; Egg(s); Embryogeny; Embryology; Evolution; Fertilization; Hermaphrodites; individual species, organisms; Morphogenesis; Ovulation; Parthenogensis; Sexuality; Spermatozoa; Spontaneous generation
Reproductive system
anatomy of
ancient Greek study, **1**:265a; **6**:318a
medieval Latin study, **9**:468b
Renaissance study, **1**:619b; **4**:521a; **8**: 204a–b; **9**:165b; **14**:10a
seventeenth-century study, **4**:509b; **5**: 484b; **6**:509a; **7**:612b; **13**:33b, 170b–171a
eighteenth-century study, **6**:566a–b; **9**:482b, 511b; **13**:563b
nineteenth-century study, **8**:302b
comparative, **8**:254a
embryogeny of, **9**:570b; **11**:308a
see also Uterus
Repsold, Adolf (1806–1871), 11:376–378; **2**:102a; **13**:109b
meridian circle, **4**:27a*; **13**:113a
Repsold, Johann Adolf (1838–1919), 11: 376–378; **6**:241a
Repsold, Johann Georg (1770–1830), 11: 376–378; **5**:301b; **12**:234a
A. Repsold and Sons (Hamburg), **5**:405a
Reptiles, **11**:503a; **12**:216a
adaptation of, **4**:147b
anatomy, **13**:402b
fossil, **2**:489a,b, 504b–505a; **3**:526b, 608b; **4**:147b, 549a; **5**:296a, 357b; **6**:594b–595a; **7**:60a, 61b; **8**:43b, 104a, 572a; **9**:134b; **10**:40b, 262a; **11**:336a; **14**: 410a
hearing, **5**:355a; **7**:516b
morphology
and evolution, **2**:504a–b
neurology, **14**:57b
taxonomy of, **2**:493b; **7**:547b; **9**:572b; **13**:25b; **14**:410a
winged, **2**:489b
see also Herpetology; specific reptiles
Repulsion, electrostatic, **6**:170b
theories of, **4**:215b; **5**:132b–133a, 135b–136b; **7**:371b
Repulsion, mechanical
theories of, **7**:228a–b; **15**:377a
Résal, Henri-Amé, **8**:382b; **11**:78b
Research Corporation, **3**:436b; **14**:121a, 393a–b
Research Institute for Chemotherapy
see Ehrlich (Paul) Institute
Resection, **2**:130a; **8**:34a–b
Reseda, **7**:471a
Reshᶜayna
see Sarjiyūs al-Raʾs ᶜAynī
Residue(s) (math.), **3**:135b, 136b, 139b–140a; **4**:474b; **5**:450a; **8**:62a
biquadratic, **4**:341b; **5**:305b; **7**:52a; **9**:175a
calculus of, **3**:252a–b
concept of, **11**:54b
cubic, **4**:341b; **7**:52a

definition of, **3**:139b
quadratic, **4**:341b, 573b; **6**:390b
reciprocity of, **8**:135b, 137b, 138b
theory of, **7**:54b; **8**:367a; **10**:247b
foundations of, **12**:512b
Residue theorem, **3**:138b, 139a,b
Resins, **6**:166b; **14**:349b
analysis of, **3**:279a; **10**:497b; **14**:156b
Bakelite, **1**:385b
decomposition products, **10**:498a
origin of, **14**:350b
Resistance, electrical
and dimensions, **11**:474a
measurement of, **11**:150b*; **14**:290a
standardization of, **1**:612b; **7**:93b, 182b; **8**:502a; **9**:178a, 209b; **14**:206b–207a, 330a
in metallic powders, **11**:93b
and temperature, **1**:612b; **3**:19b, 312b; **4**:80b; **7**:221b
theories of, **14**:208a–b
Resistance, mechanical
medieval study, **13**:186a, 201b–205b, 206b–207a
seventeenth-century study, **6**:607b; **9**:119a; **10**:74b–76b
eighteenth-century study, **1**:110b; **2**:39a–b
nineteenth-century study, **14**:620b
degrees of, **13**:207a
and density, **10**:75a
elastic bodies, **2**:49a
in fluids, **2**:299b; **5**:200a–b; **11**:400b
and body shape, **4**:208a
boundary, **4**:208a
frictional, **3**:441b
law of, **2**:54a; **11**:400b
measurement of, **10**:75b
proportionality to time, **2**:39b
and velocity, **10**:75a
Resolution (ship), **3**:397a; **5**:75a,b, 76b–77a
Resolvents, **3**:221b
Malfatti resolvent, **9**:55a
Resonance
acoustic, **6**:245b; **7**:137b; **14**:206a
resonant vases, **3**:153a
theory of, **3**:437b; **13**:101b
chemical, **6**:108b; **12**:419a–b
electrical, **4**:191b; **8**:510b
in oscillatory circuits, **2**:167b; **9**:215b
magnetic
ferromagnetism, **1**:284a; **7**:617a
light, **4**:578a
nuclear, **6**:105a
mechanical, **8**:596a
optical, **11**:582b
quantum mechanical, **8**:474b–475a
in radiation, **11**:10b–13a
Resonator, electric, **11**:213a
cavity, **6**:104b
Resorcinol
derivatives of, **7**:471a
Respighi, Lorenzo (1824–1889), 11:378–379; **12**:268a
Respiration
Renaissance study, **11**:528a
seventeenth-century study, **2**:312a; **5**:418b; **7**:578a; **13**:169a, 169b–170a
eighteenth-century study, **2**:509a; **6**:44b; **11**:499a; **12**:24b–25a, 286b, 287a, 557b, 562b–563a
early nineteenth-century study, **2**:336a, 422b; **11**:173b
mid nineteenth-century study, **8**:348b, 540b, 541b; **12**:466b, 467a, 502a–b
late nineteenth-century study, **5**:149b–150a; **6**:23b–25a; **10**:579b; **11**:425b
twentieth-century study, **1**:453a–455a; **6**:21b–22b; **7**:501b–503b; **8**:614a–b; **11**:138b–139a
and animal heat, **2**:483b–484a; **14**:407b; **15**:95a–b
and blood circulation, **2**:599a; **6**:64a
and blood pressure, **11**:63a
cellular, **4**:76a; **10**:579b; **12**:242b; **13**:393b–394a; **14**:175a–b
chemistry of, **1**:362b–363a; **6**:42b, 264b–265a, 505a–b; **7**:273a–b; **8**:83a–b; **12**:271a
in cold-blooded vertebrates, **4**:268a
and combustion, **6**:484b, 485a; **15**:95a
comparative studies, **2**:61b
cutaneous, **7**:501b
enzymatic systems, **2**:87a; **6**:501b; **14**:173b, 175a–b, 176b
evolution of, **1**:454a
fetal, **1**:455a; **9**:244b, 569b
at high altitude, **9**:547a
inhibition of, **14**:175a
measurement of, **5**:149b–150a; **10**:558a; **12**:466b, 467a; **14**:64b–66a
mechanics of, **2**:37a; **6**:64a, 575a–576a; **8**:203b–204a, 541b; **9**:101b, 243a; **10**:267b; **13**:32b; **15**:108a
neurophysiology of, **8**:536a; **12**:271a; **13**:530a
oxygen deprivation, **4**:285b
parenchymatous, **12**:563a
pathology of, **11**:482a
pharmacology of, **15**:103b
in plants
see Plants, respiration
pressure studies, **13**:557a
regulation of, **8**:614a
age, effects of, **8**:134a
chemical, **6**:24a
gaseous, **10**:579b; **11**:138b, 298a
nervous, **2**:61b; **8**:506a
reflex, **2**:446a–b; **6**:300a
in sleep, **9**:547a; **14**:322a
theories of
ancient Greek, **1**:262a, 267a; **4**:106b, 368b, 384b–385a; **6**:318a; **11**:128a
ancient Roman, **5**:235a
seventeenth-century, **4**:377a; **6**:183a, 484b, 485a; **8**:525b–526b; **9**:243a–246a; **13**:169b–170a, 336a–b
eighteenth-century, **2**:3a; **6**:64a; **8**:75b, 83a–84a; **12**:563a; **15**:95a–b
nineteenth-century, **14**:351a
twentieth-century, **6**:24b; **7**:501b–502a
in tissue, **2**:61b; **8**:541b; **13**:394a
see also Artificial respiration, Oxidation, Resuscitation
Respirators, **10**:609a
liquid-oxygen, **3**:12a
underwater, **12**:244b; **13**:233b
Respiratory quotient, **8**:555a–b, 615a; **10**:579b
Respiratory system
anatomy of, **12**:302b
comparative, **8**:254a; **11**:307b, 380a
embryogeny, **11**:307b
Restionaceae, **10**:13b
Resultants
see Equations, resultant
Resumptio genetica, **8**:513a
Resuscitation
ancient Greek theories of, **4**:32b–33a
nineteenth-century, **3**:6b; **9**:153a; **12**:502a; **13**:53b
Marshall Hall method, **6**:60b
microbiological, **12**:557b–558a
mouth-to-mouth, **11**:100a
see also Artificial respiration
Retaining walls, **3**:441b
Rete mirabile, **1**:620a; **4**:199a–b; **9**:166a; **10**:21a; **14**:10b
Reticular formation (anatomy), **1**:580a; **8**:552b
Reticular ordering (crystallography), **2**:431a
Reticularia, **3**:44a
Reticuloendothelial system, **8**:9a; **9**:296a; **15**:480a
defensive function, **4**:154a; **5**:317a
Reticulum
cells and fibers of, **2**:165a
Retina, **9**:10a
chemistry of, **7**:520a–b
color sensitivity
variations in, **9**:201b–202a
cones of
shifting of, **4**:372a
electrical response to light, **6**:476a–b
embryogeny of, **5**:514b
fatigue, **6**:247a,b
inversion of images, **3**:116a
length of impression, **11**:21a
light-sensitive layer, **11**:214b
rods and cones, **2**:124b; **12**:232a
sensitivity, **10**:482a
sequence of excitation, **4**:615b
structure of, **3**:424b; **4**:266b; **6**:287b; **10**:267b; **13**:556b
Retinography, **5**:66a*
Retrogradation
see Planets, retrogradation of
Retzius, Anders Adolf (1796–1860), 11:379–381; **8**:519a; **9**:570b, 573a; **11**:217b*, 382b; **13**:274a
Retzius' cavity, **11**:380b
Retzius' ligament, **11**:380b
Retzius' striae, **11**:380b
and Scheele, **12**:143b–148b *passim*
Retzius, Anders Johan, **11**:379b, 381b
Retzius, Magnus Gustaf (1842–1919), 11:381–383; **1**:562b; **5**:328b
Reuchlin, Johann, **1**:80a; **12**:150b; **15**:160b
De verbo mirifico, **1**:79b
Reuleaux, Franz (1829–1905), 11:383–385; **3**:303a; **8**:208b, 365a; **14**:404a
Theoretische Kinematik
translations of, **7**:287b
Réunion (Island), **3**:365b
Réunion des Sciences
founding of, **7**:581b
Reuss, August Emanuel (1811–1873), 11:385–387
Reuss, Franz Ambrosius (1761–1830), 11:387–388; **6**:397b*
Revegetation
after fire, **7**:102b
Revelation
see Epistemology, revelational
Reverberations, acoustical, **12**:53b, 54b
Reversibility
concept of, **3**:73b, 74b, 81a–b, 82b; **7**:217b; **12**:497b; **13**:436b
paradox of, **9**:227a
see also Irreversibility in physical systems; Reactions, chemical, reversibility of
Reversion, genetic, **1**:505b
Review of American Chemical Research
founding of, **10**:157a
Review of Scientific Instruments
editors of, **11**:441b
Revista argentina de historia natural, **1**:130b
Revista matemática hispanoamericana, **11**:394b
Revolution of 1830, **13**:128a
Revolution of 1848, **3**:336b–337a, 467a; **4**:342a; **7**:22b; **9**:538a, 560a, 569a;

Revolution of 1848 (cont.) **11**:78b–79a, 111a, 301b; **12**:295b; **13**:143a–b, 404a; **14**:40b–41a, 57b, 205a, 220b; **15**:193b, 526b
Revue d'anthropologie
founding of, **2**:478a
Revue de géographie alpine
founding of, **2**:191a
Revue de géologie
editors of, **8**:30b
Revue de l'Oise
founding of, **4**:74a
Revue de métallurgie
editors of, **11**:100b
founding of, **8**:119b
Revue de métaphysique et de morale
founding of, **2**:544b
Revue de synthèse, **11**:388b
Revue du mois, **2**:303a
Revue encyclopédique, **3**:80a,b
Revue générale de botanique, **2**:290b
Revue générale d'histologie
founding of, **11**:374a
Revue neurologique
founding of, **9**:109a
Revue scientifique, **2**:60b, 203a
founding of, **2**:439a
Revue scientifique et industrielle, **5**:370a
Revue scientifique italienne
editors of, **9**:538b
Revue semestrielle des publications mathématiques
editors of, **7**:466a; **12**:213b
Rey, Abel (1873–1940), 11:388–389; 4: 233b*; **5**:541a*; **9**:378a
philosophy, **15**:423a,b
Rey, Jean (*ca.* 1582–*ca.* 1645), 11:389; 1: 529b; **8**:211b
Rey Pastor, Julio (1888–1962), 11:394–395
Reye, Theodor (1838–1919), 11:389–390; **13**:155a; **15**:256a
apolarity theory, **9**:358a
Reyna, Francisco de la (*b. ca.* 1520), 11: 390–391
Reyneau, Charles René (1656–1728), 11:392; **1**:110b; **3**:282b; **9**:50b–51a; **11**:157b
Analyse démontrée, **9**:51a
Reynolds, Doris Livesey, **1**:394b; **6**:474b
Reynolds, James Emerson, **3**:478b; **14**:561a
Reynolds, John Russell
System of Medicine, **11**:462b
Reynolds, Osborne (1842–1912), 11:392–394; 3:477a; **9**:224a–225a; **12**:612b; **13**:362a, 365a
Reynolds' number, **5**:502a; **11**:393a
Rezbanyite, **5**:162a
Rgveda, **15**:194b–195a, 534b, 535a
Rhaetian (stratig. stage), **9**:617b
Rham, Georges de, **4**:337a
Rhazes
see al-Rāzī
Rhea, **3**:569b
Rhea (satellite), **9**:115b
Rheims
cathedral school, **5**:365a
Rheinhold, Erasmus, **2**:404a
Rheita, Anton Maria Schyrlaeus de (Antonín Maria širek z Vrajtu) (1597–1660), 11:395; 5:287b
Rhenish Mountains
geology, **11**:500a
Rhenium
atomic weight, **6**:481a
complex compounds of, **8**:109a
discovery of, **10**:136a–b
spectra, **14**:599a
Rheobase, **8**:29a
Rheology, **9**:221a
of blood, **1**:534a–b
Rheostat
invention of, **14**:290a
Rheticus, George Joachim (1514–1574), 11: 395–398; 6:144a; **11**:365b; **14**:274a,b, 471a
and Copernicus, **10**:245b
correspondence of, **11**:288b
library of, **3**:262a
Narratio prima, **3**:406a; **10**:245a
Opus Palatinum de triangulis, **3**:262a
revision of, **11**:4a–b
tables of, **4**:619a
trigonometry, **14**:18b
Rhetoric
ancient Greek, **4**:367a
ancient Roman, **2**:230b, 232b; **3**:109b
eighteenth-century, **8**:469b
nineteenth-century, **14**:288a
Rhetorius, **10**:419a; **15**:125a, 545b, 550b
Rheumatic fever, **2**:35b; **8**:7a; **14**:40a
etiology of, **14**:624a
Rheumatism, **1**:399b–400a; **5**:418b
chronic, **3**:205a
treatment of
steroid, **3**:353a
Rhine River
history of, **11**:277a
Rhinencephalon, **1**:545a
Rhinoceros, **3**:38a
classification of, **5**:46a
fossils of, **3**:264a
Rhinology, **13**:567b
Rhinoscopy, **2**:305b
dorsal, **3**:531a
Rhizobium, **6**:237b; **15**:14a
Rhizocarps
anatomy of, **9**:601a
Rhizocrinus, **12**:107a
Rhizoctonia, **2**:34b
Rhizophoraceae, **2**:519b
Rhizopods, **2**:627a; **3**:337a; **12**:231a
naming of, **4**:234b–235a
Rhizopus, **6**:73a
Rho, Giacomo, **14**:162b
Rhodamine B, **2**:59b
Rhode, Johan, **1**:482a; **4**:377a
Rhode Island
Geological Survey, **7**:45a
geology, **4**:361b
Rhodes (island), **5**:345a
Rhodes (star), **15**:210b
Rhodesian redwater fever, **7**:428b
Rhodium, **2**:589b; **3**:301b, 617b
as catalyst, **4**:241b; **5**:595b; **13**:313a
complex compounds of, **8**:108b–109a
coordination compounds of, **7**:179a
discovery of, **3**:232b; **14**:487b
refining of, **3**:235a; **8**:109a
Rhodizonic acid, **5**:431b
Rhododendron, **1**:422b; **6**:489b, 494a; **8**: 178a; **10**:164b
Rhodophyta, **7**:535b
Rhodora, **4**:584a
Rhodotorula, **12**:208a
Rhombencephalon, **1**:545b–546a
Rhomboids
area of, **4**:609b
Rhonchi, **7**:557a
Rhone (river)
basin, **6**:168b
Rhône-Poulenc Chemical Company, **5**:100a
Rhubarb, **1**:411b
chemical analysis, **8**:329a; **12**:148a
in pharmacy, **1**:125a; **12**:146b
Rhumb lines, **6**:124b, 125a; **12**:501a
Rhynchocephalian, **4**:147b
Rhynia, **8**:5b
Rhyolites, **4**:564a; **9**:367b
Rhythm
ancient theories of, **1**:283a
medieval theories of, **4**:526a
chronos protos, **4**:526a
poetic
isosyllabic, **1**:564b
Rhytina, **2**:423a
Ribaucour, Albert (1845–1893), 11:398; 3: 96a, 559b
congruences of, **4**:338b*
transformations of, **4**:338a*
Ribaud, Gustave Marcel, **5**:187a
Ribeiro de Castro, Carlos, **3**:185a
Ribeiro Santos, Carlos (1813–1882), 11:398–399
Ribit, Jean, **1**:571a; **13**:507a–b
Riboflavin
see Vitamin B_2
Ribonucleic acid
see RNA
Ribose, **8**:275b–276a
Ribs
resection of, **8**:34a–b
Ricardo, David, **9**:67b, 69b; **10**:486a; **15**:405a
Ricardus
see Richard
Riccati, Jacopo Francesco (1676–1754), 11: 399–401; 1:164a; **4**:515b, 516b; **9**:55a
hydrodynamics, **2**:39b
Riccati equations, **2**:37b, 57a; **4**:478b; **5**:414a, 449b; **7**:561b; **8**:384a; **11**:400b
Riccati, Vincenzo (1707–1775), 11:401–402
Ricchebach, Giacomo, **3**:13b, 14a
Ricci, Gregorio
see Ricci-Curbastro
Ricci, Matteo (1552–1610), 11:402–403; 7: 19b; **10**:103b
cosmology, **14**:162a
translations by, **3**:311b; **4**:457b*; **14**:164b
Ricci, Michelangelo (1619–1682), 11:404–405; 2:309a; **3**:182b; **4**:506a
correspondence of, **2**:308b, 311b; **10**:332b; **12**:460a; **13**:436a, 438b, 439a
mathematics, **5**:527a; **10**:336b
Ricci, Ostilio (1540–1603), 11:405–406; 5: 238a
Ricci-Curbastro, Gregorio (1853–1925), 11: 406–411; 8:284a
calculus, **3**:263b; **4**:280b, 336b; **8**:389b; **12**:214a
geometry, **5**:555a; **11**:455a
Ricci's tensor, **11**:407b
Ricciocarpus, **9**:601a
Riccioli, Giambattista (1598–1671), 11:411–412; 1:483b; **3**:100a; **4**:166a; **5**:527b; **7**:376a; **8**:26a; **10**:153b
Almagestum novum, **1**:164b, 508a, 513a; **3**:528b; **5**:542b; **6**:362a; **7**:308a
Astronomia Reformata, **5**:543a
Geographiae et Hydrographiae Reformatae, **5**:543a
and Grimaldi, **5**:542b–543a
optics, **10**:475b
star catalog, **6**:364a
Riccò, Annibale (1844–1919), 11:412; 13: 232b, 233a*
Rice, **7**:89a; **13**:460a
and beriberi, **4**:311a–b; **5**:209a; **14**:392b–393a
Richard of Fournival, **6**:273a
Biblionomia, **1**:63b; **4**:447b; **5**:360a; **7**: 171b
Richard of Salerno, **12**:83a
Richard Swineshead
see Swineshead, Richard

Richard of Wallingford (*ca.* 1292–1336), 11: 414–416; 2:435a; 4:164b
Albion, 7:38b, 119b
De sectore, 7:38b
Richard, Jules Antoine (1862–1956), 11: 413–414
Richard's paradox, **11**:413b
Richard, Louis Claude Marie, **15**:267a
Démonstrations botaniques, ou Analyse du Fruit, **8**:371a
Richard, Louis Paul Émile (1795–1849), 11: 414; **5**:260b; **6**:307a
Richards, Alfred Newton, **1**:533b; **5**:43a; **7**:502b; **12**:470b; **15**:105b
Richards, Francis John (1901–1965), 11:416; **5**:523b, 524a*
Richards, Theodore William (1868–1928), 11:416–418; **8**:290a; **15**:467a(n)
atomic theory, **6**:117a
atomic weights, **3**:398a, 480b(n); **11**:173b
gas spectra, **13**:473a
influence of, **2**:458a; **6**:118a, 260a, 481a
radiochemistry, **12**:505b
students of, **8**:289a; **11**:576b
thermochemistry, **11**:577a; **15**:438b–439a
thermodynamics, **8**:289b
Richardson, Benjamin Ward (1828–1896), 11:418–419; **2**:547b, 566b; **12**:487b, 490a,b, 502a,b, 503b; **14**:495b
Richardson, George Fleming, **13**:441b
Richardson, Owen Willans (1879–1959), 11:419–423; **3**:367a, 597a; **8**:499b*
Electron Theory of Matter, **4**:576b
radiation, **3**:372b; **8**:11b; **9**:397a
Richardson's law, **11**:421a–b
Richarz, Franz, **7**:458a; **12**:423a; **14**:218a
Riche, Jean, **2**:587b
Riche de Prony, Gaspard-François-Clair-Marie
see Prony, Gaspard-François-Clair-Marie Riche de
Richelieu, Armand Jean du Plessis, Duc de, **3**:360b; **4**:49a; **6**:299a; **9**:507a
Richelot, Friedrich Julius, **1**:294a; **3**:314a; **4**:378a; **7**:51a; **10**:25a; **12**:216b, 255b; **14**:158a, 202a
Richer of Saint-Remy, **5**:365a,b
Richer, Jean (1630–1696), 11:423–425
Cayenne expedition, **3**:101b, 103a; **6**:606b
experiments of, **2**:592b; **15**:271a
horology, **10**:91b(n)
Richer de Belleval, Pierre
see Belleval, Pierre Richer de
Richet, Charles Robert (1850–1935), 11: 425–432; **2**:439a; **5**:411a*; **8**:90a*; **9**:103a*; **11**:101b; **15**:469a
Richmann, Georg Wilhelm (1711–1753), 11: 432–434; **1**:67a; **4**:129a; **11**:609a
atmospheric electricity, **5**:134b; **8**:470a
cooling, **3**:495a; **13**:10b
Richmann's law, **14**:353b, 354a–b
Richter, Charles Francis
Richter scale, **5**:596b
Richter, Jeremias Benjamin (1762–1807), 11:434–438; **3**:465b; **7**:388b
chemistry, **2**:93a; **5**:532b
Richter, (Hieronymus) Theodor, **5**:162a, 455a
Richter, Viktor Yulevich, **9**:290b
Richthofen, Ferdinand von (1833–1905), 11: 438–441; **11**:540a
biographies of, **10**:503b
critics of, **10**:167b
Führer für Forschungsreisende
influence of, **10**:504b
geomorphology, **11**:210a
influence of, **1**:174a; **5**:336b
loess, **3**:462b
stratigraphy, **7**:370b
students of, **4**:193a,b; **6**:215a; **9**:149b; **11**:524b
Richtmyer, Floyd Karker (1881–1939), 11: 441–442
Ricin, **4**:297a–b, 298a
protein nature of, **9**:284b
Ricinus, **12**:59a; **15**:67b
Rickets, **5**:209a, 425b; **6**:136a, 183b, 232a*; **9**:243a; **15**:418a–b
etiology of, **5**:37a; **10**:243b; **15**:418a
treatment of, **5**:209a; **14**:444b; **15**:418b, 419a
Ricketts, Howard Taylor (1871–1910), 11: 442–443; **11**:175a
editions of works, **6**:233a*
Rickettsia, **5**:317a; **11**:175a, 442a–443a; **14**:623b
cultures of, **7**:624b
Rickettsiosis
prevention of, **15**:454b
Ricochet, **15**:490b
Ricord, Philippe, **2**:574b; **11**:515b
Rideal, Eric Keightley, **10**:329b
Ridgway, Robert (1850–1929), 11:443–444; **1**:331b, 405b, 406a; **13**:25a
Ibn Riḍwān, Abu'l-Ḥasan ᶜAlī ibn ᶜAlī ibn Jaᶜafar al-Miṣrī (998–*ca.* 1069), 11: 444–445; **1**:513b; **2**:619a,b
commentaries by, **15**:173b, 185b, 188a, 239b
Riḍwān al-Khurāsānī, **1**:532b; **15**:254a
Riḍwān ibn al-Razzāz, **12**:362b
Riecke, Eduard (1845–1915), 11:445–447; **10**:347a; **11**:475b, 480b(n); **12**:613b, 615a; **14**:209a*
electron theory, **4**:191b; **14**:205a–b, 208b
Riecke's rule, **1**:551b
Rieger, Johann Christoph, **12**:40b
Riemann, Georg Friedrich Bernhard (1826–1866), 11:447–456
acquaintances of, **2**:105a; **4**:2a, 342b; **5**: 472a; **7**:506b
analysis, **2**:465a; **3**:263b, 315a; **5**:204a, 263b; **6**:308b; **7**:398b; **8**:558b; **10**: 139b; **11**:407b, 409a; **14**:86a, 572b; **15**:386a
Cauchy-Riemann equations, **3**:137a,b, 138a, 139a; **4**:477b; **11**:449a
critics of, **14**:222b
and Dedekind, **4**:1b, 2b, 3a–b
editions of works, **4**:4b; **14**:202a,b
equation of, **3**:309b
geometry, **3**:96a; **4**:417b, 374a; **5**:307b; **6**:391a, 522a; **11**:22a; **12**:246a; **13**: 264b; **15**:198a
influence of, **1**:7a, 599b; **2**:105b; **3**:263a, 322a; **4**:205b; **6**:75b, 95a; **7**:621b; **11**:54b, 124a, 407a, 475b; **12**:213a, 330b
influences on, **4**:342a, 482a; **5**:97b, 301a, 307a
mathematical physics, **4**:306b; **6**:26a, 344a; **8**:490b–491a, 599b; **10**:27b
number theory, **6**:214a, 522a; **9**:173b
Riemann functions, **8**:367a; **11**:454a–b
Riemann hypothesis, **1**:306b; **6**:114a, 393b; **11**:453a
Riemann integral, **6**:95b; **8**:110a–b, 111a,b, 562b; **11**:454b
Riemann problem, **2**:144b; **11**:407a, 452b–453a
Riemann spaces, **3**:96a–b, 323a; **7**:207b; **8**:284a, 434a; **11**:454b–455a; **12**:609b
Riemann surfaces
see Surface(s), Riemann
Riemann-Christoffel symbols, **3**:263b
Riemann-Roch theorem, **4**:617b*; **6**:308a; **7**:621b; **8**:585a; **12**:331b
theorems of, **10**:244b, 248b
topology, **4**:174a; **8**:389b; **11**:60a
Über die Hypothesen welche der Geometrie zu Grunde liegen, **1**:600a; **6**:249a
translation of, **3**:322b
zeta function, **1**:306b; **2**:238b; **4**:475a; **6**:4a, 113b, 214b; **7**:102a; **13**:424a, 498a, 561b
Ries (or Risz, Riesz, Ris, or Riese), Adam (1492–1559), 11:456–458; **7**:419a; **13**:60b; **14**:325b
Ries, Emil, **14**:278b
Riesengebirge (Germany), **12**:187a
botany, **10**:13b
Riess, Peter, **6**:553a; **11**:539b
Riesz, Frigyes (Frédéric) (1880–1956), 11: 458–460; **3**:230b; **4**:548a; **8**:112a; **9**: 126b
acquaintances of, **4**:561b; **5**:608a
analysis, **4**:562a; **5**:97b; **6**:5a; **11**:157a; **14**:89a
Riesz representation theorem, **11**:459a
Riesz-Fischer theorem, **6**:392a; **11**:459a
students of, **11**:259b, 260a
Riesz, Marcel, **4**:548a, 561b; **6**:114a
Riffat Bey, **10**:15b
Rift valleys, **3**:325b; **7**:144b; **14**:402b
Rigaud, Stephen Peter, **6**:128a, 511b; **7**:146a, 163a; **14**:485a
Rigel (star)
spectrum, **14**:56a
Righi, Augusto (1850–1920), 11:460–461; **13**:370a(n)
comets, **11**:412b
electrodynamics, **6**:51b; **9**:98b
Right ascension, **2**:99a–b; **14**:577b; **15**:552a, 571a, 573a, 578b, 592a, 612a, 615a, 616b, 621a
definition of, **5**:614a
Right-handedness
see Left- and Right-handedness
Rigor mortis, **2**:31a
Riley, Charles Valentine, **6**:524b; **10**:273a; **15**:130b
Riley, Dennis Parker, **4**:522a
Rima, Tommaso (1775–1843), 11:461–462
Rinde, Herman, **13**:160a,b
Rinderpest
see Cattle plague
Rindfleisch, Daniel (Bucretius), **3**:99a; **12**: 577b
Ringer, Sydney (1835–1910), 11:462–463
Ringer's solution, **11**:463a
Rings (math.)
Artin rings, **1**:307b
associative, **1**:307b
concept of, **4**:5a; **10**:138b
of invariants, **6**:389a–b
noncommutative, **10**:139a
polynomial, **6**:389b
sections of, **3**:182b
theory of, **5**:108a
Ringworm
see Tinea
Rinman, Sven (1720–1792), 11:463–464; **2**: 5a; **3**:473a; **7**:254b
Rinne, Friedrich, **10**:124b, 125b
Rinow, Willy, **6**:497a
Río, Andrés Manuel del (1764–1849), 11: 464–465; **3**:346b
Rio de Janeiro
sanitation, **15**:96b–97a
Rio Grande
geology of, **2**:549a
Riocreux, Alfred, **13**:394b, 395b

Río-Hortega, Pío del (1882–1945), 11:465–466; 11:274a
Riolan, Jean, Sr., **4**:209a; **11**:466a; **13**:507b
Riolan, Jean, Jr. (1580–1657), 11:466–468; 8:310a; **10**:426a
anatomy, **10**:477b
on contemporary scientists, **1**:482b; **4**:209a; **12**:577b
and Harvey, **6**:157a,b; **10**:477a; **15**:5a*
Ripensis, Christiernus Johannis, **2**:409a
Ripensis, Johannes Franciscus, **2**:401b
Ripiphorus, **2**:321b
Risner, Friedrich (*d. ca.* 1580), 11:468; 1:81a; **11**:289a(n)
Optica libri quatuor, **12**:501a
Opticae thesaurus Alhazeni, **5**:525a; **6**:197a, 200a; **11**:97b; **15**:180a
Ristoro (Restoro) d'Arezzo (*fl. ca.* 1210–*ca.* 1282), 11:468–469
Composizione del mondo, **10**:421a
Ritchey, George Willis (1864–1945), 11:470; 3:509b; **4**:249a; **9**:45b; **10**:473a
Ritchie, Edward Samuel, **11**:604a
Ritchie, John, **3**:194a
Ritt, Joseph Fels (1893–1951), 11:470–471
Rittener, Théophile, **8**:543a
Rittenhouse, David (1732–1796), 11:471–473; 1:485a; **5**:143b; 7:89b
hygrometer, **1**:65b
Ritter, August, **3**:496a; **4**:279a; **8**:1b; **12**:19a
Ritter, Heinrich, **9**:578b; **15**:526a
Ritter, Johann Wilhelm (1776–1810), 11:473–475; 9:570a; **10**:182b, 183a, 184a, 191b; **12**:281a
Ritter, Karl, **1**:173b–174a; **10**:168b; **11**:180b, 337a,b; **14**:54a
Die Erdkunde von Asien, **3**:532a; **12**:300a–b, 301b
Ritter, Robert, **11**:581b
Ritter, Wilhelm, **3**:496a
Ritter von Reuss, A. E.
see Reuss, August Emanuel
Ritthausen, Heinrich, **10**:242a,b
Ritz, Walter (1878–1909), 11:475–481; 7:515a; **10**:347a–b
Ritz combination principle, **12**:43b
Riva-Rocci, Scipione (1863–1937), 11:481–482
Rivault, David, **1**:229b
Riverius, Stephanus
see La Riviere, Estienne de
Rivers
action of, **7**:185b–186a; **8**:564a,b, 566a, 568a
annual flow, **14**:53a–b
base level, **5**:395b; **7**:144b
classification of, **3**:593b; **8**:467a; **14**:53a
currents, **8**:466a,b; **9**:370a; **14**:53a
history of, **8**:33b; **11**:277a
ice on, **12**:168a
motion of, **3**:115b–116a
mouths of, **9**:43a
pollution of, **5**:125a, 395b; **12**:478b; **14**:409b
profiles of, **8**:466a,b
regulation of, **2**:470a; **4**:501b; **5**:195b; **7**:499b; **9**:369b–370a
riverbeds, **14**:589a
formation of, **8**:466a–467a; **14**:621a; **15**:153b*
silting of, **13**:85a
stability of, **8**:467a
theories of, **5**:401a; **10**:280b, 521b–522a; **12**:310a; **14**:53a
see also Hydrology; specific rivers by name; Streams; Valleys
Rivett, Albert Cherbury David (1885–1961), 11:482–483
Rivière de Précourt, Émile Valère (1835–1922), 11:483–484
Rivinus, Augustus Quirinus
see Bachmann, Augustus Quirinus
Rivista di matematica
founding of, **10**:442b
Rivista di storia delle scienze mediche e naturali
editors of, **9**:378a
Rivista di studi sessuali e di eugenetica, **9**:378a
RNA, **10**:277a; **15**:252b
distribution of, **4**:603b–604a
structure of, **5**:589b
in viruses, **5**:141a
Roach
meiosis, **2**:626b
Robartes, Francis, **12**:128b
Roberg, Lars, **1**:305b; **3**:210b
Robert of Chester, **1**:22b, 512a; **4**:611b; **7**:362b; **11**:458a, 504a, 591a; **15**:175a,b, 186a, 256a
Robert the Englishman, **1**:382a; **3**:27a; **7**:120a, 127a(n); **12**:62a; **13**:400b
Robert of Lincoln
see Grosseteste, Robert
Robert Retinensis
see Robert of Chester
Robert de' Rossi, **1**:274a
Robert Dawson Evans Memorial for Clinical Research and Preventive Medicine, **11**:576b
Roberti, Johannes, **6**:254a
Roberts, Isaac (1829–1904), 11:484–485
Roberts, William Chandler
see Roberts-Austen, William Chandler
Roberts, Dale & Co., **3**:84b
Roberts-Austen, William Chandler (1843–1902), 11:485–486; 7:529a; **10**:246b
"iron-carbon" diagram, **3**:233b
metals, **13**:245b
Robertus
see Robert
Roberval, Gilles Personne (Personier) de (1602–1675), 11:486–491; 7:578a; **8**:162a; **10**:104a*
acquaintances of, **2**:353b; **3**:63b, 64b; **4**:46a,b; **6**:112b, 445b, 598b; **9**:316b, 498a, 598a, 599a,b; **10**:332a, 342a
correspondence of, **4**:567b; **13**:434a, 435b; **14**:49b
cosmology, **9**:320b(n)
critics of, **12**:85b
and Descartes, **3**:64a; **4**:571a
editions of works, **7**:577b, 578b
experiments of, **11**:506b
geometry, **1**:245b; **4**:47b; **5**:527a; **10**:115b, 336a,b, 342a,b, 495b; **11**:402a; **13**:434b, 436a; **15**:118a
quadrature, **3**:152a; **10**:103b
influence of, **1**:474b; **5**:259b*
logic, **9**:120a
manuscripts of, **13**:479b
physics, **1**:341b, 541b; **10**:332b–333a, 334b, 342a–b
priority disputes, **5**:522b*; **7**:583b; **9**:16a; **13**:434b
Roberval's problem, **12**:206b
Robespierre, Maximilien de, **3**:71a, 72a; **5**:91b, 92a, 93b, 603a
Robin, Charles-Phillipe (1821–1885), 11:491–492; 4:266b; **9**:422a; **10**:32a; **11**:425b; **12**:174b, 551b*
Dictionnaire de médecine, **3**:184b
Robin, Jean, 7:536b; **13**:450a
Robin, Vespasien, 7:537a
Robinet, Jean-Baptiste René (1735–1820), 11:492–493; 7:592b
Robinia, **3**:173a*; **6**:219b
Robinow, Carl Franz, **13**:3a
Robins, Benjamin (1707–1751), 11:493–494; 13:350b
New Principles of Gunnery, **4**:470b
Robinson, Abraham, **9**:71b
Robinson, Coleman Townsend, **5**:555b
Robinson, Elihu, **3**:537b, 538a,b
Robinson, George, **14**:318a
Robinson, Harold, **12**:30a
Robinson, Robert, **5**:589a; **8**:32a, 292a; **14**:412a
Robinson, Thomas Romney, **10**:328b; **14**:292b
Robiquet, Pierre-Jean (1780–1840), 11:494–495; 1:471a; **3**:159b; **4**:242b; **5**:432a; **10**:497b; **13**:499a
chemistry, **8**:333a; **14**:477a
Robison, John (1739–1805), 11:495–498
acquaintances of, **2**:174a, 179a; **14**:196b
on contemporary scientists, **2**:175a, 180b, 181a; **10**:36b; **14**:197b
editions of works, **14**:198a*
optics, **9**:267a
structural mechanics, **6**:452a
students of, **6**:54a; **8**:261b
Robison, Robert, **6**:111a
Robot, **13**:431b
Rocca, Gian Antonio, **3**:152a
Roccamonfina volcano, **2**:439b
Rocha-Lima, Henrique da, **11**:175a, 442a
Roche, Édouard Albert (1820–1883), 11:498; 7:85b
Roche's limit, **11**:498a
Rochefort (town), **2**:201a
Rochelle salt, **3**:505a
electrical characteristics, **7**:527a
Rochleder, Friedrich, **12**:452a
Rochon, Abbé Alexis Marie de, **2**:327b; **3**:440b
Rock, John, **10**:611a; **13**:97b
Rock crystal, **1**:5b
Rock salt, **5**:68b; **14**:61b
optical properties, **11**:582a–b
Rockefeller Foundation, **2**:345b; **3**:190b; **5**:37a, 40b, 208b, 378a; **7**:503a; **10**:601a
and astronomy, **12**:349a
International Education Board, **4**:485b; **6**:32a
National Research Fellowships, **6**:31b
Rockefeller Institute for Medical Research (New York City), **5**:316b; **8**:275b, 446b; **9**:265b
animal pathology, **9**:586a; **12**:482b
founding of, **5**:40a–b; **11**:176b
Rockefeller Sanitary Commission, **13**:63a
Rockets, **5**:433a–b; **7**:465a–b; **9**:404a, 504b; **11**:494a
ascending motion, **9**:324b
Congreve, **3**:388b; **9**:491b
fuel, **3**:511a, 578b; **7**:379a
historiography of, **9**:491b
long-distance, **7**:465a–b
motion of, **9**:504b; **13**:483b–484a
multistage, **13**:483b
rotating, **6**:35a
space, **7**:247b, 465b; **13**:483b–484a
underwater, **6**:35a; **9**:491b
see also Artillery, Missiles
Rocks
abyssal, **13**:148a
alkaline, **6**:475b; **7**:249b
genesis of, **3**:548a*
analysis of, **3**:292b, 293a; **4**:564a; **5**:563a, 577b; **7**:5b; **9**:61b, 531a; **11**:270b, 560b
calcareous, **1**:623a

carbonate
  genesis of, **3**:161b; **14**:132b
 chemical composition of, **5**:212b
  classification of, **10**:126a
  oxygen coefficient, **2**:159a
  variation diagram, **8**:42a
 classification of, **2**:496b–497a; **3**:313b, 484a; **5**:212b, 445a; **6**:580a; **8**:246a, 416a, 594b, 615b; **9**:367a, 620b; **10**: 430a; **11**:304a, 560b; **12**:514a; **13**: 330b–331a; **14**:260a, 261a; **15**:55a
  chemical, **8**:147a, 285b
  graphical, **14**:440b
  mineralogical, **3**:474a; **7**:5a–b, 249b; **11**:547b
 clastic, **6**:228a, 229a
 crystalline, **2**:496a; **12**:433a
 dating of, **2**:259b–260a; **5**:205b; **6**:14a, 474b–475a; **7**:45a; **10**:288b, 289a; **14**:588b
  radioactive methods, **6**:16a; **7**:161a–b; **8**:42a–b; **13**:107b, 617b
 evolution of, **12**:274a
 fabric patterns, **10**:126a
 hybrid, **6**:116a
 igneous, **2**:439b; **3**:161a, 328a; **4**:361b; **5**:39a; **13**:283a
  Atlantic group, **1**:551b
  basic, **2**:289b
  chemical analysis of, **1**:20a; **2**:588a; **3**: 293a–b, 484a; **14**:183a–b
  classification of, **3**:484a; **5**:89a; **6**:582b; **7**:112b–113a; **8**:285b; **9**:367a; **11**: 547b; **14**:183b
  crystallization of, **14**:58b–59a
  dating of, **8**:42a–b
  differentiation of, **2**:369b–370a, 486a; **5**:89b; **6**:116a; **7**:5a
  distribution of, **6**:116b; **7**:5b
  evolution of, **2**:370a
  intrusive, **3**:547b; **4**:366b; **6**:582b; **8**: 594b
  nomenclature of, **3**:484a; **8**:511b
  nucleus theory, **11**:547b
  and ore deposits, **14**:440b
  origin of, **1**:234a–b; **3**:567b; **6**:54a–b, 475b, 580a, 582b–583b; **8**:285b, 594a–b; **9**:367b; **14**:132b, 312a
  Pacific group, **1**:551b
  phase diagrams, **2**:369b
  phosphate in, **5**:104a
  succession of, **7**:370b–371a; **11**:439a
  synthesis of, **5**:89a–b; **9**:367a–b
  ultrabasic, **2**:289b
  volcanic-plutonic-hypabyssal cycle, **6**: 116a
 see also Basalt, Lava, Magma
 intrusive, **7**:62b
 intrusive vs. extrusive, **3**:593a
 magmatic
  see Rocks, igneous
 magnetization of, **9**:181b
 metamorphic, **2**:289b; **3**:161a; **5**:39a, 562b–563a; **7**:249a, 474a; **8**:511b; **9**:173a; **11**:561a; **13**:148b
  classification of, **1**:551b–552a; **4**:409a; **5**:563a; **12**:271b
  composition of, **5**:563a; **10**:126b
  foliation of, **1**:551b
  origin of, **2**:439b; **3**:568a; **4**:409b; **5**: 563a; **7**:249b; **9**:26b; **12**:262b; **15**:55a
  orthorocks, **1**:552a
  pararocks, **1**:552a
 meteoric, **1**:528a
 mineralogical composition of, **10**:126a
 mixed, **12**:273a
 phase changes, **4**:262a
 phase diagrams, **10**:124b
 phosphate, **3**:161b
 plasticity, **1**:52a
 porosity, **9**:257b
 primitive
  origins of, **8**:616a
 pyroxenic, **2**:588a
 remelting of, **4**:141a
 sedimentary, **2**:347b; **3**:161a–162a, 328a; **4**:361b
  classification of, **5**:486b–487a; **10**:126b
  composition of, **3**:161b, 293b; **10**:126b
  naming of, **5**:486b–487a
  origin of, **1**:623a–b; **4**:309a; **5**:444a–445a, 486b–487a; **6**:580b–581a; **8**: 616a; **10**:281a; **12**:545a; **14**:132b
  sequence of, **3**:594b; **10**:32b
 siliceous, **3**:161a
  genesis of, **3**:161b
 subalkaline
  reaction series, **2**:370a
 succession, **8**:594a
  order of, **2**:553a–b, 569a; **14**:259b–263a
 trachytic, **2**:588a
 volcanic
  see Basalt; Lava; Magma; Rocks, igneous
 see also Geology; particular rocks, as Granite, etc.; Petrography; Petrology; Sedimentology; Weathering
Rocky Mountain spotted fever
 etiology of, **11**:442b–443a
Rocky Mountains
 exploration of, **11**:118b
 geology, **6**:186b–187b; **11**:439a
 paleontology, **5**:500b
 stratigraphy, **11**:321a
 surveying of, **4**:265b; **5**:395b
Rocour, Georges, **8**:463a
Rod (unit of length), **12**:500b
Rod cells, **5**:36a*
Rodents, **3**:439a; **10**:16b; **11**:234a
 body temperature, **4**:285b
 caudal autotomy, **3**:492b
 see also specific rodents
Roderique, Johann Ignaz, **2**:15b
Rodier, Georges Paul Eugène, **5**:149b
Rodonea (curve), **5**:499b
Rodoxanthin, **13**:487b
Rodrigues, João
 see Lusitanus, Amatus
Rodriguez de Lima, Juan Manuel, **3**:97a
Roebuck, John (1718–1794), 11:499–500; **2**:174b; **14**:196b
Roemer, Ferdinand (1818–1891), 11:500–501; **6**:57b; **7**:61a; **11**:209a
Roemer, Friedrich Adolph (1809–1869), 11: 501–502; **5**:547b
Roemer, Hermann, **11**:501b; **12**:236b
Roemer
 see also Römer
Roentgen, Wilhelm
 see Röntgen, Wilhelm Conrad
Roentgenology, **2**:597b
 see also Radiology, X rays
Roesel von Rosenhof, August Johann (1705–1759), 11:502–503
Roger of Hereford (*fl.* second half of twelfth century), 11:503–504
Rogers, Arthur William, **4**:262a
Rogers, Eric, **3**:369b
Rogers, Henry Darwin (1808–1866), 11: 504–506; **3**:553b; **4**:73b, 364a,b; **6**: 503a; **8**:260b; **10**:258a
Rogers, James, **11**:504b
Rogers, Robert, **11**:504b
Rogers, William Augustus, **9**:531a; **14**:449a
Rogers, William Barton (1804–1882), 11: 504–506; **3**:553b; **4**:364a,b; **9**:100b; **10**:599b
Rognley, Oswald, **3**:367b
Rogoff, Julius Moses, **13**:53b; **15**:73a
Rogowski, Walter, **12**:527b
Rohault, Jacques (1620–1675), 11:506–509; **9**:51a, 498a,b; **13**:335a
 meteorology, **9**:117b
 Oeuvres postumes, **3**:320b
 students of, **12**:127a
 Traité de physique, **6**:36a
  translations of, **3**:294b
Rohdewald, Margarete, **14**:411b, 412b
Rohn, Karl (1855–1920), 11:509–510
Rohon, Josef Victor, **3**:530a
Rohr, Louis Otto Moritz von, **5**:590b; **7**: 448a
Roiti, Antonio, **14**:85b, 86a
Rojas, Juan de, **14**:400b, 594a
Rokitansky, Karl von, **3**:463a,b; **5**:565b; **6**: 434a; **12**:294b, 295a,b, 450a; **14**:40b
Rolando, Luigi (1773–1831), 11:510–511
 anatomy, **14**:15a
Rolfinck, Guerner (1599–1673), 11:511; **14**:212b
Roll, Peter Guy, **5**:273a
Rolle, Michel (1652–1719), 11:511–513; **5**: 259b
 analysis, **9**:51a; **11**:392a; **12**:118a; **13**:129a
 critics of, **12**:117b; **13**:586a
 theorem of, **8**:385a; **11**:512b
Rolleston, George (1829–1881), 11:513–515; **13**:341b
Rollet, Joseph-Pierre-Martin (1824–1894), 11:515–516
 Rollet's chancre, **11**:515b
Romain, Pierre Ange, **10**:609b
Romakasiddhānta, **8**:46b; **13**:582a; **15**:542b–550b *passim,* 556a, 588a
Roman balance
 see Steelyard
Roman de la Rose, **1**:91b
Romanes, George John (1848–1894), 11: 516–520; **2**:599a; **3**:90a; **5**:80b, 82b; **9**:512b; **12**:355b; **14**:239a*
Romanovsky, Gennadii Danilovich, **9**:592b
Romans, Berthelemy de, **3**:273a
Romanticism
 eighteenth-century, **4**:88a
 early nineteenth-century, **1**:69a–b, 489b; **2**:223b, 595a; **9**:568a; **10**:306b; **11**: 473b; **12**:154a, 155b, 156a; **15**:82b
 mid nineteenth-century, **2**:596b; **3**:600a, 604a; **9**:383a–b; **13**:238a; **15**:50b
 nineteenth-century critics of, **4**:616b; **5**: 193a
 see also *Naturphilosophie*
Romanus
 see Roomen, Adriaan van
Romberg, Moritz Heinrich, **6**:60a, 440a
Rome, Adolphe, **9**:298b; **10**:294a
Rome (Italy)
 archaeology, **2**:570b
 architecture, **15**:515b
 Collegio Romano
  observatory, **3**:13b
 geology, **2**:481b
 Observatory of the Campidoglio, **3**:14a
 Policlinico Umberto I, **1**:360a
 St. Peter's basilica, **11**:65b
 Santo Spirito Hospital
  library, **7**:614b
 Technology Office, **5**:407a
Rome, University of
 chemistry at, **9**:527a
 physics at, **4**:578a–580b

Romé de l'Isle (or Delisle), Jean-Baptiste Louis (1736–1790), 11:520–524; 2:431a; 6:181a; 10:496b; 12:65b; 14:258a
crystallography, 6:179a, 181b
Essai de cristallographie, 3:61b
mineralogy, 6:182a; 11:167b
Romer, Eugeniusz Mikołaj (1871–1954), 11:524–525
Rómer, István, 7:22b
Römer, Ole Christensen (or Roemer, Olaus) (1644–1710), 11:525–527; 1:481a; 2:593a; 6:513a
astronomy, 9:90a,b; 10:596a; 15:325a
speed of light, 3:101b, 103b; 12:448b
ballistics, 2:202a
instruments, 13:99b
observations of, 5:257a; 6:512b, 513b
thermometry, 4:517a–b
Römer
see also Roemer
Romieu, Jean-Baptiste, 6:245a
Romme, Gilbert, 1:206b; 15:335b
Ronchi, Vasco, 13:437b
Ronchitti, Cecco de (pseudonym), 3:60a
Rondeau, Pierre, 11:427b
Rondelet, Guillaume (1507–1566), 11:527–528
acquaintances of, 1:108b, 596a; 3:342a; 8:435b; 10:152a
ichthyology, 9:573a
influence of, 13:451a; 14:413b
De piscibus marinis, 8:120b
students of, 1:525b; 3:533b; 11:311b
Rondoni, Pietro, 8:557a
Ronsard, Pierre de, 10:116b, 152b, 494a; 12:135a
Röntgen (Roentgen), Wilhelm Conrad (1845–1923), 11:529–531; 3:479b; 6:350a*; 12:527b; 14:618a
acquaintances of, 14:170b; 15:251b
roentgen current, 4:306a; 11:530a–b
spectroscopy, 7:526a
students of, 4:100b; 7:468b, 553a; 15:251a
X rays, 1:558b; 3:498b; 8:181a; 13:366a
Rood, Ogden Nicholas (1831–1902), 11:531–532; 12:37a
Roofing, 5:63b
Rooke, Lawrence (1622–1662), 11:532; 3:482a; 14:369a
Roomen, Adriaan van (1561–1615), 11:532–534; 2:526a; 12:499b
Ideae mathematicae, 3:181b; 4:8b
mathematics, 7:258b; 14:22a–b
Roosevelt, Cornelius, 2:439a
Roosevelt, Franklin Delano, 2:374b; 3:373a; 4:313a; 9:399a; 13:227a
Roosevelt Hospital (N.Y.C.), 6:77b
Roots (bot.), 3:581a–b; 5:535b; 12:167b, 259a; 13:405b
adventitious, 9:78b
corrosive action of, 12:59a
Roots (math.)
medieval discussion, 4:607b–608a, 609a; 7:257a; 13:510a
Renaissance discussion, 3:274a; 11:590a; 13:59b, 60a
bounds on, 2:574a
characteristic, 2:197b
complex, 3:136b; 5:97a; 8:385a
cube, 1:227a; 2:618b; 3:273b; 4:607b–608a, 609a; 5:408b–409a; 8:420b
of binomials, 5:409b
graphic extraction of, 3:277b
method for, 6:315a, 511a; 7:325b–326a, 533a; 8:234b, 238a–b; 9:614a–b; 10:540a; 12:63a, 610b; 13:540a, 545a,b
tables of, 15:668a, 670a
expansion of, 8:139a
imaginary, 4:58a, 475a; 5:98b, 409b; 7:567a; 8:385a; 9:610a; 10:51b
negative, 4:58a; 14:544a
notation of, 8:41a; 13:60a; 14:21a
primitive, 7:54b; 11:61b
real, 2:573a–574a; 5:98b
number of, 3:136b; 5:98a
separation of, 8:139a
square, 1:222a; 2:618b; 3:133a, 273b; 4:491a, 607b, 609a; 5:298b–299a; 8:198a; 11:533b; 13:315a; 15:693a
classification of, 13:302a
method for, 1:437b; 3:126a–128a, 268a; 6:314b, 511a; 7:118a, 129a, 258a, 533a, 559a*; 8:234b, 316a, 420b–421a; 9:614a–b; 11:229b, 230a; 12:63a; 13:323b, 540a, 544b, 545b; 15:690b–691a, 691b, 692a–b
of polynomials, 7:241b; 12:92b
tables of, 11:457b; 15:668a
values for, 5:366a, 409a; 10:238b
Rooymans, C. J. M., 13:611a
Roozeboom, Hendrik Willem Bakhuis (1854–1907), 11:534–535; 7:59a; 12:495b; 13:243b
physical chemistry, 8:119a; 13:244b
Ropes
stretching of, 3:183a
Rore, Cipriano da, 1:606b; 5:249b
Rosa, Daniele (1857–1944), 11:535
Rosa, Edward Bennett, 1:325b
Rosaceae, 1:412b; 8:371b; 9:253b
Rosanes, Jakob (1842–1922), 11:535–536; 10:343a; 15:42a
Rosaniline, 3:85a; 5:2a; 6:463a; 7:283a; 8:329b
Rosati, Giovanni, 1:611b
Rosati (society), 3:71a
Roscelinus
school of, 1:1a–b
Roscoe, Henry Enfield (1833–1915), 11:536–539; 11:37a*
acquaintances of, 4:127b; 6:111b; 7:380a; 9:347b; 12:208b–209a
actinometer, 4:182b–183a
chemistry, 7:462a; 9:290b; 12:209a*; 15:84b*
photochemical law, 1:21b; 2:589a; 12:248b
students of, 2:428a; 6:110b; 12:238a, 493b; 13:362a, 389a; 14:560b
Roscoff (France)
marine zoological laboratory, 4:12a,b; 7:546a
Rose, Gustav (1798–1873), 11:539–540; 3:70a; 9:426a*; 11:541a
acquaintances of, 2:91b, 556b, 586b; 4:289b; 6:552a; 9:18b; 11:540b; 14:240a, 241a; 15:526a
mineralogy, 15:526b
students of, 7:498b; 8:511b; 11:270b, 560b
Rose, Heinrich (1795–1864), 11:540–542; 3:70b; 5:393b; 6:166b; 14:476a
acquaintances of, 2:91b, 586b; 9:18b; 11:34a
chemistry, 5:164a; 14:529b
students of, 7:619b; 9:445a; 11:270b, 560b; 12:236a; 14:315b, 329b; 15:457a
Rose, Joseph Nelson, 10:285b
Rose, Valentin (the elder), 7:394a; 9:106b; 15:205b
Rose's metal, 11:539a
Rose, Valentin (the younger), 5:202a; 7:394b; 9:106b; 14:239b; 15:171b, 206a
Rose (flower), 12:175a
medicinal use of, 9:466b*
Rosellus of Arezzo, 9:435a
Rosen, Nathan, 8:474b
Rosenberg, Hans Oswald (1879–1940), 11:542–543
Rosenberger, Johann Karl Ferdinand (1845–1899), 11:543–544
Rosenberger, Otto August (1800–1890), 11:544–545; 7:108b
Rosenblueth, Arturo (1900–1970), 11:545–547; 5:65b; 15:74a–b, 76b
Rosenblum, Salomon, 3:502a
Rosenbusch, Harry (Karl Heinrich Ferdinand) (1836–1914), 11:547–548; 5:456a
crystallization series, 4:141b
influence of, 2:289b; 6:475b; 13:273a
petrography, 3:161a; 8:42a
students of, 1:51a; 4:597b; 5:455a; 7:5a, 62b; 11:551a; 12:83b, 140a, 271b; 14:515a; 15:427a
Rosenfeld, Léon, 2:250b; 15:43a
Rosenhain, Johann Georg (1816–1887), 11:548; 7:51a, 54b
analysis, 5:471b; 6:308b; 11:451a
Rosenhain, Walter (1875–1934), 11:548–550
metals, 13:246a
Rosenheim, Arthur (1865–1942), 11:550–552; 4:485a; 14:445a
Rosenheim, (Sigmund) Otto, 14:335a
Rosenhof, August Johann Roesel von
see Roesel von Rosenhof, August Johann
Rosenmüller, Johann Christian, 4:288a; 9:483b
Handbuch der Anatomie, 14:200a
Rosenthal, Artur, 2:265a
Rosenthal, David, 11:216b
Rosetta Stone, 7:615a; 14:569b; 15:681a
Rosicrucianism, 1:159b–160a; 2:542b; 3:579a; 4:550b; 6:145b, 306a; 9:23b; 12:307b; 14:209b, 226a, 367b–368b
opposition to, 5:47a; 8:310b; 14:370a
Röslin, Eucharius, 8:483b, 484a
Rosolic acid, 5:489a
Ross, Alexander
Commentum de terrae motu circulari, 14:365a
Ross, Andrew, 1:135b; 8:414b
Ross, Frank Elmore (1874–1960), 11:552–554
Ross, James Clark (1800–1862), 11:554–555; 7:70b
Antarctic expedition, 3:557b; 4:291a; 6:327a, 489a
Arctic expedition, 6:263a; 14:491b
geomagnetism, 12:49a, 50b, 52a
nebulae, 12:266b
Ross, John, 11:554a,b; 12:49a
Ross, Ronald (1857–1932), 11:555–557; 6:12b; 9:83b*
malaria, 5:503b–504a; 7:428a; 8:66a; 9:82b; 12:482a
Ross, William Horace, 12:504b
Ross Institute of Tropical Hygiene, 11:556b
Rossby, Carl-Gustaf Arvid (1898–1957), 11:557–559; 2:168b; 12:366b
Rossby diagram, 11:558a
Rossby equation, 11:558a
Rossby waves, 11:558a
Rosse, William
see Parsons, William, Third Earl of Rosse
Rosser, Barkley, 5:375b
Rossetti, Francesco (1833–1885), 11:559
Rossignol, Hippolyte, 10:395a,b
Rössler, Adolf, 1:502b
Rostan, Léon Louis (1790–1866), 11:559–560

Rota, Francesco, **9**:11b, 12a
Rotation, **2**:606b; **7**:478a
of asymmetric body, **7**:479a
axes of, **12**:283b
center of, **2**:38b
instantaneous, **2**:38b–39a
of liquids, **1**:598b; **7**:54b–55a; **8**:384b
mechanics of
eighteenth-century, **2**:38b–39a; **7**:564b; **8**:135a; **12**:283b; **15**:348a–b
nineteenth-century, **5**:111a; **11**:62a; **12**:534b
twentieth-century, **7**:453b, 492b
physiological effects of, **2**:515b
of superfluids, **8**:481a–482a
of viscous fluids, **9**:107b
see also Motion, rotational
Rotatoria, **4**:290a
Roth, Justus Ludwig Adolph (1818–1892), 11:560–561; 8:511b
Roth, Klaus Friedrich, **13**:391a
Rothamsted Experimental Station, **5**:7b; **8**: 92b–93b; **15**:492b–493a
Rothe, Heinrich August, **2**:158a
Rothe, Hermann, **5**:122b
Rothe, Peter, **4**:55b; **5**:409b
Rothman, Johan, **8**:374b, 376b
Rothmann, Christoph (*d. ca.* 1608), 11:561–562; 2:406b
astronomy, **14**:359a
Rotifera, **6**:167a; **8**:127b; **9**:112b; **12**:557b
classification of, **7**:98b
morphogenesis, **7**:98a
reproduction, **9**:185a
sexuality, **9**:185a–b
Rotliegende (stratig. stage), **5**:445a
Rotogravure, **4**:283a*
Rotors (math.), **2**:594a
Rotterdam
Collegium Mechanicum, **1**:566b
medical school, **9**:557b–558a
Rouelle, Guillaume-François (1703–1770), 11:562–564; 2:174a; **13**:565b
chemistry, **3**:561a*; **8**:76a, 619a
geology, **4**:71a; **8**:68b
influence of, **3**:419a–b; **4**:70a, 88a; **13**: 603a
students of, **1**:58b, 529a; **3**:5a, 560b; **5**: 355a; **6**:578a; **8**:68a, 69a, 72b, 618b; **9**:53b, 478a; **10**:325b; **12**:63b; **13**: 494a, 602a
Rouelle, Hilaire-Marin (1718–1779), 11: 564; 4:346a; **8**:619b; **10**:496a; **11**: 166a; **12**:64b
chemistry, **5**:92b–93a; **8**:74b
pharmacy, **11**:562a
Rouelle, Jean (*b. ca.* 1751), 11:564–565
Rouen
Academy of Sciences, **10**:614b
water supply, **8**:71a
Rouffignac (cave), **2**:450b
Rouget, Charles Marie Benjamin (1824–1904), 11:565–567
cells of Rouget, **11**:566a
Rouillier, Karl Frantsovich (1814–1858), 11:567–568
Round, Henry Joseph, **9**:99a
Roundworm, **13**:527b
Rous, Francis Peyton, **5**:40a; **9**:586a; **12**: 483b
Rousseau, Jean Jacques, **1**:115b; **3**:462a*; **4**:27b; **7**:611a; **13**:565b
acquaintances of, **3**:380a,b; **4**:85a,b; **8**:68a; **9**:67a; **11**:563a; **13**:390a
critics of, **3**:119b; **8**:115a
Émile, **1**:139a; **3**:485a; **8**:81b; **12**:120b; **15**:272a
influence of, **1**:139b; **2**:321b; **3**:172b, 284b; **9**:7a; **10**:611b; **15**:82b
Roussel de Vauzème, **2**:443a
Roussy, Gustave, **9**:108b
Routh, Edward John (1831–1907), 11:568–569; 4:35b; **9**:199a; **13**:363b, 427b*
students of, **2**:397b; **10**:447a,b; **13**:101a, 362b
Treatise on the Dynamics of a System of Rigid Bodies, **9**:211a
Roux, Augustin, **2**:572a; **3**:560b; **11**:616b
Roux, Philibert Joseph, **5**:565b
Roux, Pierre Paul Émile (1853–1933), 11: 569; 8:274a*; **10**:415b*
acquaintances of, **1**:576b; **3**:501b; **8**:66a; **10**:385b; **14**:36b, 551a–b
anthrax, **3**:588a; **10**:392b–397a *passim*
critics of, **11**:597a
diphtheria, **1**:576a; **8**:449b
on Duclaux, **4**:212a
influence of, **10**:355b; **15**:97b
at Institut Pasteur, **4**:211b; **5**:99b; **6**:11b, 298a
rabies, **8**:313a; **10**:401b, 403b, 405a; **14**: 551a–b
septicemia, **10**:407b
students of, **2**:584a; **3**:22a, 203b; **12**: 315a,b; **15**:454a
syphilis, **4**:300a; **9**:334b; **10**:18b
tuberculosis, **7**:426b
Roux, Wilhelm (1850–1924), 11:570–575; 3:114a*; **6**:224a; **7**:402a
critics of, **14**:428a–b
cytology, **13**:89a; **14**:235a
embryology, **4**:187a,b; **5**:447a; **6**:132b, 435b; **9**:517a,b; **14**:427b–428b, 429b
Entwicklungsmechanik, **8**:445b
heredity, **8**:587a
influence of, **14**:236b
Rovereto, Gaetano (1870–1952), 11:575–576
Rovida, Giulio C., **8**:557a
Rowe, Allan Winter (1879–1934), 11:576–577; 11:418a
Rowland, Henry Augustus (1848–1901), 11: 577–579; 1:133a, 166b; **2**:424a; **6**: 51a,b; **14**:497a
electromagnetism, **4**:306a; **6**:252a; **7**:56a; **11**:530a
spectroscopy, **12**:21b, 73a; **14**:55b
Rowland grating
see Diffraction grating
Rowning, John (*ca.* 1701–1771), 11:579–580
Compendious System of Natural Philosophy, **11**:140a, 142b–143a
Roy, Charles Smart, **15**:100a
Roy, William, **11**:285a
Royal Academy of Brussels
see Académie Royale des sciences . . . de Belgique, Brussels
Royal Agricultural Society of England (London), **3**:163a
Royal Astronomical Society
see Astronomical Society of London
Royal Astronomical Society of New Zealand
founding of, **12**:532b
Royal Botanic Garden (Edinburgh)
in nineteenth century, **1**:423a
in twentieth century, **1**:422b
Royal Botanic Gardens (Kew)
in eighteenth century, **1**:88a–89b, 435b; **3**:172b; **12**:516b
in nineteenth century, **1**:614b, 615a; **2**: 477a; **6**:491a–b; **8**:371b; **13**:342a–343a
Britten and, **2**:476a,b
establishment of, **6**:494a–b
herbarium, **1**:412b; **6**:491a
Jodrell Laboratory, **6**:491a
Palm House, **6**:494a
Royal Botanical Expedition to New Spain, **9**:432b, 433a
Royal College of Chemistry (London), **6**: 461b
Royal College of Physicians
see London, College of Physicians
Royal College of Science (Dublin), **7**:224a
Royal College of Surgeons
see London, College of Surgeons
Royal Danish Academy of Sciences
competitions of, **9**:504b
Royal Dublin Society, **6**:385b; **7**:389a
radium institute, **7**:161b
Royal Geographical Society, **9**:123b
founding of, **5**:519a; **9**:585a
Royal Horticultural Society (England), **1**: 615a; **13**:341b
founding of, **1**:89a
Royal Institution of Great Britain, London, **2**:399b; **3**:601a, 604b; **4**:79a–b; **13**: 521b, 522b; **14**:562b
and Accum, **1**:43b, 44a
chemistry, **6**:384a
Davy-Faraday Laboratory, **9**:467b
Faraday and, **4**:529a, 532a, 534b, 539b
founding of, **1**:435a; **5**:276a,b; **13**:350b, 351b–352a
journal of, **3**:556b
Royal Institution of the Netherlands, **13**: 184a
Royal Irish Academy, **7**:388a
Royal jelly, **10**:277a
Royal Medical Society (Edinburgh)
founding of, **3**:495a
Royal Meteorological Society
founding of, **5**:413b
Royal Microscopical Society
standard thread, **9**:141b
Royal Observatory, Greenwich
see Greenwich, Royal Observatory
Royal Prussian Institute for Experimental Therapy
see Ehrlich (Paul) Institute
Royal School of Mines (G.B.), **11**:485b–486a
Royal Scottish Geographical Society
founding of, **5**:339a
Royal Scottish Society of Arts
founding of, **2**:451b
Royal Society of Arts (London), **3**:350b
founding of, **6**:44b
Royal Society of Canada
founding of, **6**:564b
Royal Society of Edinburgh, **6**:579b, 585b; **9**:480b; **14**:132a
founding of, **3**:495a; **6**:579a; **11**:496a
Royal Society of London
in seventeenth century, **1**:376a; **2**:349b, 507a; **3**:362b; **4**:96a, 495a; **5**:427a, 484b, 534b; **6**:68b, 69b; **8**:130b, 524a–b; **9**:63a, 509a,b; **10**:200a, 201a–202b; **11**:121b; **13**:528a; **14**:209b–210a, 451b; **15**:29b, 744b, 748b
in early eighteenth century, **2**:325a; **3**: 458a; **4**:43b–44a, 518a; **5**:53b–54a, 434a, 435b, 516a–b; **6**:6a, 36b–37a, 176a; **10**:84a,b
in late eighteenth century, **1**:435a; **2**:186a; **3**:349b–350a; **4**:132a, 149a; **5**:76b, 468b; **6**:577a; **13**:39b, 350b; **14**:453a
in early nineteenth century, **2**:483a, 232b; **3**:545a, 603b; **5**:518b; **6**:292a–b, 293a; **7**:262b; **14**:486a–b
in mid nineteenth century, **6**:325b; **14**: 185a
in late nineteenth century, **3**:475b; **5**: 83a–b; **6**:488b; **9**:512b; **12**:544b

Royal Society of London (cont.)
in twentieth century, **3**:480a; **5**:335a; **13**:370a; **15**:201b
astronomy, **10**:584b; **12**:414a
Bakerian Lectures, **1**:411b; **14**:508b
blood circulation experiments, **4**:37b
Catalogue of Scientific Papers, **6**:279b–280a; **9**:467b
Commercium epistolicum, **2**:54a; **10**:49b, 51a–b, 68a, 84b
criticism of, **1**:354b; **5**:415a, 416a–b; **6**:401a; **12**:552a
Donation Fund, **14**:486b
electricity, **14**:419a
epidemiology, **2**:528b
Evolution Committee, **1**:505b; **14**:251b
founding of, **1**:317b; **2**:377b; **3**:482a–b; **4**:377a, 495b; **5**:606b–607a; **6**:483a; **9**:313a, 507a–b; **10**:200a, 200b–201a, 565a–b; **12**:582a; **14**:151b, 370b–371b, 413a, 510b
geomagnetism, **12**:49b–52b *passim*
and Hauksbee, **6**:169a–173a *passim*
historiography of, **12**:580b–586b
and Hooke, **6**:483a–b, 487a
language, **14**:362a, 373b
and Leeuwenhoek, **8**:127b, 129a
and Leibniz, **8**:165a; **9**:452b
library of, **3**:348a; **4**:496a
mathematics, **7**:163a
medicine, **11**:2b
meteorology, **3**:158a; **5**:17b; **11**:174a
microscopy, **1**:411a
motto, **14**:363b
museum, **5**:535a
and Pearson, **10**:463a
Philosophical Transactions, **5**:54a; **6**:38b, 68b, 577a; **10**:201b–202a; **12**:85a; **13**:425b
editors of, **11**:40b
physiology, **8**:524b–526a
Population Study Group, **5**:43a
precursors of, **3**:482a; **4**:96b, 495b; **5**:606b–607a; **6**:482a, 483a; **10**:200b–201a, 564b, 565a–b; **11**:532a; **12**:583b–584a; 585a–b; **14**:151a–b, 364a, 367b–368a, 368b, 369a, 371a
Proceedings, **7**:84a
reform movement, **5**:560b; **12**:52a
Water Research Committee, **5**:128b
weights and measures, **5**:491b
Royal Society of New South Wales
founding of, **3**:298a
Royal Society of Sciences (Madrid), **7**:183b
Royal Society of Tropical Medicine
founding of, **9**:83a
Royce, Josiah, **3**:335a; **4**:384a; **6**:261a; **12**:609b
Royds, Thomas, **5**:331a; **12**:30a,b
Royer, Louis, **5**:186b
Rozental, Iosif Sergeevich, **10**:434b
Rozental, Isidor, **4**:201a,b
Rozet, Claude, **2**:341b
Rozhdestvensky, Dmitry Sergeevich (1876–1940), 11:580–581; 7:554b
Rozier, Abbé François, **2**:321b; **5**:411a; **8**:72a; **10**:326a
Ṛtus (seasons)
definitions of, **10**:426a; **15**:535a, 538b
Ruangsiri, Chanai, **4**:155a
Rubber, **10**:436a; **13**:342a,b; **14**:349b
chemistry of, **9**:376a
crystal, **14**:182b
hydrogenation of, **13**:2a
natural, **8**:621a
discovery of, **15**:271a
production of, **4**:362a; **5**:604b
reclaiming, **8**:113a
structure of, **13**:1b–2a
synthesis of, **1**:415a; **2**:324a, 624b; **3**:86a,b; **4**:553a; **6**:118b; **7**:21a,b, 323b, 457a–b; **8**:108a–b; **10**:122a; **13**:1a–b, 66b, 163a; **14**:247b
vulcanization of, **6**:517a
see also Latex
Rubber tree, **3**:177a*; **13**:459b
Rubens, Heinrich (Henri Leopold) (1865–1922), 11:581–585; 15:448b
acquaintances of, **1**:519a; **5**:117a; **6**:16a, 28b; **7**:528b
correspondence, **6**:350a*
critics of, **11**:150b
radiation, **7**:528a; **8**:551b–552a; **10**:346a; **11**:12a, 13a
spectroscopy, **10**:345b
students of, **11**:347b
Rubens, Otto, **15**:451a
Rubens, Peter Paul, **1**:81a,b; **8**:204b; **13**:508a
Rubiadin, **9**:95b
Rubian, **12**:236b
Rubidium
discovery of, **2**:589b; **3**:476a; **7**:381b
in plant nutrition, **11**:416a
radioactivity of, **6**:16a
in solar spectrum, **12**:20b
spectrum, **11**:580b
Rubin, Edgar, **9**:562b
Rubin, Leon, **5**:6a
Rubner, Max (1854–1932), 11:585–586; 1:325b; **4**:373a*; **5**:196a; **10**:563a*; **14**:66a
physiology, **2**:561b; **8**:29b; **11**:428a
Rubus, **1**:396a,b; **9**:403b; **10**:13b; **15**:268a,b
Ruby, **3**:232b; **13**:511b
heat, effects of, **3**:560b
synthesis of, **5**:157b, 295a
Rücker, Arthur William, **13**:517b
Rudall, Kenneth Maclaurin, **6**:133b
Rudbeck, Olof (1630–1702), 11:586–588; **6**:380a; **8**:375b; **10**:132a
lymphatics, **1**:482b; **10**:477b
Rudbeck, Olof (the younger), **1**:305b; **8**:374b, 375a; **11**:587b, 588a
Rüdenberg, Reinhold (1883–1961), 11:588–589; **2**:318a
Rudio, Ferdinand (1856–1929), 11:589
Rudolf II, Holy Roman emperor, **2**:293a, 404a, 412b, 541b, 602a; **3**:471b; **4**:184a, 393b; **5**:588a; **6**:254a; **7**:297a,b, 299b–300a; **8**:309a,b, 310a; **9**:23a,b; **11**:532b, 606a,b; **12**:306b, 307a; **13**:613a,b
Rudolf of Bruges
De astrolabia, **1**:23a
Rudolff (or Rudolf), Christoff (*fl.* first half of sixteenth century), 11:589–592; 11:457a,b
Coss, **4**:468a; **13**:59b, 60b
editions of, **12**:205a; **13**:59a
Rudolfinerhaus (Vienna)
founding of, **2**:130a
Rudolphi, Karl Asmund (1771–1832), 11:592–593; **4**:289b, 290a; **9**:568a,b, 573b; **11**:214a
students of, **2**:422b; **4**:288b; **8**:519a; **9**:344b; **12**:354a, 420b
Rudolphine Tables
see Kepler, Johann, *Tabulae Rudolphinae*
Ruedemann, Rudolf (1864–1956), 11:593–594; 13:534a*
Ruel (Ruellius), Jean (1474–1537), 11:594–595; **4**:121b; **14**:507b; **15**:162b*
Ruete, Christian Georg Theodor, **4**:162b
Ruffer, Marc Armand (1859–1917), 11:595–596; **8**:409a
Ruffin, Edmund, **4**:182a
Ruffini, Angelo (1864–1929), 11:596–598
Ruffini's corpuscle, **11**:596b
Ruffini, Paolo (1765–1822), 11:598–600; 2:105a
Abel-Ruffini theorem, **1**:15a; **11**:598b
on algebraic equations, **1**:591b; **3**:254b; **5**:263a; **15**:226a
editions of works, **2**:320a
Ruffini-Horner method
see Horner, William George, Horner's method
Sopra la determinazione delle radici . . . , **6**:511a
theorem of, **3**:132a, 142b
Ruffo, Giordano (*fl.* middle of thirteenth century), 11:600–601
Rufigallic acid, **11**:495a
Rufinus (*fl.* second half of thirteenth century), 11:601
Rufinus, Lucius Cuspius, **5**:227b–228a
Rufus of Ephesus (*fl. ca.* first century A.D.), 11:601–603; **6**:317a,b, 318a; **11**:444b; **12**:540b, 541b(n); **15**:244a(n)
Rugan, Henry Fisher, **3**:87a
Ruhe, Jacob, **9**:188a
Ruhland, Wilhelm, **2**:561b
Rühlmann, Richard, **14**:330a
Rühmkorff, Heinrich Daniel (1803–1877), 11:603–604; **5**:340b, 341a
induction coils, **2**:439a; **3**:18a; **5**:87a; **9**:98b
Ruini, Carlo (*ca.* 1530–1598), 11:604–605
Ruiz, Hipólito (1754–1816), 11:605–606; **3**:154b; **4**:156b
and Pavón, **15**:470a–471a
Rukh, Shāh, **7**:255b
Ruland, Martin (1569–1611), 11:606–607; **7**:150b; **8**:309a, 310b; **12**:307a
Rule of three, **3**:273b; **4**:611a; **9**:22b; **11**:457b, 590a
Rülein, Ulrich (Ulrich Rülein von Calw) (*ca.* 1465–1523), 11:607–609
Rulie, Karl Frantsovich
see Rouillier, Karl Frantsovich
Ruling engine, **11**:578b; **12**:37a
Rum
manufacture of, **6**:383a–b
Rumford
see Thompson, Benjamin
Rumford Chemical Company, **6**:517b
al-Rūmī
see Qāḍī Zāda al-Rūmī
Ruminants
anatomy, **10**:568a
dentition, **5**:469b
Rümker, Georg Friedrich Wilhelm, **8**:539a
Rümker, Karl Ludwig Christian, **2**:472a; **11**:82b, 114a, 115a; **12**:192a
Rumovsky, Stepan Yakovlevich (1734–1812), 11:609–610; **4**:471a; **8**:252a
Rumpf, George Eberhard, **8**:394b
Rumpf, Theodor, **14**:278b
Rune(s), **14**:505b
Runge, Carl David Tolmé (1856–1927), 11:610–615; **6**:253a*; **10**:347a; **11**:481a(n); **14**:63a*; **15**:40a
acquaintances of, **2**:191b; **11**:8a, 476b; **12**:249b
critics of, **11**:479a
and Kayser, **7**:267b–268a
mathematics, **4**:562a
Runge formula, **11**:480b(n); **12**:43a
Runge-Kutta procedure, **11**:613a
Runge's rule, **11**:612b
spectroscopy, **1**:425b; **10**:346a,b, 347b; **11**:479a

Runge, Friedlieb Ferdinand (1794–1867), 11:615–616; 2:586b; 3:159b; 11:50a
chemistry, 5:197b; 11:494b
Runkle, John Daniel, 4:592a; 6:399a
Runs, theory of, 2:319a
Rupelian epoch, 2:108b
Rupp, Henrietta Maria, 11:422b
Ruppius, Heinrich Bernhard, 13:450b
Ruprecht, Anton von, 9:559a
Ruscelli, Girolamo (or Alexis of Piedmont [?]) (*d. ca.* 1565), 15:491–492
Rusconi, Mauro, 3:424b
Rush, Benjamin (1746–1813), 11:616–618; 2:174a; 6:521a; 8:612b; 11:148b*, 263a; 13:215a; 14:456a
critics of, 13:46b
Ibn Rushd, Abu'l-Walīd Muḥammad ibn Aḥmad ibn Muḥammad (Averroës) (1126–1198), 12:1–9
acquaintances of, 13:488b; 14:637b
Aristotelian commentaries, 1:198a, 267b, 272b, 273b, 276a, 409a; 2:394a, 590a; 3:470a
translations of, 9:362a, 362b–363a
astronomy, 4:442a; 15:33a,b, 34b
chemical theory, 12:311b
commentaries on, 2:609b; 10:122b
critics of, 1:279b; 5:105a
influence of, 3:218b; 5:403a; 8:280b; 9:364a; 10:555b; 12:429a; 14:388a, 457b, 458b; 15:45b
influences on, 4:524a; 14:593b
al-Kulliyyāt, 14:637b
mathematics, 7:83a
medicine, 1:290b; 11:251b; 15:498b
optics, 10:475a
physics, 1:100b, 410b; 6:378a
translations of works, 13:400b, 401b
Rusinov, L. I., 7:527a
Ruska, Ernst, 2:318a
Ruska, Heinrich, 13:3b
Russ, Sidney, 6:16a; 9:550a
Russell, Alexander Smith, 9:543a; 12:30b, 32a, 507a(n)
Russell, Bertrand Arthur William (1872–1970), 12:9–17; 9:502b; 14:469b*
acquaintances of, 3:90a; 9:383b; 10:442b
antinomy of, 2:593b; 3:57b, 278a; 5:154b; 12:13a
causation, 7:234a
on contemporary scientists, 7:317a–b; 10:606a,b
critics of, 2:512b; 3:111a; 11:60b, 61a
epistemology, 11:356a
Essay on the Foundations of Geometry, 12:609b
and Frege, 5:152b, 153a
historiography, 4:4b; 8:156b
Human Knowledge, Its Scope and Limits, 11:357b
influences on, 10:444a
logic, 5:350b; 8:154b, 159a,b, 160a, 262b, 473a, 590b
pacifism, 4:313a; 6:114b*
Principia mathematica, 5:153b; 11:107a, 285b; 14:303b–304b
editions of, 3:456a
Principles of Mathematics, 14:304a
students of, 14:344a, 468a
Russell, Edward John (1872–1965), 15:492–493
Russell, Elias Harlow, 6:53a
Russell, Harry Luman, 1:357a
Russell, Henry Chamberlaine (1836–1907), 15:493–494
Russell, Henry Norris (1877–1957), 12:17–24; 4:218b, 219b*; 12:349b; 13:96a*; 14:558a
Adams-Russell effect, 12:21b
astronomy, 2:196b; 4:219a; 8:442b; 10:290a
Hertzsprung-Russell diagram, 4:280a; 5:272b; 6:351a; 8:2a; 10:317b; 12:18a, 19a, 22b
influence of, 3:367a; 13:117b
and Shapley, 12:346a, 348a–b
spectroscopy, 9:256b*; 12:71a
Vogt-Russell theorem, 12:22b
Russell, Israel Cook, 3:594a; 7:148b, 149a
Russell, John, sixth duke of Bedford, 6:494a
Russell, John Scott, 4:167b; 5:200a; 6:503a; 15:200a
Russell Sage Institute of Pathology, 8:555b
Russia
agriculture, 7:32b–33a; 15:505a–511b
anthropology, 1:173a, 174b, 387b–388a; 3:198a
astronomy, 7:6a–b; 10:158b–159a; 13:138b–139a
biological sciences
botany, 9:45a; 10:165a; 13:248b–249a, 450a
ecology, 10:17a
forestry, 9:535a–b; 14:107a
genetics, 9:565a
microbiology, 6:298b
physiology, 12:270a
zoology, 7:32b–33a; 10:283b–284a
celestial mechanics, 9:449a
census, 12:301a
Central Aerohydrodynamic Institute, 3:196a
chemistry, 1:361a–363a; 2:623b; 6:354b; 7:203b; 8:469b; 9:287b, 305a; 14:125a
petrochemistry, 9:608a–b
education, 5:210a; 7:320a; 9:324a; 10:248a, 428b, 620b; 14:125a
ethnography, 1:175a, 387b
exploration of, 10:283a–b; 11:180b–182a; 12:300b; 14:50a
geodesy, 7:497a; 10:231b–232a; 13:412b–413a
geography, 1:173a–174b, 388a; 4:23a; 5:75a, 76b, 428a–429a; 7:510a–511b, 512b–513a; 8:251a–b; 12:300a–301a; 13:249a
cartography, 7:265b; 9:231b
paleogeography, 7:250a–253a
Geological Committee, 6:238a
Geological Survey, 10:128a
geology, 1:161b–163b, 388a; 2:555b; 3:236a–237a, 529b–530a; 4:143a–144b, 308b–309b, 401a, 597b–599a; 5:212b; 6:238a–239a; 7:249a–253b, 319b–320a; 8:147b, 285b, 470b; 9:592a–594a; 10:265b, 428a–430a; 12:514a–b; 15:427a–428a
glaciology, 9:593b; 15:427b
gravimetry, 10:159a–b; 13:45b
hydrology, 1:162a; 7:410b
meteorology, 14:52a–53b
mineralogy, 5:212b; 12:329a–b
oceanography, 4:42a–43a; 9:331a–b; 12:411b; 14:634a–635b
paleontology, 4:308b–309b; 10:222a
seismology, 9:593a, 594a
stratigraphy, 9:584a,b; 10:428b; 12:514b; 13:620b–621a
history, 7:6a; 8:470b, 471a
intellectual history, 7:482b, 485a,b
logic, 11:94b–95a
mathematics, 4:474a; 8:558a; 9:17a; 11:327a; 13:549a
medicine, 5:269a
epidemiology, 5:270a–271a
medieval visitors to, 1:516a, 517a
natural history, 1:29b, 387b–388a; 4:144a, 289b, 308a–b; 7:475b, 495a–b
Neurosurgical Institute, 2:597b
philosophy, 14:636a
physics, 5:188b; 7:617a; 8:102b, 107b, 469a; 11:580b, 581a
atomic energy, 7:527b
ballistics, 10:250b
politics, 15:75a–b
sociology, 4:401a
Soviet science, 3:235a
technology, 8:184b; 12:425a; 14:106a
aviation, 14:620a–b, 621a–b
civil engineering, 3:287a
industry, 7:253b; 9:291b, 292b
metallurgy, 3:271a
railroads, 6:131b
space, 7:465b
weights and measures, 9:292b
Russian Academy of Sciences
see Akademiia nauk USSR
Russian Astronomical Society, 7:472b
competitions of, 13:114a
Russian Chemical Society
founding of, 9:287b
Russian Entomological Society
see Vsesoiuznoe entomologicheskoe obshchestvo
Russian Geographical Society
see Geograficheskoe obshchestvo USSR
Russian language
orthography, 12:300b
reform of, 8:470b
in science, 8:471a; 15:65b, 66b
Russian Physics and Chemistry Society
see Russkoe fizicheskoe i khimicheskoe obshchestvo
Russian Physiological Society, 10:435a
Russian revolution, 1:361a–b
Russian Scientific and Technical Institute of Nutrition, 3:271a
Russkoe fizicheskoe i khimicheskoe obshchestvo, 2:620a; 9:305a
Russo-Japanese War, 4:6b
Russula, 9:35a
Rust
prevention of, 6:381a
Rust fungi, 2:583b; 3:612a,b, 613a–b; 5:56b
sexuality in, 2:583a; 6:121b
Rutgers Medical College, 6:521b
Ruthenium, 2:589b
compounds of, 6:525b
discovery of, 2:620b; 3:301a
Rutherford, Daniel (1749–1819), 12:24–25; 2:174a, 177a; 9:484a
translations of works, 8:85b(n)
Rutherford, Ernest (1871–1937), 12:25–36; 3:503a*; 6:118a; 13:370a; 15:61a
acquaintances of, 2:397b, 399a; 5:103a; 6:14b; 8:10a; 11:419b; 12:238b, 419a; 14:420b; 15:25b, 104b
alpha particles, 2:338a, 398a,b; 5:272a; 9:543b; 13:107b, 369a; 15:42b
atomic model, 2:240b–242b, 600b; 7:469a; 8:181a; 11:281a; 12:43b
and Bohr, 2:240b, 241a, 244a,b; 12:31b
and Boltwood, 2:259a
on contemporary scientists, 4:356b
correspondence of, 13:446b
electricity, 13:366a
and Geiger, 5:330b–331a
influence of, 1:490b; 3:328b–330a *passim;* 6:117a,b; 11:117a, 347b
influences on, 13:52b
ionization, 14:421a
Manchester Laboratory, 9:354a, 542b; 10:288a

Rutherford, Ernest (cont.)
radioactivity, **1**:500a; **2**:258b; **3**:480a, 499b, 500a, 507b–508a; **5**:394b; **6**:365a; **8**:575b; **11**:280a; **12**:191b, 504b; **15**:464a
thorium X, **1**:560a; **12**:504b
students of, **3**:368a, 564a; **6**:15b
Rutherford, William, **12**:344b, 345a
Rutherford Laboratory, **3**:330a
Rutherfurd, Lewis Morris (1816–1892), 12:36–37; 4:352a*; **5**:480a; **6**:541b; **9**:567b*; **11**:578b; **12**:176a
Rütimeyer, Karl Ludwig (1825–1895), 12:37–39; 6:436a; **13**:124b*
Rutty, John, **7**:389a
Ruysch, Frederik (1638–1731), 12:39–42; 6:508b, 618a; **13**:169a
anatomy, **2**:226a; **4**:590a; **9**:153a; **13**:170b
critics of, **2**:226a; **15**:29a
Dilucidatio valvularum in vasis lymphaticis et lacteis, **6**:509a
students of, **4**:352b; **6**:231a; **15**:28b
translations of works, **15**:28b
Ruzicka, Francis Charles Julian, **4**:91a
Rydberg, Johannes (Janne) Robert (1854–1919), 12:42–45
Rydberg formula, **1**:425b; **2**:243a–244a; **7**:267b, 268a; **11**:476a–b, 478b, 479a
Rydberg-Schuster law, **12**:43a, 238b
Rydberg's constant, **4**:281a; **5**:609a–b; **11**:422a; **12**:43a, 528b
spectroscopy, **11**:479a, 612a,b
Rye, **5**:523b
ergotism, **14**:93b
Marchfeld, **13**:478b
origin of, **15**:508b
Rymers, Nicolas, **5**:397b
Rymsdyck, Jan van, **6**:566b, 569b
Rythmomachia, **1**:468a; **6**:302a

S

al-Saᶜātī, Muḥammad al-Khurāsānī, **15**:254a
Sabaeans
see Ḥarrānian science
Sabatier, Armand (1834–1910), 12:45–46
Sabatier, Paul (1854–1941), 12:46–47; 5:540b; **7**:21b; **11**:283b*
Sabatier, Raphael Bienvenu, **1**:333a; **2**:64a; **5**:251b
al-Ṣabbāḥ, Muḥammad ibn, **9**:84a–b
Sabbia, Emilio Francesco, **3**:118a; **9**:268a
Ṣabians
see Ḥarrānian science
Sabin, Florence Rena (1871–1953), 12:48–49; 5:28a*; **9**:56b, 58a*; **15**:75b
Sabin (acoustical unit), **12**:54b
Sabine, Edward (1788–1883), 12:49–53; 3:94a*, 215a*; **5**:293a*; **6**:277b
geodesy, **6**:325b
magnetism, **1**:364a, 521b; **8**:426b; **11**:555a
sunspots, **14**:480b
Sabine, Paul Earls (1879–1958), 12:53–54
Sabine, Wallace Clement Ware (1868–1919), 12:54; 2:458a; **12**:53a–b; **13**:473a
Saccardo, Pier' Andrea, **10**:531a; **15**:507b
Saccharic acids, **13**:577b
Saccharimeter, **3**:341b; **12**:519a
Saccharin
in food, **11**:371a; **14**:357b
Saccharinic acids, **10**:15a
Saccharomyces
life cycle, **6**:100b
spores, **5**:584b; **6**:101a
Saccheri, (Giovanni) Girolamo (1667–1733), 12:55–57; 1:600a
Euclides ab omni naevo vindicatus, **4**:417a, 451a; **6**:76b
geometry, **7**:330a; **8**:429b
Logica demonstrativa, 6:76b
Neo-Statica, **1**:541b
Saccheri quadrilateral, **6**:202a; **7**:329b, 600a; **12**:56a; **13**:510b
translations of works, **6**:76b
Sacco, Luigi (1769–1836), 12:57–58
Saccorhiza, **12**:125b
Sacculina, **4**:12a, 207a; **13**:355b
Sachariassen, Johannes, **7**:73a–b
Sachs, Carl, **4**:201a, 202b, 204b
Sachs, Hans, **4**:298b, 299b, 302a*; **5**:179a
Sachs, Julius von (1832–1897), 12:58–60
acquaintances of, **3**:581a; **5**:437b; **7**:395b; **8**:445a
on contemporary scientists, **9**:442b; **12**:175a
historiography, **1**:70b; **4**:265a; **5**:216b; **8**:378b; **9**:180a; **15**:692a
History of Botany, **2**:188b, 427a*
influences on, **6**:40a
Lehrbuch der Botanik, **2**:103a, 371a; **5**:438a, 623b; **14**:95b
translations of, **13**:341b
plant physiology, **6**:39b; **14**:97a
students of, **10**:575b; **12**:166a, 259a; **13**:87b; **14**:96a
Sackur, Otto, **5**:394b, 622a; **6**:15b; **13**:40b, 41b; **15**:444a
Sacramento River, **5**:395b
Sacrobosco, Johannes de (John of Holywood) (*d. ca.* 1256), 12:60–63; 4:605a; **15**:256a
De algorismo, **7**:176b, 362b, 418b
commentaries on, **3**:512a
astronomy, **7**:120a
commentaries on, **7**:112a; **10**:540a
influence of, **5**:22b, 239b; **10**:270b; **11**:339a; **12**:439a; **13**:62a(n)
De sphaera, **1**:84a, 94a, 382a, 468a; **2**:401b, 540b; **3**:26a; **4**:542b; **5**:361a; **10**:474b
commentaries on, **2**:20b; **3**:311b; **4**:157a; **9**:363a, 364a
editions of, **2**:601b; **3**:280b; **11**:395b; **15**:154a, 475b
translations of, **3**:558b
Saᶜd al-Dīn al-Farghānī, **11**:249b
al-Ṣadafī
see Ibn Yūnus, Abu'l-Ḥasan ᶜAlī ibn ᶜAbd al-Raḥmān ibn Aḥmad ibn Yūnus al-Ṣadafī
Ṣaḍaśītimukhas (arcs), **10**:426a; **15**:546b
Sadīd al-Dīn Muḥammad ibn Masᶜūd al-Kāzarūnī, **9**:603b, 604a
Sadler, Charles Albert, **1**:458b
Šafarevič, Igor Rostislavovich, **1**:307a; **6**:393b
Safety
in diving, **6**:24b
in electric technology, **3**:299a
in mining, **1**:573a; **6**:24b, 549b
Ibn al-Ṣaffār (Abu'l-Qāsim Aḥmad), **7**:361a; **9**:39b; **11**:31b, 32a
translations of works, **13**:400b
Saffron plant
diseases of, **4**:223b
Safranine, **2**:59b
Safrole, **3**:279b
Sage, Balthazar-Georges (1740–1824), 12:63–69; 2:572a; **3**:5a,b, 199a, 473b; **8**:623b*; **11**:520b, 523a
Saggi dell'esperienze naturali fatte nell'Accademia del Cimento, **9**:498b
Sagittarius (constellation), **5**:363b
nebulae, **3**:528b
Sagnac, Georges M. M. (1869–1928), 12:69–70; 1:457a; **8**:10a
Saguens, Jean, **9**:25b
Saha, Meghnad (1894–1956), 12:70–71; 10:290a; **11**:265a; **13**:96a*; **15**:47b*,b, 48a
gas theory, **15**:48b
ionization theory, **1**:56a; **9**:405a; **12**:20a–b
Saha-Boltzmann relationship, **12**:21b
Sahara, **13**:552a
travel routes, **1**:414a
Ṣāḥib al-Shurṭa
see Abraham bar Ḥiyya ha-Nasi
Sahl ibn Abān, **9**:36a
Ṣāᶜid (Abū al-Qāsim) ibn Aḥmad ibn ᶜAbd al-Raḥmān ibn Muḥammad ibn Ṣāᶜid al-Qurṭubī al-Andalusī (=Qāḍī Ṣāᶜid), **1**:5b; **4**:542a; **6**:189b; **7**:223a; **9**:39b; **13**:538a; **14**:112b, 592b, 593a
Ibn Saᶜīd al-Maghribī, **9**:36a
Kitāb Basṭ al-arḍ fi'l-ṭūl wa 'l-ᶜard, **1**:29a
Ibn al-Ṣāᵓigh
see Ibn Bājja, Abū Bakr Muḥammad ibn Yaḥyā ibn al-Ṣāᵓigh
Saijyo school, **1**:83b
Sainfoin, **12**:317a
Sainsbury, Harrington, **11**:462b
St. Andrews University (Scotland)
in seventeenth century, **5**:525a
observatory, **5**:183a–b, 184a, 525a
Saint-Ange, Gaspard Martin, **1**:518a,b
St. Anthony's fire, **6**:123b
Saint-Cloud, William of
see William of Saint-Cloud
Sainte-Croix, Gaudin de, **5**:417b
Sainte-Mesme, Marquis de
see L'Hospital (L'Hôpital), Guillaume-François-Antoine de
Saint-Étienne
School of Mines, **5**:185b
Saint-Georges-Chatelaison coal mines, **3**:411b
St.-Germain-en-Laye
Musée des Antiquités Nationales, **9**:538b
Saint-Gobain, Compagnie de, **3**:316b
St. Helena (island)
geology, **3**:547b
observatory, **7**:146a
Saint-Hilaire, Augustin François César Prouvençal (Auguste de) (1779–1853), 12:72; 13:489a
Saint-Hilaire, Geoffroy
see Geoffroy Saint-Hilaire, Étienne
St. Ignatius's bean, **9**:8a
St. John, Charles Edward (1857–1935), 12:72–73
St. John's College, Cambridge
in mid-nineteenth century, **9**:392b
St. Louis, Mo.
Louisiana Purchase Exposition, **4**:91b; **6**:30b
Missouri Botanical Garden, **13**:456b; **15**:130a, 224a
World's Fair, **2**:267a; **4**:220a
St. Martin, Alexis, **1**:542b–543a
Saint-Martin, Louis-Claude de, **2**:223b
St. Paul's Rocks, **3**:567b
St. Petersburg
Institute of Experimental Medicine, **14**:36b
Polytechnical Institute, **15**:251a
see also Petersburg mathematical school, Petersburg paradox
St. Petersburg, Imperial Academy of Science
in eighteenth century, **5**:428a; **6**:305a; **8**:469b; **11**:432a

in nineteenth century, **1**:387b; **2**:624a; **8**:418a; **9**:292a; **10**:247b–248a
in twentieth century, **13**:26a–b
"Academic expeditions," **10**:283a
chemistry, **7**:529a
competitions of, **1**:207a; **3**:284a; **4**:481a–b; **7**:404a–b
and Euler, **3**:223b; **4**:468b–471a, 472a–473a
founding of, **5**:448b–449a
mathematics, **10**:247b–248a
Medical-Surgical Academy, **3**:488a; **10**:620a
Military Medical Academy, **5**:269a
observatory, **2**:141a; **4**:23a
Polytechnic Institute, **9**:324a
zoological museum, **2**:422b–423a
St. Petersburg, University of
in nineteenth century, **8**:187b; **9**:292a
in twentieth century, **13**:26a
biochemistry at, **10**:22a
chemistry at, **9**:305a
geography at, **14**:52b
mathematics at, **3**:226a; **8**:560b; **10**:248a; **14**:630a
Saint-Pierre, Charles Castel, Abbé de, **13**:584b
Saint-Simon, Claude Henri de Rouvroy, Comte de, **2**:187b; **3**:287b(n), 326b, 375a,b, 379a; **5**:250b; **11**:301b; **13**:591a, 593b, 594b(n); **15**:132b
Saint-Simonianism, **10**:2b, 4b; **15**:50b, 489a
Saint-Venant, Adhémar Jean Claude Barré de (1797–1886), 12:73–74; **2**:356a; **4**:208b*; **10**:4b*
algebra, **15**:194a
hydrodynamics, **3**:195a; **4**:591b; **13**:75a
mechanics, **3**:146a; **11**:123b
St. Victor, Hugh of
see Hugh of St. Victor
Saint Vincent, Gregorius (1584–1667), 12:74–76; **6**:598b; **7**:557b; **9**:97a, 311a; **13**:235b
critics of, **6**:600a
geometry, **5**:527a; **13**:480a
influences on, **10**:336a,b; **13**:561a
number theory, **8**:161a
Opus geometricum, **14**:148a
St. Vincent Island, **5**:39a
Sajnovics, Johannes, **6**:234a
Sakabe, Hiroyasu, **1**:90a
Sakai, S., **15**:469b
Sākalyasaṃhita, **15**:613b
Sakharov, Andrei Dmitrievich, **13**:241a
Sakharov, Vladimir Vladimirovich (1902–1969), 12:76–77
Saks, Stanisław (1897–1942), 12:77–78
Sal ammoniac
in alchemical theory, **5**:421a; **7**:42b
manufacture of, **1**:527a; **3**:201a; **4**:223b; **6**:578a
synthesis of, **12**:79b
transitions of, **6**:519b–520a
Sal benzoes, **12**:146b
Sal microcosmicum, **12**:146b
Sal mirabile
see Glauber, Johann Rudolph, Glauber's salt
Sal viperinum, **13**:234b
Sala, Angelo (Angelus) (1576–1637), 12:78–80; **7**:194b
Saladini, Girolamo, **11**:401b–402b
Ṣalāḥ al-Dīn al-Ayyūbī, **9**:602b
Salamanca, University of
in Renaissance, **1**:573b
Salamander, **5**:518a; **10**:520a; **12**:216a
brain, **6**:321a
embryology, **3**:245b; **6**:132b–133b
endocrinology, **7**:101a
regeneration of, **5**:447b*; **12**:555b
speciation, **3**:572b
Salcher, Peter, **8**:597a
Salem, Massachusetts
Peabody Museum, **9**:536b
Salerno, school of, 12:80–83; **3**:393a; **6**:123a; **11**:601b; **13**:38b, 142a
anatomy, **12**:80b–83a
medicine, **3**:394a; **10**:21b
Salicylates
in foods, **14**:357b
Salicylic acid, **1**:578a; **3**:448a
derivatives of, **3**:341b
formula for, **3**:448b, 449a
structure of, **9**:355b
synthesis of, **7**:452b
al-Ṣāliḥī, **12**:362a
Salimbene, **5**:147a; **9**:361b
Salisbury, Richard Anthony, **2**:518b–519a
Salisbury, Rollin Daniel (1858–1922), 12:83–84; **3**:191b*; **4**:562b; **14**:403a*
Saliva
chemistry of, **5**:431b; **10**:557b
lysozyme in, **5**:29b
microbes in, **10**:398b, 399a,b, 401a
nature of, **9**:63b; **13**:31b
oxygen in, **1**:453b
role of, **14**:287a
secretion of
see Secretion, salivary
Salix, **5**:191a; **9**:282a; **12**:471b
phylogeny, **4**:401b
Sallo, Denys de (1626–1669), 12:84–86; **5**:259a
Salmon, Daniel Elmer, **12**:480b, 481a
Salmon, George (1819–1904), 12:86–87; **3**:163b, 164a, 314b; **7**:522b; **13**:219b
algebra, **3**:164a–b; **7**:163b
geometry, **3**:167b
Higher Plane Curves, **3**:167a
Treatise on the Analytic Geometry of Three Dimensions, **3**:168a
Salmon, Thomas, **14**:152b
Salmon (fish), **6**:267b
diseases of, **9**:587b
metabolism, **8**:452a; **9**:381a
symbiosis, **1**:622b
Salmonella, **8**:450a; **12**:480b
identification of, **4**:276a; **6**:432b; **12**:482a
Salmonidae, **14**:57b
Salomo ibn Pater, **6**:197b
Salomonsen, Carl Julius (1847–1924), 12:87–89; **4**:296a; **6**:100b; **7**:100a, 423b, 502b; **10**:17b; **14**:250b*
Salpa, **2**:502a; **15**:82b
life cycle, **6**:590b
Salpêtrière
see Paris, Salpêtrière
Salt (NaCl), **5**:351b; **12**:175a
in alchemical theory, **1**:549b; **3**:596b; **4**:209b, 387b; **5**:353b, 422a; **8**:131a, 174a; **10**:309b, 310a; **12**:311b, 312a
analysis of, **6**:236b
crystallography of, **14**:491a; **15**:62b
deposits, origin of, **4**:8a; **8**:173b; **10**:171a
as fertilizer, **9**:609b
manufacture of, **3**:348b*; **4**:164a, 554b; **6**:165a*; **11**:354b; **12**:520b
from sea water, **2**:524a*; **3**:561a*
metabolism of, **4**:621a
mining of, **1**:95b; **3**:41b, 210a; **6**:369b; **7**:253a; **10**:128b
nutritional role, **2**:585b
physiological effects, **11**:462b–463a
specific gravity, **7**:346a
thermal dissociation, **3**:176b
see also *Tria prima*
Salt of tartar
see Potassium carbonate
Salter, John William, **6**:591b
Saltpeter
see Niter
Salts
acid, **8**:520b
alchemical theory, **5**:421b–422b
analysis of, **2**:93a,b, 292a; **3**:289b; **4**:223b; **5**:323a–b; **6**:439a; **7**:388b; **8**:171b
classification of, **2**:7b–8a, 381a; **3**:495a; **10**:281a; **11**:562a–b
complex, **3**:271b
solutions of, **4**:229a
crystal forms, **8**:520a
decomposition of, **3**:419a
double, **3**:333b, 495a; **5**:412b*
deposits of, **7**:529b–530a; **10**:170b
oceanic, **13**:580a
origin of, **10**:171a
double, **2**:258a; **7**:529b; **8**:340b, 373b; **11**:534b; **13**:580a
electrical conductivity, **14**:268b–269a
extraction of, **12**:329b
in fermentation, **6**:111b
Fischer's, **5**:7a
fixed, **7**:524b
formulas for, **13**:373a
fused, **8**:503a
hydrated
water-vapor pressure, **3**:617b
impurities in, **3**:201b
isomorphous, **13**:517b
metal, **5**:323a–b, 420b; **11**:541a
conductivity, **11**:142a, 541a
decomposition of, **11**:435b
electrochemistry of, **3**:557a
nature of, **6**:478a; **8**:173b
origin of, **8**:173b
physiological effects, **11**:462b–463a
production of, **2**:292a
purification of, **2**:143a
regrouping in, **3**:246b; **5**:324b
solubility, **1**:616b; **4**:238b–239a, 352b; **5**:324a; **7**:208a; **8**:618b
triple, **5**:7a
vapor pressure, **14**:330a
vapors, **8**:47a; **12**:191b
vegetable, **8**:173b–174a
volatility, **2**:589a
see also Solutions; specific elements, salts of
Salusbury, Thomas, **1**:484a; **7**:293a
Mathematical Collections, **3**:348b
Saluzzo di Menusiglio, Giuseppe Angelo, **1**:548a; **7**:561a
Salvarsan, **5**:29b; **7**:392b; **10**:18b; **13**:228b–229a; **14**:115b
invention of, **4**:300b–301b; **15**:252b
Salvia, **3**:422a; **5**:383b
Salviani, Ippolito (1514–1572), 12:89–90; **1**:596a
Salvin, Osbert, **12**:258a
Salvioli, Gaetano, **2**:164b, 165a
Sāmānid dynasty, **2**:148a–b; **4**:523a–b
Samarium, **3**:321b, 478a
atomic weight, **2**:428b
discovery of, **2**:254b
Samarkand
medieval university, **11**:227a
observatory, **7**:255b–256b, 353b; **13**:535b–536a
al-Samarqandī, Najīb al-Dīn Abū Ḥāmid Muḥammad ibn ᶜAlī ibn ᶜUmar (*d.* 1222), 12:90–91
al-Samarqandī, Shams al-Dīn Muḥammad ibn Ashraf al-Ḥusaynī (*fl.* 1276), 12:91; **4**:455a*; **14**:575b

Samarskite, **3**:479a; **9**:110b
al-Samaw᾿al, ibn Yaḥyā al-Maghribī (*d. ca.* 1180), 12:91–95; 7:241b–245a *passim*
Sambon, Louis Westenra, **9**:83a
Samelson, Hans, **6**:497a
Ibn al-Samḥ (Abu 'l-Qāsim Aṣbagh), **3**:25b; **9**:39b; **14**:594b
Sammlung Kombinatorisch-analytischer Abhandlungen, **6**:404a
Samoylov, Aleksandr Filippovich (1867–1930), 12:95; **12**:271a
Sampatkumāra
  commentaries of, **2**:416b
Sampling, statistical, **1**:155a; **5**:507b; **14**:382b
  theories of, **5**:10a, 477a; **10**:461a; **14**:552b
Sampo kokon tsuran, **15**:750b
Sampson, Ralph Allen (1866–1939), 12:95–96; **4**:279a; **9**:613a*; **13**:501a*
Samuel ben Judah, **7**:38a, 82b
Saṃvatsara (year)
  definition of, **15**:535a
San chiao to, **3**:270a
San Francisco
  fire, **3**:87a
San Francisco Bay, **5**:395b
  geology, **8**:99b
San Juan Mountains, **3**:483b–484a; **8**:41b–42a
San Salvador, **15**:88b
Sanad ibn ᶜAlī, **1**:35b; **4**:438a–b, 541b; **7**: 343a; **14**:537b, 538a
Sanarelli, Giuseppe (1864–1940), 12:96–97; **3**:94b; **4**:620a; **11**:346b
  Sanarelli's phenomenon, **12**:96b
Sancassani, Dionisio Andrea, **9**:4b
Sanchez (Sanches), Francisco (*ca.* 1550–1623), 12:97–98
Sanctorius
  see Santorio, Santorio
Sand, Gerhard Alexis Cappelen, **7**:100a,b
Sand, Knud, **10**:570a
Sand (geol.)
  black, **13**:519a
  ferruginous, **5**:14a–b
  magnetic, **5**:515b
  origin of, **10**:166b, 209b
  water level in, **14**:620b
Sand dunes
  migration of, **8**:242b; **10**:166b
Sandage, Allan Rex, **1**:353a; **6**:30b, 530b; **13**:475a
Sandalwoods, **3**:43a
Sandell, Ernest Birger, **13**:226b
Sandemanians, **4**:527a; **13**:413b
Sander, Bruno, **3**:325b; **13**:148b
Sanders, William, **5**:525a, 528a,b
Sanderson, Ezra Dwight (1878–1944), 12: 98–99
Sanderson, George, **14**:462b
Sanderson, John S. B.
  see Burdon-Sanderson, John Scott
Sandford, Kenneth Stuart, **1**:286a
Sandglass
  see Hourglass
Sandmeyer, Traugott, **9**:356a–b
Sandpipers, **3**:438b
Sandstede, Heinrich, **15**:150a
Sandstone
  fossils in, **11**:340b
  origin of, **6**:56a
  petroliferous, **6**:229a
  variegated (Bunter sandstone), **1**:95b; **4**: 540b; **5**:155b, 445a
Sandström, Johan Wilhelm, **2**:168b, 557b
Sandwich Islands, **1**:594b
Sangaku tablets, **1**:83a
Sanger, Frederick, **1**:11b
Sanger, Margaret, **10**:611a
Sanio, Karl Gustav (1832–1891), 12:99–100
Sanitary Commission, U. S., **5**:480a
Sanitation
  see Hygiene
Sanjar ibn Malikshāh, **7**:325b, 336a,b, 337a,b, 339a
Śaṅkara, **10**:129a
  commentaries of, **2**:115a, 116a, 119a
  Karaṇavaiṣṇava, **15**:589a–b
  sine table, **15**:589a
Śaṅkara Vāriyar, **10**:129b
Śaṅkaranārāyaṇa
  commentaries of, **2**:115a
Saṅkrānti (astron. period), **3**:584b; **10**:426a
  definition of, **15**:546b
Saṅku (sine), **15**:552b, 559b
Sano Antei
  Hi zoshi, **15**:740b
Sano, Fritz, **4**:397b
Sansan (France), **8**:43b
Sanskrit intellectual tradition
  see India
Sanskrit language, **15**:194b–195a
Sanson, Louis, **11**:215a
Sansone, Giovanni, **14**:46b
Santa Cruz, Alonso de, **3**:424a
Santalaceae, **5**:539b
Santalum, **6**:266b
Santi, Renato, **9**:296a
Santiago (Cape Verde island)
  geology, **3**:567b
Santini, Vincenzo, **9**:369b
Santo Domingo
  geography, **5**:214b
  geology, **5**:214a
Santone, Elia Vineto, **4**:451a
Santonin
  structure of, **3**:46b
  synthesis of, **3**:246b
Santorin (Greece), **5**:89a; **11**:336b; **12**:180a
Santorini, Giovanni Domenico (1681–1737), 12:100–101; **9**:511a
Santorio, Santorio (Sanctorius) (1561–1636), 12:101–104; **2**:306a; **4**:383b; **13**:30b
  "static" experiments, **4**:136a
  De statica medicina, **1**:392a; **2**:308a; **6**:37b
Sao hirsuta, **1**:469a
Sap
  flavor changes, **1**:101b
  movement, theories of, **2**:286b; **4**:264b, 402a; **6**:136b; **7**:16a*; **8**:128b–129a; **9**:485b; **10**:520a–b; **11**:314b; **13**:89b, 462b; **14**:413a
    circulation theory, **2**:390a; **5**:535a–b; **6**:36b, 38a–40a; **7**:409a; **8**:205b–206a; **9**:114b
    tension theory, **4**:131a
  pressure of, **6**:36b, 38a–39b
Saphea (astron. instrum.), **7**:124b–125a, 126a; **11**:415a; **14**:273b, 400a–b, 594a–b
Saponification, **2**:66a; **13**:310a; **14**:603a
  elucidation of, **3**:242a–b
Saponins, **14**:445b
  structure of, **14**:444b
Saporta, Louis Charles Joseph Gaston de (1823–1896), 12:104–105; **9**:112a; **11**:270a
  critics of, **14**:600a
Sapphire, **3**:232b
  medical use of, **6**:397a
Sappho (asteroid), **5**:405a
Saprolegnia, **11**:152b, 153b
Sarabhai, Vikram, **15**:27b
Sarasa, Alfonso Antonio de, **9**:311a; **12**:75b
Sarcina, **13**:154b
Sarcocele
  venereal, **8**:125b
Sarcode, **3**:337a, 612b; **4**:235a–236a; **12**:231b
  description of, **4**:235b
Sarcolemma, **2**:376a
Sarcoma, **4**:299b; **9**:586a
  Ewing's, **4**:499b
  Jensen's, **9**:550a
Sarcoplasm
  function of, **2**:339b
Sarcoptes, **3**:97a
Sarcosporidiosis, **12**:482a
Śārdūlakarṇāvadāna, **15**:538a, 584b
Sarek Mountains, **6**:78b
Sargasso Sea, **12**:190b
Sargon II of Assyria, **15**:636b, 640a,b
Sarjiyūs al-Ra᾿s ᶜAynī (Sergius of Resaina), **1**:275a,b; **15**:235b–236a, 238b
Sarlet, Henri, **5**:149b
Sarmiento, Martin (Friar), **3**:97a
Saros (astron. period), **8**:143b, 179a; **13**: 295b–296a; **15**:795a, 799a, 804a
Sarotti, Ambrose, **10**:292b–293a
Sarpi, Paolo (1552–1623), 12:105–106; **2**: 606b; **4**:508a; **5**:239b, 240a, 241a, 382b; **11**:96a; **12**:101b
Ibn al-Sarrāj, **4**:523b; **12**:357b
Sarrau, Jacques Rose Ferdinand, **1**:128b; **3**:486a
Sarrus, Pierre Frédéric, **8**:62b; **12**:291b
Sars, George Ossian, **8**:324a; **9**:443a; **12**:107a
Sars, Michael (1805–1869), 12:106–107; **13**:360a; **15**:430b
Sarsaparilla, **10**:622a
Sarsi, Lotario (pseudonym)
  see Grassi, Orazio
Sarton, George Alfred Léon (1884–1956), 12:107–114; **2**:358b
  acquaintances of, **6**:260a; **9**:378a
  critics of, **15**:176a–b, 179a
  historiography, **1**:279b; **3**:380a*; **13**: 251b*, 254b; **14**:35b; **15**:173a, 180a, 183b, 402a*
    of anatomy, **9**:468a
    of astronomy, **7**:223b
    of geography, **7**:19b
    of mathematics, **3**:249a,b, 254a, 265b, 266b; **5**:265a*; **7**:212b, 572b*; **8**: 313b; **9**:501b*; **15**:199a*, 466a*
    of medicine, **3**:243b*; **5**:227b; **15**: 266b(n)
    of optics, **6**:93b*, 210a*; **9**:192a
    of physics, **4**:431a
    of physiology, **6**:48a*
  influences on, **9**:342a
Sartorius von Waltershausen, Wolfgang, **2**:588a; **4**:65b
Sartorius (muscle), **7**:520a
Sarychev, Gavriil Andreevich (1763–1831), 12:114–115; **8**:101a
Sarzosa, Francisco, **15**:154a
Sascerides, Gellius, **2**:406a, 412a
Sassafras, **10**:622a
  oil of, **3**:279b
Ṣassanian culture, **1**:32b; **4**:556a
Śatānanda (*fl.* 1099), 12:115–116
  Bhāsvatī, **15**:608a–b
Satellites, artificial, **2**:511b; **3**:221b*; **4**: 408a*; **7**:465b
  and gravimetry, **13**:609a
Satellites, astronomical
  dimensions of, **1**:466b
  discovery of, **1**:466b
  distances of
    and diameter, **11**:498a
  orbits of, **13**:422b
  perturbations of, **12**:448b
  theories of, **4**:547b; **12**:280b

tidal deformation, **12**:247b
see also specific planets (satellites of) and specific satellites by name
Satō, Masaoki
Sanpô Kongenki, **12**:290b
Satō Nobuhiro (1769–1850), 12:116–117
Satō Nobukage (1674–1732), 12:116–117
Satō Nobusue (1724–1784), 12:116–117
Saturation, chemical, **2**:68a, 79b
Saturday Review, **13**:522b
Saturn (planet)
atmosphere of, **12**:267a
brightness, **11**:542a
conjunctions of, **7**:374a
distance of, **2**:403b; **3**:409a
equator, **13**:113a
figure of, **6**:361b; **12**:267a
and Halley's comet, **3**:283a–b
illumination of, **12**:282b
and Jupiter, **2**:401b; **4**:481b; **7**:290b, 297b, 374a, 376a; **8**:179a; **10**:64b, 76b, 77b; **14**:389b; **15**:250a, 323a–325a, 329a–331b, 351a, 353b
motion of, **7**:598b; **10**:34a
irregularities of, **2**:43a, 328b; **6**:515b; **8**:144b*, 179a; **15**:289b, 329a–331b
observations of, **2**:97b, 285b; **5**:241b; **11**:411b; **12**:162a; **15**:218b
occultation of, **3**:14a
oppositions of, **8**:144b*, 178b; **15**:356a
orbit of, **5**:224a; **7**:563a; **8**:277a, 279a
inequalities in, **7**:562a
perturbations of, **6**:103b, 399b; **9**:174b
period of, **10**:180b
photography of, **3**:366a; **7**:472a; **12**:36b, 186b, 454a; **14**:521b
rings of, **2**:349b; **3**:23a, 606a; **4**:128a; **6**:432a; **10**:176b; **12**:266b
Cassini's division, **3**:102b; **7**:385a
concept of, **4**:506a; **5**:287b; **6**:603b–604a, 609b
crepe ring, **2**:284b; **3**:605b; **5**:257a
differential rotation, **9**:203b
eccentricity, **12**:239b, 267a
gap between, **6**:335a
inclination, **2**:592b–593a
mathematical analysis of, **7**:477b, 479a; **10**:479b; **15**:296a–b, 321b–322b, 348b
micrometric measurement of, **6**:471a; **13**:115a
nature of, **1**:465b, 597b; **3**:102b–103a; **6**:335a; **7**:270b; **9**:202a–203b, 217b, 219a; **12**:226b; **15**:322a–b
origin of, **7**:385b; **11**:21b
radial velocities, **1**:597b
rotation of, **4**:69b; **6**:335a; **11**:114b
spectroscopy of, **1**:597b; **7**:270b; **12**:368a
stability, **12**:187b
structure, **3**:105a; **8**:522a
varying appearance, **2**:388b; **4**:107b; **8**:145a*; **13**:113a–b
rotation
period of, **3**:357b; **6**:49a, 335a; **9**:504a; **12**:455a
satellites of, **6**:575a*; **11**:162b; **13**:113b; **15**:354b
discovery of, **2**:284b; **3**:102b; **6**:329b, 334b–335a, 603b; **8**:46a; **9**:115b
micrometric measurement of, **13**:115a
orbits of, **6**:49b; **7**:608a; **10**:35a, 77a, 602a; **11**:552a–b; **13**:113a
size
diameter, **2**:388b
mass, **15**:356a
spectrum, **12**:267a
surface, **3**:102b; **6**:335a
white spots, **3**:357b; **6**:49a
synodic period, **4**:467a; **15**:219a
tables of, **2**:359b; **4**:14b; **7**:86b; **15**:331a,b, 356a
zodiacal period, **5**:345b
Satya, **12**:576b; **13**:582a; **14**:549b
Satyrus, **5**:227b, 228a, 229a–b
Sauerkraut
and scurvy, **8**:363a
Saunders, Bernard Charles, **9**:402b; **11**:479b
Saunders, Edith Rebecca, **1**:505b; **13**:158a
Saunders, Frederick Albert, **12**:18b, 20b, 21a
Saunderson, Nicholas, **3**:459b; **4**:85b; **6**:139a
Saunier, Pierre Paul, **9**:366a
Saura days, **4**:556a
Saurapakṣa (school of astronomy), **9**:175b, 580a; **11**:264a; **15**:534a, 600a, 602b, 608b–618a, 618b, 620a, 622b, 624b
Saurians, **2**:489a, 569b–570a
Saurin, Joseph (1659–1737), 12:117–119; 11:512b, 513a
Sauropsida, **6**:595a
Sauropus primaevus, **8**:104a
Saussure, Horace Bénédict de (1740–1799), 12:119–123; 10:8b
acquaintances of, **3**:43b; **10**:602b; **12**:560b; **14**:72a
botany, **5**:277b
correspondence of, **14**:72a–b
and Deluc, **4**:28b–29a
electricity, **14**:73a–b
geology, **15**:56a
granite, **6**:583b
theories, **2**:568a, 571a; **3**:210b
hygrometer, **4**:28b
influence of, **4**:152b; **11**:521a; **14**:259b
pneumatic chemistry, **14**:76a
students of, **11**:134b
terrestrial magnetism, **2**:135a
Voyages dans les Alpes, **6**:582b
Saussure, Nicolas-Théodore de (1767–1845), 12:123–124; 4:152b; **5**:325a; **8**:345b; **12**:121a, 309a
Sauter, Fritz, **8**:476a, 477b
Electromagnetic Fields and Interactions, **1**:24a
Sauton, Benjamin, **8**:273b
Sauvage, Henri Émile, **1**:491a
Sauvageau, Camille-François (1861–1936), 12:124–126
Sauvages de la Croix, François Boissier de, **3**:579a; **13**:215a
Sauveur, Albert (1863–1939), 12:126–127; 9:139b*; **11**:101a
Sauveur, Joseph (1653–1716), 12:127–129; 2:49a; **3**:258b; **4**:88b; **5**:59a; **10**:319a
Savage, Leonard Jimmie, **3**:461b
Foundations of Statistics, **1**:532a
Savage, Thomas Staughton, **14**:533b
Savannah (biogeog.), **14**:181a
Savart, Félix (1791–1841), 12:129–130
acoustics, **3**:258b; **14**:492b
electromagnetism, **2**:137a
Savary, Félix, **6**:278b, 279a; **8**:36b
Savasorda
see Abraham bar Ḥiyya ha-Nasi
Save-the-Redwoods League
founding of, **7**:102b; **9**:314a
Savelev, Aleksandr Stepanovich, **8**:187b, 188b
Savery, Thomas
steam engine, **4**:44a; **10**:36b, 293a
Savić, Pavle Petar, **4**:580a; **7**:158b; **9**:262a
Savich, Aleksei Nikolaevich
Practical Astronomy, **10**:542a
Savigny, Marie-Jules-César Lelorgne de (1777–1851), 12:130–131; 1:328b; **11**:234a
Savile, Henry, **2**:462b; **4**:212b, 451a; **5**:343a–b; **7**:39a; **10**:489a
Savory, William, **8**:407b
Savot, Louis
L'architecture françoise des bastimens particuliers, **2**:201b
Sawaguchi, Kazuyuki
Kokon sanpōki, **12**:290b
Saxifraga, **15**:147a
Saxony
geology, **3**:433b, 462b; **5**:155b–156a
mining, **6**:369b; **13**:455a–b
natural history, **13**:425b
Saxton, Joseph (1799–1873), 12:131–132; 4:183a
Say, Thomas (1787–1834), 12:132–134; 1:489b; **3**:391b; **8**:266b, 615b; **10**:164b, 439b; **13**:467a; **14**:328b
American Entomology, **10**:439b
Sayf al-Dawla, **4**:524b; **6**:186a; **11**:226a,b; **15**:250b
al-Ṣaymarī, Abu'l-ᶜAnbas, **1**:38a; **9**:161a
Saytzeff reaction, **5**:540a
Sazerac, Robert, **8**:273b
Scabies, **9**:75a; **14**:638a
etiology of, **2**:291a–b; **5**:570b; **8**:125b; **10**:384a; **11**:301a, 342b
transmission of, **5**:107a
treatment of, **3**:97a; **13**:229b
Scabiosa, **7**:166a
Scala naturae
ancient Greek discussion, **1**:262a
ancient Roman discussion, **11**:42a
medieval discussion, **7**:27b; **8**:549b; **11**:232a–b
seventeenth-century discussion, **11**:493a; **13**:526b, 528a
eighteenth-century discussion, **2**:286b; **3**:526a–b; **5**:356a; **7**:367b, 585b, 604a; **9**:533b; **13**:564a; **15**:54b–55a
nineteenth-century discussion, **2**:187a–b, 570b; **3**:395b, 522b–523b, 526a–b; **4**:290b–291a; **5**:296a, 357b, 370b; **7**:592a; **13**:506b
twentieth-century discussion, **8**:517b
Chinese discussion, **8**:391b
mineralogical, **7**:587b
Scale insect, **6**:524b
Scales, musical, **4**:526a; **15**:265b
chromatic, **1**:232b, 282b; **11**:223a
diatonic, **1**:232b, 282b
enharmonic, **1**:232b, 282b; **11**:223a
tempered, **5**:249b
see also Harmonics
Scaliger, Joseph Justus, **1**:126b; **3**:533b; **7**:306a; **11**:533b; **12**:134a,b; **14**:21b, 22a
Cyclometrica elementa, **2**:360b; **11**:533b
Scaliger (Bordonius), Julius Caesar (1484–1558), 12:134–136; 1:495a; **2**:590b; **3**:312a; **5**:426b; **9**:338a; **10**:152a; **13**:210b
commentaries of, **10**:112a
critics of, **3**:262b
natural philosophy, **12**:312a,b
Scallop
muscle
glycogen and glycine in, **3**:256b
Scalp
diseases of, **3**:596b
Scandia, **9**:110b
Scandinavia
archaeology, **9**:489a–490a
botany, **14**:116b–117a
ethnology, **10**:197a–b
geography, **10**:197a–b
geology, **3**:547b; **4**:408b–409b
geophysics, **13**:610a
paleontology, **13**:9b

Scandinavia (cont.)
stratigraphy, **2**:555b–556a; **5**:329b–330a; **9**:584b; **14**:59b–60a
medicine, **6**:123b
paleogeography, **8**:519a–b
see also individual countries and regions
Scandium, **3**:475a
atomic weight, **3**:321b
discovery of, **2**:589b; **9**:290b
spectrum, **12**:22a
Scaphe (astron. instrum.), **4**:390a
Scarification, **15**:645b
Scarlet fever, **2**:575b; **4**:405a; **6**:232b; **8**:449a; **9**:95a; **10**:613a; **11**:37b
edema, **14**:253a, 464a
treatment of, **4**:404b; **14**:464a
Scarpa, Antonio (1752–1832), 12:136–139; **3**:437b; **12**:561b, 562b
anatomy, **5**:267a
students of, **1**:492b; **4**:146a; **11**:98b, 494b
Scatchard, George, **3**:336a
Scattering, **6**:615a; **7**:492a
of alpha particles, **2**:240b, 337b–338a; **5**:331a; **9**:543b; **12**:30b–31b; **13**:369a; **15**:42b
of beta particles, **2**:337b–338a; **12**:31b; **13**:368b–369a
of electrons, **2**:338a,b; **3**:597b; **8**:181a
of gamma rays, **3**:368a; **9**:261b; **13**:369a
of light, **11**:265a, 265b–266b; **12**:464b; **13**:101b, 103b, 104b, 368b; **14**:617a
anomalous, **7**:186b
in atmosphere, **13**:522a
combination scattering, **7**:622a–b; **9**:76b
by crystals, **7**:622a; **9**:76b
fluctuation scattering, **9**:76b
by liquids, **11**:265b–266a; **15**:6b
quantum theory, **13**:240a
in solutions, **3**:620a–b
of neutrons, **4**:582a; **8**:482a
of nucleons, **4**:583a
of photons, **14**:420b
of pions, **4**:583a
of positrons, **15**:26b
of protons, **7**:617b
small-angle, **2**:338a
of sound, **13**:104b
statistical theories, **2**:338a; **5**:331a, 473b; **13**:241b; **15**:42b–43a
of sunlight, **8**:501b
theory of, **12**:31a–b
of X rays, **1**:457a–458b; **2**:398b; **3**:367a, 430a, 619b; **4**:196b; **13**:368b–369a
Compton effect, **1**:459a; **2**:338a, 488a; **3**:367b–370a; **4**:317b; **5**:331b; **9**:261b; **12**:528a
Schaaffhausen, Hermann, **5**:206b–207a
Schaeberle, John Martin (1853–1924), 12:139; **1**:466a; **3**:35a
Schaefer, Vincent Julius, **8**:24b
Schäfer, Clemens, **5**:306b, 310a; **8**:552a
Schafer, Edward Albert
see Sharpey-Schäfer, Edward Albert
Schaffhausen (Switzerland)
fortification of, **4**:260b
Schaffhausen trio, **10**:567b
Schardt, Hans (1858–1931), 12:139–141; **1**:236b; **11**:375a; **13**:146a
geology, **1**:235b; **5**:221b; **6**:228a; **8**:543b; **13**:285a
Schaskolskii, Marianna, **5**:186b
Schaub, Jean-Jacques, **13**:127a
Schauder, Pawel Juljusz, **1**:427a; **2**:145b, 513a; **6**:472a
Schaudinn, Fritz Richard (1871–1906), 12:141–143; **4**:300a; **8**:273b; **10**:18b, 141b–142a; **11**:174b–175a; **12**:423b–424a
Scheeffer, Ludwig, **8**:111b
Scheele, Carl Wilhelm (1742–1786), 12:143–150; **2**:421b; **10**:26a, 148b
acquaintances of, **4**:346b; **13**:280b
correspondence of, **14**:355b*
editions of works, **8**:247b*
and Gahn, **5**:222b–223a
influence of, **2**:92a; **9**:5b; **12**:67b(n)
inorganic chemistry, **5**:602a
acids, **8**:76a, 618b; **9**:106b
chlorine, **10**:496b
nitrogen, **12**:25a
oxygen, **8**:86a(n)
phosphorus, **9**:105b
Mémoires de chymie, **5**:601b
mineral chemistry, **2**:92b
organic chemistry, **3**:10b; **12**:320b
acids, **8**:76a
calculi, **5**:92b
glycerol, **3**:242b
Prussian blue, **8**:619a
photosynthesis, **7**:12a,b
pneumatic chemistry, **14**:75a
Scheele's green, **12**:146b
theoretical chemistry, **2**:6a
translations of works, **1**:563b; **15**:207b*
Scheelite, **12**:148a
Scheffer, Henrik Theophilus, **2**:421b, 524a; **3**:473b; **14**:225a
Scheffers, Georg (1866–1945), 12:150; **4**:205b; **8**:324a
Schegk, Jakob (or Schegkius or Scheggius; sometimes called Jakob Degen) (1511–1587), 12:150–151
Scheiding, Fritz, **8**:329a
Scheiner, Christoph (1573–1650), 12:151–152; **2**:526a; **3**:528a; **5**:242a; **7**:375a
instruments, **5**:278b; **6**:361a,b
optics, **1**:475a; **9**:97b
Rosa ursina, **5**:515b
on sunspots, **3**:179b; **7**:193b; **14**:415a
Scheiner, Julius (1858–1913), 12:152–153; **4**:168a*; **6**:28b
astrophysics, **1**:21b; **11**:542b; **13**:481b; **14**:56a,b
Die Spectralanalyse der Gestirne
translation of, **5**:199a
Schellbach, Karl Heinrich, **1**:370a; **3**:314a; **4**:340b, 341a; **5**:472a; **6**:286a; **10**:24a
Schelling, Frederick Wilhelm Joseph (later von Schelling) (1775–1854), 12:153–159; **1**:70a; **2**:160a,b, 552b; **4**:200a; **5**:446a; **7**:234a, 608a
acquaintances of, **11**:473b, 474a; **12**:197a; **14**:351a
critics of, **2**:545a, 595a; **4**:616b; **5**:193a; **6**:81a; **7**:367a; **8**:373b; **11**:592b; **12**:175a, 198b; **15**:127b, 132b
influence of, **1**:73a; **2**:425b; **6**:395b; **10**:194b; **12**:198b, 281a, 607a; **14**:241b; **15**:86b
Naturphilosophie, **1**:69b; **4**:146a,b, 530b; **9**:572a, 578b; **10**:182b
students of, **7**:505b; **10**:11a; **13**:402a, 462b
Schelver, Franz Josef, **5**:218a
Schenck, [Johann] Heinrich [Rudolf], **13**:90a
Schenk, August, **10**:502a; **11**:439b
Scherer, Alexander Nicolaus von, **3**:465b; **5**:558b; **7**:378a; **14**:508a; **15**:171b
Scherer's Journal
see *Allgemeines Journal der Chemie*
Schering, Ernst Christian Julius, **5**:310b*; **6**:470a; **7**:506b; **11**:448a
Scherk, Heinrich Ferdinand, **7**:521b, 522a; **11**:544b
Scherrer, Paul, **3**:618a, 619b; **9**:262a
Schetelig, Jacob Grubbe Cock, **2**:486a
Scheubel, Johann, **4**:449b; **11**:339a
Scheuchzer, Johann Jakob (1672–1733), 12:159; **7**:412a; **8**:4a; **14**:501b–502a; **15**:55b
correspondence of, **6**:305a; **15**:52b, 53b
geology, **8**:543b
Herbarium diluvianum, **7**:197b
paleontology, **9**:152b; **15**:54b, 55a
Physica sacra, **7**:411b
Sciagraphia lithologica curiosa, seu lapidum nomenclator, **7**:401a
Scheutz, Georg, **1**:355b
Schiaparelli, Giovanni Virginio (1835–1910), 12:159–162; **12**:266b; **13**:232a
astronomy, **1**:172b; **8**:504a, 521b, 522a; **10**:41b; **13**:474b
critics of, **14**:242b
and Dembowski, **4**:30a
historiography, **1**:247a, 509b; **10**:529b; **11**:203b*; **15**:203b–204a
Precursori di Copernico nell'antichità, **3**:512a
Schick, Béla, **4**:404b
Schickard, Wilhelm (1592–1635), 12:162–163; **9**:169b*; **10**:332a
Schiefergebirge, **2**:108a
Schiff, Hugo Josef (1834–1915), 12:163–164
Schiff bases, **12**:164a
Schiff's reagent, **4**:603b
Schiff, Leonard Issac, **8**:478a
Schiff, Moritz (1823–1896), 12:164–165; **2**:30a; **6**:11b, 225a; **8**:7a; **9**:546b; **12**:163b
Schild, Alfred, **7**:10b
Schiller, Johann Christoph Friedrich von, **6**:327a, 550a; **7**:366b; **10**:183b, 482b; **12**:178b, 607a
Schiller, Nikolai Nikolaevich, **6**:252a
Schilling, Johann Jacob, **2**:324b
Schilling, Nikolai Gustavovich, **7**:511a
Schilling, Pavel, **7**:56a
Schilt, Heinz, **6**:497a
Schimmelbusch, Curt, **4**:277a; **8**:408a
Schimper, Andreas Franz Wilhelm (1856–1901), 12:165–167; **13**:90a
Schimper, Karl Friedrich (1803–1867), 12:167–168; **2**:425b, 426a; **5**:599a; **9**:44b, 149a; **15**:130a
Schimper, Wilhelm Philipp (Guillaume Philippe) (1808–1880), 12:168–169; **6**:491a; **8**:264a
Schistosoma, **2**:127b; **11**:596a
Schists, **3**:567b; **5**:548b, 556b
bituminous, **12**:182a
crystalline, **1**:551b; **5**:562b–563a; **13**:283b–284a, 285a
Dalradian, **1**:394a
lineation, **1**:153b–154a
origin of, **9**:620b; **12**:593b
Schizophrenia, **10**:611a
treatment of, **4**:287a(n); **7**:190a, 191b
Schizophytes, **4**:207a
Schizosaccharomyces, **5**:584b
Schizotrypanum cruzi, **3**:185b
Schjellerup, Hans Carl Frederik Christian (1827–1887), 12:169–170
historiography, **15**:208a
star catalog, **2**:146b
Schläfli, Ludwig (1814–1895), 12:170–173; **3**:167b; **7**:51b; **13**:20b
Schläfli function, **12**:172a
Schläfli symbols, **12**:171b
Schlechtendal, Diederich Franz Leonhard von, **3**:611b; **7**:533b; **15**:82a,b
Schlegel, Paul Marquard, **13**:474a
Schlegel, Victor, **15**:195a–b
Schleicher, August, **6**:7b
Schleiden, Jacob Mathias (1804–1881), 12:173–176; **6**:73a; **11**:152a
cell formation, **9**:600b

cell theory, **2**:160a; **3**:29b, 337a,b; **6**:266a; **11**:152b–153a
on contemporary scientists, **11**:300b
critics of, **2**:426b, 532a; **6**:465a–b, 466a; **11**:153a; **12**:230b; **13**:542a
on fertilization, **5**:539b; **6**:266a–b
Grundzüge der wissenschaftlichen Botanik, **6**:465a; **9**:600a; **11**:152a
influences on, **5**:195a
and Naegeli, **9**:600a,b, 601a
Die Pflanze und ihr Leben, **6**:6b
philosophy, **7**:235a; **15**:167a
and Schwann, **12**:243a
students of, **4**:200a, 371b; **5**:381a; **7**:401b; **14**:350a
works translated, **6**:267a*
Schleip, Waldemar, **11**:597a
Schlemm, Friedrich, **2**:124a; **11**:360b
Schlemm's canal, **11**:380a
Schlenk, Fritz, **4**:486a; **6**:111a
Schlenk, Wilhelm, **5**:465a; **8**:52b
Schlenk, Wilhelm, Jr., **6**:304a
Schlesinger, Frank (1871–1943), 12:176–177; **2**:511a; **3**:510b*; **4**:351a; **6**:400b*; **12**:18b, 37b*; **13**:474b
Schlesische Gesellschaft für Vaterländische Kultur, **5**:441a
Schleyer, Johann Martin, **10**:443b
Schlick, Ernst Otto, **12**:527a, 575a
Schlick, (Friedrich Albert) Moritz (1882–1936), 12:177–179; **14**:470a*
Raum und Zeit in der gegenwärtigen Physik, **6**:26a–b
Schliemann, Heinrich (1822–1890), 12:179–182; **1**:162a; **14**:42a–b
Schlieper, Adolph, **5**:125b
Schloezer, August Ludwig, **1**:554a
Schlotheim, Ernst Friedrich, Baron von (1765–1832), 12:182–183; **2**:491b; **5**:155b; **6**:456a
paleobotany, **13**:44a
Schlumberger, Charles (1825–1905), 12:183
Schmalhausen, Ivan Fedorovič (Johannes Theodor), **3**:530a
Schmeidler, Werner, **10**:139a
Schmerling, Philippe-Charles (1791–1836), 12:183–184; **5**:109b; **15**:51a
Schmidel (or Schmiedel), Casimir Christoph (1718–1792), 12:185–186; **3**:415a*; **5**:379a; **6**:219a,b
Schmidt, Bernhard Voldemar (1879–1935), 12:186
telescope, **1**:352b; **3**:63a
Schmidt, Carl August von (1840–1929), 12:186–187; **7**:186b
Schmidt's law, **12**:187a
Schmidt, Carl Ernst Heinrich, **2**:124a, 162a, 585a; **13**:243a,b; **14**:63b; **15**:109a, 455b, 457a
Schmidt, Erhard (1876–1959), 12:187–190; **3**:62b; **4**:560b; **12**:247a*
analysis, **6**:392a; **12**:246b
Gram-Schmidt process, **12**:188b, 189b
Hilbert-Schmidt theorem, **12**:189a
students of, **6**:496a; **11**:256a; **14**:613b, 614a
Schmidt, Ernst Johannes (1877–1933), 12:190–191; **13**:148b
Schmidt, Friedrich, **3**:529a,b, 530a
Schmidt, Gerhard Carl Nathaniel (1865–1949), 12:191–192; **11**:283a*
radioactivity, **1**:559b; **2**:258b; **3**:499a; **9**:261b; **11**:280b; **12**:26b
Schmidt, Hans Karl Gustav, **4**:155b
Schmidt, Heinrich Willy, **9**:260b
Schmidt, Johann Friedrich Julius (1825–1884), 12:192; **1**:616a; **12**:226b; **13**:475b
Charte der Gebirge des Mondes, **1**:569a
Schmidt, Karl, **6**:227b
Schmidt, Oscar, **5**:446b–447a; **9**:112a; **14**:626b
Schmidt-Ott, Friedrich Gustav Adolf Edward Ludwig, **4**:269a; **8**:52b; **12**:615b
Schmiedeberg, Oswald, **2**:586a; **8**:451a, 452a; **9**:381b; **10**:175b; **14**:495b; **15**:99b, 100a
Schmiedel
see Schmidel, Casimir Christoph
Schmit-Jensen, Hans Oluf, **7**:503a
Schmitt, Francis Otto, **4**:398a; **8**:283a
Schmitz, Friedrich, **7**:535b
Schnee, Walter, **6**:472b
Schneegans, Daniel, **5**:221b
Schneider, Friedrich Anton (1831–1890), 12:192–194; **14**:235a
Schneiderhöhn, Hans (1887–1962), 12:194–195; **8**:371a*; **10**:127b*
Schneller, Heribert, **9**:564a
Schober, C. G., **14**:495a
Schoenberg, Mario, **5**:272b
Schoenflies, Arthur Moritz (1853–1928), 12:195–196; **2**:217a, 513a
on contemporary scientists, **3**:53b; **7**:400a*; **9**:427a*
crystallography, **1**:460b, 461b; **5**:212a; **6**:303a; **12**:512a; **14**:525b
Einführung in die mathematische Behandlung der Naturwissenschaften, **15**:435a
geometry, **11**:79b
influences on, **2**:431b; **6**:180b; **7**:522b
Schoenflies-Federov groups, **9**:183b
set theory, **2**:303b, 513b; **3**:56a; **6**:230b
topology, **7**:312a
Scholasticism
ancient Roman, **2**:231a
medieval, **1**:196b–199b, 378b; **2**:391b–395b, 590b, 609a–611a; **3**:15a, 513a; **4**:129b; **6**:544a; **7**:117b–118a; **9**:31b; **10**:171b–174b, 223b; **11**:469b; **15**:189a
Renaissance, **2**:401a, 539a; **3**:172a, 516a; **4**:585a–b; **5**:105a, 119a; **9**:32b, 132b–133b; **11**:71a, 73b; **12**:105b, 151a, 548a; **14**:47a, 586b
seventeenth-century, **2**:13a, 14a, 601b; **5**:426b, 427a; **9**:25b; **10**:43b; **14**:482b, 582a
eighteenth-century, **8**:378b; **9**:140a; **13**:563a; **14**:483a–b
Arab influence on, **1**:410b; **7**:23a
kinematics, **6**:378a–379a; **9**:133a–b
mathematics, **1**:226b, 381a; **3**:150a; **4**:93a, 447a
and medicine, **1**:289b, 290a; **5**:425b–426a; **11**:605a; **12**:82a, 83a
opposition to, **2**:377b–378b, 545a; **4**:96a; **5**:285a–288a *passim;* **10**:334a; **11**:286b; **12**:97b; **13**:564b
Parisian, **1**:228a; **2**:603a–608a; **3**:470a; **14**:388a
rediscovery of, **4**:225a, 227b–228a
spread to Continent, **1**:565b; **2**:193a, 194b
statics, **8**:215a, 223b
Schönbein, Christian Friedrich (1799–1868), 12:196–199; **1**:160b, 161a, 362a; **2**:340b; **4**:599b; **7**:45a; **9**:380b; **13**:618a; **14**:329b
lamp, **2**:340a
Schönemann, Theodor, **5**:264a
Schöner, Johannes (1477–1547), 12:199–200; **6**:144a, 145a*; **11**:395b, 415b; **13**:62a(n); **14**:275b; **15**:477b
astronomy, **15**:474b
cartography, **15**:156a
Schönfeld, Eduard (1828–1891), 12:200
observations of, **1**:243a,b; **11**:545a; **12**:153a
Schönherr, Carl Johan (1772–1848), 12:200–201; **5**:605b
Schonland, Basil Ferdinand Jamieson (1896–1972), 12:201–202; **15**:60a
Schönlein, Johann Lucas (1793–1864), 12:202–203; **4**:146a; **11**:368b–369a
pathology, **1**:493b; **5**:566a
Schönlein's disease, **12**:203b*
students of, **2**:129b; **4**:200a; **6**:241b; **9**:259a; **14**:40a
Schönrock, Otto, **7**:65b
School and Society
founding of, **3**:130b
Schoolcraft, Henry Rowe (1793–1864), 12:203–205
Schooten, Frans van, the Elder, **12**:205a, 610a
Schooten, Frans van (*ca.* 1615–1660), 12:205–207; **1**:474b, 541a; **3**:615a; **4**:576a(n); **9**:599a; **12**:610b; **14**:20b–21a, 465b
correspondence of, **6**:359b, 598b
Descartes, editions of, **2**:47a; **3**:614b; **8**:161b; **10**:43b
Exercitationes mathematicae, **2**:50a; **6**:537a
influence of, **1**:481a; **10**:45b
optics, **9**:159a
students of, **6**:359a, 536a–b, 538a, 598a; **14**:465a
Schooten, Pieter van, **12**:206b; **13**:479b
Schopenhauer, Arthur, **2**:123a; **5**:445b; **6**:26a, 247b; **7**:189b, 505b; **8**:159b, 445a; **12**:218a
Schopfer, William-Henri (1900–1962), 12:207–208
Schorlemmer, Carl (1834–1892), 12:208–209; **2**:623b; **10**:517a; **11**:537b, 538a; **12**:493b; **14**:560b; **15**:84b*
Schott, Charles Anthony (1826–1901), 12:209–210; **6**:188a
Schott, Eduard, **9**:139a
Schott, Gaspar (1608–1666), 12:210–211; **7**:374b, 375a; **13**:30b
Mechanica hydraulico-pneumatica, **5**:575a
Technica curiosa, **2**:83b; **5**:575a
Schott, Otto Friedrich (1851–1935), 12:211–212; **1**:9a; **5**:144a
Schott Glass Manufacturing Company, **14**:633a
Schottky, Friedrich Hermann (1851–1935), 12:212–213; **4**:2b; **14**:222a
Schottky's theorem, **12**:212b–213a
Schottky, Walter, **6**:546b; **8**:475b
Schoute, Pieter Hendrik (1846–1923), 12:213; **7**:465b; **12**:172b
Schouten, Jan Arnoldus (1883–1971), 12:214
geometry, **11**:455a
Ricci Calculus, **4**:336b
Schouten, Jodocus, **13**:583b
Schouw, Joakim Frederik (1789–1852), 12:214–215; **6**:267a*
Schrader, Franz, **9**:104a
Schrader, Heinrich Adolph, **1**:46a; **9**:456a
Schrader, Johann Gottlieb Friedrich, **12**:226a
Schram, Robert [Gustav], **10**:219b
Schramm, Gerhard, **5**:141a
Schreber, Johann Christian Daniel, **6**:218a–b; **8**:379b; **12**:185b
Schreck, Johann, **9**:147a; **14**:162a
Schreiber, Heinrich, **11**:457a,b, 589b; **14**:471b(n)

Schreibers, Karl (or Carl) Franz Anton von (1775–1852), 12:215–216; 12:543b; 13:245b; 14:326b
Schreier, Otto, 1:307a–b
Schreinemakers, Frans Antoon Hubert, 11:534b, 535a
Schreve, Forrest, 8:425b
Schrieffer, John Robert, 7:221b; 8:477b, 480b
Schröder, Friedrich Wilhelm Karl Ernst (1841–1902), 12:216–217; 8:262b, 542b; 12:228b; 15:195a
logic, 2:297a,b; 10:442b; 11:94b, 107a; 12:12a
Vorlesungen über die Algebra der Logik, 10:485a
Schröder, Heinrich Georg Friedrich, 7:464a; 8:159b; 10:367a
Schröder, Johann Christian, 8:130b
Schroder, Wilhelm von, 7:284a
Schrödinger, Erwin (1887–1961), 12:217–223; 8:473a; 14:341b*
acquaintances of, 4:312b; 6:163b; 9:260b; 15:448a
field theory, 4:330b
influences on, 8:490a
quantum theory, 2:466a; 3:368b; 8:52a, 292a; 10:423a
Schrödinger equations, 4:339b*; 7:492a–b, 574b; 8:284b, 473b, 475a, 498b
validity of, 6:614b–615a
wave mechanics, 2:248b; 4:280a, 318a–b; 5:271b, 473a; 6:89b; 10:213b; 12:529b; 15:42b
Schroeder van der Kolk, Jacobus Ludovicus Conradus (1797–1862), 12:223–225; 4:163a,b; 6:137b
Schroeter, Heinrich Eduard (1829–1892), 12:225–226; 7:109b; 10:343a; 13:132b
Schröter, Adam, 1:49b; 10:306b
Schröter, Johann Hieronymus (1745–1816), 12:226; 2:97b; 4:169a; 6:112a, 334a; 8:321b; 10:198b, 199a
Schröter, Johann Samuel, 5:205b; 14:119b
Schrötter, Anton von (1802–1875), 12:227; 2:340b; 4:168a*; 5:341a; 8:507b; 14:349a
Schubert, Friedrich Theodor, 4:477b; 5:209b; 8:300a
Schubert, Hermann Cäsar Hannibal (1848–1911), 12:227–229; 6:570b; 13:125a
calculus, 6:394a
geometry, 3:213b; 7:164a; 12:331a; 13:133a
Schubert sets, 12:228b
Schuchert, Charles (1858–1942), 12:229–230; 6:57b; 13:518b, 531b
on contemporary scientists, 1:470a; 9:134b*, 316a*; 14:299a*, 392b*, 627b*
geology, 3:520b*; 5:486b; 13:532a
Schuler, Wolfgang, 2:404a
Schultes, Joseph August, 13:392b
Schultz, Gustav, 7:283a
Schultz, Jack, 13:134b, 137a
Schultze, Max Johann Sigismund (1825–1874), 12:230–233; 2:366b; 4:291a
cytology, 1:540a; 2:165b, 532a; 3:337b; 4:235b, 236a; 9:442a; 11:296a
Schultz-Lupitz, Albert, 6:237b
Schulze, Franz Ferdinand (1815–1873), 12:233–234; 5:34b; 10:367a; 12:141b, 142a,b; 14:233b
Schulze, Johann Karl, 7:569a
Schulze, Walter, 1:441a
Schumacher, Heinrich Christian (1780–1850), 12:234–235; 5:303a; 10:110b, 542b; 11:377a, 378b*
acquaintances of, 3:302a; 5:301b; 6:103a–b, 326b; 8:469b; 13:109b
astronomy, 10:207a
on Challis, 3:187a–b
correspondence of, 2:98a; 5:302b, 308a; 6:522a; 8:433a; 11:544b; 13:110a
Schumann, Friedrich, 9:562a
Schumann, Victor (1841–1913), 12:235–236; 8:578b
Schunck, Charles, 12:236b
Schunck, Henry Edward (1820–1903), 12:236–237; 9:95b; 12:478a, 479a*
Schur, Friedrich Heinrich, 8:326a
Schur, Issai (1875–1941), 12:237; 2:217a; 5:198a; 11:256a; 14:211b
algebra, 6:572a; 14:90b
group theory, 5:198b; 14:283b–284a
Schur index, 12:237b
Schuster, Arthur (1851–1934), 12:237–239; 4:278a, 280a; 5:330b; 11:538a; 13:365a, 370a
on contemporary scientists, 11:242a*; 13:53a*, 107a*, 364a
correspondence of, 3:481b*; 13:371b*
electricity, 13:106a*
at Manchester, 12:29b
radiation, 3:477a; 9:404b; 13:366b, 367a
Rydberg-Schuster law, 12:43a, 238b
Schuster-Schwarzschild model, 12:250a
spectroscopy, 1:425b
Schut, Willem, 3:334a
Schütz, Wilhelm, 7:428b
Schütze, Albert, 15:522a
Schützenberger, Paul, 1:361a, 362a; 3:498b; 8:9b, 329a; 15:422b
Schwab, Leonhardt Conrad, 13:578a
Schwabe, Samuel Heinrich (1789–1875), 12:239–240
sunspots, 7:609a; 12:51a; 14:480b
Schwägrichen, Christian Friedrich, 4:288a; 6:219b
Schwalbe, Gustav, 7:504b, 519b; 11:570a,b, 571a
Schwann, Theodor Ambrose Hubert (1810–1882), 12:240–245; 5:149a
acquaintances of, 1:476b; 4:200a; 12:203a
biliary fistula, 1:544a
cell formation, 2:32b; 6:270a
cell theory, 3:29b; 4:264b; 8:302a; 9:572b; 11:491b
on contemporary scientists, 11:300b
critics of, 2:426b, 532a; 7:438b; 11:369b; 12:230b
fermentation, 2:561a; 3:9a–b; 8:402b; 10:361b
histology, 2:376a
influences on, 1:543a; 11:217a
pepsin, 1:543b; 8:342b
philosophy of science, 4:203b; 7:234b–235a; 8:515b
and Schleiden, 12:173b, 174a
spontaneous generation, 10:367a,b
yeast, nature of, 13:506b
Schwartz, Robert S., 4:522b
Schwarz, Hermann Amandus (1843–1921), 12:245–247; 3:62b; 4:2b, 174a; 14:222b, 223a
acquaintances of, 4:617a; 6:230b; 7:507a–b, 522b; 14:222a
analysis, 8:367a; 11:58b, 59a–b; 13:26b
and Cantor, 3:53a
geometry, 7:398a; 10:442a; 11:22a; 14:605a
influence of, 4:561b; 11:394b
Schwarz-Bunyakovsky inequality, 6:472b; 12:246b; 15:66b
Schwarz's lemma, 3:63a; 12:246b
students of, 2:217a; 10:275a; 14:613b
topology, 11:451b–452a
Schwarzert, Philipp
see Melanchthon
Schwarzschild, Karl (1873–1916), 12:247–253; 4:280a; 7:236b; 9:564a
acquaintances of, 7:238b; 11:542a,b
astrophysics, 4:279a; 9:404b
correspondence of, 6:350a
influences on, 12:282b
optics, 9:46a; 11:583a
relativity, 8:278a
Schuster-Schwarzschild model, 12:250a
Schwarzschild radius, 12:251b
Schwarzschild's law, 12:248b
stellar statistics, 4:278b; 7:238a
students of, 1:428b
Schweidler, Egon von, 5:395a*; 6:355a; 12:218a
Schweigger, Johann Salomo Christoph (1779–1857), 12:253–255; 2:340a; 3:497a; 11:50a, 437a; 14:489a; 15:172b
multiplier, electrical, 4:201b
students of, 6:96a; 14:204a
Schweikart, Ferdinand Karl (1780–1859), 12:255; 2:274a; 13:264b
Schweins, Ferdinand, 13:13a, 18a
Schweitzer, Albert, 8:447a; 11:9a,b; 12:222b
Schweizerische Mineralogische und Petrographische Gesellschaft, 10:127a
Schweizerische mineralogische und petrographische Mitteilungen
founding of, 5:562b
Schweizerische Zeitschrift für Heilkunde
editors of, 12:165a
Schwendener, Simon (1829–1919), 12:255–256; 3:421b; 10:166a
Das mechanische Prinzip im anatomischen Bau der Monokotylen, 5:623b
Schwenter, Daniel, 5:409a; 8:160b; 10:264a
Geometria practica, 3:128a
Schwerd, Friedrich Magnus, 8:51b
Schwers, Frédéric, 15:442a
Sciagraphy, 9:25b
Sciatic nerve, 9:505b
Sciatic notch, 1:400a
Sciatica
arthritic, 1:400a; 3:437b
nervous, 3:437b
Science(s)
and art, 7:402b; 15:515a, 744b–745a, 749a, 753a
classification of
ancient Greek, 1:252a, 256a–b, 2:231a–b
medieval Arabic, 4:440a; 5:592a–b; 7:42a–b; 11:249a; 13:292b; 15:249b, 496b–498a
medieval Latin, 5:551b; 6:544a; 9:364a; 12:429a
seventeenth-century, 1:374a
nineteenth-century, 1:142a; 3:376a, 377a–378a, 452b–453b; 10:482b; 15:137b–138a
control of, 7:108a
definition of, 3:452a–b; 8:195b; 15:731a
financial support of
governmental, 6:280b, 384b–386a; 7:89b; 11:353a; 12:544b; 13:52a, 459b; 14:454a; 15:124a, 447b, 755b
private, 1:71b, 582b–583a; 2:471b–472b, 486b; 4:194b, 559b; 5:37a, 65a; 10:601a; 13:522b; 15:2b, 123b, 755b
and history, 3:452b–453b; 4:203b; 7:604b; 8:183b; 11:14a; 15:730b, 732b
and humanities, 4:203a
institutionalization of, 1:375b–376a; 6:279b–280b, 380b; 10:546b; 13:598b; 14:367b–371b; 15:123a–125a, 341b–342b, 346b, 744a–b, 748b, 754b

international cooperation, **7**:2a, 10b, 439b, 598b; **9**:320b(n); **15**:465b
aeronautics, **7**:248a
agriculture, **15**:493a
astronomy, **3**:428b; **4**:251a, 359a; **5**: 316a, 405a–b, 561a; **6**:30b–31a; **7**: 236a, 238a–b, 481a, 580b–581a; **9**: 164a, 551b; **10**:35a, 119b, 159a, 615a; **11**:236b; **13**:99a–b, 118b–119a, 121a, 232b, 500b; **14**:178b, 589b
atomic energy, **15**:26a–b
biology, **4**:216b; **15**:124a–125a
chemistry, **6**:481a–b; **9**:287a
geodesy, **2**:329b; **6**:239b
geography, **10**:506a; **13**:605a
geology, **13**:617b
geophysics, **5**:306a, 309b; **11**:236b
historiography, **13**:253b, 254b–255a
medicine, **9**:265b; **15**:74b–76a
meteorology, **2**:628b; **4**:174–175a; **9**: 196a–b; **11**:236b, 558a; **12**:595b; **14**:357a
military technology, **3**:330a
nutrition, **15**:420a
oceanography, **12**:191a
parasitology, **12**:453b
pharmacy, **15**:107a
physics, **2**:252b; **3**:330a; **8**:489b; **11**: 117b; **12**:521a
polar exploration, **4**:193b
and politics, **3**:330b; **7**:240a
publications, **6**:279b–280a
statistics, **11**:236b
technology, **13**:505b
see also *Carte du Ciel;* entries under International . . .
interrelations of, **1**:231b; **3**:6b; **5**:61a; **11**:26a–27a, 28a; **12**:524b; **15**:742a
and medicine, **3**:244a–245a; **7**:520b; **9**: 325b
and morality, **2**:63b–64a, 65a, 596b; **5**: 474b*; **14**:343b
philosophy of
ancient Greek, **1**:252a, 334b; **11**:26b–27a; **13**:330a
ancient Roman, **5**:228a, 232a
medieval Arabic, **4**:524b–525b; **7**:340a; **9**:29b, 171b; **11**:323b–324a; **15**:496b–498a
medieval Latin, **1**:381b; **2**:604a, 604b–605a, 607b–608a
Renaissance, **1**:97b–98a; **5**:106a–107a; **8**:195a–196a; **12**:97b
early seventeenth-century, **7**:194a–b, 294b, 539a; **9**:316a–318b, 369a–b; **10**:489b, 491b; **14**:362a–363b, 367a–b
late seventeenth-century, **5**:59a–62b, 415a–417a; **8**:153a–154a, 155b–156b, 437b; **10**:202a–b; **14**:509a
early eighteenth-century, **2**:17a–18a; **6**:557b–559a; **9**:490b, 532b, **10**:120b–121a; **13**:564b; **14**:483b
late eighteenth-century, **2**:577a–578a; **3**:380a–383a; **4**:88b–89b, 178a*; **6**: 468b–469b, 586b–587a; **7**:224b–235a, 587b, 597b–598a, 604a–b; **8**: 81a–b; **9**:188b; **10**:278b; **13**:495b; **15**:341b
early nineteenth-century, **1**:142a, 143b–144a, 178a, 354b; **2**:595a–596b; **5**: 167a, 376b; **6**:552a–553a; **8**:373b–374a; **9**:569b; **10**:183b; **15**:127b–128a
mid nineteenth-century, **2**:453b; **3**: 375b–379a, 434b, 451a–453b; **4**: 203a–204a; **5**:261b, 264a; **6**:243b–244b, 252b, 279b–280b, 573a; **8**: 515a–b; **9**:566b, 600a; **10**:355a; **12**: 570a–b; **14**:293b–294b; **15**:135a–145b, 406a–415b
late nineteenth-century, **2**:64b–65a, 277b, 563b–564a; **3**:456a; **4**:616b; **7**:105b–107a; **8**:600b–605b; **9**:291a, 332a, 382a–b; **10**:448b–449a; **11**:9a, 543b–544a; **12**:607b–610a; **13**:550b; **14**:629b; **15**:79a–80a, 422b
early twentieth-century, **2**:460a–461a; **3**:31b–34b; **4**:100b–102a, 225a–227b, 374a–b; **7**:190a; **8**:183b–184b, 257a–b, 286a–b; **9**:341a–342a, 378a, 419a; **10**:594b–595a; **11**:14a–b, 355a–358a, 388b–389a, 413a; **12**:177b–178b; **13**:345b; **14**:3a, 124a, 173b–174a, 306b–308b, 435a–b, 469b; **15**:422b–424b, 449a–b, 466b
mid twentieth-century, **1**:365b–366a; **2**:253a–b; **3**:359b; **5**:122a–123a; **6**: 548a; **7**:193a; **10**:217a–b, 424a–b; **11**:546a; **13**:98b*; **15**:455a
Chinese, **12**:385a–b
and history, **14**:294a
popular enthusiasm for, **9**:326b
practical applications of
medieval discussion, **1**:380a–b
Renaissance discussion, **11**:289a
seventeenth-century discussion, **1**: 375b–376a; **6**:141a–b, 380b–381a; **10**:565b
eighteenth-century discussion, **2**:74b, 77a–b, 175a, 179b; **3**:384b, 435b; **5**: 135a, 602a, 603a,b; **6**:43b–44a, 383b, 468b; **9**:152b–153a, 477a; **11**:141a; **12**:462b; **13**:494b–496a; **14**:354b; **15**:113a
nineteenth-century discussion, **1**:201a; **3**:557b; **6**:1b, 24a, 110a, 293a, 494b; **8**:613b; **10**:351b–352a, 361a–b; **11**: 293b–294a; **12**:433b; **13**:105a
twentieth-century discussion, **2**:169b; **6**:31a; **9**:399a; **12**:438a, 493b, 613a; **13**:505a–b
industrial, **1**:44a–45b, 304a; **2**:587a; **3**:19b, 89a; **4**:87b, 257a, 414a, 554b; **5**:583b; **6**:25a, 368b, 463a; **7**:389a–b; **8**:116a–120a; **9**:286b, 467a–b; **13**: 387a, 547b; **14**:213b–214a
see also Industrial research; specific industries; specific sciences
professionalization of, **3**:539a, 544b–546b, 553b, 556b, 557b–558a; **6**: 382b; **10**:108b; **13**:433a; **15**:728a,b, 747a, 748b, 753b
and society
eighteenth-century discussion, **3**:200a; **15**:341b
nineteenth-century discussion, **3**:389b; **4**:163b; **8**:443a; **9**:392a; **11**:219b; **12**:278a–b; **14**:114a; **15**:38a, 375b
twentieth-century discussion, **1**:506a; **2**:400a; **3**:564b; **5**:123b*, 457a; **6**:21b; **8**:184b, 440b; **10**:217a, 236b; **11**: 117b; **12**:71b, 504a–b, 505b–506a; **13**:73b, 419a–b; **15**:19a, 107a, 134a–142a, 406a–415b
Japanese discussion, **15**:746a–b, 747a–b, 748a, 752a, 755a,b
theory vs. technology, **3**:417a; **4**:87b; **9**: 461b; **10**:127a; **11**:344a, 385a, 496b, 549b, 579a; **12**:425b; **14**:197a–198a; **15**:138b, 449a–b
see also Education; Explanation, scientific; Government, and science; Historiography; Method, scientific; Philosophy, and science; Popularization of science; Religion, and science
Science, **2**:103a; **3**:130a,b; **11**:388b
founding of, **1**:583a
Science Advisory Board (U.S.), **2**:374b; **3**: 373a; **9**:399a; **15**:2b
Science and Culture
founding of, **12**:71b
Science fiction, **1**:584b; **7**:306b–307a
Science Observer Code, **3**:194a
Science Press (Lancaster, Pa.)
founding of, **3**:130b
Science Service
founding of, **3**:131a
Scientia
editors of, **9**:378a
Scientia generalis, **8**:153b–154a
Scientific American, **12**:23b
Scientific Computing Service, **3**:373b
Scientific Delimitation of Constellations, **4**: 27a
Scientific Monthly
founding of, **3**:130b
Scientific Research and Development, Office of (U.S.A.), **3**:336a; **13**:570a
Scientific Revolution, **5**:237b–238a; **6**:482a; **7**:487a,b; **11**:289a; **14**:210a
Scientism, **5**:255b; **14**:69b
Scilla, Agostino (1629–1700), 12:256–257
Scinaia, **13**:165b
Scintillation method, **3**:480a; **4**:356a; **5**:331a
Sclater, Philip Lutley (1829–1913), 12:257–258; **6**:594b; **14**:138b
Sclater, William Lutley, **12**:257b
Sclerosis
amyotrophic, **3**:205b
cerebral, **3**:490a
coronary, **9**:95a
patchy, **3**:205b
syphilitic dissipating, **1**:580a
Sclerotinia, **10**:1a; **14**:93b
Scolecite, **5**:497a
Scolia, **4**:504b
Scopoli, Giovanni Antonio, **3**:473b; **12**:558b, 561b–562a
Scorpio (constellation), **9**:297a
novae, **15**:221b(n)
Scorpion, **8**:27a
Scot
see Michael Scot
Scotia (ship), **9**:78b
Scotism, **3**:172a; **4**:255a; **10**:485a
Scotland
agriculture, **6**:578b
education, **3**:265a; **5**:69a
ethnology, **1**:562a–b
exploration of, **5**:337a
geology, **1**:394a–b; **2**:341a; **3**:191b, 569a; **5**:333b–338a, 338b–339a, 518b; **6**: 55a, 116a, 582b; **7**:69a–b; **8**:564a, 565a, 593b–594b; **9**:583a; **10**:9a; **13**:273a; **14**:132a
mineralogy, **11**:304b
paleobotany, **8**:5b; **14**:462b
volcanism, **4**:549a; **5**:336a–b
lochs, **9**:590a
oceanography, **9**:589b–590a
religion, **9**:388b
Scott, Alexander, **3**:480b(n)
Scott, David Aylmer, **1**:443a
Scott, Dukinfield Henry (1854–1934), 12: 258–260; **2**:371a; **3**:614a*; **5**:604b; **8**:5a; **13**:342a; **14**:399a
botany, **5**:498a
on contemporary scientists, **5**:605b*; **8**: 5b; **9**:602a*; **11**:155b*, 373b*; **14**: 463a*, 600a*
Scott, Ernest Lyman, **1**:441a, 442b; **6**:285a
Scott, Gladwys M., **9**:550a
Scott, Howard, **9**:312a
Scott, Leon, **7**:445b
Scott, Oliver, **5**:498a

Scott, Robert Falcon, **3**:211b
  Antarctic Expedition, 1910–1913, **3**:616b; **9**:124a
Scott, William Berryman (1858–1947), 12:260–261; **1**:130b; **10**:241a
  Introduction to Geology, **3**:595b
Scott Polar Research Institute (Cambridge, Eng.), **3**:617a
Scott-Couper
  see Couper, Archibald Scott
Scotus, John Duns
  see Duns Scotus, John
Scotus, Theophrastus (pseudonym), **3**:597a*
Screw
  advancing device, **8**:170b
  Archimedes', **1**:213b; **2**:40a; **3**:8b, 9a; **6**:34b; **8**:212b; **14**:366a; **15**:519a
  hydraulic, **9**:609b
  surface of, **13**:435a
  see also Spiral
Screw-cutter, **6**:311a, 312a
Scripps Institution of Oceanography
  founding of, **7**:447b
Scripta mathematica
  editors of, **15**:256b
Scriptorium, **3**:109b
Scripture
  see Bible
Scrofula, **4**:405a
Scrolls (math.)
  theory of, **3**:168a
Scrope, George Julius Poulett (1797–1876), 12:261–264; **3**:552a; **7**:185b
  Memoir on the Geology of Central France, **8**:566a
Scrophulariaceae, **2**:519a; **8**:513a
Scudder, Samuel Hubbard (1837–1911), 12:264; **10**:272b; **13**:456a
Sculptor (constellation), **12**:349a
Scultetus, Bartholomaeus, **2**:402a
Scurvy, **2**:617a; **3**:397a; **10**:426b*
  experimental study, **8**:362a–b
  infant, **6**:111b
  theories of, **14**:407b
  treatment of, **1**:26a, 98b, 434b; **5**:209a; **6**:111b; **7**:15a*; **8**:362b–363a, 585b; **12**:457a
Scutellum, **5**:217a
Scutum (constellation), **7**:608a
  R Scuti
    variability, **10**:608a
Sea anemone, **3**:550a
Sea cow, **13**:28b
Sea level
  atmospheric pressure effects, **11**:555a*
  changes in, **3**:193a*, 569a, 595a; **7**:144a; **8**:31b*, 187a; **10**:564a; **11**:131a
    eustatic, **5**:329b, 486b, 487a–b; **10**:503b; **13**:147a–b
Sea urchin
  anatomy, **1**:264b; **13**:556b
  classification of, **7**:401a
  eggs, **1**:264b; **2**:363a–b, 364b; **5**:51b; **6**:337a, 338a; **8**:446a; **9**:517a
  embryology, **3**:184b; **4**:187a–b, 188b; **9**:517b
  fertilization, **2**:626a; **5**:51b, 52b; **8**:357b
    artificial, **4**:12b
  fossil, **11**:336a
  physiology, **5**:148b
Sea water
  absorption of gases, **4**:127b
  chemical composition, **2**:557b; **4**:127b; **5**:72a
    bromine, **1**:416a–b; **3**:437a*; **13**:67a*
    and depth, **13**:360b
    gold, **5**:622b
    iodine, **1**:416a–b; **3**:437a*; **5**:297a
  corrosion by, **5**:190a
  distillation of, **3**:317a*; **6**:44a; **8**:363a; **9**:342b
  extraction of salts from, **1**:416b, 557b
  origin of, **13**:148a
  physical properties, **2**:557b; **7**:416b; **9**:607b
    color, **11**:265a; **15**:83a
    density, **12**:412a; **15**:431b
    equation of state, **4**:344b
    luminescence, **3**:52a; **4**:157a*, 290a; **7**:375b; **12**:560b
    penetration of light, **5**:52a; **10**:564a
    specific gravity, **8**:188b; **9**:43a
  radioactivity in, **10**:564a
  see also Hydrography, Hydrology, Oceans
Seaborg, Glenn Theodore, **3**:371a; **4**:581b; **8**:47b, 96a
Seal (animal), **13**:25b
  fisheries, **13**:352b
  intelligence, **3**:521a
  physiology, **12**:470a
Seares, Frederick Hanley (1873–1964), 12:264–266; **6**:34a*; **8**:583a*; **12**:345b–346a
Seas
  see Oceans; Sea water; individual seas by name
Seasons, **9**:141a; **15**:716b
  alternation of, **1**:260a
  astronomical, **3**:319b; **5**:106a, 345b, 346a
  dating of, **12**:447b
  descriptions of, **2**:118a
  and humors, **6**:426b
  Indian, **15**:535a
  inequality of, **1**:247b; **3**:405b; **4**:463b; **11**:190a
  lengths of, **1**:510b; **3**:21b; **4**:460a; **9**:338b; **11**:190b; **15**:211a, 677b, 706b–707a, 708a
Seaweed, **12**:457a
  iodine in, **5**:297a
  in nutrition, **8**:29b
Seba, Albert
  Thesaurus, **1**:305b, 306a
Sebacic acid, **13**:309b
Sébert, Hippolyte, **6**:545a,b
Sebokht, Severus, **7**:362a; **13**:323a
Secant, **5**:527b, 528b
  addition of, **5**:594a*
  first use of, **9**:191a
  naming of, **11**:4a
  tables of, **11**:397a
Secchi, (Pietro) Angelo (1818–1878), 12:266–270; **2**:433a; **8**:503b; **11**:412a; **12**:160b; **13**:232a,b; **14**:55b
  astronomy, **3**:13b; **6**:350b; **12**:160a,b
  spectra, classification of, **4**:180a; **5**:34a; **9**:391b
Sechenov, Ivan Mikhaylovich (1829–1905), 12:270–271; **6**:60a; **10**:431a, 433a
  acquaintances of, **2**:27a, 316a, 531a; **8**:559b; **12**:95a
  blood pump, **8**:540b
  critics of, **15**:22a*
  neurophysiology, **12**:397a; **13**:530a
  students of, **7**:34b; **12**:336b; **14**:105a
Secretin, **12**:617b
Secretion, **2**:596b
  biliary, **2**:124a, 237b, 335b; **15**:108b
  brain and, **2**:484a
  digestive juices
    hormonal control of, **5**:43a
    nervous control of, **2**:124a
  electrical phenomena, **1**:536b
  gastric, **1**:613b
    gastric pouch technique, **6**:225b–226a
    influences upon, **1**:543b; **15**:108a
    localization of, **2**:28a
    nervous control of, **2**:28a; **10**:582b; **11**:426b
    see also Gastric juice
  hepatic, **6**:226a
  internal, **2**:312a; **8**:447b; **10**:252b
    concept of, **2**:26b, 525b
  intestinal, **6**:226a; **10**:433a
  measurement of, **12**:240b
  mucous, **5**:43a
  nature of, **2**:301b; **4**:203a; **5**:469b; **6**:225b; **7**:274b; **9**:64a–b, 571b; **14**:199b
    neurological aspects, **8**:15a
  pancreatic, **1**:537a; **2**:28a, 237b; **4**:604a; **5**:484a; **6**:226a; **7**:520a; **9**:9a; **13**:171a, 223b, 403b
  in plants, **5**:582a; **9**:345b
  renal
    theories of, **2**:376a; **8**:541b
  salivary, **6**:225b; **8**:14b–15a, 452b; **10**:433a
    induced, **11**:296b
    nervous control of, **2**:30a,b; **8**:540b, 541b
    psychic, **10**:433b
  sebaceous, **1**:545a
  see also Glands
Sector (math. instrument), **5**:593b; **8**:22a
Sedative salt
  see Boric acid
Sederholm, Johannes Jakob (1863–1934), 12:271–275; **1**:237a; **4**:408b
Sedgwick, Adam (1785–1873), 12:275–279; **2**:288b, 289a, 569a; **4**:365a
  acquaintances of, **2**:567a; **3**:395b; **5**:80b; **6**:288b, 502a; **8**:104a; **12**:254a, 261b; **13**:620b; **14**:397b; **15**:151b
  on contemporary scientists, **12**:488b–489a, 491a; **13**:547b–548a
  correspondence of, **3**:297b
  critics of, **3**:567a,b
  Discourse on the Studies of the University, **5**:67b
  influence of, **3**:192a, 297a, 565b; **5**:436b; **7**:45a; **8**:461a; **11**:501b; **12**:52a
  influences on, **2**:570a; **8**:486b
  and Murchison, **9**:583a, 583b–584b; **12**:276b, 277a–b
  stratigraphy, **13**:620b–621a
  students of, **7**:185b; **13**:272b
Sedgwick, William Thompson, **7**:170b; **9**:142b; **14**:251a, 424a, 425b–426a
Sédillot, Charles, **10**:389b
Sédillot, Jean Jacques Emmanuel, **14**:575a
Sédillot, Louis Pierre Eugène Amélie, **1**:42a; **6**:203b; **7**:259a; **11**:227a
Sedimentation, **3**:568a; **5**:39a, 156a, 577b; **8**:463b
  biological contribution to, **2**:556a; **3**:293b; **5**:533b
  cycle of, **2**:89b; **5**:72a; **6**:582a
  and tectonics, **6**:168a; **10**:507a
  total amount of, **3**:293b
Sedimentology
  Renaissance, **8**:242a
  seventeenth-century, **13**:34a
  eighteenth-century, **14**:132b
  nineteenth-century, **2**:289b; **3**:293b; **8**:565a; **10**:171a, 502a; **12**:514b, 543a; **14**:398a
  twentieth-century, **1**:236b, 470a, 623a–b; **6**:229a; **10**:126b, 507a; **11**:321a; **13**:518a–519a
  marine, **11**:372a
  see also Geology; Rocks, sedimentary

Sediments, **3**:434a
deep-sea
composition of, **10**:564a
oceanic, **11**:113a–b
radioactivity in, **10**:564a
Sedov, Georgy Yakovlevich (1877–1914), 12:279–280; **12**:301a; **14**:50b
Sée, Marc, **3**:490a
See, Thomas Jefferson Jackson (1866–1962), 12:280–281; 10:35b*
Seebeck, Thomas (1770–1831), 12:281–282; **3**:497a; **12**:49b
Seeds
anatomy, **9**:418a; **13**:165b
classification of, **5**:217a
composition of, **1**:293a–b; **12**:124a; **13**:405b
proteins, **9**:284a,b; **10**:242a–243b
dispersing of, **9**:60b
evolution of, **12**:260a
fossil, **5**:498a; **11**:373a
morphology, **12**:357b
nature of
ancient Greek study, **13**:332b
Renaissance study, **5**:379a; **8**:206b
seventeenth-century study, **4**:135b; **8**:128a
nineteenth-century study, **2**:492a
twentieth-century study, **1**:561a–b
physiology, **2**:290b
absorption of water, **3**:44a
development, **5**:582a; **13**:405b–406a
metabolism, **12**:59a
resistance, **4**:131a
self-burying, **3**:582a*
as taxonomic criterion, **11**:314b
temperature, effects of, **12**:557a; **13**:405b
see also Germination, Pollination
Seeley, Harry Govier, **2**:504a,b
Seeliger, Hugo von (1849–1924), 12:282–283; **4**:273a; **7**:236b; **8**:599a; **12**:247b, 248b, 250b
Sefström, Nils Gabriel, **2**:94b; **3**:346b; **11**:464b; **14**:447a
Segner, János-András (Johann Andreas von) (1704–1777), 12:283–284; **4**:471a, 481a; **10**:120a
theorem of, **2**:574a
Segre, Corrado (1863–1924), 12:284–286; **3**:167b; **8**:285a*; **11**:390b*; **12**:57a*; **13**:6b*, 623b*
acquaintances of, **3**:117a
critics of, **13**:133a
geometry, **5**:569a; **7**:185a; **8**:504b; **11**:46a; **13**:125a
influences on, **10**:139b
Segre lines, **12**:285a
Segre varieties, **12**:284b
students of, **3**:399a; **4**:373a,b, 522b; **12**:330b
Zeuthen-Segre invariant, **12**:284b
Segrè, Emilio, **4**:578b–581b *passim;* **8**:96a; **9**:37b; **10**:136b
Segregation (biol.); **8**:589a; **9**:618b; **14**:34a
laws of, **9**:280a; **14**:96b, 99b
and reduction division, **3**:422b
somatic, **1**:505a
Segregation (mineral.), **4**:141a; **5**:576b
Séguin, Armand (1767–1835), 12:286–287; **3**:159b, 455a; **8**:84a; **10**:497a; **14**:79b
Seguin, Marc (Seguin Aîné) (1786–1875), 12:287–289
Seiches, **3**:265a; **6**:480a; **12**:342a; **15**:158b
Seide, Oskar, **3**:246b
Seidel, Ludwig, **10**:559a
Seidel, Philipp Ludwig von (1821–1896), 12:289–290; **2**:430a; **5**:144a; **7**:51a, 522b; **10**:194a*; **11**:7b; **13**:22b
Seignetto-electricity, **7**:527a
Seismograph, **4**:500a; **7**:9b; **8**:388a; **9**:61a, 407a, 444b
galvanometric, **5**:462a
horizontal-pendulum, **10**:210b
inverted-pendulum, **14**:327a–b
portable, **14**:327b
reflection, **4**:8a
refraction, **4**:8a
Seismological Society of America
organization of, **8**:99b
Seismological Society of Japan, **9**:407a
Seismology
eighteenth-century, **14**:453a
mid nineteenth-century, **9**:61a; **10**:204a
late nineteenth-century, **1**:364b; **4**:266a, 500a; **5**:89b; **7**:31a; **9**:406b–407b, 593a, 594a; **10**:203a–b, 210b–211a; **12**:187a, 192a; **15**:158b
early twentieth-century, **3**:357b; **4**:515a*; **5**:462a–b; **6**:147a; **7**:9a–b, 413b, 474a; **8**:99b, 388a; **9**:181b, 443b–445a; **10**:231a–232a; **11**:361b–362a; **13**:500b; **14**:327a–b, 402b
mid twentieth-century, **5**:596a–597a; **12**:341b–342a
experimental, **9**:407a; **10**:211a
historical, **7**:9b; **9**:407a
naming of, **9**:61a
theoretical, **7**:595a
in volcanology, **7**:63a
see also Earthquakes
Seismometry, **5**:462a
Séjour, Achille-Pierre
see Dionis du Séjour, Achille-Pierre
Seki Kowa
see Seki Takakazu
Seki, Takakazu (*ca.* 1642–1708), 12:290–292; **1**:90a–b; **3**:270b; **8**:314b; **14**:540a; **15**:745a–752b *passim*
Byodai meichi, **15**:751a
Daijutsu bengi, **15**:751a
Selaginellales, **2**:160a; **10**:575b
Selander, Nils Haqvin, **13**:111a
Select Society (Edinburgh), **9**:480b
Selective breeding, **3**:120b–122a; **4**:270b–271b; **5**:10b; **6**:594a; **15**:509b–510b
see also Hybridization
Selenates
crystals, **9**:424b
Selenic acid, **9**:424b
Selenides, **12**:46b
Selenium
allotropy, **6**:438b
compounds of, **3**:476a
optical activity, **11**:86a
as dehydrating agent, **4**:90b–91a
discovery of, **2**:94b
preparation of, **5**:431b, 595b
reactions of, **5**:7b
spectrum, **3**:125a
vapor density, **13**:468a
Selenococcidium, **4**:207a
Selenography
Renaissance, **5**:400a
seventeenth-century, **3**:103a; **4**:128a; **5**:542b; **6**:126a, 361b–362a; **8**:25b; **9**:485b; **11**:395a, 411b
eighteenth-century, **4**:166a; **8**:178b, 321b; **9**:233b, 234a; **12**:226b; **14**:419a–b
nineteenth-century, **1**:568b–569a; **2**:147a–b; **9**:2a; **10**:584a–b; **12**:192a, 267a
twentieth-century, **9**:415a
see also Moon
Seleucus of Seleucia, **1**:247b; **6**:305b; **15**:678b
Sell, Ernest, **8**:349a
Selmi, Francesco, **12**:163b
Selwyn, Alfred Richard Cecil (1824–1902), 12:292–294; **1**:51a; **6**:564a; **13**:525a
Selye, Hans, **3**:352b
Selys-Longchamps, Marc de, **1**:601a–b
Semanario del nuevo reino, **3**:16b
Semang race, **9**:144a
Semantics, **2**:277b; **14**:391a
rational, **3**:278b
see also Language, Linguistics
Semen
formative power, **12**:151a
nature of, **9**:482b; **10**:598b
Semenov, Nikolai Nikolaevich, **6**:404a; **15**:251a,b
Semenov-Tian-Shanskii
see Semyonov-Tyan-Shansky
Semeology, **14**:365b
see also Symptomatology
Semicircle
center of gravity, **8**:224b
Semicirculus (geod. instrum.), **2**:419a
Semiconductors, **15**:251b, 449b
electric breakdown in, **5**:161a
manufacture of, **15**:251b
Semimetals, **2**:421b
Semiminerals, **2**:143a
Semina, **6**:256b, 257b; **10**:309a–b; **12**:335b
Seminaria, **5**:107a
Seminvariants, **3**:164b
Semiotics, **14**:365b
Semmelweis, Ignaz Philipp (1818–1865), 12:294–297; **8**:409a; **10**:389b; **14**:42b, 297a
Semnopithecus, **2**:110a
Semon, Richard Wolfgang (1859–1918), 12:297–298; **12**:218a
Semper, Carl Gottfried (1832–1893), 12:299; **1**:421a; **5**:34b
Sempervivum, **7**:166a
Semple, John Greenlees, **12**:228b
Semyonov-Tyan-Shanskу, Petr Petrovich (1827–1914), 12:299–302
Senac, Jean-Baptiste (*ca.* 1693–1770), 12:302–303; **4**:267b; **8**:352a; **11**:99b, 100a
Nouveau cours de chymie, **8**:619a
Traité de la structure du coeur, **8**:526b
Sénarmont, Henri Hureau de (1808–1862), 12:303–306; **4**:66a, 222b; **9**:392b
Senckenbergische Naturforschende Gesellschaft (Frankfurt am Main), **2**:625b
Senderens, Jean Baptiste, **12**:47a
Sendivogius (Sędzimir or Sędziwoj), Michael (1566–1636), 12:306–308; **5**:420a; **8**:311a
Novum lumen chymicum, **1**:571a
Senebier, Jean (1742–1809), 12:308–309; **2**:286b; **12**:123b*, 561b
acquaintances of, **3**:43b; **11**:132b; **12**:560b; **14**:72a, 79b
influence of, **3**:45a; **8**:82b
plant physiology, **7**:12b; **8**:345b
translations by, **12**:560a, 563a
zoology, **12**:555b, 562b
Seneca, Lucius Annaeus (*ca.* 4 B.C.–A.D. 65), 12:309–310
editions of works, **3**:534a
geography, **15**:88a
historiography, **3**:391a; **13**:297a
influence of, **1**:381b; **5**:121a; **7**:27b; **13**:92a; **14**:377a(n)
influences on, **11**:105b(n)
meteorology, **13**:93b
philosophy, **14**:375a(n); **15**:641b
Senecio
alkaloids, **15**:2b, 103b
Senegal
natural history, **1**:58b

Senescence
see Aging
Sengers, Jan Vincent, **4**:376b
Senn, François Louis, **13**:493a
Sennert, Daniel (1572–1637), 12:310–313; **1**:550a; **7**:164b; **11**:511a
De chymicorum cum Aristotelicis et Galenicis consensu et dissensu, **10**: 306a; **12**:335b
Hypomnemata physica, **9**:15a
Senoi tribe, **9**:144a
Sens, Council of (1140), **1**:2a
Sensation
electrical excitation, **11**:474a; **14**:77a–b, 78b
of heat, **7**:599a
laws of, **1**:584a
of motion, **8**:597a,b; **11**:215a–b
physiology of, **3**:531a; **4**:63a–b, 64a; **5**: 184b–185a; **7**:516b; **8**:596a–598a; **11**:135b, 214b–215a, 425b–426a; **12**:270b; **13**:393a; **14**:113b, 200b–201b, 527a
of pressure, **10**:267b; **14**:200b
psychology of, **8**:596a–597a, 603b–605a; **10**:485a–b
quantitative study, **4**:557b–558a; **14**: 200b–201b
of temperature, **14**:200b
theories of
ancient Greek, **1**:260b, 262a; **4**:33a, 368b; **13**:94a, 333b
ancient Roman, **11**:42a
medieval Arabic, **13**:407b
Renaissance, **5**:106b; **8**:197a–b, 203a–b
seventeenth-century, **3**:381a; **4**:62a, 63a–b, 64a; **6**:158a–159a, 445b; **7**: 612b; **14**:406b, 408a
eighteenth-century, **3**:1b, 381a–382a; **8**:115a
nineteenth-century, **1**:404a; **4**:557a–558a; **5**:44b, 445b; **6**:242b, 243a, 247b–249a, 300a–b; **9**:561b; **11**:173b; **12**:270b
twentieth-century, **3**:335a; **14**:200b–201b
Japanese, **15**:738a
specific nerve energies, **9**:569b–570a
threshold, concept of, **14**:200b–201a
time sense, **3**:531a
unconscious, **6**:159a
see also Esthesiology, Hearing, Olfaction, Pain, Perception, Taste, Touch, Vision
Sensationalism
see Epistemology, sensationalist
Sensibility
seventeenth-century discussion, **6**:158b
eighteenth-century discussion, **5**:55b–56b; **6**:64b–65a; **8**:115a; **14**:321b
twentieth-century discussion, **3**:247b
deep, **5**:185a
differential, **3**:15a
opposition to
eighteenth-century discussion, **3**:15a–b
recurrent, **9**:10a
universal, **4**:89a–b
vegetable, **1**:260b; **3**:422a; **5**:624a; **7**:538a; **11**:315b
Sensilla, **14**:314b
Sensitive plant
see *Mimosa pudica*
Sensitizer, immunological, **2**:300b
Sensitometry (photographic), **1**:21b; **4**:282b–283a
Sensorium commune, **11**:159a,b; **12**:510a
Sentences
see Peter Lombard, *Sentences*
Sentiment intérieur, **7**:591a–b
Sepsis
experimental study, **1**:574b; **9**:10b
generalized, **12**:296a–b
Septicemia, **1**:323a; **2**:598b; **5**:219a; **9**:95a; **10**:389a–b, 397b, 407b
and anthrax, **10**:387a–b, 397a
bovine, **3**:588a
Sequences (math.)
divergent, **13**:428b
fundamental, **3**:53b–54a; **12**:189b
see also Series
Sequential analysis, **14**:121b–122a
Sequents
midsequent theorem, **5**:350b–351a
Serasin, Eduard, **5**:52a
Serenus (*fl. ca.* fourth century A.D.), 12: 313–315; **1**:191b; **4**:491a; **6**:71a; **9**: 269a
translations of works, **3**:364b
Sergent, Edmond (1876–1969), 12:315; **8**: 66b*
Sergent, Étienne, **12**:315a
Sergius of Theodosiopolis (Reshᶜayna)
see Sarjiyūs al-Raᵓs ᶜAynī
Seri Indians, **7**:149a
Sericulture, **11**:311a; **12**:316b–317a; **15**: 661a(n)
Series, **3**:53a, 255b; **4**:117a; **12**:375b
algebraic, **14**:540a–b
anticipation of, **3**:615b
arc tangent, **8**:163a
arithmetical, **3**:268b; **4**:550a; **6**:403b
summation of, **4**:606b; **11**:590a; **15**: 693a–694a
vs. geometric, **14**:471a
asymptotic behavior, **2**:320a; **13**:55b
Bernoulli, **2**:52b
binomial, **1**:15a; **2**:48b, 507b
convergent, **9**:453a–b
Bohr-Fourier, **2**:238b
Cantor, **3**:54a
composition, **7**:168a
conditionally convergent, **3**:136b
convergent, **1**:427a; **2**:50a; **3**:126b, 146b; **4**:125a, 125b–126a, 342b; **5**:204b; **6**:113b; **7**:351b, 514a, 567b; **8**:62a, 62b–63a, 163a, 558b; **9**:303b; **10**: 228b; **13**:81b, 153a, 250b; **15**:305a–b, 306a–308a, 364a,b, 486a
d'Alembert's theorem, **1**:116a
rapidity principle, **8**:253b
sum of, **2**:38a; **3**:137a
testing of, **2**:287b; **6**:577b*; **15**:98b
theory of, **1**:207a; **3**:136a; **5**:526b; **14**: 610b
development of, **6**:403b; **11**:599a; **12**:93a
Dirichlet, **1**:306b; **2**:238b; **4**:103a, 125a; **6**:114b*, 214b; **7**:408a; **11**:256b; **12**: 451b
transformations, **6**:571b
divergent, **2**:38a, 57a, 304a; **3**:178a; **4**: 475a; **5**:450a; **6**:114b*; **7**:575b; **8**: 558b; **9**:303b; **10**:573a; **11**:599a; **12**: 237b; **13**:58a
theory of, **1**:116a; **2**:49b, 303b; **3**:136a; **4**:476a–b, 561b; **11**:268a; **14**:122b
double
convergence of, **3**:136a; **13**:81b
Eisenstein, **4**:342b; **6**:214b, 571b
expansion in, **2**:48b; **3**:222b; **5**:527b; **8**: 162b; **10**:111a; **13**:78a, 480a; **14**:500a; **15**:484b, 486a
exponential, **2**:48b, 49a–b
derivation of, **2**:50a
factorial, **10**:117b
Farey, **14**:95a
Fibonacci, **4**:607a, 611a; **5**:409a; **8**:531b
Fourier, **1**:147b; **4**:103a, 196b, 476a; **5**: 95a,b, 97a,b, 122b; **6**:572a; **7**:408a; **8**:110b, 367a, 382b, 389a; **9**:202b; **11**:156b, 448a, 454b; **13**:78a, 374b, 424a, 428a; **14**:221b, 282a; **15**:384a
convergent, **1**:427a; **4**:125b, 205b, 561b; **5**:96a; **6**:472b; **7**:167b, 514a; **10**:248b
divergent, **4**:206a
integrability of, **4**:206a
singularities of, **4**:561b
summability of, **4**:561b; **7**:54a
theory of, **2**:238b; **3**:54a, 178b; **4**:126a, 205b–206a, 561b; **6**:114b*; **13**:27a; **14**:573a
fundamental, **3**:53b–54a
gamma, **13**:69a
geometrical, **3**:136a, 268b; **13**:60a
summation of, **4**:572a, 606b; **11**:590a; **12**:75a
Gregory's, **5**:527b; **15**:633a(n)
harmonic, **8**:162a
divergence of, **2**:49b; **9**:303a
infinite, **2**:53a
hypergeometric, **5**:204a, 302a, 482a; **7**: 521b, 523a; **9**:127a; **11**:268a, 448a; **12**:247a
indeterminate, **11**:599a
infinite
medieval Latin discussion, **10**:228b; **13**:192b–193b
Renaissance discussion, **3**:126a; **13**: 210a
late seventeenth-century discussion, **2**:507a,b; **5**:521a, 528a–529a; **14**:150b
early eighteenth-century discussion, **2**:56b; **3**:460a; **4**:475b–476b, 477a; **13**:265a–266b
late eighteenth-century discussion, **1**: 90b, 207b; **6**:577b*; **7**:599b
early nineteenth-century discussion, **1**:239a; **2**:276b; **5**:302a
mid nineteenth-century discussion, **5**: 572b–573a; **8**:429a
late nineteenth-century discussion, **6**: 95b; **8**:61b; **13**:263b–264a
early twentieth-century discussion, **1**: 307a, 407b; **2**:490b; **3**:146b; **7**:574b; **8**:253b
bounds of, **2**:50b
convergence of, **4**:35b; **7**:508a; **8**:611a; **10**:48a, 51a, 67b; **15**:65b
expansions, **4**:126b; **5**:299a; **10**:46a–47a, 48a, 51a, 66a,b, 67b
general term of, **5**:450a
integration of, **8**:110b–111a
sum of, **3**:136a, 227b–228a; **5**:449b; **7**:54a; **9**:303a, 500a; **10**:327a–b; **13**: 69a
theory of, **7**:102a
transformation of, **4**:476a; **5**:449b–450a; **13**:68b–69a
integral, **3**:269a–b; **6**:309a
interpolating, **5**:342b; **13**:69a
inversion of, **6**:403b
L, **1**:306b–307a; **6**:214b
lacunary, **6**:3b
Lagrange, **3**:140a; **7**:566b–567a
Lambert's, **7**:599b
Laplace, **4**:561b
Laurent, **3**:138b; **8**:62a
Leibniz, **8**:163a
limits of, **5**:201b
logarithmic, **2**:49b; **8**:161a,b, 163a; **9**: 304a; **10**:47a
Maclaurin, **8**:611a; **13**:266a
Malmsténian, **8**:253b
Mercator, **5**:528a; **8**:161b; **10**:87a(n)

orthogonal, **1**:427b; **11**:156b; **14**:573a
convergence of, **6**:451b
p-adic, **3**:222a
Poincaré, **7**:399a
power, **2**:37b–38a; **3**:139a, 223a; **4**:474b, 476a,b, 478a, 561b; **5**:152a, 204a; **7**:570b; **8**:164a, 558b; **9**:308a, 503a*; **10**:47a, 48a; **11**:149a; **13**:78a, 81b; **14**:222b; **15**:370a, 384b
convergent, **3**:136a, 140a; **14**:220b
summation of, **1**:15a; **3**:270b; **11**:229b–230a
recurrent, **2**:37b; **4**:607a, 611a; **5**:98b; **7**:561a; **9**:454a; **13**:68b, 69a; **15**:278a–279b
recurro-recurrent, **15**:278b–279b, 280b, 282a, 285b, 306b
semiconvergent, **3**:136b
Stirling, **8**:367a; **13**:69a
summation, **4**:561b; **8**:62a; **9**:499b–500a; **10**:228b; **13**:267a, 350a; **15**:10a
Taylor, **3**:229a; **4**:477b; **5**:204a, 528b; **7**:570b; **8**:367a; **13**:266a
continuation of, **6**:3b
convergence of, **3**:136a, 223a; **4**:547b; **5**:204b; **6**:3b
development of, **5**:201a
singularities in, **4**:548a
theory of
seventeenth-century, **6**:304b
eighteenth-century, **5**:209b, 450a; **7**:162b; **15**:278a–279b, 306a–308a
nineteenth-century, **2**:288a; **12**:511b; **13**:81b, 263b
twentieth-century, **6**:113b; **13**:391a
time, **5**:10b
trigonometric, **2**:238b; **3**:53b; **4**:124a, 476a, 478b; **5**:95a, 96a, 528b; **6**:113b; **8**:110b; **10**:327b; **13**:562a; **14**:221b
convergence of, **4**:125b–126a, 548a*; **8**:429a, 558b; **13**:139a
summation of, **4**:126a
theory of, **8**:112a, 432b, 558a; **13**:562b
see also Progressions, mathematical; Sequences; Singularities, of series; Transformations (math.), of series
Seringe, Nicolas, **7**:165a–b
Serology
nineteenth-century, **2**:300a–b; **5**:564a–b; **7**:391b–392a; **11**:429a–b; **12**:88a–b; **14**:218b–219a
twentieth-century, **2**:300a–b; **6**:432b–433a; **7**:100a–b, 623a–624b; **9**:545a–b; **10**:18b, 142a; **12**:481b–482a; **13**:228b–229a; **15**:522a–523a
and anthropology, **6**:433a
see also Immunology
Serotherapy, **1**:575a–577a; **3**:22a–b; **4**:297a–299b, 404a,b; **7**:391b–392a; **10**:18b; **11**:271b; **15**:96b–97a
reactions to, **5**:317a
see also Antitoxins; Serums, therapeutic
Serotonins, **1**:534a
Serpentine (rock), **2**:289b
origin of, **1**:394b
Serranidae, **1**:264b
Serres, Antoine Étienne Reynaud Augustin (1786–1868), 12:315–316; 1:518a; **5**:357a; **8**:254b; **12**:318a
Serres-Meckel law, **12**:316a
Serres, Olivier de (or des) (1539–1619), 12: 316–317
Serres de Mesplès, Marcel Pierre Toussaint de (1780–1862), 12:317–318; 3:264a
Serret, Joseph Alfred (1819–1885), 12:318–319; 2:349b; **3**:178a; **7**:564a; **8**:382b, 385a; **11**:414b
differential geometry, **2**:288a, 350a*
Frenet-Serret formulas, **5**:158a
group theory, **4**:83a; **5**:264a; **7**:167b
Lehrbuch der Differential- und Integralrechnung
editions of, **12**:150a
Serret, Paul Joseph, **12**:318b
Sertoli, Enrico (1842–1910), 12:319–320; 5:459b
Sertoli cells, **12**:319b
Sertularia, **6**:590a
Sertürner, Friedrich Wilhelm Adam Ferdinand (1783–1841), 12:320–321; 3:159b; **11**:495a*
Serullas, Georges-Simon (1774–1832), 12: 321–322; 8:331a, 333b; **9**:91b
Serum, blood
see Blood, serum
Serums, therapeutic
anticholeraic, **2**:300b
antidiphtheric, **12**:88a; **15**:454a
antimeningitis, **5**:40b; **12**:482b; **15**:522b
antiplague, **8**:556b
antipneumococcus, **12**:482b
antistreptococcus, **4**:404b
antitoxic, **4**:297b
antivenin, **3**:22b
artificial, **4**:398b
hemolytic, **2**:300b
hypersensitivity, **12**:482a
polyvalent, **15**:522a
precipitating, **2**:300b
standardization of, **1**:576a; **11**:271b
see also Antitoxins, Serotherapy
Servetus, Michael (*ca.* 1511–1553), 12:322–325; 3:356a; **5**:119a, 121a
Christianismi restitutio, **3**:356a
Servite Order, **12**:105b
Servois, François-Joseph (1767–1847), 12: 325–326; 1:207b, 239b; **5**:111b, 367b; **7**:566b
Servomechanisms, **1**:570b; **6**:223a
Sesbania, **12**:259a
Seshagiri Rao, K., **11**:265a
Sessé y Lacasta, Martín de (*ca.* 1751–1808), 12:326–328; 3:177a; **9**:432b, 433a
collections of, **15**:471a
Sestini, Aldo, **3**:533a
Sestini, Benedict, **4**:168a
Setchell, William Albert (1864–1943), 12: 328–329
Setchenov
see Sechenov
Sète, Station Zoologique de, **12**:45b
Seth, Andrew, **6**:26a
Seti I, **15**:706a
Seton, Alexander, **4**:96a; **8**:311a; **12**:307b
Sets, **3**:150b
A, **8**:559a
analytic, **13**:562a
B, **8**:558b–559a
Borel
cardinality of, **6**:177a
Borel-measurable, **2**:304a
calculus of, **2**:297a–b
Cantor, **3**:55a
closed, **9**:249a
compact, **1**:407b
dimension of, **9**:248b–249a
convex, **9**:414a
countable, **2**:303b–304a; **3**:55a, 57b; **8**:112a; **14**:222b
duality principle, **2**:297b
equivalent, **3**:55b–56a; **7**:168b
equivalence theorem, **2**:58a; **3**:57a
exterior measure, **7**:167b
finiteness of, **4**:4a
infinite, **2**:277a–b; **3**:56a; **8**:111a–b; **12**:364b
mapping of, **3**:54a–55a
measurable, **8**:558b
null, **11**:58a
ordered, **5**:108a; **6**:177a
partially ordered, **6**:177a
perfect, **1**:407b
perfect discontinuous, **3**:55a
power of, **3**:57b
projective, **8**:559a
Schubert, **12**:228b
theory of
nineteenth-century, **2**:277a–b, 303b–304a; **3**:54a–57b; **6**:95b; **10**:443a; **11**:60b; **12**:196a, 217a, 599b
twentieth-century, **1**:407b; **2**:351b, 513a; **6**:176b–177a, 570a; **7**:71b–72a, 444b; **8**:112a, 558b–559a; **11**:413b; **12**:13b, 189b, 364b, 426b, 451b; **13**:562a; **14**:89b, 304a, 613b–615a
applications of, **2**:58a
axiomatic, **1**:427b; **5**:108a; **14**:89b, 614b–615a
cardinality, **2**:58a
definite property, **5**:108a; **14**:614b–615a
δs-operations, **6**:177a
descriptive, **1**:427b, 428a; **6**:177a
foundations of, **2**:513b; **5**:108a–b
maximal principle, **6**:177a
sieving process, **8**:559a
well-ordering theorem, **14**:614a
transfinite, **1**:407b, 428a; **3**:55b
Settala, (Senator) Lodovico, **1**:316a; **3**:60b; **9**:4b
Seubert, Karl, **9**:348a, 351a,b
Seurat, Georges, **3**:241b
Seurat, Léon Gaston, **9**:185b
Seven wives, problem of, **4**:606b, 612a
Seven Years War, **10**:615a
medicine, **8**:362a–b
Severgin, Vasily Mikhaylovich (1765–1826), 12:329–330; 12:514a
Severi, Francesco (1879–1961), 12:330–332; 3:117a; **4**:373b; **5**:569a; **10**:139b; **12**:285a
geometry, **11**:45b, 55a; **12**:228b
Severin, Christian (also known as Longomontanus) (1562–1647), 12:332; 2:406a; **6**:520b; **7**:294b; **13**:433b; **14**:471a
astronomy, **2**:412a, 413b; **7**:303b
correspondence of, **2**:413a; **11**:561b
critics of, **3**:63b; **10**:495b
tables, **6**:515b; **9**:611a
Severino, Marco Aurelio (1580–1656), 12: 332–334; 1:482a; **2**:308a, 309a; **6**:508b; **10**:567b; **13**:474a
Severinus, Petrus (or Peder Sørenson) (*ca.* 1542–1602), 12:334–336; 4:208b; **7**:539a; **9**:440a
Idea medicinae philosophicae
commentaries on, **3**:596b
Severtsov, Aleksey Nikolaevich (1866–1936), 12:336–339; 7:454a; **12**:409b
Severus Sebokht, **7**:362a; **13**:323a
Sèvres porcelain works, **2**:493b; **3**:5a, 560b; **6**:236b; **8**:54b, 621b; **11**:353a; **13**:310a
Sewage, **3**:6b; **8**:450b
deodorizing of, **12**:478b
purification of, **3**:22b; **8**:403a
utilization of, **3**:475a; **8**:93a
Seward, Albert Charles (1863–1941), 12: 339–340; 1:205b; **3**:582b*; **6**:116b*; **9**:618a*; **13**:90a*, 344a*, 345a,b
Sewell, William, **14**:187b
Sewerage, **10**:559a–b
Sex
determination of
dietetic, **11**:211b

Sex (cont.)
environmental, 6:337a, 518a; 8:588a–b; 9:518a
genetic, 2:456a; 3:121a, 422b, 492b–493a; 5:453b; 8:586b, 588a–589a; 9:282b, 518a–b; 10:570a–b; 11:212a–b; 12:416b, 417a; 13:90a, 135b, 136a, 137b; 14:431a–432a, 595b
hormonal, 8:358a–359a
theories of, 9:496b, 518a–519a, 522a–b
inversion, 10:570a
physiology of
see Physiology, sexual
see also Sexual characteristics, Sexuality
Sex cells
differentiation of, 1:545b
Sex education, 9:86a
Sex ratio, 3:120b; 8:588b, 589a; 9:518a–b, 520a
Sextant, 2:148b, 153a; 3:262a, 528a; 5:24b, 434a; 6:6a, 362b–363a; 7:81b, 338b; 8:22a; 9:5b; 14:491b
Brahe's, 2:402b, 403b, 406a, 411a
bubble, 12:249b
differential, 5:466a
Fakhrī, 7:260a, 353a–b; 13:536a–b
portable, 2:471b; 6:363a
scale division, 2:141a
Sextius Niger, Quintus, 4:119b, 120b
Sextus Empiricus (*fl. ca.* A.D. 200), 12:340–341; 1:150b; 4:32a–b; 5:347a; 10:181a; 11:105b(n); 14:612b
Sexual characteristics
intersexuality, 5:453b
secondary, 2:344b; 3:575a; 14:595b
hormonal influences on, 10:570a–b
physiological conditioning of, 10:570a
Sexual selection, 3:574b–575a; 13:135a
Sexuality, 5:385b
in animals, 3:521a; 7:64b
of cryptogams, 3:337b; 5:217a, 539b; 11:152a, 153a–154b
infantile, 5:176b–177a
development of, 5:177b–178a, 179a, 180b
in protozoa, 7:99b
in psychoanalysis, 2:448b; 5:172b, 176a, 179a; 7:190b
in yeasts, 5:584b
see also Plants, sexuality
Seyffer, Karl Felix von, 2:268a; 12:518b
Seymouria, 14:410a
Sezawa, Katsutada (1895–1944), 12:341–342
Sforza, Ludovico (il Moro), 8:193b, 206b, 208a, 211b, 214a, 242a; 10:269b, 271a
Shackleton Expedition (1907–1909), 3:591a; 9:198a
Shādhān, Abū Saᶜīd, 9:160b; 13:538a
Mudhākarāt, 1:35a, 39b*
Shadows
medieval discussion, 2:153a–b; 5:612b; 6:195a–b, 544b; 7:2b
Renaissance discussion, 4:260a
seventeenth-century discussion, 10:52a–b
eighteenth-century discussion, 9:471a, 475b
nineteenth-century discussion, 2:473a; 9:474a; 14:343a
colored, 13:351b
Indian discussion, 9:22b
see also Gnomonics
Shadwell, Thomas, 6:483b; 8:415b
Shafarevich, Igor Rostislavovich, 4:474b
Shaffer, Philip Anderson, 1:442b; 8:447b
Shaftesbury, Earl of
see Cooper, Anthony Ashley
Shain
see Shayn
Shakerley, Jeremy (1626–*ca.* 1655), 12:342–343; 14:446b
Shale
black, 11:594a; 13:519a
copper, 12:195a
fossils in, 8:32b; 14:120b
Shaler, Nathaniel Southgate (1841–1906), 12:343–344; 1:74b
acquaintances of, 3:592b, 594a; 5:49b; 14:1a
geology, 3:593a
paleontology, 5:50b*; 14:517b
students of, 2:500a; 7:62b, 144a
Shāmila (astron. instrum.), 13:536a
Shamlī, 15:236b
Shane, Charles Donald, 14:521b
Shang yuan (Grand Cycle), 3:250b–251a
Shankland, Robert Sherwood, 15:8a
Shanks, William (1812–1882), 12:344–345
Shannon, Claude Elwood, 5:9a
Shanyavsky People's University, Russia
biology laboratory, 7:454b, 456a
Shapiro, Arnold, 14:311b
Shapley, Harlow (1885–1972), 12:345–352; 1:352b; 5:363a; 9:330b; 12:24a*, 265b
astronomical distances, 1:398a; 3:509b; 6:30b, 529b; 8:106a, 582b; 13:98b, 475a
binary stars, 2:196b; 12:19b–20a
and Curtis, 6:529b; 8:582b; 12:347b–348a
spiral nebulae, 6:529b; 8:582b
Shapley-Ames Catalogue, 12:349a
Sharaf al-Dawla, 7:353b; 11:240a; 13:149a
al-Sharīf al-Idrīsī
see al-Idrīsī, Abū ᶜAbd Allāh Muḥammad ibn Muḥammad ibn ᶜAbd Allāh ibn Idrīs, al-Sharīf al-Idrīsī
Shark
anatomy, 9:572b–573a; 13:33b
embryology, 3:610b; 13:527b
eye, 6:326b
fossils
teeth, 1:482a, 491b; 7:252a–b; 13:33b–34a
reproduction, 13:33b
Sharonov, Vsevolod Vasilievich (1901–1964), 12:352–354; 13:409a*
Sharp and Roberts, manufacturing engineers, 3:287a
Sharpe, Richard Bowdler, 5:481a; 12:258b
Sharpey, William (1802–1880), 12:354; 2:598b; 5:80a; 12:355a; 13:274a
correspondence of, 8:401b
embryology, 1:477a
Sharpey's bone fibers, 6:183b; 12:354a
students of, 5:81a; 8:400b, 401a; 10:204b; 11:517a; 14:397b
Sharpey-Schäfer, Edward Albert (1850–1935), 12:355–357; 1:535b; 5:80b; 6:60b; 9:19b; 12:617b
endocrinology, 1:442b; 10:205a; 15:106b
histology, 11:517b
neurophysiology, 8:17b(n)
Text Book of Physiology, 12:397b
Sharrock, Robert (1630–1684), 12:357; 6:41a
Ibn al-Shāṭir, ᶜAlāᵓ al-Dīn Abu'l-Ḥasan ᶜAlī ibn Ibrāhīm (*ca.* 1305–*ca.* 1375), 12:357–364; 13:511a; 15:259a
cosmology, 3:406b; 11:202a, 251a–b
Shatsky, Nikolai Sergeevich, 7:251a,b
Shatunovsky, Samuil Osipovich (1859–1929), 12:364–365
Shaver, John Rodney, 1:498b
Shaw, George, 6:577a
Shaw, George Bernard, 2:12a; 12:108a, 109a; 14:512a
Shaw, Henry, 13:456b; 15:130a
Shaw, John, 1:584a; 9:9b
Shaw, Peter (1694–1764), 12:365–366; 2:177b, 227b; 6:176a, 460b*
Shaw, William Napier (1854–1945), 12:366–367; 5:423b; 11:558a; 13:102a
Shayn, Grigory Abramovich (1892–1956), 12:367–369; 6:105b; 10:31b*; 13:118a, 119a
Shchubnikov, Aleksei Vasilevich, 5:186b
Shear, Cornelius Lott, 3:318a
Shear, physical, 3:441b–442a; 10:319b
coefficient of, 3:442a
internal stress, 3:145b, 496b
Shée, Henri, 8:113b
Sheep, 3:200b
breeding of, 1:582b; 7:33a; 9:278a; 14:595b–596a; 15:113a
diseases of, 2:575b; 7:100b; 9:185b
Indian, 2:205a
Karakul, 14:596a
Merino, 1:435b, 492b; 7:89a; 15:113a
phylogeny, 10:16b
raising of, 1:436a; 15:113a
Sheffer, Henry Maurice, 12:14b, 15a
Sheffield Scientific School
see Yale University
Sheffield University
technology, 13:505a
Shelburne, 2d Earl of (William Petty), 11:140b–141a
Shell shock, 4:353b
Shellac
bleaching of, 7:378a
Shellfish
dyestuffs from, 4:223b; 11:331a–b
Shells, 5:518a
analysis of, 2:627b; 6:166b
classification of, 2:591b; 7:588b
collecting of, 13:443b
crystallization of, 8:416a
fossil, 3:313a, 568a–b; 7:588b–589b, 592b; 8:566a–571a
geographical distribution, 5:478a
growth lines, 8:242a
growth of, 11:328a, 331a
see also Conchology, Malacology
Shen Ch'in-p'ei, 3:250a, 266a
Shen Kua (1031–1095), 12:369–393
Shen-nung pen-ts'ao, 8:392b, 396a*
Shenstone, Allen Goodrich, 12:21a
Shepovalnikov, Nikolai Petrovich, 10:433a
Sheppard, Philip Macdonald, 3:572b
Sherard, James, 4:99a; 12:394a,b
Sherard, William (1659–1728), 12:394–395; 2:390a, 516b; 3:129a–b; 9:17b; 13:442b, 449a
Pinax, 4:98b
Sherrington, Charles Scott (1857–1952), 12:395–403; 13:520a; 14:27a
acquaintances of, 5:81a, 207b
on contemporary scientists, 4:595a*; 5:255a, 256a*; 8:19a*, 412b*, 535b*; 11:276a*
historiography, 4:584b, 586a; 5:446b*; 6:61a*
influence of, 5:42a, 208a; 9:19b, 20a; 11:597b
influences on, 6:60a; 11:517b–518a
neurophysiology, 5:65a, 463b; 6:59a; 8:15b; 9:266a; 13:530b
students of, 5:41a, 64b
Shewhart, Walter Andrew, 14:382a–b, 383a, 552b, 554a
Shibukawa, Harumi (1639–1715), 12:403–404; 15:733a, 734a, 756a

    astronomy, **7**:20a; **15**:731a–732a
      Shunju jutsureki, **15**:731a–b
      Tenmon keito, **15**:731b
  Shibukawa Kagesuke, **15**:734b–737a
  Shida, Toshi, **9**:180b, 181b
  Shield (geol.), **10**:168a, 169a; **13**:148a
  Shields, John, **11**:278b–279a
  Shiga, Kiyoshi, **4**:298b, 299b; **7**:392b
  Shigella, **7**:392b
  Shih Chung-yung, **14**:539a,b, 543a
  Shih Hsin-tao, **3**:266b
  Shihāb al-Dīn Muḥammad al-Ījī al-Bulbulī, **9**:604a
  Ibn al-Shiḥna al-Ḥalabī, **1**:28b
  Shīʿism, **9**:171a; **13**:510a, 513a; **15**:249a–250b
  Shikata, Masuzo, **6**:372b
  Shikimic acid
    structure of, **5**:6a
  Shilov, Nikolay Aleksandrovich (1872–1930), 12:404–405
  Shimada, Junichi, **8**:577a
  Shimpen jinkoki, **15**:749b
  Shipbuilding
    see Architecture, naval
  Shiphaulers' argument, **1**:253a
  Shippen, William, **1**:485a; **6**:567b
  Ships
    compartmentalization, **2**:84b
    copper sheathing, **3**:557b, 603b
    distance measurement, **11**:65b
    gauging, **10**:572a
    icebreakers, **9**:42b, 43a–b, 292b; **15**:431a
    iron-armored, **10**:329a
    lightning and, **3**:557b
    masts, **2**:343a; **3**:39a; **4**:468b
    for ocean research, **4**:43a
    performance tests, **7**:383a
    resistance
      laws of, **5**:200a–b
      scale-model studies, **5**:200a
    rigging, **4**:224a
    roll and pitch, **2**:43b, 84b; **7**:437a; **11**:293b
      control of, **5**:200a
    rudder forces, **2**:43a
    screw propulsion, **5**:200b
    seaworthiness tables, **7**:514a
    speed measurement, **5**:593b; **11**:5a
    speed vs. length, **11**:345a
    stability, **2**:430b
    submerged, raising of, **13**:260a
    theory of, **3**:230a; **7**:514a; **14**:620a
    two-hulled, **10**:565b
    ventilation in, **2**:84a; **6**:24b; **9**:153a; **14**:354b
    vibration in, **12**:527a
    waterlines, **11**:293b
    woods used for, **4**:156b
    see also Architecture, naval; Submarines
  Shipworms, **7**:447b
    protection against, **9**:459a
  Shirakatsi, Anania (also known as Ananias of Shirak) (*ca.* 620–*ca.* 685), 12:405–406; **10**:300a, 419a
  al-Shīrāzī, ʿAbd al-Malik ibn Muḥammad, **11**:249b
  al-Shīrāzī, Quṭb al-Dīn
    see Quṭb al-Dīn al-Shīrāzī
  Shivering, **11**:428a–b
  Shizuki, Tadao (1760–1806), 12:406–409
    Tenmon kanki, **1**:314a
  Shmalhauzen, Ivan Ivanovich (1884–1963), 12:409–410; **12**:338b,b*, 339a*
  Shmatko, Mikhail Kapitonovich, **7**:530a
  Shnirelman, Lev Genrikhovich (1905–1938), 12:410–411; **6**:393b
  Shōchō (astron. law), **1**:312a–b
  Shock, electric, **4**:216a
    perception of, **11**:425b–426a
    recovery from, **1**:304a
  Shock, hydraulic, **14**:620b
  Shock, medical, **10**:620b
    cause of, **6**:264b
    spinal, **6**:60a; **12**:397b
    surgical, **8**:614b
    traumatic, **5**:291a; **15**:73a–b
      experimental, **3**:353a
    treatment of, **3**:336a; **4**:398b; **6**:265a; **15**:73b, 76b
    wound, **1**:537b; **4**:398b; **13**:530b
    see also Shell shock
  Shoenberg, David, **8**:480b, 483a
  Shogi (game), **15**:749a, 753a
  Shohat, James Alexander, **4**:562a
  Shokalsky, Yuly Mikhaylovich (1856–1940), 12:411–413; **12**:301a
  Shooting stars
    see Meteors
  Shoreline, **4**:563a
    displaced, **5**:395b
    elevation of, **5**:329b
    formation of, **7**:143b, 144a
      cyclic theory, **3**:595b; **7**:144a
    protection of, **7**:144a
    types of, **13**:148a
  Short, James (1710–1768), 12:413–414; **4**:148b, 149a; **6**:6a; **11**:609b; **14**:419a–b, 453a
    telescopes, **8**:178b; **11**:284b; **14**:452b
  Short, Thomas
    Medicina Britannica, **1**:487a
  Shorthand, **3**:20a
  Shōsa (finite differences), **3**:270b
  Shoulder
    morphology of, **2**:504b
  Shrimp
    biometrics, **14**:251a
  Shtokman, Vladimir Borisovich (1909–1968), 12:414–416
  Shubert, Fedor Fedorovich, **13**:111b
  Shubin, Semen Petrovich, **13**:240b
  Ibn Shuhdī al-Karkhī, **15**:236b
  Shujāʿ ibn Aslam, al-Miṣri
    see Abū Kāmil Shujāʿ ibn Aslam ibn Muḥammad ibn Shujāʿ
  Shull, Aaron Franklin (1881–1961), 12:416–418
  Shull, George Harrison, **4**:271b
  Shumova-Simanovskaia, Ekaterina Olimpievna, **10**:433a
  Sial (geol.), **13**:148a
  Siam
    geography, **13**:583b
  Siberia
    biology, **9**:374b–375a
    ecology, **10**:17a
    ethnography, **1**:173a; **13**:166b
    exploration of, **3**:532a–b; **5**:428a–429a; **6**:552a; **8**:147b; **10**:283a–b; **11**:108b
    geography, **7**:495a, 510b; **10**:169a
    geology, **3**:529b–530a, 532a–b; **4**:598b; **10**:167a–169a
    glaciation, **7**:510b–511a; **10**:169a
    mineralogy, **6**:354a
    natural history, **4**:289b
    permafrost, **9**:375a
  Sibirskii matematicheskii zhurnal
    editors of, **9**:72a
  Siboglinum, **3**:148b
  Sibthorp, John
    Flora Graeca, **1**:520a–b; **8**:372b
  Siciliano, Luigi, **15**:469b
  Sicily
    medieval scholarship, **1**:270b; **5**:147a–b; **7**:8a
    nineteenth-century politics, **3**:45b–46b
    botany, **11**:130a
    geography, **9**:192a
    geology, **8**:544a, 567a–568a, 573a–b; **12**:257a, 261b
    volcanoes, **12**:562a
  Sickle cell
    see Erythrocytes
  Ibn Sīda, **11**:246b
  Siddhānta, **1**:511b, 512a; **15**:221a
  Siddhāntadarpaṇasiddhaparyayādayaḥ, **15**:622b
  Siderism, **11**:474b
  Siderite
    composition of, **3**:326b
  Siderostat, **5**:86a, 87a, 316a
  Sidgwick, Nevil Vincent (1873–1952), 12:418–420; **7**:469b; **9**:402a–b; **11**:482b; **12**:199a*
  Sīdī ʿAlī Reʾīs, **9**:36a
  Siebeck, Friedrich Herrmann, **4**:551a
  Siebold, Carl Theodor Ernst von (1804–1885), 12:420–422; **7**:439b; **8**:271a; **11**:360b; **12**:612b*
    botany, **3**:337a; **4**:291a
    helminthology, **1**:602b
    influences on, **11**:593a
    physiology, **2**:596a
    students of, **2**:127b; **9**:259a, 332a; **10**:427a; **14**:233a; **15**:78a
  Siebold, Philipp Franz van, **7**:20a
  Siedentopf, Henry Friedrich Wilhelm (1872–1940), 12:422–423; **10**:525a; **12**:593b; **14**:633b
  Siedlecki, Michał (1873–1940), 12:423–424; **4**:133b*
  Sieff Research Institute (Israel), **5**:623a
  Siegbahn, Manne, **2**:488a; **3**:429b; **4**:195b; **9**:262b
  Siegel, Carl Ludwig, **5**:342b; **6**:214b, 393b, 394a, 571b; **7**:398b; **11**:53b; **13**:391a
  Siegel, John, **12**:142a
  Siemens, Charles William (Carl Wilhelm) (1823–1883), 12:424; **8**:20a, 116a; **9**:226b; **12**:425b
  Siemens, Ernst Werner von (1816–1892), 12:424–426; **6**:223a–b, 243b, 343b; **7**:526b*; **11**:8b; **14**:360a; **15**:124b
    engineering, **3**:18a, 19b; **7**:56a
  Siemens Bros. (Siemens and Halske), **2**:318a; **6**:223a; **12**:425a–b
  Siena (Italy)
    water supply, **9**:146b
  Sierpiński, Wacław (1882–1969), 12:426–427; **7**:71b, 73a*; **9**:248a; **12**:78a; **14**:592a
    number theory, **2**:236b; **14**:282b
    problems of, **9**:249a
    topology, **9**:248b, 249b
  Sierra Club, **7**:148b–149a
  Sierra Nevada
    geology, **1**:552b; **11**:439a
    stratigraphy, **1**:472a
    surveying, **7**:148b, 370b
    zoology, **5**:545b
  Sigaud de Lafond, Joseph-Aignan (1730–1810), 12:427–428; **10**:147a
  Sigault, Jean René, **8**:353a
  Siger of Brabant (*ca.* 1240–*ca.* 1284), 12:428–429; **1**:197a
  Śīghra anomalies, **15**:563b, 569b, 584a, 596a, 601a, 610a, 612a, 613a, 620a–b
  Sight
    see Vision
  Sigillaria, **2**:492a; **14**:600a
  Sigismund III, King of Poland, **9**:15b, 308b; **12**:306b, 307a
  Signac, Paul, **3**:241b

Signal Corps, U.S., **3**:367a
Weather Service
see Weather Bureau, U.S.
Signatures, doctrine of, **4**:209b; **7**:459a, 485a; **10**:307b–308a; **11**:97a,b; **12**:335a
critics of, **7**:537b
Signer, Rudolf, **13**:2b
Significance, statistical, **11**:237a
analysis of variance, **5**:8b–9a, 477a; **14**: 382b
t test, **5**:8a–b, 10a, 477a; **10**:454b–455a, 457b
Signs, mathematical
Descartes' rule of, **2**:573b; **5**:97b–98a, 567b; **7**:573b; **12**:283a
Signs, philosophical, **1**:335a
Sigorgne, Pierre (1719–1809), 12:429–430; **11**:158a; **13**:494a
Sigüenza y Góngora, Carlos de (1645–1700), 12:431
al-Sijzī, Abū Saʿīd Aḥmad ibn Muḥammad ibn ʿAbd al-Jalīl (*ca.* 945–*ca.* 1020), 12:431–432; **10**:300a; **11**:240a
algebra, **12**:92a
al-Jāmiʿ al-Shāhī, **1**:36b, 37a,b, 39a
Risāla fī qismat al-zāwiya, **11**:240b
Silanes, **13**:70b
Silber, Paolo, **3**:279b
Silbermann, Johann Theobald, **2**:69a; **4**:554b
Silberstein, Ludwig, **2**:168a
Silenium
occurrence of, **7**:45a
Silesia
flora, **3**:337a
geology, **2**:108a–109a, 552b, 553b–554a; **3**:70b, 325a–b; **11**:500b, 539b
paleobotany, **5**:441b–442a
Sileto, Gerard of
see Gerard of Silteo
Silica, **14**:633b
crystallization of, **3**:8b; **6**:178a
formula of, **9**:110a
in minerals, **6**:564b
polymorphs, **4**:563b
Silica gels
structure of, **2**:627b
Silicates, **3**:293b; **4**:409b*; **13**:618a
analysis of, **8**:54b; **13**:476a
biodeposition of, **9**:589b
in cements, **8**:117a–b
crystallization of, **14**:58a–59a
decomposition of, **7**:394b
formulas of, **5**:369b
fused, **4**:141a–b
high-temperature studies, **7**:59a–b
hydrated, **14**:588a
phase diagrams, **13**:246b
phase-equilibrium studies, **2**:369b
solubility, **4**:141a
structure, **15**:63a–b
Silíceo, Juan Martínez, **4**:237a
Silicide, **6**:480b–481a
Silicon
atomic weight, **13**:389b
compounds of, **9**:287a; **13**:70b, 71a, 468a
organic, **7**:372b, 552a; **14**:478a
in earth's crust, **3**:293b
and iron, **13**:468a
isolation of, **3**:602a; **4**:77b; **14**:475b, 478a
isotopes, **7**:154b
in sea water, **5**:72a
spectra, **8**:441b
Silicon hydride, **14**:478a
Silicon tetrafluoride
discovery of, **11**:145b
Silicones, **7**:372b
Silicosis, **10**:305b, 307a
Silin, Victor Pavlovich, **13**:241a,b
Silk
artificial
Chardonnet silk, **3**:207a
dyeing of, **8**:621b
electrical properties, **14**:70a
fibers of, **1**:319b
spiders', **11**:331b
Silkworm, **8**:536a; **9**:440a
anatomy, **9**:65a; **13**:169b, 172a
diseases of, **1**:417b–418a; **4**:211a; **10**: 372b–376b; **11**:234a; **15**:11b
muscardine, **1**:329a, 492b–493b; **10**: 373a; **12**:202b–203a
genetics, **7**:286a, 455b
life cycle, **12**:317a
reproduction, **12**:559a
see also Sericulture
Silliman, Benjamin (1779–1864), 12:432–434; **3**:45b; **4**:291b; **8**:613a; **9**:88a*; **10**:586a
acquaintances of, **3**:397b, 546a; **10**:446a; **11**:218b
correspondence of, **6**:57b; **8**:103b; **12**:204b
gastric juice, **1**:542b
geology, **1**:413a
influence of, **3**:549b; **6**:564a
students of, **1**:43b; **4**:273b; **6**:437a; **10**: 150a; **14**:315b
Silliman, Benjamin, Jr. (1816–1885), 12: 434–437; **6**:564a; **10**:150a; **12**:433b
Sillimanite, **13**:617a
Sills (geol.), **2**:554b
Silphidae, **6**:507a
Silt
origin of, **8**:30b
Silteo, Gerard of
see Gerard of Silteo
Silurian period, **1**:494b; **3**:236b; **5**:476a; **6**: 187a; **7**:45a; **8**:486a–b
coral reefs, **3**:190a
naming of, **9**:583b
paleobotany, **8**:6a
paleogeography, **7**:250b
paleontology, **1**:494b; **2**:128b; **3**:392a, 529a; **4**:308b; **5**:50a,b, 488a*, 538a; **6**:58b*; **7**:320a; **11**:500b
stratigraphy, **1**:210a, 468b–469a; **2**:555b–556a; **4**:262b, 365a(n); **8**:32b, 33a; **9**:583b–585a; **10**:265b; **11**:386a; **12**: 277a–b, 514b; **13**:518b, 620b–621a
Silurian System
see Silurian period, stratigraphy
Silva theory, **3**:15a
Silver, **5**:595b; **15**:651b
alloys, **5**:576b; **6**:561a; **8**:312a; **11**:485b
assaying, **8**:621a; **11**:485b
atomic weight, **3**:398a, 480b(n); **4**:241a; **9**:110a; **10**:510b; **11**:417b
as catalyst, **13**:313a
colloidal, **9**:295b
economics of, **13**:145a
electrochemical equivalent, **7**:449a; **9**: 155a,b
extraction, **5**:421a; **10**:511b
metallurgy, **3**:473a; **4**:345b, 394a; **6**:79b–80a; **13**:327a, 411b
mining, **1**:448a–b; **4**:346b; **5**:155b; **7**:200b; **8**:150a; **10**:555b; **11**:338a, 607b; **13**: 455b, 525b; **14**:60b
ores, **1**:557b
amalgamation treatment, **1**:448a,b; **4**: 345b, 346b
deposits of, **7**:370b; **11**:608a
refining of, **11**:499a
salts, **3**:271b; **12**:79b
organic, **8**:339a–b
in photography, **4**:282b–283a; **13**:238b
tree of Diana, **6**:478a
separation from copper, **2**:565a
separation from gold, **2**:143a; **8**:213b; **10**:557b; **13**:245b, 327a
tests for, **2**:381b
quantitative, **5**:325b; **9**:558b
Silver fluoride, **5**:474a
Silver fulminate, **4**:185a
Silver sulfide, **13**:327b
Silvester, Henry Robert, **6**:60b
Sima (geol.), **13**:148a
Simeiz observatory, **6**:105b; **10**:30a–b
Siṃhatilaka Sūri, **12**:598b
Similarity of figures, **4**:419a, 420a–421b
Simirenko, Vladimir Lvovich, **15**:425b
Simmonds, George Harvey, **3**:92b
Simms, William, **3**:92b, 605b; **13**:470b
Simon Bredon
see Bredon, Simon
Simon of Lendenaria, **4**:237b
Simon de Phares (*b. ca.* 1450; *d.* after 1499), 12:439–440
astrology, **15**:35b
historiography, **3**:28a(n); **7**:125b, 141a(n); **14**:401a(n); **15**:125a, 191a(n)
Simon, Alfred Walter, **2**:247b
Simon, Franz Eugen (Francis) (1893–1956), 12:437–439; **8**:368b, 475b, 479b; **15**:450a,b, 451a*
thermodynamics, **3**:369b; **15**:432b, 440a, 443b, 444b–447a
Simon, Sir John, **12**:466a,b, 467a, 503a
Simplexes
decomposition of, **12**:172a
Simplicity, principle of
in crystallography, **6**:180b–181a
in logic, **4**:101a
in physics, **3**:309b; **4**:320a; **5**:167a; **11**:12b
Simplicius (*b. ca.* 500; *d.* after 533), 12:440–443; **1**:279b; **4**:367a, 460b, 461b, 465a*, 467a; **5**:346b, 403a; **10**:112a; **14**:582a
commentaries on Aristotle, **1**:272a, 276a, 279a; **3**:21b; **4**:104a, 463a; **6**:411b, 412b; **7**:1a, 134b, 138b; **9**:437b, 438b; **10**:324a; **11**:30a(n); **13**:91a; **14**: 605a(n)
commentaries on Euclid, **4**:415b, 437b; **10**:5b–6a
cosmology, **1**:137b; **4**:463b; **7**:135a,b; **15**:204a
geometry, **1**:171a–b; **4**:33b, 433a(n); **7**: 79a–b, 80a; **11**:589a
influences on, **1**:137a; **14**:609b
Simplon Tunnel, **2**:466a; **6**:228a; **8**:543b; **11**:375a; **12**:140b
Simpson, Edward, **8**:105a
Simpson, George, **6**:463a
Simpson, George Gaylord, **3**:572b–573a
Simpson, Sir James Young, **1**:476b; **4**:163a; **8**:402a, 404a–b, 405a; **12**:502b
Simpson, Thomas (1710–1761), 12:443–445; **6**:452a
The Doctrine of Annuities and Reversions Deduced From General and Evident Principles, **9**:454b
on observational errors, **2**:42b
Simpson's rule, **5**:528a; **10**:443a; **12**:444b–445a
Simson, Robert (1687–1768), 12:445–447; **3**:78a, 214b*; **9**:159a
acquaintances of, **8**:609b
critics of, **12**:444a
geometry, **10**:297a
influences on, **4**:426b
Simson line, **14**:141a
and Stewart, **13**:54a–b
students of, **3**:494a

works translated, 4:433b(n), 434a(n), 450b; 10:303a*
Simultaneity
concept of, 2:460b; 4:321b–322a, 331b; 12:177b
Ibn Sīnā, Abū ᶜAlī al-Ḥusayn ibn ᶜAbdallāh (Avicenna) (980–1037), 15:494–501; 1:276a; 3:249a; 4:441b, 447b; 9:83b
alchemy, 5:580a; 9:23b
anatomy, 1:620a; 4:285a
on Aristotle, 10:173b
chemical theory, 12:311b
commentaries on, 1:107a; 11:249b; 12: 7Bb; 13:510a
correspondence of, 2:149a
cosmology, 12:6b
critics of, 8:248b, 249a; 12:5b, 6b, 7Ba
influence of, 1:382a, 538b; 2:23a, 546b; 3:218b, 548b; 4:121b, 440b; 6:123a; 9:364a; 10:555b; 11:231a, 247b–251b *passim*, 311b, 312a; 13:509a; 14:35b, 388a, 457b, 458b
influences on, 1:267b; 4:524a; 7:343a
music, 11:249b
natural philosophy, 1:28a, 100b, 410a
time, 1:27b
optics, 1:380a
rainbow, 7:213b, 215b
philosophy, 1:276b, 408b–409a; 7:330b; 9:30a–b, 84a, 111b; 11:249b; 13:488b, 511b–512a
psychology, 1:26b–27a; 5:592a
on al-Rāzī, 11:326a–b
theology, 9:28b, 29a, 30a
translations of works, 4:486b; 9:361b, 362b; 13:401b
Aḥkam al-adwiya al-qalbiyya (De viribus cordis), 1:290a
De anima
translation of, 5:591b–592a
al-Hidāya (Guidance)
commentaries on, 9:604b
al-Ishārāt wa'l-tanbīhāt, 11:248a, 249b
commentaries on, 11:238b; 13:513a
Metaphysics
translations of, 5:592a
De mineralibus, 13:511b
Posterior Analytics, 5:592b
translations of, 5:592a
al-Qānūn, 1:484b; 2:435b; 10:305a; 12: 7Bb; 13:509a; 15:246b(n)
commentaries on, 7:213a, 216a–b; 9: 603b; 11:239a*, 248a, 249a, 251b; 12:102a, 103b; 13:415b*, 511b
translations of, 15:185b
Salāmān wa-Absāl, 15:232a
al-Shifā' (Sufficientia), 1:26b; 4:439b
translations of, 5:592a
Sinai, Mount, 9:406b
Sinān ibn Thābit ibn Qurra, Abū Saᶜīd (*ca.* 880–943), 12:447–448; 2:152b; 7:2b
Sinanthropus pekinensis, 2:171b–172a; 5: 486b, 500b; 13:275a
Sinclair, George, 5:525a
Tyrocinia Mathematica in Novem Tractatus, 12:445b
Sinclair, John, 3:579b, 580a; 7:12b; 11:304b
Sinclair, Robert, 12:445b
Sindhind (Sind Hind)
see *Zīj al-Sindhind*
Sine, 1:511b; 2:118a; 3:432a; 5:528a; 7:118a; 11:202a; 15:208b
calculation of, 1:249a; 2:419a; 7:256b, 258b–259a; 9:555b–556a; 11:227b–228b
definition of, 5:612b
integral, 5:414a
integration of, 11:487a,b
powers of, 11:472a
rule of, 11:4a, 350a; 14:577b
discovery of, 9:84b
in a polygon, 7:352b–353a, 500a,b; 8: 27b
tables
ancient Greek, 15:209a
ancient Roman, 9:297b
medieval Arabic, 1:42a, 513b; 5:612a–b; 7:256b, 361a, 531b; 13:536b; 14: 577a, 578a
medieval Latin, 1:62b; 7:123b, 129a; 15:163b
Renaissance, 1:179a; 2:602b; 3:181b; 9:611a–b; 11:350b, 396a–b, 397b; 15:155b, 478b
seventeenth-century, 5:408b
eighteenth-century, 8:137a
Indian, 15:550b–551a, 552a, 559a, 570a–b, 579b–592a *passim,* 596a, 600a,b, 602a, 612b–620a *passim,* 626a–b
Sine condition (optics), 1:7b–8a; 3:101a
Sine theorem, 8:280a; 9:555b; 13:291b, 510b; 15:33b
Singapore, 11:261b
Singer, Charles, 1:611b; 9:378a, 468a; 12: 111a,b
Singleton, E. (pseudonym)
see Holden, Edward Singleton
Singularities (math.), 3:138a–b, 139a, 141a, 460b; 5:608b; 8:366b
of algebraic curves, 2:465a; 3:315a, 468b; 5:567a, 569a; 6:547b; 11:45b
classification of, 6:75b
of algebraic surfaces, 5:569a; 11:450a; 12:86b
composition of, 6:3b
condensation of, 6:95b
of differential equations, 2:144b, 350a, 358a; 3:221a; 5:203b, 204b
of field equations, 4:329b, 330b, 331a; 7:10a
of Fourier series, 4:561b
of plane curves, 3:167a; 5:368b
polar, 6:3b, 4a; 11:450a
of series, 4:547b, 548a; 6:3b
types of, 11:207a
Sinology, 11:402b–403a
Sino-Swedish Expedition, 6:216b–217a
Sinuses
sphenoid, 14:456b
Sinusoid, 4:506a
Siphon
ancient Greek study, 6:311b
medieval study, 5:402b
seventeenth-century study, 1:424a; 2: 311a, 377b; 9:13b; 13:336a, 438b
Siphonophores, 6:7b, 9a; 14:57b
Siphunculata, 3:492b
Sipunculoidea, 11:113b
Siren, 6:566b
Siren, acoustical, 3:9a
Sirenia, 2:423a–b
Sirius
editors of, 7:400b
Sirius (star), 15:659a(n), 677a, 680b, 721b
astrological use of, 10:547b, 548a
as decanal star, 15:712b, 715a, 716a
distance, 5:527b
heliacal risings, 14:578a; 15:707a,b, 708b, 711a, 725a
observations of, 9:162b
orbit, 1:339b
parallax, 6:71a, 512b
photometry of, 11:542b
proper motion, 3:14a; 6:71a; 10:542b
radial velocity, 6:542b
Sothic cycle, 15:709a
spectrum, 5:143b; 6:542b; 9:392a; 14:56a
variation in motion, 2:99b
Sirius, companion of (Sirius B)
density, 4:279b
discovery of, 3:288a
spectra, 1:56b–57a; 4:279b, 280b; 9:504a
Sirk, Hugo, 4:141b
Sirturi, Hieronymus, 13:437b
Sisson, Jonathan, 2:140b; 5:491b, 492a; 8: 178b; 11:284b
Sītārāma Jhā, 7:315a,b
commentaries of, 5:274b
Sitter, Willem de (1872–1934), 12:448–450; 2:511a; 4:280b, 281a; 5:405b; 7:18b*
cosmology, 6:353a, 531a; 7:237a
earth's rotation, 10:34a
Śiva, 9:607a
Siwalik Hills, 4:518b
Sixtus V, Pope, 3:559a; 9:168a, 308b; 10: 512b, 598a; 12:105b
Size
inheritance of, 4:271a
and velocity, 5:15a
see also Mass
Skaergaard (Greenland), 4:564b
Skaphē (astron. instrum.), 1:246b
Skeat, Walter William, 5:604b; 9:160b
Skeletal system, 4:333b; 5:581a
branchial, 6:573b
fetal, 3:342b
infantile, 3:342b
mechanics of, 8:202a–b
morphogenesis, 7:61b; 11:255b
pubic bones, 7:17a
tarsus, 1:46b
visceral
embryogeny, 6:573a
see also Bones, Joints (anat.), Osteology, Spine
Skepticism
ancient Greek, 5:347a; 10:581b
ancient Roman, 12:340a–341a
medieval Latin, 10:225b
Renaissance, 1:80a; 12:97a–b
seventeenth-century, 5:415b–416a; 6: 253b–254a, 257b, 448a; 13:214b
opposition to, 9:316b, 317a; 14:301b
eighteenth-century, 4:85a–b; 6:555b, 557b–559b
nineteenth-century, 6:592a, 596b
twentieth-century, 4:225b; 8:603b
see also Doubt, methodical
Skin
blood vessels
injury response, 8:295a
diseases of, 2:314b; 5:566a, 570b; 10:418a
classification of, 8:506b
fungal, 5:566a; 10:384a
neurotropic, 8:273b
histology, 8:9a; 11:216b
injury reactions, 15:106a
innervation of, 8:8b, 9a; 9:259a, 422a; 11:566a–b, 596b–597a; 12:396b; 14: 113b
muscles, 8:401a
sensibility, 3:531a; 5:184b–185a; 13:393a, 493a; 14:200b–201a
spatial localization, 3:531a
transplantation, 8:447a
tumors, 10:19a
see also Dermatology; Epidermis
Skippon, Philip, 14:413a,b
Škoda, Josef (1805–1881), 12:450–451; 12: 294b, 295b, 296b
Skolem, Albert Thoralf (1887–1963), 12: 451–452; 5:108a,b*; 11:108a*; 13: 216b*

Skolem, Albert Thoralf (cont.)
Löwenheim-Skolem paradox, **6**:393a
set theory, **5**:108a; **14**:615a
Skotodynamics, **6**:90a
Skottsberg, Carl, 7:210b, 535a
Skraup, Zdenko Hans (1850–1910), 12:452; **5**:489a
Skryabin, Konstantin Ivanovich (1878–1972), 12:452–453
Skull
anatomy, **1**:619b; **14**:505a
comparative, **2**:504a–b; **5**:442b–443a; **6**:591b–592a; **15**:112b
classification of, **11**:514b
embryogeny, **6**:592a
evolution of, **7**:480a
formation of, **7**:438a
fossils, **2**:172a, 189a–b; **3**:328b
fractures of
Renaissance discussion, **1**:618b
injuries to
Renaissance discussion, **2**:336b
nineteenth-century discussion, **11**:515b
lesions of
Renaissance discussion, **1**:618b
sutures, **1**:264b; **7**:17a
veins, **2**:443a
vertebral theory of, **5**:443b–444a; **10**:195b; **15**:126a–b
see also Craniology, Head
Sky
airglow, **13**:107b
blue color of, **3**:303b, 600a; **4**:506b; **8**:12b*; **11**:265a; **12**:353a, 497b; **13**:101b, 522a; **14**:499b*
illumination of, **1**:429b
night glow, **13**:107b
polarization in, **1**:358a
see also Atmosphere, Clouds
Skye, Isle of
geology, **1**:394a
Slags, **10**:511b
analysis of, **14**:478a
mineralogy, **14**:58a
properties of, **8**:118b
Slare, Frederick, **6**:41a; **7**:525b
Slater, John Clarke, **2**:247b, 248a, 338a; **3**:369a–b; **4**:282a; **5**:331b; **7**:492a; **8**:186a; **12**:419a
Slates, **9**:61a; **11**:547b
cleavage, **3**:567b; **12**:542b–543a
origin of, **15**:427b
pressure effects, **13**:521b
Slavery
Abolitionists, **14**:456b
legislation against, **6**:553a
opposition to, **6**:551a; **10**:277b
rationalization of, **1**:73b
Slavs
ethnography, **7**:4a–b
Slawik, Antonius (pseudonym), **11**:159a
Sleep
physiology of, **5**:137b; **15**:262a
and respiration, **9**:547a; **14**:322a
theories of, **9**:41a; **11**:135b; **13**:94a, 543a; **14**:408a
Sleeping sickness, **2**:528b–529a, 533b; **7**:429a
treatment of, **4**:300a; **15**:252b
Sleeswijk, Jan Gerard, **2**:300b
Slide rule, **6**:271b; **9**:80a
chemical, **14**:488b, 489a
circular
invention of, **4**:13a–b; **10**:255a
invention of, **4**:13a–b; **10**:255a
precursors of, **5**:593b
Slime mold, **14**:93b
Slip (physics), **9**:224b, 225a; **14**:170b
Slipher, Earl C. (1883–1964), 12:453–454
Slipher, Vesto Melvin (1875–1969), 12:454–456; **6**:530b–531a; **8**:522a; **10**:473b
Sloan, David Harold, **6**:104a
Sloan Foundation, **15**:2b
Sloane, Sir Hans (1660–1753), 12:456–459; **6**:44b
acquaintances of, **1**:305b; **3**:349b, 459b; **4**:172b, 295a; **5**:352a; **6**:139a; **8**:375a
botanical gardens, **1**:486b; **5**:539a; **9**:390b
collections, **1**:432b
correspondence of, **7**:274b; **9**:141b, 465b*; **13**:449a
herbarium, **2**:476a, 517a; **12**:394b
influence of, **2**:188b; **14**:193a
library, **5**:344a*; **7**:205a; **13**:508a*
patronage of, **3**:129b; **5**:516a; **8**:307b
at Royal Society, **5**:53b; **10**:84a; **15**:29b
Sloan-Kettering Institute for Cancer Research, **7**:316b
Sloth, **2**:570a
fossil, **7**:89b; **13**:72a
Slovenia, **5**:366b
Slug, **15**:505b
Sluse, René-François de (1622–1685), 12:459–461; **6**:359b
correspondence of, **3**:348b; **5**:524b; **6**:598b; **8**:250b; **10**:48b
geometry, **6**:537a, 600b; **10**:336b
Miscellanea, **5**:521a
optics, **1**:475a
pearls of Sluse, **5**:521a; **6**:359a; **9**:599b
Slutsky, Evgeny Evgenievich (1880–1948), 12:461
Slye, Maude, **14**:252b
Slyke, Donald Dexter van
see Van Slyke, Donald Dexter
Small, John Kunkel, **3**:196b; **6**:122a
Small, William, **7**:88a
Smallpox, **3**:437b; **5**:157a; **8**:390b; **11**:37b, 325b; **13**:292b
confluent, **7**:11b
description of, **1**:399b; **2**:444b
epidemics, **11**:514b; **13**:214b; **14**:405b
etiology of, **1**:493b; **6**:73a; **11**:175a
light, effects of, **4**:621a
mortality rate, **2**:42a–b
pathology of, **3**:447b; **14**:228b
prevention of, **15**:96b–97a
transmission of, **5**:107a
treatment of, **13**:215a; **14**:501a
vaccination
early eighteenth-century, **1**:322b; **12**:458b; **13**:257a; **14**:83a; **15**:271b–272a
late eighteenth-century, **1**:116b; **2**:42b; **5**:137b; **7**:11b–12a, 96a–97b; **8**:352b, 506a; **9**:481b, 500b; **10**:326a, 445b; **12**:57a, 215b; **13**:181b, 465a; **14**:194b
nineteenth-century, **7**:425b; **10**:390a, 391b, 395b; **12**:57b
twentieth-century, **5**:270a
opposition to, **15**:429b
statistical studies, **15**:373b
Smeaton, John (1724–1792), 12:461–463; **3**:51b; **4**:149a; **8**:209b; **11**:499a; **14**:418b
Smeatonian Society, **12**:462a
Smekal, Adolf Gustav Stephan (1895–1959), 12:463–465; **4**:196b; **7**:492a; **11**:265b
Smellie, William, **3**:37b, 38a; **6**:568b
Smelting, **3**:454b
chemistry of, **11**:34a
reverberatory furnace, **1**:448b
Smilax, **10**:622a
Smit, Maurits (pseudonym), **7**:59b
Smith, Adam, **2**:174b, 175a; **6**:556b, 579a; **9**:384b; **11**:135a; **15**:405a
An Inquiry Into the Nature and Causes of the Wealth of Nations, **11**:164a
Smith, Archibald, **4**:492a; **9**:225b
Smith, David Eugene, **3**:134b; **8**:100b, 271b; **10**:480b; **12**:108b; **15**:256a
Smith, Edgar Fahs (1854–1928), 12:465; **5**:350a*, 518b*; **6**:115a*
Smith, Edward (*ca.* 1818–1874), 12:465–467
Smith, Erwin Frink (1854–1927), 12:467–468; **7**:101a; **10**:415b*
Smith, Grafton Elliot
see Elliot Smith, Grafton
Smith, Henry John Stephen (1826–1883), 12:468–469; **5**:413b–414a; **7**:506b; **12**:244a–b; **13**:218a, 341b
number theory, **9**:173b, 411b
Smith, Homer William (1895–1962), 12:470–471
Smith, Hugh, **10**:321b
Smith, James (optics), **8**:414b
Smith, James Edward (1759–1828), 12:471–472; **2**:454b*; **6**:539a; **8**:98a, 380a*; **11**:588a*
acquaintances of, **2**:518b; **6**:521a; **10**:531a
botany, **6**:492b, 493a; **7**:198b; **12**:589b(n)
English Botany, **1**:358b; **12**:552b
The English Flora, **2**:19a
Smith, John (math., seventeenth cent.), **3**:430b; **14**:147b
Smith, John (bot., nineteenth cent.), **1**:89a, 521a; **9**:340b
Smith, John Lawrence, **2**:254b
Smith, Kirstine, **10**:456a, 457a
Smith, Levi B., **12**:345a
Smith, Miles H., **14**:169b
Smith, Paul Althaus, **2**:145a
Smith, Philip Edward (1884–1970), 12:472–477
Smith, Robert (1689–1768), 12:477–478; **3**:430b, 431b
A Compleat System of Opticks in Four Books, viz. A Popular, a Mathematical, a Mechanical, and a Philosophical Treatise, **3**:454b; **4**:45b; **6**:328a
Harmonics, or the Philosophy of Musical Sounds, **6**:328a
Smith-Helmholtz formula, **12**:477b
translations of works, **10**:572a*
Smith, Robert Angus (1817–1884), 12:478–479; **11**:536b
Smith, Sidney Irving (1843–1926), 12:479–480; **11**:176a; **14**:1b, 423b
Smith, Theobald (1859–1934), 12:480–486; **7**:424a, 428a, 429b, 430a; **10**:144b*; **11**:442b
Smith, William (1769–1839), 12:486–492; **2**:481a; **5**:578b; **14**:486b
acquaintances of, **2**:566b; **13**:448b; **14**:396b; **15**:151b
Farey and, **4**:540a
geological map, **5**:519a; **9**:87a; **11**:35b
influence of, **2**:567b; **4**:365a; **5**:14a; **10**:583b, 586a
paleontology, **1**:413a; **2**:495a; **4**:541a; **10**:322b; **12**:182a; **13**:147b
Strata Identified by Organized Fossils, **12**:552b
stratigraphy, **5**:533a; **11**:505a
Smith, William Robertson, **5**:144b–145a
Smith, Wilson (1897–1965), 12:492–493
Smithells, Arthur (1860–1939), 12:493–494; **14**:319a
Smithson, James Louis Macie (1765–1829), 12:494–495; **2**:569b; **6**:279b
Smithsonian Institution
in nineteenth century, **1**:405a, 406a; **3**:292b; **5**:214b; **7**:92a; **8**:19b; **13**:25a
in twentieth century, **5**:433a,b; **14**:121a
archaeology, **2**:549b
Astrophysical Observatory, **8**:19b

astrophysics, **1**:583a
aviation, **8**:20a
ethnology, **11**:119a–b
founding of, **6**:279b–280b; **10**:480a; **12**: 494b–495a
geophysics, **1**:552b
Harriman fund, **9**:313b
library, **5**:406b
meteorology, **5**:599b
National Museum, **1**:405a, 406a; **3**:292b, 392b*, 520a, 536a,b; **5**:49b; **8**:264b; **11**:443a
geology, **9**:315a
paleontology, **9**:255a–b
Smithsonite, **12**:494b
Smits, Andreas (1870–1948), 12:495–496; 10:125b
Smoke, **14**:319a
Smoke rings, **13**:237a
Smokeless powder, **9**:292a; **10**:133a
Smoke-screen research, **8**:24b
Smollett, Tobias, **6**:569b
Smoluchowski, Marian (1872–1917), 12: 496–498; 2:120b; **10**:207a*; **14**:523a*
Brownian movement, **4**:315a; **8**:11a; **9**: 129b; **10**:525a–b; **13**:159b
Einstein-Smoluchowski theory, **11**:265a; **14**:345a; **15**:464a
gas theory, **14**:171a
physical chemistry, **14**:319a
statistical mechanics, **9**:76b; **10**:235b
Smorodinskii, Iakov Abramovich, **7**:618a
Smut fungi, **2**:437b, 438a; **3**:612a; **6**:121b; **13**:411b; **14**:93b
Smyrna
university, **3**:62b
Smyth, Charles Piazzi (1819–1900), 12:498–499; 10:549b
Smyth, Henry DeWolf, **10**:217b
Smyth Report, **3**:371b
Smyth, Sir James Carmichael, **5**:603b
Snail, **3**:572b
anatomy, **1**:432a; **12**:298a, 299a
eggs, **2**:626a
fossil, **3**:608b; **10**:29b
generation of, **2**:591b; **13**:174a
genetics, **5**:10b; **13**:136a
heart, **2**:591b; **5**:82a
hermaphroditism, **10**:568a; **13**:174a; **15**: 67b
land, **3**:536b*; **5**:477b; **9**:536b; **13**:65b
regeneration, **12**:555b
respiration, **12**:563a
Snakebite
effects of, **11**:341b
remedies for, **9**:12a; **11**:341b
Snakes, **15**:654b(n)
scent glands, **14**:13a
tongues, **3**:116a
see also Herpetology, Reptiles, Venoms, Viper
Sneezing, **3**:437b
auguries from, **9**:364a
reflex nature of, **6**:60a; **9**:571a
Snel (Snellius or Snel van Royen), Willebrord (1580–1626), 12:499–502; 11: 561b*
acquaintances of, **1**:566b, 607b
critics of, **3**:104b
Cyclometricus, **3**:181a
Eratosthenes batavus, **9**:447b
geodesy, **9**:336a; **10**:595b, 596a
influence of, **1**:567a; **5**:409a, 593b; **11**: 288b; **13**:30b
influences on, **10**:153b; **11**:468a; **14**:461a
logarithms, **9**:159a
navigation, **10**:161a
optics, **4**:59b; **6**:197a; **10**:475b
problem of, **12**:500b
Snel-Huygens inequalities, **5**:526b
Snel's law, **6**:129a*; **12**:501a–b; **15**:362b
students of, **5**:408b; **6**:520a
Snell, Karl, **1**:7a; **7**:401b
Snelus, George James, **13**:346b–347a
Snider-Pellegrini, Antonio, **14**:216a
Snow, Charles Percy, **1**:322a*; **5**:272b
Snow, Francis Huntington, **7**:285b
Snow, John (1813–1858), 12:502–503; 10: 559a
Mode of Communication of Cholera, **2**: 575a
Snow, **9**:466b
density, **11**:309a
geological effects, **10**:126b
and landscape formation, **9**:150b
measurement of, **12**:595b
nature of, **5**:535a
radioactivity of, **7**:527a
red, **1**:521a
snow cover
and climate, **14**:53b
vaporization of, **10**:553a
Snowflakes
shape of, **7**:299b
Soap
analysis of, **2**:74b; **3**:242a–b, 419b
gels, **14**:633b
mechanism of, **12**:593b–594a
medical use of, **14**:320b
production of, **3**:561a*; **10**:497a; **15**:651a
see also Saponification
Soap bubbles, **15**:59a
Sober, Herbert Alexander, **13**:421b
Sobolev, Leonid Vasilevich, **1**:440b, 441a
Sobrero, Ascanio, **10**:132b; **11**:85a
Sobrerol, **11**:85a
Social contract, **6**:447b–448a
Social criticism
Renaissance, **1**:79b–80a
Social instincts
see Instinct, social
Social physics, **3**:378a; **11**:236b, 237a
Social reform
scientists and, **1**:9a, 134a, 201a; **5**:74a; **6**:60b, 267b, 509b, 519a; **8**:84b; **11**: 301a–b, 534b; **12**:466b–467a; **13**: 144a–b; **14**:552a
Social sciences
mathematical models in, **3**:386b; **11**:237a
Social work
scientists and, **1**:77a, 471b; **2**:124a; **3**: 133a; **7**:72b
Socialism
nineteenth-century, **3**:184a; **9**:385a; **10**: 448a; **12**:209a; **13**:24a; **14**:133b, 134a, 223b; **15**:132b–133b, 144b, 405a
twentieth-century, **7**:151b, 154b, 157b; **8**:184b; **9**:565a; **10**:441b, 525b; **12**: 108a, 109a–b; **13**:551a; **14**:339a; **15**:38a
Sociedad de Amigos del País, **15**:429b
Sociedad Económica de Amigos del País, **4**:346a
Società Botanica Fiorentina
founding of, **9**:368a; **13**:449b
Società degli Spettroscopisti Italiani, **13**: 232b
founding of, **8**:503b; **11**:412a
Società dei Quaranta, **11**:401b, 599a
see also Società Italiana delle Scienze
Società Filotecnica
founding of, **11**:95b
Società Italiana delle Scienze, **14**:80a
see also Società dei Quaranta
Societas Ereunetica, **7**:193b
Societas Meteorologica Palatina, **4**:174b
Societas Spinozana
founding of, **2**:544b
Société Alfred Binet, **2**:131b
Société Anatomique (Paris)
founding of, **3**:489b
Société Botanique de France (Paris), **3**:359a
Société Chimique de France (Paris), **2**:621a; **5**:100a; **14**:530a
Société d'Acclimatation, Paris
founding of, **5**:359b
Société d'Agriculture de Paris, **13**:497a
Société d'Anthropologie (Paris), **5**:506a
Société d'Arcueil
see Arcueil, Société d'
Société d'Émulation et d'Agriculture de l'Ain, **7**:581a
Société d'Émulation pour les Sciences Pharmaceutiques
founding of, **5**:72b
Société d'Encouragement pour l'Industrie Nationale, **3**:199b, 358b, 498a–b; **5**: 101a; **9**:493a; **10**:326a; **11**:111a; **13**: 309a
founding of, **9**:474a
Société d'Histoire Naturelle de Paris, **7**:586b
founding of, **1**:328b
Société de Biologie
founding of, **11**:491b
Société de l'Harmonie Universelle, **9**:327a,b
Société de Micrographie
founding of, **1**:417b
Société de Pathologie Exotique
founding of, **8**:66a; **9**:328b
Société de Pathologie Végétale et d'Entomologie Agricole de France, **9**:78b
Société de Physique (Zurich)
founding of, **5**:380a
Société de Physique et d'Histoire Naturelle de Genève, **13**:595b
Société de Recherches et de Perfectionnements Industriels
founding of, **4**:253b
Société des Amis des Sciences, **10**:594a
Société des Arts (Paris), **3**:281a; **5**:87b, 567a; **10**:145a
Société des Ingénieurs du Vide, **4**:254a
Société Entomologique de France (Paris)
founding of, **1**:329a
Société Ethnologique de Paris
founding of, **4**:286a
see also *Mémoires de la Société ethnologique de Paris*
Société Française de Physiologie Végétale, **3**:359a
Société Géologique de France, **4**:349b; **14**: 88b
founding of, **2**:341b; **4**:67b; **11**:133b; **15**: 114b
see also *Bulletin de la Société Géologique de France*
Société Géologique Suisse
founding of, **11**:375a
Société Helvétique des Sciences Naturelles, **3**:42b
Société Linnéenne (Paris), **2**:321b
Société Mathématique de France
founding of, **15**:31a
Société Médicale d'Émulation
founding of, **2**:122a
Société Positiviste (Paris), **3**:375a
Société Royale de Médecine (Paris), **5**:89b–90a, 90b, 91b
founding of, **8**:353a, 506b, 621a; **13**:495b
Société Royale des Sciences de Montpellier, **3**:199a–b, 201a
Societies, scientific
seventeenth-century, **1**:375b–376a
eighteenth-century, **3**:544b

Societies, scientific (cont.)
nineteenth-century, 3:603b–604a; 4:182a
twentieth-century, 4:220b
see also individual societies by name, and list at end of Index
Society
esprit général, 9:491a
evolution of, 12:571b; 15:136b, 144b
genetics and, 4:271b–272b
philosophy of, 11:43b, 323b; 14:361b–362a
statistical description of, 3:386b
see also Culture; Science(s), and society; Sociology
Society for Experimental Biology and Medicine
founding of, 8:556a; 9:265b
Society for Mathematics (Hungary)
founding of, 4:380b
Society for Natural Sciences (Hungary)
see Természettudományi Társulat
Society for Philosophical Experiments and Conversations, 6:383a, 384a; 14:562b
Society for Promoting Useful Knowledge
founding of, 3:557b
Society for Psychical Research (G.B.), 3:476a
Society for the Diffusion of Useful Knowledge, 1:459b; 4:35b; 12:524a
Society for the Encouragement of Arts, Manufactures and Commerce, 1:411b; 12:365b, 488a
founding of, 12:413a
Society for the Improvement of Medical Knowledge (Edinburgh), 9:480b
Society for the Promotion of the Natural Sciences in the Netherlands Indies, 13:459b
Society for the Propagation of Christian Knowledge, 3:238b; 12:542a
Society of American Bacteriologists
founding of, 7:170b; 10:155a
Society of Chemical Industry, 11:538a
founding of, 12:236b
Society of Civil Engineers
founding of, 12:462a
Society of Dermatology and Syphilography, 11:516a
Society of German Scientists and Physicians
reorganization, 6:434b
Society of Glass Technology, 13:505b
Society of Jesus
see Jesuit order
Society of Polish Mathematicians, 4:84a
Society of Telegraphic Engineers and Electricians
founding of, 3:288b
Society Islands, 3:396b
Socinianism, 2:164a; 4:212b; 11:140b
Sociology
medieval Arabic, 7:321b–322b; 11:233a
eighteenth-century, 9:490b–491a; 14:68b–69a
nineteenth-century, 2:172b; 3:376a, 378a–b; 4:401a; 5:255a, 506a; 6:551a,b; 9:68a–70a; 11:237a–b; 12:570b–571b; 14:44b; 15:133a–142a, 406a–415b
twentieth-century, 2:12a; 6:122b, 262a; 10:503a; 12:98b
historical method, 3:378a–b
and medicine, 5:452b
Mesopotamian, 15:637a, 646b, 650b
rural, 12:98b, 99a
theoretical, 5:178b; 9:490b–491a
Socotra
flora of, 1:422a
Socrates, 1:282a; 2:550a; 4:30b; 6:405b, 406b; 7:187a; 11:325a,b; 13:290a, 301b, 314b; 14:606a
in Plato, 11:23b–29b *passim*
Socrates (the younger), 13:302a, 314b
Soda, 5:351b
compounds of, 13:310b
manufacture of, 3:202a; 5:602a; 6:383b; 8:113b–114a, 553b; 9:106a, 467a; 10:497a; 13:358b
from salt, 4:250a; 7:277b; 12:520b–521a
Soddy, Frederick (1877–1956), 12:504–509; 2:257b; 6:14b, 15a, 117a; 11:283b*; 13:159b–160a
acquaintances of, 3:87a; 10:288a; 11:420a
isotopes, concept of, 1:321a
radioactive decay, 1:560a; 2:258b, 259a,b; 3:480a, 499b, 500a, 508a; 5:394b; 11:280b; 12:27b–28b
radioactive displacement rule, 6:16a, 365a; 12:32a
radioactivity, 11:280a; 15:464a
translations by, 10:526a*
Söderbäck, Erik Gustaf, 2:171a
Soḍhala
commentaries of, 2:119a
Sodium
alloys, 12:321b
atomic weight, 10:510b; 11:417b
as catalyst, 7:457b
compound nature of, 3:602b; 13:311a
discovery of, 3:601b–602a
fused, 6:365a
isolation of, 2:588b; 5:320b–321a
oxides, 13:312a
production of, 4:77b; 13:311a
reaction with organic compounds, 5:13a
salts of, 3:202a, 455a; 9:106a
crystals, 5:124a; 11:85b
decomposition of, 10:446a
in sea water, 5:72a
spectrum, 7:553b; 9:267a; 11:580b; 14:598a
D line, 3:420a; 5:86b; 9:391b
in tissues, 9:549b
vapor, 14:497b
fluorescence of, 4:253b
monatomicity, 4:79a
Sodium benzoate, 3:257a
Sodium pyrophosphate
discovery of, 3:289b
Sodium sulfate, 5:420a
Sodium thiosulfate, 6:325a; 9:296a
composition of, 10:178a
Soemmerring, Samuel Thomas (1755–1830), 12:509–511; 2:597a*; 4:146b; 5:75b; 6:454a; 11:593a; 13:402a
anatomy, 5:442b; 8:352b, 506a
Sohncke, Leonhard (1842–1897), 12:511–512; 5:212a; 6:180b; 14:344a*; 15:195a
crystallography, 1:460b, 461b; 5:557a; 6:358a
Sohotski, Yulyan Vasilevitch, 3:226a
Soil(s)
nineteenth-century study, 4:143b–145b; 10:242a; 13:249a
twentieth-century study, 1:623a; 5:458a; 13:518a; 14:106b–107a
chemistry of, 8:147b, 346b; 10:150b; 14:145a; 15:172b
acids, 9:558a–b
nitrogen fixation, 2:356b–357a
oxygen, 8:425b
pH, 2:170b
phosphates, 5:104a
salts, 14:107a
vitamins, 12:208a
classification of, 4:145a
as climatic indicators, 2:549a
conservation, 13:519a
and disease, 10:258b, 558b–560a
freezing of, 5:428b
fungi in, 12:468a
heat conduction, 5:68b
mapping of, 4:143b–145b *passim;* 13:249a
microbes in, 4:211b; 14:37b; 15:492b
mixing of, 14:145a
moisture in, 8:425b
origin of, 1:623a; 4:144b–145a; 9:315b; 10:430b
and plants, 2:357a; 3:45b, 586a; 5:72b; 6:256b; 7:358a; 12:124a; 14:117a
fertility measurement, 2:356b–357a
sandy
improvement of, 6:237b
sterilization of
and productiveness, 15:492b
temperature of, 14:67b, 117a
topsoil, 4:143b
zonality of, 4:144b, 145a
Soil mechanics, 2:565b; 3:441a, 444a; 11:164b, 294a
sliding wedge theory, 3:442a
Soil poisons, 3:463b
Soil science, 8:285a; 9:421a, 535a; 10:430b; 14:36b
Soissons, Council of (1121), 1:1b
Sokhotsky, Yulian-Karl Vasilievich (1842–1927), 12:512–513
Sokolov, Aleksei Petrovich, 5:462a
Sokolov, Dmitry Ivanovich (1788–1852), 12:513–515
Sol (chem.), 5:494b
coagulation, 14:319a
Solanaceae, 10:13b
Solander, Daniel Carl (1733–1782), 12:515–517; 1:89b; 3:351a, 365b*; 5:75b, 77a; 6:566b
and Banks, 1:88b, 434a–b; 3:397a; 8:379b
Solander case, 12:516a
Solanine
discovery of, 14:112a
Solano, Francisco, 13:601a
Solanum, 7:376a
Solar constant
see Radiation, solar, solar constant
Solar equation, 13:538b
Solar system
absolute velocity of, 9:374a
evolution of, 12:280a–b
flattening of, 7:232a
motion of, 13:114a
origin of, 7:85b; 8:522a; 9:552b; 10:42a; 12:409a; 13:270b
stability of, 4:126b; 8:277a, 560b; 10:78a; 15:322b–333a
see also Cosmology, Planets
Solar wind, 2:434a; 11:412b
Solberg, Halvor Skappel, 2:168b
Solbrig, August, 5:570b
Soldani, Ambrogio (or Baldo Maria) (1736–1808), 12:517–518
Solder, 13:327a
Soldering iron, 5:203a
Soldner, Johann Georg von (1776–1833), 12:518–519; 7:608a; 9:156a
Soleil, Jean-Baptiste-François (1798–1878), 12:519
Solenoids, 1:203a
Solidification
of gases, 4:79b
of liquids, 1:129a
Solidism, 1:315a

Solids, geometrical
Archimedean, **4**:97b
development of, **4**:260a
cochlea, **1**:164b
generation of, **7**:1a
of least resistance, **3**:459a; **10**:68a
in nature, **3**:596b
non-Platonic, **3**:169a
Platonic
see Solids, geometrical, regular
regular, **1**:141b, 245b, 518a–b; **3**:181a, 596b; **4**:97b, 424a, 609b; **5**:113a; **7**:290b–292a, 398a; **9**:191b; **10**:271a; **11**:533a–b; **15**:671a
angle of inclination, **7**:29b
construction of, **4**:423a–b; **11**:28a, 222b; **13**:303b–304a
development of, **4**:260a
inscribed, **6**:415a, 616b; **10**:294b
surface vs. volume, **10**:295b
rotational, **1**:228b; **3**:183a; **5**:527a; **10**:296b; **12**:460a
classification of, **13**:435a
surface areas, **6**:600a
volumes, **1**:164b; **3**:152a; **4**:571b; **5**:588b; **7**:583b; **13**:435a–436a
semiregular, **10**:295b
space-filling, **8**:617a
see also Polyhedra; specific solids
Solids, physical
atomic theory of, **5**:160b
chemical reactions of, **8**:373b; **13**:247a
compressibility, **11**:530a
elasticity, **1**:128b; **3**:303b; **6**:342a; **7**:601b; **9**:166b
electrolytic conduction, **15**:449b
Kelvin-Kirchoff-Neumann bodies, **4**:230a, 230b–231a
kinetic theory of, **2**:466a
pressure effects, **12**:593b
stability, **4**:230b
structures of, **5**:597b
theory of, **2**:431b
mechanical properties, **5**:161a
thermal motion in, **5**:161a; **15**:363a
Solifluction, **13**:147b
Solinus, Gaius Julius, **7**:27b; **9**:140b
Sollas, William Johnson (1849–1936), 12:519–520; **2**:288b; **12**:545b*
Solliolis, Barthélémy de, **15**:156a
Solms-Laubach, Hermann zu (Count), **3**:31a
Solomon, king of Israel, **15**:634b
temple, **14**:29a–b
Solomon ben Abraham ben David, **12**:7Bb
Solomon ben Ayyub ben Joseph, **12**:7Bb
Solomon Islands, **2**:342b
Solstices
and calendar, **5**:553a
observation of, **1**:246b; **2**:148a, 150b; **9**:337a, 338b; **11**:188a, 189b, 240a; **12**:403a; **15**:218b, 270b
position of, **3**:101a, 108a; **13**:320a, 440b; **15**:219b, 677a, 680b
summer, **4**:390a, 459b; **15**:673a
Solubility, **2**:499a; **3**:334a; **4**:229a; **7**:162a
and chemical structure, **12**:418b
of gases, **2**:170a; **12**:271a
of glass, **4**:214b
of oils, **8**:618b
of salts, **1**:616b; **4**:229a, 238b–239a, 352b; **5**:324a; **7**:208a; **8**:618b
effects of gas on, **2**:524a
Solutions, **8**:373b
absorption spectra, **7**:162a
analogy to gases, **1**:299a–b; **13**:579a–b
boiling point, **11**:298b
colligative properties, **11**:279a; **14**:182b; **15**:460a
concentrated
boiling points, **3**:176b
cooperative phenomena, **5**:103b
critical points, **3**:620b
density, **3**:211b; **13**:547b
distribution in, **12**:404b; **15**:434a
electrical conduction in, **3**:334a; **6**:439a,b; **7**:203b, 449a–b, 462a
electrified systems, **4**:229a
electrolytic, **7**:162a
fractional solution, **3**:242b
freezing point, **11**:298a–299b; **15**:459a
and concentration, **14**:191a
heat of solution, **4**:228b, 554b
ionic coefficients, **12**:495b
ionic strength
and activity coefficients, **2**:499a; **8**:290b
isohydric, **1**:300a
nonaqueous, **3**:176b; **14**:125a
optical properties, **9**:376b
and concentration, **2**:139a–b
critical opalescence, **3**:620b
partial molal properties, **4**:229a
partition coefficient, **2**:69b–70a; **11**:426a; **12**:404b; **15**:434a
physiological
Cushing's solution, **3**:518a
in wound treatment, **5**:29b
pressure effects, **13**:244b
pseudo-solutions, **11**:279a
solid, **1**:461a; **5**:187a; **7**:529a–b; **8**:119a; **12**:593b; **13**:580a; **14**:241a, 588a
and atomic size, **1**:462a; **6**:561a
heterogeneity of, **3**:239a
theory of, **6**:561a; **11**:550a; **13**:617b
solvent-solute relations, **7**:162a; **13**:244b
specific gravity, **9**:291a–b
specific heats
and concentration, **9**:110a
standard, **5**:325b–326a
strength of, **13**:547b
supercooling, **8**:520a
supersaturation, **2**:254b; **8**:520a; **15**:461b
surface tension, **14**:67a
temperature effects, **13**:244b
theories of
seventeenth-century, **12**:312b
nineteenth-century, **2**:186b; **4**:203a; **7**:203b, 462a; **8**:510b; **10**:577b; **11**:8b; **13**:244a–b
twentieth-century, **7**:162a, 208b, 387a; **14**:125a
chemical, **9**:287b, 291b
turbidity
and molecular weight, **3**:620b
vapor pressures, **5**:324a; **11**:298a–299b; **13**:243a
viscosity, **7**:518b
volume changes, **4**:554b; **14**:191a
see also Concentration, chemical; Electrolytes; Saturation
Solvation, **14**:125a
Solvay, Ernest (1838–1922), 12:520–521; **9**:467a
Solvay Congresses, **2**:488b; **4**:318b, 326a, 579b; **5**:272a; **7**:154a; **8**:11a, 489b, 497b; **10**:425a, 525b; **12**:521a, 528b; **15**:442b, 449b
Solvents
ionization, **14**:125a
universal, **7**:525a; **10**:309b; **11**:40b
Solvolysis, **14**:125a
Somaliland, **13**:552a
Somasiddhānta, **15**:612a
Somatology, **2**:208a
Somerset Coal Canal, **12**:487a
Somersetshire
geology, **4**:11a; **12**:487a–b, 490a–b
Somerville, Mary Fairfax Greig (1780–1872), 12:521–525
Somerville, William, **12**:523a–524a
Somerville College, Oxford, **12**:525a
Someśvara
commentaries of, **1**:308a; **2**:417b
Sommerfeld, Arnold (Johannes Wilhelm) (1868–1951), 12:525–532; **4**:357a; **7**:400a*; **12**:196b*, 221b
acquaintances of, **2**:466a; **3**:617b; **4**:577b; **7**:264a; **8**:51a; **11**:422a; **12**:71a; **14**:68a; **15**:451a
Atombau und Spektrallinien, **6**:614b
atomic theory, **7**:469b; **10**:422b
on contemporary scientists, **6**:395a*; **8**:598a; **12**:223a, 498b*
correspondence of, **10**:422b; **11**:9a, 584b*; **12**:614a
critics of, **12**:219a, 615b
electron theory, **5**:473b; **11**:447a
influence of, **6**:615a; **12**:614b; **14**:336b
mathematical physics, **1**:456b; **7**:399a; **11**:421b, 476b
quantum theory, **2**:244b, 245b; **3**:125a; **4**:317a; **7**:26b, 528b; **12**:251b; **13**:42a; **15**:48b, 442b
Sommerfeld-Debye theory, **12**:464a
spectroscopy, **4**:195b; **9**:397b; **10**:348a; **12**:21b
fine structure constant, **10**:423b
students of, **7**:26b, 468b; **8**:473a; **10**:422a; **11**:355a; **13**:40b
Sommering
see Soemmerring, Samuel Thomas
Sommerville, Duncan McLaren Young (1879–1934), 12:532–534
Somogyi, Zoltán, **13**:226b
Somov, Osip Ivanovich (1815–1876), 12:534; **2**:425a
Sonar, **10**:605a; **11**:422a
Sonder, Otto Wilhelm, **6**:163a; **7**:534a
Sonin, Nikolay Yakovlevich (1849–1915), 12:535; **3**:226a, 230b; **10**:544b; **15**:65b
Sonnerat, Pierre (1748–1814), 12:535–538
Sonneschmidt, Friedrich, **4**:345b
Sonnini de Manoncourt, Charles Nicolas Sigisbert, **12**:536b
Sophia Charlotte, electress of Brandenburg, **8**:150b, 151b
Sophisms, **6**:376b; **7**:66a; **13**:193b, 194b, 207b, 208a,b; **14**:391b
Sophists, **1**:170b, 250a; **2**:549b; **6**:405a–407a; **10**:230a
Second Sophistic, **5**:228a, 229a–b
Sophonias, **13**:308a
Sopp, O.
see Olsen, O. J.
Soranus of Ephesus (*fl.* second century A.D.), 12:538–542; **6**:318a–b, 418b
Sorbière, Samuel-Joseph, **5**:419b; **6**:446b, 598b; **9**:498a,b; **13**:335a
Relation d'un voyage en Angleterre, **12**:584b–585b
Sorbite, **10**:246b
Sorbitol, **2**:86b
Sorbose, **2**:86b
synthesis of, **5**:3a
Sorby, Henry Clifton (1826–1908), 12:542–546; **3**:434a, 435a*; **4**:532a; **10**:246b; **13**:8a; **14**:625a
metallurgy, **11**:549a; **13**:246a
petrography, **3**:161a; **5**:89a; **8**:42a; **10**:110a; **13**:273a
Sorcery, **14**:118a
scientists accused of, **1**:468a
see also Magic

Sørensen, Søren Peter Lauritz (1868–1939), 12:546–547; 8:335b; 7:180a*; 8:607b; 13:228b
Sørenson, Peder
see Severinus, Petrus
Soret, Charles, 5:597b; 13:101b
Soret, Jacques-Louis, 1:161a; 5:597b; 8:122a; 9:110b
Sorrel
chemical analysis of, 12:148a
Sosigenes (*fl.* first century B.C.), 12:547; 1:117b; 3:558b; 4:463b
Soto, Domingo de (*ca.* 1494–1560), 12:547–548; 1:198b; 3:172a, 280b, 420b; 6:379a; 13:350a
Soul
in animals, 1:61b; 11:325b; 13:528a, 543b, 564a
and body, 3:245a, 321a*; 6:59b, 573b; 7:137a–b; 8:115a, 514b–515a; 12:224b; 13:179b; 15:262a
separability of, 2:193a
and brain, 2:162a; 11:326a; 13:528a, 543b
continuity of, 6:290a,b
definition of, 11:71b; 14:535b(n)
divisibility of, 9:572a
and form, 1:260a–b; 6:152a, 459a
immateriality of, 11:599a
immortality of, 5:107b, 592a; 6:556b; 7:520b, 612b; 8:436b; 10:122b; 11:444b; 13:52b
intellect and, 1:27a, 260b
location of
ancient Greek discussion, 11:128a
medieval Latin discussion, 6:396b
Renaissance discussion, 14:10b
seventeenth-century discussion, 4:63a; 6:152b; 14:407a
eighteenth-century discussion, 2:3a; 6:64a; 8:506a; 12:510a; 13:179b–180a
nineteenth-century discussion, 9:572a
and matter, 11:325a
mortality of, 1:118b; 3:295a–b; 4:81b; 8:515a; 9:30a; 11:71b, 74a
phylogeny of, 2:596b
in plants, 7:537b, 538a; 12:175b
theories of
ancient Greek, 1:235a, 324b; 13:94a
Byzantine, 12:441b–442a
medieval Arabic, 1:26b–27a, 33b–34a; 7:23a; 11:325a–b, 326a; 15:249a–b, 262a, 494b, 495b
Renaissance, 4:585a,b; 13:279b
seventeenth-century, 14:407b–408a
eighteenth-century, 13:179b; 15:525a–b
nineteenth-century, 9:572a; 11:592b; 14:114a
atomistic, 4:32b–33b; 8:538a
materialist, 7:605a, 607a; 13:179b
transmigration of, 11:325b
tripartite, 5:235b
see also Animism, Psychology
Soulavie, Jean-Louis Giraud (1752–1813), 12:549–550; 2:204b; 3:526a; 12:489b
Souleyet, Louis-François-Auguste (1811–1852), 12:550–551
Sound, 2:324b; 3:459a
absorption of, 12:53b, 54b
amplification of
resonant vases, 3:153a
audibility limits, 7:445b; 13:103b
filtration of, 13:104b
frequency of vibration, 3:9a; 7:445a
determination of, 12:127b–128a
graphic recording of, 7:445b
intensity of
and distance, 9:319b
night, effects of, 6:553b
laws of, 9:230b
and light, 14:565a–566a
mathematical study, 4:577b
and motion, 1:233a
Doppler effect, 4:167b; 5:20a; 8:596a
overtones, 4:222b
photographic recording, 9:386b
in pipes
see Pipes, sound in
propagation of, 4:376a, 481a; 6:165a*; 7:565a, 578a
in air, 6:171b; 8:510b; 13:522a; 14:39a
in hydrogen, 13:76b
mechanical, 1:374a; 14:289b
in vacuum, 2:83b, 377b; 13:439a
in vapors, 2:135b; 15:357a
quanta of, 13:240a
refraction of, 11:393b
reproduction of, 10:604b
electrical, 1:293a; 14:290a
scattering of, 13:104b
shadows, 8:122a
speed of, 1:203a; 5:78b, 435b, 516a, 530b; 8:122a; 15:378a
in air, 10:76a; 15:356b–358a, 487a
in cast iron, 2:135b
in gases, 1:129a; 3:258b; 7:526a; 15:486a
measurement of, 2:135b; 3:108b; 5:288b; 6:292b–293a; 7:183a; 9:90b, 319b, 459b; 14:39a
in solids, 3:258b
in water, 13:127a–b, 130b
theories of
ancient Greek, 1:233a; 13:94a
medieval, 1:101a; 4:525b–526a; 5:550b, 551b
Renaissance, 3:99a
early seventeenth-century, 5:245a; 9:318a–b, 319b
late seventeenth-century, 6:609a; 10:76a, 520b
eighteenth-century, 7:587a; 8:468b; 9:33b
nineteenth-century, 2:88b; 6:245b; 13:76a–b; 14:186a
mathematical, 8:384b; 13:76a–b
tone quality, 7:445b
transmission of
see Sound, propagation of
wind, effect of, 13:76a
see also Acoustics; Hearing; Timbre; Waves, sound
Sound ranging, 3:220b, 372b, 564a; 4:407b; 7:553a; 9:386b, 404a; 14:244a; 15:61a
Sound waves
see Waves, sound
Sounding (naut.), 8:101a; 9:508a*
deep-sea lead, 4:291b
electronic, 3:505a
Sounding gauge, 4:44a
Soundproofing
advocacy of, 7:77b
Sousa, Martim Alfonso de, 10:161a, 237a
South, James (1785–1867), 12:551–552; 6:263a, 324a, 325a–b
South Africa
anthropology, 2:505a; 5:196a
botany, 6:162b–163a; 13:342b, 391b–392b
coastline, 4:262b
geology, 3:547b; 4:261b–263a
natural history, 13:282a–b
paleontology, 2:505a
surveying of, 5:405b
South America
and Africa, similarities between, 4:262b, 310a; 14:215a, 216a
and Australia, possible former connection, 6:594b
botany, 11:47b, 130a; 12:594a–b; 14:386b; 15:429b–430a
ethnography, 1:48a
exploration of, 4:602b–603a; 6:550b–551b; 7:200a–b; 9:149a; 12:578a–b; 14:133b–134b, 187a
geodesy, 10:523b–524a
geography, 1:351a–b; 2:374a; 10:619a
geology, 3:567b–568b; 4:262b; 8:570b; 13:63b
hydrography, 5:16a, 17a
natural history
Renaissance, 1:48a
eighteenth-century, 1:351b; 4:156b; 7:57b–58a
nineteenth-century, 1:501a–b; 3:569b–570a; 9:344b; 10:221a,b
paleontology, 10:261b, 262a
volcanoes, 11:336b
zoology, 12:211a, 258a; 13:552a
see also specific countries
South Carolina
medicine, 15:226b–227a
South Georgia Island, 3:397a
South Pacific
charting of, 4:256a
ethnology, 5:277a
exploration of, 1:434a–b; 2:342b–343a; 3:365b, 549b; 5:224b
natural history, 3:549b–550b
South Pole, exploration of, 4:193b
South Sandwich Islands, 3:397a
South Seas
see South Pacific
Southard, William Freeman, 2:367a
South-West Africa
geology, 12:194b
Sovereignty, 6:445a
absolute, 6:447b–448a, 448b
Sovetov, Aleksandr Vasilievich, 4:144b
Soviet Academy of Sciences
see Akademiia nauk USSR
Soviet Union
see Russia
Sowden, John Clinton, 5:6a
Sowerby, George, 9:87a
Sowerby, James (1757–1822), 12:552–553; 2:567b; 8:564a
British Botany, 2:183b
Mineral Conchology, 4:541a
Sowerby, James de Carle, 12:552b
Space
ancient Greek concept of, 13:93a
Renaissance concept of, 13:279b
seventeenth-century concept of, 2:13b–14a; 5:574b
eighteenth-century concept of, 1:111b; 8:159b; 14:84b, 484a
nineteenth-century concept of, 1:140b–141a
twentieth-century concept of, 1:24a; 7:488a; 14:283b, 306b–308b
absolute, 3:90a, 296a; 5:188a, 287a–b; 7:226a,b, 230b–231a; 9:510a; 10:70a, 416b; 11:324a; 12:608b–609a
arguments against, 2:17b; 8:158b, 599a
continuity of, 1:115b; 4:4a, 417a
curvature of, 3:323a; 5:189a*; 6:532a; 8:599b; 12:248b–249a
divisibility of, 7:229b
homogeneity of, 5:188a
infinity of, 1:28a; 2:221a; 4:416b–417a; 5:114a, 575b–576a
and matter, 2:326a; 3:322b–323a; 11:489b
multiplicity of, 2:331a
quantization of, 3:369b
relative, 2:460b; 4:322a, 323b; 7:225b–226b, 229b–231a; 8:11b, 158b; 11:324a

space-time, **5**:188a; **8**:292b; **9**:405b–406a, 414b
geometrization of, **4**:328a, 329a–331b; **8**:434a
uniformity of, **14**:305a–306a
subjective, **15**:504b
see also Place
Space(s) (math.)
absolute, **2**:270a, 271a; **8**:141b–142a
algebraic representation of, **6**:90a–b, 91b
Banach, **1**:428a; **2**:513a; **11**:459a
bicompact, **3**:170b–171a
closure, **6**:177a
of constant curvature, **3**:179a; **5**:201a,b; **8**:434a; **11**:455a
construction of, **3**:428a; **6**:394a
continuous, **3**:171a
curvature of, **11**:454b–455b; **12**:609a–b
curved
parallelism in, **8**:284a
cutting of, **14**:591a–b
differential, **14**:345a–b
division of, **13**:16b
elliptical, **3**:323a, 399a; **8**:434a
parallelism in, **5**:200b
Euclidean, **2**:465a; **4**:419a; **5**:8a
complex, **8**:431a–b
fiber, **4**:336b
Finsler, **3**:96b*; **4**:336b
four-dimensional, **9**:430b
geometric, **2**:214b–215b; **10**:416b
harmonic, **4**:337a
Hilbert, **1**:427b; **5**:150b, 152a; **6**:392a; **8**:457b; **12**:187b, 189a–190a; **13**:58a, 549a; **14**:46b, 90a–91b, 282a, 346b, 454a
homogeneity of, **4**:417a
hyperbolic, **8**:434a
infinite, **8**:141b
infinite-dimensional, **13**:428b
linear, **1**:427b; **6**:392b; **11**:408b, 450a
Lobachevskian, **8**:434a; **10**:486a; **12**:609a,b; **13**:414b
locally-convex, **13**:428b
metric structure, **1**:428a; **6**:177a; **11**:454b–455b; **13**:549a; **14**:122b
multi-dimensional, **1**:500a; **2**:465a, 512a; **3**:178b–179a, 221a; **5**:271b; **7**:577a; **11**:390a
multiplicity of, **3**:111b
n-dimensional, **3**:166b; **4**:523a; **5**:8a; **8**:112a; **11**:394b; **13**:549a
non-Euclidean, **2**:356a; **5**:201b; **12**:11b
normed, **13**:549a
numerical, **2**:513a
projective, **3**:171a; **13**:133a
complex, **11**:79b; **13**:5b
pseudo-Euclidean, **8**:431b, 433a,b
pseudospherical, **2**:465a
Riemann, **3**:96a–b, 323a; **7**:207b; **8**:284a, 434a; **11**:454b–455a; **12**:609b
spherical, **3**:399a
subprojective, **7**:207b
symmetric, **11**:53b
tangent, **4**:336b, 337a
theories of, **2**:274a; **5**:154b; **6**:176b–177a; **7**:207a–b; **8**:284b, 434a; **11**:454b–455b; **12**:609a–b
topological, **6**:177a, 391b; **12**:16b; **13**:549a
vector, **2**:351b
Space groups (crystallog.), **10**:125a
Space probes, **5**:405a, 433a–b
Space technology, **7**:247b–248a, 465a–b
Space travel, **4**:600a; **13**:483a–484a; **14**:363a, 364b
Spain
botany, **3**:154b; **8**:121a
economics, **13**:448a
geography, **6**:550b; **7**:2a
geology, **13**:621a
industry, **9**:140a
natural history, **3**:97a, 154b*; **14**:585a
politics, **15**:75b–76a
Spallanzani, Lazzaro (1729–1799), 12:553–567; 1:492b; **3**:425b
correspondence of, **13**:465a
critics of, **9**:140a; **11**:159a
editions of works, **2**:340a
influence of, **5**:56b; **7**:31b; **9**:86a
microbiology, **1**:586a; **5**:56b; **10**:365a, 370b
physiology
digestion, **1**:543a–b; **13**:46b, 47a
reproduction, **1**:561a; **2**:287a; **8**:160a; **13**:563b
on spontaneous generation, **7**:111a; **10**:10b, 367a
translations of works, **1**:563b; **12**:308b
Spallitta, Francesco, **15**:469a
Spanish-American War, **3**:518a
Spanish Civil War, **6**:22b
Spärck, Hakon Ragnar Gisiko, **7**:503a
Spark, electric, **4**:216a; **7**:416a
temperature of, **4**:78b
see also Discharge, electric, spark
Sparrman, Anders, **5**:76b; **8**:379b
Sparta, **4**:81b
Spasm
pharyngeal, **8**:506b
pyloric, **8**:506b
respiratory, **11**:425b
Spasmolytics, **5**:100a
Spasticity, **5**:207b, 208a
Speaking machine
eighteenth-century, **3**:578b
Speaking trumpet, **7**:376b; **9**:486a, 529b
Speaking tube, **7**:599a
Spearman, Charles Edward, **10**:453b; **14**:382a, 383a
Speciation, **3**:569b–570a, 572b
Species, biological
competition of, **8**:512a, 568b; **9**:69a; **11**:567b
concept of
ancient Greek, **1**:260b–261a, 278a
eighteenth-century, **1**:306a; **2**:579b–580a; **8**:377a; **9**:368a
nineteenth-century, **5**:67b; **7**:165b–166b; **9**:375a; **14**:99b–100a
twentieth-century, **3**:318a; **5**:222a*; **15**:508a
definition of, **1**:251a,b; **4**:68a; **6**:296a, 594b; **8**:415b; **11**:315b; **13**:443a
degradation of, **7**:585b, 587b, 588b; **11**:234b; **12**:551a
extinction of, **2**:481a, 492a; **3**:569b; **6**:310a
theories of, **4**:68a; **5**:32a; **6**:614a; **7**:588b, 589a, 592b, 604a; **8**:568b–569a, 594b; **9**:533b; **10**:222a; **11**:535b, 567b; **12**:549b; **14**:387a
fixity of, **1**:337a, 469a; **2**:286b, 492a, 580b; **3**:378a, 526a; **4**:288b, 504b; **5**:32a, 218b, 356b, 429a, 577b; **6**:120b, 266a; **7**:102b, 165b–166b, 376b, 461a, 585b, 592a; **8**:377a–b; **10**:212a, 284a; **11**:314b; **12**:318a, 357a–b; **14**:44b; **15**:344b
geographical distribution of
see Biogeography
mutability of, **2**:426b; **3**:318a, 523b, 526b, 527b, 553a, 569b–570a, 579b; **4**:290b; **5**:356a, 443a, 498a; **6**:8a, 65b, 275b, 486b, 592a; **7**:604a; **8**:416a; **9**:280b, 375a, 536a, 618b; **10**:209b, 391b–382a; **11**:235b, 315b, 567b; **14**:189b–190a, 313a; **15**:127b
origin of
ancient Roman theories, **4**:368a; **8**:538a
medieval theories, **6**:275b
early eighteenth-century theories, **5**:429a; **8**:377b–378a
late eighteenth-century theories, **3**:523b, 43b; **5**:356a; **6**:469a; **9**:188a
early nineteenth-century theories, **3**:570a–573b; **5**:192a; **6**:10a; **11**:137a–138a
mid nineteenth-century theories, **1**:501a; **2**:161b, 205b–206a; **3**:43a, 192b; **5**:67b, 71a; **6**:266a; **7**:320a, 439a, 546a; **8**:572b–574b; **10**:262b; **12**:277b–278a; **14**:134b–139a
late nineteenth-century theories, **9**:432a; **11**:519a; **14**:99b–100b, 237a–238a
early twentieth-century theories, **1**:623b; **3**:493a; **11**:535b; **13**:151a
perfection of, **8**:464a; **12**:318b
quantitative definition, **14**:251a–b
replacement of, **5**:67b; **8**:569a
splitting of, **3**:572b; **7**:166a–b; **9**:313b
survival of, **2**:257a
see also Evolution; Natural selection; Variation, biological
Species, mathematical, **14**:19b–20b
Specific heat
see Heat, specific
Specific volume, **9**:286b, 289a,b; **10**:518b
and chemical affinity, **15**:457b
Specificity, biological
DNA, role of, **1**:342b–343a
immunological, **2**:300b
alteration of, **1**:342b–343a
protein theory of, **1**:343a; **7**:467b
Spectacles
see Eyeglasses
Spectra
combination principle, **12**:43b
dispersion formula, **6**:146b
displacement law, **3**:124b; **10**:346a
enhanced lines, **1**:56a
fine structure, **4**:281b; **10**:348a; **12**:528b–529a
Fraunhofer lines, **1**:166a, 556a; **2**:453a; **4**:183a; **5**:86b, 143a–144a; **7**:381b; **9**:566a–b; **12**:269a; **14**:491b
wavelengths of, **1**:166b; **7**:528a
high multiplicities, **1**:370b–371a
hyperfine structure, **1**:371a; **4**:579a; **14**:599a
intensity of lines, **2**:600b; **10**:236a
line broadening, **2**:170a; **5**:118a; **8**:496a–b; **14**:598a
line splitting
electric field, **2**:246b; **12**:529a, 614b
magnetic, **1**:370b–371a; **2**:245a, 247b; **6**:29b; **10**:236a, 347b; **11**:612b; **14**:598a–b
theories of, **11**:612b
Lyman ghosts, **8**:578b
multiple lines, **3**:124b–125a
multiplets, **9**:154b; **10**:348a
theory of, **3**:125a; **7**:492b; **9**:397b; **13**:82a
photography of, **9**:391b–392a
quantum theory, **7**:27a; **15**:48b
series, **3**:124a; **5**:102b*; **10**:346a,b, 347a–348a
Balmer series, **4**:498a; **11**:476a, 478b–479a
formulas for, **1**:425a–b; **2**:243a–244a; **6**:164a; **11**:476a–477a, 478b–479a, 611b–612b; **12**:21a–b, 42b–43b, 238a–b; **13**:339a
laws of, **1**:370b; **14**:598b
Paschen series, **1**:425b; **10**:347b
theories of, **7**:267b–268a; **11**:475b–477a, 478a–479b

Spectra (cont.)
sharp lines, **3**:369b; **10**:347b–348a
and temperature, **14**:479b
types
absorption, **2**:488a; **3**:327a, 430a; **4**:79a, 253b, 497b; **5**:143b; **6**:108b–109a; **7**:273a, 382a, 553b; **9**:391b, 405a; **12**:236b; **14**:175b
arc, **3**:124b; **4**:79a; **7**:469b
atomic, **1**:370b–371a; **2**:243a–248b; **3**:124a–125b; **5**:102a–b; **11**:580b–581a; **12**:22a; **14**:497b
band, **3**:430a; **4**:79a; **5**:102a, 118a; **8**:442a–b, 473a, 474a; **9**:606b; **10**:214a
complex, **3**:124b
continuous, **2**:488a; **4**:497b, 498a; **10**:214a
discontinuous, **3**:478a
emission, **1**:166a; **3**:36a; **4**:79a; **5**:143b; **8**:293b; **9**:154b
flame, **4**:79a; **5**:143a, 483b; **9**:267a, 391b
molecular, **1**:21b; **4**:69a; **5**:118a
monochromatic, **6**:94b
Raman, **3**:336a; **4**:82b
rotational, **5**:609b
spark, **1**:166a; **2**:589b; **3**:124b; **4**:79a; **5**:497a; **7**:469b; **9**:397a; **13**:77b
vibration, **1**:519b
ultimate lines, **5**:497a
variation of, **5**:102a
wavelengths of, **4**:498a
weak lines, **9**:415a
Zeeman effect
see Zeeman, Pieter, Zeeman effect
see also individual elements; Light; Radiation; Spectroscopy; Spectrum, visual
Spectral analysis
late eighteenth-century, **9**:267a
mid nineteenth-century, **2**:589a; **3**:476a; **4**:179a; **5**:341a; **7**:379b, 381b; **9**:223a, 391b; **13**:77b
late nineteenth-century, **3**:420a; **8**:503b; **10**:519a; **11**:538a; **12**:238a; **13**:82a
early twentieth-century, **1**:56a; **2**:600b; **5**:102a, 497a–b; **7**:622a–b; **9**:96a–b; **13**:454a
gas impurities, **1**:519b
Spectrobolometer, **8**:20b
Spectrochemistry, **1**:341a; **9**:256a–b
Spectrocomparator, **6**:146b
Spectrogram, **11**:612a
Spectrograph, **4**:69a, 498a; **5**:405b; **7**:270b; **12**:454b
Bruce, **5**:199b
longitudinal, **13**:408b
Mills, **3**:35a, 36a
nebular, **13**:117b
solar, **9**:415a; **12**:73a
stellar, **2**:196b; **3**:510a; **4**:179b; **6**:146b; **11**:18b; **12**:153a, 251a
vacuum, **12**:235b
Spectroheliograms
hydrogen α, **4**:498a
Spectroheliograph, **2**:424a; **4**:498a; **6**:28b, 32a; **8**:503b; **11**:297a, 412a
invention of, **4**:69b; **6**:28a
Spectrohelioscope
invention of, **6**:32a; **7**:75a–b
Spectrometer
Bragg, **3**:368a, 369a; **15**:62b
echelon, **11**:422b
X-ray, **2**:399a–b; **3**:367b; **4**:195b; **11**:531a
Spectrophotometry, **1**:598a; **2**:382b; **4**:283a; **7**:458a; **9**:415a, 563b; **11**:150b, 542a–b; **12**:96a, 353a, 368a; **14**:55b, 174a
Spectropolariscope, **2**:382b
Spectroscope, **1**:366a, 370b; **2**:424a; **4**:161b–162a, 251a; **5**:292b, 561b*; **6**:541b; **7**:381b; **8**:551b; **12**:37a, 60a, 269a; **13**:344a
direct-view, **14**:479b
optical properties, **13**:101b
reversion, **14**:54b, 629a
Spectroscopy
early nineteenth-century, **2**:452b–453a; **5**:143a–144a; **6**:324b
mid nineteenth-century, **1**:166a–b, 556a; **2**:589a–b; **3**:476a–b; **4**:183a, 497b; **5**:19a, 86b, 126b; **6**:138a, 293b–294a, 327a, 505a, 541b–543a; **7**:379b; **8**:596a; **9**:391b, 566a–b; **11**:47a; **12**:132a, 544a–b; **13**:77b, 558a; **15**:110a
late nineteenth-century, **1**:19a, 21b, 133a, 425a–b; **2**:254b; **3**:234a, 279a, 478a, 499a; **4**:69a, 79a, 179b–180a, 513b–514a; **5**:483b; **6**:94b; **7**:267b–268b, 497a; **8**:20b; **9**:531a; **10**:345b–346b; **11**:578b–569a, 611b–612b; **12**:42a–44a, 235b–236a, 236b, 238a–b, 499a; **13**:473a
early twentieth-century, **1**:370b–371a; **2**:169b–170a, 243a–248b, 600b; **3**:124a–125b, 327a–b, 329a, 429b–430a, 502a–b; **4**:253b, 513b–514a; **5**:101b–102a, 497a–b; **7**:553b–554b; **8**:578b–579a; **9**:256a–b, 397a–b, 606b; **10**:347a–348b; **11**:420a,b, 422a, 580b–581a; **12**:20b–23a, 528b–529b, 614b; **14**:497b
mid twentieth-century, **6**:615a; **8**:473a; **11**:422b
alpha-ray, **3**:329a
in astronomy
early nineteenth-century, **5**:143a–b; **6**:237a
mid nineteenth-century, **1**:166a–b; **2**:433a,b, 485b; **4**:161b–162a, 168a; **5**:86b, 126b; **6**:541b–543a; **7**:74b–75b, 381b–382a; **8**:440b–442b; **9**:154b, 391b–392a; **11**:379a; **12**:37a, 267a–269b; **14**:479b
late nineteenth-century, **1**:19a, 21b, 295b, 597a–598b; **2**:146b; **3**:35a, 36a–b, 50a–b, 420a; **4**:69a–b, 179b–180a, 251a, 497b–498a; **5**:33b–34a, 199a–b, 405b; **7**:270b–271a, 461a, 528a; **8**:20b, 465a–b, 503b, 539a–b; **9**:154b, 184b, 194b–195a, 373a, 564a; **10**:528b, 600a–b; **11**:319b–320a, 412a–b; **12**:499a; **13**:232a–b, 344a–b; **14**:55a–56b, 449a, 557b, 629a–b
early twentieth-century, **1**:55b–57a, 352a; **2**:196a–b; **3**:510a; **4**:270a, 279b, 280a–b, 513b–514a; **5**:101b–102a, 182a–184a; **6**:146b–147a, 350b–353a, 508a, 531a–b; **7**:186b, 268a–b; **9**:405a, 503b–504a; **10**:473b; **11**:18b–19a; **12**:20b–23a, 71a, 153a, 176b, 251a–b, 454b–456a; **13**:116a–b, 117b–118a; **14**:482a, 521a–b
mid twentieth-century, **8**:364a–b; **9**:256b*; **12**:367b–368b; **13**:117b–118b
atomic, **10**:348a
beta-ray, **2**:338b; **3**:329a; **12**:32a, 34a
in biology, **7**:273a; **9**:549b
in chemistry, **5**:589b; **13**:238b; **14**:175b
electromagnetic, **1**:284a
fluorescent, **8**:293b; **13**:77b
gamma-ray, **2**:338b; **12**:32a, 34a
infrared, **1**:558b; **3**:327a–b, 336a, 430a; **4**:183a; **9**:566b; **10**:347a–b; **11**:477a, 479b
L-S coupling, **12**:21a
luminescent, **1**:556b
mass spectrograph, **1**:321a–b; **5**:459a; **6**:117b; **8**:95a,b; **10**:289a; **14**:218a–b
monochromatic, **8**:503b
notation, **12**:21a
phosphorescent, **2**:254b–255a; **3**:478a; **8**:293b
in quantitative analysis, **11**:485b
standardization of, **1**:166b; **8**:465b
theories of, **3**:476a; **7**:381b–382a; **8**:367b, 510b; **9**:38a, 223a, 397b; **11**:476a, 478a–479b; **12**:20b–21b, 22a, 42a–44a, 529a–b; **13**:82a, 104a, 368a, 369a
combination principle, **11**:478b–479b
dissociation hypothesis, **8**:441a–b
electric-doublet, **13**:156a
ultraviolet, **4**:68b, 79a, 183a; **8**:578b; **9**:154b, 397a–b, 566a–b; **12**:235b–236a
X-ray, **2**:245a, 487b–488a; **3**:329a, 367b–370a, 429b–430a; **4**:195b, 196a–197a, 578b; **11**:441a–b; **12**:528b
focusing effect, **2**:487b
rotating-crystal method, **2**:487b
ultrasoft, **3**:430a
see also Diffraction grating; Spectra; specific bodies and materials
Spectrum, mathematical, **6**:392a; **14**:281b–282a, 454a
Spectrum, visual
brightness distribution of colors, **7**:458a
number of colors, **2**:453a; **4**:94b; **9**:200a; **14**:491b
shape of, **9**:118b
theories of, **10**:176b
Speculation (phil.), **1**:366b; **2**:453b; **9**:129b
Speculum (metal), **4**:179a; **5**:422b, 562a; **6**:5b, 324b; **8**:46a
Speech
control of, **1**:496a
cerebral localization, **2**:478a; **5**:173a–b
disorders of, **1**:496a
stammering, **6**:563a
harmonic analysis of, **4**:500a
and intelligence, **5**:505a–b
mechanical imitation of, **9**:322a (n); **14**:366a; **15**:194a
see also Language; Phonation; Physiology, linguistic; Voice
Speleology, **2**:566b; **11**:483b–484a
see also Caves
Spemann, Hans (1869–1941), 12:567–569; 10:427b; **15**:171a*
on contemporary scientists, **4**:188a; **6**:132b; **11**:575b*
embryology, **4**:187a; **11**:597b
Spence, William (entomol.), **1**:502b
Spencer, Adam, **5**:130a,b
Spencer, Baldwin, **5**:145b
Spencer, Herbert (1820–1903), 12:569–572; 1:496a; **5**:252a; **13**:353a*, 522b, 524a*
critics of, **5**:213a; **7**:384a; **14**:96b
on evolution, **6**:140a; **7**:47b, 48b; **11**:518b; **14**:237a
influence of, **2**:10b; **3**:322b, 573a; **5**:290b, 438b; **6**:53a; **7**:48a; **12**:109a; **13**:9a
influences on, **3**:89a
Principles of Biology, **5**:70a; **15**:92b
sociology, **5**:255a,b
Study in Sociology, **8**:362b
Spencer, James, **11**:281a
Spencer, Leonard James (1870–1959), 12:572–573; 4:142b*; **9**:380a*; **13**:286a*, 518a*
Spencer Jones, Harold (1890–1960), 12:573–574; 4:352a*; **10**:34a; **12**:450a*
Spencer Lens Company (Hamburg, New York), **4**:563b

Spengel, Leonhard, **4**:461a
Spenner, Fridolin Karl Leopold, **12**:167b
Sperm whale, **11**:528b
Spermaceti, **3**:414b; **5**:91a
Spermatia, **13**:395b
Spermatogenesis, **2**:344b; **3**:188b*; **5**:35b; **6**:338b; **12**:45b, 319b; **14**:235b
  in birds, **5**:598b
  in *Bufo,* **1**:545b
  chromosomes, **8**:587b–588a, 589a
  in myriapods, **2**:345a
  in *Sacculina,* **4**:207a
Spermatogonia, **13**:489b
Spermatozoa
  ancient Roman study, **12**:539b
  seventeenth-century study, **8**:127b
  eighteenth-century study, **12**:557a–b, 559a–b; **15**:52b
  nineteenth-century study, **1**:386b; **5**:35b; **11**:132b
  twentieth-century study, **9**:496a
  chemical characterization of, **7**:467a
  composition of, **9**:380b, 381a
  differentiation, **9**:496a, 497a
  in female genital tract, **5**:43a
  midpiece, **2**:363a
  motility, **7**:438b
  naming of, **1**:386b
  nature of, **2**:579b; **7**:437b; **9**:482b; **12**:557a–b; **13**:563b
  nonmotile, **7**:455a
  nucleus, **6**:338a–b
  origin of, **11**:132b
  in ovary, **2**:161a–b
  penetration into ovum, **1**:477a; **2**:161b; **5**:51b; **6**:337b–338a; **7**:455a; **8**:127b
  preservation of, **7**:32a; **9**:86a
  role of, **12**:559a–b; **15**:54a
  structure, **7**:455a; **11**:383a; **14**:113a
Spermatozoids, **10**:13a, 576a; **13**:394b–395a
Spermism, **7**:441a
Spermophilus, **3**:521a
Sperry, Elmer Ambrose (1860–1930), 12: 574–575
Sperry Gyroscope Company, **6**:104b
Speusippus (*ca.* 408–339 B.C.), 12:575–576; **1**:250a, 259a; **9**:269b; **11**:23b, 27b, 29a; **14**:534b, 535a; **15**:202a
Speyer, Georg, **4**:298b
Sphacelariaceae, **11**:154a; **12**:125a–b
Sphaerobacteria, **3**:338b
Sphaeroplea, **3**:337b
Sphagna, **12**:169a; **15**:149a
Sphene, **11**:539a
Sphenodon, **2**:504b
Sphenophyllum, **12**:259b; **14**:600a
Sphere, **3**:399a
  area of, **1**:219b, 220a, 221b, 227a,b, 445a; **12**:247a
  characterization of, **4**:336a–b
  circular sections of, **1**:339a
  and cone, **1**:220b–221b
  cubature of, **7**:559a*
  cuttings of, **14**:591a
  and cylinder, **1**:213a–b, 219b, 220b–221b
  development of, **4**:260a
  division of, **4**:105a, 108b–109b; **9**:21b; **14**:274b; **15**:116b
  functional equation of, **3**:428a
  homotopy, **6**:497a
  intersecting, **13**:14b
  motion of, **1**:338b–339a
   in incompressible fluid, **4**:126b
  n-dimensional, **2**:513a
  polygonometry of, **1**:429a
  as a polyhedron, **4**:33b
  segments of, **4**:490a; **11**:240a
  tangent, **5**:111a; **12**:227b–228a; **13**:14b
  volume of, **1**:219b, 220b–221b, 227a,b, 444b; **3**:152a, 277a; **4**:466b; **13**:319b
Sphere, armillary
  see Armillary sphere
Sphere, celestial
  circles of, **3**:319b; **5**:345b
  division of, **1**:508b
  rotation of, **4**:429b
Spheres, astronomical doctrine of
  ancient Greek, **1**:257b–258a, 338b–339a; **3**:21b–22a; **4**:463b, 466b–467a; **9**: 269b; **10**:590b; **11**:187b–188b; **13**: 326b
  ancient Roman, **9**:1b; **11**:197a–b
  Byzantine, **7**:138b
  medieval Arabic, **1**:33b–34a, 511a–b; **6**: 198a; **12**:3a–5b; **13**:292a; **15**:496a–b
  medieval Hebrew, **8**:281a
  medieval Latin, **1**:199a; **2**:24a, 551a, 606b; **3**:25b; **5**:115b; **7**:27b, 112b, 124b, 142a; **9**:136b, 362a; **13**:534b; **14**:388b, 390a–b, 400a, 401a
  Renaissance, **2**:407b, 408a, 410a; **3**:409b; **11**:533a; **15**:475b
  critics of, **1**:338b; **5**:399b, 403a
  Egyptian, **4**:622a
  harmony of
   see Harmony, universal
  Indian, **10**:129a
  music of, **6**:617b; **11**:203a, 223b, 469a
Spherics, **13**:319b–320a
Spheroid
  attraction of, **5**:158a; **7**:37b; **8**:135a, 136a; **10**:61a–b, 63b, 68b, 74a; **15**:296a–b, 316a–322a, 337b, 351a,b, 356b, 380a
  prolate, **4**:429a
  viscous
   tides in, **3**:583a
Spherometer, **2**:137b; **3**:366a
Sphex, **4**:504b
Sphingomyelin, **8**:276a
Sphinx, **10**:40a*
Sphujidhvaja (*fl.* A.D. 269), 12:576–577; **10**:425b; **13**:582a; **14**:549a
  Yavanajātaka, **15**:538b–541a *passim,* 543a
Sphygmogram, **15**:101a
Sphygmology, **8**:390b
  see also Pulse
Sphygmomanometer, **4**:398a; **10**:205a; **11**: 482a
Spica (star), **2**:404a; **9**:296b; **15**:218a, 676b
  double nature of, **14**:56a
Spice plants
  cultivation of, **11**:64b
Spider, **8**:27a; **11**:317a
  nomenclature, **3**:320a
  taxonomy, **5**:502b; **8**:415a–b
  see also Arachnology
Spiegel, Adriaan van den (also Spieghel, Spigelius, Spiegelius, Adriano Spigeli) (1578–1625), 12:577–578; **4**:508a; **6**:153a
  De humani corporis fabrica, **3**:99a
Spiegeleisen (metallurgy), **15**:25a
Spiegelius
  see Spiegel, Adriaan van den
Spielmann, Jacob Reinbold, **2**:6b; **7**:394a; **9**:104b
  Institutiones chemiae, **3**:7a
Spigelius
  see Spiegel, Adriaan van den
Spilite, **5**:39a
  see also Basalt
Spiller, John, **3**:476a
Spina, Bartolomeo de, **11**:74a
Spinach, **1**:101b; **15**:67b
Spinal animal, **15**:105b
Spinal cord, **3**:487a; **5**:44b
  anatomy, **1**:496b, 580a; **5**:196a, 251a; **6**: 533b; **10**:252b; **12**:165a, 396a–b; **13**:493a, 587b
   comparative, **14**:126b
   myelogenesis technique, **5**:27a–b
  conduction in
   action potentials, **5**:291a
  crossing of nervous pathways, **2**:525a
  cutting of, **1**:496b; **2**:525a; **3**:220a
   hemisection, **2**:525a
  dorsal roots, **3**:205b; **14**:229a
  ganglia, **3**:343a; **8**:282b; **11**:296a; **12**:137b
  histology, **2**:124b; **5**:172a; **7**:438a, 439b; **13**:171a; **14**:229a
  hyaline degeneration, **2**:584b*
  irritation of, **13**:493a
  metameric organization, **8**:134a
  motor functions, **12**:397b, 399b–400a
  motor pathways
   cerebral influence, **5**:208a
  neural arc
   concept of, **6**:59b–60a
  pathology of, **8**:7b; **9**:108b
  psychic function, **10**:579a
  puncture of, **8**:506a
  pyramidal tract, **5**:27a–b
  reflex function, **6**:59b–60a; **10**:579a; **12**: 396a–401a; **14**:321a,b
  secondary degeneration of, **3**:205b; **8**:7a
  surgery of, **3**:517b
  tumors, **6**:518b–519a
Spinal pressure
  and blood pressure, **3**:518a
Spindle, mitotic, **5**:51a–b; **9**:524a
Spine
  evolution of, **7**:170a
  intervertebral joints, **14**:450a
  spinal canal, **4**:586a
  veins of, **2**:443a
Spinels
  crystal form, **1**:20a
Spinola, Daniele, **2**:308a
Spinors, **3**:95b; **6**:615a; **7**:10a; **13**:600a
Spinoza, Baruch (or Benedict), **6**:448a; **10**: 447b; **12**:157a*
  acquaintances of, **6**:538a; **8**:149b; **9**:31b; **10**:201a; **13**:33a, 479b
  critics of, **2**:379a; **3**:381b
  influence of, **2**:545a,b; **3**:56b; **4**:85a; **7**: 505b; **9**:510a, 572a; **12**:218a
  influences on, **3**:470a; **4**:64b, 502a; **9**:98a; **15**:69b
  philosophy, **3**:295a, 492a; **7**:482b
Spinozism, **2**:55a, 224b; **3**:322b; **5**:442b
Spinthariscope, **3**:480a; **6**:117b
Spiral, **2**:349a; **5**:527b; **11**:404a
  Archimedes', **1**:164b, 219b; **3**:391a; **4**: 104a, 259b; **5**:588b; **10**:295a; **12**:75b; **13**:436a; **14**:22a
   tangent to, **14**:21b
  construction of, **10**:52a
  general, **1**:164b
  infinite, **5**:527a
  logarithmic, **2**:47b; **4**:58a, 259b; **5**:528a; **6**:124b–125a; **8**:505a; **13**:436a
  parabola, comparison with, **3**:183a; **11**: 487b; **12**:75b
  parabolic, **2**:47b
  quadrature of, **1**:219b; **4**:571b
  tangential, **4**:259b
  tangents to, **13**:436b
  transformation of, **5**:527a
  see also Screw
Spiric sections, **10**:529b–530a
Spirillosis
  treatment of, **4**:300a,b

Spirillum
Obermeier's, **15**:454b
Spirit, animal
see Animal spirits
Spirit lamp, **5**:203a
Spiritualism
Renaissance, **5**:119a–121b; **10**:308b–311a
seventeenth-century, **4**:96a; **5**:416b–417a; **9**:509a–510a; **14**:210a
nineteenth-century, **2**:625a; **3**:240b, 378a, 475a–476a; **5**:531a; **6**:115a, 541a; **8**: 515a; **9**:292a; **10**:355a, 371b; **13**:52b; **14**:134a, 137a–138a, 629b
twentieth-century, **11**:431a; **14**:139b
critics of, **7**:498b
and religion, **3**:475a–b; **7**:68a; **13**:52b; **14**:137a–138a, 629b
see also Animism, "Occult" phenomena, Parapsychology, Psychic research, Theosophy
Spirobacteria, **3**:338b
Spirochetes, **4**:300a; **5**:317a; **7**:422a, 624a; **10**:155b; **11**:174b–175a; **12**:141b, 142a–b; **15**:454b
in brain, **10**:142a
culture of, **10**:142a
intestinal, **4**:207a
sanguicolous, **4**:207a; **8**:296b
Spirochetosis, **7**:429a
Spirogyra, **11**:153b; **14**:98b
reproduction, **3**:612b; **5**:217a
Spirometer, **7**:502b; **12**:467b*
invention of, **6**:575b
Spitsbergen Islands
exploration of, **4**:251a; **5**:225a; **10**:148a
geology, **3**:237a; **6**:78b
latitude of, **3**:237a
Spitzer, Lyman, Jr., **1**:353b
Spix, Johann Baptist von (1781–1826), 12: 578–579; **1**:501b; **5**:442b; **9**:149a
Spleen, **4**:178b
anatomy, **2**:478a; **5**:514b; **14**:286b
and blood volume, **1**:454b–455a
embryogeny, **5**:514b
enlargement of, **1**:439a
hemocatatonistic function, **2**:339a–b
and hemolysis, **1**:439a–b
histology, **3**:113b*; **9**:56b–57a
metabolism, **6**:506b
pathology, **1**:439a–b; **8**:352b
physiology, **5**:514b
role of, **1**:259a, 318b; **5**:230b; **6**:151a, 367b; **9**:57a; **13**:223b, 403a; **15**:479b
size, **1**:454b; **9**:165b
and stomach contraction, **8**:352b
Splenectomy, **5**:366b; **13**:567b; **14**:587a
for Banti's disease, **1**:439a
effects of, **15**:479b
for hemolytic jaundice, **1**:439a–b
Spoerer, Gustav Friedrich Wilhelm (1822–1895), 12:579
Sponges, **6**:7b, 8b, 613a; **13**:554b
animal vs. vegetable nature, **4**:236a; **12**: 560b
embryogeny, **4**:12a; **6**:9a
fossil, **3**:608b; **6**:58b*; **9**:87a; **12**:520a; **13**:531b; **14**:626a–b
freshwater, **14**:203a
metamorphism, **4**:12a
reproduction, **4**:207a
Spongilla, **14**:203a
Spontaneous generation
ancient Greek discussion, **1**:261b, 262a, 263b, 265a–b; **2**:591a–b; **13**:332b
medieval Arabic discussion, **7**:64b
medieval Latin discussion, **10**:555b
Renaissance discussion, **4**:387a; **5**:107a; **12**:317a; **14**:507b
seventeenth-century discussion, **1**:293b, 550a; **2**:591a–b; **3**:180b, 332b; **4**:509a; **5**:439b; **6**:161b, 603b; **8**:127b, 128a, 416a; **11**:317a, 341b–342b; **12**:616a; **13**:173b, 174a, 563b
eighteenth-century discussion, **1**:392a, 492b; **2**:579b; **3**:579b; **7**:111a, 604a; **8**:580a; **10**:10a–b; **12**:554b–555a, 556b, 558a
early nineteenth-century discussion, **2**: 321a; **6**:120b; **7**:587b, 588b, 592b
mid nineteenth-century discussion, **1**: 417a, 493b; **3**:192b, 378a, 612a,b, 614a; **4**:210b–211a, 236b, 288b, 290b–291a, 291b; **6**:8a; **7**:421a; **8**: 169b, 254b, 403a; **9**:85b–86a, 601b; **10**:355a, 366b–372a; **11**:110a–b; **12**: 233a–b, 241a; **14**:350b, 534a; **15**:110a
late nineteenth-century discussion, **1**: 496a,b; **2**:33a, 575b; **3**:188b, 338b, 339a–b; **10**:380b–383b, 389b; **13**: 522a
twentieth-century discussion, **1**:496b, 497a–b
explanations of, **7**:590a–b
Spores, **2**:437a, 583a; **5**:217a; **6**:465b
colors of, **5**:192b
discovery of, **3**:180a
formation of, **4**:546a; **6**:101a, 121b; **9**: 600a–b; **12**:99b
germination, **3**:30a; **4**:220a
heat-resistant, **12**:558a
monosperm cultures, **13**:405a
pathogenic, **10**:386a–389a, 393a, 394b, 397b
pressure, effects of, **8**:608a
role of, **6**:219a; **7**:197b
swarm, **3**:612b; **11**:154a
reactions of, **13**:88b
thermoresistant, **3**:339a–b; **10**:370b, 372a–b, 383a
Spörlin, Johann Jacob, **5**:418b
Sporochnales, **12**:125b
Sporogenesis, **7**:421b–422a
tetrads, **2**:492a
Sporophytes, **2**:371b; **13**:165b
origin of, **3**:30a; **6**:465b
antithetic theory, **3**:30b
homologous theory, **3**:30b
Sporotrichum, **9**:176a
Sporozoa, **8**:145b, 270b
development, **4**:207a
life histories, **12**:423b
taxonomy, **4**:207a
Sporus of Nicaea (*fl.* second half of third century A.D.), 12:579–580; **4**:103b, 104b, 463b, 489b, 491a
Sprat, Thomas (1635–1713), 12:580–587; **2**:507a; **14**:369a, 379a(n)
History of the Royal Society, **5**:416a, 606b; **10**:43b; **14**:363b, 371b
Spratt, Thomas Abel Bremage, **5**:67a
Spread (math.), **2**:513b, 514a
Sprengel, Christian Konrad (1750–1816), 12:587–591; **2**:257a; **7**:442a
Das entdeckte Geheimniss der Natur . . . , **2**:517b; **7**:417b
on reproduction, **7**:441a; **13**:506b; **15**:67b
Sprengel, Herman Johann Philipp
Sprengel pump, **2**:258a
Sprengel, Joachim Friedrich, **12**:587b
Sprengel, Kurt Polycarp Joachim (1766–1833), 12:591–592; **7**:553b; **9**:252a; **12**:587b, 588a, 589a(n),b(n)
Historia rei herbariae, **11**:312b
historiography, **4**:122a*; **6**:459b; **11**:37b; **13**:178b*
soil analysis, **8**:346b
Spring, Frédéric-Antoine, **12**:244a; **15**:51b
Spring, Walthère Victor (1848–1911), 12: 592–594; **5**:149a; **13**:577b
Springs (mech.), **6**:485a; **8**:212b
force of, **8**:197b–198a
leaf, **3**:118b
spiral, **6**:482b
torsion, **3**:118b
Springs (water), **6**:228a, 502b; **12**:140b
explanations of, **2**:158a; **4**:71a; **5**:401a; **6**:14a, 502b; **10**:521b; **11**:41a, 387b; **12**:135b, 210b, 554a–b; **13**:564a
thermal, **4**:218a; **6**:460a, 480a; **11**:469a; **12**:306a
Sprinkler system, **3**:388b
Spruce, Richard (1817–1893), 12:594
Sprung, Adolf Friedrich Wichard (1848–1909), 12:594–596
Spurway, Helen, **6**:22b
Spurzheim, Johann Christoph (1776–1832), 12:596–597; **5**:251a–b, 252a, 253b, 255a; **8**:272a; **14**:15a
Squalermo, Luigi
see Anguillara, Luigi
Square
area of, **1**:31a; **4**:609b
center of gravity, **8**:224a
diagonal of, **1**:94a; **2**:394b, 395b; **11**: 222a
of proportionality, **8**:239b (fig.), 240a
transformation of, **1**:42a
Square, magic
see Magic square
Squid, **14**:1b
anatomy, **10**:9b
Squire, William, **12**:502b
Śrīdatta
commentaries of, **2**:417b
Śrīdhara (*fl.* ninth century), 12:597–598; **1**:309b; **2**:115b
commentaries of, **2**:116a; **12**:599a
Laghukhecarasiddhi, **15**:588a
mathematics, **12**:598b
Pāṭīgaṇita, **9**:614b
Śrīkaṇṭha
commentaries of, **2**:114b, 115a, 116a
Śrīnivāsa Yajvan, **2**:119b
commentaries of, **2**:416b
Śrīpati (*fl.* 1039–1056), 12:598–599; **15**:586a, 619a, 625b
commentaries on, **10**:314a*
Jātakapaddhati, **1**:49a
Jyotiṣaratnamālā, **7**:583a
Siddhāntaśekhara, **15**:581a–583a, 596b, 614b
sine table, **15**:579a, 582a, 612b
Ssobolew
see Sobolev
Ssu-k'u ch'üan-shu, **8**:314a; **14**:159b
Ssu-ma Ch'ien
Shih-chi, **15**:730b
Ssu-yüan shu, **3**:265b, 266a–267a; **8**:314b
Stability, mathematical
theory of, **6**:572a; **13**:73a
trigonometric, **2**:145a
Stability, physical
of celestial bodies, **4**:126b
of floating bodies, **1**:326b; **4**:480b
rule of metacenters, **4**:230b
of mechanical systems, **3**:418b; **8**:560a–562b; **11**:568b
of solids, **4**:230b
theory of, **9**:449b
of thermodynamic systems, **4**:231a
see also Equilibrium, physical
Stabius, Johannes, **9**:162a; **14**:273a, 275b; **15**:179a

Stäckel, Paul Gustav (1862–1919), 12:599; **3**:135b; **4**:339b*, 371a; **5**:311a*, 313b*; **7**:507a; **10**:544b
Theorie der Parallellinien von Euklid bis auf Gauss, **7**:600a; **12**:255a; **13**:264a–b
Stackhousiaceae, **2**:519b
Stade (unit of distance), **4**:390b
Städeler, Georg Andreas Karl, **9**:131a
Staderini, Rutilio
Staderini's nucleus, **8**:552b
Stadius, Joannes, **13**:49b
Ephemerides, **2**:401b
Staël, Auguste de, **15**:82b
Staël-Holstein, Anne-Louise-Germaine Necker, Baronne de, **13**:127a; **15**:82a,b
Staff, Hans von, **11**:336a
Staffa (island), **4**:549a
Staffordshire
natural history, **11**:40b
Stages (geol.)
concept of, **1**:285b–286a; **10**:222a–b
Stahl, Georg Ernst (*ca.* 1660–1734), 12:599–606; **6**:259a–b
acquaintances of, **2**:2b; **10**:25b; **11**:109a
chemistry, **2**:74b–77a; **7**:188a–b; **9**:106a; **10**:26a
phlogiston theory, **3**:454b; **5**:353a; **6**:458b; **8**:531a; **11**:562b
phosphorus, **9**:105b
critics of, **2**:226b; **8**:76b, 80a, 298a; **11**:257b, 592b; **14**:321a; **15**:206a, 525a
influence of, **1**:451a; **2**:122b, 292a; **4**:352b; **6**:460a; **7**:188a; **12**:365b; **13**:543a, 602b; **14**:145a, 333a; **15**:94a
influences on, **10**:311a; **14**:212b
philosophy, **1**:479a; **7**:188a; **14**:321a
opposition to, **2**:3a; **6**:63b
Specimen Beccherianum, **1**:550b
translations of works, **6**:468b; **11**:562b; **12**:365b
Staining, microbiological, **5**:36a; **8**:8b, 9a, 272a, 589b; **9**:64a, 79a, 93b; **10**:17b; **11**:369a; **13**:590a; **14**:228a,b, 229a, 350a; **15**:479b
aniline dyes, **6**:224a; **7**:422a; **10**:577a
carmine, **3**:424a, 425a; **12**:193a–b
chromic-acid, **2**:124b; **3**:424a
counterstaining, **5**:35a; **7**:423b
differential, **1**:613b; **8**:272a; **12**:545a; **14**:228b
Gram's method, **5**:495a–b
discovery of, **11**:69b
fuchsine, **12**:87b
gold sublimate, **11**:275a
Heidenhain's stain, **6**:224a
methylene blue, **4**:296b; **8**:450b; **11**:274b, 382b
nucleic acids, **4**:603b
osmic acid, **8**:9a; **11**:296a
phagocytic, **5**:424b
silver carbonate, **11**:465b
silver nitrate, **5**:460a–b; **11**:274a,b, 296a, 382b
standardization, **8**:589b
theories of, **9**:381a
vital, **1**:540a; **3**:30a; **4**:296b; **10**:105a, 577a; **12**:48b
zinc sulfate, **13**:465a
Staircases, **4**:49a
Blondel's rule, **2**:201b
Stalactites
structure of, **15**:53b, 54a
Stallo, Johann Bernhard (1823–1900), 12:606–610; **2**:461a; **3**:32a, 33a, 379a; **6**:564b; **8**:597b, 599a
Stamen, **2**:219a; **11**:315b; **14**:586a
Stampioen, Jan Jansz, de Jonge (*b.*1610, *d.* after 1689), 12:610–611
Standard deviation, **2**:319a; **5**:8a, 10a; **9**:453b; **10**:451b, 452a, 458a–459a
naming of, **10**:450a
Standardization of measurement
see Measurement, standardization of
Standards, Bureau of (U.S.), **2**:565b; **5**:407b; **9**:462b
naval radiotelegraphic laboratory, **1**:338a
Stanford University
biology at, **7**:286a; **13**:269a
Stangeria, **8**:5b
Stanhope, Philip, **6**:556b; **12**:446b
Stanley, Wendell Meredith, **7**:35b; **13**:152b, 420a; **15**:14b
Stannius, Hermann Friedrich (1808–1883), 12:611–612; **8**:536a
Stanovoy Range, **7**:511a
Stanton, Thomas Ernest (1865–1931), 12:612–613
Staphylococcus, **5**:30a, 42b; **10**:389b; **11**:429a
chemotherapy, **4**:154a
discovery of, **8**:406a
phagotherapy, **6**:298b
Star catalogs
ancient Greek, **15**:217a–b
ancient Roman, **1**:511a, 530a; **9**:297a; **11**:195a; **15**:421a
medieval Arabic, **1**:511a; **2**:155a; **4**:542b–543a; **7**:259b, 324b, 337b; **9**:160b; **11**:202b; **13**:149a–b, 537a; **14**:400a
medieval Latin, **1**:122a; **7**:119a,b, 142a–b; **14**:391a*
Renaissance, **2**:411a, 412b; **11**:561b; **14**:359a; **15**:478a
seventeenth-century, **1**:169a; **5**:24b–25a; **6**:69b, 363b–364a; **7**:305b, 373b, 543b; **11**:411b
eighteenth-century, **1**:403b*; **6**:323a; **7**:81a, 87a, 543b, 544b–545a; **8**:179a; **9**:90a, 164b, 234a; **10**:591b; **14**:485a,b, 583a, 589b
early nineteenth-century, **1**:402b–403a; **2**:472a; **4**:16b, 269b; **10**:592b; **11**:545a; **12**:234a, 552a
mid nineteenth-century, **1**:241a–b, 242a–243a; **3**:14a, 92b; **4**:369b; **5**:480a–b; **6**:263b; **7**:145b–146a; **8**:613b; **12**:170a, 200b; **13**:110a–b, 120a
late nineteenth-century, **1**:339b–340a; **2**:99a, 146b, 332b–333a, 614b–615a; **3**:194a, 261b*; **4**:185a–b, 251a; **5**:223b; **6**:146b; **7**:236a–b, 481a; **8**:504a, 539b; **9**:194b, 551b; **10**:35a, 279b, 543b; **11**:379a; **12**:200b; **13**:114a; **14**:449a
twentieth-century, **1**:87b; **2**:126b; **3**:50a; **4**:278a, 359a; **5**:405a–b; **6**:353a; **7**:472b; **9**:504a, 514b, 564a; **10**:30b; **12**:265a,b; **13**:139a, 153b; **14**:242b*
analysis of, **10**:159a
of bright stars, **12**:176b–177a
colorimetric, **13**:408b
of faint stars, **10**:30b, 159a; **13**:139a
Indian, **15**:613a, 621b, 625a, 628a–b
limited-magnitude, **1**:242b–243a
Mesopotamian, **15**:657b(n), 676a
N30, **9**:514b
photometric, **13**:482a
revisions of, **1**:403a,b*; **4**:175b
southern hemisphere, **7**:236a–b, 543b
spectroscopic binaries, **9**:504a
zonal, **7**:608a; **11**:545a; **12**:176b
Star clusters, **6**:239b, 351b–352a; **10**:473b, 526b
ages of, **6**:351b; **7**:86a
catalogs of, **2**:221a; **4**:185a–b; **6**:325b, 326b, 327b, 471a; **9**:330a; **14**:479b
color-magnitude diagrams, **6**:351b
distances of, **12**:346b–347a
distribution within, **13**:118b
galactic, **13**:474b–475a
distances of, **13**:475a
globular, **4**:279a; **5**:183a; **6**:507b, 529b, 532b; **9**:410b–411a; **10**:318a*; **11**:49a
distribution of, **5**:182a; **12**:347a
radial velocities of, **12**:456a
variable stars in, **1**:352–353a, 398a; **12**:346b–347a
kinetic theory, **7**:86a
measurement of, **5**:199b; **7**:608a
open, **4**:279a
photography of, **7**:472a; **8**:465a; **11**:484b; **12**:37a
proper motions of, **8**:582b; **13**:474b
southern, **5**:480b; **6**:69b
see also Pleiades
Starch, **12**:233b
animal, **11**:565a
composition of, **10**:436a–b
conversion to sugar, **4**:134a; **13**:403b
digestion of, **15**:108a
cellular, **9**:25a
fermentation of, **14**:247b
formation of, **9**:25a
grains
micellar theory, **9**:601a
origin of, **12**:166a
hydrolysis of, **7**:378b–379a; **10**:436a, 532a
structure of, **15**:257a
tests for, **15**:108a
Starfish
distribution of, **5**:66b
embryology, **9**:333b
fossil, **12**:230a
life cycle, **12**:107a
Stark, Johannes (1874–1957), 12:613–616; **2**:398b; **4**:332a(n); **8**:182b
critics of, **15**:451a
and Nazis, **10**:348b; **12**:530b–531a
and Prussian Academy, **8**:52b
radiation, **2**:245a; **5**:117b, 459a; **12**:528a; **14**:340b
radioactivity, **3**:501a
Stark effect, **7**:491b; **8**:181b; **12**:529a, 614b; **14**:172a–b
inverse of, **14**:340b
Starkey, George (1628–1665), 12:616–617; **6**:141b; **14**:451b
Starlight
atmospheric extinction, **9**:189b, 563b
Starling, Ernest Henry (1866–1927), 12:617–619; **2**:446a; **6**:226b; **8**:294b, 451b; **9**:335a*; **15**:104b
and Bayliss (Wm.), **1**:533b, 534a, 536b–537a; **12**:617b, 618a
biochemistry, **1**:453a
medicine, **15**:73b
Starling (bird)
anatomy of, **3**:342b
Stars
ages of, **1**:353a; **4**:280a
Algol type, **4**:219b*; **12**:18b; **13**:408b
atmospheres of, **2**:266a; **9**:404b–405a; **12**:349b
capture phenomena, **4**:547b
composition of, **10**:290a
escape of molecules, **9**:405a
gaseous rings, **13**:118a
ionization in, **12**:20a–b, 21b–22a
isotopes in, **12**:368a
perturbations in, **1**:598a
physical processes in, **13**:117b–118a

Stars (cont.)
atmospheres of (cont.)
physical properties, **10**:290a
radiative equilibrium, **4**:279a
axial rotation of, **1**:21b
speed of, **12**:367b–368a; **13**:118a,b
B type
see Stars, helium
Barnard's, **1**:467a
of Bethlehem, **7**:298a
binary, **1**:87b; **3**:510a; **6**:49b, 325a, 326b; **11**:18b
eclipsing, **1**:429b, 430b*; **2**:196b; **4**:219b*, 251a; **12**:18a,b, 19b–20a, 22b, 23a, 176b, 346a; **14**:56a–b
ephemerides, **1**:430b*
formation of, **12**:19a; **13**:118b
mass vs. luminosity, **6**:352b–353a
orbits, **7**:18a; **8**:539a–b; **9**:504a; **12**:346a, 368a; **13**:339a
parallaxes, **6**:352b
short-period, **13**:118a
spectroscopic, **1**:352a–b, 598a; **3**:36b; **4**:249b; **5**:199b, 468b; **8**:539a–b; **9**:194b–195a, 504a; **10**:232a, 601a; **11**:18b, 49a; **12**:176b, 368a; **13**:118a; **14**:56a–b
visual, **1**:598a; **2**:614b; **12**:20a, 23a
see also Stars, double
bright, **5**:480a; **8**:582b
classification of, **9**:194b
carbon, **12**:368a
catalogs of, *see* Star catalogs
chemical composition, **9**:392a
circumpolar, **1**:283b; **3**:92b, 261b*; **5**:224a*; **7**:146a; **13**:114a, 482a; **14**:485b; **15**:711a, 718b
clusters of
see Star clusters
collapse of, **9**:405b
collisions of, **2**:123b
colors of, **1**:353a; **4**:167b, 168a; **5**:490a; **6**:351b; **7**:135a; **8**:50a, 364a; **11**:162b; **12**:170a; **13**:408a,b
Hertzsprung gap, **6**:353a
indices of, **7**:239b; **10**:321a; **12**:249b, 265b*
and luminosity, **6**:353a; **9**:563b; **12**:368a
and temperature, **11**:542a–b
cool, **5**:102a; **8**:442b
dark companions, **2**:99b; **8**:50a
daylight observations, **10**:596b
decanal
see Decans
densities, **1**:56a–57a; **4**:279b, 280a; **12**:18b, 19a,b, 20a, 22a,b
radial variation in, **4**:219a
distances of, **1**:56a–b; **2**:100a, 101a, 328b, 403b, 468b; **4**:280b; **5**:243b, 399b, 400b, 405a; **6**:71a, 332b; **8**:281a; **9**:370b–371a; **12**:346b; **13**:110b
double-star method, **6**:330b
and magnitude, **2**:100b–101a; **6**:70b, 330b–331a; **13**:110b
distribution of, **4**:272b–273a, 279a; **6**:330a, 331a, 332b, 352b, 507b; **7**:235b–240a, 303a; **8**:364b, 617b; **10**:35a; **12**:282b; **13**:110b; **14**:92a
as divine beings, **7**:42a; **15**:659a(n)
double, **1**:209a; **2**:126b; **3**:374b, 428a, 583b; **5**:22a*; **6**:512a; **8**:613b; **9**:2b; **10**:585a; **11**:162b, 411b; **13**:12a
angular distances, **2**:100b, 101a, 614b; **5**:257b
catalogs of, **1**:87b; **2**:221a, 614b–615a; **3**:605b; **4**:29b–30a; **6**:325a–b, 327b, 329b, 330b; **7**:18a; **9**:231b; **12**:552a; **13**:110a–b, 120a
classification of, **13**:110a
discovery of, **1**:465a, 466a; **2**:614b; **3**:288a; **6**:328b, 522b, 523a, 574b; **7**:17b–18a; **9**:231b, 370b; **10**:608b*
and Doppler effect, **4**:167b, 168a
evolution of, **12**:367b
interferometry, **12**:248a
lengthening periods of revolution, **4**:219a
micrometric observation, **3**:261b; **4**:251a, 547b; **5**:554b; **6**:326b, 471a, 574b; **8**:50a; **10**:30b; **13**:110a, 115a, 120a
orbits, **1**:339b; **6**:49b; **7**:481b; **11**:156a; **12**:162a, 247b
origin of, **2**:123b; **4**:547b; **7**:85b; **8**:562b
photography of, **6**:352b; **8**:465a; **11**:554a*
reflection effect, **4**:219a
southern, **12**:280a
tidal distortion, **4**:219a
see also Stars, binary
dwarf, **1**:56b; **12**:19a–b
discovery of, **6**:351a; **9**:194b
structure of, **4**:279a,b
see also Stars, white dwarf
eclipsing
see Stars, binary, eclipsing; Stars, variable, eclipsing
emission, **3**:510a; **5**:363a
energy of
see Energy, stellar
evolution of, **1**:353a; **2**:221a; **3**:470b; **4**:280a; **5**:272b; **6**:30a; **7**:86a; **8**:2a, 442a–b; **12**:18a,b, 19a–b, 20a, 22b, 280a, 367b; **13**:118b, 120b, 475a; **15**:449a
early stars, **10**:290a
faint, **3**:374b; **5**:480a; **8**:50a, 364a–b, 523a*, 582a–b; **10**:321a; **12**:170a
proper motions, **8**:465a; **13**:95a
velocity distributions, **6**:539b
gaseous
stability of, **9**:405a
giant, **1**:56b; **12**:19a–b
discovery of, **6**:351a; **9**:194b
structure of, **4**:279a,b; **10**:290a
heat radiated from, **3**:327a; **5**:364b*; **10**:105b
helium (B-type), **1**:55b; **7**:239b; **11**:18b
classification of, **8**:539b
streaming of, **6**:74b
Hertzsprung-Russell diagram, **4**:280a; **5**:272b; **6**:351a; **8**:2a; **10**:317b; **12**:18a, 19a, 22b
high-density, **5**:160b
interiors of, **5**:103a; **7**:86a; **15**:444a
radiation pressure, **2**:120b; **4**:279a–b
reaction rates, **5**:272a
light curves, **2**:196a,b; **6**:353a; **10**:585a
lumāšu (dating) stars, **15**:640b
luminosity
and color, **6**:353a; **9**:563b–564a; **12**:368a
and mass, **4**:279b–280a; **5**:272b, 363b; **6**:74b; **12**:23a
and spectra, **1**:56b; **6**:351a; **8**:364a–b; **12**:367b
upper limit, **6**:530b
magnitudes, **1**:398a; **2**:155a; **3**:174a; **4**:557b; **5**:490a, 527b; **6**:331a; **7**:238b; **10**:483a, 600a; **12**:170a; **13**:149b
absolute, **7**:239b; **8**:106a, 364a; **13**:98a
and distances, **2**:100b–101a; **6**:70b, 330b–331a; **13**:110b
estimation by steps, **1**:242a–243a
and motion, **13**:98a
photoelectric measurement, **1**:353a
photographic, **7**:239a; **10**:321a; **12**:248a, 249b, 265a–b
sequences of, **6**:324b, 331b; **7**:146b*; **8**:106a; **11**:156a
and spectra, **10**:321a; **12**:18b–19a
and temperature, **6**:351a; **12**:18b–19a
motion of, **1**:466b; **2**:98b, 389a; **3**:36b, 357b; **7**:235b–240a; **9**:555b; **11**:49a; **12**:518b; **14**:519a; **15**:496a
apparent motion, **15**:711a
asymmetric, **8**:364b; **13**:98b
ellipsoidal hypothesis, **4**:278b–279a; **9**:77b
and magnitude, **13**:98a
periodic, **2**:100a; **11**:223b
preferential, **13**:98a–b
radial, **9**:184a
and spectra, **7**:239b
velocities, **6**:331a–b; **7**:238a; **14**:499a*, 577a
see also Stars, proper motions of; Stars, radial velocities; Trepidation
multiple, **13**:110a, 120a
nature of, **6**:195b; **7**:134b–135a; **11**:223a–b; **13**:92a; **15**:202b
nomenclature of, **1**:242a; **9**:23a; **13**:149b; **15**:673a
number of, **6**:70a–b, 331b
O type, **7**:268a; **11**:18b
occultations of
see Occultations, of stars
opacity, **4**:279b, 280a; **9**:404b–405a
orbits of, **5**:199b; **10**:317b
origin of, **8**:618a
photography of, **1**:597a; **2**:284b–285a, 285b; **5**:86a; **6**:507b–508a; **8**:465a; **10**:600b; **11**:484b–485a, 542b; **12**:36b, 139a, 248a–b, 249b; **13**:153b
populations of, **1**:353a
positions of, **2**:98b–100a, 405b, 410b–411a, 511a; **3**:194b, 298b; **4**:26b, 351b, 369b, 547b; **5**:405a, 516a; **6**:149b; **7**:124a, 209b; **9**:460a, 463a; **10**:33b; **11**:188a; **14**:359b; **15**:216b–217b
absolute, **3**:92b; **5**:24b–25a; **15**:564b
altitudes, **2**:343a, 405b–406a; **5**:274b
corrections for, **2**:388a–b
declinations, **2**:332b; **5**:302a; **7**:239a–b; **15**:216b–217a, 217b
galactic altitudes, **5**:363b
photographic determination, **2**:616a; **5**:405a; **11**:156a; **13**:45b
polar longitudes, **15**:217a,b
prediction of, **2**:100a
right ascensions, **5**:24b–25a; **10**:592b; **15**:217a
tables of, **11**:114b; **14**:538a, 593a
zenith distances, **2**:100b
proper motions of, **1**:241a–242a, 339b–340a, 467a; **2**:99a, 99b–100a, 332b–333a, 615a; **3**:105b, 428b; **4**:30a, 269b–270a, 278b–279a; **6**:513b, 539b–540a; **7**:18a, 239b, 472a,b, 481b; **8**:582a–b; **9**:234a, 514b; **10**:30b, 212b, 608a; **11**:163a
discovery of, **6**:70b–71a; **11**:553a
and distances, **2**:100b–101a; **7**:236b–238a; **12**:250b
variation of, **2**:99b; **9**:410b
pulsation of, **1**:352b, 598b; **12**:346a
mathematical theory, **4**:280a
radial velocities, **3**:35b, 261a, 509a; **4**:278b–279a; **6**:542a–b; **9**:215a; **10**:30a; **14**:56b
changes in, **1**:598a–b
and solar parallax, **5**:405b
spectroscopic determination, **1**:55b; **3**:510a; **5**:20a, 199b; **6**:350b; **8**:539a,b; **9**:503b–504a; **11**:18b; **12**:153a, 251a, 368a; **14**:55a, 56a, 521b
systematic observation, **3**:36b

red, 2:146b
energy production, 5:272b
spectra, 4:251a; 6:28b
types of, 12:19a
rising and setting of, 4:429b; 11:223a; 15:210b, 211a, 219b
heliacal, 1:339a; 2:416a,b, 417b; 3:584b; 5:274a, 345b–346a; 11:195a, 197b; 15:640b, 673a, 711a–b, 714b
scintillation, 1:202b; 5:527b; 11:469a; 13:409a
shape of, 3:282b; 9:186b
sizes of, 8:281a; 12:346a
diameters, 2:403b; 4:279b; 6:94b; 10:473a; 13:36b–37a, 409a; 14:414b
and luminosity, 4:279b–280a; 5:272b; 6:74b; 12:23a
mass, 5:182a, 363b; 7:239a; 8:539b; 13:118a
range of, 9:212b
of southern hemisphere, 2:472a; 3:35b; 5:480a; 6:69b, 263b, 326b; 7:145b, 146a
brightness, 6:553b
parallaxes, 12:176b
spectra, 1:21b; 4:69a, 161b; 5:102a, 143b, 199b, 405b; 6:146b, 541b–543a; 10:320b–321a; 11:379a; 12:348b; 14:55a
absorption lines, 4:280a–b; 9:405a; 12:368a
changes in, 13:117b
classification of, 3:50a–b; 4:180a; 5:34a; 6:350b; 9:194b, 391b–392a; 10:600a–b; 12:37a, 269a–b; 14:55b–56a
and Doppler principle, 1:598a; 3:510a; 5:20a; 14:56a–b, 629a
emission lines, 4:280b; 11:320a
interpretation of, 5:103a; 10:601a
and luminosity, 1:56b; 6:351a; 8:364a–b; 12:367b
and magnitude, 10:321a; 12:18b–19a
and motion, 7:239b
photography of, 4:179b; 6:29a, 30a, 542b; 14:521a–b
photometry, 11:542a; 14:414b
quantitative studies, 12:20b, 21b–22a
red shift, 4:279b, 280b; 5:182a
and rotation, 13:118a
temperature scale, 9:405a
type W, 14:479b
statistical studies of, 1:242b; 5:363b; 6:74b, 332b; 7:236b–239b; 8:364b, 539b; 9:410b; 10:212b; 12:250b, 282a–b, 346b; 14:482a
streams of, 4:269b, 278b–279a; 7:238a, 239a, 481b; 11:163a; 12:250b, 348a
structure of
Eddington's model, 4:279a–280b; 12:250a
neutron cores, 10:214b
theory of, 8:2a; 9:405a–b; 12:22b, 23a
subdwarf, 4:279b
supergiant, 12:349a
temperatures, 4:280a; 6:540a; 7:268b; 10:107b; 12:153a
distribution of, 9:404b
and ionization, 12:21b–22a
and magnitude, 6:351a; 12:18b–19a
surface, 12:248a, 249b; 14:414b
terrestrial influence, 10:548a
thermodynamics of, 2:120b; 9:405a
transits of, 7:608b; 15:714b, 715a
triple, 4:30a
variable, 1:430a, 530b; 3:194a; 5:363a; 6:149b–150a, 326b, 331b, 507b, 540a; 7:18a, 194a; 8:50a; 9:90a; 10:320b, 321a, 600b–601a; 12:192a; 13:12a, 99b, 118a; 14:482a
Algol, 12:18b; 13:408b
atlases, 13:12a
brightness, 1:242b; 10:232a
catalogs of, 6:150a; 9:486a, 564a; 10:317b
Cepheid, 1:598a; 3:510a; 4:249a, 280a, 282a; 5:363b; 6:30b, 352a–b, 529a–b, 530a–b; 8:106a, 582b; 9:411a; 10:107b, 290a; 12:248b, 346a–347a, 349a
classification of, 8:539b
clusters of, 1:398a
and cosmology, 9:486a
discovery of, 1:466b; 3:50a; 4:219a; 5:33b–34a, 364b*, 468a–b; 6:402b, 529a; 7:373b; 8:106a; 10:30b, 608a; 11:553a; 13:481b; 14:589b–590a
distances of, 8:582b; 12:346b
eclipsing, 1:429b; 2:196b; 4:218b, 219a; 5:468b
ephemerides, 1:169a; 6:150a
fluctuations, 4:167b
irregular, 10:608a
light curves, 2:196a,b; 6:353a
long-period, 6:70a, 150a; 8:539b; 11:542a; 12:368a
motion of, 10:317a
periodicity of, 2:349a; 5:468b; 6:352a; 7:373b; 12:346b–347a
period-luminosity relation, 8:106a; 9:411a
photography of, 12:248b
photometry of, 11:542a, 543a
semi-regular, 5:363b
short-period, 2:196a; 5:468b; 10:30b
spectra, 7:268b; 12:368a
theory of, 1:352b, 598b; 6:431b; 9:486a; 12:346a
visibilities of, 15:561b–564b, 676a
white dwarf, 4:279b, 280a; 5:103a; 8:582b; 9:405a; 13:481b
Wolf-Rayet, 3:36a; 11:320a; 14:479b
zodiacal, 7:545a; 8:179a; 9:234a; 13:114a
see also Atlases, stellar; Comets; Constellations; Meteors; Novae; Parallax, stellar; Photometry; Polestar; Pulsars; Quasars; specific stars and constellations; Supernovae
Stas, Jean-Servais (1813–1891), 12:619–620; 4:242b; 7:280b, 281a; 10:510b; 11:485b; 12:593a; 13:500a
atomic weights, 3:476b; 4:244a, 395a; 9:110a; 11:173b
errors in, 11:417a–b
Stasis, 7:503a
Staszic, Stanisław Wawrzyniec (1755–1826), 15:501–503
States of matter
see Matter, states of
Statics
ancient Greek, 6:310b, 312a–b; 15:220a
medieval, 1:94a, 228b–229a; 2:193b–194a; 7:172a–175a, 336b–337a, 339a–348a; 8:215b–216a; 13:292a
Renaissance, 3:515b; 5:238b, 247a; 8:212b, 215a–225b; 9:487b–488a; 13:48b–49a
seventeenth-century, 1:541b; 4:60b; 5:239a; 9:119b; 11:488a–b; 12:74b, 75a; 14:150a
eighteenth-century, 3:441b; 9:156a, 446b, 472a, 477a; 10:319b; 12:55a
nineteenth-century, 2:215b, 424b, 470a; 3:555a; 6:244b–245a; 9:226a–b, 430a–b; 11:294a
twentieth-century, 2:594a; 11:123a,b
architectural, 2:326b; 5:195b*
in astronomy, 15:350a–b
chemical, 4:229b, 231a, 248a*
dynamical approach, 8:215b
engineering, 5:63b; 9:577b–578a; 10:3a–b; 12:74a; 14:473a–b
and geometry, 1:214b, 219b–223b, 228b–229a; 3:150a, 182b; 5:369a*; 6:599b–600a, 601a; 10:336b–337a; 11:164a
graphic, 3:468a, 495b–496b; 8:287b; 9:578a; 13:132b
method of sections, 3:496a
reciprocal figure, 3:468b–469a
in medicine, 12:102b–103a, 103b
physiological, 6:36b–40b
reciprocal figures, 9:226a–b
static moment, principle of, 1:222b; 7:174a–b; 8:217b, 219b; 14:150a
use of term, 6:37a–b
Statilius Attalus, 12:540b
Statistical Society of London
founding of, 1:354b; 9:67b
Statistics
seventeenth-century, 5:507a–b; 6:71b; 10:565b–566a; 13:593b
eighteenth-century, 4:38b–39a, 275b; 9:453b
early nineteenth-century, 5:98b, 302a
mid nineteenth-century, 3:45b; 5:266a, 307a, 309b; 9:203b; 11:236b–237b; 13:616a–b; 15:31a–32b
late nineteenth-century, 3:231a; 10:449a–453a, 461a–462b; 11:123a, 568b; 12:301a
twentieth-century, 2:58a; 3:17a; 5:7b–10b, 476b–477a; 6:267b, 402a*, 442a; 8:512a; 9:39a*, 420a; 10:450a–465a; 12:461a–b; 14:230b–231a, 382a–384a, 436b–437b, 573b–574a
in agriculture, 9:279a; 14:552b–553b, 574a
in astronomy, 1:242b; 5:363b; 6:74b, 332b, 352b–353a, 431a, 530b, 532a, 539b–540a; 7:236b–239b; 8:364b, 539b; 9:410b; 10:41b–42a, 212b, 317a; 12:250b, 282a–b, 346b, 461b; 13:98a–b, 475b*; 14:178b, 482a
in biology, 14:437a–b
anatomy, 8:125b
anthropology, 2:211b; 9:144a
botany, 1:59a; 5:513a; 14:99a
entomology, 12:98b–99a
evolution, 3:590b; 10:450a, 461a–462a; 14:251a–252a, 574a
genetics, 12:461b
physiology, 2:32b; 4:311b; 11:416a
zoology, 14:138b
see also Biometrics
"continental direction," 15:31b
in economics, 1:155a; 2:319a; 7:104b–105a, 318b; 8:61b, 62a; 10:566a
and experimental design, 5:8b–9a; 14:553a–554b
Fermi-Dirac, 11:421b
in geology, 9:367a
geophysics, 1:478a
meteorology, 6:98a,b; 8:487a; 9:279b
mineralogy, 3:609b
oceanography, 12:414b
in industry, 5:476b–477a; 14:382b, 383a,b
legal, 3:452a; 11:237b
mathematical, 2:319a; 5:8a–10a, 304a; 11:237a–b; 12:532b; 14:121b–122b, 382a–384a, 552a–554b, 618a; 15:49b
origin of, 1:208b–209a
in medicine, 1:563b; 2:42a–b; 3:2b, 6b; 5:137b; 6:59a; 7:44b; 8:122a; 10:404b, 405b; 11:33b, 237a; 12:294b–295a, 296b, 450b, 503a; 14:139a, 230b–231a
methodology
ascertainment methods, 14:230b
block designs, 14:553a, 553b–554a

Statistics (cont.)
decision functions, **14**:122a
graphical representation, **10**:450a, 450b–451a, 452a–b, 455b, 456a
interval estimation, **14**:437a–b
law of small numbers, **2**:319a
nonparametric methods, **1**:155b; **14**:383b
sequential estimation, **10**:461a
small samples, **5**:477a; **10**:451a, 456b; **14**:382b
stochastic equations, **14**:122b
tolerance limits, **14**:383b
variance-difference method, **1**:154b
military, **11**:180b; **14**:553b
order, **2**:319a; **10**:461a; **14**:383b
origin of word, **11**:236b
in physics, **2**:247b, 248b, 250a, 251a,b, 261a–262b; **3**:564a; **4**:577b–578a, 578b, 583a; **5**:103a–b, 161a, 330b, 331a,b, 391b–392b, 473a–b; **7**:493a, 617b, 618a; **8**:185b–186b, 473b, 478a–479a; **9**:129b, 218a–225b; **11**:10a, 12b; **12**:218b, 219a, 221b, 222b, 238b, 464a, 496b, 497b, 528a, 529b; **13**:241b; **14**:91b, 245a, 554b, 617a; **15**:6b, 43a, 432b, 442a, 444a
population, **12**:461b; **14**:179a–b
in psychiatry, **13**:396a
in psychology, **3**:130a; **5**:266a; **10**:454a; **14**:382b
on scientific productivity, **3**:130b
social, **11**:236b, 237a–b; **13**:616a–b; **15**:304a
theoretical, **2**:318b–319a; **5**:8a–10a
see also Correlation, statistical; Demography; Distribution, statistical; Likelihood, statistical; Population, statistical studies; Probability; Regression; specific tests; Terminology, statistical; Vital statistics
Statolith, **5**:624a
Status lymphaticus, **4**:405a
Staudinger, Hermann (1881–1965), 13:1–4; **7**:518b; **9**:354a, 377a
Staudinger's law, **13**:2b
Staudinger, Magda, **13**:3b
Staudt, Karl Georg Christian von (1798–1867), 13:4–6; **10**:605b; **11**:413a; **12**:285a; **13**:132b
geometry, **1**:591a; **3**:165b, 314a; **5**:309a; **6**:391a,b; **7**:185a, 397a; **8**:542b; **9**:173b; **10**:343b–344a; **11**:390a
Staudt-Clausen theorem, **3**:302b; **13**:6a–b
Stauropteris, **8**:354b; **12**:260a
Stauroscope, **5**:557a
Stay, Benedict, **2**:327b, 330a
Steacie, Edgar William Richard (1900–1962), 13:6–7
Stead, John Edward (1851–1923), 13:7–8
Steam
condensation of, **3**:19a
in disinfection, **5**:219a; **7**:423a; **8**:408a
equation for, **3**:20a
expansion of, **3**:19b; **13**:381a; **14**:196b
flow of, **3**:19a
as heat source, **13**:352a
in irrigation, **2**:390a
latent heat, **2**:179a–b; **5**:216a
motive power, **3**:80b; **6**:311b; **8**:211b–212a; **9**:530a; **15**:86b
and temperature, **12**:288b
pressure of, **1**:203a; **3**:541a; **15**:487a
and temperature, **2**:104a; **3**:80b; **4**:241b
properties of, **9**:462b; **13**:73b
saturated
specific heat, **13**:381a
specific heat, **3**:20a
apparent, **11**:292a
supersaturated state, **3**:19b
total heat, **3**:20a; **15**:487a
volume of, **8**:211b
Steam baths, medical use of, **9**:153a
Steam engine
applications of, **4**:348a; **6**:370a; **7**:22a, 89a; **11**:345a*
construction of, **2**:354b–355a, 586b; **3**:388b; **4**:44a; **5**:84b; **13**:73a
double action, **2**:104a; **14**:196b
economy of, **6**:431b–432a; **9**:237a; **11**:165a
governors for, **5**:87a, 387a; **14**:106a
high-pressure, **11**:355a
historiography of, **6**:2b; **13**:398b
indicator, **3**:287a; **4**:39b; **5**:388b; **7**:514a; **8**:124a; **14**:196b
invention of, **10**:36a–b
Perkins, **3**:82b
precursors of, **8**:212a; **9**:530a; **10**:293a
regenerative, **12**:424b
regulation of, **3**:358a
reversible, **3**:308a
theory of
mechanical, **3**:224a, 228b, 417b, 418b; **14**:618a
thermodynamic, **3**:20a, 79b–82b *passim,* 287a, 316a; **4**:501a*; **10**:286a–b; **11**:306a, 353a; **13**:73b, 381a–b; **14**:618a
valves, **3**:287a; **4**:39b; **10**:292b; **14**:618a
Watt's, **2**:179a; **14**:196b–197a
see also Boilers, steam; Locomotive, steam
Steam hammer
see Hammer, steam
Steam turbine
see Turbine, steam
Steamboat, **3**:411b; **9**:491b, 615b; **11**:340b; **15**:472a
Stearic acid, **8**:58a
Stearin
synthesis of, **2**:66a
Stearns, Joyce Clennam, **3**:367b
Stebbing, Thomas Roscoe Rede (1835–1926), 13:8–9; **13**:356a*
Stebbins, Joel, **6**:30a, 529b; **12**:19b
Stedman, Thomas Lathrop
Stedman's Practical Medical Dictionary, **4**:252b*
Steel
chemistry of, **11**:329a; **12**:543b–544a
carbon content, **2**:421b; **3**:233b; **5**:601b; **10**:246b; **11**:463b
corrosion, **5**:189b, 190a; **9**:60b; **11**:100b
phase rule, **11**:534b
phosphorus, **13**:8a
spectroscopy, **3**:234a; **7**:622a
magnetization of, **11**:578a; **14**:330a
metallurgy, **1**:609a; **2**:467a; **3**:87a, 233a–234b; **4**:531b–532a; **6**:164b, 319b, 368b, 479b, 480a; **7**:383a; **9**:472b, 477b; **11**:100b–101a, 463b; **12**:126a–b; **13**:8a, 246b, 327a, 346a, 346b–347a, 571a; **15**:25a
casting, **3**:234a, 326b; **6**:5a; **9**:591b
cementation, **7**:78a; **11**:328b
cold-working, **12**:382b
hardening, **10**:246a–b; **11**:486a, 549a–b
heat treatment, **13**:272a
ore, direct production from, **3**:234a
quenching, **3**:212a; **8**:119a; **12**:126b
smelting, **3**:234a
softening, **11**:100b
tempering, **1**:581a; **6**:480a; **10**:246a; **11**:101a, 329a
transformation, **6**:480a; **12**:544a
physical properties of, **1**:609b; **2**:459a; **7**:57a; **12**:613a
brittleness, **3**:212a; **7**:57a
compressibility, **2**:458b
dilatation, **3**:212b*; **11**:472a
flexibility, **2**:459a; **11**:329a
hardness, **10**:246a–b
magnetism, **3**:212a, 498a; **6**:5b, 504b; **12**:281b
plastic flow, **2**:460a
testing of, **14**:473b
structure of, **1**:581a; **3**:212a; **11**:550a; **12**:543b–544a
crystallization, **3**:234a
fracture surfaces, **2**:467a
grain-size, **12**:126b
types of
aluminum, **3**:212a
antimony, **3**:212a
boron, **3**:212a*
chrome, **11**:100b
Damascus, **1**:581a; **3**:326b
high-speed, **13**:272a
Indian, **10**:446a
magnetic, **6**:479b, 480a
manganese, **6**:5a; **9**:591b
silicon, **3**:212a; **6**:5b
special, **5**:583b
stainless, **11**:100b
tungsten, **10**:246a
Steele, William John, **13**:236b, 378a
Steelyard, **7**:341a, 342b, 346a; **8**:215b, 223a; **12**:595b; **14**:542a
Steenbock, Harry, **1**:357a; **4**:357b; **6**:135b, 136a; **14**:444b
Steenstrup, (Johannes) Japetus Smith (1813–1897), 13:9–10; **2**:617b*; **5**:495a; **6**:99b–100a; **12**:421a
Generationswechsel, **12**:107a
Steepest descents, method of, **3**:564a
Stefan, Josef (1835–1893), 13:10–11; **8**:507b, 511a
cooling, **4**:240a
Stefan-Boltzmann law of radiation, **2**:266a; **11**:150a,b; **13**:11a; **14**:339b, 340a
Stefan's constant, **1**:519b; **14**:38b
Stefan's law, **13**:11b
Stefan's number, **13**:11b
students of, **2**:261a; **10**:212b; **12**:496a
Steffens, Henrik, **6**:395b; **9**:578b; **10**:182b; **12**:155b, 156b; **14**:257a; **15**:526a
Stegocephalia, **3**:462b
Stegosaurus, **4**:147b
Stein, Johan Willem Jakob Antoon, S.J. (1871–1951), 13:12; **4**:273a*
Stein, William Howard, **15**:15b
Steinach, Eugen, **2**:344b
Steiner, André Louis, **4**:504b
Steiner, Jakob (1796–1863), 13:12–22; **2**:214b; **3**:466b; **4**:341a, 342b; **5**:367b; **13**:217b; **15**:192b
acquaintances of, **7**:51b; **11**:41b, 44b; **14**:222a
correspondence of, **5**:339b; **12**:172a,b
editions of works, **14**:223a
geometry, **3**:153a, 213b, 314a; **4**:2a; **6**:357b; **7**:599b; **9**:55b, 157a; **12**:247a; **13**:132b
influence of, **2**:191b, 192a; **5**:471b; **7**:109b; **13**:132b
influences on, **1**:42a; **8**:306a
lectures of, **12**:225b
Steiner curves, **13**:20a
Steiner relationship, **13**:18a
Steiner surface, **3**:117a; **13**:20b–21a, 623b

students of, **1**:294b; **2**:299a; **3**:59a; **6**:230a; **7**:108b, 505b; **11**:447b; **12**:170b, 225b; **14**:480b
Systematische Entwicklung der Abhängigkeit geometrischer Gestalten voneinander, **13**:5a
theorem of, **8**:250b; **13**:17a
translations of works, **12**:170b; **14**:472a
Steinhaus, Hugo, **1**:427a
Banach-Steinhaus theorem, **1**:428a
Steinheil, Karl August (1801–1870), 13:22–23; **3**:302a; **4**:557b; **5**:144a; **12**:289b; **13**:109b
instrumentation, **5**:305b; **7**:209b; **11**:377b
Steinitz, Ernst (1871–1928), 13:23; **7**:530b
theorem of, **14**:122b
Steinmetz, Charles Proteus (1865–1923), 13:24–25; **7**:288b; **14**:495a
Steinschneider, Moritz, **7**:223b; **10**:111b; **11**:31a, 240b
Stejneger, Leonhard Hess (1851–1943), 13:25; **4**:310b*
Steklov, Vladimir Andreevich (1864–1926), 13:25–28; **3**:230b; **4**:288a*; **5**:189a*; **8**:563a*; **9**:125b, 128a; **10**:30a, 251a*
analysis, **10**:248b
influences on, **3**:226a; **8**:562b
mechanics, **3**:195a
Stelar theory, **3**:30a
Steller, Georg Wilhelm (1709–1746), 13:28–29; **5**:428b, 429a
Stellio, **2**:322b
Stelluti, Francesco (1577–1652), 13:29–30; **3**:179b; **12**:257a
Stems (bot.), **12**:259a
polystely, **5**:604b–605a
stele
segmentation of, **5**:605a
structure of, **6**:401a; **12**:256a; **13**:542a
vascular bundles, **8**:354a
woody, **9**:455b
Stendhal (pseudonym of Marie Henri Beyle), **3**:133b, 523b, 524a; **5**:250b; **7**:196b; **11**:6b
Stenflo, Jan Olof, **8**:582b
Stenhammar, Christian, **5**:191a
Steno, Nicolaus
see Stensen, Niels
Stenosis, **2**:601a
aortic, **14**:408a
mitral, **3**:518a; **14**:26a
pyloric, **2**:188b
Stensen, Niels (Nicolaus Steno) (1638–1686), 13:30–35; **2**:309b; **14**:449b
acquaintances of, **10**:477a; **13**:169a, 170a,b, 335a
anatomy, **9**:64a; **13**:170b, 171a; **14**:15a
Cartesianism, **4**:65a
correspondence of, **13**:169a–b
crystallography, **6**:486b; **11**:523a
Discours sur l'anatomie du cerveau, **4**:65a
Elementorum myologiae specimen, **4**:65a
embryology, **2**:591b
geology, **7**:412a; **9**:533a; **12**:257a
stratigraphy, **2**:578a; **8**:242b
influence of, **11**:507b; **14**:25b, 259b, 450b
influences on, **1**:482b; **6**:508b; **13**:223b
De musculis et glandulis, **12**:85a
physiology, **3**:483a; **9**:63b, 244b
Prodromus, **2**:431a
editions of, **10**:202a
Stensen's duct, **13**:31a
theology, **2**:306b, 310a
zoology, **9**:573a
Stensiö, Erik Helge Oswald (Andersson), **3**:610b
Stentor, **8**:357a
Stepanov, Vyacheslav Vassilievich (1889–1950), 13:35–36; **2**:239a; **4**:287b; **5**:342a; **11**:157b*
Stephan, Édouard Jean Marie (1837–1923), 13:36–37; **10**:529a*; **11**:321a*
Stephan's quintet, **13**:36b
Stephanus (or Stephen) of Alexandria (*fl.* first half of seventh century A.D.), 13:37–38; **13**:323b
Stephanus of Athens, **13**:37b
Stephanus Philosophus, **13**:38b
Stephanus, Carolus
see Estienne (Stephanus), Charles
Stephen of Antioch (*fl.* first half of twelfth century), 13:38–39
anatomy, **12**:81b
translations by, **3**:393b; **9**:41b*; **12**:81a
Stephen of Byzantium, **6**:212b, 305b
Stephen of Provins, **9**:361b, 362b
Stephens, Joanna, **6**:44a, 139a; **12**:365b; **14**:320a–b
Stephens, Walter, **5**:13b
Stephenson, Marjorie, **6**:499a–501b *passim*
Stepling, Joseph (1716–1778), 13:39–40; **7**:490b
Steppes, **4**:144a; **7**:496a–b
botany, **13**:248b–249a
ecology, **10**:17a
forestry, **14**:106b–107a
Sterculiaceae, **2**:521a
Stereochemistry
mid nineteenth-century, **1**:518b, 578a; **2**:485b, 622b; **4**:247a
late nineteenth-century, **1**:320a, 340b–341a, 389b, 390a–b; **5**:3b–4a; **6**:108a; **8**:109b; **9**:356b–357a, 360a–b; **10**:359a; **13**:575b–577b; **14**:124b, 265a–272a, 454b–455a
early twentieth-century, **1**:106a, 319a–b; **3**:271b; **5**:5b, 128b; **8**:291b–292a; **9**:402b–403a; **11**:85a–88b; **14**:445b; **15**:2b
mid twentieth-century, **3**:235b; **4**:21a; **5**:100a; **6**:538b; **7**:517b–518a; **10**:125a; **12**:419a–b
and atomic theory, **2**:169b
critics of, **3**:300a
geometric configuration theory, **3**:235b
naming of, **9**:356b
planetary model, **4**:247a
rotary dispersion, **3**:271b
"rule of distance," **3**:271b
see also Compounds, organic; Molecules; Optical activity; Stereoisomerism
Stereocomparator, **7**:472a; **8**:582a; **11**:208a; **14**:482a
Stereogoniometer, **3**:357b
Stereoisomerism, **7**:517b; **9**:402b–403a; **14**:270a–271b, 601b; **15**:2b
in crystals, **14**:588a–b
of inorganic compounds, **14**:271a–b
inversion, **14**:124b
and pharmacological action, **15**:103a–b
theories of, **8**:109b; **9**:356b; **13**:576b–577b; **14**:455a
see also Carbon, structure of; Compounds, organic, optical properties; Optical activity
Stereometry, **2**:473a; **4**:608b; **5**:588b; **7**:300a–b; **8**:237b–238a; **9**:463a
Stereophotogrammetry, **11**:208a
Stereoscope, **6**:248b; **9**:226a
invention of, **14**:290a, 493a
Stereoscopy
see Vision, stereoscopic
Stereotaxy, **6**:518b
Stereotypy, **3**:560b; **13**:413b
Steric hindrance, **1**:340b–341a; **9**:356b, 357a, 360b, 403a
Sterigmatocystis, **9**:461a
Sterility, sexual, **4**:67a*, 106a
Sterilization, **4**:405a
by boiling, **1**:493a–b, 496b; **12**:555a, 557a
in carbon dioxide, **1**:303a–b
of culture media, **3**:188b–189a
by heat, **2**:437a; **12**:563b; **13**:522a
of liquids
porcelain filter, **3**:188b–189a
see also Antisepsis, Disinfection
Stern, Curt, **9**:523b; **14**:432b
Stern, Moritz Abraham, **3**:59a; **4**:1b, 2a, 341b; **6**:230a, 570b; **13**:557a
Stern, Otto (1888–1969), 13:40–43; **3**:329a; **4**:253b, 579b; **10**:424b; **15**:41a, 444a
Stern-Gerlach experiment, **13**:42b
Stern Foundation, **4**:299b
Sternberg, Charles, **2**:489a
Sternberg, George, **9**:142b; **11**:346a,b
Sternberg, Kaspar Maria von (1761–1838), 13:43–44; **2**:491b; **11**:130a
Sternberg, Pavel Karlovich (1865–1920), 13:45–46
Sternbergite, **13**:44b
Sterne, Carus
see Krause, Ernst Ludwig
Sterneck, Robert von, **6**:240b
Steroids, **15**:252b, 258b
physiological effects, **15**:258b
structure of, **4**:522a; **14**:445b; **15**:258b
Sterols
structure of, **14**:334b, 444b–445b
Steudel, Hermann, **4**:603a,b
Stevens, Edward (*ca.* 1755–1834), 13:46–47
Stevens, Nettie Maria, **8**:588b; **9**:519a, 520b; **14**:431a–b
Stevens Institute of Technology (Hoboken, N.J.), **13**:398b
Stevenson, John, **6**:578a
Stevenson, Thomas, **6**:498b
Stevenson, Walter, **7**:161b
Stevin, Simon (1548–1620), 13:47–51; **1**:367b, 566b; **2**:602b; **3**:181b; **4**:117b; **5**:246b, 409b; **8**:208b; **9**:610b; **12**:446b
Arithmetic, **2**:281a; **14**:24b
critics of, **8**:267a–b
editions of, **5**:408b
and Benedetti, **1**:607b
and De Groot, **4**:8b
La Disme, **7**:533a; **10**:255a
editions of works, **5**:408b; **12**:205a
geodesy, **12**:500b
hydrostatic paradox, **1**:223a, 607b; **13**:49a
hydrostatics, **10**:334a
influence of, **1**:567a; **12**:499b
influences on, **1**:229b; **14**:461a
mathematics, **1**:607b; **9**:598b; **10**:335b, 336b; **14**:51b
notation, **5**:408b–409a
mechanics, **1**:606a; **5**:239b; **8**:211a; **11**:488a
method of, **8**:602a
music theory, **5**:249b
Oeuvres, **2**:47b
Thiende, **7**:121a(n); **14**:51a
translations of works, **12**:500a
Waterwicht, **4**:8b
Steward, Frederick Campion, **6**:443b
Stewart, Balfour (1828–1887), 13:51–53; **11**:122a; **12**:51b, 238a; **13**:362a
Stewart, Dugald, **2**:174b; **8**:261b; **11**:135a, 136b; **14**:288b
Stewart, George Neil (1860–1930), 13:53–54; **15**:73a
Stewart, John (medicine), **8**:407a

Stewart, John Quincy, **12**:20b, 23a; **14**:558a
Stewart, Matthew (1717–1785), 13:54–55; **3**:78a; **8**:611b; **12**:446a
Stewart's theorem, **10**:297a; **13**:54b
Stewart, Thomas Dale, **9**:211b; **13**:429a
Stewart, William (senator), **11**:119b
Stibbs, Douglas Walter Noble, **5**:184a
Stibnite
optical properties, **12**:304b
Stiborius Boius, Andreas, **15**:475a
Stichotropismus, **11**:597a
Stickelberger, Ludwig, **5**:198a,b; **14**:222a
Stiefel, Eduard Ludwig, **6**:497a
Stieltjes, Thomas Jan (1856–1894), 13:55–58; **3**:428b*; **6**:308a
analysis, **3**:230b, 251a; **11**:261a
and Markov, **9**:126b
problem of moments, **9**:126b
Stieltjes integrals, **6**:392a; **8**:112a, 562b; **9**:126b; **13**:55a, 57b–58a, 562a; **14**:573a
Stieve, Hermann, **2**:344b
Stifel (Styfel, Styffel, Stieffell, Stifelius, or Stieffel), Michael (*ca.* 1487–1567), 13:58–62; **4**:550b; **11**:591a; **12**:205a
algebra, **2**:618a,b; **3**:312a; **11**:339a, 590b; **12**:92b, 206a
arithmetic, **10**:338b(n)
Arithmetica integra, **2**:602b; **9**:611b; **10**:494a
Deutsche Arithmetica, **11**:458b(n)
influence of, **11**:457b, 590a
trigonometry, **4**:550a, 551a
Stigmaria, **8**:461a
Stilbene, **10**:578b
discovery of, **8**:59a
Stilbestrol
synthesis of, **5**:208b
Stilbospora, **3**:45a
Stiles, Charles Wardell (1867–1941), 13:62–63
Stille, Wilhelm Hans (1876–1966), 13:63–65; **3**:434b; **10**:169a
Stillingfleet, Edward, **14**:361a, 374b
Stilwell, George R., **7**:36b
Stimpson, William (1832–1872), 13:65–66; **5**:406b, 478b
Stimulants, dietetic, **1**:543b
Stine, Charles Milton Altland (1882–1954), 13:66–67
Stirk, George
see Starkey, George
Stirling, James (1692–1770), 13:67–70; **3**:167a, 459b, 460a; **4**:475b; **5**:567a; **7**:494b; **13**:54b
Stirling series, **8**:367a; **13**:69a
Stirling's formula, **9**:453a–b
Stirling's theorem, **13**:69b; **15**:307b, 308a, 364a
Stjerneborg (Brahe's observatory), **2**:405a
Stobaeus, Joannes, **4**:415a,b; **6**:306a; **9**:269a; **13**:308a
Stobaeus, Kilian, **8**:374b; **14**:144a
Stochastics, **3**:142a; **14**:345b–347a
Stochiology, **5**:193b
Stock, Alfred (1876–1946), 13:70–71; **9**:452a*
Stock, Chester (1892–1950), 13:71–72; **9**:314b*
Stockard, Charles Rupert, **10**:291b
Stöckhardt, Adolf, **6**:237a,b
Stockholm
Museum of Natural History, **5**:329a; **10**:437a
Stockholm, University of
geology at, **5**:329b–330a
Institute of Biochemistry, **4**:485b
mineralogical institute, **2**:486a
Vitamin Institute, **4**:485b
Stodart, James, **4**:531b; **10**:446a; **14**:475b
Stodola, Aurel Boleslav (1859–1942), 13:72–74; **11**:478a; **15**:7a
Stöffler, Johann, **1**:383a; **7**:418b; **15**:154b
Stöhrer, Emil, **8**:188a
Stoichiometry, **2**:93a, 158a; **4**:134b; **5**:431a; **11**:434b–438a; **15**:462b, 463b, 464b
laws of, **5**:203a
Stoics, **1**:75a, 248a; **3**:319b; **11**:103b–105b; **13**:83b
cosmology, **5**:347a; **13**:92a; **14**:606a–607a
critics of, **1**:119a; **4**:385b; **12**:340b; **14**:613a
ethics, **14**:606a, 607b
historiography, **13**:333a–b
influence of, **1**:205a, 212a–b, 235a, 315a, 333b, 495a–b; **6**:305b; **7**:137b; **11**:39b, 201b; **12**:310a; **14**:377a(n); **15**:262b
logic, **12**:340b; **14**:607a–b, 613a
natural philosophy, **1**:255b; **7**:1a; **11**:42a; **14**:375a(n), 606a–607b
pneumatic theory, **1**:261b, 324a–b
Stokes, Adrian, **10**:142b
Stokes, George Gabriel (1819–1903), 13:74–79; **8**:39a; **10**:94a*; **13**:104b
acquaintances of, **6**:293b, 343b; **10**:483a; **11**:241b; **13**:217b, 365a
analysis, **5**:97b
astrophysics, **13**:383a
chemistry, **6**:505b
correspondence of, **9**:201a, 219b
and Crookes, **3**:474b, 480a
elasticity theory, **9**:199b, 220a–b
equations, **8**:502a, 516b; **9**:206b; **13**:377a
gas theory, **9**:217b, 219b
geodesy, **13**:609a
hydrodynamics, **4**:591b; **8**:516b; **11**:63b
influence of, **2**:491a; **3**:583b; **8**:592a; **9**:199a
influences on, **6**:502a
optics
aberration, **9**:214b, 220b, 372a
ether, **8**:490b, 493a; **9**:215a, 372b
refraction theory, **8**:592b; **9**:213a–b; **11**:292a
and Royal Society, **5**:126b(n), 413b
spectroscopy, **7**:382a–b; **12**:235b, 544a
Stokes lines, **13**:78a
Stokes parameters, **13**:77a
Stokes phenomenon, **13**:78a
Stokes's law, **1**:254b; **9**:395b, 396a; **10**:525b; **13**:446a
Stokes's theorem, **9**:207a
students of, **1**:456b; **7**:594b; **10**:447a; **13**:101a
Stokes, Pringle, **5**:16a
Stokes, William, **12**:450a
Stokes-Adams syndrome, **4**:398a; **5**:366b
Stoletov, Aleksandr Grigorievich (1839–1896), 13:79–81; **5**:462a; **8**:106b; **13**:446b
Stoll, Maximilian, **1**:333a; **3**:427a
Stolpenite, **4**:597b
Stolz, Otto (1842–1905), 13:81; **13**:562a; **14**:222a
Stomach
achalasia, **14**:408a
acid in, **2**:176a; **3**:351b; **4**:210b*; **6**:255a; **11**:616b; **13**:403b
anatomy
ancient Greek discussion, **4**:384a
Renaissance discussion, **1**:619a
seventeenth-century discussion, **14**:408a
nineteenth-century discussion, **11**:380b
bacteria in, **11**:428b
cancer of, **1**:612a
contraction of, **1**:544a
and nerves, **13**:556b
and spleen, **8**:352b
defense mechanism, **2**:28a
innervation of, **3**:69a
mucosa
replenishment of cells, **1**:613b
pathology, **8**:352b
physiology, **3**:69a
surgery of, **2**:130a
wall
digestion of, **6**:566b; **12**:560a
see also Digestion; Digestive system; Gastric juice; Secretion, gastric
Stomata
see Leaves, stomata
Stomatitis
mercurial, **8**:125b
Stomodaeum, **8**:27a
Stone, Edward James, **11**:163a
Stone, Marshall Harvey, **2**:144b; **6**:392a; **14**:90b
Stone
for building, **9**:315a; **10**:126b; **12**:489a; **15**:515a
coloring of, **15**:649b, 650b
cutting of, **2**:333b
graphical method, **4**:47a–49a *passim;* **7**:577a, 578a
strength of, **10**:3b
Stone Age, **2**:189b, 190a, 341a; **14**:505b, 506a,b
subdivision of, **9**:489a–b
see also Neolithic, Paleolithic
Stone skipping, **12**:555a
Stonecastle, Henry
see Baker, Henry
Stonehenge, **10**:549b
origins of, **3**:209a
Stoner, Edmund Clifton, **2**:247b; **10**:423a
Stoney, George Johnstone (1826–1911), 13:82; **3**:477a, 479a; **4**:175b; **8**:510a; **9**:222b, 224a,b; **11**:281b(n); **12**:238a–b
Stoppani, Antonio, **9**:538a
Storer, David Humphreys, **6**:56b; **7**:169b
Storm(s)
barometry, **2**:147a
forecasting of, **11**:320a
laws of, **3**:297b
motion of, **4**:410b
direction of, **5**:137a
surface wind pattern, **8**:487a
theories of, **4**:592a–b; **8**:487a; **9**:107b–108a; **10**:500b; **11**:340a–b; **12**:366b; **13**:93b
warning systems, **8**:278b
see also Cyclone, Thunderstorms, Wind, etc.
Størmer, Fredrik Carl Mülertz (1874–1957), 13:82–83; **13**:216b*; **14**:32a
Stott, Alice Boole, **12**:213b
Stovaine, **5**:99b, 100a
Stovarsol, **5**:100a; **8**:273b
Stoves, **3**:7a; **5**:130a,b; **11**:471b; **13**:412a*
kitchen range, **13**:351b
Strabo (*ca.* 63 B.C.–*ca.* A.D. 25), 13:83–86; **11**:312a
on Eratosthenes, **4**:388a–391a *passim*
geography, **9**:532b; **11**:225b; **15**:88a
Geography, **5**:347a
historiography, **4**:108b, 109a, 460b; **6**:212b, 213a; **11**:105b; **13**:319a; **15**:678b
influences on, **11**:105b(n); **15**:210a
Strachey, John (1671–1743), 13:86–87; **11**:35a
Strain (physics), **3**:146a; **10**:3a
definition of, **11**:294a
intramolecular, **3**:443b
see also Stress

Strain energy, **3**:118a; **6**:534b; **9**:578a
Strain theory, chemical, **1**:390a–b
Strakhov, Pyotr Ivanovitch, **2**:474b
Strange, John, **3**:210b
Strangles (disease), **7**:100b
Strangulation
  anatomical and physiological effects of, **1**:266b
Strasbourg
  fortification of, **4**:260b
  museums, **14**:88a
Strasbourg, University of, **3**:263a; **5**:185b, 463a; **14**:126a
  botany at, **3**:612a
  chemistry at, **13**:337b
  library, **14**:324a
  physics at, **14**:244a
Strasburger, Eduard Adolf (1844–1912), 13:87–90; **5**:52a; **12**:589a(n), 591a*
  acquaintances of, **7**:498b; **10**:576b; **11**:152b
  cell theory, **2**:626a; **3**:29b; **5**:582a; **9**:601a
  critics of, **3**:422a; **6**:467b(n)
  embryology, **5**:35b; **6**:338b, 465b
  heredity, **14**:235a
  influence of, **4**:131a; **11**:573a
  students of, **3**:30a, 187b; **4**:130b; **6**:121a,b; **12**:166a
Strasser, Hans, **11**:571b
Strassman, Fritz, **2**:251b; **4**:580a; **5**:161b; **6**:16b–17a; **9**:262a,b; **13**:227a
Strata
  compensation of, **2**:329a
  condensation of, **6**:229b
  correlation of, **4**:274a, 361b, 363b; **5**:32b, 155b, 156a, 205a–b; **8**:486b; **10**:258b, 428b; **11**:131a, 133b; **12**:276a; **13**:448b; **14**:392a
    intercontinental, **5**:330a; **10**:164a; **13**:581b, 620b–621a; **15**:152a, 153a
    Jurassic system, **1**:285a–286a
  dating of, **5**:296a
  dip, **13**:87a
  inclination of, **10**:209a
Strategy, games of, **2**:304a; **14**:92b
Stratigraphy
  late seventeenth-century, **11**:41a; **12**:257a; **13**:34a; **14**:501b–502a
  early eighteenth-century, **9**:135a, 534a; **13**:87a
  late eighteenth-century, **2**:347b–348a; **4**:493a; **5**:205a–b, 578b; **8**:68b–69a, 147a; **11**:35a–b; **12**:487a–b, 517a–b, 549a; **14**:260a, 311b–312a
  early nineteenth-century, **2**:481a, 494a–497a, 567b, 568b; **3**:313b, 395a–b; **4**:11a*, 274a, 540b; **5**:14a–b, 155b–156a; **6**:456b; **8**:564b–568a; **10**:9a, 209a–b, 586a; **11**:133b; **12**:182a–b, 487b–491b; **14**:88a, 210b–211a
  mid nineteenth-century, **1**:20a, 95b, 210a–211a, 468b–469a, 572a; **2**:108b–110a, 129a, 341b, 497b–498a, 556a; **3**:297b, 530a, 608a; **4**:68a, 308b–309a, 363b–365a; **5**:67b–68a, 295b, 436b–437a, 533a–534a, 538a; **6**:57b, 186b–187b, 238b, 437b; **7**:460b–461a; **8**:486a–b; **9**:99b–101a, 240b–241a, 255a–b, 346a, 583a–585a, 617b; **10**:258a–b, 265b, 584a; **11**:130b–131a, 235b, 269b, 276b–277a, 375a, 385b–386b, 399a, 439a, 501b–502a, 505a, 567a; **12**:276a–277b, 514b; **13**:145b, 581b, 620a–621a; **14**:439a; **15**:152a–b
  late nineteenth-century, **1**:131a,b, 163a, 397a, 471b–473a, 491a; **2**:346b–347a, 486b; **3**:236b, 321b–322a, 591a; **4**:361b; **5**:89b, 335a, 336b–337a, 475b–476a, 497b–498a; **6**:168a–b; **7**:253a; **8**:30b–31a, 32b–33b, 463a–b, 543a–b; **9**:579a–b, 593b; **10**:29a, 128a, 428a–429b; **11**:372a; **12**:140a–b; **14**:120b, 392a–b
  twentieth-century, **1**:394a; **2**:189b, 505a; **3**:520a, 533a; **4**:39a, 262a–263a, 409b; **5**:39a, 50a, 221b, 329b–330a, 486a–487b, 500b; **6**:228b–229b; **7**:267a; **11**:321a, 594a; **13**:63b, 274b, 518b–519a, 531b–532b; **14**:59b, 60a
  and archaeology, **12**:181b; **14**:506a
  and bionomics, **5**:486b
  classification systems, **1**:285b–286a; **11**:375b; **13**:581b
  criteria for
    current-bedding, **1**:394a
  dry-land theory, **1**:470a
  nomenclature, **1**:285b–286a; **4**:274a
  and paleobotany, **14**:298a
  type locating, **13**:581b
  type sections, **2**:497a
  see also Biostratigraphy; Geology, time scale; specific countries and regions; specific geological time divisions
Strato of Lampsacus (*d. ca.* 268 B.C.), 13:91–95; **1**:246a,b, 252b; **4**:105b, 461a; **6**:310b; **13**:85a
  pneumatics, **3**:491a; **4**:383b
Stratosphere
  temperature inversion, **5**:187b
Stratton, Frederick John Marrion (1881–1960), 13:95–96; **3**:584b*; **4**:69b, 498a*; **6**:34a*
Straus, Ernst Gabor, **4**:331a
Strawberry, **1**:411b; **13**:450a
Stream gauge, **8**:541a
Streams, **7**:144b
  antecedent, **3**:593b, 594a; **11**:118b
  capacity of, **3**:66b
  classification of, **3**:595b
  graded
    concept of, **5**:395b
  meanders, **3**:595b
  pressure against banks, **3**:66b
  self-purification, **7**:171a
  source of water in, **3**:66b; **10**:280b
  velocity of, **7**:500b
    stratification of, **3**:66b
  see also Geomorphology, Rivers, Valleys
Strecker, Adolph Friedrich Ludwig, **5**:1b, 214b; **6**:505b; **15**:108b
Strecker, Otto, **7**:282b, 283b
Streete, Thomas (1622–1689), 13:96; **5**:23b; **7**:308a; **10**:43b
  Astronomia Carolina, **7**:308a; **10**:60b
  translation of works, **4**:166a
Streeter, George Linius (1873–1948), 13:96–98; **9**:56b, 57a
Strehlke, Friedrich, **3**:258b
Strength
  concept of, **1**:253a–b
Strength of materials
  Renaissance study, **8**:212b
  seventeenth-century study, **5**:245a,b; **9**:119b
  eighteenth-century study, **2**:576b; **3**:441b–442a; **5**:410b; **10**:319a; **11**:496a,b
  early nineteenth-century study, **1**:459b; **4**:257a; **10**:2b–3b
  mid nineteenth-century study, **6**:452a–b; **7**:383a–b; **9**:60b–61a; **14**:473a–b
  late nineteenth-century study, **4**:501a*; **5**:64a, 501b; **6**:342a; **9**:131a, 139a, 446b; **11**:384a; **13**:398b; **14**:620a
  twentieth-century study, **6**:534a,b; **8**:301a, 500b; **12**:613a
Streptococcus, **2**:528b; **8**:273a; **14**:624a
  chemotherapy, **4**:154a–b; **5**:42b, 344b
  hemolytic infection, **5**:317a
  phagotherapy, **6**:298b
  pyogenic, **7**:100b; **12**:88a
  and scarlet fever, **4**:404b; **8**:449a
  toxins, **14**:624a
Streptomyces, **4**:220b
Streptomycin, **4**:155b
Stress, geological
  horizontal, **6**:187a
Stress (physics), **3**:145b, 146a; **7**:453b; **10**:3a
  electrical, **8**:40b
  frameworks, **9**:226a–b
  in gases, **9**:220a–221a, 224a–225a
  internal, **8**:212b
  Mohr circle, **3**:496b; **9**:446b
  tangent, **8**:301a; **9**:224b–225a
  and thermoelectricity, **4**:500b
  see also Strain
Striae (geol.), **2**:496b, 556b
Stricker, Pierre, **2**:345b
Strigelius, Victorinus, **10**:8a
Strobili, **12**:259b
Stroboscope, **11**:21a
Strömberg, Gustaf Benjamin (1882–1962), 13:98; **10**:473b*
Stromboli (volcano), **12**:261b, 562a
  eruptions of, **7**:376a
Stromeyer, Friedrich, **2**:586b; **5**:203a, 429b, 430a; **9**:423a; **13**:499a; **14**:111b
Strömgren, Bengt, **2**:616a; **13**:99b
Strömgren, Svante Elis (1870–1947), 13:98–99; **2**:616a; **4**:280a; **11**:527a*; **14**:453b, 454a
Strömholm, Daniel, **12**:504b; **13**:159b
Strong, John Donovan, **6**:30a
Strong, Martin (pseudonym), **7**:276b*
Strong, Theodore, **6**:399a
Strongyles, **9**:185b
Strongylocentrotus, **8**:357b
Strontianite, **6**:495b; **14**:132a
Strontium
  atomic weight, **13**:389b
  compounds of, **2**:565a; **8**:520a; **12**:546a
  discovery of, **6**:495b; **7**:394b; **15**:96a*
  isolation of, **2**:588b; **3**:602a; **8**:520a
  preparation of, **9**:178a
Strophoid (curve), **5**:499b; **11**:402a
Struck, Heinrich, **7**:423a
Structure, mathematical, **2**:352a–b; **11**:79b
  algebraic, **4**:9a
  topological, **4**:9a
Structures, geologic, **1**:236a
  igneous, **1**:153b
  see also Faulting, Folding, Geosynclines, Nappes, Tectonics
Struss (Struthius), Józef (1510–1568), 13:100
Strutt, John William, Third Baron Rayleigh (1842–1919), 13:100–107; **10**:91b(n); **13**:370a
  acoustics, **8**:399b, 473b
  acquaintances of, **5**:423b; **6**:343b; **9**:372b; **11**:241b, 612b; **12**:238a; **13**:365a
  capillarity, **4**:76b
  on contemporary scientists, **5**:390b; **13**:79b*, 524b*; **14**:184a, 186a
  correspondence of, **5**:391b
  electrical standards, **3**:19b; **8**:502a
  electron theory, **8**:496b
  gases, **9**:229a(n); **11**:279a–b
    atomic weight, **3**:398a
    Avogadro's number, **8**:501b
    noble, **3**:479b; **4**:80a
  hydrodynamics, **13**:76a
  influence of, **3**:620b; **8**:510a; **9**:211b, 228a(n); **11**:264b; **13**:619a; **14**:617a

Strutt, John William (cont.)
optics, **5**:483a; **9**:213a; **11**:265a, 266b; **12**:497b; **13**:78a
radiation, **4**:315b; **7**:528a; **11**:476a, 582b
Rayleigh lines, **9**:76b
Rayleigh scattering, **7**:622a; **13**:522a
Rayleigh waves, **8**:516b
Rayleigh-Jeans law, **7**:84b–85a; **13**:102b, 104a
Rayleigh-Ritz method, **11**:480b(n)
relativity, **2**:382b
statistics, **10**:461a
students of, **15**:46b
technology, **8**:11b
Strutt, Robert [Robin] John, Fourth Baron Rayleigh (1875–1947), 13:107–108; **5**:102a; **6**:474b; **13**:370b*; **15**:61a*
radioactivity, **2**:260a; **6**:475a; **12**:29a; **14**:588b
Strutton, William Russell, **7**:624b
Struve, Friedrich Georg Wilhelm (or Vasily Yakovlevich) (1793–1864), 13:108–113; **3**:302a, 374b; **5**:144a; **7**:480b; **11**:609b
parallax determination, **2**:101b; **3**:404b; **5**:404a; **6**:263b; **9**:2a, 371a
at Pulkovo Observatory, **10**:198a, 542b
star catalog, **4**:30a
Struve, Georg Otto Hermann (1886–1933), 13:113
Struve, Gustav Wilhelm Ludwig (or Ludwig Ottovich), (1858–1920), 13:113–114; **5**:363a; **7**:265a
Struve, Karl Hermann (or Hermann Ottovich) (1854–1920), 13:114–115
Struve, Kirill Vasil'evich, **11**:108a
Struve, Otto (1897–1963), 13:115–120; **7**:482b*; **10**:199b*; **12**:24b*, 367b
acquaintances of, **5**:363a
astronomy, **1**:56b
on contemporary scientists, **5**:364b*; **12**:346a
historiography, **6**:34b
spectroscopy, **4**:280b; **5**:199b*
Struve, Otto Wilhelm (or Otton Vasilievich) (1819–1905), 13:120–121; **2**:434a; **13**:110a, 112b*, 412b
on contemporary scientists, **3**:224b; **4**:30a; **8**:279b*; **14**:629b
observations of, **13**:339a
at Pulkovo Observatory, **1**:6a, 371b; **2**:433a
Struve family, **13**:109a
Strychnine, **2**:31a; **5**:421a, 589a; **9**:9b
biological test for, **6**:60b
detection of, **6**:294a*
discovery of, **10**:497b
effects of, **12**:612a
existence predicted, **9**:8a
isolation of, **3**:159b; **10**:498a
structure of, **10**:517b
Strychnos, **12**:259a
Stuart, Alexander (1673–1742), 13:121–123
Stuart, John, Third Earl of Bute, **1**:88a
Stubbs, George, **6**:567b
"Student" (pseudonym)
see Gossett, William Sealy
Studer, Bernhard (1794–1887), 13:123–124; **2**:108a; **4**:73b*; **12**:38a
Studia mathematica
founding of, **1**:427b
Studien zur Geschichte der Medizin, **13**:142a
Study, Eduard (1862–1930), 13:124–126; **2**:191b; **3**:399a; **4**:370b, 618b; **5**:472b
algebra, **11**:55b
geometry, **12**:228b
Stukeley, William, **5**:53b; **6**:36a,b; **10**:84a
Stupor, **10**:613b
Sturgeon, William (1783–1850), 13:126; **6**:278a; **7**:180b, 181a
Sturgeon (fish)
neuroanatomy, **12**:612a
Sturli, Adriano, **7**:623a
Sturm, (Jacques) Charles-François (1803–1855), 13:126–132; **8**:382b
acquaintances of, **5**:94b; **8**:305b; **10**:247b; **13**:375a; **15**:200b
algebra, **2**:217b, 574a; **3**:136b; **13**:218b
analysis, **5**:96b; **7**:109b
students of, **11**:206a
Sturm-Liouville theory, **2**:144a–b; **5**:608a; **7**:408a; **8**:384a–385a; **11**:59a; **14**:281b–282a
Sturm's functions, **2**:299a; **13**:129b, 217a, 218b
Sturm's theorem, **5**:98a; **13**:129a–b, 218b
translations of works, **8**:458b
Sturm, Friedrich Otto Rudolf (1841–1919), 13:132–133; **7**:110a*; **12**:226a, 226a*; **13**:21a(n); **15**:195a
Synthetische Untersuchungen über Flächen dritter Ordnung, **13**:20b
Sturm, Jacob, **10**:13b
Sturm, Johann Christoph, **1**:229b; **4**:166a,b
Sturm, Richard, **3**:467b
Sturtevant, Alfred Henry (1891–1970), 13:133–138; **3**:122a; **9**:515b–516a, 521b–523b *passim,* 564b; **14**:101a,b, 432a
Sturz, Georg, **7**:419a; **11**:457a, 458a; **14**:325b
Stylocephalus, **8**:146a
Styrene
synthesis of, **1**:415b
Su Shih, **12**:391b*
Su Sung, **12**:384b
Suadicani, Carl Ferdinand, **6**:287a
Suan-ching shih-shu, **8**:419a
Subba Row, Yellapragada, **4**:155b
Subbotin, Mikhail Fedorovich (1893–1966), 13:138–140
Subconscious, **8**:322a
Subdeterminants, **3**:142b
Subjectivism, **4**:281a; **7**:106b; **14**:226b
Sublimation (chem.)
of arsenic, **2**:314a
theories of, **12**:312b
Submarines
medieval, **13**:568a
seventeenth-century, **4**:184a,b; **14**:366a, 367a
eighteenth-century, **9**:343b; **13**:178b
nineteenth-century, **9**:491b
defense measures, **8**:24b; **14**:384a
detection, **8**:11b; **10**:605a
physiological studies, **6**:22b
sound-detection systems, **2**:460a; **3**:505a; **10**:525b
Subsets, **3**:55a–b
axiom of, **5**:108a
Subsidence, geological, **3**:568a, 595a; **4**:265b; **5**:476b, 498b; **13**:85a
of islands, **3**:551a–552a, 567b
and oceans, **3**:568b; **6**:168a; **13**:147a
recent, **5**:476b; **8**:567a
theories of, **1**:552b; **8**:570b
Substance, concept of, **1**:277b, 278a; **4**:52a; **8**:156b–157a, 158b, 159a; **13**:510a
Substitution, logical, **3**:278b; **8**:155b
Substitution theory, chemical, **1**:518b, 578a; **2**:624b; **4**:245a–b, 246a,b; **5**:325a, 370a–b, 381a; **8**:56b, 57a–b; **14**:156b, 601a
critics of, **14**:478a
Substitutions (math.), **3**:221b, 268a; **4**:478a, 548a*
conjugated, **1**:307a
geometric, **3**:213b
groups of, **5**:472b
linear, **1**:294a–b; **5**:198b; **7**:562a; **12**:237b; **14**:13b, 86a
theory of, **2**:105a; **5**:262b; **8**:61b; **9**:175a
Subtraction
medieval techniques, **7**:532b; **9**:614b; **12**:92b–93a; **13**:540a
algebraic, **7**:241b; **12**:92b–93a
borrowing, **9**:614b
Egyptian techniques, **15**:683b–684a
Indian techniques, **9**:614b
logical, **2**:296a, 297b; **7**:105a; **8**:154b
Succinate
biological oxidation of, **13**:393b–394a
Succinic acid, **4**:76a; **13**:388b
metabolism of, **7**:421a, 520a
Succow, Friedrich Wilhelm Ludwig, **12**:167a
Succulents, **3**:44a; **6**:184b
Suchten (or Zuchta), Alexander (*ca.* 1520–*ca.* 1590), 13:140–141; **4**:394a; **12**:307b
Sucrose, **6**:185b, 538b
Sudan
agriculture, **5**:523b
Sūdhakara Dvivedin, **7**:219b
commentaries of, **2**:416b, 417a; **5**:274b
Sudhoff, Karl Friedrich Jakob (1853–1938), 13:141–143; **5**:609b; **9**:378a; **14**:324a
historiography, **3**:393a, 395a*, 415a*; **4**:170b*; **11**:128b*, 365b*; **12**:82a,b; **15**:161b*, 183a, 190a(n)
Sudhoffs Archiv, **13**:142a
Sudhoffs Klassiker der Medizin, **13**:142a
Suess, Eduard (1831–1914), 13:143–149; **10**:220a*
acquaintances of, **1**:18a; **10**:29a; **13**:285a; **14**:626a
Das Antlitz der Erde, **1**:236b, 393b; **8**:544a
influence of, **10**:504b; **13**:284b–285a
translations of, **9**:103b; **12**:520a; **13**:285a
on contemporary scientists, **14**:523a*
critics of, **10**:169a
geomorphology, **10**:167b
influence of, **2**:89b; **3**:499a; **10**:503b; **13**:271a
students of, **1**:162a, 491a; **4**:140b; **6**:97a; **10**:428a
tectonics, **1**:472a, 551b; **4**:39a; **6**:228a; **7**:250b–251a; **8**:543b; **10**:168a,b
Suez, Gulf of, **13**:552a
Suez canal, **3**:287a; **13**:147b
Sufficiency, statistical, **15**:31b
al-Ṣūfī, Ibn Abi'l Fatḥ, **12**:362b
al-Ṣūfī, Abu'l-Ḥusayn ᶜAbd al-Raḥmān ibn ᶜUmar al-Rāzī (903–986), 13:149–150; **9**:35b, 297a; **12**:431b
astronomy, **11**:240a, 247a; **14**:575b
star magnitudes, **2**:155a; **12**:170a
on calendar, **1**:382b
star catalog, **11**:202b; **13**:537a
translations of works, **13**:510a
Ṣūfism, **11**:248a, 249b, 250a; **12**:7Aa; **13**:513b
Sugar(s)
in albumen, **1**:471a
analysis of
reducing capacities, **7**:393b
in blood, **8**:614b; **14**:492b, 523b
in body fluids, **2**:29a
cane sugar
extraction of, **15**:24b
inversion of, **11**:298a; **15**:458a
reactions of, **14**:360a–b
similarity to beet sugar, **2**:139b
structure of, **8**:508a
substitutes for, **10**:325b; **12**:321b

composition of, **4**:244b
consumption of, harmful effects, **2**:585b
degradation of, **14**:603a
detection of, **15**:108a, 109b
biochemical, **7**:405b
cupropotassic solution, **1**:471a
fermentation of, **4**:485b; **6**:110b; **9**:425a–b
formation of, **3**:347b
identification of, **5**:1b
invert, **9**:425b
isolation of, **2**:66a; **9**:105a–b
in liver, **1**:471a; **2**:29a; **15**:109b
manufacture of, **14**:357a–b
from starch, **7**:378b–379a
metabolism of, **1**:471a; **2**:28b–29a; **4**:360a; **7**:519a; **8**:614b; **11**:51b; **15**:74a, 109b
in molasses, **2**:139b
in nucleic acids, **4**:603b; **7**:466b; **15**:252b
polarimetry of, **2**:139b; **9**:425b
reactions of, **10**:15a; **14**:360a–b, 445b
refining of, **3**:556b; **4**:41b; **6**:383a–b; **9**:425b; **10**:37a, 436a
structures of, **5**:2b–4b; **6**:185b, 538b; **8**:276a, 508a; **14**:603a
synthesis of, **5**:3a–4a; **8**:275b
in urine, **14**:408a, 523b
see also Beet sugar; specific sugars, *e.g.,* Glucose
Sugar beet, **14**:96a
breeding of, **14**:34a
Sugar cane, **11**:565a*; **13**:354b
physiology of, **14**:255a
Suggestion, **2**:35b
Sugita Genpaku, **15**:740a, 741a
Kaitai shinsho, **15**:738a, 741b, 742a
al-Suhrawardī, Shihāb al-Dīn, **11**:247b, 249a,b, 250a; **13**:513a
Suicide, **6**:556b; **8**:506b
of scientists, **2**:267b; **4**:294b; **6**:176b, 285b, 481b; **7**:194b; **8**:560b, 593a; **9**:37b, 142a, 178a, 355a, 389b, 569a; **10**:560b; **15**:157b
Suisset
see Swineshead, Richard
Suite de la clef
see *Journal de Verdun*
Suiyan-Sinkiang Highway Expedition, **6**:216b, 217a
Sulaymān al-Mahrī, **9**:36a
Sulcometer, **1**:97b
Sulfa drugs, **4**:154b–156a; **5**:344b
pharmacological action
and molecular structure, **9**:296a
Sulfamides, **5**:100a; **8**:273b
Sulfanilamide, **7**:273b
Sulfanilic acid, **4**:297a
Sulfates
composition of, **8**:339b; **14**:488a
crystals, **5**:90b; **9**:423b, 424b
Sulfides, **4**:141a; **14**:600b–601a
composition of, **11**:168a–169b; **12**:46b; **13**:310b
deposits of, **2**:159a; **14**:59b–60a
formation of, **14**:58b, 59a
reactions of, **8**:312a
Sulfites, **5**:92a
Sulfocamphoric acid, **14**:156b
Sulfonamides
antibacterial action, **4**:154b, 155b–156a; **5**:30a, 344b
synthesis of, **5**:344b
Sulfonanilamide
synthesis of, **5**:344b
Sulfones
and leprosy, **4**:155a
Sulfosalts, **1**:529a; **4**:141a
Sulfovinic acid, **8**:333a–335a; **12**:320b
Sulfur
in alchemical theory, **1**:549b; **3**:596b; **4**:209b, 387b; **5**:353a–b, 422b; **7**:524b; **8**:131a, 174a; **10**:309b–310a, 555a; **12**:311b, 312a; **13**:511b
atomic weight, **3**:398a; **4**:241a; **13**:499b
boiling point, **3**:20a
calcination of, **2**:75a; **12**:605a
coagulated, structure of, **2**:627b
combustion of, **3**:202b*; **6**:42b; **8**:74b, 76b, 77a; **13**:469a
compounds of, **2**:565a; **5**:100a, 323a–b; **11**:541a; **13**:389b
inorganic, **5**:73a; **12**:593a
metallic, **4**:345b
optical activity, **11**:86a
organic, **4**:20b; **6**:501b; **14**:600b
polysulfides, **12**:147b
crystals, **8**:107a; **9**:424a
deposits of, **10**:170b; **15**:502b
origin of, **5**:558b; **10**:265b
detection of, **2**:381b
determination of, **9**:95b
doubly-bound, **4**:20b
electrical properties, **5**:575b
in enzymes, **6**:501b
isomerism, **12**:197b
medical use of
Renaissance, **4**:209a
seventeenth-century, **2**:308b; **5**:421a
eighteenth-century, **3**:97a
twentieth-century, **5**:100a
metabolism of, **1**:10b
mining of, **2**:439b; **4**:598b
oxidation of, **2**:75a
in sea water, **5**:72a
tetravalent, **3**:10b
toxicity, **6**:443a
valence, **2**:622b; **8**:508a
vapor density, **3**:47b; **13**:468a
see also Tria prima
Sulfur dioxide
conversion of, **11**:34a
discovery of, **6**:255a; **11**:145a
liquefaction of, **3**:326b; **9**:471b, 477b
liquid, **3**:176a,b
Sulfur nitride, **5**:73a
Sulfur tetrachloride, **1**:553b
Sulfuration, **6**:586b
Sulfuric acid, **5**:323a; **6**:237a*
crystallization of, **7**:278a*
dibasicity, **5**:372a–b
effects of
on organic materials, **2**:385b, 386a; **3**:194a*; **5**:92a–b, 374b*; **7**:282a; **8**:373b
formula for, **8**:508a
hydrates of, **7**:162a
naming of, **5**:602b
preparation of, **5**:420b; **6**:255a; **11**:34a
industrial, **1**:471b; **3**:200b, 201a, 202a, 316a; **4**:185a; **5**:223a; **8**:553b
lead chamber process, **5**:157b, 326a; **11**:499a; **14**:447a
reactions of, **14**:395a–b, 396a
salts of, **8**:520b
see also Oleum; Vitriol, oil of
Sulfuric ether
see Ether, sulfuric
Sullivan, Eugene, **3**:292b
Sullivan, John Henry, **15**:37b
Sullivant, William Starling, **5**:512b; **8**:264a
Sulphur (ship), **5**:481a
Sulzer, Johann Georg, **7**:412a, 597a; **14**:77b
Sumatiharṣa Gaṇi
commentaries of, **2**:119a; **12**:599a
Sumerian language, **15**:635a,b, 673a
Sumerians, **14**:504b
mathematics, **2**:320a; **15**:667b, 668a
Summa philosophiae, **11**:31b
Summability, **2**:238b
theory of, **9**:426b
Summation, **1**:218a; **2**:38a, 49b, 53a, 238b, 320a; **3**:178a; **4**:476b, 561b, 606b–607a; **6**:176b, 472a–b; **7**:619a; **8**:279b, 367a; **9**:127a; **13**:562b
Euler-Maclaurin formula, **4**:475b–476a; **10**:249b; **12**:535a
Gaussian, **4**:125a; **7**:621b
geometrical, **1**:219b–220a, 221b; **7**:583a; **13**:289b, 290a
infinite, **3**:140a
number theory, **4**:125b
of reciprocal squares, **2**:57a
topological, **9**:249a
Sumner, Francis Bertody (1874–1945), 13:150–151
Sumner, James Batcheller (1887–1955), 13:152–153; 1:11b
critics of, **14**:412b
Sumner, John Bird, **2**:568b–569a
Sumner Line, **3**:374b
Sun
age of, **13**:383a
in alchemical theory, **5**:421b, 422b
atmosphere of, **2**:328b; **4**:69b; **7**:75a; **10**:519a; **11**:320a; **15**:344b–345b
atoms ejected from, **9**:405a
chemical composition, **1**:166b; **6**:615a; **12**:22a, 23a
inversion layer, **12**:268a
ionization in, **12**:20a–b
radiative transfer, **8**:364a
refraction in, **12**:187b
reversing layer, **14**:557b
selective absorption by, **8**:20b
see also Sun, chromosphere
brightness of, **2**:284b
distribution of, **7**:186b
chemical composition of, **6**:333a–b; **7**:382a
chromosphere
discovery of, **8**:441a
ionization in, **12**:71a
refraction in, **12**:187b
spectrum of, **2**:433a, 434a; **4**:270a, 498a; **5**:102b; **12**:268a; **13**:95b; **14**:557b
comets and, **1**:301a; **2**:433a–b; **10**:234a–b
cooling of, **13**:382b–383a
corona, **6**:29b; **7**:271a*; **14**:498b*
brightness of, **4**:180a; **10**:321a
movement of matter in, **2**:434a; **5**:364a
photography of, **6**:28b, 105b, 106a, 508a; **7**:472a; **8**:581a–b; **9**:184a; **11**:412a; **14**:449a
photometry of, **12**:353a
polarization of, **8**:581a–b; **10**:31b
rays, **2**:434a; **6**:106a
and solar activity, **6**:105b–106a
spectrum of, **1**:598a; **5**:183b, 199a–b; **6**:119a, 542b; **8**:539b; **9**:504a; **10**:31b, 106a–b; **11**:412b; **12**:368a; **13**:95b
structure of, **13**:408b
theories of, **1**:301a; **12**:139a
thickness of, **12**:268a
cosmological place of, **3**:408a–b
distance from earth, **1**:248a, 248b–249b, 311a, 340a; **3**:25a–b, 103a, 409a, 457b; **4**:161b, 391b; **5**:404b; **6**:69a–b, 334a; **7**:303a–b; **8**:423a, 613b; **10**:299a; **11**:104a, 188a, 194a–b, 197b; **12**:406a, 574a; **13**:55a; **14**:452b, 637a;

Sun (cont.)
15:214b–215b, 561a, 594a, 604b, 605b
see also Astronomical unit
effects on earth
electric, **1**:300b
heat, **5**:551a
magnetism, **3**:260a
electromagnetic properties, **11**:393a
energy of
see Energy, solar
energy transfer, **12**:250a
faculae, **1**:597b; **2**:434a; **12**:151b, 267b
figure of, **13**:482a
filaments, **11**:412b
flares, **3**:93a; **12**:267b
flash spectrum, **1**:56a; **5**:199a; **6**:508a; **11**:412b; **12**:72b–73a, 249b–250a
quantitative analysis of, **10**:290a
flocculi, **6**:28a,b, 29b
heat of
see Heat, solar
hydrogen in, **12**:22a
life on, **6**:333b
magnetism in, **1**:56a; **6**:29b–30a, 32a
magnetically active (*M*) regions, **1**:478a
measurement of, **6**:30a; **8**:582b–583a
and moon
conjunctions of, **7**:130a–b, 139b; **10**:299a; **15**:590b
mutual attraction of, **15**:332a, 380b–381b
motion of, **1**:241b–242a; **2**:24a; **5**:24a; **7**:2b, 143a; **10**:78a; **12**:359a–b; **13**:98a; **14**:519a, 521b, 593b; **15**:35a, 540b
irregularities of, **11**:190a
maximum equation, **12**:359b
mean motion, **5**:614b–616a; **6**:515b; **7**:130a, 139b; **10**:5b; **11**:190a, 195a, 202a; **15**:611a–b, 613a–b, 623a–b, 624a–b, 625a–b
speed of, **3**:101a; **6**:331a–b; **13**:291b; **15**:539b, 679a
nature of, **4**:368a, 555a; **6**:445a; **7**:134b–135a; **10**:590b; **13**:279b; **14**:415a
opacity, **9**:405a
orbit, **2**:413b; **3**:35b, 36b; **7**:481b; **8**:178b; **9**:141a; **10**:180b, 317b; **15**:676a,b
apogee, **1**:510b–511a; **2**:411a–b; **3**:405b–406a, 409b; **5**:24a; **8**:280b; **11**:190b–191a, 202a, 250b; **14**:576b, 593b; **15**:211b, 545a
apsis, **11**:321b
dimensions of, **15**:586a, 608a–b
eccentricity, **1**:312b, 510b, 511a; **2**:411a–b; **5**:24a, 345b, 614b; **6**:515b; **11**:191a, 202a; **12**:359b; **14**:576b; **15**:211b
elongation, **15**:592b
maximum equation, **4**:556a; **14**:576b
perturbations of, **3**:285b*; **7**:544b
zodiacal period, **5**:345b
photography of, **1**:597a; **2**:196a; **5**:18b, 86a; **6**:28a; **7**:76a, 77a; **9**:184a; **12**:36b; **13**:481b; **14**:449a
photosphere, **9**:616a; **10**:107b
motion of, **8**:1b; **12**:267b
nature of, **12**:267b
pressure at, **12**:20b
physics of, **5**:364a; **9**:415a; **11**:149b; **12**:187b; **13**:232a–b
polarimetry of, **1**:202a
position of, **3**:364a; **15**:536b, 678b, 729b
altitude, **2**:411a; **5**:612b, 613a–b; **6**:126a, 302a; **7**:118a, 119a; **12**:518a; **14**:577b; **15**:219b, 552a, 559b, 571a–572a
declination, **1**:573b; **2**:435b; **5**:613a; **7**:123b; **8**:65a; **11**:189a–b; **14**:275a
galactic altitude, **3**:563b
longitude, **5**:274a; **7**:362a, 512b; **9**:607a; **10**:419a; **11**:553b*; **15**:537a, 539a–b, 543a, 545a, 557b–558a, 601b, 614a
prominences, **1**:202b; **6**:402b; **11**:379a; **13**:472b; **14**:629a
cycles of, **11**:412a–b
dynamics of, **12**:268b
movements of, **8**:581b
photography of, **6**:28a, 508a; **14**:557b
spectra, **4**:498a; **6**:542a; **7**:75a; **8**:440b–441a, 503b; **9**:184a; **11**:320a; **12**:268a–b
theories of, **7**:186b; **13**:120b
white, **13**:232b
rising and setting, **15**:679a
rotation, **1**:597a; **2**:147a, 328a; **10**:31b, 129a; **12**:95b; **15**:557a
axis of, **3**:93a; **9**:184a
and geomagnetism, **1**:478a; **9**:184a
period of, **2**:196a; **6**:362a; **13**:139b*
spectroscopic determination of, **4**:251a; **13**:344b; **14**:55a, 557b, 629a
variation in, **1**:56a; **3**:93a; **4**:251a; **6**:74b; **12**:579a–b; **14**:414a
shadow of, **15**:576a
size of, **4**:391b; **10**:299a; **12**:406a
angular diameter, **1**:248b, 249a; **6**:250b
apparent diameter, **15**:219a
circumference, **15**:608a–b
diameter, **1**:249a–b, 511a; **2**:403b; **3**:101a, 238b, 528b; **4**:25a*; **7**:298b–299a; **8**:145a*; **9**:554b; **10**:595b; **11**:104a; **15**:215a–b, 544b, 550a, 560b, 579a, 594a–b, 605a, 608a, 798a
radius, **15**:608a–b
volume, **1**:248b
spectrum, **4**:161b; **8**:20b; **9**:154b; **10**:107b; **11**:412a–b; **12**:37a, 267b–268a; **13**:344a–b; **14**:55a, 557b
absolute wavelengths, **9**:564a
absorption bands, **7**:74b–75b; **9**:256*; **12**:499a
Balmer lines, **4**:497b–498a; **7**:553b; **13**:95b
cyanogen bands, **5**:182b
D lines, **5**:86b; **7**:381b
displacement, **12**:72b–73a
Fraunhofer lines, **5**:143a–b; **6**:541b; **7**:528a
helium lines, **5**:126b; **11**:279b
mapping of, **1**:166a–b; **5**:143a–b; **10**:519a; **11**:579a; **14**:55b
photography of, **1**:21b; **4**:69b, 179b–180a; **6**:327a; **11**:542a; **12**:37a
quantitative studies, **12**:21b–22a
relativistic deflection, **12**:73a, 251b
resonance lines, **7**:186b
separation from terrestrial spectra, **3**:420a
tables of, **12**:37a
telluric rays, **7**:74b
theories of, **5**:102a–b; **12**:71a
wavelength variations, **5**:182b
spiculae, **12**:268b
and stellar light, **3**:36a–b
stellar magnitude, **13**:482a
structure of, **4**:161b; **8**:2a; **11**:320a; **12**:250a
density, **8**:1b
pressure distribution, **8**:1b
surface activity, **1**:478a, 597b; **6**:28b, 106a, 471a; **8**:581a
and cosmic radiation, **11**:348a
optical theory of, **7**:186b
periodicity of, **12**:461b
theories of, **12**:368a
surface features
see Sun, photosphere
tables of
ancient Greek, **1**:191a
ancient Roman, **11**:189b, 190a
medieval Arabic, **5**:614b, 616a; **7**:259b; **14**:578a,b; **15**:259b
medieval Latin, **2**:435b; **7**:118b, 119a,b, 123a,b, 130b; **10**:540b; **14**:390a,b
Renaissance, **1**:573b; **14**:583b; **15**:477b
seventeenth-century, **3**:101a–b; **5**:24a; **6**:115b; **7**:577b; **8**:65a; **9**:336b*
eighteenth-century, **3**:430b; **4**:14b, 22b; **7**:544b; **8**:179a; **9**:233b, 234b; **14**:583a
nineteenth-century, **4**:16b; **10**:207b; **11**:114a,b
twentieth-century, **7**:6a
temperature of, **4**:78b; **8**:1b; **11**:559b; **12**:95b; **13**:11a, 481b; **14**:38b–39a, 185b
distribution of, **12**:266b
see also Eclipses, solar; Heat, solar; Midnight sun; Parallax, solar; Radiation, solar; Solar wind; Spectroheliograms; Sunlight; Sunspots
Sun Ssu-kung, **12**:380b
Sundararāja, **10**:129b; **15**:602b
Sundew
see *Drosera*
Sundials
ancient Greek discussion, **1**:246b; **4**:467a; **13**:319a, 320a
medieval Arabic discussion, **1**:509a; **4**:542a; **6**:197b; **7**:362b; **13**:291b
medieval Latin discussion, **5**:365b
Renaissance discussion, **1**:605a, 606b; **4**:260a; **6**:144a; **7**:418b; **9**:335b; **14**:273b; **15**:478a
early seventeenth-century discussion, **4**:13b; **5**:343b, 593b*; **7**:376a; **9**:25b; **10**:255a*; **13**:614a
late seventeenth-century discussion, **3**:348a; **9**:485a, 530a*
early eighteenth-century discussion, **2**:326b; **3**:51a; **4**:38b, 166a; **8**:170b; **10**:255b
late eighteenth-century discussion, **2**:451a; **10**:614b; **14**:589b
nineteenth-century discussion, **11**:320a
twentieth-century discussion, **6**:74b
altitude, **6**:301b
Butterfield type, **2**:133a
calibration of, **3**:364a
conical, **4**:110a
construction of, **11**:197b; **15**:116a
graphic method, **4**:47b
cylindrical, **6**:301b; **7**:119b
hemispherical, **4**:110a; **10**:7a*
historiography of, **14**:471b
horizontal, **12**:361a–b; **15**:163b
mural, **6**:362b
navicula, **15**:154b
portable, **6**:301b–302a; **15**:518b
reflexive, **6**:362b
telescopic, **9**:465a
vertical, **14**:578a; **15**:478a
see also Gnomonics
Sundman, Karl Frithiof (1873–1949), 13:153–154; 8:284b
Sunflower
rust disease, **14**:93b
Sung Ching-ch'ang, **3**:250a; **14**:540a
Sung dynasty, **3**:256a, 265b; **12**:370a–371a, 373a, 377b–378a, 383b, 384a–391a
Sunlight
colors of
submarine, **6**:68b
scattering of, **8**:501b

tenuity of, 2:328b
see also Radiation, solar
Sunset
green rays, 12:512a; 13:409a
Sunspots
seventeenth-century study, 3:179b, 528a; 5:515b; 6:124b, 126b, 361b; 7:193b, 374b; 9:336a; 11:411b
eighteenth-century study, 2:328a; 7:373b; 9:330b; 11:83a*
nineteenth-century study, 2:147a, 434a; 12:192a, 510b; 13:482a
twentieth-century study, 10:233a*; 12:352b
charting of, 3:115b; 10:543a; 13:472b
and cosmology, 1:588a; 5:244a
cycle of, 1:372a; 6:74b, 106a; 7:386b*, 609a; 10:213a*; 11:412a–b; 13:52a; 14:480b–481a
discovery of, 12:239b
and economics, 7:104b
and meteorology, 8:441b; 13:232b
polarity law, 6:29b
theories of, 8:441b
development of, 1:478a
discovery of, 5:241b–242a; 7:299b; 12:151b
distribution of, 3:93a; 9:184a; 12:579b
Evershed effect, 4:498a
magnetic fields of, 4:498a; 6:29b
motion of, 1:597a; 3:92b–93a; 6:362a; 12:267b
nature of, 4:19a, 555a; 6:333b, 362a; 7:186b; 10:490a; 13:256b; 14:415a
photography of, 6:106a
relative radiation of, 6:278a
rotation of
axis of, 12:152a
and latitude, 9:184a; 12:579a–b
spectra, 1:55b–56a; 4:498a; 5:102a; 6:28b; 8:441b; 10:31b; 12:20b, 71a, 72b–73a, 268a; 14:557b
intensity measurements, 9:415a
line splitting, 11:379a
photography of, 6:29a–b
statistical studies, 14:574a
systematic observations, 3:92b–93a, 261a; 6:512b; 9:184a; 12:239b
temperature of, 1:598a; 6:29b
and terrestrial magnetism, 4:354b; 7:609a; 11:412b; 12:51a–b, 210a; 13:52a; 14:480b
and weather, 9:279b
Sunstroke, 14:496a
Sun-tzu suan ching, 3:250a–b; 14:545b
Superconductivity
see Conduction, electrical
Supercooling, 2:177b, 186b; 15:315a, 461b
Superego, 5:178a–b
Superfluidity, 7:272b, 617b; 8:477a, 478a–479a
and rotation, 8:481a–482a
and superconductivity, 8:478b–479a, 482a
Superior, Lake
land district survey, 4:73b
Supernovae, 1:352b; 4:249a
discovery of, 2:403a; 6:150a
photography of, 11:470a
reactions in, 5:272b
Superphosphate
manufacture of, 8:93a
Superposition, principle of, 2:41b, 42a, 395a, 553a; 3:144a; 4:222b; 5:205a; 8:399b; 10:3a; 14:612b
opposition to, 10:494a
Supersaturation, 2:254b; 3:19b; 8:520a; 15:461b
Supersonics, 8:597a; 14:498a
Superstition, 9:40a
Supposition, theory of, 1:3a; 2:233a, 604b; 6:376b–377b
Suramin sodium, 5:100a
Surd, 13:302a; 15:503a–b
Surface(s)
algebraic, 4:523a; 5:569a; 6:547a–b; 8:250a–b, 562a; 10:553b; 11:60a, 509b
generation of, 15:193a
irreducible, 3:117a–b
theory of, 3:315b; 4:373b; 7:163b–164a, 446b; 9:553b; 10:594a; 11:390a; 12:330b–332a
transformation of, 10:275a
anallagmatic, 7:573b; 9:553b
applicable, 2:288a; 3:331a, 559b; 4:102b, 479b; 9:409b
couples of, 5:201b
area of, 10:442a, 443a
bending of, 10:544b–545a
binding with cloth, 3:225a
branching, 11:449a
caustic, 13:130b
Clifford, 3:323a
complete, 6:497a
complex, 11:46b
concept of, 7:72a
conical
definition of, 3:152b
of constant curvature, 1:599b; 2:288a; 5:202a; 11:398b
deformations of, 8:385b
convex
congruence theorem, 6:497a
rigidity of, 14:283a
cubic, 3:467b, 468a; 4:523a; 7:397a; 9:358b; 14:336a
classification of, 3:167b
construction of, 8:250a–b; 13:20b
division of, 12:172b
lines on, 3:167b; 5:339b; 12:86b, 172b
models of, 14:343a
theory of, 3:314b, 315b; 13:132b
curvature of, 4:102b, 479b; 5:376a–b; 11:455a
cylindrical
definition of, 3:152b
deformation of, 2:351a; 3:225a, 428a, 559b; 4:335b; 6:497b; 14:231b
infinitesimal, 14:283a
theory of, 4:336a
developable, 3:467b, 468a; 4:336a, 479b; 7:615a; 9:471a, 475b, 476b; 12:246b
division of, 4:609b
elastic, 5:375b–376a; 7:601b; 15:363a
equipotential, 6:240a
existence theorem, 2:288a
families of, 9:475b, 476a–b; 11:408a
focal
concept of, 9:476a
helicoid, 4:102b
hyperelliptical, 4:373b
intersection of, 7:109b, 601a
isometric, 14:231b–232a
Joachimsthal, 7:109b
Kummer, 5:471b; 6:547b; 7:397a, 523b; 11:509b, 589a
of least area
see Surface(s), minimal
mapping of, 5:568b
mechanics of, 8:124a
minimal, 2:288a, 351a; 3:559b; 4:174a; 5:339b, 376b; 7:561b; 8:136a; 9:476b; 11:22a, 398a–b, 454b; 12:246b; 13:130a; 15:23a
n-dimensional, 3:63a
of negative curvature, 6:4b
one-sided, 9:431a
open
classification of, 7:312b
orthogonal, 2:350a; 3:559b, 560a; 4:257b
osculatory, 11:45b
oval, 7:185a
Peterson, 10:545a
plane sections of, 3:282a
potential, 4:287b
quadric, 2:454b–455a; 3:282a, 399a; 4:479b; 5:339b; 6:2a; 12:213b, 225b; 13:19a
normals to, 7:109b
quartic, 3:468a; 9:358b; 11:509b–510a; 12:284b; 13:20b–21a, 250b
rational, 3:468b; 4:375a*; 5:568b
reciprocal, 12:86b
Riemann, 3:141a–b, 263a, 315a, 322b, 323a, 468b; 4:269a*; 6:571b, 572a; 7:397b–398a, 399a; 11:207a, 447b, 449a–450b, 452a; 14:283a
construction of, 8:542b
mapping of, 11:54a
of rotation, 1:600a; 2:351a; 3:432a; 4:429a; 5:527a; 9:127a, 409b; 10:337a; 13:130a; 14:158b, 231b
ruled, 2:351a; 3:117a; 4:102b, 350a; 9:471a, 475b, 476b; 11:46a; 12:86b; 13:418a
singularities of, 5:569a; 11:450a; 12:86b
spiric, 10:529a–530a
Steiner, 3:117a; 13:20b–21a, 623b
systems of, 4:336a
theory of, 1:600a; 2:88b, 288a, 470a, 577a; 3:178b–179a, 225a, 331a, 559b–560a; 4:102b–103a, 479b; 5:482a; 7:109a–b, 185a; 8:385b; 9:80b, 342b, 343a–b, 358b, 476a; 10:544b–545a; 11:408b; 12:285a; 13:132b, 418a; 14:231b–232a
transformation of, 6:547b
translation, 3:222a; 12:150a
Veronese, 13:623b
W surfaces, 14:231b
wave, 4:177b; 9:80b; 11:45b, 46b
see also Area
Surface tension, 4:203a, 211b, 228b
and evaporation, 13:11a–b
measurement of, 2:239b; 6:118a–b; 14:67a
theories of, 11:22a
Surgeons, Royal College of
see London, College of Surgeons
Surgery
ancient Roman, 3:174b; 12:540a
opposition to, 5:232a
Byzantine, 10:418b
medieval, 2:314a–315a; 3:218a–b, 393b; 6:276a–b; 9:41a, 603a; 11:238b; 14:584b; 15:499a
Renaissance, 1:611b, 618a–b; 2:336b, 546a–547a; 3:40a, 98b, 343a, 354a–355a, 534a; 4:511b, 521a; 5:348a, 361b, 580b, 581a; 7:16b; 8:554b; 10:315b–316b; 14:27b–28a
early seventeenth-century, 1:315b; 4:511b; 7:147b; 9:4a–b; 12:333a–b
late seventeenth-century, 4:495a; 5:419a; 6:508b–509a; 7:284a; 9:322b–323a, 581b–582a; 15:28b–29a
early eighteenth-century, 4:172b, 173a; 6:44a; 8:114b; 11:112a; 13:567b
late eighteenth-century, 2:122a; 3:37b, 38a, 67b; 6:367a, 567a–568a; 8:353a; 9:483b; 12:137a, 427b–428a; 13:448b; 14:296b–297a
early nineteenth-century, 2:483a–b; 4:22a*; 6:478b–479a, 521a; 8:96b–97b; 11:461b–462a

Surgery (cont.)
mid nineteenth-century, 2:376b, 386b, 478a; 5:45b; 6:120b; 8:34a–b, 399b–407b; 10:619b–621a
late nineteenth-century, 2:129b–130a; 6:77a–78a, 518a–519a; 8:408a–b; 14:278b–279a
twentieth-century, 1:438b; 2:597b; 3:518a–b; 7:502b–503a; 8:552b; 13:158a; 14:279a; 15:65a
aseptic techniques
medieval, 2:314b–315a
nineteenth-century, 8:403a–409a; 10:384b, 389b
twentieth-century, 3:91a; 4:277a
dressings, 10:385b, 386a
experimental
late seventeenth-century, 14:587a
nineteenth-century, 1:417b; 10:432a, 433a; 11:98b–99a
twentieth-century, 3:91a–b, 121a–b, 517b–518a
hypothermia, 7:502b–503a
and medicine, 8:114b; 10:304b
Mesopotamian, 15:645b–646a
microscopic
nineteenth-century, 1:417b
twentieth-century, 12:568a
military
Renaissance, 9:11b–12a; 10:304b, 315b–316a; 14:5b–6a
seventeenth-century, 14:209b
eighteenth-century, 6:231a, 566b
nineteenth-century, 2:386b; 6:479a; 10:620a,b; 13:353b–354a
twentieth-century, 2:597b; 3:91b, 518b
pathological, 11:98b–99a
plastic
Renaissance, 1:204a
nineteenth-century, 2:130a
thoracic, 9:266a
vascular, 2:122a; 3:90b–91a; 11:461b–462a
see also Amputation; Apparatus, surgical; Dissection; Education, surgical; Instruments, surgical; Neurosurgery; Psychosurgery; Resection; specific operations and organs; Transplantation; Trephination; Vivisection
Suri seiran, 1:90b
Surian, Joseph, 4:67a; 11:47a
Suringar, Willem Frederik Reinier (1832–1898), 13:154–155; 10:254a*; 13:459a; 14:95b
Surra, 7:428a
Surrey (county)
geology, 5:437a
Surveying
ancient Greek, 6:312b
ancient Roman, 1:418a–b; 3:59a
medieval Arabic, 1:40a, 41b
medieval Latin, 4:609b, 612a; 5:365a; 6:544b
Renaissance, 4:97a, 586b; 5:349a; 7:419a; 9:581a, 609b; 11:406a; 13:260a–b
seventeenth-century, 1:126b; 5:593b; 6:483b; 7:577b; 8:65b, 132a; 9:115b, 336a; 10:151a, 264b*, 565a, 596a–b
eighteenth-century, 2:299b, 329b; 3:344b; 4:131b–132a; 6:576b; 7:89a; 8:351a; 9:234a–b; 10:572a; 11:285a, 471b; 12:486b–487a; 13:68a; 14:279b, 583a
nineteenth-century, 1:615b; 2:102a, 332b; 4:273b; 5:303a–304a, 405b, 537b; 6:489b; 7:19a–b, 148b–149b, 370b, 615a; 11:95a–b, 291b, 399a; 12:518a–b; 13:521a; 14:232b
twentieth-century, 3:616b–617a; 5:489b
cadastral, 8:351a; 15:335b
Chinese, 12:380a; 14:539b
Japanese, 15:744a, 745b, 746b, 747b
methods, 7:149b
nautical
see Hydrography
photography in, 11:95b
plane-table, 9:581a
rapid, 11:95a
standardization of measurement, 1:612b
traverse method, 7:19b
underground, 6:213b
see also Geodesy, Hydrography, Triangulation
Survival of the fittest, 3:573a; 8:538a; 12:570b; 15:426b
Sūryadāsa
commentaries of, 2:115b, 117a
Sūryadeva Yajvan of Kerala
commentaries of, 1:308a,b; 2:114b; 9:580b; 12:599a
Sūryamaṇi
commentaries of, 2:116a
Sūryasiddhānta, 5:612a,b, 618b; 8:46b; 12:115b; 15:586b, 597a, 598b, 603a, 605a, 607b–618a *passim,* 629a
commentaries on, 11:264a, 290a–b
see also Lāṭadeva, *Sūryasiddhānta*
Susa tablets, 4:610b; 15:667a, 671a
Suslin, Mikhail Yakovlevich, 8:558a, 559a
Suslyga, Lawrence, 7:298a
Sussex (county)
geology, 5:14a–b
paleontology, 9:87a–b
Suter, Heinrich (1848–1922), 13:155
historiography, 1:30b, 224a; 4:454a*, 455a*, 543b; 7:3b*, 223b, 256a, 337b, 352b, 531b; 9:615a*; 10:299b; 13:294b*; 15:179b, 186b
translations by, 6:206a*,b*, 208a*; 11:241a*, 245a*
Sutherland, A. (inoculation), 7:11b
Sutherland, William (1859–1911), 13:155–156
Sutherland constant, 11:291a; 13:156a
Sutlej (river)
source of, 6:216a
Sutton, Daniel, 7:11b; 8:352b
Sutton, Robert, 7:96a; 8:352b
Sutton, Thomas, 9:202a
Sutton, Walter Stanborough (1877–1916), 13:156–158; 2:364a; 3:422b; 8:588b, 589a; 9:496b; 14:424b, 431a
Sutures (surgical), 7:208b
catgut, 8:404a; 14:584b
silver-wire, 8:402a
Suvorov atoll, 8:101a
Svedberg, The (Theodor) (1884–1971), 13:158–164; 10:525b; 14:633a
biochemistry, 13:2a–b, 419b
gases, 12:497a
radiochemistry, 12:504b
students of, 13:418b
Svedelius, Nils Eberhard (1873–1960), 13:164–166; 13:392b*
Sverdrup, Harald Ulrik (1888–1957), 13:166–167; 2:168b; 12:415b
Sverdrup, Otto, 13:166a
Svetlov, Pavel Grigorevich, 7:34a
Swab, Anton, 2:5a
Swabia
stratigraphy, 11:235b
Swain, Lorna, 7:595a
Swainson, William (1789–1855), 13:167–168; 1:330b; 14:133b, 187a,b
anthropology, 11:137b
Swallow (bird), 14:299b
hibernation, 3:129b, 350a; 8:379b
migration, 3:350a
Swallowing
mechanism of, 15:71b, 72b
reflex nature of, 6:60a; 9:265b–266a
Swammerdam, Jan (1637–1680), 13:168–175; 4:65a; 9:595a
acquaintances of, 2:317a; 5:484a; 6:508b; 13:31a, 33b
anatomy, 12:40b; 15:29b
Biblia naturae, 2:227b
editions of, 6:401a
publication of, 4:268a
embryology, 11:133a, 333a
entomology, 5:439b, 440b; 8:127a; 11:331b
Ephemeri vita
translations of, 13:527a
influence of, 5:328b; 8:157b, 579b; 13:179b
influences on, 13:223b
muscle contraction, 3:483a
on reproduction, 1:293b; 4:291a; 5:484b; 6:509a; 8:128a, 580a; 15:67b
Tractatus . . . de respiratione usuque pulmonum, 9:243a
Swamp cypress, 9:54a
Swamps, 4:145b
drainage, 12:344a, 371b
Swann, William Francis Gray (1884–1962), 13:175–177; 8:94a–b
Swarts, Frédéric Jean Edmond (1866–1940), 13:177
Swarts reaction, 13:177a
Swarts, Théodore, 7:281a; 13:177a
Swartz, Olof (1760–1818), 13:177–178
Swasey, Ambrose, 3:35b; 6:28a
Sweating
see Perspiration
Sweating sickness, 3:13a, 413a; 15:160b
Sweden
archaeology, 14:506a
astronomy, 8:364b
biology, 11:379b
botany, 5:190b–191a
calendar, 14:178b
chemistry, 5:222b–223a; 6:380b
entomology, 5:605b
ethnology, 10:197a–b
geography, 10:197a–b
Geological Survey, 9:617b
geology, 2:496b; 5:329b–330a; 6:436b–437a; 8:570b; 13:409b
history, 11:587b
iron industry, 11:463b
mining, 3:473a; 5:223a; 13:409b
natural history
eighteenth-century, 1:305a–306a; 2:4b; 8:376a
paleontology, 2:490a; 9:617b
politics, 6:217a
population, 14:179a–b
research councils, 13:159a
Science Advisory Council, 13:419a
Swedenborg, Emanuel (1688–1772), 13:178–181; 5:340b
biographies of, 11:125b, 382b
geology, 3:210b; 8:519a
Minerale de ferro
translation of, 3:454b
Swedish Academy of Sciences, 5:328a; 14:178b, 355a
founding of, 8:375b; 13:409b
Swedish Institute of Metallography, 1:609a
Swedish Metal Research Institute, 1:609b
Swedish Seed Association, 10:130a

Sweet pea
genetics, **11**:212b
Swelling, **4**:203a
Swieten, Gerard van (1700–1772), 13:181–183; **1**:332b, 333a; **2**:225b; **3**:2a; **7**:57b; **8**:327b; **9**:325b
Świętokrzyskie Mountains, **15**:502b, 527a
Swinden, Jan Hendrik van (1746–1823), 13:183–184; 7:499b, 500a
Swine cholera
see Hog cholera
Swine diphtheria
see Hog cholera
Swine erysipelas, **8**:449b
bacillus, **7**:425b
vaccination, **7**:100b; **10**:399b–400b
Swine fever
see Hog cholera
Swine plague, **8**:449b; **12**:480b
bacteria, **15**:522a
Swineshead, John, **13**:184b–185a
Swineshead (Swyneshed, Suicet, etc.), Richard (*fl. ca.* 1340–1355), 13:184–213; **2**:590b; **3**:420b; **4**:237b; **6**:379a; **8**:198a; **9**:32b; **13**:349a–350a
Liber calculationum, **2**:393b; **9**:611b
Swineshead, Roger, **13**:184b–187b, 189a, 190a, 195a, 208a
Swiss Mathematical Society
founding of, **5**:206a
Switches, electrical, **4**:202b
Switching, electronic, **1**:455b
Switzerland
climate, **9**:189b
education, **5**:339b; **8**:35b
flora, **6**:61b, 65b–66a
geography, **4**:217b; **12**:159b
Geological Commission, **6**:227b, 228a; **13**:124a
geology, **1**:235b–237a; **3**:210a–211a; **5**:221a–b; **6**:222a, 227a–228a, 228b–229b; **7**:43b–44a; **10**:124a, 127a; **11**:375a; **13**:123b–124a
Geotechnical Commission, **5**:563a
mapping, **4**:73b; **5**:221b; **6**:227b, 229a; **7**:43b–44a
guide books, **4**:275a–b
limnology, **15**:158b
natural history, **11**:272a
paleontology, **6**:221a–222a; **8**:4a; **12**:159b
Sycosis, **5**:566a
Sydenham, Thomas (1624–1689), 13:213–215; **1**:399b; **2**:224b, 225b; **5**:367a; **8**:437a,b; **9**:306b, 400b; **11**:617a; **12**:365b, 457b
Sydenham's chorea, **13**:215a
Sydenstricker, Edgar, **5**:452b
Sydney, Australia
observatory, **15**:493b
Sydney Exhibition, **11**:384b
Syenites, **3**:433b
nepheline, **7**:249b
Syllogism, **1**:94b, 278b, 404a; **2**:230b, 231b–233a *passim,* 296b, 601b, 604b, 609b; **3**:60a; **4**:36a, 138a, 433b(n); **6**:83a; **8**:546b; **14**:483b, 581a,b
conditional, **13**:510a
hypothetical, **2**:232a
in mathematics, **3**:585a; **4**:226b
moods of, **4**:462a
Stoic, **2**:232a
Theophrastian, **2**:232a
unfigured, **6**:82a
Syllogistics, **8**:154a
Sylow, Peter Ludvig Mejdell (1832–1918), 13:215–216; **2**:615b; **3**:142b, 168b; **8**:323b, 324a; **12**:452a
Sylvanite
analysis of, **9**:559a
Sylvester II, Pope
see Gerbert (Gerbert d'Aurillac)
Sylvester, James Joseph (1814–1897), 13:216–222; **2**:469b
acquaintances of, **4**:35b; **6**:451a; **7**:506b; **10**:483b
algebra, **10**:484a; **13**:129b
equation theory, **2**:113a, 114a
invariant theory, **1**:294a; **3**:164b, 169b, 314b; **5**:472b; **6**:309a, 388b; **8**:617a
matrix theory, **3**:166a
and Cayley, **3**:163a, 164a, 167a, 170a; **13**:217a, 218a–b, 219b, 221a
on contemporary scientists, **5**:568b
correspondence of, **12**:86b
formula of, **15**:686a, 699b
historiography, **15**:687a,b, 703b*
mechanics, **11**:62a
poetry, **7**:479a
students of, **6**:76a
Sylvestrene, **13**:410b
Sylvius, Franciscus dele Boë (1614–1672), 13:222–223
acquaintances of, **1**:482a; **2**:317a; **13**:31a
anatomy, **13**:31b
digestion, **2**:237b; **13**:171a
influence of, **13**:33a, 169b; **14**:26a
influences on, **10**:311a
students of, **5**:484a; **9**:581b; **12**:40a; **13**:169a, 282a, 479b
Sylvius, Jacobus
see Dubois, Jacques
Symachus, **14**:35b
Symbiosis, **1**:622b; **2**:34b, 290b; **3**:259b; **6**:237b; **9**:176a
artificial, **12**:207b–208a
mycorrhizal, **3**:429a
naming of, **3**:613b
Symbols
in geometry, **1**:474a
in psychology, **7**:191b, 193a
Syme, James, **5**:469a, 470a; **8**:401a,b, 404b
Symmachus, Q. Aurelius Memmius, **1**:269b; **2**:228b
Symmedian point, **8**:175b, 176a
Symmer, Robert (*ca.* 1707–1763), 13:224–225; **14**:70a
electricity, **1**:547b; **10**:147a; **14**:71a, 353a
Symmetry
biological, **9**:573b, 620b
botanical, **13**:405b
embryogeny of, **1**:153a; **6**:133a–b; **7**:95a
philosophy of, **1**:261b; **10**:278b
plane of, **10**:39b
and zoological classification, **1**:387a
and causation, **3**:505b
crystallographic, **2**:441a; **3**:505a; **5**:211a, 557a; **6**:179b, 180a; **7**:59b; **13**:2b–3a; **14**:525b
centrosymmetries, **5**:186b
classes of, **5**:124a; **6**:358a; **9**:379b; **12**:195b–196a, 511b–512a; **14**:240b
effects of, **6**:304a
enantiomorphous, **1**:461b
etch figures, **1**:528b
Friedel's law, **5**:186b
incomplete, **9**:620a
internal, **12**:511b–512a; **15**:114b
mathematics of, **1**:460b–462a
and piezoelectricity, **3**:504b
point groups, **1**:552a; **5**:212a
polar, **1**:461b
pseudosymmetry, **1**:461a; **9**:58b; **14**:241a, 525b
space groups, **1**:461b–462a; **5**:212a; **6**:303a–b; **12**:512a; **15**:17a
surface, **5**:456a
geometrical, **1**:222a; **5**:482a; **7**:167b
mathematics of, **2**:431b
in physics, **9**:211a–b; **10**:425a; **11**:117b
laws of, **3**:505a–b
statistical, **6**:304a
Symons, John, **6**:569a
Sympathetic medicine, **5**:48a; **6**:254a
Sympathy, powder of, **4**:95b
Sympathy, principle of, **5**:106b–107a; **10**:308b, 309a; **11**:42a; **14**:606b
Sympathy and antipathy, doctrine of, **5**:47b–48a
Symphysiotomy, **8**:353a; **12**:427b–428a
experimental, **3**:38a
Symptomata, **1**:180b–185b passim; **4**:568a
Symptomatology
seventeenth-century, **10**:567b
eighteenth-century, **2**:301b–302a
Synaeresis, **1**:561b
Synapses, **8**:92a; **12**:396b, 399b
concept of, **12**:397a–b
Synchrocyclotron, **8**:95b
Synchronicity, theory of, **7**:190a, 192b
Synchrotron, **3**:330a
Synchytrium, **14**:93b
Synclines, **10**:430a; **12**:277a
Syncoryne, **8**:519a
Syncrisis, **4**:569b; **12**:312b
Synedra, **13**:458b
Synesius of Cyrene (*ca.* 370-*ca.* 414), 13:225–226; **6**:616a; **10**:208b; **14**:631a; **15**:219b
Syng, Philip, **5**:131a
Syngameon, **8**:513a
Syngamy, **6**:337a
Synge, John Lighton, **14**:305b, 306a
Syngnathus acus, **1**:265a
Synkaryon, **9**:34b
Synovia
naming of, **10**:310b
Synthèse, **5**:123a
Synthesis, chemical
concept of, **2**:65b
mineral substances, **1**:557b
organic
catalytic, **1**:415a–b
petrochemical, **1**:415b
photochemical, **3**:279b
see also Chemistry, organic, synthetic; specific compounds
Synthesis, logical, **1**:188a; **2**:545a; **4**:101a,b, 416a–b; **6**:203a, 424b; **8**:159a; **10**:51a–52a, 85b; **14**:483b
Syōsahō, **12**:291b
Syphilis, **8**:273a
Renaissance discussion, **1**:204a; **2**:546b; **4**:521a; **5**:105b–106a; **8**:248b; **9**:178b
congenital, **8**:273b; **10**:307a; **11**:515b
diagnosis of, **6**:433a
Wassermann test, **2**:300b, 584a; **4**:299b, 301a; **7**:624a; **8**:273b; **10**:18b, 142a; **15**:522b–523a
etiology of, **1**:493b; **6**:73a; **8**:273b; **9**:75a; **10**:18a–b, 153b; **11**:174b–175a; **12**:141b, 142a–b
geographic origin of, **5**:411b
of heart, **7**:614a
historiography of, **13**:142b
host response to, **14**:624a
infantile, **3**:463b
naming of, **5**:105b
paralytic, **4**:300b
symptomatology, **9**:179a; **14**:28a
transmission of, **10**:18b; **11**:515b
to fetus, **1**:612a
to monkeys, **9**:334b

Syphilis (cont.)
  treatment of, **4**:585a; **14**:6a
    arsenical compounds, **4**:300a–b; **10**:18b
    fever-therapy, **14**:115b
    guaiac wood, **1**:620b; **5**:105b; **9**:75a, 179a; **10**:305a, 306b
    mercury, **1**:60a, 620b, 621a; **5**:105b; **9**:334b; **10**:305a–b, 307a; **13**:181b; **14**:28a
    metallotherapy, **8**:273b
    Salvarsan, **5**:29b
  visceral, **10**:307a
  see also Neurosyphilis
Syphon
  see Siphon
Syria, **15**:723b
  botany, **6**:492a; **11**:311b
Syriac language, **15**:232a
Syrianus, **2**:229b; **4**:159a–b; **11**:160b, 161b
Syringe, **3**:588a; **12**:103a
Syrups (pharm.), **12**:323b
Systems, mathematical
  algebraic
    theory of, **9**:71a–b
  analytic, **2**:144b–145a
  covariant, **11**:408a–b
  differential, **3**:95a; **14**:13b
    theory of, **3**:95b–96a
  elliptical, **10**:554a
  Fourier, **5**:608a
  fundamental, **5**:204b
  hyperbolic, **10**:554a
  hypercomplex, **9**:457b; **11**:55b
  infinite, **4**:4b; **7**:435a–436b; **11**:471a
  invariant, **11**:407a–b
  linear, **4**:478a
  orthogonal, **4**:287b; **5**:608a–b; **11**:398b
  parabolic, **10**:554a
  Pfaffian, **2**:145a; **3**:314a; **12**:214a
Systems, physics
  dynamical
    theory of, **2**:145a
  ensembles of, **5**:392a–b
  inertial, **4**:321a–322b, 323b
Systole, **3**:220a; **4**:586a; **8**:204a, 525b; **14**:4b, 5a
  shape of the heart during, **4**:589b
Sytinskaya, N. N., **12**:353a
Syzygies, astronomical, **2**:117b, 119a; **7**: 219b, 582b; **9**:607a; **11**:194b; **15**: 340b, 380b–381a, 794a–b, 795a–b, 798a, 803b
Syzygies, mathematical, **3**:165a; **6**:389a–b
Syzygites megalocarpus, **4**:288b
Szász, Otto, **4**:561b
Szebellédy, László (1901–1944), 13:226; 14:448b*
Szegö, Gabor, **4**:548a, 561b, 562a
Szent-Györgyi, Albert, **6**:185b; **15**:258a
Szilard, Leo (1898–1964), 13:226–228; 4: 313a, 580b–581a
  Szilard-Chalmers reaction, **13**:227a
Szily, Pál (1878–1945), 13:228–229
Szily von Nagy-Szigeth, Coloman, **2**:261a

T

t test, **5**:8a–b, 10a, 477a; **10**:454b–455a, 457b
Ta yen (Great Extension), **3**:251b, 253b; **4**:606b, 612a
al-Ṭabarī, Abu'l-Ḥasan Aḥmad ibn Muḥammad (*fl.* tenth century), 13:229–230
al-Ṭabarī, Abu'l-Ḥasan ᶜAlī ibn Sahl Rabbān (*ca.* 808-*ca.* 861), 13:230–231; 7:358a–b
Tabasheer, **12**:494b
Tabernaemontanus, Johann Teodor, **5**:362a
Tabes dorsualis, **3**:204b*; **14**:229a
Tables, actuarial
  eighteenth-century, **2**:418b
  nineteenth-century, **1**:402b
Tables, anatomical
  Renaissance, **3**:342a
Tables, astrological
  medieval, **7**:259b, 338a; **12**:432a
  Renaissance, **14**:273b
Tables, astronomical
  ancient Greek, **1**:191a; **13**:322a; **15**:218b–219a
  ancient Roman, **11**:189b–197a *passim*
  Byzantine, **12**:406a
  medieval Arabic, **1**:34b, 36a, 38b, 61a, 62b, 122a, 508a–b, 510a, 512b; **2**: 154b, 155a; **3**:26b; **4**:556a; **5**:612a–618b *passim;* **6**:274a; **7**:79a, 82b, 124a, 141b–142a, 259b, 324b, 334a–b, 336b–338b, 360b–361b, 363a, 531b, 532a; **9**:39a,b; **10**:5b; **11**:227a, 249a, 250b; **12**:358a–b, 360b–361a, 362a–b; **13**:149b, 509b, 536b–537a; **14**:537b–538b, 546a–b, 575b–579a, 593a–b; **15**:259a–260b
  medieval Hebrew, **1**:383a; **8**:280b, 281b
  medieval Latin, **1**:122a, 382b, 383a; **2**:24a, 401b–402a, 435a,b; **3**:26a–b, 218a; **4**:185b; **5**:553a; **6**:273a, 273b–274a, 301b; **7**:118b–120a, 122b–126a, 129b–131a, 139a–141a; **10**:540b, 541a–b; **11**:321b–322a, 415a, 504a; **14**:390a,b; **15**:187b
  Renaissance, **1**:573b; **2**:401b; **11**:350b–351a, 366a; **14**:273b, 359a, 583b; **15**:474b, 476a–b, 477b, 478a
  early seventeenth-century, **2**:131a; **3**: 101a–b; **6**:514b; **7**:304a–306a, 308a; **8**:25b, 28a, 65a; **9**:247a,b, 336b*, 527b
  late seventeenth-century, **3**:103a; **5**:23b–24a, 25b, 543a; **6**:115b, 516a; **7**:308a, 577b; **8**:132a; **9**:485a; **13**:96a
  early eighteenth-century, **2**:388b, 389a; **3**:105a; **4**:22b–23a; **5**:88a, 435a; **8**: 178b–179a; **9**:77b; **14**:178b–179a
  late eighteenth-century, **1**:400b–401a; **3**: 197b, 283a, 430b–431a; **4**:14b, 481b; **7**:86b, 544b, 580a–b, 598b; **8**:144b*; **9**:90b, 162b–163a, 164b, 233b, 234b; **14**:178b–179a, 583a; **15**:325a,b, 326b, 327a, 331a,b, 337b, 349a,b, 353a–b, 354b
  early nineteenth-century, **2**:359b, 471b, 600a; **4**:16b; **8**:369b; **11**:7a
  mid nineteenth-century, **3**:14a; **6**:103b; **8**:278a; **10**:207b; **11**:114a–115a
  late nineteenth-century, **2**:332b, 441b; **4**:176a; **6**:403a*; **7**:481b; **9**:563b; **10**:34a–35a, 219b, 480b*–481a*; **12**:170a
  early twentieth-century, **1**:429a; **2**:516a–b; **4**:359a; **5**:224a; **7**:6a–b, 237a; **12**: 95b; **15**:129a
  Alfonsine
    see Alfonso of Castile, tables of
  Chinese, **14**:162a
  Egyptian, **15**:715a–b
  ephemerides
    ancient Greek, **13**:323b
    medieval, **1**:437b–438a; **7**:140b; **13**:401a
    Renaissance, **1**:165b; **2**:401b; **9**:12b; **11**:351a; **15**:474b, 478b
    early seventeenth-century, **1**:244b; **4**: 551b; **7**:306b, 308a
    late seventeenth-century, **1**:169a; **3**: 100a, 101b, 248a, 534b; **5**:22b; **9**: 485a; **13**:96a
    eighteenth-century, **2**:220b, 600a; **5**: 435a; **6**:234a,b; **7**:542b, 576a; **9**:77b, 163a, 234b; **14**:589b
    nineteenth-century, **2**:99b; **4**:176a, 514b; **5**:256b–257a; **6**:150a, 399b, 402b; **8**:557b; **9**:421b; **10**:35a, 219a; **11**:114b–115a, 545a; **12**:234b; **13**:4a
    twentieth-century, **1**:156b, 157a, 430b*; **2**:516b; **3**:472a; **5**:224a; **7**:6b; **10**:30b, 159a
    calculation of, **1**:437b–438a; **6**:514b; **7**:373b; **8**:132a; **12**:163a; **15**:679a, 729a,b
    Chinese, **12**:375a
    Japanese, **1**:310a
    medical, **1**:399b
    and occultation, **1**:429b
    theory of, **7**:7a, 481a
  Indian, **1**:5b, 33a, 34b; **2**:119a; **4**:100a, 556a; **5**:274a–276a; **6**:115b; **7**:81a,b, 315a, 512b; **9**:21a, 42a, 175b, 607a; **11**:264a; **15**:535a–628b *passim*
  Marseilles Tables, **14**:593a,b
  Mayan, **15**:774b, 778a–b, 779a, 784a–802b, 811a,b, 812a,b, 814a
  Mesopotamian, **15**:676a, 679a–b
  Prutenic (Prussian) Tables
    see Reinhold, Erasmus, *Prussian Tables*
  Rudolphine Tables
    see Kepler, Johann, *Tabulae Rudolphinae*
  Sindhind
    see *Zīj al-Sindhind*
  Toledan
    see al-Zarqālī, Toledan tables
  Toulouse tables, **7**:129b
    critics of, **14**:390a
  see also Moon; specific stars and planets; Sun
Tables, ballistic
  seventeenth-century, **13**:437a
  eighteenth-century, **1**:581b; **2**:202a
  nineteenth-century, **3**:224b; **10**:250b
  twentieth-century, **2**:198a
Tables, calendric
  medieval Arabic, **1**:62b; **2**:152b; **4**:556a; **7**:142a, 337a; **14**:576b
  medieval Latin, **1**:382b, 383a, 565a–b
  Alfonsine
    see Alfonso of Castile, tables of
  Indian, **9**:42a
  Mayan, **15**:772a, 773a, 778a–784a
Tables, chemical
  eighteenth-century, **2**:6b–7a; **5**:353b–354a; **9**:5b; **11**:109a–b
  nineteenth-century, **2**:93b–94a; **3**:289a, 292b; **7**:620a; **9**:95b; **10**:37b–38a; **11**:437a; **14**:112a
  twentieth-century, **7**:323b; **8**:485a; **14**: 182a
Tables, chronological
  medieval Arabic, **2**:152b
Tables, chronometric
  eighteenth-century, **2**:389a
Tables, crystallographic, **6**:303a
Tables, engineering, **3**:20a, 442b; **9**:462b
Tables, geodetic
  nineteenth-century, **6**:240b
Tables, geographical
  medieval Arabic, **1**:29a; **2**:154b; **7**:259b, 337a–b; **9**:89a; **14**:547b
  seventeenth-century, **7**:305b; **11**:411b
  eighteenth-century, **11**:609b
Tables, geological
  nineteenth-century, **9**:241a; **12**:514b
  twentieth-century, **8**:42a; **13**:500b
Tables, geophysical, **14**:327b
Tables, logarithmic
  Renaissance, **13**:60a

seventeenth-century, **2**:411a, 462b, 602b–603a; **3**:152b; **4**:551a; **5**:344a, 593b; **7**:304b; **9**:17a, 310b, 610a–613a; **10**:255a, 264b*; **14**:51a
eighteenth-century, **4**:38b; **11**:164a
nineteenth-century, **2**:441b; **6**:522a; **8**:137a, 140b; **9**:463a
twentieth-century, **13**:140a*
antilogarithms, **1**:90b–91a
eight-place, **1**:355a
Oriental, **1**:84a, 90b–91a
seven-place, **1**:90b; **4**:16a, 551a
typography of, **1**:356b*
Tables, mathematical
Byzantine, **12**:406a
medieval Arabic, **7**:533a
medieval Latin, **4**:606a, 609b
Renaissance, **3**:181b; **9**:12b; **15**:155a
seventeenth-century, **5**:159a; **9**:554b
eighteenth-century, **6**:577a; **7**:559a, 599b; **8**:351b*
nineteenth-century, **1**:459b; **2**:441b; **3**:466b; **5**:414a, 572b; **7**:494b; **8**:137a,b, 138b, 140b; **9**:127a; **10**:482a
twentieth-century, **2**:90a; **3**:373b, 374a*; **6**:570a
Egyptian, **15**:681a–b, 683b–689b, 704b*–705b*
Japanese, **15**:752a
Mesopotamian, **15**:668a, 670a
multiplication, **2**:233b; **7**:533a; **10**:540a; **15**:155a, 668a
see also Tables, logarithmic; Tables, statistical; Tables, trigonometric
Tables, meteorological
seventeenth-century, **5**:25b
eighteenth-century, **3**:540a
nineteenth-century, **5**:413a, 600a*
Tables, mineralogical, **9**:367a
Tables, oceanographic, **14**:635b*
Tables, optical
seventeenth-century, **6**:125b, 128a
Tables, physics
seventeenth-century, **5**:382a
Tables, statistical, **5**:8b, 10a,b; **10**:451a, 454b–455b, 464b
Tables, trigonometric
ancient Greek, **15**:208b–209b
ancient Roman, **9**:297b; **11**:188b
medieval Arabic, **1**:42a–b, 513b; **2**:154b; **5**:612a–b; **7**:257b–259a, 259b, 338a, 361a, 531b, 532a; **13**:536b; **14**:577a, 578a–b, 593b
medieval Hebrew, **8**:280a
medieval Latin, **1**:62b; **2**:435a; **11**:415b; **15**:163a–b
Renaissance, **1**:179a; **3**:181b, 262a–b; **4**:619a; **8**:27b; **9**:190b, 611a–b; **11**:350b, 396a–b, 397a–b, 561b; **14**:18b–19a
seventeenth-century, **3**:152b; **4**:551a; **5**:344a, 408b, 409a,b; **9**:12b, 554b; **10**:255a, 264b*; **11**:3b, 4a–b, 533b; **13**:437a; **14**:51b
eighteenth-century, **2**:300a; **4**:38b; **7**:559a, 598b; **11**:164a
twentieth-century, **1**:157a
centesimal, **4**:16a
Tabor, John (*b.* 1667), 13:231–232
Tabrīzī, Yār-Aḥmed, **7**:324a
Tabula Smaragdina
commentaries on, **2**:317b; **7**:41a; **11**:32a
Tabulae biologicae, **5**:584b
Tacchini, Pietro (1838–1905), 13:232–233; **1**:19a; **2**:433a; **8**:503b; **11**:412a; **12**:269b
Taccola, Mariano di Jacomo (1381–*ca.* 1458), 13:233–234; **8**:208a; **9**:146b, 147a
critics of, **14**:275a
geography, **14**:275a
Tachenius, Otto (*fl. ca.* 1640–1670), 13:234–235
Hippocrates chimicus, **14**:639b
Tachymetry, **11**:95a,b; **13**:260a
Tachytes, **4**:504b
Tacke, Ida, **10**:136a
Tacoma
see Rainier, Mount
Taconic System, **1**:469b; **4**:363b–365a; **6**:57a; **8**:462a; **9**:100b, 173a; **11**:594a; **14**:120b
Tacquet, Andreas (1612–1660), 13:235–236; **1**:474b; **13**:561a
Euclidean commentaries, **1**:474a
mathematics, **1**:164a; **4**:451b; **10**:336a,b
Taenia, **1**:602b; **8**:270b
life cycle, **5**:502b–503a
Taeniarhynchosis
control of, **12**:453a
Tagesansicht, **4**:556b, 557a
Ibn Taghrībirdī, **4**:541b
Tagliacozzi, Gaspare, **1**:204a; **4**:507b
Tagore, Rabindranath, **2**:325b; **15**:47a, 48a, 49b–50a
Ibn Ṭāhir
see al-Baghdādī, Abū Manṣūr ᶜAbd al-Qāhir ibn Ṭāhir ibn Muḥammad ibn ᶜAbdallah, al-Tamīmī, al-Shafiᶜī
Tahiti, **3**:396b
botany, **10**:324a
discovery of, **2**:342b
geology, **3**:551a,b, 552b
language, **10**:324a
Tai Chen, **8**:419a,b
Taine, John (pseudonym), **1**:584b
Taisnier, Jean
Opusculum . . . de natura magnetis, **1**:606a; **7**:295a
Tait, Peter Guthrie (1831–1901), 13:236–237
acquaintances of, **3**:308a; **5**:389a; **6**:502a; **7**:413a; **9**:199a; **11**:241b
and Cayley, **3**:163b, 166a
and Clausius, **3**:306a–b; **5**:388a
on contemporary scientists, **3**:497a; **5**:69a*; **11**:294b*
correspondence of, **13**:218b
critics of, **3**:477a; **13**:523a
gas theory, **1**:128b; **9**:222b, 224a
geology, **12**:593b
influence of, **3**:32a; **4**:500a; **10**:483b
mathematics
quaternions, **5**:391a; **9**:215b
topology, **3**:169a
mechanics, **14**:569a
hydrodynamics, **11**:58a; **13**:244b
ozone, **1**:161a
Quaternions, **3**:163b; **6**:92a
Sketch of Thermodynamics, **5**:388a
spiritualism, **13**:52b
students of, **9**:588b; **13**:53a
thermodynamics, **5**:388a; **9**:239b
Treatise on Natural Philosophy, **9**:211a; **13**:386a
Takagi Teiji, **1**:307a; **6**:297b
Takahashi, Kageyasu, **7**:20a; **15**:734b
Takahashi Yoshitoki, **1**:311b, 312a, 313b; **7**:18b, 19a
Zōshū shōchō hō, **1**:312a–b, 313b
Takaki, Tomoe, **15**:522a
Takamine, Jokichi, **1**:10b–11a; **10**:205b
Takebe, Katahiro, **1**:84a, 90b; **3**:266a; **15**:750a, 751a,b, 752b, 756a
Hatubi sanpō endan genkai, **12**:291a
Taklamakan desert, **6**:215b
Talbot, William Henry Fox (1800–1877), 13:237–239; **13**:273a
Talbot's bands, **13**:77a
Talbotypy, **4**:283a*
Talc
mining of, **6**:229a
Talcott, Andrew, **6**:513b
Talent
inheritance of, **5**:266a
Talismans, **1**:34a, 37b; **7**:42a; **15**:497b
Talleyrand-Périgord, Charles-Maurice de, **4**:14b; **15**:271b, 334a
Talmud
commentaries on, **13**:401b
Talmudic law, **9**:28a
Talus (geol.), **10**:429a
Tamarkin, Jacob David, **13**:26a; **14**:345b
Tamm, Igor Evgenievich (1895–1971), 13:239–242; **5**:161b*; **9**:77a; **13**:599a
Tamm-Dankov method, **13**:241a
Tammann, Gustav Heinrich Johann Apollon (1861–1938), 13:242–248; **2**:431b; **5**:576a; **6**:480a; **15**:433a, 435a
Tammann oven, **13**:247b
Tana, Lake, **3**:533a
Tanfilev, Gavriil Ivanovich (1857–1928), 13:248–249
Tang, Fei-fang, **14**:624a
T'ang Shen-wei, **8**:391a–b, 396b*
Tangents (geom.), **3**:460a,b; **4**:33b; **5**:368a,b, 526b, 567a; **7**:570b; **10**:49b, 494b
of algebraic curves, **6**:359a, 537a; **12**:460b; **13**:480b
conjugate, **12**:285a
construction of, **1**:474b; **5**:528a; **10**:52a; **11**:487b; **12**:206b, 460b
Descartes' method for, **3**:615b; **4**:57b
determination of, **8**:304b; **11**:404a
double, **3**:167a, 315a; **5**:339b; **6**:357b; **13**:20a,b
inflectional, **3**:167a
inverse, **2**:52b; **3**:615b–616a; **6**:607b; **10**:48b
method of, **4**:570a–571a; **5**:527a; **10**:66b; **11**:609a; **12**:118a
problem of, **4**:46b
rule of, **4**:58a, 575a, 619a; **6**:537b–538a, 600b
Tangents (trig.), **5**:528b; **8**:163a
medieval Arabic use of, **1**:511b–512a; **5**:612b–613a; **10**:6a
inverse, **8**:164b
naming of, **11**:4a
tables of, **9**:611a; **11**:350b, 397a; **13**:536b; **14**:578a
Tani, Jinzan, **7**:20a
Tannery, Jules (1848–1910), 13:249–251; **2**:350a; **3**:456a; **4**:177a,b; **5**:262a; **8**:385a; **11**:207b*
Tannery, Marie, **13**:255a
Tannery, Paul (1843–1904), 13:251–256
biographies of, **8**:505a
critics of, **6**:417b(n); **13**:305b(n), 306b(n)
historiography, **1**:232b; **3**:380a*, 614b–615a, 615b; **4**:105a*, 429a,b, 430b, 463b, 488b, 491b*; **6**:382a, 408a, 410b, 416a–b; **7**:584a*; **9**:300a,b; **10**:7b*, 180a,b, 302b(n), 304a*, 344a, 529b; **11**:185b; **12**:315a*, 580b*; **14**:611a; **15**:204a
Diophantus, **4**:111a, 113a, 117b, 159b
Fermat, **4**:567b
influence of, **2**:358b; **7**:30a*; **12**:108a
students of, **8**:124b; **11**:388a
Tanning, **3**:601a; **10**:497a; **12**:286b; **14**:349b
artificial agents, **6**:166b
natural agents, **9**:576b; **15**:653a–b
Tannins, **5**:6a; **8**:339a; **14**:350a
composition of, **5**:4b; **12**:164a
extraction of, **8**:585b
Tansley, Arthur George, **2**:185a

Tanstetter Collimitius, Georg, 2:537a; 7:118a, 120a; 15:475a, 476a, 478a
Tantalum, 6:166b; 11:541a; 13:468a
  discovery of, 4:343b–344a
  isotopes of, 12:505b
al-Ṭanṭāwī, Muḥammad ibn Muṣṭafā, 12:361a; 15:259b
Tanton, Thomas Leslie, 1:394a
T'ao Hung-ching, 8:396a*
Taoism, 12:385b, 386a, 387a,b; 14:163a
Tapes, magnetic, 4:7a
Tapeworms, 7:414b; 12:577b; 13:527b; 14:2a
  treatment for, 8:623a*
  see also Cysticerci, Taenias
Tapir, 11:262a
Tappeiner, Hermann Edler von, 7:423b
Tar
  beechwood, 11:359a–b
Tar water, 3:344b; 6:47b*
Tarde, Jean (*ca.* 1561–1636), 13:256–257
Tardenoisian (archaeol.), 10:607a
Tardigrada, 3:492b; 9:575b; 12:557b
  anabiosis, 3:588b
Tarentum
  Pythagoreans at, 1:231b
Targioni Tozzetti, Giovanni (1712–1783), 13:257–258; 9:368a
Tarim basin, 6:215b
Ibn Ṭāriq
  see Yaᶜqūb ibn Ṭāriq
Tarsier
  anatomy, 12:224a
  taxonomy of, 3:523a; 5:356a; 6:535b
Tarski, Alfred, 2:277b; 8:546a,b; 10:274a
Tartaglia (Tartalea, Tartaia), Niccolò (*ca.* 1499–1557), 13:258–262; 2:320a; 5:238a, 246b; 11:405b; 12:548b
  algebra, 1:368a; 3:65b; 4:586b–587b, 595b; 5:409b
  on Archimedes, 1:229a–b
  and Benedetti, 1:604b, 605a
  and Cardano, 3:64b–65a; 4:586b–587b; 13:259a
  critics of, 9:488a
  and Ferrari, 2:280a, 284a; 4:586b–587b; 13:259a–b
  General trattato di numeri et misure, 2:361a; 3:65a, 311b; 4:587b; 10:270b
  geometry, 4:596b; 9:157a; 10:338b(n)
  mechanics, 1:94b; 5:247a
  Quesiti et inventioni diverse, 1:607a; 2:280a; 3:65a; 4:587a, 595b; 7:174b; 9:487b
  Tartaglia's triangle, 13:259b
  translations by, 1:605b; 3:149b; 4:449b
Tartar, 10:305b, 309a,b
  solution of, 4:223b
Tartaric acid, 10:516b
  isolation of, 12:145b, 148a
  manufacture of, 3:202a; 8:93b
  optical properties, 2:139b; 10:357a–360b
  and racemic acid, 5:325a; 8:332a; 10:357a–360b
  reactions of, 4:76a; 8:339a–341a *passim*
  structure of, 1:319a, 320a
Tartrates, 13:580a
Taschenbuch für die gesammte Mineralogie
  founding of, 8:245b
Ṭāsh Köprü Zādeh, 7:366a, 338b; 13:514b
Tashiro, Shiro (Tashiro Shirosuke) (1883–1963), 13:262–263
Tashkent
  observatory, 13:138b
Tasman, Abel Janszoon, 3:396b; 13:583a
Tasmania
  botany, 6:489a
  paleobotany, 6:489b
Tassoni, Alessandro, 1:293a; 10:521b
Taste
  ancient Greek study, 4:33a
  seventeenth-century study, 1:593a
  brain center for, 1:580a
  and facial nerves, 2:29b
  neurophysiology of, 13:556b, 557a
  theories of, 8:102b; 9:63b; 11:173b
Taste buds, 6:224a; 13:590a
Tasyīr (prorogation), 7:79a, 335a
Tata Institute for Fundamental Research, 15:26a
al-Ṭaṭāwī, Muḥyī al-Dīn, 3:356a
Tate, John Torrence, 1:307a; 3:367a; 6:214b
Tatra Mountains, 15:502a
Tatum, John, 4:528b
Tauber, Alfred (1866–*ca.* 1942), 13:263–264
  Abel-Tauber theorem, 6:113b; 13:263b
  Tauberian theorems, 14:345b–346a
Taurine, 5:431b; 9:284b
Taurinus, Franz Adolf (1794–1874), 13:264–265; 12:255a
Taurus (constellation)
  nebulae, 6:402b; 11:553b*
Tautochrone, 5:54b; 7:109b, 446a, 560b; 11:207a
Tautology, 5:351a; 14:468b
Tautomerism, 4:553b; 6:108b; 11:534b
  amino-imino, 3:246b
  concept of, 1:390b; 2:624b
  intraannular, 13:388b
  keto-enol, 1:341a; 3:286b, 347b; 5:5b, 381b; 9:354a; 13:388b
Tavel, Franz von, 2:437b
Taverner, Leonard, 3:87a
al-Tawḥīdī, 15:250a,b, 263b
Taxation
  reform of, 13:593a–b
Taxidermy, 14:187a
Taxodiaceae, 8:354b
Taxonomy
  circular system, 3:192b
  and idealism, 1:70b–71a; 5:192a, 443a
  "natural," 4:512b; 6:8a,b; 10:9a
  philosophy of, 2:109b–110a; 4:512a–b; 5:191b–192a; 11:521a; 12:578b*
  see also Classification; Species, biological
Taxonomy, botanical
  medieval Latin, 1:101b
  Renaissance, 2:218b, 219b; 3:413a, 414b; 4:139a; 5:379a–b; 8:435b; 12:135a; 14:586a; 15:81a
  seventeenth-century, 1:369a–b, 524a; 3:180a; 5:535b; 7:194b; 9:18a, 528b–529a; 11:314a–316a; 13:29b–30a, 442b–443b
  early eighteenth-century, 1:306a; 2:225a; 3:344a, 351a, 535a; 4:98b–99b; 6:65b–66a, 232a*; 7:197a–b, 199a–b; 8:460a; 9:368a–b, 539b; 11:83a; 13:178a
  late eighteenth-century, 1:46a, 58b; 5:74b, 217a, 443a; 6:219b, 400b–401a, 539a; 7:198a–b, 585a–b, 586a; 9:254a, 390a, 434a; 12:471b; 13:392b, 425a–b; 14:463b
  early nineteenth-century, 1:69a–b, 329a; 2:517a–521a; 3:44a; 5:191a–192b; 6:493b; 7:610a–b; 8:371b, 373b; 9:456a; 10:13a–b, 163b, 164b, 531a–b; 11:262b; 13:354b, 506b, 596a; 14:386b–387a; 15:82b, 267a–b
  mid nineteenth-century, 1:70b, 71a, 358b–359a, 615a; 2:19a, 491b–492b; 3:42b, 612a–b; 4:306b, 560a; 5:512a–513a, 539b, 546a–b; 6:296a, 465a, 489a, 491b; 7:165b–166b, 196b, 320a, 498b; 8:372a–b; 9:340b; 11:154a; 13:432b; 14:33b; 15:130a,b
  late nineteenth-century, 1:412b, 422a, 615a; 2:103a–b, 185a, 477a; 3:338b–339b, 613b; 4:401b; 6:100a, 490a; 7:395b; 8:354b, 512b; 9:587a; 10:1a–b, 130a, 253b; 13:154a, 405b, 406a; 14:181b, 192b; 15:147a–148b, 513b
  early twentieth-century, 1:422b; 2:371a–b, 583a; 3:318a; 4:583b; 6:101a; 7:90b–91a, 102b, 535a–b; 8:513a–b; 12:125b; 15:149b–150b, 224a–b, 507a, 510a–b
  anatomy and, 13:405b
  artificial, 1:369a–b
  and chemistry, 8:60a; 15:68a
  Darwinism and, 3:338a
  ecological, 15:510a–b
  experimental, 7:166a–b; 8:513a–b
  historiography of, 2:160a
  Linnaean, 1:143a, 262a, 358b; 3:154b*, 177a, 344a, 351a, 535a; 4:295a; 5:328a, 378b, 380a; 6:539a; 7:585a–b; 8:375a–378b; 9:433a, 456a; 11:605b; 12:515b; 13:178a; 14:116b
    overthrow of, 1:520b; 5:512b
    reform of, 3:579b
  methodology, 8:377a–b; 15:268b
  microchemical methods, 15:150a
  natural
    seventeenth-century, 11:314b; 13:443b
    early eighteenth-century, 3:344a; 7:197a, 199b; 8:377a
    late eighteenth-century, 1:58b; 5:217a; 7:198b, 585a,b, 586a
    early nineteenth-century, 1:69b, 358b–359a; 6:296a; 11:262b
    mid nineteenth-century, 2:246b; 5:512b; 6:491b; 7:498b; 8:372a–b; 13:432b
  photosynthesis, as criterion, 7:12b
  statistical methods, 1:59a
  type specimens, 15:224a
  see also individual plants and groups of plants
Taxonomy, microbiological, 7:406a–b; 10:364a, 389b–390a
Taxonomy, paleontological
  eighteenth-century, 14:119b, 502a–b; 15:54b–55a
  early nineteenth-century, 10:583b; 12:182a–b; 13:44a
  mid nineteenth-century, 2:109b–110a; 10:212a; 11:386a–b
  late nineteenth-century, 7:252a; 9:579a; 10:29b, 429b; 12:264b; 13:531b; 14:313a, 626a–b
  early twentieth-century, 3:519b–520b; 12:337a; 13:345b; 14:410a
  philosophy of, 14:626b
Taxonomy, zoological
  ancient Greek, 1:259b, 262a–b; 12:576a
  medieval, 1:102a; 5:147b; 7:64b
  Renaissance, 1:596a; 5:379b
  seventeenth-century, 7:459a; 8:415a–b; 11:316b–317a
  early eighteenth-century, 1:306a; 7:401a; 8:378b–379a; 11:331b–332a, 503a; 13:174a; 15:53b–54a, 111a
  late eighteenth-century, 1:58b, 139a; 2:4b, 474a, 509b; 3:523a; 4:512a–b; 5:355a; 7:586a, 588b; 9:575a–b; 10:283a, 284a, 437a, 509b; 11:67a; 13:425a–b; 15:112a–b
  early nineteenth-century, 1:331b, 387a; 2:187b, 493b; 3:521a, 525a; 4:308a; 5:31b, 356a; 6:120a–b; 7:547b, 588b,

592a; **8**:48b–49a; **10**:13a; **11**:592b; **12**:133b; **13**:355b, 402a; **15**:126b
  mid nineteenth-century, **1**:405b, 502a, 504a; **2**:282a, 422b, 498a; **3**:89a, 378a, 550a–b, 552b–553a; **4**:234b–235a, 235b, 290a–291a; **5**:359a–b, 480b; **6**:454a, 589b–591a, 592b–595a; **7**:546a; **8**:270a–b, 519a; **9**:286a, 572b–573b; **10**:221b, 261a–263a, 439b–440a; **11**:234b; **12**:231a, 232a, 421b, 550b–551a; **13**:9b, 168a, 554a–b; **15**:78b, 227a–b
  late nineteenth-century, **3**:438b, 536a–b; **5**:46a–b, 73b–74a, 109a–b, 385b, 406b, 555b; **6**:8b, 167a–b, 336b, 507a; **7**:98b, 169b–170a, 475a; **9**:313b, 496a; **10**:273a,b, 522b; **11**:443b; **12**:264b; **13**:8b–9a, 551b–552a; **14**:1b–2a, 410b
  early twentieth-century, **1**:507a; **2**:504b–505a; **3**:17a, 519b–520b; **4**:42a, 142b, 207a, 310a, 494a–b; **5**:467b; **6**:524b; **7**:286a; **9**:328a–b; **12**:453a; **13**:134b, 286b; **14**:203b
  mid twentieth-century, **13**:138a
  binary (dichotomous) system, **5**:31b
  circular system, **13**:168a
  and embryology, **1**:603a; **7**:475a; **11**:234b
  and genetics, **4**:308a
  Linnaean, **2**:204a; **9**:575a–b
  morphological approach, **8**:270b
  "natural," **4**:308a; **7**:588b; **8**:48b–49b
  philosophy of, **12**:576a
  quinary system, **13**:168a
  resistance to, **2**:580a–b
  terminology, **1**:306a; **4**:512a
  transitional types, **15**:112b
  see also individual animals and groups of animals
al-Tayfūrī (Zakariyyā ibn ᶜAbd Allāh), **15**:236a, 238b
Taylor, Alfred Swaine, **2**:420b*; **10**:177a
Taylor, Brook (1685–1731), 13:265–268
  acoustics, **4**:88b
  acquaintances of, **5**:516a; **6**:174b(n)
  analysis, **3**:282b; **14**:180a
  and Bernoulli, **2**:48b, 57b; **13**:265b, 266b, 267a
  correspondence of, **9**:500a
  influence of, **2**:418a; **3**:460a; **5**:567a; **6**:173a; **7**:561b
  Madras Catalogue, **4**:176a
  Methodus incrementorum, **2**:53b, 54a
  Taylor's expansion, **2**:275a, 276b; **5**:521a; **8**:140b, 366b; **10**:573b; **13**:265b–266b; **15**:198b, 384b
  Taylor's series, **3**:136a, 223a, 229a; **4**:477b, 547b, 548a; **5**:201a, 204a,b, 528b; **6**:3b; **7**:570b; **8**:367a; **13**:266a
  Taylor's theorem, **2**:275a; **10**:51a, 442a
    generalization of, **11**:498b
Taylor, Charles Vincent (1885–1946), 13:268–269
Taylor, Frank Bursley (1860–1938), 13:269–271; 5:487a; **14**:216a
Taylor, Frederick Winslow (1856–1915), 13:271–272; 3:442b; **8**:119b; **12**:274b
  Taylor-White process, **10**:246b
Taylor, Harry Francis West, **5**:589b
Ibn al-Ṭayyib, Abu'l-Faraj, **2**:619a; **11**:445a
Tbilisi University, **11**:327a
Tchebycheff, Pafnuty Lvovich
  see Chebyshev, Pafnuty Lvovich
Tea, **7**:205a; **15**:430a
  as beverage, **10**:426b*
    effects of, **13**:282b
  cultivation of, **4**:518b; **13**:282b
Teall, Jethro Justinian Harris (1849–1924), 13:272–273; 2:288b; **5**:39a; **8**:34a*
Tears
  of blood, **6**:183b; **13**:262b
  lysozyme in, **5**:29b
Tebb, William, **14**:139a
Technetium, **10**:136b
Technogenesis, **4**:600a
Technology
  ancient Greek, **1**:213b–214b; **6**:311a–313a; **13**:331a
  ancient Roman, **15**:518a–520b
  medieval Arabic, **15**:265b
  medieval Latin, **6**:544a; **13**:327a–b, 348a–b
  Renaissance, **1**:97b; **8**:206b–214b; **9**:147a–b; **10**:23a–b
  early seventeenth-century, **1**:566b–567a; **4**:183b–185a; **14**:362b–363a, 366a–b
  late seventeenth-century, **6**:141b; **7**:578a–b; **9**:118a, 507b, 529b–530a; **10**:292b–293a; **11**:41a; **12**:211a; **13**:480a; **14**:369b
  early eighteenth-century, **3**:155b; **4**:469b, 471a; **11**:328a–331a; **13**:178b–179a
  late eighteenth-century, **1**:554b; **4**:71a; **6**:2a; **7**:89a, 490b, 581a; **8**:147b; **9**:342b–343b; **10**:108b–109a; **12**:286b; **13**:351a–352a, 413b; **15**:471b–472a
  early nineteenth-century, **3**:388b; **4**:41b; **11**:80a; **12**:317b; **14**:326b; **15**:472a
  mid nineteenth-century, **1**:201a; **2**:354b–355a, 438b–439a; **3**:224a, 288b; **4**:18b–19a; **5**:87a, 340a–341a; **10**:557b–558a; **11**:603b–604a; **13**:22b–23a; **15**:24a–25a
  late nineteenth-century, **2**:200a–b; **3**:224b, 299a; **4**:39b–40a, 283b–284b; **11**:384b–385a; **12**:574b–575a
  early twentieth-century, **3**:299a–b, 436b; **5**:407a; **8**:184b; **9**:458b–459a; **10**:236b; **12**:575a
  mid twentieth-century, **15**:8b
  chemical
    see Chemistry, technological
  Chinese, **12**:382a–383a, 384b; **15**:730b
  and fire, **15**:651a, 652b
  mechanical, **6**:368b
  Mesopotamian, **15**:646b–653b
  military
    ancient Greek, **6**:312b; **10**:587a–588b
    medieval, **13**:568a
    Renaissance, **13**:567b–568b
    seventeenth-century, **2**:305b; **13**:592b
    eighteenth-century, **5**:603a; **6**:1a; **9**:472b; **11**:472a; **13**:351a
    nineteenth-century, **1**:128b; **3**:388b; **5**:561a; **6**:5a,b; **9**:491b–492a
    twentieth-century, **3**:372b, 373a; **5**:433a; **6**:22b; **7**:247a, 465a; **8**:11b, 12a, 24b, 102b; **9**:386b
    see also Artillery; Ballistics; Engineering, military; Gunpowder; Missiles; Rockets; Weapons
  and progress, **9**:477a
  and society, **15**:133a, 136a, 139a, 406b–409a
  see also Education, technical; Electrotechnology; Engineering; Historiography, of technology; Industrial research; Instrumentation; Machines; Science(s), practical applications of; specific industries
Tecnomasio italiano
  founding of, **11**:95b
Tectology, **15**:38b–39a
Tectonics
  early nineteenth-century, **3**:395b; **4**:402b
  mid nineteenth-century, **1**:20a–b; **2**:108b; **4**:309a, 349b–350a; **5**:533a; **9**:585a, 620b; **10**:265b–266a; **11**:439a; **12**:167b–168a, 301a
  late nineteenth-century, **1**:20a–b, 163b, 174b; **2**:486b; **5**:476a–b, 548b; **6**:168a–b, 227a–228b; **7**:250b–251b; **8**:30b, 463a–464a; **9**:103b, 593b–594b; **10**:168a, 429b–430a; **12**:140a–b; **13**:145b–148b; **14**:402b
  early twentieth-century, **1**:236a–237a, 394a–b, 472b; **3**:325a–326a, 434b, 533a; **5**:221a–b; **6**:228b–229b; **7**:267a, 474a; **8**:543b–544b; **10**:168a–169a, 507a–508b; **11**:575b; **13**:63b–64b, 283a–285b
  mid twentieth-century, **2**:558b
  comparative, **7**:474a
  fracture lines, **7**:474a
  granite, **3**:325a–b
  historical, **13**:64a–b
  historiography of, **1**:395a
  magmatic, **3**:325b
  mapping, **1**:236b, 237a
  and paleogeography, **6**:168b; **7**:251a–b
  and sedimentation, **6**:168a; **10**:507a
  theoretical, **10**:430a
  see also Geology, structural; Mountains
Teesdale, Christopher, **3**:399b
Teeth
  abrasion of, **11**:159a
  canine, **10**:16a–b
  cetacean, **3**:184b
  Czermak spaces, **3**:530b
  dentition, **4**:487a
    primary, **4**:520a
    secondary, **4**:520a
  enamel
    analysis of, **6**:166b; **9**:527a
    development of, **11**:514b
  evolution of, **7**:480a
  fossil, **1**:482a, 491b; **7**:252a–b; **9**:527a; **10**:16a–b, 261b; **13**:33b–34a
  and horns, compensation principle, **5**:443a–b
  human, **6**:567a
  ontogeny of, **2**:255b; **4**:520a; **5**:469b; **7**:61b; **9**:253a; **12**:315b
  pathology of, **11**:596a
  sensitivity, **4**:487a
  structure of, **3**:530b; **4**:487a, 522b; **10**:261b; **11**:216b, 380b
  as taxonomic tool, **3**:521a
  transplantation of, **6**:567a
  used as medicine, **1**:482a
Teichmann, Ludwik Karol (1823–1895), 13:273–274
Teilhard de Chardin, Pierre (1881–1955), 13:274–277; 2:171b; **3**:606b; **8**:257a
Teit, James, **2**:212a–b
Tekhnologichesky zhurnal
  founding of, **12**:329b
Tektites
  origin of, **12**:572b–573a
Tela cellulosa, **6**:62a–b
Telecardiography, **4**:334a
Telecommunication, **9**:486a
Telefunkengesellschaft, **1**:24a
Telegraphy
  electric, **1**:203a, 460a; **2**:439a; **3**:529a; **6**:223a, 280b; **7**:56a; **9**:166b; **10**:519a; **11**:213a; **12**:425a–b, 510b; **13**:22b; **14**:204b, 290a
    anticipation of, **14**:80b
    Atlantic cable, **3**:535b–536a; **4**:248b, 500a, 559a–b; **5**:479b; **12**:424b, 425b; **13**:386b–387a; **14**:145b

Telegraphy (cont.)
electric (cont.)
effect of sun, **4**:248b
grounded circuits, **4**:248b
indicator, **12**:425a
inductance effects, **6**:211b
invention of, **5**:305b, 309b; **7**:45a–b
retardation effects, **3**:288b
telegraphers' equation, **6**:212a; **7**:381a; **10**:594b
theory of, **9**:210b
in geophysics, **14**:204b
Japanese, **15**:744a
in length measurement, **6**:240b
in longitude measurement, **1**:364a; **5**:479b
mathematical checking system, **6**:402a*
multiplex, **4**:194b, 284a; **11**:579a
optical, **1**:138a; **2**:104a; **8**:64b; **10**:588b
fire signals, **8**:191a
printing, **4**:248b
in seismology, **8**:388a
stock tickers, **4**:283b, 284a
submarine, **3**:288b; **7**:94a*, 288a; **9**:196a; **10**:605a; **14**:290a
see also Telegraphy, electric, Atlantic cable
weather service, **4**:601a; **5**:17b
wireless, **2**:427b, 487b; **4**:189b; **5**:33a; **8**: 40b; **9**:98b–99a; **11**:93b–94a, 420b, 461a; **15**:46b–47a
prediction of, **13**:287a
see also Radio
Telekino, **13**:431b
Telemeter, **2**:327b; **11**:95a
Telemetry, **6**:223a
astronomical, **1**:136a
Teleology
ancient Greek, **1**:149b, 255b, 259a–261a, 315a; **2**:234a; **4**:106a; **8**:537b; **11**:29b; **13**:329b, 332b
ancient Roman, **5**:230b, 232b; **8**:538a
medieval, **11**:230b; **12**:82a
Renaissance, **4**:508a,b; **11**:528a, 604b; **13**:587a; **14**:8b
seventeenth-century, **4**:510a; **7**:303a; **8**: 156a; **9**:509b; **12**:333a; **14**:286b
eighteenth-century, **4**:471b; **5**:443b; **6**: 581b, 586b; **7**:232b–233a, 368a, 440b; **10**:120b, 278a–b; **11**:43a–b, 616a; **12**:601a–604a; **14**:484a; **15**:54a
nineteenth-century, **1**:388b; **2**:290a, 426b; **3**:553a; **5**:447a; **7**:366a; **8**:514b, 515a; **9**:601b; **10**:273b, 427b, 580b; **11**: 173a; **12**:224b, 242b–243a, 278a; **14**:44b, 529b; **15**:408b
twentieth-century, **6**:261a–b; **11**:430b–431a; **12**:250b; **13**:275b–276b
and anatomy, **5**:624a
critics of, **13**:92a; **15**:139b
immanent, **1**:255b
and physiology, **1**:479b; **2**:240a; **5**:230b, 232b
Teleosaurus, **3**:526b; **5**:357b
Teleost
embryogeny, **10**:105a
Telephone, **13**:103b
automatic dialing, **4**:7a
invention of, **1**:582b; **4**:248a–b; **13**:357a
loudspeaking receiver, **4**:284a
Telephony, **1**:292b–293a; **2**:439a; **3**:207a, 298b; **4**:248a–b; **7**:101b; **10**:604b–605a
biophysical application of, **1**:303b
carrier, **7**:108a
inductance coil, **11**:213a
interference, **4**:588b
transatlantic, **7**:108a
transcontinental, **4**:7a; **7**:107b–108a
Telephotography, **2**:559b; **7**:108a
Telescope
seventeenth-century, **3**:101b, 102b, 248b*, 262a; **6**:126a, 361a–b, 363a–b, 487a, 520b; **9**:485b; **12**:105b, 151b; **14**:451b, 637a
eighteenth-century, **9**:330a; **14**:485a, 589a
nineteenth-century, **12**:289b
twentieth-century, **9**:46b; **10**:158b
antecedents of, **5**:106a
construction, **2**:503a, 541a
card stops, **6**:361b
cross hairs, **5**:278b; **11**:471b; **14**:415b*
diaphragms, **3**:116a; **7**:543b
eyepiece, **3**:23a; **5**:542b; **6**:361b; **11**:395a
floating collimators, **7**:263a
focal point, **1**:341a–b
housings, **6**:363a–b
lenses, **13**:437b
levels, **9**:485b
micrometer, **5**:278b; **6**:333b–334a
mounting, **6**:361a–b, 603a–b; **11**:284b
objective, **11**:95b, 379a
reticle, **9**:485b
sights, **1**:341a,b
terminology, **11**:395a
invention of, **2**:305b; **5**:241a; **7**:73a–b; **9**:336a–b; **11**:97b–98a
sizes
forty-foot, **6**:322b, 327a, 329b, 334b
15-inch, **2**:284b, 285b; **13**:109b
18½-inch, **3**:288a
25-inch, **10**:31b
26-inch, **6**:49a
30-inch, **10**:528b
36-inch, **3**:366a–b; **6**:28a; **7**:270b
40-inch, **1**:55a; **3**:288a
60-inch, **1**:56a; **3**:366a; **6**:29a, 30a; **11**:470a
72-inch, **2**:424a; **10**:328b
82-inch, **13**:116b
100-inch, **1**:55a, 56b; **6**:30a–b, 529a; **11**:470a
200-inch, **1**:55a, 353a; **6**:32a–b, 532b; **10**:473a
theory of, **4**:158b; **6**:603b; **7**:299a
types
achromatic, **2**:216b, 328b, 451b; **3**:284a; **4**:148b–149a, 482a; **5**:143b; **6**:119a; **7**:87b*; **9**:330a; **11**:285a; **13**:22b
astrographic, **4**:26b; **5**:315b, 316a, 405b
Bruce, **1**:398a, 466b
Campani, **2**:310b; **3**:23a
Cassegrain, **3**:97b–98a; **5**:525b, 561a, 562a; **7**:263a; **8**:582a
Clark, **3**:509a
clockwork, **4**:180b
collimating, **11**:471b
Cooke, **4**:26b, 27a
Crossley, **3**:327a; **7**:271a
double-image, **7**:87a
equatorial, **1**:19a; **2**:147b; **3**:288a, 605b; **5**:144a, 561a, 562a; **6**:282a, 523b*; **7**:209b; **8**:46a; **9**:607b*
floating zenith, **4**:270a
Fraunhofer, **1**:568b; **3**:238b; **7**:608a; **9**:2a
Galilean, **5**:241a
Gregorian, **7**:263a; **9**:330a
Hale, **1**:55a
long-focus, **4**:128a; **14**:449a
meridian, **7**:239a; **9**:551b
Merz, **12**:161a, 267a, 269a
Northumberland equatorial, **3**:187a
photographic, **5**:561a, 562a; **6**:282a; **10**:600b
photoheliographic, **4**:19a
reflecting, **1**:55b, 56a; **2**:141a, 216b; **3**:152b, 284a, 366a; **4**:27a, 45b–46a*, 148b, 179a,b; **5**:87a, 525b, 529a, 561b, 562a; **6**:5b–6a, 176a, 324b, 328b–330a, 335b, 508a; **8**:170b; **9**: 464a, 615b; **10**:54a, 56b, 60a, 328b; **11**:470a; **12**:186a–b, 226a, 413a; **14**:637a
refracting, **1**:55a, 465a, 466a, 568b; **2**: 284b, 285b; **3**:97b–98a, 261b, 284a, 288a, 509a, 605b; **4**:148b; **5**:144a, 315b, 316a; **12**:413b
Schmidt, **1**:352b; **3**:63a; **5**:183a, 184a
Snow, **1**:55b, 56a; **6**:29a–b
solar, **1**:55b; **5**:182b; **6**:29a; **10**:31b
stereogonic, **11**:95a
tower, **1**:55b; **5**:182b; **6**:29b
wide-angle, **1**:352b
zenith, **2**:332b; **4**:555a; **8**:388a; **10**:159a, 231a
see also Instrumentation, astronomical; Radiotelescope
Telesio, Bernardino (1509–1588), 13:277–280; **1**:447b; **9**:316b
philosophy, **12**:333a; **13**:438a; **15**:69b
De rerum natura, **15**:68b, 69a
Teletypewriter, **6**:223a
Teleutospores, **3**:613a,b; **4**:220a; **13**:489b
Television, **1**:287b; **2**:427b, 428a; **5**:33a; **7**: 36b
color, **8**:96a
Telford, Thomas, **1**:459b; **11**:291a, 496b; **12**:131b, 132a
Teller, Edward, **4**:313a, 579a, 582b; **5**:118a, 273a; **8**:95b, 96a; **10**:215a; **15**:43a
Gamow-Teller selection rule, **5**:272b
Telluric acid
determination of, **11**:551a
Telluric rays, **7**:74b
Telluric screw, **1**:572a–b
Tellurium, **5**:595a,b; **9**:18b
atomic weight, **2**:428b
determination of, **11**:551a
discovery of, **7**:391a, 394b; **9**:559a
naming of, **9**:559a–b
occurrence of, **7**:45a
reactions of, **5**:7b
vapor density, **13**:468a
Tellus, **11**:558a
Telome theory, **8**:354b
Telpherage, **7**:93b
Temperament (music), **1**:282b, 606b; **5**: 450b*; **12**:127b; **14**:152b
of fretted instruments, **3**:154a; **5**:249b
Temperaments, medical theory of, **10**:417b, 434b; **15**:240a, 498b
Temperance movement, **6**:519a
Temperature
atmospheric, **3**:261b*; **4**:175a; **5**:428b; **10**:585a
and altitude, **5**:187b; **8**:262a; **9**:266b, 444b; **15**:355a–b
determination of without thermometer, **5**:199a
and disease, **3**:436a
and life, **4**:285b; **12**:328b
concept of, **2**:169b
critical, **1**:161a; **5**:462a; **6**:470a; **9**:287a
and density, **7**:346a
distribution of, **4**:126a, 222b; **13**:75a
and electrical resistance, **1**:612b; **3**:19b, 312b; **4**:80b; **7**:221b
vs. heat, **2**:177b–178a; **3**:304a, 308b; **9**: 132b–133a; **14**:155b
of mixtures, **11**:432a–b
physical interpretation of, **4**:314a
thermodynamic scale
see Absolute temperature

universal function of, **3**:305a–b
see also Absolute temperature, Absolute zero, Body temperature, Heat, Pyrometry, Thermometry
Tempier, Étienne, **1**:378b, 382b; **2**:605a; **5**: 401b, 402b; **6**:274a; **12**:428b
Temple, William, **8**:309b
Temple Coffee House Botany Club, **1**: 432a
Temporalism, **8**:517a
Tencin, Claudine Alexandrine Guérin de, **1**:323a; **3**:380b; **5**:58a
Tendons
repair of, **14**:322a
structure of, **2**:165a
Tenerife, Peak of, **12**:498b–499a
Tengen jutsu, **3**:254a; **8**:314b
Tengnagel von Camp, Franz Gansneb, **2**: 401a, 412b, 413b; **7**:297a
Tennant, Charles, **3**:289a
Tennant, Smithson (1761–1815), 13:280–281; 14:486a
chemistry, **3**:346b, 603a; **5**:92a; **10**:446a; **14**:486b–487a
Tennent, David Hilt (1873–1941), 13:281–282
Tenner, Karl Ivanovich, **13**:111a
Tennessee (river), **7**:144b
Tennessee (state)
geology, **13**:467a
Tenon, Jacques René, **3**:2b, 436b*, 440b; **5**:251b
Ten Rhyne, Willem (1647–1700), 13:282–283
Tension (phys.), **3**:441b–442a
coefficient of, **3**:442a
electrical, **7**:380b; **10**:188b, 189b, 190b, 191b, 192a; **14**:70b, 72b, 73a, 79b
fundamental unit, **14**:73b–74a
in strings, **8**:219b–222a
Tensor, **4**:280b, 337a
classes of, **9**:215b
concept of, **3**:146a
curvature, **3**:263b; **4**:328a
energy-momentum, **7**:26b
gravitation, **4**:327b
naming of, **14**:62a
Ricci's tensor, **11**:407b
Riemann-Christoffel, **3**:263b
strain, **3**:146a
stress, **3**:146a
Tensor analysis, **4**:577b; **7**:207b; **12**:214a–b
Tentacularia, **2**:322a
Tenzan zyutu, **12**:291a,b
Teratogenesis
experimental, **1**:153a; **3**:184b; **8**:254b; **13**:458a, 555b
theories of, **6**:62b–63a; **11**:159a
Teratology
Renaissance, **1**:109b
seventeenth-century, **1**:523b; **9**:66a; **11**: 41a; **13**:527a
eighteenth-century, **6**:62b–63a; **8**:171b; **14**:450b; **15**:524b
nineteenth-century, **3**:45a, 184b, 259a, 588b; **4**:267a; **5**:357a, 358b, 359a; **8**:254b, 269b; **9**:252a, 253a
botanical, **2**:286b; **9**:24b; **13**:506b; **14**:99a, 350a; **15**:130b
naming of, **5**:359a
theories of, **8**:171b
Terbium, **3**:478b
discovery of, **2**:254b, 255a; **9**:541b
Teredinidae
geographic distribution of, **9**:459a
Teredos, **6**:138a
Természettudományi Társulat, **4**:379a
Termier, Pierre (1859–1930), 13:283–286; 1:235b; **8**:544a; **9**:58b; **11**:375a; **13**: 146a, 149a*
geology, **5**:187a; **6**:168b
Terminists, Paris, **1**:198a, 199b; **2**:590b, 604a–b; **3**:172a, 420b; **4**:237b; **10**: 419b, 420b; **12**:547b, 548a; **13**:349a; **14**:391a
Terminology
anatomical
ancient Greek, **6**:317a–b
ancient Roman, **12**:540a–b
medieval Latin, **9**:468b; **12**:81b, 82b
Renaissance, **1**:620a–b; **4**:511a; **9**:165b; **14**:4b, 7b–8a
seventeenth-century, **5**:484b; **13**:32b, 33a
eighteenth-century, **12**:510a; **14**:450b
Mesopotamian, **15**:642b, 643a
Aristotelian, **1**:250b–262b *passim,* 277a–279b
astronomical, **1**:339a, 511b; **2**:349a; **4**: 429b; **5**:614b; **15**:475b, 643b
biochemical, **2**:86b, 95b
biological
eighteenth-century, **1**:306a
early nineteenth-century, **1**:386b; **2**: 123a; **7**:589a; **9**:496b; **10**:287a
mid nineteenth-century, **1**:540a; **2**: 127b, 160a; **6**:136b; **8**:270a–b; **9**:442a; **11**:491b; **13**:356a
late nineteenth-century, **5**:35a–b, 386a; **8**:16a, 27a; **14**:100b, 126b
twentieth-century, **7**:114a
botanical, **2**:160a; **5**:217a, 535b–536a; **8**: 377b; **12**:100a; **13**:88b, 332b; **15**: 162a*–b*
chemical
medieval, **6**:80a
seventeenth-century, **6**:257a; **12**:312b
eighteenth-century, **4**:343b; **6**:46b(n)
early nineteenth-century, **1**:345a, 346b; **5**:323b, 325a,b, 431b
mid nineteenth-century, **2**:65b, 66a, 69a, 95b; **4**:554b; **5**:126a, 494b, 536b; **7**:281a; **10**:178a
late nineteenth-century, **1**:298b; **5**:489a; **15**:460a
twentieth-century, **2**:429a; **8**:32a; **12**: 404b; **13**:394a
crystallographic, **1**:463a*; **10**:125a
geographical
medieval Arabic, **9**:89a
nineteenth-century, **12**:300b
geological
eighteenth-century, **11**:35b; **13**:410a
nineteenth-century, **7**:45a, 474a; **8**: 596b, 615b; **9**:61a, 103b; **10**:258b; **13**:146b–148a; **14**:292b
twentieth-century, **1**:285b–286a, 394b, 551b–552a; **2**:549a; **4**:598a; **8**:544a; **9**:257b; **10**:505a; **13**:64a
mathematical
ancient Greek, **1**:180b, 183b–184a, 185a, 215a, 245a; **4**:421a; **5**:346b; **8**:189b; **9**:269a, 270b; **13**:305b(n); **15**:503b
medieval, **1**:511b–512a; **4**:609a,b, 611b; **5**:612a–b; **7**:359b; **10**:224a–b
Renaissance, **2**:618b; **3**:272b, 273a, 277a; **4**:619a
seventeenth-century, **2**:47b; **5**:409a, 526b, 593b; **9**:12b–13a, 303b
eighteenth-century, **5**:450b(n); **15**:99a
nineteenth-century, **1**:238a,b; **3**:142b, 164b; **5**:204b, 368a; **6**:92a; **7**:444b; **9**:215b, 307a–b; **13**:220b–221a; **15**: 65b, 66b
twentieth-century, **1**:307a–b; **2**:352a; **6**:5a; **7**:408a
Chinese, **8**:315b–319a *passim*
Mesopotamian, **15**:668b
medical, **1**:11a, 323a; **2**:443a; **6**:60a, 318a; **11**:254a; **15**:645a–b
mineralogical, **2**:155a; **11**:539b; **12**:329b
musical, **1**:282a–b; **9**:141a
neurological, **12**:395b–396a
optical, **1**:7b, 8b; **10**:89b(n); **11**:395a
pharmaceutical, **2**:155a, 547a; **13**:38b
philosophical, **2**:163b; **7**:485a; **15**:261a–b, 263a
physics
ancient Greek, **1**:253a, 277a–279b
medieval Latin, **13**:190a
Renaissance, **5**:397b, 398b; **8**:215a
seventeenth-century, **2**:523a
eighteenth-century, **5**:129a; **12**:128b
nineteenth-century, **3**:417a–b, 499a; **6**:212a; **9**:201a, 236b; **11**:293a; **14**: 62a, 292b, 522b(n)
twentieth-century, **3**:501a; **8**:148b; **15**: 491a
Japanese, **12**:407b
physiological, **6**:59b; **12**:356a; **15**:106b–107a
statistical, **5**:9a; **10**:450a, 451b, 465b(n), 466b(n)
see also Dictionaries, Nomenclature
Termite, **5**:502b; **7**:447b
parasites of, **4**:207a
Terpenes, **5**:541a; **6**:185a,b; **13**:406b
analysis of, **14**:142a
chemistry of, **9**:608b
classification of, **13**:550a
derivatives of, **13**:410b
discovery of, **5**:412b
reactions of, **1**:288b; **13**:550a
structure of, **4**:91b; **10**:517b; **13**:410b, 549b–550a; **14**:142a
synthesis of, **1**:106a–b; **3**:271b
thermal decomposition of, **13**:410b
Terquem, Olry (geol., 1797–1887), **12**:183a
Terquem, Olry (math., 1782–1862), **13**:129b
Terra, Helmut de, **13**:275b
Terra Nova (ship), **9**:124a
Terra pinguis, **1**:549b–550a
Terraces (geol.), **6**:438a; **9**:150b; **13**:147b
marine, **7**:144a
Quaternary, **4**:39a
Terrella, **5**:398a,b, 399a
Terrestrial magnetism
Renaissance study, **13**:49b
seventeenth-century study, **2**:462a; **6**:71b
eighteenth-century study, **2**:389a; **3**:436a; **6**:234a; **14**:354b
early nineteenth-century study, **1**:460a; **4**:256a; **5**:78b–79a, 187b, 318a–b, 319a; **6**:106b–107a, 550a,b, 551a, 552a,b
mid nineteenth-century study, **1**:166a, 364a; **2**:430b, 431a, 472a; **3**:260a; **4**:174b, 400b; **5**:305a–306a; **6**:119a, 277b–278a, 279a; **7**:608b–609b; **8**: 426b, 487a–b; **9**:430a; **11**:111b, 554b–555a; **12**:49a–52a, 209b–210a; **14**: 204a–205a, 206a–b
late nineteenth-century study, **1**:521b–522a; **3**:261a; **6**:470a; **12**:238b; **13**: 52a–b, 413a
twentieth-century study, **3**:204a, 565a*; **4**:270a, 577a; **5**:161b, 187b; **7**:413b; **8**:102b; **12**:187a; **13**:166b, 175b–176a
and aurora borealis, **3**:260a; **5**:258b; **6**: 277b, 553b; **8**:487b
and eclipses, **1**:521b
magnetic equator, **4**:256a; **7**:74a

Terrestrial magnetism (cont.)
magnetic storms, **5**:306a
mapping of, **1**:521b; **4**:492a; **6**:107a; **7**:413b, 480b, 609a; **8**:426b; **9**:155a; **11**:412b, 554b; **12**:187a; **13**:517b; **14**:206b, 354b
measurement of, **6**:107a; **7**:450b; **11**:150b
observation of, **5**:258b, 305b–306a; **6**:327a; **7**:577b; **8**:187a, 426b, 487a–b; **12**:50a; **14**:452b–453a
and solar rotation, **1**:478a; **9**:184a
and statistics, **1**:478a
and sunspots, **4**:354b; **11**:412b; **12**:51a–b, 210a; **13**:52a; **14**:480b
theories of, **3**:260a; **5**:397b–401b; **6**:107a, 144a; **7**:609a; **8**:107b; **12**:49b; **13**:156a
mathematical, **1**:521b; **5**:306a; **7**:595a; **10**:213a*
variations in, **1**:478a, 521b; **4**:256a; **7**:609a; **8**:426b; **12**:50a
altitude effects, **2**:135a; **5**:318a–b
diurnal, **3**:260a; **5**:79a, 319a; **7**:595a; **12**:51b, 238b; **13**:184a
Terrestrial Magnetism
founding of, **1**:521b–522a
Tertian fever, **9**:95a
Tertiary period, **1**:131b; **2**:556b, 569a; **4**:265b, 401a; **10**:167b, 507a; **13**:147a, 274b
climate, **6**:221b
paleobotany, **12**:104b; **15**:147b
paleogeography, **7**:250b
paleontology, **1**:163a–b, 491b; **2**:109b, 110a, 491b–492a; **3**:264a, 391b, 392a, 536a–b, 608a; **4**:39a, 234a, 518b; **5**:214b*, 295b–296a, 422a, 500b; **6**:8b, 221a–222a; **7**:43b, 91b, 92a; **8**:43b, 103b, 566b–571a; **9**:87a, 240b, 314a, 617b; **10**:29a–b, 32b, 221b, 222a; **11**:269b, 386a,b, 500a–b, 502a; **13**:145b, 275a; **15**:91b, 152a
rocks, **3**:161b; **5**:214b; **8**:463a, 566a
stratigraphy, **1**:131a, 210a, 394b; **2**:108b–109a, 495b, 497b; **3**:567b; **4**:68a–b; **5**:68a, 437a; **6**:187a; **8**:566b, 567b–568a, 569b, 570b–571a; **9**:240b; **10**:428a, 429a; **11**:130b–131a, 133b, 386a; **13**:620b; **14**:210b–211a
tectonics, **9**:367b; **10**:168b; **13**:271a
volcanic activity, **4**:548b, 549a; **5**:39a, 336a–b; **6**:116a; **11**:336b; **13**:283b
see also Eocene epoch, Miocene epoch, Neocene epoch, Oligocene epoch, Pliocene epoch,
Tertullian (Quintus Septimius Florens Tertullianus), **3**:238b; **4**:383a; **6**:405b; **7**:27b; **12**:80b, 539a, 540a,b
Tesla, Nikola (1856–1943), 13:286–287; 4:589a
Tesselation, **12**:533a–b
Tessier, Henri Alexandre, **2**:321b; **3**:523b; **5**:355b, 356a
Testacea, **5**:66b; **12**:560b
reproduction, **1**:265a; **4**:509a
Testicles
anatomy, **6**:65a
degeneration in, **2**:344b
descent of, **11**:565b
dual function of, **2**:344b
extracts of, **2**:525b; **10**:353b
histology, **8**:302b; **12**:319b
periodic sexual activity in, **2**:345b
transplantation of, **2**:73a; **9**:86a; **10**:570a
Testing, psychological, **14**:550a
intelligence, **2**:132a–b; **4**:272a; **5**:377b, 378b
Testosterone, **8**:302b
Testudo, **2**:322b
Testut, Jean Léo (1849–1925), 13:288; 3:90b
Tetanus (contraction), **14**:105b
induced, **9**:505b
negative fluctuation of injury current, **4**:202a
strychnine, **4**:201b
Tetanus (disease), **2**:529a
bacillus, **7**:391b
and respiration, **11**:425b
serotherapy, **1**:575a–b; **7**:391b–392a
treatment of, **14**:187b, 559a*
vaccination, **1**:577a; **8**:273a; **11**:271b
Tetanus toxin, **1**:11b; **4**:297b; **7**:392a; **11**:569b; **15**:522a
Tetany, **8**:295a
in infants
cause of, **4**:404a
parathyroid, **3**:352b; **13**:589b
Tetractinellida, **12**:520a
Tetractys, **11**:221a
Tetragonometry, **7**:599b
Tetrahedron, **5**:568a*
center of gravity, **8**:200a, 224b–225b
equality of, **6**:393b
Möbius, **9**:430b
volume of, **2**:270b, 271a; **8**:421b; **13**:259b
see also Pyramid
Tetrahedron
editors of, **7**:323a
Tetrahymena, **13**:268b
Tetraphenylmethane, **5**:464b
Tetrapods, **6**:594b
Tetrarhynchus, **1**:602b
Tetraspores, **13**:165a–b
Tetrathionate, **9**:296a
Tetrode, Otto, **13**:41b; **15**:444a
Tetrode (thermionics)
invention of, **6**:546b
Texas
geology, **11**:500a
paleontology, **14**:410a
Texas, University of
astronomy at, **13**:116b
Texas cattle fever, **7**:428a,b
etiology of, **12**:480a, 480b–481a
Texas tick fever, **11**:442b
Textile machinery, **1**:356a; **8**:213a
Textiles
bleaching of, **2**:77b; **6**:58b, 283a, 385b; **14**:197a
dyeing of, **2**:77b
manufacture of
Renaissance, **12**:317a
seventeenth-century, **1**:549a
eighteenth-century, **2**:292a; **4**:71a, 250a
late nineteenth-century, **3**:207a; **15**:24b
Mesopotamian, **15**:649b, 664a(n)
physics of, **1**:319a
printing of, **3**:400b*, 481a*, 488b; **4**:250a; **6**:385a; **10**:532a
structure of, **1**:319a–b
synthetic, **3**:207a
waterproof, **3**:289a
see also Sericulture; specific textiles
Teyler's Museum, **8**:489a; **9**:152a,b; **13**:183b
Teyler's Society, **9**:456b
Teysmannia, **13**:460a
Thābit ibn Qurra, al-Ṣābiʾ al-Ḥarrānī (836–901), 13:288–295; 1:507b
acquaintances of, **11**:244a
astronomy, **14**:593b
calendar, **1**:382b; **5**:553a
cosmology, **7**:124b, 142a; **12**:3a
tables, **1**:510b; **7**:338a; **14**:593a
trepidation, **2**:20b, 24a; **13**:324a
biographies of, **7**:187a
commentaries by, **11**:249b
critics of, **14**:390b
De expositione nominum Almagesti, **15**:179b
De figura alchata
translations of, **15**:177b
influence of, **3**:26b; **9**:21b; **15**:475b, 477a
influences on, **1**:508b; **14**:538b
Liber karastonis, **1**:224b, 228b–229a; **8**:215b, 216a, 222b
translations of, **15**:178a
mathematics
geometry, **6**:199b, 202a, 206a*; **7**:2b–3a, 329a; **8**:429b; **11**:240b; **13**:510b
trigonometry, **7**:38b; **11**:415a
On the Measurement of the . . . Parabola, **1**:445b
On the Measurement of the Parabolic Solids, **1**:445b
De motu accessionis et recessionis
translations of, **15**:175a, 179b
statics, **7**:341a, 342b, 345b
translations by, **1**:224a, 226a, 444a, 446a; **4**:438a–447b *passim;* **7**:25a; **9**:297a, 299b, 615a; **10**:300a; **11**:244a; **13**:320a, 540a; **15**:236b, 238b
translations of works, **1**:62a
Thackeray, William Grasett, **4**:269b
Thaddaeus Florentinus
see Alderotti, Taddeo
Thalazia, **2**:322a
Thalén, Tobias Robert, **1**:166b, 296b; **12**:42b
Thales (*ca.* 625-*ca.* 547), 13:295–298; 6:406a, 410b; **10**:208a; **14**:537a
astronomy, **1**:214a; **4**:463b; **6**:381b
geometry, **4**:463a
Thalès, **11**:388b
Thalline, **4**:297a
Thallium
atomic weight, **3**:476b
discovery of, **3**:476b
optical activity, **7**:518b
Thallophytes
classification of, **9**:368b
Thames River, **12**:545a
tunnel, **15**:151b
Than, Károly (1834–1908), 13:298–299; 4:378a; **14**:447b, 448b
Than's salt, **13**:299a
Thaumaturgy, **15**:202b
Thaw, William, **2**:423b; **8**:19b
Thaxter, Roland (1858–1932), 13:299–300; 4:220a
Thayer, William Sydney (1864–1932), 13:300–301; 8:447a
Theaetetus (*ca.* 417–369), 13:301–307; 4:414b, 415a, 423b; **10**:300a
acquaintances of, **11**:22b, 27b; **13**:314b
geometry, **4**:424a; **8**:192a,b; **11**:26b, 28a, 29b, 222b; **13**:317b
Theaters, anatomical
Renaissance, **1**:604a
Thebaine
discovery of, **10**:498a
Thege, Miklós von Konkoly
see Konkoly Thege, Miklós von
Theine, **9**:558b
Theis, Charles Vernon, **9**:258a
Theism, **3**:470b; **7**:231b–232a
Thelycke (chem.), **10**:183a, 184a
Thema mundi, **4**:622a
Themison of Laodicea, **1**:315a
Themistius (*ca.* 317-*ca.* 388), 13:307–309
Aristotelian paraphrases, **1**:269b, 271a, 272a, 276a
commentaries by, **7**:25a; **9**:437a, 438b; **14**:535b(n); **15**:176b
geometry, **1**:171b

influence of, **2**:229b, 230a, 232b, 619b; **11**:444b; **14**:582a
translations of works, **2**:283b; **13**:401b
Thenard, Louis Jacques (1777–1857), 13: 309–314; **1**:416a
acquaintances of, **3**:455a, 545b; **4**:238b; **9**:424b; **10**:511b; **11**:494b; **15**:347b
analytical method, **8**:331b
catalysis, **4**:241a–b
chemical classification, **3**:159a
and Gay-Lussac, **4**:239b; **5**:320a–322a, 324b; **13**:311a–b, 312a
influence of, **2**:356b; **3**:202a; **4**:234a
inorganic chemistry, **5**:92a
chlorine, **3**:602b; **6**:284b
elements, isolation of, **9**:451a
potassium, **3**:400a, 602b
organic chemistry, **7**:378b
photochemistry, **4**:182b
students of, **4**:77b; **5**:430a; **8**:330a; **10**: 532a; **12**:197a
Thenard's blue, **13**:310a
Traité de chimie élémentaire, théorique et pratique, **11**:171a
Theobromine
industrial production of, **5**:2b
Theodolite, **2**:133a, 372b; **5**:213a, 271b; **6**: 312b; **7**:609a; **9**:551b; **11**:291b; **14**: 216a
Theodore (Abū Qurra?), **1**:275b
Theodore of Gaza, **1**:274b; **9**:362b
Theodoric Borgognoni of Lucca
see Borgognoni of Lucca, Theodoric
Theodoric of Freiberg
see Dietrich von Freiberg
Theodorus, Master, **4**:605a,b, 610a, 611a; **5**:147b; **9**:362b
Theodorus of Cyrene (*ca.* 465-*ca.* 399), 13: 314–319; 15:503, **6**:410b; **11**:22b; **13**:301b, 302a, 304b
Theodosius of Bithynia (*b.* second half of second century B.C.), 13:319–321; **1**:22b, 41b, 224a, 381b; **13**:433b
astronomy, **10**:295b; **15**:205a(n)
editions of works, **3**:585a; **9**:190b; **13**:509b
De habitationibus
editions of, **9**:191a
translations of, **9**:190b; **15**:177b, 179b, 180a
manuscripts of, **14**:273a
Sphaerics, **4**:429b; **6**:203b(n), 271b; **9**: 298a, 364a
commentaries on, **1**:63b; **3**:312a
editions of, **1**:471a; **9**:555b; **11**:18a
translations of, **1**:23a; **7**:81a; **11**:32a,b, 244a; **13**:401b; **15**:175a, 177a, 628b
translations of works, **11**:245a
Theodotion, **14**:35b
Theolite, **1**:394b
Theology
ancient Greek, **11**:29b; **14**:536a–537a
ancient Roman, **1**:333b–337b; **2**:231a, 232b–233a; **12**:310b
Byzantine, **7**:134b–135b; **11**:161a–b, 183a–b
medieval Arabic, **1**:27a; **7**:40a, 41b, 64a; **9**:172a, 604b; **11**:249a–250a; **12**:1a, 7Ab–7Bb, 91a; **13**:510a, 513a; **15**:9b, 232a–b, 261b, 262b–263a, 494b
medieval Hebrew, **3**:469b–470a; **9**:28a–31b
medieval Latin, **1**:1a–2b, 91a–b, 197a–b, 289b, 290b; **2**:395b, 603b–605b *passim*, 609a; **4**:92a–b, 254b–255b, 396a–b; **5**:113b, 115a–b, 401b, 402b, 459a–550a; **6**:275a, 396a, 543b–544a; **7**:117b; **8**:548a–550a; **9**:136b, 363a–b; **10**:171b–174b, 223a, 474a; **12**:428b; **13**:340a–b, 347b–348a; **14**:35a, 388a–b
Renaissance, **1**:587b–590a; **2**:535b, 541a–542b; **3**:171b, 280a,b, 512b–516a *passim*, **4**:387a; **5**:119a–121b, 378b–379a; **6**:306a; **9**:609a–610a; **10**:245a, 310b; **11**:71b–74a, 402b, 561b; **12**: 151a, 322b–323a, 324a, 547b–548a; **13**:58b–59a, 59b, 491b, 501b; **14**:47b, 225a–226b
early seventeenth-century, **1**:133b–134b, 158b–160a, 480a–b; **2**:163a–164a, 222a–223b, 381b–382a; **3**:359b–362b; **4**:95b; **5**:588a, 606a–b; **6**:127b; **7**:300a, 306b, 307b; **8**:309b; **9**:316b; **12**:105b; **13**:573a–b; **14**:301b, 367a
late seventeenth-century, **1**:292a, 474a, 549a; **2**:381b–382a, 613b; **3**:294b, 484b; **4**:158a–b, 221a–b, 505a–b; **5**: 414b–417b; **6**:129b; **8**:436b; **9**:47a–49b, 509a–510a; **14**:152b–153a, 361a–362a, 370b, 371b–372a
early eighteenth-century, **1**:531a; **2**:382a; **3**:244b, 294b–296a; **4**:85a, 471b; **8**: 379b; **10**:81b–82a, 614a; **13**:179b–180a; **14**:295b–296b
late eighteenth-century, **3**:199a; **6**:66b; **11**:139a–141b, 143a; **12**:153b–154a; **14**:119a–b, 191a
nineteenth-century, **2**:568b–569a, 570a–b; **3**:192b, 378b–379a, 470b; **6**:9b; **7**:509b; **11**:116a; **12**:86a–b, 240a, 244a,b, 466a; **13**:9a; **14**:294a,b, 516b
twentieth-century, **2**:11b–12a; **7**:483b–484b, 488a–b
astral, **1**:507b
atomism and, **3**:209a,b
Ḥarrānian, **1**:33b–34a, 507b
and language, **1**:274b
and mathematics, **1**:91a
Mesopotamian, **1**:507b
natural
ancient Greek, **14**:607a
medieval, **4**:255a–b; **10**:172b
seventeenth-century, **1**:159a–b; **2**:381b–382a; **5**:416b; **7**:275b; **8**:416a; **10**:85a; **11**:313b, 317b–318a; **14**:361a–362a
early eighteenth-century, **3**:244b, 295a, 296a; **4**:40b–41a; **6**:139b, 259b; **8**: 580a–b; **10**:120a; **11**:43a–44a
late eighteenth-century, **1**:487b; **3**:465a; **6**:586b; **9**:575a; **10**:277b–279a, 509b
nineteenth-century, **2**:568b, 570a–b; **3**:573b–574a; **6**:8b–9a, 437b, 438a, 453b; **7**:366a; **9**:196b; **11**:116a; **12**: 262b, 278a; **13**:78b; **14**:529b
twentieth-century, **4**:188a; **9**:341b; **14**: 318a
critics of, **6**:556b
negative, **9**:28b–29a, 31a
and philosophy, **2**:163b, 232b; **6**:448a–449b; **8**:436b; **9**:30a, 48a–49b; **11**: 71a–74a, 257b, 258b; **12**:2b, 7Ab, 8a
rational, **2**:232b; **11**:141a
and science
see Religion, and science
see also Calvinism; Historiography, of religion; Religion; specific theologies
Theon of Alexandria (*fl.* second half of fourth century A.D.), 13:321–325; **4**:489a, 491a
astronomy, **15**:33b, 34a,b
commentaries of, **10**:294a, 299a; **11**:97b, 202a; **13**:289a, 325b; **14**:603b, 604a–b
critics of, **12**:446b
edition of Euclid, **4**:416a, 430a,b, 437b–438a, 444a, 447a, 448b, 450b
influence of, **1**:33a, 42a; **8**:191b; **13**: 62a(n); **14**:458b
influences on, **1**:223b
Manual Tables, **1**:511b
trigonometry, **7**:361a,b; **9**:297a, 298b; **11**:196b
Theon of Smyrna (*fl.* early second century A.D.), 13:325–326; **9**:276a(n)
Arithmetica, **2**:349a
commentaries of, **3**:14b
cosmology, **15**:203a
historiography, **2**:618b; **4**:391b; **9**:269b; **10**:179b; **12**:314a; **15**:215b
influence of, **3**:319b; **4**:160a; **14**:538a
philosophy, **3**:127b–128a
Theophanes, **3**:21a
Theophilus, Byzantine emperor, **8**:191a
Theophilus (Theophilus Presbyter or Rugerus) (*fl.* early twelfth century), 13: 326–328
Theophrastus (*ca.* 371-*ca.* 287), 13:328–334; **1**:152a, 205b, 255b, 258b, 268a, 350b; **4**:30b, 33a, 82a, 105b
botany, **1**:101b, 261a; **4**:106b–107a; **11**: 595a; **13**:343a; **15**:67b
commentaries on, **5**:232a; **12**:134b
on Democritus, **4**:32a
and Eudemus, **4**:460b–463b *passim*
historiography, **6**:382a; **8**:269a–b; **13**: 303b, 297a; **14**:536b
influence of, **1**:150b–151a, 250a; **7**:539b; **11**:288b, 312a; **14**:586a; **15**:80b, 515a
natural philosophy, **13**:91b–94b *passim*
students of, **13**:91a
translations of works, **3**:534a; **12**:135a
Characters, **6**:405b
Historia plantarum
editions of, **13**:343a
De lapidibus
translations of, **6**:401a
Metaphysics, **1**:272a; **13**:91b–92a
De sensibus, **4**:32a
De signis, **2**:283a; **4**:460a
De ventis, **2**:283a
Theophylline, **5**:2a,b
Theopompus, **2**:550a
Theorica lignea (astron. instrum.), **7**:119a
Theory
evaluation of, **4**:227b; **8**:55b
explanatory power of, **3**:33b–34a
mathematical vs. mechanical, **3**:33a–34b
nature of, **4**:374a
and observation, **5**:102b
role of, **4**:225a, 226b–227a; **10**:78b; **11**: 14a–b
structure of, **3**:32b–34b
see also Hypothesis; Science, theory vs. technology
Theosophy, **3**:476a; **6**:253b; **7**:355b–356a; **8**:515b; **11**:249b, 250a
see also Spiritualism
Therapeutics
ancient, **1**:212b, 315a; **4**:386a; **5**:231b; **6**:422a–b
medieval Arabic, **15**:498b–499a
Renaissance, **3**:471b; **5**:233b; **10**:307b–308a
seventeenth-century, **6**:256a; **13**:214b–215a, 223b
eighteenth-century, **1**:563b; **2**:83a, 301a,b; **5**:57a, 157a; **10**:613a
nineteenth-century, **1**:10a, 360a–b, 574b–577a; **2**:91a, 547b–548a; **4**:252a, 621a; **12**:450b
twentieth-century, **1**:304a–b
experimental, **14**:495b–496b
Therapsida, **2**:504b

Therapy
dietary, **10**:417b
electric, **1**:304a–b; **2**:83a, 91a
endocrine, **3**:353a; **12**:384a
gaseous, **1**:563b; **3**:600a–b; **6**:283a
heat, **1**:304a–b
intravenous, **12**:503a
low-fluid, **4**:621a
low-salt, **4**:621a
magnetic
see Animal magnetism
occupational, **11**:365a
ultraviolet, **3**:327b
see also Electrotherapy; Hydrotherapy; Injection; Organotherapy; Oxygen, therapeutic use; Physiotherapy; Radiotherapy; Serotherapy
Theriac, **1**:120b, 167a; **3**:413a; **6**:427b; **7**:187b; **11**:341b; **12**:8a, 92a; **14**:637b
Thermalism, **2**:301b
Thermionics, **3**:597b; **4**:355b; **5**:33a; **11**:420a–421b
naming of, **11**:420a
Thermobalance, **3**:239b; **6**:480a
Thermochemistry
early nineteenth-century, **4**:240b–241a
mid nineteenth-century, **1**:160b–161a, 579b; **2**:69a–b; **4**:554b; **6**:354a–b; **11**:298a; **13**:358b–359a, 467b–468a
late nineteenth-century, **4**:20b; **6**:177b–178a, 244b; **7**:203b; **8**:545b; **9**:619b; **11**:8b; **12**:46b; **14**:265b; **15**:437b–438a
early twentieth-century, **3**:294a*; **11**:418a, 576b, 577a; **13**:246b; **15**:441a–442a
laws of, **2**:69a; **6**:354a–b
maximum work principle, **2**:69a
critics of, **4**:225b, 232a
Thermocline, **2**:142a, 558a
Thermocouple, **1**:490b; **2**:142a; **3**:308a; **8**:117b; **10**:107b; **11**:486a, 559b; **14**:339a
Thermodynamics
early nineteenth-century, **3**:8b, 80b–83b
mid nineteenth-century, **2**:88a–b; **3**:303b–310a; **4**:258a, 554b; **6**:431b–432a; **7**:182b, 601b; **8**:510b; **9**:210a, 227a–b, 445b; **11**:292a–293b, 353a–b; **12**:288a–289a; **13**:377b–384a, 522a, 614b–615a; **14**:618a; **15**:85a–86b
late nineteenth-century, **1**:600a–b; **2**:427a, 559a; **3**:19b–20a; **4**:20b, 225b, 228a–232a, 500a; **5**:387a–392b, 502a, 587b; **6**:243a, 245a, 342a; **7**:462a, 466a; **8**:118b, 257b*, 500a, 598b–599a; **9**:394a, 461b–462a, 616b; **11**:8a–12b, 149b, 446a; **13**:73b, 102b, 237a; **14**:109b–111b, 339a–340a; **15**:462b–463a
early twentieth-century, **1**:365b; **2**:169b, 565b; **3**:299a–b, 334a, 564b*; **4**:293b–294a, 314a–317b, 320b; **6**:470a; **8**:11a–b, 52a, 289a–293b, 503a; **9**:462a–b; **10**:162a–163a; **11**:12b–13b, 420a; **13**:40b–41b, 176a, 244b–245a; **15**:436a–449a, 463b–464b
mid twentieth-century, **3**:63a; **7**:617a; **8**:478b; **12**:437b–439a
adiabatic invariants, **2**:245b–246a
applications of, **5**:390b; **7**:446b; **11**:345a; **15**:438a
and astrophysics, **12**:70b–71a
atmospheric, **2**:168a; **4**:361a; **9**:107b–108a; **11**:558a; **12**:187a–b, 366b; **14**:216a
of celestial bodies, **4**:360b–361a
characteristic functions, **4**:228b
chemical
mid nineteenth-century, **8**:510b
late nineteenth-century, **4**:228b–230a; **6**:243a, 244b, 519b–520a; **7**:65a–b; **8**:118b; **9**:360a, 394a, 619b; **13**:578a–579b; **14**:123b, 560b–561a; **15**:432b, 433b, 457a–b, 458b
early twentieth-century, **8**:47a, 289a–293b; **9**:354b; **14**:182b; **15**:435b–449a
mid twentieth-century, **6**:404b
computational systems, **9**:462b
diagrams, **9**:461b, 462a–b
and electromagnetism, **2**:266a; **8**:387b
and ferromagnetism, **14**:245b
first law of, **3**:83b, 304a–b, 305a; **14**:123b; **15**:437b, 447a
axiomatization of, **4**:229b
as work principle, **2**:499b
foundations of, **2**:460b; **6**:245a
generalization of, **4**:228a–b
and geochemistry, **3**:293a
geometric methods, **5**:388a–389a
of heterogeneous mixtures, **5**:389a–390b, 391b
and kinetic theory of gases, **3**:307a; **5**:391b; **12**:497a–b; **15**:443b–444a
laws of, **2**:499a–b; **5**:387b
in biochemistry, **6**:22a
Clausius' couplet, **3**:306a
and universe, **1**:301a
see also Thermodynamics, first law of, etc.
of liquids, **2**:466a; **3**:620a
quasi-static processes, **4**:229a,b
and rational mechanics, **2**:471a
relativistic, **13**:429b
second law of, **2**:499b; **3**:83a, 287b, 305a, 480a; **5**:387b, 390b, 391b; **6**:245a; **8**:510b, 598b; **9**:227a, 360a, 462a; **10**:604a; **11**:8a, 9a,b; **12**:570b–571a; **13**:226b–227a, 275b, 382b, 383a; **14**:123b; **15**:437b, 438a, 441a–b, 443a, 447a, 464b, 465b
and fluctuations, **12**:497a–b; **13**:226b–227a
mechanical analogies, **2**:261a, 265a; **3**:308a–309a
statistical foundation, **4**:293b; **12**:497b
of solids, **2**:466a; **7**:601b
statistical, **5**:103b, 598a; **9**:227a–b; **11**:420a; **12**:218b; **13**:41a–b
technical, **10**:163a
third law of, **4**:316b; **11**:418a; **12**:437b, 438b; **13**:41b; **15**:432b, 436a–447b
Thermoelectric pile, **10**:192a, 500a
electromotive force, **10**:188b–189b
Thermoelectricity
eighteenth-century study, **1**:66b–67a, 67b
nineteenth-century study, **1**:557a; **3**:260b; **4**:192a, 228b, 264b; **5**:98b; **6**:96b; **7**:526b; **10**:500a; **11**:8b, 474a; **13**:237a; **15**:433a
twentieth-century study, **1**:609b; **4**:7a; **6**:51b; **8**:47a; **14**:617a
discovery of, **3**:497a; **12**:281b
inversion effect, **3**:497a
and morphology, **11**:539b
and stress, **4**:500b
thermoelectric series, **12**:281b
Thomson effect, **3**:497a
Thermograph, **11**:550a
Thermoluminescence, **15**:49b
Thermometer, **1**:556a; **2**:141a, 444b; **3**:534b, 540a; **7**:578a; **11**:66a*
air, **1**:138a,b; **3**:19b; **5**:353a; **6**:255a; **7**:160a; **11**:530a; **12**:102b–103a; **13**:425a
comparison with mercury, **4**:240a
automatic, **1**:586a
calibration, **3**:534b; **6**:487a; **14**:553b
clinical, **8**:102b
Delisle's, **4**:23b
double-scale, **1**:586a
electrical, **1**:547a; **2**:438b; **7**:371b
gas, **10**:545b; **14**:39a
hypsometric, **4**:518a
invention of, **12**:102b–103a
linseed oil, **2**:178a
liquid, **13**:437b
manufacture of, **1**:586a; **5**:340b
maximum, **5**:340b
maximum-minimum, **12**:25a*, 132a
mercury, **1**:553b; **3**:158a; **5**:226a, 582b
metal, **2**:439a; **13**:425a
platinum, **3**:19b
resistance, **3**:19b; **11**:581b
spirit, **4**:517a
standard, **5**:340b; **11**:46b
theory of, **5**:97b
vapor-tension, **13**:71a
volumetric, **11**:330b–331a
see also Pyrometer, Thermometry
Thermometergraph, **1**:586a
Thermometry
seventeenth-century, **1**:138a,b; **2**:379b; **3**:534b; **6**:487a; **7**:376a, 577b; **9**:14a
early eighteenth-century, **3**:174a; **4**:26a, 517a–b; **5**:353a; **9**:33b; **11**:330b–331a, 526b; **13**:265b
late eighteenth-century, **2**:177b–179b; **9**:234a; **13**:425a; **15**:355b
early nineteenth-century, **1**:240a, 586a; **4**:240a
mid nineteenth-century, **1**:449a,b; **5**:19a; **7**:181b; **8**:1a, 2a; **11**:111b; **13**:377b–378a
late nineteenth-century, **1**:612b; **3**:19b; **5**:582b; **8**:117b; **14**:330b, 339a
early twentieth-century, **4**:253b; **6**:470a
Celsius' scale, **3**:174a
clinical, **11**:462b
electrical, **1**:557a; **6**:225a–b
Fahrenheit scale, **4**:517a–b; **11**:330b, 432b
geological, **8**:42a
Kelvin scale
see Absolute temperature
physiological, **1**:303a
Réaumur scale, **11**:330b–331a
Rømer scale, **4**:517a–b; **11**:526b
standardization of, **3**:19b
thermoelectric scale, **14**:339a
zero point, **1**:586a; **6**:487a
see also Absolute zero; Calorimetry; Oceans, thermometry; Pyrometry; Temperature; Thermocouple; Thermometer
Thermoneutrality, law of, **6**:354b
Thermopile, **2**:587a; **3**:327a; **5**:68b; **7**:74a; **9**:566b; **11**:50b, 582b
Thermoscope, **4**:81a; **5**:47b; **10**:500a; **12**:103a; **13**:437b
Thermostat, **4**:184b; **12**:60a
Thermotaxis, **5**:270b
Therocephalia, **2**:504b
Thessalus of Tralles, **12**:538b
Thetis (submarine), **6**:22b
Theudius of Magnesia (*fl.* fourth century B.C.), 13:334; **4**:415b; **6**:414b
Theurgy, **1**:34a; **2**:539b; **11**:160b, 183a,b, 184b; **15**:497b
Thévenot, Melchisédech (*ca.* 1620–1692), 13:334–337; **6**:598b; **13**:30a, 33a, 169a, 172a
capillarity, **9**:498a–b
Thiamine
see Vitamin(s), B_1

Thiazole, **5**:215a
synthesis of, **6**:108a
Thiboumery, **10**:498a
Thiele, F. K. Johannes (1865–1918), 13:337–338; **3**:511b; **4**:91a, 485a; **13**:1a; **14**:334a
Thiele, Thorvald Nicolai (1838–1910), 13:338–339; **2**:616a; **6**:513a(n); **7**:185a; **14**:280b
Thiele transformation, **13**:339a
Thiele-Innes constants, **7**:18a; **13**:339a
Thiem, Gunter, **9**:257b
Thierry of Chartres (Magister Theodoricus Carnotensis) (*d. ca.* 1155), 13:339–341; **1**:91b; **2**:19b, 21a; **5**:592b; **15**:181a
Thiersch, Karl, **8**:406b
Thiesen, Max, **11**:150b
Thilenius, R., **4**:171a
Thio acids, **11**:551b
Thioalcohols, **14**:600b
Thiocarbamide, **3**:272a
Thiocyanates, **2**:171a; **11**:551a–b
Thiocyanic acid, **5**:558b
Thiodiglycol
synthesis of, **3**:246b
Thionuric acid, **8**:343b
Thionyl chloride
synthesis of, **10**:532a; **12**:164a
Thiophanes, **9**:608b
Thiophene, **14**:601b
discovery of, **9**:356b
synthesis of, **3**:246b
Thiosemicarbazones, **4**:155b
Third-man argument, **2**:550a
Thirty Years' War, **3**:360a
Thiry, Ludwig, **1**:544a
Thiry, Paul Henri
see Holbach, Paul Henri Thiry, Baron d'
Thiselton-Dyer, William Turner (1843–1928), 13:341–344; **6**:163a*
Thixotropy, **15**:160a
Thiyūfīl al-Ruhāwī, **15**:236a–b
Thölde, Johann, **13**:559a
Tholeiite, **1**:394b
Thollon, Louis (1829–1887), 13:344
Tholoids, **2**:554b
Thom, Charles, **5**:30a
Thomas of Bologna, **2**:22a–b
Thomas of Cantimpré (Thomas Brabantinus) (*ca.* 1186–*ca.* 1294), 13:347–349; **14**:35a
Thomas of Erfurt
De disciplina scholarium, **2**:231a
Thomas of Prague, **2**:397a*
Thomas, Cyrus, **11**:119b
Thomas, Harold Wolfstan, **4**:300a
Thomas, Herbert Henry, **1**:394b
Thomas, Hugh Hamshaw (1885–1962), 13:345–346; **13**:273b*; **14**:503a*
Thomas, Llewellyn Hilleth, **4**:578b
Thomas, Sidney Gilchrist (1850–1885), 13:346–347; **15**:25a
Thomas, Tracy Yerkes, **4**:336b; **13**:600a
Thomas, W. (quantum mech.), **7**:518b
Thomas, William (physiol.), **2**:3a
Thomaz, Alvaro (*b.* second half of fifteenth century), 13:349–350; **3**:171b, 420b; **6**:127a
Liber de triplici motu, **13**:210a–b
Thomism
medieval, **2**:20b, 590b; **5**:401b
Renaissance, **3**:172a; **12**:548a
twentieth-century, **15**:424b
Thompson, Benjamin (Count Rumford) (1753–1814), 13:350–352; **3**:83b, 599a; **5**:276b; **6**:55a; **11**:354a; **14**:562b
Thompson, D'Arcy Wentworth (1860–1948), 13:352–353; **4**:132b, 133b*; **5**:80b; **8**:39b; **13**:344a*; **14**:87b*, 203b*
Thompson, Harold Warris, **6**:404b
Thompson, Sir Henry, **3**:261b
Thompson, John Griggs, **2**:615b
Thompson, John Vaughan (1779–1847), 13:353–356
Thompson, Keith, **8**:483a
Thompson, Silvanus Phillips (1851–1916), 13:356–357; **3**:481b*; **4**:540a*; **5**:407b; **10**:537b(n), 539a*; **13**:126b*, 388a*
Dynamic Electric Machinery, **9**:395a
Thompson effect, **11**:446b
Thompson Indians, **2**:212b
Thomsen, Christian Jürgensen (1788–1865), 13:357–358; **9**:489a; **14**:506a,b
Thomsen, Hans Peter Jörgen Julius (1826–1909), 13:358–359; **2**:499a; **6**:350a
thermochemistry, **2**:69a; **6**:244b, 354b; **15**:437a–438a, 441b, 457a–b
Thomson, Albert, **5**:501a
Thomson, Sir Charles Wyville (1830–1882), 13:360–361; **7**:92b*; **9**:588a; **11**:113b; **13**:236b, 352b; **15**:78a
Challenger expedition, **2**:557b; **3**:89a; **9**:588b, 589a
entomology, **2**:321b
Thomson ridge, **9**:589a
Thomson, David Landsborough, **3**:352b
Thomson, Elihu (1853–1937), 13:361–362; **4**:601b
Thomson, George
Galeno-pale, **7**:150b
Thomson, George Paget, **3**:598a; **13**:447a
Thomson, James (chem.), **13**:236b, 374a, 380b–381a, 381b; **14**:560b
Thomson, James (nat. phil.), **13**:224a
Thomson, John (medicine), **2**:172b
Thomson, Joseph John (1856–1940), 13:362–372; **3**:328b; **4**:191b; **5**:459a; **8**:39a; **12**:32b; **13**:52b; **15**:47b*
acquaintances of, **1**:320a; **2**:240b, 397b; **4**:194b; **9**:616a; **10**:31b; **11**:123a; **12**:238a; **15**:61a
chemical theories, **8**:291a–b
atomic structure, **9**:230b–231a; **10**:106b; **12**:31a–b
ionization, **14**:421a
and Compton, A. H., **3**:368a
Conduction of Electricity Through Gases, **1**:320b
on contemporary scientists, **13**:156a; **14**:421a
Corpuscular Theory of Matter, **7**:469a; **8**:291a
correspondence of, **4**:197b*
critics of, **2**:399a
electricity, **3**:310b(n); **5**:133b; **11**:420a, 421a; **13**:82b
Electricity and Matter, **5**:609a
electrodynamics, **1**:456b
on electrons, **2**:241b; **3**:478a, 499b; **4**:355b; **7**:264b; **8**:496b; **13**:446b, 447a(n); **15**:464a,b
radiation
cathode rays, **1**:321a, 559b; **3**:477b; **6**:439b; **8**:180b–181a
X rays, **1**:457a; **3**:367b, 368b–369a; **4**:355a
and Rutherford, **12**:26a–b, 27a
students of, **3**:31b; **7**:84b; **8**:10a; **11**:419b; **13**:445b; **15**:40a
Thomson, Robert Dundas, **13**:372b
Thomson, Thomas (1773–1852), 13:372–374; **1**:156a; **2**:336a; **13**:374a
acquaintances of, **2**:335b; **4**:65b; **13**:499a
An Attempt to Establish the First Principles of Chemistry by Experiment, **11**:172b; **13**:499b
chemistry, **9**:245b; **11**:167a
atomic theory, **3**:543a; **5**:320a; **6**:386a
atomic weight, **11**:173a,b
correspondence of, **3**:291b, 542b
historiography, **2**:175a; **6**:166b, 586a; **8**:88a*
influence of, **13**:313a–b
influences on, **2**:174a
students of, **3**:488b; **5**:492b
System of Chemistry, **3**:541b; **11**:170b, 437a
see also *Annals of Philosophy*
Thomson, Thomas (botanist), **6**:490a
Thomson, Sir William (Baron Kelvin of Largs) (1824–1907), 13:374–388; **4**:410b; **9**:230b, 239b; **13**:101b–102a
acquaintances of, **4**:500a; **5**:412a, 413b; **6**:242a, 343b, 502a; **7**:93b, 382a; **9**:372b, 588b; **11**:241b; **12**:49b, 496a; **13**:75a, 347a, 365a
biographies of, **8**:40b; **13**:357a
on contemporary science, **3**:471a; **5**:462a; **6**:51b; **11**:293a–b; **12**:28b; **13**:74b, 77b, 79b*
correspondence of, **9**:206a, 209a; **13**:78b
critics of, **3**:306a, 479a; **8**:1b; **9**:606b; **13**:499b, 523a; **14**:503b
and Curie (Pierre), **3**:504a, 506b
electricity, **6**:349a*; **9**:155a, 205b, 209b, 210a,b, 211a, 216b; **13**:80a, 374b–377b, 384a–b; **14**:32b, 73b
electromotive force, **10**:481b; **14**:32b
and Faraday, **4**:538a
geology, **6**:503a; **15**:158b
earth, age of, **2**:260a; **3**:191a; **6**:475a; **7**:161a, 370b; **8**:575a; **13**:107b, 383a–b
tides, **4**:591a–b
heat theory, **3**:304b–305a, 306a; **5**:387b; **11**:292b; **13**:378a–384b
hydrodynamics, **11**:58a; **13**:76a
influence of, **3**:32a, 583a; **8**:385b, 592a; **13**:237b, 614b; **14**:569a, 598a
influences on, **13**:237a; **14**:185b; **15**:200b
and Joule, **7**:182a–b; **13**:378b–382a *passim*
Joule-Thomson effect, **4**:79b; **6**:94a; **7**:182b; **8**:365b
lectures of, **2**:466a
magnetism, **9**:206b–207a, 207b, 211b; **13**:363a, 377a–b, 384a–b
"Mathematical Theory of Magnetism," **9**:206b
mathematics, **11**:449b
matter theory, **3**:599b; **4**:230a; **12**:31a; **13**:380a–b; **15**:464a
atom size, **8**:510a
ether, **3**:146b; **5**:391a; **8**:40a; **9**:213a,b, 214a; **13**:384a–386b
vortex theory, **2**:168a; **6**:250a; **8**:40a; **13**:52b, 362b, 385a–386a
and Maxwell, **9**:202b, 205b–211b *passim,* 220b, 221a, 222b, 224b; **13**:375a, 377b, 384b, 385b–386a
mechanics, **14**:569a
meteorology, **9**:107b
philosophy, **12**:609a
scientific explanation, **13**:375b–377b, 384b
solar heat, **9**:239a
students of, **2**:503b; **7**:313a; **11**:277b; **12**:493b
technology
instrumentation, **9**:227a; **13**:377b
telegraphy, **4**:559a–b; **7**:381a; **9**:220b

Thomson, Sir William (cont.)
thermodynamics, 2:263b; 3:82a, 83a, 287b; 5:388a; 9:227a; 13:377b–384a
thermometry, 13:377b–378a
Thomson effect, 3:497a; 6:51b
Treatise on Natural Philosophy, 9:211a; 13:237b*, 386a
Thomson-Houston Company, 13:361b
Thoracic duct
see Lymphatic system
Thorax
paracentesis of, 6:367a; 9:482b, 483a; 12:103a
Thore, Jules, 3:475b
Thoria, 8:24a
Thorium, 2:258b; 4:563b; 9:262a; 11:279b
atomic weight, 2:429a; 9:290b
bombardment of, 4:580a
complex compounds of, 3:235b
determination of, 11:551a
discovery of, 2:94b
extraction of, 12:505b
fission, 7:554b
radioactivity, 1:559b; 3:499a,b, 507a; 7:159b*; 12:27a–b, 28a, 504b
decay series, 2:259b; 12:505b
discovery of, 12:191b
as spectroscopic source, 9:256a
thorium X, 1:560a; 12:504b; 13:159b
Thorndike, Edward Lee, 7:98b; 8:446a; 9:512b–513a; 14:549b
Thorndike, Lynn, 1:5a, 49b, 381b, 468a; 2:20b; 3:130a; 5:361a, 402a; 7:112a,b, 140b; 9:364a, 378a; 10:281a; 12:311a–b; 15:492a
Thoron, 11:280b
Thorpe, Jocelyn Field (1872–1940), 13:388–389; 10:517b*
Thorpe, Thomas Edward (1845–1925), 13:389–390; 3:158b*; 5:216b*; 6:185a; 12:493b; 13:410b, 517b
chemistry, 7:282b; 11:537b
on contemporary scientists, 3:280a*; 7:464b*; 12:479a*
Dictionary of Applied Chemistry
editions of, 13:389a*
Thorpe, W. V., 15:106a
Thought
economy of, 8:601a,b, 604b
Thought experiments, 1:567b, 607a; 3:306a; 6:37b; 7:232b, 487a; 8:606a(n); 10:226a; 11:14a, 15a; 14:341a; 15:443b
Thought process, 2:132a
experimental analysis of, 14:528b
Thouin, André (1747–1824), 13:390; 2:321b, 322a; 3:173a*; 7:199a; 9:366a; 10:326a; 13:495b
Thrasyllus, 10:112b; 13:325b, 326a
Three, rule of
see Rule of three
Three-body problem
seventeenth-century study, 10:73b, 78a–b
eighteenth-century study, 1:113b, 400b–401a; 7:563b–564a, 580a; 8:179a
nineteenth-century study, 2:350b, 351a; 3:583b; 5:508b; 6:88b, 399b; 8:384b; 9:174b; 10:212b; 11:57a–b, 58a, 60a; 12:248a; 13:339a, 422b; 14:85b–86a
twentieth-century study, 1:156b–157a, 429a; 2:144a, 145a, 516a; 3:58a*, 220b–221a; 8:284b; 9:450a*; 13:153a
and comets, 3:283a–b
generalization of, 9:553b*
and moon, 3:283a
periodic orbits, 2:616a; 13:99a
variation of elements, 4:481b
Thrombidium, 6:267a
Thrombin
isolation of, 6:526a–b
Thrombocytes, 2:165b; 4:277a; 9:417a; 12:232a
formation of, 4:277a; 6:526b
Thromboplastin
isolation of, 6:526b
Thrombosis, 3:489b; 4:277a; 14:40a, 229a, 249b, 250a
Throop Polytechnic Institute, 6:31b
Throscidae, 6:507a
Thrush (disease), 5:566a; 11:491b
Thrust (geol.), 8:33a; 11:594a
Thruston, Malachi, 4:377b; 9:243b–244a
Thue, Axel (1863–1922), 13:390–391; 6:393b; 11:107b; 12:452a
Thuillier, Louis, 7:425a; 10:385b, 395a, 399b–400a; 11:569a
Thuja, 13:487b
Thule
discovery of, 11:225a–b
Thulium
discovery of, 3:321b
Thulium II, 9:544a
Thunberg, Carl Peter (1743–1828), 13:391–393; 5:216b–217a, 218a; 9:152b; 13:178a; 14:116a–b
botony, 1:88b; 7:205a; 8:379b
Thunberg, Thorsten Ludvig (1873–1952), 13:393–394; 7:273a
Thunder
divination by, 8:191b
theories of, 3:535a; 8:174a–b; 11:66b–67b, 104b
Thunderfish, 2:127b
Thunderstorms, 1:547b; 3:593a; 8:278b, 469a; 9:443a, 464a*
mechanism of, 12:187b
moon's influence on, 1:300b
theories of, 12:512a; 14:422b
Thuret, Gustave Adolphe (1817–1875), 13:394–396; 3:337b; 9:601a, 618a; 11:153a,b; 14:93a
Thüringer Wald
geology, 5:205a–b, 444a,b, 445a; 6:456b
Thuringia
geology, 2:109a; 3:434a; 5:155b–156a; 6:456b
Thurmann, Jules, 5:533a,b; 9:101a; 14:88a
Thurnam, John (1810–1873), 13:396
Thurneysser, Leonhard (Thurnyser, Lienhart) (1531–1596), 13:396–398; 6:255a
Thurston, Robert Henry (1839–1903), 13:398–399; 10:286a
Thymaridas (*fl.* first half of fourth century B.C.), 13:399–400
Epanthema, 4:117a, 607a
Thyme
oil of, 5:351b
Thymic acid
see Apurinic acid
Thymine, 4:604a
discovery of, 7:466b
Thymonucleic acid
see DNA
Thymus
Renaissance description of, 1:619b
role of, 6:367b
Thyratron
invention of, 6:546b
Thyroid hormone
preparation of, 7:101a; 15:11a
therapeutic use of, 3:415b
Thyroidectomy, 6:78a; 14:115a
Thyrotrophic hormone (TSH), 3:352b
Thyroxine
structure of, 15:258a
Ibn Tibbon, Jacob ben Machir (*ca.* 1236–1305), 13:400–401; 7:119b, 124b
Quadrans novus, 7:119a
translations by, 4:442a; 6:197b; 7:38a, 125a; 9:302a*; 14:400b, 593b, 594a
works translated, 10:540b
Ibn Tibbon, Judah ben Moses, 13:401b
Ibn Tibbon, Judah ben Saul, 13:400a–b
Ibn Tibbon, Moses ben Samuel (*fl.* 1240–1283), 13:401–402
translations by, 4:442a; 6:201b; 7:38a; 8:280a,b; 9:28b, 31b; 12:7Bb; 15:33a
Ibn Tibbon, Samuel ben Judah, 13:400b
Tiberius Claudius Menecrates, 3:174b
Tibet
exploration of, 6:215b–217a; 11:181a–b
geology, 6:216a–b; 9:104a
Ibn Tibūgā, 14:594a–b
Tic
facial, 1:580a
Tichonow
see Tikhonov
Tick
and disease, 7:429a; 11:442b; 12:315a, 481a; 15:454b
Tickle, Thomas, 1:390b
Tickling, 5:185a
Tidal wave
mechanism of, 4:159a
Tides, atmospheric, 1:113b, 196a; 9:326a; 15:301a
and barometric pressure, 15:372b, 376b, 380b–381b
lunar, 1:478a; 15:301a, 381a,b, 382a
solar, 3:583a; 5:301a
Tides, geodynamic, 8:516b; 12:187a; 14:306a
Tides, marine
Renaissance study, 10:513b
seventeenth-century study, 4:377b*; 6:71b, 514b; 11:423a
eighteenth-century study, 3:105a; 8:144b
nineteenth-century study, 1:364b; 3:583b; 6:326b; 14:293a–b
twentieth-century study, 3:204a; 13:166b; 14:634b
calculation of, 1:459b; 5:547a; 8:530a; 13:167a
mechanical, 4:591b
cotidal lines, 14:293a–b
establishment of the port, 8:530a
friction of, 3:583b; 4:20a, 591a–b; 12:449a; 14:569a; 15:301b
high
theory of, 14:634b
industrial utilization of, 4:614a
inequality of, 3:248b; 14:293b; 15:297a–301a *passim,* 340b–341a
loading of earth's surface, 7:595a
mathematical study, 2:466b; 4:591b; 15:192b, 296b–301b, 320a–b, 339b–341b, 352b
motive force, 13:614a
phase differences, 7:595a
recording of, 12:132a
tables of, 5:25b; 15:341b
theories of
ancient Greek, 11:104a
ancient Roman, 11:75b
medieval, 1:36a, 101a, 198b, 565b; 5:551a, 552b; 9:363a; 15:33b, 35b
Renaissance, 3:423b; 4:158b–159a; 5:399b, 400a, 401a, 547a; 13:49b; 15:45a
seventeenth-century, 1:424a, 568a; 3:248b; 4:506a–b; 5:239b–240a, 244a, 288a; 7:301b; 10:71a, 73b, 78a
eighteenth-century, 2:43a; 3:282b; 8:611a; 15:296b–301b, 340a–341a

nineteenth-century, **4**:19b, 591a–b; **7**: 595a; **14**:293a, 569a
twentieth-century, **4**:614a; **6**:615a
Tidskrift for Mathematik
editors of, **14**:618b
Tiedemann, Friedrich (1781–1861), 13:402–404; **2**:162a; **13**:461b, 463a
acquaintances of, **1**:476a; **2**:160b; **5**: 430a,b; **6**:268b; **9**:568b, 570b
biochemistry, **5**:431b
students of, **9**:456b; **12**:164b, 202a, 354a
van Tieghem, Philippe (1839–1914), 13:405–406; **2**:290b; **3**:429a; **5**:605a; **10**:385b; **11**:428b; **12**:125a
van Tieghem cell, **2**:437b
Tiemann, Johann Carl Wilhelm Ferdinand (1848–1899), 13:406–407; **11**:551a
Reimer-Tiemann reaction, **13**:406b
T'ien Shan, **12**:300a–301a
geography, **6**:216b
T'ien-yuan (celestial monad), **3**:252a, 253a, 254a, 266b
T'ien-yüan-shu method, **3**:254a; **8**:314b–320a; **12**:290b
Tierra del Fuego, **3**:397a
Tiessen, Ernst, **11**:439b–440a
Tietz, J.D.
see Titius, Johann Daniel
al-Tīfāshī, Shihāb al-Dīn Abu'l-ʿAbbās Aḥmad ibn Yūsuf (1184–*ca.* 1254), 13: 407–408; **7**:343b
Tifernate
see Gregory of Cittá di Castello
Tiglath-Pileser I, **15**:637b
Tigris (river), **15**:639a
Tikhonov, Andrei Nikolaevich, **3**:171a
Tikhov, Gavriil Adrianovich (1875–1960), 13:408–409; **6**:106b*
Tikhov-Nordmann effect, **13**:408b
Tilas, Daniel (1712–1772), 13:409–410; **3**: 210b, 473a
Tilden, Sir William Augustus (1842–1926), 13:410–411; **1**:320a; **3**:48b; **5**:104a*, 413a*; **10**:38b*; **11**:284a*
Tillamook Indians, **2**:212a
Tillet, Mathieu (1714–1791), 13:411–412; **3**:387b(n); **6**:237a*; **8**:621a; **11**:132a
and Académie des sciences, **15**:333a,b, 334a,b
Tilley, Cecil Edgar, **1**:394b; **5**:39a
Tillite, **4**:262a,b; **7**:267a; **10**:504a; **14**:132b
Tillo, Aleksey Andreevich (1839–1900), 13: 412–413; **13**:112b*
Tilloch, Alexander (1759–1825), 13:413–414; **10**:108b
Tillotson, John, **1**:474a; **2**:612b; **14**:295b, 361a, 370a, 374a,b, 374b(n), 375a(n)
Tilly, Joseph-Marie de (1837–1906), 13: 414–415
Ibn al-Tilmīdh, Amīn al-Dawla Abu'l-Ḥasan Hibat Allāh ibn Ṣaʿīd (*ca.*1073–1165), 13:415–416
Tiltmeter, **7**:265b; **10**:210b
Timaeus of Locri, **11**:25a, 225b
Timan (mountains), **3**:236b
Timber
see Wood
Timbre
theories of, **6**:245b; **12**:128b
Time
absolute, **3**:90a, 296a; **5**:287a–b; **10**:70a; **11**:324a
opposition to, **2**:17b; **8**:158b, 599a
and algebra, **6**:90b–91a
astronomical
definition of, **8**:40b
daylight saving time
concept of, **5**:130b
definition of, **1**:253a; **3**:26b, 470a; **6**:257a–b
determination of
ancient Greek, **6**:617a; **15**:216b
medieval Arabic, **4**:556a; **5**:613a–614a; **14**:577b, 578b–579a
medieval Latin, **1**:564b–565a; **2**:24a; **6**:301b–302a
Renaissance, **15**:478a
seventeenth-century, **11**:532b
eighteenth-century, **2**:389a; **3**:109a*
nineteenth-century, **1**:19a, 402b–403b; **2**:97b, 471b; **3**:565b; **6**:107a; **7**:209b
twentieth-century, **2**:126b; **4**:270a, 407b; **7**:266a; **11**:552b; **12**:96a
astronomical, **10**:279b; **15**:557a, 563a,b, 569b, 577a, 577b–578a, 591b, 604a, 679a
Egyptian, **11**:706a; **15**:707b–716a
Indian, **1**:308a; **15**:536b–538b, 593b, 616a, 624b
lunar six intervals, **15**:677a,b, 680b
Mesopotamian, **15**:644a, 674b
radioactive methods, **3**:508a
at sea, **2**:43a
from shadows, **2**:153b
direction of, **2**:264b; **11**:358a
and energy, **13**:241b
Ephemeris Time, **2**:511b
equation of, **2**:412a; **5**:23b; **7**:304a, 361a; **10**:540b; **11**:191a; **15**:582b, 583b–584a, 586a
geographical differences in, **3**:319b
Greenwich mean time, **3**:288b
mathematical, **10**:49b–50a
mean time, **5**:87b; **15**:536a
as measure of motion, **1**:474b
multiplicity of, **2**:11b, 331a
relative, **4**:322a, 323b
space-time, **5**:188a; **8**:292b; **9**:405b–406a, 414b
geometrization of, **4**:328a, 329a–331b; **8**:434a
specificity of, **15**:423b
subjective, **15**:504b
theories of
ancient Greek, **13**:93a–b; **14**:608a–609a
Byzantine, **12**:441b–442a
medieval Arabic, **1**:27b; **11**:231b, 324a–b; **15**:496a
Renaissance, **13**:279b
seventeenth-century, **2**:13b–14a; **8**:158b
eighteenth-century, **1**:111b; **7**:225b–226b, 367a; **8**:159b; **14**:84b, 484a
nineteenth-century, **1**:140b–141a; **2**: 9a–b; **8**:40b; **10**:484a–b; **14**:527a
twentieth-century, **1**:24a; **2**:11a–b, 460b; **4**:321b–322a; **5**:188a; **7**:192b; **8**:11b, 498a; **9**:406a; **11**:358a; **14**: 306b–308b, 608b–609a
Egyptian, **15**:657a(n)
Indian, **15**:555a–b
relativistic, **9**:212b
time service; **4**:26b, 270a, 354a, 407b; **5**:404a; **7**:270b; **8**:19b, 613b
units of, **2**:154b; **3**:175b; **7**:219b; **8**:198a; **10**:426a
Universal Time, **2**:511b
see also Calendar; Chronology; Clock; Congresses, scientific; Eternity; Geology, time scale; Horology; Simultaneity; Space; Sundial; Year, etc.
Time series, **1**:154b
oscillatory, **14**:574a
Time-and-motion study, **13**:272a
Timerding, Heinrich Carl Franz Emil, **2**: 594a
Timiryazev, Kliment Arkadievich (1843–1920), 13:416–417; **9**:534b; **10**:1a; **12**:336b; **15**:509b
Timocharis, **1**:283b; **2**:405a; **9**:297a; **11**:188a; **15**:217a,b, 218a
Timofeev-Ressovsky, Nikolai Vladimirovich , **9**:565a
Timor (island)
geology, **6**:229b
paleontology, **2**:110a
Timoshenko, Stepan Prokofevich, **5**:64b; **6**:534a; **11**:478a
Timothy I (Catholicos), **10**:21a
Tin, **5**:595b; **9**:62a; **15**:651b
allotropy, **3**:333b
alloys, **3**:560b; **5**:91a–b; **8**:321a; **9**:60b, 106b; **11**:534b; **13**:310b
in antiquity, **1**:388a
arsenic in, **1**:529b; **9**:106a
atomic weight, **2**:429a
calcination of, **11**:389a–b
compounds of, **2**:199b, 565a; **3**:201b, 202a; **10**:530b, 578b
optical activity, **11**:86a
organic, **5**:125b–126a; **7**:552a
oxides, **5**:157b; **11**:169b
sulfides, **2**:8a; **14**:508b
deposits of, **11**:225b
dimorphism, **5**:197b
electrical resistance, **8**:480b
gray, **5**:197b
mining of, **3**:586b
placers, **2**:500b
quantitative determination of, **10**:511a
specific gravity, **7**:346a
"Tin disease," **3**:333b
Tinea, **5**:566a
Ting Chieh, **8**:314a
Ting Ch'ü-chung, **3**:266a
Ting-mu (definite base numbers), **3**:252a, 253a
Ting-yung-shu (definite use numbers), **3**: 253b
Tinplate
manufacture of, **5**:101a; **11**:330a
Tinseau d'Amondans, Charles de (1748–1822), 13:417–418; **9**:476a
Tintinnids, **7**:447a
Tiryns, Greece, **12**:180b–181a
Tisber
see Heytesbury, William
Tiselius, Arne Wilhelm Kaurin (1902–1971), 13:418–422; **13**:164b*
Tisserand, François Félix (1845–1896), 13: 422–424; **2**:88b, 126b; **8**:279b*, 383a; **10**:528b; **11**:498b*
Tisserand's criterion, **13**:422b–423a
Traité de méchanique céleste, **3**:18b–19a
Tissues, animal
aging of, **8**:283a
analysis of, **8**:584a
autolysis, **14**:252b
as basis of anatomy, **2**:122b–123a; **6**: 62b
cardiac, **5**:82a–b, 280b–281a
classification of, **2**:122b; **6**:270a; **7**:438b
connective, **2**:165a, 301b; **5**:34b; **11**:361a, 374a; **15**:524b
embryogeny, **6**:435a
histology, **5**:36a; **9**:56a; **11**:296b
in vitro cultivation of, **3**:91a; **8**:283a
and infection, **2**:165b
mobile cells of, **2**:165a–b
structure of, **14**:596b
culture of, **3**:91a–b; **6**:131b–132a; **8**:447b; **14**:623b
environmental effects on, **2**:61a

Tissues, animal (cont.)
formation of, **12**:242a
metabolism, **13**:262b
nutrition, **11**:159b
oxidation, **4**:296b
mucous, **2**:301b
nature of, **1**:261b; **4**:383b; **7**:438b, 440a; **13**:556a–b
physiology of, **6**:270a
properties of, **2**:122b
immunity, **5**:317a
lability, **14**:105b
specific vitality, **2**:61a
regeneration, **4**:299a; **14**:228b–229a
respiration, **2**:61b; **8**:541b; **13**:394a
and injury, **13**:262b
water balance, **9**:19b
see also Histology
Tissues, vegetable
classification of, **5**:624a; **9**:345a, 600b
respiration
and injury, **13**:262b
Tisza, Laszlo, **8**:478b, 481a
Titan (satellite), **10**:35a
Titania (satellite), **13**:113b
Titanic acid, **11**:541a
Titanite, **5**:556b; **11**:539a
Titanium, **11**:541a; **14**:478a
atomic weight, **13**:389b
compounds of, **11**:540b
discovery of, **7**:394b
isolation of, **4**:77b; **9**:451b
spectrum, **12**:22a
Titanotheres, **10**:241b
Titchener, Edward Bradford, **9**:562b; **14**: 527a, 528b, 550b
Titchmarsh, Edward Charles (1899–1963), 13:424; **6**:114a
Tithis (time units), **2**:416a, 417b; **3**:584b; **4**:100a; **5**:275a; **6**:115b; **7**:315a, 512b; **9**:607a; **10**:426a; **15**:588b
definition of, **15**:536a, 538b
Tithonian (stratig. stage), **10**:212a
paleontology, **5**:533b; **14**:626a
Tithonometer, **4**:182b
Titius (Tietz), Johann Daniel (1729–1796), 13:424–426; **2**:221a
Titius-Bode series
see Bode, Johann Elert, Bode's law
Titmouse
speciation, **3**:572b
Titration, **8**:520b; **9**:445b; **13**:299a
colorimetric, **14**:448a
concept of, **5**:325a–b
coulometric, **13**:226b
electrometric, **5**:589b; **7**:208a
indicators, **2**:170b; **3**:290b; **5**:325b; **10**: 440b; **13**:228a–b
potentiometric, **7**:208a
theory of, **2**:170b
Titus, **11**:38b, 39a
Tizard, Henry, **8**:368b; **11**:482b
Tizard, Thomas, **9**:589a
Tizard mission, **3**:330a
Tlingit Indians, **2**:212a
Toad
larvae
digestive organs, **3**:424b
poisons, **14**:334b
Toadstones, **4**:549a; **14**:312a, 463b
Toaldo, Giuseppe, **3**:426a, 436a
Tobacco
Renaissance discussion, **5**:448a; **10**: 116b
seventeenth-century discussion, **9**:466b*; **10**:426b*
cultivation of, **3**:7b
selective breeding, **4**:271a,b
diseases of
mosaic disease, **3**:94b; **4**:220a–b; **5**: 141a, 317a; **7**:34b–35b; **13**:152b; **14**:553a; **15**:14b
wildfire, **7**:34b
dry distillation of, **14**:601a
historiography of, **13**:404a
smoking of, **6**:124a
opposition to, **6**:519a
Todd, John Lancelot, **7**:429a
Todd, Robert Bentley, **1**:539b; **2**:375a, 376a
Cyclopaedia of Anatomy and Physiology, **2**:375b; **10**:261a
Todd, T. Wingate, **2**:171b
Todhunter, Isaac (1820–1884), 13:426–428; **4**:35b; **6**:502a
historiography, **3**:135b, 386b; **15**:282a, 320a, 322b, 368a
History of the Mathematical Theory of Probability, **8**:306b
History of the Theory of Elasticity, **10**: 447b, 448b
Toepler, August, **13**:420b
Toeplitz, Otto (1881–1940), 13:428; **5**:152a*; **6**:235b, 394b*; **11**:256b; **13**:23a, 318b(n)
Toeplitz conditions, **13**:428b
Tohoku Imperial University, **6**:479b
Toise (unit of length), **8**:351a
standardization of, **2**:102a; **5**:491b; **8**:22a
Tokamak (thermonuclear installation), **15**: 9a
Tokugawa shogunate, **1**:310b; **15**:728b, 730a, 737a, 739b, 742a–749a *passim,* 755b, 756a,b
Tokyo Imperial University, **15**:743b
seismology, **7**:9a
Toland, John, **2**:542b
Toledan tables
see al-Zarqālī, Toledan tables
Toledo (Spain)
school of translators, **5**:591b
Tollens, Bernhard Christian Gottfried, **3**: 286b; **5**:4a; **7**:281b
Tolman, Cyrus Fisher, Jr., **14**:442a
Tolman, Richard Chace (1881–1948), 13: 429–430; **6**:531b–532a; **8**:290b
electrodynamics, **9**:211b; **11**:421b
Toluene, **5**:541a
constitution of, **8**:508a
discovery of, **3**:10b; **10**:498a; **14**:156a–b
isomerism, **14**:141b
oxidation of, **11**:370b
Tolyl red, **2**:59b
Tomato
red pigment formation, **4**:220a
Tombaugh, Clyde William, **8**:522b; **12**:456a
Tomeo, Nicolò Leonico, **5**:104b; **8**:248a
Tomkeieff, Sergei Ivanovich, **1**:394b; **13**: 441b
Tompion, Thomas, **5**:491a; **6**:482b
Tomsk Technological Institute, **10**:168a
Tone, musical, **1**:282a,b
Tonga Island, **3**:397a
Tongrian (stratig. stage), **2**:108b; **9**:240b
Tongue, **13**:31b, 32b
anatomy, **1**:593a
histology, **9**:63b; **14**:143a
innervation of, **13**:590a
Tònolo, Angelo, **14**:46b
Tonometer, **7**:502a
Tooke, Christopher, **6**:126a
Top, spinning
stability of, **8**:228b
Topaz, **5**:89a
refraction in, **2**:452b
thermoelectricity, **6**:96b*
Topics, logical, **2**:229b
Topography
see Altitude, determination of; Geography; Geomorphology; Landforms; Maps; Physiography; Surveying
Topology
eighteenth-century, **4**:479b; **13**:571b
mid nineteenth-century, **2**:274b; **3**:141a, 315a; **7**:384a; **9**:215b, 430b–431a; **11**:449a–452a; **13**:237a; **15**:193a
late nineteenth-century, **2**:105b, 236b, 274b, 320a; **3**:54b–55a, 178a, 322b, 323a; **6**:4a–5a, 176b–177a; **7**:397b; **10**:483b, 484b; **12**:196a
early twentieth-century, **1**:307b, 427b, 428a; **2**:145b, 274b, 512b, 513a; **3**: 95b, 96a, 170b–171a; **4**:9a, 269a, 336b, 560b; **5**:609a; **6**:394a; **7**:72a, 312a–313a; **9**:248a–249b; **11**:459b; **12**:411a; **13**:548b–549b, 599b–600a; **14**:283a–b
mid twentieth-century, **2**:351b; **3**:95b, 96a, 170b–171a; **8**:326b, 457b; **10**: 553b; **11**:260a; **12**:78a; **14**:90a–91b, 310b–311b, 590b–592a
abstract, **13**:549a
and algebra, **8**:326b
algebraic, **6**:496a–497a; **11**:54b, 59b–60b; **12**:451b; **14**:311a
combinatorial, **3**:169a, 170b, 171a; **7**: 167b; **11**:362b; **14**:311a
foundations of, **11**:362b
point-set, **12**:426b
relative distance, **9**:248b
Topsell, Edward
History of Four-Footed Beasts and Serpents, **9**:440b; **14**:508a
Topsøe, Haldor, **15**:83b
Torch
atomic-hydrogen, **8**:23b
oxyacetylene, **8**:118a
Torell, Otto [Martin], **10**:148a, 502a
Torelli, Joseph, **1**:230a
Tornado, **9**:444b, 486b; **12**:380b
theories of, **2**:329a; **4**:592b; **9**:279b
Toronto, University of, **7**:445a
Banting Institute, **1**:443b; **5**:6a
botany at, **7**:90b
Connaught Laboratories, **1**:442b, 443a
Torpedo fish, **1**:304a; **2**:129b; **3**:157b; **4**:202b; **5**:268b; **9**:177a, 259a; **11**:566b; **12**: 560b; **14**:113b
Torque
conservation of, **3**:74b
in thin cylinders, **3**:443b
Torquetum (astron. instrum.), **2**:24a, 404a; **7**:38b, 119b; **14**:401a
Torre, Marcantonio della (1481–1511), 13: 430–431
Torres Quevedo, Leonardo (1852–1936), 13:431–432
Torres Strait, **3**:397a; **4**:492a; **7**:185b
Torrey, John (1796–1873), 13:432–433; **3**: 196b, 197a; **4**:273b; **10**:164b
Flora of North America, **5**:512a–b
contributors to, **10**:164a
Torreya, **3**:196b
Torriano, Gianello, **4**:164b
Torricelli, Evangelista (1608–1647), 13:433–440; **2**:306b
acquaintances of, **2**:307b, 309a,b; **3**:115b; **5**:246b, 247b; **11**:404a
ballistics, **2**:202a
barometric experiment, **2**:83b, 84a, 311a, 377b; **5**:574b; **6**:36a; **9**:13b, 16a–b; **10**:332b, 333a, 334a, 342b, 546b; **12**:210b; **13**:438a–439b, 445a
correspondence of, **2**:308b; **11**:488a; **14**: 49b

critics of, **11**:507a
editions of works, **8**:505a; **14**:49b
geometry, **1**:164b, 474b; **5**:499b, 527a; **11**:487b; **14**:49b, 148a
indivisibles, **1**:164a; **2**:320a; **3**:150a, 152a–b; **10**:336a,b
hydrodynamics, **1**:165a, 567b; **3**:116a; **9**:13b–14a, 369b
influence of, **2**:13a; **4**:505b; **9**:596a; **12**: 459b
influences on, **1**:229b, 424b; **9**:25a; **13**: 561a
manuscripts of, **11**:404b
Opera geometrica, **14**:148a
Torricelli's parabola, **13**:437a
Torricelli's point, **13**:436a
Torricelli's principle, **13**:436b
see also Vacuum, Torricellian
Torridonian (stratig. division), **8**:33a,b
Torsion, **3**:440a, 443a–444a; **14**:330a
force of, **3**:443a
theory of, **2**:288a; **3**:443b; **8**:301a; **11**: 123b; **15**:164b
in wires, **11**:123a
Torsion balance
see Balance, torsion
Tortelli, Giovanni, **1**:274a
Torticollis, **8**:506b
Tortoise
heart, **4**:268a; **5**:280b–281a
speciation, **3**:569b
Tortona, Carlo Antonio, **2**:591a
Tortonian (stratig. stage), **9**:240b
Torula, **6**:100b
Torus, **10**:529a–530a
volume of, **4**:109b–110a; **7**:300b
Toscanelli dal Pozzo, Paolo (1397–1482), 13:440–441; **1**:381b; **2**:534b, 535b; **3**:513a; **15**:87b–88a
Totemism, **5**:145a, 178b
Toucasia, **9**:579a
Touch, sense of, **3**:382a; **5**:463a; **10**:267b; **14**:200b–201a
Toulmin, George Hoggart (1754–1817), 13: 441–442; **6**:588b*
Toulouse, University of, **12**:46b
observatory, **3**:428a,b; **13**:422b
Tourist industry
early nineteenth-century, **4**:275b
Tourmaline, **5**:457b; **8**:463a; **9**:559b
electrical properties, **3**:51b–52a; **14**:353b
piezoelectricity, **1**:557a; **3**:504b
pyroelectricity, **2**:5a; **3**:505b
thermoelectricity, **1**:66b–67a, 67b
magnetic properties, **11**:47a
refraction in
and polarization, **2**:138b–139a; **12**:281b
and thermal radiation, **5**:68b
Tournaisian (stratig. stage), **7**:460b
Tournal, Paul, **12**:184a; **15**:51a
Tournaments, mathematics of, **4**:137b; **14**: 614a
Tournefort, Joseph Pitton de (1656–1708), 13:442–444; **5**:352b; **13**:553b
acquaintances of, **6**:130a; **9**:17b; **11**:47a
botany, **4**:135b
on contemporary scientists, **9**:529a
critics of, **4**:99a; **11**:315a
Éléments de botanique, **11**:83b, 315a
herbarium, **12**:394b; **15**:111b
Institutiones rei herbariae, **9**:390a
editions of, **7**:197b
students of, **5**:351b; **7**:197a; **8**:171a; **12**: 394a, 457b; **13**:553a
taxonomic system, **1**:58b; **2**:580a; **7**:585a; **8**:374b, 376b, 377a; **9**:254a, 368a, 434a; **11**:83a
Tournière, Charles Penot de, **3**:284a
Tourtia, **1**:210b
Toussaint, Jean-Joseph Henry, **9**:505b; **10**: 390b, 392b–393a, 394a–b, 398a, 407b
Towarzystwo Przyjaciół Nauk, **15**:502a
Tower of Hanoi, math. problem of, **8**:531b
Town planning
ancient Roman, **15**:514b
Renaissance, **1**:97b
Towneley, Christopher, **3**:457b; **12**:342b; **13**:445a
Towneley, Richard (1629–1707), 13:444–445; **5**:278b, 279a(n)
Townsend, John Sealy Edward (1868–1957), 13:445–447; **8**:10a; **9**:544a; **13**:367b; **14**:420b
physics, **5**:331a; **12**:30a; **13**:80b, 366a
Ramsauer-Townsend effect, **13**:446b–447a; **15**:491a
Townsend discharge, **13**:446b
Townsend, Joseph (1739–1816), 13:447–449; **2**:566b, 568a; **12**:487b, 490a
The Character of Moses Established for Veracity as an Historian, **12**:490b
Toxalbumine, **1**:575b
Toxicology
ancient Greek, **10**:589a
Byzantine, **10**:418a–b
medieval Arabic, **7**:42b; **9**:41a; **11**:244b; **12**:90b; **13**:230b; **14**:118a–b, 637b
Renaissance, **9**:466b
seventeenth-century, **14**:256a–b
early nineteenth-century, **3**:159a–160a, 237b; **4**:22a*; **14**:187a–b
mid nineteenth-century, **2**:31a, 316a; **4**: 252a; **5**:297a,b; **9**:422a, 425b; **10**: 511a; **12**:620a
late nineteenth-century, **3**:22a–b; **5**:464a; **10**:252b*; **11**:425b, 429b; **13**:558a
twentieth-century, **1**:11a,b, 92b; **3**:257a; **5**:42b; **9**:35a, 506b*; **11**:101b, 429b–431a; **13**:70b–71a; **15**:228b
experimental
seventeenth-century, **10**:568a; **11**:341b
eighteenth-century, **5**:56a–b; **6**:17b
nineteenth-century, **1**:575a; **7**:438b
twentieth century, **9**:295b–296a
see also Poisoning, Venom
Toxins, **11**:428b
action of, **1**:301b; **11**:569a–b
instability of, **4**:298a
reabsorption of, **2**:161a
side-chain theory, **4**:298b–299b
spectrum theory, **4**:298a
test dose, **4**:298a
titration of, **11**:271b
toxin-antitoxin mixtures, **12**:482a
Toxites, Michael, **1**:49b; **4**:169b; **10**:305a, 306b; **13**:141a
Toxodon, **10**:261b, 262a
Toxoids, **4**:298a
Toxophore, **4**:298b–299a
Toxoplasmosis, **15**:454b
Toys
ancient Greek, **3**:491a; **6**:311b, 312b–313a
seventeenth-century, **6**:477b
hydrostatic, **13**:438b
hygrometric, **6**:477b
mechanical, **6**:481b
pneumatic, **10**:588b
Tozzetti, Giovanni Targioni, **2**:309b
Tozzi, Don Bruno (1656–1743), 13:449; **1**: 391b
Traccia, Accademia della
see Accademia della Traccia
Trace elements, **2**:87a; **4**:357b; **11**:558b
Trachea, **8**:128b
anatomy of, **9**:511b
Tracheids, **12**:100a
Tracheotomy
medieval Arabic, **14**:638a
Renaissance, **5**:581a
seventeenth-century, **12**:103a
nineteenth-century, **2**:445a; **14**:187b
Trachoma, **11**:175a
etiology of, **10**:142a,b; **15**:454b
treatment of, **11**:177b
Trachytes, **2**:554b; **3**:484a; **9**:367b
Tracks, fossil
see Ichnology
Tractarianism, **2**:569a, 570a,b
Tractrix, **1**:600a; **5**:111a; **6**:600b–601a; **8**: 162b, 164a; **11**:402a
Tradescant, John (*ca.* 1570–1638), 13:449–451; **8**:307b
Tradescant, John (1608–1662), 13:451; **8**: 307b
collections, **14**:287a
Traditionalism
opposition to, **9**:171b
Tragus, Heironymus
see Bock, Jerome
Traill, Henry Duff
Social England, **3**:463b
Trains
see Railroads
Trajectories, **2**:54a, 57b; **6**:305a
almost periodic, **13**:36a
orthogonal, **2**:57a; **4**:477b; **8**:430b
parabolic, **1**:424a; **2**:202a; **3**:66b; **5**:246a; **6**:127a
reciprocal, **4**:468b
Transactions of the American Mathematical Society
editors of, **14**:356a
founding of, **9**:502a
Transactions of the Connecticut Academy of Arts and Sciences, **5**:389a
Transactions of the Royal Society of Edinburgh
editors of, **11**:34a
Transbaikalia
geology, **10**:167b–168a, 265b
Transcaspian Region
geology, **10**:166b
Transcendentals
see Functions, transcendental; Numbers, transcendental
Transference (psychol.), **5**:174a–b
Transformationism (geol.), **2**:159a
Transformations (math.), **3**:263a; **5**:572b; **7**:53a–b, 54b; **8**:166a; **15**:307b, 386a
affine, **5**:202a; **7**:3a
algebraic
see Equations, transformation of
Bertini, **3**:468b
birational, **2**:465a; **3**:467b, 468a–b, 469a; **4**:523a; **7**:164a; **10**:140a; **11**:79b; **12**:284b
biuniform, **10**:275a
Bring-Jerrard, **2**:467b–468a
canonical, **6**:89b; **10**:250a
Cayley, **14**:90a
circular, **9**:431a
Clebsch, **2**:198a
contact, **4**:205b, 370b; **8**:323b, 324a, 325a–b; **9**:476b; **12**:150a; **13**:422b
continuous, groups of, **2**:121b; **7**:397b
coordinate, **5**:111a, 567a
Cremona, **2**:85b; **3**:323a, 467b, 468b; **4**: 523a; **5**:569a; **7**:164a; **8**:505a; **11**: 536a; **12**:330b; **13**:133a
definition of, **3**:468b
equiaffine, **13**:291a
Eulerian, **2**:50a; **4**:479b

Transformations (cont.)
Fourier, **3**:136b, 137a, 142a, 143b–144a, 145a; **4**:196b; **6**:211b; **14**:345b–346a; **15**:384b–385a
fourth-degree, **2**:271b
fractional, **2**:615a
geometric, **1**:590b–591a; **2**:85b, 192a; **3**: 128b*, 213a, 225a, 515a–b; **5**:259b*, 527a–b; **6**:606a; **8**:176a, 385b; **9**:598b; **14**:23b–24a, 466a
groups of, **2**:197a; **4**:337a; **8**:434a; **10**: 594b; **12**:533a
homographic, **3**:169a, 468b
hyperalgebraic, **12**:284b–285a
integral, **1**:500a; **5**:96b; **8**:161b, 163a
inversive, **8**:433b
isographic, **7**:164a
Laguerre, **7**:185b*; **8**:324a
Landen, **7**:619a–b; **8**:139b, 140a
Laplace, **1**:500a; **2**:58a, 144b; **6**:211b; **10**:610b; **15**:382b–386b
Legendre, **7**:530b; **8**:136a, 140b–141a
linear, **2**:294b, 594a, 615a,b; **3**:65b, 164a, 166a,b; **6**:308b, 357a, 402a; **12**:284a–b; **13**:220a–b; **14**:13b
groups of, **7**:398a, 398b–399a
Lorentz, **2**:594a; **4**:322a,b, 323b, 331b; **8**:40b, 495b, 496a, 498a; **11**:58b; **14**:62a, 305a, 317b
Mellin, **15**:383b
Möbius, **9**:431a
noninvolutory, **3**:468a
orthogonal, **3**:142b
of plane, **2**:198a; **5**:569a
on principal axes, **6**:392a
projective, **4**:49b; **10**:140a
quadratic, **3**:323a; **5**:569a; **8**:140a–b; **10**: 140a
Radon, **11**:261a
rational, **5**:482a; **10**:275a
by reciprocal polars, **2**:214b; **3**:213a; **9**: 80a–b, 475a, 476a
secant, **5**:159b
of series, **5**:449b–450a; **9**:127a
surface, **2**:145b; **4**:336a–337a; **9**:199b
theory of, **2**:470a; **3**:167b; **6**:308b, 309a; **8**:458a*
Thiele, **13**:339a
third-degree, **2**:271b
Tschirnhaus, **2**:467b; **7**:103a, 211a
Watson, **14**:189a
see also Invariants, Projection
Transformer, electric, **4**:588b; **5**:33a; **6**:5b; **13**:24a, 361b, 570a–b
air-core, **13**:287a
Transformism
eighteenth-century, **2**:580a, 581b; **6**:469a; **9**:26b–27a, 93a
nineteenth-century, **1**:130a,b, 387a–b; **4**: 266b, 309a; **5**:385b; **6**:10a; **9**:112a; **10**:287b; **11**:178b*; **12**:514b
twentieth-century, **3**:429b*; **4**:409b; **8**: 257b; **10**:124b, 126b
opposition to, **8**:254b
precursors of, **11**:493b
Transfusion
see Blood, transfusion of
Transgression, marine, **7**:250b; **13**:147b
Trans-Himalaya Range, **6**:216a
Transients (phys.), **6**:211b
Transistors, **4**:7a
Transit instrument, **2**:140b, 196b, 388b, 471b, 472a; **3**:14a, 92b, 261a, 366b*, 605b; **4**:175b; **5**:316a, 405a, 491b; **6**:69b, 103b, 263b; **7**:543a; **8**:179a; **9**:514a; **10**:608a; **11**:377b, 526b; **14**: 485a,b, 589a
axis of rotation, **4**:407b
photographic, **10**:159a
portable, **5**:404a; **11**:377a
repeating circle, **9**:250b
universal, **10**:159a
Transition strata, **2**:496a; **9**:583b, 584a
Transitivity (math.), **2**:145a–b
Transits (astron.), **2**:155a
calculation of, **7**:87b*
observation of, **2**:149b, 150b; **3**:52a; **5**: 404b; **6**:263b, 513b, 515b
times of, **5**:404b
"stellar magnitude equation," **2**:196b
see also specific celestial bodies
Translation
philosophy of, **1**:274a; **9**:436a; **15**:175b–176a
see also Transmission of science, via translation
Transmission of science
Arabic tradition, **4**:611b–612a; **6**:301a–302a; **14**:547a–548a; **15**:176a–b
Armenian tradition, **1**:274b
Chinese tradition, **4**:612a
Greek tradition, **1**:268a–269a; **2**:21a; **4**: 611b–612a; **7**:24b–25a; **8**:191a–b, 248a–249b; **10**:417a, 418b; **14**:619a; **15**:176a–b, 474b, 494b–495a, 533b
Indian tradition, **4**:502b, 612a; **14**:546a–b; **15**:533b–534b
Latin tradition, **1**:269a–274b; **7**:28a; **9**: 140b–141a
Mesopotamian tradition, **15**:536a
Syriac tradition, **1**:275a–b
translation institutes
Abbasid, **1**:32b, 444a
via commentaries, **1**:272b, 273b, 275b–276b; **7**:212b–213a; **15**:534a
via copying of manuscripts, **3**:109b–110a; **6**:189b; **15**:534a
via correspondence
seventeenth-century, **2**:349a–b; **3**:179b, 348a–b, 457b; **4**:36b; **5**:524b–525a, 606b–607a; **6**:141a–b; **9**:507b, 599a–b; **10**:44b, 200b–202b, 546b; **12**:210b, 460a; **13**:335b–336b, 434a–b
eighteenth-century, **3**:349b–351a, 459b, 461b, 486a; **4**:23b, 469b; **5**:75b–76a, 352a, 375a–b, 380a; **14**:178b; **15**:53a
nineteenth-century, **6**:307b, 308a, 553b; **10**:601b; **12**:552b
via *quaestiones,* **1**:275b–276b; **4**:237b
via translation
Arabic to Armenian, **4**:442a–b
Arabic to Hebrew, **4**:442a, 502a–b; **6**: 197b; **7**:38a; **10**:111b; **13**:400a–b, 401b
Arabic to Latin, **1**:22b–23a, 35b–39b *passim,* 61a–63b *passim,* 82a–b, 107a, 122a–b, 226a–227b, 266a, 270a, 271a–b, 272b, 290a, 378b, 512a–513a, 620a; **2**:308b–309a; **3**: 262a, 364a,b, 393a–394a; **4**:444b–448a, 542a; **5**:591b–592a; **6**:197a,b, 272b–273b; **7**:23a, 38a, 82b, 133b, 363a; **8**:547b–548a; **9**:361a–364b, 604a; **10**:111b; **11**:31b–32b, 504a; **13**:38a–b; **15**:173a–190a, 499a–b, 580b
Arabic to Persian, **4**:442a
Arabic to Sanskrit, **2**:155b; **7**:62a, 81a–b; **15**:628b
Arabic to Syriac, **4**:442a–b; **15**:237a
Arabic to Turkish, **9**:604a
Babylonian to Hittite, **15**:644b
Dutch to Japanese, **1**:311b–312a, 314a
English to Sanskrit, **15**:629a
Greek to Arabic, **1**:31a, 35a, 121a, 188a, 191b, 224a–226a, 275a–b, 339a, 444a; **4**:117a–b, 121b, 438a–443a; **5**:233b; **7**:25a; **9**:39b, 85a, 297b; **11**:202a, 244a, 244b–245a; **13**:289a; **15**:125b, 230b–242b *passim,* 261a–b, 263a
Greek to Armenian, **1**:268b
Greek to Hebrew, **1**:339a
Greek to Latin, **1**:22b, 61a–b, 99a–b, 107a, 199b, 205a, 223b, 227b–229b, 268b, 269a–274b, 339a, 378b–379a, 468a; **2**:228b–234b *passim;* **3**:13b, 14b, 149b, 363b–364b, 534a; **4**:105b, 117b, 121a–b, 396a–b, 443a–444b; **5**:233b, 549a–b, 586a; **6**:306a; **7**:65b–66b; **8**:360b–361a; **9**:434b–438b; **10**:21b; **11**:18a, 42a
Greek to Sanskrit, **14**:549a–b; **15**:533b, 538b, 540b, 541b, 545a, 555a
Greek to Syriac, **1**:268b, 275a–b; **5**: 233b; **7**:25a; **10**:111b; **15**:230b–242b *passim*
Hebrew to Arabic, **1**:22b–23a
Hebrew to Latin, **4**:609a; **11**:31b–32a
Indian to Arabic, **1**:5b, 30b; **7**:223a–b
Latin to Arabic, **7**:187a; **15**:420b
Pahlavi to Arabic, **1**:36b
Persian to Sanskrit, **15**:628b
Sanskrit to Arabic, **2**:155b; **4**:555b–556b; **15**:611a–b
Syriac to Arabic, **1**:275a–b, 508b; **7**: 187b; **10**:111b; **15**:237a–238b
Transmitters
wireless, **5**:33a; **11**:213a
range of, **2**:427b
Transmutation of metals, **1**:101b, 550a, 571b; **2**:227a; **5**:421a–b, 550b; **6**:255a, 257a; **7**:41b, 342b, 376b, 525a; **8**: 310a, 311a, 620a–b; **9**:106b, 336a, 364b; **11**:606b; **12**:79b; **13**:348a, 559a; **14**:508a; **15**:650a
and atomism, **2**:13b
denial of, **5**:580a; **13**:141a–b
explanations of, **1**:49b, 451a; **9**:24a; **10**: 555a–b
universal, **5**:421b; **10**:83a
Transpiration, **2**:286b, 290b; **3**:581b–582a; **4**:285b; **5**:277b, 523b; **6**:38b; **8**:425b; **14**:349b
demonstration of, **6**:39a–b; **7**:409a; **14**: 501b
mechanism of, **4**:131a,b
thermal, **11**:393b
Transplantation, surgical, **8**:447a,b
in animals, **2**:61a, 73a; **3**:91a; **9**:86a
compatibility reactions, **8**:447b
see also specific organs
Transportation, geological, **2**:549a; **3**:162a, 210b–211a
Transportation, mechanical
ancient Greek, **6**:312a
medieval, **7**:4b
nineteenth-century, **11**:340b
twentieth-century, **13**:484a
Transportation, public, **10**:337a–b
Transposition (math.), **7**:351b; **10**:484a
Trans-Siberian railway, **10**:167b
Transubstantiation, doctrine of, **10**:172b
Transuranic elements
see Elements, chemical, transuranic
Transversals, **2**:402a, 405b, 412b, 455a; **3**: 78b*; **4**:427a, 611b; **12**:325b, 431b, 432a
Ceva's theorem, **3**:182a
Trap (rock), **5**:519a; **9**:61a
Trapezium, **6**:414b
area of, **15**:670b, 671a
center of gravity, **8**:224a
Trapezoid
area of, **4**:609b
dimensions of, **15**:691a–693a
Trapham, Thomas, **14**:287a

Traube, Ludwig, **2**:129b; **5**:196a; **7**:504a; **13**:451b–452a
Traube-Hering curves, **5**:150a
Traube, Moritz (1826–1894), 13:451–453; **1**:362b; **2**:561b; **10**:377a, 577a; **14**:64b
Trauma, **11**:99a
and local hysteria, **3**:205b
neurological, **9**:109a
Traut, Herbert, **10**:291b–292a
Travailleur (ship), **7**:92a
Travelers
medieval, **1**:516a–517b; **2**:619a; **7**:4a, 8a; **9**:171a; **14**:547a,b
Renaissance, **1**:596a; **10**:616a–b; **11**:311b–312a; **14**:335b
seventeenth century, **7**:204a–205a
eighteenth-century, **2**:56b, 327a; **3**:291a; **7**:390b, 581a; **9**:135a; **13**:442b, 448a; **14**:583a
nineteenth-century, **1**:501b–502a, 562b; **2**:463b, 555a; **7**:496a; **8**:521a; **11**:385a; **12**:72a; **14**:135a; **15**:268a
twentieth-century, **3**:188a; **15**:506a
Travers, Morris William (1872–1961), 13:453–455; 14:318b, 319b*
and Ramsay, **11**:279b–280a
Treadmill
efficiency of, **8**:211a
in physiological research, **12**:466b
Treasury Islands, **2**:342b
Treasury of Analysis, **4**:392b, 416a, 425b, 426b, 429a; **10**:295b
Treasury of Astronomy, **4**:429b, 430a; **6**:617a; **8**:191a; **10**:295b; **13**:320a
Trebra, Friedrich Wilhelm Heinrich von (1740–1819), 13:455–456
Trécul, Auguste, **10**:376b, 380b
Tree frog, **2**:322a
color mechanism, **3**:588b
Tree of Diana (mineral.), **6**:478a
Tree snail, **6**:613a
Trees
age of, **3**:43a
bark
tracheae, **15**:113b
budding, **7**:503b
ecology, **5**:441a
fertilization, **10**:1b
fossil, **2**:553b; **3**:568a
growth, **8**:206a; **12**:167b
environmental influences, **1**:101b; **3**:45a
statistical studies, **7**:239b
identification of, **9**:24b
pathology, **5**:441a
phyllotaxis, **8**:206a–b
ring formation, **8**:206a; **12**:100a
roots, **8**:512b
rot, protection from, **6**:381a
unisexuality of, **3**:575a
see also Arboriculture; Dendrology; Forestry; Sap; specific trees; Wood
Tréfuël, Jacques, **4**:154b; **5**:100a; **8**:273b
Tréfuël, Thérèse, **4**:154b; **5**:100a; **8**:273b
Trelease, William (1857–1945), 13:456; 3:197a*; **15**:224a
Trematodes, **1**:602b; **2**:127b, 617a; **5**:109a, 453b; **12**:453b; **14**:203a
Trembley, Abraham (1710–1784), 13:457–458; 8:579b; **12**:308b, 555b, 560b
Mémoires pour servir à l'histoire d'un genre de polypes d'eau douce, **8**:580a
zoology, **1**:411a; **2**:286a; **10**:569a; **11**:333b–334a
Tremella, **3**:426a
Trémery, Jean-Louis, **15**:360a, 361a
Trench fever, **2**:529a
Trendelenburg, Friedrich, **11**:462a
Trent, Council of, **4**:212b; **5**:105a; **10**:197a; **12**:548a–b; **13**:573a
Trepanation
see Trephination
Trephination
Renaissance use of, **2**:546b; **14**:28a
twentieth-century use of, **2**:1a
Trepidation theory, **1**:312b, 313b, 511a, 512b; **2**:20b, 24a; **3**:26b; **7**:112b, 304a; **13**:292a, 324a; **14**:273b, 593a, 593b–594a; **15**:475b, 546a–b, 586b, 602a, 610a
opposition to, **8**:280b; **15**:34a
see also Precession of equinoxes
Treponema, **4**:300b; **8**:273b; **11**:515b; **12**:141b
Treub, Melchior (1851–1910), 13:458–460; **7**:16a*; **8**:512b; **10**:1b; **13**:154b
Trevelyan, Walter Calverly, **5**:71b
Treviranus, Gottfried Reinhold (1776–1837), 13:460–462; **2**:123a, 443a; **13**:404b*, 463a
Treviranus, Ludolph Christian (1779–1864), 13:462–463; **10**:11b; **12**:592a; **13**:404b*
Trevisanus
see Bernard of Trevisan
Trew, Christoph Jacob, **3**:350b; **4**:295a
Tria prima, **1**:549b, 571b; **5**:47b, 422b; **7**:165a, 356a, 538b; **8**:82b, 130b–131a; **10**:309b–310a; **11**:606b; **12**:311b–312a, 335a–b, 604b; **13**:559b
opposition to, **6**:478a
Triacetic acid; **3**:347b
Triana, José Gerónimo (Jerónimo) (1826–1890), 13:463–464
Triangle
altitude of, **6**:208a*
analytic, **5**:567a
angles, sum of, **4**:9a; **7**:600a; **8**:141b–142a, 430a
area of, **1**:226a, 227a, 445a; **3**:254b, 277a; **4**:609b; **5**:365b; **6**:314b; **7**:175a; **12**:314a; **15**:670b, 671a
as basic form of nature, **5**:470b
Brocard circle of, **2**:479a, 480a
Brocard points of, **2**:479b–480a
center of gravity; **8**:223b–224a
differential, **12**:501a
division of, **4**:426a
equiangular, **6**:414b
equilateral, **4**:610a
geodetic, **1**:591a
geometry of, **2**:479b–480a; **4**:515a, 516a, 602a; **8**:175b; **9**:358b; **10**:24b; **12**:228b
Heronian, **5**:528a; **12**:206b
homological, **11**:413a
inscribed, **5**:209b; **10**:297a; **13**:18b
isosceles, **4**:415a, 417b; **6**:414b; **10**:299b
construction of, **7**:29b
movement on a plane, **2**:215b
nine-point circle of, **4**:602a
number theory of, **5**:159b; **11**:222a
pedal line, **14**:141a
in perspective, **4**:48b
polar, **11**:4a; **12**:501a; **14**:22a
Pythagorean, **8**:163a; **11**:220b
quadrisection of, **2**:47a
as representation of motion, **8**:232a–233b
right, **1**:225b–226a; **5**:409a; **6**:414b
rational, **5**:159b
sides of, **6**:414b
spherical, **5**:209b; **6**:144a, 271b; **8**:135b–136a, 137b; **9**:297b; **10**:151b, 161b; **12**:501a, 518a–b; **14**:274a
area of, **5**:409a; **6**:125a
graphic solution of, **1**:591a; **3**:312a
right, **7**:405a; **9**:610b; **13**:320a, 510b
sine theorem, **9**:555b; **11**:4a; **13**:510b
supplementary, **5**:409a
symmedians
intersection of, **8**:175b, 176a–b
theory of, **2**:274a; **4**:117b
see also Trigonometry
Triangulation
astronomical, **2**:100a; **8**:432a; **12**:518a
chains of, **6**:240a
double, **8**:423a–424a
geographical, **2**:102a; **3**:108a,b, 250a; **4**:15a–b, 16a, 217b; **5**:78b, 303a–b, 542b–543a; **7**:2a, 497a, 499b–500b, 542b; **8**:22a, 351a; **9**:250b, 336a; **10**:231b; **12**:209b, 500a–b; **13**:111a; **14**:315b; **15**:376a
North American Datum, **2**:372b
instrumentation, **2**:419a
principle of, **5**:349a
skeleton method, **10**:595b
Trianon, botanical garden of, **7**:199b
Triarylmethyls, **3**:246b
Triassic period, **5**:476a
paleobotany, **3**:187b; **12**:169a; **13**:345b; **14**:600a
paleoclimatology, **12**:168a
paleogeography, **7**:250a,b
paleontology, **1**:95b, 491b; **2**:109b, 110a, 504b; **3**:530a; **5**:214b*; **6**:221b, 438a; **7**:60a, 61b; **9**:100b, 314a, 346a; **10**:40b, 262a; **11**:340b, 386a
stratigraphy, **2**:89b, 109a; **4**:262b, 361b, 540b; **5**:155b, 156a; **6**:229b; **7**:253a; **8**:33b; **11**:439a; **12**:487a; **13**:87a; **14**:88a
volcanic sequence, **3**:593a
Triatoma, **2**:533b; **3**:185b, 566a
Tribe, Alfred, **5**:412b
Triboelectricity, **5**:136a; **6**:171a; **9**:471b; **14**:70a, 353a
Triboluminescence, **2**:55a; **3**:271b
Triceratops, **4**:147b
Trichina, **8**:270b
Trichinella, **8**:169a
Trichinosis, **10**:261a, 384a; **13**:62b–63a
Trichloroacetic acid, **4**:246a
Trichocephalus, **5**:502b, 503a
mode of development, **3**:588a
Trichogyne, **13**:395b
Trichomycetes, **8**:146a
Trichophyton, **5**:566a; **9**:176a
Trichosurus, **2**:504a
Trichuris, **1**:26a
Tridacna, **13**:551b, 552a
Tridymite, **4**:563b
Triesnecker, Franz, **2**:599b, 600a; **6**:234b
Trifluoroacetic acid, **13**:177a
Trigonocarpus, **12**:260a
Trigonometry
ancient Greek, **1**:249a–b; **15**:208b–209b
ancient Roman, **9**:297a–300b; **11**:188b–189a
medieval Arabic, **1**:22b, 42a–b, 122a, 508b, 511b–512a; **2**:153a–b, 154b; **5**:612a–613a; **7**:257b–259a, 532a; **9**:84b–85a, 555a–556a; **10**:6a; **11**:227b–228b; **12**:361b; **13**:510b, 536b; **14**:538a, 577a–b, 578a–b, 593b; **15**:260a
medieval Hebrew, **8**:280a
medieval Latin, **2**:435a; **6**:544b; **7**:118a, 120a, 123b, 128b–129a, 140a, 141b–142b; **11**:32b, 415a
Renaissance, **2**:406a–b; **3**:181a–b, 312a; **4**:6a*, 619a; **8**:27b–28a; **9**:170a*, 191a,b, 611a; **11**:350a–b; **14**:18b–19b, 22a, 24a–b, 470b–471a, 514a; **15**:155b–156a, 478b
early seventeenth-century, **2**:419a–b, 462b–463a; **3**:152b; **4**:551a; **5**:344a, 408b–409a, 593b; **6**:272a*; **9**:13a,

Trigonometry (cont.)
611a–613a; **10**:151b; **11**:3b–4b, 533b; **12**:501a
late seventeenth-century, **3**:348a; **5**:528a–b; **10**:255a; **14**:162b
eighteenth-century, **4**:38b; **5**:209b; **7**:87b*, 599b; **8**:299b–300a; **9**:17a, 454b; **14**:280b
nineteenth-century, **2**:441b; **3**:263b; **6**:502b; **8**:431a; **9**:429a, 463a; **14**:500a
early twentieth-century, **12**:364b
absolute, **2**:271a
analytical, **7**:405a
in astronomy, **1**:314a; **7**:140a, 142a
Chinese, **12**:378a
of chords, **7**:258a–b
complex numbers used in, **1**:239a
hyperbolic, **12**:255a
Indian, **14**:471a(n)
naming of, **11**:3b
non-Euclidean, **13**:264b
notation, **10**:151b, 255a; **11**:533b
parabolic, **5**:573b*
prosthaphaeretic method, **2**:406a–b, 413a, 602b; **3**:262a–b, 312a; **9**:611a; **11**:561b; **14**:274a–b, 470b–471a
spherical
ancient Greek, **6**:617a; **15**:209b
ancient Roman, **9**:297b–300b; **11**:189a
medieval Arabic, **1**:512a; **2**:118a, 154b; **7**:38a–39a, 83a, 335a, 352b–353a; **9**:84b–85a; **13**:291b; **15**:259b–260a
Renaissance, **2**:407b; **4**:619a; **8**:27b; **11**:350a; **14**:19a–b, 273a–275a
seventeenth-century, **2**:462a; **3**:621b; **5**:409a; **9**:310a,b, 527b, 610b; **12**:501a, 610a
eighteenth-century, **2**:328a; **4**:166a, 479b; **5**:567b; **7**:405a; **8**:135b–136a, 300a
nineteenth-century, **1**:449b; **8**:431a, 504a; **9**:410a, 431a; **13**:81a
anharmonic property, **9**:299a–b
foundations of, **13**:125a
Indian, **15**:548a–b, 550b–553b, 559a, 570a–573a, 582b–583a, 592a
theorem of sines, **1**:42b; **15**:568b, 580a, 586b
theorem of tangents, **1**:42b
see also Menelaus of Alexandria, theorem of
see also Astronomy, spherical; Historiography, of trigonometry; Sine, etc.; Tables, trigonometric; Triangulation
Trihedral
moving, **3**:428a, 559b
Trilia, Bernard of
see Bernard of Le Treille
Trillion, **3**:273a; **5**:409a
Trilobites, **2**:490a, 496a; **5**:518a, 578b; **8**:33b, 379a; **11**:321a, 336a; **13**:44a; **14**:120a
metamorphosis of, **1**:469a
Trilogarithm, **7**:619b
Trimen, Henry, **6**:490a; **13**:341b, 342a
Trimethyl acetic acid, **2**:624a
Trinidad (W.I.); **10**:617a
discovery of, **15**:89b
Trinitrotoluene
derivatives of, **13**:406b
Trinity, dogma of, **1**:1b; **2**:163b, 231a, 232b; **3**:295b; **5**:550a; **7**:303a, 307b; **10**:81b, 474b; **12**:322b–323a; **14**:152b–153a
Trinity College, Cambridge
in seventeenth century, **14**:370b
in nineteenth century, **12**:9b–10a
in twentieth century, **12**:10a–b
observatory, **3**:430b
physical sciences at, **3**:431a
physiology at, **5**:80a
Trinity College, Dublin
in nineteenth century, **12**:86a–b
mathematics at, **6**:86a
observatory, **8**:426b
science at, **7**:161b
Trinity College, Oxford
in seventeenth century, **6**:129b
Trinity House, Corporation of, **13**:105a
Triode, **7**:108a; **10**:519a
feedback in, **6**:546b
invention of, **1**:287a, 292b, 304a; **4**:6b–7a; **9**:99a
Trioen, Cornelius, **3**:37b
Trioses, **5**:6a
Triphenylmethane
compounds, **3**:246b
synthesis of, **7**:283a
Triphenylmethyl radical, **5**:464b–465a
Tripod, optical, **3**:239a
Tripoli, **11**:312b
paleontology, **2**:110a
Triquetrum (astron. instrum.), **1**:509a; **2**:404a; **7**:259b, 338b
Triremes, **2**:312a
Trisection
see Angle, trisection of; Arc, trisection of
Trismegistus
see Hermes Trismegistus
Trithemius, Johannes, **1**:79b; **5**:115a; **6**:396b, 397a; **10**:304b
Triticum, **15**:507a
Tritium, **12**:34a
Triton (satellite), **8**:46a
Trituration, **8**:373b
Trivium, **1**:104b; **2**:229a, 401a; **3**:109b; **5**:592b; **8**:548a; **9**:140b; **13**:339b, 588b
Trjitzinsky, Waldemar Joseph, **2**:144b
Trocar
invention of, **12**:103a
Trochiliscids, **7**:252b
Trochophore, **12**:107a
Trochophore theory, **6**:167a–b
Trochozoa, **5**:385b
Troil, Uno von (Bishop), **4**:549a
Trois-Frères, Les (cave), **2**:450b
Troja, Michele (1747–1827), 13:464–465; **11**:67a
Trojanus, Curtius, **7**:174b; **13**:259b–260a
Trommsdorff, Johann Bartholomäus (1770–1837), 13:465–466; **2**:565a; **3**:465b; **5**:533a*; **9**:559b; **15**:172a,b, 205b
chemistry, **5**:532b; **11**:437b
Trommsdorff, Wilhelm Bernhard, **5**:532a; **13**:465b; **15**:205b
Tronchin, Théodore, **2**:73b
Troost, Gerard (1776–1850), 13:466–467; **7**:44b; **8**:103b, 266b; **10**:258a
Troost, Louis Joseph (1825–1911), 13:467–468; **6**:177b; **9**:451b; **10**:246b
Troostite, **10**:246b
Troostwijk, Adriaan Paets van (1752–1837), 13:468–469; **7**:12a, 499b; **9**:152b; **10**:446a
Tropfke, Johannes (1866–1939), 13:469–470; **3**:312b*
critics of, **15**:221a(n)
Trophoblast, **6**:536a
naming of, **6**:535b
Trophocoel theory, **8**:3b
Tropical medicine
see Medicine
Tropics (cartog.), **1**:574a
Tropidoleptus, **14**:392a
Tropisms, animal, **7**:98b; **8**:445b–446a, 446b; **10**:611a
Troposphere, **11**:558a
Trotter, Alexander Pelham, **6**:211b
Trotter, Coutts, **5**:81b
Trotter, Thomas, **8**:363a
Troughton, Edward (1753–1836), 13:470–471
instrumentation, **2**:471b, 472a; **9**:231b; **12**:552a; **14**:485a
Troughton, John, **13**:470b
Troughton and Simms (instrument co.), **5**:405a; **14**:448b
Trousseau, Armand, **11**:51b; **13**:493a
Trousseau's phenomenon, **8**:295a
Trout (fish), **6**:267b
rainbow, **8**:146a
Trout Lake Limnological Laboratory, **2**:142a
Trouton, Frederick Thomas (1863–1922), 13:471–472; **8**:496b; **11**:290b; **13**:82b*
Trouton's law, **8**:545b; **13**:471b
Trouvelot, Étienne Léopold (1827–1895), 13:472–473
Trowbridge, John (1843–1923), 13:473; **3**:372b; **4**:194b, 195a; **6**:51a; **10**:481b
Troy (Ilium), **12**:179b–181b
Trudaine, Daniel-Charles, **4**:71a; **9**:478a; **12**:65a,b; **13**:495a
Trudaine de Montigny, Jean Charles Philibert, **8**:75b, 83b; **9**:5b; **13**:495a
Trudeau, Edward Livingston, **7**:424a, 426b
Trudy po prikladnoi botanike, genetike i selektsii, **15**:507a
True, Alfred Charles, **1**:325b
Truesdell, Clifford Ambrose III, **3**:135b; **8**:227a,b; **9**:223b–224a, 225b
Truffles
cultivation of, **9**:176a
Trulli, Giovanni (1598–1661), 13:473–474
Truman, Harry S., **4**:582a; **13**:227b
Trump, John George, **13**:570a
Trumpler, Robert Julius (1886–1956), 13:474–475; **3**:35b, 36b; **7**:237b; **9**:504b*; **12**:455a
Truth
definition of, **8**:439a
scientific
temporal nature of, **11**:544a
Truth-table, **7**:105b; **11**:107a
Trypan red, **4**:299b–300a
Trypanosoma, **12**:142a
cultivation of, **10**:155b
identification of, **5**:566a; **9**:328b
life cycles, **2**:533b; **7**:429a; **11**:175a
resistance of, **9**:328b
T. cruzi, **3**:185b
T. lewisi, **8**:296b
Trypanosomiasis, **2**:528a–529a; **3**:185b–186a; **7**:428a
chemotherapy, **4**:299b–300b; **7**:429a–b; **9**:328b
control of, **7**:429b
Tryparsamide, **5**:100a; **15**:252b
Trypsin, **7**:520a
activation of, **1**:537a–b
Tryptophan
isolation of, **6**:500b
metabolism of, **4**:358a
nutritional role, **6**:501a; **9**:285a; **10**:243a
Tsang (anat.), **15**:740b
Ts'ao Hsiao-chung, **8**:397a*
Tschebytschew, Pafnuty Lvovich
see Chebyshev, Pafnuty Lvovich
Tschermak, Gustav (1836–1927), 13:475–477; **1**:551a–b; **4**:140b; **6**:358b
Tschermak von Seysenegg, Erich (1871–1962), 13:477–479; **3**:421b; **7**:623b; **9**:277a; **14**:104b*
genetics, **14**:96b, 99b
Tschermaks mineralogische und petrographische Mitteilungen
see *Mineralogische Mitteilungen*

Tschichatschew
see Chikhachev
Tschirnhaus, Ehrenfried Walther (1651–1708), 13:479–481; **9**:447a
algebra, **7**:103a; **8**:162a
correspondence of, **3**:348b
critics of, **3**:458b; **4**:516b
influence of, **2**:47a; **7**:567a
and Leibniz, **8**:160a, 162a,b; **13**:479b–480a, 480b
technology, **6**:477b; **11**:330a
Tschirnhaus transformation, **2**:467b; **7**:103a, 211a
Tschuprow, Alexander Alexandrovitch, **1**:154b; **2**:319a; **15**:31b
Tserasky (or Cerasky), Vitold Karlovich (1849–1925), 13:481–482; **2**:196a; **7**:471b; **10**:233b; **13**:45a
Zöllner-Tserasky photometer, **13**:481b
Tsetse fly
see *Glossina*
TSH
see Thyrotrophic hormone
Tsien, Hsue-shen, **7**:247b, 248a
Tsimshian Indians, **2**:210a, 212a
Tsingtao, **11**:439b
Tsiolkovsky, Konstantin Eduardovich (1857–1935), 13:482–484; **5**:433a; **7**:465a
Tsu Cheng-chih, **8**:420a
Tsu Ch'ung-chih (*ca.* 429–*ca.* 500), 13:484–485
pi, value of, **7**:258b; **8**:420a; **9**:335b
Tsu I, **3**:266b
Tsu Keng, **12**:377a; **13**:484b
Tsuchimikado family, **15**:733a,b, 735a,b, 745b
Tsumeb mine, **12**:194b
Tsunami, **7**:9b; **9**:181a; **12**:342a
geomorphological theory, **6**:55b
Tsung shu, **3**:253a
Tsung teng, **3**:253a
Tsutsugamushi disease, **13**:263a
Tsvet (or Tswett), Mikhail Semenovich (1872–1919), 13:486–488; **9**:96a; **15**:257b
Tu Shih-jan, **3**:266a,b, 269a; **8**:419a
Tuamotu Archipelago, **2**:342b
Tuber(s)
as food source, **9**:18a
see also Tuberization
Tuberculin, **4**:297a,b, 404a; **7**:392a, 426b–427a, 430a; **8**:409a; **15**:522b
hypersensitivity, **14**:624a
Stichreaktion, **4**:405a
Tuberculosis
seventeenth-century discussion, **6**:256a
eighteenth-century discussion, **1**:563b
nineteenth-century discussion, **2**:575b; **3**:463b; **4**:21b
acute vs. chronic, **14**:408a
animal vs. human, **7**:100b; **9**:545b*; **12**:481b
bacillus, **3**:22b; **4**:296b; **5**:220a*, 270b; **7**:423b, 424a; **8**:273a, 275b, 608a; **12**:481a,b, 482b; **14**:551b
of bone, **5**:220b; **11**:596a
bovine, **1**:576b; **3**:463a; **7**:423b–424a, 427a, 428b, 429a,b; **8**:409a; **10**:527a; **12**:481b
in children, **4**:404a
and climate, **14**:464a
control of, **7**:430a; **8**:313a, 557a
diagnosis of, **11**:462b
bacteriological, **7**:424a
chemical, **3**:160a
immunological, **7**:427a
X-ray, **4**:405a
etiology of, **5**:270b; **7**:423b–424a; **8**:9a; **12**:87b
experimental study, **2**:598b; **3**:220a; **5**:219a
heredity and, **10**:464a; **12**:96b
induced, **14**:115a
of joints, **8**:402b
laryngeal
treatment of, **4**:297b
lesions, **2**:444b; **7**:556b
miliary, **14**:218b, 229a
occupational, **10**:305b, 307a
of paracentral region, **3**:203b
pulmonary, **3**:437b; **6**:576a; **11**:510a
histology of, **9**:95a; **11**:374a; **12**:224a
treatment of, **4**:297a,b; **7**:425b–427a; **9**:540a; **11**:482a
serology, **6**:433a; **11**:429a–b; **12**:48b; **15**:523a
spread of, **1**:576b; **8**:7a; **9**:86a
treatment of
dietary, **11**:429b
pharmaceutical, **4**:155b; **11**:263b
vaccination, **1**:576b; **2**:584a; **3**:22b; **9**:545b; **14**:512b
BCG, **3**:22b
see also Lupus, Phthisis
Tuberization
of orchids, **2**:35a
of potatoes, **2**:35a; **3**:429a; **9**:461a
Tübingen, University of
chemistry at, **5**:429b
Faculty of Sciences, **9**:442a
geology at, **11**:235b
and Kepler, **7**:293a
medicine at, **15**:160b
Physical Institute, **2**:427b
physics at, **1**:370b, 371a; **10**:347a, 348a
Tuc d'Audoubert (cave), **2**:450b
Tucker, Robert, **8**:175b
Tufa, **12**:182a
origin of, **2**:439b
Ibn Ṭufayl, Abū Bakr Muḥammad (Abubacer) (*ca.* 1110–1185), 13:488–489; **12**:1b, 2a–b, 8a; **15**:33a
Ḥayy ibn Yaqẓān, **9**:604b
Tuff, **4**:548b, 563b; **8**:511b; **12**:168a
Tulasne, Charles, **3**:612a, 613a; **11**:132a; **13**:489a
Tulasne, Louis-René (1815–1885), 13:489–490; **3**:612a, 613a; **11**:132a
Tull, Jethro
Horse-Hoeing Husbandry, **4**:224a; **6**:578b
Tulley, William, **1**:135b; **8**:414a
Tulp, Nicolaas (1593–1674), 13:490–491; **10**:621b; **13**:222b
Tulp's valve, **13**:490b
Tumors, **2**:478a; **3**:218b; **11**:445a
abdominal, **2**:464a
of bladder, **1**:322b
of blood vessels, **3**:518b
classification of, **11**:465b
cytology, **15**:479b
diagnosis of, **3**:490a; **9**:572b
fermentation in, **14**:173b, 174b
fibrous, **1**:612a
of heart, **1**:612a
histology, **9**:572b; **11**:491b
histolysis, **15**:480a
intracranial, **1**:569b; **3**:490a, 517b, 518b; **8**:7a
location techniques, **4**:286b
malignant
chromosome theory, **2**:364b
transplantation of, **4**:299b
mediastinal, **14**:638a
morphology, **4**:499b
ovarian, **2**:386b
of pituitary gland, **3**:518b; **6**:518b
spinal, **6**:518b–519a
theory of, **11**:601a
transplantation of, **8**:447b; **9**:586a
treatment of, **1**:322b
radiotherapy, **9**:550a; **13**:570a
see also Gliomas, Meningiomas, Oncology
Tumulus, **14**:398a
Tundra, **10**:17a; **13**:249a
Tung Yu-ch'eng, **3**:269a
Tunga penetrans, **1**:26a
Tungstates, **14**:588b
Tungsten, **3**:473b
compounds of, **2**:565a; **3**:400b–401a, 617b; **5**:394a; **10**:532b
deposits of, **3**:609b
discovery of, **4**:345b, 346b; **8**:147a
electric current in, **11**:420b
in electric lamps, **7**:92b; **8**:23a–b
evaporation of, **8**:23b
filaments, **8**:23a–b, 24a
isolation of, **9**:451b
preparation of, **4**:21a
work function, **3**:598a*
Tungstic acid, **5**:595b
discovery of, **12**:148a
Tunica vaginalis, **3**:587b
Tunicates
embryogeny, **2**:502a; **3**:120b, 148b; **7**:475a
morphology, **1**:601a–b
physiology, **6**:590b
taxonomy, **1**:601a; **6**:590b
Tuning (music)
see Instruments, musical, tuning of
Tuning fork, **7**:445b; **8**:399a; **13**:11b
clock, **7**:445a
distribution of sound, **14**:206a
Tunnels, **11**:375a; **13**:143b; **15**:151b
ventilation of, **6**:265a
Tunstall, Cuthbert (1474–1559), 13:491–492; **10**:494a
Tupaia, **6**:535b
Tupolev, Andrey Nikolaevich (1888–1972), 13:492; **7**:465a; **14**:621b
Turán, Paul, **4**:562a; **14**:591b
Turba philosophorum, **1**:571b; **7**:41a; **10**:555b; **14**:632a
Turbellaria, **3**:148b; **5**:385b; **9**:328b; **12**:231a
Turbidimeter, **11**:208a
Turbidity currents, **1**:394b; **3**:548a
Turbines, **11**:393b
gas, **13**:73b
hydraulic, **4**:481a; **5**:101a; **10**:286b*; **11**:344a
naming of, **5**:101a
reaction, **8**:211b; **12**:283a
steam, **2**:565b; **3**:20a, 239a; **8**:212a; **11**:306a; **13**:73a–b
water, **14**:618a
wind, **13**:614a
Turbulence (aerodyn.), **9**:419b
law of, **7**:247a
theory of, **11**:124b
Turbulence (hydrodyn.), **11**:393a; **12**:527a
Türck, Ludwig (1810–1868), 13:492–493; **3**:351a; **5**:27a, 571a
Turgor, **14**:97b–98a
and respiration, **9**:24b
Turgot, Anne-Robert-Jacques (1727–1781), 13:494–497; **3**:383a; **15**:271b
acquaintances of, **3**:1a, 383b; **4**:71a; **8**:68a, 84b; **9**:500b; **11**:563a
administration, **3**:440a; **8**:85b(n); **15**:304b, 333b
on contemporary scientists, **12**:64a–b
correspondence of, **3**:286a*
and education, **2**:335a; **9**:471b
"Expansibilité," **8**:73a
medicine, **14**:14b, 15b

Turgot, Étienne-François (1721–1788), 13: 497
Turin
  Academy of Sciences, 7:561a, 567b
  Electricity Exposition, 4:588b
  Royal Geological Museum, 1:587a
Turin, University of
  astronomical observatory, 11:6b
  mathematics at, 10:441a
  parasitology at, 10:527a
  physics at, 1:343b
  physiology at, 9:546b
  veterinary school, 10:397a
Turing, Alan Mathison (1912–1954), 13: 497–498; 11:107a
Turkestan
  see Turkistan
Turkey (country)
  archaeology, 12:180a–181b; 14:504b–505a
  geography, 2:341b; 6:215a
  geology, 10:507a; 13:620b
  natural history, 12:561a–b
  medicine, 1:125a
  paleontology, 1:210b
Turkey (bird)
  diseases of, 12:481a, 482a,b
Turkey vulture, 1:331b
Türkheim, Ludwig Freiherr von, 13:493a
Turkistan
  exploration of, 11:210a–b
  geography, 6:215b
  geology, 9:592a,b
Turner, Dawson, 6:493a; 7:610a; 8:564a
  Historia fucorum, 6:492b
Turner, Edward (1796–1837), 13:499–500; 4:181b; 5:492b; 10:510b; 14:489b
  atomic weights, 11:173b
Turner, Frederick Jackson, 2:374a; 4:409a
Turner, Herbert Hall (1861–1930), 13:500–501; 4:498a; 5:414b*; 6:32b(n), 34a*; 8:388b*, 535b*; 10:36a*, 585a
  formulas of, 1:429a–b
  Zöppritz-Turner tables, 13:500b
Turner, Louis Alexander, 12:21a
Turner, Matthew, 11:143b
Turner, Peter (1586–1652), 13:501; 14:147a
Turner, R. N., 3:370b
Turner, Richard, 6:494a
Turner, Thomas, 3:546a
Turner, William (1508–1568), 13:501–503; 5:384a; 11:316a; 13:352b; 14:507b
  The Names of Herbes, 2:476a
Turner, William (1832–1916), 13:503–504; 9:588b; 11:515a*
Turner, William Ernest Stephen (1881–1963), 13:504–506; 9:313b*; 12: 212b*
Turnip
  tumors in, 7:101a
Turnip fly, 10:39b
Turpentine, 1:345b; 4:77b; 10:530b
  derivatives of, 11:85a; 13:410b
  isomerism, 2:65b
  rotation of light, 2:138b, 139a
Turpeth mineral, 4:209a
Turpin, Pierre Jean François (1775–1840), 13:506–507; 4:67a; 12:243b
Turquet (astron. instrum.)
  see Torquetum (astron. instrum.)
Turquet de Mayerne, Theodore (1573–1655), 13:507–508; 1:571a; 4:209a,b, 388a; 8:310a; 9:440b; 10:310b; 14: 508a
Turquoise, 11:328b
Turriano, Jannello, 3:66a
Turtle, 2:322b
  fossil, 4:147b; 12:38b
Tuscany
  geology, 3:328a; 13:34a
  natural history, 13:257b–258a
al-Ṭūsī, Muḥammad ibn Muḥammad ibn al-Ḥasan (Naṣīr al-Dīn) (1201–1274), 13:508–514
  arithmetic, 10:338b(n); 13:545b
    ratios, 7:327b
  astronomical tables, 7:324b; 11:227a
  astronomy, 7:81a, 336b; 11:202a; 13: 140b*; 15:34b
    at Marāgha Observatory, 7:260a, 337a; 9:555a; 11:202a
  calendar, 7:259b
  commentaries on, 11:249b, 251b
  editions by, 13:289a
    Archimedes, 1:224a; 9:615a; 11:240a
    Banū Mūsā, 1:444a; 7:335a
    Euclid, 4:440a–445b *passim;* 7:62a
    Menelaus, 9:22a, 301b*
    Ptolemy, 15:628b
  geometry, 6:202a; 7:330a, 335a; 8:429b; 12:56b
    proportions, 7:83a
  Īlkhānī zīj, 12:362a; 14:576b, 578a
  influence of, 9:35b; 12:91a, 358b
  Jawāmiʿ al-ḥisāb bi'l-takht wa'l-turāb, 7:257a, 326a
  optics, 11:250a
  al-Risāla al-Shāfiya . . . , 6:202b; 7:79a,b, 80a,b
  students of, 11:248a; 13:514b
  Tadhkira, 6:198b, 199a–b; 11:251a
    commentaries on, 11:249a; 15:628b–629a
  Taḥrīr al-Majisṭī
    commentaries on, 15:629a
  translations by, 13:150a*
  translations of works, 11:248b
  trigonometry, 7:352b; 9:84b
al-Ṭūsī, Sharaf al-Dīn al-Muẓaffar ibn Muḥammad ibn al-Muẓaffar (*d. ca.* 1214), 13:514–517; 7:247a
  al-Ṭūsī's staff, 13:514b
Tutton, Alfred Edwin Howard (1864–1938), 13:517–518
Tuve, Merle Antony, 8:94a
Tuzet, Odette, 4:207a
Twenhofel, William Henry (1875–1957), 13:518–519
Twilight
  medieval Arabic study, 7:82b; 14:579a
  medieval Latin study, 5:401b
  Renaissance study, 5:400b
  eighteenth-century study, 2:4b
  twentieth-century study, 1:519b
  duration of, 10:161b; 12:360b
  Egyptian study, 15:713a
  second twilight, 1:519b; 10:161b
Twining, Alexander Catlin, 10:41a
Twins, 3:149a*, 245b
  bovine, 8:358a–b
  genetics of, 3:555b
  joined, 6:62b–63a
  statistics of, 14:230b
Twisting
  of rods, 4:230b
  of shells, 4:230b
Two-body problem, 10:73a–b
  inverse, 2:54b
Twort, Frederick William (1877–1950), 13: 519–521; 6:298a; 15:418a
Tylor, Edward Burnett, 5:145a
Tyndall, John (1820–1893), 13:521–524; 1: 288a; 3:475b, 476a; 11:537a
  acquaintances of, 3:303b; 4:527a; 5:125a, 126a; 8:528a; 11:241b, 277a; 12:52b
  biology, 3:339b; 10:372a, 382a, 383a
  on contemporary scientists, 2:575a; 4: 540a*; 5:69a*
  correspondence of, 3:470b; 9:239b; 13: 384a
  critics of, 1:496b, 497a
  "On Dust and Disease," 8:405a
  geology, 12:593b
  Heat Considered as a Mode of Motion, 11:10a–b
  influence of, 6:53a; 7:162a; 8:122a, 443b
  and Joule-Mayer dispute, 3:306a
  lectures of, 3:292a
  physics, 8:399b; 12:593b; 13:11a
  Tyndall effect, 13:522a
  Tyndallization, 13:522a
Type, casting of, 2:143a; 15:24b
Type theory, chemical, 1:518b; 2:199b, 621a–b; 3:193b–194a; 4:246b–247a; 5:295a, 315a, 372b–374a, 431a; 6: 462a–b, 463b, 564b; 7:280a–b; 9: 286b, 288a; 10:177b–178a; 12:619b; 14:156b, 395a–396a, 530b, 531b
  critics of, 7:452a, 453a
  mixed types, 10:178a
Type theory, mathematical, 3:278a–b; 5: 154b; 12:13a,b, 16a; 14:304a
Type theory, psychological, 7:190b–193a
Typhoid fever
  Renaissance discussion, 1:399b
  seventeenth-century discussion, 1:391b; 14:405b
  nineteenth-century discussion, 1:438b; 2:575a–b; 3:589b*; 4:263b; 7:421b; 8:7a; 10:252a; 11:516a*; 12:316a*
  twentieth-century discussion, 5:317a
  bacillus, 5:128b
    cultivation of, 4:276b; 5:219a–b; 8:450b
    discovery of, 4:276a,b; 7:425b
  control of, 6:433a; 7:428b; 12:451a
    carriers, 11:346a
  diagnosis of, 2:300b; 5:564b; 6:298b
  epidemics, 11:176b; 14:405b
  etiology of, 6:73a; 10:559a–b; 11:148a; 12:96b; 14:249b
  individualization of, 2:445a
  lesions, 2:445a
  mouse, 8:450a
  muscles, effects on, 14:126a
  perforation of gut, 3:518a
  prognostic test, 4:297a
  transmission of, 2:445a; 11:346a
  urine, crystals in, 12:202b
  vaccination, 6:433a; 10:104b; 14:511b, 512b
  see also Fowl typhoid, Horse typhoid
Typhus
  seventeenth-century study, 14:405b
  eighteenth-century study, 11:148a
  nineteenth-century study, 2:35b; 14:40b
  twentieth-century study, 5:270a, 451a; 6:432b
  endemic, 14:623b
  epidemics, 6:432b; 14:40b, 42a, 623a–b
  etiology of, 7:624b; 11:175a, 442b–443a; 14:42a, 623b
  exanthematous, 2:445a, 534a; 5:366b; 6: 73a; 11:559b
  immunity, 15:522a
  murine, 15:454b
  transmission of, 5:270b, 451b; 15:454a–b
  treatment of, 11:177b, 599b; 13:229a
  vaccination, 6:433a; 14:623b; 15:454b
Typography
  Renaissance, 4:260a
  eighteenth-century, 14:414b
  for the blind, 4:252a
Tyramine, 15:10b
Tyrén, Helge, 13:163b

Tyrocidin, **5**:42b
Tyroff, Hermann Jakob, **5**:217a
Tyrosinase, **2**:86b
Tyrosine, **4**:400a; **6**:505b; **7**:467a
Tyrothricin, **5**:42b
Tyrrell, Joseph Burr (1858–1957), 13:524–526
Tyrrhenian Stage, **4**:39a; **7**:30b
Tyson, Edward (*ca.* 1650–1708), 13:526–528; **13**:175a*
Tytler, James, **4**:528a–b
Tyzzer, Ernest Edward, **12**:482b

U

al'Ubaid (ruins), **14**:504a–b
Ubaldo, Guido
  see Monte, Guidobaldo, Marchese del
Ubisch, Leopold von, **2**:363b
Ucello, Paolo, **2**:535a
Udall, Nicholas, **5**:348a
Udayadivākara
  commentaries of, **2**:115a
al-ᶜUdhrī, **1**:413b, 414a; **7**:4a
Uexküll, Jakob Johann von (1864–1944), 15:503–505; **7**:519b; **9**:19b, 20a
Uganda
  Geological Survey, **6**:475b
Uhlenbeck, George Eugene, **2**:248a; **8**:478b; **9**:397b; **11**:420b; **15**:44a
Uhlenhuth, Paul, **4**:300a; **15**:522a
Uhlig, Victor Carl, **8**:544a
Uhlworm, Oscar, **8**:449b
Uintatheres, **9**:134b
Ujjayinī (India), **15**:538a,b
Ukhtomsky, Alexei Alexeivich (1875–1942), 13:529–530; **12**:271a
Ukraine
  flora, **5**:216b, 218a
Ukrainian S. S. R. Academy of Sciences, **13**:617a
Ulam, Stanislaw Marcin, **1**:427a; **5**:273a; **7**:530b
Ulcers, **1**:315b; **3**:98b
  biochemistry of, **5**:209a; **13**:262b
  endocardial
    bacterial nature of, **9**:95a
  gastric, **3**:490a; **13**:262b, 263a
  peptic, **8**:555a*
  treatment of, **6**:479a; **10**:306a,b; **14**:323b*
  tubercular, **4**:198a
  typhic, **4**:198a
Ulloa y de la Torre Giral, Antonio de (1716–1795), 13:530–531; **5**:435b; **6**:512b; **7**:183a
Ulmic acid, **9**:558a–b
Ulmin, **2**:385b
Ulmus, **13**:154b
Ulrich of Strasbourg (or Ulricus de Argentina or Ulrich Engelberti) (*d. ca.* 1278), 13:534; **1**:99a
Ulrich, Edward Oscar (1857–1944), 13:531–534; **1**:494b; **5**:50a; **12**:229b, 230a
  critics of, **5**:486b; **13**:518b–519a
Ulrich, Georg, **4**:1b
Ulstad, Philipp (*fl.* early sixteenth century), 13:534–535; **10**:308a
Ultracentrifuge, **13**:160b–161a, 162b, 419b; **14**:617a
Ultrafiltration, **14**:633b
Ultramarine, **7**:59b*
  analysis of, **3**:316a
  synthesis of, **3**:316a
Ultrametamorphism, geological, **3**:434a
Ultramicroscope, **12**:593b
  cardioid, **12**:423a
  invention of, **14**:633a–b
  slit, **10**:525a; **12**:423a
Ultravirus, **8**:273b
Ulugh Beg (1394–1449), 13:535–537; **7**:62a, 219b, 255a, 337a; **9**:36a; **13**:149b
  acquaintances of, **11**:227a
  algebra, **7**:259a
  astronomical instruments, **7**:353b
  and al-Kāshī, **7**:255b–256b
  star catalog, **1**:403b*; **6**:364a
  Zīj i Ulugh Beg, **7**:81a, 256a,b, 259b, 324b; **12**:362a,b; **14**:578a–b; **15**:616a–b, 626a–627b, 628b, 629a
Ulyanov, Vladimir Ilyich
  see Lenin, Vladimir Ilyich
ᶜUmar ibn ᶜAbd al-ᶜAzīz, **7**:187b
ᶜUmar ibn al-Farrukhān al-Ṭabarī (*fl.* 762–812), 13:538–539; **1**:33a, 34b; **4**:555b; **15**:125a,b
ᶜUmar al-Khayyāmī
  see al-Khayyāmī
Umāśaṅkara Miśra
  commentaries of, **7**:315a,b
al-Umawī, Abū ᶜAbdallāh Yaᶜīsh ibn Ibrāhīm ibn Yūsuf ibn Simāk al-Andalusī (*fl.* fourteenth century), 13:539–540; **11**:230a
Umbelliferae
  classification of, **1**:306a; **9**:528b
  poisonous, **14**:256a
Umbgrove, Johannes Hermann Frederik, **13**:611a
Umbilical cord
  anatomy
    Renaissance discussion, **1**:619b
    seventeenth-century discussion, **4**:510a–b
  jelly, **14**:286b–287a
Umbriel (satellite), **8**:46a
Umov, Nikolai Alekseevich, **1**:283b, 284a, 598a–b
Umwelt theory, **15**:504a–b
Unanue, José Hipólito (1755–1833), 13:541
Uncertainty principle
  see Indeterminacy
Unconformity (geol.), **11**:35b, 439b; **12**:272b, 276b
  angular, **13**:87a
Unconscious, the, **2**:447b–448a; **5**:146a, 174a–b, 177b, 178a, 180b–181a; **8**:157b
  collective, **7**:191b–192a
  historical structures in, **7**:190b–191a
  type theory, **7**:191a–b
Underflow, **4**:208a
Underhill, Anthony Lispenard, **2**:198a
Undermining
  explosive, **9**:146b; **13**:233b
Undulant fever, **2**:528a; **9**:546a*
Unemployment, **7**:318a
Unger, Franz (1800–1870), 13:542–543; **9**:278a–b, 279b, 600b; **14**:349a
  Die Exantheme der Pflanzen, **12**:203a
Ungulates
  dentition, **12**:38a
  phylogeny, **7**:480a; **12**:38a
Uniformitarianism
  seventeenth-century, **12**:257a
  eighteenth-century, **4**:72a; **5**:205a; **6**:582b; **9**:26b; **12**:122b; **13**:441b–442a
  early nineteenth-century, **1**:413b; **2**:556a; **5**:31b–32b; **7**:589b; **8**:565b–575b *passim;* **10**:9a; **11**:133b, 134a; **12**:182b, 262a–263b
  mid nineteenth-century, **2**:570b; **3**:392b, 434b, 552b; **6**:187a, 437b; **9**:420b; **12**:276b, 514b
  late nineteenth-century, **1**:130a; **3**:162a, 594a; **8**:260b; **10**:32b; **12**:272b
  twentieth-century, **12**:273a, 274a
  opposition to, **1**:470a–b; **9**:583b; **10**:209b; **13**:382b–383b
  vs. actualism, **6**:457a–b
Uniformization (math.), **6**:394a
Union College
  engineering at, **13**:24b
Union Mondiale des Académies, **13**:145a
Union of Biological Sciences, **8**:356b
Unionicola, **7**:111a
Unionidae, **8**:103a–b, 104a, 357a
Unitarianism, **3**:295b; **6**:140a; **11**:140b, 141a
  opposition to, **9**:537a; **14**:152b
Unitary problem (physics), **7**:211a–b
United Company of Barber-Surgeons (London), **5**:348a
United Nations, **2**:374b
  Atomic Energy Commission, **13**:429b
  Educational, Scientific, and Cultural Organization (UNESCO), **12**:350a
United States
  agriculture, **7**:89a
  archaeology, **2**:549b; **11**:177a; **14**:443a, 517a
  astronomy, **5**:479b; **9**:195b
  biological sciences
    botany, **1**:484b–485a, 485b; **3**:196b–197a, 562b–563a; **4**:273b, 583b–584a; **5**:512a–514a; **9**:365b–366b; **13**:432b; **14**:192b
    ethnology, **11**:119a–120a; **14**:328b–329a
    marine biology, **13**:66a
    physiology, **9**:142b; **15**:107b, 110a–b
    zoology, **5**:12a, 478a–b; **7**:169b–170a; **12**:133a,b; **14**:410b, 416b–418a; **15**:227a
  chemistry, **5**:621b; **6**:517a; **8**:290b–291a, 612b; **10**:156b–157a; **11**:371a, 616b; **12**:435b; **15**:1a–3a
  Declaration of Independence, **5**:138a; **7**:88b
  ecology, **14**:1b
  education, **2**:217b, 272a–b; **9**:56b, 57b–58a; **10**:14b; **12**:435a; **15**:108a
  exploration, **7**:89b; **10**:32a; **11**:118b; **12**:133a, 204a; **15**:91b, 130a
    fortieth parallel, **4**:365b, 366a; **6**:13b; **7**:370a,b; **14**:192a–b, 312b
  geography, **2**:372b–373a; **9**:537a–b; **10**:480a
    surveying, **11**:471b; **15**:152b
  geology
    eighteenth-century, **9**:537b; **14**:68b
    early nineteenth-century, **8**:615a–616a
    mid nineteenth-century, **3**:392a, 553b; **4**:363b–365a; **6**:524a; **9**:255a–b; **10**:258a–b; **11**:118b–120a; **14**:439a–b; **15**:151b–152b
    late nineteenth-century, **2**:500a; **3**:593a–594a; **4**:265b–266a, 365b–366b; **5**:49b, 395a–b; **6**:13b–14a, 186b–187b; **8**:123a, 576b; **9**:100a–101a; **10**:32b; **14**:132b; **15**:91b–92a
    twentieth-century, **5**:49b
    geomorphology, **2**:374a, 549a–b; **4**:562b–563a
    geophysics, **6**:165b–166a
    glaciers, **6**:187b; **14**:443a, 516b–517b
    mineralogy, **3**:313a–b; **12**:204b; **14**:441b
    paleontology, **9**:134a, 314a, 540b; **13**:72a; **14**:313a, 409b; **15**:91b–92a
    stratigraphy, **13**:621a

United States (cont.)
historiography, **7**:486a–b; **12**:110a, 113a–b
mathematics, **1**:65b–66a; **2**:217b; **9**:501b–502b
medicine, **2**:172b; **6**:526b; **8**:555b–556a
pathology, **11**:176a; **14**:249a–250b
pharmacology, **14**:496a
psychoanalysis, **5**:179b–181a
psychology, **11**:617a–b; **14**:550a
public health, **14**:249a–250b
meteorology, **9**:196a–197a
natural history, **1**:73b–74a; **5**:518a; **10**:439b; **11**:263a,b
physics, **8**:2a–b; **10**:599b
politics, **7**:89a
scientists, statistical data on, **3**:130b
settling of, **11**:119a, 119b–120a
technology
bridge engineering, **3**:495b–496a
industry, **7**:462b
metallurgy, **11**:549b
mining, **4**:366b; **14**:441b
tourism, **8**:264a
see also America; individual states; North America; specific governmental agencies, *e.g.,* Geological Survey, U.S.
United States Congress
apportionment of representatives, **6**:570b
United States National Museum
see Smithsonian Institution
United States South Seas Surveying and Exploring Expedition
see Wilkes Expedition
United States Weather Bureau
see Weather Bureau, U.S.
Units (math.)
complex, **1**:370a; **7**:505b–506a, 508a
concept of, **4**:5a
divisibility of, **8**:267a–b
theory of, **4**:125a–b; **5**:97a–b
Unity
metaphysics of, **3**:514b
Universal Brotherhood of Friends of Truth, **5**:469b, 470b
Universal joint, **6**:487a; **8**:214a; **12**:211a
Universal Spectator and Weekly Journal
founding of, **1**:411a
Universalism, **5**:120a
Universals
ancient discussion, **1**:2b, 118a
medieval discussion, **1**:2b–3a, 61b; **2**:231b; **6**:377a; **10**:174a
Renaissance discussion, **5**:106b, 107a
seventeenth-century discussion, **5**:286a; **6**:447a
twentieth-century discussion, **3**:335b
Universe
age of, **6**:531b; **9**:405b
center of, **3**:410a; **9**:137b; **10**:420b, 590a, 591a; **11**:224a; **15**:496a–b
chemistry of, **4**:599b–600a; **5**:457b
creation of, **1**:549b; **2**:21a; **3**:14b, 527b; **6**:306a; **7**:134b, 135a–b, 231b–232a; **12**:441b
continual, **3**:479a; **8**:617b–618a
design in, **5**:104a
and entropy, **1**:301a; **2**:264a–b; **15**:448b–449a
evolution of, **7**:232a
expanding universe theory, **4**:281b, 330a; **5**:272b; **6**:531a–b; **9**:405b–406a
finitude of, **15**:261b
five-dimensional theory, **7**:26b, 211b
infinity of, **1**:567a; **2**:540a, 541a, 542b, 543a; **3**:514a; **4**:97b; **6**:70a–b; **8**:537b, 538b; **9**:317a
multiplicity of, **2**:331a
order of, **2**:234a; **15**:54a, 333a, 638b, 644a, 645a
origin of
Big Bang theory, **5**:273a
rotation of, **5**:273a; **14**:415a
size of
ancient estimates, **1**:247a, 248a, 257a
medieval estimates, **8**:281a
Renaissance estimates, **3**:403a–b
eighteenth-century estimates, **6**:70a–b
Mesopotamian estimates, **15**:638b
modern estimates, **1**:353a; **4**:282a; **6**:30b; **9**:411a; **12**:249a, 347b–348a
steady state, **8**:618a; **15**:344b, 449a
structure of, **6**:30b; **7**:235b; **11**:393b; **13**:120b, 429b
unity of, **5**:357b, 358a; **14**:364b, 388b
see also Cosmogony, Cosmology
University College, London, **5**:266a; **14**:397b
Birkbeck Laboratory, **5**:104a
chemistry at, **11**:278a; **13**:499a; **14**:394b
eugenics at, **10**:464a–465a
founding of, **6**:509b
Galton Laboratory, **5**:11a
geology at, **14**:210b
hospital, **11**:462a
pharmacology at, **15**:99b–100a
physiology at, **5**:80a
University College, Sheffield, **12**:544b
Unmoved Mover, **1**:259b, 260a; **12**:6b
Unsöld, Albrecht Otto Johannes, **7**:186b; **12**:22a, 530a
Untersuchungen zur Naturlehre des Menschen und der Thiere, **9**:456b
Unverdorben, Otto, **5**:197b
Unzer, Johann August (1727–1799), 13:543–544; 14:321a–b
Upas, **5**:82a–b; **9**:8a; **10**:498a
Upham, Warren, **14**:517b
Uppsala University
in seventeenth century, **11**:587a–b
eighteenth-century curriculum, **1**:305b
astronomical observatory, **3**:173b, 174a
botanical garden, **8**:375b
botany at, **13**:392a
chemistry at, **14**:144a–b
Medical Faculty, **11**:587a–b
physiology at, **6**:476b
al-Uqlīdisī, Abu'l-Ḥasan Aḥmad ibn Ibrāhīm (*fl.* 952–953), 13:544–546
arithmetic, **7**:533a; **9**:614b, 615a
Kitāb al-fuṣūl fi'l-ḥisāb al-hindī, **7**:257a; **13**:539b
Ur (city), **14**:504a–b
Urachus, **5**:586a; **11**:99b
Uracil
discovery of, **7**:466b
Ural Mountains
botany, **8**:251b
geography, **7**:480b
geology, **3**:236a–237a; **5**:211a; **6**:238a,b, 552a; **7**:249a–b, 251b; **8**:252a; **9**:592a
mining, **5**:428b
paleontology, **10**:222a
stratigraphy, **7**:253a; **9**:584a
Uramil, **8**:343b
Urania Observatory, Berlin, **9**:2b
Uraniborg (Brahe's castle), **2**:404b–406a; **11**:525b
coordinates of, **10**:596a
Uranie (ship), **5**:224b; **11**:242b
Uraninite, **3**:479b; **11**:281b(n)
radioactivity, **3**:507a
Uranium, **2**:338b
atomic weight, **9**:289a, 290b
bombardment of
with neutrons, **4**:580a; **6**:16b–17a; **7**:158b; **9**:261b–263a
compounds of, **4**:554a
complex, **2**:235b
oxides, **15**:7b
compressibility, **2**:460a
decay products, **9**:262a
decay series, **2**:258b–260a; **6**:16b; **12**:28b–29a, 505b
discovery of, **7**:394b
fission, **2**:251b–252a; **6**:17a; **7**:527a, 554b; **9**:262b–263a
chain reaction, **3**:371a; **4**:581a–582a; **7**:155a
discovery of, **7**:158b
isolation of, **2**:565a; **4**:563b; **9**:451b
isotopes, **9**:263a; **12**:505b
separation of, **5**:493a; **8**:482a; **10**:215a
nuclear isomerism, **6**:16b
ores, **3**:499a; **5**:610b
deposits of, **7**:155b
see also Pitchblende
radioactivity, **1**:559a–b, 560a; **3**:499a,b, 507a; **4**:355b; **12**:27a, 28a
ratio with radium in minerals, **2**:258b–259a; **12**:29a
salts
purification of, **7**:498a
and uranium X, **3**:480a; **12**:27b
Uranus (planet), **2**:516b
discovery of, **2**:216b; **6**:322a, 324b, 328b–329a, 335a; **8**:299b; **15**:311b–312a
distance of, **13**:425a
motion of, **1**:53a–b; **2**:359b; **3**:186b; **4**:14b, 19b; **15**:353b
and Neptune, discovery of, **8**:277a–b
perturbations of, **5**:257a,b; **10**:479b
naming of, **2**:221a
observations of, **9**:330b; **11**:472a; **13**:113b
pre-discovery, **2**:359b; **5**:24b; **8**:179b
orbit of, **2**:216b; **4**:14b; **5**:224a; **8**:277a, 522b; **15**:316b
photography of, **7**:472a
rotation, **4**:69b
period of, **8**:522a; **9**:504a; **12**:455a
satellites of, **6**:49b, 335b, 471a; **7**:608a; **8**:46a; **10**:35a; **13**:113b; **15**:354b
size
mass, **6**:335b; **7**:608a
spectrum, **12**:267a; **14**:55a
tables of, **2**:359b; **10**:34a
Urbain, Georges (1872–1938), 13:546–547; 5:295b*; **14**:532b*
chemistry, **9**:110b, 544a
spectroscopy, **3**:478a
Urban VIII, Pope, **3**:115b, 180a; **5**:241b, 243a, 244b, 246b; **7**:375a; **9**:15b; **13**:474a; **15**:69b
Urban, Friedrich Maria, **4**:557b
al-ʿUrḍī, Muʾayyad, **12**:358b
Ure, Alexander, **5**:492b
Ure, Andrew (1778–1857), 13:547–548; 10:258a
Chemical Dictionary, **14**:489a
Urea, **5**:149b
adduct products, **6**:304a
in blood, **5**:432b
composition of, **4**:244b
derivatives of, **14**:329b
determination of, **2**:316b; **9**:401b; **10**:493a
dry distillation of, **14**:476a
excretion of, **1**:471a; **5**:432b; **13**:575a; **14**:63b–66a
isolation of, **2**:227a; **5**:93a; **11**:172b
medical use of, **4**:209a
metabolism of, **13**:152a
theories of, **2**:162a; **10**:22a–b; **11**:428b; **14**:63b–66a
naming of, **5**:92b
nutritional role, **6**:136a
synthesis of, **2**:324a; **14**:474b–475a

Urease
isolation of, **13**:152a–b; **14**:412b
Uredinales, **2**:583b; **3**:612a,b; **9**:34b, 35a
life cycles, **13**:489b
Urediospores, **3**:613a
Ureides, **9**:183a
Ureter
neurophysiology of, **4**:371b
Urethanes, **5**:53a
Urethra
strictures, **6**:479a*
Urey, Harold Clayton, **4**:582b; **8**:95a, 292b; **14**:182b
Urgebirge (geol. theory), **5**:444a,b
Uric acid, **5**:92b, 197b; **6**:284a; **10**:22a
composition of, **8**:331b, 343a
crystalline forms, **9**:422a
derivatives of, **1**:389b, 390b; **8**:343a–b; **14**:477a
determination of, **6**:500b
formation of, **9**:284b
isolation of, **12**:148b
metabolism of, **6**:506b; **7**:467a
structure of, **5**:2a–b
synthesis of, **5**:2a, 431b; **6**:506b; **13**:453a; **15**:13a
Uridil, Joseph, **15**:73b
Urinary system
anatomy, **4**:520b; **5**:216b; **9**:511b
disorders of, **10**:385b–386a, 582a; **11**:602b
evolution of, **1**:421a–b
see also Bladder; Calculi; Kidney(s); Urology
Urine
medieval study, **3**:393b; **7**:23a; **9**:364a
Renaissance study, **3**:413a; **11**:339b
seventeenth-century study, **6**:255a; **14**:405a
eighteenth-century study, **5**:353a
nineteenth-century study, **11**:172b–173a
acidity vs. alkalinity, **2**:28a
analysis of
Renaissance, **4**:170b*
eighteenth-century, **5**:92b–93a; **6**:237a*; **8**:174b
nineteenth-century, **2**:336a; **5**:218a; **6**:505a; **8**:331b; **10**:204b; **11**:172b
twentieth-century, **5**:53a; **6**:22a; **11**:577a
testing papers, **10**:204b
chylous, **7**:24b*
composition of
albumin, **2**:464a,b
ammonia, **13**:575a
bilirubin, **4**:296b–297a
carbon-nitrogen ratio, **11**:129a
and diet, **10**:557a
metabolites, **1**:10b; **10**:557a; **14**:477a
phosphorus, **9**:105b–106a
pigments, **10**:493a
sugar, **14**:408a, 523b
uric acid, **6**:500b
concentration
mechanism of, **7**:519a
fermentation of, **10**:378b
formation of
seventeenth-century theory, **1**:592b–593a
nineteenth-century theory, **6**:226a–b
twentieth-century theory, **15**:101b–103a
germs in, **10**:382a–383a
pharmaceutical use of, **5**:421a; **12**:383b–384a
preservation of, **10**:370a, 371b
prognostic use of, **2**:435b; **5**:231b, 233a
retention of, **1**:322b
secretion of
diagnostic use of, **6**:287b
see also Kidney(s)
Urodela
cytology, **12**:423b
Urogenital system, **14**:126b
evolution of, **5**:467b
see also Bladder; Calculi; Kidney(s); Reproductive system; Urinary system
Urology
Renaissance, **8**:204a
eighteenth-century, **13**:465a
nineteenth-century, **6**:284a; **11**:99a
Uropoda, **8**:528b
Uroscopy, **1**:290a; **4**:285a
Urotropine, **2**:621a
Urpflanze, **5**:443a
Ursa Major (constellation)
alpha Ursae Majoris, **10**:232a
delta Ursae Majoris, **13**:110b
eta Ursae Majoris
parallax, **2**:100b
zeta Ursae Majoris, **10**:601a; **13**:110a; **14**:56b
star cluster, **6**:352a; **8**:539b
Ursa Minor (constellation)
alpha Ursae Minoris
see Polaris
Ursinus, Benjamin, **7**:304b
Cursus mathematici practici, **9**:613a
Urso of Calabria, **3**:394a; **12**:82b–83a
Ursus (Bär), Nicolai Reymers, **2**:406b, 410a, 413a,b; **7**:294b; **11**:533b; **14**:470b
Urthiere, **10**:195a
Urtica, **5**:216b
Urticaria, **10**:19a
Uryson, Pavel Samuilovich (1898–1924), 13:548–549; **3**:170b; **9**:249a; **12**:410b; **14**:591a
Uryson's lemma, **13**:549a
Uspekhi fizicheskikh nauk
founding of, **8**:103a
Uspekhi sovremennoy biologii
founding of, **7**:456a
U.S.S.R.
see Russia
Usṭāt, **15**:263a
Ustilaginales, **2**:437b; **3**:612a; **9**:34b; **14**:93b
life cycles, **13**:489b
Usury, **5**:401b
Utah
geology, **4**:265b; **11**:118b
oil shale, **3**:610a
paleontology, **14**:312b
Uterus
anatomy
ancient Roman discussion, **5**:228a
medieval Latin discussion, **9**:468b; **12**:81b
Renaissance discussion, **1**:619b; **6**:562b; **8**:204a; **9**:166a; **13**:569a; **14**:10a
seventeenth-century discussion, **4**:511a; **13**:169a
eighteenth-century discussion, **4**:173a; **6**:533b; **12**:41a
nineteenth-century discussion, **7**:421a
comparative, **1**:264b
blood volume in, **1**:455a
diseases of, **12**:92a; **14**:278b–279a
experimental use of, **15**:105b, 106a
function of
seventeenth-century discussion, **4**:509b
histology, **11**:216b, 492a
metabolism of, **1**:455a
prolapse of, **6**:318b; **14**:279a
Utilitarianism
medieval, **1**:380a–381b
seventeenth-century, **13**:335b
eighteenth-century, **1**:233b; **3**:7b, 382a; **11**:43a–b, 140a–b, 332a; **13**:601b; **14**:144b
nineteenth-century, **1**:404a; **3**:378b; **6**:140a; **9**:383a, 385a; **12**:591b; **15**:502a
twentieth-century, **9**:189a
critics of, **12**:278a
Utility (economic)
concept of, **7**:104b–105a
marginal, **7**:105a
Utility (math.)
concept of, **3**:460a, 461b
Utility (phil.)
concept of, **1**:115a
Utopias, **1**:159b, 375b–376a; **2**:65a, 187b; **5**:266b*; **6**:141a
Utpala, **13**:582b, 583a
commentaries of, **2**:417b; **9**:580b
Utrecht, University of
in nineteenth century, **9**:459b
botany at, **14**:255a–b
chemistry at, **4**:162b
Donders and, **4**:163a
herbarium, **9**:417b
observatory, **6**:149a; **9**:415a
physics at, **10**:236a
Uttarakhaṇḍakhādyaka, **15**:581a
Utzschneider, Joseph von, **3**:302a; **5**:142b, 143a, 144a, 302a; **11**:354b, 355a
Utzschneider-Fraunhofer Institute (Munich), **3**:302a; **4**:146b; **5**:142b–143a, 144a; **11**:354b

V

Vācaspati Miśra
commentaries of, **2**:118a
Vacca, Giovanni, **4**:596a
Vaccination
controversy over, **2**:73b; **3**:38a, 463b; **7**:11b; **14**:139a–b; **15**:271b–272a, 429b
see also Immunization; specific diseases
Vaccination, preventive
early eighteenth-century, **1**:322b; **12**:458b; **13**:257a; **14**:83a; **15**:271b–272a
late eighteenth-century, **1**:116b; **2**:42b; **3**:577b; **5**:137b; **7**:11b–12a, 96a–97b; **8**:352b, 506a; **9**:481b, 500b; **10**:198a, 326a, 445b, 613a; **12**:57a, 215b; **13**:181b, 465a; **14**:194b; **15**:429b
early nineteenth-century, **12**:57b; **14**:456a
late nineteenth-century, **1**:511a, 576b, 577a; **2**:575a, 584a; **3**:189a, 220a; **4**:211a; **5**:269a–b, 270a,b; **6**:11b–12b, 518b; **7**:100b, 286b, 424b, 425b, 427b; **8**:7b, 313a, 556b; **10**:390a–408a; **12**:88b; **14**:511b, 512b
early twentieth-century, **2**:300b; **3**:22b, 336a; **5**:270a; **6**:433a; **7**:286b; **8**:273a; **9**:545b; **10**:104b; **11**:271b–272a; **12**:96b, 482b; **14**:623b; **15**:96b, 454b
experimental, **3**:588a; **4**:276b; **10**:15b, 18a–b
hazards of, **11**:515b
intensive method, **5**:269b, 270a
nasal, **12**:96b
Vaccination, therapeutic, **2**:584a; **5**:29b; **13**:229a
Vaccines
antirabies, **9**:334a
antituberculosis, **8**:608a
antiviral, **11**:271b
chemical, **10**:407b
combination of, **11**:271b–272a
contamination of, **6**:12a–b
deterioration of, **10**:396b
manufacture of, **10**:396a
polyvalent, **15**:522a

Vaccines (cont.)
preparation of, 3:91b, 189a; 5:40b; 7:97a; 10:353a; 11:569a; 12:480b
see also Immunization; specific diseases; Vaccination
Vaccinia
see Cowpox
Vacuoles, 5:584b; 14:98b; 433a
Vacuum
ancient Greek discussion, 6:311a; 13:92a, 92b–93a; 14:606a
medieval discussion, 1:27b–28a, 410a; 2:194b; 3:470a; 4:525a; 5:402b
Renaissance discussion, 3:66b; 4:237b; 10:416b; 13:279b
early seventeenth-century discussion, 2:354a, 419a; 5:245b–246a, 288b; 9:13b, 16a–b, 25a; 10:332b–334b, 342b; 11:489a–b; 13:438a–439a; 15:69a
late seventeenth-century discussion, 2:377b–378a; 5:574b–575a; 6:450a; 7:612a; 11:507a–b; 12:210b; 13:335a
eighteenth-century discussion, 3:295a, 296a; 9:471b; 12:407b
and atomism, 1:373a–b, 495b, 567a; 4:31b–32a
denial of, 1:254a; 3:431b; 4:60a; 6:446a; 9:15a, 52a, 137b; 10:137a; 14:610a; 15:496a
electrical attraction in, 1:547b
electrical induction in, 1:347b
electromagnetic effects in, 6:279a
evaporation in, 8:256a
extramundane, 2:395b; 5:400a, 401b
high, 1:292b
intercorpuscular, 6:608b
life in, 13:439a
light in, 3:599a; 9:16a
measurement of, 11:531b
motion through, 1:198a, 254a–b, 255a, 410a, 424b, 606a, 607a; 2:606b–607a; 5:402b; 7:135b; 9:16a
sound in, 2:83b, 377b; 13:439a
Torricellian, 5:340b; 9:319b; 10:137a, 546b
electric discharge in, 5:292b
within a vacuum, 1:341a; 10:137a, 333b; 11:489a, 506b
see also *Horror vacui;* Pumps
Vacuum technology, 3:329b, 475a, 477a, 495a, 502a; 4:78b–79a, 80a, 253a–254a; 5:340b–341a; 8:24a; 13:70b, 71a
Vacuum tubes, 1:195b; 5:340b–341a; 7:263b; 14:340b
see also Electron tubes
Vagner (or Wagner), Egor Egorovich (1849–1903), 13:549–550; 2:623b
Vagner oxidation, 11:92b; 13:549b
Vagus nerve, 2:124b; 4:413a
accessory, 12:137b–138a
and blood pressure, 10:432b
compound nature of, 5:281–282
development of, 3:245b
and digestion, 10:582b
function of, 6:225a; 13:556b
and heartbeat, 2:111a; 5:82a–b, 280a, 463a; 6:225a; 8:453a
and hunger mechanism, 3:69a
and intestine, 10:579a
and respiration, 2:446a; 6:300a
section of, 2:3a, 29b, 30a, 525b; 10:220a, 433a
Vaidhṛta
definition of, 15:546a, 564b
Vaijā Paṇḍita, 12:599a
Vailati, Giovanni (1863–1909), 13:550–551; 8:599a; 12:55a,b
Vaillant, Léon-Louis (1834–1914), 13:551–553
Vaillant, Sébastien (1669–1722), 13:553–554; 7:199a; 8:376b; 11:83a; 12:394b; 13:442b–443a
Botanicon Parisiense, 2:227b
Vaillard, Louis, 1:440b
Vaiśeṣika Sūtra, 4:31a
Vākya (table), 2:119b; 15:596a, 601b, 602a
Vākyakaraṇa, 15:601b–602b, 606a
Valangian (stratig. stage), 6:228b, 229a
Valcarenghi, Paolo, 5:384b
Valden, Pavel Ivanovich
see Walden, Paul
Valence, 8:502b
in alloys, 6:561a
and chemical properties, 3:479a; 7:469a–b
concept of, 1:345b, 578a; 2:621b, 622b–623a; 3:48b; 4:400a; 5:126a, 373b; 6:463b; 10:177b–178a; 14:531a–b
direction of bonds, 1:390b; 13:388b
and ionic volumes, 1:462a
partial, 1:415a; 6:108b; 13:338a
theories of, 5:190a; 8:32a, 47a, 276a, 289b–290a, 291a–292a, 293b; 9:289b; 10:158a; 12:593a; 14:265b–272a
electronic, 12:419a–b
mathematical, 5:472b
octet theory, 8:23b
variable, 2:199b; 3:448b, 449a; 5:126a; 8:508a
see also Bonding, chemical; specific elements
Valenciennes, Achille (1794–1865), 13:554–555; 1:602b; 3:525a; 7:547a; 11:243b
Valens, Vettius, 1:33a, 38a, 39a, 191a; 10:299b, 548a–b; 11:226a; 15:125b, 211b
Valentin, Gabriel Gustav (1810–1883), 13:555–558; 2:596a; 3:424b, 425a; 4:73b*; 11:368a; 14:57a
anatomical theory, 12:241b
physiology, 6:269a; 8:302a; 11:216b; 14:113b
Valentine, Basil (Basilius Valentinus) (n.d.), 13:558–560; 7:150b
Triumphant Chariot of Antimony, 11:48b
Valentini, Michael Bernhard, 15:67b, 68a
Valeral
isolation of, 3:193b
Valerianus, Magnus
see Magni, Valeriano
Valerio (Valeri), Luca (1552–1618), 13:560–561; 1:229b; 3:179b; 5:238b, 241a; 9:303b; 10:335b; 13:235b
De centro gravitatis solidorum, 7:558a
Valerone
isolation of, 3:193b
Valine
discovery of, 5:4b
Valk, Gerhard, 6:364a
Valk, Leonhard, 6:364a
Valla, Giorgio, 1:274a, 603b; 8:248a; 9:132b; 14:274b
De expetendis et fugiendis rebus opus, 1:229a; 4:448b; 8:236b–237a, 238a
Vallarta, Manuel Sandoval, 3:370b
Vallée-Poussin, Charles-Jean-Gustave-Nicolas de la (1866–1962), 13:561–562
analysis, 1:408a; 5:306b; 15:23b
number theory, 3:228a; 4:475a; 6:4a; 7:616a; 8:139a; 9:173b; 11:453a
Valleriol, François, 1:526a
Valleys
below sea level, 7:30b
cross valleys, 3:593b, 594a
formation of, 2:568a, 569a, 570a; 3:395b; 4:143b; 5:519a; 6:55b, 502b; 8:241b, 616a; 9:27a, 150a–b; 10:586a; 11:36a, 131a, 269b, 277a; 12:263b; 13:564a
cycles of development, 3:593b–594a
erosion theory, 4:402b; 12:38b; 13:257b
glacial effects, 10:502a
history of, 6:456b
rift valleys, 3:325b; 7:144b; 14:402b
see also Rivers, Streams
Valli, Eusebio, 14:77b–78a
Vallisnieri (Vallisneri), Antonio (1661–1730), 13:562–565; 3:180b, 332a; 4:511b; 9:4a, 511a
geology, 9:531b, 532a,b, 533a; 12:554b; 15:52b
taxonomy, 11:331b
Valmont de Bomare, Jacques-Christophe (1731–1807), 13:565–566
Dictionnaire raisonné universel d'histoire naturelle, 8:619b
Valorous (ship), 7:92a
Valsalva, Anton Maria (1666–1723), 13:566–567; 3:437b; 5:226a; 8:204a; 9:66b, 511a,b
editions of works, 9:512a
Valsalva's test, 13:567b
Valson, Claude Alfons, 3:133a,b, 135a–b; 4:554b
Valturio, Roberto (1405–1475), 13:567–568; 13:233b
Valuations, theory of, 7:530b
Value (economic)
concept of, 7:104b–105a
Valverde, Juan de (*ca.* 1520-*ca.* 1588), 13:568–569; 3:356a
Historia de la composición del cuerpo humano, 3:355a,b
Valves (eng.)
conical, 15:254a–b, 255a
Vályi, Julius, 7:530b
Van—
some names with this prefix have been listed under the next part of the name
Vanadic acid
determination of, 11:551a
Vanadium, 5:197b
atomic weight, 11:538a; 13:389a
compounds of, 5:394a; 11:537b–538b
discovery of, 2:94b; 3:346b; 11:464b
isolation of, 9:451b; 11:538a
Vanadium black, 11:538b
Van Allen, James Alfred
Van Allen belts, 13:83a
Van Alphen-de Haas effect, 5:610b
Van Beneden
see Beneden
Van Biesbroeck, George, 1:467a; 13:116a
Van Buren, William Holme, 7:23b, 24a
Van Cittert, Pieter Hendrik, 2:600b
Van Dale, Antonius, 5:60a
Van de Graaff, Robert Jemison (1901–1967), 13:569–571
Van de Graaff accelerator, 2:338b; 11:460a; 13:570a
Van den Broek, Antonius, 9:543a; 12:505a
Van den Hove, Maarten
see Hortensius, Martinus
Vanderbilt University, 14:439a
Van der Eycke, Simon, 3:181a; 11:533b
Vandermonde, Alexandre-Théophile (also known as Alexis, Abnit, and Charles-Auguste Vandermonde) (1735–1796), 13:571–572; 5:55a, 601b; 7:494b, 567a, 570a; 15:276a, 310a
and Monge, 9:471a,b, 472a,b, 477b
Vandermonde determinant, 3:142b; 13:571a–b
Van der Waals, Johannes Diderik
see Waals, Johannes Diderik van der

Van der Waerden, Bartel Leendert, **6**:473a; **7**:444b; **10**:139a; **13**:23b
historiography, **6**:407a, 411b; **10**:590b–591a; **11**:221b, 222b, 223b, 224a,b; **13**:296b, 303a, 317a; **14**:593b, 611b; **15**:204a, 697b
mathematics, **7**:10a; **12**:228b, 331a
Van der Willigen, Volkert Simon Maarten, **5**:340a, 341a; **10**:134a
van der Wyk, Antoine J. A., **9**:354b
Vandevelde, Guillaume, **5**:149a
Van Deyl, Harmanus, **1**:135b
Vandiver, Harry Shultz, **2**:145b
van Dorp, Willem Anne, **5**:215a; **8**:329a
Van Engers, Edwin Maurits, **7**:492b
Vanessa, **10**:489a
Van Hise, Charles Richard (1857–1918), 13:572–573; 4:562b; **5**:563a, 598b; **11**:210a; **13**:518b
Vanillin, **5**:432b
synthesis of, **13**:406b
Vanini, Giulio Cesare (*ca.* 1585–1619), 13: 573–574
Vanishing points, **2**:535b
Van Mons, Jean Baptiste, **1**:45a; **12**:619a
van Niel, Cornelis Bernardus, **7**:406a
Vannucci, Dino, **3**:113a
Vanoise
geology, **13**:283b, 285a
van Reede tot Drakestein, Hendrik Adriaan , **13**:282b
Van Rensselaer, Stephen, **4**:273b, 274a; **6**: 56b; **15**:151b
Van Rhijn, Pieter Johannes, **7**:237a, 472a
Van Slyke, Donald Dexter (1883–1971), 13: 574–575; 8:276b*; **14**:523b
physiology, **12**:470b
Van Slyke, Lucius Lincoln, **6**:135b
van't Hoff, Jacobus Henricus (1852–1911), 13:575–581; 11:534a
acquaintances of, **1**:297b; **13**:243b; **14**: 530a; **15**:433a, 456a, 458b
La chimie dans l'espace, **7**:282b
translations of, **14**:455a
on contemporary scientists, **7**:620b*; **11**:299b*; **15**:468b*
correspondence of, **1**:297a
critics of, **1**:288b; **3**:300a; **7**:452b
Études de dynamique chimique, **15**:458b
editions of, **3**:334a
influence of, **2**:69a, 265b; **11**:298a; **15**: 435b
influences on, **11**:299b
oceanic salt deposits, **10**:171a
physical chemistry, **1**:362b; **3**:333b
electrolytic dissociation, **1**:299a–300a; **15**:460a,b
equilibria, **8**:118a; **15**:459a, 463b
kinetics, **5**:587a–b; **6**:110a; **9**:305a; **14**:109a
osmotic pressures, **10**:577a; **13**:453a; **14**:98a
solutions, **2**:186b; **11**:299a; **13**:244a
thermodynamics, **8**:289b; **15**:438b–439a
stereochemistry, **1**:340b; **2**:485b, 622b; **5**:3b; **7**:517b; **8**:109b; **10**:359a; **14**: 265a,b, 269b, 455a
students of, **1**:286b, 431a; **3**:333a; **4**: 165a,b, 485a; **6**:538b; **7**:162a
van't Hoff isochor, **13**:578b
van't Hoff-Le Châtelier principle, **13**:578b
van't Hoff's laws, **7**:503a; **10**:156a
Van Tieghem
see under Tieghem
Vanuxem, Lardner (1792–1848), 13:581; 4:364a; **8**:103a; **11**:505a; **15**:153a
Van Velden, Martin-Étienne, **14**:254a
Van Vleck, Edward Burr, **3**:4a
van Woerkom, Adrianus Jan Jasper, **13**:99a
Vapor(s)
nineteenth-century study, **3**:80a–82b; **9**: 19a
condensation of, **13**:471b
dielectric constants, **8**:106b
entropy, **9**:461b; **13**:73b
heat in, **3**:316a
metallic, **4**:253b
nature of, **8**:72b, 73a
optical properties, **14**:497b
salts, **8**:47a; **12**:191b
saturated, **4**:228b
solutions, analogy with, **8**:255b–256b
specific volumes, **10**:518b
thermodynamic properties, **3**:20a; **11**: 292a
see also Gases; Steam
Vapor density
and atomic weight, **3**:47b, 48b
and chemical composition, **3**:10b; **5**:372b
determination of, **3**:47b; **4**:77b–78a, 243b–244a; **5**:320b, 323b; **9**:355b–356a, 424b–425a, 619b; **13**:299a, 389b
and temperature, **13**:468a
see also specific compounds
Vapor pressure, **4**:229a; **11**:278b; **15**:459a
of binary liquid mixtures, **9**:107b
and boiling point, **12**:495b
determination of, **1**:349a; **5**:340b; **12**:595b
of liquid systems, **7**:462a; **14**:560b–561a
of salts, **14**:330a
of solutions, **5**:324a; **11**:298a–299b; **13**: 243a
tables of, **3**:80b
and temperature, **3**:287a, 304a; **6**:520a; **14**:560b–561a
of water, **3**:540b; **4**:28b; **9**:425a
see also Molecular weight, determination of; specific compounds
Vapor tension, **1**:300a
Vaporimeter, **5**:340b
Vaporization
heat of, **2**:178b, 179a; **3**:287a; **6**:520a; **8**:545b
of liquids, **3**:8b–9a
of metals, **9**:451b
and pressure, **14**:560b
and temperature drop, **8**:73a
theories of, **7**:587a; **8**:79b, 80a; **13**:494b; **15**:312a
in a vacuum, **4**:253b
see also Evaporation
Varadarāja, **9**:22b
Varāhamihira (*fl.* sixth century), 13:581–583; 1:33a; **12**:576b, 599a; **15**:540b, 541b, 543a, 608b
cosmology, **15**:555a, 590b
Pañcasiddhāntikā, **8**:46b; **10**:419b, 426a; **12**:115b; **14**:28b; **15**:537a, 541a, 542a, 545a–554a *passim*, 573a, 581a, 603a, 604a–605a
Vararuci
Candravākyas, **15**:601b
Varenius, Bernhardus (Bernhard Varen) (1622–1650), 13:583–584; 13:30b
Geographia generalis
editions of, **10**:44a
Variables
complex, **1**:408a; **2**:121b, 615a; **3**:265a; **4**:617b*; **5**:96a, 152a; **9**:128a; **11**: 459b; **13**:424a, 428a, 562b; **14**:188a
concept of, **9**:303b
continuous, **9**:611b
convergent, **9**:307b
dimensionless, **2**:565b
interdependence of, **7**:136b
logical, **3**:278b
nonnumerical, **2**:295a
p-adic, **5**:342b
progressive, **9**:307a–b
rate of change, **7**:136b
real, **1**:407a–408a; **2**:351b; **11**:260a; **13**: 562b
separation of, **5**:96a; **10**:248a; **12**:220a
sequences of, **9**:128b–129b
Varian, Russell Harrison, **6**:104b
Varian, Sigurd Furgus, **6**:104b
Variance, statistical, **5**:9b
analysis of, **5**:8b–9a, 477a; **14**:382b
asymptotic, **10**:458a
reciprocal, **5**:9a
Variation, biological, **2**:204b; **4**:487a; **8**:415b, 528a; **9**:75b; **10**:283b
eighteenth-century theories, **9**:188a
nineteenth-century theories, **2**:206a; **5**: 385b; **6**:613b; **7**:395b–396a; **8**:98a, 125b; **9**:278b, 279b–282a, 408b; **10**: 400b; **14**:99a–b, 235a–238a *passim*; **15**:92b
twentieth-century theories, **1**:505b–506a; **2**:10b–11a; **7**:99a; **11**:535b; **14**:314 **15**:507a–508a
and anthropology, **2**:208a–212b *passim*
geographic, **5**:453b
and heredity, **7**:114b
homologous series, law of, **15**:507a–508a
and statistics, **5**:266a; **10**:450a; **14**:99a, 251a
Variation, mathematical
bounded, **13**:562a
coefficient of, **10**:451b
concept of, **4**:479a
Variations, calculus of
see Calculus of variations
Variegation
inheritance of, **3**:422b, 493a
see also Pigmentation
Varieties, mathematical theory of, **3**:170b, 171a; **5**:201b; **6**:389b; **10**:140a–141a; **11**:408b; **12**:284a, 330b–331b
Cantorian, **13**:549a
curvature of, **11**:407b
n-dimensional, **11**:409a
Riemannian, **11**:407b, 409b
Segre varieties, **12**:284b
Variety, taxonomical
concept of, **1**:306a
Varignon, Pierre (1654–1722), 13:584–587; 11:157b
acquaintances of, **2**:52a, 53b; **4**:221b; **5**: 60b, 259b
analysis, **8**:304b; **11**:512b
correspondence of, **1**:206b
influence of, **1**:110b; **5**:61a
mechanics, **10**:66a
Nouvelle mécanique, **2**:54b
parallelogram of forces, **2**:38b; **7**:611a
students of, **3**:39a, 104b, 281a; **11**:67b, 328a
Variograph, **10**:159a
Variola
see Smallpox
Variolisateurs, **1**:322b
Varley, Alfred, **12**:425b
Varley, Cromwell Fleetwood, **3**:475b
Varnish, **5**:203a; **12**:394b*; **13**:327b
Varolio, Costanzo (1543–1575), 13:587–588; 1:46b; **8**:54a; **10**:21b
Varrentrapp, Franz, **7**:393a
Varro, Marcus Terentius (116–27 B.C.), 13:-588–589; 1:350b; **3**:174b, 175b, 176a; **4**:81b; **7**:27b; **9**:140b, 141a
Varuṇa, **2**:417b

Varves, **12**:272b
chronology of, **5**:329b–330a
Vasari, Giorgio, **1**:621a; **8**:193b, 204b; **10**: 271a; **11**:406a
Lives of the Painters, Sculptors and Architects, **2**:534b–535a; **8**:207a–b; **13**:430b
Vascularization, **9**:418b
Vasculose
see Cellulose
Vasilev, Aleksandr Vasilevich, **3**:226a; **9**: 127b, 128a; **11**:94b
Vasilev, G. A., **10**:435a
Vasiṣṭha, **15**:538b, 540b–542b *passim,* 545a, 601b
Vṛddhavasiṣṭhasiddhānta, **15**:540b, 612b–613a, 617b
Vasomotor system, **8**:7b, 452b; **9**:505a–506a; **12**:165a
acid-alkaline theory, **5**:280a
antidromic reactions, **1**:537a; **8**:17a
depression of, **2**:367a
discovery of, **2**:30a–b; **6**:38a
in muscles, **5**:279b–280a
temperature reactions, **1**:450a
theories of, **9**:506a
Vasopressors, **15**:106b
Vassaeus, Ludovicus, **5**:348a
Vassale, Giulio (1862–1913), 13:589–590
Vassalli-Eandi, Antonio Maria, **14**:78a
Vassar College
in nineteenth century, **10**:240a
Vassiliev
see Vasilev
Vassura, Giuseppe, **8**:505a
Vastarini-Cresi, Giovanni (1870–1924), 13: 590
Vater, Abraham, **6**:169b; **8**:146b; **10**:267a
Vaṭeśvara (*b.* 880), 13:590; 15:596b, 599b
Karaṇasāra, **15**:598b
sine table, **15**:600a
Vaṭeśvarasiddhānta, **15**:598b
Vatican, **3**:558b
observatory, **13**:12a
Vauban, Sébastien le Prestre de (1633–1707), 13:590–595; **3**:71a; **12**:127b; **13**:50a
Vaucanson, Jacques de, **8**:115a
Vaucher, Jean Pierre Étienne (1763–1841), 13:595–596; **2**:492a; **3**:43b; **5**:217a; **10**:603b*; **11**:132b
Vaucheria, **3**:337b; **11**:153b; **13**:595b
Vaughan, Thomas, **9**:507b
Vaughan, Victor Clarence, **1**:10a; **9**:56a; **10**:154b, 155a
Vaulezard, J. L., **4**:48a; **14**:20a
Cinq livres des zététiques de Fr. Viette, **1**:541a
Vaults
resistance of, **11**:80a
Vauquelin, Nicolas Louis (1763–1829), 13: 596–598; **3**:346b; **12**:67a, 321b
acquaintances of, **3**:237b; **5**:202b; **11**: 494b, 540b
chemistry, **2**:385b; **3**:201b; **5**:322b; **11**: 437b, 494b; **12**:287a; **13**:281a; **14**: 492a
on contemporary scientists, **3**:317a*; **14**:487b
critics of, **15**:172a
and Fourcroy, **5**:91a, 92a–93a; **8**:331b; **13**:596b, 597a–b
influence of, **3**:240b; **13**:309b
students of, **3**:240a, 346a; **5**:430a, 558a; **10**:436a; **13**:309a
Vautier, Théodore, **14**:39a
Vaux, Clothilde de, **3**:375a
Vavilov, Nikolay Ivanovich (1887–1943), 15:505–513; **9**:565a; **15**:426b*
Vavilov, Sergey Ivanovich (1891–1951), 13: 598–599; **8**:107b, 471b*; **11**:180a*; **13**:241a
Vavilov-Cherenkov effect, **13**:599a
Veblen, Oswald (1880–1960), 13:599–600; **2**:198b; **6**:114a
acquaintances of, **4**:618b; **14**:310b
on contemporary scientists, **4**:619a*
geometry, **4**:339a*; **12**:533b; **14**:212a*, 311a, 559b
influence of, **4**:336b
influences on, **9**:502a
Projective Geometry, **9**:173b
topology, **6**:496a; **7**:312a
Vectocardiography, **2**:601a
Vector analysis
eighteenth-century, **3**:73b–75a; **9**:475a
nineteenth-century, **2**:559a; **3**:166a; **4**: 579b; **5**:63b; **6**:91b–92b; **7**:473a–b; **8**:501b, 502a, 542b; **9**:207a, 215b; **10**:248b, 443a, 483b–484a; **12**:74a; **15**:195b, 196a–198b
twentieth-century, **2**:169a, 593b; **7**:207b, 437a; **10**:20a, 444a; **12**:189b; **14**:437a
foundations of, **2**:594a
opposition to, **13**:237a
parallel, **4**:337a
and philosophy, **15**:79b
theory of, **1**:591a
projective, **7**:473a
Vector notation, **1**:24a, 238b; **2**:594a; **6**:92a, 211b–212a; **8**:495a; **9**:430a
Vector quantities
addition of, **14**:280a
classes of, **9**:206b, 215b
curvature, **8**:389b
derivatives of, **2**:594a
linear transformations of, **2**:594a
magnetic, **9**:206b
multiplication of, **14**:280a–b
Vectors, biological, **4**:620a–b; **12**:315a; **15**: 454b
concept of, **9**:82a–b
Vedas, **15**:534b
Vedel, Anders Sörensen, **2**:401b, 402a
Veezaerdt, Paulus, **5**:439b
Vega (ship), **10**:148b; **13**:165a
Vega (star), **6**:512b; **13**:118a–b
distance to, **9**:371a
parallax, **2**:101b, 468b; **3**:14a; **5**:404a; **6**: 263b; **9**:2a; **13**:110b, 120b
photography of, **2**:285a; **4**:179b
spectrum, **4**:180a
Vegetation
aerial survey of, **13**:345a
and climate, **5**:546b; **6**:551b; **14**:117a
stages of, **9**:34b
types of, **6**:551b
zonation of, **10**:168a; **12**:215a, 301a; **13**: 249a
Veinberg, Boris Petrovich, **4**:597b
Veins (anat.), **1**:46b
azygos, **3**:41a; **4**:199b, 487a
function of, **6**:153a, 154b–155a, 156b–157a
hepatic, **3**:99b
portal, **3**:99b; **5**:426a; **9**:56b, 165b
renal, **3**:41a
sacral, **3**:41a
structure of, **11**:216b
suturing of, **3**:91a
valves, **1**:98a–b, 523a,b; **3**:40a–b, 41a; **4**:508b–509a, 520a–b; **5**:327b; **6**:153a, 155a, 156b–157a; **8**:554a; **10**:598b; **12**:106a
varicose, **4**:508b; **11**:461b–462a
see also Capillaries; Cardiovascular system; Pulmonary vein
Veins (geol.)
metalliferous, **4**:350a
origin of, **14**:262a–263a
Veitch, James, **2**:451a
Vejdovský, František (1849–1939), 15:513–514; **11**:257a
Veksler, Vladimir Iosifovich (1907–1966), 13:600–601
Velaria, **2**:53a
Veliger, **12**:107a
Vellozo, José Mariano da Conceição (1742–1811), 13:601–602
Velocity
actual vs. potential, **2**:330a
addition of, **4**:322a–b
angular, **4**:480a
composition of, **2**:551a
Bradwardine's law, **1**:290b; **2**:194b, 392a–394a; **7**:116a; **9**:133a–b; **10**:224a; **13**:186a, 194a, 197a–207a *passim;* **14**:156a; **15**:265a
calculation of, **2**:590b; **13**:186a–187a, 201a–205b, 207a
curvilinear, **5**:360a
distribution of, **4**:208a; **7**:86a
group, **11**:393b
group vs. phase, **6**:90a
instantaneous, **1**:256a; **2**:606b; **6**:379b; **9**:611b; **10**:47b
limitation of, **1**:198a
linear vs. angular, **2**:551a
and mass, **7**:264a; **8**:496a–b, 498a
measurement of, **4**:208a
medieval concept of, **1**:197b
radial, **1**:55b; **6**:94b
and resistance of medium, **1**:410a–b; **10**:75a; **13**:203a–204a
and size, **5**:15a
of translation, **5**:360a
uniform, **1**:253a–b; **5**:360b
in a vacuum, **1**:410a
vectorial nature of, **9**:33b
see also Acceleration, Motion, Virtual velocities
Velpeau, Alfred, **3**:490a, 587b
Veltheim, August, **14**:60b
Velum palatinum, **5**:384b
Vena cava, **1**:267a
Venel, Gabriel François (1723–1775), 13: 602–604; **1**:529a; **9**:306b*; **11**:167a, 563a
Venereology
Renaissance, **1**:620b, 621a; **2**:337a; **8**:125b
seventeenth-century, **10**:426b*
eighteenth-century, **1**:323a; **5**:411b; **6**:18a, 567a; **7**:605b; **8**:362a; **13**:181b
nineteenth-century, **8**:97b; **10**:17b–18b; **11**:515b–516a; **15**:454a
twentieth-century, **7**:624a
see also specific diseases
Venesection
see Phlebotomy
Venetian Republic, **1**:233b–234a; **4**:164a
Venetz, Ignatz (1788–1859), 13:604–605; **3**:210a, 211a
Venice (Italy)
hydraulic engineering, **9**:486b
Library of St. Mark, **1**:274a
theology, **12**:105b
Venice Exposition, **1**:573a
Vening Meinesz, Felix Andries (1887–1966), 13:605–611; **2**:373a; **14**:515b
pendulum, **9**:181a
Vening Meinesz belts, **13**:609b–610b
Venkateswaran, S., **11**:265b

Venn, John (1834–1923), 13:611–613; 2: 297b; **6**:130a*; **7**:105b
probability, **7**:106b; **9**:419b
Venn diagrams, **4**:137b; **13**:612b–613a
Venoms, **10**:141b
cobra, **5**:100a
Heloderma, **8**:447b
immunization to, **3**:22a–b
rattlesnake, **9**:422a
toad, **14**:445b
viper, **3**:560b; **5**:56a–b; **11**:341b
Ventasso (lake), **12**:554b
Ventenat, Étienne, **3**:172b
Ventilation, **12**:366b
experimental studies, **10**:558a; **11**:536b
in factories, **6**:576a; **9**:153a
in mines, **3**:358a; **7**:78b*; **11**:305b
in ships, **2**:84a; **6**:24b; **9**:153a; **14**:354b
in tunnels, **6**:265a
Ventilators
centrifugal, **3**:358a; **4**:44a
encased-blade, **3**:358b
invention of, **6**:44b
Ventriculites, **9**:87a
Venturi, Giovanni Battista, **8**:242b; **9**:485b; **11**:164b
Venturi tube, **2**:354b
Venus (planet), **5**:22a*; **13**:472b
apparent diameter, **6**:515b; **15**:219a
and astrological prediction, **15**:672b, 673b–674a
atmosphere, **1**:57a; **6**:334a; **8**:577b; **10**: 107b; **11**:553a
discovery of, **2**:4b
conjunctions of, **7**:374a; **12**:151b; **14**:489b; **15**:777a
distance of
from earth, **8**:281a; **11**:197b
from sun, **3**:409a
irradiation of, **13**:113b
life on, **1**:301a
motion of, **2**:410b–411a, 551a; **3**:15a; **11**:224a; **15**:353b
retrogradation, **3**:408b
observations of, **11**:223a; **13**:325b
orbit of, **3**:457b; **6**:515b–516a; **15**:202b–204a
eccentricity, **7**:139b; **12**:360a; **14**:576b
elongation, **3**:408a,b, 409a; **13**:326a
parallax, **7**:543b
phases of, **2**:410a; **3**:115b; **5**:243b; **10**:489a
and cosmology, **1**:588a; **5**:241b; **7**:302b
photography of, **7**:77a; **11**:553a; **12**:454a; **14**:521b
position of, **3**:407b–409a; **4**:359b*
latitude, **5**:618a–b
rising and setting of
heliacal, **3**:101b; **15**:777b, 778a, 781a–b, 786a, 787a–b
rotation
period of, **3**:101b; **6**:334a; **12**:162a, 226b, 455a
size
diameter, **2**:388b; **3**:457b; **13**:113b
mass, **15**:290a
and solar distance, **6**:69a–b
and solar parallax, **1**:157b–158a, 340a; **4**:351b; **5**:258a, 479b; **10**:107b
space probes, **7**:465b
spectrum, **4**:181a*; **12**:73a; **14**:55a
surface features, **11**:163a; **12**:226b
synodic period, **4**:467a; **15**:219a, 760b, 777a
tables of, **7**:123a, 124a; **8**:369b; **9**:421b; **14**:576b, 577a; **15**:778a–b, 779a, 784a–b, 785a–b, 788a–b
transits of, **9**:330b; **11**:114b, 115a; **12**:37a, 290a; **13**:40a*; **14**:589b
in 1639, **3**:457b; **5**:526a; **6**:515b–516a; **12**:343a
in 1761, **2**:4b–5a, 327a; **3**:197b, 198a; **4**:23b–24a, 131b; **5**:88a; **6**:69b, 512b; **7**:87b*; **8**:143b–144a, 178b; **9**:162b, 164a, 231b; **10**:256a, 615a; **11**:609b; **12**:413b; **14**:178b, 452b
in 1769, **1**:434a; **2**:327b; **3**:106b*, 197b, 198a, 396b; **4**:132a; **5**:88a; **6**:69b, 234a–b; **7**:564b; **8**:299b, 321b; **9**: 164b, 231b; **10**:34b, 607b, 615a–b; **11**:114b, 472a, 609b; **12**:413b–414a; **14**:178b, 415a*, 452b
in 1874, **1**:19a, 21b, 157b, 340a, 422a; **4**:179b; **5**:196a, 404b; **6**:49a, 119a; **7**:77a; **9**:551b; **10**:34b, 219a, 480a, 543b; **11**:114b, 163a, 207b; **12**:282a; **13**:115a, 232b, 481b; **14**:242b, 449a, 557b; **15**:493b
in 1882, **1**:340a; **2**:333a; **6**:49a, 119a, 149b, 282b*; **7**:271a*; **9**:564a; **10**:34b, 119b; **11**:207b
and solar parallax, **6**:234b, 511b; **7**: 580b–581a; **11**:207b
Venus calendar (Maya), **15**:776b–789a, 812a–b
Venus year, **15**:760b, 777a
visibility of, **15**:777a, 781a
zodiacal period, **5**:345b
Vera causa, **3**:600a
Verantius, Faustus (Fausto Vrančić or Veranzio) (1551–1617), 13:613–614
Veratrine, **2**:111a; **9**:9b; **15**:252b
discovery of, **10**:497b
isolation of, **3**:159b
Verbiest, Ferdinand, **14**:160a, 162a
Verdeil, François, **11**:492a
Verdet, Marcel Émile (1824–1866), 13:614–615; 3:135b, 307a; **9**:154a
Verdet's constant, **13**:615a
Verdigris
medical use of, **14**:585a
Ver Eecke, Paul (1867–1959), 13:615–616; 4:605b; **5**:589a*; **10**:300b; **11**:162a*
Vereeniging tot Bevordering van de Opleiding tot Instrumentmaker
founding of, **7**:222a
Verein Deutscher Ingenieure
founding of, **5**:501b
Verein für Landwirtschaft
founding of, **6**:221a
Vereinigte Astronomische Gesellschaft, **6**: 112a
Vergil
see Virgil
Verhandlungen der Deutschen physikalischen Gesellschaft
editors of, **7**:458a
Verhulst, Pierre-François (1804–1849), 13: 616; 9:69a
Verification, **3**:376b
historical, **2**:545a–b
Vermont
Geological Survey, **6**:564a
geology, **3**:547b
Vernadsky, Vladímir Ivanovich (1863–1945), 13:616–620; 1:162b; **4**:145b, 597b, 599a,b; **8**:285a, 472b*; **13**:275a; **14**:525a
Vernalization, **5**:523b
Verne, Jules, **8**:11b; **14**:85b
Verneuil, Aristide, **11**:425b, 426a
Verneuil, Philippe Édouard Poulletier de (1805–1873), 13:620–621; 1:210a–b; **4**:67b; **6**:57b; **7**:320a; **8**:44b, 45a*; **9**:585b*
Verney, Ernest Basil, **12**:355b
Vernia, Nicoletto, **10**:122b; **11**:71a; **13**:209b
Vernier, Pierre (1584–1638), 13:621–622; 2:412b; **10**:161a
vernier scale, **5**:491a,b; **6**:363a
Vernier (instrum.), **10**:572a; **13**:622a
Veronal, **5**:2b
Veronese, Giuseppe (1854–1917), 13:623; 3:117a, 166b; **8**:284a; **12**:284a
Veronese's surface, **13**:623b
Verrill, Addison Emery (1839–1926), 14: 1–2; 10:272b; **11**:176a; **12**:479a,b, 480a*; **14**:423b
Verrocchio, Andrea del, **8**:193b, 208a, 241a
Verruga peruana, **10**:142b
Versed sine
Arabic use of, **1**:511b; **5**:612b
Versiera, **4**:572b
definition of, **1**:76a–b; **5**:499a–b
Versorium (magn.), **5**:398a, 399b
Vertebrates
anatomy
comparative, **8**:254a; **9**:572b–573a; **12**:45a; **14**:331a; **15**:166a–170a
archetype, **5**:443a–444a
brain
quantitative studies, **14**:14b
cold-blooded, **15**:91a–b
development of, **1**:387a; **7**:454a
segmentation, **6**:167b
embryogeny, **5**:356b–357a; **6**:435a–436a, 591b–592a; **7**:95a; **9**:416a; **11**:307b–308a; **12**:336b; **13**:96b–97b
evolution of, **2**:505b; **5**:283a–b, 467b; **6**: 535b–536a; **7**:61a–b; **9**:346a; **10**:241b; **11**:361a; **12**:337a, 409b; **15**:123a, 513b
fossil, **2**:494b; **7**:60b; **8**:565b; **9**:87a–b, 346a
and invertebrates, transition hypothesis, **3**:526b; **5**:357a
psychology of, **14**:550b
reproductive organs, **13**:461b
terrestrial
origin of, **12**:409b
unity of plan, **5**:356b; **6**:9b
see also Embryology; Paleontology; specific classes, orders and families of vertebrates; Zoology
Vertigo, **1**:446b–447a; **2**:515b; **3**:530b; **7**: 605b; **11**:215a–b; **14**:408a
galvanic, **11**:215b
Verulam, Baron
see Bacon, Francis
Verworn, Max (1863–1921), 14:2–3; 4: 373a*; **5**:37b; **6**:338b; **7**:98a,b; **8**:534a; **14**:430a
Vesalius, Andreas (1514–1564), 14:3–12; 3:64b, 99a; **4**:199a, 488a; **9**:75b
acquaintances of, **12**:323b; **13**:100a
anatomy, **2**:337a; **3**:41a, 342b, 356a; **5**: 585b; **7**:17a; **8**:204a
and Colombo, **3**:354b–355a
critics of, **3**:355a; **4**:198b, 486b, 508b
drawings of, **1**:523b; **4**:87a
editions of works, **2**:226a; **15**:4a
embryology, **4**:510b
influence of, **2**:224b; **4**:487a,b; **6**:509a; **8**:54a; **11**:605a; **12**:97b; **15**:162a*
psychology, **10**:21b
students of, **3**:12b; **13**:569a
surgery, **7**:16b; **14**:5b–6a
works plagiarized, **5**:348a–b, 580b; **14**:11b
Anatomicarum Gabrielis Falloppii observationum examen, **4**:521a
China Root Letter, **3**:355a; **4**:138b
Epitome, **5**:348a; **14**:11a–b
Examen, **3**:355a

Vesalius, Andreas (cont.)
De humani corporis fabrica, **3**:40a,b, 354b, 355a,b; **4**:413a, 586a; **5**:348a; **6**:203b(n); **7**:17a; **11**:33a
commentaries on, **4**:519b–521a
corrections to, **13**:569a
Tabulae anatomicae, **1**:523a
editions of, **13**:142a
Vesicles
embryonic, **2**:161b
Vesling, Johann (1598–1649), 14:12–13; 1: 482a,b; **6**:508b; **10**:477b; **12**:333a
Vespasian, **11**:38b
Vespidae, **4**:504a
Vespucci, Amerigo, **1**:530b; **10**:617a, 619a; **14**:127b; **15**:89b
Vessiot, Ernest (1865–1952), 14:13–14; 4: 177b; **8**:326a
Vesta (asteroid), **6**:94b, 334b
discovery of, **10**:199a
motion of, **10**:528b
Vestea, Alfonso di, **10**:15b
Vestnik estestvennykh nauk
editors of, **11**:567a
Vestnik opytnoi fiziki i elementarnoi matematiki
editors of, **7**:207a
Vesuvius (volcano), **2**:554a; **3**:549b, 603b; **4**:149b; **6**:84a, 551b; **7**:62b, 63a; **8**: 285b, 567a; **11**:336b; **12**:120a, 300a; **13**:621a
eruptions, **2**:481a; **3**:291a; **4**:26a; **5**:89a; **6**:227b; **7**:376a, 549a; **8**:321a; **11**:38b, 303b; **12**:261b, 562a
evolution of, **2**:439b
mineralogy, **14**:588a
structure, **3**:552a; **4**:350a; **8**:568a, 573a
Veterinary science
medieval, **2**:314a; **11**:600b–601a; **13**:292b
Renaissance, **5**:379b; **11**:595a, 604b–605a
eighteenth-century, **13**:495a; **14**:15a–16a; **15**:113a
nineteenth-century, **3**:219b; **7**:100a–b; **12**:480a–481b
twentieth-century, **7**:32b, 286b–287a; **13**:66b
Mesopotamian, **15**:646b
Viala, Pierre, **9**:78b
Viatkin, Vasilii Lavrentevich, **13**:536a
Viau, Théophile de, **7**:537b
Viault, François, **2**:62b
Vibrating strings
medieval study, **2**:606a
seventeenth-century study, **4**:52b; **6**:608a; **9**:114b, 319a,b
eighteenth-century study, **1**:113b; **2**:40b–42a; **4**:125b, 476b–477a, 478b; **7**: 561b; **12**:128a–b; **13**:267a; **15**:368b
nineteenth-century study, **2**:427a; **4**:222b; **6**:242b; **8**:385b, 399b
twentieth-century study, **11**:265a
curves of vibration, **2**:41b
equation of, **2**:41a; **4**:478b; **8**:384a; **9**: 471a
frequencies of, **1**:606b; **2**:40b–41a; **6**:245a; **9**:319b
isochronism, **3**:615b
nodal points, **3**:258b; **12**:128a–b
simple modes, **2**:40b–41a
uneven thickness, **2**:41b
weighted, **4**:222b
Vibration(s), **2**:311a
of bells, **9**:174b
composite, **2**:41a
of coupled systems, **12**:129b
of cylindrical rods, **3**:258b
of elastic media, **2**:427a; **3**:145b; **13**:102b
energy of, **8**:473b
mathematical theory of, **3**:258b–259a; **7**:514a–b; **9**:174a; **11**:460a–b; **12**: 341b, 342a; **15**:487a–b
of membranes, **4**:478b; **5**:151a, 202a; **7**: 235b; **11**:59a
of metallic ellipsoid, **2**:466a
molecular, **5**:117b–118a; **8**:596a; **11**:266a
multivibrators, **9**:76b
nonlinear, **9**:76b–77a
optical observation of, **8**:398a–399b; **14**: 289b
of prismatic rods, **3**:258b
sensation of, **5**:185a
simple, **2**:41a
of surfaces, **5**:375b–376b; **12**:129b–130a
sand pattern technique, **3**:258a–259a; **5**:375b
see also Oscillation, Vibrating strings
Vibrionia, **3**:337b; **10**:364a,b
Vibrios, **3**:589a*; **10**:370a, 375b
cholera, **2**:300a–b; **5**:564a; **10**:267b
fetus, **12**:482b
septic, **10**:387a–b, 389a–b, 390b, 397a, 399a
Vicary, Thomas, **5**:348b
Anatomie of Mans Bodie, **5**:348a
Vico, Francesco de, **12**:266a,b
Vicq d'Azyr, Félix (1748–1794), 14:14–17; 5:536b; **8**:353a, 506b, 622a(n); **11**:99b
acquaintances of, **3**:6a; **5**:89b; **10**:609a; **12**:136b
on contemporary scientists, **11**:563a
influences on, **15**:112a
medicine, **3**:2b; **13**:495b
Victor, Paul-Émile, **3**:204a
Victoria, Queen of England, **5**:255a; **6**:12a, 322b; **8**:405b, 407a, 409b, 462b; **12**: 502b; **13**:361a
Victoria, Australia
Geological Survey, **12**:293a
Victoria (asteroid), **5**:405a
and solar parallax, **1**:340a
Victoria Institute of London, **13**:74b
Victoria Land (Antarctica), **11**:554b
Victorinus, Gaius Marius, **2**:229a–b; **7**:27b
De definitionibus, **1**:269b; **2**:230b
Victorium, **3**:478b
Victorius of Aquitaine, **1**:565b
Vidal de la Blache, Paul, **1**:208a; **2**:538b; **9**:150a
Videsott, P., **3**:533a
Vidius, Vidus
see Guidi, Guido
Vidyādhara
Pañcāṅgavidyādharī, **15**:588b
Vieille, Paul Marie Eugène, **1**:128b; **2**:69b, 70a
Vielle, Edmond, **10**:607a
Vienna (Austria)
House of the Society of Physicians, **2**:130a
latitude, **8**:350b
Royal Academy of Sciences, **12**:227a; **13**:145a
St. Anna Children's Hospital, **4**:404a
Vienna Exposition, **1**:147b; **5**:341a; **11**: 384b
water supply, **5**:564b; **13**:144b
Vienna, University of
in Middle Ages, **6**:275b; **7**:117b, 119b–120a; **12**:62b
in eighteenth century, **6**:233b–234a
in nineteenth century, **13**:145a
in early twentieth century, **9**:260a
anatomy at, **6**:618b
botany at, **14**:349a–b
founding of, **1**:93b
medicine at, **1**:332b; **2**:225b; **5**:172b, 564a; **9**:325b; **12**:294b–295b; **13**:181b; **14**: 40b
museum, **12**:215b–216a
observatory, **5**:489b, 561a; **6**:233b–234a; **10**:279b; **14**:242b
paleobiology at, **1**:18a
physics at, **12**:217b–218a
Vienna Basin
paleontology, **10**:221b
Vienna Circle, **2**:461a; **12**:177b; **14**:468b
Vienne, Council of, **8**:547a
Vierteljahrsschrift der Naturforschenden Gesellschaft in Zürich
founding of, **14**:481a
Viète, François (1540–1603), 14:18–25; 3: 181b; **7**:193b
acquaintances of, **5**:382a; **11**:532b
algebra, **4**:566b–569b *passim,* 572a, 612a; **5**:409b; **6**:511a; **7**:103a; **9**:158b; **11**: 288a, 488a, 533a–b
analysis, **9**:50a
calendar, **3**:312a
editions of works, **12**:205b
geometry, **1**:188b, 191b; **6**:599b; **10**:297a
influence of, **1**:541b; **2**:88b; **4**:60b(n), 75a, 566b; **5**:408b, 409a, 410a; **8**:267b; **10**:45b; **13**:15a; **14**:619a
influences on, **11**:289a(n)
manuscripts of, **12**:205b
tables, **8**:27b
trigonometry, **8**:28a
De aequationum recognitione et emendatione, **3**:64a
Apollonius Gallus, **11**:533b; **12**:500a
In artem analyticem isagoge, **1**:541a
translations of, **6**:112b
Canon mathematicus, **14**:471a
Munimen adversus nova Cyclometrica, **2**:360b
Pseudo-mesolabium, **2**:360b
Supplementum geometriae, **10**:115b
Tactionum, **5**:382a
Variorum de rebus mathematicis responsorum, **3**:311b
Zetetica, **4**:117b
translations of, **6**:112b
Vieussens, Raymond (*ca.* 1635–1715), 14: 25–26; 4:589a; **6**:63a; **8**:506a; **10**: 567b; **13**:179b
critics of, **1**:323b*; **12**:41a; **14**:15a
Vigani, John Francis (*ca.* 1650–1713), 14: 26–27; 6:35b–36a; **10**:44b
Vigo, Giovanni da (1450–1525), 14:27–28
Vigors, Nicholas, **5**:480b
Vijayananda (or Vijayanandin) (*fl.* 966), 14: 28–29
Karaṇatilaka, **15**:611a–612a
sine table, **15**:612b
Vikings
in America, **6**:517b
Vilant, Nicholas, **8**:261a; **9**:159a
Villalpando, Juan Bautista (1552–1608), 14: 29–30
Villanova
see Arnald of Villanova
Villard de Honnecourt (*b. ca.* 1190), 14:30–31; 10:536b
Villard, Paul (1860–1934), 14:31–32; 14: 39b*
Villari, Emilio (1836–1904), 14:32–33
Villefranche
see La Roche, Estienne de
Villefranchian (stratig. stage), **3**:607a
Villemin, Jean-Antoine, **2**:598b; **7**:423b
Villeneuve, Michel de (pseudonym), **12**:323a
Villiger, Walter, **12**:250a

Villon, Antoine de, **9**:527b
Vilmorin, Pierre Louis François Leveque de (1816–1860), 14:33–34
Vimtrup, Bjovolf, **7**:502b
Vinçard, Nicolas, **3**:61b
Vincent of Beauvais (*ca.* 1190–*ca.* 1264), 14: 34–36; 5:592b; **8**:242a
Speculum doctrinale, **9**:364a
Speculum naturale, **15**:35a
Vinci, Leonardo da
see Leonardo da Vinci
Vinci, Pierino da, **8**:193b
Vindicianus, **12**:81b, 539b
Schema anatomica, **12**:81a
Vinegar
manufacture of, **2**:227a; **8**:251a; **10**:365a–366a
preservation of, **10**:326a; **12**:148b
and radical vinegar, **1**:65a
wood, **5**:420b
Vinegar-eels, **7**:110b
Vinen, William Frank, **8**:482a
Vines, Sydney Howard, **1**:422b; **2**:183b, 371a; **5**:80b; **12**:58b; **13**:341b; **15**:46b
Vines (bot.)
diseases of, **1**:587a; **3**:259b; **5**:504a; **9**:78b, 112b; **10**:406a; **13**:596a
see also Viticulture
Viniculture, **3**:425b; **4**:139a; **9**:445b; **15**:648b
see also Oenology, Wine
Vinogradov, Aleksandr Pavlovich, **13**:619b
Vinogradov, Ivan Matveevich, **4**:83a; **6**: 393b; **14**:95a, 282b
Vinogradsky, Sergey Nikolaevich (1856–1953), 14:36–38; 15:14a
Viola, **7**:165b
Violent inclination
see Inclination, violent
Violet, Fabius, **4**:210b*; **6**:255a; **10**:307b
Violle, Jules Louis Gabriel (1841–1923), 14: 38–39; 13:11a
Viper, **1**:263b
circulatory system, **4**:268a
sexual organs, **14**:13a
see also Venoms
Viperine salt, **14**:639b
Viquesnel, Auguste, **1**:210b
Virchow, Rudolf Carl (1821–1902), 14:39–44; 10:563b*; **11**:368b; **12**:203b*; **14**:242a*
acquaintances of, **1**:562b; **2**:207a; **3**:425a; **4**:201b; **6**:434b; **8**:7b; **9**:333b, 574b*; **11**:255a; **12**:87b, 203a
anthropology, **5**:206b
archaeology, **12**:180b, 181a
on cell formation, **2**:165b
cell theory, **2**:32b; **12**:242a
Cellularpathologie, **5**:470a; **6**:7a
critics of, **11**:296b, 491b, 492a
disease theory, **1**:575b; **3**:463b; **9**:512a
influence of, **2**:165b; **4**:277a
influences on, **3**:490a; **12**:156b
and Koch, **7**:421a–426b *passim*
neurohistology, **11**:275a
pathology, **3**:489b; **9**:572b; **10**:562b*
tuberculosis, **7**:435a*; **8**:9a
pathology institute, **6**:505a; **7**:519b
philosophy, **6**:7a; **7**:235a; **8**:445a
public health, **8**:270b
students of, **2**:563b; **3**:463a; **4**:275b; **6**: 287a, 434a, 440a; **8**:8a,b, 9a; **9**:347b; **11**:570a; **12**:395b; **14**:2b, 228a; **15**: 166b
Vīreśvara
commentaries of, **2**:116a, 117a
Virey, Julien-Joseph (1775–1846), 14:44–45; 6:120b; **7**:572b*; **12**:322a*
Virgil, **1**:350b; **2**:21a; **3**:391a, 413a; **4**:393a, 560a; **7**:27b; **8**:536b; **11**:288b; **13**: 343a; **15**:660a
Virginia (state)
education, **7**:89b
exploration of, **6**:124a
geography, **7**:88b–89a
geology, **3**:593b; **11**:504b, 505a
Mineralogical Society, **4**:182a
Virginia, University of
founding of, **7**:90a
medical school, **4**:252a
Virginia Company, **6**:20b
Virginia creeper, **3**:575b
Virgo (constellation)
galactic cluster, **6**:530b, 531a; **9**:330a
gamma Virginis, **6**:325a
nebulae, **9**:330a
Virial equation, **3**:308b
Virial theorem, **5**:160b
Virly, Charles André Hector Grossart de, **5**:602b
Virology
eighteenth-century, **7**:96b–97b
nineteenth-century, **5**:270b; **10**:390a–408a
twentieth-century, **8**:273b; **9**:586a; **12**: 96b, 492a–493a; **14**:624a
techniques, **12**:96b
Virtanen, Artturi Ilmari (1895–1973), 14: 45–46; 7:518a
Virtual displacement, **7**:172b; **8**:215b; **9**: 488a; **10**:250a–b
Virtual velocities, **2**:54b, 214a, 215b; **3**:417b, 454a; **5**:239a, 241b; **7**:562a, 568b; **8**:215b; **10**:250a; **12**:325b; **13**:391a, 585b–586a
Virtual work, principle of, **4**:228b; **5**:98b; **9**:268a, 488a; **11**:488b
Virtus impressa, **2**:605b, 607b
Virūpākṣa Sūri, **1**:308b
Viruses
attenuation of, **3**:189a; **7**:97b, 286b; **10**: 390b, 402b, 403b–404a; **11**:569a
chemical action of, **7**:518a
cowpox, **7**:96b; **10**:390a; **12**:492b
crystalline, **7**:35b
cultures of
in animal cells, **3**:91b
in vivo, **5**:40b
discovery of, **3**:189a; **7**:34b–35b; **8**:450a–b; **15**:14b
evolution of, **13**:520b
filtrable, **5**:270b; **7**:35a–b; **10**:142b, 401b; **11**:347a
ultrafiltration, **12**:96b
herpes, **12**:492b; **14**:624a
influenza, **7**:518a; **12**:492b; **15**:454b
measles, **5**:451b
myxomatosis, **12**:96b
naming of, **7**:96b
parasitic nature of, **7**:35b
plant, **4**:522a
poliomyelitis, **5**:40b; **7**:624a; **8**:273b
pressure, effects of, **8**:608a
size, measurement of, **14**:624a
smallpox, **7**:97a
structure of, **4**:522a; **5**:141a
tobacco mosaic, **7**:34b–35b; **13**:152b
transmission of, **7**:286b–287a; **12**:492b
tubercular, **12**:96b
ultramicroscopic, **13**:520a–b
yellow fever, **3**:94b; **4**:620a–b
see also Bacteriophages, Ultravirus
Vis insita, **10**:70a
Vis nervosa, **11**:159a
Vis plastica, **5**:426b–427a
Vis seminalis, **12**:311b
Vis viva
seventeenth-century discussion, **6**:608a, 611b; **8**:152a, 152b–153a, 159a–b, 162b
early eighteenth-century discussion, **1**: 112b, 113a; **2**:54b, 330a; **3**:215b–216a, 296a, 454a; **4**:44a; **5**:509b; **9**: 33b–34a; **11**:494a; **14**:84a
late eighteenth-century discussion, **2**: 299b; **3**:304a; **7**:561b, 562a; **12**:462b
nineteenth-century discussion, **3**:417b, 418b; **7**:182a; **9**:236a; **10**:3b; **13**:378b–379a, 381b, 383b; **14**:568b–569a
conservation of, **1**:112a; **2**:39a,b, 43b; **3**: 73a–74b; **6**:244a; **8**:153a, 163b; **12**: 287b; **15**:200b, 313a, 339b
and heat theory, **3**:156a; **15**:313a
loss of, **13**:130b
variation in, **13**:128b
Viscosity, **2**:466a; **3**:334a; **7**:553a; **13**:75a
of blood, **1**:534a–b
and chemical composition, **7**:529b
and elasticity, **9**:220b
of gases, **2**:263a; **3**:477b; **4**:376a; **8**:478b; **9**:217b–222b *passim;* **11**:290b–291a; **13**:156a
of liquids, **3**:334a; **13**:389b
magnetic, **1**:284b
measurement of, **4**:141b; **13**:2b; **14**:182b
mechanical, **12**:613a
of minerals, **4**:141a
and molecular weight, **13**:2b
and pressure, **9**:217b, 219b
radiative, **7**:85b–86a
of solutions, **7**:518b
and temperature, **9**:220a, 221b; **11**:291a; **13**:471b
theories of, **5**:161a; **7**:617b; **9**:219a–b, 220b
Visean (stratig. stage), **7**:460b; **8**:463b
Visher, Stephen
Scientists Starred, **3**:130b
Visibility
indexes of, **12**:352b
Visio intellectualis, **3**:513b, 515b, 516a
Vision
afterimages, **3**:116a, 562a; **4**:557a; **6**: 247a,b; **7**:375b; **8**:399a; **10**:60a, 491a; **11**:21a
astigmatism, **1**:86b; **4**:614b, 615b; **5**:590b; **6**:508a; **13**:77a; **14**:564b
binocular, **6**:194a, 300a; **8**:123a; **9**:319a; **10**:60a; **14**:253a
theories of, **6**:248a–249a; **11**:460b; **14**: 492b–493a
color, **3**:327b*, 539a–b; **4**:557a; **5**:86b; **7**:516b; **8**:529a; **9**:226a; **11**:214b; **13**:53b
accidental colors, **11**:21a
color blindness, **3**:539b; **6**:22a, 247a, 278b, 327a, 476b; **7**:458a; **9**:200a–201a, 521a; **12**:218b
Talbot-Plateau law, **11**:21a
theories of, **1**:21a; **4**:615b; **5**:445b; **6**: 246a–247a, 300a–b; **7**:458a; **9**:200a–202a, 562b; **11**:267a; **12**:218b, 232a; **14**:83b, 564b–565a
contrast, **6**:247b
distortion of, **4**:60a
errors of, **6**:194a
and health, **9**:75a
interior stimuli, **9**:570a
irradiation, **11**:21a
and learning, **8**:45b
myopia, **2**:72b
nyctalopia, **5**:137b
perimetry, **11**:215a

tension of
and pitch, **4**:264a
see also Physiology, linguistic
Voelckers, (J.) Carl, **6**:287b
Voeykov, Aleksandr Ivanovich (1842–1916), 14:52–54; 11:182a*
Voeykov axis, **14**:53a
Vogel, Eduard, **6**:553a; **14**:54a
Vogel, Heinrich August, **10**:497b
Vogel, Hermann Carl (1841–1907), 14:54–57; 8:465a,b
binary stars, **5**:468b
comets, **1**:429a
nebulae, **12**:266b
at Potsdam Observatory, **5**:199a; **6**:28b; **9**:563b; **12**:153a, 251a
solar rotation, **14**:557b, 629a
Spectroskopische Beobachtungen der Sterne, **9**:564a
Vogel, Rudolf August, **1**:332b–333a; **14**:224b, 332b
Vögelin, Johann, **11**:591a
Vogelsang, Hermann Peter Joseph, **5**:340a, 341a; **6**:457a; **10**:246b
Vogt, Carl (1817–1895), 14:57–58; 5:533b; **6**:7a, 8b; **8**:3b; **9**:456b; **12**:607b; **14**:114a
Lectures on Man
editions of, **6**:563b
physiology, **14**:64b
Vogt, Heinrich, **14**:482a; **15**:208a, 217a, 421a
Vogt-Russell theorem, **12**:22b
Vogt, Johan Hermann Lie (1858–1932), 14:58–59; 2:369b; **4**:141a,b, 408b
Vogt, Marthe Louise, **15**:106b
Vogt, Oskar, **5**:26b, 27b; **9**:565a; **11**:274b
Vogt, Thorolf (1888–1958), 14:59–60; 1:394a; **4**:408b; **8**:42a
Voice
mechanism of production, **3**:8b; **9**:571b; **14**:565a
nature of, **3**:99a
see also Phonation; Physiology, linguistic; Speech; Vocal cords
Void
see Vacuum
Voigt, Johann Carl Wilhelm (1752–1821), 14:60–61; 5:205b; **6**:456a; **8**:245b; **14**:261b
Voigt, Woldemar (1850–1919), 14:61–63; 4:193a*; **7**:383a*; **10**:29a*; **11**:446a, 447b*
acquaintances of, **11**:612b, 613a; **12**:614b
optics, **7**:553b; **14**:598b
students of, **2**:217a; **4**:100b, 189b, 190a; **8**:50b; **11**:475b, 476b; **12**:422b; **14**:67a; **15**:40a
vectors, **9**:215b
Voit, Carl von (1831–1908), 14:63–67
acquaintances of, **1**:325b; **2**:161a, 162a; **4**:403b; **10**:558a–b
on contemporary scientists, **7**:521a*; **10**:29a*, 563a*; **11**:179a*; **12**:512b*
critics of, **10**:580a
metabolism, **8**:349a; **13**:452b
students of, **2**:366b; **5**:563b; **6**:264a; **8**:555a; **11**:585a
Voit standard, **3**:257a
Voit, Kurt, **4**:604a
Volapük (language), **10**:443b
Volatiles (geol.), **4**:563b–564b
Volatility, **6**:478a
Volatilization
see Vaporization
Volcanic activity, **4**:406b
in Carboniferous period, **5**:336a
cyclic nature of, **7**:63a
in Devonian period, **5**:336a
and elevation of land, **3**:567a–b, 568a; **4**:265b; **8**:568a
and gravity, **14**:515b
groundwater, role of, **7**:63a–b
nuée ardente, **7**:548b–549a
oceans, influence on, **5**:72a
in Paleozoic era, **4**:549a
phreatic eruptions, **7**:63a
prediction of, **7**:63a
in Primary era, **1**:472b
protection from, **7**:63b
and radioactivity, **4**:266a
recent, **6**:187a; **11**:336b
in Tertiary period, **4**:548b, 549a; **5**:39a, 336a–b; **6**:116a; **11**:336b; **13**:283b
theories of
seventeenth-century, **7**:376a
eighteenth-century, **2**:554b; **4**:72a; **11**:302b–304a; **12**:549a–b; **14**:261b
nineteenth-century, **2**:158a–b, 555b; **3**:412a, 552a–b, 586a, 604a; **4**:349b–350a; **6**:503a, 551b; **8**:567b, 570a, 573a–b, 575a–b; **9**:61a–b, 425b; **12**:261b, 262b, 263a; **13**:382b–383a
twentieth-century, **4**:563b–564a; **8**:286a
Volcanistic philosophy
see Vulcanism
Volcanoes, **3**:365b
ash flow, **4**:564a
chemistry of, **5**:89a, 315b
classification of, **7**:63a; **12**:262a
"embryonic," **3**:325b
extinct, **9**:368b
formation of, **11**:134a
fossil, **11**:560b
gases emitted from, **2**:588a
geographical distribution of, **3**:551a–b, 569a
heat, origin of, **4**:152a
interdependence of, **2**:555b
linear eruptions, **11**:336b
lunar, **6**:334a; **7**:263a, 400a
mineralogy, **14**:588a
vegetation, **12**:166b
see also Craters of elevation, Mud volcanoes
Volcanology
ancient Greek, **11**:104b
medieval Latin, **9**:363b
seventeenth-century, **2**:311a, 312a
eighteenth-century, **2**:83a, 439b, 554a–b; **3**:291a; **4**:26a, 70a–72a, 150b–152a, 548b–549a; **5**:77a, 578a; **6**:54a, 84a–b; **7**:495a; **8**:321a; **11**:303b; **12**:549a, 562a
early nineteenth-century, **2**:158a, 481a–b, 554b–555a, 568a; **3**:586a, 603b–604a; **6**:551b; **8**:566a, 567a–b, 568a, 615b, 616a–b; **10**:9a; **12**:261b
mid nineteenth-century, **1**:19b, 20a–b; **2**:588a; **3**:549b, 552a–b; **5**:88b–89b, 207a*, 611a; **7**:510b; **8**:573a–b; **9**:61a–b, 425b; **11**:133b–134a, 269b–270a, 560b; **12**:263a–b, 300a; **13**:621a
late nineteenth-century, **1**:491a; **4**:140b, 266a; **5**:21b, 336a–b; **6**:14a, 215b, 227b; **7**:413b; **10**:211a; **12**:192a; **13**:148a, 283b; **14**:183a
early twentieth-century, **1**:153b, 300b, 394b; **2**:370a; **3**:325a, 533a, 548a*; **4**:262a, 563b–564a; **5**:39a; **7**:5b, 62b–63b, 474a, 548b–549a; **11**:336a,b, 412b
mid twentieth-century, **3**:325b; **8**:285b
Volga (river), **8**:465b, 466b
geology, **10**:428a,b
Volkmann, Alfred Wilhelm, **2**:124b; **4**:556b, 557b; **5**:463a; **6**:224b–225a; **8**:514a; **13**:556b; **14**:113b
Volkmann, Paul Oskar Eduard (1856–1938), 14:67–68; 10:29a*; **12**:526a
Volkmann, Richard von, **8**:406b
Volkoff, George Michael, **10**:214b
Volmar (monk), **6**:396b
Volmer, Max, **13**:41b, 42b; **15**:448a, 450b
Volney, Constantin-François Chasseboeuf, Comte de (1757–1820), 14:68–69; 3:1a, 383a; **8**:615a,b
Volt
standardization of, **9**:155b
Volta, Alessandro Giuseppe Antonio Anastasio (1745–1827), 14:69–82
acquaintances of, **5**:138a; **8**:321a; **12**:67b(n), 137a; **15**:312b
correspondence of, **11**:142a
critics of, **1**:586a; **8**:36a; **10**:135b; **12**:254a
electricity, **1**:144b; **14**:353b
animal, **5**:268b; **14**:76b–80a
electrochemical series, **1**:347b; **14**:78a–80a
electrostatics, **7**:371b; **14**:70a–74a
influence of, **1**:343b; **5**:320b; **12**:155a
influences on, **1**:68a; **13**:225a
inventions
condensator, **14**:71a, 73a, 74a, 78b
electrometer, **14**:73b, 78a,b
electrophorus, **1**:548a; **7**:13a; **14**:71a–b, 72b
eudiometer, **5**:318b; **7**:620b–621a; **14**:75a
meteorology, **1**:586a; **14**:73a–74a
pneumatics, **14**:72a, 74a–76b
students of, **1**:492b; **11**:494b
Volta effect, **11**:460a
see also Voltaic cell, Voltaic pile
Volta, Giovanni Serafino, **12**:560a, 561b
Volta Laboratory Association (Volta Bureau), **1**:583a
Voltaelectricity, **3**:260b
Voltaic cell, **1**:557a–b; **4**:536a; **5**:7a, 474b
decline in power, **3**:557a
electromotive force, **11**:297b, 298a
heat effects, **4**:229a
heat evolved, **11**:298a
Voltaic pile, **1**:107b; **2**:91a; **3**:68a; **4**:74a; **5**:7a, 268b; **13**:311a
chemical effects of, **2**:92a; **3**:68a; **5**:320b; **10**:182b
design of, **8**:36b
electromotive force of, **1**:555b; **10**:187b–189b
factors affecting, **1**:555b
invention of, **14**:76b–80a
naming of, **9**:152a
theories of, **2**:136b; **4**:29a
Voltaire, François Marie Arouet de (1694–1778), 14:82–85; 1:117a; **4**:85a, 203b; **10**:145b
acquaintances of, **1**:115a; **3**:281b, 365a, 383b; **4**:217b; **7**:442b, 562b, 581b; **8**:115a, 506a; **11**:272a
biology, **13**:457b
and Mme. du Châtelet, **3**:215a–216b
critics of, **1**:402a*; **10**:10a
Diatribe du docteur Akakia . . ., **4**:471b; **9**:188b
Éléments de la philosophie de Newton, **3**:215b, 284a; **5**:510b
historiography, **3**:296b(n); **9**:47a; **13**:593b, 594a(n)
Koenig dispute, **4**:471b; **7**:443a
Lettres philosophiques, **3**:215a
and Maupertuis, **9**:186b–189a *passim*; **14**:83a–b
metaphysics, **3**:381b, 382a
Micromégas, **9**:188b
on Newton, **3**:115a

Voltaire, François Marie Arouet de (cont.)
philosophy, **4**:85b; **5**:600b
translations of works, **15**:502a
Voltameter, **11**:298a
Volterra, Vito (1860–1940), 14:85–88; 12:286a*
acquaintances of, **1**:406b; **4**:577b; **13**:550b
analysis, **5**:151a; **6**:5a; **8**:111a; **10**:610b; **14**:13b
correspondence of, **8**:563b*
critics of, **10**:443b
historiography, **2**:359a*; **4**:516b*
influence of, **8**:512a; **10**:514a
influences on, **2**:105b; **3**:145a
students of, **4**:373b; **10**:513b
Volterra equations, **1**:427b; **6**:392a
Voltolini, Friedrich, **3**:531a
Voltz, Philippe Louis (1785–1840), 14:88; 4:347b; **5**:533a
Volume
of barrel, **3**:152b; **7**:300b, 418b; **10**:572a; **13**:435b
Cavalieri principle, **3**:150b–152a
of cone, **1**:220a, 227a,b; **4**:424a, 466b; **8**:421b; **15**:670b
of cube, **3**:277a
of cupola, **13**:290a
of curved figures, **1**:214b, 215a
of cylinder, **1**:227b; **8**:421b; **15**:670b
of frustum, **8**:421b; **15**:670b
of hyperboloid, **3**:182b; **13**:435b
mixed, **9**:414a
of paraboloidal solids, **1**:445b; **3**:182b
of polyhedra, **1**:591a; **4**:609b
of prism, **8**:421b; **15**:670b
of pyramid, **1**:220a; **3**:277a; **4**:424a, 466b; **8**:421b; **12**:375b; **15**:670b–671a, 697a–b
of sphere, **1**:219b, 220b–221b, 227a,b, 444b; **3**:152a, 277a; **4**:466b; **13**:319b
of tetrahedron, **2**:270b, 271a; **8**:421b; **13**:259b
theory of, **12**:364b
of torus, **4**:109b–110a; **7**:300b
of vault, **3**:152b
see also Solids, geometric, rotational
Volume, atomic
see Atomic volume
Volumeter, **10**:301a
Volvelle (astron. instrum.), **10**:540b
Volvox, **3**:337b; **12**:557a; **14**:398a–b
Vomer, **2**:504a
Vomiting
mechanism of, **9**:9a
reflex nature of, **9**:571a
Von Neumann, Johann (or John) (1903–1957), 14:88–92
acquaintances of, **6**:297a; **7**:531a; **15**:43a
analysis, **14**:282a
on contemporary scientists, **5**:108b*
ergodic theory, **2**:145b
game theory, **2**:304b
group theory, **2**:239a, 513a; **5**:609a
mathematical physics, **6**:392a
number theory, **3**:56b
von Neumann algebras, **14**:90b–91b
Voronin, Mikhail Stepanovich (1838–1903), 14:93–94; 10:1a
Voronoy, Georgy Fedoseevich (1868–1908), 14:94–95; 3:178a, 226a; **12**:426a
Vortex
circular, **2**:168b
conservation of, **2**:168a; **6**:249b
theory of, **10**:514a; **14**:206a
see also Motion, vortical
Vortex theory, Cartesian, **2**:38b; **3**:105b; **5**:62a; **6**:604a, 609b, 610a–b; **9**:52a–b; **11**:158a; **13**:96a
critics of, **2**:54b; **3**:431a–b, 444a; **10**:76a–b; **12**:430a
Voskresenskii, Aleksandr Abramovich, **9**:286a, 287a, 304b
Voss (Vossius), Gerhard Johann, **9**:168b, 501a; **10**:43b, 200b; **12**:61a
Vossius, Isaac, **11**:312b; **12**:501a–b
Votes, **12**:532b
degrees of preference, **4**:137b
theory of, **3**:384b, 386b–387b; **4**:137b; **15**:371a
Votoček, Emil, **2**:429a; **6**:375a
Vowels
physical characteristics of, **7**:445b
Vrain-Lucas, Denis, **3**:214a
Vrasskii, Vladimir Pavlovich, **7**:31b
Vries, Hugo de (1848–1935), 14:95–105; 10:253b
acquaintances of, **3**:430a; **10**:577a; **12**:58b; **13**:154b
botany, **13**:579b
on contemporary scientists, **8**:513a
genetics
critics of, **8**:513b
Mendelism, **3**:422b; **7**:623b; **9**:277a; **13**:477b; **14**:431a
mutation theory, **3**:421b; **5**:293b; **7**:114b, 115a, 166b, 439a; **9**:520b, 523b; **13**:137a; **14**:238a; **15**:507b
influence of, **3**:121a, 422a
Mutationstheorie, **4**:546a; **9**:520a
students of, **14**:203a, 255a
Vṛndāvana
commentaries of, **2**:116a; **12**:116a, 598a
Vrolik, Willem, **12**:224a
Vsesoiuznoe entomologicheskoe obshchestvo
founding of, **1**:388a
Vulcanism (theory), **1**:20a, 327a, 413b; **2**:158a–b, 481a, 554b, 568a; **3**:433b; **4**:72a, 151a; **6**:549a; **8**:246a; **9**:532b–534a; **10**:284a; **11**:304a; **14**:60b
Vulcano (volcano), **12**:562a
Vulf, Yuri Viktorovich
see Wulff, Georg
Vulpian, (Edme-Félix) Alfred, **2**:366b; **3**:588a; **10**:205b, 404a,b; **11**:275a, 425b, 426b, 428b
Vultures
food, locating of, **5**:147b; **14**:187b
Vvedenskii, Boris Alekseevich, **7**:622b
Vvedenskii, Nicolai Egorovich, **7**:505b; **13**:529a,b, 530a
Vvedensky, Nikolay Evgenievich (1852–1922), 14:105; 7:34b; **10**:434b; **12**:271a
Vyatipāta
definition of, **15**:546a, 564b, 592a
Vyeñkaṭa
commentaries of, **5**:275a
Vyshnegradsky, Ivan Alekseevich (1831–1895), 14:105–106; 9:286a; **10**:248a, 551a
O regulyatorakh pryamogo deystvia, **15**:7a
Vysotsky, Georgy Nikolaevich (1865–1940), 14:106–108
Vyvyan, Richard Norman, **9**:99a

W

Waage, Peter (1833–1900), 14:108–109; 5:587a; **15**:458a
law of mass action, **2**:68b, 69a; **4**:412a; **9**:110a, 352b; **10**:440b
see also Mass action, law of
Waals, Johannes Diderik van der (1837–1923), 14:109–111; 1:128b; **7**:220a, 271b; **8**:488b; **11**:534a; **13**:81a; **14**:597b
equation of state, **2**:40a; **3**:308b, 309a–b; **7**:221a; **8**:503a; **11**:417b; **14**:561a; **15**:441a
law of corresponding states, **1**:129a; **7**:220b
Over de continuiteit van den gas- en vloeistoftoestand, **7**:515b
physical chemistry, **2**:265b, 267a; **7**:462a, 466a; **8**:475a; **9**:222b; **13**:244b
van der Waals forces, **14**:111b
Waard, Cornelis de, **2**:83b–84a; **5**:439a; **8**:267a; **13**:253b, 255a
Wackenroder, Heinrich Wilhelm Ferdinand (1798–1854), 14:111–112; 5:381a
Wackenroder's solution, **14**:112a
Wada Yasushi, **15**:746a, 752a
Wadham College, Oxford, **12**:580b, 582a,b, 584a, 585b; **14**:364a, 369a–b
gardens, **14**:369b
Waessenaer, Jacob van, **12**:610b
Ibn Wāfid, Abū al-Muṭarrif ᶜAbd al-Raḥmān (Abenguefit, Abenguéfith, Albenguéfith, Abel Nufit) (*fl. ca.* 1008–1075), 14:112–113
Liber medicinarum simplicium et ciborum
translations of, **15**:185a
Wager, Lawrence Rickard, **1**:394b
Wagner, Artur, **6**:98b
Wagner, August, **1**:597b
Wagner, Carl, **13**:247a
Wagner, Egor Egorovich (or Georg G.)
see Vagner, Egor Egorovich
Wagner, Ernst, **5**:26b, 27a, 463a; **6**:299b; **7**:468b
Wagner, Moritz, **6**:553a; **11**:308b
Wagner, Rudolph (1805–1864), 14:113–114; 8:514a; **11**:216a, 217a; **12**:421a
acquaintances of, **1**:476b; **8**:269b–270a; **13**:557a
Handwörterbuch der Physiologie, **2**:73a; **4**:557b; **8**:270b; **9**:441b; **12**:612a
Icones physiologicae, **9**:259a
and Meissner, **9**:259a
students of, **2**:129b; **7**:519a; **12**:164b
zoology, **12**:421b
Wagner von Jauregg (Wagner-Jauregg), Julius (1857–1940), 14:114–116
Wahl, Arthur Charles, **4**:581b; **8**:47b
Wahlenberg, Göran (Georg) (1780–1851), 14:116–117; 12:215a
Ibn Waḥshiyya, Abū Bakr Aḥmad ibn ᶜAli ibn al-Mukhtār (*ca.* 860–*ca.* 935), 14:117–119; 7:40a; **11**:231a
Nabatean Agriculture, **1**:350b; **14**:118a
Wakeman, Alfred John, **10**:243b
Wakley, Thomas, **6**:59a, 59b–60a
Walachia
geography, **9**:150a
Walaeus, Johannes, **10**:307b; **13**:222b
Waland, Robert Louis, **5**:184a
Walch, Johann Ernst Immanuel (1725–1778), 14:119–120; 2:203b; **5**:205b; **7**:412a,b; **12**:182b
Walcott, Charles Doolittle (1850–1927), 14:120–121; 4:365a(n); **6**:13b, 57b; **7**:5a, 62b; **8**:21a*; **12**:229b
Wald, Abraham (1902–1950), 14:121–123
Wald, František (1861–1930), 14:123–124; 8:599a; **15**:463a, 464a
Wald, Georgius an und von
see Amwald, Georgius
Walden, George Ballard, **1**:442b

Walden, Paul (Pavel Ivanovich Valden) (1863–1957), 14:124–125; 3:176a; 5:5a
Walden's inversion, 8:276a; 14:124b
Waldeyer-Hartz, Wilhelm von (1836–1921), 14:125–127; 5:196a; 9:495b; 11: 584a*; 15:525b*
on contemporary scientists, 1:421b; 7: 439b; 11:383a
critics of, 1:177a
cytology, 6:338b; 13:89a; 14:236a
students of, 4:296a; 5:148b; 14:228a
Waldschmidt-Leitz, Ernst, 14:412b
Waldseemüller, Martin (1470–*ca.* 1518), 14: 127–128; 1:179a
Wale, Johannes de, 1:482a; 6:508b; 11:587a
Wales, William, 5:75a
Wales (G.B.)
botany, 8:307b
geology, 4:10b; 5:518b; 6:116a
stratigraphy, 9:583b; 10:586a; 12:277a
mining, 4:349a
natural history, 8:308a
population, 5:507b
Walgensten, Thomas Rasmussen, 7:375b; 13:30b
Walker, Alexander (1779–1852), 14:128–131
Walker, Arthur Geoffrey, 9:405b
Walker, John (1731–1803), 14:131–133; 6: 579a; 7:69a,b; 8:593b
Walker, John (chem.), 7:22a
Walker, Miles, 3:329a
Walker, Sears, 1:364a; 10:479b
Walker Chemical Company, 4:250a
Wall rock
alterations in, 5:548b
Wallace, Alfred Russel (1823–1913), 14: 133–140; 3:576b; 5:255a,b; 8:98a; 11:337b; 12:594b*; 13:343b*
and Bates, 1:500b–501a, 502a–b
on contemporary scientists, 2:205b; 12: 570a
and Darwin, 5:513a; 14:135a–136a, 137a
on evolution, 1:502b; 3:573b; 5:67b; 6: 490b; 8:573a,b, 574a; 9:69a
"On the Law Which Has Regulated the Introduction of New Species," 2: 206a; 8:572b
Malay Archipelago, 3:30a
on mimicry, 1:503b–504a; 9:561a
Wallace's line, 14:138b
Wallace, Robert James, 6:29b
Wallace, William (1768–1843), 14:140–141; 6:263a; 12:523a
Wallach, Otto (1847–1931), 14:141–142; 4:194b; 6:185a; 8:329b*, 502b; 11: 87a; 13:244a; 14:444a, 479a*
Wallenburg Foundation, 4:485b
Wallenstein, Albrecht von, 1:244b; 7:306a–b, 307a
Waller, Augustus Désiré, 4:333b
Waller, Augustus Volney (1816–1870), 14: 142–144; 5:571a; 11:275a, 296a
Wallerian degeneration, 14:142b–143a
Waller, John, 6:36a
Waller, Richard, 4:40b; 6:482a, 487b
Wallerant, Frédéric, 9:182b, 341a
Wallerius, Johan Gottschalk (1709–1785), 14:144–145; 2:4b, 5a; 3:473a; 7:412b; 12:515a
evaporation, 8:72b
Minéralogie
translations of, 6:468b
mineralogy, 13:565b
translations of works, 14:225a
Wallich, George Charles (1815–1899), 14: 145–146
Wallich, Nathaniel, 5:539a; 11:262a
Wallingford
see Richard of Wallingford
Wallis, John (1616–1703), 14:146–155; 5: 521b*, 606b; 6:128a; 9:303a; 10: 87b(n); 13:501a
acoustics, 12:128b
acquaintances of, 2:377b; 5:425b; 6:482a; 10:254b; 12:584a; 13:479b; 14:369a, 405a
astronomy, 5:25b
correspondence of, 3:348b; 4:95b; 5: 159a,b; 6:359b, 598b; 7:583a; 8:250b; 10:44b; 12:460a
deaf, teaching of, 1:411a
editions of works, 13:253b
and Hobbes, 6:449b–450a; 14:152a
influence of, 1:239a; 2:47a; 5:247b, 520b; 10:60b; 15:368a
mathematics
algebra, 14:279b
analysis, 4:474b, 478a; 7:52a; 9:50a, 304a; 14:180a
combinatorial analysis, 6:403b; 9:159b*
geometry, 7:330a, 583b; 8:429b; 10:46b, 336b; 12:56b
interpolation, 1:474b; 2:48b, 50a, 56b
number theory, 3:128a; 4:573a
mechanics, 6:602a; 9:116b; 10:62a
and Newton-Leibniz dispute, 8:164b–165a
physiology, 8:525a
and Royal Society, 5:607a; 12:583b; 14: 364a, 368a,b, 510b
on tides, 3:248b
works edited by, 5:23b; 6:516b*; 10:300b, 303a*
Arithmetica infinitorum, 1:474a; 2:349a, 507a; 3:152a; 6:450a
influence of, 10:43b, 45b, 46a, 47a
Arithmetica universalis, 3:128a
De cycloide, 1:474b
Mechanics
publication of, 3:348b
Tractatus de sectionibus conicis, 14:466a–b
Treatise of Algebra, 1:238a; 9:49b
publication of, 3:348b
Wallis, Samuel, 2:342b
Wallroth, Christian Friedrich Wilhelm, 1: 46a
Walnut tree
leaves
antibacterial action, 3:588a
Walpole, Horace, Fourth Earl of Orford, 6:45a; 11:304b; 14:320a
Walsh, Don, 10:598a
Walsh, Joseph, 5:227a–b, 230a
Waltenberg, Romaine George, 9:312a
Walter of Brittany, 9:435b
Walter Burley
see Burley, Walter
Walter of Evesham
see Walter of Odington
Walter of Odington (*fl. ca.* 1280–1330), 14: 155–156
Walter, Philippe (1810–1847), 14:156–157; 10:498a
Walters, Geoffrey King, 8:482b
Walther, Bernhard, 4:260a; 8:281b; 11:351b; 14:273a
Walther, Johannes, 5:486b; 6:457b; 10:171a; 11:593b
Walton, Ernest Thomas Sinton, 3:329b, 330a,b; 5:272a; 8:95a; 12:33b; 15:26b
Walton, Izaak (1593–1683), 14:157–158; 6:541a; 14:287a
Wan (Wang) Fan, 8:419b; 13:484b
Wandelaar, Jan, 15:4b
Wang An-shih, 12:370b–371a, 372a,b, 373a, 381a, 386a–390b *passim*
Wang Cheng, 9:147a
Wang Chien, 3:266a
Wang, Hsi, 8:357a, 359a
Wang Hsi-shan (1628–1682), 14:159–168
Wang Lai, 3:269a
Wang, Shou Chin, 8:186a, 474b, 475a
Wang Yang-ming, 14:163a, 164a
Wang Ying-lin, 3:250a
Wangerin, Albert (1844–1933), 14:158–159; 5:311b*; 6:230b*; 11:545b*
Wanklyn, James Alfred (1834–1906), 14: 168–170; 5:126b, 540b; 7:393b; 13: 237a
Wannier, Gregory Hugh, 7:493a
Wantzell, Pierre Laurent, 7:328a
War Department (U.S.), 14:520b
Warburg, Aby, 3:110b
Warburg, Emil Gabriel (1846–1931), 14: 170–172; 5:117a; 15:448b
acquaintances of, 6:16a; 12:496a
on contemporary scientists, 9:567b*
correspondence of, 6:350a*; 11:531a
critics of, 5:118b
elasticity, 5:557a
gas theory, 7:526a; 9:224b, 225a; 12: 496a–b
hydrodynamics, 11:63b
magnetism, 4:500b–501a; 11:460b
optics, 13:107b
students of, 4:194b, 485a; 7:59a; 10:105b; 11:347b
Warburg, Otto Heinrich (1883–1970), 14: 172–177; 15:448a
on contemporary scientists, 4:156b*, 304b*
enzymes, 1:363a; 4:358a, 486a; 7:273a
fermentation, 5:6a
oxidation, 13:393b, 394a
photosynthesis, 2:184b; 4:362b
students of, 4:362a; 9:359a
Warburg Institute (London), 3:110b
Warburton, Peter Egerton
Journey Across the Western Interior of Australia, 1:502a
Ward, Harry Marshall, 5:80b, 128b
Ward, James Clifton, 13:273a
Ward, Lester Frank, 10:32b; 11:120a; 14: 297b
Ward, Robert DeCourcy, 3:595b
Ward, Seth (1617–1689), 14:177–178; 3: 248b; 4:40a; 9:310a; 13:501a; 14: 369a, 379a(n)
acquaintances of, 2:377b; 6:482a; 11:532a; 12:580b; 14:405a
Astronomia geometrica, 5:525b, 526a; 14:446b
astronomy, 2:349a; 4:166a; 13:96a
cosmology, 9:310b; 14:446b
Idea, 5:528a
on language, 14:373a, 380a(n)
on scientific method, 4:54b–55a
Vindiciae academiarum, 14:209b, 370a, 373a
Warden, Adam, 3:531a
Ibn al-Wardī, 1:28b
Ware, Louise, 12:73a
Warfare
aeronautic, 7:613b
biological, 7:287a
chemical, 1:431a; 5:128a, 190a, 291a, 465b, 541a, 622a; 6:16a, 21b, 265a, 501b; 7:323a; 8:47a, 292a; 9:354a; 10:121b; 13:429a; 14:319a, 334a, 393a; 15:444a
operations analysis, 7:613b
underwater, 9:491b

Warfare (cont.)
see also Engineering, military; Military science; specific wars; Technology, military; Weapons
Wargentin, Pehr Wilhelm (1717–1783), 14: 178–179; **8**:299a; **12**:147a; **14**:355a
tables of, **15**:325a,b, 326b, 329b–330a, 337b
Waring, Edward (1736–1798), 14:179–181; **3**:555a; **5**:450b; **7**:570a
Meditationes algebraicae, **7**:565b; **14**:438a
Miscellanea analytica, **10**:277b; **14**:438a
theorem of, **4**:83a; **8**:386b(n); **12**:411a
proof of, **6**:392b
Warington, Robert, **8**:93b
Warltire, John, **3**:157b
Wärmetod, **1**:301a; **2**:264a–b; **3**:479a; **14**: 123b; **15**:448b–449a
Warming, Johannes Eugenius Bülow (1841–1924), 14:181–182; **3**:317b
Warner, Richard
History of Bath, **12**:490a–b
Warner, Walter, **6**:127b, 128a, 445b, 446a; **9**:319a
Warner Observatory (Rochester, N.Y.), **1**: 464a,b
Warren, Ernest Henry, **9**:403a
Warren, Gouverneur Kemble, **6**:186b
Warren, John
A Treatise on the Geometrical Representation of the Square Roots of Negative Quantities, **1**:240a; **6**:90a
Warren, Joseph Weatherhead, **2**:367a
Warrington Academy, **5**:75a, 76b; **11**:140a,b, 143b
Warsaw
Nencki Institute of Experimental Biology, **10**:23a
Radium Institute, **3**:502a
Royal Botanical Garden, **3**:596b
serum institute, **6**:432b
Warsaw, University of
physics at, **7**:10b
reorganizations of, **4**:84a; **7**:71b
Warsaw Scientific Society, **4**:84a
Warthog, **13**:527b
Wartmann, Élie, **10**:193b(n)
Warts, **9**:66a
Washburn, Edward Wight (1881–1934), 14: 182–183
Washington, George, **7**:89a; **14**:453a
Washington, Henry Stephens (1867–1934), 14:183; **3**:293a, 484a; **4**:564b; **7**:5a–b
Washington, D.C.
surveying of, **7**:89a
Washington Academy of Sciences
founding of, **9**:313b
Washington Philosophical Society, **4**:265b
Washington University (St. Louis)
Compton and, **3**:371b
al-Wasīṭī, Maymūn ibn Najīb, **12**:91b
Wasps
ichneumon
see Ichneumonidae
life cycles, **8**:415b
paralyzing instinct, **4**:504a,b
sexes of, **4**:40b
Wassermann, August Paul von (1866–1925), 15:521–524; **4**:297b, 304a*, 305a*; **7**:427a
syphilis, **2**:300b
Wassermann test, **2**:300b, 584a; **4**:299b, 301a; **7**:624a; **8**:273b; **10**:18b, 142a; **15**:522b–523a
mechanism of, **7**:624a
Wasserstein, Abraham, **13**:315a
Watches
eighteenth-century, **5**:491a
escapement
cylinder, **5**:491a
see also Chronometers, Clock, Horology
Water
acidic nature of, **1**:300b
air in, **9**:117b
analysis of, **5**:126b, 128a–b; **6**:237a*; **7**: 184a; **8**:68b, 70a–b, 71b; **10**:177a; **12**:478a–b; **14**:448a
geochemical, **3**:292b; **9**:258a
boiling point
and altitude, **3**:16b; **9**:443a
and atmospheric pressure, **4**:518a
carbon dioxide in, **8**:312a
carbonation of
eighteenth-century, **2**:5b; **7**:14b*; **11**: 144a–b
in chemical reactions, **5**:371b; **11**:541a
color of, **12**:593b
composition of, **3**:326b; **5**:318b; **7**:603b; **8**:78b, 79b, 620b; **9**:477a–b; **11**:146a, 416b; **12**:64b–65a, 619b; **14**:75a–b
gravimetric, **4**:127b, 239b; **12**:287a
priority dispute, **2**:186a; **14**:197a
compressibility, **2**:592b; **3**:52a; **9**:14a, 119b
as conductor, **8**:177a
contamination of
bacteria, **8**:556b; **10**:382b
metals, **14**:448a
mixing, **6**:427b
of crystallization, **5**:186a; **8**:69b, 73b; **13**:420a
decomposition of, **8**:79a–b; **9**:343a,b, 472a; **10**:446a; **12**:197b; **13**:469a
theories of, **10**:135a–b
density, **3**:66b, 482b; **4**:577a; **6**:495b; **7**: 341b
dielectric constant, **9**:376b
dissociation, heat of, **11**:577a
distillation of, **8**:71b–72a
electrolysis of, **3**:68a, 600b, 601b; **5**:558a, 560a; **10**:108b; **11**:474a; **12**:197b
as element, **1**:549b; **5**:353a–b; **7**:538b; **8**: 71b–73b, 131a, 174a, 201a; **13**:297a; **14**:157b
conversion to air, **1**:567b
conversion to earth, **1**:549b; **2**:179b; **8**:71b–72a; **10**:80a; **12**:146a
natural place of, **1**:336b
evaporation of, **1**:495a; **2**:178b; **3**:304b; **8**:72b, 256a
freezing of, **2**:177b, 186b, 311a, 593a; **6**:171b, 426b–427a; **7**:269b
effects of dissolved substances, **2**:186b
expansion measurement, **5**:341a
pressure effects, **13**:380b–381a
and temperature, **14**:354a
generation of air, **2**:592b
hardness, **2**:170b; **10**:430a
scale of, **3**:289b
"soap-test," **3**:289b
heavy, **3**:329b; **4**:582a; **7**:155a–b, 519a; **8**:47b*; **12**:34a
lethal effects of, **8**:293a
manufacture of, **4**:545a; **14**:182b
ratio to light water, **4**:545b
ionic dissociation of, **6**:538b
isotopic composition, **14**:182b
juvenile, **13**:148a
mass of, **10**:519a
in metabolism, **1**:357a; **4**:621a
in minerals, **14**:588a
molecular weight, **1**:344b
molecules
size of, **8**:510a; **14**:185b
structure of, **1**:519b
motive force, **9**:114b
natural cycle of, **3**:66b, 541a; **7**:376a; **8**: 201a, 242a, 252a; **9**:119a, 258a; **10**: 522a; **14**:53a, 107a
nitrogen in, **5**:126b
nutritional value, **13**:351a
optical properties, **6**:125b; **11**:265a
physical properties, **7**:578a; **9**:578b
and plant nutrition, **7**:538a
purification of, **3**:189a; **6**:283b*, 397a; **8**:450b; **10**:329b; **11**:177a; **13**:406b
radioactivity, effects of, **5**:394b
radioactivity of, **4**:356b
softening of
Clark's process, **3**:289b
natural, **9**:258a
solvent properties, **1**:288b
specific gravity, **7**:346a; **15**:334a
specific heat, **2**:589b; **3**:20a; **6**:271a; **10**: 27b; **11**:393a, 578b
spectra, **2**:170a
sterile, **13**:453a
subterranean, **1**:550a; **14**:620b
chemical action of, **3**:587a
supercooling of, **2**:177b
supplies of
Renaissance, **9**:146b; **10**:280b
seventeenth-century, **4**:46a; **10**:596a
eighteenth-century, **3**:440b; **4**:38b; **5**: 410b; **6**:44b; **8**:71a–b
nineteenth-century, **1**:528a; **3**:481a*; **4**:144a; **5**:564b; **7**:44a, 149a; **8**:7b, 464a; **9**:60b, 392a; **10**:436a, 511a*, 559a; **11**:131a, 355a, 399a; **12**:140b, 451a; **13**:144b; **14**:96b
twentieth-century, **2**:374a, 548b; **10**: 128a
analysis of, **5**:52a, 125a; **7**:427b; **8**:70b, 71a–b; **10**:326a; **11**:51a–b, 176b; **12**:233a; **14**:169b
contamination, **5**:73a, 128b; **7**:425a; **8**:556b; **10**:382b; **12**:481a, 503a; **14**: 448a
piping, **6**:517a; **10**:109a; **14**:620b
storage, **6**:44a
transport, **15**:514b
surface tension, **2**:239b; **14**:67a
synthesis of, **3**:157b; **8**:78a–b; **9**:140a, 152b, 471b, 472a, 477a,b; **12**:287a, 427b; **13**:469a
thermal expansion, **13**:351a
vadose, **13**:148a
vapor, **13**:522a
in atmosphere, **3**:540a–541a
condensation, **14**:421a–422a
spectra, **7**:74b; **10**:345b
vapor pressure, **3**:540b; **4**:28b; **9**:425a
and temperature, **14**:76a
see also Flow, Ground water, Hydraulics, Hydrology, Ice, Irrigation, Lakes, Limnology, Mineral water, Neptunism, Oceans, Rivers, Sea water, Springs, Steam, Streams, Wells
Water clock
see Clock
Water flea, **2**:142a
Water gas, **9**:427a
Water glass (chem.), **5**:203a
Water hammer, **3**:118b
Water level, measurement of, **6**:223a
Water meter, **12**:424b; **14**:495a
Water mill, **13**:233b
Water snail (mech.), **1**:213b
Waterhouse, Benjamin, **7**:97a
Waterspouts, **3**:344b; **12**:561a; **14**:353b, 451b
Waterston, John James (1811–1883), 14: 184–186; **7**:509b; **9**:219b; **13**:104b
equipartition theorem, **13**:237a
solar heat, **9**:239a

Waterton, Charles (1782–1865), 14:186–188
Watertown Arsenal, **2**:460a
Waterwheel, **3**:8a; **4**:45b; **5**:101a; **8**:211a–b; **11**:77b, 80a
efficiency of, **12**:462b
al-Wāthiq, **7**:358b; **13**:230b; **15**:231a
Watson, George Neville (1886–1965), 14: 188–189; 7:619b*; **8**:584b; **14**:317b
Watson transforms, **14**:189a
Watson, Henry William, **9**:223b; **11**:568b
Watson, Hewett Cottrell (1804–1881), 14: 189–191
Watson, James Craig, **3**:374a
Theoretical Astronomy, **3**:35a, 36a
Watson, James Dewey, **5**:140b–141a, 273a, 589b; **15**:18a
Watson, John Broadus, **4**:161a; **8**:446b; **14**:549b
Watson, Morison, **9**:588b
Watson, Richard (1737–1816), 14:191–192; **2**:186b; **11**:298a
Watson, Sereno (1826–1892), 14:192–193; **4**:584a; **5**:514a*; **8**:264b*; **13**:456a
Watson, Thomas John, Sr., **15**:128b, 129a
Watson, William (1715–1787), 14:193–196; **13**:450b*
acquaintances of, **3**:51a; **4**:149a; **6**:328b, 329a,b; **7**:11b; **11**:140b; **14**:418b
on contemporary scientists, **2**:524a*; **5**: 130b
correspondence of, **10**:147a, 569a
electricity, **5**:131a, 135a
platinum, **13**:530b
Watson-Watt, Robert Alexander, **1**:196a
Watt, Gregory, **3**:598b–599a; **6**:579a
Watt, James (1736–1819), 14:196–199; 1: 564a*; **2**:175a; **3**:400a, 579a; **5**:87a; **9**:615b; **15**:136a
acquaintances of, **6**:579a; **7**:277a; **11**:141a, 354a, 496a; **14**:463a
and Black, **2**:179a–b, 180a
indicator diagram, **3**:287a; **5**:388b
law of, **3**:316a
parallel mechanism, **3**:228b
steam engine, **10**:36b; **11**:499b
translations of works, **3**:80b
water controversy, **2**:186a; **11**:146a
Watt, Joachim de, **10**:305b
Watteville, Charles de, **5**:497a
Watts, Henry
Dictionary of Chemistry, **9**:557b
Watts, Isaac, **4**:527b–528a
Logick, or the Right Use of Reason, **14**: 288a
Watts, John Isaac, **2**:428b
Waulsortian facies, **7**:460b–461a
Wave(s)
atmospheric, **2**:147a
Cauchy-Poisson, **7**:436b
earthquake
see Wave(s), seismic
electrical
optical behavior, **11**:581b–582a
phase-displaced, **4**:588b
theory of, **8**:516a
velocity of, **11**:582a; **13**:473a
electromagnetic, **4**:189b, 329b; **6**:278b, 342b, 343a, 344a; **7**:381a; **8**:493b; **9**:98b–99a, 210a; **10**:518b
detection of, **8**:444a; **11**:93b
existence proved, **6**:345b; **9**:212b; **11**: 581b–582a
optical behavior, **6**:346a; **11**:460b–461a
ponderomotive action, **8**:107a
practical application of, **6**:348b–349a
propagation of, **4**:305b; **9**:155a; **12**:527a
refraction of, **8**:107a
short, **8**:107a
velocity of, **1**:456b; **6**:345a–346a
wavelengths, **14**:348a
see also Microwaves; Waves, Hertzian
explosive, **2**:70a
theory of, **7**:387a
fields of
propagation law, **2**:248b
filtration of, **13**:102b
fluid, **2**:356a; **3**:132a, 143a,b
canal, **8**:284b; **15**:200a
Fresnel, **11**:413a
gravitational, **4**:329b; **12**:280b
height of, **15**:200a
Hertzian
centimeter, **15**:46b–47a
measurement of, **10**:604b; **14**:348a
quasi-optical properties, **2**:325a
reflection of, **4**:6b
solar, **4**:69b
velocity of, **4**:194b; **9**:77a
Laval, **5**:64a
light, **14**:205b–206a
interference phenomena, **5**:19a
ponderomotive action, **8**:107a
standing, **14**:347b, 348a–b
transverse nature of, **5**:170a–b
Love, **8**:516b
matter, **4**:318a; **11**:421b
oscillatory, **13**:75b
plane harmonic, **3**:144a
ponderomotive action of, **8**:106b–107a
propagation of, **3**:263a; **8**:597a; **14**:13b, 204a
in elastic systems, **4**:230b; **7**:466a; **10**: 27b; **14**:200a
in fluids, **4**:230a–b; **10**:75b–76a; **14**: 205b–206a
quasi waves, **4**:230a–b
radio
see Wave(s), Hertzian
seismic, **1**:500a; **4**:71a; **5**:89b, 596b; **7**: 413b, 595a; **10**:211a
components of, **7**:413b; **12**:187a
curved paths, **12**:187a
surface, **8**:516b; **12**:341b–342a
theory of, **8**:147a
time-distance curve, **12**:187a
travel times, **5**:596b; **7**:9a; **9**:61a, 407a, 444b; **14**:327b
types of, **9**:61a; **10**:203b
velocity of, **5**:89b; **8**:388a; **9**:444a; **12**: 187a; **14**:327b
shock, **11**:454a; **13**:73b, 76a, 104b
angle of, **8**:597a
in compressed liquids, **7**:436b
photography of, **15**:60a
propagation of, **4**:230a–b; **5**:89b, 273a; **11**:293a
theory of, **3**:263a; **4**:376a; **6**:545a,b; **7**: 387a; **14**:92a
sound, **1**:483b; **7**:526a; **8**:398a–399b; **11**: 214a; **13**:104b; **14**:204a, 205b–206a, 216a
high-frequency, **3**:505a
large-amplitude, **4**:306a–b
optical grating, **3**:620a; **11**:266b
photography of, **14**:497a
propagation of, **4**:230a–b; **5**:202a; **11**: 454a; **14**:327b
quantification of, **13**:240a
refraction of, **7**:407b
sensitive flame, **8**:122a
see also Sound
spherical, **3**:144a; **5**:483a–b
spin, **7**:493a
stationary, **7**:466a
surface, **8**:516b
theory of, **10**:20a
two-dimensional motion, **7**:595a
ultrasonic, **14**:497b–498a
amplification of, **8**:11b
see also Light, theories of, wave theory; Oceans, waves
Wave equation, **1**:499b; **3**:144a, 565a*; **6**: 251b; **8**:473b, 492a,b, 501b, 592b; **9**:175a; **10**:554a; **11**:476a; **12**:220a, 221a; **13**:241a; **14**:317a; **15**:27b
current-density vector, **5**:473a,b
derivation of, **4**:281b
λ, value of, **4**:281a
possible states, **5**:473b
precursors of, **1**:113b; **4**:478b
Wave fronts, **3**:144a, 145a
Wave function, **6**:147b, 148a, 245b; **8**:473b–474b, 477a, 480b; **12**:221a–b
normalization of, **10**:214a
Wave mechanics, **4**:280a, 318a–b; **5**:122b, 160b, 271b, 272a, 473a–b; **6**:89b, 545b, 615a; **8**:186a–b; **9**:616b; **10**: 214a, 423b; **11**:422a; **12**:33a, 218b–221b, 222b, 529b; **15**:42b, 49a, 63a
validity of, **6**:614b
see also Quantum mechanics
Wave theory
see Light, theories of, wave theory
Wavelengths
and frequency, **10**:76a
measurement of, **1**:166b, 357b; **5**:143b; **8**:578b; **9**:154b; **10**:519a; **11**:149b, 579a, 582a; **14**:566a–567b
Wax
analysis of, **2**:484b
Boltwax, **2**:258a–b
manufacture of, **11**:332a
Wayjan ibn Rustam
see al-Qūhī (or al-Kūhī), Abū Sahl Wayjan ibn Rustam
al-Wazzān al-Zayyātī al-Gharnāṭī, al-Ḥasan ibn Muḥammad
see Leo the African
Wealden formations, **3**:606a–b; **5**:14a–b, 437a; **6**:502b; **9**:87b; **12**:339b
freshwater origin of, **9**:87a
Weapon salve, **5**:47b–48a; **6**:254a
Weapons, **15**:652b
incendiary, **3**:21a
metallurgy of, **3**:234a; **5**:583b
see also Artillery; Atomic bomb; Ballistics; Cannon; Hydrogen bomb; Rockets; Technology, military; Warfare
Weather
asymmetry in, **4**:256a
control of, **8**:24b
daily records of
Renaissance, **2**:410b
seventeenth-century, **1**:566b; **7**:306b; **9**:119a
eighteenth-century, **3**:435b, 538a, 540a; **4**:224b*
nineteenth-century, **3**:593a; **4**:410a
forecasting
ancient Greek, **4**:172a; **11**:197b; **15**: 220a
Renaissance, **2**:410a,b; **13**:260b; **14**: 275b
seventeenth-century, **5**:575a
eighteenth-century, **7**:389a
nineteenth-century, **2**:533a; **4**:592a; **5**: 17b, 258b; **6**:402b; **8**:487a; **9**:279b; **12**:187b, 411b
twentieth-century, **4**:601a; **11**:557b; **14**:50b
long-range, **6**:98b; **11**:558a
nineteen-year cycle, **3**:436a

Weather (cont.)
  simultaneous observations, **2**:533a, 628a–b; **5**:575a
  and sunspots, **9**:279b
  thirty-five-year cycle, **3**:248a
  see also Climatology, Meteorology, Rain, Storms, Wind
Weather Bureau, U.S., **1**:6a; **3**:593a; **4**:601a; **11**:557b
Weathering, **1**:623a; **5**:72b, 338a; **6**:228a; **7**:376a; **9**:315a; **10**:126b, 429b
  chemical, **7**:267a
  see also Erosion
Weaving
  see Textiles
Webb, Philip Barker, **15**:471a
  Histoire naturelle des îles Canaries, **10**:221b
Weber, Anna van Bosse, **14**:203a
Weber, Eduard Friedrich Wilhelm, **5**:26b; **6**:505a; **14**:199a, 200a, 204b, 205a, 206a
Weber, Ernst Heinrich (1795–1878), 14:199–202; **6**:434b; **9**:101b; **12**:137a; **14**:205a
  physiology, **3**:531a; **14**:113b
  students of, **5**:26b; **6**:299b, 505a; **8**:514a
  wave motion, **3**:258b; **14**:204a, 206a
  Weber fraction, **4**:557b; **14**:201a
  Weber-Fechner law, **4**:557b–558a; **7**:458a; **9**:561b; **14**:201a
  Weber's ossicles, **14**:199b
Weber, Heinrich (1842–1913), 14:202–203; **4**:2b, 5a; **6**:388a; **13**:23b; **14**:209a*
  acquaintances of, **9**:358a; **11**:448a,b
  algebra, **6**:297b
  analysis, **7**:621b; **11**:449a, 451b
  on contemporary scientists, **4**:206b*; **7**:509a*
  correspondence of, **4**:4b
  Lehrbuch der Algebra, **2**:615b
  number theory, **6**:214a
  students of, **14**:67a; **15**:256a, 432b
Weber, Heinrich Friedrich, **10**:346a; **11**:480a(n)
Weber, Joseph, **4**:329b
Weber, Leonhard, **1**:529a
Weber, Max Wilhelm Carl (1852–1937), 14:203
Weber, Wilhelm Eduard (1804–1891), 14:203–209; **5**:307a; **12**:254a; **14**:32b, 199a
  acquaintances of, **4**:1b; **6**:570b; **11**:448a
  electrodynamics, **6**:343b–344a, 344b; **7**:380b, 381a; **8**:10b, 188a, 490b–491a, 492a, 494b; **9**:205a,b, 209a,b; **10**:27b
  electromagnetism, **1**:146a; **7**:450a–b
  and Gauss, **1**:7a; **5**:305a–306b, 308a, 309b; **7**:450b, 608b; **14**:204a–b
  induction coil, **14**:330a
  law of electrical force, **3**:309b; **8**:510b; **14**:205a, 207a–208a, 208b–209a
    critics of, **6**:250b–251a; **14**:205b
  measurement, **5**:407b
  students of, **1**:7a; **6**:505a; **7**:449a; **11**:445b–446a, 447b; **13**:79b
  terrestrial magnetism, **10**:192b; **12**:50a
  translations of works, **12**:53a*
  wave motion, **3**:258b; **14**:200a
Weber (magnetic unit), **14**:205b
Webster, David Locke, **4**:195b
Webster, John (1610–1682), 14:209–210; **5**:48a; **13**:560a*
  Academiarum examen, **14**:370a, 378b(n)
  critics of, **14**:365b, 367a–b, 373a
Webster, John White, **6**:517a; **7**:44b
Webster, Thomas (1773–1844), 14:210–211; **5**:14a; **8**:564b; **9**:87a
Wechsel, André, **5**:348b
Wedderburn, Joseph Henry Maclagan (1882–1948), 14:211–212; **1**:307b; **4**:83a, 618b; **9**:457b; **11**:55b; **13**:600a
Wedekind, Edgar [Leon Waldemar Otto], **11**:85b
Wedel, Georg Wolfgang (1645–1721), 14:212–213; **2**:2b; **6**:259a, 458a; **7**:458b; **11**:511b*
  Physiologia medica, **2**:2b
Wedenskij
  see Vvedenskii
Wedge, **14**:366b
Wedgwood, John, **1**:89a
Wedgwood, Josiah (1730–1795), 14:213–214; **3**:565a, 578b, 579a; **6**:383a; **7**:277a; **8**:261b; **10**:107b; **11**:141a, 145a; **14**:463a
Wedgwood, Thomas, **4**:182a; **8**:261b
Weeds, **6**:122b; **15**:60b
Week, **15**:713a
Weevil, **12**:201a; **13**:29b
Weffring, Blasius, **1**:78a
Wegener, Alfred Lothar (1880–1930), 14:214–217; **5**:487a; **13**:148b
  continental drift, **1**:236b, 623b; **6**:475a–b; **13**:271a
  critics of, **3**:325b; **13**:285b, 610b
Wegmann, César Eugène, **12**:274a
Wehnelt, Arthur Rudolph Berthold (1871–1944), 14:217–218
  Wehnelt cylinder, **14**:217b
Wei P'u, **12**:372a, 377b, 378a, 389a
Weichselbaum, Anton (1845–1920), 14:218–219; **7**:623a
Weiditz, Hans, **2**:536a, 536b–537a
Weierstrass, Karl Theodor Wilhelm (1815–1897), 14:219–224; **3**:143b, 466b; **4**:3b, 205b, 341a; **7**:55b*; **12**:246b; **13**:21b*
  acquaintances of, **1**:294a; **4**:2b; **5**:572a, 573a; **6**:230a; **7**:522a, 523a; **9**:427a
  algebra, **12**:284a,b
  analysis, **2**:105a, 275b, 276a, 304a, 358a, 470a; **3**:53a, 139b, 169a; **4**:103a, 206a; **6**:3b, 286b, 391b; **7**:479a, 621b; **8**:111b; **10**:594a; **11**:54b, 451a; **12**:513a; **13**:81b, 250b, 562a; **14**:614a; **15**:23b
  approximation theorems, **4**:561b
  Bolzano-Weierstrass theorem, **2**:275b
  calculus of variations, **2**:192a, 198a; **9**:232b
  on contemporary scientists, **4**:4b, 342b
  critics of, **7**:398a
  geometry, **4**:174a; **8**:433b; **11**:22a; **13**:20b
  influence of, **4**:4a; **7**:101b, 396b; **9**:426b; **10**:610a,b; **11**:54b, 149a, 327a; **12**:245b, 246a, 599b; **15**:755a
  and Kronecker, **7**:506b, 507a–508a, 523a; **14**:222b
  number theory, **2**:277a; **4**:419b, 466a; **9**:307b
  students of, **2**:271b, 272a; **3**:53a, 54a; **5**:203b; **6**:95a, 286a, 472a, 570b; **7**:407b, 459b, 477b; **8**:253a; **9**:358a, 426a, 502a, 563b; **10**:24a, 343a; **11**:8a, 611a; **12**:212b, 213a; **13**:81a, 155a; **14**:630a; **15**:65b
  theorems of
    generalization of, **5**:201b
    proof of, **6**:472b, 572a
  translations of works, **14**:472a
  Weierstrass coordinates, **8**:433b
  Weierstrass functions, **6**:214b; **14**:343a
  Weierstrass' integral, **5**:201b
Weigel, Christian Ehrenfried (1748–1831), 14:224–225; **4**:343b; **8**:87a*; **11**:523b*
Weigel, Valentin (1533–1588), 14:225–227; **1**:159a; **2**:222b; **5**:120a, 121b; **7**:485a
Weigert, Carl (1845–1904), 14:227–230; **4**:295b, 302b*; **12**:87b
  influence of, **4**:296a
  pathology, **4**:299a; **5**:495b; **7**:422a; **10**:131a
  students of, **1**:569a; **3**:447a; **10**:17b
Weight
  ancient Greek concept of, **1**:223a–b; **13**:92a
  ancient Roman concept of, **9**:300b
  medieval concept of, **7**:343b–345b
  Renaissance concept of, **8**:199b, 228a
  atomistic explanation of, **4**:32a; **5**:287a
  conservation of, **7**:194b
  and density of medium, **7**:344a–b
  vs. mass, **6**:608b; **10**:62a, 75b, 77b
  mathematical problem of, **1**:368a; **4**:606b
  psychophysical perception of, **9**:562a
  specific, **13**:260b
  standardization of units, **9**:393a; **13**:23a
  theories of
    seventeenth-century, **9**:115a
    eighteenth-century, **15**:288a–b
  variation of
    geographical, **2**:136a; **10**:76b; **15**:271a
  see also Congresses, scientific; Gravimetry; Measurement, standardization of
Weight function, **3**:229b
Weights and Measures, Office of (U. S.), **1**:364b; **2**:333a; **6**:165b
Weil, Adolf
  Weil's disease, **10**:142a
Weil, André, **1**:306b; **2**:351b; **6**:394a; **10**:140a; **11**:53b, 55b
Weimar, duchy of
  geology, **14**:60b
Weinbaum, Sidney, **8**:474b
Weinberg, Arthur, **4**:299b
Weinberg, Wilhelm (1862–1937), 14:230–231
  Hardy-Weinberg law, **14**:230b
Weiner, Joseph Sidney, **3**:606b, 607a
Weingarten, Julius (1836–1910), 14:231–232
Weinstein, Max Bernhard, **11**:8b
Weiprecht, Karl, **7**:511a
Weisbach, Julius Ludwig (1806–1871), 14:232; **11**:209a; **14**:617b
Weiser, Conrad, **1**:486b; **4**:493a
Weis-Fogh, Torkel, **7**:503b
Weismann, August Friedrich Leopold (1834–1914), 14:232–239; **3**:121a; **6**:9b; **9**:112a; **12**:567b
  critics of, **5**:447a; **6**:338b; **9**:520b; **14**:96b
  cytology, **13**:89a, 157b
  embryology, **3**:121b; **4**:187b; **9**:600b
  evolution, **6**:9a; **11**:20a
  Germ-Plasm, **15**:93a
  heredity, **8**:587a, 589a; **9**:601b; **11**:518b; **12**:298b; **14**:433b
  influences on, **2**:625b; **5**:52a; **13**:460b
  students of, **4**:186b; **7**:516a; **8**:447a; **9**:143b
  zoology, **15**:158b
Weiss, Christian Samuel (1780–1856), 14:239–242; **2**:586b; **4**:289b; **11**:524a*
  acquaintances of, **5**:202b; **6**:142b
  crystallography, **1**:460b; **2**:482a; **3**:232b; **9**:392b–393a, 448b; **11**:271a
  law of zones, **9**:620a; **14**:241a
  mineralogy, **7**:254a
  students of, **2**:108a; **10**:27a; **11**:235b, 270b, 539a
  Weiss indices, **14**:241a
Weiss, Edmund (1837–1917), 14:242–243; **2**:126a; **10**:212b; **12**:160b–161a

Weiss, Pierre (1865–1940), 14:243–247; 11:481b*
acquaintances of, **1**:519a; **3**:3b; **11**:477a, 480b(n)
Cotton-Weiss method, **14**:244a
Curie-Weiss law, **3**:4a; **14**:245b
physics, **3**:4a; **8**:11a; **11**:481a(n)
Weiss domains, **14**:245a–b
Weisse, Maximilian, **13**:110b
Weizmann, Chaim (1874–1952), 14:247–248; 4:313a
Weizsäcker, Carl Friedrich von, **7**:232a, 235a
Weizsäcker-Williams method, **4**:578a
Welch, William Henry (1850–1934), 14:248–250; 5:40b; **6**:232b; **15**:65a
acquaintances of, **3**:447a; **4**:296a; **6**:77b, 526a; **8**:447a; **9**:56a, 57b, 58a, 265b, 422b; **11**:176b; **12**:87b, 481a, 483b; **15**:64b
on contemporary scientists, **7**:435a*; **10**:144b*
correspondence of, **15**:71b
historiography, **7**:45b*
influence of, **1**:610a; **3**:517a
influences on, **7**:425b, 429b
pathology, **8**:450b
students of, **5**:40a; **11**:345b
Welcker, Hermann, **6**:225a
Welding
electric, **13**:361b
Weldon, Walter Frank Raphael (1860–1906), 14:251–252; 1:505a, 506a; **7**: 94b; **10**:469b*
evolution, **10**:463a–b
statistics, **10**:449a,b, 451a, 462b
Well problems (math.), **4**:606a, 612a
Wellcome, Henry, **11**:120b; **15**:105a
Wellcome Chemical Research Laboratories, **11**:120b, 121a
Wellesley, 1st Marquis of
see Colley, Richard
Wellmann, Max, **3**:174b, 175a; **11**:602a
Wells, Alfred Kingsley, **1**:394b
Wells, Harry Gideon (1875–1943), 14:252–253; 10:243a
Wells, Maurice Kingsley, **1**:394b
Wells, Thomas Spencer, **8**:406b, 407b
Wells, William Charles (1757–1817), 14: 253–254; 11:134b
Wells (water), **7**:240b
artesian, **3**:191b*; **10**:128a, 280b
disinfection of, **3**:7a
recharging of, **10**:280b
testing of, **9**:257b
see also Water
Welsbach, Carl Auer von, **3**:321b, 478a
Welser, Mark, **3**:179b; **5**:241b–242a; **10**: 240a; **12**:151b
Welter, Jean Joseph, **5**:323a, 325b; **15**:357a–b, 358a, 378a
Weltzien, Carl, **3**:48b–49a; **9**:348a
Welwitschia, **2**:371a; **6**:492a
Wen shu (original number), **3**:253a
Wendelin (Vendelinus), Gottfried (1580–1667), 14:254–255; **7**:126a(n); **8**:26a, 250b
Went, Friedrich August Ferdinand Christian (1863–1935), 14:255; 4:335b*; **13**: 460b*; **14**:96b, 105a*
Went, Frits Warmolt, **3**:581b
Wentzel, Gregor, **4**:196b; **7**:469a; **10**:423b–424a; **12**:21a, 464a, 529b; **15**:42b
Wenzel, Karl Friedrich, **2**:6a; **5**:601a; **7**: 389b(n); **11**:437a; **13**:310a
Wenzel, Leland Keith, **9**:257b, 258a
Wepfer, Johann-Jakob (1620–1695), 14: 255–256; 10:567b, 568a; **14**:407a
Werneck asylum, **5**:570a
Werner, Abraham Gottlob (1749–1817), 14: 256–264
acquaintances of, **4**:152a,b; **5**:202b, 518b; **6**:18b; **8**:246a
on contemporary scientists, **2**:440b; **6**: 587a(n)
correspondence of, **8**:245b
critics of, **1**:413b; **3**:433b; **9**:448a; **10**:209a; **14**:240a
influence of, **2**:552b–557a *passim;* **5**:578b; **7**:604a; **11**:387b, 521a; **12**:122a
influences on, **3**:474a; **5**:205b; **9**:534a
minerals, classification of, **1**:359b; **2**:482a, 552b; **3**:232b; **4**:274a; **14**:463b
orogenesis, **11**:273a
petrography, **3**:434a; **8**:594b
stratigraphy, **3**:313b; **5**:533a; **10**:209a
students of, **1**:327a; **2**:440a; **3**:210a; **4**: 346b, 347b; **5**:155a–b, 156a; **6**:549b; **7**:69b; **9**:447b, 620a; **11**:387a, 464a; **12**:182a; **13**:466b; **14**:60a, 239b
theory of the earth, **5**:13b; **10**:209b; **14**: 60b
veins, origin of, **8**:616b
Werner, Alfred (1866–1919), 14:264–272; 4:20b; **7**:180a*
chemistry, **5**:381a; **6**:108b; **8**:329b
coordination theory, **2**:199b; **3**:272a, 301b; **4**:20b–21a; **5**:350a*; **7**:179b; **11**:551b; **13**:546b; **14**:265b–272a
and Jørgensen, **7**:179b
stereochemistry, **1**:340b; **6**:108a; **9**:356b, 402b; **11**:87b
students of, **2**:64a, 169b; **5**:595a; **10**:578a; **15**:257a
Werner, Johann(es) (1468–1522), 14:272–277; 4:259a
astronomy, **13**:324a
De elementis conicis, **7**:133b
manuscripts of, **6**:144a
De triangulis sphaericis libri quatuor, **14**:470b–471a
trigonometry, **3**:262b; **14**:470b
Werner Institute of Nuclear Chemistry, **13**:163b–164a
Wernerian Natural History Society
founding of, **7**:69b; **13**:372b
Wernerism, **5**:31b, 537b; **6**:437b, 584b; **7**:45a, 69a–70b; **10**:8b; **15**:151b
opposition to, **2**:568a; **8**:616a; **10**:586a
Wernicke, Carl (1848–1905), 14:277–278
Wernicke's aphasia, **1**:496a; **14**:277a
Wernicke's encephalopathy, **14**:277b
Wernicke, Erich Arthur Emmanuel, **1**:576a
Wertheim, Ernst (1864–1920), 14:278–279
Wertverlauf, **5**:154a–b
Wesleyanism, **12**:465b
Wessel, Caspar (1745–1818), 14:279–281; 1:237b, 239a; **5**:111b; **15**:196a
Wessely, Franz Xaver, **8**:595b–596a
West Indies, **4**:213b, 214a
botany, **5**:546a; **13**:177b–178a, 354b
exploration of, **9**:587b
geology, **8**:615b
mapping of, **2**:201a
natural history, **7**:57b–58a; **12**:457b; **13**: 353b–354a
zoology, **5**:406b; **8**:266b
West Point, U.S. Military Academy, **6**:471b; **7**:445a
West Riding Lunatic Asylum (Wakefield), **4**:594a
West Virginia, geology, **14**:301a
Westermann, George, **12**:175a
Western Electric Company Laboratories, **3**:597a–b
Western Ontario, University of
Collip Medical Research Laboratory, **3**: 353a
Western Union International Telegraph Expedition, **3**:535b–536a
Westfeld, Christian Friedrich Gotthard Henning, **2**:6a
Westgren, Arne, **6**:561a
Westheimer, Frank, **7**:323b
Westhoff, Hermann, **12**:79a
Westinghouse, George, **13**:286b
Westinghouse Electric & Manufacturing Company, **1**:287b; **3**:367a–b
Westminster Assembly, **14**:364a
Weston, Edward
Weston cell, **3**:334a
Westphal, Carl Friedrich Otto, **1**:580a; **6**: 440a; **14**:277a
Westphalian (stratig. stage), **5**:476a
Westrumb, Johann Friedrich, **5**:532b
Westwood, John Obadiah, **13**:355b
Weyl, Hermann (1885–1955), 14:281–285; 5:350b; **6**:496a; **8**:326b
acquaintances of, **10**:138a, 139a; **12**:222a
analysis, **2**:238b; **6**:572b; **7**:492b; **11**:452b; **12**:13b; **14**:89a, 90a
on contemporary scientists, **3**:96b; **5**: 473a*; **6**:394b; **10**:423b
foundations of mathematics, **6**:393a, 473a; **14**:612a*
geometry, **4**:336b, 337a; **11**:455a
Die Idee der Riemannschen Fläche, **7**: 398a
influence of, **3**:95b; **12**:214a
influences on, **12**:426a
mathematical physics, **2**:236b; **4**:281a, 330b–331a; **12**:220a,b
Raum, Zeit, Materie, **10**:422a
Weyl chambers, **14**:284b
Weyl group, **14**:284b
Whale, **9**:432a
anatomy, **14**:203a
fisheries, **3**:203b; **6**:442b
mammalian nature of, **8**:378b
see also Cetacea
Whaling industry, **6**:267b; **9**:492a
Wharton, George (1617–1681), 14:285–286; 1:317a
Wharton, Thomas (1614–1673), 14:286–287; 13:451a
Adenographia, **5**:425b
anatomy, **1**:46b; **9**:64a; **10**:478a; **13**:31b
Whateley, Richard (1787–1863), 14:287–289; 8:528b
Wheat, **3**:459b, 479b; **15**:648b
analysis of, **3**:7b; **9**:401b; **10**:325b
classification of, **14**:33b
diseases and pests, **4**:220a; **8**:177a; **9**:78b; **13**:596a
mildew, **6**:219b
rust fungi, **3**:613a–b
smut, **2**:438a; **11**:131b, 132a; **13**:411b
wheat worm, **3**:588b
germination, **12**:124a
immunity, **15**:505b, 507a
milling of, **9**:62a
preservation of, **3**:7a
Wheatstone, Charles (1802–1875), 14:289–291; 6:278a; **7**:93b
acoustics, **3**:258b; **4**:534b
acquaintances of, **3**:474b; **6**:211a; **12**: 131b; **13**:238a
electricity, **12**:425b
speed of, **1**:202a; **4**:559a; **5**:19a, 85a
influence of, **3**:476a; **6**:402b
stereoscope, **6**:248b; **14**:493a
telegraph, **12**:425a
Wheatstone bridge, **3**:260b; **8**:20a; **14**: 290a
Wheatstone dial plate, **2**:439a
Wheel, **14**:366b; **15**:666a(n)
theory of, **6**:312a

Wheel (cont.)
toothed, **10**:298a
water-raising, **15**:519a
Wheelchair, **3**:388b
Wheeler, George Montague, **4**:265b; **5**:395b
Wheeler, John Archibald, **2**:251b; **5**:161b
Wheeler, Mortimer, **14**:505a
Wheeler, William Morton (1865–1937), 14: 291–292; **6**:262a; **7**:593b*; **8**:529b*, 587a, 588a; **11**:334b(n), 575a*; **15**: 525b*
Wheler, Granville, **5**:516b–517a
Whelk, **1**:265a
Whewell, William (1794–1866), 14:292–295
acquaintances of, **6**:324a; **8**:530b*; **10**: 260a; **12**:524a
causation, **7**:234a; **8**:159b
on contemporary scientists, **13**:500a*
correspondence of, **6**:82b–83a
critics of, **1**:178a, 364b; **2**:441a; **7**:106a; **11**:116a
crystallography, **9**:392b; **15**:115a*
influence of, **1**:404a; **4**:546b; **9**:199a
influences on, **3**:238b
philosophy of science, **3**:375b; **8**:569b; **9**:213b
scientific terminology, **4**:537a
Whiddington, Richard, **9**:543a–b
Whig party, **8**:524a
Whin sill, **13**:273a
Whipworms, **4**:275b
Whirlpools, **2**:356a
theory of, **10**:20a
Whistler, Daniel, **5**:425b
Whiston, William (1667–1752), 14:295–296; **1**:234a; **5**:510b
acquaintances of, **3**:294b, 431a; **6**:35b; **10**:45a, 83b, 84a
on contemporary scientists, **5**:520b; **6**: 175b; **10**:61a, 84b, 93b*
cosmogony, **11**:317b
critics of, **7**:275b
mathematics, **13**:235b
Praelectiones astronomicae, **5**:24a
students of, **6**:36a; **9**:479b
theology, **2**:382a
Whitbread, Samuel Charles, **5**:413b; **9**:69b
Whitby, Lionel Ernest Howard, **4**:154b
White, Charles (1728–1813), 14:296–297
White, Charles David (1862–1935), 14:297–299; **14**:312b
White, Gilbert (1720–1793), 14:299–300; **1**:331b; **6**:43b; **14**:187b
The Natural History and Antiquities of Selborne, **2**:205a; **6**:295b; **10**:509b
White, Israel Charles (1848–1927), 14:300–301
White, J. Maunsel, **13**:272a
White, Paul Dudley, **8**:295a
White, Thomas (1593–1676), 14:301–302; **5**:416a
White, William Alanson, **5**:180a
White lead, **7**:78a
manufacture of, **3**:201a, 317a*; **4**:134a, 250a; **13**:331a
medical use of, **14**:585a
White leg, **14**:297a
White Sea, **7**:410b
geology, **15**:428a
hydrology, **4**:42b
level of, **15**:427b
Whitehead, Alfred North (1861–1947), 14: 302–310
The Concept of Nature, **6**:26b
An Enquiry Concerning the Principles of Natural Knowledge, **12**:16b
logic, **8**:154b; **12**:13b
philosophy, **6**:260b; **14**:609a
Principia mathematica, **5**:153b; **11**:107a, 285b; **12**:14a–15a; **14**:303b–304b
and Russell, **12**:10a, 12a
students of, **14**:310b
Whitehead, John Henry Constantine (1904–1960), 14:310–311; **5**:153b; **8**:473a
The Foundations of Differential Geometry, **13**:600a
Whitehead, Richard, **9**:465a
Whitehouse, Edward Orange Wildman, **13**:387a
Whitehurst, John (1713–1788), 14:311–312; **4**:549a; **11**:35a
An Inquiry into the Original State and Formation of the Earth, **13**:441b
Whitfield, Robert Parr (1828–1910), 14: 312–313; **6**:57a
Whitford, Albert Edward, **6**:30a
Whitman, Charles Otis (1842–1910), 14: 313–315; **3**:590a; **5**:598b; **7**:170b; **8**: 355a,b, 356a; **14**:291a
Whitney, Hassler, **13**:600a
Whitney, Josiah Dwight (1819–1896), 14: 315–316; **4**:73b, 364b; **5**:214a; **7**:45a
and Silliman, **12**:436a
Whitrow, Gerald James, **9**:405b–406a
Whittaker, Edmund Taylor (1873–1956), 14:316–318; **2**:144a; **3**:457b*; **4**: 282b*; **11**:49b*
acquaintances of, **3**:564a; **6**:548a; **12**:10a
astronomy, **4**:282a
on contemporary scientists, **5**:77b; **6**: 212b*; **14**:87b*
A Course of Modern Analysis, **14**:188b
historiography, **3**:135b; **15**:401b*
History of the Theories of Aether and Electricity, **5**:15b
mathematical physics, **1**:499b; **6**:92b, 93a(n)
Whittaker's functions, **14**:317a
Whooping cough, **1**:399b
vaccination, **2**:300b
Whytlaw-Gray, Robert (1877–1958), 14: 318–319; **11**:280b; **13**:454b
Whytt, Robert (1714–1766), 14:319–324; **2**:173b, 174a; **3**:494b; **9**:482a,b; **13**: 122b, 123a, 448a
calculi, **2**:175a–b
neurology, **6**:59a, 60a
Whytt's reflex, **14**:321b
Wiadomości matematyczne
founding of, **4**:84a
Wichelhaus, Hermann, **7**:281a, 282b, 283a
Wickersheimer, Ernest (1880–1965), 14: 324–325; **6**:124a*, 430b*; **10**:311b*
Widal, Ferdinand
Gruber-Widal reaction, **5**:564b
Widenmann, Johann Friedrich Wilhelm, **14**:261b
Wideröe, Rolf, **8**:95a
Widman (or Weideman or Wideman), Johannes (*ca.* 1462–*ca.* 1498), 14:325–326; **7**:419a; **11**:457a,b, 591a; **13**: 62a(n)
Widmannstätten (or Widmanstetter), Aloys Joseph Beck Edler von (1754–1849), 14:326–327; **12**:543b; **13**:245b
Widmannstätten figures, **1**:581a; **12**:543b; **14**:326b
Wiechert, Emil (1861–1928), 14:327–328; **4**:357b*; **5**:596a; **12**:187a, 526a; **13**: 246b
electron theory, **8**:494a; **9**:217a; **13**:367a
Wied, Maximilian zu (1782–1867), 14:328–329
Wiedemann, Eilhard [Ernst Gustav], **5**:330b; **7**:212b, 334a, 336a, 338b, 339a, 343b, 346b; **11**:248b; **12**:191a; **14**:217b, 329b, 330a
Wiedemann, Gustav Heinrich (1826–1899), 14:329–331; **4**:189b; **14**:348a; **15**:456a
acquaintances of, **5**:63b; **6**:73b, 271a, 343b; **14**:629a
batteries, **10**:481b
on contemporary scientists, **3**:310a
correspondence of, **6**:350a*
physical chemistry, **3**:617b
students of, **10**:481b; **12**:595a; **14**:628b
Wiedemann-Franz ratio, **8**:502a; **11**: 446b–447a; **14**:330a
Wiedemanns Annalen
see *Annalen der Physik und Chemie*
Wiedersheim, Robert (1848–1923), 14:331; **9**:143b; **15**:169a
Wiegleb, Johann Christian (1732–1800), 14: 332–333; **9**:347a; **14**:386b; **15**:205b
Wieland, Heinrich Otto (1877–1957), 14: 334–335; **3**:512a*; **14**:411b; **15**:157a
biochemistry, **6**:500b; **7**:273a; **10**:282a; **13**:393b; **14**:444b, 445b
Wieland (or Guilandinus), Melchior (*ca.* 1520–1589), 14:335–336; **1**:124a, 522b; **9**:309a
Wieleitner, Heinrich (1874–1931), 14:336–337; **2**:430b*; **3**:59a; **5**:383a*, 574a*; **13**:213a*, 470a, 561a*; **14**:155a*
Wien, Max, **3**:620a; **12**:218b, 526a
Wien, Wilhelm Carl Werner Otto Fritz Franz (1864–1928), 14:337–342; **12**:527b, 615a
acquaintances of, **11**:8b; **12**:526a
correspondence of, **12**:219a–b
electrodynamics, **4**:313b, 319b; **6**:348a; **8**:497a
electronics, **8**:51b
energy, **5**:609a; **7**:264b
Kanalstrahlen und Ionizations . . . , **11**: 420b
philosophy, **11**:14a
radiation, **5**:459a; **8**:180b, 289b; **11**:11a–b, 12a, 150a; **12**:613b; **13**:369a–b; **14**: 31b–32a
thermometry, **6**:470a
Wien's laws, **7**:528a; **8**:20b, 551b; **11**: 150b, 582b
displacement law, **4**:294a, 314b; **10**: 346a; **14**:338a–b, 339b
distribution law, **4**:315b; **10**:346b; **11**: 11b, 12a; **14**:338b
Wiener, Alexander, **7**:624b
Wiener, Ludwig Christian (1826–1896), 14: 342–344; **2**:465a
Wiener, Norbert (1894–1964), 14:344–347; **5**:65b; **6**:350a*
acquaintances of, **15**:42a–b, 43a
analysis, **2**:238b, 239a; **9**:129b; **13**:264a
cybernetics, **1**:428a; **2**:325b; **9**:227a; **11**: 546a
philosophy, **15**:39a
probability, **1**:366b; **7**:599b; **14**:454a
Wiener problem, **14**:346a
Wiener-Hopf equation, **14**:346a,b
Wiener, Otto (1862–1927), 14:347–349; **6**: 74a*; **12**:236a*
Wiener, Philip, **8**:517a
Wiersma, Eliza Cornelis, **5**:610b
Wiesbaden Agricultural College, **5**:164b
Wiesner, Julius von (1838–1916), 14:349–350; **3**:581b; **5**:623b; **7**:16a*
Wiesner, K., **6**:373a
Wigand, Albert Julius Wilhelm (1821–1886), 14:350; **4**:307a; **10**:575b
Wight, Robert, **5**:539a; **6**:491a
Wight, Isle of
geology, **5**:14a–b, 68a; **6**:288b; **8**:564b; **14**:210b–211a
Wigner, Eugene Paul, **3**:371a; **4**:313a; **6**: 561b; **7**:555a; **15**:43a

Wilamowitz-Moellendorff, Ulrich von, **4**: 463b; **5**:229b; **6**:418b; **11**:25a
Wilberforce, Lionel Robert, **1**:457b
Wilberforce, Samuel, **3**:574a; **4**:183b
and Huxley, **6**:592b–593a
Wilbrand, Johann Bernhard (1779–1846), 14:351–352; 1:578a
Wilcke, Johan Carl (1732–1796), 14:352–356; 12:149a*; **13**:183b
and Aepinus, **1**:66b, 67a
electricity, **1**:67a; **2**:5a; **3**:51b; **5**:136a; **14**:70a, 71b
Wilczynski, Ernest Julius (1876–1932), 14: 356; 5:204b*
Wild, Heinrich (1833–1902), 14:356–357; 10:28b
Wild, Wilhelm, **1**:362a
Wild flowers, **14**:300a
identification manuals, **3**:318a
Wildt, Rupert, **12**:455a
Wiley, Harvey Washington (1844–1930), 14:357–358; 3:294a
Wilfarth, Hermann, **6**:237b
Wilhelm
see also William
Wilhelm IV, Landgrave of Hesse (1532–1592), 14:358–359; 2:403b–404a, 406b, 602a; **6**:145b; **11**:378b*, 561b
star catalog, **6**:364a
Wilhelmsuhr, **14**:359a
Wilhelm, Richard, **7**:192a
Wilhelm Roux' Archiv für Entwicklungsmechanik der Organismen, **4**:188a
Wilhelmy, Ludwig Ferdinand (1812–1864), 14:359–360
Wilkes, Charles, **3**:549b; **9**:195b; **10**:439b
Wilkes Expedition (1838–1842), **3**:549b; **10**:439b; **13**:433a
Wilkins, George, **13**:167a
Wilkins, John (1614–1672), 14:361–381
acquaintances of, **2**:377b; **3**:208b; **4**:495b; **5**:415a, 606b; **10**:200b, 201a; **11**:532a; **12**:580b; **14**:405a, 509a
on contemporary scientists, **14**:510a
Discovery of a New World, **5**:59b
Essay Towards a Real Character and a Philosophical Language, **3**:362a; **11**:314a
influence of, **6**:130a; **8**:153b
influences on, **5**:247b
and Royal Society, **5**:534b, 607a; **6**:482a; **10**:564b; **12**:582a–585b *passim,* **14**: 151b, 510b
translations of works, **4**:166a
Vindiciae academiarum, **14**:177b, 209b
Wilkins, Maurice Hugh Frederick, **5**:140a
Wilkinson, David Todd, **5**:273a
Wilkinson, Lancelot, **7**:81b; **15**:629a
Wilks, Samuel, **1**:60b
Wilks, Samuel Stanley (1906–1964), 14: 381–386; 10:472b*
Will, Clifford Martin, **14**:305b–306a
Will, Heinrich, **2**:340b; **7**:279b, 393a, 463a; **8**:331a; **9**:619b; **12**:208b
Will (philos.), **1**:334a
definition of, **7**:67b
as energy, **1**:404a
experimental analysis of, **8**:445a–446a
and intellect, **5**:549b
see also Free will
Willcock, Edith Gertrude, **5**:451b
Willdenow, Karl Ludwig (1765–1812), 14: 386–388; 6:549a; **12**:587b, 588b; **14**:116b, 332b; **15**:267a
Grundriss der Kräuterkunde, **14**:117b
Willemite, **8**:289a
Willemoes-Suhm, Rudolph von, **9**:588b
William
see also Wilhelm
William III, King of England, **2**:612b, 613a; **9**:400b; **15**:28b
William of Auvergne (Guilielmus Arvernus or Alvernus) (*ca.* 1180–1249), 14: 388–389; 1:62a, 377b; **5**:592a
William of Champeaux, **1**:1b
William of Conches, **2**:19b, 21b
William the Englishman (*fl.* thirteenth century), 14:399–402; 14:390b(n)
astronomy, **7**:125a; **15**:35a
translations by, **14**:594a
William Heytesbury
see Heytesbury, William
William of Luna, **1**:272b
William of Moerbeke
see Moerbeke, William of
William of Ockham
see Ockham, William of
William of Paris
see William of Auvergne
William of Saint-Cloud (*fl.* end of thirteenth century), 14:389–391; 7:139a; **10**: 541a
translations by, **7**:129a; **14**:593b
William of Sherwood (Shyreswood, Shirewode) (*fl.* thirteenth century), 14: 391–392; 2:604b; **6**:377a
William and Mary, College of
establishment of, **1**:432a
Williams, Albert, **3**:609b
Williams, Evan James, **2**:242a
Williams, George Huntington, **7**:5a–b
Williams, Henry Shaler (1847–1918), 14: 392; 14:297b
Williams, John (geol.), **11**:35a
Williams, John Lloyd, **13**:165a
Williams, Robert Runnels (1886–1965), 14: 392–394; 14:446a
Williams, Roger John, **14**:393b
Williams, William Ewart, **11**:422b
Williamson, Alexander William (1824–1904), 14:394–396; 10:177a–b; **11**: 278a, 536b
on acids, **5**:372b
atomic weights, **10**:37b
on contemporary scientists, **2**:515b; **4**: 395b*; **9**:19a*, 426a*; **10**:511b*; **14**:532a*
etherification theory, **1**:298b; **2**:70b; **5**: 373a
influences on, **6**:462b
and Kekulé, **7**:280a
type theory, **3**:193b; **6**:564b; **8**:59b
Williamson's synthesis, **14**:395a
Williamson, William Crawford (1816–1895), 14:396–399; 8:5a; **12**:259a,b, 339b
Willis, Bailey (1857–1949), 14:402–403; 3: 587a; **6**:528a*; **7**:62b, 63a; **11**:209a,b, 210a
Willis, Robert (1800–1875), 14:403–404; 13:78b; **14**:31a*, 295a*
Willis, Thomas (1621–1675), 14:404–409
acquaintances of, **6**:482a; **9**:400a; **14**:369a
anatomy, **11**:112a; **12**:137b
Cerebri anatome, **12**:85a; **14**:509b
circle of Willis, **14**:407a
critics of, **6**:64a, 387b
Diatribae duae medico-philosophicae, **8**: 525a
De fermentatione sive De motu intestino particularum in quovis corpore, **6**: 46b(n)
influence of, **1**:593a; **2**:226a; **9**:244a; **11**: 40b; **14**:25b
influences on, **4**:65a; **10**:311a
and Lower, **8**:524a, 525a–b, 526b
physiology, **6**:64a; **8**:526b; **9**:243a,b
Williston, Samuel Wendell (1851–1918), 14: 409–411; 7:285b; **8**:587a
Willow tree
seeds, **2**:219a
Willstätter, Richard (1872–1942), 14:411–412; 4:90b; **5**:622a, 623a; **9**:96a; **13**: 1a; **14**:334a
enzymes, **13**:152b
students of, **5**:99b; **7**:517a; **9**:354a; **15**:1a
Willughby, Francis (1635–1672), 14:412–414; 9:508a*; **10**:509b; **13**:450a; **14**: 379a(n)
botany, **6**:38b
History of Fishes, **13**:527b
and Ray, **11**:313a–b; **14**:412b–413b
zoology, **11**:316a–317a; **14**:374a
Wilsing, Johannes (1856–1943), 14:414; 11:542b; **12**:153a
Wilson, Alexander (1714–1786), 14:414–415; 4:19a; **9**:266b
Wilson, Alexander (1766–1813), 14:415–418; 1:331a, 489b; **2**:282a; **10**:438b
American Ornithology, **1**:330a
editions of, **7**:70a
Wilson, Alexander Philips
see Philip, Alexander Philips Wilson
Wilson, Benjamin (1721–1788), 14:418–420; 2:5a; **3**:51b, 52a; **7**:13a; **10**:147a; **12**:413b; **13**:224b
Wilson, Bertram Martin, **14**:188b
Wilson, Charles Thomson Rees (1869–1959), 14:420–423; 8:10a; **11**:117a, 419b; **12**:201b; **15**:60b
cloud chamber, **1**:366a, 490b; **2**:398b; **6**:117b; **7**:152b, 153b, 158a; **9**:261b; **11**:117a; **12**:33a; **13**:367b; **14**:421a–422a
Wilson, Edmund Beecher (1856–1939), 14: 423–436; 9:515b, 523a; **15**:124a
on contemporary scientists, **2**:361a, 363a, 364b
cytology, **5**:52b; **8**:588b; **9**:519a, 521a
embryology, **8**:357a; **13**:158a
cell lineage, **3**:389b; **8**:355a
influences on, **2**:502a; **9**:496b
students of, **8**:587a, 589a; **9**:521b, 564b; **13**:151a, 156b, 157a–b
Wilson, Edwin Bidwell (1879–1964), 14: 436–438; 2:594a; **5**:391a; **12**:416b; **14**:623a
Wilson, Ernest Dana, **6**:117a–b; **9**:110a
Wilson, George, **5**:333b, 335b; **6**:517b; **8**: 298a
On Colour Blindness, **9**:200b
A Compleat Course of Chemistry, **6**:41b; **8**:297a
Wilson, Harold Albert, **8**:94b; **9**:395b; **11**: 419b, 420a,b
Wilson, James (agric.), **14**:357b
Wilson, James Maurice, **4**:137b
Wilson, John (1741–1793), 14:438; 7:565b
Wilson's theorem, **14**:438a
Wilson, John (philos.), **6**:81a
Wilson, Patrick, **9**:267a
Wilson, Robert Erastus, **9**:375b
Wilson, Robert Woodward, **5**:273a
Wilson, William, **12**:31b
Wilson, Woodrow, **3**:389a; **4**:618b; **6**:31a
Wiman, Anders, **6**:572a
Wimereux (France)
biological station, **5**:385a
Winch, **2**:215b
heat-driven, **3**:8a
Winchell, Alexander (1824–1891), 14:439–440; 3:189b; **6**:524b*; **14**:441a, 442b, 517b
Winchell, Alexander Newton (1874–1958), 14:440–441; 13:518b; **14**:443a
Winchell, Horace, **14**:440b
Winchell, Horace Vaughn (1865–1923), 14: 441–442; 14:443a

Winchell, Newton Horace (1839–1914), 14:442–443; 10:259b*; 13:533b*; 14:439b, 441a,b, 517b
Winchell family, 14:438b
Wind(s), 3:436a; 6:585b; 12:121a; 14:52b
and atmospheric pressure (Buys Ballot's law), 2:628b; 12:595a
average direction, 7:599a
charting of, 5:520a; 6:71b; 9:196a
force of, 2:410b
as geological agent, 10:166b; 11:181b
and ground, friction of, 12:595a
laws of, 3:297b
remote recording of, 12:595b
sea breeze, 3:593a
and sun, effects of, 12:595a
theories of
ancient Greek, 2:283a; 11:104b; 13:93b
ancient Roman, 15:518a–b
Renaissance, 1:607b; 5:401a
seventeenth-century, 6:71b; 9:117b, 119a; 13:438a
eighteenth-century, 1:113b; 3:540a
nineteenth-century, 4:175a, 592a–b; 6:97b–98b; 9:196b; 13:382b
trade, 14:275a; 15:289a, 301a
dust of, 4:289a
velocity
measurement of, 6:126b; 14:494b–495a
see also Storms
Wind tunnel, 8:301a; 11:124a, 306a; 12:613a; 13:483a, 492a; 14:520b
Windaus, Adolf Otto Reinhold (1876–1959), 14:443–446; 4:90b, 91b; 14:334b; 15:157a
Windler, S. C. H. (pseudonym), 14:478a
Windmill, 4:8b; 9:153a
construction of, 13:50a–b
design of, 3:442b; 8:211b; 13:233b
horizontal, 3:578b
Windmill Street School of Anatomy
see Great Windmill Street School of Anatomy
Window (geol.), 8:544a; 13:285a–b
Windrows, 8:24b
Wine, 9:558b
aging of, 10:366a
alcohol content, 2:420b; 5:340b
analysis of, 5:165a
concentrates of, 5:421a
diseases of, 10:366a–b
distillation of, 1:527a
medical use of, 1:315a
preservation of, 10:326a, 366b
production of, 5:422b
chaptalization, 3:200b–201a
see also Oenology, Viniculture, Viticulture
Wine, oil of, 8:333b
Wine press, 6:312a
Wing, Vincent (1619–1668), 14:446–447; 5:23b; 10:43b, 60b; 12:342b–343a
Wing (aeronaut.)
angle of attack, 14:621a
cantilever, 9:578a
longitudinal moment on
angle of attack, 3:196a
profile, 14:621b
rate of circulation around, 3:196a; 14:621b
stream pressure on, 3:196a
theory of, 3:195b–196a; 9:419b; 10:20a; 11:124b
see also Aerodynamics; Airplane; Drag, aerodynamic
Wingate, Edmund, 6:271b
Winiwarter, Hans von, 2:130a; 14:431a
Winkler, Clemens (1838–1904), 14:447; 4:311a; 9:290b, 291a; 10:493a; 11:34a
Winkler gas burette, 14:447a
Winkler, Lajos Wilhelm (1863–1939), 14:447–448; 13:226a
Winkler method, 14:448a
Winlock, Anna, 14:449a
Winlock, Joseph (1826–1875), 14:448–449; 10:483a
Winnecke, Friedrich August Theodor, 1:157b; 4:351a
Winsløw, Jacob (or Jacques-Bénigne) (1669–1760), 14:449–451; 4:267b, 589b; 6:62a; 8:114b, 171b; 9:323b; 14:320a
Winsor, Frederick Albert, 1:44a
Winterl, Jacob Joseph, 3:232b; 7:390b; 10:183a, 184a
Winthrop, John (1606–1676), 14:451–452; 1:134a; 5:606b; 6:141b; 12:616b; 14:452a
Winthrop, John (1714–1779), 14:452–453; 5:520a
Wintner, Aurel (1903–1958), 14:453–454; 2:239a
Winzer, F. A.
see Winsor, Frederick Albert
Wires
elasticity of, 3:443b
magnetized, 3:445b
Wirsung, Johann Georg, 9:64a; 10:567b; 13:31a
Wirsung duct, 10:477b; 13:589a
Wirtinger, Wilhelm, 11:448b; 12:218a
Wirtz, Carl, 6:531a; 11:542a; 12:455b
Wisconsin
agriculture, 8:591a
Geological Survey, 6:57b
geology, 3:189b–190a; 4:562b, 563a; 14:315b
glaciation, 13:270a–b
limnology, 7:184a–b
Wisconsin, University of, 3:190a
chemistry at, 7:208b; 13:160a–b
geology at, 13:518b
pharmacy at, 11:120b
Washburn Observatory, 6:471a
Wiseman, Richard, 2:315a
Wishart, John, 14:382b, 383a
Wish-fulfillment theory, 5:176a–b, 177a
Wislicenus, Johannes (1835–1902), 14:454–455; 5:127b*; 6:107b; 9:360a; 14:601b
physical chemistry, 15:463a
physiology, 5:126b; 14:65b
stereochemistry, 13:576a–b
students of, 1:553a; 3:347a; 4:165a; 5:128a,b; 10:517a
Wistar, Caspar (1761–1818), 14:456–457; 11:616b
Wistar Institute of Anatomy and Biology (Phila.), 4:161a
Wisteria, 14:457a(n)
Witelo (Vitello) (*ca.* 1230–*ca.* 1275), 14:457–462; 1:383a; 9:435a
critics of, 2:435a
influence of, 4:8b, 157b; 6:361a; 7:269b; 10:474a; 11:288b; 12:501a
influences on, 1:170a
optics, 1:81a; 4:53a; 5:28b, 409a, 525a,b; 6:125b, 194b; 7:298a,b
Perspectiva, 6:197a; 9:435a,b; 11:203a
editions of, 11:468a
Witham, Henry (1779–1844), 14:462–463; 2:492a
Observations on Fossil Vegetables, 10:110a,b
Withering, William (1741–1799), 14:463–465; 11:141a
An Account of the Foxglove, 3:578a
A Botanical Arrangement of all the Vegetables naturally growing in Great Britain, 3:578a
and Darwin (Erasmus), 3:578a
Witherite, 14:463b
Witmer, Lightner, 5:377b
Witt, Ernst, 11:56a
Witt, Jan de (1625–1672), 14:465–467; 1:566b; 3:384b; 6:538a; 10:43b, 45b; 12:205b, 206b; 13:17a
Elementa curvarum linearum, 14:148a
Witt, Otto, 3:85a
Wittgenstein, Ludwig (Josef Johann) (1889–1951), 14:467–470; 5:153b; 11:285b; 12:15a, 177b; 14:304a
Wittich (Wittichius), Paul (*ca.* 1555–1587), 14:470–471; 2:406a–b, 413a; 9:611a
Wleügel, Severin, 6:271a
Woepcke, Franz (1826–1864), 14:471–473; 4:426a, 431a–b, 454a*, 544b*, 613a(n),b*; 7:240b–241a, 329a, 331a; 9:557a*, 614a; 10:299b; 11:229b; 13:294b*
Wöhler, August (1819–1914), 14:473–474; 11:464b
Wöhler's laws, 14:473b
Wöhler, Friedrich (1800–1882), 14:474–479; 1:389b; 2:587a; 3:49a; 7:451a
acquaintances of, 3:286b; 5:12b, 381a, 430b; 9:18b, 91a, 541b; 11:540b
and Berzelius, 2:91b, 93a, 96b
biochemistry, 7:519b; 12:243b
correspondence of, 10:436a
Grundriss der Organischen Chemie, 5:13a
translations of, 11:370b
inorganic chemistry, 13:597a
and Liebig, 8:331a–335a *passim,* 339a, 342b–343b; 11:495a; 14:474b, 476a–477b
organic chemistry, 5:371a
cyanates, 5:325a
urea preparation, 2:67a
radical theory, 2:96a
students of, 1:578a; 3:300a; 7:450b, 519a; 12:163b, 465a; 13:243a; 14:63a, 141b, 233a; 15:457a
Wojciech, Oczok, 1:204a
Wolbach, Simeon Burt, 11:442a, 443a
Wolcott, Alexander Simon, 4:182b
Wolf, Charles Joseph Étienne (1827–1918), 14:479–480; 11:319b
Wolf-Rayet stars, 3:36a
Wolf, Emil, 15:43b
Wolf, Johann Rudolf (1816–1893), 14:480–481; 5:490b*; 9:2b; 11:561b; 12:239b; 13:21b*, 22b*, 155a
Wolf, Maximilian Franz Joseph Cornelius (1863–1932), 14:481–482; 2:424a; 4:219a; 10:279b
Wolf (animal)
fossil, 12:184a
Wolfers, Jacob Philipp, 4:370a; 11:114b
Wolff, Caspar Friedrich (1734–1794), 15:524–526; 6:65a; 11:159a
Wolffian bodies, 11:307b–308a; 15:525a
Wolffian duct, 1:421a–b; 15:525a
Wolff, Christian (1679–1754), 14:482–484; 3:216a; 8:253a; 13:39b, 480b
Anfangsgründe aller mathematischen Wissenschaften, 9:233a
commentaries on, 2:467b
Elementa
editions of, 3:460a
influence of, 2:274a; 7:225b, 442b, 596a, 597b

Philosophia sive ontologia, 8:159b
philosophy, 4:471b; 7:598a; 8:159a, 160a; 15:53a
universal language, 13:180a
students of, 7:490a,b; 8:467b
translations of works, 8:467b
Wolff, Gustaf, 12:567b
Wolfsohn, Günther, 7:554a
Wollan, Ernest Omar, 3:370a
Wollaston, Francis (1731–1815), 14:484–486; 14:486a
Wollaston, William Hyde (1766–1828), 14:486–494
acquaintances of, 3:291b, 545a; 5:71b; 6:324a; 8:288b; 9:10a, 424b; 12:523b; 13:281a
biographies of, 6:285b*
chemistry, 3:603a; 4:535b, 536a
analytical, 3:232b; 6:166b
atomic theory, 3:543a; 13:373a
equivalent weights, 3:47b; 5:320a
physiological, 5:92b
crystallography, 6:180b, 182a
optics, 2:135a
color theory, 2:453a
instrumentation, 13:238b
spectra, 5:143a; 9:267a
wave theory, 9:73a; 15:361b–362a
The Religion of Nature Delineated, 5:129b
and Royal Society, 3:603b
Wollastonite, 14:491a
Woltman, Reinhard (1757–1837), 14:494–495; 11:377a
instrumentation, 3:358a
Woman's Field Army, 9:586b
Women
education of
eighteenth-century, 1:75b; 3:578a; 11:617b
nineteenth-century, 2:60b, 316a, 623b; 3:608a; 7:477b; 10:128a, 240a, 620b; 11:382a; 12:525a; 13:45a, 481b, 504a; 14:529b; 15:450b
twentieth-century, 9:260a–b; 15:417b
objection to, 2:162b
emancipation of, 7:477b; 12:525a; 14:134a, 223b
presumed inferiority of, 6:425b–426a; 14:512a
as scientists
astronomy, 3:49b–50b; 5:33b–34a; 8:105b–106b; 9:194a–195a, 421b
biochemistry, 3:415b–416b; 5:139b–141b
biology, 15:124a
botany, 1:205b–206a
chemistry, 9:340b–342a
mathematics, 1:75b–77a; 5:375a–376b, 472b–473a; 6:615b–616a; 7:477a–479b; 10:137b–139a; 12:521b
medicine, 12:48a–b
natural philosophy, 3:215a–216b
Nobel Prize winners, 3:416a, 500a, 501a
physics, 3:497b–502b; 7:157b–159a; 8:484b–485b; 9:260a–263a
status of, 14:44b
suffrage, 12:10b
Wood, Alexander, 3:31b
Wood, Anthony à, 2:550b; 3:208b, 248b; 5:344a; 6:127b; 8:524a; 9:507b; 14:363b, 374b, 377b(n), 378a(n), 379b(n)
Wood, Benjamin DeKalbe, 15:128b
Wood, Ethel, 8:33a
Wood, Horatio C (1841–1920), 14:495–497
Wood, John (surg.), 8:407b
Wood, Robert Williams (1868–1955), 14:497–499; 4:253b; 7:36a, 554b; 11:266a, 583a; 15:60a
Wood
distillation of, 5:420b
expansion of, 11:472a
formation of, 12:100a
fossil, 3:433b; 6:489b; 8:354b; 13:29b
structure of, 10:110b
hydrolysis of, 2:4a
identification of, 12:100a
lamination of, 15:253b
microscopy of, 5:535b; 12:99b–100a
preservation of, 9:458b–459a
structural properties, 4:224a; 8:128b–129a; 9:418b; 14:349b
sulfuric acid, effects of, 2:385b
tensile strength, 1:459b; 2:576b; 5:410b; 10:3b
transformation into coal, 2:3b
see also Trees
Wood borer, 3:549a
Woodhouse, James, 6:114b; 8:613a; 11:616b; 15:471b
Woodhouse, Robert (1773–1827), 14:500; 12:445a(n)
Woodpecker
anatomy, 3:343a, 570b
Woodruff, Elmer Grant, 3:610a; 4:7b
Woods Hole, Marine Biological Laboratory, 3:17a, 590a; 4:161a; 8:356a
founding of, 6:613a; 12:479b; 14:314a
Woods Hole, Oceanographic Institution, 8:356b
Woodville, William, 7:97a
Woodward, Arthur Smith, 2:171b; 3:606b
Woodward, John (1665–1728), 14:500–503; 13:178b
botany, 6:38b
cosmogony, 11:317b
critics of, 6:129b; 9:532a, 533b; 10:84a
geology, 7:412a; 14:296a; 15:55a,b
influence of, 2:614a*; 6:130a; 8:4b; 9:539b; 11:43a
lectures of, 2:382a
paleontology, 9:533a
Woodward, Robert Simpson (1849–1924), 14:503–504; 6:29b, 399a–b; 9:231a*
and Sarton, 12:110a–111a
Woodyard, John Robert, 6:104b
Wool
dyeing of, 3:241a; 6:236b
improvement of, 15:113a
merino, 1:435b
structure of, 1:319a–b
sulfuric acid, effects of, 2:386a
Woolley, Charles Leonard (1880–1960), 14:504–505; 10:550a
Woolwich
Royal Military Academy, 1:459b; 3:260a
Wooster, William Alfred, 9:261a
Wooster, College of, 3:366b
Worcester Foundation for Experimental Biology
founding of, 10:611a
Wordsworth, William, 1:489b; 3:538a; 6:86b; 9:383a, 385a; 11:140a; 12:276b
Work
analysis of
elementary motions, 13:272a
concept of, 1:253a–b; 3:417b; 4:229b; 8:209a; 10:3b
conservation principle, 3:74b, 118a, 454a; 8:230a–b; 10:3b
definition of, 3:73a
and heat, 3:303b, 304a–306a; 5:86b–87a, 387b; 12:288b; 13:378a; 14:123b; 15:85b
conversion coefficient, 3:83b
internal, 3:417b
vs. external, 3:304a–b
maximum by a horse, 4:44a
maximum by a man, 2:37b, 43a; 3:442b; 4:44a
measurement of, 13:272a
physiology of, 2:37a–b; 12:271a
principle of, 7:173a–174b
transmission of, 3:417b, 418a
units of measurement, 3:417b
virtual work principle, 4:228b; 5:98b; 9:268a, 488a; 11:488b
see also Least work; Maximum work
Work functions, 7:36b
Work Projects Administration (W.P.A.)
scientific translations, 2:319b*
World War I, 2:272b; 4:262a, 485a, 500b, 614a; 5:100a, 187a–188a, 220a, 453a, 486a, 489b, 561b; 6:25b, 31a, 394b; 9:544b; 10:275a
anthropology, 3:555b; 4:271b, 272a
chemical warfare, 1:454a–b; 5:128a, 190a, 465a–b, 541a, 622a–b; 6:16a, 21b; 10:121b; 11:139a; 12:404b; 14:334a, 393a, 601b; 15:2a, 159b, 444a
chemistry, 6:185a, 265a; 11:84b, 87a–b, 482b; 14:247b; 15:1b
communications, 1:24a; 2:488a
geology, 2:500b; 4:598a; 7:144b
industry, 3:401a; 4:165b; 5:1a
interruption of research, 2:347a, 399b, 427b, 488a, 560b–561a; 3:622b; 4:301a, 312b; 5:117b; 6:355b; 7:94b, 240a; 8:295a; 9:144b, 260b, 314b, 335a; 10:347b; 11:257b, 420b; 12:32b, 110a, 398b–399a, 412a; 13:95b, 486a–b, 520b; 14:283b, 412a; 15:61a, 157a, 521b
medicine, 1:537b; 2:384b; 3:91b, 501b, 518b; 4:353b, 398b, 618a; 5:29b, 291a; 6:432b, 519a; 8:102b; 11:556b, 577a–b; 13:229a, 300b–301a; 14:511b, 512a; 15:73a–b, 454b
nutrition, 3:257a–b; 5:37a, 225b; 6:499b; 8:29b, 556a; 9:144b; 10:493b; 11:586a; 15:201b, 418b
photography, 13:345a, 408a
postwar boundaries, 2:374a
psychology, 14:550a
relief work, 3:69a; 15:431b–432a
reparations, 7:318a
technology, 2:303a, 370a, 399b, 460a; 3:220b, 299a, 372b, 501b, 505a, 564a, 611a; 4:407b, 563b; 5:103a, 128a, 202a, 433a, 465a–b, 583b; 6:5b; 7:247a, 446a, 468b; 8:11b, 24b, 119b, 532b; 9:373b, 386b, 403b, 419b, 550a, 552b; 10:525b, 605a; 11:470a, 580b; 12:54a, 575a; 13:361b, 454b, 505a, 598a; 14:86b, 244a, 515b
World War II, 1:443b; 2:303b, 345b, 400a; 3:91b, 197b, 299b, 564b; 5:103b, 208b; 6:371a; 7:155b, 486a
astronomy, 1:352b–353a; 5:183a; 15:129a
biochemistry, 1:361b; 9:577a
chemical warfare, 5:190a
chemistry, 13:163a
cryptography, 6:402a
economics, 7:318b
geology, 4:599a
interruption of research, 1:427b, 428b, 519b; 2:238b, 544b, 582b; 3:170b, 176b; 4:84a, 102a, 561a; 5:206a, 350b, 458a; 6:481b; 7:59a–b, 155a,b, 555a; 8:12a, 480b, 546a; 9:248b, 359a, 565a; 10:30b, 236b; 11:258a,

World War II (cont.)
525a; 13:498a, 520b, 530b, 608b; 15:25b, 157b
medicine, 5:30a, 42b–43a, 208a, 209a; 15:76b, 420a
meteorology, 11:557b, 558b
nuclear research, 2:252a–b, 338b; 3:371a–b; 4:313a, 581b–582a; 5:118a–b, 610b; 6:17a; 8:95b; 10:215a–216a; 13:429b; 14:89a
nutrition, 14:393b
oceanography, 13:167a
operations analysis, 7:613b
photography, 13:345a–b
physics, 6:147b
psychology, 14:551a
tactics, 15:19a
technology, 1:534a; 2:460a; 3:373a, 511a; 4:90b, 362a; 5:272b–273a, 433a; 6:22b, 104b, 534a–b; 7:10b, 465a, 527a–b, 622b; 8:24b; 12:201b; 13:454b, 492b, 505b, 570a, 598b; 14:92a, 346b, 553a–b; 15:8b
see also Nazism
World-Soul, 1:80a, 259b; 2:553a; 3:14b; 9:98a; 12:155b; 13:340b; 14:388b
Worm, Ole (Olaus Wormius) (1588–1654), 14:505; 2:317a; 4:507b; 12:333a; 14:449b
Wormian bones, 14:505a
Worms
developmental conditions, 14:595b
embryogeny, 7:475a
hematophagous, 5:502b
hematotrophic, 4:198a
intestinal, 1:602b; 3:588a; 4:198a, 236a; 7:15a*, 557a; 10:418a; 11:592a
origin of, 13:563b
life cycles, 8:415b; 11:592b
parasitic, 15:430a
regeneration, 2:286a; 7:33b; 12:555b
reproduction, 8:128a; 12:421a
taxonomy, 1:603a; 9:575b; 10:283a; 12:453a
see also Helminthology; specific worms
Worsaae, Jens Jacob (1821–1885), 14:506–507; 9:489a,b; 12:87b; 13:9b
Worshipful Society of Apothecaries, 13:507b
Wort (malt infusion), 8:585b
Worthen, Amos Henry, 9:255b; 14:315b
Wotton, Edward (1492–1555), 14:507–508; 9:440b
Wotton, Henry, 7:306a; 14:157a,b
Woulfe, Peter (*ca.* 1727–1803), 14:508–509
Woulfe's bottle, 14:508a
Wound fever, 2:129b
Wound shock, 1:537b; 4:398b; 13:530b
Wounds, 3:98b
biology of, 4:618a
chest, 7:16b
cleaning of
medieval techniques, 2:314b
drainage, 8:405b
festering, 7:284a
gunshot, 9:422a
treatment of, 2:336b, 337a, 547a; 3:91a; 4:209a; 6:566b; 9:12a; 10:315b–316a; 14:6a, 28a
head, 3:518b
healing of, 8:403b, 447a
hormone theory, 1:303b; 5:624a
infected, 8:273a; 14:250a
bacteriology of, 7:422b
theories of, 9:4b, 12a
treatment of
medieval, 3:219a; 6:276b
Renaissance, 2:546b; 9:12a; 10:306b–307a
seventeenth-century, 9:4a–b
nineteenth-century, 1:574b; 8:402b–408a *passim;* 10:620b
twentieth-century, 3:91b; 5:29b; 14:512a
alchemical, 4:95b
dietary, 10:621a
magnetic, 6:254a
Wourali poison, 14:187a
Wray, John
see Ray, John
Wren, Christopher (1632–1723), 14:509–511; 6:482a
acquaintances of, 3:482a; 4:495b, 496a; 6:483b, 487b; 9:400a; 11:532a; 12:580b, 584b; 14:369a, 405a
anatomical illustration, 14:406a, 407a
architecture, 13:451b
celestial mechanics, 10:64a, 72a
influences on, 14:377a(n)
intravenous injection, 1:360a; 4:37b; 8:524b; 14:407a
mathematics, 1:474b; 5:527a; 9:599b; 10:336b
mechanics, 6:602a; 8:151a, 152a; 9:116b; 10:62a; 14:150a
Wrench, John William, Jr., 12:345a
Wright, Almroth Edward (1861–1947), 14:511–513; 2:584a; 5:29a,b; 9:334b
Wright, Charles, 5:512b
Wright, Edward (1561–1615), 14:513–515; 2:462a; 5:593a; 9:610a; 10:151b; 14:365a, 375a(n)
cartography, 4:214a; 9:309b
Certaine Errors in Navigation, 5:343b, 344a
Wright, Frederick Bennett, 14:517a
Wright, Frederick Eugene (1877–1953), 14:515–516; 4:564b; 8:42a; 14:503b*
Wright, George Frederick (1838–1921), 14:516–518; 5:513b
Wright, James Homer, 3:447b
Wright, Orville (1871–1948), 14:520–521; 8:20a
Wright, Samuel, 9:610a
Wright, Sewall, 3:123a, 572b; 4:272a; 6:22a; 9:282b
Wright, Thomas (1711–1786), 14:518–520; 6:331b; 7:598b; 13:470b; 15:754a
An Original Theory or New Hypothesis of the Universe, 6:332a–b; 7:231b
Wright, Wilbur (1867–1912), 14:520–521; 8:20a; 10:275a
Wright, William Hammond (1871–1959), 14:521–522; 9:504b*
Wróblewski, Zygmunt Florenty von (1845–1888), 14:522–523; 1:128b; 4:79b; 10:206a–b
Wroński, Józef Maria
see Hoëné-Wroński (Hoehne), Józef Maria
Wu Chia-shan, 3:266a–b
Wu Ch'i-chün, 8:394b
Wu Ch'ien, 3:249a,b
Wu Ching, 14:539b, 540b
Wu, Hsien (1893–1959), 14:523–524; 5:53a
Wucherer, Otto Eduard Heinrich, 4:198a
Wu-hsing ta-i, 15:733b
see also Five Phases (Chinese science)
Wulf, Theodor, 6:355a; 11:583a
Wulff, Georg (Yuri Viktorovich) (1863–1925), 14:525–526; 9:293b*
Curie-Wulff principle, 14:525b
Wulff construction, 14:525b
Wülfing, Ernst Anton, 11:547b
Wüllner, Adolph, 8:181b; 11:299a; 13:10b–11a, 243a
Wundt, Wilhelm (1832–1920), 14:526–529; 1:580a; 2:1b; 3:130a, 247b; 4:558b*; 7:504a; 9:562a,b; 15:460a
Grundzüge der physiologischen Psychologie, 6:52b
logic, 12:217a
psychophysics, 4:558a
Wünsch, Christian Ernst, 9:227b(n)
Wurdemann, William, 12:132a
Würm stage, 5:109b
Wurster dyes, 14:411b
Württemberg (Germany)
Karlsschule, 7:366b
volcanoes, 3:325b
Wurtz, Charles-Adolphe (1817–1884), 14:529–532; 5:315b
acids, 2:66a
acquaintances of, 2:254b, 621a; 3:448a–b; 5:315a; 7:281b; 8:109b; 9:287a; 14:394a
amines, 1:156a
on contemporary scientists, 1:416a; 3:243a; 5:370a; 13:577b
influence of, 3:48b; 6:462a
radicals, 2:621b
students of, 1:578a; 2:598a; 7:551b; 9:360a; 11:135b; 13:298b, 575b
Théorie atomique, 9:290b
type theory, 5:374a
Wurtz-Fittig reaction, 5:13a
Würzburg (Germany)
Julius Hospital, 12:202b
Würzburg, University of
in nineteenth century, 4:146b; 7:439a
medicine at, 12:202a
physics at, 12:615a
Wüstenfeld, (Heinrich) Ferdinand, 15:173b–177a *passim,* 180b, 183b–190a *passim*
Wyckoff, Ralph Walter Graystone, 8:485a
Wylie, Alexander, 3:251a; 11:403a
Wyman, Jeffries (1814–1874), 14:532–534; 2:366a; 5:478b*, 479a*; 8:103b; 11:219a
Wyman, Jeffries (*b.* 1901), 3:336a
Wynne, William Palmer, 1:288b; 13:505a
Wynn-Williams, Charles Eryl, 12:34a
Wyoming
geology, 4:564a; 6:187a
paleontology, 5:500b; 14:409b
Wyrouboff, Grégoire, 13:254b

X

X Club, 5:125a; 6:596b; 12:570a; 13:522b, 523a
X rays, 3:372b; 4:231b; 5:199a, 458b; 6:104a; 9:542a; 10:525a; 11:420a; 13:473a; 14:217b, 338b; 15:251b
absorption, 1:457a, 458a–b; 13:367b; 14:218a
and wave length, 4:195a
biological effects, 14:174b; 15:76b
cell division, 9:549b–550a
genetic, 6:338b; 9:565a
characteristic radiation, 1:458b; 4:195b; 7:27a; 9:543a–544b
diffraction of, 4:196a
Compton effect, 1:459a; 2:338a, 488a; 3:367b–370a; 4:317b; 5:331b; 9:261b; 12:528a
and condensation, 1:490b; 14:421a
Coolidge tube, 4:195a
diagnostic use of
see Radiology

diffraction of, 2:398b, 431b; 3:367a, 564a; 7:272a, 468b; 9:542b–543a; 12:464b, 527b; 14:617a
magnetization, effects of, 3:367b
mathematical analysis of, 12:527a; 14:617a
theories of, 1:319a; 8:51a–b; 9:543a; 12:464b; 15:61b–63a
discovery of, 1:558b; 3:475a, 479b, 498b; 8:181a; 11:530b–531b
dispersion of, 3:430a
ejection of electrons by, 2:338a; 8:10a
fluorescence analysis
discovery of, 6:366a
generation of, 2:487b; 13:570a
grating spectra, 3:370a
hardness of
see X rays, absorption
homogeneous, 4:195b
industrial applications of, 13:570a
interference patterns, 3:619b; 8:51a–52a
ionization, 8:10a; 12:26b, 69a; 13:366a–b
J-Phenomenon, 1:459a
and light, 3:502a
of long wavelength, 3:502a
and luminescence, 1:558b–559a; 11:58b
photographic quantum sensitivity, 10:136b
polarization of, 1:457b–458a; 2:203a; 3:369b
reflection of, 3:370a, 564b*; 4:196a–b; 9:543a; 15:8a, 62a–b
scattering of, 1:457a–458b; 2:398b; 3:367a, 430a, 619b; 4:196b; 13:368b–369a
secondary, 1:456b–458b; 3:372a*; 8:10a; 12:69a
soft, 11:422a
spectrometry of, 2:399a–b; 7:27a, 468b–469b, 491b
theories of, 12:464a
tertiary, 4:197a
theories of, 9:542b–543a, 544b; 10:347a; 12:238b; 14:32a; 15:61b–62b
ether pulse, 1:457a–458a; 2:398a–b
Maxwell-Lorentz, 1:458a
neutral-pair, 2:398a–b
quantum, 2:398b; 3:368a–370a; 7:468b–469a; 10:214a; 12:614a; 14:341a
transmission of, 8:369a
wave nature of, 3:369a; 12:31b
wavelength of, 2:399a; 3:368a–b; 8:51b; 12:32a; 14:340b–341a; 15:63a
and absorption, 4:195a
Zeeman effect, 11:461a
see also Angiography; Crystallography, X-ray; Radiation; Radiology; Roentgenology; Spectroscopy, X-ray
Xanthine
structure of, 5:2a–b
synthesis of, 5:2b
Xanthogen method, 3:271b
Xanthogenates, 14:600b
Xanthophyll, 7:35a; 13:487b
Xanthorhamnin, 8:329b
Xenarchus, 7:134b; 13:83b
Xenia phenomenon, 3:421b, 422a; 13:478a–b
Xenocoeloma, 9:328b
Xenocrates of Chalcedon (*ca.* 396–313 B.C.), 14:534–536; 12:575b; 15:202b
Xenoliths, 7:62b
Xenon
density, 11:280b
discovery of, 11:280a; 13:454a
penetrability, 15:491a
Xenophanes (*ca.* 580-*ca.* 478), 14:536–537; 10:208a, 280b
critics of, 6:291a
Xenophilus, 1:282a
Xenophon, 1:259a, 263b, 266a; 6:405b, 406b, 407a; 11:23b, 128b; 13:317b
Xerography
discovery of, 8:322a
Ximenes, Leonardo, 13:440b
Ximénez, Francisco, 6:309b
Xiphidium, 8:587b, 588a
Xylander
see Holzmann, Wilhelm
Xylem, 5:605a
Xyloïdine, 2:386a

Y

Y alloy, 11:550a
Yādava, 5:274b; 9:607a; 15:625a
Yahūda ibn Solomon Kohen, 15:33a
Yaḥyā ibn Abī Manṣūr (*d.* 832), 14:537–538; 1:275b; 14:592b
Judicia seu propositiones . . . , 1:23a
Zīj al-Mumtaḥan, 14:575b, 576a,b; 15:624a
Yaḥyā ibn ᶜAdī, 4:524a; 15:261b, 263a
Yaḥyā ibn al-Biṭrīq, 1:275b; 9:362b; 15:236b, 263a
Yaḥyā al-Kāshī, 13:515a
Yaḥyā ibn Sarāfyūn (Serapion maior), 4:121b
Yaḥyā ibn Sarāfyūn (Filius Serapionis)
Practica Serapionis, 15:185a
Yaᶜīsh ibn Ibrāhīm
see al-Umawī, Abū ᶜAbdallāh Yaᶜīsh Ibrāhīm ibn Yūsuf ibn Simāk al-Andalusī
Yajñeśvara
commentaries of, 5:275a
Yak, 6:567b
Yale University
in nineteenth century, 12:432b, 433a–b
astronomy at, 10:42a
biochemistry at, 9:284a
botany at, 15:149a
chemistry at, 2:260b; 10:150a–b; 12:435a
child development studies, 5:378a
geology at, 12:230a
Medical School, 8:555b
history of medicine, 5:208a–b
physiology at, 5:207b–208a
observatory, 4:351a
Peabody Museum, 14:1a
Sheffield Scientific School, 3:256b; 5:394a; 8:578a; 10:150a; 11:176a; 12:433b
Yale Zone Catalogue, 3:50a
Yallaya, 15:618b
commentaries of, 1:308b; 9:580b
Yamagiwa, Eigo, 8:577a
Yamaji, Nushizumi, 1:83b, 90a
Yamanouchi, Shigeo, 13:165a
Yāmaṭa
commentaries of, 2:417b
Yamawaki Toyo, 15:739b–740a, 740b
Yang, Chen Ning, 4:582b; 9:38b; 10:425a
Yang Hui (*fl. ca.* 1261–1275), 14:538–546; 3:254b, 255a, 265b, 266b, 268a; 8:419b; 12:375b
Ibn Yaᶜqūb
see Ibrāhīm ibn Yaᶜqūb al-Isrāᵓīlī al-Turṭushi
Yaᶜqūb ibn Ayyūb, 11:230a
Yaᶜqūb ibn Isḥāq
see al-Kindī, Abū Yūsuf Yaᶜqūb ibn Isḥāq al-Ṣabbāḥ
Yaᶜqūb al-Manṣūr (caliph 1184–1199), 1:279b
Yaᶜqūb ibn Ṭāriq (*fl.* second half of eighth century), 14:546; 4:555b; 7:223a, 360b, 363b(n)
al-Yaᶜqūbī, 4:453a*, 542a; 13:323a
Yāqūt al-Ḥamawī al-Rūmī, Shihāb al-Dīn Abū ᶜAbdallāh Yāqūt ibn ᶜAbd Allāh (1179–1229), 14:546–548; 2:619a; 4:555b, 556a; 6:79b; 9:35b; 13:415a; 15:242b(n)
Yard (measure)
standardization of, 1:403a; 2:141a; 5:491b–492a; 13:517b
Yare River, 8:564a
Yarrell, William, 11:380a
Ibn al-Yāsmīnī, 11:229b
Yativṛṣabha (*fl.* sixth century A.D.), 14:548–549
Yavaneśvara (*fl. ca.* A.D. 150), 14:549; 10:425b; 12:576b; 15:538b
Yaws (disease)
treatment of, 4:301a; 9:546a*; 10:621b
Yazdijird III, 15:605a
Year
Arabic, 3:26b; 4:542b, 556a; 5:553a
civil, 15:708a–709b
Coptic, 14:578a
Egyptian, 4:542b; 15:706a
great, 11:223b–224a
Hindu, 3:158a
Indian, 15:535a, 603b
Julian, 1:512a; 2:152a; 3:26b, 558b; 4:542b; 5:224a*; 15:540b
leap, 15:708a, 709b
length of, 15:556a, 583b, 584b, 590b, 600a, 601b, 603b, 609a, 619b–625b *passim,* 707b, 708a
lunar, 2:152a,b; 3:21b; 5:553a; 15:708a–b, 709b, 723b
lunisolar, 1:383a; 2:152a
Mayan
chronological year, 15:765a, 766a–b
computing year, 15:765a, 773a
division of, 15:765a
lunar half-year, 15:760a, 774a–775a, 810b
Venus year, 15:760b, 777a, 780a–789a *passim*
Mesopotamian, 15:723b
Persian, 2:152a; 4:542b
regnal, 15:708b
sidereal, 4:556a
length of, 10:180a; 15:218a–b, 545a
solar, 2:152a; 3:21b, 26b; 15:218b, 590b, 707a
length of, 4:459b–460a; 5:553a; 7:304a; 10:180a–b; 12:378b, 403a, 547a; 15:211b, 536a–b, 538a,b, 539a, 540b, 543a, 545a
Syrian, 4:542b
tropical, 3:250b; 15:543a, 760b
divisions of, 12:379a; 15:774b
length of, 1:312b–313b, 382b, 511a; 3:402b; 11:189b; 12:378b, 403b; 13:536a; 14:162a; 15:218a–b, 732a,b, 735a, 765a
see also Calendar
Yeasts, 3:259b
in beer, 6:100a–101a
in biochemical measurement, 7:405b
brewer's, 10:243b, 362a,b, 363b, 364a
chemical composition
enzymes, 6:111a,b
nucleic acids, 4:603b; 6:505b
culture methods, 6:100b
extraction of fluid from, 2:561b
fermentation, 6:501b; 12:241a–b, 243b–244a; 13:312a

Yeasts (cont.)
growth of, **9**:394a; **12**:241a–b
life cycle, **6**:100b; **10**:377a–b
mitosis, **5**:584b
milk-sugar, **2**:562b
mutations, **6**:101a
nature of, **2**:437b; **3**:9a–b; **5**:584b; **9**:425a; **10**:362a–b, 380b–382a; **13**:312a, 506b
nutrition, **11**:310b–311a
phylogeny, **5**:584b
sexuality in, **5**:584b
specificity of, **10**:362a–b, 371b, 381a,b
spores, **6**:101a
taxonomy, **6**:101a
see also Fermentation
Yeats, Grant David, **9**:245b
Yellow fever
eighteenth-century study, **3**:344b; **14**:323a*
nineteenth-century study, **8**:124b, 313a
twentieth-century study, **3**:185a; **5**:451a
diagnosis of, **4**:620a
epidemics, **14**:456a
etiology of, **3**:94b; **4**:620a–b; **8**:450b; **10**:142a–b; **11**:346b–347a; **12**:96b
naming of, **1**:26a
prevention of, **15**:96b–97a
transmission of, **6**:521a, 524b; **8**:133a; **9**:10b; **11**:346a–347a
treatment of, **13**:46b; **14**:456a
vaccination, **10**:142a
Yellow Fever Commission (U.S.), **4**:620a
Yellow River (China), **11**:181a
Yellowstone National Park, **6**:187a
geology, **7**:5a
surveying of, **6**:13b–14a
Yemen
geography, **10**:117a
Yen Chen-ch'ing, **12**:380b
Yen Su, **12**:384b
Yen-mu (operation modulus), **3**:253a
Yenri method, **1**:90b
Yen-shu (operation numbers), **3**:252a, 253a
Yeo, Gerald, **4**:593b–594a; **12**:355b
Yerkes, Charles Tyson, **6**:28b
Yerkes, Robert Mearns (1876–1956), 14:549–551; **6**:320b; **15**:77b*
Yerkes Observatory, **1**:466b; **3**:288a; **5**:199a–b; **10**:320b–321a; **13**:116a
early development, **1**:55a,b
founding of, **6**:28b
Yersin, Alexandre (1863–1943), 14:551–552; **7**:392a, 427b; **8**:449b; **11**:569a
Yew, **5**:442a
Yin-yang, **12**:374a, 390b, 407b, 409a; **15**:729a–b, 739a, 740a, 742a, 745b, 746b
Ying-yung suan-fa, **14**:545b
Ylem, **5**:273a
Yo jutsu problems, **1**:83b, 91a
Yoder, Hatten Schuyler, Jr., **4**:409a
Yoga (period), **3**:584b; **4**:100a; **5**:275a; **15**:588b
definition of, **15**:537a–b
Yogatārās (junction stars), **15**:554a, 564b, 578a,b, 593b, 595b, 598b, 607b, 617b, 625a
Yogīndra, **12**:115b
Yogurt, **9**:335a
Yojana (unit of length), **15**:556b, 591a, 605b
Yoredale Series, **10**:584a
York (England)
cathedral school, **1**:104b
Yorkshire (England)
geology, **10**:584a; **12**:491a
Yorkshire Philosophical Society, **10**:583b
Yosemite National Park
founding of, **14**:315b
Yoshida Koyu, **15**:745b
Jinkoki, **15**:747b, 754b
Yoshimasu Todo, **15**:740a, 741a,b
Youden, William John (1900–1971), 14:552–557
Youden plot, **14**:554a
Youden squares, **14**:553a
Young, Alfred (math.), **14**:283b
Young, Arthur (agric.), **6**:219b; **13**:448a
Young, Charles Augustus (1834–1908), 14:557–558; **4**:181b*; **5**:199a; **6**:28a, 29a,b; **8**:503b; **9**:386a; **12**:23a
Young, Grace Chisholm, **14**:572b
Young, James (chem.), **10**:511a
Young, John Richardson (*ca.* 1782–1804), 14:558–559; **6**:510b; **11**:616b; **12**:558b; **13**:46b
Young, John Wesley (1879–1932), 14:559–560; **12**:285b*
Projective Geometry, **9**:173b; **13**:599b
Young, Stewart (chem.), **14**:624a
Young, Sydney (1857–1937), 14:560–562; **11**:278b; **13**:454b
Young, Thomas (1773–1829), 14:562–572; **2**:335b; **9**:72b
acoustics, **3**:258b; **6**:245a; **14**:565a
acquaintances of, **1**:201a; **6**:383a; **12**:523b; **13**:352a
capillarity, **15**:360a
on contemporary scientists, **11**:496a; **14**:253b; **15**:362b
correspondence of, **14**:493a; **15**:388a*
critics of, **3**:539b; **9**:73a
editions of works, **10**:437b
error theory, **11**:237a
influence of, **5**:86b, 445b; **15**:199b
influences on, **2**:174a; **3**:441b; **8**:262a; **14**:491b
linguistics, **14**:569b; **15**:704b*
matter theory, **8**:510a; **14**:185b
mechanics, **4**:591b; **14**:568b–569a
medicine, **14**:562b–563a
Nautical Almanac, **6**:263a
optics
physical, **1**:201b, 202a, 357b; **3**:145b; **5**:19a, 137a, 143b, 167b, 168a, 169a; **6**:277b, 278b, 333b; **9**:74a, 214a, 264b; **10**:59a; **12**:519a; **14**:565a–568b
physiological, **9**:200a,b; **14**:563b–565a
tides, **14**:293a, 569a
Young-Helmholtz theory, **6**:246b–247b; **7**:458a
Young, William Alexander, **10**:142b
Young, William Henry (1863–1942), 14:572–573; **4**:206b; **8**:110b
Theory of Sets of Points, **6**:451b
Young, William John, **2**:562a; **4**:485b; **6**:111a; **9**:359b
Ytterbia, **9**:110b
Ytterbium, **3**:321b; **13**:546a
discovery of, **9**:110b
Yttria, **4**:343b
discovery of, **5**:216a
fractionation of, **3**:478a,b
Yttrium, **3**:478a
atomic weight, **9**:290b
discovery of, **9**:541b
isolation of, **14**:476a
Yu I, **12**:404a
T'ien-ching huo-wen, **15**:733a
Yu Sung-nien, **14**:540a
Yu shu (math.), **3**:253a
Yuan Hao-wen, **3**:266b; **8**:315a
Yuan-shu (original number), **3**:252a, 253a
Yucca, **12**:259a; **15**:130b
Yuga (period), **1**:34b; **15**:534b–535a, 536a,b, 537a, 538b, 555a–b, 590a, 597a
Yugeśvara Jhâ
commentaries of, **5**:274b
Yūḥannā ibn Māsawayh
see Ibn Māsawayh (Mesuë), Yūḥannā
Yūḥannā al-Qass, **4**:438a
Yūḥannā ibn Yūsuf, **7**:343a
Yukawa, Hideki, **4**:580a; **9**:606b; **11**:117a; **15**:27a
Yukon
geology, **13**:525b
Yule, George Udny (1871–1951), 14:573–574; **10**:449b–450a, 453a, 454a, 472a*; **14**:155a*
Yung shu (use number), **3**:253a
Yung-lo ta-tien, **8**:419a
Ibn Yūnus, Abu'l-Ḥasan ᶜAlī ibn ᶜAbd al-Raḥmān ibn Aḥmad ibn Yūnus al-Ṣadafī (*d.* 1009), 14:574–580; **2**:360a; **4**:117b; **6**:189a; **14**:538b
cosmology, **12**:3a
historiography, **1**:510b; **4**:543b(n); **5**:612a; **7**:363b(n); **10**:5b
trigonometry, **14**:471a(n)
al-Zīj al-Ḥākimī al-kabīr, **9**:21b; **12**:362b; **13**:293b*; **15**:259b
Yūsuf al-Fāsī, **6**:189a–b
Yūsuf ibn Ibrāhīm, **1**:82a
Yvon, Paul, **6**:112b

Z

z test, **5**:477a
Zabarella, Jacopo (1533–1589), 14:580–582; **1**:495a
De methodis, **7**:269a
natural philosophy, **1**:495b; **12**:312a; **13**:278b, 279a
Zaccagna, Domenico, **7**:31a
Zach, Franz Xaver von (1754–1832), 14:582–583; **10**:199b*, 592b; **11**:82b; **14**:359b*
acquaintances of, **5**:303a; **6**:112a, 456a; **8**:369a
astronomy, **6**:128a; **10**:592a; **15**:392b*
correspondence of, **9**:251b; **10**:198b, 592a; **11**:83a*
Zacharias Scholasticus
Ammonius, **1**:137a
Zacuto (or Zacut), Abraham bar Samuel bar Abraham (*ca.* 1450–*ca.* 1522), 14:583–584; **8**:281b; **13**:401a
Almanach perpetuum, **1**:573b
Zag, Isaac ibn Sid (Rabi), **1**:122b
Zagreb
Geophysical Institute, **9**:443b, 444b
Zahn, Friedrich Wilhelm, **4**:277a
al-Zahrāwī, Abu'l-Qāsim Khalaf ibn ᶜAbbās (Abulcasis) (*ca.* 936–*ca.* 1013), 14:584–585; **1**:350b, 538b; **2**:546b; **3**:218b; **14**:638a
De cirurgia
translations of, **15**:185a
influences on, **9**:39b; **10**:418b; **15**:265b
Zaitsev, Aleksandr Mikhailovich, **2**:623b; **13**:549b
Zakariyā ibn Muḥammad ibn Maḥmūd
see al-Qazwīnī, Zakariyā ibn Muḥammad ibn Maḥmūd, Abū Yaḥyā
Zalužanský ze Zalužan, Adam (*ca.* 1558–1613), 14:586
Zambeccari, Giuseppe (1655–1728), 14:586–587
Zamberti, Bartolomeo, **4**:448a–451a *passim;* **10**:271a
Zamboni, Giuseppe, **4**:29a, 74a
Zambonini, Ferruccio (1880–1932), 14:587–589; **5**:556b
Zambonini's rule, **14**:588a
Zammit, Themistocles, **2**:528a
Zänger, Christian, **2**:15b(n)

Zanotti, Eustachio (1709–1782), 14:589–590; 9:77b
Zante (island)
geology, 7:30b
Zarankiewicz, Kazimierz (1902–1959), 14:590–592
Zarlino, Gioseffo, 1:606b; 5:238a,b, 249a–b
al-Zarqālī (Azarquiel), Abū Isḥāq Ibrāhīm ibn Yaḥyā al-Naqqāsh (*d.* 1100), 14:592–595; 1:509b; 11:321b–322a; 14:538b
Almanac, 13:401a
astronomy, 1:511a; 3:25b, 405b–406a; 11:202a; 12:3a; 15:34a
Canones sive regulae super tabulas astronomiae, 14:19a
translations of, 15:175b, 187b
saphea, 1:438a; 7:125a; 14:400a–b
Toledan tables, 1:222a, 381b; 3:26b; 6:274a; 7:124a, 141b–142a, 363a; 9:363b; 11:321b, 322a, 415a, 504a; 13:401a; 14:400a,b; 15:187b, 477a
commentaries on, 14:401a
critics of, 14:390a
translations of works, 12:199b; 13:400b
Treatise on the Movement of the Fixed Stars, 15:35a
trigonometry, 15:478b
Zavadovsky, Mikhail Mikhaylovich (1891–1957), 14:595–596; 7:454b
Zavarzin, Aleksey Alekseevich (1886–1945), 14:596–597
Zawadski, Alexander Johann Anton, 9:278b
Zayd ibn Rifāᶜa, 15:250a
Zayn al-ᶜArab al-Miṣrī, ᶜAlī ibn ᶜAbdallāh, 9:603b
al-Zayyāt, Aḥmad ibn al-Ḥusayn ibn ᶜAlī . . ., 14:118a
al-Zayyātī al-Gharnāṭī, al-Ḥasan ibn Muḥammad al-Wazzān
see Leo the African
Zea, Francisco Antonio, 11:605b
Zea, 3:422a; 9:281b; 15:67b, 68a, 161a
Zechmeister, Laszlo, 6:366a; 13:487b
Zeeman, Pieter (1865–1943), 14:597–599; 3:481b*; 7:186b; 8:489b, 499a*; 12:497a; 14:499b*
electron theory, 13:368a
Zeeman effect, 1:370b, 371a; 2:247b, 248a; 3:420a; 4:498a; 6:29b; 8:40b, 496a; 10:347b, 348a; 11:379a, 461a, 478b; 14:61b
measurement of, 14:599a
on molecular lines, 4:69a
theory of, 10:422b–423a; 11:612b; 12:528b, 529a
Zeidler, Othmar, 9:576b
Zeijlstra, Hein Hidde, 14:97a, 101a
Zeiller, Otto, 5:332a
Zeiller, René Charles (1847–1915), 14:599–600; 12:105a*
Zeipel, Hugo von, 4:19b; 8:364a
Zeise, William Christopher (1789–1847), 14:600–601; 8:334b, 336a–b
Zeise's salt, 9:18b
Zeishold, Herman, 15:465b
Zeiss, Carl, 1:6b–9a *passim;* 5:144a; 7:422b, 448a; 12:211b
Zeiss, Roderich, 1:9a; 12:211b
Zeiss (Carl) Foundation, 1:6b, 9a; 12:212a
Zeiss Optical Works, 1:6b–9a *passim;* 5:590b; 7:448a; 11:208a; 12:174b, 423a; 14:633b
Zeitschrift der Oesterreichischen Gesellschaft für Meteorologie, 6:97a
Zeitschrift für allgemeine Physiologie
founding of, 14:2b
Zeitschrift für analytische Chemie
founding of, 5:164b–165a
Zeitschrift für angewandte Mathematik und Mechanik
founding of, 9:419a
Zeitschrift für anorganische Chemie, 11:552a
Zeitschrift für anorganische und allgemeine Chemie
editors of, 8:503a; 13:248a*
Zeitschrift für Astronomie und verwandte Wissenschaften
founding of, 8:369b
Zeitschrift für Biologie
founding of, 10:558b; 14:66b*
Zeitschrift für Chemie, 1:578a; 5:12b
Zeitschrift für Chemie und Industrie der Kolloide
editors of, 10:251b
Zeitschrift für Chemie und Pharmacie
see *Zeitschrift für Chemie*
Zeitschrift für die gesamte Kälteindustrie
founding of, 8:500a
Zeitschrift für Elektrochemie
editors of, 15:450a
Zeitschrift für Hygiene und Infektionskrankheiten, 5:220a
founding of, 7:425b; 10:559b
Zeitschrift für Kristallographie
editors of, 10:125a; 14:588b
Zeitschrift für Kristallographie und Mineralogie
founding of, 5:557b
Zeitschrift für Parasitenkunde
founding of, 6:73a
Zeitschrift für Physik, 5:332a
Zeitschrift für physikalische Chemie, 15:433a
editors of, 7:162a
founding of, 1:536b; 13:576a; 15:456a, 459b–460a
Zeitschrift für Physiologie
founding of, 13:461b, 463a
Zeitschrift für physiologische Chemie
editors of, 7:467b*
founding of, 1:536b; 6:505b
Zeitschrift für praktische Geologie
editors of, 14:442a
Zeitschrift für Psychologie und Physiologie der Sinnesorgane
editors of, 7:458a
Zeitschrift für rationelle Medicin, 6:268b
Zeitschrift für Vermessungswesen
editors of, 6:239b
Zeitschrift für Vulkanologie
editors of, 11:336b
Zeitschrift für wissenschaftliche Botanik, 9:600a
Zeitschrift für wissenschaftliche Zoologie
founding of, 7:439b; 12:421a
Zejszner (or Zeuschner), Ludwik (1805–1871), 15:526–527
Zeleny, Charles, 13:136a
Zeleny, John, 13:366a
Zelinsky, Nikolay Dmitrievich (1861–1953), 14:601–603; 1:415b; 3:271a; 4:601a*; 9:130b; 12:404b
Zell, Heinrich, 11:396a
Zemlevedenie
editors of, 7:512b
founding of, 1:173b
Zemplén, Géza (1883–1956), 14:603
Zendrini, Bernardino, 11:400a,b; 15:52b
Zenith sector (astron. instrum.), 2:387b, 388a; 5:491b; 8:350b; 9:463b–464a
Zenker, Friedrich Albert von, 8:270b
Zenneck, Jonathan Adolf Wilhelm, 7:468b
Zeno of Citium (*ca.* 335–263), 14:605–607; 4:388b; 13:307b
Zeno of Elea (*ca.* 490–*ca.* 425), 14:607–612; 8:269a; 9:382b
paradoxes, 1:149b, 255b; 7:485b; 8:151b; 12:75a
Zeno of Sidon (*ca.* 150–*ca.* 70); 14:612–613; 11:106a*
Zeno, Niccolò, 10:239a
Zenodorus (*fl.* early second century B.C.), 14:603–605; 10:181a, 295b; 13:321b; 15:115b, 116a
Zentralblatt für Bakteriologie, Parasitenkunde, Infektionskrankheiten und Hygiene
founding of, 8:449b–450a
Zeolite, 2:572b*; 3:473b; 4:141a, 565a*; 11:372a; 14:588a
cation exchange, 5:203a
water in, 5:186a; 13:420a
Zeppelin, Ferdinand von, 9:578a
Zerbi, Gabriele, 1:47a, 617b, 619a–b
Zermelo, Ernst Friedrich Ferdinand (1871–1953), 14:613–616
analysis, 9:502b
axiomatization, 12:15b–16a
physics, 2:264a; 4:293b
set theory, 3:57a; 5:108a; 6:393b; 8:559a; 12:13b
students of, 7:71b
Zermelo-Fraenkel system, 3:57b; 5:108a
Zermelo's axiom, 5:108a
Zernike, Frits (1888–1966), 14:616–617; 1:7a–b; 3:430a; 10:235b, 236b*
Zernike polynomials, 14:617a
Zero, 4:502b, 611b
addition or subtraction of, 3:273b; 4:607b
definition of, 5:153b
division by, 2:274b–275a
symbol, 13:340a
Chinese use of, 3:254a; 8:315a–b
Zervas, Leonidas, 15:15b
Zeta function
see Functions (math.), zeta
Zeuner, Gustav Anton (1828–1907), 14:617–618; 3:303a; 8:365a, 500a; 9:462a
Zeuner diagram, 14:618a
Zeuschner
see Zejszner
Zeuthen, Hieronymus Georg (1839–1920), 14:618–619; 1:192a; 2:238a; 12:228a; 13:252b
geometry, 3:168a, 213b; 4:427a; 9:273a
historiography, 1:41a, 245a; 4:428a, 429a, 433b(n), 436a*; 7:326b; 10:544a*; 13:255a, 303a–b, 304b, 306b(n), 316a–b
Zeuthen-Segre invariant, 12:284b
Zharkov, Gelii Frolovich, 13:241b
Zhemchuzhnii, Sergei Feodorovich, 7:529a, 530a
Zhukovsky, Nikolay Egorovich (1847–1921), 14:619–622; 5:188b; 8:300b–301a, 558a, 560a; 10:20a; 13:492a
and Chaplygin, 3:194b–195a, 196a
Chaplygin-Zhukovsky postulate, 3:196a; 14:621b
editions of works, 7:473b
Zhurnal eksperimentalnoy biologii
founding of, 7:456a
Zhurnal fizicheskoi khimii, 1:361b
Ziegler, Jacob, 10:197a
Ziemssen, Hugo Wilhelm von, 10:559b
Zīj, 2:154b; 12:358a
etymology of, 1:508b
see also Tables, astronomical, medieval Arabic
Zīj al-Arkand, 1:33a, 34b; 15:605a, 607b
Zīj al-Harqan, 15:596b
Zīj al-Mumtaḥan, 10:5b; 11:202b; 13:149b; 14:575b, 576a,b; 15:624a
Zīj al-Shāh, 1:33a, 34b; 4:556a; 7:360b, 361a; 14:546b; 15:581a, 605a

Zīj al-Sindhind, **1**:5b, 33a, 34b, 61a, 62b; **3**:26b; **4**:542a, 543b, 556a; **5**:612a, 613b; **6**:79b; **7**:82b, 223a, 338a, 360b–361b, 363a; **9**:39a,b; **10**:5b; **11**:322a; **12**:361a; **13**:401a; **14**:546a–b, 576b, 593a; **15**:580b–581a
Zilva, Sylvester Solomon, **6**:111b
Zimmermann, Eberhard August Wilhelm von, **5**:298b, 301a
Zimmermann, Robert, **7**:597b
Zinc, **9**:62a
  chemistry of, **11**:109a
  compounds of, **4**:554a–b; **6**:236b; **11**:109a
    crystallography of, **15**:61b
    mineral, **8**:289a
    organic, **2**:623a; **5**:125b–126a
  copper-zinc couple, **5**:412b
  equivalent weight, **4**:554a
  medical use of, **11**:177b
  metallurgy of, **4**:394a; **7**:254b
  nutritional role, **6**:136a
  ores, **3**:190a, 606b; **12**:494b
    deposits of, **11**:608a; **14**:315b
Zinc oxide
  manufacture of, **5**:602a
Zincke, Hans, **4**:1b, 3b
Zincke, Theodor, **6**:15a; **7**:282b; **11**:92b; **15**:157a
Zingiber, **7**:58b
Zinin, Nikolay Nikolaevich (1812–1880), 14:622; **1**:579a; **2**:316a, 620b, 624a; **9**:287a; **10**:132b
  chemistry, **5**:197b; **9**:290a
Zinn, Walter, **3**:371a; **4**:580b
Zinsser, Hans (1878–1940), 14:622–625; **12**:483b
  Brill's disease (Brill-Zinsser's disease), **14**:623b
Zintl, Eduard, **6**:561a
Zionism, **4**:313a; **14**:247b–248a; **15**:422b
Zippin, Leo, **2**:513a
Zircon, **7**:161a
  radioactivity of, **8**:42b; **12**:520a
Zirconium
  discovery of, **7**:394b
  isolation of, **9**:451b
  spectrum, **3**:429b–430a
Zirkel, Ferdinand (1838–1912), 14:625; **2**:289b; **3**:462b; **11**:135b; **12**:543b; **14**:183a
  petrology, **3**:161a; **7**:370b; **8**:42a
  students of, **3**:483b; **7**:473b; **10**:502a; **14**:183a
Zittel, Karl Alfred von (1839–1904), 14:626–627; **1**:130b; **7**:252a, 412a
  acquaintances of, **2**:625b; **10**:502a; **11**:308b
  Handbuch der Palaeontologie, **1**:473a; **12**:230a; **13**:531b
  historiography, **3**:345a*, 624b*; **4**:28a; **9**:346b*; **11**:439a
  influences on, **2**:481a; **12**:38b
  paleontology, **10**:212a
  students of, **1**:162a, 491a; **2**:489a,b; **7**:61a, 62b; **8**:543a
Živa, **11**:217a
Ziyārid dynasty, **2**:148b, 149a
Zoantharians, **9**:112b
Zoarces, **12**:190b
Zodiac, **15**:723a
  astrological use, **5**:345b; **7**:118b, 125b, 299b, 301b; **15**:674a, 725a–b
  division of, **7**:118a, 123a, 124a; **15**:676a–b, 677b–678a, 721a, 725a–b
  as frame of reference
    chronology, **5**:616a; **15**:640b, 679b, 680b, 724a
    position, **1**:339a; **15**:541a, 581b, 673a
    size, **1**:248b, 249a
  obliquity of, **3**:319b; **15**:676b
  origin of, **8**:145a*; **15**:706a, 719b–720b
  representations of, **5**:348a; **15**:721a–723a
  rising of signs, **1**:509a; **4**:542b; **5**:345b; **6**:617a; **15**:209b–210a, 539b, 573a
  setting of signs, **9**:297a
Zodiacal light, **1**:301a; **2**:433a; **3**:103a; **6**:105b; **9**:90a; **10**:35a; **12**:192a, 282b, 451b
  spectrum, **12**:499a
Zoea, **13**:355a–b
Zollman, Philip Henry, **9**:534a
Zöllner, Johann Karl Friedrich (1834–1882), 14:627–630; **9**:563b; **14**:54b, 55a, 205b, 527b
  Die Principien einer elektrodynamischen Theorie der Materie, **14**:209a
  Zöllner-Tserasky photometer, **13**:481b
Zolotarev, Egor Ivanovich (1847–1878), 14:630–631; **3**:226a, 230b; **5**:508b; **9**:124b; **12**:534b*
  number theory, **3**:228b; **9**:126a; **12**:513a; **14**:94b
Zomotherapy, **11**:429b, 431a
Zone plate, **13**:101b
Zones, biogeographical, **1**:405b; **9**:313b; **14**:138b
  oceanic, **5**:66b, 67a; **7**:91b–92a; **13**:360a–b; **14**:145b–146a
Zones, biostratigraphical, **10**:212a
Zones, geographical, **1**:622b; **3**:319b, 423b; **4**:145a, 391a, 392b; **5**:345b; **7**:334b, 361b; **10**:474b; **11**:104a; **12**:61b; **13**:85a
  displacement of, **1**:623a
Zones, geological, **1**:285b–286a; **3**:520a; **10**:128a
Zoochlorellae, **15**:14a
Zoogeography
  eighteenth-century, **8**:251b
  early nineteenth-century, **2**:320b; **4**:285b; **8**:49a; **10**:283b–284a; **11**:137b, 243a
  mid nineteenth-century, **1**:504a; **5**:478a, 481a; **6**:594b; **8**:519a–b, 572b; **10**:221a; **11**:567a; **12**:257b–258a, 479b; **14**:134b–135b, 138a–139a
  late nineteenth-century, **5**:109b; **6**:222a, 613a; **9**:313b, 432a, 496a; **10**:16b–17a; **11**:108b, 443b; **12**:298a; **13**:552a; **14**:203b
  twentieth-century, **1**:623b; **5**:70b; **9**:459a; **13**:151a
  and evolution, **14**:135a,b, 138a–139a
  laws of, **7**:169b–170a
  statistical, **14**:138b
  see also Biogeography, Chorology
Zoogloeae, **13**:456a
Zoological Society of London, **5**:480b; **12**:258a
  founding of, **11**:261b
Zoologische Beiträge
  founding of, **12**:193a
Zoologischer Anzeiger, **15**:78b
Zoology
  ancient Greek, **1**:120b, 258b–266a; **4**:463b
  medieval Arabic, **3**:548b–549a; **7**:64a–b
  medieval Latin, **1**:102a; **9**:364a; **14**:35b
  Renaissance, **3**:13a; **5**:379b; **11**:254a; **14**:507a–508a
  early seventeenth-century, **7**:164b; **9**:122a–b; **10**:176b
  late seventeenth-century, **4**:267b–268a; **7**:578a; **9**:313a; **10**:519b–520b; **11**:316a–317a; **14**:413a–b
  early eighteenth-century, **2**:579a–581a; **7**:401a; **11**:332b–334a
  late eighteenth-century, **1**:351b; **2**:342b; **3**:522b–523b; **4**:178a; **5**:76b, 77a, 355a; **6**:401a; **7**:391a; **8**:251b; **9**:575a–576a; **10**:436b–437a, 509b; **11**:503a; **12**:562a–b; **14**:299b–300a; **15**:111a–113a
  early nineteenth-century, **1**:331a; **2**:106a, 422b–423b, 481a; **3**:521a–b; **4**:274a, 308a–b; **5**:224b, 276b*, 355b–358a; **6**:120a; **8**:265b, 266a–b, 373b; **10**:518a; **11**:242a–243b, 510a–b; **12**:215b–216a, 578a–b; **13**:167b–168a, 354b–356a, 402a–b, 554a; **15**:82b–83a
  mid nineteenth-century, **1**:71b–72a, 177a, 405a–406a; **2**:282a, 617a–b, 625b–627b; **3**:524b–527b; **4**:234a–236b, 288b–291b; **5**:358b–359b; **6**:7a–10b, 137b–138a, 453b–454b; **7**:437b; **8**:269b–271a, 518b–519b, 528a; **9**:344b, 431b–432a, 559b–561a; **11**:178a, 513b–515a; **12**:164b–165a, 169a, 318a–b; **13**:9b–10a, 554a–555a
  late nineteenth-century, **1**:600b–601b, 602a–603a; **2**:493b, 501b–502b; **3**:247b, 492b–493a, 532a, 569b–575a, 590a–b; **4**:12a–b, 186b–188b; **5**:45b–46b, 109a–b, 385a–386a, 406b, 446b–447a; **6**:11a–b, 131b, 167a–b, 336b, 442a, 613a–614a; **7**:401b–402b; **8**:3a–4a, 26b–27a; **9**:185a–b, 313b; **10**:272b–273b, 285b; **12**:297b–298a; **14**:203a, 233a–238b, 313b–315a; **15**:92a–93a, 430a–b, 513a–b
  twentieth-century, **2**:221b; **3**:148a–b, 204a, 247b–248a; **5**:453a–454a, 598b–599a; **6**:319b; **7**:98a–99b, 314a–b, 454a–456a; **8**:355a–359a, 586b–589b; **10**:251b; **11**:535a–b; **13**:63a; **15**:503b–504b
  agricultural, **13**:460a
  applied, **10**:427a
  comparative, **2**:339b; **10**:240a
  as experimental science, **7**:546a; **13**:457a–458b
  geology and, **1**:73a
  invertebrate
    seventeenth-century, **8**:415a–416b; **11**:317a
    eighteenth-century, **11**:331a–332b; **13**:457a–458b, 465a
    early nineteenth-century, **2**:187a; **5**:31b; **7**:588b–589b; **8**:48b–49b; **12**:130b–131a; **13**:461a,b
    mid nineteenth-century, **3**:608b; **5**:477b–478b; **9**:407b–409a; **11**:234a–b; **12**:299a–b, 420b–422a, 550a–551a; **15**:165b–166b
    late nineteenth-century, **1**:328b–329a; **5**:51a–52b; **6**:355a–b; **9**:495b, 536b; **10**:273b, 522b–523b; **13**:8b–9a; **15**:158a–b
    twentieth-century, **4**:142b; **10**:149a; **11**:19a–20a, 257a–b; **14**:291b
  marine
    ancient Greek, **1**:263a, 264a–b
    Renaissance, **11**:527b–528b
    eighteenth-century, **10**:283a; **12**:560b–561a
    early nineteenth-century, **1**:328b–329a; **4**:308b; **7**:414a, 610b; **8**:266b; **13**:355a–356a
    mid nineteenth-century, **3**:549b–550b; **4**:289a–291b; **5**:66b–68a, 469b; **6**:7a–b, 8b, 589b–591a; **7**:91b–92a, 545b–546a; **8**:270a, 519a–b; **9**:332a, 408a–b, 432a, 573a–b; **11**:113a–b, 234a–b, 308a, 491a–492a; **12**:106b–107b, 231a, 232a; **13**:65b–66a; **14**:145b–146a
    late nineteenth-century, **1**:72a, 162a; **2**:501b–502b; **3**:89a–b; **5**:51a–52b,

148b–149b; **6**:9a, 10b, 287b–288a; **8**:3b, 609b; **9**:112a–113a, 515a–518a; **10**:272b; **12**:336b, 479a–b; **13**:552a; **14**:1a–2a, 203a–b, 251a–b, 424a–b; **15**:78a, 123a–125a
twentieth-century, **3**:148a–b; **4**:42a–b; **5**:467a–b; **7**:503a–b; **9**:328a–b; **11**: 101b–102a, 211b; **13**:151a
applied, **9**:113a
mathematical, **1**:131a
medical, **2**:422b; **13**:63a
and paleontology, **1**:73a, 131a; **10**:29b; **12**:517b; **13**:552a
philosophy of
seventeenth-century, **13**:526b–528a
eighteenth-century, **13**:543b
nineteenth-century, **5**:31b–32b; **6**:7a–10b, 454a, 591a; **10**:427b; **11**:234b; **15**:78b
preservation techniques, **9**:313b
vertebrate
eighteenth-century, **7**:547a–b
nineteenth-century, **1**:405a–b; **5**:277a; **6**:591b–594b; **7**:319b; **10**:16a–17a; **11**:234a; **14**:57a–58a; **15**:91a–b, 126a–127a
twentieth-century, **5**:545a–b; **13**:151a
see also Anatomy, comparative; Animals; Classification, zoological; Congresses, scientific; Entomology; Evolution; Ichthyology; Illustration, zoological; individual species, genera, families, orders, etc.; Malacology; Ornithology; Physiology; Protozoology; Taxonomy, zoological
Zoonic acid, **13**:309b
Zoophilus (pseudonym), **2**:205b
Zoophytes, **3**:550a–b; **4**:236b*; **5**:216b; **7**: 412b, 545b–546a, 610b*; **8**:373b, 414b; **11**:243a; **12**:131a; **13**:355b, 554b
opposition to, **15**:112b
Zoospores, **12**:125b; **13**:395a
Zoothamnium, **7**:455a
Zopf, Friedrich, **14**:37a
Zöppritz, Karl, **14**:327a,b
Zöppritz, Karl Jacob
Zöppritz-Turner tables, **13**:500b
Zoroaster, **12**:432a; **14**:632a; **15**:674a
Zoroastrianism, **11**:324a
Zosimus of Panopolis (*fl. ca.* A.D. 300), 14: 631–632; 10:208a
Zotterman, Yngve, **13**:393a
Zsigmondy, Richard Adolf (1865–1929), 14: 632–634; 10:525a; **12**:423a, 593b
Zur Erkenntnis der Kolloide, **13**:159a
Zubler, Leonhard, **2**:419a
Zubov, Nikolay Nikolaevich (1885–1960), 14:634–636; 8:418b*; **12**:412b
Zubov, Vasily Pavlovich (1899–1963), 14: 636; 11:404b*
Zucchi, Niccolò (1586–1670), 14:636–637
Zuchta
see Suchten
Zuelzer, Georg Ludwig, **1**:441a
Ibn Zuhr, Abū Marwān ᶜAbd al-Malik ibn Abi'l-ᶜAlāᵓ (Abhomeron or Avenzoar) (*ca.* 1092–1162), 14:637–639; 1:290a; **12**:1b; **15**:498b
al-Taisīr fi'l-mudāwāt wa'l-tadbīr, **12**: 7Bb
Zuider Zee, **6**:137b; **8**:489a
Zuntz, Nathan, **1**:454a; **12**:319b
Ibn Zurayq, **12**:362a
Zurich
Eidgenössisches Polytechnikum
see Zurich, Eidgenössische Technische Hochschule
Eidgenössische Technische Hochschule, **4**:2b; **5**:555a; **8**:503a; **10**:423b–424a; **11**:344b; **13**:73a; **14**:244a, 481a
medicine, **9**:582a
Mineralogical and Petrographical Institute, **5**:563a
observatory, **14**:480b
surgical hospital, **2**:129b
Zurich, University of
in nineteenth century, **10**:195b
in twentieth century, **8**:3b
anthropology at, **9**:144a
geology at, **12**:140a
psychology at, **14**:527b
zoology at, **8**:3a–b
Zwelfer, Johann (1618–1668), 14:639–640; 8:130b
Pharmacopoeia Augustana, **13**:234b
Zwicky, Fritz, **1**:352b
Zwingli, Huldreich (or Ulrich), **5**:121a, 378b
Zygmund, Antoni, **12**:78a
Zygomycetes, **2**:437a
Zygoneura, **6**:167b
Zygosaccharomyces, **5**:584b
Zygospore
formation of, **5**:217a
Zygotes, **3**:30b
Zymase, **2**:561b–562a; **6**:111a
Zymin, **2**:562a
Zymogens, **7**:520a
Zymohexase, **14**:174b

Contributors

HANS AARSLEFF
Princeton University
ANDREAE; BISTERFELD; BOEHME; COMENIUS; S. FRANCK; HAAK; SPRAT; V. WEIGEL; WILKINS

GIORGIO ABETTI
Istituto Nazionale di Ottica · Osservatorio Astrofisico di Arcetri · University of Florence
ANTONIO ABETTI; ANTONIADI; G. CALANDRELLI; I. CALANDRELLI; HORN D'ARTURO; LORENZONI; ODIERNA; PIAZZI; RICCÒ; SCHIAPARELLI; SECCHI; TACCHINI; TOSCANELLI DAL POZZO

JOHN W. ABRAMS
University of Toronto
W.W. CAMPBELL

H.B. ACTON
University of Edinburgh
CARR; R.B. HALDANE

MARK B. ADAMS
University of Pennsylvania
A.O. KOVALEVSKY; VAVILOV

FREDERIC J. AGATE, JR.
Columbia University
P.E. SMITH

L.R.C. AGNEW
University of California, Los Angeles
BLAIR

S. MAQBUL AHMAD
Aligarh Muslim University
YĀQŪT AL-ḤAMAWĪ; AL-IDRĪSĪ; IBN KHURRADĀDHBIH; IBN MĀJID; AL-MAQDISĪ; AL-MASʿŪDĪ; AL-QAZWĪNĪ

KATHLEEN AHONEN
University of Michigan
GLAUBER

ERIC J. AITON
Didsbury College of Education
LEIBNIZ

LUÍS DE ALBURQUERQUE
University of Coimbra
LAVANHA; LISBOA; D.P. PEREIRA; PIRES; ZACUTO

MICHELE ALDRICH
Smith College · Smithsonian Institution · Aaron Burr Papers
HAYDEN; HITCHCOCK; KING; NEWBERRY; PEALE; F.B. TAYLOR

A. F. O'D. ALEXANDER
BOUVARD; J. BRADLEY; DREYER

MAX ALFERT
University of California, Berkeley
M. HEIDENHAIN

D.J. ALLAN
University of Glasgow
PLATO

MEA ALLAN
W.J. HOOKER

DEAN C. ALLARD
Department of the Navy
BAIRD

GARLAND ALLEN
Washington University
CASTLE; CONKLIN; FOL; MCCLUNG; T.H. MORGAN; PAINTER; SHULL; E.B. WILSON

FEDERICO ALLODI
MASCAGNI

TORSTEN ALTHIN
Royal Institute of Technology, Stockholm
BENEDICKS; NOBEL; RINMAN

PETER AMACHER
University of California, Los Angeles
C. BELL; FREUD

EDOARDO AMALDI
University of Rome
MAJORANA

PIETRO AMBROSIONI
University of Bologna
SANARELLI

G.C. AMSTUTZ
University of Heidelberg
BISCHOF; ESKOLA; V. GOLDSCHMIDT; GRODDECK; NIGGLI; OCHSENIUS; POSĚPNÝ; PUMPELLY; ROTH; ZIRKEL

G.C. ANAWATI
Institut Dominicain d'Études Orientales, Cairo
ḤUNAYN IBN ISḤĀQ; IBN SĪNĀ

ADEL ANBOUBA
Institut Moderne de Liban
AL-SAMAWʾAL; AL-ṬŪSĪ

DAVID L. ANDERSON
Oberlin College
GOLDSTEIN

R. CHRISTIAN ANDERSON
Brookhaven National Laboratory
DESSAIGNES

HENRY N. ANDREWS
University of Connecticut
W.C. WILLIAMSON

JEAN ANTHONY
Muséum National d'Histoire Naturelle
MILNE-EDWARDS

TOBY A. APPEL
Kirkland College · Johns Hopkins University
LACAZE-DUTHIERS; LACÉPÈDE; LEREBOULLET; QUOY; RANVIER; SOULEYET; VALENCIENNES

WILBUR APPLEBAUM
University of Illinois · Illinois Institute of Technology
CRABTREE; HORROCKS; ROOKE; SCHICKARD; STREETE; WING

ROGER ARNALDEZ
University of Paris
IBN BUṬLĀN; IBN RIDWĀN; IBN RUSHD

RICHARD P. AULIE
Loyola University · Encyclopaedia Britannica
BOUSSINGAULT; GAINES; E.J. RUSSELL

CORTLAND P. AUSER
Bronx Community College · City University of New York
BOWEN; BOWIE; LARSEN

OLIVER L. AUSTIN, JR.
University of Florida
SCLATER

WILLIAM H. AUSTIN
University of Houston
GLANVILL; MORE

COLETTE AVIGNON
University of Orléans-Tours
EVELYN

ROBERT AYCOCK
North Carolina State University
E.F. SMITH

JOSEF BABICZ
Polish Academy of Sciences
E.M. ROMER

VASSILY BABKOFF
Academy of Sciences of the U.S.S.R.
SAKHAROV

Contributors

LAWRENCE BADASH
University of California, Santa Barbara
BOLTWOOD; HAHN; E. RUTHERFORD

D.L. BAILEY
University of Toronto
BULLER

A. ALBERT BAKER, JR.
Grand Valley State College · University of Oklahoma · California State University, Fullerton
CLAISEN; CURTIUS; H. FISCHER; FRITZSCHE; GABRIEL; GRIGNARD; A. LADENBURG; TIEMANN

JOHN R. BAKER
University of Oxford
TREMBLEY

NANDOR L. BALAZS
State University of New York at Stony Brook
EINSTEIN

ERNEST BALDWIN
University College, London
F.G. HOPKINS

D.M. BALME
Queen Mary College
ARISTOTLE

GEORGE B. BARBOUR
University of Cincinnati
FENNEMAN

A. CLIFFORD BARGER
Harvard Medical School
CANNON

WILLIAM A. BARKER
University of Santa Clara
A.N. WHITEHEAD

MARGARET E. BARON
W.G. HORNER; C. HUTTON; IVORY; W. JONES; NAPIER; TODHUNTER

WALTER BARON
University of Hamburg
BLUMENBACH

GEORGE BASALLA
University of Delaware
FITZROY

ISABELLA G. BASHMAKOVA
Academy of Sciences of the U.S.S.R.
MOLIN; VORONOY; ZOLOTAREV

DONALD G. BATES
McGill University
SYDENHAM; J.R. YOUNG

EDWIN A. BATTISON
Smithsonian Institution
G. GRAHAM; J. HARRISON

IRINA V. BATYUSHKOVA
Academy of Sciences of the U.S.S.R.
KARPINSKY; MUSHKETOV; NIKITIN; A.P. PAVLOV

HANS BAUMGÄRTEL
OTHENIO ABEL; F.A. VON ALBERTI; ALBRECHT; BECKE; BROILI; RULEIN; TSCHERMAK; J.C.W. VOIGT

J.C. BEAGLEHOLE
Victoria University of Wellington
COOK

WILLIAM B. BEAN
University of Iowa · University of Texas Medical Branch
FINLAY; REED

JOSEPH BEAUDE
Centre National de la Recherche Scientifique
CLERSELIER; B. LAMY

HEINRICH BECK
Pädagogische Hochschule Bamberg der Universität Würzburg
HERBART

STANLEY L. BECKER
Bethany College
MCCOLLUM

ROBERT P. BECKINSALE
University of Oxford
DELUC; W. HOPKINS; MARTONNE; A. PENCK; W. PHILLIPS; A.C. RAMSAY; RATZEL; RICHTHOFEN; K.M. VON STERNBERG

SILVIO A. BEDINI
Smithsonian Institution · National Museum of History and Technology
BIRD; CAMPANI; DIVINI; JEFFERSON

JOHN J. BEER
University of Delaware
BESSEMER; CARO; MITTASCH; WEIZMANN

HEINRICH BEHNKE
University of Münster/Westphalia
RADEMACHER

YVON BELAVAL
L. BRUNSCHVICG

WHITFIELD J. BELL, JR.
American Philosophical Society Library
J. BARTRAM; W. BARTRAM; J. DIXON; HARLAN; LEA; S.G. MORTON; C.W. PEALE; WISTAR

LUIGI BELLONI
University of Milan
BONOMO; BOTAZZI; CESTONI; COGROSSI; DUBINI; FONTANA; LANDRIANI; LARGHI; MALPIGHI; MARCHI; MENGHINI; MORGAGNI; MORICHINI; ODDI; PACCHIONI; REDI; RIMA; RIVA–ROCCI; SACCO; TRULLI

ENRIQUE BELTRÁN
Mexican Society of History of Science and Technology · Instituto Mexicano de Recursos Naturales
ALZATE Y RAMÍREZ; HERRERA; LICEAGA; MAST; SIEDLECKI; SIGÜENZA Y GÓNGORA

OTTO THEODOR BENFEY
Earlham College · Guilford College
COUPER; J.L. MEYER; F.K.J. THIELE; WASHBURN

SAUL BENISON
University of Cincinnati
CANNON

JOHN A. BENJAMIN
ADDISON

RICHARD BERENDZEN
Boston University · American University
MAANEN; H.A. NEWTON; PARKHURST; PEASE; PERRINE; POOR; RITCHEY; V.M. SLIPHER; C.A. YOUNG

WALTER L. BERG
Central Washington State College
SHALER

JAMES D. BERGER
Indiana University
BÜTSCHLI; HENKING

ALEX BERMAN
University of Cincinnati
CADET; CADET DE GASSICOURT; CADET DE VAUX; CAVENTOU; J.B.A. CHEVALLIER; DEROSNE; FÉE; FORDOS; GAULTIER DE CLAUBRY; GOBLEY; GUIGNARD; LESSON; MILLON; MOISSAN; PARMENTIER; P.-J. PELLETIER; PELOUZE; PERSONNE; PERSOZ; POGGIALE; ROBIQUET; VIREY

MICHAEL BERNKOPF
Pace College
FREDHOLM; HALPHEN; H. VON KOCH; LAGUERRE; E. SCHMIDT; STIELTJES

PIERRE BERTHON
Archives, Académie des Sciences, Paris
DUNOYER DE SEGONZAC

A.E. BEST
University of Edinburgh
POURFOUR DU PETIT; ROUGET

RICHARD BIEBL
University of Vienna
TSCHERMAK VON SEYSENEGG; WIESNER

KURT R. BIERMANN
Akademie der Wissenschaften der DDR
CLAUSEN; DEDEKIND; EISENSTEIN; HUMBOLDT; JOACHIMSTHAL; L. KRONECKER; KUMMER; NETTO; WEIERSTRASS

OLEXA MYRON BILANIUK
Swarthmore College
LICHTENBERG

FRANCIS BIRCH
BRIDGMAN

ARTHUR BIREMBAUT
D'ARCHIAC; AUBUISSON DE VOISINS; BERTIN; BLONDEL; BOUÉ; BOULLANGER; BOURDELIN; BRAVAIS; CARANGEOT; C. COMBES; ÉLIE DE BEAUMONT; HASSENFRATZ

P.W. BISHOP
Smithsonian Institution
JARS; MUSHET; OSMOND; STANTON

ASIT K. BISWAS
Department of Environment, Ottawa
DUFOUR; FICHOT; FORTIN; GAMBEY; LAUSSEDAT; PALISSY; P. PERRAULT; WILD

MARGARET R. BISWAS
McGill University · Department of Environment, Ottawa
FORTIN; GAMBEY; LAUSSEDAT; PALISSY; P. PERRAULT; WILD

A. BLAAUW
European Southern Observatory
DE SITTER; EASTON; KAPTEYN

L.J. BLACHER
Academy of Sciences of the U.S.S.R.
DOGEL; ESCHSCHOLTZ; GURVICH; P.P. IVANOV; V.O. KOVALEVSKY; LAVRENTIEV; SCHMALGAUSEN; ZAVADOVSKY; ZAVARZIN

MAX BLACK
Cornell University
WITTGENSTEIN

WILLIAM A. BLANPIED
American Association for the Advancement of Science
BHABHA; JAGADISCHANDRA BOSE; SATYENDRANATH BOSE

JOSEPH L. BLAU
Columbia University
M.R. COHEN

GEORGE BOAS
Johns Hopkins University
LOVEJOY

R.P. BOAS, JR.
Northwestern University
BOURBAKI; HUNTINGTON; TITCHMARSH

HERMANN BOERNER
University of Giessen · University of Göttingen
CARATHÉODORY; ENGEL; KNESER; SCHUR; SCHWARZ

WALTER BÖHM
LOSCHMIDT; STEFAN

UNO BOKLUND
Royal Pharmaceutical Institute, Stockholm
G. BRANT; CLEVE; CRONSTEDT; EKEBERG; GAHN; HIÄRNE; SCHEELE; WALLERIUS

BRUNO A. BOLEY
Northwestern University
CASTIGLIANO; MENABREA; POLENI; PORRO

MARTIN BOPP
Botanical Institute, University of Heidelberg
KLEBS; SACHS

ALFRED M. BORK
University of California, Irvine
FITZGERALD

O. BORŮVKA
Czechoslovak Academy of Sciences
LERCH

H.J.M. BOS
State University of Utrecht
HUYGENS

WALLACE A. BOTHNER
University of New Hampshire
JAGGAR; A. JOHANNSEN

FRANK BOURDIER
École Pratique des Hautes Études
BARROIS; CHRISTOL; F. CUVIER; G. CUVIER; DÉCHELETTE; DEPÉRET; DUMÉRIL; GAUDRY; E. GEOFFROY SAINT-HILAIRE; I. GEOFFROY SAINT-HILAIRE; SERRES DE MESPLÈS

E.J. BOWEN
University of Oxford
D.L. CHAPMAN

CARL B. BOYER
Brooklyn College, City University of New York
BOULLIAU

MARJORIE NICE BOYER
York College, City University of New York
BOROCIUS

MARY A.B. BRAZIER
University of California, Los Angeles
UKHTOMSKY

GERT H. BRIEGER
Johns Hopkins University
EBERTH; HEKTOEN; KELSER; LOEFFLER; METCHNIKOFF; SABIN; THAYER; WELCH

G.E. BRIGGS
University of Cambridge
BLACKMAN

J. MORTON BRIGGS, JR.
University of Rhode Island
D'ALEMBERT; PARENT; ROBINS

L. BRILLOUIN
BRILLOUIN

JAMES E. BRITTAIN
Georgia Institute of Technology
PACINOTTI

T.A.A. BROADBENT
BOOLE; LANDEN; MACAULAY; MCCOLL; MACMAHON; MATHEWS; RAMSEY; B.A.W. RUSSELL; SHANKS; VENN; W. WALLACE; J. WILSON

W.H. BROCK
University of Leicester
ASTON; CROOKES; ERDMANN; E. FRANKLAND; HOFMANN; T.S. HUNT; MARCHAND; MARIGNAC; ODLING; PENNY; PROUT; TILLOCH; E. TURNER; WANKLYN; A.W. WILLIAMSON

BARUCH A. BRODY
Massachusetts Institute of Technology
Rice University
W. HAMILTON; W.E. JOHNSON; WHATELY

JOAN BROMBERG
Niels Bohr Institute
FÖPPL

JOHN HEDLEY BROOKE
University of Lancaster
GERHARDT; WURTZ

HARCOURT BROWN
PEIRESC; SALLO

SANBORN C. BROWN
Massachusetts Institute of Technology
B. THOMPSON

THEODORE M. BROWN
Princeton University · City College, City University of New York
BELLINI; CHEYNE; DESCARTES; GALVANI; LOWER; MAYOW; PITCAIRN; STUART

VIGGO BRUN
THUE

STEPHEN G. BRUSH
University of Maryland
BOLTZMANN; ENSKOG; R. FOWLER; J. HERAPATH; LENNARD-JONES; WATERSTON

GERD BUCHDAHL
Whipple Science Museum
BERKELEY; C. WOLFF

JED Z. BUCHWALD
Harvard University · University of Toronto
MELLONI; MOSOTTI; NOBILI; W. THOMSON; VILLARI

K.E. BULLEN
University of Sydney
BARTELS; DAVID; DEBENHAM; IMAMURA; KNOTT; LAMB; LOVE; J. MILNE; MOHOROVIČIĆ; R.D. OLDHAM; OMORI; SEZAWA; WEGENER; WIECHERT

Contributors

VERN L. BULLOUGH
San Fernando Valley State College
ALDEROTTI; BAILLIE; CHAULIAC; HENRY OF MONDEVILLE; HUNDT; KNOX; LIND; MACLEAN; MONDINO DE' LUZZI; PANDER; RATHKE; VARENIUS

O.M.B. BULMAN
Sedgwick Museum, Cambridge
BONNEY

IVOR BULMER–THOMAS
CONON OF SAMOS, DINOSTRATUS; DIONYSODORUS; DOMNINUS OF LARISSA; EUCLID; EUDEMUS OF RHODES; EUTOCIUS OF ASCALON; HIPPIAS OF ELIS; HIPPOCRATES OF CHIOS; HYPSICLES OF ALEXANDRIA; ISIDORUS OF MILETUS; LEO; LEODAMUS OF THASOS; MENAECHMUS; MENELAUS OF ALEXANDRIA; OENOPIDES OF CHIOS; PAPPUS OF ALEXANDRIA; PERSEUS; SERENUS; THEAETETUS; THEODORUS OF CYRENE; THEODOSIUS OF BITHYNIA; ZENODORUS

WERNER BURAU
University of Hamburg
CLEBSCH; GÖPEL; GRASSMANN; C.F. KLEIN; J. KOENIG: KÖNIGSBERGER; LUEROTH; W.F. MEYER; PLÜCKER; REYE; ROHN; ROSANES; J.G. ROSENHAIN; SCHEFFERS; SCHROETER; SCHUBERT; SCHWEIKART; STAUDT; STUDY; F.O.R. STURM; WANGERIN; WEINGARTEN

JOHANN JAKOB BURCKHARDT
University of Zurich
FUETER; GEISER; GRÄFFE; M. GROSSMANN; RUDIO; SCHLÄFLI; STEINER; J.R. WOLF

E.H.S. BURHOP
University College, London
C.F. POWELL

DEAN BURK
National Cancer Institutes
R. KUHN; O.H. WARBURG

JOHN G. BURKE
University of California, Los Angeles
BOURNON; BROCHANT DE VILLIERS; CLEAVELAND; CORDIER; DUFRÉNOY; J. FORBES; FOUQUÉ; HESSEL; O. LEHMANN; LEONHARD; MICHEL–LÉVY; MOHS; K.F. NAUMANN; F.E. NEUMANN; P. PREVOST; QUENSTEDT; SOHNCKE; VALMONT DE BOMARE

RICHARD W. BURKHARDT, JR.
University of Illinois
LATREILLE

J.C. BURKILL
University of Cambridge
G.H. HARDY; HOBSON; VALLÉE–POUSSIN; W.H. YOUNG

LESLIE J. BURLINGAME
Mount Holyoke College
LAMARCK

JOHN C. BURNHAM
Ohio State University
CALKINS; CARROLL; CHILD; JENNINGS; YERKES

CONRAD BURRI
Swiss Federal Institute of Technology
GRUBENMANN

E. ALFRED BURRILL
VAN DE GRAAFF

HAROLD L. BURSTYN
Carnegie–Mellon University · William Paterson College
BUYS BALLOT; CROLL; FERREL; FOUCAULT; M.F. MAURY; J. MURRAY; POURTALÈS; REDFIELD

H.L.L. BUSARD
State University of Leiden
BOUVELLES; BUOT; CARCAVI; CLAVIUS; DEPARCIEUX; DESPAGNET; FRENICLE; GULDIN; C. HARDY; HENRY OF HESSE; LA FAILLE; VAN LANSBERGE; LE PAIGE; PITISCUS; ROOMEN; VER EECKE; VIÈTE

ROBERT E. BUTTS
University of Western Ontario
WHEWELL

PERRY BYERLY
University of California, Berkeley
REID

G.V. BYKOV
Academy of Sciences of the U.S.S.R.
BEKETOV; BUTLEROV; MARKOVNIKOV; A.N. POPOV; ZININ

JEROME J. BYLEBYL
University of Chicago
COLOMBO; W. HARVEY; LAURENS; LEONICENO; PICCOLOMINI; RIOLAN; WILLDENOW

WILLIAM F. BYNUM
University College, London
BRUNTON; DALE; S.W. MITCHELL; B.W. RICHARDSON; RINGER; H.W. SMITH; WERNICKE; WITHERING; ZINSSER

R.W. CAHN
University of Sussex
W. ROSENHAIN

ANDRÉ CAILLEUX
Laval University
BLANC; LAPPARENT; E.A.I.H. LARTET; L. LARTET

RONALD S. CALINGER
Rensselaer Polytechnic Institute
BLISS; BOLZA; BOUTROUX; CASTILLON; DICKSON; E.H. MOORE

JOHN T. CAMPBELL
KAUFMANN

W.A. CAMPBELL
University of Newcastle Upon Tyne
MOND; SOLVAY; C. WINKLER

LUIGI CAMPEDELLI
University of Florence
FRISI; GHETALDI; MAGALOTTI; MAGINI; M. RICCI; RICCIOLI; ZUCCHI

KENNETH L. CANEVA
University of Utah
A.–A. DE LA RIVE; C.–G. DE LA RIVE; OHM; SCHWEIGGER

GEORGES CANGUILHEM
University of Paris
BICHAT; CABANIS

WALTER F. CANNON
Smithsonian Institution
BUCKLAND

ROBERT CANTWELL
Time Inc.
A. WILSON

MILIČ ČAPEK
Boston University
STALLO

J.B. CARLES
Institut Catholique de Toulouse
MOLLIARD; RAULIN

ELOF AXEL CARLSON
State University of New York at Stony Brook
H.J. MULLER

ERIC T. CARLSON
Cornell University Medical Center
RUSH

ALBERT V. CAROZZI
University of Illinois
J.F. BRANDT; BELLARDI; CAYEUX; DESOR; GUYOT; W. HAMILTON; HAUG; MAILLET; MUNIER–CHALMAS; RASPE; RENARD; H.B. DE SAUSSURE; SCHLUMBERGER

JEFFREY CARR
University of Leeds
M. LISTER

ETTORE CARRUCCIO
Universities of Bologna and Turin
ANGELI; BELLAVITIS; BERTINI; BETTI; BIANCHI; BONCAMPAGNI; BORTOLOTTI; CASTELNUOVO; CATALDI; CAVALIERI; LORIA; P. RUFFINI

H.B.G. CASIMIR
University of Leiden
KRAMERS

JAMES H. CASSEDY
National Library of Medicine
E.O. JORDAN; STILES

CARLO CASTELLANI
University of Parma
ALDROVANDI; GAGLIARDI; GHISI; LANCISI; MANTEGAZZA; MARCHIAFAVA; MOSSO; PERRONCITO; POLI; ROLANDO; SALVIANI; STRUSS; TROJA; ZAMBECCARI

W.B. CASTLE
Harvard University
G.R. MINOT

MARÍA ASUNCIÓN CATALÁ
University of Barcelona
ORTEGA

BERNARDO J. CAYCEDO
F. D'ELHUYAR; J. D'ELHUYAR

PIERRE CHABBERT
Hôpitaux de Castres
PINEL; PORTAL

CARLOS CHAGAS
UNESCO, Brazilian Delegation
CHAGAS

JAMES F. CHALLEY
Princeton University · Vassar College
S. CARNOT; M. SEGUIN

JOHN CHALLINOR
University College of Wales
FALCONER; FAUJAS DE SAINT-FOND; A. GEIKIE; J. GEIKIE; HARKER; C. LAPWORTH; MANTELL; E.T. NEWTON; G. OWEN; J. PLAYFAIR; PRESTWICH; T. WEBSTER; WHITEHURST

M.C. CHANG
Worcester Foundation for Experimental Biology
PINCUS

SEYMOUR L. CHAPIN
California State University, Los Angeles
BAILLY; CAMUS; J.-N. DELISLE; FOUCHY; GODIN; JEAURAT; LE GENTIL DE LA GALAISIÈRE; PONS

ALLAN CHAPMAN
University of Oxford
SHAKERLEY; G. WHARTON

CARLETON B. CHAPMAN
The Commonwealth Fund, New York City · Dartmouth Medical School
J.S. HALDANE; E. SMITH; STARLING

JEAN CHÂTILLON
Catholic Institute of Paris
GILES OF ROME; HUGH OF ST. VICTOR

GEORGES CHAUDRON
University of Paris · Laboratoire de Recherches Métallurgiques
CHARPY; CHEVENARD; GUILLAUME; GUILLET; HÉROULT; PORTEVIN

H. CHIARI
University of Vienna
WEICHSELBAUM

ROBERT A. CHIPMAN
University of Toledo
DEPREZ; HEFNER-ALTENECK; JENKIN; MARCONI; C.W. SIEMENS; STEINMETZ

RICHARD CHORLEY
University of Cambridge
DAUBRÉE; A. HEIM; D.W. JOHNSON; W.D. JOHNSON; W. PENCK

FREDERICK B. CHURCHILL
Indiana University
F.M. BALFOUR; W. ROUX

J.G. VAN CITTERT-EYMERS
BURGER; HARTING; HARTSOEKER

STIG CLAESSON
University of Uppsala
SVEDBERG

MARSHALL CLAGETT
Institute for Advanced Study, Princeton
ADELARD OF BATH; ARCHIMEDES; GERARD OF BRUSSELS; JOHN OF PALERMO; LEONARDO DA VINCI; ORESME

RONALD W. CLARK
J.B.S. HALDANE

THOMAS H. CLARK
McGill University
F.D. ADAMS; R. BELL; BILLINGS; J.W. DAWSON; W.E. LOGAN

EDWIN CLARKE
University College, London · Wellcome Institute for the History of Medicine
BASTIAN; BEEVOR; BROCA; FERRIER; FLECHSIG; FRITSCH; M. HALL; HITZIG; HORSLEY; J.J. HUBER; HUTCHINSON; J.H. JACKSON; C.L. MORGAN; NEWPORT; PHILIP; TWORT; C. WHITE

M.J. CLARKSON
Liverpool School of Tropical Medicine
T.R. LEWIS; MANSON

ARCHIBALD CLOW
British Broadcasting Corporation
DUNDONALD; ROEBUCK

F.A.L. CLOWES
University of Oxford
SHARROCK

N.H. CLULEE
Frostburg State College
RULAND

BRUCE C. COGAN
Johns Hopkins University
H.N. RUSSELL

I. BERNARD COHEN
Harvard University
DELAMBRE; B. FRANKLIN; A.M. MAYER; I. NEWTON

LORD COHEN OF BIRKENHEAD
University of Hull
E. DARWIN

ROBERT S. COHEN
Boston University
ENGELS; FRANK; MARX

EDWIN H. COLBERT
Museum of Northern Arizona
OSBORN; W.B. SCOTT

ERIC M. COLE
ATWOOD

WILLIAM COLEMAN
Johns Hopkins University · Northwestern University
BATESON; BLAINVILLE; GAIMARD; GEGENBAUR; GRATIOLET; KIELMEYER; MENURET DE CHAMBAUD; NAUDIN

RUNAR COLLANDER
OVERTON

D.E. COOMBE
University of Cambridge
T. JOHNSON

ANDRÉ COPAUX
École Supérieure de Physique et de Chimie, Paris
COPAUX

GEORGE W. CORNER
American Philosophical Society
BRACHET; CARREL; FLEXNER; MALL; C.S. MINOT; STREETER

CARL W. CORRENS
University of Göttingen
FRANKENHEIM

DAVID W. CORSON
University of Arizona
POLNIÈRE; MASSON

ALBERT B. COSTA
University of Notre Dame · Duquesne University
ARFVEDSON; K.F. VON AUWERS; BEHREND; CHANCEL; CHEVREUL; CIAMICIAN; A.C.L. CLAUS; L. COHN; COURTOIS; CRUM; DEWAR; H.B. DIXON; ERLENMEYER; FOWNES; GELMO; HANTZSCH; HILDITCH; V. MEYER; MICHAEL; H.J. SCHIFF; SWARTS; J.F. THORPE; WALTER; WISLICENUS

Contributors

PIERRE COSTABEL
École Pratique des Hautes Études
BAIRE; BEGHIN; CHAZY; CORIOLIS; DEBEAUNE; J.B. DU HAMEL; GALLOIS; M.-G. HUMBERT; P. HUMBERT; LAGNY; MALEBRANCHE; MILHAUD; MORIN; MYLON; PARDIES; PÉRÈS; POISSON; REYNEAU; VARIGNON

JULIANA HILL COTTON
MANARDO

E. COUMET
E. LE ROY

CHARLES COURY
University of Paris
DUVAL; TESTUT

C.F. COWAN
W.H. EVANS

RUTH SCHWARTZ COWAN
State University of New York at Stony Brook
ALLEN; BARTHEZ; BONNIER; GATES; WELDON

E. HORNE CRAIGIE
University of Toronto
TÜRCK

PAUL F. CRANEFIELD
The Rockefeller University
BREUER

MAURICE CRANSTON
London School of Economics
LOCKE; MILL

J.K. CRELLIN
Wellcome Institute for the History of Medicine
JONSTON; POUCHET; J. REY; R. ROSS; THURNAM

ERIKA CREMER
BODENSTEIN

F.A.E. CREW
PUNNETT

A.C. CROMBIE
University of Oxford
R. BACON; DESCARTES; GROSSETESTE; MERCENNE

M.P. CROSLAND
University of Kent at Canterbury
AVOGADRO; J.E. BÉRARD; BERTHELOT; BIOT; CHAPTAL; DULONG; GAY-LUSSAC; GERHARDT; THENARD

MICHAEL J. CROWE
University of Notre Dame
BROMWICH; H. HANKEL; HOÜEL; A.F. MÖBIUS

HOWARD A. CRUM
University of Michigan Herbarium
EVANS

CHARLES A. CULOTTA
ARSONVAL; F.G. BENEDICT; BIDDER

L.W. CURRIER
B.K. EMERSON

STANISŁAW CZARNIECKI
Polish Academy of Sciences
CZEKANOWSKI; CZERSKI; STASZIC; ZEJSZNER

AL-DABBAGH
BANŪ MŪSĀ

ŽARKO DADIĆ
Yugoslav Academy of Sciences and Arts
DOMINIS

F. DAGOGNET
BACHELARD

PER F. DAHL
Brookhaven National Laboratory
COLDING

W.S. DALLAS
A.E. REUSS

EUGENIO DALL'OSSO
ARANZIO

JOHN F. DALY, S.J.
St. Louis University
SACROBOSCO

J.M.A. DANBY
North Carolina State University
H.C. PLUMMER

GLYN DANIEL
University of Cambridge
BEDDOE; BREUIL; LEAKEY; MONTELIUS; MORTILLET; PETRIE; PIETTE; PITT-RIVERS; SCHLIEMANN; WOOLLEY; WORM; WORSAAE

KARL H. DANNENFELDT
Arizona State University
CALLINICOS OF HELIOPOLIS; DIOCLES; DIOCLES OF CARYSTUS; HERMES TRISMEGISTUS; OLYMPIODORUS; RAUWOLF; STEPHANUS OF ALEXANDRIA; SYNESIUS OF CYRENE

UMBERTO MARIA D'ANTINI
SARPI

ROBERT DARNTON
Princeton University
MESMER

EDWARD E. DAUB
University of Wisconsin
CLAUSIUS; HARCOURT; KRÖNIG

MAURICE DAUMAS
Conservatoire National des Arts et Métiers
LANGLOIS

HERBERT A. DAVIDSON
University of California at Los Angeles
CRESCAS

AUDREY B. DAVIS
Smithsonian Institution
LIEBERKÜHN; MOSS; MURPHY

HALLOWELL DAVIS
Central Institute for the Deaf
A. FORBES

STACEY B. DAY
Memorial Sloan-Kettering Cancer Center
STEVENS; TASHIRO

G.E.R. DEACON
National Institute of Oceanography, U.K.
PETTERSSON

MARGARET DEACON
National Institute of Oceanography
DITTMAR

GAVIN DE BEER
CHARPENTIER; C.R. DARWIN; GOODRICH; LANKESTER; SLOANE; VENETZ

KAREL L. DE BOUVÈRE, S.C.J.
University of Santa Clara
A.N. WHITEHEAD

ALLEN G. DEBUS
University of Chicago
BECHER; DUCHESNE; FLUDD; SEVERINUS; VALENTINE; J. WEBSTER

RONALD K. DEFORD
University of Texas
G. GILBERT

CLAUDE K. DEISCHER
University of Pennsylvania
L. GMELIN; W.B. HERAPATH

ALBERT DELAUNAY
Institut Pasteur, Paris
CALMETTE; CHAMBERLAND; DUCLAUX; RAMON; P.P.E. ROUX; SERGENT

SUZANNE DELORME
Centre Internationale de Synthèse
FONTENELLE; METZGER; MONTMOR; A. REY

S. DEMIDOV
Academy of Sciences of the U.S.S.R.
PETROVSKY

Y.A. DEMIDOVICH
Academy of Sciences of the U.S.S.R.
POTANIN

ALFRED R. DESAUTELS, S.J.
Holy Cross College
CASTEL

A. DE SMET
Bibliothèque Royale de Belgique
VAN LANGREN

R.G.C. DESMOND
India Office Library of Records, London
C. BABINGTON; J.G. BAKER; I.B. BALFOUR; J.H. BALFOUR; J.D. HOOKER; SPRUCE

SOLOMON DIAMOND
California State University, Los Angeles
WUNDT

BERN DIBNER
Burndy Library
ALDINI; DU MONCEL; FERRARIS; HOPKINSON

D.R. DICKS
University of London
CLEOMEDES; DOSITHEUS; ERATOSTHENES; EUCTEMON; GEMINUS; HECATAEUS OF MILETUS; HICETAS OF SYRACUSE

SALLY H. DIEKE
Johns Hopkins University
ABNEY; ANDOYER; ANDRÉ; AVEST; BAADE; BACKLUND; S.I. BAILEY; BAILY; D. BROUWER; CURTISS; DUGAN; DUNCAN; ELKIN; HOUGH; M.J. JOHNSON; KEELER; H.J. KLEIN; I.W. LUBBOCK; C.S. LYMAN; G. MOLL; SCHWARZSCHILD; SEARES; STRÖMBERG; TRUMPLER; W. H. WRIGHT

J. DIEUDONNÉ
University of Nice
CARTAN; C. JORDAN; MINKOWSKI; POINCARÉ; VON NEUMANN; WEYL

HÂMIT DILGAN
Istanbul University
AL-SAMARQANDĪ; QĀḌĪ ZĀDA AL-RŪMĪ

AUBREY DILLER
Indiana University
DICAEARCHUS; PYTHEAS OF MASSALIA

HERBERT DINGLE
University of London
A. FOWLER; HUGGINS; LOCKYER

JERZY DOBRZYCKI
Institute for the History of Science, Polish Academy of Sciences, Warsaw
HOËNÉ-WROŃSKI

JESSIE DOBSON
Royal College of Surgeons Hunterian Museum
CRUIKSHANK; J. HUNTER; W. HUNTER

WILLIAM DOCK
State University of New York at Brooklyn
W.C. WELLS

YVONNE DOLD-SAMPLONIUS
AL-JAYYĀNĪ; AL-KHĀZIN; AL-MĀHĀNĪ; AL-SIJZĪ; SINĀN IBN THĀBIT IBN QURRA; AL-QŪHĪ

CLAUDE E. DOLMAN
University of British Columbia
BREFELD; D. BRUCE; BUDD; BULLOCH; CREIGHTON; DOBELL; EHRLICH; ESCHERICH; A. FLEMING; H.H.R. KOCH; J. LISTER; NOGUCHI; PETTENKOFER; T. SMITH; SPALLANZANI

M.A. DONK
Rijksherbarium, Leiden
PERSOON

J.D.H. DONNAY
McGill University
FANKUCHEN; FRIEDEL

ARTHUR DONOVAN
West Virginia University
CRAWFORD

J.G. DORFMAN
Institute for the History of Science and Technology, Moscow
ARKADIEV; ARTSIMOVICH; FRENKEL; GOLITSYN; KURCHATOV; G.S. LANDSBERG; P.P. LAZAREV; P.N. LEBEDEV; MANDELSHTAM; A.S. POPOV; ROZHDESTVENSKY; STOLETOV; TAMM; VAVILOV

HAROLD DORN
Stevens Institute of Technology
O. GREGORY; HODGKINSON; KATER; NEWCOMEN; ROBISON; SMEATON; S.G. THOMAS; WATT; WEDGWOOD

H. DÖRRIE
University of Münster/Westphalia
XENOCRATES OF CHALCEDON

SIGALIA C. DOSTROVSKY
Worcester Polytechnic Institute · Barnard College
BOTHE; CHLADNI; J.-M. DUHAMEL; EUCKEN; J.A. EWING; C. FABRY; H. FOSTER; LISSAJOUS; MAIRAN; PÉROT; SAGNAC; SAUVEUR; SAVART; VILLARD; VIOLLE; WHEATSTONE

A. VIBERT DOUGLAS
Queen's University, Ontario
EDDINGTON

A.G. DRACHMANN
CTESIBIUS; HERO OF ALEXANDRIA; PHILO OF BYZANTIUM

STILLMAN DRAKE
University of Toronto
BALDI; BALIANI; BARTOLI; G.B. BENEDETTI; BERTI; CASTELLI; CESI; G. GALILEI; V. GALILEI; LE TENNEUR

OLLIN J. DRENNAN
Western Michigan University
HITTORF; HORSTMANN; F.W.G. KOHLRAUSCH; R.H.A. KOHLRAUSCH

JOHN M. DUBBEY
University College, London
BROUNCKER; COTES; DE MORGAN

LOUIS DULIEU
University of Montpellier
BELLEVAL; BERTHOLON; BORDEU

G.S. DUNBAR
University of California, Los Angeles
RECLUS

RAYNOR L. DUNCOMBE
Nautical Almanac Office, U.S. Naval Observatory
H.R. MORGAN

K.C. DUNHAM
Institute of Geological Sciences, London
F.J.O. EVANS; FLETT; GODWIN-AUSTEN; HOLMES; TEALL; TYRRELL

L.C. DUNN
Columbia University
W.L. JOHANNSEN

A. HUNTER DUPREE
Brown University
A. AGASSIZ; A. GRAY; TORREY; WYMAN

DAVID R. DYCK
University of Winnipeg
ELLER VON BROCKHAUSEN

JOY B. EASTON
West Virginia University
DEE; L. DIGGES; T. DIGGES; RECORDE; TUNSTALL; P. TURNER; WITT

BRUCE S. EASTWOOD
Kansas State University
GRIMALDI

M.V. EDDS, JR.
Brown University
W.K. BROOKS

SIDNEY EDELSTEIN
Dexter Chemical Corporation
DREBBEL; W.H. PERKIN

THE EDITORS
COLUMBUS

J.M. EDMONDS
University of Oxford Museum
ARKELL; KIDD; LHWYD; J. PHILLIPS; SOLLAS

JOHN T. EDSALL
Harvard University
E.J. COHN

FRANK N. EGERTON III
University of Wisconsin–Parkside
E. FORBES; GRAUNT; FOREL; FUCHS; LOVELL; PETTY; J. TOWNSEND; H.C. WATSON

OLIN J. EGGEN
Australian National University
AIRY; CHALLIS

DAVID E. EICHHOLZ
University of Bristol
PLINY

CAROLYN EISELE
Hunter College, City University of New York
BÔCHER; COUTURAT; ENRIQUES; G.W. HILL; G.A. MILLER; B. PEIRCE; B.O. PEIRCE II; C.S. PEIRCE

Contributors

CHURCHILL EISENHART
U.S. Department of Commerce, National Bureau of Standards
K. PEARSON; WILKS; YOUDEN

VĚRA EISNEROVÁ
VEJDOVSKÝ; ZALUŽANSKÝ

JON EKLUND
Smithsonian Institution
DUHAMEL DU MONCEAU; W. LEWIS

FRANÇOIS ELLENBERGER
University of Paris–South
BOURGUET

JAMES ELLINGTON
University of Connecticut
PAUL CARUS; KANT

H. ENGEL
Zoölogisch Museum, Amsterdam
ARTEDI; BILLROTH; BOLK

ANN MARIE ERDMAN
Florida State University
PEKELHARING

GUNNAR ERIKSSON
University of Umeå
ACHARIUS; C.A. AGARDH; J.G. AGARDH; FORSSKÅL; E. FRIES; PONTEDERA; SWARTZ; C.P. THUNBERG; WAHLENBERG

VASILIY A. ESAKOV
Academy of Sciences of the U.S.S.R.
ANUCHIN; DOKUCHAEV; KRUBER; M.P. LAZAREV; PALLAS; PRZHEVALSKY; SARYCHEV

I. ESTERMANN
Israel Institute of Technology
STERN

FOCKO EULEN
Ruhr University, Bochum
KELLNER

P.P. EVALD
Polytechnic Institute of New York
C.H. HERMANN

CHARLES L. EVANS
W.M. BAYLISS

D.G. EVANS
National Institute for Biological Standards and Control, London
W. SMITH

DAVID S. EVANS
University of Texas
J.F.W. HERSCHEL

C.W.F. EVERITT
Stanford University
F. LONDON; H. LONDON; MAXWELL

JOSEPH EWAN
Tulane University
BANISTER; BARTON; BESSEY; A.W. CHAPMAN; CLEMENTS; DARLINGTON; ENGELMANN; FERNALD; J. GAERTNER; R.M. HARPER; HITCHCOCK; HOSACK; JEPSON; A. KELLOGG; LESQUEREUX; MERRILL; MICHAUX; ORTON; PURSH; RAFINESQUE; SETCHELL; J. TRADESCANT, SR.; J. TRADESCANT, JR.; TRELEASE

JOAN M. EYLES
FAREY; FEATHERSTONHAUGH; FITTON; R. GRIFFITH; JAMESON; JUKES; T. OLDHAM; RAMOND DE CARBONNIÈRES; RENNELL; W. SMITH; SOWERBY

VICTOR A. EYLES
W. BABINGTON; DE LA BECHE; FEATHERSTONHAUGH; GREENOUGH; J. HALL; J. HUTTON; JOLY; MACCULLOCH; STRACHEY; TOULMIN; J. WOODWARD

J. FABER
Hubrecht Laboratory, Utrecht
HUBRECHT

W.M. FAIRBANK
Stanford University
F. LONDON; H. LONDON

EDUARD FARBER
BAEKELAND; P. BOREL; G. CLAUDE; DÖBEREINER; EDER; E. FISCHER; FRÉMY

KATHLEEN FARRAR
University of Manchester
GULLAND

W.V. FARRAR
University of Manchester
DONNAN; HAMPSON; A. LAPWORTH; LAWES AND GILBERT; K.H. MEYER; NERI; NIEUWLAND; PAYEN; W.H. PERKIN, JR.; L. PLAYFAIR; K.L. REICHENBACH; SCHUNK

SISTER MAUREEN FARRELL, F.C.J.
University of Manchester
SCHUMACHER; SCHWABE; SOLDNER; STEIN

GIOVANNI FAVILLI
University of Bologna
LUSTIG

IGNAZIO FAZZARI
University of Florence
LUNA

VERA N. FEDCHINA
Academy of Sciences of the U.S.S.R.
MIKLUKHO–MAKLAY; SEDOV; SEMYONOV–TYAN–SHANSKY; VYSOTSKY

A.S. FEDOROV
Academy of Sciences of the U.S.S.R.
CHERNOV; KRASHENINNIKOV; VIZE

I. FEDOSEYEV
Academy of Sciences of the U.S.S.R.
BELLINGSGAUZEN; KRASNOV; LEPEKHIN; LOKHTIN; MAKAROV; TILLO; VERNADSKY; VOEYKOV

BERNARD T. FELD
Massachusetts Institute of Technology
SZILARD

JEAN FELDMANN
Pierre and Marie Curie University
LAMOUROUX; SAUVAGEAU; THURET

LUCIENNE FÉLIX
University of Paris
BACHELIER; BOUSSINESQ; BRIOT; DRACH; PAINLEVÉ; C.E. PICARD; VESSIOT

E. A. FELLMANN
Institut Platonaeum, Basel
JOHANN I. BERNOULLI; FABRI; J. HERMANN; J.S. KOENIG

EDWIN FELS
DRYGALSKI

JAMES W. FELT, S.J.
University of Santa Clara
A.N. WHITEHEAD

FRANK FENNER
Australian National University
FLOREY

EUGENE S. FERGUSON
University of Delaware
BOURDON; BRINELL; KENNEDY; THURSTON

KONRADIN FERRARI D'OCCHIEPPO
University of Vienna
ANGELUS; GRAFF; HARTWIG; HELL; OPPENHEIM; OPPOLZER; PALISA; E. WEISS

SARAH FERRELL
JAMES

MARTIN FICHMAN
York University
GUIBERT; JUNCKER; E. KÖNIG; J.–B. LE ROY; MAGNENUS; MALOUIN; P. PETIT; PRIVAT DE MOLIÈRES; SIGORGNE; TABOR; ULSTAD

M. FIERZ
Federal Institute of Technology, Zurich
PAULI

KARIN FIGALA
Deutches Museum · Technische Hochschule, Munich
LONICERUS; PAULLI

N. FIGUROVSKY
Academy of Sciences of the U.S.S.R.
LOVITS

C. P. FINLAYSON
University of Edinburgh Library
MONRO (PRIMUS); MONRO (SECUNDUS)

BERNARD S. FINN
Smithsonian Institution
A. G. BELL; J.L. CLARK; CUMMING; EDISON; GLAZEBROOK; E. H. HALL; IVES; M.H. VON JACOBI; J. KERR; PELTIER; RÜHMKORFF; STURGEON

W.S. FINSEN
INNES

WALTHER FISCHER
DOELTER; GROTH; A.A. HEIM; HELMERT; LIESGANIG; MALLET; SÉNARMONT

C.S. FISHER
Brandeis University
BATEMAN; DEHN; FINE; GORDAN

DONALD W. FISHER
New York State Museum and Science Service
J. HALL, JR.; RUEDEMANN

HEINZ FLAMM
University of Vienna
GRUBER

J.O. FLECKENSTEIN
University of Basel
JACOB II BERNOULLI; JOHANN I BERNOULLI; JOHANN II BERNOULLI; JOHANN III BERNOULLI; NIKOLAUS I BERNOULLI; NIKOLAUS II BERNOULLI; EMDEN

DONALD FLEMING
Harvard University
J. DRAPER; J. LOEB

MARCEL FLORKIN
University of Liège
BENEDEN; DODOENS; FREDERICQ; GRAMME; SCHWANN; SLUSE; STAS

MENSO FOLKERTS
Technische Universität, Berlin
BALBUS; CENSORINUS; TITIUS; J. WERNER

JAROSLAV FOLTA
Czechoslovak Academy of Science
ČECH; CESÁRO; JONQUIÈRES; KLÜGEL

GEORGE A. FOOTE
BANKS

ERIC G. FORBES
University of Edinburgh
BIRT; CARRINGTON; FREUNDLICH; MASKELYNE; J.T. MAYER

GEORGE S. FORBES
Harvard University
COOKE

MICHAEL FORDHAM
JUNG

PAUL FORMAN
Smithsonian Institution
BACK; BARKLA; W.H. BRAGG; DUANE; ORNSTEIN; PASCHEN; RITZ; C.D.T. RUNGE; SMEKAL; SOMMERFELD

FREDERICK M. FOWKES
Lehigh University
HARKINS

DEAN R. FOWLER
Center for Process Studies, Claremont, California
A.N. WHITEHEAD

ROBERT FOX
University of Lancaster
DUPRÉ; GAUDIN; LAPLACE; N. LEBLANC; A.T. PETIT; M.–A. PICTET; B. POWELL; REGNAULT

ATTILIO FRAJESE
University of Rome
CABEO

VINCENZO FRANCANI
Polytechnique of Milan
BREISLAK

PIETRO FRANCESCHINI
University of Florence
ACHILLINI; AROMATARI; BACCELLI; BANTI; BECCARI; BENIVIENI; BIZZOZERO; BUONANNI; CASTALDI; CHIARUGI; G. DELLA TORRE; GALEAZZI; GRASSI; MENEGHETTI; PACINI; A. RUFFINI; SCARPA

ROBERT G. FRANK, JR.
University of California, Los Angeles
T. WILLIS

EUGENE FRANKEL
Trinity College, Hartford
BABINET; BENOIT; BRACE; H. LLOYD; NICOL; SEEBECK; VERDET

V. FRANK-KAMENETSKY
Leningrad State University
WULFF

FRITZ FRAUNBERGER
QUINCKE; RAMSAUER; REGENER; SCHUMANN; O. WIENER

ARTHUR H. FRAZIER
U.S. Geological Survey
SAXTON; WOLTMAN

H.C. FREIESLEBEN
BRUHNS; DORNO; ENCKE; GALLE; HARDING; J.F. HARTMANN; KAYSER; KONKOLY THEGE; LOHSE; LUDENDORFF; LUTHER; C. MAYER; MOLLWEIDE; C.A.F. PETERS; C.F.W. PETERS; SPOERER; STEINHEIL; M.F.J.C. WOLF

RICHARD D. FRENCH
Science Council of Canada
OLIVER; J.G. PRIESTLEY

HANS FREUDENTHAL
State University of Utrecht
ARBUTHNOT; CAUCHY; HAAR; HEINE; HERMITE; HILBERT; HOPF; HURWITZ; KERÉKJÁRTÓ; KNOPP; LIE; LOEWNER; NIEUWENTIJT; A. PRINGSHEIM; QUETELET; RIEMANN; SCHOENFLIES; SCHOTTKY; SYLOW; N. WIENER

JOHN E. FREY
Northern Michigan University
A. STOCK

B. VON FREYBERG
University of Erlangen–Nuremberg
BAIER; HOFF; J.G. LEHMANN; SCHLOTHEIM; WALCH

GEORGE F. FRICK
University of Delaware
CATESBY; COLLINSON

WALTER FRICKE
Astronomisches Rechen-Institut, Heidelberg
BESSEL

O.R. FRISCH
University of Cambridge
MEITNER

KURT VON FRITZ
University of Munich
ARCHYTAS OF TARENTUM; PHILOLAUS OF CROTONA; PYTHAGORAS OF SAMOS; ZENO OF ELEA; ZENO OF SIDON

JOSEPH S. FRUTON
Yale University
BERGMANN; CORI; HARDY; HOPPE–SEYLER; KEILIN; MEYERHOF; J.R. SUMNER; WILLSTÄTTER

TSUNESABURO FUJINO
Osaka University
KITASATO

J.Z. FULLMER
Ohio State University
GARNETT; REMSEN

DAVID J. FURLEY
Princeton University
EPICURUS; HERACLITUS OF EPHESUS; HERODOTUS OF HALICARNASSUS; LUCRETIUS; SEXTUS EMPIRICUS; ZENO OF CITIUM

L.K. GABUNA
Academy of Sciences of the U.S.S.R.
DOLLO

A.E. GAISSINOVITCH
Academy of Sciences of the U.S.S.R.
WOLFF

JEAN–CLAUDE GALL
Université Louis Pasteur, Strasbourg
VOLTZ

Contributors

HENRI GALLIARD
Brumpt

PAUL GANIÈRE
Corvisart

JUSTO GARATE
Ameghino

ELIZABETH B. GASKING
University of Melbourne
A.E.R.A. Serres

GERALD L. GEISON
University of Minnesota · Princeton University
Beale; A.C.H. Braun; Burdon–Sanderson; F. Cohn; Cushny; Dujardin; Fletcher; M. Foster; P. Frankland; Gaskell; Henfrey; J.N. Langley; Loewi; Lucas; Mellanby; Pasteur; N. Pringsheim; Rolleston; Schultze; Thiselton–Dyer

A.M. GEIST–HOFMAN
Moleschott

WILMA GEORGE
University of Oxford
Bering; Bory de Saint–Vincent; F. Darwin

WALTHER GERLACH
University of Munich
Elster; Geitel

PATSY A. GERSTNER
Dittrick Museum of Historical Medicine, Cleveland
J. Hill; J.T. Klein; K.N. Lang; J. Morton; J. Parkinson

A. GEUS
University of Marburg
Preyer; Roesel von Rosenhof; Schmidel; Siebold

RUTH ANN GIENAPP
Baeyer; Bamberger; E.O. Beckmann; Beilstein; Bernthsen

ARTHUR C. GIESE
Stanford University
C.V. Taylor

GEORGE E. GIFFORD, JR.
Harvard University
A. Gould; C.T. Jackson

BERTRAND GILLE
L.B. Alberti; Brunelleschi

KARL–ERNST GILLERT
Robert Koch Institut des Bundesgesundheitsamtes
Wassermann

R.J. GILLINGS
University of New South Wales
The Mathematics of Ancient Egypt

JEAN GILLIS
University of Ghent
Anschütz; Kekule von Stradonitz

CHARLES C. GILLISPIE
Princeton University
Bélidor; L. Carnot; Condillac; Diderot; Koyré; Laplace; Voltaire

STEWART GILLMOR
Wesleyan University
Borda; Bossut; Coulomb; D'Arcy; Thévenot

OWEN GINGERICH
Smithsonian Astrophysical Observatory
Argoli; G.P. Bond; W.C. Bond; A.J. Cannon; W. Fleming; A. Hall; Kepler; Lacaille; Leavitt; A.C. Maury; Méchain; Messier; Reinhold; Shapley

BENTLEY GLASS
State University of New York at Stony Brook
Maupertuis

PAUL GLEES
University of Göttingen
Leydig; Nissl; Waldeyer–Hartz; Wiedersheim

THOMAS F. GLICK
Boston University
Leo the African; Markham; Martínez; Río–Hortega; Ruiz; Sanchez; Teilhard de Chardin; Unanue; Vellozo

MARIO GLIOZZI
University of Turin
Cardano; Fabbroni; Levi–Civita; Torricelli

MARTHA TEACH GNUDI
University of California, Los Angeles
Dorn; Torre

HARRY GODWIN
University of Cambridge
Arber; Bower

GEORGE GOE
New School for Social Research
Kaestner; Lukasiewicz

EDWARD D. GOLDBERG
Scripps Institution of Oceanography
F.W. Clarke; Forchhammer; V.M. Goldschmidt; Washington

STANLEY GOLDBERG
Hampshire College
Max Abraham; Bucherer; Drude; Riecke; Trouton; W. Voigt

G.J. GOODFIELD
B.C. Brodie

D.C. GOODMAN
Open University
B.C. Brodie, Jr.; A.C. Brown; Tennant; W.H. Wollaston

JUDITH R. GOODSTEIN
California Institute of Technology
Tolman

MORRIS GORAN
Roosevelt University
Haber

J. ELSIE GORDON
University of Oxford
Highmore

H.B. GOTTSCHALK
University of Leeds
Strato of Lampsacus

T.A. GOUDGE
Bergson

A. GOUGENHEIM
Académie des Sciences de l'Institut de France
G. Perrier

J.B. GOUGH
Washington State University
Achard; A.E. Becquerel; Blondlot; I. Born; Bucholz; Charles; Fahrenheit; Fizeau; Gouy; C. Le Roy; Lasage; Réaumur; Sigaud De Lafond

STEPHEN JAY GOULD
Museum of Comparative Zoology, Harvard University
Hyatt

I. DE GRAAF BIERBRAUWER–WÜRTZ
L.W. Winkler

JUDITH V. GRABINER
Bezout; Budan de Boislaurent

LOREN GRAHAM
Columbia University
Bogdanov

RAGNAR GRANIT
Karolinska Institutet, Stockholm · National Institutes of Health · Nobel Institute for Neurophysiology
Fernel; Holmgren; Kalm

EDWARD GRANT
University of Indiana, Bloomington
Blasius of Parma; Jordanus De Nemore; Peter Peregrinus; Pomponius Mela; Vives

GILLES GRANGER
University of Aix–Marseilles
Condorcet; Cournot

IVOR GRATTAN-GUINESS
Middlesex Polytechnic of Enfield
FOURIER; LAPLACE; LAURENT; MATHIEU; RIESZ; STÄCKEL

FRANK GREENAWAY
Science Museum, London
H.C.H. CARPENTER; GRIESS; HADFIELD; KIRKALDY; PERCY; PLATTNER; POTT; ROBERTS–AUSTEN; STEAD; W.E.S. TURNER

JOHN C. GREENE
University of Connecticut
B. SILLIMAN; VOLNEY

JOSEPH T. GREGORY
University of California, Berkeley
BASSIANI; BROCCHI; BROOM; GRANGER; G.F. JAEGER; C. STOCK

SAMUEL L. GREITZER
Rutgers University
BRIANCHON; CREMONA; LAMÉ; LEMOINE

NORMAN T. GRIDGEMAN
National Research Council of Canada
BABBAGE; DODGSON; FISHER; GALTON; JEVONS; KIRKMAN; LOTKA; F.–E.–A. LUCAS; VON MISES; E.B. WILSON

REESE E. GRIFFIN, JR.
University of Pennsylvania
HOLBROOK

A.T. GRIGORIAN
Institute of the History of Science and Technology, Moscow · Academy of Sciences of the U.S.S.R.
ANDRONOV; BRASHMAN; BUNYAKOVSKY; CHAPLYGIN; DAVIDOV; FRIEDMANN; GALERKIN; IOFFE; KOCHIN; KOLOSOV; KOTELNIKOV; A.N. KRYLOV; N.M. KRYLOV; L.D. LANDAU; LEXELL; LEYBENZON; LYAPUNOV; MESHCHERSKY; NEKRASOV; PETERSON; N.P. PETROV; PRIVALOV; SEGNER; SHATUNOVSKY; SOMOV; THĀBIT IBN QURRA; TSIOLKOVSKY; TUPOLEV; VEKSLER; VYSHNEGRADSKY; ZHUKOVSKY; V.P. ZUBOV

N.A. GRIGORIAN
Academy of Sciences of the U.S.S.R.
BEKHTEREV; BURDENKO; ORBELI; I.P. PAVLOV; SAMOYLOV; VVEDENSKY

M.D. GRMEK
Archives Internationales d'Histoire des Sciences
BARRESWIL; C. BERNARD; BAGLIVI; BROWN-SÉQUARD; DODART; ESTIENNE; FERREIN; GERBEZIUS; GRISOGONO; GUIDI; MAGENDIE; ROBIN; ROSTAN; SANTORIO; VIEUSSENS; VERANTIUS

MICHAEL GROSS
Hampshire College
MAREY; MORAT

MORTON GROSSER
JOHN COUCH ADAMS

ERIC C. GROVES
British Museum
G. WHITE

JACOB W. GRUBER
Temple University
BOULE; MIVART; PUTNAM

J. GRUNOW
SPRUNG

JEAN–CLAUDE GUÉDON
University of Montreal
BÉCHAMP; CORNETTE

HENRY GUERLAC
Cornell University
JOSEPH BLACK; S. HALES; THE HAUKSBEES; LA BROSSE; LAVOISIER; SAGE; VAUBAN

FRANCISCO GUERRA
Laboratorios Abelló · Wellcome Historical Medical Library
ABREU; CRISTOBAL ACOSTA; AZARA; CASAL JULIAN; CERVANTES; DESCOURTILZ; DOMBEY; FEUILLÉE; MOLINA; MONARDES; SAINT–HILAIRE; VALVERDE

LAURA GUGGENBUHL
Hunter College, City University of New York
BROCARD; FEUERBACH

HEINRICH GUGGENHEIMER
Polytechnic Institute of New York
EISENHART

ALICE A. GUIMOND
Holyoke Community College
HERBERT

NORBERT GÜNTHER
ERNST ABBE

DOUGLAS GUTHRIE
BARCLAY

V. GUTINA
Academy of Sciences of the U.S.S.R. · Institute for the History of Science and Technology, Moscow
GAMALEYA; IVANOVSKY; VINOGRADSKY

KARLHEINZ HAAS
HESSE; HINDENBURG; HUDDE; TAURINUS; ZEUTHEN

IAN HACKING
University of Cambridge
BAYES; DE MOIVRE; GOSSET; MONTMOR

WILLEM D. HACKMAN
Museum of the History of Science
SWINDEN

ROGER HAHN
University of California, Berkeley
ARAGO

H.R. HAHNLOSER
VILLARD DE HONNECOURT

K. HAJNIS
Charles University
HRDLIČKA

A. RUPERT HALL
Imperial College of Science and Technology
E. BARTHOLIN; DESAGULIERS; 'SGRAVESANDE; OLDENBURG; VIGANI

JOHN S. HALL
Lowell Observatory, Flagstaff
E.C. SLIPHER

MARIE BOAS HALL
Imperial College of Science and Technology
BOYLE; DIGBY; FREIND; HARTLIB; HELLOT; HOMBERG; KUNCKEL; P. SHAW; TACHENIUS

ROBERT E. HALL
The Queen's University of Belfast
AL-KHĀZINĪ

JOHN HALLER
Harvard University
RENEVIER; TERMIER

PIERRE G. HAMAMDJIAN
University of Paris
DALIBARD

SAMI K. HAMARNEH
Smithsonian Institution
AL-MAJŪSĪ; ABU'L-ḤASAN AḤMAD IBN MUḤAMMAD AL-ṬABARĪ; ABU'L-ḤASAN ʿALI IBN SAHL RABBAN AL-ṬABARĪ; IBN AL-TILMĪDH; IBN AL-QUFF; IBN WAḤSHIYYA; AL-ZAHRĀWĪ; IBN ZUHR

WALLACE P. HAMBY, M.D.
PARÉ

J. VAN DEN HANDEL
KAMERLINGH ONNES

THOMAS L. HANKINS
University of Washington
LALANDE; P.–C. LE MONNIER; W.R. HAMILTON

OWEN HANNAWAY
Johns Hopkins University
BARCHUSEN; DAVISON; C. GLASER; GOHORY; LE FEBVRE; L. LEMERY; N. LEMERY; TURQUET DE MAYERNE

Contributors

BERT HANSEN
Fordham University
L.W. BAILEY; BAKEWELL; BARRANDE; BATHER; C. E. BERTRAND; M.A. BERTRAND; I. BOWMAN; BRONN; J. BRUCE; FÜCHSEL

KOKITI HARA
ROBERVAL

ROBERT H. HARDIE
BARNARD

NIKOLAUS M. HÄRING
Pontifical Institute of Mediaeval Studies, Toronto
THIERRY OF CHARTRES

THOMAS M. HARRIS
University of Reading
SEWARD

ROY FORBES HARROD
Royal Economic Society, London
KEYNES

RICHARD HART
National Academy of Sciences
PEASE; POOR; V.M. SLIPHER; C.A. YOUNG

R.S. HARTENBERG
Northwestern University
CULMANN; EYTELWEIN; GRASHOF; HIRN

HAROLD HARTLEY
PARTINGTON

WILLY HARTNER
AL-BATTĀNĪ

E. RUTH HARVEY
University of Toronto
QUSṬĀ IBN LŪQĀ

HELMUT HASSE
Journal für die Reine und Angewandte Mathematik
HENSEL

MELVILLE H. HATCH
University of Washington
HORN; HOWARD; SCUDDER

JAGDISH N. HATTIANGADI
BAIN

RALPH E. HAUPT
United States Naval Observatory
F.E. ROSS

THOMAS HAWKINS
Boston University
BERWICK; BURNSIDE; CALLANDREAU; LEBESGUE; SAKS

FREDERICK HEAF
Welsh National School of Medicine
LAENNEC

JEAN VAN HEIJENOORT
Brandeis University
HERBRAND

J.L. HEILBRON
University of California, Berkeley
AEPINUS; BECCARIA; BOSE; BROGLIE; CANTON; CAVALLO; DUFAY; S. GRAY; KINNERSLEY; KLEIST; MOSELEY; NOLLET; RICHMANN; SYMMER; J.J. THOMPSON; VOLTA; W. WATSON; WILCKE

ROGER HEIM
Muséum National d'Histoire Naturelle
LE DANTEC; MAIRE

FRANZ HEIN
German Academy of Sciences
GUTBIER

H.M. HEINE
University of Oxford
SENECA

KARL HEINIG
University of Berlin
SCHORLEMMER

C. DORIS HELLMAN
Queens College, City University of New York
BRAHE; DÖRFFEL; PEURBACH

STERLING B. HENDRICKS
United States Department of Agriculture
COTTRELL

JOHANNES HENIGER
State University of Utrecht
LEEUWENHOEK

DAVID HEPPELL
Royal Scottish Museum
GOODSIR; JEFFREYS

JOHN W. HERIVEL
Queen's University, Belfast
CORNU

ARMIN HERMANN
University of Stuttgart
BORN; A. HAAS; W. KUHN; KURLBAUM; LAUE; LENARD; LUMMER; SCHRÖDINGER; SOMMERFELD; STARK

HEINRICH HERMELINK
AL-JAYYĀNĪ

DIETER B. HERRMANN
Archenhold Observatory, Berlin
LINDENAU; LAMONT; MADLER; G. MÜLLER; NICOLAI; VOGEL; WILHELM IV; ZOLLNER

MAYO DYER HERSEY
BUCKINGHAM

MAXIMILIAN HERZBERGER
Louisiana State University
GULLSTRAND

MARY HESSE
F. BACON

RICHARD G. HEWLETT
United States Atomic Energy Commission
K.T. COMPTON

C.C. HEYDE
Scientific and Industrial Research Organization
BIENAYMÉ

ERWIN N. HIEBERT
Harvard University
BJERRUM; MACH; NERNST; OSTWALD

JOEL H. HILDEBRAND
University of California, Berkeley
LATIMER

DONALD R. HILL
AL-JAZARĪ

JULIAN W. HILL
The Crystal Trust
CAROTHERS

BROOKE HINDLE
Smithsonian Institution
COLDEN; FARRAR; GODFREY; GREENWOOD; RITTENHOUSE

EDWARD HINDLE
The Royal Society
J.G. KERR

H.M. HINE
University of Oxford
SENECA

ERICH HINTZSCHE
University of Bern
CALDANI; A. CORTI; HALLER; HENLE; KOELLIKER; REMAK; SOEMMERRING; VALENTIN; VESLING

TETU HIROSIGE
Nihon University
HONDA; ISHIWARA

ANN M. HIRSCH-KIRCHANSKI
University of California, Berkeley
W. GRIFFITH

E. HLAWKA
University of Vienna
RADON; TAUBER

MICHAEL E. HOARE
Australian National University
G. FORSTER; J. FORSTER

M. HOCQUETTE
MAIGE; MANGIN; TURPIN

HEBBLE E. HOFF
Baylor University
DENIS

E. DORRIT HOFFLEIT
Yale University · Maria Mitchell Observatory

M. MITCHELL; SCHLESINGER; TROUVELOT

J.E. HOFMANN
University of Tübingen
JAKOB BERNOULLI I; BRAUNMÜHL; M.B. CANTOR; CURTZE; CUSA; DYCK; GUENTHER; LEIBNIZ; SAINT VINCENT; SCHOOTEN; SUTER; TROPFKE; TSCHIRNHAUS

ERNST HÖLDER
University of Mainz
HÖLDER

HELMUT HÖLDER
University of Münster/Westphalia
ZITTEL

A. HOLLMAN
University College Hospital, London
T. LEWIS

FREDERIC L. HOLMES
University of Western Ontario
BARCROFT; DALTON; LIEBIG; RICHET; VOIT

WILLIAM T. HOLSER
University of Oregon
W. BARLOW; MALLARD; MIERS; RÍO; C.S. WEISS

OLAF HOLTEDAHL
BRØGGER

GERALD HOLTON
Harvard University
BRIDGMAN; FRANK

DORA HOOD
D. BLACK

S. HOOGERWERF
EINTHOVEN

MARJORIE HOOKER
United States Geological Survey
A. LACROIX

R. HOOYKAAS
State University of Utrecht
BEECKMAN; HAÜY; ROMÉ DE L'ISLE

HO PENG-YOKE
University of Malaya · Griffith University
CH'IN CHIU-SHAO; CHU SHIH-CHIEH; LI CHIH; LIU HUI; YANG HUI

DAVID HOPKINS
A.H. BROOKS

I.B. HOPLEY
Clifton College
LIPPMANN

BRIGITTE HOPPE
Deutches Museum
K. GAERTNER

ZDENĚK HORNOF
ŠKODA

MICHAEL A. HOSKIN
University of Cambridge
AITKEN; BIRMINGHAM; CURTIS; DAWES; C.L. HERSCHEL; W. HERSCHEL; MACLEAR; ROBERTS; T. WRIGHT

GEORGE F. HOURANI
State University of New York at Buffalo
IBN ṬUFAYL

PIERRE HUARD
René Descartes University
ASTRUC; BAILLOW; J.F. BÉRARD; BERNHEIM; BRESCHET; BROUSSAIS; CRUVEILHIER; LE DOUBLE; LIEUTAUD; LORRY; PECQUET; PEYER; RABELAIS; VICQ D'AZYR

WŁODZIMIERZ HUBICKI
Marie Curie–Skłodowska University
BOODT; CENTNERSZWER; ERCKER; KOSTANECKI; LIBAVIUS; MAIER; MARCHLEWSKI; SENDIVOGIUS; SUCHTEN; THURNEYSSER

KARL HUFBAUER
University of California, Irvine
CRELL; GREN; HENCKEL; KLAPROTH; C. NEUMANN; TROMMSDORFF; WEDEL; C.E. WEIGEL

SALLY SMITH HUGHES
University of North Carolina
BEIJERINCK

THOMAS PARKE HUGHES
University of Pennsylvania
C.M. HALL; MIDGLEY; E.W. VON SIEMENS; SPERRY

G.L. HUXLEY
Queen's University of Belfast
ANTHEMIUS OF TRALLES; AUTOLYCUS OF PITANE; BRIGGS; EUDOXUS OF CNIDUS; SOSIGENES; THEON OF SMYRNA; THEUDIUS OF MAGNESIA; THYMARIDAS

AARON J. IHDE
University of Wisconsin
ALDER; BABCOCK; ELVEHJEM; EULER–CHELPIN; FUNK; GOMBERG; HARDEN; HART; KAHLENBERG; LEVENE; WILEY; R.R. WILLIAMS

TATYANA D. ILYINA
Academy of Sciences of the U.S.S.R.
MIKLUKHO-MAKLAY

MARIE-JOSÉ IMBAULT-HUARD
René Descartes University
LE DOUBLE; LIEUTAUD; LORRY; PECQUET; PEYER; VICQ D'AZYR

ALBERT Z. ISKANDAR
The Wellcome Institute for the History of Medicine
ḤUNAYN IBN ISḤĀQ; IBN SĪNĀ; IBN AL-NAFĪS; IBN RUSHD

JEAN ITARD
Lycée Henry IV
ARBOGAST; BILLY; BOBILLIER; BOUQUET; BRET; CHUQUET; CLAIRAUT; A. GIRARD; HENRION; KRAMP; S.F. LACROIX; LAGRANGE; P.A. LAURENT; LEGENDRE; OCAGNE; J.A. RICHARD; L.P.E. RICHARD; ROLLE; SAINT–VENANT

C. DE JAGER
Astronomical Institute at Utrecht
MINNAERT

ILSE JAHN
University of Berlin
EHRENBERG

MELVIN E. JAHN
University of California, Berkeley
BERINGER

W.O. JAMES
Imperial College of Science and Technology
GOEBEL

S.A. JAYAWARDENE
Science Museum, London
BOMBELLI; FERRARI; PACIOLI

JULIAN JAYNES
Princeton University
FECHNER; G.E. MÜLLER

EDOUARD JEAUNEAU
BERNARD OF CHARTRES; BERNARD SILVESTRE

REESE V. JENKINS
Case Western Reserve University
FRAUNHOFER; TALBOT

A.C. JERMY
British Museum (Natural History)
PRESL

BØRGE JESSEN
University of Copenhagen
H. BOHR; J.L.W.V. JENSEN; PETERSEN

RICHARD I. JOHNSON
Museum of Comparative Zoology
STIMPSON

JEAN JOLIVET
École Pratique des Hautes Études
AL–KINDĪ

ROBERT JOLY
Free University of Brussels
HIPPOCRATES OF COS

CHARLES W. JONES
University of California, Berkeley
BEDE

DANIEL JONES
Oregon State University

GORE; W. GREGORY; HAHNEMANN; MATTHIESSEN; F.F. RUNGE; SMITHSON; WIELAND

PHILLIP S. JONES
University of Michigan
J.E. ARGAND; CRAMER; J.W. GLAISHER; KARPINSKI; B. TAYLOR; VANDERMONDE; WESSEL

R.V. JONES
University of Aberdeen
BOYS; F.A. LINDEMANN

DAVID JORAVSKY
Northwestern University
MICHURIN

P. JOVET
Centre National de Floristique
L'ÉCLUSE; L'OBEL; MAGNOL; PÉRON; PLUMIER; POIVRE; RUEL; THOUIN; S. VAILLANT

SHELDON JUDSON
Princeton University
BRYAN; DAVIS

GEORGE KAHLSON
University of Lund
T.L. THUNBERG

GISELA KANGRO
SALA

HANS KANGRO
University of Hamburg
GEISSLER; JUNGIUS; C. KHUNRATH; H. KHUNRATH; KIRCHER; J.H.J. MÜLLER; PLANCK; E. PRINGSHEIM; RUBENS; SENNERT; WEHNELT; WIEN

SATISH C. KAPOOR
University of Saskatchewan
BERTHOLLET; DUMAS; AUGUSTE LAURENT

ROBERT H. KARGON
Johns Hopkins University
CHARLETON; HARRIS; W. MOLYNEAUX; REYNOLDS; ROSCOE; SCHUSTER; R.A. SMITH

T.N. KARI–NIAZOV
Academy of Sciences of the U.S.S.R.
ULUGH BEG

MIROSLAV KATĚTOV
BANACH; HAUSDORFF

GEORGE B. KAUFFMAN
California State University, Fresno
BLOMSTRAND; C.E. CLAUS; DELÉPINE; FRIEND; GENTH; T. GRAHAM; GULDBERG; HOWE; JØRGENSEN; W. KOSSEL; H.G. MAGNUS; PFEIFFER; WAAGE; A. WERNER

MORRIS KAUFMAN
Rubber and Plastics Processing Industry Training Board
CHARDONNET

ALAN S. KAY
AVERY; BARBOUR; P. BECQUEREL; N. BERNARD; G. BERTRAND; BEXON; BURDACH

MARSHALL KAY
Columbia University
GRABAU; STILLE

B.M. KEDROV
Academy of Sciences of the U.S.S.R.
LENIN; LOMONOSOV; MENDELEEV

KENNETH D. KEELE
Research Fellow, Wellcome Institute
LEONARDO DA VINCI

ROBIN KEEN
Gillingham Technical High School
F. WÖHLER

BRIAN B. KELHAM
KANE; STONEY

ALEX G. KELLER
University of Leicester
GHINI; LUSITANUS; D'ORTA; C. PERRAULT; PLOT; RONDELET; G. SCHOTT; O. DE SERRES; STELLUTI

SUZANNE KELLY, O.S.B.
Carroll College · Stonehill College
BURNET; W. GILBERT; NORMAN

EDWIN C. KEMBLE
Harvard University
BRIDGMAN

MARTHA B. KENDALL
Vassar College
BASSLER; BECKER; BÉGUYER DE CHANCOURTOIS; BARRELL; BEUDANT; BUCHER; C. DAWSON; FUHLROTT; LESLEY; RIVIÈRE DE PRÉCOURT

E.S. KENNEDY
Brown University
AL-BĪRŪNĪ

HUBERT C. KENNEDY
Providence College
BURALI-FORTI; CESARE; PADOA; PARSEVAL DES CHÊNES; PEANO; PIERI; POST; VAILATI

G.B. KERFERD
University of Manchester
ANTIPHON; DEMOCRITUS; LEUCIPPUS

MILTON KERKER
Clarkson College of Technology
CLAPEYRON; PAMBOUR; ZSIGMONDY

GÜNTHER KERSTEIN
BERGIUS; BOSCH; CRAMER; GEHLEN; HERMBSTAEDT; WIEGLEB

DANIEL J. KEVLES
California Institute of Technology
J.S. AMES; BARUS; D.C. MILLER; MILLIKAN; NICHOLS; ROOD; ROWLAND; W.C.W. SABINE

GEOFFREY KEYNES
T. BROWNE

PEARL KIBRE
Hunter College, City University of New York
ADAM OF BODENSTEIN; BERNARD OF TREVISAN; PETRUS BONUS; THOMAS OF CANTIMPRÉ

JOHN S. KIEFFER
St. John's College, Annapolis
ANATOLIUS OF ALEXANDRIA; ATHENAEUS OF ATTALIA; CALLIPPUS

C.G. KING
Columbia University
WU

DAVID A. KING
Smithsonian Institution Project in Medieval Islamic Astronomy, Cairo · New York University
AL-KHALĪLĪ; IBN AL-SHĀṬIR; IBN YŪNUS

H.C. KING
Royal Ontario Museum
H. GRUBB; T. GRUBB; HADLEY; LASSELL

LAWRENCE J. KING
State University of New York at Geneseo
C.K. SPRENGEL; A. WALKER

LESTER S. KING
American Medical Association
BRIGHT; STAHL

STEFAN J. KIRCHANSKI
University of California, Berkeley
W. GRIFFITH

PAUL A. KIRCHVOGEL
Landesmuseum, Kassel
BRAMER; FAULHABER; REPSOLD FAMILY

GEORGE KISH
University of Michigan
JOSÉ DE ACOSTA; APIAN; CORONELLI; G. DELISLE; DUDLEY; GEMMA FRISIUS; GERMANUS; MAURO; G. MERCATOR; MÜNSTER; A.E. NORDENSKIÖLD; WALDSEEMÜLLER

MARC KLEIN
Louis Pasteur University · Faculté de Médecine, Strasbourg
ANCEL; BALBIANI; BERTHOLD; BOUIN; GRAAF; LAVERAN; MOHL; OKEN; RASPAIL; SCHLEIDEN; WICKERSHEIMER

MARTIN J. KLEIN
Yale University
J. GIBBS; EHRENFEST; EINSTEIN

F. KLEIN–FRANKE
AL–TĪFASHĪ

FRIEDRICH KLEMM
Deutsches Museum
J. BECKMANN; LINDE; POGGENDORFF; VALTURIO

HENRY S. VAN KLOOSTER
Rensselaer Polytechnic Institute
F.M. JAEGER

BRONISLAW KNASTER
BROŻEK; JANISZEWSKI; MASURKIEWICZ; ZARANKIEWICZ

DAVID KNIGHT
University of Durham
A.C. BECQUEREL; BEDDOES; H. DAVY; DERHAM; GIRTANNER; SCHONLAND

OLE KNUDSEN
University of Aarhus
A.O. RANKINE

HIDEO KOBAYASHI
Hokkaido University Shimane University
KOTŌ; B.S. LYMAN; SATŌ FAMILY

AKIRA KOBORI
Sangyo University, Kyoto
SEKI; TSU CH'UNG-CHIH

MANFRED KOCH
Bergbau Bücherei, Essen
CANCRIN; DECHEN; C.F.A. HARTMANN; HECHT; HEYNITZ; JUSTI; KARSTEN; LÖHNEYSS; TREBRA

HULDRYCH M. KOELBING
A.Q. BACHMANN

ROBERT E. KOHLER
University of Pennsylvania
ADAMS; KENDALL; G.N. LEWIS

KENKICHIRO KOIZUMI
University of Pennsylvania
DAVISSON

L. KOLDITZ
Humboldt University of Berlin
ROSENHEIM

ZDENĚK KOPAL
University of Manchester
BEER; BICKERTON; G. DARWIN; GOODRICKE; P.A. HANSEN; MICHELL; E. PIGOTT; N. PIGOTT; RHEITA; O.C. RÖMER; SPENCER JONES

ELAINE KOPPELMANN
Goucher College
CHASLES; D.F. GREGORY; M. LÉVY; MANNHEIM; MOUTARD; PEACOCK; PLATEAU; WOODHOUSE

SHELDON J. KOPPERL
Grand Valley State College
GADOLIN; W.N. HAWORTH; KHARASCH; KIPPING; KRAUS; M.J.L. LE BLANC; MOSANDER; T.W. RICHARDS; ROWE; SCHRÖTTER; SERULLAS; T.E. THORPE; URBAIN; WILHELMY

HANS-GÜNTHER KÖRBER
Zentralbibliothek des Meteorologischen Dienstes, Potsdam
GUERTLER; HALLWACHS; W.G. HANKEL; HOLBORN; JOLLY; A. KÖNIG; KUNDT; OLSZEWSKI; C.W.W. OSTWALD; F.W. OSTWALD; PULFRICH; O.F. SCHOTT; SIEDENTOPF; WIEDEMANN; L.C. WIENER; WRÓBLEWSKI

T.W. KORZYBSKI
Polish Academy of Sciences
PARNAS

A.N. KOST
University of Moscow
CHICHIBABIN

J. KOVALEVSKY
Bureau des Longitudes
DELAUNAY; FAYE

FRITZ KRAFFT
University of Mainz
GUERICKE

EDNA E. KRAMER
Polytechnic Institute of New York
AGNESI; GERMAIN; HYPATIA; S. KOVALEVSKY; A.E. NOETHER; M. NOETHER; SOMMERVILLE

FREDERICK KREILING
Polytechnic Institute of New York
LEIBNIZ

JAN KREJČI
Purkyne University
STERNBERG

CLAUDIA KREN
University of Missouri
AILLY; ALAIN DE LILLE; BERNARD OF VERDUN; DOMINICUS DE CLAVASIO; GUNDISSALINUS; HERMANN THE LAME; ROGER OF HEREFORD

A.D. KRIKORIAN
State University of New York at Stony Brook
F. GREGORY; HOAGLAND; LIVINGSTON; F.J. RICHARDS

VLADISLAV KRUTA
Purkyně University, Brno
CZERMAK; DUTROCHET; EDWARDS; FLOURENS; J. GMELIN; GRUBY; R.P.H. HEIDENHAIN; HERING; LEGALLOIS; J.G. MENDEL; PLENČIČ; PROCHÁSKA; PROWAZEK; PURKYNĚ; REICHERT; A.A. RETZIUS; RUDOLPHI; TEICHMANN; TIEDEMANN; UNZER; WAGNER; E.H. WEBER

DAVID KUBRIN
Liberation School, San Francisco
JOHN KEILL

FRIDOLF KUDLIEN
University of Kiel
AËTIUS OF AMIDA; AGATHINUS; ALCMAEON OF CROTONA; ALEXANDER OF TRALLES; ARETAEUS OF CAPPADOCIA; CELSUS; GALEN; ORIBASIUS; PHILINUS OF COS; RUFUS OF EPHESUS

H.G. KUHN
University of Oxford
J. FRANCK

EMIL KUHN-SCHNYDER
University of Zurich
GAGNEBIN

P.G. KULIKOVSKY
Moscow University
BELOPOLSKY; BLAZHKO; BREDIKHIN; GERASIMOVICH; HANSKY; IDELSON; KAVRAYSKY; KOSTINSKY; KOVALSKY; MAKSUTOV; MOISEEV; NEUYMIN; NUMEROV; A.Y. ORLOV; S.V. ORLOV; PARENAGO; RUMOVSKY; SHARONOV; SHAYN; SHIRAKATSI; P.K. STERNBERG; SUBBOTIN; TIKHOV; TSERASKY

NAOITI KUMAGAI
MATUYAMA

PAUL KUNITZSCH
University of Munich
IBN QUTAYBA; AL-ṬŪSĪ

KAZIMIERZ KURATOWSKI
Polish Academy of Sciences
SIERPINSKI

G.D. KUROCHKIN
Academy of Sciences of the U.S.S.R.
CHERNYSHEV; SEVERGIN

LOUIS I. KUSLAN
Southern Connecticut State College
BERTHIER; BÖTTGER; FAVRE; NORTON; RAOULT

GISELA KUTZBACH
University of Wisconsin
BUCHANAN; HANN; J. LECONTE; LOOMIS; MARGULES; ROSSBY; W.N. SHAW

V.I. KUZNETSOV
Academy of Sciences of the U.S.S.R.
BALANDIN; FAVORSKY; IPATIEV; KONDAKOV; VAGNER

RODOLPHINE J. CH. V. TER LAAGE
State University of Utrecht
DONDERS

YVES LAISSUS
Muséum National d'Histoire Naturelle
D'ARGENVILLE; CELS; COMMERSON; LA CONDAMINE; J. DE JUSSIEU; J. MARCHANT; N. MARCHANT

Contributors

I.M. LAMB
Harvard University
THAXTER

BENGT–OLOF LANDIN
University of Lund
CLERCK; FABRICIUS; C. GEER; GYLLENHAAL; PAYKULL; SCHÖNHERR

M.G. LAROSHEVSKY
Academy of Sciences of the U.S.S.R.
SECHENOV

LAURENS LAUDAN
University of Pittsburgh
APELT; COMTE; FERGUSON

DELFINO LAURI
PAGANO

P.S. LAURIE
Royal Greenwich Observatory
ELLIS; EVERSHED; HIND; J.C. ROSS

JOHN LAW
University of Keele
BRAGG

GEORGE H.M. LAWRENCE
Hunt Institute for Botanical Documentation
F. NYLANDER; W. NYLANDER

EDWIN LAYTON
University of Minnesota
F.W. TAYLOR

CHAUNCEY D. LEAKE
University of California, San Francisco
SANTORINI

WILLIAM LEFANU
Royal College of Surgeons
CLIFT; HAVERS; HEWSON; HOME; KEITH; MAYO; T. WHARTON

GORDON LEFF
University of York
DUNS SCOTUS

HENRY M. LEICESTER
University of the Pacific
BERZELIUS; CANNIZZARO; W.M. CLARK; DEVILLE; FISCHER; FITTIG; FOLIN; G.H. HESS; H.C. JONES; KARRER: KOLBE; KOPP; S.V. LEBEDEV; LE BEL; LE CHÂTELIER; MUIR; P. MÜLLER; SØRENSEN; VIRTANEN; WALLACH; WINDAUS

CZESŁAW LEJEWSKI
University of Manchester
LEŚNIEWSKI

RICHARD LEMAY
The Graduate School and University Center of the City University of New York
GERARD OF CREMONA

DONALD J. LE ROY
National Research Council of Canada
W.L. MILLER

JEAN F. LEROY
Muséum National d'Histoire Naturelle
BOSC; A.-T. BRONGNIART; TOURNEFORT

JOHN E. LESCH
Princeton University
ROMANES

ERNA LESKY
University of Vienna
BRÜCKE; WAGNER VON JAUREGG; WERTHEIM

MARTIN LEVEY
ABRAHAM BAR ḤIYYA; ABŪ KĀMIL; IBN EZRA; AL–SAMARQANDĪ; IBRĀHĪM IBN YAʿQŪB

JACQUES R. LÉVY
Paris Observatory
BIGOURDAN; F.A. CLAUDE; COSSERAT; ESCLANGON; L. FABRY; GAILLOT; P. GAUTIER; HAMY; P.M. HENRY; P.P. HENRY; JANSSEN; LE VERRIER; LYOT; MINEUR; PERROTIN; RAYET; ROCHE; STEPHAN; THOLLON; TISSERAND; C.J.E. WOLF

E.B. LEWIS
California Institute of Technology
STURTEVANT

O.A. LEZHNEVA
Academy of Sciences of the U.S.S.R.
EICHENWALD; LENZ; V.V. PETROV

JOHN H. LIENHARD
University of Kentucky
NUSSELT; PRANDTL

CAMILLE LIMOGES
University of Montreal
DAUBENTON; E. PERRIER; PLUCHE; QUATREFAGES DE BRÉAU

DAVID C. LINDBERG
University of Wisconsin-Madison
FLEISCHER; PECHAM; RISNER; WITELO

G.A. LINDEBOOM
Free University, Amsterdam
BOERHAAVE; CAMPER; EIJKMAN; GRIJNS; HORNE; RUYSCH; SPIEGEL; F.D.B. SYLVIUS

STEN LINDROTH
University of Uppsala
BROMELL; CELSIUS; LINNAEUS; N.E. NORDENSKIÖLD; OLAUS MAGNUS; RUDBECK; STELLER; SWEDENBORG; WARGENTIN

R.B. LINDSAY
Brown University
C.G. DARWIN; HARTREE; RICHTMYER; J.W. STRUTT; SWANN; R.W. WOOD

A.C. LLOYD
University of Liverpool
IAMBLICHUS; PLOTINUS

DAVID P.C. LLOYD
The Rockefeller University
GASSER

J.A. LOHNE
Municipal Gymnasium, Flekkefjord
HARRIOT

ALBERT G. LONG
Hancock Museum, University of Newcastle Upon Tyne
W.H. LANG; WITHAM

ESMOND R. LONG
University of Pennsylvania
COUNCILMAN; H.G. WELLS

JAMES LONGRIGG
University of Newcastle Upon Tyne
ANAXAGORAS; ERASISTRATUS; HEROPHILUS; NICOLAUS OF DAMASCUS; PRAXAGORAS OF COS; THALES

J. M. LÓPEZ DE AZCONA
Comisión Nacional de Geologia, Madrid
CORTÉS DE ALBACAR; IBÁÑEZ; JUAN Y SANTACILLA; MEDINA; NUÑEZ SALACIENSE; RIBEIRO SANTOS

EDGAR R. LORCH
Columbia University
RITT

JACOB LORCH
Hebrew University
BODENHEIMER

RICHARD P. LORCH
University of Manchester
JĀBIR IBN AFLAḤ

MARIO LORIA
GIORGI

FLOYD G. LOUNSBURY
Yale University
MAYA NUMERATION, COMPUTATION, AND CALENDRICAL ASTRONOMY

D.J. LOVELL
Massachusetts College of Optometry
COBLENZ

EDWARD LURIE
University of Delaware
J.L.R. AGASSIZ; MARCOU

EDYTHE LUTZKER
HAFFKINE

AVERIL M. LYSAGHT
S. PARKINSON

MADELEINE LY–TIO–FANE
Sugar Industry Research Institute, Mauritius
SONNERAT

A.J. McCONNELL
University of Dublin
SALMON

RUSSELL McCORMMACH
Johns Hopkins University
CAVENDISH; HERTZ; LORENTZ; J.W. NICHOLSON

JOHN B. McDIARMID
University of Washington
THEOPHRASTUS

ERIC McDONALD
ADET; BAUMÉ; BAYER; BUCQUET; D'ARCET

MARVIN W. McFARLAND
Library of Congress
LANCHESTER; PICCARD; WRIGHT BROTHERS

A.G. MacGREGOR
E.M. ANDERSON; E.B. BAILEY

WILLIAM McGUCKEN
University of Akron
J. SCHEINER

SUSAN M.P. McKENNA
University of Michigan
BRINKLEY; BRISBANE; DOWNING

LUDOLF VON MACKENSON
Astronomisch–Physikalisches Kabinett
A. WÖHLER

ROBERT M. McKEON
Babson College
ANTHELME; AUZOUT; BOCHART DE SARON; COURTIVRON; LE FÈVRE; J. LEMAIRE; P. LEMAIRE; NAVIER; PRONY; VERNIER

DUNCAN McKIE
University of Cambridge
E.D. CLARKE; W.H. MILLER; L.J. SPENCER; TUTTON

H. LEWIS McKINNEY
University of Kansas
BATES; BLYTH; W. LAWRENCE; JOSEPH LECONTE; F. MÜLLER; A.R. WALLACE

VICTOR A. McKUSICK
Johns Hopkins Hospital
BRÖDEL; CHAUVEAU; SUTTON

PATRICIA P. MacLACHLAN
College of DuPage
W. JOHNSON; PAPIN

SAUNDERS MAC LANE
University of Chicago
VEBLEN; E.B. WILSON

PATRICK J. McLAUGHLIN
St. Patrick's College, Maynooth
CALLAN

ROY MacLEOD
University of Sussex
TYNDALL

NORA F. McMILLAN
Merseyside County Museums
SWAINSON

ERNAN McMULLIN
AUGUSTINE OF HIPPO; BELLARMINE

ROBERT J. McRAE
Eastern Montana College
RITTER

MICHAEL McVAUGH
University of North Carolina
ARNALD OF VILLANOVA; CONSTANTINE THE AFRICAN; FREDERIC II; RUFINUS; RUSCELLI

ROGERS McVAUGH
University of Michigan Herbarium
MOCIÑO; PALMER; SESSÉ Y LACASTA

FRANCIS R. MADDISON
Museum of the History of Science, Oxford
DONDI

KARL MÄGDEFRAU
University of Munich
CAMERARIUS; CESALPINO; SANIO

WILHELM MAGNUS
New York University
HELLINGER

MUHSIN MAHDI
Harvard University
AL-FĀRĀBĪ

EDWARD P. MAHONEY
Duke University
NIFO

MICHAEL S. MAHONEY
Princeton University
AMSLER; DESCARTES; FERMAT; GOLDBACH; HERO OF ALEXANDRIA; LE POIVRE; MARIOTTE; NICERON; RAMUS; J.K.F. ROSENBERGER; SAURIN; STAMPIOEN

CLIFFORD L. MAIER
Wayne State University
ÅNGSTRÖM; BALMER; RYDBERG

E.H. MAJER
BÁRÁNY

CARLTON MALEY
Wayne State University
P.E. SABINE

JOSEPH M. MALINE
COPE

J.C. MALLET
Centre National de Floristique
L'ÉCLUSE; L'OBEL; MAGNOL; PÉRON; PLUMIER; POIVRE; RUEL

M. MALLET
Centre National de Floristique
THOUIN; S. VAILLANT

ERNST M. MANASSE
North Carolina Central University
SPEUSIPPUS

S. MANDELBROJT
Collège de France
HADAMARD

JEROME H. MANHEIM
California State University, Long Beach
I. FUCHS; J.H. PRATT

NIKOLAUS MANI
University of Bonn
BERT; LANGERHANS; SUDHOFF

FREDERICK GEORGE MANN
University Chemical Laboratory, Cambridge
MILLS; POPE

AUGUSTO MARINONI
LEONARDO DA VINCI

F. MARKGRAF
University of Zurich Botanical Garden
MARKGRAF

ŽELJKO MARKOVIĆ
BOŠKOVIĆ

YVES MARQUET
University of Dakar
IKHWĀN AL-ṢAFĀʾ

BRIAN G. MARSDEN
Smithsonian Astrophysical Observatory
CHANDLER; CROMMELIN; FROST; B. GOULD; D. KIRKWOOD; LOWELL; J.H. MOORE; NEWCOMB; W.H. PICKERING; ST. JOHN

DANIEL MARTIN
University of Glasgow
WHITTAKER

L. MARTON
National Bureau of Standards
EÖTVÖS

A. HUGHLETT MASON
University of Virginia
MASON

ARNALDO MASOTTI
Polytechnic of Milan
FERRO; MAUROLICO; M. RICCI; O. RICCI; TARTAGLIA

DANIEL MASSIGNON
E. BAUER

KIRTLEY F. MATHER
Harvard University
CHAMBERLIN; DALY; S.F. EMMONS; FOERSTE; LINDGREN; MERRILL; SALISBURY; VAN HISE

M.V. MATHEW
Royal Botanic Garden Library, Edinburgh
HENSLOW

Contributors

OTAKAR MATOUŠEK
Charles University, Prague
RÁDL

KURT MAUEL
Verein Deutscher Ingenieure
F.R.H.C. MOLL

ALEXANDER P.D. MAURELATOS
University of Texas
EMPEDOCLES OF ACRAGAS

SEYMOUR MAUSKOPF
Duke University
BALARD; BAUDRIMONT; BRACONNOT; PROUST

KENNETH O. MAY
University of Toronto
APPELL; E.T. BELL; E. BOREL; GAUSS

JOSEF MAYERHÖFER
BIELA; BÜRG; HASENÖHRL; V.F. HESS

ERNST MAYR
Harvard University
RIDGWAY; SARS; SEMPER; WHITMAN

OTTO MAYR
Smithsonian Institution
LECORNU; H. LORENZ; C.O. MOHR; MOLLIER; MÜLLER-BRESLAU; PERRONET; R.-P. PICTET; PITOT; REDTENBACHER; G.F. VON REICHENBACH; REULEAUX; STODOLA; ZEUNER

A. J. MEADOWS
University of Leicester
NEWALL; PRITCHARD; RANYARD

JAGDISH MEHRA
Université Libre de Bruxelles
MIE; RAMAN

EVERETT MENDELSOHN
Harvard University
H.P. BOWDITCH

KURT MENDELSSOHN
University of Oxford
SIMON

ROBERT M. MENGEL
University of Kansas
AUDUBON

A.A. MENIAILOV
Academy of Sciences of the U.S.S.R.
FERSMAN; FYODOROV; LEVINSON-LESSING

N.M. MERKOULOVA
Academy of Sciences of the U.S.S.R.
HUGONIOT

PHILIP MERLAN
Scripps College
ALEXANDER OF APHRODISIAS; AMMONIUS; BRYSON OF HERACLEA

DANIEL MERRIMAN
Yale University
HJORT; M'INTOSH; WALLICH

ROBERT K. MERTON
Columbia University
SARTON

HERBERT MESCHKOWSKI
Free University, Berlin
G. CANTOR

JEAN MESNARD
University of Paris
NOEL

CHARLES R. METCALF
Royal Botanic Gardens, Kew
GREW

R. MICHARD
Paris Observatory
DESLANDRES

MARKWART MICHLER
University of Giessen
SORANUS OF EPHESUS

W.E.K. MIDDLETON
University of British Columbia
BOUGUER; J. GLAISHER

MIKLÓS MIKOLÁS
Technical University of Budapest
FEJÉR

S.R. MIKULINSKY
Institute of the History of Science, Moscow
BOLOTOV; DYADKOVSKY; MOROZOV; PIROGOV; ROUILLIER

WYNDHAM D. MILES
National Institutes of Health
BANCROFT; GREEN; HARE; E.F. SMITH

DONALD G. MILLER
Lawrence Radiation Laboratory
DUHEM

G.H. MILLER
Edinboro State College
BLICHFELDT

GENEVIEVE MILLER
Case Western Reserve University
G.N. STEWART

ERIC L. MILLS
Dalhousie University
STEBBING

LORENZO MINIO-PALUELLO
University of Oxford
ABAILARD; ARISTOTLE; BOETHIUS; JAMES OF VENICE; MICHAEL SCOT; MOERBEKE; PLATO OF TIVOLI

MARCEL MINNAERT
HOEK; HORTENSIUS; JULIUS; KAISER; PANNEKOEK; STEVIN

SAMUEL I. MINTZ
City College of the City University of New York
HOBBES

E. MIRZOYAN
Academy of Sciences of the U.S.S.R.
SEVERTSOF

MICHAEL E. MITCHELL
University College, Galway
GLEICHEN-RUSSWORM; W. H. HARVEY

JÜRGEN MITTELSTRASS
University of Konstanz
LEIBNIZ

KR. PEDER MOESGAARD
University of Aarhus
C. HORREBOW

A.G. MOLLAND
University of Aberdeen
JOHN OF DUMBLETON

A.M. MONNIER
University of Paris
DEVAUX; ERLANGER; FIESSINGER; LAPICQUE; MACHEBOEUF; PEZARD; PORTIER; ROSENBLUETH; A. SABATIER

A.M. MONSEIGNY
Muséum National d'Histoire Naturelle
E.A.I.H. LARTET; L. LARTET

GIUSEPPE MONTALENTI
University of Rome
VALLISNIERI

ERNEST A. MOODY
ALBERT OF SAXONY; BURIDAN; OCKHAM

CHARLOTTE E. MOORE
National Bureau of Standards
MEGGERS

ELLEN J. MOORE
U.S. Geological Survey · Natural History Museum, San Diego
CONRAD; E. MITCHELL; G. TROOST

J.E. MORÈRE
Institut d'Histoire des Sciences
AGUILON

WILLIAM J. MORISON
University of Louisville
G.F. WRIGHT

J.B. MORRELL
University of Bradford
T. THOMSON

GLENN R. MORROW
PROCLUS

EDGAR W. MORSE
California State College, Sonoma · University of California, Davis
BREWSTER; S.H. CHRISTIE; GASSIOT; R. SMITH; T. YOUNG

MARSTON MORSE
BIRKHOFF

GIUSEPPE MORUZZI
University of Pisa
MATTEUCCI

JEAN MOTTE
University of Montpellier
BROUSSONET; DELILE; DRAPARNAUD

DON F. MOYER
Ripon College
S.P. LANGLEY; MACCULLAGH

ANN MOZLEY
New South Wales Institute of Technology
W.B. CLARKE; RIVETT

DALE M. J. MUELLER
Texas A.&M. University
TOZZI

PIERCE C. MULLEN
Montana State University
KOFOID; RICKETTS

D. MÜLLER
University of Copenhagen
STEENSTRUP; WARMING

LETTIE S. MULTHAUF
KIRCH FAMILY; NANSEN; NIESTEN; NORWOOD; OLBERS; ROTHMANN; SAMPSON; SCHRÖTER; STRATTON; ZACH

ROBERT P. MULTHAUF
Smithsonian Institution
ROLFINCK; ZWELFER

ALIDA M. MUNTENDAM
MARUM

ARNE MÜNTZING
University of Lund
NILSSON–EHLE

GIULIO MURATORI
University of Ferrara
CANANO

JOHN E. MURDOCH
Harvard University
BRADWARDINE; BURLEY; EUCLID; SWINESHEAD

SHIGERU NAKAYAMA
University of Tokyo
ASADA GŌRYŪ; HIRAYAMA; INŌ; KIMURA; JAPANESE SCIENTIFIC THOUGHT; SHIBUKAWA; SHIZUKI

J.A. NANNFELDT
Institute for Taxonomic Botany, Uppsala
SVEDELIUS

GERALD D. NASH
University of New Mexico
WHITNEY

SEYYED HOSSEIN NASR
University of Teheran
AL–TŪSI; QUTB AL–DIN

HENRY NATHAN
OECD, Directorate for Scientific Affairs
ARNAULD; BEAUGRAND; F. BERNSTEIN; FATOU; WEDDERBURN

A. NATUCCI
University of Genoa
BUONO; G.C. FAGNANO DEI TOSCHI; G.F. FAGNANO DEI TOSCHI; GRANDI; MALFATTI; MENGOLI; J. RICCATI; V. RICCATI; VIVIANI

G. NAUMOV
Academy of Sciences of the U.S.S.R.
KROPOTKIN; MIDDENDORF; OBRUCHEV

CLIFFORD M. NELSON
University of California, Berkeley
RÜTIMEYER; ULRICH; WHITFIELD

JOHN NICHOLAS
University of Pittsburgh
N.R. CAMPBELL

J.P. NICOLAS
ADANSON

AXEL V. NIELSEN
BURRAU; DUNÉR; P.N. HORREBOW; LAU; OLUFSEN; SCHJELLERUP; STRÖMGREN; T.N. THIELE

WŁODZIMIERZ NIEMIERKO
Nencki Institute of Experimental Biology
NENCKI

W.A. NIEUWENKAMP
State University of Utrecht
C.L. VON BUCH; KRAYENHOFF; VENING MEINESZ

ALBERT NIJENHUIS
University of Pennsylvania
SCHOUTEN

H.M. NOBIS
Deutsches Museum
J. FRIES

ROBERT LAING NOBLE
University of British Columbia
COLLIP

CALVERT E. NOLAND
San Diego State University
PACKARD

LOWELL E. NOLAND
University of Wisconsin
BIRGE; GUYER; JUDAY

ERIK NORIN
University of Uppsala
HEDIN

CALVERT E. NORLAND
San Diego State College
SANDERSON

J.D. NORTH
Museum of the History of Science, Oxford
R. BACON; CAYLEY; T. CHEVALLIER; CHILDREY; W.H.M. CHRISTIE; CLIFFORD; GELLIBRAND; T. HENDERSON; HEVELIUS; HORNSBY; JERRARD; MACMILLAN; MELVILL; W.A. MILLER; NASMYTH; PARSONS; T.E.R. PHILLIPS; PLASKETT; PROCTOR; RICHARD OF WALLINGFORD; HENRY CHAMBERLAINE RUSSELL; H.J.S. SMITH; SYLVESTER; TAIT; J.H.C. WHITEHEAD; F. WOLLASTON; YULE

A. NOUGARÈDE
University of Paris
MIRBEL; VAN TIEGHEM

LUBOŠ NOVÝ
Czechoslovak Academy of Sciences
BÜRGI; ČECH; CESÁRO; DICKSTEIN; P.D.G. DU BOIS–REYMOND; GUDERMANN; JONQUIÈRES; MARCI DE KRONLAND

MARY JO NYE
University of Oklahoma
A. GAUTIER; HAUTEFEUILLE; P. SABATIER; L.J. TROOST

KARL–GEORG NYHOLM
University of Uppsala
LOVÉN

WILFRIED OBERHUMMER
Austrian Academy of Sciences
JACQUIN

GORDON W. O'BRIEN
University of Minnesota
H. POWER

HERBERT H. ODOM
Sir George Williams University
PRICHARD

HERBERT OETTEL
ARONHOLD; G. CEVA; T. CEVA; JUEL; LALOUVÈRE; LINDELÖF; NIELSEN; SKOLEM

CHRISTOFFER OFTEDAHL
Technical University of Norway
J.H.L. VOGT; T. VOGT

YNGVE ÖHMAN
Stockholm Observatory
LINDBLAD

ROBERT OLBY
University of Leeds
ASTBURY; F.A. BAUER; F.L. BAUER; D. CAMPBELL; CORRENS; DILLENIUS; FEULGEN; W. FLEMMING; R. FRANKLIN; W.A.O. HERTWIG; KOELREUTER; KÖHLER; K.M.L.A. KOSSEL; MIESCHER; NAEGELI; STAUDINGER; UNGER

Contributors

PETER D. OLCH
National Institutes of Health
W.S. HALSTED

OLIVIERO M. OLIVO
University of Bologna
LEVI

RICHARD G. OLSON
University of California, Santa Cruz
LESLIE

EUGENIUSZ OLSZEWSKI
Polish Academy of Sciences
M.T. HUBER

C.D. O'MALLEY
SALOMON ALBERTI; T. BARTHOLIN; BERENGARIO DA CARPI; BOTALLO; CAIUS; DUBOIS; EDWARDES; EUSTACHI; FALLOPPIO; T. GEMINUS; GUINTER; INGRASSIA; LINACRE; MASSA; NEMESIUS; VAROLIO; VESALIUS

CHARLES O'NEILL
BRADFORD

A. LEO OPPENHEIM
MAN AND NATURE IN MESOPOTAMIAN CIVILIZATION

JANE OPPENHEIMER
Bryn Mawr College
BAER; BOVERI; DOHRN; DRIESCH; R.G. HARRISON; K.W.T.R. VON HERTWIG; NICHOLAS; TENNENT

OYSTEIN ORE
NIELS HENRIK ABEL; P.G.H. BACHMANN; DIRICHLET; HOLMBOE; RAMANUJAN

V. OREL
Mendelianum, Moravian Museum
J.G. MENDEL

ALEXANDER M. OSPOVAT
Oklahoma State University
A.G. WERNER

G.E.L. OWEN
University of Oxford
ARISTOTLE

SHIN'ICHI OYA
AJIMA NAONOBU

A. PABST
University of California, Berkeley
H.A. BAUMHAUER; BREITHAUPT; RAMMELSBERG; G. ROSE

LEROY E. PAGE
Kansas State University
J. FLEMING; SCROPE

WALTER PAGEL
Wellcome Institute of the History of Medicine
ERASTUS; HELMONT; HILDEGARD OF BINGEN; PARACELSUS

GEORGE F. PAPENFUSS
University of California, Berkeley
KÜTZING; KYLIN

A. PAPLAUSCAS
Academy of Sciences of the U.S.S.R.
EGOROV; LUZIN; URYSON

JOHN PARASCANDOLA
University of Wisconsin
L.J. HENDERSON; Y. HENDERSON; J.U. LLOYD; MELTZER; NOVY; F.B. POWER; VAN SLYKE

FRANKLIN PARKER
West Virginia University
L. LOEB; PEARL; A.E. WRIGHT

RICHARD A. PARKER
Brown University
EGYPTIAN ASTRONOMY, ASTROLOGY, AND CALENDRICAL RECKONING

E.M. PARKINSON
Worcester Polytechnic Institute
W.J.M. RANKINE; STOKES

JOHN PASSMORE
Australian National University
HUME

ELIZABETH C. PATTERSON
Albertus Magnus College
SOMMERVILLE

H.W. PAUL
University of Florida
MEYERSON

LINUS PAULING
Stanford University
DICKINSON; A.A. NOYES

JACQUES PAYEN
École Pratique des Hautes Études
Conservatoire National des Arts et Métiers
AMAGAT; AMONTONS; BETANCOURT Y MOLINA; BION; BLONDEL; BREGUET; CAGNIARD DE LA TOUR; CAHOURS; CAILLETET; CLÉMENT; DESORMES; GRAMONT

LEONARD M. PAYNE
Royal College of Physicians, London
CROONE; MERRETT

KAI O. PEDERSEN
University of Uppsala
SVEDBERG; TISELIUS

KURT MØLLER PEDERSEN
University of Aarhus
DOVE; HANSTEEN; HENRICHSEN; KRAFT; MALUS; POISEUILLE

OLAF PEDERSEN
University of Aarhus
BRYTTE; HARPESTRAENG; JOHANNES LAURATIUS DE FUNDIS; JOHN SIMONIS OF SELANDIA; MOHN; ORTELIUS; PETER PHILOMENA OF DACIA

GIORGIO PEDROCCO
BELLANI; BONVICINO

J.D.Y. PEEL
University of Liverpool
H. SPENCER

RUDOLF PEIERLS
University of Oxford
OPPENHEIMER

MARGARET R. PEITSCH
Department of Energy, Mines and Resources, Canada
FICHOT

JEAN PELSENEER
University of Brussels
MANSION; NEUBERG; TILLY; VERHULST; WENDELIN

JON V. PEPPER
Polytechnic of the South Bank
GUNTER

ENRIQUE PÉREZ ARBELÁEZ
Jardín Botánico José Celestino Mutis, Bogotá
CALDAS; TRIANA

FRANCIS PERRIN
Académie des Sciences, Paris · Collège de France
JOLIOT; JOLIOT–CURIE; P. WEISS

G. PETIT
University of Paris
ALBERT I OF MONACO; BONAPARTE; MARION

MOGENS PIHL
University of Copenhagen
C.A. BJERKNES; V.F.K. BJERKNES

STUART PIERSON
Memorial University of Newfoundland
LEONHARDI; LYONET; MAGELLAN; J.F. MEYER; H. ROSE; TILLET; A. SÉGUIN

VICENTE R. PILAPIL
California State University, Los Angeles
COMAS SOLÁ; PÉREZ DE VARGAS; SERVETUS

P.E. PILET
University of Lausanne
BONNET; A. CANDOLLE; A.–P. CANDOLLE; CHODAT; R. COMBES; FOREL; J. GESNER; K. GESNER; J. GLASER; ERRERA; PLATTER; J.–L. PRÉVOST; N.T. DE SAUSSURE; SCHEUCHZER; SCHOPFER; SENEBIER; C. VOGT; WEPFER; YERSIN

MARTIN PINE
Queens College, City University of New York
POMPONAZZI

SHLOMO PINES
The Hebrew University
ABU'L-BARAKĀT AL-BAGHDĀDĪ; IBN BĀJJA; MAIMONIDES; AL-RĀZĪ

DAVID PINGREE
Brown University · University of Chicago
ABŪ MAᶜSHAR; ACYUTA PIṢĀRAṬI; ĀRYABHAṬA I; ĀRYABHAṬA II; BHĀSKARA I; BHĀSKARA II; BRAHMADEVA; BRAHMAGUPTA; DÁSABALA; DINAKARA; DOROTHEUS OF SIDON; AL-FAZĀRĪ; GAṆEŚA; HARIDATTA I; HARIDATTA II; IBN HIBINTĀ; HISTORY OF MATHEMATICAL ASTRONOMY IN INDIA; JAGANNĀTHA; JAYASIṂHA; KAMALĀKARA; KANAKA; KEŚAVA; KṚṢṆA; LALLA; LĀṬADEVA; LEO THE MATHEMATICIAN; MAHĀDEVA; MAHĀVĪRA; MAHENDRA SŪRI; MAKARANDA; MANILIUS; MĀSHĀᵓALLĀH; MATHURĀNĀTHA ŚARMAN; MUNĪŚVARA VIŚVARŪPA; MUÑJĀLA; NĀGEŚA; NĀRĀYAṆA NĪLAKAṆṬHA; PARAMEŚVARA; PAULIŚA; PAUL OF ALEXANDRIA; PETOSIRIS; PLANUDES; AL-QABĪṢĪ; RĀGHAVĀNANDA ŚARMAN; RAṄGANĀTHA; ŚATĀNANDA; SPHUJIDHVAJA; ŚRĪDHARA; ŚRĪPATI; ᶜUMAR IBN AL-FARRUKHĀN AL-ṬABARĪ; VARAHĀMIHIRA VATEŚVARA; VIJAYANANDA; YAᶜQUB IBN ṬĀRIQ; YATIVṚṢABHA; YAVANEŚVARA

D. ANTON PINSKER, S.J.
Archivar des Österreichischen Provinz des Jesuitenordens, Vienna
STEPLING

JACQUES PIQUEMAL
Université Paul Valéry, Montpellier
A. JORDAN

A.F. PLAKHOTNIK
Academy of Sciences of the U.S.S.R.
DERYUGIN; KNIPOVICH; LITKE; MESYATSEV; SHOKALSKY; SHTOKMAN; N.N. ZUBOV

LUCIEN PLANTEFOL
University of Paris
COSTANTIN; GUILLIERMOND; G. LAMY; L.-G. LE MONNIER; NICOT; PEYSSONNEL

A.F. PLATÉ
Academy of Sciences of the U.S.S.R.
NAMETKIN; ZELINSKY

M. PLESSNER
BAYLAK AL-QIBJĀKĪ; JĀBIR IBN ḤAYYĀN; AL-JĀḤIẒ; AL-TĪFASHĪ; ZOSIMUS OF PANOPOLIS

HOWARD PLOTKIN
University of Western Ontario
S. MOLYNEAUX; E.C. PICKERING; A. WILSON

S. PLOTKIN
Academy of Sciences of the U.S.S.R.
MAGNITSKY

DENISE MADELEINE PLOUX, S.N.J.M.
College of the Holy Names
MATRUCHOT

JESSIE POESCH
Newcomb College · Tulane University
T.R. PEALE

J.B. POGREBYSSKY
BRILL; BRIOSCHI; KOROLEV

ARTHUR W. POLLISTER
Columbia University
T.H. MONTGOMERY

GIANLUIGI PORTA
LUCIANI

ERICH POSNER
Birmingham Regional Hospital Board
DOMAGK

JAMES H. POTTER
Stevens Institute of Technology
RATEAU

EMMANUEL POULLE
École Nationale des Chartes
FINE; FUSORIS; HENRY BATE OF MALINES; JOHN OF LIGNÈRES; JOHN OF MURS; JOHN OF SAXONY; JOHN OF SICILY; RAYMOND DE MARSEILLES; SIMON DE PHARES; WILLIAM OF SAINT-CLOUD; WILLIAM THE ENGLISHMAN

YVONNE POULLE-DRIEUX
RUFFO

LORIS PREMUDA
University of Padua
ABANO; ASELLI; ASSALTI; BARTOLOTTI; CASSERIO; CORTI; MAGATI; MAGGI; MERCATI; RUINI; VALSALVA; WIELAND, M.

HANS PRESCHER
Staatliches Museum für Mineralogie und Geologie, Dresden
COTTA

DEREK J. DE SOLLA PRICE
Yale University
CHAUCER; VITRUVIUS POLLIO

PAUL H. PRICE
West Virginia Geological and Economic Survey
I.C. WHITE

R.D.F. PRING-MILL
University of Oxford
LULL

J.A. PRINS
Technological University of Delft
CLAY; COSTER; W. HAAS; KEESOM; WAALS; ZERNIKE

JOHANNES PROSKAUER
HOFMEISTER; NEES VON ESENBECK

WILLIAM B. PROVINE
Cornell University
EAST

MANUEL PUIGCERVER
University of Barcelona
CABRERA

CARROLL PURSELL
University of California, Santa Barbara
DANFORTH; J.C. MERRIAM

HANS QUERNER
Institut für Geschichte der Medizin, University of Heidelberg
GOETTE; HIS; KLEINENBERG; KÜHN; MEYEN; K.A. MÖBIUS; ROSA; M.W.C. WEBER

EUGENE RABINOWITCH
State University of New York at Albany
R. EMERSON

SAMUEL X. RADBILL
College of Physicians of Philadelphia
College of Physicians of New Jersey
DUNGLISON; GESELL; ISAACS; OTT; A. PLUMMER; PRINGLE; WHYTT

JOHN B. RAE
Harvey Mudd College
KETTERING

VARADARAJA V. RAMAN
Rochester Institute of Technology
CATALÁN; KALUZA; P.C. RAY; SAHA

PAUL RAMDOHR
University of Heidelberg
CARNALL; CREDNER; LOSSEN; ROSENBUSCH; SCHNEIDERHÖHN

HANS RAMSER
VOLKMANN; E.G. WARBURG

R.A. RANKIN
University of Glasgow
G.N. WATSON

RHODA RAPPAPORT
Vassar College
GUETTARD; MALESHERBES; MONNET; G.-F. ROUELLE; H.-M. ROUELLE; J. ROUELLE; SOULAVIE; A.-R.-J. TURGOT; É.-F. TURGOT

ROSHDI RASHED
Centre National de la Recherche Scientifique
IBRĀHĪM IBN SINĀN; KAMĀL AL-DĪN; AL-KARAJĪ; AL-KINDĪ

Contributors

P.M. RATTANSI
University of Cambridge
BASSO; BEGUIN; BORRICHIUS; CUDWORTH

ROY A. RAUSCHENBERG
Ohio University
A. HAWORTH; W. HUDSON; SOLANDER

J.R. RAVETZ
University of Leeds
FOURIER

GERHARD REGNÉLL
University of Lund
HISINGER; NATHORST

NATHAN REINGOLD
Smithsonian Institution
CLEVELAND ABBE; BACHE; L.A. BAUER; N. BOWDITCH; CATTELL; ESPY; O. GIBBS; HARKNESS; HASSLER; HAYFORD; J. HENRY; LANE; RUTHERFURD; E. SABINE; C.A. SCHOTT; R.S. WOODWARD

LADISLAO RETI
MARTINI; LEONARDO DA VINCI

SAMUEL REZNECK
Rensselaer Polytechnic Institute
EATON; FITCH; HORSFORD; HOUGHTON

P.W. RICHARDS
University College of North Wales
HEDWIG; W.P. SCHIMPER

R.A. RICHARDSON
University of Western Ontario
MACALLUM

JOHN M. RIDDLE
North Carolina State University
DIOSCORIDES

ARNULF RIEBER
Pädagogische Hochschule Bamberg der Universität Würzburg
HERBART

M. HOWARD RIENSTRA
Calvin College
G. DELLA PORTA

GUGLIELMO RIGHINI
Osservatorio Astrofisico di Arcetri
DEMBOWSKI; EMANUELLI; RESPIGHI; ROSSETTI

MARIA LUISA RIGHINI–BONELLI
Istituto e Museo di Storia della Scienza, Florence
BARANZANO; DANTI; DONATI; MIELI

L.M. DE RIJK
Filosofisch Institut, Leiden
WILLIAM OF SHERWOOD

RUTH GIENAPP RINARD
Kirkland College
GRAEBE; LANDOLT; LIEBERMANN

S. DILLON RIPLEY
Smithsonian Institution
LEVAILLANT

GUENTHER B. RISSE
University of Wisconsin–Madison
DÖLLINGER; EHRET; EICHLER; HOFFMANN; HOUSSAY; MECKEL; REIL; SCHAUDINN; M. SCHIFF; SEMMELWEIS; K.P.J. SPRENGEL; VIRCHOW; WILBRAND

PHILIP C. RITTERBUSH
Archives of Institutional Change, Washington, D.C.
LEIDY

LUCILLE B. RITVO
Albertus Magnus College
G. HARTMANN

ANDRÉ RIVIER
XENOPHANES

J.M. ROBERTSON
University of Glasgow
K. LONSDALE

ABRAHAM ROBINSON
Yale University
L'HOSPITAL; MÉRAY; MITTAG–LEFFLER; STOLZ; TOEPLITZ

GLORIA ROBINSON
Yale University School of Medicine
BASSI; JULIUS VICTOR CARUS; CHAMBERLAIN; DE BARY; GAFFKY; JENKINSON; PAULY; PFEFFER; PLATE; F.H. PRATT; I.–B. PRÉVOST; PRUDDEN; RABL; SCHNEIDER; SCHÖNLEIN; STRASBURGER; WEISMAN

BERNARD ROCHOT
École Pratique des Hautes Études
BÉRIGARD; GASSENDI

ALAN J. ROCKE
University of Wisconsin, Madison
BARGER; JACOBS

ANNE CLARK RODMAN
Johns Hopkins University
HOWELL

JOEL M. RODNEY
Widener College · Elmira College
S. CLARKE; FOLKES; PALEY

FRANCESCO RODOLICO
University of Florence
ARDUINO; COCCHI; DAINELLI; ISSEL; MARSILI; MICHELI; RISTORO D'AREZZO; ROVERETO; SCILLA; SOLDANI; TARGIONI; TOZZETTI

JACQUES ROGER
University of Paris
BOUCHER DE PERTHES; BUFFON; MONTESQUIEU; ROBINET; WHISTON

ANDREW DENNY ROGERS III
L.H. BAILEY, JR.

JOHN ROGERS
Yale University
H.D. ROGERS; W.B. ROGERS

HANS ROHRBACH
University of Mainz
FEIGL

COLIN A. RONAN
Journal of the British Astronomical Association
R. BRADLEY; COMMON; DYSON; HALLEY; PARSONS; PINGRÉ

ALFRED ROMER
A.H. BECQUEREL

VASCO RONCHI
Istituto Nazionale di Ottica
GIOVAN BATISTA AMICI

CONRAD E. RONENBERG
JOHN JACOB ABEL; E.F. ARMSTRONG; H.E. ARMSTRONG

GRETE RONGE
GEUTHER; HÖNIGSCHMID; WACKENRODER

PAUL G. ROOFE
University of Kansas
BENSLEY; COGHILL; C.J. HERRICK; C.L. HERRICK; LASHLEY

B. VAN ROOTSELAAR
State Agricultural University, Wageningen
BOLZANS; L.E.J. BROUWER; CHWISTEK; FRAENKEL; FREGE; TURING; ZERMELO

PAUL LAWRENCE ROSE
James Cook University · University of Cambridge · New York University · St. John's University, New York
DUDITH; KECKERMAN; MAGIOTTI; MAGNI; MARLIANI; MONTE; B. PEREIRA; SCALIGER; TACCOLA

EDWARD ROSEN
City University of New York
J. BAYER; COMMANDINO; COPERNICUS; JANSEN; MÄSTLIN; MAYR; NOSTRADAMUS; NOVARA; OSIANDER; REGIOMONTANUS; RHETICUS; RICHER; SCHÖNER; TARDE; VANINI

GEORGE ROSEN
Yale University School of Medicine
BEAUMONT; LUDWIG

CHARLES E. ROSENBERG
University of Pennsylvania
ATWATER; GOLDBERGER; LUSK; H.N. MARTIN

B.A. ROSENFELD
Academy of Sciences of the U.S.S.R.
AL–KĀSHĪ; AL–KHAYYĀMĪ;

LOBACHEVSKY; MALTSEV; THĀBIT IBN QURRĀ

L. ROSENFELD
Nordic Institute for Theoretical Atomic Physics, Copenhagen
NIELS BOHR; JOULE; G.R. KIRCHHOFF

FRANZ ROSENTHAL
Yale University
IBN BAṬṬŪṬA; IBN KHALDŪN

BARBARA ROSS
University of Massachusetts
MORAY

JOHN ROSS
Massachusetts Institute of Technology
J.G. KIRKWOOD

SYDNEY ROSS
Rensselaer Polytechnic Institute
FREUNDLICH; SPRING

BERNARD ROTH
Stanford University
R. WILLIS

K.E. ROTHSCHUH
University of Münster/Westphalia
J.G. BERGER; T.L.W. BISCHOFF; BOHN; E. DU BOIS-REYMOND; ENGELMANN; FICK; FREY; GOLTZ; HENSEN; H. KRONECKER; KÜHNE; LOTZE; MEISSNER; PFLÜGER; RUBNER; STANNIUS; VERWORN

HUNTER ROUSE
University of Iowa
DU BUAT; FOURNEYRON; FROUDE; P. GIRARD; REECH; WEISBACH

ALAIN ROUSSEAU
BRETONNEAU

DOROTHEA RUDNICK
Yale University
CHAMISSO

G. RUDOLPH
University of Kiel
JULIUS BERNSTEIN; M.G. RETZIUS; TRAUBE; WEIGERT

MARTIN J. RUDWICK
Free University, Amsterdam
University of Cambridge
A. BRONGNIART; CONYBEARE; L. HORNER; W. LONSDALE; H. MILLER; MURCHISON; L.-C. PRÉVOST; SEDGWICK

MICHAEL T. RYAN
University of Chicago
VILLALPANDO

EUGENIUSZ RYBKA
Jagiellonian University
HUFNAGEL

DAVID RYNIN
University of California, Berkeley
SCHLICK

A. I. SABRA
Harvard University
AL-FARGHĀNĪ; IBN AL-HAYTHAM; AL-JAWHARĪ; AL-NAYRĪZĪ

DONALD HARRY SADLER
Royal Greenwich Observatory
COMRIE; COWELL

MORRIS H. SAFFRON
New Jersey College of Medicine and Dentistry, Rutgers University
DONALDSON; DUGGAR; GAY; SALERNITAN ANATOMISTS

A.S. SAIDAN
University of Jordan
AL-BAGHDĀDĪ; KUSHYĀR; AL-NASAWĪ; AL-QALSĀDĪ; AL-UMAWĪ; AL-UGLĪDISĪ

JOSEPH SAJNER
University of Brno
APÁTHY

S. SAMBURSKY
Hebrew University
JOHN PHILOPONUS

JULIO SAMSÓ
Universidad Autónoma de Barcelona
AL-BIṬRŪJĪ; LEVI BEN GERSON; MANṢŪR IBN ᶜALĪ IBN ᶜIRĀQ

A. P. M. SANDERS
State University, Utrecht · Biohistorisch Institut der Rijksuniversität, Utrecht
GARNOT; GARREAU; MARTIUS; A. F. W. SCHIMPER; SCHOUW; SCHWENDENER; SPIX

A. T. SANDISON
University of Glasgow
RUFFER

BETTINA F. SARGEANT
E. I. du Pont de Nemours and Company
STINE

R. SATTLER
McGill University
HABERLANDT

RALPH A. SAWYER
University of Michigan
T. LYMAN

WILLIAM L. SCHAAF
Brooklyn College, City University of New York · Florida Atlantic University
BACHET DE MÉZIRIAC; DECHALES; LEURECHON; MASERES; OZANAM

SUSAN SCHACHER
Yale University
BOISBAUDRAN; BRAUNER; BUNSEN

HANS SCHADEWALDT
University of Düsseldorf
BEHRING; BILHARZ; HELLRIEGEL; HIRSZFELD; LEUCKART; R. MARTIN; NEISSER; POLLENDER

CARL SCHALÉN
University of Lund
MÖLLER; SUNDMAN

NORMAN SCHAUMBERGER
Bronx Community College, City University of New York
JESSE DOUGLAS

GUSTAV SCHERZ
STENSEN

ROBERT SCHLAPP
University of Edinburgh
CHRYSTAL

EBERHARD SCHMAUDERER
Deutsches Museum
DIELS; EMBDEN; J. FUCHS; LUNGE; SERTÜRNER

F. SCHMEIDLER
University of Munich Observatory
E. GROSSMANN; HALM; MAURER; SCHÖNFELD; SEELIGER; WILSING

CHARLES B. SCHMITT
Warburg Institute
BONAVENTURA; BORRO; CAMPANELLA; DALÉCHAMPS; MAIGNAN; PATRIZI; SCHEGK; SEVERINO; ZABARELLA

RUDOLPH SCHMITZ
University of Marburg
AGRIPPA; H. BRUNSCHWIG; E. CORDUS; V. CORDUS; HARTIG; J. HARTMANN

CECIL J. SCHNEER
University of New Hampshire
E. EMMONS; GRESSLY; KNORR; LEONARDO DA VINCI

IVO SCHNEIDER
University of Munich
NEANDER

BRUNO SCHOENEBERG
University of Hamburg
ARTIN; HECKE; C. F. KLEIN; E. LANDAU; G. LANDSBERG; LIPSCHITZ; STEINITZ; H. WEBER

ROBERT E. SCHOFIELD
Case Western Reserve University
J. PRIESTLEY; ROWNING

RONALD SCHORN
SEE

DOROTHY V. SCHRADER
AḤMAD IBN YŪSUF

HERBERT SCHRIEFERS
University of Bonn
E. BUCHNER; BUNGE

Contributors

GERALD SCHRODER
Technische Hochschule Carolo Wilhelmina
CROLLIUS

HANS HENNING SCHROTH
University of Bonn
GUDDEN

DOROTHY M. SCHULLIAN
Cornell University Library
A. BENEDETTI; COITER; COTUGNO

B. P. M. SCHULTE
Netherlands Society for the History of Medicine, Mathematics, and Exact Sciences
H. BERGER

CYNTHIA A. SCHUSTER
University of Montana
H. REICHENBACH

JOHN A. SCHUSTER
University of Leeds
ROHAULT

E. L. SCOTT
Stamford High School, Lincolnshire
T. ANDERSON; ANDREWS; BLAGDEN; BROWNRIGG; T. CLARK; COOPER; GLADSTONE; GROVE; HATCHETT; T. HENRY; W. HENRY; HOPE; KEIR; KIRWAN; R. LUBBOCK; D. MACBRIDE; NEWLANDS; G. PEARSON; D. RUTHERFORD; URE; R. WATSON; WOULFE

HAROLD W. SCOTT
Michigan State University
J. WALKER

J.F. SCOTT
St. Mary's College of Education, Middlesex
BRAIKENRIDGE; CRAIG; DELAMAIN; HEATH; MACLAURIN; OUGHTRED; WARING; WREN

T. K. SCOTT, JR.
Purdue University
PAUL OF VENICE

CHRISTOPH J. SCRIBA
University of Hamburg
BLASCHKE; BORCHARDT; CRELLE; GRASSMANN; C. G. J. JACOBI; LAMBERT; REIDEMEISTER; WALLIS; WIELEITNER

JAN SEBESTIK
Centre Universitaire International
BINET

DANIEL SEELEY
Boston University
PERRINE

BENIAMINO SEGRE
Academia Nazionale dei Lincei
SEVERI

EMILIO SEGRÈ
University of California, Berkeley
FERMI

A. SEIDENBERG
University of California, Berkeley
MASCHERONI; G. MOHR; PASCH; WILCZYNSKI

EDITH SELOW
Justus Liebig-Universität
CASSIRER; DINGLER; SCHELLING

E. M. SENCHENKOVA
Academy of Sciences of the U.S.S.R.
MAKSIMOV; NAVASHIN; PALLADIN; PRYANISHNIKOV; TIMIRYAZEV; TSVET; VORONIN

E. SENETA
Australian National University, Canberra
BIENAYMÉ

PAUL SENTEIN
University of Montpellier
BATAILLON

ROGER SERVAJEAN
Paris Observatory
FLAMMARION

JACQUES SESIANO
Brown University
DIOPHANTUS OF ALEXANDRIA

THOMAS B. SETTLE
Polytechnic Institute of Brooklyn
BORELLI

CAROL SHAMIEH
Boston University
MAANEN

A.N. SHAMIN
Institute for the History of Science and Technology, Moscow
BACH; K. S. KIRCHHOF

C. D. SHANE
University of California, Santa Cruz
HUSSEY

ROBERT S. SHANKLAND
Case Western Reserve University
A. H. COMPTON; K. R. KOENIG

HAROLD I. SHARLIN
Iowa State University
P. BARLOW; F. BRAUN

WILLIAM D. SHARPE
Seton Hall University
ISIDORE OF SEVILLE

WILLIAM R. SHEA
McGill University
C. SCHEINER

NABIL SHEHABY
Warburg Institute
ISḤĀQ IBN ḤUNAYN

I. P. SHELDON-WILLIAMS
University College, Dublin
ERIUGENA

A. G. SHENSTONE
Princeton University
R. W. LADENBURG

OSCAR S. SHEYNIN
Academy of Sciences of the U.S.S.R.
O.J.V. ANDERSON; BORTKIEWICZ; KRASOVSKY

N.P. SHIKHOBALOVA
Academy of Sciences of the U.S.S.R.
SKRYABIN

KAZUO SHIMODAIRA
AIDA YASUAKI

ELIZABETH NOBLE SHOR
Scripps Institution of Oceanography
COUES; DALL; DAVENPORT; DAY; DEAN; DEGOLYER; EIGENMANN; GABB; T. GILL; GRINNELL; GUTENBERG; G. S. HALL: D. S. JORDAN; V. L. KELLOGG; MARSH; C. H. MERRIAM; E.S. MORSE; SAY; SCHOOLCRAFT; STEJNEGER; F.B. SUMNER; VERRILL; S. WATSON; WHEELER; C. D. WHITE; WILLISTON

GEORGE G. SHOR, JR.
GUTENBERG

ROBERT R. SHROCK
Massachusetts Institute of Technology
TWENHOFEL

DANIEL M. SIEGEL
University of Wisconsin
B. STEWART

ROBERT SIEGFRIED
University of Wisconsin
J. DAVY

ROBERT H. SILLIMAN
Emory University
FRESNEL

DIANA M. SIMPKINS
Polytechnic of North London
Northwestern Polytechnic, London
JAMES BLACK; H.H. DIXON; FARMER; FRAZER; FRERE; J. GOULD; GWYNNE–VAUGHAN; R.A. HARPER; J. HUNT; KNIGHT; E. W. MACBRIDE; MALTHUS; MOFFETT; MOTTRAM; G.R.M. MURRAY; RAFFLES; ELLIOT SMITH; J.E. SMITH; S.I. SMITH; H.H. THOMAS; VAUCHER; VILMORIN

NATHAN SIVIN
Massachussetts Institute of Technology
LI SHIH–CHEN; SHEN KUA; WANG HSI–SHAN

P.N. SKATKIN
I.I. IVANOV

W.A. SMEATON
University College, London
BERGMAN; CHARDENON; COLLET–DESCOTILS; FOURCROY; C. GEOFFROY; E.F. GEOFFROY; GUYTON DE MORVEAU; MACQUER; E.J. DE MONTGOLFIER; M.J. DE MONTGOLFIER; B. PELLETIER; PILATRE DE ROZIER; SENAC; VAUQUELIN; VENEL

PIETER SMIT
Catholic University, Nijmegen
HOEVEN; HUSCHKE; KLUYVER; G. R. TREVIRANUS; L. C. TREVIRANUS; WIGAND

JOSEF SMOLKA
Czechoslovak Academy of Sciences
DIVIŠ

CYRIL STANLEY SMITH
Massachusetts Institute of Technology
BARBA; BELAIEW; BIRINGUCCIO; HUME–ROTHERY; JEFFRIES; MERICA; A. SAUVEUR; SORBY; THEOPHILUS

CHARLES P. SMYTH
Princeton University
DEBYE

I. SNAPPER
Veterans Administration Hospital, Brooklyn
HEURNE

IAN N. SNEDDON
University of Glasgow
ROUTH; SIMSON; M. STEWART

H. A. M. SNELDERS
University of Utrecht
ARRHENIUS; E. H. VON BAUMHAUER; CALLENDAR; CHENEVIX; E.J. COHEN; HILDEBRANDT; H.F. LINK; MULDER; MUNCKE; RIGHTER; ROOZEBOOM; SCHÖNBEIN; SCHROEDER VAN DER KOLK; SMITS; TEN RHYNE; TROOSTWIJK; VAN'T HOFF

E. SNORRASON
Rigshospitalet, Copenhagen
C. BARTHOLIN; GRAM; E.C. HANSEN; C.O. JENSEN; KROGH; LANGE; O.F. MÜLLER; SALOMONSEN; WINSLØW

C.P. SNOW
BERNAL

T.A. SOFIANO
Academy of Sciences of the U.S.S.R.
FRENZEL

Z.K. SOKOLOVSKAYA
Academy of Sciences of the U.S.S.R.
STRUVE FAMILY

Y.I. SOLOVIEV
Academy of Sciences of the U.S.S.R. Institute of the History of Natural Science and Engineering, Moscow
BORODIN; CHUGAEV; KABLUKOV; KONOVALOV; KURNAKOV; LUGININ; MENSHUTKIN; SHILOV; TANFILEV

FRED SOMKIN
Cornell University
J. LUBBOCK

GLENN SONNEDECKER
University of Wisconsin–Madison
H.C. WOOD

ROBERT SOULARD
Musée du Conservatoire National des Arts et Métiers
NIEPCE

HAROLD SPEERT
Columbia University
PAPANICOLAOU

PAUL SPEISER
University of Vienna
LANDSTEINER

ROBERT SPENCE
University of Kent
COCKCROFT

JAMES BROOKES SPENCER
Oregon State University
ZEEMAN

LARRY T. SPENCER
Plymouth State College
E.D. MONTGOMERY

PIERRE SPEZIALI
University of Geneva
CROUSAZ; DINI; FUBINI; GUCCIA; L'HUILLIER; MOUTON; MYDORGE; RICCI–CURBASTRO; SEGRE; J.C.F. STURM; J. TANNERY

J.W.T. SPINKS
University of Saskatchewan
STEACIE

ERNEST G. SPITTLER, S.J.
John Carroll University
HINSHELWOOD; MORLEY; PANETH

NILS SPJELDNAES
Aarhus University
DE GEER; EBEL; ESCHOLT; HAMBERG; NIEBUHR; RENAULT; E.J. SCHMIDT; SCHREIBERS; STØRMER; SVERDRUP; C.J. THOMSEN; TILAS

FRANS A. STAFLEU
State University of Utrecht
ENGLER; L'HÉRITIER DE BRUTELLE; A.H.L. DE JUSSIEU; A. DE JUSSIEU; A.-L. DE JUSSIEU; B. DE JUSSIEU; KUNTH; KUNTZE; MIQUEL; REDOUTÉ

WILLIAM H. STAHL
ARISTARCHUS OF SAMOS; CALCIDIUS; FIRMICUS MATERNUS; MACROBIUS; MARTIANUS CAPELLA

JERRY STANNARD
ALEXANDER OF MYNDOS; ALPINI; ANGUILLARA; ARCHIGENES; ASCLEPIADES; BOCK; BOLOS OF MENDES; BRUNFELS

WILLIAM STANTON
University of Pittsburgh
DANA

MARTIN S. STAUM
University of Calgary
MARGGRAF

WILLIAM T. STEARN
British Museum (Natural History)
BRITTON; R. BROWN; GERARD; LINDLEY; MEDICUS; METTENIUS; P. MILLER; MOENCH

MAX STECK
University of Munich
DÜRER

ARTHUR R. STEELE
University of Toledo
PAVÓN Y JIMÉNEZ

C.G.G.J. VAN STEENIS
University of Leiden
LOTSY; TREUB

WALLACE STEGNER
Stanford University
DUTTON; J.W. POWELL

NANCY STEPAN
Yale University
CRUZ

CURT STERN
University of California, Berkeley
WEINBERG

S.M. STERN
ISAAC ISRAELI

SHLOMO STERNBERG
Harvard University
WINTNER

JOHANNES STEUDEL
Medizinhistorisches Institut, University of Bonn
AUENBRUGGER; BEZOLD; G.W. BISCHOFF; HYRTL; J.P. MÜLLER

LLOYD G. STEVENSON
Johns Hopkins University
BANTING; MACLEOD

BERNHARD STICKER
ARGELANDER; A.J.G.F. VON AUWERS; BENZENBERG; BODE; BRANDES; BREMIKER; BRENDEL; BUSCH

FRANS STOCKMANS
University of Brussels
GOSSELET; GRAND'EURY; LIGNIER; LOHEST; RAMES; SAPORTA; ZEILLER

Contributors

R.H. STOY
Royal Observatory, Edinburgh
D. GILL

JANIS STRADINŠ
Academy of Sciences of the U.S.S.R.
GROTTHUSS; WALDEN

K. AA. STRAND
United States Naval Observatory
HERTZSPRUNG

V.T. STRINGFIELD
United States Geological Survey
MEINZER

PER STRØMHOLM
University of Oslo
HERIGONE; HYLLERAAS; TACQUET; VALERIO

DIRK J. STRUIK
Massachussetts Institute of Technology
ADRAIN; BELTRAMI; J.L.F. BERTRAND; BLAEU; F. BOLYAI; J. BOLYAI; P.-O. BONNET; CEULEN; CHRISTOFFEL; CODAZZI; COOLIDGE; DANDELIN; DARBOUX; DE GROOT; DUPIN; FANO; FRENET; GERBERT; GERGONNE; HEURAET; KORTEWEG; METIUS FAMILY; MEUSNIER DE LA PLACE; MUSSCHENBROEK; RIBAUCOUR; SACCHERI; SCHOUTE; SERRET; SNEL; VLACQ

ROGER H. STUEWER
University of Minnesota
GAMOW; MASCART; PERRIN; G.C.N. SCHMIDT

A.H. STURTEVANT
BRIDGES

CHARLES SÜSSKIND
University of California, Berkeley
APPLETON; E.H. ARMSTRONG; ARNOLD; AUSTIN; BARKHAUSEN; BORRIES; J.C. BOSE; DE FOREST; FEDDERSEN; FESSENDEN; J.A. FLEMING; W.W. HANSEN; HEAVISIDE; HULL; JEWETT; KÁRMÁN; KENNELLY; LANGMUIR; E.O. LAWRENCE; LODGE; PIERCE; PUPIN; S.P. THOMPSON; E. THOMSON; SIDGWICK

JUDITH P. SWAZEY
Boston University · Harvard University
R. MAGNUS; MARIE; EGAS MONIZ; SHERRINGTON; WALLER

KENNETH M. SWEEZEY
TESLA

LOYD S. SWENSON, JR.
University of Houston
GODDARD; GUYE; MICHELSON; O.W. RICHARDSON

NOEL M. SWERDLOW
University of Chicago
PEURBACH

EDITH DUDLEY SYLLA
North Carolina State University
BURLEY; SWINESHEAD

FERENC SZABADVÁRY
Technical University, Budapest
CLASSEN; DEBRAY; ESSON; FARKAS; N.W. FISCHER; FRESENIUS; GÖRGEY; HEVESY; IRINYI; JAHN; KITAIBEL; KJELDAHL; R. LORENZ; MARTINOVICS; MITSCHERLICH; C.F. MOHR; F. MÜLLER; A. NAUMANN; NODDACK; PÉAN DE SAINT-GILLES; PREGL; SZEBELLÉDY; SZILY; THAN; ZEMPLÉN

MANFRED E. SZABO
Concordia University
GENTZEN; SPORUS OF NICAEA

GEORGIO TABARRONI
Universities of Modena and Bologna
CAPRA; MANFREDI; MICHELINI; MONTANARI; RIGHI; ZANOTTI

CHARLES H. TALBOT
Wellcome Institute for the History of Medicine
BREDON; STEPHEN OF ANTIOCH; VIGO

G.A. TAMMANN
University of Basel
TAMMANN

LEONARDO TARÁN
Columbia University
ANAXILAUS OF LARISSA; ANAXIMANDER; ANAXIMENES OF MILETUS; ARATUS OF SOLI; ARISTYLLUS; NICOMACHUS OF GERASA; PARMENIDES OF ELEA

JULIETTE TATON
Centre International de Synthèse, Paris
D'ANVILLE; ARBOS; BLANCHARD; BRUNHES; BUACHE; MOUCHEZ; PEZENAS; J. PICARD

RENÉ TATON
École Pratique des Hautes Études
BARBIER; BOOSE; BOUR; BOURDELOT; B. BRISSON; M.J. BRISSION; CASSINIS (I,II,III,IV); CHÂTELET; CLOUET; DALENCÉ; DESARGUES; DIONIS DU SÉJOUR; FONTAINE; F. FRANÇAIS; J.F. FRANÇAIS; GALOIS; GUA DE MALVES; HACHETTE; KOENIGS; G.-P. LA HIRE; P. DE LA HIRE; LANCRET; LIOUVILLE; MARALDI FAMILY; MONGE; B. PASCAL; É. PASCAL; J. PICARD; POINSOT; PONCELET; POUILLET; PUISEUX; SERVOIS; P. TANNERY; TINSEAU D'AMONDANS

DOUGLASS W. TAYLOR
University of Otago Medical School
RAMÓN Y CAJAL; SHARPEY; SHARPEY-SCHÄFER

GEORGE TAYLOR
Royal Botanic Gardens, Kew
WILLIAM AITON; WILLIAM TOWNSEND AITON; G. BENTHAM; M.J. BERKELEY; BRITTEN; BRUNKES

KENNETH L. TAYLOR
University of Oklahoma
COTTE; DELAFOSSE; DES CLOIZEAUX; DESMAREST; DOLOMIEU; DUPERREY; LAMÉTHERIE; NECKER;

MIKULÁŠ TEICH
University of Cambridge · University of Oxford
HEYROVSKÝ; HORBACZEWSKI; SKRAUP; F. WALD

SEVIM TEKELI
Ankara University
ḤABASH AL-ḤĀSIB; AL-KHUJANDĪ; MUḤYI'L-DĪN AL-MAGHRIBĪ; PIRĪ RAIS

OWSEI TEMKIN
Johns Hopkins University
GLISSON

ANDRZEJ A. TESKE
BIALOBRZESKI; NATANSON; SMOLUCHOWSKI

ANDRÉE TÉTRY
École Pratique des Hautes Études
CAULLERY; CHABRY; J.-B. CHARCOT; J.-M. CHARCOT; CUÉNOT; DELAGE; DUBOSCQ; GIARD; R. GOLDSCHMIDT; MESNIL

ARNOLD THACKRAY
University of Pennsylvania
ACCUM; BOSTOCK; CARLISLE; DALTON; DANIELL; DAUBENY; B. HIGGINS; W. HIGGINS; W. NICHOLSON; SARTON

JEAN THÉODORIDÈS
Centre National de la Recherche Scientifique
AUDOUIN; DAVAINE; FABRE; É.-L. GEOFFROY; HALLIER; HÉRELLE; LÉGER; MAUPAS

JOACHIM THIELE
F.K.C.L. BÜCHNER

K. BRYN THOMAS
Royal Berkshire Hospital · Reading Pathological Society, Library
W. BOWMAN; BUSK; W.B. CARPENTER; JAMES DOUGLAS; SNOW; W. TURNER; WATERTON

PHILLIP DRENNON THOMAS
Wichita State University
ALCUIN OF YORK; ALFONSO EL SABIO; CASSIODORUS; NUTTALL; PAUL OF AEGINA; C.W. THOMPSON; VARRO; WALTER OF ODINGTON; T. WHITE; WIED

ROSE THOMASIAN
University of New Hampshire
MORO

RUTH D'ARCY THOMPSON
D.A.W. THOMPSON

ELIZABETH H. THOMSON
Yale University, School of Medicine
CUSHING; B. SILLIMAN, JR.

VICTOR E. THOREN
Indiana University
CASSEGRAIN; FLAMSTEED; GASCOIGNE; SEVERIN; WARD

V.V. TIKHOMIROV
Academy of Sciences of the U.S.S.R.
ABICH; ANDRUSOV; L.S. BERG; EICHWALD; FRENZEL; GOEPPERT; HELMERSON; KEYSERLING; OZERSKY; SOKOLOV

RONALD C. TOBEY
University of California, Riverside
OMALIUS D'HALLOY; VERNEUIL

HEINZ TOBIEN
University of Mainz
BEYRICH; CLOOS; DESHAYES; ERMAN; ESCHER VON DER LINTH; FREIESLEBEN; HEER; JACCARD; JAEKEL; MARGERIE; MAYER-EYMAR; C.E.H. VON MEYER; NEHRING; NEUMAYR; OPPEL; D'ORBIGNY; RECK; A.E. REUSS; F.A. REUSS; F. ROEMER; F.A. ROEMER; K.F. SCHIMPER; STUDER

RUTH TODD
Smithsonian Institution
CUSHMAN

CHRISTOPHER TOLL
University of Uppsala
AL-HAMDĀNĪ

GERALD JAMES TOOMER
Brown University
APOLLONIUS OF PERGA; CAMPANUS OF NOVARA; DIOCLES; HERACLIDES PONTICUS; HIPPARCHUS; AL-KHWĀRIZMĪ; METON; NICOMEDES; PTOLEMY; THEON OF ALEXANDRIA; VITRUVIUS POLLIO

CAROLYN TOROSIAN
University of California, Berkeley
SHERARD

ANDRZEJ TRAUTMAN
University of Warsaw
INFELD

THADDEUS J. TRENN
University of Regensburg
GEIGER; GIESEL; W. RAMSAY; SMITHELLS; SODDY; R.J. STRUITT; SUTHERLAND; J.S.E. TOWNSEND; TRAVERS; TROWBRIDGE; WHYTLAW-GRAY; S. YOUNG

F.G. TRICOMI
Academy of Sciences of Turin
PINCHERLE; PLANA; VERONESE; VITALI

D.N. TRIFONOV
Academy of Sciences of the U.S.S.R.
CHERNYAEV; LEBEDINSKY

VICTOR A. TRIOLO
Temple University
J. EWING; FINSEN

HENRY S. TROPP
Humboldt State University · Smithsonian Institution · University of Toronto
ECKERT; FIELDS; GOMPERTZ; GOURSAT; G.B. HALSTED; MOULTON; RADÓ; J.W. YOUNG

A.J. TURNER
MILLINGTON

G. L'E. TURNER
University of Oxford · Museum of the History of Science, Oxford
H. BAKER; DOLLOND; J.J. LISTER; B. MARTIN; MORLAND; NOBERT; RÖNTGEN; SHORT; SOLEIL; B. WILSON; C.T.R. WILSON; J. WINTHROP

R. STEVEN TURNER
University of New Brunswick
HELMHOLTZ; J.R. MAYER

SHERWOOD D. TUTTLE
University of Iowa
KAY

G. UBAGHS
University of Liège
FRAIPONT; L.-G. DE KONINCK; SCHMERLING

T. VON UEXKÜLL
University of Ulm
UEXKÜLL

CAROL URNESS
University of Minnesota
PENNANT

GEORG USCHMANN
Deutsche Akademie der Naturforscher Leopoldina · University of Jena
HAECKEL; HATSCHEK; KRAUSE; A. LANG; SEMON

F.M. VALADEZ
University of California, Los Angeles
JAMES KEILL

H.L. VANDERLINDEN
University of Ghent
DELPORTE

PETER W. VAN DER PAS
BERNARD ALBINUS; BERNARD SIEGFRIED ALBINUS; CHRISTIAN B. ALBINUS; FREDERIK B. ALBINUS; BIDLOO; GOEDAERT; HEISTER; INGEN-HOUSZ; JOBLOT; KAEMPFER; KNUTH; KUENEN; OUDEMANS; PISO; SURINGAR; SWIETEN; TULP; VRIES

B.L. VAN DER WAERDEN
University of Zurich
MATHEMATICS AND ASTRONOMY IN MESOPOTAMIA

GERALD R. VAN HECKE
Harvey Mudd College
COLLIE; W.A. NOYES; TILDEN

F. VAN STEENBERGHEN
University of Louvain
SIGER OF BRABANT

ARAM VARTANIAN
New York University
D'HOLBACH; LA METTRIE

FRANCIS E. VAUGHAN
VEMCO Corporation
LAWSON

STIG VEIBEL
Technical University of Denmark
BRØNSTED; H.P.J.J. THOMSEN; ZEISE

L. VEKERDI
Library, Hungarian Academy of Sciences
KÜRSCHÁK

G. VERBEKE
University of Louvain
SIMPLICIUS; THEMISTIUS

J.-J. VERDONK
BUTEO; CHRISTMANN; DASYPODIUS; FINK; LA ROCHE; PELETIER

JUAN VERNET
University of Barcelona
ᶜABBĀS IBN FIRNĀS; ABU'L-FIDĀ˒; ABŪ ḤĀMID; IBN AL-ᶜAWWĀM; AL-BAKRĪ; IBN AL-BANNĀ˒; IBN AL-BAYṬĀR BEHAIM; CARAMUEL; CAVANILLES; AL-DAMĪRĪ; IBN HAWQAL; HERNÁNDEZ; IBN JULJUL; AL-KHUWĀRIZMĪ; AL-MAJRĪṬĪ; MARTÍ FRANQUÉS; MELLO; MUTIS Y BOSSIO; REY PASTOR; JACOB BEN MACHIR IBN TIBBON; MOSES BEN SAMUEL IBN TIBBON; TORRES QUEVEDO; ULLOA Y DE LA TORRE GIRAL; IBN WAFĪD; YAḤYĀ IBN ABĪ MANṢŪR; AL-ZARQĀLI

GRAZIELLA FEDERICI VESCOVINI
University of Turin
FRANCIS OF MARCHIA; FRANCIS OF MEYRONNES; MARSILIUS OF INGHEN

THÉODORE VETTER
Société Française d'Histoire de la Médecine
LE CAT

Contributors

HUBERT BRADFORD VICKERY
Connecticut Agricultural Experiment Station
CHITTENDEN; L.B. MENDEL; OSBORNE

G. VIENNOT-BOURGIN
Institut National Agronomique
TULASNE

JEAN VIEUCHANGE
Institut Pasteur
BORDET

MAURICE B. VISSCHER
University of Minnesota
CARLSON

KURT VOGEL
University of Munich
ARISTAEUS; DIOPHANTUS; FIBONACCI; JOHN OF GMUNDEN; KÖBEL; MONTUCLA; RIES; RUDOLFF; STIFEL; WIDMAN; WITTICH; WOEPCKE

T.M. VOGELSANG
G.H.A. HANSEN

FRED W. VOGET
Southern Illinois University
BOAS

ALEKSANDR VOLODARSKY
Institute for the History of Science and Technology, Moscow · Academy of Sciences of the U.S.S.R.
BUGAEV; FUSS; GRAVE; K.S. KIRCHHOF

E. VOLTERRA
University of Texas
VOLTERRA

C.H. WADDINGTON
University of Edinburgh
SPEMANN

GERHARD WAGENITZ
Systematisch-Geobotanisches Institut, Berne
GRISEBACH

EARL WALKER
Johns Hopkins Hospital
FULTON

WILLIAM A. WALLACE, O.P.
Catholic University of America
ALBERTUS MAGNUS; AQUINAS; BERNARD OF LE TREILLE; BORGOGNONI OF LUCCA; BUONAMICI; CELAYA; CIRUELO; CORONEL; DIETRICH VON FREIBERG; DULLAERT OF GHENT; GERARD OF SILTEO; GILES OF LESSINES; LAX; MAIOR; SOTO; THOMAZ; ULRICH OF STRASBOURG; VINCENT OF BEAUVAIS; WILLIAM OF AUVERGNE

HELEN WALLIS
British Museum
HAKLUYT

P.J. WALLIS
University of Newcastle Upon Tyne
GREEN; PELL; SIMPSON; STIRLING; E. WRIGHT

ANTHONY A. WALSH
Dickinson College
SPURZHEIM

J.L. WALSH
OSGOOD

R. WALZER
University of Oxford
MARINUS

C.W. WARDLAW
D.H. SCOTT

JOHN WARD-PERKINS
VITRUVIUS POLLIO

E.H. WARMINGTON
University of London
POSIDONIUS; STRABO

DEBORAH JEAN WARNER
Smithsonian Institution
BOSS; BRASHEAR; W.R. BROOKS; E.W. BROWN; BURNHAM; A. CLARK; COMSTOCK; MAUNDER; S.B. NICHOLSON; OUTHIER; C.H.F. PETERS; SCHAEBERLE; SMYTH; SOUTH; WINLOCK

J.B. WATERHOUSE
University of Toronto
HAAST; MAWSON

AARON C. WATERS
University of California, Santa Cruz
B. WILLIS

DIEDRICH WATTENBERG
Archenhold Observatory
POWALKY; ROSENBERG; O.A. ROSENBERGER; C.A. VON SCHMIDT

RAY L. WATTERSON
University of Illinois
LILLIE

CHARLES WEBSTER
University of Oxford
ALSTED; W. AMES; ASHMOLE; J. BAUHIN; BURGERSDIJK; ENT; MORISON; J. RAY; TOWNELEY; W. TURNER; WALTON

RODERICK S. WEBSTER
Adler Planetarium, Chicago
NAIRNE; RAMSDEN; TROUGHTON

C.E. WEGMANN
E. ARGAND

E. WEGMANN
LUGEON; SCHARDT; SEDERHOLM; SUESS

SIGFRID VON WEIHER
Sammlung von Weiher zur Geschichte der Technik
RÜDENBERG

DORA B. WEINER
Manhattanville College
LEURET; RENAUT; ROLLET; L.-L. VAILLANT

ADRIENNE R. WEILL-BRUNSCHVICG
DE BROGLIE; M. CURIE; JACQUET; LANGEVIN

PIERRE WELANDER
Massachusetts Institute of Technology
EKMAN

MARY A. WELCH
University of Nottingham
WILLUGHBY

GEORGE A. WELLS
University of London
GOETHE

JOHN W. WELLS
Cornell University
LESUEUR; MATHER; RAYMOND; VANUXEM; H.S. WILLIAMS

F.W. WENT
University of Nevada
WENT

RACHEL HORWITZ WESTBROOK
NEEDHAM

RICHARD S. WESTFALL
Indiana University
HOOKE; PEMBERTON

RALPH H. WETMORE
JEFFREY

FRANZ WEVER
Max-Planck-Institut, Düsseldorf
HEYN; MARTENS

JOYCE WEVERS
State University of Utrecht · Vening Meinesz Laboratory, Utrecht
HAIDINGER; WIDMANNSTÄTTEN

ALWYNE WHEELER
British Museum (Natural History)
J.V. THOMPSON; WOTTON

GEORGE W. WHITE
University of Illinois
L. EVANS; MACLURE; J. MORSE; D.D. OWEN

D.T. WHITESIDE
Whipple Science Museum
BARROW; COLLINS; D. GREGORY; J. GREGORY; N. MERCATOR

CHARLES A. WHITNEY
Smithsonian Astrophysical Laboratory
H. DRAPER; HOLDEN

GERALD J. WHITROW
Imperial College of Science and Technology
A. FORSYTH; HUBBLE; E.A. MILNE

GWENETH WHITTERIDGE
G. BAUHIN

THOMAS WIDORN
BIELA; BÜRG

W.P.D. WIGHTMAN
King's College, Aberdeen
CULLEN; FRANCESCA

RONALD S. WILKINSON
Library of Congress
GROTE; REYNA; STARKEY; J. WINTHROP

MARY E. WILLIAMS
Skidmore College
COLE; L.S. HILL

L. PEARCE WILLIAMS
Cornell University
AMPÈRE; DE LA RUE; FARADAY; OERSTED

WESLEY C. WILLIAMS
Case Western Reserve University
BARRY; CHAMBERS; DUVERNEY; FLOWER; GAYANT; H. GRAY; HUXLEY; MÉRY; R. OWEN; TYSON

HELMUT M. WILSDORF
AGRICOLA

CURTIS A. WILSON
University of California, San Diego
CYSAT; DOPPELMAYR; HEYTESBURY

J.T. WILSON
University of Toronto
DU TOIT

LEONARD G. WILSON
University of Minnesota
ARISTOTLE; JENNER; LYELL

C. GORDON WINDER
University of Western Ontario
SELWYN

R.P. WINNINGTON–INGRAM
ARISTOXENUS

MARY P. WINSOR
University of Toronto
S. FORBES; SAVIGNY; SWAMMERDAM

FRANK H. WINTER
Smithsonian Institution
CONGREVE; W. HALE; MONTGÉRY; W. MOORE

F.R. WINTON
L.E. BAYLISS

J. WITKOWSKI
BANACHIEWICZ

EDWIN WOLF II
Library Company of Philadelphia
J. LOGAN

J.H. WOLF
University of Munich
MOLDENHAWER; MURALT

J. WOLFOWITZ
University of Illinois
A. WALD

M.L. WOLFROM
C.S. HUDSON; NEF

M. WONG
BELON

A.E. WOODRUFF
Yeshiva University
DOPPLER; JEANS; LARMOR; POYNTING; W.E. WEBER

HARRY WOOLF
Johns Hopkins University · The Institute for Advanced Study
CHAPPE D'AUTEROCHE

HELEN WRIGHT
WALTER SYDNEY ADAMS; G.E. HALE

O. WRIGHT
University of London
AL–FĀRĀBĪ

DENISE WROTNOWSKA
Institut Pasteur
FOURNEAU; LEVADITI; PRUNER BEY

HANS WUSSING
Karl Marx University
FROBENIUS; C.L.F. LINDEMANN; C.G.A. MAYER; C.G. NEUMANN; PFAFF; SCHRÖDER

JEAN WYART
University of Paris
P. CURIE; S.–D.A. LÉVY; MAUGUIN

ERI YAGI
Toyo University
NAGAOKA

FRANCES A. YATES
University of London
BRUNO

ELLIS L. YOCHELSON
United States Geological Survey
MEEK; SCHUCHERT; WALCOTT

H.S. YODER, JR.
Carnegie Institution of Washington
CROSS; FENNER; HAGUE; IDDINGS; WINCHELL FAMILY; F.E. WRIGHT

L.D.G. YOUNG
SEE

ROBERT M. YOUNG
University of Cambridge
GALL; HARTLEY

ALEXANDER A. YOUSCHKEVITCH
Academy of Sciences of the U.S.S.R. · Institute for the History of Science and Technology, Moscow
KHINCHIN; MARKOV; MINDING; SLUTSKY

A.P. YOUSCHKEVITCH
Institute for the History of Science and Technology, Moscow · Academy of Sciences of the U.S.S.R.
ABŪ'L–WAFĀ'; SERGEY NATANOVICH BERNSTEIN; BOHL; BRING; CHEBOTARYOV; CHEBYSHEV; DA CUNHA; EULER; GELFOND; KAGAN; AL–KĀSHĪ; AL–KHAYYĀMĪ; LEXELL; OSTROGRADSKY; PETERSON; PORETSKY; PRIVALOV; RAZMADZE; SEGNER; SHATUNOVSKY; SHNIRELMAN; SOKHOTSKY; SONIN; STEKLOV; STEPANOV; V.P. ZUBOV

S.Y. ZALKIND
Institute for the History of Science and Technology, Moscow
KOLTZOFF

BRUNO ZANOBIO
University of Pavia
FABRICI; FRACASTORO; GOLGI; LUCIANI; MATTIOLI; NEGRI; PAGANO; PENSA; PIANESE; L. PORTA; SERTOLI; VASTARINI–CRESI

Societies Indexed

Academia Caesarea Leopoldina Carolina Naturae Curiosorum
Academia Leopoldina Naturae Curiosorum
Academia Naturae Curiosorum Sacra
Academia Operosorum
Academia Parisiensis
Academia Secretorum Naturae
Académie Bourdelot
Académie de Calvin
Académie de Genève
Académie de Médecine, Paris
Académie des Sciences, Paris
Académie des Sciences, Belles-Lettres et Arts (Rouen)
Académie des Sciences et Beaux-Arts des États-Unis
Académie Française
Académie Internationale d'Histoire des Sciences
Académie Montmor
Académie Royale (Paris)
Académie Royale de l'Architecture (Paris)
Académie Royale de Peinture et de Sculpture (Paris)
Académie Royale des Sciences, des Lettres et des Beaux-Arts de Belgique
Académie Royale des Sciences et des Belles Lettres de Berlin
Academy of Natural Sciences
Academy of Sciences, Turin
Academy of Sciences (USSR)
Accademia Cosmografica degli Argonauti
Accademia degli Inquieti
Accademia degli Investiganti
Accademia degli Oziosi
Accademia dei Fisiocritici
Accademia dei Lincei
Accademia dei Percossi
Accademia dei Segreti
Accademia del Cimento
Accademia della Crusca
Accademia della Fucina
Accademia della Traccia
Accademia dell' Esperienza
Accademia delle Scienze dell'Istituto di Bologna
Accademia d'Italia
Accademia Fisica-matematica
Accademia Reale (Rome)
Accademia Scientifica-Litteraria (Milan)
Aerial Experimental Association
African Association
Agriculture Society of Edinburgh
Akademi Internasional de Lingu Universal
Akademie der Wissenschaften (Berlin)
Akademiia nauk (USSR)
Allgemeine Elektrizitäts-Gesellschaft
All-Union Association of Scientists and Technicians
Altomare, Academy of
American Academy of Arts and Sciences
American Association for Cancer Research
American Association for the Advancement of Science
American Association of Anatomists
American Association of Economic Entomologists
American Association of Geologists
American Association of Neurological Surgeons
American Association of Pathologists and Bacteriologists
American Association of Physical Anthropologists
American Association of University Professors
American Association of University Women
American Association of Variable Star Observers
American Astronomical Society
American Breeders Association
American Cancer Society
American Chemical Society
American Crystallographic Association
American Epidemiological Society
American Ethnological Society
American Genetic Association
American Geographical Society
American Geological Society
American Mathematical Society
American Medical Association
American Meteorological Society
American Metrological Society
American Ornithologists' Union
American Philosophical Society
American Photographic Society
American Physical Society
American Physiological Society
American Phytopathological Society
American Psychoanalytic Association
American Psychological Association
American Public Health Association
American Society for Horticultural Science
American Society for Pharmacology and Experimental Therapeutics
American Society of Agronomy
American Society of Biological Chemists
American Society of Mammalogists
American Society of Mechanical Engineers
American Society of Naturalists
American Society of Zoologists
American-Soviet Medical Society
American Union Academy of Literature, Science and Art
Analytical Society (Cambridge Univ.)
Anatomical Society of Great Britain and Ireland
Anthropological Society of London
Antiquaries, Society of
Apothecaries, Society of
Arcueil, Société d'
Aristotelian Society for the Systematic Study of Philosophy
Askesian Society
Association for Medical Instruction
Association Française pour l'Avancement des Sciences
Association Internationale de Botanistes
Association Internationale de Géodésie
Association of American Geographers
Association of American Geologists
Association of German Mathematicians
Association Polytechnique
Astronomical Society of Berlin
Astronomical Society of London
Astronomical Society of the Atlantic
Astronomical Society of the Pacific
Astronomische Gesellschaft
Athénée Médical
Aurelian Society
Australian Academy of Science
Austrian Academy of Sciences
Bansha (Association of Foreign Learning)
Batavian Club
Bath Philosophical Society
Bayerische Akademie der Wissenschaften
Berlin Academy
Berlin Geographical Society
Berlin Medical Society
Biological Society of Washington (D.C.)
Bohemian Scientific Society (Prague)
Botanical Exchange Club
Botanical Society of America
Botanical Society of Edinburgh
British Association (for the Advancement of Science)
British Astronomical Association
British Medical Association
British Meteorological Society
British Mineralogical Society
British Ornithologists' Club
British Physiological Society
British Science Guild
California Academy of Sciences
California Botanical Society
Cambridge Antiquarian Society
Cambridge Philosophical Society
Canterbury Philosophical and Literary Institution
Cape Philosophical Society
Chapter Coffee House Society
Chemical Institute of Canada
Chemical Society, London
Chemical-Physical Society (Vienna)
Chicago, Academy of Sciences
Cimento, Accademia del
Circolo Matematico di Palermo
City Philosophical Society (London)
Collegium Curiosum sive Experimentale
Colorado Scientific Society
Copenhagen Academy
Cracow Academy of Sciences
Crusca, Accademia della
Derby Philosophical Society
Deutsche Botanische Gesellschaft
Deutsche Chemische Gesellschaft

Societies Indexed

Deutsche Elektrochemische Gesellschaft
Deutsche Geologische Gesellschaft, Berlin
Deutsche Gesellschaft für Metallkunde
Deutsche Gesellschaft zur Bekämpfung der Geschlechtskrankheiten
Deutsche Mathematiker-Vereinigung
Deutsche Physikalische Gesellschaft, Berlin
Dijon Academy
Dublin Philosophical Society
Edinburgh Botanical Club
Edinburgh Mathematical Society
Edinburgh Philosophical Society
Egypt Exploration Society
Entomological Society (Cambridge, Eng.)
Entomological Society (of London)
Epidemiological Society
Esperienza, Accademia dell'
Fachgemeinschaft der deutschen Hochschullehrer der Physik
Florence, Academy of
Fraternitas Medicorum
Free Association for Microbiology
French Astronomical Society
Fucina, Accademia della
Galvanic Society
Genetical Society (Gr. Brit.)
Genetics Society of America
Gentleman's Society of Spalding
Geograficheskoe obshchestvo (USSR)
Geographical Society of Finland
Geological Institute, Berlin
Geological Society of America
Geological Society of London
German Anthropological Society
German Botanical Society
German Geological Society, Berlin
German Metallographic Society
German Research Association
German Society for Electron Microscopy
German Society for Technical Röntgenology
German Society for the History of Medicine and Science
Gesellschaft Deutscher Naturforscher und Aerzte
Haarlem Academy
Hakluyt Society
Harvey Cushing Society
Helmholtz-Gesellschaft
History of Science Society (Cambridge, Mass.)
Horticultural Society of London
Hungarian Academy of Sciences
Indian Academy of Sciences
Indian Association for the Cultivation of Science
Inquieti, Accademia degli
Institut de France
Institut International d'Embryologie
Institut National des Sciences et des Arts, Paris
Institute of Chemistry
Institution of Electrical Engineers (G.B.)
Institution of Engineers
International Academy for the History of Science
International Association of Academies
International Association of Chemical Societies
International Association of Seismology
International Association of Seismology and Physics of the Earth's Interior
International Astronomical Society
International Federation of Electron Microscope Societies
International Geodesic Association
International Psychoanalytic Association
International Working Men's Association
Investiganti, Accademia degli
"Invisible College"
Ionian Academy
Italian Anatomical Society
Italian Geological Society
Italian Mathematical Union
Italian Society of the History of Medicine
Japanese Association for the Advancement of Science
Kaiser Wilhelm Gesellschaft zur Förderung der Wissenschaften
Kolloid Gesellschaft
Lazzaroni
Lebedev Physics Society, Moscow
Leopoldina
Lichfield Botanical Society
Lincei, Accademia dei
Lincoln (England), Mechanics Institution
Linnean Society of London
London Geological Society
London Mathematical Society
Lunar Society (Birmingham)
Lycée des Arts (Paris)
Lyceum Medicum Londinense
Magnetischer Verein
Magyar Tudomanyos Akademia, Budapest
Manchester Literary and Philosophical Society (England)
Marine Biological Association (U.K.)
Mathematical and Physical Society (Hungary)
Mathematical Association of America
Mathematical Society (Berlin)
Mathematischer Verein (Berlin)
Max Planck Gesellschaft zur Förderung der Wissenschaften
Medical Research Society (London)
Melun, Agricultural Society
Metaphysical Club (Cambridge)
Metaphysical Society (London)
Mineralogical Society (St. Petersburg)
Minnesota Academy of Sciences
Moscow Mathematical Society
Moscow Physics Society
Moskovskoe fizicheskoe obshchestvo imeni I.I. Lebedeva
Moskovskoe matematicheskoe obshchestvo
National Academy of Sciences (U.S.)
National Geographic Society (Washington)
National Society for Medical Research (U.S.)
Natural History Society of Edinburgh
Natural Science Society (Switzerland)
Nederlandsche Vereeniging voor Koeltechniek
Netherlands Botanical Society
Netherlands Society of Sciences (Haarlem)
New York Academy of Medicine
New York Academy of Sciences
New York Mathematical Society
New York Pathological Society
Newcastle Chemical Society
Notgemeinschaft der deutschen Wissenschaften
Nuremberg, Cosmographical Society
Oxford Philosophical Society
Paläontologische Gesellschaft (Berlin)
Paleontological Society (U.S.)
Paris Chemical Society
Petrologists' Club of Washington
Pharmazeutischer Verein (Baiern)
Philosophical and Literary Society (Liverpool)
Philosophical Society (London)
Philosophical Society of Edinburgh
Philosophical Society of Glasgow
Philosophical Society of Washington (D.C.)
Philosophy of Science Association
Physical Society of London
Physikalisch-Medizinische Gesellschaft, Würzburg
Physikalische Gesellschaft (Berlin)
Physikalische-technische Reichsanstalt
Physikalischer Verein (Frankfurt am Main)
Physiological Society (London)
Polish Astronautical Society
Polish Physical Society
Private College of Amsterdam
Prussian Academy
Prussian Geodetic Institute
Ray Society
Real Sociedad Económica Vascongada de Amigos del País
Réunion des Sciences
Rosati
Rouen, Academy of Sciences
Royal Academy of Brussels
Royal Agricultural Society of England
Royal Astronomical Society (London)
Royal Astronomical Society of New Zealand
Royal Danish Academy of Sciences
Royal Dublin Society
Royal Geographical Society (London)
Royal Horticultural Society (England)
Royal Institution of Great Britain, London
Royal Institution of the Netherlands
Royal Irish Academy
Royal Medical Society (Edinburgh)
Royal Meteorological Society (Gr. Brit.)
Royal Microscopical Society (London)
Royal Scottish Geographical Society
Royal Scottish Society of Arts
Royal Society of Arts (England)
Royal Society of Canada
Royal Society of Edinburgh
Royal Society of London
Royal Society of New South Wales
Royal Society of Sciences (Madrid)
Royal Society of Tropical Medicine (London)
Russian Academy of Sciences
Russian Astronomical Society
Russian Chemical Society
Russian Entomological Society
Russian Geographical Society
Russian Physics and Chemistry Society
Russian Physiological Society
Russkoe fizicheskoe i khimicheskoe obshchestvo
St. Petersburg, Imperial Academy of Science
Save-the-Redwoods League
Schlesische Gesellschaft für Vaterländische Kultur
Schweizerische Mineralogische und Petrographische Gesellschaft
Seismological Society of America
Seismological Society of Japan
Select Society (Edinburgh)
Senckenbergische Naturforschende Gesellschaft (Frankfurt am Main)
Sierra Club
Smeatonian Society
Sociedad de Amigos del País
Sociedad Económica de Amigos del País
Società Botanica Fiorentina
Società degli Spettroscopisti Italiani
Società dei Quaranta
Società Filotecnica
Società Italiana delle Scienze
Societas Ereunetica
Societas Meteorologica Palatina
Societas Spinozana
Société Alfred Binet
Société Anatomique (Paris)
Société Botanique de France (Paris)
Société Chimique de France (Paris)
Société d'Acclimatation, Paris

Société d'Agriculture de Paris
Société d'Anthropologie (Paris)
Société d'Arcueil
Société d'Émulation et d'Agriculture de l'Ain
Société d'Émulation pour les Sciences Pharmaceutiques
Société d'Encouragement pour l'Industrie Nationale
Société d'Histoire Naturelle de Paris
Société de Biologie
Société de Lecture
Société de l'Harmonie Universelle
Société de Micrographie
Société de Pathologie Exotique
Société de Pathologie Végétale et d'Entomologie Agricole de France
Société de Physique (Zurich)
Société de Physique et d'Histoire Naturelle de Genève
Société de Recherches et de Perfectionnements Industriels
Société des Amis des Sciences (Paris)
Société des Arts (Paris)
Société des Ingénieurs du Vide
Société Entomologique de France (Paris)
Société Ethnologique de Paris
Société Française de Physiologie Végétale
Société Géologique de France
Société Géologique Suisse
Société Helvétique des Sciences Naturelles
Société Linnéenne (Paris)
Société Mathématique de France
Société Médicale d'Émulation (Paris)
Société Positiviste (Paris)
Société Royale de Médecine (Paris)
Société Royale des Sciences de Montpellier
Society for Experimental Biology and Medicine (U.S.)
Society for Mathematics (Hungary)
Society for Natural Sciences (Hungary)
Society for Philosophical Experiments and Conversations
Society for Promoting Useful Knowledge
Society for Psychical Research (G.B.)
Society for the Diffusion of Useful Knowledge
Society for the Encouragement of Arts, Manufactures and Commerce (England)
Society for the Improvement of Medical Knowledge (Edinburgh)
Society for the Promotion of the Natural Sciences in the Netherlands Indies
Society for the Propagation of Christian Knowledge
Society of American Bacteriologists
Society of Chemical Industry (London)
Society of Civil Engineers (Gr. Brit.)
Society of Dermatology and Syphilography
Society of German Scientists and Physicians
Society of Glass Technology
Society of Polish Mathematicians
Society of Telegraphic Engineers and Electricians
Soviet Academy of Sciences
Sozietät der Wissenschaften
Statistical Society of London
Surgeons, Royal College of
Swedish Academy of Sciences
Swedish Seed Association
Swiss Mathematical Society
Temple Coffee House Botany Club
Természettudományi Társulat
Teyler's Society
Towarzystwo Przyjaciół Nauk
Traccia, Accademia della
Ukrainian S. S. R. Academy of Sciences
Union Mondiale des Académies
Union of Biological Sciences
Universal Brotherhood of Friends of Truth
Vereeniging tot Bevordering van de Opleiding tot Instrumentmaker
Verein Deutscher Ingenieure
Verein für Landwirtschaft
Vereinigte Astronomische Gesellschaft
Victoria Institute of London
Virginia, Mineralogical Society
Vsesoiuznoe entomologicheskoe obshchestvo
Warsaw Scientific Society
Washington Academy of Sciences
Washington Philosophical Society
Wernerian Natural History Society
Worshipful Society of Apothecaries
Yorkshire Philosophical Society
Zoological Society of London

Periodicals Indexed

Abhandlungen aus dem mathematischen Seminar der Hamburgischen Universität
Acta arithmetica
Acta astronomica
Acta crystallographica
Acta eruditorum
Acta litterarum ac scientiarum Regiae Universitatis hungarica . . .
Acta Mathematica
Acta medica et philosophica hafniensa
Acta physicochimica URSS
Acta scientiarum mathematicarum
Addisonia
Afhandlingar i fysik, kemi och mineralogi
Akademische Sternkarten
Album der Natuur
Algebra i logika. Seminar
Allgemeine berg- und hüttenmännische Zeitung
Allgemeine nordische Annalen der Chemie
Allgemeines Journal der Chemie
Allgemeines Magazin
Allgemeines Repertorium der Mineralogie
American Breeders' Magazine
American Chemical Journal
American Chemical Society Scientific Monographs
American Ephemeris and Nautical Almanac
American Geologist
American Journal of Anatomy
American Journal of Mathematics
American Journal of Physical Anthropology
American Journal of Physiology
American Journal of Psychology
American Journal of Science
American Journal of Science and Arts
American Mathematical Monthly
American Midland Naturalist
American Naturalist
Anales de ciencias naturales
Anales de historia natural
Analyst
Annalen der Chemie
Annalen der Naturphilosophie
Annalen der Pharmacie und Chemie
Annalen der Physik
Annalen der Physik und Chemie
Annalen für Meteorologie und Erdmagnetismus
Annales de chimie (et de physique)
Annales de Gergonne
Annales de l'agriculture française
Annales de l'Institut Océanographique
Annales de l'Institut Pasteur
Annales de la médecine physiologique
Annales de mathématiques pures et appliquées
Annales de parasitologie humaine et comparée
Annales des mines
Annales des sciences naturelles
Annales d'hygiène publique et de médecine légale
Annales scientifiques de l'École normale supérieure
Annali di matematica pura ed applicata
Annals and Magazine of Natural History, including Zoology, Botany, and Geology
Annals and Magazine of Zoology and Botany
Annals of Anatomy and Physiology
Annals of Botany
Annals of Electricity
Annals of Eugenics
Annals of Mathematical Statistics
Annals of Mathematics
Annals of Medicine and Surgery
Annals of Philosophical Discovery and Monthly Reporter of the Progress of Science and Art
Annals of Philosophy
Annals of Philosophy, Natural History, Chemistry, Literature, Agriculture, and the Mechanical and Fine Arts
Annals of the New York Academy of Sciences
Année biologique
Année psychologique
Annuaire astronomique et météorologique
Annuaire de chimie
Annual Record of Science and Industry
Anthropological Review
Anthropologie
L'anthropologie
Anthropologischer Anzeiger
Applied Mechanics Reviews
Archaeologia
Archeion
Archiv der Mathematik and Physik
Archiv der Pharmazie
Archiv der reinen und angewandten Mathematik
Archiv för mathematik og naturvidenskab
Archiv für Anatomie, Physiologie und wissenschaftliche Medizin
Archiv für Anatomie und Physiologie
Archiv für Anthropologie
Archiv für Bergbau und Hüttenwesen
Archiv für die gesamte Physiologie des Menschen und der Tiere
Archiv für die Physiologie
Archiv für Entwicklungsmechanik der Organismen
Archiv für Geschichte der Medizin und der Naturwissenschaften
Archiv für Hygiene
Archiv für Hygiene und Bakteriologie
Archiv für Mikrobiologie
Archiv für mikroskopische Anatomie und Entwicklungsmechanik
Archiv für Mineralogie, Geognosie, Bergbau- und Hüttenkunde
Archiv für Naturgeschichte
Archiv für pathologische Anatomie und Physiologie, und für die klinische Medizin
Archiv für praktische Geologie
Archiv für Psychiatrie und Nervenkrankheiten
Archiv für Rassen- und Gesellschaftsbiologie
Archiv für wissenschaftliche Kunde von Russland
Archiv für Zellforschung
Archives d'anatomie microscopique
Archives de Genève
Archives de zoologie expérimentale et générale
Archives des sciences physiques et naturelles. Supplément à la Bibliothèque universelle de Genève
Archives internationales d'histoire des sciences
Archives of Pathology
Archives of Universal Science
Archivio di storia della scienza
Archivio italiano di anatomia e di embriologia
Archivio italiano di biologia
Archivio per l'antropologia e la etnografia
Archivio per le scienze mediche
Archivos de histología normal y patólogica
Arsberättelser över vetenskapernas framsteg
Arzt, Der
Astronomical Journal
Astronomical Newsletter
Astronomichesky ezhegodnik
Astronomische Nachrichten
Astronomisches Jahrbuch, Berlin
Astronomy and Astro-Physics
Astrophysical Journal
Asuntos varios sobre ciencias y artes
Baileya
Beiblätter zu den Annalen der Physik
Beiträge zur Biologie der Pflanzen
Beiträge zur Krystallographie und Mineralogie
Beiträge zur Morphologie und Physiologie der Pilze
Beiträge zur Physiologie der Sinnesorgane
Berg- und hüttenmännische Zeitung, mit besonderer Berücksichtigung der Mineralogie und Geologie
Berichte der Deutschen chemischen Gesellschaft
Berichte über die Leistungen der niederen Thiere
Berliner astronomisches Jahrbuch
Bibliographia genetica
Bibliotheca sacra
Bibliothèque britannique

Periodicals Indexed

Bibliothèque italique
Bibliothèque universelle de Genève
Biochemica et biophysica acta
Biochemical Journal
Biokhimiia
Biologia Centrali-Americana
Biological Abstracts
Biologicheskii zhurnal
Biologische Untersuchungen
Biometrika
Bolshaia meditsinskaia entsiklopediia
Bolshaia sovetskaia entsiklopediia
Bon jardinier
Botanical Magazine
Botanical Miscellany
Botanical Register
Botanische Jahrbücher für Systematik, Pflanzengeschichte und Pflanzengeographie
Botanische Zeitung
Botanisches Zentralblatt
Brain
British Magazine
British Medical Journal
Brittonia
Bryologist
Bulletin astronomique
Bulletin biologique de la France et de la Belgique
Bulletin . . . de Férussac
Bulletin de la Société Géologique de France
Bulletin de la Société philomathique de Paris
Bulletin de l'Institut Océanographique
Bulletin de l'Institut Pasteur
Bulletin de pharmacie
Bulletin des sciences et de l'industrie
Bulletin des sciences mathématiques
Bulletin des sciences médicales
Bulletin des sciences naturelles et de géologie
Bulletin du jardin botanique de Buitenzorg
Bulletin du Musée Océanographique
Bulletin hydrografique
Bulletin of the American Mathematical Society
Bulletin of the Astronomical Institutes of the Netherlands
Bulletin of the New York Mathematical Society
Bulletin of the Torrey Botanical Club
Bulletin scientifique de la France et de la Belgique
Bulletin scientifique du département du Nord
Bulletin statistique
Bullettino Boncompagni
Bullettino di bibliografia e di storia delle scienze matematiche e fisiche
Bullettino di paletnologia italiana
Cambridge Mathematical Journal
Canadian Naturalist & Geologist
Cancer Research
Chemical Abstracts
Chemical News
Chemical Reviews
Chemische Annalen für die Freunde der Naturlehre, Arzneygelahrtheit, Haushaltungskunst und Manufacturen
Chemisches Journal für die Freunde der Naturlehre, Arzneygelahrtheit, Haushaltungskunst und Manufacturen
Ciel et terre
Circulaire de l'Observatoire de Cracovie
Der Civil-Ingenieur
Clinical Science
Coelum
Collection of Czechoslovak Chemical Communications
Comptes rendus hebdomadaires des séances de l'Académie des sciences
Comptes rendus mensuels des travaux chimiques
Connaissance des temps
Contributions, Cushman Laboratory for Foraminiferal Research
Correspondance astronomique, géographique, hydrographique et statistique
Correspondance mathématique et physique
Cosmos
Crelle's Journal
Curtis's Botanical Magazine
Cytologia
Daedalus hyperboreus
Deutsche Wochenschrift für Staat, Kirche und Volksleben
Diario literario de México
Dinglers polytechnisches Journal
Dublin Journal of Medical and Chemical Science
Economic Geology
Economic Journal
Economist
Edinburgh Journal of Science
Edinburgh Philosophical Journal
Edinburgh Review
The Electrician
Endocrinology
Engler's Jahrbücher
Enseignement mathématique
Enzymologia
Ephemerides of Eclipsing Binaries
Erkenntnis
Erythea
Essais et recherches de mathématiques et de physique
Excerpta medica
Exercices de mathématiques
Férussac's Bulletin
Flora
Fortschritte der Mathematik
Die Fortschritte der Physik
Fundamenta mathematicae
Gazeta de literatura
Gazette de santé
Gazzetta chimica italiana
Gemeinnützige Abhandlungen
The General Magazine of Arts and Sciences
Genetica
Genetics
Gentes herbarum
Gentleman's Diary
Gentleman's Magazine
Gentleman's Mathematical Companion
Geographical Journal
Geographical Review
Geologicheski vestnik
Gergonne, Annales
Giornale dei Letterati
Giornale di geologia pratica
Good Housekeeping
Gorny zhurnal
Göttingische Zeitungen von gelehrten Sachen
Göttingisches Magazin der Wissenschaften und Litteratur
Guy's Hospital Reports
Hamburger mathematische Einzelschriften
Handbuch der physiologischen Botanik
Heart
Helminthologia
Helvetica physica acta
Hemel en dampkring
Histologische und Histopathologische Arbeiten über die Grosshirnrinde
Holländische Beiträge zu den anatomischen und physiologischen Wissenschaften
Homme
Hospitalstidende
Human Biology
Ibis
Icones plantarum
Indian Journal of Physics
Indicateur de l'archéologue
Intermédiare des mathématiciens
International Catalogue of Scientific Literature
International Journal of American Linguistics
International Seismological Summary
Internationale photographische Monatsschrift für Medizin und Naturwissenschaften
Internationale Zeitschrift für Metallographie
Internationale Zeitschrift für Vitaminforschung
Isis
Isis oder enzyklopädische Zeitung von Oken
Izvestiya Akademii nauk SSSR
Jahrbuch der Radioaktivität und Elektronik
Jahrbuch für Elektrochemie
Jahrbuch für psychoanalytische und psychopathologische Forschungen
Jahrbuch für wissenschaftliche Botanik
Jahrbuch über die Fortschritte der Mathematik
Jahrbücher für wissenschaftliche Botanik
Jahresbericht über die Fortschritte der Chemie
Jahresberichte über die Fortschritte der gesammten Medicin in allen Ländern
Jenaische Zeitschrift für Medizin und Naturwissenschaft
Johns Hopkins University, Studies from the Biological Laboratory
Journal de chimie physique
Journal de Chirurgie
Journal de Genève
Journal de géologie
Journal de l'anatomie et de la physiologie normales et pathologiques de l'homme et des animaux
Journal de l'École polytechnique
Journal de Liouville
Journal de mathématiques pures et appliquées
Journal de médecine
Journal de médecine, chirurgie et pharmacie
Journal de médecine et de chirugie de Toulouse
Journal de Paris
Journal de pharmacie et de chimie
Journal de physiologie expérimentale et de pathologie
Journal de physique
Journal de Trévoux
Journal de Verdun
Journal der Pharmacie
Journal des sçavans
Journal d'histoire naturelle
Journal encyclopédique
Journal für Chemie, Physik und Mineralogie
Journal für Chemie und Physik
Journal für die Baukunst
Journal für die reine und angewandte Mathematik
Journal für makromolekulare Chemie
Journal für Physik und physikalische Chemie des Auslandes
Journal für praktische Chemie
Journal für technische und ökonomische Chemie
Journal helvétique
Journal littéraire de la Haye
Journal of Anatomy

Periodicals Indexed

Journal of Anatomy and Physiology
Journal of Animal Behavior
Journal of Applied Psychology
Journal of Atmospheric Research
Journal of Biological Chemistry
Journal of Botany
Journal of Botany and Kew Gardens Miscellany
Journal of Comparative Neurology
Journal of Economic Entomology
Journal of Experimental Medicine
Journal of Experimental Zoology
Journal of Genetics
Journal of Geology
Journal of Geophysical Research
Journal of Heredity
Journal of Infectious Diseases
Journal of Meteorology
Journal of Morphology
Journal of Natural Philosophy, Chemistry and the Arts
Journal of Neurophysiology
Journal of Organic Chemistry
Journal of Pharmacology and Experimental Therapeutics
Journal of Physical Chemistry
Journal of Physiology
Journal of Preventive Medicine
Journal of Religious Psychology
Journal of Sedimentary Petrology
Journal of the American Chemical Society
Journal of the American Medical Association
Journal of the Bombay Natural History Society
Journal of the British Astronomical Association
Journal of the Chemical Society
Journal of the Franklin Institute
Journal of the History of Ideas
Journal of the Scientific Laboratories of Denison University
Justus Liebig's Annalen der Chemie
Kew Bulletin
Knowledge and Illustrated Scientific News
Kolloidchemische Beihefte
Kosmos
Ladies' Diary
Lancet
Leipziger Magazin für Naturkunde, Mathematik und Ökonomie
Leipziger Magazin für reine & angewandte Mathematik
Leonardo
Lesnoy zhurnal
Liebig, Jahresbericht
Linnaea
Liouville's Journal
Literary Gazette
London Journal of Botany
Maanedsskrift for Dyrlaeger
Madroño
Magazin for Naturvidenskaben
Magazin für die gesamte Mineralogie, Geognosie und mineralogische Erdbeschreibung
Magazin für die Oryktographie von Sachsen
Magazin für die Technologie
Magazin von merkwürdigen neuen Reisebeschreibungen
Magazin vor Aerzte
Makromolekulare Chemie
Manchester Memoirs
Mankind Quarterly
Matematicheskii sbornik
Matematisk Tidsskrift
Matematyka
Matériaux pour l'histoire positive et philosophique de l'homme
Mathematics Teacher
Mathematische Annalen
Mathematische Nachrichten
Mathematische Zeitschrift
Mathesis
Médecine éclairée par les sciences physiques
Mededeelingen
Medical Essays and Observations
Die medizinische Reform
Mémoires de la Société ethnologique de Paris
Mémoires de la Société géologique du Nord
Mémoires pour servir à l'histoire des plantes
Mémoires pour servir à l'histoire naturelle des animaux
Mémorial de l'officier du génie
Memorie del Regio osservatorio di Firenze ad Arcetri
Memorie della Società degli spettroscopisti italiani
Mercure suisse
Mercurio Peruano
Messenger of Mathematics
Metallographist
Meteorologicheskii vestnik
Meteorologische Zeitschrift
Mind
Mineralogical Abstracts
Mineralogical Magazine
Mineralogische Mitteilungen
Miscellanea curiosa mathematica
Mitteilungen zur Geschichte der Medizin und der Naturwissenschaften
Mittheilungen aus dem Kaiserlichen Gesundheitsamt
Monatliche Correspondenz zur Beförderung der Erd- und Himmelskunde
Monatshefte
Monist
Moniteur scientifique
Monografie matematyczne
Monthly American Journal of Geology and Natural Science
Monthly Notices of the Royal Astronomical Society
Morphologisches Jahrbuch
National Geographic Magazine
Nature
Naturforscher
Nautical Almanac (G.B.)
Nautical Almanac (U.S.)
Nederlandsch kruidkundig archief
Nederlandsch Lancet
Nemophila
Neue Gesellschaftliche Erzählungen
Neues allgemeines Journal der Chemie
Neues Berliner Jahrbuch für die Pharmazie
Die neuesten Entdeckungen in der Chemie
Die neuesten Forschungen im Gebiete der technischen und ökonomischen Chemie
Nevrologicheskii vestnik
New York Medical Journal
Nicholson's Journal
Nieuw archief voor wiskunde
Norddeutsche Zeitung
Nordisk astronomisk tidsskrift
North American Entomologist
Notes From the Royal Botanic Garden, Edinburgh
Nouvelle correspondance mathématique
Nouvelles annales de mathématiques
Nova acta Academiae Caesareae Leopoldino Carolinae germanicae naturae curiosorum
Nuovo cimento
Nyt Tidsskrift for Fysik og Kemi
Obozrenie psikhiatrii, nevrologii i eksperimentalnoy psikhologii
Observaciones sobre la fisica, historia natural y artes utiles
Observations sur la physique, sur l'histoire naturelle, et sur les arts
Observatory
The Open Court
Osiris
Osservazioni e memorie dell'osservatorio astrofisico di Arcetri
Palaeobiologica
Palaeontographica
Palaeontologia Indica
Paläontologische Zeitschrift
Parasitology
Periodico di matematiche
Petrus Camper
Pflüger's Archiv
Philadelphia Medical and Physical Journal
Philosophical Collections
Philosophical Magazine
Philosophical Transactions
Philosophische Studien
Philosophisk repertorium for faedrelandets nyeste litteratur
Phrenological Journal
Physical Review
Physikalische Blätter
Physikalische Zeitschrift
Physiological Abstracts
Physiological Zoology
Pochvovedenie
Policlinico
Politecnico
Polski przegląd kartograficzny
Polytechnisches Notizblatt
Polyteknisk tidsskrift
Poor Richard's Almanack
Popular Science Monthly
Prace matematyczno-fizyczne
Priroda
Proceedings of the Edinburgh Mathematical Society
Proceedings of the Geological Society of London
Proceedings of the National Academy of Sciences
Proceedings of the Royal Society of London
Proceedings of the Zoological Society of London
Progressus rei botanicae
Psyche
Psychological Review
Quarterly Journal of Experimental Physiology
Quarterly Journal of Mathematics
Quarterly Journal of Microscopical Science
Quarterly Journal of Pure and Applied Mathematics
Quarterly Journal of Science
Quarterly Journal of Science, Literature and the Arts
Quarterly Review of Biology
Quaternaria
Railway Magazine and Annals of Science
The Reader
Recent Progress in Hormone Research
Recherches physicochimiques
Records of the Past
Recueil zoologique suisse
Reil's journal
Rendiconti del Circolo matematico di Palermo
Repertorium der Physik
Repertorium für Anatomie und Physiologie
Repertorium für die Pharmacie
Resumptio genetica
Review of American Chemical Research
Review of Scientific Instruments
Revista argentina de historia natural

Periodicals Indexed

Revista matemática hispanoamericana
Revue d'anthropologie
Revue de géographie alpine
Revue de géologie
Revue de l'Oise
Revue de métallurgie
Revue de métaphysique et de morale
Revue de synthèse
Revue du mois
Revue encyclopédique
Revue générale de botanique
Revue générale d'histologie
Revue neurologique
Revue scientifique
Revue scientifique et industrielle
Revue scientifique italienne
Revue semestrielle des publications mathématiques
Rhodora
Rivista di matematica
Rivista di storia delle scienze mediche e naturali
Rivista di studi sessuali e di eugenetica
Saggi dell'esperienze naturali fatte nell'Accademia del Cimento
Sammlung Kombinatorisch-analytischer Abhandlungen
Saturday Review
Scherer's Journal
School and Society
Schweizerische mineralogische und petrographische Mitteilungen
Schweizerische Zeitschrift für Heilkunde
Science
Science and Culture
Scientia
Scientific American
Scientific Monthly
Scripta mathematica
Semanario del nuevo reino
Sibirskii matematicheskii zhurnal
Sirius
Studia mathematica
Studien zur Geschichte der Medizin
Sudhoffs Archiv
Sudhoffs Klassiker der Medizin
Suite de la clef
Synthèse
Taschenbuch für die gesammte Mineralogie
Tecnomasio italiano
Tekhnologichesky zhurnal
Tellus
Terrestrial Magnetism
Tetrahedron
Teysmannia
Thalès
Tidskrift for Mathematik
Transactions of the American Mathematical Society
Transactions of the Connecticut Academy of Arts and Sciences
Transactions of the Royal Society of Edinburgh
Trudy po prikladnoi botanike, genetike i selektsii
Tschermaks mineralogische und petrographische Mitteilungen
Universal Spectator and Weekly Journal
Untersuchungen zur Naturlehre des Menschen und der Thiere
Uspekhi fizicheskikh nauk
Uspekhi sovremennoy biologii
Verhandlungen der Deutschen physikalischen Gesellschaft
Vestnik estestvennykh nauk
Vestnik opytnoi fiziki i elementarnoi matematiki
Vierteljahrsschrift der Naturforschenden Gesellschaft in Zürich
Wiadomości matematyczne
Wiedemanns Annalen
Wilhelm Roux' Archiv für Entwicklungsmechanik der Organismen
Zeitschrift der Oesterreichischen Gesellschaft für Meteorologie
Zeitschrift für allgemeine Physiologie
Zeitschrift für analytische Chemie
Zeitschrift für angewandte Mathematik und Mechanik
Zeitschrift für anorganische Chemie
Zeitschrift für anorganische und allgemeine Chemie
Zeitschrift für Astronomie und verwandte Wissenschaften
Zeitschrift für Biologie
Zeitschrift für Chemie
Zeitschrift für Chemie und Industrie der Kolloide
Zeitschrift für Chemie und Pharmacie
Zeitschrift für die gesamte Kälteindustrie
Zeitschrift für Elektrochemie
Zeitschrift für Hygiene und Infektionskrankheiten
Zeitschrift für Kristallographie
Zeitschrift für Kristallographie und Mineralogie
Zeitschrift für Parasitenkunde
Zeitschrift für Physik
Zeitschrift für physikalische Chemie
Zeitschrift für Physiologie
Zeitschrift für physiologische Chemie
Zeitschrift für praktische Geologie
Zeitschrift für Psychologie und Physiologie der Sinnesorgane
Zeitschrift für rationelle Medicin
Zeitschrift für Vermessungswesen
Zeitschrift für Vulkanologie
Zeitschrift für wissenschaftliche Botanik
Zeitschrift für wissenschaftliche Zoologie
Zemlevedenie
Zentralblatt für Bakteriologie, Parasitenkunde, Infektionskrankheiten und Hygiene
Zhurnal eksperimentalnoy biologii
Žhurnal fizicheskoi khimii
Živa
Zoologische Beiträge
Zoologischer Anzeiger

*Lists of Scientists By Field**

ASTRONOMY

Abetti
Abney
Abraham Bar Ḥiyya
Abū Ma'Shar
Abu'l-Wafā'
Acuta Pisāraṭi
Adams, J. C.
Adams, W. S.
Adelard of Bath
Ailly
Airy
Aitken
Albertus Magnus
Albrecht
Alfonso El Sabio
Alzate y Ramírez
Anaximander
Andoyer
André
Angelus
Ångström
Anthelme
Antoniadi
Apian
Arago
Aratus of Soli
Argelander
Argoli
Aristarchus of Samos
Aristotle
Aristyllus
Arrest
Āryabhaṭa I
Āryabhaṭa II
Asada Gōryū
Autolycus of Pitane
Auwers, A. J. G. F. von
Auzout

Baade
Backlund
Bacon, R.
Bailey, S. I.
Bailly
Baily
Banachiewicz
Banū Mūsā
Barbier
Barnard
Barocius
Al-Battānī
Bayer
Beer
Benzenberg
Bergman
Bernard of Le Treille
Bernard of Verdun
Bernoulli, Jakob I
Bernoulli, Johann (Jean) III
Berti
Bessel
Bhāskara I
Bhāskara II
Bickerton
Biela
Bigourdan
Billy
Birmingham
Birt
Al-Bīrūnī
Al-Biṭrūjī
Blazhko
Bochart de Saron
Bode
Boguslavsky
Bond, G. P.
Bond, W. C.
Borelli
Bošković
Boss
Boulliau
Bour
Bouvard
Bowditch, N.
Bradley, J.
Brahe
Brahmadeva
Brahmagupta
Brandes
Bravais
Bredikhin
Bredon
Bremiker
Brendel
Brinkley
Brisbane
Brooks, W. R.
Brouwer, D.
Brown, E. W.
Bruhns
Brytte
Buot
Bürg
Bürgi
Burnham
Burrau
Busch

Calandrelli, G.
Calandrelli, I.
Caldas
Callandreau
Callippus
Campani
Campanus of Novara
Campbell, W. W.
Camus
Cannon, A. J.
Capra
Carrington
Cassini, G. D.
Cassini, J.
Cassini, J. D.
Cassini de Thury
Castelli
Cauchy
Cayley
Celcius
Challis
Chandler
Chappe d'Auteroche
Chaucer
Chazy
Chevallier, T.
Christie, W. H. M.
Christmann
Clairaut
Claude, F. A.
Clausen
Clavius
Cleomedes
Comas Solá
Common
Comrie
Comstock
Conon of Samos
Copernicus
Cosserat
Cotes
Cowell
Crabtree
Crommelin
Curtis
Curtiss
Cysat

Dalencé
D'Arcy
Darwin, G. H.
Daśabala
Dasypodius
Dawes
Delambre
De La Rue
Delaunay
Delisle, J. N.
Delporte
Dembowski
Deslandres
Dinakara
Dionis du Séjour
Dixon, J.
Dominicus de Clavasio
Donati
Dondi
Doppelmayer
Doppler
Dörffel
Dorno
Dorotheus of Sidon
Dositheus
Downing
Draper, H.
Dreyer
Dudith
Dugan
Duncan
Dunér
Dyson

Easton
Eckert
Eddington
Elkin
Ellis
Emanuelli
Embden, R.
Encke
Esclangon
Euclid
Euctemon
Eudoxus of Cnidus
Euler
Evershed
Ibn Ezra

Fabry, L.
Al-Farghānī
Faye
Al-Fazārī
Ferguson
Ferrel
Feuillée
Fine, O.
Fink
Firmicus Maternus
Flammarion
Flamsteed
Fleming, W. P.
Fontenelle
Foster H.
Fouchy
Fowler, A.
Frenicle de Bessy
Freundlich, E. F.
Frisi
Frost
Fusoris
Fuss

*Initials, first names, or dates are given for those scientists with the same surname.

Gaillot
Galilei, G.
Galle
Gaṇeśa
Gascoigne
Gassendi
Gauss
Geminus
Gerard of Silteo
Gerasimovich
Giles of Lessines
Giles of Rome
Gill, D.
Glaisher, J. W. L.
Godin
Goodricke
Gould, B. A.
Graff
Gregory D.
Gregory J.
Grimaldi
Grisogono
Grossmann, E. A. F. W.

Ḥabash al-Ḥāsib
Hale, G. E.
Hall, A.
Halley
Halm
Hamy
Hansen, P. A.
Hansky
Hansteen
Harding
Haridatta I
Haridatta II
Harkness
Harriot
Hartmann, J. F.
Hartwig
Ibn al-Haytham
Hell
Helmert
Henderson, T.
Henry Bate of Malines
Henry of Hesse
Henry, P. M.
Henry, P. P.
Heraclides Ponticus
Hermann the Lame
Hermes Trismegistus
Herschel, C. L.
Herschel, J. F. W.
Herschel, W.
Hertzsprung
Hevelius
Ibn Hibintā
Hicetas of Syracuse
Hill, G. W.
Hind
Hipparchus of Rhodes
Hippocrates of Chios
Hirayama
Hoek
Holden
Horn d'Arturo
Hornsby
Horrebow, C.
Horrebow, P. N.
Horrocks
Hortensius
Hoüel
Hough
Hubble
Hufnagel
Humboldt
Huggins
Hussey
Huygens
Hypsicles of Alexandria

Ibrāhīm Ibn Sinān
Idelson
Innes
Inō

Jābir ibn Aflaḥ
Jagannātha
Janssen
Al-Jawharī
Jayasiṃha
Al-Jayyānī
Jeans
Jeaurat
Johannes Lauratius de Fundis
John of Gmunden
John of Lignères
John of Murs
John of Saxony
John of Sicily
John Simonis of Selandia
Johnson, M. J.

Kaiser
Kamalākara
Kanaka
Kapteyn
Al-Kāshī
Kavraysky
Keckermann
Keeler
Kepler
Keśava
Al-Khalīlī
Al-Khayyāmī
Al-Khāzin
Al-Khāzinī
Al-Khujandī
Al-Khwārizmī
Kimura
Kirch, G.
Kirch, M. M. W.
Kirch, Christfried
Kirch, Christine
Kirkwood, D.
Klein, H. J.
Konkoly Thege
Kostinsky
Kovalsky
Kramp
Kṛṣṇa
Kushyār

Lacaille
Lagrange
La Hire, G. P. de
La Hire, P. de
Lalande
Lalla
Lambert
Lamont
Langley
Lansberge
Laplace
Lassell
Lāṭadeva
Lau, H. E.
Lavanha
Leavitt
Le Fèvre
Le Gentil
Le Monnier, P. C.
Leo the Mathematician
Le Verrier
Levi ben Gerson
Lexell
Liesganig
Lindblad
Lindenau
Lippmann
Lockyer
Lohse
Loomis
Lorenzoni
Lowell
Ludendorff
Luther
Lyman, C. S.
Lyot

Maanen
Maclear
Macmillan
Mädler
Magini
Mahtādeva
Al-Māhānī
Mahendra Sūri
Al-Majrīṭī
Makaranda
Maksutov
Manfredi
Manilius
Manṣūr
Maraldi, G. F.
Maraldi, G. D.
Markgraf
Martianus Capella
Māshā'allāh
Maskelyne
Mason
Mästlin
Mathurānātha Sárman
Maunder
Maurer
Maurolico
Maury, A.
Mayer, C.
Mayer, J. T.
Mayr
Méchain
Melvill
Menelaus of Alexandria
Mercator
Messier
Meton
Michael Scot
Michell
Miller, W. A.
Milne, E. A.
Mineur
Minnaert
Mitchell, M.
Möbius, A. F.
Moiseev
Moll, G.
Möller
Mollweide
Molyneux, S.
Molyneux, W.
Montanari
Monte
Moore, J. H.
Morgan, H. R.
Morin
Mouchez
Moulton
Mouton
Muḥyi 'l-Dīn
Müller, G.
Munīśvara Viśvarūpa
Muñjāla
Mutis y Bossio

Nāgeśa
Nasmyth
Al-Nayrīzī
Neuymin
Newall
Newcomb
Newton, H. A.
Newton, I.
Nicholson, J. W.
Nicholson, S. B.
Nicolai
Niesten
Nīlakaṇṭha
Nostradamus
Novara
Numerov
Nuñez Salaciense

Odierna
Oenopedes of Chios
Olbers
Olufsen
Oppenheim
Oppolzer
Orlov, A. Y.
Orlov, S. V.
Outhier

Palisa
Pannekoek
Pappus of Alexander
Parameśvara
Parenago
Parkhurst
Parsons
Paul of Alexandria
Pauliśa
Pease
Peirce, B.
Peiresc
Pereira, B.
Pérez de Vargas
Perrine
Perrotin
Peter Philomena of Dacia
Peters, C. F. W.
Peters, C. A. F.
Peters, C. H. F.
Petit, P.
Petosiris
Peurbach

Pezenas
Phillips, T. E. R.
Philolaus of Crotona
Piazzi
Picard, J.
Pickering, E. C.
Pickering, W. H.
Pigott, E.
Pigott, N.
Pingré
Plana
Plaskett
Plato of Tivoli
Plummer, H. C.
Poincaré
Poleni
Pons
Poor
Poretsky
Powalky
Pritchard
Proclus
Proctor
Ptolemy
Puiseux
Pythagoras of Samos

Al-Qabīṣī
Qāḍī Zāda
Al-Qūhī
Quṭb al-Dīn

Rāghavānanda Śarman
Ramus
Raṅganātha
Ranyard
Rayet
Raymond of Marseilles
Regiomontanus
Reinhold
Respighi
Rheita
Rheticus
Ricci, Matteo
Riccioli
Riccò
Richard of Wallingford
Richer
Ritchey
Rittenhouse
Roberts
Roche
Roger of Hereford
Römer
Rooke
Rosenberg
Rosenberger, O. A.
Ross, F. E.
Rosse (Parsons)
Rothmann
Rumford (Thompson, B.)
Rumovsky
Ibn Rushd
Russell, H. N.
Russell, H. C.
Rutherfurd

Sacrobosco
St. John
Saint Vincent
Al-Samarqandī (*fl.* 1276)
Ṣampson
Śatānanda
Schaeberle
Scheiner, C.
Scheiner, J.
Schiaparelli
Schickard
Schjellerup
Schlesinger
Schmidt, C. A. von
Schmidt, J. F. J.
Schöner
Schönfeld
Schröter
Schumacher
Schwabe
Schwarzschild
Seares
Secchi
See
Seeliger
Seidel
Severin
Shakerley
Shapley
Sharonov
Ibn al-Shāṭir
Shayn
Shen Kua
Shibukawa
Shirakatsí
Sigüenza y Góngora
Al-Sijzī
Simon de Phares
Ibn Sīnā
Sinān
Sitter
Slipher, E.
Slipher, V. M.
Smyth
Snel
Soldner
Somerville
Sosigenes
South
Spencer, J. H.
Sphujidhvaja
Ṣpoerer
Śrīpati
Stein
Steinheil
Stephan
Stephanus of Alexandria
Stepling
Sternberg, P. K.
Stewart, M.
Stratton
Streete
Strömberg
Strömgren
Struve, F. G. W.
Struve, G. O. H.
Struve, G. W. L.
Struve, K. H.
Struve, O.
Struve, O. W.
Subbotin
Al-Ṣūfī
Sundman
Synesius of Cyrene

Tacchini
Tarde
Thābit ibn Qurra
Theodosius of Bithynia
Theon of Alexandria
Theon of Smyrna
Thiele, T. N.
Thollon
Ibn Tibbon, J.
Tikhov
Tisserand
Titius
Toscanelli dal Pozzo
Trouvelot
Trumpler
Tserasky
Turner, H. H.
Al-Ṭūsī (Naṣir al-Dīn)
Al-Ṭūsī (Sharaf al-Dīn)

Ulugh Beg
'Umar ibn al-Farrukhān

Varāhamihira
Vaṭeśvara
Vijayananda
Vogel

Ibn Waḥshiyya
Wang Hsi-Shan
Ward
Wargentin
Waterston
Wendelin
Werner, J.
Wharton, G.
Whewell
Wilhelm IV
William of Saint-Cloud
William the Englishman
Wilsing
Wilson, Alexander (*d.* 1786)
Wing
Winlock
Winthrop, J. (*d.* 1676)
Winthrop, J. (*d.* 1779)
Wolf, C. J. E.
Wolf, J. R.
Wolf, M. F. J. C.
Wollaston, F.
Wright, T.
Wright, W. H.

Yaḥyā ibn Abī Manṣūr
Yaʿqūb ibn Ṭāriq
Yativṛṣabha
Yavaneśvara
Young, C. A.
Ibn Yūnus

Zach
Zacuto
Zanotti
Al-Zarqālī
Zöllner

CHEMISTRY

Abel, J. J.
Accum
Achard
Adam of Bodenstein
Adams, R.
Adet
Agrippa
Albertus Magnus
Alder
Ampère
Anaxilaus of Larissa
Anderson, T.
Andrews
Anschütz
Arfvedson
Armstrong, E. F.
Armstrong, H. E.
Arrhenius
Aston
Atwater
Auwers, K. F. von
Avogadro
Babcock
Bach
Baekeland
Baeyer
Balandin
Balard
Bamberger
Bancroft
Barchusen
Barger
Barlow, W.
Barreswil
Baudrimont
Bauer, E.
Baumé
Baumhauer, E. H. von
Baumhauer, H. A.
Bayen
Ibn Al-Bayṭār
Béchamp
Becher
Beckmann, E. O.
Becquerel, A. C.
Beddoes

Beguin
Behrend
Beilstein
Beketov
Bellani
Benedict
Bérard, J. E.
Bergius
Bergman
Bergmann
Bernard of Trevesan
Bernthsen
Berthelot
Berthier
Berthollet
Bertrand, G.
Berzelius
Bischof
Bjerrum
Black, Joseph
Blagden
Blomstrand
Bodenstein
Boerhaave
Boisbaudran
Boltwood
Bonvicino
Borodin
Borrichius
Bosch
Bostock
Böttger
Bourdelin
Boussingault
Boyle
Braconnot
Brande
Brandt, G.
Brauner
Brodie, B. C. Jr.
Brønsted
Brown, A. C.
Brownrigg
Brunton
Buchanan
Buchner, E.
Bucholz
Bucquet
Bunsen
Butlerov
Buys Ballot

Cadet de Gassicourt, L. C.
Cadet de Gassicourt, C. L.
Cadet de Vaux
Cahours
Callinicos of Heliopolis
Cannizzaro
Caro
Carothers
Caventou
Centnerszwer
Cervantes
Chancel
Chapman, D. L.
Chaptal
Chardenon
Chardonnet
Charpy
Chenevix
Chernyaev
Chernov
Chevenard
Chevallier, J. B. A.
Chevreul
Chichibabin
Chittenden
Chugaev
Ciamician
Claisen
Clark T.
Clark, W. M.
Clarke, F. W.
Classen
Claus, A. C. L.
Claus, C. E.
Clément
Cleve
Clouet
Cohen, E. J.
Cohn, E. J.
Cohn, L.
Collet-Descotils
Collie
Cooke
Cooper
Copaux
Cordus, V.
Cori
Cornette
Cottrell
Couper
Courtois
Cramer
Crawford
Crell
Crollius
Cronstedt
Crookes
Crum
Cullen
Curtius
Cushny

Dale
Dalton, J.
Daniell
D'Arcet
Daubeny
Davison
Davy, H.
Davy, J.
Day
Debray
Debye
De La Rue
Delépine
Derosne
Desormes
Despagnet
Dessaignes
Deville
Dewar
Dickinson
Diels
Dioscorides
Dittmar
Dixon, H. B.
Döbereiner
Doelter
Domagk
Donnan
Dorn
Draper, J. W.
Duchesne
Duclaux
Duhamel du Monceau
Duhem
Dulong
Dumas
Dundonald

Eder
Ekeberg
Elhuyar, F. D'
Elhuyar, J. J. D'
Eller von Brockhausen
Elvehjem
Embden, G.
Ercker
Erdmann
Erlenmeyer
Esson
Eucken
Euler-Chelpin

Fankuchen
Faraday
Farkas
Favorsky
Favre
Fersman
Feulgen
Fischer E. H.
Fischer, H.
Fischer, H. O. L.
Fischer, N. W.
Fittig
Folin
Forchhammer
Fordos
Fourcroy
Fourneau
Fownes
Frankenheim
Frankland, E.
Frankland, P. F.
Franklin, R. E.
Freind
Frémy
Fresenius
Freundlich, H. M. F.
Friedel
Friend
Fritzsche
Fuchs, J. N. von
Funk
Fyodorov

Gabriel
Gadolin
Gahn
Galeazzi
Gaudin
Gaultier de Claubry
Gautier, A. E. J.
Gay-Lussac
Gehlen
Gelmo
Genth
Geoffroy, C. J.
Geoffroy, E. F.
Gerhardt
Geuther
Gibbs, O. W.
Giesel
Girtanner
Gladstone
Glaser, C.
Glauber
Gmelin, L.
Gobley
Gohory
Goldschmidt, V.
Goldschmidt, V. M.
Gomberg
Gore
Görgey
Graebe
Graham, T.
Green, J.
Gregory, W.
Gren
Griess
Grignard
Groth
Grotthuss
Grove
Guertler
Guibert
Guillaume
Guillet
Guldberg
Gulland
Gutbier
Guyton de Morveau

Haber
Hadfield
Hahn
Hahnemann
Haldane, John B. S.
Hall, C. M.
Hall, J.
Hampson
Hankel, W. G.
Hantzsch
Harcourt
Harden
Hardy
Hare
Harkins
Hartmann, C. F. A.
Hartmann, J.
Hassenfratz
Hatchett
Hautefeuille
Haüy
Haworth, W. N.
Hellot
Hellriegel
Helmont
Henckel
Henderson, L. J.
Henry, T.
Henry, W.
Herapath, W. B.
Hermann, C. H.
Hermbstaedt
Herschel, J. F. W.
Hess, G. H.
Hessel
Hevesy
Heyrovský
Hiärne
Higgins, B.
Higgins, W.
Hildebrandt
Hilditch
Hinshelwood
Hisinger
Hittorf
Hoffmann, F.

Hofmann, A. W. von
Homberg
Hönigschmid
Hope
Hopkins, F. G.
Hoppe-Seyler
Horbaczewski
Horsford
Horstmann
Houssay
Howe, J. L.
Hudson, C. S.
Hume-Rothery
Hunt, T. S.

Ipatiev
Irinyi

Jābir ibn Ḥayyān
Jackson, C. T.
Jacobs, W. A.
Jacquet
Jacquin
Jaeger, F. M.
Jahn
Jeffries
Johnson, W.
Joly
Jones, H. C.
Jørgensen
Juncker

Kablukov
Kahlenberg
Kane
Karrer
Karsten
Keilin
Keir
Kekule von Stradonitz
Kellner
Kendall
Kharasch
Khunrath, C.
Khunrath, H.
Kidd
Kielmeyer
Kipping
Kirchhof, K. S.
Kirkaldy
Kirkwood, J. G.
Kirwan
Kitaibel
Kjeldahl
Klaproth
Kluyver
Kohlrausch, F. W. G.
Kolbe
Kondakov
Koninck
Konovalov
Kopp
Kossel, K. M. L.
Kostanecki
Kraus
Kuhn, R.
Kuhn, W.
Kühne
Kunckel
Kurnakov

La Brosse
Ladenburg, A.
Lamétherie
Landolt
Langmuir
Lapworth
La Rive, C. G. de
Latimer
Laurent, A.
Lavoisier
Lawes, J. B.
Lawes, G. J. H.
Lebedev, S. V.
Lebedinsky
Le Bel
Le Blanc M. J. L.
Leblanc, N.
Le Châtelier
Le Febvre
Lehmann, J. G.
Lehmann, O.
Lemery, L.
Lemery, N
Lennard-Jones
Leonhardi
Levene
Lewis, G. N.
Lewis, W.
Libavius
Liebermann
Liebig
Link
Lloyd, J. U.
Lomonosov
London, F.
Lonsdale, K.
Lorenz, R.
Loschmidt
Lovits
Lubbock, R.
Luginin
Lunge

Macallum
Macbride, D.
McCollum
Macculloch
Macheboeuf
Maclean
Macquer
Magellan
Magnus H. G.
Maier
Malouin
Marchand
Marchlewski
Marggraf
Marignac
Markovnikov
Martí Franqués
Martinovics
Matthiessen
Mayow
Mellanby
Meltzer
Mendel, L. B.
Mendeleev
Meneghetti
Menghini
Menshutkin
Merica
Metzger
Meyer, J. F.
Meyer, J. L.
Meyer, K. H.
Meyer, V.
Meyerhof
Michael
Midgley
Mielli
Miescher
Miller, W. A.
Miller, W. L.
Millon
Mills
Mitscherlich
Mittasch
Mohr, C. F.
Moissan
Mond
Monge
Monnet
Moray
Morichini
Morley
Mosander
Muir
Mulder
Müller, Franz
Müller, P.
Mushet

Nametkin
Naumann, A.
Nef
Nencki
Neri
Nernst
Neumann, C.
Newlands
Nicholson, W.
Nieuwland
Niggli
Nobel
Noddack
Norton
Noyes, A. A.
Noyes, W. A.

Odling
Olszewski
Olympiodorus
Osborne
Osmond
Ostwald, C. W. W.
Ostwald, F. W.
Overton

Palladin
Paneth
Paracelsus
Parmentier
Parnas
Partington
Pasteur
Payen
Péan de Saint-Gilles
Pearson, G.
Pekelharing
Pelletier, B.
Pelletier, P. J.
Pelouze
Penny
Percy
Pérez de Vargas
Perkin, W. H.
Perkin, W. H. Jr.
Perrin
Personne
Persoz
Petrov, V. V.
Petrus Bonus
Pettenkofer
Pfeffer
Pfeiffer, P.
Plattner
Playfair, L.
Plot
Poggiale
Pope
Popov, A. N.
Pott
Power, F. B.
Pregl
Prévost, I. B.
Priestley, J.
Proust
Prout
Pryanishnikov

Rammelsberg
Ramsay, W.
Raoult
Ray, P. C.
Al-Rāzī
Regnault
Reichenbach, K. L.
Remsen
Rey, J.
Richards, T. W.
Richardson, B. W.
Richter
Rinman
Ritter
Rivett
Roberts-Austen
Robiquet
Roebuck
Rolfinck
Romé de l'Isle
Roozeboom
Roscoe
Rose, G.
Rose, H.
Rosenhain, W.
Rosenheim
Rouelle, G. F.
Rouelle, H.-M.
Rouelle, J.
Rowe
Ruland
Runge, F. F.
Ruscelli
Russell, E. J.
Rush
Rutherford, D.
Rutherford, E.

Sabatier, P.
Sage
Sala
Saussure, N. T. de
Scheele
Schiff, H. J.
Schmidt, G. C. N.
Schoenflies
Schönbein
Schopfer

Schorlemmer
Schott, O. F.
Schrötter
Schulze
Schunck
Schweigger
Sechenov
Séguin, A.
Sendivogius
Sennert
Sertürner
Serullas
Severgin
Severinus
Shaw, P.
Shilov
Sidgwick
Sigaud de Lafond
Silliman, B.
Silliman, B. Jr.
Skraup
Smith, Edgar F.
Smith, R. A.
Smithells
Smithson
Smits
Soddy
Solvay
Sørensen
Spring
Stahl
Starkey
Stas
Staudinger
Steacie
Stead
Stephanus of Alexandria
Stine
Stock, A.
Suchten
Sumner, J. B.
Sutherland
Svedberg
Swarts
Szebellédy
Szily

Tachenius
Talbot
Tammann
Tashiro
Tennant
Than
Thenard
Theophilus
Thiele, F. K. J.
Thomsen, H. P. J. J.
Thomson, T.
Thorpe, J. F.
Thorpe, T. E.
Thurneysser
Tiemann
Tilden
Tillet
Tiselius
Tolman
Traube
Travers
Trommsdorf
Troost, L. J.
Troostwijk
Tsvet
Turner, E.
Turquet de Mayerne
Tutton

Ulstad
Urbain
Ure

Vagner
Valentine
Van Slyke
Van't Hoff
Vauquelin
Venel
Vernadsky
Vigani
Virtanen

Waage
Wackenroder
Ibn Waḥshiyya
Wald, F.
Walden
Wallach
Wallerius
Walter of Odington
Walter
Wanklyn
Warburg, O. H.
Washburn
Waterston
Watson, R.
Watt
Webster, J.
Wedel
Wedgwood
Weigel, C. E.
Weizmann
Werner, A.
Whytlaw-Gray
Wiedemann
Wiegleb
Wieland, H. O.
Wiley
Wilhelmy
Williams, R. R.
Williamson, A. W.
Willstätter
Windaus
Winkler, C.
Winkler, L. W.
Wislicenus
Wöhler, F.
Wollaston, W. H.
Wood, H.
Woulfe
Wu
Wurtz

Young, S.

Al-Zahrāwī
Zambonini
Zeise
Zelinsky
Zemplén
Zinn
Zosimus of Panopolis
Zsigmondy
Zwelfer

EARTH SCIENCES

Abbe, C.
Abich
Abu'l Fidā'
Abū Ḥāmid al-Gharnāṭī
Acosta, J.
Adams, F. D.
Agassiz, A.
Agassiz, J. L. R.
Agricola
Albert I of Monaco
Alberti F. A. von
Albertus Magnus
Alzate y Ramírez
Ameghino
Anderson, E. M.
André
Andrusov
Anuchin
Anville
Apian
Arbos
Archiac
Arduino
Argand, E.
Aristotle
Arkell
Astbury
Aubuisson De Voisins
Azara
Babinet
Babington, W.
Baer
Baier
Bailey, E. B.
Bailey, L. W.
Bailey, S. I.
Baily, F.
Bakewell
Al-Bakrī
Barba
Barlow, W.
Barrande
Barrell
Barrois
Bassani
Bassler
Bather
Bauer
Baumhauer
Baylak Al-Qibjāqī
Becke
Becker
Béguyer De Chancourtois
Behaim
Belaiew
Bell, R.
Bellinsgauzen
Benzenberg
Berg
Bergman
Bering
Bernal
Berthier
Bertrand, M. A.
Bessel
Beudant
Beyrich
Billings
Birge
Al-Bīrūnī
Bischof
Bjerknes, V. F. K.
Black, James
Blaeu
Blanchard
Blomstrand
Bonaventura
Bonney
Boodt
Borelli
Born, I. E. von
Bosc
Bošković
Boucher
Boué
Bougainville
Bouguer
Boule
Boullanger
Bourguet
Bournon
Bowen
Bowie
Bowman, I.
Brandt, G.
Bravais
Breislak
Breithaupt
Brocard
Brocchi
Brochant de Villiers
Brøgger
Broili
Bromell
Brongniart, A.
Bronn
Brooks, A. H.
Broom
Bruce, J.
Brunhes
Bryan
Buache
Buch
Buchanan
Bucher
Buckland
Burnet
Bütschli
Buys Ballot

Cabeo
Caldas
Campanella
Cancrin
Carangeot
Carnall
Carpenter, H. C. H.
Cassini, G. D.
Cassini, J.
Cassini, J. D.
Cassini de Thury
Cayeux
Chamberlin
Chambers
Charcot, J. B.
Charpentier
Charpy
Chenevix
Chernov
Chernyshev
Chevenard
Childrey
Christol
Clarke, E. D.
Clarke, F. W.
Clarke, W. B.
Cleaveland
Cleve
Clift
Cloos
Clouet
Cocchi
Columbus
Conon of Samos
Conrad
Conybeare
Cook
Cordier
Coronelli
Cortés de Albacar
Cotta
Cotte
Credner
Croll
Cronstedt
Cross
Cushman
Cuvier, G.
Czekanowski
Czerski

Dainelli
Dalton, J.
Daly
Dana
Daniell
Danti
Darwin, C. R.
Daubeny
Daubenton
Daubrée
David
Davis
Dawson, J.
Dawson, J. W.
Day
Debenham
Dechen
DeGolyer
De La Beche
Delafosse
Delambre
Delisle, G.
Delisle, J.-N.
Deluc, J. A.
Deperet
Deryugin
Des Cloizeaux
Deshayes
Desmarest
Desor
Dicaearchus of Messina
Doelter
Dokuchaev
Dollo
Dolomieu
Dorno
Dove
Drygalski
Dudley
Dufrénoy
Duperrey
Du Toit
Dutton

Easton
Eaton
Ebel
Ehrenberg
Eichwald
Ekeberg
Ekman
Elhuyar, F. D'
Elhuyar, J. J. D'
Élie de Beaumont
Ellis
Emerson, B. K.
Emmons, E.
Emmons, S. F.
Eratosthenes
Ercker
Erman
Escher von der Linth
Escholt
Eskola
Espy
Evans, F. J. O.
Evans, L.

Falconer
Fankuchen
Farey
Faujas de Saint-Fond
Faye
Featherstonhaugh
Fenneman
Fenner
Ferrel
Fersman
Fichot
Fitton
Fitzroy
Fleming, J.
Flett
Foerste
Forbes, E.
Forbes, J. D.
Forchhammer
Forel
Forster, G. A.
Forster, J. R.
Foster, H.
Fouqué
Fraipont
Frankenheim
Franklin, B.
Freiesleben
Frenzel
Frere
Friedel
Fuchs, J. N. von
Füchsel
Fyodorov

Gabb
Gadolin
Gagnebin
Gahn
Gaimard
Gaudry
Geer, G. J. de
Geikie, A.
Geikie, J.
Gellibrand
Gemma Frisius
Genth
Germanus
Gessner
Gilbert, G. K.
Glaisher, J.
Glauber
Gmelin, J. G.
Godwin-Austen
Goethe
Goldschmidt, V.
Goldschmidt, V. M.
Golitsyn
Goodrich
Gosselet
Grabau
Gramont
Grand'eury
Granger
Greenough
Gressly
Griffith, R. J.
Groddeck
Groth
Grubenmann
Gua de Malves
Guenther
Guertler
Guettard
Guillaume
Guillet
Gunter
Gutenberg
Guyot

Haast
Hadfield
Hague
Haidinger
Hakluyt
Hall, J.
Hall, J. Jr.
Halley
Hamberg
Al-Hamdānī
Hamilton, W.
Hann
Harker
Harper, R. M.
Hartmann, C. F. A.
Hassler
Haug
Haüy
Ibn Hawqal
Hayden
Hayford
Hecataeus of Miletus
Hedin
Heer
Heim, A.
Heim, A. A.
Helmersen
Helmert
Henckel
Hermann, C. H.
Héroult
Hessel
Heyn
Heynitz
Hiärne
Hisinger
Hitchcock
Hoff
Holmes
Hopkins, W.
Horner, L.
Hough
Houghton
Humboldt
Hume-Rothery
Hunt, T. S.
Hutton, J.
Huxley, T. H.
Hyatt

Ibáñez e Ibáñez de Ibero
Ibrāhīm ibn Yaʿqūb
Iddings
Al-Idrīsī
Imamura
Inō
Issel

Jaccard
Jackson, C. T.
Jaeger, F. M.
Jaeger, G. F.
Jaekel
Jaggar
Jameson
Jars
Jefferson
Jeffries
Johannsen, A.
Johnson, D. W.
Johnson, W. D.
Joly
Juan y Santacilla
Juday
Jukes
Jussieu, A. de
Justi

Kaempfer
Kalm
Karpinski
Karpinsky
Karsten
Kater
Kavraysky
Kay
Keyserling
Ibn Khurradādbih
Al-Khwārizmī

King
Kirkaldy
Kirwan
Kitaibel
Klein, H. J.
Knipovich
Knorr
Knott
Knudsen
Koninck
Konkoly Thege
Kotō
Kovalevsky, V. O.
Krasheninnikov
Krasnov
Krasovsky
Krayenhoff
Kropotkin
Kruber

Lacaille
Lacroix, A.
La Hire, G.P. de
La Hire, P. de
Lamarck
Lamb
Lamétherie
Lang, K. N.
Langren
Lapparent
Lapworth
Larsen
Lartet, E. A. I. H.
Lartet, L.
Laussedat
Lavanha
Lavoisier
Lawson
Lazarev, M.P.
Lazarev, P.P.
Leakey
Le Châtelier
Leconte, Joseph
Lehmann, J. G.
Lenz
Leo the African
Leonardo da Vinci
Leonhard
Lepekhin
Le Roy, C.
Lesley
Lesquereux
Lesson
Le Verrier
Levinson-Lessing
Lévy, S. D. A.
Leybenzon
Lhwyd
Liesganig
Lindgren
Link
Linnaeus
Lisboa
Lister, M.
Litke
Logan, W. E.
Lohest
Löhneyss
Lokhtin
Lomonosov
Lonsdale, W.
Loomis
Lorenzoni
Lossen
Love
Lugeon
Lyell
Lyman, B. S.
Lyman, C. S.

Macculloch
Maclure
Magini
Maillet
Ibn Mājid
Makarov
Mallard
Mallet
Mantell
Al-Maqdisī
Maraldi, G. D.
Maraldi, G. F.
Marcou
Margerie
Margules
Marion
Mariotte
Markham
Martí Franqués
Martonne
Mason
Al-Masʿūdī
Mather
Matuyama
Mauguin
Maurer
Mauro
Maury, M. F.
Mawson
Mayer, J. T.
Mayer-Eymar
Méchain
Medina
Meek
Meinzer
Mendel, J. G.
Mercator
Merica
Merriam, J. C.
Merrill
Mesyatsev
Metius, A. A.
Meyer, C. E. H. von
Michel-Lévy
Middendorf
Miers
Miklukho-Maklay
Miller, H.
Miller, W. H.
Milne, J.
Mitchell, E.
Mitscherlich
Mohn
Mohorovičič
Mohr, C. F.
Mohs
Molina
Monnet
Montanari
Montelius
Moray
Moro
Morozov
Morse, J.
Mortillet
Mosander
Mosso
Mouchez
Munier-Chalmas
Münster
Murchison
Murray, J.
Mushet
Mushketov

Nansen
Nathorst
Naumann, K. F.
Necker
Neumann, F. E.
Neumayr
Newberry
Nicol
Niebuhr
Niggli
Nikitin
Nordenskiöld, N. A. E.
Norman
Norwood
Numerov

Obruchev
Ochsenius
Odierna
Olaus Magnus
Oldham, R. D.
Oldham, T.
Omalius d'Halloy
Omori
Oppel
Oppolzer
Orbigny
Orlov, A. Y.
Ortelius
Orton
Osborn
Osmond
Outhier
Owen, D. D.
Owen, G.
Owen, R.
Ozersky

Palissy
Pallas
Pander
Pappus of Alexander
Parkinson, J.
Pavlov, A. A.
Peirce, C. S.
Penck, A.
Penck, W.
Percy
Pereira, D. P.
Pérez de Vargas
Perrier, G.
Peters, C. F. W.
Pettersson
Petty
Pezenas
Phillips, J.
Phillips, W.
Picard, J.
Piatte
Pirī Rais
Plattner
Playfair, J.
Poleni
Pomponius Mela
Porro
Portevin
Pošepný
Potanin
Pourtalès
Powell, J. W.
Prestwich
Prévost, L. C.
Przhevalsky
Pumpelly
Pythias of Massalia

Al-Qazwīnī
Quenstedt

Rames
Ramond de Carbonnières
Ramsay, A. C.
Rankine, A. O.
Raspe
Ratzel
Raymond
Reck
Reclus
Redfield
Reid
Renard
Renault
Renevier
Rennell
Reuss, A. E.
Reuss, F. A.
Ribeiro Santos
Ricci, Mateo
Riccioli
Riccò
Richthofen
Ries
Rinman
Río
Rittenhouse
Rivière de Précourt
Roberts-Austen
Roche
Roemer F.
Roemer, F. A.
Rogers, H. D.
Rogers, W. B.
Romé de l'Isle
Romer
Rose, G.
Rosenberger
Rosenbusch
Rosenhain, W.
Ross, J. C.
Rossby
Roth
Rouelle, G. F.
Rouillier
Rovereto
Ruedemann
Rülein
Rumovsky
Russell, H. C.
Rütimeyer

Sabine, E.
Sage
Salisbury

Saporta
Sarychev
Satō Nobuhiro
Satō Nobukage
Satō Nobusue
Saussure, H. B. de
Sauveur, A.
Schardt
Scheuchzer
Schimper, K. F.
Schlotheim
Schlumberger
Schmerling
Schmidt, C. A. von
Schmidt, J. F. S.
Schneiderhöhn
Schöner
Schott, C. A.
Schouw
Schuchert
Schumacher
Scilla
Scott, D. H.
Scott, W. B.
Scrope
Sederholm
Sedgwick
Sedov
Selwyn
Semyonov-Tyan-Shansky
Sénarmont
Serres de Mesplès
Setchell
Severgin
Seward
Sezawa
Shaler
Sharonov
Shaw, W. N.
Shirakatsí
Shokalsky
Shtokman
Silliman B.
Silliman, B. Jr.
Smith, William
Smyth
Sohncke
Sokolov
Soldani
Soldner
Sollas
Sorby
Soulavie
Sowerby
Spencer, L. J.
Sprung
Staszic
Stead
Steller
Stensen
Sternberg, K. M. von
Sternberg, P. K.
Stewart, B.
Stille
Stock, C.
Størmer
Strabo
Strachey
Struve F. G. W.
Struve, G. W. L.
Struve, O. W.
Studer
Suess
Sverdrup
Swedenborg
Swinden

Tachini
Tanfilev
Tarde
Taylor, F. B.
Teall
Teilhard de Chardin
Termier
Theophilus
Theophrastus
Thomas, H. H.
Thomas, S. G.
Thompson, D. W.
Thomson, C. W.
Al-Tīfāshī
Tilas
Tillo
Toscanelli dal Pozzo
Toulmin
Townsend, J.
Trebra
Troost, G.
Tschermak
Turner, H. H.
Al-Ṭūsī (Nasir al-Dīn)
Tutton
Twenhofel
Tyrrell

Ulrich, E. O.
Urbain

Valmont de Bomare
Van Hise
Vanuxem
Varenius
Venetz
Vening Meinesz
Vernadsky
Verneuil
Vize
Voeykov
Vogt, J. H. L.
Vogt, T.
Voigt, J. C. W.
Volney
Voltz
Vysotsky

Waage
Walch
Walcott
Waldseemüller
Walker, J.
Wallerius
Washington
Webster, T.
Wegener
Weiss, C. S.
Wells, W. C.
Wendelin
Werner, A. G.
Werner, J.
White, C. D.
White, I. C.
Whitehurst
Whitfield
Whitney
Widmannstätten
Wiechert
Wild
Williams, H. S.
Williamson, W. C.
Willis, B.
Williston
Winchell, A.
Winchell, A. N.
Winchell, H. V.
Winchell, N. H.
Witham
Woltmann
Woodward, J.
Woodward, R. S.
Wright, E.
Wright, F. E.
Wright, G. F.
Wulff

Yāqūt al-Ḥamawī al-Rūmī
Young, T.

Zach
Zambonini
Zeiller
Zejszner
Zirkel
Zittel
Zubov

LIFE SCIENCES

Abano
Abel, J. J.
Abel, O.
Abreu
Acharius
Achillini
Acosta, C.
Adam of Bodenstein
Adanson
Addison
Aëtius of Amida
Agardh, C. A.
Agardh, J. G.
Agassiz, A.
Agassiz, J. L. R.
Agathinus
Agrippa
Aiton, W.
Aiton, W. T.
Alberti, L. B.
Alberti, S.
Albertus Magnus
Albinus, B.
Albinus, B. S.
Albinus, C. B.
Albinus, F. B.
Alcmaeon of Crotona
Alderotti, Taddeo
Aldrovandi
Alexander of Myndos
Alexander of Tralles
Allen
Alpini
Alzate Y Ramírez
Ameghino
Amici
Ancel
Anguillara
Anuchin
Apáthy
Aranzio
Arber
Archiac
Archigenes
Aretaeus of Cappadocia
Argenville
Aristotle
Arkell
Arnald of Villanova
Aromatari
Arsonval
Artedi
Asclepiades
Aselli
Assalti
Astbury
Astruc
Athenaeus of Attalia
Atwater
Audouin
Audubon
Auenbrugger
Avery
Ibn al-ᶜAwwām
Azara
Babington, C. C.
Baccelli
Bachmann, A. Q.
Baer
Baier
Baglivi
Bailey, L. H.
Baillie
Baillou
Baird
Baker, J. G.
Balbiani
Balfour, F. M.
Balfour, I. B.
Balfour, J. H.
Banister
Banks
Banti
Banting
Bárány
Barbour
Barchusen

Barclay
Barcroft
Barrande
Barrois
Barry
Barthez
Bartholin, C.
Bartholin, T.
Bartolotti
Barton
Bartram, J.
Bartram, W.
Bassani
Bassi
Bassler
Bastian
Bataillon
Bates
Bateson
Bather
Baudrimont
Bauer, F. L.
Bauer, F. A.
Bauhin, G.
Bauhin, J.
Bayliss, L. E.
Bayliss, W. M.
Ibn al-Bayṭār
Beale
Beaumont
Beccari
Becquerel, P.
Beddoe
Beddoes
Beevor
Behring
Beijerinck
Bekhterev
Bell, C.
Bellardi
Belleval
Bellini
Belon
Beneden, E. van
Beneden, P. J. van
Benedetti, A.
Benedict
Benivieni
Bensley
Bentham
Bérard, J. F.
Berengario Da Carpi
Berg
Berger, H.
Berger, J. G.
Bergman
Bérigard
Beringer
Berkeley, M. J.
Bernal
Bernard, C.
Bernard, N.
Bernheim
Bernoulli, D.
Bernstein, J.
Bert
Berthold
Bertrand, C. E.
Bertrand, G.
Bessey
Bexon
Beyrich
Bezold
Bichat
Bidder
Bidloo
Bilharz
Billings
Billroth
Binet
Birge
Bischoff, G. W.
Bischoff, T. L. W.
Bizzozero
Black, D.
Black, James
Black, Joseph
Blackman
Blainville
Blair
Blanc
Blumenbach
Blyth
Boas
Bock
Bodenheimer
Boerhaave
Bogdanov
Bohn
Bolk
Bolos of Mendes
Bolotov
Bonaparte
Bonnet, C.
Bonnier
Bonomo
Bordet
Bordeu
Borel, P.
Borelli
Borgognoni of Lucca
Bory de Saint-Vincent
Bosc
Bose, J.
Bostock
Botallo
Bottazzi
Boucher
Bouin
Boule
Bourdelot
Bourguet
Boveri
Bowditch, H. P.
Bower
Bowman, W.
Brachet
Bradford
Bradley, R.
Brandt, J. F.
Braun, A. C. H.
Bravais
Bredon
Brefeld
Breschet
Bretonneau
Breuer
Breuil
Bridges
Bright
Brisson, M. J.
Britten
Britton
Broca
Brödel
Brodie, B. C.
Broili
Brongniart, A. T.
Bronn
Brooks, W. K.
Broom
Broussais
Broussonet
Brown A. C.
Brown, R.
Brown, T.
Brown-Séquard
Bruce, D.
Brücke
Brumpt
Brunfels
Brunschwig
Brunton
Bucholz
Buckland
Budd
Buffon
Buller
Bulloch
Bunge
Buonanni
Burdach
Burdenko
Burdon-Sanderson
Burger
Busk
Ibn Buṭlān
Bütschli

Cabanis
Cadet de Gassicourt, C. L.
Cadet de Vaux
Caius
Caldani
Caldas
Calkins
Calmette
Camerarius
Campbell, D. H.
Camper
Canano
Candolle, A. de
Candolle, A. P. de
Cannon, W. B.
Carangeot
Cardano
Carlisle
Carlson
Carpenter, W. B.
Carrel
Carroll
Carus, J. V.
Casal Julian
Casseri
Castaldi
Castle
Catesby
Cattell
Caullery
Cavanilles
Cels
Cervantes
Cesalpino
Cesi
Cestoni
Chabry
Chagas
Chamberlain
Chamberland
Chambers
Chamisso
Chaplin
Chapman, A. W.
Charcot, J. B.
Charcot, J. M.
Chardenon
Charleton
Chauliac
Chauveau
Cheyne
Chiarugi
Child
Childrey
Chittenden
Chodat
Christol
Claus, C. E.
Clements
Clerck
Clift
Coghill
Cogrossi
Cohn, E. J.
Cohn, F. J.
Coiter
Colden
Collinson
Collip
Colombo
Combes, R.
Commerson
Conklin
Conrad
Constantine the African
Cope
Cordus, E.
Cordus, V.
Cori
Cornette
Correns
Corti, A.
Corti, B.
Corvisart
Costantin
Cotugno
Coues
Councilman
Crawford
Creighton
Crollius
Croone
Cruikshank
Cruveilhier
Cruz
Cuénot
Cullen
Cushing
Cushman
Cushny
Cuvier, F.
Cuvier, G.
Czermak
Czerski

Dale
Daléchamps
Dalibard
Dall
Dalton, J. C.
Al-Damīrī
Danforth
Darlington
Darwin, C. R.
Darwin, E.
Darwin, F.
Daubenton
Davaine
Davenport
Davison
Davy, J.
Dawson, J.
Dean
De Bary
Déchellete
Delage

Delile
Denis
Deperet
Derham
Deryugin
Descartes
Descourtilz
Deshayes
Desor
Devaux
Digby
Dillenius
Diocles of Carystus
Dionis du Séjour
Dioscorides
Dixon, H. H.
Dobell
Dodart
Dodoens
Dogel
Dohrn
Dokuchaev
Döllinger
Dollo
Domagk
Dombey
Dominicus de Clavasio
Donaldson
Donders
Dondi
Dorn
Douglas, James
Draparnaud
Driesch
Dubini
Dubois
Dubois-Reymond, E. H.
Duboscq
Duchesne
Duclaux
Duggar
Du Hamel
Duhamel du Monceau
Dujardin
Duméril
Dunglison
Dutrochet
Duval
Duverney
Dyadkovsky

East
Eaton
Ebel
Eberth
Edwardes
Edwards
Egas Moniz
Ehrenberg
Ehret
Ehrlich
Eichler
Eichwald
Eigenmann
Eijkman
Einthoven
Eller von Brockhausen
Elliot Smith
Elvehjem
Embden
Emerson, R.
Engelmann, G.
Englemann, T. W.
Engler
Ent

Erasistratus
Erastus
Erlanger
Erman
Errera
Escherich
Eschscholtz
Estienne
Euler-Chelpin
Eustachi
Evans, A. W.
Evans, W. H.
Evelyn
Ewing, J.

Fabre
Fabrici
Fabricius
Falconer
Falloppio
Farmer
Fechner
Fée
Fernald
Fernel
Ferrein
Ferrier
Feuillée
Feulgen
Fick
Fiessinger
Fink
Finlay
Finsen
Fischer, H. O. L.
Fisher, R. A.
Fitch
Flechsig
Fleming, A.
Fleming, J.
Flemming
Fletcher
Flexner
Florey
Flourens
Flower
Fludd
Foerste
Fol
Folin
Fontana
Forbes, A.
Forbes, E.
Forbes, S. A.
Forel
Forsskål
Foster, M.
Fourcroy
Fracastoro
Fraipont
Frankland, P. F.
Franklin, R. E.
Frazer
Frederick II of Hohenstaufen
Fredericq
Freind
Freud
Frey
Fries, E. M.
Fritsch
Fuchs, L.
Fuhlrott
Fulton
Funk

Gaertner, J.
Gaertner, K. F. von
Gaffky
Gagliardi
Gaimard
Gaines
Galeazzi
Galen
Gall
Galton
Galvani
Gamaleya
Garnett
Garnot
Garreau
Gaskell
Gasser
Gates
Gaudry
Gaultier de Claubry
Gay
Gayant
Geer, C. de
Gegenbaur
Geminus, T.
Geoffroy, C. J.
Geoffroy, E. F.
Geoffroy, E. L.
Geoffroy Saint-Hilaire, E.
Geoffroy Saint-Hilaire, I.
Gerard, J.
Gerbezius
Gesell
Gesner
Gessner
Ghini
Ghisi
Giard
Giles of Rome
Gill, T. N.
Girtanner
Glaser, C.
Glaser, J. H.
Glauber
Gleichen-Russworm
Glisson
Gmelin, J. G.
Goebel
Goedaert
Goeppert
Goethe
Goette
Gohory
Goldberger
Goldschmidt, R. B.
Golgi
Goltz
Goodrich
Goodsir
Gosselet
Gould, A. A.
Gould, J.
Graaf
Grabau
Gram
Grand'eury
Granger
Grassi
Gratiolet
Gray, A.
Gray, H.
Green, J.
Gregory, F. G.
Gregory, W.
Gressly
Grew
Griffith, W.
Grijns

Grinnell
Grisebach
Grote
Gruber
Gruby
Gudden
Guettard
Guidi
Guignard
Guilliermond
Guinter
Gulland
Gullstrand
Gurvich
Guyer
Gwynne-Vaughan
Gyllenhaal

Haberlandt
Haeckel
Haffkine
Hahnemann
Haldane, John B. S.
Haldane, John Scott
Hales
Hall, G. S.
Hall, J. Jr.
Hall, M.
Haller
Hallier
Halsted, W. S.
Al-Hamdānī
Hansen, E. C.
Hansen, G. H. A.
Harden
Hardy
Harlan
Harper, R. A.
Harper, R. M.
Harpestraeng
Harrison, R. G.
Hart
Hartig
Harting
Hartley
Hartmann, J.
Harvey, W.
Harvey, W. H.
Hatschek
Haug
Havers
Haworth, A. H.
Hedwig
Heer
Heidenhain, M.
Heidenhain, R. P. H.
Heister
Hektoen
Helmholtz
Helmont
Henderson, L. J.
Henderson, Y.
Henfrey
Henking
Henle
Henry of Mondeville
Hensen
Henslow
Herapath, W. B.
Herbart
Herbert
Hérelle
Hering
Hernández
Herophilus

Herrera
Herrick, C. J.
Herrick, C. L.
Hertwig, K. W. T. R. von
Hertwig, W. A. O.
Heurne
Hewson
Hiärne
Highmore
Hill, J.
Hippocrates of Cos
Hirszfeld
His
Hitchcock
Hitzig
Hjort
Hoagland
Hoeven
Hoffmann, F.
Hofmeister
Holbrook
Holmgren
Home
Hooker, J. D.
Hooker, W. J.
Hopkins, F. G.
Hoppe-Seyler
Horbaczewski
Horn
Horne
Horsley
Hosack
Houghton
Houssay
Howard
Howell
Hrdlička
Huber, J. J.
Hubrecht
Hudson, W.
Humboldt
Ḥunayn ibn Isḥāq
Hundt
Hunt, J.
Hunter, J.
Hunter, W.
Huschke
Hutchinson
Hutton, J.
Huxley, T. H.
Hyatt
Hyrtl

Ingen-Housz
Ingrassia
Isaac Israeli
Isaacs
Isḥāq ibn Ḥunayn
Ivanov, I. I.
Ivanov, P. P.
Ivanovsky

Jaccard
Jackson, C. T.
Jackson, J. H.
Jacquin
Jaeger, G. F.
Jaekel
Al-Jāḥiẓ
James, W.
Jameson
Jefferson
Jeffrey
Jeffreys
Jenkinson
Jenner
Jennings
Jensen, C. O.
Jepson
Joblot
Johannsen, W. L.
Johnson, T.
Jonston
Jordan, C. T. A.
Jordan, D. S.
Jordan, E. O.
Juday
Ibn Juljul
Juncker
Jung
Jungius
Jussieu, A. de
Jussieu, A. H. L. de
Jussieu, A. L. de
Jussieu, B. de
Jussieu, J. de

Kaempfer
Kalm
Keilin
Keill, James
Keith
Kellner
Kellogg, A.
Kellogg, V. L.
Kelser
Kendall
Kerr, J. G.
Keyserling
Khunrath, C.
Khunrath, H.
Kidd
Kielmeyer
Kitaibel
Kitasato
Klebs
Klein, J. T.
Kleinenberg
Kluyver
Knight
Knipovich
Knorr
Knox
Knuth
Koch, H. H. R.
Koelliker
Koelreuter
Kofoid
Köhler
Koltzoff
König, E.
Koninck
Kossel, K. M. L.
Kovalevsky, A. O.
Kovalevsky, V.
Kraft
Krasheninnikov
Krasnov
Krogh
Kronecker, H.
Kühn, A.
Kühne
Kunth
Kuntze
Kützing
Kylin

La Brosse
Lacaze-Duthiers
Lacépède
La Condamine
Laennec
Lamarck
Lamétherie
La Mettrie
Lamouroux
Lamy, G.
Lancisi
Landsteiner
Lang, A.
Lang, K. N.
Lang, W. V.
Lange
Langerhans
Langley
Lankester
Lapicque
Larghi
La Rive, C. G. de
Lartet, E. A. I. H.
Lartet, L.
Lashley
Latreille
Laurens
Laveran
Lavoisier
Lavrentiev
Lawrence, W.
Lazarev, P. P.
Lea
Leakey
Le Cat
L'Écluse
Leconte, John
Leconte, Joseph
Le Dantec
Le Double
Leeuwenhoek
Le Febvre
Legallois
Leger
Lehmann, J. G.
Leidy
Lemery, L.
Lemery, N.
Le Monnier, L. G.
Leonardo da Vinci
Leonhardi
Leoniceno
Lepekhin
Lereboullet
Le Roy, C.
Lesquereux
Lesson
Lesueur
Leuckart
Leuret
Levaditi
Levaillant
Levene
Levi
Lewis, Thomas
Lewis, Timothy R.
Lewis, W.
Leydig
L'Héritier de Brutelle
Lhwyd
Libavius
Liceaga
Lieberkühn
Liebig
Lieutaud
Lignier
Lillie
Linacre
Lind
Lindley
Link
Linnaeus
Li Shih-Chen
Lister, J.
Lister, M.
Livingston
Lloyd, J. U.
L'Obel
Loeb, J.
Loeb, L.
Loeffler
Loewi
Lohest
Lonicerus
Lorry
Lotsy
Lotze
Lovell
Lovén
Lower
Lubbock, J.
Lucas, K.
Luciani
Ludwig
Luna
Lusitanus
Lusk
Lustig
Lyell
Lyonet

Macallum
Macbride, D.
MacBride, E. W.
McClung
McCollum
Mach
Macheboeuf
McIntosh
Macleod
Magati
Magendie
Maggi
Magnenus
Magnol
Magnus, R.
Maimonides
Maire
Al-Majūsī
Maksimov
Malesherbes
Mall
Malouin
Malpighi
Manardo
Mangin
Manson
Mantegazza
Marchant, J.
Marchant, N.
Marchi
Marchia Fava
Marci of Kronland
Marcou
Marey
Marie
Marion
Mariotte
Markgraf
Marliani
Marsh
Marsili
Marsilius of Inghen

Martí Franqués
Martin, H. N.
Martin, R.
Martínez
Martius
Marum
Mascagni
Massa
Mast
Matruchot
Matteucci
Mattioli
Maupas
Maupertuis
Mayer, J. R.
Mayer-Eymar
Mayo
Mayow
Meckel
Medicus
Meek
Meissner
Mellanby
Meltzer
Mendel, J. G.
Mendel, L. B.
Meneghetti
Menghini
Menuret de Chambaud
Mercati
Merriam, C. H.
Merriam, J. C.
Merrill
Mersenne
Méry
Mesmer
Mesnil
Metchnikoff
Mettenius
Meyen
Meyer, C. E. H. von
Meyerhof
Michaux
Micheli
Michelini
Michurin
Middendorf
Miescher
Miller, P.
Millington
Millon
Milne-Edwards
Minot, C. S.
Minot, G. R.
Miquel
Mirbel
Mitchell, E.
Mitchell, S. W.
Mivart
Möbius, K. A.
Mociño
Moench
Moerbeke
Moffett
Mohl
Moldenhawer
Moleschott
Molina
Molliard
Monardes
Mondino de' Luzzi
Monro, A. (I)
Monro, A. (II)
Montanari
Montgomery, E. D.
Montgomery, T. H.
Morat
Moray
Morgagni
Morgan, C. L.
Morgan, T. H.
Morichini
Morin
Morison
Morozov
Morse, E. S.
Mortillet
Morton, J.
Morton, S. G.
Moss
Mosso
Mottram
Müller, Fritz
Müller, G. E.
Muller, H. J.
Müller, J. P.
Müller, O. F.
Munier-Chalmas
Muralt
Murphy
Murray, G. R. M.
Mutis y Bossio

Naegeli
Ibn al-Nafīs
Nansen
Nathorst
Naudin
Navashin
Neander
Necker
Needham
Nees von Esenbeck
Negri
Nehring
Neisser
Nemesius
Nencki
Neumayr
Newberry
Newport
Newton, E. T.
Nicholas
Nicol
Nicolaus of Damascus
Nicolle
Nicot
Nifo
Nilsson-Ehle
Nissl
Noguchi
Nordenskiöld, N. E.
Norton
Nostradamus
Novy
Nuttall
Nylander, F.
Nylander, W.

Oddi
Odierna
Oken
Olaus Magnus
Olbers
Oliver
Oppel
Orbeli
Orbigny
Oribasius,
Orta
Orton
Osborn
Ostwald, C. W. W.
Ott
Oudemans
Overton
Owen, R.

Pacchioni
Pacini
Packard
Pagano
Painter
Palissy
Palladin
Pallas
Palmer
Pander
Papanicolaou
Paracelsus
Paré
Parkinson, J.
Parkinson, S.
Parmentier
Parnas
Pasteur
Paul of Aegina
Paulli
Pauly
Pavlov, I. P.
Pavón y Jiménez
Paykull
Peale, R.
Peale, T. R.
Pearl
Pearson, K.
Pecquet
Pekelharing
Peletier, J.
Pelletier, P. J.
Pembertom
Pennant
Pensa
Pérez de Vargas
Péron
Perrault, C.
Perrault, P.
Perrier, E.
Perroncito
Personne
Persoon
Petrie
Pettenkofer
Peyer
Peyssonnel
Pezard
Pfeffer
Pflüger
Philinus of Cos
Philip
Phillips J.
Philolaus of Crotona
Pianese
Piccolomini
Piette
Pincus
Pinel
Pires
Pirogov
Piso
Pitcairn
Pitt-Rivers
Plate
Plato of Tivoli
Platter
Plenčič
Pliny
Plot
Plumier
Plummer, A.
Poggiale
Poiseuille
Poivre
Poli
Pollender
Pontedera
Porta, G. della
Porta, L.
Portal
Portier
Potanin
Pouchet
Pourfour du Petit
Powell, J. W.
Power, F. B.
Power, H.
Pratt, F. H.
Praxagoras of Cos
Presl
Prevost, I. B.
Prevost, J. L.
Preyer
Prichard
Priestley, J. G.
Pringle
Pringsheim, N.
Procháska
Prout
Prowazek
Prudden
Pruner Bey
Pryanishnikov
Przhevalsky
Pumpelly
Punnett
Purkyně
Pursh
Putnam

Quatrefages de Bréau
Quenstedt
Ibn al-Quff
Quoy
Quṣṭa ibn Lūqā
Quṭb al-Dīn

Rabelais
Rabl
Rádl
Raffles
Rafinesque
Raman
Rames
Ramon
Ramond de Carbonnières
Ramón y Cajal
Ranvier
Raspail
Rathke
Ratzel
Raulin
Rauwolf
Ray, J.
Raymond
Al-Rāzī
Réaumur
Reck
Redfield
Redi

Redouté
Reed
Reichert
Reil
Remak
Renault
Renaut
Renevier
Retzius, A. A.
Retzius, M. G.
Reuss, A. E.
Reyna
Richards, F. J.
Richardson, B. W.
Richet
Ricketts
Ridgway
Ibn Riḍwān
Rima
Ringer
Río-Hortega
Riolan
Ristoro
Rittenhouse
Ritter
Riva-Rocci
Rivière de Précourt
Robin
Robinet
Robiquet
Roemer, F.
Roemer, F. A.
Roesel von Rosenhof
Rolando
Rolfinck
Rolleston
Rollet
Romanes
Rondelet
Roomen
Rosa
Rosenblueth
Ross, R.
Rostan
Rouelle, J.
Rouget
Rouillier
Roux, P. P. E.
Roux, W.
Rowe
Rubner
Rudbeck
Rudolphi
Ruedemann
Ruel
Ruffer
Ruffini, A.
Ruffini, P.
Ruffo
Rufinus
Rufus of Ephesus
Ruini
Ruiz
Ruland
Rülein
Ruscelli
Rush
Ibn Rushd
Russell, E. J.
Rütimeyer
Ruysch

Sabatier, A.
Sabin
Sacco
Sachs
Saint-Hilaire
Sakharov
Sala
Salernitan Anatomists
Salomonsen
Salviani
Al-Samarqandi (*d.* 1222)
Al-Samaw'al
Samoylov
Sanarelli
Sanchez
Sanderson
Sanio
Santorini
Santorio
Saporta
Sars
Saussure, H. B. de
Saussure, N. T. de
Sauvageau
Savigny
Say
Scaliger
Scarpa
Schaudinn
Scheele
Schegk
Scheuchzer
Schiff, M.
Schimper, A. F. W.
Schimper, K. F.
Schimper, W. P.
Schleiden
Schlotheim
Schlumberger
Schmerling
Schmidel
Schmidt, E. J.
Schneider
Schönherr
Schönlein
Schoolcraft
Schopfer
Schouw
Schreibers
Schroeder van der Kolk
Schuchert
Schultze
Schulze
Schwann
Schwendener
Sclater
Scott, D. H.
Scott, W. B.
Scudder
Sechenov
Séguin, A.
Semmelweis
Semon
Semper
Senac
Senebier
Sennert
Sergent
Serres, A. E. R. A.
Serres, O. de
Serres de Mesplès
Sertoli
Sertürner
Serullas
Servetus
Sessé y Lacasta
Setchell
Severino
Severinus
Severtsov
Seward
Sharpey
Sharpey-Schäfer
Sharrock
Sherard
Sherrington
Shmalhauzen
Shull
Siebold
Siedlecki
Sigaud de Lafond
Sigüenza y Góngora
Ibn Sīnā
Ṣinān
Škoda
Skryabin
Sloane
Smith, Edward
Smith, Erwin F.
Smith, H. W.
Smith, J. E.
Smith, P. E.
Smith, S. I.
Smith, T.
Smith, Wilson
Snow
Soemmerring
Solander
Sollas
Sonnerat
Soranus of Ephesus
Sorby
Souleyet
Sowerby
Spallanzani
Spemann
Spencer, H.
Spiegel
Spix
Sprengel, C. K.
Sprengel, K. P. J.
Spruce
Spurzheim
Stahl
Stannius
Starkey
Starling
Stebbing
Steenstrup
Stejneger
Steller
Stelluti
Stensen
Sternberg, K. M. von
Stevens
Stewart, G. N.
Stiles
Stimpson
Stock, C.
Strasburger
Streeter
Struss
Stuart
Sturtevant
Suchten
Sudhoff
Sumner, F. B.
Sumner, J.
Suringar
Sutton
Svedelius
Swainson
Swammerdam
Swartz
Swedenborg
Swieten
Sydenham
Sylvius
Szilard

Al-Ṭabarī (11th cent.)
Al-Ṭabarī (9th cent.)
Tabor
Tachenius
Targioni Tozzetti
Tashiro
Taylor, C. V.
Teichmann
Teilhard de Chardin
Tennent
Ten Rhyne
Testut
Thābit ibn Qurra
Thaxter
Thayer
Theophrastus
Thiselton-Dyer
Thomas, H. H.
Thompson, D. W.
Thompson, J. V.
Thouin
Thunberg, T. L.
Thuret
Thurnam
Ibn Tibbon, M.
Tiedemann
Van Tieghem
Al-Tīfāshī
Tikhov
Tillet
Ibn al-Tilmīdh
Timiryazev
Tiselius
Titius
Torre, M. della
Torrey
Toscanelli dal Pozzo
Tournefort
Townsend, J.
Tozzi
Tradescant, J. I
Tradescant, J. II
Traube
Trelease
Trembley
Treub
Treviranus, G. R.
Treviranus, L. C.
Triana
Troja
Trommsdorf
Troost, G.
Trouvelot
Trulli
Tschermak von Seysenegg
Tsvet
Ibn Ṭufayl
Tulasne
Tulp
Türck
Turgo, E. F.
Turner, W. (*d* 1568)
Turner, W. (*d.* 1916)
Turpin
Turquet de Mayerne
Twort
Tyndall
Tyson

Ukhtomsky
Ulloa
Ulrich
Ulstad
Unanue
Unger

Unzer
Uexküll

Vaillant, L. L.
Vaillant, S.
Valenciennes
Valentin
Vallisnieri
Valmont de Bomare
Valsalva
Valverde
Van Slyke
Varolio
Varro
Vassale
Vastarini-Cresi
Vaucher
Vavilov, N. I.
Vejdovský
Vellozo
Venel
Verneuil
Verrill
Verworn
Vesalius
Vesling
Vicq d'Azyr
Vieussens
Vigani
Vigo
Vilmorin
Vinogradsky
Virchow
Virey
Virtanen
Vogt, C.
Voit, C. von
Voronin
Vries
Vvedensky

Wackenroder
Ibn Wāfid
Wagner
Wagner von Jauregg
Wahlenberg
Ibn Waḥshiyya
Walch
Walcott
Waldeyer-Hartz
Walker, A.
Walker, J.
Wallace, A. R.
Waller
Wallich
Walton
Wanklyn
Warburg, O. H.
Warming
Wassermann
Waterton
Watson, H. C.
Watson, S.
Watson, W.
Weber, E. H.
Weber, M. W. C.
Webster, J.
Wedel
Weichselbaum
Weigert
Weinberg
Weismann
Weizmann
Welch
Weldon
Wells, H. G.
Wells, W. C.
Went
Wepfer
Wernicke
Wertheim
Wharton, T.
Wheeler
White, C.
White, G.
Whitfield
Whitman
Whytt
Wickersheimer
Wied
Wiedersheim
Wiegleb
Wieland, M.
Wiesner
Wigand
Wilbrand
Willdenow
Williams, H. S.
Williams, R. R.
Williamson, N. C.
Willis, T.
Williston
Willughby
Wilson, Alexander (*d.* 1813)
Wilson, Edmund B.
Wilson, Edwin B.
Winsløw
Winthrop
Wistar
Witham
Withering
Wolff, C. F.
Wollaston, W. H.
Wood, H.
Woodward, J.
Worm
Wotton
Wright, A. E.
Wu
Wundt
Wyman

Yerkes
Yersin
Young, T.

Al-Zahrāwī
Zalužanský
Zambeccari
Zavadovsky
Zavarzin
Zeiller
Zinsser
Zittel
Ibn Zuhr
Zwelfer

MATHEMATICS

Abel, N. H.
Abraham Bar Hiyya
Abū Kāmil
Abul'-Wafā'
Adams, J. C.
Adelard of Bath
Adrain
Aepinus
Agnesi
Aguilon
Ahmad Ibn Yūsuf
Aida Yasuaki
Ajima Naonobu
Albert of Saxony
Alberti, L. B.
Albertus Magnus
Alembert
Alzate y Ramírez
Ampère
Amsler
Anatolius of Alexandria
Anderson, O. J. V.
Andoyer
Angeli
Anthemius of Tralles
Antiphon
Apollonius of Perga
Apell
Arbogast
Arbuthnot
Archimedes
Archytas of Tarentum
Argand, J. R.
Aristaeus
Aristarchus of Samos
Aristotle
Aristoxenus
Arnauld
Aronhold
Artin
Āryabhaṭa I
Āryabhaṭa II
Atwood
Autolycus of Pitane
Auzout
Azara

Babbage
Bachelier
Bachet de Méziriac
Bachmann, P. G. H.
Bacon, R.
Al-Baghdādī
Baire
Balbus
Balmer
Banach
Ibn Al-Bannā'
Banū Mūsā
Barbier
Barlow, P.
Barocius
Barrow
Bartholin, E.
Bateman
Al-Battānī
Bayes
Beaugrand
Bell, E. T.
Bellavitis
Beltrami
Benedetti, G. B.
Bernoulli, D.
Bernoulli, Jakob I
Bernoulli, Jakob II
Bernoulli, Johann I
Bernoulli, Johann II
Bernoulli, Johann III
Bernoulli, Nikolaus I
Bernoulli, Nikolaus II
Bernstein F.
Bernstein, S. N.
Bertini
Bertrand, J. L. F.
Berwick
Bessel
Betti
Bezout
Bhāskara II
Bianchi
Bienaymé
Billy
Birkhoff
Al-Bīrūnī
Bjerknes, C. A.
Blaschke
Blasius of Parma
Blichfeldt
Bliss
Bobillier
Bôcher
Boethius
Bohl
Bohr, H.
Bolyai, F.
Bolyai, J.
Bolza
Bolzano
Bombelli
Bonnet, P. O.
Boole
Borchardt
Borda
Borel, E.
Borelli
Bortkiewicz
Bortolotti
Bošković
Bosse
Bossut
Bougainville
Boulliau
Bouquet
Bour
Bourbaki
Boussinesq
Boutroux
Bouvelles
Bradwardine

Braikenridge
Bramer
Brashman
Braunmühl
Bredon
Bret
Brianchon
Briggs
Brill
Brillouin
Bring
Brinkley
Brioschi
Briot
Brisson
Brocard
Bromwich
Brouncker
Brouwer, L. E. J.
Brożek
Bryson of Heraclea
Budan de Boislaurent
Bugaev
Bunyakovsky
Buot
Burali-Forti
Bürgi
Burnside
Burrau
Buteo

Cabeo
Calandrelli, I.
Callippus
Campanus of Novara
Camus
Cantor, G.
Cantor, M. B.
Caramuel y Lobkowitz
Carathéodory
Carcavi
Cardano
Carnot, L. N. M.
Cartan
Castel
Castelnuovo
Castillon
Cataldi
Cauchy
Cavalieri
Cayley
Čech
Cesàro
Ceulen
Ceva, G.
Ceva, T.
Chaplygin
Chasles
Chebotaryov
Chebyshev
Cheyne
Ch'in Chiu-Shao
Christmann
Christoffel
Chrystal
Chu Shih-Chieh
Chuquet
Ciruelo
Clairaut
Clarke, S.
Clausen
Clavius
Clebsch
Clifford
Codazzi
Cole
Collins
Commandino
Comte
Condorcet
Conon of Samos
Coolidge
Cosserat
Cotes
Cournot
Couturat
Craig
Cramer
Crelle
Cremona
Crousaz
Culmann
Cunha
Curtze
Cusa

Dandelin
Danti
Darboux
D'Arcy
Darwin, C. G.
Darwin, G. H.
Dasypodius
Davidov
Debeaune
Dechales
Dedekind
Dee
De Groot
Dehn
Delamain
Democritus
De Morgan
Deparcieux
Desargues
Descartes
Dickson
Dickstein
Digges, L.
Digges, T.
Dini
Dinostratus
Diocles
Dionis du Séjour
Dionysodorus
Diophantus of Alexandria
Dirichlet
Dodgson
Dominicus de Clavasio
Domninus of Larissa
Doppelmayr
Doppler
Dositheus
Douglas, Jesse
Drach
Du Bois-Reymond, P. D. G.
Dudith
Duhamel
Dupin
Dupré
Dürer
Dyck

Egorov
Eisenhart
Eisenstein
Engel
Enriques
Eratosthenes
Esclangon
Euclid
Eudoxus of Cnidus
Euler
Eutocius of Ascalon
Ibn Ezra

Fabri
Fagnano dei Toschi, G. F.
Fagnano dei Toschi, G. C.
Fano
Farrar
Fatou
Faulhaber
Feigl
Fejér
Fermat
Ferrari
Ferrel
Ferro
Feuerbach
Fibonacci
Fields
Fine, H. B.
Fine, O.
Fink
Fisher, R. A.
Fontaine
Fontenelle
Forsyth
Fourier
Fraenkel
Français, F. J.
Français, J. F.
Francesca
Frank, P.
Fredholm
Frege
Frenet
Frenicle de Bessy
Friedmann
Fries, J. F.
Frisi
Frobenius
Fubini
Fuchs, I. L.
Fueter
Fuss
Fyodorov

Galerkin
Galois
Galton
Gauss
Geiser
Gelfond
Gellibrand
Geminus
Gemma Frisius
Gentzen
Gerard of Brussels
Gerbert
Gergonne
Germain
Ghetaldi
Giorgi
Girard, A.
Glaisher, J. W. L.
Goldbach
Gompertz
Göpel
Gordan
Gossett
Goursat
Gräffe
Grandi
Grassman
Graunt
Grave
'sGravesande
Green, G.
Gregory, D.
Gregory, D. F.
Gregory, J.
Gregory, O. G.
Grossmann, M.
Gua de Malves
Guccia
Gudermann
Guenther
Guldin
Gunter

Haar
Ḥabash al-Ḥāsib
Hachette
Hadamard
Halley
Halphen
Halsted, G. B.
Hamilton, W.
Hamilton, W. R.
Hankel, H.
Hardy, C.
Hardy, G. H.
Harriot
Hartmann, G.
Hartmann, J.
Hartree
Hausdorff
Ibn al-Haytham
Heath
Hecht
Hecke
Heine
Hellinger
Henrion
Hensel
Heraclides Ponticus
Herbrand
Hérigone
Hermann the Lame
Hermann, J.
Hermite
Hero of Alexandria
Hesse
Heuraet
Heytesbury
Hilbert
Hill, G. W.
Hill, L. S.
Hindenburg
Hipparchus
Hippias of Elis
Hippocrates of Chios
Hobbes
Hobson
Hodgkinson
Hoëné-Wroński
Hölder
Holmboe
Hopf
Hopkins, W.
Horner, W. G.
Hoüel

Hudde
Hugh of St. Victor
Humbert, M. G.
Humbert, P.
Huntington
Hurwitz
Hutton, C.
Huygens
Hypatia
Hypsicles of Alexandria

Ibrāhīm ibn Sinān
Isidorus of Miletus
Ivory

Jābir ibn Aflah
Jacobi, C. G. J.
Jagannātha
Janiszewski
Al-Jawharī
Al-Jayyānī
Jensen, J. L. W. V.
Jerrard
Jevons
Joachimsthal
John of Gmunden
John of Lignères
John of Murs
Johnson, W. E.
Jones, W.
Jonquières
Jordan, C.
Jordanus de Nemore
Juel
Jungius

Kaestner
Kagan
Kaluza
Kamāl al-Dīn
Al-Karajī
Al-Kāshī
Keckermann
Keill, John
Kerékjártó
Keynes
Al-Khalīlī
Al-Khayyāmī
Al-Khāzin
Khinchin
Al-Khujandī
Al-Khwārizmī
Kirkman
Klein, C. F.
Klügel
Kneser
Knopp
Köbel
Koch, H. von
Kochin
Koenig, J.
Koenig, J. S.
Koenigs
Kolosov
Königsberger
Korteweg
Kotelnikov
Kovalevsky, S.
Kraft
Kramp
Krasovsky
Kronecker, L.
Krylov, A. N.
Krylov, N. M.
Kummer
Kürschák
Kushyār

La Condamine
Lacroix, S. F.
La Faille
Lagny
Lagrange
Laguerre
La Hire, P. de
Lalouvère
Lamb
Lambert
Lamé, G.
Lamy, B.
Lancret
Landau, E.
Landen
Landsberg, G.
Lansberge
Laplace
La Roche
Laurent, M. P. H.
Laurent, P. A.
Lavanha
Lax
Lebesgue
Legendre
Leibniz
Lemoine
Leo
Leo the Mathematician
Leodamas of Thasos
Leonardo da Vinci
Le Paige
Le Poivre
Lerch
Le Roy, E.
Lesniewski
Le Tenneur
Leurechon
Levi ben Gerson
Levi-Civita
Lévy, M.
Lexell
L'Hospital
L'Huillier
Li Chih
Lie
Lindelöf
Lindemann, C. L. F.
Liouville
Lipschitz
Liu Hui
Lobachevsky
Loewner
Loewy
Loomis
Loria
Lotka
Love
Lucas, F. E. A.
Lueroth
Łukasiewicz
Lull
Luzin
Lyapunov

Macaulay
McColl
Maclaurin
MacMahon
Macmillan
Magini
Magnitsky
Al-Māhānī
Mahāvīra
Maior
Malebranche
Malfatti
Maltsev
Mannheim
Mansion
Manṣūr
Marci of Kronland
Markov
Martianus Capella
Mascheroni
Maseres
Matthews
Mathieu
Maupertuis
Maurolico
Mayer, C. G. A.
Mazurkiewicz
Mello
Menabrea
Menaechmus
Menelaus of Alexandria
Mengoli
Méray
Mercator
Mersenne
Meshchersky
Metius, A.
Metius, A. A.
Metius, J.
Meusnier de la Place
Meyer, W. V.
Milhaud
Miller, G. A.
Miller, W. H.
Minding
Mineur
Minkowski
Mises
Mittag-Leffler
Möbius, A. F.
Moerbeke
Mohr, G.
Moiseev
Moivre
Molin
Mollweide
Monge
Monte
Montmort
Montucla
Moore, E. H.
Morland
Moulton
Moutard
Mouton
Muḥyi 'l-Dīn
Munīśvara Viśvarūpa
Mydorge
Mylon

Nairne
Napier
Nārāyaṇa
Al-Nasawī
Al-Nayrīzī
Neander
Nekrasov
Netto
Neuberg
Neumann, C. G.
Neumann, F. E.
Newton, H. A.
Newton, I.
Nicholson, J. W.
Nicomachus of Gerasa
Nicomedes
Nielsen
Nieuwentijt
Noether, A. E.
Noether, M.
Norwood
Nuñez Salaciense

Ocagne
Oenopedes of Chios
Oresme
Ortega
Osgood
Ostrogradsky
Oughtred
Ozanam

Pacioli
Padoa
Painlevé
Pappus of Alexander
Parseval des Chênes
Pascal, B.
Pascal, E.
Pasch
Patrizi
Peacock
Peano
Pearson, K.
Pecham
Peirce, B.
Peirce, B. O. II
Peirce, C. S.
Peletier, J.
Pell
Pemberton
Pérès
Perseus
Peter Philomena of Dacia
Petersen
Peterson
Petrovsky
Peurbach
Pfaff
Picard, C. E.
Pieri
Pincherle
Pitiscus
Plana
Planudes
Plato of Tivoli
Playfair, J.
Plücker
Poincaré
Poisson
Poinsot
Poleni
Poncelet
Poretsky
Porta, G. della
Post
Pratt, J. H.

Lists of Scientists By Field

Prévost, I. B.
Pringsheim, A.
Privalov
Privat de Molières
Proclus
Ptolemy
Puiseux
Pythagoras of Samos

Qāḍī Zāda
Al-Qalaṣādī
Quetelet
Al-Qūhī

Rademacher
Radó
Radon
Ramanujan
Ramsden
Ramsey
Ramus
Razmadze
Réaumur
Recorde
Regiomantanus
Reichenbach, H.
Reidemeister
Reye
Reyneau
Rey Pastor
Rheticus
Ribaucour
Ricatti, J. F.
Ricatti, V.
Ricci, Matteu
Ricci, Michelangelo
Ricci, O.
Ricci-Curbastro, G.
Richard, J. A.
Richard, L. P. E.
Richard of Wallingford
Riemann
Ries
Riesz
Risner
Ritt
Roberval
Robins
Rohn
Rolle
Roomen
Rosanes
Rosenhain, J. G.
Rowning
Rudio
Rudolff
Ruffini, P.
Rumovsky
Runge, C. D. T.
Russell, B. A. W.
Rydberg

Saccheri
Sacrobosco
Saint-Venant
Saint-Vincent
Saks
Salmon
Al-Samarqandī (*fl.* 1276)
Al-Samaw'al
Saurin
Scheffers
Scheuchzer
Schickard
Schläfli
Schmidt, E.
Schoenflies
Schooten
Schott, G.
Schottky
Schoute
Schouten
Schröder
Schroeter
Schubert
Schur
Schuster
Schwarz
Schweikart
Segner
Segre
Seidel
Seki
Semyonov-Tyan-Shansky
Serenus
Serret
Servois
Severi
Sezawa
Shanks
Shatunovsky
Shen Kua
Shirakatsí
Shnirelman
Sierpiński
Sigüenza y Góngora
Al-Sijzī
Simpson
Simson
Sinān
Skolem
Sluse
Slutsky
Smith, H. J. S.
Snel
Sokhotsky
Somerville
Sommerville
Somov
Sonin
Ṣporus of Nicaea
Ṣrīdhara
Srīpati
Stäckel
Stampioen
Staudt
Steiner
Steinitz
Steklov
Stepanov
Stephanus of Alexandria
Stepling
Stevin
Stewart, M.
Stieltjes
Stifel
Stirling
Stokes
Stolz
Stoney
Størmer
Study
Sturm, C.-F.
Sturm, F. O. R.
Subbotin
Suter
Swineshead
Sylow
Sylvester

Tacquet
Tait
Talbot
Tannery, J.
Tannery, P.
Tartaglia
Tauber
Taurinus
Taylor, B.
Thābit ibn Qurra
Theaetetus
Theodorus of Cyrene
Theodosius of Bithynia
Theon of Alexandria
Theon of Smyrna
Theudius of Magnesia
Thiele, T. N.
Thomaz
Thompson, D. W.
Thue
Thunberg, C. P.
Thymaridas
Tilly
Tinseau d'Amondans
Titchmarsh
Todhunter
Toeplitz
Tolman
Torricelli
Tropfke
Troughton
Tschirnhaus
Tsu Ch'ung-Chih
Tunstall
Turing
Turner, P.
Al-Ṭūsī (Naṣir al-Dīn)
Al-Ṭūsī (Sharaf al-Dīn)

Al-Umawī
Al-Uqlīdīsī
Uryson

Valerio
Vallée-Poussin
Vandermonde
Varignon
Veblen
Venn
Ver Eecke
Verhulst
Veronese
Vessiot
Viète
Villalpando
Vitali
Viviani
Vlacq
Volterra
Von Neumann
Voronov

Wald, A.
Wallace, W.
Wallis
Wangerin
Waring
Watson, G. N.
Weber, H.
Wedderburn
Weierstrass
Weingarten
Werner, J.
Wessel
Weyl
Whiston
Whitehead, A. N.
Whitehead, J. H. C.
Whittaker
Widman
Wieleitner
Wiener, L. C.
Wiener, N.
Wilczynski
Wilks
Wilson, Edwin B.
Wilson, J.
Winlock
Winthrop, J. (*d.* 1779)
Wintner
Witt
Wittich
Woepcke
Woodhouse
Woodward, R. S.
Wren
Wright, E.

Xenocrates of Chalcedon

Yang Hui
Yativṛṣabha
Youden
Young, J. W.
Young, W. H.
Yule
Ibn Yūnus

Zanotti
Zarankiewicz
Zenodorus
Zeno of Elea
Zeno of Sidon
Zermelo
Zeuthen
Zhukovsky
Zolotarev
Zucchi

HISTORY, PHILOSOPHY, DISSEMINATION of KNOWLEDGE

Abailard
Abano
Abu'l-Barakāt
Abu'l-Fidā'
Adanson
Agrippa
Ailly
Alain de Lille
Albert of Saxony
Albertus Magnus
Alcuin of York
Alexander of Aphrodisias
Alfonso el Sabio
Alsted
Ames, W.
Ammonius
Anatolius of Alexandria
Anaxagoras
Anaximander
Anaximenes of Miletus
Andreae
Antiphon
Apelt
Aquinas
Arago
Archytas of Tarentum
Aristotle
Armstrong, H. E.
Ashmole
Atwater
Augustine of Hippo

Bachelard
Bacon, F.
Bacon, R.
Bain
Baird
Ibn Bājja
Baranzano
Bartholin, C.
Basso
Ibn Baṭṭūṭa
Bede
Bellarmine
Bergson
Bérigard
Berkeley, G.
Bernard of Chartres
Bernard of Le Treille
Bernard Silvestre
Al-Bīrūnī
Bisterfeld
Al-Biṭrūjī
Blasius of Parma
Boehme
Boethius
Bogdanov
Bolzano
Boncompagni
Borro
Bošković
Bossut
Bourdelot
Bourguet
Boutroux
Boyle
Bradwardine
Braunmühl
Bridgman
Bruno
Brunschvicg
Büchner, F. K. C. L.
Buonamici
Burgersdijk
Buridan
Burley

Cabanis
Calcidius
Campanella
Campbell, N. R.
Cardano
Carr
Carus, P.
Cassiodorus
Cassirer
Cattell
Celaya
Celsus
Cesalpino
Censorinus
Cesi
Charleton
Châtelet
Christmaan
Chrystal
Chwistek
Ciruelo
Clarke, S.
Clerselier
Cohen, M. R.
Collinson
Comenius
Comte
Condillac
Constantine the African
Coronel
Cournot
Couturat
Creighton
Crell
Crelle
Crescas
Crousaz
Cudworth
Cusa
Cuvier, G.

Darwin, C. R.
Della Torre
Democritus
Derham
Descartes
Dicaearchus of Messina
Dickstein
Diderot
Dietrich von Freiberg
Digby
Dingler
Domninus of Larissa
Driesch
Du Hamel
Duhem
Dullaert of Ghent
Duns Scotus
Dyadkovsky

Eaton
Empedocles of Acragas
Engels
Enriques
Epicurus
Erastus
Eriugena
Eudemus of Rhodes

Fabri
Al-Fārābī
Flourens
Folkes
Fontenelle
Forster, G. A.
Forster, J. R.
Fracastoro
Francis of Marchia
Francis of Meyronnes
Franck, S.
Frank, P.
Franklin, B.
Frederick II Hohenstaufen
Fries, J. F.

Galen
Gallois
Garnett
Gassendi
Geminus
Gerard of Cremona
Giles of Lessines
Giles of Rome
Glanvill
Glisson
'sGravesande
Green, J.
Greenwood
Grosseteste
Gundissalinus

Haak
Haldane, R. B.
Hamilton, W.
Harris
Hartlib
Heath
Helmholtz
Helmont
Heraclides Ponticus
Heraclitus of Ephesus
Herbart
Hermes Trismegistus
Herodotus of Halicarnassus
Heytesbury
Hildegard of Bingen
Hipparchus of Rhodes
Hippias of Elis
Hobbes
Höené-Wroński
Holbach
Hugh of St. Victor
Humbert, P.
Humboldt
Hume
Ḥunayn ibn Isḥāq
Hutton, J.
Huxley, T. H.
Huygens
Hypatia

Iamblichus
Ikhwān al-Ṣafā'
Isaac Israeli
Ishaq ibn Hunayn
Isidore of Seville

James of Venice
James, W.
Jevons
John of Dumbleton
John of Gmunden
John of Palermo
John Philoponus
Johnson, W. E.
Jungius

Kant
Karpinski
Ibn Khaldūn
Al-Khayyāmī
Khunrath, H.
Al-Khwārizmī
Al-Kindī
Kircher
Köbel
Koyré
Kraft
Krause

La Condamine
Lamarck
Lambert
Lamétherie
La Mettrie
Lamy, G.
Lax
Leibniz
Lenin
Le Roy, E.
Lesniewski
Leucippus
Levi ben Gerson
Libavius
Linacre
Locke
Logan, J.
Lotze
Lovejoy
Lucretius
Łukasiewicz
Lull
Lyell

McColl
Mach
Macrobius
Magalotti
Magnenus
Maimonides
Maior
Malebranche
Malthus
Mansion
Marinus
Marsilius of Inghen
Martianus Capella
Martin, B.
Marum
Marx
Mersenne
Metzger
Meyerson
Michael Scot
Mieli
Milhaud
Mill
Moerbeke
Montesquieu
Montgomery, E. D.
Montmor
Montucla
More
Morgan, C. W.

Ibn al-Nafīs
Newton, I.
Nieuwentijt
Nifo

Ockham
Oken
Oldenburg
Olympiodorus
Oresme
Oribasius
Osiander

Padoa
Paley
Pappus of Alexandria
Paracelsus
Parmenides of Elea
Partington
Pascal, B.
Patrizi
Paul of Aegina
Paul of Venice
Peale, C. W.
Peano
Pecham
Peirce, C. S.
Peiresc
Petty
Philolaus of Crotona
Planck
Planudes
Plato
Plotinus
Pluche
Poggendorff
Polinière
Pomponazzi
Porta, G. della
Posidonius
Post
Prevost, P.
Priestley, J.
Proclus
Psellus
Purkyně
Pythogoras of Samos

Al-Qazwīnī
Ibn al-Quff
Qusṭā ibn Lūqā
Ibn Qutayba
Quṭb al-Dīn

Rádl
Ramus
Ray, P. C.
Al-Rāzī
Reichenbach, H.
Rey, A.
Rittenhouse
Robinet
Roger of Hereford
Rohault
Rosenberger, J. K. F.
Rowning
Rudio
Ruffini, P.
Ibn Rushd
Russell, B. A. W.

Sabatier, A.
Sacrobosco
Sallo
Al-Samarqandī (*fl.* 1276)
Sanchez
Sarpi
Sarton
Scaliger
Schegk
Schelling
Schleiden
Schlick
Schliemann
Seneca
Servetus
Sextus Empiricus
Shen Kua
Shizuki
Siger of Brabant
Sigorgne
Simplicius
Ibn Sīnā
Smithson
Solvay
Soto
Spencer, H.
Speusippus
Sprat
Stallo
Staszic
Stephen of Antioch
Strabo
Strato of Lampsacus
Sudhoff
Swedenborg
Swineshead

Al-Ṭabarī (11th cent.)
Al-Ṭabarī (9th cent.)
Tannery, P.
Teilhard de Chardin
Telesio
Thābit ibn Qurra
Thales
Themistius
Theophrastus
Thévenot
Thierry of Chartres
Thomas of Cantimpré
Thompson, D. W.
Thompson, S. P.
Thomsen, C. J.
Ibn Tibbon, J.
Ibn Tíbbon, M.
Tilloch
Towneley
Ibn Ṭufayl
Tunstall
Turgot, A. R. J.
Al-Ṭūsī (Naṣir al-Dīn)
Tyndall

Ulrich of Strasbourg

Vailati
Vanini
Varro
Venn
Vincent of Beauvais
Vives
Vlacq
Volkmann
Volney
Voltaire

Wallace, A. R.
Weigel, V.
Wendelin
Whately
Whewell
Whiston
White, T.
Whitehead, A. N.
Whittaker
Wickersheimer
Wieleitner
Wien
Wiener, L. C.
Wilkins
William of Auvergne
William of Sherwood
Winthrop, J. (*d.* 1676)
Witelo
Wittgenstein
Wolf, J. R.
Wolff, Casper F.
Wolff, Christian
Wooley
Worsaae

Xenocrates of Chalcedon
Xenophanes

Yāqūt al-Ḥamawī al-Rūmī
Young, T.

Zabarella
Zeno of Citium
Zeno of Elea
Zeno of Sidon
Zittel

PHYSICS

Abbe, E.
Abraham
Abu'l-Barakāt
Achard
Adams, W. S.
Aepinus
Aguilon
Albert of Saxony
Alberti, L. B.
Albertus Magnus
Aldini
Alembert
Amagat
Ames, J. S.
Amici
Amontons
Ampère
Andronov
Angeli
Ångström
Apell
Appleton
Arago
Archimedes
Atchytas of Tarentum
Aristotle
Arkadiev
Arnold

Lists of Scientists By Field

Arrhenius
Arsonval
Artsimovich
Astbury
Aston
Atwood
Aubuisson de Voisins
Austin
Auzout
Avogadro

Babinet
Bache
Back
Bacon, R.
Baker, H.
Baldi
Baliani
Balmer
Barkhausen
Barkla
Barlow, P.
Barrow
Bartels
Bartholin, E.
Bartoli
Barus
Bateman
Bauer, E.
Bauer, L. A.
Beccaria
Becquerel, A. E.
Becquerel, A. C.
Becquerel, A. H.
Beeckman
Béghin
Bélidor
Bellani
Belopolsky
Benedetti, G. B.
Benedicks
Benoit
Benzenberg
Bergman
Bérigard
Bernoulli, D.
Bernoulli, Jakob I.
Bertholon
Berti
Bhabha
Białobrzeski
Biot
Bjerknes, C. A.
Bjerknes, V. F. K.
Bjerrum
Black, Joseph
Blagden
Blasius of Parma
Blondlot
Bohr, N. H. D.
Boltwood
Boltzmann
Borda
Borelli
Born
Borries
Bose, G. M.
Bose, J.
Bose, S.
Bošković
Bothe
Bouguer
Bour
Bourguet
Boussinesq
Boyle
Boys
Brace
Bragg, W. H.
Bragg, W. L.
Brandes
Braun, F.
Bravais
Brewster
Bridgman
Brillouin
Briot
Brisson, M. J.
Broglie
Bucherer
Buckingham
Buono
Buot
Burger
Buridan
Burley

Cabrera
Cagniard de la tour, C.
Cailletet
Callan
Callendar
Campbell, N. R.
Camus
Canton
Cardano
Carlisle
Carnot, N. L. S.
Cassegrain
Castel
Castelli
Catalán
Cauchy
Cavallo
Cavendish
Charles
Chladni
Christiansen
Christie, S. H.
Chrystal
Clairaut
Clausius
Clay
Coblentz
Cockcroft
Colden
Colding
Combes, C. P. M.
Compton, A. H.
Compton, K. T.
Coriolis
Cornu
Corti, B.
Cosserat
Coster
Coulomb
Courtivron
Crawford
Crookes
Cumming
Curie, M.
Curie, P.

Dalencé
Dalton, J.
Daniell
Darwin, C. G.
Davidov
Davisson
Debye
De Forest
De Groot
Delaunay
Deluc
Democritus
Deprez
Desaguliers
Descartes
Des Cloizeaux
Deslandres
Devaux
Dewar
Dickinson
Dietrich von Freiberg
Diocles
Diviš
Dollond
Dominis
Donnan
Doppelmayer
Doppler
Dove
Drebbel
Drude
Duane
Dubois-Reymond, E. H.
Du Buat
Dufay
Duhamel
Duhem
Dullaert of Ghent
Dulong
Du Moncel
Dunoyer de Segonzac
Duperrey
Dupré
Dutrochet

Eddington
Ehrenfest
Eichenwald
Einstein
Ellis
Elster
Emden, R.
Enskog
Eötvös
Erman
Esclangon
Eucken
Euclid
Euler
Evershed
Ewing, J. A.

Fabbroni
Fabry, C.
Fahrenheit
Fankuchen
Faraday
Farkas
Farrar
Feddersen
Fermi
Ferraris
Fessenden
Fitzgerald
Fizeau
Fleischer
Fleming, J. A.
Föppl
Forbes, J. D.
Foucault
Fourier
Fowler, A.
Fowler, R. H.
Franck, J.
Frank, P.
Frankenheim
Franklin, B.
Franklin, R. E.
Fraunhofer
Frenicle de Bessy
Frenkel
Fresnel
Friedel
Friedmann
Fries, J. F.
Frisi
Froude
Fyodorov

Galerkin
Galilei, G.
Galilei, V.
Galvani
Gamow
Gascoigne
Gassiot
Gauss
Gautier, P. F.
Gay-Lussac
Geiger
Geitel
Gerasimovich
Gibbs, J. W.
Gilbert, N.
Giles of Rome
Giorgi
Glazebrook
Goddard
Goethe
Goldschmidt, V.
Goldstein
Golitsyn
Gordon
Gore
Gouy
Graham, T.
Gramont
Grashof
'sGravesande
Gray, S.
Green, G.
Gregory, D.
Gregory, J.
Gren
Grimaldi
Grosseteste
Groth
Grotthuss
Grove
Guericke
Guillaume
Guldberg
Gullstrand
Guye
Guyton de Morveau

Haas A. E.
Haas, W. J. de

Hachette
Hahn
Hale, G. E.
Hall, E. H.
Halley
Hallwachs
Hamilton, W. R.
Hamy
Hankel, W. G.
Hansen, W. W.
Hansteen
Harkins
Harriot
Hartree
Hartsoeker
Hasenöhrl
Hauksbee, Francis (*d.* 1713)
Hauksbee, Francis (*d.* 1763)
Ibn al-Haytham
Heaviside
Helmholtz
Henrichsen
Henry of Hesse
Henry, J.
Henry, P. M.
Henry, P. P.
Herapath, J.
Hermann, C. H.
Hero of Alexandria
Herschel, J. F. W.
Hertz
Hess, V. F.
Hessel
Hevesy
Heytesburg
Heyrovsky
Hirn
Hittorf
Hobbes
Hodgkinson
Holborn
Honda
Hooke
Hopkinson
Horstmann
Huber, M. T.
Huggins
Hugoniot
Hull
Humboldt
Hume-Rothery
Hutton, J.
Huygens
Hyleraas

Infeld
Ingen-Housz
Ioffe
Ishiwara
Ives

Jacobi, M. H. von
Jaeger, F. M.
Jahn
Janssen
Jeans
Joblot
Joliot
Joliot-Curie
Jolly
Joly
Jones, H. C.
Jordanus de Nemore
Joule
Julius

Kaluza
Kamāl al-Dīn
Kamerlingh Onnes
Kármán
Kaufmann
Kayser
Keesom
Keill, John
Kelvin (W. Thomson)
Kennedy
Kennelly
Kepler
Kerr, J.
Al-Khāzinī
Kinnersley
Kirchhof, G. R.
Kleist
Klingenstierna
Klügel
Knudsen
Kochin
Koenig, J. S.
Koenig, K. R.
Koenigs
Kohlrausch, F. W. G.
Kohlrausch, R. H. A.
Kolosov
König, A.
Konkoly Thege
Korolev
Kossel, W. L. J. P. H.
Kotelnikov
Kraft
Kramers
Kramp
Krönig
Krylov, A. N.
Kuenen
Kuhn, W.
Kundt
Kurchatov
Kurlbaum

Ladenburg, R. W.
Lagrange
La Hire, P. de
Lamb
Lambert
Lamont
Lamy, B.
Landau L. D.
Landriani
Landsberg, G. S.
Lane
Langevin
Langley
Langmuir
Laplace
La Rive, A. A. de
La Rive, C. G. de
Larmor
Laue
Laurent, P. A.
Lawrence, E. O.
Lazarev, P. P.
Lebedev, P. N.
Le Cat
Le Châtelier
Leconte, John
Lecornu
Leeuwenhoek
Lehmann, O.
Leibniz
Le Monnier, L. G.
Lenard
Lennard-Jones
Lenz
Leonardo da Vinci
Le Roy, C.
Le Roy, J. B.
Lesage
Leslie
Le Tenneur
Le Verrier
Levi ben Gerson
Levi-Civita
Lewis, G. N.
Leybenzon
Lichtenberg
Linde
Lindemann, F. A. (Lord Cherwell)
Link
Lippmann
Lissajous
Lister, J. J.
Lloyd, H.
Lockyer
Lodge
Lomonosov
London, F.
London, H.
Lonsdale, K.
Lorentz, H. A.
Lorenz, H.
Lorenz, L. V.
Lorenz, R.
Lorenzoni
Loschmidt
Love
Lummer
Lyapunov
Lyman, T.
Lyot

MacCullagh
Mach
Magellan
Magiotti
Magnenus
Magni
Magnus, H. G.
Maignan
Mairan
Majorana
Maksutov
Malebranche
Mallet
Malus
Mandelshtam
Manfredi
Marci of Kronland
Marconi
Margules
Mariotte
Marliani
Marsilius of Inghen
Martianus Capella
Mascart
Masson
Mathieu
Matteucci
Matuyama
Mauguin
Maupertuis
Maurolico
Maxwell
Mayer, A. M.
Mayer, J. R.
Meggers
Meitner
Melloni
Melvill
Merica
Mersenne
Meshchersky
Meusnier de la Place
Michelini
Michelson
Mie
Miller, D. C.
Miller, W. H.
Miller, W. A.
Millikan
Milne, E. A.
Mineur
Mises
Mittasch
Mohr, C. F.
Moiseev
Moll, G.
Mollier
Molyneux, S.
Molyneux, W.
Monge
Montanari
Monte
Moore, W.
Morley
Moseley
Mossotti
Müller, G.
Müller, J. H. J.
Muncke
Musschenbroek
Mydorge

Nagaoka
Nairne
Natanson
Navier
Nekrasov
Nernst
Neumann, C. G.
Neumann, F. E.
Newall
Newton, I.
Niceron
Nichols
Nicholson, J. W.
Nicol
Niggli
Nobili
Noel
Nollet
Norman
Numerov
Nusselt

Oersted
Ohm
Olszewski
Oppenheimer
Orlov, S. V.

Ornstein
Osmond
Ostrogradsky
Ostwald, F. W.

Pacinotti
Paneth
Pardies
Parent
Pascal, B.
Paschen
Pasteur
Pauli
Pecham
Peirce, B. O. II
Peltier, J. C. A.
Pemberton
Pereira, B.
Pérès
Pérot
Perrin
Peter Peregrinus
Petit, A. T.
Petit, P.
Petrov, N. P.
Petrov, V. V.
Pezenas
Philo of Byzantium
Piccard
Pictet, M.-A.
Pictet, R. -P.
Pierce
Planck
Plateau
Playfair, J.
Plücker
Poggendorff
Poincaré
Poinsot
Poiseuille
Poisson
Poleni
Poli
Polinière
Poncelet
Popov, A. S.
Porro
Pouillet
Powell, B.
Powell, C. F.
Power, H.
Poynting
Prandtl
Prévost, I.-B.
Prevost, P.
Priestley, J.
Pringsheim, E.
Pritchard
Privat de Molières
Puiseux
Pulfrich
Pupin

Quincke

Raman
Ramsauer
Ramsay, W.
Ramsden
Ramus
Rankine, A. O.
Rankine, W. J. M.
Ranyard
Rateau
Rayleigh (Strutt)
Réaumur
Redtenbacher
Reech
Regener
Regnault
Reichenbach, G. F. von
Reid
Reuleaux
Reynolds
Ricci-Curbastro
Ricco
Richardson, O. W.
Richer
Richmann
Richtmyer
Riecke
Riemann
Righi
Risner
Ritter
Ritz
Roberval
Robison
Roche
Röntgen
Rood
Roozeboom
Rosenberg
Rosenberger, J. K. F.
Ross
Rossetti
Routh
Rowland
Rozhdestvensky
Rubens
Rüdenberg
Runge, C. D. T.
Russell, H. N.
Rutherford E.
Rutherfurd
Rydberg

Sabine, E.
Sabine, P. E.
Sabine, W. C. W.
Sagnac
Saha
Saint-Venant
Saurin
Sauveur, J.
Savart
Scheiner, J.
Schmidt, B. V.
Schmidt, C. H. von
Schmidt, G. C. N.
Schoenflies
Schönbein
Schonland
Schrödinger
Schuster
Schweigger
Secchi
Sechenov
Seebeck
Segner
Seguin, M.
Sénarmont
Seneca
Sharanov
Shaw, W. N.
Shayn
Short
Siedentopf
Sigaud de Lafond
Sigorgne
Simon
Smeaton
Smekal
Smith, R.
Smits
Smoluchowski
Snel
Soddy
Sohncke
Somerville
Sommerfeld
Somov
Spring
Stark
Steacie
Stefan
Steinheil
Steinmetz
Steklov
Stepling
Stern
Stewart, B.
Stokes
Stoletov
Stoney
Störmer
Strutt, J. W.
Strutt, R. J.
Sturgeon
Sturm, J. C. F.
Sutherland
Svedberg
Sverdrup
Swann
Swinden
Swineshead
Symmer
Synesius of Cyrene
Szilard

Taccola
Tait
Tamm
Tammann
Tartaglia
Tesla
Thābit ibn Qurra
Thollon
Thomaz
Thompson B.
Thompson, S. P.
Thomson, J. J.
Thomson, W.
Tikhov
Tiselius
Tisserand
Titius
Tolman
Torricelli
Towneley
Townsend, J. S. E.
Travers
Troughton
Trouton
Trowbridge
Tschirnhaus
Tsiolskovsky
Tupolev

Van de Graaff
Van't Hoff
Varignon
Vavilov, S. I.
Veksler
Verdet
Villalpando
Villard
Villari
Violle
Vogel
Voigt, W.
Volkmann
Volta
Von Neumann
Vyshnegradsky

Waals
Warburg, E. G.
Washburn
Waterston
Watson, W.
Watt
Weber, W. E.
Wehnelt
Wegener
Weiss, C. S.
Weiss, P.
Weyl
Wheatstone
Whitehead, A. N.
Whittaker
Whytlaw-Gray
Wiechert
Wiedemann
Wien
Wiener, L. C.
Wiener, O.
Wilcke
Wilhelmy
Wilson, B.
Wilson, C. T. R.
Wilson, Edwin B.
Witelo
Wollaston, W. H.
Wood, R. W.
Woodward, R. S.
Wren
Wróblewski
Wulff

Young, S.
Young, T.

Zeeman
Zernike
Zeuner
Zeuthen
Zhukovsky
Zöllner

TECHNOLOGY, ENGINEERING

Amsler
Armstrong, E. H.

Beckmann, J.
Bell, A. G.
Berti
Bertin
Bessemer
Betancourt y Molina
Bion
Bird
Blondel, A. E.
Blondel, N. F.
Biringuccio
Bourdon
Brashear
Breguet
Brinell
Brioschi
Brisson, B.
Brunelleschi

Campani
Carnot, L. N. M.
Castigliano
Chaplygin
Chardonnet
Clapeyron
Clark, A.
Clark, A. G.
Clark, G. B.
Clark, J. L.
Claude, G.
Congreve
Cottrell
Courtivron
Ctesibius

De La Rue
Deprez
Derosne
Diderot
Divini
Drebbel
Dufour
Duhamel du Monceau

Edison
Eichenwald
Eytelwein

Ferguson
Ferraris
Fessenden
Fleming, J. A.
Föppl
Fortin
Foucault
Fourneyron
Fraunhofer
Froude

Gambey
Gautier, P. F.
Gay-Lussac
Geissler
Giorgi
Girard, P. S.
Goddard
Godfrey
Graham, G.
Gramme
Grashof
Gregory, O. G.
Grubb, H.
Grubb, T.
Guericke

Hadley
Hale, W.
Hampson
Harrison, J.
Harting
Hartmann, G.
Hartsoeker
Hauksbee, F. (*d.* 1713)
Hauksbee, F. (*d.* 1763)
Hefner-Alteneck
Hellot
Hermbstaedt
Héroult
Hevelius
Heyn
Hodgkinson
Huygens

Ioffe

Jacquet
Jansen
Jars
Al-Jazarī
Jefferson
Jeffries
Jenkin
Jewett
Joblot

Kennedy
Kennelly
Kettering
Al-Khāzinī
Kirkaldy
Köhler
Korolev
Krayenhoff
Krylov, A. N.

Lanchester
Landriani
Langlois
Langren
Le Châtelier
Lecornu
Leeuwenhoek
Lemaire, J.
Lemaire, P.
Leonardo da Vinci
Le Roy, J. B.
Lévy, M.
Leybenzon
Linde
Lippmann
Lorenz, H.
Lunge

Magellan
Mallard
Mallet
Marconi
Martens
Martin, B.
Martini
Medina
Menabrea
Merett
Metius, A.
Metius, A. A.
Metius, J.
Meusnier de la Place
Mohr, C. O.
Moll, F. R. H. C.
Montgéry
Montgolfier, E. J. de
Montgolfier, M. J. de
Morland
Moutard
Müller-Breslau
Mushet

Nasmyth
Navier
Neri
Newcomen
Nicholson, W.
Niepce
Nobel
Nobert

Pambour
Papin
Percy
Perrault, C.
Perronet
Petrov, N. P.
Pilatre de Rozier
Pitot
Plattner
Poleni
Popov, A. S.
Portevin
Pouillet
Prony

Rankine, W. J. M.
Rateau
Réaumur
Redtenbacher
Reech
Repsold, A.
Repsold, J. A.
Repsold, J. G.
Reuleaux
Ribaucour
Ribeiro Santos
Ricci, O.
Rittenhouse
Robins
Roebuck
Rosenhain, W.
Rüdenberg
Rühmkorff

Santorio
Sauveur, A.
Saxton
Schott, G.
Schott, O. F.
Schumann
Seguin, M.
Siemens, C. W.
Siemens, E. W. von
Smeaton
Soleil
Solvay
Sperry
Stanton
Stead
Steinmetz
Stevin
Stodola
Swedenborg

Taylor, F. W.
Tesla
Thomas, H. H.
Thompson, S. P.
Thomson, E.
Thurston
Torres Quevedo
Towneley
Travers
Tupolev
Turner, W. E. S.

Valturio
Vauban

Venetz
Verantius
Vernier
Villard de Honnecourt
Vitruvius
Vyshnegradsky
Watt
Wedgwood
Weisbach
Willis, R.
Wöhler, A.
Woltman
Wright, O.
Wright, W.
Zucchi